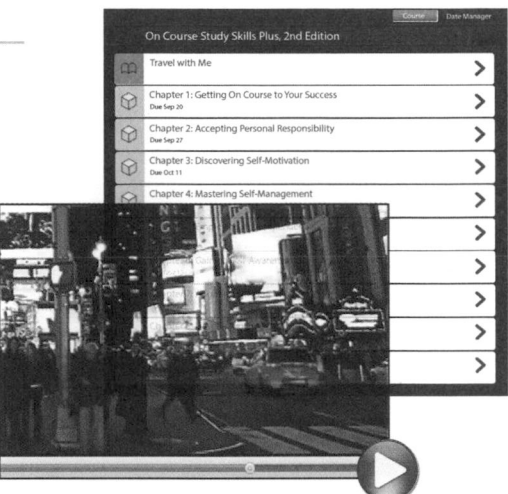

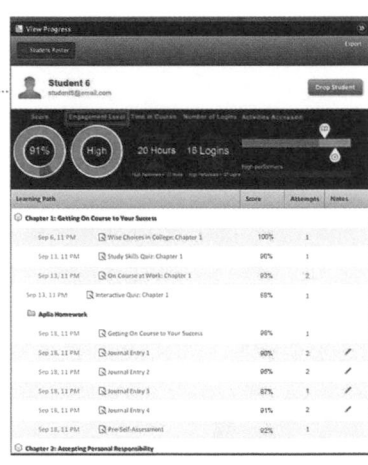

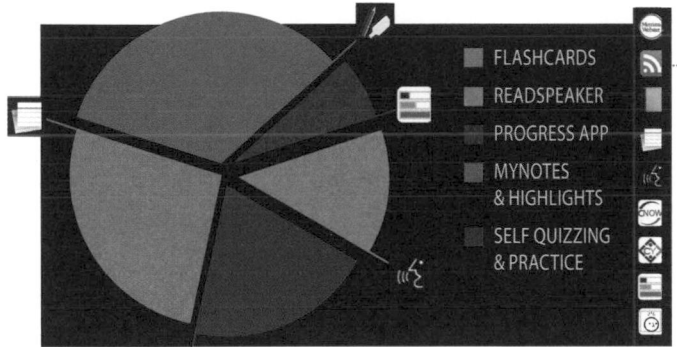

Third Canadian Edition

AUTOMOTIVE TECHNOLOGY

A Systems Approach

Third Canadian Edition

AUTOMOTIVE TECHNOLOGY

A Systems Approach

JACK ERJAVEC

MARTIN RESTOULE
ALGONQUIN COLLEGE

STEPHEN LEROUX
CENTENNIAL COLLEGE

ROB THOMPSON

NELSON

NELSON

Automotive Technology: A Systems Approach, Third Canadian Edition
by Jack Erjavec, Martin Restoule, Stephen Leroux, and Rob Thompson

Vice President, Editorial Higher Education:
Anne Williams

Publisher:
Jackie Wood

Marketing Manager:
Cara Cortese

Technical Reviewer:
Jeff Oakes

Developmental Editor:
Courtney Thorne

Photo Researcher and Permissions Coordinator:
Natalie Barrington

Production Project Manager:
Jennifer Hare

Production Service:
Integra Software Services Pvt. Ltd.

Copy Editor:
Michael Kelly

Proofreader:
Jill Pellarin

Indexer:
Gaile Brazys

Design Director:
Ken Phipps

Managing Designer:
Franca Amore

Interior Design:
Peter Papayanakis

Cover Design:
Peter Papayanakis

Cover Image:
General Motors LLC. Used with permission, GM Media Archives.

Compositor:
Integra Software Services Pvt. Ltd.

Printed and bound in Canada
5 6 7 20 19 18

For more information contact Nelson Education Ltd., 1120 Birchmount Road, Toronto, Ontario, M1K 5G4. Or you can visit our Internet site at nelson.com

Additional credits for recurring images: automotive diagnostic line art (chapter openers), VoodooDot/Shutterstock; hazard warning sign (Warning! boxes), NEGOVURA/Shutterstock

Library and Archives Canada Cataloguing in Publication Data

Erjavec, Jack, author
Automotive technology : a systems approach / Jack Erjavec, Martin Restoule (Algonquin College), Stephen Leroux (Centennial College), Rob Thompson. — Third Canadian edition.

Includes index.
ISBN 978-0-17-653152-2 (bound)

1. Automobiles—Maintenance and repair—Textbooks. I. Restoule, Martin, author II. Leroux, Stephen, author III. Thompson, Rob (Rob D.) author IV. Title.

TL152.E75 2015
629.28'7 C2014-906605-8

ISBN-13: 978-0-17-653152-2
ISBN-10: 0-17-653152-1

BRIEF CONTENTS

CONTENTS

PHOTO SEQUENCES

PREFACE

ABOUT THE BOOK

Advancing technology continues to improve the operation and integration of the various systems of the automobile. These changes present ongoing challenges for the student to become a successful automotive technician. The third Canadian edition of *Automotive Technology: A Systems Approach* was designed and written to continue to prepare students for those challenges. This book concentrates on the need-to-know essentials of the various automotive systems—and how they have changed from the vehicles of yesterday—the operation of today's vehicles, and what to expect in the near future. New technology is addressed throughout the book, but some older technology remains in this edition as technicians will still see this technology in older vehicles. This does not mean the pages are filled with fact after fact. Rather, each topic is explained in a logical way, slowly but surely. Many years of teaching have provided us with a good sense of how students read and study technical material. We also know what draws their interest to a topic and keeps it there. This knowledge has been incorporated in the writing and the features of this book.

NEW TO THIS EDITION

This new edition is the largest revision to date. This third Canadian edition builds upon the strengths and success of the market-leading second Canadian edition. The new design has a cleaner look with an easy-to-read font and clear objectives at the beginning of each chapter. The new "Go To" feature allows for easier navigation, and the "Performance Tip" feature introduces performance-enhancing techniques.

There are a total of four new chapters in this edition: Chapter 8, Preventive Maintenance and Basic Services; Chapter 36, Hybrid Vehicles; Chapter 37, Electric Vehicles; and Chapter 43, Electronic Automatic Transmissions. Additionally, each existing chapter has been updated in response to the changing industry. Many of the photos and photo sequences have been revised and updated for clarity or to present updated technology. We've expanded our coverage to ensure that the most recent technological advancements in automotive technology have been incorporated, as well as the many elements unique to the role of a Canadian professional automotive technician.

Finally, through the invaluable assistance of automotive instructors and contributors across Canada, we've focused on continuing to ensure that the text is technically sound and up-to-date. We are confident that the changes throughout the third Canadian edition will continue to meet the unique needs of Canadian students and instructors and will keep *Automotive Technology: A Systems Approach* the preferred resource on the market.

Section 1 gives an overview of the automotive industry, careers, working as a technician, tools, diagnostic equipment, and basic automotive systems. Chapter 1 explores the career opportunities in the automotive industry and includes coverage of alternative career options. Updated features include current vehicle and industry information. Chapter 2 discusses workplace skills. This chapter goes through the process of getting a job and keeping it in today's market. It also covers some of the duties common to all automotive technicians. Chapter 6 covers the special and diagnostic tools required for use in all eight system sections in this textbook. Chapter 7 presents both the science and math principles that govern the operation of the automobile. Too often, we as educators assume that our students know these basics. We have included this chapter to serve as a reference for those students who want to be good technicians and need a better understanding of why things happen the way they do. Chapter 8 is a new chapter in Section 1, covering preventive maintenance and basic vehicle services. Preventive maintenance is becoming a greater part of the day-to-day tasks of today's technicians.

Section 2 has thorough coverage and updated photos of the latest engine designs, materials, and technologies. There is also coverage of theory, diagnosis, and service to in-block and overhead camshaft engines. There are discussions on the latest trends, including the most current variable valve timing and lift systems.

Section 3 covers basic electricity and electronics, with a separate chapter covering troubleshooting and repairing electrical and electronic systems. The coverage of all major electrical systems includes new technologies such as adaptive systems (e.g., cruise control), many new accessories, dashboard displays, and servicing current electrical systems. The rest of the electricity section includes up-to-date coverage on body computers and the use of lab scopes and graphing meters.

Engine performance is covered in Section 4 and includes a thorough discussion of fuels and alternative energy sources. The fuels chapter (Chapter 27) includes discussion about gasoline, diesel, and alternate fuels. Chapter 29 has been updated with new photos and current information, including injector service and electronic throttle control. Chapter 33 has been updated to cover the latest boost controls for turbocharged engines. Chapters 34 and 35 contain the latest information on emission control system operation, testing, and servicing. Chapter 36 is another new chapter to this third Canadian edition and is dedicated to hybrid vehicles, including all the relevant information on safe working procedures, operation, and the various hybrid vehicle systems. Chapter 37 is also a new chapter, covering the most current advancements in electric vehicle technologies.

Sections 5 and 6 cover transmissions and drivelines, including comprehensive discussions of new clutch technologies, six-speed transmissions, automatic manual transmissions, and differential designs. Chapter 43 is another new chapter that offers information all about electronic automatic transmission and transaxle controls, including operation, diagnosis, and servicing. This section (Chapter 44) also includes two new photo sequences: Checking Transmission Fluid Level on a Vehicle without a Dipstick and Changing Automatic Transmission Fluid and Filter.

Many topics in Section 7 have been updated, such as tire and wheel servicing, tire pressure monitoring systems, current suspension and steering systems, and four-wheel alignment procedures.

Section 8 covers brakes and provides current coverage of the latest technologies, inspection and servicing of drum, disc and antilock brakes, and stability control and traction control systems. Heating and air-conditioning systems are discussed in the final two chapters in Section 9. These chapters have been updated with new refrigerant information and current inspection and service procedures.

ORGANIZATION AND GOALS OF THIS EDITION

The third Canadian edition of *Automotive Technology: A Systems Approach* remains a comprehensive guide to the service and repair of contemporary automobiles. The book is divided into nine sections, the first being an introduction to the automotive industry and general service work and procedures. The remaining eight relate to specific automotive systems and include many diagnostic procedures. With the complexity of today's vehicles, being able to break down the vehicle into subsystems or individual components allows the student to focus on that particular component or system operation. Diagnostic and service procedures can be generic as well as unique to different automobile manufacturers. Because many automotive systems are integrated, we identify these important relationships between systems and also flag any potential safety issues. While this edition has a new chapter dedicated to hybrid vehicles, there are also references to hybrid technologies throughout the book to identify how the systems are different in a hybrid vehicle.

Effective diagnostic skills begin with learning to isolate the problem. By identifying the system that contains the problem, the exact cause is easier to pinpoint.

Learning to think logically about troubleshooting problems is crucial to mastering this essential skill. Therefore, logical troubleshooting techniques are discussed throughout this text. Each chapter describes ways to isolate both the problem system and the individual components of that system.

This systems approach provides the student with important preparation for the Certificate of Qualification (Red Seal) certification exams. The book's sections are outlined to match the Red Seal test specifications and address many of the identified National Occupational Analysis tasks. Further, the review questions at the end of every chapter give students practice answering Red Seal–style questions. The single most important goal of this text continues to be giving students a better understanding of the total vehicle. This will give students the very best chance to become skilled automotive technicians with unlimited career potential.

ACKNOWLEDGEMENTS

We would like to gratefully acknowledge and thank these knowledgeable educators for their comments and suggestions throughout the development of this book.

CONTRIBUTOR AND TECHNICAL REVIEWER

Jeff Oakes
Conestoga College

Jeff was instrumental in the development of this edition. His tireless effort, extremely thorough review, and true passion for the automotive trade have made this a better book.

CONTENT AND TECHNICAL REVIEWERS

Winston Bertie
Centennial College

Greg Buerk
British Columbia Institute of Technology

Ross Dunn
British Columbia Institute of Technology

Mubasher Faruki
British Columbia Institute of Technology

Greg Henderson
Vancouver Community College

Mike Howells
British Columbia Institute of Technology

Russ Hunter
British Columbia Institute of Technology

Robert Hyde
Algonquin College

Scott McLaughlin
Canadore College

Hans Reimer
Fanshawe College

Brian Southgate
Saskatchewan Institute of Applied Science and Technology

ABOUT THE AUTHORS

Jack Erjavec has become a fixture in the automotive textbook publishing world. He has many years of experience as a technician, educator, author, and editor, and has authored or co-authored more than 40 automotive textbooks and training manuals. Mr. Erjavec holds a Master of Arts degree in vocational and technical education from Ohio State University. He spent 20 years at Columbus State Community College as an instructor and administrator and has also been a long-time affiliate of the North American Council of Automotive Teachers, including serving on the board of directors and as executive vice-president. Jack was also associated with ATMC, SAE, ASA, ATRA, AERA, and other automotive professional associations.

Martin Restoule has 30 years of experience as a Canadian Interprovincial Standards 310S and 310T technician. Martin is an automotive and truck transport technician professor and coordinator at Algonquin College in Ottawa and is very active in the various Ontario college motive power educators committees. He is currently a member of both the local and provincial motive power apprenticeship advisory committees as well as the Ontario Motive Power Curriculum Delivery Advisory Committee. Martin is also very active with Skills Ontario and Skills Canada automotive competitions. In addition to chairing the Ontario and national skills competitions, he also chairs the National Technical Committee for Skills Canada and represents Canada at World Skills as the Canadian expert. He has authored and co-authored a number of ASE test prep manuals, as well as the Canadian Truck and Transport Technician Interprovincial Certificate of Qualification Test Preparation manual.

Stephen Leroux has over 25 years of experience in the field as an automotive service technician. He possesses a 310S Canadian Interprovincial Standard along with many other industry training certificates obtained while working in various dealerships across the Greater Toronto Area. He is a professor and coordinator at Centennial College, as well as an avid supporter of and participant at Skills Ontario and Skills Canada events. Stephen is also involved with the Ontario Youth Apprenticeship Program and participates with the advisory committee of his local school board. He has written a number of articles on vehicle safety and maintenance for national newspapers and has also participated in various interviews on the subjects. Stephen earned a college diploma in automotive technology from Centennial College, a certificate in adult learning, and is also an associate member of the Society of Automotive Engineers.

Rob Thompson has about 20 years of teaching experience and worked in the industry for 10 years prior to being an educator. He is an ASE master automobile technician with A9 and L1 certification. He is also currently an officer of the North American Council of Automotive Teachers (NACAT) and has served on NACAT's board.

Learning how to maintain and repair today's automobiles can be a daunting endeavour. To guide the readers through this complex material, we have built in a series of features that will ease the teaching and learning processes.

LEARNING OUTCOMES

Each chapter begins with the purpose of the chapter, stated in a list of learning outcomes. Both cognitive and performance outcomes are included in the lists. The outcomes state the expected outcome that will result from completing a thorough study of the contents in the chapters.

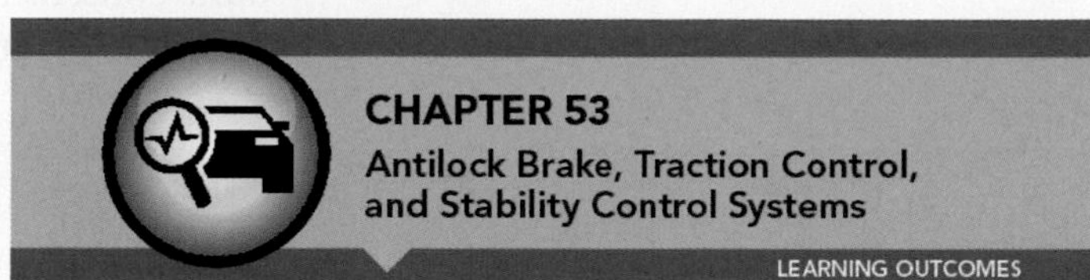

CHAPTER 53
Antilock Brake, Traction Control, and Stability Control Systems

LEARNING OUTCOMES

- Explain how antilock brake systems work to bring a vehicle to a controlled stop.
- Describe the differences between an integrated and a nonintegrated antilock brake system.
- Briefly describe the major components of a two-wheel antilock brake system.
- Briefly describe the major components of a four-wheel antilock brake system.
- Describe the operation of the major components of an antilock brake system.
- Describe the operation of the major components of automatic traction and stability control systems.
- Explain the best procedure for finding ABS faults.
- List the precautions that should be followed whenever working on an antilock brake system.

WARNING! AND CAUTION!

Instructors often tell us that shop safety is their most important concern. Cautions and warnings appear frequently to alert students to important safety concerns.

WARNING!

The use of a standard camshaft bearing driver and hammer is not recommended for aluminum heads because the aluminum bearing supports are very easily damaged or broken, which can result in expensive head replacement.

CAUTION!

Always wear eye protection when operating any type of grinding equipment.

SHOP TALK

These features appear throughout the book to give practical, common-sense advice on service and maintenance procedures.

SHOP TALK

Never change spark plugs when the cylinder head is hot. The bores for the plugs can take on an oval shape as the cylinder head

CUSTOMER CARE

Creating a professional image is an important part of shaping a successful career in automotive technology. The customer care tips were written to encourage professional integrity. They give advice on educating customers and keeping them satisfied.

CUSTOMER CARE

When attempting to diagnose the cause of abnormal engine noise, it may be necessary to temper the enthusiasm of customers

USING SERVICE INFORMATION

Learning to use available service information is critical to becoming a successful technician. The source of information varies from printed to online material. The gathering of information can be a time-consuming task but nonetheless is extremely important. We have included a feature that points the student in the right direction to find the right information.

USING SERVICE INFORMATION

Crankshaft specifications can be found in the engine specification section of a service information system.

PERFORMANCE TIP

This feature introduces students to the ideas and theories behind many performance-enhancing techniques used by professionals.

PERFORMANCE TIP

Adding nitrous oxide to the air/fuel mixture is not something done by oil refineries. Rather, it is commonly done by those seeking more instantaneous power from their engines. Nitrous oxide is

GO TO ▶ Chapter 14 for information on servicing and installation of oil pumps.

GO TO

This feature tells the student where to look in the text for prerequisite and additional information on the relevant topic.

HYBRID VEHICLES

In addition to a new dedicated chapter on hybrid vehicles, abundant content on hybrid technologies is presented throughout the textbook within the specific system areas.

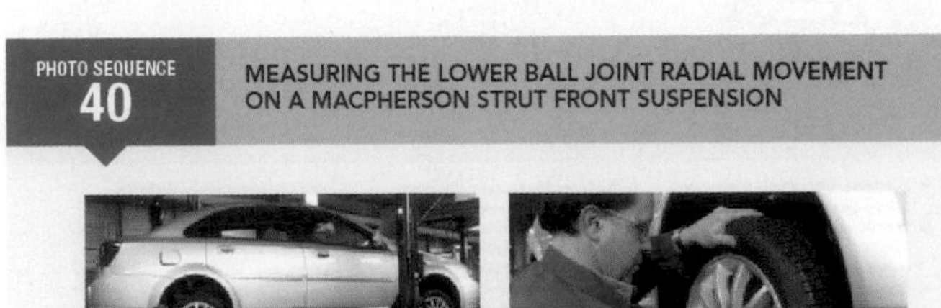

PHOTO SEQUENCE
40
MEASURING THE LOWER BALL JOINT RADIAL MOVEMENT ON A MACPHERSON STRUT FRONT SUSPENSION

P40-1 Position the car correctly on a chassis hoist.

P40-2 Grasp the front tire at the top and bottom, and rock the tire inward and outward while a co-worker visually checks for movement in the front wheel bearing. If there is movement, adjust or replace the wheel bearing.

PHOTO SEQUENCES

There are 45 step-by-step photo sequences scattered throughout the text that illustrate practical shop techniques. The photo sequences focus on techniques that are common, need-to-know service and maintenance procedures. These photo sequences give students a clean, detailed image of what to look for when they perform these procedures.

PROCEDURE

Insert Valve Seat Removal and Replacement

STEP 1 To remove the damaged insert, use a puller or a pry bar (Figure 12–80).

STEP 2 After removal, clean up the counterbore or recut it to accommodate oversized inserts.

PROCEDURES

This feature gives detailed, step-by-step instructions for important service and maintenance procedures. These hands-on procedures appear frequently and are given in great detail because they help to develop good shop skills.

KEY TERMS

KEY TERMS

Each chapter ends with a list of the terms that were introduced in the chapter. These terms are highlighted in the text when first used, and are also defined in the Glossary.

SUMMARY

SUMMARY

Highlights and key bits of information from the chapter are listed at the end of each chapter. This listing is designed to serve as a refresher for the reader.

REVIEW QUESTIONS

REVIEW QUESTIONS

A combination of short-answer essay and interprovincial examination–style multiple-choice questions makes up the end-of-chapter questions. These questions are used to challenge the reader's understanding of the chapter's content. The chapter objectives and content are used as the basis for the review questions.

Undersize bearings are available in .0254 mm (0.001 in.), or .0508 mm (0.002 in.)sizes for shafts that are uniformly worn by that amount. Undersize bearings are also available in thicker sizes, such as .2540 mm (0.010 in.), .5080 mm (0.020 in.), and .7620 mm (0.030 in.), for use with crankshafts that have been refinished (or reground) to one of these standard undersizes. The difference in thickness of the bearing is normally stamped onto the backside of the bearing. Bearings may also be colour-coded to indicate their size.>

METRIC EQUIVALENTS

Throughout the text, measurements are given in metric and UCS increments.

ABOUT THE NELSON EDUCATION TEACHING ADVANTAGE (NETA)

The **Nelson Education Teaching Advantage (NETA)** program delivers research-based instructor resources that promote student engagement and higher-order thinking to enable the success of Canadian students and educators. Be sure to visit Nelson Education's **Inspired Instruction** website at **www.nelson.com/inspired/** to find out more about NETA. Don't miss the testimonials of instructors who have used NETA supplements and have seen student engagement increase!

INSTRUCTOR RESOURCES

All NETA and other key instructor ancillaries are provided on the **Instructor Companion Site** at **www.nelson.com/automotivetechnology3Ce**, giving instructors the ultimate tool for customizing lectures and presentations. Instructor materials can also be accessed through **www.nelson.com/instructor.**

- **NETA Test Bank:** This resource was written by Krystyna Lagowski. It includes over 1000 multiple-choice questions written according to NETA guidelines for effective construction and development of higher-order questions. Also included are 500 true/false questions, 500 completion questions, and 430 short answer questions.

The NETA Test Bank is available in a new, cloud-based platform. **Testing Powered by Cognero®** is a secure online testing system that allows instructors to author, edit, and manage test bank content from any place with Internet access. Nelson Testing Powered by Cognero for *Automotive Technology* can be accessed through **www.nelson.com/instructor.** Printable versions of the Test Bank in Word and PDF formats are available upon request.

- **NETA PowerPoint:** Microsoft® PowerPoint® lecture slides for every chapter have been created by Martin Restoule of Algonquin College. There is an average of 50 slides per chapter, many featuring key figures, tables, and photographs from *Automotive Technology: A Systems Approach*, Third Canadian Edition.
- **Image Library:** This resource consists of digital copies of figures, short tables, and photographs used in the book. Instructors may use these jpegs to customize the NETA PowerPoint or create their own PowerPoint presentations.
- **Instructor's Manual:** This resource was written by Dean Key of the University of Fraser Valley. It is organized according to the textbook chapters and includes instructional outlines, engagement strategies, and answers to review questions from the textbook.
- **Day One:** Day One—Prof In Class is a PowerPoint presentation that instructors can customize to orient students to the class and their text at the beginning of the course.

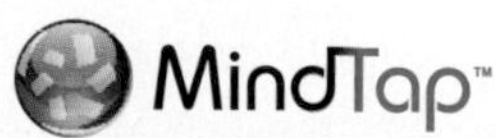

- **MindTap** for *Automotive Technology* is a personalized teaching experience with relevant assignments that guide students to analyze, apply, and elevate thinking, allowing instructors to measure skills and promote better outcomes with ease. A fully online learning solution, MindTap combines all student learning tools—readings, multimedia, activities, and assessments—into a single Learning Path that guides the student through the curriculum. Instructors personalize the experience by customizing the presentation of these learning tools to their students, even seamlessly introducing their own content into the Learning Path. Instructors can access MindTap for *Automotive Technology* at **www.nelson.com/instructor.**

STUDENT RESOURCES

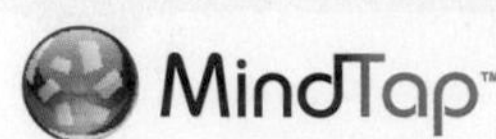

Stay organized and efficient with **MindTap**—a single destination with all the course material and study aids you need to succeed. Built-in apps leverage social media and the latest learning technology. For example:

- ReadSpeaker will read the text to you.
- Flashcards are pre-populated to provide you with a jump start for review—or you can create your own.
- You can highlight text and make notes in your MindTap Reader. Your notes will flow into Evernote, the electronic notebook app that you can access anywhere when it's time to study for the exam.
- Self-quizzing allows you to assess your understanding.

Visit **www.nelson.com/student** to start using MindTap. Enter the Online Access Code from the card included with your text. If a code card is *not* provided, you can purchase instant access at **NELSONbrain.com**.

Canadian Automotive Service Technician: Certificate of Qualification Test Preparation, 1e

- Through an exclusive arrangement with Centennial College Press, Nelson is proud to represent this outstanding exam preparation guide by Sean Bennett, Centennial College, and Dave Weatherhead, Centennial College (published by Centennial College Press; ISBN: 0-91-985261-0). This book is an essential Certificate of Qualification (C of Q) exam preparation guide for automotive service technicians in Canada. It features study strategies, scheduling, and tips for automotive service technicians ready to challenge provincial licence examinations. The provincial training standards and test reportable subjects are used as the basis for the test questions that appear in the book, and there are two complete sample tests formatted identically to the C of Q. Each test contains the exact number of questions in each reportable subject to those in a C of Q test, and a detailed answer analysis for each question is provided in the back of the book. A vital tool for any automotive service technician student in Canada, this book provides students with all they need to study effectively, strategize, and pass the Certificate of Qualification exam.

CHAPTER 1
Careers in the Automotive Industry

LEARNING OUTCOMES

- Describe the reasons why today's automotive industry is considered a global industry.
- Explain how computer technology has changed the way vehicles are built and serviced.
- Explain why the need for qualified automotive technicians is increasing.
- Describe the major types of businesses that employ automotive technicians.
- List some of the many job opportunities available to people with a background in automotive technology.
- Describe the different ways a student can gain work experience while attending classes.
- Describe the requirements for Red Seal certification of automotive technicians.

WE SUPPORT THE CANADIAN INTERPROVINCIAL STANDARDS PROGRAM

SERVICING TODAY'S VEHICLES

When the first automobile rolled down a street over 100 years ago, life changed. Only the elite owned one of these early horseless carriages, which were a sign of wealth and status. Today, an automobile is a necessity. Most Canadians would have a difficult time surviving without an automobile. We need our cars and we need the automotive industry. Each year, millions of new cars and light trucks are produced and sold in North America **(Figure 1–1)**. The automotive industry's part in the total economy of both Canada and the United States is second only to the food industry. Manufacturing, selling, and servicing these vehicles is an incredibly large, diverse, and expanding industry.

FIGURE 1–1
Ford's F-Series pickup has been the best-selling vehicle in Canada for many years.
Martin Restoule

Thirty years ago, North America's "big three" automakers—General Motors Corporation, Ford Motor Company, and Chrysler Corporation—dominated the auto industry. That is no longer true. The industry is now a global industry **(Table 1–1)**. Automakers from Japan, Korea, Germany, Sweden, and other European and Asian countries compete with North American companies for domestic and foreign sales; in fact, the two best-selling cars in North America are from Japanese car companies.

Several foreign manufacturers, such as Honda, Toyota, and BMW, operate assembly plants in Canada and the United States. Chrysler LLC declared bankruptcy in 2009 and was taken over by Fiat S.p.A., both the Canadian and American governments, and the United Auto Workers (UAW) union. This move made Chrysler a global company once again because its merger with Daimler dissolved. Fiat has since bought out the other partners. Many smaller auto manufacturers have been bought by larger companies to form larger global automobile companies. Most often, the ownership of a car company is not readily identifiable by the brand name, an example of which is Ford Motor Company; Ford brands include Ford, Lincoln, and Mazda. Many more mergers and acquisitions in

TABLE 1–1 FACTS ABOUT THE PASSENGER CARS AND LIGHT- AND MEDIUM-DUTY TRUCKS SOLD IN NORTH AMERICA (ALL FIGURES ARE APPROXIMATE)

MANUFACTURER	OWNED BY	COMMON BRANDS	COUNTRY OF ORIGIN	ANNUAL SALES (UNITS SOLD)
BMW AG	Shareholders 53% and family 47%	BMW, Mini, and Rolls-Royce	Germany	330 thousand
Chrysler Group	Fiat	Chrysler, Dodge, and Jeep	Italy and North America	1.6 million
Daimler AG	Aabar Investments 8%, Kuwait Investments 7%, Renault-Nissan 3%, and shareholders 81%	Bentley, Daimler Trucks & Buses, Mercedes Benz, and Smart	Germany	288 thousand
Fiat S.p.A.	Family 30% and shareholders 70%	Abarth, Alfa Romeo, Chrysler, Ferrari, Fiat, Lancia, and Maserati	Italy	22 thousand
Ford Motor Company	Family 40% and shareholders 60%	Ford, Lincoln, and Mazda	North America	2.2 million
Fuji Heavy Ind. Ltd.	Shareholders 81%, Toyota 16%, Suzuki 2%, and Fuji 1%	Subaru	Japan	320 thousand
Geely Automotive	Li Shu Fu 50% and shareholders 50%	Volvo	China	71 thousand
General Motors	U.S. Treasury 32%, UAW Trust 10%, Canada DIC 9%, and shareholders 49%	Buick, Chevrolet, GMC, Holden, Opel, and Vauxhall	North America	2.8 million
Honda Motor Co.	Shareholders 80% and various banks 20%	Acura and Honda	North America and Japan	1.5 million
Hyundai Motor Co.	Shareholders 74%, Hyundai Mobis 21%, and Chung Mong Koo 5%	Hyundai and Kia	Korea	1.3 million
Mazda Motor Co.	Shareholders 80%, various banks and Ford Motor Co. 20%	Mazda	Japan	270 thousand
Mitsubishi Motors	Shareholders 71% and Mitsubishi Corp. 29%	Mitsubishi	Japan	79 thousand
Nissan Motor Corp.	Shareholders 52%, Renault SA 44%, Daimler 3%, and Nissan 1%,	Nissan and Infinity	Japan	1.2 million
Porsche Auto Holding	Family 90% and Qatar Holding 10%	Porsche	Germany	29 thousand
Suzuki Motor Corp.	Shareholders 79%, Volkswagen 20%, and Fuji HD 1%	Suzuki	Japan	27 thousand
Tata Motors	Tata 35%, Indian banks 14%, and shareholders 50%	Jaguar, Rover, and Tata	India	60 thousand
Toyota Motor Corp.	Shareholders 85%, Toyota 9%, and others 6%	Daihatsu, Isuzu, Lexus, Scion, Tesla, and Toyota	Japan	1.9 million
Volkswagen AG	Porsche 54%, Lower Saxony 20%, Qatar Holding 17%, and shareholders 10%	Audi, Bentley, Bugatti, Lamborghini, and Volkswagen	Germany	500 thousand

the future will continue to create additional global automobile manufacturers. A number of vehicles are built jointly by North American and foreign manufacturers. These vehicles are built in North America to be sold here or exported to other countries. Some of these joint ventures manufacture automobiles overseas and import the vehicles into North America.

This cooperation between manufacturers and the public acceptance of imported vehicles has resulted in an extremely wide selection of vehicles from which customers may choose. This variety has also created new challenges for automotive technicians based on one simple fact: Along with the different models come variations in their systems.

The Importance of Automotive Technicians

The automobile started out as a simple mechanical beast. It moved people and things with little regard to the environment, safety, and comfort. Through the years, these concerns have been the impetus for design changes. A technical area that has affected vehicle design the most is the same one that has greatly influenced the rest of our lives—electronics. Today's automobiles are sophisticated, electronically controlled machines. To provide comfort and safety while still being friendly to the environment, these new machines use the latest developments of many different technologies—mechanical and chemical engineering, hydraulics, refrigeration, pneumatics, physics, and, of course, electronics.

Because electronics play an important part in the operation of all automotive systems, an understanding of electronics is a must for all automotive technicians. The needed level of understanding is not that of an engineer; instead, technicians need a practical understanding of electronics. In addition to mastering the mechanical skills needed to remove, repair, and replace faulty or damaged components, today's technician also must be able to diagnose and service electronic systems.

Computers and electronic devices are used to control the engine and its support systems. Because of these controls, today's automobiles use less fuel, perform better, and run more cleanly than those in the past **(Figure 1–2)**.

FIGURE 1–2

Today's cars use less fuel, perform better, and run more cleanly than they did in the past.

Martin Restoule

Electronic controls also are used in nearly all systems of an automobile. The number of electronically controlled systems on cars and trucks increases each year. There are many reasons for the heavy insurgence of electronics into automobiles. Electronics are based on electricity, and electricity moves at the speed of light. This means the operation of the various systems can be monitored and changed very quickly. Electronic components have no moving parts, are durable, do not require periodic adjustments, and are very light. All of these factors allow today's automobiles to be cleaner, safer, more efficient, and better performing than vehicles of the past.

The application of electronics has also led to the success of hybrid vehicles **(Figure 1–3)**. A hybrid vehicle has two separate sources of power. Those power sources can work together to move the vehicle, or each source can power the vehicle on its own. Today's hybrid vehicles are moved by electric motors and/or a gasoline engine. Hybrid vehicles are complex machines, and all who work on them must be properly trained.

FIGURE 1–3

The Toyota Prius is the best-selling hybrid vehicle.

Martin Restoule

The design of today's automobiles is also influenced by legislation. Throughout history, automobile manufacturers have been required to respond to new laws designed to make automobiles safer and run more cleanly. In response to these laws, new systems and components are introduced. Anyone desiring to be a good technician must regularly update his or her skills to keep up with the technology.

Legislation has not only influenced the design of gasoline-powered vehicles, it has also led to a wider use of diesel engines in passenger vehicles. By mandating cleaner diesel fuels, the laws have opened the door for clean-burning and highly efficient diesel engines. These new engines have also been fitted with electronic controls.

Some provinces have passed laws that require vehicle owners to have their cars' exhaust tested on an annual or bi-annual basis. These provinces require the vehicles to pass some form of an **Inspection/Maintenance (I/M)** test. These tests may require a shop to perform a computerized emission system scan or operate the vehicle on a chassis dynamometer through a range of loads and speeds to mimic regular driving conditions, followed by an idling period. A vehicle not achieving the specified emission limits will require maintenance of its emission control devices. That is why it is called an Inspection/Maintenance (I/M) test. This test, and the laws that tell the owners their vehicles must pass the test, affects the work of a technician. The cause of test failures must be found and corrected.

The Need for Quality Service

Vehicles will continue to become more complex; therefore, the need for good technicians will continue to grow. Currently there is a great shortage of qualified automotive technicians. This means there are, and will be, excellent career opportunities for good technicians. Good technicians are able to diagnose problems in both the simple and complex systems of today's automobiles. Of course, after the cause of a problem has been identified, the system must be properly serviced or repaired **(Figure 1–4)**.

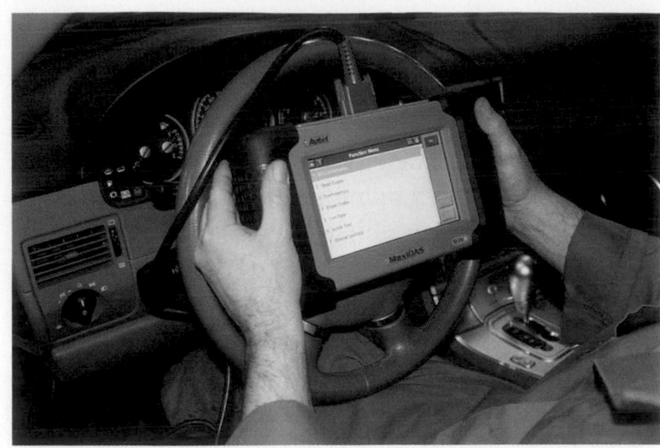

FIGURE 1–4

Good technicians are able to follow a specific manufacturer's diagnostic charts and interpret the results of diagnostic tests.

Martin Restoule

With the increase in the price of new vehicles came increased public demand for very reliable vehicles. The public also demands that when things do go wrong, they should be corrected the first time owners take the vehicle back to the dealership—they expect the problem to be "repaired right the first time." This feeling also carries through to older vehicles, those out of warranty, and those serviced by someone other than the dealership. Paying for repairs and parts that do not fix the problem is not something consumers want—nor should we expect them to. It is also not something that helps the reputation of technicians or the manufacturer of the vehicle.

The primary reason some technicians are unable to fix a particular problem is simply that they cannot find the cause of the problem. Today's vehicles are complex, which means that a great amount of knowledge and understanding is required to diagnose them. Today's technicians must have good **diagnostic skills**. Individuals who can identify and solve problems the first time a vehicle is brought into a shop are wanted by the industry. For these technicians, there are many excellent opportunities.

The high cost of electronic components and many mechanical parts has made the hit-or-miss method of repair too expensive. Too often, mechanics who do not understand how to properly troubleshoot an electronic system automatically replace its most expensive component—the computer, which often results in a very expensive wrong guess. Computers are very reliable. Normally, the cause of a problem in a computer system is the failure of an inexpensive switch or sensor, a corroded wire or poor electrical connection, or a bad mechanical part within the system.

The Need for Ongoing Service

The use of electronic controls has not eliminated the need for routine service and scheduled maintenance **(Figure 1–5)**. In fact, it has made it more important than ever. Although the computer systems can make adjustments to cover up some problems, a computer

FIGURE 1–5

Regular preventive maintenance is important for keeping electronic control systems operating correctly. A common part of preventive maintenance is changing the engine's oil and filter.

Martin Restoule

cannot replace parts that wear. A computer cannot tighten loose belts, change weak or dirty coolant, or change dirty engine oil. Simple problems such as these can set off a chain of unwanted events in an engine control system. Electronic controls are designed to help a well-maintained vehicle operate efficiently. They are not designed to repair systems.

The computer, through its control devices, may attempt to compensate for a problem by making adjustments to the engine's systems. As a result, the engine will run reasonably well, but its overall performance and efficiency will be lowered.

Various maintenance procedures usually are performed according to a schedule recommended by the vehicle's manufacturer. These maintenance procedures are referred to as **preventive maintenance (PM)** because they are designed to prevent problems. Scheduled preventive maintenance normally includes oil and filter changes, coolant and lubrication services, replacement of belts and hoses, and replacement of spark plugs, filters, and worn electrical parts **(Figure 1–6)**.

If the vehicle's owner fails to follow the recommended maintenance schedule, the vehicle's warranty might not cover problems that result. For example, if the engine fails during the period of time covered by the warranty, the warranty may not cover the engine if the owner does not have proof that the engine's oil was changed according to the recommended schedule.

WARRANTIES A **warranty** is an agreement by the auto manufacturer to have its authorized dealers repair, replace, or adjust certain parts if they become defective. This agreement normally lasts until the vehicle has been driven a certain number of kilometres, typically 60 000 km (37 000 mi.), and/or until the vehicle has been owned for a certain length of time, typically three years. In order for the warranty to cover the cost of the repair, the problem must occur within the time or kilometres covered by the warranty. There are basically two types of warranties: those offered by the manufacturer and those mandated by federal and provincial laws.

The details of most manufacturer warranties normally vary, depending on the manufacturer, vehicle model, and year. Most manufacturers provide several levels of warranty coverage. There is often a basic warranty that covers the complete vehicle (including wear items such as brakes, clutches, and even bulbs) for the first year or first 20 000 km (12 500 mi.), whichever comes first. Additional warranties may cover the engine, transmission, drive axles, powertrain, battery, safety restraint systems, the body, or other parts of the vehicle. These warranties extend the warranty time for these items. Sometimes on these warranties, the owner must pay a certain amount of money, called the **deductible**. The manufacturer pays for all repair costs over the deductible amount. Battery warranties are often prorated, which means that the amount of the repair bill covered by the warranty decreases over time. Some of these warranties are held by a third party, such as the manufacturer of tires. Although the manufacturer sold the vehicle with the tires already installed, the warranty of the tires is the responsibility of the tire manufacturer.

The two government-mandated warranties are the Federal Emissions Defect Warranty and the Federal Emissions Performance Warranty. The Federal Emissions Defect Warranty ensures that the vehicle meets all required emissions regulations and that the vehicle's emission control system works as designed and will continue to do so for two years or 40 000 km (25 000 mi.). Typically covered by this warranty are the following systems:

- Air induction
- Fuel metering
- Ignition
- Exhaust
- Positive crankcase ventilation
- Fuel evaporative control
- Emission control system sensors

The warranty does not cover malfunctions caused by accidents, floods, misuse, modifications,

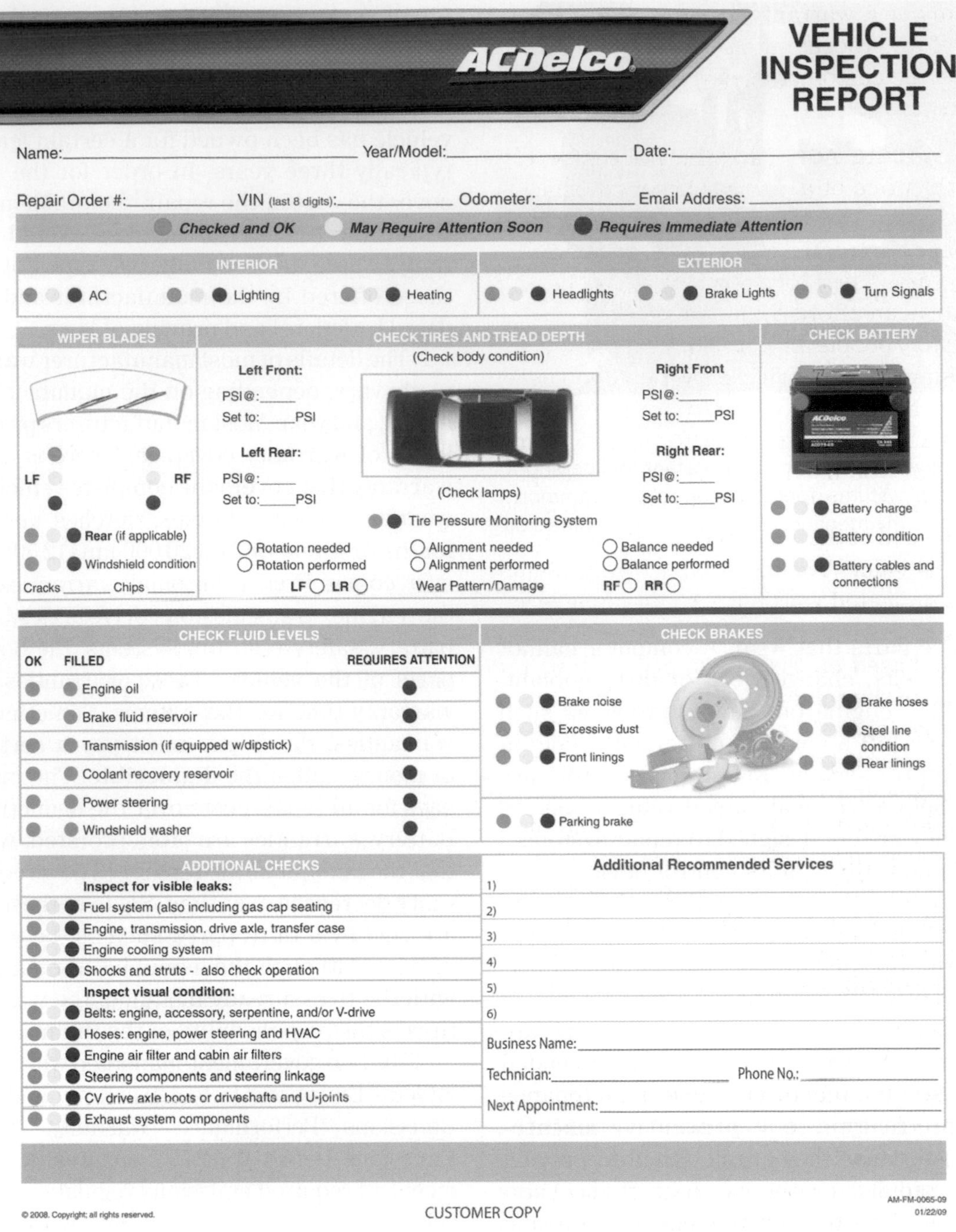

ACDelco

VEHICLE INSPECTION REPORT

Name:________ Year/Model:________ Date:________

Repair Order #:________ VIN (last 8 digits):________ Odometer:________ Email Address:________

Checked and OK | May Require Attention Soon | Requires Immediate Attention

INTERIOR: AC | Lighting | Heating

EXTERIOR: Headlights | Brake Lights | Turn Signals

WIPER BLADES

LF RF

Rear (if applicable)

Windshield condition

Cracks ______ Chips ______

CHECK TIRES AND TREAD DEPTH

(Check body condition)

Left Front: PSI@:_____ Set to:_____PSI

Right Front: PSI@:_____ Set to:_____PSI

Left Rear: PSI@:_____ Set to:_____PSI

Right Rear: PSI@:_____ Set to:_____PSI

(Check lamps)

Tire Pressure Monitoring System

Rotation needed | Alignment needed | Balance needed

Rotation performed | Alignment performed | Balance performed

LF LR | Wear Pattern/Damage | RF RR

CHECK BATTERY

Battery charge

Battery condition

Battery cables and connections

CHECK FLUID LEVELS

OK	FILLED	REQUIRES ATTENTION
	Engine oil	
	Brake fluid reservoir	
	Transmission (if equipped w/dipstick)	
	Coolant recovery reservoir	
	Power steering	
	Windshield washer	

CHECK BRAKES

Brake noise | Brake hoses

Excessive dust | Steel line condition

Front linings | Rear linings

Parking brake

ADDITIONAL CHECKS

Inspect for visible leaks:

- Fuel system (also including gas cap seating)
- Engine, transmission. drive axle, transfer case
- Engine cooling system
- Shocks and struts - also check operation

Inspect visual condition:

- Belts: engine, accessory, serpentine, and/or V-drive
- Hoses: engine, power steering and HVAC
- Engine air filter and cabin air filters
- Steering components and steering linkage
- CV drive axle boots or driveshafts and U-joints
- Exhaust system components

Additional Recommended Services

1)

2)

3)

4)

5)

6)

Business Name:________

Technician:________ Phone No.:________

Next Appointment:________

© 2008. Copyright; all rights reserved. CUSTOMER COPY AM-FM-0065-09 01/22/09

FIGURE 1–6

A typical preventive maintenance schedule.

Portions of materials contained herein have been reprinted with permission of General Motors under License Agreement #1410912.

poor maintenance, or the use of leaded fuels. The Federal Emissions Performance Warranty covers the catalytic converter and engine control module for a period of eight years or 130 000 km (80 000 mi.). The manufacturer's dealer will repair these emission-related parts covered by the warranty, free of charge. If the owner properly maintains the vehicle, and all of the emission control devices are working correctly, the vehicle should pass any emission test approved by a provincial environment ministry.

All warranty information can be found in the vehicle's Warranty Information Booklet. Whenever there are questions about the warranties, carefully read that section in the owner's manual. If you are working on a vehicle and know the part or system is covered under a warranty, make sure to tell the customer before proceeding with your work. Doing this will save the customer money, and you will earn his or her trust.

INCREASED VEHICLE AGE Like the price of everything else, the price of a new car has risen sharply. To purchase a new car, many people have taken out loans and have contracted to make car payments for up to eight years. These long-term loans are the only way that many can afford a new vehicle.

Another way people are able to afford a new car is through leasing. When you lease a vehicle, you never really own it. Normally you use the vehicle for two to five years and then give it back to the dealership. During the time you use it, you have a regular payment and are responsible for the maintenance of the vehicle. At the end of the lease, the car can be bought for its residual value. The **residual value** is simply the vehicle's projected worth at the end of the lease. Leasing is attractive to some because the monthly payments are based on the selling price minus the residual value of the vehicle.

The average age of on-the-road automobiles is approximately seven years. There are no signs that the trend of keeping cars longer will stop. Older vehicles provide most of the major repair and overhaul work performed in dealerships and independent garages. However, this does not mean that these cars are not complex. Most of the cars on the road have electronic controls, and servicing them requires training and specialized equipment.

Career Opportunities

Automotive service technicians can enjoy careers in many different types of automotive businesses. Because of the skills required to be a qualified technician, there are also career opportunities for those who do not want to repair automobiles the rest of their lives. There are also many opportunities for good technicians who want to change careers. The knowledge required to be a good service technician can open many doors of opportunity.

DEALERSHIPS New car dealerships **(Figure 1–7)** serve as the link between the vehicle manufacturer and the customer. They are privately owned businesses. Most dealerships are franchised operations, which means the owners have signed a contract with particular auto manufacturers and have agreed to sell and service their vehicles.

The manufacturer usually sets the sales and service policies of the dealership. Most warranty repair work is done at the dealership. The manufacturer then pays the dealership for making the repair. The manufacturer also provides the service department at the dealership with the training, special tools, equipment, and information needed to repair its vehicles. The manufacturer also helps the dealerships get service business. Often, manufacturers' commercials stress the importance of using their replacement parts and promote their technicians as the most qualified to work on their products.

FIGURE 1–7

Dealerships sell and service vehicles made by specific auto manufacturers.

Martin Restoule

Working for a new car dealership can have many advantages. Technical support, equipment, and the opportunity for ongoing training are usually excellent. At a dealership, you have a chance to become very skillful in working on the vehicles you service. However, working on one or two types of vehicles does not appeal to everyone. Some technicians want diversity.

INDEPENDENT SERVICE SHOPS **Independent shops** (**Figure 1–8**) may service all types of vehicles or may specialize in particular types of cars and trucks or specific systems of a car. Independent shops outnumber dealerships by 6:1. As the name states, an independent service shop is not associated with any particular automobile manufacturer. Many independent shops are started by technicians eager to be their own boss and run their own business.

An independent shop may range in size from a two-bay garage with two to four technicians, to a multiple-bay service centre with 20 to 30 technicians. A **bay** is simply a work area for a complete vehicle. The amount of equipment in an independent shop varies; however, most are well-equipped to do the work they do best. Working in an independent shop may help you develop into a well-rounded technician.

FIGURE 1–8

Full-service gasoline stations are not as common as they used to be, but they are a good example of an independent service shop.

Martin Restoule

Specialty shops specialize in areas such as engine rebuilding; transmission/transaxle overhauling; and air conditioning, brake, exhaust, cooling, emissions, and electrical work. A popular type of specialty shop is the "quick lube" shop, which takes care of the preventive maintenance of vehicles. It hires lubrication specialists who change fluids, belts, and hoses, in addition to checking certain safety items on the vehicle.

The number of specialty shops that service and repair only one or two systems of the automobile has steadily increased over the past 10 to 20 years. Technicians employed by these shops have the opportunity to become very skillful in one particular area of service.

FRANCHISE REPAIR SHOP A great number of jobs are available at service shops that are run by large companies, such as Firestone, Goodyear, Midas, and Speedy. These shops do not normally service and repair all of the systems of the automobile. However, their customers do come in with a variety of service needs. Technicians employed by these shops have the opportunity to become very proficient in many areas of service and repair.

Some independent shops may look like they are part of a franchise but are actually independent. Good examples of this type of shop are the NAPA service centres. These centres are not controlled by NAPA, nor are they franchises of NAPA. They are called NAPA service centres because the facility has met NAPA's standards of quality and the owner has agreed to use NAPA as his primary source of parts and equipment.

STORE-ASSOCIATED SHOPS Other major employers of auto technicians are the service departments of department stores **(Figure 1–9)**. Many large stores that sell automotive parts often offer certain types of automotive services, such as brake, exhaust system, and wheel and tire work.

FLEET SERVICE AND MAINTENANCE Any company that relies on several vehicles to do its business faces an ongoing vehicle service and preventive maintenance problem. Small fleets often send their vehicles to an independent shop for maintenance and repair. Large fleets, however, usually have their own preventive maintenance and repair facilities and technicians.

Utility companies (such as hydro, telephone, or cable TV), car rental companies, overnight delivery services, and taxicab companies are good examples of businesses that usually have their own service departments. These companies normally purchase their vehicles from one manufacturer. Technicians who work on these fleets have the same opportunities and benefits as technicians in a dealership. In fact, the technicians of some large fleets are authorized to do warranty work for the manufacturer. Many good career opportunities are available in this segment of the auto service industry.

FIGURE 1–9

Canadian Tire stores are a good example of large stores that sell automotive parts and offer a broad range of automotive services.

Martin Restoule

JOB CLASSIFICATIONS

The automotive industry offers numerous types of employment for people with a good understanding of automotive systems.

Service Technician

A **service technician** assesses vehicle problems, performs all necessary diagnostic tests, and competently repairs or replaces faulty components. The skills to do this job are based on a sound understanding of auto technology, on-the-job experience, and continuous training in new technology as it is introduced by auto manufacturers.

Individuals skilled in automotive service are called technicians, not mechanics. There is a good reason for this. *Mechanic* stresses the ability to repair and service mechanical systems. While this skill is still very much needed, it is only part of the technician's overall job. Today's vehicles require mechanical knowledge plus an understanding of other technologies, such as electronics, hydraulics, and pneumatics.

A technician may work on all systems of the car or may become specialized. Specialty technicians concentrate on servicing one system of the automobile, such as electrical, brakes **(Figure 1–10)**, or transmission. These specialties require advanced and continuous training in that particular field.

Shop Foreman

The **shop foreman** is the one who helps technicians with more difficult tasks and serves as the quality control expert. In some shops, this is the role of the **lead tech**. For the most part, both jobs are the same. Some shops have technician teams. On these teams, there are several technicians, each with a different level of expertise. The lead tech is sort of the shop foreman of the team. Lead techs and shop foremen have a good deal of experience and excellent diagnostic skills.

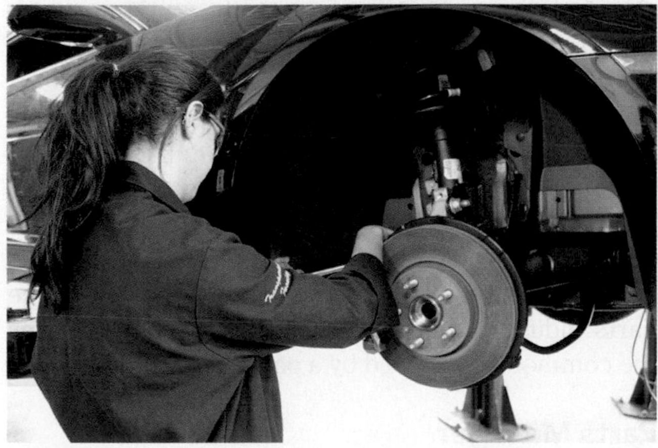

FIGURE 1–10

Specialty technicians work on only one vehicle system, such as brakes.

Martin Restoule

Service Adviser

The person who greets customers at a service centre is the **service adviser**, sometimes called a service writer or consultant. Service advisers need to have an understanding of all major systems of an automobile and be able to identify all major components and their locations. They also must be able to describe the function of each of those components and be able to identify related components. A good understanding of the recommended service and maintenance intervals and procedures is also required. With this knowledge, they are able to explain the importance and complexity of each service and are able to recommend other services.

A thorough understanding of warranty policies and procedures is also a must. Service advisers must be able to explain and verify the applicability of warranties, service contracts, service bulletins, and campaign/recalls procedures.

In most dealerships, service advisers also serve as the liaison between the customer and the technician. They have responsibility for explaining the customer's concerns and/or requests to the technician plus keeping track of the progress made by the technician so that the customer can be informed. This monitoring is also important because it impacts the completion of service on the vehicles of other customers.

Often, automotive technicians, apprentices, or students of automotive service programs realize a need to change career choices but desire to stay in the service industry. Becoming a service writer, adviser, or consultant is a good alternative. This job is good for those who have the technical knowledge but lack the desire or physical abilities to repair automobiles.

Many of the requirements for being a successful technician apply to being a successful service consultant. However, being a service consultant requires greater skill levels in customer relations, internal communication and relations, and sales. Service consultants must communicate well with customers, over the telephone or in person, in order to satisfy their needs or concerns. Most often this satisfaction involves the completion of a repair order, which contains customer information, instructions to the technicians, and a cost estimate.

Accurate estimates are not only highly appreciated by the customer, but they are also required by law in most provinces. Writing an accurate estimate requires a solid understanding of the automobile, good communication with the customers and technicians, and good reading and math skills.

Most shops use computers to generate the repair orders and estimates and to schedule the shop's workload. Therefore, having solid computer skills is an asset for service advisers.

Service Manager

The **service manager** is responsible for the operation of the entire service department at a large dealership or independent shop. Normally, customer concerns and complaints are handled by the service manager. Therefore, a good service manager has good people skills in addition to organizational skills and a solid automotive background.

In a dealership, the service manager makes sure the manufacturers' policies on warranties, service procedures, and customer relations are carried out. The service manager also arranges for technician training and keeps all other shop persons informed and working together.

Service Director

Large new car dealerships often have a **service director** who oversees the operation of the service and parts departments as well as the body shop. The service director has the main responsibility of keeping the three departments profitable. The service director coordinates the activities of these separate departments to ensure efficiency.

Many service directors begin their career as technicians. As technicians, they demonstrate a solid knowledge of the automotive field and have outstanding customer-relations skills and good business sense. The transition from technician to director typically involves promotion to various other managerial positions first.

Parts Counterperson

A **parts counterperson (Figure 1–11)** can have several different duties and is commonly called a parts person or specialist. Parts specialists are found in nearly all automotive dealerships and auto parts retail and wholesale stores. They sell auto parts directly to customers and issue materials and supplies to auto repair specialists working in automotive services facilities and body shops. A parts counterperson must be friendly, professional, and efficient when working with all customers, both on the phone and in person.

FIGURE 1–11

A parts counterperson has an important role in the operation of a store or dealership.

Martin Restoule

Depending on the parts store or department, duties may also include delivering parts, purchasing a variety of automotive parts, maintaining inventory levels, and issuing parts to customers and technicians. Responsibilities include preparing purchase orders, scheduling deliveries, assisting in the receipt and storage of parts and supplies, and maintaining contact with vendors. An understanding of automotive terminology and systems is a must for a good parts counterperson.

This career is an excellent alternative for those who know about cars but would rather not work on them. Much of the knowledge required to be a technician is also required for a parts person. However, a parts specialist requires a different set of skills. Most automotive parts specialists acquire the sales and customer service skills needed to be successful, primarily through on-the-job experience and training. They may also gain the necessary technical knowledge on the job or through educational programs and/or experience. Some provinces now offer parts counterperson apprenticeships. To better understand the world of the parts industry, refer to **Figure 1–12**, which defines the common terms used by a parts person.

Parts Manager

The **parts manager** is in charge of ordering all replacement parts for the repairs the shop performs. The ordering and timely delivery of parts is extremely important for the smooth operation of the

ACCOUNTS RECEIVABLE Money due from a customer.

ALPHANUMERIC A numbering system commonly used in parts catalogues and price listings. This system uses a combination of letters and numbers. They are placed in order starting from the left digit and working across to the right.

BACK ORDER Parts ordered from a supplier that have not been shipped to the store or shop because supplier has none in its inventory.

BILL OF LADING A shipping document acknowledging receipt of goods and stating terms of delivery.

CATALOGUING The process of looking up the needed parts in a parts catalogue.

CORE CHARGE A charge that is added when a customer buys a remanufactured part. Core charges are refunded to the customer when he or she returns a rebuildable part.

CORRECTION BULLETIN A bulletin that corrects catalogue errors due to printing errors or inaccurately assigned part numbers.

CUSTOMER RELATIONS A description of how a salesperson interacts with the customer.

DEALERS The jobber's wholesale customers, such as service stations, garages, and vehicle dealers, who install parts in their customers' vehicles.

DISCOUNT The amount of savings being offered to a customer, normally expressed as a percentage.

DISTRIBUTOR A large-volume parts-stocking business that sells to wholesalers.

FREIGHT CHARGE A charge added to special order parts to cover their transportation to the store.

GROSS PROFIT The selling price of a part minus its cost.

HIGH-VOLUME Describes a popular item, which is sold in large numbers.

INDIVIDUALLY PRICED The condition of having each part of a display priced for customer convenience.

INVENTORY The parts a store or shop has in its possession for resale.

INVENTORY CONTROL A method of determining amounts of merchandise to order based on supplies on hand and past sales of the item.

INVOICE The record of a sale to a customer.

JOBBER The owner or operator of an auto parts store usually wholesaling products to volume purchasers such as dealers, fleet owners, and businesses. They also may sell retail to do-it-yourselfers.

LIST PRICE The suggested selling price for an item.

MARGIN Same as gross profit.

MARKUP The amount a business charges for a part above the actual cost of the part.

NET PRICE A business's profit after deducting the cost of all of its merchandise and all expenses involved in operating the business.

NO-RETURN POLICY A store policy that certain parts cannot be returned after purchase. It is common to have a no-return policy on electrical and electronic parts.

ON HAND The quantity of an item that the store or shop has in its possession.

PERPETUAL INVENTORY A method of keeping a continuous record of stock on hand through sales receipts and/or invoices.

PHYSICAL INVENTORY The process whereby each part is manually counted and the number on hand is written on a form or entered into a computer.

PROFIT The amount received for goods or services above the shop's or store's cost for the part or service.

PURCHASE ORDER A form giving someone the authority to purchase goods or services for a company.

REMANUFACTURED PART A part that has been reconditioned to its original specifications and standards.

RESTOCKING FEE The fee charged by a store or supplier for having to handle a returned part.

RETAIL Selling merchandise to walk-in trade (do-it-yourselfers).

RETURN POLICY A policy regarding the return of unwanted and unneeded parts. Return policies may include restocking fees or prohibit the return of certain parts.

SELLING PRICE The price at which a part is sold. This price will vary according to the type of customer (retail or wholesale) that is purchasing the part.

SPECIAL ORDER An order placed whenever a customer purchases an item not normally kept in stock.

STOCK ORDER A process by which the store orders more stock from its suppliers in order to maintain its inventory.

STOCK ROTATION Selling the older stock on hand before selling the newer stock.

SUPERSESSION BULLETIN A bulletin sent by the parts supplier that lists part numbers that now replace (supersede) previous part numbers.

TURNOVER The number of times each year that a business buys, sells, and replaces a part.

VENDOR The supplier.

WARRANTY RETURN A defective part returned to the supplier due to failure during its warranty period.

WAREHOUSE DISTRIBUTOR The jobber's supplier who is the link between the manufacturer and the jobber.

WHOLESALE The business's price to large-volume customers.

FIGURE 1–12

Some of the common terms used by parts personnel.

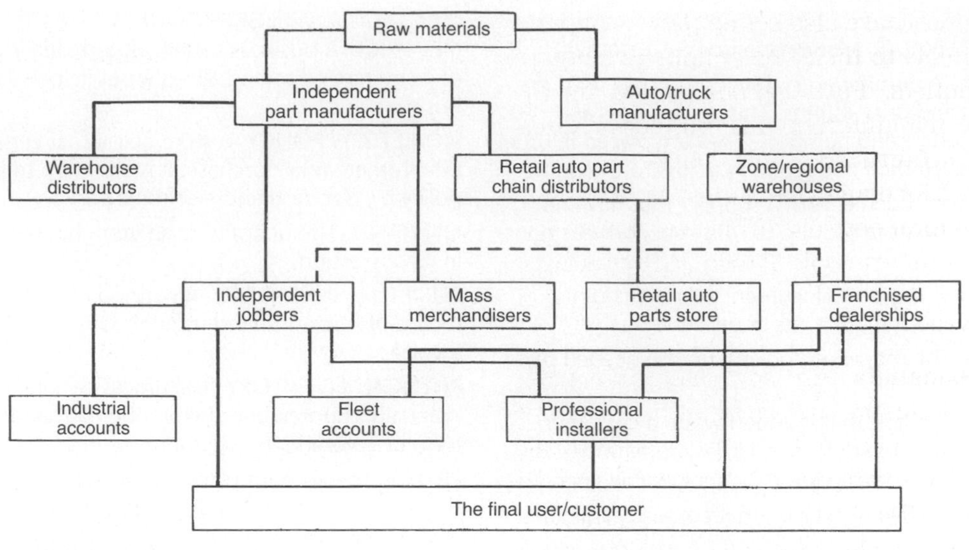

FIGURE 1–13

The auto parts supply network.

shop. Delays in obtaining parts or omitting a small but crucial part from the initial parts order can cause frustrating holdups for both the service technicians and customers.

Most dealerships and large independent shops keep an inventory of commonly used parts, such as filters, belts, hoses, and gaskets. The parts manager is responsible for maintaining this inventory.

An understanding of automotive systems and their parts, thoroughness, attention to detail, and the ability to work with people face-to-face and over the phone are essential for a parts manager.

RELATED CAREER OPPORTUNITIES

In addition to careers in automotive service, there are many other job opportunities directly related to the automotive industry.

Parts Distribution

The **aftermarket** refers to the network of businesses **(Figure 1–13)** that supplies replacement parts to independent service shops, car and truck dealerships, fleet operations, and the general public.

Vehicle manufacturers and independent parts manufacturers sell and supply parts to many warehouse distributors throughout Canada. These **warehouse distributors (WDs)** carry substantial inventories of many part lines.

Warehouse distributors serve as large distribution centres. WDs sell and supply parts to parts wholesalers, commonly known as jobbers.

Jobbers sell parts and supplies to shops and do-it-yourselfers. Jobbers often have a delivery service that gets the desired parts to a shop shortly after the shop has ordered them. Some parts stores focus on individual or walk-in customers. These businesses offer the do-it-yourselfers repair advice, and some even offer testing of old components. Selling good parts at a reasonable price and offering extra services to their customers are the characteristics of successful parts stores. Many jobbers operate machine shops that offer another source of employment for skilled technicians. Jobbers or parts stores can be independently owned and operated. They can also be part of a larger national chain **(Figure 1–14)**.

FIGURE 1–14

Many parts stores are part of a national corporation with stores located across the country.

Auto manufacturers have also set up their own parts distribution systems to their dealerships and authorized service outlets. Parts manufactured by the original vehicle manufacturer are called **original equipment manufacturer (OEM)** parts.

Opportunities for employment exist at all levels in the parts distribution network, from warehouse distributors to the counterpeople at local jobber outlets.

Marketing and Sales

Companies that manufacture equipment and parts for the service industry are constantly searching for knowledgeable people to represent and sell their products. For example, a sales representative working for an aftermarket parts manufacturer should have good knowledge of the company's products. The sales representative also works with WDs, jobbers, and service shops to make sure the parts are being sold and installed correctly. They also help coordinate training and supply information so that everyone using their products is properly trained and informed.

Other Opportunities

Other career possibilities for those trained in automotive service include automobile and truck recyclers, insurance company claims adjusters, auto body shop technicians, and trainers for the various manufacturers or instructors for an automotive program. The latter two careers require solid experience and a thorough understanding of the automobile. It is not easy being an instructor or trainer; however, passing on knowledge can be very rewarding. Undoubtedly, there is no other career that can have as much impact on the automotive service industry as that of a trainer or instructor.

TRAINING FOR A CAREER IN AUTOMOTIVE SERVICE

Those interested in a career in auto service can receive training in formal school settings—secondary, post-secondary, and vocational schools, and technical or community colleges, both private and public.

Student Work Experience

There are many ways to gain work experience while you are a student. You may already be involved in one of the following; if not, consider becoming involved in one of these possibilities.

JOB SHADOWING PROGRAM In this program, you follow an experienced technician or service writer. The primary program objective is to expose you to the "real world," to see what it takes to be a successful technician or service writer. By job shadowing, you will also become familiar with the total operation of a service department.

MENTORING PROGRAM This program has the lowest participation rate of all these programs but can be one of the most valuable. In a mentoring program, you have someone who is successful to use as an expert. Your mentor has agreed to stay in contact with you, to answer questions, and to encourage you. When you have a good mentor, you have someone who may be able to explain things a little differently than the way things were explained in class. A mentor may also be able to give real-life examples of why some of the things you need to learn are important.

COOPERATIVE EDUCATION This type of program can vary in length from approximately four months to as long as two years. Depending on the school or program offered, the work placement can be a paid or non-paid, partial- or full-day placement. Secondary-school cooperative-education programs are generally non-paid placements that are designed to expose the student to actual shop experience while obtaining school credits. The students that enrol in these programs have generally completed a number of automotive shop classes. Cooperative education programs offered by post-secondary institutions usually include paid work placements. Not only do you earn an hourly wage while you are working, you also earn credit toward your degree or diploma. While at work, you get a chance to practise and perfect what you learned in school. Your experiences at work are carefully coordinated with your experiences at school; therefore, it is called a cooperative program—industry cooperates with education.

PART-TIME EMPLOYMENT The success of this experience depends on you and your drive to learn. Working part-time will bring you good experience, some income, and a good start in getting a great full-time position after you have completed school. The best way to approach this is to find a position and service facility that will allow you to grow. You need to start at a right level and be able to take on more difficult tasks when you are ready. The most difficult challenge when working part-time is to keep up with your education while you are working. Many times, work may get in the way, but if you truly want to learn, you will find a way to fit your educational needs around your work schedule. It should be noted that in some provinces, most repairs in an automotive repair facility require the worker to be a registered apprentice or a qualified Automotive Service Technician.

Canada's Automotive Apprenticeship Program

Canada has an extremely well-defined apprenticeship program for nearly 50 trades, one of which is automotive service. The programs are based on the tasks performed by certified workers in the trade (journeyperson). The apprenticeship programs are usually four to five years in length with 85 percent of the time working in the occupation and the rest of the time spent at a school. Because most of Canada (and other countries) requires that automotive technicians be certified or licensed in the area in which they are working, participating in an apprenticeship program is mandatory and is one way that individuals can get a start in many automotive-related careers. This certification is unique to Canada and does not include ASE certification.

How **apprenticeship** works is quite simple. The apprentice is employed at a dealership or independent shop and is registered with either provincial or territorial ministries of education, labour, training, or college of trades (Ontario). Once registered, the apprentice will gain work experience and skills on the job and will be sent periodically to school for training. When the necessary on-the-job tasks are acquired and the in-school programs completed, the apprentice can challenge the Canadian Automotive Service Technician Interprovincial Standards Examination.

Provincial and territorial apprenticeship acts provide the standards and conditions of training, such as the methods of registering apprentices, curriculum, accreditation, and certification. One of the benefits of this program is that you are in the workforce earning while learning. The in-school training sessions can be offered in a number of ways. Traditional training sessions are usually in blocks of six to ten weeks in duration and are attended by the apprentice once a year for three or four years. Alternative delivery methods are offered in some provinces that allow apprentices to attend school as little as a day a week, in the evenings, on the weekends, or online. Many automobile manufacturers also present special tailored programs that offer all the mandatory apprenticeship training along with special product training on their own product line. These programs are generally referred to as modified apprenticeship programs (MAP) and usually require the apprentice to alternate between the school and the sponsoring dealer every 8 weeks for a total of 64 weeks. Examples of these programs are the Acura/Honda AHAP, Chrysler CAPS, Ford ASSET, General Motors ASEP, and Toyota T-TEP programs. Whatever apprenticeship program you choose, you start the program as a helper to an experienced technician and begin to work more on your own and on more challenging repairs as you progress through the program.

Canada's **Red Seal Program (Figure 1–15)** was designed to make it easier for skilled workers in many different trades to move across the country and obtain jobs. Through the program, apprentices who have completed their training and achieve an average of 70 percent or better on their Interprovincial Standards examination will obtain a Red Seal endorsement on their Certificate of Qualification. The Red Seal allows qualified automotive technicians to work in the field in most provinces and territories in Canada without needing to take further examinations. The Red Seal is a mark of high achievement and, in most cases, is not a requirement for employment.

For more information, go to www.red-seal.ca.

The Need for Continuous Learning

Training in automotive technology and service does not end with your certificate of qualification. Nor does the *need to read* end. A professional technician constantly learns and keeps up to date. In order to maintain your image as a professional and to keep your knowledge and skills up to date, you need to do what you can to learn new things. You need to commit yourself to lifelong learning. There are many ways you can keep up with the changing times and

We support the Red Seal Program of Canada

FIGURE 1–15
Canada's Red Seal Program is a mark of high achievement.

technology. Short courses on specific systems or changes are often available from manufacturers, training groups, associations, and schools. It is wise to participate in these classes as soon as you can. If you wait too long, other changes will have occurred, and you may have a difficult time catching up.

Besides attending classes, you can also keep updated by reading automotive magazines or the latest editions of automotive textbooks. As soon as you realize what subjects you need refreshing or updating on, respond as soon as you can. A good technician takes advantage of every opportunity to learn.

ASE CERTIFICATION

The United States–based National Institute for **Automotive Service Excellence (ASE)** has established a voluntary certification program for automotive, heavy-duty truck, auto body repair, and engine-machine-shop technicians. In addition to these programs, ASE also offers individual testing in the areas of automotive and heavy-duty truck parts, service consultant, alternate fuels, advanced engine performance, and a variety of other areas. This certification system combines voluntary testing with on-the-job experience to confirm that technicians have the skills needed to work on today's more complex vehicles. ASE recognizes two distinct levels of service capability—the automotive technician and the master automotive technician. The master automotive technician is certified by ASE in all major automotive systems. The automotive technician may have certification in only several areas.

To become ASE-certified, a technician must pass one or more tests that stress system diagnosis and repair procedures. The eight basic certification areas in automotive repair follow:

1. Engine repair
2. Automatic transmission/transaxle
3. Manual transmissions and drive axles
4. Suspension and steering
5. Brakes
6. Electrical systems
7. Heating and air conditioning
8. Engine performance (driveability)

After passing at least one exam and providing proof of two years of hands-on work experience, the technician becomes ASE-certified. Retesting is necessary every five years to remain certified. A technician who passes one examination receives an automotive technician shoulder patch. The master automotive technician patch is awarded to technicians who pass all eight of the basic automotive certification exams **(Figure 1–16)**.

FIGURE 1–16

ASE certification shoulder patches worn by (left) automobile technicians and (right) master automobile technicians.

ASE also offers advanced-level certification in some areas. The most commonly sought advanced certification for automobile technicians is the L1, or Advanced Engine Performance. Individuals seeking this certification must be certified in Electricity and Engine Performance before taking this exam.

ASE also offers specialist certifications. For example, to become a certified Undercar Specialist, you must have certification in Suspension and Steering, Brake, and Exhaust Systems (a specialty test). Certification is also available for Parts Counterperson and Service Consultants.

As mentioned, ASE certification requires that you have two years of full-time, hands-on working experience as an automotive technician. You may receive credit toward this two-year experience requirement by completing formal training in one or a combination of the following:

- High school training
- Post–high school training
- Short courses
- Apprenticeship programs

Each certification test consists of 40 to 80 multiple-choice questions. The questions are written by a panel of technical service experts, including domestic and import vehicle manufacturers, repair and test equipment and parts manufacturers, working automotive technicians, and automotive instructors. All questions are pretested and quality-checked on a national sample of technicians before they are included in the actual test. Many test questions force the student to choose between two distinct repair or diagnostic methods.

In Canada, the ASE certification program is not recognized by any ministry or level of government and is considered only as additional training to the mandatory apprenticeship training. However, some employers encourage their employees to attain ASE certification.

For further information on the ASE certification program, go to www.ase.com.

KEY TERMS

Aftermarket (p. 12)
Apprenticeship (p. 14)
Automotive Service Excellence (ASE) (p. 15)
Bay (p. 7)
Deductible (p. 5)
Diagnostic skills (p. 4)
Independent shops (p. 7)
Inspection/Maintenance (I/M) (p. 4)
Jobbers (p. 12)
Lead tech (p. 9)
Original equipment manufacturer (OEM) (p. 13)
Parts counterperson (p. 10)
Parts manager (p. 10)
Preventive maintenance (PM) (p. 5)
Red Seal Program (p. 14)
Residual value (p. 7)
Service adviser (p. 9)
Service director (p. 10)
Service manager (p. 10)
Service technician (p. 9)
Shop foreman (p. 9)
Warehouse distributors (WDs) (p. 12)
Warranty (p. 5)

SUMMARY

- The modern auto industry is a global industry involving vehicle and parts manufacturers from many countries.
- Electronic computer controls are found on many auto systems, such as engines, ignition systems, transmissions, steering systems, and suspensions. The use of electronics in automobiles is increasing rapidly.
- The increasing complexity of vehicles, the increasing age of vehicles on the road, and the need to comply with federal laws concerning emission control and mileage are three reasons the need for quality service technicians is increasing.
- Preventive maintenance is extremely important in keeping today's vehicles in good working order.
- New car dealerships, independent service shops, specialty service shops, fleet operators, and many other businesses are in great need of qualified service technicians.
- A solid background in auto technology may be the basis for many other types of careers within the industry. Some examples are parts management, sales, and marketing positions.
- Training in auto technology is available from many types of secondary, vocational, and technical schools.
- An apprenticeship is an agreement between an apprentice, an employer, and the government. Apprenticeship is an industry-based, on-the-job learning system with technical training that produces a certified journeyperson. When the specified training period is complete, the apprentice who successfully writes the Interprovincial Standards examination will then obtain his or her Certificate of Qualification.
- The Red Seal Program provides greater mobility across Canada for certified technicians. Apprentices who have completed their training and become certified technicians are able to obtain a Red Seal endorsement on their Certificates of Qualification by achieving 70 percent or higher on their Interprovincial Standards Exam.
- The National Institute for Automotive Service Excellence (ASE) certification process involves both written tests and credit for on-the-job experience. Testing is available in many areas of auto technology.

REVIEW QUESTIONS

1. Give a brief explanation of why electronics are so widely used on today's vehicles.
2. Explain the basic requirements for becoming a successful automotive technician.
3. List at least five different types of businesses that hire service technicians. Describe the types of work these businesses handle and the advantages and disadvantages of working for them.
4. Name four ways that you can gain work experience while you are a student.
5. What must an automotive service apprentice successfully complete during the on-the-job portion of his or her apprenticeship?
   a. a prescribed set of engine repair tasks
   b. a prescribed set of brake system repair tasks
   c. a prescribed set of steering system repair tasks
   d. a prescribed set of complete vehicle repair tasks

6. Where is repair work usually performed on vehicles that are still under the manufacturer's warranty?
   a. independent service shops
   b. dealerships
   c. specialty shops
   d. fleet service departments
7. Which of the following businesses perform work on only one or two automotive systems?
   a. dealerships
   b. independent service shops
   c. specialty shops
   d. fleet service departments
8. Normally, whose job is it to prepare a repair cost estimate for a customer?
   a. service manager
   b. parts manager
   c. master automotive technician
   d. service adviser
9. What automobile system has produced the greatest changes in today's automobile?
   a. the vehicle electronics
   b. the engine system
   c. the drivetrain
   d. the transmission
10. What must today's automotive technician have to become successful?
    a. an understanding of electronics
    b. the ability to repair and service mechanical systems
    c. the dedication to always be willing to learn new technology
    d. all of the above
11. Which of the following would be performed under scheduled preventive maintenance?
    a. water pump replacement
    b. transmission overhaul
    c. engine oil replacement
    d. tire rotation
12. Who determines the new car warranty conditions for the automobile's engine and drive-train systems?
    a. the selling dealership
    b. the automobile manufacturer
    c. the manufacturer's engine supplier
    d. the government
13. Who determines the new car warranty conditions for the automobile's emission control systems?
    a. the selling dealership
    b. the automobile manufacturer
    c. the manufacturer's engine supplier
    d. the government
14. What must a person do to become a certified automotive technician in Canada?
    a. complete a post-secondary automotive program
    b. complete a cooperative education program
    c. complete a secondary school automotive program
    d. complete an automotive service technician apprenticeship
15. What minimum grade must be achieved on the Certificate of Qualification Exam to obtain the Red Seal endorsement?
    a. 60%
    b. 70%
    c. 80%
    d. 90%
16. What does the term *prorated* mean in regards to the warranty of an automotive part?
    a. a full warranty for the life of the vehicle
    b. a full warranty for a set number of years
    c. a warranty that decreases over time
    d. a warranty that is determined by the government
17. Which of the following components would likely be covered under a prorated warranty?
    a. tires and batteries
    b. engine and drivetrain components
    c. steering and brake system components
    d. climate control system components
18. What are auto parts stores that sell aftermarket parts and supplies to service shops and the general public called?
    a. warehouse distributors
    b. mass merchandisers
    c. jobbers
    d. freelancers
19. What advantage does a Red Seal endorsement provide?
    a. the ability to work in most provinces without further testing
    b. the ability to work in many other countries
    c. the ability to work with automotive refrigerants
    d. the ability to work on vehicle electronics
20. What is ASE testing in Canada?
    a. Certificate of Qualification testing
    b. Interprovincial Qualification testing
    c. mandatory testing in various automotive technology areas
    d. voluntary testing in various automotive technology areas

CHAPTER 2
Workplace Skills

LEARNING OUTCOMES

- Develop a personal employment plan.
- Seek and apply for employment.
- Prepare a resumé and cover letter.
- Prepare for an employment interview.
- Accept employment.
- Understand how automotive technicians are compensated.
- Understand the proper relationship between an employer and an employee.
- Explain the key elements of on-the-job communications.
- Be able to use critical-thinking and problem-solving skills.
- Explain how you should look and act to be regarded as a professional.
- Explain how fellow workers and customers should be treated.

This chapter gives an overview of what you should do to get a job and how to keep it. The basis for this discussion is respect: respect for yourself, your employer, fellow employees, your customers, and everyone else. Also included in this discussion are the key personal characteristics required of all who seek to be successful automotive technicians and employees.

SEEKING AND APPLYING FOR EMPLOYMENT

Becoming employed, especially in the field in which you want a career, involves many steps. As with many things in life, you must be adequately prepared before taking the next step toward employment. This discussion suggests ways you can prepare and what to expect while taking these steps.

Employment Plan

An **employment plan** is nothing more than an honest appraisal of yourself and your career hopes. It includes your specific job goals, when you expect to reach them, and how your attitudes, interests, aptitudes, and skills match the requirements of your target job. An employment plan should contain your short-term goals (four to six months) and long-term goals, as well as a prioritized list of potential employers or types of employers. You may need to present your employment plan to someone while you are seeking employment, so make sure it is complete. Even if no one else will see it, you should be thorough because it will help you make good career choices and will serve as the basis for your personal marketing tools.

Think about the type of job you want and do some research to determine the requirements for that job. Evaluate yourself against those requirements and determine if you are ready for the job. Also consider the conditions in which you would work with that type of job. Ask yourself if you are willing and able to be a productive worker in those conditions. If not, find a job that is similar to your desire and pursue that type of job.

To begin the self-appraisal part of your employment plan, think about what your interests and skills are. Ask yourself the following questions:

- Why am I looking for a job?
- What specifically do I hope to gain by having a job?
- What do I like to do?
- What am I good at?
- Which of my skills would I like to use in my job?

By honestly answering these questions, you should be able to identify the jobs that will help you meet your goals. If you are just seeking a job to pay bills or buy a car and have no intention of turning this job into a career, be honest with yourself and your potential employer. If you are hoping to begin a successful career, realize you will probably start at the bottom of the ladder to success. You must also realize that how quickly you climb the ladder is your responsibility. An employer's responsibility is merely to give you a fair chance to climb it.

Honestly evaluate yourself, and your life, to determine what skills you have. Even if you have never had a job, you still have skills and talents that can be offered to an employer. Think about your life and make a list of all of the things you have learned from your school, friends, and family, and through television, volunteering, books, hobbies, and so on. You may be surprised by the number of skills you really have. Identify these skills as being either technical or personal skills.

Technical skills include things you can do well and enjoy, such as the following:

- Using a computer
- Working with tools, machines, or equipment
- Doing math problems
- Maintaining or fixing things
- Figuring out how things work
- Making things with your hands
- Working with ideas and information
- Solving puzzles or problems
- Studying or reading
- Doing experiments or researching a topic
- Expressing yourself through writing

Personal skills are also called **soft skills** and are things that are part of your personality. These are things you are good at or enjoy doing, such as the following:

- Working with people
- Caring for or helping people
- Working as a member of a team and independently
- Leading or supervising others
- Following orders or instructions
- Persuading people
- Negotiating with others

By identifying these skills, you will have created your personal skills inventory. From the inventory, you should match your skills and personality to the needs and desires of potential employers. The inventory will also come in handy when marketing yourself for a job, such as when preparing your resumé and cover letter and during an interview.

Identifying Job Possibilities

One of the things you identified in your employment plan was your preferred place to work. This may have been a specific business but probably was

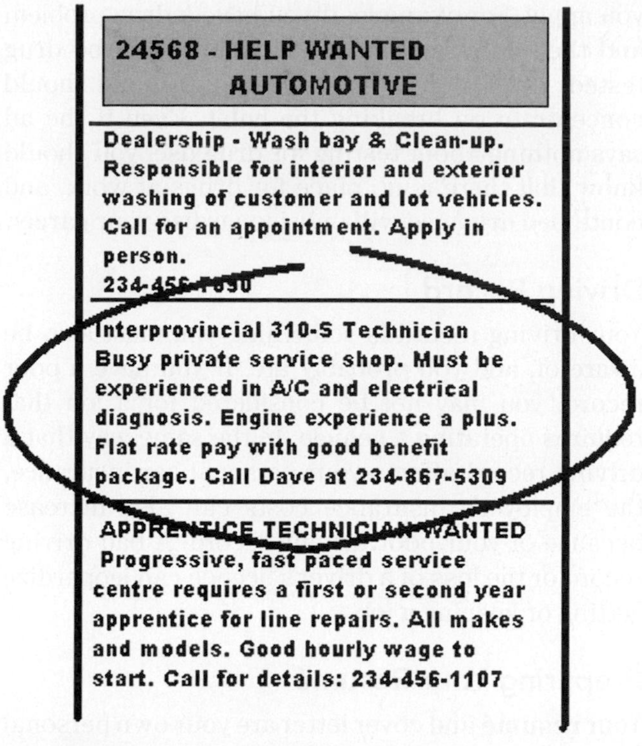

24568 - HELP WANTED AUTOMOTIVE

Dealership - Wash bay & Clean-up. Responsible for interior and exterior washing of customer and lot vehicles. Call for an appointment. Apply in person. 234-456-7890

Interprovincial 310-S Technician Busy private service shop. Must be experienced in A/C and electrical diagnosis. Engine experience a plus. Flat rate pay with good benefit package. Call Dave at 234-867-5309

APPRENTICE TECHNICIAN WANTED Progressive, fast paced service centre requires a first or second year apprentice for line repairs. All makes and models. Good hourly wage to start. Call for details: 234-456-1107

FIGURE 2–1

Check the help-wanted ads in your local newspaper for businesses that are looking for technicians.

a type of business, such as a new car dealership or independent shop. Now your task is to identify the companies in that type of business that are looking for someone. To do this, you can look through the help-wanted section in the newspaper **(Figure 2–1)**. You can also check your school's job-posting board or ask people you know who already work in the business. If there is nothing available in the business you prefer, look for openings in the type of business that was second on your priority list.

There are also job websites (such as www.workopolis.com, www.careerbuilder.ca, www.apprenticesearch.com, and www.jobbankone.com) that may have job postings you may be interested in. These sites also will guide you through the application process.

Do not limit your job search to just looking at help-wanted ads or websites. If you are interested in working in a particular shop, visit the shop and talk to people that work there. Speak with the manager about being in an automotive program and about current or forthcoming job openings. If a job is not currently available, you may be able to intern in the shop, without pay, as a way to gain experience and to be ready for when the next opening occurs.

Carefully look at the description of the job. Make sure you meet the qualifications for the job before you apply. For example, if you have a drug problem and the ad states that all applicants will be drug tested, you should not bother applying and should concentrate on breaking the habit. Even if the ad says nothing about testing for drug use, you should know that there is no place for drugs at work, and continued drug use will only jeopardize your career.

Driving Record

Your driving record is something you must also be aware of, and you probably are. If you have a poor record, you may not be considered for a job that requires operating a vehicle. In the same way that a driving record affects your personal car insurance, the employer's insurance costs can also increase because of your poor driving record. A bad driving record or the loss of a driver's licence can jeopardize getting or keeping a job.

Preparing Your Resumé

Your **resumé** and cover letter are your own personal marketing tools and may be an employer's first look at you. Although not all employers require a resumé, you should prepare one for those that do. Preparing a resumé also forces you to look at your qualifications for a job. That alone justifies having a resumé.

A resumé must be neatly written. If you do not have access to a computer, your local library probably has one available for public use.

Keep in mind that although you may spend hours writing and refining your resumé, an employer may take only a minute or two from his or her busy schedule to look it over. With this in mind, put together a resumé that tells the employer who you are in such a way that he or she wants to interview you.

A resumé normally includes your contact information, career objective, skills and/or accomplishments, work experience, and statement about references. There are different formats you can follow when designing your resumé. If you have limited work experience, make sure the resumé emphasizes your skills and accomplishments rather than work history. Even if you have no work experience, you can sell yourself by highlighting some of the skills and attributes you identified in your employment plan.

When listing or mentioning your attributes and skills, express them in a way that shows how they relate to the job you are seeking. For instance, if you practise every day at your favourite sport so that you can make the team, you may want to describe yourself as being persistent, determined, motivated, and goal-oriented. If you have ever pulled an all-nighter to get an assignment done on time, it can mean that you work well under pressure and always get the job done. Another example would be if you keep your promises and do what you say you will do, you may want to describe yourself as reliable, a person who takes commitment seriously.

Identifying your skills may be a difficult task, so have your family and/or friends help you. Keep in mind that you have qualities and skills that employers want. You need to recognize them, put them in a resumé, and tell them to your potential employer. Do not put the responsibility of figuring out who you are on the employers—tell them.

Figure 2–2 is an example of a basic resumé for an individual seeking an entry-level position as a technician.

Here are some guidelines to follow when you are designing your resumé:

- Make sure your resumé is neat, uncluttered, and easy to read.
- Use quality white paper.
- Keep it short—a maximum of two pages.
- Let the resumé tell your story, but do not try to oversell yourself.
- Use dynamic words to describe your skills and experience, such as accomplished, achieved, communicated, completed, created, delivered, designed, developed, directed, established, founded, instructed, managed, operated, organized, participated, prepared, produced, provided, repaired, and supervised.
- Choose your words carefully; remember that the resumé is a look at you.
- Make sure all information is accurate.
- Make sure the information you think is the most important stands out and is positioned near the top of the page.
- Design your resumé with a clean letter type (font) and wide margins (1.5 in. on both sides is good) so that it is easy on the eyes.
- Only list the "odd" jobs you had if they are related to the job you are applying for.
- Do not repeat information.
- Proofread the entire resumé to catch spelling and grammatical errors. If you find them, fix them and print a new, clean copy.

References

A **reference** is someone who will be glad to tell a potential employer about you and your work habits. A reference can be anyone who knows you other than a family member or close friend. Employers contact

Jack Erjavec
1234 My Street
Somewhere, NS A1B 2C3
123-456-7890

Performance-oriented student, with an excellent reputation as a responsible and hard-working achiever, seeking a position as an entry-level automotive technician in a new car dealership.

Skills and Attributes

- People oriented
- Motivated
- Committed
- Strong communication and teamwork skills
- Honest
- Reliable
- Organized
- Methodical
- Creative problem-solver
- Good hand skills

Work Experience

2012–2014 Somewhere Soccer Association (Assistant coach)

- Instructed and supervised junior team
- Performed administrative tasks as the coach required

2010–2012 Carried out various odd jobs within the community

- Washing and waxing cars, picking up children from school, raking leaves, cutting grass

Education

Somewhere Senior High School, graduated in 2011
Somewhere Community College, currently enrolled in the Automotive Technology Program

Extracurricular Activities

2012–2014 Active member of the college hockey team
2012–2014 Member of the college soccer team

Hobbies and Activities

Reading auto-related magazines and websites, going to races, video games, working on cars with family and friends.

References

Available upon request.

FIGURE 2–2

A sample of a resumé for someone who has little work experience.

references to verify or complete their picture of who you are. Make a list of three to five people, with their contact information, that you can use as references. If you do not present references to a potential employer, he or she may assume that you cannot find anyone who has anything nice to say about you. You probably will not be considered for the job.

Choose your references wisely. Teachers (past and present), coaches, and administrators of your school are good examples of who you can ask to be a reference. People you have worked for or have helped are also good references. Try also to get someone whose opinion is respected, such as a priest, minister, or elder or someone you know well who is in a high position.

Always talk to your references first, if possible, and get permission to give their names and telephone numbers to an employer. If they do not seem comfortable with giving you a reference, take the hint and move on to someone else. If someone is willing to provide you with a written reference, make several copies of the recommendation so that you can attach it to your resumé and/or the applications you fill out.

Application forms often have a section for personal references. Make sure you have your list with you when applying for a job. Your list of references does not need to be included in your resumé; merely state that references are available on request. Make sure you give copies of your resumé to your references.

Preparing Your Cover Letter

A cover letter **(Figure 2–3)** should be presented with every resumé you mail, email, fax, or personally deliver. A cover letter gives you a chance to point out exactly why you are perfect for the job. You should not send out the same cover letter to all potential employers. Adjust the letter to match the company and position you are applying for. Yes, this means a little more work, but it will be worth it.

Address the letter to the person doing the hiring. Do ***not*** use "Dear Sir," Dear Madam," or "To whom it may concern." If the job posting does not give the hiring person's name, you can normally find out to whom to send the resumé by calling the employer and asking so that you can address the cover letter correctly. You can also try checking LinkedIn (www.linkedin.com) or other business information sites to determine the name of the person conducting the hiring.

Typically, a cover letter has three paragraphs, each with a purpose. In the first one, you tell the employer you are interested in working for that company, the position you are interested in, and why. This paragraph also includes how you found out about the open position, which could be a reference to a help-wanted ad, a job posting at school, and/or a referral by someone who works for the company. Make sure this paragraph shows that you know about the company and what the job involves.

In the second paragraph, sell yourself by addressing one or two of your qualifications for the job and describing them in more detail than you do in your resumé. Make sure you expand on the material in your resumé rather than simply repeat it. Point out any special training or experience you have that directly relates to the job. When doing this, give a summary without listing the places and dates. This information is listed in your resumé, so simply refer to the resumé for details. This summary is another opportunity for you to let the employers know you understand what they do, what the job involves, and how you can help them.

The third paragraph is typically the end or closing. Make sure you thank the employer for taking the time to review your resumé, and ask him or her to contact you to make an appointment for an interview. Make sure you give a phone number where you can be reached. If you have particular times when it is best to contact you, put those times in this paragraph.

Make sure you have a clear and understandable message on your telephone's answering machine, just in case you miss a potential employer's call. Also have an organized work area around the phone so that you accurately schedule any interview appointments.

GUIDELINES FOR WRITING AN EFFECTIVE COVER LETTER Follow these guidelines while preparing and writing your cover letter:

- Address the letter to a person, not just a title. If you do not know the person's name, call the company and ask for the correct spelling of the person's name and his or her title.
- Make sure the words you use in the letter are upbeat.
- Use a natural writing style, keeping it professional but friendly.
- Try hard not to start every sentence with "I"; make some "you" statements.
- Check the letter for spelling and grammatical errors. This is a critical step!
- Type the letter on quality paper and make sure it is neat and clean.
- Make sure you sign the letter before sending it.

Contacting Potential Employers

Unless the help-wanted ad or job posting tells you otherwise, it is best to drop off your resumé and cover letter in person (preferably to the person who does the hiring). When you are doing this, make sure the employer knows who you are and the job you want. Make sure you are prepared for what happens next. You may be given an interview right then. You may be asked to fill out an application. If so, fill it out.

Before you leave, thank the employer and ask if you can call back in a few days if you do not hear from him. If you do not hear back within a week, call to make sure the employer received your resumé. If you are told that the job is filled or that no jobs are available, politely thank him for considering you. Ask if it is okay for you to stay in touch in case there is a future job opening.

Jack Erjavec
1234 My Street
Somewhere, NS A1B 2C3

March 24, 2014

Mrs. Need Someone
Service Manager, Exciting New Cars
56789 Big Dealer Avenue
Somewhere, NS A1B 2C4

Re: application for an entry-level automotive technician position

Dear Mrs. Someone:

Your ad in the March 14 edition of the *Dogpatch* for an automotive technician greatly interested me, as this position is very much in line with my immediate career objective—a career position as an automotive technician in a new car dealership. Because of the people and cars featured at your dealership, I know working there would be exciting.

I have tinkered with cars for most of my life and am currently enrolled in the Automotive Technology program at Somewhere Community College. I chose this program because you are on the advisory council and I knew it must be a good program. I have good hand skills and I work hard to be successful. My being on the varsity soccer team for four years should attest to that. I also enjoy working with people and have developed excellent communication skills. The position you have open is a perfect fit for me. A resumé detailing my skills and work experience is attached for your review.

I would appreciate an opportunity to meet with you to further discuss my qualifications. In the meantime, many thanks for your consideration, and I look forward to hearing from you soon. I can be reached by phone at 123-456-7890, most weekdays after 2:30 PM. If I am unable to answer the phone when you call, please leave a message and I will return your call as soon as I can. Thanks again.

Sincerely,

Jack Erjavec

encl.

FIGURE 2–3

An example of a cover letter that can be sent with the resumé in Figure 2–2.

Applications

An application form is a legal document that summarizes who you are. It is also another marketing tool for you. Filling out the application is the first task the employer has asked you to do, so do it thoroughly and carefully. Make sure you are prepared to fill out an application before you go. Take your own pen and a paper clip so that you can attach your resumé to the application. Make sure you have your reference list with you. When filling out the application, neatly print your answers.

Make sure you follow the directions carefully. Read through the application before you fill in the blanks; this gives you a better chance of filling it out neatly and correctly. A messy application or one with crossed-out or poorly erased information tells employers you may not care about the quality of your work.

By following the directions on the application and providing the employer with the information asked for, you are demonstrating that you have the ability to read, understand, and follow written instructions, rules, and procedures. When answering the questions in the application, be honest.

When you have completed the application, sign it and attach your cover letter and resumé to it.

Many companies use electronic applications, which may be only electronic versions of a paper application. Depending on the company, you may complete one on your own from anywhere or you may be required to complete it at the place of employment. Some companies use online applications that can be linked with Facebook, Google+, and LinkedIn accounts. This form of application may also ask you to upload a current resumé. In some cases, online applications also serve as a type of personality test, asking you value or judgment questions designed to determine what type of person you are. It is important to note that many employers will check applicants out on Facebook and other social media websites as part of their decision-making process. If you have questionable content on your Facebook page or elsewhere, you should consider removing it before beginning your job search.

The Interview

Typically, if employers are interested in you, you will be contacted to come in for an interview. This is a good sign. If they were not impressed with what they know of you so far, they probably would not ask for an interview. Knowing this should give you some confidence as you prepare for the interview.

Although an interview does not last very long, it is a time when you can either get the job or lose it. Whether or not you realize it, you have a variety of qualities you can sell.

Get ready for the interview by taking some time to learn as much as you can about the company. Think of some of the reasons the company should hire you. When doing this, think of how both of you would benefit. Think of questions you might ask the interviewer to show you are interested in the job and the business. Then make a list of questions that you think the employer might ask. Think about how you should answer each of them and practise the answers with your family and friends. Some of the more common interview questions include the following:

- What can you tell me about yourself?
- Why are you interested in the job?
- What are your strengths and weaknesses?
- If we offer you a job, what can you offer us?
- Do you have any questions about the job?

Make sure you are on time (early is good) for the interview. If you are not exactly sure how to get to the business, what the travel time is, or what types of problems you may face getting there (such as traffic jams or construction), make a trip there a couple of days before at the time of the interview. If you must be late, or if you cannot make it to the interview, call the employer as soon as possible and explain why. Ask if you can arrange for a new interview time.

Determine the days and hours you can work and when you can start to work before you go to the interview. Make sure you take your social insurance number card (SIN card), extra copies of your resumé, a list of your references and their contact information, as well as copies of any letters of recommendation you have. Do not be surprised if your interviewer takes notes during the interview. You should also take paper and a pen so that you can take notes as well.

Here are some things you should do to have a successful interview:

- Show up looking neat and professional. Wear something more formal than what you would wear on the job.
- Try to relax before the interview.
- Turn off your cellphone or leave it in your car.
- When you are greeted by the interviewer, look him or her in the eyes, introduce yourself and be ready to shake hands. Do it firmly but do not show how strong your grip is!
- Listen closely to the interviewer and look at the interviewer while he or she talks.
- Answer all questions carefully and honestly. If you do not have an immediate answer, think about it before you open your mouth. If you do not understand the question, restate the question in the way you understand it. The interviewer will then know what question you are answering.
- Never answer questions with a simple yes or no. Answer all questions with examples or explanations that show your qualities or skills.
- Market yourself but do not lie about or exaggerate your abilities.
- Show your desire and enthusiasm for the job, but try to be yourself—not too shy, not too aggressive.
- Never say anything negative about other people or past employers.
- Do not be overly familiar with the interviewer, and do not use slang during the interview, even if the interviewer does.
- Restate your interest in the job and summarize your good points at the end of the interview.

- Ask the interviewer if you can call back in a few days.

After the Interview

When you have left the business after the interview, go to a quiet place and reflect on what just took place. Think about what you did well and what you could have done better. Write these down so that you can refer to them when you are preparing for your next interview.

Within three days after the interview, contact the interviewer thanking her for her time. Make sure you remind her of your interest and qualifications. Take advantage of this additional chance to market yourself, but do not be overly aggressive when doing this. And do not beg for the job!

Remember, finding a job takes time and seldom do you land a job on your first attempt. If you do not get a job offer as a result of a first interview, do not give up. Do your best not to feel depressed or dejected. Simply realize that, although you are qualified, someone with more experience was chosen. Send a thank-you letter anyway; this may prompt the interviewer to think of you the next time a similar job becomes available.

Review your cover letter, resumé, and interview experience. Identify anything that can improve your marketing tools. Do not feel shy about asking the employer who did not hire you what you could have done better. Discuss your job hunt with family and friends who will provide support and encouragement. Explore other options. Do not rule out volunteering or job shadowing as a means of connecting with the workplace.

If you do get a job offer, do not be afraid to discuss the terms and conditions before accepting. Find out, or confirm, things such as what you will be doing, the hours you will be working, how you will be paid, and what to do when you report to work the first day. If you have any concerns, do not hesitate to share them with someone whose opinion you respect before committing yourself to the job. Do not commit to the job and then change your mind a few days later. Think seriously about the job before you accept or decline it.

ACCEPTING EMPLOYMENT

When you accept the job, you are entering into an agreement with the employer. That agreement needs to be honoured. Make sure you are ready to start to work. You need to have transportation to and from work and the required tools and clothes for the job.

You will also need a social insurance number. If you do not already have one, you need one quickly. In Canada, you can apply for a social insurance number if you are a legal citizen or if you are a permanent or temporary resident with the proper documentation and have permission to work in Canada.

To apply for a social insurance number, you must first obtain all of the required documents. A list of these documents and the application form (NAS 2120-05-04) is available at Service Canada Centres or on the Service Canada website (www.servicecanada.gc.ca). The completed forms can be delivered in person to Employment and Social Development Canada (ESDC) buildings or to Service Canada Centres or by mail. Once the forms are completed and submitted, it may take more than two weeks for you to receive a card with your number on it.

Typically before you begin to work, or at least before you get paid, you will fill out federal and provincial income tax forms. These forms give the company authorization to deduct income taxes from your wages. When you are an employee, the company must deduct those taxes. One form you will fill out is the personal exemption form. This form tells the employer how much, according to a scale, should be deducted from your pay for taxes. Depending on whether the employer offers benefits, you may have to complete a number of other related forms and possibly have your doctor complete insurance-related health forms.

Compensation

Part of your agreement with an employer is compensation. Basically, the employer agrees to pay you in exchange for your work. Along with the pay, the employer may offer benefits it pays for or provides and some that you pay part of. Make sure you understand the benefits and seek help in choosing which you should participate in.

Keep in mind that when you accept employment, you accept the terms of compensation offered to you. Do not show up on the first day of work demanding more. After you have worked, progressed on the job, and made the company money, you can ask for more.

HOURLY WAGES Most often, new or apprentice technicians are paid a fixed wage for every hour they work. The amount of pay per hour depends on the business, your skill level, and the work you will be doing. While collecting an hourly wage, you have a chance to learn the trade and the business. Time is usually spent working with a master technician or doing low-skilled jobs. As you learn more and become more productive, you can earn more. Many shops pay a good hourly rate to their productive technicians. Some have bonus plans that allow technicians to make more when they are highly productive. Nearly all service facilities for fleets pay their technicians an hourly wage.

COMMISSION When technicians are paid on a **commission** basis, they receive a minimum hourly wage plus a percentage of what the shop receives for performing various services. This pay system can work well for technicians who are employed in a shop whose business fluctuates through the year. This system, along with the *flat-rate* system (below), is often referred to as an incentive-pay system.

FLAT RATE **Flat rate** is a pay system in which a technician is paid for the amount of work, meaning labour hours he or she does. The flat-rate system favours technicians who work in a shop that has a large volume of work. Although this pay plan offers excellent wages, it is not recommended for new and inexperienced technicians.

Every conceivable service to every different model of vehicle has a flat-rate time. These times are assigned by the automobile manufacturers. The times are based on the average time it takes a team of technicians to perform the service on new vehicle models. Flat-rate times are listed in a **labour guide**, which can be a manual or be available on a computer. When you are paid on a flat-rate basis, your pay is based on that time, regardless of how long it took to complete the job.

To explain how this system works, suppose a technician is paid $20 per hour flat rate. If a job has a flat-rate time of three hours, the technician will be paid $60 for the job, regardless of how long it took to complete it. Experienced technicians beat the flat-rate time, nearly all of the time. Their weekly pay is based on the time turned in, not on the time spent. If the technician turns in 60 hours of work in a 40-hour workweek, he or she actually earned $30 each hour worked. However, if he or she turned in only 30 hours in the 40-hour week, the hourly pay is $15.

The flat-rate times from manufacturers are also used for warranty repairs. Once a vehicle gets a little older, it takes a little longer to service it. This is because dirt, rust, and other conditions make the service more difficult. Because of this, the flat-rate times for older vehicles are longer. Because non-dealership service facilities normally work on "out-of-warranty" vehicles (therefore, older vehicles), the flat-rate times are about 20 percent higher than those for a newer vehicle. These flat-rate times are given in most service information systems and flat-rate manuals published by Chilton, Motor, and Mitchell.

At times, a flat-rate technician will be paid for the amount of time spent on the job. This is commonly referred to as *straight* or *clock time*. Straight time is paid when a service procedure is not listed in the flat-rate manual and when the customer's concern requires more than normal diagnostic time.

TEAM SYSTEM The team system is a variation of the flat-rate system. The technicians on a team are paid according to the total hours the team completes. The team is comprised of A, B, and C technicians; their designations are based on their skill levels. There is normally one A technician who does advanced diagnostics. An A technician receives the highest compensation on the team. One or two B technicians are also on the team. These are technicians that can handle somewhat difficult diagnostics and most repairs. They are paid at a lower flat rate wage than the A technician. Normally, there are also two C technicians. These technicians are typically apprentices and are capable of doing normal services and minor repairs. As these technicians' skills improve, they can move up the ladder and improve their pay.

To give an example of compensation, suppose that a four-technician team turned in a total of 144 hours for the week. That means each technician will be paid for 36 hours. If the A technician earns $28 per hour, his total compensation would be $1,008. The B technician earns $22 per hour and would earn $792 dollars for the week. The two C technicians earn $18 per hour, and each would make $648.

BENEFITS Along with the pay, the employer may offer benefits, sometimes called fringe benefits. The cost of the benefits may be paid for by the business, or you may need to pay a share or all of the costs. There is no common benefit package for automotive technicians. Common benefits include the following:

- Dental, medical, and vision insurance
- Retirement plan
- Paid vacation days
- Paid sick days
- Uniforms and uniform cleaning services
- Update training

When accepting employment, make sure you understand the benefits, and seek help in choosing which benefits you should participate in.

TOTAL EARNINGS Depending on the business, you may be paid weekly, every two weeks, or twice a month. The total amount of what you earn is called your **gross pay**. This is not your take-home pay, or **net pay**. Your net pay is the result of subtracting all taxes and benefit costs from your gross pay. These deductions may include the following:

- Federal income taxes
- Provincial income taxes
- Employment insurance deduction
- Your contribution toward the Canada Pension Plan
- Uniform costs

WORKING AS AN AUTOMOTIVE TECHNICIAN

Landing a job is only the beginning of your career. Once you have the job, you will have to keep it. Your performance during the first few weeks will determine how long you will stay employed and how soon you will get a raise or a promotion. During the first weeks on the job and the rest of the time you have the job, make sure you arrive on time to work. If you are going to be late or absent, call the employer as soon as you can. Once you are at work, be cheerful and cooperative with those around you, but do not spend a lot of time talking or texting when you should be working.

Find out what is expected of you, such as how many hours you are expected to work and when breaks are allowed. Show that you are willing to learn and to help out in emergencies. Make sure you ask about anything you are not sure of, but try to think things out for yourself whenever you can.

A successful automotive technician has good training, a desire to succeed, and a commitment to be a good technician and employee. A good employee works well with others and strives to make the business successful. The required training is not just in the automotive field. Because good technicians spend a great deal of time working with electronic service information systems, good reading skills are a must. Technicians must also be able to accurately describe what is wrong to customers and the service adviser. Often these descriptions are done in writing; therefore, a technician also needs to be able to write well.

Employer–Employee Relationships

Being a good employee requires more than job skills. When you become an employee, you sell your time, skills, and efforts. In return, your employer pays you for these resources.

As part of the employment agreement, your employer also has certain responsibilities:

- *Instruction and supervision.* You should be told what is expected of you. A supervisor should observe your work and tell you if it is satisfactory and offer ways to improve your performance.
- *Clean, safe place to work.* An employer should provide a clean and safe work area and a place for personal cleanup.
- *Wages.* You should know how much you are to be paid before accepting a job. You should understand what your pay will be based on. Will you be paid by the hour, by the amount of work completed, or by a combination of these two?
- *Benefits.* When you are hired, you should be told what benefits, in addition to wages, you can expect. Fringe benefits usually include paid vacation days and employer contributions to dental, vision, drug, and retirement plans.
- *Opportunity and fair treatment.* Opportunity means you are given a chance to succeed and possibly advance within the company. Fair treatment means all employees are treated equally, without prejudice or favouritism.

On the other side of this business transaction, employees have responsibilities to their employers. Your obligations as an employee include the following:

- *Regular attendance.* A good employee is reliable. Businesses cannot operate successfully unless their workers are on the job.
- *Following directions.* As an employee, you are part of a team. Doing things your way may not serve the best interests of the company.
- *Team Membership.* A good employee works well with others and strives to make the business successful.
- *Responsibility.* Be willing to answer for your behaviour and work habits.
- *Productivity.* Remember that you are paid for your time as well as your skills, knowledge, and effort. You have a duty to be as effective as possible when you are at work.
- *Loyalty.* Loyalty is expected by any employee. Being loyal means that you act in the best interests of your employer, both on and off the job.

Communications

Employers value employees who can communicate. Effective communications include listening, reading, speaking, and writing. Communication is a two-way process. It involves sending a message and receiving a response, and possibly clarifying that the response has been received and understood. Because communication is based on a message, the role of listening and reading is receiving that message.

You should carefully follow all oral and written directions that pertain to your job. If you do not fully understand them, ask for clarification. You also need to be a good listener. Like other things in life, messages can appear to be good, bad, or have little worth to you. Regardless of how you rate the message, you should show respect to the person giving the message. Look at the person while she speaks and listen to her message before you respond. Try to fully comprehend the message by asking questions about it and gathering as many details as possible. Do not try to control the conversation; give listeners a chance to speak. ***Hint:*** Try to put yourself in the other person's shoes and listen without bias.

Obviously, when you read something, you are receiving a message without the advantage of seeing the message sender. Therefore, you must take what you read at face value. This is important because being able to read and understand the information and specifications given in service information systems is a must for automotive technicians **(Figure 2–4)**.

The purpose of speaking and writing is to send a message. Do your best to think through the words you use to convey the message. Pay attention to how the intended receiver of the information is listening and adjust your words and mannerisms accordingly. This consideration is also important when you write out your message. Think about who the message is going to, and adjust your words to match the abilities and attitudes of the reader. Also, keep in mind that more than one person may read it, so think of others' needs as well.

Proper telephone etiquette is also important. Most businesses will tell you how to answer the phone, typically involving the name of the company followed by your name. Make sure you listen carefully to the person calling. When you are the one making the call, make sure you introduce yourself and state the overall purpose of the phone call. Again, the key to proper phone etiquette is respect.

NONVERBAL COMMUNICATION In any communication, some of the true meaning is lost. In many cases, the heard message is often far different from the one intended because the words spoken are not always understood or are interpreted wrongly because of personal feelings. Therefore, it is important to realize that a major part of communication is nonverbal.

GENERAL SPECIFICATIONS	
DISPLACEMENT	1.9L
NUMBER OF CYLINDERS	1–4
BORE AND STROKE	
1 9L	82 × 88 (3.23 × 3.46)
FIRING ORDER	1-3–4–2
OIL PRESSURE (HOT 2000 RPM)	240–450 kPa (35–65 psi)
DRIVE BELT TENSION	178–311 (40–70 Lb-Ft)
CYLINDER HEAD AND VALVE TRAIN 1 2	
COMBUSTION CHAMBER VOLUME (oc)	EFI-HO 55 ± 1.6
VALUE GUIDE BORE DIAMETER	EFI 39.9 ± 0.8
Intake	13.481-13 519 mm (0.531-0.5324 in.)
Exhaust	13.481-13.519 mm (0.531-0.532 in.)
VALVE GUIDE I.D.	
Intake and Exhaust	8.063–8.094 mm (.3174–.3187 in.)
Width — Intake & Exhaust	1.75–2.32 mm (0.069–0.091 in.)
Angle	45°
Runout (T.I.R.)	0.076 mm (0.003 in.) MAX.
Bore Diameter (Insert Counterbore Diameter)	
Intake	(EFI-HO) 43.763 mm (1.723 in.) MIN.
	43.788 mm (1.724 in.) MAX.
	(EFI) 39.940 mm (1.572 in.) MIN.
	39.965 mm (1.573 in.) MAX.
Exhaust	(EFI-HO) 38.263 mm (1.506 in.) MIN.
	38.288 mm (1.507 in.) MAX.
	(EFI) 34.940 mm (1.375 in.) MIN.
	39.965 mm (1.573 in) MAX.
GASKETS SURFACE FLATNESS	0.04 mm (0.0016 in.)/26 mm (1 in.)
	0.08 mm (0.003 in.)/156 mm (6 in.)
	0.15 mm (0.006 in.) Total
HEAD FACE SURFACE FINISH	0.7/2.5 0.8 (28/100 .030)
VALVE STEM TO GUIDE CLEARANCE	
Intake	0.020–0.069 mm (0.0008–0.0027 in.)
Exhaust	0.046–0.095 mm (0.0018–0.0037 in.)
VALVE HEAD DIAMETER	
Intake	42.1–41.9 mm (1.66–1.65 in.)
Exhaust	37.1–36.9 mm (1.50–1.42 in.)
VALVE FACE RUNOUT	
LIMIT	Intake & Exhaust 05 mm (0.002 in.)
VALVE FACE ANGLE	45.6°
VALVE STEM DIAMETER (Std.)	
Intake	8.043–8.025 mm (0.3167–0.3159 in.)

FIGURE 2–4

Being able to read and understand the information and specifications given in service information systems is a must for automotive technicians.

Ford Motor Company

Nonverbal communication is a key part of sending and receiving a message. Pay attention to your nonverbal communication as well as that of others.

Nonverbal communication includes such things as body language and tone. Body language includes facial expression, eye movement, posture, and gestures. All of us read people's faces for ways to interpret what they say or feel, such as looking for a nod of a head. We also look at posture to provide insights about how the other person feels about the message. Posture can indicate self-confidence, aggressiveness, fear, guilt, or anxiety. Similarly, we look at gestures such as how they place their hands or give a handshake.

Many scholars have studied and classified body language and have defined which certain behaviours they indicate. Some divide postures into two basic groups: open/closed and forward/back.

- Open/closed is the most obvious. People with their arms folded, legs crossed, and bodies turned away are signalling that they are rejecting or are closed to messages. People fully facing you with open hands and both feet planted on the ground are saying they are open to and accepting the message.
- Forward/back indicates whether people are actively or passively reacting to the message. When they are leaning forward and pointing toward you, they are actively accepting or rejecting the message. When they are leaning back, looking at the ceiling, doodling on a pad, or cleaning their glasses, they are either passively absorbing or ignoring the message.

You can alter the meaning of words significantly by changing the tone of your voice. Think of how many ways you can say "no"; you can express mild doubt, terror, amazement, anger, and other emotions.

Solving Problems and Critical Thinking

Employers value someone who can think critically and act logically to evaluate situations. They also value employees with the ability to solve problems and make decisions. **Critical thinking** is the art of being able to judge or evaluate something without bias or prejudice. When diagnosing a problem, critical thinkers are able to locate the cause of the problem because they respond to what is known, not what is supposed.

Good critical thinkers begin their process of problem solving by careful observation of what is or what is not happening. Based on these observations, they declare something as a fact. For example, it is a fact that the right headlight does not light, and it is a fact that the left headlight does light. Based on these facts, a critical thinker is quite sure that the source of the problem is related to the right headlight and not the left one. Therefore, the focus of any testing of the system is centred on the right headlight. A critical thinker then studies the circuit and determines the test points. Prior to conducting any test, he or she knows what to test for and what the possible results are. Further, he or she knows what those results would indicate.

Critical thinkers solve problems in an orderly way and do not depend on chance or guesswork. They come to conclusions based on a sound reason. They also understand that if a specific problem exists only during certain conditions, there are a limited number of solutions. They further understand the relationship between how often the problem occurs and the probability of accurately predicting the problem. Also, they understand that one problem may cause other problems and know how to identify the connection between the problems.

Solving problems is something we do every day of our lives. Often, the problems are trivial, such as deciding what to watch on television. Other times, they are critical and demand much thought. At these times, thinking critically will really pay off. It is impossible to guarantee that critical thinking will lead to the correct decision; however, it will lead to good decisions and solutions.

DIAGNOSIS The word **diagnosis** is commonly used to define a primary duty of an automotive technician. Diagnosis is not guessing, and it is more than following a series of interrelated steps in order to find the solution to a specific problem. Diagnosis is a way of looking at systems that are not functioning properly and finding out why. Through an understanding of the purpose and operation of the system, you can accurately diagnose problems.

Service information systems and service manuals give diagnostic aids for many different systems. These are either symptom-based aids or flow charts. **Flow charts** or decision trees **(Figure 2–5)** guide you through a step-by-step process. As you answer the questions given at each step, you are told what your next step should be. Symptom-based diagnostic charts **(Figure 2–6)** focus on a solid definition of the problem and offer a list of possible causes of the problem. Sometimes the diagnostic aids are a combination of the two, a flow chart based on clearly defined symptoms.

When these diagnostic aids are not available or prove to be ineffective, good technicians conduct a visual inspection and then take a logical approach to solving the problem. This approach relies on critical thinking skills, as well as system knowledge. Logical diagnosis follows these steps:

1. **Verify that the problem exists.** After interviewing the customer, take the vehicle for a road test and try to duplicate the problem, if possible.
2. **Do some preliminary checks.** Research all available information to determine the possible causes of the problem. Try to match the exact problem with a symptoms chart, or think about what is happening and match a system or some components to the problem.

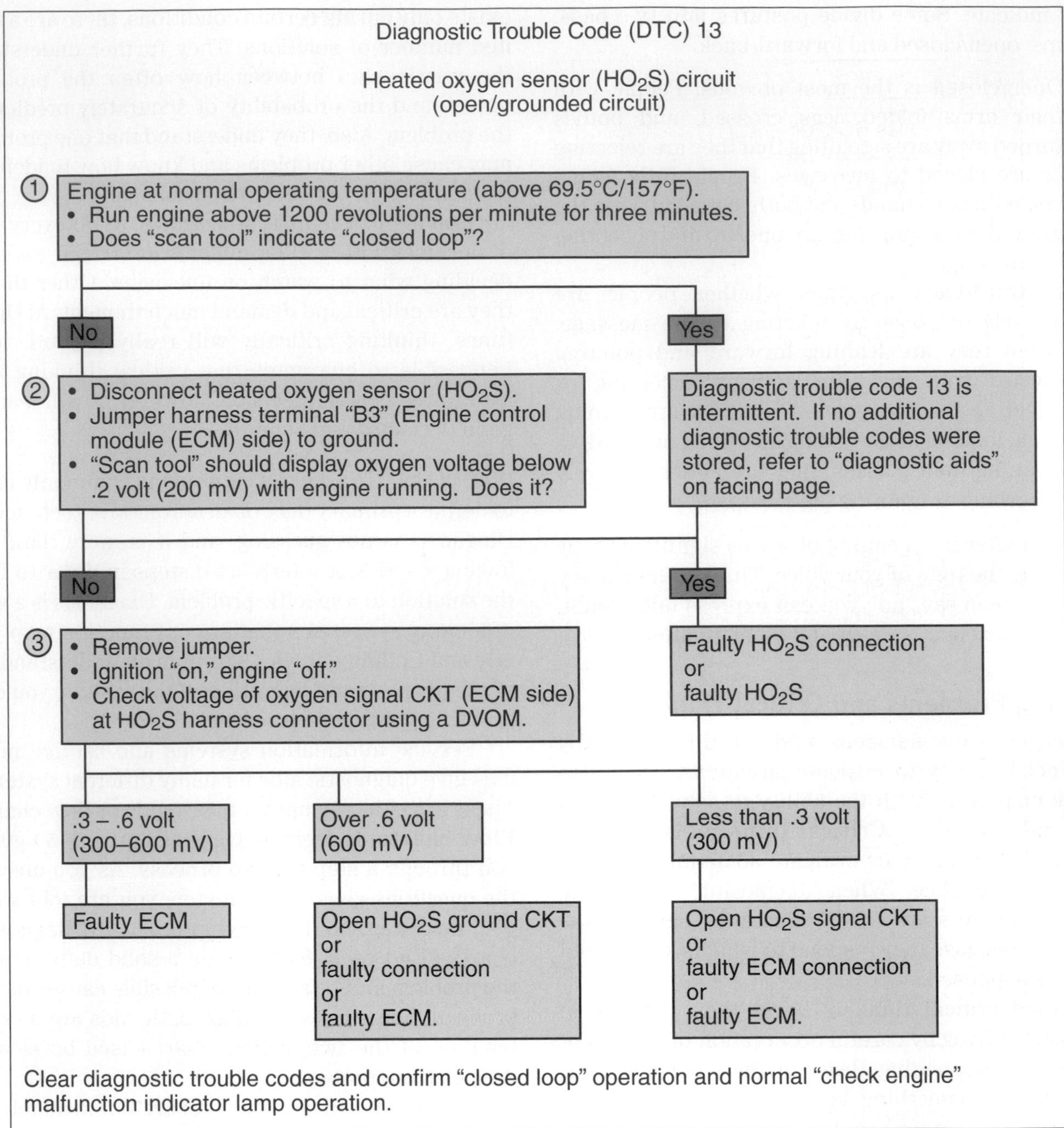

Diagnostic Trouble Code (DTC) 13
Heated oxygen sensor (HO2S) circuit
(open/grounded circuit)
① Engine at normal operating temperature (above 69.5°C/157°F).
• Run engine above 1200 revolutions per minute for three minutes.
• Does "scan tool" indicate "closed loop"?
No
Yes
② • Disconnect heated oxygen sensor (HO2S).
• Jumper harness terminal "B3" (Engine control module (ECM) side) to ground.
• "Scan tool" should display oxygen voltage below .2 volt (200 mV) with engine running. Does it?
Diagnostic trouble code 13 is intermittent. If no additional diagnostic trouble codes were stored, refer to "diagnostic aids" on facing page.
No
Yes
③ • Remove jumper.
• Ignition "on," engine "off."
• Check voltage of oxygen signal CKT (ECM side) at HO2S harness connector using a DVOM.
Faulty HO2S connection
or
faulty HO2S
.3 – .6 volt
(300–600 mV)
Over .6 volt
(600 mV)
Less than .3 volt
(300 mV)
Faulty ECM
Open HO2S ground CKT
or
faulty connection
or
faulty ECM.
Open HO2S signal CKT
or
faulty ECM connection
or
faulty ECM.
Clear diagnostic trouble codes and confirm "closed loop" operation and normal "check engine" malfunction indicator lamp operation.

FIGURE 2–5

A typical decision tree for diagnostics.

American Isuzu Motors Inc.

3. **Thoroughly define what the problem is and when it occurs.** Pay strict attention to the conditions present when the problem happens. Also pay attention to the entire vehicle; another problem may be evident to you that was not evident to the customer.
4. **Conduct a visual inspection.** Look at all possible sources of the concern.
5. **Diagnose the concerned systems.** Conduct the necessary tests to determine which components are good and which ones are not operating normally.
6. **Locate and repair the problem.** Once you have identified the cause of the concern, follow the recommended procedures for making the repair.
7. **Verify the repair.** Never assume that your work solved the original problem. Go back to step 2 and see if the problem or concern still exists, before returning the vehicle to the customer.

DTC B3138 DOOR LOCK CIRCUIT (HIGH)

STEP	ACTION	YES	NO
1	Was BCM diagnostic check performed?	Go to step 2	See BCM diagnostics
2	* Check for current DTCs with scan tool. * Disconnect power to LH door lock switch. * Does scan tool display B3138 as a current code?	Go to step 3	Go to step 6
3	* Check for current DTCs with scan tool. * Disconnect power to RH door lock switch. * Does scan tool display B3138 as a current code?	Go to step 4	Go to step 7
4	* Disconnect the brown BCM (C!) connector. * Backprobe connectors with a digital multimeter. * Measure voltage between A4 (LT BLU) and ground. * Does multimeter show battery voltage?	Go to step 5	Go to step 8
5	Locate and repair short to battery voltage in CKT 195 (LT BLU) between the BCM and the LOCK relay, or the left or right front door switches.	Go to step 9	—
6	Replace the LH power door lock switch. Is the repair complete?	Go to step 9	—
7	Replace the RH power door lock switch. Is the repair complete?	Go to step 9	—
8	* Replace the BCM. * Program the BCM with proper calibrations. * Perform the learn procedure. Is the repair complete?	Go to step 9	—
9	* Reconnect all disconnected components. * Clear the DTCs. Is the action complete?	System okay	—

FIGURE 2–6

A symptom-based diagnostic chart.

Professionalism

The key to effective communication is respect. You should respect others and others should respect you. Respect cannot be commanded; it must be earned. As a technician, you can earn the respect of others in many ways. All of these are the result of the amount of professionalism you display. Professionalism carries through all aspects of life and is best shown by having a positive attitude, displaying good behaviour, and accepting responsibility.

A successful automotive technician is a highly skilled and knowledgeable individual. A professional automotive technician has the same skills and knowledge but also dresses and acts appropriately. A professional demonstrates positive behaviours, attitudes, and responsibility.

Demonstrations of positive behaviours and attitudes include evidence of the following:

- Self-esteem, pride, and confidence
- Honesty, integrity, and personal ethics
- A positive attitude toward learning, growth, and personal health
- Initiative, energy, and persistence to get the job done
- Respect for others
- A display of initiative and assertiveness

You show your employer and the rest of the world that you are responsible by displaying these traits:

- The ability to set goals and priorities in work and personal life
- The ability to plan and manage time, money, and other resources to achieve goals
- The willingness to follow rules, regulations, and policies
- The willingness to fulfill the responsibilities of your job
- Responsibility and accountability for your decisions and actions
- The ability to apply ethical reasoning

COPING WITH CHANGE Demonstrations of your professionalism are also evident by how you react to change. Unfortunately, work environments never stay the same. New rules and regulations, supervisors, fellow employees, vehicle systems, and vehicles are all potential sources of stress. Rather than focusing on the negatives of the changes, you should identify the positives. Doing this will help you minimize stress. If you feel stress, do what you can to relieve it. Activities such as walking, running, or playing sports help reduce stress. Sharing the stress with others also helps as long as you show respect to the person you want to listen. When you are stressed, it is difficult to be a productive worker. Therefore, do your best to put things in perspective, and do some critical thinking to identify what you can do to change the situation that is causing the stress.

When the source of stress is related to your job, you may be in a situation that will not change or one that you are no longer willing to cope with. In these cases, it may be wise to find employment elsewhere.

If you decide that leaving your job is the best solution to the problem, do it professionally. Do not simply stop showing up for work or walk up to the employer and say, "I quit!" The best way to quit a job is to write a letter of resignation and present it personally to the employer. The letter should state your reason for leaving the company. When doing this, be careful not to attack the business, the employer, or fellow workers. You can simply say you are looking at other opportunities or have found another job. Badmouthing the business is a sure way of losing a good work reference, one that you may need to land your next job. The letter should also include the last day you intend to work. Your last day should be approximately two weeks after you notify the employer. At the end of your letter of resignation, thank the employer for the opportunity to work for him or her and for the experiences provided for your personal growth.

Interpersonal Relationships

As an employee, you have certain responsibilities toward your fellow workers. You will be a member of a team. Teamwork means cooperating with and caring about other workers. All members of the team (the business that employs you) should understand and contribute to the goals of the business. Keep in mind that if the business does not make money, you may not have a job in the future. Your responsibility is more than simply doing your job. You should also consider the following:

- Suggest improvements that may make the business more successful.
- Display a positive attitude.
- Work with team members to achieve common goals.
- Respect the thoughts and opinions of your fellow workers and your employer.
- Exercise give-and-take for the benefit of the business.
- Value individual diversity.
- Respond to praise or criticism in a professional way.
- Provide constructive praise or criticism.
- Channel and control emotional reactions to situations.
- Resolve conflicts in a professional way.
- Identify and react to any intimidation or harassment.

CUSTOMER RELATIONS Good customer relations begin at the technician level. Learn to listen and communicate clearly. Be polite and organized, particularly when dealing with customers on the telephone. Always be as honest as you possibly can.

Look like and present yourself as a professional, which is what automotive technicians are. Professionals are proud of what they do and they show it. Always dress and act appropriately and watch your language, even when you think no one is near.

Respect the vehicles on which you work. They are important to your customers. Always return the vehicle to the owner in a clean, undamaged condition. Remember, a car is the second-largest expense a customer has. Treat it that way. It does not matter if you like the car. It belongs to the customer; treat it respectfully.

Explain the repair process to the customer in understandable terms. Whenever you are explaining something to a customer, make sure you do this in a simple way without making the customer feel stupid. Always show customers respect and be courteous to them. Not only is this the right thing to do, but it also leads to loyal customers. Make repair estimates as precise as possible. No one likes surprises, particularly when substantial amounts of money are involved.

To help develop your customer relations skills, special Customer Care tips appear throughout this text. They contain sound advice you can share with customers on personal vehicle care. They also give you advice on how to conduct business in a courteous and professional manner.

KEY TERMS

Commission (p. 26)
Critical thinking (p. 29)
Diagnosis (p. 29)
Employment plan (p. 18)
Flat rate (p. 26)
Flow charts (p. 29)
Gross pay (p. 27)
Labour guide (p. 26)
Net pay (p. 27)
Nonverbal communication (p. 28)
Reference (p. 20)
Resumé (p. 20)
Soft skills (p. 19)

SUMMARY

- An employment plan is an honest appraisal of yourself and your career hopes.
- A reference is someone who will be glad to tell a potential employer about you and your work habits.
- A resumé and cover letter are personal marketing tools and may be the first look at you an employer has.
- A resumé normally includes your contact information, career objective, skills and/or accomplishments, work experience, education, and statement about references.
- A cover letter gives you a chance to point out exactly why you are perfect for a particular job.
- An application form is a legal document that summarizes who you are.
- Good preparation for an employment interview will result in a good experience.
- Automotive technicians are typically paid an hourly wage or on the flat-rate system.
- A successful automotive technician has good training, a desire to succeed, and a commitment to be a good technician and a good employee.
- As part of an employment agreement, your employer has certain responsibilities to you, and you have responsibilities to the employer.
- Effective communications include listening, reading, speaking, and writing.
- Nonverbal communication is a key part of sending and receiving a message and includes such things as body language and tone.
- Employers value someone who can think critically and act logically to evaluate situations and who has the ability to solve problems and make decisions.
- Diagnosis means finding the cause or causes of a problem. It requires a thorough understanding of the purpose and operation of the various automotive systems.
- Diagnostic charts found in service manuals can aid in diagnostics.
- Professionalism is best displayed by having a positive attitude, displaying good behaviour, and accepting responsibility.
- New rules and regulations, supervisors, fellow employees, vehicle systems, and vehicles are all potential sources of stress, and your professionalism will be measured by how well you cope.
- Teamwork means cooperating with and caring about other workers.
- Good customer relations is a quality of good technicians and is based on respect.

REVIEW QUESTIONS

1. What type of information should go into your employment plan?
2. What does it mean to be paid based on a flat rate?
3. What should be included in the three main paragraphs of a cover letter?
4. What is the first step in accurately diagnosing a problem?
   a. Test to pinpoint the cause of the problem.
   b. Gather as much information as you can about the problem.
   c. Isolate the problem by testing.
   d. Replace components to resolve the problem through elimination.
5. Which of the following is a desired characteristic of a good resumé?
   a. It is neat, uncluttered, and easy to read.
   b. It has a list of all jobs you have done, whether for pay or just to help someone out.
   c. It is five to six pages long.
   d. Important information is sprinkled throughout the paper.

6. Which of the following would make a good reference on a resumé?
   a. a close friend
   b. a past or present teacher
   c. a close family member
   d. your spouse
7. When your workplace becomes too stressful, how should you inform your employer that you are leaving?
   a. Stop showing up.
   b. Present a letter of resignation in person to your employer.
   c. Walk up to your employer and tell him or her, "I quit."
   d. Mail a letter of resignation to your employer.
8. When filling out a job application form, what should you do when you encounter a question that does not pertain to you?
   a. Fill it out the best you can.
   b. Leave the answer space blank.
   c. Make up something that will help you get the position.
   d. Write N/A (not applicable) in the space.
9. Which of the following is the right thing to do when you are being interviewed for a job?
   a. Show up looking neat and in the clothing you would wear on the job.
   b. Hesitate when answering questions to show that you always think things over well.
   c. To avoid saying too much or offending the interviewer, answer as many questions as you can with a simple yes or no.
   d. Move about constantly while being interviewed to show that you are eager to work.
10. Which of the following shows good nonverbal communication during an interview?
    a. folded arms or crossed legs
    b. leaning back and looking at the ceiling
    c. doodling on a pad
    d. leaning forward and nodding
11. What should you do after a job interview?
    a. Visit the business daily to check up on your job prospect.
    b. Write a letter in a couple of days, thanking the interviewer for his or her time.
    c. Phone for another interview.
    d. Sit by the phone and wait.
12. When a customer is describing a problem, what should you do to show you are interested?
    a. Sit with your arms crossed and listen.
    b. Clean your glasses to show your attention to detail.
    c. Control the conversation to show your determination.
    d. Look at the customer and listen to his or her message before responding.
13. Which one of the following is considered a soft skill?
    a. solving math problems
    b. playing video games
    c. supervising others
    d. reading
14. Which of the following is considered a technical skill?
    a. enjoying solving puzzles or problems
    b. caring for or helping people
    c. having the ability to work independently
    d. taking care to follow orders or instructions
15. If a flat-rate technician, who is paid $20 per hour, takes three hours to complete a four-hour job, what should he expect to be paid?
    a. $20 **b.** $40
    c. $60 **d.** $80
16. Unless otherwise stated, what is the best way of submitting your resumé to a prospective employer?
    a. by mail **b.** in person
    c. by email **d.** by fax
17. What government department issues social insurance numbers (SIN)?
    a. Immigration Canada
    b. Consumer and Corporate Affairs
    c. Employment and Social Development Canada
    d. Ministry of Training, Colleges and Universities
18. What is the recommended pay plan for new and inexperienced technicians?
    a. straight time **b.** piece work
    c. flat rate **d.** commission

CHAPTER 3
Working Safely in the Shop

LEARNING OUTCOMES

- Understand the importance of safety and accident prevention in an automotive shop.
- Explain the basic principles of personal safety, including protective eye wear, clothing, gloves, shoes, and hearing protection.
- Explain the procedures and precautions for safely using tools and equipment.
- Explain the precautions that need to be followed to safely raise a vehicle on a lift.
- Explain what should be done to maintain a safe working area in a shop, including running the engines of vehicles in the shop and venting the exhaust gases.
- Describe the purpose of the laws concerning hazardous wastes and materials, including the WHMIS right-to-know legislation.
- Describe your rights, as an employee and/or student, to have a safe place to work.

Working on automobiles can be dangerous. It can also be fun and very rewarding. To keep the fun and rewards rolling in, you need to prevent accidents by working safely. In an automotive repair shop, there is great potential for serious accidents, simply because of the nature of the business and the equipment used. When there is carelessness, the automotive repair industry can be one of the most dangerous occupations.

Shop accidents can cause serious injury, temporary or permanent disability, and death. Think about these facts:

- Vehicles, equipment, and many parts are very heavy; their weight can cause severe injuries.
- Fan blades, drive belts, and rotating components pose severe cutting and pinching hazards.
- Many parts of a car become very hot and can cause severe burns.
- High fluid pressures can build up inside the cooling system, fuel system, or battery; these can spray dangerous fluids on you and especially into your eyes.
- Batteries contain highly corrosive acids and potentially explosive gases; these can cause bad skin burns or blindness.
- Fuels and commonly used cleaning solvents are flammable.
- Exhaust fumes are poisonous and can be deadly.
- During some repairs, technicians can be exposed to harmful dust particles and vapours that can cause chronic or terminal diseases.

All of these can be enough to scare you away from working on cars. However, the chances of you being injured while working on a car are very small if you learn to work safely and use common sense. Shop safety is the responsibility of everyone in the shop—you, your fellow students or employees, and your employer or instructor. Everyone must work together to protect the health and welfare of all who work in the shop.

This chapter contains basic safety guidelines concerning personal, work area, tool and equipment, and hazardous material safety. In addition to what is in this chapter, special Warning boxes are included throughout this book to alert you to situations in which carelessness could result in personal injury. Finally, when working on cars, always follow safety guidelines given in electronic service information systems, service manuals, and other technical literature. They are there for your protection.

PERSONAL SAFETY

Personal safety simply involves those precautions you take to protect yourself from injury. Wear personal protective equipment (PPE), dress appropriately, work professionally, and handle tools and equipment correctly. PPE are items such as safety glasses, uniforms, and safety boots or shoes that help protect you while working in the shop.

Personal Safety Precautions

EYE PROTECTION Your eyes can become infected or permanently damaged by many things in a shop.

Some procedures, such as grinding, result in tiny particles of metal and dust that are thrown off at very high speeds. These metal and dirt particles can easily get into your eyes, causing scratches or cuts on your eyeball. Pressurized gases and liquids escaping a ruptured hose or hose fitting can spray a great distance. If these chemicals get into your eyes, they can cause blindness. Dirt and sharp bits of corroded metal can easily fall down into your eyes while you are working under a vehicle.

Eye protection should be worn whenever you are exposed to these risks. To be safe, you should wear **safety glasses** whenever you are working in the shop. There are many types of eye protection available **(Figure 3–1)**. To provide adequate eye protection, safety glasses have lenses made of safety glass or plastic and are usually identified by a "CSA Z94.3," "CSA," or the American "ANSI Z87" rating. They also offer some sort of side protection. Regular prescription glasses do not offer sufficient protection and should not be worn as a substitute for safety glasses. When prescription glasses are worn in a shop, they should be fitted with side shields.

Wearing safety glasses at all times is a good habit to get into. To help develop this habit, wear safety glasses that fit well and feel comfortable.

Some procedures may require that you wear other eye protection in addition to safety glasses. For example, when you are working around air-conditioning systems, you should wear splash goggles, and when cleaning parts with a pressurized spray, you should wear a face shield. The face shield not only gives added protection to your eyes, but it also protects the rest of your face.

EYE FIRST AID If chemicals such as battery acid, fuel, or solvents get into your eyes, flush them immediately and continuously with clean water. Have someone call a doctor and get medical help immediately.

Many shops have eye wash stations **(Figure 3–2)** or safety showers that should be used whenever you

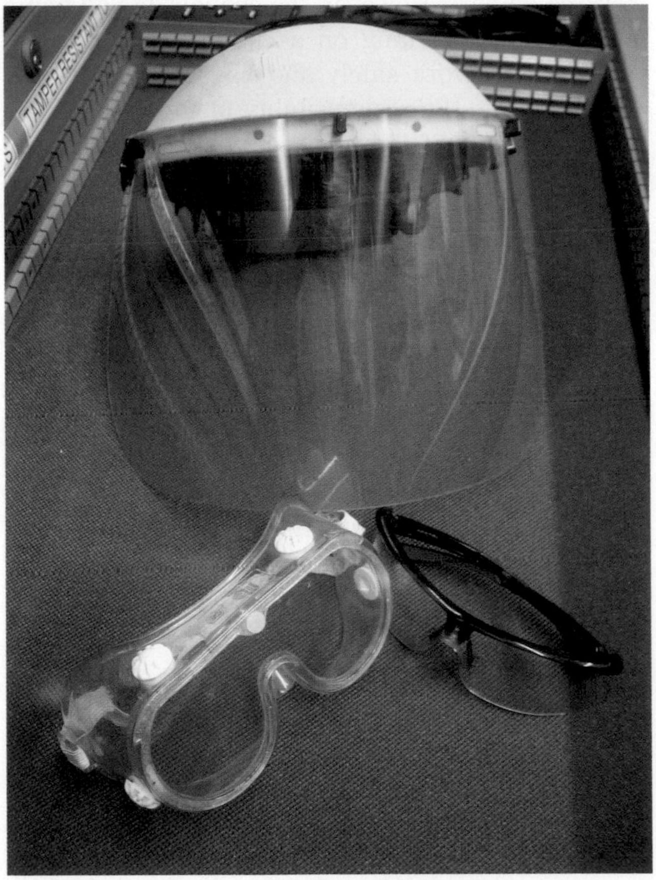

FIGURE 3–1

Types of eye protection include a face shield (top), safety or splash goggles (left), and safety glasses (right).

Martin Restoule

FIGURE 3–2

A typical eye wash station.

Martin Restoule

or someone else has been sprayed or splashed with a chemical.

CLOTHING Clothing that hangs out freely, such as shirt tails, can create a safety hazard and cause serious injury. Nothing you wear should be allowed to dangle in the engine compartment or around equipment. Shirts should be tucked in and buttoned, and long sleeves buttoned or carefully rolled up. Your clothing should be well fitted and comfortable but made with strong material. Some technicians prefer to wear coveralls or shop coats to protect their personal clothing. Your work clothes should offer you some protection but should not restrict your movement.

Keep your clothing clean. If you spill gasoline or oil on yourself, change that item of clothing immediately. Oil against your skin for a prolonged period can produce rashes or other allergic reactions. Gasoline can irritate cuts and sores as well as make you a flammable hazard.

HAIR AND JEWELLERY Long hair and loose, hanging jewellery can create the same type of hazard as loose-fitting clothing. They can get caught in moving engine parts and machinery. If you have long hair, tie it back or tuck it under a cap.

Never wear rings, watches, bracelets, or neck chains. These can easily get caught in moving parts and cause serious injury. Metal items worn by automotive technicians can also short an electronic circuit, can become very hot, and may cause severe burns to skin and body parts to which they are attached.

SHOES Automotive work involves the handling of many heavy objects, which can be accidentally dropped on your feet or toes. Always wear steel or nonmetallic composite, safety-toed shoes that are made of leather or a similar material, or boots with nonslip soles. Safety shoes can give added protection to your feet. Sport shoes, street shoes, and sandals are inappropriate in the shop.

GLOVES Good hand protection is often overlooked. Gloves are worn to protect your hands from disease and injury, and to keep them clean. Many different types of gloves are worn by technicians. A well-fitted pair of heavy work gloves or "mechanics" gloves should be worn during operations such as grinding and welding or when handling high-temperature components. Polyurethane or vinyl gloves should be worn when handling strong and dangerous caustic chemicals. These chemicals can easily burn your skin. Latex surgical-type and nitrile gloves are worn as protection against disease to keep grease from building up under and around your fingernails. Latex gloves are more comfortable to wear but weaken when they are exposed to gas, oil, and solvents. Nitrile gloves are not as comfortable as latex gloves, but they are not affected by gas, oil, and solvents. Your choice of hand protection should be based on what you are doing.

DISEASE PREVENTION When you are ill with something that may be contagious, see a doctor and do not go to work or school until the doctor says there is little chance of someone else contracting the illness from you. Doing this will protect others, and if others do this, you will be protected.

You should also be concerned with and protect yourself and others from blood-borne pathogens. **Blood-borne pathogens** are pathogenic microorganisms that are present in human blood and can cause disease in humans. These pathogens include, but are not limited to, hepatitis B virus (HBV) and human immunodeficiency virus (HIV). For everyone's protection, any injury that causes bleeding should be dealt with as a threat to others. You should avoid contact with the blood of another. If you need to administer some form of first aid, make sure you wear hand protection before you do so. You should also wear gloves and other protection when handling the item that caused the cut. Most importantly, like all injuries, report the accident to your instructor or supervisor.

EAR PROTECTION Exposure to very loud noise levels for extended periods of time can lead to hearing loss. Air wrenches, engines running under a load, and vehicles running in enclosed areas can all generate annoying and harmful levels of noise. Simple earplugs or earphone-type protectors should be worn in environments that are constantly noisy.

RESPIRATORY PROTECTION It is not uncommon for a technician to work with chemicals that have toxic fumes. **Respirators** or respiratory masks should be worn whenever you are exposed to toxic fumes or excessive amounts of dust. Cleaning parts with solvents and painting are examples of when respiratory masks should be worn.

High-efficiency particulate air (HEPA) filtration masks should also be worn when handling parts that have asbestos dust on them or when handling hazardous materials. The proper handling of these materials is covered in great detail later in this chapter.

LIFTING AND CARRYING Knowing the proper way to lift heavy objects is important. You should also use back-protection devices when you are lifting a heavy object. Always lift and work within your ability and ask others to help when you are not sure whether you can handle the size or weight of an object. Even small, compact parts can be surprisingly heavy or unbalanced. Think about how you are going to lift something before beginning. When lifting any object, follow these steps:

1. Place your feet close to the object. Position your feet so that you will be able to maintain a good balance.
2. Keep your back and elbows as straight as possible. Bend your knees until your hands reach the best place to get a strong grip on the object **(Figure 3–3)**.
3. If the part is in a cardboard box, make sure the box is in good condition. Old, damp, or poorly sealed boxes will tear, and the part will fall out.
4. Firmly grasp the object or container. Never try to change your grip as you move the load.
5. Keep the object close to your body and lift it up by straightening your legs. Use your leg muscles, not your back muscles.
6. If you must change your direction of travel, never twist your body. Turn your whole body, including your feet.
7. When placing the object on a shelf or counter, do not bend forward. Place the edge of the load on the shelf and slide it forward. Be careful not to pinch your fingers.
8. When setting down a load, bend your knees and keep your back straight. Never bend forward. This strains the back muscles.
9. When lowering something heavy to the floor, set the object on blocks of wood to protect your fingers.

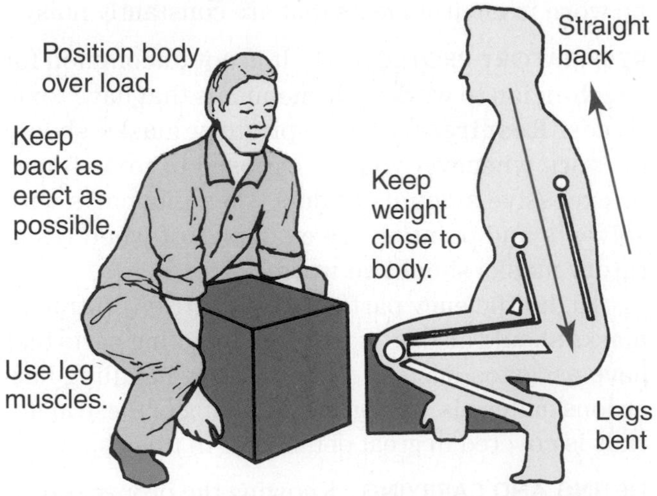

FIGURE 3–3
Use your leg muscles—*never* your back—to lift heavy objects.

Professional Behaviour

Accidents can be prevented simply by the way you act. The following are some guidelines to follow while working in a shop. This list does not include everything you should or should not do; it merely gives you some things to think about.

- Never smoke while working on a vehicle or while working in the shop.
- To prevent serious burns, keep your skin away from hot metal parts such as the radiator, exhaust manifold, tailpipe, catalytic converter, and muffler.
- Always disconnect electric engine cooling fans when working around the radiator. Many of these will turn on without warning and can easily chop off a finger or hand. Make sure you reconnect the fan after you have completed your repairs.
- When working with a hydraulic press, make sure the pressure is applied in a safe manner. It is generally wise to stand to the side when operating the press.
- Properly store all parts and tools by putting them away in a place where people will not trip over them. This practice not only cuts down on injuries but also reduces time wasted looking for a misplaced part or tool.
- Keep your work area clean and uncluttered. Make sure you clean up all spills before continuing to work.

TOOL AND EQUIPMENT SAFETY

An automotive technician must adhere to the following shop safety guidelines when working with tools and equipment.

Hand Tool Safety

Careless use of simple hand tools, such as wrenches, screwdrivers, and hammers, causes many shop accidents that could be prevented. Keep in mind the following tips when using hand tools:

- Keep all tools grease-free. Oily tools can slip out of your hand, causing broken fingers or at least cut or skinned knuckles.
- Inspect your tools for cracks, broken parts, or other dangerous conditions before you use them. Never use broken or damaged tools.
- Hand tools should be used only for the purpose they were designed for. Use the right tool for the job. Never use a wrench or pliers as a hammer; also never use screwdrivers as chisels.

- Make sure the tool is of professional quality.
- When using a wrench, always pull it toward you; do not push it away from you. When using an adjustable wrench; pull the wrench so that the force of the pull is on the nonadjustable jaw.
- Always use the correct size of wrench.
- Use a box-end or socket wrench whenever possible.
- Do not use deep-well sockets when a regular size socket will work. The longer socket develops more twist torque and tends to slip off the fastener.
- When using an air impact wrench, always use impact sockets.
- Never use pliers to loosen or tighten a nut; use the correct wrench.
- Always be sure to strike an object with the full face of the hammerhead.
- Always wear safety glasses when using a hammer and/or chisel.
- Never strike two hammer heads together.
- Be careful when using sharp or pointed tools.
- Do not place sharp tools or other sharp objects in your pockets.
- If a tool is supposed to be sharp, make sure it is sharp. Dull tools can be more dangerous than sharp tools.
- Use knives, chisels, and scrapers in a motion that will keep the point or blade moving away from your body.
- Always hand a pointed or sharp tool to someone else with the handle toward the person receiving the tool.

Power Tool Safety

Many shops have areas specifically marked as special safety areas **(Figure 3–4)**. These safety areas or zones often house equipment such as bench grinders, solvent tanks, welding equipment, and drill presses, which present special hazards in the shop. Working within or even near these areas often requires additional PPE over and above standard safety glasses, uniforms, and boots. For example, when using a bench grinder, a full-face shield should be used to prevent debris from flying into your face. Be sure to identify what forms of PPE are necessary when working in and around these areas.

Power tools are operated by an outside source of power, such as electricity, compressed air, or hydraulic pressure. Safety around power tools is very important. Serious injury can result from carelessness. Always wear safety glasses when using power tools.

FIGURE 3–4

Some areas of the shop require additional safety precautions.

If the tool is electrically powered, make sure it is properly grounded. Check the wiring for cracks in the insulation, as well as for bare wires, before using it. Also, when using electrical power tools, never stand on a wet or damp floor. Disconnect the power source before doing any work on the machine or tool. Before plugging in any electric tool, make sure its switch is in the OFF position. When you have finished using the tool, turn it off and unplug it. Never leave a running power tool unattended.

When using power equipment on a small part, never hold the part in your hand. Always mount the part in a bench vise or use vise grip pliers. Never try to use a machine or tool beyond its stated capacity or for operations requiring more than the rated power of the tool.

When working with larger power tools, such as a bench or floor grinding wheel, check the machine and the grinding wheels for signs of damage before using them. If the wheels are damaged, they should be replaced and not used. Check the speed rating of the wheel to make sure it matches the speed of the machine. Never spin a grinding wheel at a speed higher than it is rated for. Be sure to place all safety guards in position **(Figure 3–5)**. A safety guard is a protective cover that is placed over a moving part. Although the safety guards are designed to prevent injury, you should still wear safety glasses and/or a face shield while using the machine. Make sure there are no people or parts around the machine before starting it. Keep your hands and clothing away from the moving parts. Maintain a balanced stance while using the machine.

FIGURE 3–5

A bench grinder with its safety shields in place.

Martin Restoule

Compressed Air Equipment Safety

Tools that use compressed air are called **pneumatic tools**. Compressed air is used to inflate tires, apply paint, and drive tools. Compressed air can be dangerous when it is not used properly.

When using compressed air, safety glasses and/or a face shield should be worn. Particles of dirt and pieces of metal, blown by the high-pressure air, can penetrate your skin or get into your eyes. Some provinces require air lines to be regulated down to a maximum of 210 kPa (30 psi) when connected to an air gun. Only specially designed or vented air guns can be connected to an unregulated air source.

Before using a pneumatic tool, check all hose connections for leaks and the air line for damage. Always hold an air nozzle securely when starting or shutting off the compressed air. A loose nozzle can whip suddenly and cause serious injury. Never point an air nozzle at anyone. Never use compressed air to blow dirt from your clothes or hair or use compressed air to clean the floor or workbench. Also, never spin bearings with compressed air. If the bearing is damaged, one of the steel balls or rollers might fly out and cause serious injury.

Finally, pneumatic tools must always be operated at the pressure recommended by the manufacturer.

Lift Safety

Always be careful when raising a vehicle on a lift or a hoist. Adapters and hoist plates must be positioned correctly on lifts to prevent damage to the underbody

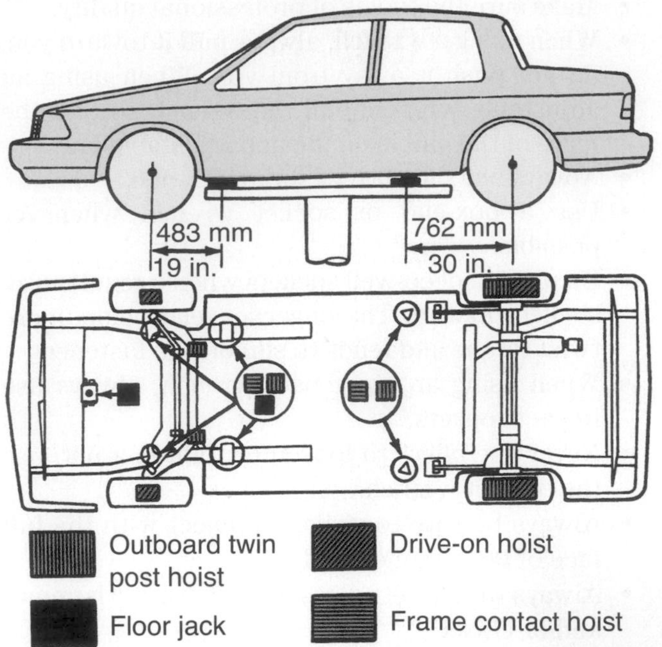

FIGURE 3–6

Hoisting and lifting points for a typical unibody car.

Chrysler Group LLC

of the vehicle. There are specific lift points. These points allow the weight of the vehicle to be evenly supported by the adapters or hoist plates. The correct lift points can be found in the vehicle's service information. **Figure 3–6** shows typical locations for unibody and frame cars. These diagrams are for illustration only. Also, always follow the recommended instructions for operating a particular lift.

Once the lift supports are properly positioned under the vehicle, raise the lift until the supports contact the vehicle. Then, check the supports to make sure they are in full contact with the vehicle. Shake the vehicle to make sure it is securely balanced on the lift, then raise the lift to the desired working height.

WARNING!

Before working under a car, make sure the lift's locking device is engaged.

The Automotive Lift Institute (ALI) is an association concerned with the design, construction, installation, operation, maintenance, and repair of automotive lifts. The organization's primary concern is safety. Every lift approved by ALI has the label shown in **Figure 3–7**. It is a good idea to read through the safety tips included on that label before using a lift.

AUTOMOTIVE LIFT

SAFETY TIPS

Post these safety tips where they will be a constant reminder to your lift operator. For information specific to the lift, always refer to the lift manufacturer's manual.

1. Inspect your lift daily. Never operate if it malfunctions or if it has broken or damaged parts. Repairs should be made with original equipment parts.
2. Operating controls are designed to close when released. Do not block open or override them.
3. Never overload your lift. Manufacturer's rated capacity is shown on nameplate affixed to the lift.
4. Positioning of vehicle and operation of the lift should be done only by trained and authorized personnel.
5. Never raise vehicle with anyone inside it. Customers or by-standers should not be in the lift area during operation.
6. Always keep lift area free of obstructions, grease, oil, trash, and other debris.
7. Before driving vehicle over lift, position arms and supports to provide unobstructed clearance. Do not hit or run over lift arms, adapters, or axle supports. This could damage lift or vehicle.
8. Load vehicle on lift carefully. Position lift supports to contact at the vehicle manufacturer's recommended lifting points. Raise lift until supports contact vehicle. Check supports for secure contact with vehicle. Raise lift to desired working height. CAUTION: If you are working under vehicle, lift should be raised high enough for locking device to be engaged.
9. Note that with some vehicles, the removal (or installation) of components may cause a critical shift in the center of gravity and result in raised vehicle instability. Refer to the vehicle manufacturer's service manual for recommended procedures when vehicle components are removed.
10. Before lowering lift, be sure tool trays, stands, etc. are removed from under vehicle. Release locking devices before attempting to lower lift.
11. Before removing vehicle from lift area, position lift arms and supports to provide an unobstructed exit (See Item #7).

These "Safety Tips," along with "Lifting it Right," a general lift safety manual, are presented as an industry service by the Automotive Lift Institute. For more information on this material, write to: ALI, P.O. Box 1519, New York, NY 10101.

Look For This Label on all Automotive Service Lifts.

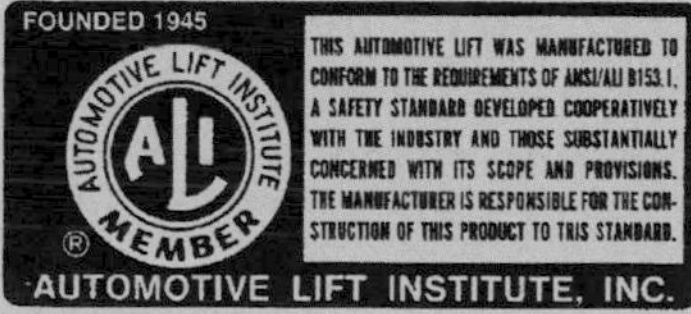

FOUNDED 1945
AUTOMOTIVE LIFT INSTITUTE
ALI
MEMBER
THIS AUTOMOTIVE LIFT WAS MANUFACTURED TO CONFORM TO THE REQUIREMENTS OF ANSI/ALI B153.1, A SAFETY STANDARD DEVELOPED COOPERATIVELY WITH THE INDUSTRY AND THOSE SUBSTANTIALLY CONCERNED WITH ITS SCOPE AND PROVISIONS. THE MANUFACTURER IS RESPONSIBLE FOR THE CONSTRUCTION OF THIS PRODUCT TO THIS STANDARD.
AUTOMOTIVE LIFT INSTITUTE, INC.

FIGURE 3–7

Automotive lift safety tips.

Automotive Lift Institute

Jack and Jack Stand Safety

A vehicle can be raised by a hydraulic jack **(Figure 3–8)**. A handle on the jack is moved up and down to raise a vehicle, and a valve is turned to release the hydraulic pressure in the jack to lower it. The jack has a lifting pad, which must be positioned under an area of the vehicle's frame or at one of the manufacturer's recommended lift points. Never place the pad under the floor pan or under steering and suspension components because it can be damaged by the weight of the vehicle. Always position the jack so that the wheels of the vehicle can roll as the vehicle is being raised.

FIGURE 3–8

Typical hydraulic jack.

Lincoln Automotive Company

WARNING!

Never use a lift or jack to move something heavier than it is designed for. Always check the rating before using a lift or jack. If a jack is rated for 2 tonnes, do not attempt to use it for a job that requires a 5-tonne jack. It is dangerous for you and the vehicle.

Safety stands, also called **jack stands**, are supports of various heights that sit on the floor. They are placed under a sturdy chassis member, such as the frame or axle housing, to support the vehicle. Once the safety stands are in position, the hydraulic pressure in the jack should be slowly released until the weight of the vehicle is on the stands. Like jacks, jack stands also have a capacity rating. Always use a jack stand of the correct rating.

Never move under a vehicle when it is supported by only a hydraulic jack. Rest the vehicle on the safety stands before moving under the vehicle.

The jack should be removed after the jack stands are set in place. This eliminates a hazard, such as a jack handle sticking out into a walkway. A jack handle that is bumped or kicked can cause a tripping accident or cause the vehicle to fall.

FIGURE 3–9

A heavy-duty hydraulic crane.

Lincoln Automotive Company

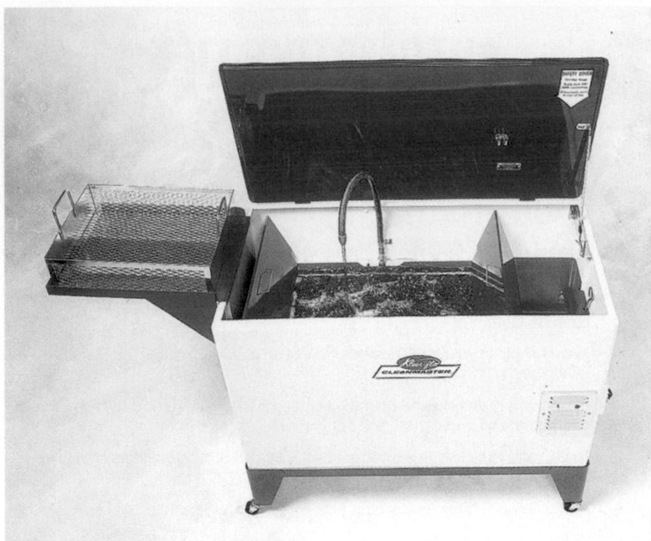

FIGURE 3–10

An automotive parts washer.

KLEENTEC/KLEER-FLO

Chain Hoist and Crane Safety

Heavy parts of the automobile, such as engines, are removed by using chain hoists **(Figure 3–9)** or hydraulic cranes. Another term for a chain hoist is chain fall. Cranes often are called cherry pickers.

To prevent serious injury, chain hoists and cranes must be properly attached to the parts being lifted. Always use bolts with enough strength to support the object being lifted. Place the chain hoist or crane directly over the assembly. Then, attach the chain or cable to the hoist.

Cleaning Equipment Safety

Parts cleaning is a necessary step in most repair procedures. Cleaning automotive parts can be divided into three basic categories.

Chemical cleaning relies on chemical action to remove dirt, grease, scale, paint, or rust. A combination of heat, agitation, mechanical scrubbing, or washing may be used to remove dirt. Chemical cleaning equipment includes small parts washers **(Figure 3–10)**, hot/cold tanks, pressure washers, spray washers, and salt baths.

Thermal cleaning relies on heat, which bakes off or oxidizes the dirt. Thermal cleaning leaves an ash residue on the surface that must be removed by an additional cleaning process, such as airless shot blasting or spray washing.

Abrasive cleaning relies on physical abrasion to clean the surface. This includes everything from a wire brush to glass bead blasting, airless steel shot blasting, abrasive tumbling, and vibratory cleaning.

Vehicle Operation

When a customer brings a vehicle in for service, certain driving rules should be followed to ensure your safety and the safety of those around you. For example, before moving a car into the shop, buckle your safety belt. Make sure no one is near, the way is clear, and there are no tools or parts under the car before you start the engine.

Check the brakes before putting the vehicle in gear. Then, drive slowly and carefully in and around the shop.

When road testing the car, obey all traffic laws. Drive only as far as is necessary to check the automobile and verify the customer's complaint. Never make excessively quick starts, turn corners too quickly, or drive faster than conditions or laws allow.

If the engine must be kept running while you are working on the car, block the wheels to prevent the vehicle from moving. Place the transmission in PARK for automatic transmissions or in NEUTRAL for manual transmissions. Set the parking (emergency) brake. Never stand directly in front of or behind a running vehicle.

When parking a vehicle in the shop, always roll the windows down. This allows access inside if the doors accidentally lock.

VENTING THE ENGINE'S EXHAUST Whenever you need to have the engine running for diagnosis or

FIGURE 3–11

When running an engine in a shop, make sure the vehicle's exhaust is connected to the shop's exhaust ventilation system.

Martin Restoule

service, the engine's exhaust must be vented to the outside. **Carbon monoxide (CO)** is present in the exhaust. CO is an odourless, tasteless, and colourless deadly gas. Inhaling CO can cause brain damage and, in severe cases, death. Early symptoms of CO poisoning include headaches, nausea, and fatigue. Most shops have an exhaust ventilation system **(Figure 3–11)**; always use it. Connect the hose from the vehicle's tailpipe to the intake for the vent system. Make sure the vent system is turned on before running the engine. If the work area does not have an exhaust venting system, use a hose to direct the exhaust out of the building.

Electrical Safety

To prevent personal injury or damage to the vehicle, you should always take the necessary precautions before working on or around a vehicle's electrical system. You should disconnect the battery before disconnecting any electrical wire or component. This prevents the possibility of a fire or electrical shock. It also eliminates the possibility of an accidental short, which can ruin the car's electrical system. Disconnect the negative or ground cable first **(Figure 3–12)**, then disconnect the positive cable.

Because electrical circuits require a ground to be complete, by removing the ground cable you eliminate the possibility of a circuit accidentally becoming completed. When reconnecting the battery, connect the positive cable first, then the negative.

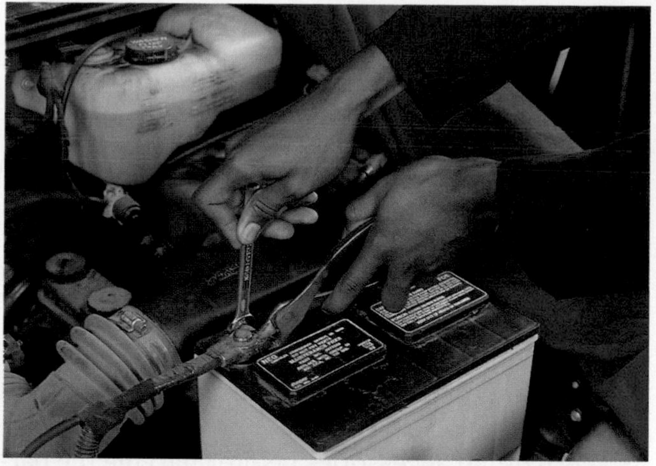

FIGURE 3–12

Before doing any electrical work or working around the battery, disconnect the negative lead of the battery.

Also, remove wristwatches and rings before servicing any part of the electrical system. This helps prevent the possibility of electrical arcing and burns. When disconnecting electrical connectors, do not pull on the wires. When reconnecting the connectors, make sure they are securely connected.

BATTERY PRECAUTIONS Because the vehicle's electrical power is stored in a battery or battery pack, special handling precautions must be followed when working with or near batteries. Hybrid and other electric vehicles have very high voltages; therefore, special precautions that apply to these vehicles are given below, following the general precautions for batteries.

- Make sure you are wearing safety glasses (preferably a face shield) and protective clothing when working around and with batteries.
- Keep all flames, sparks, and excessive heat away from the battery at all times, especially when it is being charged.
- Never lay metal tools or other objects on the battery because a short circuit across the terminals can result.
- All batteries have an electrolyte, which is very corrosive. It can cause severe injuries if it comes in contact with your skin or eyes. If an electrolyte gets on you, immediately wash with baking soda and water. If the acid gets in your eyes, immediately flush them with cool water for a minimum of 15 minutes and get immediate medical attention.
- Acid from a battery can damage a vehicle's paint and metal surfaces, and it can harm shop equipment. Neutralize any electrolyte spills during servicing.

- The most dangerous battery is one that has been overcharged. It is hot and has been, or still may be, producing large amounts of hydrogen. Allow the battery to cool before working with or around it. Also, never use or charge a battery that has a frozen electrolyte.
- Always use a battery carrier or lifting strap to make moving and handling batteries easier and safer.
- Always charge a battery in well-ventilated areas.
- Never connect or disconnect charger leads when the charger is turned on. This generates a dangerous spark.
- Turn off all accessories before charging the battery and correct any parasitic drain problems.
- Make sure the charger's power switch is off when you are connecting or disconnecting the charger cables to the battery.
- Always double-check the polarity of the battery charger's connections before turning the charger on. Incorrect polarity can damage the battery or cause it to explode.
- Never attempt to use a charger as a boost to start the engine.

HIGH-VOLTAGE SYSTEMS Electric-drive vehicles (battery-operated, hybrid, and fuel-cell electric vehicles) have high-voltage electrical systems (from 42 volts to 650 volts). These high voltages can kill you! Fortunately, most high-voltage circuits are identifiable by size and colour. The cables have thicker insulation and are typically coloured orange **(Figure 3–13)**. The connectors are also orange. On most vehicles, the high-voltage cables are enclosed in an orange shielding or casing; again, the orange indicates high voltage. In addition, the high-voltage battery pack and most high-voltage components have HIGH VOLTAGE caution labels **(Figure 3–14)**. Be careful not to touch these wires and parts. Here are some other safety precautions that should be adhered to when working on an electric-drive vehicle:

FIGURE 3–13

The high-voltage cables on this Honda Civic hybrid are coloured orange and are enclosed in orange casing.

FIGURE 3–14

Most high-voltage components in a hybrid vehicle have HIGH VOLTAGE caution labels.

- Always follow the safety guidelines given by the vehicle's manufacturer.
- Obtain the necessary training before working on these vehicles.
- Be sure to perform each repair operation correctly.
- Disable or disconnect the high-voltage system before performing services to those systems. Do this according to the procedures given by the manufacturer.
- Any time the engine is running in a hybrid vehicle, the generator is producing high voltage. Take care to prevent being shocked.
- Before doing any service to an electric-drive vehicle, make sure the power to the electric motor is disconnected or disabled.
- Systems may have a high-voltage capacitor that must be discharged after the high-voltage system has been isolated. Make sure to wait the prescribed amount of time (normally about 10 minutes) before working on or around the high-voltage system.
- After removing a high-voltage cable, cover the terminal with vinyl electrical tape.
- Always use insulated tools.
- Use only the tools and test equipment specified by the manufacturer, and follow the test procedures defined by the equipment manufacturer.

- Alert other technicians that you are working on the high-voltage systems by posting a warning sign such as HIGH-VOLTAGE WORK: DO NOT TOUCH.
- Wear insulating gloves when working on or around the high-voltage system. Make sure they have no tears, holes, or cracks and that they are dry. The integrity of the gloves should be checked before using them.
- Always install the correct type of circuit protection device into a high-voltage circuit.
- Many electric motors have a strong permanent magnet in them; do not handle these parts if you have a pacemaker.
- When an electric-drive vehicle needs to be towed into the shop for repairs, make sure it is not towed on its drive wheels. Doing this will drive the generator(s), which can overcharge the batteries and cause them to explode. Always tow these vehicles with the drive wheels off the ground or move them on a flatbed.

Rotating Pulleys and Belts

Be careful around belts, pulleys, wheels, chains, or any other rotating component. When working around these, make sure your hands, shop towels, or loose clothing do not come in contact with the moving parts. Hands and fingers can be quickly pulled into a revolving belt or pulley even at engine idle speeds.

WORK AREA SAFETY

Your work area should be kept clean and safe. The floor and bench tops should be kept clean, dry, and orderly. Any oil, coolant, or grease on the floor can make it slippery. Slips can result in serious injuries. To clean up oil, use a commercial oil absorbent. Keep all water off the floor. Water causes smooth floors to become slippery, and it readily conducts electricity. Aisles and walkways should be kept clean and wide enough for you to easily move through them. Make sure the work areas around machines are large enough for you to safely operate the machines.

Make sure all drain covers are snugly in place. Open drains or covers that are not flush to the floor can cause toe, ankle, and leg injuries.

Keep an up-to-date list of emergency telephone numbers clearly posted next to the telephone. The numbers of a doctor, a hospital, and fire and police departments should be included. Also, the work area should have a first-aid kit for treating minor injuries and eye-flushing kits readily available. You should know where these items are kept.

Flammable Liquids

GASOLINE Gasoline is a highly **flammable** volatile liquid. Something that is flammable catches fire and burns easily. A **volatile liquid** is one that vaporizes very quickly. Flammable volatile liquids are potential firebombs. Always keep gasoline, ethanol, or diesel fuel in an approved safety can **(Figure 3–15)**, and never use gasoline to clean your hands or tools.

The presence of gasoline is so common that its dangers are often forgotten. A slight spark or an increase in heat can cause a fire or explosion. Gasoline fumes are heavier than air. Therefore, when an open container of gasoline is sitting about, the fumes spill out over the sides of the container. These fumes are more flammable than liquid gasoline and can easily explode.

Never smoke around gasoline or in a shop filled with gasoline fumes. If the vehicle has a gasoline leak or you have caused a leak by disconnecting a fuel line, wipe it up immediately and stop the leak. Make sure that any grinding or welding that may be taking place in the area is stopped until the spill is totally cleaned up and the floor has been flushed with water. The rags used to wipe up the gasoline should be taken outside to dry, then stored in an approved dirty rag container. If vapours are present in the shop, keep the doors open and turn on the ventilating system. It takes only a small amount of fuel mixed with air to cause combustion.

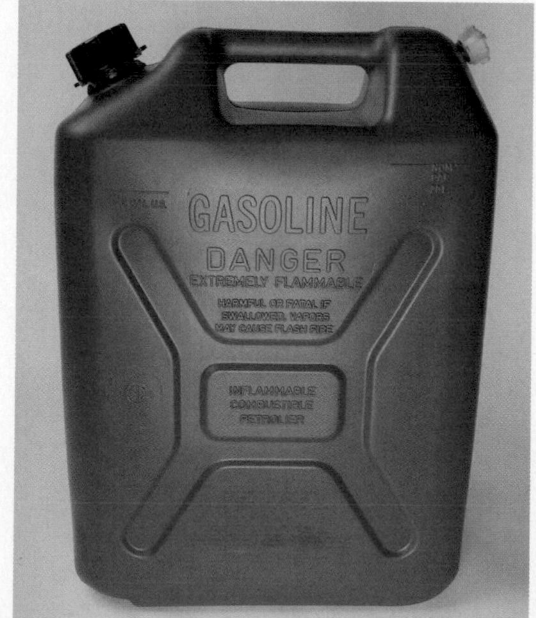

FIGURE 3–15
Flammable liquids should be stored in approved safety containers.

ETHANOL Most commonly found as E85 (15 percent gasoline mixed with 85 percent ethanol), ethanol is a very volatile liquid. Ethanol is a non-petroleum-based fuel and is used as an alternative fuel to gasoline. Ethanol is also used as an additive to increase the octane rating of gasoline. Handle and store E85 in the same way as gasoline.

DIESEL FUEL Diesel fuel is not as volatile as gasoline but should be stored and handled in the same way. It is also not as refined as gasoline and normally contains many impurities, including active microscopic organisms that can be highly infectious. If diesel fuel happens to get on an open cut or sore, thoroughly wash it immediately.

Handle all solvents (or any liquids) with care to avoid spillage. Keep all solvent containers closed, except when pouring. Proper ventilation is very important in areas where volatile solvents and chemicals are used. Solvents and other combustible materials must be stored in approved and designated storage cabinets **(Figure 3–16)** or rooms. Be extra careful when transferring flammable materials from bulk storage. Static electricity can build up to the point where it creates a spark that could cause an explosion. A grounding strap or wire is recommended between containers when transferring flammable liquids to discharge any static electricity without creating potentially dangerous sparks. Discard or clean all empty solvent containers. Solvent fumes in the bottom of these containers are very flammable. Never light matches or smoke near flammable solvents and chemicals, including battery acids.

FIGURE 3–16

Store combustible materials in approved safety cabinets.

Martin Restoule

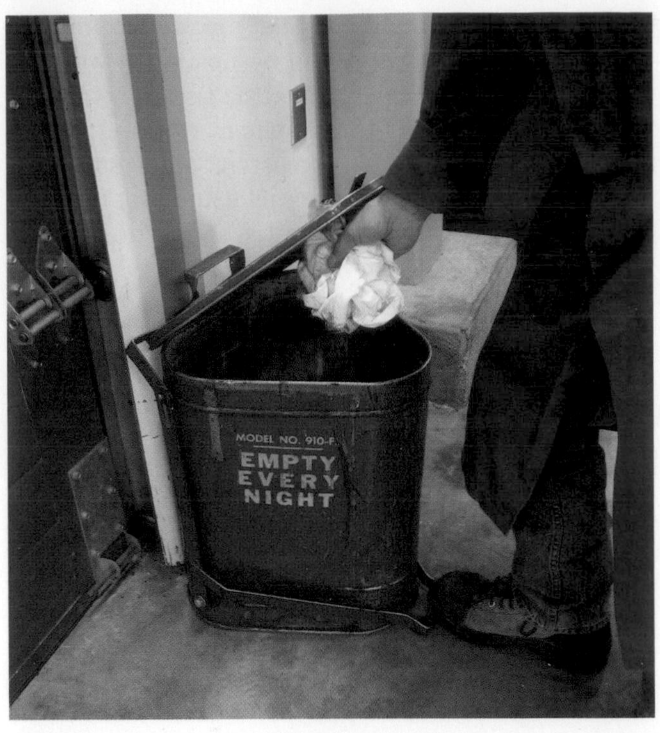

FIGURE 3–17

Dirty rags and towels should be kept in an approved container.

Martin Restoule

Oily rags should also be stored in an approved metal container **(Figure 3–17)**. When these oily, greasy, or paint-soaked rags are left lying about or are not stored properly, they can cause spontaneous combustion. Spontaneous combustion results in a fire that starts by itself, without being set off by a match or some other source of ignition.

Fire Extinguishers

Know where all of the shop's fire extinguishers are located and what types of fires they put out **(Table 3–1)**. Fire extinguishers are clearly labelled as to what type they are and what types of fires they should be used on. Make sure you use the correct type of extinguisher for the type of fire you are dealing with. A multipurpose dry chemical fire extinguisher will put out ordinary combustibles, flammable liquids, and electrical fires. Never put water on a gasoline fire. The

TABLE 3–1 GUIDE TO EXTINGUISHER SELECTION

	CLASS OF FIRE	TYPICAL FUEL INVOLVED	TYPE OF EXTINGUISHER
Class A Fires (green)	**For Ordinary Combustibles** Put out a class A fire by lowering its temperature or by coating the burning combustibles.	Wood Paper Cloth Rubber Plastics Rubbish Upholstery	Water*[1] Foam* Multipurpose dry chemical[4]
Class B Fires (red)	**For Flammable Liquids** Put out a class B fire by smothering it. Use an extinguisher that gives a blanketing, flame-interrupting effect; cover whole flaming liquid surface.	Gasoline Oil Grease Paint Lighter fluid	Foam* Carbon dioxide[5] Halogenated agent[6] Standard dry chemical[2] Purple K dry chemical[3] Multipurpose dry chemical[4]
Class C Fires (blue)	**For Electrical Equipment** Put out a class C fire by shutting off power as quickly as possible and by always using a nonconducting extinguishing agent to prevent electric shock.	Motors Appliances Wiring Fuse boxes Switchboards	Carbon dioxide[5] Halogenated agent[6] Standard dry chemical[2] Purple K dry chemical[3] Multipurpose dry chemical[4]
Class D Fires (yellow)	**For Combustible Metals** Put out a class D fire of metal chips, turnings, or shavings by smothering or coating with a specially designed extinguishing agent.	Aluminum Magnesium Potassium Sodium Titanium Zirconium	Dry powder extinguishers and agents only

*Cartridge-operated water, foam, and soda-acid types of extinguishers are no longer manufactured. These extinguishers should be removed from service when they become due for their next hydrostatic pressure test.

Notes:

(1) Freezes in low temperatures unless treated with antifreeze solution, usually weighs more than 9 kg (20 lb.), and is heavier than any other extinguisher mentioned.

(2) Also called ordinary or regular dry chemical (sodium bicarbonate).

(3) Has the greatest initial fire-stopping power of the extinguishers mentioned for class B fires. Be sure to clean residue immediately after using the extinguisher so that sprayed surfaces will not be damaged (potassium bicarbonate).

(4) The only extinguishers that fight A, B, and C classes of fires. However, they should not be used on fires in liquefied fat or oil of appreciable depth. Be sure to clean residue immediately after using the extinguisher so that sprayed surfaces will not be damaged (ammonium phosphates).

(5) Use with caution in unventilated, confined spaces.

(6) May cause injury to the operator if the extinguishing agent (a gas) or the gases produced when the agent is applied to a fire are inhaled.

water will just spread the fire. The proper fire extinguisher will smother the flames. Remember, during a fire never open doors or windows unless it is absolutely necessary; the extra draft will only make the fire worse. Make sure the fire department is contacted before or during your attempt to extinguish a fire.

In the event a fire becomes more than can be safely handled with a fire extinguisher, you and everyone else in the shop will need to evacuate the area. Every school and public building should have evacuation routes posted. These routes should clearly identify the nearest emergency exits. During the first days of class or at the start of a new job, make sure you locate the evacuation routes and understand exactly where you are to go in the event of an emergency.

To extinguish a fire, stand 2 to 3 m (6 to 10 ft.) from the fire. Hold the extinguisher firmly in an upright position. Aim the nozzle at the base of the fire and use a side-to-side motion, sweeping the entire width of the fire. Stay low to avoid inhaling the smoke. If it gets too hot or too smoky, get out. Remember, never go back into a burning building for anything. To help remember how to use an extinguisher, remember the word "PASS."

Pull the pin from the handle of the extinguisher.

Aim the extinguisher's nozzle at the base of the fire.

Squeeze the handle.

Sweep the entire width of the fire with the contents of the extinguisher.

MANUFACTURER'S WARNINGS AND GOVERNMENT REGULATIONS

As you work on cars, there are many parts and fluids that you do not want to keep. Some of the common wastes, as well as how to dispose of them, are given here.

A typical shop contains many potential health hazards for those working in it. These hazards can cause injury, sickness, health impairments, discomfort, and even death. Hazards can be classified as chemical, wastes, physical, and ergonomic:

- **Chemical hazards** are caused by high concentrations of vapours, gases, or solids in the form of dust.
- **Hazardous wastes** are those substances that are the result of a service.
- **Physical hazards** include excessive noise, vibration, pressures, and temperatures.
- **Ergonomic hazards** are conditions that impede normal and/or proper body position and motion.

Most Canadian workers are covered by provincial or territorial occupational health and safety legislation. Workplace/workers' compensation boards, workplace health and safety commissions, departments or ministries of labour, or ministries of skills, development and labour administer the various legislations. These regulations must be understood and followed to ensure a safe workplace. Depending on the size of the workplace, many shops must have a safety committee in place to ensure that the shop and the workers are adhering to all of the safety concerns. Regardless of how many work in the shop, everyone in a shop has the responsibility for adhering to these regulations.

In Canada, Environment Canada—under the *Canadian Environmental Protection Act* (CEPA)—and provincial environment ministries identify and regulate the use of many potentially hazardous materials and also regulate their disposal. As a result, these regulations are best treated as adjoining laws that deal with hazardous materials and waste. There are strict rules and regulations that help to promote safety in the auto shop. These are described throughout this text whenever they are applicable. Maintaining a vehicle involves handling and managing a wide variety of materials and wastes. Some of these wastes can be toxic to fish, wildlife, and humans when improperly managed. No matter the amount of waste produced, it is to the shop's legal and financial advantage to manage the wastes properly and, even more importantly, to prevent the pollution of our natural resources.

WHMIS RIGHT-TO-KNOW LEGISLATION

An important part of a safe work environment is the employees' knowledge of potential hazards. Every employee in the shop is protected by **WHMIS right-to-know legislation**. These laws were put into effect when the federal government in 1987 passed the **Workplace Hazardous Materials Information System (WHMIS)** legislation (Bill C-70). Although this legislation is federal, all provinces and territories have included this legislation in their own health and safety acts. These regulations apply to all companies, including auto repair shops.

The general intent of WHMIS right-to-know legislation is for employers to provide their employees with a safe working place when hazardous materials are involved. Specifically, there are three areas of employer responsibility.

Primarily, all employees must be trained about their rights under the legislation, the nature of the hazardous chemicals in their workplace, and the contents of the labels on the chemicals. All of the information about each chemical must be posted on **material safety data sheets (MSDSs)** and must be accessible by all employees. The manufacturer of the chemical must give these sheets to its customers, if they are requested to do so **(Figure 3–18)**. They detail the chemical composition and precautionary information for all products that can present a health or safety hazard. An MSDS must include the following information about the product:

- The trade and chemical name of the product
- The manufacturer of the product
- All of the ingredients of the product
- Health hazards such as headaches, skin rashes, nausea, and dizziness
- The product's physical description, which may include the product's colour, odour, permissible exposure limit (PEL), threshold limit value (TLV), specific gravity, boiling point, freezing point, evaporation data, and volatility rating
- The product's explosion and fire data, such as flash point
- The reactivity and stability data
- The product's weight compared to air
- Protection data, including first aid and proper handling

Employees must become familiar with the general uses, protective equipment, accident or spill

HEXANE

MSDS Safety Information

Ingredients

Name: HEXANE (N_HEXANE)
% Wt: >97
OSHA PEL: 500 PPM
ACGIH TLV: 50 PPM
EPA Rpt Qty: 1 LB
DOT Rpt Qty: 1 LB

Health Hazards Data

LD50 LC50 Mixture: LD50:(ORAL,RAT) 28.7 KG/MG
Route Of Entry Inds _ Inhalation: YES
Skin: YES
Ingestion: YES
Carcinogenicity Inds _ NTP: NO
IARC: NO
OSHA: NO
Effects of Exposure: ACUTE:INHALATION AND INGESTION ARE HARMFUL AND MAY BE FATAL. INHALATION AND INGESTION MAY CAUSE HEADACHE, NAUSEA, VOMITING, DIZZINESS, IRRITATION OF RESPIRATORY TRACT, GASTROINTESTINAL IRRITATION AND UNCONSCIOUSNESS. CONTACT W/SKIN AND EYES MAY CAUSE IRRITATION. PROLONGED SKIN MAY RESULT IN DERMATITIS (EFTS OF OVEREXP)
Signs And Symptoms Of Overexposure: HLTH HAZ:CHRONIC:MAY INCLUDE CENTRAL NERVOUS SYSTEM DEPRESSION.
Medical Cond Aggravated By Exposure: NONE IDENTIFIED.
First Aid: CALL A PHYSICIAN. INGEST:DO NOT INDUCE VOMITING. INHAL:REMOVE TO FRESH AIR. IF NOT BREATHING, GIVE ARTIFICIAL RESPIRATION. IF BREATHING IS DIFFICULT, GIVE OXYGEN. EYES:IMMED FLUSH W/PLENTY OF WATER FOR AT LEAST 15 MINS. SKIN:IMMED FLUSH W/PLENTY OF WATER FOR AT LEAST 15 MINS WHILE REMOVING CONTAMD CLTHG & SHOES. WASH CLOTHING BEFORE REUSE.

Handling and Disposal

Spill Release Procedures: WEAR NIOSH/MSHA SCBA & FULL PROT CLTHG. SHUT OFF IGNIT SOURCES:NO FLAMES, SMKNG/FLAMES IN AREA. STOP LEAK IF YOU CAN DO SO W/OUT HARM. USE WATER SPRAY TO REDUCE VAPS. TAKE UP W/SAND OR OTHER NON-COMBUST MATL & PLACE INTO CNTNR FOR LATER (SU PDAT)
Neutralizing Agent: NONE SPECIFIED BY MANUFACTURER.
Waste Disposal Methods: DISPOSE IN ACCORDANCE WITH ALL APPLICABLE FEDERAL, STATE AND LOCAL ENVIRONMENTAL REGULATIONS. EPA HAZARDOUS WASTE NUMBER:D001 (IGNITABLE WASTE).
Handling And Storage Precautions: BOND AND GROUND CONTAINERS WHEN TRANSFERRING LIQUID. KEEP CONTAINER TIGHTLY CLOSED.
Other Precautions: USE GENERAL OR LOCAL EXHAUST VENTILATION TO MEET TLV REQUIREMENTS. STORAGE COLOR CODE RED (FLAMMABLE).

Fire and Explosion Hazard Information

Flash Point Method: CC
Flash Point Text: _9F,_23C
Lower Limits: 1.2%
Upper Limits: 77.7%
Extinguishing Media: USE ALCOHOL FOAM, DRY CHEMICAL OR CARBON DIOXIDE. (WATER MAY BE INEFFECTIVE).
Fire Fighting Procedures: USE NIOSH/MSHA APPROVED SCBA & FULL PROTECTIVE EQUIPMENT (FP N).
Unusual Fire/Explosion Hazard: VAP MAY FORM ALONG SURFS TO DIST IGNIT SOURCES & FLASH BACK. CONT W/STRONG OXIDIZERS MAY CAUSE FIRE. TOX GASES PRDCED MAY INCL:CARBON MONOXIDE, CARBON DIOXIDE.

FIGURE 3–18

Material safety data sheets are an important part of employee training and should be readily accessible.

procedures, and any other information regarding the safe handling of the hazardous material. This training must be given to employees annually and provided to new employees as part of their job orientation.

Furthermore, all hazardous material must be properly labelled, indicating what health, fire, or reactivity hazard it poses and what protective equipment is necessary when handling each chemical. The manufacturer of the hazardous waste materials must provide all warnings and precautionary information, which must be read and understood by the user before application. Attention to all label precautions is essential for the proper use of the chemical and for prevention of hazardous conditions. A list of all hazardous materials used in the shop must be posted for the employees to see.

Finally, shops must maintain documentation on the hazardous chemicals in the workplace, proof of training programs, records of accidents or spill incidents, satisfaction of employee requests for specific chemical information via the MSDS, and a general right-to-know compliance procedure manual utilized within the shop.

HAZARDOUS MATERIALS

WARNING!

When handling any hazardous waste material, be sure to wear the proper safety equipment covered under the WHMIS right-to-know legislation. Follow all required procedures correctly. This includes the use of approved respirator equipment.

As mentioned before, some of the materials used in auto repair shops can be dangerous. The solvents and other chemical products used in an auto shop carry warnings and caution information that must be read and understood by all who use them.

Many repair and service procedures generate hazardous wastes. Dirty solvents and cleaners are good examples of hazardous wastes. Something that is classified as a hazardous waste by Environment Canada or the various provincial environment ministries will be identified and regulated through various federal and provincial acts or bills. It should be noted that no material is considered a hazardous waste until the shop is finished using it and is ready to dispose of it.

Regulations on hazardous waste handling and generation have led to the development of equipment now commonly found in shops. Examples of these are thermal cleaning units, close-loop steam cleaners, waste oil furnaces, oil filter crushers, refrigerant recycling machines, engine coolant recycling machines, and highly absorbent cloths.

WARNING!

The shop is ultimately responsible for the safe disposal of hazardous wastes, even after the waste leaves the shop. Only licensed waste removal companies should be used to dispose of the waste. Make sure you know what the company is planning to do with the waste. Make sure you have a written contract stating what is supposed to happen to the waste. Leave nothing to chance.

Many shops use full-service haulers to remove hazardous waste from the property. Besides hauling the hazardous waste away, the hauler also takes care of all the paperwork, deals with the various government agencies, and advises the shop on how to recover the disposal costs.

Guidelines for Handling Shop Wastes

To protect yourself, you should consider the following.

OIL Recycle oil. Set up equipment, such as a drip table or screen table with a used-oil collection bucket, to collect oil that drips off parts. Place drip pans underneath vehicles that are leaking fluids. Do not mix other wastes with used oil, except as allowed by your recycler. Used oil generated by a shop (and/or oil received from household do-it-yourself generators) may be burned on-site in a commercial space heater. Also, used oil may be burned for energy recovery. Contact provincial and local authorities to determine requirements and to obtain necessary permits.

OIL FILTERS Drain for at least 24 hours; crush **(Figure 3–19)** and recycle used oil filters.

BATTERIES Recycle batteries by sending them to a reclaimer or back to the distributor. Keep shipping receipts to demonstrate that you have recycled. Store batteries in a watertight, acid-resistant container. Inspect batteries for cracks and leaks when they come in. Treat a dropped battery as if it were cracked. Acid residue is hazardous because it is corrosive and may contain lead and other toxins. Neutralize spilled acid by covering it with baking soda or lime, and dispose of all hazardous material.

METAL RESIDUE FROM MACHINING Collect metal filings when machining metal parts and recycle if possible. Prevent metal filings from falling into a storm sewer drain.

FIGURE 3–19

A hydraulic single oil filter crusher.

SPX/OTC Service Solutions

REFRIGERANTS Recover and/or recycle refrigerants during the servicing of air-conditioning systems. It is not allowable to knowingly vent refrigerants into the atmosphere. Recovery and/or recycling during servicing must be performed in most provinces by a Ministry of Environment or CEPA-certified technician using certified equipment and following specified procedures.

SOLVENTS Replace hazardous chemicals with less toxic alternatives that have equal performance. For example, substitute water-based cleaning solvents for petroleum-based solvent degreasers. To reduce the amount of solvent used when cleaning parts, use a two-stage process: dirty solvent followed by fresh solvent. Hire a hazardous waste management service to clean and recycle solvents. Store solvents in closed containers to prevent evaporation. Evaporation of solvents contributes to ozone depletion and smog formation. In addition, the residue from evaporation must be treated as a hazardous waste.

CONTAINERS Cap, label, cover, and properly store all liquid containers and small tanks within a diked area and on a paved impermeable surface to prevent spills from running into surface or ground water.

OTHER SOLIDS Store materials such as scrap metal, old machine parts, and worn tires under a roof or tarpaulin to protect them from the elements and to prevent potential contaminated runoff. Consider recycling tires by retreading them.

LIQUID RECYCLING Collect and recycle coolants from radiators. Store transmission fluids, brake fluids, and solvents containing chlorinated hydrocarbons separately, and recycle or dispose of them properly.

SHOP TOWELS/RAGS Keep waste towels in a closed container marked CONTAMINATED SHOP TOWELS ONLY. To reduce costs and liabilities associated with disposal of used towels, investigate using a laundry service that is able to treat the wastewater generated from cleaning the towels.

HIRING A HAULER Hire a reputable, financially stable, and provincially approved hauler who will dispose of your shop wastes legally. If hazardous waste is dumped illegally, your shop may be held responsible.

WASTE STORAGE Always keep hazardous waste separate, properly labelled and sealed in the recommended containers. The storage area should be covered and may need to be fenced and locked if vandalism could be a problem. Select a licensed hazardous waste hauler after seeking recommendations and reviewing the firm's permits and authorizations.

Asbestos

Asbestos has been identified as a health hazard. **Asbestos** is a term used to describe a number of naturally occurring fibrous materials. It has been identified as a carcinogen and has been shown to cause a number of diseases, including cancer. Asbestos-caused cancer, or mesothelioma, is a form of lung cancer. When breathed in, the asbestos fibres cause scarring of the lungs and/or cause damage to the lung's air passages. The injuries and scars in the lung become an effective holding place for the asbestos. Obviously, you want to avoid breathing in asbestos dust and fibres. Face masks equipped with HEPA filters will protect your lungs from asbestos exposure. When working with asbestos materials, such as brake pads, clutch discs, and some engine gaskets, there are certain guidelines you should follow. All asbestos waste must be disposed of in accordance with federal or provincial asbestos regulations.

Generally, these regulations do not regulate the removal of asbestos brakes unless debonding or grinding of asbestos brake pads constitutes large volumes of the shop's work. At such facilities, the asbestos materials are regulated as a hazardous waste and handled accordingly; they are stored in an enclosed container and sent to a hazardous waste hauler. However, even when asbestos wastes are not regulated as hazardous wastes, the CEPA recommends that shops capture asbestos from brake shoes in a separate container. Use a low-pressure/wet-cleaning method. *Never* blow brake dust and never use an air hose for cleaning.

One asbestos cleaning method is the use of a **high-efficiency particulate air (HEPA)** vacuum cleaner. This vacuum cleaner captures the asbestos in a special filter. When used on a brake system, the vacuum cleaner completely houses the brake assembly. Built-in gloves allow a technician to clean the assembly with compressed air through a window and without direct contact with the assembly. The dust is drawn in by the vacuum cleaner.

WARNING!

Make sure the enclosure of the vacuum cleaner fits tightly around the brake or clutch assembly before using compressed air to clean it.

Once the filter in the HEPA vacuum cleaner is full, it must be wetted with a mist of water before it is removed from the cleaner. The filters must be placed in an impermeable container, labelled, and disposed of in the manner prescribed by local or provincial laws.

Another approved asbestos cleaning method is to use water mixed with an organic solvent or wetting agent. It is important that the wetting agent be allowed to flow through the brake drum or around the brake disc before the brakes are disassembled for further cleaning. Position a catch basin under the brake assembly to capture any contaminated liquid. As the assembly is being disassembled, the parts should be misted with the wetting agent.

To minimize the risks of working around asbestos, follow these simple personal hygiene guidelines:

- Do not smoke while or after working with the materials.
- Thoroughly wash yourself before eating.
- Shower after work.
- Change into work clothes when you arrive at work, and change out of your work clothes after work. Do not take work clothing home.

KEY TERMS

Abrasive cleaning (p. 42)
Asbestos (p. 51)
Blood-borne pathogens (p. 37)
Carbon monoxide (CO) (p. 43)
Chemical cleaning (p. 42)
Chemical hazards (p. 48)
Ergonomic hazards (p. 48)
Flammable (p. 45)
Hazardous wastes (p. 48)
High-efficiency particulate air (HEPA) (p. 51)
Jack stands (p. 41)
Material safety data sheets (MSDSs) (p. 48)
Physical hazards (p. 48)
Pneumatic tools (p. 40)
Power tools (p. 39)
Respirators (p. 37)
Safety glasses (p. 36)
Safety stands (p. 41)
Thermal cleaning (p. 42)
Volatile liquid (p. 45)
WHMIS right-to-know legislation (p. 48)
Workplace Hazardous Materials Information System (WHMIS) (p. 48)

SUMMARY

- Dressing safely for work is very important. Wear snug-fitting clothing, eye and ear protection, protective gloves, steel-toed shoes, and caps to cover long hair.
- When choosing eye protection, make sure it has safety glass and offers side protection.
- A respirator should be worn whenever you are working around toxic fumes or excessive dust.

- When shop noise exceeds safe levels, protect your ears by wearing earplugs or earmuffs.
- Safety while using any tool is essential, and even more so when using power tools. Before plugging in a power tool, make sure the power switch is off. Disconnect the power before servicing the tool.
- Always observe all relevant safety rules when operating a vehicle lift or hoist. Jacks, jack stands, chain hoists, and cranes can also cause injury if not operated safely.
- Use care while moving a vehicle in the shop. Carelessness and playing around can lead to a damaged vehicle and serious injury.
- Carbon monoxide (CO) is a poisonous gas present in engine exhaust fumes. Exhaust must be properly vented from the shop using tailpipe hoses or other reliable methods.
- Adequate ventilation is also necessary when working with any volatile solvent or material.
- Much of the work on an automobile is around or with the vehicle's battery and electrical system. To prevent personal injury or damage to the vehicle, always take necessary precautions before working.
- Gasoline and diesel fuel are highly flammable and should be kept in approved safety cans.
- Never light matches or smoke near any combustible materials.
- It is important to know when to use each of the various types of fire extinguishers. When fighting a fire, aim the nozzle at the base and use a side-to-side sweeping motion.
- Right-to-know laws came into effect in 1987 and are designed to protect employees who must handle hazardous materials and wastes on the job.
- Material safety data sheets (MSDSs) contain important chemical information and must be furnished to all employees annually. New employees should be given the sheets as part of their job orientation.
- All hazardous and asbestos waste should be disposed of according to CEPA regulations.

REVIEW QUESTIONS

1. What is the correct way to dispose of used oil filters?
2. Where in the shop should a list of emergency telephone numbers be posted?
3. When should eye protection be worn?
4. How should a class B fire be extinguished?
5. What is the correct procedure for using a fire extinguisher to put out a fire?
6. What type of masks should be worn when working with brake shoes or clutch discs?
   a. masks equipped with CEPA filters
   b. masks equipped with HEPA filters
   c. masks equipped with OHSA filters
   d. masks equipped with CSA filters
7. Which of the following statements about safety glasses is correct?
   a. They should be worn only when performing grinding operations.
   b. Any glasses with plastic lenses will provide sufficient protection.
   c. They should be worn any time you are working in the shop.
   d. Any glasses with glass lenses will provide sufficient protection.
8. What type of fire extinguisher should be used for an electrical fire?
   a. foam
   b. water
   c. dry powder
   d. multipurpose dry chemical
9. Which of the following shop wastes do not require approved waste disposal collection?
   a. brake shoes
   b. used engine oil
   c. used engine coolant
   d. contaminated gasoline
10. Which of the following relates to the right-to-know legislation?
    a. OHSA **b.** CEPA
    c. HEPA **d.** WHMIS
11. Which method for cleaning parts may leave a residue that must be removed by further cleaning?
    a. cold chemical **b.** abrasive
    c. thermal **d.** hot chemical
12. What does WHMIS right-to-know legislation refer to?
    a. auto emission standards
    b. hazards associated with chemicals used in the workplace
    c. employee benefits
    d. hiring practices
13. Which of the following is a poisonous gas that requires proper ventilation?
    a. CO_2 **b.** CO
    c. HC **d.** H_20
14. What term refers to a material that reacts violently with water or other materials?
    a. corrosivity **b.** volatility
    c. ignitability **d.** reactivity

15. When extinguishing a fire, how far away from the fire should you be?
    a. 1 to 2 m (3.3 to 6.6 ft.)
    b. 2 to 3 m (6.6 to 10 ft.)
    c. 3 to 4 m (10 to 13 ft.)
    d. 4 to 5 m (13 to 16.5 ft.)
16. What should you do when lifting any object?
    a. Place your feet far apart.
    b. Turn with your hips.
    c. Keep your back as straight as possible.
    d. Bend forward to place the object on a bench.
17. What is the recommended way to store gasoline?
    a. in an approved safety container under a workbench
    b. in a sealed container under a workbench
    c. in a sealed container in a designated storage cabinet
    d. in an approved safety container in a designated storage cabinet
18. What can compressed air be used to blow off?
    a. dirt from bolt holes
    b. dirt from a work bench
    c. dust from brake shoes
    d. dust from your coveralls
19. When should nitrile protective gloves be worn?
    a. when grinding
    b. when welding
    c. when washing parts
    d. when handling hot components
20. To properly dispose of an oil filter, how long should it be allowed to drain?
    a. 6 hours **b.** 12 hours
    c. 18 hours **d.** 24 hours

CHAPTER 4
Automotive Systems

LEARNING OUTCOMES

- Explain the major events that have influenced the development of the automobile in the recent past.
- Explain the difference between unitized and body-over-frame vehicles.
- Describe the manufacturing process used in a modern automated automobile assembly plant.
- List the basic systems that make up an automobile, and name their major components and functions.

HISTORICAL BACKGROUND

The automobile has changed quite a bit since the first horseless carriage went down a North American road. In 1896, both Henry Ford and Ransom Eli Olds test drove their first gasoline-powered vehicles. This same year is credited as the beginning of the automotive industry, not because of what Ford or Olds did, but because the Duryea Brothers by 1896 had made 13 cars in their factory, which was the first to make cars for customers. The introduction of the Ford Model T was a turning point in the auto industry because it was the first car to be built on an assembly line and was very affordable.

In the beginning, the automobile looked like the horse-drawn carriage it was designed to replace **(Figure 4–1)**. In fact, for many years most cars looked like carriages. In 1919, 90 percent of the cars had carriage-like open bodies. Although body styles changed, cars continued to have carriage-like features. It was not until 1939 that running boards began disappearing.

These early cars had rear-mounted engines and very tall tires. They were designed to move people down dirt roads. The automobile changed to meet new conditions: Roads were improved and became paved; more people owned cars; manufacturers tried to sell more cars; concern for safety and the environment grew; and new technology was developed. Because of all of these changes, the automobile became more practical, more affordable, safer, more comfortable, more dependable, and faster. Although many improvements have been made to the original design, the basic structure of the automobile has changed very little.

FIGURE 4–1

The 1886 Benz Patent Motor Wagen, one of the first automobiles made.

Mercedes-Benz of N A., Inc.

Nearly all of today's cars still use gasoline engines to drive two or more wheels. A steering system is used to control the direction of the car. A brake system is used to slow down and stop the car. A suspension system is used to absorb road shocks and help the driver maintain control on bumpy roads. The parts of these major systems are mounted on steel frames, and the frame is covered with body panels. These panels give the car its shape and protect those inside from the weather and dirt. The body panels also offer some protection for the passengers if the automobile is in an accident.

Although the basics of an automobile have changed little in the past 100 years, the parts and

FIGURE 4–2

A cutaway of a late-model car showing some of the technology of today's cars.

Chrysler Group LLC

the control systems have changed greatly. The entire automobile is technologically light-years ahead of Ford's and Olds's early models **(Figure 4–2)**. The use of new technology has changed the slow, unreliable, user hostile vehicles of the early 1900s into vehicles that travel at very high speeds, operate trouble-free for thousands of kilometres, and provide comforts that even the rich had not dreamed of in 1896.

The most dramatic changes have occurred during the past 40 years. In 1965, the United States passed legislation limiting the levels of exhaust emissions. Although there was little immediate effect on the industry because of this legislation, automobile manufacturers needed to focus on the future. They needed to build cleaner-burning engines. In the following years, stricter emissions laws were passed, and manufacturers were required to develop systems to control emissions.

World events in the 1970s continued to shape the development of today's automobile. An oil embargo by Arab nations in 1973 caused the price of gasoline to quickly increase to four times its normal price. This event caused most North Americans to realize that the supply of gasoline and other non-renewable resources was limited. Buyers wanted cars that were not only kind to the environment but that also used a smaller amount of fuel.

In 1977, Transport Canada began setting annual **Company Average Fuel Consumption (CAFC)** goals for the vehicle manufacturers and importers. This voluntary program is similar to the mandatory **Corporate Average Fuel Economy (CAFE)** program in the United States. This program requires automakers to not only manufacture clean-burning engines but also equip their vehicles with engines that burn gasoline efficiently. Under the CAFC/CAFE standards, different models from each manufacturer are tested for the number of litres used to travel 100 km. The fuel efficiencies of these vehicles are averaged together to arrive at a corporate average. The CAFC/CAFE standards have increased many times since they were first established. In Canada, manufacturers are encouraged to meet the CAFC standards, and additional incentives encourage them to increase the production of vehicles that operate on alternative fuels. In the United States, a manufacturer that does not meet CAFE standards for a given model year faces heavy fines. The current proposed CAFC/CAFE standards will increase fuel economy to 4.32 L/100 km or 54.5 mpg (US) between the years 2017 and 2025.

MODERN POWER PLANTS

In trying to produce more efficient vehicles, North American manufacturers put four-cylinder and other small engines into their vehicles, instead of large eight-cylinder engines. Some basic engine systems such as carburetors and ignition breaker points were replaced by electronic fuel injection and electronic ignition systems.

By the mid-1980s, the North American automobile had gained a measure of self-control over emissions and fuel efficiency through the use of computers and other electronics. Fuel and air were carefully monitored and consumed in proportions that maximized the performance of the smaller engines while minimizing the production of harmful pollutants.

After a prolonged period of economic growth in the 1980s, the demand for good performance was once again a shaping force in automotive design. Electronic sensors are now used to monitor engine functions. Computerized engine control systems control air and fuel delivery, ignition timing, emission systems operation, and a host of other related operations. The result is a clean-burning, fuel-efficient, and powerful engine **(Figure 4–3)**.

DESIGN EVOLUTION

Not too long ago, nearly every car and truck built in North America was built with body-over-frame construction, rear-wheel drive, and symmetrical designs. Today, most vehicles do not have a separate frame; instead, the frame and body are built as a single unit,

FIGURE 4–3

A late-model lightweight V-12 engine.

Chrysler Group LLC

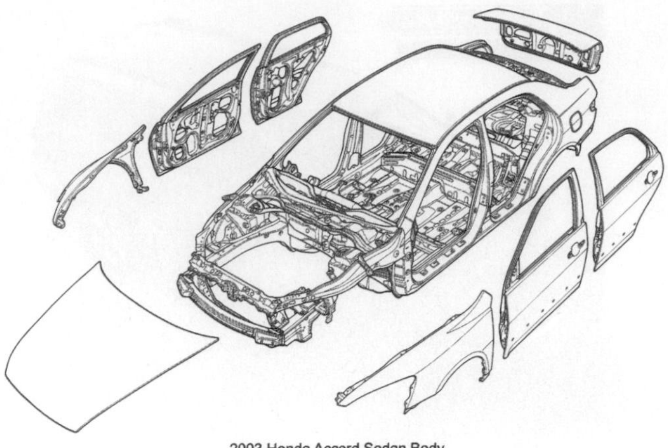

FIGURE 4–4

The structure of a unibody car.

American Honda Motor Co., Inc.

called a unibody. In 1977, when most cars were built on a frame, the average weight of a car was 2041 kg (4500 lbs.). Today, because of unibody construction and changes in materials, the average weight is 1360 kg (3000 lbs.). Most trucks are still built on a frame.

Another major influence on design was the switch from rear-wheel drive to front-wheel drive. Making this switch accomplished many things. The most notable benefits of front-wheel drive are improved traction for the drive wheels, increased interior space, shorter hood lines, and a very compact driveline. Because of the weight and loads that pickup trucks are designed to move, they remain rear-wheel drive.

Perhaps the most obvious design change through the years has been in body styles. Body styles have changed in response to the other design considerations and to trends of the day. For example, in the 1950s, North Americans had a strange preoccupation with the unknown, outer space; this led to cars that had rocketlike fins. Since then, fins have disappeared and body styles have become more rounded to reduce air drag.

Unitized Construction

A **unibody** has no separate frame **(Figure 4–4)**. It is a stressed hull structure in which each of the body parts supplies structural support and strength to the entire vehicle.

The major advantage of unibody vehicles is that they tend to be more tightly constructed because the major parts are all welded together. This design characteristic helps protect the occupants during a collision. However, it causes damage patterns that differ from those of body-over-frame vehicles. Rather than localized damage, the stiffer sections used in unibody design tend to transmit and distribute impact energy throughout more of the vehicle.

Nearly all unibodies are constructed from steel. A few cars, such as the Audi A8, use aluminum instead. An aluminum car body and frame can weigh up to 40 percent less than an identical body made of steel.

Body-over-Frame Construction

In body-over-frame construction, the frame is the vehicle's foundation. The body and all major parts of the vehicle are attached to the frame. The frame must also be strong enough to keep the other parts in alignment should a collision occur.

The frame is an independent, separate component that is not welded to any of the major units of the body. The body is generally bolted to the frame **(Figure 4–5)**. Large, specially designed rubber mounts are placed between the frame and body structure to reduce noise and vibration from entering the passenger compartment. Quite often, two layers of rubber are used in the mounting pads to provide a quieter and smoother ride. Body-over-frame designs are still used on many of today's pickup trucks, full-size vans, and a few full-size passenger cars.

The frame rails are made of stamped steel and are welded together. Some frames are made by a **hydroforming** process, which uses high-pressure water to shape the steel into the desired shape.

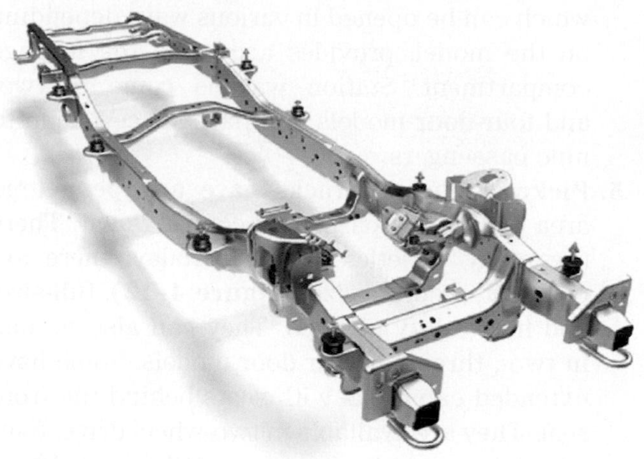

FIGURE 4–5

A typical hydroformed frame for a pickup truck.

Ford Motor Company

FIGURE 4–7

This Hyundai Genesis is a high-horsepower, rear-wheel-drive coupe.

Martin Restoule

BODY DESIGNS

Various methods of classifying vehicles exist. Vehicles may be classified by engine type, body/frame construction, fuel consumption structure, type of drive, or the classifications most common to consumers, which are body shape, seat arrangement, and number of doors. Eight basic body shapes are used today:

1. **Sedan**. A vehicle with front and back seats that accommodates four to six persons is classified as either a two- or four-door sedan **(Figure 4–6)**. Often, a two-door sedan is called a coupe **(Figure 4–7)**.
2. **Convertible**. Although they were absent from the domestic market for several years, manufacturers began offering convertible cars again in 1985 and have continued ever since. Convertibles usually have vinyl or cloth roofs that can be raised or lowered. A few late-model convertibles feature a folding metal roof that tucks away in the trunk when it is down **(Figure 4–8)**. Convertibles are available in two- and four-door models. Some convertibles have both front and rear seats. Those without rear seats are commonly referred to as sports cars **(Figure 4–9)**.
3. **Liftback** or **hatchback**. The distinguishing feature of this vehicle is its rear luggage compartment, which is an extension of the passenger compartment. Access to the luggage compartment is gained through an upward opening hatch-type door **(Figure 4–10)**. A car of this design can be a three- or five-door model. The third or fifth door is the rear hatch.
4. **Station wagon**. A station wagon **(Figure 4–11)** is characterized by its roof, which extends straight back, allowing a spacious interior luggage compartment in the rear. The rear door,

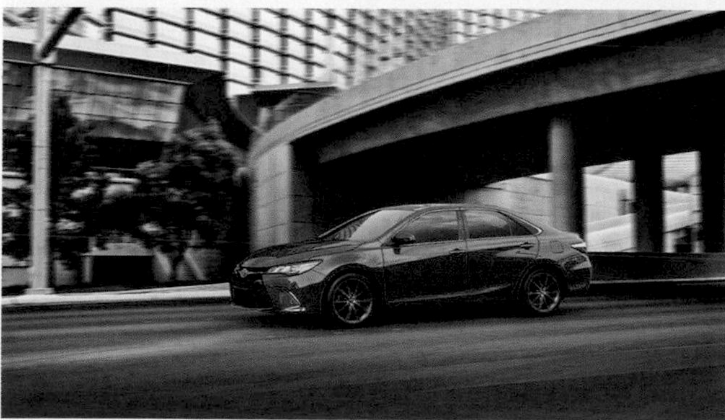

FIGURE 4–6

This Toyota Camry is an example of a typical late-model sedan.

Toyota Canada

FIGURE 4–8

This BMW 435i Cabriolet features a retractable steel hardtop.

Martin Restoule

which can be opened in various ways depending on the model, provides access to the luggage compartment. Station wagons come in two- and four-door models and have space for up to nine passengers.

5. **Pickups**. Pickup trucks have an open cargo area behind the passenger compartment. There are many varieties available today; there are compact, medium-size **(Figure 4–12)**, full-size, and heavy-duty pickups. They can also be had in two-, three-, or four-door models. Some have extended cab areas with seats behind the front seat. They are available in two-wheel drive, four-wheel drive, or all-wheel drive **(Figure 4–13)**.
6. **Vans**. The van body design **(Figure 4–14)** has a tall roof and a totally enclosed large cargo or passenger area. Vans can seat from two to twelve passengers, depending on size and design. Basically there are two sizes of vans: minivans and full size.

FIGURE 4–9

Two-passenger convertibles are called sports cars. The Mercedes SLK 350 is the third generation of the popular SLK model line.

Martin Restoule

FIGURE 4–10

The distinguishing feature of the Hyundai Veloster hatchback is its rear luggage compartment, which is an extension of the passenger compartment. It also features three passenger doors.

Martin Restoule

FIGURE 4–11

A late-model Subaru Outback station wagon.

Subaru Canada Inc.

FIGURE 4–12

The Toyota Tacoma is an example of a medium-size pickup truck with four doors.

Martin Restoule

FIGURE 4–13

A full-sized Dodge Ram pickup with four-wheel drive.

Martin Restoule

FIGURE 4–14

This Chrysler Town and Country is an example of a late-model minivan. Full-size vans are also available.

Chrysler Group LLC

7. **Sport utility vehicles (SUVs)**. This classification of vehicles covers a range of body designs. SUVs are best described as multipurpose vehicles and, depending on their size and design, can carry a wide range of passengers. A good majority of SUVs have four-wheel drive, although some do not. The classification of SUV implies that the vehicles are designed to do well off the road. This is not always the case. Buyers may choose SUVs for status, size, utility, and/or off-road play. Most small SUVs are based on an automobile platform and take on many different looks and features **(Figure 4–15)**. Mid-size SUVs are larger and typically offer more features and comfort **(Figure 4–16)**. There are many large SUVs available **(Figure 4–17)**. These vehicles can seat up to nine adults and tow up to six tonnes.
8. **Crossover vehicles**. Crossover vehicles are a new trend in automotive offerings. These vehicles are a mixture of a station wagon and an SUV. They have SUV features but are not quite the size of an SUV **(Figure 4–18)**.

FIGURE 4–15

This Toyota RAV4 is considered a small SUV and has many unique features, including all-wheel drive.

Martin Restoule

FIGURE 4–16

This Lexus GX460 is a mid-size SUV with a V-8 engine and many luxuries.

Toyota Canada

FIGURE 4–17

The Mercedes ML 350 is a good example of a large SUV.

Martin Restoule

FIGURE 4–18

The Honda Crosstour is an example of a crossover vehicle. It features the benefits and comfort of a station wagon with four-wheel drive like an SUV.

Martin Restoule

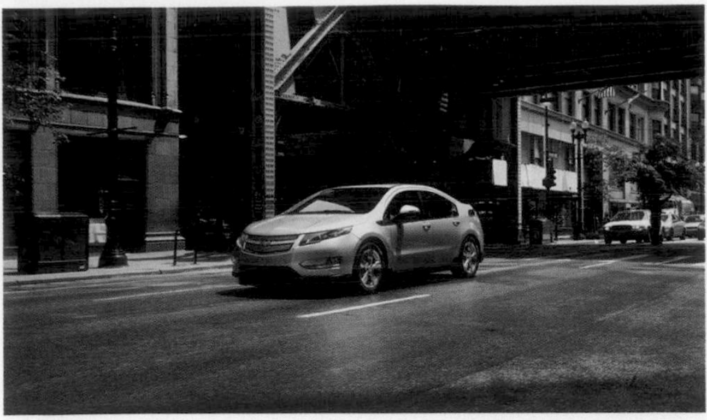

FIGURE 4–19

The Chevrolet Volt is a series hybrid and is referred to as an extended range electric vehicle.

General Motors

In addition to the eight body classifications listed above, there are two additional classes of vehicles defined by how they are powered:

- Hybrid vehicles. There are many hybrid vehicles available today. Each of these incorporates different technologies that combine two power sources to achieve low emissions and excellent fuel mileage **(Figure 4–19)**.
- Electric vehicles. As time passes, more manufacturers are introducing pure electric vehicles. One of the first to be available is the Nissan Leaf **(Figure 4–20)**.

TECHNOLOGICAL ADVANCES

Perhaps the thing that has brought about the greatest change in the automotive industry is the computer. Not only are engine support systems controlled by computers, but nearly every other major system on a car has some sort of electronic control. Initially, electronic controls were added to help maintain low emission levels from vehicles. As these controls became more sophisticated, they improved engine performance. Electronic controls have done so much for engine efficiency that some of the early emission control systems have been eliminated.

FIGURE 4–20

The Nissan Leaf is a pure electric vehicle. This vehicle is not assisted by any type of engine.

Nissan Canada

The use of electronics has and will continue to change the automobile, as will advances made in other technologies. New composite materials are being used for engine parts. Soon, most of the engine may be made with plastic-based or ceramic materials. Steel body panels are being replaced with aluminum and plastic parts that are bonded to a frame with special adhesives. The remaining steel body parts are thin, high-strength steel. Steel, aluminum, and composites are being used to reduce vehicle weight. Reduced weight results in better performance and fuel economy, especially when the weight of moving engine parts is reduced. A loss of weight here reduces the frictional drag of the engine, resulting in great increases in power output.

Vehicles powered by electric motors have also been introduced by the manufacturers. Some use a battery as the power source for the motor; others use a small automotive-type engine to charge a battery and to act as a supplemental power source to the electric motor. There are even vehicles that use hydrogen to feed a fuel cell that is capable of generating electricity for the electric motor.

The way automobiles of today and tomorrow look and run is being shaped by the constant influx of new technology. Automotive technicians must stay abreast of these changes and be able to diagnose and service the new systems.

THE BASIC ENGINE

The engine provides the power to drive the wheels of the vehicle. All automobile engines, both gasoline and diesel, are classified as internal combustion engines because the combustion or burning that provides the power to move the vehicle takes place inside the engine. **Combustion** is the burning of an air/fuel mixture. As a result of combustion, large amounts of pressure are generated in the engine. This pressure or energy is used to power the car. The engine must be built strong enough to hold the pressure and temperatures formed by combustion.

Diesel engines have been around a long time and are found mostly in big heavy-duty trucks. However, they are also used in some pickup trucks and are becoming more common in automobiles due

FIGURE 4–21

A four-cylinder automotive diesel engine.

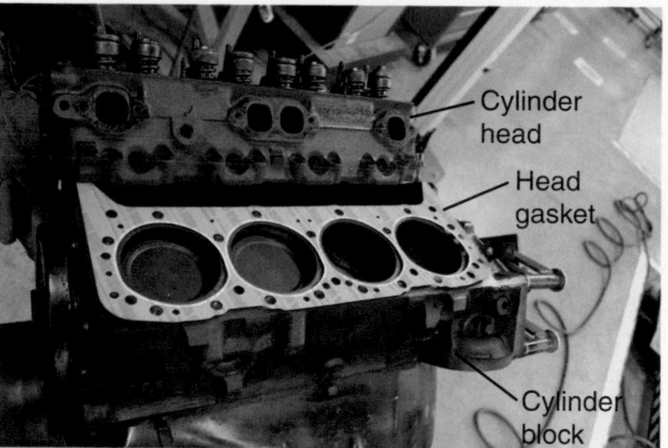

FIGURE 4–22

The two major units of an engine, the cylinder block and cylinder head, are sealed together with a gasket and bolted together.

Martin Restoule

to their fuel economy **(Figure 4–21)**. Although the construction of gasoline and diesel engines is similar, their operation is quite different. Diesel engines achieve better fuel economy than gasoline engines of the same size. With new technologies and the cleaner fuels, their emissions levels can be comparable to the best gasoline engines.

A gasoline engine relies on a mixture of fuel and air that is ignited by a spark to produce power. A diesel engine also uses fuel and air, but it does not need a spark to cause ignition. A diesel engine is often called a compression ignition engine. This is because its intake air is tightly compressed, which greatly raises its temperature. The fuel is then injected into the compressed air. The heat of the compressed air ignites the fuel, and combustion takes place. The following sections cover the basic parts and the major systems of a gasoline engine.

Cylinder Block

The biggest part of the engine is the **cylinder block**, which is also called an **engine block (Figure 4–22)**. The cylinder block is a large casting of metal (cast iron or aluminum) that is cast and drilled with holes to allow for the passage of lubricants and coolant within the block and to provide spaces for the movement of mechanical parts. The block contains the cylinders, which are round passageways fitted with pistons. The block houses or holds the major mechanical parts of the engine.

Cylinder Head

The **cylinder head** fits on top of the cylinder block to close off and seal the top of the cylinders. The **combustion chamber** is an area into which the air/fuel mixture is compressed and burned. The cylinder head contains all or most of the combustion chamber. The cylinder head also contains **ports**, which are passageways through which the air/fuel mixture enters and burned gases exit the cylinder. A cylinder head can be made of cast iron or aluminum.

Piston

The combustion of air and fuel takes place between the cylinder head and the top of the piston. The **piston** is a can-shaped part closely fitted inside the cylinder **(Figure 4–23)**. In a four-stroke cycle engine, the piston moves through four different movements or strokes to

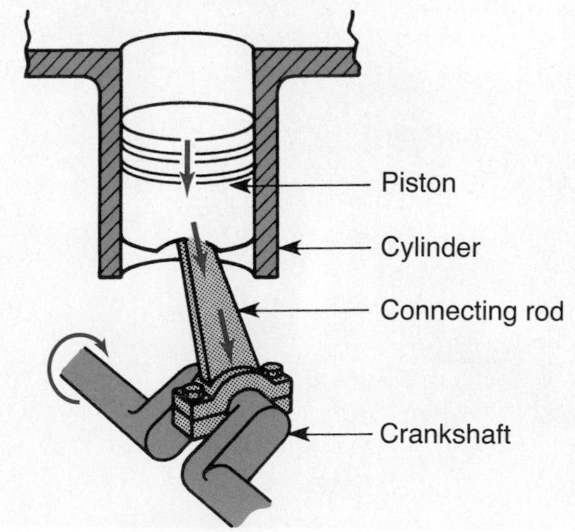

FIGURE 4–23

The engine's pistons fit tightly in the cylinders and are connected to the engine's crankshaft with connecting rods.

complete one cycle. These four are the intake, compression, power, and exhaust strokes. On the intake stroke, the piston moves downward, and a charge of air/fuel mixture is introduced into the cylinder. As the piston travels upward, the air/fuel mixture is compressed in preparation for burning. Just before the piston reaches the top of the cylinder, ignition occurs and combustion starts. The pressure of expanding gases forces the piston downward on its power stroke. When it reciprocates, or moves upward again, the piston is on the exhaust stroke. During the exhaust stroke, the piston pushes the burned gases out of the cylinder.

Connecting Rods and Crankshaft

The reciprocating motion of the pistons must be converted to rotary motion before it can drive the wheels of a vehicle. This conversion is achieved by linking the piston to the **crankshaft** with a **connecting rod**. As the piston is forced down on the power stroke, the connecting rod transfers this force to the crankshaft, causing it to rotate. The end of the crankshaft is connected to the transmission to continue the power flow through the drivetrain and to the wheels.

Valve Train

A **valve train** is a series of parts used to open and close the intake and exhaust ports. A valve is a movable part that opens and closes a passageway. A camshaft controls the movement of the valves **(Figure 4–24)**, causing them to open and close at the proper time. Springs are used to close the valves.

FIGURE 4–24

The valve train for one cylinder. Notice that this engine uses two intake and two exhaust valves in each cylinder.

FIGURE 4–25

The blue manifold is the intake manifold and the red manifold is for the exhaust.

Ford Motor Company

Manifolds

A **manifold** is a ductwork assembly used to direct the flow of gases to or from the combustion chambers. Two separate manifolds are attached to the cylinder head **(Figure 4–25)**. The **intake manifold** delivers a mixture of air and fuel to the intake ports. The **exhaust manifold** mounts over the exhaust ports and carries exhaust gases away from the cylinders.

ENGINE SYSTEMS

The following sections present a brief explanation of the systems that help an engine run and keep running.

Lubrication System

The moving parts of an engine need constant lubrication. Lubrication limits the amount of wear and reduces the amount of **friction** in the engine.

Motor or engine oil is the fluid used to lubricate the engine. Several litres of oil are stored in an **oil pan** bolted to the bottom of the engine block. The oil pan is also called the crankcase or **oil sump**. When the engine is running, an oil pump draws oil from the pan and forces it through oil galleries, which are small passageways that direct the oil to the moving parts of the engine.

Oil from the pan passes through an oil filter before moving through the engine **(Figure 4–26)**. The filter removes dirt and metal particles from the oil. Premature wear and damage to parts can result from dirt in the oil. Regular replacement of the oil filter and oil is an important step in a preventive maintenance program.

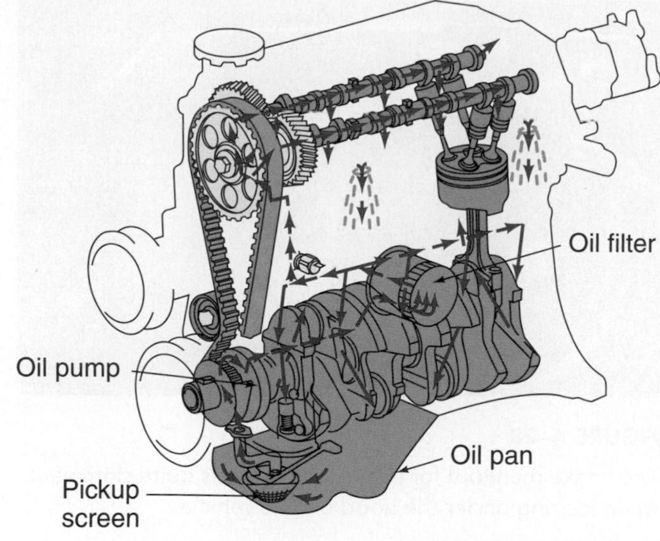

FIGURE 4–26

Oil flow in a typical engine's lubrication system.

Cooling System

The combustion of the air/fuel mixture produces large amounts of heat in the engine. This heat must not be allowed to exceed safe limits and must be maintained at proper operating temperatures, or it can easily damage and warp engine components. To prevent this, engines have a cooling system **(Figure 4–27)**.

The most common way to remove excess heat from an engine is to circulate a liquid coolant through passages in the engine block and cylinder head. An engine can also be cooled by passing air over and around the engine. Air-cooled engines are no longer installed in automobiles today because it is very difficult to maintain a constant temperature at the cylinders. If the engine is kept at a constant temperature, it will run more efficiently. A liquid cooling system also has a supply of hot coolant available to operate a heater for the passenger compartment. The cooling

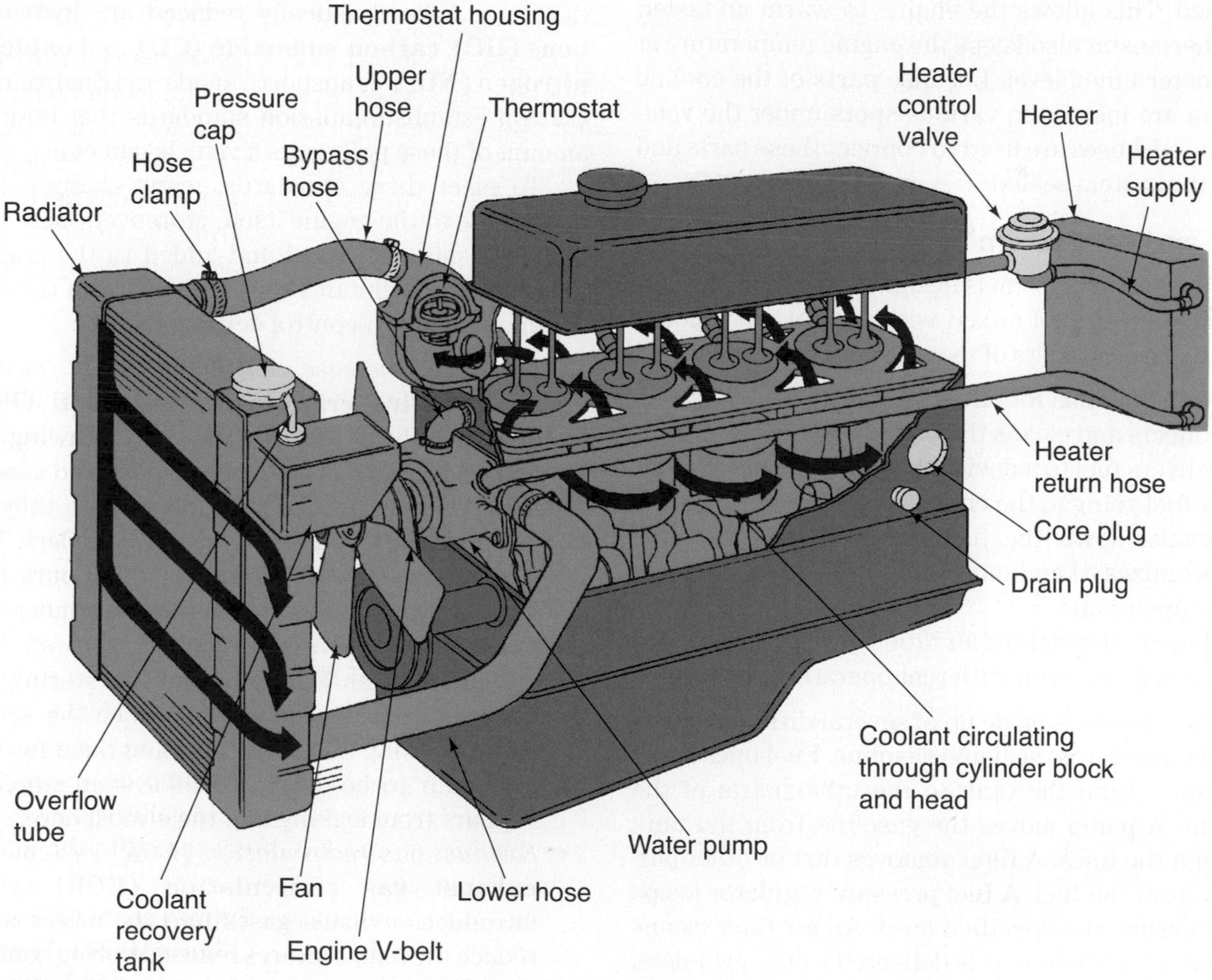

FIGURE 4–27

A typical cooling system.

Gates Corporation

system is designed to cool the engine, not the passengers inside the car. Cooling the passengers is the responsibility of the air-conditioning system.

A typical cooling system relies on a **water pump** that circulates the coolant through the system. The pump is driven by the engine or an electric motor. The coolant, a mixture of water and antifreeze, is pushed through passages called **water jackets** in the cylinder block and head to remove excess heat from the area around the cylinders. The heat picked up by the coolant is sent to the **radiator**, which transfers the coolant's heat to the outside air as the coolant flows through its tubes. To help remove the heat from the coolant, a cooling fan is used to pull cool outside air in through the fins of the radiator.

To raise the boiling point of the coolant, the cooling system is pressurized. To maintain this pressure, a radiator or **pressure cap** is fitted to the radiator. A **thermostat** is used to block off circulation in the system until a preset temperature is reached. This allows the engine to warm up faster. The thermostat also keeps the engine temperature at a predetermined level. Because parts of the cooling system are located in various spots under the vehicle's hood, hoses are used to connect these parts and keep the system sealed.

Fuel and Air System

The fuel and air system is designed to supply the correct amount of fuel mixed with the correct amount of air to the cylinders of the engine. This system also

- stores the fuel for later use;
- collects and cleans the outside air;
- delivers fuel to a device that controls the amount of fuel going to the engine;
- breaks down the fuel into very fine droplets (atomizes it) and mixes the fuel with air to form a vapour; and
- changes the fuel and air ratios to meet the needs of the engine during different operating conditions.

The fuel system is made up of several different parts. A fuel tank stores the liquid gasoline. Fuel lines carry the liquid from the tank to the other parts of the system. A pump moves the gasoline from the tank through the lines. A filter removes dirt or other particles from the fuel. A fuel pressure regulator keeps the pressure at a specified level. An air filter cleans the outside air before it is delivered to the cylinders. Fuel injectors or a carburetor mix the liquid gasoline with air for delivery to the cylinders. An intake manifold directs the air/fuel mixture to each of the cylinders **(Figure 4–28)**.

FIGURE 4–28

The intake manifold for this four-cylinder is quite dominant when looking under the hood on this vehicle.

Chrysler Group LLC

Emission Control System

Today's engines have been engineered to emit very low amounts of certain pollutants. The pollutants that have been drastically reduced are **hydrocarbons (HC)**, **carbon monoxide (CO)**, and **oxides of nitrogen (NO_x)**. Transport Canada and Environment Canada establish emission standards that limit the amount of these pollutants a vehicle can emit.

To meet these standards, many changes have been made to the engine itself. Moreover, there have been systems developed and added to the engines to reduce the pollutants they emit. A list of the most common pollution-control devices follows:

- *Positive crankcase ventilation (PCV) system.* The **positive crankcase ventilation (PCV)** system reduces HC emissions by drawing fuel and oil vapours from the crankcase and sending them into the intake manifold, where they are delivered to and burned in the cylinders. This system prevents the pressurized vapours from escaping the engine and entering the atmosphere.
- *Evaporative emission control system.* This system reduces HC emissions by storing fuel vapours from the fuel system when the vehicle is not running and then releasing them into the intake air to be burned. This system stops the vapours from leaking into the atmosphere.
- *Exhaust gas recirculation (EGR) system.* The **exhaust gas recirculation (EGR)** system introduces exhaust gases into the intake air to reduce the temperatures reached during combustion. This reduces the chances of NO_x forming during combustion.
- *Catalytic converter.* Located in the exhaust system, the catalytic converter allows for the

burning or converting of HC, CO, and NO_x into harmless substances, such as water.

- *Air injection system*. This system reduces HC emissions by introducing fresh air into the exhaust stream to cause minor combustion of the HC in the engine's exhaust. Most newer engines do not need this system to achieve acceptable emissions levels.

Diesel Emission Controls

Many of the systems used on gasoline engines are also used on diesel engines to reduce their emissions. In the past, emissions have always been an obstacle to having diesel cars. Today, that problem is rapidly disappearing. Many new diesel vehicles have an assortment of traps and filters to clean the exhaust before it leaves the tailpipe; others use selective catalytic reduction (SCR) systems. SCR is a process in which a reductant is injected into the exhaust stream and then absorbed onto a catalyst. A reductant removes oxygen from a substance and combines the oxygen with another substance to form another compound. In this case, oxygen is separated from the NO_x and is combined with hydrogen to form water.

Exhaust System

During the exhaust stroke, the engine's pistons push the burned air/fuel mixture, or exhaust, out of the cylinder and into the exhaust manifold. From the manifold, the gases travel through the exhaust system until they are expelled into the atmosphere **(Figure 4–29)**. The exhaust system is designed to

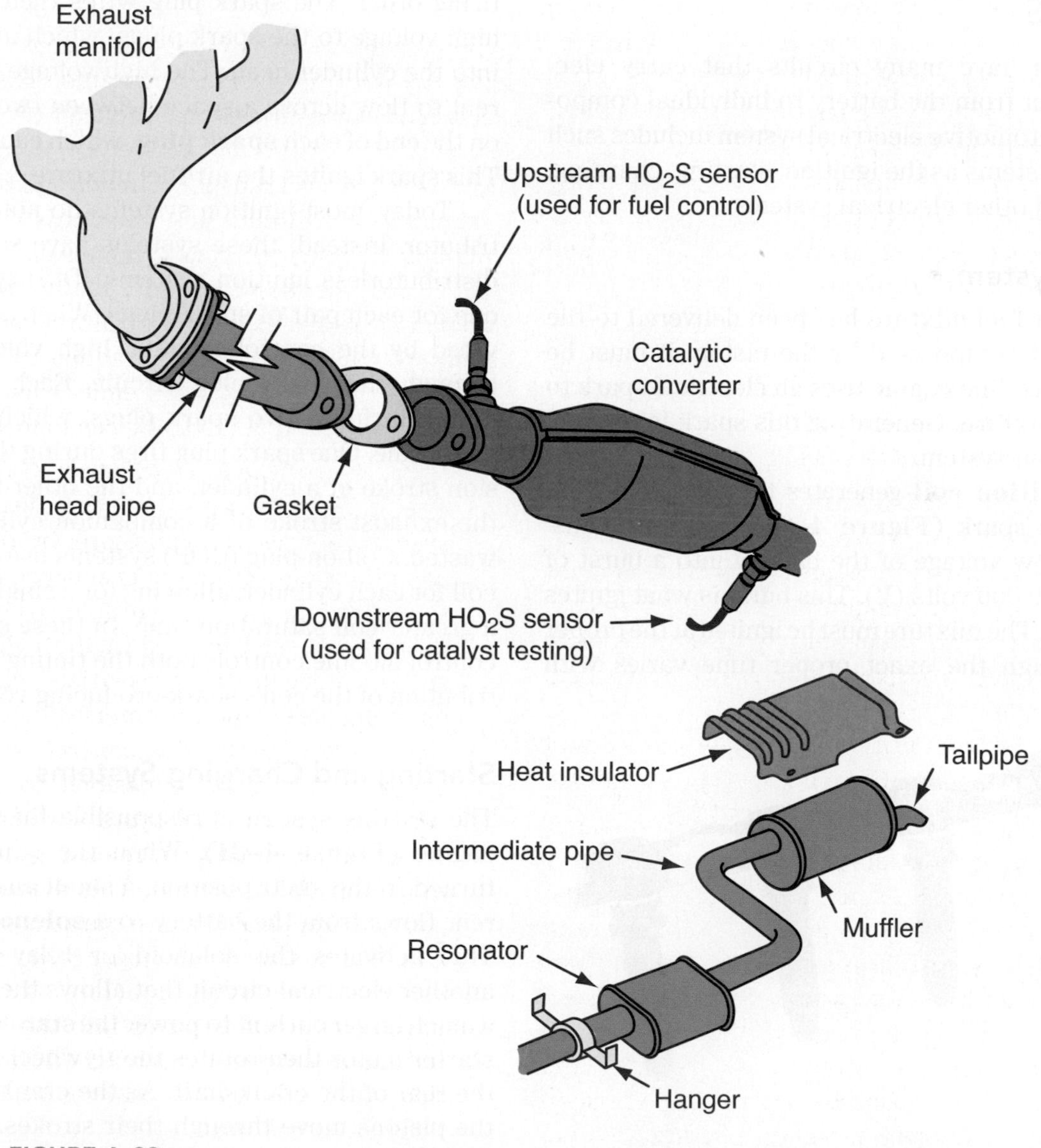

FIGURE 4–29

A typical exhaust system on a late-model car.

direct toxic exhaust fumes away from the passenger compartment, to quiet the sound of the exhaust pulses, and to burn or catalyze pollutants in the exhaust. A typical exhaust system contains the following components:

- Exhaust manifold and gasket
- Exhaust pipe, seal, and connector pipe
- Intermediate pipes
- Catalytic converter(s)
- Muffler
- Resonator
- Tailpipe
- Heat shields
- Clamps, gaskets, and hangers

ELECTRICAL AND ELECTRONIC SYSTEMS

Automobiles have many circuits that carry electrical current from the battery to individual components. An automotive electrical system includes such major subsystems as the ignition, starting, charging, lighting, and other electrical systems.

Ignition System

After the air/fuel mixture has been delivered to the cylinder and compressed by the piston, it must be ignited. A gasoline engine uses an electrical spark to ignite the mixture. Generating this spark is the role of the ignition system.

The **ignition coil** generates the electricity that creates this spark **(Figure 4–30)**. The coil transforms the low voltage of the battery into a burst of 30 000 to 100 000 volts (V). This burst is what ignites the mixture. The mixture must be ignited at the proper time, although the exact proper time varies with engine design; ignition must occur at a point before the piston has completed its compression stroke.

FIGURE 4–30

An ignition module and coil assembly for four cylinders.

In most engines, the motion of the piston and the rotation of the crankshaft are monitored by a **crankshaft position (CKP) sensor**. The sensor electronically tracks the position of the crankshaft and relays that information to a control module. Based on input from the crankshaft position sensor, an electronic engine control computer turns battery current to the coil on and off at precisely the right time so that the voltage surge arrives at the cylinder at the right time.

The voltage surge from the coil must be distributed to the correct cylinder, because only one cylinder is fired at a time. In earlier systems, this was the job of the **distributor**. A distributor is driven by the camshaft. It transfers the high-voltage surges from the coil to the spark plug wires in the correct firing order. The spark plug wires then deliver the high voltage to the spark plugs, which are threaded into the cylinder head. The high voltage causes current to flow across a space between two electrodes on the end of each **spark plug**, which causes a spark. This spark ignites the air/fuel mixture.

Today, most ignition systems do not have a distributor. Instead, these systems have several coils; distributorless ignition systems (DIS) typically use one for each pair of spark plugs. When a coil is activated by the control module, high voltage is sent through the spark plug circuit. Each spark plug circuit includes two spark plugs, which fire at the same time. One spark plug fires during the compression stroke of a cylinder, and the other fires during the exhaust stroke of a companion cylinder and is wasted. Coil-on-plug (COP) systems use a dedicated coil for each cylinder, allowing for a single spark and a greater coil saturation time. In these systems, the control module controls both the timing and the distribution of the coil's spark-producing voltage.

Starting and Charging Systems

The starting system is responsible for starting the engine **(Figure 4–31)**. When the ignition key is turned to the START position, a small amount of current flows from the battery to a **solenoid**, or relay. This activates the solenoid or relay and closes another electrical circuit that allows the full battery a much larger current to power the starter motor. The starter motor then rotates the flywheel mounted on the rear of the crankshaft. As the crankshaft turns, the pistons move through their strokes. At the correct time for each cylinder, the ignition system provides the spark to ignite the air/fuel mixture. If good

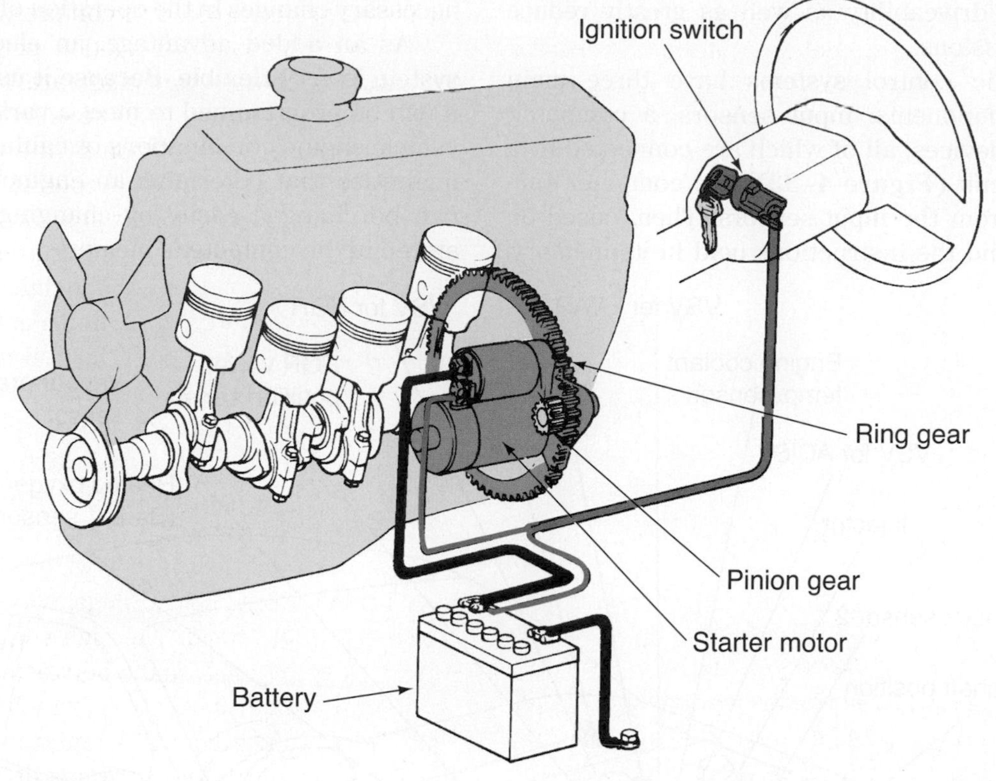

FIGURE 4–31

A typical starting system.

combustion takes place, the engine will now rotate on its own without the need of the starter motor. The ignition key is now allowed to return to the ON position. From this point on, the engine will continue to run until the ignition key is turned off.

The power to operate the starter circuit comes from the battery. While the starter is rotating the crankshaft, it uses a lot of electricity. This tends to lower the amount of stored electrical power in the battery. Therefore, a system is needed to recharge the battery so that engine starts can be made in the future.

The charging system is designed to recharge and maintain the battery's state of charge. It also provides electrical power for the ignition system, air conditioner, heater, lights, radio, and all electrical accessories when the engine is running.

The charging system includes an AC generator (alternator), voltage regulator, indicator light, and the necessary wiring **(Figure 4–32)**. Rotated by the engine's crankshaft through a drive belt, the **AC generator** converts mechanical energy into electrical energy. When the output or electrical current from the charging system flows back to the battery, the battery is being charged. When the current flows out of the battery, the battery is said to be discharging.

Electronic Engine Controls

Nearly all vehicles have an electronic engine control system. This is a system composed of many electronic and electromechanical parts. It is designed to continuously monitor the engine and to make adjustments that will allow the engine to run more efficiently. Electronic engine control systems

FIGURE 4–32

The major components of a late-model AC generator.

Robert Bosch GmnH

dramatically improve fuel efficiency, engine performance, and driveability as well as greatly reduce exhaust emissions.

Electronic control systems have three main types of components: input sensors, a computer, and output devices, all of which are connected by a network circuit **(Figure 4–33)**. The computer analyzes data from the input sensors. Then, based on the inputs and the instructions held in its memory, the computer directs the output devices to make the necessary changes in the operation of some systems.

As an added advantage, an electronic control system is very flexible. Because it uses a computer, it can be programmed to meet a variety of different vehicle engine combinations or calibrations. Critical quantities that determine an engine's performance can be changed easily by changing data that are stored in the computer's memory.

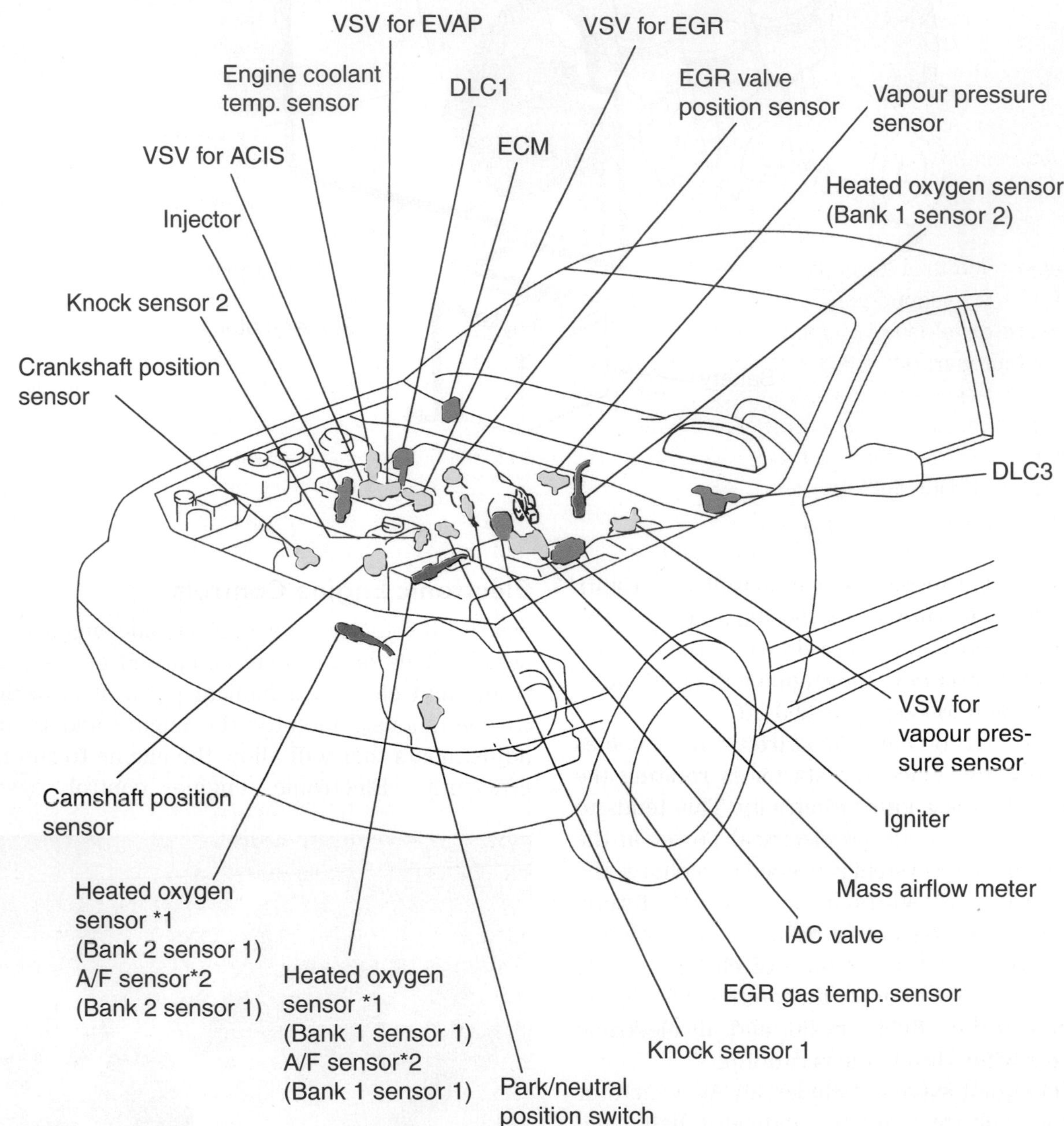

FIGURE 4–33

Late-model electronic engine control systems are made up of many different sensors and actuators and a central computer or control module.

Reprinted with permission

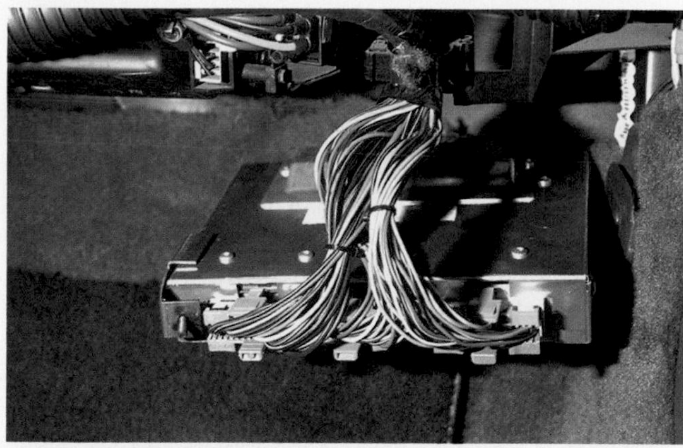

FIGURE 4–34

A typical automotive computer.

On-Board Diagnostics

Today's engine control systems are **on-board diagnostic (OBD-II)** second-generation systems. These systems were developed to ensure proper emission control system operation for the vehicle's lifetime by monitoring emission-related components and systems for deterioration and malfunction. The OBD system consists of engine and transmission control modules, their sensors and actuators, along with the diagnostic software.

The computer **(Figure 4–34)** can detect system problems even before the driver notices a driveability problem, because many problems that affect emissions can be electrical or even chemical in nature.

When the OBD system determines that a problem exists, a corresponding diagnostic trouble code is stored in the computer's memory. The computer also illuminates a yellow dashboard light or MIL (malfunction indicator lamp) displaying CHECK ENGINE, SERVICE ENGINE SOON, or an engine symbol. This light informs the driver of the need for service, not of the need to stop the vehicle.

A blinking or flashing dashboard lamp indicates a rather severe level of engine misfire. When this occurs, the driver should reduce engine speed and load and have the vehicle serviced as soon as possible. After the problem has been fixed, the dashboard lamp will turn off automatically or must be reset.

HEATING AND AIR-CONDITIONING SYSTEMS

Heating and air-conditioning systems do little for the operation of a vehicle; they merely provide comfort for the vehicle's passengers. The heating system basically adds heat to the vehicle's interior, whereas air-conditioning removes heat. To do this, the systems rely on many parts to put basic theories to work.

Heating Systems

To meet federal safety standards, all vehicles must be equipped with passenger compartment heating and windshield defrosting systems. The main components of an automotive heating system are the heater core, the heater control valve, the blower motor and fan, and the heater and defroster ducts. The heating system works with the engine's cooling system and converts the heat from the coolant circulating inside the engine to hot air, which is blown into the passenger compartment. A heater hose transfers hot coolant from the engine to the heater control valve and then to the heater core inlet. As the coolant circulates through the core, heat is transferred from the coolant to the tubes and fins of the core. Air blown through the core by the blower motor and fan then picks up the heat from the surfaces of the core and transfers it into the passenger compartment. After giving up its heat, the coolant is then pumped out through the heater core outlet, where it is returned to the engine's cooling system to be heated again.

Transferring heated air from the heater core to the passenger compartment is the job of the heater and defroster ducts. The ducts are typically part of a large plastic shell that connects to the necessary inside and outside vents. Contained inside the duct are also the doors required to direct air to the floor, dash, and/or windshield.

Air-Conditioning Systems

In an automotive air-conditioning (A/C) system, heat is removed from the passenger compartment and moved to the outside of the vehicle.

The substance used to remove heat is called the **refrigerant**. To understand how a refrigerant is used to cool the interior of a vehicle, the effects of pressure and temperature on it must be first understood. If the pressure of the refrigerant is high, so is its temperature. Likewise, if the pressure is low, so is its temperature. Therefore, changing its pressure can change the refrigerant's temperature.

To absorb heat, the temperature and pressure of the refrigerant are kept low. To get rid of the heat, the temperature and pressure are high. As the refrigerant absorbs heat, it evaporates or changes from a liquid to a vapour. As it dissipates heat, it condenses and changes from a vapour to a liquid. These two changes of state occur continuously as the refrigerant circulates through the system.

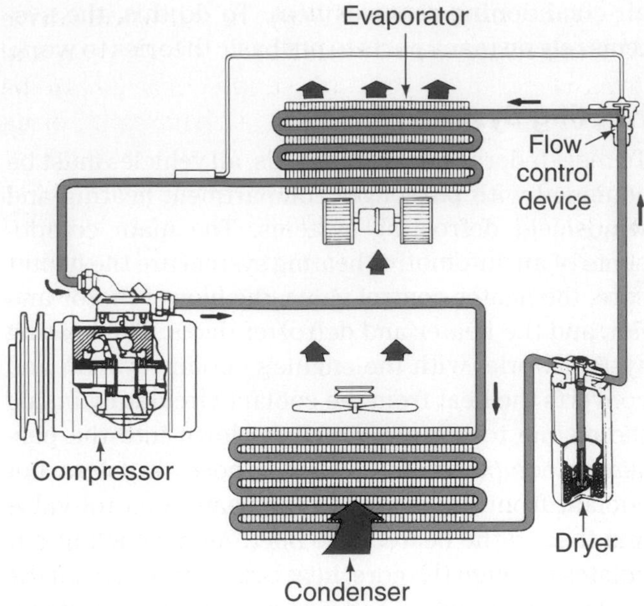

FIGURE 4–35

A simple look at an air-conditioning system. The blue signifies low pressure and the red is high pressure.

An A/C system is a closed, pressurized system. It consists of a compressor, a condenser, a receiver/dryer or accumulator, an expansion valve or orifice tube, and an evaporator. The best way to understand the purpose of the components is to divide the system into two sides: the high side and the low side. High side refers to the side of the system that is under high pressure and high temperature. Low side refers to the low-pressure, low-temperature side **(Figure 4–35)**.

COMPRESSOR The **compressor** separates the high and low sides of the system. Its primary purpose is to draw the low-pressure and low-temperature vapour from the evaporator and compress this vapour into high-temperature, high-pressure vapour. The compressor also circulates or pumps the refrigerant through the system. The compressor is located on the engine and is driven by the engine's crankshaft via a drive belt.

Compressors are equipped with an electromagnetic clutch as part of the compressor pulley assembly. The clutch is designed to engage the pulley to the compressor shaft when the clutch coil is energized. When the clutch is not engaged, the compressor shaft does not rotate, and the pulley freewheels. The clutch provides a way for turning the compressor on or off.

CONDENSER The **condenser** consists of coiled tubing mounted in a series of thin cooling fins to provide maximum heat transfer in a minimum amount of space. The condenser is normally mounted just in front of the vehicle's radiator. It receives the full flow of ram, or forced, air from the movement of the vehicle or airflow from the radiator fan when the vehicle is standing still.

The condenser condenses or liquefies the high-pressure, high-temperature vapour coming from the compressor. To do so, it must give up its heat. Very hot, high-pressure refrigerant vapour enters the inlet at the top of the condenser, and as the hot vapour passes down through the condenser coils, heat moves from the refrigerant into the cooler air that flows across the coils and fins. This loss of heat causes the refrigerant to change from a high-pressure hot vapour to a high-pressure warm liquid. The high-pressure warm liquid flows from the bottom of the condenser to the receiver/dryer or to the refrigerant metering device if an accumulator is used.

RECEIVER/DRYER The **receiver/dryer** is a storage tank for the liquid refrigerant from the condenser. The refrigerant flows into the receiver tank, which contains a bag of desiccant (moisture-absorbing material). The desiccant absorbs unwanted water and moisture in the refrigerant.

ACCUMULATOR Most late-model systems have an **accumulator** rather than a receiver/dryer. The accumulator is connected into the low side at the outlet of the evaporator. The accumulator contains a desiccant and is designed to store excess refrigerant. If liquid refrigerant flows out of the evaporator, it will be collected by and stored in the accumulator. The main purpose of an accumulator is to prevent liquid from entering the compressor.

THERMOSTATIC EXPANSION VALVE/ORIFICE TUBE The refrigerant flow to the evaporator must be controlled to obtain maximum cooling while ensuring complete evaporation of the liquid refrigerant within the evaporator. This is the job of a **thermostatic expansion valve (TEV or TXV)** or a fixed orifice tube. The TEV is mounted at the inlet to the evaporator and separates the high-pressure side of the system from the low-pressure side. The TEV regulates the refrigerant flow to the evaporator by balancing the inlet flow to the outlet temperature.

Like the TEV, the **orifice tube** is the dividing point between the high- and low-pressure sides of the system. However, its metering or flow-rate control does not depend on comparing evaporator pressure and temperature. It is a fixed orifice. The flow rate is determined by pressure difference across the orifice and by the additional cooling of the refrigerant in the bottom of the condenser after it has changed from vapour to liquid.

EVAPORATOR The **evaporator** is made up of tubes that are mounted in a series of thin cooling fins. The evaporator is usually located beneath the dashboard or instrument panel.

The low-pressure, low-temperature liquid refrigerant from the TEV or orifice tube enters the evaporator as a spray. The heat at the evaporator causes the refrigerant to boil and change into a vapour. The transfer of heat from the evaporator to the refrigerant causes the evaporator to get cold. The hot air from inside the vehicle is sent past the evaporator to provide heat for refrigerant evaporation. As the process of heat loss from the air to the evaporator core surface is taking place, any moisture in the air condenses on the outside of the evaporator core and is drained off as water. This dehumidification of air adds to passenger comfort.

REFRIGERANT LINES There are three major refrigerant lines. Suction lines are located between the outlet side of the evaporator and the inlet side or suction side of the compressor. They carry the low-pressure, low-temperature vapour to the compressor. The discharge or high-pressure line connects the compressor to the condenser. The liquid lines connect the condenser to the receiver/dryer and the receiver/dryer to the inlet side of the expansion valve. Through these lines, the refrigerant travels in its path from a gas state (compressor outlet) to a liquid state (condenser outlet) and then to the inlet side of the expansion valve, where it vaporizes at the evaporator.

DRIVETRAIN

The **drivetrain** is made up of all components that transfer power from the engine to the driving wheels of the vehicle. The exact components used in a vehicle's drivetrain depend on whether the vehicle is equipped with rear-wheel drive, front-wheel drive, or four-wheel drive.

Today, most cars are front-wheel drive (FWD). Some larger luxury and performance cars are rear-wheel drive (RWD). Most pickup trucks and large SUVs are also RWD vehicles. Power flow in RWD vehicles passes through the **clutch** or **torque converter**, manual or automatic transmission, and the driveline (driveshaft assembly). Then it goes through the rear final drive assembly, the rear-driving axles, and onto the rear wheels.

Power flow through the drivetrain of FWD vehicles passes through the clutch or torque converter, then moves through the transaxle, the driving axles, and onto the front wheels.

Four-wheel drive (4WD) or all-wheel drive (AWD) vehicles combine features of both rear- and front-wheel-drive systems so that power can be delivered to all wheels either on a permanent or an on-demand basis. Typically, if a truck, pickup, or large SUV has 4WD, the system is based on RWD, and a front drive axle is added. When a car has AWD or 4WD, the drivetrain is a modified FWD system. Modifications include a rear drive axle and an assembly that transfers some of the power to the rear axle.

Transmissions and Transaxles

A transmission is found in RWD vehicles. Transaxles are used in FWD vehicles. Both provide various drive gears that allow the vehicle to move forward and in reverse. The primary difference between a transmission and a transaxle is that a transaxle unit also contains the final drive gears. The final drive gears for RWD vehicles are contained in a separate axle housing. Transmissions and transaxles can be either manually shifted or shifted automatically.

Clutch

A clutch is used with manual transmissions/transaxles. It mechanically connects the engine's flywheel to the transmission/transaxle input shaft **(Figure 4–36)**. This is accomplished by a special friction disc that is splined to the input shaft of the transmission. When the clutch is engaged, the friction disc contacts the flywheel and transfers power to the input shaft.

When stopping, starting, and shifting from one gear to the next, the clutch is disengaged by pushing down on the clutch pedal. This moves the clutch disc away from the flywheel, stopping the power flow to the transmission. The driver can then shift gears without damaging the transmission or transaxle. Releasing the clutch pedal re-engages the clutch and allows power to flow from the engine to the transmission.

Manual Transmission

A manual or standard transmission is one in which the driver manually selects the gear of choice. Proper gear selection allows for good driveability and requires some driver education.

Whenever two or three gears have their teeth meshed together, a gearset is formed. The movement of one gear in the set will cause the others to move. If any of the gears in the set are a different size than the others, the gears will move at different speeds.

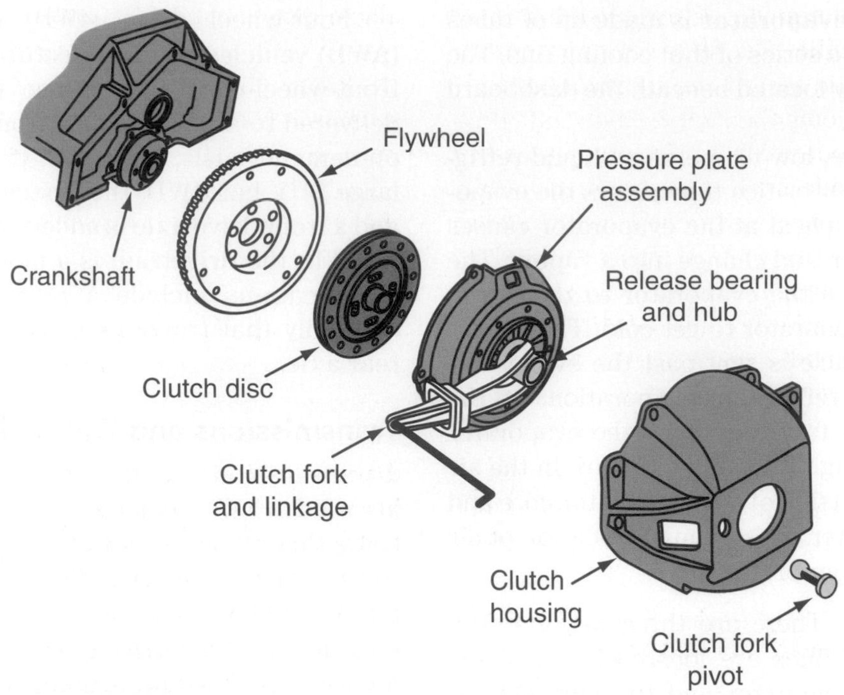

FIGURE 4–36

The major components of a clutch assembly for a manual transmission.

The size ratio of a gearset is called the **gear ratio** of that gearset.

A manual transmission or transaxle houses a number of individual gearsets, which produce different gear ratios **(Figure 4–37)**. The driver selects the desired operating gear or gear ratio. A typical manual transmission can have three to six forward gear ratios, and neutral and reverse.

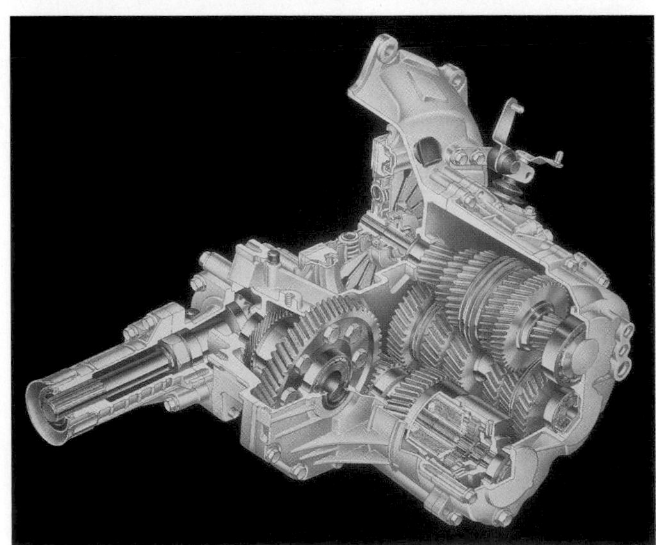

FIGURE 4–37

A typical manual transaxle.

Chrysler Group LLC

Automatic Transmission

An automatic transmission does not need a clutch pedal and shifts through the forward gears without the control of the driver. Instead of a clutch, it uses a flex-plate and a torque converter to transfer power from the engine's crankshaft to the transmission input shaft. The torque converter allows for smooth transfer of power at all engine speeds **(Figure 4–38)**.

FIGURE 4–38

Cutaway of a six-speed automatic transmission shown with the torque converter in the housing.

BMW NA, LLC

Shifting in an automatic transmission is controlled by a hydraulic and/or electronic control system. In a hydraulic system, an intricate network of valves and other components uses hydraulic pressure to control the operation of planetary gearsets. These gearsets provide from three to nine forward speeds, and neutral, park, and reverse gears. Newer electronic shifting systems use electric solenoids to control shifting mechanisms. Electronic shifting is precise and can be varied to suit certain operating conditions. All late-model automatic transmission–equipped vehicles have electronic shifting.

Dual-Clutch (Shaft) Transmissions

One of the latest trends to save fuel and reduce emissions is the use of dual-clutch or shaft transmissions **(Figure 4–39)**. These transmissions are a combination of automatic and manual transmissions. Shifts can be made on demand or automatically.

Continuously Variable Transmissions (CVTs)

Continuously variable transmissions are a type of automatic transmission but instead of using sets of gears to change speeds, two variable diameter pulleys and a metal belt are used. By changing the diameter of the input and output pulleys, the CVT can keep the engine speed within its most efficient range of operation, improving fuel economy.

Driveline

Drivelines are used on RWD vehicles and 4WD vehicles. They connect the output shaft of the transmission to the gearing in the rear axle housing. They are also used to connect the output shaft to the front and rear drive axles on a 4WD vehicle.

A driveline consists of a hollow drive or propeller shaft that is connected to the transmission and drive axle by universal joints (U-joints). These U-joints allow the driveshaft to move with the movement of the rear suspension, preventing damage to the shaft.

Rear Axle/Final Drive

On RWD vehicles, the driveshaft turns perpendicular to the forward motion of the vehicle. The bevelled gearing in the rear axle housing is designed to turn the direction of the power so that it can be used to drive the wheels of the vehicle. The power flows into bevelled drive gears, where it changes direction, then flows to the rear axles and wheels **(Figure 4–40)**.

The gearing in the final drive also multiplies the torque it receives from the driveshaft by providing a final gear reduction. The **differential** divides the torque between the left and right driving axles and wheels so that a differential wheel speed is possible. This means one wheel can turn faster than the other when going around turns.

Driving Axles

Driving axles are solid steel shafts that transfer the torque from the differential to the driving wheels. A separate axle shaft is used for each driving wheel. In RWD vehicles, the driving axles, final drive, and differential are enclosed in an axle housing that protects and supports these parts. Some RWD axle units are mounted to an independent suspension, and the drive axle assembly is similar to that of FWD vehicles.

Each drive axle is connected to the side gears in the differential. The inner ends of the axles are

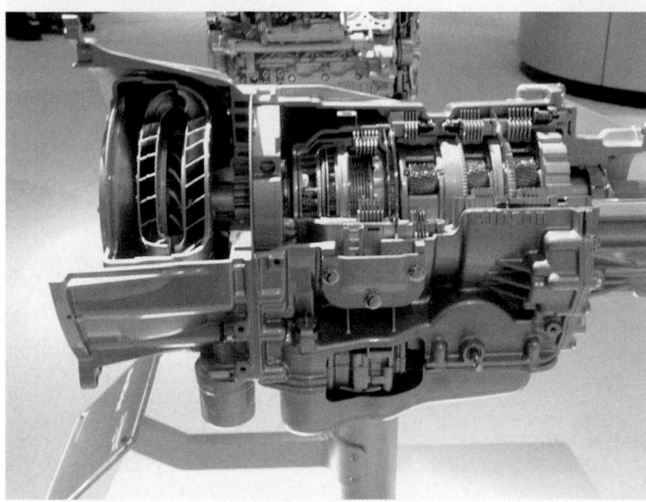

FIGURE 4–39

A cutaway of a dual-clutch transmission.

©Cengage Learning 2015

FIGURE 4–40

The driveline connects the output from the transmission to the differential unit and drive axles.

BMW NA, LLC

splined to fit into the side gears. As the side gears are turned, the axles to which they are splined turn at the same speed.

The drive wheels are attached to the outer ends of the axles. The outer end of each axle has a flange mounted to it. A **flange** is a rim for attaching one part to another part. The flange, fitted with studs, at the end of an axle holds the wheel in place. **Studs** are threaded shafts, resembling bolts without heads. One end of the stud is screwed or pressed into the flange. The wheel fits over the studs and a nut, called the **lug nut**, is tightened over the open end of the stud. This holds the wheel in place.

The final drive carrier supports the differential case and the inner end of each axle. A bearing inside the axle housing supports the outer end of the axle shaft. This bearing, called the axle bearing, allows the axle to rotate smoothly inside the axle housing.

Transaxle

A **transaxle** is used on FWD vehicles. It is made up of a transmission and final drive housed in a single unit **(Figure 4–41)**. The gearsets in the transaxle provide the required gear ratios and direct the power flow into the final drive. The final drive gearing provides the final gear reduction and splits the power flow between the left and right drive axles.

The drive axles extend from the sides of the transaxle. The outer ends of the axles are fitted to the hubs of the drive wheels. **Constant velocity (CV) joints** mounted on each end of the drive axles allow for changes in length and angle without affecting the power flow to the wheels.

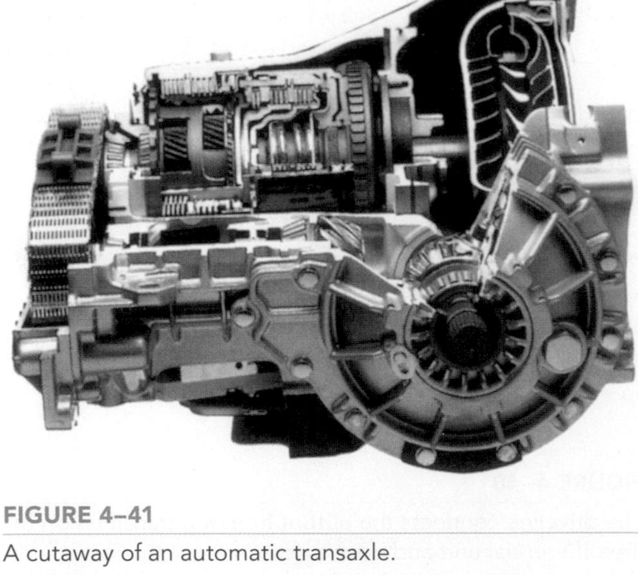

FIGURE 4–41

A cutaway of an automatic transaxle.

Chrysler Group LLC

Four-Wheel-Drive System

Four-wheel-drive or all-wheel-drive vehicles combine the features of rear-wheel-drive transmissions and front-wheel-drive transaxles. Additional **transfer case** gearing splits the power flow between a differential driving the front wheels and a rear differential that drives the rear wheels. This transfer case can be a housing bolted directly to the transmission/transaxle, or it can be a separate housing mounted somewhere in the driveline. Most RWD-based four-wheel-drive vehicles have a driveshaft connecting the output of the transmission to the rear axle and another connecting the output of the transfer case to the front drive axle. Typically, AWD cars have a centre differential that splits the torque between the front and rear drive axles.

RUNNING GEAR

The **running gear (Figure 4–42)** of a vehicle includes those parts that are used to control the vehicle, which includes the wheels and tires and the suspension, steering, and brake systems.

Suspension System

The suspension system **(Figure 4–43)** includes such components as the springs, shock absorbers, MacPherson struts, torsion bars, axles, and

FIGURE 4–42

The running gear in a typical late-model FWD car.

Chrysler Group LLC

FIGURE 4–43

A strut assembly of a typical suspension system.

Ford Motor Company

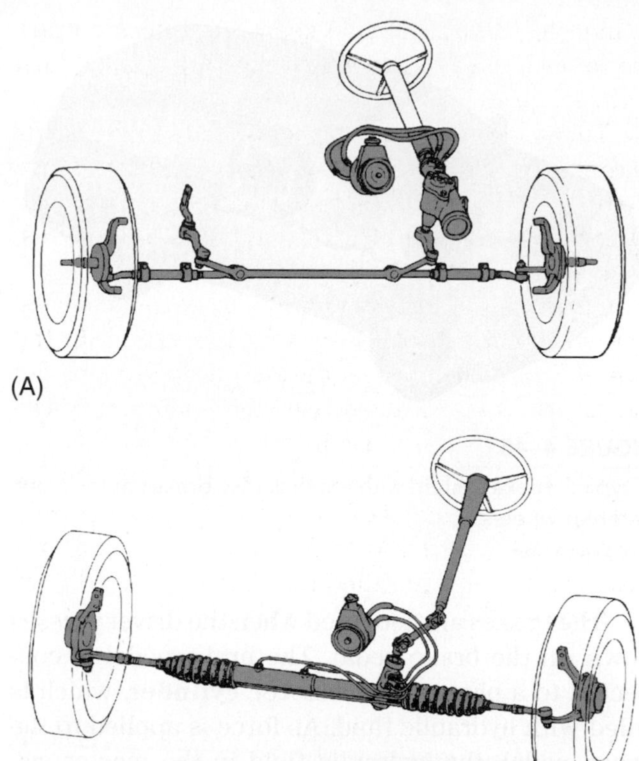

FIGURE 4–44

(A) A parallelogram-type steering system. (B) A rack-and-pinion steering system.

Federal Mogul Corporation

connecting linkages. These are designed to support the body and frame, the engine, and the drivelines. Without these systems, the comfort and ease of driving the vehicle would be reduced.

Springs or **torsion bars** are used to support the axles of the vehicle. The two types of springs commonly used are the coil spring and the leaf spring. Torsion bars, which are long spring steel rods, are also used. One end of the rod is connected to the frame, while the other end is connected to the movable parts of the axles. As the axles move up and down, the rod twists and acts as a spring.

Shock absorbers dampen the upward and downward movement of the springs. This is necessary to limit the vehicle's reaction to a bump in the road.

Steering System

The steering system allows the driver to control the direction of the vehicle. It includes the steering wheel, steering gear, steering shaft, and steering linkage.

Two basic types of steering systems are used today: the **rack-and-pinion** and **recirculating ball** (parallelogram) systems **(Figure 4–44)**. The rack-and-pinion system is commonly used in passenger cars. The recirculating ball system is normally used only on larger pickup trucks and SUVs.

Steering gears provide a gear reduction to make changing the direction of the wheels easier. On most vehicles, the steering gear is also power assisted to ease the effort of turning the wheels. In a power-assisted system, a pump provides hydraulic fluid under pressure to the steering gear. Pressurized fluid is directed to one side or the other of the steering gear to make it easier to turn the wheels. Some newer power-assisted steering systems use electric motors to provide the required assist rather than using hydraulics.

Some vehicles are equipped with speed-sensitive power-steering systems. These systems change the amount of power assist according to vehicle speed. The greatest amount of power assist occurs when the vehicle is moving slowly, and it decreases as speed increases.

Brakes

Obviously, the brake system is used to slow down and stop a vehicle **(Figure 4–45)**. Brakes, located at each wheel, use friction to slow and stop a vehicle.

FIGURE 4–45

A typical brake system with antilock disc brakes at the front and rear wheels.

Toyota Motor Sales

The brakes are activated when the driver presses down on the brake pedal. The brake pedal is connected to a plunger in a **master cylinder**, which is filled with hydraulic fluid. As force is applied to the brake pedal, the hydraulic fluid in the master cylinder is pressurized and transferred through brake hoses and lines to the four brake assemblies.

Two types of brakes are used—**disc brakes** and **drum brakes**. Many vehicles use a combination of the two types: disc brakes at the front wheels **(Figure 4–46)** and drum brakes at the rear wheels; others have disc brakes at all wheels.

Most vehicles have power-assisted brakes. Many vehicles use a vacuum **brake booster** to increase the force applied to the plunger in the master cylinder. Others use hydraulic pressure from the power-steering pump to increase the pressure on the brake fluid. Both of these systems lessen the amount of force that must be applied to the brake pedal and increase the responsiveness of the brake system.

Nearly all late-model vehicles have an **antilock brake system (ABS)**. The purpose of ABS is to prevent skidding during hard braking and braking on slippery winter roads. This gives the driver more directional control of the vehicle during most braking conditions.

Wheels and Tires

The only contact a vehicle has with the road is through its tires and wheels. Tires are made of rubber and other materials to give them strength, and they are filled with air. Wheels are made of metal and are bolted to the axles or spindles **(Figure 4–47)**. Wheels hold the tires in place. Wheels and tires come in many different sizes. Their sizes must be matched to one another and to the vehicle.

FIGURE 4–46

A disc brake unit with a wheel speed sensor for ABS.

Chrysler Group LLC

FIGURE 4–47

An alloy wheel with high-performance tires.

Mercedes-Benz of N.A., Inc.

HYBRID VEHICLES

A **hybrid electric vehicle (HEV)** uses one or more electric motors and an engine to propel the vehicle **(Figure 4–48)**. Depending on the design of the system, the engine may move the vehicle by

FIGURE 4–48

The electric motor in this hybrid arrangement fits between the engine and the transmission.

itself, assist the electric motor while it is moving the vehicle, or it may drive a generator to charge the vehicle's batteries. The electric motor may power the vehicle by itself or assist the engine while it is propelling the vehicle. Many hybrids rely exclusively on the electric motor(s) during slow speed operation, on the engine at higher speeds, and on both during some driving conditions. Complex electronic controls monitor the operation of the vehicle. Based on the operating conditions, electronics control the engine, electric motor, and generator.

A hybrid's electric motor is powered by high-voltage batteries, which are recharged by a generator driven by the engine and through **regenerative braking**. The engines used in hybrids are specially designed for the vehicle and electric assist. They operate more efficiently, resulting in very good fuel economy and very low tailpipe emissions. HEVs can provide the same performance as, if not better than, a comparable vehicle equipped with a larger engine.

There are primarily two types of hybrids: the parallel design and the series design. A parallel HEV uses either the electric motor or the gas engine to propel the vehicle, or both. The engine in a true series HEV is used only to drive the generator that keeps the batteries charged. The vehicle is powered only by the electric motor(s). Most current HEVs are considered as having a series/parallel configuration because they have the features of both designs.

Recently, some true series hybrids have been released. These are commonly called range-extending hybrids because they are capable of driving farther before recharging the batteries than a pure electric vehicle. Although most current hybrids are focused on fuel economy, the same ideas can be used to create high-performance vehicles. Hybrid technology is also influencing off-the-road performance. By using individual motors at the front and rear drive axles, additional power can be applied to certain drive wheels when needed.

Electric Vehicles

Because of the current fuel economy and emission regulations set by the government, auto manufacturers are introducing a variety of battery-operated electric vehicles.

ALTERNATIVE FUELS

There are several ways to reduce fuel consumption, other than using electric drive. Much research has and is being conducted on the use of alternative fuels in internal combustion engines. By using alternative fuels, we not only reduce our reliance on oil but also reduce emissions and the effects an automobile's exhaust has on global warming. Common alternative fuels are as follows:

- Propane, also referred to as liquefied petroleum gas (LPG)
- Methyl alcohol, normally called methanol
- Ethyl alcohol, also referred to as ethanol
- Compressed natural gas (CNG)

KEY TERMS

AC generator (p. 67)
Accumulator (p. 70)
Antilock brake system (ABS) (p. 76)
Brake booster (p. 76)
Carbon monoxide (CO) (p. 64)
Clutch (p. 71)
Combustion (p. 60)
Combustion chamber (p. 61)
Company Average Fuel Consumption (CAFC) (p. 55)
Compressor (p. 70)
Condenser (p. 70)
Connecting rod (p. 62)
Constant velocity (CV) joints (p. 74)
Convertible (p. 57)
Corporate Average Fuel Economy (CAFE) (p. 55)
Crankshaft (p. 62)
Crankshaft position (CKP) sensor (p. 66)
Crossover vehicle (p. 59)
Cylinder block (p. 61)
Cylinder head (p. 61)
Differential (p. 73)

Disc brakes (p. 76)
Distributor (p. 66)
Drivetrain (p. 71)
Drum brakes (p. 76)
Engine block (p. 61)
Evaporator (p. 71)
Exhaust gas recirculation (EGR) (p. 64)
Exhaust manifold (p. 62)
Flange (p. 74)
Friction (p. 62)
Gear ratio (p. 72)
Hatchback (p. 57)
Hybrid electric vehicle (HEV) (p. 76)
Hydrocarbons (HC) (p. 64)
Hydroforming (p. 56)
Ignition coil (p. 66)
Intake manifold (p. 62)
Liftback (p. 57)
Lug nut (p. 74)
Manifold (p. 62)
Master cylinder (p. 76)
Oil pan (p. 62)
Oil sump (p. 62)
On-board diagnostic (OBD-II) (p. 69)
Orifice tube (p. 70)
Oxides of nitrogen (NO_x) (p. 64)
Pickup (p. 58)
Piston (p. 61)
Ports (p. 61)
Positive crankcase ventilation (PCV) (p. 64)
Pressure cap (p. 64)
Rack-and-pinion (p. 75)
Radiator (p. 64)
Receiver/dryer (p. 70)
Recirculating ball (p. 75)
Refrigerant (p. 69)
Regenerative braking (p. 77)
Running gear (p. 74)
Sedan (p. 57)
Shock absorbers (p. 75)
Solenoid (p. 66)
Spark plug (p. 66)
Sport utility vehicle (SUV) (p. 59)
Springs (p. 75)
Station wagon (p. 57)
Steering gears (p. 75)
Studs (p. 74)
Thermostat (p. 64)
Thermostatic expansion valve (TEV or TXV) (p. 70)
Torque converter (p. 71)
Torsion bars (p. 75)
Transaxle (p. 74)
Transfer case (p. 74)
Unibody (p. 56)
Valve train (p. 62)
Van (p. 58)
Water jackets (p. 64)
Water pump (p. 64)

SUMMARY

- Dramatic changes to the automobile have occurred over the last 40 years, including the addition of emission control systems, more fuel-efficient and cleaner-burning engines, and lighter body weight.
- In addition to being lighter than body-over-frame vehicles, unibodies offer better occupant protection by distributing impact forces throughout the vehicle.
- Today's computerized engine control systems regulate such things as air and fuel delivery, ignition timing, and emissions. The result is an increase in overall efficiency.
- All automotive engines are classified as internal combustion, because the burning of the fuel and air occurs inside the engine. Diesel engines share the same major parts as gasoline engines, but they do not use a spark to ignite the air/fuel mixture.
- The cooling system maintains proper engine operating temperatures. Liquid cooling is more efficient than air cooling and more commonly used.
- The lubrication system distributes motor oil throughout the engine. This system also contains the oil filter necessary to remove dirt and other foreign matter from the oil.
- The fuel system is responsible not only for fuel storage and delivery, but also for atomizing and mixing it with the air in the correct proportion.
- The exhaust system has three primary purposes: to channel toxic exhaust away from the passenger compartment, to quiet the exhaust pulses, and to reduce the emissions in the exhaust.
- The electrical system of an automobile includes the ignition, starting, charging, and lighting systems. Electronic engine controls regulate these systems very accurately through the use of computers.
- The heating and air-conditioning systems of a vehicle provide comfort for the passengers of the vehicle. Both systems are dependent on the proper operation of the engine. The heating system basically adds heat to the vehicle's interior, whereas air conditioning removes heat and moisture/humidity from the cabin air.
- Modern automatic transmissions use a computer to match the demand for acceleration with engine speed, wheel speed, and load conditions. It then chooses the proper gear ratio and, if necessary, initiates a gear change.
- The running gear is critical to controlling the vehicle. It consists of the suspension system, braking system, steering system, and wheels and tires.
- Hybrid electric vehicles use one or more electric motors and an engine to propel the vehicle. Depending on the design of the system, the engine may move the vehicle by itself, may assist the electric motor while it is moving the vehicle, or may drive a generator to charge the vehicle's batteries. The electric motor may power the vehicle by itself or assist the engine while it is propelling the vehicle.

REVIEW QUESTIONS

1. Under the CAFC/CAFE standards, for what are vehicles tested?
2. List the benefits of the design switch from rear-wheel-drive to front-wheel-drive vehicles.
3. Define internal combustion.
4. In addition to the battery, what does the charging system include?
5. What are the differences between the parallel design and the series design hybrid vehicle?
6. Which of the following is a typical emission control system?
   a. API **b.** SAE
   c. CEPA **d.** EFE
7. What is used in an automatic transmission to transfer power from the flywheel to the transmission's input shaft?
   a. differential **b.** U-joint
   c. torque converter **d.** constant velocity joint
8. How is a four-stroke gasoline engine classified?
   a. external combustion
   b. internal combustion
   c. continuous combustion
   d. alternate combustion
9. Which of the following systems is designed to reduce hydrocarbon (HC) by routing fuel and engine oil vapours to the intake manifold?
   a. evaporative emission control system
   b. positive crankcase ventilation (PCV) system
   c. exhaust gas recirculation (EGR) system
   d. air injection system
10. What two vehicle types are crossover vehicles based on?
    a. a station wagon and a hatchback
    b. a station wagon and an SUV
    c. a station wagon and a convertible
    d. a hatchback and an SUV
11. What is the order of strokes in a four-stroke cycle gasoline engine?
    a. intake, compression, power, and exhaust
    b. intake, power, compression, and exhaust
    c. intake, power, exhaust, and compression
    d. intake, compression, exhaust, and power
12. What does the valve train do?
    a. It delivers fuel to a device that controls the amount of fuel going to the engine.
    b. It houses the major parts of the engine.
    c. It converts a reciprocating motion to rotary motion.
    d. It opens and closes the intake and exhaust ports of each cylinder.
13. What engine cooling system component is responsible for raising the boiling point of the engine's coolant?
    a. the thermostat
    b. the pressure cap
    c. the water pump
    d. the radiator
14. Which of the following vehicles would most likely utilize body-over-frame construction?
    a. a pickup truck
    b. a minivan
    c. a four-door hardtop
    d. a station wagon
15. Which emission control system introduces exhaust gases into the intake air to reduce the formation of NO_x in the combustion chamber?
    a. evaporative emission controls
    b. exhaust gas recirculation
    c. air injection
    d. early fuel evaporation
16. What component is part of both the charging and the starting systems?
    a. the starter
    b. the alternator
    c. the battery
    d. the AC generator
17. What vehicle commonly uses a transaxle?
    a. a rear-wheel-drive car
    b. a two-wheel-drive pickup
    c. a front-wheel-drive car
    d. a four-wheel-drive pickup
18. Which of the following is part of the drivetrain?
    a. differential
    b. steering
    c. suspension
    d. brakes
19. What air-conditioning system component works with the compressor to divide the A/C system into high and low sides?
    a. the receiver/dryer
    b. the accumulator
    c. the orifice tube
    d. the evaporator
20. What does the reductant do in a diesel exhaust system?
    a. Oxygen is separated from NO_x and combined with hydrogen to form water.
    b. Oxygen is separated from NO_x and combined with carbon to form CO.
    c. Oxygen is added to nitrogen to form NO_x.
    d. Oxygen is removed from NO_x to form nitrogen and oxygen.

CHAPTER 5
Hand Tools and Shop Equipment

LEARNING OUTCOMES

- List the basic units of measurement for length, volume, and mass in the two measuring systems.
- Describe the different types of fasteners used in the automotive industry.
- List the various mechanical measuring tools used in the automotive shop.
- Describe the proper procedure for measuring with a micrometer.
- List some of the common hand tools used in auto repair.
- List the common types of shop equipment, and state their purpose.
- Describe the use of common pneumatic, electrical, and hydraulic power tools found in an automotive service department.
- Describe the different sources for service information that are available to technicians.

Repairing the modern automobile requires the use of various tools. Many of these tools are common hand and power tools used every day by a technician. Other tools are very specialized and are only for specific repairs on specific systems and/or vehicles. This chapter presents some of the more commonly used hand and power tools with which every technician must be familiar. Because units of measurement play such an important part in tool selection and in diagnosing automotive problems, this chapter begins with a presentation of measuring systems. Prior to the discussion on tools, there is a discussion on another topic that relates very much to tools, measuring systems, and fasteners.

MEASURING SYSTEMS

In 1970, the metric system (the *Système International d'Unités*, or SI) became the official measurement system in Canada. However, most regulations have been repealed or have been no longer enforced since 1984. Today's cars are totally metric, but many aftermarket components and accessories may still use imperial measurement, as do some older Canadian and American cars that are still on the road.

The basic unit of linear measurement in the imperial system is the inch. The basic unit of linear measurement in the metric system is the metre. The metre (m) is easily broken down into smaller units, such as the centimetre ($\frac{1}{100}$ m) and millimetre ($\frac{1}{1000}$ m).

All units of measurement in the metric system are related to each other by a factor of 10. Every metric unit can be multiplied or divided by the factor of 10 to get larger units (multiples) or smaller units (submultiples). This makes the metric system much easier to use, with less chance of math errors, than when using the imperial system **(Figure 5–1)**.

The United States passed the *Metric Conversion Act* in 1975 in an attempt to get American industry and the general public to use the metric system, as the rest of the world does. Although the general American public has been slow to drop the customary measuring system of inches, gallons, and pounds, many industries, led by the automotive industry, have adopted the metric system for the most part.

Due to the United States's conversion, nearly all vehicles are now built to metric standards. Technicians must be able to measure and work with

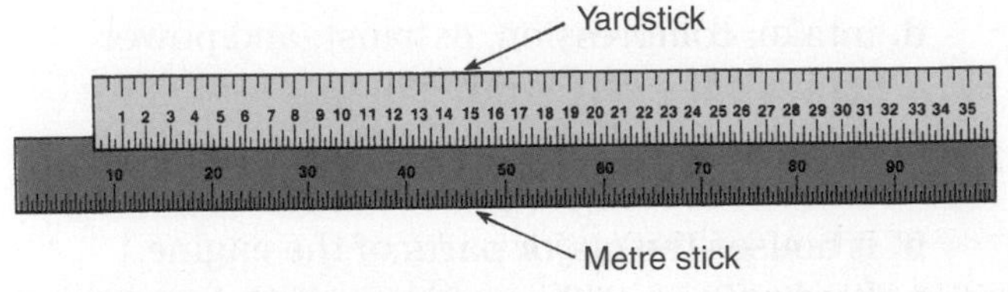

FIGURE 5–1

A metre stick is made of 1000 increments known as millimetres and is slightly longer than a yardstick.

both systems of measurement. The following are some common equivalents in the two systems:

Linear Measurements

1 metre (m) = 39.37 inches (in.)
1 centimetre (cm) = 0.3937 inch
1 millimetre (mm) = 0.03937 inch
1 inch = 2.54 centimetres
1 inch = 25.4 millimetres
1 mile (mi.) = 1.6093 kilometres (km)

Square Measurements

1 square inch = 6.452 square centimetres
1 square centimetre = 0.155 square inch

Volume Measurements

1 cubic inch = 16.387 cubic centimetres
1000 cubic centimetres = 1 litre (l)
1 litre = 61.02 cubic inches
1 gallon (gal.) = 3.78541 litres

Weight Measurements

1 ounce (oz.) = 28.3495 grams (g)
1 pound (lb.) = 453.59 grams
1000 grams = 1 kilogram (kg)
1 kilogram = 2.2046 pounds

Temperature Measurements

1 Fahrenheit (F) = $\frac{9}{5}$ C + 32°
1 Celsius (C) = $\frac{5}{9}$ (F − 32°)

Pressure Measurements

1 pound per square inch (psi) = 0.07031 kilogram per square centimetre (kg/cm^2)
1 kilogram per square centimetre = 14.22334 pounds per square inch
1 pound per square inch = 6.8948 kilopascals (kPa)
1 kilopascal = 0.1450 pound per square inch
1 kilogram per square centimetre = 98.0665 kilopascals
1 kilopascal = 0.0102 kilogram per square centimetre
1 bar = 14.504 pounds per square inch
1 pound per square inch = 0.06895 bar
1 atmosphere = 14.7 pounds per square inch

Torque Measurements

10 foot-pounds (ft·lb) = 13.558 Newton metres (N·m)
1 N·m = 0.7375 ft·lb
1 ft·lb = 0.138 m·kg
1 cm kg = 7.233 ft·lb
10 cm kg = 0.98 N·m

FASTENERS

Fasteners are those things used to secure or hold parts of something together. Many types and sizes of fasteners are used by the automotive industry. Each fastener is designed for a specific purpose and condition. One type of fastener most commonly used is the threaded fastener. Threaded fasteners include bolts, nuts, screws, and similar items that allow a technician to install or remove parts easily **(Figure 5–2)**.

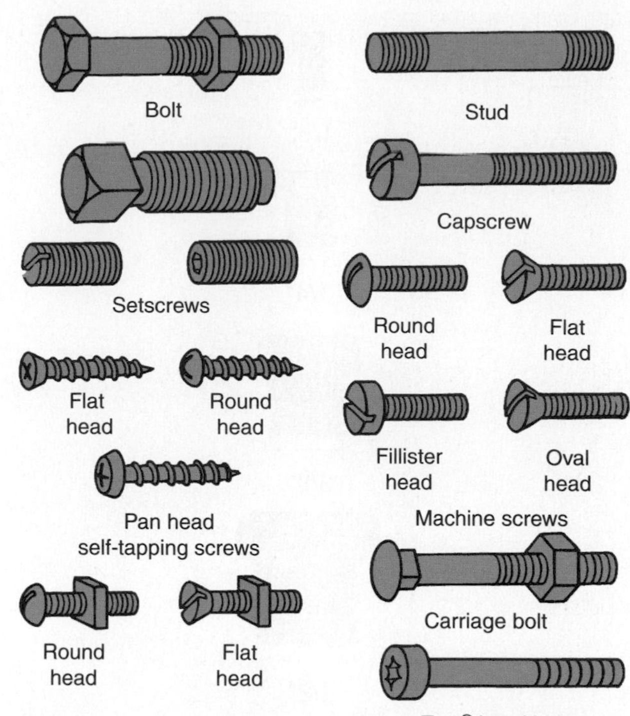

FIGURE 5–2

Common automotive threaded fasteners.

Threaded fasteners are available in many sizes, designs, and threads. The threads can be either cut or rolled into the fastener. Rolled threads are 30 percent stronger than cut threads. They also offer better fatigue resistance because there are no sharp notches to create stress points. Fasteners are made to metric or imperial measurements. There are four classifications for the threads of imperial fasteners: Unified National Coarse (UNC), Unified National Fine (UNF), Unified National Extrafine (UNEF), and Unified National Pipe Thread (UNPT or NPT). Metric fasteners are also available in fine and coarse threads.

NPT is the standard thread design for joining pipes and fittings. There are two basic designs: tapered and straight-cut threads. Straight-cut pipe thread is used to join pipes, but it does not provide a good seal at the joining point. Tapered pipe threads provide a good seal because the internal and external threads compress against each other as the joint is tightened. Most often, a sealant is used on pipe threads to provide a better seal. Pipe threads are commonly used at the ends of hoses and lines that carry a liquid or gas **(Figure 5–3)**.

Coarse threads are used for general-purpose work, especially where rapid assembly and disassembly is required. Fine-threaded fasteners are used

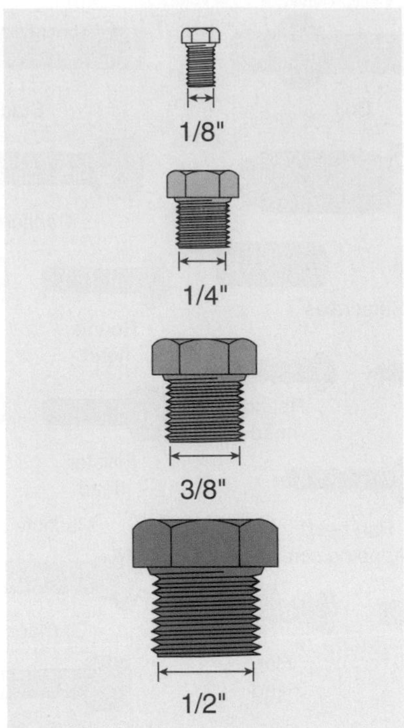

FIGURE 5–3

Various sizes of pipe fittings used with lines and hoses.

where greater holding force is necessary. They are also used where greater resistance to vibration is desired.

Bolts have a head on one end and threads on the other. Bolts are identified by defining the head size, shank diameter, thread pitch, length, and grade **(Figure 5–4)**. Bolts have a shoulder below the head, and the threads do not travel all the way from the head to the end of the bolt.

Cap screws are similar to bolts; however, cap screws have no shoulder. The threads travel from the head to the end of the bolt. It is important that you never use a cap screw in place of a bolt.

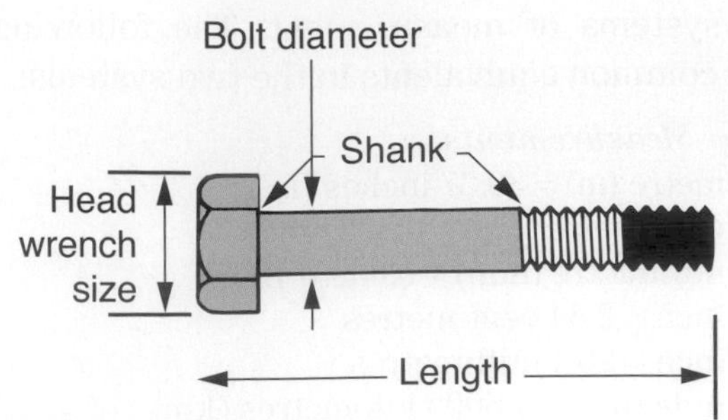

FIGURE 5–4

Basic terminology for bolt identification.

Studs are rods with threads on both ends. Most often, the threads on one end are coarse, and the other end has fine threads. One end of the stud is threaded into a threaded bore. A hole in the part to be secured is fitted over the stud and held in place with a nut that is threaded over the stud. Studs are used when the clamping pressures of a fine thread are needed and a bolt will not work. If the material the stud is being threaded into is soft (such as aluminum) or granular (such as cast iron), fine threads will not withstand a great amount of pulling force on the stud. Therefore, a coarse thread is used to secure the stud in the work piece, and a fine-threaded nut is used to secure the other part to it. Doing this results in having the clamping force of fine threads and the holding power of coarse threads.

Nuts are used with other threaded fasteners when the fastener is not threaded into a piece of work. Nuts of many different designs are found on today's cars **(Figure 5–5)**. The most common one is

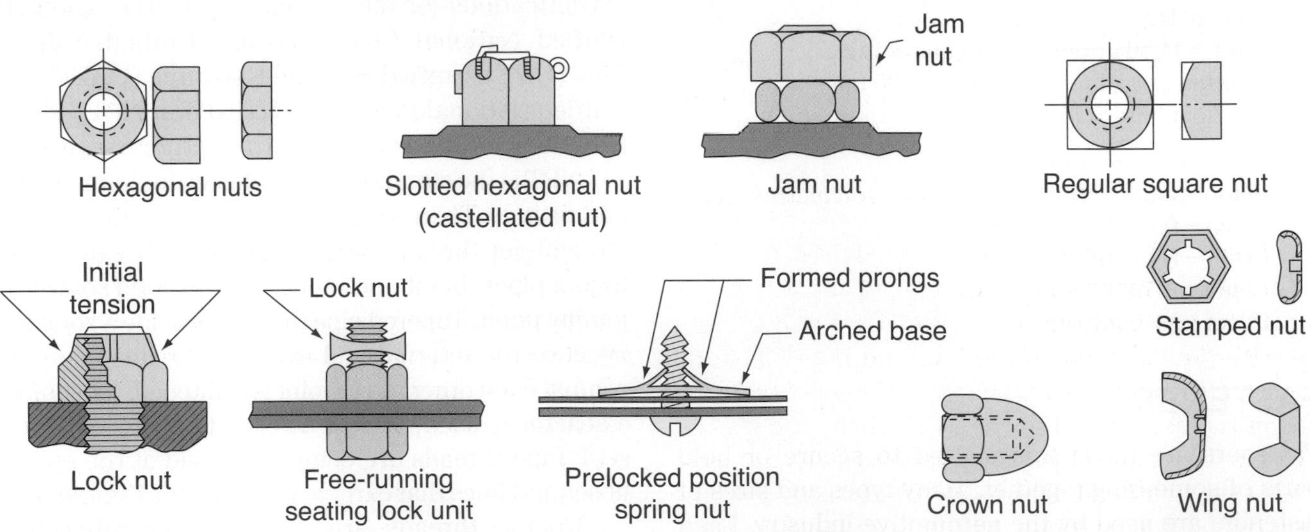

FIGURE 5–5

Many different types of nuts are used on automobiles. Each type has a specific purpose.

FIGURE 5–6

Two common types of locknuts: distorted thread (Stover) and nyloc nuts.

Martin Restoule

the hex nut, which is used with studs and bolts and is tightened with a wrench.

LOCKNUTS Locknuts are often used in places where vibration may tend to loosen a nut. Many locking nuts are standard nuts with nylon inserted into a section of the threads. The nylon cushions the vibrations. Others types of locknuts are regular nuts that are distorted to provide resistance to thread movement **(Figure 5–6)**.

Setscrews are used to prevent rotary motion between two parts, such as a pulley and shaft. Setscrews are either headless or have a square head. Headless setscrews require an Allen wrench or screwdriver to loosen and tighten them.

MACHINE SCREWS The length of machine screws is entirely threaded. These screws have a head on one end and a flat bottom on the other. Machine screws are used to mount one component to another that has a threaded bore. They are also used with a nut to hold parts together. Machine screws can have a round, flat, Torx, oval, or fillister head.

Self-tapping screws are used to fasten sheet-metal parts or to join light metal, wood, or plastic parts together. These screws form their own threads in the material they are screwed into.

Bolt Identification

The **bolt head** is used to loosen and tighten the bolt. A socket or wrench fits over the head and is used to screw the bolt in or out. The size of the bolt head varies with the diameter of the bolt and is available in metric and imperial wrench sizes. Many confuse the size of the head with the size of the bolt. The size of a bolt is determined by the diameter of its shank. The size of the bolt head determines what size wrench is required to turn or rotate it with. **Table 5–1** lists the most common bolt head sizes. Notice that the sizes are listed as millimetres or as fractions of an inch.

TABLE 5–1 STANDARD BOLT HEAD SIZES

COMMON ENGLISH (SAE) HEAD SIZES	COMMON METRIC (ISO) HEAD SIZES
Wrench Size (inches)*	Wrench Size (millimetres)*
3/8	9
7/16	10
1/2	11
9/16	12
5/8	13
11/16	14
3/4	15
13/16	16
7/8	17
15/16	18
1	19
1 1/16	20
1 1/8	21
1 3/16	22
1 1/4	23
1 5/16	24
1 3/8	26
1 7/16	27
1 1/2	29
	30
	32

*This does not suggest equivalency.

Bolt diameter is the measurement across the major diameter of the threaded area or across the **bolt shank**. The length of a bolt is measured from the bottom surface of the head to the end of the threads.

The **thread pitch** of a bolt in the imperial system is determined by the number of threads that are in one inch of the threaded bolt length and is expressed in number of threads per inch. A UNF bolt with a 3/8 -inch (9.54 mm) diameter would be a 3/8 × 24 bolt. It would have 24 threads per inch. Likewise, a 3/8-inch (9.54 mm) UNC bolt would be called a 3/8 × 16.

The distance, in millimetres, between two adjacent threads determines the thread pitch in the metric system. This distance will vary between 1.0 and 2.0, and depends on the diameter of the bolt. The lower the number, the closer the threads are placed and the finer the threads are.

The bolt's tensile strength, or grade, is the amount of stress or stretch it is able to withstand before it breaks. The type of material the bolt is made of and the diameter of the bolt determines its grade. In the imperial system, the tensile strength of a bolt is

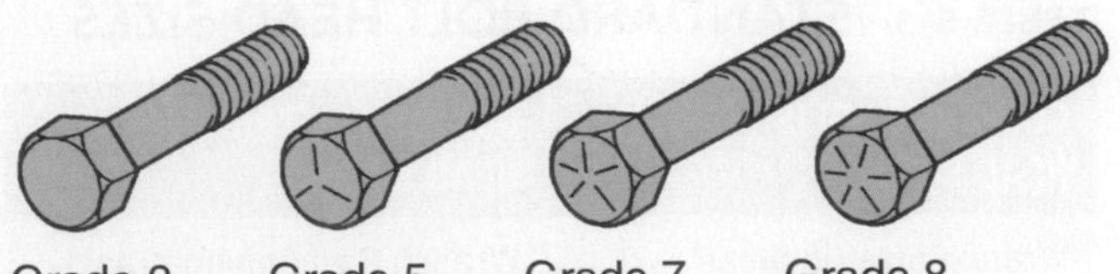

Customary (inch) bolts—SAE identification marks correspond to bolt strength—increasing numbers represent increasing strength.

Metric bolts—ISO identification class numbers correspond to bolt strength—increasing numbers represent increasing strength.

FIGURE 5–7

Bolt grade markings.

identified by the Society of Automotive Engineers (SAE), which uses a number of radial lines (**grade marks**) on the bolt's head. More lines mean higher tensile strength **(Figure 5–7)**. Count the number of lines and add two to determine the grade of a bolt.

A property class number on the bolt head identifies the grade of metric bolts. This numerical identification is from the International Organization for Standardization (ISO) and comprises two numbers. The first number represents the tensile strength of the bolt—the higher the number, the greater the tensile strength. The second number represents the yield strength of the bolt. This number represents how much stress the bolt can take before it is unable to return to its original shape without damage. The second number represents a percentage rating. For example, a 10.9 bolt has a tensile strength of 1000 MPa (145 000 psi) and a yield strength of 900 MPa (90 percent of 1000). A 10.9 metric bolt is similar in strength to an SAE (Society of Automotive Engineers) grade 8 bolt.

Nuts are graded to match their respective bolts **(Table 5–2)**. For example, a grade 8 nut must be used with a grade 8 bolt. If a grade 5 nut were used, a grade 5 connection would result. Grade 8 and critical applications require the use of fully hardened flat washers. These will not dish out when torqued, as soft washers will.

Bolt heads can pop off because of fillet damage. The **fillet** is the smooth curve where the shank flows into the bolt head **(Figure 5–8)**. Scratches in this

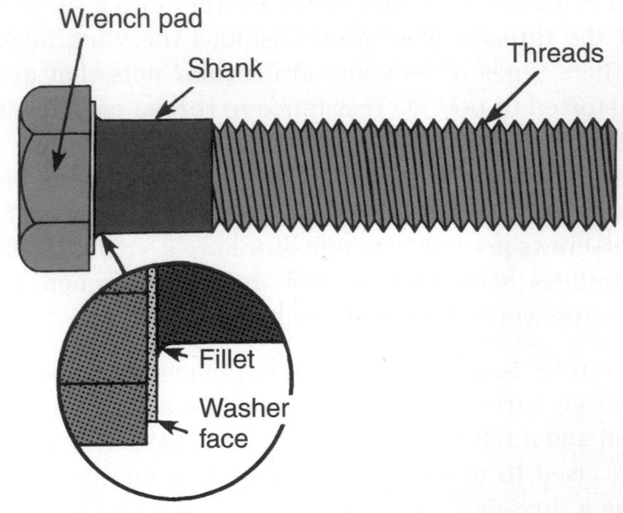

FIGURE 5–8

Bolt fillet detail.

TABLE 5–2 **STANDARD NUT STRENGTH MARKINGS**

INCH SYSTEM		METRIC SYSTEM	
Grade	Identification	Class	Identification
Hex Nut Grade 5	3 Dots	Hex Nut Property Class 9	Arabic 9
Hex Nut Grade 8	6 Dots	Hex Nut Property Class 10	Arabic 10
Increasing dots represent increasing strength.		Can also have blue finish or paint dab on hex flat. Increasing numbers represent increasing strength.	

area introduce stress to the bolt head, causing failure. Removing any burrs around the edges of holes can protect the bolt head. It is also a good practice to place flat washers with their rounded, punched side against the bolt head and their sharp side to the work surface.

Fatigue breaks are the most common type of bolt failure. A bolt becomes fatigued from working back and forth when it is too loose. Undertightening the bolt causes this problem. Bolts can also be broken or damaged by overtightening, being forced into a nonmatching thread, or bottoming out, which happens when the bolt is too long.

Tightening Bolts

Any fastener is near worthless if it is not as tight as it should be. When a bolt is properly tightened, it will be "spring loaded" against the part it is holding. This spring effect is caused by the stretch of the bolt when it is tightened. Normally, a properly tightened bolt is stretched to 70 percent of its elastic limit. The elastic limit of a bolt is that point of stretch from which the bolt will not return to its original shape when it is loosened. Not only will an overtightened or stretched bolt have insufficient clamping force, it will also have distorted threads. The stretched threads will make it more difficult to tighten and loosen the bolt or a nut on the bolt **(Figure 5–9)**.

Washers

Many different types of washers are used with fasteners. The type of washer it is defines the purpose of the washer. Flat washers are used to spread out the load of tightening a nut or bolt. This stops the bolt head or nut from digging into the surface as it is tightened. Always place flat washers with their rounded, punched side against the bolt head. Soft, flat washers, sometimes called compression washers, are also used to spread the load of tightening and help seal one component to another. Copper washers are often used with oil pan bolts to help seal the pan to the engine block.

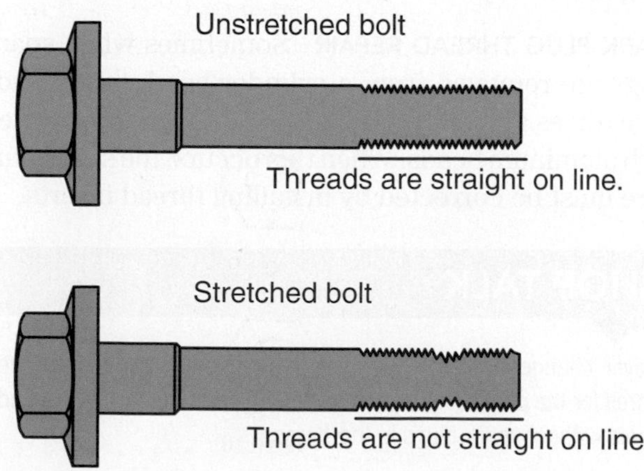
Unstretched bolt
Threads are straight on line.
Stretched bolt
Threads are not straight on line.

FIGURE 5–9
A comparison of an unstretched and a stretched bolt.

Thread Lubricants and Sealants

Often, manufacturers recommend that the threads of a bolt or stud be coated with a sealant or lubricant. The most commonly used lubricant is antiseize compound. Antiseize compound is used where a bolt might become difficult to remove after a period of time, for example, in an aluminum engine block. Thread lubricants introduce the possibility of a hydrostatic lock, where oil is trapped in a blind hole. When the bolt contacts the oil, it cannot compress it; therefore, the bolt cannot be properly tightened and a fractured part may result.

Thread sealants are used on bolts that are tightened into an oil cavity or coolant passage. The sealant prevents the liquid from seeping past the threads. Another commonly used thread chemical, called threadlocker, prevents a bolt from working loose as the engine or another part vibrates.

Thread Pitch Gauge

The use of a **thread pitch gauge (Figure 5–10)** provides a quick and accurate method of checking the thread pitch of a fastener. The leaves of this measuring tool are marked with the various pitches. To check the pitch of threads, simply match the teeth of the gauge with the threads of the fastener. Then, read the pitch from the leaf.

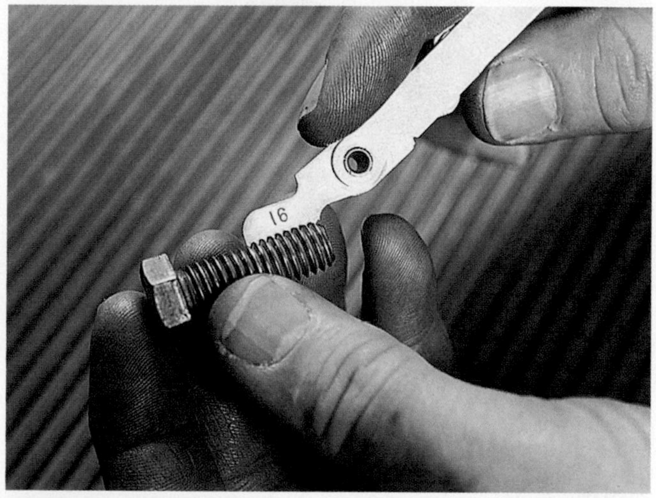
16

FIGURE 5–10
Using a thread pitch gauge to check the thread pitch on a bolt.

Thread pitch gauges are available for the various types of fastener threads used by the automotive industry: metric threads, International Standard threads, Unified National Coarse and Fine threads, and Whitworth threads.

Taps and Dies

The hand **tap** is a small tool used for hand-cutting internal threads **(Figure 5–11)**. An internal thread is cut on the inside of a part, such as a thread on the inside of a nut. Thread chaser taps are also available that only clean and restore threads that were previously cut. Taps are selected by size and thread pitch. Photo Sequence 1 goes through the correct procedure for repairing damaged threads with a tap.

When tapping a bore, rotate the tap in a clockwise direction. Then, turn the tap counter-clockwise about a quarter turn to break off any metal chips that may have accumulated in the threads. These small metal pieces can damage the threads as you continue to tap. These metal chips are gathered in the tap's flutes, which are recessed areas between the cutting teeth of the tap **(Figure 5–12)**. After backing off the tap, continue rotating the tap clockwise. Remember to back off the tap periodically and make sure all of the existing threads in the bore have been recut by the tap.

Hand-threading **dies** are the opposite of taps because they cut external (outside) threads on bolts, rods, and pipes rather than internal threads. Dies are made in various sizes and shapes, depending on the particular work for which they are intended. Dies may be solid (fixed size), be split on one side to permit adjustment, or have two halves held together in a collet that provides for individual adjustments. Dies fit into holders called die stocks.

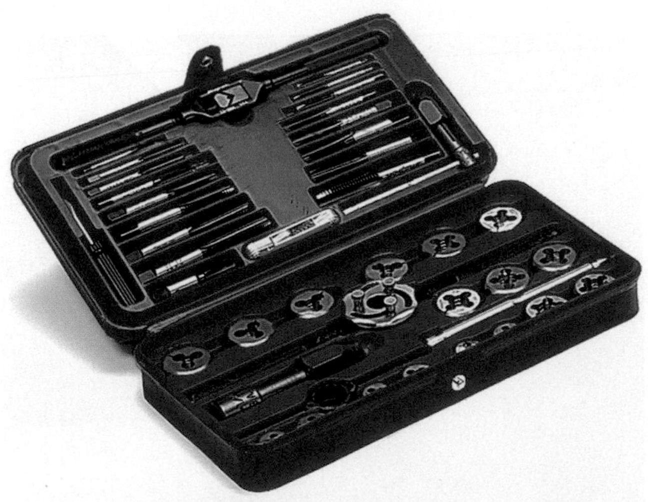

FIGURE 5–11

A tap and die set.

Snap-on Tools Company

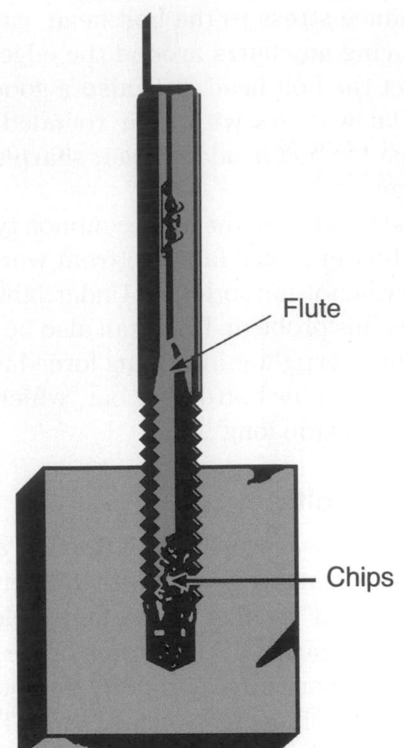

FIGURE 5–12

Metal chips are gathered into the flutes of a tap.

Threaded Inserts

When the threads in a bore are excessively damaged, it is better to replace them than try to tap them. A thread insert can be used to restore the original threads. Inserts require drilling the bore to a larger diameter and tapping that bore to allow the insert to be screwed into it. The inner threaded diameter of the insert will provide fresh threads for the bolt **(Figure 5–13)**.

SPARK PLUG THREAD REPAIR Sometimes when spark plugs are removed from a cylinder head, the threads have traces of metal on them. This happens more often with aluminum heads. When this occurs, the spark plug bore must be corrected by installing thread inserts.

SHOP TALK

Never change spark plugs when the cylinder head is hot. The bores for the plugs can take on an oval shape as the cylinder head cools without spark plugs in the bores.

When installing spark plugs, if the plugs cannot be installed easily by hand, the threads in the cylinder

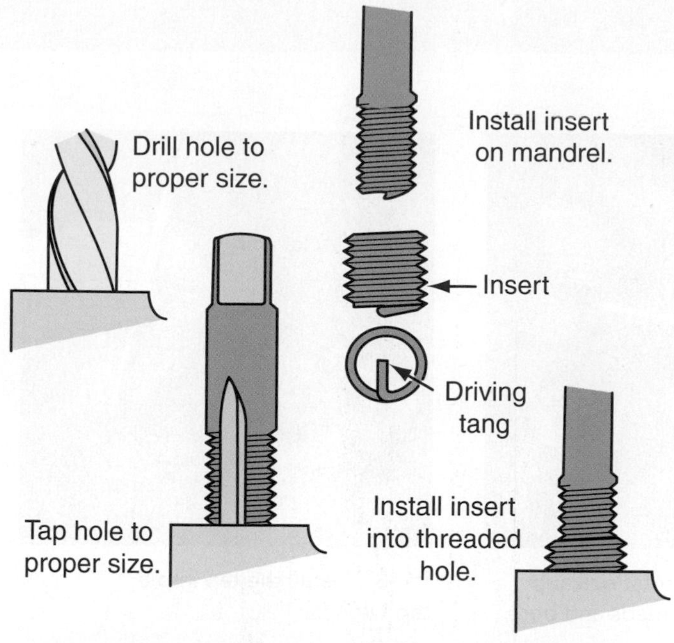

FIGURE 5–13

Using a threaded insert (heli-coil®) to repair damaged threads.

Emhart Fastening Teknologies

head may need to be cleaned with a thread-chasing tap. There are special taps for spark plug bores, simply called **spark plug thread taps**. Be especially careful not to cross-thread the plugs when working with aluminum heads. Always tighten the plugs with a torque wrench and the correct spark plug socket, following the vehicle manufacturer's specifications. Also, when changing spark plugs in aluminum heads, the temperature of the heads should be an ambient temperature before attempting to remove the plugs.

MEASURING TOOLS

Some service work, such as engine repair, requires very exact measurements, often in thousandths (0.001) of a millimetre or ten-thousandths (0.0001) of an inch. Accurate measurements with this kind of precision can be made only by using precise measuring devices.

Measuring tools are precise and delicate instruments. In fact, the more precise they are, the more delicate they are. They should be handled with great care. Never pry, strike, drop, or force these instruments. They may be permanently damaged.

Precision measuring instruments, especially micrometers, are extremely sensitive to rough handling. Clean them before and after every use. All measuring should be performed on parts that are at room temperature to eliminate the chance of measuring something that has contracted because it was cold or has expanded because it was hot.

SHOP TALK

Check measuring instruments regularly against known good equipment to ensure that they are operating properly and are capable of accurate measurement. Always refer to the appropriate material for the correct specifications before performing any service or diagnostic procedures. The close tolerances required for the proper operation of some automotive parts make using the correct specifications and taking accurate measurements very important. Even the slightest error in measurement can be critical to the durability and operation of an engine and other systems.

Machinist's Rule

The **machinist's rule** looks very much like an ordinary ruler. Each edge of this basic measuring tool is divided into increments based on a different scale. As shown in **Figure 5–14**, a typical metric machinist's rule is usually divided into 0.5 mm and 1 mm increments. Of course, imperial system machinist's rules are also available. Imperial system rules may have scales based on ⅛, 1⁄16, 1⁄32, and 1⁄64 in. (3.18, 1.59, .79, and .40 mm) intervals.

Some machinist's rules may be based on decimal intervals. These are typically divided into increments of 0.1, 0.03, and 0.01 mm (1⁄10, 1⁄50, and 1⁄1000 in.). Decimal machinist's rules are very helpful when measuring dimensions specified in decimals; they make such measurements much easier.

Vernier Caliper

A **vernier caliper** is a measuring tool that can make inside, outside, or depth measurements. It is marked in both metric and British imperial divisions called a

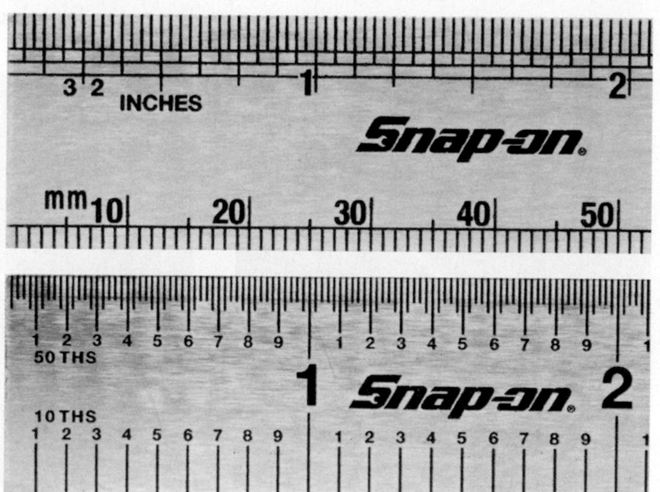

FIGURE 5–14

Graduations on a typical machinist's rule.

PHOTO SEQUENCE 1

REPAIRING DAMAGED THREADS WITH A TAP

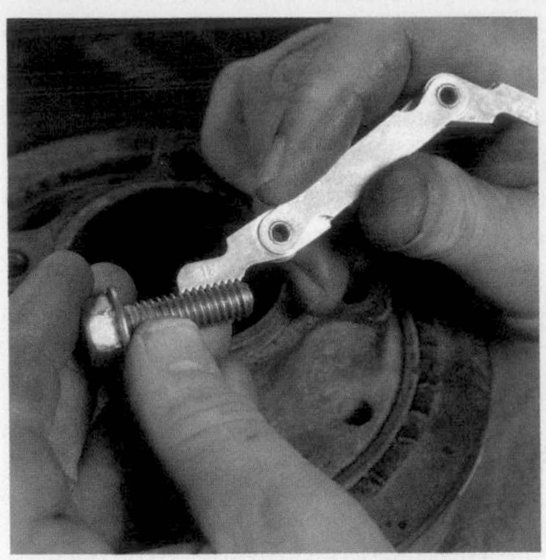

P1–1 Using a thread pitch gauge, determine the thread size of the fastener that should fit into the damaged internal threads.

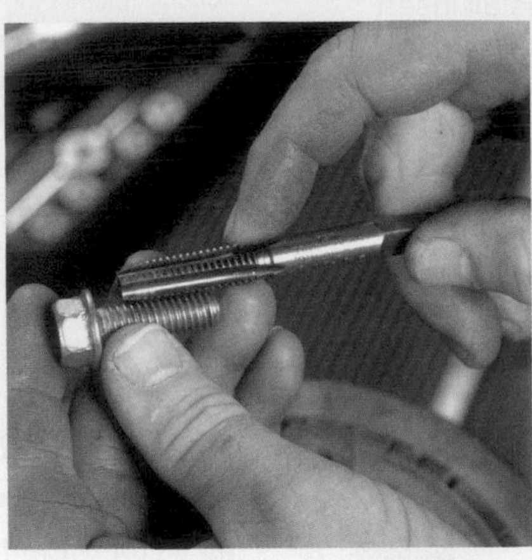

P1–2 Select the correct size and type of tap for the threads and bore to be repaired.

P1–3 Install the tap into a tap wrench.

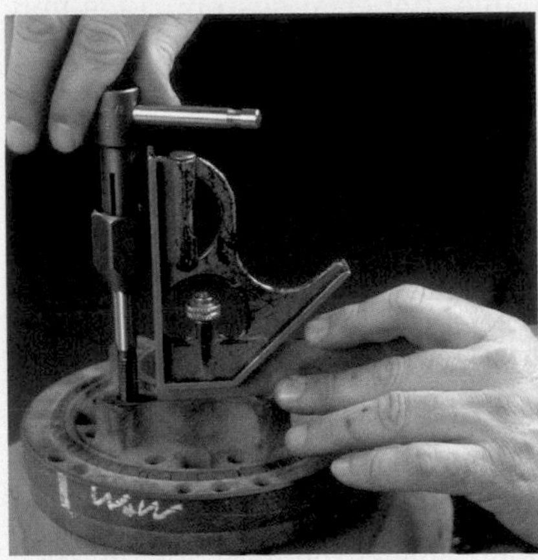

P1–4 Start the tap squarely in the threaded hole using a machinist square as a guide.

P1–5 Rotate the tap clockwise into the bore until the tap has run through the entire length of the threads.

P1–6 Drive the tap back out of the hole by turning it counter-clockwise.

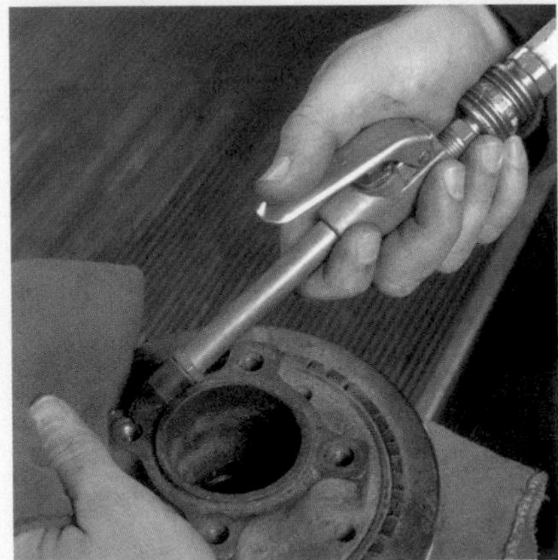

P1–7 Clean the metal chips left by the tap-out of the hole.

P1–8 Inspect the threads left by the tap to be sure they are acceptable.

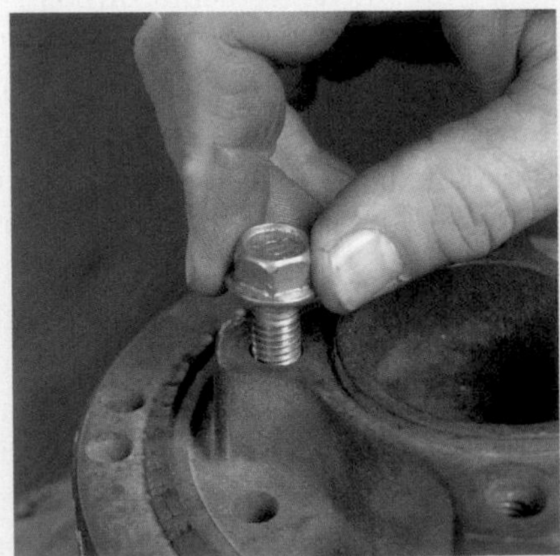

P1–9 Test the threads by threading the correct fastener into the threaded hole.

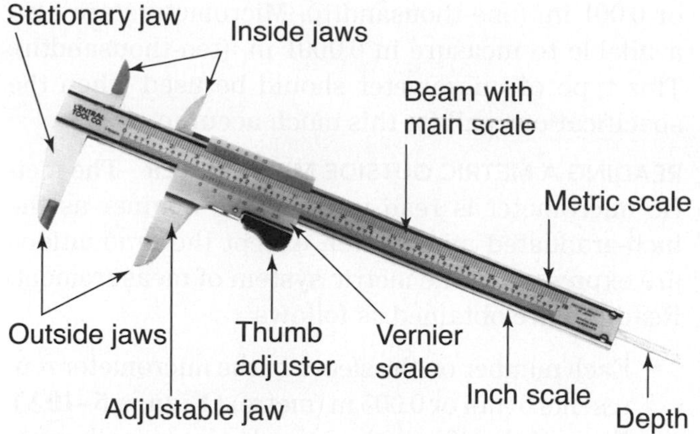

FIGURE 5–15

A vernier caliper.

Central Tools, Inc.

vernier scale. A vernier scale consists of a stationary scale and a movable scale, in this case, the vernier bar to the vernier plate. The length is read from the vernier scale.

A vernier caliper has a movable scale that is parallel to a fixed scale **(Figure 5–15)**. These precision measuring instruments are capable of measuring outside and inside diameters, and most will even measure depth. Vernier calipers are available in both metric and imperial scales. The main scale of the caliper is divided into centimetres, and inches; most measure up to 15 cm or 6 in. The metric main scale is divided into centimetres, with each centimetre divided by 10 mm graduations. The imperial main scale is divided into inches, with each inch divided into 10 parts, each equal to 0.100 in. The area between the 0.100 marks is divided into four. Each of these divisions is equal to 0.025 inch **(Figure 5–16)**.

The metric vernier scale is divided into 50 equal divisions, each representing 0.02 mm. The imperial vernier scale has 25 divisions, each one representing 0.001 in. Measurement readings are taken by combining the main and vernier scales. At all times, only one division line on the main scale will line up with a line on the vernier scale **(Figure 5–17)**. This is the basis for accurate measurements.

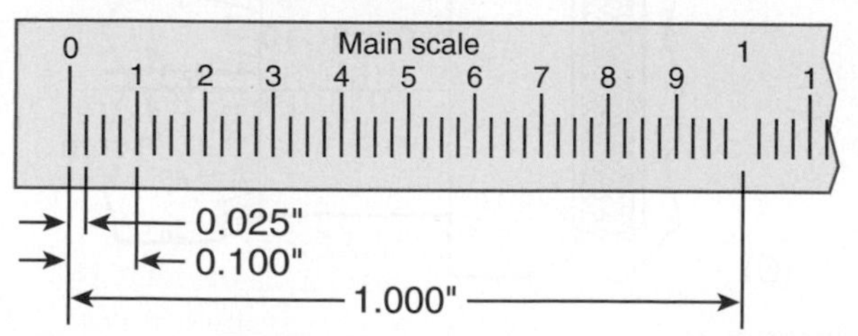

FIGURE 5–16

Each line of the main scale equals 0.025 in.

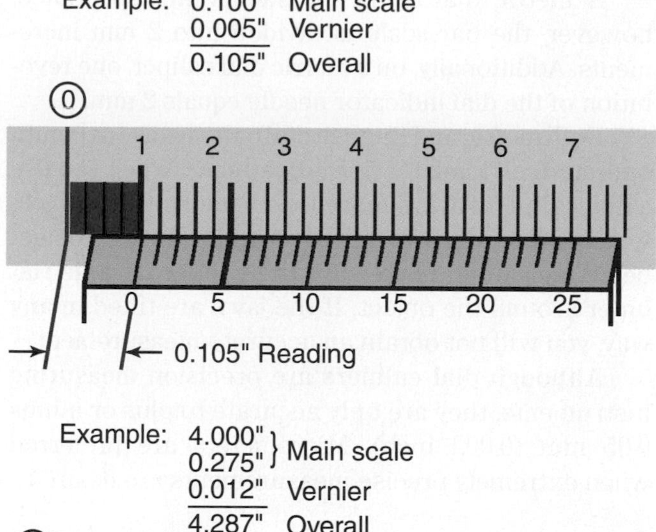

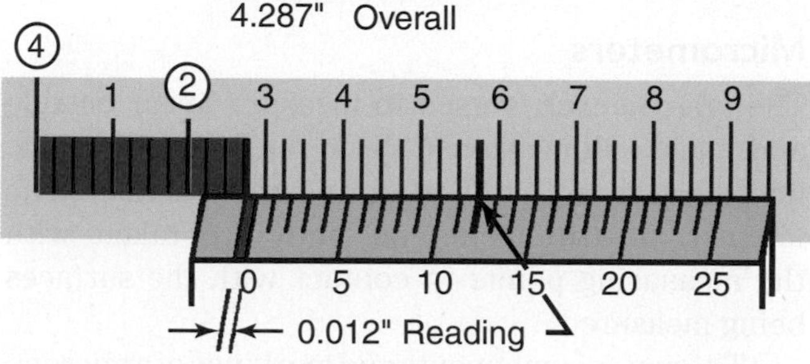

FIGURE 5–17

To get a final measurement, line up the vernier scale line that is exactly aligned with any line on the main scale.

To read the caliper, locate the line on the main scale that lines up with the zero (0) on the vernier scale. If the zero lined up with the 1 on the main scale, the reading would be 1 cm or 0.100 in., depending on the scale used. If the zero on the vernier scale does not line up exactly with a line on the main scale, then look for a line on the vernier scale that does line up with a line on the main scale.

Dial Caliper

The **dial caliper** is an easier-to-use version of the vernier caliper. Metric dial calipers typically measure from 0 to 150 mm in increments of 0.02 mm. Imperial calipers commonly measure dimensions from 0 to 6 in. (0 to 150 mm). The dial caliper features a depth scale, a bar scale, a dial indicator, inside measurement jaws, and outside measurement jaws.

The main scale of a British imperial dial caliper is divided into graduations of one-tenth (0.1) inch. The dial indicator is divided into graduations of one-thousandth (0.001) inch. Therefore, one revolution of the dial indicator needle equals one-tenth inch on the bar scale.

A metric dial caliper is similar in appearance; however, the bar scale is divided into 2 mm increments. Additionally, on a metric dial caliper, one revolution of the dial indicator needle equals 2 mm.

Both metric and English dial calipers use a thumb-operated roll knob for fine adjustment. When you use a dial caliper, always move the measuring jaws backward and forward to centre the jaws on the object being measured. Make sure the caliper jaws lie flat on or around the object. If the jaws are tilted in any way, you will not obtain an accurate measurement.

Although dial calipers are precision measuring instruments, they are only accurate to plus or minus 0.05 mm (0.002 inch). Micrometers are preferred when extremely precise measurements are desired.

Micrometers

The **micrometer** is used to measure linear outside and inside dimensions. Both outside and inside micrometers are calibrated and read in the same manner. Measurements on both are taken with the measuring points in contact with the surfaces being measured.

The major components and markings of a micrometer include the frame, anvil, spindle, locknut, sleeve, sleeve numbers, sleeve long line, thimble marks, thimble, and ratchet **(Figure 5–18)**. Micrometers are calibrated in either metric or inch graduations and are available in a range of sizes. The proper procedure for measuring with an inch-graduated outside micrometer is outlined in Photo Sequence 2.

Most micrometers are designed to measure objects with accuracy to 0.01 mm (one-hundredth) or 0.001 in. (one-thousandth). Micrometers are also available to measure in 0.0001 in. (ten-thousandth). This type of micrometer should be used when the specifications call for this much accuracy.

READING A METRIC OUTSIDE MICROMETER The metric micrometer is read in the same manner as the inch-graduated micrometer, except the graduations are expressed in the metric system of measurement. Readings are obtained as follows:

- Each number on the sleeve of the micrometer represents 5 mm or 0.005 m (metre) **(Figure 5–19A)**.
- Each of the 10 equal spaces between each number, with index lines alternating above and below the horizontal line, represents 0.5 mm, or five-tenths of a millimetre. One revolution of the thimble changes the reading one space on the sleeve scale, or 0.5 mm **(Figure 5–19B)**.
- The bevelled edge of the thimble is divided into 50 equal divisions with every fifth line numbered: 0, 5, 10, ... 45. Because one complete revolution of the thimble advances the spindle 0.5 mm, each graduation on the thimble is equal to one-hundredth of a millimetre **(Figure 5–19C)**.

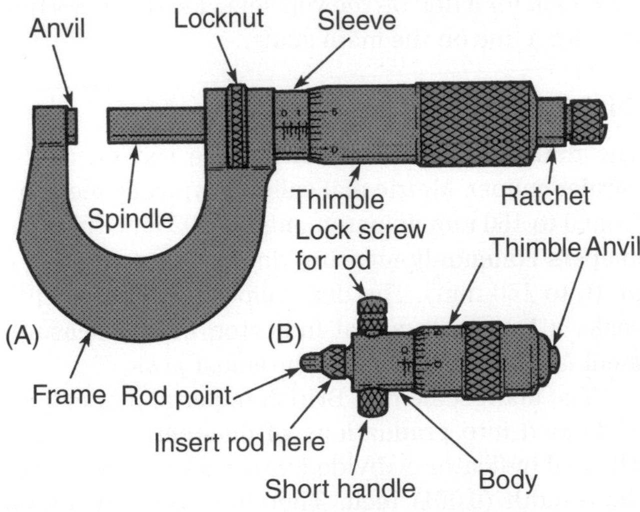

FIGURE 5–18

Major components of (A) an outside and (B) an inside micrometer.

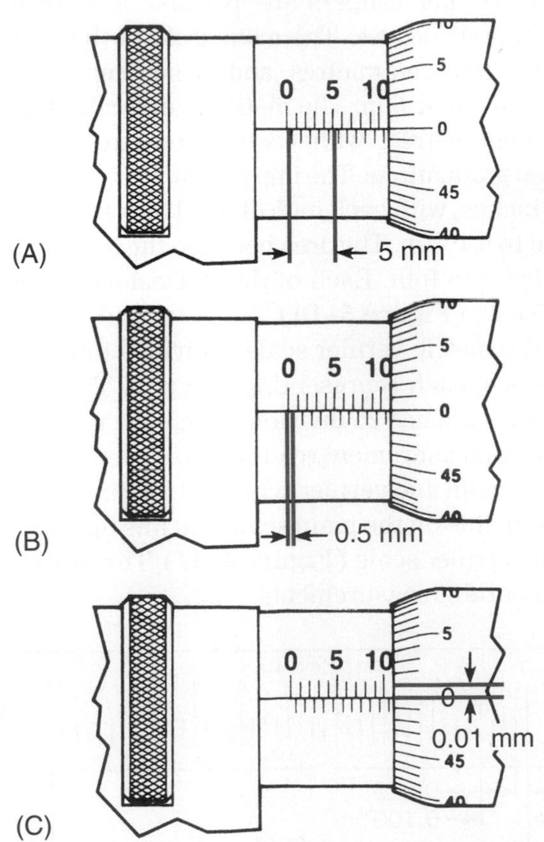

FIGURE 5–19

Reading a metric micrometer: (A) 5 mm plus (B) 0.5 mm plus (C) 0.01 mm equals 5.51 mm.

USING A MICROMETER

PHOTO SEQUENCE 2

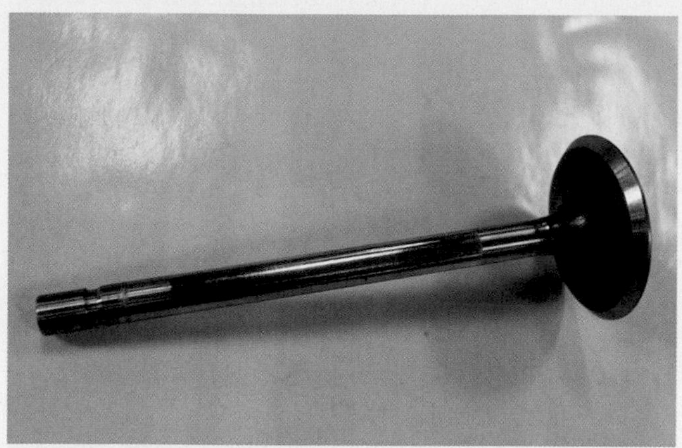

P2–1 Micrometers can be used to measure the diameter of many different objects. By measuring the diameter of a valve stem in two places, the wear of the stem can be determined.

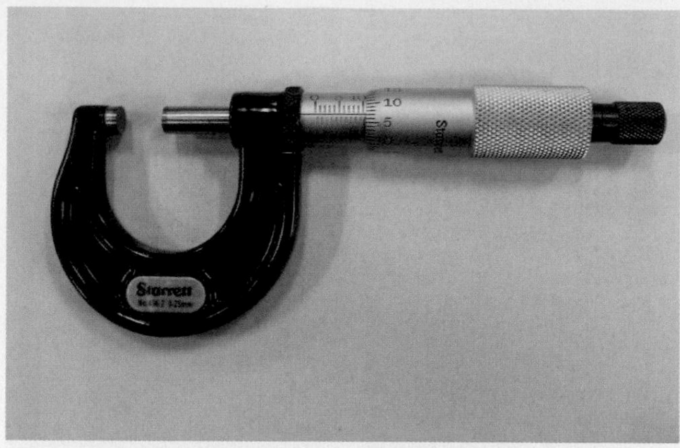

P2–2 Because the diameter of a valve stem is less than 25 mm, a 0 to 25 mm outside micrometer is used.

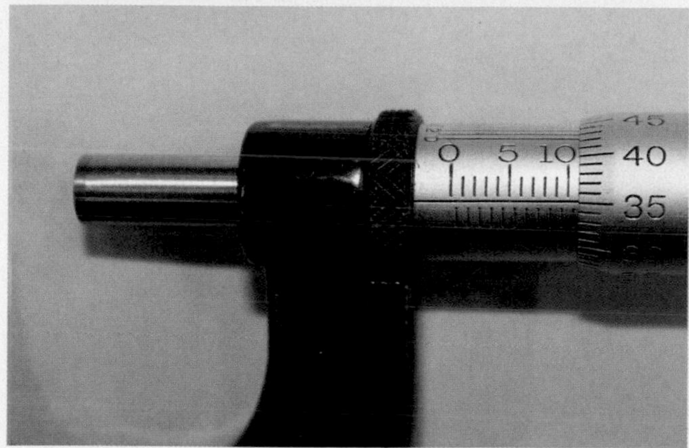

P2–3 The graduations on the sleeve each represent 0.5 mm, with larger lines representing whole millimetres. Most micrometers have every fifth millimetre identified numerically. To read a measurement on a micrometer, begin by counting whole millimetre lines uncovered by the sleeve and possibly an additional 0.5 line.

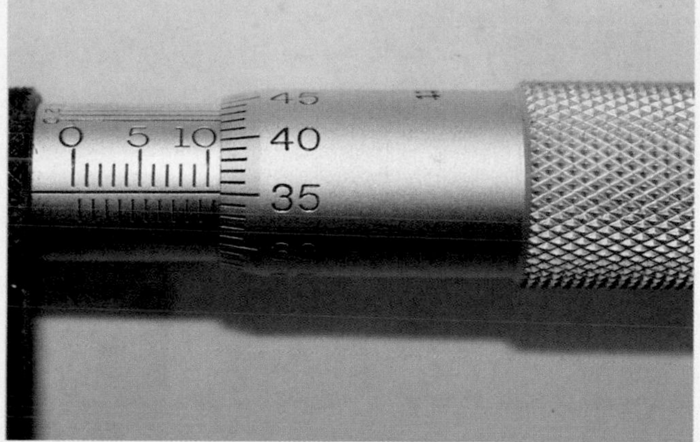

P2–4 The graduations on the thimble assembly are in 1⁄100 of a millimetre and define the area between the lines on the sleeve. The number indicated on the thimble is added to the measurement shown on the sleeve.

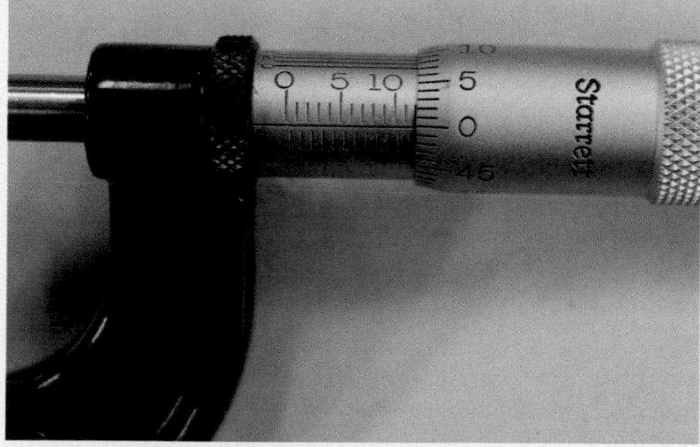

P2–5 A micrometer reading of 12.0 mm.

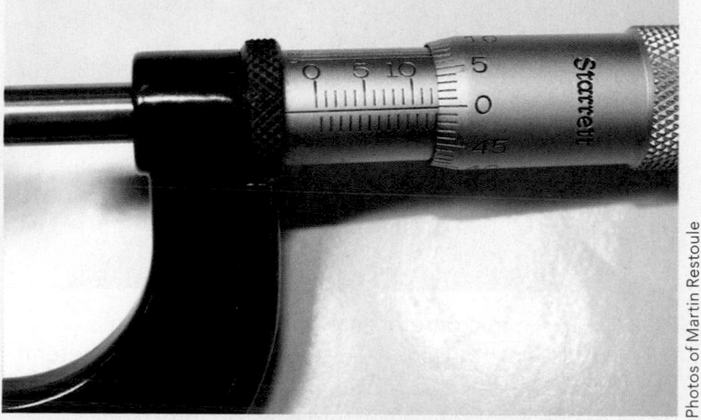

P2–6 A micrometer reading of 12.50 mm.

Photos of Martin Restoule

USING A MICROMETER *(continued)*

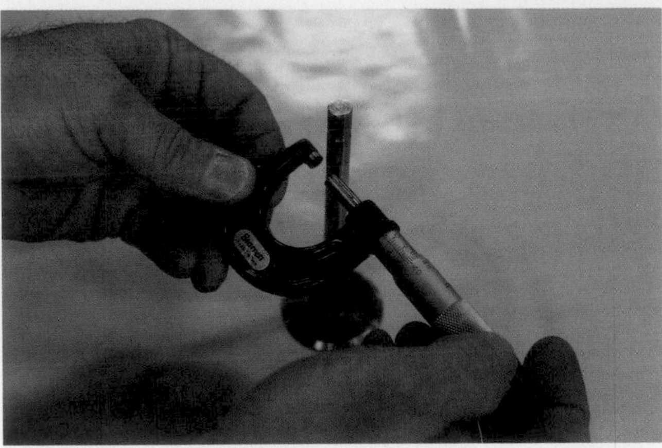

P2–7 Normally, little stem wear is evident directly below the keeper grooves. To measure the diameter of the stem at that point, close the micrometer around the stem.

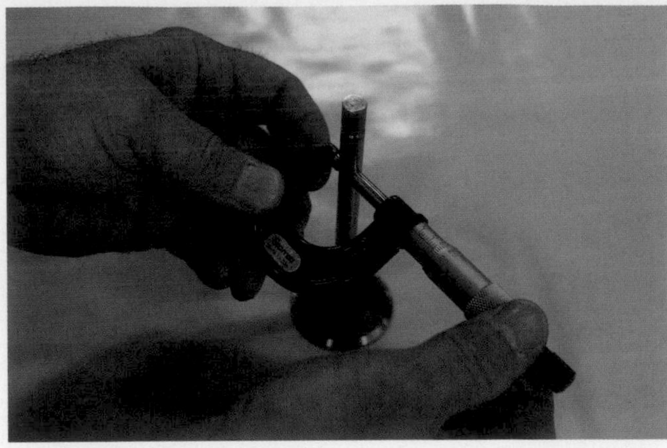

P2–8 To get an accurate reading, slowly close the micrometer until a slight drag is felt while passing the valve in and out of the micrometer.

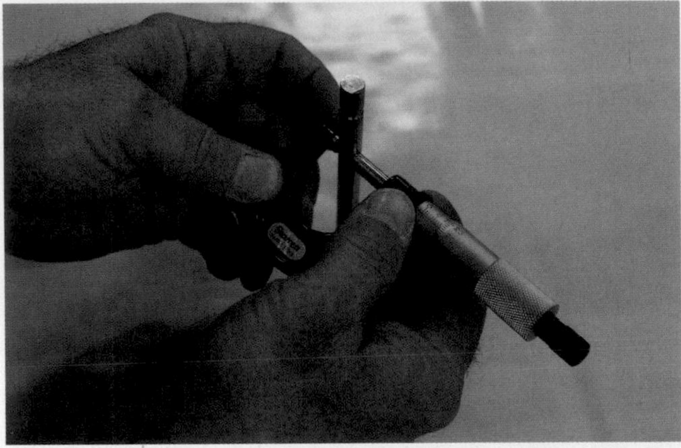

P2–9 To prevent the reading from changing while you move the micrometer away from the stem, use your thumb to roll the lock collar to activate a lock lever.

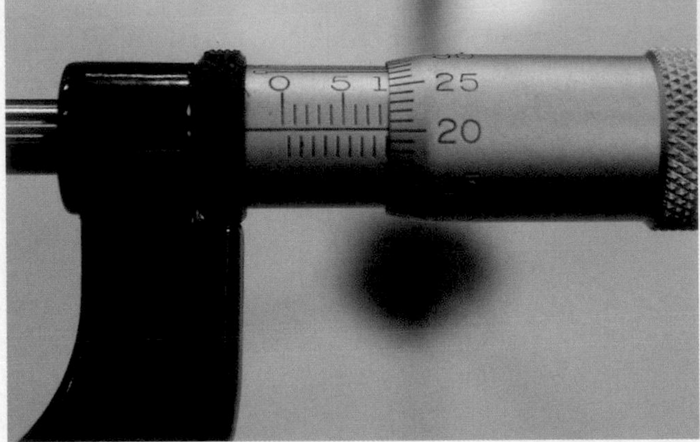

P2–10 This reading of 8.70 mm represents the diameter of the valve stem at the top of the wear area.

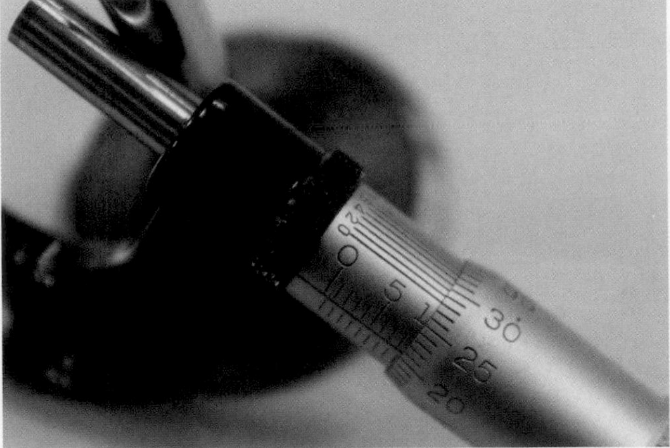

P2–11 Some micrometers are able to measure in 0.001 mm (one-thousandth). Use this type of micrometer if the specifications call for this much accuracy. Note that the exact diameter of the valve stem is 8.700 mm.

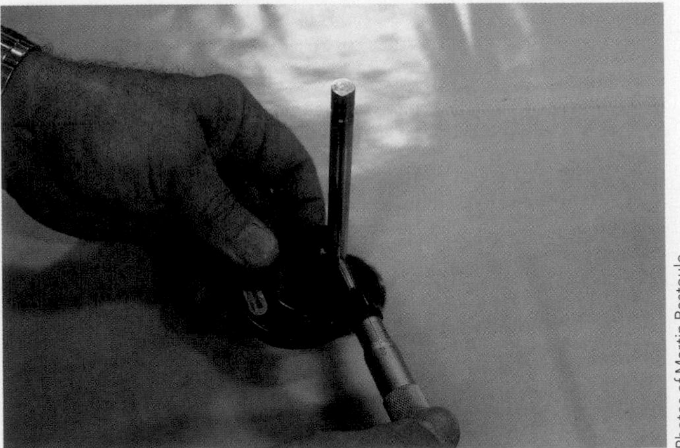

P2–12 Most valve stem wear occurs above the valve head. The diameter here should also be measured. The difference between the diameter of the valve stem just below the keepers and just above the valve head represents the amount of valve stem wear.

Photos of Martin Restoule

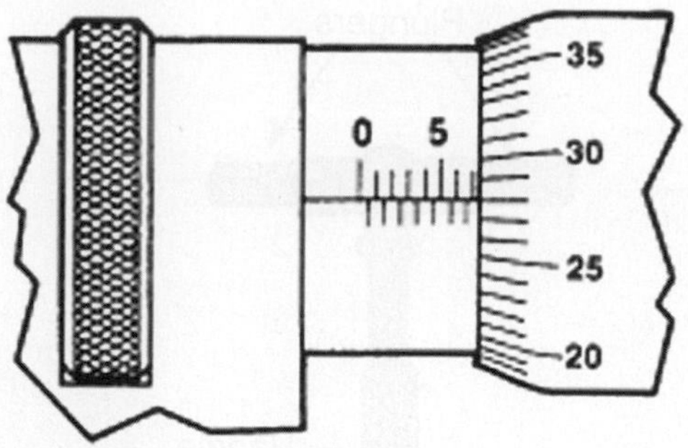

FIGURE 5–20

The total reading on this micrometer is 7.28 mm.

- As with the inch-graduated micrometer, the three separate readings are added together to obtain the total reading **(Figure 5–20)**.

USING AN OUTSIDE MICROMETER To measure small objects using an outside micrometer, open the jaws of the tool and slip the object between the spindle and anvil. While holding the object against the anvil, turn the thimble using your thumb and forefinger until the spindle contacts the object. Never clamp the micrometer tightly. Use only enough tightening force on the thimble to allow the work to just fit between the anvil and spindle. To get accurate readings, you should slip the micrometer back and forth over the object until you feel a very light resistance, while at the same time rocking the tool from side to side to make certain the spindle cannot be closed any further **(Figure 5–21)**. When a satisfactory adjustment

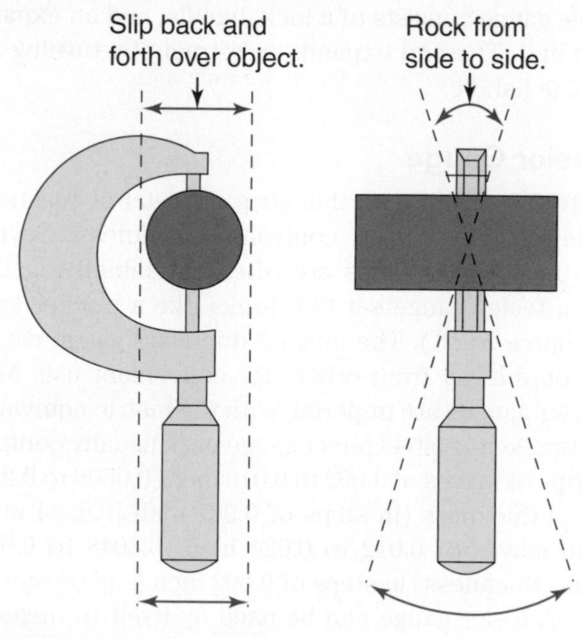

FIGURE 5–21

Slip the micrometer over the object and rock it from side to side.

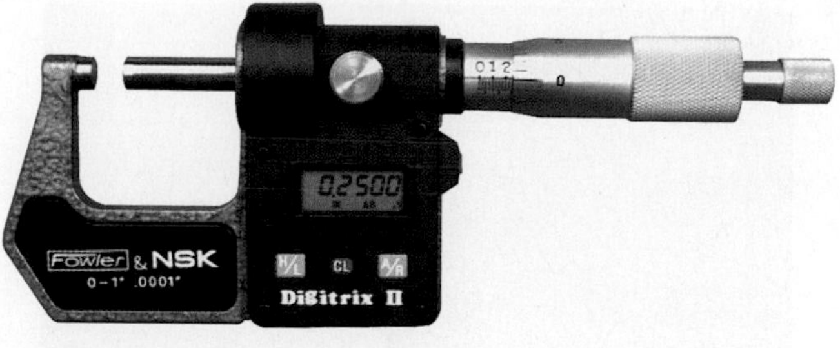

FIGURE 5–22

A digital micrometer.

Fred V. Fowler Co., Inc.

has been made, lock the micrometer. Read the measurement scale.

Some technicians use a digital micrometer, which is easier to read. Because these tools do not have the various scales, the measurement is displayed and read directly off the micrometer. **Figure 5–22** shows a digital outside micrometer.

To measure a larger object such as a piston, select a micrometer of the proper size. Micrometers are available in a number of different sizes. The size is dictated by the smallest measurement it can make to the largest. Examples of these micrometer sizes are 0 to 25 mm (0 to 1 in.), 25 to 50 mm (1 to 2 in.), 50 to 75 mm (2 to 3 in.), and 75 to 100 mm (3 to 4 in.).

READING AN INSIDE MICROMETER Inside micrometers are used to measure the inside diameter of a bore or hole. To do this, place the tool inside the bore and extend the measuring surfaces until each end touches the bore's surface. If the bore is large, it might be necessary to use an extension rod to increase the micrometer's range. These extension rods come in various lengths. The inside micrometer is read in the same manner as an outside micrometer.

To obtain a precise measurement, keep the anvil firmly against one side of the bore and rock the micrometer back and forth and side to side to ensure that the micrometer is in the centre of the bore. As with the outside micrometer, this procedure will require a little practice until you get the feel of the correct resistance on both ends of the tool.

READING A DEPTH MICROMETER A depth micrometer **(Figure 5–23)** is used to measure the distance between two parallel surfaces. The sleeves, thimbles, and ratchet screws operate in the same way as other micrometers. Likewise, depth micrometers are read in the same way as other micrometers.

If a depth micrometer is used with a gauge bar, it is important to keep both the bar and the micrometer from rocking. Any movement of either part will result in an inaccurate measurement.

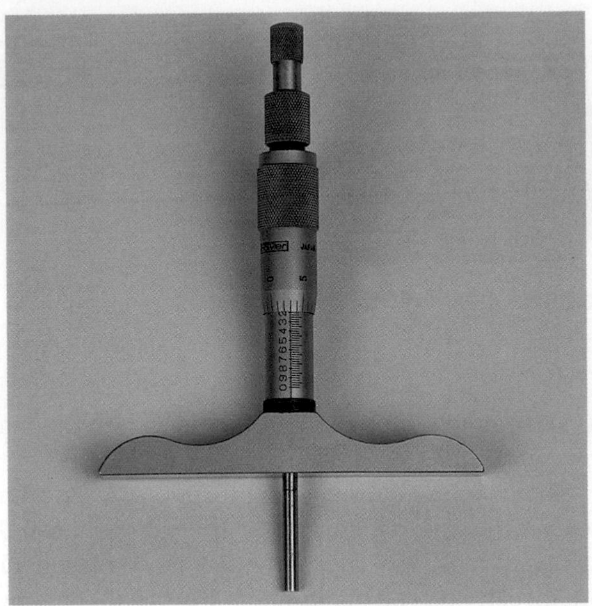

FIGURE 5–23

A depth micrometer.

Central Tools, Inc.

SHOP TALK

Measurements with any micrometer will be reliable only if the micrometer is calibrated correctly. To calibrate a micrometer, close the micrometer over a micrometer standard. If the reading differs from that of the standard, the micrometer should be adjusted according to the instructions provided by the tool manufacturer. Proper care of a micrometer is also important to ensure accurate measurements. This care includes the following:

- Always clean the micrometer before using it.
- Do not touch the measuring surfaces.
- Store the tool properly. The spindle face should not touch the anvil face; a change in temperature might spring the micrometer.
- Clean the micrometer after use. Wipe it clean of any oil, dirt, or dust, using a lint-free cloth.
- Never use the tool as a clamp or tighten the jaws too tightly around an object.
- Do not drop the tool.
- Check the calibration weekly.

Telescoping Gauge

Telescoping gauges (Figure 5–24) are used for measuring bore diameters and other clearances. They may also be called **snap gauges**. They are available in sizes ranging from approximately 7 mm (1/4 in.) through 150 mm (6 in.). Each gauge consists of two telescoping plungers, a handle, and a lock screw. Snap gauges are normally used with an outside micrometer.

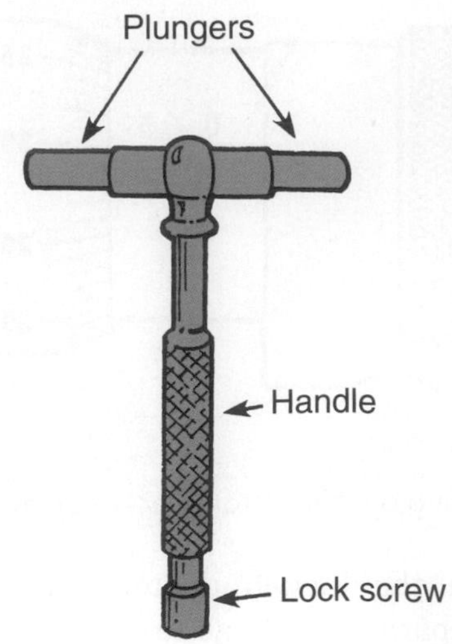

FIGURE 5–24

Parts of a telescoping gauge.

To use the telescoping gauge, insert it into the bore and loosen the lock screw. This will allow the plungers to snap against the bore. Once the plungers have expanded, tighten the lock screw. Then, remove the gauge and measure the expanse with a micrometer.

Small Hole Gauge

A small hole or **ball gauge** works just like a telescoping gauge. However, it is designed to be used on small bores. After it is placed into the bore and expanded, it is removed and measured with a micrometer **(Figure 5–25)**. Like the telescoping gauge, the small hole gauge consists of a lock, handle, and an expanding end. The end expands or retracts by turning the gauge handle.

Feeler Gauge

A **feeler gauge** is a thin strip of metal or plastic of known and closely controlled thickness. Several of these metal strips are often assembled together as a feeler gauge set that looks like a pocket knife **(Figure 5–26)**. The desired thickness gauge can be pivoted away from others for convenient use. Most feeler gauges are imperial, with the metric equivalent in brackets. A steel feeler gauge pack usually contains strips or leaves of 0.002 to 0.010 inch (0.0508 to 0.2540 mm) thickness (in steps of 0.001 inch [0.0254 mm]) and leaves of 0.012 to 0.024 inch (0.3048 to 0.6096 mm) thickness (in steps of 0.002 inch [0.0508 mm]).

A feeler gauge can be used by itself to measure piston ring side clearance, piston ring end gap, connecting rod side clearance, crankshaft endplay, and other distances.

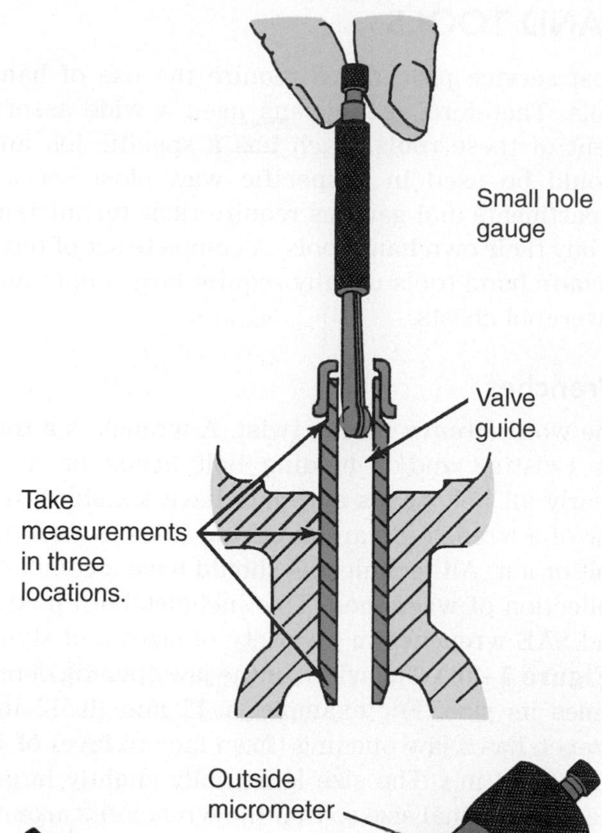

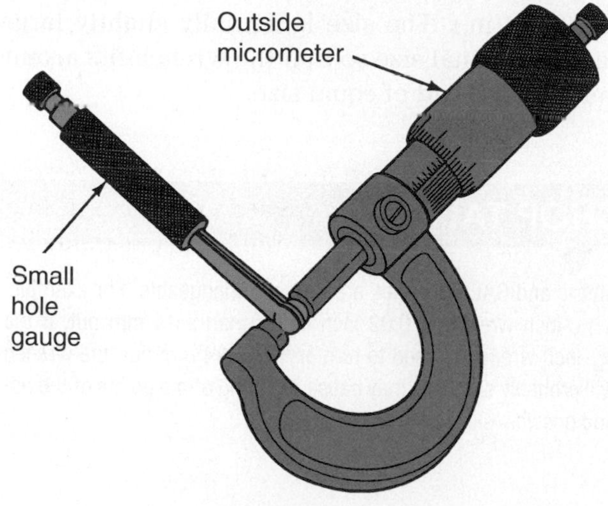

FIGURE 5–25

Insert the ball gauge into the bore to be measured. Then expand it, lock it, and remove it. Now measure it with an outside micrometer.

Ford Motor Company

FIGURE 5–26

Typical feeler gauge set.

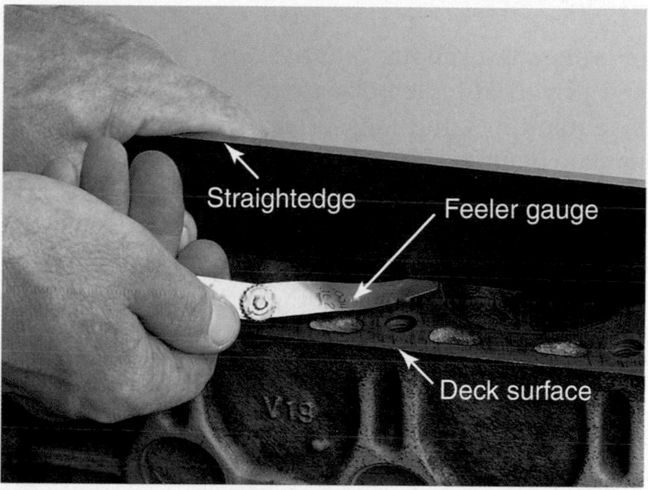

FIGURE 5–27

Using a feeler gauge and precision straightedge to check for warpage.

Round wire feeler gauges are often used to measure spark plug gap. The round gauges are designed to give a better feel for the fit of the gauge in the gap.

Straightedge

A **straightedge** is no more than a flat bar machined to be totally flat and straight, and to be effective it must be flat and straight. Any surface that should be flat can be checked with a straightedge and feeler gauge set. The straightedge is placed across and at angles on the surface. At any low points on the surface, a feeler gauge can be placed between the straightedge and the surface **(Figure 5–27)**. The size gauge that fills in the gap indicates the amount of warpage or distortion.

Dial Indicator

The **dial indicator (Figure 5–28)** is used to measure movement. Although dial indicators are available in a number of different ranges of accuracy, most metric dial indicators are graduated in 0.02 mm (two-hundredths of a millimetre) increments and imperial in 0.001 in. (one-thousandth inch) increments. Common uses of the dial indicator include measuring valve lift, journal concentricity, flywheel

FIGURE 5–28

A dial indicator with a highly adaptive holding fixture.

Federal-Mogul Corporation

or brake rotor runout, gear backlash, and crankshaft end play. Dial indicators are available with various face markings and measurement ranges to accommodate many measuring tasks.

To use a dial indicator, position the indicator rod against the object to be measured. Then, push the indicator toward the work until the indicator needle travels far enough around the gauge face to permit movement to be read in either direction **(Figure 5–29)**. Zero the indicator needle on the gauge. Always be sure the range of the dial indicator is sufficient to allow the amount of movement required by the measuring procedure. For example, never use a 25 mm (1 in.) indicator on a component that will move 50 mm (2 in.).

FIGURE 5–29

This dial indicator setup will measure the amount this axle can move in and out.

HAND TOOLS

Most service procedures require the use of hand tools. Therefore, technicians need a wide assortment of these tools. Each has a specific job and should be used in a specific way. Most service departments and garages require their technicians to buy their own hand tools. A complete set of technician's hand tools usually require large upper and lower tool chests.

Wrenches

The word *wrench* means twist. A wrench is a tool for twisting and/or holding bolt heads or nuts. Nearly all bolt heads and nuts have six sides; the jaw of a wrench fits around these sides to turn the bolt or nut. All technicians should have a complete collection of **wrenches**. This includes both metric and SAE wrenches in a variety of sizes and styles **(Figure 5–30)**. The width of the jaw opening determines its size. For example, a 13 mm (0.512 in.) wrench has a jaw opening (from face to face) of 13 mm (0.512 in.). The size is actually slightly larger than its nominal size so that the wrench fits around a nut or bolt head of equal size.

SHOP TALK

Metric and SAE wrenches are not interchangeable. For example, a 9/16-inch wrench is 0.02 inch larger than a 14 mm nut. If the 9/16-inch wrench is used to turn or hold a 14 mm nut, the wrench will probably slip. This may cause rounding of the points of the nut and possibly skinned knuckles as well.

FIGURE 5–30

A technician needs many different sets of wrenches.

Martin Restoule

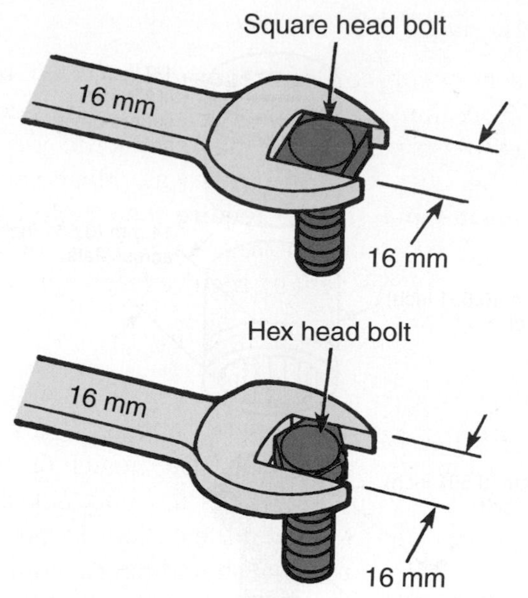

FIGURE 5–31

An open-end wrench grips only two sides of a fastener.

The following is a brief discussion of the types of wrenches used by automotive technicians.

OPEN-END WRENCH The jaws of the open-end wrench **(Figure 5–31)** allow the wrench to slide around two sides of a bolt or nut head where there might be insufficient clearance above or on one side of the nut to accept a box wrench.

BOX-END WRENCH The end of the box-end wrench is boxed or closed rather than open. The jaws of the wrench fit completely around a bolt or nut, gripping each point on the fastener. The box-end wrench is not likely to slip off a nut or bolt. It is safer than an open-end wrench. Box-end wrenches are available in 6 point and 12 point **(Figure 5–32)**. The 6-point box end grips the bolt head more securely than a 12-point box-end wrench can and avoids damage to the bolt head.

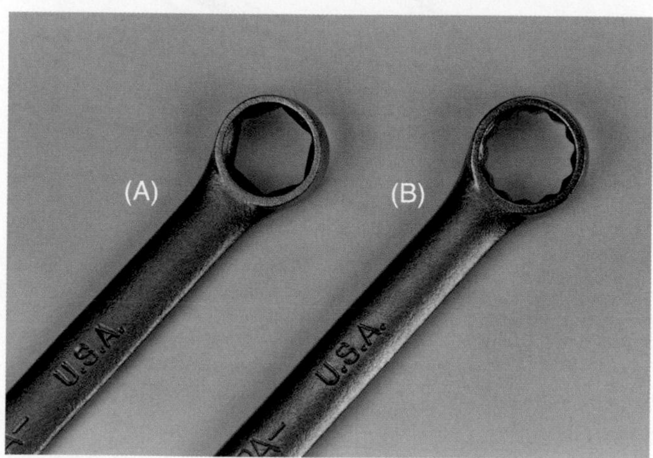

FIGURE 5–32

(A) Six-point and (B) twelve-point box-end wrenches are available.

COMBINATION WRENCH The combination wrench has an open-end jaw on one end and a box end on the other. Both ends are the same size. Every auto technician should have two sets of wrenches: one for holding and one for turning. The combination wrench is probably the best choice for the second set. It can be used with either open-end or box-end wrench sets and can be used as an open-end or box-end wrench.

FLARE NUT (LINE) WRENCHES Flare nut or line wrenches should be used to loosen or tighten brake line or tubing fittings. Using open-ended wrenches on these fittings tends to round the corners of the nut, which are typically made of soft metal and can distort easily. Flare nut wrenches surround the nut and provide a better grip on the fitting. They have a section cut out so that the wrench can be slipped around the brake or fuel line and dropped over the flare nut.

ALLEN WRENCH Setscrews are used to fasten door handles, instrument panel knobs, engine parts, and even brake calipers. A set of metric and fractional hex head wrenches, or Allen wrenches **(Figure 5–33)**, should be in every technician's toolbox. An Allen wrench can be L-shaped or can be mounted in a socket driver and used with a ratchet.

ADJUSTABLE-END WRENCH An adjustable-end wrench (commonly called a crescent wrench) has one fixed jaw and one movable jaw. The wrench opening can be adjusted by rotating a helical adjusting screw that is mated to teeth in the lower jaw. Because this type of wrench does not firmly grip a bolt's head, it is likely to slip. Adjustable wrenches should be used carefully and *only* when it is absolutely necessary. Be sure to put all of the turning force or effort on the fixed jaw.

Sockets and Ratchets

A set of metric and imperial sockets combined with a ratchet handle and a few extensions should be included in your tool set. The ratchet allows you to turn the socket in one direction with force and in the other direction without force, which allows you to tighten or loosen a bolt without removing and resetting the wrench after you have turned it. In many situations, a socket wrench is much safer, faster, and easier to use than any other wrench. In fact, sometimes it is the only wrench that will work.

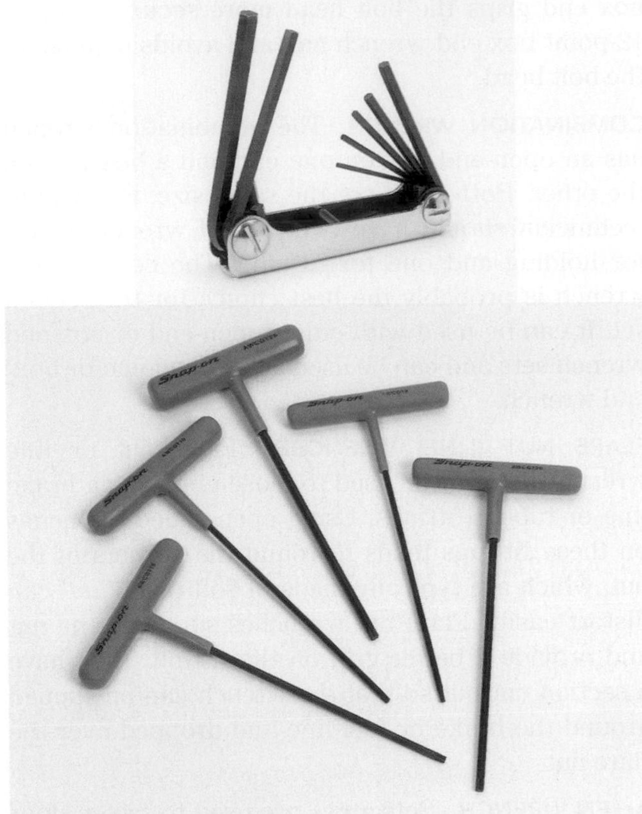

FIGURE 5–33

Top: A handy tool containing many different Allen wrenches. Bottom: T-handle Allen wrenches designed for better gripping and easier torque application.

Snap-on Tools Company

The basic socket wrench set consists of a ratchet handle and several barrel-shaped sockets. The socket fits over and around a bolt or nut **(Figure 5–34)**. Inside, it is shaped like a box-end wrench. Sockets are available in 6, 8, or 12 points. A 6-point socket has stronger walls and improved grip on a bolt compared to a normal 12-point socket. However, 6-point sockets have half the positions of a 12-point socket. Six-point sockets are mostly used on fasteners that are rusted or rounded. Eight-point sockets are available to use on square nuts or square-headed bolts. Some axle and transmission assemblies use square-headed plugs in the fluid reservoir.

The top side of a socket has a square hole that accepts a square lug on the socket handle. This square hole is the drive hole. The size of the hole and handle lug (6.35 mm, 9.54 mm, 12.70 mm [¼ in., ⅜ in., ½ in.], and so on) indicates the drive size of the socket wrench. One handle fits all the sockets in a set. On better-quality handles, a spring-loaded ball in the square drive lug fits into a depression in the socket. This ball holds the socket to the handle. An

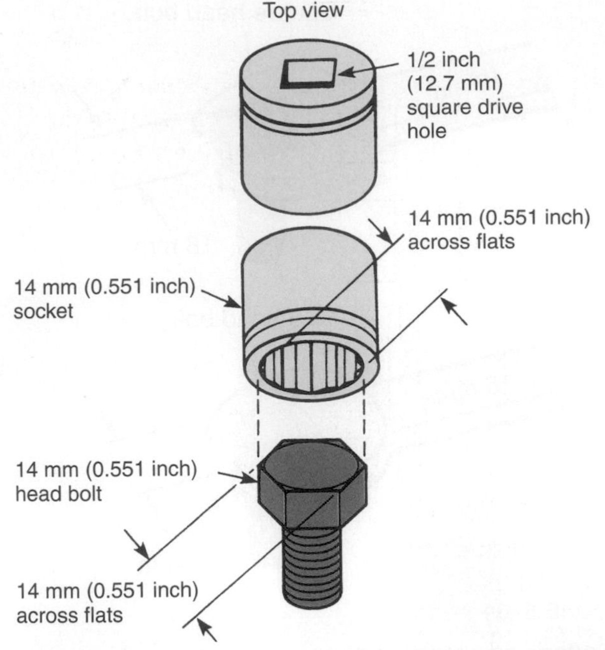

FIGURE 5–34

The size of the correct socket is the same size as the size of the bolt head or nut.

assortment of socket (ratchet) handles is shown in **Figure 5–35**.

Not all socket handles are ratcheting. Some, called breaker bars, are simply long arms with a swivel drive used to provide extra torque onto a bolt to help loosen it. These are available in a variety of lengths and drive sizes. Sometimes, nut drivers are used. These handles

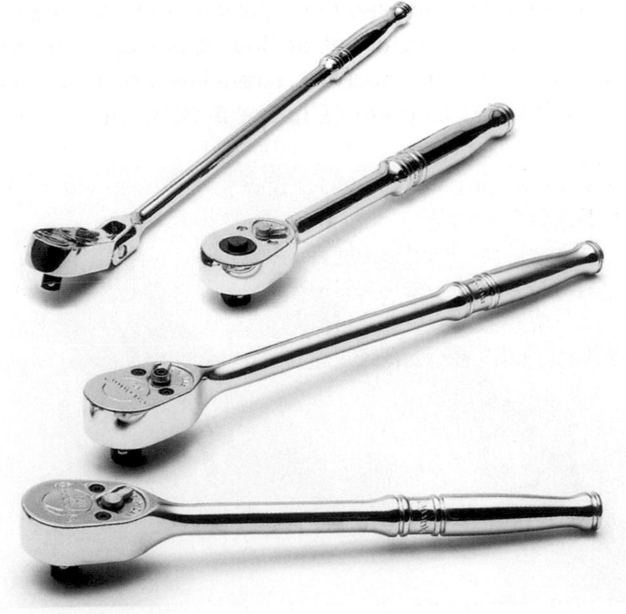

FIGURE 5–35

An assortment of ratchets.

Snap-on Tools Company

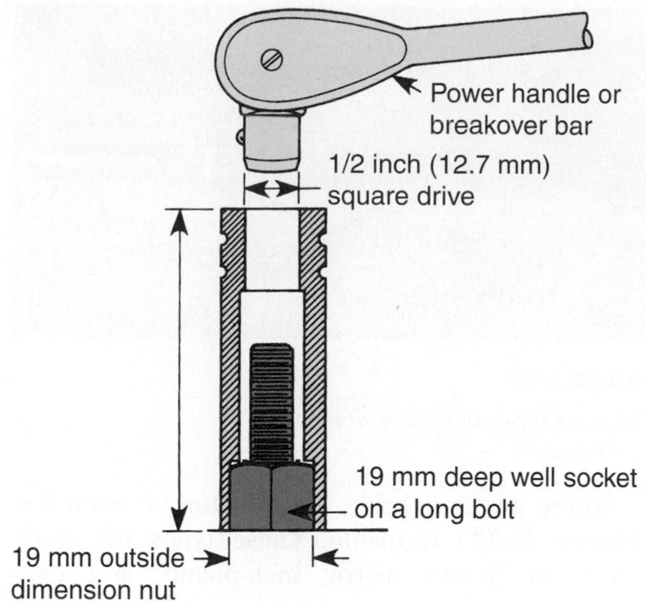

FIGURE 5–36

Deep-well sockets fit over the ends of bolts and studs.

look like screwdrivers and have a driveshaft on the end of the shaft. Sockets and/or various attachments are inserted on the drive lug. These drivers are used only when bolt tightness is low.

Sockets are available in various sizes, lengths, and bore depths. Both metric and SAE socket wrench sets are necessary for automotive service. Normally, the larger the socket size, the longer the socket or the deeper the well. Deep-well sockets **(Figure 5–36)** are made extra long to fit over bolt ends or studs. A spark plug socket is an example of a special purpose deep-well socket. Deep-well sockets are also good for reaching nuts or bolts in limited-access areas. Deep-well sockets should not be used when a regular-size socket will do the job. The longer socket develops more twist torque and tends to slip off the fastener.

Heavier-walled sockets are designed for use with an impact wrench and are called impact sockets. Most sockets are chrome-plated, except for impact sockets, which are not.

WARNING!

Never use a nonimpact socket with an impact wrench.

SPECIAL SOCKETS Screwdriver (including Torx® driver) and Allen wrench attachments are also available for use with a socket wrench. **Figure 5–37** shows a typical set of screwdriver attachments and

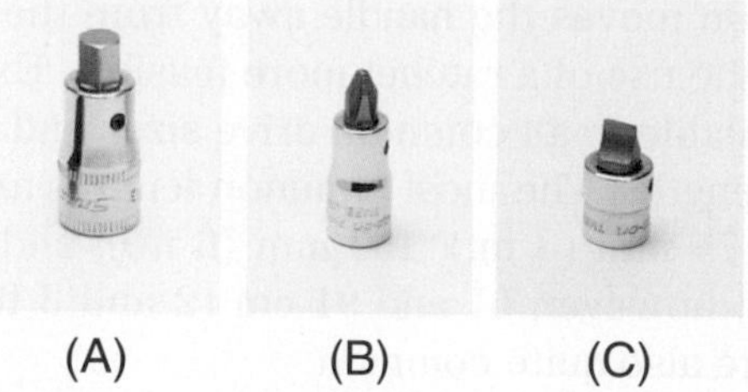

FIGURE 5–37

Typical screwdriver attachment set including (A) a hex driver, (B) a Phillips driver, and (C) a slot tip driver.

Snap-on Tools Company

three specialty sockets. These socket wrench attachments are very handy when a fastener cannot be loosened with a regular screwdriver. The leverage given by the ratchet handle is often just what it takes to break a stubborn screw loose.

Swivel sockets are also available. These sockets are fitted with a flexible joint that accommodates odd angles between the socket and the ratchet handle. These sockets are often used to work bolts that are difficult to reach.

Although crowfoot sockets are not really sockets, they are used with a ratchet or breaker bar. These sockets are actually the end of an open-end or line wrench made with a drive bore, which allows a ratchet to move the socket.

CROWFOOT WRENCH ADAPTERS Crowfoot wrenches are typically used when a hex fitting is used on a line or fitting that is in a restricted or shrouded area that prevents the use of a regular socket or wrench. These wrench adapters are available in open-end and box-end 6- and 12-point, as well as flare-nut, configurations **(Figure 5–38)**.

EXTENSIONS An extension is commonly used to separate the socket from the ratchet or handle. The

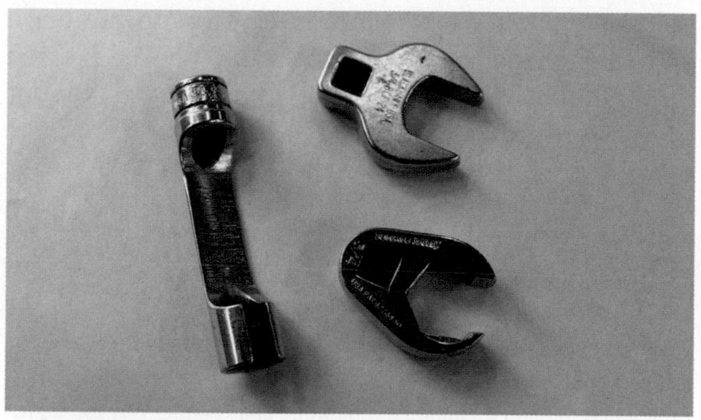

FIGURE 5–38

Types of crowfoot wrench adapters.

Martin Restoule

extension moves the handle away from the bolt and makes the use of a ratchet more feasible. Extensions are available in all common drive sizes and in a variety of lengths. The most common lengths are 25 mm (1 in.), 75 mm (3 in.), 150 mm (6 in.), and 250 mm (10 in.); however, 61 and 91 cm (2 and 3 ft.) extensions are also quite common.

Wobble extensions allow a socket to pivot slightly at the drive connection. This type of extension provides for a more positive connection to the socket than swivel joints but allows only approximately 16° of flexibility.

SOCKET ADAPTERS When sockets of a different drive size must be used with a particular ratchet or handle, an adapter can be inserted between the socket and the drive on the handle. An example of a common adapter is one that allows for the use of a ¼-inch drive socket on a ⅜-inch drive ratchet.

Torque Wrenches

Torque wrenches measure how tight a nut or bolt is. Nearly all of a car's nuts and bolts should be tightened to a certain amount and have a torque specification that is expressed in Newton-metres (N·m) (metric) or pound-feet (lb·ft) (USCS) or a combination of pound-feet or Newton-metres plus an angle. A Newton-metre is the work or torque accomplished by a force of one Newton through a distance of one metre. A pound-foot is the work or torque accomplished by a force of one pound through a distance of one foot. When comparing a Newton-metre to a pound-foot, 1 N·m is equal to 0.7375621 lb·ft.

A torque wrench is basically a ratchet or breaker bar with some means of displaying the amount of torque exerted on a bolt when force is applied to the handle. Torque wrenches are available with the various drive sizes. Sockets are inserted onto the drive and then placed over the bolt. As force is exerted on the bolt, the torque wrench indicates the amount of torque.

> **SHOP TALK**
>
> Following torque specifications is critical. However, there is a possibility that the torque spec is wrong as printed. (In other words, someone made a mistake.) If the torque spec seems way too tight or loose for the size of bolt, find the torque spec in a different source. If the two specs are the same, use it. If they are different, use the one that seems right.

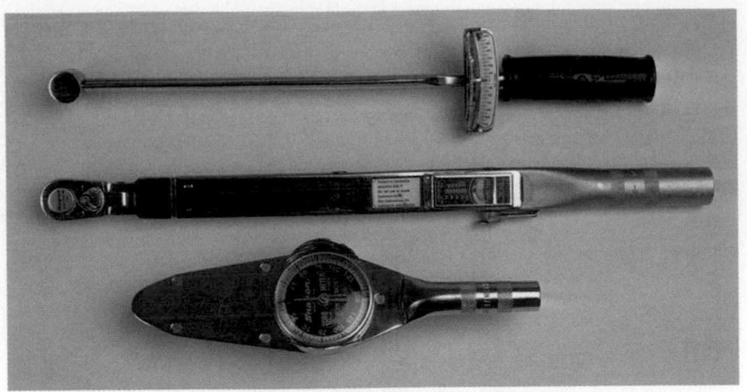

FIGURE 5–39

The basic types of torque wrenches.

Martin Restoule

There are four basic types of torque wrenches **(Figure 5–39)** available. These types are available with Newton-metre, inch-pound, and foot-pound increments.

- *A beam torque wrench.* On this type, the beam points to the torque reading. This torque wrench is not highly accurate.
- *A click-type torque wrench.* On this type of wrench, the desired torque reading is set on the handle. When the torque reaches that level, the wrench clicks.
- *A dial torque wrench.* This type of torque wrench has a dial that indicates the torque exerted on the wrench. Some designs of this type have a light or buzzer that turns on when the desired torque is reached.
- *A digital readout torque wrench.* This style of torque wrench displays the torque digitally and is commonly used to measure turning effort as well as to tighten bolts. Some designs of this type have a light or buzzer that turns on when the desired torque is reached. Any good torque wrench allows a technician to tighten a bolt or nut to the correct torque.

The correct torque provides the tightness and stress that the manufacturer has found to be the most desirable and reliable. For example, engine-bearing caps that are too tight distort the bearings, causing excessive wear and incorrect oil clearance. This often results in rapid wear of other engine parts due to decreased oil flow. Insufficient torque can result in out-of-round bores and subsequent failure of the parts.

When using a torque wrench, follow these steps to get an accurate reading:

1. Locate the torque specs and procedures in a service manual.
2. Mentally divide the torque specification by three.

3. Hold the wrench so that it is at a 90° angle from the fastener being tightened.
4. Tighten the bolt or nut to one-third of the specification.
5. Then, tighten the bolt to two-thirds of the spec.
6. Now tighten the bolt to within 15 Newton-metres (11 pound-feet) of the spec.
7. Tighten the bolt to the specified torque.
8. Recheck the torque.

Screwdrivers

A screwdriver drives a variety of threaded fasteners used in the automotive industry. Each fastener requires a specific kind of screwdriver, and a well-equipped technician has several sizes of each.

SHOP TALK

A screwdriver should not be used as a chisel, punch, or pry bar. Screwdrivers were not made to withstand blows or bending forces. When misused in such a fashion, the tips will wear, become rounded, and tend to slip out of the fastener. At that point, a screwdriver becomes unusable. Remember, a defective tool is a dangerous tool.

Screwdrivers are defined by their sizes, their tips **(Figure 5–40)**, and the types of fasteners they should be used with. Your tool set should include both blade and Phillips drivers in a variety of lengths from 5 cm (2 in.) "stubbies" to 30 cm (12 in.) screwdrivers. You also should have an assortment of special screwdrivers, such as those with a Torx® head design.

- *Standard tip screwdriver.* A slotted screw accepts a screwdriver with a standard or blade-type tip. The standard tip screwdriver is probably the most common type **(Figure 5–41)**. It is useful for turning carriage bolts, machine screws,

FIGURE 5–40

The various screwdriver tips that are available.

Snap-on Tools Company

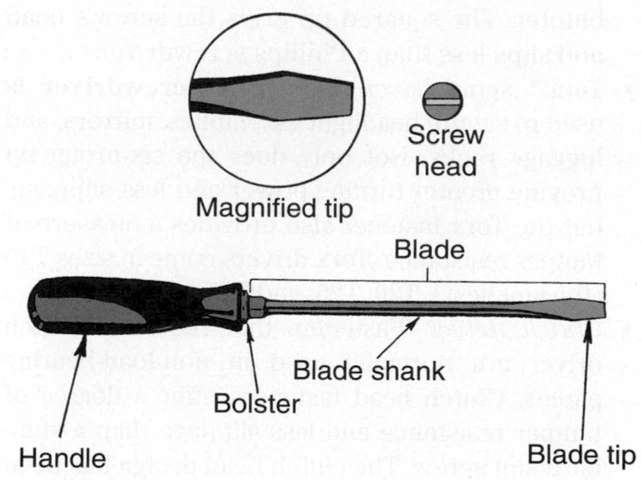

FIGURE 5–41

The standard tip screwdriver is used with slotted head fasteners.

Snap-on Tools www.snapon.com

and sheet metal screws. The width and thickness of the blade determine the size of a standard screwdriver. Always use a blade that fills the slot in the fastener.

- *Phillips screwdriver.* The tip of a **Phillips screwdriver** has four prongs that fit the four slots in a Phillips head screw **(Figure 5–42)**. The four surfaces enclose the screwdriver tip, so it is less likely that the screwdriver will slip out of the fastener. Phillips screwdrivers come in sizes #0 (the smallest), #1, #2, #3, and #4 (the largest).
- *Reed and Prince screwdriver.* The tip of a Reed and Prince screwdriver is like a Phillips except that the prongs come to a point rather than to a blunt end.
- *Pozidriv® screwdriver.* The **Pozidriv screwdriver** is like a Phillips, but its tip is flatter and

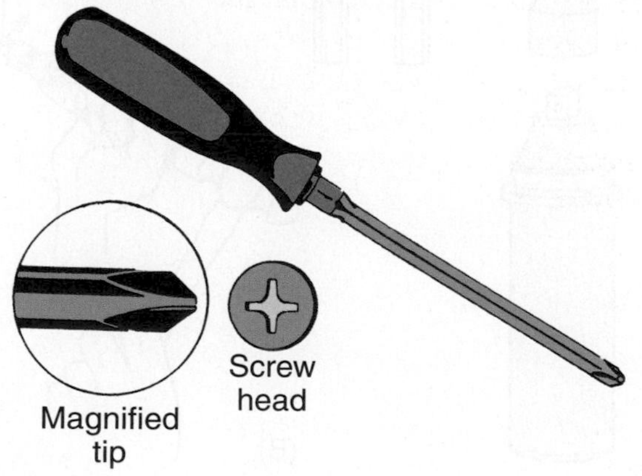

FIGURE 5–42

The tip of a Phillips screwdriver has four prongs that provide a good grip on the fastener.

Snap-on Tools www.snapon.com

blunter. The squared tip grips the screw's head and slips less than a Phillips screwdriver.

- *Torx® screwdriver.* The **Torx screwdriver** is used to secure headlight assemblies, mirrors, and luggage racks. Not only does the six-prong tip provide greater turning power and less slippage, but the Torx fastener also provides a measure of tamper resistance. Torx drivers come in sizes T15 (the smallest), T20, T25, and T27 (the largest).
- *Clutch driver.* Fasteners that require a clutch driver are normally used in non-load-bearing places. Clutch head fasteners offer a degree of tamper resistance and less slippage than a standard slot screw. The clutch head design has been called a butterfly or figure-eight. Automotive technicians do not often use these drivers.
- *Robertson® or Scrulox® screwdriver.* The Robertson or Scrulox screwdriver has a square tip. The tip fits into a square recess in the top of a fastener. This type of fastener is commonly used on truck bodies, campers, and boats.

Impact Screwdriver

An impact screwdriver is used to loosen stubborn screws. Impact screwdrivers have interchangeable heads and bits that allow the handles of the tools to be used with various screw head designs.

To use an impact screwdriver **(Figure 5–43)**, select the correct bit and insert it into the driver's head. Then hold the bit against the screw slot while firmly twisting the handle in the desired direction. Strike the handle with a hammer. The force of the hammer will exert a downward force on the screw and, at the same time, exert a twisting force on the screw.

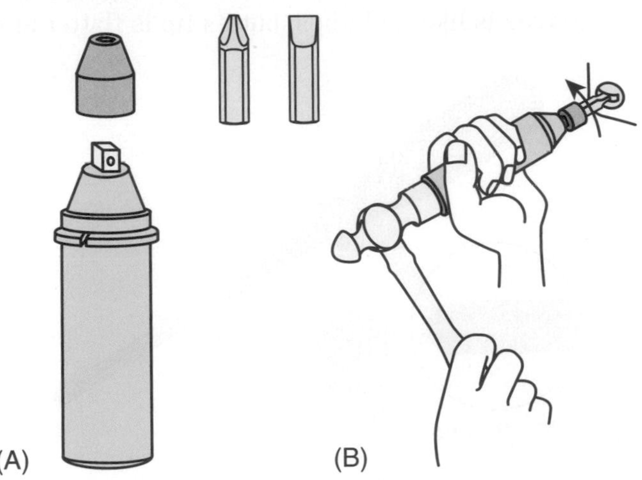

FIGURE 5–43

(A) An impact screwdriver set. (B) An impact screwdriver automatically tries to rotate the screw when it is struck with a hammer.

Pliers

Pliers are gripping tools used for working with wires, clips, and pins. At a minimum, an auto technician should own several types: standard pliers for common parts and wires, needle nose for small parts, and large, adjustable pliers for large items and heavy-duty work. A brief discussion on the different types of pliers follows:

- *Combination pliers* **(Figure 5–44)** are the most common type of pliers and are frequently used in many kinds of automotive repair. The jaws have both flat and curved surfaces for holding flat or round objects. Also called slip-joint pliers, the combination pliers have many jaw-opening sizes. One jaw can be moved up or down on a pin

FIGURE 5–44

Various combination pliers.

Snap-on Tools Company

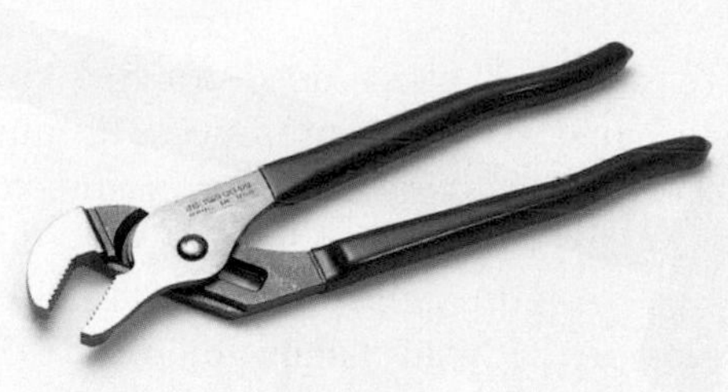

FIGURE 5–45

Typical adjustable pliers.

Snap-on Tools Company

attached to the other jaw to change the size of the opening.

- *Adjustable pliers*, commonly called *channel locks* **(Figure 5–45)**, have a multiposition slip joint that allows for many jaw-opening sizes.
- *Needle nose pliers* have long, tapered jaws **(Figure 5–46)**. They are great for holding small parts or for reaching into tight spots. Many needle nose pliers also have wire-cutting edges and a wire stripper. Curved needle nose pliers allow you to work on a small object around a corner.
- *Locking pliers*, or *vise grips*, are similar to the standard pliers, except they can be tightly locked around an object **(Figure 5–47)**. They are extremely useful for holding parts together. They are also useful for getting a firm grip on a badly rounded fastener that is impossible to turn with a wrench or socket. Locking pliers come in several sizes and jaw configurations for use in many auto repair jobs.
- *Diagonal-cutting pliers*, or cutters, are used to cut electrical connections, cotter pins, and wires on a vehicle. Jaws on these pliers have extra-hard cutting edges that are squeezed around the item to be cut **(Figure 5–48)**.

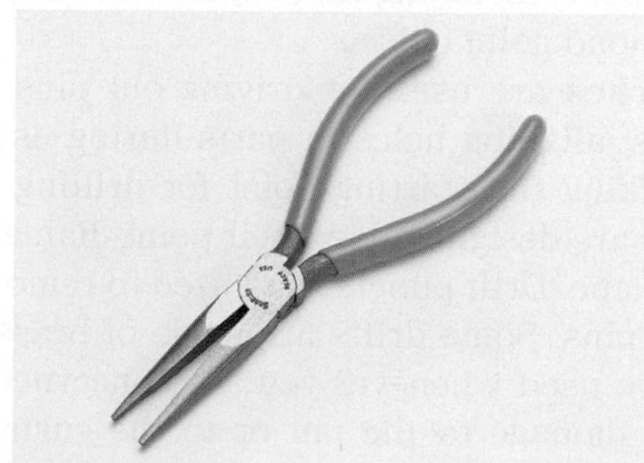

FIGURE 5–46

Typical needle nose pliers.

Snap-on Tools Company

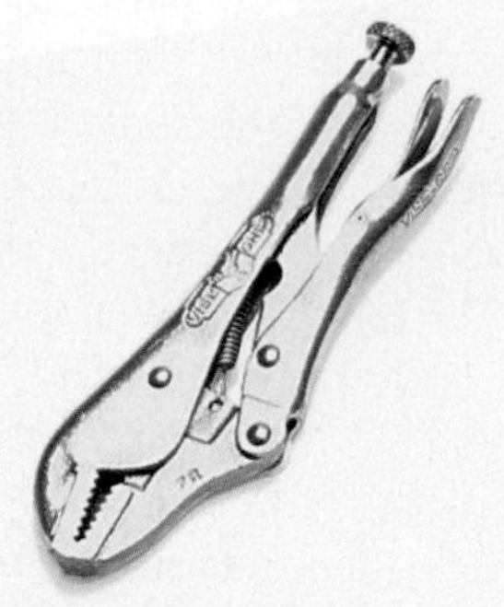

FIGURE 5–47

Locking pliers, commonly called vise grips.

Snap-on Tools Company

- *Snap or lock ring pliers* are made with a linkage that allows the movable jaw to stay parallel throughout the range of opening **(Figure 5–49)**. The jaw surface is usually notched or toothed to prevent slipping.
- *Retaining ring pliers* are identified by their pointed tips that fit into holes in retaining rings.

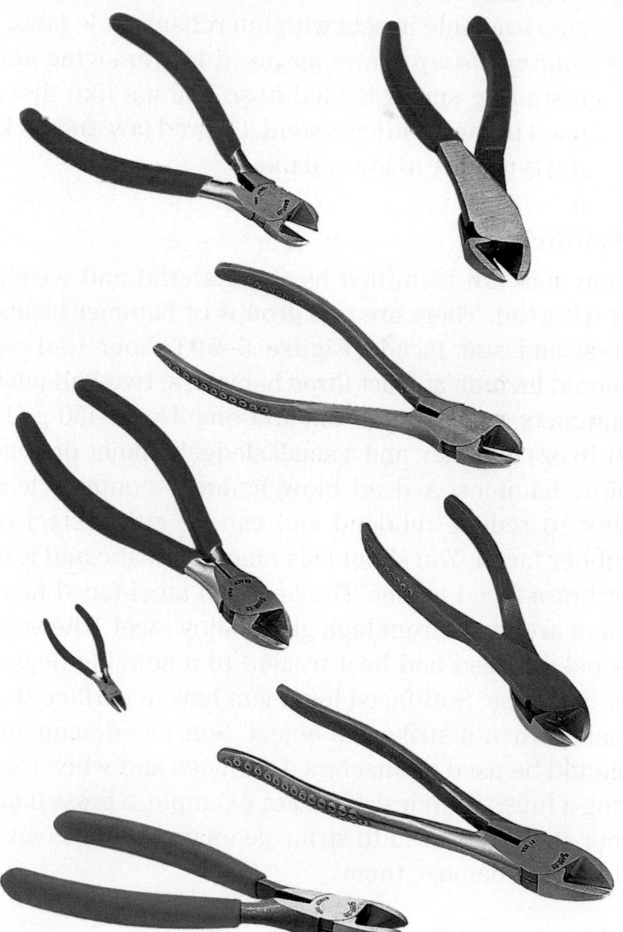

FIGURE 5–48

Various diagonal-cutting pliers.

Snap-on Tools Company

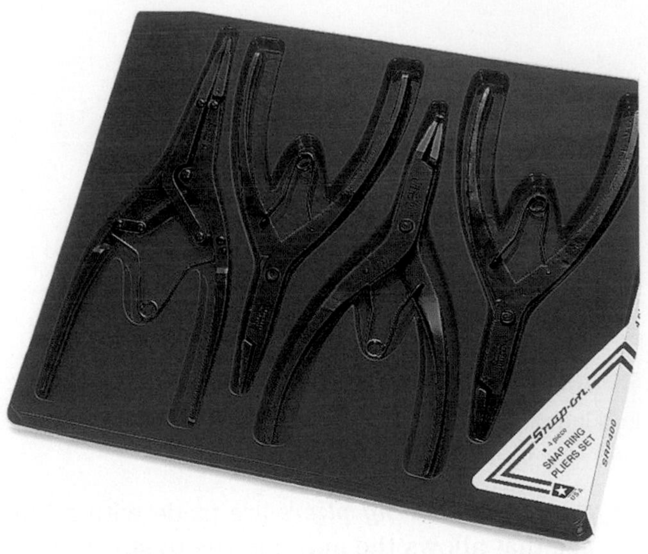

FIGURE 5–49

Snap ring pliers.

Snap-on Tools Company

Retaining ring pliers come in fixed sizes but are also available in sets with interchangeable jaws.

- *Spring clamp pliers* are used for removing and installing spring-loaded hose clamps like those used in the cooling system. Curved jaw and locking types are also available.

Hammers

Hammers are identified by the material and weight of the head. There are two groups of hammer heads: steel and soft faced **(Figure 5–50)**. Your tool set should include at least three hammers: two ball-peen hammers, one 225 g (8 oz) and one 340 to 450 g (12 to 16 oz) hammer, and a small sledgehammer or dead blow hammer. A dead blow hammer contains lead shot to reduce rebound and can be either steel or rubber faced. You should also have a plastic and lead or brass-faced mallet. The heads of steel-faced hammers are made from high-grade alloy steel. The steel is deep forged and heat treated to a suitable degree of hardness. Soft-faced hammers have a surface that yields when it strikes an object. Soft-faced hammers should be used on machined surfaces and when marring a finish is undesirable. For example, a brass hammer should be used to strike gears or shafts because it will not damage them.

Chisels and Punches

Chisels are used to cut metal by driving them with a hammer. Automotive technicians use a variety of chisels for cutting sheet metal, shearing off rivet and

FIGURE 5–50

Soft-faced hammers.

Snap-on Tools Company

bolt heads, splitting rusted nuts, and chipping metal. A variety of chisels are available, each with a specific purpose, including flat, cape, round-nose cape, and diamond point chisels.

Punches are used for driving out pins, rivets, or shafts; aligning holes in parts during assembly; and marking the starting point for drilling a hole. Punches are designated by their point diameter and punch shape. Drift punches are used to remove drift and roll pins. Some drifts are made of brass; these should be used whenever you are concerned about possible damage to the pin or to the surface surrounding the pin. Tapered punches are used to line up bolt holes. Starter or centre punches are used to make an indentation before drilling to prevent the drill bit from wandering.

FIGURE 5–51

Stud installation/removal tool.

Removers

Rust, corrosion, and prolonged heat can cause automotive fasteners, such as cap screws and studs, to become stuck. A box wrench or socket is used to loosen cap screws. A special gripping tool is designed to remove studs. However, if the fastener breaks off, special extracting tools and procedures must be employed.

One type of stud remover is shown in **Figure 5–51**. These tools are also used to install studs. Stud removers have hardened, knurled, or grooved eccentric rollers or jaws that grip the stud tightly when operated. Stud removers/installers are turned by a socket wrench drive handle, a socket, or a wrench.

Extractors are used on screws and bolts that are broken off below the surface. Twist drills, fluted extractors, and hex nuts are included in a screw extractor set **(Figure 5–52)**. This type of extractor

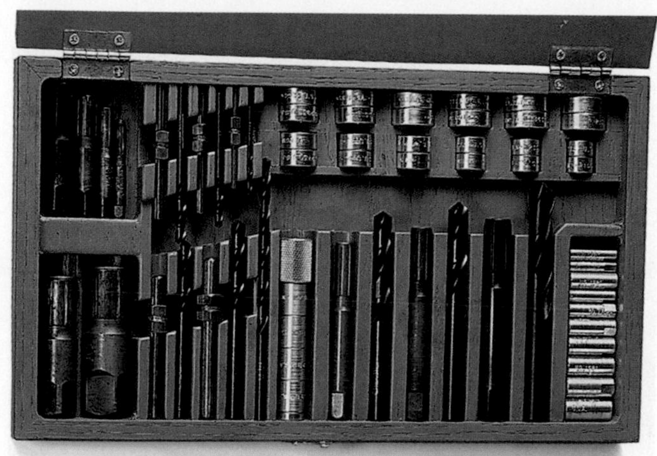

FIGURE 5–52

Screw extractor set.

Snap-on Tools Company

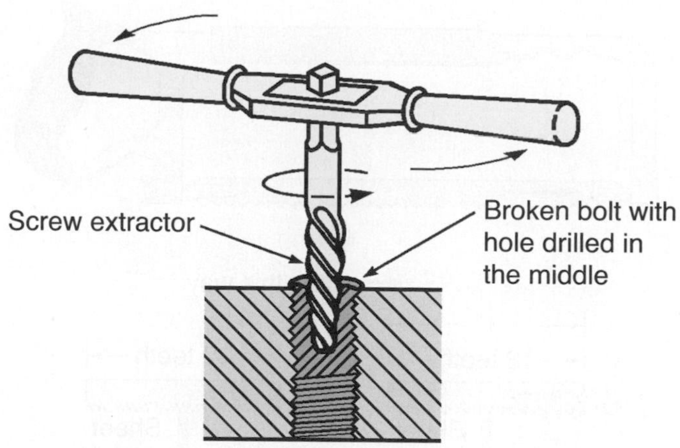

FIGURE 5–53

Using a screw extractor to remove a broken bolt.

lessens the tendency to expand the screw or stud that has been drilled out by providing gripping power along the full length of the stud.

Screw extractors are often called easy outs. To use an extractor, the bolt must be drilled and the extractor forced into that bore. The teeth of the extractor grip the inside of the drilled bore and allow the bolt to be turned out **(Figure 5–53)**. Easy outs typically have the size of the required drill bit stamped on one side.

At times, a broken bolt can be loosened and removed from its bore by driving it in a counterclockwise direction with a chisel and hammer. A bolt broken off above the surface may be able to be removed with locking pliers.

Hacksaws

A hacksaw is used to cut metal **(Figure 5–54)**. The blade only cuts on the forward stroke. The teeth of the blade should always face away from the saw's handle. The number of teeth on the blade determines the type of metal the saw can be used on. A fine-toothed blade is best for thin sheet metal, whereas a coarse blade is used on thicker metals.

When using a hacksaw, never bear down on the blade while pulling it toward you; this will dull the blade. Use the entire blade while cutting.

Files

Files are commonly used to shape or smooth metal edges. Files typically have square, triangular, rectangular (flat), round, or half-round shapes **(Figure 5–55)**. They also vary in size and coarseness. The most commonly used files are the half-round and flat with either single-cut or double-cut designs.

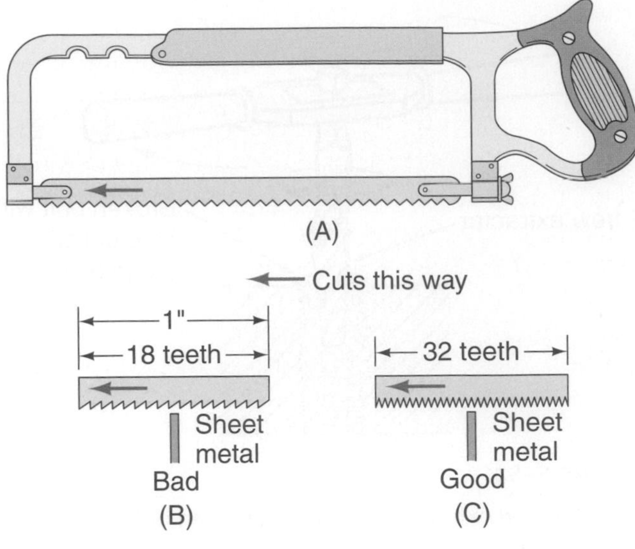

FIGURE 5–54

(A) The teeth on the blade in a hacksaw should face forward. (B) A coarse blade should not be used with sheet metal. (C) A fine blade will work well with sheet metal.

A single-cut file has its cutting grooves lined up diagonally across the face of the file. The cutting grooves of a double-cut file run diagonally in both directions across the face. Double-cut files are considered first cut or roughening files because they can remove large amounts of metal. Single-cut files are considered finishing files because they remove small amounts of metal.

To avoid personal injury, files should always be used with a plastic or wooden handle. Like hacksaws, files only cut on the forward stroke. Coarse files are used for soft metals, and smoother, or finer, files are used to work steel and other hard metals.

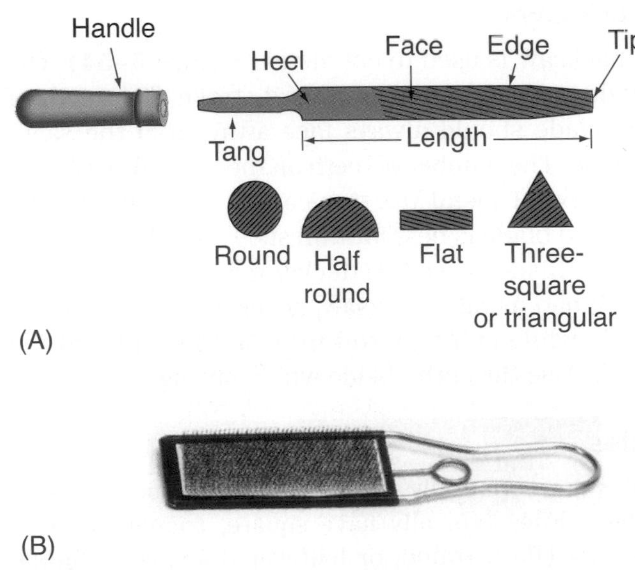

FIGURE 5–55

(A) Files come in a variety of shapes. (B) A file card.

Ford Motor Company; Snap-on Tools Company

Keep files clean, dry, and free of oil and grease. To clean filings from the teeth of a file, use a special tool called a file card.

Gear and Bearing Pullers

Many precision gears and bearings have a slight interference fit (**press fit**) when installed on a shaft or housing. For example, the inside diameter of a bore is 0.025 mm (0.001 in.) smaller than the outside diameter of a shaft. When the shaft is fitted into the bore, it must be pressed in to overcome the 0.025 mm (0.001 in.) interference fit. This press fit prevents the parts from moving on each other. The removal of gears and bearings must be done carefully. Prying or hammering can break or bind the parts. A puller with the proper jaws and adapters should be used when applying force to remove gears and bearings. Using proper tools, the force can be applied with a slight and steady motion.

Many pullers are designed to accommodate inside and outside pulls. These pullers may come with various jaw lengths and shapes to allow them to work in a number of different situations.

Some pullers are fitted to the end of a slide hammer **(Figure 5–56)** to remove slightly press-fit items. After the jaws of the puller are secure in or around the object to be removed, the weight on the tool's hammer is slid back with force against the handle of the tool, generating a pulling force and jerking the object out of its bore.

To pull something out of a bore, the puller must be designed to expand its jaws outward. The jaws also must be small enough to reach into the bore

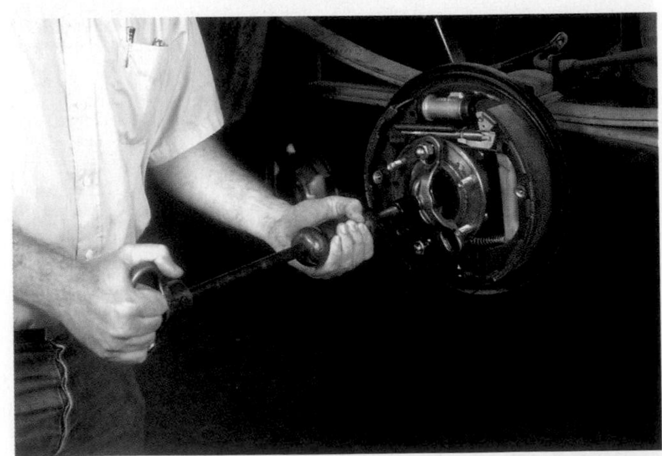

FIGURE 5–56

Using a slide hammer-type puller to remove a drive axle.

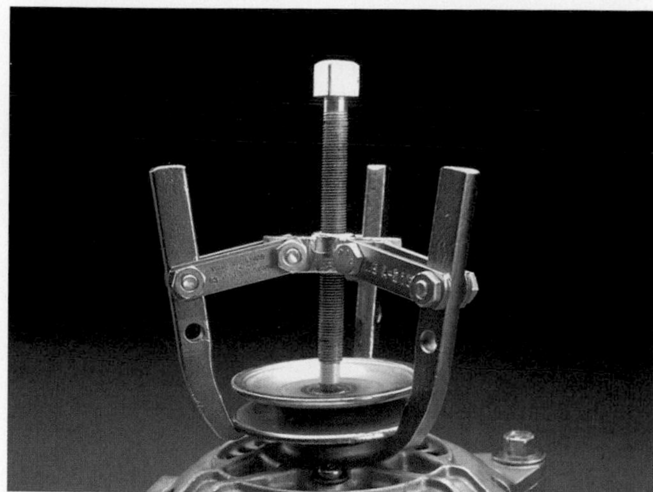

FIGURE 5–57

The jaws on this puller are reversible to allow for inside and outside pulls.

without damaging the bore while still firmly gripping the object that is being removed. This type puller is commonly used to remove seals, bushings, and bearing cups.

Jaw-type pullers are used to pull an object off a shaft. These pullers are available with two or three jaws **(Figure 5–57)**. Jaw-type pullers are commonly used to remove bearings, pulleys, and gears.

Some pullers are actually pushers. A push-puller is used to push a shaft out of its bore in a housing. It is often difficult to grip the end of the shaft with a puller, so a push-puller is used to move the shaft out of the bore.

Bearing, Bushing and Seal Drivers

Another commonly used group of special tools includes the various designs of bearing, bushing, and seal drivers. Auto manufacturers supply their dealerships with drivers for specific components. However, universal sets of drivers are also available. These sets include a variety of driver plates, each of a different diameter. The plates are often reversible. The flat side of the plate is used to install seals, and the tapered side is used to install tapered bearing races. A driver handle is threaded into the appropriate plate. The bearing or seal is driven into place by tapping on the driver hammer.

Always make sure you use the correct tool for the job; bushings and seals are easily damaged if the wrong tool or procedure is used. Car manufacturers and specialty tool companies work closely together to design and manufacture special tools required to repair cars.

Trouble Light

Adequate light is necessary when working under and around automobiles. A **trouble light** may be battery powered (like a flashlight) or may need to be plugged into a wall socket. Some shops have trouble lights that pull down from a reel suspended from the ceiling. Trouble lights should have LED or fluorescent bulbs. Incandescent bulbs should not be used because they can pop and burn. Take extra care when using a trouble light. Make sure the cord does not get caught in a rotating object. The bulb or tube is surrounded by a cage or enclosed in clear plastic to prevent accidental breaking and burning.

Creeper

Rather than crawl on your back to work under a vehicle, use a creeper. A creeper is a platform with small wheels. It allows you to slide under a vehicle and easily manoeuvre while working. To protect yourself and others around you, never have the creeper lying on the floor when you are not using it. Accidentally stepping on it can result in a serious fall. Always keep the creeper standing on its end when it is not being used.

SHOP EQUIPMENT

Some tools and equipment are supplied by the service facility, and few technicians have these as part of their tool assortment. These tools are commonly used, but there is no need for each technician to own them. Many shops have one or two of each.

Bench Vises

Often, repair work is completed with a part or assembly removed from the vehicle. Typically, the repairs are safely and quickly made by securing the assembly. Small parts are usually secured with a bench vise. The vise is bolted to a workbench to give it security. The object to be held is placed into the tool's jaws, and the jaws are tightened around the object. If the object could be damaged or marred by the jaws, brass jaw caps are installed over the jaws before the object is placed between them.

Bench Grinder

This electric power tool is generally bolted to a workbench. The grinder should have safety shields and guards. Always wear face protection when using a grinder. A bench grinder is classified by wheel size; 15 to 25 cm (6 to 10 in.) wheels are the most common in auto repair shops. Three types of wheels are available with this bench tool.

1. Grinding wheel, for a wide variety of grinding jobs from sharpening cutting tools to deburring
2. Wire wheel brush, for general cleaning and buffing, removing rust, scale, and paint, deburring, and so forth

   Note: For safety reasons, when using a wire wheel on a bench grinder, the tool rest must be completely removed.
3. Buffing wheel, for general purpose buffing, polishing, and light cutting

Presses

Many automotive jobs require the use of powerful force to assemble or disassemble parts that are press-fit together. Removing and installing piston pins, servicing rear axle bearings, pressing brake drum and rotor studs, and performing transmission assembly work are just a few examples. Presses can be hydraulic, electric, air, or hand-driven. Capacities range up to 150 tonnes of pressing force, depending on the size and design of the press. Smaller arbor and C-frame presses can be bench or pedestal mounted, while high-capacity units are freestanding or floor mounted **(Figure 5–58)**.

WARNING!
Always wear safety glasses when using a press.

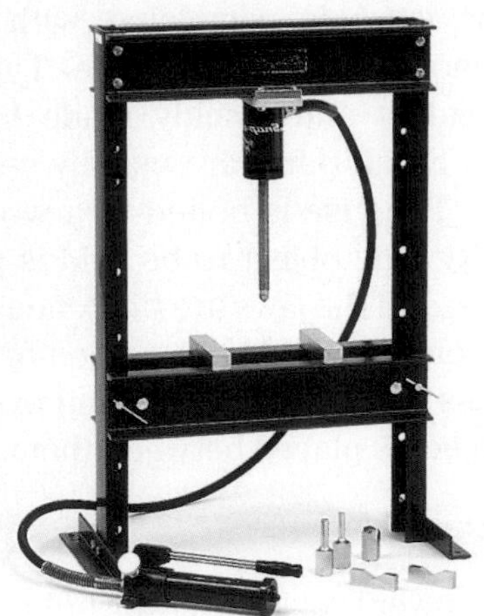

FIGURE 5–58

A floor-mounted hydraulic press.

Snap-on Tools Company

Grease Guns

Some shops are equipped with air-powered grease guns; in others, technicians use a manually operated grease gun. Both types can force grease into a grease fitting. Hand-operated grease guns are often preferred because the pressure of the grease can be controlled by the technician. However, many shops use low air pressure to activate a pneumatic grease gun. The suspension and steering system may have several grease or zerk fittings.

Oxyacetylene Torches

Oxyacetylene torches **(Figure 5–59)** have many purposes. In the automotive service industry, they are used to heat metal when two parts are difficult to separate, to cut metal (such as when replacing exhaust system parts), and to weld or connect two metal parts together.

Oxyacetylene welding and cutting equipment uses the combustion of acetylene in oxygen to produce a flame temperature of about 3100°C (5600°F). Acetylene is used as the fuel and oxygen is used to aid in the combustion of the fuel.

The equipment includes cylinders of oxygen and acetylene, two pressure regulators, two flexible hoses (one for each cylinder), and a torch. The torches are selected for the job being done—welding torch for welding, brazing, soldering, and heating; and cutting torch for cutting metal. There are three sets of valves for each gas: the tank valve, the regulator valve, and the torch valve. The oxygen hose is coloured green, and the acetylene hose is red. The acetylene connections have left-hand threads, and the oxygen connectors have right-hand threads.

WELDING AND HEATING TORCH The hoses connect the cylinders to the torch. The torch has separate valves for each gas. The torch is comprised of the valves, a handle, a mixing chamber (where the fuel and oxygen mix), and a tip (where the flame forms). Many different tips can be used with a welding torch. Always select the correct size for the job.

CUTTING TORCH A cutting torch is used to cut metal. It is similar to a welding torch. However, the cutting torch has a third tube from the valves to the mixing chamber. It carries high-pressure oxygen, which is controlled by a large lever on the torch. During cutting, the metal is heated until it glows orange, and then a lever on the torch is pressed to pass a stream of oxygen through the heated metal to burn it away where the cut is desired.

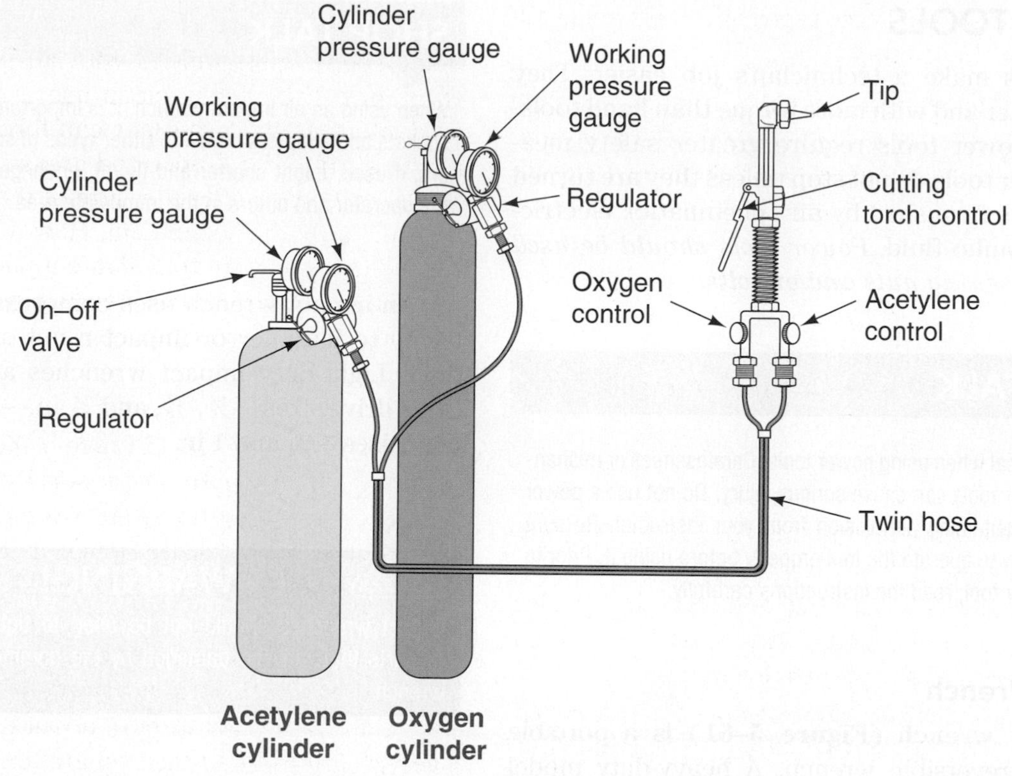

FIGURE 5–59
Oxyacetylene welding equipment shown with a cutting torch.

PRECAUTIONS Never use oxyacetylene equipment unless you have been properly trained to do so. Also, adhere to all safety precautions, including the following:

- While using a torch, severe and fatal burns and violent explosions can result from inattention and carelessness.
- Before using an oxyacetylene torch, make sure that all flammable materials, such as grease, oil, paint, sawdust, and so on, are cleared from the area.
- Keep oxygen away from all combustibles.
- Wear approved shaded goggles with enclosed sides, or a shield with a shaded lens to protect your eyes from glare and sparks.
- Wear leather gloves to protect your hands from burns.
- Wear clothes and shoes/boots appropriate for welding.
- Make sure the gas cylinders are securely fastened upright to a wall, post, or portable cart.
- Never move an oxygen tank around without its valve cap screwed in place.
- Never lay an acetylene tank on its side while being used.
- Never oil an oxygen regulator.

MINI-DUCTOR® This tool uses the principle of electrical induction to create heat **(Figure 5–60)**. It is faster and safer than a torch; inductive heating heats and removes seized nuts, bearings, pulleys, and other metal or mechanical hardware and parts from corrosion and/or threadlock compounds. It relies on a high-frequency electromagnet that heats only metal objects or objects containing metal when they are exposed to the tool's magnetic field. It provides immense heat without the danger of an open flame or damage to nearby plastics. The magnetic field moves through non-metals, without heating them, and heats the metal underneath the non-metals.

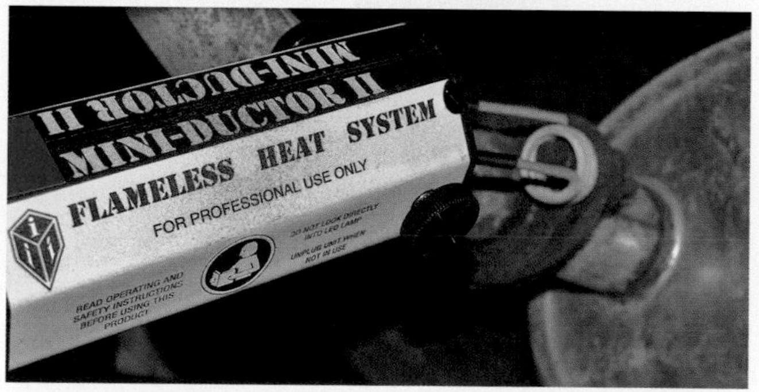

FIGURE 5–60
A Mini-Ductor®

POWER TOOLS

Power tools make a technician's job easier. They operate faster and with more torque than hand tools. However, power tools require greater safety measures. Power tools do not stop unless they are turned off. Power is furnished by air (pneumatic), electricity, or hydraulic fluid. *Power tools should be used only for loosening nuts and/or bolts.*

SHOP TALK

Safety is critical when using power tools. Carelessness or mishandling of power tools can cause serious injury. Do not use a power tool without obtaining permission from your instructor. Be sure you know how to operate the tool properly before using it. Prior to using a power tool, read the instructions carefully.

Impact Wrench

An **impact wrench (Figure 5–61)** is a portable hand held reversible wrench. A heavy-duty model can deliver up to 600 N·m (450 ft·lb) of torque. When triggered, the output shaft, onto which the impact socket is fastened, spins freely at 2000 to 14 000 rpm, depending on the wrench's make and model. When the impact wrench meets resistance, a small spring-loaded hammer situated near the end of the tool strikes an anvil attached to the driveshaft onto which the socket is mounted. Each impact moves the socket around a little until torque equilibrium is reached, the fastener breaks, or the trigger is released. Torque equilibrium occurs when the torque of the bolt equals the output torque of the wrench. Impact wrenches can be powered either by air or by electricity.

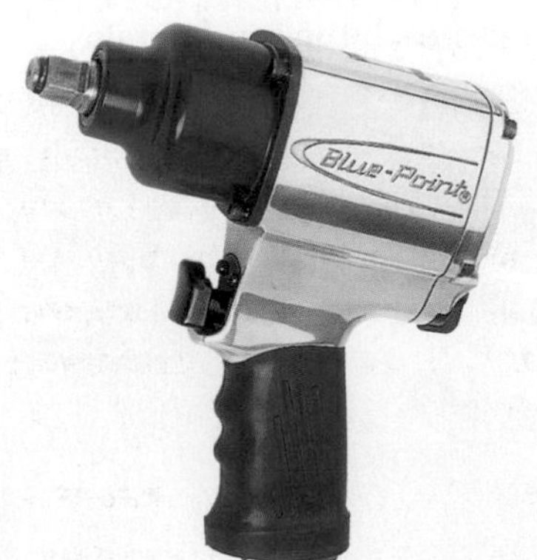

FIGURE 5–61

A typical air impact wrench.

Snap-on Tools Company

SHOP TALK

When using an air impact wrench, it is important that only impact sockets and adapters be used. Other types of sockets and adapters, if used, might shatter and fly off, endangering the safety of the operator and others in the immediate area.

An impact wrench uses compressed air or electricity to hammer or impact a nut or bolt loose or tight. Light-duty impact wrenches are available in three drive sizes—¼, ⅜, and ½ in.—and two heavy-duty sizes—¾ and 1 in.

WARNING!

Impact wrenches should not be used to tighten critical parts or parts that may be damaged by the hammering force of the wrench.

Many technicians are using battery-powered electric impact wrenches. The tools currently on the market can provide several hundred foot-pounds of torque, making them more than adequate for removing lug nuts and other fasteners.

Air Ratchet

An air ratchet, like the hand ratchet, has a special ability to work in hard-to-reach places. Its angle drive reaches in and loosens or tightens where other hand or power wrenches just cannot work **(Figure 5–62)**. The air ratchet looks like an ordinary ratchet but has a fat handgrip that contains the air vane motor and drive mechanism. Air ratchets usually have a ⅜-inch drive. Air ratchets are not torque sensitive; therefore, a torque wrench should be used on all fasteners after snugging them up with an air ratchet.

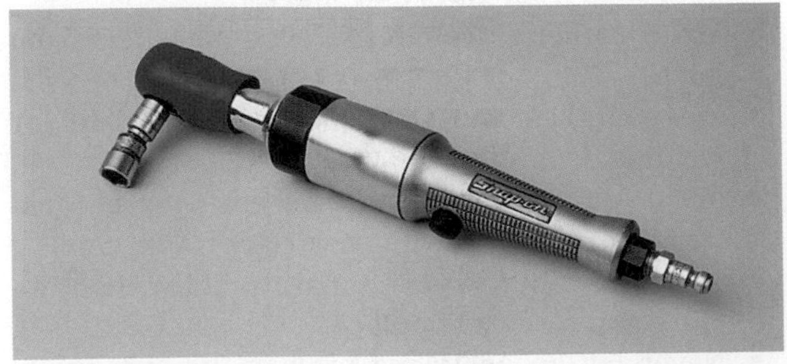

FIGURE 5–62

An air ratchet.

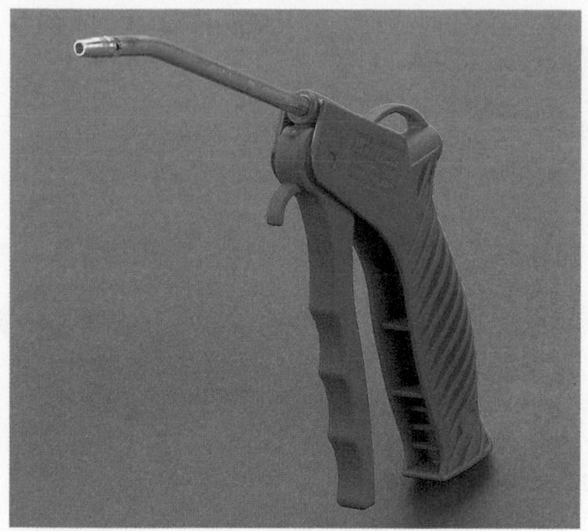

FIGURE 5–63

An OSHA-approved air blowgun.

ITW DeVilbiss Co.

Air Drill

Air drills are usually available in ¼-, ⅜-, and ½-inch chuck sizes. They operate in much the same manner as an electric drill, but are smaller and lighter. This compactness makes them a great deal easier to use for drilling operations in auto work.

Blowgun

Blowguns are used for blowing off parts during cleaning. Never point a blowgun at yourself or anyone else. A blowgun **(Figure 5–63)** snaps into one end of an air hose and directs airflow when a button is pressed. Always use an air blowgun that meets the American OHSA (Occupational Health and Safety Act) standards. Before using a blowgun, be sure it has not been modified to eliminate air-bleed holes on the side.

JACKS AND LIFTS

Jacks are used to raise a vehicle off the ground and are available in a variety of sizes. The most common one is a hydraulic floor jack, which is classified by the weights it can lift: 1½, 2, and 2½ tonnes, and so on. The other design of portable floor jacks uses compressed air. Pneumatic jacks are operated by controlling air pressure at the jack.

CAUTION!

Before lifting a vehicle with air suspension, turn off the system. The switch is usually in the trunk.

FIGURE 5–64

Whenever you have raised a vehicle with a floor jack, the vehicle should be supported with jack stands.

Martin Restoule

When a vehicle is raised by a jack, it should be supported by safety stands **(Figure 5–64)**. Never work under a car with only a jack supporting it; always use safety stands that are correctly rated for the weight of the vehicle being supported. Hydraulic seals in the jack can let go and allow the vehicle to drop.

Floor Jack

A floor jack is a portable unit mounted on wheels. The lifting pad on the jack is placed under the appropriate lift points of the vehicle, and the jack handle is operated with a pumping action. This forces fluid into a hydraulic cylinder in the jack, and the cylinder extends to force the jack lift pad upward and to lift the vehicle. Always be sure that the lift pad is positioned securely under one of the car manufacturer's recommended lifting points. To release the hydraulic pressure and lower the vehicle, the handle or release lever must be turned slowly.

The maximum lifting capacity of the floor jack is usually written on the jack decal. Never lift a vehicle that exceeds the jack lifting capacity. This action may cause the jack to break or collapse, resulting in vehicle damage or personal injury.

Lift

The hydraulic floor lift is the safest lifting tool and is able to raise the vehicle high enough to allow you to walk and work under it. Various safety features prevent a hydraulic lift from dropping if a seal does leak or if air pressure is lost. Before lifting a vehicle, make sure the lift is correctly positioned. The lift arms must be placed under the car manufacturer's

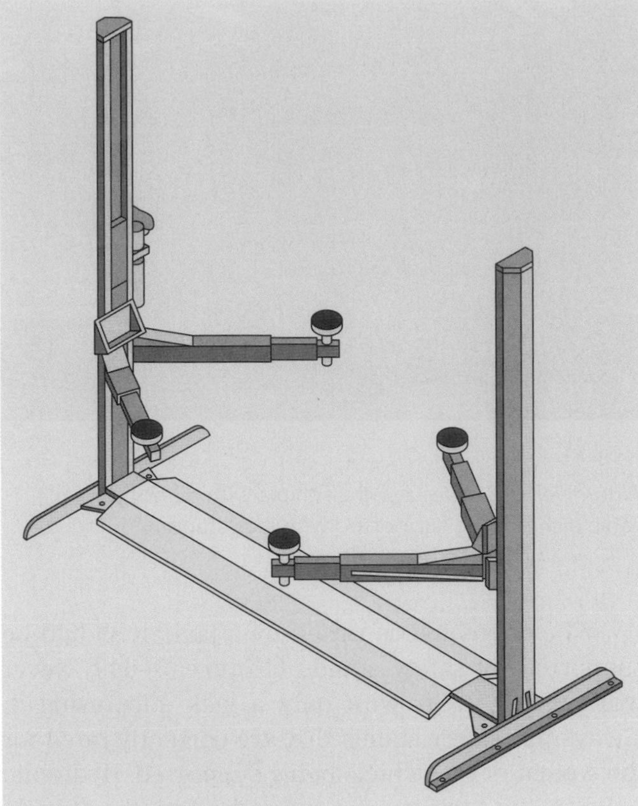

FIGURE 5–65

An above-ground or surface-mount frame-contact lift.

Automotive Lift Institute

recommended lifting points prior to raising a vehicle. There are three basic types of lifts: frame contact **(Figure 5–65)**, wheel contact, and axle engaging. These categories define where the frame contact points align with the vehicle.

Some older shops use in-floor single or twin post lifts. Some lifts have an electric motor, which drives a hydraulic pump to create fluid pressure and force the lift upward. Other lifts use air pressure from the shop air supply to force the lift upward. If shop air pressure is used for this purpose, the air pressure is applied to fluid in the lift cylinder. A control lever or switch is placed near the lift. The control lever supplies shop air pressure to the lift cylinder, and the switch turns on the lift pump motor. Always be sure that the safety locks are engaged after the lift is raised. When the safety lock is released, a release lever is operated slowly to lower the vehicle.

The arms of a lift are fitted with **foot pads (Figure 5–66)** or adapters that can be lifted up to contact the vehicle's lift points to add clearance between the arms and the vehicle. This clearance allows for secure lifting without damaging any part of the body or underbody of the vehicle.

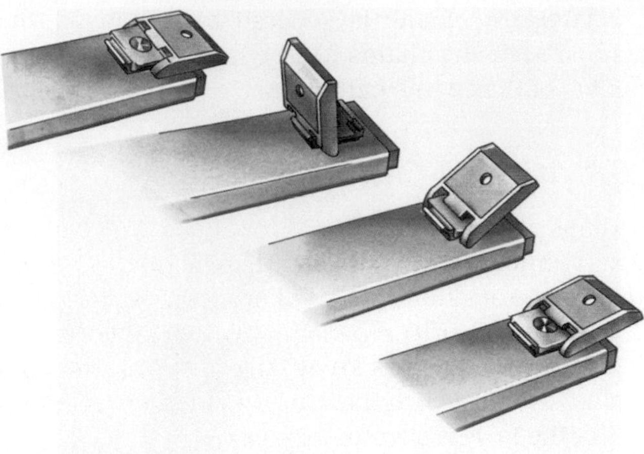

FIGURE 5–66

Foot pads on the arms of a lift.

Automotive Lift Institute

Portable Crane

To remove and install an engine, a portable crane, frequently called a cherry picker, is used. To lift an engine, attach a pulling sling or chain to the engine. Some engines have eye plates for use in lifting. If they are not available, the sling must be bolted to the engine. The sling attaching bolts must be large enough to support the engine and must thread into the block a minimum of 1½ times the bolt diameter. Connect the crane to the chain. Raise the engine slightly and make sure the sling attachments are secure. Carefully lift the engine out of its compartment.

Lower the engine close to the floor so that the transmission and torque converter or clutch can be removed from the engine, if necessary.

Engine Stands/Benches

After the engine has been removed, use the crane to raise the engine. Position the engine next to an engine stand. Most stands use a plate with several holes or adjustable arms. The engine must be supported by at least four bolts that fit solidly into the engine. The engine should be positioned so that its centre is in the middle of the engine's stand adapter plate. The adapter plate can swivel in the stand. By centring the engine, the engine can be easily turned to the desired working positions.

Some shops have engine mounts bolted to the top of workbenches. The engine is suspended off the side of the workbench. These have the advantage of a good working space next to the engine, but they are not mobile and all engine work must be done at that location.

After the engine is secured to its mount, the crane and lifting chains can be removed, and disassembly of the engine can begin.

SERVICE INFORMATION

Perhaps the most important tools you will use are service information systems. There is no way a technician can remember all of the procedures and specifications needed to repair an automobile correctly. Good information plus knowledge allow a technician to fix a problem with the least amount of frustration and at the lowest cost to the customer.

Auto Manufacturers' Service Information

The primary source for repair and specification information for any car, van, or truck is the manufacturer. The manufacturer publishes appropriate service information each year, for every vehicle built.

The information may be divided into sections such as chassis, suspension, steering, emission control, fuel systems, brakes, basic maintenance, engine, transmission, body, and so on **(Figure 5–67)**. Each major section is divided into subsystems **(Figure 5–68)**. These cover all repairs, adjustments, specifications, diagnostic procedures, and any required special tools.

- General information & identification
- Body
- Brakes
- Chassis electrical
- Diagnostics
- Driveline
- Engine cooling
- Engine electrical
- Engine mechanical
- Engine performance & emission controls
- Fuel systems
- Heating, ventilation, & air conditioning
- Steering
- Suspension
- Transmissions
- Wiring diagrams

FIGURE 5–67

The main index of a factory service manual showing that the manual is divided by major vehicle systems.

Reprinted with permission

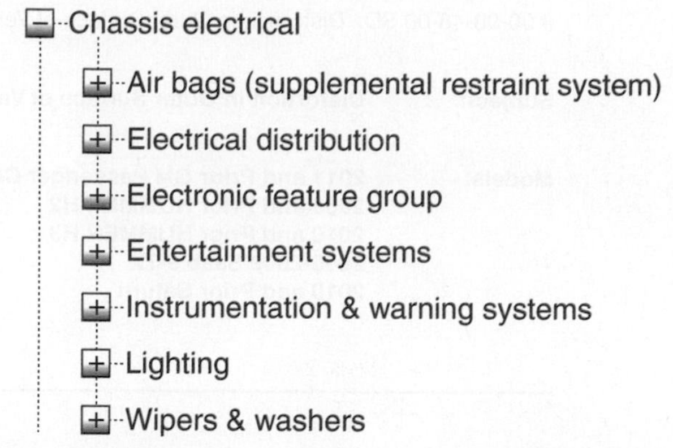

FIGURE 5–68

After the basic system is selected on the main screen, the program moves to the subsystems of that main system.

Because many technical changes occur on specific vehicles each year, the manufacturers' service information needs to be constantly updated. Updates are published as service bulletins (often referred to as technical service bulletins or TSBs) that show the changes in specifications and repair procedures during the model year **(Figure 5–69)**. These changes do not appear in the service information until the next year. The car manufacturer provides these bulletins to dealers and repair facilities on a regular basis.

Automotive manufacturers also publish a series of technician reference books. The publications provide general instructions about the service and repair of the manufacturers' vehicles and also indicate their recommended techniques.

General and Specialty Repair Manuals

Service information is also available from independent companies rather than the manufacturers. These companies pay for and get most of their information from the car makers. Examples of these companies are Chilton, Mitchell 1 (Snap-on ShopKey), and AllData **(Figure 5–70)**.

Finding Information

Although the information from different publishers varies in presentation and arrangement of topics, all are easy to use after you become familiar with their organization. Most are divided into a number of sections, each covering different aspects of the vehicle. The beginning sections commonly provide vehicle identification and basic maintenance information. The remaining sections deal with each different vehicle system in detail, including diagnostic, service,

00-08-48-00 SD: Distortion in Outer Surface of Vehicle Glass - (Sep 10, 2010)

Subject: **Distortion in Outer Surface of Vehicle Glass**

Models: **2011 and Prior GM Passenger Cars and Trucks**
2009 and Prior HUMMER H2
2010 and Prior HUMMER H3
2005-2009 Saab 9-7X
2010 and Prior Saturn

This bulletin is being revised to add model years. Please discard Corporate Bulletin Number 00-08-48-005 C (Section 08 - Body and Accessories).

Distortion in the outer surface of the windshield glass, door glass or backlite glass may appear after the vehicle has:

- Accumulated some mileage.
- Been frequently washed in automatic car washes, particularly "touchless" car washes.

This distortion may look like a subtle orange peel pattern, or may look like a drip or sag etched into the surface of the glass.

Some car wash solutions contain a buffered solution of hydrofluoric acid which is used to clean the glass. This should not cause a problem if used in the correct concentration. However, if not used correctly, hydrofluoric acid with attack the glass, and over time, will cause visual distortion in the outer surface of the glass which cannot be removed by scraping or polishing.

If this condition is suspected, look at the area of the windshield under the wipers or below the belt seal on the side glass. The area of the glass below the wipers or belt seal will not be affected and what looks like a drip or sag may be apparent at the edge of the wiper or belt seal. You may also see a line on the glass where the wiper blade or the belt seal contacts the glass.

Important: The repair will require replacing the affected glass and is not a result of a defect in material or workmanship. Therefore, is not covered by New Vehicle Warranty.

FIGURE 5–69

A technical service bulletin (TSB).

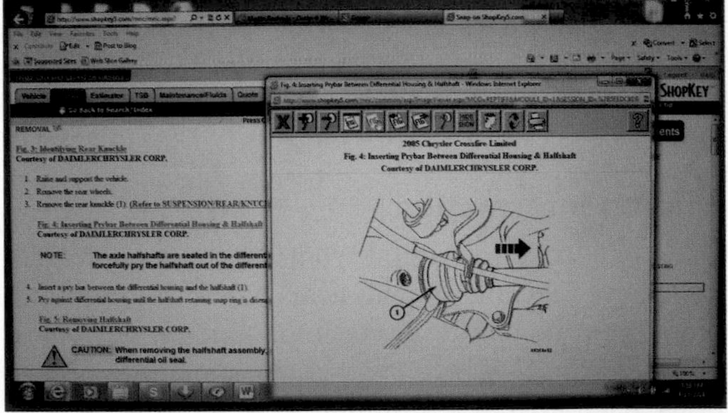

FIGURE 5–70

All the relevant information to perform a service is displayed on the computer screen.

Martin Restoule

and overhaul procedures. Each section has an index indicating more specific areas of information.

To obtain the correct system specifications and other information, you must first identify the exact system you are working on. The best source for vehicle identification is the VIN. Service information may also help you identify the system through identification of key components or other identification numbers and/or markings.

Aftermarket Suppliers' Guides and Catalogues

Many of the larger parts manufacturers have excellent guides on the various parts they manufacture or supply. They also provide updated service bulletins on their products. Other sources for up-to-date technical information are trade magazines and trade associations.

Lubrication Guides

These specially designed service manuals contain information on lubrication, maintenance, capacities, and underhood service. The lubrication guide includes lube and maintenance instructions, lubrication diagrams and specifications, vehicle lift points, and preventive maintenance mileage/time intervals. The capacities listed include cooling system, air-conditioning, cooling system air bleed locations, wheel and tire specifications, and wheel lug torque specifications. The underhood information includes specifications for tune-up; mechanical, electrical, and fuel systems; diagrams; and belt tension.

Owner's Manuals

An owner's manual comes with the vehicle when it is new. It contains operating instructions for the vehicle and its accessories. It also contains valuable information about checking and adding fluids, safety precautions, a complete list of capacities, and the specifications for the various fluids and lubricants for the vehicle.

Hotline Services

Hotline services provide answers to service concerns by telephone. Manufacturers provide help by telephone for technicians in their dealerships. There are subscription services for independents to be able to get repair information by phone. Some manufacturers also have a phone modem system that can transmit computer information from the car to another location. The vehicle's diagnostic link is connected to the modem. The technician in the service bay runs a test sequence on the vehicle. The system downloads the latest updated repair information on that particular model of car. If that does not repair the problem, a technical specialist at the manufacturer's location will review the data and propose a repair.

iATN

The International Automotive Technicians Network (iATN) is a group of thousands of professional automotive technicians from around the world. The technicians in this group exchange technical knowledge and information with other members. The website address for this group is www.iatn.net.

KEY TERMS

Ball gauge (p. 94)
Blowguns (p. 111)
Bolt diameter (p. 83)
Bolt head (p. 83)
Bolt shank (p. 83)
Chisels (p. 104)
Dial caliper (p. 89)
Dial indicator (p. 95)
Dies (p. 86)
Extractors (p. 105)
Feeler gauge (p. 94)
Fillet (p. 84)
Foot pads (p. 112)
Grade marks (p. 84)
Impact wrench (p. 110)
Machinist's rule (p. 87)
Micrometer (p. 90)
Phillips screwdriver (p. 101)
Pliers (p. 102)
Pozidriv screwdriver (p. 101)
Press fit (p. 106)
Punches (p. 104)
Snap gauges (p. 94)
Spark plug thread taps (p. 87)
Straightedge (p. 95)
Tap (p. 86)
Telescoping gauges (p. 94)
Thread pitch (p. 83)
Thread pitch gauge (p. 85)
Torque wrenches (p. 100)
Torx screwdriver (p. 102)
Trouble light (p. 107)
Vernier caliper (p. 87)
Wrench (p. 96)

SUMMARY

- Repairing the modern automobile requires the use of many different hand and power tools. Units of measurement play a major role in tool selection. Therefore, it is important to be knowledgeable about the metric and imperial systems of measurement.
- Measuring tools must be able to measure objects to a high degree of precision. They should be handled with care at all times and cleaned before and after every use.
- A micrometer can be used to measure the outside diameter of shafts and the inside diameter of holes. It is calibrated in either metric or inch graduations.
- Telescoping gauges are designed to measure bore diameters and other clearances. They usually are used with an outside micrometer. Small hole gauges are used in the same manner as the telescoping gauge, usually to determine valve guide diameter.
- The thread pitch gauge provides a fast and accurate method of measuring the distance between threads in millimetres or threads per inch (pitch) of fasteners. This is done by matching the teeth of the gauge with the fastener threads and reading the pitch directly from the leaf of the gauge.
- It is crucial to use the proper amount of torque when tightening nuts or cap screws on any part of a vehicle, particularly the engine. A torque-indicating wrench makes it possible to duplicate the conditions of tightness and stress recommended by the manufacturer.
- Metric and SAE size wrenches are not interchangeable. An auto technician should have a variety of both types.

- A screwdriver, no matter what type, should never be used as a chisel, punch, or pry bar.
- The hand tap is used for hand-cutting internal threads and for cleaning and restoring previously cut threads. Hand-threading dies cut external threads and fit into holders called die stocks.
- Carelessness or mishandling of power tools can cause serious injury. Safety measures are needed when working with such tools as impact and air ratchet wrenches, blowguns, bench grinders, lifts, hoists, and hydraulic presses.
- The primary source of repair and specification information for any vehicle is the manufacturer's service manual. Updates are published as service bulletins and include changes made during the model year, which will not appear in the manual until the following year.

REVIEW QUESTIONS

1. How often should the calibration of a micrometer be checked?
2. List some common uses of the dial indicator.
3. What determines the size of a wrench?
4. *True or False?* The same information available in service manuals and bulletins is also available electronically: on compact discs (CD-ROMs), digital video discs (DVDs), and the Internet.
5. Describe the grading methods for both imperial and metric bolts.
6. Which of the following wrenches is the best choice for turning a bolt?
   a. open end **b.** box end
   c. combination **d.** line
7. How many graduations are found on the thimble of a metric micrometer?
   a. 10 **b.** 25
   c. 50 **d.** 100
8. What does each thimble graduation on a metric micrometer equal?
   a. one-tenth of a millimetre
   b. one-hundredth of a millimetre
   c. one-thousandth of a millimetre
   d. one-ten-thousandth of a millimetre
9. Which of the following screwdrivers is like a Phillips but has a flatter and blunter tip?
   a. standard **b.** Torx®
   c. Pozidriv® **d.** clutch head
10. Which of the following types of pliers is best for grasping small parts?
    a. adjustable **b.** needle nose
    c. retaining ring **d.** snap ring
11. What tool should be used to drive a roll pin from an aluminum component?
    a. centre punch **b.** brass drift
    c. tapered punch **d.** diamond point chisel
12. What is an extractor designed to remove?
    a. worn seals **b.** worn bushings
    c. pistons **d.** broken bolts
13. What tool should be used to measure a valve guide bore?
    a. a ball gauge
    b. a telescoping gauge
    c. an inside micrometer
    d. a vernier caliper
14. What tool should be used when the measurement must be within one-hundredth of a millimetre?
    a. a machinist's rule **b.** a vernier caliper
    c. a micrometer **d.** a dial caliper
15. What type of bolt damage could cause a bolt head to pop off?
    a. shank damage
    b. damaged threads
    c. rounded head hex corners
    d. fillet damage
16. Which of the following can a hand tap thread?
    a. a nut **b.** a bolt
    c. a pipe **d.** a steel rod
17. What tool can be used to measure inside, outside, and depth measurements?
    a. a micrometer **b.** a dial indicator
    c. a vernier caliper **d.** a telescoping gauge
18. What type of fastener is generally used to secure headlamp assemblies, mirrors, and luggage racks?
    a. Torx® fasteners
    b. clutch tip fasteners
    c. slotted tip fasteners
    d. Robertson® fasteners
19. When using a hydraulic vehicle lift, what should be done before working under the vehicle?
    a. Make sure the trouble light is working.
    b. Make sure the lift's locking devices are fully engaged.
    c. Release all power supplies to the lift.
    d. Select all of the tools required to perform the repair.
20. What do the vehicle manufacturers publish when new service procedures or component updates are needed?
    a. service manuals
    b. specialty repair manuals
    c. technical service bulletins
    d. owner's manuals

CHAPTER 6
Diagnostic Equipment and Special Tools

LEARNING OUTCOMES

- Describe the various diagnostic and service tools used to check and repair an engine and its related systems.
- Describe the various diagnostic and service tools used to check and repair electrical and electronic systems.
- Describe the various diagnostic and service tools used to check and repair a vehicle's drivetrain.
- Describe the various diagnostic and service tools used to check and repair a vehicle's running gear for wear and damage.
- Describe the various diagnostic and service tools used to check and repair a vehicle's heating and air-conditioning system.

Diagnosing and servicing the various systems of an automobile require many different tools. Tools that are used to check the performance of a system or component are commonly referred to as diagnostic tools. Tools designed for a particular purpose or system are referred to as special tools. This chapter looks at the common diagnostic and special tools required for the different systems of a vehicle.

ENGINE REPAIR TOOLS

Engine repair, service, and diagnostic tools are discussed in this chapter. This discussion does not cover all of the tools you may need; only the most commonly used are discussed. Details of when and how to use these tools are presented in Section 2 of this book.

Compression Testers

Engines depend on the compression of the air/fuel mixture for power output. The compression stroke of the piston compresses the air/fuel mixture within the combustion chamber. If the combustion chamber leaks, some of the air/fuel mixture will escape while it is being compressed, resulting in a loss of power and a waste of fuel. The leaks can be caused by burned valves, a blown head gasket, worn rings, a slipped timing belt or chain, worn valve seats, a cracked head, and more.

If a symptom suggests that the cause of a problem may be poor compression, a compression test is performed.

A **compression gauge** is used to check cylinder compression. The dial face on the typical compression gauge indicates pressure in both **kilopascals (kPa)** and **pounds per square inch (psi)**. The range is usually 0 to 2100 kPa and 0 to 300 psi. There are two basic types of compression gauges: the push-in gauge and a thread-in gauge.

The push-in type has a short stem that is either straight or bent at a 45° angle. The stem ends with a tapered rubber tip that fits any size spark plug hole. After the spark plugs have been removed, the rubber tip is placed in the spark plug hole and held there while the engine is cranked through several compression cycles. Although simple to use, the push-in gauge may give inaccurate readings if it is not held tightly in the hole.

The thread-in gauge has a long, flexible hose that ends in a threaded adapter **(Figure 6–1)**. This type of compression tester is often used because its flexible hose can reach into areas that are difficult to reach with a push-in-type tester. The threaded adapters are changeable and come in several thread sizes to fit 10 mm, 12 mm, 14 mm, and 18 mm diameter holes. The adapters thread into the spark plug holes.

Most compression gauges have a vent valve that holds the highest pressure reading on its meter. Opening the valve releases the pressure when the test is complete.

Many technicians are using pressure transducers instead of mechanical compression testers. The transducer connects to a compression tester

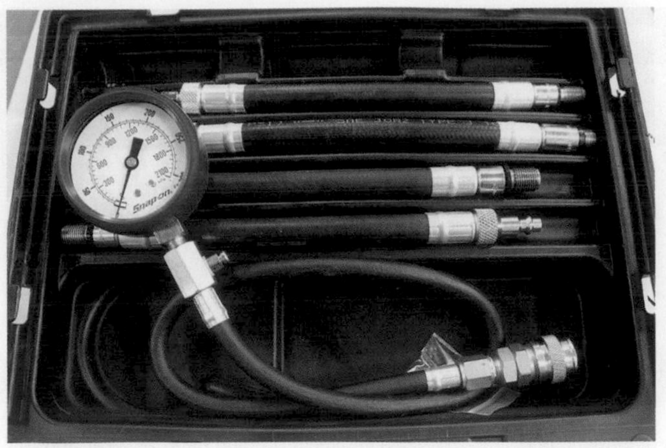

FIGURE 6–1

A thread-in compression gauge set.

Martin Restoule

hose and plugs into a scope. Using a pressure transducer provides much greater detail about what is happening in the cylinder than a standard compression gauge.

Cylinder Leakage Tester

If a compression test shows that any of the cylinders are leaking, a cylinder leakage test can be performed to measure the percentage of compression loss and help locate the source of leakage.

A **cylinder leakage tester (Figure 6–2)** applies compressed air through the spark plug hole and into a cylinder, with the piston at the top of its bore. A threaded adapter on the end of the air pressure hose threads into the spark plug hole. A pressure regulator in the tester controls the pressure applied to the cylinder. A gauge registers the percentage of air pressure lost from the cylinder when the compressed air is applied. The scale on the dial face reads 0 to 100 percent.

A zero reading means there is no leakage in the cylinder. A reading of 100 percent indicates that the cylinder will not hold any pressure. The location of the compression leak can be found by listening and feeling around various parts of the engine.

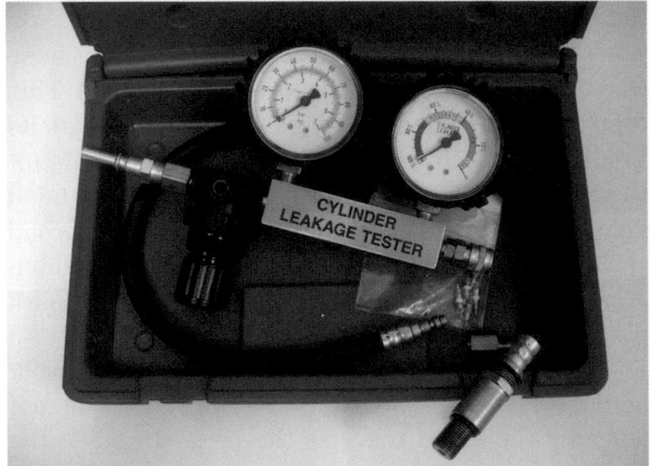

FIGURE 6–2

A cylinder leakage tester.

Martin Restoule

Oil Pressure Gauge

Checking the engine's oil pressure gives information about the condition of the oil pump, the pressure regulator, and the entire lubrication system. Lower than normal oil pressures can be caused by excessive engine bearing clearances. Oil pressure is checked at the sending unit passage with an externally mounted mechanical oil pressure gauge. Various fittings are usually supplied with the oil pressure gauge to fit different openings in the lubrication system.

Stethoscope

A **stethoscope** is used to locate the source of engine and other noises. The stethoscope pickup is placed on the suspected component, and the stethoscope receptacles are placed in the technician's ears **(Figure 6–3)**. Some sounds can be heard easily without using a listening device, but others are impossible to hear unless amplified, which is what a stethoscope does. It can also help you distinguish between normal and abnormal noise.

ELECTRONIC STETHOSCOPE While trying to locate the source of a noise, the best results are obtained with an electronic listening device. With this tool, you can tune into the noise, which allows you to eliminate all other noises that might distract or mislead you. Also, many electronic stethoscopes can record the sounds they amplify. The recordings can be played back on the stethoscope or audio equipment.

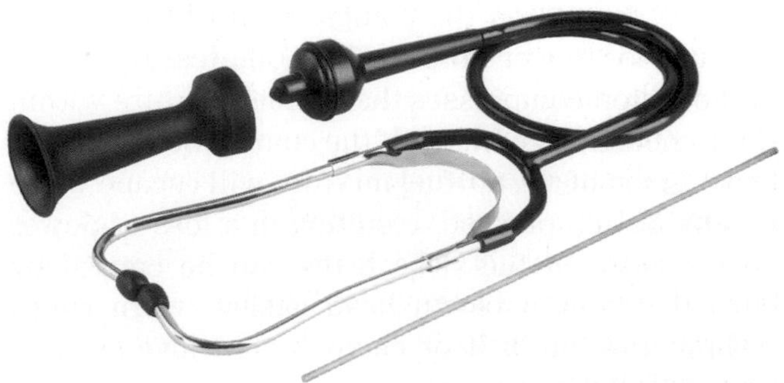

FIGURE 6–3

A stethoscope.

SPX-OTC Automotive Electronic Diagnostic Tools

Electronic stethoscopes digitize the sound waves. Once the sound waves are digitized, they can be amplified and carefully listened to.

Engine Removal and Installation Equipment

The engines of most RWD and some FWD vehicles are removed by lifting them from the top. Others must be removed from the bottom, and the procedure requires different equipment. Make sure you follow the instructions given by the manufacturer and use the appropriate tools and equipment. The required equipment varies with manufacturer and vehicle model; however, most accomplish the same thing.

To remove the engine from under the vehicle, the vehicle must be raised. A crane and/or support fixture is used to hold the engine and transaxle assembly in place while the engine is being readied for removal. Once the engine is ready, the crane is used to lower the engine onto an engine cradle. The cradle is similar to a hydraulic floor jack and is used to lower the engine further so that it can be rolled out from under the vehicle.

Often, a transverse-mounted engine is removed with the transaxle as a unit. The transaxle can be separated from the engine once it has been lifted out of the vehicle. In this case, the drive axles must be disconnected from the transaxle before removing the unit.

Ridge Reamer

After many kilometres of use, a ridge forms at the top of the engine's cylinders. Because the top piston ring stops travelling before it reaches the top of the cylinder, a ridge of unworn metal is left. This ridge may have to be removed to allow the pistons to be pushed out of the block without damaging them. If the ridge is great enough to prevent piston removal, it can be removed with a **ridge reamer (Figure 6–4)**. The tool is adjusted for the bore and then inserted into it. Rotate the tool clockwise with a wrench to remove the ridge. Remove just enough metal to allow the piston assembly to slip out of the bore without causing damage to the surface of the bore or to the piston. If the ridge is too large, the top rings will hit it and possibly break the ring lands. Remember, if a ridge is noticeable, the cylinders will probably require machining to fit suitable new oversize pistons and rings.

After the ridge-removing operation, wipe all the metal cuttings out of the cylinder. Use an oily rag to wipe the cylinder. The cuttings will stick to it.

FIGURE 6–4

A ridge reamer.

Ring Compressor

A **ring compressor (Figure 6–5)** is used to install a piston with piston rings into a cylinder bore. The compressor wraps around the piston rings to make their outside diameter smaller than the inside diameter of the bore. With the compressor tool adjusted properly, the piston assembly can be pushed easily into the bore without damaging the bore or piston.

There are a number of different types of ring compressors. One style has an adjustable band with a ratchet mechanism to tighten it around the piston. Another style uses ratcheting pliers to tighten a steel band around the piston. The bands are available in a variety of sizes. A third type has a single band that is wrinkled. The band is tightened by moving a lever. Once the rings are totally compressed into the piston, a thumbscrew is tightened to hold the band in position. Another style uses a tapered collar that is sized for a specific cylinder or adjustable to fit a range of cylinders. These collars are placed above the cylinder, and the piston assembly is simply inserted through the collar and into the cylinder.

Ring Expander

To prevent damage to the piston rings during removal and installation, a **ring expander** should be used. While installing a piston ring, the ring must be made

FIGURE 6–5

Various piston ring compressors.

Martin Restoule

large enough to fit over the piston. To do this, the piston rings are placed into the jaws of the expander, and the handle of the tool is squeezed to expand the ring **(Figure 6–6)**. The rings should be expanded only enough to fit over the piston.

FIGURE 6–6

A piston ring expander.

Ring Groove Cleaner

Before installing piston rings onto a piston, the ring grooves should be cleaned. The carbon and other debris that may be present in the back of the groove will not allow the new piston rings to compress evenly and completely into the grooves. Piston ring grooves are best cleaned with a **ring groove cleaner**. This tool is adjustable to fit the width and depth of the groove. Make sure it is properly adjusted before using it.

Dial Bore Indicator

Cylinder bore taper and out-of-roundness can be measured with a micrometer and a telescoping gauge. However, most shops use a **dial bore gauge (Figure 6–7)**. This gauge typically consists of a handle, guide blocks, a lock, an indicator contact, and an indicator. These gauges also come with extensions that make them adaptable to various size bores. As the dial bore gauge is moved inside the bore, the indicator will show any change in the bore's diameter.

Cylinder Deglazer

The proper surface finish on a cylinder wall acts as a reservoir for oil to lubricate the piston rings and prevent piston and ring scuffing. If the inspection and measurements of the cylinder wall show that surface conditions, taper, and out-of-roundness are within acceptable limits, the cylinder walls only need to be deglazed. Combustion heat, engine oil, and piston movement combine to form a thin residue on the cylinder walls that is commonly called glaze.

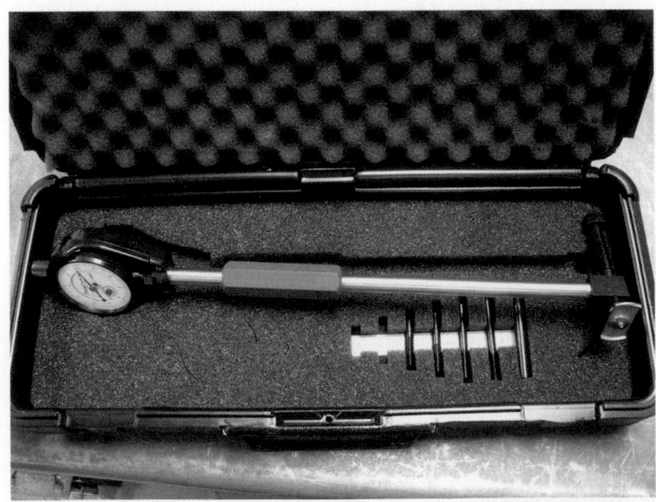

FIGURE 6–7

A dial (cylinder) bore gauge.

Martin Restoule

The common types of cylinder deglazers or **glaze breakers** use an abrasive with about 220 or 280 grit. The glaze breaker is installed in a slow-moving electric drill or in a honing machine. Many deglazers use round stones that extend on coiled wire from the centre shaft. This type of deglazer may also be used to lightly hone the bore. Various sizes of resilient-based hone-type brushes are available for honing and deglazing.

Cylinder Hone

A cylinder should be honed whenever there are minor problems with the bore. Honing sands the walls to remove imperfections. A **cylinder hone** usually consists of two or three stones. The hone rotates at a selected speed and is moved up and down the cylinder's bore. The stones have outward force applied to them and remove some metal from the bore as they rotate within it. Honing oil flows over the stones and onto the cylinder wall to control the temperature and flush out any metallic and abrasive residue. The correct stones should be used to ensure that the finished walls have the correct surface finish. Honing stones are classified by grit size; usually, the lower the grit number, the coarser the stone.

Cylinder honing machines are available in manual and automatic models. The major advantage of the automatic type is that it allows the technician to dial in the exact crosshatch angle needed.

When cylinder surfaces are badly worn or excessively scored or tapered, a **boring bar** is used to cut the cylinders for oversize pistons or sleeves. A boring bar leaves a pattern on the cylinder wall similar to uneven screw threads; therefore, you should hone the bore to the correct finish after it has been bored. A rigid hone should only be used for this procedure.

Cam Bearing Driver Set

The camshaft is supported by several friction-type bearings, or bushings. They may be designed as a one-piece unit, which is pressed into the camshaft bore in the cylinder head or block. The bearings are press fit into the block or head, using a bushing driver and hammer. Many overhead camshaft (OHC) engines use split bearings to support the camshaft. Camshaft bearings are normally replaced during engine rebuilding.

V-Blocks

The various shafts in an engine must be straight. It is impossible to see any distortions unless the shaft is severely damaged. Warped or distorted shafts will cause many problems, including premature wear of the bearings they ride on. The best way to check a shaft is to place the ends of the shaft onto V-blocks. These blocks will support the shaft and allow you to rotate the shaft. Place the plunger of a dial indicator on the journals of the shaft, and rotate the shaft. Any movement of the indicator's needle suggests a problem.

Valve and Valve Seat Resurfacing Equipment

The most critical sealing surface in the valve train is between the face of the valve and its seat in the cylinder head. Leakage between these surfaces reduces the engine's compression and power and can lead to valve burning. To ensure proper seating of the valve, the contact or sealing area on the valve face and seat must be the correct width, at the correct location, and concentric with the guide. These conditions are accomplished by renewing the surface of the valve face and seat.

Valve and valve seat grinding or refacing is done by using a grinding stone or metal cutters to achieve a fresh, smooth surface on the valve faces and stem tips. Valve faces suffer from burning, pitting, and wear caused by opening and closing millions of times during the life of an engine. Valve stem tips wear because of friction from the rocker arms or actuators.

Valve Guide Repair Tools

The amount of valve guide wear can be measured with a ball gauge and micrometer. If wear or taper is excessive, the guide must be machined or replaced. If the original guide can be removed and a new one inserted, press out the old valve guide by placing the properly sized driver into the guide. Then press out the guide. To install a new guide, use a press and the same driver that was used to remove the old guide. Align the new guide and press straight down, not at an angle.

Knurling is one of the fastest ways to restore the inside diameter (ID) dimensions of a worn valve guide. The process raises the surface of the guide ID by plowing tiny furrows through the surface of the metal. As the knurling tool cuts into the guide, metal is raised or pushed up on either side of the indentation, effectively decreasing the ID of the guide hole. A burnisher is used to make the ridges flat and produce the proper-sized hole to restore the correct guide-to-stem clearance.

Reaming is used to repair worn guides by increasing the ID of a guide to take an oversize valve stem or by restoring the guide to its original diameter after installing inserts. When reaming a guide, limit the amount of metal removed and always reface the valve seat after the valve guide has been reamed. Some valve guide liners or inserts are not precut to length, and the excess must be milled off before finishing.

Valve Spring Compressor

To remove the valves from a cylinder head, first the valve spring assemblies must be removed. To do this, the valve spring must be compressed enough to remove the valve keepers and then the retainer. Many types of **valve spring compressors** are available. Some designs allow valve spring removal while the cylinder head is still on the engine block; other designs are used only when the cylinder head is removed.

The pry bar–type compressor is used when installing valve oil seals when the cylinder head is still mounted to the block. With the cylinder's piston at top dead centre (TDC), shop air is fed into the cylinder to hold the valve up and prevent it from falling into the cylinder. The pry bar is then used to compress the valve spring so that the valve keepers can be removed.

Some OHC engines require the use of a special spring compressor **(Figure 6–8)**. Often, these special tools can be used when the cylinder head is attached to the block or when it is on a bench. These compressors bolt to the cylinder head and have a threaded plunger that fits onto the retainer. As the plunger is tightened down on the retainer, the spring compresses.

C-clamp–type valve compressors can be used on cylinder heads only after they have been removed. This type of compressor usually is a universal tool with interchangeable jaws. The spring is compressed either pneumatically or manually after the compressor is in place. One end of the clamp is positioned on the valve head and the other on the valve's retainer. After the compressor is adjusted, it will squeeze down on the spring. Once the spring is compressed, the valve keepers can be removed. Then the tension of the compressor is slowly released, and the valve retainer and spring can be removed.

FIGURE 6–8

A typical spring compressor for OHC valves.

Valve Spring Tester

Before valve springs are reused, they should be checked to make sure they are within specifications. This checking should include their freestanding height and squareness. If those two dimensions are good, the spring should be checked with a **valve spring tester**. A valve spring tester checks the spring's open-and-close tension. Correct spring tension on a closed valve guarantees a tight seal and proper valve head cooling. The opening spring tension overcomes valve train inertia and closes the valve when it should close. The tester's gauge reflects the tension of the spring when it is compressed to the installed or valve-closed height. Compare the tension readings on the tester to specifications. Any tension readings outside the specifications given indicate the valve spring should be replaced.

Torque Angle Gauge

Most manufacturers recommend the torque-angle method for tightening cylinder head bolts, which requires the use of a **torque angle gauge**. Typically, two steps are involved: tighten the bolt to the specified torque, and then tighten the bolt an additional amount. The latter is expressed in degrees. To accurately measure the number of degrees added to the bolt, a torque angle gauge **(Figure 6–9)** is attached

FIGURE 6–9

A torque angle gauge is attached to the drive lug of a torque wrench.

SPX-OTC Automotive Electronic Diagnostic Tools

to the wrench. The additional tightening will stretch the bolt and produce a very reliable clamp force or load that is much more accurate than can be achieved just by torquing.

Oil Priming Tool

Prior to starting a freshly rebuilt engine, the oil pump must be primed. There are several ways to pre-lubricate, or prime, an engine. Some older engines allow the oil pump to be driven by an electric drill. The oil pump is driven for several minutes; the valve cover should be removed to see whether there is any oil flow to the rocker arms. If oil has reached the cylinder head, the engine's lubrication system is full of oil and is operating properly. If no oil has reached the cylinder head, there is a problem either with the pump, with the alignment of an oil hole in a bearing, or perhaps with a plugged gallery.

Using a pre-lubricator **(Figure 6–10)**, which consists of an oil reservoir attached to a continuous air supply, is the best method of pre-lubricating an engine without running it. When the reservoir is attached to the engine and the air pressure is turned on, the pre-lubricator will supply the engine's lubrication system with oil under pressure.

Cooling System Pressure Tester

A cooling system pressure tester **(Figure 6–11)** contains a hand pump and a pressure gauge. A hose is connected from the hand pump to an adapter that fits on the radiator filler neck. This tester is used to pressurize the cooling system and check for coolant leaks. Additional adapters are available to connect the tester to the radiator cap. This test checks the pressure relief action of the cap.

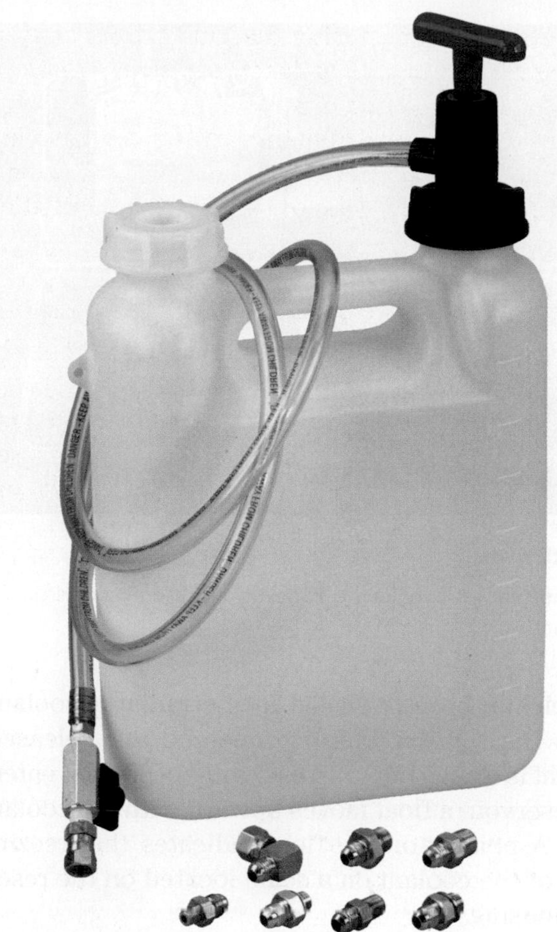

FIGURE 6–10

An engine pre-luber kit.

SPX-OTC Automotive Electronic Diagnostic Tools

Coolant Hydrometer

A coolant **hydrometer** is used to check the amount of antifreeze in the coolant. This tester contains a pickup hose, a coolant reservoir, and a squeeze bulb.

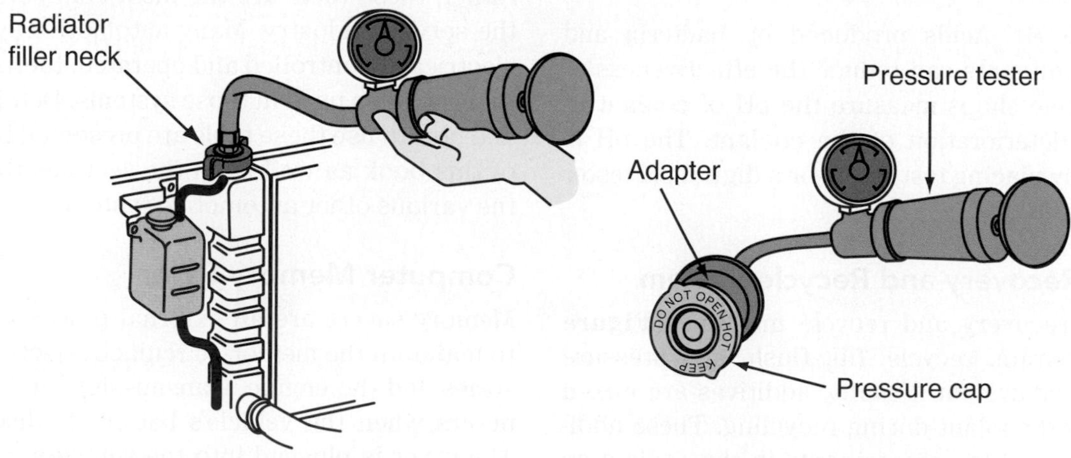

FIGURE 6–11

A cooling system pressure tester.

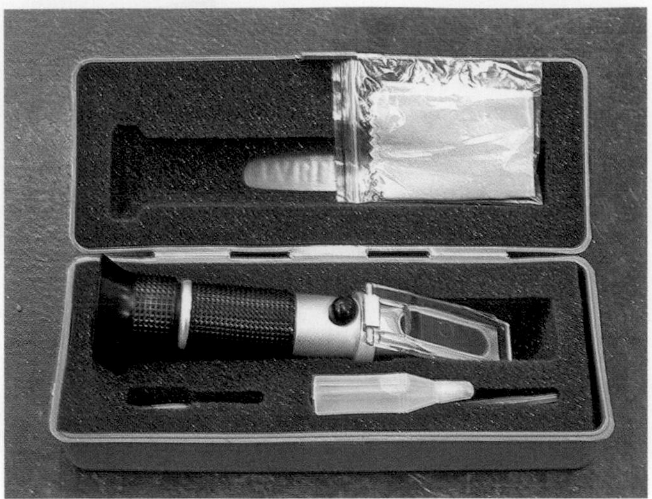

FIGURE 6–12

A refractometer checks the condition of the engine's coolant.

The pickup hose is placed in the radiator coolant. When the squeeze bulb is squeezed and released, coolant is drawn into the reservoir. As coolant enters the reservoir, a float moves upward with the coolant level. A pointer on the float indicates the freezing point of the coolant on a scale located on the reservoir housing.

REFRACTORY TESTERS For many shops, the preferred way to check coolant is with a **refractometer (Figure 6–12)**. This tester works on the principle that light bends as it passes through a liquid. A sample of the coolant is placed in the tester. As light passes through the sample of coolant, it bends and shines on a scale in the tester. A reading is taken at the point on the scale where there is a separation of light and dark. Most refractory coolant testers can also check the electrolyte in a battery.

MEASURING pH Acids produced by bacteria and other contaminants can reduce the effectiveness of coolant. Some shops measure the pH of coolant to determine deterioration of the coolant. The pH is measured by placing test strips or a digital pH tester into the coolant.

Coolant Recovery and Recycle System

A coolant recovery and recycle machine **(Figure 6–13)** can drain, recycle, fill, flush, and pressure test a cooling system. Usually, additives are mixed into the used coolant during recycling. These additives either bind to contaminants in the coolant so that they can be easily removed, or they restore some of the chemical properties in the coolant.

FIGURE 6–13

A coolant recycling machine that drains, back-flushes, and fills the cooling system.

SPX/OTC Service Solutions

ELECTRICAL/ELECTRONIC SYSTEM TOOLS

Electrical system service and diagnostic tools are discussed in the following paragraphs. This discussion does not cover all of the tools you may need; rather, these tools are the most commonly used by the service industry. Many automotive systems are electrically controlled and operated; therefore, these tools are also used in those systems. Details of when and how to use these tools are presented in Section 3 of this book as well as in the sections that discuss the various other automotive systems.

Computer Memory Saver

Memory savers are an external power source used to maintain the memory circuits in electronic accessories and the engine, transmission, and body computers when the vehicle's battery is disconnected. The saver is plugged into the vehicle's cigar lighter outlet. It can be powered by a 9- or 12-volt battery (**Figure 6–14**).

FIGURE 6–14

A 12-volt memory saver.

Snap-on Tools Company

Some have the following features:

- The capability to connect to a vehicle's cigarette lighter receptacle, a standard 120 VAC power outlet, or the OBDII port
- Use with 9-volt battery (not included)
- Maintains memory of the on-board vehicle computer when the power source is disconnected
- LED indication of state of internal battery charge
- LED confirmation of circuit between the cigarette lighter and the vehicle's battery with the ignition on or off
- 4.5 amp hour sealed lead acid battery
- Can scan engine codes while also acting as a memory saver, which stores the vehicle data for review and analysis

Circuit Tester

Circuit testers **(Figure 6–15)** are used to check for voltage in an electrical circuit. Low-voltage testers are used to troubleshoot 6- to 12-volt circuits. High-voltage circuit testers diagnose primary and secondary ignition circuits.

A circuit tester, commonly called a **test light**, looks like a stubby ice pick. Its handle is transparent and contains a light bulb. A probe extends from one end of the handle and a ground clip and wire from the other end. When the ground clip is attached to a good ground and the probe touched to a live connector, the bulb in the handle will light up. If the bulb does not light, voltage is not available at the connector.

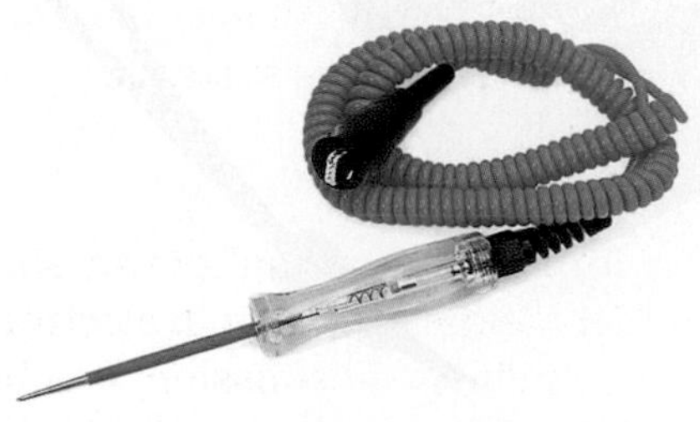

FIGURE 6–15

A typical circuit tester, commonly called a test light.

Snap-on Tools Company

WARNING!

Do not use a conventional 12-volt test light to diagnose components and wires in computer systems. The current draw of test lights may damage computers and computer system components. High-impedance test lights are available for diagnosing computer systems.

A self-powered test light is called a **continuity tester**. Used on open circuits, it looks like a regular test light, except that it has a small internal battery. When the ground clip is attached to one end of the wire or circuit and the probe is touched to the other end, the lamp will light if there is continuity in the circuit. If an open circuit exists, the light will not be illuminated.

WARNING!

Do not use any type of test light or circuit tester to diagnose automotive air bag systems. Use only the vehicle manufacturer's recommended equipment on these systems.

Multimeter

A **multimeter** is a must for diagnosing the individual components of an electrical system. Multimeters have different names, depending on what they measure and how they function. A volt-ohm-milliamp meter is referred to as a VOM, or DVOM if it is digital. A **digital multimeter (DMM)** can measure many more things than volts, ohms, and low current.

Most multimeters **(Figure 6–16)** measure direct current (DC) and alternating current (AC) amperes, volts, and ohms. More advanced multimeters may also measure diode continuity, frequency, temperature, engine speed, and dwell and/or duty cycle.

Multimeters are available with either digital or analog displays. DMMs provide great accuracy by measuring in tenths, hundredths, or thousandths of a unit. Several test ranges are usually provided for each of these functions. Some meters have multiple test ranges that must be manually selected; others are auto-ranging.

Analog meters use a sweeping needle against a scale to display readings and are not as precise as digital meters. Analog meters have low input impedance and should not be used on sensitive electronic

FIGURE 6–16

A typical multifunctional low-impedance multimeter.

circuits or components. Digital meters have high impedance and can be used on electronic circuits as well as electrical circuits.

Voltmeter

A voltmeter measures the voltage available at any point in an electrical system. A voltmeter can also be used to check the voltage drop across an electrical circuit, component, switch, or connector. A voltmeter can also be used to check for proper circuit grounding.

A **voltmeter** relies on two leads: a red positive lead and a black negative lead. The red lead should be connected to the positive side of the circuit or component. The black should be connected to the ground or to the negative side of the component. Voltmeters should be connected in parallel with the circuit being tested.

Ohmmeter

An **ohmmeter** measures the resistance in a circuit. In contrast to the voltmeter, which operates by the voltage available in the circuit, an ohmmeter is battery powered. The circuit being tested must have no power applied. If the power is on in the circuit, the ohmmeter will be damaged.

The two leads of the ohmmeter are placed across the circuit or component being tested. The red lead is placed on the positive side of the circuit and the black lead is placed on the negative side. The meter sends current through the component and determines the amount of resistance based on the voltage dropped across the load. The scale of an ohmmeter reads from 0 to infinity (∞). A 0 reading means there is no resistance in the circuit and may indicate a short in a component that should show a specific resistance. An infinity reading indicates a number higher than the meter can measure, which usually indicates an open circuit.

Ammeter

An **ammeter** measures current flow in a circuit. Current is measured in amperes. Unlike the voltmeter and ohmmeter, the ammeter must be placed into the circuit or in series with the circuit being tested. Normally, this requires disconnecting a wire or connector from a component and connecting the ammeter in series between the wire or connector and the component. The red lead of the ammeter should always be connected to the side of the connector closest to the positive side of the battery, and the black lead should be connected to the other side.

It is much easier to test current using an ammeter with an inductive pickup **(Figure 6–17)**. The pickup clamps around the wire or cable being tested. The ammeter determines amperage based on the magnetic field created by the current flowing through the wire. This type of pickup eliminates the need to separate the circuit to insert the meter.

FIGURE 6–17

An ammeter with an inductive pickup is often called a current probe.

SPX-OTC Automotive Electronic Diagnostic Tools

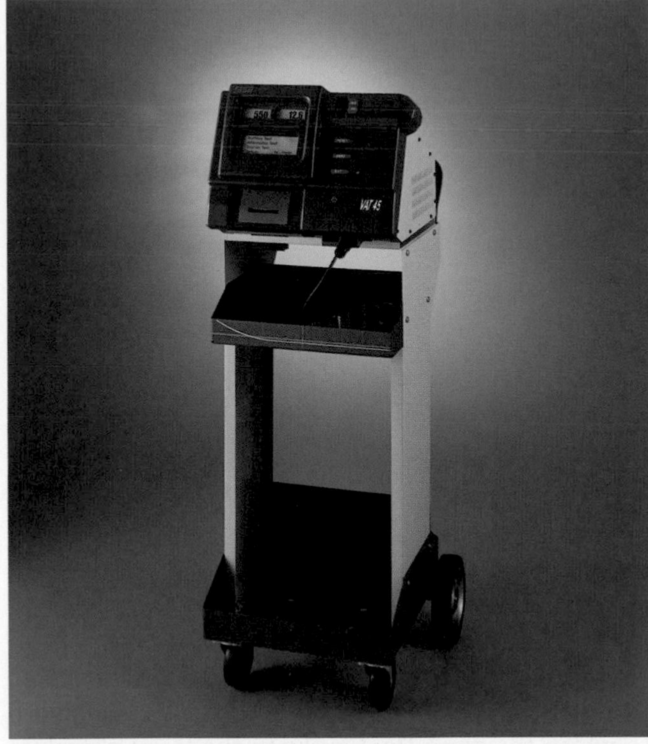

FIGURE 6–18

The volt/ampere tester is used for testing batteries and the starting and charging systems.

Snap-on Tools Company

Volt/Ampere Tester

A **volt/ampere tester (VAT)**, shown in **Figure 6–18**, is used to test batteries, starting systems, and charging systems. The tester contains a voltmeter, ammeter, and carbon pile. The carbon pile is a variable resistor. When the tester is attached to the battery and turned on, the carbon pile draws current out of the battery. The ammeter will read the amount of current draw. The maximum current draw from the battery, with acceptable voltage, is compared to the rating of the battery to see if the battery is okay. A VAT also measures the current draw of the starter and the current output from the charging system.

Battery Capacitance Tester

Many manufacturers recommend that a capacitance or conductance test be performed on batteries **(Figure 6–19)**. Conductance describes a battery's ability to conduct current. It is a measurement of the plate surface available in a battery for chemical reaction. Measuring conductance provides a reliable indication of a battery's condition and is correlated to battery capacity. Conductance can be used to detect cell defects, shorts, normal aging, and open circuits, which can cause the battery to fail.

FIGURE 6–19

A battery capacitance tester.

To measure conductance, the tester creates a small signal that is sent through the battery, and then measures a portion of the AC current response. The tester displays the service condition of the battery. The tester will indicate that the battery is good, needs to be recharged and tested again, has failed, or will fail shortly.

Lab Scopes

An oscilloscope, or **lab scope**, is a visual voltmeter **(Figure 6–20)**. A lab scope converts electrical signals to a visual image representing voltage changes over a period of time. This information is displayed in the form of a continuous voltage line called a **waveform**, or **trace**. A scope displays any change in voltage as it occurs.

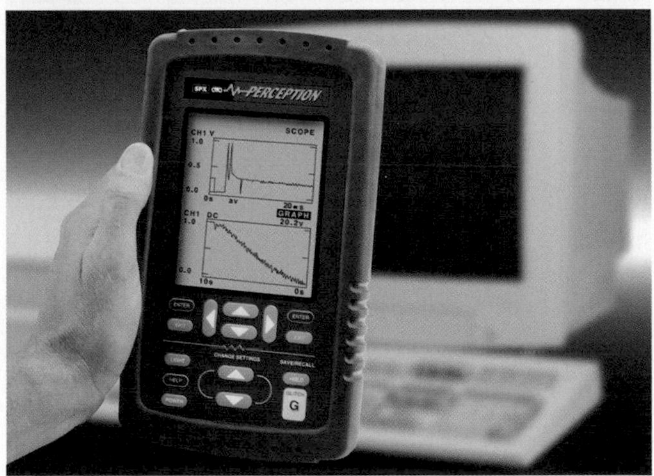

FIGURE 6–20

A hand held dual-trace lab scope.

OTC Tool and Equipment, Division of SPX Corporation

An upward movement of the trace on an oscilloscope indicates an increase in voltage, and a downward movement represents a decrease in voltage. As the voltage trace moves across an oscilloscope screen, it represents a specific length of time.

The size and clarity of the displayed waveform is dependent on the voltage scale and the time reference selected. Most scopes are equipped with controls that allow voltage and time-interval selection.

Dual-trace scopes can display two different waveform patterns at the same time. This type of scope is especially important for diagnosing intermittent problems.

Graphing Multimeter

A graphing digital multimeter displays readings over time, similar to a lab scope. The graph shows the minimum and maximum readings on a graph, as well as displaying the current reading **(Figure 6–21)**. By observing the graph, a technician can detect any undesirable changes during the transition from a low reading to a high reading, or vice versa. These glitches are some of the more difficult problems to identify without a graphing meter or a lab scope.

Battery Hydrometer

On unsealed batteries, the specific gravity of the electrolyte can be measured to give a fairly good indication of the battery's state of charge. A hydrometer **(Figure 6–22)** is used to perform this test. A battery hydrometer consists of a glass tube or barrel, rubber bulb, rubber tube, and a glass float or hydrometer with a scale built into its upper stem. The glass tube encases the float and forms a reservoir for the test electrolyte. Squeezing the bulb pulls electrolyte into the reservoir.

When filled with test electrolyte, the sealed hydrometer float bobs in the electrolyte. The depth to which the glass float sinks in the test electrolyte indicates its relative weight compared to water. The reading is taken off the scale by sighting along the level of the electrolyte.

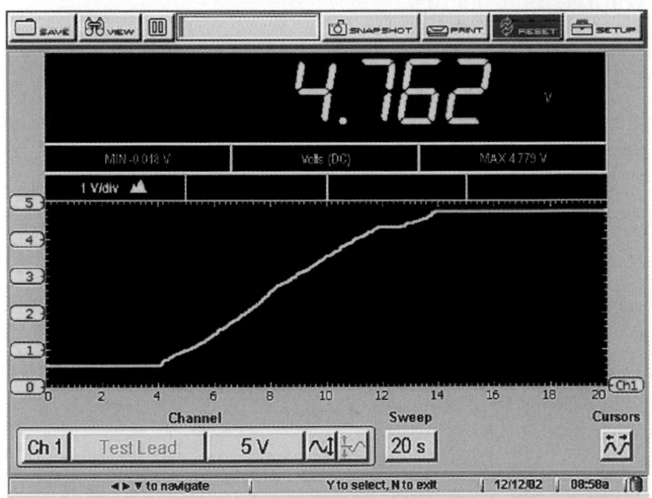

FIGURE 6–21

The screen of a graphing multimeter.

Reproduced under licence from Snap-on Incorporated. All of the marks are marks of their owners.

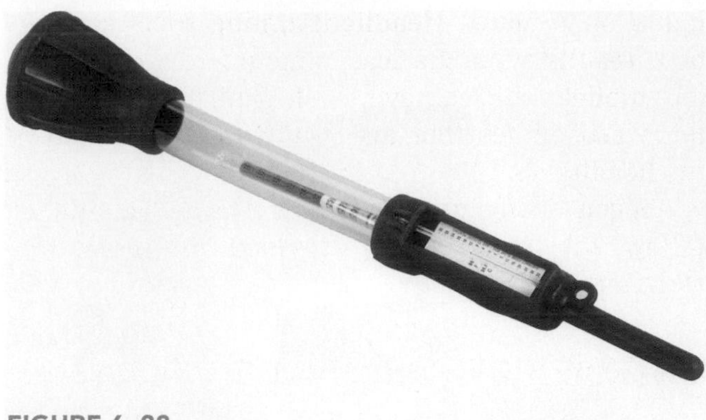

FIGURE 6–22

A battery hydrometer is used to measure the specific gravity of a battery's electrolyte.

SPX-OTC Automotive Electronic Diagnostic Tools

Wire and Terminal Repair Tools

Many automotive electrical problems can be traced to faulty wiring. Loose or corroded terminals; frayed, broken, or oil-soaked wires; and faulty insulation are the most common causes.

Wires and connectors are often repaired or replaced. Sometimes an entire length of wire is replaced; other times only a section is. In either case, the wire must have the correct terminal or connector to work properly in the circuit. Wire cutters, stripping tools, terminal crimpers, and **connector picks** are the most commonly used tools for wire repair. Also, soldering equipment is used to provide the best electrical connection for a wire to another wire and for a wire to a connector.

Connector picks are designed to reach into electrical connectors and release the locking tabs on the terminals. Doing this allows the terminal and the attaching wire to be removed from the terminal.

Headlight Aimers

Headlights must be kept in adjustment to obtain maximum illumination. Properly adjusted headlights will cover the correct range and afford the driver a proper nighttime view. Headlights can be adjusted using headlamp-adjusting tools or by shining the lights on a chart. Headlight-aiming tools give the best results with the least amount of work. Many late-model vehicles have levels built into the headlamp assemblies that are used to correctly adjust the headlights.

When using any headlight-aiming equipment, follow the instructions provided by the equipment manufacturer.

ENGINE PERFORMANCE TOOLS

Diagnostic and special tools for the air, fuel, ignition, emission, and engine-control systems are discussed in the following paragraphs. This discussion does not cover all of the tools you may need; rather, these tools are the most commonly used by the service industry. Some are also used when diagnosing or servicing the controls of other automotive systems. Details of when and how to use these tools are covered in Section 4 of this book as well as in other sections where necessary.

Scan Tools

The introduction of computer-controlled ignition and fuel systems brought with it the need for tools capable of troubleshooting electronic engine control systems. A variety of computer scan tools are available today that do just that. A **scan tool (Figure 6–23)** is a microprocessor designed to communicate with the vehicle's computer. Connected to the electronic control system through diagnostic connectors, a scan tool can access trouble codes, run tests to check system operations, and monitor the activity of the system. Trouble codes and test results are displayed on an LCD (liquid crystal display) screen or printed out on the scan tool's printer.

The scan tool is connected to specific diagnostic connectors on the vehicle. Some manufacturers have one diagnostic connector that connects the data wire from each computer to a specific terminal in this connector. Other vehicle manufacturers have several different diagnostic connectors on each vehicle, and each of these connectors may be connected to one or more computers. The scan tool must be programmed for the model year, make of vehicle, and type of engine.

With OBD-II, the diagnostic connectors are located in the same place on all vehicles. Also, any scan tool designed for OBD-II will work on all OBD-II systems. The OBD-II scan tool has the ability to run diagnostic tests on all systems and has freeze frame capabilities.

As automotive computer systems become more complex, the diagnostic capabilities of scan testers continue to expand. Many scan testers now have the capability to store, or "freeze," data into the tester during a road test **(Figure 6–24)** and then play back these data when the vehicle is returned to the shop.

There are many different scan tools available. Some are a combination of other diagnostic tools, such as a lab scope and a graphing multimeter. These may have the following capabilities:

- Retrieve diagnostic trouble codes (DTCs).
- Monitor system operational data.
- Reprogram the vehicle's electronic control modules.
- Perform systems diagnostic tests.
- Display appropriate service information, including electrical diagrams.

FIGURE 6–23

A typical scan tool shown with various test leads, adapters, and memory card.

SPX/OTC Service Solutions

FIGURE 6–24

Using a scan tool during a road test.

OTC Tool and Equipment, Division of SPX Corporation

- Display technical service bulletins (TSBs).
- Display troubleshooting instructions.
- Perform easy tool updating through a personal computer (PC).

Some scan tools work directly with a PC through uncabled communication links, such as Bluetooth. Others use a personal digital assistant (PDA). These small hand held units allow you to read DTCs, monitor the activity of sensors, and view Inspection/Maintenance (I/M) system test results to quickly determine what service the vehicle requires. Most of these scan tools also have the ability to do the following:

- Perform system and component tests.
- Report test results of monitored systems.
- Print DTC/freeze frame and display full diagnostic code descriptions.
- Exchange files between a PC and a PDA.
- View and print files on a PC.
- Generate emissions reports.
- View IM/Mode 6 information.
- Display relative TSBs.

Engine Analyzers

When performing a complete engine performance analysis, an engine analyzer is used. Although the term engine analyzer is often loosely applied to any multipurpose test meter, a complete engine analyzer incorporates most, if not all, of the test instruments needed to diagnose driveability problems. Most engine analyzers have the following diagnostic tools: a compression gauge, an exhaust analyzer, pressure and vacuum gauges, a voltmeter, an ohmmeter, a vacuum pump, an ammeter, a tachometer, an oscilloscope, a timing light and probe, and, of course, a scan tool.

With an engine analyzer, you can perform tests on the battery, starting system, charging system, primary and secondary ignition circuits, electronic control systems, fuel system, emissions system, and engine assembly. The analyzer is connected to these systems by a variety of leads, inductive clamps, probes, and connectors. The data received from these connections are processed by several computers within the analyzer.

Most engine analyzers have both manual and automatic test modes. In the manual modes, any single test, such as cylinder compression or generator output, can be performed. The manual test mode is useful when there is little need to do a complete test. The automatic test mode is useful when looking at general performance. When the automatic test mode is selected, specific tests are automatically performed in a specific sequence.

The analyzer may compare all the test results to the vehicle manufacturer's specifications. When the test series is completed, the analyzer prints a report indicating those readings that were not within specifications. Many analyzers also provide diagnostic assistance for the problems indicated by the readings that were not within specifications.

Fuel Pressure Gauge

A **fuel pressure gauge** is essential for diagnosing fuel injection systems **(Figure 6–25)**. These systems rely on fuel pressures from 240 to 480 kPa (35 to 70 psi). A drop in fuel pressure reduces the amount of fuel delivered to the injectors and results in a lean air/fuel mixture.

A fuel pressure gauge is used to check the discharge pressure of fuel pumps, the regulated pressure of fuel injection systems, and injector pressure drop. These tests can identify faulty pumps, regulators, or injectors and restrictions in the fuel delivery system. Restrictions are typically caused by a dirty fuel filter, collapsed hoses, or damaged fuel lines.

Some fuel pressure gauges have a valve and outlet hose for testing fuel pump discharge volume. The specification for discharge volume is given as a number of litres or pints of fuel that should be delivered in a certain number of seconds.

CAUTION!

While testing fuel pressure, be careful not to spill gasoline. Gasoline spills may cause explosions and fires, resulting in serious personal injury and property damage.

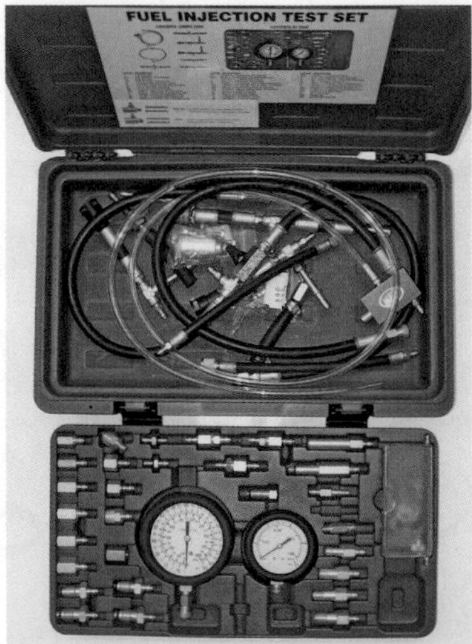

FIGURE 6–25

A fuel pressure gauge and adapters.

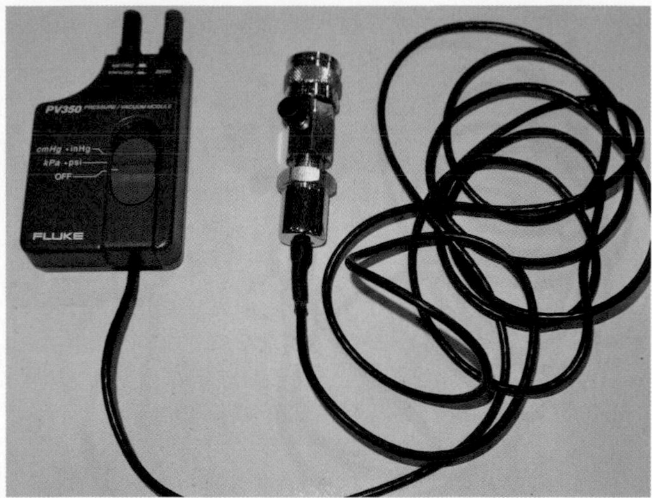

FIGURE 6–26

A pressure transducer.

Pressure Transducer

A pressure transducer **(Figure 6–26)** measures pressure and changes it into an electrical signal so that it can be displayed on a lab scope of a graphing DMM. A pressure transducer can measure such things as fuel pressure, oil pressure, exhaust pressure, intake pressure (vacuum), crankcase pressure, radiator pressure, and cylinder compression.

A fuel pressure transducer connects to the Schrader valve on the fuel rail. The resultant trace on the lab scope shows the changes in fuel pressure at the fuel rail. If the pattern is uniform and consistent, a normal operating fuel system is noted.

Injector Balance Tester

The injector balance tester **(Figure 6–27)** tests the injectors in a port fuel-injected engine. A fuel pressure gauge is used during the injector balance test. The injector balance tester contains a timing circuit, and some injector balance testers have an off–on switch. A pair of tester leads must be connected to the battery with the correct polarity. The injector terminals are disconnected, and the other leads on the tester are attached to the injector terminals.

The fuel pressure gauge is connected to the fuel rail, and the ignition switch is cycled two or three times until the specified fuel pressure is indicated on the pressure gauge. When the tester push button is depressed, the tester energizes the injector for a specific length of time, and the technician records the pressure decrease on the fuel pressure gauge. This procedure is repeated on each injector.

If the pressure drops little, or if there is no pressure drop, the injector's orifice is restricted or the injector is faulty. If there is an excessive amount of pressure drop, the injector plunger is sticking open, which results in a rich air/fuel mixture.

WARNING!

Electronic fuel injection systems are pressurized, and these systems require depressurizing prior to fuel pressure testing and other service procedures.

FIGURE 6–27

A fuel injection balance tester.

Injector Circuit Test Light

A special test light called a **noid light** can be used to determine if a fuel injector is receiving its proper voltage pulse from the computer **(Figure 6–28)**. The wiring harness connector is disconnected from

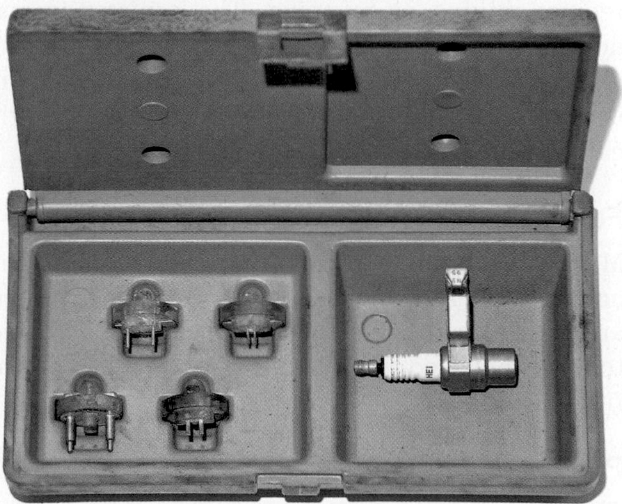

FIGURE 6–28

A set of noid lights and a test spark plug.

the injector, and the noid light is plugged into the connector. After disabling the ignition to prevent starting, the engine is turned over by the starter motor. The noid light will flash rapidly if the voltage signal is present. No flash usually indicates an open in either the power feed or ground circuit to the injector.

Fuel Injector Cleaners

Fuel injectors spray a certain amount of fuel into the intake system. If the fuel pressure is low, not enough fuel will be sprayed. Low pressure may also occur if the fuel injector is dirty. Normally, clogged injectors are the result of inconsistencies in gasoline detergent levels and the high sulphur content of gasoline. When these sensitive fuel injectors become partially clogged, fuel flow is restricted. Spray patterns are altered, causing poor performance and reduced fuel economy.

The solution to a sulphated and/or plugged fuel injector is to clean it, not replace it. There are two kinds of fuel injector cleaners. One is a pressure tank. A mixture of solvent and unleaded gasoline is placed in the pressure tank. The vehicle's fuel pump is disabled, and on some vehicles, the fuel line must be blocked between the pressure regulator and the return line. Then the hose on the pressure tank is connected to the service port in the fuel system. The in-line valve is then partially opened and the engine is started. It should run at approximately 2000 rpm for about 10 minutes to clean the injectors thoroughly.

An alternative to the pressure tank is a pressurized canister **(Figure 6–29)** in which the solvent solution is premixed. Use of the canister-type cleaner is similar to this procedure but does not require mixing or pumping. The canister is connected to the injection system's servicing fitting, and the valve on the canister is opened. The engine is started and allowed to run until it dies. Then the canister is discarded.

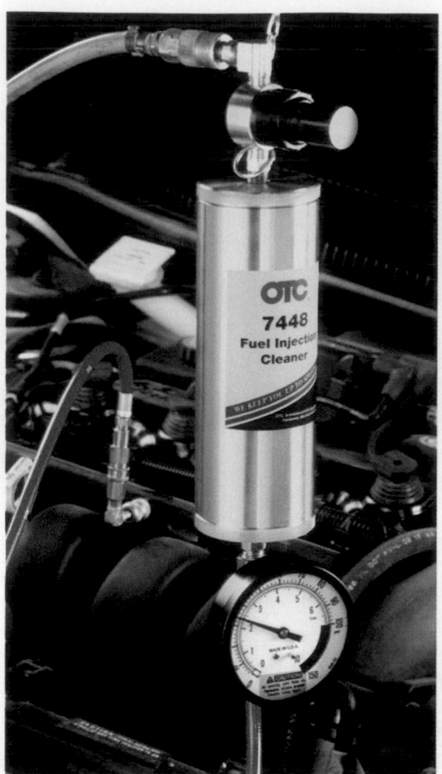

FIGURE 6–29

A fuel injector cleaner using the shop's compressed air.

OTC Tool and Equipment, Division of SPX Corporation

Fuel Line Tools

Many vehicles are equipped with quick-connect line couplers. These work well to seal the connection but are nearly impossible to disconnect if the correct tools are not used. There is a variety of quick-connect fittings and tools.

Pinch-Off Pliers

The need to temporarily close off a rubber hose is common during diagnostics and service. Special pliers are designed to do this without damaging the hose **(Figure 6–30)**. These pliers are much like vise grip pliers in that they hold their position until they are released. The jaws of the pliers are flat and close in a parallel motion, which prevents damage to the hose.

Vacuum Gauge

Measuring intake manifold vacuum can diagnose the condition of an engine. Manifold vacuum is tested with a vacuum gauge. **Vacuum** is formed on a piston's

FIGURE 6–30

Pinch-off pliers closing off a vacuum hose.

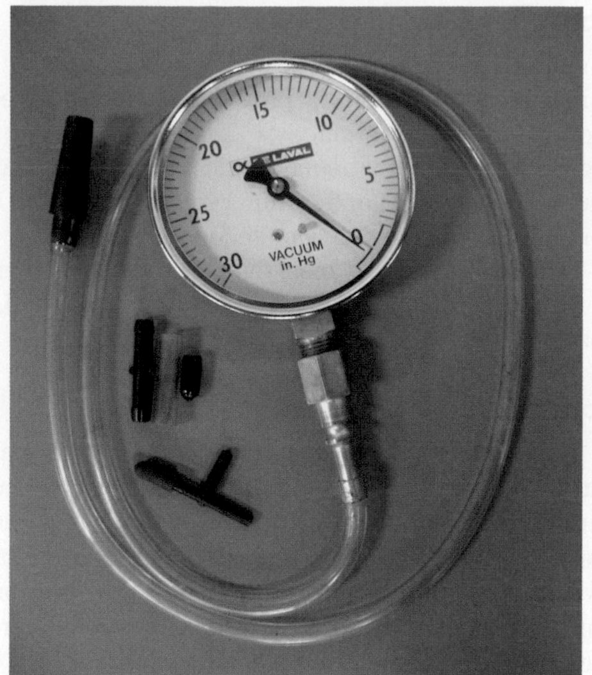

FIGURE 6–31

A vacuum gauge used to measure engine vacuum.

Martin Restoule

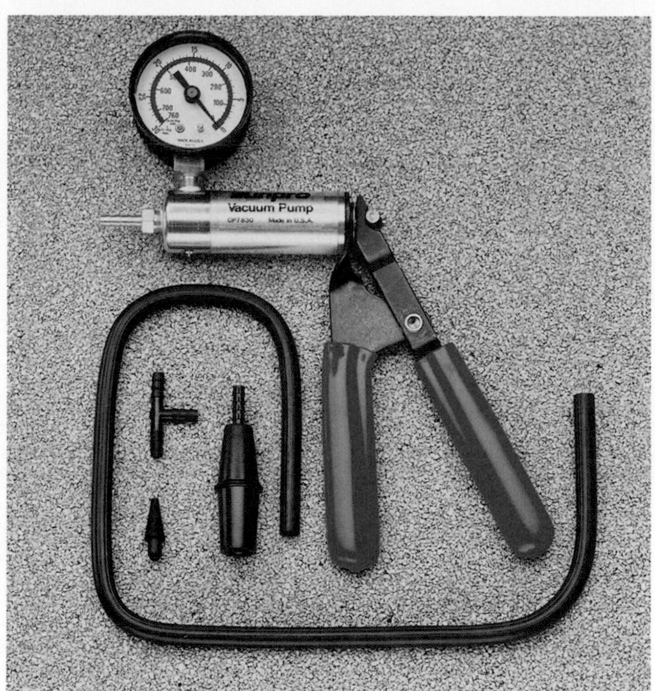

FIGURE 6–32

A typical vacuum pump with accessories.

Actron Manufacturing Company

intake stroke. As the piston moves down, it lowers the pressure of the air in the cylinder—if the cylinder is sealed. This lower cylinder pressure is called engine vacuum. Vacuum is measured in millimetres of mercury (mm Hg), centimetres of mercury (cm Hg), and inches of mercury (in. Hg).

To measure vacuum, a flexible hose on the vacuum gauge **(Figure 6–31)** is connected to a source of manifold vacuum, either on the manifold or at a point below the throttle plates. The test is made with the engine cranking or running. A good vacuum reading is typically at least 40 cm Hg (16 in. Hg). However, a reading of 38 to 50 cm Hg (15 to 20 in. Hg) is normally acceptable. Because the intake stroke of each cylinder occurs at a different time, the production of vacuum occurs in pulses. If the amount of vacuum produced by each cylinder is the same, the vacuum gauge will show a steady reading. If one or more cylinders are producing different amounts of vacuum, the gauge will show a fluctuating reading.

Vacuum Pump

There are many vacuum-operated devices and vacuum switches on cars. These devices use engine vacuum to cause a mechanical action or to switch something on or off. The tool used to test vacuum-actuated components is the vacuum pump **(Figure 6–32)**. There are two types of vacuum pumps: an electrical-operated pump and a hand held pump. The hand held pump is most often used for diagnostics. A hand held vacuum pump consists of a hand pump, a vacuum gauge, and a length of rubber hose used to attach the pump to the component being tested. Tests with the vacuum pump can usually be performed without removing the component from the vehicle.

The vacuum pump is also used to locate vacuum leaks by connecting the vacuum pump to a suspect vacuum hose or component and applying vacuum. If the needle on the vacuum gauge begins to drop after the vacuum is applied, a leak exists somewhere in the system.

Vacuum Leak Detector

A vacuum or compression leak might be revealed by a compression check, a cylinder leakage test, or a manifold vacuum test. Finding the location of the leak can be very difficult.

A simple but time-consuming way to find leaks in a vacuum system is to check each component and vacuum hose with a vacuum pump. Simply apply vacuum to the suspected area and watch the gauge for any loss of vacuum. A good vacuum component holds the vacuum applied to it.

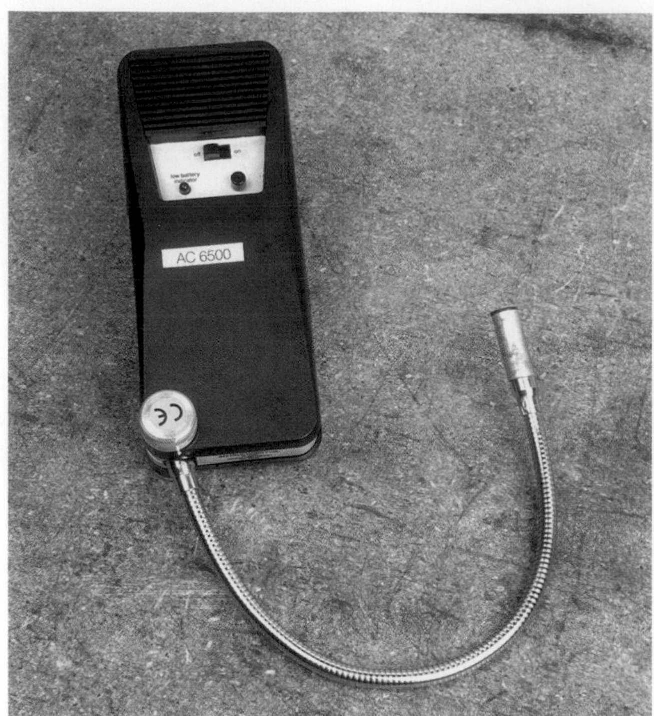

FIGURE 6–33

An ultrasonic vacuum leak detector.

Another method of leak detection uses an ultrasonic leak detector **(Figure 6–33)**. Air rushing through a vacuum leak creates a high-frequency sound higher than the range of human hearing. An ultrasonic leak detector is designed to hear the frequencies of the leak. When the tool is passed over a leak, the detector responds to the high-frequency sound by emitting a warning beep. Some detectors also have a series of light-emitting diodes (LEDs) that light up as the frequencies are received. The closer the detector is moved to the leak, the more LEDs light up or the faster the beeping occurs, allowing the technician to zero in on the leak. An ultrasonic leak detector can sense leaks as small as 0.05 mm (1/500 in.) and accurately locate the leak to within 1.59 mm (1/16 in.).

An ultrasonic leak detector can also be used to detect the source of compression leaks, bearing wear, and electrical arcing. It can also be used to diagnose fuel injector operation.

Tachometer

A **tachometer** is used to measure engine speed. Like other meters, tachometers are available in analog and digital types. Tachometers are connected to the ignition system to monitor ignition pulses, which are then converted to engine speed by the meter.

Spark Tester

A **spark tester** is a fake spark plug. The tester is constructed like a spark plug but does not have a ground electrode. In place of the electrode, there is a grounding clamp. Using spark testers is an easy way to determine if an ignition problem is caused by something in the primary or secondary circuit.

The spark tester is inserted in the spark plug end of an ignition cable. When the engine is cranked, a spark should be seen from the tester to ground.

Logic Probes

In some circuits, pulsed or digital signals pass through the wires. These on–off digital signals either carry information or provide power to drive a component. Many sensors, used in a computer-control circuit, send digital information back to the computer. To check the continuity of the wires that carry digital signals, a logic probe can be used.

A **logic probe** is similar in appearance to a test light. It contains three different-coloured LEDs. A red LED lights when there is high voltage at the point being probed. A green LED lights to indicate low voltage. A yellow LED indicates the presence of a voltage pulse. The logic probe is powered by the circuit and reflects only the activity at the point being probed. When the probe's test leads are attached to a circuit, the LEDs display the activity.

Sensor Tools

Oxygen sensors are replaced as part of the preventive maintenance program and when they are faulty. Because they are shaped much like a spark plug with wires or a connector coming out of the top, ordinary sockets do not fit well. For this reason, tool manufacturers provide special sockets for these sensors **(Figure 6–34)**.

FIGURE 6–34

A heated oxygen sensor socket.

SPX OTC Automotive Electronic Diagnostic Tools

Special sockets are also available for other sending units and sensors.

Static Strap

Because electronic components are sensitive to voltage, static electricity can destroy them. Static straps can be worn while working on or around electronic components. These straps typically are worn around a wrist and connected to a known good ground on the vehicle. The straps discharge all static electricity to the ground of the vehicle, thereby eliminating the chance of static electricity moving to the electronic components.

Pyrometer

A **pyrometer** is an electronic device that measures temperature. These are used to measure the temperature of the cooling system and the effectiveness of a catalytic converter. A converter should be checked for its ability to convert CO and HC into CO_2 and water by doing a comparison or delta temperature test. To conduct this test, a hand held digital pyrometer is used **(Figure 6–35)**. By touching the pyrometer probe or placing it near to the exhaust pipe just ahead of and just behind the converter, there should be an increase of at least 38°C (100°F) or 8 percent above the inlet temperature reading as the exhaust gases pass through the converter. If the outlet temperature is the same or lower, nothing is happening inside the converter. A pyrometer can also be used to measure the temperature of the coolant at various stages of its travel.

FIGURE 6–35

A hand held digital infrared temperature sensor.

SPX-OTC Automotive Electronic Diagnostic Tools

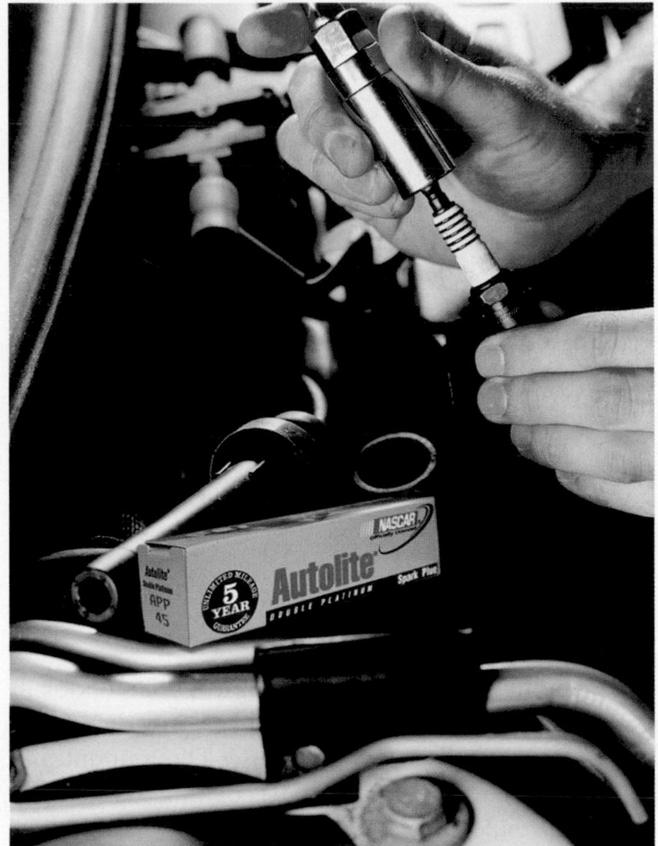

FIGURE 6–36

A spark plug socket.

Honeywell International Inc.

Spark Plug Sockets

Special sockets are available for the installation and removal of spark plugs **(Figure 6–36)**. These deep sockets have a hex nut drive at the end to allow a technician to turn them with a ratchet or an open-end wrench. The sockets are available in the common sizes of spark plugs (5/8 in. and 13/16 in.) and have a 3/8-inch drive. The socket is built with a rubber sleeve that surrounds the insulator part of the spark plug to prevent cracking or other damage to the plug while it is being removed or installed.

Exhaust Analyzers

Federal laws require that cars and trucks meet specific emission levels. Provincial governments have also passed laws requiring that owners maintain their vehicles so that the emissions remain below an acceptable level. Some provinces require an annual or bi-annual emissions inspection to meet that goal. Many shops have an exhaust analyzer for inspection purposes.

Exhaust analyzers **(Figure 6–37)** are also very valuable diagnostic tools. By looking at the quality

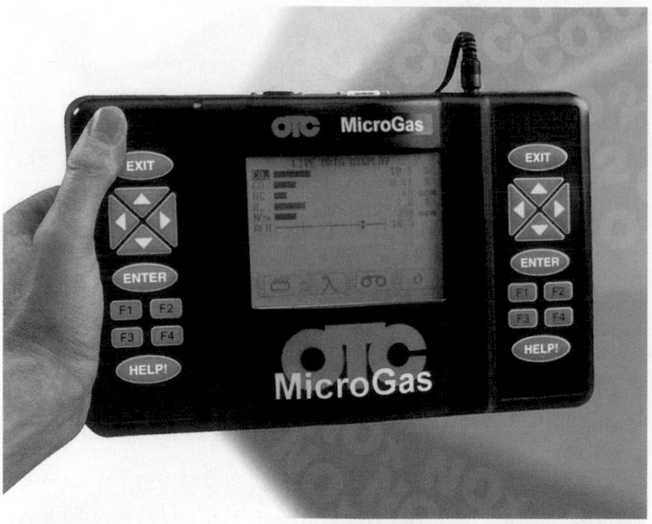

FIGURE 6–37

A MicroGas five-gas exhaust analyzer.

SPX/OTC Service Solutions

of an engine's exhaust, a technician is able to look at the effects of the combustion process. Any defect can cause a change in exhaust quality. The amount and type of change serves as the basis of diagnostic work.

Early emission analyzers measured the amount of hydrocarbons (HC) and carbon monoxides (CO). Exhaust analyzers normally measure HC in parts per million (ppm) or grams per kilometre (g/km). CO is measured as a percent of the total exhaust.

Many of the emission control devices that have been added to vehicles over the past 30 years—especially catalytic converters—have decreased the amount of HC and CO in the exhaust. These devices alter the contents of the exhaust. Therefore, checking the HC and CO contents in the exhaust may not be a true indication of the operation of an engine.

The manufacturers of exhaust analyzers have altered their machines so that they can look at the efficiency of an engine in spite of the effectiveness of the emission controls. These machines are four-gas exhaust analyzers. In addition to measuring HC and CO levels, a four-gas exhaust analyzer also monitors carbon dioxide (CO_2) and oxygen (O_2) levels in the exhaust. The latter two gases are changed only slightly by the emission controls and therefore can be used to check engine efficiency. Many exhaust analyzers are also available that measure a fifth gas, oxides of nitrogen (NO_x).

By measuring NO_x, CO_2, and O_2, in addition to HC and CO, a technician gets a look at the efficiency of the engine. Keep in mind that an exhaust analyzer is an excellent diagnostic tool and is not just for comparing emission levels against standards. There is a desired relationship among the five gases. Any deviation from this relationship can be used to diagnose a driveability problem.

Chassis Dynamometer

A chassis **dynamometer**, commonly called a dyno, is used to simulate a road test. A vehicle can be driven through a wide assortment of operating conditions without leaving the shop. Because the vehicle is stationary, test equipment can be connected and monitored while the vehicle is driven under various loads. This is extremely valuable when diagnosing a problem. A chassis dyno can also be used for performance tuning.

The vehicle's drive wheels are positioned on large rollers. The electronically controlled rollers offer rotational resistance to simulate the various loads a vehicle may face.

Some performance shops have an engine dynamometer that directly measures the output from an engine. A chassis dynamometer measures the engine's output after it has passed through the driveline.

Hybrid Tools

A hybrid vehicle is an automobile and as such is subject to many of the same problems as a conventional vehicle. Most systems in a hybrid vehicle are diagnosed in the same way as well. However, a hybrid vehicle has unique systems that require special procedures and test equipment. Also, make sure you follow all test procedures precisely as they are given.

GLOVES Always wear safety gloves when working on or around the high-voltage systems. These gloves must be class 0 rubber insulating gloves, rated at 1000 volts (these are commonly called **lineman's gloves**). Also, to protect the integrity of the insulating gloves, as well as you, wear leather gloves over the insulating gloves while doing a service.

TEST EQUIPMENT An important diagnostic tool is a DMM. However, the meter used on hybrids, electric vehicles (EVs), and fuel-cell electric vehicles (FCEVs) should be classified as a category III meter. There are basically four categories for low-voltage electrical meters, each built for specific purposes and to meet certain standards. Low voltage, in this case, means voltages less than 1000 volts. The categories define how safe a meter is when measuring certain circuits. The standards for the various categories are defined by the American National Standards Institute (ANSI), the International Electrotechnical

FIGURE 6–38

Only meters with this symbol should be used on the high-voltage systems in a hybrid vehicle.

Commission (IEC), and the Canadian Standards Association (CSA). A CAT III meter **(Figure 6–38)** is required for testing hybrid vehicles because of the high voltages, three-phase current, and the potential for high transient voltages. **Transient voltages** are voltage surges that occur in AC circuits. To be safe, you need a CAT III 1000 V meter. A meter's voltage rating reflects its ability to withstand transient voltages. Therefore, a CAT III 1000 V meter offers more protection than a CAT III meter rated at 600 volts.

The probes for the meter's leads should have safety ridges or finger positioners. These help prevent contact between your fingertips and the meter's test leads.

Another important tool is an **insulation resistance tester**. This can check for voltage leakage through the insulation of the high-voltage cables. Obviously, no leakage is desired as any leakage can cause a safety hazard as well as damage to the vehicle. Minor leakage can also cause hybrid system–related driveability problems. This meter should be used by anyone who might service a damaged hybrid vehicle, such as someone doing body repair. This should also be a CAT III meter.

TRANSMISSION AND DRIVELINE TOOLS

The following paragraphs discuss the commonly used repair and diagnostic tools for manual and automatic transmissions as well as those required for driveline service. Details of when and how to use these tools are covered in Sections 5 and 6.

Transaxle Removal and Installation Equipment

The removal and replacement (R&R) of transaxles mounted to transversely mounted engines require different tools than those needed to remove a transmission from a rear-wheel-drive vehicle.

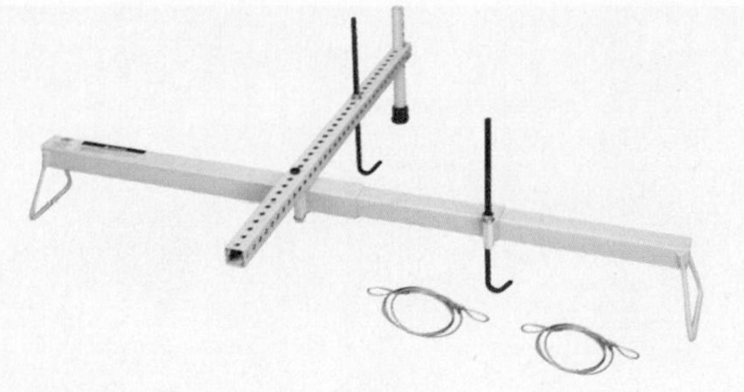

FIGURE 6–39

An engine support is used on many FWD vehicles to hold the engine in place while the transaxle is removed.

SPX-OTC Automotive Electronic Diagnostic Tools

To remove the engine and transmission from under the vehicle, the vehicle must be raised. A crane and/or support fixture is used to hold the engine and transaxle assembly in place while the assembly is being readied for removal. When everything is set for removal of the assembly, the crane is used to lower the assembly onto a cradle. The cradle is similar to a hydraulic floor jack and is used to lower the assembly so that it can be rolled out from under the vehicle. The transaxle can be separated from the engine once it has been removed from the vehicle.

When the transaxle is removed as a single unit, the engine must be supported while it is in the vehicle before, during, and after transaxle removal. Special fixtures (**Figure 6–39**) mount to the vehicle's upper frame or suspension parts. These supports are attached to the engine and support it so that the transmission can be removed.

Transmission/Transaxle Holding Fixtures

Special holding fixtures should be used to support the transmission or transaxle after it has been removed from the vehicle. These holding fixtures may be stand-alone units or may be bench mounted. They allow the transmission to be easily repositioned during repair work.

Transmission Jack

A transmission jack **(Figure 6–40)** is designed to help you while removing and reinstalling a transmission from under the vehicle. The weight of the transmission makes it difficult and unsafe to remove it without assistance and/or a transmission jack. These jacks fit under the transmission and are usually equipped with hold-down chains, which are used to secure the transmission to the jack. The transmission's weight rests on the jack's saddle.

Transmission jacks are available in two basic styles. One is used when the vehicle is raised by a

FIGURE 6–40

A typical transmission jack.

hydraulic jack and sitting on jack stands. The other style is used when the vehicle is raised on a lift.

Axle Pullers

Axle pullers are used to pull rear axles in RWD vehicles. Most rear axle pullers are slide hammer type.

Special Tool Sets

Vehicle manufacturers and specialty tool companies work closely together to design and manufacture special tools required to repair transmissions. Most of these special tools are listed in the appropriate service information and are part of each manufacturer's "essential tool kit."

Clutch Alignment Tool

To keep the clutch disc centred on the flywheel while assembling the clutch, a clutch alignment tool is used. The tool is inserted through the input shaft opening of the pressure plate and is passed through the clutch disc. The tool then is inserted into the pilot bushing or bearing. The outside diameter (OD) of the alignment tool that goes into the pilot must be only slightly smaller than the inside diameter (ID) of the pilot bushing. The OD of the tool that holds the disc in

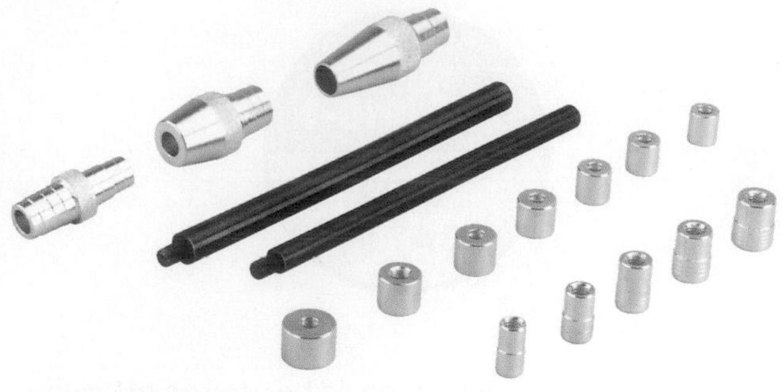

FIGURE 6–41

A clutch alignment tool set with various sizes of pilots, adapters, and alignment cones.

SPX-OTC Automotive Electronic Diagnostic Tools

place must likewise be only slightly smaller than the ID of the disc's splined bore. The effectiveness of this tool depends on its diameter, so it is best to have various sizes of clutch alignment tools **(Figure 6–41)**.

Clutch Pilot Bearing/Bushing Puller/Installer

To remove and install a clutch pilot bearing or bushing, special tools are needed. These tools not only make the job easier but also prevent damage to the bore in the flywheel.

Universal Joint Tools

Although servicing universal joints can be done with hand tools and a vise, many technicians prefer the use of specifically designed tools. One such tool is a C-clamp modified to include a bore that allows the joint's caps to slide in while tightening the clamp over an assembled joint to remove it **(Figure 6–42)**. Other tools are the various drivers used with a press to press the joint in and out of its yoke.

Driveshaft Angle Gauge

Critical to the durability of universal joints and vibration-free vehicle operation is the angle of the driveshaft. The angle of the driveshaft at the

FIGURE 6–42

A universal joint bearing press with adapters.

Snap-on Tools Company

FIGURE 6–43

An angle gauge used to check the angle of the driveshaft.

Snap-on Tools Company

transmission should equal its angle at the drive axle. There are many ways to measure the angle; one way involves the use of an **inclinometer**, or driveshaft angle gauge **(Figure 6–43)**.

Hydraulic Pressure Gauge Set

A common diagnostic tool for automatic transmissions is a hydraulic pressure gauge. A pressure gauge measures pressure in kilopascals (kPa) and/or pounds per square inch (psi). The gauge is normally part of a kit that contains various fittings and adapters.

SUSPENSION AND STEERING TOOLS

Common suspension and steering repair and diagnostic tools as well as wheel alignment tools and equipment are discussed in the following paragraphs. Details of when and how to use these tools are covered in Section 7.

Tire Tread Depth Gauge

A tire tread depth gauge measures the depth of a tire's tread to determine the wear of the tire. This measurement should be taken at three or four locations around the tire's circumference to obtain an average tread depth. This gauge is used to determine the remaining life of a tire as well as for comparing wear of one tire to the other tires. It is also used when making tire warranty adjustments.

Tire Pressure Monitoring Sensor (TPMS) Tester

These tools are required for checking and resetting tire pressure monitoring sensor (TPMS) warning lights on vehicles after an under-inflated tire has been repaired or replaced. The tool **(Figure 6–44)**

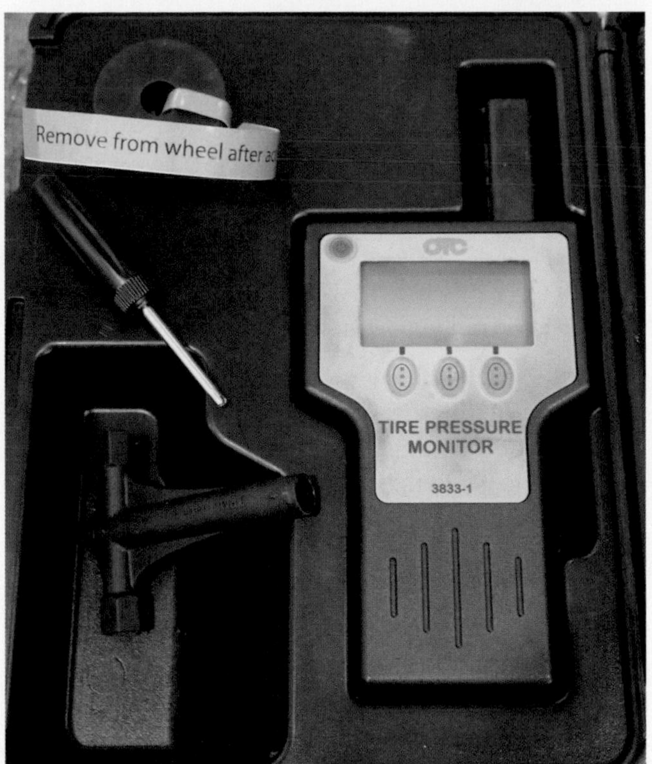

FIGURE 6–44

A tire pressure monitoring system tester.

©Cengage Learning 2015

is also used to program a replacement TPMS. It does this using the relearn option, if available, in the vehicle's engine management system.

Power-Steering Pressure Gauge

A power-steering pressure gauge is used to check the power-steering pump. The pressure gauge, with a shutoff valve, is installed between the pump and the steering gear. This tester is also used when checking hydraulic boost brake systems. Because the power-steering pump delivers extremely high pressure during this test, the manufacturer's recommended procedure must be followed.

Control Arm Bushing Tools

A variety of tools are available to remove and replace control arm bushings. Old bushings are pressed out of the control arm. A C-clamp tool can be used to remove the bushing. The C-clamp is installed over the bushing. An adapter is selected to fit on the bushing and push the bushing through the control arm. Turning the handle on the C-clamp pushes the bushing out of the control arm.

New bushings can be installed by driving or pressing them in place. Adapters are available for the C-clamp tool to press the new bushings into

FIGURE 6–45

A ball joint/tie rod removal tool.

Snap-on Tools Company

their bore. After the correct adapters are selected, position the bushing and tool on the control arm. Turning the C-clamp handle pushes the bushing into the control arm.

Tie-Rod End and Ball Joint Puller

A tie-rod end and ball joint puller can be used to remove tie-rod ends and pull ball joint studs from the steering knuckle. A tie-rod end remover is a safer and easier way of separating ball joints than a pickle fork **(Figure 6–45)**.

Ball joint removal and installation tools are designed to remove and replace pressed-in ball joints in steering knuckles and control arms on front and rear suspension systems. Often these tools are used in conjunction with a hydraulic press. The size of the removal and pressing tool must match the size of the ball joint.

Front Bearing Hub Tool

Front bearing hub tools are designed to remove and install front-wheel bearings on FWD cars. These bearing hub tools are usually designed for a specific make of vehicle, and the correct tools must be used for each application. Failure to do so may result in damage to the steering knuckle or hub. Also, the use of the wrong tool will waste quite a bit of your time.

Pitman Arm Puller

A pitman arm puller is a heavy-duty puller designed to remove the pitman arm from the pitman shaft **(Figure 6–46)**. These pullers can also be used to separate tie-rod ends and ball joints.

Tie-Rod Sleeve-Adjusting Tool

A tie-rod sleeve-adjusting tool **(Figure 6–47)** is required to rotate the tie-rod sleeves to make wheel alignment adjustments. Never use anything except a tie-rod adjusting tool to adjust the tie-rod sleeves. Tools such as pipe wrenches will damage the sleeves.

FIGURE 6–46

A pitman arm puller is designed to remove the pitman arm from the pitman shaft.

SPX-OTC Automotive Electronic Diagnostic Tools

Steering Column Special Tool Set

A wheel puller is used to remove the steering wheel from its shaft. Mount the puller over the wheel's hub after the horn button and air bag have been removed. Install the bolts into the threaded bores in the steering wheel. Then tighten the puller's centre bolt against the steering wheel shaft until the steering wheel is free.

Special tools are also required to service the lock mechanism and ignition switch.

Shock Absorber Tools

Often, shock absorbers can be removed with regular hand tools, but there are times when special tools may be necessary. The shocks are under the vehicle and are subject to dirt and moisture, which may make it difficult to loosen the mounting nut from the stud of the shock. Wrenches are available to hold the stud while attempting to loosen the nut. There are also tools for pneumatic chisels that help to work off the nut.

FIGURE 6–47

A tie-rod sleeve-adjusting tool.

SPX-OTC Automotive Electronic Diagnostic Tools

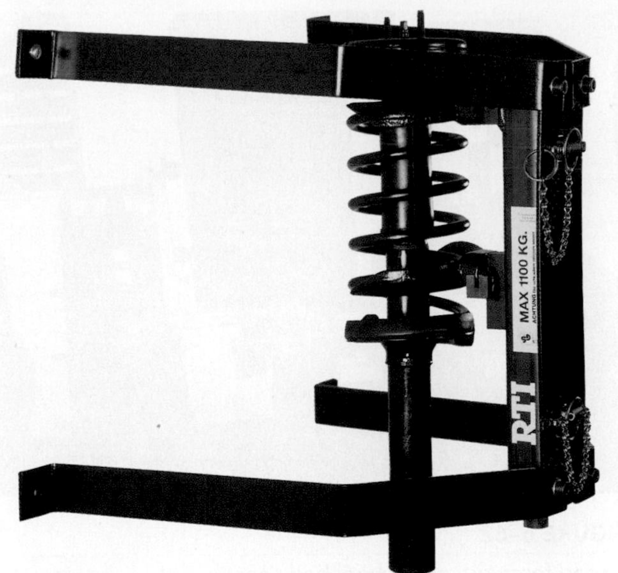

FIGURE 6–48

A cradle-type coil spring compressor.

RTI Technologies, Inc.

Spring/Strut Compressor Tool

Many types of coil spring compressor tools are available to the automotive service industry. These tools are designed to compress the coil spring and hold it in the compressed position while removing the strut from the coil spring **(Figure 6–48)**, removing the spring from a short-long arm (SLA) suspension, or performing other suspension work. Various types of spring compressor tools are required on different types of front suspension systems.

One type of spring compressor uses a threaded rod that fits through two plates, an upper and lower ball nut, a thrust washer, and a forcing nut. The two plates are positioned at either end of the spring. The rod fits through the plates with a ball nut at either end. The upper ball nut is pinned to the rod. The thrust washer and forcing nut are threaded onto the end of the rod. Turning the forcing nut draws the two plates together and compresses the spring.

Power Steering Pump Pulley Special Tool Set

When a power steering pump pulley must be replaced, it should never be hammered off or on. Doing so will cause internal damage to the pump. Normally, the pulley can be removed with a gear puller, although special pullers are available **(Figure 6–49)**. To install a pulley, a special tool is used to press it in place.

FIGURE 6–49

A power steering pump pulley service kit.

SPX-OTC Automotive Electronic Diagnostic Tools

Brake Pedal Depressor

A brake pedal depressor must be installed between the front seat and the brake pedal to apply the brakes while checking some front-wheel alignment angles to prevent the vehicle from moving.

Wheel Alignment Equipment—Four Wheel

Many automotive shops are equipped with a computerized four-wheel alignment machine **(Figure 6–50)** that can check all front- and rear-wheel alignment angles quickly and accurately.

FIGURE 6–50

A computerized four-wheel alignment setup.

RTI Technologies, Inc.

Once the machine is set up and the vehicle information is keyed into the machine, the alignment measurements are instantly displayed. Also displayed are the specifications for that vehicle. In addition to the normal alignment specifications, the screen may display asymmetric tolerances, different left- and right-side specifications, and cross-specifications. (A difference is allowed between left and right sides.) Graphics and text on the screen show the technician where and how to make adjustments. As the adjustments are made, the technician can observe the changes.

Tire Changer

Tire changers are used to demount and mount tires. A wide variety of tire changers are available, and each one has somewhat different operating procedures. Always follow the procedure in the equipment operator's manual and the directions provided by your instructor.

Wheel Balancer—Electronic Type

The most commonly used wheel balancer requires that the tire/wheel assembly be mounted on the balancer's spindle. Weights are added to balance the tire/wheel assembly. The wheel assembly is rotated at high or low speed, depending on the machine, and the machine indicates the amount of weight to be added and the location where the weights should be placed.

Several electronic dynamic/static balancer units are available that permit balancing while the wheel and tire are on the vehicle. Often, a strobe light flashes at the heavy point of the tire and wheel assembly.

Wheel Weight Pliers

Wheel weight pliers are combination tools designed to install and remove clip-on lead wheel weights **(Figure 6–51)**. The jaws of the pliers are designed to hook into a hole in the weight's bracket. The pliers are then moved toward the outside of the wheel, and the weight is pried off. On one side of the pliers is a plastic hammer head used to tap the weights onto the rim.

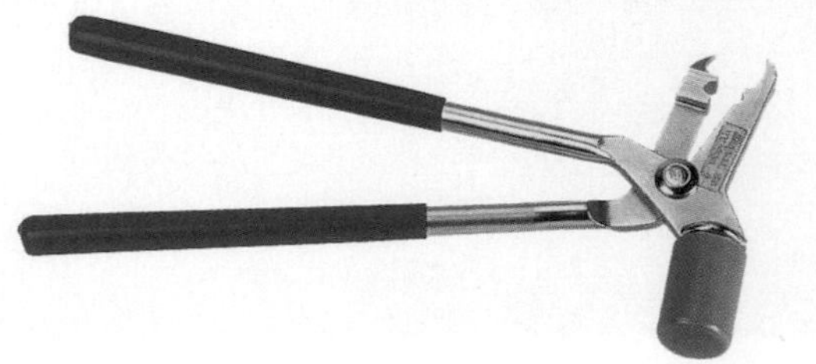

FIGURE 6–51

Wheel weight pliers.

Snap-on Tools Company

FIGURE 6–52

A road force tire and wheel balancer.

Road Force Balancer

A fairly new tire and wheel balancing machine is a road force balancer **(Figure 6–52)**. The primary purpose of the machine is to see if a wheel and tire assembly is round while it is rolling down the road. To do this, the wheel/tire assembly is mounted to the machine and a roller is pressed against the tire as the shaft slowly rotates. The roller presses with a constant force. If the tire is round, the roller will simply roll with the tire. If the tire is not round, the roller will move back and forth to measure the loaded runout of the tire.

Once the loaded runout is determined, the machine calculates how to minimize the loaded runout of the assembly. This normally requires the dismounting and remounting of the tire onto the wheel, but in a different location. After road force balance is rechecked, the assembly is rebalanced conventionally with weights.

BRAKE SYSTEM TOOLS

Common repair and diagnostic tools for brake service are discussed in the following paragraphs. Additional details of when and how to use these tools are presented in Section 8.

Cleaning Equipment and Containment Systems

The following systems and methods are used to safely contain harmful brake dust that may contain asbestos in the workplace; one or more of these should be used whenever you are doing brake work.

With negative-pressure enclosure and HEPA vacuum systems, cleaning and inspecting brake assemblies is performed inside a tightly sealed protective enclosure that covers and contains the entire brake assembly. The enclosure prevents the release of asbestos fibres into the air and is designed so that you can clearly see the work in progress. Low-pressure wet cleaning systems wash dirt from the brake assembly and catch the contaminated cleaning agent in a basin. The reservoir contains water with an organic solvent or wetting agent. To prevent any asbestos-containing brake dust from becoming airborne, the flow of liquid should be controlled so that the brake assembly is gently flooded.

Hold-Down Spring and Return Spring Tools

Drum brake shoe return springs are very strong and require special tools for removal and installation. Most return spring tools have special sockets and hooks to release and install the spring ends. Some are built like pliers **(Figure 6–53)**.

Brake shoe hold-down springs are much lighter than return springs, and many such springs can be released and installed by hand. A hold-down spring tool **(Figure 6–54)** looks like a cross between a screwdriver and a nut driver. A specially shaped end grips and rotates the spring-retaining washer.

Boot Drivers, Rings, and Pliers

Dust boots attach between the caliper bodies and pistons of disc brakes to keep dirt and moisture out of the caliper bores. A special driver is used to install a dust boot with a metal ring that fits tightly on the caliper body. The circular driver is centred on the boot placed against the caliper and then hit with a hammer to drive the boot into place. Other kinds of dust boots fit into a groove in the caliper bore before the piston is installed. Special rings or pliers are then needed to expand the opening in the dust boot and let the piston slide through it for installation.

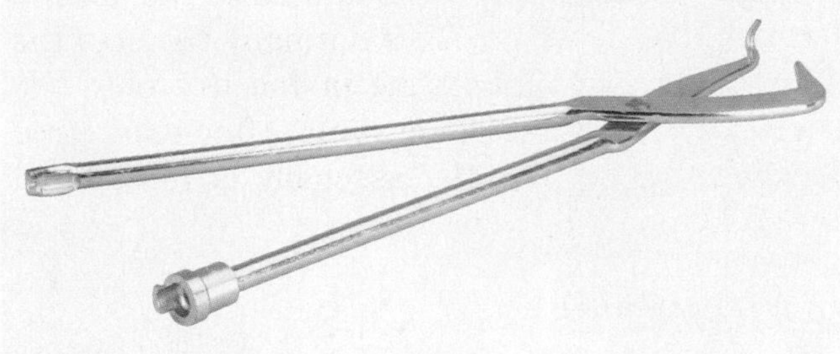

FIGURE 6–53

Brake spring pliers.

SPX-OTC Automotive Electronic Diagnostic Tools

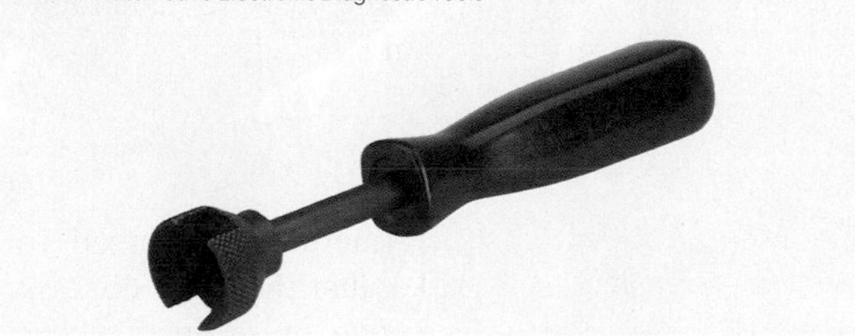

FIGURE 6–54

A hold-down spring compressor tool.

SPX-OTC Automotive Electronic Diagnostic Tools

Caliper Piston Removal Tools

A caliper piston can usually be slid or twisted out of its bore by hand. Rust and corrosion (especially where road salt is used in the winter) can make piston removal difficult. One simple tool that will help with the job is a set of special pliers that grips the inside of the piston and allows you to move it by hand with more force. These pliers work well on pistons that are only mildly stuck.

For a severely stuck caliper piston, a hydraulic piston remover can be used. This tool requires that the caliper be removed from the car and installed in a holding fixture. A hydraulic line is connected to the caliper inlet, and a hand-operated pump is used to apply up to 6895 kPa (1000 psi) of pressure to loosen the piston. Because of the danger of spraying brake fluid, always wear eye protection when using this equipment.

Drum Brake Adjusting Tools

Although drum brakes built during the past 30 years have some kind of self-adjuster, brake shoes still require an initial adjustment after they are installed. The star wheel adjusters of many drum brakes can be adjusted with a flat-blade screwdriver. Brake adjusting spoons **(Figure 6–55)** and wire hooks designed for this specific purpose can make the job faster and easier.

Brake Cylinder Hones

Cylinder hones are used to clean light rust, corrosion, pits, and built-up residue from the bores of master cylinders, wheel cylinders, and calipers. A hone can be a very useful—sometimes necessary—tool when you need to overhaul a cylinder. A hone will not, however, save a cylinder with severe rust or corrosion.

The most common cylinder hones have two or three replaceable abrasive stones at the ends of

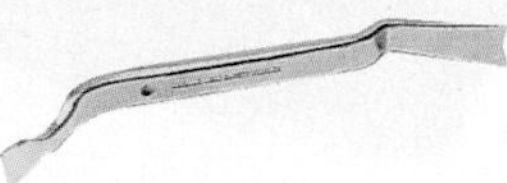

FIGURE 6–55

A drum brake adjustment tool.

Snap-on Tools Company

spring-loaded arms. Spring tension is usually adjustable to maintain proper stone apply force against the cylinder walls. The other end of the hone is mounted in a drill motor.

Another kind of hone is the **brush hone**, or **ball hone**. It has abrasive balls attached to flexible metal brushes that are, in turn, mounted on the hone's flexible shaft. In use, centrifugal force moves the abrasive balls outward against the cylinder walls. A brush hone provides a superior surface finish and is less likely to remove too much metal than a stone hone.

Tubing Tools

The rigid brake lines, or pipes, of the hydraulic system are made of steel tubing to withstand high pressure and to resist damage from vibration, corrosion, and work hardening. Brake lines often can be purchased in preformed lengths to fit specific locations on specific vehicles. Straight brake lines can also be purchased in many lengths and several diameters and bent to fit specific vehicle locations. The tools **(Figure 6–56)** you should have are as follows:

- A tubing cutter and reamer
- Tube benders
- A double flaring tool for SAE flares
- An International Standards Organization (ISO) flaring tool for European-style ISO flares

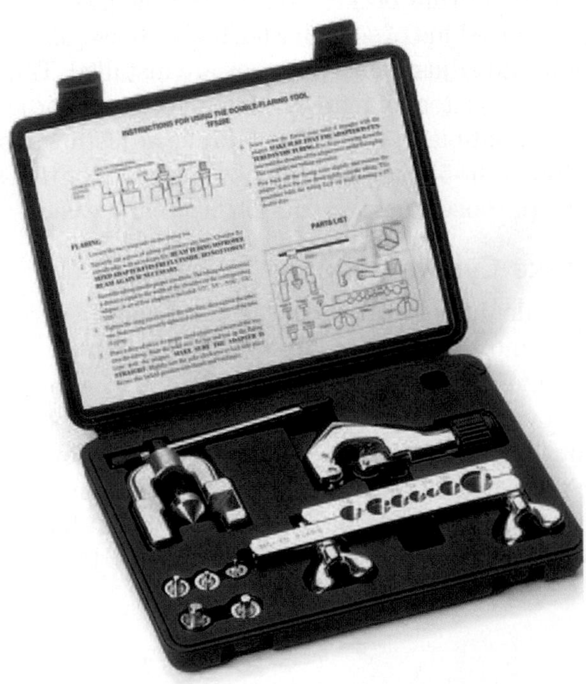

FIGURE 6–56

A typical tubing tool set.

Snap-on Tools Company

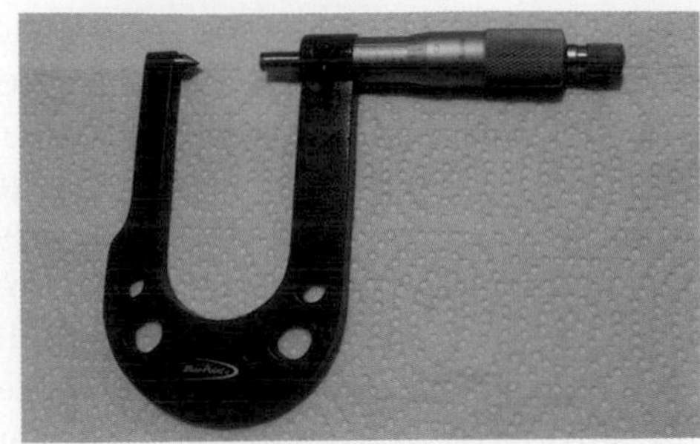

FIGURE 6–57

A brake disc micrometer.

Snap-on Tools Company

Brake Disc Micrometer

A special micrometer is used to accurately check the thickness of a brake rotor. A brake disc micrometer **(Figure 6–57)** has pointed anvils that allow the tip to fit into grooves worn on the rotor. This type of micrometer is read in the same way as other micrometers but is made with a range from 7 to 35 mm (0.300 to 1.300 in.). Digital calipers are also used to measure disc brake thickness **(Figure 6–58)**.

Drum Micrometer

A drum micrometer is a single-purpose instrument used to measure the inside diameter of a brake drum. A drum micrometer has two movable arms on a shaft. One arm has a precision dial indicator; the other arm has an outside anvil that fits against the inside of the drum. In use, the arms are secured on

FIGURE 6–58

A digital caliper for measuring brake disc thickness.

Bendix Brakes by Allied Signal

the shaft by lock screws that fit into grooves every 2 mm on the shaft. The shaft is graduated in 1 cm major increments, and the dial indicator is graduated in 0.02 mm increments. The imperial drum micrometer shaft has 1-inch major graduations, and the lock screws fit in notches every 1/8 in. (0.125). The dial indicator is graduated in 0.005-inch increments.

Brake Shoe Adjusting Gauge (Calipers)

A brake shoe adjusting gauge is an inside-outside measuring device **(Figure 6–59)**. This gauge is often called a brake shoe caliper. During drum brake service, the inside part of the gauge is placed inside a new or resurfaced drum and expanded to fit the drum diameter. The lock screw is then tightened and the gauge moved to the brake shoes installed on the backing plate. The brake shoes are then adjusted until the outside part of the gauge just slips over them. This action provides a rough adjustment of the brake shoes. Final adjustment must still be done after the drum is installed, but the brake shoe gauge makes the job faster.

Brake Lathes

Brake lathes are special power tools used only for brake service. They are used to resurface brake rotors and drums. This involves cutting away very small amounts of metal to restore the surface of the rotor or drum. The traditional brake lathe is an assembly mounted on a stand or workbench. The bench lathe requires that the drum or rotor be removed from the vehicle and mounted on the lathe for service.

As the drum or rotor is turned on the lathe spindle, a carbide steel-cutting bit is passed over the drum or rotor friction surface to remove a small amount of metal. The cutting bit is mounted rigidly on a lathe fixture for precise control as it passes across the friction surface.

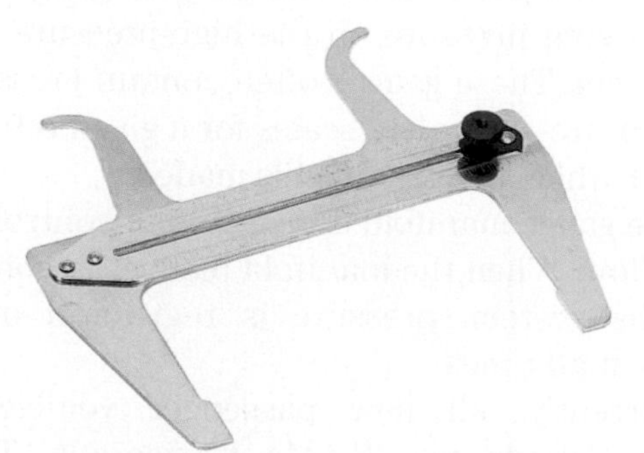

FIGURE 6–59

A drum brake shoe adjusting gauge.

Snap-on Tools Company

FIGURE 6–60

An on-vehicle disc brake lathe.

RTI Technologies, Inc.

An on-car lathe **(Figure 6–60)** is bolted to the vehicle suspension or mounted on a rigid stand to provide a stable mounting point for the cutting tool. The rotor may be turned by the vehicle's engine or drivetrain (for a FWD vehicle) or by an electric motor and drive attachment on the lathe. As the rotor is turned, the lathe cutting tool is moved across both surfaces of the rotor to refinish it. An on-car lathe not only has the obvious advantage of speed, but it also rotates the rotor on the vehicle wheel bearings and hub so that these sources of runout, or wobble, are compensated for during the refinishing operation.

Bleeder Screw Wrenches

Special bleeder screw wrenches are often used to open bleeder screws. Bleeder screw wrenches are small, 6-point box wrenches with strangely offset handles for access to bleeder screws in awkward locations. The 6-point box end grips the screw more securely than a 12-point box wrench can and avoids damage to the screw.

Pressure Bleeders

Removing the air from the closed hydraulic brake system is very important. This removal is done by bleeding the system. Bleeding can be done manually with a vacuum pump or a pressure bleeder **(Figure 6–61)**. The latter is preferred because it is quick and very efficient.

Pressure bleeding is a fast and efficient way to bleed a brake system for two reasons. First, the

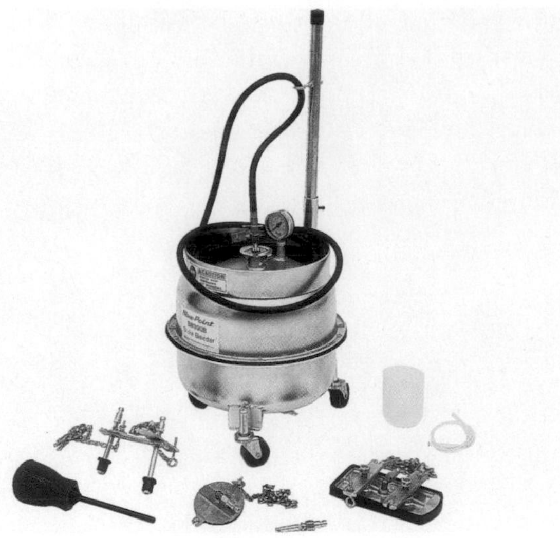

FIGURE 6–61

A diaphragm-type pressure bleeder.

Snap-on Tools Company

master cylinder does not have to be refilled several times, and second, the job can be done by one person. Pressurized brake fluid flows into the master cylinder and out through the brake lines, quickly forcing air out of the lines.

Other tools used in brake bleeding operations include a large rubber syringe, used to remove fluid from the master cylinder on some systems; master cylinder bleeder tubes, used to return fluid to the master cylinder reservoir from the outlet ports during bench bleeding; and assorted line and port plugs, used to close lines and valves temporarily during service and keep out dirt and moisture.

HEATING AND AIR-CONDITIONING TOOLS

Common repair and diagnostic tools for the heating, ventilation, and air-conditioning systems are discussed in the following paragraphs. Details of when and how to use these tools are covered in Section 9.

Manifold Gauge Set

A **manifold gauge set (Figure 6–62)** is used when recovering, charging, evacuating, and for diagnosing trouble in an air-conditioning (A/C) system. With the legislation on handling refrigerants, all gauge sets are required to have a valve device to close off the end of the hose so that the fitting not in use is automatically shut.

Depending on the gauge set manufacturer, the low-pressure gauge can be graduated into kilopascals,

FIGURE 6–62

An R-134a manifold gauge set.

Robinair, SPX Corporation

from 0 to 825 (with a cushion to 1725) in 0.1 kPa increments; **bars**, from 0 to 8.25 (with a cushion to 17.25) in 0.1-bar increments; or in pounds of pressure, from 1 to 120 (with a cushion to 250) in 1-pound graduations. In the opposite direction (vacuum range), the gauge can be graduated in centimetres of mercury (cm Hg) from 0 to 76, millimetres of mercury (mm Hg) from 0 to 760, kilopascals from 0 to –100, bar from 0 to –1, and inches of mercury (in Hg) from 0 to 30. This is the gauge that should always be used in checking pressure on the low-pressure side of the system. The high-pressure gauge can be graduated in kilopascals from 0 to 3500 in 50 kPa graduations, bars from 0 to 35 in 0.5 bar graduations, or in pounds of pressure from 0 to 500 in 5-pound graduations. This gauge is used for checking pressure on the high-pressure side of the system. These gauges often contain pressure- to temperature-equivalent scales for a given refrigerant type for which the gauge set is made.

The gauge manifold is designed to control refrigerant flow. When the manifold test set is connected into the system, pressure is registered on both gauges at all times.

Currently, all new passenger vehicles sold in the Canada use R-134a refrigerant. The US Environmental Protection Agency (EPA) recently approved HFO-1234yf (hydrofluoro-olefin) refrigerant for use in mobile air-conditioning systems (also adopted into Canadian regulations). Because 1234yf, R-134a, and R-12 are not interchangeable, separate sets of hoses, gauges, and other equipment are required to service vehicles. All equipment used to service R-134a and R-12 systems must meet SAE standard J1991. The service hoses on the manifold gauge set must have manual or automatic backflow valves at the service port connector ends to prevent the refrigerant from being released into the atmosphere during connection and disconnection. Manifold gauge sets for R-134a can be identified by labels on the gauges and/or have a light blue colour on the face of the gauges.

R-134a service hoses have a black stripe along their length and are clearly labelled. The low-pressure hose is blue with a black stripe; the high-pressure hose is red with a black stripe; and the centre service hose is yellow with a black stripe.

Electronic Leak Detector

An electronic leak detector **(Figure 6–63)** is safe and effective and can be used with all types of refrigerants. A hand held, battery-operated electronic leak detector has a test probe that is moved close to the areas of suspected leaks. Because refrigerant is heavier than air, the probe should be positioned below the test point. An alarm or a buzzer on the detector indicates the presence of a leak. On some models, a light flashes when refrigerant is detected.

Fluorescent Leak Tracer

To find a refrigerant leak using the fluorescent tracer system, a fluorescent dye is introduced into the air-conditioning system with a special infuser included with the detector equipment. Run the air conditioner for a few minutes, giving the tracer dye fluid time to circulate and penetrate. Wear protective goggles and scan the system with a black-light glow gun. Leaks in the system will shine under the black light as a luminous yellow-green.

Refrigerant Identifier

A refrigerant identifier **(Figure 6–64)** is used to identify the type of refrigerant present in a system. This test should be done before any service work. The tester is used to identify the purity and quality of the refrigerant sample taken from the system.

Refrigerant Charging Station

A charging station removes, recycles, evacuates, and recharges an air-conditioning system. Other equipment may be used to remove and recycle a system's refrigerant **(Figure 6–65)**. These machines draw the old refrigerant out of the system, filter it, separate the oil, remove moisture and air, and store the refrigerant for future use.

All recycled refrigerant must be safely stored in Canadian Transport Commission (CTC) specified or CSA-approved containers. Containers specifically made for R-134a should be so marked. Before any container of recycled refrigerant can be used, it must be checked for noncondensable gases.

FIGURE 6–63

An electronic leak detector.

Robinair, SPX Corporation

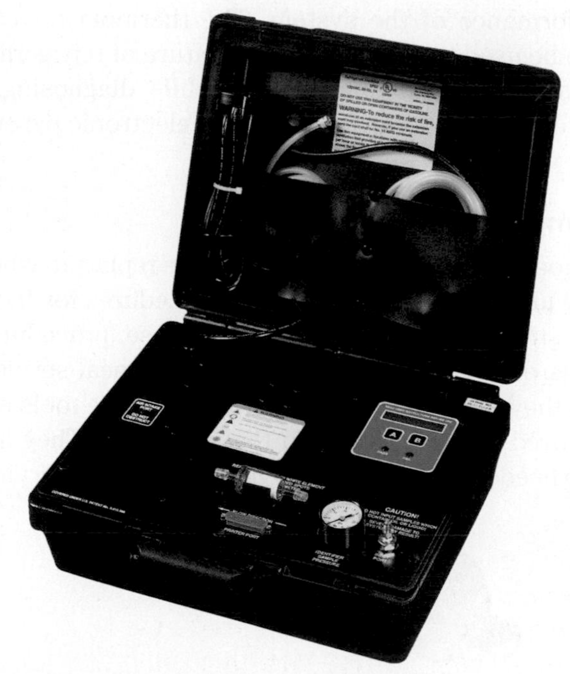

FIGURE 6–64

A refrigerant identifier.

RTI Technologies, Inc.

FIGURE 6–65

A single-pass refrigerant recovery and recycling machine.

RTI Technologies, Inc.

Thermometer

A digital readout or dial-type **thermometer (Figure 6–66)** is used to measure the air temperature at the vent outlets, which indicates the overall performance of the system. The thermometer can also be used to check the temperature of refrigerant lines, hoses, and components while diagnosing a system. While doing the latter, an electronic pyrometer works best.

Compressor Tools

Although compressors are usually replaced when they are faulty, certain service procedures for them are standard practice. Most of these procedures focus on compressor clutch and shaft seal service, and they require special tools. Clutch plate tools are required to gain access to the shaft seal. They are also needed to reinstall the clutch plate after service.

FIGURE 6–66

A dial-type thermometer used to check A/C operation.

Robinair, SPX Corporation

FIGURE 6–67

An A/C clutch spanner wrench.

SPX-OTC Automotive Electronic Diagnostic Tools

Typically to replace a shaft seal, you will need an adjustable or fixed spanner wrench **(Figure 6–67)**, clutch plate installer/remover, ceramic seal installer/remover, seal assembly installer/remover, seal seat installer/remover, shaft seal protector, snap ring pliers, O-ring remover, and O-ring installer. Some of these tools are for a specific model compressor; others are universal fit or have interchangeable parts to allow them to work on a variety of compressors.

Hose and Fitting Tools

An A/C system is a closed system, meaning outside air should never enter the system, and the refrigerant in the system should never exit to the outside. To maintain this closed system, special fittings and hoses are used. Often, special tools, such as the spring-lock coupling tool set, are required when servicing the system's fittings and hoses. Without this tool, it is impossible to separate the connector and not damage it.

KEY TERMS

Ammeter (p. 126)
Ball hone (p. 144)
Bars (p. 146)
Boring bar (p. 121)
Brush hone (p. 144)
Compression gauge (p. 117)
Connector picks (p. 128)
Continuity tester (p. 125)
Cylinder hone (p. 121)
Cylinder leakage tester (p. 118)
Dial bore gauge (p. 120)
Digital multimeter (DMM) (p. 125)
Dual-trace (p. 128)
Dynamometer (p. 136)
Fuel pressure gauge (p. 130)
Glaze breakers (p. 121)
Hydrometer (p. 123)
Inclinometer (p. 139)
Insulation resistance tester (p. 137)
Kilopascals (kPa) (p. 117)
Lab scope (p. 127)
Lineman's gloves (p. 136)
Logic probe (p. 134)
Manifold gauge set (p. 146)
Multimeter (p. 125)
Noid light (p. 131)
Ohmmeter (p. 126)
Pounds per square inch (psi) (p. 117)

Pyrometer (p. 135)	Torque angle gauge (p. 122)
Refractometer (p. 124)	Trace (p. 127)
Ridge reamer (p. 119)	Transient voltages (p. 137)
Ring compressor (p. 119)	Vacuum (p. 132)
Ring expander (p. 119)	Valve spring compressors (p. 122)
Ring groove cleaner (p. 120)	Valve spring tester (p. 122)
Scan tool (p. 129)	Volt/ampere tester (VAT) (p. 127)
Spark tester (p. 134)	Voltmeter (p. 126)
Stethoscope (p. 118)	Waveform (p. 127)
Tachometer (p. 134)	
Test light (p. 125)	
Thermometer (p. 148)	

SUMMARY

- Common diagnostic tools used to check an engine and its related systems include compression gauge, cylinder leakage tester, oil pressure gauge, stethoscope, dial bore indicator, valve spring tester, cooling system pressure tester, coolant hydrometer, engine analyzer, fuel pressure gauge, injector balance tester, injector circuit test light, vacuum gauge, vacuum pump, vacuum leak detector, spark tester, logic probe, pyrometer, exhaust analyzer, and chassis dynamometer.
- Common tools used to service an engine and its related systems include transaxle removal and installation equipment, ridge reamer, ring compressor, ring expander, ring groove cleaner, cylinder deglazer, cylinder hone, boring bar, cam bearing driver set, V-blocks, valve and valve seat resurfacing equipment, valve guide repair tools, valve spring compressor, torque angle gauge, oil priming tool, a coolant recovery and recycle system, fuel injector cleaners, fuel line tools, pinch-off pliers, timing light, and spark plug sockets.
- Some of the common diagnostic tools for electronic and electrical systems include test light, continuity tester, voltmeter, ohmmeter, ammeter, volt/ampere tester, digital multimeter (DMM), lab scope, scan tool, graphing multimeter, and battery hydrometer.
- Common electrical and electronic system service tools include computer memory saver, wire and terminal repair tools, headlight aimers, static straps, and sensor tools.
- Diagnostic tools for a vehicle's drivetrain include a driveshaft angle gauge and hydraulic pressure gauge set.
- Tools required to service the drivetrain include transaxle removal and installation equipment, transmission/transaxle holding fixtures, transmission jack, axle pullers, special tool sets, clutch alignment tool, clutch pilot bearing/bushing puller/installer, and universal joint tools.
- The various diagnostic tools used on a vehicle's running gear include tire tread depth gauge, power steering pressure gauge, wheel alignment equipment, brake disc micrometer, and drum micrometer.
- Some of the common tools used to service a vehicle's running gear include control arm bushing tools, tie-rod end and ball joint pullers, front bearing hub tool, pitman arm puller, tie-rod sleeve adjusting tool, steering column special tool set, shock absorber tools, spring/strut compressor tool, power steering pump pulley special tool set, brake pedal depressor, tire changer, wheel balancer, wheel weight pliers, brake cleaning equipment and containment systems, hold-down spring and return spring tools, boot drivers and pliers, caliper piston removal tools, drum brake adjusting tools, brake cylinder hones, tubing tools, brake shoe adjusting gauge, brake lathes, bleeder screw wrenches, and pressure bleeders.
- Common tools used to check a vehicle's heating and air-conditioning system include manifold gauge set, service port adapter set, electronic leak detector, fluorescent leak tracer, and thermometer.
- Tools used to service air-conditioning systems include refrigerant identifier, refrigerant charging station, refrigerant recovery and recycling system, compressor tools, and hose and fitting tools.

REVIEW QUESTIONS

1. What are the two types of test lights and how do they differ?
2. How is knurling used to repair worn valve guides?
3. Name the two basic types of compression gauges.
4. What tool is used to test engine manifold vacuum?
5. Why is a lab scope used when diagnosing a circuit?
6. Which of the following brake shoe adjusting gauge statements is true?
   - **a.** an inside-outside precision measuring tool graduated in millimetres
   - **b.** provides a quick, rough adjustment of the brake shoes
   - **c.** an inside-outside precision measuring tool graduated in 0.001 in.
   - **d.** resembles an inside micrometer and is read like a standard micrometer

7. Which of the following circuits cannot be tested with a conventional 12-volt test light?
   a. headlight circuit
   b. charging system circuit
   c. engine control module circuit
   d. starting system circuit
8. Which of the following tools measures resistance to current flow in a circuit?
   a. an ohmmeter **b.** a voltmeter
   c. an ammeter **d.** a test light
9. What vacuum gauge reading should a properly operating engine produce?
   a. 200 mm Hg (7.9 in. Hg)
   b. 400 mm Hg (15.7 in. Hg)
   c. 600 mm Hg (23.6 in. Hg)
   d. 800 mm Hg (31.5 in. Hg)
10. When using an oscilloscope, what would an upward movement of the trace indicate?
   a. an increase in resistance
   b. a decrease in resistance
   c. an increase in voltage
   d. a decrease in voltage
11. What is the first step in performing a four-wheel alignment after the wheel units are installed?
   a. Perform wheel runout compensation.
   b. Read camber angles.
   c. Read caster angles.
   d. Adjust front-wheel toe.
12. Which of the following statements about manifold gauge sets is true?
   a. An adapter is required for using R-12 gauges on an R-134a system.
   b. A manifold gauge set is used when charging and evacuating, and for diagnosing trouble in an A/C system.
   c. The gauge manifold is designed to control refrigerant flow. When the manifold test set is connected into the system, pressure is registered on both gauges when the valves are opened.
   d. R-134a service hoses have a black stripe along their length; the low-pressure hose is blue, and the high-pressure hose is yellow.
13. What is of major concern when removing a steering wheel?
   a. damaging the horn controls
   b. damaging the end of the steering wheel shaft
   c. stripping the bolt holes of the steering wheel
   d. deploying the air bag
14. What can be determined when using a cylinder leakage tester?
   a. an oil pressure leak
   b. cylinder compression pressure leakage in kPa or psi
   c. where a leak from the cylinder is located
   d. cylinder combustion pressure leakage in kPa or psi
15. Which of the following conditions can be revealed by fuel pressure readings?
   a. faulty fuel pump
   b. restricted fuel return line
   c. restricted fuel delivery system
   d. all of the above
16. What does a compression tester read?
   a. cylinder pressure during the intake stroke
   b. cylinder pressure during the compression stroke
   c. cylinder pressure during the power stroke
   d. cylinder pressure during the exhaust stroke
17. When should a cylinder be ridge reamed?
   a. before the pistons are removed
   b. after the pistons have been removed
   c. after the cylinder has been bored
   d. after the cylinder has been deglazed
18. Which vehicle usually has the engine removed by lowering it from the engine compartment?
   a. rear-wheel-drive cars
   b. pickup trucks
   c. front-wheel-drive cars
   d. sport utility vehicles
19. When should the torque angle gauge be used?
   a. after the bolt is tightened to the specified torque
   b. after the preliminary torque step and before the specified torque is reached
   c. before any torque steps are performed
   d. to check the bolt torque after the engine has operated
20. Which of the following multimeter statements is correct?
   a. Digital multimeters have high impedance and should not be used on sensitive electronic circuits.
   b. Digital multimeters have low impedance and should not be used on sensitive electronic circuits.
   c. Analog multimeters have high impedance and should not be used on sensitive electronic circuits.
   d. Analog multimeters have low impedance and should not be used on sensitive electronic circuits.

CHAPTER 7
Basic Theories and Math

LEARNING OUTCOMES

- Describe how all matter exists.
- Explain what energy is and how energy is converted.
- Calculate the volume of a cylinder.
- Explain the forces that influence the design and operation of an automobile.
- Describe and apply Newton's laws of motion to an automobile.
- Define friction and describe how it can be minimized.
- Describe the various types of simple machines.
- Differentiate between torque and horsepower.
- Interpret the difference between a vibration and a sound.
- Define Pascal's law and give examples of how it applies to an automobile.
- Explain the behaviour of gases.
- Describe the effects of heat on matter.
- Describe what is meant by the chemical properties of a substance.
- Explain the difference between oxidation and reduction.
- Describe the origin and practical applications of electromagnetism.

This chapter contains many of the subjects you have learned or will learn in other courses. This material is not intended to take the place of those other courses; rather, it serves to emphasize the knowledge you will need to gain employment and be successful in an automotive career. Many of the facts presented in this chapter are addressed later in greater detail according to the topic. Make sure you understand the contents of this chapter. If you have difficulty answering the chapter review questions, study the appropriate content in the chapter until you clearly understand and are able to answer the questions correctly.

MATTER

Matter is anything that occupies space. All matter exists as a gas, liquid, or solid. Gases and liquids tend to flow and are influenced by and respond to pressure. A gas has neither shape nor volume of its own and tends to expand without limits. A liquid takes a shape and has volume. Both liquids and gases will tend to take the shape of the container or vessel that they are stored in. (As an example, think of the shape of a balloon that is inflated.) A solid is matter that does not flow.

Atoms and Molecules

All matter consists of countless tiny particles called **atoms** and **molecules**. An atom may be defined as the smallest particle of an element in which all the chemical characteristics of the element are present. Atoms are so small they cannot be seen with an electron microscope, which magnifies millions of times. A substance with only one type of atom is referred to as an **element**. More than 100 elements are known to exist at present, and of the known elements, 92 occur naturally. The remaining elements have been manufactured in laboratories **(Figure 7–1)**.

Small, positively charged particles called protons are located in the centre, or nucleus, of each atom. In most atoms, neutrons are also located in the nucleus. Neutrons have no electrical charge, but they add weight to the atom. The positively charged protons tend to repel each other, and this repelling force could destroy the nucleus. The presence of the neutrons with the protons in the nucleus cancels out the repelling action of the protons and keeps the nucleus together. Electrons are small, very light particles with a negative electrical charge. Electrons move in orbits around the nucleus of an atom.

A proton is approximately 1840 times heavier than an electron, and this makes the electron much

Group:	1	2	3	4	5	6	7	8	9	10	11	12	13	14	15	16	17	18
Period	1A	2A	3B	4B	5B	6B	7B		8B		1B	2B	3A	4A	5A	6A	7A	8A
1	1 H																	2 He
2	3 Li	4 Be											5 B	6 C	7 N	8 O	9 F	10 Ne
3	11 Na	12 Mg											13 Al	14 Si	15 P	16 S	17 Cl	18 Ar
4	19 K	20 Ca	21 Sc	22 Ti	23 V	24 Cr	25 Mn	26 Fe	27 Co	28 Ni	29 Cu	30 Zn	31 Ga	32 Ge	33 As	34 Se	35 Br	36 Kr
5	37 Rb	38 Sr	39 Y	40 Zr	41 Nb	42 Mo	[43] Tc	44 Ru	45 Rh	46 Pd	47 Ag	48 Cd	49 In	50 Sn	51 Sb	52 Te	53 I	54 Xe
6	55 Cs	56 Ba	*	72 Hf	73 Ta	74 W	75 Re	76 Os	77 Ir	78 Pt	79 Au	80 Hg	81 Tl	82 Pb	83 Bi	84 Po	85 At	86 Rn
7	87 Fr	88 Ra	**	[104] Unq	[105] Unp	[106] Unh	[107] Uns	[108] Uno	[109] Une	[110] Uun	[111] Uuu	[112] Uub	[113] Uut	[114] Uuq	[115] Uup	[116] Uuh	[117] Uus	[118] Uuo

***Lanthanides:**	57 La	58 Ce	59 Pr	60 Nd	[61] Pm	62 Sm	63 Eu	64 Gd	65 Tb	66 Dy	67 Ho	68 Er	69 Tm	70 Yb	71 Lu
****Actinides:**	89 Ac	90 Th	91 Pa	92 U	[93] Np	[94] Pu	[95] Am	[96] Cm	[97] Bk	[98] Cf	[99] Es	[100] Fm	[101] Md	102 No	103 Lr

Legend:

Alkali metals	Noble gases
Alkaline earth metals	Halogens
Other metals	Other nonmetals
Semiconductors	No data available

FIGURE 7–1

The periodic table of the elements, with each element's natural state shown.

easier to move than a proton. Because the electrons are orbiting around the nucleus, centrifugal force tends to move the electrons away from the nucleus. However, the attraction between the positively charged protons and the negatively charged electrons holds the electrons in their orbits. The atoms of the different elements have different numbers of protons, electrons, and neutrons. Some of the lighter elements have the same number of protons and neutrons in the nucleus, but many of the heavier elements have more neutrons than protons.

The simplest atom is the hydrogen (H) atom, which has one proton in the nucleus and one electron orbiting around the nucleus **(Figure 7–2)**. The nucleus of a copper (Cu) atom contains 29 protons and 34 neutrons, and 29 electrons orbit in 4 different rings around the nucleus. Because 2, 8, and 18 electrons are the maximum number of electrons on the first 3 electron rings next to the nucleus, the fourth ring must have 1 electron. The outer ring of an atom is called the valence ring, and the number of electrons on this ring determines the electrical characteristics of the element. Elements are listed on the atomic scale, or periodic table, according to their number of protons and electrons. For example, hydrogen is number 1 on this scale and copper is number 29.

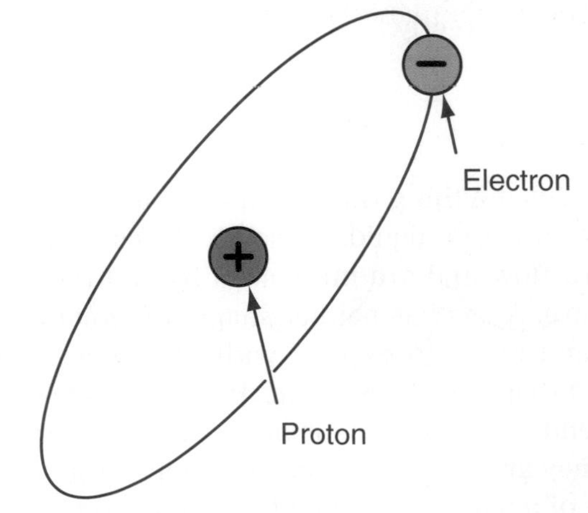

FIGURE 7–2

A hydrogen atom.

For some elements, a single atom does not exist. An example of this is oxygen, which has a chemical symbol of O. Pure oxygen exists as pairs of oxygen atoms and has a symbol of O_2, which is a molecule of oxygen. A molecule is the smallest particle of an element or compound that can exist and still retain the characteristics of the element or compound. Some materials contain only one type of atom, whereas a compound may be described as a liquid, solid, or gas that contains two or more types of atoms. An oxygen atom readily combines with another oxygen atom or atoms of many other elements to form a compound. Many other atoms also have this characteristic.

Water is a compound that contains oxygen and hydrogen atoms. The chemical symbol for water is H_2O. This chemical symbol indicates that each molecule of water contains two atoms of hydrogen and one oxygen atom **(Figure 7–3)**.

Ions

An **ion** is an atom or molecule that has lost or gained one or more electrons. As a result, it has a negative or positive electrical charge. A negatively charged ion has more electrons than it has protons. The opposite is true of positively charged ions, which have fewer electrons than protons. Ions are denoted in the same way as other atoms and molecules except for a superscript symbol or number that shows the electrical charge and the number of electrons gained or lost. For example, hydrogen with a positive charge (H^+) and oxygen with a negative charge (O^{2-}) is called an **oxide**.

PLASMA Considered by scientists as the fourth state of matter, **plasma** refers to an ionized gas that has about an equal amount of positive ions and electrons. The electrons travel with the nucleus of the atoms but can move freely and are not bound to it. The gas at this point no longer behaves as a gas. It now has electrical properties and creates a magnetic field, which radiates light and other forms of electromagnetic energy. It typically takes the form of gas-like clouds and is the basis of most stars. In fact, our sun is really just a large piece of plasma. Plasma is the most common form of matter in the universe. Plasma in the stars and in the space between them occupies nearly 99 percent of the visible universe. Plasma does not exist as a solid, liquid, or gas; it is different and has a much different temperature range. Plasma is more dense than other states of matter.

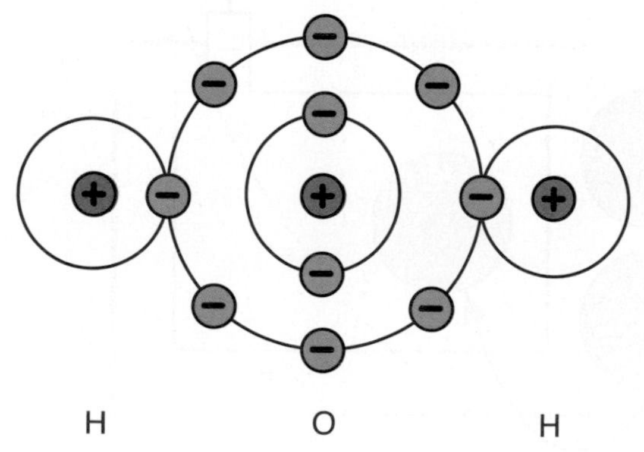

FIGURE 7–3
A molecule of water.

States of Matter

For most everyday uses, we need to be concerned with three states of matter: solid, liquid, and gas. The particles of a solid are held together in a rigid structure. When a solid dissolves into a liquid, its particles break away from this structure and mix evenly in the liquid, forming a **solution**. When heated, most liquids **evaporate**, which means that the atoms or molecules of which they are made break free from the body of the liquid to become gas particles. If all of the liquid in a solution has evaporated, the solid is left behind. The particles of the solid normally arrange in a structure called a crystal.

ABSORPTION AND ADSORPTION Not all solids dissolve in a liquid; rather, the liquid will be either absorbed or adsorbed. The action of a sponge serves as the best example of absorption. When a dry sponge is put into water, the water is absorbed by the sponge. The sponge does not dissolve; the water merely penetrates into the sponge, and the sponge becomes filled with water. There is no change to the atomic structure of the sponge, nor does the structure of the water change. If we put a glass into water, the glass does not absorb the water. However, the glass still gets wet, as a thin layer of water adheres to the glass. This is adsorption. Materials that *absorb* fluids are **permeable** substances. **Impermeable** substances do *not* absorb fluids. Some materials are impermeable to most fluids, whereas others are impermeable to just a few.

ENERGY

Energy can be produced by many different methods and may also provide the ability to perform work. All matter consists of atoms and molecules that are in constant motion, and therefore all matter has different amounts of energy. Energy is not matter, but it affects the behaviour of matter. Everything that happens requires energy, and energy comes in many forms.

Each form of energy can change into other forms. However, the total amount of energy never changes; it can only be transferred from one form to another, not created or destroyed. This is known as the principle of the conservation of energy.

As an example, **engine efficiency** is a measure of the relationship between the amount of energy put into or applied to the engine and the amount of output energy available from the engine. Engine efficiency is expressed as a percentage. The formula for determining efficiency is as follows: (output energy ÷ input energy) × 100.

Other aspects of the engine are expressed in efficiencies. They include mechanical efficiency, volumetric efficiency, and thermal efficiency. They are expressed as a ratio of input (actual) to output (maximum or theoretical). Efficiencies are always less than 100 percent. The difference between the efficiency and 100 percent is the percentage lost during the process. For example, if there were 100 units of energy put into the engine and 28 units were available on the output to power the vehicle, then the efficiency would be equal to 28 percent. This would mean that 72 percent of the energy received was wasted or lost.

Kinetic and Potential Energy

When energy is released to do work, it is called **kinetic energy**. This type of energy may also be referred to as energy in motion **(Figure 7–4)**. Stored energy may be called **potential energy**.

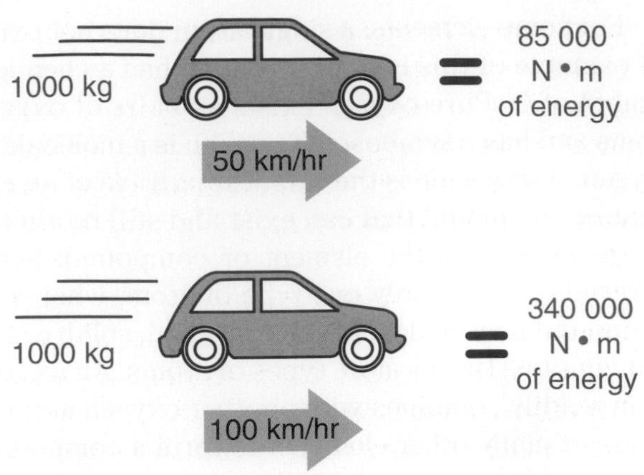

FIGURE 7–4

The kinetic energy of a moving vehicle increases exponentially with its speed.

Many components and systems have potential energy and, at times, kinetic energy. The ignition system is a source for high electrical energy. The heart of the ignition system is the ignition coil, which has much potential energy. When it is time to fire a spark plug, that energy is released and becomes kinetic energy as it creates a spark across the gap of a spark plug.

Energy Conversion

Energy conversion occurs when one form of energy is changed to another form. Because energy is not always in the desired form, it must be converted to a form we can use. Some of the most common automotive energy conversions are listed here.

- *Chemical to thermal energy.* Chemical energy in gasoline or diesel fuel is converted to thermal energy when the fuel burns in the engine cylinders.
- *Chemical to electrical energy.* The chemical energy in a battery **(Figure 7–5)** is converted

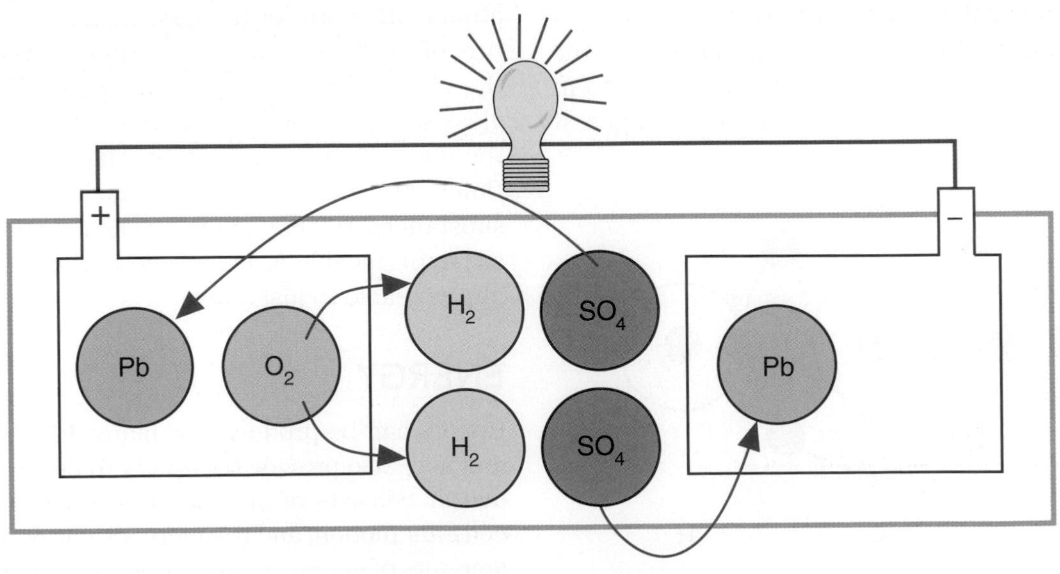

FIGURE 7–5

Chemical energy is converted to electrical energy in a battery.

to electrical energy to power many of the accessories on an automobile.

- *Electrical to mechanical energy.* In the automobile, the battery supplies electrical energy to the starting motor, and this motor converts the electrical energy to mechanical energy to crank the engine.
- *Thermal to mechanical energy.* The thermal energy that results from the burning of the fuel in the engine is converted to mechanical energy, which is used to move the vehicle.
- *Mechanical to electrical energy.* The generator is driven by mechanical energy from the engine. The generator converts this energy to electrical energy, which powers the electrical accessories on the vehicle and recharges the battery.
- *Electrical to radiant energy.* Radiant energy is light energy. In the automobile, electrical energy is converted to thermal energy, which heats up the inside of light bulbs so that they illuminate and release radiant energy.
- *Kinetic to mechanical to electrical energy.* Hybrid vehicles have a system, called regenerative braking, that uses the energy of the moving vehicle (kinetic) to rotate a generator. The mechanical energy used to operate the generator is used to provide electrical energy to charge the batteries (**Figure 7–6**) or power the electric-drive motor.

Mass and Weight

Mass is the amount of matter in an object. **Weight** is a force and is measured in kilograms or pounds. Gravitational force gives the mass its weight. As an example, a spacecraft can weigh 500 tonnes (1 102 311 pounds) here on earth where it is affected by the earth's gravitational pull. In outer space,

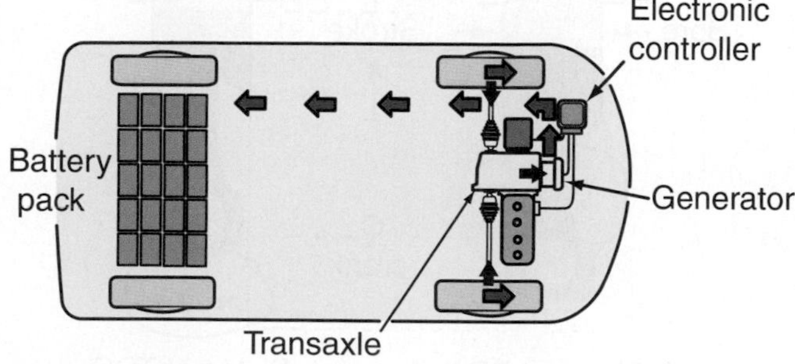

FIGURE 7–6

Regenerative braking captures some of the vehicle's kinetic energy to charge the batteries or power the electric-drive motor.

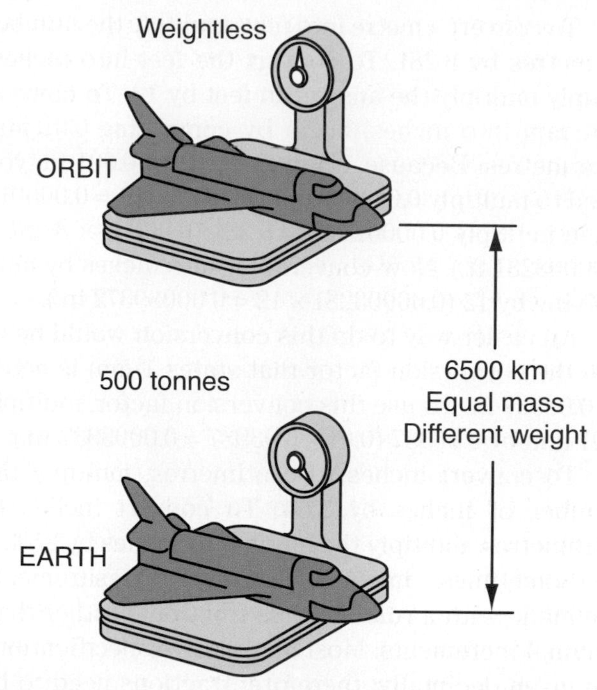

FIGURE 7–7

The difference in weight of a space shuttle on Earth and in space.

beyond the earth's gravity and atmosphere, the spacecraft is nearly weightless (**Figure 7–7**).

Automobile specifications list the weight of a vehicle primarily in two ways. **Gross weight** is the total weight of the vehicle when it is fully loaded with passengers and cargo. **Curb weight** is the weight of the vehicle when it is not loaded with passengers or cargo.

To convert kilograms into pounds, simply multiply the weight in kilograms by 2.2046. For example, if something weighs 5 kg, it weighs 11.023 lbs. (5 × 2.2046). To express the answer in pounds and ounces, convert the 0.023 lb. into ounces. Because there are 16 oz. in a pound, multiply 16 by 0.023 (16 × 0.023 = 0.368). Therefore, 5 kg is equal to 11 lb. 0.368 oz.

Size

The size of something is related to its mass. The size of an object defines how much space it occupies. Size dimensions are typically stated in terms of length, width, and height. Length is a measurement of how long something is from one end to another. Width is a measurement of how wide something is from one side to another. Obviously, height is a measurement of the distance from something's bottom to its top. All three of these dimensions are measured in metres, centimetres, and millimetres in the metric system (yards, feet, and inches in the imperial system).

To convert a metre into feet, multiply the number of metres by 3.281. To convert the feet into inches, simply multiply the answer in feet by 12. To convert 0.01 mm into inches, begin by converting 0.01 mm into metres. Because 1 mm is equal to 0.001 m, you need to multiply 0.01 by 0.001 (0.001 × 0.01 = 0.00001). Then multiply 0.00001 m by 3.281 (0.00001 × 3.281 = 0.00003281 ft.). Now convert feet into inches by multiplying by 12 (0.00003281 × 12 = 0.00039372 in.).

An easier way to do this conversion would be to use the conversion factor that states 1 mm is equal to 0.03937 in. To use this conversion factor, multiply 0.01 mm by 0.03937 (0.01 × 0.03937 = 0.0003937 in.).

To convert inches to centimetres, multiply the number of inches by 2.54. To convert inches to millimetres, multiply the number of inches by 25.4.

Sometimes, imperial distance measurements are made with a rule that has fractional rather than decimal increments. Most automotive specifications are given decimally; therefore, fractions need to be converted into decimals. It is also easier to add and subtract dimensions if they are expressed in decimal form rather than in fractions. For example, suppose you want to find the rolling circumference of a tire, and you know the diameter of the tire to be 60 cm (23⅝ in.). The distance around the tire is the circumference, and it is equal to the diameter multiplied by a constant called pi (π). Pi is equal to approximately 3.14; therefore, the circumference of the tire is equal to 60 cm (23⅝ in.) multiplied by 3.14 (60 × 3.14 = 188.4). This equals 188.4 cm (74.183 in.). To calculate the imperial measurement, the measurement must be changed to a whole number and the decimal equivalent of the fraction. To convert the ⅝ to a decimal, divide the 5 by the 8 (5 ÷ 8 = 0.625). Therefore, the imperial tire diameter measurement is 23.625 in. Now multiply the diameter by (23.625 × 3.14 = 74.183). The circumference of the tire is slightly more than 74 in.

VOLUME

Volume is also a measurement of size and is related to mass and weight. Volume is the amount of space occupied by an object in three dimensions: length, width, and height. For example, a kilogram of gold and a kilogram of feathers both have the same weight, but the kilogram of feathers occupies a much larger volume. The measurement for volume in the metric system is cubic centimetres or litres **(Figure 7–8)**. In the imperial system, volume is measured in cubic inches, cubic feet, cubic yards, or gallons.

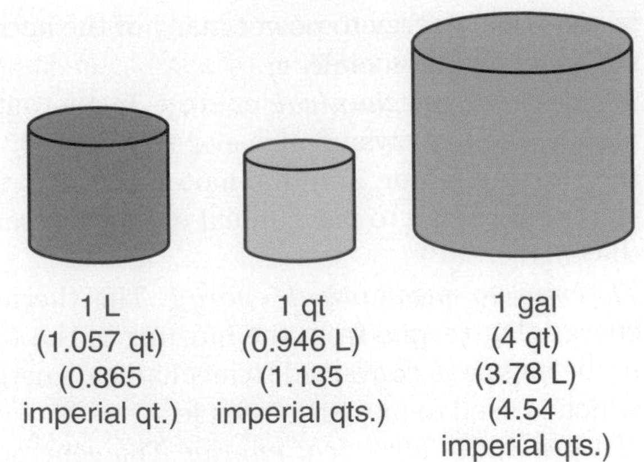

FIGURE 7–8

A comparison of metric and imperial units of volume.

The volume of a container is basically calculated by multiplying the measured length, width, and height of an object. For example, if a box has a length of 5 cm (2 in.), a width of 8 cm (3 in.), and a height of 10 cm (4 in.), it has a volume of 400 cc (5 × 8 × 10) (24 cu. in. [2 × 3 × 4 = 24]). Different shapes have different formulas for calculating volume, but all consider the three basic dimensions of the object.

The volume of an engine's cylinders determines its size, expressed as displacement. This size does *not* reflect the external dimensions (length, width, and height) of the engine. Cylinder **displacement** is the volume of a cylinder between when the cylinder's piston is at its lowest point of travel, or bottom dead centre (BDC), and its highest point of travel, or top dead centre (TDC). This is called the stroke of the piston **(Figure 7–9)**. Displacement is usually measured in cubic centimetres, litres, or cubic inches. The total displacement of an engine (including all

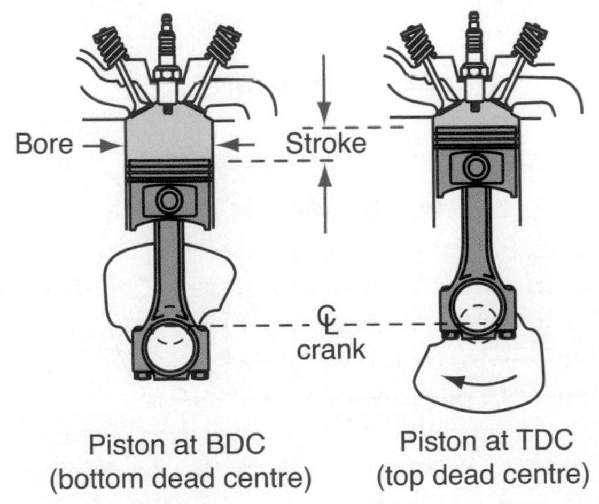

FIGURE 7–9

The bore and stroke of an engine.

cylinders) is a rough indicator of its power output. Total displacement is the sum of displacements for all cylinders in an engine. Engine displacement may be calculated as follows:

$$Displacement = \pi \times R^2 \times L \times N$$

Where $\pi = 3.1416$

R = radius of the cylinder opening or the diameter (bore) ÷ 2

L = length of stroke

N = number of cylinders in the engine

Most of today's engines are described by their metric displacement. Cubic centimetres and litres are determined by using metric measurements in the displacement formula.

Example: Calculate the metric displacement of a four-cylinder engine with a 78.9 mm stroke and a 100 mm bore. Before you use the formula to find the displacement in cubic centimetres, you must convert the millimetre measurements to centimetres: 78.9 mm = 7.89 cm and 100 mm = 10 cm.

Displacement = $3.1416 \times 5^2 \times 7.89 \times 4$

Displacement = 2479 cubic centimetres (cc) or approximately 2.5 litres (L)

Example: Calculate the cubic inch displacement (CID) of a six-cylinder engine with a 3.7 in. bore and 3.4 in. stroke.

$CID = 3.1416 \times 1.85^2 \times 3.4 \times 6$

$CID = 219.66$

Engine displacement can also be calculated by using this formula:

0.7854 × Bore × Bore × Stroke × Number of cylinders = Displacement

Ratios

Often, automotive features are expressed as ratios. A ratio expresses the relationship between two things. If something is twice as large as some other thing, the ratio is 2:1. Sometimes, ratios are used to compare the movement of an object. For example, if a gear with a 5 cm (2 in.) diameter drives a gear with a 10 cm (4 in.) diameter, the ratio of the gears is 5:1, but the drive ratio would be 2:1 (two turns of the input gear to one turn of the output gear)

The **compression ratio** of an engine expresses how much the air/fuel mixture will be compressed as the piston in a cylinder moves from BDC to TDC of the cylinder. The compression ratio is defined as the ratio of the volume in the cylinder above the piston when the piston is at the bottom of its travel to the volume in the cylinder above the piston when the piston is at its uppermost position **(Figure 7–10)**. The formula for calculating the compression ratio is as follows:

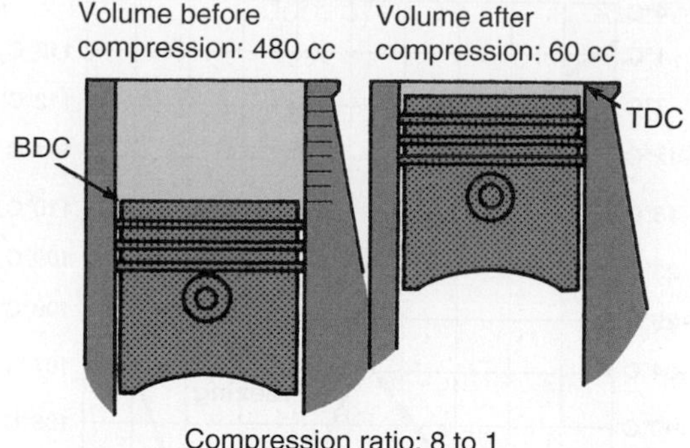

FIGURE 7–10

An engine's compression ratio indicates the amount the air/fuel mixture is compressed during the compression stroke.

volume above the piston at BDC ÷
volume above the piston at TDC

or

total cylinder volume÷
total combustion chamber volume

In many engines, the top of the piston is even, or level, with the top of the cylinder block at TDC. The combustion chamber is in the cavity in the cylinder head above the piston. This is modified slightly by the shape of the top of the piston. The volume of the combustion chamber must be added to each volume in the formula to get an accurate calculation of compression ratio.

Example: Calculate the compression ratio if the total piston displacement is 800 cc (48.82 cu. in.) and the combustion chamber volume is 100 cc (6.10 cu. in.).

800 + 100 ÷ 100 = 9.1

Therefore, the compression ratio is 9.1 to 1, or 9.1:1.

Proportions

Ratios can also be used to express the correct mixture for something. An example of this would be the amount of engine coolant that should be mixed with water before the engine's cooling system is refilled **(Figure 7–11)**. Typically, specifications call for 50 percent coolant and 50 percent water, or a ratio of 1:1. This mixture allows for maximum hot and cold protection. To apply this ratio, suppose a cooling system has a capacity of 9.5 L. Most engine coolant is sold in 4 L containers. To determine how much coolant or antifreeze to put in the system, we divide

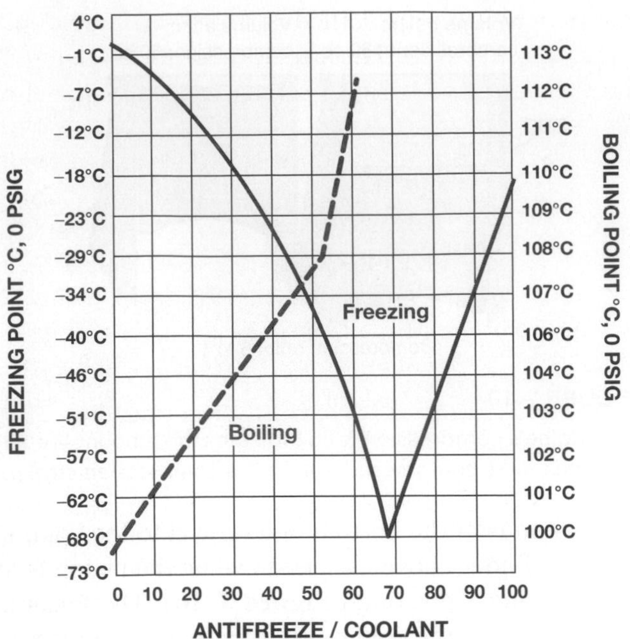

FIGURE 7–11

The relationship of the percentage of antifreeze to the freezing and boiling points of the engine's coolant.

the total capacity by 2, which gives the quantity equal to 50 percent of the capacity (9.5 ÷ 2 = 4.75). Therefore, to obtain the correct mixture, 4.75 L of coolant should be mixed with 4.75 L of water.

FORCE

A **force** is a push or pull and can be large or small. Force can be applied to objects by direct contact or from a distance. Gravity and electromagnetism are examples of forces that are applied from a distance. Forces can be applied from any direction and with any intensity. For example, if a pulling force on an object is twice that of the pushing force, the object will be pulled at one-half the pulling force. When two or more forces are applied to an object, the combined force is called the resultant. The resultant is the sum of the size and direction of the forces. For example, when a mass is suspended by two lengths of wire, each wire should carry half the weight of the mass. If we move the attachment of the wires so that they are at an angle to the mass, the wires now carry more force. The wires carry the force of the mass plus a force that pulls against the other wire.

Automotive Forces

When a vehicle is at rest, gravity exerts a downward force on the vehicle. The ground exerts an equal and opposite upward force and supports the vehicle. When the engine is running and its power output is transferred to the vehicle's drive wheels, the wheels exert a force against the ground in a horizontal direction. This force causes the vehicle to move but is opposed by the mass of the vehicle **(Figure 7–12)**. To move the vehicle faster, the force supplied by the wheels must increase beyond the opposing forces. As the vehicle moves faster, it pushes against the air as it travels. This push becomes a growing opposing force, and the force at the drive wheels must overcome the force in order for the vehicle to increase speed. After the vehicle has achieved the desired speed, no additional force is required at the drive wheels. The study of the effects of forces on vehicles is generally referred to as vehicle dynamics.

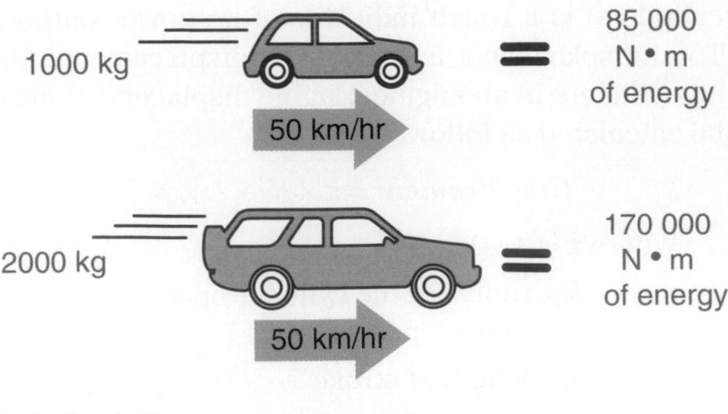

FIGURE 7–12

The amount of energy required to move a vehicle depends on its mass.

BALANCED AND UNBALANCED FORCE When the applied forces are balanced and there is no overall resultant force, the object is said to be in **equilibrium**. An object sitting on a solid flat surface is in equilibrium because its weight is supported by the surface and there is no resultant force. If the surface is put on an angle, the object will tend to slide down the surface. If the surface is at a slight angle, the force will cause the object to slowly slide down the surface. If the surface is at a severe angle, the downward force will cause the object to quickly slide down the slope. In both cases, the surface is still supplying the force needed to support the object, but the pull of gravity is greater, and the resultant force causes the object to slide down the slope.

TURNING FORCES Forces can cause rotation as well as straight-line motion. A force acting on an object that is free to rotate has a turning effect, or turning force. This force is equal to the size of the force multiplied by the distance of the force from the turning point around which it acts.

FIGURE 7–13

A tire at an angle will roll in the same way as a cone would.

Forces on Tires and Wheels

If you roll a cone-shaped piece of metal on a smooth surface, the cone does not roll in a straight line. Rather, it moves toward the direction of the tilt on the cone. The weight of the cone is applied to the surface, but part of the weight at the large end of the cone is applied at an angle to the small end of the cone **(Figure 7–13)**. Riding a bicycle is another example. When you want to make a left turn, it is easier if you tilt the bicycle to the left. The reason for this action is that a tilted, rolling wheel tends to move in the direction of the tilt. Similarly, if a tire and wheel on an automobile are tilted, the tire and wheel will tend to move in the direction of the tilt.

While riding a bicycle, the force applied to the bicycle is projected through the bicycle's front fork to the road surface by your weight. The centreline of the front fork is tilted rearward in relation to the vertical centreline of the wheel. When the handle bars are turned, the tire pivots on the vertical centreline of the wheel. Because the tire's pivot point is behind the front fork centreline where your weight is projected against the surface of the road, the front wheel tends to return to the straight-ahead position after a turn. The wheel also tends to remain in the straight-ahead position as the bicycle is driven. This principle of resultant forces is also the basis for the theories applied during front-wheel alignment.

Centrifugal/Centripetal Forces

When an object moves in a circle, its direction is continuously changing, and all changes in direction require a force. The forces required to maintain circular motion are called **centripetal force** and **centrifugal force**. The size of these forces depends on the size of the circle and the mass and speed of the object.

Centripetal force tends to pull the object toward the centre of the circle, whereas centrifugal force tends to push the object away from the centre. The centripetal force that keeps an object whirling around on the end of a string is caused by **tension** in the string. If the string breaks, there is no longer string tension and centripetal force, and the object will fly off in a straight line because of the centrifugal force on it. Gravity is the centripetal force that keeps the planets orbiting around the sun. Without this centripetal force, Earth would move in a straight line through space.

Wheel and Tire Balance

When the weight of a wheel and tire assembly is distributed equally around the centre of wheel rotation, the wheel and tire have proper static balance. Being statically balanced, the wheel and tire assembly will not tend to rotate by itself, regardless of the wheel position. If the weight is not distributed equally, the wheel and tire assembly is statically unbalanced. As the wheel and tire rotate, centrifugal force acts on this static unbalance and causes the wheel to "tramp" or "hop" **(Figure 7–14)**.

Dynamic balance exists when the weight thrown to the sides of the tire and wheel assembly is equal when the assembly is rotating **(Figure 7–15)**. To illustrate this balance, assume we have a bar with a ball attached by string to both ends of the bar. If we cause the bar to rotate, the balls will turn with the bar, and centripetal and centrifugal force will keep the balls in an orbit around the rotating bar. If the two balls weigh the same and are at an equal distance from the bar, the bar will rotate smoothly. However, if one of the

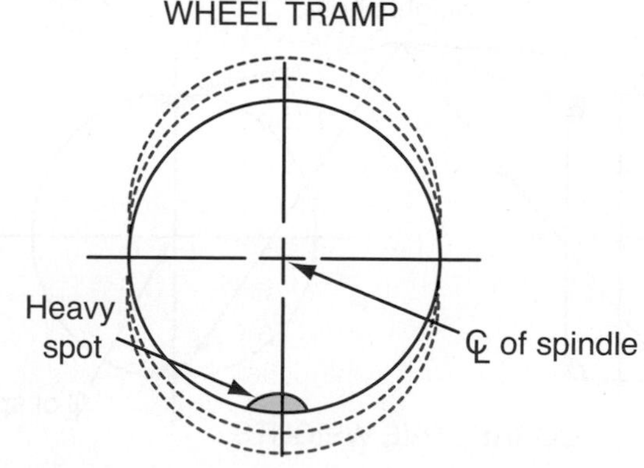

FIGURE 7–14

Wheel tramp is the result of a tire and wheel assembly being statically unbalanced.

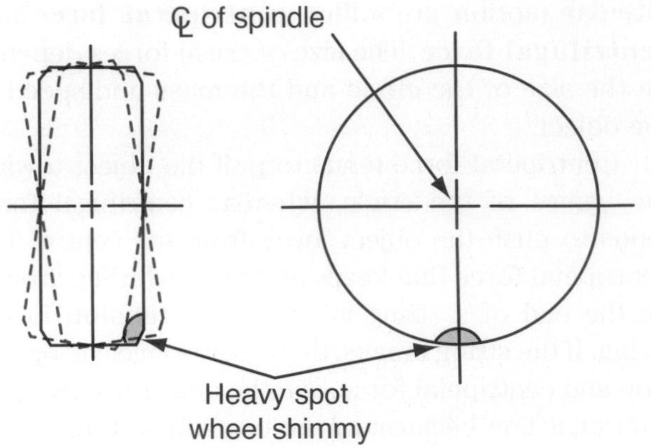

FIGURE 7–15

Dynamic imbalance causes wheel shimmy.

balls is heavier than the other, the bar will wobble as it rotates. The greater the difference in the weight of the balls, the greater the wobble. The wobble can eventually destroy the mechanism used to rotate the bar.

Now, if we add some weight to the end of the bar that has the lighter ball, the weights and forces can be equalized and the wobble removed. This principle illustrates how we dynamically balance a wheel and tire assembly **(Figure 7–16)**.

When we think of all the parts of an automobile that rotate, it is easy to see why proper balance is important. Improper balance can cause premature wear or destruction of parts.

Pressure

Pressure is a force applied against an object and is measured in units of force per unit of surface area (kilograms per square centimetre or pounds per square inch). Mathematically, pressure is equal to the applied force divided by the area over which the force acts. Consider two 10 kilogram (kg) weights sitting on a table; one occupies an area of 1 square centimetre (cm^2), and the other an area of 4 cm^2. The pressure exerted by the first weight would be 10 kilograms per square centimetre or 10 kg/cm^2. The other weight, although it weighs the same, will exert only 2.5 kg/cm^2 ($10\ kg \div 4\ cm^2 = 2.5\ kg/cm^2$). An imperial example would be two 10 lb. weights sitting on a table; one occupies an area of 1 sq. in., and the other an area of 4 sq. in. The pressure exerted by the first weight would be 10 lb. per 1 sq. in., or 10 psi. The other weight, although it weighs the same, will exert only 2.5 psi (10 lb. per 4 sq. in. = $10 \div 4 = 2.5$). This illustrates an important concept: A force acting over a large area exerts less **pressure** than the same force acting over a small area.

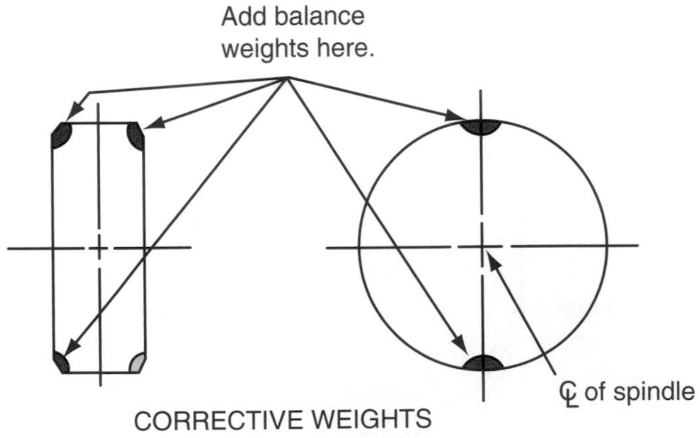

FIGURE 7–16

Adding a weight to counteract the heavy spot of a tire and wheel assembly.

Because pressure is a force, all principles of force apply to pressure. If more than one pressure is applied to an object, the object will respond to the resultant force. Also, all matter (liquids, gases, and solids) tends to move from an area of high pressure to a low-pressure area.

TIME

The word **time** is used to mean many things. For our look into science, time will be defined as a measurement of the duration of an event that has happened, is happening, or will happen. In everyday events, time is measured by the increments of a clock: seconds, minutes, and hours. For automotive purposes, depending on the application or system in use, time may be measured in minutes, seconds, milliseconds, degrees, or even revolutions per a given unit of time measurement. Often, an automotive technician is concerned with how long something occurs, such as the length of time a spark plug fires to cause combustion in an engine's cylinder. This time, called spark duration, is typically about 3 milliseconds (0.003 seconds) and is measured with a lab scope because it would be very difficult to measure that short a time period with a clock.

Technicians also monitor how many times a cycle is repeated within a period of time. An example of this would be with a tachometer, which measures engine revolutions per minute and is an often-used diagnostic tool.

MOTION

When the forces being applied to an object become unbalanced (where the opposing forces are not equal), they will change the motion of the object. The object's speed, direction of motion, or both will change. The greater the mass of an object, the greater the force needed to change its motion. This resistance to change in motion is called **inertia**. Inertia is the tendency of an object at rest to remain at rest, or the tendency of an object in motion to stay in motion. The inertia of an object at rest is called static inertia, whereas dynamic inertia refers to the inertia of an object in motion. Inertia exists in liquids, solids, and gases. When you push and move a parked vehicle, you overcome the static inertia of the vehicle. If you catch a ball in motion, you overcome the dynamic inertia of the ball.

When a force overcomes static inertia and moves an object, the object gains momentum. **Momentum** is the product of an object's weight times its speed. Momentum is a type of mechanical energy. An object loses momentum if another force overcomes the dynamic inertia of the moving object.

Rates

Speed is the distance an object travels in a set amount of time. It is calculated by dividing distance covered by time taken. We refer to the speed of a vehicle in terms of kilometres per hour (km/h) or miles per hour (mph). **Velocity** is the speed of an object in a particular direction. **Acceleration**, which occurs only when a force is applied, is the rate of increase in speed. Acceleration is calculated by dividing the change in speed by the time it took for that change. **Deceleration** is the reverse of acceleration, as it is the rate of a decrease in speed.

Newton's Laws of Motion

How forces change the motion of objects was first explained three centuries ago by Sir Isaac Newton. These explanations are known as Newton's laws. Newton's first law of motion is called the law of inertia. It states that an object at rest tends to remain at rest and an object in motion tends to remain in motion unless some force acts on it. When a car is parked on a level street, it remains stationary unless it is driven or pushed.

Newton's second law states that when a force acts on an object, the motion of the object will change. This change in motion is equal to the size of the force divided by the mass of the object on which it acts. Trucks have a greater mass than cars. Because a large mass requires a larger force to produce a given acceleration, a truck needs a larger engine than a car.

Newton's third law says that for every action there is an equal and opposite reaction. A practical application of this law occurs when the wheel on a vehicle strikes a bump in the road surface. This action drives the wheel and suspension upward with a certain force, and a specific amount of energy is stored in the spring. After this action occurs, the spring forces the wheel and suspension downward with a force equal to the initial upward force caused by the bump.

Friction

Friction is a force that slows or prevents motion in two moving objects or surfaces that touch. Friction may occur in solids, liquids, and gases. The joining or bonding of the atoms at each of the surfaces causes the friction. When you attempt to pull an object across a surface, the object will not move until these bonds have been overcome. Smooth surfaces produce little friction; therefore, only a small amount of force is needed to break the bonds between the atoms. Rougher surfaces produce a larger friction force because stronger bonds are made between the two surfaces **(Figure 7–17)**. To move an object over a rough surface, such as sandpaper, a great amount of force is required.

Friction is put to good use in disc brakes **(Figure 7–18)**. The friction force between the disc and brake pad slows down the rotation of the wheel, reducing the vehicle's speed. In doing so, it converts the kinetic energy of the vehicle into heat.

LUBRICATION Friction can be reduced in two main ways: by lubrication or by the use of rollers or bearings. The presence of oil or another fluid between two surfaces keeps the surfaces apart. Because fluids (liquids and gases) flow, they allow movement between surfaces. The fluid keeps the surfaces

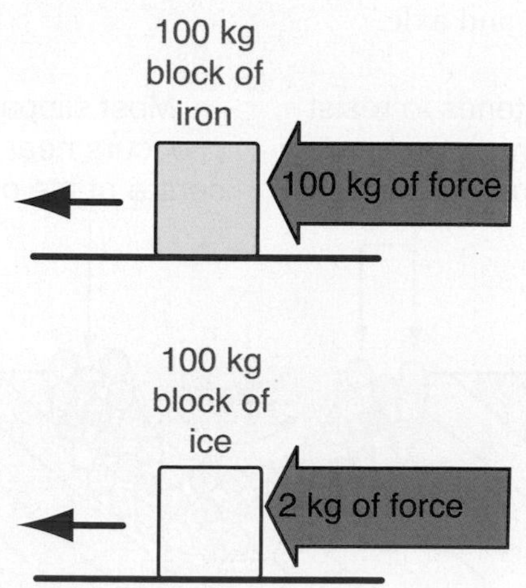

FIGURE 7–17

Sliding ice across a surface produces less friction than sliding a rougher material, such as iron, across a surface.

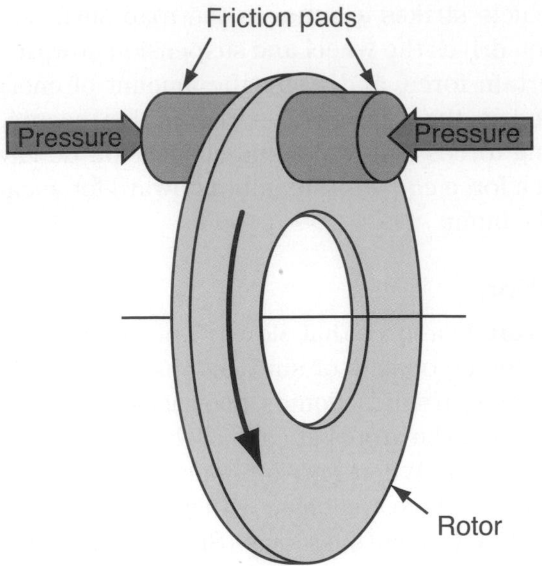

FIGURE 7–18

As pressure is applied to the friction pads, the pads attempt to stop the rotor to which the tire and wheel are attached.

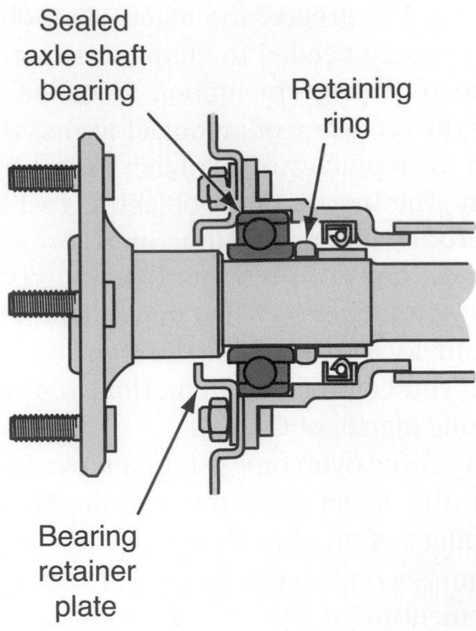

FIGURE 7–20

A typical ball bearing assembly for an axle shaft.

apart, allowing them to move smoothly past one another **(Figure 7–19)**.

ROLLERS Rollers (or bearings) placed between two surfaces keep the surfaces apart. An object placed on rollers will move smoothly if pushed or pulled. Rollers actually use friction to grip the surfaces and produce rotation. Instead of sliding against one another, the surfaces produce turning forces, which cause each roller to spin, leaving very little friction to oppose motion. Bearings are a type of roller used to reduce the friction between moving parts such as a wheel and its axle **(Figure 7–20)**. As the wheel turns on the axle, the balls in the bearing roll around inside the bearing, drastically reducing the friction between the wheel and axle.

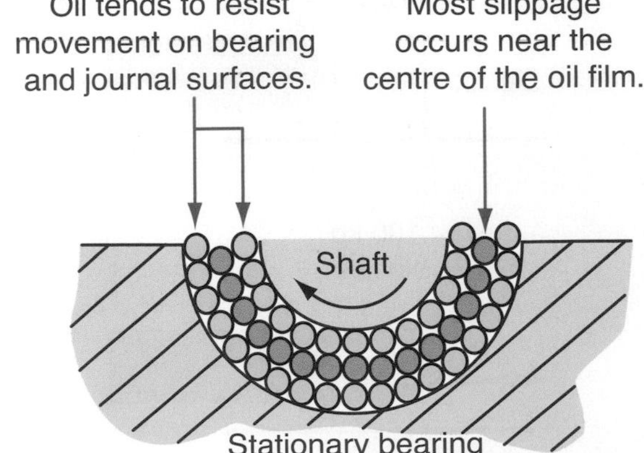

FIGURE 7–19

Oil separates the rotating shaft from the stationary bearing.

Air Resistance

When a car is driven down the road, resistance occurs between the air and the car's surface. This resistance or friction opposes the momentum, or mechanical energy, of the moving vehicle. The mechanical energy from the engine must overcome the vehicle's inertia and the friction of the air striking the vehicle. The faster an object moves, the greater the air resistance.

Body design, obviously, affects the amount of friction developed by the air striking the vehicle. The total resistance to motion caused by friction between a moving vehicle and the air is referred to as coefficient of drag (Cd). At 70 km/h (44 mph), half of the engine's mechanical energy can be used to overcome air resistance. Therefore, reducing a vehicle's Cd can be a very effective method of improving fuel economy. Coefficient of drag is also called aerodynamic drag.

Aerodynamics is the study of the effects of air on a moving object **(Figure 7–21)**. The basics of this science are fairly easy to understand. The larger the air area facing the moving air, the more force will be put on that area by the air. The air tends to hold back or resist the forward motion of the object moving against it.

The less air a vehicle pushes out of its way as it moves, the less power it will need to move at a given speed. By studying a vehicle's aerodynamics, engineers strive to get the air flowing over the vehicle in a manner that will reduce the amount of effort required

FIGURE 7–21

The movement of air as it goes over a car.

DaimlerChrysler Corporation

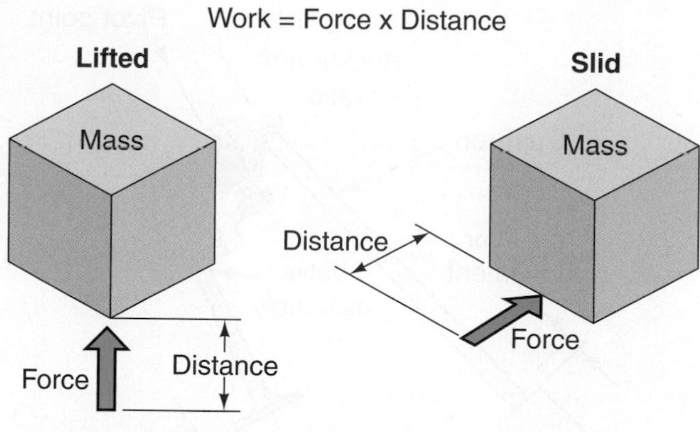

FIGURE 7–22

When work is performed, a mass is moved a certain distance.

by the vehicle's engine to propel the vehicle through the air. Spoilers and air diffusers are used to direct the air over vehicle surfaces to help reduce the amount of air friction a vehicle will encounter when moving.

Most aerodynamic design work is done initially on a computer; then the design is checked and modified by placing a vehicle with that design in a wind tunnel. A wind tunnel is a carefully constructed facility with a large fan at one end. Inside the tunnel, the movement of air over, under, and around the vehicle is studied.

Ideally, the air that is moved by the vehicle will follow the contours of the vehicle, which prevents the air from doing funny things as it is pushed away. If the air that moves under the vehicle has a place to push up, causing what is known as "lift," this situation creates poor handling, which can be very unsafe. Air can also be trapped under the vehicle, which increases the air drag of the vehicle. If the air moving over the top pushes against the vehicle, there is an increase in air drag as it creates what is known as "down-force." To aid in vehicle handling characteristics, engineers will use some of the above mentioned devices to minimize the effects of wind resistance on a moving vehicle.

WORK

When a force moves a certain mass a specific distance, **work** is done. When work is accomplished, the mass may be lifted or slid on a surface against a resistance or opposing force **(Figure 7–22)**. Work is equal to the applied force multiplied by the distance the object moved (force × distance = work) and is measured in **Newton**-metres **(Figure 7–23)**, watts, or foot-pounds. For example, the force needed to move a 1500 kg car 15 m has to be converted to Newtons to produce a Newton-metre reading: 1 kilogram equals 9.8067 Newtons, so 1500 kg-force × 9.8067 = 14 710 N. Then to get a Newton-metre reading, 14 710 × 15 = 220 650 Nm of work being produced. Imperial system calculations would have a 3300 lb. car moving 50 ft., equalling 165 000 foot-pounds of work being produced.

During work, a force acts on an object to start, stop, or change the direction of the object. It is possible to apply a force to an object and not move the object. For example, you may push with all your strength on a car stuck in a ditch and not move the car. Under this condition, no work is done. Work is only accomplished when an object is started, stopped, or redirected by a force.

Simple Machines

A machine is any device that can be used to transmit a force and, in doing so, change the amount of force and/or its direction. A common example of a simple machine that does both is a valve rocker arm. One end of a rocker arm is pushed up by the action of the engine's camshaft. When this happens, the other end of the rocker arm pushes down on a valve to open it. A rocker arm also is designed to change the size of the force applied to it. Rocker arms provide more movement on the valve side or output than on the input side, a condition referred to as the ratio of

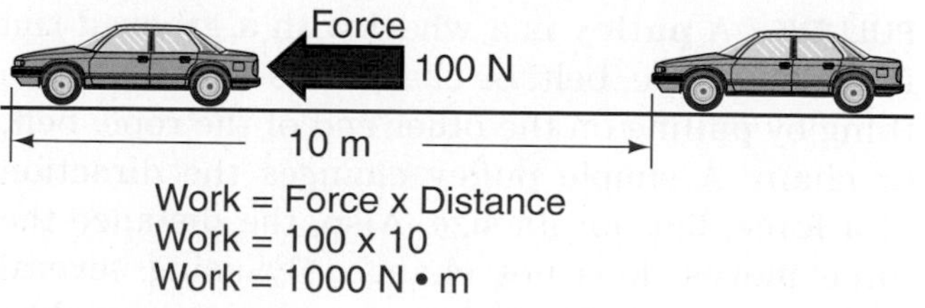

FIGURE 7–23

One thousand Newton-metres of work.

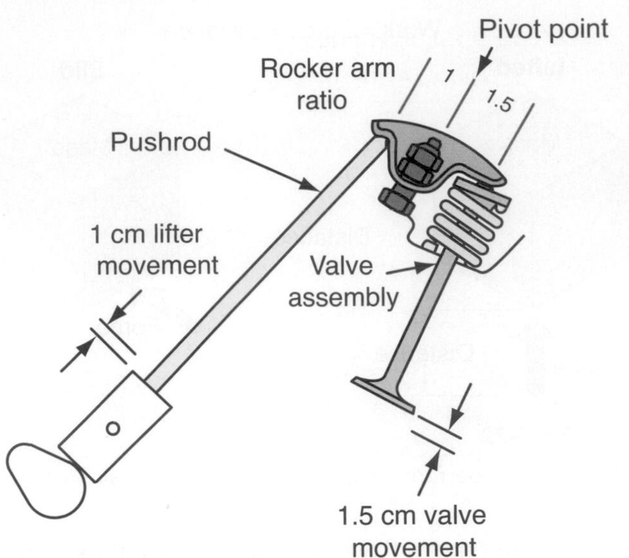

FIGURE 7–24

A rocker arm with a ratio of 1.5:1.

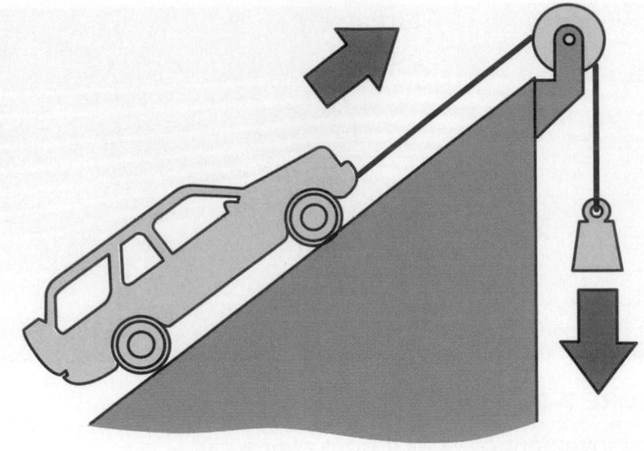

FIGURE 7–25

It takes less energy to pull a mass up an inclined plane than to lift the mass vertically.

the rocker arm. If a rocker arm has a ratio of 1.5:1, one end will move 1.5 times more than the other **(Figure 7–24)**. For example, if the camshaft causes one end of the rocker arm to move 1 cm (0.4 in.), the other end will move 1.5 cm (0.6 in.).

The force applied to a machine is called the effort, and the force it overcomes is called the **load**. The effort is often smaller than the load, because a small effort can overcome a heavy load if the effort is moved a larger distance. The machine is then said to give a mechanical advantage. Although the effort will be smaller when using a machine, the amount of work done, or energy used, will be equal to or greater than that without the machine.

INCLINED PLANE The force required to drag an object up a slope **(Figure 7–25)** is less than that required to lift it vertically. However, the overall distance moved by the object is greater when pulled up the slope than if it were lifted vertically. A screw is like an inclined plane wrapped around a shaft. The force that turns the screw is converted to a larger one, which moves a shorter distance and drives the screw in.

PULLEYS A **pulley** is a wheel with a grooved rim in which a rope, belt, or chain runs to raise something by pulling on the other end of the rope, belt, or chain. A simple pulley changes the direction of a force, but not its size. Also, the distance the force moves does not change. By using several pulleys connected together as a block and tackle, the size of the force can be changed too so that a heavy load can be lifted using a small force. With a double pulley, the applied force required to move an object can be reduced by one-half, but the distance the force must be moved is doubled. A quadruple pulley can reduce the force by four times, but the distance will be increased by four times. Pulleys of different sizes can change the amount of required applied force as well as the speed or distance the pulley needs to travel to accomplish work **(Figure 7–26)**.

LEVERS A **lever** is a device made up of a bar turning about a fixed pivot point called the fulcrum that uses a force applied at one point to move a mass on the other end of the bar. Types of levers are divided into classes. In a class one lever, the fulcrum is between the effort and the load **(Figure 7–27)**. The load is larger than the effort, but it moves through a smaller distance. A pair of pliers is an example of a class one lever. In a class two lever, the load is between the fulcrum and the effort. A wheelbarrow is an example of a class two lever. Here again, the load is greater than the effort and moves through a smaller distance **(Figure 7–28)**. In a class three lever, the effort is between the fulcrum and the load, and the load moves in the same direction as the effort. In this case, the load is less than the effort but it moves through a greater distance.

GEARS A **gear** is a toothed wheel that becomes a machine when it is meshed with another gear. The action of one gear is that of a rotating lever and moves the other gear meshed with it. Based on the size of the gears in mesh, the amount of force applied from one gear to the other can be changed. Keep in mind that this change in gear size does not change the amount of work performed by the gears, although as the force changes so does the distance of

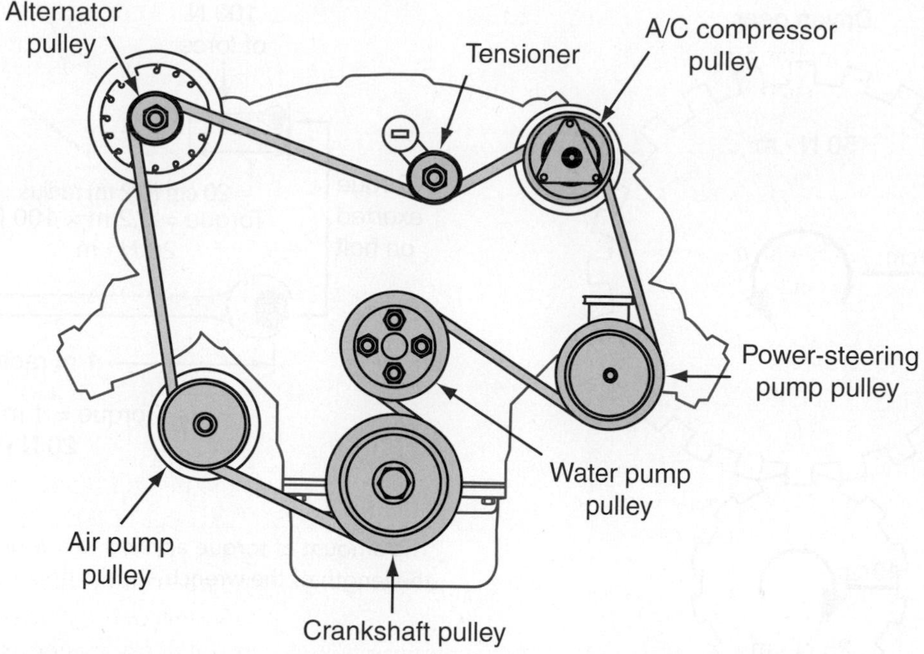

FIGURE 7–26

Accessories are driven by a common drive belt, but they rotate at different speeds because of the differences in pulley size.

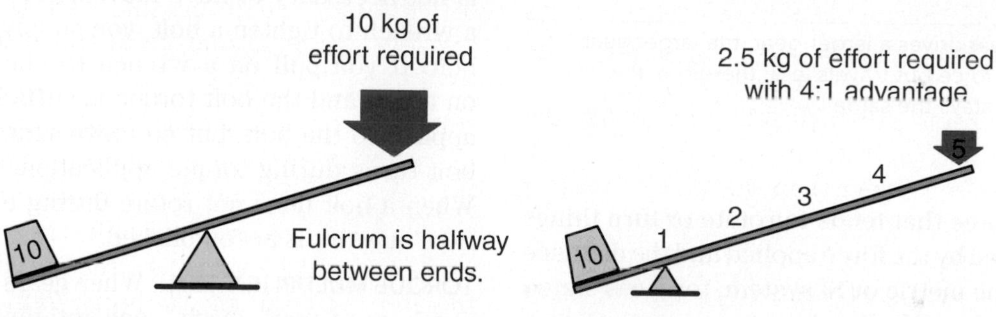

FIGURE 7–27

A mechanical advantage can be gained with a class one lever.

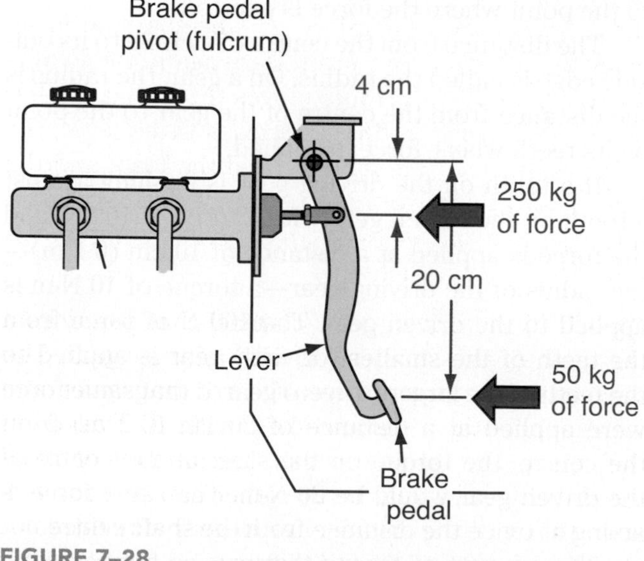

FIGURE 7–28

A brake pedal assembly is an example of a class two lever.

travel **(Figure 7–29)**. The relationship of force and distance is inverse. Gear ratios express the mathematical relationship (diameter and number of teeth) of one gear to another.

WHEELS AND AXLES The most obvious application of a wheel and axle is a vehicle's tires and wheels. These units revolve around an axle and limit the amount of area of a vehicle that contacts the road. Wheels function as rollers to reduce the amount of friction between a vehicle and the road. Basically, the larger the wheel, the less force is required to turn it. However, the wheel must move farther as it gets larger. An example of this principle is a steering wheel. A steering wheel that is twice the size of another will require one-half the force to turn it but will also require twice the distance to accomplish the same work.

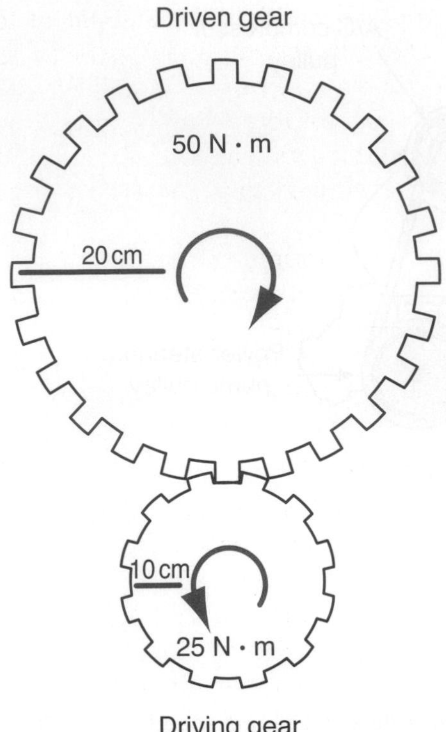

FIGURE 7–29

When a small gear drives a larger gear, the larger gear turns with more force but travels less; therefore, the amount of work stays the same.

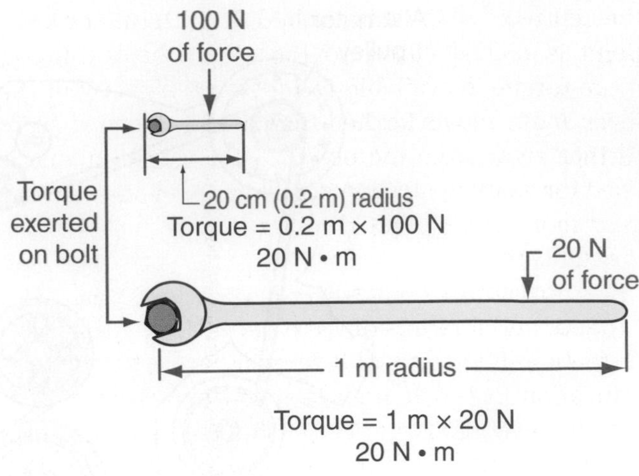

FIGURE 7–30

The amount of torque applied to a wrench is changed by the length of the wrench.

Torque

Torque is a force that tends to rotate or turn things and is measured by the force applied and the distance travelled. In the metric or SI system, torque is stated in Newton-metres (N·m) or kilogram-metres (kg·m). The technically correct imperial unit of measurement for torque is pounds per foot (lb·ft). However, it is common to see torque stated in terms of foot-pounds (ft·lb).

An engine creates torque and uses it to rotate the crankshaft. The combustion of gasoline and air in the cylinder creates pressure against the top of a piston. That pressure creates a force on the piston and pushes it down. The force is transmitted from the piston to the connecting rod and from the connecting rod to the crankshaft. The engine's crankshaft rotates with a torque that is transmitted through the drivetrain to turn the drive wheels of the vehicle.

Torque is force times leverage, the distance from a pivot point to an applied force. Torque is generated any time a wrench is turned with force. If the wrench is 20 cm (0.2 m) long and you put a force of 100 N on it, 20 N·m (0.2 × 100) is being generated. To generate the same amount of torque while only exerting 20 N of force, the wrench would have to be 1 m long **(Figure 7–30)**. An imperial system example would have a 1-ft.-long wrench with 20 lb. of force applied to it, generating 20 lb·ft of torque. To have torque, it is not necessary to have movement. When you pull a wrench to tighten a bolt, you supply torque to the bolt. If you pull on a wrench to check the torque on a bolt and the bolt torque is sufficient, torque is applied to the bolt, but no movement occurs. If the bolt turns during torque application, work is done. When a bolt does not rotate during torque application, no work is accomplished.

TORQUE MULTIPLICATION When gears with different numbers of teeth mesh, each rotates at a different speed and force. Torque is calculated by multiplying the force by the distance from the centre of the shaft to the point where the force is exerted.

The distance from the centre of a circle to its outside edge is called the radius. On a gear, the radius is the distance from the centre of the gear to the point on its teeth where force is applied.

If a tooth on the driving gear is pushing against a tooth on the driven gear with a force of 100 N, and the force is applied at a distance of 10 cm (0.1 m)—the radius of the driving gear—a torque of 10 N·m is applied to the driven gear. The 100 N of force from the teeth of the smaller (driving) gear is applied to the teeth of the larger (driven) gear. If that same force were applied at a distance of 20 cm (0.2 m) from the centre, the torque on the shaft at the centre of the driven gear would be 20 N·m. The same force is acting at twice the distance from the shaft centre.

The amount of torque that can be applied from a power source is proportional to the distance from the centre at which it is applied. If a fulcrum or pivot point is placed closer to the object being moved, more torque is available to move the object, but the lever must move farther than if the fulcrum were further away from the object. The same principle is used for gears in mesh: A small gear will drive a large gear more slowly, but the larger gear will produce greater torque.

A drivetrain consisting of a driving gear with 11 teeth and a radius of 2 cm and a driven gear with 44 teeth and a radius of 8 cm will have a torque multiplication factor of 4 and a speed reduction of one-fourth. Thus, the larger gear will turn with four times the torque but one-fourth the speed **(Figure 7–31)**. The radii between the teeth of a gear act as levers; therefore, a gear that is twice the size of another has twice the lever arm length of the other.

Gear ratios express the mathematical relationship of one gear to another. Gear ratios can be varied by changing the diameter and number of teeth of the gears in mesh. A gear ratio also expresses the amount of torque multiplication between two gears. The ratio is obtained by dividing the diameter or number of teeth of the driven gear by the diameter or teeth of the drive gear. If the smaller driving gear has 11 teeth and the larger gear has 44 teeth, the ratio is 4:1.

Power

Power is a measurement for the rate, or speed, at which work is done. The metric unit for power is the watt. A watt is equal to one Newton-metre per second. You can multiply the amount of torque in Newton-metres by the rotational speed to determine the power in watts. Power is a unit of speed combined with a unit of force. For example, if you were pushing something with a force of 1 N and it moved at a speed of 1 m/s, the power output would be 1 watt (W).

In electrical terms, one watt is equal to the amount of electrical power produced by a current of one ampere across a potential difference of one volt. Mathematically, this is expressed as Power (P) = Voltage (E) × Current (I) or $P = E \times I$.

Horsepower

Horsepower is the rate at which torque is produced. James Watt is credited with being the first person to calculate horsepower and power. He measured the amount of work that a horse could do in a specific time. He found that a horse could move 330 lb. (150 kg) 100 ft. (30.5 m) in 1 minute **(Figure 7–32)**. Therefore, he determined that one horse could do 33 000 ft·lb of work in 1 minute. Thus, 1 horsepower (hp) is equal to 33 000 ft·lb per minute, or 550 ft·lb per second. Two horsepower could do this same amount of work in a half-minute. If you push a 3000 lb. (1360 kg) car for 11 ft. (3.3 m) in one-quarter minute, you produce 4 horsepower.

An engine that produces 300 lb·ft of torque at 4000 rpm produces 228 hp at 4000 rpm. This is based on the formula that horsepower is equal to torque multiplied by engine speed and that sum divided by 5252 ([torque × engine speed] ÷ 5252 = horsepower). The constant, 5252, is used to convert the rpm for torque and horsepower into revolutions per second.

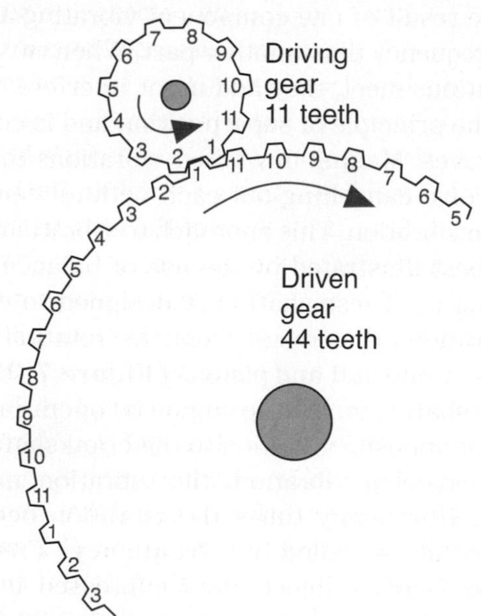

FIGURE 7–31
The driving gear must rotate four times to rotate the driven gear once.

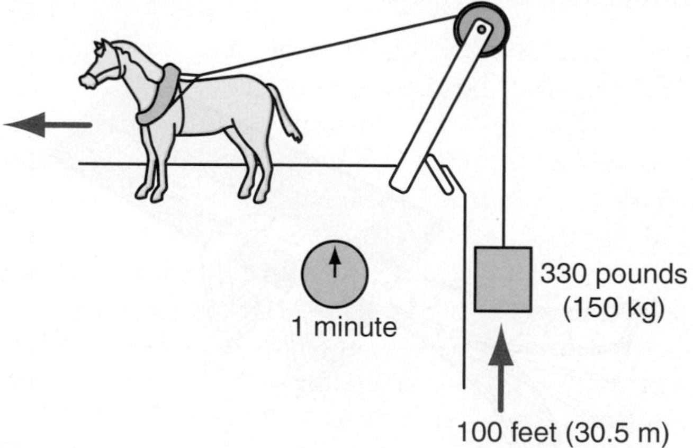

FIGURE 7–32
James Watt defined 1 horsepower as the power a horse exerts pulling 330 lb. (150 kg) 100 ft. (30.5 m) in 1 minute.

WAVES AND OSCILLATIONS

An **oscillation** is any single swing of an object back and forth between the extremes of its travel. When that motion travels through matter or space, it becomes a **wave**. A mass suspended by a spring, for example, is acted upon by two forces: gravity and the tension in the spring. At the point of equilibrium, the resultant of these forces is zero; they cancel each other out. When the mass is given an initial downward push, the tension of the spring exceeds the weight of the mass. The resultant upward force accelerates the mass back up toward its original position, by which time it has momentum, carrying it farther upward. When the weight exceeds the tension in the spring, the mass is pulled down again, and the oscillation repeats itself until the mass is at equilibrium. As the mass oscillates toward the equilibrium position, the size of the oscillation decreases unless it is again pushed downward. The air around the mass is upset as the mass oscillates and becomes an air wave.

Vibrations

When an object oscillates, it vibrates **(Figure 7–33)**. To prevent the vibration of one mass from causing a vibration in other masses, the oscillating mass must be isolated from other objects. This task is often difficult. For example, consider the engine in an automobile. Even the best-running engine vibrates as it runs. To reduce the transfer of engine vibrations to the vehicle, the engine is held in place by special mounts. The materials used in the mounts must keep the engine in its location but also be elastic enough to absorb the engine's vibrations **(Figure 7–34)**. If the engine were mounted solidly to the vehicle, the vibrations would be felt throughout the vehicle.

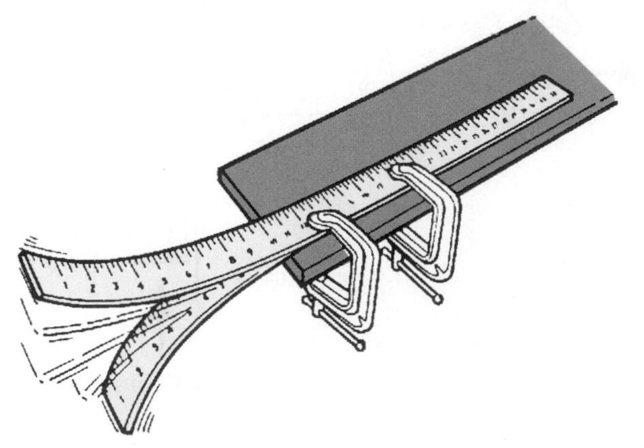

FIGURE 7–33

Vibrations happen in cycles.

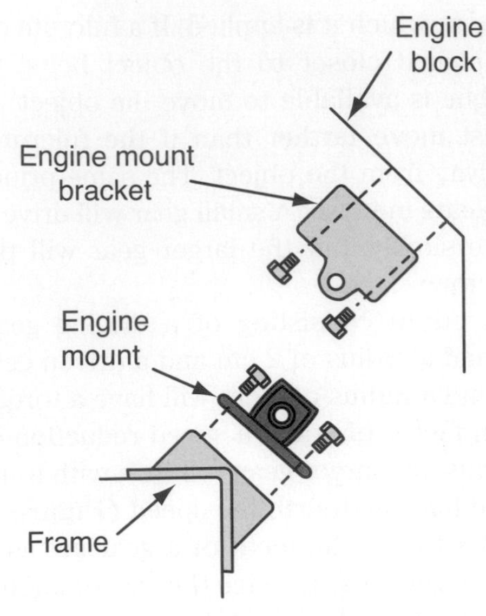

FIGURE 7–34

An engine mount holds the engine in place and isolates engine vibrations from the rest of the vehicle.

In an automobile, many parts vibrate as they operate. Through careful engineering, these vibrations can be isolated or insulated, thereby reducing the chances of the vibrations moving through the vehicle. Vibration control is also important for the reliability of components. If the vibrations are not controlled, the object could shake itself to destruction. Vibration control is the best justification for always mounting components in the way they were designed to be mounted.

Unwanted and uncontrolled vibrations are typically the result of one component vibrating at a different frequency than another part. When two waves or vibrations meet, they add up or interfere. This is called the principle of superposition and is common to all waves. Making unwanted vibrations tolerable may involve cancelling out each with an equal and opposite vibration. This approach to vibration reduction is best illustrated by the use of balance shafts in an engine. These shafts are designed to counter the vibrations that result from the rotation of the engine's crankshaft and pistons **(Figure 7–35)**. The balance shaft smooths the engine by operating at an equal but opposite vibration to the crankshaft.

To cancel a vibration, the vibration must be defined. How many times the vibration occurs in one second is called its **frequency**. Frequency **(Figure 7–36)** is most often expressed in **hertz (Hz)**. One hertz is equal to one cycle per second. The hertz is named in honor of Heinrich Hertz, an early German investigator of radio wave transmission. The **amplitude** and velocity must also be

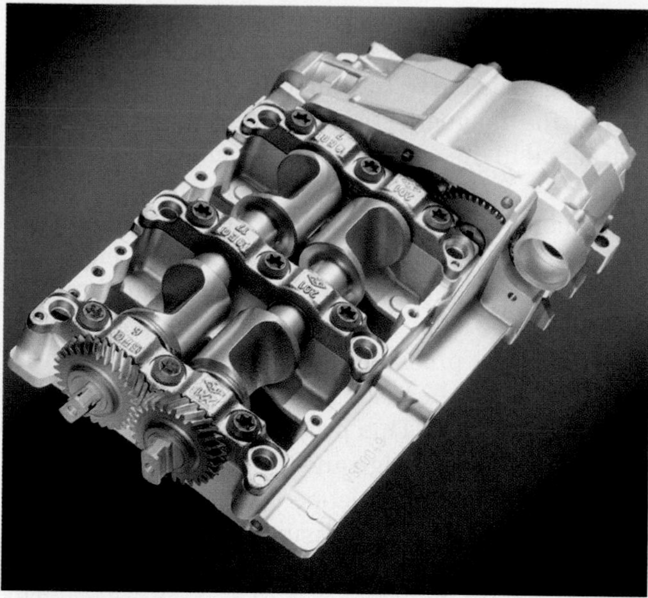

FIGURE 7–35

Balance shafts are driven by the crankshaft and work to counter crankshaft pulses and vibrations by acting with an equal force but in the opposite direction.

BMW of North America, Inc

defined. The amplitude of a vibration is its intensity or strength **(Figure 7–37)**. The velocity of a vibration is the result of its amplitude and its frequency. Because every material has a unique resonant or natural vibration frequency, everything on an automobile can vibrate at its own frequency.

Sound

Vibration is very common and results in the phenomenon of sound. In air, the vibrations that cause sound are transmitted as a wave between air molecules; many other substances transmit sound in a similar way. A vibrating object causes variations in pressure in the surrounding air. Areas of high and low pressure, known as compressions and rarefactions, move through the air as sound waves. Compression makes the sound waves denser, whereas rarefaction makes them less dense. The distance between each compression of a sound wave is called its **wavelength**. Sound waves with a short wavelength have a high frequency and a high-pitched sound.

When the rapid variations in pressure occur between about 20 Hz and 20 kHz, sound is potentially audible. Audible sound is the sensation (as detected by the ear) of very small rapid changes in the air pressure above and below atmospheric pressure.

Certain terms are used to describe sound:

- The pitch of a sound is based on its frequency. The greater the frequency, the higher the pitch.
- A decibel is a numerical expression of the relative loudness of a sound.

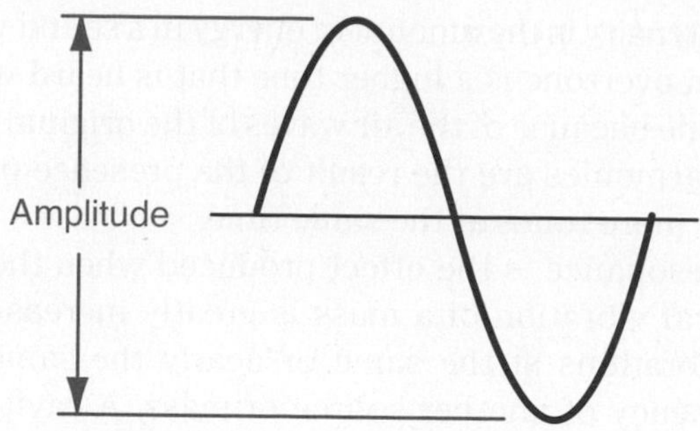

FIGURE 7–37

Amplitude is a measurement of a vibration's intensity.

Portions of materials contained herein have been reprinted with permission of General Motors under License Agreement #1410912.

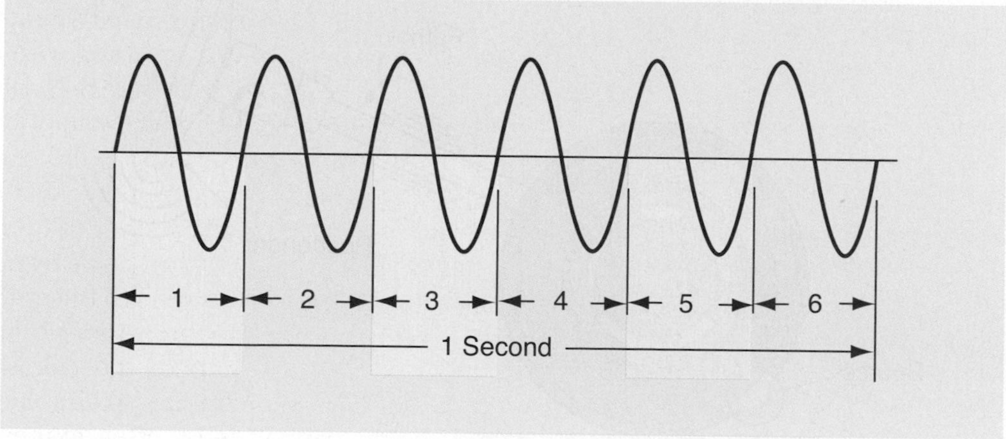

FIGURE 7–36

Frequency is a statement of how many cycles occur in a second. This is an example of 6 Hz frequency.

- Intensity is the amount of energy in a sound wave.
- An overtone is a higher tone that is heard with a tone because of the air waves of the original tone.
- Harmonics are the result of the presence of two or more tones at the same time.
- Resonance is the effect produced when the natural vibration of a mass is greatly increased by vibrations at the same or nearly the same frequency of another source or mass. A cavity has certain resonant frequencies. These frequencies depend on the shape and size of the cavity and the velocity of sound within the cavity.

When diagnosing automotive systems, you will often be told to listen to the sound of something. Mostly you will be paying attention to the type of sound and its intensity and frequency. The tone of the sound usually indicates the type of material causing the noise. If there is a high pitch, you know that the source of the sound is something that can vibrate quickly. This means the source is less rigid than something that vibrates at a low pitch. Although pitch depends on the frequency of a sound, the frequency itself can identify the possible sources of the sound. For example, if a sound from an engine increases with an increase in engine speed, you know the source of the sound must be something that is moving faster as a result of the increase in engine speed. If the frequency of the sound appears to be at one-half the speed of the engine, you know the source of the sound is something that is rotating at that speed, such as the camshaft.

SPEAKERS A speaker converts electrical energy into sound energy or waves. A constantly changing electrical signal is fed to the coil of a speaker, which lies within the magnetic field of a permanent magnet. The signal in the coil causes it to behave like an electromagnet, making it push against the field of the permanent magnet. The speaker cone is then pushed in and out by the coil in time with the electrical signal. As the cone moves forward, the air immediately in front of it is compressed, causing a slight increase in air pressure. It then moves back past its rest position and causes a reduction in the air pressure (rarefaction). This process continues so that a wave of alternating high and low pressure is radiated away from the speaker cone at the speed of sound. These changes in air pressure are actually sound. The sound from a speaker may be amplified by the space or cavity that surrounds the speaker cone. The room or area that the speaker sits in also works to amplify the sound.

Noise

Noise is any unwanted signal or sound. It can be random or periodic. To identify the source of a noise, it is important to remember that sound or noise is a vibration and that the vibration may be travelling through other components. Therefore, the source of the noise is not always where it may appear **(Figure 7–38)**. There are many potential sources of noise in an automobile, and manufacturers work hard to prevent noise.

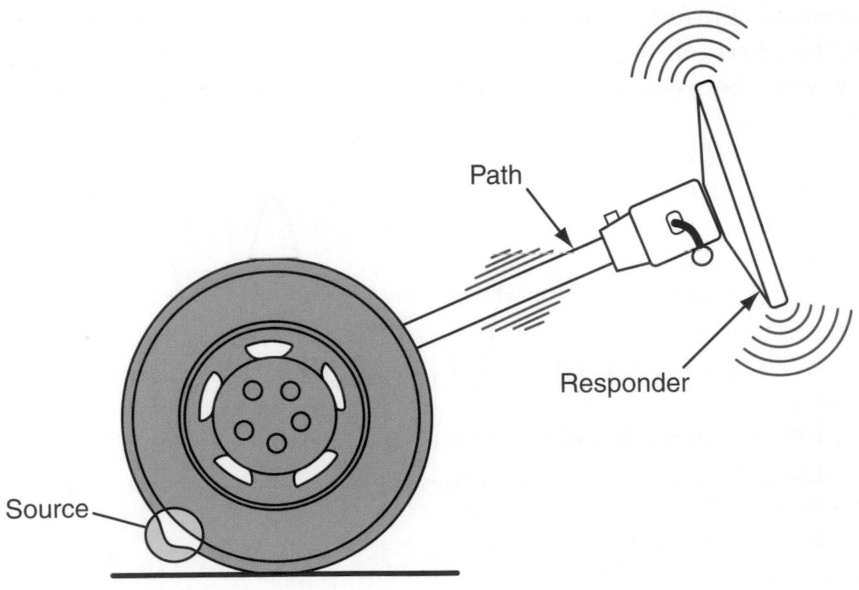

FIGURE 7–38

A vibration and/or noise will easily move through components so that it appears that the responder (in this case, the steering wheel) is the cause of the noise or vibration.

Noise can be prevented or reduced by three different approaches. The most effective way is to intervene at the design stage to redesign a noisy component so that it produces less noise. A relatively new technique for noise reduction is antinoise, or active noise control, which involves producing a sound that is similar to, but out of phase with, the noise. This sound effectively cancels the original noise. More obvious methods of noise reduction, or passive noise control, involve the use of filters, insulation, and noise barriers.

A filter is an electrical circuit that allows signals in certain frequency ranges to pass through and that blocks all other frequencies. Sound insulation prevents sound from travelling from one place to another. Heavy materials such as concrete are the most effective materials for sound insulation. Sound insulation or deadening materials are placed strategically throughout a modern vehicle. Some sound-deadening materials actually absorb sounds. These materials must be able to vibrate without creating sound. Sound insulators rarely absorb sound.

LIGHT

Light is a form of electromagnetic radiation. In free shape, it travels in a straight line at 300 million metres per second. When a beam of light meets an object, a proportion of the rays may be reflected. Some light may be absorbed and some transmitted. Without reflection, we would only be able to see objects that give out their own light. Light always reflects from a surface at the same angle at which it strikes. Therefore, parallel rays of light reflecting off a very flat surface remain parallel. A beam of light reflecting from an irregular surface scatters in all directions. Light that passes through an object is bent or refracted. The angle of refraction depends on the angle at which the light meets the object and the material it passes through. Lenses and mirrors can cause light rays to diverge or converge. When light rays converge, they can reach a point of focus.

These principles are the basis for fibre-optic lighting. With fibre optics, the light from a single lamp moves through one or more fibre cables to illuminate a point remote from the source lamp. The fibre cables are designed to allow the light to travel without losing intensity because of reflection. The light beam is bent through refraction as it travels through the cable and can be delivered to many locations at the same time.

Photo Cells

Radiation is produced in the sun's core during its nuclear reactions and is the source of most of Earth's energy. A transfer of energy from electromagnetic radiation to electrical energy takes place in a photovoltaic (photo) cell, or solar cell. When no sunlight falls on it, it can supply no electricity.

LIQUIDS

A fluid is something that does not have a definite shape. Fluids and gases share some similar properties, one being that fluids and gases tend to conform to the shape of their container. A major difference between a gas and a liquid is that a gas always fills a sealed container, whereas a liquid may not. A gas also readily expands or compresses according to the pressure exerted on it. Liquids are basically incompressible, which gives them the ability to transmit force **(Figure 7–39)**. The pressure applied to a liquid in a sealed container is transmitted equally in all directions and to all areas of the system and acts with equal pressure on all areas. As a result, liquids can provide great increases in the force available to do work. They also always seek a common level. A liquid under pressure may also change from a liquid to a gas in response to temperature changes.

Liquids exert pressure on immersed objects, resulting in an upward force called upthrust (buoyancy). The upthrust is equal to the weight of the liquid displaced by the immersed object. If the upthrust on

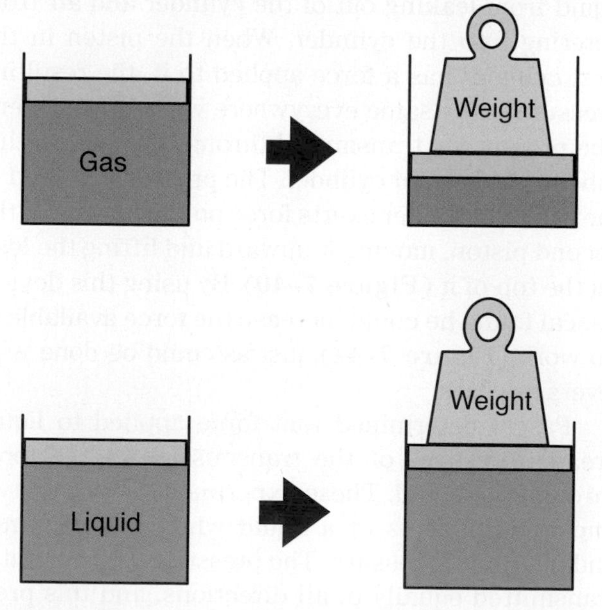

FIGURE 7–39
Gases compress; liquids (fluids) do not.

an object is greater than the weight of the object, then the object will float. Large ships float because they displace huge amounts of water, producing a large upthrust.

Laws of Hydraulics

Hydraulics is the study of liquids in motion. Liquids predictably respond to pressures exerted on them. Their reaction to pressure is the basis of all hydraulic applications. This fact allows hydraulics to do work. A simple hydraulic system has liquid, a pump, lines to carry the liquid, control valves, and an output device. The liquid must be available from a continuous source, such as an oil pan or sump. An oil pump is used to move the liquid through the system. The lines that carry the liquid may be pipes, hoses, or a network of internal bores or passages in a single housing. Control valves are used to regulate hydraulic pressure and direct the flow of the liquid. The output device is the unit that uses the pressurized liquid to do work.

More than 300 years ago, a French scientist, Blaise Pascal, determined that if you had a liquid-filled container with only one opening and applied force to the liquid through that opening, the force would be evenly distributed throughout the liquid. This explains how pressurized liquid is used to operate and control systems, such as the brake and automatic transmission systems.

Pascal constructed the first known hydraulic device, which consisted of two sealed containers connected by a tube. The pistons inside the cylinders seal against the walls of each cylinder, preventing the liquid from leaking out of the cylinder and air from entering into the cylinder. When the piston in the first cylinder has a force applied to it, the resulting pressure is the same everywhere within the system. The pressure is transmitted through the connecting tube to the second cylinder. The pressurized fluid in the second cylinder exerts force on the bottom of the second piston, moving it upward and lifting the load on the top of it **(Figure 7–40)**. By using this device, Pascal found he could increase the force available to do work **(Figure 7–41)**, just as could be done with levers or gears.

Pascal determined that force applied to liquid creates pressure or the transmission of the force through the liquid. These experiments revealed two important aspects of a liquid when it is confined and put under pressure. The pressure applied to it is transmitted equally in all directions, and this pressure acts with equal force at every point in the container. If a liquid is confined and a force is applied,

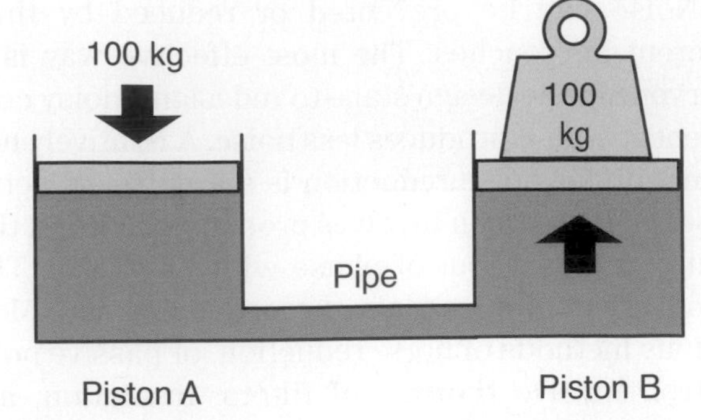

FIGURE 7–40

In a hydraulic circuit, pressure is transferred equally throughout the system.

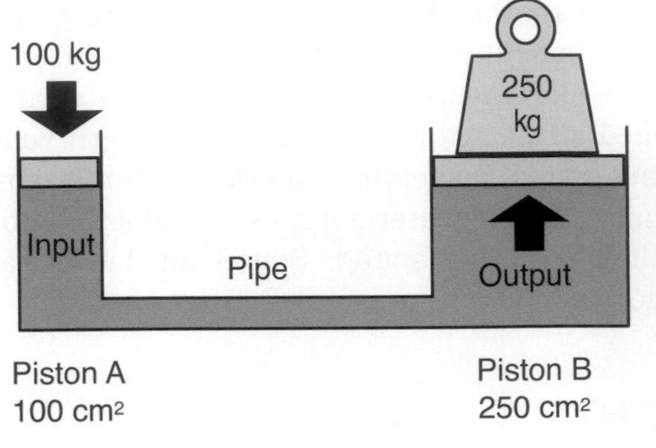

FIGURE 7–41

The force available to do work can be increased by increasing the size of the piston doing the work.

pressure is produced. In order to pressurize a liquid, the liquid must be in a sealed container. Any leak in the container will decrease the pressure.

Mechanical Advantage with Hydraulics

Hydraulics are used to do work in the same way a lever or gear does. All of these systems transmit energy. Because energy cannot be created or destroyed, these systems only redirect energy to perform work and do not create more energy. If a hydraulic pump provides 7 kg/cm^2 (100 psi), there will be 7 kg of force on every square centimetre (100 lb. force on every square inch) of the system **(Figure 7–42)**. If the system includes a piston with an area of 322 cm^2 (50 sq. in.), each square centimetre receives 7 kg of force (each square inch receives 100 lb. of force). This means there will be 2254 kg (5000 lb.) of force exerted from this piston **(Figure 7–43)**, but the output's travel will decrease proportionally. The use of the larger piston gives the system a **mechanical advantage**, or increase in the force available to do work. The multiplication of force through a hydraulic

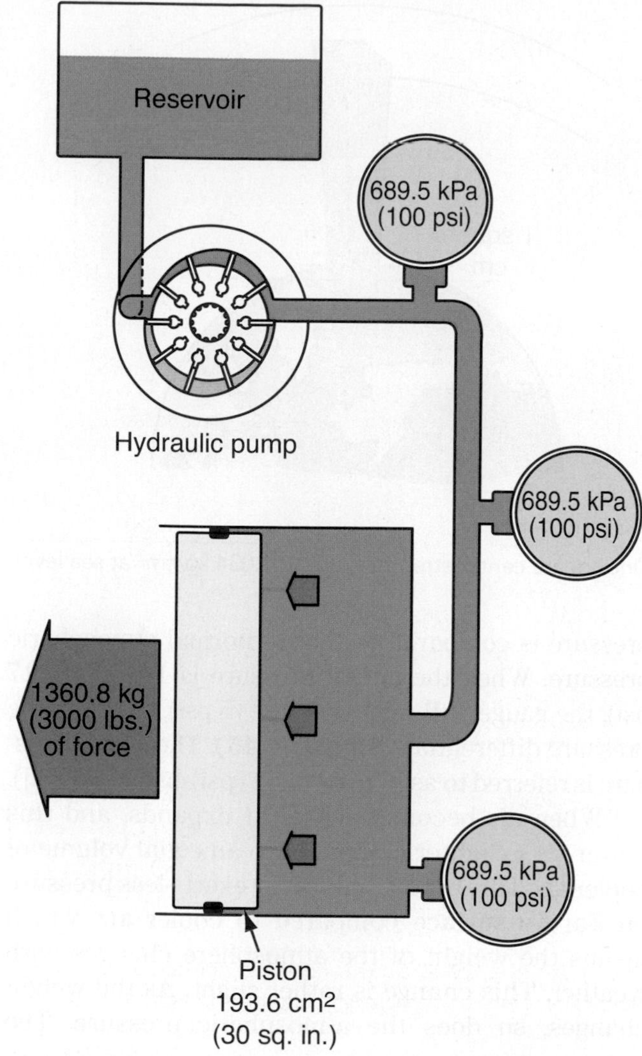

FIGURE 7–42

A pressure applied to a liquid is transmitted equally and acts with equal force at every point within the hydraulic circuit.

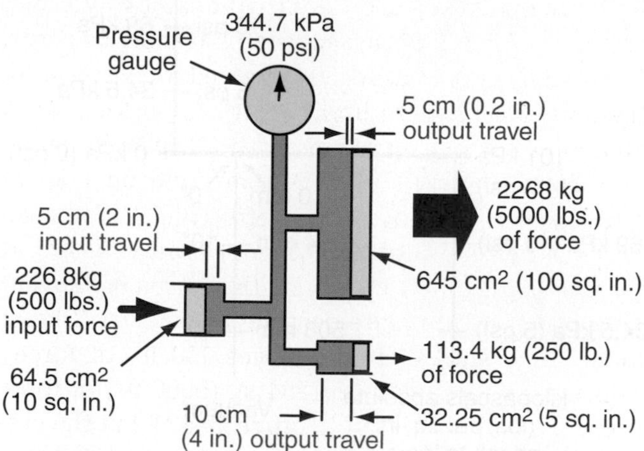

FIGURE 7–43

Hydraulic systems can provide an increase in force (mechanical advantage), but the output's travel will decrease proportionally.

system is directly proportional to the difference in the piston sizes throughout the system.

By changing the size of the pistons in a hydraulic system, force is multiplied, and as a result, low amounts of force can be used to move heavy objects. The mechanical advantage of a hydraulic system can be further increased by the use of levers to increase the force applied to a piston.

Although the force available to do work is increased by using a larger piston in one cylinder, the total movement of the larger piston is less than that of the smaller one. A hydraulic system with two cylinders, one with a 20 cm^2 (3.10 sq. in.) piston and the other with a 10 cm^2 (1.55 sq. in.) piston, will double the force at the second piston; however, the total movement of the larger piston will be half the distance of the smaller one.

The use of hydraulics to gain a mechanical advantage is similar to the use of levers or gears. Hydraulics is preferred when the size and shape of the system is of concern. In hydraulics, the force applied to one piston transmits through the fluid, and the opposite piston has the same force on it. The distance between the two pistons in a hydraulic system does not affect the force in a static system. Therefore, the force applied to one piston can be transmitted without change to another piston located somewhere else.

A hydraulic system responds to the pressure or force applied to it. The mere presence of different-sized pistons does not always result in fluid power. Either the pressure applied to the pistons must be different, or the size of the pistons must be different in order to cause fluid power. If an equal amount of pressure is exerted onto both pistons in a system and both pistons are the same size, neither piston will move, and the system is balanced or is at equilibrium. The pressure inside the hydraulic system is called **static pressure** because there is no fluid motion.

When an unequal amount of pressure is exerted on the pistons, the piston with the least amount of pressure on it will move in response to the difference between the two pressures. Likewise, if the size of the two pistons is different and an equal amount of pressure is exerted on the pistons, the fluid will move. The pressure of the fluid while it is in motion is called **dynamic pressure**.

GASES

A gas comprises independent particles—atoms or molecules—in random motion. This means that a gas will fill any container into which it is placed. The random movement of gas particles also ensures that any two gases sharing the same container will totally mix. This phenomenon is called **diffusion**.

The kinetic energy in atoms and molecules increases as the temperature increases. A decrease in temperature reduces the kinetic energy. Molecules in solids move slowly compared to those in liquids or gases. Gas molecules move quickly compared to liquid molecules. Because gas molecules are in constant motion, they spread out to fill all the space available. At higher temperatures, gas molecules spread out more, whereas at lower temperatures, gas molecules move closer together. The bombardment of particles against the sides of the container produces pressure.

Behaviour of Gases

Three simple laws describe the predictable behaviour of gases: Boyle's law, the pressure law, and Charles's law. Each of these laws describes a relationship among the pressure, volume, and temperature of a gas.

Boyle's law states that the volume and pressure of a mass of gas at a fixed temperature is inversely proportional. If the pressure on a gas increases, its volume will decrease; likewise, if the volume is increased, the pressure will decrease.

The pressure law (Gay-Lussac's law) states that the pressure exerted by a gas held at a constant volume is proportional to its temperature. This means that as the gas temperature increases, so does its pressure.

Charles's law states that the volume of a given mass of gas is directly proportional to its temperature. Simply put, this means the higher the temperature, the greater the volume. If the volume cannot change, the pressure of the gas will. Therefore, the pressure and temperature of a gas are also directly related. If you increase one, you also increase the other. This explains why cold air is denser than warm air.

Air Pressure

Because air is gaseous matter with mass and weight, it exerts pressure on Earth's surface. A one-square-centimetre column of air extending from Earth's surface to the outer edge of the atmosphere weighs 1.034 kg (1 sq. in. = 14.7 lb.) at sea level. Therefore, atmospheric pressure is 1.034 kg/cm^2 (14.7 psi), or 101 kPa at sea level **(Figure 7–44)**. **Atmospheric pressure** may be defined as the total weight of Earth's atmosphere. Pressure greater than atmospheric pressure may be measured in kPa (psi gauge [psig]). Using a standard pressure gauge, air

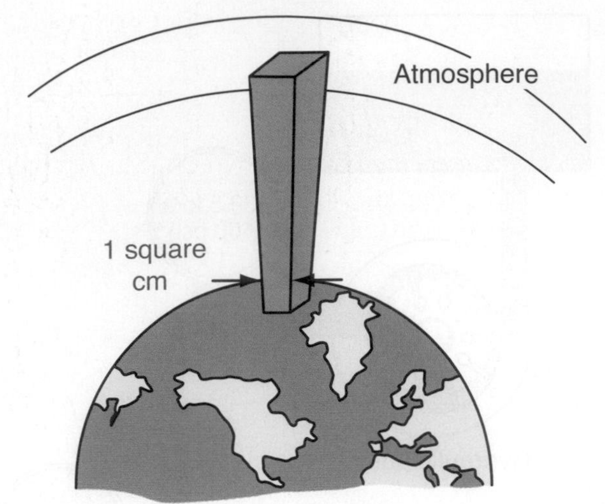

FIGURE 7–44

One square centimetre of air equals 1.034 kg/cm^2 at sea level.

pressure is compared to that of normal atmospheric pressure. When the actual pressure is 136 kPa (19.7 psi), the gauge will read 34.5 kPa (5 psi), showing the pressure differential **(Figure 7–45)**. The actual pressure is referred to as kPa absolute (psi absolute [psia]).

When air becomes hotter, it expands, and this hotter air is lighter compared to an equal volume of cooler air. This hotter, lighter air exerts less pressure on Earth's surface compared to cooler air, which means the weight of the atmosphere changes with weather. This change is rather slight. As the weight changes, so does the atmospheric pressure. The

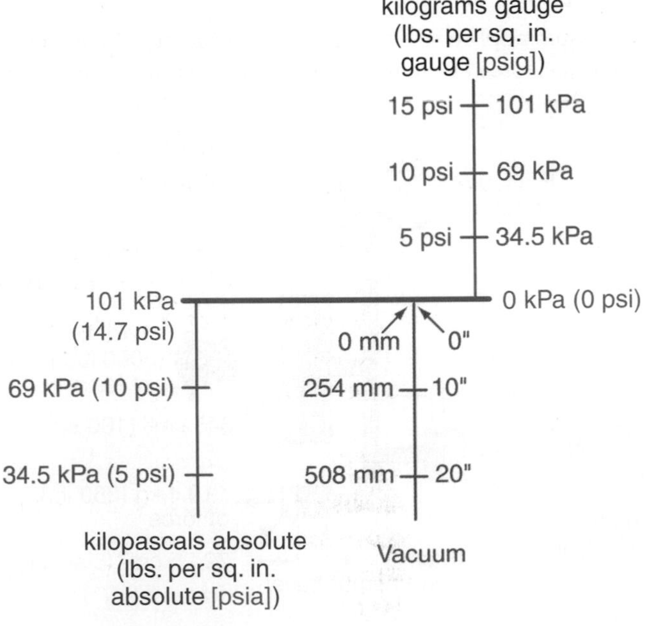

FIGURE 7–45

The relationship between psia and psig.

change in atmospheric pressure is measured with a barometer and is called **barometric pressure**. Barometric pressure at normal atmospheric pressure is 760 mm Hg (29.92 in. of mercury). The increments for measuring barometric pressure are based on the increments of a barometer, which are either 1 mm (metric) or 0.05 in. (imperial). A barometer is a J-shaped tube with mercury in it. One end of the tube is exposed to normal atmospheric pressure and the other end to current atmospheric pressure. When the current atmospheric pressure equals normal atmospheric pressure, the level of the mercury will rise 760 mm (29.92 in.) up the tall side of the J. When the current atmospheric pressure is lower than normal, the normal atmospheric pressure pushes the mercury down. Likewise, when current atmospheric pressure is higher than normal, it pushes the mercury up the tube. The amount of mercury movement reflects the difference in the two pressures. This corresponds with a universal law that states that a high pressure always moves toward a lower pressure.

Although the pressure of the atmosphere only changes slightly, the impact of these changes can be critical to the overall operation of an engine. The combustion process depends on having the correct amount of air enter into the cylinders. If the calibrations for the air and the accompanying amount of fuel did not consider the changes in atmospheric pressure, the engine would most often not receive the correct mixture of air and fuel. Today's engines are equipped with a sensor to monitor barometric pressure.

To further consider the law that states that a high pressure always moves to a lower pressure, consider what happens when a nail punctures an automotive tire. The high-pressure air in the tire leaks out until the pressure inside the tire is equal to atmospheric pressure outside the tire. When the tire is repaired and inflated, air with a pressure higher than atmospheric pressure is forced into the tire.

When you climb above sea level, atmospheric pressure decreases. The weight of a column of air is less at an elevation of 1524 m (5000 ft) than it is at sea level. As altitude continues to increase, atmospheric pressure and weight continue to decrease. At an altitude of several hundred kilometres above sea level, Earth's atmosphere ends, and there is no pressure beyond that point.

VACUUM Scientifically, **vacuum** is defined as the absence of atmospheric pressure. However, it is commonly used to refer to any pressure less than atmospheric pressure. Vacuum may also be referred to as low pressure simply because it is a pressure less than atmospheric pressure.

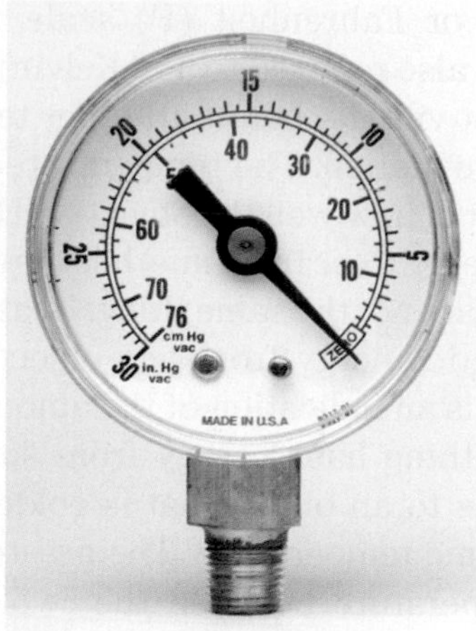

FIGURE 7–46

A vacuum gauge measures pressures below atmospheric pressure in units of inches of mercury.

Vacuum could be measured with pressure gauges, but millimetres of mercury (mm Hg, or inches of mercury [in. Hg]) are most commonly used for this measurement **(Figure 7–46)**. Let us assume that a plastic U-tube is partially filled with mercury, and atmospheric pressure is allowed to enter one end of the tube. If vacuum is supplied to the other end of the U-tube, the mercury is forced downward by the atmospheric pressure. When this movement occurs, the mercury also moves upward on the side where the vacuum is supplied. If the mercury moves downward 254 mm (10 in.), where the atmospheric pressure is supplied, and upward 254 mm (10 in.), where the vacuum is supplied, 508 mm Hg (20 in. Hg) is supplied to the U-tube. The highest possible, or perfect, vacuum is approximately 759.46 mm Hg (29.9 in. Hg).

HEAT

Heat is a form of energy and is used in many ways. The main sources of heat are the sun, Earth, chemical reactions, electricity, friction, and nuclear energy. Heat is the result of the kinetic energy present in all matter; therefore, everything has heat. Cold objects have low kinetic energy because their atoms and molecules are moving very slowly, whereas hot objects have more kinetic energy because their atoms and molecules are moving fast.

Temperature is an indication of an object's kinetic energy. Temperature is measured with a thermometer, which has either a Celsius (C)

(Centigrade) or Fahrenheit (F) scale. At absolute zero (–273°C, also referred to as 0 Kelvin), particles of matter do not vibrate, but at all other temperatures, particles have motion. The temperature of an object is a statement of how well the object will transfer heat or kinetic energy to or from another object. Heat and temperature are not the same thing. Heat is the movement of kinetic energy from one object to another; temperature is an indication of the amount of kinetic energy something has. Energy from something hot always moves to an object that is colder, until both are at the same temperature. The greater the difference in temperature between the two objects, the faster the heat will flow from one to the other.

Heat is measured in **calories** and British thermal units (BTUs). One calorie is equal to the amount of heat needed to raise the temperature of 1 g of water 1°C. One BTU is the amount of heat required to heat 1 lb. of water by 1°F.

Heat Transfer

Heat transfers between two substances at different temperatures through convection, conduction, or radiation. **Convection** is the transfer of heat by the movement of a heated object. Convection can be easily seen by watching a pot of water on a hot stove. The water on the bottom of the pot is the first to be heated by the stove. As the water at the bottom becomes hotter, it expands and becomes lighter than the water at the top of the pan, causing the heavier water to sink toward the bottom and push the warmer water up. This action continues until all of the water in the pan is the same temperature.

Conduction is the movement of heat through a material. Much of the heat generated by a running engine is not used to drive the vehicle. Rather, it is "wasted" as it heats the parts of the engine. Some of the heat produced during combustion is absorbed by the surfaces exposed to it, such as the piston top, combustion chamber of the cylinder head, valves, and cylinder walls. This heat must be controlled at safe limits to prevent damage to the engine components, breakdown in lubrication, scuffing, or uncontrolled expansion of pistons and rings. The combustion process superheats the gases present in the combustion chamber, causing them to rapidly expand, resulting in huge pressures. It is these expanding gases that drive the pistons downward, ultimately driving the crankshaft. The engine's cooling system uses conduction to move the heat off the parts to help cool the engine. Because the energy from something hot moves toward something colder, the heat from the engine moves to the engine's coolant circulating throughout the system. The object that receives or absorbs conducted heat can be liquid, gas, or solid.

Radiation does not rely on another material to transfer heat. The moving atoms and molecules within an object create waves of radiant energy called infrared rays. Hot objects give off more infrared rays than colder objects; therefore, a hot object gives off infrared rays to anything around it that is colder. No movement is necessary to transfer this heat. You can feel radiation in action by simply putting your hand near something hot. This radiation can heat up other objects. The hot object cools as it radiates its heat energy. In an engine's cooling system, the radiator uses radiation to transfer heat from the coolant into the surrounding air.

The Effects of Temperature Change

Whenever the temperature of an object has changed, a transfer of heat has occurred. A transfer of heat may also cause the object to change size or its state of matter. The amount of heat required to raise the temperature of 1 g of mass 1°C is called the specific heat capacity. Every substance has its own specific heat capacity, and this factor is assigned to material based on its difference from water, which has a specific heat capacity of 1. For example, the temperature of 1 g of water will increase by 10°C if 10 cal of heat are transferred to it. But if 10 cal of heat were added to 1 g of copper, the temperature would increase by 111°C because copper has a specific heat capacity of only 0.09 as compared to the 1.0 specific heat capacity of water.

As heat moves in and out of a mass, the movement of atoms and molecules in that mass increases or slows down. With an increase in motion, the size of the mass tends to get bigger or expand through a process called **thermal expansion**. **Thermal contraction** takes place when a mass has heat removed from it and the atoms and molecules slow down. All gases and most liquids and solids expand when heated, with gases expanding the most. Solids, because they are not fluid, expand and contract at a much lower rate. It is important to realize that all materials do not expand and contract at the same rate. For example, an aluminum component will expand at a faster rate than a similar iron component, which explains why aluminum cylinder heads have unique service requirements and procedures when compared to iron cylinder heads.

Thermal expansion takes place every time fuel and air are burned in an engine's cylinders. The

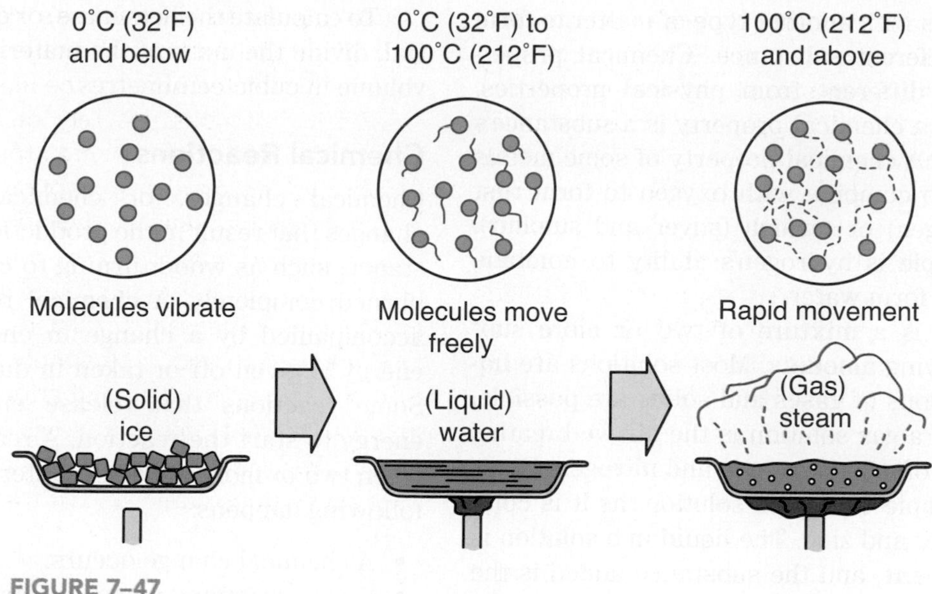

FIGURE 7–47

Water can exist in three different states of matter.

sudden temperature increase inside the cylinder causes a rapid expansion of the gases, which pushes the piston downward and causes engine rotation.

Typically, when heat is added to a mass, the temperature of the mass increases. This does not always happen, however. In some cases, the additional heat causes no increase in temperature but causes the mass to change its state (solid to liquid or liquid to gas). For example, if we heat an ice cube to 0°C (32°F), it will begin to melt **(Figure 7–47)**. As heat is added to the ice cube, the temperature of the ice cube will not increase until it becomes a liquid. The heat added to the ice cube that did not raise its temperature but caused it to melt is called **latent heat** or the heat of fusion. Each gram of ice at 0°C requires 80 cal of heat to melt it to water at 0°C. As more heat is added to the 0°C water, the water's temperature will once again increase. This process continues until the temperature of the water reaches 100°C (212°F), the boiling temperature of water. At this point, any additional heat applied to the water is latent heat, causing the water to change its state to that of a gas. This added heat is called the heat of evaporation.

To change the water gas back to liquid water, the same amount of heat required to change the liquid to a gas must be removed from the gas. At that point, the gas condenses to a liquid. As additional heat is removed, the temperature will drop until enough heat is removed to bring its temperature back down to freezing (melting in reverse) point. At that time, latent heat must be removed from the liquid before the water turns to ice again.

Controlling Heat

There is a particular temperature range in which all parts of an automobile will operate best. Engineers strive to control those temperatures to ensure reliable and efficient operation. This is a major task, as parts that do not perform well when hot are often mounted close to something that is very hot. High heat could transfer to the heat-sensitive parts if insulation or passing outside air were not present. Although some parts tolerate extreme temperatures, they must still be protected from overheating. The combustion inside a cylinder can generate temperatures as great as 1370°C (2500°F). If that heat transferred uncontrollably to the metal parts of an engine, those parts would expand to the point where they could no longer move or could melt. This is why the engine's cooling system is so important. The cooling system is a controlled heat-transfer system designed to protect the engine and allow it to run more efficiently.

CHEMICAL PROPERTIES

The properties of something are those characteristics used to identify or describe it. Physical properties are readily observable characteristics, such as colour, size, lustre, and smell. Physical changes are those changes that do not result in the production of a new substance, such as melting an ice cube.

Chemical properties are only observable during a chemical reaction, and they describe how one type of matter reacts with another type of matter to form a new and different substance. Chemical properties are quite different from physical properties. An example of a chemical property is a substance's ability to burn. A chemical property of some metals is the ability to combine with oxygen to form rust (iron and oxygen) or tarnish (silver and sulphur). Another example is hydrogen's ability to combine with oxygen to form water.

A solution is a mixture of two or more substances in varying amounts. Most solutions are liquids, but solutions of gases and solids are possible. An example of a gas solution is the air we breathe; it is composed of mostly oxygen and nitrogen. Brass is a good example of a solid solution, as it is composed of copper and zinc. The liquid in a solution is called the **solvent**, and the substance added is the solute. If both are liquids, the one present in the smaller amount is usually considered the solute. Solutions can vary widely in terms of how much of the dissolved substance is actually present. A heavily diluted (much water) acid solution has very little acid and may not be harmful or even noticeably acidic.

Specific Gravity

Specific gravity is the heaviness or relative density of a substance compared to that of water. If something is 3.5 times as heavy as an equal volume of water, its specific gravity is 3.5. Its density is 3.5 g/cm^3 or 3.5 kg/L. Because it is a ratio of two quantities that have the same dimensions (mass per unit volume), specific gravity has no dimension.

Specific gravity checks of a battery's electrolyte are an indication of the battery's state of charge **(Figure 7–48)**.

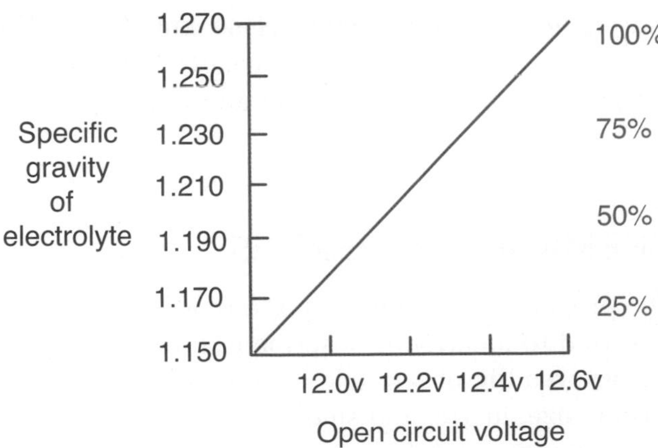

FIGURE 7–48

Specific gravity checks of a battery's electrolyte are an indication of the battery's state of charge.

To calculate the heaviness, or **density**, of a material, divide the mass of the material in grams by its volume in cubic centimetres.

Chemical Reactions

Chemical changes, or chemical reactions, are changes that result in the production of another substance, such as wood turning to carbon after it has burned completely. A chemical reaction is always accompanied by a change in energy. This means energy is given off or taken in during the reaction. Some reactions that release energy need some energy to start the reaction. A reaction takes place when two or more molecules interact and one of the following happens:

- A chemical change occurs.
- Single reactions occur as part of a large series of reactions.
- Ions, molecules, or pure atoms are formed.

Catalysts and Inhibitors

Reactions need a certain amount of energy to happen. If they do not have it, the reaction probably cannot happen. A **catalyst** lowers the amount of energy needed to make a reaction happen. A catalyst is any substance that affects the speed of a chemical reaction without itself being consumed or changed. Catalysts tend to be highly specific, reacting with one substance or a small set of substances. In a car's catalytic converter, the platinum catalyst converts unburned hydrocarbons and nitrogen compounds to products that are harmless to the environment **(Figure 7–49)**. Water, especially salt water, catalyzes oxidation and corrosion. An inhibitor is the opposite of a catalyst and stops or slows the rate of reaction.

Acids/Bases

An ion is an atom or group of atoms with one or more positive or negative electric charges. Ions are formed when electrons are added to or removed

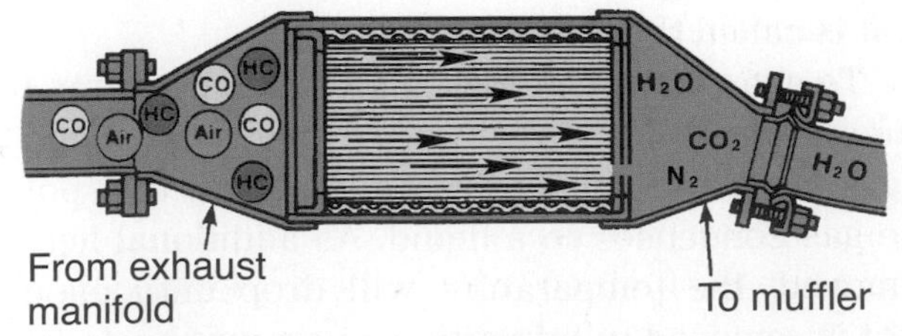

FIGURE 7–49

A basic catalytic converter that changes pollutants into chemicals that are good for the environment.

from neutral molecules or other ions. Many crystalline substances are composed of ions held in regular geometric patterns by the attraction of oppositely charged particles for each other. Ions are what make something an **acid** or a **base**.

Acids are compounds that break into hydrogen (H^+) ions and another compound when placed in an aqueous (water) solution. They have a sour taste, are corrosive, react with some metals to produce hydrogen, react with carbonates to produce carbon dioxide, change the colour of litmus from blue to red, and become less acidic when combined with alkalis. Most acids are slow-reacting, especially if they are weak acids. Acids also react with bases to form salts.

Alkalis (bases) are compounds that release hydroxide ions (OH^-) and react with hydrogen ions to produce water, thus neutralizing each other. Most substances are neutral (not an acid or a base). Alkalis feel slippery, change the colour of litmus from red to blue, and become less alkaline when they are combined with acids.

A hydroxide is any compound made up of one atom each of hydrogen and oxygen, bonded together and acting as the hydroxyl group or hydroxide anion (OH^-). An oxide is any chemical compound in which oxygen is combined with another element. Metal oxides typically react with water to form bases or with acids to form salts. Oxides of nonmetallic elements react with water to form acids or with bases to form salts.

A salt is a chemical compound formed when the hydrogen of an acid is replaced by a metal. Typically, an acid and a base react to form a salt and water.

pH The **pH scale** is used to measure how acidic or basic a solution is. Its name comes from the fact that pH is the absolute value of the power of the hydrogen ion concentration. The scale goes from 0 to 14. Distilled (pure) water is 7. Acids are from 0 and 7 and bases are from 7 to 14. When the pH of a substance is low, the substance has many H^+ ions. When the pH is high, the substance has many OH^- ions. The pH value helps inform scientists and technicians of the nature, composition, or extent of reactivity of substances.

The pH of something is typically checked with litmus paper. Litmus is a mixture of coloured organic compounds obtained from several species of lichen. Lichen is a type of plant that is actually a combination of a fungus and an algae. Litmus test strips can be used to check the condition of the engine's coolant **(Figure 7–50)**.

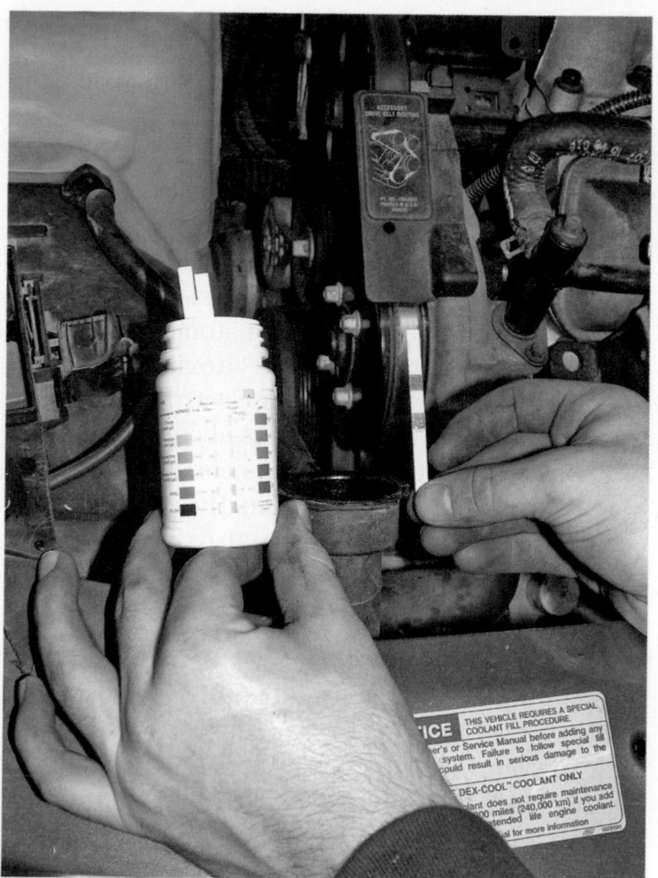

FIGURE 7–50

Litmus test strips can be used to check the condition of an engine's coolant.

Reduction and Oxidation

Oxidation is a chemical reaction in which a substance combines with oxygen. Rapid oxidation produces heat fast enough to cause a flame. When fuel burns, substances in the fuel combine with oxygen to form other compounds. This chemical reaction is combustion, which produces heat and fire. The rusting of iron is also an example of oxidation. Unlike fire, rusting occurs so slowly that little heat and no flames are produced.

The addition of hydrogen atoms or electrons is **reduction**. Oxidation and reduction always occur simultaneously: One substance is oxidized by the other, which it reduces. During oxidation, a molecule provides electrons. During reduction, a molecule accepts electrons. Oxidation and reduction reactions are usually called redox reactions. Redox is any chemical reaction in which electrons are transferred. Batteries, also known as voltaic cells, produce an electrical current at a constant voltage through redox reactions.

An oxidizing agent is a substance that accepts electrons and oxidizes something else while being reduced in the process. A reducing agent is a substance that provides electrons and reduces something else being oxidized in the process.

Every atom or ion has a prescribed oxidation number. This value compares the number of protons in an atom and the number of electrons assigned to that atom. In many cases, the oxidation number reflects the actual charge on the atom, but there are many cases in which it does not. The oxidation number is reduced in reduction by adding electrons. The oxidation number of an atom is increased during oxidation by removing electrons. All free, uncombined elements have an oxidation number of zero. Hydrogen, in all its compounds except hydrides, has an oxidation number of +1. Oxygen, in all its compounds except peroxides, has an oxidation number of –2.

RUST AND CORROSION The rusting of iron is an example of oxidation. Unlike fire, rusting occurs so slowly that little heat is produced. Iron combines with oxygen to form rust. The rate at which this occurs depends on several factors: temperature, surface area (more iron exposed for oxygen to reach), and catalysts (speed up a reaction but do not react and change themselves).

Corrosion is the wearing away of a substance due to chemical reactions. It occurs whenever a gas or liquid chemically attacks an exposed surface. This action is accelerated by heat, acids, and salts. Some materials naturally resist corrosion; others can be protected by painting, coating, galvanizing, or anodizing.

Galvanizing involves the coating of zinc onto iron or steel to protect it against exposure to the atmosphere. If galvanizing is properly applied, it can protect the metals for 15 to 30 years or more.

Metals can be anodized for corrosion resistance, electrical insulation, thermal control, abrasion resistance, sealing, improving paint adhesion, and decorative finishing. Anodizing is a process that electrically deposits an oxide film from an aqueous solution onto the surface of a metal, often aluminum. During the process, dyes can be added to the process to give the material a coloured surface.

Metallurgy

Metallurgy is the art and science of extracting metals from their ores and modifying the metals for a particular use. It also concerns the chemical, physical, and atomic properties and structures of metals and the principles by which metals are combined to form alloys. An **alloy** is a mixture of two or more metals. Steel, for example, is an alloy of iron plus carbon and other elements.

A metal can be defined as any substance that has one or more of the following properties:

- Good heat and electric conduction
- Malleability—can be hammered, pounded, or pressed into a shape without breaking
- Ductility—can be stretched, drawn, or hammered without breaking
- High light reflectivity—can make light bounce off its surface
- The capacity to form positive ions in a solution and hydroxides rather than acids when its oxides meet water

About three-quarters of the elements are metals. The most abundant metals are aluminum, iron, calcium, sodium, potassium, and magnesium.

HARDNESS The hardness of something is a statement of its resistance to scratching. **Hardening** is a process that increases the hardness of a metal, deliberately or accidentally, by hammering, rolling, carburizing, heat treating, tempering, or other physical processes. All of these methods deform the metal but also compact the atoms or molecules to make the material denser. The first few deformations imposed by such treatment weaken the metal, but because of the crystalline structure of metal, its strength increases with continued deformations.

Carburizing is a method used to surface-harden steel by heat or mechanical means to increase the hardness of the outer surface while leaving the core relatively soft. This combination of a hard surface and soft interior withstands very high stress and fatigue and also offers low cost and superior flexibility in manufacturing. To carburize, the steel parts are placed in a carbonaceous environment (with charcoal, coke, and carbonates, or carbon dioxide, carbon monoxide, methane, or propane) at a high temperature for several hours. The carbon diffuses into the surface of the steel, altering the crystal structure of the metal. Gears, ball and roller bearings, and piston pins are often carburized.

Heat treating is the changing of the properties of a metal (including iron, steel, aluminum, copper, and titanium) by using heat. It is used to harden metals that have different crystal structures at low and high temperatures. The metal is heated and then quenched (cooled rapidly) to retain the high-temperature constituent. Mid-heating (tempering) may then be used to attain the desired hardness.

Heating followed by slow cooling (annealing) is used to soften metals.

Tempering is the heat treating of metal alloys, particularly steel, to result in specific properties. For example, raising the temperature of hardened steel from 400°C (752°F) and holding it for a time before quenching it in oil decreases its hardness and brittleness and produces strong steel. Quench and temper heat treating is applied at many different cooling rates, holding times, and temperatures and is a very important means of controlling the properties of steel.

Solids under Tension

The atoms of a solid are closely packed, giving it a greater density than most liquids and gases. A solid's rigidity is the result of the strong attraction between its atoms. A force pulling on a solid moves these atoms further apart, creating an opposing force called tension. If a force pushes on a solid, the atoms move closer together, creating compression. These principles explain how springs function. Springs are used in many automotive systems, most obviously in suspension systems **(Figure 7–51)**.

An elastic substance is a solid that gets larger under tension, gets smaller under compression, and returns to its original size when no force is acting on it. Most solids show some elastic behaviour, but there is usually a limit to the force from which recovery is possible. Stresses beyond its elastic limit cause a material to yield, or flow, and the result is permanent distortion or breakage. This limit depends on the material's internal structure; for example, steel, although strong, has a low elastic limit and can be extended only about 1 percent of its length, whereas rubber can be elastically extended to about 1000 percent. Another factor involved in elasticity is the cross-sectional area of the material involved.

Tensile strength is the ratio of the maximum load a material can support without breaking when being stretched and is dependent on the cross-sectional area of the material. When stresses less than the tensile strength are removed, the material completely or partially returns to its original size and shape. As the stress approaches that of the tensile strength, the material forms a narrow, constricted region that is easily broken. Tensile strengths are measured in units of force per unit area.

Electrochemistry

Electrochemistry is a branch of chemistry concerned with the relationship between electricity and chemical change. Many spontaneous chemical reactions release electrical energy, and some of these reactions are used in batteries and fuel cells to produce electric power. The basis for electricity is the movement of electrons from one atom to another.

Electrolysis is an electrochemical process in which electric current is passed through a substance, causing a chemical change, usually the gaining or losing of electrons. It is carried out in an electrolytic cell consisting of separated positive and negative electrodes immersed in an electrolyte solution containing ions.

An **electrolyte** is a substance or compound that conducts electric current as a result of dissociation of its molecules into positive and negative ions. Dissociation is the breaking up of a chemical compound into simpler parts as a result of added energy. The most familiar electrolytes are acids, bases, and salts, which ionize when dissolved in solvents such

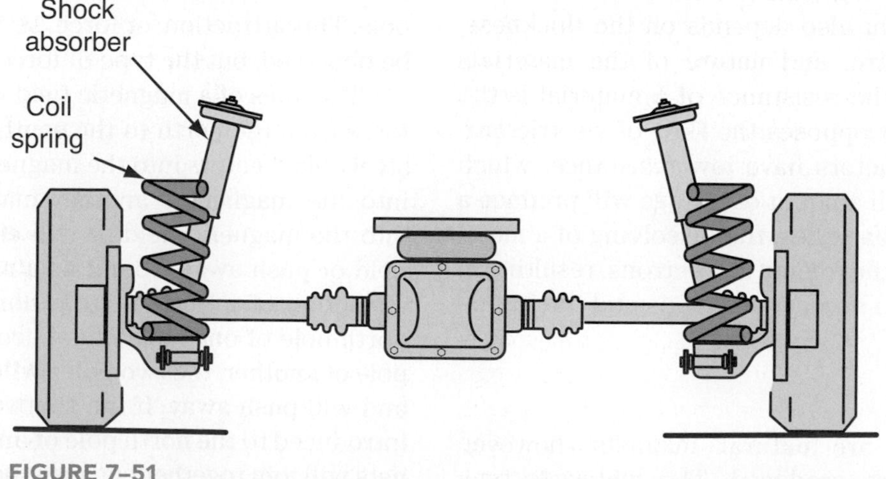

FIGURE 7–51

The placement of springs in an independent rear suspension.

as water and alcohol. Ions drift to the electrode of opposite charge in an electric field and are the conductors of current in electrolytic cells.

ELECTRICITY AND ELECTROMAGNETISM

All electrical effects are caused by electric charges. There are two types of electric charge: positive and negative. These charges exert electrostatic forces on each other due to the strong attraction of electrons to protons. An electric field is the area on which these forces have an effect. In atoms, protons carry positive charge, and electrons carry negative charge. Atoms are normally neutral, having an equal number of protons and electrons. However, an atom can gain or lose electrons, for example, by being rubbed. It then becomes a charged atom, or ion. Electricity has many similarities with magnetism. For example, the lines of the electric fields between charges take the same form as the lines of magnetic force, so magnetic fields can be said to be equivalent to electric fields. Charges of the same type repel, whereas charges of a different type attract **(Figure 7–52)**.

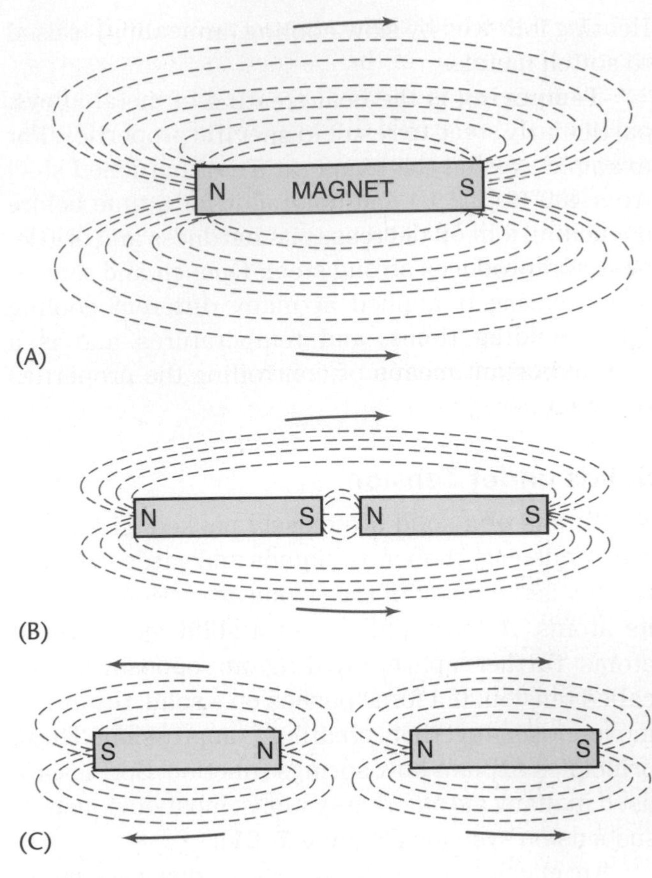

FIGURE 7–52

(A) In a magnet, lines of force emerge from the north pole and travel to the south pole before passing through the magnet back to the north pole. (B) Unlike poles attract, whereas (C) similar poles repel each other.

Electricity

An electric circuit is simply the path along which an electric current flows. Electrons carry a negative charge and can be moved around a circuit by electrostatic forces. A circuit usually consists of a conductive material, such as a metal, in which the electrons are held very loosely to their atoms, thus making movement possible. The strength of the electrostatic force is the voltage. The resulting movement of the electric charge is called an electric current. The higher the voltage, the greater the current will be, but the current also depends on the thickness, length, temperature, and nature of the materials that conduct it. The resistance of a material is the extent to which it opposes the flow of electric current. Good conductors have low resistance, which means that a small amount of voltage will produce a large current. In batteries, the dissolving of a metal electrode causes the freeing of electrons, resulting in their movement to another electrode and the formation of a current.

Magnets

Some materials are natural magnets; however, most magnets are produced. The materials typically used to make a permanent magnet are called ferromagnetic materials. These materials consist of mostly iron compounds that are heated. The heat causes the atoms to shift direction, and once they all point in the same direction the metal becomes a magnet. Two distinct poles, called the north and south poles, are at the ends of the magnet, and there is an attraction between the north pole and the south pole. This attraction, or force, set up by a magnet can be observed, but the type of force is not known.

The lines of a magnetic field form closed lines of force from the north to the south. If another iron or steel object enters into the magnetic field, it is pulled into the magnet. If another magnet is introduced into the magnetic field, it will either move into the field or push away from it as a result of the natural attraction of a magnet from north to south. If the north pole of one magnet is introduced to the north pole of another, the two poles will oppose each other and will push away. If the south pole of a magnet is introduced to the north pole of another, the two magnets will join together because the opposite poles are attracted to each other.

The strength of the magnetic force is uniform all around the outside of the magnet. The force is strongest at the surface of the magnet and weakens with distance. If you double the distance of an object from a magnet, the force is reduced by ⅛.

The strength of a magnetic field is typically measured with devices known as magnetometers and in units of Gauss (G).

Electromagnetism

Any electrical current produces magnetism that affects other objects in the same way as permanent magnets. The arrangement of force lines around a current-carrying conductor, its magnetic field, is circular. The magnetic effect of electrical current is increased by making the current-carrying wire into a coil **(Figure 7–53)**.

When a coil of wire is wrapped around an iron bar, it is called an **electromagnet**. The magnetic field produced by the coil magnetizes the iron bar, strengthening the overall effect. A field like that of a bar magnet is formed by the magnetic fields of the wires in the coil. The strength of the magnetism produced depends on the number of coils and the size of the current flowing in the wires. Electromagnetic coils and permanent magnets are arranged inside an electric motor so that the forces of electromagnetism create rotation of the armature.

Producing Electrical Energy

There are many ways to generate electricity. The most common way is to use coils of wire and magnets in a generator. Whenever a wire and magnet are moved relative to each other, a voltage is produced **(Figure 7–54)**. In a generator, the wire is wound into a coil. The more turns in the coil and the faster the coil moves, the greater the voltage. The coils or magnets spin around at high speed, typically turned by steam pressure. The steam is usually generated by burning coal or oil, a process that creates pollution. Renewable sources of electricity, such as hydroelectric power, wind power, solar energy, and geothermal power, produce only heat as a pollutant. In automobiles, the generator is spun by a belt driven by the engine's crankshaft. In a generator, the kinetic energy of a spinning object is converted into electrical energy.

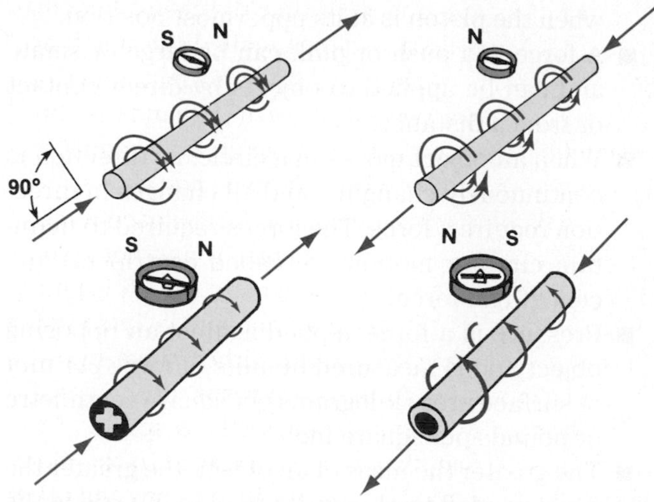

FIGURE 7–53

When current is passed through a conductor such as a wire, magnetic lines of force are generated round the wire at right angles to the direction of the current flow.

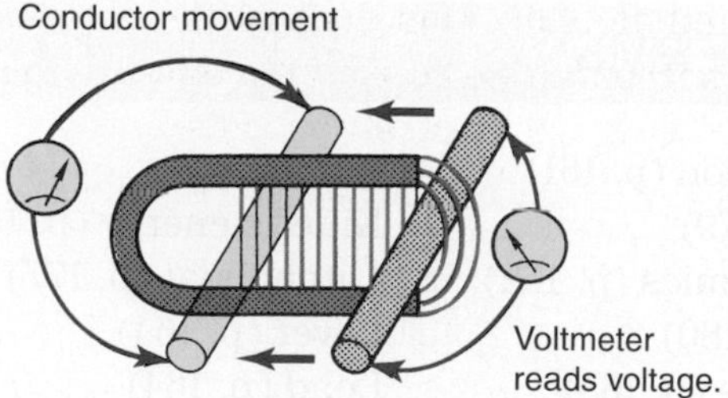

FIGURE 7–54

Moving a conductor across magnetic lines of force induces a voltage in the conductor.

A solar cell converts the energy of sunlight directly into electrical energy using layers of semiconductors. Electricity is produced by causing electrons to leave the atoms in the semiconductor material. Each electron leaves behind a hole or gap. Other electrons move into the hole, leaving holes in their atoms. This process continues all the way around a circuit. The moving chain of electrons is an electrical current.

Radio Waves

Electricity and magnetism are directly related. A changing electric field produces a changing magnetic field, and vice versa. Whenever an electric charge, such as that carried by an electron, accelerates, it gives out energy in the form of electromagnetic radiation. For example, electrons moving up and down a radio antenna produce a type of radiation known as radio waves. Electromagnetic radiation consists of oscillating electric and magnetic fields. There is a wide range of different types of electromagnetic radiation, called the electromagnetic spectrum, extending from low-energy radio waves to high-energy, short wavelength gamma rays, including visible light and X-rays.

KEY TERMS

Acceleration (p. 161)
Acid (p. 179)
Aerodynamics (p. 162)
Alloy (p. 180)
Amplitude (p. 168)
Atmospheric pressure (p. 174)
Atoms (p. 151)
Barometric pressure (p. 175)
Base (p. 179)
Calories (p. 176)
Carburizing (p. 180)
Catalyst (p. 178)
Centrifugal force (p. 159)
Centripetal force (p. 159)
Compression ratio (p. 157)
Conduction (p. 176)
Convection (p. 176)
Curb weight (p. 155)
Deceleration (p. 161)
Density (p. 178)
Diffusion (p. 174)
Displacement (p. 156)
Dynamic pressure (p. 173)
Electrolysis (p. 181)
Electrolyte (p. 181)
Electromagnet (p. 183)
Element (p. 151)
Engine efficiency (p. 154)
Equilibrium (p. 158)
Evaporate (p. 153)
Force (p. 158)
Frequency (p. 168)
Friction (p. 161)
Gear (p. 164)
Gross weight (p. 155)
Hardening (p. 180)
Heat (p. 175)
Heat treating (p. 180)
Hertz (Hz) (p. 168)
Horsepower (p. 167)
Impermeable (p. 153)
Inertia (p. 161)
Ion (p. 153)
Kinetic energy (p. 154)
Latent heat (p. 177)
Lever (p. 164)
Load (p. 164)
Mass (p. 155)
Matter (p. 151)
Mechanical advantage (p. 172)
Molecules (p. 151)
Momentum (p. 161)
Newton (p. 163)
Oscillation (p. 168)
Oxidation (p. 179)
Oxide (p. 153)
Permeable (p. 153)
pH scale (p. 179)
Plasma (p. 153)
Potential energy (p. 154)
Power (p. 167)
Pressure (p. 160)
Pulley (p. 164)
Radiation (p. 176)
Reduction (p. 179)
Solution (p. 153)
Solvent (p. 178)
Specific gravity (p. 178)
Speed (p. 161)
Static pressure (p. 173)
Temperature (p. 175)
Tempering (p. 181)
Tensile strength (p. 181)
Tension (p. 159)
Thermal contraction (p. 176)
Thermal expansion (p. 176)
Time (p. 160)
Torque (p. 166)
Vacuum (p. 175)
Velocity (p. 161)
Volume (p. 156)
Wave (p. 168)
Wavelength (p. 169)
Weight (p. 155)
Work (p. 163)

SUMMARY

- Matter is anything that occupies space, and it exists as a gas, liquid, or solid.
- All matter consists of countless tiny particles called atoms and molecules.
- When a solid dissolves into a liquid, a solution is formed. Not all solids dissolve into a liquid; rather, the liquid is either absorbed or adsorbed.
- Materials that *absorb* fluids are permeable substances. Impermeable substances do *not* absorb fluids.
- Energy is the ability to do work, and all matter has energy.
- The total amount of energy never changes; it can only be transferred from one form to another, not created or destroyed.
- When energy is released to do work, it is called kinetic energy. Stored energy may be called potential energy.
- Energy conversion occurs when one form of energy is changed to another form.
- Mass is the amount of matter in an object. Weight is a force and is measured in kilograms or pounds. Gravitational force gives the mass its weight.
- Volume is the amount of space occupied by an object.
- The volume of an engine's cylinders determines its size, expressed as displacement.
- The compression ratio of an engine is defined as the ratio of the volume in the cylinder above the piston when the piston is at the bottom of its travel to the volume in the cylinder above the piston when the piston is at its uppermost position.
- A force is a push or pull, can be large or small, and can be applied to objects by direct contact or from a distance.
- When an object moves in a circle, its direction is continuously changing, and all changes in direction require a force. The forces required to maintain circular motion are called centripetal and centrifugal force.
- Pressure is a force applied against an opposing object and is measured in units of force per unit of surface area (kilograms per square centimetre or pounds per square inch).
- The greater the mass of an object, the greater the force needed to change its motion. Inertia is the tendency of an object at rest to remain at rest, or the tendency of an object in motion to stay in motion.

- When a force overcomes static inertia and moves an object, the object gains momentum. Momentum is the product of an object's weight times its speed.
- Speed is the distance an object travels in a set amount of time. Velocity is the speed of an object in a particular direction. Acceleration, which occurs only when a force is applied, is the rate of increase in speed. Deceleration, the rate of decrease in speed, is the reverse of acceleration.
- Newton's laws of motion state that an object at rest tends to remain at rest and an object in motion tends to remain in motion unless some force acts on it; when a force acts on an object, the motion of the object will change; and for every action there is an equal and opposite reaction.
- Friction is a force that slows or prevents motion of two moving objects that touch.
- Friction can be reduced in two main ways: by lubrication or by the use of rollers or bearings.
- Aerodynamics is the study of the effects of air on a moving object.
- When a force moves a certain mass a specific distance, work is done.
- A machine is any device that can be used to transmit a force and, in doing so, change the amount of force and/or its direction. Examples of simple machines are inclined planes, pulleys, levers, gears, and wheels and axles.
- Torque is a force that tends to rotate or turn things and is measured by the force applied and the distance travelled.
- Gear ratios express the mathematical relationship of one gear to another.
- Power is a measurement of the rate at which work is done and is measured in watts.
- Horsepower is the rate at which torque is produced.
- An oscillation is any single swing of an object back and forth between the extremes of its travel. When that motion travels through matter or space, it becomes a wave.
- How many times a vibration occurs in one second is called frequency. Frequency is most often expressed in hertz (Hz), which is equal to one cycle per second. The amplitude of a vibration is its intensity or strength.
- Noise is any unwanted signal or sound and can be random or periodic.
- Light is a form of electromagnetic radiation, which travels in a straight line at 300 million metres per second.
- A gas always fills a sealed container, whereas a liquid may not. A gas also readily expands or compresses according to the pressure exerted on it. Liquids are basically incompressible, which gives them the ability to transmit force.
- Hydraulics is the study of liquids in motion.
- Pascal constructed the first known hydraulic device and established what is known as Pascal's law of hydraulics.
- The pressure inside the hydraulic system is called static pressure because there is no fluid motion. The pressure of the fluid while it is in motion is called dynamic pressure.
- Boyle's law states that the volume and pressure of a mass of gas at a fixed temperature are inversely proportional.
- The pressure law states that the pressure exerted by a gas at constant volume increases as the temperature of the gas is increased.
- Charles's law states that the volume of a mass of gas depends on its temperature.
- Atmospheric pressure is the total weight of Earth's atmosphere. Pressure greater than atmospheric pressure may be measured in kPa gauge (kPag) or psi gauge (psig); actual pressure is measured in kPa absolute (kPaa) or psi absolute (psia).
- Scientifically, vacuum is defined as the absence of atmospheric pressure; however, it is commonly used to refer to any pressure less than atmospheric pressure.
- Heat is a form of energy caused by the movement of atoms and molecules and is measured in calories or British thermal units (BTUs).
- Temperature is an indication of an object's kinetic energy and is measured with a thermometer that has either a Celsius (or Centigrade, C) or Fahrenheit (F) scale.
- Convection is the transfer of heat by the movement of a heated object.
- Conduction is the movement of heat through a material.
- Through radiation, heat is transferred by radiant energy.
- As heat moves in and out of a mass, the size of the mass tends to change.
- Sometimes, additional heat causes no increase in temperature but causes the mass to change its state; this heat is called latent heat.
- The liquid in a solution is called the solvent, and the substance added is the solute.
- Specific gravity is the heaviness or density of a substance compared to that of water.

- A catalyst is any substance that affects a chemical reaction without itself being consumed or changed.
- An ion is an atom or group of atoms with one or more positive or negative electric charges. Ions are formed when electrons are added to or removed from neutral molecules or other ions. Ions are what make something an acid or a base.
- The pH scale is used to measure how acidic or basic a solution is.
- Oxidation is a chemical reaction in which a substance combines with oxygen. The addition of hydrogen atoms or electrons is reduction.
- Hardening is a process that increases the hardness of a metal, deliberately or accidentally, by hammering, rolling, carburizing, heat treating, tempering, or other physical processes.
- An elastic substance is a solid that gets larger under tension, gets smaller under compression, and returns to its original size when no force is acting on it.
- Tensile strength is the ratio of the maximum load a material can support without breaking when being stretched and is dependent on the cross-sectional area of the material.
- Electrolysis is an electrochemical process in which electric current is passed through a substance, causing a chemical change, usually the gaining or losing of electrons.
- An electrolyte is a substance or compound that conducts electric current as a result of dissociation of its molecules into positive and negative ions.
- Any electrical current produces magnetism. When a coil of wire is wrapped around an iron bar, it is called an electromagnet.
- The most common way to produce electricity is to use coils of wire and magnets in a generator.

REVIEW QUESTIONS

1. Which of the following best describes Newton's first law of motion?
   a. An object at rest tends to remain at rest.
   b. The change in motion of an object is equal to the amount of force divided by the mass.
   c. The force acting on an object is equal to its area multiplied by the amount of pressure applied.
   d. For every action there is a reaction.
2. In which automotive system would Pascal's law most likely be applied?
   a. cooling system **b.** brake system
   c. suspension system **d.** electrical system
3. Describe six different forms of energy.
4. Describe four different types of energy conversion.
5. When energy is released to do work, it is described as which type of energy?
   a. static **b.** potential
   c. kinetic **d.** mechanical
6. Which of the following would produce a dynamic imbalance?
   a. weight evenly placed around the tire centreline
   b. weight evenly placed around the outer edge of the tire
   c. weight evenly placed around the inside edge of the tire
   d. weight evenly placed around the tire diagonally to the tire centreline
7. Which of the following best describes tire tramp?
   a. a tire hopping up and down
   b. a tire shimmying
   c. a tire wobbling
   d. a tire that is dynamically imbalanced
8. What can be found in the nucleus of an atom?
   a. protons and electrons
   b. protons and neutrons
   c. neutrons and electrons
   d. only neutrons
9. How is work calculated?
   a. pressure × time **b.** force ÷ by time
   c. force × distance **d.** pressure × distance
10. Which of the following statements about energy is correct?
    a. Energy is matter.
    b. Energy is the ability to do work.
    c. Energy is created by pressure.
    d. Energy is destroyed as it changes form.
11. Which of the following automotive components is a class two lever?
    a. a rocker arm **b.** the brake pedal
    c. the crankshaft **d.** a window crank
12. What is a hertz a measurement of?
    a. frequency **b.** amplitude
    c. wave length **d.** intensity
13. What is torque defined as?
    a. a pushing force **b.** pressure
    c. a pulling force **d.** a twisting force

14. Which of the following is a measurement term for torque?
    a. kilopascals (kPa)
    b. cubic centimetres (cc)
    c. Newton-metres (N·m)
    d. hertz (Hz)
15. Which of the following laws refers to hydraulics?
    a. Charles's law
    b. Pascal's law
    c. Boyle's law
    d. Ohm's law
16. What must occur in a hydraulic system to increase the output force?
    a. increase the size of the input piston
    b. have both input and output pistons of equal size
    c. decrease the size of the input piston
    d. connect both pistons through large passageways
17. What type of energy does the automobile brake system convert the vehicle's kinetic energy to?
    a. mechanical energy
    b. thermal energy
    c. electrical energy
    d. chemical energy
18. What is the correct formula for calculating cylinder volume?
    a. 3.1416 × bore × stroke
    b. 3.1416 × bore × bore × stroke
    c. 3.1416 × radius × stroke
    d. 3.1416 × radius × radius × stroke
19. How is a torque increase achieved through gearing?
    a. having a large gear drive a smaller gear
    b. having a small gear drive a large gear
    c. having two equal-sized gears
    d. having a small gear placed between two larger gears
20. When calculating the specific gravity of a substance, what is its relative density compared to?
    a. water
    b. mercury
    c. battery acid
    d. air

CHAPTER 8
Preventive Maintenance and Basic Services

LEARNING OUTCOMES

- Describe the information that should be included on a repair order.
- Explain how repair costs can be estimated.
- Explain how the vehicle and its systems can be defined by deciphering its VIN.
- Explain the importance of preventive maintenance, and list at least six examples of typical preventive maintenance services.
- Understand the differences between the types of fluids required for preventive maintenance and know how to select the correct one for a particular vehicle.
- Explain how the design of a vehicle determines what preventive maintenance procedures must be followed.

Preventive maintenance services are those services performed not to correct problems but rather to prevent them. These and other basic services are covered in this chapter. All of these services may be performed by technicians in many different types of service facilities—dealerships, independents, and specialty shops. Regardless of what type of shop, the first thing a tech needs to worry about is the repair order.

REPAIR ORDERS

A **repair order (RO)** is written for every vehicle brought into the shop. ROs may also be called service or work orders. ROs contain information about the customer, the vehicle, the customer's concern or request, an estimate of the cost for the services, and the time the services should be completed. ROs are legal documents that are used for many other purposes, such as payroll and general recordkeeping **(Figure 8–1)**. Legally, an RO protects the shop and the customer.

Although every shop may enter different information onto the original RO, most ROs contain the following information:

- Complete customer information
- Complete vehicle identification
- The service history of the vehicle
- The customer's concern
- The preliminary diagnosis of the problem
- An estimate of the amount of time and costs required for the service
- The time the services should be completed
- The name or other identification of the technician assigned to perform the services
- The actual services performed with their cost
- The parts replaced during the services and their cost
- Recommendations for future services

An RO is signed by the customer, who in doing so authorizes the services and accepts the terms noted on the RO. The customer also agrees to pay for the services when they are completed. Many provinces require a customer signature to begin repair work and for a change in the original estimate. If a signature is not required for changes in the original estimate, all phone conversations concerning the estimate should be noted on the RO.

In most cases when a customer signs the RO, he or she acknowledges the shop's right to impose a mechanic's lien. This lien basically says that the shop may gain possession of the vehicle if the customer does not pay for the agreed-upon services and the vehicle remains at the shop for a period of 90 or more days. This clause ensures that the shop will receive some compensation for the work performed, whether or not the customer pays the bill.

Service records are kept by the shop to maintain the vehicle's service history and for legal purposes. Evidence of repairs and recommended repairs is very important for settling potential legal disputes with the vehicle's owner.

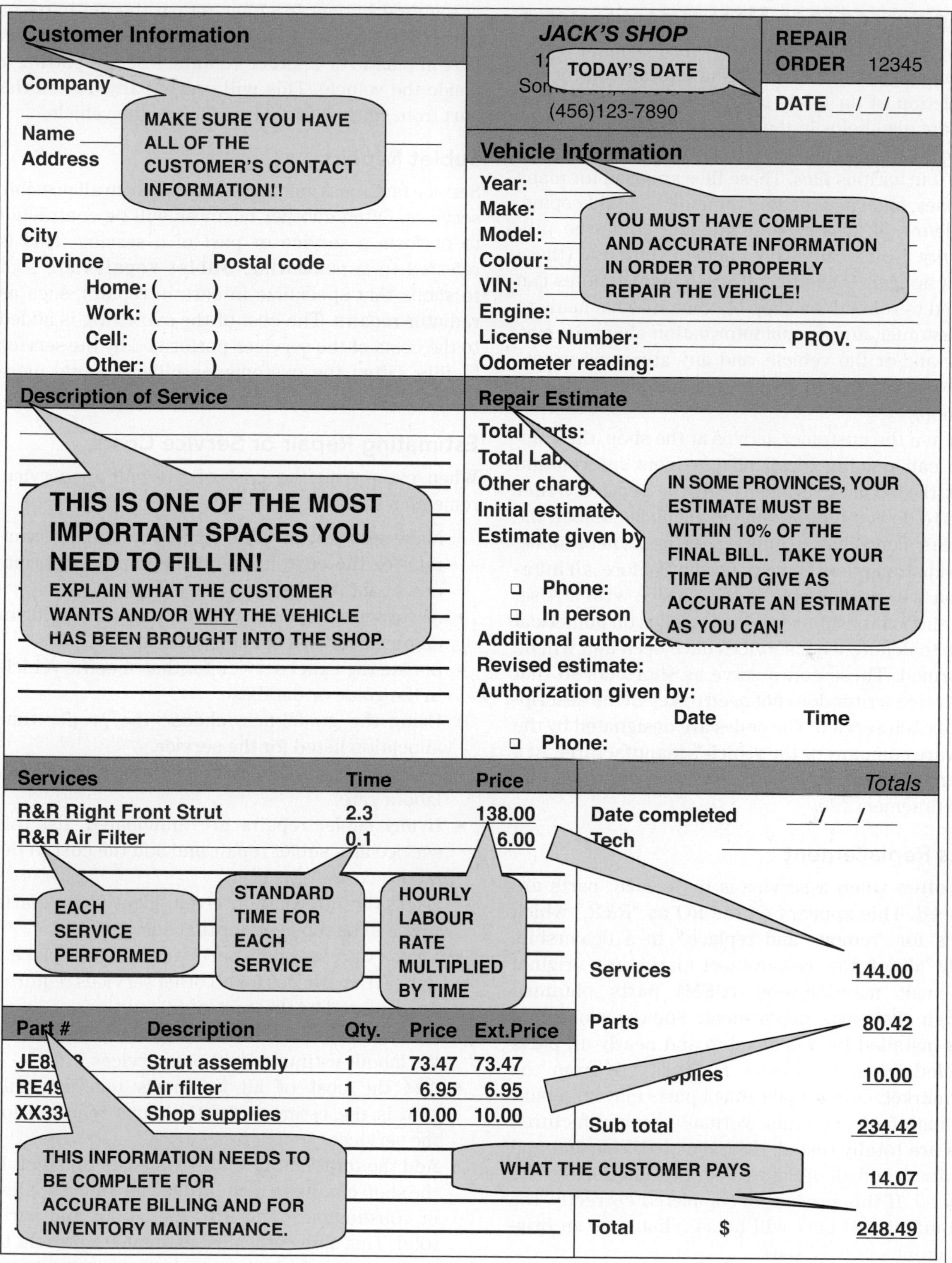

Customer Information

Company

Name

Address

City

Province Postal code

Home: ( )

Work: ( )

Cell: ( )

Other: ( )

JACK'S SHOP

(456)123-7890

REPAIR ORDER 12345

DATE / /

Vehicle Information

Year:

Make:

Model:

Colour:

VIN:

Engine:

License Number: PROV.

Odometer reading:

Description of Service

Repair Estimate

Total Parts:

Total Lab

Other charg

Initial estimate

Estimate given by

❑ Phone:

❑ In person

Additional authorized

Revised estimate:

Authorization given by:

Date Time

❑ Phone:

Services	Time	Price
R&R Right Front Strut	2.3	138.00
R&R Air Filter	0.1	6.00

Part #	Description	Qty.	Price	Ext.Price
JE8	Strut assembly	1	73.47	73.47
RE49	Air filter	1	6.95	6.95
XX334	Shop supplies	1	10.00	10.00

Totals

Date completed / /

Tech

Services	144.00
Parts	80.42
S pplies	10.00
Sub total	234.42
	14.07
Total $	248.49

FIGURE 8–1

A completed repair order.

Computerized Shop Management Systems

Today, most service facilities use computerized shop management software. The information for the completion of an RO is input into a computer. The software also helps in the estimation of repair costs. The software also takes information from the RO and saves it in various files. These files are used for many purposes, such as schedule reminders, bookkeeping, vehicle/owner history, and tracking employee productivity. Notes can also be added to the RO (these do not appear on the RO). These personal notes can be used to remind the shop of commitments made to the customer, any special information about the customer and/or the vehicle, and any abnormal events that took place during the customer's last visit to the shop.

When the customer arrives at the shop, the computer can quickly recall all pertinent information about the vehicle. Typically, all the service writer needs to do is input the vehicle's licence number, the vehicle's identification number, or the owner's name. If the customer has been to the shop before, all information will be available to the service writer. Also, most shop management software relies on numerical codes to denote what services have been and will be performed. These codes serve as shortcuts so that the service writer does not need to key in the description of each service. The codes are designated by the software company or the vehicle's manufacturer. At a dealership, these codes link directly to the warranty reimbursement file.

Parts Replacement

Very often when a service is performed, parts are replaced. This appears on the RO as "R&R," which stands for "remove and replace." In a dealership, nearly all of the replacement parts are original equipment manufacturer (OEM) parts obtained through the parts department. Some replacement parts installed by a dealership and nearly all parts installed by other service facilities are from the aftermarket. Other replacement parts may be rebuilt or remanufactured units. Normally, remanufactured parts are totally tested, disassembled, cleaned, and machined, and all of the weak or dysfunctional parts replaced. If this process is completed correctly, the remanufactured part will be as reliable as an original equipment (OE) part.

If the replaced part has no core value, the shop disposes of the part. However, many shops offer the part to the customer as proof that the part was removed and a new one installed. In many provinces, the shop is required by law to either return the old parts or allow the customer to inspect them. Always place the part in plastic or another container before putting it inside the vehicle. This will prevent any dirt on the part from getting on anything inside the vehicle.

Sublet Repairs

Service facilities typically do not perform all possible services. Often another business will be contracted to perform a service or part of a service. This is referred to as subletting. **Sublet repairs** are sent to shops that specialize in certain repairs, such as radiator repairs. The cost of the subletting is added to the costs of the services performed by the service facility. Often the customer is billed slightly more than the actual cost of the sublet repair.

Estimating Repair or Service Costs

When estimating the cost of a repair or service, remember the following:

- Make sure you have the correct contact information for the customer and the correct information about the vehicle.
- Always use the correct labour and parts guide or database for that specific vehicle.
- Locate the exact service for that specific vehicle in the guide or database.
- Using the guidelines, choose the proper time allocation listed for the service.
- Multiply the allocated time by the shop's hourly labour rate.
- If any sublet repairs are anticipated, list this service as a sublet repair and add the cost to the labour costs.
- Using the information given, identify the parts that will be replaced for that service.
- Locate the cost of the parts that will be replaced.
- Repeat the process for all other services required or requested by the customer.
- Add all of the labour costs together; this sum is the labour estimate for those services.
- Add the cost of all the parts together; this sum is the estimate for the parts required for the services.
- Add the total labour and parts costs together. If the shop charges a standard fee for shop supplies or consumables, add it to the labour and parts total. This sum is the cost estimate to present to the customer.

The customer is protected against being charged more than the estimate given on the RO, unless he or she later authorizes a higher amount. Some provinces

When estimating the cost of a repair or service:

- Make sure you have the correct contact information for the customer.
- Make sure you have the correct information about the vehicle.
- Always use the correct labour and parts guide or database for that specific vehicle.
- Locate the exact service for that specific vehicle in the guide or database.
- Using the guidelines provided in the guide or database, choose the proper time allocation listed for the service.
- Multiply the allocated time by the shop's hourly labour rate.
- If any sublet repairs are anticipated, list this service as a sublet repair and add the cost to the labour costs.
- Using the information given in the guide or database, identify the parts that will be replaced for that service.
- Locate the cost of the parts in the guide or database or in the catalogues used by the shop.
  - Repeat the process for all other services required or requested by the customer.
  - Multiply the time allocations by the shop's hourly flat rate.
  - Add all of the labour costs together; this sum is the labour estimate for those services.
  - Add the cost of all the parts together; this sum is the estimate for the parts required for the services.
  - Add the total labour and parts costs together. If the shop charges a standard fee for shop supplies, add it to the labour and parts total. This sum is the cost estimate to present to the customer.

FIGURE 8–2

Guidelines for estimating the cost of repairs.

allow shops to be within 10 percent of the estimate, whereas others hold the shop to the amount that was estimated. **Figure 8–2** lists some things to follow when estimating the cost of services and repairs.

VEHICLE IDENTIFICATION

Before any service is done to a vehicle, it is important to know exactly what type of vehicle you are working on. The best way to do this is to refer to the **vehicle identification number (VIN)**. The VIN is on a plate behind the lower corner of the driver's side of the windshield as well as other locations on the vehicle. The VIN is made up of 17 characters and contains all pertinent information about the vehicle. The use of the 17 number and letter code became mandatory in 1981 and is used by all manufacturers of vehicles both domestic and foreign. Most new vehicles have a scan code below the VIN **(Figure 8–3)**.

Each character of a VIN has a particular purpose. The first character identifies the country where the vehicle was manufactured, as in the following examples:

- 1 or 4 – U.S.A.
- 2 – Canada
- 3 – Mexico
- J – Japan
- K – Korea
- S – England
- W – Germany

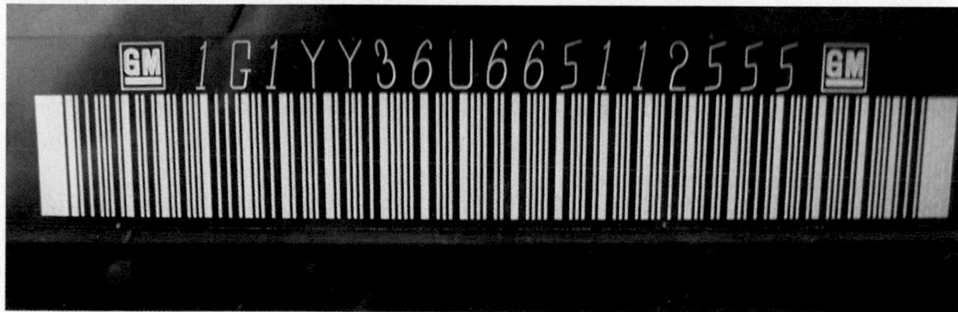

FIGURE 8–3

A vehicle identification plate.

The second character identifies the manufacturer:

- A – Audi
- B – BMW
- C – Chrysler
- D – Mercedes Benz
- F – Ford
- G – General Motors
- H – Honda
- N – Nissan
- T – Toyota

The third character identifies the vehicle type or manufacturing division (passenger car, truck, bus, and so on). The fourth through eighth characters identify the features of the vehicle, such as the body style, vehicle model, and engine type.

The ninth character is used to identify the accuracy of the VIN and is a check digit. The tenth character identifies the model year:

- 8 – 2008
- 9 – 2009
- A – 2010
- B – 2011
- C – 2012
- D – 2013
- E – 2014
- F – 2015
- G – 2016

The eleventh character identifies the plant where the vehicle was assembled, and the twelfth to seventeenth characters identify the production sequence of the vehicle as it rolled off the manufacturer's assembly line.

PREVENTIVE MAINTENANCE

Preventive maintenance (PM) involves performing certain services to a vehicle on a regularly scheduled basis before there is any sign of trouble. Regular inspection and routine maintenance can prevent major breakdowns and expensive repairs. It also keeps cars and trucks running efficiently and safely.

Once the vehicle's warranty expires, repairs will have to be paid for by the customer. A 2012 study found that the average age of passenger cars in the United States is over 11 years old. The number of vehicles on the road, average vehicle age, and length of ownership have been increasing for years. To keep older vehicles running properly, the PM program is vitally important.

A survey of 2375 vehicles conducted during National Car Care Month found that more than 90 percent of the cars lacked some form of service. The cars were inspected for exhaust emissions, fluid levels, tire pressure, and other safety features. The results indicated that 34 percent of the cars had restricted air filters, 27 percent had worn belts, 25 percent had clogged PCV filters, 14 percent had worn hoses, and 20 percent had bad batteries, battery cables, or terminals.

During the fluid and cooling system inspection, 39 percent failed due to bad or contaminated transmission or power-steering fluid, 36 percent had worn-out or dirty engine oil, 28 percent had inadequate cooling system protection, and 8 percent had a faulty radiator cap.

In the safety category, 50 percent failed due to worn or improperly inflated tires, 32 percent had inoperative headlights or brake lights, and 14 percent had worn wipers.

Maintenance Schedules and Reminders

A typical PM schedule recommends particular service at mileage or time intervals. Driving habits and conditions should also be used to determine the frequency of PM service intervals. For example, vehicles that frequently are driven for short distances in city traffic often require more frequent oil changes due to the more rapid accumulation of condensation and unburned fuel in the oil. Most manufacturers also specify more frequent service intervals for vehicles that are used to tow a trailer or those that operate in extremely dusty or unusual conditions.

It is important that your customer understands the differences between what the vehicle manufacturer considers normal and severe service. In many cases, if a vehicle is driven in any of the conditions listed in the severe service category, the maintenance intervals increase to twice as often as for normal service. Because operating conditions and manufacturer's maintenance recommendations vary, many shops suggest using the manufacturer's PM schedule as the minimum requirement. The shop may then tailor the vehicle's PM schedule based on the actual operating conditions.

Many vehicles now use maintenance reminder systems. These systems often show a message, such as the engine oil life percentage, in a driver information centre. Other vehicles show codes for the next required service **(Figure 8–4)**. To determine what services are specified by the code, check the vehicle's service information.

FIGURE 8–4

A maintenance indicator and service identifier.

Safety Inspections

Some provinces require annual or biennial vehicle safety inspections. The intent of these inspections is to improve road safety. Research shows that provinces with annual safety inspection programs have 20 percent fewer accidents than those without safety inspections. These inspections consist of a series of safety-related checks of various systems and areas of a vehicle. For example, some common checks are shown in **Figure 8–5**. The exact systems and subsystems that are inspected vary. The inspections are part of the vehicle registration process. Often automobile dealers are required to complete a safety inspection on all used vehicles before they are sold and then report the results to the customer.

BASIC SERVICES

Often while performing PM on a vehicle, a technician notices the need for a minor repair. Both PM and those basic minor services are covered in the rest of this chapter.

CUSTOMER CARE

Whenever you do any service to a vehicle, use fender covers **(Figure 8–6)** and do not leave fingerprints on the exterior or interior of the car. Use floor, seat, and steering wheel covers to protect the interior. If oil or grease gets on the car, clean it off.

Engine Oil

Engine oil is a clean or refined form of **crude oil**. Crude oil, when taken out of the ground, is dirty and does not work well as a lubricant for engines. Crude oil must be refined to meet industry standards. Engine oil (often called motor oil) is just one of the many products that come from crude oil. Engine oil is specially formulated to do the following:

- Flow easily through the engine
- Provide lubrication without foaming
- Reduce friction and wear
- Prevent the formation of rust and corrosion
- Cool the engine parts it flows over
- Keep internal engine parts clean

Engine oil contains many additives, each intended to improve the effectiveness of the oil. The **American Petroleum Institute (API)** classifies engine oil as standard or S-class for passenger cars and light trucks and as commercial or C-class for heavy-duty commercial applications. The various types of oil within each class are further rated according to their ability to meet the engine manufacturers' warranty specifications **(Table 8–1)**. Engine oils can be classified as **resource-conserving** (fuel-saving) **oils**. These are designed to reduce friction, which in turn reduces fuel consumption. Friction modifiers and other additives are used to achieve this.

In addition to the API rating, oil **viscosity** is important in selecting engine oil. The ability of oil to resist flowing is its viscosity. The thicker the oil, the higher its viscosity rating. Viscosity is affected by temperature; hot oil flows faster than cold oil. Oil flow is important to the life of an engine. Because an engine operates under a wide range of temperatures, selecting the correct viscosity is very important.

TABLE 8–1 **ENGINE OIL SERVICE RATINGS**

RATING	COMMENTS
SA	Straight mineral oil (no additives), not suitable for use in any engine.
SB	Non-detergent oil with additives to control wear and oil oxidation
SC	Obsolete since 1964
SD	Obsolete since 1968
SE	Obsolete since 1972
SF	Obsolete since 1980
SG	Obsolete since 1988
SH	Obsolete since 1993
SJ	Started in 1997
SL	Started in 2001
SM	Started in 2005
SN	Started in 2011

INSPECT WINDSHIELD AND OTHER GLASS FOR:

Cloudiness, distortion, or other obstruction to vision.

Cracked, scratched, or broken glass.

Window tinting.

Operation of front door glass.

INSPECT WINDSHIELD WIPER/WASHER FOR:

Operating condition.

Condition of blade.

INSPECT WINDSHIELD DEFROSTER FOR:

Operating condition.

INSPECT MIRRORS FOR:

Rigidity of mounting.

Condition of reflecting surface.

View of road to rear.

INSPECT HORN FOR:

Electrical connections, mounting, and horn button.

Emits a sound audible for a minimum of 200 feet.

INSPECT DRIVER'S SEAT FOR:

Anchorage.

Location.

Condition.

INSPECT SEAT BELTS FOR:

Condition.

INSPECT HEADLIGHTS FOR:

Approved type, aim, and output.

Condition of wiring and switch.

Operation of beam indicator.

INSPECT OTHER LIGHTS FOR:

Operation of all lamps, lens colour, and condition of lens.

Aim of fog and driving lamps.

INSPECT SIGNAL DEVICE FOR:

Correct operation of indicators (visual or audible).

Illumination of all lamps, lens colour, and condition of lens.

INSPECT FRONT DOORS FOR:

Handle or opening device permits the opening of the door from the outside and inside of the vehicle.

Latching system that holds door in its proper closed position.

INSPECT HOOD FOR:

Operating condition of hood latch.

INSPECT FLUIDS FOR:

Levels that are below the proper level.

INSPECT BELTS AND HOSES FOR:

Belt tension, wear, or absence.

Hose damage.

INSPECT POLLUTION CONTROL SYSTEM FOR:

Presence of emissions system—evidence that no essential parts have been removed, rendered inoperative, or disconnected.

INSPECT BATTERY FOR:

Proper anchorage.

Loose or damaged connections.

INSPECT FUEL SYSTEM FOR:

Any part that is not securely fastened.

Liquid fuel leakage.

Fuel tank filler cap for presence.

INSPECT EXHAUST SYSTEM FOR:

Damaged exhaust—manifold, gaskets, pipes, mufflers, connections, etc.

Leakage of gases at any point from motor to point discharged from system.

INSPECT STEERING AND SUSPENSION FOR:

Play in steering wheel.

Wear in bushings, kingpins, ball joints, wheel bearings, and tie-rod ends.

Looseness of gear box on frame, condition of drag link, and steering arm.

Wheel alignment and axle alignment.

Broken spring leaves and worn shackles.

Shock absorbers.

Broken frame.

Broken or missing engine mounts.

Lift blocks.

INSPECT FLOOR PAN FOR:

Holes that allow exhaust gases to enter occupant compartment.

Conditions that create a hazard to the occupants.

INSPECT BRAKES FOR:

Worn, damaged, or missing parts.

Worn, contaminated, or defective linings or drums.

Leaks in system and proper fluid level.

Worn, contaminated, or defective disc pads or discs.

Excessive pedal play.

INSPECT PARKING BRAKE FOR:

Proper adjustment.

INSPECT TIRES, WHEELS, AND RIMS FOR:

Proper inflation.

Loose or missing lug nuts.

Condition of tires, including tread depth.

Mixing radials and bias ply tires.

Wheels that are cracked or damaged so as to cause unsafe operation.

FIGURE 8–5

A safety inspection may include these items.

Because of changes in engine technology such as variable valve timing, placing pistons rings closer to the top of the piston, and the adoption of hybrid powertrains, using the correct viscosity oil is more critical than ever. Using the incorrect weight oil can cause excessive oil consumption, increased fuel consumption, variable valve timing (VVT) system faults, and other concerns.

The **Society of Automotive Engineers (SAE)** has established an oil viscosity classification system that is accepted throughout the industry. This system is a numeric rating in which the higher viscosity, or heavier weight, oils receive the higher numbers. For example, oil classified as SAE 50 weight oil is heavier and flows slower than SAE 10 weight oil.

Modern engine oils are **multiviscosity oils**. These oils carry a combined classification such as 5W-30. This rating says the oil has the viscosity of both a 5- and a 30-weight oil. The "W" after the 5 notes that the oil's viscosity was tested at –18°C (0°F). This is commonly referred to as the "winter grade." Therefore, the 5W means the oil has a viscosity of 5 when cold. The 30 rating is the hot rating. This rating was the result of testing the oil's viscosity at 100°C (212°F). To formulate multiviscosity oils, polymers are blended into the oil. Polymers expand when heated. With the polymers, the oil maintains its viscosity to the point where it is equal to 30-weight oil. The SAE classification and the API rating are displayed on the container of oil **(Figure 8–7)**.

FIGURE 8–7

The SAE classification and the API rating are displayed in this way on a container of oil.

Reproduced courtesy of the American Petroleum Institute.

ILSAC Oil Ratings

International Lubricant Specification Advisory Committee (ILSAC), formed by both North American and Japanese vehicle manufacturers, has developed an oil rating that combines SAE viscosity ratings and the API service rating. If engine oil meets the standards, a "starburst" symbol is displayed on the container **(Figure 8–8)**. ILSAC standards require that an oil meet specific standards for low- and high-temperature operation, deposit control, sludge control, and more.

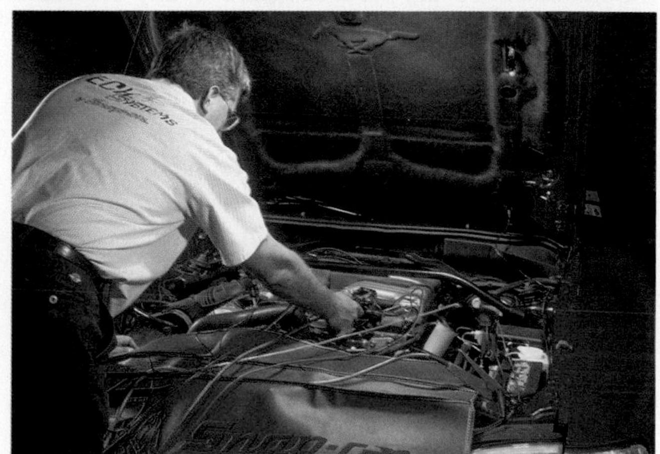

FIGURE 8–6

Fender covers should be used when working under the hood.

©Cengage Learning 2015

SHOP TALK

Many engines have very specific requirements. Always install the type of oil specified by the manufacturer. Never assume that a particular type of oil can be used in an engine.

ACEA Oil Ratings

ACEA stands for the Association of Constructors of European Automobiles. There are ACEA oil ratings for both gasoline and diesel engines. Requirements include meeting standards for sulphated ash, phosphorus, sulphur, and high-temperature shear.

Manufacturers' Oil Ratings

The vehicle manufacturers themselves also have specific oil ratings. These ratings are a result of

FIGURE 8–8

The ILSAC certification mark, commonly referred to as "the starburst."

Reproduced courtesy of the American Petroleum Institute.

changes in engine technology, seal and gasket technology, and emission standards. One of the newest oil ratings is General Motors' Dexos standard. This new oil standard is required for all late-model GM engines.

SYNTHETIC OILS Synthetic oils are made through a chemical, not natural, process. The introduction of synthetic oils dates back to World War II. Synthetic oils have many advantages over mineral oils, including better fuel economy and engine efficiency by reducing friction; they have low viscosity in low temperatures and a higher viscosity in warm temperatures, and they tend to have a longer useful life. Synthetic oils cost more than mineral oils, which is the biggest drawback for using them. Engine oils that are blends of mineral oils and synthetics to keep the cost down are available and offer many of the advantages of synthetic oil.

MAINTENANCE Perhaps the PM service that is best known to the public is changing the engine's oil and filter. Because oil is the lifeblood of an engine, it is critical that the oil and filter are changed on a regular basis. Photo Sequence 3 shows the steps involved in changing the engine oil and oil filter. Whenever doing this, make sure the oil is the correct rating for the vehicle.

In between oil and filter changes, the level of the oil should be periodically checked. When doing this, make sure the vehicle is parked on level ground. Locate and remove the oil **dipstick**. With a clean rag, wipe the oil from the dipstick and reinsert it all the way in its tube. Remove it again and check the level of the oil. If the level is at the FULL mark, the level is okay. If the level is at the ADD mark, this means the level is about one litre low. Regardless of the level, examine the oil for evidence of dirt. If the oil is contaminated, it must be changed.

OIL FILTER The oil pumped through the system passes through an oil filter. This filter is normally changed along with the oil. There are many shapes and sizes of filters and each may require a special tool to remove and install one **(Figure 8–9)**.

SHOP TALK

Most late-model vehicles have a maintenance reminder lamp in the instrument panel. This lamp comes on when maintenance is due and will stay on until it is turned off by a technician. The method for doing this varies with the manufacturer, so always check the service information for the correct procedure.

FIGURE 8–9

A variety of oil filter tools.

Cooling System

Whenever you change an engine's oil, you should also do a visual inspection of the different systems under the hood, including the cooling system. Inspect all cooling system hoses for signs of leakage and/or damage. Replace all hoses that are swollen, are cracked, or show signs of leakage. The radiator should also be checked for signs of leaks; if any are evident, the radiator should be repaired or replaced. Also check the front of the radiator for any buildup of dirt and bugs **(Figure 8–10)**. This buildup can restrict airflow through the radiator and should be removed by thorough cleaning.

The level and condition of the engine's coolant should also be checked. Check the coolant's level at the coolant recovery tank **(Figure 8–11)**. It should be between the LOW and FULL lines. If the level is too low, more coolant should be added through the cap of the tank, not the radiator. Bring the level up to the FULL line. Always use the correct type of coolant when topping off or replacing it. Look at the colour of the coolant when checking the level. It should not look rusty, crusty, or cloudy. If the coolant looks contaminated, the cooling system should be flushed and new coolant put into the system.

SHOP TALK

Recycle all used antifreeze/coolant or take it to an authorized collection point. Do not dump old coolant into a sewage drain, the ground, or any body of water.

FIGURE 8–10

A buildup of dirt and bugs can restrict airflow through the radiator.

> **CAUTION!**
>
> Never remove the radiator cap when the coolant is hot. Because the system is pressurized, the coolant can be hotter than boiling water and will cause severe burns. Wait until the top radiator hose is not too hot to touch. Then press down on the cap and slowly turn it until it hits the first stop. Now slowly let go of the cap. If there is any built-up pressure in the system, it will be released when the cap is let up. After all pressure has been exhausted, turn the radiator cap to remove it.

FIGURE 8–11

The level of coolant in the cooling system should be checked at the coolant recovery tank.

COOLANT Engine **coolant** is a mixture of water and antifreeze. Water alone has a boiling point of 100°C (212°F) and a freezing point of 0°C (32°F) at sea level. A mixture of 70 percent antifreeze and 30 percent water will raise the boiling point of the mixture to 136°C (276°F) under 15 psi of pressure and lower the freezing point to –64°C (–84°F). Normally, the recommended mixture is a 50/50 solution of water and antifreeze.

The antifreeze concentration must always be a minimum of 44 percent all year and in all climates. If the percentage is lower than 44 percent, engine parts may be eroded by cavitation, and cooling system components may be severely damaged by corrosion. **Cavitation** is erosion or pitting caused by the formation and rapid collapse of vapour bubbles formed in fast-moving or boiling coolant. These imploding bubbles strike components under tremendous pressure, creating surface pitting.

Five types of coolant are commonly available:

- **Ethylene glycol:** This was once the most commonly used antifreeze. It uses inorganic acid technology (IAT), is green in colour, and provides good protection regardless of climate, but it is poisonous. IAT coolant is not compatible with newer long-life coolants.

> **CAUTION!**
>
> Never leave ethylene glycol or propylene glycol coolant out and lying around. Both children and animals will drink it because of its sweet taste. The coolant is poisonous and can cause death.

- **Propylene glycol:** This type has the same basic characteristics as ethylene glycol–based coolant but is not sweet tasting and is less harmful to animals and children. Propylene glycol–based coolants are not used as factory-fill coolants and should not be mixed with ethylene glycol.
- **Phosphate-free:** This is ethylene glycol–based coolant with zero phosphates, which makes it more environmentally friendly. Phosphate-free coolant is recommended by some auto manufacturers.
- **Organic acid technology (OAT):** This coolant is also environmentally friendly and contains zero

PHOTO SEQUENCE 3

CHANGING THE OIL AND OIL FILTER

P3–1 Always make sure the vehicle is positioned safely on a lift or supported by jack stands before working under it. Before raising the vehicle, allow the engine to run until it reaches normal operating temperature. After it is warm, turn off the engine.

P3–2 The tools and other items needed to change the engine's oil and oil filter are rags, a funnel, an oil filter wrench, safety glasses, and a wrench for the drain plug.

P3–3 Place the oil drain pan under the drain plug before beginning to drain the oil.

P3–4 Loosen the drain plug with the appropriate wrench. After the drain plug is loosened, quickly remove it so that the oil can freely drain from the oil pan.

P3–5 Make sure the drain pan is positioned so that it can catch all of the oil.

P3–6 While the oil is draining, use an oil filter wrench to loosen and remove the oil filter.

P3–7 Make sure the oil filter seal comes off with the filter. Then place the filter into the drain pan so that it can drain. After it has completely drained, discard the filter according to local regulations.

P3–8 Wipe off the oil filter sealing area on the engine block. Then apply a coat of clean engine oil onto the new filter's seal.

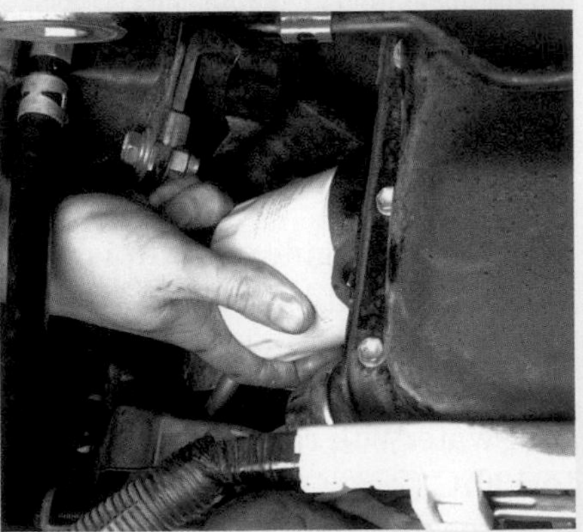

P3–9 Install the new filter and hand-tighten it. Oil filters should be tightened according to the directions given on the filter.

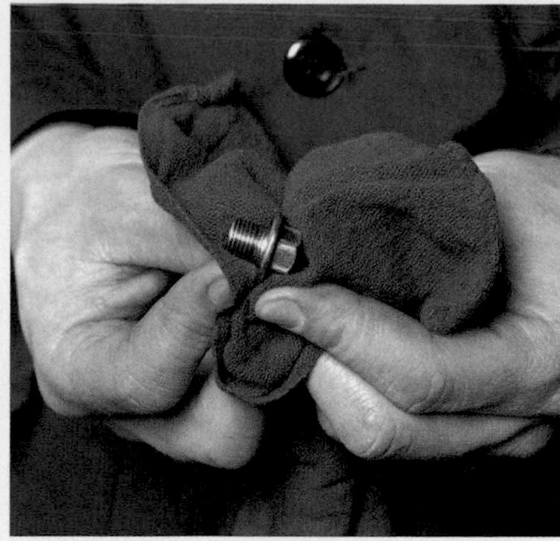

P3–10 Prior to installing the drain plug, wipe off its threads and sealing surface with a clean rag.

P3–11 Tighten the drain plug according to the manufacturer's recommendations. Over-tightening can cause thread damage, whereas under-tightening can cause an oil leak.

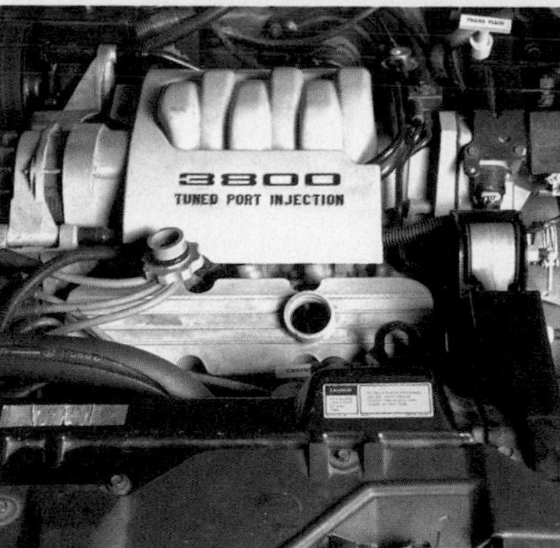

P3–12 With the oil filter and drain plug installed, lower the vehicle and remove the oil filter cap.

P3–13 Carefully pour the oil into the engine. The use of a funnel usually keeps oil from spilling on the engine.

P3–14 After the recommended amount of oil has been put in the engine, check the oil level.

P3–15 Start the engine and allow it to reach normal operating temperature. While the engine is running, check the engine for oil leaks, especially around the oil filter and drain plug. If there is a leak, shut down the engine and correct the problem.

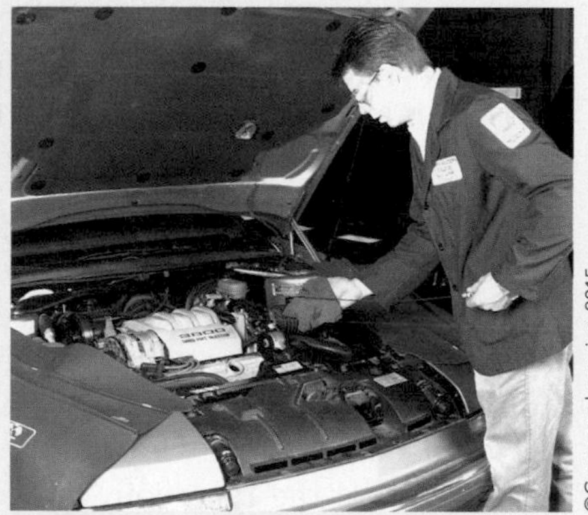

P3–16 After the engine has been turned off, recheck the oil level and correct it as necessary.

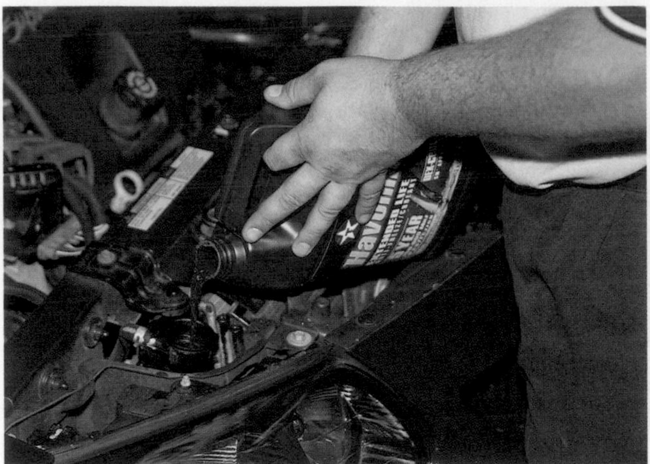

FIGURE 8–12

Ethylene glycol is the most commonly used antifreeze/coolant and is green in colour. OAT coolant is orange and is often referred to by the brand name DEX-COOL.

©Cengage Learning 2015

phosphates or silicates. This orange coolant is often referred to by the brand name DEX-COOL and is used in all late-model GM vehicles **(Figure 8–12)**.

- **Hybrid organic acid technology (HOAT):** This is similar to OAT coolant but has additives that make the coolant less abrasive to water pumps. This type of coolant is used by Ford and Chrysler and is not compatible with IAT or OAT coolants.

In addition to the five commonly available types of coolant, many manufacturers require a specific coolant for their vehicles. These coolants may be phosphate free, silicate free, borate free, or all of the above. Factory coolants are usually sold premixed 50/50 coolant and pure water.

When inspecting and servicing the coolant, it is important to note that colour alone does not really determine what coolant is appropriate for the vehicle. Several colours of orange, pink, yellow, and blue are used in modern vehicles, and each vehicle manufacturer has specific coolant use specifications. Before pouring in any coolant, first check the service information to determine exactly which coolant is required.

COOLANT CONDITION A coolant hydrometer is used to check the amount of antifreeze in the coolant. This tester contains a pickup hose, coolant reservoir, and squeeze bulb. The pickup hose is placed in the radiator coolant. When the squeeze bulb is squeezed and released, coolant is drawn into the reservoir. As coolant enters the reservoir, a pivoted float moves upward with the coolant level. A pointer on the float indicates the freezing point of the coolant on a scale located on the reservoir housing.

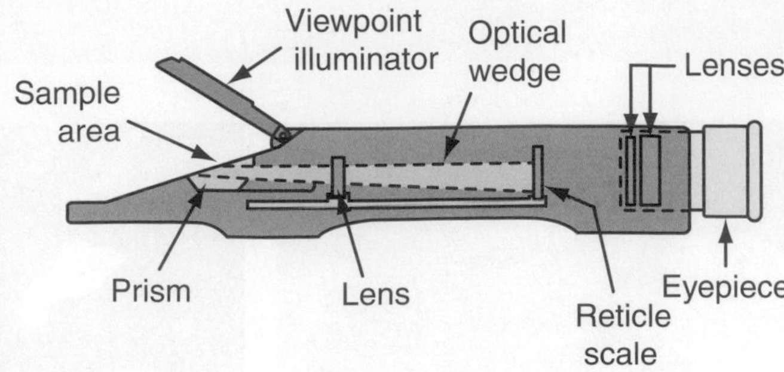

FIGURE 8–13

A refractometer that tests coolant condition and battery electrolyte.

©Cengage Learning 2015

A refractometer **(Figure 8–13)** offers a precise way to check coolant condition. Most refractometers can also measure the specific gravity of battery electrolyte and test the condition of brake fluid. A sample of the fluid is placed in the sample area of the meter, and as light passes through the sample, a line is cast on the meter's scale. The line shows the concentration of the antifreeze in the coolant **(Figure 8–14)**.

Litmus strips are also used to evaluate coolant. The test strips are immersed into a sample of coolant. After about one to five minutes, the strip will change colour. The colour of the strip is then compared to a scale on the container of strips. Matching the colours will indicate the freeze protection level and the acidity of the coolant **(Figure 8–15)**.

Drive Belts

V-belts and **V-ribbed (serpentine) belts** are used to drive water pumps, power-steering pumps, air-conditioning compressors, generators, and emission control pumps. Heat has adverse effects on drive belts; it can cause the belts to harden and crack. Excessive heat normally comes from slippage. Slippage can be caused by improper belt tension or oily conditions. When there is slippage, the support bearing of the component driven by the belt can be damaged.

The angled sides of V-belts contact the inside of the pulleys' grooves **(Figure 8–16)**. This point of contact is where motion is transferred. As a V-belt wears, it begins to ride deeper in the groove. This reduces its tension and promotes slippage. Because this is a normal occurrence, periodic adjustment of belt tension is necessary.

Drive belts can be used to drive a single part or a combination of parts. An engine can have three or more V-belts.

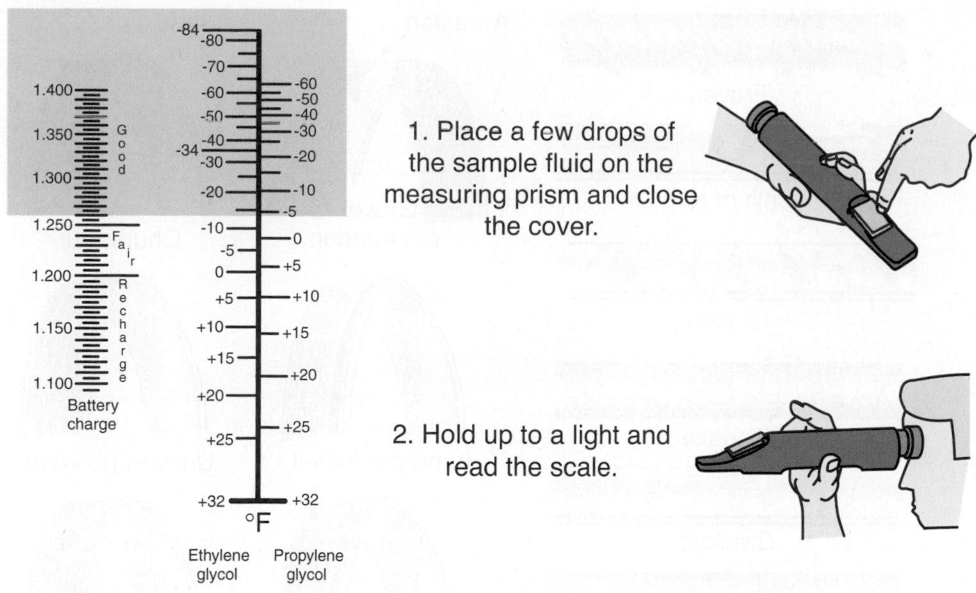

FIGURE 8–14

Measuring antifreeze and battery electrolyte levels with a refractometer.

©Cengage Learning 2015

V-BELT INSPECTION Even the best V-belts last only an average of four years. That time can be shortened by things that can be found by inspecting the belts. Check the condition of all of the drive belts on the engine. Carefully look to see if they have worn or glazed edges, tears, splits, and signs of oil soaking **(Figure 8–17)**. If these conditions exist, the belt should be replaced. Also inspect the grooves of the pulleys for rust, oil, wear, and other damage. If a pulley is damaged, it should be replaced. Rust, dirt, and oil should be cleaned off the pulley before installing a new belt.

Misalignment of the pulleys reduces the belt's service life and brings about rapid pulley wear, which causes thrown belts and noise. Undesirable side or end thrust loads can also be imposed on pulley or pump shaft bearings. Check alignment with a straightedge. Pulleys should be in alignment within 1.59 mm (1/16 inch) per foot (30 cm) of the distance across the face of the pulleys.

A quick check of a belt's tension can be made by locating the longest span of the belt between two pulleys. With the engine off, press on the belt midway through that distance. If the belt moves more than ½ inch per foot of free span, the belt should be adjusted. Keep in mind that different belts require

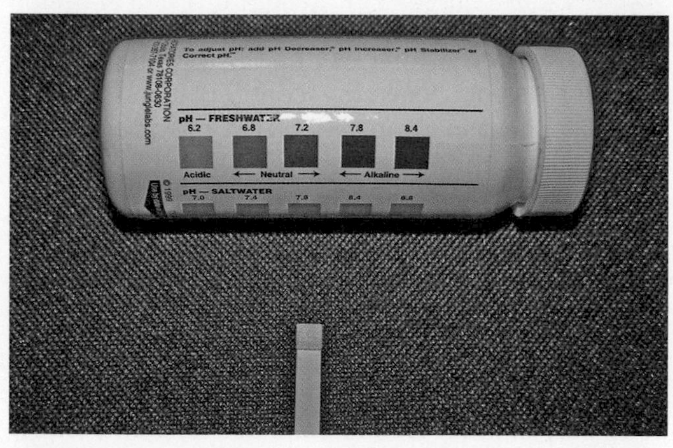

FIGURE 8–15

Matching the colour of a test strip to the scale on its container will indicate the freeze protection level and the acidity of the coolant.

©Cengage Learning 2015

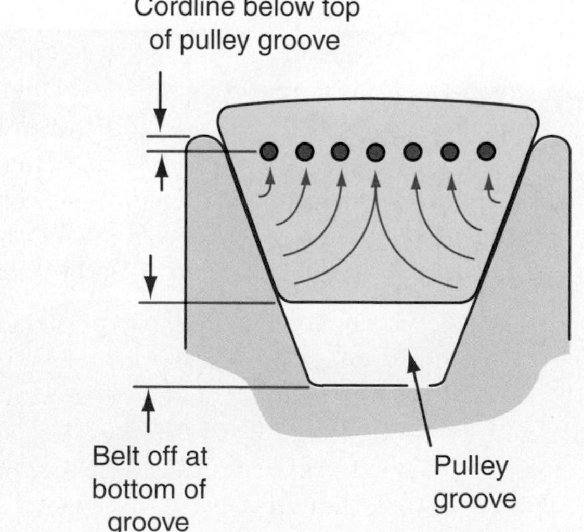

FIGURE 8–16

The sides of a V-belt contact the grooves of the pulleys.

©Cengage Learning 2015

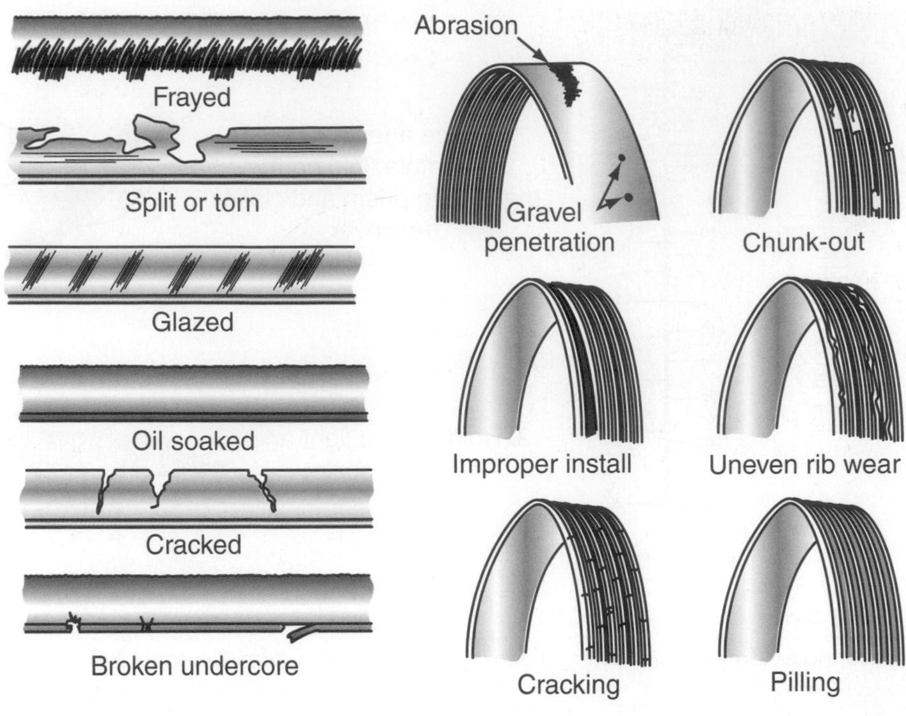

FIGURE 8–17

Drive belts should be inspected.

©Cengage Learning 2015

different tensions. The belt's tension should be checked with a belt tension gauge **(Figure 8–18)**.

USING SERVICE INFORMATION

Proper belt tightening procedures are given in the specification section of the service information.

FIGURE 8–18

Check the tension of a drive belt with a belt tension gauge.

©Cengage Learning 2015

The exact procedure for adjusting belt tension depends on what the belt is driving. Normally, the mounting bracket for the component driven by the belt and/or its tension adjusting bolt is loosened. The mounting brackets on generators, power-steering pumps, and air compressors are designed to be adjustable. Some brackets have a hole or slot to allow the use of a prybar. Other brackets have a ½-inch square opening in which a breaker bar can be installed to move the component and tighten the belt. Other engines have an adjusting bolt, sometimes called a jackscrew, that can be tightened to correct the belt tension. Loosen the mounting bolts and hold the component in the position that provides for the correct tension. Be careful not to damage the part you are prying against. Then tighten the mounting bolts or tension adjusting bolt to keep the tension on the belt. Once tightened, recheck the belt tension with the tension gauge.

V-RIBBED BELTS Most late-model vehicles use a V-ribbed or multi-ribbed belt to drive all or most accessories. Multi-ribbed belts are long and follow a complex path that weaves around the various pulleys **(Figure 8–19)** and are also called serpentine belts. Proper tension is critical due to the complex routing. Serpentine belts are flat on the outside and have a series of continuous ribs on the inside.

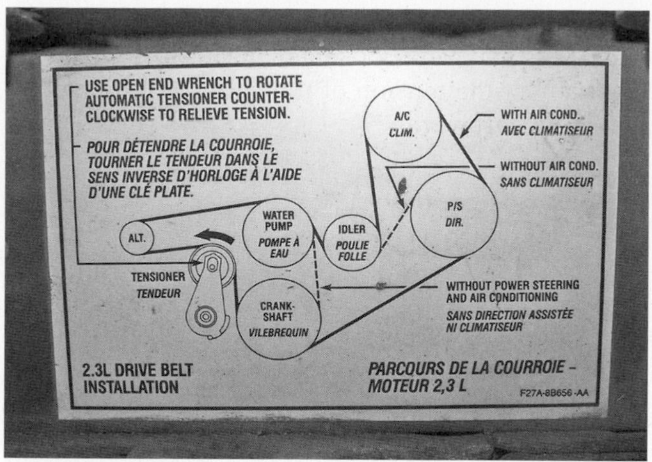

FIGURE 8–19

The routing of a typical serpentine belt drive.

These ribs fit into matching grooves in the pulleys. Both the ribbed side and the flat side of the belt can be used to transfer power. Over time, the belts will stretch and lose their tension. To compensate for this and to keep a proper amount of belt tension, serpentine belt systems have an automatic belt tensioner pulley. This pulley is a spring-loaded pulley **(Figure 8–20)** that exerts a predetermined amount of pressure on the belt to keep it at the desired tension.

V-ribbed belts are exposed to dirt, rocks, salt, and water; these along with pulley shape and slight misalignments result in rib-surface wear. Belts are designed to have a clearance between the belt and the pulley. When rib material is lost, that clearance is lost, which eliminates a way for dirt and water to pass through the system.

A belt that has stretched in length can cause slippage and the bottoming out of the belt tensioner, which can cause damage to mating parts.

V-ribbed belts can be made of neoprene (polychloroprene) or EPDM (ethylene propylene diene monomer) rubber. Both of these are types of synthetic rubber, and each has its own advantages and disadvantages.

V-RIBBED BELT INSPECTION The procedures for inspecting these belts are similar but vary with the belt's construction. Although neoprene belts are dependable for up to 100 000 km (60 000 miles), through use they begin to develop cracks, uneven rib wear, edge and backside wear, glazing, and noise **(Figure 8–21)**. If any of these are evident during an inspection of the belt, the belt should be replaced.

EPDM belts resist cracking; therefore, a look for cracks may not be the best way to check these belts. It is better to check these belts for a loss of material. A loss as little as 5 percent of material can cause a loss of tension and/or belt slip, which will affect the operation of components and lead to their failure.

If a belt does not have the proper tension, it may squeal or chirp, it may roll off a pulley, or it may slip.

FIGURE 8–20

A belt tensioner for a serpentine belt.

Gates Corporation

FIGURE 8–21

A damaged V-ribbed belt.

Excessive tension may put unwanted forces on the pulleys and the shafts they are attached to, leading to noise, belt breakage, glazing, and damage to the bearings and bushings in the driven components. Improper tension is normally caused by belt stretch or a faulty tensioner. Do not assume that the belt tensioner is working properly. Measure the belt tension when performing a belt inspection and after a new belt is installed.

BELT REPLACEMENT If a drive belt is damaged, it should be replaced. If there is more than one drive belt, all should be replaced even if only one is bad. Always use an exact replacement belt. The size of a new belt is typically given, along with the part number, on the belt container **(Figure 8–22)**. You can verify that the new belt is a replacement for the old one by physically comparing the two. This, however, does not account for any belt stretch that may have occurred. Therefore, only use this comparison as verification. The best way to select the correct replacement belt is with a parts catalogue and/or by matching the numbers on the old belt to the numbers on the new belt.

To replace a V-belt on some engines, it may be necessary to remove the fan, fan pulley, and other accessory drive belts to gain access to belts needing replacement. Also, before removing the old drive belt, disconnect the electric cooling fan at the radiator, if the vehicle has one. Remove the old belt by loosening the components that have adjusting slots for belt tension. Then slip the old belt off. Check the condition and alignment of the pulleys. Correct any problems before installing the new belt. Place the new belt around the pulleys. Once in place, loosely tighten the bolts that were loosened during belt removal. Then adjust the tension of the belt and retighten all mounting hardware.

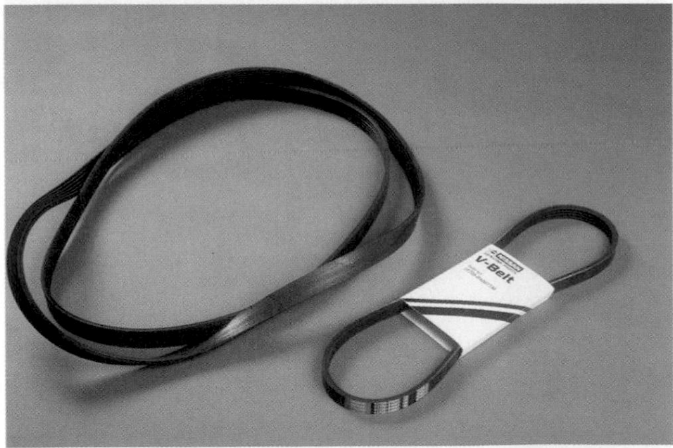

FIGURE 8–22

The size and part number of a new belt are given on the belt container. The size can be verified by physically comparing the old with the new belt.

SHOP TALK

It is never advisable to pry a belt onto a pulley. Obtain enough slack so that the belt can be slipped on without damaging either the V-belts or a pulley. Some power-steering pumps have a ½-inch drive socket to aid in adjusting belts to the proper tension without prying against any accessory.

Photo Sequence 4 shows the correct procedure for inspecting, removing, replacing, and adjusting a V-ribbed belt. Before removing a serpentine belt, locate a belt routing diagram in the service information or on an underhood decal. Compare the diagram with the routing of the old belt. If the actual routing is different from the diagram, draw the existing routing on a piece of paper.

After installation of a new belt, the engine should be run for 10 to 15 minutes to allow belts to seat and reach their initial stretch condition. Modern steel-strengthened V-belts do not stretch much after the initial run-in, but it is often recommended that the tension of the belt be rechecked after 8000 km (5000 miles).

Stretch-Fit Belts

Many newer cars and light trucks have stretch-fit belts that require no tensioner. This type of belt is typically used to drive one accessory, such as the A/C compressor. The belt is installed using a special tool that attaches to the drive pulley. Rotating the pulley gently stretches the belt over the pulleys and into the grooves. Once installed, there is no provision for adjusting belt tension. The old belt is removed by cutting it off.

Air Filters

If an air filter is doing its job, it will get dirty. Most air filters are made of pleated paper to increase the filtering area. As a filter gets dirty, the amount of air that can flow through it is reduced. This is not a problem until less air than what the engine needs can get through the filter. Without the proper amount of air, the engine will not be able to produce the power it should, nor will it be as fuel efficient as it should be.

TYPICAL PROCEDURE FOR INSPECTING, REMOVING, REPLACING, AND ADJUSTING A DRIVE BELT

PHOTO SEQUENCE **4**

P4–1 Inspect the belt by looking at both sides.

P4–2 Look for signs of glazing.

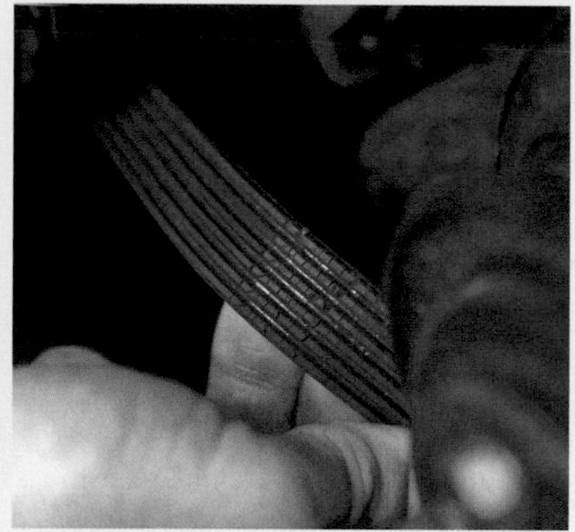

P4–3 Look for signs of tearing or cracking.

P4–4 To replace a worn belt, locate the tensioner or generator pulley.

P4–5 Loosen the hold-down fastener for the tensioner or generator pulley.

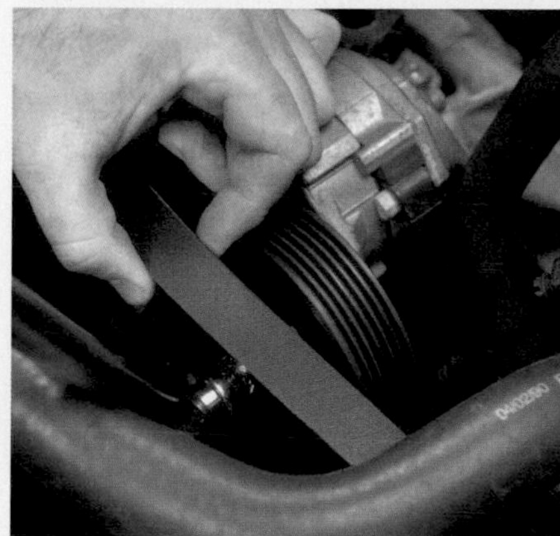

P4–6 Pry the tensioner or generator pulley inward to release the belt tension and remove the belt.

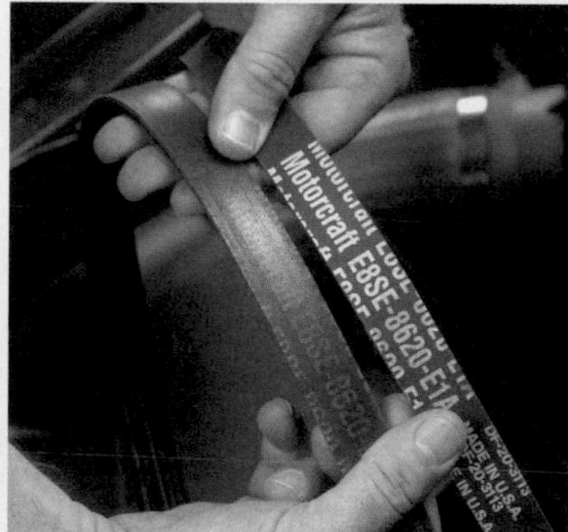

P4–7 Match the old belt up for size with the new replacement belt.

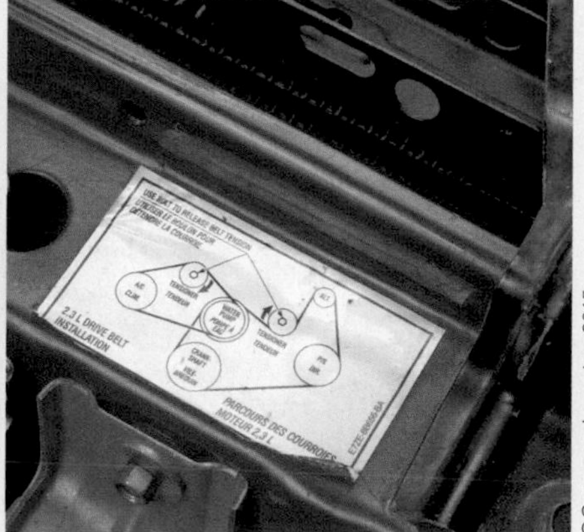

P4–8 Observe the belt routing diagram in the engine department.

TYPICAL PROCEDURE FOR INSPECTING, REMOVING, REPLACING, AND ADJUSTING A DRIVE BELT *(continued)*

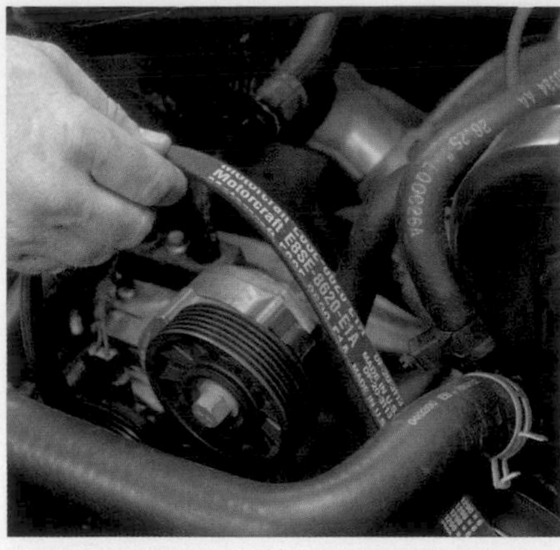

P4–9 Install the new belt over each of the drive pulleys. Often the manufacturer recommends a sequence for feeding the belt around the pulleys.

P4–10 Pry out the tensioner or generator pulley to put tension on the belt.

P4–11 Install the belt squarely in the grooves of each pulley.

P4–12 Measure the belt deflection in its longest span. If a belt tension gauge is available, use it and compare the tension.

P4–13 Pry the tensioner or generator pulley to adjust the belt to specifications.

P4–14 Tighten the tensioner or generator pulley fastener.

P4–15 Start the engine and check the belt for proper operation.

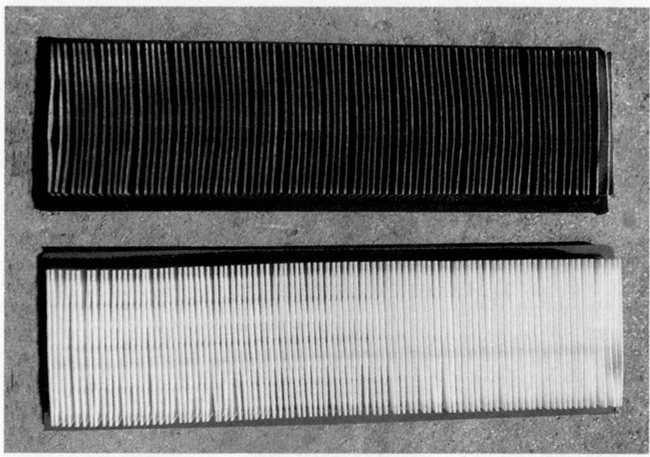

FIGURE 8–23

A dirty and a clean air filter.

Included in the PM plan for all vehicles is the replacement of the air filter. This mileage or time interval is based on normal operation. If the vehicle is used, or has been used, in heavy dust, the life of the filter is shorter. Always use a replacement filter that is the same size and shape as the original. An air filter should also be periodically checked for excessive dirt or blockage **(Figure 8–23)**.

When replacing the filter element, carefully remove all dirt from the inside of the housing. Large pieces of dirt and stones accumulate here. It would be disastrous if that dirt got into the cylinders. Also make sure that the air cleaner housing is properly aligned and closed around the filter to ensure good airflow of clean air. If the filter does not seal well in the housing, dirt and dust can be pulled into the air stream to the cylinders. The shape and size of the air filter element depends on its housing; the filter must be the correct size for the housing, or dirt will be drawn into the engine.

Battery

The battery is the main source of electrical energy for the vehicle. It is very important that it is inspected and checked on a regular basis.

SHOP TALK

It should be noted that disconnecting the battery on late-model cars removes some memory from the engine's computer and the car's accessories. Besides losing the correct time on its clock or the programmed stations on the radio, the car might run roughly, the airbag light may turn on, and transmission shifting may be affected.

PROCEDURE

STEP 1 Visually inspect the battery cover and case for dirt and grease.

STEP 2 Check the electrolyte level (if possible).

STEP 3 Inspect the battery for cracks, loose terminal posts, and other signs of damage.

STEP 4 Check for missing cell plug covers and caps.

STEP 5 Inspect all cables for broken or corroded wires, frayed insulation, or loose or damaged connectors.

STEP 6 Check the battery terminals, cable connectors, metal parts, hold-downs, and trays for corrosion damage or buildup—a bad connection can cause reduced current flow.

STEP 7 Check the heat shield for proper installation on vehicles so equipped.

If the battery or any of the associated parts are dirty **(Figure 8–24)** or corroded, they should be removed and cleaned. Photo Sequence 5 shows the correct procedure for cleaning a battery, a battery tray, and battery cables.

SHOP TALK

When removing or installing a battery, always use the built-in battery strap or a battery lifting tool to lift the battery in or out of its tray.

Transmission Fluid

The oil used in automatic transmissions is called **automatic transmission fluid (ATF)**. This special fluid is dyed red so that it is not easily confused

FIGURE 8–24

A really dirty battery.

PHOTO SEQUENCE
5

TYPICAL PROCEDURE FOR CLEANING A BATTERY CASE, TRAY, AND CABLES

P5–1 Loosen the battery negative terminal clamp.

P5–2 Use a terminal clamp puller to remove the negative cable.

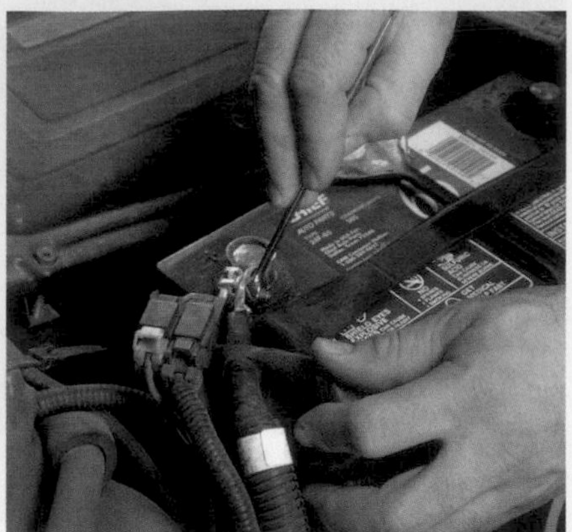

P5–3 Loosen the battery positive terminal clamp.

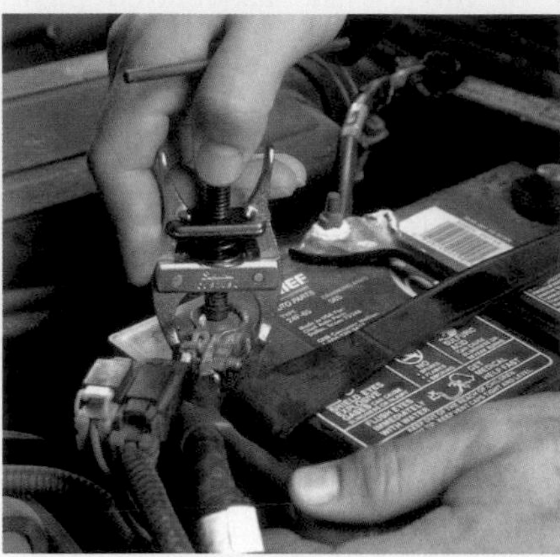

P5–4 Use a terminal clamp puller to remove the positive clamp.

P5–5 Remove the battery hold-down hardware and any heat shields.

P5–6 Remove the battery from the tray.

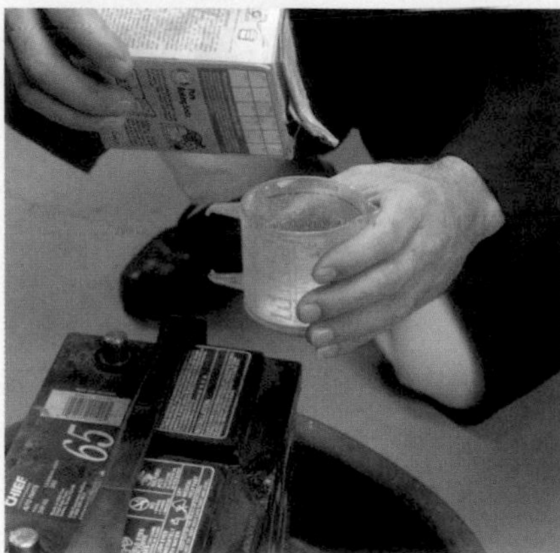

P5–7 Mix a solution of baking soda and water.

P5–8 Brush the baking soda solution over the battery case, but do not allow the solution to enter the cells of the battery.

P5–9 Flush the baking soda off with water.

P5–10 Use the scraper and wire brush to remove corrosion from the hold-down hardware.

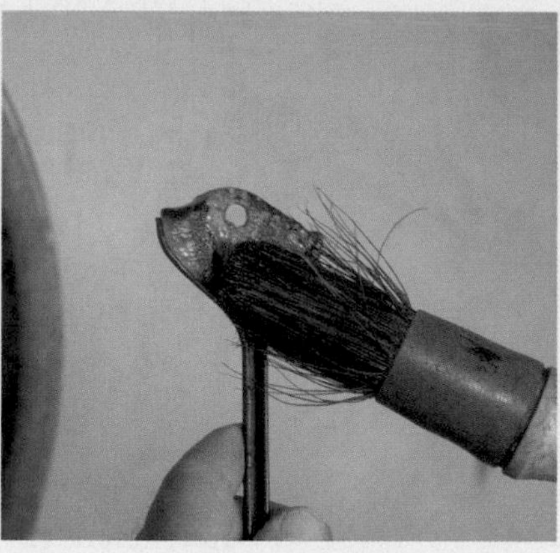

P5–11 Brush the baking soda solution over the hold-down hardware and then flush with water.

P5–12 Allow the hardware to dry. Then paint it with corrosion-proof paint.

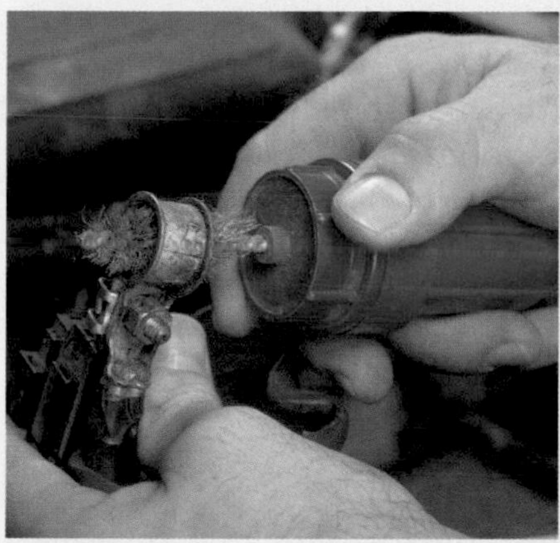

P5–13 Use a terminal cleaner brush to clean the battery cables.

P5–14 Use a terminal cleaner brush to clean the battery posts.

P5–15 Install the battery back into the tray. Also install the hold-down hardware.

P5–16 Install the positive battery cable. Then install the negative cable.

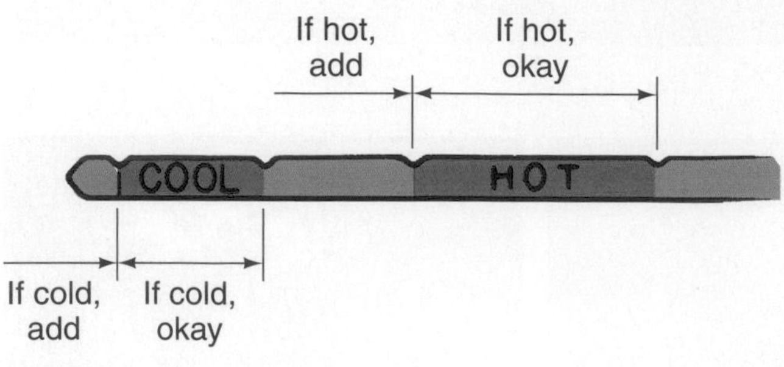

FIGURE 8–25

Automatic transmission fluid should be checked regularly. Normally, the level is checked when the engine is warm. The normal cold level is well below the normal hot level.

with engine oil. Before checking the fluid, make sure the engine is warm and the vehicle is level. Then set the parking brake and allow the engine to idle. Sometimes it is recommended that the ATF level be checked when the transmission is placed into park; however, some may require some other gear. Make sure you follow those requirements. Locate the fluid dipstick (normally located to the rear of the engine) and pull it out of its tube. Check the level of the fluid on the dipstick **(Figure 8–25)**. If the level is low, add only enough to bring the level to FULL. Make sure you use only the fluid recommended by the manufacturer.

SHOP TALK

Many late-model vehicles do not have a transmission dipstick and require special procedures to check the fluid. Refer to the vehicle's service information to find out how to check the fluid.

The condition of the fluid should be checked while checking the fluid level. The normal colour of ATF is pink or red. If the fluid has a dark brownish or blackish colour and/or a burned odour, the fluid has been overheated. A milky colour indicates that engine coolant has been leaking into the transmission's cooler in the radiator.

After checking the ATF level and colour, wipe the dipstick on absorbent white paper and look at the stain left by the fluid. Dark particles are normally band and/or clutch material, whereas silvery metal particles are normally caused by the wearing of the transmission's metal parts. If the dipstick cannot be wiped clean, it is probably covered with varnish, which results from fluid oxidation. Varnish will cause the transmission's valves to stick, causing improper shifting speeds. Varnish or other heavy deposits indicate the need to change the transmission's fluid and filter.

The exact fluid that should be used in an automatic transmission depends on the transmission design and the year the transmission was built. It is very important that the correct type of ATF be used. Each manufacturer has a fluid specification that must be followed. Always refer to the service or owner's manual for the correct type of fluid to use. Some transmission dipsticks are also marked with the type of ATF required.

Some transmissions require the use of unique fluids. Continuously variable transmissions (CVTs) require a fluid that is much different from that used in automatic transmissions. Always use the fluid recommended by the manufacturer. The use of the wrong fluid may cause the transmission to operate improperly and/or damage the transmission.

MANUAL TRANSMISSIONS Manual transmissions, transaxles, and drive axle units require the use of specific lubricants or oils, and the levels should be checked according to the recommended time frames. Some manufacturers recommend that the fluids be periodically changed. Most repair shops have an air-operated dispenser for these fluids; others rely on a hand-operated pump.

Power-Steering Fluid

Normally, the power-steering fluid level is checked with the engine warm but turned off. If the fluid is cold, it will read lower than normal. The filler cap on the power-steering pump normally has a dipstick. Unscrew the cap and check the level **(Figure 8–26)**. Add fluid as necessary. Sometimes the fluid used in these systems is ATF; check the service information for the proper fluid type before adding fluid.

Brake Fluid

Brake fluid levels are checked at the master cylinder. Older master cylinders are made of cast iron or aluminum and have a metal bail that snaps over the master cylinder cover to hold it in place. Normally, the bail can be moved in only one direction. Once moved out of the way, the master cylinder cover can be removed. Once removed, the fluid levels can be checked.

Newer master cylinders have a metal or plastic reservoir mounted above the cylinder. The reservoir will have one or two caps. To check the fluid level

FIGURE 8–26

The filler cap on a power-steering pump normally has a dipstick to check the fluid level.

©Cengage Learning 2015

FIGURE 8–27

Translucent brake fluid reservoirs allow the fluid level to be observed from the outside.

©Cengage Learning 2015

in a metal reservoir, the cap must be removed. Most often, the caps are screwed on. The caps on some plastic reservoirs have snaps to hold them. Unsnap the cap to check the fluid. It is important to clean the area around the caps before removing them. This prevents dirt from falling into the reservoir. A rubber diaphragm attached to the inside of the caps is designed to stop dirt, moisture, and air from entering into the reservoir. Make sure the diaphragm is not damaged.

Most new plastic reservoirs are translucent and allow the fluid level to be observed from the outside **(Figure 8–27)**. Do not open the master cylinder reservoir unnecessarily as this allows air and moisture to contact the brake fluid.

While checking the fluid level, look at the colour of the fluid. Brake fluid tends to absorb moisture, and its colour gives clues as to the moisture content of the fluid. Dark- or brown-coloured fluid indicates contamination, and the system must be flushed and the fluid replaced.

When it is necessary to add brake fluid, make sure the fluid is the correct type and is fresh and clean. There are basically four types of brake fluids: DOT 3, DOT 4, DOT 5, and DOT 5.1.

DOT 3, DOT 4, and DOT 5.1 fluids are polyalkylene-glycol-ether mixtures, called **polyglycol** for short. The colour of both DOT 3 and DOT 4 fluid ranges from clear to light amber. DOT 5 fluids are all silicone-based because only silicone fluid—so far—can meet the DOT 5 specifications. Currently, few vehicle manufacturers recommend DOT 5 fluid for use in their brake systems. Although the other three fluid grades are compatible, they do not combine well if mixed together in a system. Therefore, it is best to use the fluid type recommended by the manufacturer and never mix fluid types in a system.

Clutch Fluid

On some vehicles with a manual transmission, there is another, but smaller, master cylinder close to the brake master cylinder. This is the clutch master cylinder. Its fluid level needs to be checked, which is done in the same way as brake fluid. In most cases, the clutch master cylinder uses the same type of fluid as the brake master cylinder. However, check this out before adding any fluid.

Diesel Exhaust Fluid

Some newer diesel engine vehicles use a special diesel exhaust fluid (DEF) made of urea and pure water. This mixture is stored in a tank and is injected into the diesel exhaust after-treatment system to reduce NO_x emissions. How often the tank needs to be filled depends on the vehicle's operating conditions, but customers should expect to refill the tank about every 8000 km (5000 miles). If the tank runs out, the on-board computer system will limit engine and vehicle speed until the tank is refilled and the system operation returns to normal.

Windshield Wipers

Check the condition of the windshield wipers. Wiper blades can become dull, torn, or brittle. If they are, they should be replaced. Also check the condition of the wiper arms. Look for signs of distortion or damage. Also check the spring on the arm. This spring is designed to keep the wiper blade fairly tight against the windshield. If the spring is weak or damaged, the blade will not do a respectable job cleaning the glass.

Most wiper blade assemblies have replaceable blades or inserts. To replace the blades, grab hold of the assembly and pivot it away from the windshield. Once the arm is moved to its maximum position, it should stay there until it is pivoted back to the windshield. Doing this will allow you to easily replace the wiper blades without damaging the vehicle's paint or glass.

Although most wiper blades are replaced as an assembly, there are three basic types of wiper blade inserts or refills. Look carefully at the old blade to determine which one to install. Remove the old insert and install the new one. After installation, pull on the insert to make sure it is properly secured. If the insert comes loose while the wipers are moving across the windshield, the wiper arm could scratch the glass.

Most often, wiper blades are replaced as an assembly. There are several methods used to secure the blades to the wiper arm **(Figure 8–28)**. Most replacement blades come with the necessary adapters to secure the blade to the arm.

Wiper arms are either mounted onto a threaded shaft and held in place by a nut, or they are pressed over a splined shaft. Some shaft-mounted arms are held in place by a clip that must be released before the arm can be pulled off. When installing wiper arms, make sure they are positioned so that the blades do not hit the frame of the windshield while they are operating. When checking the placement and operation of the wipers, wet the windshield before turning on the wipers. The water will serve as a lubricant for the wipers.

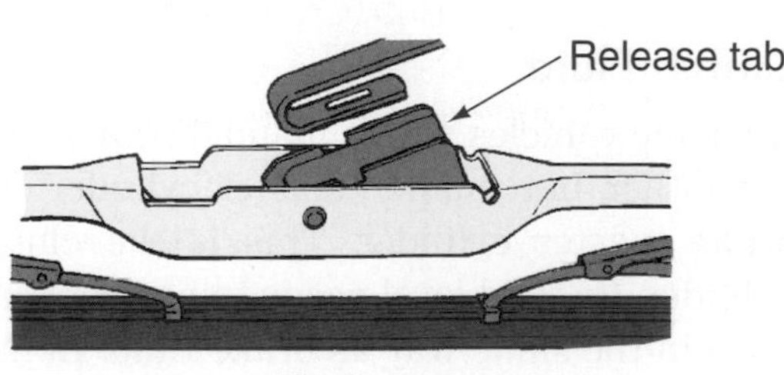

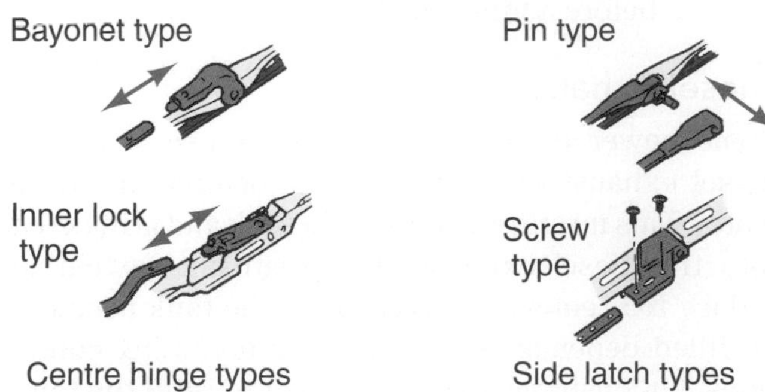

FIGURE 8–28

Examples of the different ways that wiper blades are secured to the wiper arm.

©Cengage Learning 2015

FIGURE 8–29

Check the level of the windshield washer fluid reservoir.

©Cengage Learning 2015

WINDSHIELD WASHER FLUID The last fluid level to check is the windshield washer fluid **(Figure 8–29)**. Visually check the level and add as necessary. Always use windshield washer fluid, and never add water to the washer fluid, especially in cold weather. Water can freeze and crack the tank or clog the washer hoses and nozzles.

Tires

The vehicle's tires should be checked for damage and wear. Tires should have at least 1.6 mm (1/16") of tread remaining. Any less and the tire should be replaced. Tires have tread wear indicators moulded into them. When the wear bar shows across the width of the tread, the tire is excessively worn. A tire wear gauge gives an accurate measurement of the tread depth **(Figure 8–30)**. Also check the tires for bulges, nails, tears, and other damage. All of these indicate the tire should be replaced.

FIGURE 8–30

A tire tread depth gauge.

INFLATION Check the inflation of the tires. To do this, use a tire pressure gauge **(Figure 8–31)**. Press the gauge firmly onto the tire's valve stem. The air pressure in the tire will push the scale out of the tool. The highest number shown on the scale is the air pressure of the tire. Compare this reading with the specifications for the tire.

FIGURE 8–31

Check the tires and wheels for damage and proper inflation.

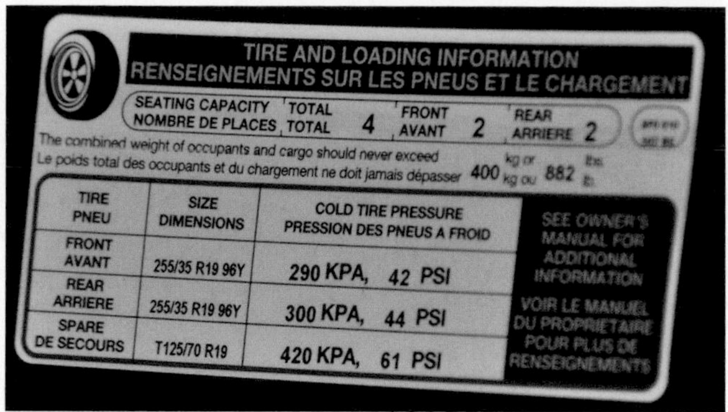

FIGURE 8–32

The tire placard gives the recommended cold tire pressure for that vehicle.

The correct tire pressure is listed in the vehicle's owner's manual or on a decal (placard) stuck on the driver's doorjamb **(Figure 8–32)**. The air pressure rating on the tire is not the amount of pressure the tire should have. Rather, this rating is the maximum pressure the tire should ever have when it is cold.

New vehicles are fit with tire pressure monitoring systems. These systems either have an air pressure sensor inside of each wheel or use wheel speed sensor data to determine tire pressure. When the pressure is below or above a specified range, the vehicle's computer causes a warning light on the dash to illuminate. This alerts the driver of a problem. For vehicles that have pressure sensors, the sensors can be, and should be, checked as part of regular maintenance.

TIRE ROTATION To equalize tire wear, most car and tire manufacturers recommend that the tires be rotated. Front and rear tires perform different jobs and can wear differently, depending on driving habits and the type of vehicle. In RWD vehicles, for instance, the front tires usually wear along the outer edges, primarily because of the scuffing and slippage encountered in cornering. The rear tires wear in the centre because of acceleration thrusts. To equalize wear, it is recommended that tires be rotated as recommended for the type of vehicle being serviced (see Chapter 46). Radial tires should be initially rotated at 12 000 kilometres (7500 miles) and then at least every 24 000 kilometres (15 000 miles) thereafter. It is important that directional tires are kept rotating in the direction they are designed for. This means the tires may need to be dismounted from the wheel, flipped, and reinstalled on the rim before being put on the other side of the car.

FIGURE 8–33

Wheel lugs should be tightened to the specified torque.

FIGURE 8–34

Torque sticks are colour-coded to indicate torque setting.

LUG NUT TORQUE Obviously, to rotate the tires you must remove the tire/wheel assemblies and then reinstall them. Before reinstalling a tire/wheel assembly on a vehicle, make sure the wheel studs are clean and not damaged; then clean the axle/rotor flange and wheel bore with a wire brush or steel wool. Coat the axle pilot flange with disc brake caliper slide grease or an equivalent. Place the wheel on the hub. Install the lug nuts and tighten them alternately to draw the wheel evenly against the hub. They should be tightened to a specified torque **(Figure 8–33)** and sequence to avoid distortion. Many tire technicians snug up the lug nuts, and then when the car is lowered to the floor, they use a torque wrench for the final tightening.

WARNING!

Over-torquing of the lug nuts is a common cause of disc brake rotor distortion. Also, an over-torqued lug distorts the threads of the lug and could lead to failure.

Some technicians use a torque absorbing adapter, also called a torque stick **(Figure 8–34)**, to tighten the lug nuts. Follow the torque stick manufacturer's instructions for setting and checking torque accuracy to ensure the lugs are not over- or under-torqued. Make sure to use the correct stick for the required torque. Then check the actual torque of the lug nuts with a torque wrench.

Chassis Lubrication

A PM procedure that is becoming less common because of changing technology is chassis lubrication. However, all technicians should know how to do this. During the lubrication procedure, grease is forced between two surfaces that move or rub against each other. The grease reduces the friction produced by the movement of the parts. During a chassis lube, grease is forced into a pivot point or joint through a grease fitting. Grease fittings are found on steering and suspension parts, which need lubrication to prevent wear and noise caused by their action during vehicle operation.

Grease fittings are called **zerk fittings** and are threaded into the part that should be lubricated. A fitting at the end of a manual or pneumatic grease gun fits over the zerk to inject the lubricant. Older vehicles have zerk fittings in many locations, whereas newer vehicles use permanently lubricated joints. Some of these joints have threaded plugs that can be removed to lubricate the joint. A special adapter is threaded onto the grease gun and into the plug's bore to lubricate the joint. After grease has been injected into the joint, the plug should be reinstalled or a zerk fitting installed. On some vehicles, rubber or plastic plugs are installed at the factory; they should never be reused.

Carefully look at the joints to see if the joint boots are sealed or not. Some joints, such as tie-rod ends and ball joints, are sealed with rubber boots. If the boots are good, wipe any old grease or dirt from the zerk fittings, push the grease gun's nozzle straight onto a zerk fitting, and pump grease slowly into the joint **(Figure 8–35)**. If the joint has a sealed boot, put just enough grease into the joint to cause the boot to slightly expand. If the boot is not sealed, put in enough grease to push the old grease out. Then wipe off the old grease and any excess grease. Repeat this at all lubrication points.

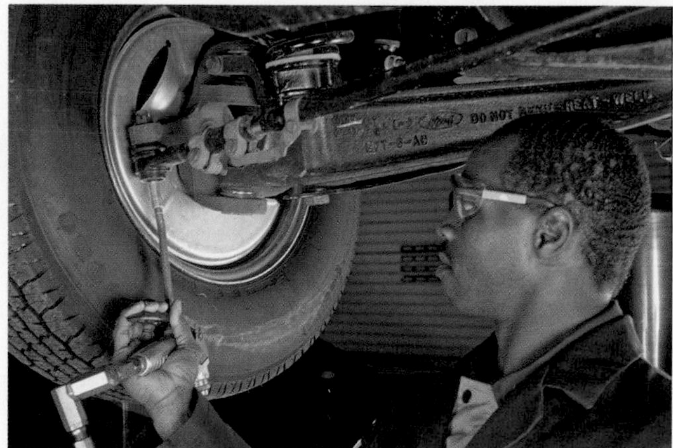

FIGURE 8–35

A grease gun forces lubrication into a joint through a zerk fitting.

NLGI grade	Worked Penetration after 60 Strokes at 77°F (25°C)	Appearance
000	44.5–47.5 mm	fluid
00	4.00–4.30 mm	fluid
0	3.55–3.85 mm	very soft
1	3.10–3.40 mm	soft
2	2.65–2.95 mm	moderately soft
3	2.20–2.50 mm	semifluid
4	1.75–2.05 mm	semihard
5	1.30–1.60 mm	hard
6	0.85–1.15 mm	very hard

FIGURE 8–36

The table shows the NLGI grades and the worked penetration ranges.

Class	Purpose
GA	Mild duty—wheel bearings
GB	Mild to moderate duty—wheel bearings
GC	Mild to severe duty—wheel bearings
LA	Mild duty—chassis parts and universal joints
LB	Mild to severe duty—chassis parts and universal joints

FIGURE 8–37

ASTM grease designation guide.

GREASES Greases are made from oil blended with thickening agents. There are a few synthetic greases available that meet the same standards as petroleum greases. The thickening agent increases the viscosity of the grease. Greases are categorized by a **National Lubricating Grease Institute (NLGI)** number and by the thickeners and additives that are in the grease. Some greases are also labeled with an "EP," which means they have extreme pressure additives. The number assigned by the NLGI is based on test results and the specifications set by the American Society for Testing Materials (ASTM).

The ASTM specifies the consistency of grease using a penetration test. During this test, the grease is heated and placed below the tip of a test cone. The cone is dropped into the grease. The distance the cone is able to penetrate the grease is measured. The cone will penetrate deeper into soft grease. The NLGI number represents the amount of penetration **(Figure 8–36)**. The higher the NLGI number, the thicker the grease is. NLGI #2 is typically specified for wheel bearings and chassis lubrication.

The NLGI also specifies grease by its use and has established two categories for automotive use. Chassis lubricants are identified with the prefix "L," and wheel bearing lubricants have a prefix of "G." Greases are further defined within those groups by their overall performance. Chassis greases are classified as either LA or LB, and there are three classifications for wheel bearing greases (GA, GB, and GC). Many types of greases are labelled as both GC and LB and are acceptable for both. These are often referred to as multipurpose greases **(Figure 8–37)**. The NLGI certification mark is included on the grease's container **(Figure 8–38)**.

HYBRID VEHICLES

Hybrid vehicles are maintained and serviced in the same way as conventional vehicles, except for the hybrid components. The latter include the high-voltage battery pack and circuits, which must be respected when doing any service on the vehicles. Other services to hybrid vehicles are normal services that must be completed in a different way.

For the most part, service to the hybrid system is not something that is done by technicians unless they are certified to do so by the automobile manufacturer. Keep in mind that a hybrid has nearly all of the basic systems as a conventional vehicle, and these are diagnosed and serviced in the same way.

NATIONAL LUBRICATING GREASE INSTITUTE
NLGI
AUTOMOTIVE WHEEL BEARING LUBRICANT
GC

NATIONAL LUBRICATING GREASE INSTITUTE
NLGI
AUTOMOTIVE WHEEL BEARING & CHASSIS LUBRICANT
GC-LB

NATIONAL LUBRICATING GREASE INSTITUTE
NLGI
AUTOMOTIVE CHASSIS LUBRICANT
LB

FIGURE 8–38

NLGI identification symbols.

Through an understanding of how the hybrid vehicle operates, you can safely service them.

One of the things to pay attention to is the stop-start feature. You need to know when the engine will normally shut down and restart. Without this knowledge, or the knowledge of how to prevent this, the engine may start on its own when you are working under the hood. Needless to say, this can create a safety hazard. There is a possibility that your hands or something else could be trapped in the rotating belts or hit by a cooling fan. Unless the system is totally shut down, the engine may start at any time when its control system senses that the battery needs to be recharged.

In addition, there is a possibility that the system will decide to power the vehicle by electric only. When it does this, there is no noise, just a sudden movement of the vehicle. This can scare you and can be dangerous. To prevent both of these, always remove the key from the ignition. Make sure the "READY" lamp in the instrument cluster is off; this lets you know the system is also off. Hybrids from Toyota and Hyundai use a "READY" light, but other manufacturers use other indicators to let you know the vehicle is powered up and ready to be driven. Make sure you know how to tell when the vehicle is completely powered down before beginning any work.

Maintenance

Maintenance of a hybrid vehicle is much the same as a conventional one. Care needs to be taken to avoid anything orange while carrying out the maintenance procedures.

The computer-controlled systems are extremely complex and are very sensitive to voltage changes. This is why the manufacturers recommend a thorough inspection of the auxiliary battery and connections every six months.

The engines used in hybrids are modified versions of engines found in other models. Other than fluid checks and changes, there is little maintenance required on these engines. However, there is less freedom in deciding the types of fluids that can be used and the parts that can replace the original equipment. Hybrids are not very forgiving. Always use the exact replacement parts and the specified fluids.

Typically, the engine oil used in a hybrid is very light, for example 0W-20 oil. If heavier oil is used, the computer may see this as a problem and prevent the engine from starting. The heavier oil may cause an increase in the current required to crank the engine. If the computer senses very high current draw while attempting to crank the engine, it will open the circuit in response.

Special attention is necessary for the cooling systems in hybrid vehicles. Most hybrids have two cooling systems, one for the engine and one for the hybrid drive systems and/or battery packs. These cooling systems do not intermix, and different coolants are used in each. If servicing a hybrid vehicle cooling system, first make sure you are working on the correct cooling system. Cooling the hybrid system is important, and checking the coolant's condition and level is an additional PM check. The cooling system used in some hybrids features electric pumps and storage tanks **(Figure 8–39)**. The tanks store heated coolant and could cause injury if you are not aware of how to check them.

Many hybrids use passenger compartment air for the high-voltage battery cooling system. These air supply systems often contain air filters that may also need to be serviced at regular intervals. There is a filter in the ductwork from the outside of the vehicle to the battery box. This filter needs to be periodically

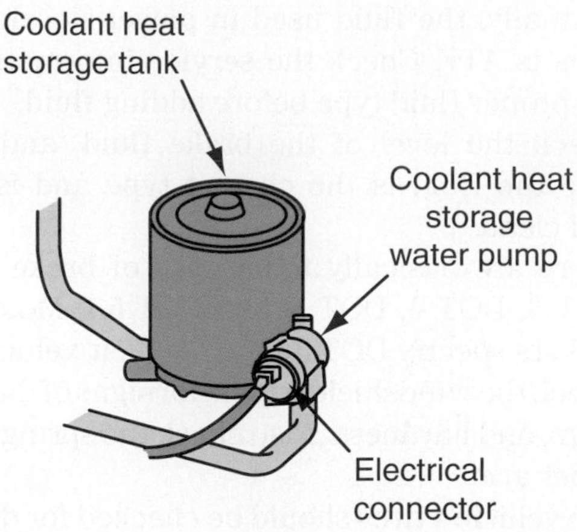

FIGURE 8–39

The hot coolant storage tank for Toyota hybrids.

changed. If the filter becomes plugged, the temperature of the battery will rise to dangerous levels. In fact, if the computer senses high temperatures, it may shut down the system.

A normal part of PM is checking power-steering and brake fluids. The power-steering systems used by the manufacturers vary; some have a belt-driven pump, some have an electrically driven pump, and others have a pure electric and mechanical steering gear. Each variety requires different care; therefore, always check the service information for the specific model before doing anything to these systems. Also keep in mind that some hybrids use the power-steering pump as the power booster for the brake system.

Hybrids are all about fuel economy and reduced emissions. Everything that would affect these should be checked on a regular basis. Items such as tires, brakes, and wheel alignment can have a negative effect, and owners of hybrids will notice the difference. These owners are constantly aware of their fuel mileage due to the displays on the instrument panel.

ADDITIONAL PM CHECKS

The following PM checks are tasks that should be performed at these suggested time intervals to help ensure safe and dependable vehicle operation.

Time: While operating the vehicle

- Pay attention to and note any changes in the sound of the exhaust or any smell of exhaust fumes in the vehicle.
- Check for vibrations in the steering wheel. Notice any increased steering effort or looseness in the steering wheel.
- Notice if the vehicle constantly turns slightly or pulls to one side of the road.
- When stopping, listen and check for strange sounds, pulling to one side, increased brake pedal travel, or hard-to-push brake pedal.
- If any slipping or changes in the operation of the transmission occur, check the transmission fluid level.
- Check for fluid leaks under the vehicle. (Water dripping from the air-conditioning system after use is normal.)
- Check the automatic transmission's park function.
- Check the parking brake.

Time: At least monthly

- Check the operation of all exterior lights, including the brake lights, turn signals, and hazard warning flashers.

Time: At least twice a year

- Check the pressure in the spare tire.
- Check headlight alignment.
- Check the muffler, exhaust pipes, and clamps.
- Inspect the lap/shoulder belts for wear.
- Check the radiator, heater, and air-conditioning hoses for leaks or damage.

Time: At least once a year

- Lubricate all hinges and all outside key locks.
- Lubricate the rubber weather strips for the doors.
- Clean the body's water drain holes.
- Lubricate the transmission controls and linkage.

KEY TERMS

American Petroleum Institute (API) (p. 193)
Automatic transmission fluid (ATF) (p. 207)
Cavitation (p. 197)
Coolant (p. 197)
Crude oil (p. 193)
Dipstick (p. 196)
Multiviscosity oils (p. 195)
National Lubricating Grease Institute (NLGI) (p. 215)
Polyglycol (p. 211)
Repair order (RO) (p. 188)
Resource-conserving oils (p. 193)
Serpentine belt (p. 200)
Society of Automotive Engineers (SAE) (p. 195)
Sublet repairs (p. 190)
V-belts (p. 200)
Vehicle identification number (VIN) (p. 191)
Viscosity (p. 193)
V-ribbed belts (p. 200)
Zerk fittings (p. 214)

SUMMARY

- A repair order (RO) is a legal document used for many purposes.
- An RO includes a cost estimate for the repairs. By law, this estimate must be quite accurate.
- Preventive maintenance (PM) involves regularly scheduled vehicle service to keep it operating efficiently and safely. Technicians should stress the importance of PM to their customers.
- Engine oil is a clean or refined form of crude oil. It contains many additives, each intended to improve the effectiveness of the oil.
- The American Petroleum Institute (API) classifies engine oil according to its ability to meet the engine manufacturers' warranty specifications.
- The Society of Automotive Engineers (SAE) has established an oil viscosity classification system that has a numeric rating in which the higher viscosity, or heavier weight, oils receive the higher numbers.
- Changing the engine's oil and filter should be done on a regular basis.
- Whenever the engine's oil is changed, a thorough inspection of the cooling systems should be done.
- Normally, the recommended mixture for engine coolant is a 50/50 solution of water and antifreeze.
- V-belts and V-ribbed (serpentine) belts are used to drive water pumps, power-steering pumps, air-conditioning compressors, generators, and emission control pumps.
- If a belt does not have the proper tension, it may produce squealing and chirping noises; allow the belt to roll off a pulley; or slip, which reduces the power that drives a component.
- Excessive belt tension may put unwanted forces on the pulleys and the shafts they are attached to, leading to noise, belt breakage, glazing, and damage to the bearings and bushings in water pumps, generators, and power-steering pumps.
- The air filter should be periodically checked for excessive dirt or blockage, and a replacement filter should be the same size and shape as the original.
- The battery is the main source of electrical energy for the vehicle. It is very important that it is checked on a regular basis.
- If the battery or any of the associated parts are dirty or corroded, remove the battery and clean them.
- The condition of the automatic transmission fluid (ATF) should be checked while checking the fluid level.
- Normally, the fluid used in power-steering systems is ATF. Check the service information for the proper fluid type before adding fluid.
- Check the level of the brake fluid, and make sure the fluid is the correct type and is fresh and clean.
- There are basically four types of brake fluids: DOT 3, DOT 4, DOT 5, and DOT 5.1. Most automakers specify DOT 3 fluid for their vehicles.
- Check the windshield wipers for signs of dullness, tears, and hardness. Also check the spring on the wiper arm.
- The vehicle's tires should be checked for damage and wear as well as for proper inflation.
- To equalize tire wear, most car and tire manufacturers recommend that the tires be rotated after a specified mileage interval.
- Several parts of a vehicle may need periodic lubrication; always use the correct type of grease when doing this.
- Some PM procedures are unique to hybrid vehicles; always follow the recommendations of the manufacturer.
- When servicing a hybrid vehicle, always respect its high-voltage system.

REVIEW QUESTIONS

1. Describe the information found in a VIN.
2. List at least five things that should be checked while inspecting a vehicle's battery.
3. List five different types of oil rating systems.
4. Why should you wipe off the outside of a zerk fitting before injecting grease into it?
5. How does a technician determine the proper inflation of a vehicle's tires?
6. Which of the following is a true statement about a mechanic's lien?
   a. This lien states that the shop may gain possession of the vehicle if the customer does not pay for all services performed.
   b. The right to impose a mechanic's lien can be exercised by a shop within 30 days after the services have been completed.
   c. A shop's right to impose a lien on the vehicle being serviced does not require acknowledgement by the customer prior to beginning any services to the vehicle.
   d. The right to impose a mechanic's lien can be exercised by a shop 90 days after the completion of the agreed-upon services.

7. Which of the following greases are best suited for lubricating automotive wheel bearings?
   a. LA b. LB
   c. GA d. GC
8. What engine oil viscosity is bested suited for use in a hybrid electric vehicle?
   a. 0W-20 b. 5W-20
   c. 10W-30 d. 15W-40
9. When presenting a repair estimate to a customer, what should be included in the estimate?
   a. only the total cost of the repair parts
   b. only the total cost of the shop labour
   c. the total cost of the repair parts and shop labour
   d. the total cost of the repair parts and shop labour, plus any shop consumables
10. What is an additional procedure during a PM service on a hybrid vehicle?
   a. inspecting the high voltage output
   b. checking the hybrid system coolant's condition and level
   c. addition of a thicker oil for summer engine operation
   d. auxiliary battery replacement
11. Which of the following statements about oil's viscosity is correct?
   a. It measures the ability of oil to resist flow.
   b. Viscosity is affected by temperature; hot oil flows slower than cold oil.
   c. In the API system of oil viscosity classification, the lighter oils receive a higher number.
   d. Heavyweight oils are best suited for use in low-temperature regions.
12. Which of the following VINs would represent a 2014 Canadian-made Dodge vehicle?
   a. 1C3AN65L9EXO29415
   b. 2C3AN65L9EXO29415
   c. 2C3AN65L94XO29415
   d. 2D3AN65L94XO29415
13. What is the viscosity rating of an engine oil with a classification of 5W-30?
   a. 5 at 0°C and 30 at 100°C
   b. 5 at 100°C and 30 at 0°C
   c. 5 at –18°C and 30 at 100°C
   d. 5 at 100°C and 30 at –18°C
14. What type of anti-freeze is usually green in colour?
   a. propylene glycol
   b. ethylene glycol
   c. organic acid technology (OAT)
   d. hybrid organic acid technology (HOAT)
15. When should V-belt tension be rechecked after the initial run-in?
   a. 2000 km b. 4000 km
   c. 6000 km d. 8000 km
16. What would be indicated by dark particles left on the rag after checking the automatic transmissions fluid level?
   a. worn clutch discs
   b. worn transmission gearing
   c. engine coolant contamination
   d. a varnish buildup
17. When replacing a pleated paper air filter element, what else should be done?
   a. Wash and reuse the old paper element.
   b. Tap the filter element to remove excess dirt.
   c. Use compressed air to blow out excess dirt.
   d. Clean out the air filter housing.
18. What is the reason for newer manufacturer oil ratings?
   a. colder temperature operation
   b. changes to seal and gasket technology
   c. extended oil-change intervals
   d. use of turbochargers and superchargers
19. Which of the following PM checks should be made most often?
   a. coolant in the coolant recovery reservoir
   b. operation of all exterior lights
   c. engine oil level
   d. spare tire pressure
20. What is the minimum tire tread depth allowed before recommended replacement?
   a. 1.6 mm (1/16 inch)
   b. 3.2 mm (1/8 inch)
   c. 4.8 mm (3/16 inch)
   d. 6.4 mm (1/4 inch)

CHAPTER 9
Automotive Engine Designs and Diagnosis

LEARNING OUTCOMES

- Describe the various ways in which engines can be classified.
- Explain what takes place during each stroke of the four-stroke cycle.
- Outline the advantages and disadvantages of the in-line and V-type engine designs.
- Define important engine measurements and performance characteristics, including bore and stroke, displacement, compression ratio, engine efficiency, torque, and horsepower.
- Outline the basics of diesel, stratified, and Miller cycle engine operation.
- Explain how to evaluate the condition of an engine.
- List and describe abnormal engine noises.

WE SUPPORT THE CANADIAN INTERPROVINCIAL STANDARDS PROGRAM

INTRODUCTION TO ENGINES

The engine **(Figure 9–1)** provides the power to drive the vehicle's wheels. All automobile engines, both gasoline and diesel, are classified as internal-combustion engines because the combustion or burning that creates energy takes place inside the engine.

The biggest part of the engine is the cylinder block **(Figure 9–2)**. The cylinder block is a large casting of metal that is cast and drilled with holes to allow for the passage of lubricants and coolant through the block and provide spaces for movement of mechanical parts. The block contains the cylinders, which are round passageways fitted with pistons. The block houses or holds the major mechanical parts of the engine.

The cylinder head fits on top of the cylinder block to close off and seal the top of the cylinder **(Figure 9–3)**. The combustion chamber is an area into which the air/fuel mixture is compressed and burned. The cylinder head contains all or most of the combustion chamber. The cylinder head also contains ports through which the air/fuel mixture enters and burned gases exit the cylinder and the bore for the spark plug.

FIGURE 9–1
Today's engines are complex, efficient machines.
Martin Restoule

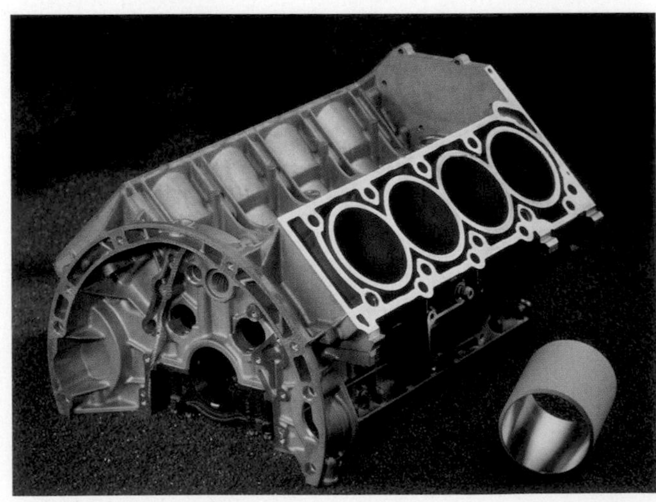

FIGURE 9–2
A cylinder block and cylinder liner for a late-model aluminum V-8 engine.
DaimlerChrysler Corporation

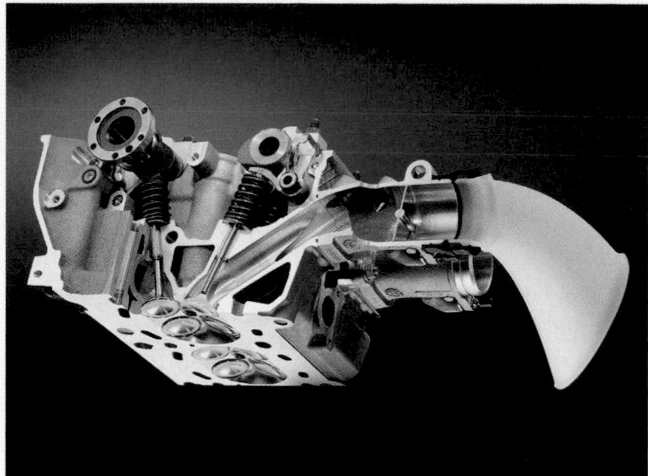

FIGURE 9–3

A cylinder head assembly for a late-model in-line six-cylinder engine.

BMW of North America, Inc.

The valve train is a series of components used to open and close the intake and exhaust ports. A valve is a movable part that opens and closes the ports. A camshaft controls the movement of the valves. Springs are used to help close the valves.

The up-and-down motion of the pistons must be converted to rotary motion before it can drive the wheels of a vehicle. To do this, the pistons are connected to a crankshaft by connecting rods **(Figure 9–4)**. The upper end of the connecting rod moves with the piston. The lower end is attached to the crankshaft and moves in a circle. The end of the crankshaft is connected to the flywheel or flexplate. Through the flywheel or flexplate, the power from the engine is indirectly sent to the drive wheels.

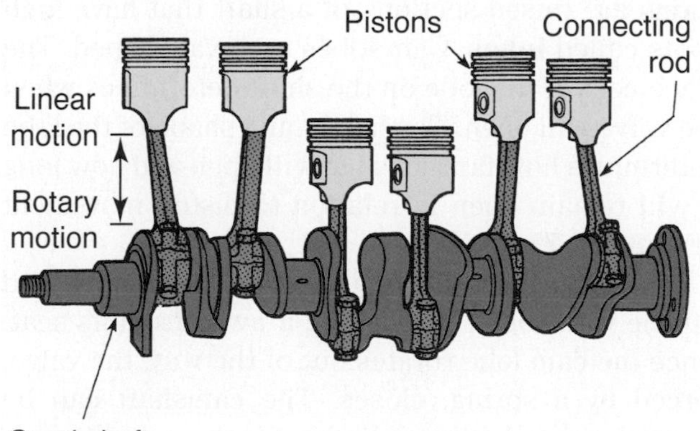

FIGURE 9–4

The reciprocating motion of the pistons is converted to rotary motion by the crankshaft.

FIGURE 9–5

A typical late-model engine.

American Honda Motor Co., Inc.

Engine Construction

Modern engines are designed to meet the performance and fuel efficiency demands of the public. Most are made of lightweight engine castings and stampings; non-iron materials (e.g., aluminum, magnesium, fibre-reinforced plastics); and fewer and smaller fasteners to hold things together. These fasteners are made possible through computerized joint designs that optimize loading patterns. Each of these newer engine designs has its own distinct personality based on construction materials, casting configurations, and design **(Figure 9–5)**.

ENGINE CLASSIFICATIONS

Today's automotive engines can be classified in several ways depending on the following design features:

- *Operational cycles.* Most technicians will come in contact with only four-stroke cycle engines. A few older cars have used, and some in the future may use, a two-stroke engine.
- *Number of cylinders.* Current engine designs include 3-, 4-, 5-, 6-, 8-, 10-, 12-, and 16-cylinder engines.
- *Cylinder arrangement.* The cylinders of an engine can be arranged in line with each other, in a "V" with an equal number of cylinders on each side, or directly across from each other, or horizontally opposed **(Figure 9–6)**.

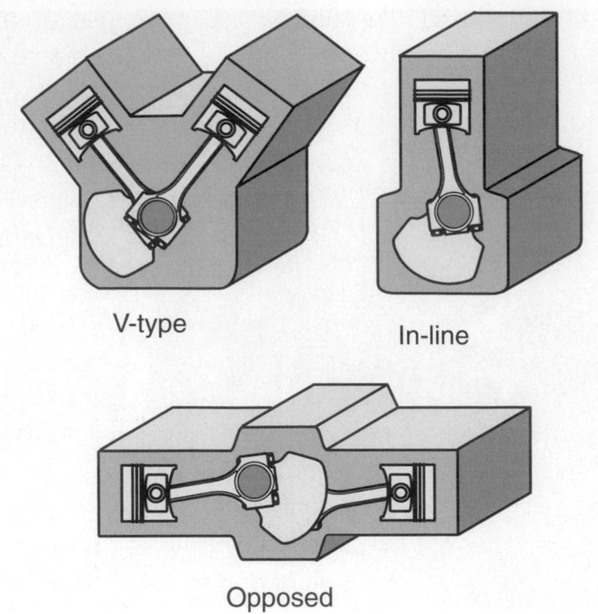

FIGURE 9–6

Basic engine configurations.

- *Valve train type.* Valve trains can be either the **overhead camshaft (OHC)** or the camshaft in-block **overhead valve (OHV)** design. Some engines have separate camshafts for the intake and exhaust valves. These are based on the OHC design and are called **dual overhead camshaft (DOHC)** engines. V-type DOHC engines have four camshafts—two on each side.
- *Ignition type.* There are two types of ignition systems: spark and compression. Gasoline engines use a spark ignition system. In a spark ignition system, the air/fuel mixture is ignited by an electrical spark. Diesel engines, or compression ignition engines, have no spark plugs. An automotive diesel engine relies on the heat generated as air is compressed in the cylinder to ignite the air/fuel.
- *Cooling systems.* There are both air-cooled and liquid-cooled engines in use. All current engines have liquid-cooling systems.
- *Fuel type.* The fuels currently used in automobiles include gasoline, natural gas, methanol diesel, ethanol, and propane. The most commonly used is gasoline, although most gasoline is blended with ethanol.

Four-Stroke Gasoline Engine

The engine provides the rotating power to drive the wheels through the transmission and driving axles. All vehicle engines, both gasoline and diesel, are classified as internal combustion because the combustion or burning of the fuel takes place inside the engine. They require an engine constructed to withstand the high temperatures and pressures created by the burning of vaporized fuel mixed with air.

The **combustion chamber** is the space between the top of the piston and the cylinder head. It is an enclosed area in which the air/fuel mixture is burned. If all of the fuel in the chamber is burned, complete combustion has taken place.

In order to have complete combustion, the right amount of fuel must be mixed with the right amount of air. This mixture must be compressed in a sealed container, then ignited by the right amount of heat at the right time. When these conditions exist, nearly all the fuel that enters a cylinder is burned. A percentage (about 30 percent) of the released energy is converted to useful power that moves the vehicle. Automotive engines have more than one cylinder. Each cylinder should receive the same amount of air, fuel, and heat, if the engine is to run efficiently.

Although the combustion must occur in a sealed cylinder, the cylinder must also have a way to allow heat, fuel, and air into it. There must also be a way to allow the burnt air/fuel mixture out so that a fresh mixture can enter and the engine can continue to run. To meet these requirements, engines are fitted with valves.

There are at least two valves at the top of each cylinder. The air/fuel mixture enters the combustion chamber through an intake valve and leaves (after having been burned) through an exhaust valve. The valves are accurately machined plugs that fit into machined openings. A valve is said to be seated or closed when it rests in its opening or seat. When the valve is pushed off its seat, it opens.

A rotating camshaft, driven by the crankshaft, opens and closes the intake and exhaust valves. **Cams** are raised sections of a shaft that have high spots called **lobes**. Cam lobes are oval shaped. The placement of the lobe on the shaft determines when the valve will open. The height and shape of the lobe determines how far the valve will open and how long it will remain open in relation to piston movement **(Figure 9–7)**.

As the camshaft rotates, the lobes rotate and lift the valve open by pushing it away from its seat. Once the cam lobe rotates out of the way, the valve, forced by a spring, closes. The camshaft can be located either in the cylinder block or in the cylinder head. The camshaft is driven by the crankshaft through gears, or sprockets, and a cogged belt, or timing chain.

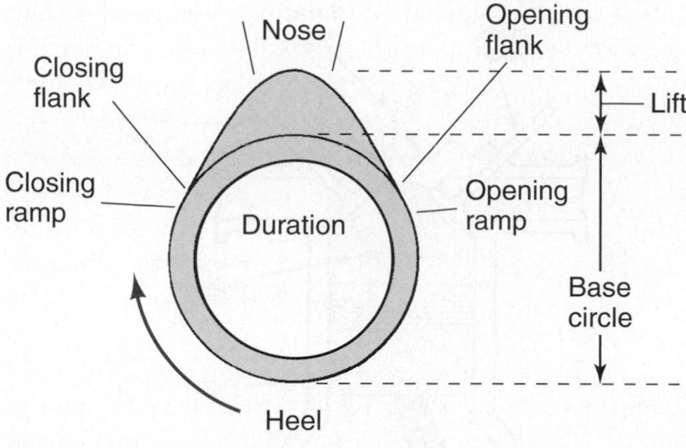

FIGURE 9–7

The height and width of a cam lobe determine when and for how long a valve will be open.

When the action of the valves and the spark plug is properly timed to the movement of the piston, the combustion cycle takes place in four strokes of the piston: the intake stroke, the compression stroke, the power stroke, and the exhaust stroke. The camshaft turns at half the crankshaft speed and rotates one complete turn during each complete four-stroke cycle.

FOUR-STROKE CYCLE A **stroke** is the full travel of the piston either up or down in a cylinder's bore. The reciprocal movement of the piston during the four strokes is converted to a rotary motion by the crankshaft. It takes two full revolutions of the crankshaft to complete the four-stroke cycle. One complete revolution of the crankshaft is equal to 360°; therefore, it takes 720° to complete the four-stroke cycle. During one piston stroke, the crankshaft rotates 180°.

The piston moves down by the pressure produced during combustion, but this moves the piston only enough to complete one stroke. In order to keep the engine running, the piston must travel through the other three strokes. The inertia of a flywheel attached to the end of the crankshaft keeps the crankshaft rotating and allows the piston to complete the rest of the four-stroke cycle. This stored flywheel inertia is also used to compress the air/fuel mixture just before combustion. A heavy flywheel is found only on engines equipped with a manual transmission. Engines with automatic transmissions have a flexplate and a torque converter. The weight and motion of the fluid inside the torque converter serve as a flywheel.

INTAKE STROKE The first stroke of the cycle is the intake stroke. As the piston moves away from **top dead centre (TDC)**, the intake valve opens **(Figure 9–8A)**. The downward movement of the piston increases the volume of the cylinder above it, reducing the pressure in the cylinder. This reduced pressure, commonly referred to as engine vacuum, causes the atmospheric pressure to push air or an air/fuel mixture through the open intake valve. (Some engines are equipped with a super- or turbocharger that pushes more air past the valve.) In most engines, the intake valve closes after the piston has reached **bottom dead centre (BDC)**, the lowest point of piston travel in the cylinder. This delayed closing of the valve increases the volumetric efficiency of the cylinder by packing as much air or air/fuel mixture into it as possible.

COMPRESSION STROKE The compression stroke begins once the intake valve closes as the piston is moving upward from BDC. The closed intake valve traps the air/fuel mixture in the cylinder **(Figure 9–8B)**. The upward movement of the piston compresses the air or air/fuel mixture, thus heating it up. The amount of pressure and heat formed by the compression stroke depends on the amount of air in the cylinder and the compression ratio of the engine. The volume of the cylinder with the piston at BDC compared to the volume of the cylinder with the piston at TDC determines the compression ratio of the engine. In most modern engines, fuel is injected into the cylinder (direct injection) sometime during the compression stroke.

POWER STROKE The power stroke begins as the piston moves away from TDC with the compressed fuel mixture ignited **(Figure 9–8C)**. With the valves still closed, an electrical spark across the electrodes of a spark plug ignites the air/fuel mixture. The burning mixture rapidly expands, creating a very high pressure against the top of the piston. This drives the piston down toward BDC. The downward movement of the piston is transmitted through the connecting rod to the crankshaft.

EXHAUST STROKE The exhaust valve opens before the piston reaches BDC on the power stroke **(Figure 9–8D)**. Pressure within the cylinder causes the exhaust gas to rush past the open valve and into the system. Movement of the piston from BDC pushes most of the remaining exhaust gas from the cylinder. As the piston nears TDC, the exhaust valve begins to close as the intake valve starts to open. The exhaust stroke completes the four-stroke cycle. The opening of the intake valve begins the cycle again. This cycle occurs in each cylinder and is repeated over and over, as long as the engine is running.

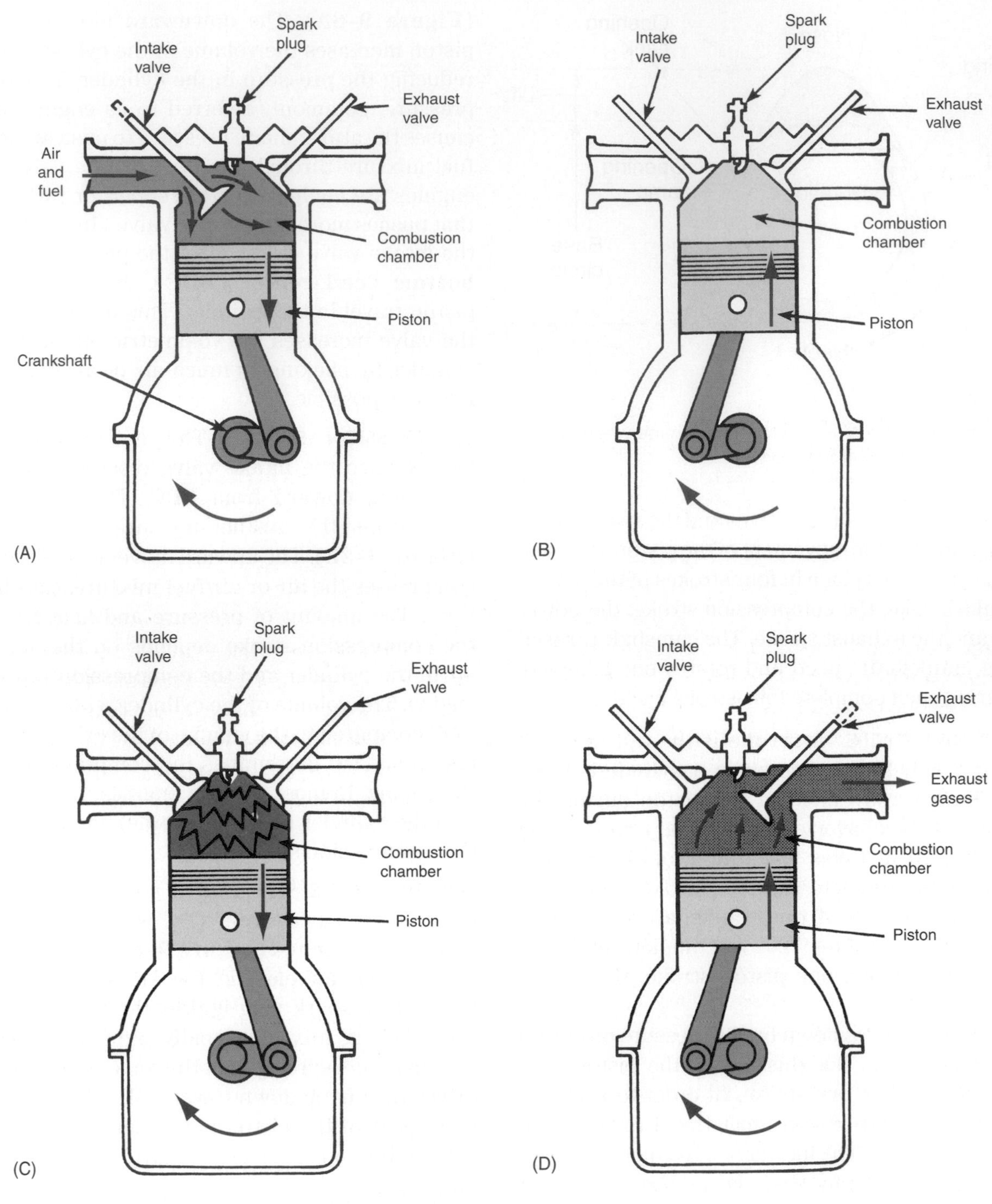

FIGURE 9–8

(A) Intake stroke, (B) compression stroke, (C) power stroke, and (D) exhaust stroke.

FOUR-STROKE OPERATING DYNAMICS When all of the above stroke descriptions are tied together, a better understanding of the four-stroke cycle operation will result.

To begin the cycle, the intake valve must begin to open before the piston reaches TDC (intake stroke) to allow the valve to be partially open when the piston begins to move downward from TDC. This is a very important point to understand, because as the pistons travel from TDC to approximately halfway down the cylinder, the pistons accelerate, and speed and travel is the greatest. The acceleration of the piston is what establishes the flow and inertia of the intake air/fuel charge into the cylinder. As the piston continues to travel down to BDC, it will progressively slow down until it stops at BDC. With

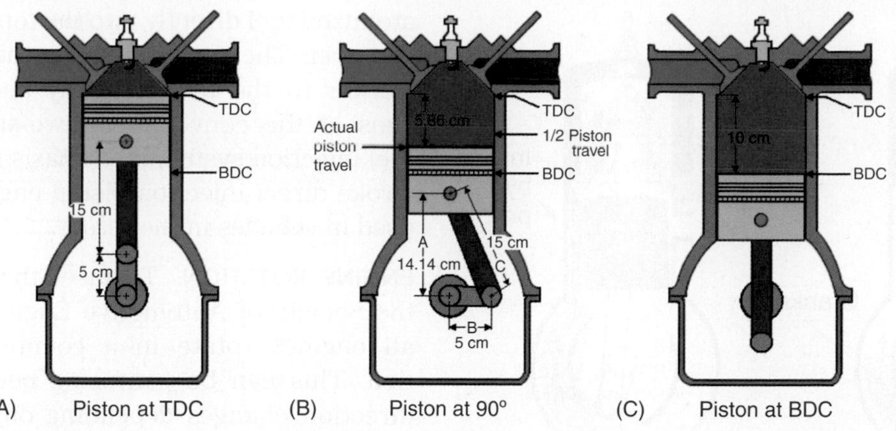

FIGURE 9–9

(A) This figure shows the piston at TDC. The rod length is 15 cm and the throw is 5 cm. This produces a stroke of 10 cm. (B) This figure shows the crankshaft at the 90° position from TDC. The three axes (wrist pin, connecting rod, and main bearing journals) form a triangle. The calculations were arrived at by using the Pythagorean theorem, which states: $A^2 + B^2 = C^2$ or $(A \times A) + (B \times B) = (C \times C)$ or $(14.14 \times 14.14) + (5 \times 5) = (15 \times 15)$. If the stroke is 10 cm and the piston travelled 5.68 cm in the first 90°, the travel in the next 90° would be only 4.14 cm. (C) This figure shows the piston at BDC, a full 10 cm from TDC.

the intake valve still open at BDC, the inertia of the intake charge will continue to fill the cylinder.

The crankshaft continues to rotate past BDC and positions the connecting rod for the compression upstroke. Depending on the engine rpm operating range, the intake valve will generally be closed midway between BDC and 90° before TDC. When the intake valve closes, compression of the intake charge can begin. Before the piston reaches TDC on the compression stroke, the spark will occur, igniting the air/fuel mixture. Depending on the engine rpm, the spark will be timed to allow the ignition process to exert pressure on the piston when it's at the correct position.

The power stroke begins as the piston travels downward from TDC. During the first 90° of crankshaft rotation from TDC, the piston will travel a greater distance down the cylinder than during the second 90° of rotation to BDC. This is due to the travel of the crankshaft away from the cylinder centreline plus its travel downward toward BDC. This is best shown by using the Pythagorean theorem **(Figure 9–9)**. During the power stroke, from the midway point down, the piston is actually slowing down until it comes to a stop at BDC.

Depending on the engines rpm operating range, the exhaust valve will generally be opened midway between 90° after TDC and BDC, signalling the beginning of the exhaust event. This establishes exhaust gas flow out past the opened valve while also relieving cylinder pressure and the pressure against the top of the piston. This pressure reduction places less force on the connecting rod and crankshaft when it's at or approaching an ineffective crank angle. This increases engine-bearing life.

The exhaust stroke continues as the piston moves up from BDC to slightly after TDC, which allows the valve to remain at least partially open to TDC.

At this point of crankshaft travel, both the intake and exhaust valves are open. This is known as valve overlap. The flow of outgoing exhaust gases past the closing exhaust valve will develop a low pressure in the cylinder, which allows atmospheric pressure to force the intake charge past the opening intake valve. A new cycle can now begin.

Valve opening times and distances are controlled by the camshaft lobes and are known as valve timing. This is revisited in the camshaft section of Chapter 11.

Two-Stroke Gasoline Engine

In the past, several imported vehicles have used two-stroke engines. As the name implies, this engine requires only two strokes of the piston to complete all four operations: intake, compression, power, and exhaust **(Figure 9–10)**. This is accomplished as follows:

1. Movement of the piston from BDC to TDC completes both intake and compression.
2. When the piston nears TDC, the compressed air/fuel mixture is ignited, causing an expansion of the gases. During this time, the intake and exhaust ports are closed.

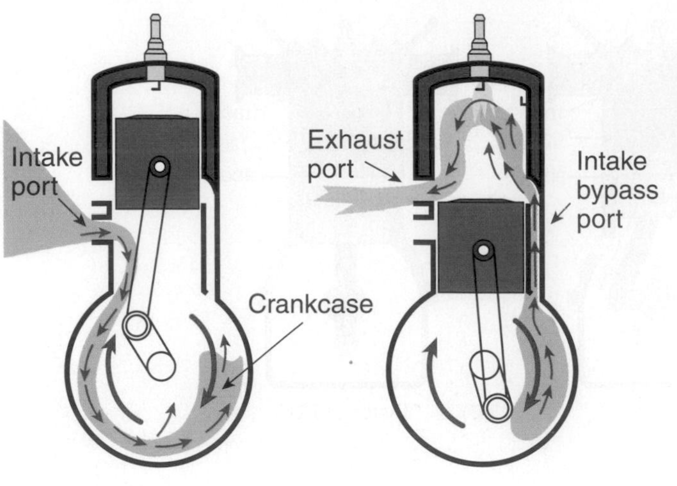

FIGURE 9–10

A two-stroke cycle.

3. Expanding gases in the cylinder force the piston down, rotating the crankshaft.
4. With the piston at BDC, the intake and exhaust ports are both open, allowing exhaust gases to leave the cylinder and the air/fuel mixture to enter.

Although the two-stroke-cycle engine is simple in design and lightweight because it lacks a valve train, it has not been widely used in automobiles. It tends to be less fuel efficient and releases more pollutants in the exhaust than four-stroke engines. Oil is often present in the exhaust because the engines require a constant oil delivery to the cylinders to keep the piston lubricated. Some of these engines require a certain amount of oil to be mixed with the fuel.

In recent years, however, thanks to a revolutionary pneumatic fuel injection system, there has been some interest in the two-stroke engine. The injection system uses compressed air to flow highly atomized fuel directly into the top of the combustion chamber. The system may be the long-sought-after answer to the fuel economy and emissions problems of the conventional two-stroke engine. This fuel injection system is the basis for the orbital two-stroke direct-injection piston engine, which may be used in vehicles in the future.

ENGINE ROTATION To meet the standards set by the Society of Automotive Engineers (SAE), nearly all engines rotate in a counter-clockwise direction. This can be confusing because its apparent direction changes depending on which end of the engine you look at. From the front of the engine, it rotates in a clockwise direction. The standards are based on the rotation of the flywheel, which is at the rear of the engine, and there the engine rotates counter-clockwise.

Combustion

Although many different things can affect combustion in the engine's cylinders, the ignition system has the responsibility for beginning and maintaining the combustion process. Obviously, when combustion does not occur in any of the cylinders, the engine will not run. If combustion occurs in all but one or two cylinders, the engine may start and run but will run poorly. Lack of or poor combustion is not always caused by the ignition system. It can also be caused by problems in the engine, the air/fuel system, or the exhaust system.

When normal combustion occurs, the burning process moves from the gap of the spark plug across the compressed air/fuel mixture. The movement of this flame front should be rapid and steady and should end when all of the air/fuel mixture has been burned **(Figure 9–11)**. During normal combustion,

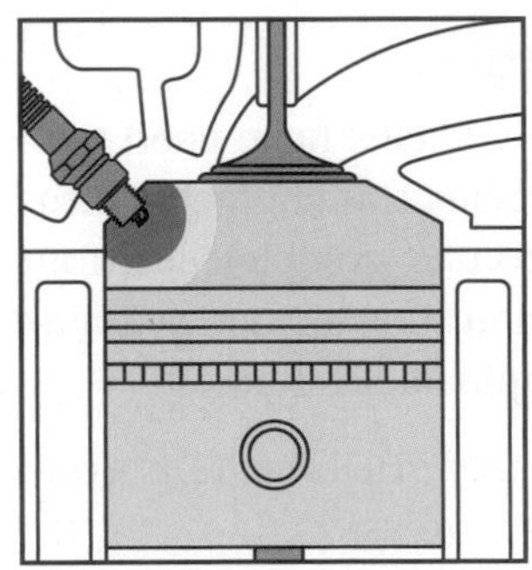

1. Spark occurs

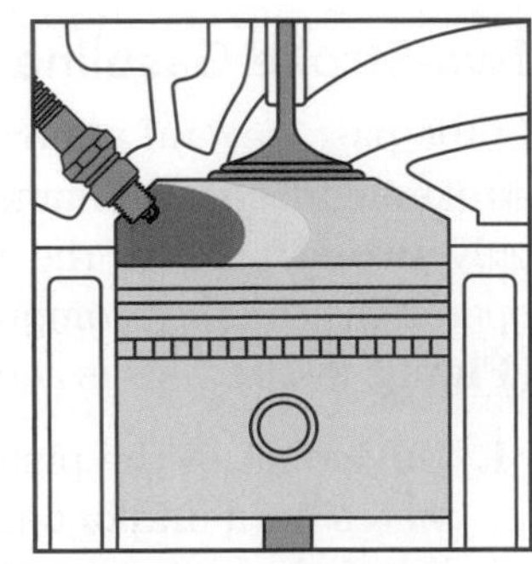

2. Combustion starts

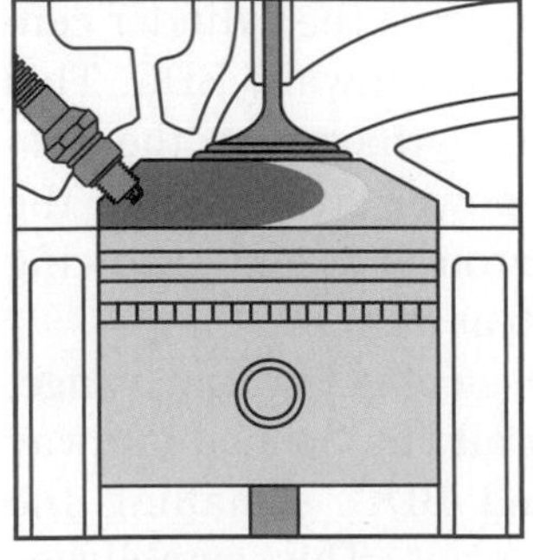

3. Continues rapidly

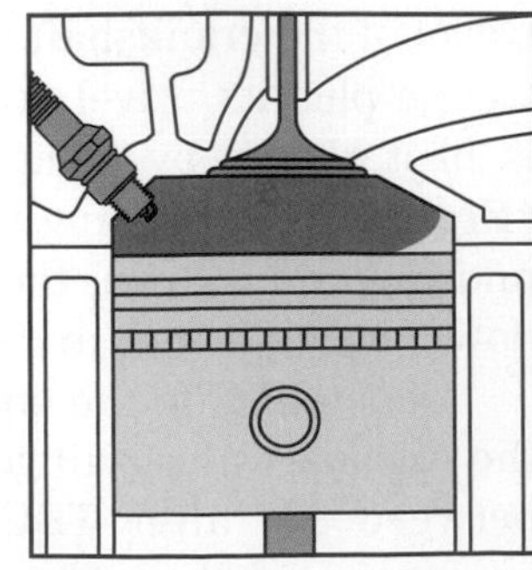

4. Complete

FIGURE 9–11

Normal combustion.

the rapidly expanding gases push down on the piston with a powerful but constant force.

When all of the air and fuel in the cylinder are involved in the combustion process, complete combustion has occurred. When something prevents this, the engine will misfire or experience incomplete combustion. Misfires cause a variety of driveability problems, such as a lack of power, poor gas mileage, excessive exhaust emissions, and a rough running engine.

Engine Configurations

Depending on the vehicle, either an in-line, V-type, slant, or opposed cylinder design can be used. The most popular designs are in-line and V-type engines.

IN-LINE ENGINE In the in-line engine design **(Figure 9–12)**, the cylinders are all placed in a single row. There is one crankshaft and one cylinder head for all of the cylinders. The block is cast so that all cylinders are located in an upright position.

In-line engine designs have certain advantages and disadvantages. They are easy to manufacture and service. However, because the cylinders are positioned vertically, the front of the vehicle must be higher. This affects the aerodynamic design of the car.

V-TYPE ENGINE A V-type engine design has two rows of cylinders **(Figure 9–13)** located 60 to 90° away from each other. A V-type engine uses one crankshaft, which is connected to the pistons on both sides of the V. This type of engine has two cylinder heads, one over each row of cylinders.

FIGURE 9–12

The cylinder block for an in-line engine.

Chrysler LLC

FIGURE 9–13

A V-type engine.

Chrysler LLC

One advantage of using a V-configuration is that the engine is not as high or as long as one with an in-line configuration. If eight cylinders are needed for power, a V-configuration makes the engine much shorter, lighter, and more compact. Many years ago, some vehicles had an in-line eight-cylinder engine. The engine was very long, and its long crankshaft also caused increased torsional vibrations in the engine.

A variation of the V-type engine is the W-type engine. These engines are basically two V-type engines joined together at the crankshaft. This design makes the engine more compact. They are commonly found in late-model Volkswagens, Bentleys, and the Bugatti Veyron.

SLANT CYLINDER ENGINE Another way of arranging the cylinders is in a slant configuration. This arrangement is much like an in-line engine, except the entire block has been placed at a slant. The slant engine was designed to reduce the distance from the top to the bottom of the engine. Vehicles using the slant engine can be designed more aerodynamically.

OPPOSED CYLINDER ENGINE In this design, two rows of cylinders are located opposite the crankshaft **(Figure 9–14)**. These engines have a common crankshaft and a cylinder head on each bank of cylinders. Porsches and Subarus use this style of engine, commonly called a boxer engine. Boxer engines have a low centre of gravity and tend to run smoothly during all operating conditions.

FIGURE 9–14

A horizontally opposed cylinder engine, commonly called a boxer engine.

Chrysler LLC

FIGURE 9–16

An engine block with two camshafts.

Portions of materials contained herein have been reprinted with permission of General Motors under License Agreement #1410912.

Camshaft and Valve Location

Two basic valve and camshaft placement configurations of the four-stroke gasoline engines are used in vehicles.

OVERHEAD VALVE (OHV) As the name implies, the intake and exhaust valves on an overhead valve engine are mounted in the cylinder head and are operated by a camshaft located in the cylinder block. This arrangement requires the use of valve lifters, pushrods, and rocker arms to transfer camshaft rotation to valve movement **(Figure 9–15)**. The intake and exhaust manifolds are attached to the cylinder head.

A recent development by GM incorporates two camshafts in the engine block **(Figure 9–16)**. There is a separate camshaft for the intake valves and another for the exhaust valves. This setup allows for variable intake valve operation while keeping the exhaust valves dictated by their camshaft. Having the camshafts placed in the engine block eliminates the need for long valve timing belts and/or chains.

OVERHEAD CAM (OHC) An overhead cam engine also has the intake and exhaust valves located in the cylinder head. But as the name implies, the cam is located in the cylinder head. In an overhead cam engine, the valves are operated directly by the camshaft or through cam followers or tappets **(Figure 9–17)**. Some engines have separate camshafts for the intake and the exhaust valves. These are called dual overhead camshaft (DOHC) engines.

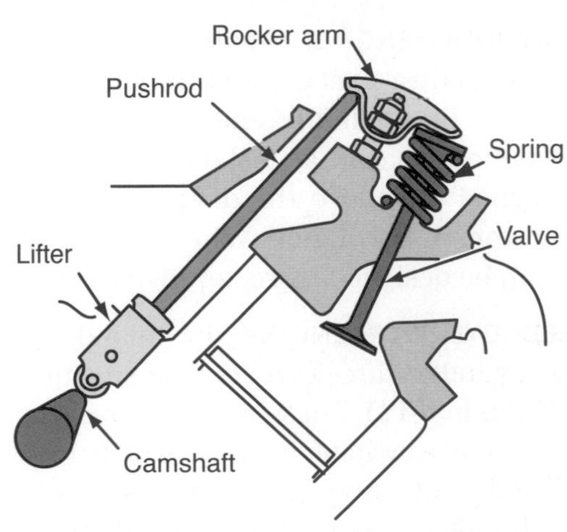

FIGURE 9–15

The basic valve train for an overhead valve engine.

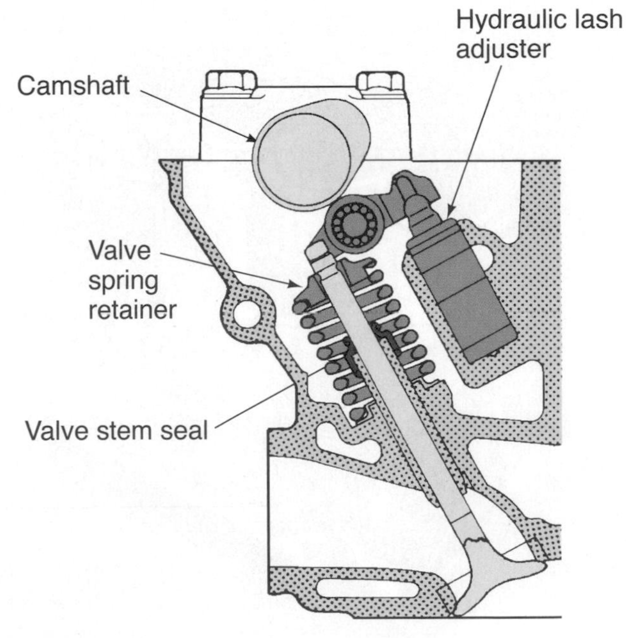

FIGURE 9–17

Basic valve and camshaft placement in an overhead camshaft engine.

Hyundai Motor America

Valve and Camshaft Operation

In OHV engines with the camshaft in the block **(Figure 9–18)**, the valves are operated by valve lifters and pushrods that are actuated by the camshaft. On overhead cam engines, the cam lobes operate the valves directly, and there is no need for pushrods. (Lifters may be used between the camshaft and valves.)

The camshaft is driven by the crankshaft through gears, or sprockets, and a cogged belt, or timing chain. The camshaft turns at half the crankshaft speed and rotates one complete turn during each complete four-stroke cycle.

Engine Location

The engine is placed in one of three locations. In most vehicles, it is in front of the passenger compartment. Front-mounted engines can be positioned either longitudinally or transversely with respect to the vehicle.

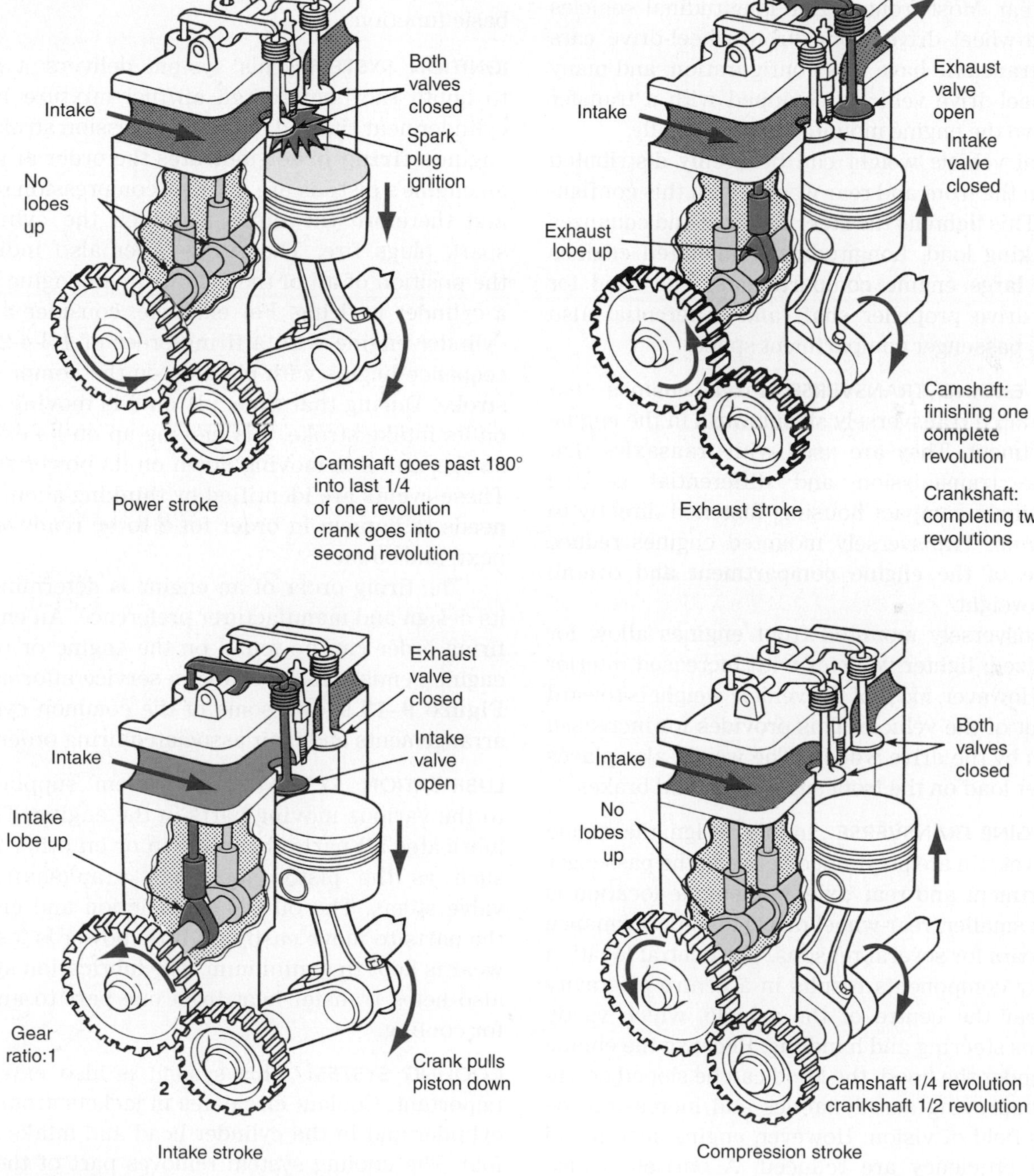

FIGURE 9–18

Valve operation in an overhead valve engine.

The second engine location is a mid-mount position between the passenger compartment and rear suspension. The third, and least common, engine location is the rear of the vehicle. The engines are typically opposed-type engines.

Each of these engine locations offers advantages and disadvantages.

FRONT ENGINE LONGITUDINAL With this arrangement, the engine, transmission, front suspension, and steering system are located in the front of the vehicle, and the differential and rear suspension are in the rear. Most front engine longitudinal vehicles are rear-wheel drive. Some front-wheel-drive cars with a transaxle have this configuration, and many four-wheel-drive vehicles equipped with a transfer case have the engine mounted longitudinally.

Total vehicle weight can be evenly distributed between the front and rear wheels with this configuration. This lightens the steering force and equalizes the braking load. Longitudinally mounted engines require large engine compartments. The need for a rear drive propeller shaft and differential also reduces passenger compartment space.

FRONT ENGINE TRANSVERSE Front engines that are mounted transversely sit sideways in the engine compartment. They are used with transaxles that combine transmission and differential gearing into a single compact housing, fastened directly to the engine. Transversely mounted engines reduce the size of the engine compartment and overall vehicle weight.

Transversely mounted front engines allow for down-sized, lighter vehicles with increased interior space. However, most of the vehicle weight is toward the front of the vehicle. This provides for increased traction by the drive wheels. The weight also places a greater load on the front suspension and brakes.

MID-ENGINE TRANSVERSE In this design, the engine and drivetrain are positioned between the passenger compartment and rear axle. Mid-engine location is used in smaller, rear-wheel-drive, high-performance sports cars for several reasons. The central location of heavy components results in a centre of gravity very near the centre of the vehicle, which vastly improves steering and handling. Because the engine is not under the hood, the hood can be sloped downward, improving aerodynamics and increasing the driver's field of vision. However, engine access and cooling efficiency are reduced. A barrier is also needed to reduce the transfer of noise, heat, and vibration to the passenger compartment.

Gasoline Engine Systems

The operation of an engine relies on several other systems. The efficiency of these systems affects the overall operation of the engine.

AIR/FUEL SYSTEM This system makes sure the engine gets the right amount of both air and fuel needed for efficient operation. For many years, air and fuel were mixed in a carburetor, which supplied the resulting mixture to the cylinder. Today, most late-model automobiles have a fuel injection system, which replaces the carburetor but performs the same basic function.

IGNITION SYSTEM This system delivers a spark to ignite the compressed air/fuel mixture in the cylinder near the end of the compression stroke. An engine's **firing order** indicates the order at which an engine's pistons are on their compression stroke and therefore the order in which the cylinders' spark plugs fire. The firing order also indicates the position of all of the pistons in an engine when a cylinder is firing. For example, consider a four-cylinder engine with a firing order of 1-3-4-2. The sequence begins with piston 1 on the compression stroke. During that time, piston 3 is moving down on its intake stroke, 4 is moving up on its exhaust stroke, and 2 is moving down on its power stroke. These events are identified by thinking about what needs to happen in order for 3 to be ready to fire next, and so on.

The firing order of an engine is determined by its design and manufacturer preference. An engine's firing order can be found on the engine or on the engine's emissions label and in service information. **Figure 9–19** shows some of the common cylinder arrangements and their associated firing orders.

LUBRICATION SYSTEM This system supplies oil to the various moving parts in the engine. The oil lubricates all parts that slide in or on other parts, such as the piston, bearings, crankshaft, and valve stems. The oil reduces friction and enables the parts to move easily, so little power is lost and wear is kept to a minimum. The lubrication system also helps transfer heat from one part to another for cooling.

COOLING SYSTEM This system is also extremely important. Coolant circulates in jackets around the cylinder and in the cylinder head and intake manifold. The cooling system removes part of the heat produced by combustion and prevents the engine from being damaged by overheating.

①②③④⑤⑥
FIRING ORDER 1-5-3-6-2-4
6 CYLINDER

②④⑥⑧ **RIGHT BANK**
①③⑤⑦ **LEFT BANK**
FIRING ORDER 1-8-4-3-6-5-7-2
V-8

⑤③① **RIGHT BANK**
⑥④② **LEFT BANK**
FIRING ORDER 1-4-5-2-3-6
V-6

②④⑥⑧ **RIGHT BANK**
①③⑤⑦ **LEFT BANK**
FIRING ORDER 1-8-7-2-6-5-4-3
V-8

①②③④ **RIGHT BANK**
⑤⑥⑦⑧ **LEFT BANK**
FIRING ORDER 1-5-4-8-6-3-7-2
V-8

①②③④
FIRING ORDER 1-3-4-2
1-2-4-3
4 CYLINDER

②④⑥ **RIGHT BANK**
①③⑤ **LEFT BANK**
FIRING ORDER 1-6-5-4-3-2
V-6

①②③④ **RIGHT BANK**
⑤⑥⑦⑧ **LEFT BANK**
FIRING ORDER 1-5-4-2-6-3-7-8
V-8

FIGURE 9–19

Common cylinder firing orders.

EXHAUST SYSTEM This system removes the burned gases from the combustion chamber and limits the noise produced by the engine. It also carries deadly carbon monoxide (CO) away from the passenger compartment to the rear of the vehicle. Both gasoline and diesel engines utilize catalytic converters in their exhaust systems to reduce harmful emissions produced during combustion. Catalytic converters rely on catalysts, usually precious metals to chemically react with harmful gases like NO_x to reduce it to nitrogen and oxygen. Many newer "clean" diesels also include an exhaust fluid injection system to reduce the NO_x produced.

EMISSION CONTROL SYSTEM Several control devices designed to reduce the amount of pollutants released by the engine have been added to the engine. Engine design changes, such as reshaped combustion chambers and altered valve timing, have also been part of the manufacturers' attempt to reduce emission levels.

ENGINE MEASUREMENT AND PERFORMANCE

Many of the engine measurements and performance characteristics a technician should be familiar with were discussed in Chapter 7. What follows are some of the important facts of each.

Bore and Stroke

The **bore** of a cylinder is simply its diameter measured in millimetres (mm) or inches (in.). The stroke is the distance the piston travels between TDC and BDC. The bore and stroke determine the displacement of the cylinders. When the bore and stroke are of equal size, the engine is called a **square** engine. Engines that have a larger bore than stroke are called oversquare, and engines with a larger stroke than bore are referred to as being undersquare. **Oversquare** engines can be fit with larger valves in the combustion chamber and use shorter connecting rods, which means these engines are capable of running at higher engine speeds. But because of the size of the bore, the engines tend to be physically larger than undersquare engines. **Undersquare** engines have long connecting rods that aid in the production of more power at lower engine speeds. A square engine is a compromise between the two designs.

The **crank throw** is the distance from the crankshaft's main bearing centreline to the crankshaft connecting rod journal or throw centreline. An engine's stroke is twice the crank throw **(Figure 9–20)**.

Displacement

A cylinder's displacement is the swept volume of the cylinder. This is the volume of the cylinder from BDC to TDC. The trend in recent years has been towards a smaller displacement engine fitted with turbo- or superchargers. Many manufacturers have moved from eight-cylinder to six-cylinder or from six-cylinder to four-cylinder engines to improve fuel economy. Using a turbo- or supercharger maintains high levels of performance, and the smaller engine improves economy. As an example, Ford recently announced it will be offering its 1.0 litre turbocharged three-cylinder EcoBoost engine in North America.

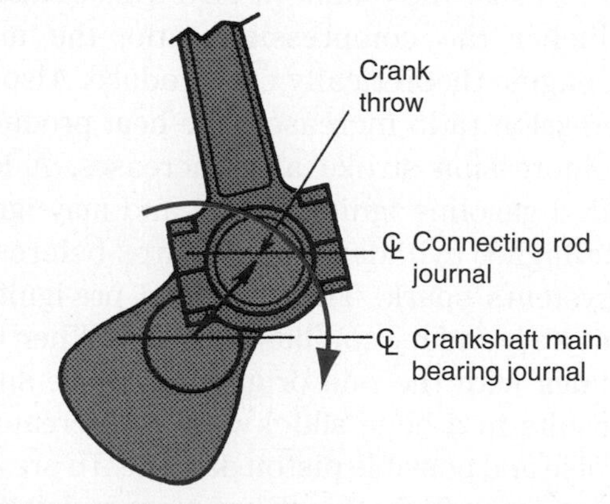

FIGURE 9–20

The stroke of an engine is equal to twice the crank throw.

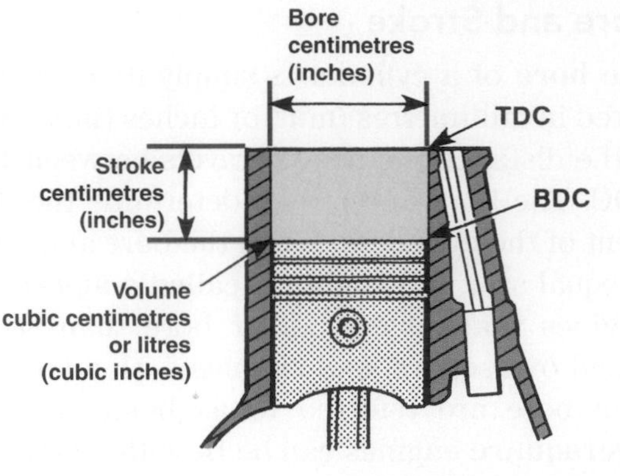

FIGURE 9–21

Displacement is the volume the cylinder holds between TDC and BDC.

An engine's displacement is the sum of the displacements of each of the engine's cylinders **(Figure 9–21)**. Typically, an engine with a larger displacement produces more torque than a smaller displacement engine; however, many other factors influence an engine's power output. Engine displacement can be changed by changing the size of the bore and stroke of an engine.

Compression Ratio

An engine's stated compression ratio is a comparison of a cylinder's volume when the piston is at BDC to the cylinder's volume when the piston is at TDC. The compression ratio is a statement of how much the air/fuel mixture is compressed during the compression stroke. It is important to keep in mind that this ratio can change through wear and carbon and dirt buildup in the cylinders. For example, if a great amount of carbon collects on the top of the piston and around the combustion chamber, the volume of the cylinder changes. This buildup of carbon will cause the compression ratio to increase because the volume at TDC will be smaller.

The higher the compression ratio, the more power an engine theoretically can produce. Also, as the compression ratio increases, the heat produced by the compression stroke also increases. A low-octane-rated gasoline ignites easily and may ignite due to the higher cylinder temperatures before the ignition system's spark. This is called pre-ignition and causes a secondary combustion front. When this front collides with the one originated at the spark plug, it results in a huge shockwave that creates a pinging noise and possible piston damage. To prevent this, higher octane fuels, which are more resistant to self-ignition under high pressures and temperature, should be used in higher-compression engines.

Engine Efficiency

One of the dominating trends in automotive design is increasing an engine's efficiency. **Efficiency** is simply a measure of the relationship between the amount of energy put into an engine and the amount of energy available from the engine. Other factors, or efficiencies, affect the overall efficiency of an engine.

VOLUMETRIC EFFICIENCY Volumetric efficiency describes the engine's ability to fill its cylinders with the air/fuel mixture. If the engine's cylinders are able to be completely filled with the air/fuel mixture during its intake stroke, the engine has a volumetric efficiency of 100 percent. Typically, engines running at low speed and a wide-open throttle (under heavy loads) have a volumetric efficiency of 80 to 100 percent if they are not equipped with a turbo- or supercharger. Basically, an engine becomes more efficient as its volumetric efficiency is increased.

Turbochargers and superchargers force more air into the cylinders and therefore increase the volumetric efficiency of the engine. In fact, anything that is done to an engine to increase the intake air volume will increase its volumetric efficiency.

THERMAL EFFICIENCY Thermal efficiency is a measure of how much of the heat formed during combustion is available as power from the engine. Typically, only one-third of the heat is used to power the vehicle. The rest is lost to the exhaust system, surrounding air, engine parts, and engine's coolant. Obviously, when less heat is lost, the engine is more efficient.

MECHANICAL EFFICIENCY Mechanical efficiency is a measure of how much power is available once it leaves the engine compared to the amount of power that was exerted on the pistons during the power stroke. Power losses occur because of the friction generated by the moving parts. Minimizing friction increases mechanical efficiency.

Torque versus Horsepower

Torque is a twisting or turning force. Horsepower is the rate at which torque is produced. When using pound-feet (ft·lb) as units of torque, revolutions per minute (rpm) for engine speed, and horsepower (hp) for power, power can be calculated with the following formula:

$$\text{hp} = (\text{ft·lb} \times \text{rpm}) \div 5252$$

A graph of the relationship between the horsepower and torque of an engine is shown in **Figure 9–22**. The constant 5252 represents the fact that torque and horsepower will be equal when the engine is running

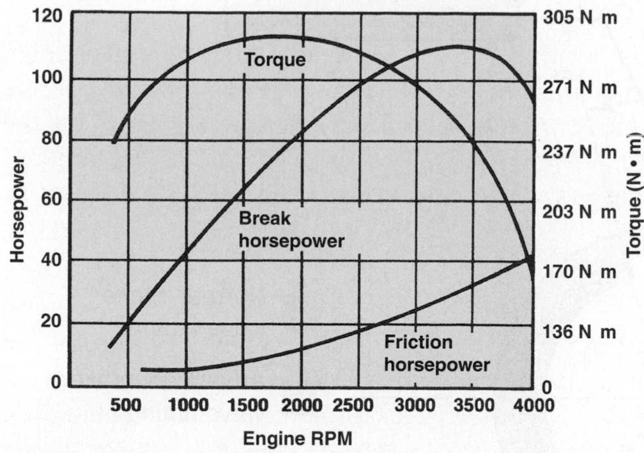

FIGURE 9–22

The relationship between horsepower and torque.

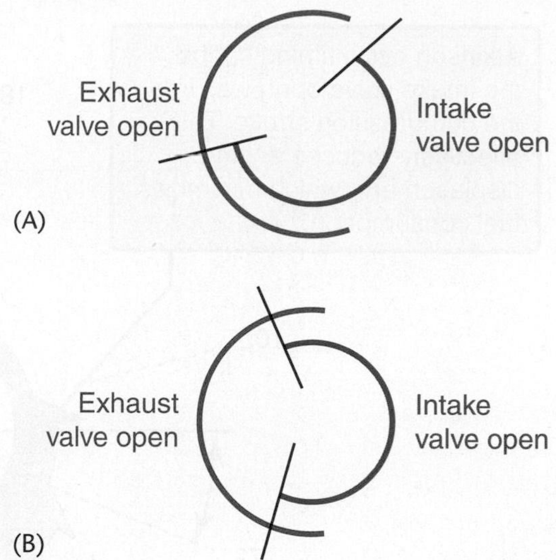

FIGURE 9–23

(A) Typical valve timing for an Atkinson cycle engine.
(B) Typical valve timing for a conventional four-stroke cycle engine. Notice that the intake valve in the Atkinson cycle engine opens and closes later.

at 5252 rpm. This constant applies only to calculations using the English system of measurement. If the calculated power is expressed in kilowatts (kW) and torque is expressed in Newton-meters (N·m), the constant should be 9549. Therefore, the formula for calculating power is as follows:

$$\text{kW} = (\text{N·m} \times \text{rpm}) \div 9549$$

This means that a graph of power and torque based on rpm using metric units would show that torque and power would be equal at 9549 rpm.

Atkinson Cycle Engines

An **Atkinson cycle** engine is a four-stroke cycle engine in which the intake valve is held open longer than normal during the compression stroke **(Figure 9–23)**. As the piston is moving up, the mixture is being compressed and some of it pushed back into the intake manifold. As a result, the amount of mixture in the cylinder and the engine's effective displacement and compression ratio are reduced. Often, the Atkinson cycle is referred to as a five-stroke cycle because there are two distinct cycles during the compression stroke. The first is while the intake valve is open and the second is when the intake valve is closed.

In a conventional engine, much engine power is lost due to the energy required to compress the mixture during the compression stroke. The Atkinson cycle reduces this power loss, and this leads to greater engine efficiency. The Atkinson cycle also effectively changes the length of time the mixture is being compressed. Most Atkinson cycle engines have a long piston stroke. Keeping the intake valve open during compression effectively shortens the stroke. However, because the valves are closed during the power stroke, that stroke is long. The longer power stroke allows the combustion gases to expand more and reduces the amount of heat that is lost during the exhaust stroke. As a result, the engine runs more efficiently than a conventional engine.

Although these engines provide improved fuel economy and lower emissions, they also produce less power. The lower power results from the lower operating displacement and compression ratio. Power also is lower because these engines take in less air than a conventional engine.

HYBRID ENGINES Many hybrid vehicles have Atkinson cycle engines. The low-power output from the engine is supplemented with the power from the electric motors. This combination offers good fuel economy, low emissions, and normal acceleration.

Most Atkinson cycle engines use variable valve timing to allow the engine to run with low displacement (Atkinson cycle) or normal displacement. The opening and closing of the intake valves is controlled by the engine control system **(Figure 9–24)**. While the valve is open during the compression stroke, the effective displacement of the engine is reduced. When the displacement is low, fuel consumption is minimized, as are exhaust emissions. The engine runs with normal displacement when the intake valves close earlier. This action provides for more power output. The control unit adjusts valve timing according to engine speed, intake air volume,

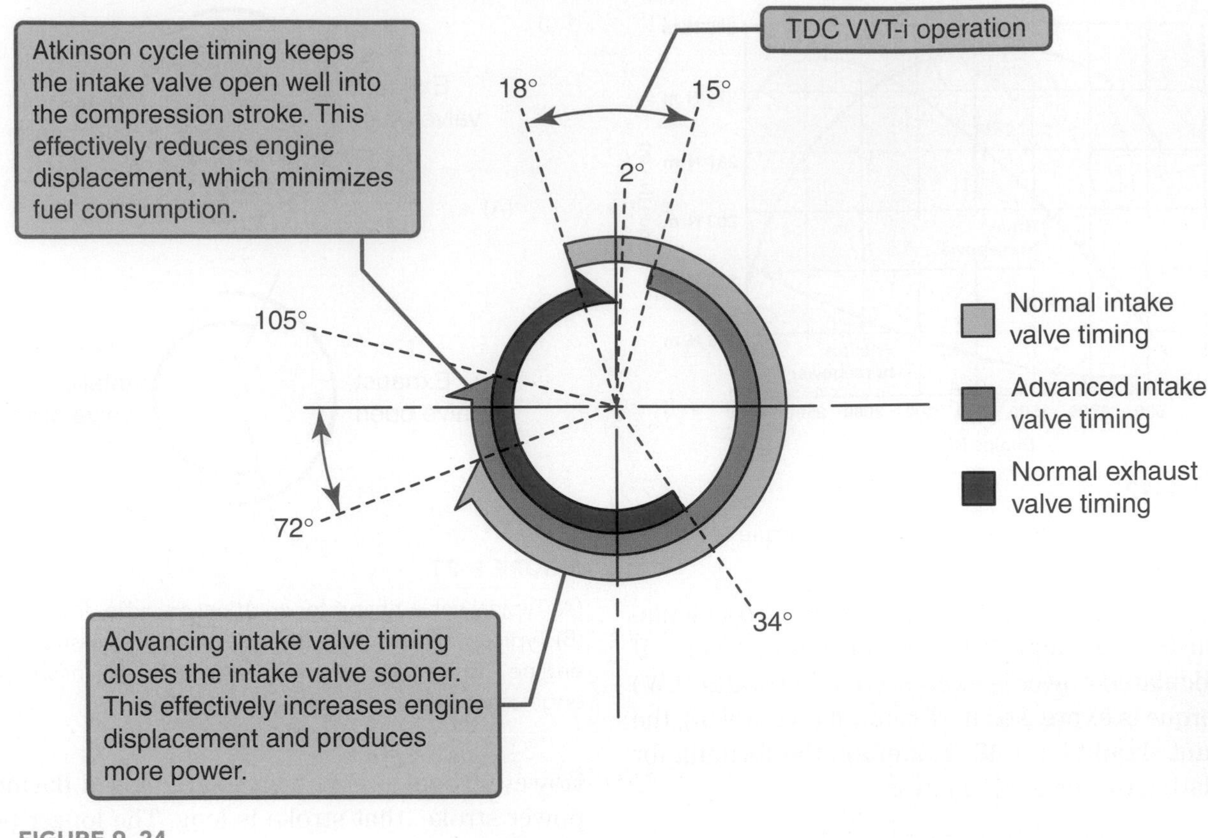
Atkinson cycle timing keeps the intake valve open well into the compression stroke. This effectively reduces engine displacement, which minimizes fuel consumption.
TDC VVT-i operation
18°
15°
2°
105°
72°
34°
Normal intake valve timing
Advanced intake valve timing
Normal exhaust valve timing
Advancing intake valve timing closes the intake valve sooner. This effectively increases engine displacement and produces more power.

FIGURE 9–24

Toyota's VVT-i (variable valve timing with intelligence) changes the engine from a conventional four-stroke cycle to an Atkinson cycle according to the vehicle's operating conditions.

throttle position, and coolant temperature. Because this system responds to operating conditions, the displacement of the engine changes accordingly.

In typical systems, the control unit sends commands to the camshaft timing oil control valve. A phaser at the end of the camshaft is driven by the crankshaft. The control unit regulates the oil pressure sent to the phaser. A change in oil pressure changes the position of the camshaft and the timing of the valves. An advance in timing results when oil pressure is applied to the timing advance chamber. When the oil control valve is moved and the oil pressure is applied to the timing retard side vane chamber **(Figure 9–25)**, the timing is retarded.

MILLER CYCLE ENGINES An Atkinson cycle engine with forced induction (supercharging) is called a Miller cycle engine. The decrease of intake air and resulting low power is compensated for by the supercharger. The supercharger forces air into the cylinder during the compression stroke. Keep in mind that the actual compression stroke in an Atkinson cycle engine does not begin until the intake valve closes. The supercharger in a Miller cycle engine forces more air past the valve, and therefore there is more air in the cylinder when the intake closes.

OTHER AUTOMOTIVE POWER PLANTS

The gasoline-powered, internal-combustion piston engine has been the primary automotive power plant for many years and probably will remain so for years to come. Present-day social requirements and new technological developments, however, have necessitated searches for ways to modify or replace this time-proven workhorse. In an attempt to reduce fuel consumption and harmful exhaust emissions, many manufacturers are supplementing or modifying the basic internal combustion engine. Due to the advancements made in electronic controls, they are becoming a viable alternative to the conventional gasoline engine.

Hybrids

A hybrid vehicle has at least two different types of power or propulsion systems. Today's hybrid vehicles have an internal combustion engine and an

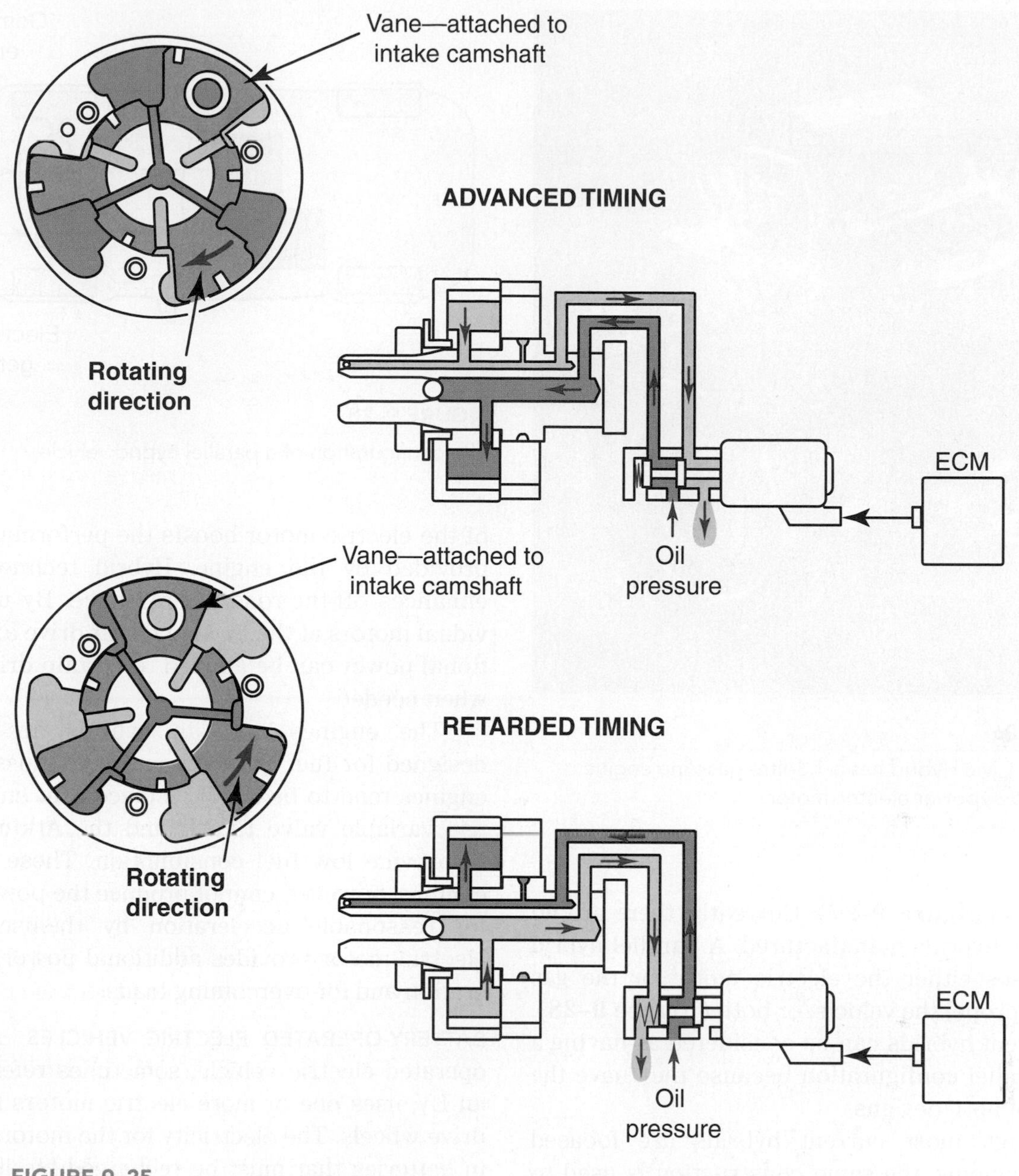

FIGURE 9–25
Oil flow for the VVT-i as it advances and retards the valve timing.

electric motor (some vehicles have more than one electric motor). A hybrid's electric motor is powered by batteries and/or ultracapacitors, which are recharged by a generator that is driven by the engine **(Figure 9–26)**. They are also recharged through regenerative braking. The engine may use gasoline, diesel, or an alternative fuel. Complex electronic controls monitor the operation of the vehicle. Based on the current operating conditions, electronics control the engine, electric motor, and generator.

Depending on the design of the hybrid vehicle, the engine may power the vehicle, assist the electric motor while it is propelling the vehicle, or drive a generator to charge the vehicle's batteries. The electric motor may propel the vehicle by itself, assist the engine while it is propelling the vehicle, or act as a generator to charge the batteries. Many hybrids rely exclusively on the electric motor(s) during slow-speed operation, on the engine at higher speeds, and on both during some certain driving conditions.

Often, hybrids are categorized as series or parallel designs. In a series hybrid, the engine never directly powers the vehicle. Instead, it drives a generator, and the generator either charges the batteries or directly powers the electric motor that drives

FIGURE 9–26

The Honda Civic Hybrid has a 1.3-litre gasoline engine and a 20-horsepower electric motor.

American Honda Motor Co., Inc.

the wheels **(Figure 9–27)**. Currently, there are no true series hybrids manufactured. A parallel hybrid vehicle uses either the electric motor or the gas engine to propel the vehicle, or both **(Figure 9–28)**. Most current hybrids can be considered as having a series/parallel configuration because they have the features of both designs.

Although most current hybrids are focused on fuel economy, the same construction is used to create high-performance vehicles. The added power

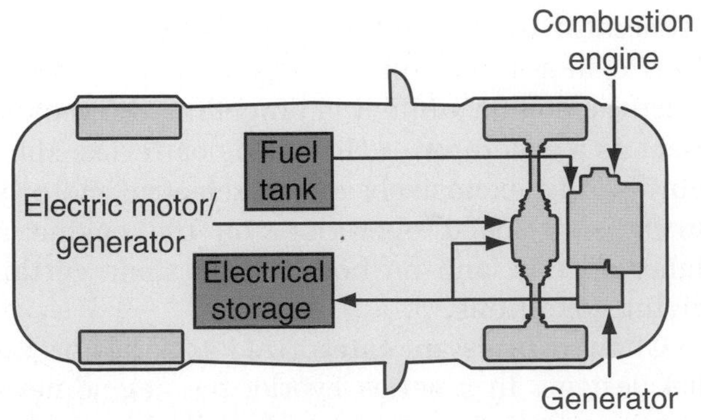

FIGURE 9–27

The configuration of a series hybrid vehicle.

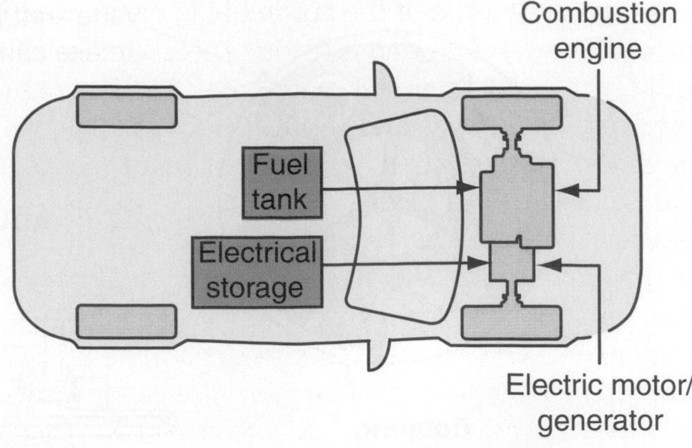

FIGURE 9–28

The configuration of a parallel hybrid vehicle.

of the electric motor boosts the performance levels provided by the engine. Hybrid technology also enhances off-the-road performance. By using individual motors at the front and rear drive axles, additional power can be applied to certain drive wheels when needed.

The engines used in hybrids are specially designed for fuel economy and low emissions. The engines tend to be small displacement engines that use variable valve timing and the Atkinson cycle to provide low fuel consumption. These advanced engines, however, cannot produce the power needed for reasonable acceleration by themselves. The electric motor provides additional power for acceleration and for overcoming loads.

BATTERY-OPERATED ELECTRIC VEHICLES A battery-operated electric vehicle, sometimes referred to as an EV, uses one or more electric motors to turn its drive wheels. The electricity for the motors is stored in batteries that must be recharged by an external electrical power source. Normally, they are recharged by plugging them into an outlet at home or other locations. The recharging time varies with the type of charger, the size and type of battery, and other factors. Normal recharge time is four to eight hours.

An electric motor is quiet and has few moving parts. It starts well in the cold, is simple to maintain, and does not burn petroleum products to run. The disadvantages of an EV are limited speed, power, and range as well as the need for heavy, costly batteries. However, an EV is much more efficient than a conventional gasoline-fuelled vehicle. EVs are considered zero emissions vehicles because they do not directly pollute the air. The only pollution associated with them is the result of creating the electricity to charge their batteries.

In the early days of the automobile, electric cars outnumbered gasoline cars. Today, there are few EVs on the road, but they are commonly used in manufacturing, shipping, and other industrial plants, where the exhaust of an internal combustion engine could cause illness or discomfort to the workers in the area. They are also used on golf courses, where the quiet operation adds to the relaxing atmosphere. Some auto manufacturers are still studying their use. Whether battery-operated EVs return to the market really depends on the development of new batteries and motors. To be practical, EVs need to have much longer driving ranges between recharges and must be able to sustain highway speeds for great distances.

FUEL CELL ELECTRIC VEHICLES Although just experimental at this time, there is much promise for fuel cell EVs. These vehicles are powered solely by electric motors, but the energy for the motors is produced by fuel cells. Fuel cells rely on hydrogen to produce the electricity. A fuel cell generates electrical power through a chemical reaction. A fuel cell EV uses the electricity produced by the fuel cell to power motors that drive the vehicle's wheels **(Figure 9–29)**. The batteries in these vehicles do not need to be charged by an external source.

Fuel cells convert chemical energy to electrical energy by combining hydrogen with oxygen. The hydrogen can be supplied directly as pure hydrogen gas or through a "fuel reformer" that pulls hydrogen from hydrocarbon fuels such as methanol, natural gas, or gasoline. Simply put, a fuel cell is comprised of two electrodes (the anode and the cathode) located on either side of an electrolyte. As the hydrogen enters the fuel cell, the hydrogen atoms give up electrons at the anode and become hydrogen ions in the electrolyte. The electrons that were released at the anode move through an external circuit to the cathode. As the electrons move toward the cathode, they can be diverted and used to power the vehicle. When the electrons and hydrogen ions combine with oxygen molecules at the cathode, water and heat are formed. There are no smog-producing or greenhouse gases produced. Although vehicles equipped with reformers emit some pollutants, those that run on pure hydrogen are true zero-emission vehicles.

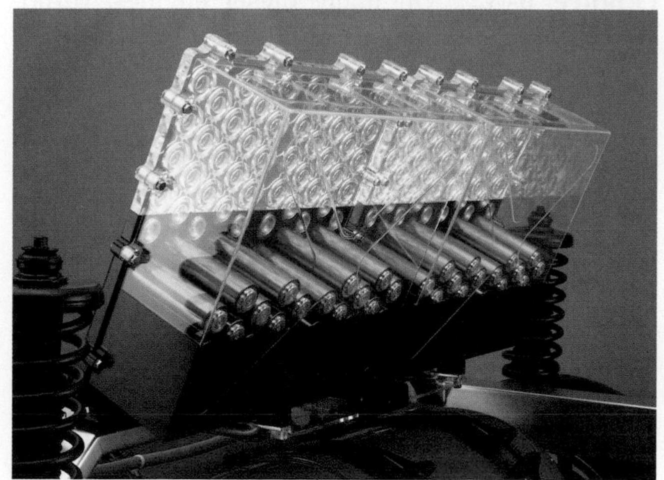

FIGURE 9–29

The sources of power for a fuel cell electric vehicle: fuel cell stack (left), power control unit (centre), and lithium ion battery pack (right).

American Honda Motor Co., Inc.

Rotary Engines

The **rotary engine**, or **Wankel engine**, is somewhat similar to the standard piston engine in that it is a spark-ignition, internal-combustion engine. Its design, however, is quite different. For one thing, the rotary engine uses a rotating motion rather than a reciprocating motion. In addition, it uses ports rather than valves for controlling the intake of the air/fuel mixture and the exhaust of the combusted charge. This design has been used in a few cars through the years, but the public has not accepted them because they are not very fuel efficient and produce high power only at high engine speeds. The main part of a rotary engine is a roughly triangular rotor that rotates within an oval-shaped housing. The rotor has three convex faces, and each face has a recess in it. These recesses increase the overall displacement of the engine. The tips of the rotor are always in contact with the walls of the housing as the rotor moves to seal the sides (chambers) to the walls. As the rotor rotates, it creates three separate chambers of gas. Also as it rotates, the volume between the sides of the rotor and the housing continuously changes. During rotor rotation, the volume of the gas in each chamber alternately expands and contracts. It is how a rotary engine rotates through the basic four-stroke cycle.

DIESEL ENGINES

Diesel engines were invented by Dr. Rudolph Diesel, a German engineer, and first marketed in 1897. The diesel engine is now the dominant power plant in heavy-duty trucks, construction equipment, farm equipment, buses, marine, and some automotive applications. Diesel vehicles are very common in

FIGURE 9–30

A European four-cylinder passenger car diesel engine.

Europe and other places where cleaner fuels are available **(Figure 9–30)**.

The operation of a **diesel engine** is comparable to a gasoline engine. They also have a number of components in common, such as the crankshaft, pistons, valves, camshaft, and water and oil pumps. They both are available as two- or four-stroke combustion cycle engines. However, diesel engines rely on compression ignition **(Figure 9–31)**. A diesel engine uses the heat produced by compressing air in the combustion chamber to ignite the fuel. The compression ratio of diesel engines can be three times (as high as 25:1) that of a gasoline engine, though newer engine technologies allow diesel engines to use lower compression ratios. As intake air is compressed, its temperature rises to 700°C to 900°C (1300°F to 1650°F). Just before the air is fully compressed, a fuel injector sprays a small amount of diesel fuel into the cylinder. The high temperature of the compressed air instantly ignites the fuel. The combustion causes increased heat, and the resulting high pressure moves the piston down on its power stroke.

FIGURE 9–32

A high-output turbo diesel engine.

Construction

A diesel engine must be made stronger to contain the extremely high compression and combustion pressures. A diesel engine produces less horsepower than a same-sized gasoline engine. Therefore, to provide the required power, the displacement of the engine is increased. This results in a physically larger engine. Diesels have high torque outputs at very low engine speeds but do not run well at high engine speeds **(Figure 9–32)**. On many diesel engines, turbochargers and intercoolers are used to increase their power output.

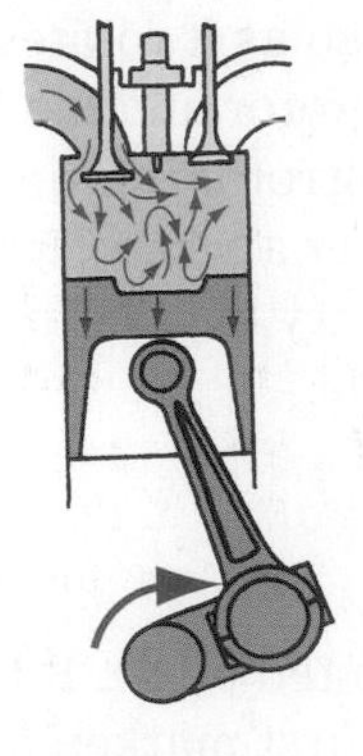

Compression

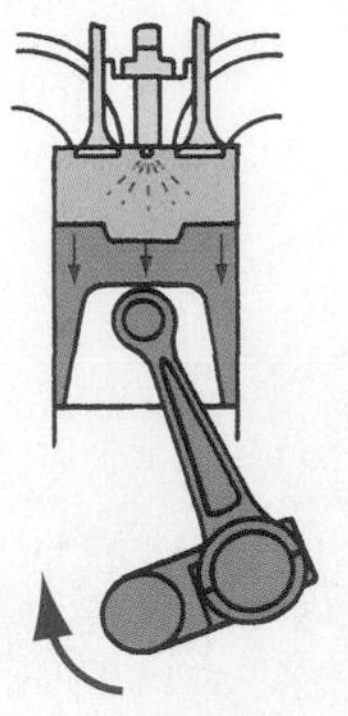

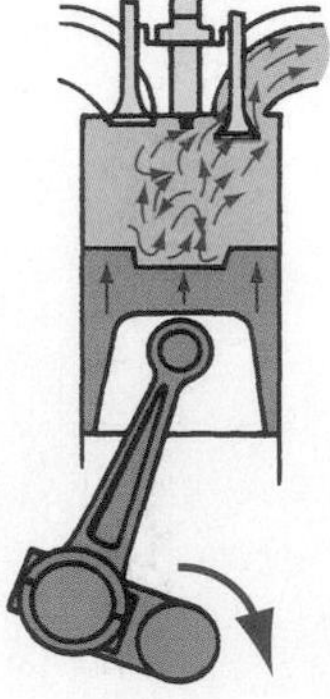

FIGURE 9–31

A four-stroke diesel engine cycle.

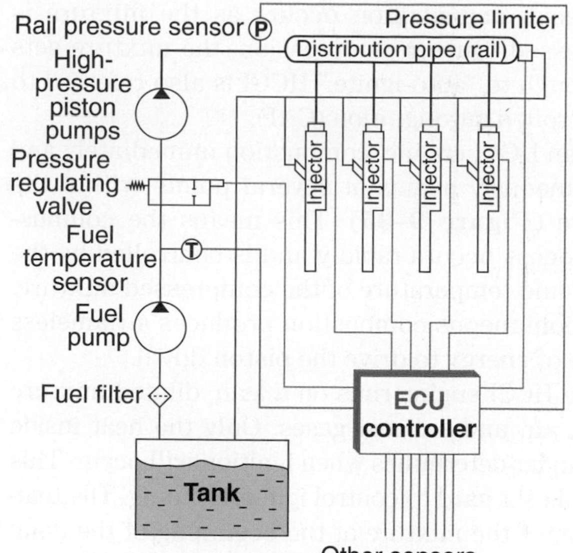

FIGURE 9–33

A common rail fuel injection system.

Fuel injection is used on all diesel engines. Older diesel engines had a distributor-type injection pump driven and regulated by the engine. The pump supplied fuel to injectors that sprayed fuel into the engine's combustion chamber. Newer diesel engines are equipped with common rail systems **(Figure 9–33)**. Common rail systems use **direct injection (DI)**. The injectors' nozzles are placed inside the combustion chamber. The top of the pistons has a depression where the initial combustion takes place. The injector must be able to withstand the temperature and pressure inside the cylinder and be able to deliver a fine spray of fuel into those conditions. These systems have a high-pressure fuel rail connected to individual solenoid-type injectors. Depending on the system, the injection pressure can range from 150 to 250 mPa (megapascal; 21 756 to 36 260 psi).

These injectors are computer controlled and attempt to match injector operation to the operating conditions of the engine. Newer diesel fuel injectors rely on stacked piezoelectric crystals rather than solenoids. Piezo crystals expand quickly when electrical current is applied to them. The crystals allow the injectors to respond quickly to the needs of the engine. With this style of injector, diesel engines are quieter, more fuel efficient, and cleaner, and they have more power.

When compared to gasoline engines, diesel engines offer many advantages. They are more efficient and use less fuel than a gasoline engine of the same size. Diesel engines are very durable. This is due to stronger construction and the fact that diesel fuel is a better lubricant than gasoline. This means that the fuel is less likely to remove the desired film of oil on the cylinder walls and piston rings of the engine. Diesel engines are also better suited for moving heavy loads at low speeds. Many diesel engines are fitted with a turbocharger to increase their power. Combining turbochargers with common rail injection systems has resulted in more horsepower.

STARTING In cold weather, diesel engines can be difficult to start because the cold air is unable to reach temperatures high enough to cause combustion. This problem is compounded by the fact that the cold metal of the cylinder block and head absorbs the heat generated during the compression stroke. Therefore, many diesel engines use **glow plugs** to help ignite the fuel during cold starting. These small electrical heaters are placed inside the cylinder and are used only when the engine is cold. Other diesels have a resistive grid heater in the intake manifold to warm the air until the engine reaches operating temperature.

SOUND A characteristic of a diesel engine is its sound. This noise, knock, or clatter is caused by the sudden ignition of fuel as it is injected into the combustion chamber. Through the use of electronically controlled common rail injector systems, this noise has been reduced.

EMISSIONS This has been an obstacle for diesel cars, and new stricter emissions standards go into effect shortly. Cleaner, low-sulphur diesel fuel has been available in Canada since 2007. With new technologies and the cleaner fuel, a diesel engine is able to run as cleanly as most gasoline engines. Many diesel vehicles have an assortment of traps and filters to clean the exhaust before it enters the atmosphere. Also, diesel engines produce very little CO because they run with an abundance of air.

Diesel vehicles may be equipped with particulate filters and catalytic converters **(Figure 9–34)**. Particulate filters catch the black soot (unburned carbon compounds) that is expelled from a typical diesel's exhaust. Most diesel cars have **selective catalytic reduction (SCR)** systems to reduce NO_x emissions. SCR is a system that has a substance injected into the exhaust stream and then absorbed in a catalyst. This breaks down the exhaust's NO_x to form H_2O and N_2. Others use NO_x traps.

FIGURE 9–34

A catalytic converter and particulate trap for a diesel engine.

Martin Restoule

Homogeneous Charge Compression Ignition Engines

Within the next few years, some automobiles will be equipped with **homogeneous charge compression ignition (HCCI)** engines. HCCI engines offer the high efficiency and torque of a diesel engine while providing the low emissions and power of a gasoline engine. Basically, these engines have a combustion process that allows a gasoline or diesel engine to operate with either compression ignition or spark ignition. With spark ignition, the air and fuel are mixed (homogenized) before ignition, and ignition is caused by a spark. In a diesel engine, the air and fuel are never mixed. The air is compressed, and ignition occurs when fuel is sprayed into the high-temperature air. In an HCCI engine, the air and fuel are mixed, and ignition occurs as the mixture is compressed. During compression, the mixture gets hot enough to "auto-ignite." HCCI is also referred to as controlled auto-ignition (CAI).

In an HCCI engine, combustion immediately and simultaneously begins at several points within the mixture **(Figure 9–35)**. This means the combustion process occurs rapidly and is controlled by the quality and temperature of the compressed mixture. This spontaneous combustion produces a flameless release of energy to drive the piston down.

The HCCI engine runs on a lean, diluted mixture of fuel, air, and exhaust gases. Only the heat inside the cylinder determines when ignition will occur. This fact makes it hard to control ignition timing. The temperature of the mixture at the beginning of the compression stroke must be increased to auto-ignition temperatures at the end of the compression stroke. Auto-ignition usually occurs when the temperature reaches 777°C to 827°C (1430°F to 1520°F) for gasoline. The engine's control unit must supply the correct amount of fuel mixed with the correct amount of air in order for combustion to occur at the right time. In addition, the control unit must provide a mixture that is hot enough to be able to auto-ignite at the end of the compression stroke. Therefore, it must be able to vary the compression ratio, the temperature of the intake air, the pressure of the intake air, or the amount of retained or reinducted exhaust gas. The role of the control unit is extremely important for proper operation.

DUAL MODE A practical application of an HCCI engine would be one with dual-mode capabilities. The spark ignition mode could be used when high

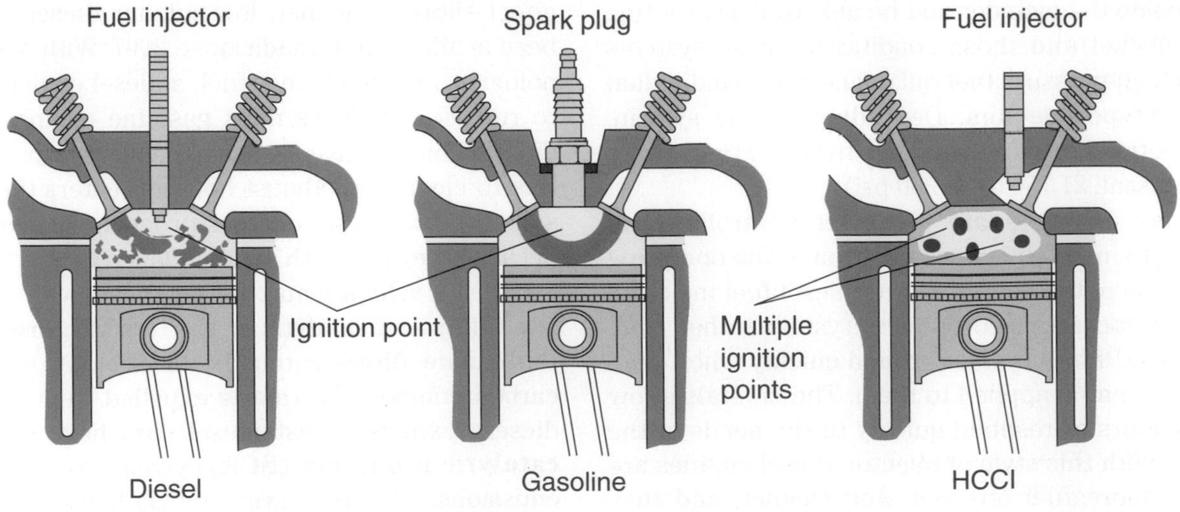

FIGURE 9–35

A comparative look at the ignition of a diesel, gasoline, and HCCI engine.

©Cengage Learning 2015

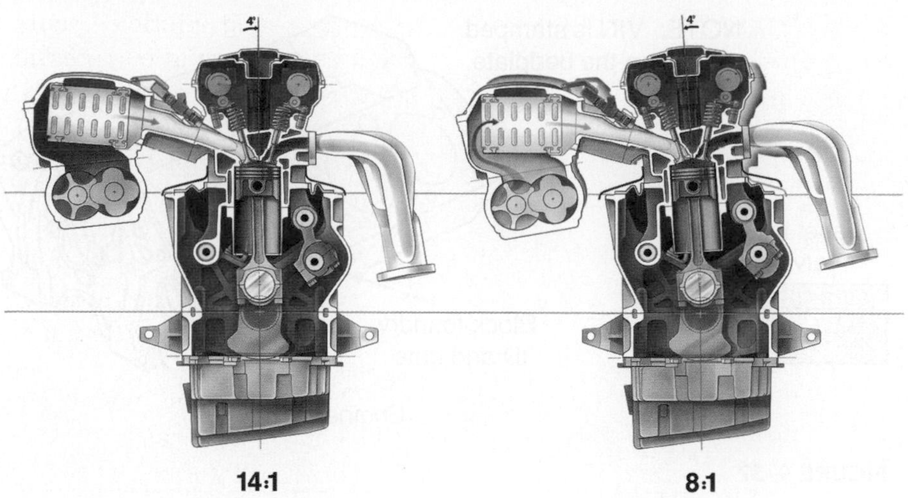

FIGURE 9–36

The Saab variable compression (SVC) engine can vary the engine's compression ratio from 8:1 to 14:1.

Saab Cars North America

power is required, and the compression ignition mode would be used during steady loads and speeds. To do this, the engine must be able to smoothly switch from the HCCI mode to the spark ignition mode from one cylinder firing to the next. This would require precise control of valve timing, air and fuel metering, and spark plug timing.

BENEFITS A gasoline HCCI engine could deliver almost the same fuel economy as a diesel engine and at a much lower cost. GM estimates that HCCI could improve gasoline engine fuel efficiency by 15 percent while emitting near-zero amounts of NO_x and particulate matter. In fact, HCCI engines emit extremely low levels of NO_x without a catalytic converter.

However, a gasoline engine running in HCCI mode produces more noise and vibrations than a conventional engine. Also, it tends to experience incomplete combustion, which leads to hydrocarbon and carbon monoxide emissions. To rectify this, HCCI engines are fitted with typical emission control systems, including an oxidizing catalytic converter.

Variable Compression Ratio Engines

Variable compression engines are being explored, not only for use with HCCI but for use in conventional engines. Changing the compression ratio is one way to provide power when needed and minimize fuel consumption. One way to do this is through changes in valve timing. The process is similar to the modifications made for the Atkinson cycle. Another way is to change the volume of the combustion chamber in response to the engine's operating conditions.

In some engines, the compression ratio is altered by changing the slope of the cylinder head in relation to the engine block. This changes the volume of the combustion chamber **(Figure 9–36)**. The cylinder head is pivoted at the crankshaft by a hydraulic actuator, and the angle can be as much as 4°. The engine management system adjusts the angle in response to engine speed, engine load, and fuel quality. The cylinder head is sealed to the engine block by a rubber bellows.

ENGINE IDENTIFICATION

Using Service Information

Normally, information used to identify the size of an engine is given in the service information for a vehicle at the beginning of the section covering a particular vehicle.

By referring to the VIN, much information about the vehicle can be determined. Identification numbers are also found on the engine. Engine blocks often have a serial number stamped into them **(Figure 9–37)**. An engine code is generally found beside the serial number. A typical engine code might be DZ or MO. These letters indicate the horsepower rating of the engine, whether it was built for an automatic or manual transmission, and other important details. The engine code will help you determine the correct specifications for that particular engine.

Casting numbers are often mistaken for serial numbers and engine codes. Manufacturers use a casting number to identify major engine parts on the assembly line. Casting numbers seldom can be

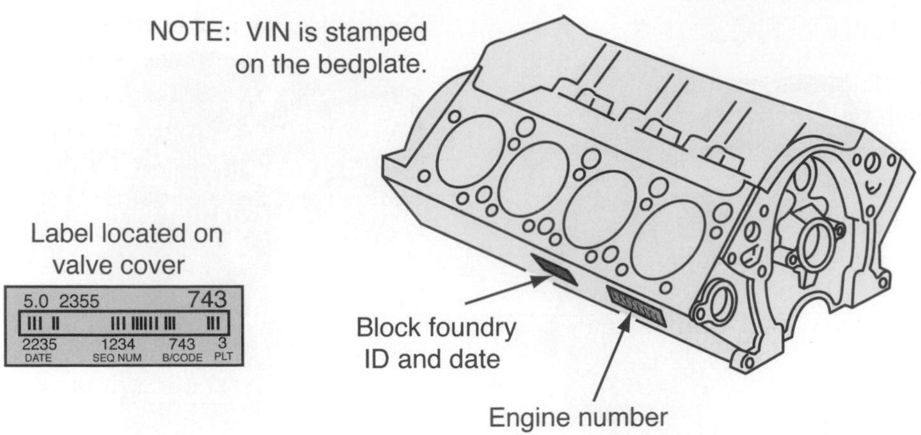

FIGURE 9–37

Examples of the various identification numbers found on an engine.

used to identify the type of engine. Normally, they are raised from the metal, whereas ID numbers are usually stamped.

Engine ID Tags

Many engines have ID tags or stickers attached to various places on the engine, such as the valve cover or oil pan. The tags include the displacement, assembly plant, model year, change level, engine code, and date of production. The location of these stickers or tags on a particular engine may be given in the service information.

Casting Numbers

Whenever an engine part such as an engine block or head is cast, a number is put into the mould to identify the casting and the date when the part was made. This date does not indicate when the engine was assembled or placed into the vehicle. A part made during one year may be installed in the vehicle in the following year; therefore, the casting date may not match the model year of the vehicle. Casting numbers should not be used for identifying the displacement of an engine. They only indicate the basic design of an engine. The same block or head can be used with a variety of different displacement engines.

Underhood Label

All vehicles produced since 1972 have an underhood label called the Vehicle Emission Control Information (VECI) label **(Figure 9–38)**. This gives some useful information regarding the emissions' rating of the vehicle and, at times, information necessary to perform maintenance **(Figure 9–39)** and an emissions inspection or to order engine and engine management parts.

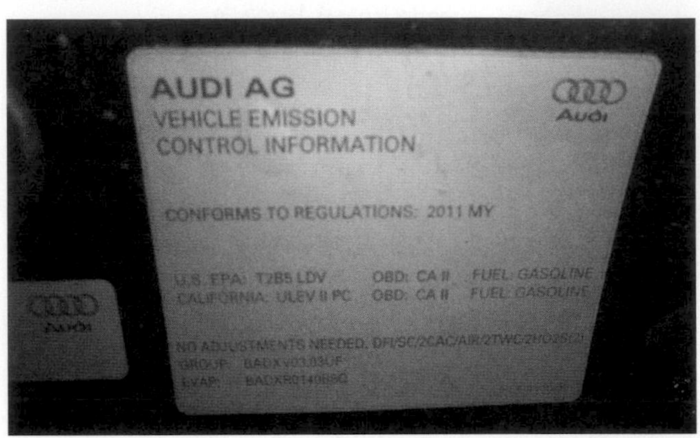

FIGURE 9–38

A current VECI label.

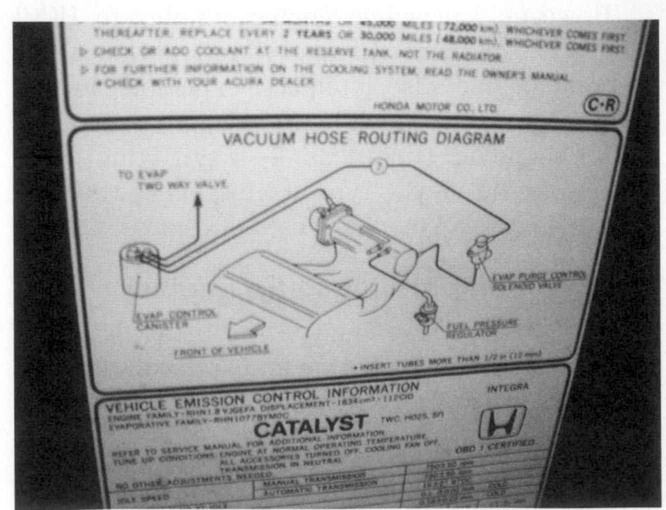

FIGURE 9–39

Some information that may be included on a VECI label.

ENGINE DIAGNOSTICS

As the trend toward the integration of ignition, fuel, and emission systems progresses, new diagnostic tools and techniques are constantly being developed to diagnose electronic engine control systems. However, not all engine performance problems are related to electronic controls; therefore, technicians still need to understand basic engine tests. These tests are an important part of modern engine diagnosis.

Compression Test

Internal combustion engines depend on compression of the air/fuel mixture to maximize the power produced by the engine. The upward movement of the piston on the compression stroke compresses the air/fuel mixture within the combustion chamber. The air/fuel mixture gets hotter as it is compressed. The hot mixture is easier to ignite, and when ignited, it generates much more power than the same mixture at a lower temperature.

If the combustion chamber leaks, some of the air/fuel mixture will escape when it is compressed, resulting in a loss of power and a waste of fuel. The leaks can be caused by burned valves, a blown head gasket, worn rings, a slipped timing belt or chain, worn valve seats, a cracked head, and more.

An engine with poor compression (lower compression pressure due to leaks in the cylinder) will not run correctly and cannot be tuned to factory specifications. If a symptom suggests that the cause of a problem may be poor compression, a compression test is performed.

A compression gauge is used to check cylinder compression. The dial face on the typical compression gauge indicates pressure in both kilopascals (kPa) and pounds per square inch (psi). Most compression gauges have a vent valve that holds the highest pressure reading on its meter. Opening the valve releases the pressure when the test is complete. The steps for conducting a cylinder compression test are shown in Photo Sequence 6.

Ford, Toyota, and other hybrids have Atkinson cycle engines. These engines delay the closing of the intake valve, which means that the overall compression ratio and displacement of the engine are reduced. Therefore, when conducting a compression test on these engines, expect a slightly lower reading than what you would expect from a conventional engine.

To conduct a compression test on a Ford hybrid, you must use a scan tool, and the one from Ford is preferred. The scan tool allows you to enter into the engine cranking diagnostic mode. This mode allows the engine to crank with the fuel injection system disabled. It also makes sure that the starter motor/generator is not activated (except for activating the starter motor to crank the engine), which is good for safety purposes but also because the load of the generator cannot affect the test results because it is not energized. Always follow the sequence as stated in the service information. Failure to do so will result in bad readings.

WET COMPRESSION TEST Because many things can cause low compression, it is advisable to conduct a wet compression test on the low cylinders. This test allows you to identify if it is caused by worn or damaged piston rings. To conduct this test, add two squirts of oil into the low cylinders. Then measure the compression of that cylinder. If the readings are higher, it is very likely that the piston rings are the cause of the problem. The oil temporarily seals the piston to the cylinder walls, which is why the readings increased. If the readings do not increase, or increase only slightly, the cause of the low readings is probably the valves.

Running Compression Test

Some engine problems, such as worn camshaft lobes, are not easily detected using a cranking compression test. When diagnosing a cylinder that is not producing as much power as other cylinders yet shows normal cranking compression, a running compression test may be needed. Using a standard compression test kit, remove the Schrader valve from the tester's adapter hose, and thread it into the spark plug hole for the cylinder being tested. Start the engine and note the gauge. Note that because the gauge will show both pressure and vacuum, the needle will sweep up and down. This can cause the needle to impact the needle stop on some gauges. With the engine running, note the maximum pressure. During running compression tests, cylinder pressures will likely be in the 70 to 80 psi range. To more accurately see what is happening in the cylinder while the engine is running, many technicians perform this test using electronic pressure transducers and a scope.

Cylinder Leakage Test

If a compression test shows that any of the cylinders are leaking, a cylinder leakage test can be performed to measure the percentage of compression lost and help locate the source of leakage. A cylinder leakage tester applies compressed air to a cylinder through the spark plug hole. Before the air is applied to the cylinder, the piston of that cylinder must be at TDC on its compression stroke to ensure that the valves of that cylinder are closed.

PHOTO SEQUENCE 6

CONDUCTING A CYLINDER COMPRESSION TEST

P6–1 Before conducting a compression test, disable the ignition and the fuel injection system (if the engine is so equipped).

P6–2 Prop the throttle plate into a wide-open position to allow an unrestricted amount of air to enter the cylinders during the test.

P6–3 Remove all of the engine's spark plugs.

P6–4 Connect a remote starter button to the starter system.

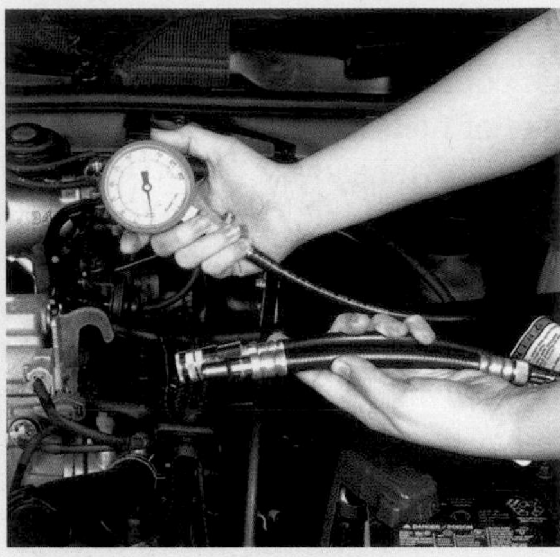

P6–5 Many types of compression gauges are available. The screw-in type tends to be the most accurate and easiest to use.

P6–6 Carefully install the gauge into the spark plug hole of the first cylinder.

P6–7 Connect a battery charger to the car to allow the engine to crank at consistent and normal speeds needed for accurate test results.

P6–8 Depress the remote starter button, and observe the gauge's reading after the first engine revolution.

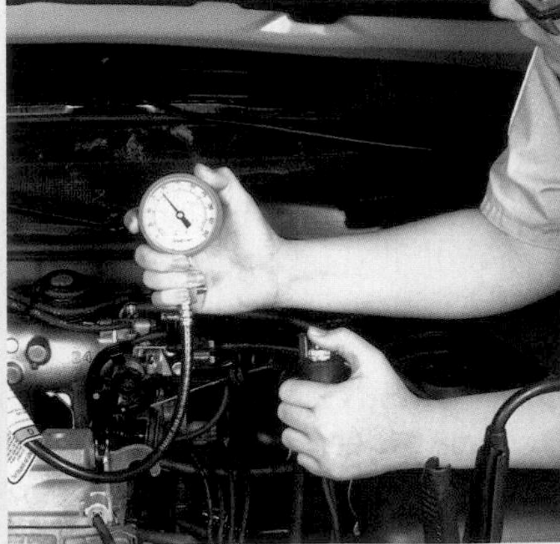

P6–9 Allow the engine to turn through four revolutions, and observe the reading after the fourth. The reading should increase with each revolution.

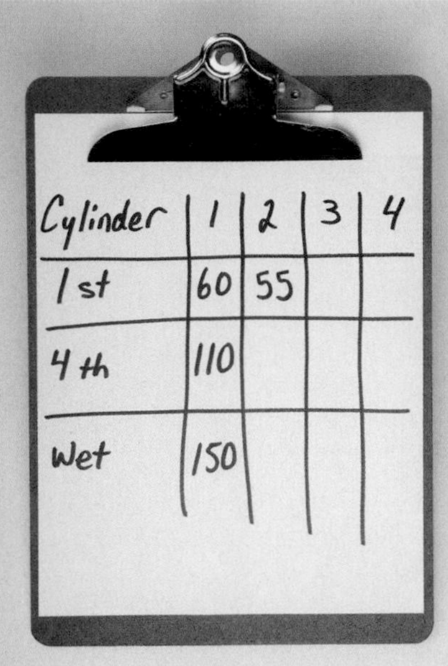

P6–10 Readings observed should be recorded. After all cylinders have been tested, a comparison of cylinders can be made.

P6–11 Before removing the gauge from the cylinder, release the pressure from it, using the release valve on the gauge.

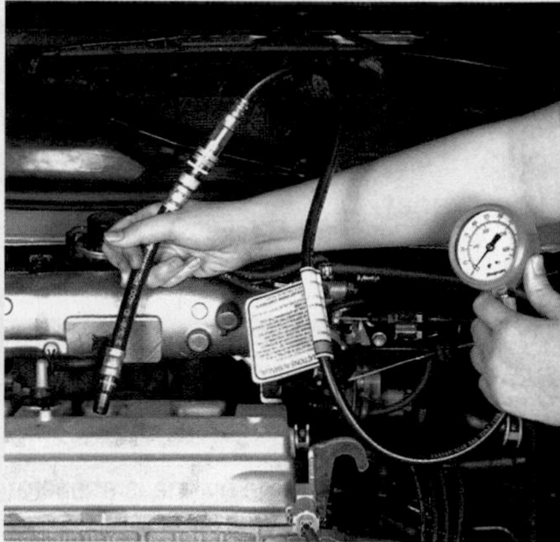

P6–12 Each cylinder should be tested in the same way.

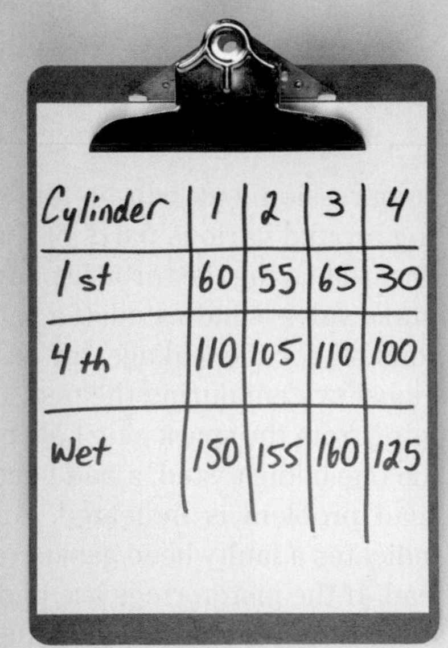

Cylinder 1 2 3 4
1st 60 55 65 30
4th 110 105 110 100
Wet 150 155 160 125

P6–13 After completing the test on all cylinders, compare them. If one or more cylinders is much lower than the others, continue testing those cylinders with the wet test.

P6–14 Squirt a small amount of oil into the weak cylinder(s).

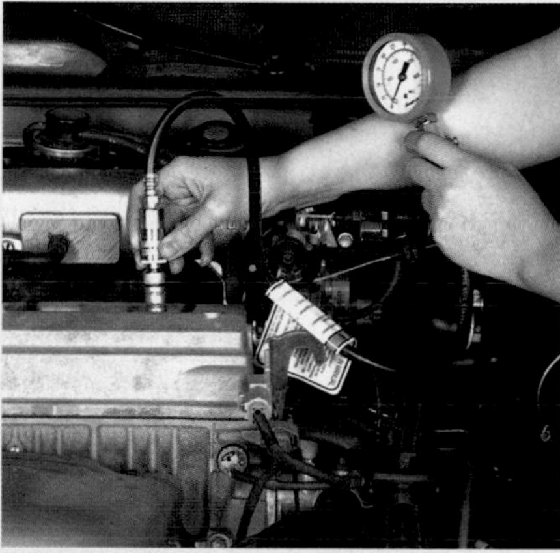

P6–15 Reinstall the compression gauge into that cylinder, and conduct the test.

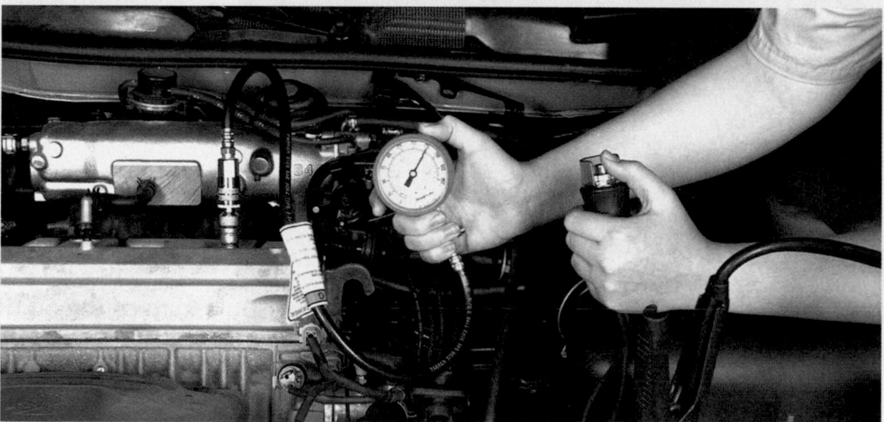

P6–16 If the reading increases with the presence of oil in the cylinder, the most likely cause of the original low readings was poor piston ring sealing. Using oil during a compression test is normally referred to as a wet test.

A threaded adapter on the end of the air-pressure hose screws into the spark plug hole. The source of the compressed air is normally the shop's compressed-air system. A pressure regulator in the tester controls the pressure applied to the cylinder. A gauge registers the percentage of air pressure lost from the cylinder when the compressed air is applied. The scale on the gauge typically reads 0 to 100 percent.

PROCEDURE

STEP 1 Make sure the engine is at operating condition.

STEP 2 Remove the radiator cap, oil filler cap, dipstick tube, air filter cover, and all spark plugs.

STEP 3 Rotate the crankshaft with a remote starter button so that the piston of the tested cylinder is at TDC on its compression stroke **(Figure 9–40)**. This ensures that the valves of that cylinder are closed.

STEP 4 Insert the threaded adapter on the end of the tester's air-pressure hose into the spark plug hole.

STEP 5 Allow the compressed air to enter the cylinder.

STEP 6 Observe the gauge reading **(Figure 9–41)**.

STEP 7 Listen and feel to identify the source of any escaping air.

A zero reading means there is no leakage in the cylinder. A reading of 100 percent indicates that the cylinder will not hold any pressure. Any reading that is more than 0 percent indicates some leakage.

FIGURE 9–40

Rotate the engine so that the piston of the cylinder that will be tested is at TDC (compression) before checking leakage.

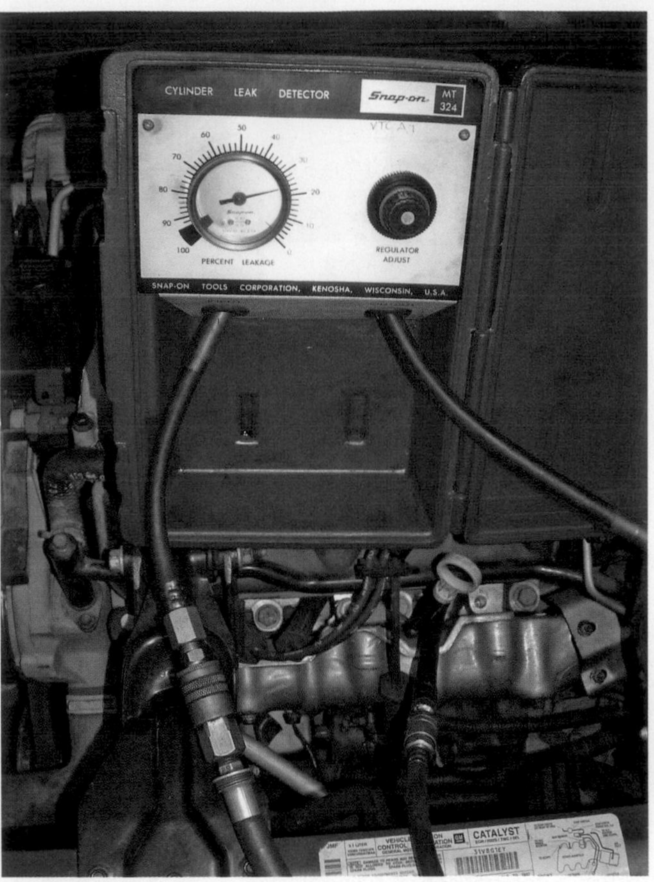

FIGURE 9–41

The reading on the tester is the percentage of air that leaked out during the test.

The location of the compression leak can be found by listening and feeling around various parts of the engine. If air is felt or heard leaving the throttle plate assembly, a leaking intake valve is indicated. If a bad exhaust valve is responsible for the leakage, air can be felt leaving the exhaust system during the test. If air is felt or heard coming from the spark plug hole of the cylinder next to the one being tested, a bad head gasket or cylinder head problem is indicated. Air leaving the radiator indicates a faulty head gasket or a cracked block or head. If the piston rings are bad, air will be heard leaving the valve cover's breather cap or the oil dipstick tube.

Most engines, even new ones, experience some leakage around the rings. Up to 20 percent is considered acceptable during the leakage test. When the engine is actually running, the rings will seal much better, and the actual percent of leakage will be lower. However, there should be no leakage around the valves or the head gasket.

Cylinder Power Balance Test

The cylinder power balance test is used to see if all of the engine's cylinders are producing the same amount of power. Ideally, all cylinders will produce the same amount. To check an engine's power balance, the spark plugs for individual cylinders are shorted out one at a time, and the change in engine speed is recorded. Little or no decrease in speed indicates a weak cylinder. If all of the readings are fairly close to each other, the engine is in good condition. If the readings from one or more cylinders differ from the rest, there is a problem. Further testing may be required to identify the exact cause of the problem. If all of the cylinders are producing the same amount of power, engine speed will drop the same amount as each plug is shorted. Unequal cylinder power balance can be caused by the following problems:

- Defective ignition coil
- Defective spark plug wire
- Defective or worn spark plug
- Damaged head gasket
- Worn piston rings
- Damaged piston
- Damaged or burned valves
- Broken valve spring
- Worn camshaft
- Defective lifters, pushrods, and/or rocker arms
- Leaking intake manifold
- Faulty fuel injector

A power balance test is performed quickly and easily using an engine analyzer, because the firing of the spark plugs can be automatically or manually controlled by pushing a button. Some vehicles have a power balance test built into the engine control computer. This test is either part of a routine self-diagnostic operating mode or must be activated by the technician.

WARNING!

On some computer-controlled engines, certain components must be disconnected before attempting the power balance test. Always check the service manual for appropriate procedures. Be careful not to run the engine with a shorted cylinder for more than 15 seconds. The unburned fuel in the exhaust can cause overheating and damage to the catalytic converter and create an unsafe situation. Also, run the engine for at least 10 seconds between cylinder shortings.

Override the controls of the electric cooling fan by using jumper wires to make the fan run constantly. If the fan control cannot be bypassed, disconnect the fan.

Connect the engine analyzer's leads, then turn the engine on and allow it to reach normal operating temperature. Set the engine speed at 1000 rpm and connect a vacuum gauge to the intake manifold. As each cylinder is shorted, note and record the rpm drop and the change in vacuum.

As each cylinder is shorted, a noticeable drop in engine speed should be noted. Little or no decrease in speed indicates a weak cylinder. If all of the readings are fairly close to each other, the engine is in good condition. If the readings from one or more cylinders differ from the rest, there is a problem. Further testing may be required to identify the exact cause of the problem.

Nearly all late-model engines can go through a cylinder balance test as ordered by a scan tool. During this automatic test, the system will attempt to run the engine at a fixed speed. Then once that speed can be maintained, the control system will shut off spark and fuel to one cylinder and the decrease in engine speed will be measured. The system will then activate the fuel and spark to the tested cylinder and wait for several seconds until the idle stabilizes again. Then the system will repeat the process for the next cylinder, repeating until all cylinders have been tested. After all of the cylinders have been initially checked, the system will perform the test again. The results of both tests will be displayed on the scan tool.

Vacuum Tests

Measuring intake manifold vacuum is another way to diagnose the condition of an engine. Manifold vacuum is tested with a vacuum gauge. Vacuum is formed by the downward movement of the pistons during their intake stroke. If the cylinder is sealed, a maximum amount will be formed. This test is most important on engines that are not totally computer controlled. Keep in mind that the amount of manifold vacuum that is created on the intake stroke is directly related to engine speed and throttle plate position. Also, the faster the engine spins, the higher the vacuum will be if the throttle plates are closed. When the throttle plates open, vacuum will decrease.

Manifold vacuum is measured with a vacuum gauge. The gauge's hose is connected to a vacuum fitting on the intake manifold **(Figure 9–42)**. Normally, a "tee" fitting and short piece of vacuum hose are used to connect the gauge.

FIGURE 9–42

The vacuum gauge is connected to the intake manifold, where it reads engine vacuum.

Vacuum gauge readings **(Figure 9–43)** can be interpreted to identify many engine conditions, including the ability of the cylinder to seal, the timing of the opening and closing of the engine's valves, and ignition timing.

Ideally, each cylinder of an engine will produce the same amount of vacuum; therefore, the vacuum gauge reading should be steady and give a reading of at least 430 mm of mercury (mm Hg), or 17 in. of mercury (in. Hg). If one or more cylinders produces more or less vacuum than the others, the needle of the gauge will fluctuate. The intensity of the fluctuation indicates the severity of the problem. For example, if the reading on the vacuum gauge fluctuates between 255 and 430 mm Hg (10 and 17 in. hg), you should look at the rhythm of the needle. If the needle seems to stay at 430 mm Hg (17 in. Hg) most of the time but drops to 255 mm Hg (10 in. Hg) and quickly rises, you know the reading is probably caused by a problem in one cylinder. Fluctuating or low readings can indicate many different problems. For example, a low, steady reading might be caused by retarded ignition timing or incorrect valve timing. A sharp vacuum drop at regular intervals might be caused by a burned intake valve. Other conditions that can be revealed by vacuum readings follow:

- Stuck or burned valves
- Improper valve or ignition timing
- Weak valve springs
- Faulty positive crankcase ventilation (PCV), exhaust gas recirculation (EGR), or other emission-related system
- Uneven compression
- Worn rings or cylinder walls
- Leaking head gaskets
- Vacuum leaks
- Restricted exhaust system
- Ignition defects

VACUUM TRANSDUCERS A vacuum or pressure transducer is an excellent way to check an engine's

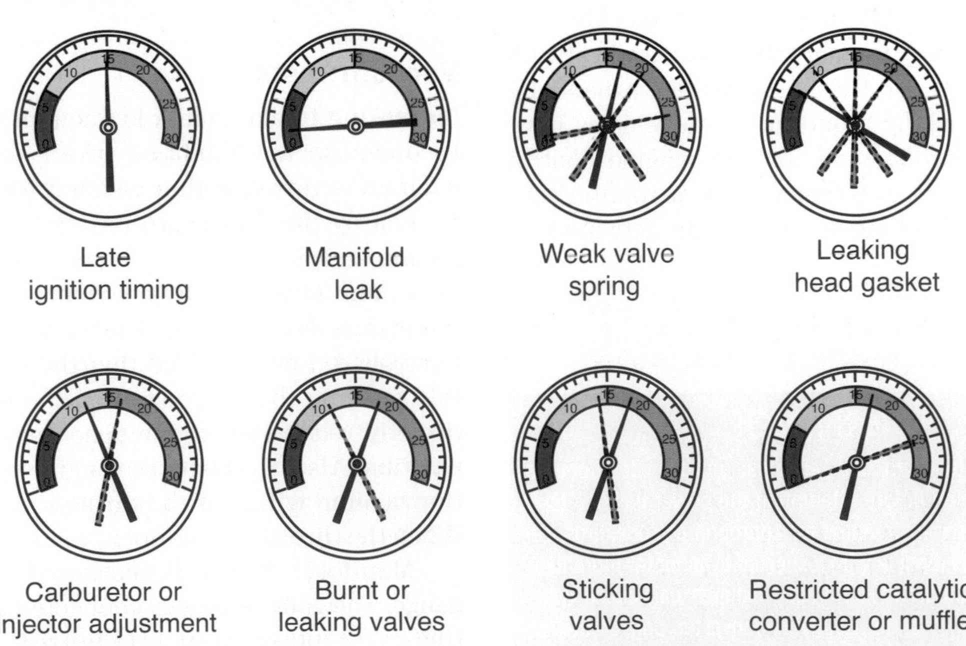

FIGURE 9–43

Vacuum gauge readings and the engine condition indicated by each.

vacuum. A transducer can transform a condition and convert it into an electronic signal. This has many advantages; the signal, as viewed on a scope, can allow you to observe other engine activities, and that may help you discern what is going on in the engine.

The connectors for the transducer are connected to a scope and to a source of manifold vacuum. On the displayed pattern on the scope, keep in mind, the vacuum waveform will be inverted. An increase in pressure, or decrease in vacuum, moves the waveform lower.

Oil Pressure Testing

An oil pressure test is used to determine the wear of an engine's parts. The test is performed with an oil pressure gauge, which measures the pressure of the oil as it circulates through the engine. Basically, the pressure of the oil depends on the efficiency of the oil pump and the clearances through which the oil flows. Excessive clearances, most often caused by wear between a shaft and its bearings, will cause a decrease in oil pressure. Excessive bearing clearances are not the only possible causes for low oil pressure readings; others are oil pump–related problems, a plugged oil pickup screen, weak or broken oil pressure relief valve, low oil level, contaminated oil, or low oil viscosity. Higher than normal pressures can be caused by too much oil, cold oil, high oil viscosity, restricted oil passages, or a faulty pressure regulator. A loss of performance, excessive engine noise, and poor starting can be caused by abnormal oil pressure. When the engine's oil pressure is too low, premature wear of its parts will result.

An oil pressure tester is a gauge with a high-pressure hose attached to it. The scale of the gauge typically reads from 0 to 690 kPa (0 to 100 psi). Using the correct fittings and adapters, the hose is connected to an oil passage in the engine block. Normally, the engine's oil pressure sensor is removed, and the hose is connected to that port **(Figure 9–44)**.

To conduct the test, simply follow the guidelines given in the service information and observe the gauge. The pressure is read when the engine is at normal operating temperatures and at a fast idle speed. To get accurate results from the test, make sure you follow the manufacturer's recommendations and compare your findings to specifications. Excessive bearing clearances are not the only possible causes for low oil pressure readings; others are

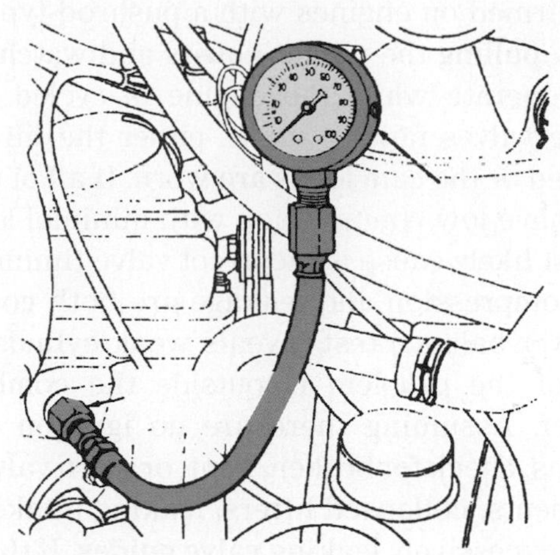

FIGURE 9–44

The oil pressure gauge is installed into the oil pressure sending unit's bore in the engine block.

Chrysler Group LLC

pump-related problems, a plugged oil pickup screen, a weak or broken oil pressure relief valve, low oil level, contaminated oil, or low oil viscosity. Too much oil, cold oil, high oil viscosity, restricted oil passages, and a faulty pressure regulator can cause high oil pressure readings.

EVALUATING THE ENGINE'S CONDITION

Once the compression, cylinder leakage, vacuum, and power balance tests are performed, a technician is ready to evaluate the engine's condition. For example, an engine with good relative compression but high cylinder leakage past the rings is typical of a high-mileage worn engine. This engine would have these symptoms: excessive blowby, lack of power, poor performance, and reduced fuel economy.

If these same compression and leakage test results are found on an engine with comparatively low mileage, the problem is probably stuck piston rings that are not expanding properly. If this is the case, try treating the engine with a combustion chamber cleaner, oil treatment, or engine flush. If this fails to correct the problem, an engine overhaul is required.

A cylinder that has poor compression but minimal leakage indicates a valve train problem. Under these circumstances, a valve might not be opening at the right time, might not be opening enough, or might not be opening at all. This condition can be confirmed on engines with a pushrod-type valve train by pulling the rocker covers and watching the valves operate while the engine is cycled. If one or more valves fails to move, either the lifters are collapsed or the cam lobes are worn. If all of the cylinders have low compression with minimal leakage, the most likely cause is incorrect valve timing.

If compression and leakage are both good, but the power balance test reveals weak cylinders, the cause of the problem is outside the combustion chamber. Assuming there are no ignition or fuel problems, check for broken, bent, or worn valve train components, collapsed lifters, leaking intake manifold, or excessively leaking valve guides. If the latter is suspected, squirt some oil on the guides. If they are leaking, blue smoke will be seen in the exhaust.

Fluid Leaks

When inspecting the engine, check it for leaks **(Figure 9–45)**. There are many different fluids under the hood of an automobile, so care must be taken to identify the type of fluid that is leaking **(Figure 9–46)**. Carefully look at the top and sides of the engine, and note any wet residue that may be present. Sometimes, road dirt will mix with the leaking fluid and create a heavy coating. Also look under the vehicle for signs of leaks or drips; make sure you have good lighting. Note the areas around the leaks, and identify the possible causes. Methods for positively identifying the source of leaks from various components are covered later in this section. All leaks should be corrected because they can result in more serious problems.

Sometimes, smell will identify the fluid. Gasoline evaporates when it leaks out and may not leave any residue, but it is easy to identify by its smell.

FIGURE 9–45
Oil leaking from around the oil pan gasket.

DESCRIPTION	PROBABLE SOURCE
Honey or dark greasy fluid	Engine oil
Honey or dark thick fluid with a chestnut smell	Gear oil
Green, sticky fluid	Engine coolant
Slippery clear or yellowish fluid	Brake fluid
Slippery red fluid	Transmission or power-steering fluid
Bluish watery fluid	Washer fluid

FIGURE 9–46
Identification of fluid leaks.

Exhaust Smoke Diagnosis

Examining and interpreting the vehicle's exhaust can give clues of potential engine problems. Basically, there should be no visible smoke coming out of the tailpipe. There is an exception to this rule, however; on a cold day after the vehicle has been idling for a while, it is normal for white smoke to come out of the tailpipe. This is nothing else but the water that has condensed in the exhaust system becoming steam. However, the steam should stop once the engine reaches normal operating temperature. If it does not, a problem is indicated. The colour of the exhaust is used to diagnose engine concerns **(Figure 9–47)**.

NOISE DIAGNOSIS

More often than not, an engine malfunction will reveal itself first as an unusual noise. This can happen well before the problem affects the driveability of the vehicle. Problems such as loose pistons, badly worn rings or ring lands, loose piston pins, worn main bearings and connecting rod bearings, loose vibration damper or flywheel, and worn or loose valve train components all produce telltale sounds. Unless the technician has experience in listening to and interpreting engine noises, it can be very hard to distinguish one from the other.

When correctly interpreted, engine noise can be a very valuable diagnostic aid. For one thing, a costly and time-consuming engine teardown might be avoided. Always make a noise analysis before doing any repair work. This way, there is a much greater likelihood that only the necessary repair procedures will be done. Careful noise diagnosis also reduces

ENGINE TYPE	VISIBLE SIGN	DIAGNOSIS	PROBABLE CAUSES
Gasoline	Grey or black smoke	Incomplete combustion or excessively rich air/fuel mixture	• Clogged air filter • Faulty fuel injection system • Faulty emission control system • Ignition problem • Restricted intake manifold
Diesel	Grey or black smoke	Incomplete combustion	• Clogged air filter • Faulty fuel injection system • Faulty emission control system • Wrong grade of fuel • Engine overheating
Gasoline and diesel	Blue smoke	Burning engine oil	• Oil leaking into combustion chamber • Worn piston rings, cylinder walls, valve guides, or valve stem seals • Oil level too high
Gasoline	White smoke	Coolant/water is burning in the combustion chamber	• Leaking head gasket • Cracked cylinder head or block
Diesel	White smoke	Fuel is not burning	• Faulty injection system • Engine overheating

FIGURE 9–47
Exhaust analysis

CUSTOMER CARE

When attempting to diagnose the cause of abnormal engine noise, it may be necessary to temper the enthusiasm of customers who think they have pinpointed the exact cause of the noise using nothing more than their own two ears. Although the owner's description may be helpful (and should always be asked for), it must be stressed that one person's "rattle" can be another person's "thump." You are the professional. The final diagnosis is up to you. If customers have been proved correct in their diagnosis, make it a point to tell them so. Everyone feels better about dealing with an automotive technician who listens to them.

the chances of ruining the engine by continuing to use the vehicle despite the problem.

Using a Stethoscope

Some engine sounds can be easily heard without using a listening device, but others are impossible to hear unless amplified. A stethoscope **(Figure 9–48)** is very helpful in locating engine noise by amplifying the sound waves. It can also distinguish between normal and abnormal noise. The procedure for using a stethoscope is simple. Use the metal prod to trace the sound until it reaches its maximum intensity.

FIGURE 9–48
Using a stethoscope helps to identify the source of an abnormal noise.

Once the precise location has been discovered, the sound can be better evaluated. A rubber hose or a sounding stick, which is nothing more than a long, hollow tube, works on the same principle, though a stethoscope gives much clearer results.

The best results, however, are obtained with an electronic listening device. With this tool you can tune into the noise. Doing this allows you to eliminate all other noises that might distract or mislead you.

WARNING!

Be very careful when listening for noises around moving belts and pulleys at the front of the engine. Keep the end of the hose or stethoscope probe away from the moving parts. Physical injury can result if the hose or stethoscope is pulled inward or flung outward by moving parts.

Common Noises

Following are examples of abnormal engine noises, including a description of the sound, its likely cause, and ways of eliminating it **(Figure 9–49)**. An important point to keep in mind is that insufficient lubrication is the most common cause of engine noise. For this reason, always check the oil level first before moving on to other areas of the vehicle. Some noises are more pronounced on a cold engine because clearances are greater when parts are not expanded by heat. Remember that aluminum and iron expand at different rates as temperatures rise. For example, a knock that disappears as the engine warms up is probably piston slap or knock. An aluminum piston expands more than the iron block, allowing the piston to fit more closely as engine temperature rises.

RING NOISE This sound can be heard during acceleration as a high-pitched rattling or clicking in the upper part of the cylinder. It can be caused by worn rings or cylinders, broken piston ring lands, or

Type	Sound	Mostly Heard During	Possible Causes
Ring noise	High-pitched rattle or clicking	Acceleration	■ Worn piston rings ■ Worn cylinder walls ■ Broken piston ring lands ■ Insufficient ring tension
Piston slap	Hollow, bell-like	Cold engine operation and is louder during acceleration	■ Worn piston rings ■ Worn cylinder walls ■ Collapsed piston skirts ■ Misaligned connecting rods ■ Worn bearings ■ Excessive piston to wall clearance ■ Poor lubrication
Piston pin knock	Sharp, metallic rap	Hot engine operation at idle	■ Worn piston pin ■ Worn piston pin boss ■ Worn piston pin bushing ■ Lack of lubrication
Main bearing noise	Dull, steady knock	Louder during acceleration	■ Worn bearings ■ Worn crankshaft
Rod bearing noise	Light tap to heavy knocking or pounding	Idle speeds and low-load higher speeds	■ Worn bearings ■ Worn crankshaft ■ Misaligned connecting rod ■ Lack of lubrication
Thrust bearing noise	Heavy thumping	Irregular sound, may be heard only during acceleration	■ Worn thrust bearing ■ Worn crankshaft ■ Worn engine saddles
Tappet noise	Light regular clicking	Idle	■ Improper valve adjustment ■ Worn or damaged valve train ■ Dirty hydraulic lifters ■ Lack of lubrication
Timing chain noise	Severe knocking	Increases with increase in engine speed	■ Loose timing chain

FIGURE 9–49

Common engine noises.

insufficient ring tension against the cylinder walls. Ring noise is corrected by replacing the rings, pistons, or sleeves or reboring the cylinders.

PISTON SLAP This sound is commonly heard when the engine is cold and often gets louder when the vehicle accelerates. When a piston slaps against the cylinder wall, the result is a hollow, bell-like sound. Piston slap is caused by worn pistons or cylinders, collapsed piston skirts, misaligned connecting rods, excessive piston-to-cylinder wall clearance, or lack of lubrication, resulting in worn bearings. Correction requires either replacing the pistons, reboring the cylinder, replacing or realigning the rods, or replacing the bearings. Shorting out the spark plug of the affected cylinder might quiet the noise.

PISTON PIN KNOCK Piston pin knock is a sharp, metallic rap that can sound more like a rattle if all the pins are loose. It originates in the upper portion of the engine and is most noticeable when the engine is idling and the engine is hot. Piston pin knock sounds like a double knock at idle speeds. It is caused by a worn piston pin, piston pin boss, or piston pin bushing, or lack of lubrication, resulting in worn bearings. To correct it, either install oversized pins, replace the boss or bushings, or replace the piston.

RIDGE NOISE This noise is less common but very distinct. As a piston ring strikes the ridge at the top of the cylinder, the result is a high-pitched rapping or clicking noise that becomes louder during deceleration **(Figure 9–50)**.

There can be more than one reason for the ridge interfering with the ring's travel. For one thing, if new rings are installed without removing the old ridge, the new rings will contact the ridge and make a noise. Also, if the piston pin is very loose or the connecting rod has a loose or burned-out bearing, the piston will go high enough in the cylinder for the top ring to contact the ridge. Thus, in order to eliminate ridge noise, remove the old ring ridge and replace the piston pin or piston.

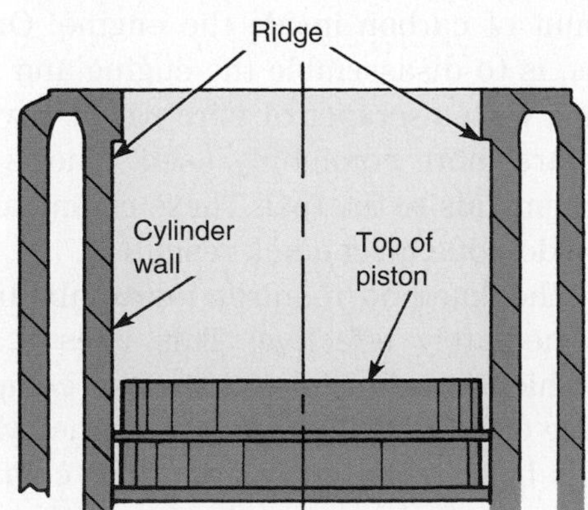

FIGURE 9–50
When the piston strikes the ridge at the top of the cylinder, a high-pitched rapping or clicking sound is heard.

ROD-BEARING NOISE The result of worn or loose connecting rod bearings, this noise is heard at idle as well as at speeds over 60 km/h (35 mph). Depending on how badly the bearings are worn, the noise can range from a light tap to a heavy knock or pound. Shorting out the spark plug of the affected cylinder can lessen the noise unless the bearing is extremely worn. In this case, shorting out the plug will have no effect. Rod-bearing noise is caused by a worn bearing or crankpin, a misaligned rod, or lack of lubrication, resulting in worn bearings. To correct it, service or replace the crankshaft, realign or replace the connecting rods, and replace the bearings.

MAIN OR THRUST BEARING NOISE A loose crankshaft main bearing produces a dull, steady knock, whereas a loose crankshaft thrust bearing produces a heavy thump at irregular intervals. The thrust bearing noise might be audible only on very hard acceleration. Both of these bearing noises are usually caused by worn bearings or crankshaft journals. To correct the problem, replace the bearings, and service or replace the crankshaft.

TAPPET NOISE Tappet noise is characterized by a light, regular clicking sound that is more noticeable when the engine is idling. It is the result of excessive clearance in the valve train. The clearance problem area is located by inserting a feeler gauge between each lifter and valve, or between each rocker arm and valve tip, until the noise subsides. Tappet noise can be caused by improper valve adjustment, worn or damaged parts, dirty hydraulic lifters, or lack of lubrication. To correct the noise, adjust the valves, replace any worn or damaged parts, or clean or replace the lifters.

ABNORMAL COMBUSTION NOISES **Preignition** and detonation noises are caused by abnormal engine combustion. For instance, **detonation** knock or **ping** **(Figure 9–51)** is a noise most noticeable during acceleration with the engine under load and running at normal temperature. Excessive detonation can be very harmful to the engine. It is often caused by advanced ignition timing or substantial carbon buildup in the combustion chambers that increases combustion pressure. Carbon deposits that get so

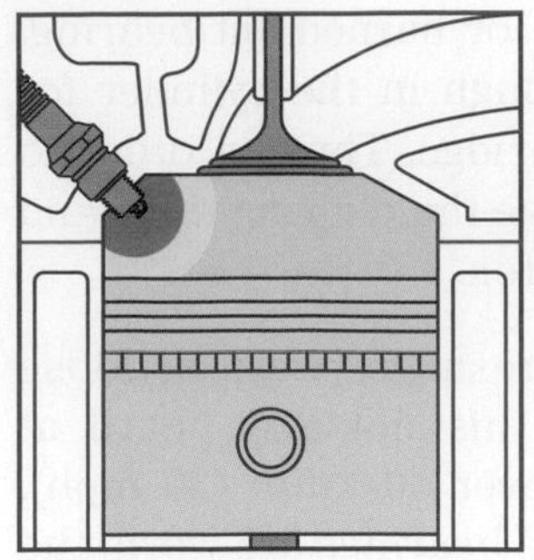
1. Spark occurs
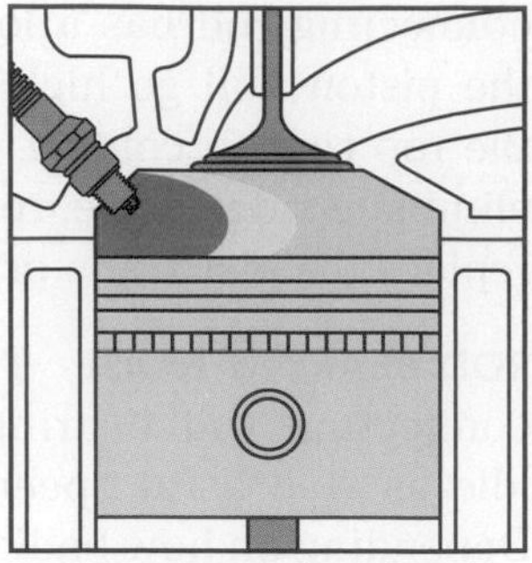
2. Combustion starts
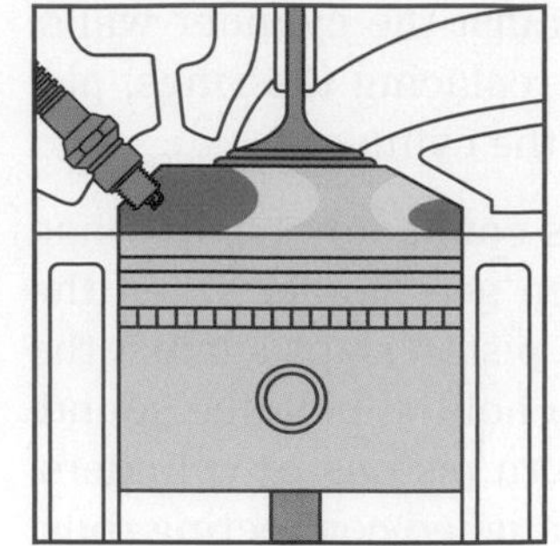
3. Combustion continues
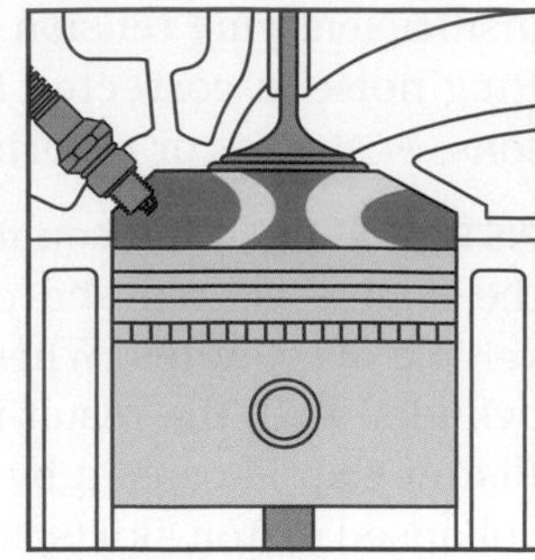
4. Detonation occurs

FIGURE 9–51

Detonation.

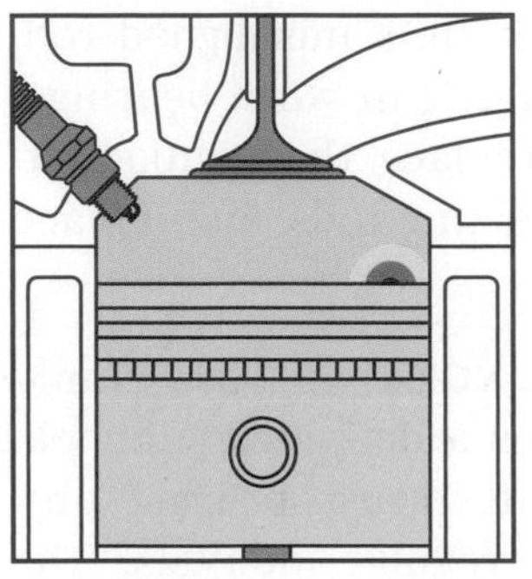
1. Ignited by hot carbon
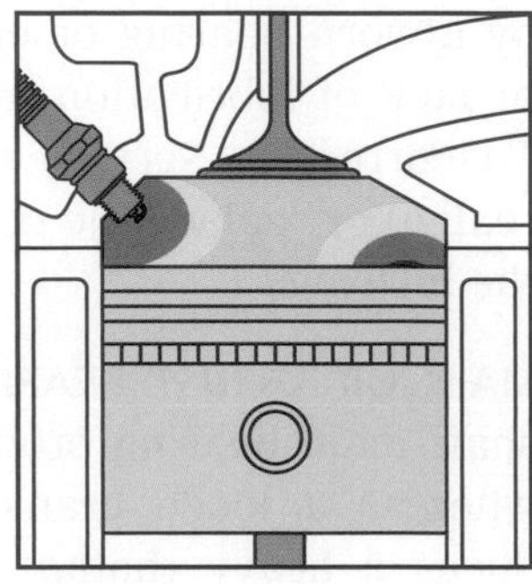
2. Ignition spark
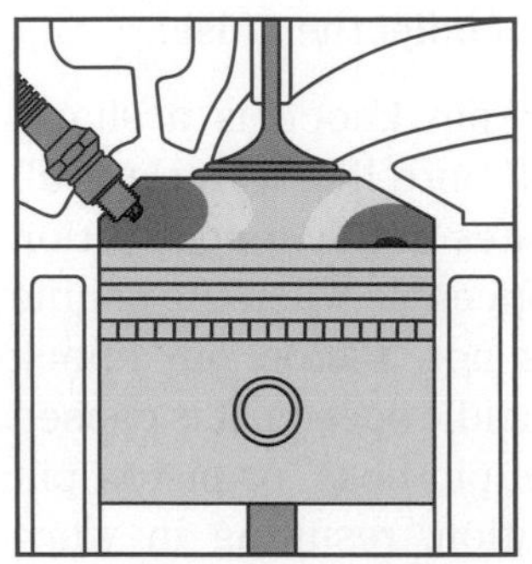
3. Flame fronts collide
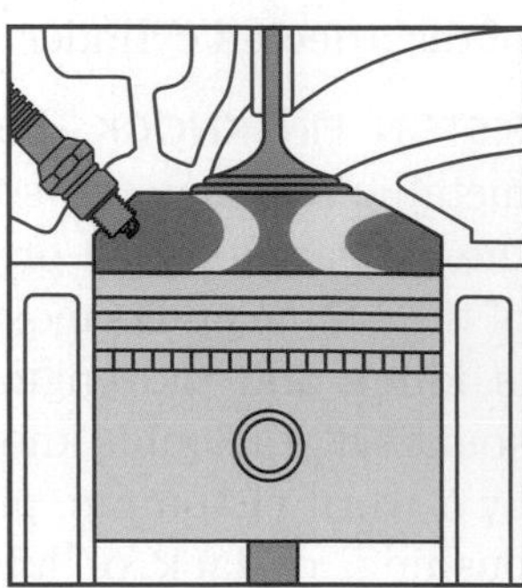
4. Ignites remaining fuel

FIGURE 9–52

Preignition.

hot they glow will also preignite the air/fuel mixture, causing preignition **(Figure 9–52)**. Another possible cause is fuel with octane that is too low. Detonation knock can usually be cured by removing carbon deposits from the combustion chambers with a rotary wire brush as well as by recommending the use of a higher octane gasoline. A malfunctioning EGR valve can also cause detonation.

Sometimes, abnormal combustion combines with other engine parts to cause noise. For example, rumble is a term used to describe the knock or noise resulting from another form of abnormal ignition. Rumble is a vibration of the crankshaft and connecting rods that is caused by multisurface ignition. Rumble is a form of preignition in which several flame fronts occur simultaneously from overheated deposit particles. A loose vibration damper causes a heavy rumble or thump in the front of the engine that is more apparent when the vehicle is accelerating from idle under load or is idling unevenly. A loose flywheel causes a heavy thump or light knock at the back of the engine, depending on the amount of play and the type of engine. Both of these problems are corrected either by tightening or replacing the damper or flywheel.

CLEANING CARBON DEPOSITS A buildup of carbon on the top of the piston or intake valve, or in the combustion chamber **(Figure 9–53)**, can cause a number of driveability concerns, including preignition. There are a number of techniques used to remove or reduce the amount of carbon inside the engine. One way, of course, is to disassemble the engine and remove the carbon with a scraper or wire wheel. Two other methods are more commonly used. One is simply adding chemicals to the fuel. These chemicals work slowly, so do not expect quick results.

The other method requires more labour but is more immediately effective. This uses a carbon blaster, which is a machine that uses compressed air to force crushed walnut shells into the cylinders. The shells beat on the piston top and combustion chamber walls to loosen and remove the carbon. Basically, to use a carbon blaster, the intake manifold and spark plugs are removed. The output hose of the blaster is attached to a cylinder's intake port

FIGURE 9–53

Carbon buildup on the intake and exhaust valves.

or inserted into the bore for the fuel injector. A hose is inserted into the spark plug bore; this is where the shells and carbon exit the cylinder. Once connected to the cylinder, the blaster forces a small amount of shells in and out of the cylinder. Hopefully, the carbon deposits leave with the shells. To help remove any remaining bits of shells, compressed air is applied to the cylinder. This operation is done at each cylinder. It is important to note that any remaining shell bits will be burned once the engine is run again.

KEY TERMS

Atkinson cycle (p. 233)
Bore (p. 231)
Bottom dead centre (BDC) (p. 223)
Cams (p. 222)
Combustion chamber (p. 222)
Crank throw (p. 231)
Detonation (p. 253)
Diesel engine (p. 238)
Direct injection (DI) (p. 239)
Dual overhead camshaft (DOHC) (p. 222)
Efficiency (p. 232)
Firing order (p. 230)
Glow plugs (p. 239)
Homogeneous charge compression ignition (HCCI) (p. 240)
Lobes (p. 222)
Overhead camshaft (OHC) (p. 222)
Overhead valve (OHV) (p. 222)
Oversquare (p. 231)
Ping (p. 253)
Preignition (p. 253)
Rotary engine (p. 237)
Selective catalytic reduction (SCR) (p. 239)
Square (p. 231)
Stroke (p. 223)
Top dead centre (TDC) (p. 223)
Torque (p. 232)
Undersquare (p. 231)
Wankel engine (p. 237)

SUMMARY

- Automotive engines are classified by several different design features such as operational cycles, number of cylinders, cylinder arrangement, valve train type, valve arrangement, ignition type, cooling system, and fuel system.
- The basis of automotive gasoline engine operation is the four-stroke cycle. This includes the intake stroke, compression stroke, power stroke, and exhaust stroke. The four strokes require two full crankshaft revolutions.
- The most popular engine designs are the in-line (in which all the cylinders are placed in a single row) and V-type (which features two rows of cylinders). The slant design is much like the in-line, but the entire block is placed at a slant. Opposed cylinder engines use two rows of cylinders located opposite the crankshaft.

- The two basic valve and camshaft placement configurations currently in use on four-stroke engines are the overhead valve and overhead cam.
- Bore is the diameter of a cylinder, and stroke is the length of piston travel between top dead centre (TDC) and bottom dead centre (BDC). Together, these two measurements determine the displacement of the cylinder.
- Compression ratio is a measure of how much the air and fuel are compressed during the compression stroke. The higher the compression ratio is, the more power an engine can produce.
- Horsepower is the rate at which torque is produced by an engine. The torque is then transmitted through the drivetrain to turn the driving wheels of the vehicle.
- Instead of relying on a spark for ignition, diesel engines use the heat produced by compressing air in the combustion chamber to ignite the fuel.
- Features of both the gasoline and the diesel engine are found in the stratified charge engine. Its major advantages are good part-load fuel economy, low exhaust emissions, and an ability to operate on low-octane fuel.
- In addition to the diesel and the stratified charge, other automotive engines include the rotary or Wankel, the Miller cycle, and the Atkinson cycle. The future may bring many electric-powered and hybrid vehicles.
- The vehicle identification number, or VIN, is used to identify correct engine specifications. It is stamped on a metal tab, which is riveted to the top of the instrument panel.
- A compression test is conducted to check a cylinder's ability to seal and therefore its ability to compress the air/fuel mixture inside the cylinder.
- A cylinder leakage test is performed to measure the percentage of compression lost and to help locate the source of leakage.
- A cylinder power balance test reveals whether all of an engine's cylinders are producing the same amount of power.
- Vacuum gauge readings can be interpreted to identify many engine conditions, including the ability of the engine's cylinders to seal, the timing of the opening and closing of the engine's valves, and ignition timing.
- An oil pressure test measures the pressure of the engine's oil as it circulates throughout the engine. This test is very important because abnormal oil pressures can cause a host of problems, including poor performance and premature wear.
- An engine malfunction often reveals itself as an unusual noise. When correctly interpreted, engine noise can be a very helpful diagnostic aid.

REVIEW QUESTIONS

1. What occurs in the combustion chamber of a four-stroke engine?
2. Name the four strokes of a four-stroke-cycle engine.
3. As an engine's compression ratio increases, what should happen to the octane rating of the gasoline?
4. What test can be performed to check the efficiency of individual cylinders?
5. Describe tappet noise.
6. Which of the following statements about engines is true?
   a. The engine provides the rotating power to drive the wheels through the transmission and driving axle.
   b. Only gasoline engines are classified as internal combustion.
   c. The combustion chamber is the space in the cylinder between top dead centre and bottom dead centre.
   d. Only diesel engines are classified as internal combustion.
7. Which stroke in the four-stroke cycle begins as the compressed fuel mixture is ignited in the combustion chamber?
   a. power stroke
   b. exhaust stroke
   c. intake stroke
   d. compression stroke
8. What is compression ratio?
   a. diameter of the cylinder
   b. cylinder arrangement
   c. the amount the air/fuel mixture is compressed
   d. none of the above
9. What engine component is responsible for opening and closing the intake and exhaust valves?
   a. the crankshaft
   b. the camshaft
   c. the intake manifold
   d. the exhaust manifold

10. Which engine design would have the highest compression ratio?
   - **a.** diesel engine
   - **b.** Miller cycle engine
   - **c.** two-stroke-cycle engine
   - **d.** stratified charge engine

11. What engine test can be used to pinpoint the cause of a low cylinder compression test reading?
   - **a.** a cylinder leakage test
   - **b.** a cylinder power balance test
   - **c.** a chassis dynamometer test
   - **d.** a vacuum test

12. What is piston slap?
   - **a.** grooves on the side of the piston
   - **b.** force applied to the piston
   - **c.** noise made by the piston when it contacts the cylinder wall
   - **d.** a high-pitched clicking that becomes louder during deceleration

13. Which engine system removes burned gases and limits noise produced by the engine?
   - **a.** exhaust system
   - **b.** emission control system
   - **c.** ignition system
   - **d.** air/fuel system

14. Which of the following would best describe the noise caused by a piston slap?
   - **a.** a sharp metallic noise
   - **b.** a hollow bell-like noise
   - **c.** a dull knocking noise
   - **d.** a high-pitched ticking noise

15. What is the stroke of the piston equal to?
   - **a.** half of the crankshaft throw
   - **b.** the crankshaft throw
   - **c.** two times the crankshaft throw
   - **d.** four times the crankshaft throw

16. During a cylinder leakage test, where would you expect to find air leaking from if the rings were worn?
   - **a.** at the exhaust tailpipe
   - **b.** at the throttle plate
   - **c.** at the oil fill port
   - **d.** at the radiator

17. What could cause a higher than normal oil pressure test reading?
   - **a.** a defective oil pressure regulator valve
   - **b.** worn connecting rod bearings
   - **c.** worn crankshaft main bearings
   - **d.** using a lower viscosity engine oil than specified

18. What would cause an engine to produce a sharp, metallic, rapping sound that originates in the upper portion of the engine and is most pronounced during idle?
   - **a.** piston slap
   - **b.** worn piston pins
   - **c.** worn connecting rod bearings
   - **d.** worn crankshaft main bearings

19. Which of the following engine problems may be indicated by good results from a compression test and a cylinder leakage test but poor results from a cylinder power balance test?
   - **a.** incorrect valve timing
   - **b.** collapsed lifter
   - **c.** burnt intake valve
   - **d.** worn piston rings

20. Which of the following is an expression of how much of the heat formed during the combustion process is available as power from the engine?
   - **a.** mechanical efficiency
   - **b.** engine efficiency
   - **c.** volumetric efficiency
   - **d.** thermal efficiency

CHAPTER 10
Engine Disassembly

LEARNING OUTCOMES

- Prepare an engine for removal.
- Remove an engine from an FWD and an RWD vehicle.
- Describe how to disassemble and inspect an engine.
- Name the three basic cleaning processes.
- Identify the types of cleaning equipment.
- Describe the common ways to repair cylinder head cracks.

Careful diagnostics will determine if a starting or running problem is caused by the engine itself or by one of its systems, such as the ignition or air/fuel system. When the engine is the source of the problem, or its parts are broken or excessively worn, it normally needs to be rebuilt or overhauled. Although some engine repairs can be made with the engine still in the vehicle, most require its removal.

Before removing the engine, clean it and the area around it. Protect the fuel injection and ignition systems while cleaning. Cover them with plastic bags and try not to spray directly at the systems. Also check the service information for the correct procedure for removing the engine from a particular vehicle. Make sure you adhere to all precautions as they vary from model to model.

Complete disassembly and assembly of the engine block and cylinder head are covered in Chapters 11 and 12.

REMOVING AN ENGINE

Make sure you have the tools and equipment required for the job before you begin. In addition to hand tools and some special tools, you will need a crane **(Figure 10–1)** and a jack.

The basic procedures for engine removal vary depending on whether the engine is removed from the bottom of the vehicle or through the hood opening. Many FWD vehicles require removal of the engine from the bottom, whereas most RWD vehicles require the engine to come out from the hood opening. The engine exit point is something to keep in mind while you are disconnecting and removing items in preparation for engine removal.

FIGURE 10–1

To pull an engine out of a vehicle, the chain on the lifting crane is attached to another chain secured to the engine.

OTC Tool and Equipment, Division of SPX Corporation

General Procedures

When removing an engine, setting the vehicle on a frame contact lift is recommended. Make sure to block the wheels so that the vehicle does not move while you are working. Open the hood and put fender covers on both front fenders **(Figure 10–2)**. Once the vehicle is in position, relieve the pressure in the fuel system, using the procedures outlined by the manufacturer.

CUSTOMER CARE

Make sure your hands, shoes, and clothing are clean before getting into a customer's car. Disposable seat and floor coverings should be used to help protect the interior.

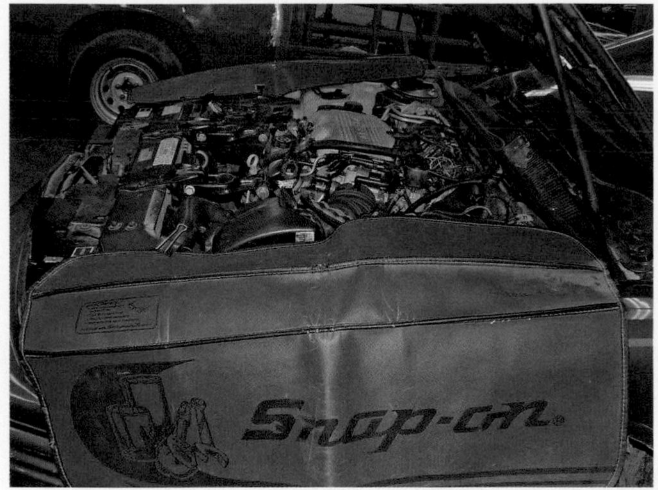

FIGURE 10–2

Before doing anything, put covers on the fenders.

BATTERY Disconnect all negative battery cables and place them away from the battery. Then remove the positive cable and the battery. Remember to install a memory saver before you disconnect the battery to prevent the vehicle's computers and other devices from losing what they have stored in their memory.

HOOD The vehicle's hood may get in the way during engine removal. If the hinges allow the hood to be set straight up above the engine compartment, prop it in that position with a proper hood prop or safely with wood or a broom. If the hood must be removed, mark the location of the hinges on the hood. Then unbolt and remove the hood with the help of someone else, and place it in a safe place. Make sure not to damage the vehicle's paint while doing this.

FLUIDS Drain the engine's oil and its coolant from the radiator and engine block, if possible. To increase the flow of the coolant out of the system, remove the radiator cap. Make sure the engine is cool before opening the coolant drain and before removing the radiator cap. If the transmission will be removed with the engine, drain its fluid.

UNDERBODY CONNECTIONS While you are under the vehicle, disconnect the exhaust system, shift linkage, transmission cooling lines, all electrical connections, vacuum hoses, and clutch linkages from the transmission. If the clutch is hydraulically operated, unbolt the slave cylinder and set it aside, if possible. If this is not possible, disconnect and plug the line to the cylinder **(Figure 10–3)**.

AIR-FUEL SYSTEM Remove the air intake ducts and air cleaner assembly. Disconnect and plug the fuel line at the fuel rail. If the engine is equipped with a return fuel line from the fuel pressure regulator, disconnect that as well **(Figure 10–4)**. Make sure all fuel lines are closed off with pinch pliers or the appropriate plug or cap. Most late-model fuel lines have quick-connect fittings that are separated by squeezing the retainer tabs together and pulling the fitting off the fuel line nipple.

Disconnect all vacuum lines at the engine. Make sure these are labelled before disconnecting them. Late-model vehicles have a VECI decal

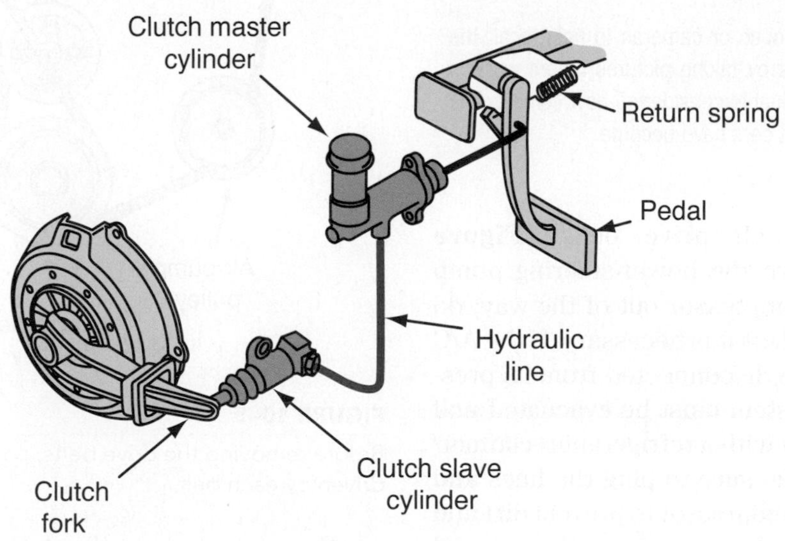

FIGURE 10–3

If the transmission will be removed with the engine, disconnect or remove the clutch slave cylinder.

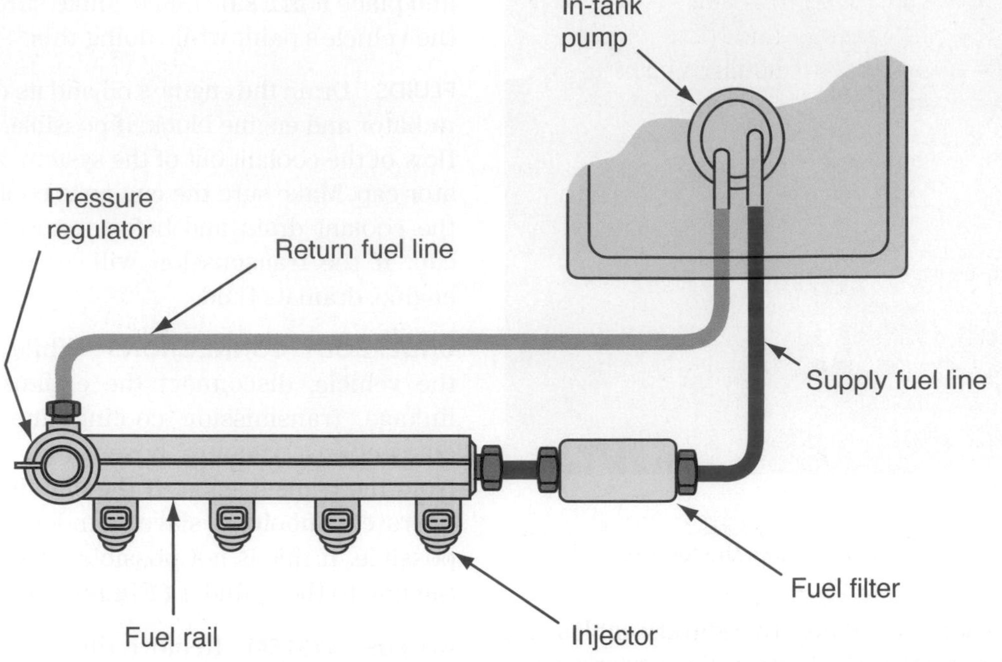

FIGURE 10–4

Disconnect and plug the fuel lines at the fuel rail and the pressure regulator.

under the hood. The diagram and labels will make it easier to reconnect the hoses when the engine is reinstalled.

Now disconnect the throttle linkage at the throttle body and the electrical connector to the throttle position (TP) sensor.

SHOP TALK

Some technicians use cellphones or cameras to help recall the locations of underhood items by taking pictures before work is started. This can be quite valuable considering how complex the underhood systems of current cars have become.

ACCESSORIES Remove all drive belts **(Figure 10–5)**. Unbolt and move the power-steering pump and air-conditioning compressor out of the way; do not disconnect lines unless it is necessary. If the A/C compressor needs to be disconnected from its pressure hoses, first the system must be evacuated and the refrigerant captured with a refrigerant reclaimer/recycling machine. Make sure to plug the lines and the connections at the compressor to prevent dirt and moisture from entering. In some provinces, special certification or an ozone depletion prevention (ODP) card is required by law if the air-conditioning system is required to be opened or serviced in any way.

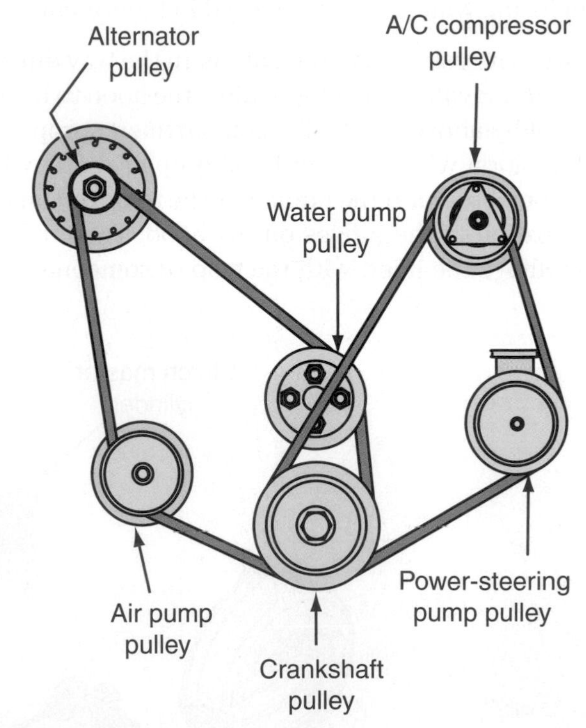

FIGURE 10–5

Before removing the drive belts, pay attention to what is driven by each belt.

Remove or move the A/C compressor bracket, power-steering pump, air pump, and any other components attached to the engine. Disconnect and plug all transmission and oil cooler lines.

ELECTRICAL CONNECTIONS Unplug all electrical wires between the engine and the vehicle. Use masking tape as a label to identify all wires and hoses that are disconnected, and cap all hoses.

SHOP TALK

When removing the fasteners, pay close attention to their size and type. Many brackets used to secure the engine or accessories use several different size fasteners. Mark and organize the fasteners so that their proper location can be easily determined during reassembly. It is a good idea to store the fasteners in several different containers, one for each system or section of the engine.

Some engines have a crankshaft position sensor attached above the flywheel or flex plate. This sensor must be removed before separating the engine from the bell housing. Make sure the engine ground strap is disconnected, preferably at the engine.

COOLING SYSTEM Disconnect the heater inlet and outlet hoses. Then disconnect the upper and lower radiator hoses and any vacuum hose that may get in the way during engine removal. Disconnect the exhaust system; attempt to do this at the exhaust manifold. Now carefully check under the hood to find and remove anything that may interfere with engine removal.

If the radiator is fitted with a fan shroud **(Figure 10–6)**, carefully remove it along with the cooling fan. If the vehicle is equipped with an electric

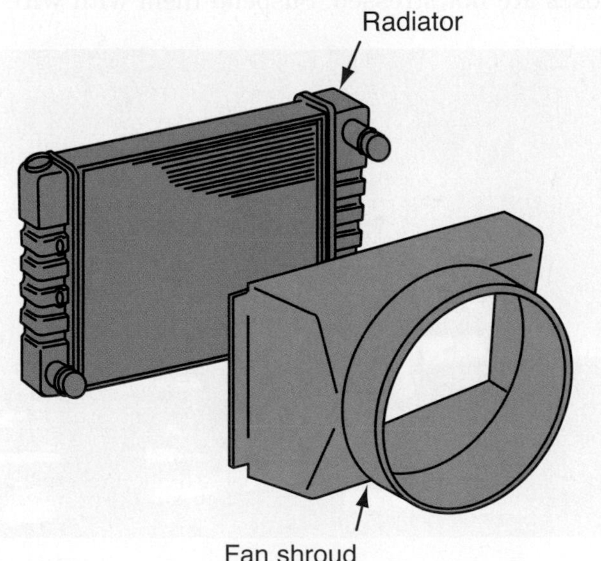

FIGURE 10–6

If the radiator is fitted with a fan shroud, remove it before attempting to remove the radiator.

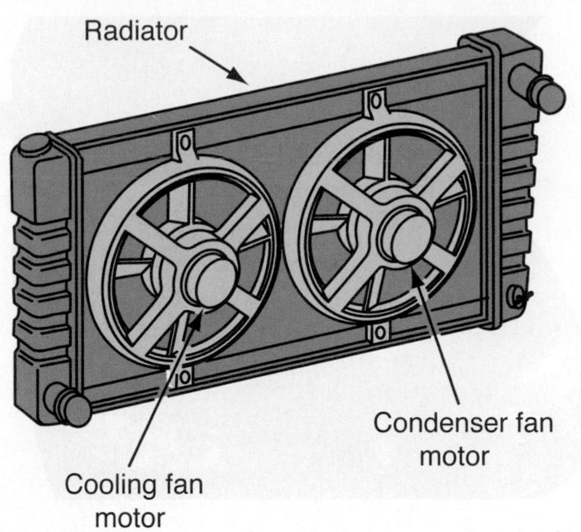

FIGURE 10–7

The electric cooling fans usually can be removed as a unit with the radiator.

cooling fan, disconnect the wiring to the cooling fan. Unbolt and remove the radiator mounting brackets and remove the radiator. Normally, the electric cooling fan assembly and radiator can be removed as a unit **(Figure 10–7)**.

Removing the engine from an RWD vehicle is generally more straightforward than removing one from an FWD model, as there is usually easy access to the cables, wiring, and bell housing bolts. Engines in FWD cars, because of their limited space, can be more difficult to remove as you may need to disassemble or remove large assemblies such as engine cradles, suspension components, brake components, splash shields, or other pieces that would not usually affect RWD engine removal.

FWD Vehicles

Before removing the engine, identify any special tool needs and precautions recommended by the manufacturer. Most often, the engine in an FWD vehicle is removed through the bottom of the vehicle. Special tools may be required to hold the transaxle and/or engine in place as it is being disconnected from the vehicle **(Figure 10–8)**. Always refer to the service information before proceeding to remove the transaxle. You will waste much time and energy if you do not check the information first.

When the engine is removed through the bottom of the vehicle, use an engine cradle and dolly to support the engine. If the manufacturer recommends engine removal through the hood opening, use an engine hoist. Regardless of the method of removal,

FIGURE 10–8

A transverse engine support bar provides the necessary support when removing an engine from an FWD vehicle.

OTC Tool and Equipment, Division of SPX Corporation

the engine and transaxle are usually removed as a unit. The transaxle can be separated from the engine once it has been lifted out of the vehicle.

DRIVE AXLES Using a large breaker bar, loosen and remove the axle shaft hub nuts. It is recommended that these nuts be loosened with the vehicle on the floor and the brakes applied. Doing so makes the job easier and reduces the chance of damaging the constant velocity (CV) joints and wheel bearings.

Raise the vehicle so that you can comfortably work under it. Then remove the wheel and tire assemblies for the front wheels. Tap the splined CV joint shaft with a soft-faced hammer to see if it is loose. Most will come loose with a few taps. Many Ford FWD cars use an interference fit spline at the hub, and you will need a special puller for this type of CV joint **(Figure 10–9)**. The tool pushes the shaft out, and on installation pulls the shaft back into the hub.

Disconnect all suspension and steering parts that need to be removed according to the service information. Index the parts so that wheel alignment will be close after reassembly. Normally, the lower ball joint must be separated from the steering knuckle. The ball joint will either be bolted to the lower control arm or be held into the knuckle with a pinch bolt **(Figure 10–10)**. Once the ball joint is loose, the control arm can be pulled down and the knuckle can be pushed outward to allow the CV joint shaft to slide out of the hub.

The inboard joint either can then be pried out or will slide out. Some transaxles have retaining clips that must be removed before the inner joint can be removed. Others have a flange-type mounting. These

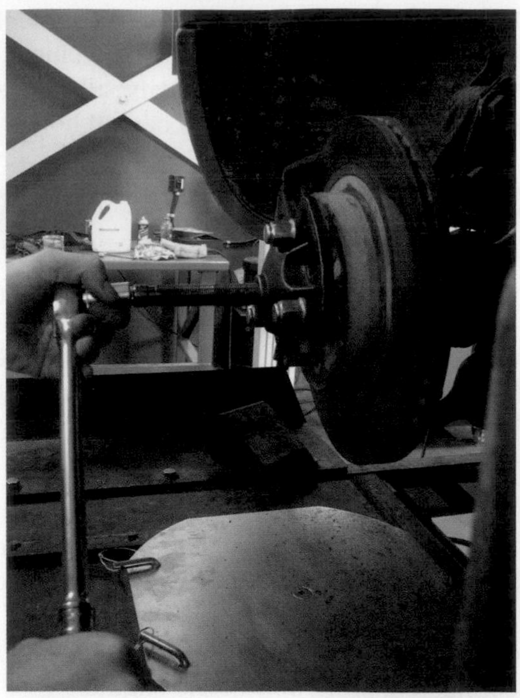

FIGURE 10–9

Sometimes a puller is required to separate the axle shaft from the hub.

Martin Restoule

must be unbolted for removal of the shafts. In some cases, the flange-mounted driveshafts may be left attached to the wheel and hub assembly and unbolted only at the transmission flange. The free end of the shafts should be supported and placed out of the way.

Pull the drive axles out of the transaxle. While removing the axles, make sure the brake lines and hoses are not stressed. Suspend them with wire to

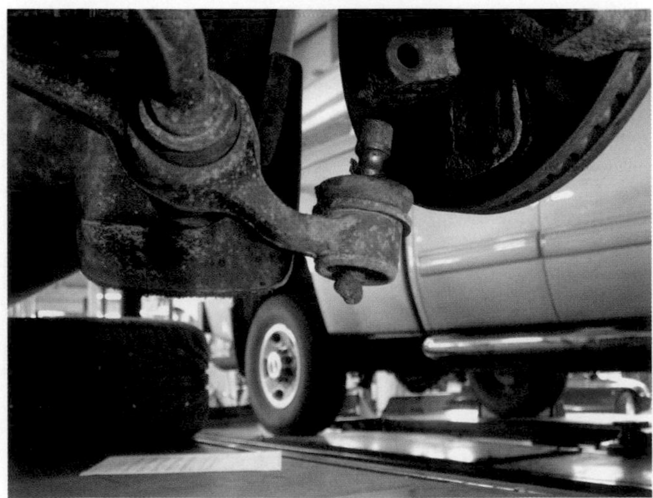

FIGURE 10–10

Before the axle shaft can be removed, the lower ball joint must be separated from the steering knuckle.

Martin Restoule

relieve the weight on the hoses and to keep them out of the way.

TRANSAXLE CONNECTIONS Disconnect all electrical connectors and the speedometer cable or connector at the transaxle. Working under the hood, disconnect the shift linkage or cables and the clutch cable. Now the shift linkages, electrical connections, and speedometer cables should be disconnected.

The exhaust system may also need to be lowered or partially removed. Remove any heat shields that may be in the way.

STARTER Now remove the starter. The starter wiring may be left connected, or you can completely remove the starter from the vehicle to get it totally out of the way. The starter should never be left to hang by the wires attached to it. The weight of the starter can damage the wires or, worse, break the wires and allow the motor to fall, possibly on you or someone else. Always securely support the starter and position it out of the way after you have unbolted it from the engine.

REMOVING THE ENGINE THROUGH THE HOOD OPENING Connect the engine sling or lifting chains to the engine. Use the lifting hooks on the engine **(Figure 10–11)**, or fasten the sling to the points given in the service information system. Connect the sling to the crane and raise the crane just enough to support the engine.

From under the vehicle, remove the cross member. Then remove the mounting bolts for the engine at the engine and transmission mounts. With the transmission jack supporting the transmission, remove the transaxle mounts.

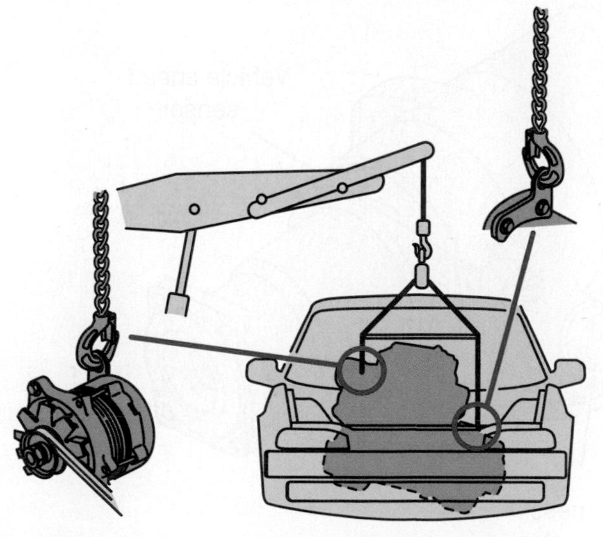

FIGURE 10–11

Some engines are equipped with eye plates to which the hoist can be safely attached.

From under the hood, remove all remaining mounts. Raise the engine slightly to free the engine from the mounts. Then slowly raise the engine from the engine compartment. Guide the engine around all wires and hoses to make sure nothing gets damaged. Once the engine is cleared from the vehicle, prepare to separate it from the transaxle.

REMOVING THE ENGINE FROM UNDER THE VEHICLE Position the engine cradle and dolly under the engine. Adjust the pegs of the cradle so that they fit into the recesses on the bottom of the engine and secure the engine.

Remove all engine and transmission mount bolts. If required, remove the frame member from the vehicle. It may also be necessary to disconnect the steering gear from the frame. Double-check to ensure all wires and hoses are disconnected from the engine. With the transmission jack supporting the transmission, remove the transaxle mounts.

Slowly raise the vehicle, lifting it slightly away from the engine. As the vehicle is lifted, the engine remains on the cradle. During this process, continually check for interference with the engine and the body of the vehicle. Also watch for any wires and hoses that may still be attached to the engine. Once the engine is clear of the vehicle, prepare to separate the engine from the transaxle.

RWD Vehicles

On most RWD vehicles, the engine is removed through the hood opening, requiring the use of an engine hoist. Refer to the service information to determine the proper engine lift points. Never lift the engine by intake manifold bolts. If the transmission is being removed with the engine, it may be easier if you locate the hook of the engine hoist to the chain in such a manner that the engine tips a little toward the transmission. Remember to lift the engine enough to clear the vehicle; it may be necessary to adjust the length of the hoist boom and legs. Before beginning the final steps for removing the engine, check for anything behind and under the engine that should be disconnected.

TRANSMISSION If the engine and transmission must be separated before engine removal, loosen and remove all clutch (bell) housing bolts. If the vehicle has an automatic transmission, remove the torque converter mounting bolts.

If the transmission is being removed with the engine, place a drain pan under the transmission and drain the fluid from the transmission. Once the fluid is out, move the drain pan under the rear of

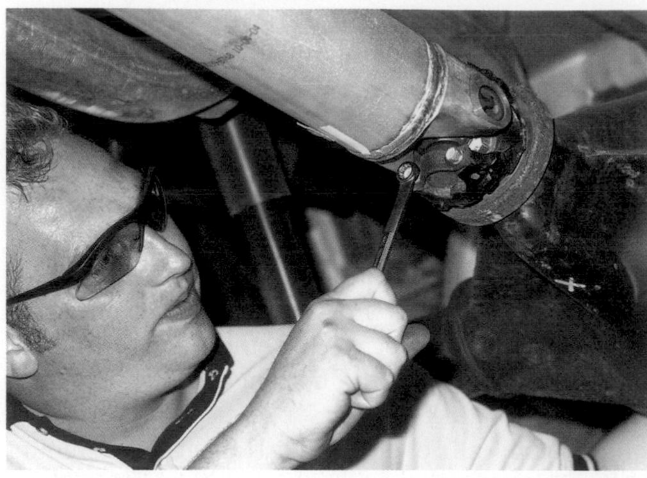

FIGURE 10–12

Mark the alignment of the rear U-joint and the pinion flange/yoke before removing the driveshaft.

the transmission. Use chalk or a paint stick to index the alignment of the rear U-joint and the pinion flange **(Figure 10–12)**. Then remove the driveshaft. Disconnect the parts of the exhaust system that may get in the way. Disconnect all electrical connections **(Figure 10–13)** and the speedometer cable or connector at the transmission. Make sure you place these away from the transmission so that they are not damaged during transmission removal or installation. Disconnect and remove the transmission and clutch linkage. It is best to do this by disconnecting as little as possible.

Use a transmission jack (preferred) or a floor jack to support the transmission **(Figure 10–14)** and unbolt the engine mounts. If the engine is removed with its transmission, the front of the engine must come straight up as the transmission moves away from the bottom of the vehicle. Remove the transmission mount and cross member **(Figure 10–15)**.

REMOVING THE ENGINE To lift the engine out of the vehicle, attach a pulling sling or chain (see Figure 10-1) to the engine. Some engines have eye plates for use in lifting. If they are not available, the sling must be bolted to the engine at the

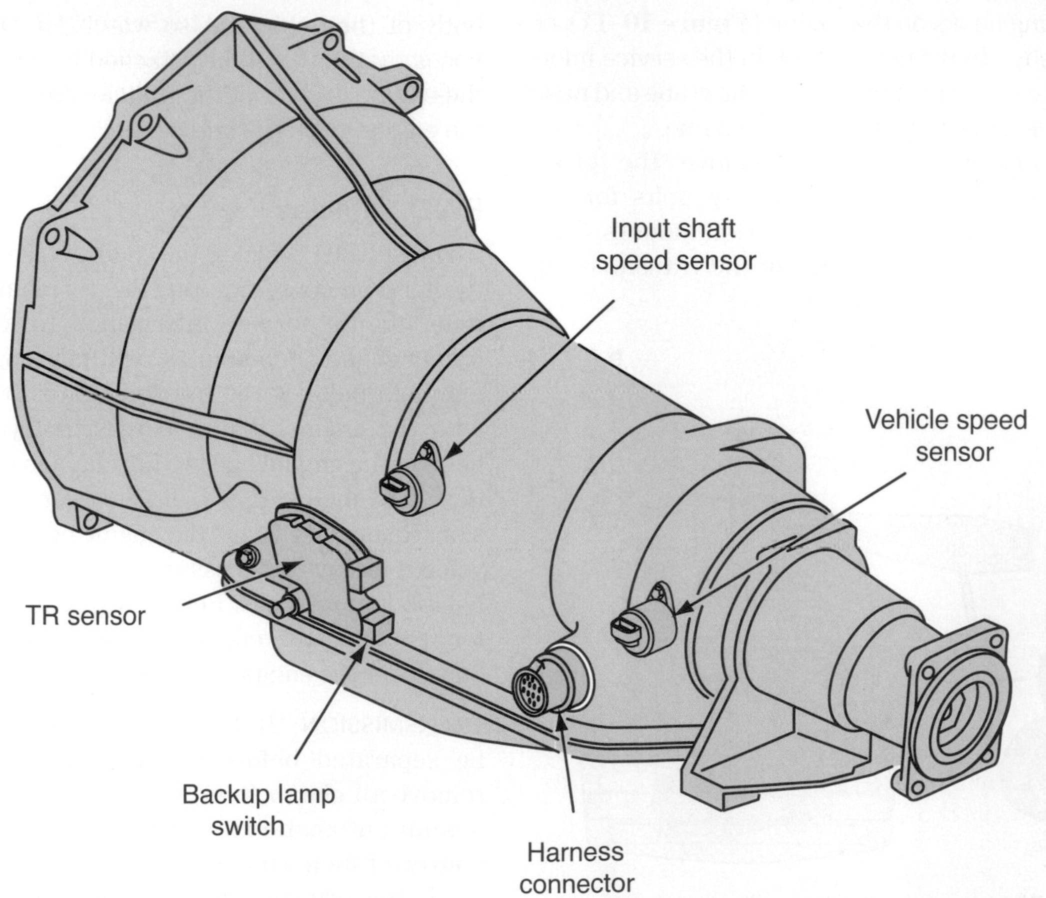

FIGURE 10–13

If the transmission will be removed with the engine, make sure all electrical connections are disconnected. Move the wiring harnesses to the side so that they are not damaged during removal.

FIGURE 10–14

Use a transmission jack to securely support the transmission before removing the motor mounts.

©Cengage Learning 2015

FIGURE 10–15

If the engine is removed with its transmission, remove the transmission mount and cross member.

©Cengage Learning 2015

recommended lifting points. The sling-attaching bolts must be large enough to support the engine and must thread into the block a minimum of 1½ times the bolt diameter.

Centre the boom of the crane directly over the engine, and raise the engine slightly. Make sure the engine is securely fastened to the chain and that nothing else is still attached to the engine. Continue raising the engine while pulling it forward. Make sure that the engine does not bind or damage any compartment component during this procedure. When the engine is high enough to clear the radiator, roll the crane and engine straight out and away from the vehicle.

Lower the engine close to the floor so that it can be transported to the desired location. If the transmission was removed with the engine, remove the bell housing bolts and inspection plate bolts. On vehicles with an automatic transmission, remove the torque converter-to-flex plate bolts. Use a C-clamp or other brace to prevent the torque converter from falling. Also mark the location of the torque converter in relation to the flex plate for reference during installation.

ENGINE DISASSEMBLY AND INSPECTION

It is a good habit to remove the flex plate or flywheel before mounting the engine onto the engine stand. Once the engine is mounted on the stand, it may be difficult, if not impossible, to remove the flex plate or flywheel.

Raise the engine and position it next to an engine stand **(Figure 10–16)**. Mount the engine to the engine stand with bolts. Most stands use a plate with several holes or adjustable arms. The engine must

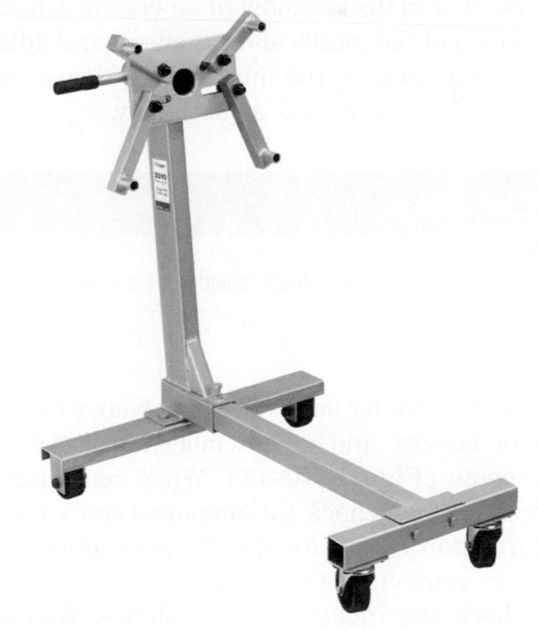

FIGURE 10–16

A typical engine stand.

SPX Corporation, Aftermarket Tool and Equipment Group

be supported by at least four bolts that fit solidly into the engine. The engine should be positioned so that its centre is in the middle of the engine stand's adapter plate to ensure that the engine is not too heavy when rotated on the engine stand.

Once the engine is securely mounted to the engine stand, remove the sling or lifting chain. The engine can now be disassembled and cleaned. Always refer to the service information before you start to disassemble an engine.

Slowly disassemble the engine and visually inspect each part for any signs of damage. Look for excessive wear on the moving parts. Check all parts for signs of overheating, unusual wear, and chips. Look for signs of gasket and seal leakage.

The following engine teardown of both cylinder head and block can be considered typical. Exact details will vary slightly depending on the style and type of engine. For instance, in some engines, the overhead camshaft is mounted directly in the cylinder head. In other engines, it is located in a separate housing that is mounted on the cylinder head.

USING SERVICE INFORMATION

Look up the specific model car and engine prior to disassembling the engine.

Cylinder Head Removal

The first step in disassembly of an engine is usually the removal of the intake and exhaust manifolds. On some in-line engines, the intake and exhaust manifolds are often removed as an assembly.

SHOP TALK

It is important to let an aluminum cylinder head cool completely before removing it.

To start cylinder head removal, remove the valve cover or covers, and disassemble the rocker arm components **(Figure 10–17)**. When removing the rocker assembly, check the manufacturer's manual or information system for specific instructions.

After removing the rocker arm and pushrods, check the rocker area for sludge. Excessive buildup can indicate a poor oil change schedule and is a signal to look for similar wear patterns on other components.

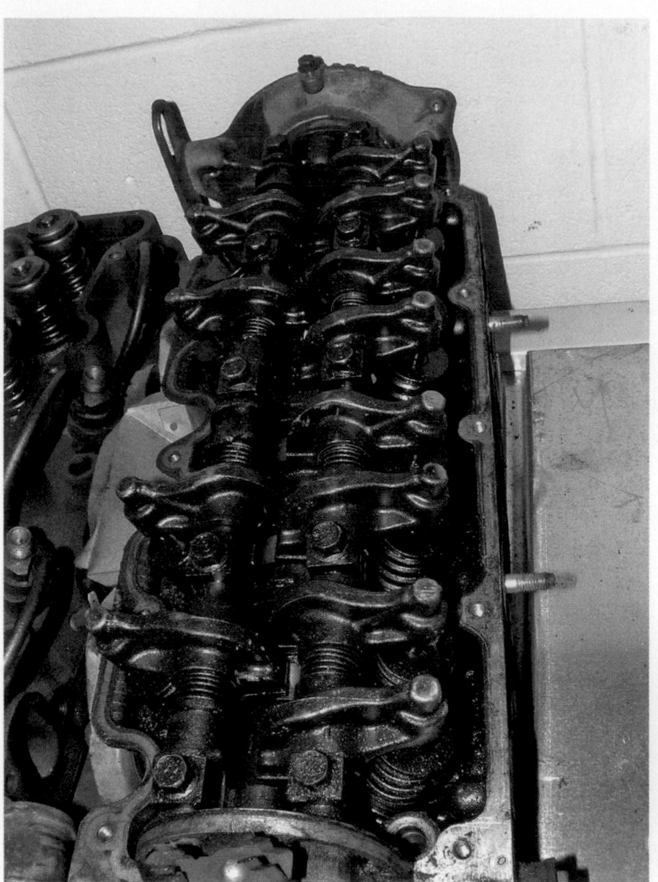

FIGURE 10–17

Remove the valve cover and disassemble the rocker arm components. Check the rocker area for sludge. Keep the rocker arms or rocker arm assemblies in the order they were installed.

On overhead camshaft (OHC) engines, the timing belt cover must be removed. Under the cover is the timing belt or chain and sprockets. In the service information, there will be a description of the type and location of the timing marks on the crankshaft and camshaft sprockets. If possible, rotate the crankshaft to check the alignment of the sprockets. If the shafts are not aligned, make note of this for later reference. The valves will hit the pistons on some engines when the timing belt or chain slips, skips, or breaks. These engines are commonly called **interference engines**. When the valves hit the piston, they will bend. The valves in **freewheeling engines** will not hit the piston when valve timing is off. However, the keys and keyways in the camshaft sprocket may be damaged.

Interference engines typically have a decal on the cam cover that states the belt must be changed at a particular mileage interval. Potential valve and/or piston damage is the reason why timing belt replacement is recommended.

FIGURE 10–18

Two examples of wear that results when a lifter fails to rotate on the camshaft lobe.

TRW, Incorporated

The belt or chain must be removed before removing the cylinder head. Locate and move the belt's tensioner pulley to remove its tension on the belt. Slip the belt off the camshaft and crankshaft sprockets, if possible.

When removing the cylinder head, keep the pushrods and rocker arms or rocker arm assemblies in exact order. Use an organizing tray or label the parts with a felt-tipped marker to keep them together and labelled accurately. This type of organization aids greatly in diagnosing valve-related problems. Also, carefully check the lifters for dished bottoms or scratches, which indicate poor rotation **(Figure 10–18)**.

The cylinder head bolts are loosened one or two turns each, working from the outside of the cylinder head inward **(Figure 10–19)**. This procedure prevents the distortion that can occur if bolts are all loosened at once. The bolts are then removed, again following the outside-to-centre sequence. With the bolts removed, the cylinder head can be lifted off. The cylinder head gasket should be inspected and saved to compare with the new head gasket during reassembly.

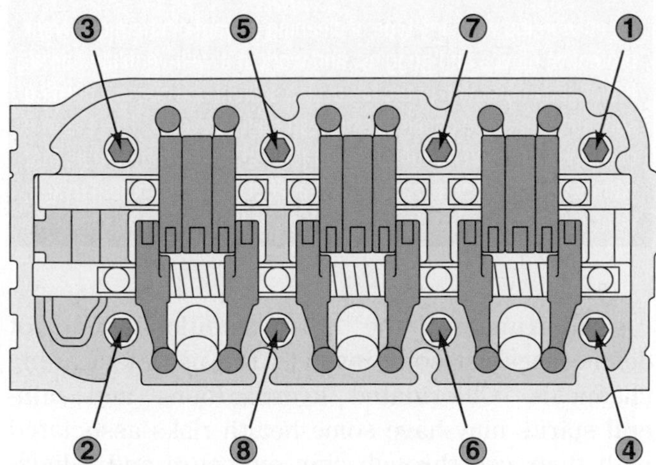

FIGURE 10–19

When loosening cylinder head bolts, follow the sequence given by the manufacturer.

©Cengage Learning 2015

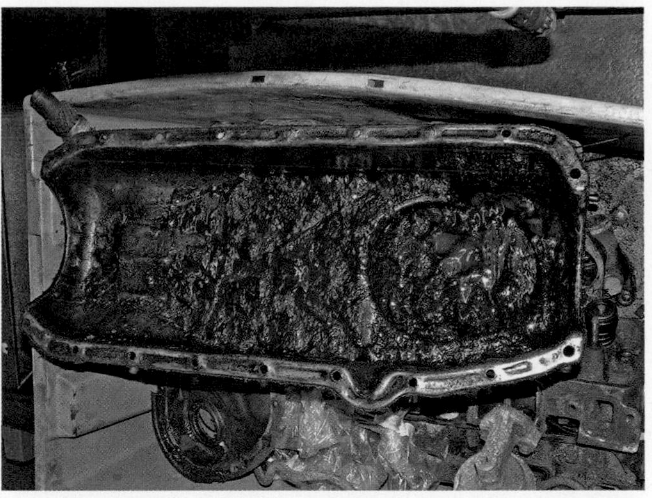

FIGURE 10–20

A major buildup of sludge inside this oil pan.

©Cengage Learning 2015

GO TO ▶ Chapter 12 for the procedures for disassembling and servicing cylinder heads.

The water pump should be removed from the front of the engine. This is usually a simple unbolt and removal procedure. The engine can now be rotated on its stand so that the oil pan is up. Remove the oil pan and look inside for metal shavings and sludge **(Figure 10–20)**. Both of these are indications of problems.

The timing cover can now be removed. On some engines, this may require removing the harmonic balancer or vibration damper. This normally requires a special puller.

GO TO ▶ Chapter 11 for the procedures for disassembling and servicing engine short blocks.

CLEANING ENGINE PARTS

When the block or cylinder head parts have been removed, they must be thoroughly cleaned **(Figure 10–21)**. The cleaning method depends on the component to be cleaned and the type of equipment available. An incorrect cleaning method or agent can often be more harmful than no cleaning at all. For example, using **caustic soda** to clean an aluminum part will dissolve the part.

FIGURE 10–21
From (A) grime to (B) shine.

Only after all components have been thoroughly and properly cleaned can an effective inspection be made or proper machining be done.

Types of Soil Contaminants

Being able to recognize the type of dirt you are to clean will save you time and effort. Basically, there are four types of dirt.

WATER-SOLUBLE SOILS The easiest dirt to clean is **water-soluble soil**, which includes dirt, dust, and mud.

ORGANIC SOILS **Organic soils** contain carbon and cannot be effectively removed with plain water. There are three distinct groupings of organic soils:

- Petroleum by-products derived from crude oil, including tar, road oil, engine oil, gasoline, diesel fuel, grease, and engine oil additives
- By-products of combustion, including carbon, varnish, gum, and sludge **(Figure 10–22)**
- Coatings, including such items as rust-proofing materials, gasket sealers and cements, paints, waxes, and sound-deadener coatings

RUST **Rust** is the result of a chemical reaction that takes place when iron and steel are exposed to oxygen and moisture. Corrosion, like rust, results from a similar chemical reaction between oxygen and metal containing aluminum. If left unchecked, both rust and corrosion can physically destroy metal parts quite rapidly. In addition to metal destruction, rust also acts to insulate and prevent proper heat transfer inside the cooling system.

FIGURE 10–22
Water and by-products of combustion combine with engine oil to form sludge.
Chevron Texaco Lubricants

SCALE When water containing minerals and deposits is heated, suspended minerals and impurities tend to dissolve, settle out, and attach to the surrounding hot metal surfaces. This buildup of minerals and deposits inside the cooling system is known as **scale**. Over a period of time, scale can accumulate to the extent that passages become blocked, cooling efficiency is compromised, and metal parts start to deteriorate.

Cleaning with Chemicals

There are three basic processes for cleaning automotive engine parts. The first process discussed is chemical cleaning.

This method of cleaning uses chemical action to remove dirt, grease, scale, paint, and/or rust.

CAUTION!

When working with any type of cleaning solvent or chemical, wear protective gloves and goggles, and work in a well-ventilated area. Prolonged immersion of the hands in a solvent can cause a burning sensation. In some cases, a skin rash might develop. There is one caution to mention about all manufactured cleaning materials that cannot be overemphasized: Read the labels carefully before mixing or using. Before using, learn how to operate the equipment.

Unfortunately, the most traditional line of defence against soils involves the use of cleaning chemicals. Chlorinated hydrocarbons and mineral spirits may have some health risks associated with their use through skin exposure and inhalation of vapours. Hydrocarbon cleaning solvents are also flammable. The use of water-based nontoxic chemicals can eliminate such risks.

WARNING!

Prior to using any chemical, read through all the information given on the Workplace Hazardous Materials Information System's (WHMIS) material safety data sheets (MSDSs) for that chemical. Become aware of the health hazards presented by the various chemicals.

Hydrocarbon solvents are labelled hazardous or toxic and require special handling and disposal procedures. The makers of many water-based cleaning solutions claim that their products are biodegradable. Once the cleaning solution has become contaminated with grease and grime, it too becomes a hazardous or toxic waste that can be subject to the same disposal rules as a hydrocarbon solvent.

Some manufacturers offer waste-handling and solvent-recycling services. The old solvent is recycled by a distillation process to separate the sludge and contaminants. The solvent is then returned to service and the contaminants disposed of. Independent services for maintaining hot tanks and spray washers are also available.

CHEMICAL CLEANING MACHINES *Parts Washers.* Parts washers (often called **solvent tanks**) are one of the most widely used and inexpensive methods of removing grease, oil, and dirt from the metal surfaces of a seemingly infinite variety of automotive components and engine parts. A typical washer setup **(Figure 10–23)** might consist of a tank to hold a given volume of cleaning solution and some method of applying the solution. These methods include soaking, soaking and agitation, solvent streams, and spray gun applicators.

Soak Tanks. There are two types of soak tanks: cold and hot. Cold soak tanks are commonly used to clean carburetors, throttle bodies, and aluminum parts. A typical cold soak unit consists of a tank to hold the cleaner and a basket to hold the parts to be cleaned. After soaking with or without gentle agitation is complete, the parts are removed, flushed with water, and blown dry with compressed air.

Cleaning time is short, about 20 to 30 minutes, when the chemical cleaner is new. The time becomes progressively longer as the chemical ages. Agitation by raising and lowering the basket (usually done by hand) will reduce the soak period to about 10 minutes. Some more-elaborate tanks are agitated automatically.

FIGURE 10–23

A typical parts washer.

Martin Restoule

Hot soak tanks are actually heated cold tanks. The source of heat is either electricity, natural gas, or propane. The solution inside the hot tanks usually ranges from 71°C to 93°C (160°F to 200°F). Most tanks are generally large enough to hold an entire engine block and its related parts.

Hot tanks use a simple immersion process that relies on a heated chemical to lift the grease and grime off the surface. Liquid or parts agitation may also be used to speed up the job. Agitation helps shake the grime loose and also helps the liquid penetrate blind passageways and crevices in the part **(Figure 10–24)**. Generally speaking, it takes one to several hours to soak most parts clean.

Hot Spray Tanks. Because of the *Canadian Environmental Protection Act (CEPA)* regulations against open steam cleaning, the spray tank has become more popular. The hot spray tank resembles a large automatic dishwasher and is designed to remove organic and rust soils from a variety of automotive parts **(Figure 10–25)**. In addition to parts

FIGURE 10–24

A hot soak tank.

FIGURE 10–25

A compact hot spray tank.

being bathed and soaked as in the hot soak method, spray washers add the benefit of moderate pressure cleaning.

Using a hot jet-spray washer can cut cleaning time to less than 10 minutes. Citrus-based non-toxic detergents or chemical-based strong soap solutions can be used as a cleaning agent. The speed of this system, along with lower operating costs, makes it popular with many machine shop owners.

SHOP TALK

Caustic soda, also known as sodium hydroxide, has been used as a cleaner and can be a very dangerous irritant to the eyes, skin, and mucous membranes. These chemicals should be used and handled with care. Because of the accumulation of heavy metals, it is considered a hazardous waste material and must be disposed of in accordance with CEPA guidelines.

Spray washers are often used to pre-clean engine parts prior to disassembly. A pass-through spray washer is fully automatic once the parts have been loaded, and the cabinet prevents the runoff from going down the drain or onto the ground (which is not permitted in many areas because of local waste disposal regulations). Spray washers are also useful for post-machining cleanup to remove machine oils and metal chips.

Thermal Cleaning

The second basic process for cleaning engine parts is thermal cleaning. This process relies on heat to bake off or oxidize dirt and other contaminants.

Thermal cleaning ovens **(Figure 10–26)**, especially the pyrolytic type, have become increasingly popular. The main advantage of thermal cleaning is a total reduction of all oils and grease on and in blocks, heads, and other parts. The high temperature inside the oven (generally 343°C to 426°C [650°F to 800°F]) oxidizes all the grease and oil, leaving behind a dry, powdery ash on the parts. The ash must then be removed by shot blasting or washing. The parts come out dry, which makes subsequent cleanup with shot blast or glass beads easier because the shot will not stick.

SHOP TALK

A slow cooling rate is recommended to prevent distortion that could be caused by unequal cooling rates within complex castings.

FIGURE 10–26

A cleaning furnace.

Pollution Control Products Co.

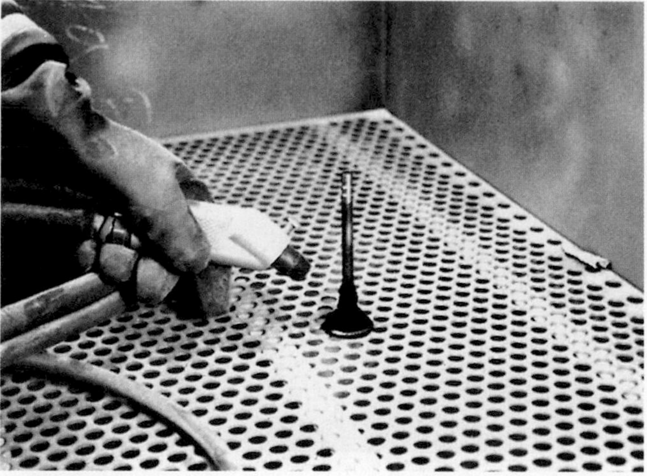

FIGURE 10–27

Using a blast nozzle to clean the back side of a valve.

One of the major attractions of cleaning ovens is that they offer a more environmentally acceptable process than chemical cleaning. But, although there is no solvent or sludge to worry about with an oven, the ash residue that comes off the cleaned parts must still be handled according to local disposal regulations.

The maintenance procedure given in the owner's manual must be followed if the ovens are to operate properly.

Abrasive Cleaners

The third process used to clean engine parts involves the use of abrasives. Most abrasive cleaning machines are used in conjunction with other cleaning processes rather than as a primary cleaning process itself. Parts must be dry and grease-free when they go into an abrasive blast machine. Otherwise, the shot or beads will stick.

ABRASIVE CLEANING METHODS *Abrasive Blaster.* Airless shot and grit blasters are best used on parts that will be machined after they have been cleaned **(Figure 10–27)**. Two basic types of media are available: shot and grit. Shot is round; grit is angular in shape. Steel shot and glass beads are used primarily for cleaning operations where etching or material removal is not desired. Steel shot and glass beads are also used for peening the surfaces of certain parts. **Peening** is a process of hammering on the surface of a part. This packs the molecules tightly, thereby increasing the part's resistance to fatigue and stress.

Grit is used primarily for aggressive cleaning jobs or where the surface of the material needs to be etched to improve paint adhesion. Because grit cuts into the surface as it cleans, it removes dirt and scale faster than shot blasting or glass beading. But it also removes metal, leading to some change in tolerances. The beneficial effect of grit blasting is that it roughens the surface, leaving a matte finish to which paint or other surface treatments will stick better than a peened or polished surface. However, grit blasting is an abrasive process that chews out pits in the surface, into which pollutants and blast residue can settle. This leads to stress corrosion unless the surface is painted or treated. The tiny crevices also focus surface stresses in the metal, which can lead to cracking in highly loaded parts. Because of that, grit should never be used for peening.

The type of media used for a given job depends on the job itself and the type of equipment. Steel shot is normally used with airless wheel blast equipment, which hurls the shot at the part with the centrifugal force of the spinning wheel. Used with air blast equipment, glass beads are blown through a nozzle by compressed air.

Parts Tumbler. A cleaning alternative that can save considerable labour when cleaning small parts such as engine valves is a tumbler. Various cleaning media can be used in a tumbler to scrub the parts clean. This saves considerable hand labour and eliminates dust. In some tumblers, all parts are rotated and tilted at the same time.

Vibratory Cleaning. Shakers, as they are frequently called, use a vibrating tub filled with ceramic steel, porcelain, or aluminum abrasive to scrub parts

FIGURE 10–28

A vibratory parts cleaner.

C & M TOPLINE

FIGURE 10–30

Using a power scraper pad will prevent any metal from being removed.

Martin Restoule

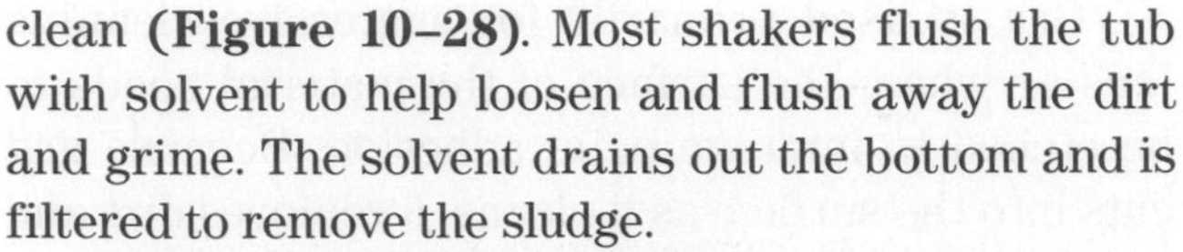

clean **(Figure 10–28)**. Most shakers flush the tub with solvent to help loosen and flush away the dirt and grime. The solvent drains out the bottom and is filtered to remove the sludge.

Cleaning by Hand. Some hand cleaning is inevitable. Regardless of the cleaning process used, it is usually necessary to remove gallery plugs and hand clean the oil galleries **(Figure 10–29)**. Another often-neglected area is between the heat shield and the bottom of the intake manifold, where carbon and oil deposits collect. This shield should be removed before cleaning the manifold. Any residual dirt left in the engine after cleaning can lead to failure. Therefore, proper cleaning of all engine components is vital in the rebuilding process. Remove any surface irregularities with very fine, emery cloth. Make sure

FIGURE 10–29

It is often necessary to remove the gallery plugs and hand clean the oil galleries.

Martin Restoule

to keep any dirt out of the cylinder bores. The special power scraper pad shown in action in **Figure 10–30** is guaranteed not to remove any metal.

Carbon can be removed from parts using a twist-type wire brush driven by an electric or air drill motor. Using brushes can often be a time-consuming job. Some shops use a wire brush in addition to another cleaning method. Moving the drill motor in a light circular motion against the carbon helps to crack and dislodge the carbon for easier wire brush cleaning.

ALTERNATIVE CLEANING METHODS Three of the most popular alternatives to traditional chemical cleaning systems are ultrasonic cleaning, citrus chemicals, and salt baths.

Ultrasonic Cleaning. This cleaning process has been used for a number of years to clean small parts such as jewellery, dentures, and medical instruments. Recently, however, the use of larger ultrasonic units has expanded into small engine parts cleaning. **Ultrasonic cleaning (Figure 10–31)** utilizes high-frequency sound waves to create microscopic bubbles that burst into energy to loosen soil from parts. Because the tiny bubbles do all the work, the chemical content of the cleaning solution is minimized, making waste disposal less of a problem.

Citrus Chemicals. Some chemical producers have developed citrus-based cleaning chemicals as a replacement for the hazardous solvent and alkaline-based chemicals. Because of their citrus origin, these chemicals are safer to handle and easier to dispose of, and they even smell good.

FIGURE 10–31

An ultrasonic parts cleaner.

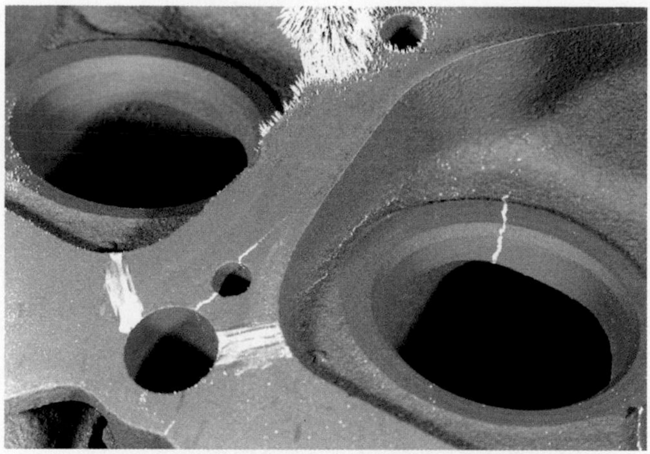

FIGURE 10–32

Examples of stress cracks.

Salt Bath. The **salt bath** is a unique process that uses high-temperature molten salt to dissolve organic materials, including carbon, grease, oil, dirt, paint, and some gaskets. For cast iron and steel, the salt bath operates at about 371°C to 454°C (700°F to 850°F). For aluminum or combinations of aluminum and iron, a different salt solution is used at a lower temperature (about 315°C [600°F]). The contaminants precipitate out of the solution and sink to the bottom of the tank, from which they must be removed periodically. The salt bath itself lasts indefinitely as long as the salt is maintained properly. Like a hot tank, the temperature of the salt bath is maintained continuously.

CRACK DETECTION AND REPAIR

Once the engine parts have been cleaned and given a thorough visual inspection, actual repair work begins. If cracks in the metal casting were discovered during the inspection, they should be repaired or the part replaced.

Cracks in metal castings are the result of stress or strain in a section of the casting. This stress or strain finds a weak point in that section of the casting and causes it to distort or separate at that point **(Figure 10–32)**. Such stresses or strains in castings can develop from the following:

- Pressure or temperature changes during the casting procedure may cause internal material structure defects, inclusion, or voids.
- Fatigue may result from fluctuating or repeated stress cycles. It might begin as a small crack and progress to a larger one under the action of the stress.
- Flexing of the metal may result due to its lack of rigidity.
- Impact damage may occur by a solid, hard object hitting a component.
- Constant impacting of a valve against a hardened seat may produce vibrations that could possibly lead to fracturing a thin-walled casting.
- Chilling of a hot engine by a sudden rush of cold water or air over the surface may happen.
- Excessive overheating is possible due to improper operation of an engine system.

Cracks can be found by visual inspection; however, many are not easily seen. Therefore, engine rebuilders use special equipment to detect cracks, especially if there is reason to suspect a crack.

Magnetic particle inspection (MPI) uses a permanent magnet or an electromagnet to create a magnetic field in a cast-iron unit **(Figure 10–33)**. When the legs of the detector tool are placed on the metal, the magnetic field travels through the metal. Iron filings are sprinkled on the surface to detect a secondary magnetic field resulting from a crack **(Figure 10–34)**. Because the secondary magnetic field will not form if the crack is in the same direction as the magnet, the magnet must be rotated and the metal checked in both directions.

Another common way to detect cracks is called magnetic fluorescent crack detection. This method does not rely on a magnet; rather, it uses a magnetic, fluorescent paste. The paste is spread over an area. The magnetic paste will flow over the metal. A black light is used to look at the paste. Wherever the paste has created a line, it has filled in a crack. This method is great for finding small cracks.

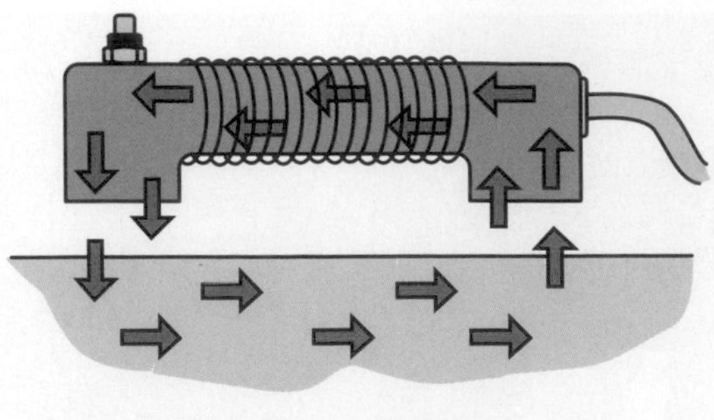

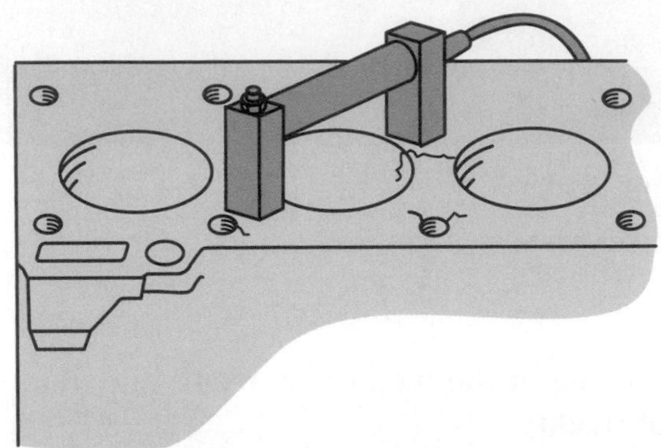

FIGURE 10–33

MPI testing passes a magnetic field through the iron item being checked.

Dye Penetrant

Another common way to detect cracks is by using three separate chemicals: penetrant, cleaner, and developer. The part to be checked must be clean and dry. This check must be done according to the following sequence:

1. Spray or brush the penetrant onto the surface.
2. Wait five minutes.
3. Spray the cleaner onto a clean cloth.
4. Wipe off all visible penetrant.

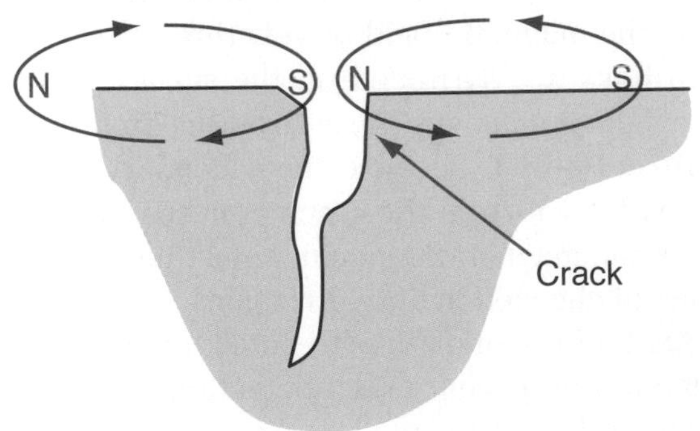

FIGURE 10–34

A crack causes two opposing magnetic poles to form on each side. The iron filings used with the magnet will show these fields.

FIGURE 10–35

Cracks appear as red lines when a dye penetrant is used.

LOCK-N-STITCH Inc.

5. Spray the developer on the tested area.
6. Wait until the developer is totally dry.
7. Inspect the area. Cracks will appear as a red line **(Figure 10–35)**.

No matter what caused the crack, it is important to relieve the stress at the point of distortion or cracking and then add more metal and close the crack. This can be accomplished by the cold process of pinning or the hot process of welding.

Furnace Welding Crack Repairs

Furnace welding is considered by many people to be the best way to repair cracks in a cast-iron head. By preheating the entire casting, the problem of stress cracks forming during the cooling-off period is eliminated. Heat welding, however, requires a good heat source and proficient welding skills. As a rule, this repair is conducted only by a specialist.

Repairing Aluminum Heads

Aluminum heads have become popular in recent years, primarily because of the weight savings they offer. However, there are many problems associated with an aluminum cylinder head. The typical shop is most likely to encounter the following problems:

- Cracks in the aluminum between the valve seat rings **(Figure 10–36)**. These cracks, usually quite small, require close inspection to find. They very seldom leak and can be closed by a light peening. Some shops make no repairs to them.
- Bottomside cracks coming from the coolant passages. These cracks can be repaired by veeing

FIGURE 10–36

Cracks between the centre two valve seats are common with aluminum heads.

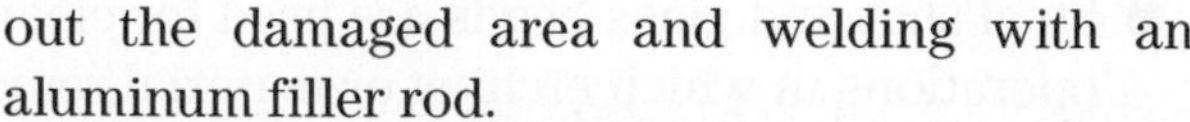

out the damaged area and welding with an aluminum filler rod.

- Topside cracks across the main oil artery. These cracks, although not too common, are usually very visible **(Figure 10–37)**. Most authorities recommend replacing the head completely if such a crack is found. The length of time required to make the repair is not reasonable. The labour cost is not worth the risk of possible failure and is higher than the cost of purchasing an uncracked core.
- Detonation damage can occur on any cylinder **(Figure 10–38)**. Repairs can be made by welding and freehand machining with a rotary burr in a die grinder.
- Meltdown damage is a somewhat common occurrence on the high-swirl combustion chamber heads. Again, repairs can be made by welding and freehand machining.
- Coolant-related metal erosion. If damage around coolant passages is excessive, if the side of any valve seat has been exposed, or if the combustion chamber shows erosion, the head must be repaired or replaced **(Figure 10–39)**. Coolant erosion can be easily fixed by welding and resurfacing.

FIGURE 10–37

The topside oil artery crack appeared when an oxyacetylene flame was passed over the casting. Carbon in the flame was trapped in the crack, highlighting it.

FIGURE 10–38

The effect of detonation on a combustion chamber.

Tungsten inert gas (TIG) welding or **gas tungsten arc welding (GTAW)** is the preferred repair technique for aluminum heads **(Figure 10–40)**. Welding aluminum is often considered difficult because it welds differently than iron or steel. When exposed to air, aluminum forms an oxide coating on the surface that helps protect the metal against further corrosion. The oxide layer makes welding difficult because it interferes with fusing and weakens

FIGURE 10–39

Severe coolant-related damage can sometimes be repaired, but normally a damaged cylinder head is replaced.

FIGURE 10–40

TIG-welded aluminum head repair.

the weld. Cleaning the surface can remove the oxide. However, as soon as the metal is heated, oxide reforms (unless the weld is bathed in a constant supply of inert gas).

KEY TERMS

Caustic soda (p. 267)
Freewheeling engines (p. 266)
Gas tungsten arc welding (GTAW) (p. 275)
Interference engine (p. 266)
Organic soils (p. 268)
Peening (p. 271)
Rust (p. 268)
Salt bath (p. 273)
Scale (p. 268)
Solvent tanks (p. 269)
Tungsten inert gas (TIG) welding (p. 275)
Ultrasonic cleaning (p. 272)
Water-soluble soil (p. 268)

SUMMARY

- When preparing an engine for removal and disassembly, always follow the specific service information system procedures for the particular vehicle being worked on.
- A hoist and chain or crane are needed to lift an engine out of its compartment. Mount the engine to an engine stand with a minimum of four bolts, or set it securely on blocks.
- Although an engine teardown of both the cylinder head and block is a relatively standard procedure, exact details vary among engine types and styles. The vehicle's service information should be considered to be the final word.
- An understanding of specific soil types can save time and effort during the engine cleaning process. The main categories of contaminants include water-soluble and organic soils, rust, and scale.
- Protective gloves and goggles should be worn when working with any type of cleaning solvent or chemical. Read the label carefully before using, as well as all of the information provided on material safety data sheets.
- Parts washers, or solvent tanks, are a popular and inexpensive means of cleaning the metal surfaces of many automotive components and engine parts. Regardless of the type of solvent used, it usually requires some brushing, scraping, or agitation to increase the cleaning effectiveness.
- Alternatives to chemical cleaning have emerged in recent years, including thermal cleaning, ultrasonic cleaning, salt baths, and citrus chemical cleaning.
- Steel shot and glass beads are used for cleaning operations in which etching or material removal is not desired. Grit, the other type of abrasive blaster, is used for more aggressive cleaning jobs.
- Some degree of manual cleaning is necessary in any engine rebuilding job. Very fine, abrasive paper should be used to remove surface irregularities. A twist-type wire brush driven by an electric or air drill motor is also helpful, though it can be time-consuming to work with.
- TIG welding is the preferred repair technique for aluminum cylinder heads. Because it reacts differently to heat than iron or steel, aluminum is considered a challenge to repair by welding.

REVIEW QUESTIONS

1. What precautions should be taken when cleaning an engine prior to its removal?
2. What should be worn when working with any type of cleaning solvent or chemical?
3. What is the best way to repair cracks in a cast-iron head?
4. What are the typical steps for removing an engine from a rear-wheel-drive vehicle?
5. What is the most common method of removing an engine from a front-wheel-drive vehicle?
   a. Lower the engine and transaxle as an assembly with the cradle.
   b. Separate the transaxle from the engine and lower the engine with the cradle.
   c. Raise the engine and transaxle as an assembly with a crane.
   d. Raise only the engine with a crane.

6. What is usually the first step in disassembling an engine?
   a. removing the intake and exhaust manifold(s)
   b. disassembling the cylinder heads
   c. removing the piston assemblies
   d. removing the oil pan and crankshaft
7. What is the best bay arrangement for lowering an engine from a front-wheel-drive vehicle?
   a. a flat floor bay equipped with a crane
   b. a flat floor bay equipped with a floor jack and stands
   c. a drive-on hoist bay
   d. a frame-contact hoist bay
8. When removing an engine from a vehicle, after lifting the hood and installing fender covers, what is the next step?
   a. Pressure-test the cooling system.
   b. Relieve the pressure in the fuel system.
   c. Drain all engine fluids.
   d. Remove all engine accessories.
9. What is the recommended procedure for removing a cylinder head?
   a. Use an impact wrench to swiftly remove each head bolt, starting in the centre and working outward.
   b. Use an impact wrench to swiftly remove each head bolt, starting from the outside of the cylinder head and working inward.
   c. Using a breaker bar, loosen each head bolt one or two turns, working from the centre outward.
   d. Using a breaker bar, loosen each head bolt one or two turns, starting from the outside of the cylinder head and working inward.
10. What is the buildup of minerals and deposits inside the cooling system called?
    a. organic soil   b. scale
    c. rust   d. grime
11. Which of the following is true of hydrocarbon solvents?
    a. They are toxic.
    b. They are nonflammable.
    c. They are nontoxic.
    d. They are biodegradable.
12. Which of the following abrasive cleaning methods can be used to etch the surface of a material?
    a. shot media blasting
    b. grit media blasting
    c. steel shot blasting
    d. glass bead blasting
13. Which of the following crack repair methods can only be used on aluminum engine components?
    a. furnace welding
    b. pinning
    c. metal inert gas (MIG) welding
    d. tungsten inert gas (TIG) welding
14. Which cleaning method uses high-frequency sound waves to create microscopic bubbles that loosen dirt from parts?
    a. ultrasonic   b. salt bath
    c. thermal   d. caustic
15. Which of the following parts cleaning machines can be used to pre-clean engine components prior to disassembly?
    a. pyrolytic ovens
    b. abrasive blasters
    c. spray washers
    d. vibratory cleaners
16. Which of the following bolts do not require a special loosening procedure to prevent distortion?
    a. cylinder head bolts
    b. main bearing cap bolts
    c. camshaft bearing cap bolts
    d. connecting rod cap bolts
17. What is the minimum number of bolts used to mount an engine to an engine stand?
    a. two   b. three
    c. four   d. five
18. What is the next step in engine removal once the vehicle is in place and the fuel system is depressurized?
    a. Remove the exhaust system.
    b. Raise the vehicle.
    c. Disconnect and remove the starter.
    d. Disconnect the negative battery cable.
19. When removing an engine from a vehicle, what should be done only after the crane or supports are in place?
    a. removal of the drive axles or driveshaft
    b. removal of the torque converter mounting bolts
    c. removal of the engine mount bolts
    d. removal of the shift linkage and cables
20. What method of repair should be used on an aluminum cylinder head with a crack coming from a coolant passage on the deck surface?
    a. Vee out the damaged area and weld with an aluminum filler rod.
    b. Lightly peen over the crack.
    c. Weld over the area and freehand machine with a rotary burr.
    d. Drill and pin the crack with threaded rod.

CHAPTER 11
Lower End Theory and Service

LEARNING OUTCOMES

- Disassemble and inspect an engine's cylinder block.
- List the parts that make up a short block, and briefly describe their operation.
- Describe the major service and rebuilding procedures performed on cylinder blocks.
- Describe the purpose, operation, and location of the camshaft.
- Describe the four types of camshaft drives.
- Inspect the camshaft and timing components.
- Describe how to install a camshaft and its bearings.
- Explain crankshaft construction, inspection, and rebuilding procedures.
- Explain the function of engine bearings, flywheels, and harmonic balancers.
- Explain the common service and assembly techniques used in connecting rod and piston servicing.
- Explain the purpose and design of the different types of piston rings.
- Describe the procedure for installing pistons in their cylinder bores.
- Inspect, service, and install an oil pump.

The lower end of an engine is the cylinder block. This includes the block, crankshaft, bearings, pistons, piston rings, and oil pump (and sometimes the camshaft). Many of these parts are made by casting or forging. To cast is to form molten metal into a particular shape by pouring it into a mould. To **forge** is to form metal into a shape by heating it and pressing it into a mould. Some forging is done with cold metals. These manufactured parts then undergo a number of machining operations, such as the following:

- The top of the block must be smooth so that the cylinder head can seal it.
- The bottom of the block is machined to allow for proper sealing of the oil pan.
- The cylinder bores are machined smooth and have the correct diameter to accept the pistons.
- The main bearing area of the block must be cut in a series of bores that are in a straight line and match the diameter of the journals on the crankshaft. Camshaft bearing bores must also be aligned.

When there is a major engine failure, shops either rebuild or replace the engine. Most often the **short block** is repaired or replaced as an assembly **(Figure 11–1)**. A basic short block assembly consists of a cylinder block, crankshaft, crankshaft bearings, connecting rods, pistons and rings, and oil gallery and core plugs. Parts related to the short block, but not necessarily included with it, are the flywheel and **harmonic balancer**. A short block may also include the engine's camshaft and timing gear. A **long block** is basically a short block with cylinder heads.

FIGURE 11–1

A cutaway showing the fit of the piston assemblies and crankshaft in an engine block.

SHORT BLOCK DISASSEMBLY

This chapter begins with the assumption that the engine is on an engine stand, and the cylinder head(s) are removed from the cylinder block. If the oil pan and water pump are still attached to the block, remove them before proceeding.

FIGURE 11–2

The construction of a harmonic balancer.

FIGURE 11–3

The timing gears and chain on a typical OHV engine.

Remove the harmonic balancer, also called a vibration damper. The harmonic balancer has an inner hub bonded with rubber to an outer ring. Its purpose is to absorb the torsional vibrations of the crankshaft **(Figure 11–2)**. Removal of the balancer often requires the correct type puller. If an unsuitable puller is used, it is likely the rubber bonding will be damaged, which would make the balancer useless, causing engine vibrations and crankshaft damage. Once the balancer is removed, carefully check the rubber for tears or other damage. If there are any faults, replace the balancer.

On OHV engines, remove the timing cover. Under the cover are the timing gears. The crankshaft's timing sprocket has a slight interference fit and can normally be pulled off by hand, as can the camshaft sprockets. The camshaft sprocket and chain must be removed with the crankshaft sprocket **(Figure 11–3)**.

Before removing the gears and chain, check the deflection of the timing chain. Depress the chain at its midway point between the gears, and measure the amount that the chain can be deflected. If the deflection exceeds specifications, the timing chain and gears should be replaced.

Loosen the camshaft sprocket, and pull the timing gears and chain from the engine. Be careful not to lose the keys in each shaft or any shims that may be behind the sprockets.

Often, the timing chain assembly has tensioners and guides. Normally, the timing gear assembly is replaced during an engine overhaul, as are its tensioners and guides.

The timing belt or chain on OHC engines was already loosened before the cylinder head was removed. Remove it and unbolt the timing chain or belt cover, and gently pry it away from the block and cylinder head. Remove the crankshaft position sensor, timing chain guide, chain tensioner slipper, and chain. On some engines, the oil pump is driven by the crankshaft at the front of the engine **(Figure 11–4)**. The pump should be unbolted and removed. Rotate the crankshaft counter-clockwise to align the timing marks on the oil pump sprocket with the mark on the oil pump. Remove the bolt for the sprocket and the chain tensioner plate and spring. Then remove the oil pump sprockets and chain. Then remove the oil screen from the oil pump.

If the lifters have not been removed, do so now. Place them on a bench in the order they were removed. Carefully pull the camshaft out of the block. Support the camshaft to avoid dragging its lobes over the surfaces of the camshaft bearings. This can damage the bearings and lobes.

Cylinder Block Disassembly

Rotate the engine on its stand so that the bottom is facing up. Remove the oil pan if it was not previously removed. Then remove the oil pump. Be careful not

FIGURE 11–4

On most engines, the oil pump is driven by the crankshaft at the front of the engine.

to lose the driveshaft while pulling the pump off the engine.

If the engine has balance shafts, check the thrust clearance of the shafts before removing the assembly. Set a dial indicator so that it can read the back-and-forth movement (end play) of the shaft. Measure the total distance that the shaft is able to move in the housing. Compare that reading to specifications. If the reading is more than the specified maximum, the balance shaft housing and bearings should be replaced. Unbolt the housing, following the sequence given in the specifications **(Figure 11–5)**.

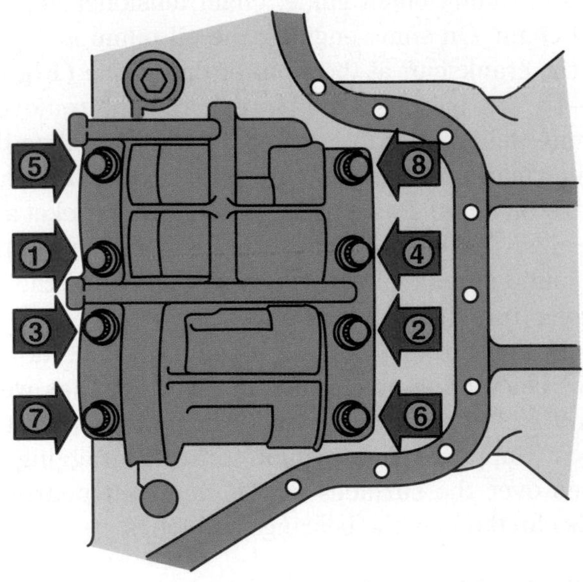

FIGURE 11–5

The balance shaft housing must be unbolted in the prescribed order to prevent shaft and housing warpage.

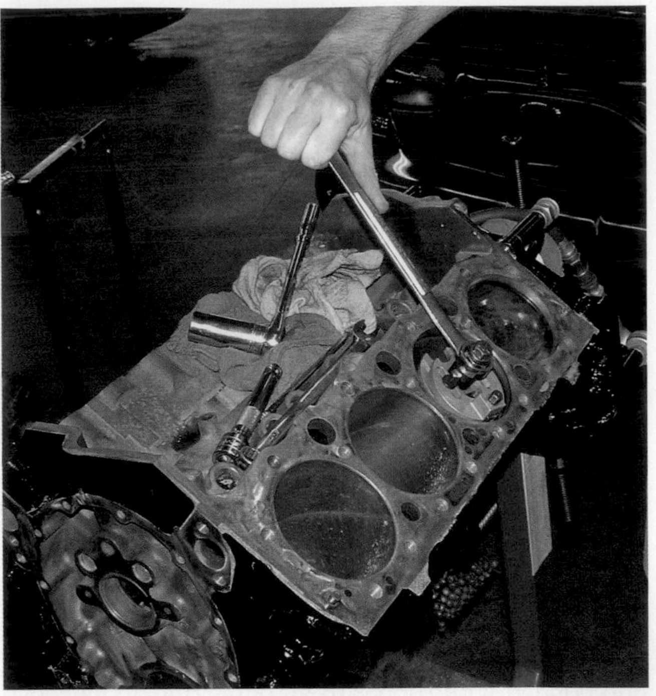

FIGURE 11–6

A ridge reamer can be used in cylinders before removing the pistons.

Lift the balance shaft(s) out of the housing. Inspect the bearings for unusual wear or damage. Keep the bearings in their original location. Check the journals on the balance shafts for scratches, pitting, and other damage. If a bearing or journal is damaged, replace the bearings and/or balance shaft.

Rotate the engine so that the bottom is at the bottom again. Rotate the crankshaft to put a cylinder at BDC. Carefully remove the cylinder ridge with a ridge reamer tool. Rotate the tool clockwise to remove the ridge **(Figure 11–6)**. Do not cut too deeply because an indentation may be left in the bore. Remove just enough metal to allow the piston to slip out of the bore without causing damage to the bore.

The ridge is formed at the top of the cylinder. Because the top ring stops travelling before it reaches the top of the cylinder, a ridge of unworn metal is left **(Figure 11–7)**. Carbon also builds up above this ridge, adding to the problem. If the ridge is not removed, the piston's rings may be damaged as the piston is driven up in its bore.

Repeat the process on all cylinders. After removing the ridges, use an oily rag to wipe the metal cuttings out of the cylinder.

Rotate the engine to put the bottom side up. Check all connecting rods and main bearing caps for correct position and numbering. If the numbers are not visible,

FIGURE 11–7

Normal cylinder wear.

©Cengage Learning 2015

use a centre punch or number stamp to number them **(Figure 11–8)**. Caps and rods should be stamped on the external flat surface. If the rods are already numbered or marked, make sure the marks designate the cylinder where the rods should be installed. If not, remark them to show their current location.

To remove the piston and rod assemblies, position the throw of the crankshaft at the bottom of its stroke. Remove the connecting rod nuts and cap. Remember that the caps and rods must remain as a set. To help remove the cap, tap the cap lightly with a soft hammer or wood block. Cover the rod bolts with protectors to avoid damage to the crankshaft journals. Carefully push the piston and rod assembly out with the wooden hammer handle or wooden drift, and support the piston as it comes out of the cylinder. Be sure that the connecting rod does not damage the cylinder wall during removal.

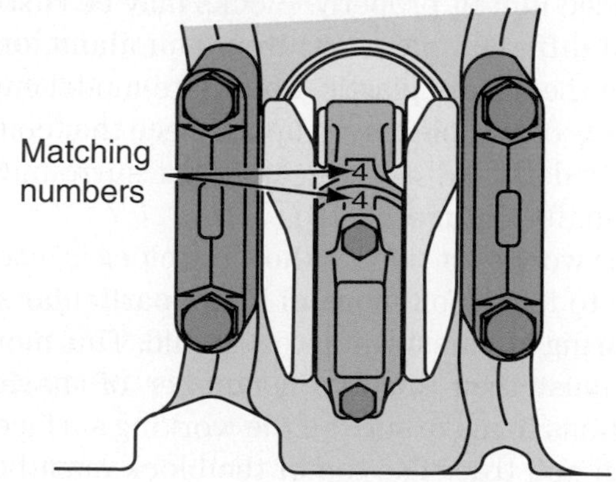

FIGURE 11–8

Check all connecting rods and main bearing caps for correct position and numbering. If the numbers are not visible, use a number stamp or centre punch to number them.

©Cengage Learning 2015

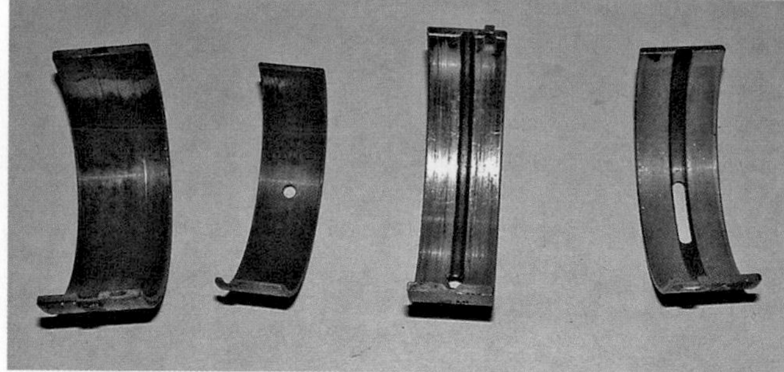

FIGURE 11–9

The bearings on the left have no babbitt left, and the ones on the right are slightly worn and scored.

©Cengage Learning 2015

Loosen and remove the main bearing cap bolts and main bearing caps. Keeping the main bearing caps in order is very important. The location and position of each cap should be marked. Many engines have a main bearing girdle or bearing support. These use at least four bolts at each main bearing. It is important that you follow the recommended bolt-loosening sequence.

After removing the main bearing caps, carefully remove the crankshaft. Then store the crankshaft in a vertical position to avoid damage.

Remove the rear main oil seal and main bearings from the block and caps. Examine the bearing inserts for signs of unusual wear, embedded metal particles, lack of lubrication, antifreeze contamination, oil dilution, and uneven wear **(Figure 11–9)**. Then carefully inspect the main journals on the crankshaft for damage **(Figure 11–10)**.

Engine blocks have **core plugs**, also called expansion plugs. When the block is made, sand cores are used to provide the various passages inside the block. The sand is partly broken and dissolved when the hot metal is poured into the mould.

FIGURE 11–10

Inspect each crankshaft journal for damage and wear.

©Cengage Learning 2015

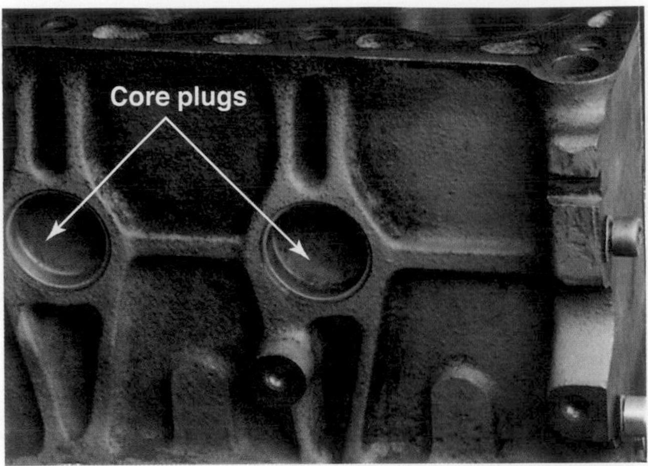

FIGURE 11–11

Core plugs in an engine block.

To remove the remaining sand, holes are bored into the block. These core holes are machined and core plugs installed to seal them **(Figure 11–11)**. Blocks are also made with passageways for oil. These are machined in the block and are sealed with plugs.

The block cannot be thoroughly cleaned unless all core plugs and oil plugs are removed. To remove cup-type "freeze"/core plugs, drive them in on a slant, and use channel lock pliers to pull them out. A fat-type plug can be removed by drilling a hole near the centre and inserting a slide hammer to pull it out. On some engines, the cup-type plug can be removed by driving the plug out from the backside with a long rod.

Sometimes, removing threaded front and rear oil gallery plugs can be difficult. Using a drill and screw extractor can help.

After cleaning, the block and its parts must be visually checked for cracks or other damage.

SHOP TALK

Using heat to melt paraffin into the threads of oil plugs will make removal much easier. As the part is heated, it will expand and the paraffin will leak down between the threads. Because the paraffin serves as a lubricant, you will be able to loosen the plug. Hot paraffin burns, so wear gloves when handling it.

CYLINDER BLOCK

The cylinder block makes up the lower section of the engine. It houses the areas where combustion of the air/fuel mixture takes place. The upper section of the engine, known as the cylinder head, bolts to the top of the cylinder block. The head is also part of the combustion chamber and contains the valve train components.

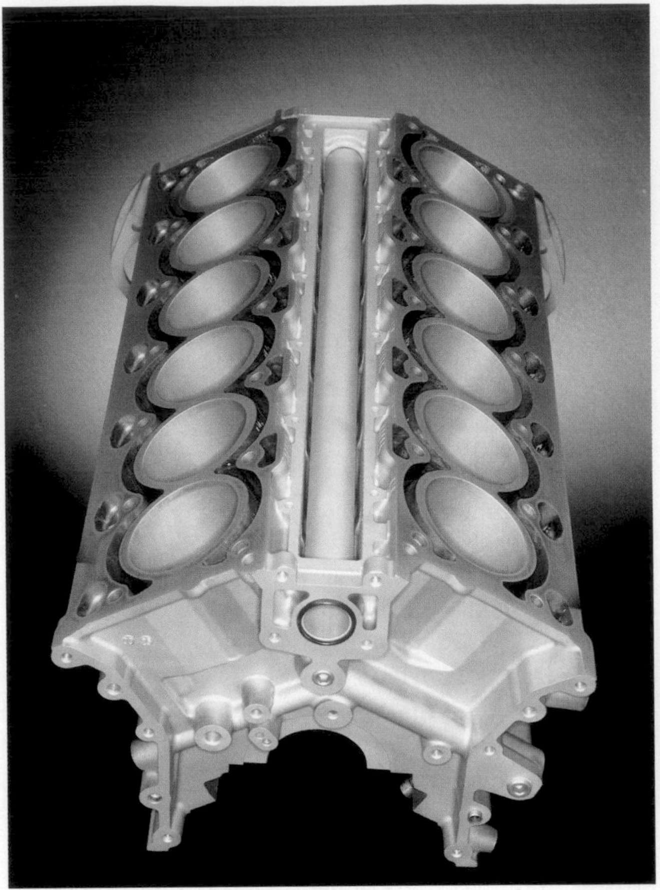

FIGURE 11–12

An engine block for a 12-cylinder engine.

The cylinder block **(Figure 11–12)** is normally a one-piece casting, machined so that all the parts contained in it fit properly. Blocks may be cast from several different materials: iron, aluminum, or possibly in the future, plastic. Some late-model engines are made of two pieces, an upper piece that contains the cylinders and a lower unit that surrounds the crankshaft **(Figure 11–13)**.

The word **cast** refers to how the block is made. To cast is to form molten metal into a particular shape by pouring or pressing it into a mould. This moulded piece must then undergo a number of machining operations to make sure all the working surfaces are smooth and true. The top of the block must be perfectly smooth so that the cylinder head can seal it. The base or bottom of the block is also machined to allow for proper sealing of the oil pan. The cylinder bores must be smooth and have the correct diameter to accept the pistons.

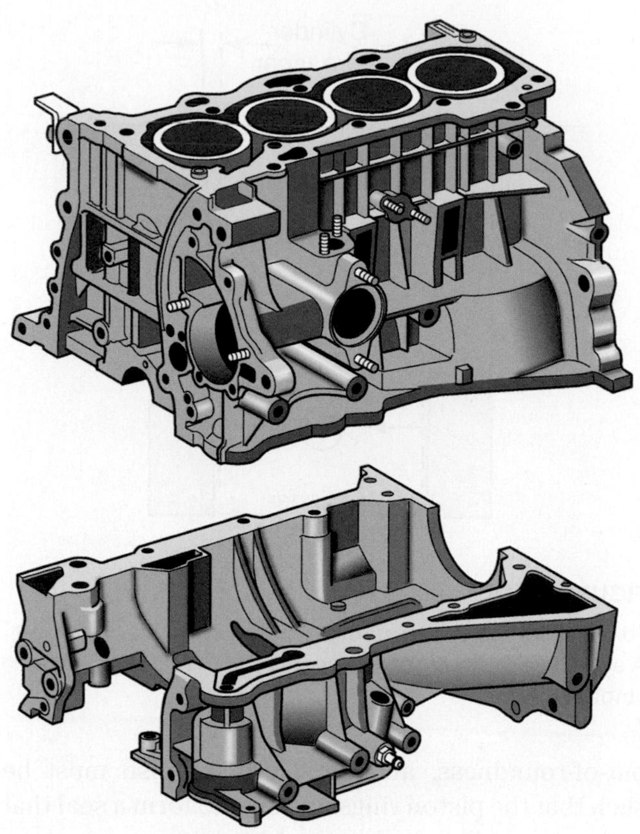

FIGURE 11–13

Aluminum engine blocks are often two-piece units.

Toyota Motor Sales

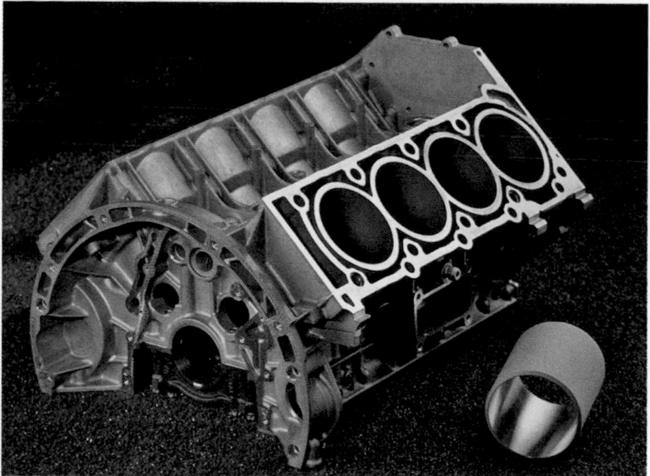

FIGURE 11–14

A cylinder block and cylinder liner for a late-model aluminum V8 engine.

Chrysler LLC

The main bearing area of the block must be align bored to a diameter that will accept the crankshaft. Camshaft bearing surfaces must also be aligned. The word *bore* means to drill or machine a hole. Align boring cuts a series of holes in a straight line.

Cast-iron blocks offer great strength and controlled warpage. With the increased concern for improved gasoline mileage, however, car manufacturers are trying to make the vehicle lighter. One way to do this is to reduce the weight of the block. Aluminum is often used to reduce this weight. Certain materials are added to aluminum to make the aluminum stronger and less likely to warp from the heat of combustion. The cylinder walls of aluminum blocks may be treated with a special coating, but most have a sleeve or liner to serve as cylinder walls **(Figure 11–14)**. The liners are normally made of a cast-iron alloy. Most are very thin and cannot be serviced, and the block must be replaced if the walls are damaged. Liners are pressed into the block or are placed in the mould before the block is cast. After the metal is poured, the liner cannot be removed. The liners have ribs on their outside diameter. These ribs hold the liner in place and increase its ability to dissipate heat to the block.

Lubrication and Cooling

A cylinder block contains a series of oil passages that allows engine oil to be pumped through the block and crankshaft and on to the cylinder head. The oil lubricates, cools, seals, and cleans engine components. Water jackets are also cast in the block around the cylinder bores. Coolant circulates through these jackets to transfer heat away.

Some engine blocks are cast with a plastic spacer for the water jackets. The spacers provide a uniform distribution of heat throughout the cylinders by directing the flow of coolant toward the normally hotter areas.

A cylinder block has machined areas on which to mount other parts. These include threaded bores. Brackets and housings may also be cast onto the basic block.

CYLINDER BLOCK RECONDITIONING

Before any reconditioning or rebuilding work is started, threaded holes should be cleaned with the correct-size tap to remove any and all burrs or dirt to allow for proper bolt torquing. Use a bottoming tap in any blind holes. **Chamfering** or counterboring will eliminate thread pulls and jagged edges. If there is damage to the threads, they should be repaired. To restore damaged threads in an aluminum part, a threaded insert should be installed in the bore.

Check the block for cracks and other damage. Cast-iron blocks can be checked by magnafluxing. Aluminum blocks are checked for cracks with penetrant dye and a black light. Some cracks can be repaired; however, if they are in critical areas, the block should be replaced.

Deck Flatness

The top of the engine block where the cylinder head mounts, is called the **deck**. To check deck warpage, use a precision straightedge and feeler gauge. With the straightedge positioned diagonally across the deck, the amount of warpage is determined by the size of feeler gauge that fits into the gap between the deck and the straightedge.

Some engines have special deck flatness requirements. Always refer to the manufacturer's specifications. If specifications are not available, use 0.0762 mm (0.003 in.) per 152.4 mm (6 in.), and no more than 0.1524 mm (0.006 in.) maximum on any length. If the deck is warped beyond limits, the block should be decked or replaced. Decking requires a special grinder that will shave off small amounts of metal, leaving a flat surface. Some manufacturers do not recommend decking, especially if the block is aluminum. If the block has more than one deck surface (such as a V-type engine), each deck must be machined to the same height. This allows for uniform compression and manifold alignment. If the deck is warped and not corrected, coolant and combustion leakage can occur.

Cylinder Walls

Inspect the cylinder walls for scoring, roughness, or other signs of wear. Ring and cylinder wall wear can be accelerated by dirt.

Scuffed or scored pistons, rings, and cylinder walls can act as passages for oil to bypass the rings and enter the combustion chamber. Scuffing and scoring occur when the oil film on the cylinder wall is ruptured, allowing metal-to-metal contact of the piston rings on the cylinder wall. Cooling system hot spots, oil contamination, and fuel wash are typical causes of this problem.

Cylinder Bore Inspection

Normally, the most cylinder wear occurs at the top of the ring travel area. Pressure on the top ring is at a peak and lubrication at a minimum when the piston is at the top of its stroke. A ridge of unworn material will remain above the upper limit of ring travel. Below the ring travel area, wear is negligible because only the piston skirt contacts the cylinder wall.

A properly reconditioned cylinder must have the correct diameter and have no taper or out-of-roundness, and the surface finish must be such that the piston rings will seat to form a seal that will control oil and minimize blowby.

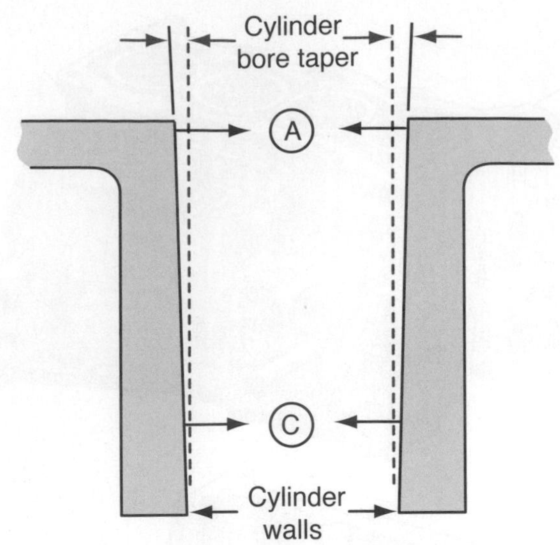

FIGURE 11–15

To check for taper, measure the diameter of the cylinder at A and C. The difference between the two readings is the amount of taper.

Taper is the difference in diameter between the bottom of the cylinder bore and the top of the bore just below the ridge **(Figure 11–15)**. Subtracting the smaller diameter from the larger one gives the cylinder taper. Some taper is permissible, but normally not more than 0.1524 mm (0.006 in.). If the taper is less than that, reboring the cylinder is not necessary.

Cylinder out-of-roundness is the difference of the cylinder's diameter when measured parallel with the crank and then perpendicular to the crank **(Figure 11–16)**. Out-of-roundness is measured at the top of the cylinder just below the ridge. Typically, the maximum allowable out-of-roundness is 0.0381 mm (0.0015 in.). Normally, a cylinder bore is checked for out-of-roundness with a dial bore gauge **(Figure 11–17)**. However, a telescoping gauge can also be used.

When using a dial bore gauge or a telescoping gauge to check a cylinder's bore, make sure the measuring arms are parallel to the plane of the crankshaft. The best way to do this is to rock the gauge until the smallest reading is obtained **(Figure 11–18)**.

Cylinder Bore Surface Finish

The proper surface finish on a cylinder wall acts as a reservoir for oil to lubricate the piston rings and prevent piston and ring scuffing. Piston ring faces

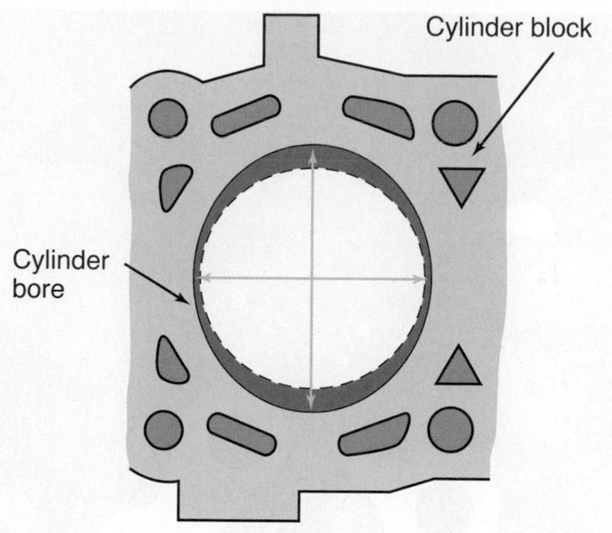

FIGURE 11–16

To check cylinder out-of-roundness, measure the bore in different locations.

Ford Motor Company

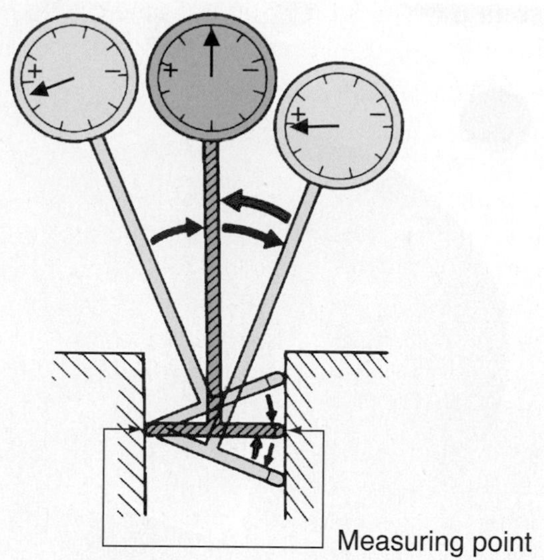

FIGURE 11–18

To get an accurate reading, rock the gauge until the smallest reading is obtained.

can be damaged and can experience premature wear if the cylinder wall is too rough. A surface that is too smooth will not hold enough oil and will not allow the rings to seat properly. Obtaining the correct cylinder wall finish is important.

The desired cylinder wall finish is a crosshatch pattern, comprised of many small crisscross grooves **(Figure 11–19)**. Ideally, these grooves cross at 50- to 60-degree angles, although anything in the 20- to 60-degree range is acceptable. This finish leaves millions of tiny diamond-shaped areas, which serve as lubricant reservoirs **(Figure 11–20)**. This finish also leaves small flat areas, or plateaux, on the surface. A film of oil adheres to these areas to act as a bearing surface for the piston rings.

If the angle of the crosshatch is too steep, the oil film will be too thin, resulting in ring and cylinder scuffing. If the angle is too flat, the pistons may hydroplane, and excessive oil consumption will result.

CYLINDER DEGLAZING If the inspection and measurements of the cylinder wall show that surface conditions, taper, and out-of-roundness are within acceptable limits, the cylinder walls may only need to be deglazed. Combustion heat, engine oil, and

FIGURE 11–17

Cylinder bore is checked for out-of-roundness with a dial bore gauge.

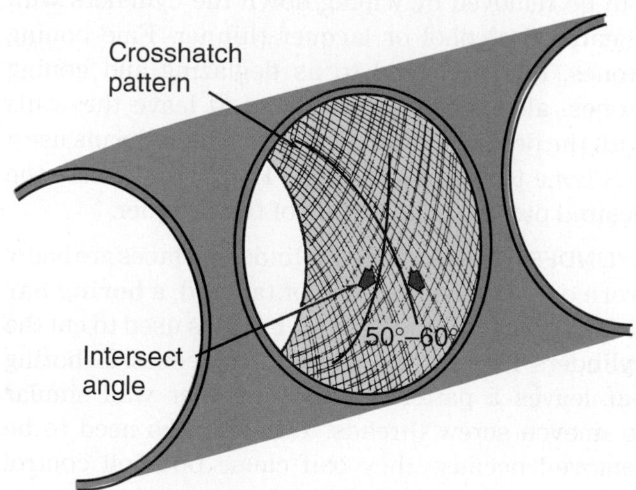

FIGURE 11–19

Ideal crosshatch pattern for cylinder walls.

Chrysler LLC

FIGURE 11–20

The desired cylinder wall finish for most types of piston rings.

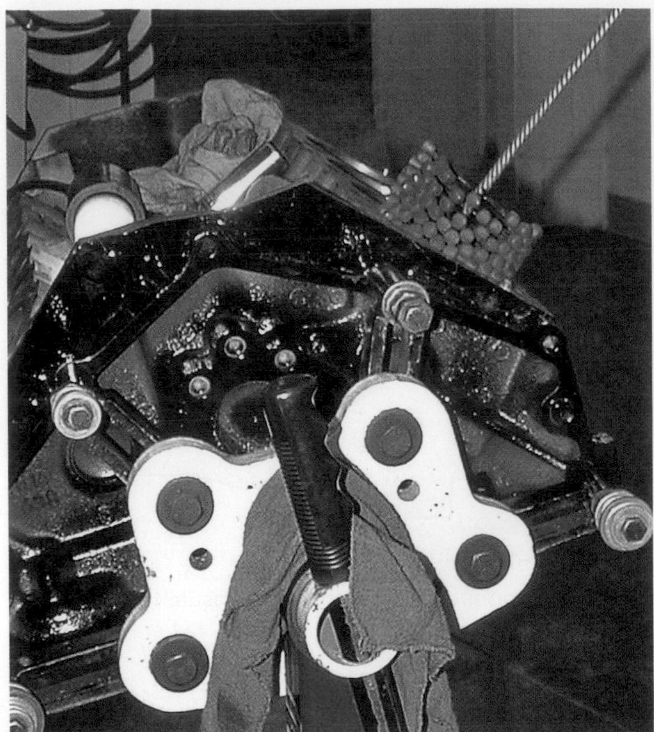

FIGURE 11–21

Using a resilient-based, hone-type brush, commonly called a ball hone.

piston movement combine to form a thin residue on the cylinder walls, commonly called **glaze**.

A glazed cylinder wall allows the piston rings to slide over the walls of the cylinder, preventing a positive seal between the two.

It is easy to confuse glaze with the polished surface that appears on the walls of the cylinder after the engine has some miles on it. Often, true glazing can be removed by wiping down the cylinders with denatured alcohol or lacquer thinner. Fine honing stones, often referred to as deglazing and honing stones, also remove the glaze and leave the walls with the desired finish. Most often, technicians use a ball hone to deglaze **(Figure 11–21)** and create the desired pattern on the walls of the cylinder.

CYLINDER BORING When cylinder surfaces are badly worn or excessively scored or tapered, a **boring bar** or boring machine **(Figure 11–22)** is used to cut the cylinders for oversize pistons or sleeves. A boring bar leaves a pattern on the cylinder wall similar to uneven screw threads. These marks need to be removed because they can cause poor oil control and excessive blowby. Therefore, you should hone the bore after it has been bored. This honing will slightly enlarge the cylinder, so the boring should be performed only to the exact size of the new pistons. The honing procedure will produce the required piston clearance.

CYLINDER HONING A cylinder hone usually consists of three or four stones **(Figure 11–23)**. The **hone** is spun and moved up and down the cylinder's bore. The stones are forced outward and remove some metal from the bore as they rotate within it. The

FIGURE 11–22

A cylinder being bored.

Jaspar Engine and Transmission Exchange, Inc.

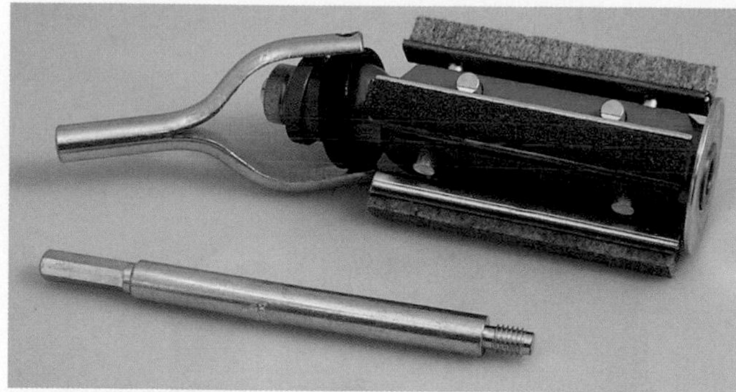

FIGURE 11–23

A cylinder hone.

Lisle Corp.

FIGURE 11–25

Torque plates are fastened to the block during cylinder boring and honing to prevent block distortion during the machining process.

Jaspar Engine and Transmission Exchange, Inc.

grade of stones is typically specified by the manufacturer of the piston rings. Honing stones are classified by grit size; usually, the lower the grit number, the coarser the stone.

Cylinder honing machines are available in manual and automatic models **(Figure 11–24)**. The major advantage of the automatic type is that it allows the technician to dial in the exact crosshatch angle needed. When honing a cylinder by hand, make sure you use a slow-speed (200 to 450 rpm) electric drill. Mount the honing tool into the drill, and insert it into the bore. Adjust the stones so that they fit snugly to the narrowest section of the cylinder. Move the drill and the hone up and down in the bore with short strokes. Never remain in one spot too long. Squirt some honing oil on the walls, and occasionally stop honing and clean the stones before restarting. Honing oil is used to control the temperature of the cylinder walls and stones and to flush out any metallic and abrasive residue. Continue until the desired results are achieved.

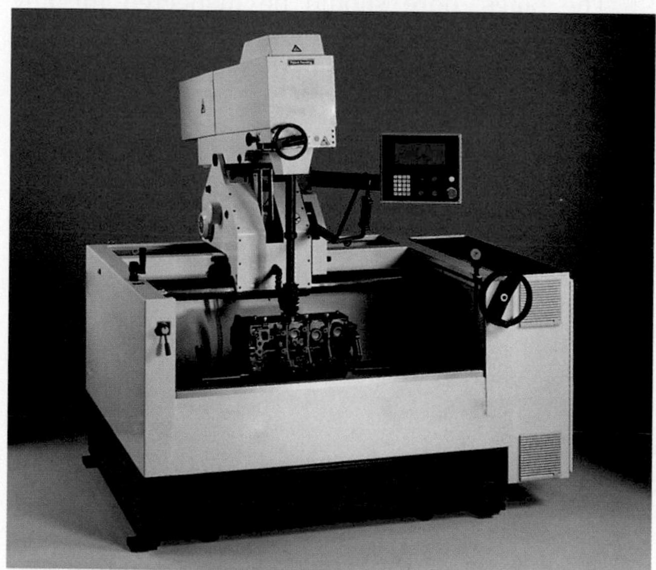

FIGURE 11–24

An automatic cylinder hone machine.

Sunnen Products Company

Torque plates simulate the weight and structure of a cylinder head. They are used by engine rebuilding shops and are fastened to the cylinder block to equalize or prevent twist and distortion when honing or boring a cylinder **(Figure 11–25)**.

After resurfacing the cylinder walls, use plenty of hot, soapy water; a stiff bristle brush; and a soft, lint-free cloth to clean the residue **(Figure 11–26)**. Then rinse the block with water and dry it thoroughly. Lightly coat the bore with clean, light engine oil to prevent rust.

CAUTION!

Always wear eye protection when operating deglazing, honing, or boring equipment.

Lifter Bores

Carefully examine each of the bores for the valve lifters. Look for cracks and evidence of excessive wear. Oblong or egg-shaped bores indicate wear. Typically, if these bores exceed allowable wear limits or are damaged, the engine block is replaced. If the bores are rusted, glazed, or have burrs and high spots, they can be honed with a brake wheel cylinder hone. Be careful not to remove more than 0.013 mm (0.0005 in.) of metal while honing.

FIGURE 11–26

After honing, clean the cylinder with soapy water.

Martin Restoule

Checking Crankshaft Saddle Alignment

Figure 11–27 is an exaggerated illustration of crankcase housing bores that are out of alignment. If the block is warped and its main bearing bores are out of alignment, the crankshaft will inflict heavy loads on one side of the main bearings. Engine blocks that are not severely warped can be repaired by an operation called **line boring**, a machining operation in which the main bearing housing bores are rebored to standard size and alignment **(Figure 11–28)**. Badly warped blocks must be replaced.

The alignment of the crankshaft saddle bore can be checked with a precisely ground arbor placed into the bearing bores. The arbor is rotated in the bores. The effort required to rotate it determines the alignment of the bores.

If a proper arbor is not available, saddle alignment can be checked with a straightedge **(Figure 11–29)**. Place the straightedge in the saddles as shown. Using a feeler gauge that is half the maximum specified oil clearance, try to slide the feeler under the straightedge. If this can be done at any saddle, the saddles are out of alignment, and the block must be line-bored. Repeat this procedure at two other parallel positions in the saddles.

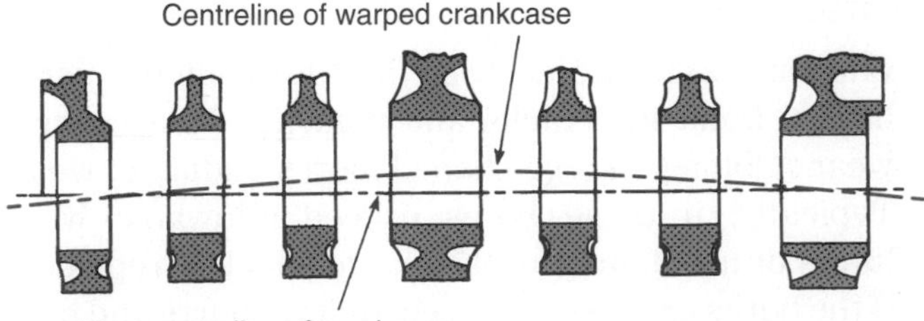

FIGURE 11–27

An exaggerated view of crankcase housing misalignment.

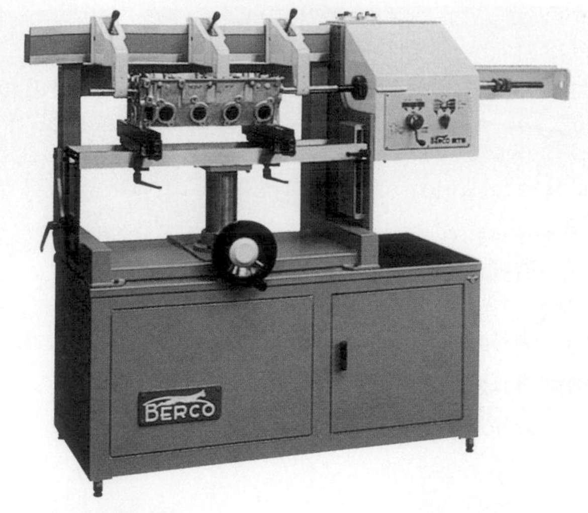

FIGURE 11–28

A line boring machine for correct crankshaft saddle alignment.

Peterson Machine Tool Inc.

Out-of-roundness of the saddles can also be checked by bolting on the main bearing caps and checking each bore with a dial bore gauge or an out-of-roundness indicator.

Installing Core Plugs

Old core and oil gallery plugs are normally removed and replaced during cylinder block reconditioning. When installing new core plugs, make sure they are the correct size and type.

The plugs' bore should be inspected for any damage that would interfere with the proper sealing. If the bore is damaged, it should be bored out for an

FIGURE 11–29

Checking bore alignment with a straightedge and feeler gauge.

Martin Restoule

oversized plug. Oversize (OS) plugs are identified by the OS stamped on the plug.

Coat the plug or bore lightly with a nonhardening oil-resistant (oil gallery) or water-resistant (cooling jacket) sealer. The three basic core plugs are installed as follows.

DISC- OR DISHED-TYPE This type fits in a recess in the engine casting with the dished side facing out **(Figure 11–30A)**. With a hammer, hit the disc in the centre of the crown, and drive the plug into the bore until just the crown becomes flat. In this way, the plug will expand properly and give a good tight fit.

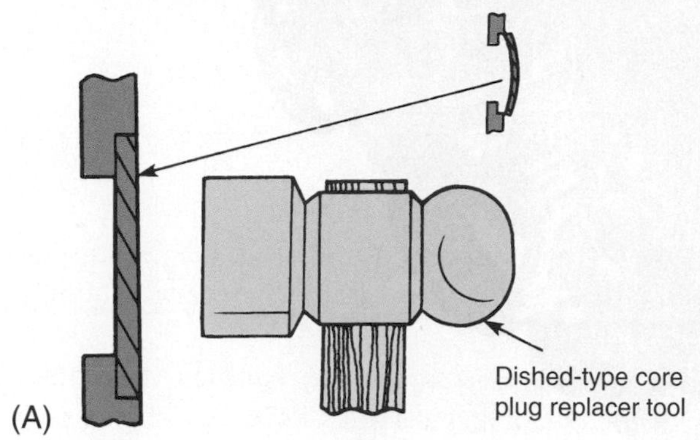

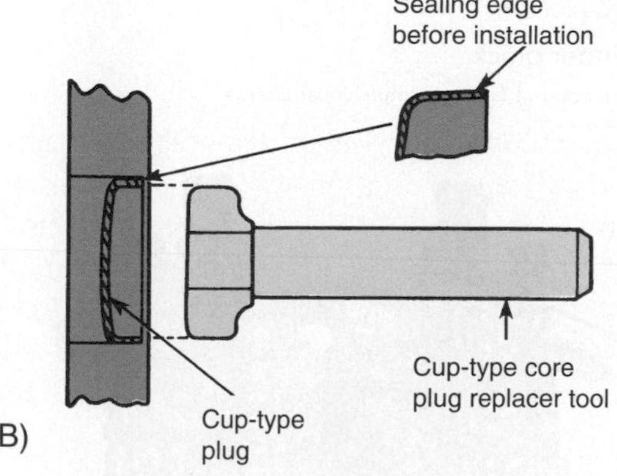

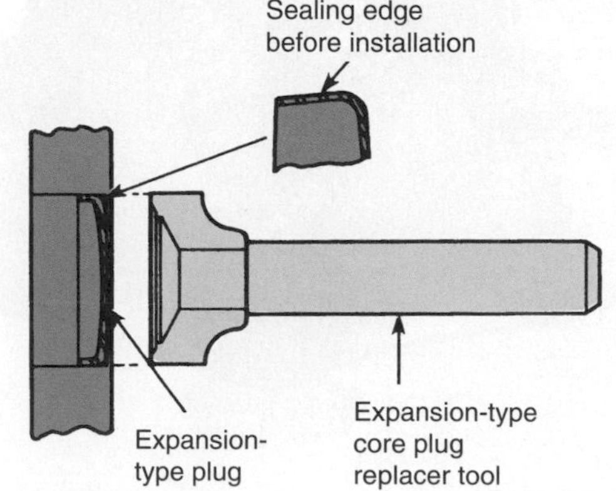

FIGURE 11–30

Core plug installation methods: (A) dished, (B) cup, and (C) expansion.

CUP-TYPE This type of plug is installed with the flanged edge outward **(Figure 11–30B)**. The flange on cup-type plugs flares outward, with the largest diameter at the outer (sealing) edge. The flanged (trailing) edge must be below the chamfered edge of the bore to effectively seal the plugged bore.

EXPANSION-TYPE This type of plug is installed with the flanged edge inward **(Figure 11–30C)**. The maximum diameter of this plug is located at the base of the flange with the flange flaring inward. When installed, the trailing (maximum) diameter must be below the chamfered edge of the bore to effectively seal.

CAMSHAFT

A camshaft is a shaft **(Figure 11–31)** with a cam for each exhaust and intake valve, each one placed to allow for the proper timing of each valve. A cam is a device that changes rotary motion into reciprocating motion. Each cam has a high spot or lobe that controls the opening of the valves. The height of the lobe is proportional to the amount the valve will open.

The camshaft can be located in either the cylinder block or cylinder head(s). The camshaft fits into a bore next to the crankshaft on most in-line engines, unless the engine has overhead camshafts. On V-type engines, the camshaft lies in bore above the crankshaft at the centre of the block. When the camshaft is in the block **(Figure 11–32)**, the valves are opened through lifters, pushrods, and rocker arms. As the cam lobe rotates, it pushes up on the lifter, which lifts up the pushrod, moving one end of the rocker arm up while the other end pushes the valve down to open it. As the cam rotates, the valve spring closes the valve and maintains the contact

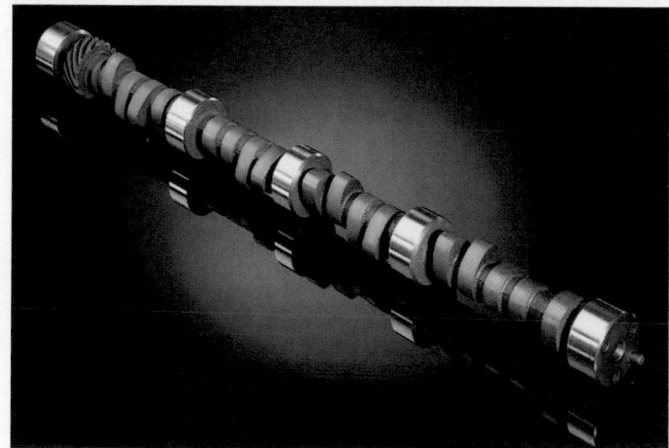

FIGURE 11–31

A camshaft for a V8 engine.

Melling Engine Parts

FIGURE 11–32

The camshaft for this engine is located in the cylinder block.

Portions of materials contained herein have been reprinted with permission of General Motors under License Agreement #1410912.

between the valve and the rocker arm, thereby keeping the pushrod and the lifter in contact with the rotating cam.

Overhead camshaft (OHC) engines have the camshaft mounted above the cylinders, in or on the cylinder head **(Figure 11–33)**. OHC engines have no need for pushrods. As the camshaft rotates, the cams ride directly above the valves. The lobes open the valves by either directly depressing the valve or by depressing the valve through the use of a cam follower, rocker arm, or bucket-type tappet. The closing of the valves is still the responsibility of the valve springs.

Service to the camshaft(s) in an OHC engine is usually done when reconditioning the engine's cylinder head. In overhead valve (OHV) engines, the camshaft and related parts are inspected and serviced while the short block is being reconditioned.

A camshaft is driven by the crankshaft at half its speed. This is accomplished through the use of a camshaft drive gear or drive sprocket that is twice as large as the crankshaft sprocket. For every two complete turns of the crankshaft, the camshaft

FIGURE 11–33

The camshaft(s) are located above the cylinders in OHC engines.

BMW of North America, Inc.

turns once. During that full rotation, the intake and exhaust valves open and close once.

To synchronize the opening and closing of the valves with the position and movement of the pistons, the camshaft is timed to the crankshaft. In the typical valve timing diagram **(Figure 11–34)**, valve action is shown in relation to crankshaft rotation. The intake valve starts to open at 21° before the piston has reached TDC and remains open until it has travelled 51° past BDC. The number of degrees between the valve's opening and closing is called intake valve duration time (252°).

The exhaust stroke begins at 57° before BDC and continues until 15° after TDC, or a total exhaust valve duration time of 252° of crankshaft rotation. The specifications for the camshaft used for the figure show the duration of the intake valve to be the same as the exhaust. This is typical; however, some camshafts are designed with different durations for the intake and exhaust valves. These camshafts are called dual-pattern cams. Different engine designs require different valve opening and closing times. Therefore, each engine design has a unique camshaft.

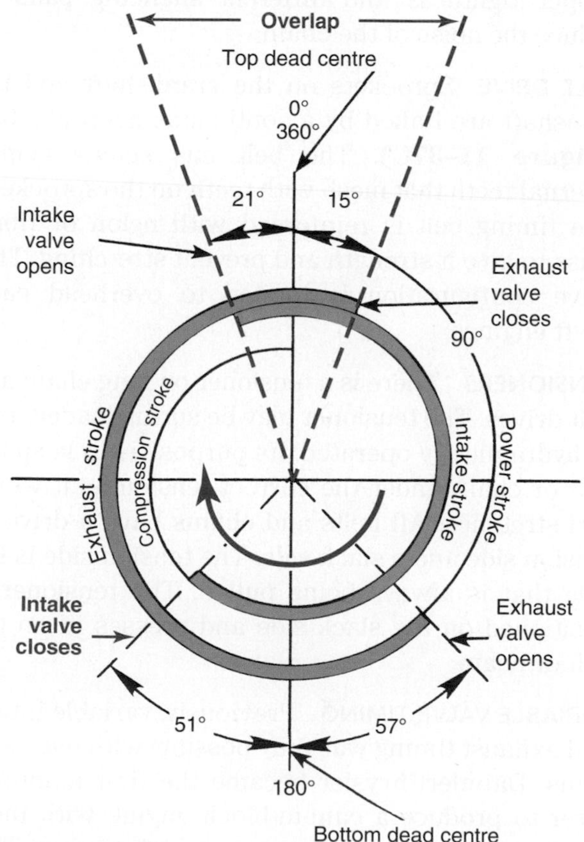

FIGURE 11–34

Typical valve timing diagram.

The actual design of the cams and lobes varies with the type of lifter or follower used in the engine. There are four distinct types of lifters: solid non-roller, hydraulic non-roller, solid roller, and hydraulic roller. Camshafts designed for solid and hydraulic non-roller cams are often called "flat-tappet" cams. A camshaft must be matched with the type of lifter for which it was designed.

Valve Timing Terminology

Many different terms are used to define the specifications of a camshaft. The actual shape of a cam lobe is called the cam profile. The profile determines the duration and lift provided by the camshaft. Valve overlap is controlled by the placement of the lobes on the camshaft.

Camshaft **duration** is how long the cam holds the valves open. It is expressed in crankshaft degrees. The width of the lobe determines the cam's duration. Lift is the distance the lobe moves the lifter or follower to open the valve. Maximum lift is determined by the height of the lobe. The valve is fully open only when the lifter is at the top of the lobe. Camshaft lift does not always express how far the valve is open. Rocker arms increase the actual amount of valve opening.

Both valves are open slightly at the end of the exhaust stroke and the beginning of the intake stroke; this is called valve **overlap**. Overlap is critical to exhaust gas scavenging. A camshaft with a long overlap helps empty the cylinders at high engine speeds for improved efficiency. However, because both valves are open for a longer period, low-rpm cylinder pressure tends to drop. Because the amount of overlap has an effect on cylinder pressure, it affects overall engine efficiency and exhaust emissions. Valve overlap also helps get the intake mixture moving into the cylinder. As the exhaust gases move out of the cylinder, a low pressure is present in the cylinder. This low pressure causes atmospheric pressure to push the intake charge into the cylinder. Less overlap provides for more pressure in the cylinder at low speeds, resulting in more torque at lower speeds.

LOBE TERMINOLOGY Other terms are used to describe a camshaft's profile **(Figure 11–35)**:

- *Base circle*—The base circle is the cam without its lobe. It is also the part of the cam where valve adjustments are made.
- *Nose*—The nose is the highest portion of the cam lobe measured from the base circle. This point provides for the maximum amount of lift.

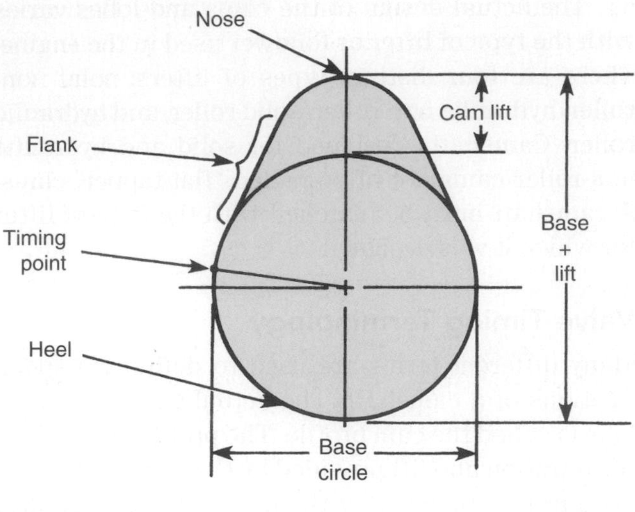

FIGURE 11–35

Cam lobe nomenclature.

- *Ramp (flank)*—The ramps are the sides of a cam lobe that lie between the nose and base circle. The ramp on one side is for valve opening and the other for valve closing. How quickly a valve will open and close depends on the steepness of the ramp.

Timing Mechanisms

The camshaft and crankshaft must always remain in the same relative position to each other. They must also be aligned to each other. The alignment is initially set by matching marks on the gears for both shafts; these are called timing marks **(Figure 11–36)**. The following are the basic configurations for driving the camshaft.

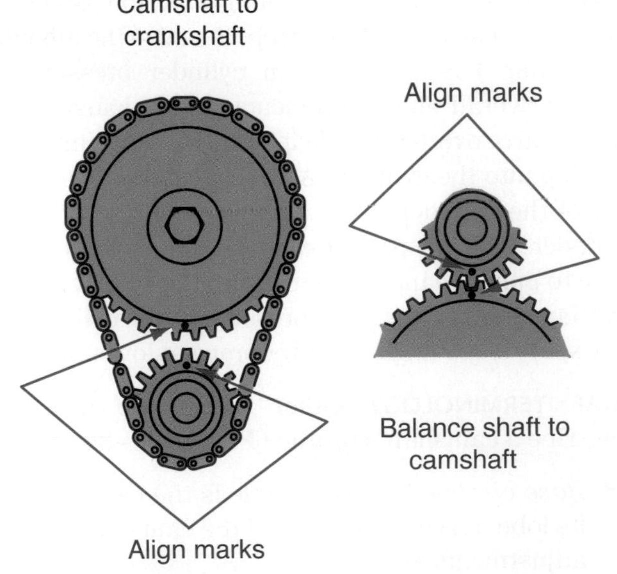

FIGURE 11–36

Camshaft-to-crankshaft and balance shaft–to-camshaft timing marks.

GEAR DRIVE A gear on the crankshaft meshes directly with another gear on the camshaft **(Figure 11–37A)**. The gear on the crankshaft is usually made of steel. The gear on the camshaft may be steel for heavy-duty applications, or it may be aluminum or pressed fibre when quiet operation is a major consideration. The gears are helical in design because helical gears are stronger and also tend to push the camshaft backward during operation to help prevent the camshaft from walking out of the block.

CHAIN DRIVE Sprockets on the camshaft and the crankshaft are linked by a continuous chain **(Figure 11–37B)**. The sprocket on the crankshaft is usually made of steel. The sprocket on the camshaft may be steel for heavy-duty applications. When quiet operation is a major consideration, an aluminum sprocket with nylon covering on the teeth is used. Nearly all OHV engines use a chain drive system. Chain drives are also used on many OHC engines, especially DOHCs. Often, multiple chains are used and arranged in an elaborate fashion. These chain arrangements use a chain tensioner to maintain proper tightness and different silencing pads to reduce the noise of the chain.

BELT DRIVE Sprockets on the crankshaft and the camshaft are linked by a continuous neoprene belt **(Figure 11–37C)**. The belt has square-shaped internal teeth that mesh with teeth on the sprockets. The timing belt is reinforced with nylon or fibreglass to give it strength and prevent stretching. This drive configuration is limited to overhead camshaft engines.

TENSIONERS There is a tensioner on long chain and belt drives. The tensioner may be spring loaded and/or hydraulically operated. Its purpose is to keep the belt or chain under the correct tension as it wears and stretches. All belts and chains have a drive or tension side and a slack side. The tension side is the side that is always being pulled. The tensioner is positioned on the slack side and presses in on the belt or chain.

VARIABLE VALVE TIMING Previously, variable intake and exhaust timing was only possible with overhead cams. DaimlerChrysler became the first manufacturer to produce a cam-in-block engine with independent control of exhaust camshaft timing. This system was introduced in the 2008 Dodge Viper SRT10's 8.4-litre engine. This is the first production

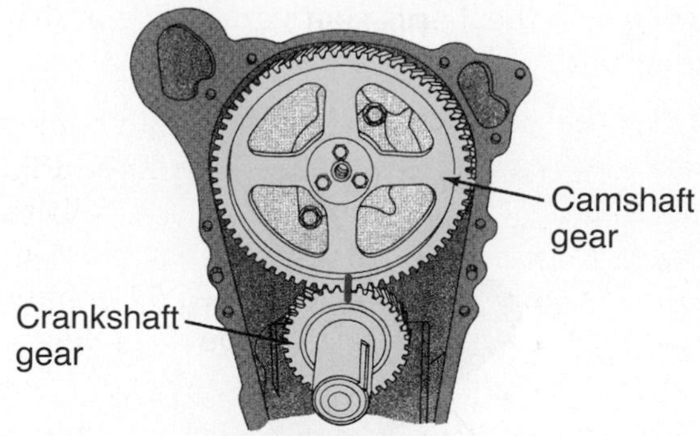

(A) OHV engine with gear-driven camshaft

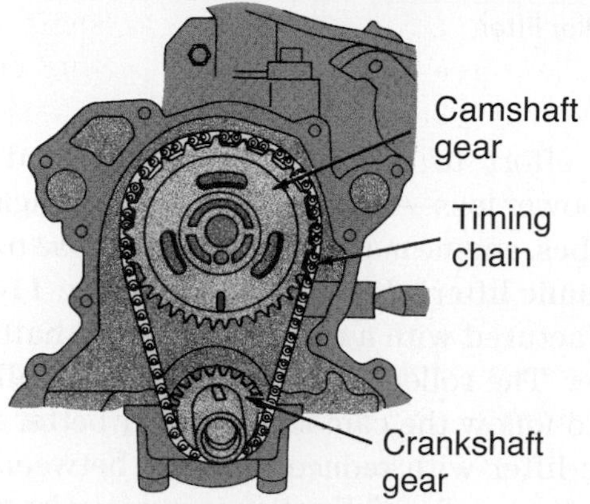

(B) OHV engine with timing chain and gears

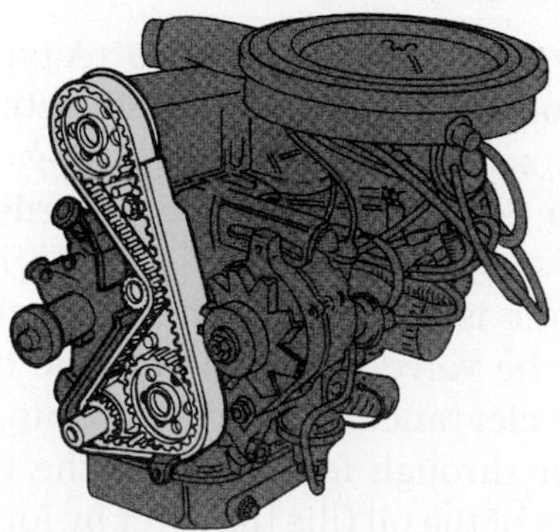

(C) OHC engine with belt-driven camshaft

FIGURE 11–37

The different timing drive mechanisms used today.

Ford Motor Company

pushrod–equipped engine with true **variable valve timing (VVT)**. The VVT system electronically adjusts valve overlap by changing exhaust valve opening times in response to engine speed and load. The system provides an increase in horsepower and torque. It also reduces fuel consumption and exhaust emissions.

This VVT system uses a special camshaft and phaser. The phaser is attached to the end of the camshaft. Inside the phaser are vanes that move within a fixed cavity inside a sealed hub. The movement of the vanes is controlled by oil pressure. The applied oil pressure is controlled by the powertrain control module (PCM). The PCM transmits a signal to a solenoid to move a valve spool that regulates the flow of oil to the phaser cavity. As the applied oil pressure increases, the vanes move against spring pressure. Each vane can rotate a total of 22.5° inside its chamber.

The camshaft is actually two camshafts: an inner shaft and an outer hollow tube–type shaft. It is a camshaft within a camshaft. The exhaust lobes are attached to the outer shaft and the intakes are pinned to the inner camshaft through slots in the outer tube **(Figure 11–38)**. Locking pins pass through the slots and are driven through the intake cam lobe assemblies. The exhaust lobes are pressed into position on the hollow outer shaft.

The phaser hub is fit with an external gear that is driven, via a chain, by the crankshaft. The vanes are connected to the outer tube and the hub. As the vanes rotate, the position of the exhaust lobes, in relationship to the intake lobes, changes. The amount the exhaust lobes can move is limited by the size of the oil cavity and the slots in the outer tube.

Valve Lifters

Valve lifters, sometimes called **cam followers** or tappets, follow the contour or shape of the cam lobe. Lifters are either mechanical (solid) or hydraulic. Solid valve lifters provide for a rigid connection between the camshaft and the valves. Hydraulic valve lifters provide the same

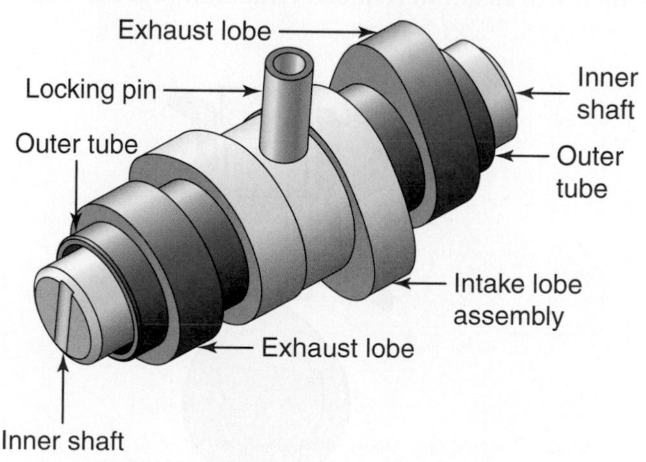

FIGURE 11–38

The basic construction of the camshaft within a camshaft used for variable valve timing on an OHV engine.

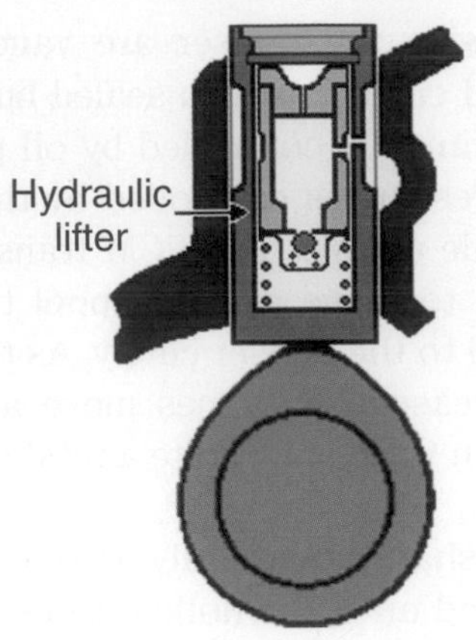

FIGURE 11–39

A hydraulic valve lifter.

connection but use oil to act as a solid link while maintaining proper valve lash of near zero, compensating for material expansion and contraction due to changing temperatures and wear of valve train components.

Hydraulic lifters **(Figure 11–39)** are designed to automatically compensate for the effects of engine temperature. Changes in temperature cause valve train components to expand and contract. Hydraulic lifters are designed to automatically maintain a direct connection between valve train parts.

Solid lifters **(Figure 11–40)** do not have this built-in feature and require a clearance between the parts of the valve train. This clearance allows for expansion of the components as the engine gets hot. Periodic adjustment of this clearance must be made. Excessive clearance might cause a clicking noise. This clicking noise is also an indication of the hammering of valve train parts against one another, which will result in reduced camshaft and lifter life.

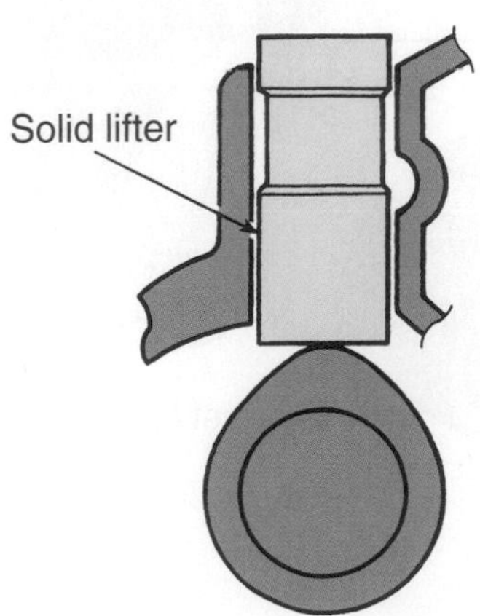

FIGURE 11–40

A solid valve lifter.

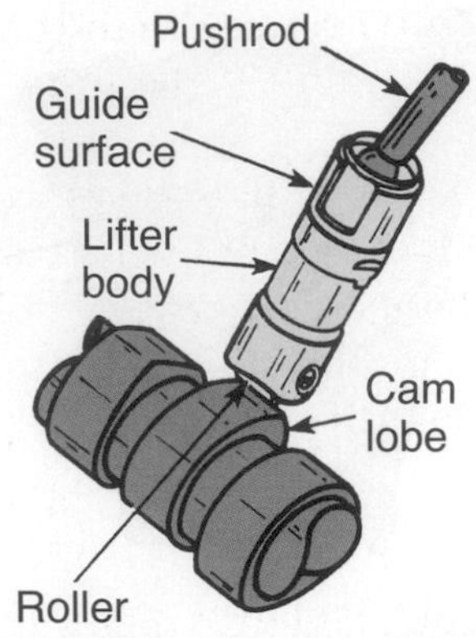

FIGURE 11–41

A typical roller lifter.

In an effort to reduce the friction—and the resulting power loss—from the lifter rubbing against the cam lobes, engine manufacturers often use roller-type hydraulic lifters. Roller lifters **(Figure 11–41)** are manufactured with a roller on the camshaft end of the lifter. The roller acts like a wheel and allows the lifter to follow the cam lobe contour better than a flat-type lifter with reduced friction between the two contacting surfaces. Friction is reduced because the lifter rolls along the surface of the cam lobe as opposed to rubbing against it.

OPERATION OF HYDRAULIC VALVE LIFTERS A typical hydraulic lifter contains a plunger, an oil-metering valve, a pushrod seat, a check valve spring, and a plunger return spring housed in a hardened iron body.

When the lifter is resting on the base circle of the cam lobe, the valve is closed, and the lifter maintains a zero clearance in the valve train. Oil is fed to the lifter through feed holes in the lifter bore. The pressure of the oil fills the lifter by forcing down the check valve inside the lifter. When the oil pressure is equal above and below the lifter plunger, the check valve will close. The oil between the plunger and the check valve forms a rigid connection between the lifter and the pushrod. Whenever there is some clearance in the valve train, a spring between the plunger and the lifter body pushes the plunger up to eliminate the clearance. This creates a lower pressure beneath the plunger, allowing more oil to flow through the check valve into the cavity below the plunger. As the cam lobe turns and opens a valve, the valve spring tension against the plunger can push it down slightly in the lifter body, which allows a small amount of oil to leak out. This leaking out of oil is called **leakdown**. Once the cam rotates and the lifter returns to the base of the cam, oil pressure will again fill the lifter, resetting the plunger position.

If a hydraulic lifter has excessive leakdown or does not fill with oil, a noise will be heard from the engine. Non-roller-type lifters must also be able to rotate in their bore when the engine is running. This prevents wear on the bottom of the lifter.

Camshaft Bearings

The camshaft and balance shafts are supported by several friction-type bearings, or bushings. They are designed as one piece and are typically pressed into the camshaft bore in the engine block **(Figure 11–42)**. The bearings are made of either aluminum or steel with a lining of babbitt. **Babbitt** is a soft material made of mostly lead and tin. Alloys of aluminum are often found in late-model engines. Aluminum bearings have a longer service life in newer engines because the engines run at higher temperatures. Aluminum bearings are harder than babbitt and therefore are more susceptible to damage from dirt and poor lubrication.

Balance shaft and camshaft bearings are normally replaced during engine rebuilding. The old bearings should be inspected for signs of unusual wear that may indicate an oiling or bore alignment problem. Some OHC camshafts use machined bores in the aluminum cylinder head as a bearing surface. Often when this bore is damaged, the cylinder head is replaced.

Balance Shafts

Many late-model engines are fitted with one or more balance (silence) shafts to smooth engine operation. An engine's crankshaft is one of the main sources of engine vibration because its shape makes it inherently out of balance as it spins. Balance shafts are designed to cancel out these vibrations.

In its basic form, a balance shaft is fitted with counterweights designed to mirror the throws of the crankshaft. These weights are rotated in the opposite direction as the crankshaft. As the engine turns, the opposing weights mutually cancel out any vibrations. To do this, the balance shafts rotate at twice the speed of the crankshaft and are synchronized or timed to the rotation of the crankshaft. If the balance shaft(s) are not timed to the crankshaft, the engine may vibrate more than it would without the balance shaft assembly.

Balance shafts are located in the engine block to the right and left side of the crankshaft **(Figure 11–43)** or in the camshaft bore of OHC engines. The shafts are supported by full-round bearings pressed into the block.

Balance shafts are typically inspected and serviced as part of reconditioning or building a short block. The service procedures for balance shafts are the same as those for servicing camshafts. The shaft journals and bearings need to be checked for wear, damage, and proper oil clearances. The runout of the shafts and drive gears should also be checked.

FIGURE 11–42

The typical camshaft bearing is a full-round design.

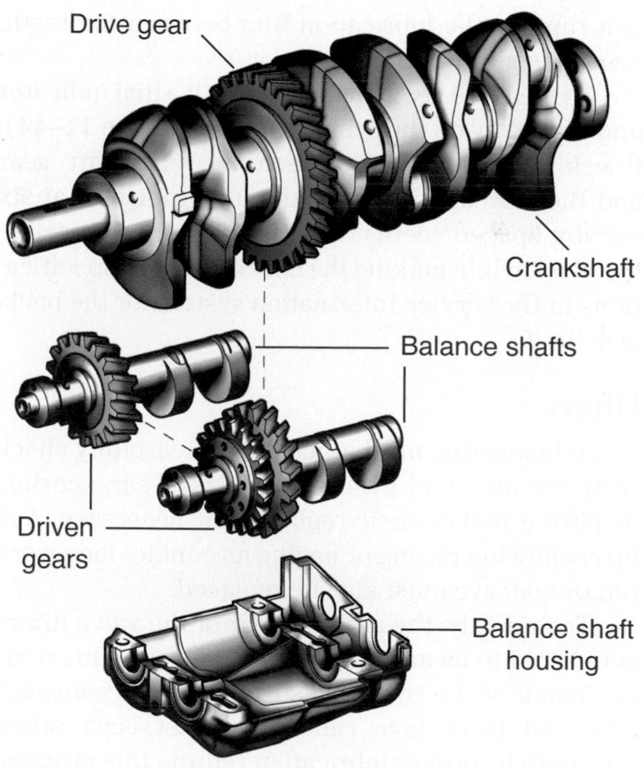

FIGURE 11–43

Balance shaft assemblies for a four-cylinder engine.

Toyota Motor Sales

INSPECTION OF CAMSHAFT AND RELATED PARTS

As the engine is being disassembled, all parts should be carefully inspected, including the camshaft and timing gears. The bearings for the camshaft should also be carefully inspected.

Timing Components

The timing belt or chain and crankshaft/camshaft gears (sprockets) should be inspected or replaced if damaged or worn. This inspection should include a timing chain deflection check. To conduct this test, simply depress the chain at its midway point between the gears, and measure the amount that the chain can be deflected. If the deflection measurement exceeds specifications, the timing chain and gears should be replaced.

A gear with cracks, spalling, or excessive wear on the tooth surface is an indication of improper **backlash** (either insufficient or excessive). With excessive backlash, operation will be noisy because the teeth will make violent impact contact. When coupled with the normal valve train loads, this overloading causes accelerated tooth wear and often breakage. Insufficient backlash places a bind on the gears. Also, it generates high-contact forces that can rupture the lubrication film between the teeth, causing spalling and wear.

To measure gear backlash, install a dial indicator and bracketry on the cylinder block **(Figure 11–44)**. Check the backlash between the camshaft gear and the crankshaft gear with a dial indicator at six equally spaced teeth. Hold the gear firmly against the block while making the check. Refer to specifications in the service information system for the backlash limits.

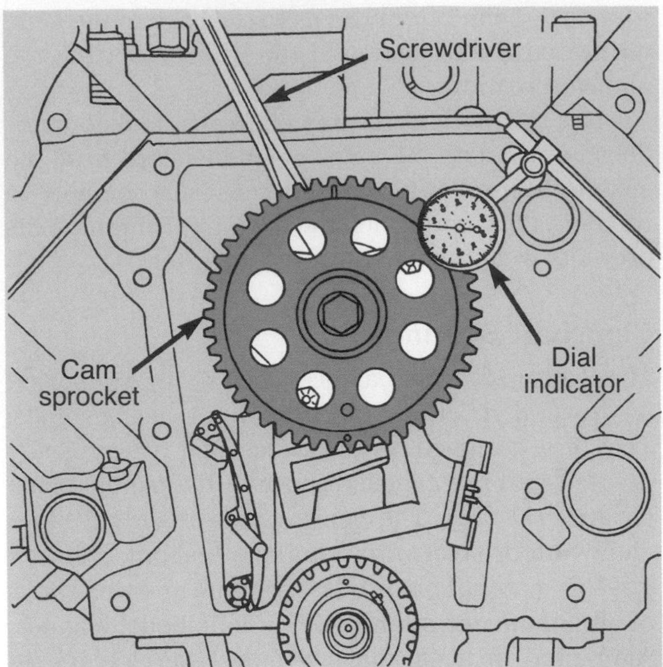

FIGURE 11–44

Checking camshaft end play with a dial indicator.

Chrysler LLC

Lifters

When inspecting mechanical lifters, carefully check their bottoms and pushrod sockets. Wear, scoring, or pitting makes their replacement necessary. Any lifter showing pitting or having its contact face worn flat or concave must also be replaced.

Technically, the normal wear of the valve lifters is referred to as adhesive or galling wear. This wear is a result of the two solid surfaces of the camshaft lobe and lifter face rubbing against each other. Fortunately, proper lubrication retards this process. However, excessive loading negates the beneficial effects of the lubricant and accelerates the wear process **(Figure 11–45)**. Examples of excessive

FIGURE 11–45

Concave lifter base wear due to long-life rotational contact with the camshaft lobe.

Martin Restoule

FIGURE 11–46

The possible wear patterns of a lifter that does not spin in its bore.

TRW, Incorporated

loading would be incorrectly matched valve springs (too much spring tension), old lifters on a new camshaft, or new lifters on an old camshaft.

If a camshaft and lifters are going to be reused, the lifters must remain with their respective lobes. Worn valve lifters and improper camshaft installation are common causes of camshaft/lifter failure.

The normal wear path is off centre with no edge contact between the lifter and the lobe. The taper on the cam lobe and the spherical radius of the lifter bottom are specifically designed to result in an offset contact pattern, causing the lifter to rotate. The spinning lifter reduces the sliding friction and equalizes the load around the lifter bottom **(Figure 11–46)**.

Whenever the valve train is disturbed, the hydraulic lifters should be removed, disassembled, cleaned, and checked. They should be kept in sequence during removal so that they can be put back in the same place. Lifters should be replaced if the bottoms are worn or pitted or if a new camshaft is installed. The bottoms of new lifters are generally spherical **(Figure 11–47)**.

Any time hydraulic lifters are removed, the varnish and deposits should be carefully removed from the lifter bores in the engine block, and the galleries should be flushed with pressurized oil to clear any dirt from the holes that feed the lifters.

After cleaning, check the lifter's leakdown with a leakdown tester. Lifter (tappet) leakdown rate is important. If the tappets leak down too quickly, noisy operation will result. When diagnosis indicates no cause for noisy tappet operation, the condition can sometimes be remedied by checking the lifter leakdown rate and replacing all lifters that are outside specifications.

Camshaft

After the camshaft has been cleaned, check each lobe for scoring, scuffing, fractured surface, pitting, and signs of abnormal wear. Also check for plugged oil passages.

Premature lobe and lifter wear is generally caused by metal-to-metal contact between the cam lobe and lifter bottom due to inadequate lubrication. The nose will be worn from the cam lobes, and the lifter bottoms will be worn to a concave shape or may be worn completely away. This type of failure usually begins within the first few minutes of operation and is the result of insufficient lubrication.

There are several ways to check cam lobes for wear, but the two most popular are the dial indicator and outside micrometer.

With the camshaft in the engine, use a dial indicator to check the lift of each cam lobe **(Figure 11–48)**. Make sure the pushrod is in the valve lifter socket. Install the dial indicator so that the cup-shaped adapter fits into the end of the pushrod and is in the same plane as the pushrod movement. Connect a remote starter switch into the starting circuit. With the ignition switch off, bump

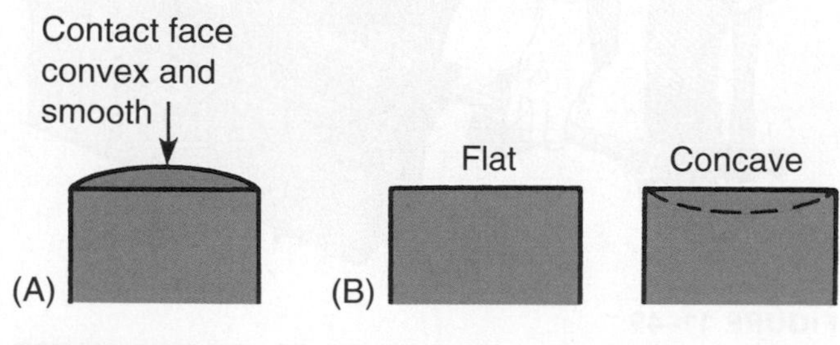

FIGURE 11–47

(A) Acceptable and (B) unacceptable valve lifter bottoms. (Drawing (A) is purposely exaggerated to show convex.)

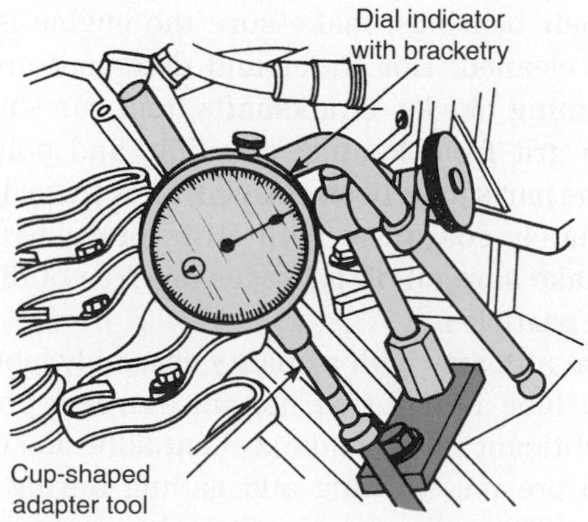

FIGURE 11–48

Checking a camshaft lobe using a dial indicator.

the crankshaft over until the lifter is on the base circle of the camshaft lobe. At this point, the pushrod will be in its lowest position. Set the dial indicator at zero. Continue to rotate the crankshaft slowly by hand until the pushrod is in its fully raised position (highest indicator reading). Compare the total lift recorded on the indicator with specifications. If the lift on the lobe is below the specified service limits, the camshaft and lifters must be replaced.

With the camshaft removed from the engine, cam lobe height can be measured with an outside micrometer. Place the micrometer so that it can measure from the heel to the nose of the lobe. Record the measurement for each intake and exhaust lobe. Any variation in height indicates wear. Also check the measurements taken against the manufacturer's cam lobe heights.

Measure each camshaft journal in several places with a micrometer to determine if it is worn excessively. If any journal is 0.0254 mm (0.001 in.) or more below the manufacturer's specifications, it should be replaced.

The camshaft should also be checked for straightness with a dial indicator. Place the camshaft on V-blocks. Position the dial indicator on the centre bearing journal and slowly rotate the camshaft. If the dial indicator shows runout (a 0.0508 mm [0.002 in.] deviation), the camshaft is not straight and must be replaced.

If the engine has a worn or damaged camshaft, identify and fix the cause of the damage before installing a new camshaft, lifters, and/or followers.

INSTALLING THE CAMSHAFT AND RELATED PARTS

Before installing the camshaft and balance shafts with their bearings, make sure the engine is thoroughly cleaned. Hot water and detergent are best for cleaning blocks, crankshafts, and camshafts to remove grit from honing, grinding, and polishing. Once the parts have been cleaned, blow them dry and immediately coat them with oil to prevent rusting. Also make sure all oil passages are free of dirt and foreign particles.

Coat all parts with a quality assembly lubricant. A good lube is one that has an extreme pressure (EP) lubricant rating and excellent adhesion quality to help prevent scuffing and galling during initial startup. The adhesion quality also prevents the lubricant from draining off the components during engine reassembly.

Camshaft Bearings

Although the installation of camshaft bearings can be done after the rest of the short block is assembled, it may be easier to align the oil holes of the bearings when the crankshaft is not yet installed. Keep in mind that any engine block that needs to have its main bearing bore alignment corrected due to distortion is likely to have camshaft bearing bore misalignment problems.

Cam bearings are normally press-fit into the block or head using a bushing driver and hammer **(Figure 11–49)**. The camshaft journals may have different diameters, with the smallest being on the rear of the block and each journal being progressively larger. Therefore, the bearing at the rear of the block should be installed first.

The new bearing is fit over the expanding mandrel of the tool, and the length of the tool is set into the block. A guide cone on the tool is used to keep the tool centred in the bore. Once the bearing is at the outside of its bore, rotate the bearing to align the oil hole in the bearing with the oil hole in the block.

On blocks with grooves behind the bearings, the bearing should be installed with the oil hole at the 2 o'clock position as viewed from the front for normal clockwise camshaft rotation. This position introduces oil into the clearance space outside the loaded area and allows shaft rotation to build an oil film ahead of the load.

While holding the centring cone against the outside bore, drive the bearing into its bore. If the cone and tool are allowed to move while inserting the bearings, the bearing can be damaged. While driving the bearings into their bore, be careful not to shave metal off the backs of the bearings. This may cause a

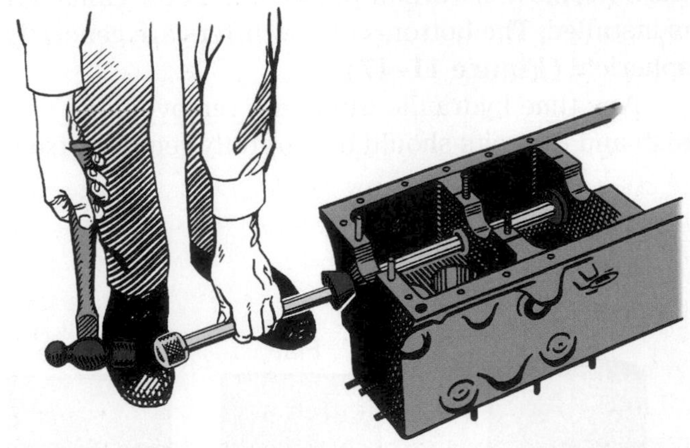

FIGURE 11–49

Cam bearings are normally press-fit into the block or head using a bushing driver and hammer.

Lisle Corp.

buildup of metal between the outside of the bearing and the housing bore, which will result in a reduction of clearance. To prevent galling, check the housing bores for proper lead-in chamfer before installing the bearings.

After the bearing is fully seated in its bore, double-check the alignment of the bearing's oil hole with the oil hole in the block by inserting a wire through the holes or by squirting oil into the holes. If the oil does not run out, the holes are misaligned. This procedure should be repeated with each bearing.

WARNING!

The use of a standard camshaft bearing driver and hammer is not recommended for aluminum heads because the aluminum bearing supports are very easily damaged or broken, which can result in expensive head replacement.

Camshaft

To install the camshaft, wipe off each cam bearing with a lint-free cloth; then thoroughly coat the camshaft lobes, bearing journals, and distributor drive gear (if there is one) with assembly lube. Also lubricate the lifters. Most premature cam wear develops within the first few minutes of operation. Prelubrication helps to prevent this when the engine is started the first time. Special prelubricants can be used only if specifically recommended by the manufacturer.

The camshaft should be carefully installed to avoid damaging the bearings with the edge of a cam lobe or journal. Be especially careful to keep it straight to prevent it from cutting or grooving the bearings. A threaded bolt in the front of the camshaft can be helpful in guiding the cam in place. Some technicians temporarily install the timing gear onto the camshaft to aid in installing the camshaft into the engine block. An alternative is to install the camshaft while the block rests on its end. When the camshaft is in place, install the thrust plate and the timing gear.

A camshaft timing gear may need to be pressed off and a replacement pressed on the camshaft prior to installing the camshaft into the block. Be sure to align the thrust plate with the woodruff key during removal to prevent damage to the thrust plate. Both the thrust plate and timing gear must then be aligned with the woodruff key for assembly. Never hammer a gear or sprocket into the shaft. Heat all metal and aluminum gears on a hot plate heated to between 90°C and 150°C (200°F and 300°F). To ensure ease of installation, install the gear while it is still hot. This step does not apply to fibre gears. These gears should be carefully installed. Press the camshaft into the gear, and be sure to keep the gear square and aligned with the keyway at all times.

FIGURE 11–50

A camshaft being ground.

Jaspar Engine and Transmission Exchange, Inc.

Once the shaft is completely in the block, the shaft should be able to be turned by hand. Binding can be caused by a damaged bearing, a nick on the cam's journal, or slight misalignment of the block journals. The cause of the problem should be identified. If the bearing clearance is too small, some technicians ream away a slight amount of the bearing; others hone the bearing. Both of these tasks need to be done carefully. The most practical way to increase the clearance is to grind down the camshaft journals **(Figure 11–50)**. Reaming or honing the inside diameter of cam bearings is not recommended because grit may become embedded in bearing surfaces, which will cause shaft wear.

CRANKSHAFT

Crankshafts **(Figure 11–51)** are generally made of cast iron, forged cast steel, or nodular iron, and then machined. At the centreline of the crankshaft are the main bearing journals **(Figure 11–52)**. These journals must be machined to a very close tolerance because the weight and movement of the crankshaft is supported at these points. The number of main bearings is determined by the design of the engine. V-block engines generally have fewer main bearings than an in-line engine with the same number of cylinders because the V-block engine uses a shorter crankshaft.

FIGURE 11–51

A crankshaft.

Offset from the crankshaft centreline are the connecting rod–bearing journals. The degree of offset and the number of journals are determined by engine design. An in-line six-cylinder engine has six connecting rod journals. A V8 engine has only four. Each journal has two connecting rods attached to it, one from each side of the V. The connecting rod journal is also called the crank pin.

Because the connecting rod journals are offset from the centreline of the crankshaft, the weight and pressure from the pistons are also offset from the centre of the crankshaft. This could create an imbalanced condition. However, to allow for smooth engine operation, counterweights are added to the crankshaft. These weights are part of the crankshaft and are positioned opposite the connecting rod journals **(Figure 11–53)**.

The machining of the main and rod bearing journals must have a very smooth surface at the bearing area. The bearings must fit tightly enough to eliminate noise but must also have enough clearance between them and their journals to allow an oil film of 0.0381 mm (0.0015 in.) and 0.0508 mm (0.002 in.) to form.

The crankshaft rotates on this film of oil. The oil is supplied by the engine's oil pump. If the crankshaft

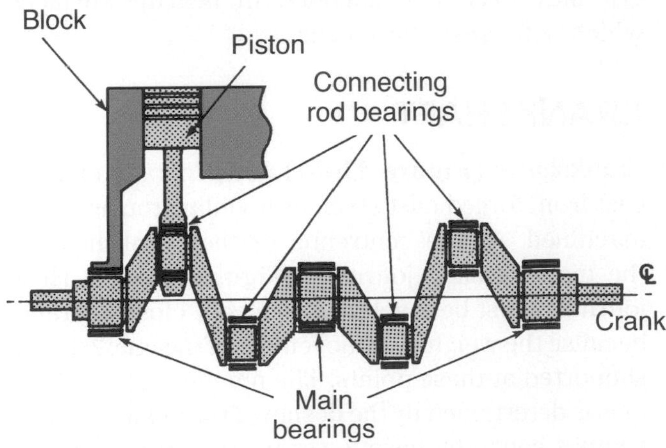

FIGURE 11–52

Crankshaft bearing and journal locations.

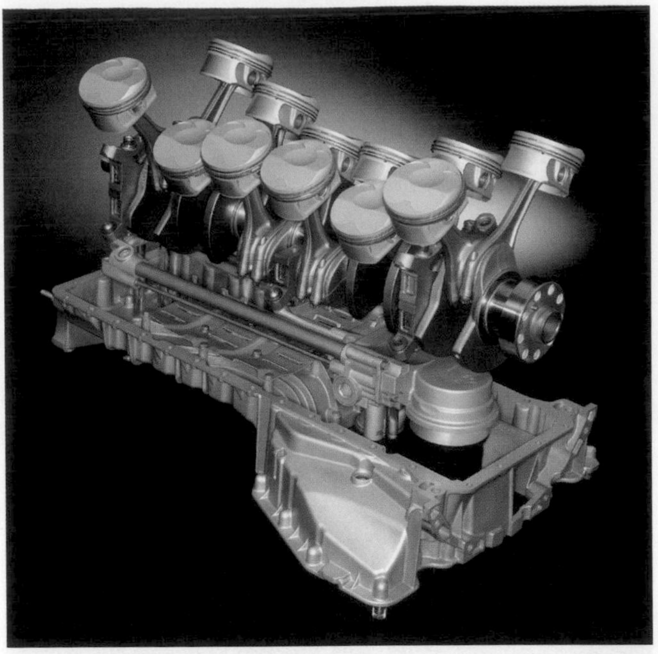

FIGURE 11–53

A crankshaft with the pistons and connecting rods attached. Notice the counterweights below the connecting rod journals.

BMW of North America, Inc.

journals become out of round, tapered, or scored, the oil film will not form properly, and the journal will contact the bearing surface, which causes early bearing or crankshaft failure. The main and rod bearings are generally made of lead-coated copper or tin and aluminum. Both of these are softer materials than that used to make the crankshaft. By using the soft material, any wear will appear first on the bearings. Early diagnosis of bearing failure most often will spare the crankshaft, and only the bearings will need to be replaced.

The bearings are fed oil under pressure. In order for the oil to reach these bearings, oil passages must be drilled into the crankshaft. Each main bearing journal has a hole drilled into it with a connecting hole or holes leading to one or more rod bearing journals. In this way, all bearing journals receive oil to protect both the bearing and the journal.

The crankshaft configuration determines the engine block design, or the positioning of the connecting rod journals around the centreline of the crankshaft **(Figure 11–54)**.

A crankshaft has two distinct ends. One is called the flywheel end, and as its name implies, this is where the flywheel is attached. The front end or belt drive end of the crankshaft contains a threaded snout or is drilled and tapped. The timing gear and damper are mounted on the snout.

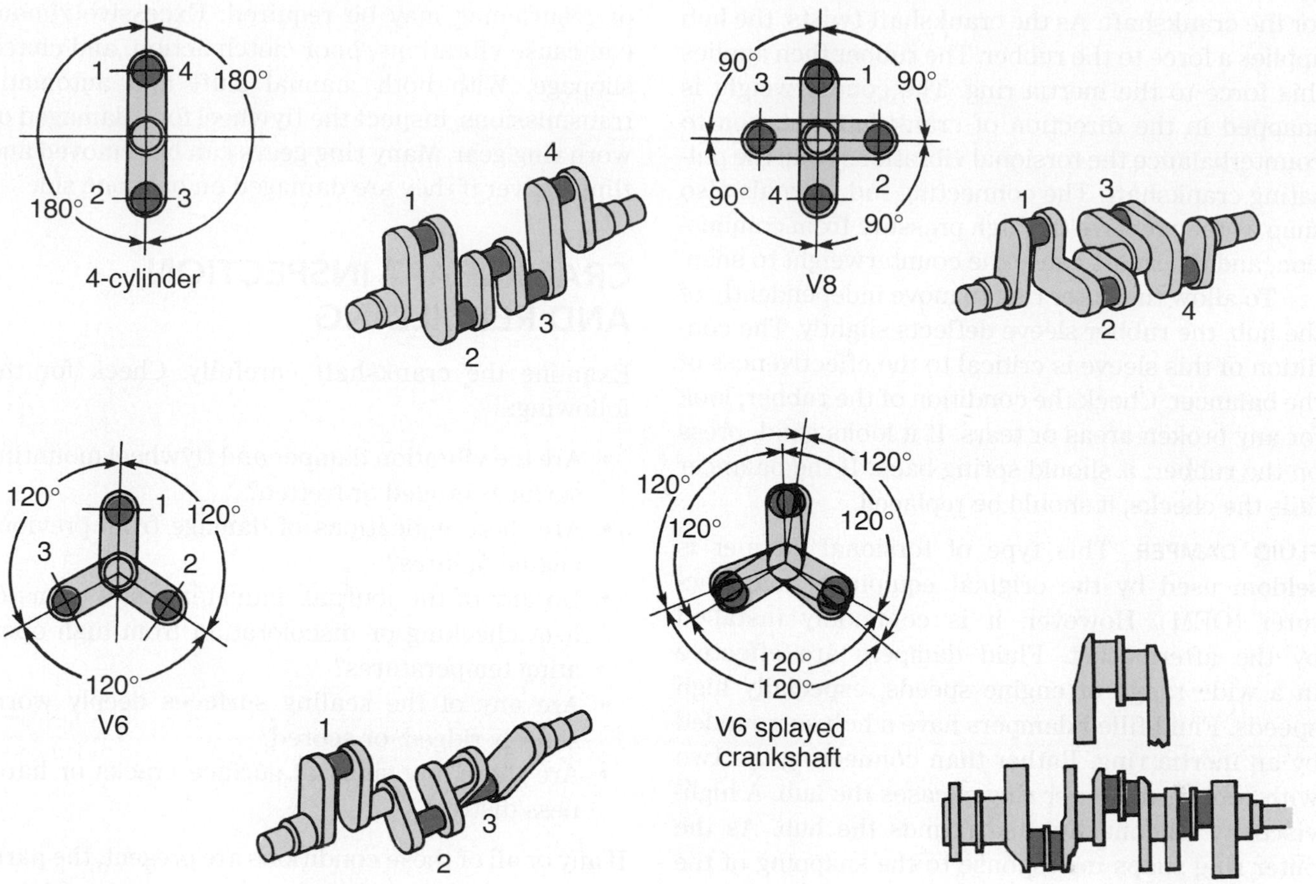

FIGURE 11–54

Various crankshaft configurations.

Crankshaft Torsional Dampers

Combustion causes an extreme amount of pressure in a cylinder (more than two tonnes each time a cylinder fires). This pressure is applied to the pistons and moves through the connecting rods to the crankshaft. This downward force causes the crankshaft to rotate. In an engine with more than one cylinder, this pressure is exerted at different places on the crankshaft and at different times. As a result, the crankshaft tends to twist and deflect, causing torsional harmonic vibrations. These vibrations constantly change, but there are specific engine speeds where these harmonics are amplified. This increase in torsional vibration can cause damage to the crankshaft, the engine, and/or any accessories that are driven by the crankshaft.

These harmful vibrations are often limited by a torsional damper located at the front of the crankshaft. There are two common types of torsional dampers: harmonic balancers and fluid dampers. Both use friction to reduce crankshaft vibrations.

HARMONIC BALANCER A harmonic balancer **(Figure 11–55)**, also called a vibration damper, is the most common. A harmonic balancer has a cast-iron hub and an outer cast-iron inertia ring that is connected to the hub with an elastomer (rubber) sleeve. The hub of the harmonic balancer is pressed onto the snout of the crankshaft. The inertia ring is heavy and is machined to serve as a counterweight

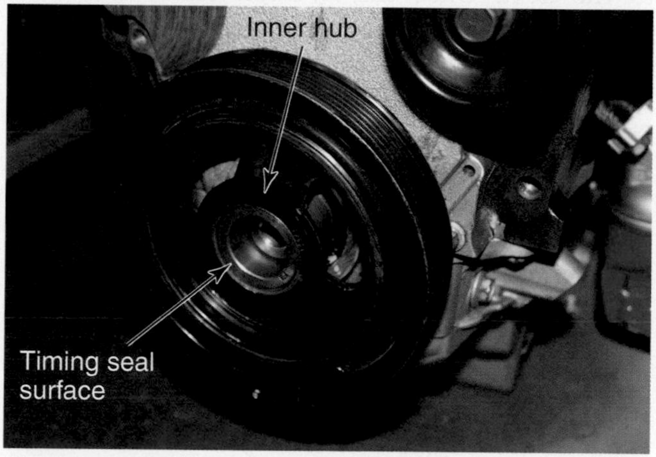

FIGURE 11–55

A vibration damper harmonic balancer.

Martin Restoule

for the crankshaft. As the crankshaft twists, the hub applies a force to the rubber. The rubber then applies this force to the inertia ring. The counterweight is snapped in the direction of crankshaft rotation to counterbalance the torsional vibrations from the pulsating crankshaft. The connecting rod journals also snap as they receive the high pressure from combustion, and the snaps cause the counterweight to snap.

To allow the outer ring to move independently of the hub, the rubber sleeve deflects slightly. The condition of this sleeve is critical to the effectiveness of the balancer. Check the condition of the rubber; look for any broken areas or tears. If it looks good, press on the rubber; it should spring back. If the balancer fails the checks, it should be replaced.

FLUID DAMPER This type of torsional damper is seldom used by the original equipment manufacturer (OEM). However, it is commonly installed by the aftermarket. Fluid dampers are effective in a wide range of engine speeds, especially high speeds. Fluid-filled dampers have a hub surrounded by an inertia ring. Rather than connecting the two with rubber, the outer ring encases the hub. A high-viscosity silicone fluid surrounds the hub. As the outer ring snaps in response to the snapping of the connecting rods and crankshaft, the outer ring rubs against the hub. This rubbing creates friction. The friction is absorbed by the fluid and turned into heat. Therefore, the vibrations are changed to heat, and the heat is dissipated from the damper.

Flywheel

The **flywheel** also helps to make the engine run more smoothly by applying a constant moving force to carry the crankshaft from one firing stroke to the next. Once the flywheel starts to rotate, its weight tends to keep it rotating. This is called inertia. The flywheel's inertia keeps the crankshaft rotating smoothly in spite of the pulses of power from the pistons.

Because of its large diameter, the flywheel also makes a convenient point for the starter to connect to the engine. The large diameter supplies good gear reduction for the starter, making it easy for the starter to turn the engine against its compression. The surface of a flywheel may be used as part of the clutch. On an engine that drives an automatic transmission, a **flex plate** is used. The automatic transmission torque converter provides the weight required to attain flywheel functions.

FLYWHEEL INSPECTION Check the runout of the flywheel, and carefully inspect its surface. Replacement or resurfacing may be required. Excessive runout can cause vibrations, poor clutch action, and clutch slippage. With both manual shift and automatic transmissions, inspect the flywheel for a damaged or worn ring gear. Many ring gears can be removed and flipped over if they are damaged on only one side.

CRANKSHAFT INSPECTION AND REBUILDING

Examine the crankshaft carefully. Check for the following:

- Are the vibration damper and flywheel mounting surfaces eroded or fretted?
- Are there indications of damage from previous engine failures?
- Do any of the journal diameters show signs of heat checking or discoloration from high operating temperatures?
- Are any of the sealing surfaces deeply worn, sharply ridged, or scored?
- Are there any signs of surface cracks or hardness distress?

If any or all of these conditions are present, the parts need to be repaired or replaced.

To measure the diameter of a rod journal, use an outside micrometer **(Figure 11–56)**. Measure the journals for size, out-of-roundness, and taper **(Figure 11–57)**. Taper is measured from one side of the journals to the other. The maximum taper and out-of-roundness is 0.0254 mm (0.001 in.).

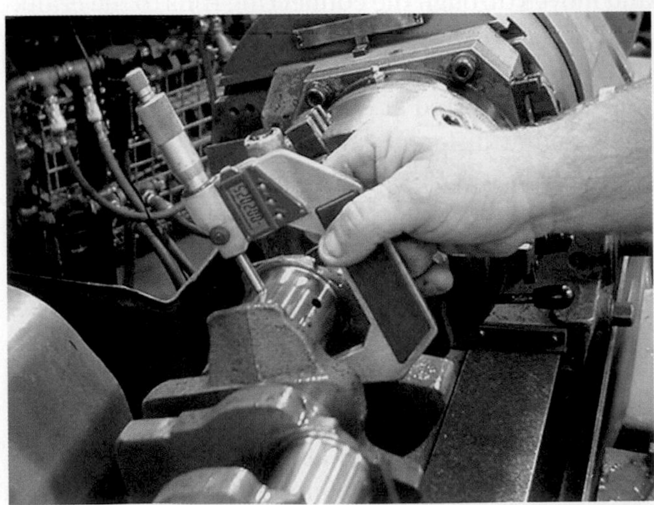

FIGURE 11–56

Measure the diameters of crank journals with an outside micrometer.

Jaspar Engine and Transmission Exchange, Inc.

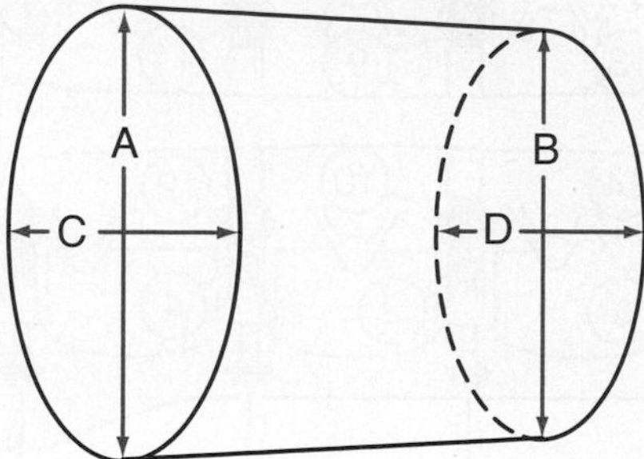

FIGURE 11–57

Checking crankshaft journals for out-of-roundness and taper.

Compare these measurements to specifications to determine if the crankshaft needs to be reground or replaced. If the journals are within specifications, the journal area needs only to be cleaned up.

USING SERVICE INFORMATION

Crankshaft specifications can be found in the engine specification section of a service information system.

Crankshaft Reconditioning

If the crankshaft is severely damaged, it should be replaced. A crankshaft that has journal taper, grooves in the journal surfaces, burnt marks, or small nicks in the journal surface may be reusable after the journals are refinished. This process grinds away some of the metal on the journals to provide an even and mar-free surface. When a crankshaft has been ground, undersize bearings are fitted to the crankshaft to provide for the proper oil clearances.

At times, minor damage to the journals can be corrected by polishing the journals with a very fine crocus cloth. A polishing tool rotates a long loop of crocus cloth against the journals as the crankshaft is rotated by a stand. The constant movement of the crocus cloth and the rotation of the crankshaft prevent the creation of any flat spots on the surface. You should also polish the journals after they have been ground.

Checking Crankshaft Straightness

To evaluate the straightness of the crankshaft, the shaft should be supported by V-blocks positioned on the end main bearing journals. Position a dial indicator at the 3 o'clock position on the centre main bearing journal.

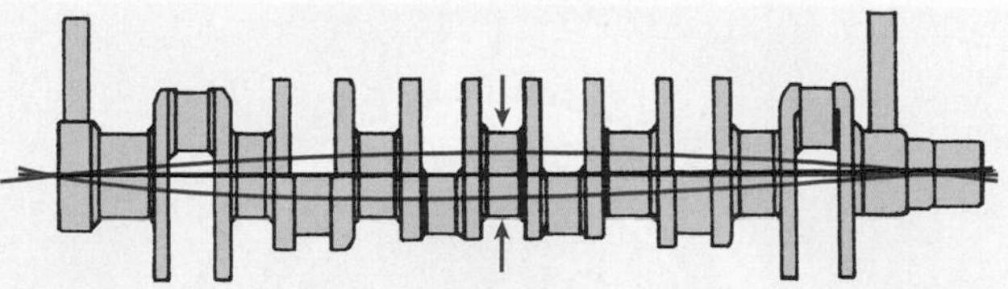

FIGURE 11–58

Evaluating alignment bow.

Set the indicator at 0 (zero), and turn the crankshaft through one complete rotation. The total deflection of the indicator, the amount greater than zero plus the amount less than zero, is the **total indicator reading (TIR)**. Bow is 50 percent of the TIR **(Figure 11–58)**. Compare the bow of the crankshaft to the acceptable alignment/bow specifications.

A special machine is designed to straighten crankshafts but is found only in serious engine rebuilding shops. In most cases, if the crankshaft is warped, it is replaced.

Crankshaft Bearings

Bearings are used to carry the critical loads created by crankshaft movement. They are a major wear item in the engine and require close inspection. Main bearings support the crankshaft journals. Connecting rod bearings are installed between the crankshaft and connecting rods.

Modern crankshaft bearings are known as insert bearings. There are two basic designs of insert bearings **(Figure 11–59)**. A **full-round** (one-piece) **bearing** is used in bores that allow the shaft's journals to be inserted into the bearing, such as a camshaft. A **split** (two halves) **bearing** is used where

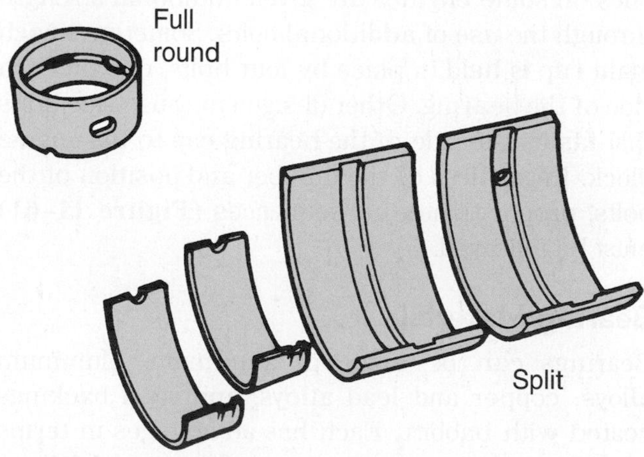

FIGURE 11–59

Full-round and split insert bearings.

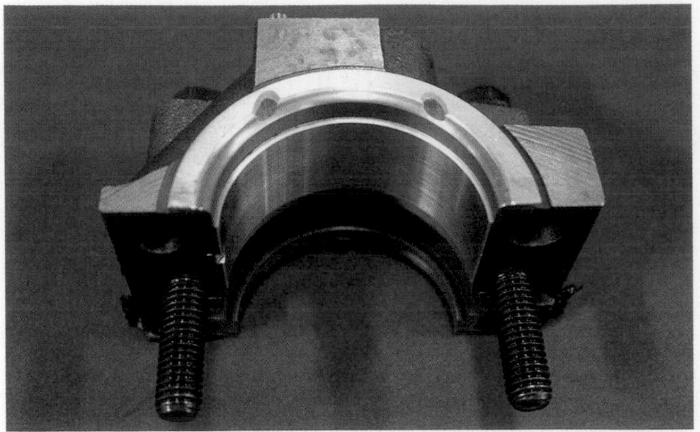

FIGURE 11–60

A thrust bearing with grooves cut into its flange to provide for better lubrication.

Martin Restoule

the bearing must be assembled around the journal with the bearing housing being of two parts also, including a cap that holds the assembly together. Crankshaft bearings are typically the split type.

Many crankshafts are fitted with a main bearing that has flanged sides. This type of bearing is typically called a thrust bearing and is used to control any horizontal movement or end play of the shaft. The flange bearing is used in the thrust position of the block. Most thrust main bearings are double-flanged **(Figure 11–60)**.

Some late-model engines do not use separate main bearing caps; instead, they are fitted with a lower engine block assembly called a bed plate. This assembly works like a bridge and contains the lower half of the bore for the main bearings. The assembly is torqued to the engine block and holds the crankshaft in place.

The main bearing caps and lower block assemblies on some engines are given additional strength through the use of additional bolts. Sometimes each main cap is held in place by four bolts, two on each side of the bearing. Other designs may use side bolts that fasten the side of the bearing cap to the engine block. Regardless of the number and position of the bolts, proper tightening sequences **(Figure 11–61)** must be followed.

Bearing Materials

Bearings can be made of aluminum, aluminum alloys, copper and lead alloys, and steel backings coated with babbitt. Each has advantages in terms of resistance to corrosion, rate of wear, and fatigue strength. Two-layer, aluminum alloy bearings are the most commonly used. These bearings contain

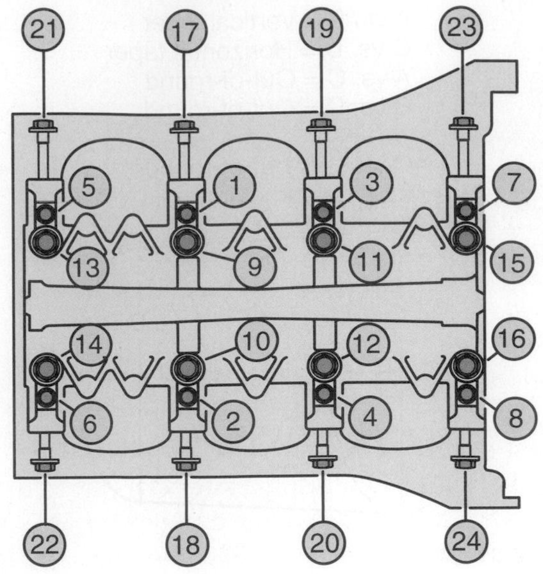

FIGURE 11–61

Six bolts secure each main cap in this engine. Each bolt must be tightened in correct sequence and to the correct torque.

©Cengage Learning 2015

silicon, which helps to reduce wear. Some bearings use a combination of three layers of metals, such as a layer of copper-lead alloy on a steel backing, followed by a thin coating of babbitt **(Figure 11–62)**. This design takes advantage of the excellent properties of each metal.

Bearing Spread

Most main and connecting rod bearings are manufactured with spread. Bearing spread means that the distance across the outside parting edges of the

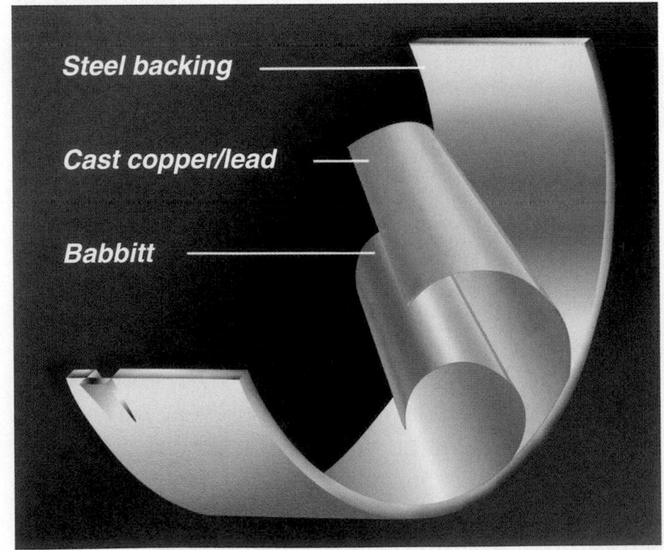

FIGURE 11–62

The basic construction of a bearing composed of three metals.

Dana Corporation

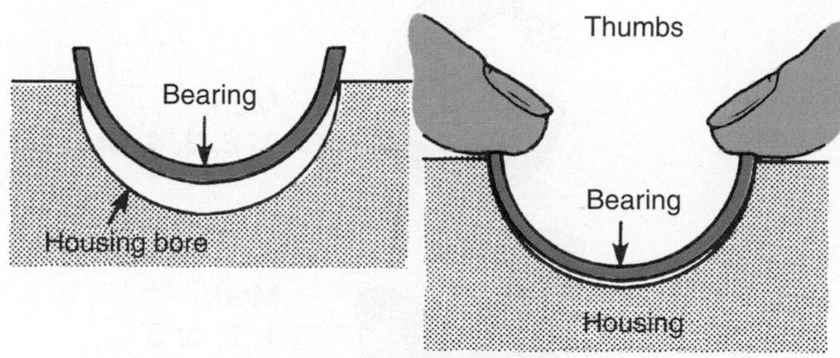

FIGURE 11–63

Spread requires a bearing to be lightly snapped into place.

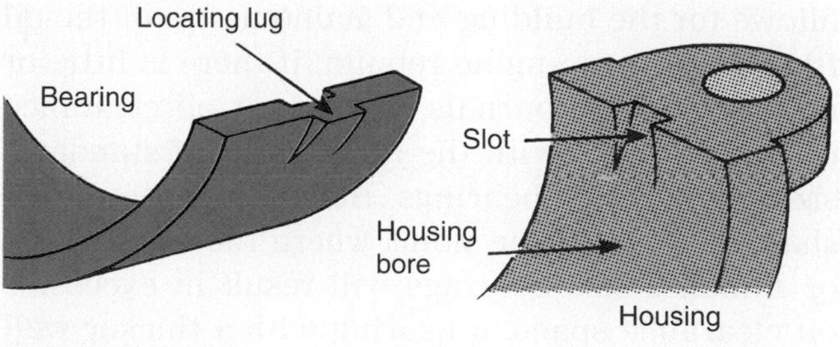

FIGURE 11–65

The locating lug fits into the slot in the housing.

bearing insert is slightly greater than the diameter of the housing bore. To position a bearing half that has spread, it must be snapped into place by a light forcing action **(Figure 11–63)**. This ensures positive positioning against the inside of the bore and helps to keep the bearings in place during assembly.

Bearing Crush

Each half of a split bearing is made so that it is slightly greater than an exact half. This can be seen quite easily when a half is snapped into place in its housing. The parting faces extend a little beyond the seat **(Figure 11–64)**. This extension is called **crush**.

When the two bearing halves are assembled and the housing cap tightened, the crush sets up a radial pressure on the bearing halves so that they are forced tightly into the housing bore.

Bearing crush increases the surface contact between the bearing and its bore, allowing for better heat transfer, compensation for slight bore distortion, and prevention of the bearings rotating during operation.

Bearing Locating Devices

Engine bearings must be provided with some means to keep them from shifting sideways in their housings during installation. Many different methods have been used by manufacturers to keep the bearings in

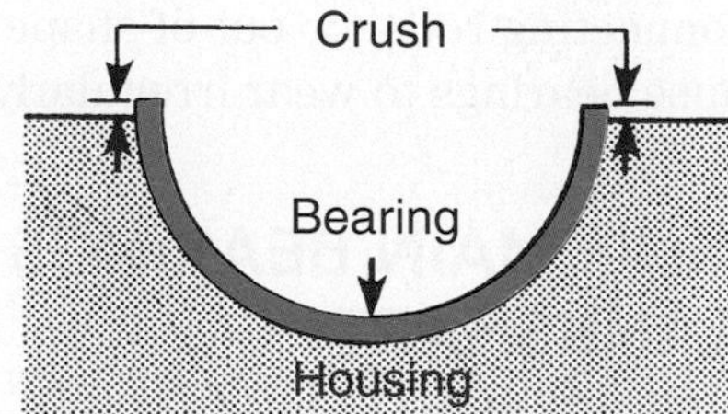

FIGURE 11–64

Crush ensures good contact between the bearing and the housing.

place. The most common way is the use of a locating lug. As shown in **Figure 11–65**, this consists of a protrusion at the parting face of the bearing. The lug fits into a slot in the bearing's bore.

Oil Grooves

Providing an adequate oil supply to all parts of the bearing surface, particularly in the load area, is an absolute necessity. In many cases, this is accomplished by the oil flow through the bearing oil clearance. In other cases, however, engine operating conditions are such that this oil distribution method is inadequate. When this occurs, some type of oil groove must be added to the bearing. Some oil grooves are used to ensure an adequate supply of oil to adjacent engine parts by means of oil throw-off.

Most OEM bearings have a full groove around the entire circumference of the bearing, and others have a half groove in which only the upper bearing half is grooved.

Oil Holes

Oil holes allow for oil flow through the engine block galleries and into the bearing oil clearance space. Connecting rod bearings receive oil from the main bearings by means of oilways in the crankshaft. Oil holes are also used to meter the amount of oil supplied to other parts of the engine. For example, oil squirt holes in connecting rods are often used to spray oil onto the cylinder walls. When the bearing has an oil groove, the oil hole normally is in line with the groove.

The size and location of oil holes is critical. Therefore, when installing bearings, you must make sure the oil holes in the block line up with holes in the bearings.

Oil Clearance

There must be a gap or clearance between the outside diameter of the crankshaft journals and the inside diameter of its bearings. This clearance allows for the building and maintenance of the oil film. During an engine rebuild, if there is little or no wear on the journals, the proper oil clearance can be restored with the installation of standard-size replacement bearings. However, if the crankshaft is worn to the point where the installation of standard-size bearings will result in excessive oil clearance space, a bearing with a thicker wall must be used. Although these bearings are thicker, they are known as undersize because the journals and crank pins of the crankshaft are smaller in diameter. In other words, they are under the standard size.

Undersize bearings are available in 0.0254 mm (0.001 in.) or 0.0508 mm (0.002 in.) sizes for shafts that are uniformly worn by that amount. Undersize bearings are also available in thicker sizes, such as 0.2540 mm (0.010 in.), 0.5080 mm (0.020 in.), and 0.7620 mm (0.030 in.), for use with crankshafts that have been refinished (or reground) to one of these standard undersizes. The difference in thickness of the bearing is normally stamped onto the backside of the bearing. Bearings may also be colour-coded to indicate their size **(Figure 11–66)**.

Often, engines are manufactured with other-than-standard journal sizes. The manufacturer uses colour codes or stamped numbers to indicate which bearing size to use **(Figure 11–67)**.

STANDARD
0.026 MM/0.001 IN.
0.052 MM/0.002 IN.
0.25 MM/0.010 IN.
0.50 MM/0.020 IN.
0.625 MM/0.025 IN.
0.75 MM/0.030 IN.
1.00 MM/0.040 IN.
1.25 MM/0.050 IN.
1.50 MM/0.060 IN.

FIGURE 11–66

Some bearings are colour-coded to indicate their size.

Dana Corporation

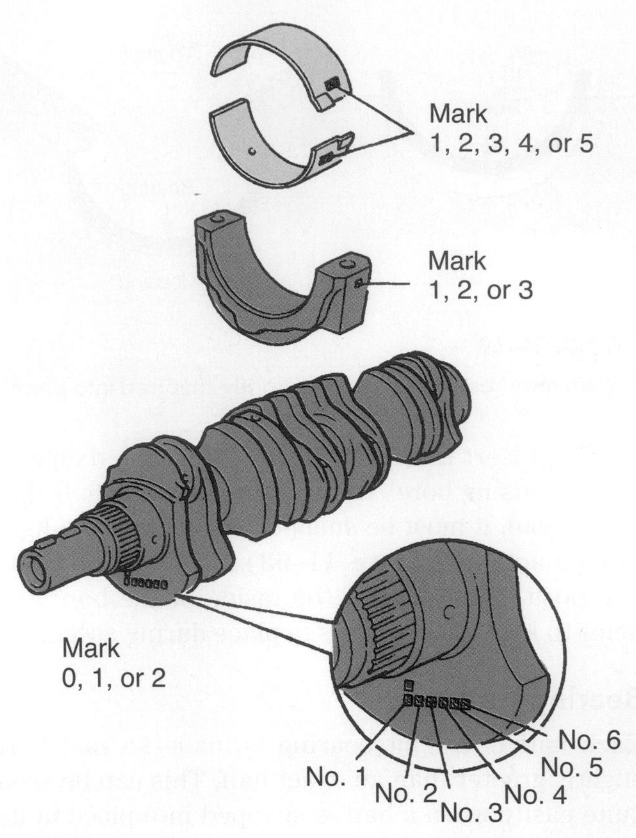

FIGURE 11–67

Size marking on a crankshaft, connecting rod, and rod bearing.

SHOP TALK

If the journals measure within specifications, but pitting and gouges exist, polish the worst journal to determine whether grinding is necessary. If polishing the journal achieves smoothness, then grinding is probably not necessary. If the crankshaft does not have to be reground, check it for straightness.

Bearing Failure and Inspection

As shown in **Figure 11–68**, bearings can fail for a variety of reasons. Oil starvation and dirt are the major reasons for bearing failure. Problems in other engine components, such as bent or twisted crankshafts or connecting rods, or out-of-shape journals, can also cause bearings to wear irregularly.

INSTALLING MAIN BEARINGS AND CRANKSHAFT

When selecting new main bearings, make sure they match the crankshaft journal diameters and main bearing bores. If the crankshaft has been ground undersize, the main bearings must also be undersize.

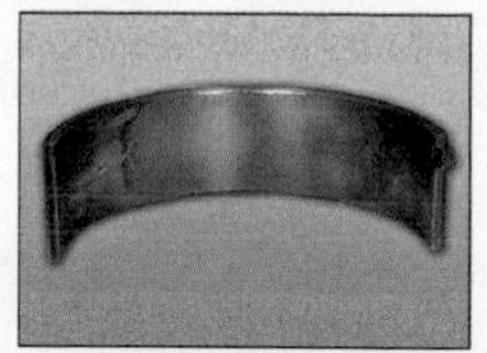
Normal wear

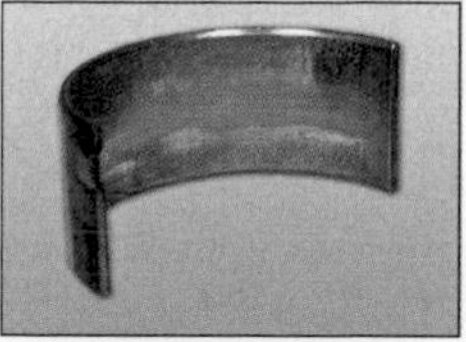
Overlay fatigue

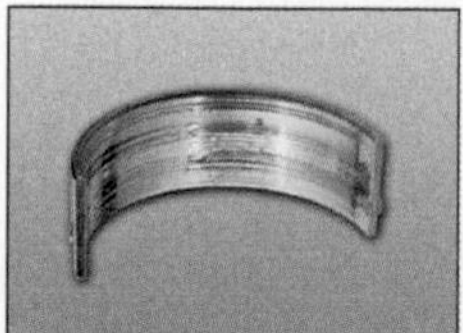
Scoring

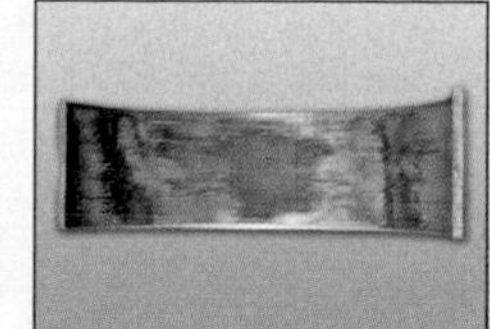
Corrosion

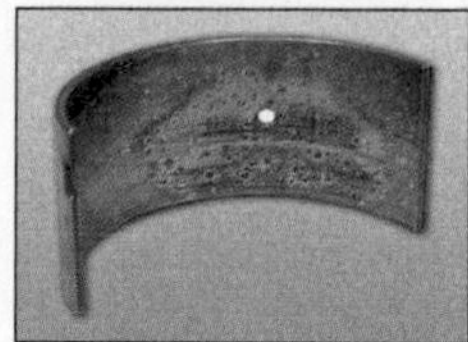
Dirt embedment

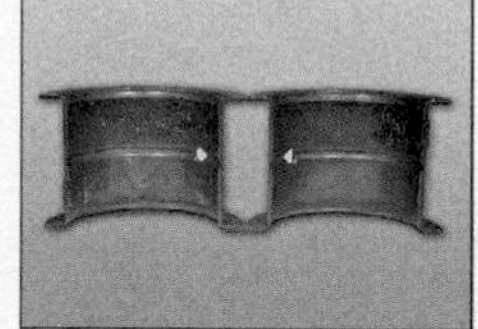
Cap shift

Distorted crankcase

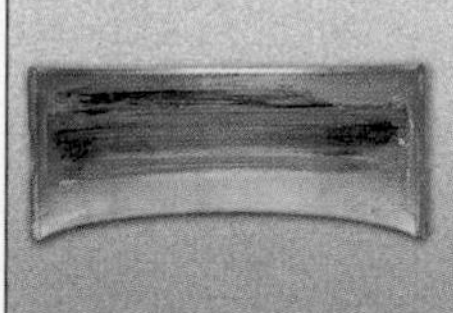
Oil starvation

Accelerated wear

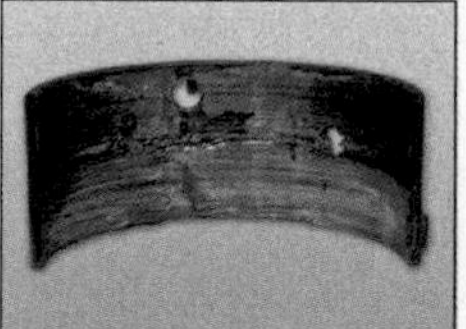
Hot short

Dirt on bearing back

Wiped

Fretting

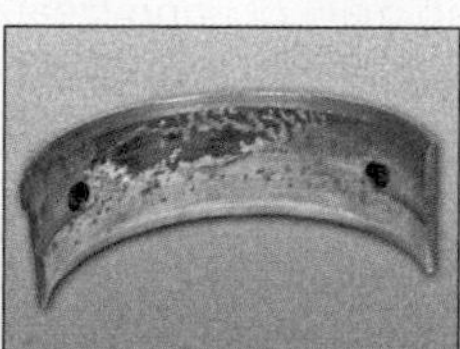
Fatigue

FIGURE 11–68

Common forms of bearing distress.

MAHLE

Similarly, if the housing bores have been machined oversize by align boring or align honing, the bearings must take up this space. Bearing size is usually marked on the bearing box and on the back of the bearing.

When the bearings are ready to be installed in the main bearing bores, make sure the bore is clean and dry before installing the bearing halves into place. Use a clean, lint-free cloth to wipe the bearing back and bore surface.

Put the new main bearing inserts into each of the main bearing caps **(Figure 11–69)** and into the bearing bores in the cylinder block housings. Make sure all holes align. The backs of the main bearing inserts should never be oiled or greased. Place the crankshaft in the block on the main bearing inserts, and arrange the main bearing caps in the correct order and direction over the crankshaft. Follow the factory markings or use those made during disassembly.

The next step is to measure the oil clearance between the crankshaft and the main bearing. Proper lubrication and cooling of the bearing depend on correct crankshaft oil clearances. Scored bearings, worn crankshaft, excessive cylinder wear, stuck piston rings, and worn pistons can result from too small an oil clearance. If the oil clearance is too great, the crankshaft might pound up and down, overheat, and weld itself to the insert bearings.

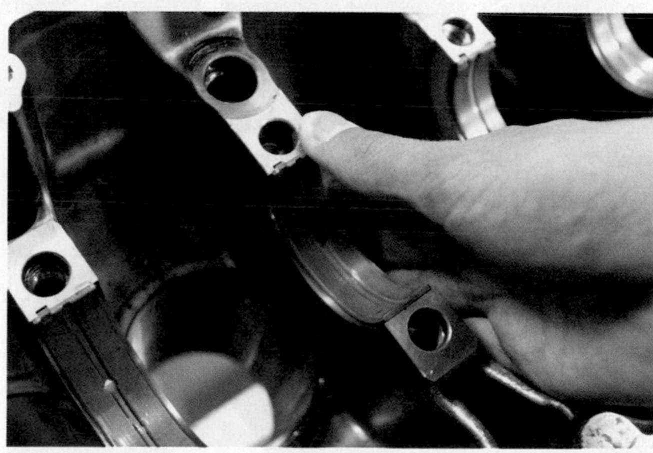
FIGURE 11–69

Place the bearing inserts into the bore, making sure the locating lugs fit into their recess.

Martin Restoule

Plastigage is fine, plastic string used to measure the oil clearance between the bearing and the crankshaft. The procedure for checking bearing clearance with Plastigage is outlined in Photo Sequence 7 included in this chapter.

PHOTO SEQUENCE 7

CHECKING MAIN BEARING CLEARANCE WITH PLASTIGAGE

P7–1 Checking main bearing clearance begins with mounting the engine block upside down on an engine stand.

P7–2 Install main bearings into bores, being careful to properly seat them. Wipe the bearings with a clean, lint-free rag.

P7–3 Carefully install the crankshaft into the bearings. Try to keep the crankshaft from moving on the bearing surfaces.

P7–4 Wipe the crankshaft journals with a clean rag.

P7–5 Place a piece of Plastigage on the journal. The piece should fit between the radii of the journal.

P7–6 Install the main caps in their proper locations and directions. Wipe the threads of the cap bolts with a clean rag.

P7–7 Install the cap bolts and tighten them according to the manufacturer's recommendations.

P7–8 Remove the main caps and observe the spread of the Plastigage. If the Plastigage did not spread, try again with a larger Plastigage.

P7–9 Compare the spread of the Plastigage with the scale given on the Plastigage container. Compare the clearance with the specifications.

P7–10 Carefully scrape the Plastigage off the journal surface.

P7–11 Wipe the journal clean with a rag.

P7–12 If the clearance was within the specifications, remove the crankshaft and apply a good coat of fresh engine oil to the bearings.

P7–13 Reinstall the crankshaft, and apply a coat of oil to the journal surfaces.

P7–14 Reinstall the main caps and tighten according to specifications.

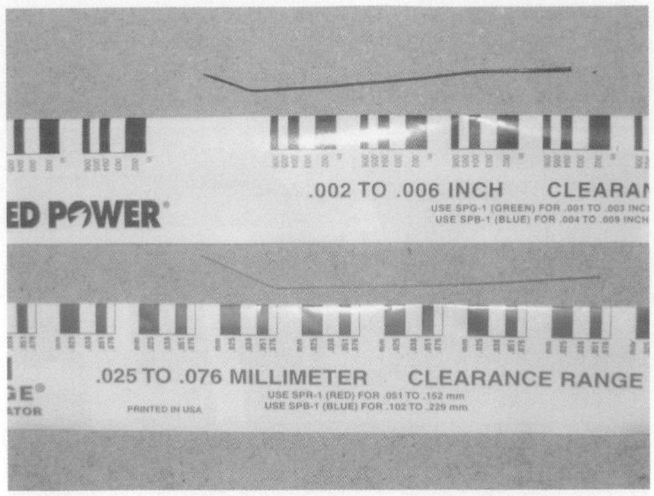

FIGURE 11–70
Plastigage is available in a variety of ranges, and the packing is colour-coded according to the range.
Martin Restoule

One side of the Plastigage's package has stripes for metric measurements, and the other side has stripes for inch measurements **(Figure 11–70)**. The string can be purchased to measure different clearance ranges. Usually, only the smallest clearance range is necessary for reassembly work.

Crankshaft End Play

Crankshaft end play can be measured with a feeler gauge by prying the crankshaft rearward and measuring the clearance between the thrust bearing

FIGURE 11–71
Crankshaft end play can be checked with a feeler gauge.
Martin Restoule

flange and a machined surface on the crankshaft. Insert the feeler gauge at several locations around the rear thrust bearing face **(Figure 11–71)**. Or position a dial indicator so that the fore and aft movement of the crankshaft can be measured **(Figure 11–72)**.

If the end play is less than or greater than the specified limits, the main bearing with the thrust surface must be exchanged for one with a thicker or thinner thrust surface. If the engine has thrust washers or shims, thicker or thinner washers or shims must be used.

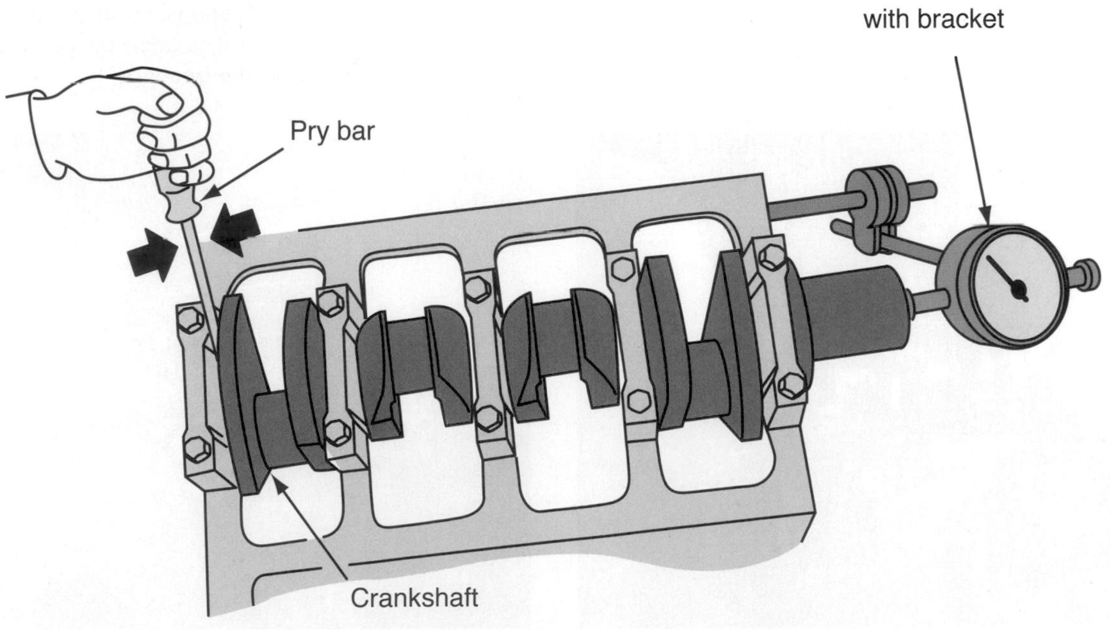

FIGURE 11–72
Crankshaft end play can be checked with a pry bar and dial indicator.

FIGURE 11–73

A typical crankshaft seal.

Most engines require the installation of main bearing seals **(Figure 11–73)** during the final installation of the crankshaft.

Connecting Rod

The connecting rod is used to transmit the pressure applied on the piston to the crankshaft **(Figure 11–74)**. Connecting rods are able to swivel at the piston and the crankshaft. This allows them to freely move the pistons up and down while they rotate around the crankshaft. A connecting rod faces great stress. The force applied during the power stroke is applied to the connecting rod as it moves through a variety of angles. The rod has great force applied to it from the top and has great resistance to movement at the bottom. The centre section of a rod is basically an I-beam. This provides maximum strength with minimum weight.

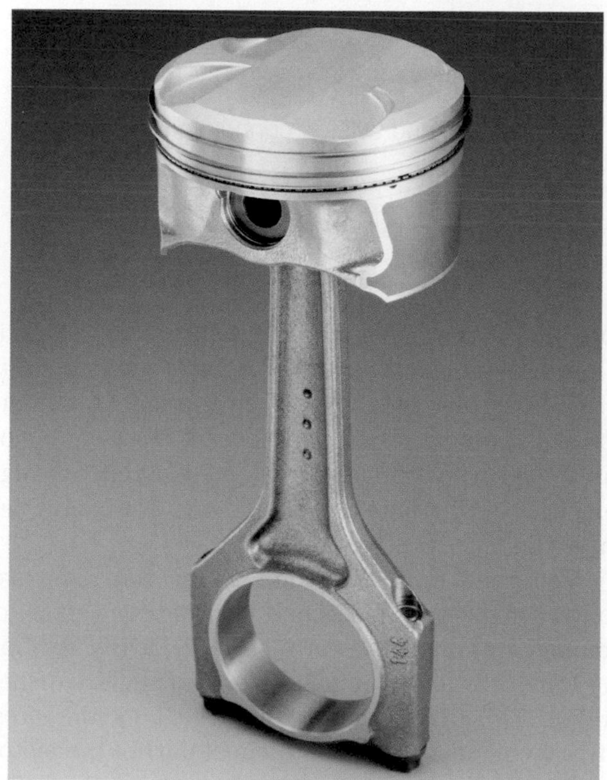

FIGURE 11–74

A piston and connecting rod assembly.

BMW of North America, Inc.

Connecting rods are kept as light as possible. They are generally forged from high-strength steel or made of nodular steel or cast iron. Cast iron is rarely used in automotive engines. Aluminum and titanium connecting rods are also used. Aluminum rods are light and have the ability to absorb high-pressure shocks, but they are not as durable as steel rods. Titanium rods are very strong and light but are rather expensive. Some late-model engines, such as the Ford 4.6-litre and the Chrysler 2.0-litre engines, have powdered (sintered) metal connecting rods. These rods are light and strong and are easily identified by their smoothness. The small end or piston pin end is made to accept the piston pin, which connects the piston to the connecting rod. The piston pin can be press-fit in the piston and free-fit in the rod. When this is the case, the small end of the rod will be fitted with a bushing. The pin can also be free-fit in the piston and press-fit in the rod. In this case, no bushings are used. The pin simply moves in the piston, using the piston hole as a bearing surface. A pin press **(Figure 11–75)** is used to separate and attach press-fit piston pins and pistons from their connecting rods. A third mounting allows the pin

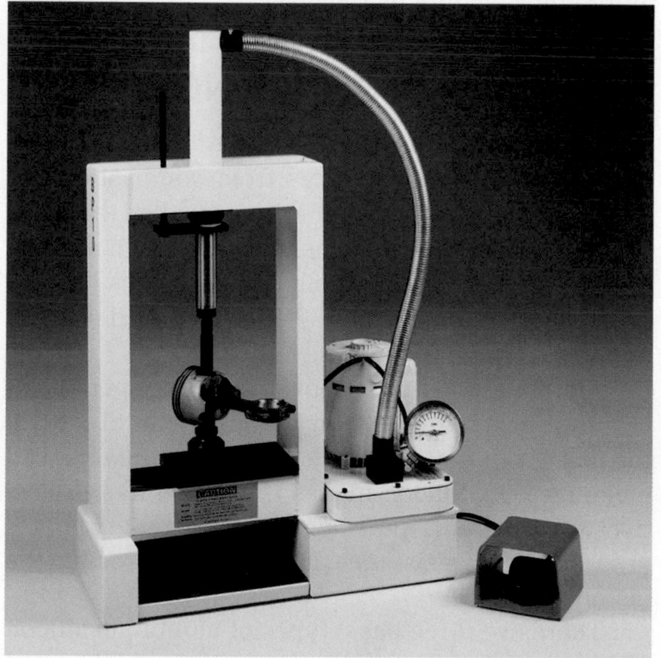

FIGURE 11–75

A pin press.

to move freely in both the piston and the rod. This design requires the use of clips or caps to prevent the pin from moving out against the cylinder walls.

The larger, or "big," end of the rod is used to attach the connecting rod to the crankshaft. This end is made in two pieces. The upper half is part of the rod. The lower half is called the rod cap and is bolted to the rod. The connecting rod and its cap are manufactured as a unit and must always be kept together. During production, the rod caps are either machined off the rod or are scribed and broken off. The big end of the rod is fitted with bearing inserts made of the same material as the main bearings. Some connecting rods have a hole drilled through the big end to the bearing area. The bearing insert might have a hole, which will align with this drilling. This hole is used to supply oil for lubricating and cooling the piston skirt. When the rod is properly installed, the oil hole should be pointing to the major thrust area of the cylinder wall.

INSPECTION Closely examine all piston skirts and bearings for unusual wear patterns that may indicate a twisted rod **(Figure 11–76)**. Rods suspected of being bent or distorted can be checked with a rod alignment checker. Normally, damaged rods are replaced, although equipment is available to straighten them and to rebore the small and big ends. Many manufacturers recommend a check of the rod bolts before reusing them. The typical procedure involves measuring the diameter of the bolt at its tension portion. If the diameter is less than the minimum, it should be replaced.

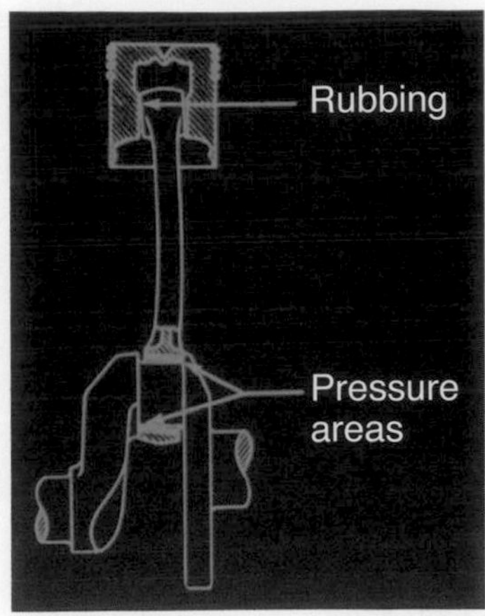

Major misalignment

Bent and twisted connecting rod

FIGURE 11–76

The effects of rod misalignment on its bearings.

MAHLE

PISTON AND PISTON RINGS

The piston forms the lower portion of the combustion chamber. The pressures from combustion are exerted against the top of the piston, called the **head** or **dome**. A piston must be strong enough to face this pressure; however, it should also be as light as possible. This is why most pistons are made of aluminum or aluminum alloys.

Aluminum pistons alloyed with copper, magnesium, nickel, and silicon are common. Silicon is the most common element mixed with aluminum to make pistons. Silicon makes the piston more resistive to corrosion and improves its strength, hardness, and wear resistance. It also helps to reduce the piston's weight.

There are three basic types of aluminum silicon alloys used in pistons: hypoeutectic, eutectic, and hypereutectic. **Hypoeutectic** pistons, common in earlier engines, have about 9 percent silicon. Most eutectic pistons have 11 to 12 percent silicon. Eutectic alloys provide good strength and are economical to make. Hypereutectic pistons have a silicon content above 12 percent. They offer low thermal expansion rates, improved groove wear, good resistance to high temperatures, greater strength, and scuff and seizure resistance.

The head of the piston can be flat, concave, convex, crowned, raised and relieved for valves, or notched for valves. Newer pistons typically are flat, are flat with valve notches, or have a slightly dished crown. The dished crown concentrates the pressure of combustion at the thickest part of the piston head, right above the top of the piston pin boss. The piston pin boss is a built-up area around the bore for the piston pin, sometimes called the **wrist pin (Figure 11–77)**. The pin bore is not always centred

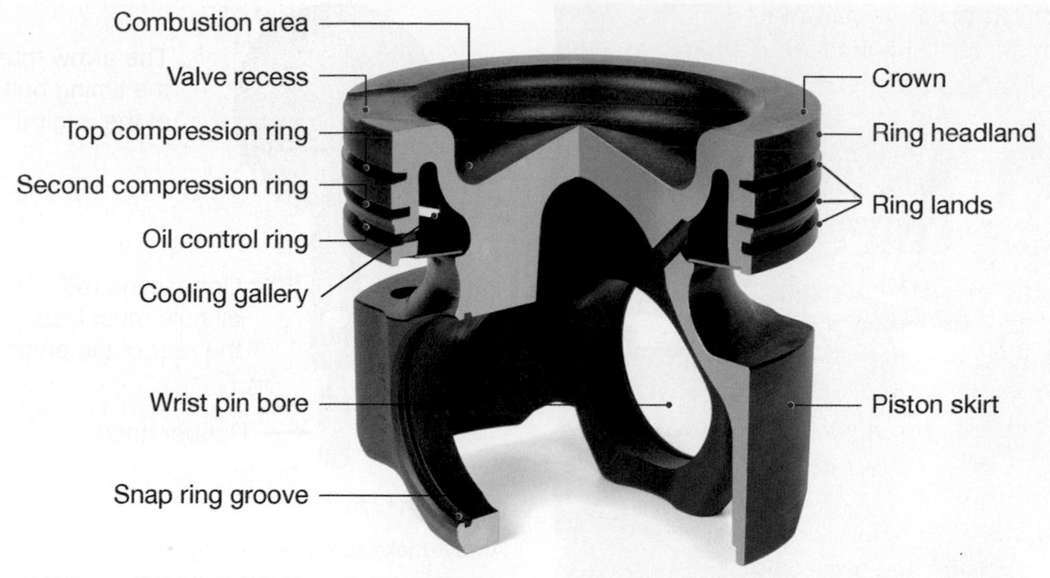

FIGURE 11–77

The features and terminology used to describe a piston.

MAHLE

in the piston. It can be offset toward the major thrust side of the piston, which is the side that will contact the cylinder wall during the power stroke.

Piston heads are often coated with hard anodizing, ceramics, or electroplating. These coatings increase the hardness and resistance to corrosion, cracking, wear, and scratching. New ceramic coatings offer nearly three times the surface hardness of traditional hard-anodized coatings. Ceramic coatings also help protect against spontaneous detonation.

Just below the dome, around the sides of the piston, is a series of grooves. The grooves are used to hold the piston rings. The high parts between the grooves are called **ring lands**. Some pistons have a ceramic coating in the top ring groove to prevent the ring from being "welded" inside the groove. Normally, there are three grooves, two compression and one oil-control. The compression grooves are located toward the top of the piston. The depth of grooves varies with the size of the piston and the type of rings used. The oil-control groove is the lowest groove on the piston. It is normally wider than the compression ring grooves and has holes or slots to allow oil to drain. The positions of the ring grooves vary with engine design. Many newer engines have the top compression ring as close as possible to the piston head. This reduces the amount of fuel that can drop down the sides of the piston before combustion. This hidden fuel is not involved in the combustion process but leaves as unburned hydrocarbons during the exhaust stroke. In this design, all rings are placed close together. On a few pistons, the piston pin bore is very close to the piston head, behind the groove for the lower oil-control ring.

The area below the piston pin is called the **piston skirt**. The area from just below the bottom ring groove to the tip of the skirt is the piston thrust surface. There are two basic types of piston skirts: the slipper and the full skirt. The full skirt is used primarily in truck and commercial engines. The slipper skirt is used in automobile engines and allows the piston enough thrust surface for normal operation. A slipper skirt also allows the piston to be lighter and reduces piston expansion because there is less material to hold heat.

Late-model engines that are capable of running to fairly high rpm use lighter pistons. These pistons have skirts only on the thrust sides. Often, the skirts are coated with molybdenum to prevent cylinder wall scuffing **(Figure 11–78)**.

SHOP TALK

The term *piston slap* is used to describe the noise made by the piston when it contacts the cylinder wall. This noise is usually heard only in older, high-mileage engines that have worn pistons or cylinder walls. The noise is most noticeable when the engine is cold or under a load.

To ensure that the piston is installed correctly and has the correct offset, the top of the piston has a

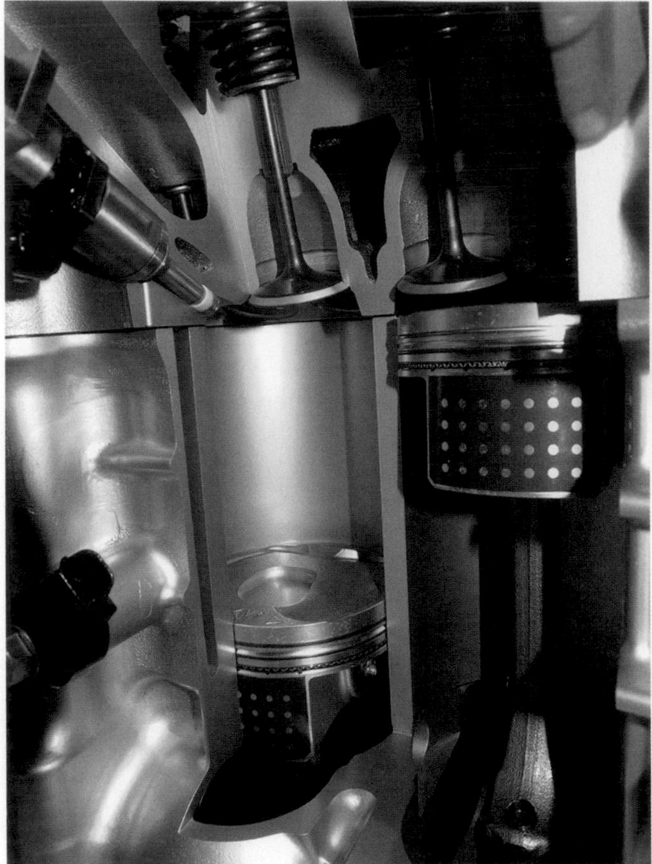

FIGURE 11–78

These pistons are coated with a special compound to reduce friction.

©Cengage Learning 2015

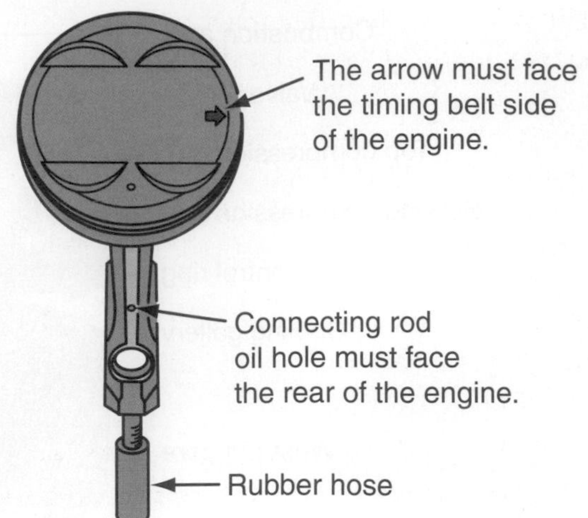
The arrow must face the timing belt side of the engine.
Connecting rod oil hole must face the rear of the engine.
Rubber hose

FIGURE 11–79

Always make sure the markings on the piston and connecting rod are in the correct relationship to each other and that they face the correct direction.

©Cengage Learning 2015

mark. The most common mark is a notch, machined into the top edge of the piston. Always refer to the appropriate service information to determine the correct direction and position of the mark. It is important that the front of the piston match the front of the connecting rod **(Figure 11–79)**.

When an engine is designed, piston expansion determines how much piston clearance will be needed in the cylinder bore. Too little clearance will cause the piston to bind at operating temperatures. Too much will cause piston slap. The normal piston clearance for an engine is about 0.0254 mm to 0.0508 mm (0.001 to 0.002 in.). This clearance is measured between the piston skirt and the cylinder wall.

Piston Terminology

Many different terms are used to describe the design of a piston; these include the following:

- *Compression distance (or height).* The distance from the centre of the piston pin bore to the top of the piston.
- *Ring belt.* The area between the top of the piston and the pin bore where the piston rings are installed.
- *Heat dam.* A narrow groove cut in some pistons to reduce heat flow to the top ring groove. During engine operation, the groove fills with carbon and absorbs the heat of combustion.
- *Land diameter.* The diameter of the ring land. On some pistons, the land diameter will be the same for each ring; on others, it will increase from top to bottom.
- *Land clearance.* The difference between the diameters of the land and the cylinder.
- *Groove root diameter.* The diameter of a piston measured at the bottom of a ring groove. The root diameter of each groove may vary with the type of ring used.
- *Groove protector.* A steel or cast-iron insert placed into the top groove of an aluminum piston to extend the life of the top compression ring.
- *Top groove spacer.* A steel spacer installed above the ring in a reconditioned groove to bring the ring's side clearance within specifications.
- *Piston pin bushing.* Found primarily in older cast-iron diesel pistons, this bushing serves as a bearing for the piston pin. It is inserted into the piston pin bore.
- *Major thrust face.* The part of the piston skirt that has the greatest thrust load. This is typically the right side when looking at the engine from the flywheel end.

- *Minor thrust face.* The part of the piston skirt that is opposite the major thrust face.
- *Skirt clearance.* The difference between the diameter of the piston skirt diameter and the diameter of the cylinder.
- *Piston skirt taper.* The difference between the diameter of the piston at the top and bottom of the skirt.
- *Piston cam.* The shape of the piston skirt area, which provides correct cylinder wall contact and clearances.

INSPECTION Each piston should be carefully checked for damage and cracks. Pay attention to the ring lands and the pin boss area. Look for scuffing on the sides of the piston **(Figure 11–80)**. Light up and down scuffing is normal. Excessive, irregular, or diagonal scuff marks indicate lubrication, cooling system, or combustion problems. Scuffing may also be caused by a bent connecting rod, a seized piston pin, or inadequate piston-to-wall clearance. If any damage is evident, the piston should be replaced.

Remove the piston rings. A ring expander should be used to remove the compression rings. Normally, the oil-control ring can be rolled off by hand. Remove the carbon from the top of the piston with a gasket scraper. Carbon and oil build up in the back part of the groove. This buildup must be removed. The dirt will prevent the rings from seating properly. Clean the grooves of the piston with a groove cleaning tool or a broken piston ring. When doing this, make sure no metal is scraped off. The oil-control ring groove has slots or holes. These should also be cleaned. Use a drill bit or small brush. Once the grooves are clean, use a brush and solvent to thoroughly clean the piston. Do not use a wire brush.

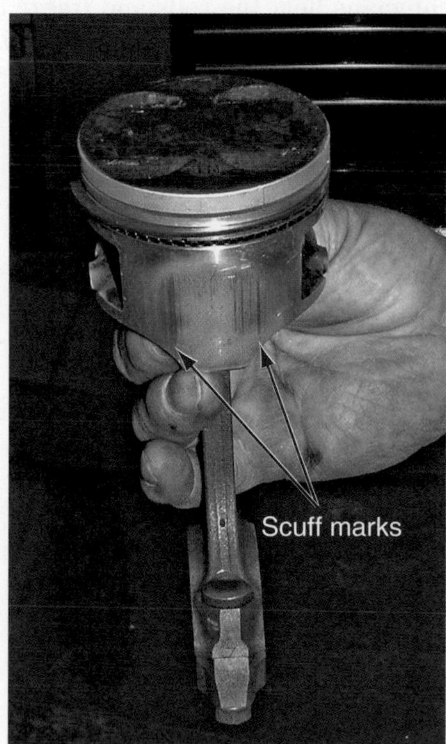

FIGURE 11–80
Each piston should be carefully checked for scuffing on the sides of the piston.

Ring side clearance should be measured. Side clearance is the difference between the thickness of the ring and the width of its groove. To measure this, place a new ring in its groove and, with a feeler gauge, measure the clearance between the ring and the top of the groove **(Figure 11–81)**. If the clearance is not within the specified range, the piston should be replaced.

The diameter of the piston should be measured. This measurement is normally taken across specific points on the skirt **(Figure 11–82)**. If the diameter is not within specifications, the piston should be replaced. Some engine rebuilders will knurl the skirts if the diameter is slightly less than specifications.

Piston Pins

Piston pins are basically thick-walled hollow tubes. Like the rest of the piston and connecting rod assembly, it is built to be strong and light. Most are made from alloy steel and are plated with chrome, carburized, and/or heat-treated to provide good wear resistance. Piston pins are lubricated by oil fed through passages in the connecting rods, oil splashing in the crankcase, or spray nozzles in the rods or pistons.

FIGURE 11–81
Ring side clearance should be checked on each piston.

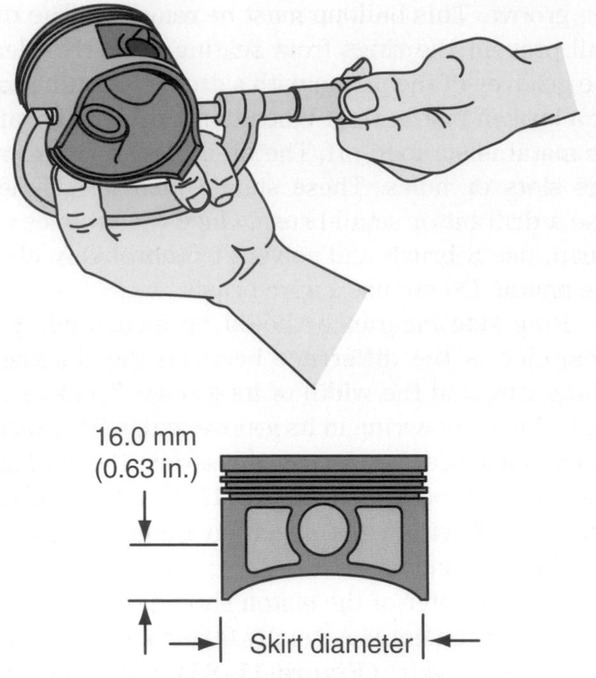

FIGURE 11–82

The diameter of the piston is measured across specific points on the skirt.

A piston pin fits through the small end of the connecting rod and the piston's pin bore. The way the pin is retained is used to describe it.

- A stationary pin is pressed into the piston. The connecting rod swivels on the pin.
- A semi-floating pin is pressed into the connecting rod. The piston swivels on the pin.
- A full-floating pin is able to move or rotate in the piston and connecting rod. The pins are retained by caps, plugs, snap rings, or spring clips inserted in the piston at the ends of the pin. Full-floating pins are the most commonly used.

Inspect the pin boss area on the piston for signs of pin wobble. Then remove the pin to inspect it. With full-floating pins, the retaining clips are removed and the pin pushed out.

A pin press is used to remove and install press-fit pins. When installing a piston pin, make sure the piston is facing the correct direction in regard to the connecting rod.

Inspect the pin closely for signs of wear. Full-floating pins should have an even wear pattern. Carefully inspect the pin bore in the piston. Because a piston is made from softer material than the pin, the piston will wear before the pin. If there are signs of uneven wear, suspect a lubrication or connecting rod problem.

Check the fit of the pin. It should move freely through the bores. Also attempt to move the pin up and down in its bores. Any movement means the piston bore or pin is worn. To determine if the bore or pin is worn, measure the diameter of the pin bore. If the bore is not within specifications, replace the piston. Then measure the diameter of the pin. If the pin is not within specifications, replace it. If the piston bore and pin meet specifications, measure the small end bore of the connecting rod **(Figure 11–83)**. If the diameter is not within specifications, replace the connecting rod.

Some manufacturers recommend a check of the pin's oil clearance. To do this, subtract the diameter of the pin from the diameter of the piston pin bore. If the oil clearance exceeds specifications, replace the piston and pin. Now subtract the diameter of the pin from the diameter of the connecting rod's small end. If the oil clearance exceeds specifications, replace the connecting rod and/or the pin.

Connecting rods may have a piston pin bushing. Measure the inside diameter of the bushings, and compare the reading to specifications. If the bushing is worn or damaged, it should be replaced. The bushing is pressed out of the rod with a pin press. Installing new bushings is also done with a press; some technicians heat the rods and freeze the pins before pressing them in. This makes installation easier. Before applying pressure on the pin, make sure it is set squarely above the bore.

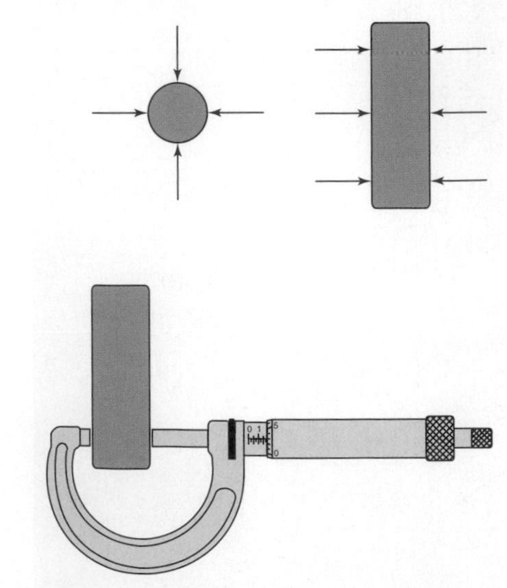

FIGURE 11–83

The piston pin is measured at a variety of spots and its diameter compared to the inside diameter (ID) of the piston's pin bore and the small end of the connecting rod.

Piston Rings

Piston rings are used to fill the gap between the piston and cylinder wall.

Piston rings must serve three functions: (1) They seal the combustion chamber at the piston; (2) they remove oil from the cylinder walls to prevent oil from entering into the combustion chamber; and (3) they carry heat from the piston to the cylinder walls to help cool the piston.

There are two basic ring families: compression rings and oil-control rings. In most engines, pistons are fitted with two compression rings and one oil-control ring. The compression rings are found in the two upper grooves closest to the piston head. The oil ring is fitted to the groove just above the wrist pin.

COMPRESSION RINGS The compression rings form the seal between the piston and cylinder walls. They are designed to use combustion pressure to force them against the cylinder wall. During the power stroke, the pressure caused by the expanding air/fuel mixture is applied between the inside of the ring and the piston groove. This forces the ring into full contact with the cylinder walls. The same force is applied to the top of the ring, forcing it against the bottom of the ring groove. These two actions help to form a tight ring seal.

Common compression rings are made of cast iron, cast iron coated with molybdenum (moly), and cast iron coated with chrome **(Figure 11–84)**. Cast iron offers a durable wear surface and costs less than a moly- or chrome-faced ring. These rings are ideal for normal driving. Moly coatings are quite porous and can hold oil. As a result, moly rings have a very high resistance to scuffing. These rings are used in engines that are run at continuous high speeds or severe load conditions. Chrome also has good resistance to scuffing but does not have the oil retention capabilities of moly. Chrome rings are recommended when driving conditions include frequent travel on dusty or unpaved roads. Chrome is very dense and hard and will push away any dirt that enters the cylinder on the intake stroke. Moly coatings, due to their porosity, will allow dirt to become embedded in a ring's face. Normally, a moly ring is used in the top ring groove, with a cast iron or chrome ring in the second groove.

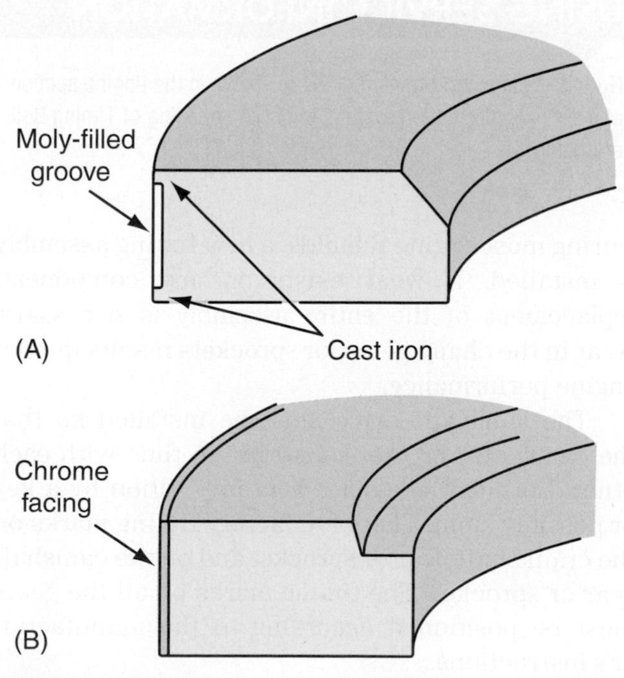

FIGURE 11–84

(A) A moly-coated compression ring. (B) A chrome-faced compression ring.

Other face coatings include ceramics, graphite, phosphate, and iron oxide. All coatings are designed to help in the wear-in process. Wear-in is the time required for the rings to conform to the shape and surface of the cylinder wall.

OIL-CONTROL RINGS Oil is constantly being applied to the cylinder walls. The oil is used to lubricate, clean the cylinder wall of carbon and dirt particles, and help cool the piston. Controlling this oil is the primary function of the oil ring. The two most common types of oil rings are the segmented oil ring and the cast-iron oil ring. Both types of rings are slotted so that excess oil from the cylinder wall can pass through the ring. The oil ring groove of the piston is also slotted. After the oil passes through the ring, it can then pass through the slots in the piston and return to the oil sump through the open section of the piston.

Segmented oil rings are made of three pieces: upper and lower scraper rails and an expander. The end gaps of the three separate pieces must be staggered to prevent oil from reaching the combustion chamber.

INSTALLING PISTONS AND CONNECTING RODS

Once the crankshaft is in place, the piston and connecting rod assemblies are installed next. Check the marks on the connecting rod caps and the connecting rods to make sure they are a match.

The insert bearings for the connecting rods must be the correct size. If the crankshaft has been machined undersize, matching rod bearing inserts must be installed. The size of the bearing inserts is printed on the box they come in and is stamped on the backs of the bearings, or they are colour-coded.

Snap the new connecting rod bearing inserts into the connecting rods and rod caps. Make sure the tang on the bearing fits snugly into the matching notch.

The piston and rod can be assembled in the block according to the procedure shown in Photo Sequence 8. Remember that connecting rods are numbered for easy identification and proper assembly. Also make sure the end gaps of the piston rings are positioned according to the manufacturer's recommendations **(Figure 11–85)** prior to installing the piston assembly in the cylinder bore.

When all the pistons and rods have been installed, connecting rod side clearance can be measured **(Figure 11–86)**. Side clearance is the amount of clearance between the crankshaft and the side of the connecting rod. Side clearance is measured with

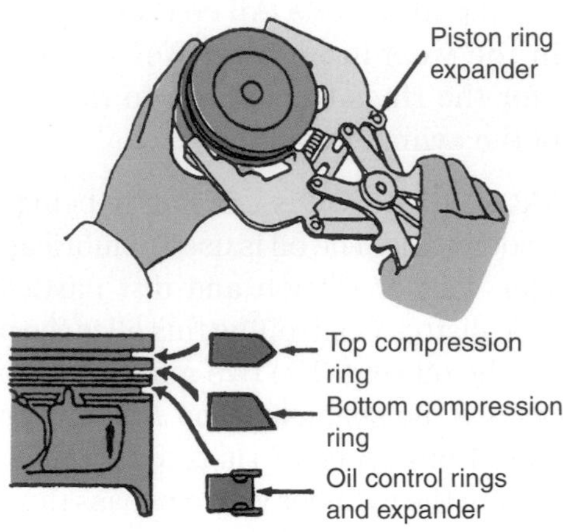

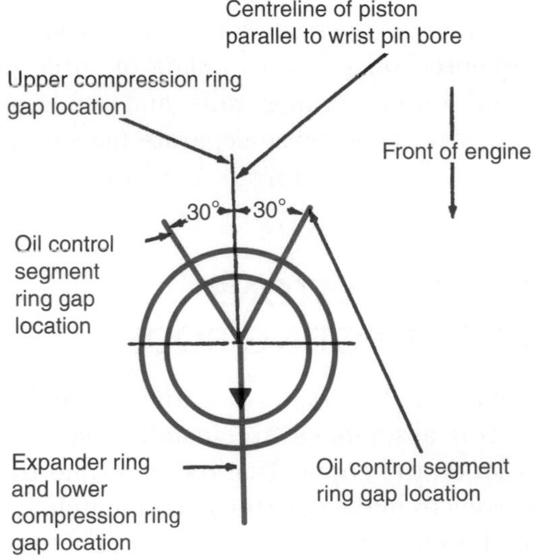

FIGURE 11–85

Install the piston rings onto the piston with a ring expander. Also make sure the ring end gaps are arranged according to specifications.

Ford Motor Company

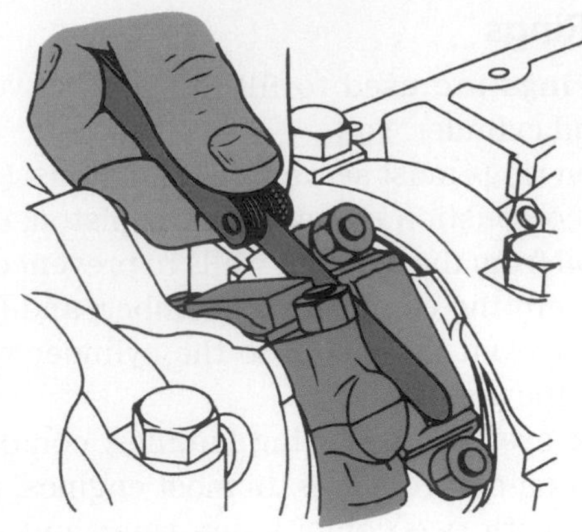

FIGURE 11–86

Measuring connecting rod side clearance.

a feeler gauge. If the clearance is not correct, the rods may need to be machined or replaced.

Be sure to coat the crankshaft assembly with clean lubricant or engine oil. After each piston assembly is installed in the block, rotate the crankshaft and check its freedom of movement. If the crankshaft is hard to rotate after a piston has been installed, remove it and look for signs of binding.

CRANKSHAFT AND CAMSHAFT TIMING

USING SERVICE INFORMATION

Normally, camshaft timing marks are shown in the engine section of a service information system, under the heading of Timing Belt or Chain R&R.

During most engine rebuilds, a new timing assembly is installed. If wear exists on any component, replacement of the entire assembly is necessary. Wear in the chain, gears, or sprockets results in poor engine performance.

The camshaft drive must be installed so that the camshaft and crankshaft are in time with each other. Both sprockets are held in position by a key or possibly a pin. There are factory timing marks on the crankshaft gear or sprocket and on the camshaft gear or sprocket. The timing marks on all the gears must be positioned according to the manufacturer's instructions.

The chain is installed on the crankshaft gear first, then around the camshaft sprocket. Never wind a chain onto the gears or use a screwdriver, pry bar,

INSTALLING A PISTON AND ROD ASSEMBLY

P8–1 Insert a new piston ring into the cylinder. Use the head of the piston to position the ring so that it is square with the cylinder wall. Use a feeler gauge to check the end gap. Compare end-gap specifications with the measured gap. Correct as needed. Normally, the gaps of the piston rings are staggered to prevent them from being in line with each other. Piston rings are installed easily with a ring expander.

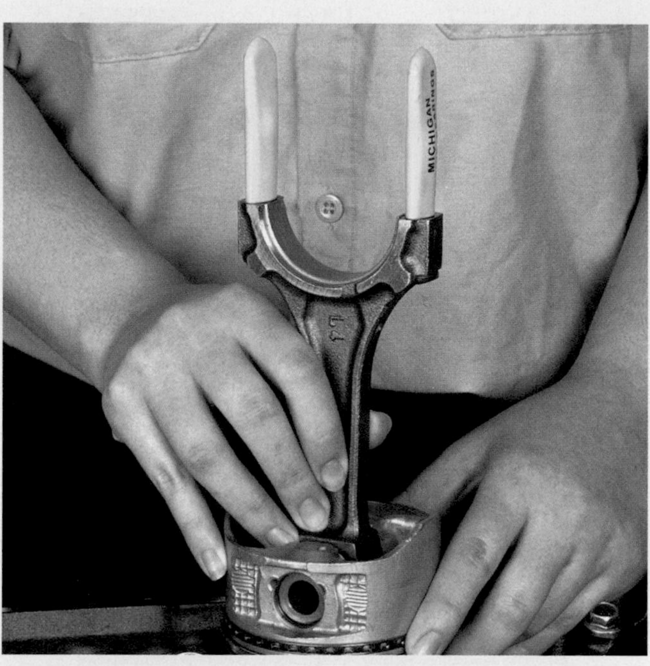

P8–2 Before attempting to install the piston and rod assembly into the cylinder bore, place rubber or aluminum protectors or boots over the threaded section of the rod bolts. This will help to prevent bore and crankpin damage.

P8–3 Lightly coat the piston, rings, rod bearings, cylinder wall, and crankpin with an approved assembly lubricant or a light engine oil. Some technicians submerge the piston in a large can of clean engine oil before it is installed.

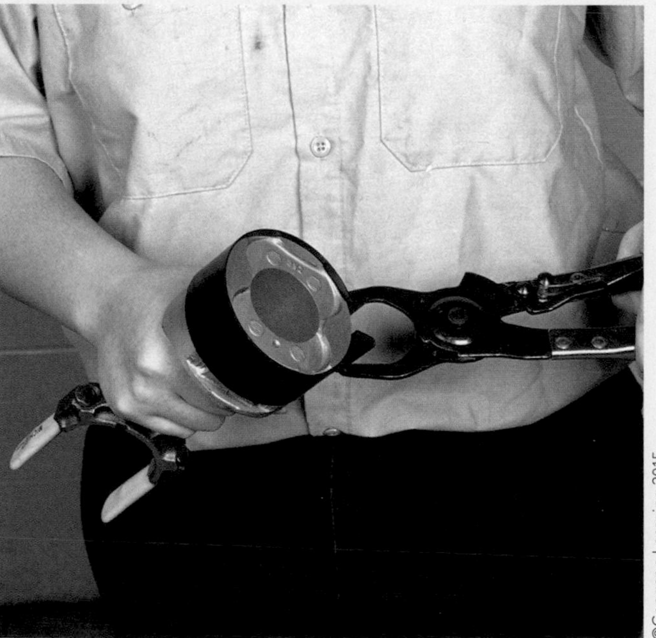

P8–4 Stagger the ring end gaps, and compress the rings with the ring compressor. This tool is expanded to fit around the piston rings. It is tightened to compress the piston rings. When the rings are fully compressed, the tool will not compress any further. The piston will fit snugly, but not tightly.

INSTALLING A PISTON AND ROD ASSEMBLY *(continued)*

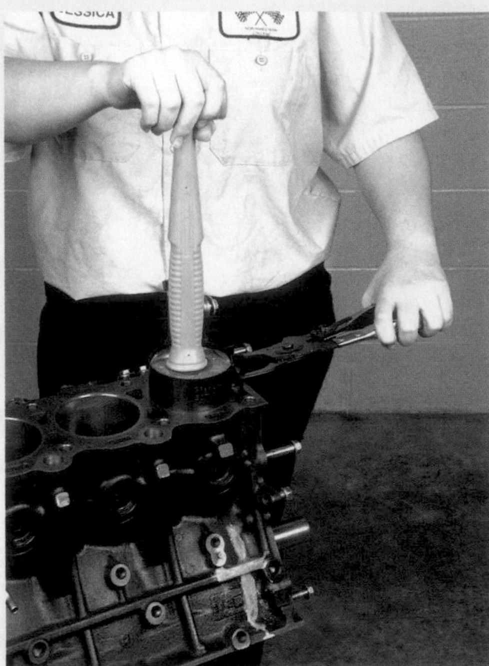

P8–5 Rotate the crankshaft until the crankpin is at its lowest level (BDC). Then place the piston/rod assembly into the cylinder bore until the ring compressor contacts the cylinder block deck. Make sure that the piston reference mark is in correct relation to the front of the engine. Also, when installing the assembly, make certain that the rod threads do not touch or damage the crankpin.

P8–6 Lightly tap on the head of the piston with a mallet handle or block of wood until the piston enters the cylinder bore. Push the piston down the bore while making sure the connecting rod fits into place on the crankpin. Remove the protective covering from the rod bolts.

P8–7 Position the matching connecting rod cap and finger-tighten the rod nuts. Make sure the connecting rod blade and cap markings are on the same side. Gently tap each cap with a plastic mallet as it is being installed to properly position and seat it. Torque the rod nuts to the specifications given in the service information. Repeat the piston/rod assembly procedure for each assembly.

or hammer to force a chain into position. Prying or pounding on the chain damages the links, causing the chain to stretch and fail. Carefully place the entire assembly as a unit onto the shafts by pressing both gears evenly, keeping the keyways aligned.

Camshaft End Play

Before proceeding to the next step in the reassembly process, check to be sure that the clearance between the camshaft boss (or gear) and the backing plate is within manufacturer's specifications. Make this check with a feeler gauge. Install shims behind the thrust plate, or reposition the camshaft gear and retest the end play. In some cases, adjustment is made by replacing the thrust plate. Some engines limit the end play of the camshaft through the use of a cam button. A nylon button or Torrington bearing sets on a spring and is installed between the camshaft timing gear and the timing chain cover. To check end play, use a dial indicator setup. Be sure the camshaft end play is not more than recommended in the service information system.

Lifters

It is important to remember that because of the relationship between the camshaft lobe and the lifter, it is extremely important that new lifters be used with a new camshaft. Before installing the lifters, prefill hydraulic lifters with oil. To do this, place them in a container of clean oil and let them sit there overnight to allow the air trapped inside to seep out. Then pump the plunger in the lifter a few times to bleed the remaining air out. Before setting the lifters into their bore in the block, coat the bores with assembly lube. After they have been installed, rotate the camshaft to check for binding or misalignment.

Oil Pump

After an engine is rebuilt, a new or rebuilt oil pump is often installed. If the old pump is to be reused, it should be carefully inspected for wear and thoroughly cleaned. The oil pump may be located in the oil pan or mounted to the front of the engine. Its installation depends on the engine's design, but it is typically driven by the crankshaft and creates suction to draw oil from the oil pan through a strainer. The pump's suction creates pressure that forces the oil through the oil filter to various passages. The oil then returns to the oil pan.

GO TO ▶ Chapter 14 for information on servicing and installation of oil pumps.

KEY TERMS

Babbitt (p. 295)
Backlash (p. 296)
Boring bar (p. 286)
Cam followers (p. 293)
Cast (p. 282)
Chamfering (p. 283)
Core plugs (p. 281)
Crush (p. 305)
Deck (p. 284)
Dome (p. 312)
Duration (p. 291)
Flex plate (p. 302)
Flywheel (p. 302)
Forge (p. 278)
Full-round bearing (p. 303)
Glaze (p. 286)
Harmonic balancer (p. 278)
Head (p. 312)
Hone (p. 286)
Hypoeutectic (p. 312)
Leakdown (p. 294)
Line boring (p. 288)
Long block (p. 278)
Overlap (p. 291)
Piston pins (p. 315)
Piston rings (p. 317)
Piston skirt (p. 313)
Plastigage (p. 307)
Ring lands (p. 313)
Short block (p. 278)
Split bearing (p. 303)
Taper (p. 284)
Total indicator reading (TIR) (p. 303)
Valve lifters (p. 293)
Variable valve timing (VVT) (p. 293)
Wrist pin (p. 312)

SUMMARY

- The basic short block assembly consists of the cylinder block, crankshaft, crankshaft bearings, connecting rods, pistons and rings, oil gallery, and core plugs. On OHV engines, the camshaft and its bearings are also included.
- The cylinder block houses the areas in which combustion occurs.
- A properly reconditioned cylinder must be of the correct diameter, have no taper or runout, and have a surface finish that allows the piston rings to seal.
- Glaze is the thin residue that forms on cylinder walls due to a combination of heat, engine oil, and piston movement.
- Core plugs and oil gallery plugs are normally removed and replaced as part of cylinder block reconditioning. The three basic core plugs are the disc or dished type, cup type, and expansion type.
- A cam changes rotary motion into reciprocating motion. The part of the cam that controls the opening of the valves is the cam lobe. The closing of the valves is the responsibility of the valve springs.
- The camshaft is supported in the cylinder block by friction-type bearings, or bushings, which are typically pressed into the camshaft bore in the

block or head. Camshaft bearings are normally replaced during engine rebuilding.

- Solid valve lifters provide for a rigid connection between the camshaft and the valves. Hydraulic valve lifters do the same but use oil to absorb the shock resulting from movement of the valve train. Roller lifters are used to reduce friction and power loss.
- The camshaft is driven by the crankshaft at half its speed.
- The most common ways to measure cam lobe wear are with a dial indicator or an outside micrometer. The dial indicator test should be conducted with the camshaft in the engine. When using an outside micrometer, the camshaft must be out of the engine.
- Most premature cam wear develops within the first few minutes of operation.
- The crankshaft turns on a film of oil trapped between the bearing surface and the journal surface. The journals must be smooth and highly polished. The flywheel adds to an engine's smooth running by applying a constant moving force to carry the crankshaft from one firing stroke to the next. The flywheel surface may be used as part of the clutch.
- Important crankshaft checks include saddle alignment, straightness, clearance, and end play.
- Bearings carry the critical loads created by crankshaft movement. Most bearings used today are insert bearings.
- Maintaining a specific oil clearance is critical to proper bearing operation. Bearings are available in a variety of undersizes.
- Today's aluminum pistons are lightweight yet strong enough to withstand combustion pressure.
- Piston rings are used to fill the gap between the piston and cylinder wall. Most of today's vehicle engines are fitted with two compression rings and one oil-control ring.
- When installing a piston and connecting rod assembly, various markings can be used to make sure the installation is correct. Always check the service information system for exact locations.
- Connecting rod side clearance determines the amount of oil throw-off from the bearings and is measured with a feeler gauge.
- During most engine rebuilds, a new timing assembly is installed. When installing the timing gears, make sure they are aligned to specifications.

REVIEW QUESTIONS

1. Name the two most common ways to measure cam lobe wear.
2. What is the deck?
3. Where does maximum cylinder bore wear occur?
4. What is the function of compression rings?
5. What type of valve lifter automatically compensates for the effects of engine temperature?
   a. hydraulic **b.** solid
   c. roller **d.** solid cam follower
6. What are most pistons used today made from?
   a. cast iron **b.** aluminum
   c. ceramic **d.** forged steel
7. What is the initial purpose of core plugs?
   a. to allow for block heater installation
   b. to allow the release of sand from the block moulding process
   c. to allow for block cleaning during rebuilding
   d. to allow for coolant expansion during freeze-up
8. What tool(s) is generally used to measure cylinder block deck warpage?
   a. vernier caliper
   b. micrometer
   c. straightedge and feeler gauge
   d. telescopic gauge and micrometer
9. What tool(s) is generally used to measure crankshaft bearing journals?
   a. vernier caliper
   b. micrometer
   c. straightedge and feeler gauge
   d. telescopic gauge and micrometer
10. Which of the following lifters will produce the least amount of camshaft wear?
    a. solid lifters **b.** hydraulic lifters
    c. roller lifters **d.** sodium-filled lifters
11. What can excessive piston clearance cause?
    a. piston slap
    b. excessive compression pressure
    c. reduced oil pressure
    d. excessive oil pressure
12. Why are cylinders deglazed?
    a. to ensure proper clearances
    b. to ensure proper ring sealing
    c. to ensure good piston fit
    d. to enlarge the cylinder
13. Each half of a split bearing is made so that it is slightly greater than an exact half. What is this extension called?
    a. spread **b.** crush
    c. locating tab **d.** embedment

14. What is the connecting rod journal also called?
    a. balancer shaft
    b. vibration damper
    c. Plastigage
    d. crank pin
15. How can main bearing bore misalignment be corrected?
    a. replacing main bearings
    b. replacing the crankshaft
    c. replacing the main bearing caps
    d. line boring
16. What is the correct procedure for boring cylinders?
    a. bore to the new piston size plus the clearance
    b. bore to the new piston size only
    c. hone to the new piston size, and bore to the clearance specification
    d. bore to the new piston size, and hone to the clearance specification
17. What type of hone should be used during a cylinder bore operation?
    a. a fly hone
    b. a ball hone
    c. a rigid hone
    d. a flex hone
18. Connecting rod bearings are to be replaced. The rod journal measures 70.866 mm (2.790 in.). The standard rod journal diameter is 71.12 mm (2.800 in.). What size bearings should you order from the parts room?
    a. U/S—0.381 mm (0.015 in.) undersize
    b. U/S—0.254 mm (0.010 in.) undersize
    c. standard size
    d. O/S—0.254 mm (0.010 in.) oversize
19. What measurements would have to be taken to calculate cylinder bore taper?
    a. take two bore readings 90° from each other at the top of ring travel in the cylinder
    b. take a bore reading at the top and bottom of ring travel along the block centreline
    c. take a bore reading at the top and bottom of ring travel at 90° to the block centreline
    d. take two bore readings 90° from each other at the bottom of the cylinder
20. What device in the valve train changes rotary motion into reciprocating motion?
    a. eccentric
    b. cam
    c. bushing
    d. mandrel

CHAPTER 12
Upper End Theory and Service

LEARNING OUTCOMES

- Describe the purpose of an engine's cylinder head, valves, and related valve parts.
- Describe the types of combustion chamber shapes found on modern engines.
- Know why there are special service procedures for aluminum and OHC heads.
- Describe the different ways that manufacturers vary valve timing.
- Perform a complete inspection on valve train components.
- Explain the procedures involved in reconditioning cylinder heads, valve guides, valve seats, and valve faces.
- Explain the steps in cylinder head reassembly.

CYLINDER HEAD

The cylinder head is made of cast iron or aluminum. On overhead valve (OHV) engines, the cylinder head contains the valves, valve seats, valve guides, valve springs, rocker arm supports, and a recessed area that makes up the top portion of the combustion chamber. On overhead cam (OHC) engines, the cylinder head contains these items plus the supports for the camshaft and camshaft bearings **(Figure 12–1)**.

Both overhead valve and overhead cam cylinder heads contain passages that match passages in the cylinder block. These passages allow coolant to circulate in the head and allow oil to drain back into the oil pan. Pressurized oil also moves through some of the passages to lube the camshaft and valve train. The coolant flows from passages in the cylinder block through the cylinder head. The coolant then flows back to other parts of the cooling system.

FIGURE 12–1

An OHC cylinder head.

DaimlerChrysler Corporation

The cylinder head also contains tapped holes in the combustion chamber to accept the spark plugs.

The sealing surface of the head must be flat and relatively smooth. However, if the surface is too smooth, the gasket may be able to squirm or move when under pressure or during expansion and contraction due to temperature change. The ideal surface for a cylinder head has a slight texture to it. This finish will grip the gasket and stop it from shifting around. This area must form a tight seal and contain the pressures formed during combustion. To aid in the sealing, a gasket is placed between the head and block. This gasket, called the head gasket, is made of special material that can withstand high temperatures, high pressures, and the expansion of the metals around it. The head also serves as the mounting point for the intake and exhaust manifolds and contains the intake and exhaust ports.

Cylinder head design is one of the most influential factors that affects the overall performance of an engine. The size and shape of the intake and exhaust ports affect the velocity and volume of the mixture entering and leaving the cylinders. Most aspects of cylinder head design are carefully tested and calibrated by the manufacturers to ensure optimal performance and fuel economy for the intended application.

Ports

Intake and exhaust ports are cast into the cylinder head. One port is normally used for each valve. However, on engines with more than two valves per cylinder, the ports for the intake or exhaust valves may be combined. These ports are called **Siamese ports** **(Figure 12–2)**. With Siamese ports, individual ports around each valve mesh together to form a larger single port that is connected to a manifold. **Cross-flow ports** are used on some engines and have intake and exhaust ports on opposite sides of the combustion chamber. Heads of this design are called cross-flow heads.

COMBUSTION CHAMBER

The performance of an engine, its fuel efficiency, and the level of pollutants in the exhaust all depend to a large extent on the shape of the combustion chamber. An efficient combustion chamber must be compact to minimize the surface area of the walls through which heat is lost to the engine's cooling system. The point of ignition (the nose of the spark plug) should be at the centre of the combustion chamber to minimize the flame path, or the distance from the spark to the furthermost point in the chamber. The shorter the flame path, the more evenly the air/fuel mixture will burn.

Manufacturers have designed several different shapes of combustion chambers. Before looking at the popular combustion chamber designs, two terms should be defined.

1. **Turbulence** is a very rapid movement of gases. Turbulence causes better combustion because the air and fuel are mixed better.
2. **Quenching** is the mixing of gases by pressing them into a thin area. The area in which gases are thinned is called the quench area.

Wedge Chamber

In the **wedge-type combustion chamber**, the spark plug is located at the wide part of the wedge

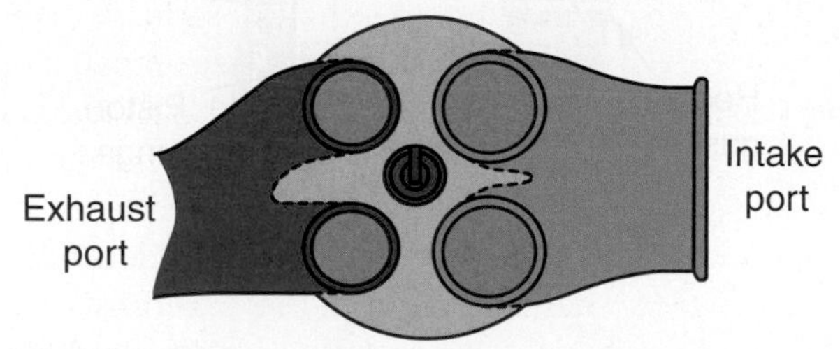

FIGURE 12–2

Siamese ports.

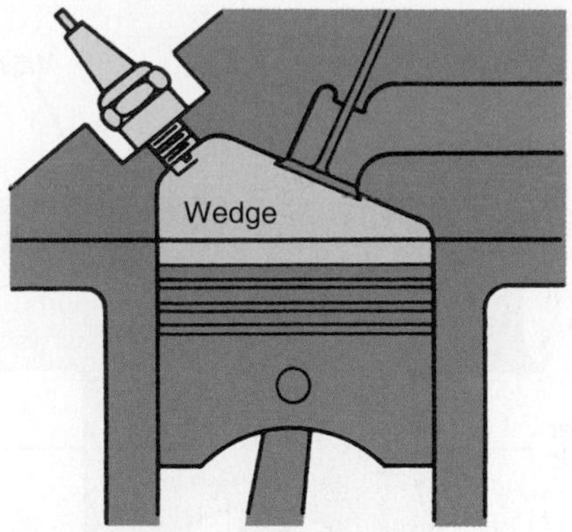

FIGURE 12–3

A typical wedge-type combustion chamber.

(Figure 12–3). As the piston comes up on the compression stroke, the air/fuel mixture is squashed in the quench area. The **quench area** causes the air and fuel to be mixed thoroughly before combustion, which helps to improve the combustion efficiency of the engine. Spark plugs are positioned to allow for rapid and even combustion. When the spark occurs, a flame front moves from the spark plug outward. The wedge-shaped combustion chamber is also called a turbulence- or swirl-type combustion chamber. On newer model cars, the quench area has been reduced, which helps reduce exhaust emissions.

Hemispherical Chamber

The **hemispherical combustion chamber** gets its name from its basic shape. Hemi is defined as half, and spherical means circle. The combustion chamber is shaped like a half-circle. This type of cylinder head is also called the **hemi-head**. The piston top forms the base of the hemisphere, and the valves are inclined at an angle of 60° to 90° to each other, with the spark plug positioned between them **(Figure 12–4)**.

This design has several advantages. The flame path from the spark plug to the piston head is short, which gives efficient burning. The cross-flow arrangement of the inlet and exhaust valves allows for a relatively free flow of gases in and out of the chamber. The result is that the engine can breathe deeply, meaning that it can draw in a large volume of mixture for the space available and give a high power output.

The hemispherical combustion chamber is considered a nonturbulence-type combustion chamber. Little turbulence is produced in this type of chamber. The air/fuel mixture is compressed evenly on the compression stroke. The spark plug is located

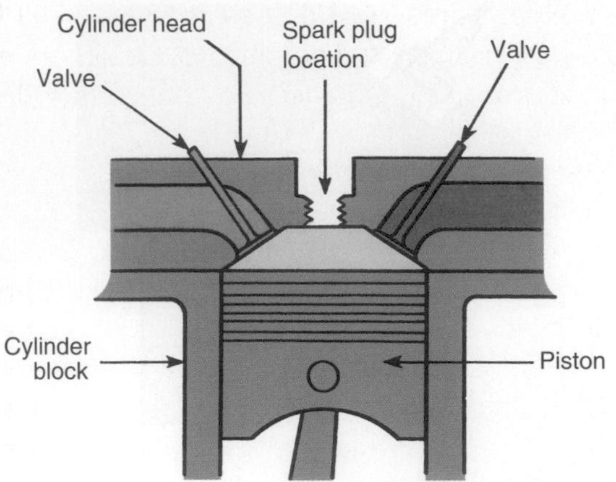

FIGURE 12–4

A typical hemispherical combustion chamber.

directly between the valves. Combustion radiates evenly from the spark plug, almost completely burning the air/fuel mixture.

One of the more important advantages of the hemispherical combustion chamber is that air and fuel can enter the chamber very easily. The wedge combustion chamber restricts the flow of air and fuel to a certain extent. This is called **shrouding**. **Figure 12–5** shows the valve very close to the side of the combustion chamber, which causes the flow of air and fuel to be restricted. Volumetric efficiency is reduced. Hemispherical combustion chambers do not have this restriction. Hemispherical combustion chambers are used on many high-performance applications, especially when large quantities of air and fuel are needed in the cylinder.

Some engines use a dome piston. This type of piston has a quench area to improve turbulence. Several variations of this design are used today.

Pentroof Chamber

Many of today's engines have a **pentroof** combustion chamber. This design is a modified hemispherical chamber. It is mostly found in engines with

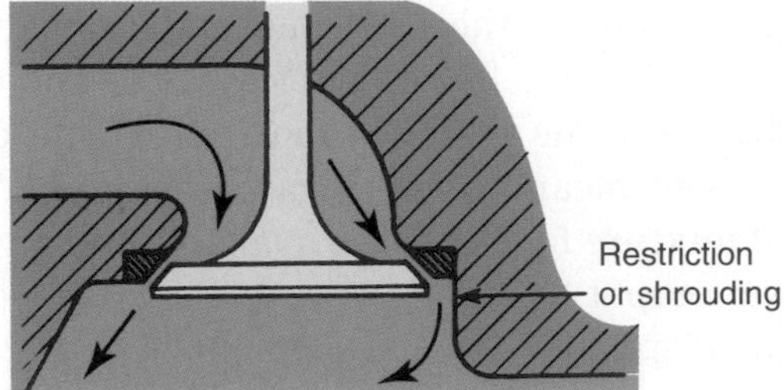

FIGURE 12–5

Shrouding is a restriction in the flow of intake gases caused by the shape of the combustion chamber.

FIGURE 12–6

A cylinder head with pentroof combustion chambers.

four valves per cylinder. The spark plug is located in the centre of the chamber, and the intake and exhaust valves are on opposite sides of the chamber **(Figure 12–6)**. Pentroof chambers have a squish area around the entire cylinder.

INTAKE AND EXHAUST VALVES

Every cylinder of a four-stroke cycle engine contains at least one intake valve to permit the air/fuel mixture to enter the cylinder and one exhaust valve to allow the burned exhaust gases to escape. The intake and exhaust valves, along with the spark plug gasket and the cylinder head gasket, must also seal the combustion chamber **(Figure 12–7)**.

The type of valve used in automotive engines is called a **poppet**. This is derived from the popping

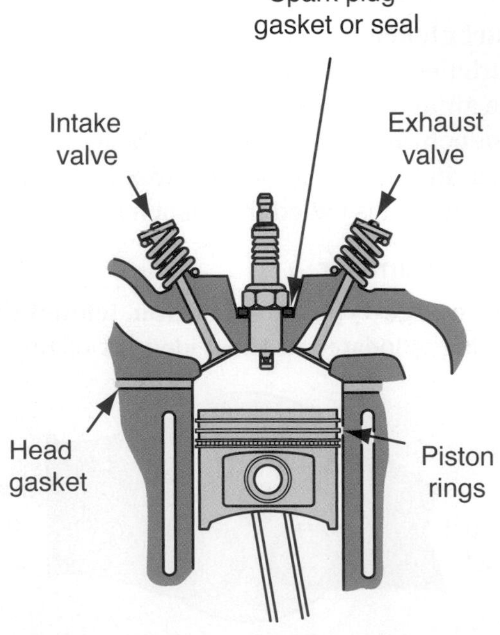

FIGURE 12–7

The sealing points of a typical engine.

action of the valve as it opens and closes. A poppet valve has a round head with a tapered face, a stem that is used to guide the valve, and a slot that is machined at the top of the stem for the valve spring retainers.

Valve Construction

Today, most valves are made from special hardened steel, steel alloys, or stainless steel. Other metals are often used in high-performance valves. Heat is an important factor in the design and construction of a valve. The material used to make a valve must be able to withstand high temperatures and be able to dissipate the heat quickly. Most of the heat is dissipated through the contact of the valve face and seat. The heat then moves through the cylinder head to its coolant passages. Heat is also transferred through the valve stem to the valve guide and again to the cylinder head **(Figure 12–8)**.

Intake and exhaust valves are typically made with different materials. Intake valves are typically low-alloy steels or heat- and corrosion-resistant high-alloy steels. Intake valves need less heat resistance because the intake air and fuel tend to cool them. Intake valves also need less corrosion protection because they are not exposed to the corrosive action of the hot exhaust gases.

The alloy used in a typical exhaust valve is chromium for oxidation resistance, with small amounts of nickel, manganese, silicon, and/or nitrogen. Heat resistance is critical for exhaust valves because they

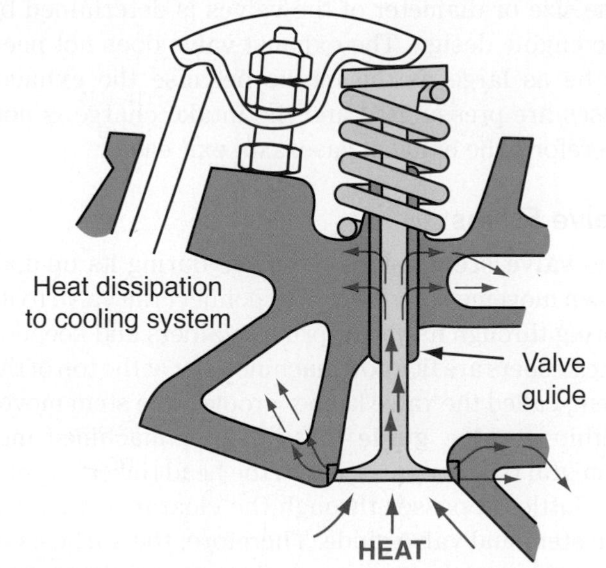

FIGURE 12–8

Exhaust valves cool by transferring heat to the liquid passages in the cylinder head.

face temperatures of 816°C to 2204°C (1500°F to 4000°F). Resistance to heat and corrosion are especially important for exhaust valves used in turbo- or supercharged engines.

A valve can be made as a single piece or two pieces. Two-piece valves allow the use of different metals for the valve head and stem. The pieces are spun welded together. These valves typically have a stainless steel head and a high-carbon steel stem. The stems are often chrome plated, so the weld is not visible. One-piece valves run cooler because the weld of a two-piece valve inhibits the flow of heat up the stem.

Today's engines require higher-quality valves that contain more nickel to withstand the heat. Most exhaust valves and some intake valves have 4 percent nickel content. The intake valves with high nickel are used in turbocharged engines. Older valves are alloyed with 2 percent nickel. The alloys used to make valves depend on the intended use and the design of the engine. Newer engines also tend to have lighter valves than what was used in the past. The lighter weight decreases the amount of power lost moving the valves and allows for higher engine speeds.

Many late-model valves have a black nitride coating to resist scuffing. Some performance valves may have their stems treated with a dry film lubricant to reduce friction and wear.

STAINLESS STEEL VALVES Stainless steel valves are commonly used. **Stainless steel** is an iron-carbon alloy with a minimum of 10.5 percent chromium content. Stainless steel does not stain, corrode, or rust as easily as ordinary steel. There are different types of stainless steels used to make valves. Austenitic stainless steels contain a maximum of 0.15 percent carbon, a minimum of 16 percent chromium, and nickel and/or manganese to give it strength and improve its heat resistance. Stainless steel is nonmagnetic.

INCONEL VALVES An alloy that is being used by many manufacturers is Inconel. Inconel has a nickel base with 15 to 16 percent chromium and 2.4 to 3.0 percent titanium. This alloy is normally used in high-temperature applications and has good oxidation and corrosion resistance. Inconel is difficult to machine; therefore, Inconel valves are replaced when they are deformed or damaged.

STELLITE VALVES Another alloy that is used in high-temperature applications is stellite. **Stellite** is an alloy of nickel, chromium, and tungsten and is nonmagnetic. Stellite is a hard-facing material that is

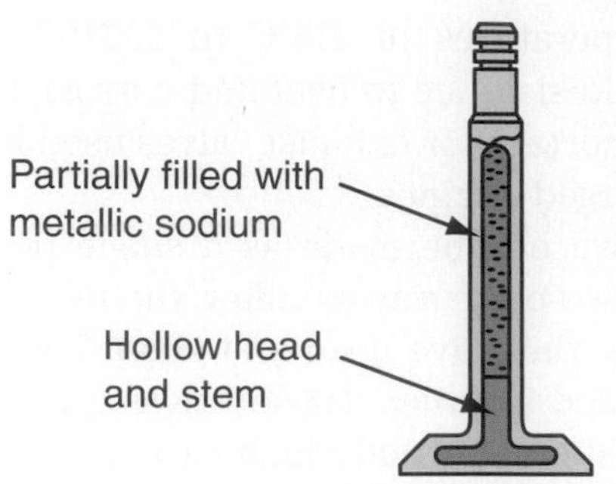

FIGURE 12–9

Some exhaust valves are partially filled with sodium to help cooling.

welded to valve faces and stems. It may also be used on the stem tip for added wear resistance. It comes in various grades depending on the mix of ingredients that are used in the alloy. This alloy has high resistance to wear, corrosion, erosion, abrasion, and galling. Stellite is available in many different grades, which are determined by the materials used in the alloy.

SODIUM FILLED Some exhaust valves have a hollow stem. The hollow section of the stem is partially filled with sodium **(Figure 12–9)**. Sodium is a silver-white alkaline metallic substance that transfers heat much better than steel. At operating temperatures, the sodium becomes a liquid. When the valve opens, the sodium splashes down toward the head and absorbs heat. Then as the valve moves up, the sodium moves away from the head and up the stem. The heat absorbed by the sodium is then transferred to the guide where it moves to the coolant passages in the head. Sodium-filled valves should not be machined.

TITANIUM VALVES Titanium alloys are added to valves to lighten them. Some high-performance engines have titanium valves. These valves dissipate heat well, are durable, and are very light. A titanium valve weighs less than half of a comparable steel valve.

CERAMIC VALVES Ceramic valves are being tested for future use. Ceramic materials weigh less than half of what a comparable-size steel valve weighs and can withstand extreme temperatures without weakening or becoming deformed.

VALVE TERMINOLOGY Valves **(Figure 12–10)** have a round head with a tapered face used to seal the intake or exhaust port. This seal is made by the **valve face** contacting the valve seat. The angle of the taper depends on the design and manufacturer of

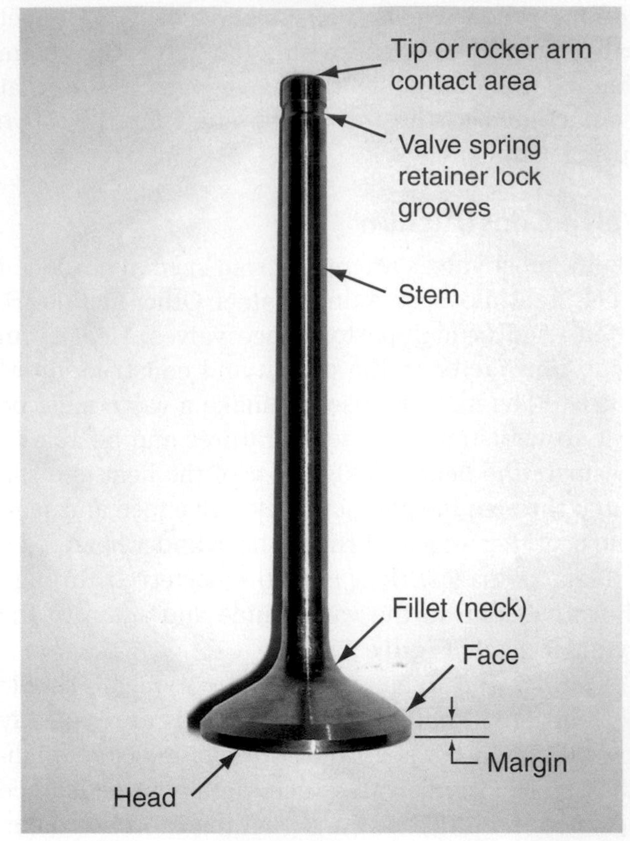

FIGURE 12–10

The parts of a typical valve.

the engine. The distance between the valve face and the head of the valve is called the **margin**. The intake and exhaust valve heads are different diameters. The intake valve is the larger of the two **(Figure 12–11)**. The size or diameter of the valves is determined by the engine design. The exhaust valve does not need to be as large as the intake because the exhaust gases are pressurized and the intake charge is not; therefore, the exhaust gases can exit easier.

Valve Stems

The **valve stem** guides the valve during its up-and-down movement and serves to connect the valve to its spring through its valve spring retainers and keepers. The keepers are fit into a machined slot at the top of the stem, called the valve keeper groove. The stem moves within a **valve guide** that is either machined into (integral type) or pressed into the head (insert type).

Little oil passes through the clearance between the stem and valve guide. Therefore, the surfaces of the guide and the stem are designed to minimize friction. Valve stems have two common types of coating to prevent wear and reduce friction: chrome plating and black nitriding. In addition to these coatings, the

FIGURE 12–11

The larger valve to the left is the intake valve, and the smaller one is the exhaust valve.

Martin Restoule

tips of the stem are hardened or stellited to resist damage from the constant hammering they face as the stems are pushed open.

Chrome-plated stems help prevent valve stem scuffing and galling and provide protection against wear during initial engine starts when no oil is present on the valve stem. Chrome plating is also widely used on high-performance valves. The thickness of the chrome plating can vary from 0.0051 to 0.0254 mm (0.0002 to 0.001 in.).

Many foreign manufacturers use a black nitride coating rather than chrome plating on the valves. Black nitride is applied to the entire valve, not just the stem. The finish of the surface is smoother than chrome; therefore, less friction is produced by the stem. The nitride coating protects the stems against scuffing and wear.

WARNING!

Never cut open any sodium-filled valves. Sodium will burn violently when it contacts moisture.

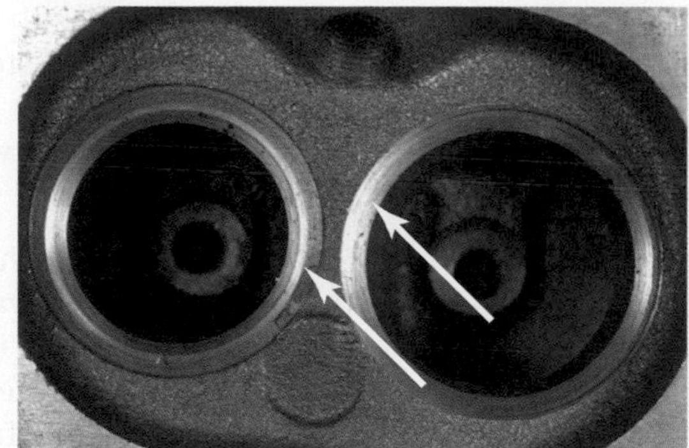

FIGURE 12–12

Valve seats.

Valve Seats

The **valve seat (Figure 12–12)** is the area of the cylinder head contacted by the valve's face. The seat may be machined in the head (integral type) or pressed into the head (insert type). Insert seats are always used in aluminum cylinder heads. They are also used to replace damaged integral seats.

Valve seats provide a sealing area for the valves. They also absorb the valve's heat and transfer it to the cylinder head. Seats must be hard enough to withstand the constant closing of the valve. Due to corrosive products found in exhaust gas, seats must be highly resistant to corrosion. When the head is made of cast iron, it has integral seats because cast iron meets those requirements. Cast iron is also used to make seat inserts. Most are induction hardened. These are hardened through electromagnetism, which heats the surface of the seat.

Many late-model engines with aluminum heads have sintered powder metal (tungsten carbide) seats. Powder metal seats are harder and more durable than cast-iron seats.

Important Valve Components of Four-Stroke Engines

VALVE GUIDES Valve guides support the valves and prevent them from moving in any direction other than up and down. They are machined to a fit of approximately 0.0508 mm (0.002 in.) clearance with the valve stem. This close clearance is important for the following reasons:

- It keeps the engine's oil from being drawn into the combustion chamber past the intake valve stem during the intake stroke and from leaking out to the exhaust port during times when the pressure in the exhaust port is lower than the pressure in the crankcase.

- It keeps exhaust gases from leaking into the crankcase area past the exhaust valve stems during the exhaust stroke.
- It keeps the valve face in perfect alignment with the valve seat.

Valve guides can be cast integrally with the head **(Figure 12–13A)**, or they can be removable **(Figure 12–13B)**. Removable valve guides usually are press-fit into the head. Aluminum heads are fitted with insert guides. Guides are made from materials that provide low friction and can transfer heat well. Cast-iron guides are mixed or coated with phosphorus and/or chrome. Bronze alloys are also used. These may contain some aluminum, silicon, nickel, and/or zinc.

VALVE SPRINGS, RETAINERS, AND SEALS The valve assembly is completed by the spring, retainer, and seal **(Figure 12–14)**. Before the spring and the retainer fit into place, a seal is placed over the valve stem. The seal acts like an umbrella to keep oil from running down the valve stem and into the combustion chamber. The spring, which keeps the valve in a normally closed position, is held in place by the retainer. The retainer locks onto the valve stem with two wedge-shaped parts that are called **valve keepers**. Some engines utilize a single valve spring per valve. Others use two or three springs. Often, the second or third spring is a flat spring called a **damper spring**, which is designed to control vibrations and is usually wound in the opposite direction to the main valve springs.

VALVE ROTATORS Many engines are equipped with mechanisms that cause the exhaust valves to rotate **(Figure 12–15)**. Their purpose is to keep carbon from building up between the valve face and seat. Carbon buildup can hold the valve partially open, causing it to burn.

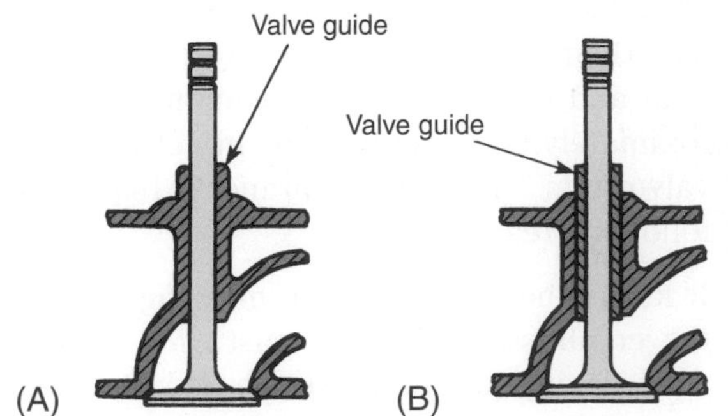

FIGURE 12–13

(A) Integral and (B) removable valve guides.

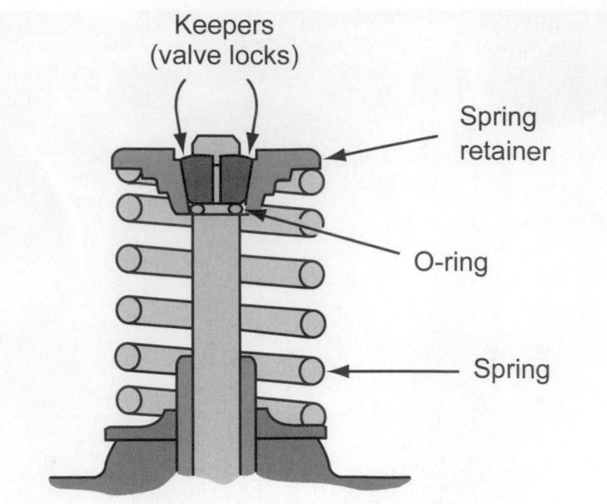

FIGURE 12–14

A valve assembly with spring, retainer, seals, and keepers.

CAMSHAFT BEARINGS The camshaft is part of the cylinder head assembly in all OHC-type engines. The unit that holds the camshaft(s) may be a separate unit bolted to the cylinder head, or the bore for the camshaft is machined into the upper part of the head. The most common design is similar to that of the crankshaft and main bearings. These cylinder heads are machined to accept one or two camshafts above the valves and have caps that secure the camshaft **(Figure 12–16)**. The camshafts are supported by split bearings. When the camshaft is held in a single structure, the camshaft is supported by one-piece insert bearings pressed into the camshaft bore.

PUSHRODS **Pushrods** are designed to be the connecting link between the rocker arm and the valve lifter; they are either solid or hollow. The pushrod fits between the valve lifter and the rocker arm to transmit cam action to the valves. Hollow pushrods allow oil to pass from the hydraulic lifter to the rocker arm assembly **(Figure 12–17)**.

PUSHROD GUIDE PLATES On some engines, **pushrod guide plates** are used to limit the side movement of

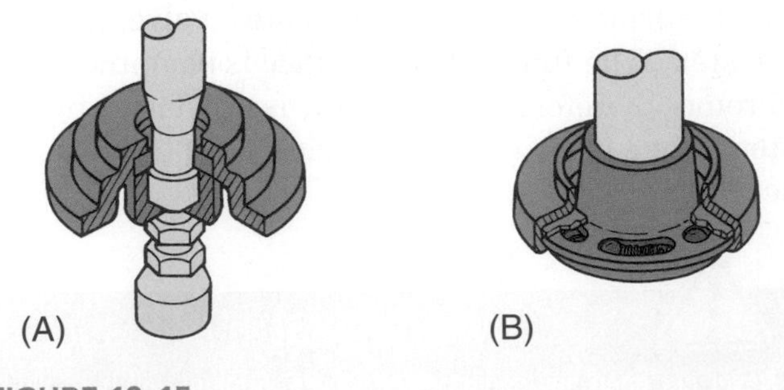

FIGURE 12–15

(A) Release and (B) positive valve rotators.

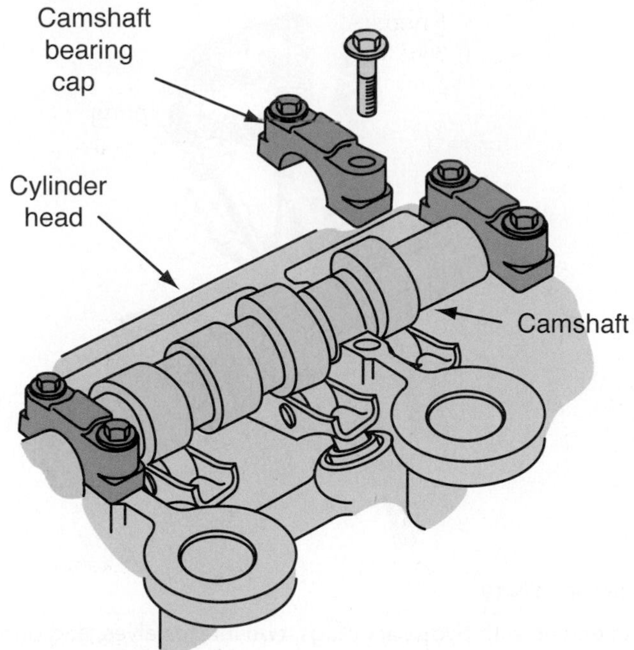

FIGURE 12–16

Many OHC cylinder heads are machined to accept one or two camshafts above the valves and have bearing caps that secure the camshaft.

the pushrods. The plates hold the pushrods in alignment with the rocker arms. When the pushrods pass through holes in the cylinder head or intake manifold, guide plates are not needed. This is true only when the holes for the pushrods are small enough to limit the sideways movement of the pushrod.

ROCKER ARMS Rocker arms are designed to change the direction of the cam's lifting force. As the lifter and pushrod move upward, the rocker arm pivots at the centre point, causing a change in direction on the valve side. This change in direction causes the valve to move downward and open. Rocker arms also permit valves to be angled.

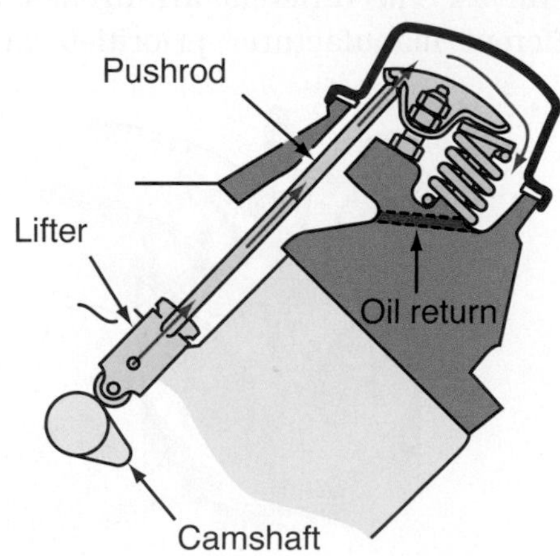

FIGURE 12–17

Most pushrods have a hole in the centre to allow oil to pass from the hydraulic lifter to the rocker arm assembly.

On some engines, the valve will open more than the actual lift of the cam lobe. This is done by changing the distance from the pivot point to the ends of the rocker arm. The difference in length from the valve end of the rocker arm and the centre of the pivot point (shaft or stud) compared to the pushrod or cam end of the rocker arm and the pivot point (shaft or stud) is expressed as a ratio. Usually, rocker arm ratios range from 1:1 to 1:1.75. A ratio larger than 1:1 produces a couple of results. First, it pushes the valves open farther than the actual lift of the cam lobe. Second, it multiplies the valve spring force through the valve train, keeping all components tight to the camshaft during operation.

Rocker arms are designed and mounted in several ways. Springs, washers, individual rocker arms, and bolts are used in this type of assembly. Other rocker arms are placed on studs that are mounted directly in the cylinder head.

Some overhead camshaft engines use rocker arms in such a way that the camshaft rides directly on top of the rocker arm. One end of the rocker arm fits over a cam follower or lifter, and the other end is directly over the valve stem **(Figure 12–18)**. Often, OHC cylinder heads have a complex arrangement of rocker arms and pushrods. Other overhead camshaft

FIGURE 12–18

The rocker arm moves with the lobes of the camshaft, and the rocker arm's movement is dampened by a spring assembly mounted next to the rocker arm.

DaimlerChrysler Corporation

engines do not use rocker arms, and the camshaft rides directly on top of the valves.

Rocker arms are made of stamped steel, cast aluminum, or cast iron. Cast adjustable rocker arms are attached to a rocker arm shaft that is mounted on the head by rocker arm brackets. Although a cast rocker arm can be resurfaced, a worn, stamped nonadjustable rocker arm must be replaced.

Cast-iron rockers are used in large, low-speed engines. They normally pivot on a common shaft. Aluminum rockers are generally used on high-performance applications and often pivot on needle bearings. The use of needle bearings reduces the friction at the pivot.

Some domestic engines are equipped with an independent stamped steel rocker arm assembly for each valve, mounted to a stud, which is either pressed or threaded into the cylinder head. On some engines, the studs are drilled to serve as an oil passage to the rocker arms. Make sure oil can pass through before installing the cylinder head on the block. Replacement press-in studs are available in standard sizes to replace damaged or worn studs and oversized to replace loose studs.

Multivalve Engines

Many newer engines use multivalve arrangements. Automotive engineers have long been obsessed with the idea of additional valves in the cylinder head. It all started in 1918 with the dual-valve Pierce Arrow, which was one of the first cars to use four valves per cylinder as a way to enhance gas flow and increase horsepower.

The basic idea behind using more than one intake and/or exhaust valve is simple—better efficiency. To improve efficiency, engineers need to improve the flow into the combustion chamber and the flow of the gases out of the chamber, which can be done in a number of ways. In the past, the common way was to make the valves larger and to change valve timing. Today, many engines use multiple valves and variable cam timing to improve the efficiency. Larger valves allowed more air in and more exhaust out, but the bigger valves weighed more than smaller ones and therefore required stronger springs to close them. The stronger springs held the valves tighter when they were closed but required more engine power to open them. This fact somewhat diminished the gains of using a larger valve. Also, when the engine runs at low speeds, the air moving past a large valve has a lower velocity than it would have if it flowed past a small valve, which reduces engine torque at low engine speeds.

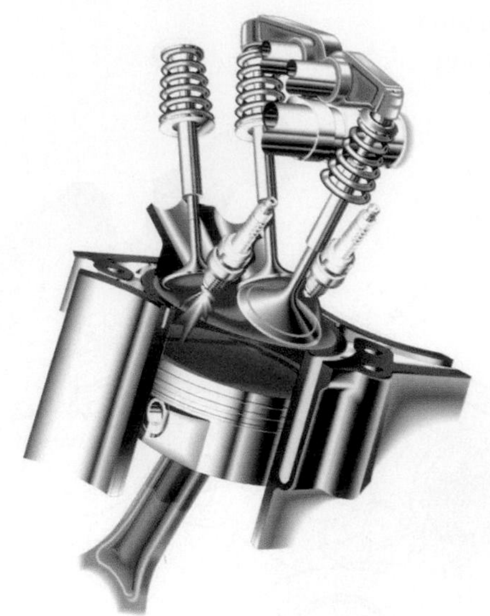

FIGURE 12–19

An engine with two spark plugs, two intake valves, and one exhaust valve for each cylinder.

Mercedes-Benz of N. A., Inc.

Although two small valves weigh as much or more than one valve, each valve weighs less and therefore the spring tension on each is less, which means the power required to open the valves is also less. Therefore, the gain in efficiency is not offset by the power required to open the valves, and the net gain from the engine is realized. Also, the velocity of the air in and out at low engine speeds is quicker than it would be with large valves.

Today, multivalve engines can have three valves per cylinder **(Figure 12–19)**, four valves per cylinder **(Figure 12–20)**, or five valves per cylinder **(Figure 12–21)**. The most common arrangement is four valves per cylinder, with two intake and two exhaust valves. The different arrangements result from different manufacturer priorities and other

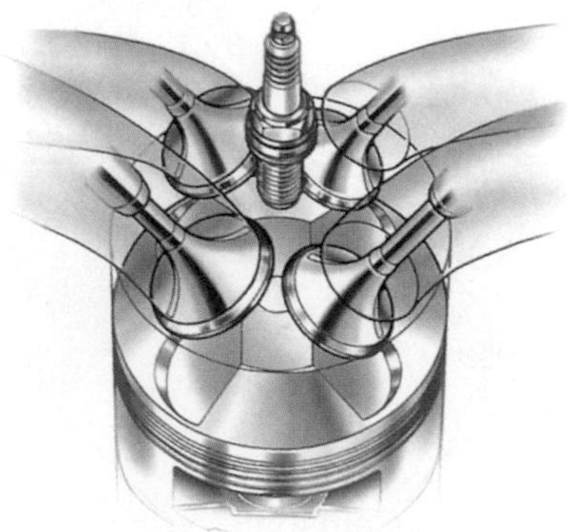

FIGURE 12–20

A typical layout for a cylinder with four valves.

American Honda Motor Co., Inc.

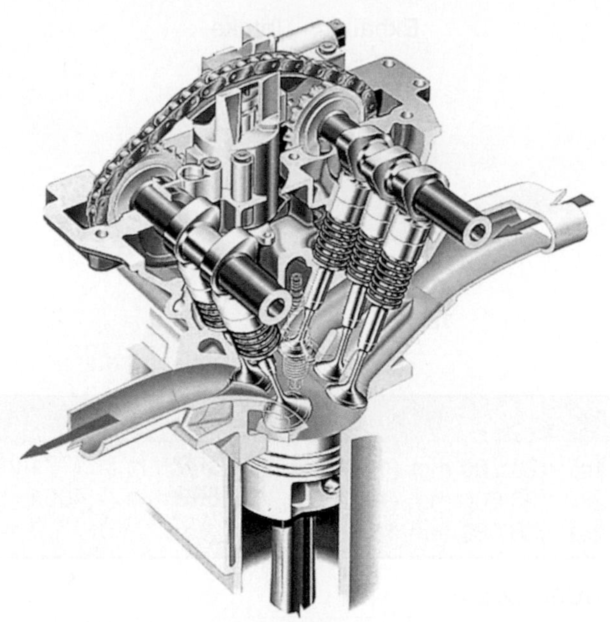

FIGURE 12–21

This five-valve-per-cylinder arrangement has three intake valves and two exhaust valves.

Audi of America, Inc.

features of the engine. In all multivalve engines, the heads are of the cross-flow design.

Using two intake and one or two exhaust valves increases the volume of the intake and exhaust ports. More mixture can enter the cylinders. Thus, multivalve engines have a more complete combustion, which reduces the chances of misfire and detonation. This results in enhanced fuel efficiency, cleaner exhaust, and increased power output. The velocity of the intake air is higher with small multiple ports than with a large single passage. The smaller valves naturally have less mass than big ones, so mechanical inertia is reduced, making a higher engine speed possible before valve float occurs. The more times the cylinder can be filled and evacuated per second, the more horsepower can be obtained.

Because increased air velocity is the main benefit of a multivalve cylinder head, the technology works best at high engine speeds.

The benefits of multivalves are offset to some extent, however, by a more complicated camshaft arrangement. The easiest way to actuate four valves per cylinder is with dual overhead camshafts. These are sometimes difficult to lubricate. The cam drive is even more complicated with V-power plants. Many use a single overhead cam per cylinder bank, with some kind of lever arm actuating the opposite bank of valves.

VARIABLE VALVE TIMING

Changing valve timing in response to driving conditions improves driveability and lowers fuel consumption and emission levels. There are many different variable valve timing (VVT) systems used on today's engines. Many systems only vary the timing of the intake or exhaust valves; some vary the timing of both valves; others vary the lift and timing of the intake or exhaust valves; and a few vary the timing and lift of both valves. Depending on the manufacturer and age of the vehicle, an engine control module (ECM) or powertrain control module (PCM) advances or retards valve timing based on operating conditions, such as engine speed and load **(Figure 12–22)**.

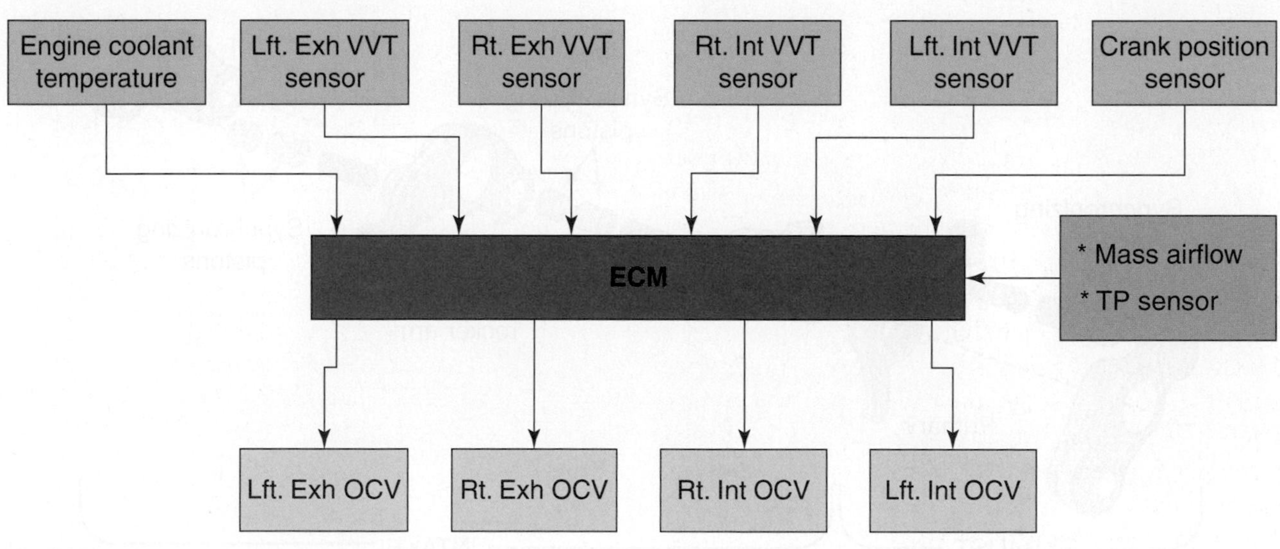

FIGURE 12–22

The information flow for a typical VVT system.

©Cengage Learning 2015

Advancing	Retarding
Begins intake event sooner	Delays intake event
Lengthens valve overlap	Shortens valve overlap
Builds more low-end torque	Builds more high-rpm power
Decreases piston-to-intake valve clearance	Increases piston-to-intake valve clearance

FIGURE 12–23

The effects of changing intake valve timing.

VVT systems are either staged or continuously variable designs. Most staged systems allow two different valve timing and lift settings. Continuously variable systems alter valve timing whenever operating conditions change. Continuously variable systems change the phasing or timing of a valve's duration **(Figure 12–23)**. These systems provide a wider torque curve, a reduction in fuel consumption, improved power at high speeds, and a reduction in hydrocarbon and NO_x emissions. On some engines, VVT has eliminated the need for an exhaust gas recirculation (EGR) valve.

Staged Valve Timing

Most staged valve timing systems switch between two or more different camshaft profiles based on operating conditions. An example of this is Honda's VTEC system used on multivalve engines. The camshaft has three lobes for each pair of intake valves. The third lobe is shaped for more valve lift and different open and close times **(Figure 12–24)**. There is a rocker arm over each of the three lobes. At low engine speeds, only the second lobe's rocker arms move the valves. At high speed, a solenoid valve sends pressurized oil through the rocker shaft to a piston in the outer rocker arms **(Figure 12–25)**. This pushes the piston partly into the centre rocker arm, locking the three rocker arms together. The valves now open according to the shape of the third lobe. When the solenoid valve closes, a spring pushes the pistons back into the outer rockers, and the engine runs with normal valve timing.

Late-model Honda i-VTEC engines switch between low-speed and high-speed lobes, similar

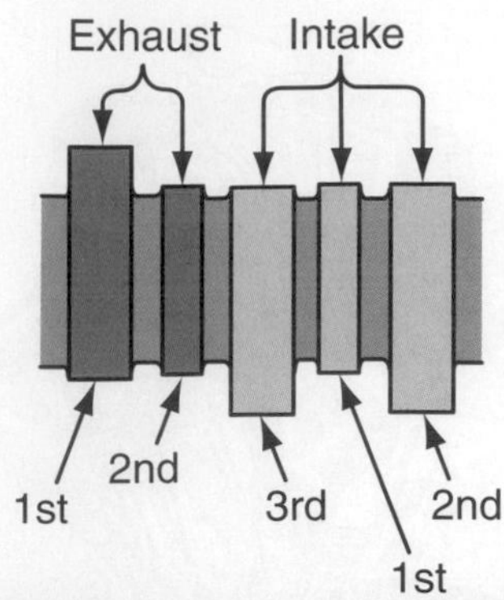

LOBE	INTAKE	EXHAUST
1st	29.700 mm (1.1692 in.)	29.900 mm (1.1771 in.)
2nd	35.568 mm (1.4003 in.)	35.699 mm (1.4054 in.)
3rd	36.060 mm(1.4196 in.)	

FIGURE 12–24

The size of the cam lobes for one cylinder for Honda's three-stage VTEC system.

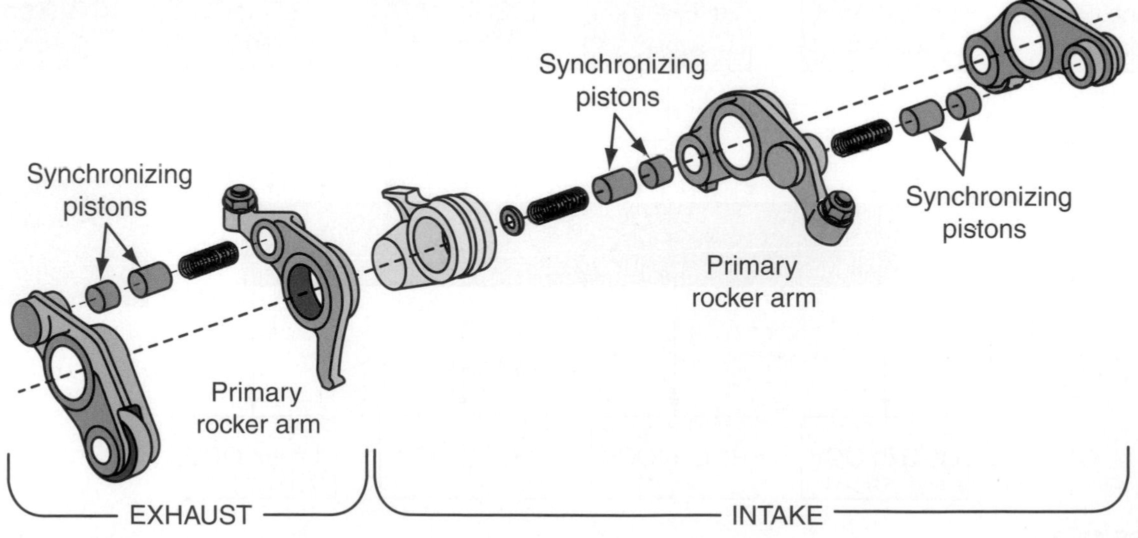

FIGURE 12–25

The synchronizing pin is controlled by oil pressure and locks the rocker arms together.

to older VTEC engine, but also change intake valve timing to reduce pumping losses at low engine speed. Under light loads, cruising, and low engine rpm, the electronic throttle control system completely opens the throttle plates to increase airflow into the engine. The i-VTEC system switches to the low-speed cam lobes and rocker arms. The cam lobe profile retards intake valve opening, allowing some of the air/fuel mixture to be pumped out of the cylinder and back into the intake manifold, as in an Atkinson cycle engine. This reduces engine output and increases fuel economy. Under high loads and rpm, the system switches to the high-speed lobes, which increases valve lift.

Continuously Variable Timing

To provide continuously variable value timing, camshafts are fitted with a **phaser**. The phaser is mounted where a timing pulley, sprocket, or gear would be **(Figure 12–26)**. The phaser allows camshaft-to-crankshaft timing to change while the engine is running. Phasers can be electronically or hydraulically controlled. In a hydraulically controlled system, oil flow is controlled by the ECM. Electronic systems rely on stepper motors **(Figure 12–27)**.

Phasers are either based on a helical gear set or vanes enclosed in a housing. A helical phaser has an outer gear driven by the timing belt, an inner gear attached to the camshaft, and a piston T placed between the outer gear and inner gear. As the hydraulically controlled piston moves, the helical splines on the piston and inner gear force the camshaft to change its position in relationship to the timing gear.

FIGURE 12–26

The phaser assemblies on a late-model engine.

FIGURE 12–27

Electric camshaft phasers that rely on stepper motors.

A vane-type phaser assembly is a sealed unit with a hub and an internal vane assembly **(Figure 12–28)**. Around the hub is the timing gear connected by a chain to the crankshaft. The vane assembly is attached to the camshaft. The vane assembly moves to change the phasing of the camshaft. At the base of the hub are oil ports. Oil from control solenoids enters and exits through these ports. When the ECM determines a need to change valve timing, oil is sent to the correct port. The pressurized oil then pushes on the vanes and causes a change in valve timing.

On a few single overhead cam (SOHC) engines, the phaser alters the timing, in equal amounts, of both the intake and exhaust camshafts. When more low-speed torque is required, the ECM orders earlier valve opening and closing. When more high-speed power is needed, the cam timing is retarded. By altering the timing of both the intake and exhaust valves, valve overlap is also changed.

TOYOTA'S VVT-I SYSTEM The engine in Toyota hybrids, like other hybrids, operates on the Atkinson cycle. However, it also runs with a conventional four-stroke cycle. The switching between the two cycles is done by valve control. Toyota's VVT-i system is reprogrammed to allow the intake valve to close later for the Atkinson cycle. The Atkinson cycle effectively reduces the displacement of the engine and is in operation when there is low engine load.

The VVT-i system is controlled by the ECM. The ECM adjusts valve timing according to engine speed, intake air volume, throttle position, and water temperature. In response to these inputs, the ECM sends commands to the camshaft timing oil control

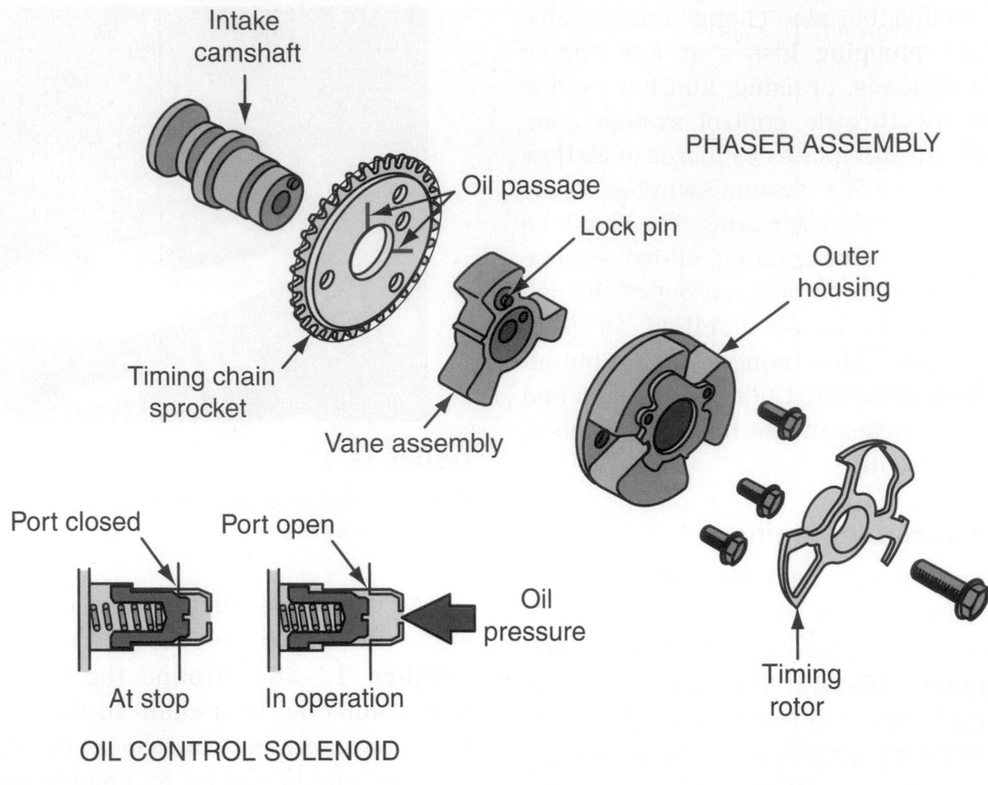

FIGURE 12–28

An exploded view of a phaser.

valve (OCV). The OCV **(Figure 12–29)** directs oil pressure to the advance or retard side of the VVT-i phaser. The position of the OCV spool valve is determined by a varying magnetic field strength, due to varying duty cycles. This magnetic field opposes a constant spring pressure in the valve. The various valve timing settings are shown in **Figure 12–30**.

The VVT-i system relies on crankshaft position sensors and camshaft position sensors to monitor camshaft position. Systems that control intake and exhaust valve timing are called dual VVT-i systems. This system allows for varying amounts of valve overlap.

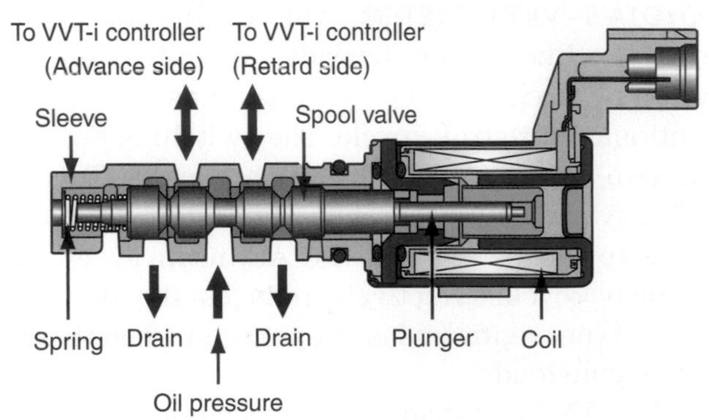

FIGURE 12–29

The spool valve in the camshaft timing oil control valve is duty-cycled by the PCM. This allows oil pressure to be applied to the advance or retard side of the phaser.

FIAT'S MULTIAIR SYSTEM Fiat's MultiAir (or UniAir) system is a variable valve lift and duration control system for gasoline or diesel engines. It is a system that relies on mechanical, hydraulic, and electronic technologies **(Figure 12–31)**.

The system controls individual intake valves through the interaction of a roller cam follower, hydraulic piston, hydraulic chamber, electronically controlled solenoid valve, and hydraulic valve actuator. As a result, the system allows for variable timing and lift for the intake valves. The camshaft directly operates the exhaust valves.

The system takes direct control of the air that passes through the intake valves. It does not rely on a throttle plate to regulate intake air; therefore, pumping losses are minimized, which accounts for many of the gains from the system. **Pumping losses** result from the work the pistons must do in order to move air in and out of the cylinders. In an engine with throttle plates, pumping losses affect overall engine performance at all times except when the throttle plates are wide open. All other times, the throttle plates restrict the air flowing to the intake valve and reduce volumetric efficiency. On its intake stroke, a

Operation state	Range	Valve timing	Objective	Effect
During idling	1	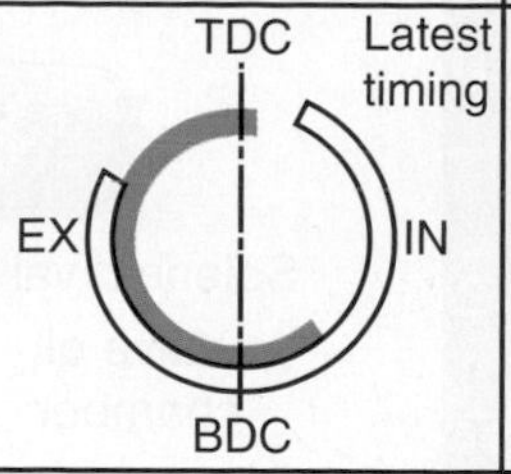 TDC, Latest timing, EX, IN, BDC	Eliminating overlap to reduce blow-back to the intake side	Stabilized idling rpm; better fuel economy
At light load	2	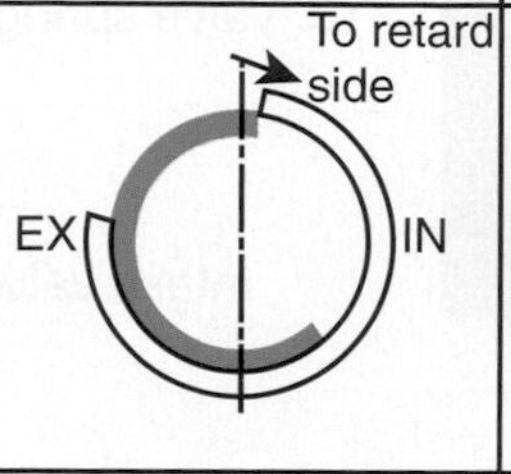 To retard side, EX, IN	Decreasing overlap to eliminate blow-back to the intake side	Ensured engine stability
At medium load	3	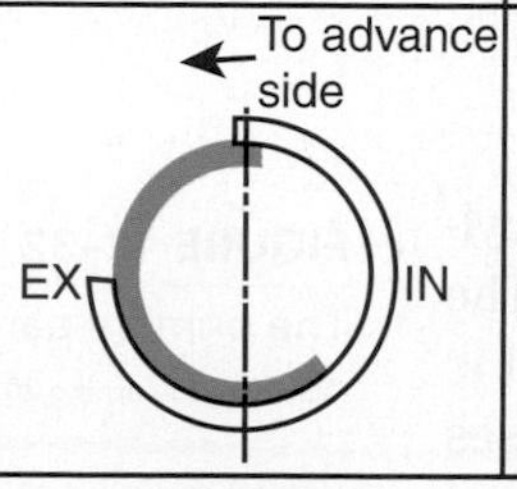 To advance side, EX, IN	Increasing overlap to increase internal EGR for pumping loss elimination	Better fuel economy; improved emission control
In low-to-medium speed range with heavy load	4	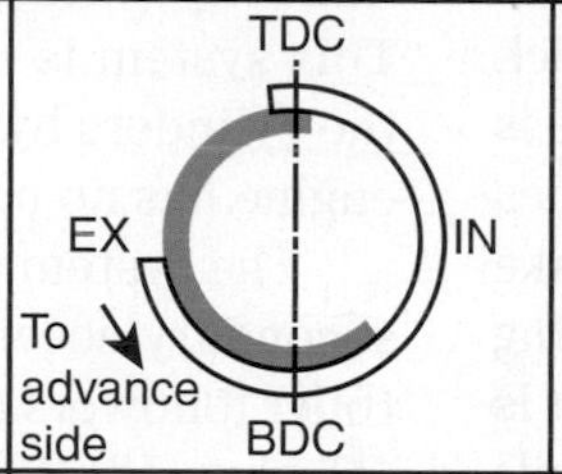 TDC, EX, IN, To advance side, BDC	Advancing the intake valve close timing for volumetric efficiency improvement	Improved torque in low-to-medium speed range
In high-speed range with heavy load	5	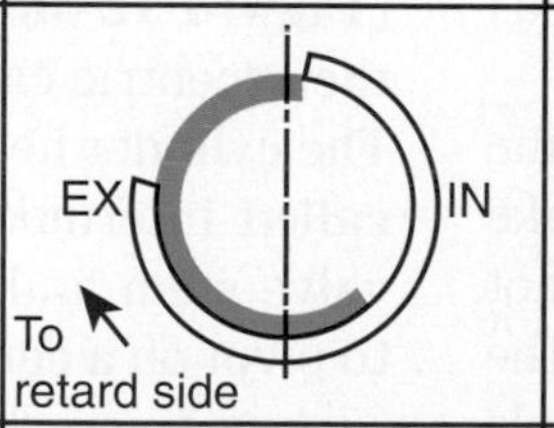 EX, IN, To retard side	Retarding the intake valve close timing for volumetric efficiency improvement	Improved output
At low temperatures	—	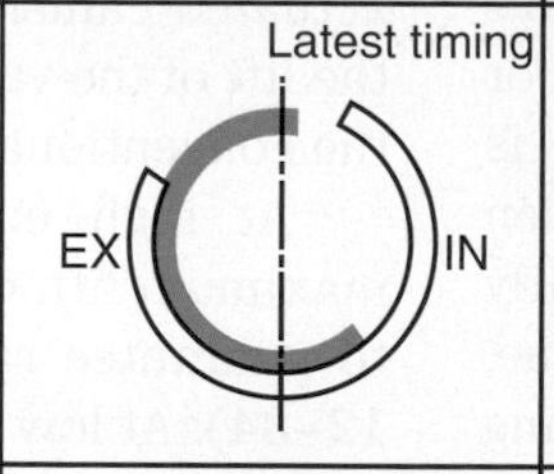 Latest timing, EX, IN	Eliminating overlap to prevent blow-back to the intake side for reduction of fuel increase at low temperatures, and stabilizing the idling rpm for decreasing fast idle rotation	Stabilized fast idle rpm; better fuel economy
Upon starting/ stopping the engine	—	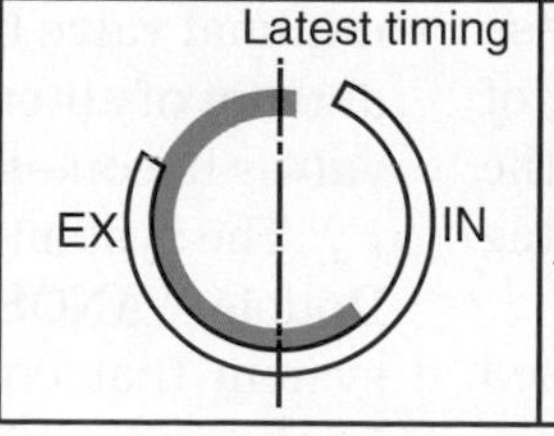 Latest timing, EX, IN	Eliminating overlap to eliminate blow-back to the intake side	Improved startability

FIGURE 12–30

The action of Toyota's VVT-i to provide for the Atkinson cycle.

FIGURE 12–31

An outside look at the valve train for Fiat's MultiAir system.

©Cengage Learning 2015

piston must use extra energy to draw air into the cylinder when the throttle plate is partially closed. The amount of lost energy depends on engine speed. It is very low when the engine is at a low rpm, but increases until the throttle plates are more than half open.

The intake valves are controlled by the action of the hydraulic solenoid valve **(Figure 12–32)**, which is controlled by an ECM. When the solenoid valve is closed, trapped oil in the hydraulic chamber provides a solid link between the cam follower and the intake valve. Therefore, the intake valves operate according to the profile of the camshaft lobes. The camshaft is designed to maximize power at high engine speeds by providing high lifts and long opening times.

When the solenoid valve is open, the hydraulic connection between the chamber and the intake valves is lost. Therefore, the intake valves do not follow the camshaft lobes, and the pressure of the valve spring closes the valve. This action controls when and how much the intake valve will open. For example, to close the valve early, the solenoid is closed at the beginning of the intake stroke and then opens partway through the intake stroke. The early closing of the intake valve provides low-speed torque.

The system reacts in an opposite way during engine startup and low engine speeds. During these times, the solenoid valve is open at the beginning of the intake stroke and then closes partway through the stroke. This action reduces emissions by increasing the speed of the incoming air.

The length of time the solenoid valve is closed not only controls the valve's duration but also controls the lift of the valve. Short solenoid closing times result in low lift. Long solenoid closing times result in high valve lift.

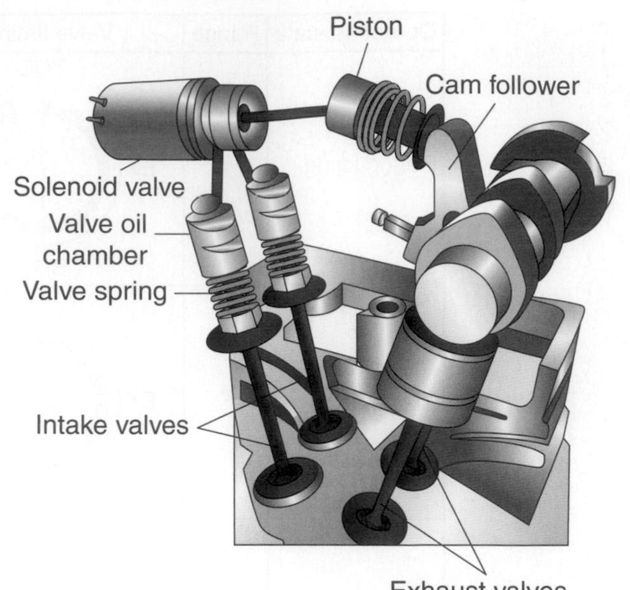

FIGURE 12–32

The primary parts of a Fiat MultiAir system.

©Cengage Learning 2015

VALVETRONIC SYSTEM Most BMW engines have infinitely variable intake valve control (Valvetronic). This system is used to regulate the flow of air into the cylinders by controlling valve lift. Therefore, the engine has no need for a throttle plate.

The system uses a conventional camshaft with a secondary eccentric shaft and a series of levers and roller followers that are activated by a stepper motor **(Figure 12–33)**. A computer changes the phase of the eccentric cam to change the action of the valves. The cylinder heads have an extra set of rocker arms, called intermediate arms, positioned between the valve stem and the camshaft. These arms are able to pivot on a central point, which is an electronically actuated camshaft. Therefore, the system can vary the lift of the valves without relying on the profile of the conventional camshaft.

At high engine speeds the system provides maximum lift, opening the ports for maximum flow to guarantee rapid filling of the cylinder **(Figure 12–34)**. At low engine speeds, the system reverts to minimal valve lift **(Figure 12–35)**. This reduces the amount of air entering the cylinder. The action of the valves becomes the engine's throttle plates.

The system is coordinated with the independent Double VANOS (variable Nockenwellensteuerung) system that continuously varies the timing of the intake and exhaust camshafts. VANOS is a hydraulic and mechanical camshaft control device. It uses engine speed and accelerator pedal position to determine ideal valve timing for the conditions.

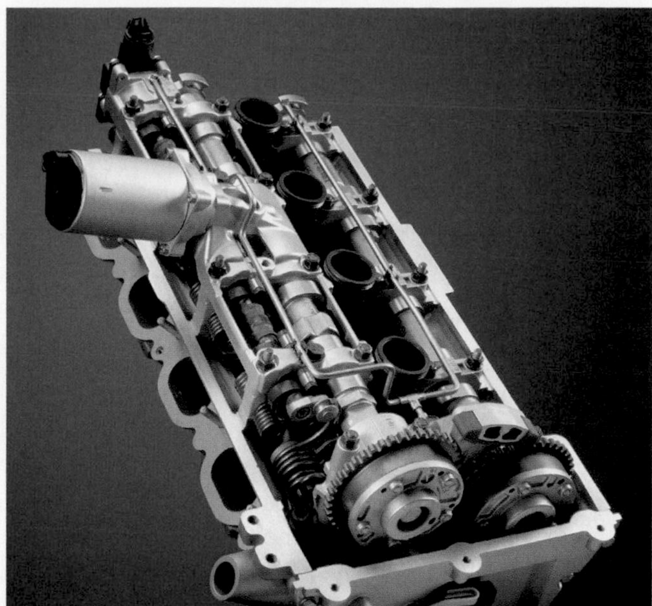

FIGURE 12–33

The secondary eccentric shaft and stepper motor assembly for the Valvetronic system.

BMW of North America, Inc.

Other VVT Systems

A unique setup for controlling intake and exhaust valve timing and lift relies on a conventional camshaft, which is ground for high performance, with

FIGURE 12–34

The position of the eccentric shaft to provide maximum lift in a Valvetronic system.

BMW of North America, Inc.

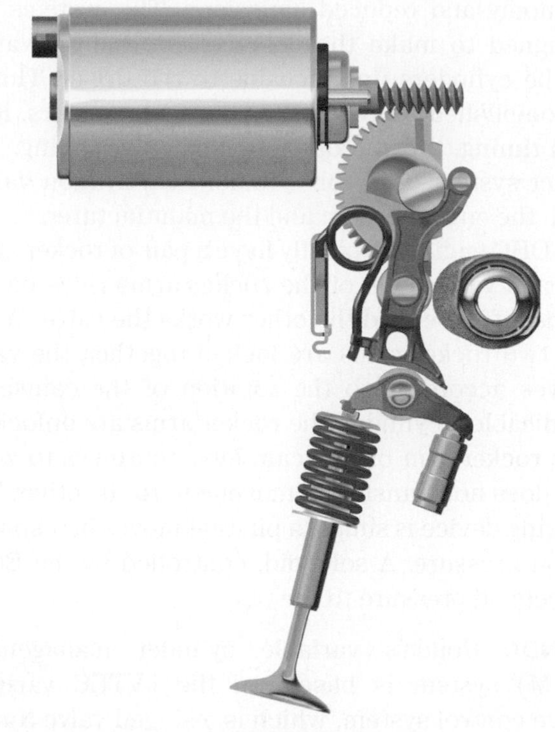

FIGURE 12–35

The position of the eccentric shaft to provide minimum lift in a Valvetronic system.

BMW of North America, Inc.

hydraulic lifters fitted with ultra-high-speed valves to bleed off the fluid. The ECM changes the pressure in the lifter to delay valve openings, change valve duration, or prevent valves from opening. A solenoid is used to control the flow of oil into a piston in each lifter that effectively determines the tappet's height.

Another design uses pressurized oil to allow a four-valve-per-cylinder engine to operate as a three-valve engine at low speeds. At high speeds, the engine uses the four valves. Below 2500 rpm, each intake valve follows a separate camshaft lobe. The primary valve opens and closes normally, whereas the secondary intake opens just enough to keep the engine running smoothly. As the engine reaches 2500 rpm, the ECM allows pressurized oil to move small pins that lock each pair of rocker arms together, causing both intake valves to follow the normal cam lobe. When the engine slows down, the pressurized oil is bled off and the pin releases, separating the two rocker arms.

Cylinder Deactivation

Cylinder deactivation works by keeping the intake and exhaust valves closed in a group of the engine's cylinders. This decreases the working displacement of the engine and provides an increase in fuel economy and reduced emissions. The systems are designed to make the deactivation and activation of the cylinders unnoticeable to the driver. This is accomplished by controlling the fuel injectors, ignition timing, throttle opening, and valve timing. The exact system used for cylinder deactivation varies with the engine design and the manufacturer.

OHC engines typically have a pair of rocker arms at each valve. One of the rocker arms rides on the camshaft lobe, and the other works the valve. When the two rocker arms are locked together, the valve moves according to the rotation of the camshaft. To disable a cylinder, the rocker arms are unlocked. The rocker arm on the cam lobe continues to work but does not transfer its movement to the other. The locking device is simply a pin that moves in response to oil pressure. A solenoid, controlled by the ECM, directs oil pressure to the pin.

HONDA Honda's variable cylinder management (VCM) system is based on the i-VTEC variable valve control system, which is a staged valve timing system. The system is primarily used on Honda's hybrid to increase the regenerative braking capabilities of the vehicle and to minimize fuel consumption. The system is called the cylinder idling system and increases the amount of energy captured during deceleration. The system also allows normal and high-output valve timing, plus cylinder idling at all or some cylinders.

Basically, the system has five rocker arms per cylinder **(Figure 12–36)**. A hydraulically controlled pin connects and disconnects the rocker arms. When there is no connection between the cam-riding rockers and the valve rocker, the cylinder is idling or deactivated. There are three separate oil passages leading to the pin. As the pressure moves through a passage, it moves the pin. The PCM controls a spool valve that directs the pressurized oil to the appropriate passage. It also controls solenoids that control the amount of pressure. With the valves closed, the pistons in those cylinders move quite freely. This, in turn, reduces the amount of engine braking or resistance that takes place during deceleration.

In a typical Honda hybrid system, when the brake pedal is released and the accelerator depressed, the vehicle moves by both electric and engine power. At this time, the engine is running in the economy mode with the valves opening by the low lift camshaft profile. When the driver is maintaining a very low cruising speed, the engine shuts off, and the electric motor powers the car by itself. During this time, the engine's rocker arms are not opening the valves. During acceleration from a low speed, the engine runs in the economy mode. During heavy acceleration, the engine runs in its high-output mode, and the electric motor assists the engine. During deceleration, the motor begins to work as a generator, and the engine's valves close and remain closed. This allows for maximum regenerative braking and reduces fuel consumption. When full engine power is needed, the system quickly gets the three idling cylinders back into action. While the cylinders are idling, the VCM system controls the ignition timing and cycles the torque converter lockup clutch to suppress any torque-induced jolting caused by the switch from six- to three-cylinder operation.

OTHER CYLINDER DEACTIVATION SYSTEMS OHV engines can also use oil pressure to deactivate the cylinders. High pressure is sent to the lifters to collapse them. The lifters then follow the cam lobes but do not move the pushrods and rocker arms. Chrysler's multi-displacement system (MDS), found

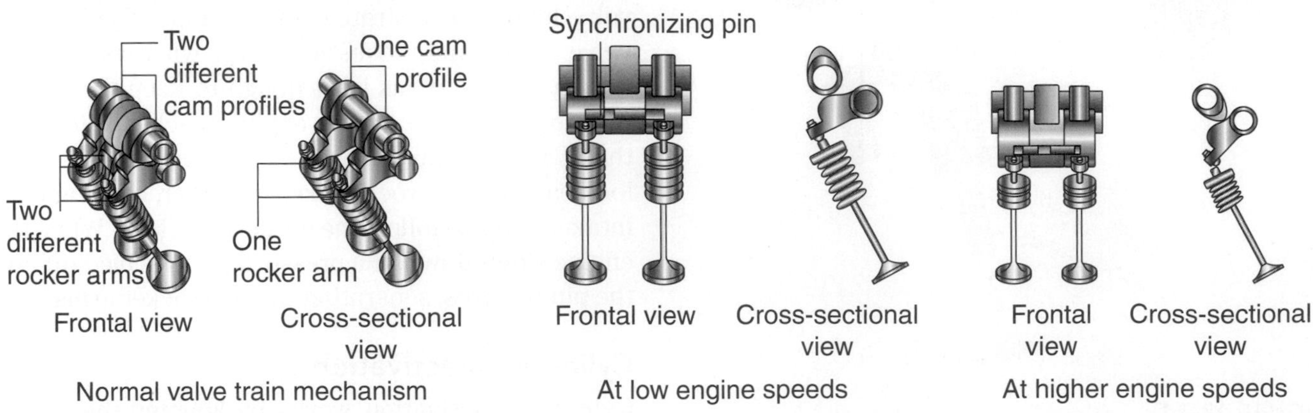

FIGURE 12–36

The activity of the rocker arms in Honda's VTEC system.

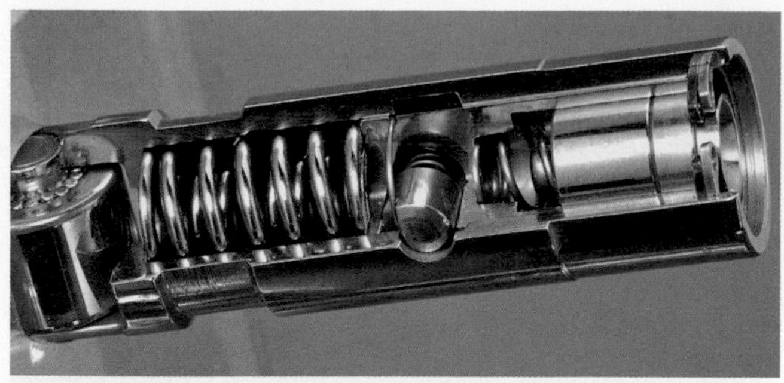

FIGURE 12–37

The lifter used in Chrysler's multi-displacement system. When the pin in the centre is moved, the piston inside the lifter is disconnected from the lifter body.

DaimlerChrysler Corporation

in some engines, has an oil circuit controlled by solenoids and unique hydraulic roller lifters. When conditions dictate that the vehicle does not need all cylinders, the ECM energizes the solenoids. Oil pressure is sent to the lifters. The pressure pushes on a small pin in the lifters **(Figure 12–37)**. As the pin moves, the piston inside the lifter is disconnected from the lifter body. The lifter body continues to move with the cam lobe, but no motion is passed on to the rocker arms.

General Motors' displacement on demand (DoD) system, now called active fuel management (AFM), uses two-stage switching lifters. The lifters have an inner and outer body connected by a spring-loaded pin. High oil pressure, sent by solenoids, collapses the spring, and the two lifter bodies disconnect **(Figure 12–38)**.

CYLINDER HEAD DISASSEMBLY

On some engines, the rocker arms must be removed before the head is disassembled. If the camshaft rides directly above the rocker arms, use the appropriate spring compressor, and depress the valve enough to pull the rocker arm out **(Figure 12–39)**. Some have the rocker arms mounted on a separate shaft. The ends of the rocker arms do not directly contact the valves; rather, a bridge rocker arm is used. The bridge rocker arm assembly is also mounted on a shaft. To remove these rocker arms, both shafts are unbolted **(Figure 12–40)**.

On all OHC engines, the camshaft must be removed before the cylinder head can be disassembled. Follow the specified order for loosening the camshaft bearing caps. Keep the caps in the order they were on the head. Also, take a picture or draw a diagram of the arrangement of the cam follower assemblies, and mark each part. This will ensure that each part is returned to the same position.

Measure the installed spring height for each valve **(Figure 12–41)**. This measurement will be needed during reassembly. To remove the valves, use a valve spring compressor. First, select a socket that fits over the valve tip and onto the retainer. Tap the socket

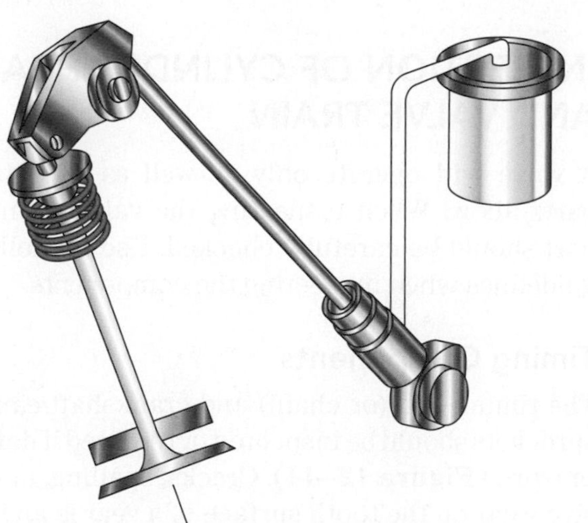

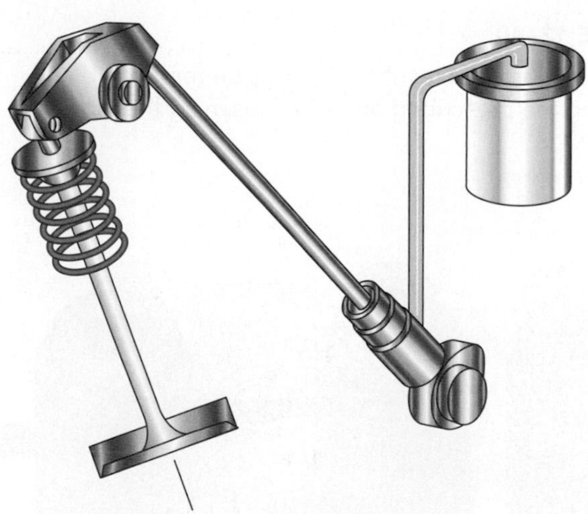

FIGURE 12–38

The action of the lifters in GM's active fuel management (AFM) system, formerly called displacement on demand (DoD).

©Cengage Learning 2015

Remove
rocker
arm.
Compress
the spring
with a special
tool or pry bar.

FIGURE 12–39

On some OHC heads, the valve springs must be slightly compressed to remove the rocker arms.

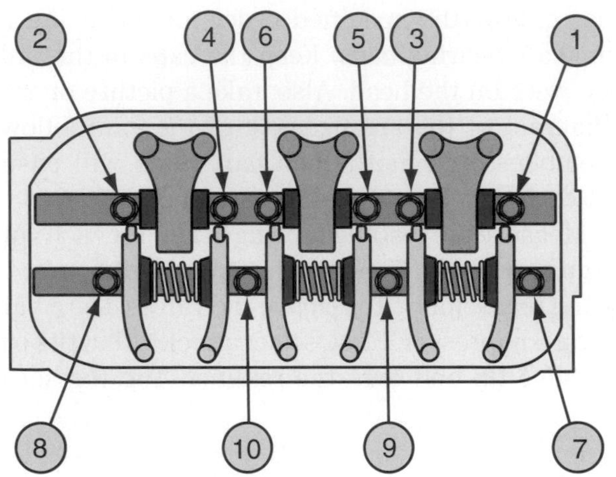
2
4
6
5
3
1
8
10
9
7

FIGURE 12–40

Rocker arm assemblies should always be removed by following the prescribed order for loosening the mounting bolts.

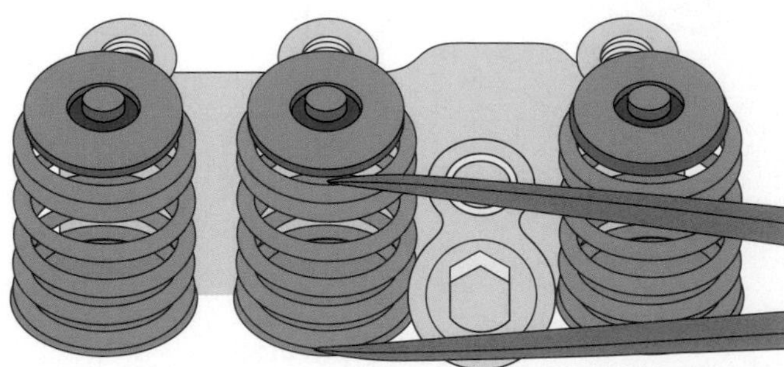

FIGURE 12–41

Measure valve spring height before disassembling the cylinder head.

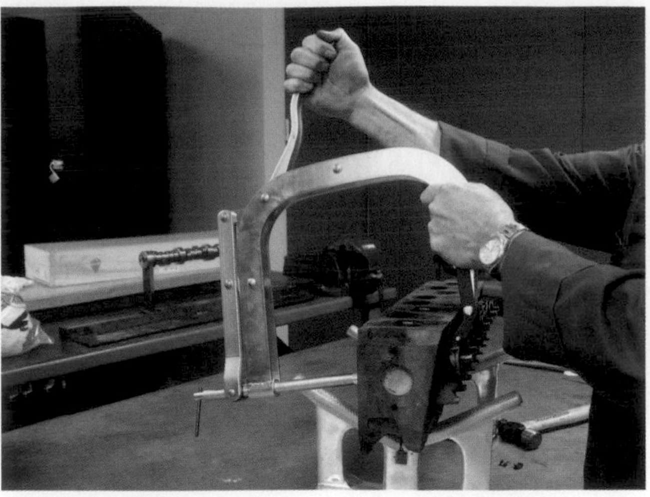

FIGURE 12–42

With a spring compressor, compress the springs just enough to remove the keepers.

with a plastic mallet to loosen the keepers. Adjust the jaws of the compressor so that they fit securely over the spring retainers. Compress the valve springs just enough to remove the keepers **(Figure 12–42)**.

Next, remove the valve oil seals and valves. Keep the assemblies together according to the cylinder they were in. If a valve cannot pass through its guide, its tip might be mushroomed or peened over. Do not force the valve through the guide. It could score or crack the valve guide or head. Raise the stem and file off the ridge until the stem slides through the guide easily **(Figure 12–43)**.

INSPECTION OF CYLINDER HEAD AND VALVE TRAIN

A valve will operate only as well as its actuating parts allow. When inspecting the valve train, each part should be carefully checked. Use the following guidelines when inspecting the components.

Timing Components

The timing belt (or chain) and crankshaft/camshaft sprockets should be inspected or replaced if damaged or worn **(Figure 12–44)**. Cracks, spalling, or excessive wear on the tooth surface of a gear is an indication of improper backlash. Stripped/broken rubber belt failure is often due to insufficient tensioning, extended service life, abusive operation, or worn tensioners. Most manufacturers recommend that a timing belt be replaced every 96 000 km (60 000 mi.). Replacing them at this interval can prevent belt

(A)

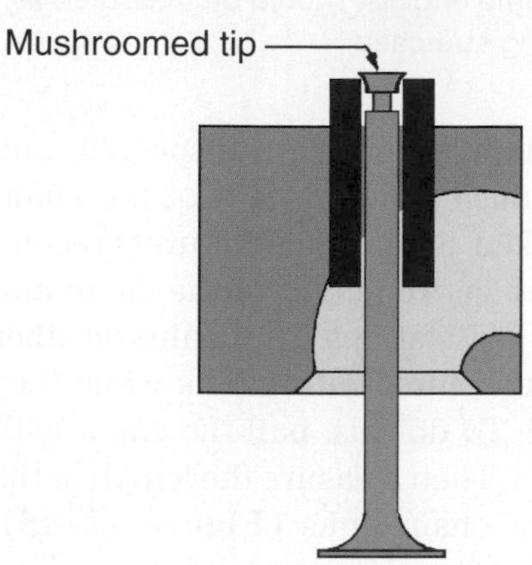

(B)

FIGURE 12–43

A mushroomed valve tip should be (A) filed before removing the valve from (B) its guide.

Martin Restoule

breakage. Loose timing belts will jump across the teeth of the timing sprockets, causing shearing of the belt teeth. Localized tensile overloads from over-revving an engine can lead to belt breakage. Also, belts on those engines equipped with adjustable tensioners should be checked for wear whenever

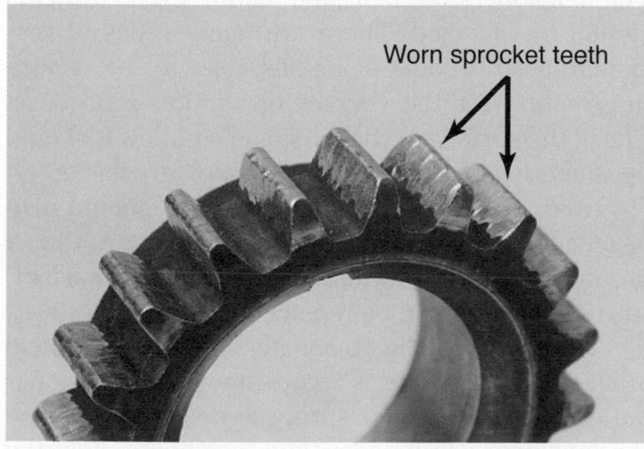

FIGURE 12–44

Worn teeth on a timing sprocket.

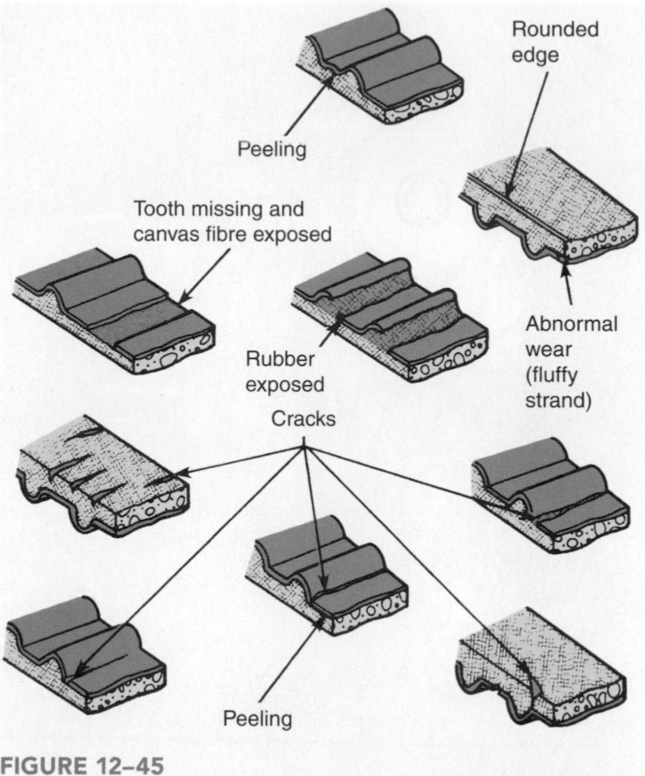

FIGURE 12–45

Various forms of timing belt wear.

a belt is retensioned. Also check the condition and operation of the tensioner. In addition, check for cord separation and cracks on all surfaces **(Figure 12–45)**. If the belts are damaged, they should be replaced. It is wise to replace the belt any time the engine is overhauled.

On many engines, severe engine damage will result from a broken timing belt. When a timing belt breaks, the camshaft no longer turns with the crankshaft. As the camshaft slows, the valves may open while the piston is moving to TDC. Because the intake valves are larger than the exhaust valves, they are more likely to hit the top of the pistons, causing damaged pistons and/or bent or broken valves. Therefore, it is absolutely necessary that the belt be replaced according to the manufacturers' recommended mileage intervals and whenever it is removed from the engine.

Timing Chains

Late-model engines may have many drive chains. Those with one chain use it to drive the camshaft(s) by the crankshaft. The engine shown in **Figure 12–46** has a long chain to drive the camshafts and short chain to drive the balance shaft. Some engines have an additional chain that aligns the intake camshaft with the exhaust camshaft. V-type engines may have

FIGURE 12–46

An engine with two separate drive chains.

a separate chain from the crankshaft to the camshafts of each cylinder bank, then additional chains to connect the intake and exhaust camshafts on each bank **(Figure 12–47)**.

FIGURE 12–47

This engine has three separate timing chains.

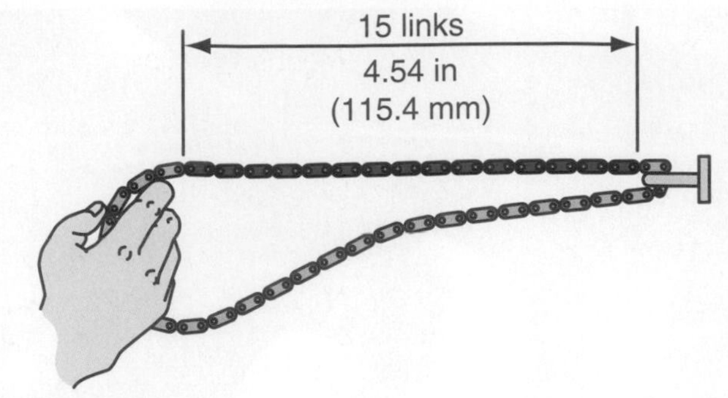

FIGURE 12–48

The timing chain on some engines should be measured in sections while it is being stretched.

Each drive chain should be inspected and replaced if it is damaged. The length of the chain should also be checked. Some manufacturers recommend measuring the entire length of the chain and comparing that to specifications. The chains on other engines should be measured in sections while they are being stretched. To do this, pull the chain with the specified tension. Then measure the length of the specified number of chain links **(Figure 12–48).** This measurement is taken at three random sections of the chain. The average length is then compared to specifications. The chain should be replaced if it is not within specifications.

Belt Idler Pulley

All idler pulleys should be rotated by hand. They are okay if they move smoothly. The pulley should also be checked for signs of lubricant leakage. Check around the seal. If leakage is evident, the idler should be replaced.

Tensioners

The tensioners of belt and chain drive systems should be checked. There are many types of tensioners used in today's engines; refer to the service information for the correct inspection procedure. Check the surface of the tensioner's pulley. It should be smooth and have no buildup of grease or oil **(Figure 12–49)**. Most belt tensioners should also be checked for signs of lubricant leakage. Check around the seal. If any damage or leakage is evident, the tensioner should be replaced.

The action of a belt tensioner should be checked. Make sure the spring is free to move the tensioner pulley. If the tensioner spring is defective, replace the tensioner. On plunger-type tensioners, hold the tensioner with both hands and push the pushrod strongly against a flat surface. The pushrod should

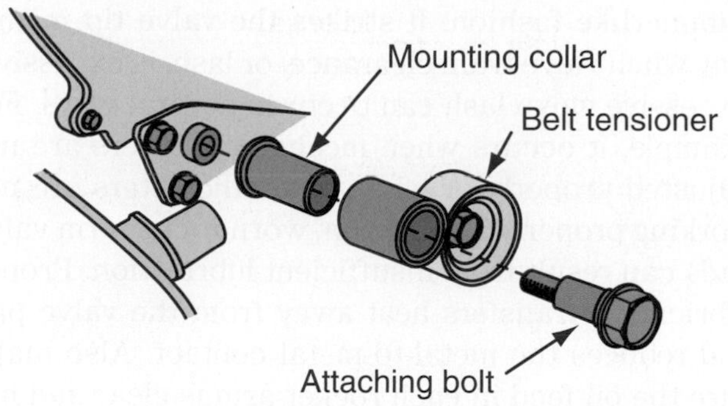

FIGURE 12–49

A timing belt tensioner assembly.

not move. If it does, replace the tensioner. Measure the distance the pushrod extends from the housing. Compare that distance to specifications. If this measurement is not within specifications, replace the tensioner.

Chain drive systems have a variety of dampers and guides in addition to a tensioner **(Figure 12–50)**. The dampers and guides should be checked for wear. In most cases, their width is measured and compared to specifications. If they are worn, they should be replaced. Again, there are different types of chain tensioners, each with a unique inspection procedure. The plunger in ratchet-type tensioners should be able to be smoothly moved out by hand but should not be able to be pushed in by hand.

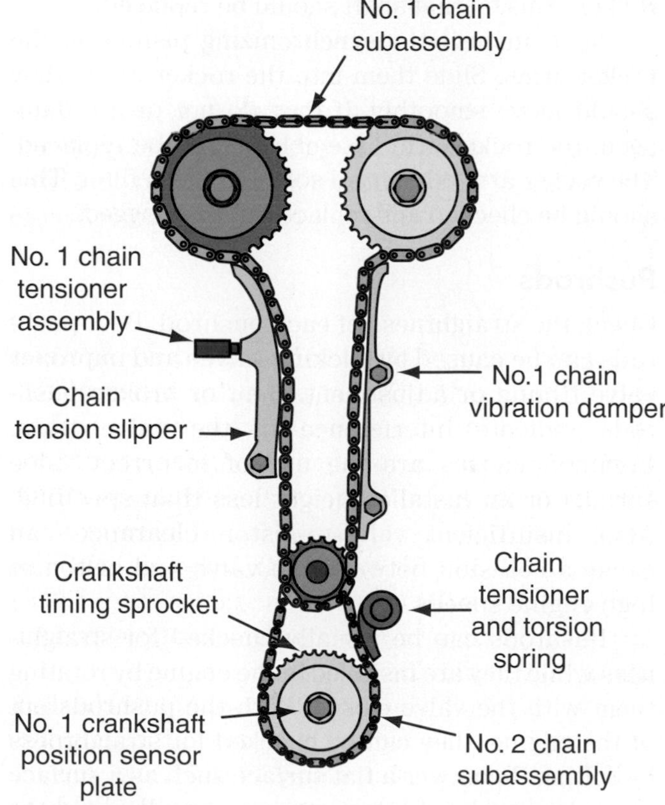

FIGURE 12–50

Timing chain components.

Gears and Sprockets

All timing gears and sprockets should be carefully inspected. A gear with cracks, spalling, or excessive wear on the tooth surface is an indication of improper backlash. All damaged or worn gears should be replaced. The oil pump, camshaft timing, crankshaft timing, and balance shaft gears and sprockets on some engines are measured with the drive chain wrapped around the individual gears. The diameter of the gear with the chain around it is measured with a vernier caliper. The caliper's jaws must connect the chain's rollers while doing this. If the diameter is less than specifications, the chain and gear or sprocket should be replaced.

Cam Phasers

Camshaft phasers are used in most VVT systems. The action of a hydraulic phaser can be checked while it is attached to the camshaft. Clamp the camshaft in a soft-jawed vise. Attempt to rotate the timing gear on the phaser. If it moves, the phaser must be replaced. Next, cover all of the oil ports on the phaser with electrical tape, except the advance port **(Figure 12–51)**. Using an air nozzle with a rubber tip, apply the specified air pressure to the exposed port **(Figure 12–52)**. The timing gear

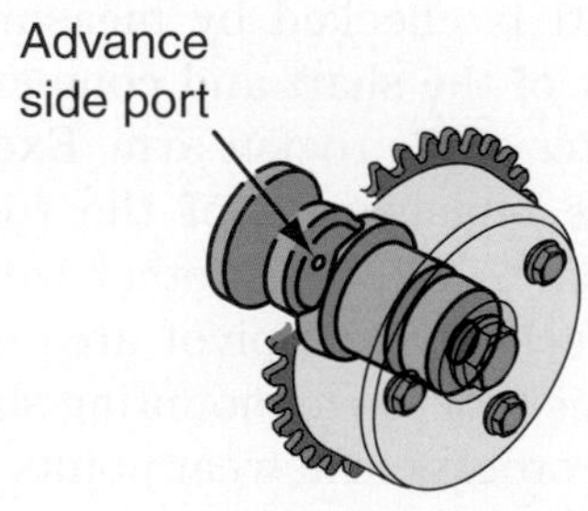

FIGURE 12–51

Location of the advance port on a camshaft phaser.

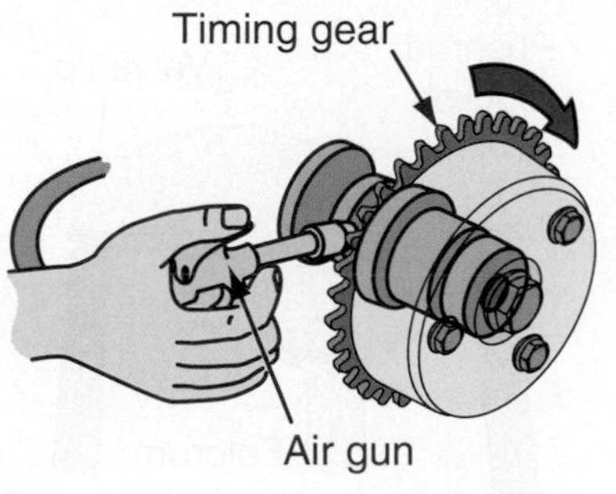

FIGURE 12–52

The action of the phaser is checked by applying air to the port and observing the timing gear.

should move counter-clockwise. When the air is released, the timing gear should move clockwise. This check should be conducted several times, and the timing gear should move smoothly. This process is then repeated at the retard port. In this case, all ports except the retard port are sealed.

SHOP TALK

The condition of a timing belt can be checked by pressing your fingernail into the hardened backside of the belt. If no impression is left, the belt is too hard and should be replaced.

Cam Followers and Lash Adjusters

Overhead cam follower arm and lash adjuster assemblies should be carefully checked for broken or severely damaged parts. If pads are used to adjust the valve lash, the cups and the shim pads must be carefully checked. A soft shim will not hold the valve at its correct lash, and therefore, each shim should be checked to make sure it is hard. To do this, place the shim on the base circle of the camshaft, and press down on the shim with your hand. You should feel no give.

Rocker Arms

Inspect the rocker shaft assembly for wear, especially at points that contact the valve stem and pushrod. The fit between a cast rocker arm and a rocker shaft is checked by measuring the outside diameter of the shaft and comparing it to the inside diameter of the rocker arm. Excessive clearance requires replacement of the rocker arm or the rocker shaft, or both. Another wear point that should be checked is the pivot area of the rocker arm. Also check for loose mounting studs and nuts or bolts. Other rocker arm wear points are shown in **Figure 12–53**.

Excessive wear on the valve pad occurs when the rocker arm repeatedly strikes the valve tip in a hammerlike fashion. It strikes the valve tip in this way when valve train clearance, or lash, is excessive. Excessive valve lash can occur in several ways. For example, it occurs when mechanical lifters are not adjusted properly or when hydraulic lifters are not working properly. In addition, worn rocker arm valve pads can result from insufficient lubrication. Proper lubrication transfers heat away from the valve pad and reduces the metal-to-metal contact. Also make sure the oil feed in each rocker arm is clear and not plugged with dirt.

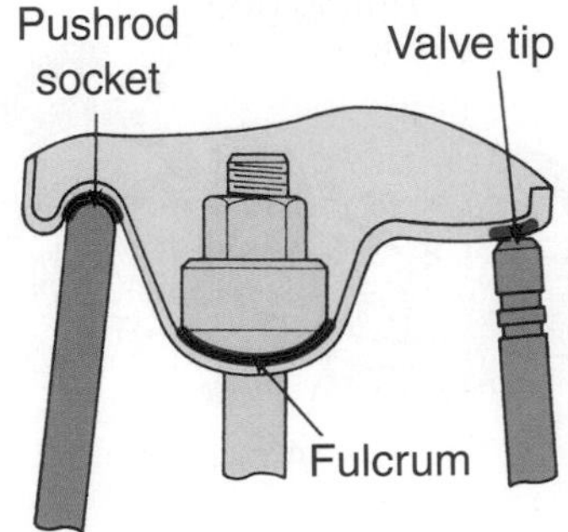

FIGURE 12–53
Rocker arm wear spots.

HONDA'S VARIABLE CYLINDER MANAGEMENT ROCKER ARMS Honda uses oil pressure and two sets of rocker arms to alter cam timing and to deactivate cylinders. This system has unique inspection and service procedures. When inspecting them, keep all parts in order so that they can be installed in their original location. Measure the outside diameter of each rocker shaft at the location of the first rocker arm. Then measure the inside diameter of the rocker arm. The difference between the two is the clearance. This clearance should be compared to specifications. Repeat this procedure for each rocker arm and shaft. If the clearance is beyond specifications, replace the shaft and all over-tolerance rocker arms. If one rocker arm needs replacement, all of the rocker arms on that shaft should be replaced.

Next, inspect the synchronizing pistons in the rocker arms. Slide them into the rocker arms; they should move smoothly. If they do not or are damaged, the rocker arm assembly should be replaced. The rocker arm oil control solenoid has a filter. This should be checked and replaced if it is clogged.

Pushrods

Check the straightness of each pushrod. Bent pushrods can be caused by sticking valves and improper valve timing or adjustment. Bent or broken pushrods indicate interference in the valve train. Common causes are the use of incorrect valve springs or an installed height less than specified. Also, insufficient valve-to-piston clearance can cause a collision between the valve and piston at high engine speeds.

Pushrods can be visually checked for straightness while they are installed in the engine by rotating them with the valve closed. With the pushrods out of the engine, they can be checked for straightness by rolling them over a flat surface such as a surface plate. If a pushrod is not straight, it will appear to hop as it is rolled. However, the most accurate way to check for straightness is by using a dial indicator.

If more than 0.0764 mm (0.003 in.) total indicated reading (TIR) is found, the pushrod is not straight and should be replaced.

Hollow pushrods should be looked through to make sure there are no blockages in the rod. No oil will get to the rocker arms if these pushrods are plugged.

The ends of the pushrods should be checked for nicks, grooves, roughness, or signs of excessive wear. Replace any damaged or worn pushrods.

When replacing pushrods, make sure they are the correct length for the application. Also, keep in mind that pushrods that are held in alignment by slots in the cylinder head or guide plates must be hardened.

Retainers and Keepers

Valve spring retainers and valve keepers hold the valve spring and valve in place. A worn retainer will allow the spring to move away from the centreline of the valve. This will affect valve operation because spring tension on the valve will not be evenly distributed. Each retainer should be carefully inspected for cracks because a cracked retainer may result in serious damage to the engine. The inside shape of most retainers is a cone that matches the outside shape of the keepers. This must be a good fit in order for the keepers to stay in their grooves on the valve stem.

The valve stem grooves should match the inside shape of the keepers. Some valves have multiple keeper grooves. Others have only one. All of the valve stem grooves should be inspected for damage and fit by inserting a keeper in them. Both the retainers and keepers should be carefully inspected for damage. If a defect is found, they should be replaced.

Valve Rotators

Most rotators impart positive rotation to the valve during each valve cycle and improve valve life two to five times, in some cases even more. In normal operation, rotators will continue to function for more than 160 000 km (100 000 mi.) and require no maintenance. However, when valves are refaced or replaced at high mileage, the rotators should be replaced, because they cannot be visually inspected accurately. Whether or not they rotate when held in the hand is no indication of their function in the engine. Although rotation can be checked only in a running engine, uneven wear patterns develop at the valve stem tip if the rotators are not functioning properly.

Valve Springs

Valve spring assemblies, including the damper springs, should be checked for correct tension, squareness, and free length as well as signs of cracks, breaks, and damage.

Cylinder Heads

Cylinder heads should be carefully inspected after they are cleaned. Any severe damage to the sealing areas indicates the head should be replaced. Also use the appropriate method for detecting cracks. Make sure the cracks are properly repaired, or replace the head. Also check the heads for dents, scratches, and corrosion around water passages, especially on aluminum heads.

Crack Repair

Common locations of cracks in a cylinder head include the following: between the spark plug bore and the valve seat, between the valve seats, around the valve guides, and in the exhaust ports. In most cases, a cracked head should be replaced. However, some cracks can be effectively repaired. Crack repair is normally done by specialty shops.

It is important to keep in mind that most cracks are caused by something other than a defect in the head. The cause of the cracking needs to be identified and corrected. No matter what caused the crack, the crack needs to be repaired if the head is reused. Crack repair is done by the cold process of pinning or the hot process of welding.

FURNACE WELDING CRACK REPAIRS Furnace welding is considered the best way to repair cracks in a cast-iron head. To furnace weld, the head is first preheated in an oven. This minimizes thermal shock when the flame of a welding torch contacts the head. After the crack has been filled with metal, the head is allowed to slowly cool before it is used.

FLAME SPRAY WELDING Flame spray (powder) welding is also used to repair cast-iron heads. This process uses nickel-based powders and a special torch to fill the crack.

REPAIRING ALUMINUM HEADS Defects in aluminum heads are commonly found as follows:

- Cracks between the valve seat rings
- Cracks in coolant passages
- Cracks across the main oil artery
- Detonation damage inside the combustion chamber
- Melted or deformed metal in the chamber
- Coolant-related metal erosion

FIGURE 12–54

TIG welding an aluminum head.

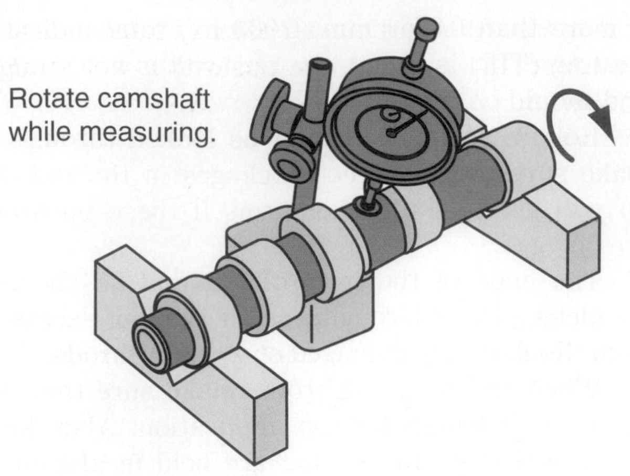

FIGURE 12–55

The camshaft should be checked for straightness.

In many cases, these problems result in head replacement. However, some heads are repairable. Tungsten inert gas (TIG) welding is the preferred way to repair aluminum heads **(Figure 12–54)**. Welding aluminum is often considered difficult because it welds differently than iron or steel. When exposed to air, aluminum forms an oxide coating on the surface that helps protect the metal against corrosion. The oxide layer makes welding difficult because it interferes with fusing and weakens the weld. A TIG welder prevents the formation of the oxide by bathing the weld with inert gas (normally argon).

PINNING CRACKS Pinning is commonly used to repair small accessible cracks in cast-iron and aluminum heads. Pins are used only when the metal is thick enough to secure the pins. Pinning is done with a drill, tap, and tapered or straight pins. Holes are drilled into the ends of the crack to keep it from spreading; then holes are drilled at various overlapping intervals along the length of the crack. After they are installed, the pins are peened to seal the surface. A sealant is not required when tapered pins are used; however, it must be applied over the repair area when straight pins are used.

Camshaft and Bearings

The old camshaft bearings should be inspected for signs of unusual wear that may indicate an oiling or bore alignment problem. On some engines, the bores in the aluminum casting serve as a bearing surface; therefore, the bore should be carefully examined.

Each lobe of the camshaft should be checked for wear, scoring, scuffing, fractured surface, pitting, and signs of abnormal wear. Also check for plugged oil passages.

The camshaft should also be checked for straightness. Place it on V-blocks. Place a dial indicator on one the middle journals. Rotate the camshaft and watch the dial indicator **(Figure 12–55)**. Compare the highest reading to specifications. If the measurement exceeds specifications, replace the camshaft. With a micrometer, measure the height of each cam lobe and the diameter of each journal. If the readings do not meet specifications, replace the camshaft.

Valves

Each valve face should be carefully checked for evidence of burning **(Figure 12–56)**, and each stem should be checked for wear **(Figure 12–57)**. Also check the stem for signs of distortion and excessive wear **(Figure 12–58)**. Replace any valves that are badly burned, worn, or bent. Reusable valves are cleaned by soaking them in solvent, which

FIGURE 12–56

A severely burnt valve.

will soften the carbon deposits **(Figure 12–59)**. The deposits are then removed with a wire buffing wheel. Once the deposits are removed, the valve can be resurfaced.

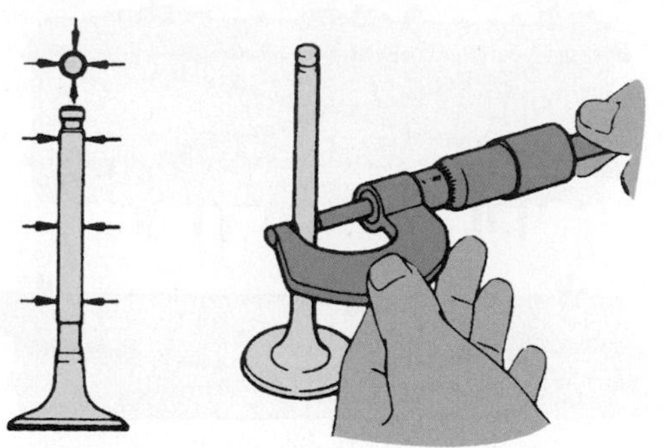

FIGURE 12–57

Checking a valve stem for wear.

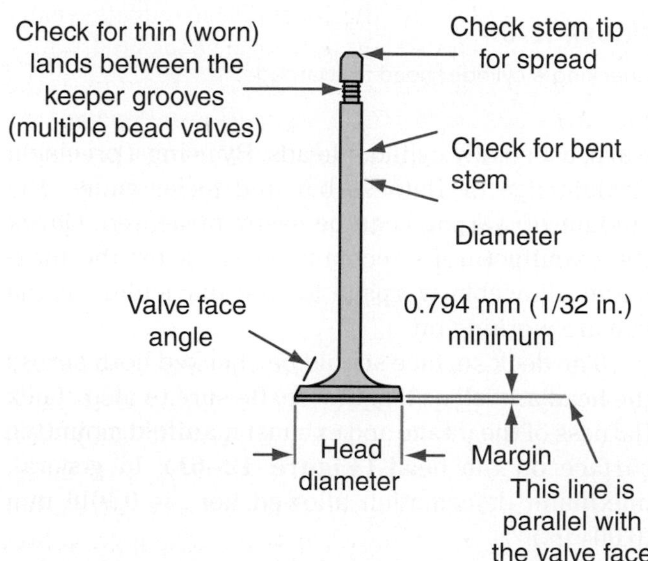
Check for thin (worn) lands between the keeper grooves (multiple bead valves)
Check stem tip for spread
Check for bent stem
Diameter
Valve face angle
0.794 mm (1/32 in.) minimum
Head diameter
Margin
This line is parallel with the valve face

FIGURE 12–58

Parts of a valve that should be checked during your inspection.

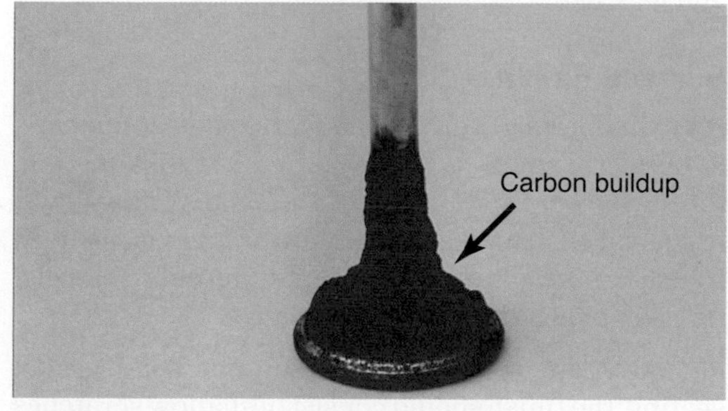
Carbon buildup

FIGURE 12–59

Carbon buildup on the back of a valve.

ALUMINUM CYLINDER HEADS

Aluminum heads are commonly used, because a typical aluminum head weighs roughly half as much as a cast-iron head. Eliminating anywhere from 9 to 18 kg (20 to 40 lb.) of weight is a plus for fuel economy, but it has its drawbacks.

Aluminum expands and contracts almost twice as much as cast iron in response to temperature changes. This creates a number of problems. When an aluminum head is mated to an iron block, the difference in thermal expansion between head and block creates a great deal of scrubbing stress on the head gasket. Unless the gasket is engineered to take such punishment, leakage and premature gasket failure can result.

Increased thermal expansion and stress can also lead to cracking. The most crack-prone areas in the head are usually the areas around the valve seats **(Figure 12–60)**. High combustion temperatures and the constant pounding of the valve against its seat often cause cracking between the intake and exhaust seats or just under the exhaust seat.

The differing rates of thermal expansion between aluminum and iron create a lot of stress throughout the head. The head wants to expand in all directions at once as it heats up, but the head bolts keep it from going sideways or lengthwise. The only place left to go is up, so the head tends to bow up in the middle.

Aluminum is not as strong as cast iron. Consequently, the head provides less top-end support for the block. This can allow more distortion in the upper cylinder bore area, affecting combustion sealing, blowby, and ring life. Using deck plates when boring the block can help minimize some of the distortion that will occur after the head is torqued down. Aluminum cylinder head bolts should never be loosened or tightened when the metal is hot. Doing so may cause the cylinder head to warp due to the torque changes.

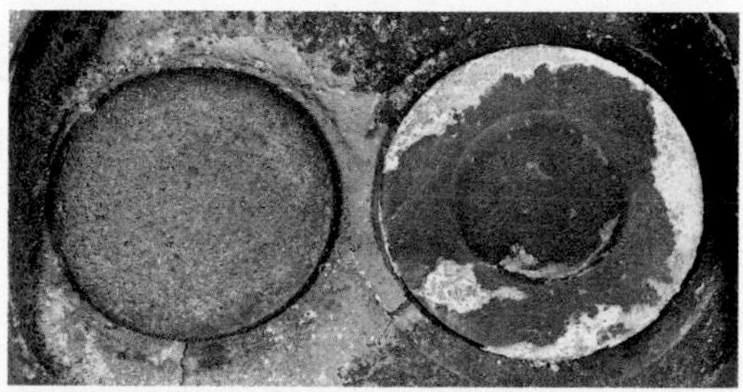

FIGURE 12–60

Lightweight aluminum heads are prone to cracking, which can lead to a recessed exhaust valve seat.

Aluminum makes a fairly good bearing material. It is soft and provides good embedability to foreign particles. But it lacks the support and rigidity of a conventional steel-backed bearing in an iron saddle. If the head overheats and warps, alignment through the cam bores is destroyed.

Aluminum has another drawback—**porosity**. The casting process sometimes leaves microscopic pores in the metal, which can weep oil or coolant. In most instances, the problem can be repaired.

Reconditioning Aluminum Cylinder Heads

Warpage in an aluminum cylinder head is usually the result of overheating (low coolant, uneven coolant circulation within the head, a too-lean fuel mixture, and incorrect ignition timing).

Alignment of the cam bores in an overhead cam head must be checked with either a straightedge and feeler gauge or with a dial indicator. If off by more than 0.0508 to 0.0762 mm (0.002 to 0.003 in.), corrective action is required.

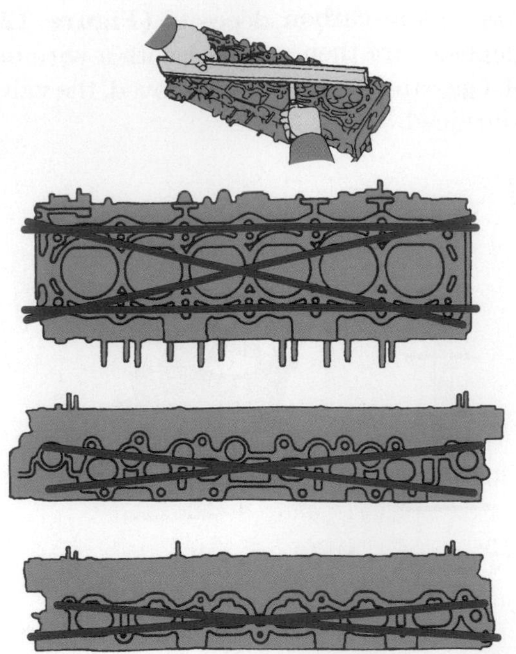

FIGURE 12–61

Checking a cylinder head for warpage.

> **SHOP TALK**
>
> Although specifications for the maximum amount of cylinder head surface warpage vary, traditionally, the maximum acceptable limit for cast-iron heads is 0.1270 mm (0.005 in.). Aluminum is not as forgiving, so 0.0508 to 0.1016 mm (0.002 in. to 0.004 in.) is a more realistic upper limit.

Removing metal from the face of any OHV head also alters valve train geometry, which limits the amount of metal that can be removed.

Aluminum cylinder heads can be straightened through the use of heat and special clamping fixtures. Some manufacturers recommend that warped heads not be straightened, but instead be replaced. Always follow the recommendations of the manufacturer.

RESURFACING CYLINDER HEADS

There are three reasons for resurfacing the deck surface of a cylinder head:

1. To make the surface flat so that the gasket seals properly
2. To raise the compression ratio
3. To square the deck to the main bores

As engines undergo heating and cooling cycles over their lifespan, certain components tend to warp, especially cylinder heads. By using a precision straightedge or flatness bar and feeler gauge, the amount of warpage can be easily measured. Check the manufacturer's recommendations for the maximum allowable warpage for the particular engine you are working on.

The deck surface should be checked both across the head as well as lengthwise. Be sure to also check flatness of the intake and exhaust manifold mounting surface on the head **(Figure 12–61)**. In general, maximum deformation allowed here is 0.1016 mm (0.004 in.).

Heads that are deformed beyond specifications must be surfaced. The finished surface, however, should not be too smooth. It must be rough enough to provide "bite," but not enough to cause a poor seal and leakage.

Surface Finish

No cylinder head surface, no matter how it appears, is perfectly smooth. When viewed in cross-section, the surface consists of a series of peaks and valleys. A special instrument, called a profilometer, is used to check surface roughness and to measure the distance between the peaks and valleys.

RESURFACING MACHINES For proper head gasket seating, the finish should consist of shallow scratches and small projections that allow for gasket support and sealing of voids.

Four different types of machines—belt surfacer, milling, broaching, and grinding—can be used in resurfacing operations.

WARNING!

Before attempting to operate any surfacing machine, be sure to become familiar with and follow all the cautions and warnings given in the machine's operation manual. Also, when operating these machines, you must wear safety glasses, goggles, or a face shield.

Belt Surfacers. Belt surfacers resemble belt sanders. These machines are easy to set up and operate. An operator merely places the part to be surfaced on the belt. A restraint rail helps keep the part positioned. Some machines have air-operated hold-down fixtures.

Resurfacing quality depends on operator skill and factors such as belt condition, machine horsepower, and the hold-down force applied.

Milling Machines. **Milling** machines cut away thin layers of metal to create a level, properly finished surface **(Figure 12–62)**. Cutters remove up to 1.27 mm (0.050 in.) per pass. Both rough and finish cuts are usually made to create the desired finish.

Broaching Machines. **Broaching** machines use an underside rotary cutter or broach. A block, cylinder head, or intake manifold is held in an inverted position as the broach passes underneath.

Surface Grinders. Surface grinders use a grinding wheel to remove metal stock **(Figure 12–63)**. They set up and operate similarly to milling machines.

STOCK REMOVAL GUIDELINES The amount of stock removed from the head gasket surface must be limited. Excessive surfacing can lead to problems in the following areas.

FIGURE 12–62

Milling an aluminum cylinder head.

FIGURE 12–63

A surface grinder.

Combustion Chamber. It might be necessary to measure and adjust the volume of an engine's combustion chambers. The combustion chamber is equal to the volume of the combustion chamber in the head plus the volume of the cylinder when the piston is at TDC.

Measuring this volume is called **cc-ing** the cylinder head. This is done with the valves and spark plugs installed. The cylinder head is mounted upside down, and a glass or plastic plate is installed over the combustion chamber. A graduated container called a burette is used to fill the combustion chamber with thin oil. The oil is poured through a hole in the plate, as shown in **Figure 12–64**. The amount of oil that enters the combustion chamber is equal to the volume (in cubic centimetres) for this cylinder.

The volume of a combustion chamber can be adjusted in several ways. It may be reduced by

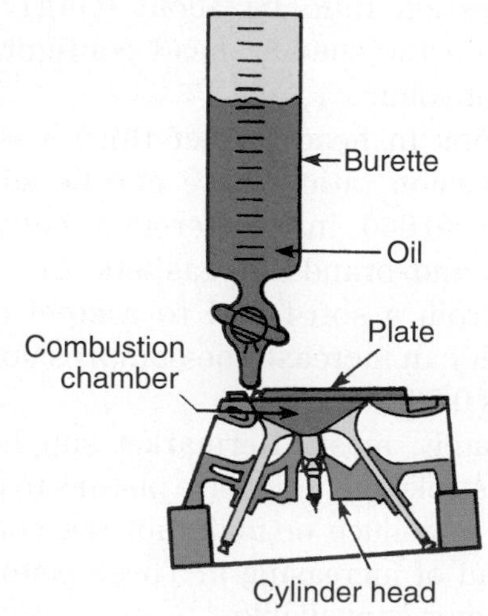

FIGURE 12–64

CC-ing a cylinder head to find the combustion volume.

surfacing the cylinder head. This, of course, reduces the volumes for all of the chambers in that cylinder head. Individual volumes can be increased by grinding the valve seats to sink the valves and by grinding and polishing metal from the combustion chamber surface. Either method can be used to equalize all the chambers and adjust them to the manufacturer's specifications.

Compression Ratio. Combustion chamber volume directly affects an engine's compression ratio. Boring the block oversize changes the swept volume and the compression ratio. Generally speaking, the compression ratio increases at the same percentage the displacement increased. Boring an engine 1.52 mm (0.060 in.) oversize will increase the displacement 147.48 cc (9 cu. in.), or slightly more than 3 percent. Assuming the replacement piston has the same compression height as the original, the initial 9.0:1 compression ratio will be increased by 3.0 percent to 9.27:1. As a rule of thumb, there is a 2 percent increase at 0.7620 mm (0.030 in.) oversize, a 4 percent increase at 1.524 mm (0.060 in.), and a 6 percent increase at 3.175 mm (0.125 in.).

Decking the block changes the compression ratio. Removing 0.2540 mm (0.010 in.) from the deck surface of an engine with a 101.6 mm (4 in.) bore, 76 cc head, 1.524 mm (0.060 in.) head gasket, and 2.032 mm (0.080 in.) deck height would raise the compression ratio by 0.14:1.

Resurfacing the head also increases the compression ratio. Though the effect varies for every type and size of chamber, a good rule of thumb is that when a head is surfaced 0.2540 mm (0.010 in.), the chamber is reduced by 1.50 cc for a 60 cc head and by 2.50 cc for a 90 cc head. This will increase the compression ratio by about 0.141:1 to 0.20:1, depending on the specific head configuration and actual swept volume.

Variations in head gasket thicknesses affect the compression ratio. There can be as much as 1.0160 mm (0.040 in.) difference between various types and brands of gaskets. For instance, changing from a soft-faced to a steel or copper shim gasket can increase the compression ratio by as much as 0.50:1.

Fortunately, most aftermarket suppliers either deck or de-stroke their oversize pistons to enable the technician to reduce or maintain the compression ratio instead of increasing it. These pistons should be used whenever available.

On many OHC engines, it is necessary to restore the distance between the camshaft gear and the crankshaft gear after material has been removed from the head. Special shims are used to raise the camshaft. If 0.7620 mm (0.030 in.) was removed from the head surface, the camshaft must be moved up 0.7620 mm (0.030 in.). If this distance is not properly restored, valve timing will be altered.

Piston/Valve Interference and Misalignment. When the block or head is surfaced, the piston-to-valve clearance during the valve overlap period becomes less. To prevent the valves from making contact with the piston, a minimum of 2.54 mm (0.100 in.) piston-to-valve clearance is recommended.

Surfacing can also cause valve tips, rocker arms, and pushrods to be dimensioned closer to the camshaft. This causes a change in rocker arm geometry and can also cause hydraulic lifters to bottom out. To correct this problem, pushrods of a different length may need to be installed.

When metal is removed from the block or heads on a V-type OHV engine, the heads will be positioned closer to the crankshaft. This downward movement causes the intake manifold to fit differently between the heads. As a result, ports might be mismatched, and manifold bolts might not line up. In order to return the intake manifold to its original alignment, corrective machining on the manifold is required.

GRINDING VALVES

Whenever the valves have been removed from the cylinder head, the valve heads and valve seats should be resurfaced. The most critical sealing surface in the valve train is between the face of the valve and its seat in the cylinder head when the valve is closed. Leakage between these surfaces reduces the engine's compression and power and can lead to valve burning. To ensure proper seating of the valve, the seat area on the valve face and seat must be the correct width, at the correct location, and concentric with the guide. These are accomplished by renewing the surface of the valve face and seat.

Valve grinding or refacing is done by machining a fresh, smooth surface on the valve faces and stem tips. Valve faces suffer from burning, pitting, and wear caused by opening and closing millions of times during the life of an engine. Valve stem tips wear because of friction from the rocker arms or actuators. Valve tips are machined after the valve face is refinished.

Valves can be refaced on either grinding **(Figure 12–65)** or cutting **(Figure 12–66)** machines.

FIGURE 12–65

Grinding a valve face.

Sioux Tools, Inc.

USING SERVICE INFORMATION

Specifications for valve angles are normally listed in the engine specifications section of a service information system.

To start the valve grinding operation, chuck the valve as close as possible to the valve head to eliminate stem flexing due to applied forces from the grinding wheel. Set the grinding angle according to the desired angle. Take light cuts, using the full grinding wheel width. Make sure coolant is striking the contact point between the valve face and grinding wheel. Remove only enough metal to clean up the valve face. A knifelike edge will heat up and burn easily or might cause preignition **(Figure 12–67)**. The width of the edge of a valve head between the top of the valve and the edge of the face is the valve margin. As a general rule, it is not advisable to grind a valve face to a point where the margin is reduced by more than 25 percent or to where it is less than 1.143 mm (0.045 in.) on the exhaust and 0.7620 mm (0.030 in.) on the intake valves.

FIGURE 12–66

A cutter-type valve resurfacer.

Neway Manufacturing, Inc.

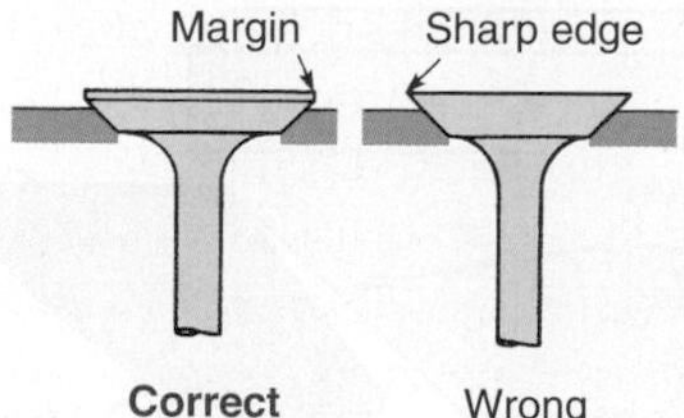

FIGURE 12–67

A sharp edge on the valve face is not recommended.

After grinding, check valve head runout. Use a dial indicator on the valve margin, and rotate the valve while it is still in the chuck. Valve runout should not exceed 0.0508 mm (0.002 in.) TIR. The face should not show any chatter marks or unground areas. After grinding, examine the valve face for cracks. Sometimes fine cracks are visible only after grinding. Sometimes they occur during grinding due to inadequate coolant flow or excessive applied wheel force.

CAUTION!

Always wear eye protection when operating any type of grinding equipment.

If an interference angle is to be used, the grinder is set ½° to 1° less than the standard 30° or 45° face angle. Always consult the manufacturer's specifications to determine whether an interference angle is to be used **(Figure 12–68)**. Grinding an interference angle produces a face angle close to 29½° or 44½°. For the valve to seat properly, the face angle cannot be less than 29° or 44°.

Removing metal from the valve face and/or seat will set the valve deeper into the port. As a result, more of the valve stem extends from the other side of the head. If the stem height is greater than that specified by the manufacturer, the valve stem tip must be ground down to bring the overall height of the stem back into specs. In some cases, if the stem height is excessive, the valve and/or valve seat must be replaced. Valve stem installed height is measured from the valve spring pad on the cylinder head to the top of the valve **(Figure 12–69)**. Never remove more material from the tip of the stem than the amount

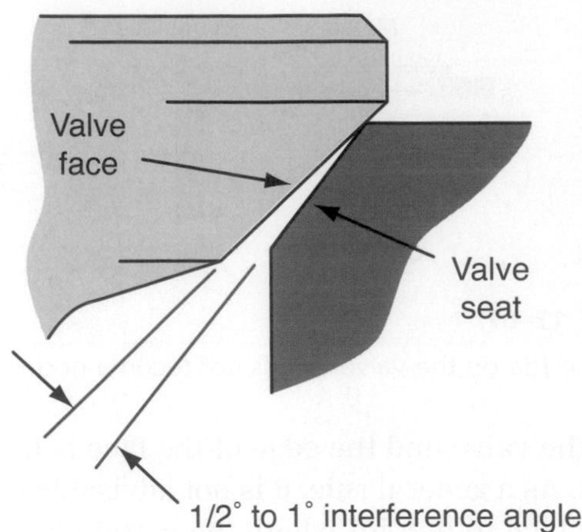

FIGURE 12–68

The difference between the seat angle and the valve face angle is known as the interference angle.

allowed by the manufacturer. When the installed height is excessive, valve train geometry can be thrown off or there can be valve lash problems.

The valve tip is ground so that it is exactly square with the stem. Because valve tips have hardened surfaces up to 0.7620 mm (0.030 in.) in depth, only 0.2540 mm (0.010 in.) can be removed during grinding. If more than 0.2540 mm (0.010 in.) is removed from the tip, the valve must be replaced. Follow the manufacturer's specifications for the allowable limits.

VALVE GUIDE RECONDITIONING

Valve guide problems can be lumped into one of three basic categories: inadequate lubrication, valve geometry problems, and wrong valve stem-to-guide clearance.

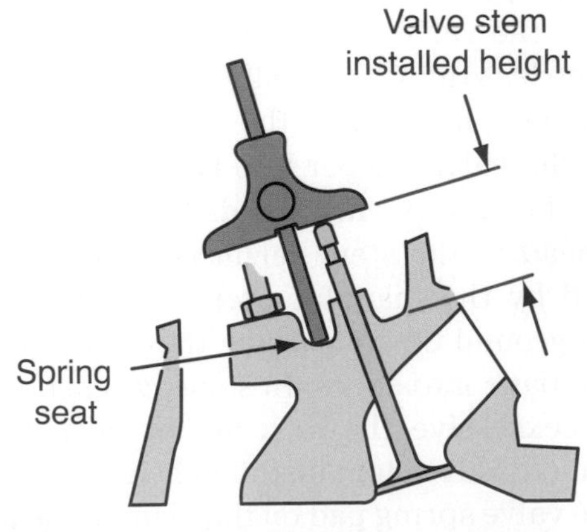

FIGURE 12–69

Measuring valve stem installed height.

Inadequate lubrication can be caused by oil starvation in the upper valve train due to low oil pressure, obstructed oil passages, improper operation of pushrods, and using the wrong type of valve seal. Insufficient lubrication results in stem scuffing, rapid stem and guide wear, possible valve sticking, and ultimately, valve failure due to poor seating and overheating.

Geometry problems include an incorrectly installed valve height, off-square springs, and rocker arm tappet screws or rocker arms that push the valve sideways every time it opens. This causes uneven guide wear, leaving an egg-shaped hole. The wear leads to increased stem-to-guide clearance, poor valve seating, and premature valve failure.

As for valve stem-to-guide clearance, a certain minimum amount is needed for lubrication and thermal expansion of the valve stem. Exhaust valves require more clearance than intakes because they run hotter. Clearance should also be close enough to prevent a buildup of varnish and carbon deposits on the stems, which could cause sticking. Insufficient clearance, however, can lead to rapid stem and guide wear, scuffing, and sticking, which prevents the valve from seating fully. This, in turn, causes the valve to run hot and burn.

The amount of valve guide wear can be measured with a ball (small-bore) gauge and micrometer. Insert and expand the ball gauge at the top of the guide. Lock it to that diameter, remove it from the guide and measure the ball gauge with an outside micrometer. Repeat this process with the ball gauge in the middle and the bottom of the guide **(Figure 12–70)**. Compare your measurements to the specifications for valve guide inside diameter. Compare these to specifications. Then compare your measurements against each other. Any difference in reading shows a taper or wear inside the guide.

Another way to check for excessive guide wear is with a dial indicator. The accuracy of this check is

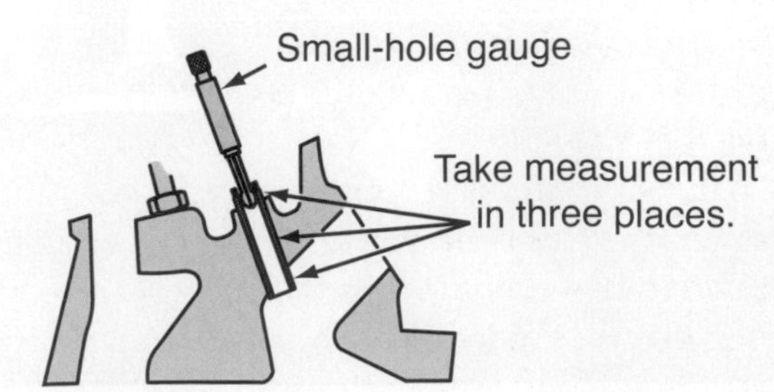

FIGURE 12–70

Valve guide wear can be measured with a small-bore gauge and a micrometer.

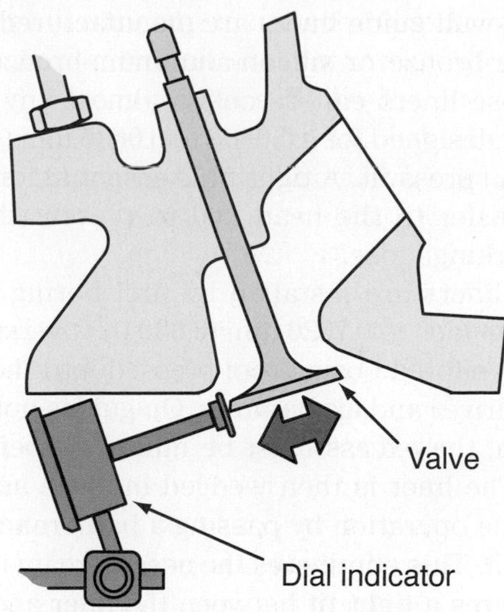

FIGURE 12–71

Checking for valve guide wear with a dial indicator.

directly dependent on the amount the valve is open during the check. Some manufacturers specify this amount or provide special spacers that are installed over the valve stem to ensure the proper height. Attach the dial indicator to the cylinder head and position it so that the plunger is at a right angle to the valve stem being measured **(Figure 12–71)**. With the plunger in contact with the valve head, move the valve toward the indicator and set the dial indicator to zero. Now move the valve from the indicator. Observe the reading on the dial while doing this. The reading on the indicator is the total movement of the valve and is indicative of the guide's wear. Compare the reading to specifications.

If the clearance is too great, oil can be drawn past both the intake and exhaust guides. Though oil consumption is more of a problem with sloppy or worn intake guides because the guides are constantly exposed to vacuum, oil can also be pulled down the exhaust guides by suction created in the exhaust port. The outflow of hot exhaust creates a Venturi effect as it exits the exhaust port, creating enough vacuum to draw oil down a worn guide **(Figure 12–72)**.

Because it retains oil well, the antiseize and antiwear characteristics of bronze allow a bronze guide to last two to five times longer than a cast-iron guide. However, bronze guides are expensive, so their use in original equipment applications is limited. Bronze is also more difficult to machine. Because of these drawbacks, bronze guides are commonly used only as aftermarket replacements.

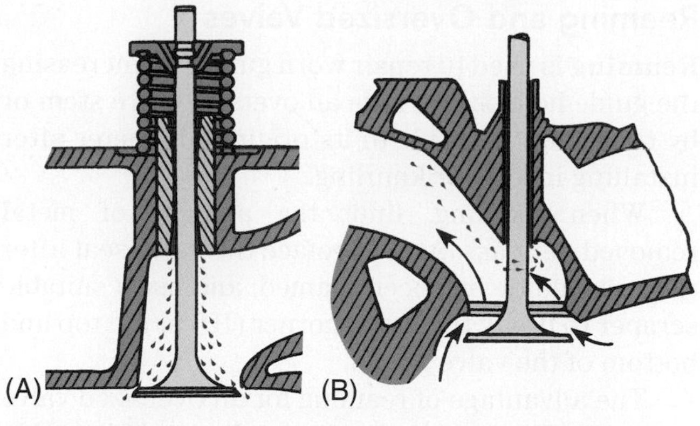

FIGURE 12–72

(A) Worn intake guides allow the intake vacuum to draw or pull oil down the guide, and (B) worn exhaust guides can do the same.

Knurling

Knurling is one of the fastest techniques for restoring the inside diameter (ID) dimensions of a worn valve guide. The process raises the surface of the guide ID by ploughing tiny furrows through the surface of the metal **(Figure 12–73)**. As the knurling tool cuts into the guide, metal is raised or pushed up on either side of the indentation. This effectively decreases the ID of the guide hole. A burnisher is used to press the ridges flat, and a reamer is then used to shave off the peaks of these ridges to produce the proper-sized hole and restore the correct guide-to-stem clearance.

One of the main advantages of knurling is that it does not change the centreline of the valve stem appreciably, so it reduces the amount of work necessary to reseat the valve. Knurling also allows a rebuilder to reuse the old valve if wear is within acceptable limits, helping to reduce rebuilding costs. In spite of its speed and simplicity, knurling is not a cure for restoring badly worn guides to their original condition.

FIGURE 12–73

Knurling restores the ID dimensions of a worn valve guide by raising the inside surface of the guide ID by ploughing tiny furrows through the surface of the metal.

Reaming and Oversized Valves

Reaming is used to repair worn guides by increasing the guide hole size to take an oversize valve stem or by restoring the guide to its original diameter after installing inserts or knurling.

When reaming, limit the amount of metal removed per pass. Always reface the valve seat after the valve guide has been reamed, and use a suitable scraper to break the sharp corner (ID) at the top and bottom of the valve guide.

The advantage of reaming for an oversized valve is that the finished product is totally new. The guide is straight, the valve is new, and the clearance is accurate. The use of oversized valve stems is generally considered to be superior to knurling. Yet, like knurling, it is relatively quick and easy. The only tool required is a reamer. The valve centreline is maintained, so the work required to finish the seat is reduced. However, because reaming requires the use of new valves, it can be more expensive on an engine with many worn guides. Its use is also limited to heads in which the guides are not worn beyond the limits of the maximum oversize valve that is available. Because of these limitations, many technicians prefer more cost-effective alternatives such as guide liners, inserts, or replacement guides.

Thin-Wall Guide Liners

The thin-wall guide liners **(Figure 12–74)** repair technique offers a number of important advantages and is also popular with many production engine rebuilders, as well as smaller shops. It provides the benefits of a bronze guide surface. It can be used with either integral or replaceable guides. It is faster, easier, and cheaper than installing new guides in heads with replaceable or integral guides, and it maintains guide centring with respect to the seats.

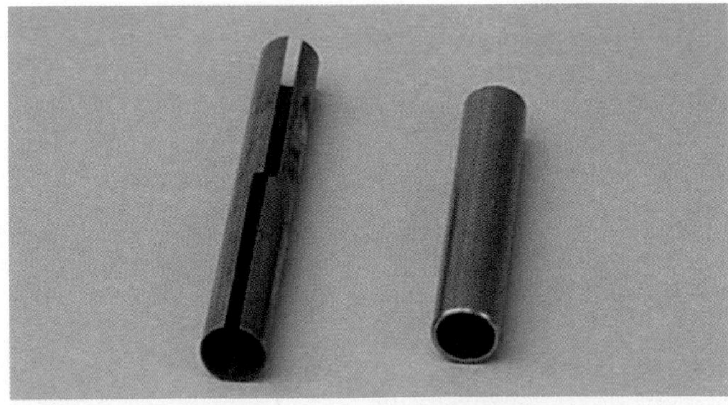

FIGURE 12–74

A thin-wall valve guide liner.

Thin-wall guide liners are manufactured from a phosphor-bronze or silicon-aluminum-bronze material. These liners can be cut to almost any length. They are designed for a 0.0508 to 0.0635 mm (0.002 to 0.0025 in.) press fit. A tight fit is essential for proper heat transfer to the head and to prevent the liner from working loose.

The liners are installed by first boring out the original guides to 0.7620 mm (0.030 in.) oversize with a special piloted boring tool pressed into the guide, using a driver and air hammer. On guides not precut to length, the excess must be milled off before finishing. The liner is then wedged in place and sized in a single operation by passing a ball broach down through it. This eliminates the need to ream it to size and ensures a tight fit between the liner and guide. If a ream finish is desired, it can be obtained by lubricating the reamer with a bronze-lube and then running it through the guide. For closer than normal stem-to-guide clearance, spiralling or knurling is suggested for added lubrication.

The only trick to using liners is to make sure the hole is round and the correct size. If the hole is distorted excessively or if it is too large, the liner will not fit properly and will cause problems.

Valve Guide Replacement

Replacing the entire valve guide is another repair option possible on cylinder heads with replaceable guides. However, pressing out the old guides and installing new ones can be difficult with some aluminum heads. Cracking the head or galling the guide hole is always a risk.

INTEGRAL GUIDES To replace integral guides, bore the old guide out and drive a thin-wall replacement guide into the hole. Many shops use a seat and guide machine for this process **(Figure 12–75)**, although it can be done with portable equipment. Drive the replacement guides in cold with approximately 0.0508 mm (0.002 in.) press fit. Use an assembly lube to prevent galling. It is necessary to keep the centreline of the guide concentric with the valve seat so that the rocker arm-to-valve stem contact area is not disturbed **(Figure 12–76)**.

Occasionally, a new guide will not be concentric with the valve seat. Install a new seat to correct the problem, and check the concentricity of the valve seat with a concentricity gauge.

INSERT GUIDES To remove an old valve guide, place a proper-sized driver so that its end fits snugly into the guide. The shoulder on the driver

FIGURE 12–75

A valve seat and guide machine.

Peterson Machine Tool Inc.

must also be slightly smaller than the outside diameter (OD) of the guide so that it will go through the cylinder head. Use a heavy ball peen hammer or air hammer to drive the pressed guide out of the cylinder head.

Pressing or driving out and installing new guides is not difficult, but there is always the danger of breaking the guide or tearing up the guide hole in the head. Cast-iron guides in particular have a tendency to gall aluminum heads. Once the hole

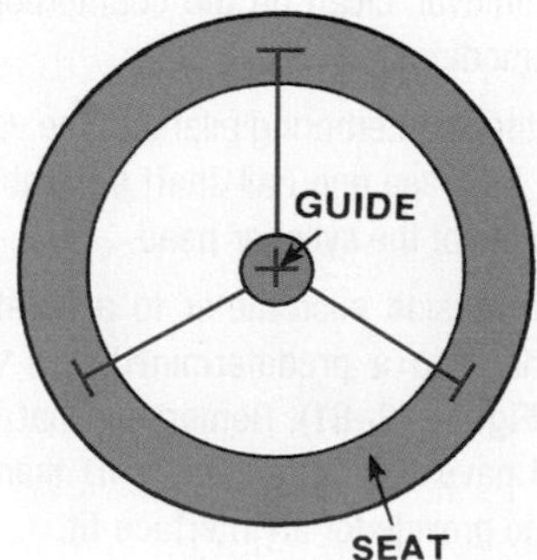

FIGURE 12–76

The centreline of the guide should be concentric with the seat.

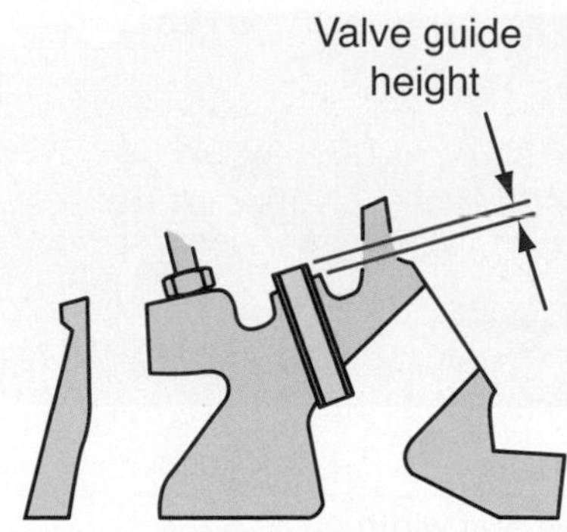

FIGURE 12–77

Make sure to check the installed height (protrusion) of the new valve guide and correct it if necessary.

is damaged, it must be bored out and an oversized guide installed—assuming one is available to fit the application. New guides should be chilled prior to installation because of the needed interference fit between the guide and head. Chilling them in a freezer or with dry ice works well. Lubricant also helps to prevent galling.

When installing new guides, be careful not to damage them. Use a press and the same driver that was used to remove the old guide. Align the new guide and press straight down, not at an angle. An air hammer and special driver can also be used to install new guides.

If the guides are cut off at an angle on the combustion chamber end or cut at an angle at one end or the other, do not press or drive against the angled end.

Find the correct amount of guide protrusion **(Figure 12–77)**. Guide height is important to avoid interference with the valve spring retainer. The guide must also fit the hole tightly or else it can work loose. The manufacturer's specifications give the correct valve guide installed height, but it is good practice to measure how far the old guides stick out of the cylinder head and to use this measurement as a reference. As each guide is installed, insert a valve. Check for any stem interference.

RECONDITIONING VALVE SEATS

The most critical sealing surface in the valve train assembly is between the face of the valve and its seat in the cylinder head when the valve is closed. Leakage between these surfaces reduces the engine's compression and power and can lead to valve burning. To ensure proper seating of the valve, the valve seat must be the correct width

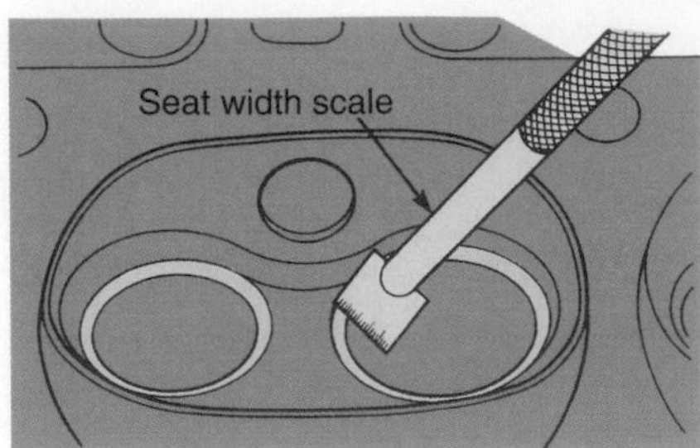
Seat width scale

FIGURE 12–78
Checking valve seat width.

(Figure 12–78), in the correct location on the valve face, and concentric with the guide (less than 0.0508 mm [0.002 in.] runout).

The ideal seat width for automotive engines can vary greatly between manufacturers. The most common intake seat width ranges between 0.762 and 1.524 mm (0.030 and 0.060 in.), and the exhaust seat width ranges between 1.524 and 2.0032 mm (0.060 and 0.080 in.). Maintaining the specified width is important to ensure proper sealing and heat transfer. However, when an existing seat is refinished to make it smooth and concentric, it also becomes wider.

Wide seats cause problems. Seating pressure drops as seat width increases. Less pressure is available to crush carbon particles that stick to the seats, and seats run cooler, allowing deposits to build up on them.

The seat should contact the valve face 0.79 mm (1/32 in.) from the margin of the valve. When the engine reaches operating temperature, the valve expands slightly more than the seat. This moves the contact area down the valve face. Seats that make too low a contact with the valve face might lose partial contact at normal operating temperatures.

Like valve guides, there are two types of valve seats—integral and insert. Integral seats are part of the casting. Insert seats are pressed into the head and are always used in aluminum cylinder heads.

Valve seats can be reconditioned or repaired by one of two methods, depending on the seat type—machining a counterbore to install an insert seat, or grinding, cutting, or machining an integral seat.

Before starting seat work, carefully check the seats for cracks **(Figure 12–79)**. Cracked integral seats sometimes can be repaired by installing inserts if the crack is not too deep. Cracked insert seats must be replaced. Check insert seats for looseness with a small pry bar. Replace them if any movement is noted.

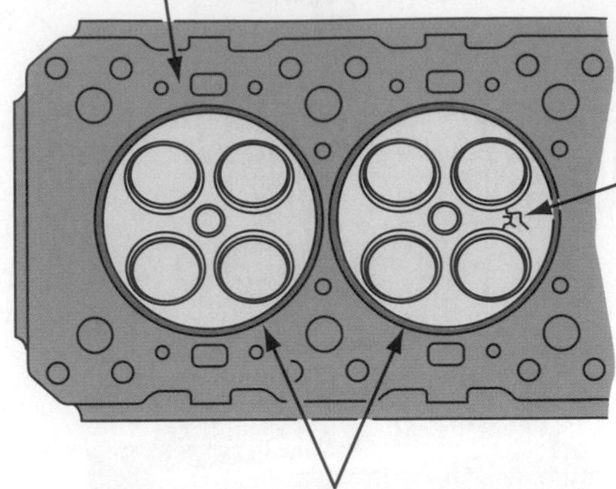
The cylinder head may be reused if corrosion is found outside the sealing area.
Replace the cylinder head if the area between the valve seats is cracked.
Replace the cylinder head if there is any corrosion within the sealing area.
IMPORTANT:
Use care when cleaning so the sealing surface is not scratched or the edges are not rounded; otherwise, the head gasket may leak.

FIGURE 12–79
A careful inspection of a cylinder head includes checking for cracks in the areas between the valve seats.

PROCEDURE

Insert Valve Seat Removal and Replacement

STEP 1 To remove the damaged insert, use a puller or a pry bar **(Figure 12–80)**.

STEP 2 After removal, clean up the counterbore or recut it to accommodate oversized inserts.

STEP 3 Insert the counterboring pilot into the valve guide. Then mount the base and ball shaft assembly to the gasket face angle of the cylinder head.

STEP 4 Use an outside micrometer to accurately expand the cutterhead to a predetermined size of the counterbore **(Figure 12–81)**. Remember that the counterbore should have a slightly smaller ID than the OD of the insert to provide for an interface fit.

STEP 5 Place the valve insert counterboring tool over the pilot and ball-shaft assembly. Preset the depth of the valve seat insert at the feed screw.

STEP 6 Cut the insert by turning the stop-collar until it reaches the present depth. Use a lubricant on the cutters for a smoother finish **(Figure 12–82)**.

STEP 7 To install the insert, heat the cylinder head in a parts cleaning oven to approximately 176°C to 204°C (350°F to 400°F). Chill the insert in a freezer or with dry ice before installation.

CAUTION!

Wear the proper gloves when handling dry ice.

STEP 8 Press the seat with the proper interference fit using a driver.

STEP 9 When the installation is complete, the edge around the outside of the insert is staked, as shown in **Figure 12–83**. By doing so, the insert will be secured more effectively in the counterbore.

FIGURE 12–82

The valve seat opening has been bored and is ready for new seat inserts.

Installing Valve Seat Inserts

The following steps outline a typical procedure for valve seat insert removal and replacement.

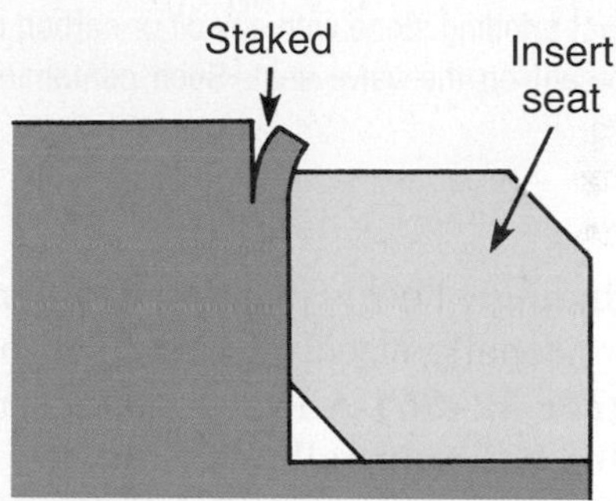

FIGURE 12–83

Staking the valve seat to the head can be done with a sharp chisel.

RECONDITIONING INTEGRAL SEATS The average valve seat width is 1.524 mm (0.060 in.), and the average seat begins 0.7620 mm (0.030 in.) from the valve margin. A properly reconditioned seat has three angles: top, 30° or 15°; seat, 45° or 30°; and throat, 60°. Typically, the 45° angle wedges tighter than the 30° seat, so it is used more often. Using three angles maintains the correct seat width and sealing position on the valve face **(Figure 12–84)**. Correct sealing

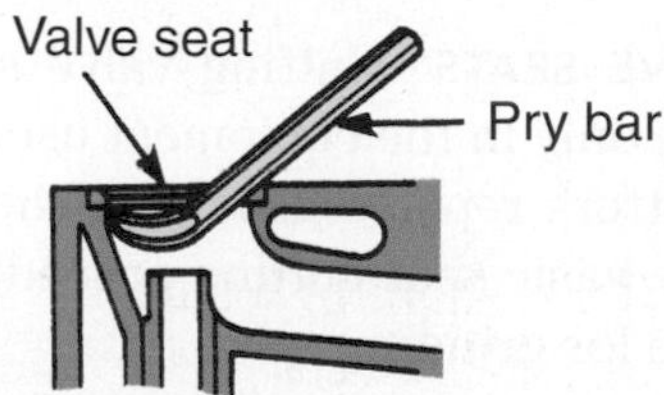

FIGURE 12–80

Using a pry bar to remove a damaged insert seal.

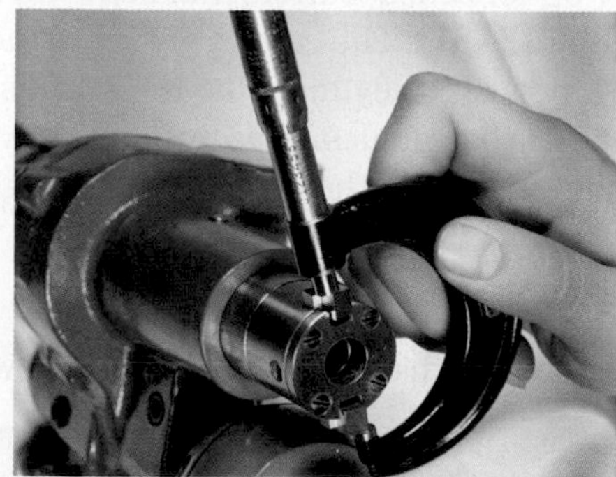

FIGURE 12–81

Using an outside micrometer to expand the cutter head to allow for an interference fit between the bore and the new valve seat.

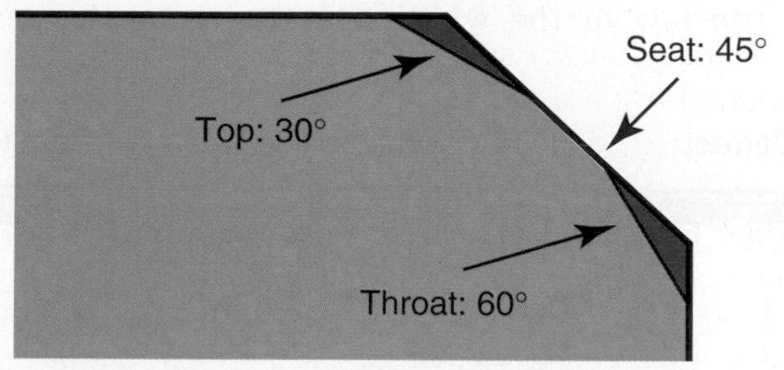

FIGURE 12–84

The three angles of a properly finished seat.

pressure and heat transfer from the valve through the seat are also affected. Always check the manufacturer's specifications for valve seat angles and valve face-to-seat contact amounts.

Integral valve seats can be reconditioned by grinding, cutting, or machining.

GRINDING VALVE SEATS When grinding a valve seat, it is very important to select and use the correct size pilot and grind stone. Hard seats use a soft stone and soft seats (cast iron) use a harder stone. The stone must be properly dressed and cutting oil used to aid in grinding.

SHOP TALK

Before grinding, many technicians clean the seats by placing a piece of fine emery cloth between the stone and the seat and giving the surface a hard rub. This will help prevent contamination of the seat grinding stone with any oil or carbon residue that might be present on the valve seat. Such contamination could cause glazing.

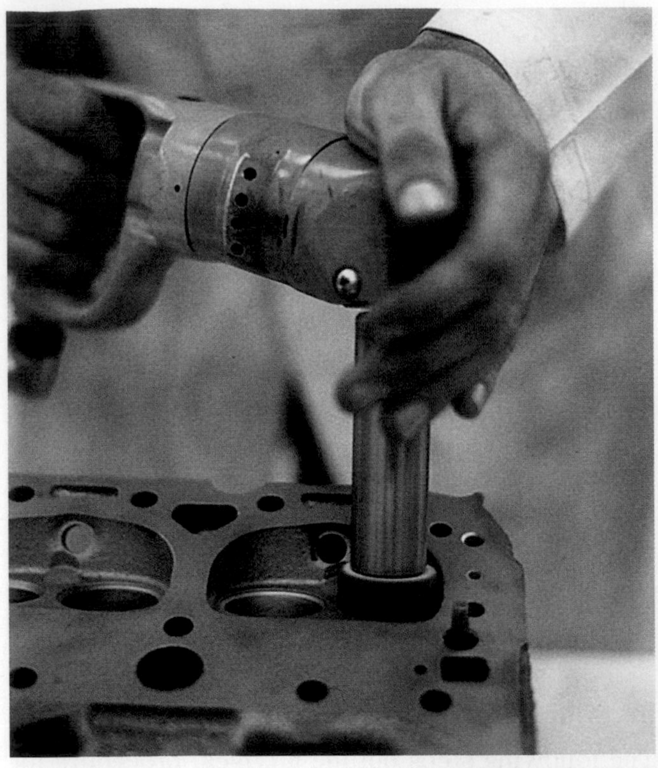

FIGURE 12–85

Grinding the seat.

The grinding wheel is positioned and centred by inserting a properly sized pilot shaft into the valve guide **(Figure 12–85)**. All valve guide service must be completed before installing the pilot.

The seat is ground by continually and quickly raising and lowering the grinder unit on and off the seat. Grinding should only continue until the seat is clean and free of defects.

After the seat is ground, valve fit is checked using machinist dye. The valve face is coated with dye, installed in its seat, and slightly rotated. The valve is then removed and the dye pattern on the valve face and valve seat inspected.

If the valve face and seat are not contacting each other evenly, or if the contact line is too high, the seat must be reground with the same stone used initially to correct the condition. If the line is too low or the width is not correct, the seat must be reground with stones of different angles **(Figure 12–86)**.

CUTTING VALVE SEATS Cutting valve seats differs from grinding only in the equipment used. Hardened valve seat cutters replace grinding wheels for seat finishing. The basic seat-cutting procedures are the same as those for grinding.

MACHINING VALVE SEATS As stated earlier in this chapter, a valve system rebuilding machine can be used to install valve guides and to machine valve seats. Some have optional seat cutters that make three-angle cuts **(Figure 12–87)**. The cutters are set to the proper diameter. Once set, they machine the seat as well as the top and throat angles. Two primary advantages of these cutters over other methods are high speed and precision.

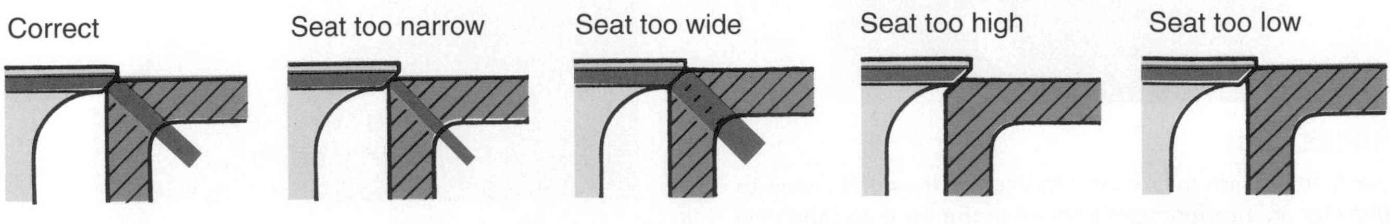

FIGURE 12–86

The fit of the valve face in the seat should be checked carefully.

FIGURE 12–87

A valve seat being cut at three angles at the same time.

Jaspar Engine and Transmission Exchange, Inc.

VALVE STEM SEALS

Valve stem seals are used on many engines to control the amount of oil allowed between the valve stem and guide. The stems and guides will scuff and wear excessively if they do not have enough lubrication. Too much oil produces heavy deposits that build up on the intake valve and hard deposits at the head end of the exhaust valve stem. Worn valve stem seals can increase the oil consumption by as much as 70 percent.

There are basically three types of seals. **Positive seals** fit tightly around the top of the stem and scrape oil off the valve as it moves up and down. Deflector, splash, or **umbrella-type seals** ride up and down on the valve stems to deflect oil away from the guides. **O-ring seals** installed over the valve stem are also used to prevent oil from moving into the guide when the valve is open.

The ultimate effectiveness of the valve stem seal depends entirely on the way it is secured to the guide. Many guides require machining to accept the stem seals. This must be done using the proper tools. A special valve guide machining tool is available for valve seal cutting. Such a tool is made up of a cutter and pilot, with sizes that vary according to the valve stem diameter and desired guide OD. The pilot is inserted into the guide, and the cutting tool machines the top of the guide.

Installing Positive Valve Seals

To install a positive valve seal **(Figure 12–88)**, place the plastic sleeve in the kit over the end of the valve stem to protect the seal as it slides over the keeper grooves. Lightly lubricate the sleeve. If it extends more than 1.59 mm (1/16 in.) below the lower keeper groove, you might want to remove the sleeve and cut off the excess length for easier removal. Carefully place the seal on the cap over the valve stem, and push the seal down until it touches the top of the valve guide. At this point, the installation cap can be removed and placed on the next valve. A special installation tool can be used to finish pushing the seal over the guide until the seal is flush with the top of the guide.

Installing Umbrella-Type Valve Seals

An umbrella-type seal is installed on the valve stem before the spring is installed. It is pushed down on

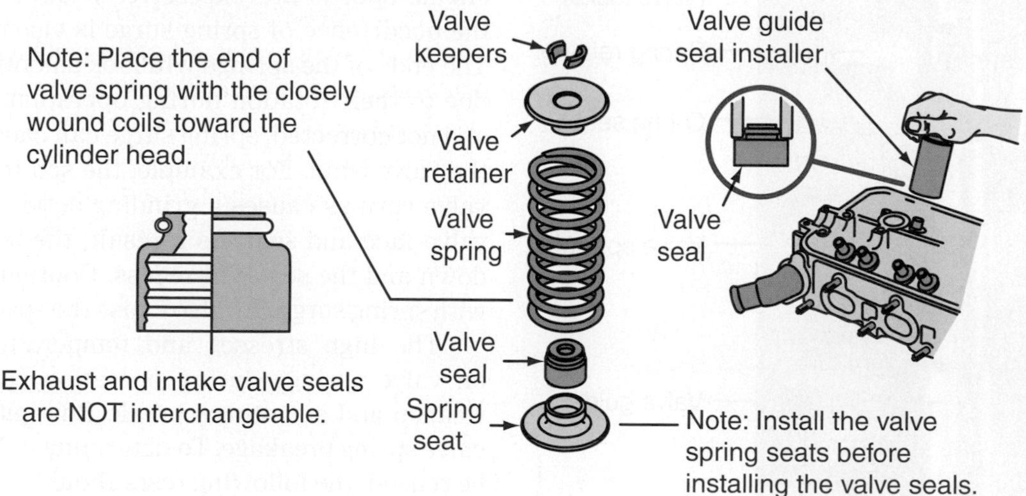
Note: Place the end of valve spring with the closely wound coils toward the cylinder head.
Exhaust and intake valve seals are NOT interchangeable.
Valve keepers
Valve retainer
Valve spring
Valve seal
Spring seat
Valve guide seal installer
Valve seal
Note: Install the valve spring seats before installing the valve seals.

FIGURE 12–88

Installation of a positive oil seal onto a valve guide.

©Cengage Learning 2015

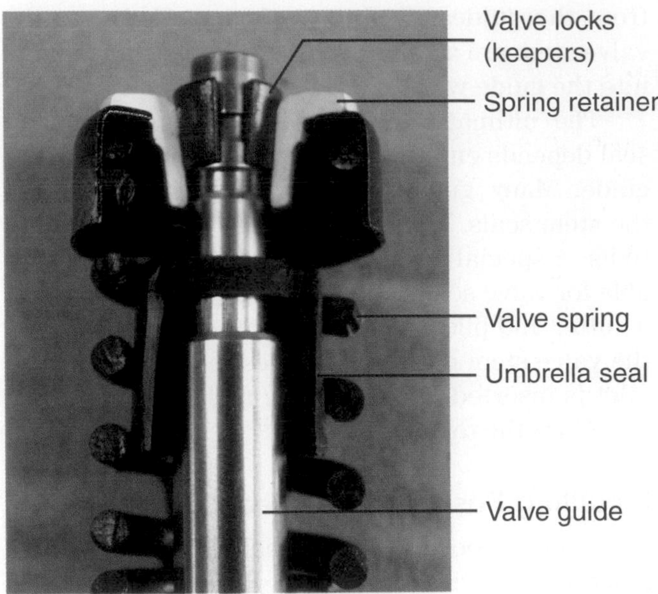

FIGURE 12–89

Valve assembly with an umbrella-type oil seal.

Martin Restoule

the valve stem until it touches the valve guide boss **(Figure 12–89)**. It will be positioned correctly when the valve first opens.

Installing O-Rings

When installing O-rings, use engine oil to lightly lubricate the O-ring. Then install it in the lower groove of the lock section of the valve stem **(Figure 12–90)**. Make sure the O-ring is not twisted.

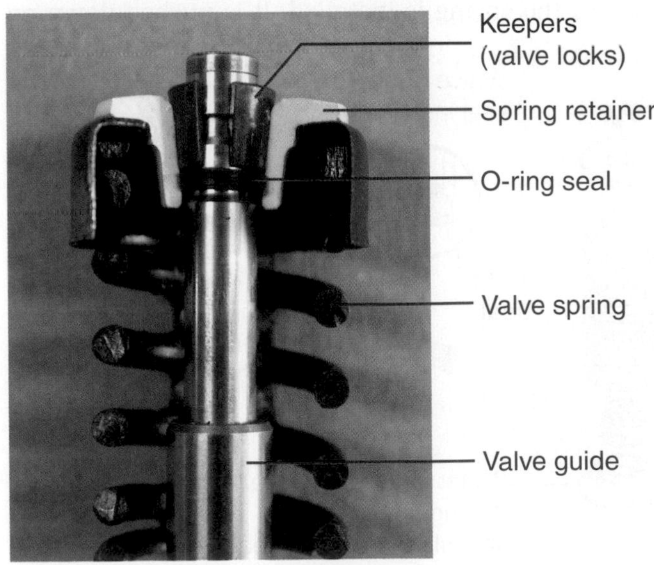

FIGURE 12–90

Valve assembly with an O-ring valve seal.

Martin Restoule

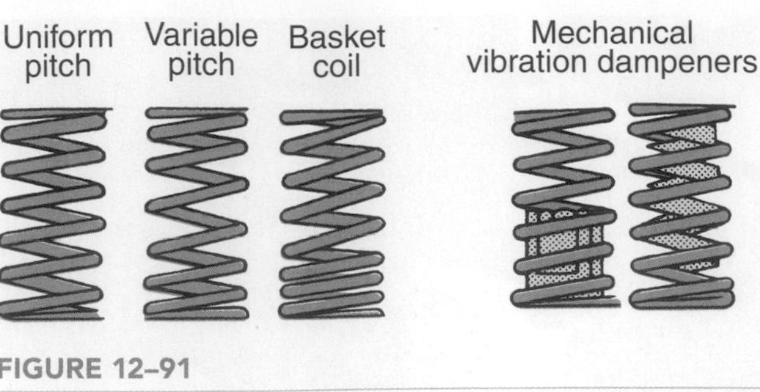

FIGURE 12–91

Common valve spring designs.

Valve Springs

The valve spring performs two functions: It closes the valve, and it maintains valve train contact during the opening and closing of the valve. Insufficient spring tension can lead to valve bounce and breakage. Too much pressure will cause premature camshaft lobe wear and can also lead to valve breakage.

The common designs of valve springs are illustrated in **Figure 12–91**. A problem that valve springs might have is **spring surge**. As the name implies, spring surge is the violent extending motion of the coils resulting in abnormal oscillation. Always install the closely wound coils of a basket coil-type spring toward the head end of the valve. Mechanical surge and vibration dampers should also be installed toward the head end of the valve. To dampen spring vibrations and increase total spring tension, some engine manufacturers use a reverse-wound secondary spring inside the main spring.

Spring surge can occur when the springs are weak, the installed spring height is improper, or engine speeds are excessive. Whatever the cause, the occurrence of spring surge is visually apparent. The ends of the springs will look smooth or polished due to their rotation during operation. If left alone and not corrected, spring surge can cause damage to the valve train. For example, the self-rotation of the valve springs causes a grinding action between the valve face and seat. As a result, the face will wear down and the seat will recess. Continued operation with spring surge can also cause the springs to break.

The high stresses and temperatures imposed on valve springs during operation cause them to weaken and sometimes break. Rust pits also cause valve spring breakage. To determine if the spring can be reused, the following tests should be performed.

FREESTANDING HEIGHT TEST Line up all the springs on a flat surface, and place a straightedge across the tops. All springs should be the same height, and free

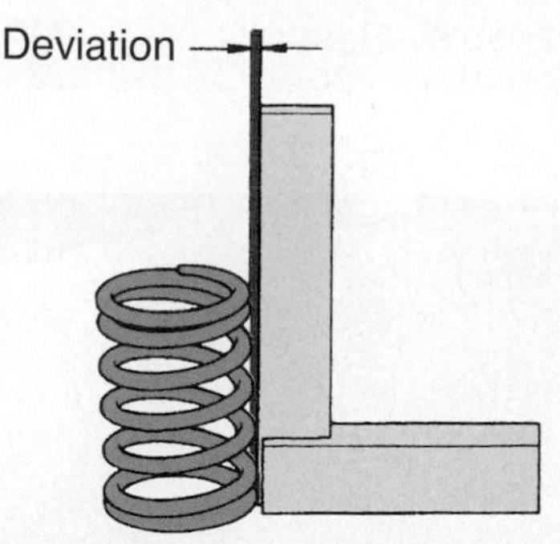

FIGURE 12–92

A spring squareness test.

length should be within 1.59 mm (1/16 in.) of OEM specifications. Throw away any spring that is not within specifications.

SPRING SQUARENESS TEST A spring that is not square will force the valve stem to one side and cause abnormal wear. To check squareness, set a spring upright against a square **(Figure 12–92)**. Turn the spring until a gap appears between the spring and the square. Measure the gap with a feeler gauge. If the gap is more than 1.524 mm (0.060 in.), the spring should be replaced.

OPEN/CLOSE SPRING TENSION TEST Use a spring tester to check for open and close spring tension. Close tension guarantees a tight seal. The open tension overcomes valve train inertia and closes the valve when it should close. Service information systems list spring tension specifications according to spring height **(Figure 12–93)**. The proper procedure for testing valve spring tension is given in Photo Sequence 9. Low spring tension may allow the valve to float during high-speed operation. Excessive spring tension may cause premature valve train wear. Any spring that does not meet specifications should be replaced.

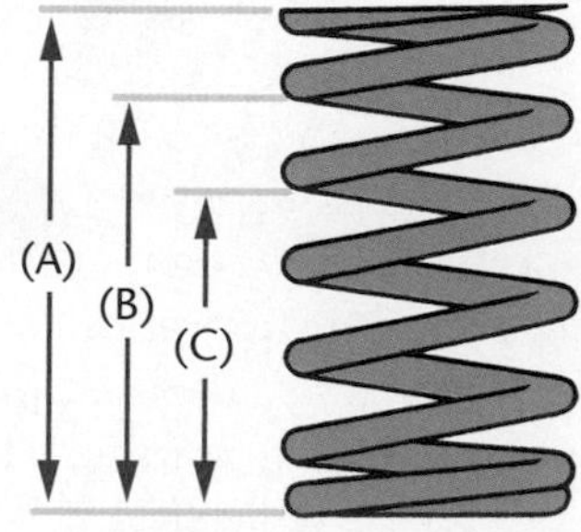

FIGURE 12–93

Valve spring height terminology: (A) free height, (B) valve closed spring height, and (C) valve open spring height.

ASSEMBLING THE CYLINDER HEAD

Before a cleaned or reconditioned cylinder head is reassembled and installed, two critical measurements must be carefully checked: the installed stem height and the installed spring height.

Installed stem height is determined by measuring the distance between the spring seat and stem tip. Because this measurement directly influences rocker arm geometry and installed spring height, accuracy and precision are important. This is especially true when the valve or valve seat has been ground. A number of tools can be used to obtain an accurate stem height reading, including a depth micrometer, vernier caliper, and telescoping gauge.

USING SERVICE INFORMATION

Stem height specifications are often unavailable in service information systems. As a guide for assembly, record the stem heights for all valves during disassembly.

Another specification can be used that corresponds directly to installed stem height, and that is installed spring height. Installed spring height is measured from the spring seat to the underside of the retainer when it is assembled with keepers and held in place. This measurement, which can be made by using a set of dividers or scales, telescoping gauge, or spring height gauge, should be taken only after valve and seat work is completed, valves are installed in their guides, and retainers and keepers are assembled.

If the spec for installed spring height for an exhaust valve is 40.64 mm (1.600 in.), and the measurement is 42.60 mm (1.677 in.), the increase in height is 1.96 mm (0.077 in.). This means the installed stem height has also been increased by 1.96 mm (0.077 in.).

Adjustments to valve spring height can be made with valve spring inserts, otherwise known as spring shims. Even though valve shims come in only three standard thicknesses—1.52, 0.7620, and 0.3810 mm (0.060, 0.030, and 0.015 in.)—using combinations of different shims gives the correct amount of compensation (within 0.1270 or 0.2540 mm [0.005 or 0.010 in.]).

SHOP TALK

Valve keepers should be replaced in pairs. If a new keeper is mated with a used one, the spring retainer may cock and break off the valve tip or allow the assembly to come apart.

PHOTO SEQUENCE
9

MEASURING AND FITTING VALVE SPRINGS

P9–1 Prior to installing the valve and fitting valve springs, all other head work should be completed.

P9–2 Install the valve into its proper valve guide.

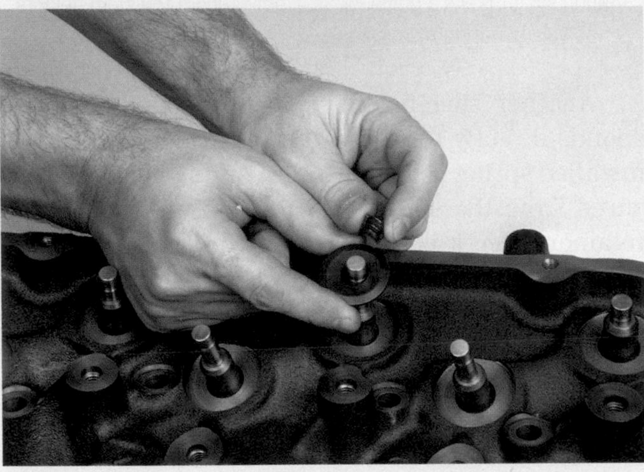

P9–3 Install the valve retainer and keepers. Without the spring, these must be held in place.

P9–4 While pulling up on the retainer, measure the distance between the bottom of the retainer and the spring pad on the cylinder head with a divider.

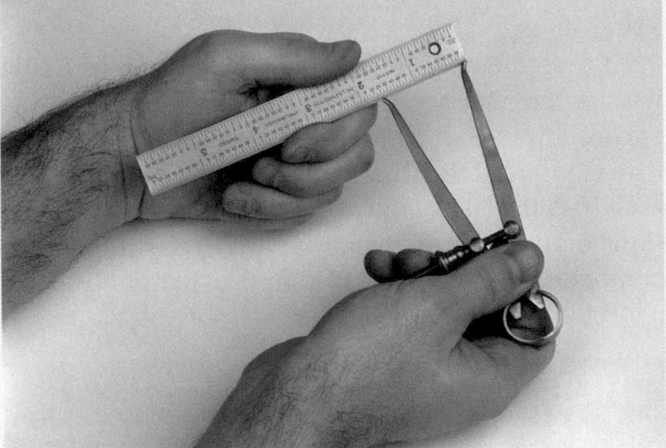

P9–5 Use a scale to determine the measurement expressed by the divider.

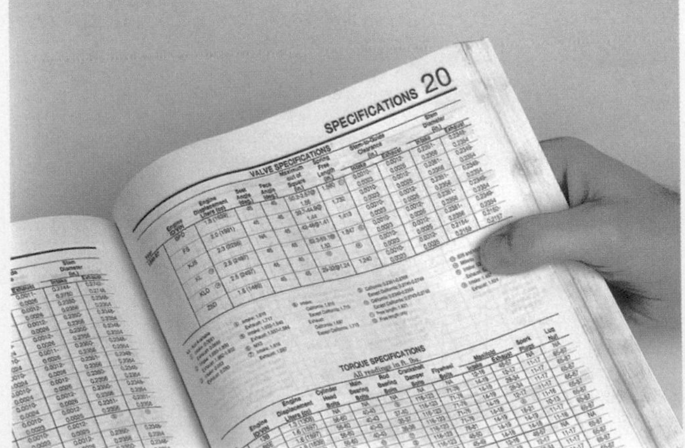

P9–6 Compare this measurement with the specifications given in the service information system for installed spring height.

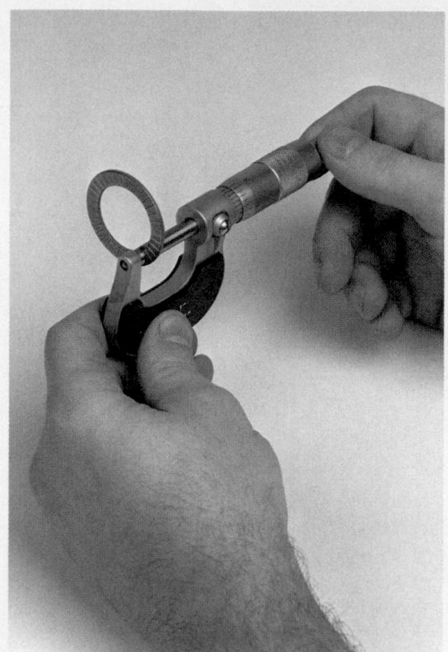

P9–7 If measured installed height is greater than the specifications, a valve shim must be placed under the spring to correct the difference.

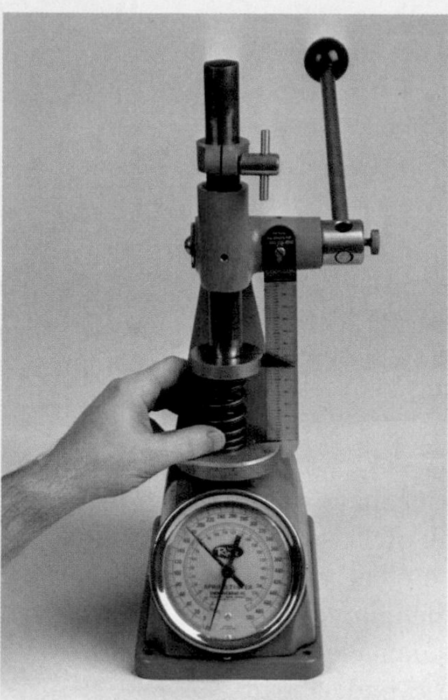

P9–8 Spring tension must be checked at the installed spring height; therefore, if a shim is to be used, insert it under the spring on the valve spring tension gauge.

P9–9 Compress the spring into the installed height by pressing down on the tester's lever.

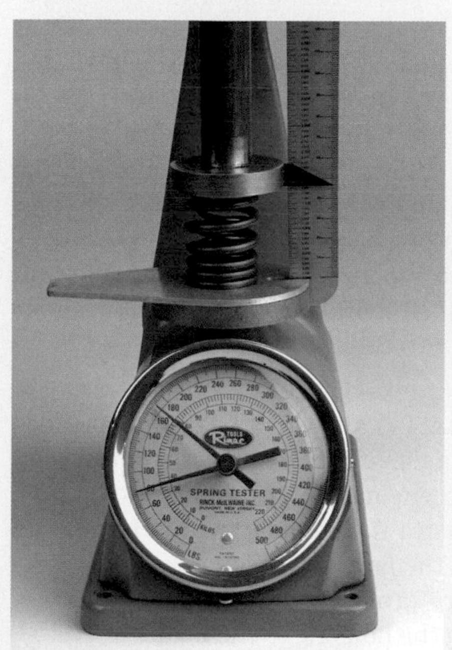

P9–10 The tension gauge will reflect the pressure of the spring when compressed to the installed or valve closed height. Compare this reading to the specifications.

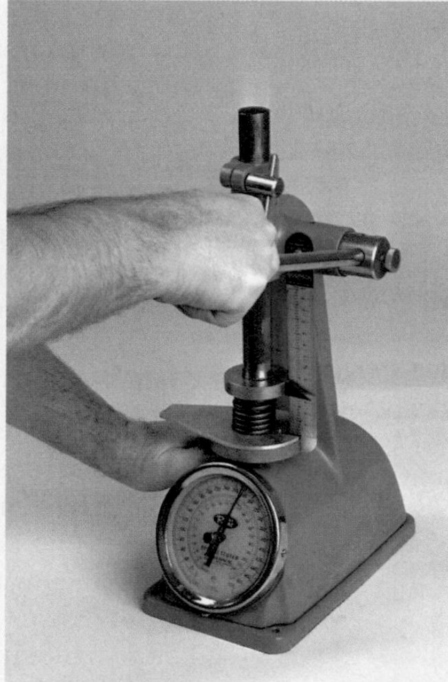

P9–11 Now compress the spring to the open height specification. Use the rule on the gauge or a scale to measure the compressed height.

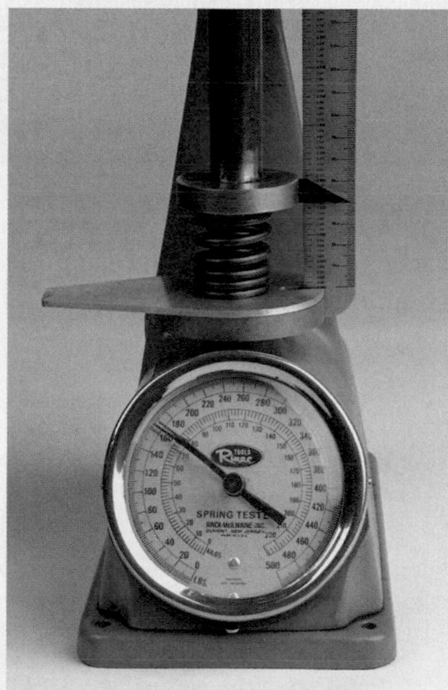

P9–12 Compare this reading to specifications. Any spring tension reading outside the tension range given in the specifications indicates the spring should be replaced. After the tension and height have been checked, the spring can be installed on the valve stem.

By comparing spring height to specifications, the desired amount of spring tension correction can be easily determined. For example, if spring height is 4.59 mm (0.180 in.) and the specifications call for 3.78 mm (0.149 in.), a 0.7620 mm (0.030 in.) shim (3.78 mm + 0.7620 mm = 4.54 mm [0.149 in. + 0.030 in. = 0.179 in.]) would be needed. If more than one shim is required, place the thickest one next to the spring, not on the head. If one side of the shim is serrated or dimpled, place that side over the valve stem and onto the spring seat.

With the valve inserted into its guide, position the valve spring inserts, valve spring, and retainer over the valve stem. Using a valve spring compressor, compress the spring just enough to install the valve keepers into their grooves **(Figure 12–94)**. Excess force applied by the spring compressor may cause the retainer to damage the oil seal. Release the spring compressor and tap the valve stem with a rubber mallet to seat the keepers. When doing this, the valve will open slightly. To prevent damage to the valves, never tap the stems with the cylinder head lying flat on the bench. Turn the head on its side or raise it off the bench.

CAUTION!

If the keepers are not fully seated, the spring assembly could fly apart and cause personal injury (or serious damage to the engine if it occurs while the engine is running). For these reasons, it is good practice to assemble the valves with the retainers facing a wall and to wear eye protection.

Valve spring compressors are available in different designs for different applications. The most commonly used compressor is hand operated. There are two different hand-operated designs: a universal tool with interchangeable jaws and a compressor specifically designed for OHC engines **(Figure 12–95)**. This compressor allows for positive contact on the retainer without removing the camshaft. Another type of spring compressor is air operated. Air-operated spring compressors are typically found in high-volume engine rebuilding shops.

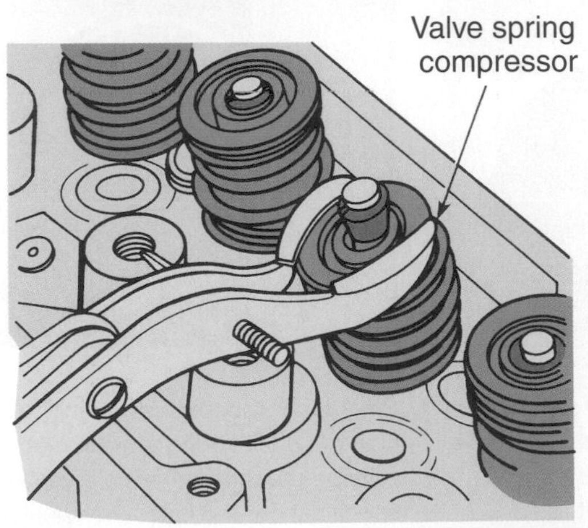

FIGURE 12–94

Compress the spring just enough to install the keepers.

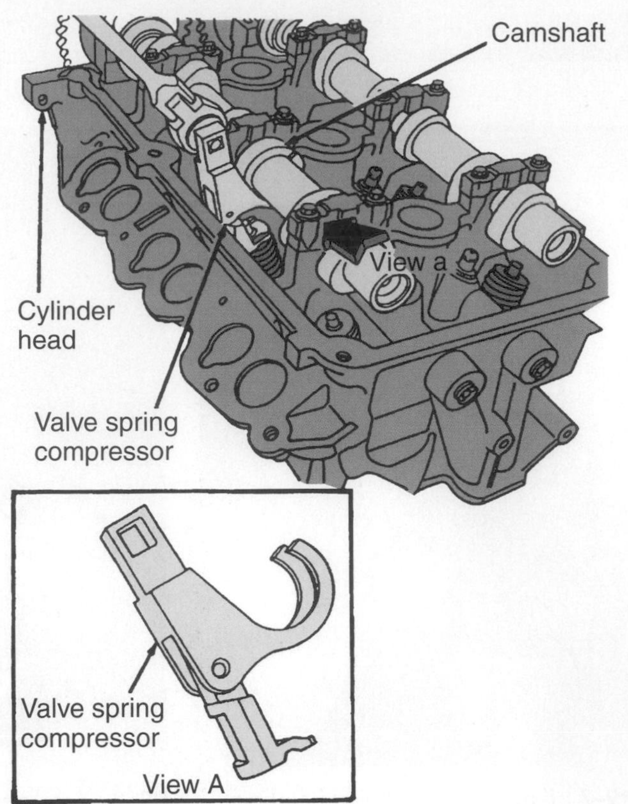

FIGURE 12–95

A special valve spring compressor for OHC engines.

Ford Motor Company

OHC Engines

After the valves are installed and the cylinder head is assembled, the camshaft can be installed in the cylinder head. Some engines have a separate camshaft housing that bolts to the cylinder head. This should be installed with the proper seals and gaskets. Make sure the seals are properly seated in their grooves before tightening the housing to the head.

Service to the camshaft bearings can now be done. Full-round insert bearings are pressed into the bores in the cylinder head. Special tools are designed to make this job easier. Never use a standard camshaft bearing driver and hammer to install these bearings. The hammering can easily break or damage the bearing supports. After each bearing is fully seated in its bore, double-check the alignment of the bearing's oil hole with the oil hole in the head.

Some overhead camshafts do not have bearing inserts. The bores in the aluminum casting serve as a bearing surface. These surfaces can be cleaned up and/or align bored if needed.

Most late-model OHC engines use split bearings and bearing caps or have a separate housing for the camshaft. Working with split camshaft bearings is like working with crankshaft main bearings. This includes checking bearing clearances with plastigage **(Figure 12–96)**.

Now gather the rocker arms, lash adjusters, pushrods, lifters, and other parts that transfer the motion of the camshaft to the valve stem. Coat all of these parts with clean engine oil. The order in which these parts should be installed is given in the service information. Some parts are installed before the camshaft is installed; others are installed after. Install the required components into their bores in the cylinder head. Make sure they fit securely and do not bind in their bores.

Before installing the camshaft, wipe off each cam bearing with a lint-free cloth, and then thoroughly coat the camshaft lobes and bearing journals with assembly lube. Special prelubricants can be used only if specifically recommended by the manufacturer.

Install the camshaft. If the assembly has bearing caps, place them in their correct position and tighten them according to specifications. Once the shaft is in place in the cylinder head, the shaft should be easily turned. On some engines, turning the camshaft causes the valves to open. If this is the case, stand the cylinder head on its end to prevent the valves from hitting the workbench while you turn the camshaft. If the cam does not turn, binding might be the cause. Binding is the result of a damaged bearing, a nick on the cam's journal, or a slight misalignment of the block journals. The cause of the problem should be identified and corrected.

Rocker arm assemblies can now be installed by turning the cam until the cam lobe for the valve faces away from a valve stem. Typically, the rocker arms can be slipped into position by depressing the valve spring slightly. Follow the same procedure for all of the valves. Install the camshaft sprockets, make sure the keyways are aligned, and then torque the retaining bolt to specifications. Now adjust the valve lash (clearance) before installing the head on the engine.

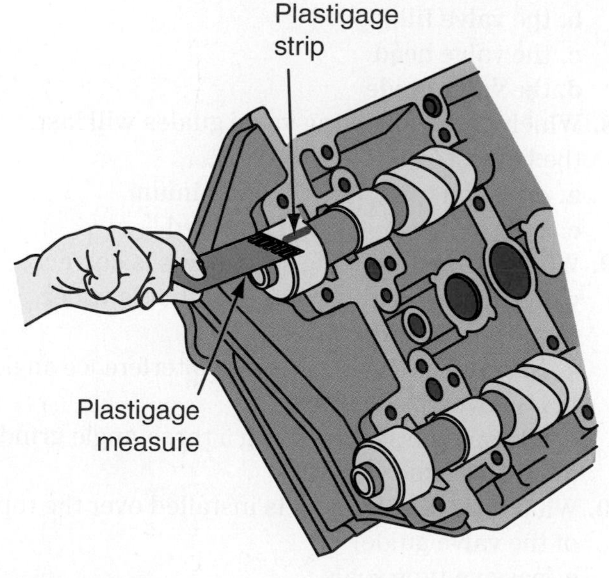

FIGURE 12–96

OHC camshaft bearing clearances can be checked with plastigage.

KEY TERMS

Broaching (p. 351)
CC-ing (p. 351)
Cross flow ports (p. 325)
Damper spring (p. 330)
Hemi-head (p. 325)
Hemispherical combustion chamber (p. 325)
Knurling (p. 355)
Margin (p. 328)
Milling (p. 351)
O-ring seals (p. 361)
Pentroof (p. 326)
Phaser (p. 335)
Poppet (p. 326)
Porosity (p. 350)
Positive seals (p. 361)
Pumping losses (p. 336)
Pushrods (p. 330)
Pushrod guide plates (p. 330)
Quench area (p. 325)
Quenching (p. 325)
Reaming (p. 356)
Shrouding (p. 326)
Siamese ports (p. 325)
Spring surge (p. 362)
Stainless steel (p. 327)
Stellite (p. 327)
Turbulence (p. 325)
Umbrella-type seals (p. 361)
Valve face (p. 328)
Valve guide (p. 328)
Valve keepers (p. 330)
Valve seat (p. 329)
Valve stem (p. 328)
Wedge-type combustion chamber (p. 325)

SUMMARY

- Pushrods are the connecting link between the rocker arm and the valve lifter.
- The rocker arm converts the upward movement of the valve lifter into a downward motion to open the valve. It also permits the valves to be angled.

- Aluminum cylinder heads are used on late-model engines because of their light weight. The thermal expansion characteristics of aluminum can lead to problems such as leaking and cracking.
- An efficient combustion chamber must be compact in order to minimize heat loss. Popular combustion chamber designs include the wedge, hemispherical, swirl, and fast-burn varieties.
- Every cylinder of a four-stroke engine contains at least one intake valve and one exhaust valve.
- Multivalve engines feature either three, four, or five valves per cylinder, which means better combustion and reduced misfire and detonation. These benefits are offset to some extent by a more complicated camshaft arrangement.
- The means of resurfacing the deck of a cylinder head include grinding, milling, belt surfacing, and broaching.
- The amount of stock removed from the cylinder head gasket surface must be limited. Excessive surfacing can create problems with the engine's compression ratio, not to mention piston/valve interference and misalignment.
- The two surfaces of a valve reconditioned by grinding are the face and the tip. Valves can be refaced on grinding or cutting machines.
- One of the fastest methods for restoring the inside diameter dimensions of a worn valve guide is knurling. Reaming repairs worn guides by increasing the guide hole size to take an oversize valve stem or by restoring the guide to its original dimension after knurling or installing inserts.
- Pressing out an old valve guide to install a new one can be difficult on some aluminum heads where the interference fit is considerable.
- To ensure proper seating of a valve, the seat must be the correct width, in the correct location on the valve face, and concentric with the guide.
- When grinding a valve seat, choosing the correct size pilot and stone is important. For soft seats such as cast iron, use a hard stone. For hard seats, a soft stone is needed.
- Valve stem seals are used to control the amount of oil between the valve stem and guide. Too much oil produces deposits, whereas insufficient lubrication leads to excessive wear.
- The valve spring closes the valve and also maintains the valve train contact during the opening and closing of the valve. To determine if a spring needs to be replaced, three tests are valuable: freestanding height, spring squareness, and open/close spring tension.
- Two critical measurements that must be made before a cylinder head is reassembled and installed are installed stem height and installed spring height. The first of these is determined by measuring the distance between the spring seat and stem tip. The latter is measured by the spring seat to the underside of the retainer when it is assembled with keepers and held in place.

REVIEW QUESTIONS

1. What happens when valve spring tension is too low?
2. Define valve margin.
3. What usually causes warpage in an aluminum cylinder head?
4. What are the two ways pushrods can be checked for straightness?
5. Why do some technicians not consider knurling a long-term repair?
6. What is the distance between the valve face and the valve head called?
   a. the stem **b.** the margin
   c. the fillet **d.** the seat
7. Which one of the following cylinder head components are worn when oil leaks into the exhaust gases or intake charge?
   a. the valve margin
   b. the valve fillet
   c. the valve head
   d. the valve guide
8. Which of the following valve guides will last the longest?
   a. cast iron **b.** aluminium
   c. bronze **d.** knurled
9. Which of the following statements is correct when dealing with a valve seat that is wider than specifications?
   a. Narrow the seat by using an interference angle.
   b. Use oversized valves.
   c. Narrow the seat by using a three-angle grind.
   d. Use ceramic valves.
10. What type of valve seal is installed over the top of the valve guide?
    a. passive-type seals
    b. umbrella-type seals
    c. O-ring-type seals
    d. positive-type seals

11. What should be done to an insert style of valve seat after it is installed in the cylinder head?
    a. Grind the insert to suit the valve.
    b. Stake around the edge to secure the seat.
    c. Heat the seat to expand the seat into its bore.
    d. Perform a three-angle cut.
12. Why is rocker arm ratio used on most valve trains?
    a. to open the valves a greater distance
    b. to reduce valve stem wear
    c. to increase oil sealing on the valve stem
    d. to decrease the force against the camshaft
13. When grinding valves and seats, what is the 1° angle difference between them called?
    a. the interference angle
    b. a narrowing angle
    c. a radius angle
    d. the throating angle
14. Why are shims placed under valve springs?
    a. to restore the correct valve spring installed height
    b. to reduce the valve spring installed height
    c. to increase the valve spring installed height
    d. to decrease valve spring tension
15. In general, what is the maximum deformation allowed in a cylinder head deck surface?
    a. 0.1 mm (0.004 in.)
    b. 0.3 mm (0.012 in.)
    c. 0.2 mm (0.008 in.)
    d. 0.4 mm (0.016 in.)
16. Which of the following is the first procedure to be performed when reconditioning a cylinder head?
    a. Grind the valve seat.
    b. Install the valve seals.
    c. Install new valve guide inserts.
    d. Install valves.
17. Which type of surfacing machine uses underside rotary cutters?
    a. milling
    b. belt
    c. broaching
    d. grinding
18. What is the ideal intake valve seat width?
    a. 1.0 mm (0.040 in.)
    b. 1.6 mm (0.063 in.)
    c. 2.4 mm (0.094 in.)
    d. 3.2 mm (0.126 in.)
19. What is the cooling of gases by pressing them into a thin area called?
    a. turbulence
    b. reaming
    c. shrouding
    d. quenching
20. Which of the following would allow a valve to be reused during cylinder head reconditioning?
    a. an exhaust valve margin less than 1.14 mm (0.045 in.)
    b. an exhaust valve with tip damage greater than 0.254 mm (0.010 in.)
    c. an exhaust valve with stem scoring
    d. an exhaust valve margin greater than 1.14 mm (0.045 in.)

CHAPTER 13
Engine Sealing and Reassembly

LEARNING OUTCOMES

- Explain the purpose of the various gaskets used to seal an engine.
- Identify the major gasket types and their uses.
- Explain general gasket installation procedures.
- Describe the methods used to seal the timing cover and rear main bearing.
- Reassemble an engine including core plugs, bearings, crankshaft, camshaft, pistons, connecting rods, timing components, cylinder head, valve train components, oil pump, oil pan, and timing covers.
- Explain the ways to prelubricate a rebuilt engine.
- Reinstall an engine and observe the correct starting and break-in procedures.

Proper sealing of an engine keeps the low-pressure liquids in the cooling system away from the cylinders and lubricating oil. It also keeps the high pressure of combustion in the cylinders. It prevents both internal and external oil leaks and suppresses and muffles noise.

TORQUE PRINCIPLES

All metals are elastic. **Elasticity** means a bolt can be stretched and compressed up to a certain point. This elastic, springlike property is what provides the clamping force when a bolt is threaded into a tapped bore or when a nut is tightened. As the bolt is tightened, it is stretched less than 0.1 mm (a few thousandths of an inch). Clamping force or holding power is created due to the bolt's natural tendency to return to its original length **(Figure 13–1)**.

Like a spring, the more a bolt is stretched, the tighter it becomes. However, a bolt can be stretched too far. This is obvious when the grip on the wrench feels "mushy." At this point, the bolt can no longer safely carry the load it was designed to support. The term **yield** represents the maximum amount of stretch a bolt can experience and still provide clamping pressure.

If a bolt is stretched into yield, it takes a permanent set and never returns to normal **(Figure 13–2)**. The bolt will continue to stretch more each time it is used, just like a piece of taffy that is stretched until it breaks. Proper use of torque will avoid this yield condition. Torque values are calculated with a 25 percent safety factor below the yield point.

FIGURE 13–1
Clamping force results from bolt stretch.
Martin Restoule

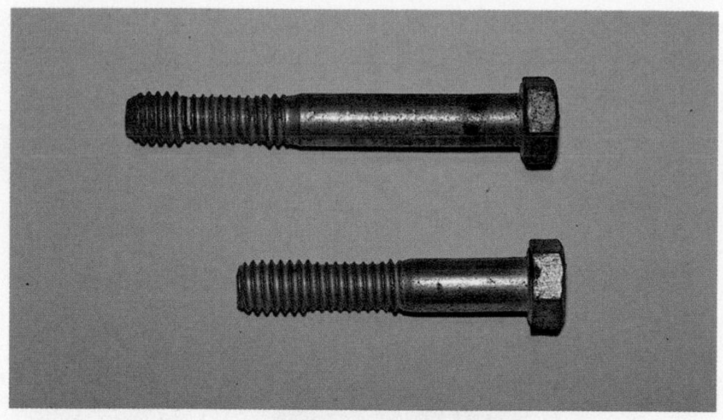

FIGURE 13–2

These bolts have been torqued past their yield points. Note the soda bottle effect.

Table 13–1 gives standard bolt and nut torque specifications. If the manufacturer's torque specifications are available, follow them precisely.

The grade and surface condition, whether plated or nonplated, dry or lubricated, oil or anti-seize, cut threads or rolled, straight shank or reduced, affects the torque/tension relationship and causes the performance of the connection to change. Because torque is actually a combination of both tension and torsion, it is also a function of friction. The bolt head or nut, whichever is being rotated, produces friction as it is turned, as do the threads when they gall together under the pulling forces of being in tension. Tests have proven that 90 percent of work energy is consumed by friction. Friction must first be overcome before any true work is done. To compensate for surface variations, the following formula may be used to approximate the required torque:

$$T = FDC$$

in which

T = The torque applied in Newton-meters
F = Friction factor (torque cofficient)
D = Bolt diameter in meters
C = Bolt clamp load in

(***Note:*** A clean, dry bolt may have a friction factor of 0.24, whereas a lightly oiled bolt may have a friction factor that is reduced to 0.15.) Examples of both bolts tightened to a torque of 50 N·m:

Dry: 50 N·m = 0.24 × 0.01 m × C   C = 20 833.3 N
Oiled: 50 N·m = 0.15 × 0.01 m × C   C = 33 333.3 N

This is a clamp load increase of over 60 percent when oil is used.

Nonplated bolts have a rougher surface than plated finishes. Therefore, it takes more torque to produce the same clamping force as on a plated bolt, even with one-third less friction. Add a lubricant and the torque might be as much as two-thirds lower.

Most printed torque values are for dry, plated bolts. Lubricants are beneficial when working with engines. They provide smoother surfaces and more consistent and evenly loaded connections. They also help reduce thread galling.

Reusing a dry nut will produce a connection with decreasing clamp force each time it is used. Nut threads are designed to collapse slightly to carry the

TABLE 13–1 **STANDARD BOLT AND NUT TORQUE SPECIFICATIONS**

SIZE NUT OR BOLT	TORQUE	SIZE NUT OR BOLT	TORQUE	SIZE NUT OR BOLT	TORQUE
1/4–20	9.45–12.15 N·m (7–9 ft·lb)	7/16–20	77–82 N·m (57–61 ft·lb)	3/4–10	324–338 N·m (240–250 ft·lb)
1/4–28	10.8–13.5 N·m (8–10 ft·lb)	1/2–13	96–101 N·m (71–75 ft·lb)	3/4–16	392–405 N·m (290–300 ft·lb)
5/16–18	17.55–22.95 N·m (13–17 ft·lb)	1/2–20	112–126 N·m (83–93 ft·lb)	7/8–9	554–567 N·m (410–420 ft·lb)
5/16–24	20.25–25.65 N·m (15–19 ft·lb)	9/16–12	122–135 N·m (90–100 ft·lb)	7/8–14	644–655 N·m (475–485 ft·lb)
3/8–16	40.5–47.25 N·m (30–35 ft·lb)	9/16–18	144–158 N·m (107–117 ft·lb)	1–8	783–797 N·m (580–590 ft·lb)
3/8–24	47.25–52.65 N·m (35–39 ft·lb)	5/8–11	185–198 N·m (137–147 ft·lb)	1–14	925–938 N·m (685–695 ft·lb)
7/16–14	62–67.5 N·m (46–50 ft·lb)	5/8–18	227–240 N·m (168–178 ft·lb)		

TABLE 13–2 FRICTION FACTORS (F) AND TORQUE REDUCTIONS FOR LUBRICATED SURFACES ON ALLOY STEEL BOLTS

LUBRICANT	FRICTION FACTOR	PERCENTAGE OF TORQUE REDUCTION REQUIRED
Colloidal copper	0.11	Reduce standard torque 45%.
Never-seize	0.11	Reduce standard torque 45%.
Grease	0.12	Reduce standard torque 40%.
Moly-cote (molybdenum disulphite)	0.12	Reduce standard torque 40%.
Heavy oils	0.12	Reduce standard torque 40%.
Graphite	0.14	Reduce standard torque 30%.
White lead	0.15	Reduce standard torque 25%.

bolt load. If dry nuts are reused, increased thread galling will result each time the nuts are reused at the same torque.

Lubrication of fasteners is recommended for consistency **(Table 13–2)**. However, be sure to lubricate all the bolts with the same lubricant. Some lubricants are more slippery than others, which affects torque values. Lubricate the bolt, never the hole. Otherwise, the bolt may merely be tightening against the oil in the hole.

If a bolt with a reduced shank diameter (e.g., a connecting rod bolt) is specified by the OEM, never replace it with a standard, straight shank bolt. A reduced shank diameter bolt looks “dog-boned.” Its function is to reduce the stress on the threads by transferring it to the shank. A standard bolt under similar conditions would break very quickly at the threads. These bolts can be checked by measuring the diameter of the threads and the shank, and comparing them to specifications **(Figure 13–3)**.

FIGURE 13–3

Bolts can be checked for stretch by measuring the shank and comparing it to specifications.

©Cengage Learning 2015

Keep the following points in mind:

1. Visually inspect the bolts.
   - Threads must be clean and undamaged. Discard all bolts that are not acceptable.
   - Run a nut over the bolt's threads by hand. Discard it if any binding occurs.
   - Clean bolt and cylinder block threads with a thread chaser or tap **(Figure 13–4)**.
2. Apply a light coat of 10W engine oil to threads and bottom face of bolt head. A sealer is required for a bolt that enters a water jacket. This will stop coolant seepage around the bolt threads. Seeping coolant could get in the oil or cause corrosion that might damage parts, resulting in engine failure.
3. Tighten bolts in the recommended sequence. This is important to prevent warpage of the cylinder head or other parts.
   - Use an accurate torque wrench.
   - Tighten bolts to the recommended torque in steps and proper sequence.
4. If bolt heads are not tight against the surface, the bolts should be removed and washers installed.
5. Make sure the bolt is the proper length (not too long).

FIGURE 13–4

Cleaning bolt holes with a tap.

Corteco

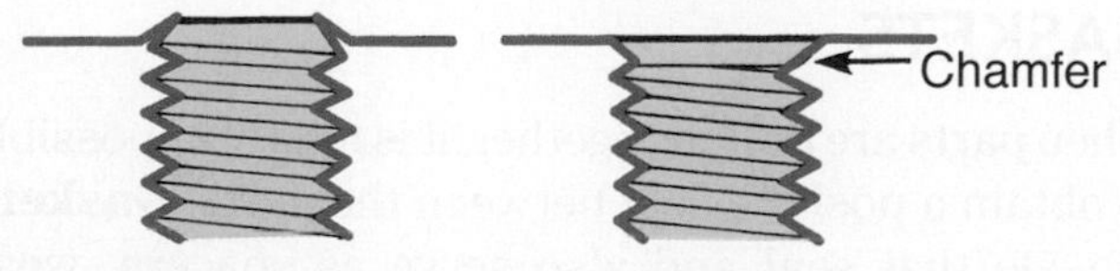

FIGURE 13–5

Bolt hole threads can pull up, leaving a raised edge. Also, if the block has been resurfaced, the threads may run up to the surface. In either case, the hole must be chamfered.

Bolt hole threads in the engine block often pull up, leaving a raised edge around the hole **(Figure 13–5)**. If the block has been resurfaced, the threads might run up to the surface. In either case, the bolt holes should be tapered at the surface by chamfering and the threads cleaned with an appropriate size bottoming tap. Always repair damaged threads to ensure proper bolt performance.

Many different nuts are used by the automotive industry. Beware of hexagon (hex) nut rotation when using power wrenches. It is deceptively easy to place a nut into a yield condition within seconds. Impact wrenches are the worst offenders. Friction is needed to prevent the nut from spinning. If the nut is lubricated, there is no friction left to stop the impact wrench from hammering the nut past the bolt's yield point or stripping the threads.

SHOP TALK

Impact wrenches should be used only to loosen nuts and bolts. Use other power or hand tools to tighten them. Final tightening should always be done with a torque wrench.

Smaller air-powered ratchets do not produce the severe force of impact wrenches and are much safer to use.

Thread Repair

A common fastening problem is threads stripping inside an engine block, cylinder head, or other structure. This problem is usually caused by overtorquing or by incorrectly threading the bolt into the hole. Rather than replacing the block or cylinder head, the threads can be replaced by the use of threaded inserts.

Torque-to-Yield (TTY) Bolts

Some fasteners are intentionally torqued into their yield condition. These fasteners, known as

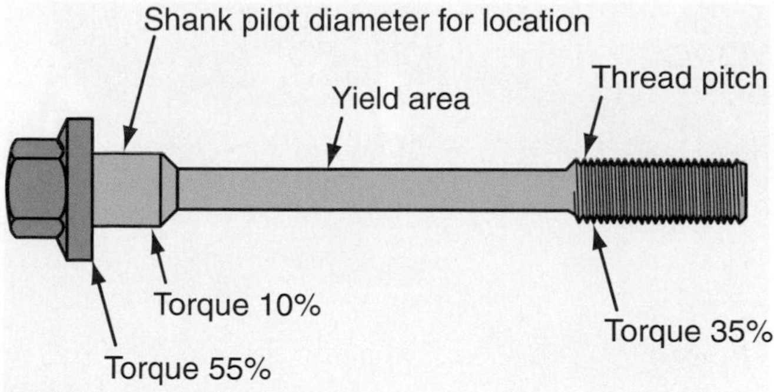

FIGURE 13–6

TTY bolts are designed with a reduced shank diameter; this is where the intended stretch occurs.

torque-to-yield (TTY) bolts, are designed to stretch when properly tightened. When a bolt is stretched to its yield point, it exerts its maximum clamping force. TTY bolts are not ordinary bolts. The bolt shank is designed to stretch **(Figure 13–6)** and spring back up to its yield point when tightened. Once at the yield point, the bolt becomes permanently stretched and will not return to its original length. Therefore, TTY bolts should not be reused.

Tightening a TTY bolt involves two distinct steps: Tighten the bolt to the specified torque, and then turn the bolt to a specified angle (measured in degrees) to load the bolt beyond its yield point **(Figure 13–7)**. A torque angle gauge is required to do this. This gauge fits between the drive of the torque wrench and the socket **(Figure 13–8)**. Once the specified torque is reached, the angle gauge is set to zero. Then the bolt is turned until the specified angle is read on the gauge.

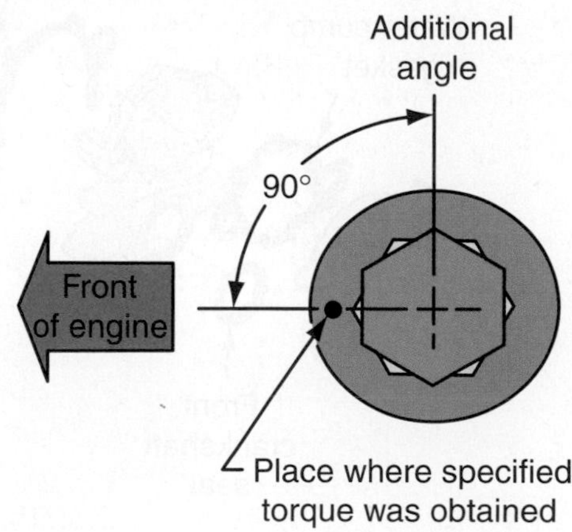

FIGURE 13–7

TTY bolts are tightened to a specified torque and then turned an additional number of degrees.

FIGURE 13–8

A TTY bolt tightened an additional number of degrees with the use of a torque angle gauge.

GASKETS

When parts are bolted together, it is nearly impossible to obtain a positive seal between the parts. **Gaskets** provide that seal and also serve as spacers, wear insulators, and vibration dampers. Gaskets are used only between two stationary parts. Seals are used if one of the parts moves. The material used to make a gasket depends on its application **(Figure 13–9)**.

Cut Gaskets

Gaskets made of paper, fibre, and cork are normally called soft, cut gaskets **(Figure 13–10)**. Each gasket is cut to the desired size and shape from a sheet of material.

PAPER/FIBRE GASKETS These are made of a fibre-reinforced paperlike material. They do a good job of sealing low-pressure, low-temperature areas. For

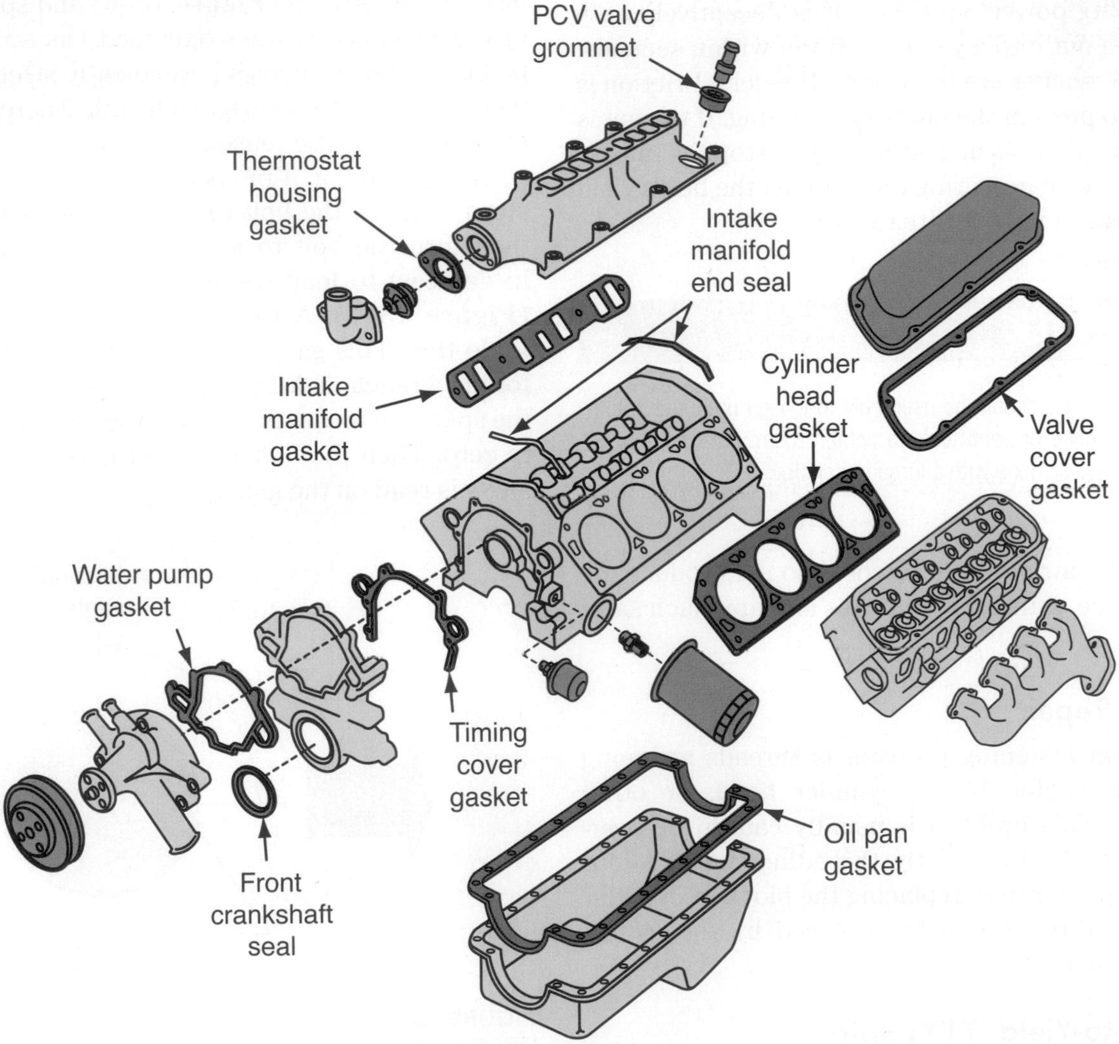

FIGURE 13–9

Typical engine gasket and seal locations.

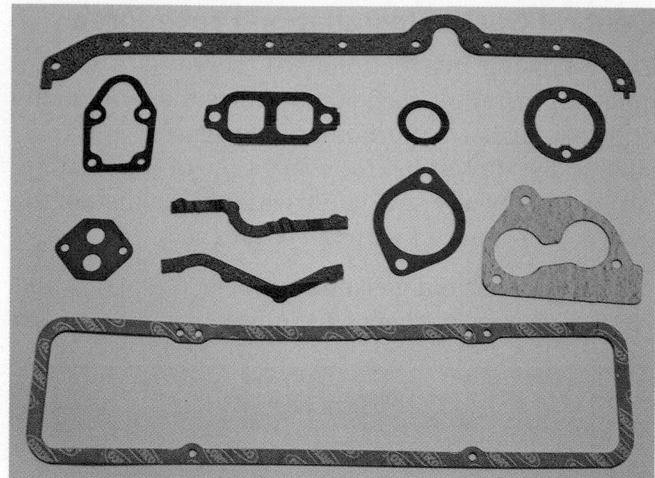

FIGURE 13–10

An assortment of cut gaskets.

some applications, paper gaskets may be relatively thick. These gaskets seldom need an additional sealant; however, a thin coating of adhesive may be used to hold the gasket in place while it is installed.

CORK GASKETS Cork gaskets are also used to seal low-pressure areas; however, they are not commonly used on today's engines. Cork gaskets are very soft, easily distorted, and absorbent and will weep some of the fluid they are sealing. They also tend to become brittle and crack over time. Most manufacturers have replaced cork gaskets with composite gaskets, typically rubberized cork **(Figure 13–11)**.

Molded Rubber Gaskets

Molded rubber gaskets provide excellent sealing and are commonly used on today's engines. These gaskets are made by injecting synthetic rubber (neoprene, nitrile, silicone, or other similar material) into a mould to form one-piece gaskets. Moulded gaskets retain their flexibility and are durable. Moulded gaskets are often used to seal intake manifolds, some thermostat or water pump housings, valve covers, and oil pans.

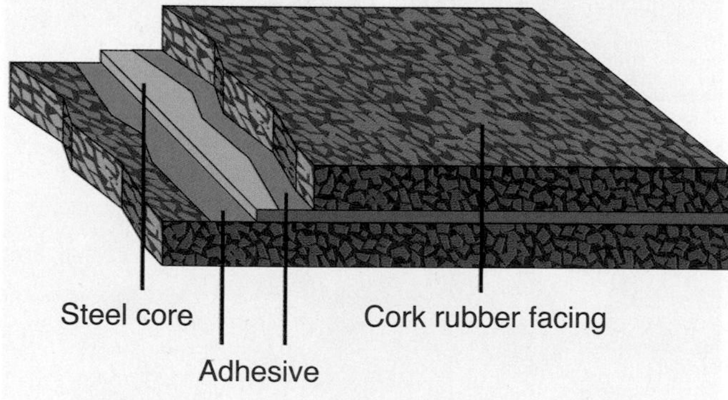

FIGURE 13–11

The composition of a cork/rubber gasket used mainly as valve cover, oil pan, and timing cover gaskets.

Dana Corporation

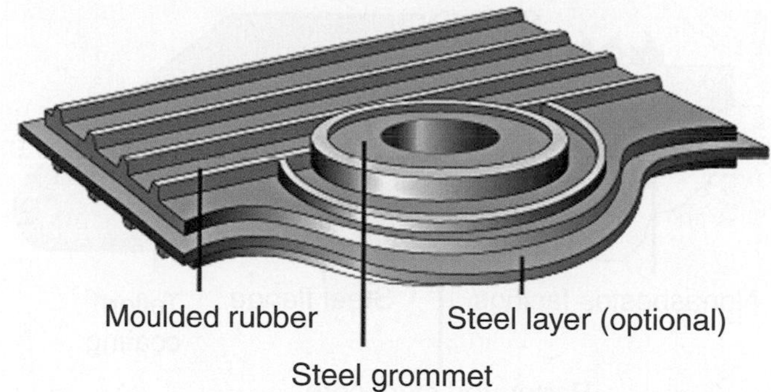

FIGURE 13–12

The composition of a moulded rubber gasket with steel grommets used mainly as valve cover and oil pan gaskets.

Dana Corporation

Some moulded gaskets have a steel insert that adds stiffness and strength to the gasket **(Figure 13–12)**. Also, some gaskets have reinforcements around the bolt holes to limit the amount of crush when the parts are tightened together.

WARNING!

Do not use sealant or adhesive on rubber gaskets; they can prevent the gasket from sealing.

Manufacturers may not use premade gaskets. Rather, they use chemical gasketing. Robotic equipment applies a bead of sealant around the sealing area. The result is called a "formed-in-place" gasket.

Hard Gaskets

Hard gaskets are made from steel, stainless steel, copper, or a combination of metals and other materials. Often, the metal is enclosed by a compressible and heat-resistant clay/fibre or Teflon® compound **(Figure 13–13)**. Hard gaskets are used for cylinder heads, exhaust manifolds, EGR valves, and some intake manifolds.

Replacement Gaskets

Gaskets can be purchased individually or in sets. Sets are often available for the service performed on the engine. The most common are timing

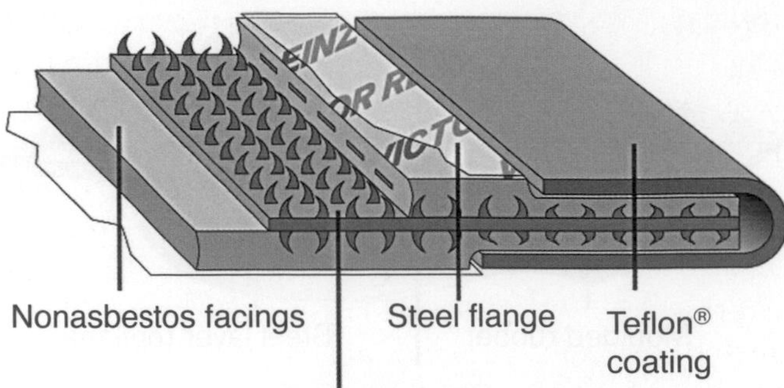

FIGURE 13–13

The composition of a Teflon-coated perforated steel core gasket used mainly as head and intake manifold gaskets.

Dana Corporation

cover, head, manifold, oil pan, and full sets. A full set contains all gaskets and seals required for rebuilding an engine **(Figure 13–14)**. Normally, there are more gaskets in a set than are needed. A particular engine may have been available with different equipment, and the extra gaskets are for those variations.

SHOP TALK

Before replacing a gasket, check for any TSBs on the engine. There may be a recommended replacement for the OEM gasket. For example, some GM engines built between 1996 and 2002 have experienced premature intake manifold gasket failures. These failures are normally caused by the corrosive effects of organic acid technology (OAT) coolant. The cure is a replacement gasket that is less susceptible to OAT.

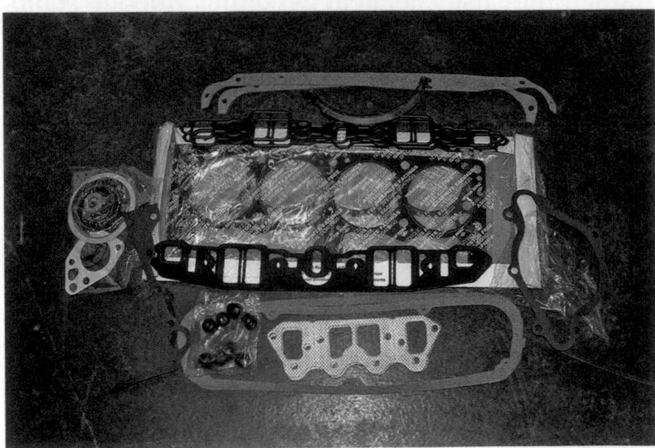

FIGURE 13–14

A full gasket set contains the gaskets and seal required for rebuilding an engine.

©Cengage Learning 2015

General Gasket Installation Procedures

The following instructions will serve as a helpful guide for installing gaskets. Because there are many different gasket materials and designs, it is impossible to list directions for every type of installation. Always follow any special directions provided in the instructions packed with the gasket sets.

1. Never reuse old gaskets.
2. Handle new gaskets carefully.

USING SERVICE INFORMATION

Always refer to the specific engine and engine part section of the service information for the recommended procedures for using sealants.

SHOP TALK

Some technicians tend to use too much sealant on gaskets. Do not make this mistake. Because sealants have less strength than gasket materials, they create weaker joints. They can also prevent gasket material from doing what it is supposed to do, which is to soak up oil and swell to make a tight seal.

1. Cleanliness is essential. New gaskets seal best when used on clean surfaces.
2. Use the right gasket in the right position. Always compare the new gasket to the mating surfaces to make sure it is the right gasket. Check that all bolt holes, dowel holes, coolant, and lubrication passages line up perfectly with the gasket. Some gaskets have directions such as "top," "front," or "this side up/down" stamped on one surface **(Figure 13–15)**. An upside-down or reversed

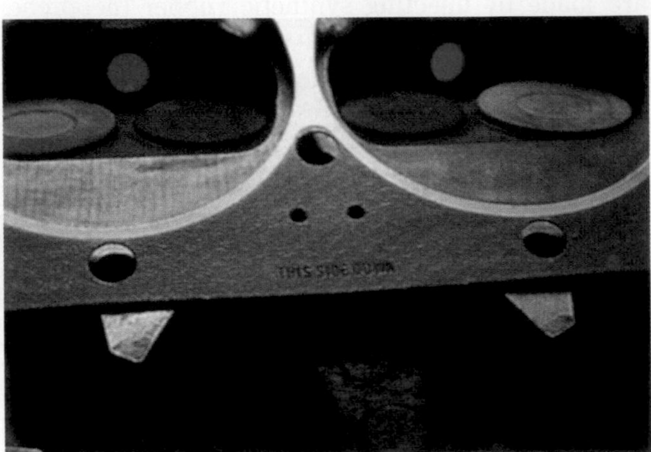

FIGURE 13–15

Some gaskets have installation directions stamped on them.

Corteco

gasket can easily cause a loss of oil pressure, overheating, and engine failure.

3. Use sealants and adhesives only when the engine or gasket manufacturer recommends their use. Some chemicals will react negatively with the gasket's coating.
4. Always start each bolt into its hole before you begin to tighten any of the bolts. Tightening bolts before all bolts are installed will cause component misalignment and prevent some bolts from threading into their holes.

SPECIFIC ENGINE GASKETS

There is a wide variety of gaskets used on engines, each with its own purpose, and each is designed for that application. The following is a discussion of the most common ones.

Cylinder Head Gaskets

Cylinder head gaskets are the most sophisticated type of gasket. They also have the most demanding job. When first starting an engine in cold weather, parts near the combustion chamber are very cold. Then, after only a few minutes of engine operation, these same parts might reach 204°C (400°F). The inner edges of the cylinder head gasket are exposed to combustion flame temperatures from 1093°C to 1648°C (2000°F to 3000°F).

Pressures inside the combustion chamber also vary tremendously. On the intake stroke, a vacuum or low pressure exists in the cylinder. Then, after combustion, pressure peaks of approximately 6895 kPa (1000 psi) occur. This extreme change from low pressure to high pressure happens in a fraction of a second.

Cylinder head gaskets, under these conditions, must simultaneously do the following:

- Seal intake stroke vacuum, combustion pressure, and the heat of combustion.
- Prevent coolant leakage, resist rust and corrosion, and in many cases, meter coolant flow.
- Seal oil passages through the block and head while resisting chemical action.
- Allow for lateral and vertical head movement as the engine heats and cools.
- Be flexible enough to seal minor surface warpage while being stiff enough to maintain adequate gasket compression.
- Fill small machining marks that could lead to serious gasket leakage and failure.
- Withstand forces produced by engine vibration.

Head gaskets for many late-model cast-iron engines have raised silicone, Viton, or fluoroelastomer sealing beads on their face to increase the clamping pressure around some areas. Most head gaskets have a steel fire ring that surrounds the top of the cylinder. These protect the gasket material used elsewhere. The durability of a head gasket can also be improved by using strong, high-temperature fibres such as aramid and kevlar and by adding reinforcements around oil passages.

BIMETAL ENGINE REQUIREMENTS Most late-model engines have aluminum cylinder heads and cast-iron blocks. When heated, aluminum expands two to three times more than steel. This creates a back-and-forth scrubbing action on a head gasket as the engine temperature changes **(Figure 13–16)**. This movement can tear a gasket apart if it is not designed to handle it. To reduce the chances of a gasket tearing, graphite or specially coated gaskets are used. **Graphite** is a relatively soft material that can withstand high temperatures; it is a natural lubricant. Teflon, molybdenum, and other similar slippery nonstick coatings are used on other gasket designs to prevent the gasket from sticking to either surface. This allows the head to expand and contract without destroying the gasket.

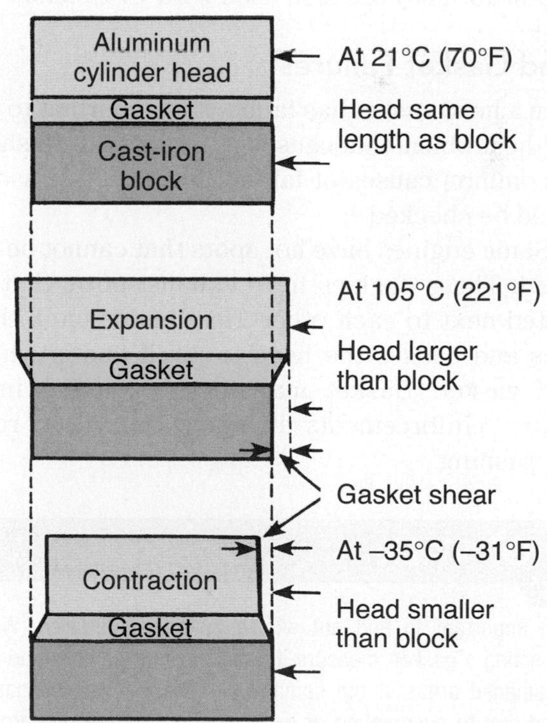

FIGURE 13–16

Thermal growth characteristics of bimetal engines.

FIGURE 13–17

The composition of an MLS head gasket.

MULTILAYER STEEL (MLS) Many engines have **multilayer steel (MLS)** head gaskets. These gaskets are comprised of three to seven layers of steel. The outer layers are embossed stainless spring steel coated with a thin layer of an antifriction coating of Teflon, nitrile rubber, or Viton **(Figure 13–17)**. The inner layers provide the necessary thickness. The use of an MLS gasket reduces the load on the head bolts and allows them to retain their shape after they are tightened in place. They are also very durable but require a smooth finish on the engine block and cylinder head. They are also used with TTY bolts.

Head Gasket Failures

When a head gasket has failed, it is important to correct the problem that caused it. **Figure 13–18** shows the common causes of failure and the systems that should be checked.

Some engines have hot spots that cannot be corrected. These engines have exhaust ports that are located next to each other. Heat builds up in these areas and causes the head to swell and crush the head gasket. Gasket manufacturers often incorporate reinforcements in those areas to resist the crushing.

SHOP TALK

It is important to find out why a head gasket failed. When inspecting a gasket, measure its thickness at the damaged and undamaged areas. If the damaged area is thinner, the gasket failed due to overheating or a hot spot. If the fire ring around the cylinder bores is cracked or burnt, preignition or detonation caused the gasket to fail.

Problem	Probable Cause
Preignition/ Detonation	Incorrect ignition timing Incorrect air-fuel mixture Vacuum leak Faulty cooling system
Overheating	Restricted radiator Cooling system leak Faulty thermostat Faulty water pump Inoperative cooling fan Faulty EGR system
Improper Installation	Wrong surface finish Incorrect bolt-tightening sequence Use of stretched or damaged bolts Improper use of sealant Use of incorrect gasket Dirty mating surfaces
Hot Spots	Use of incorrect gasket

FIGURE 13–18

Leading causes of cylinder head gasket failure.

HEAD BOLTS Installation failures are commonly caused by head bolt problems. When installing a cylinder head and bolts, do the following:

- Make sure all bolts are clean and have undamaged threads. Replace any bolts that are nicked, deformed, or worn.
- Make sure the correct length bolt is installed in the bore. Some engines use longer bolts in some locations.
- Check the length of each bolt and compare the measurements. Longer than normal bolts are stretched and should not be reused.
- Inspect the shank or top of the threaded area of the bolt for evidence of stretching.
- Never reuse TTY bolts.
- When installing an aluminum head, use hardened steel washers under the bolts. Place the washers so that their rounded edge faces up.
- Clean the thread bores in the engine block with a bottoming tap.
- Make sure the top of each thread bore is chamfered.
- If the head has been resurfaced, make sure the bolts do not bottom out in their bores.
- If they do, install hardened steel washers under the bolt to raise them.
- Lubricate the threads and the bottom of the bolt head with engine oil if so directed by the manufacturer.
- Apply the correct type of thread sealant to all bolts that go into a coolant passage.

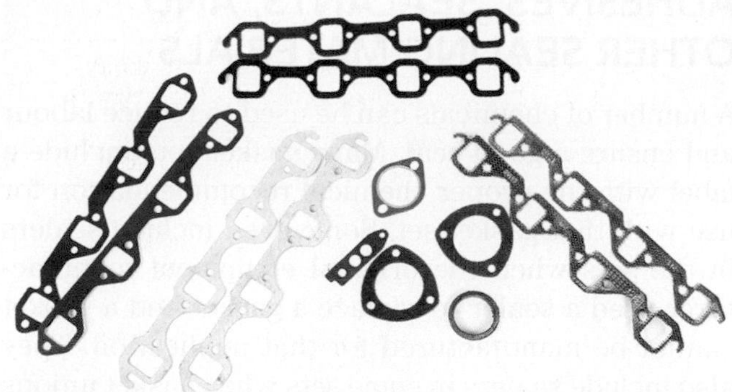

FIGURE 13–19

Assorted intake and exhaust manifold gaskets for diverse applications.

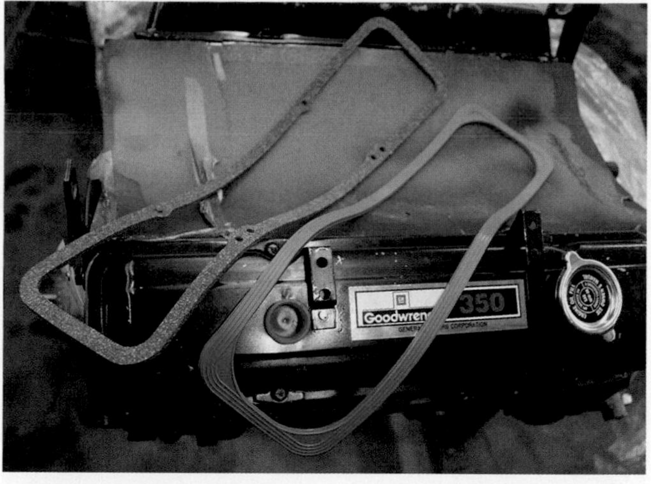

FIGURE 13–20

(left) A cork valve cover gasket. (right) A synthetic rubber valve cover gasket.

Martin Restoule

Manifold Gaskets

There are three basic types of manifold gaskets—the intake manifold, exhaust manifold, and an intake and exhaust combination. Combination gaskets are often used on in-line engines without cross-flow heads. Manifold gaskets are made of a variety of materials, depending on the application. Each type of manifold gasket has its own sealing characteristics and problems **(Figure 13–19)**. Therefore, be sure to follow the manufacturer's instructions when installing them.

Before installing a manifold and its gasket, make sure that the mating surfaces on the head and manifold are flat and free of damage. Also, always follow the recommended bolt-tightening sequence and use the exact torque specs.

Valve Cover Gaskets

Valve cover gaskets must make a seal between a steel, aluminum, magnesium, or moulded plastic cover and the cylinder head surface. On OHC engines, cam covers are normally made of die-cast aluminum. Some cylinder cover gaskets have spark plug gaskets integrated into the gasket. When installing these, make sure the gasket is perfectly aligned. Valve cover bolts are usually widely spaced so that the gasket material is able to seal without being tightly clamped **(Figure 13–20)**. Valve cover gaskets must be able to withstand high temperatures and the caustic action of acids in the oil.

Oil Pan Gaskets

An oil pan gasket seals the joint between the oil pan and the bottom of the block **(Figure 13–21)**. The oil pan gasket might also seal the bottom of the timing cover and the lower section of the rear main bearing cap.

Like valve cover gaskets, the oil pan gasket must resist hot, thin engine oil. Oil pans are usually made of stamped steel, cast iron, or cast aluminum. Because of the added weight and splash of crankcase oil,

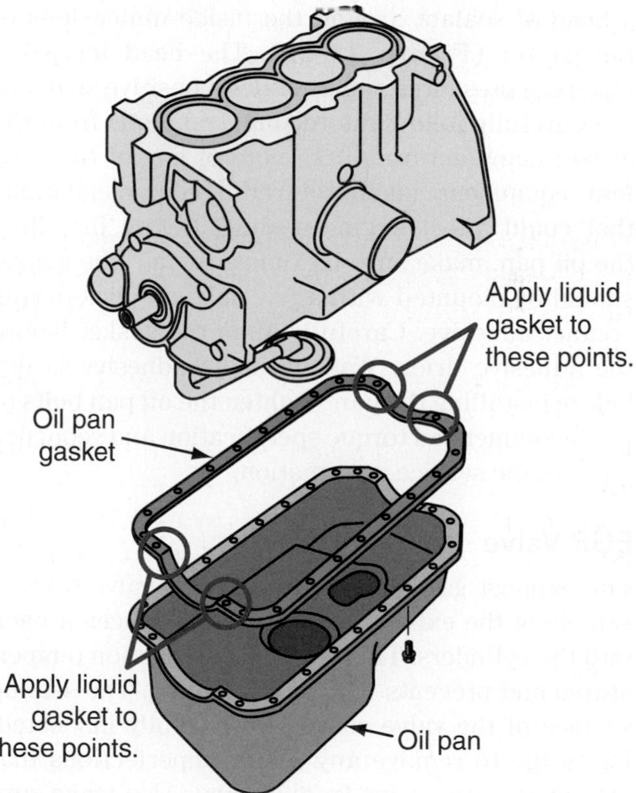

FIGURE 13–21

Oil pan gasket installation; note the points where a liquid gasket is required.

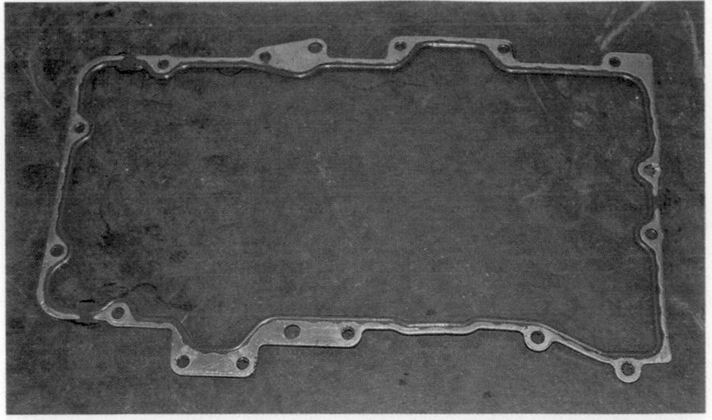

FIGURE 13–22

A hard oil pan gasket with a sealant bead.

the pan has many assembly bolts closely spaced. As a result, the clamping force on the oil pan gasket is great. The gasket is thinner and must resist crushing.

Oil pan gaskets are made of several types of material. A commonly used material is synthetic rubber, known for its long-term sealing ability. It is tough, durable, and resistant to hot engine oil. Synthetic rubber gaskets are easy to remove, so the sealing surfaces need less cleanup.

Many late-model engines use a hard gasket with a bead of sealant around the inside dimensions of the gasket **(Figure 13–22)**. The bead increases clamping pressures and provides a positive seal.

Carefully follow the recommendations from the gasket manufacturer. Take note of any of the original equipment manufacturer's recommendations that could affect engine sealing. Before installing the oil pan, make sure its flange is flat. The gasket should be mounted with a few dabs of quick-drying contact adhesive. Carefully align the gasket before the adhesive dries. Wait until the adhesive is dry before installing the pan. Tighten the oil pan bolts to the recommended torque specification and sequence given in the service information.

EGR Valve

The exhaust gas recirculation (EGR) valve takes a sample of the exhaust gases and introduces it back into the cylinders. This reduces combustion temperatures and prevents NO_x from forming. The sealing surface of the valve should be carefully inspected. Use a file to remove any minor imperfections that may prevent the valve from sealing. Also make sure that the new gasket is the correct one; some gaskets have specifically sized holes that are used to regulate exhaust flow. Using the wrong gasket could change how the engine performs.

ADHESIVES, SEALANTS, AND OTHER SEALING MATERIALS

A number of chemicals can be used to reduce labour and ensure a good seal. Many gasket sets include a label with the proper chemical recommendation for use with that gasket set. Some even include sealers in the sets when the original equipment manufacturer used a sealer to replace a gasket and a gasket cannot be manufactured for that application. They also include sealers in some sets when gasket unions need a sealant to ensure a good seal.

SHOP TALK

Chemical adhesives and sealants give added holding power and sealing ability where two parts are joined. Sealants usually are added to threads where fluid contact is frequent. Chemical thread retainers are either **aerobic** (cures in the presence of air) or **anaerobic** (cures in the absence of air). These chemical products are used in place of lock washers.

Adhesives

Gasket **adhesives** form a tough bond when used on clean, dry surfaces. Adhesives do not aid the sealing ability of the gasket. They are meant only to hold gaskets in place during component assembly. Use small dabs; they will dry quicker for fast installation. Do not assemble components until the adhesive is completely dry. Most adhesives are ideal for use on gasket applications such as valve covers, pushrod covers, manifold and manifold end seals, and oil pan and oil pan end seals **(Figure 13–23)**.

FIGURE 13–23

An adhesive is often used to hold a gasket in place during assembly.

Martin Restoule

Sealants

Manufacturers sometimes specify the use of **sealants** to assist a gasket or seal or to form a new gasket. These sealants should be used only when specified by the manufacturer. Also make sure to use the specific sealant recommended by the manufacturer.

GENERAL-PURPOSE SEALANTS General-purpose sealants come in liquid form and are available in a brush type (known as brush tack) and an aerosol type (known as spray tack). General purpose sealers **(Figure 13–24)** form a tacky, flexible seal when applied in a thin, even coat that aids in gasket sealing by helping to position the gasket during assembly.

WARNING!

Make sure every sealant you use on today's engines is oxygen-sensor safe.

WARNING!

Never use a hard-drying sealant (such as shellac) on gaskets. It will make future disassembly extremely difficult and might damage the gasket.

THREAD SEALANTS Bolts that pass through a liquid passage should be coated with Teflon thread **(Figure 13–25)** or a brush-on thread sealant. Some head bolts or water pump bolts tighten into a coolant passage and must be sealed or they will leak. These flexible sealants are nonhardening sealers that fill voids, preventing the fluid from running up the threads. They resist the chemical attack of lubricants, synthetic oils, detergents, antifreeze, gasoline, and diesel fuel.

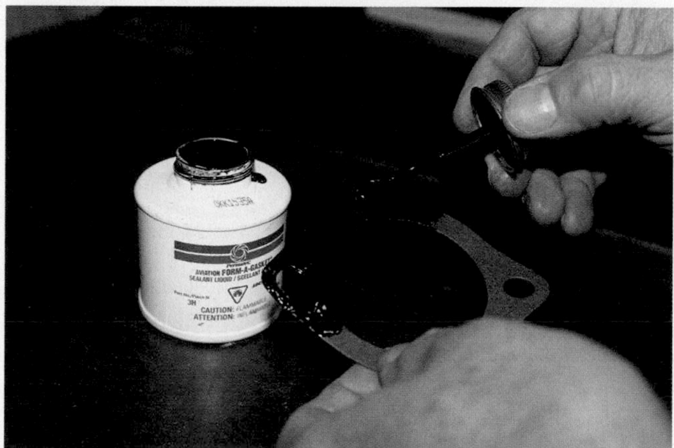

FIGURE 13–24

Applying gasket sealer with a brush.

Martin Restoule

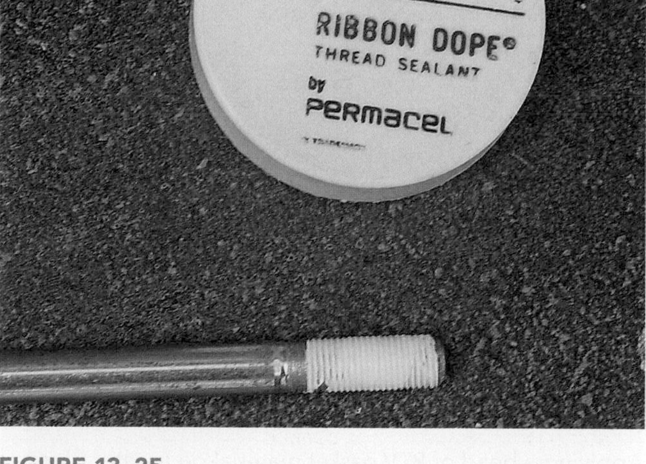

FIGURE 13–25

Bolts that pass through a liquid passage should be coated with Teflon or similar thread sealant.

SILICONE SEALANTS **Silicone** (or formed-in-place FIP) **gaskets** are a liquid sealant applied directly to mating surfaces and are allowed to cure in place. Many technicians use silicone gasketing to aid the sealing of corners, notches, or dovetails of gaskets **(Figure 13–26)**. **Room temperature vulcanizing (RTV)** silicone sealing products are the most commonly used FIP gasket products. RTV is an aerobic sealant, which means it cures or hardens in the presence of air. RTV can be used to seal two stationary parts such as water pumps **(Figure 13–27)** and oil pans. It cannot be used as a head or exhaust gasket

FIGURE 13–26

Applying a bead of RTV gasket material at the point where two gaskets meet.

FIGURE 13–27

Applying a bead of RTV gasket material on a water pump.

Loctite

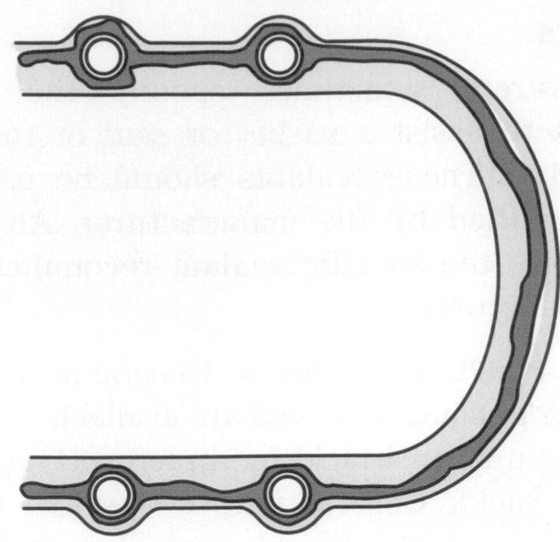

FIGURE 13–28

When applying RTV, make sure the bolt holes are encircled.

or in fuel systems. RTV comes in a variety of colours that denote the proper application. Black is for general purpose; blue is for special applications; and red is for high-temperature requirements. Always use the correct type for the application. RTV silicone sealants are impervious to most fluids, are extremely resistant to oil, have great flexibility, and adhere very well to most materials.

WARNING!

Be careful not to use excessive amounts of RTV. If too much is applied, it can loosen up and get into the oil system, where it can clog an oil or coolant passage and cause severe engine damage and/or engine overheating.

To use RTV silicone, make sure the mating surfaces are free from dirt, grease, and oil. Apply a continuous 3 mm (⅛ in.) bead on one surface only (preferably the cover side). Make sure to circle all bolt holes **(Figure 13–28)**. Adjust the shape before a skin forms (in about one minute), then remove excess RTV silicone with a dry towel or paper towel. Press the parts together. Do not slide the parts together; this will disturb the bead. Tighten all retaining bolts to the manufacturer's specified torque. Cure time is approximately one hour for metal-to-metal joints and can take up to 24 hours for 3 mm (⅛ in.) gaps.

CAUTION!

The uncured rubber contained in RTV silicone gasketing irritates the eyes. If any gets in your eyes, immediately flush with clean water or eyewash. If the irritation continues, see a doctor.

ANAEROBIC FORMED-IN-PLACE SEALANTS These materials are used for thread locking as well as gasketing. As a retaining compound, they are mostly used to hold sleeves, bearings, and locking screw nuts in place where there is a high exposure to vibration.

WARNING!

Never use a sealant or formed-in-place gasket material on exhaust manifolds.

The major difference between aerobic and anaerobic sealants, other than their method of curing, is their gap-filling ability. Typically, 1.27 mm (0.050 or 3/64 in.) is the absolute limit of any anaerobic's gap-filling ability. Some are only designed to seal 0.13 to 0.25 mm (0.005 to 0.010 in.) gaps. Anaerobic sealers are intended to be used between the machined surfaces of rigid castings, not on flexible stampings.

SHOP TALK

Once hardened, a good anaerobic bond is unbelievably tenacious and can withstand high temperatures. Therefore, care must be taken in selection. They tend to be highly specialized and not readily interchangeable. For example, there are various levels of thread-locking products that range from medium-strength antivibration agents to high-strength, weldlike retaining compounds. The inadvertent use of the wrong product could make future disassembly an impossibility. Check the label to be certain that anaerobic material will suit the purpose of the application.

Antiseize Compounds

Antiseize compounds prevent dissimilar metals from reacting with one another and seizing. This chemical-type material is used on many fasteners, especially those used with aluminum parts. Always follow the manufacturer's recommendations when using this compound.

OIL SEALS

Seals keep oil and other vital fluids from escaping around a rotating shaft. There are three basic oil seal designs: the fibre packing, the two-piece split lip design, and the one-piece radial design **(Figure 13–29)**.

SHOP TALK

Whenever installing an oil seal, make sure its lip seal is lubricated with a light coating of grease. Also make sure the lip portion of the seal is facing the direction that oil is coming against.

Timing Cover Oil Seals

An oil seal in the timing cover prevents oil from leaking around the crankshaft. The installation of this seal often requires the use of a special tool **(Figure 13–30)** or a driver. It is important that the seal be positioned squarely in the bore of the timing cover and the crankshaft be positioned in the centre of the seal.

Rear Main Bearing Seals

Rear main bearing seals keep oil from leaking at the crankshaft around the rear main bearing. There are two basic types of constructions: wick- or rope-type packing and moulded synthetic rubber.

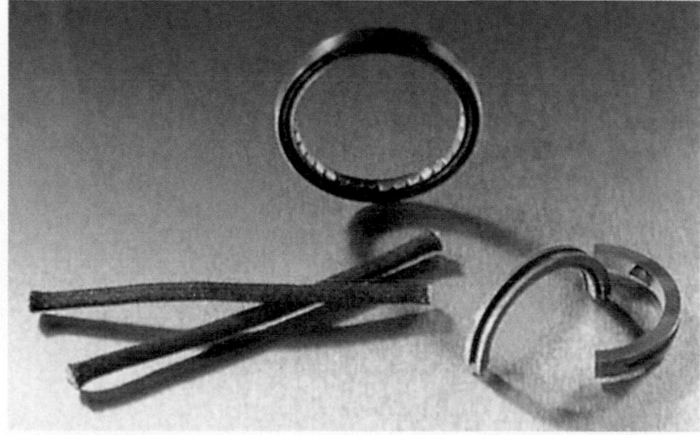

FIGURE 13–29

The three basic oil seal designs.

Corteco

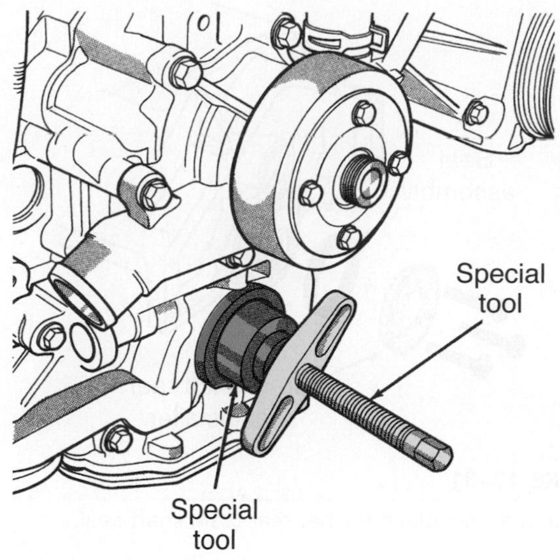

FIGURE 13–30

The installation of the timing cover oil seal requires the use of a special tool or driver.

Chrysler Group LLC

Wick- or rope-type packings are common on many older engines. Moulded synthetic rubber lip-type seals are used on many newer engines. They do a good job of sealing, even when there is some eccentricity in the shaft, as long as the surface of the shaft is very smooth. Synthetic rubber seals may sometimes be retrofitted to some older engines that have wick seals, but only if the seals are offered as an option by the sealing manufacturer.

Three types of synthetic rubber are used for rear main bearing seals. **Polyacrylate** is commonly used because it is tough and abrasion resistant, with moderate temperature resistance to 177°C (350°F). Silicone synthetic rubber has a greater temperature range, but it has less resistance to abrasion and is more fragile than polyacrylate. Silicone seals must be handled carefully during installation to avoid damage. **Viton** has the abrasion resistance of polyacrylate and the temperature range of silicone, but it is the most expensive of the synthetic types. The synthetic rubber seals may be one piece **(Figure 13–31)** or two pieces **(Figure 13–32)**.

No matter what the construction of the seal is, always check the shaft for smoothness. Shafts should be free of nicks and burrs to ensure long oil seal life. Carefully remove any roughness with a very fine emery cloth, and then clean the shaft thoroughly. The shaft should have a highly polished appearance and a smooth feel. Also be sure to check and clean the oil slinger and oil return channel in the bearing cap.

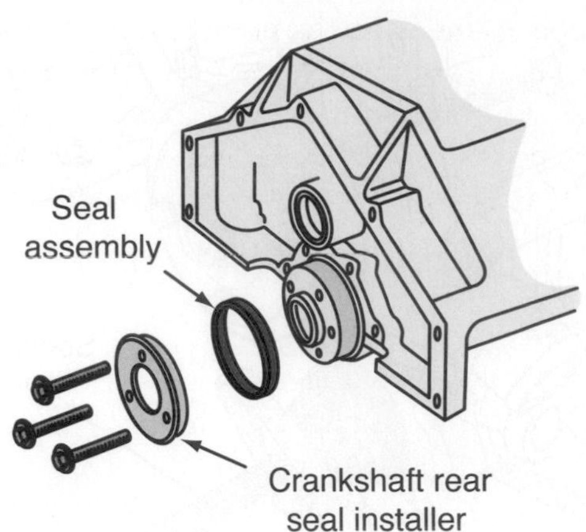

FIGURE 13–31

Installing a one-piece rubber rear crankshaft seal.

ENGINE REASSEMBLY

When reassembling an engine, the assembly sequence is essentially the reverse of the tear-down sequence outlined in a previous chapter.

Installing the Cylinder Head and Valve Train

Use a wire brush to clean the threads of the head bolts. Then check their condition and length. Many engines use head bolts of different lengths **(Figure 13–33)**, and their location is given in the service information. Lightly lubricate the threads with clean engine oil, if directed by the manufacturer.

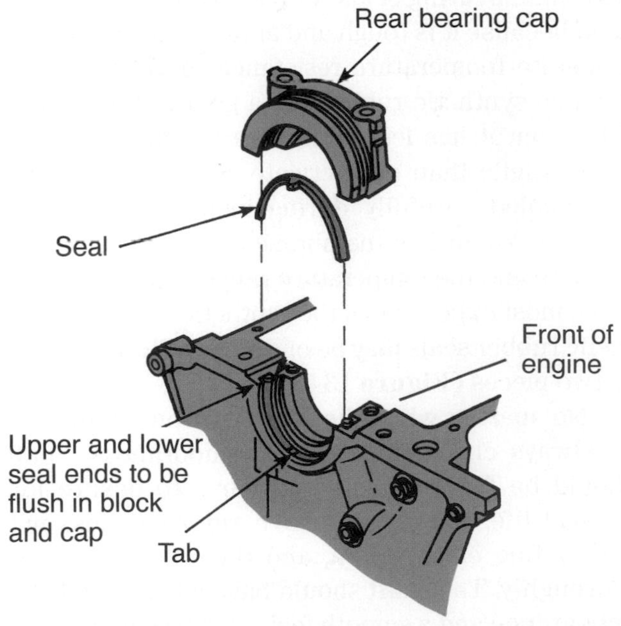

FIGURE 13–32

A typical two-piece rubber rear crankshaft seal.

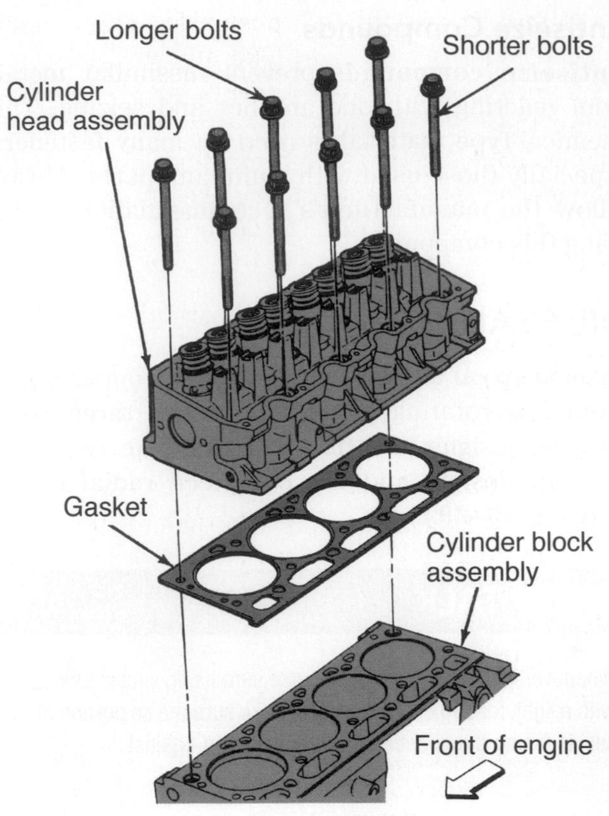

FIGURE 13–33

Head bolts are different lengths in some engines.

Ford Motor Company

Position the head gasket on the block, and make sure it matches the bores in the block. Place the cylinder head onto the block. Make sure that the dowel pins are in place and that the head and block are properly aligned.

Tighten the head bolts according to the recommended sequence **(Figure 13–34)** and to the specified torque. Most heads are tightened in a sequence that starts in the middle and moves out to the ends. The bolts are generally tightened in two or three stages.

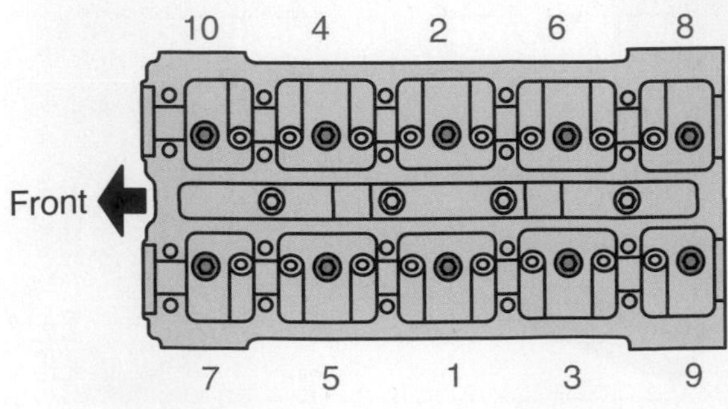

FIGURE 13–34

Always follow the specified tightening sequence when torquing cylinder head bolts.

Before inserting the pushrods, apply some assembly lube to both ends. Liberally coat the rocker arms with assembly lube or clean engine oil. Then install the pushrods and rocker arms. Many OHV engines have positive stop rocker arm adjustments. This means that when torquing the rocker arms to spec, the plunger of the hydraulic lifter is properly positioned, giving the correct lifter adjustment.

TORQUE ANGLE GAUGE Some manufacturers recommend the torque angle method for tightening cylinder head bolts, which requires the use of a torque angle gauge. Torque-to-yield bolts must be tightened according to manufacturer's recommendations. Typically, this involves two steps: tighten the bolt to the specified torque, and then tighten the bolt an additional amount. The latter is expressed in degrees. To accurately measure the number of degrees added to the bolt, a torque angle gauge is attached to the wrench. The additional tightening will stretch the bolt, producing a very reliable clamp load that is much higher than can be achieved just by torquing.

Timing Belts and Chains

The alignment of the camshaft(s) with the crankshaft is critical. The alignment marks and the correct procedure for doing this vary with the engine design. Many engines have additional chains or belts for the balance shafts and oil pump. DOHC engines may have individual chains that connect to each of the camshafts. These must be properly timed. Check the service information for the correct alignment and tension adjustment procedures **(Figure 13–35)**.

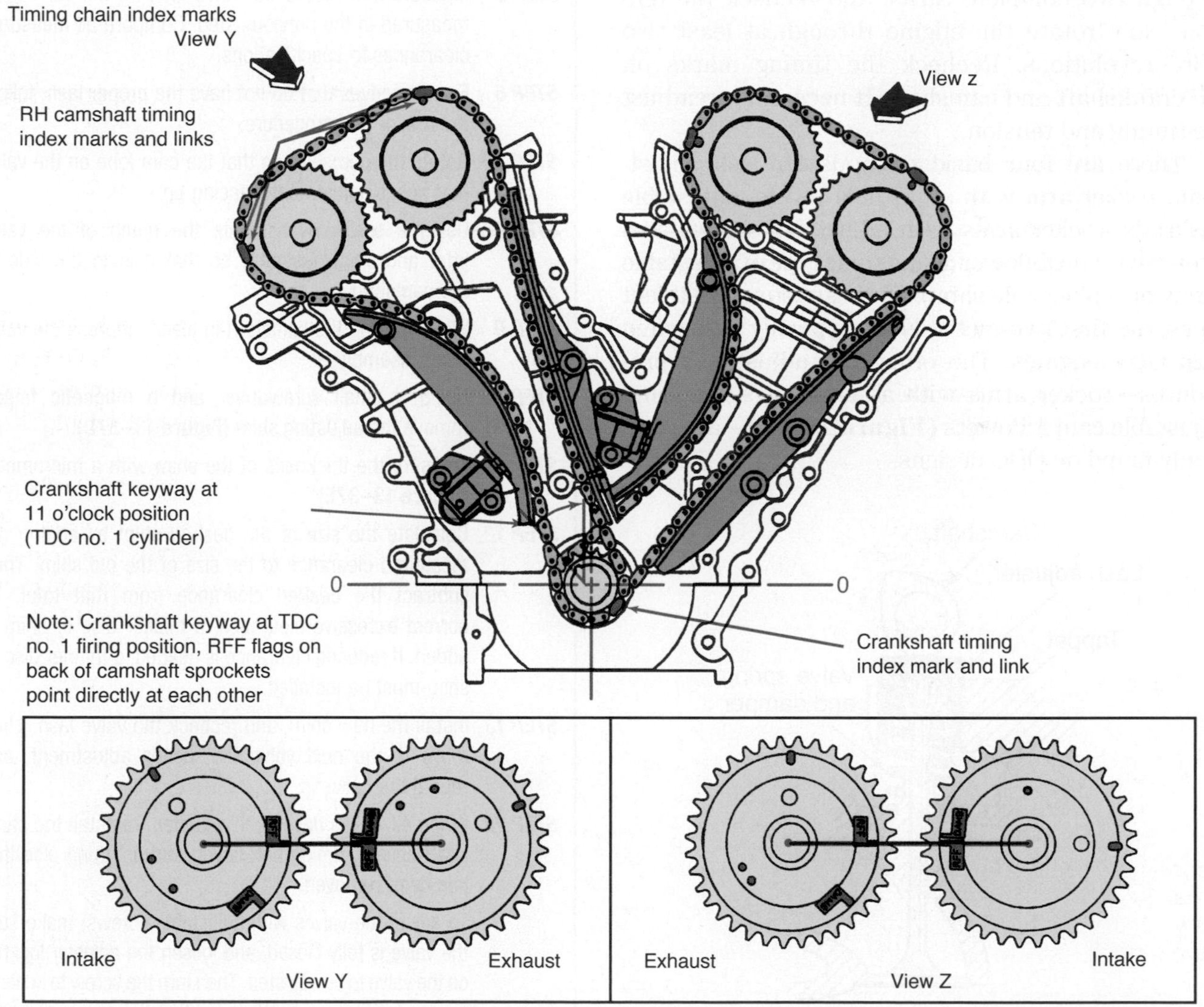

FIGURE 13–35

Note the timing marks and position of the chain tensioners on this DOHC engine.

Ford Motor Company

Make sure that the belt or chain tensioners are set properly.

USING SERVICE INFORMATION

Normally, camshaft timing marks are shown in the engine section of the service information under the heading of Timing Belt or Chain R&R.

Timing belts can be replaced with the engine in the vehicle or when the engine is on a stand. Photo Sequence 10 shows a typical procedure for changing a timing belt with the engine in a vehicle.

After the belt is installed, adjust its tension according to the manufacturer's recommendations. Then, by turning the crankshaft, rotate the engine through two complete turns, and recheck the tension. Now rotate the engine through at least two more revolutions. Recheck the timing marks on the crankshaft and camshaft. If necessary, readjust the timing and tension.

There are four basic methods for lash adjustment: rocker arm with adjustable pivots, adjustable pushrods, rocker arms with adjustable screws, and adjustable cam follower (using some type of adjustable screw or replaceable shim). Of these four adjustment types, the first two methods are typically associated with OHV engines. The other two adjustment procedures—rocker arms with adjustment screws and adjustable cam followers **(Figure 13–36)**—are commonly found on OHC designs.

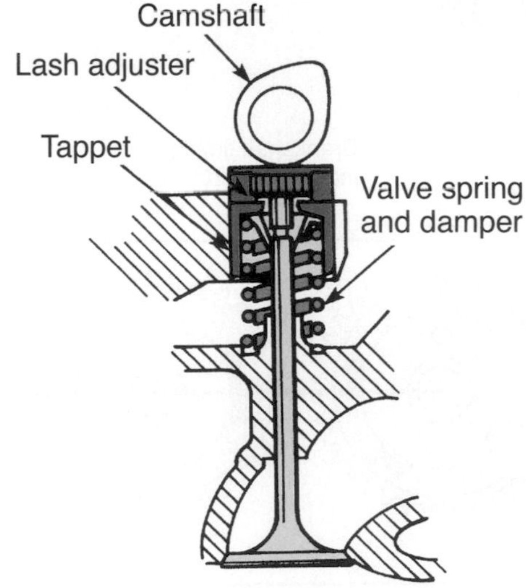

FIGURE 13–36

Some overhead cam engines feature a cam follower with an adjustment screw.

PROCEDURE

Adjusting the Valves on an Engine That Uses Shims for Valve Clearance

STEP 1 With the timing belt cover and camshaft covers removed, make sure piston 1 is at TDC **(Figure 13–37A)**.

STEP 2 Check the camshaft alignment marks. If they are not aligned, rotate the crankshaft one full turn **(Figure 13–37B)** until they are aligned.

STEP 3 Using a feeler gauge, measure and record the valve lash of the valves that are totally closed. Refer to the service information to identify the closed valves **(Figure 13–37C)**.

STEP 4 Rotate the crankshaft one full turn with piston 1 again at TDC.

STEP 5 Measure and record the valve lash on the valves not measured in the previous step. Compare all measured clearances to specifications.

STEP 6 For any valves that do not have the proper lash, follow the rest of this procedure.

STEP 7 Rotate the camshaft so that the cam lobe on the valve that needs adjustment is facing up.

STEP 8 Using a screwdriver, rotate the notch of the valve lifter and shim assembly so that it is to the side of the camshaft.

STEP 9 While holding the camshaft in place, depress the valve lifter assembly.

STEP 10 Using a small screwdriver and a magnetic finger, remove the adjusting shim **(Figure 13–37D)**.

STEP 11 Measure the thickness of the shim with a micrometer **(Figure 13–37E)**.

STEP 12 Calculate the size of the desired shim by adding the measured clearance to the size of the old shim. Then subtract the desired clearance from that total. To correct excessive clearance, a thicker disc or shim is added. If reduced clearance is needed, a thinner disc or shim must be installed.

STEP 13 Install the new shim, and recheck the valve lash. Then move to the next valve that needs adjustment, and repeat the process.

STEP 14 When all valves have been adjusted, reinstall the camshaft covers, timing belt cover, and anything else that has been removed.

To adjust the valves with adjustable screws, make sure the valve is fully closed, and loosen the adjuster locknut on the valve to be adjusted. Then turn the screw to achieve the proper clearance. While holding the screw and preventing it from turning, tighten the locknut. Follow the same procedure for each valve that needs to be adjusted.

REPLACING A TIMING BELT ON AN OHC ENGINE

PHOTO SEQUENCE 10

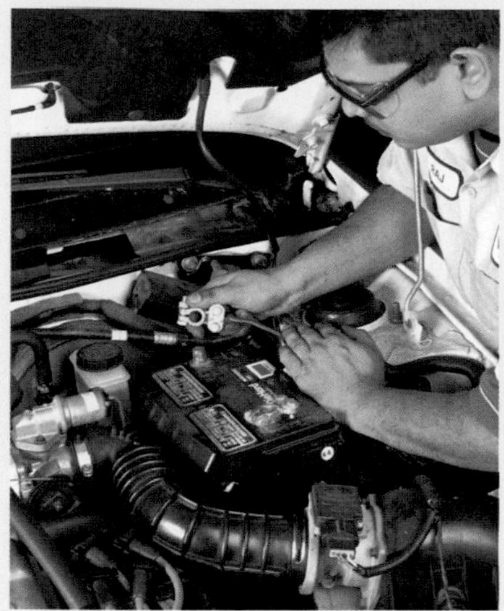

P10–1 Disconnect the negative cable from the battery prior to removing and replacing the timing belt.

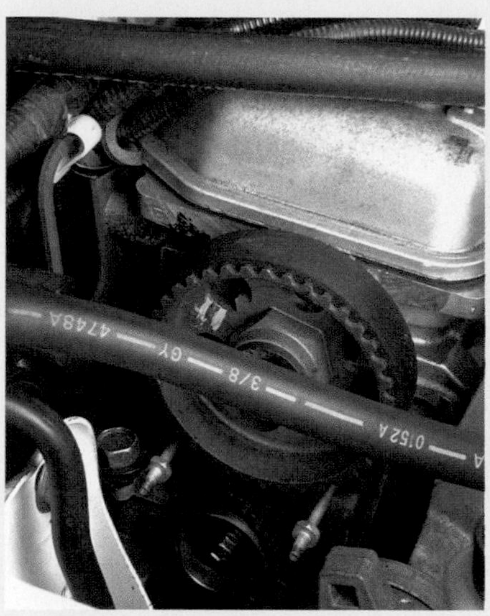

P10–2 Carefully remove the timing cover. Be careful not to distort or damage it while pulling it up. With the cover removed, check the immediate area around the belt for wires and other obstacles. If some are found, move them out of the way.

P10–3 Align the timing marks on the camshaft's sprocket with the mark on the cylinder head. If the marks are not obvious, use a paint stick or chalk to clearly mark them.

P10–4 Carefully remove the crankshaft timing sensor and probe holder.

P10–5 Loosen the adjustment bolt on the belt tensioner pulley. It is normally not necessary to remove the tensioner assembly.

P10–6 Slide the belt off the crankshaft sprocket. Do not allow the crankshaft pulley to rotate while doing this.

REPLACING A TIMING BELT ON AN OHC ENGINE *(continued)*

P10–7 To remove the belt from the engine, the crankshaft pulley must be removed. Then the belt can be slipped off the crankshaft sprocket.

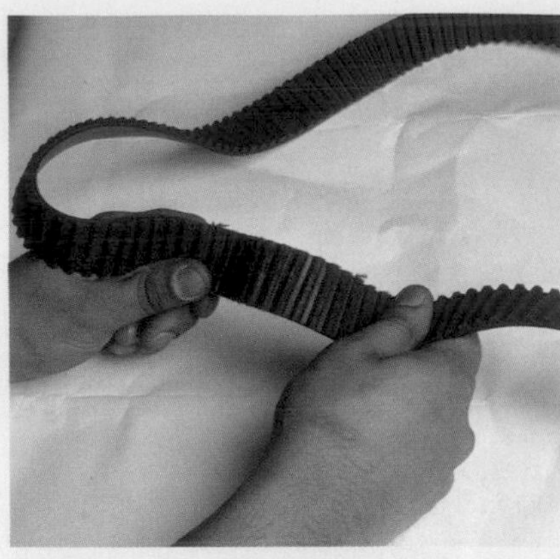

P10–8 After the belt has been removed, inspect it for cracks and other damage. Cracks will become more obvious if the belt is twisted slightly. Any defects in the belt indicate it must be replaced.

P10–9 To begin reassembly, place the belt around the crankshaft sprocket. Then reinstall the crankshaft pulley.

P10–10 Make sure the timing marks on the crankshaft pulley are lined up with the marks on the engine block. If they are not, carefully rock the crankshaft until the marks are lined up.

P10–11 With the timing belt fitted onto the crankshaft sprocket and the crankshaft pulley tightened in place, the crankshaft timing sensor and probe can be reinstalled.

P10–12 Align the camshaft sprocket with the timing marks on the cylinder head. Then wrap the timing belt around the camshaft sprocket, and allow the belt tensioner to put a slight amount of tension on the belt.

P10–13 Adjust the tension as described in the service information system. Then rotate the engine two complete turns. Recheck the tension.

P10–14 Rotate the engine through two complete turns again, then check the alignment marks on the camshaft and the crankshaft. Any deviation needs to be corrected before the timing cover is reinstalled.

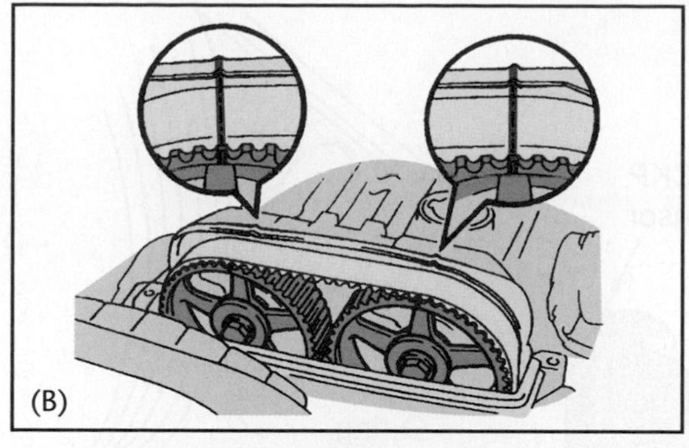

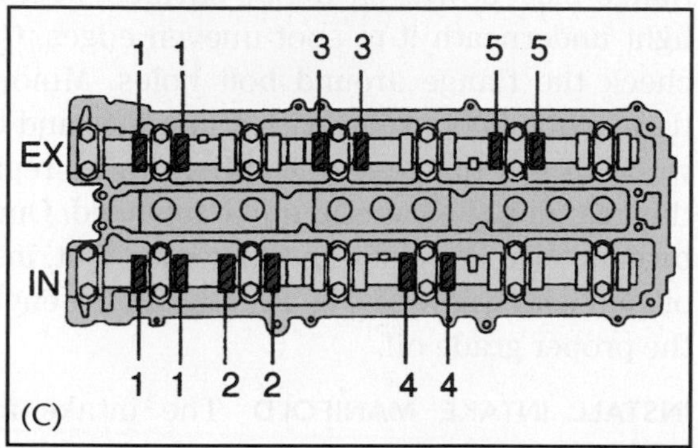

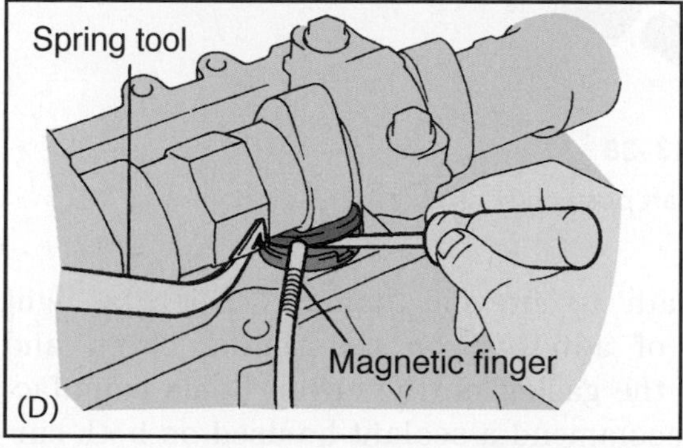

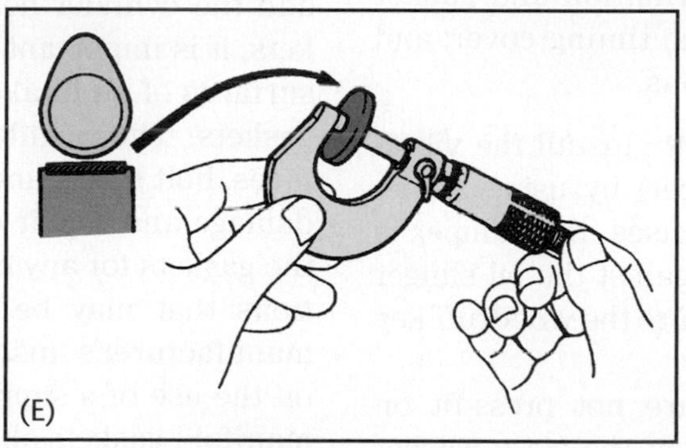

FIGURE 13–37

To adjust valve lash: (A) Rotate the crankshaft so that it is at TDC; (B) check the camshaft alignment marks; (C) measure and record the valve lash of the valves that are totally closed; (D) use a small screwdriver and a magnetic finger to remove the adjusting shim; and (E) measure the thickness of the shim with a micrometer to determine the correct shim to install.

Reprinted with permission

Final Reassembly Steps

The final steps in engine reassembly involve installing various engine covers and installing manifolds and related items that mount directly to the engine assembly.

COOLANT DRAINS AND PLUGS Make sure all coolant drains and plugs are installed in the block. Drains are normally threaded into the block with a thread sealant. This is also true for threaded plugs.

TIMING SENSORS Proper installation of the crankshaft and camshaft timing sensors is critical. Make sure to coat new O-rings with clean oil before they are installed **(Figure 13–38)**. Some sensors have a specified gap that must be set during installation. Also make sure that the trigger wheel for each sensor is properly aligned. Refer to the service information.

INSTALL THE TIMING COVER When replacing the timing cover, remove the old gaskets and seals from the timing cover and engine block.

Install the new seal using a press, seal driver, or hammer and a clean block of wood. When installing the seal, be sure to support the cover

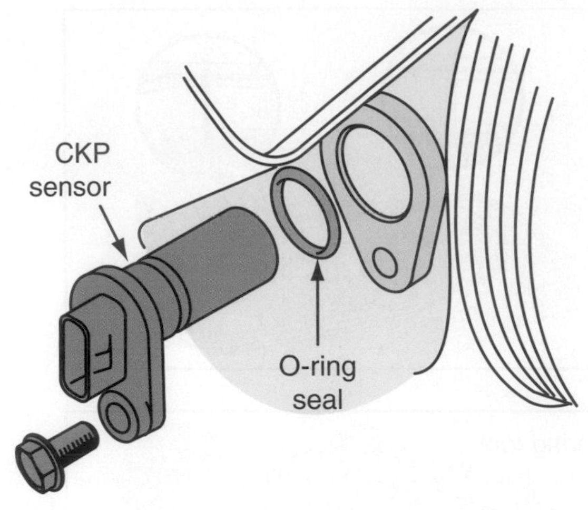

FIGURE 13–38

A crankshaft position sensor.

underneath to prevent damage. Apply a light coating of adhesive on the timing cover, and position the gasket on the cover. Some manufacturers recommend a sealant be used on both surfaces. Check the service information and gasket instructions. Finally, mount the timing cover, and torque the bolts to specifications.

INSTALL THE VIBRATION DAMPER Install the vibration damper (harmonic balancer) by using a special installation tool. In most cases, the damper is installed until it bottoms out against the oil slinger and the timing sprocket. Make sure the woodruff key is in place.

Some vibration dampers are not press-fit on the crankshaft. Be sure to install the large washer behind the damper-retaining bolt on these engines. Otherwise, the damper might fly off, causing damage and a safety hazard.

INSTALL THE VALVE COVER To install the valve cover, first make sure the cover's sealing flange is flat, and then apply contact adhesive to the valve cover's sealing surfaces in small dabs. Mount the valve cover gasket on the valve cover, and align it in position. If the gasket has mounting tabs, use them in tandem with the contact adhesive. Allow the adhesive to dry completely before mounting the valve cover on the cylinder head. Torque the mounting bolts to specifications.

SHOP TALK

Many technicians install the water pump at this point or wait until the engine is back in the vehicle. When installing a water pump, apply a coating of good waterproof sealer to a new gasket, and place it in position on the water pump. Coat the other side of the gasket with sealer, and position the pump against the engine block until it is properly seated. Install the mounting bolts and tighten them with a torque wrench in a staggered sequence to specifications. Careless tightening could cause the pump housing to crack. After tightening, check the pump to make sure it rotates freely.

INSTALL OIL PAN Before installing the oil pan gasket, check the flanges of a steel oil pan for warpage. Use a straightedge **(Figure 13–39)** or lay the pan, flange side down, on a flat surface with a flashlight underneath it to spot uneven edges. Carefully check the flange around bolt holes. Minor distortions can be corrected with a hammer and block of wood. If the flanges are too bent to be repaired in this manner, the pan should be replaced. Once it has been determined that the flanges are flat, install the oil pan with a new gasket. Finally, fill the engine with the proper grade oil.

INSTALL INTAKE MANIFOLD The intake manifold gasket seals the joint between the intake manifold and the cylinder head. As with cylinder head gaskets, it is important to properly prepare the sealing surfaces of an intake manifold before installing the gaskets. Thoroughly clean all of the sealing surfaces, bolt holes, and bolts. Inspect the surfaces for damage and repair or replace as necessary. Check the gaskets for any markings or installation instructions that may be stamped on them. Check the manufacturer's instructions for recommendations on the use of a supplementary sealant. Most intake manifold gaskets should be coated with a nonhardening sealer.

FIGURE 13–39

Checking the flatness of an oil pan flange.

Corteco

When installing an intake manifold, it is wise to use guide bolts. These guides make sure the gaskets and the manifold are perfectly aligned before tightening them in place. They also aid in setting the sealer by preventing the manifold from shifting and rupturing the sealant. When tightening the bolts or nuts, make sure you tighten them to the proper torque and in the order specified by the manufacturer.

On a V6 or V8 engine, there may be rubber or cork-rubber end seals for the front and the rear of the manifold. Before installing these seals, thoroughly clean all oil from the mating surfaces. Apply adhesive to the surface to hold the seals in place during installation. Once the intake manifold gaskets and seals are in place, apply a small bead (approximately 3 mm [⅛ in.]) of silicone RTV to the point where the seals meet the gasket.

Some V6 and V8 engines have a large one-piece combination intake manifold gasket and manifold splash guard. These are installed with the same care as other intake manifold gaskets.

Some intake gaskets purposely block off coolant passages to enable the engine coolant to flow in a predetermined path through the engine. If these ports are not blocked off, the coolant will not flow properly, and the engine can overheat.

INSTALL THE THERMOSTAT AND WATER OUTLET HOUSING Install the thermostat with the temperature sensor facing into the block. If the thermostat is installed upside down, the engine will overheat.

INSTALL EXHAUST MANIFOLD When installing the exhaust manifold(s), tighten the bolts in the centre of the manifold first to prevent cracking it. If there are dowel holes in the exhaust manifold that align with dowels in the cylinder head, make sure these holes are larger than the dowels. If the dowels do not have enough clearance because of the buildup of foreign material, the manifold will not be able to expand properly, and it may crack.

INSTALL FLYWHEEL OR FLEX PLATE Reinstall the engine sling. Raise the engine into the air on a suitable lift, and remove the engine stand mounting head. Set the assembled engine on the floor, and support it with blocks of wood while attaching the flywheel or flex plate. Be sure to use the right flywheel bolts and lock washers, and make certain the flywheel is in the correct position. These bolts have very thin heads, and the lock washers are thin. If normal bolts or washers are used, they may cause interference with the clutch disc or the torque converter. Make sure the bolts are properly torqued.

> **CAUTION!**
>
> On Hondas and some other hybrids, extreme care must be taken when installing the rotor to the rear of the engine. The rotor assembly has very strong magnets. Anyone who has a pacemaker or other magnetically sensitive medical device should not handle the rotor assembly. The rotor should only be installed with the proper tools **(Figure 13–40)**, and not with your hands. During installation, the rotor may be suddenly drawn toward the stator with great force, which can cause serious hand or finger injury.

INSTALL CLUTCH PARTS If the vehicle has a manual transmission, install the clutch **(Figure 13–41)**. Make sure the transmission's pilot bushing or bearing is in place in the rear of the crankshaft and that it is in good condition.

Using a clutch-aligning tool or an old transmission input shaft, align the clutch disc. Then tighten the disc and clutch pressure plate to the flywheel. Make sure the disc is installed in the right direction. There should be a marking on it that says "flywheel side." Then install the bell housing if it was removed from the transmission.

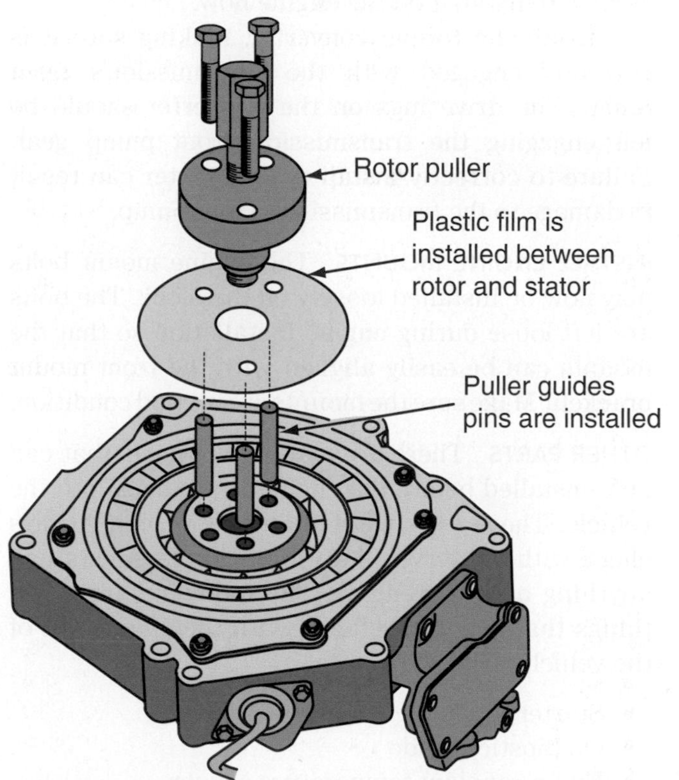
Rotor puller
Plastic film is installed between rotor and stator
Puller guides pins are installed

FIGURE 13–40

The special tools required to remove and install the rotor in a Honda hybrid.

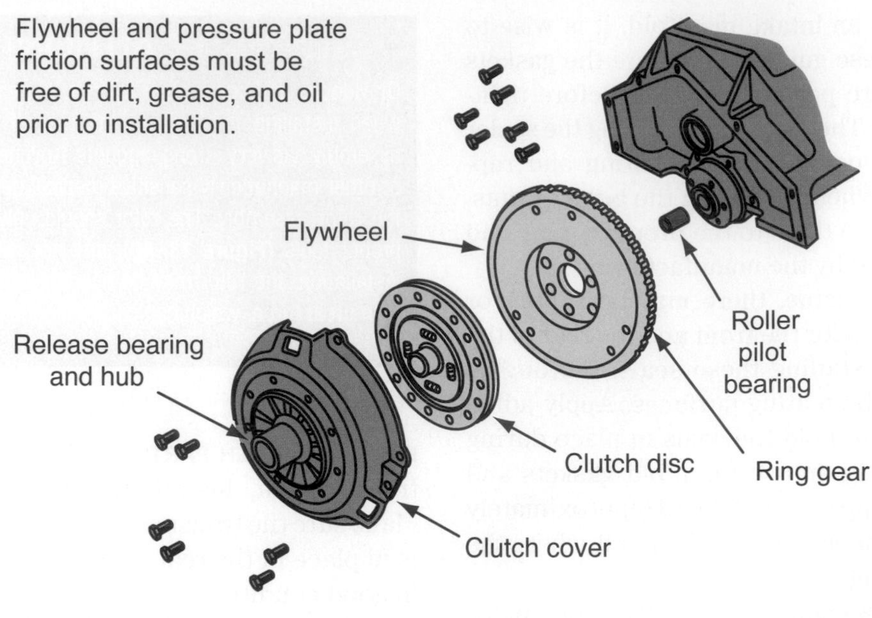

FIGURE 13–41

A typical clutch assembly.

INSTALL TORQUE CONVERTER On cars equipped with automatic transmissions, it is a good practice to replace the transmission's front pump seal. If the transmission was removed from the car with the engine, reinstall it on the engine now.

Install the torque converter, making sure it is correctly engaged with the transmission's front pump. The drive lugs on the converter should be felt engaging the transmission front pump gear. Failure to correctly install the converter can result in damage to the transmission's front pump.

INSTALL ENGINE MOUNTS The engine mount bolts may now be installed loosely on the block. The bolts are left loose during engine installation so that the mounts can be easily aligned with the front mount brackets. Make sure the mounts are in good condition.

OTHER PARTS There are many other parts that can be reinstalled before the engine is put back into the vehicle. These vary with the model of car. Always check with the service information before installing anything onto the engine. The following are a few things that may be installed with the engine out of the vehicle:

- Starter
- Oil dipstick guide
- Engine coolant temperature sensor
- Engine oil pressure switch
- Engine oil level sensor
- Knock sensor
- Fuel injectors
- Fuel rail
- PCV valve
- Camshaft timing control valve
- Drive belt tensioner
- A/C compressor
- Generator
- Drive belts
- Ignition coils
- Spark plugs

INSTALLING THE ENGINE

Installing a computer-controlled engine can be a complex task requiring special procedures. Referring to the vehicle's service information is absolutely essential for this procedure. Typically, the procedure is the reverse of the removal procedure.

Installing an Engine into an FWD Vehicle

If the engine will be installed through the top, connect the engine to a sling, and then connect the sling to the crane **(Figure 13–42)**. Slowly lower the engine into the engine compartment. Guide the engine around all wires and hoses to make sure nothing gets damaged. As the engine approaches its position in the engine compartment, align the engine and transmission mounts. Then

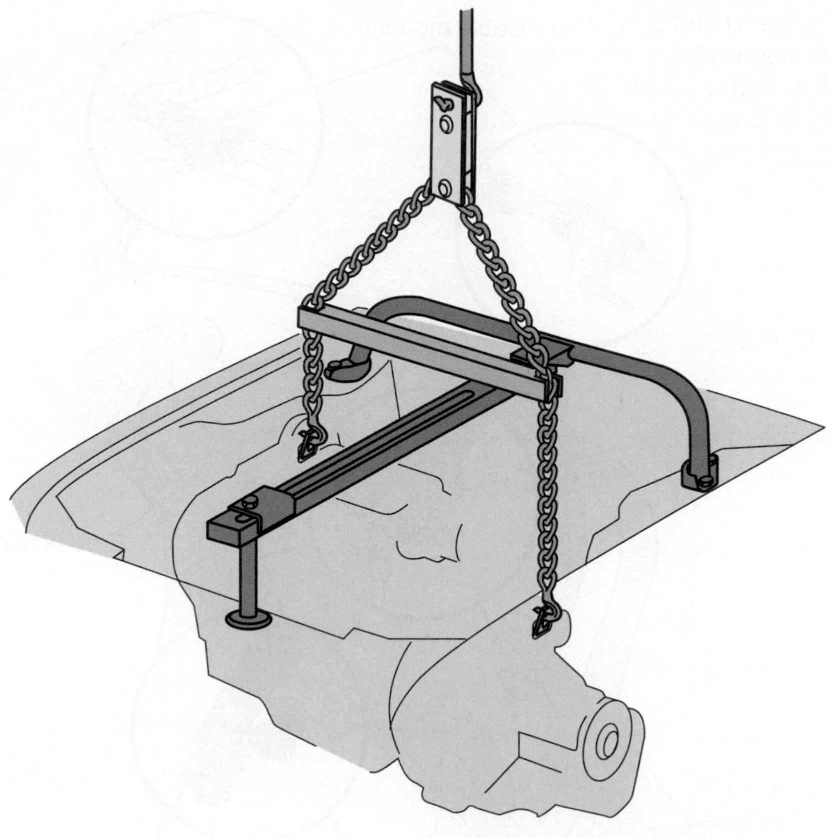

FIGURE 13–42

Equipment needed to install an engine from the top of an FWD vehicle.

lower the engine so that you can install the bolts into the mounts. Now raise the vehicle to a good working height.

If the engine will be installed from under the car, install the engine onto the engine cradle and dolly. Lift the vehicle on a hoist or lift, and position the engine under the vehicle. Slowly lower the vehicle over the engine while guiding all wires and hoses out of the way. As the vehicle gets close to the engine, align the engine and transmission mounts. Once the mounts are aligned, lower the vehicle so that you can install the bolts into the mounts. Now raise the vehicle to a good working height.

While working under the vehicle, install the axle shafts. Then install the remaining engine and transaxle mounts and braces. Connect the exhaust manifold to the exhaust system. Install any heat shields that were removed when the engine was removed. Connect all linkages, lines, hoses, and electrical wiring to the transmission **(Figure 13–43)**. Now reconnect all suspension and steering parts that were disconnected or removed, and install the wheels and tires.

SHOP TALK

Many FWD vehicles require the powertrain cradle to be properly aligned after it is reconnected. This ensures correct steering and suspension alignment. After engine installation is completed, the vehicle's wheel alignment may need to be checked and adjusted.

Lower the vehicle and remove the engine support bar or sling. Install all splash shields that were removed, and then install the tire/wheel assemblies.

Lower the vehicle to the ground, and tighten the axle hub nuts. Connect any disconnected fuel lines and heating system hoses. Now connect the engine ground strap and all electrical connectors and wires. Connect all vacuum hoses. Then connect the throttle linkage and adjust it if necessary. Install the radiator, cooling fan(s), and overflow tank. Install the upper and lower radiator hoses. Then install the heater hoses. Hybrid vehicles have coolant hoses at the inverter; make sure these are properly tightened.

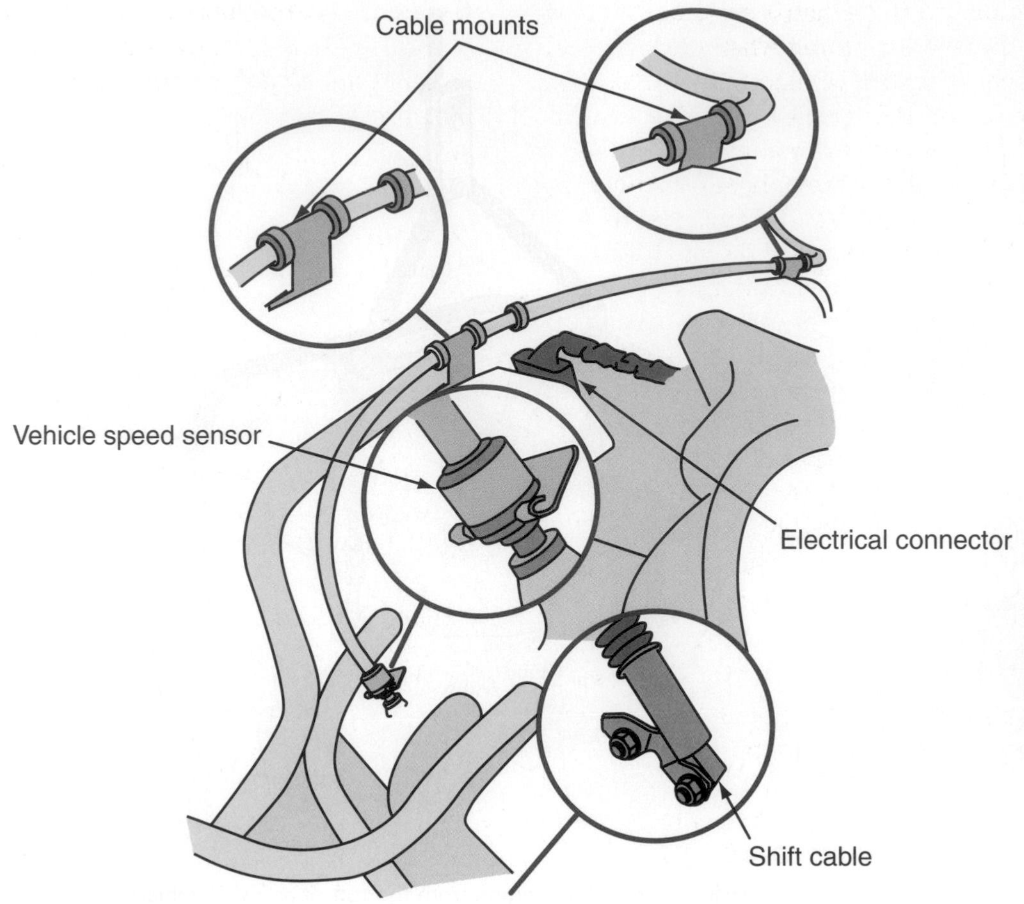

FIGURE 13–43

Various connections for a transaxle.

Install the automatic transmission fluid (ATF) cooler hoses to the transmission. Now install the A/C compressor with the drive belt, condenser fan shroud, and the electrical connectors for the fan motor and compressor clutch. Then reinstall the engine compartment support strut. Now install the air induction system, and connect any remaining items, including the battery cables. Check the engine for fuel leaks. Do this by turning the ignition switch to the ON position and allowing the fuel pump to run for a few seconds. Turn the power off, and check for signs of leaks. If there are any leaks, repair them before proceeding.

Make sure everything that was removed when the engine was pulled is reinstalled and secure. If the vehicle's hood was removed, carefully reattach it. Refill the radiator with coolant, and bleed air from the system with the heater valve open. Visually check for leaks. Prelubricate or prime the engine, and make sure the oil level is correct. On hybrid vehicles, the high-voltage system should be activated after everything is connected.

Installing an Engine in an RWD Vehicle

Connect the engine to a sling, and then connect the sling to the crane. Place a transmission jack under the transmission to hold it in position. Now slowly lower the engine into the engine compartment. Guide the engine around all wires and hoses to make sure nothing gets damaged. As the engine approaches its position in the engine compartment, align the engine to the input shaft of the transmission or the torque converter hub into the front pump. Carefully wiggle the engine until the input shaft slides through the clutch disc splines and the engine seats tightly against the transmission. Install and tighten the transmission to engine bolts. Start the engine mount bolts into their bores; you may need to wiggle the engine some to do this. Once the mount bolts are in place, tighten them, and remove the transmission jack and engine sling.

Raise the vehicle to a good working height, and install all remaining engine, torque converter, and transmission mounts and fasteners. Connect the exhaust manifold to the exhaust system. Install any heat shields that were removed when the engine was removed. Reconnect the fuel line from the fuel tank to the engine. Lower the vehicle so that you can work under the hood. Connect any remaining disconnected fuel lines and heating system hoses. Connect the engine ground strap. Connect all electrical connectors and wires. Connect all vacuum and other hoses. Now connect the throttle linkage, and adjust it if necessary.

Install the radiator and the cooling fan(s), and connect the rest of the hoses for the cooling system. Now install the air induction system, and connect any remaining items, including the battery cables.

Fill the radiator with coolant, and visually check for leaks. If the engine does not have oil in it already, add the specified amount of the proper type of oil. Prime the oil pump of the engine, and prepare the engine for startup. Install and align the hood.

Prelubrication

Not prelubing a new or rebuilt engine before starting it can cause premature bearing failure due to poor lubrication. Other parts, such as pistons, rings, and cylinder walls, need immediate lubrication to prevent scuffing, scoring, and damage. It can take as long as five minutes after the engine has started before oil is distributed through all of the vital parts of an engine. It is claimed that more than 80 percent of all engine wear occurs when an engine is first started.

These problems can be prevented by lubricating the parts as they are assembled and by forcing oil into the oil galleries. This is the purpose of prelubing. A **prelubricator (Figure 13–44)** forces oil throughout the engine before it is started. There are several ways to prelubricate, or prime, an engine. One of the most common ways is to use an air-operated prelubricator. The procedure for using this type of prelubricator follows:

FIGURE 13–44

A prelubricator.

1. Fill the oil filter with clean engine oil, and install it.
2. If the oil pressure sensor is installed, remove it.
3. Install an appropriately sized fitting for the preluber in the sensor's bore.
4. Connect the preluber hose to the fitting.
5. Open the valve on the preluber.
6. Fill the container for the preluber with at least two litres of clean oil.
7. Pump the handle on the preluber until all of the oil in it has moved into the engine.
8. Close the valve, and disconnect the hose from the fitting.
9. Remove the fitting, and install the oil pressure sensor. Be sure to tighten it to specifications.
10. Check the oil level in the engine, and add oil as needed.

SHOP TALK

While prelubing an engine, make sure there is a continuous flow of the correct type of oil. If the preluber runs out of oil during priming, an air pocket can form within the engine's lubrication system.

DISTRIBUTOR-DRIVEN OIL PUMPS On engines with a distributor-driven oil pump, the engine can be primed by driving the oil pump with an electric drill. A fabricated oil pump driveshaft is chucked in an electric drill motor and inserted through the distributor bore into the drive on the oil pump. Take the valve covers off, but loosely set them over the valves to control oil splash. Drive the oil pump with the electric drill. After running the oil pump for several minutes, remove the valve cover, and check for oil flow to the rocker arms. If oil reached the cylinder head, the engine's lubrication system is full of oil and is operating properly. If no oil reached the cylinder head, there is a problem with the pump, with an alignment of an oil hole in a bearing, or with a plugged gallery.

Starting Procedure

On engines with an ignition distributor, set the ignition timing as accurately as possible before starting the engine. The engine can be fine-tuned by using an engine analyzer or diagnostic tester after it has been started and goes through the **break-in** test.

Fill the gasoline tank with several litres of fresh gasoline. At this point, the fuel system should be primed with fuel, if possible, to ensure a prompt engine start and reduce the cranking period to a minimum. Extensive cranking of a newly rebuilt engine can damage a tappet-style camshaft, because its lobes are splash lubricated for the most part. This does not apply to roller types of lifters and camshafts. When the engine gets fuel, it will try to start. Once it does start, set the throttle to an engine speed of approximately 1500 rpm. When the engine coolant reaches normal operating temperature, turn off the engine. Look for signs of fuel, coolant, or oil leaks. After these checks, run the engine at 1200 to 1500 rpm during the warm-up period to ensure adequate initial lubrication for the piston rings, pistons, and camshaft.

Break-In Procedure

To prevent engine damage after it has been rebuilt or completely overhauled and to ensure good initial oil control and long engine life, the proper break-in procedure must be followed. Make a test run at 50 km/h (30 mph), and accelerate at full throttle to 80 km/h (50 mph). Repeat the acceleration cycle from 50 to 80 km/h (30 to 50 mph) at least 10 times. No further break-in is necessary. If traffic conditions will not permit this procedure, accelerate the engine rapidly several times through the intermediate gears during the road test. The objective is to apply a load to the engine for short periods of time and in rapid succession soon after the engine warms up. This action thrusts the piston rings against the cylinder wall with increased force and results in accelerated ring seating.

Relearn Procedures

The computer in most late-model vehicles must undergo a relearn procedure after the engine has been rebuilt. This procedure allows the computer to learn the condition of the engine and make adjustments according to the engine's restored condition. The last time the engine was run, the computer made adjustments based on the engine faults present. The relearn procedure teaches the computer that those faults were corrected. Initialization also resets the reference for the crankshaft position sensor and PCM. Always follow the manufacturer's procedures as outlined in the service information.

Initialization is also necessary to reset the operating parameters of many accessories, such as the power windows and the antitheft system. Devices such as the clock and radio will need to be manually reset.

CUSTOMER CARE

After the engine has been totally checked over, return it to the owner with the following instructions:

1. Drive the vehicle normally but avoid sustained high speed during the first 800 km (break-in period).
2. Avoid periods of extended engine idling.
3. Check the oil level frequently during the break-in period. It is not unusual to use one or two litres of oil during this time.
4. The oil and oil filter will need to be changed at the end of the break-in time.
5. The cylinder head and intake manifold bolts may need to be retorqued.
6. Some adjustments, such as valve adjustments and ignition timing, will also need to be checked.

KEY TERMS

Adhesives (p. 380)
Aerobic (p. 380)
Anaerobic (p. 380)
Antiseize compounds (p. 383)
Break-in (p. 395)
Elasticity (p. 370)
Gaskets (p. 374)
Graphite (p. 377)
Multilayer steel (MLS) (p. 378)
Polyacrylate (p. 383)
Prelubricator (p. 395)
Room temperature vulcanizing (RTV) (p. 381)
Sealants (p. 381)
Silicone gaskets (p. 381)
Torque-to-yield (TTY) (p. 373)
Viton (p. 383)
Yield (p. 370)

SUMMARY

- Elasticity means a bolt can be stretched a certain amount and, when the stretching load is reduced, return to its original size. Yield means a stretched bolt takes a permanent set and never returns to normal. Proper use of torque will prevent a yield condition.
- Gaskets serve as sealers, spacers, wear insulators, and vibration dampeners. Engine gaskets are generally classified as either hard or soft.
- General recommendations for installing gaskets include the following: Never reuse old ones; handle new ones carefully, especially the composition-type; use sealants properly; thoroughly clean all mating surfaces; and use the right gasket in the right position.

- Cylinder head gaskets on today's bimetal engines have a demanding job. The no-retorque head gasket retains a high level of clamping force.
- Contact adhesive bonds cork, rubber, fibre, and metal gaskets in place. It does not aid in sealing. Its only purpose is to hold the gasket securely during component assembly.
- General-purpose sealers aid in gasket sealing without upsetting the designed performance of most mechanical gaskets. Hard-drying sealants, such as shellac, should never be used on gaskets because they will make future disassembly very difficult.
- Flexible sealant is often used on bolt threads that go into fluid passages. Silicone gasketing, of which RTV is the best known, is used on oil pans, valve covers, thermostat housing, timing covers, and water pumps. Anaerobic formed-in-place sealants are used for both thread locking and gasketing.
- Oil seals keep oil and other vital fluids from escaping around a rotating shaft. Oil seals should always be replaced during engine rebuilding to ensure against costly do-overs.
- All engines with mechanical lifters have some method of adjustment to bring valve lash (clearance) back into specification. Rocker arms with adjustable pivots and adjustable pushrods are found on OHV engines. Rocker arms with adjustable screws and adjustable cam followers are part of OHC engines.
- The steps in final engine assembly involve installing various engine covers, prelubing the engine, and installing manifolds and related items that mount directly to the engine assembly. The best method of prelubricating an engine under pressure without running it is to use a prelubricator, which consists of an oil reservoir attached to a continuous air supply.
- A proper break-in procedure is necessary to ensure good initial oil contact and long engine life.

REVIEW QUESTIONS

1. What does it mean when a bolt is stretched into yield?
2. Name some applications of hard gaskets.
3. Where are flexible sealers most often used?
4. What are the major differences between aerobic and anaerobic sealants?
5. What is the safety factor calculated into bolt torque values?
   a. 20 percent below the bolt's yield point
   b. 25 percent below the bolt's yield point
   c. 30 percent below the bolt's yield point
   d. 35 percent below the bolt's yield point
6. What is the preferred procedure for torquing cylinder head bolts?
   a. Use a standard bolt torque chart and torque dry.
   b. Use a standard bolt torque chart and torque with engine oil.
   c. Use a standard bolt torque chart and torque with a torquing compound.
   d. Follow the manufacturer's recommendations only.
7. Which of the following is the correct procedure for tightening torque-to-yield fasteners?
   a. Tighten the bolt to a specified torque and then turn it an additional number of degrees.
   b. Tighten the bolt a specified number of degrees and then torque the bolt to specifications.
   c. Tighten the bolt to a specified torque, turn it an additional number of degrees, and re-torque to the maximum specification.
   d. Tighten the bolt to a specified torque and then re-torque to the maximum specification.
8. What should be done to an engine block bolt hole that has been pulled up?
   a. File the bolt hole flat and reuse.
   b. Chamfer the hole opening and reuse.
   c. Clean the threads with the appropriate bottoming tap.
   d. File, chamfer, and clean the bolt hole before using.
9. Unless otherwise noted, what should be done to head bolts before installing?
   a. Coat the threads and the underside of the bolt head with engine oil.
   b. Coat only the threads with engine oil.
   c. Coat the threads and the underside of the bolt head with torquing compound.
   d. Coat only the threads with torquing compound.
10. What type of sealant would be best for sealing between machined surfaces of rigid castings?
    a. black general-purpose aerobic silicone sealant
    b. anaerobic sealants
    c. blue special-application aerobic silicone sealant
    d. red high-temperature aerobic silicone sealant

11. What is the correct procedure for installing the torque converter when replacing the engine?
    a. Bolt the torque converter to the engine first.
    b. Insert the torque converter over transmission turbine shaft.
    c. Rotate the torque converter until it engages the transmission oil pump.
    d. Bolt the torque converter to the transmission housing.
12. What is the correct procedure for starting a rebuilt engine?
    a. Start and let the engine operate at 1500 rpm until operating temperature is reached.
    b. Start and let engine idle until operating temperature is reached.
    c. Start and accelerate the engine from idle to maximum rpm until operating temperature is reached.
    d. Start and road test the vehicle until operating temperature is reached.
13. During the initial test run, what should be done to properly break-in a rebuilt engine?
    a. Repeat full-throttle accelerations from 80 to 120 km/h (50 to 75 mph).
    b. Repeat part-throttle accelerations from 80 to 120 km/h (50 to 75 mph).
    c. Repeat full-throttle accelerations from 50 to 80 km/h (30 to 50 mph).
    d. Repeat part-throttle accelerations from 50 to 80 km/h (30 to 50 mph).
14. What is the purpose of using a graphite coating on a cylinder head gasket?
    a. to allow the head to slide across the gasket as it expands
    b. to seal the head to the gasket
    c. to prevent the head from moving on the gasket
    d. to prevent the need to re-torque the cylinder head
15. Which of the following lifters require valve train adjustment?
    a. hydraulic flat base lifters
    b. hydraulic roller lifters
    c. mechanical flat base lifters
    d. hydraulic cam followers
16. What gaskets can RTV silicone sealant replace?
    a. oil pan gaskets
    b. exhaust manifold gaskets
    c. intake manifold gaskets
    d. cylinder head gaskets
17. What type of rear main bearing oil seal is used on current engines?
    a. two-piece wick
    b. synthetic rubber lip
    c. two-piece rope
    d. fibre packing
18. Why do engines with cast-iron engine blocks and aluminum cylinder head engines pose problems?
    a. The cylinder block will expand two to three times more than the aluminum head.
    b. Aluminum heads compress easily when heated.
    c. The aluminum head will expand two to three times more than the cylinder block.
    d. Both materials expand tremendously during operation.
19. What is best method of circulating oil for initial startup of a rebuilt engine?
    a. Start and let the engine idle until sufficient oil pressure is obtained.
    b. Crank the engine with the ignition disabled until sufficient oil pressure is obtained.
    c. Start and run the engine at 1500 rpm until sufficient oil pressure is obtained.
    d. Use a prelubricator to prime the engine's lubrication system.
20. Which one of the following timing belt installation statements is correct?
    a. Not all timing belt tensioners require adjustment.
    b. All camshaft timing marks must be aimed directly toward the crankshaft.
    c. For the cam to be correctly timed, the mark must face straight up.
    d. All crankshaft timing marks must be aimed directly toward the camshaft.

CHAPTER 14
Lubricating and Cooling Systems

LEARNING OUTCOMES

- Name and describe the components of a typical lubricating system.
- Inspect, service, and install an oil pump.
- Describe the purpose of a crankcase ventilation system.
- List and describe the major components of the cooling system.
- Describe the operation of the cooling system.
- Describe the function of the water pump, radiator, radiator cap, and thermostat in the cooling system.
- Test and service the cooling system.

The life of an engine depends largely on its lubricating and cooling systems. If an engine does not have a supply of oil or does not cool itself, the engine will quickly be destroyed.

LUBRICATION SYSTEM

An engine's lubricating system does several important things. The main components of a typical lubricating system **(Figure 14–1)** are described here.

Engine Oil

Engine oil is specially formulated to lubricate and cool engine parts. The moving parts of an engine are fed a constant supply of oil. Engine oil is stored in the oil pan or sump. The oil pump draws the oil from the sump and passes it through a filter where dirt is removed. The oil is then moved throughout the engine via oil passages or galleries. After circulating through the engine, the oil returns to the sump.

FIGURE 14–1

Direction of oil flow through this V-10 engine.

GO TO ▶ Chapter 8 for a detailed discussion of engine oil.

OIL PUMP The oil pump is the heart of the lubricating system. The oil pump pulls oil from the oil pan through a pickup tube. The part of the tube that is in the oil pan has a filter screen, which is submerged in the oil. The screen keeps large particles from entering into the oil pump. This screen should be cleaned any time the oil pan is removed. The **oil pump pickup** may also contain a bypass valve that allows oil to enter the pump if the screen becomes totally plugged.

The oil pump may be located in the oil pan or mounted at the front of the engine **(Figure 14–2)**. Its purpose is to supply oil to cool, clean, and lubricate the various moving parts in the engine. The pump is normally driven by the crankshaft and creates suction to draw oil from the oil pan through a strainer. The pump then forces the oil through the oil filter and to various passages throughout the engine. The oil then returns to the oil pan.

An oil pump does not create oil pressure; it merely moves oil from one place to another. Oil pumps are **positive displacement pumps**; that is, the amount of oil that leaves the pump is the same amount that enters it. Output volume is proportional

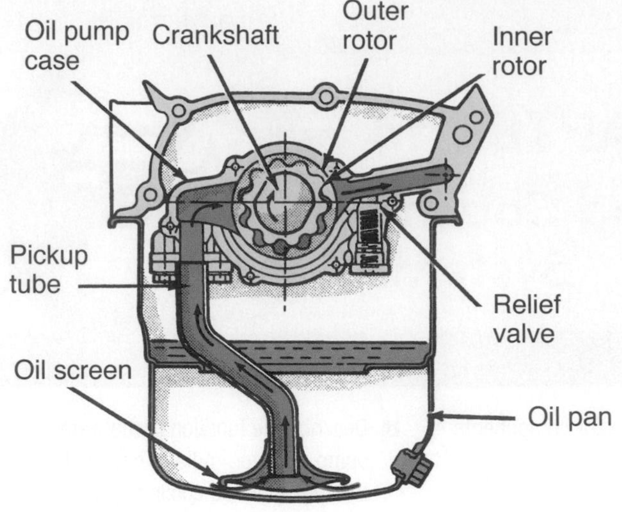

FIGURE 14–2

A crankshaft-driven rotor-type oil pump.

Chrysler Group LLC

to pump speed. As engine rpm increases, pump output also increases. As the oil leaves the pump, it passes through many passages. These passages restrict oil flow. These restrictions are what cause oil pressure. Small passages cause the pressure to increase; larger ones decrease the pressure.

This is why excessive bearing clearances will decrease oil pressure. The increased clearances reduce the resistance to oil flow and consequently increase the volume of oil circulating through the engine. This decreased resistance and increased volume lower the pressure of the oil. The ability of an oil pump to deliver more than the required volume of oil is a safety measure to ensure lubrication of vital parts as the engine wears.

Oil pressure is also determined by the viscosity and temperature of the oil. High-viscosity oil has more flow resistance than low-viscosity oil.

TYPES OF OIL PUMPS Oil pumps are driven by the camshaft or crankshaft. How the pump is driven is dictated by the location of the pump. Some oil pumps have an intermediate or driveshaft that is driven by a gear on the camshaft. Other pumps are driven by the crankshaft via a chain or gears.

Two basic types of oil pumps are used in today's engines. A **rotor-type oil pump** has an inner rotor and an outer rotor. The outer rotor is driven by the inner rotor. The outer rotor always has one more lobe than the inner rotor. When the rotors turn and the rotors' lobes unmesh, oil is drawn into that space. As the rotors continue to turn, oil becomes trapped between the lobes, cover plate, and top of the pump cavity. It is then forced out of the pump body by the meshing of the lobes. This squeezes the oil out and directs it through the engine. The amount of oil forced out of the pump depends on the diameter and thickness of the pump's rotors.

Gear-type oil pumps use a drive gear, connected to an input shaft, and a driven gear. Both gears trap oil between their teeth and the pump cavity wall. As the gears rotate, oil is forced out as the gear teeth unmesh. The output volume per revolution depends on the length and depth of the gear teeth. Another style of gear-type oil pump uses an idler gear with internal teeth that spins around the drive gear. In this style of pump, often called a crescent or trochoidal type, the gears are eccentric **(Figure 14–3)**. That is, as the larger gear turns, it walks around the smaller one, moving the oil in the space between.

The rotor type moves a greater volume of oil than the gear type because the space in the open lobe of the outer rotor is greater than the space between the teeth of the gears of a gear-type pump.

High-volume pumps are often installed by engine rebuilders. High-volume pumps have larger gears or rotors. The increase in oil volume is proportional to the increase in the size of the gears. Gears that are 20 percent larger will provide 20 percent more oil volume.

PRESSURE REGULATION Oil pumps have an oil **pressure relief valve** to prevent excessively high system pressures from occurring as engine speed increases. When the oil pressure exceeds a pre-set limit, the spring-loaded relief valve opens and allows oil to directly return to the sump. Excessive oil pressure

FIGURE 14–3

A gear-type oil pump.

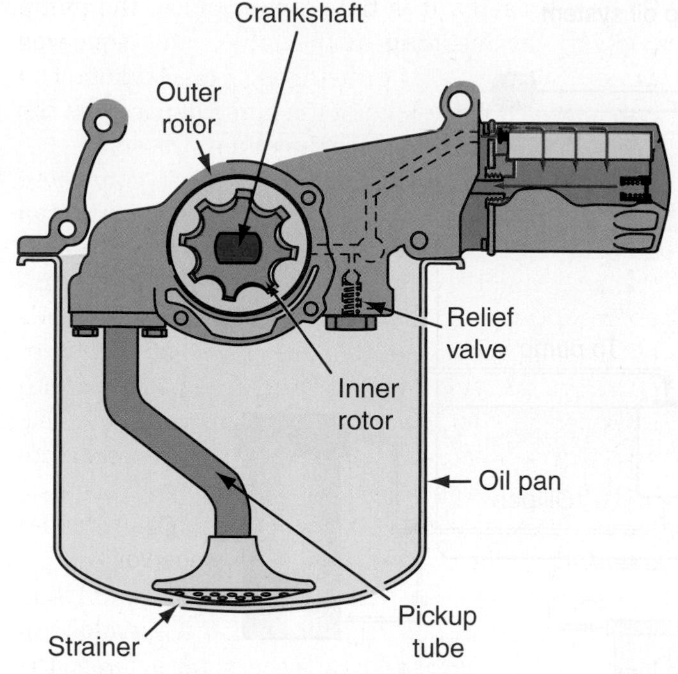

FIGURE 14–4

A rotor-type oil pump and lubrication system.

©Cengage Learning 2015

FIGURE 14–5

This oil pan was manufactured with a windage tray.

©Cengage Learning 2015

can lead to poor lubrication due to the oil blowing past parts rather than flowing over them. A pressure regulator valve is loaded with a calibrated spring that allows oil to bleed off at a given pressure **(Figure 14–4)**. When the pressure from the pump reaches a pre-set level, a check valve, ball, or plunger unseats and allows the oil to return to either the inlet side of the pump or to the crankcase.

OIL PAN OR SUMP The oil pan attaches to the crankcase or block. It serves as the reservoir for the engine's oil. It is designed to hold the amount of oil needed to lubricate the engine when it is running, plus a reserve. The oil pan helps to cool the oil through its contact with the outside air.

PAN BAFFLES In a **wet sump** system, the oil can slosh around during hard cornering or braking. During these times, it is possible for the oil to move away from the oil pump's pickup. This will cause a temporary halt in oil flow through the engine, which can destroy it. Sloshing also can affect the rotation of the crankshaft. As the crankshaft rotates through a thick puddle of oil, it meets resistance and slows down. To help prevent sloshing, many engines have baffles (windage trays) in the oil pan to limit the movement of the oil **(Figure 14–5)**.

DRY SUMP To eliminate the possibility of oil sloshing, some OEM engines are fitted with a **dry sump** oil system, as are most race engines. In a dry sump system, the oil pan does not store oil. The dry oil pan merely seals the bottom of the crankcase. The oil reservoir is a remote container set apart from the engine. Rather than having a single path for oil travel, dry sump pumps can feed oil directly to the crankshaft, valve train, and turbocharger. Normally, one external oil pump **(Figure 14–6)** is used; however, some systems have two pumps. The second pump pulls all the oil out of the sump and returns it to the reservoir. This pump also lowers the pressure inside the crankcase.

Dry sump systems provide immediate oil delivery to critical areas of the engine. They also prevent oil starvation caused by acceleration, braking, and cornering forces. Because the dry sump is smaller than a wet sump, the engine can be placed lower in the frame to improve overall handling.

OIL FILTER As the oil leaves the oil pump, it flows through an oil filter. The filter prevents the small particles of dirt and metal suspended in the oil from reaching the close-fitting engine parts. If the impurities are not filtered from the oil, the engine will wear prematurely and excessively. Filtering also increases the usable life of the oil. The filter assembly threads directly onto the main oil gallery tube. The oil from the pump enters the filter and passes through the element of the filter. From the element, the oil flows back into the engine's main oil gallery.

The oil filter is typically a disposable metal container filled with a special type of treated paper or other filter substance (cotton, felt, and the like). Some engines have a separate cartridge that fits into the block or a separate metal housing **(Figure 14–7)**.

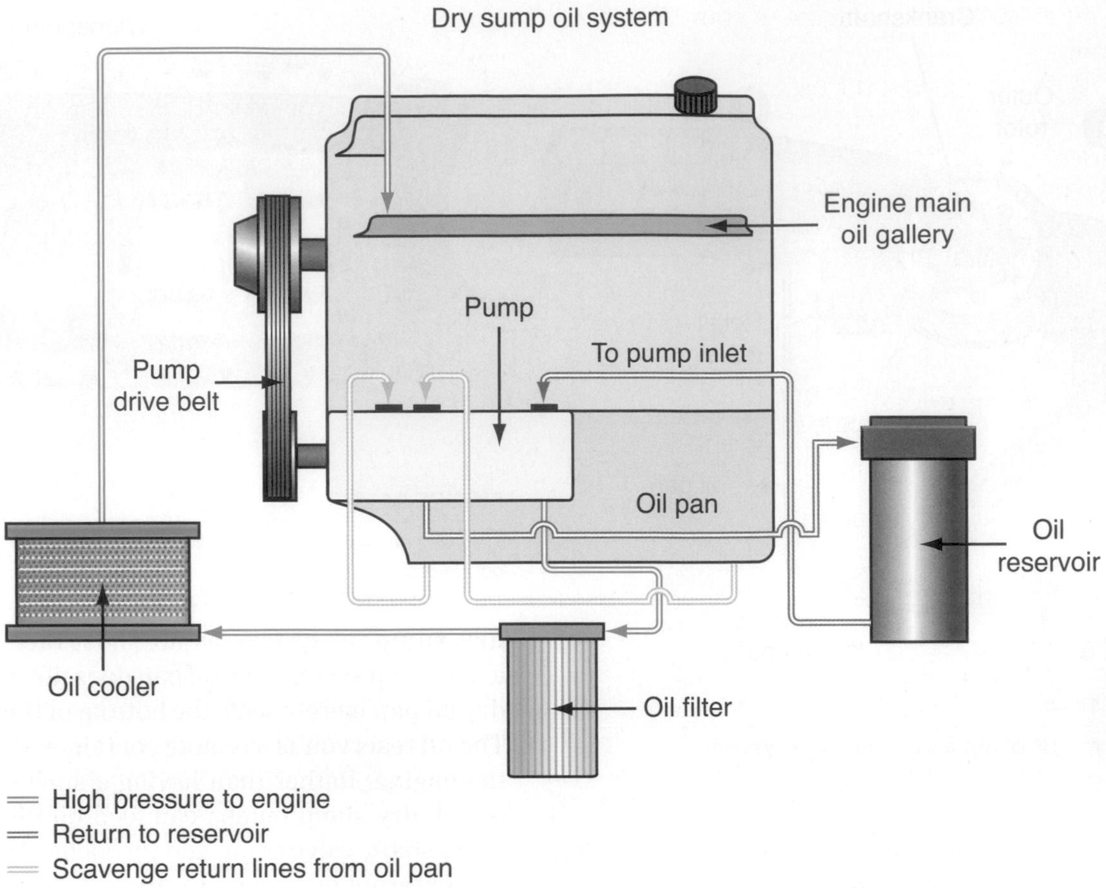

Dry sump oil system

Engine main oil gallery

Pump

To pump inlet

Pump drive belt

Oil pan

Oil reservoir

Oil cooler

Oil filter

High pressure to engine

Return to reservoir

Scavenge return lines from oil pan

FIGURE 14–6

A typical dry sump engine lubrication system.

This filter is mounted on and sealed to an adapter bolted to the block. However, it may be attached to the timing cover or remotely mounted with oil lines connecting the filter to the oil galleries in the block **(Figure 14–8)**.

FIGURE 14–7

An oil filter assembly with a replaceable element.

Martin Restoule

Oil filters may have an anti-drainback valve. This valve prevents oil drainage from the filter when the engine is not running. This allows for a supply of

FIGURE 14–8

Oil lines carry the engine oil in and out of this remote filter before it moves through the engine to lubricate parts.

BMW of North America, Inc.

filtered oil as soon as the engine is started and has oil pressure.

All of the oil going through the engine goes through the filter first. However, if the filter becomes plugged, a relief valve in the filter will open and allow oil to bypass and go directly to the engine's oil passages **(Figure 14–9)**. This provides the engine with necessary, though unfiltered, lubrication.

WARNING!

Used oil and oil filters must be disposed of properly and in accordance to local, provincial, and federal laws.

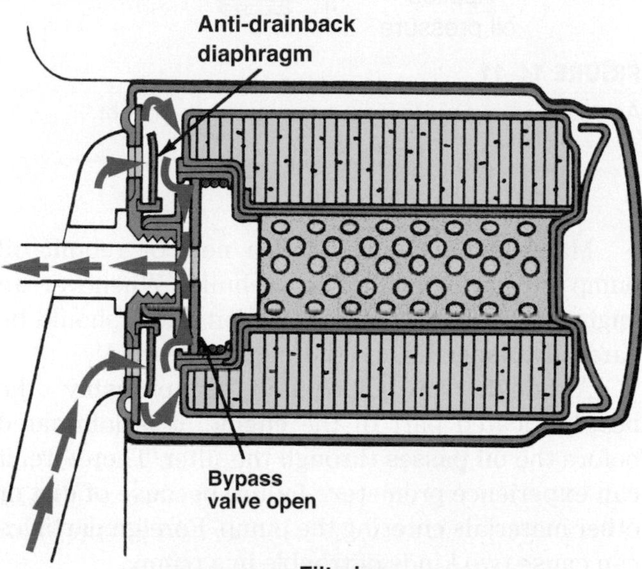

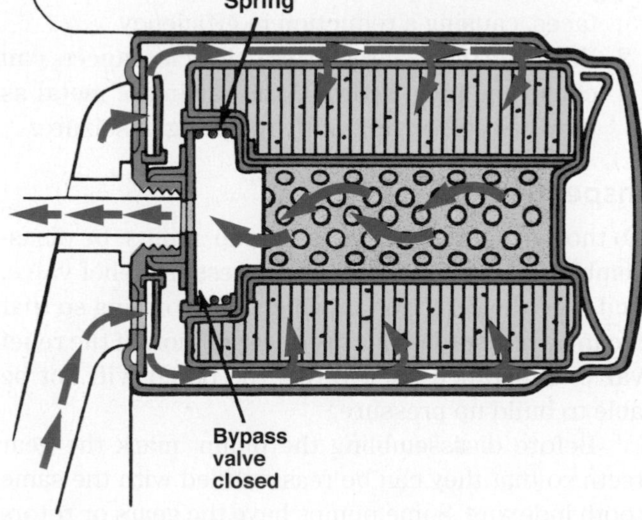

FIGURE 14–9

Oil flow through the filter.

Ford Motor Company

Oil Coolers

To control oil temperature, many diesel, high-performance, and super- or turbocharged engines have an external engine oil cooler. Hot oil mixed with oxygen breaks down (oxidizes) and forms carbon and varnish. The higher the temperature, the faster these deposits build. An oil cooler helps keep the oil at its normal operating temperature. Oil flows from the pump through the cooler and then to the engine. An oil cooler is a small radiator mounted near the front of the engine or within the radiator. Heat is removed from the oil as engine coolant flows around the cooler and air passes through it.

Engine Oil Passages or Galleries

From the filter, the oil flows into the engine's oil galleries. These galleries are interconnecting passages that were drilled into the block. The crankshaft also has oil passages (oilways) that route the oil from the main bearings to the connecting rod bearings. Engines with a remote oil filter, an oil cooler, or a dry sump system have external oil lines that move the oil to designated areas.

Dipstick

A dipstick is used to measure the level of oil in the oil pan. The end of the stick is marked to indicate where the oil level should be. Obviously, if the oil level is below that mark, oil needs to be added. Some late-model engines do not have a dipstick; instead, engine oil level is measured through the oil level sensor. The oil level is then displayed on the driver information centre.

Oil Pressure Indicator

All vehicles have an oil pressure gauge and/or a low-pressure indicator light. Oil gauges are either mechanically or electrically operated and display the actual oil pressure of the engine. The indicator light only warns the driver of low oil pressure.

In a mechanical gauge, oil travels up to the back of the gauge, where a flexible, hollow tube called a Bourdon tube uncoils as the pressure increases. A needle attached to the Bourdon tube moves over a scale to indicate the oil pressure.

Most pressure gauges are electrically controlled. An oil pressure sensor or sending unit is screwed into an oil gallery. As oil passes through an oil pressure sender **(Figure 14–10)**, it moves a diaphragm, which is connected to a variable resistor.

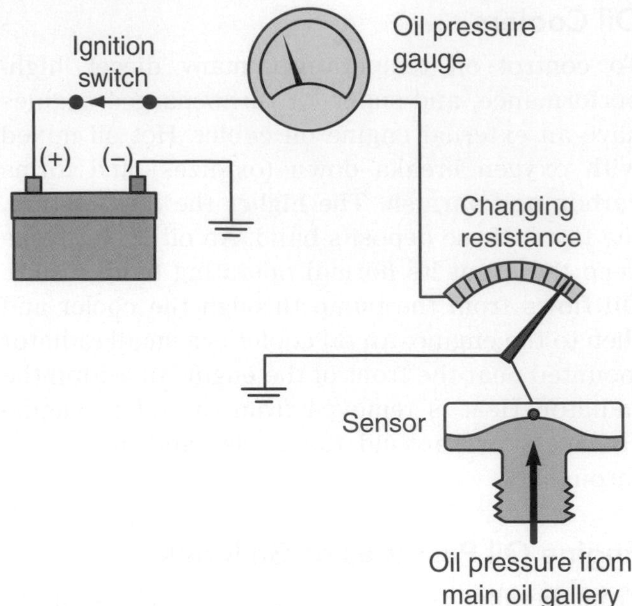

FIGURE 14–10

As oil pressure changes, the resistance in the oil pressure gauge circuit and the reading on the gauge change accordingly.

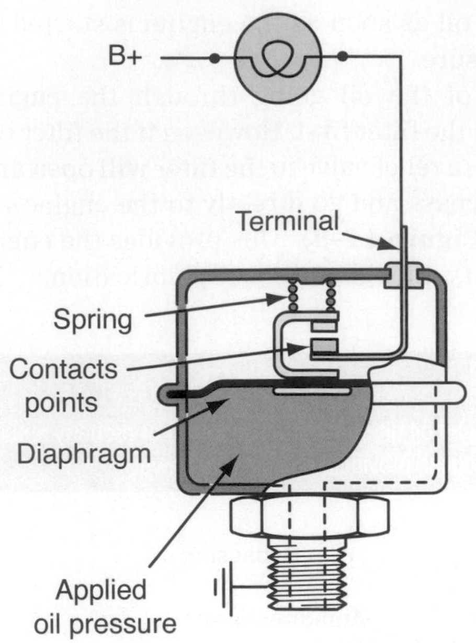

FIGURE 14–11

An oil pressure sensor for a warning switch or light.

This resistor changes the amount of current passing through the circuit. The gauge then reacts to the current and moves a needle over a scale to indicate the oil pressure, or the current is translated into a digital reading on the gauge.

Warning light systems are basically simple electrical circuits. The indicator light comes on when the circuit is completed by a sensor. This sensor has a diaphragm connected to a switch inside the sensor. Under normal conditions, the sender switch is open. When oil pressure falls below a certain level, the reduction of pressure causes the diaphragm to move and close the sender switch **(Figure 14–11)**, which completes the electrical circuit. When this happens, the warning light turns on.

GO TO Chapter 21 for further discussions on oil and other gauges.

OIL PUMP SERVICE

An oil pump is seldom the cause of lubrication problems. However, it can go bad. The best way to check an oil pump is to conduct an oil pressure test. This test will not only check the oil pump but also evaluate the rest of the lubrication system.

GO TO Chapter 9 for the correct procedure for conducting and interpreting an oil pressure test.

Many technicians install a new or rebuilt oil pump on each engine they rebuild. Whenever an engine is being overhauled, the oil pump should be carefully inspected and thoroughly cleaned.

Although the oil pump is probably the best-lubricated part of the engine, it is lubricated before the oil passes through the filter. Therefore, it can experience premature failure because of dirt or other materials entering the pump. Foreign particles can cause two kinds of trouble in a pump:

1. Fine abrasive particles gradually wear the surfaces, causing a reduction in efficiency.
2. Hard particles larger than the clearances can cause chipping, scoring, and raising of metal as they pass through, finally resulting in seizure.

Inspection

To thoroughly inspect the oil pump, it must be disassembled. Carefully remove the pressure relief valve, and note the direction in which it is pointing so that it can be reinstalled in its proper position. If the relief valve is installed backwards, the pump will not be able to build up pressure.

Before disassembling the pump, mark the gear teeth so that they can be reassembled with the same tooth indexing. Some pumps have the gears or rotors marked when they are manufactured. Once all the serviceable parts have been removed, clean them and dry them off with compressed air.

SHOP TALK

Use a paint stick or other nonviolent means to mark the gears, if possible. If a centre punch is used, make sure you file down the raised material around the indent before reassembling the pump. The raised material that results from making the indent may cause interference and wear if not removed.

USING SERVICE INFORMATION

Correct oil pump disassembly instructions are given in the oil pump unit of the engines section of a service information system.

After the pump has been disassembled and cleaned, inspect the pump gears or rotors for chipping, galling, pitting, or signs of abnormal wear. Galling is the transfer of material between mated and moving components. This can be due to the microscopic welding of the two parts together followed by the parts separating. Examine the housing bores for similar signs of wear. If any part of the housing is scored or noticeably worn, replace the pump as an entire assembly.

Check the mating surface of the pump cover for wear. If it is worn, scored, or grooved, replace the pump. Use a feeler gauge and straightedge to check its flatness **(Figure 14–12)**. There are specifications for the maximum and minimum acceptable feeler gauge thicknesses for the cover. If the cover is excessively worn, grooved, or scratched, it should be replaced.

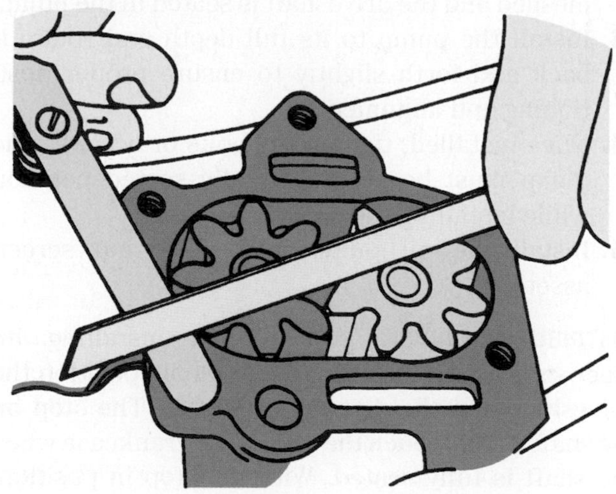

FIGURE 14–12

Use a straightedge and feeler gauge to check the flatness of the pump's sealing surface.

FIGURE 14–13

Check the clearance between the outside diameter of the outer gear and the pump body.

With gear-type pumps, reinstall the gears into the pump body. Use a feeler gauge to check the clearance between the outer gear and pump body **(Figure 14–13)**. If the housing-to-gear clearance exceeds specifications, replace the oil pump.

Use a micrometer to measure the diameter and thickness of the rotors **(Figure 14–14)**. If these dimensions are less than the specified amount, the rotors or the pump must be replaced.

With rotor pumps, reinstall the rotors into the pump body. Use a feeler gauge to check the clearance between the outer rotor and pump body. If the housing-to-rotor clearance exceeds specifications, replace the oil pump.

Position the inner and outer rotor lobes so that they face each other. Measure the clearance between

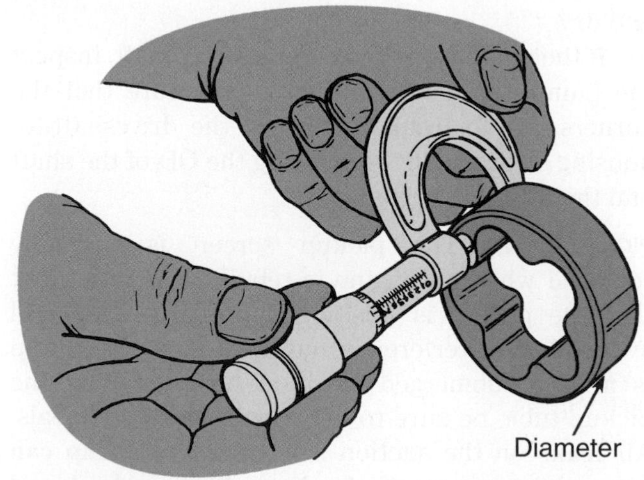

FIGURE 14–14

Measuring the outer rotor with an outside micrometer.

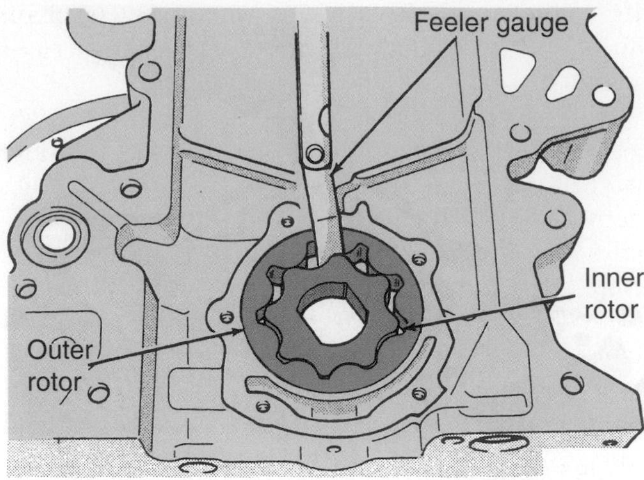

FIGURE 14–15

Measuring clearance between the inner and outer rotor lobes.
Chrysler Group LLC

them with a feeler gauge **(Figure 14–15)**. A clearance of more than 0.2540 mm (0.010 in.) is unacceptable, and the pump should be replaced. The timing case and gear thrust plate might also be worn. Excessive clearance can limit pump efficiency. Replace them as necessary.

Install the cover and tighten the bolts to specifications. The gasket used to seal the end housing is designed to provide the proper clearance between the gears and the end plate. Do not substitute another gasket or make a gasket to replace the original one. If a precut gasket was not originally used, seal the end housing with a thin bead of anaerobic sealing material. Turn the input shaft or gear by hand. It should rotate easily. If it does not, replace the pump.

Remove the old oil seal from the oil pump. Install a new seal into the pump housing. Make sure all mating surfaces of the pump are clean, undamaged, and dry.

If the pump uses a hexagonal driveshaft, inspect the pump drive and shaft to make sure that the corners are not rounded. Check the driveshaft-to-housing clearance by measuring the OD of the shaft and the ID of the drive.

PICKUP UNIT The pickup screen is normally replaced when an engine is rebuilt. It is important that the pickup is positioned properly. This will avoid oil pan interference and ensure that the pickup is always submerged in oil. When installing the pickup tube, be sure to use new gaskets and seals. Air leaks on the suction side of the oil pump can cause the pressure relief valve to hammer back and forth. Over time, this will cause the valve to fail. Air leaks can also cause oil aeration, foaming, marginal lubrication, and premature engine wear. Air leakage often comes from cracked seams in the pickup tube.

INSTALLING THE OIL PUMP

The pump should be primed before assembly. This can be done by submerging it in clean engine oil. Make sure the inlet port is fully submerged in the oil. Then turn the pump by hand until you see oil flow from the outlet of the pump. To install an older style, distributor-driven oil pump, do so in the following manner:

Crankshaft-Driven Pump

The installation of a typical crankshaft-driven oil pump is as follows:

1. Install a new oil seal into the pump.
2. Apply liquid gasket evenly to the pump's mating surfaces on the block, if so directed by the manufacturer.
3. Do not allow the gasket material to dry.
4. Coat the lip of the oil seal and the O-rings with oil.
5. Align the inner rotor with the crankshaft.
6. Install and tighten the oil pump.
7. Clean all excess grease on the snout of the crankshaft.
8. Install the oil pickup.

Cam-Driven Pumps

The installation of a typical camshaft-driven oil pump is as follows:

1. Apply a suitable sealant to the pump and block.
2. Make sure that the drive gears are properly meshed and the driveshaft is seated in the pump.
3. Install the pump to its full depth and rotate it back and forth slightly to ensure proper positioning and alignment.
4. Once installed, tighten the bolts or screws. The pump must be held in a fully seated position while installing the bolts or screws.
5. Install the oil pump inlet tube and screen assembly.

DISTRIBUTOR-DRIVEN PUMP When installing an older-style, distributor-driven oil pump, position the driveshaft into the distributor socket. The stop on the shaft should touch the roof of the crankcase when the shaft is fully seated. With the stop in position, insert the driveshaft into the oil pump. Install the pump and shaft as an assembly. Make sure that the driveshaft is seated in the pump drive. Then tighten the oil pump, attaching screws to specifications. Install the pump inlet tube and screen assembly.

BASIC LUBRICATION SYSTEM DIAGNOSIS AND SERVICE

Other than engine destruction, engine lubrication problems can cause other engine concerns, such as noise, exhaust smoke, and the need to add oil to the crankcase.

Oil Passages, Galleries, and Lines

All oil passages, galleries, and lines should be thoroughly cleaned and flushed during and after an engine has been overhauled.

Oil Consumption

Excessive oil consumption can result from external and internal leaks or worn piston rings, valve seals, or valve guides. Internal leaks allow oil to enter the combustion chamber where it is burned. Blue exhaust smoke is an indication that an engine is burning oil.

If the valve guides are worn or the valve seals are worn, cracked, or improperly installed, oil will be drawn into the cylinder during the intake stroke. If there are worn or broken piston rings or worn cylinder walls, the affected cylinders will have low compression. The oil in the cylinder also tends to foul the spark plugs, which will cause misfires, high emissions, and possible damage to the catalytic converter.

External leaks are a common cause of excessive oil consumption. These leaks can occur at the valve or cam cover gasket, oil filter, front and rear seals, oil pan gasket, and timing gear cover. Fresh oil on the clutch housing, oil pan **(Figure 14–16)**, edges of valve covers, external oil lines, or crankcase filler tube, or at the bottom of the timing gear or chain cover, usually indicates that the leak is close to or above that point.

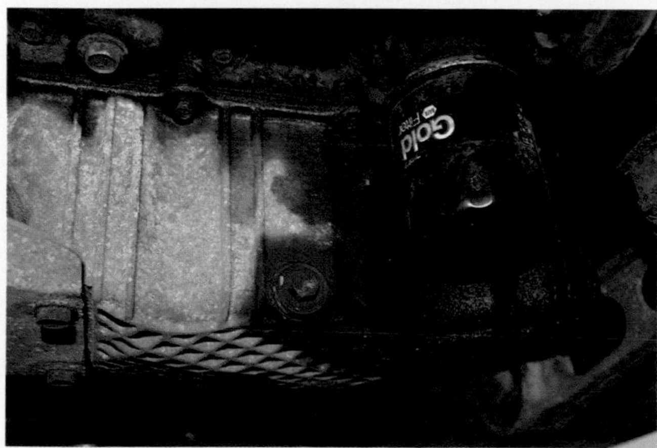

FIGURE 14–16

The presence of oil and dirt buildup around the oil pan indicates a leaky pan gasket.

Martin Restoule

When crankcase pressure is abnormally high, oil is forced out through joints that normally would not leak. Pressure develops when the positive crankcase ventilation (PCV) system is not working properly. **Blowby** is a term used for the gases that leak from the combustion chamber and enter the crankcase. Blowby gases are composed of pressurized intake gases and/or pressurized exhaust gases. The PCV system provides a continuous flow of fresh air through the crankcase to relieve the pressure and to prevent the formation of corrosive contaminants **(Figure 14–17)**.

PCV valves are designed for a particular engine's operating characteristics. Using the wrong valve can cause oil consumption as well as other problems. If the PCV valve or connecting hoses become clogged, excessive pressure will develop in the crankcase. The pressure can cause oil to leak past gaskets and seals. It might also force oil into the air cleaner or cause it to be drawn into the intake manifold.

OIL USAGE Even the smallest oil leak can cause excessive oil consumption. Losing three drops of oil every 30 m (100 feet) equals 3 litres (3 quarts) of oil lost every 1600 kilometres (1000 miles). Typically, new engines use less than a quart (0.946 L) of oil every 6000 miles (9656 km). As the engine

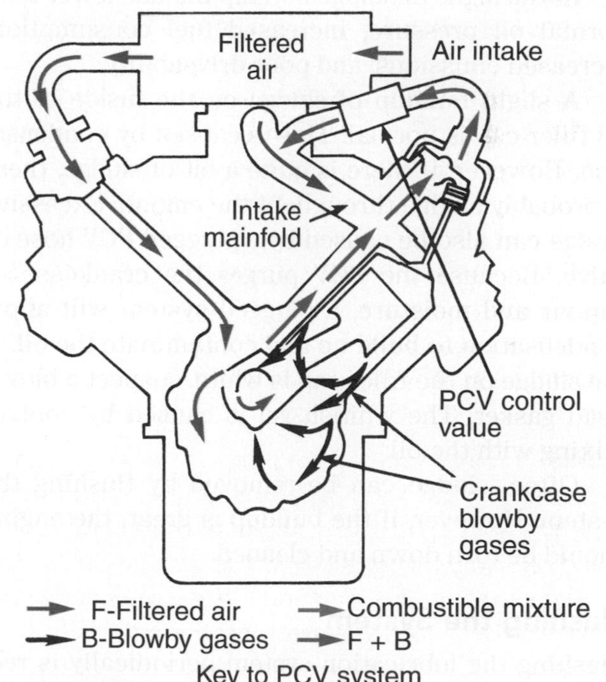

FIGURE 14–17

The operation of a PCV system.

FIGURE 14–18

Sludge buildup on the lower parts of an engine.

wears, its oil usage increases. It is not unusual for a high-mileage engine to use a litre (1 quart) of oil every 1600 kilometres (1000 miles).

SLUDGE A typical sign of poor maintenance is the buildup of yellow sludge inside the engine. **Sludge (Figure 14–18)** results from the oxidation of oil. When oil oxidizes, chemical compounds in the oil begin to break down and solidify, forming a gel substance. The gelled oil cannot circulate through the engine and collects on engine parts. This buildup of sludge can also block normal lubrication paths.

Initial signs of sludge buildup include lower than normal oil pressure, increased fuel consumption, increased emissions, and poor driveability.

A slight buildup of sludge on the inside of the oil filler cap is normal. This is caused by condensation. However, if there is quite a bit of sludge, there is probably sludge throughout the engine. Excessive sludge can also be caused by a plugged PCV hose or valve. Because the PCV purges the crankcase of vapour and moisture, a plugged system will allow condensation to build up and contaminate the oil. If the sludge on the filler cap is white, suspect a blown head gasket. The whitish gel is caused by coolant mixing with the oil.

Often, sludge can be removed by flushing the system. However, if the buildup is great, the engine should be torn down and cleaned.

Flushing the System

Flushing the lubrication system periodically is recommended by some manufacturers. However, there are others that do not recommend flushing the engine. Flushing involves running a solvent through the engine and then draining the system. The ways to do this vary, as do the solvents used. The concern of those that do not recommend flushing is simply that the solvents may loosen up some dirt or sludge that will not drain out with the oil. These contaminants can block passages and restrict oil flow.

Oil flushing solvents can be added to the engine's oil before an oil change. The engine is run for about a half-hour and then the oil and filter are changed. Flushing machines connect to the filter housing and the drain plug port after the old oil has been drained. A heated solution is pumped through the oil reservoir, passages, and oil pump, and up into the valve train. The solvent back flushes the oil pump and pickup screen and breaks up and dissolves the sludge. The remains are drawn out by a vacuum. After flushing, a new oil filter and clean oil is put into the engine.

Oil Cooler

If the engine has an external oil cooler, it and its lines should be checked for leaks. If leaks are evident, replace the lines and/or cooler. The cooler assembly should be flushed or replaced whenever sludge buildup is found in the engine. If the engine was rebuilt, the cooler should be replaced and the lines cleaned. Metal debris trapped in the cooler may become dislodged when the engine is run and cause oil starvation.

COOLING SYSTEMS

Today's internal combustion engines generate a tremendous amount of heat. This heat is created when the air/fuel mixture is ignited and expands inside the engine combustion chamber. Metal temperatures around the combustion chamber can run as high as 538°C (1000°F). To prevent the overheating of cylinder walls, pistons, valves, and other engine parts, it is necessary to dispose of excess heat.

In a liquid-cooled system **(Figure 14–19)**, heat is removed from around the combustion chambers by a heat-absorbing liquid (coolant) circulating inside the engine. This liquid is pumped through the engine and, after absorbing the heat of combustion, flows into the radiator, where the heat is transferred to the atmosphere. The cooled liquid is then returned to the engine to repeat the cycle. These systems are designed to keep engine temperatures within an ideal range where they provide peak performance with the least possible exhaust emissions.

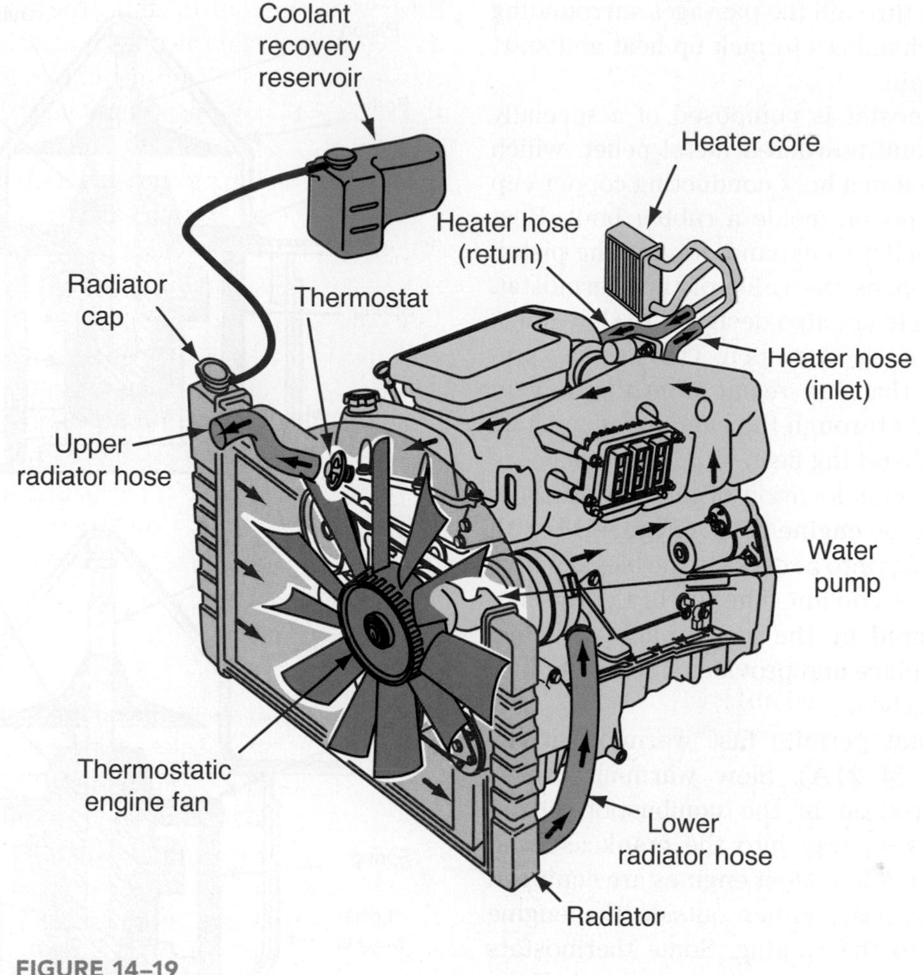

FIGURE 14–19

The major components of a liquid-cooled system. Arrows indicate the coolant flow.

Coolant

GO TO Chapter 8 for a detailed discussion of engine coolant.

Engine **coolant** is actually a mixture of water and antifreeze/coolant. Water alone has a boiling point of 100°C (212°F) and a freezing point of 0°C (32°F) at sea level. A mixture of 67 percent antifreeze and 33 percent water will raise the boiling point of the mixture to 113°C (235°F) and lower the freezing point to –69°C (–92°F). Antifreeze in excess of 67 percent will actually raise the freezing point of the mixture. The typical recommended mixture is a 50/50 solution of water and antifreeze/coolant. Some coolant suppliers offer a mixture of water and antifreeze that can be used to top off a cooling system when the level is low. It is popular in some northern locations to use a 60 percent antifreeze and 40 percent water mixture for the extra freeze protection.

Thermostat

The **thermostat** controls the minimum operating temperature of the engine. The maximum operating temperature is controlled by the amount of heat being produced by the engine at the time and the cooling system's ability to dissipate the heat.

The technical definition of a thermostat is a temperature-responsive control valve. The thermostat controls the temperature and amount of coolant entering the radiator. While the engine is cold, the thermostat remains closed, allowing coolant to circulate only inside the engine. This allows the engine to warm up uniformly and eliminates hot spots. When the coolant reaches the opening temperature of the thermostat, the thermostat begins to allow some flow of coolant to the radiator. The hotter the coolant gets, the more the thermostat opens, allowing more coolant to flow through the radiator. Once the coolant has passed through the radiator and has given up its heat, it re-enters the water pump. Here it is again pushed through the passages surrounding the combustion chambers to pick up heat and start the cycle once again.

Today's thermostat is composed of a specially formulated wax and powdered metal pellet, which is tightly contained in a heat-conducting copper cup equipped with a piston inside a rubber boot. Heat causes the wax pellet to expand, forcing the piston outward, which opens the valve of the thermostat. Today's thermostats are also designed to slow down coolant flow when they are open. This helps to prevent overheating that can result from the coolant moving too quickly through the engine, reducing its effectiveness in absorbing heat.

The most common location of the thermostat is at the front top of the engine block **(Figure 14–20)**. The heat element fits into a recess in the block, where it is exposed to hot coolant. The top of the thermostat is then covered by the water outlet housing, which holds it in place and provides a connection to the upper radiator hose.

The thermostat permits fast warm-up of the engine **(Figure 14–21A)**. Slow warm-up causes moisture condensation in the combustion chambers, which finds its way into the crankcase and causes sludge formation. Most engines are equipped with a coolant bypass, either outside the engine block or built into the casting. Some thermostats are equipped with a bypass valve that shuts off the engine bypass after warm-up, forcing all coolant to flow to the radiator.

Thermostats must start to open at a specified temperature **(Figure 14–21B)** normally 1.6°C (3°F) above or below its temperature rating. It must be fully opened at about 11°C (20°F) above the start-to-open temperature. They must also permit the passage of a specified amount of coolant when fully open and leak no more than a specified amount when fully closed.

FIGURE 14–20

A typical thermostat located in the water outlet.

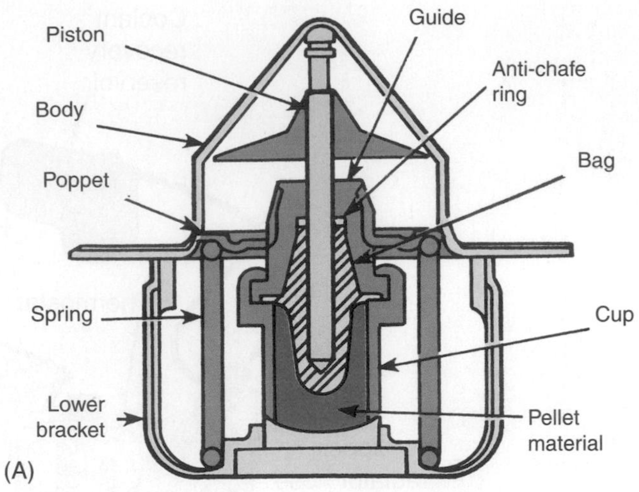

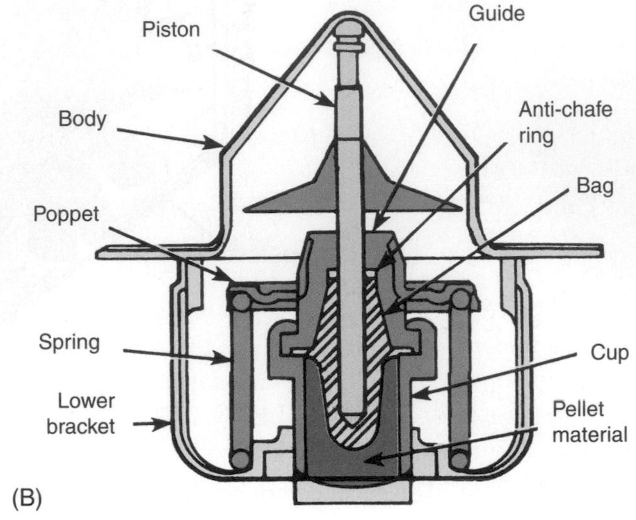

FIGURE 14–21

(A) Thermostat closed; (B) thermostat open.

Water Pump

The heart of the cooling system is the water pump. Its job is to move the coolant through the cooling system. Typically, the water pump is driven by the crankshaft through pulleys and a drive belt **(Figure 14–22)**. Some pumps may be driven off the camshaft. No matter how they are driven, they all basically work the same way. The pumps are centrifugal-type pumps **(Figure 14–23)** with a rotating paddle-wheel-type impeller to move the coolant. The shaft is mounted in the water pump housing and rotates on bearings. At the drive end, the exposed end, a pulley is mounted to accept the belt. The pulley is driven by the crankshaft. The pump housing

FIGURE 14–22

A water pump bolted to the front of an engine.

usually includes the mounting point for the lower radiator hose.

When the engine is started, the impeller pushes the water from its pumping cavity into the engine block. When the engine is cold, the thermostat is closed. This stops the coolant from reaching the top of the radiator. In order for the water pump to circulate the coolant through the engine during warm-up, a bypass passage is added below the thermostat. This passage must be kept free to eliminate hot spots in the engine. It also allows hot coolant to pass through the valve, which opens the thermostat when it reaches the proper temperature.

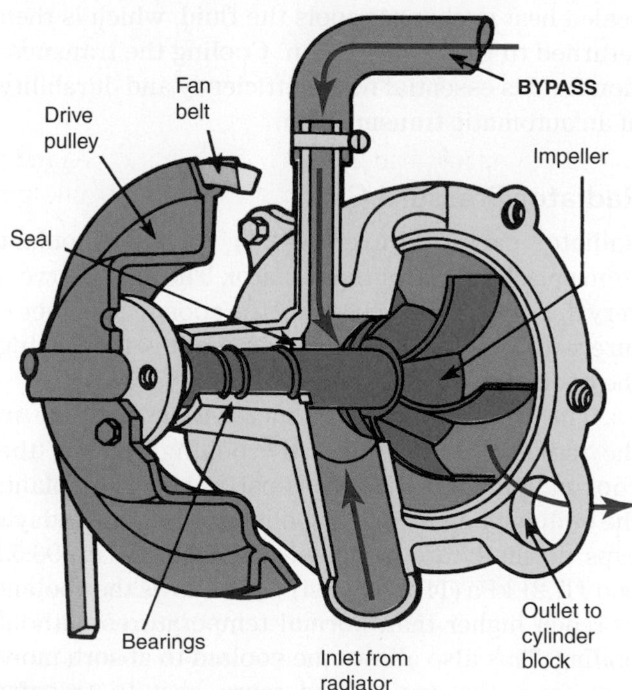

FIGURE 14–23

An impeller-type water pump.

ELECTRIC WATER PUMPS Some engines have an electric water pump for the engine cooling system. The pump is driven by a brushless DC motor that is controlled by the engine control module (ECM). The ECM regulates the amount of coolant circulating through the engine according to the operating conditions. An electric water pump provides improved fuel efficiency, efficient cooling system operation, ideal flow rates at all times, improved heater performance, and decreased engine warm-up times. The operation of these pumps can be monitored with a scan tool.

Radiator

The **radiator** is basically a liquid-to-air heat exchanger, transferring heat from the engine to the air passing through it. The radiator itself is a series of tubes and fins (collectively called the core) that expose the heat from the coolant to as much surface area as possible **(Figure 14–24)**. Attached to the sides or top and bottom of the core are plastic or aluminum tanks **(Figure 14–25)**. One tank holds hot coolant, and the other holds the cooled coolant. Cores are normally comprised of flattened aluminum tubes surrounded by thin aluminum fins. The fins conduct the heat from the tubes to the air flowing through the radiator.

FIGURE 14–24

A radiator core is made of a series of tubes and fins that expose the coolant's heat to as much surface area as possible.

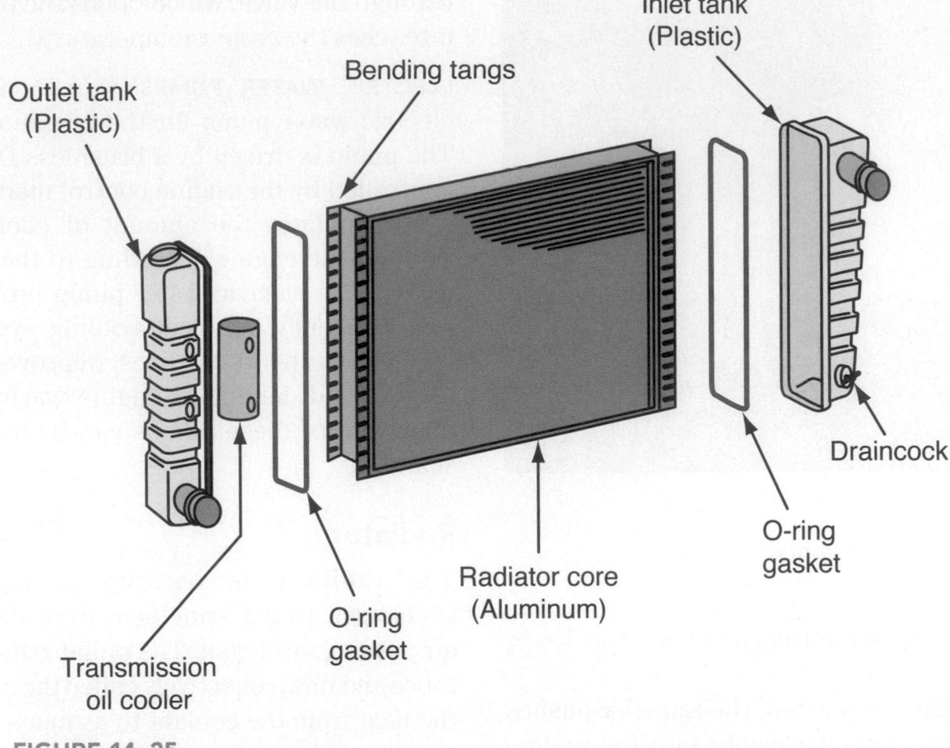

FIGURE 14–25

The core of a radiator is placed between plastic or aluminum tanks. One tank may contain the oil cooler for the automatic transmission and/or the engine.

Factors influencing the efficiency of the radiator are the basic design of the radiator, the area and thickness of the radiator core, the amount of coolant going through the radiator, and the temperature of the cooling air. It is not desirable to have an overly efficient radiator with today's engines. Over-efficiency would keep the engine at a low operating temperature, which would increase emission levels.

The radiator is usually based on one of these two designs: cross flow or down flow. In a cross-flow radiator, coolant enters on one side, travels through tubes, and collects on the opposite side. In a down-flow radiator, coolant enters the top of the radiator and exits at the bottom tank. In both radiator designs, the coolant movement is caused by the pressure differential created by the action of the water pump. Cross-flow radiators are seen most often on late-model cars because all the coolant flows through the fan airstream, and the design allows for lower hood profiles on body designs.

Most radiators feature petcocks or plugs that allow a technician to drain coolant from the system. Coolant is added to the system at the radiator cap or the recovery tank.

TRANSMISSION COOLER Radiators used in vehicles with automatic transmissions have a sealed heat exchanger, or form of radiator, located in the coolant outlet tank of the regular radiator. Metal or rubber hoses carry hot automatic transmission fluid to the heat exchanger. The coolant passing over the sealed heat exchanger cools the fluid, which is then returned to the transmission. Cooling the transmission fluid is essential to the efficiency and durability of an automatic transmission.

Radiator Pressure Cap

Radiator caps **(Figure 14–26)** keep the coolant from splashing out of the radiator. They also serve a very important role in keeping the coolant's temperature within a desired range. It does this by keeping the coolant pressurized to a specified level.

The cap allows for an increase in pressure in the radiator, which raises the boiling point of the coolant. For every 6.89 kPa (1 psi) put on the coolant, the boiling point is raised about 1.66°C (3°F). Today's caps normally are designed to hold between 96.53 and 117.21 kPa (14 and 17 psi). This allows the coolant to reach higher than normal temperatures without boiling. This also allows the coolant to absorb more heat from the engine and more heat to transfer from the radiator core to outside air. This is due to a basic law of nature that states that the greater

FIGURE 14–26

A radiator cap on a late-model engine.

the heat difference is between two objects, the faster the heat of the hotter object will move to the cooler object. When pressures exceed this level, the seal between the cap and the radiator filler neck opens and allows the excessive pressure to vent into a coolant recovery tank.

The pressure in the system is regulated by a pressure relief or vent valve in the radiator cap **(Figure 14–27)**. When the cap is tightened on the radiator's filler neck, it seals the upper and lower sealing surfaces of the neck. The pressure relief valve is compressed against the lower seal. Coolant pressure builds up as the temperature of the coolant rises. When the pressure reaches the pressure rating of the cap, it pushes up on the spring in the pressure relief valve. This opens the valve and allows excess pressure to exit the radiator through a bore between the upper and lower seals. The bore is connected by a tube to the expansion or recovery tank. When enough pressure has been released to drop system pressure below the cap's rating, the spring will close the pressure relief valve. When the coolant cools, its pressure drops. The low pressure opens the vacuum relief valve. The low pressure then draws coolant from the expansion tank to refill the radiator.

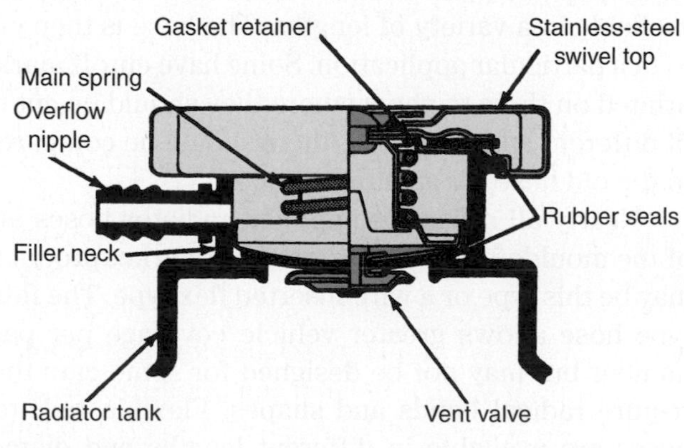

FIGURE 14–27

Parts of a radiator pressure cap assembly.

Chrysler Group LLC

All radiator caps are designed to meet SAE standards for safety. These standards specify that there shall be a detent or safety stop position, allowing pressure to escape from the system without allowing the hot coolant to blow out of the radiator's neck onto the person opening the cap. Only after all pressure has been relieved should the cap be removed from the filler neck.

Cap specifications require that the cap must not leak below the low limit of the pressure range and must open above the high limit. Pressure caps should always be tested for the proper pressure release level and checked for gasket cracking, brittleness, or deterioration each time the antifreeze is changed or when any routine cooling system maintenance is performed.

Radiator pressure caps are marked indicating the amount of pressure held in the cooling system by the pressure valve's spring. For domestic vehicles, the pressure is stated in kilopascals (kPa) and pounds per square inch (psi).

Radiator caps for older imported vehicles may be marked "0.9," which indicates that the pressure rating of the cap is 0.9 times normal atmospheric pressure. Because atmospheric pressure is 101.35 kPa (14.7 psi), a 0.9 cap has a pressure rating of about 91.22 kPa (101.35 kPa × 0.9) or 13.2 psi (14.7 psi × 0.9). Another common rating is 100. The "100" indicates that the pressure rating of the cap is 100 percent of atmospheric pressure, or 101.35 kPa (14.7 psi). Therefore, a 100 kPa (15 psi) cap would be a good substitute for a 100 cap.

SHOP TALK

Always refer to application charts or a service information system when replacing a pressure cap to make sure the new cap has the same pressure range as the original cap.

EXPANSION TANK All late-model cooling systems have an **expansion** or **recovery tank (Figure 14–28)**. Expansion tanks are designed to catch and hold any coolant that passes through the pressure cap when the engine is hot. As the engine warms up, the coolant expands. This eventually causes

FIGURE 14–28

A coolant expansion tank.

the pressure cap to release. The coolant passes to an expansion tank. When the engine is shut down, the coolant begins to shrink. Eventually, the vacuum valve inside the pressure cap opens, and coolant from the expansion tank re-enters the cooling system.

Ford vehicles often use a degas bottle, which is similar to an expansion tank. These systems do not have a separate radiator cap; instead, the degas bottle cap is the fill point and pressure cap for the system. A hose connects the degas bottle to the engine, typically near the thermostat housing. If any air is trapped in the system from a repair or service, the air will pass through the hose to the degas bottle and dissipate.

Hoses

Coolant flows from the engine to the radiator and from the radiator to the engine through radiator hoses. The radiator is solidly mounted to the vehicle, and the engine sits on rubber mounts, which means the engine can move independently of the chassis and the radiator cannot. If the engine were connected solidly to the radiator, the radiator would soon break because of the vibrations and stress. The use of butyl or neoprene rubber hoses cushions the radiator from these vibrations and prevents radiator damage.

WARNING!

When working on the cooling system, remember that at operating temperature, the coolant is extremely hot. Touching the coolant or spilling the coolant can cause serious body burns. Never remove the radiator cap when the engine is hot.

A hose is typically made up of three parts: an inner rubber tube, some reinforcement material, and an outer rubber cover. Different covers and reinforcements are used depending on the application of the hose. Hose construction differs based on where it is located and what amounts of temperature and pressure it will face. Cooling system hoses must be able to endure heavy vibrations and be resistant to oil, heat, abrasion, weathering, and pressure.

Most vehicles have at least four hoses in the cooling system; some have five or more **(Figure 14–29)**. Two small diameter hoses send hot coolant from the water pump to the heater core and back. Two larger diameter hoses move the coolant from the water pump to the radiator and back into the engine block. The fifth hose, a small diameter bypass hose, allows coolant to circulate within the engine when the thermostat is closed. This hose is not required on all engines because the bypass feature is built into the engine block or cylinder head.

Hoses are sized according to their inside diameter. For example, common heater hoses are 16 or 19 mm (⅝ or ¾ in.). Radiator hoses are larger and have reinforcements that allow them to withstand about six times the normal operating pressure of the cooling system. Lower radiator hoses are normally reinforced with wire to prevent them from collapsing due to reduced pressure caused by the action of the water pump.

Radiator hoses are seldom straight tubes. They typically must bend or curve around parts to make a good connection without kinking. Straight hoses are not used because bending causes them to collapse at the bend, causing a restriction. Most original equipment radiator hoses are moulded to a specific shape to fit specific applications. Often, moulded hoses are available in a variety of lengths. The hose is then cut to fit a particular application. Some have cutoff marks printed on them to show where they should be cut to fit different applications. Others should be compared to the old hose for a cut reference.

Nearly all original equipment radiator hoses are of the moulded, curved design. Aftermarket products may be this type or a wire-inserted flex type. The flex-type hose allows greater vehicle coverage per part number but may not be designed for some cars that require radical bends and shapes. Flexible radiator hoses are available in different lengths and diameters. This design can flex or bend into most required shapes without causing a restriction.

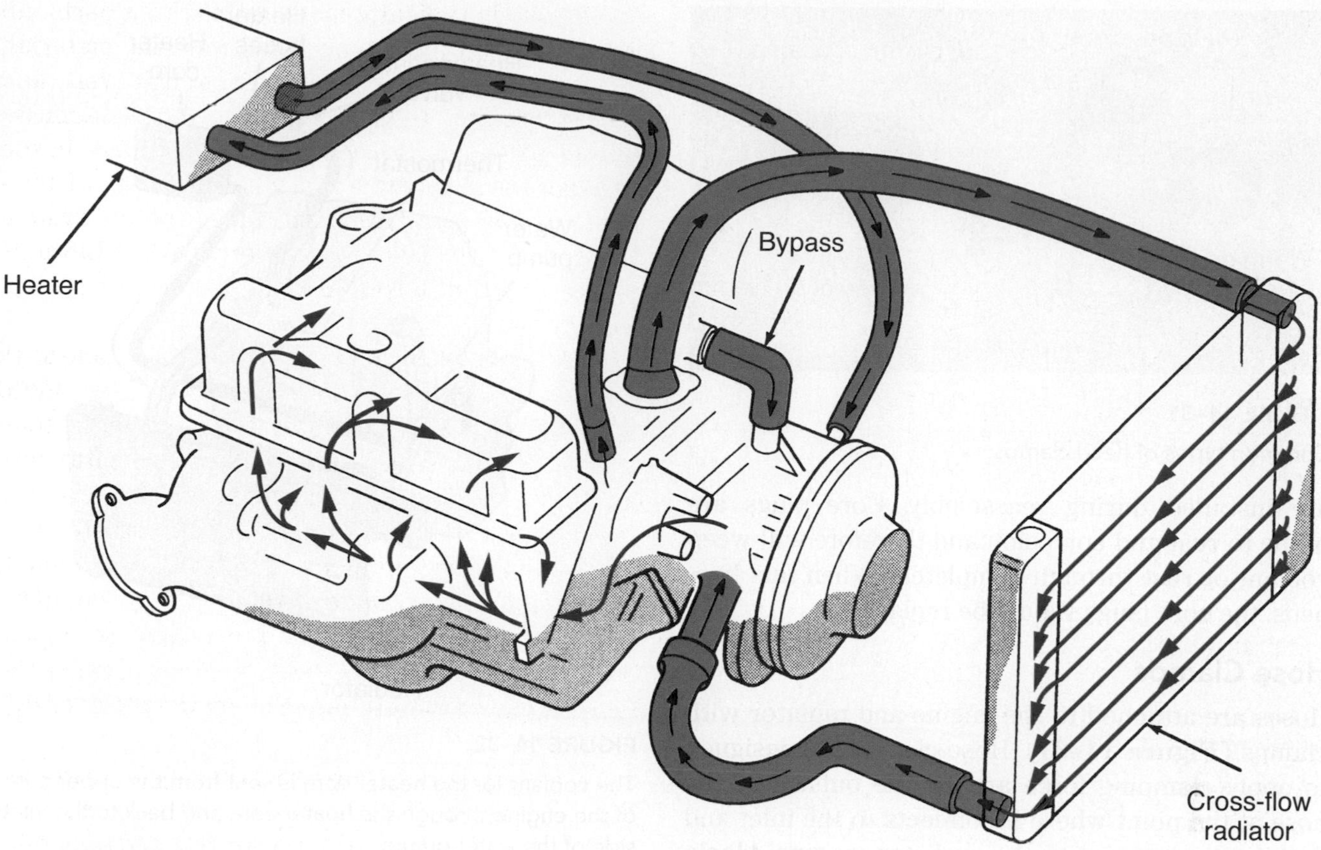

FIGURE 14–29

The routing of coolant through the cooling system hoses.

Chrysler Group LLC

Heater hoses are made with reinforcements to help keep their shape. Some applications require a moulded shape due to complex routing or curves. Rather than replacing heater hoses with specific moulded hoses, formable heater hoses are available. This hose design has a wire spine that allows the hose to bend into a curve without collapsing at the bend. Once the desired shape is obtained, the hose is cut to length and then installed.

WATER OUTLET The water outlet is the connection between the engine and the upper radiator hose. The water outlet has been called a gooseneck, elbow, inlet, outlet, or thermostat housing. Generally, it covers and seals the thermostat and in some cases includes the thermostat bypass.

Most water outlets are made of cast iron, cast aluminum, or stamped steel. Internal corrosion contributes to the failure of water outlets.

WATER JACKETS Hollow passages in the block and cylinder heads surround the areas closest to the cylinders and combustion chambers **(Figure 14–30)**.

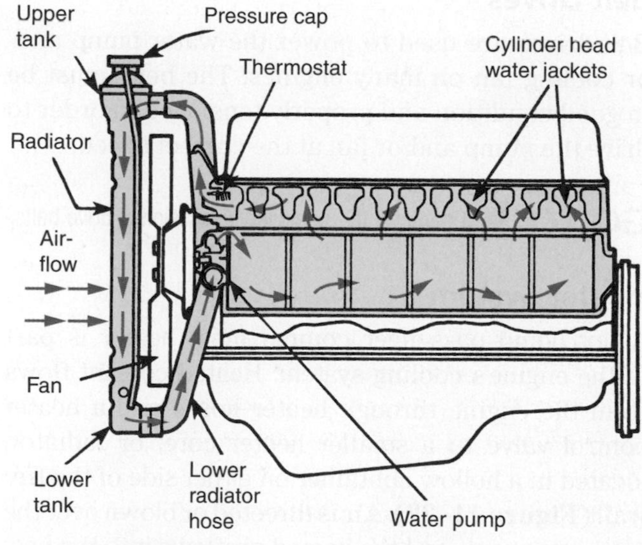

FIGURE 14–30

The cooling system circulates coolant through the engine's water jackets.

Included in the water jackets are soft (core) plugs and a block drain plug. The soft plugs and drain are usually removed during engine teardown. New ones

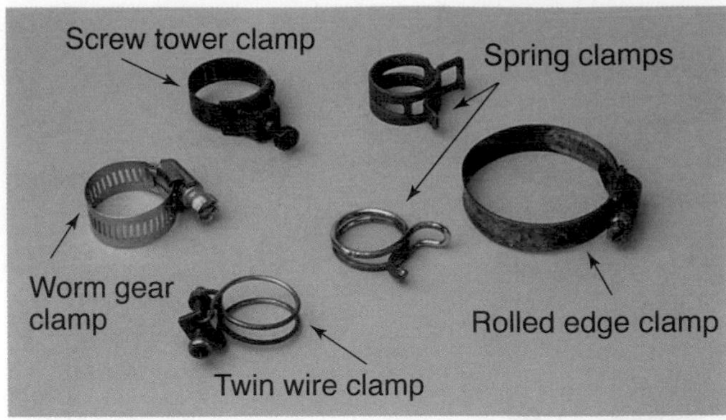

FIGURE 14–31

Common types of hose clamps.

are installed during reassembly. Core plugs are prone to rust and corrosion and therefore will weep coolant or rust through completely. When this happens, the core plugs should be replaced.

Hose Clamps

Hoses are attached to the engine and radiator with clamps **(Figure 14–31)**. Hose clamps are designed to apply clamping force around the outside of the hose at the point where it connects to the inlet and outlet connections at the radiator, engine block, water pump, or heater core. The clamping force exerted on this connection is important in making and maintaining a seal at that point.

Belt Drives

Belt drives are used to power the water pump and/or cooling fan on many engines. The belts must be in good condition and properly tensioned in order to drive the pump and/or fan at the correct speed.

GO TO ▶ Chapter 8 for a detailed discussion of drive belts.

Heater System

A hot liquid passenger compartment heater is part of the engine's cooling system. Heated coolant flows from the engine through heater hoses and a heater control valve to a smaller heater core, or radiator, located in a hollow container on either side of the fire wall **(Figure 14–32)**. Air is directed or blown over the hot heater core, and the heated air flows into the passenger compartment. Movable doors can be controlled to blend cool air with heated air for more or less heat.

Cooling Fans

The efficiency of the cooling system is based on the amount of heat that can be removed from the system and transferred to the air. The system needs air. At highway speeds, the ram air through the radiator should be sufficient to maintain proper cooling. At low speeds and idle, the system needs additional air. This air is delivered by a fan. The fan may be driven by the engine, via a belt, or driven by an electric motor.

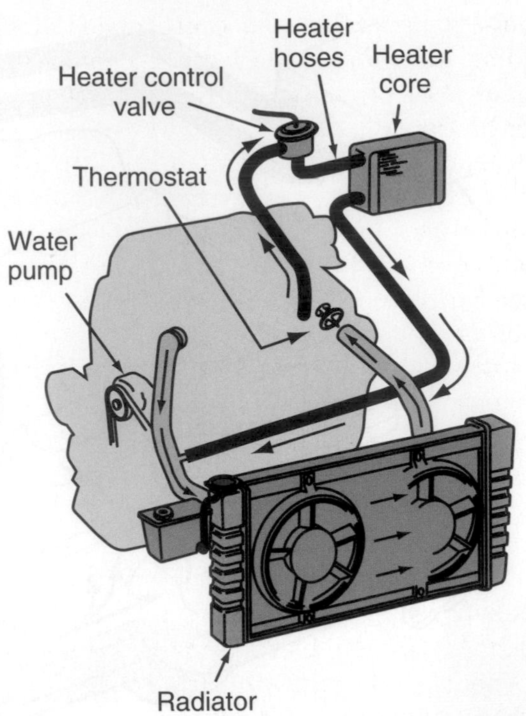

FIGURE 14–32

The coolant for the heater core is sent from the upper part of the engine through the heater core and back to the inlet side of the water pump.

The design of the fan found on a vehicle depends on the air requirements of the engine's cooling system. Diameter, pitch, and the number of blades can be varied to attain the needed flow. A fan placed more than 7.5 cm (3 in.) from the radiator becomes ineffective. It merely recirculates the hot air around the fan blades. For this reason, some radiators are equipped with shrouds. A shroud is a large, circular piece of plastic, metal, or cardboard-like material that extends outward from the radiator to enclose the fan and increase its effectiveness. These shrouds should always be kept intact and should not be modified.

A belt-driven fan is bolted to a pulley on the water pump and turns constantly with the engine. Thus, belt-driven fans always draw air through the radiator from the rear. The power pulley on the crankshaft drives the belt. The fan has several blades made of steel, nylon, or fibreglass attached to a metal hub. Any damage or distortion to the fan will cause it to be out of balance, and the fan should be replaced.

An out-of-balance fan can cause major problems, including rapid and excessive water pump bearing and seal wear or damage to the radiator if the fan blades hit the radiator.

Because fan-assisted air movement across the radiator is usually necessary at idle and low vehicle speed, various design concepts are used to limit the fan's operation at higher speeds. Horsepower is required to turn the fan. Therefore, the operation of a cooling fan reduces the available horsepower to the drive wheels, as well as the fuel economy of the vehicle. Fans are also very noisy at high speeds, adding to driver fatigue and total vehicle noise.

To eliminate this power drain during times when fan operation is not needed, many of today's belt-driven fans operate only when the engine and radiator heat up. This is accomplished by a viscous-drive **fan clutch (Figure 14–33)**. When the engine and fan clutch are cold, the fan moves independently from the fan clutch and moves little air. When the engine warms up, the fan clutch engages, locking the fan in and moving a large amount of air. The clutch unit is located between the water pump pulley and the fan. The clutch assemblies rely on a thermostatic spring or silicone fluid. In both cases, the clutch locks the fan to the fan hub when the temperature of the air passing around the fan reaches a particular point. In most cases, the clutch slips at high speeds; therefore, it is not turning at full engine speed.

Some vehicle manufacturers use flexible blades or **flex blades** that bend or change pitch based on engine speed. That is, at slower speeds, the blade pitch is at the maximum. As engine speed increases, the blade pitch decreases, as do the horsepower losses and noise levels.

FIGURE 14–33

A viscous-type fan clutch.

ELECTRIC COOLING FANS In most late-model applications, to save power and reduce the noise level, the conventional belt-driven, water-pump-mounted engine cooling fan has been replaced with an electrically driven fan **(Figure 14–34)**. This fan and motor are mounted to the radiator shroud and are not connected mechanically or physically to the engine. The 12-volt, motor-driven fan is electrically controlled by either, or both, of two methods: an engine coolant temperature switch or sensor and the air-conditioner switch.

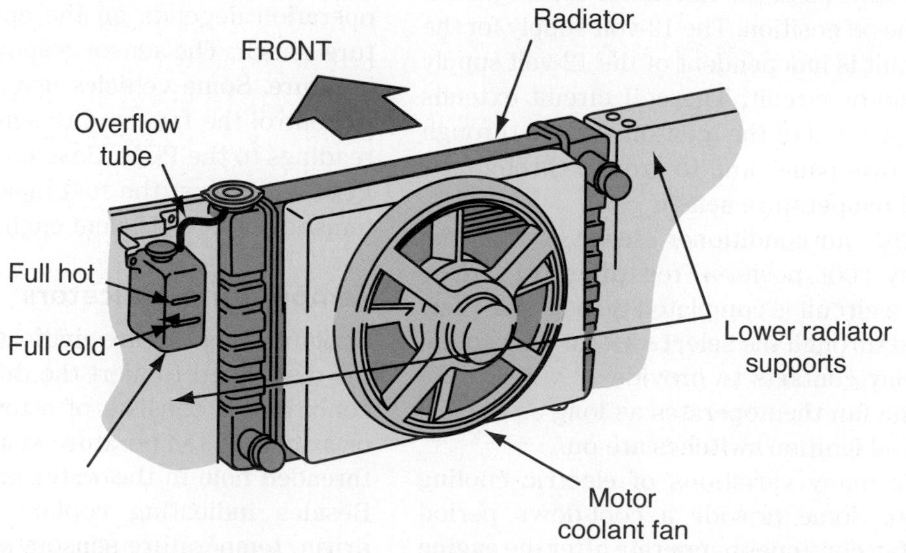

FIGURE 14–34

An electric cooling fan assembly.

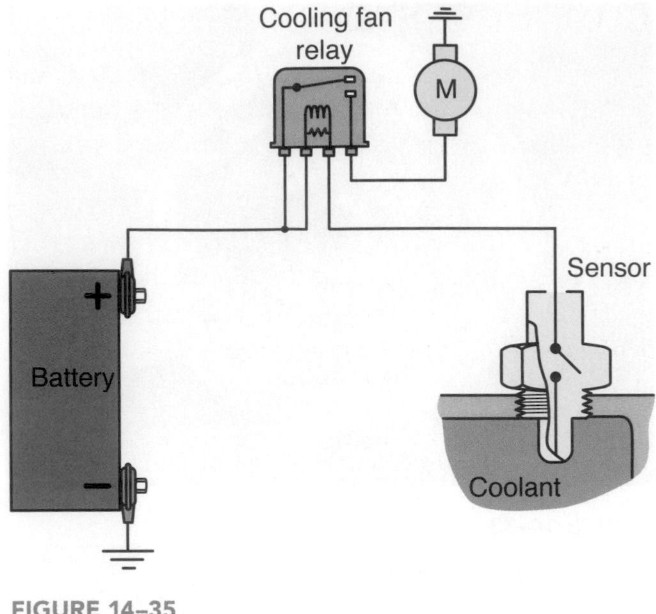

FIGURE 14–35

A simple schematic for an electric cooling fan.

As the schematic in **Figure 14–35** shows, the cooling fan motor is connected to the 12-volt battery supply through a normally open (NO) set of contacts in the cooling fan relay. During normal operation, with the air conditioner off and the engine coolant below a predetermined temperature of approximately 101.6°C (215°F), the relay contacts are open, and the fan motor does not operate.

Should the engine coolant temperature exceed approximately 110°C (230°F), the engine coolant temperature switch closes. This energizes the fan relay coil, which in turn closes the relay contacts. The contacts provide 12 volts to the fan motor if the ignition switch is in the ON position. The 12-volt supply for the relay coil circuit is independent of the 12-volt supply for the fan motor circuit. The coil circuit extends from the on terminal of the ignition switch, through a fuse in the fuse panel, and to ground through the relay coil and temperature sensor.

Should the air-conditioner select switch be turned to any COOL position, regardless of engine temperature, a circuit is completed through the relay coil to ground through the select switch. This action closes the relay contacts to provide 12 volts to the fan motor. The fan then operates as long as the air-conditioner and ignition switches are on.

There are many variations of electric cooling fan operation. Some provide a cool-down period whereby the fan continues to operate after the engine has been stopped and the ignition switch is turned off. These systems may have a second temperature sensor that controls the fan when the engine is off. The fan stops only when the engine coolant falls to a predetermined safe temperature, usually about 98.8°C (210°F). In some systems, the fan does not start when the air-conditioner select switch is turned on unless the high side of the air-conditioning (A/C) system is above a predetermined safe temperature.

Some late-model cars control the cooling fan by completing the ground through the powertrain control module (PCM). Check the service information to find out how an electric cooling fan is controlled, before working with it.

CAUTION!

The engine electric cooling fan can come on at any time without warning, even if the engine is not running. For this reason, it is always wise to remove the negative terminal from the battery or the electric cooling fan connector while working around an electric fan. Make sure you reconnect the connector before giving the car back to the customer.

Hydraulic Cooling Fans

Some Ford and many Jeep vehicles use hydraulically operated cooling fans, called the hydraulic cooling module. In these systems, the power-steering pump supplies fluid to the fan assembly. The PCM controls a solenoid that controls fan speed based on engine temperature and A/C system demand. When engine temperature reaches 105°C (220°F), the fan should be receiving full power-steering fluid flow and operating at full speed.

TEMPERATURE SENSORS Proper electric cooling fan operation depends on the operation of a temperature sensor. The sensor responds to changes in temperature. Some vehicles use more than one sensor to control the fans and to send engine temperature readings to the PCM. Based on this information, the PCM will adjust the fuel injection and ignition systems to provide efficient engine operation.

Temperature Indicators

Coolant temperature indicators are mounted in the dashboard to alert the driver of an overheating condition. It consists of a temperature gauge and/or a light. A temperature sensor is installed into a threaded hole in the water jacket **(Figure 14–36)**. Besides indicating coolant temperatures to the driver, temperature sensors supply some important information to today's computer-controlled engine control systems.

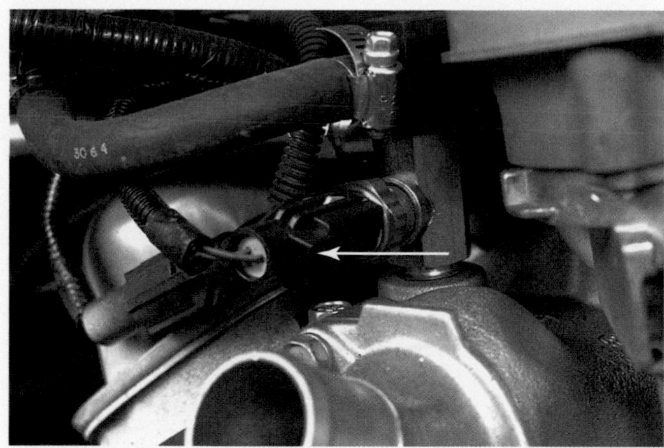

FIGURE 14–36

A coolant temperature sender or sensor.

Engine Block Heaters

This is probably the most valuable option for automobiles that are expected to start during Canadian winters. The block heater is a small heating element much like the ones found on a typical electric range. This element can be installed into a water jacket by removing a core plug **(Figure 14–37)**. Carefully place the element in a position so that it is not touching any part of the block. The removable electric cord should be connected and routed so that it is easily accessible but does not contact any hot or moving parts. These heaters should not be overused. They require only a short warm-up period to take the deep chill out of an engine that has not operated for an extended period of time.

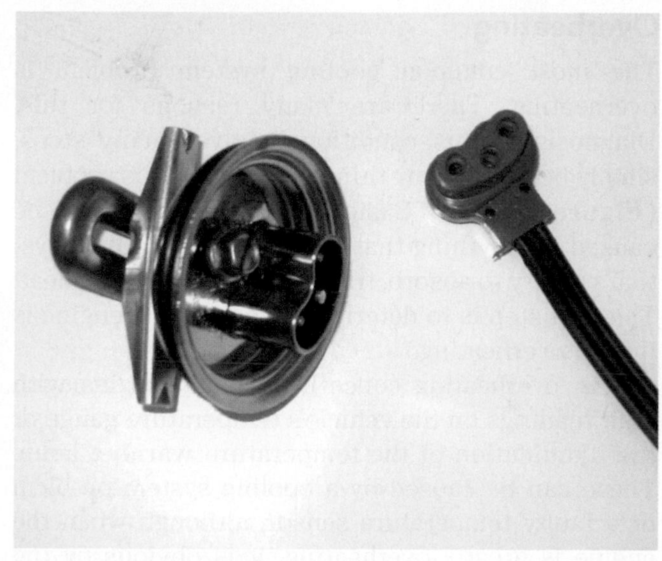

FIGURE 14–37

A block heater.

COOLING SYSTEM DIAGNOSIS

The cooling system must operate, be inspected, and be serviced as a system. Replacing one damaged part while leaving others dirty or clogged will not increase system efficiency.

Diagnosis of the system involves both a visual inspection of the parts—simple checks and tests—and leak testing. One of the first checks of the cooling system is the checking of coolant level and condition. These checks should be done during normal preventive maintenance and when there is a problem.

GO TO  Chapter 8 for a general discussion of checking the coolant level and condition.

There are marks on most expansion tanks that show where coolant levels should be when the car is hot and when it is cold. The condition or effectiveness of the coolant should be checked with a hydrometer, a refractometer, or alkaline test strips.

CAUTION!

When working on the cooling system, remember that at operating temperature, the coolant is extremely hot. Touching the coolant or spilling it can cause serious body burns. Never remove the radiator cap when the engine is hot.

The proper mixture of water and antifreeze also reduces the amount of rust and lime deposits that can form in the system. These deposits tend to insulate the walls of the water jackets. As a result, the coolant is less able to absorb the engine's heat at the points where there is scale. This causes engine hot spots that result in increased component wear and make overheating more likely **(Figure 14–38)**.

Regardless of the mixture of the coolant or the type of antifreeze used, some lime, rust, and scale will always build in a cooling system. Any deposit on the walls of the water jackets will affect engine cooling. Changes in engine temperature cause the engine parts to expand and contract. Some of these deposits then break off and become suspended in the coolant. The coolant then becomes contaminated, and the deposits may collect at a narrow passage, making the passage narrower. This restriction would further lessen the effectiveness of the cooling system. For these reasons and others, the engine's coolant should be replaced and the cooling system flushed according to the manufacturer's recommendations.

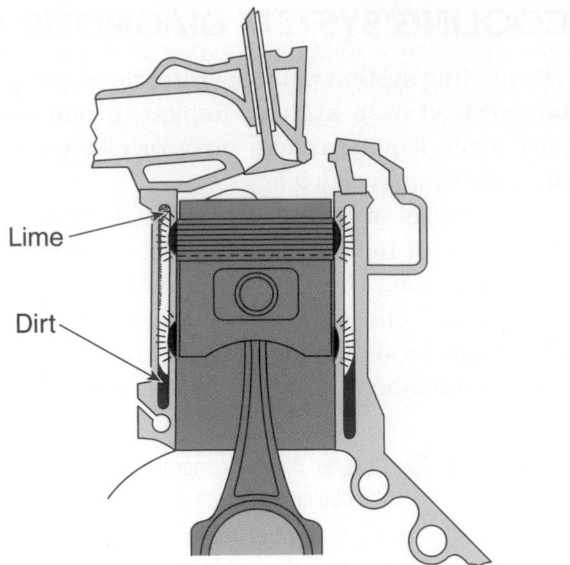

FIGURE 14–38

Lime and dirt buildup in the coolant passages tend to insulate the walls of the water jackets and can cause hot spots that result in increased wear and make overheating more likely.

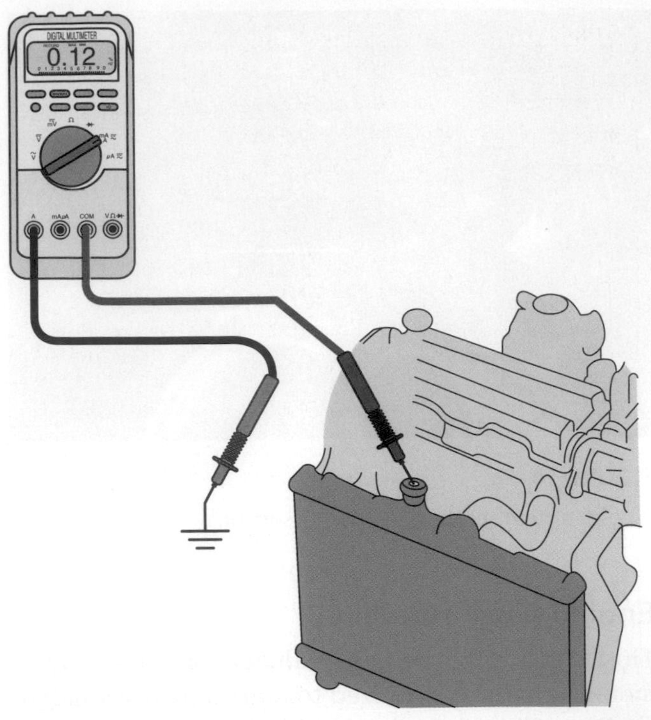

FIGURE 14–39

The setup for checking for conditions that can cause electrolysis.

Testing for Electrolysis in Cooling Systems

Electrolysis is a process in which an electrical current is passed through water, causing the separation of hydrogen and oxygen molecules. In a cooling system, electrolysis removes the protective layer on the inside of the radiator tubes. It also can cause serious engine failures. Electrolysis occurs when there is improper grounding of electrical accessories and equipment or static electricity buildup somewhere in the vehicle. Checking for these should be a part of all checks of the coolant.

To check for the conditions prone to electrolysis, use a voltmeter that is capable of measuring AC and DC voltage. Set the meter so that it can read in tenths of a volt DC. Attach the negative meter lead to a good ground. Place the positive lead into the coolant **(Figure 14–39)**. Look at the meter while the engine is cranked with the starter and record the readings. Take another reading with the engine running and all accessories turned on. Record that reading. Voltage readings of 0 to 0.3 volt are normal for a cast-iron engine; normal readings for a bimetal or aluminum engine are half that amount. Repeat the test with the DMM in the AC voltage mode. There will be AC voltage if the problem is static electricity. Any readings above the normal indicate a problem.

To isolate the problem, look at where the high voltage was measured and think about the systems that were energized during that time. If the voltage was high when all of the accessories were turned on, turn them off one at a time until the voltage drops to normal. The circuit that was turned off prior to the drop in voltage has ground problems. After the electrical problems have been corrected, flush and replace the coolant.

Overheating

The most common cooling system problem is overheating. There are many reasons for this. Diagnosis of this condition involves many steps, simply because many things can cause this problem **(Figure 14–40)**. Basically, overheating can be caused by anything that decreases the cooling system's ability to absorb, transport, and dissipate heat: The first step is to determine whether the engine is indeed overheating.

An overheating concern normally begins with high readings on the vehicle's temperature gauge or the illumination of the temperature warning lamp. These can be caused by a cooling system problem or a faulty temperature sensor, although when the engine is greatly overheating, it is obvious by the steam emitted by the system or by smell. The best way to check the accuracy of the temperature

Condition	Cause
Overheats in heavy traffic or after idling for a long time	■ Low coolant level ■ Faulty radiator cap ■ Faulty thermostat ■ Cooling fan is not turning on ■ Restricted airflow through the radiator ■ Leaking head gasket ■ Restricted exhaust ■ Water pump impeller is corroded
Overheats when driving at speed, or after repeated heavy acceleration	■ Radiator and/or block are internally clogged with rust, scale, silt, or gel ■ Restricted airflow through the radiator ■ Faulty radiator cap ■ Faulty thermostat ■ Radiator fins are corroded and falling off ■ Water pump impeller is corroded ■ Collapsed lower radiator hose ■ Dragging brakes
Overheats anytime or erratically	■ Low coolant level ■ Faulty radiator cap ■ Faulty thermostat ■ Temperature sender or related electrical problem ■ Cooling fan is not turning on
Overheats shortly after the engine is started	■ Temperature sender or related electrical problem
Seems slightly too hot all of the time; gauge nears the red zone at times	■ Radiator and/or block are internally clogged with rust, scale, silt, or gel ■ Restricted airflow through the radiator ■ Faulty radiator cap ■ Faulty thermostat ■ Radiator fins are corroded and falling off ■ Collapsed lower radiator hose ■ Cooling fan is not turning on
Bubbles in the coolant expansion tank	■ Faulty radiator cap ■ Failed head gasket
Air in the radiator but the expansion tank is full	■ Coolant leak ■ Faulty radiator cap ■ Air in the system ■ Faulty seal between the radiator cap and expansion tank ■ Failed head gasket

FIGURE 14–40

Common causes of engine overheating.

indicators is to measure the temperature of the coolant. If the indicators seem to be wrong, troubleshoot and repair the electrical circuit. Then recheck the system's temperature.

SHOP TALK

It is difficult to measure the actual temperature of the coolant on most late-model engines because the only access to the coolant is through the expansion tank. The coolant in the tank does not represent the coolant in the engine and it is not heated. Use a temperature probe or infrared sensor to measure the temperature of the inlet radiator tank. This will give an accurate measurement of system temperature.

Normal operating temperature for most engines is 91°C to 104°C (195°F to 220°F). To maintain this temperature, the coolant must circulate through the engine and radiator. Anything that will interfere with the movement of coolant, such as a faulty water pump or thermostat, or a loss of coolant, will cause overheating. Likewise, anything that interferes with the passing of air through the radiator will also cause overheating.

Effects of Overheating

Engine overheating can cause the following problems:

- Detonation
- Preignition
- Blown head gasket
- Warped cylinder head
- OHC cam seizure and breakage
- Blown hoses
- Radiator leaks
- Cylinder damage due to swelling pistons
- Sticky exhaust valve stems
- Engine bearing damage

Temperature Test

A temperature test can be performed with an infrared temperature sensor, thermometer, or temperature probe. The latter may be a feature of a digital multimeter (DMM). A temperature test monitors temperature changes through the cooling system. When a cold engine is started, the opening temperature of the thermostat can be observed as the engine warms. This can be compared to specifications to determine if the thermostat is bad. Once the engine has warmed up, the probe can be used to scan for cool spots in the radiator. These indicate an area where coolant is not flowing freely through the radiator. The cooling fan temperature switch can also be checked.

Radiator Checks

Cold spots on the radiator indicate internal restrictions. In most cases, this requires removal of the radiator so that it can be deeply flushed or replaced. Normal cooling system flushing may not remove the restrictions. The restrictions are typically caused by internal corrosion or a buildup of scale and lime.

The radiator should also be inspected for external restrictions and for evidence of leaks. Dirt, bugs, and other debris on the surface of the radiator will block airflow. These should be removed by careful cleaning. Also check for loose cooling fins. Salt and other road debris can corrode the solder used to attach the fins around the radiator's tubes. When the fins are not attached to the tubes, heat is not as easily transferred to the outside air.

Checking and Replacing Hoses

Carefully check all cooling hoses for leakage, swelling, and chafing. Also change any hose that feels mushy or extremely brittle when squeezed firmly **(Figure 14–41)**. When a hose becomes soft, it is deteriorating and should be replaced before more serious problems result. When a hose is hard, it is brittle and should be replaced. Hard hoses resist flexing and may crack rather than bend. The result is a leak.

Normally, hoses begin to deteriorate from the inside. Pieces of deteriorated hose will circulate through the system until they find a place to rest. This place is usually the radiator core, causing clogging. Deterioration can also cause leaks. Any external bulging or cracking of hoses is a definite sign of failure. When one hose fails, all of the others should be carefully inspected.

The upper radiator hose is subjected to the roughest service life of any hose in the cooling system. It must absorb more engine motion than any of the other hoses. It is exposed to the coolant at its hottest stage, and it is insulated by the hood during hot soak periods. These conditions make the upper hose the most probable to fail.

Be especially watchful for signs of splits when hoses are squeezed. These splits have a habit of bursting wide open under pressure. Also look for rust stains around the clamps. Rust stains indicate that the hose is leaking, possibly because the clamp has eaten into the hose. Loosen the clamp, slide it back, and check for cuts.

The primary cause of coolant hose failure has been identified as an electrochemical attack on the rubber compound in the hose. This is known as **electrochemical degradation (ECD)**. It occurs because the hose, engine coolant, and the engine/radiator fittings form a galvanic (battery) cell. This chemical reaction causes very small cracks in the hose, allowing the coolant to attack and weaken the reinforcement in the hose **(Figure 14–42)**. ECD can cause pinhole leaks or hose rupture under normal operating pressures. The effects of ECD are accelerated by high temperatures and vibrations.

The best way to check hoses for the effects of ECD is to squeeze the hose near the clamps or connectors. ECD occurs within 5 cm (2 in.) of the ends of the hose—not in the middle. Compare the feel of the hose between the middle and the ends. Gaps

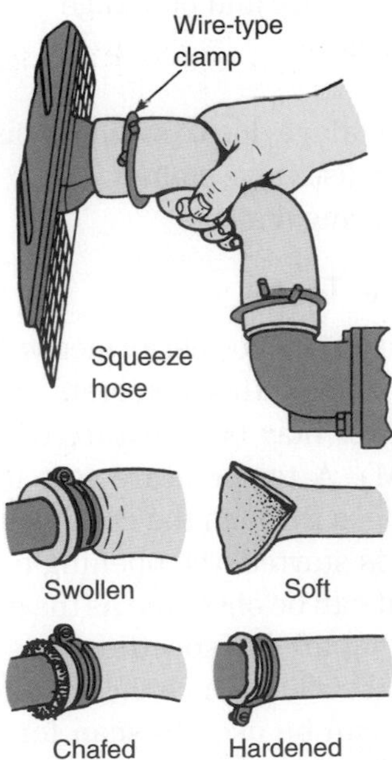

FIGURE 14–41

Defects in cooling hoses.

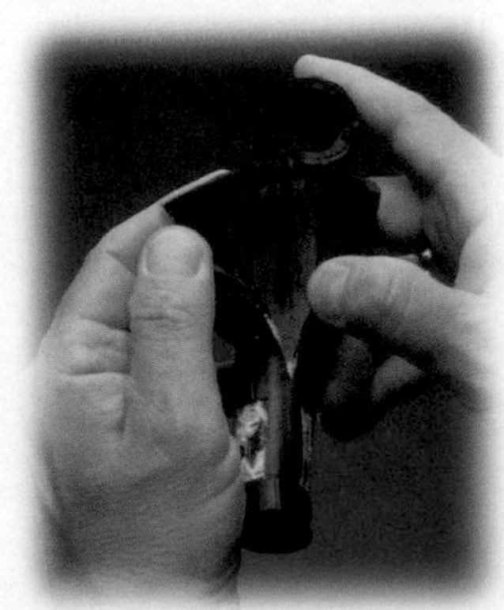

FIGURE 14–42

An ECD-damaged hose.

Gates Corporation

can be felt along the length of the hose where it has been weakened by ECD. If the ends are soft and feel mushy, chances are the hose is under attack by ECD and the hose should be replaced.

ECD can occur in any cooling system hose and will cause the most damage where the temperature is hottest and air is present with the coolant, which is why upper radiator hoses tend to fail first.

Oil is another enemy to rubber hoses. A hose damaged by oil is swollen, soft, and sticky. If the oil leak is external, eliminate the oil leak or try to reroute the hose away from the oil leak to prevent oil damage to a new hose. At times, the oil damage occurs inside the hose. This damage can be caused by transmission fluid leaking into the coolant or by an internal engine oil leak.

Belt Drives

Excessive heat, due to belt slippage, tends to cause the belts to overcure. This causes the rubber to harden and crack. Slippage can be caused by improper belt tension or oily conditions. When slippage occurs, heat not only overcures the belt, it also travels through the drive pulley and down the shaft to the support bearings. The heat can damage the bearings if the slippage is allowed to continue. As a V-belt wears, it rides deeper in the pulley groove. This reduces its tension and promotes slippage. Carefully inspect all drive belts and replace them as necessary.

GO TO ▶ Chapter 8 for the correct procedure for changing a belt and adjusting its tension.

Checking Fans and Fan Clutches

Most engine-driven fans have a clutch, and many overheating problems are caused by a defective clutch. However, the fan itself as well as the fan shroud must be thoroughly checked. The fan shroud should be securely fastened to the radiator bracket. Any damage or distortion to the fan will cause it to be out of balance. This can cause major problems, including rapid and excessive water pump bearing and seal wear or damage to the radiator if the fan blades hit the radiator. A noticeable wobble as the fan spins means that the fan should be replaced. The shroud should also be checked for cracks and other damage. A damaged shroud should be replaced.

Fan clutches are filled with silicone oil. The oil responds to speed and temperature. As the engine increases speed, the oil allows the fan to slip on its hub. This reduces the drag on the engine. If the clutch allows the fan to slip during low engine speeds, overheating can occur. A loss or deterioration of the oil will cause fan slippage. Over time, the oil begins to break down and offers less of a coupling between the fan and hub.

The clutch assembly should be carefully checked for leakage. Oily streaks radiating outward from the hub shaft mean that fluid has leaked out past the bearing seal. A leaking clutch should be replaced. The clutch should also be replaced if the fan can be spun with little or no resistance with the engine off, if the clutch wobbles when the fan is moved in or out, or if its fins are damaged or missing.

ELECTRIC COOLING FANS The action of the electric cooling fan should be observed. However, before doing so, check the mounting of the fan assembly and the condition of the fan blades. The fan should be energized when the A/C is turned on and when the coolant reaches a specified temperature. If the fan does not turn on when it should, check the motor. Do this by connecting the motor directly to the battery. If the motor runs, the problem is in the motor's control circuit. Diagnosis of this problem should follow the prescribed sequence given by the manufacturer. If the jumped motor does not run, the problem is the fan motor **(Figure 14–43)**.

Testing the Thermostat

Thermostats are often the cause of overheating or poor heater performance. They can also cause an increase in fuel consumption and poor engine

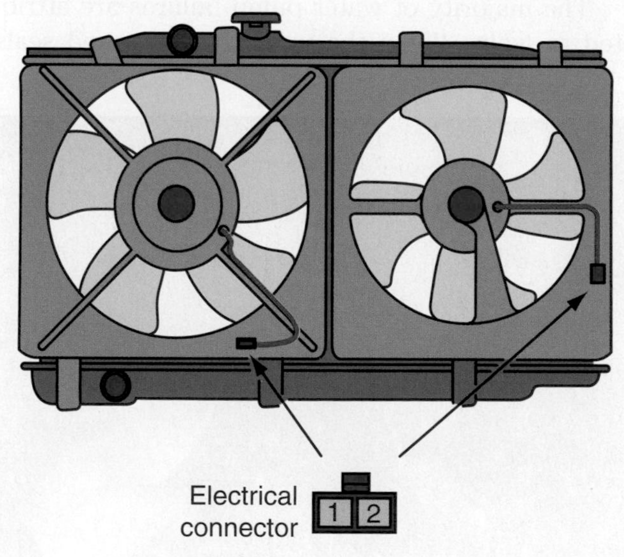

FIGURE 14–43

Test each fan motor by connecting battery power to terminal 1 and a ground to terminal 2. If either motor fails to run or does not run smoothly, replace it.

performance. Electronic engine control systems are programmed to deliver the ideal air/fuel mixture and ignition timing according to the engine's operating conditions. One of the conditions monitored by the ECM is engine temperature.

If the thermostat is stuck open, the coolant may not reach the desired temperature because it is cooled before it gets hot. If a thermostat is stuck closed because it failed or because there is a steam pocket below the thermostat, coolant will not flow between the engine and the radiator; the engine will therefore quickly overheat.

The best way to check thermostat operation is to measure engine temperature with an infrared thermometer **(Figure 14–44)**. However, you can use your hand to feel the temperature. Touch the upper and lower radiator hoses when the engine is first started. If the hoses do not become hot within several minutes, the thermostat is not opening.

Water Pump Checks

Water pumps are driven by a belt, the crankshaft or camshaft, or an electrical motor. A water pump pulls coolant from the radiator and pushes it through the engine. If the water pump is not working properly, the engine will quickly overheat. Seldom does a pump merely stop working. Problems with a pump are normally noise or leakage. An exception to this is electrically operated pumps. Electric problems can cause the pump to totally stop. The cause of these problems is identified by troubleshooting the circuit.

The majority of water pump failures are attributed to leaks. When the pump bearings and seals begin to fail, coolant will seep out of the weep hole in the casting or through the outer seal **(Figure 14–45)**. The seals may be worn out due to age or abrasives in the system, or they may have cracked because of thermal shock, such as adding cold water to an overheated engine.

FIGURE 14–44

The operation of a thermostat can be observed with an infrared thermometer.

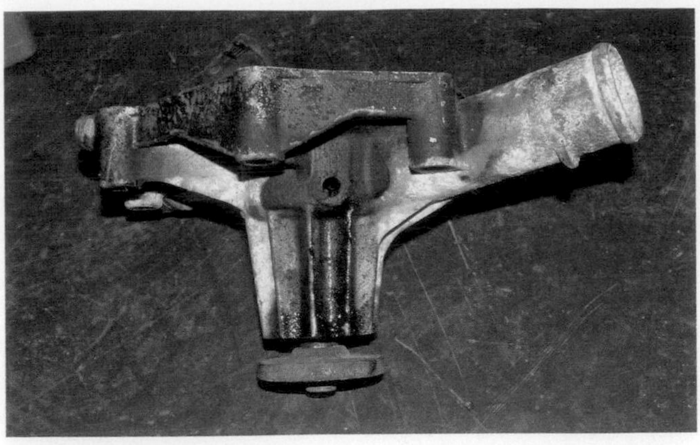

FIGURE 14–45

This pump shows signs of leakage.

CAUTION!

As soon as a water pump begins to leak, it should be replaced. Not only will the leak get worse and cause serious overheating, there is also the chance of the pump's shaft breaking and the fan moving into and destroying the radiator.

Other failures can be attributed to bearing and shaft problems and an occasional cracked casting. Water pump bearing or seal failure can be caused by surprisingly small out-of-balance conditions that are difficult to spot. Any wobble of the pump shaft or fan means the pump and/or fan should be replaced.

Through time, the impeller blades may corrode or come loose from the shaft. Both of these situations will cause the pump not to work, and it must be replaced. In less extreme cases, corrosion may cause the impeller to be loose on the shaft and not rotate all of the time. This can cause an intermittent cooling problem. Unfortunately, the only sure way of knowing the condition of the impeller is to remove the pump and inspect it.

To check the operation of the water pump, start the engine and allow it to warm up. Squeeze the upper radiator hose **(Figure 14–46)** and accelerate the engine. If a surge is felt in the hose, the pump is working.

FIGURE 14–46

The operation of the water pump can be checked by squeezing the upper radiator hose while accelerating the engine. If a surge is felt in the hose, the pump is working.

Air in the cooling system can prevent the pump from working properly. To check for the presence of air, attach one end of a small hose to the radiator overflow outlet and put the other end into a jar of water. Then make sure all hose connections are tight and the coolant level is correct. Run the engine and bring it to its normal operating temperature. Then run the engine at a fast idle. If a steady stream of bubbles appears in the jar, air is getting into the system.

Air may enter the system because of a bad head gasket. Conduct a combustion leak or compression test. If the results of the compression test show that two adjacent cylinders tested have low compression, the gasket is probably bad.

On belt-driven pumps, turn the engine off, and remove the drive belt and shroud. Grasp the fan and attempt to move it in, out, up, and down. More than 1.5 mm (1/16 in.) of movement indicates worn bearings that require water pump replacement.

If the concern is excessive noise, start the engine and listen for a bad pump bearing. Place the tip of a stethoscope on the bearing or pump shaft. If a louder-than-normal noise is heard, the bearing is defective.

CAUTION!

Whenever you are working near a running engine, keep your hands and clothing away from the moving fan, pulleys, and belts. Do not allow the stethoscope or rubber tubing to be caught by the moving parts.

Testing for Leaks

The most common cause of overheating is low coolant levels due to a leak. Leaks can occur anywhere in the system. The most common leak points include the hoses, radiator, heater core, water pump, thermostat housing, engine freeze plugs, transmission oil cooler, cylinder head(s), head gasket, and engine block. Often, only a visual inspection of the cooling system and engine is necessary to find the source of a leak. The point of the leak may be wet or have a light grey colour, the result of the coolant evaporating at that point.

Pressure Testing

The most common tool used to test a cooling system is the radiator pressure tester. A radiator pressure tester is really no more than a hand pump with a pressure gauge. The pressure tester is extremely handy for identifying the location of any leak within the cooling system. To use the tester, connect it to the radiator filler neck **(Figure 14–47)**. Run the engine until it is warm, and then pump the handle of the tester until its gauge reads the same pressure noted on the radiator cap. Watch the gauge. If the pressure drops, carefully check the hoses, radiator, heat core, and water pump for leaks. Often, the leak will initially be obvious because coolant will spray out of the leak. If the pressure drops but there are no external leaks, suspect an internal leak.

FIGURE 14–47

A cooling system pressure tester.

FIGURE 14–48

A common way to identify the source of an external leak is to use dye penetrant and a black light. Where the dyed coolant leaks, a bright or fluorescent green colour will be seen.

Tracer Products

An internal leak caused by a bad head gasket can be verified with the tester. Relieve all pressure in the cooling system. Install the pressure tester onto the radiator filler neck and start the engine. Allow the engine to idle and watch the gauge on the tester. If the pressure begins to build, it is more than likely that the head gasket is blown, and combustion gases are entering the coolant.

LEAK DETECTION WITH DYE Another common way to identify the source of a leak is to use a dye penetrant and a black light **(Figure 14–48)**. The dye is poured into the cooling system, and the engine is run until it reaches operating temperature. With the engine turned off, the engine and cooling system are inspected with the black light. Where the dyed coolant leaks, a bright or fluorescent green colour will be seen.

COMBUSTION LEAK CHECK Internal leaks are suspected when there are no visible external leaks but the engine is losing coolant or does not hold pressure during the pressure test. These leaks are typically caused by a cracked cylinder head or block or a bad head gasket that is allowing coolant to leak into the cylinders or combustion gases to leak into the cooling system. Sometimes steam, white smoke, or water in the exhaust, or coolant in the oil or oil in the coolant, will give a hint that there is a bad head gasket.

A combustion leak tester **(Figure 14–49)** is used to determine whether combustion gases are entering the cooling system. This tester is basically a glass tube with a rubber bulb. The tube is fit with a one-way valve at the bottom. To check for combustion gas leaks, follow this procedure:

FIGURE 14–49

A combustion leak tester.

Martin Restoule

PROCEDURE

STEP 1 Start the engine and allow it to reach normal operating temperature.

STEP 2 Fill the tester's glass tube with the test fluid (normally a blue liquid).

STEP 3 Carefully remove the radiator cap.

STEP 4 Make sure the coolant level in the radiator is below the lower sealing area of the filler neck.

STEP 5 Place the tester's tube into the filler neck.

STEP 6 Rapidly squeeze and release the rubber bulb. This will force air from the radiator up through the test fluid.

STEP 7 Observe the liquid. Combustion gases will change the colour of the liquid to yellow. If the liquid remains blue, combustion gases were not present.

STEP 8 Dispose of the test fluid; never return used test fluid to its original container.

A bad head gasket normally results from another problem. Make sure that problem is corrected before replacing the gasket. Bad head gaskets should be replaced as soon as possible. The excessive heat and pressures resulting from this problem are far greater than what a cooling system can handle and can lead to serious engine problems.

SHOP TALK

The O_2 sensor should be replaced after a head gasket is replaced or if there is a crack in the head or block. Coolant contains silicone and silicates that may contaminate the sensor.

Testing the Radiator Pressure Cap

Apply the proper cap-testing adapter and radiator pressure cap to the tester head. Pump the tester until the pressure valve of the cap releases pressure. The cap should hold pressure in its range as indicated on the tester gauge dial for one minute. If it does not, replace it. Remove the cap from the tester and visually inspect the condition of the cap's pressure valve and upper and lower sealing gaskets. If the gaskets are hard, brittle, or deteriorated, the cap may leak when exposed to hot, pressurized coolant **(Figure 14–50)**. It should be replaced with a new cap in the same pressure range.

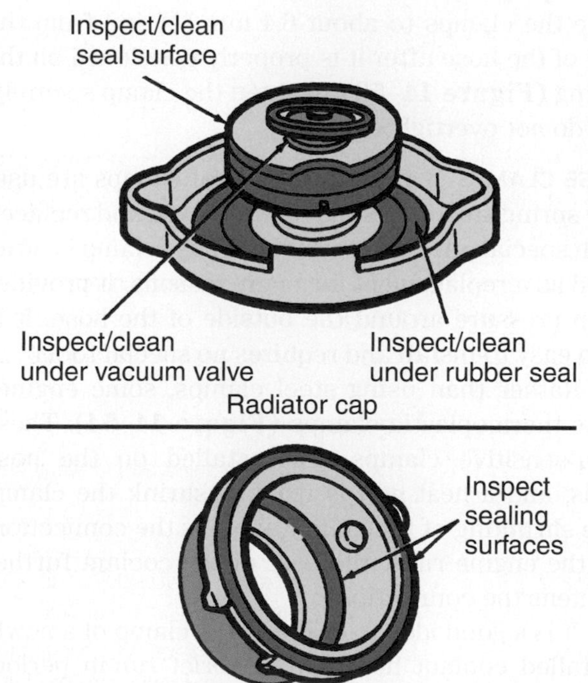

FIGURE 14–50
Radiator cap inspection.

CAUTION!

The radiator cap should never be removed when the cap or radiator is hot to the touch. When the pressure in the radiator is suddenly released, the coolant's boiling temperature is reduced, causing the coolant to immediately boil. Because coolant is a thick liquid, it will stick to your skin and cause severe burns if it is hot. The radiator should first be allowed to cool, or force cool it by lightly spraying water on the radiator core. When the cap is cool to the touch and the engine is shut off, use a cloth to cover the cap and turn it counter-clockwise one-quarter turn to the filler neck safety stop. Carefully watch for any liquid or steam loss around the ring of the cap and from the radiator overflow tube. Let the cap remain in this position until all pressure subsides. When evidence of discharge is no longer seen, use a cloth to cover the cap, press it down to pass the cap's ears over the safety stops, and continue to turn counter-clockwise to remove the cap.

Water Outlet

Internal corrosion, contributes to the failure of water outlets. Cast-iron water outlets are more resistant to corrosion than stamped steel or cast aluminum outlets. A more common cause of water outlet problems is uneven torquing of the mounting bolts, which can cause a mounting ear to crack or break off. When this happens, the outlet must be replaced.

COOLING SYSTEM SERVICE

Service to the cooling system involves replacing leaking or broken parts as well as emptying, flushing, and refilling the system. One of the common services is replacing the drive belt. Belts are a common wear item and should be installed at the correct tension. A loose belt will slip and prevent the water pump from moving coolant fast enough through the engine and/or turn the fan fast enough to cool the coolant.

GO TO ▶ Chapter 8 for a detailed discussion of drive belts and the correct procedure for changing a belt and adjusting its tension.

The water pump on many late-model engines is driven by the engine's timing belt or chain. When replacing the water pump on these engines, always replace the timing belt. Make sure all pulleys and gears are aligned according to specifications when installing the belt.

Hoses

CUSTOMER CARE

Technicians should do their customers a favour and remind them that all cooling hoses should be replaced every two to four years to prevent breakdowns.

All cooling system hoses are basically installed the same way. The hose is clamped onto inlet-outlet nipples on the radiator, water pump, and heater.

Replacement radiator hoses must be the correct diameter, length, and shape. Each has a part number, which is often printed on the old hose and on the hose package **(Figure 14–51)**.

When replacing a hose, drain the coolant system below the level being worked on. Loosen or carefully cut the old clamp. Then, using a knife, carefully cut the end of the old hose **(Figure 14–52)** so that it can slide off its fitting. If the hose is stuck, do not pry it

FIGURE 14–51

Nearly all radiator hoses have a part number painted on the outside of the hose. These numbers should be used when replacing the hose.

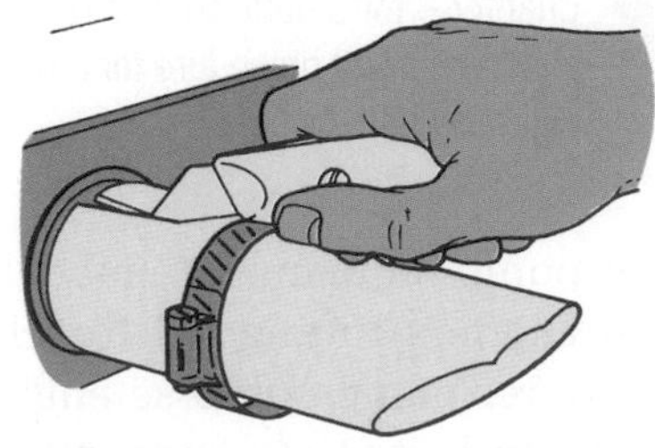

FIGURE 14–52

Cutting off an old hose.

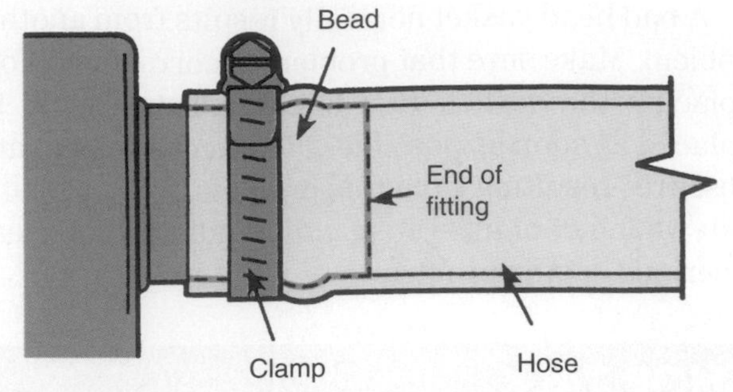

FIGURE 14–53

New clamps should be placed immediately after the bead of the fitting.

off. You could possibly damage the inlet/outlet nipple or the attachment between the end of the hose and the bead. Simply cut it more so that it can come off.

Always clean the neck of the hose fitting or nipple with a wire brush or emery cloth after the old hose has been removed. Burrs or sharp edges could cut into the hose tube and lead to premature failure, and dirt will prevent a good seal.

Dip the ends of the hose in coolant to lubricate it and slip the clamps over each end. Do not reuse old spring-type clamps, even if they look good. Slip the hose over its fittings, engine end first. In cold weather, the hose may be stiff; it can be soaked in warm water to make it more flexible. If the hose does not fit properly, remove it and reverse the ends. Then slide the clamps to about 6.4 mm (¼ in.) from the end of the hose after it is properly positioned on the fitting **(Figure 14–53)**. Tighten the clamp securely, but do not overtighten.

HOSE CLAMPS Original equipment clamps are usually spring steel wires that are removed and replaced with special pliers. A worm drive hose clamp is often used as a replacement for many reasons. It provides even pressure around the outside of the hose. It is also easy to install and requires no special tools.

Rather than using steel clamps, some engines have thermoplastic clamps **(Figure 14–54)**. These heat-sensitive clamps are installed on the hose ends, and a heat gun is used to shrink the clamp. The shrinking of the clamp tightens the connection. As the engine runs, the heat of the coolant further tightens the connection.

It is a good idea to readjust the clamp of a newly installed coolant hose after a brief run-in period. The hose end does not contract and expand at the same rate as the metal of the inlet/outlet nipple it is attached to. Rubber coolant hose, warmed by the hot coolant and hot engine, will expand. The clamp

FIGURE 14–54

Thermoplastic clamps are tightened with a heat gun.

Gates Corporation

compresses the rubber around the hose end and sets it. When the engine cools, the fitting contracts more than the rubber, and the hose will not be as secure, which can result in cold leaks of coolant at the inlet/outlet nipple when the engine is cool. Retightening the clamp eliminates the problem.

Thermostat

When replacing the thermostat, the replacement should always have the same temperature rating as the original. Using a thermostat with a different opening temperature than was originally installed on a computerized engine will affect the operation of the fuel, ignition, and emissions control systems. This is due to the fact that the wrong thermostat can prevent the system from going into closed loop.

Replacement markings on the thermostat normally indicate which end should face toward the radiator. Regardless of the markings, the sensored end must always be installed toward the engine.

When replacing the thermostat, also replace the gasket that seals the thermostat in place and is positioned between the thermostat housing and the block. Make sure the mating surfaces on the housing and block are clean and free of old gasket material **(Figure 14–55)**. Some housings are sealed with a liquid gasket. A thin line of the liquid gasket should

FIGURE 14–55

Make sure all old gasket material is removed before installing a new thermostat gasket and housing.

be applied around the water passages, and the housing installed before the liquid dries.

Normally, there are locating pins on the thermostat to position the thermostat in the block and housing **(Figure 14–56)**. Most thermostats and housings are sealed with gaskets or rubber seals. The gaskets are normally composition fibre material cut to match the thermostat opening and mounting bolt configuration of the housing. Thermostat gaskets generally come with an adhesive backing. The backing holds the thermostat securely centred in the mounting flange. This makes it easier to properly align the housing to the block.

Repairing Radiators

Most radiator leak repairs require the removal of the radiator from the vehicle. The coolant must be drained and all hoses and oil cooler lines disconnected. Bolts holding the radiator are then loosened and removed.

The actual radiator repair procedures depend on the material of which it is made and the type of damage. Most radiator repairs are made by radiator specialty shops that employ technicians with knowledge of such work. Many of today's radiators have plastic tanks, which are not repaired. If these tanks leak, they are replaced.

If the radiator is badly damaged, it should be replaced.

REPLACING THE WATER PUMP When replacing a water pump, it is necessary to drain the cooling system. Any components—belts, fan, fan shroud, shaft spacers, or viscous-drive clutch—should be removed to make the pump accessible. Most pumps

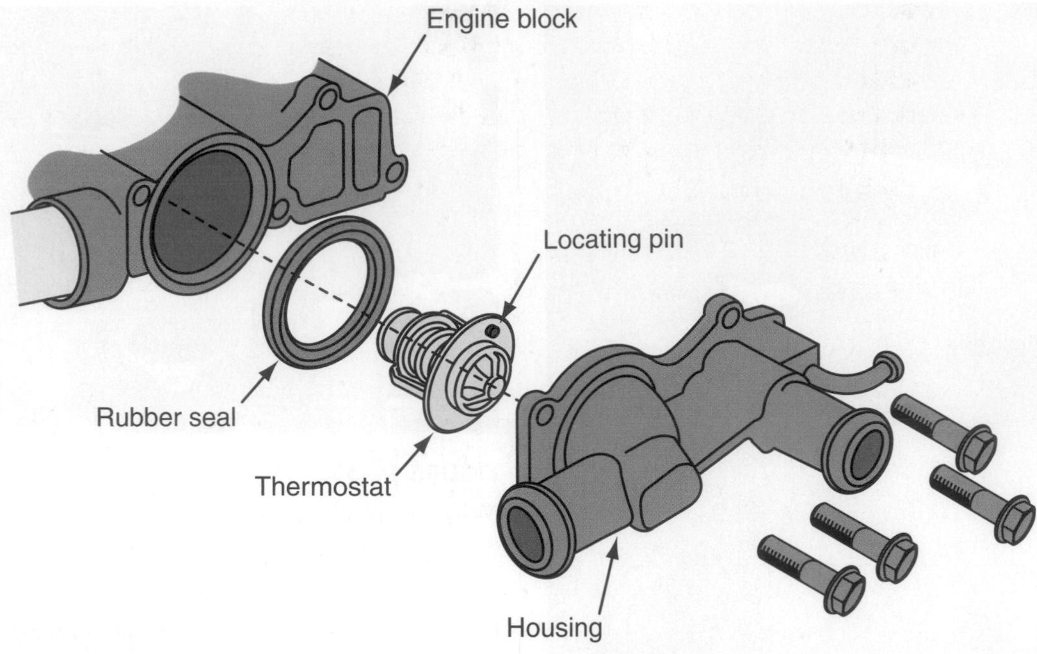

FIGURE 14–56

Normally, there are locating pins on the thermostat that are used to position the thermostat in the block and housing.

are bolted to the cylinder block. Loosen and remove the bolts in a criss-cross pattern from the centre outward. Insert a rag into the block opening and scrape off any remains of the old gasket.

CAUTION!

When working on the coolant system (e.g., replacing the water pump or thermostat), a certain amount of coolant will spill on the floor. The antifreeze in the coolant causes it to be very slippery. Always immediately wipe up any coolant that spills, to reduce or eliminate the chance of injury.

When replacing a water pump, always follow the procedures given by the manufacturer. Most often a coating of adhesive is applied to a new gasket before it is placed into position on the water pump. Some pumps are sealed to the block with an O-ring. Make sure the O-ring groove in the block is clean. When installing the O-ring, make sure it is totally inserted in the groove. Lubricate the O-ring to prevent it from tearing during installation. Position the pump against the block until it is properly seated. Install the mounting bolts and tighten them evenly in a staggered sequence to the torque specifications. Careless tightening can cause the pump housing to crack. Check the pump to make sure it rotates freely.

The water pumps on many late-model OHC engines are driven by the engine's timing belt. When replacing the water pump on these engines, always replace the timing belt. Make sure all pulleys and gears are aligned according to specifications when installing the belt.

Draining the Coolant

Part of a preventive maintenance program is changing the engine's coolant. This is done to prevent the coolant from breaking down chemically. When this happens, the coolant becomes too acidic.

Before draining the coolant, find the capacity of the system in the vehicle's specifications. This allows you to know what percentage of the coolant has been drained. Normally, 30 to 50 percent of the coolant will remain in the system.

Most radiators have a drain plug located in the lower part of a tank. Some have a petcock valve. Make sure the engine is cool before draining the coolant. Set the heater controls to the HOT position. Remove the radiator pressure cap. Remove the overflow reservoir and empty it into a catch can. Now place the catch can under the drain plug and remove the plug. If the radiator has a petcock, open it fully. If the radiator does not have a drain plug, remove the lower radiator hose. Be careful not to force the

FIGURE 14–57

Most engines have drain plugs that allow for draining coolant out of the block.

hose away from the radiator. Twist the hose back and forth while pulling it off. If the hose is stuck, slip a thin screwdriver between the hose and the radiator tube to loosen it. After the coolant stops flowing out, install the drain plug or close the petcock.

Additional coolant can be drained through drain plugs in the engine block **(Figure 14–57)**. Place the catch can below the drain plug and remove the plug. Once the coolant has drained, replace the plug. Make sure to apply sealer on the threads of the plug. Clean up any spilled coolant.

CAUTION!

Never pour engine coolant into a sewer or onto the ground. Used coolant is a hazardous waste, and its disposal should be in accordance to local laws and regulations.

CAUTION!

Coolant can be very dangerous to children and animals. It has a sweet taste and can be deadly if ingested. Never keep coolant in an open container.

Coolant Recovery and Recycle System

Whenever the coolant must be drained to service the cooling system, the used coolant should be drained and recycled by a coolant recovery and recycle machine. Typically, additives are mixed into the used coolant during recycling. These additives either bind to contaminants in the coolant so that they can be easily removed, or they restore some of the chemical properties in the coolant.

Flushing Cooling Systems

Whenever coolant is changed—and especially before a water pump is replaced—a thorough flushing should be performed. The bottom of the radiator will trap rust, dirt, and metal shavings. Draining the system will only remove the contaminants that are suspended in the drained fluid. Rust and scale will inevitably form in any cooling system. Buildups of these will affect the efficiency of the cooling system and can cause blockages inside the radiator. Flushing may not remove all debris. In fact, if the radiator is plugged, the radiator should be removed and rebuilt or replaced.

The system can be flushed in many ways. Power flushing equipment forces the old coolant and contaminants out of the system. This function is often one of the operating cycles of a coolant exchanger. Back flushing forces clean water backward through the cooling system. The discharge of fluid carries away rust, scale, corrosion, and other contaminants. Flushing guns use compressed air to back flush the system and to break loose layers of dirt and other debris **(Figure 14–58)**. This method of flushing the system is not recommended on systems that use plastic and aluminum radiators. Check the service information for the proper way of cleaning the cooling system in the vehicle being serviced.

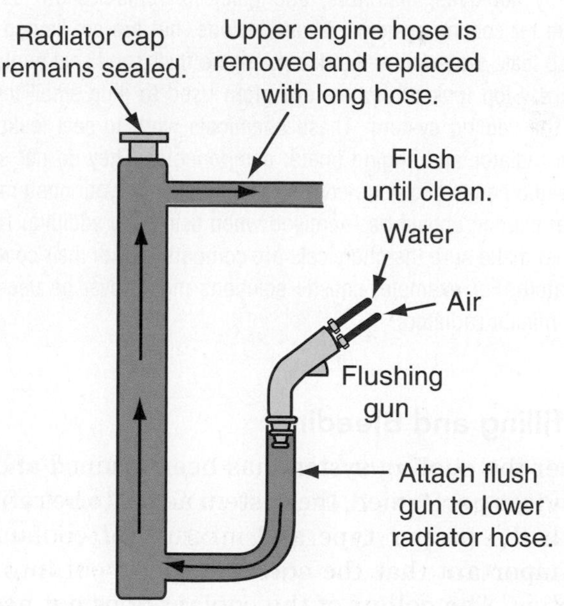
Radiator cap remains sealed.
Upper engine hose is removed and replaced with long hose.
Flush until clean.
Water
Air
Flushing gun
Attach flush gun to lower radiator hose.
REVERSE FLUSHING RADIATOR

FIGURE 14–58

The typical setup for back flushing a cooling system.

A simple way to flush the system is to drain it and fill it with water. The engine is then run and brought to its operating temperature. At this point, the engine is shut down and allowed to cool. The water is then drained. These steps are repeated until the discharged fluid is clear. The clear fluid indicates that all of the old coolant has been removed. Once this occurs, the system is drained again and refilled with coolant.

FLUSHING CHEMICALS Many different flush chemicals are available. Before using any chemical, make sure it is safe for the type of radiator. Coolant exchangers often use chemicals as part of their flushing function. The typical procedure for using flushing chemicals is to drain the system. Then pour the chemical into the radiator and top off the radiator with water. Install and tighten the radiator cap. Now start the car and set the heater control to its hottest position. Allow the engine to run until it reaches normal operating temperature. Then turn the engine off and allow it to cool. Once the engine has cooled, completely drain the radiator. If the flushing chemical requires a neutralizer, add it to the water remaining in the system, and then refill the system with fresh coolant.

CUSTOMER CARE

Many additives, inhibitors, and quick-fix remedies are available for cooling systems. These include, but are not limited to, stop leak, water pump lubricant, engine flush, and acid neutralizers. Stop leak or sealers are often used to plug small leaks in the cooling system. These chemicals work to seal leaks in the radiator and engine (metal components). They do not seal leaking hoses and hose connections. Explain to your customers that caution should be exercised when using any additive. They must make sure the chemicals are compatible with their cooling system. For example, caustic solutions must never be used in aluminum radiators.

Refilling and Bleeding

After the cooling system has been drained and all services performed, the system needs to be refilled with the proper type and mixture of coolant. It is important that the correct type is put into the system. The colour of the coolant does not necessarily indicate its purpose. Older inorganic acid coolants are green; extended-service coolants are often orange or yellow; and other coolants are often red, pink, or blue. The additives in long-life coolant are not chemically compatible with those in older green or red coolant; therefore, the system must be drained and refilled with the coolant specified by the vehicle manufacturer.

When refilling a system, determine the total capacity of the system. Half of the capacity is the amount of undiluted coolant that should be put into the system. Then top off the system with water. Loosely install the radiator cap. Run the engine until it reaches operating temperature. Then turn it off and correct the level of the coolant. Tighten the radiator cap and run the engine again and check for leaks.

When refilling the system, be sure you get it completely full. Some systems are difficult to fill without trapping air. Air pockets in the head(s), heater core, and below the thermostat can prevent proper coolant flow and cooling. If air is trapped in the engine block or cylinder head(s), it can also cause "hot spots." This can ruin the head gaskets, the cylinder walls, and the entire cooling system.

This is more of a problem on newer vehicles than older ones. In older cars, the top of the radiator sat higher than the rest of the cooling system. This positioning allowed air in the system to escape through the radiator cap. The radiator cap on many newer cars is lower than the rest of the cooling system. Because all liquids seek a natural level, that level may be above the cap and air gets trapped easily in other high places such as the block or head(s).

These new vehicles must be purged of the air after refilling. This can be done in many ways. Sometimes, jacking up the front of the car will raise the radiator cap higher than the rest of the cooling system. With the radiator cap in its first lock position, start the engine. Allow it to run until the thermostat opens and the coolant circulates. Trapped air bubbles will naturally blow out the cap. When no more air escapes, shut off the engine and correct the coolant level. Then reinstall the radiator cap and tighten it to its fully locked position.

Each vehicle may have its own specific bleeding procedure. Some engines have air bleed valves located at the high point in the system **(Figure 14–59)**. Check the service information for their location. These valves allow air to escape while the system is being filled.

To use the air bleeds, make sure the engine is warm and the heater is fully on. Connect a hose to the end of the valves and place the open end in a catch can. Open all bleed valves. Slowly put the required

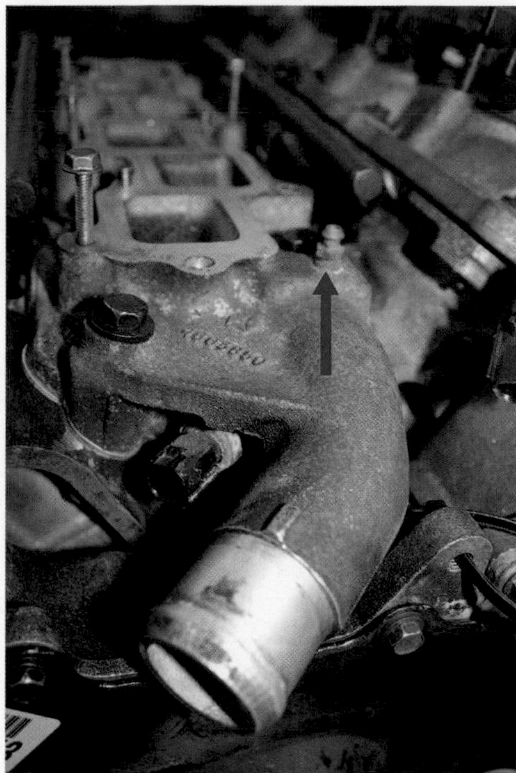

FIGURE 14–59

Some engines have a bleed screw at a high point of the system to relieve the system of trapped air.

Martin Restoule

amount of coolant into the radiator until the coolant begins to leak out of the valves. Then close the valves and top off the system.

If the system does not have a bleeder valve, disconnect a heater hose at the highest point in the system. Once fluid flows steadily out of the hose, reconnect the hose. If the system was previously fit with a flushing Tee, remove its cap to purge the system. Always recheck the fluid level and make sure that all air is purged before putting the car back into service.

Many technicians use a vacuum fill system to refill the cooling system and prevent any air from staying trapped in the system. Vacuum fill systems connect to the radiator fill neck or expansion tank and to the shop's compressed air system. Once connected, the shop air creates a vacuum in the cooling system. Within minutes, the system is purged of air and should hold a vacuum. If the vacuum bleeds off, there is a leak in the system. If the system holds a vacuum, the shop air is disconnected from the tool and a fill hose is connected and placed into a container of new coolant. Opening the valve on the vacuum fill tool forces the coolant into the system and fills it completely. Once the system is full, start the engine and make sure the coolant level is full.

Special Precautions for Hybrid Vehicles

Special coolants are required in most hybrids because the coolant cools the engine and the converter/inverter assembly. Cooling the converter/inverter is important, and checking its coolant condition and level is an additional check during preventive maintenance. The cooling systems used in some hybrids feature electric pumps and storage tanks. The tanks store heated coolant and can cause injury if the technician is not aware of how to carefully check them.

Toyota hybrids have a system that heats a cold engine with retained hot coolant to provide reduced emissions levels. Hot coolant is stored in a container. The coolant will circulate through the engine immediately after startup. The fluid also may circulate through the engine many hours after it is shut off. This fluid is under pressure and can cause serious burns to anyone who opens the system for inspection and/or repairs. To safely service this cooling system, the pump for the storage tank must be disconnected. The cooling system also is tied into the converter/inverter assembly. This also presents a potential problem, because it is easy to trap air in the cooling system due to the path of coolant flow **(Figure 14–60)**. To purge the system of air, there is a bleeder screw, and a scan tool is used to run the electrical water pump.

PROCEDURE

Recommended procedure for draining and refilling the cooling system on a Toyota hybrid includes the following steps.

STEP 1 Remove the radiator's top cover and cap.

STEP 2 Disconnect the electrical connector on the water pump to prevent circulation of the coolant.

STEP 3 Connect a drain hose to the drain port on the bottom of the coolant heat storage tank, and then loosen the yellow drain plug on the tank.

(*continued*)

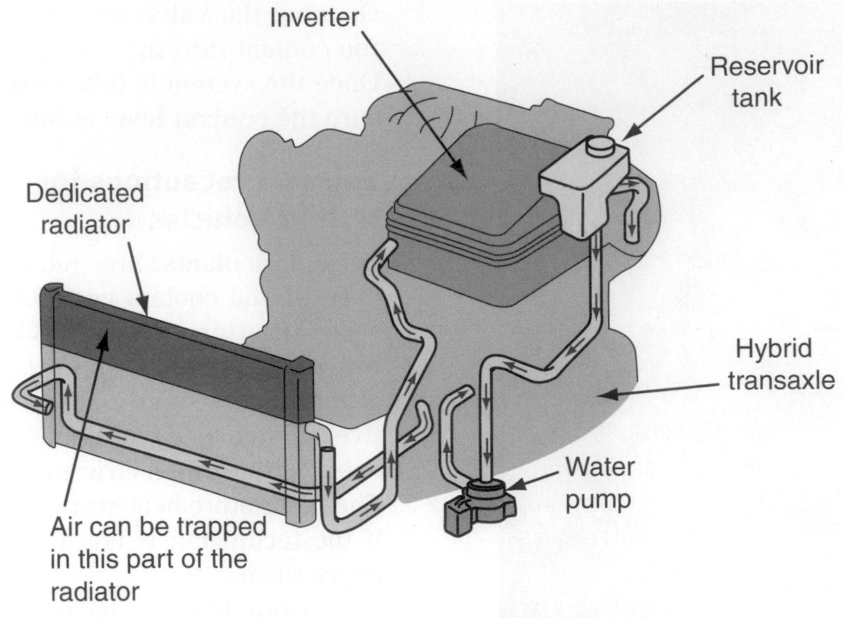

FIGURE 14–60

Because the reservoir tank and inverter are higher than the radiator cap, air is easily trapped during coolant refilling.

STEP 4 Connect a drain hose to the drain port on the lower-left corner of the radiator, and then loosen the yellow drain plug on the radiator.

STEP 5 Connect a drain hose to the drain port on the rear of the engine and loosen the drain plug.

STEP 6 After the coolant is drained, tighten the three drain plugs.

STEP 7 Reconnect the connector to the coolant heat storage tank's water pump.

STEP 8 Connect a hose to the radiator's bleeder valve port and place the other end of the hose into the coolant reservoir tank.

STEP 9 Loosen the radiator's bleeder plug.

STEP 10 Fill the radiator with the correct coolant.

STEP 11 Tighten the radiator's bleeder plug and install the radiator cap.

STEP 12 Connect the scan tool to DLC3.

STEP 13 Using the scan tool, run the water pump for the storage tank for 30 seconds, and then loosen the radiator's bleeder plug.

STEP 14 Remove the radiator cap and top off the coolant in the radiator.

STEP 15 Repeat the refilling and bleeding sequence as often as necessary. Normally when no additional coolant is needed after the sequence, the system is bled.

STEP 16 Start the engine and allow it run for one to two minutes.

STEP 17 Turn off the engine and top off the fluid, if necessary.

Ford hybrids have two separate cooling systems: One is for engine cooling and the other is for hybrid components, called the motor electronics (M/E) cooling system. The engine cooling system is conventional. The M/E cooling system uses an electric water pump **(Figure 14–61)** to move coolant through the inverter, transmission, and a separate radiator mounted next to the conventional radiator **(Figure 14–62)**. The M/E coolant reservoir is located behind the engine coolant reservoir. Although the two systems operate similarly, the M/E cooling system typically operates at lower pressures and temperatures. The fluid levels in both cooling systems must be maintained.

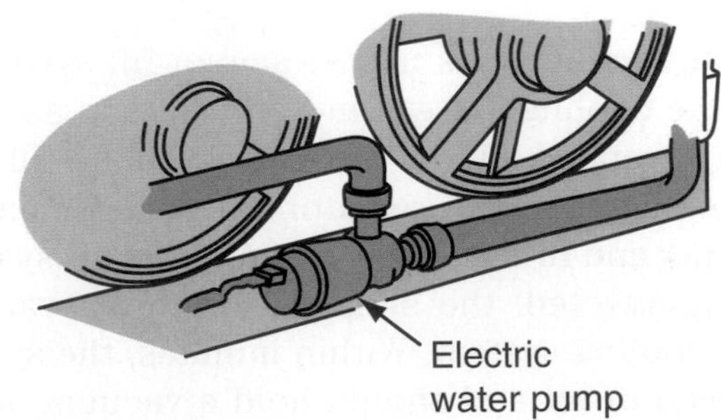

FIGURE 14–61

The electric water pump for the M/E cooling system on a Ford Escape Hybrid.

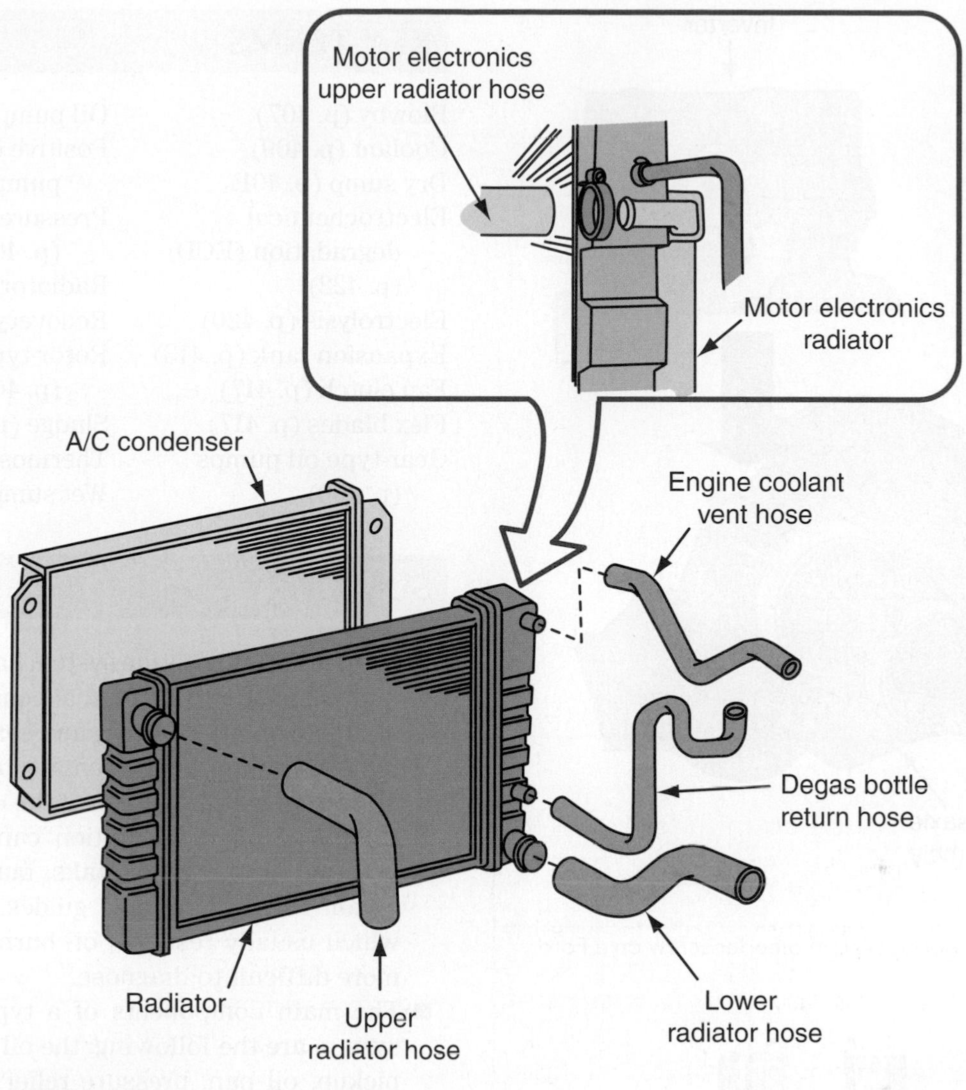

FIGURE 14–62

The radiator assembly for a Ford Escape hybrid.

It is easy to trap air in the M/E cooling system when filling the system. The system is fitted with a bleeder screw at the top of the inverter **(Figure 14–63)**. When servicing the system, make sure the high-voltage system is isolated by having the service connector in the SERVICING/SHIPPING position. Also, wear lineman's gloves because the bleeder screw is very close to the high-voltage cables.

Coolant Exchangers

A coolant exchanger **(Figure 14–64)** is used to remove old coolant from the system and to put in new coolant at the correct mixture. Some coolant exchangers will also leak test the system and flush it. Coolant exchangers are normally powered by shop air, although a few are powered by the vehicle's battery. Most exchangers move through their cycles with the engine off.

The machine is connected to the radiator filler neck, upper radiator hose, or a heater hose. The old coolant is siphoned out of the system and stored in a container. When all coolant is out, the low pressure in the cooling system siphons the fresh coolant mixture out of another container. The entire process, including flushing the system, takes only a few minutes. When complete, the system is free of air and full of coolant.

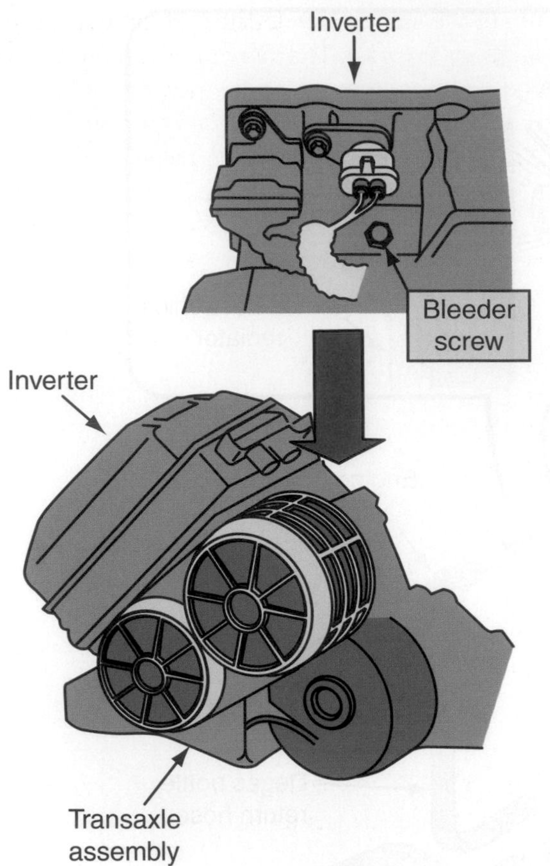

FIGURE 14–63

Location of the cooling system bleeder screw on a Ford Escape Hybrid.

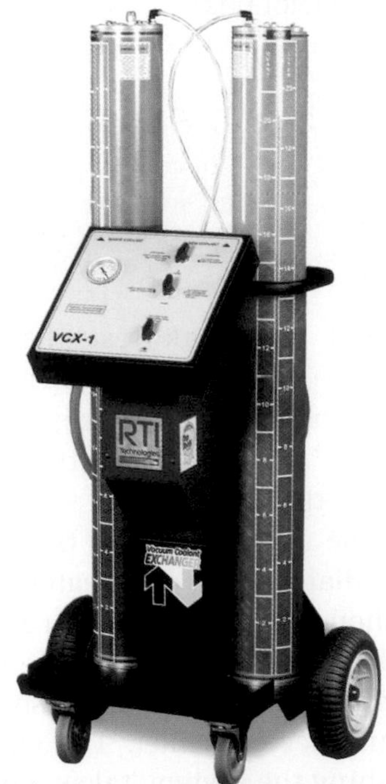

FIGURE 14–64

A coolant exchanger.

MAHLE Service Solutions

KEY TERMS

Blowby (p. 407)
Coolant (p. 409)
Dry sump (p. 401)
Electrochemical degradation (ECD) (p. 422)
Electrolysis (p. 420)
Expansion tank (p. 413)
Fan clutch (p. 417)
Flex blades (p. 417)
Gear-type oil pumps (p. 400)
Oil pump pickup (p. 399)
Positive displacement pumps (p. 399)
Pressure relief valve (p. 400)
Radiator (p. 411)
Recovery tank (p. 413)
Rotor-type oil pump (p. 400)
Sludge (p. 408)
Thermostat (p. 409)
Wet sump (p. 401)

SUMMARY

- An engine's lubrication system has several important purposes: to hold an adequate supply of oil to cool, clean, lubricate, and seal the engine; to remove contaminants from the oil; and to deliver oil to all necessary areas of the engine.
- Excessive oil consumption can be a result of external and internal leaks, faulty accessories, piston rings, and valve guides. Internal leaks, which usually result in oil burning, are usually more difficult to diagnose.
- The main components of a typical lubrication system are the following: the oil pump, oil pump pickup, oil pan, pressure relief valve, oil filter, engine oil passages, engine bearings, crankcase ventilation, oil pressure indicator, oil seals and gaskets, dipstick, and oil coolers.
- The purpose of the oil pump is to supply oil to the various moving parts in the engine. The most commonly used stock pumps are the rotor and gear type. Both are positive displacement pumps. Because the faster the pump turns, the greater the pressure becomes, a pressure-regulating valve is installed to control the maximum oil pressure.
- All automotive vehicles are equipped with either an oil pressure gauge or a low-pressure indicator light. The gauges are either mechanically or electrically operated.
- All oil leaving the oil pump is directed to the oil filter. The filter is a disposable metal container filled with a special type of treated paper or other filter substance that catches and holds the oil's impurities.
- After an engine is rebuilt, a new or rebuilt oil pump is often installed. If the old pump is to be

reused, it should be carefully inspected for wear and thoroughly cleaned.

- The fluid used as coolant today is a mixture of water and antifreeze/coolant. Closed-cooling systems are cooling systems with an expansion or recovery tank. The function of the water pump is to move the coolant efficiently through the system. The radiator transfers heat from the engine to the air passing through it. The thermostat attempts to control the engine's operating temperature by routing the coolant either to the radiator or through the bypass, or sometimes a combination of both.
- V-belts and V-ribbed belts (called serpentine belts) are used to drive water pumps and for power-steering pumps, air-conditioning compressors, generators, and emission-control pumps. The fan delivers additional air to the radiator to maintain proper cooling at low speeds and idle. Because fan air is usually only necessary at idle and low vehicle speeds, various design concepts are used to limit the fan's operation at higher speeds.
- Hollow passages in the block and cylinder heads allow coolant to flow through them. Included in the water jackets are soft plugs and a block drain plug. A temperature indicator is mounted in the dashboard to alert the driver to an overheating condition. A hot-liquid passenger heater is part of the engine cooling system. Radiators for vehicles with automatic transmissions have a sealed heat exchanger located in the coolant outlet tank.
- The basic procedure for testing a vehicle's cooling system includes inspecting the radiator filler neck, inspecting the overflow tube for dents and other obstructions, testing for external leaks, and testing for internal leaks. Most radiator leak repairs require removing the radiator from the vehicle.
- The pressure cap should hold pressure in its range, as indicated on the tester gauge dial, for one minute. A thermostat can be tested in the engine or after it is removed.
- Hoses should be checked for leakage, swelling, and chafing. Any hose that feels mushy or extremely brittle or shows signs of splitting when it is squeezed should be replaced. The majority of water pump failures are attributed to leaks of some sort. Other failures can be attributed to bearing and shaft problems and an occasional cracked casting.
- Fan operation can be checked by spinning the fan by hand. A noticeable wobble or any blade that is not in the same plane as the rest indicates replacement is in order. The fan can also be checked by removing it and laying it on a flat surface. If it is straight, all the blades will touch the surface. Never attempt to repair a damaged fan; replace it. One of the simplest ways to check a fan clutch is to visually inspect it for signs of fluid loss.
- Belt problems are easily discovered by visual inspection or by the screech of slippage. When a new belt is installed, it should be properly adjusted.
- Whenever coolant is changed, a thorough flushing should be performed. The old coolant should be captured and recycled.

REVIEW QUESTIONS

1. List some of the things that influence a radiator's efficiency.
2. What type of radiator pressure cap is found on today's vehicles?
3. Describe the simple test used to determine whether the water pump is causing good circulation.
4. What typically drives an oil pump?
5. What is the name of the component in the lubrication system that prevents excessively high system pressures from occurring as engine speed increases?
6. When should the full-flow oil filter bypass valve open?
   - **a.** when the filter element becomes clogged
   - **b.** when the oil level becomes too low
   - **c.** when the oil pressure exceeds 310 kPa (45 psi)
   - **d.** when the oil pressure is below 310 kPa (45 psi)
7. What will cause reduced oil pressure in an engine?
   - **a.** reduced engine bearing clearance
   - **b.** increased engine bearing clearance
   - **c.** decreased volume of circulating oil
   - **d.** use of a higher viscosity engine oil
8. What is the purpose of the cooling system thermostat?
   - **a.** to limit the maximum engine operating temperature
   - **b.** to maintain the coolest engine temperature possible
   - **c.** to bring the engine to operating temperature quickly
   - **d.** to control the on and off times of the cooling fans

9. What is the purpose of the vacuum valve in a radiator cap?

a. to allow coolant to escape to the radiator overflow tank

b. to allow coolant to return to the radiator from the overflow tank

c. to prevent pressure from building in the cooling system

d. to prevent coolant flow between the radiator and overflow tank

10. What determines when the electric cooling fan is turned off and on?

a. the temperature of the engine

b. the pressure of the coolant

c. the speed of the engine

d. the amount of coolant flow

11. What is the purpose of the fan shroud?

a. to protect the fan from damage

b. to prevent airflow through the radiator

c. to concentrate the airflow through the radiator

d. to protect the radiator from damage

12. Which of the following could be used as an engine oil pump?

a. a centrifugal type

b. a gear type

c. a vane type

d. an impeller type

13. How is the conventional pellet-type thermostat installed?

a. with the pellet toward the radiator

b. with the pellet toward the coolant pump

c. with the pellet toward the bypass hose

d. with the pellet toward the engine

14. For every 6.89 kPa (1 psi) of pressure placed on engine coolant, how is the coolant's boiling point affected?

a. increased by 1.11°C (2°F)

b. increased by 1.66°C (3°F)

c. increased by 2.22°C (4°F)

d. increased by 2.77°C (5°F)

15. What drives the water pump in most automotive applications?

a. gear train

b. crankshaft

c. camshaft

d. impeller

16. Which of the following is not of concern when checking a rotor-type oil pump?

a. cover flatness

b. rotor thickness

c. inner rotor to outer rotor clearance

d. inner rotor to pump housing clearance

17. Which of the following can result from excessive drive belt tension?

a. belt slippage on pulleys

b. components will be driven at reduced speeds

c. quiet operation

d. premature bearing and bushing failure in water pumps, generators, and power-steering pumps

18. When must a thermostat be fully opened?

a. 1.6°C (3°F) above its temperature rating

b. 1.6°C (3°F) above or below its temperature rating

c. 11.1°C (20°F) above the start-to-open temperature

d. 11.1°C (20°F) below the start-to-open temperature

19. What is the recommended antifreeze/water mixture for freeze-up and corrosion protection for moderate climates?

a. 50 percent antifreeze and 50 percent water

b. 60 percent antifreeze and 40 percent water

c. 70 percent antifreeze and 30 percent water

d. 80 percent antifreeze and 20 percent water

20. What could cause abnormally high engine oil pressure?

a. the wrong viscosity oil

b. worn engine bearings

c. a defective oil pressure regulator valve

d. worn oil pump gears

CHAPTER 15
Basics of Electrical Systems

LEARNING OUTCOMES

- Explain the basic principles of electricity.
- Define the terms normally used to describe electricity.
- Use Ohm's law to determine voltage, current, and resistance.
- List the basic types of electrical circuits.
- Describe the differences between a series and a parallel circuit.
- Name the various electrical components and their uses in electrical circuits.
- Describe the different kinds of automotive wiring.
- Explain the principles of magnetism and electromagnetism.

WE SUPPORT THE CANADIAN INTERPROVINCIAL STANDARDS PROGRAM

There is often confusion concerning the terms electrical and electronic. In this book, electrical and electrical systems refer to wiring and electrical parts, such as generators, lights, and voltage regulators. **Electronics** means computers; modules used to control engine and vehicle systems.

A good understanding of electrical principles is important to proper diagnosis of any system that is monitored, controlled, or operated by electricity **(Figure 15–1)**.

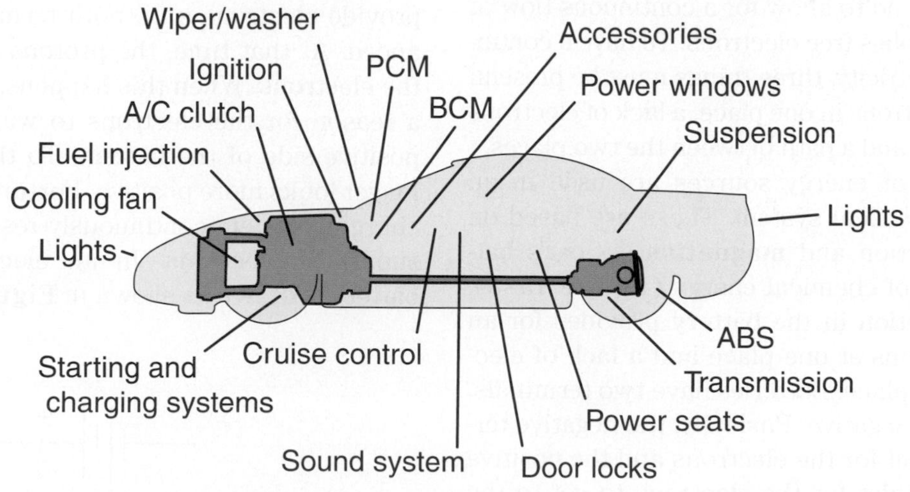

FIGURE 15–1
Basic overview of the electrical system on a late-model car.

BASICS OF ELECTRICITY

Perhaps the one reason why some people find it difficult to understand electricity is that they cannot see it. By actually knowing what it is and what it is *not*, you can easily understand it. Electricity is not magic! It is something that takes place or can take place in everything you know. It not only provides power for lights, TVs, stereos, and refrigerators, but also is the basis for the communications between our brain and the rest of our bodies. Although electricity cannot be

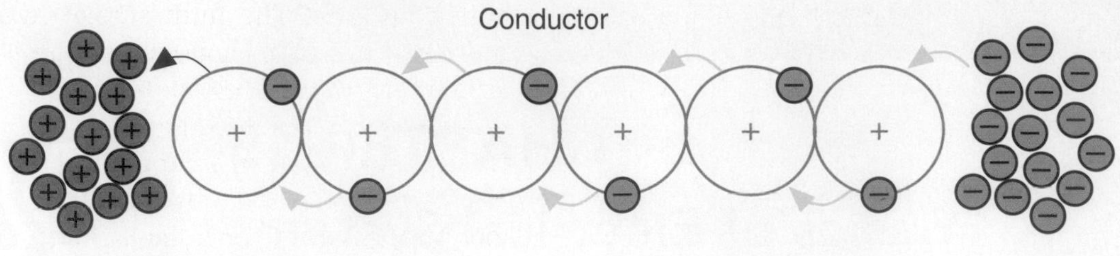

FIGURE 15–2

Electricity is the flow of electrons from one atom to another.

seen, the effects of it can be seen, felt, heard, and smelled. One of the most common displays of electricity is a lightning bolt. Lightning is electricity—a large amount of electricity. The power of lightning is incredible. Using the power from much smaller amounts of electricity to perform some work is the basis for an automobile's electrical system.

Flow of Electricity

Electricity is the flow of electrons from one atom to another **(Figure 15–2)**. The release of energy as one electron leaves the orbit of one atom and jumps into the orbit of another is electrical energy. The key to creating electricity is to provide a reason for the electrons to move to another atom.

There is a natural attraction of electrons to protons. Electrons have a negative charge and are attracted to something with a positive charge. When an electron leaves the orbit of an atom, the atom then has a positive charge. An electron moves from one atom to another because the atom next to it appears to be more positive than the one it is orbiting around.

An electrical power source provides for a more positive charge, and to allow for a continuous flow of electricity, it supplies free electrons. To have a continuous flow of electricity, three things must be present: an excess of electrons in one place, a lack of electrons in another place, and a path between the two places.

Two power or energy sources are used in an automobile's electrical system. These are based on a chemical reaction and **magnetism**. A car's battery is a source of chemical energy **(Figure 15–3)**. A chemical reaction in the battery provides for an excess of electrons at one place and a lack of electrons in another place. Batteries have two terminals, a positive and a negative. Basically, the negative terminal is the outlet for the electrons and the positive terminal is the inlet for the electrons to get to the protons. The chemical reaction in a battery causes a lack of electrons at the positive (+) terminal and an excess at the negative (–) terminal. This creates an electrical imbalance, causing the electrons to flow through the path provided by the electrical system.

FIGURE 15–3

An automotive battery.

East Penn Canada

The chemical process in the battery continues to provide electrons until both terminals become balanced. At that time, the protons are matched with the electrons. When this happens, there is no longer a reason for the electrons to want to move to the positive side of the battery. To the electrons, it no longer looks more positive. Fortunately, the vehicle's charging system continuously restores the battery's supply of electrons. In an electrical diagram, a battery is drawn as shown in **Figure 15–4**.

FIGURE 15–4

The symbol for a battery.

FIGURE 15–5

A late-model generator.

Robert Bosch Corp.

Electricity and magnetism are interrelated. One can be used to produce the other. Moving a wire (a conductor) through an already existing magnetic field (such as a permanent magnet) can produce electricity. This process of producing electricity through magnetism is called **induction**. The heart of a vehicle's charging system is the AC generator **(Figure 15–5)**. A magnetic field, driven by the crankshaft via a drive belt, rotates through a coil of wire, producing electricity. The amount of electricity produced depends on a number of things: the strength of the magnetic field, the number of wires that are passed by the magnetic field, and the speed at which the magnetic field moves past the wires.

Electricity is also produced by chemical, photoelectrical, thermoelectrical, and piezoelectrical reactions. These sources of electricity are used throughout a modern automobile. Most are the basis of operation for electronic sensors and are discussed as those sensors are introduced in this book. The two most common ways of producing electricity are through electromagnetic induction and chemical reaction. These are the main topics of the chapters in this section of the book.

ELECTRICAL TERMS

Electrical **current** describes the movement or flow of electricity. The greater the number of electrons flowing past a given point in a given amount of time, the more current the circuit has. The unit for measuring electrical current is the **ampere**, usually called an amp. The term ampere was assigned to units of current in honour of André Ampère, who studied the relationship between electricity and magnetism. The instrument used to measure electrical current flow in a circuit is called an ammeter.

When electricity flows, millions of electrons are moving past any given point at the speed of light. The electrical charge of any one electron is extremely small. It takes millions of electrons to make a charge that can be measured. In fact, one ampere of current means that 6.24 billion electrons are flowing past a given point in one second.

SHOP TALK

Current flow may be expressed in coulombs. One coulomb represents the amount of electric charge moved by one ampere of current in one second. One ampere represents 6.241 509 629 152 65×10^{18} electrons.

There are two types of current: **direct current (DC)** and **alternating current (AC)**. In direct current, the electrons flow in one direction only. In AC, the electrons change direction at a fixed rate. Typically, an automobile uses DC, whereas the current in homes and buildings is AC. Some components of the automobile generate or use AC. These are discussed in later chapters.

There are many theories about the direction of current flow. The conventional theory states that current flows from positive to negative. The electron theory says current moves from negative to positive. And the hole-flow theory basically says something is moving in both directions. Remember that a theory is not a fact. It is a concept that is yet to be proved wrong or right. Therefore, only one theory about current flow is correct and the rest are wrong. For the purposes of this book and for your own understanding of electricity, current flow **(Figure 15–6)**

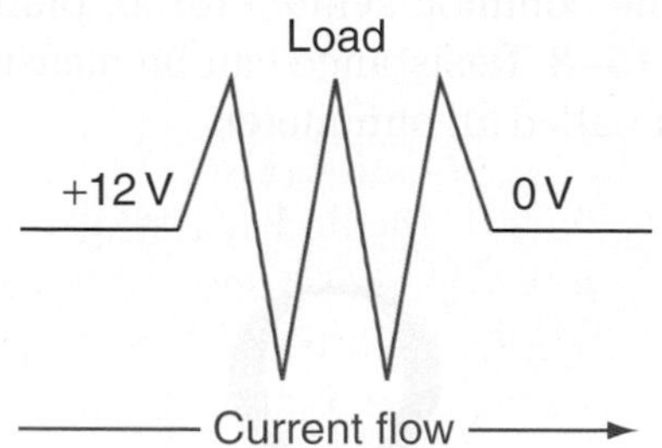

FIGURE 15–6

Current moves from a point of higher potential to a point of lower potential.

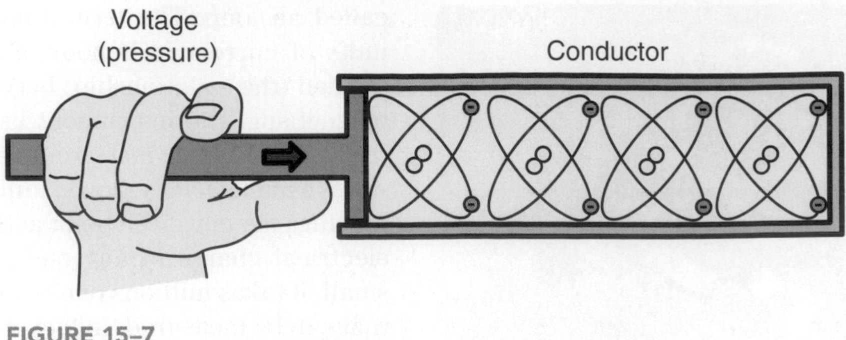

FIGURE 15–7

Voltage is electrical pressure.

is described as moving from a point of higher potential (voltage) to a point of lower potential (voltage). This statement may not be absolutely correct, but it is sound and is based on what can be observed.

Voltage is electrical pressure **(Figure 15–7)**. It is the force developed by the attraction of the electrons to protons. The more positive one side of the circuit is, the more voltage is present in the circuit. Voltage does not flow; it is the pressure that causes current flow. To have current flow, some force is needed to move the electrons between atoms. This force is the pressure that exists between a positive and negative point within an electrical circuit. This force, also called **electromotive force (EMF)**, is measured in units called volts. One volt is the amount of pressure required to move one ampere of current through a resistance of one ohm. Voltage is measured by an instrument called a voltmeter. The unit of measurement for electrical pressure was so named to honour Alessandro Volta, who, in 1800, made the first electrical battery.

When any substance flows, it meets **resistance**. The resistance to electrical flow can be measured. The resistance to current flow produces heat. This heat can be measured to determine the amount of resistance. A unit of measured resistance is called an **ohm**. The common symbol for an ohm is shown in **Figure 15–8**. Resistance can be measured by an instrument called an ohmmeter.

FIGURE 15–8

Common symbol for an ohm.

Alternating Current

AC constantly changes in voltage and direction. If a graph is used to represent the amount of DC voltage available from a battery during a fixed period, the line on the graph will be flat, which represents a constant voltage. If AC voltage is shown on a graph, it will appear as a **sine wave (Figure 15–9)**. The sine wave shows AC changing in amplitude (strength) and direction. The highest positive voltage equals the highest negative voltage. The movement of the AC from its peak at the positive side of the graph to the negative side and then back to the positive peak is commonly referred to as peak-to-peak value. This value represents the amount of voltage available at a point. During each complete cycle of AC, there are always two maximum or peak values, one for the positive half-cycle and the other for the negative half-cycle. The difference between the peak positive value and the peak negative value is used to measure AC voltages **(Figure 15–10)**.

AC does not have a constant value; therefore, as it passes through a resistance, nearly 29 percent less heat is produced when compared to DC. This is one reason that AC is preferred over DC when powering motors and other electrical devices.

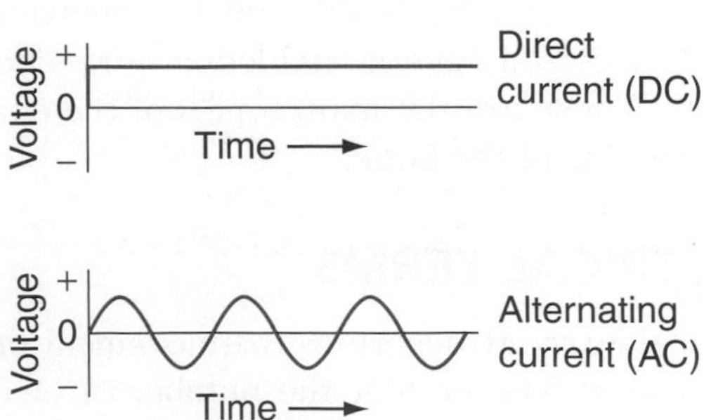

FIGURE 15–9

The difference between DC and AC.

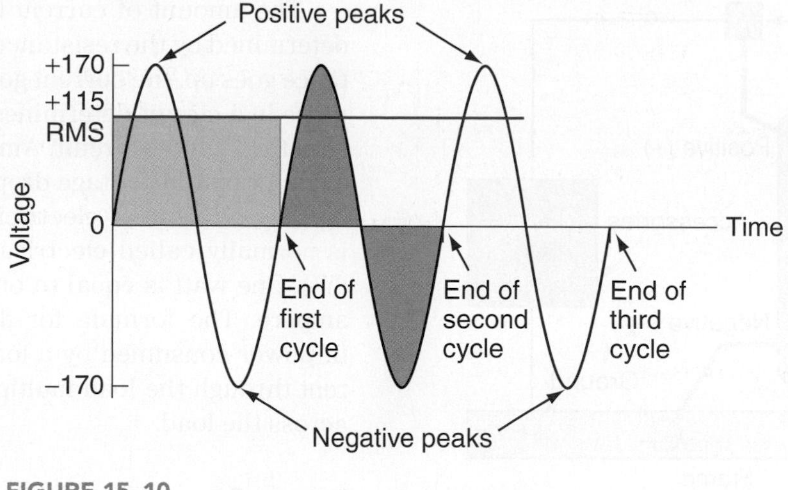

FIGURE 15–10

The action and measurement of alternating current.

The lack of heat also causes us to look at AC values differently than the same values in a DC circuit. AC has an effective value of one ampere when it produces heat in a given resistance at the same rate as does one ampere of DC. The effective value of an AC is equal to 0.707 times its maximum, or peak, current value. Because AC is caused by an alternating voltage, the ratio of the effective voltage value to the maximum voltage value is the same as the ratio of the effective current to the maximum current, or 0.707 times the maximum value. AC voltage measurements are often expressed in terms of root mean square (RMS) values.

AC voltage and current change constantly, and their values must be viewed as average or effective. When AC is applied to a resistance, as the actual voltage changes in value and direction, so does the current. In fact, the change of current is in phase with the change in voltage. An in-phase condition exists when the sine waves of voltage and current are precisely in step with one another. The two sine waves go through their maximum and minimum points at the same time and in the same direction. In some circuits, several sine waves can be in phase with each other.

If a circuit has two or more voltage pulses but each has its own sine wave that begins and ends its cycle at a different time, the waves are out of phase. If two sine waves are 180° out of phase, they will cancel each other out if they have the same voltage and current. If two or more sine waves are out of phase but do not cancel each other, the effective voltage and current are determined by the position and direction of the sine wave at a given point within the circuit **(Figure 15–11)**.

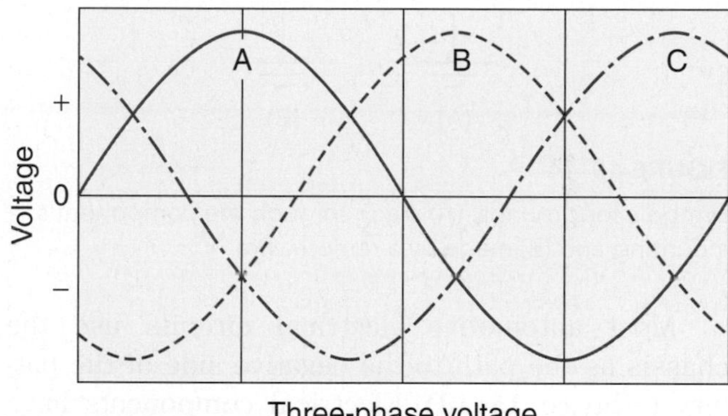

FIGURE 15–11

The sine waves of three-phase AC.

Circuit Terminology

An electrical circuit is considered complete when there is a path that connects the positive and negative terminals of the electrical power source. A completed circuit is called a **closed circuit**, whereas an incomplete circuit is called an **open circuit**. When a circuit is complete, there is **continuity**. Conductors are drawn in electrical diagrams as a line connecting two points, as shown in **Figure 15–12**.

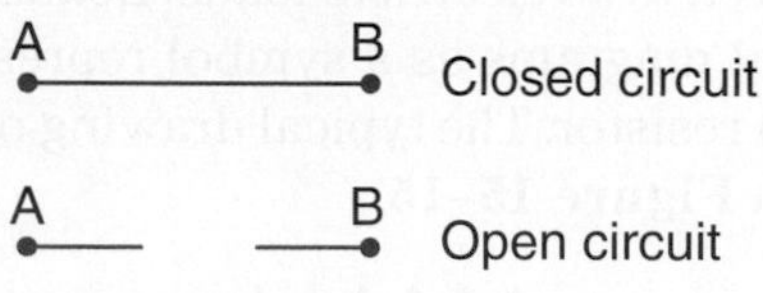

FIGURE 15–12

A simple circuit consists of a voltage source, conductors, and a resistance or load.

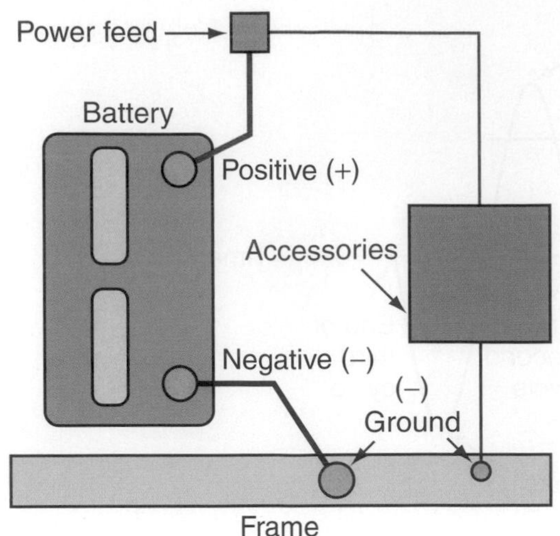

FIGURE 15–13

Most automotive electrical circuits use the chassis as the conductor for the negative side of the battery.

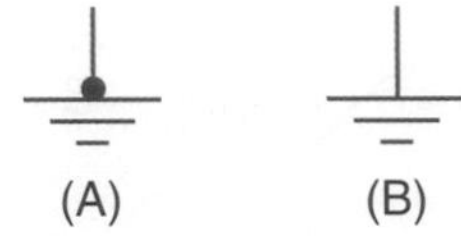

FIGURE 15–14

Symbols for grounds: (A) made through the component's mounting and (B) made by a remote wire.

Most automotive electrical circuits use the chassis as the path to the negative side of the battery **(Figure 15–13)**. Electrical components have a lead that connects them to the chassis called the **chassis ground** connections. These connections can be made through a wire or through the mounting of the component. Chassis ground connections are drawn to show which type of connection is normal for that part **(Figure 15–14)**. When the ground is made through the mounting of the component, the connection is represented with the drawing A. When the ground is made by a wire that connects to the chassis, the connection is shown as B.

In a complete circuit, electrical current can be controlled and applied to do useful work, such as light a headlight or energize a starter motor. Components that use electrical power put a load on the circuit and consume electrical energy. These components are often referred to as electrical loads. Loads are drawn in electrical diagrams as a symbol representing the part or as a resistor. The typical drawing of a resistor is shown in **Figure 15–15**.

FIGURE 15–15

The symbol for a resistor.

The amount of current that flows in a circuit is determined by the resistance in that circuit. As resistance goes up, the current goes down. The total resistance in a circuit determines how much current will flow through the circuit. Amperage stays constant in a circuit but the voltage drops as it powers a load.

The amount of electricity consumed by a load is normally called electrical power usage or watts (W). One watt is equal to one volt multiplied by one ampere. The formula for determining the amount of power consumed by a load is the amount of current through the load multiplied by the voltage drop across the load.

Power Sources

Today's vehicles operate on 12-volt electrical systems. However, the battery stores about 14 volts, although it is rated at 12 volts, and the charging system puts out 14 to 15 volts while the engine is running. The primary source of electrical power when the engine is running is the charging system, so it is fair to say that an automobile's electrical system is a 14-volt system.

Prior to 1954, vehicles had 6-volt systems. The electrical demands of accessories, such as power windows and seats, put a severe strain on the 6-volt battery. With a 12-volt system, the charging system had to work less hard, and there was plenty of electrical power for the electrical accessories.

The increase in voltage also allowed wire sizes to decrease because the amperage required to power devices was reduced. To explain this, consider an accessory that required 20 amps to operate in a 6-volt system (120 watts). When the voltage was increased to 12 volts, the system drew only 10 amps.

Today we are faced with the same situation. The use of computers and the need to keep their memories fresh has put a drain on the battery even when the engine is not running. Plus, the number of electrical accessories has grown and will continue to grow.

Today's vehicles are very sensitive to voltage change. In fact, their overall efficiency depends on a constant voltage. The demands of new technology make it difficult to maintain a constant voltage, and engineers have determined that system voltage must be increased to meet those demands. As vehicles evolve, emissions, fuel economy, comfort, convenience, and safety features will put more of a drain on the electrical system. This increased demand is the result of converting purely mechanical systems into electromechanical systems, such as steering, suspension, and braking systems. It has been estimated that in a few years the continuous electrical power demand will be 3000 to 7000 W. Current 14-volt systems are rated at 800 to 1500 W.

To meet these demands there are two possible solutions: increase the amperage capacity of the battery and charging system or increase system voltage.

Larger capacity batteries and generators are only a band-aid solution. Because the generator is driven by engine power, more power from the engine will be required to keep the higher-capacity battery charged. Therefore, overall efficiency will decrease.

By moving to a higher system voltage, the battery will need to be larger and heavier. However, because system amperage will be lower, wire size will be smaller and perhaps the weight gain at the battery will be offset by the decreased weight of the wiring.

All of the advantages of moving from 6- to 12-volt systems apply to the move from 12 to 42 volts. But why 42 volts? Engineers decided to take advantage of the fact that a 12-volt battery actually holds a 14-volt charge (3 times 14 volts equals 42 volts). Forty-two-volt systems are also desirable for safety reasons. Sixty volts can stop a person's heart from beating; therefore, 42-volt systems allow for a margin of safety.

During the transition from 14- to 42-volt systems, vehicles may be fitted with dual voltage batteries that provide 12 and 36 volts **(Figure 15–16)**. This split voltage system will provide 42 volts to high-power applications such as the power-steering, traction-control, brake, and engine-cooling systems. The 14-volt system will power low-load systems, such as lights, air conditioning, power door locks, radios, and navigation systems.

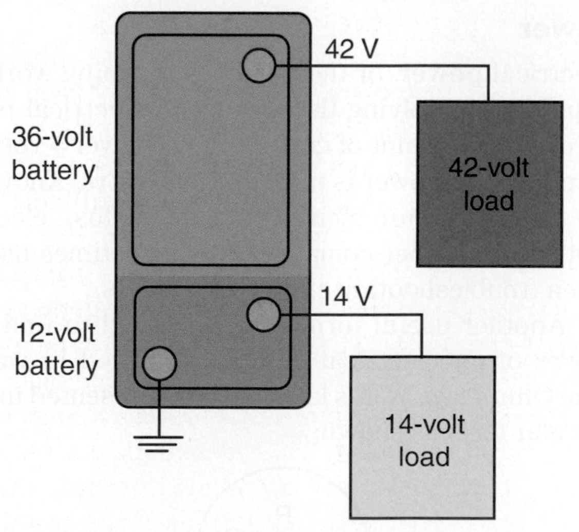

FIGURE 15–16

A split voltage arrangement providing 14 and 42 volts from a 36-volt battery.

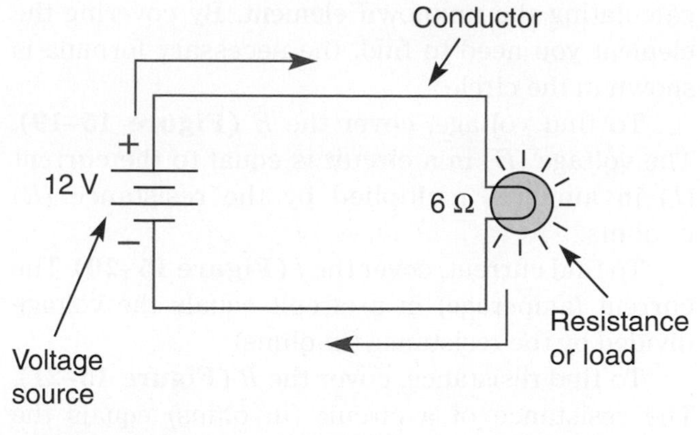

FIGURE 15–17

A simple electrical circuit.

Ohm's Law

In 1827, a German mathematics professor, Georg Ohm, published a book that included his explanation of the behaviour of electricity. His thoughts have become the basis for a true understanding of electricity. He found that it takes one volt of electrical pressure to push one amp of electrical current through one ohm of resistance. Current is directly proportional to voltage if resistance remains the same and inversely proportional to resistance if voltage remains the same. This statement is the basic law of electricity. It is known as **Ohm's law**.

A simple electrical circuit is a load connected to a voltage source by conductors. The resistor could be a fog light, the voltage source could be a battery, and the conductor could be a copper wire **(Figure 15–17)**.

In any electrical circuit, current (*I*), resistance (*R*), and voltage (*E*) are mathematically related. This relationship is expressed in a mathematical statement of Ohm's law. Ohm's law can be applied to the entire circuit or to any part of a circuit. When any two factors are known, the third factor can be found by using Ohm's law. Using the circle shown in **Figure 15–18**, you can easily find the formula for

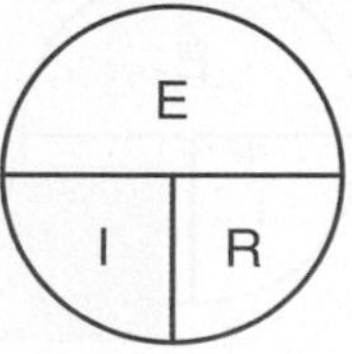

FIGURE 15–18

Ohm's law.

calculating the unknown element. By covering the element you need to find, the necessary formula is shown in the circle.

To find voltage, cover the *E* **(Figure 15–19)**. The voltage (*E*) in a circuit is equal to the current (*I*) in amperes multiplied by the resistance (*R*) in ohms.

To find current, cover the *I* **(Figure 15–20)**. The current (amperage) in a circuit equals the voltage divided by the resistance (in ohms).

To find resistance, cover the *R* **(Figure 15–21)**. The resistance of a circuit (in ohms) equals the voltage divided by the current (in amperes).

It is very important for technicians to understand Ohm's law. It explains how an increase or decrease in voltage, resistance, or current affects a circuit.

For example, if the fog light in Figure 15–17 has a 6-ohm resistance, how many amperes does it use to operate? Because cars and light trucks have a 12-volt battery and you know two of the factors in the fog light circuit, it is simple to solve for the third.

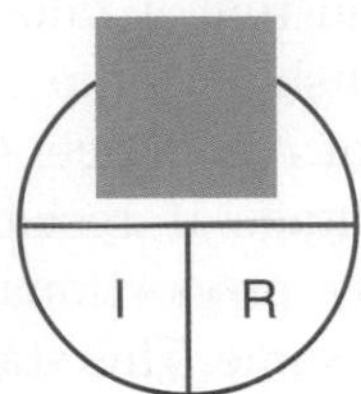

FIGURE 15–19

To find voltage, cover the *E* and use the exposed formula.

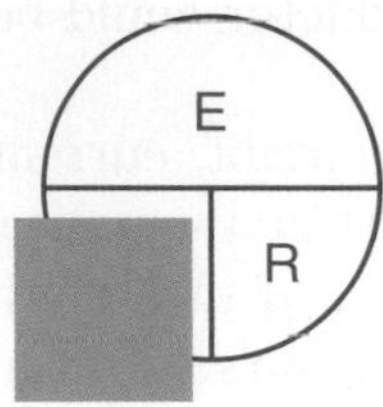

FIGURE 15–20

To find current, cover the *I* and use the exposed formula.

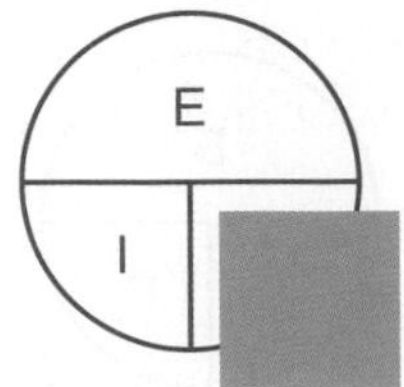

FIGURE 15–21

To find resistance, cover the *R* and use the exposed formula.

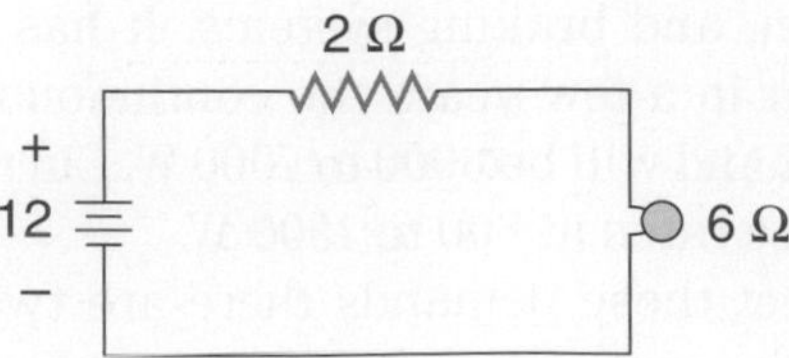

FIGURE 15–22

This one is the same circuit as Figure 15–17, but it has a corroded wire, represented by the additional resistor.

$$I\,(\text{unknown}) = \frac{E\,(12\text{ volts})}{R\,(6\text{ ohms})}$$

$$I = \frac{12}{6}$$

$$I = 2\text{ amperes}$$

In a clean, correctly wired circuit, the fog lights will draw 2 amps of current. What would happen if resistance in the circuit increases due to corroded or damaged wire or connections? If the corroded connections add 2 ohms of resistance to the fog light circuit, the total resistance is 8 ohms **(Figure 15–22)**. The amount of current flowing through the circuit for the lights decreases.

$$I = \frac{12}{6+2} = \frac{12}{8}$$

$$I = 1.5\text{ amps}$$

If the lights are designed to operate at 2 amps, this decrease to 1.5 amps causes them to burn dimly. Cleaning the corrosion away or installing new wires and connectors eliminates the unwanted resistance; the correct amount of current will flow through the circuit, allowing the lamp to burn as brightly as it should.

Power

Electrical **power**, or the rate of performing work, is found by multiplying the amount of electrical pressure by thc amount of current flow (power = voltage × amperage). Power is measured in **watts**. Knowing the power requirements of light bulbs, electric motors, and other components is sometimes useful when troubleshooting electrical systems.

Another useful formula is one used to find the power of an electrical circuit expressed in watts. Like Ohm's law, Watt's law may be represented in the circular form as shown:

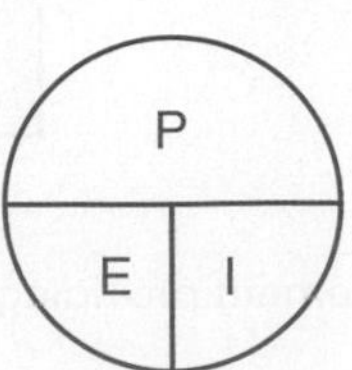

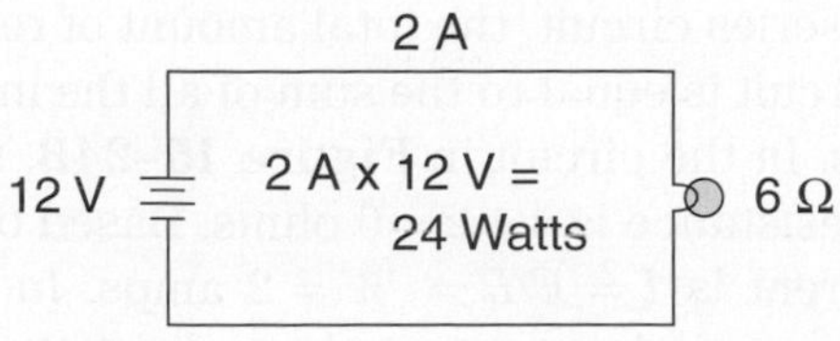

FIGURE 15–23
Electrical power is calculated by multiplying voltage by current.

That is, power (P) in watts equals the voltage (E) multiplied by the current (I) in amperes **(Figure 15–23)**. This relationship is known as **Watt's law**. Looking back at the example of the fog light circuit, we can calculate the amount of power used or heat generated by the fog light.

$$P = 12\text{ V} \times 2\text{ amps}$$
$$P = 24\text{ W}$$

The normal fog light generates 24 W of power, whereas the corroded fog light circuit produces the following.

$$P = 12\text{ V} \times 1.5\text{ amps}$$
$$P = 18\text{ W}$$

This reduction in power or heat explains the decrease in bulb brightness.

CONDUCTORS AND INSULATORS

Controlling and routing the flow of electricity requires the use of materials known as conductors and insulators. **Conductors** are materials with a low resistance to the flow of current. If the number of electrons in the outer shell or ring of an atom is less than four, the force holding them in place is weak. The voltage needed to move these electrons and create current flow is relatively small. Most metals, such as copper, silver, and aluminum, are excellent conductors.

When the number of electrons in the outer ring is greater than four, the force holding them in orbit is very strong and very high voltages are needed to move them. These materials are known as **insulators**. They resist the flow of current. Thermal plastics are the most common electrical insulators used today. They can resist heat, moisture, and corrosion without breaking down.

CAUTION!

Your body is a good conductor of electricity. Remember this when working on a vehicle's electrical system. Always observe all electrical safety rules.

Copper wire is by far the most popular conductor used in automotive electrical systems. Wire wound inside electrical units, such as ignition coils and generators, usually has a very thin baked-on insulating coating. External wiring is normally covered with a plastic-type insulating material that is highly resistant to environmental factors like heat, vibration, and moisture. Where flexibility is required, the copper wire is made of a large number of very small strands of wire woven together.

The resistance of a uniform, circular cross-section copper wire depends on the length of the wire, the diameter of the wire, and the temperature of the wire. If the length is doubled, the resistance between the wire ends is doubled. The longer the wire, the greater the resistance. If the diameter of a wire is doubled, the resistance for any given length is cut in half. The larger the wire's diameter, the lower the resistance.

In any circuit, the smallest wire that will not cause excessive voltage drop is used to minimize cost.

The other important factor affecting the resistance of a copper wire is temperature. As the temperature increases, the resistance increases. The effects of temperature are very important in the design of electrical equipment. Excessive resistance caused by normal temperature increases can hurt the performance of the equipment.

Heat is developed in any wire carrying current because of the resistance in the wire. If the heat becomes excessive, the insulation will be damaged. Resistance occurs when electrons collide as current flows through the conductor. These collisions cause friction, which in turn generates heat.

Circuits

A complete electrical circuit exists when electrons flow along a path between two points. In a complete circuit, resistance must be low enough to allow the available voltage to push electrons between the two points. Most automotive circuits contain five basic parts.

1. Power sources, such as a battery or alternator, that provide the voltage needed to cause electron flow
2. Conductors, such as copper wires, that provide a path for current flow
3. Loads, which are devices that use electricity to perform work, such as light bulbs, electric motors, or resistors

4. Controllers, such as switches or relays, that control or direct the flow of electrons
5. Circuit protection devices, such as fuses or circuit breakers

There are also three basic types of circuits used in automotive electrical systems: series circuits, parallel circuits, and series-parallel circuits. Each circuit type has its own characteristics regarding amperage, voltage, and resistance.

SERIES CIRCUITS A **series circuit** consists of one or more resistors connected to a voltage source with only one path for electron flow. For example, a simple series circuit consists of a resistor (2 ohms in this example) connected to a 12-volt battery **(Figure 15–24A)**. The current can be determined by applying Ohm's law.

$$I = \frac{E}{R} = \frac{12}{2} = 6 \text{ amperes}$$

Another series circuit may contain a 2-ohm resistor and a 4-ohm resistor connected to a 12-volt battery **(Figure 15–24B)**. The word *series* is given to a circuit in which the same amount of current is present throughout the circuit. The current that flows through one resistor also flows through other resistors in the circuit. As that amount of current leaves the battery, it flows through the conductor to the first resistor. At the resistor, some electrical pressure or voltage is dropped as the current flows through it. The decreased amount of voltage is then applied to the next resistor as current flows to it. By the time the current is flowing in the conductor leading back to the battery, all voltage has been dropped. All of the source voltage available to the circuit is dropped by the resistors in the circuit.

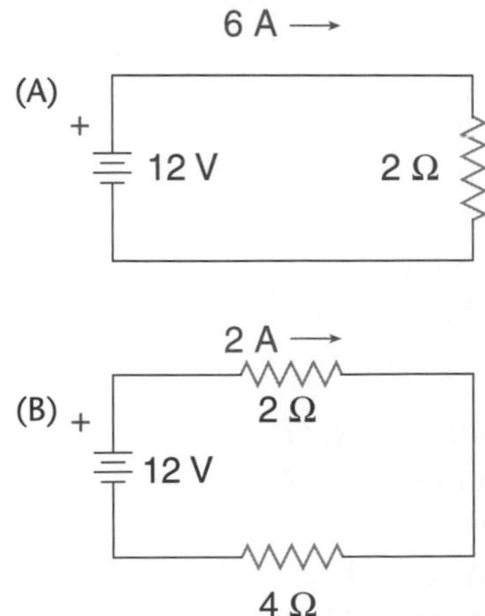

FIGURE 15–24

In a series circuit, the same amount of current flows through the entire circuit.

In a series circuit, the total amount of resistance in the circuit is equal to the sum of all the individual resistors. In the circuit in **Figure 15–24B**, the total circuit resistance is 4 + 2 = 6 ohms. Based on Ohm's law, current is $I = E/R = {}^{12}\!/\!_{6}$ = 2 amps. In a series circuit, current is constant throughout the circuit. Therefore, 2 amps of current flows through the conductors and both resistors.

Ohm's law can be used to determine the voltage drop across parts of the circuit. For the 2-ohm resistor, $E = IR = 2 \times 2 = 4$ V. For the 4-ohm resistor, $E = 2 \times 4 = 8$ V. These values are called **voltage drops**. The sum of all voltage drops in a series circuit must equal the source voltage, or 4 + 8 = 12 V.

An ammeter connected anywhere in this circuit will read 2 amps, and a voltmeter connected across each of the resistors will read 4 and 8 volts, as shown in **Figure 15–25**.

All calculations for a series circuit work in the same way no matter how many resistors there are in series. Consider the circuit in **Figure 15–26**. This circuit has four resistors in series with each other. The total resistance is 12 ohms (5 ohms + 2 ohms + 4 ohms + 1 ohm). Using Ohm's law, we can see

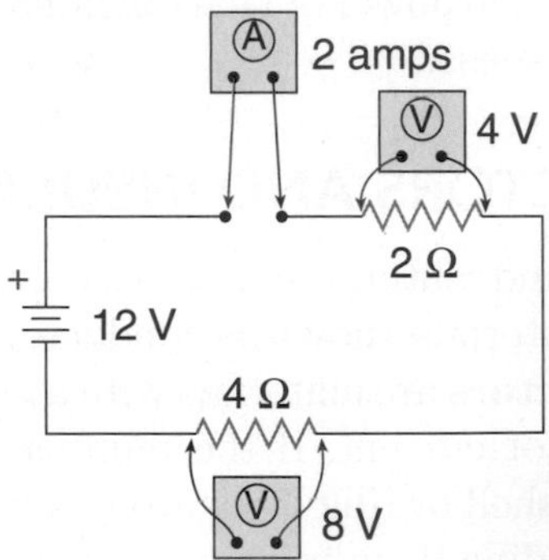

FIGURE 15–25

Measuring the current and voltage drops in the circuit.

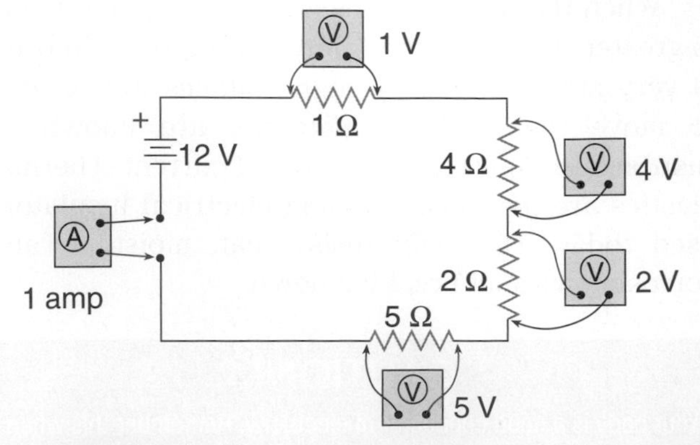

FIGURE 15–26

Values in the series circuit.

that the circuit current is 1 amp ($I = E/R = {}^{12}\!/_{12} = 1$ amp). We can also use Ohm's law to determine the voltage drop across each resistor in the circuit. For example, because the circuit current is 1 amp, 4 volts are dropped by the 4-ohm resistor ($E = I \times R = 1$ amp $\times$ 4 ohms = 4 V).

A series circuit is characterized by the following four facts:

1. There is one path for current to flow.
2. The circuit's current is determined by the total amount of resistance and source voltage in the circuit; it is constant throughout the circuit.
3. The sum of the voltage drops equals the source voltage.
4. The total resistance is equal to the sum of all resistances in the circuit.

PARALLEL CIRCUITS A **parallel circuit** provides two or more different paths for the current to flow through. Each path has separate resistors (loads) and can operate independently of the other paths. The different paths for current flow are commonly called the legs of a parallel circuit.

Consider the circuit shown in **Figure 15–27**. Two 3-ohm resistors are connected to a 12-volt battery. The resistors are in parallel with each other, because the battery voltage (12 V) is applied to each resistor and they have a common negative lead. The current through each resistor or leg can be determined by applying Ohm's law. For the 3-ohm resistors, $I = E/R = {}^{12}\!/_{3} =$ 4 amps. Therefore, the total circuit current supplied by the battery is 4 + 4 = 8 amps. Using Ohm's law, we find that 12 volts are dropped by both resistors **(Figure 15–28)**.

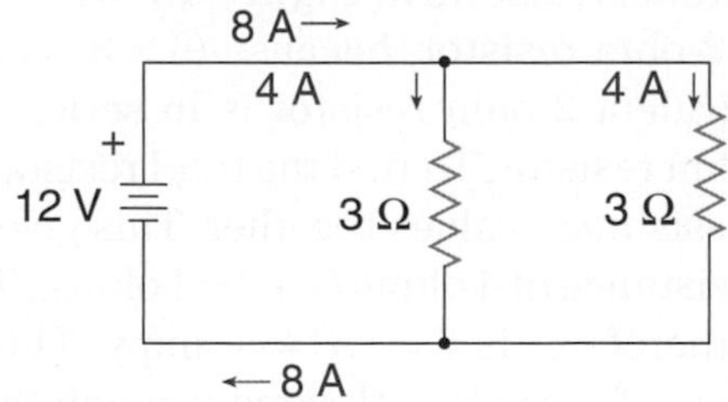

FIGURE 15–27

A simple parallel circuit.

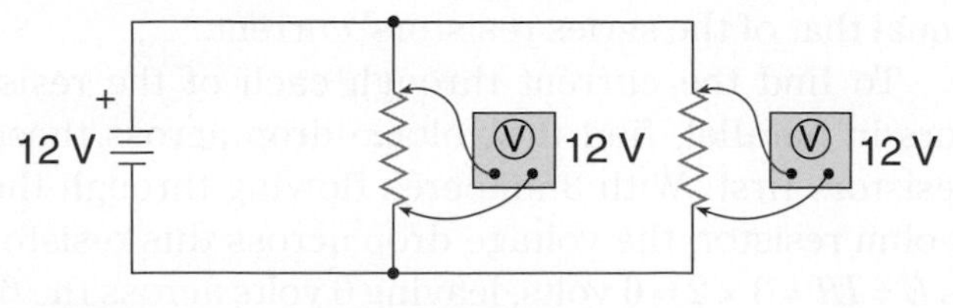

FIGURE 15–28

A parallel circuit with voltage drops shown.

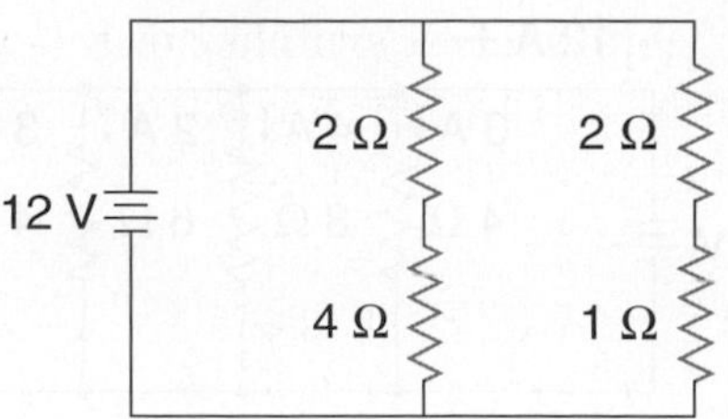

FIGURE 15–29

Series circuits within a parallel circuit.

Resistances are not added up to calculate the total resistance in a parallel circuit. Rather, they can be determined by dividing the product of their ohm values by the sum of their ohm values. This formula works only when the circuit has two parallel legs.

$$\frac{3 \text{ ohms} \times 3 \text{ ohms}}{3 \text{ ohms} + 3 \text{ ohms}} = \frac{9}{6} = 1.5 \text{ ohms}$$

Total resistance can also be calculated by using Ohm's law if you know the total circuit current and the voltage ($R = E/I = {}^{12}\!/_{8} = 1.5$ ohms).

Consider another parallel circuit, **Figure 15–29**. In this circuit, there are two legs and four resistors. Each leg has two resistors in series. One leg has a 4-ohm and a 2-ohm resistor. The total resistance on that leg is 6 ohms. The other leg has a 1-ohm and a 2-ohm resistor. The total resistance of that leg is 3 ohms. Therefore, we have 6 ohms in parallel with 3 ohms.

Current flow through the circuit can be calculated by different methods. Using Ohm's law, we know that $I = E/R$. If we take the total resistance of each leg and divide it into the voltage, we then know the current through that leg. Because total circuit current is equal to the sum of the current flows through each leg, we simply add the current across each leg together. This will give us total circuit current.

Leg1: $I = E/R = {}^{12}\!/_{6} = 2$ amps

Leg2: $I = E/R = {}^{12}\!/_{3} = 4$ amps

2 amps + 4 amps = 6 amps = total circuit current

Circuit current can also be determined by finding the total resistance of the circuit. To do this, the product-over-sum formula is used. By dividing this total into the voltage, total circuit current is known.

$$\frac{\text{Leg1} \times \text{Leg2}}{\text{Leg1} + \text{Leg2}} = \frac{6 \times 3}{6 + 3} = \frac{18}{9} = 2 \text{ ohms}$$

because $I = E/R$, $I = {}^{12}\!/_{2}$, $I = 6$ amps (total circuit current)

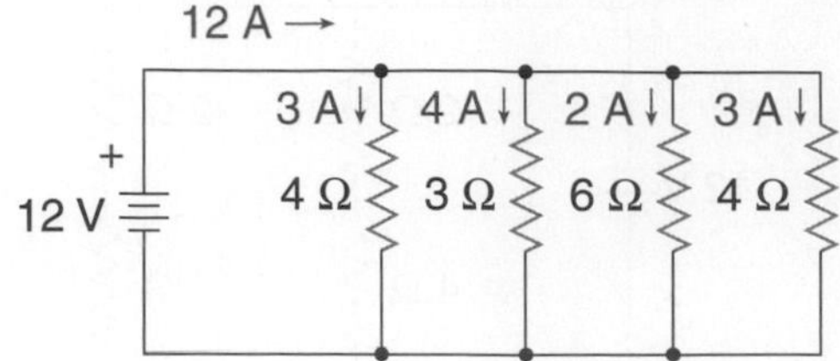

FIGURE 15–30

A parallel circuit with four legs.

When a circuit has more than two legs, the reciprocal formula should be used to determine total circuit resistance. The formula follows:

$$\frac{1}{\frac{1}{R_1}+\frac{1}{R_2}+\frac{1}{R_3}+\cdots\frac{1}{R_n}}$$

To demonstrate how to use this formula, consider the circuit in **Figure 15–30**. Here is a parallel circuit with four legs. The resistances across each leg are 4, 3, 6, and 4 ohms. Using the reciprocal formula, we will find that the total resistance of the circuit is 1 ohm. (Note that the total resistance is lower than the leg with the lowest resistance.)

$$\frac{1}{\frac{1}{4}+\frac{1}{3}+\frac{1}{6}+\frac{1}{4}}=$$

$$\frac{1}{\frac{3}{12}+\frac{4}{12}+\frac{2}{12}+\frac{3}{12}}=\frac{1}{\frac{12}{12}}=\frac{1}{1}=1$$

The total resistance of this circuit could also have been found by calculating the current across each leg and then adding them together to get the total circuit current. Using Ohm's law, if you divide the voltage by the total circuit current, you will get total resistance.

Leg1: $I = E/R = 12/4 = 3$ amps

Leg2: $I = E/R = 12/3 = 4$ amps

Leg3: $I = E/R = 12/6 = 2$ amps

Leg4: $I = E/R = 12/4 = 3$ amps

Total circuit current $= 3 + 4 + 2 + 3 = 12$ amps, then

$R = E/I = 12/12 = 1$ ohm

A parallel circuit is characterized by the following facts:

1. Current has more than one path to flow.
2. Total circuit resistance is always lower than the resistance of the branch with the lowest total resistance.
3. The sum of the current on each branch equals the total circuit current.
4. Voltage applied across each branch is the same as the applied voltage.

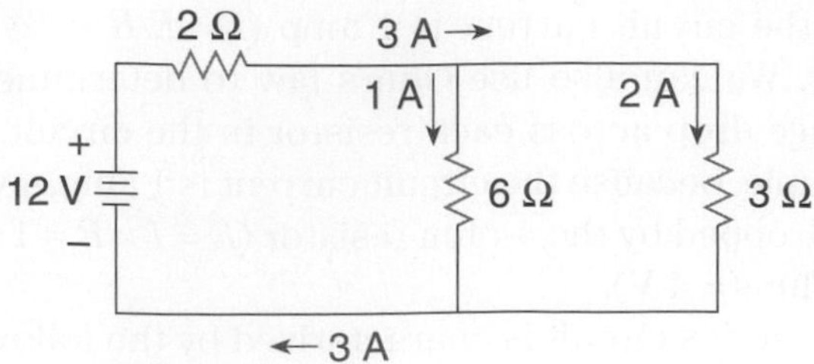

FIGURE 15–31

In a series-parallel circuit, the sum of the currents through the legs must equal the current through the series part of the circuit.

SERIES-PARALLEL CIRCUITS In a **series-parallel circuit**, both series and parallel combinations exist in the same circuit. If you are faced with the task of calculating the values in a series-parallel circuit, determine all values of the parallel circuit(s) first. By looking carefully at a series-parallel circuit, you will find that it is nothing more than one or more parallel circuits in series with each other or in series with some other resistance.

A series-parallel circuit is illustrated in **Figure 15–31**. The 6- and 3-ohm resistors are in parallel with each other and together are in series with the 2-ohm resistor.

The total current in this circuit is equal to the voltage divided by the total resistance. The total resistance can be determined as follows. The 6- and 3-ohm parallel resistors in **Figure 15–32** are equivalent to a 2-ohm resistor, because $(6 \times 3)/(6 + 3) = 2$. This equivalent 2-ohm resistor is in series with the other 2-ohm resistor. To find the total resistance, add the two resistance values together. This gives a total circuit resistance of 4 ohms ($2 + 2 = 4$ ohms). The total current, therefore, is $I = 12/4 = 3$ amps. This means that 3 amps of current is flowing through the 2-ohm resistor in series and 3 amps are divided between the resistors in parallel. In series-parallel circuits, the sum of the currents, flowing in the parallel legs, must equal that of the series resistors' current.

To find the current through each of the resistors in parallel, find the voltage drop across those resistors first. With 3 amperes flowing through the 2-ohm resistor, the voltage drop across this resistor is $E = IR = 3 \times 2 = 6$ volts, leaving 6 volts across the 6- and 3-ohm resistors. The current through the 6-ohm resistor is $I = E/R = 6/6 = 1$ amp, and through the 3-ohm

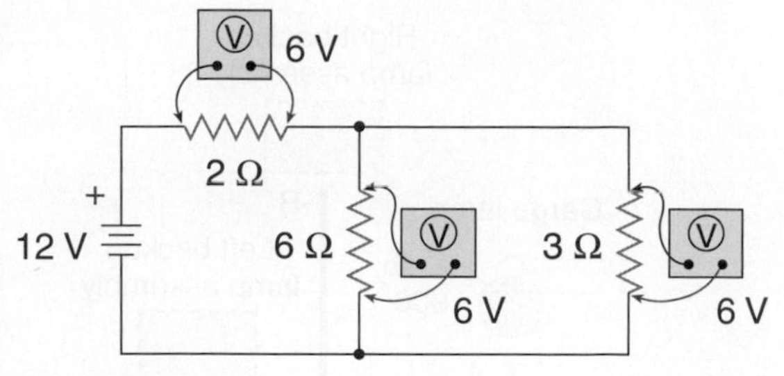

FIGURE 15–32

The circuit in Figure 15–31 with voltage drops shown.

resistor is $I = 6/3 = 2$ amps. The sum of these two current values must equal the total circuit current and it does: $1 + 2 = 3$ amps **(Figure 15–32)**. The sum of the voltage drops across the parallel part of the circuit, and the series part must also equal source voltage.

Grounding the Load

In the illustrations used to explain series, parallel, and series-parallel circuits, the return wire from the load or resistor connects directly to the negative terminal of the battery **(Figure 15–33)**. If this were the case in an actual vehicle, there would be literally hundreds of wires connected to the negative battery terminal.

To avoid this, auto manufacturers use a wiring style that involves using the vehicle's metal frame components as part of the return circuit. Using the chassis as the negative wire is often referred to as **grounding**. The wire or metal mounting that serves as the contact to the chassis is commonly called the **ground wire** or lead. As shown in **Figure 15–34**, the load is grounded directly to the metal frame. The metal frame then acts as the return wire in the circuit. Current passes from the battery, through the load, and into the frame. The frame is connected to the negative terminal of the battery through the battery's ground wire.

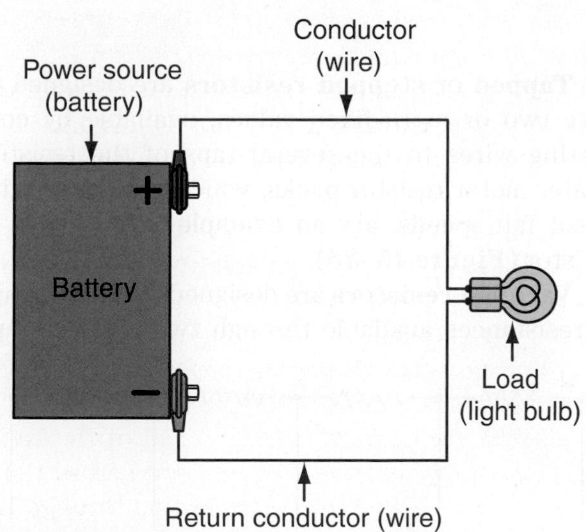

FIGURE 15–33

Electricity results from the flow of electrons from the negative side of the battery to the positive side.

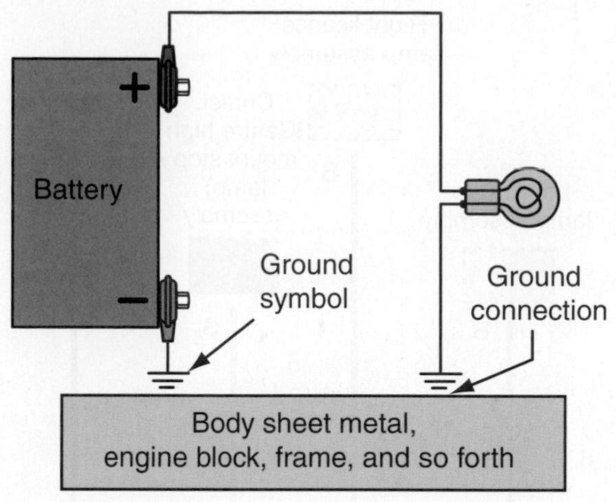

FIGURE 15–34

The same circuit as in Figure 15–33 but with a chassis ground.

An electrical component, such as an alternator, may be mounted directly to the engine block, transmission case, or frame. This direct mounting effectively grounds the component without the use of a separate ground wire. In other cases, however, a separate ground wire must be run from the component to the frame or another metal part to ensure a good connection for the return path. The increased use of plastics and other nonmetallic materials in body panels and engine parts has made electrical grounding more difficult. To ensure good grounding back to the battery, some manufacturers now use a network of common grounding terminals and wires **(Figure 15–35)**.

CIRCUIT COMPONENTS

Automotive electrical circuits contain a number of different types of electrical devices. The more common components are outlined in the following sections.

Resistors

As shown in the explanation of simple circuit design, resistors are used to limit current flow (and thereby voltage) in circuits where full current flow and voltage are not needed or desired. Resistors are devices specially constructed to apply a specific

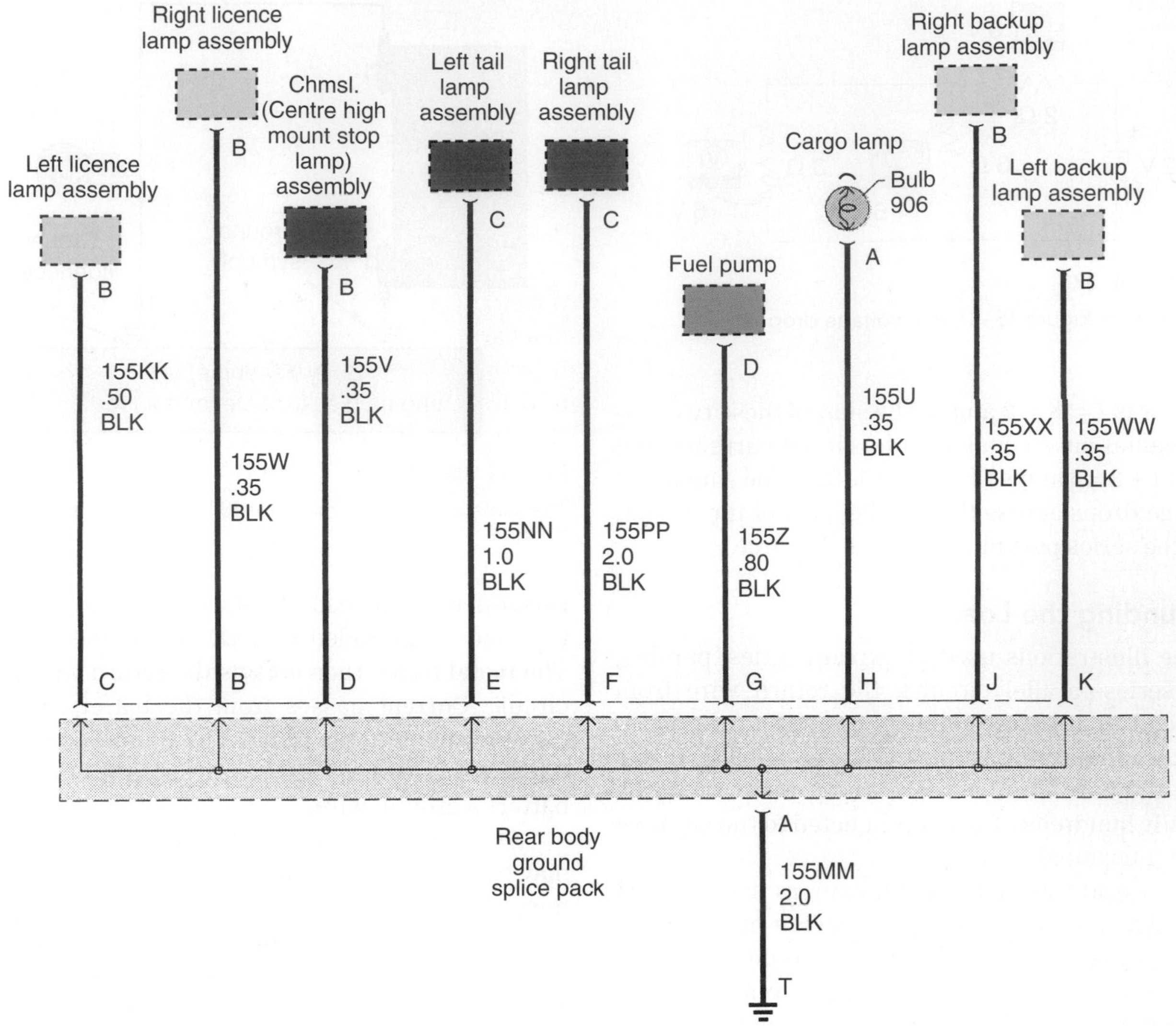

FIGURE 15–35

Some vehicles use a network of grounding wires and terminals to ensure a good ground in all electrical circuits.

amount of resistance into a circuit. In addition, some other components use resistance to produce heat and even light. An electric window defroster is a specialized type of resistor that produces heat. Electric lights are resistors that get so hot they produce light.

Automotive circuits typically contain these types of resistors: fixed value, stepped or tapped, and variable.

Fixed value resistors are designed to have only one rating, which should not change. These resistors are used to decrease the amount of voltage applied to a component, such as a ballast resistor in an ignition system. Often, manufacturers use a special wire, called **resistor wire**, to limit current flow and voltage in a circuit. This wire looks much like normal wire but is not a good conductor and is marked as a resistor.

Tapped or **stepped resistors** are designed to have two or more fixed values, available by connecting wires to the several taps of the resistor. Heater motor resistor packs, which provide for different fan speeds, are an example of this type of resistor **(Figure 15–36)**.

Variable resistors are designed to have a range of resistances available through two or more taps

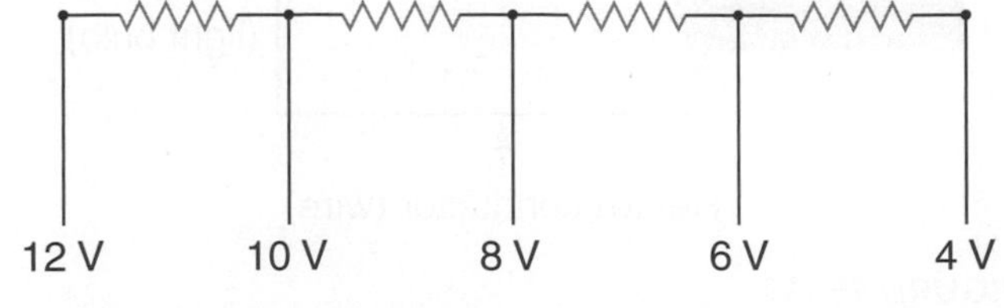

FIGURE 15–36

A stepped resistor.

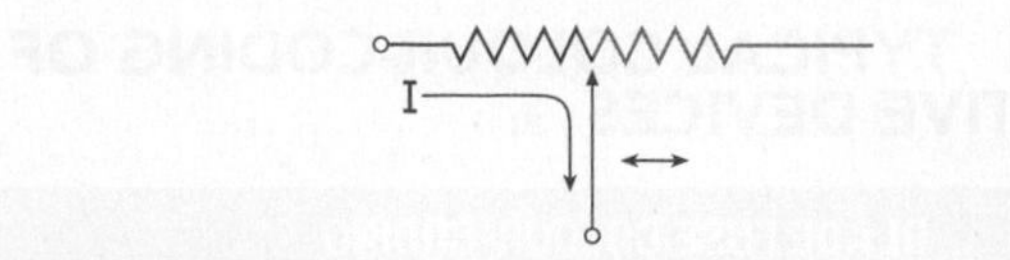

FIGURE 15–37

A rheostat.

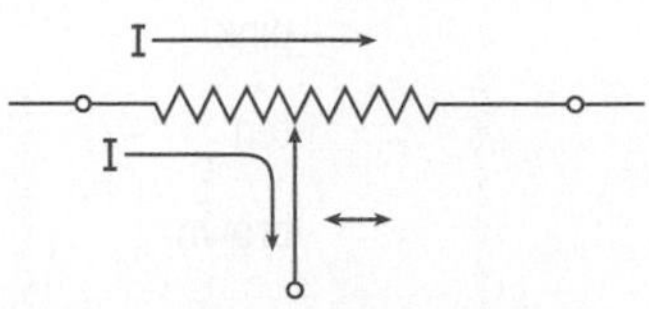

FIGURE 15–38

A potentiometer.

and a control. Two examples of this type of resistor are **rheostats** and **potentiometers**. Rheostats **(Figure 15–37)** have two connections, one to the fixed end of a resistor and one to a sliding contact with the resistor. Moving the control moves the sliding contact away from or toward the fixed end tap, increasing or decreasing the resistance. Potentiometers **(Figure 15–38)** have three connections, one at each end of the resistance and one connected to a sliding contact with the resistor. Moving the control moves the sliding contact away from one end of the resistance but toward the other end. These are called potentiometers because different amounts of potential or voltage can be applied to another circuit. As the sliding contact moves, it picks up a voltage equal to the source voltage minus the amount dropped by the resistor, so far **(Figure 15–39)**.

Another type of variable resistor is the **thermistor**. This type of resistor is designed to change its resistance value as its temperature changes. Although most resistors are carefully constructed to maintain their rating within a few ohms

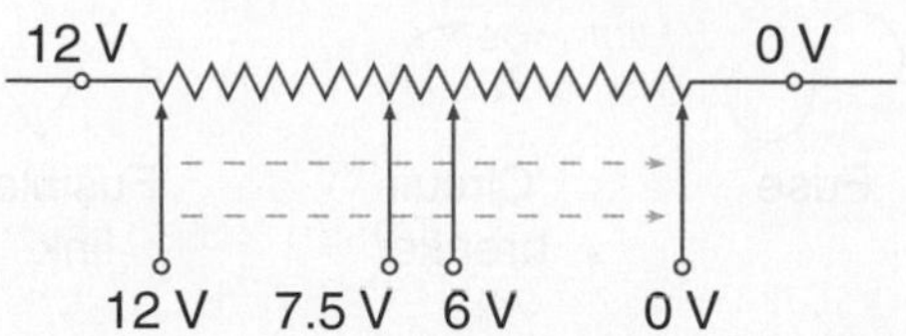

FIGURE 15–39

Voltage across a potentiometer.

through a range of temperatures, the thermistor is designed to change its rating. Thermistors are used to provide compensating voltage in components or to determine temperature. As a temperature sensor, the thermistor is connected to a voltmeter calibrated in degrees. As the temperature rises or falls, the resistance also changes, and so does the voltage across the resistor. These changes are read on the temperature gauge. Thermistors are also commonly used to sense temperature and send a signal back to a control unit. The control unit interprets the signal as a temperature value **(Figure 15–40)**.

Circuit Protective Devices

When overloads or shorts in a circuit cause too much current to flow, the wiring in the circuit heats up, the insulation melts, and a fire can result unless the circuit has some kind of protective device. Fuses, fuse links, maxi fuses, and circuit breakers are designed to provide protection from high current. They may be used singly or in combination. Typical symbols for protection devices are shown in **Figure 15–41**.

WARNING!

Fuses and other protection devices normally do not wear out. They go bad because something went wrong. Never replace a fuse or fusible link, or reset a circuit breaker, without finding out why it went bad.

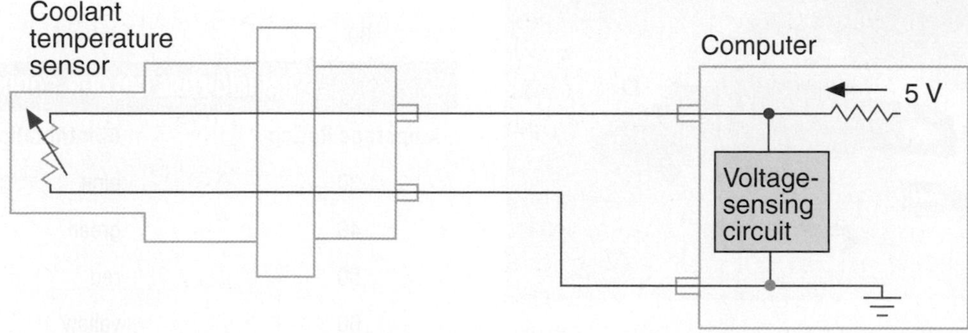

FIGURE 15–40

A thermistor is used to measure temperature. The sensing unit measures the change in resistance and translates it into a temperature value.

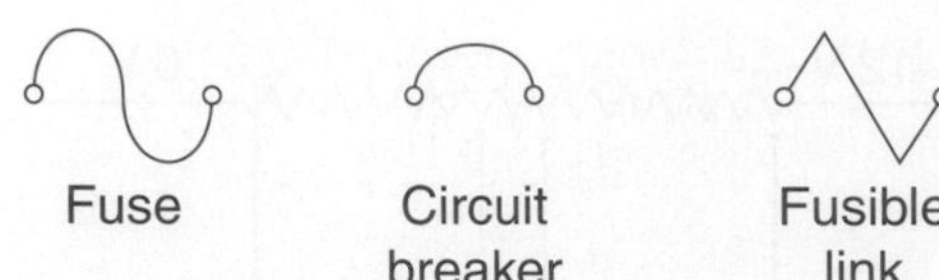

FIGURE 15–41

Electrical symbols for common circuit protection devices.

FUSES There are several types of fuses in automotive use: cartridge, blade, ceramic, mini, maxi, and Pacific **(Figure 15–42)**. The **cartridge fuse** is found on most older domestic cars and a few imports. It is composed of a strip of low-temperature melting metal enclosed in a transparent glass or plastic tube. Late-model domestic vehicles and many imports use **blade** or **spade fuses**. The **ceramic fuse** was used on many European imports. The core is a ceramic insulator with a conductive metal strip along one side.

Fuses are rated by the current at which they are designed to blow.

The current rating for blade fuses is indicated by the colour of the plastic case **(Table 15–1)**. In addition, it is usually marked on the top. The insulator on ceramic fuses is colour-coded to indicate different current ratings.

Four basic types of blade fuses are found on today's vehicles: the standard blade, the mini fuse, the maxi fuse, and the Pacific fuse. The mini fuse is the commonly used circuit protection device. Mini fuses are available in ratings from 5 to 30 amps. The maxi fuse is a serviceable replacement for a fusible link cable. It is used in circuits that have high operating current. Maxi fuses are available in 2- to 100-amp ratings; the most common is the 30 amp.

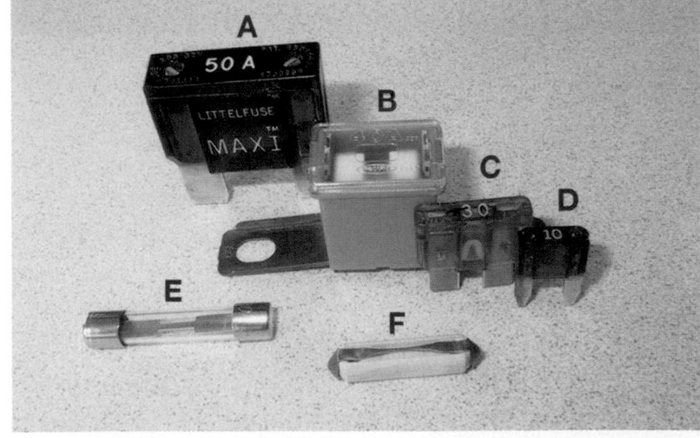

FIGURE 15–42

(A) A maxi fuse, (B) Pacific fuse, (C) blade fuse, (D) mini fuse, (E) cartridge fuse, and (F) ceramic fuse.

Gene Ozon

TABLE 15–1 TYPICAL COLOUR-CODING OF PROTECTIVE DEVICES

BLADE FUSE COLOUR-CODING	
Ampere Rating	**Housing Colour**
3	violet
4	pink
5	tan
7.5	brown
10	red
15	light blue
20	yellow
25	natural
30	light green
FUSE LINK COLOUR-CODING	
Wire Link Size	**Insulation Colour**
20 GA	blue
18 GA	brown or red
16 GA	black or orange
14 GA	green
12 GA	grey
MAXI FUSE COLOUR-CODING	
Ampere Rating	**Housing Colour**
20	yellow
30	light green
40	amber
50	red
60	blue
70	brown
80	natural
PACIFIC FUSE COLOUR-CODING	
Amperage Rating	**Identification Colour**
30	pink
40	green
50	red
60	yellow
80	black
100	blue

FIGURE 15–43

A typical fuse box or panel.

Fuses are located in a box or panel **(Figure 15–43)**, usually under the dashboard, behind a panel in the footwell, or in the engine compartment. Fuses are generally numbered, and the main components abbreviated. On late-model cars, there may be icons or symbols indicating which circuits they serve. This identification system is covered in more detail in the owner's manual and service information.

FUSE LINKS Fuse or **fusible links** are used in circuits when limiting the maximum current is not extremely critical. They are often installed in the positive battery lead to the ignition switch and other circuits that have power with the key off. Fusible links are normally found in the engine compartment near the battery. Fusible links are also used when it would be awkward to run wiring from the battery to the fuse panel and back to the load.

A fuse link **(Figure 15–44)** is a short length of small-gauge wire installed in a conductor. Because the fuse link is a lighter gauge of wire than the main conductor, it melts and opens the circuit before damage can occur in the rest of the circuit. Fuse link wire is covered with a special insulation that bubbles when it overheats, indicating that the link has melted.

FIGURE 15–44

A typical fuse link.

CAUTION!

Always disconnect the battery ground cable prior to servicing any fuse link.

MAXI FUSES AND PACIFIC FUSES Many late-model vehicles use **maxi fuses** or **Pacific fuses** instead of fusible links. Maxi fuses look and operate like two-prong, blade, or spade fuses, except they are much larger and can handle more current. (Typically, a maxi fuse is four to five times larger.) Maxi fuses are usually located in their own underhood fuse block.

Maxi fuses or Pacific fuses allow the vehicle's electrical system to be broken down into smaller circuits that are easy to diagnose and repair. For example, in some vehicles, a single fusible link controls one-half or more of all circuitry. If it burns out, many electrical systems are lost. By replacing this single fusible link with several maxi fuses or Pacific fuses, the number of systems lost due to a problem in one circuit is drastically reduced. This makes it easier to pinpoint the source of trouble.

CIRCUIT BREAKERS Some circuits are protected by **circuit breakers** (abbreviated c.b. in the fuse chart of service information). They can be fuse-panel

FIGURE 15–45

A cycling circuit breaker.

Martin Restoule

mounted or in-line. Like fuses, they are rated in amperes.

Each circuit breaker conducts current through an arm made of two types of metal bonded together (bimetal arm). If the arm starts to carry too much current, it heats up. As one metal expands faster than the other, the arm bends, opening the contacts. Current flow is broken. A circuit breaker can be cycling **(Figure 15–45)** or must be manually reset.

In the cycling type, the bimetal arm begins to cool once the current to it is stopped. Once it returns to its original shape, the contacts are closed and power is restored. If the current is still too high, the cycle of breaking the circuit is repeated.

Two types of noncycling or resettable circuit breakers are used. One is reset by removing the power from the circuit. There is a coil wrapped around a bimetal arm **(Figure 15–46A)**. When there is excessive current and the contacts open, a small current passes through the coil. This current through the coil is not enough to operate a load, but it does heat up both the coil and the bimetal arm. This keeps the arm in the open position until power is removed. The other type is reset by depressing a reset button. A spring pushes the bimetal arm down and holds the contacts together **(Figure 15–46B)**. When an overcurrent condition exists and the bimetal arm heats up, the bimetal arm bends enough to overcome the spring and the contacts snap open. The contacts stay open until the reset button is pushed, which snaps the contacts together again.

SOLID STATE CIRCUIT BREAKERS Automatic resetting solid state circuit breakers are often used to protect power-window and power-door-lock circuits.

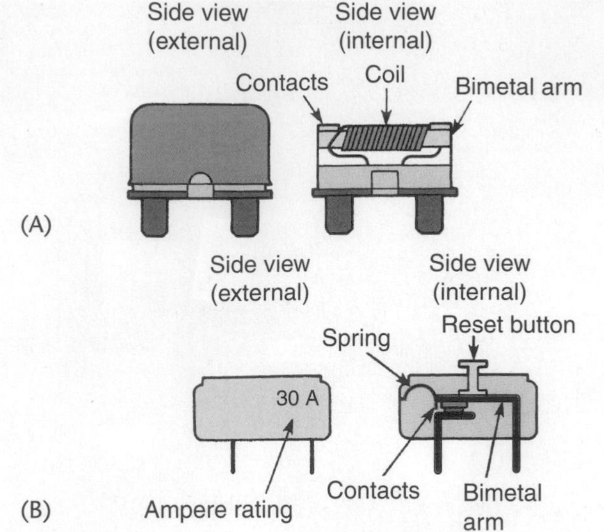

FIGURE 15–46

Resetting noncycling circuit breakers by (A) removing power from the circuit and (B) depressing a reset button.

The solid state circuit breaker contains a positive temperature coefficient (PTC) polymer crystal that creates an open circuit when excessive current is present. The device remains in its tripped position as long as voltage is applied to the circuit.

42-VOLT SYSTEMS New 42-volt systems present a unique problem to circuit protection. Most circuit protection devices used in 12-volt systems are actually rated at 32 volts. If these protection devices were used in a 42-volt system, problems such as severe damage to the vehicle's wiring and electrical components could result. The burning of the components and wiring could also cause a fire.

Protection devices for the 42-volt systems are rated at 55 volts. This rating allows protection during times of high-voltage spikes. These protection devices and their receptacles have a unique design to prevent the installation of the wrong type of fuse.

VOLTAGE LIMITER Some instrument-panel gauges are protected against voltage fluctuations that could damage the gauges or give erroneous readings. A **voltage limiter** restricts voltage to the gauges to a particular amount. The limiter contains a heating coil, a bimetal arm, and a set of contacts. When the ignition is in the on or accessory position, the heating coil heats the bimetal arm, causing it to bend and open the contacts. When the arm cools down to the point that the contacts close, the cycle is repeated. The rapid opening and closing of the contacts produces a pulsating voltage at the output terminal

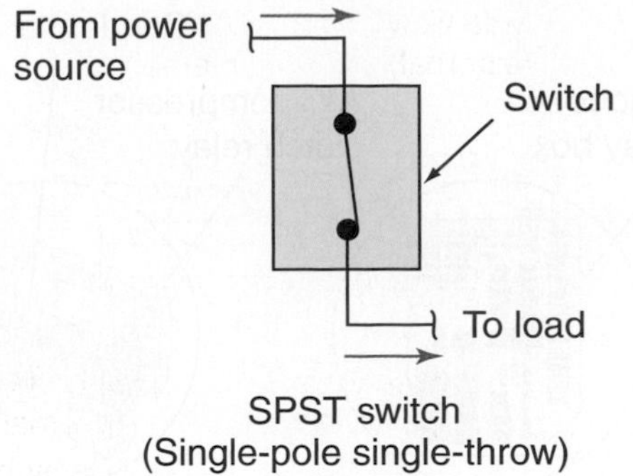

FIGURE 15–47

An SPST hinged-pawl switch diagram.

averaging about 5 volts. A voltage limiter is also called an instrument voltage regulator (IVR).

Switches

Electrical circuits are usually controlled by some type of switch. Switches do two things. They turn the circuit on or off, or they direct the flow of current in a circuit. Switches can be under the control of the driver or can be self-operating through a condition of the circuit, the vehicle, or the environment.

Contacts in a switch can be of several types, each named for the job they do or the sequence in which they work. A hinged-pawl switch **(Figure 15–47)** is the simplest type of switch. It either makes (allows for) or breaks (opens) current flow in a single conductor or circuit. This type of switch is a **single-pole single-throw (SPST)** switch. The **throw** refers to the number of output circuits, and the **pole** refers to the number of input circuits made by the switch.

Another type of SPST switch is a momentary contact switch **(Figure 15–48)**. The spring-loaded contact on this switch keeps it from closing the circuit except when pressure is applied to the button.

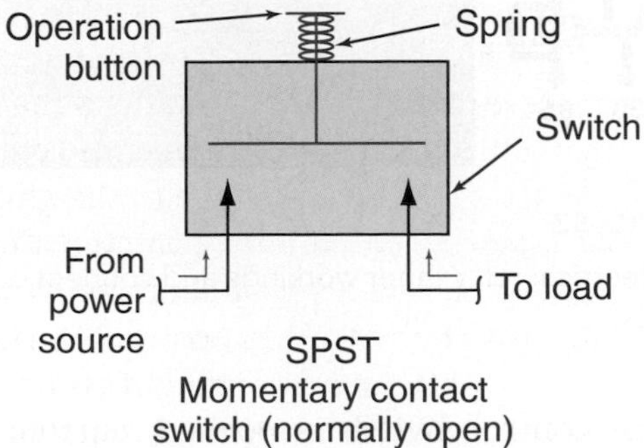

FIGURE 15–48

An SPST momentary contact switch diagram.

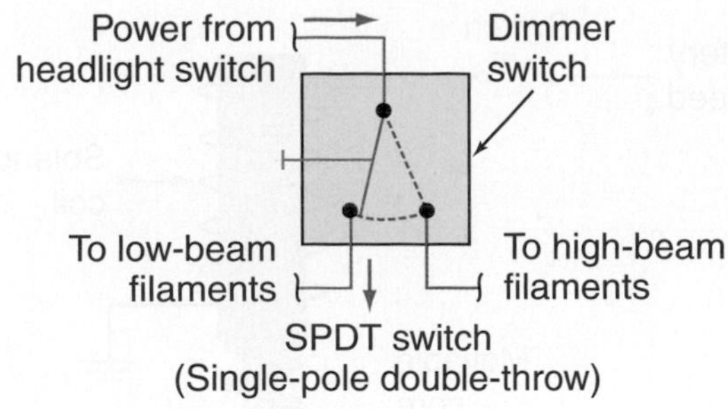

FIGURE 15–49

An SPDT headlight dimmer switch.

A horn switch is this type of switch. Because the spring holds the contacts open, the switch has a further designation: **normally open (NO)**. In the case where the contacts are held closed except when the button is pressed, the switch is designated **normally closed (NC)**. A normally closed momentary contact switch is the type of switch used to turn on the courtesy lights when one of the vehicle's doors is opened.

Single-pole double-throw (SPDT) switches have one wire in and two wires out. **Figure 15–49** shows an SPDT hinged-pawl switch that feeds either the high-beam or low-beam headlight circuit. The dotted lines in the symbol show movement of the switch pawl from one contact to the other.

Switches can be designed with a great number of poles and throws. The transmission neutral start switch shown in **Figure 15–50**, for instance, has two poles and six throws and is referred to as a **multiple-pole multiple-throw (MPMT)** switch. It contains two movable wipers that move in unison across two sets of terminals. The dotted line shows that the wipers are mechanically linked, or **ganged**. The switch closes a circuit to the starter in either P (park) or N (neutral) and to the backup lights in R (reverse).

Most switches are combinations of hinged-pawl and push-pull switches, with different numbers

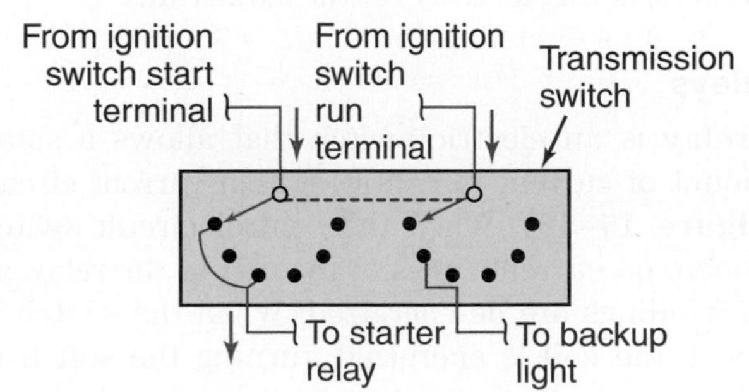

FIGURE 15–50

An MPMT neutral start safety switch.

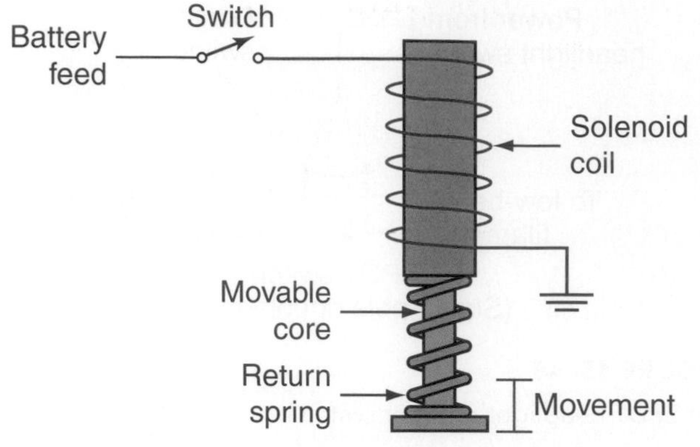

FIGURE 15–51

A solenoid is a device that has a movable electromagnetic core.

of poles and throws. Some special switches are required, however, to satisfy the circuits of automobiles. A mercury switch is sometimes used to detect motion in a component, such as the one used in the engine compartment to turn on the compartment light. Mercury switches have been banned since 2003, but many can still be found on cars that you will see in the shop.

A temperature-sensitive switch usually contains a bimetallic element heated either electrically or by some component in which the switch is used as a **sensor**. When engine coolant is below or at normal operating temperature, the engine coolant temperature switch is in its normally open condition. If the coolant exceeds the temperature limit, the bimetallic element bends the two contacts together and the switch is closed to the indicator or the instrument panel. Other applications for heat-sensitive switches are time-delay switches and flashers.

Solenoids

Solenoids (Figure 15–51) are also electromagnets with movable cores used to change electrical current flow into mechanical movement. They can also close contacts, acting as relays at the same time.

Relays

A **relay** is an electric switch that allows a small amount of current to control a high-current circuit **(Figure 15–52).** When the control circuit switch is open, no current flows to the coil of the relay, so the windings are de-energized. When the switch is closed, the coil is energized, turning the soft iron core into an electromagnet and drawing the arm down. This closes the power circuit contacts, connecting power to the load circuit **(Figure 15–53)**

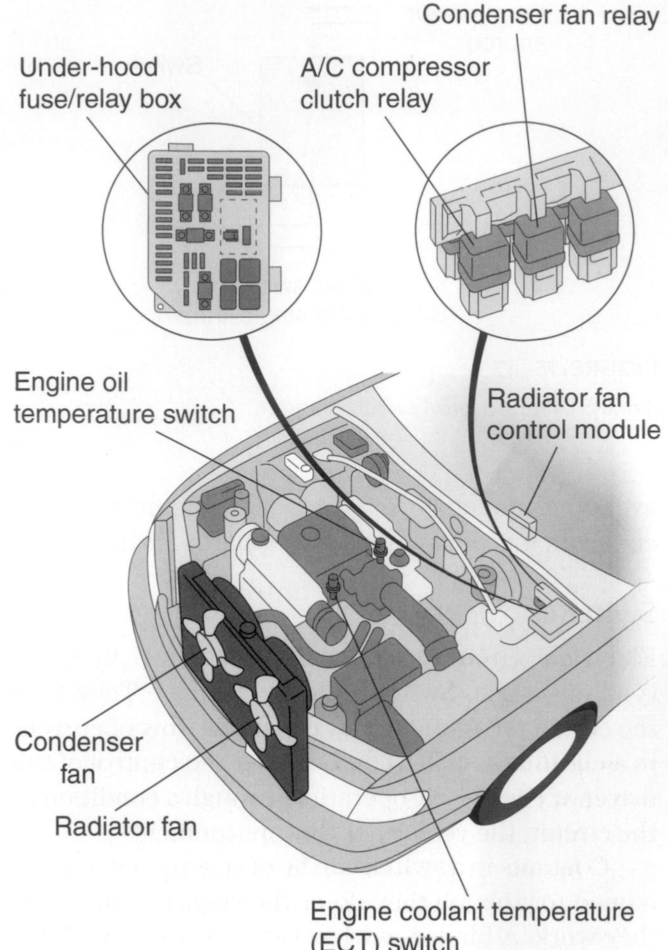

FIGURE 15–52

An electric cooling fan circuit with control modules and relays.

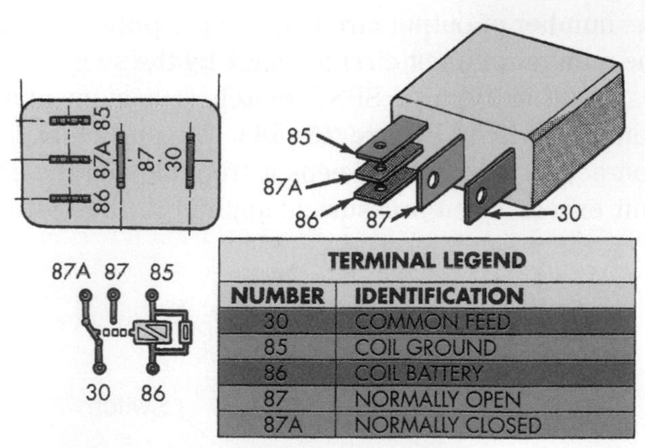

TERMINAL LEGEND	
NUMBER	**IDENTIFICATION**
30	COMMON FEED
85	COIL GROUND
86	COIL BATTERY
87	NORMALLY OPEN
87A	NORMALLY CLOSED

FIGURE 15–53

Typical electrical relay inner workings and connections.

Chrysler Group LLC

When the control switch is opened, current stops flowing in the coil, and the electromagnetism stops. This releases the arm, which breaks the power circuit contacts.

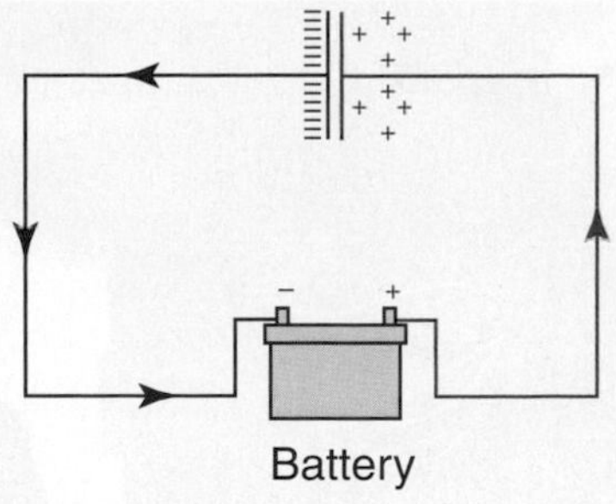

FIGURE 15–54
Charging a capacitor.

Capacitors

Capacitors (condensers) are constructed from two or more sheets of electrically conducting material with a nonconducting or **dielectric** (antielectric) material placed between them. Conductors are connected to the two sheets. Capacitors are devices that oppose a change of voltage.

If a battery is connected to a capacitor, as shown in **Figure 15–54**, the capacitor will be charged when current flows from the battery to the plates. This current flow will continue until the plates have the same voltage as the battery. At this time, the capacitor is charged.

Capacitors, known as power capacitors or stiffening capacitors, are often used with high-output car audio systems. With these large capacity units mounted close to the amplifier, the voltage fluctuations that would cause sound quality to suffer are absorbed.

Automotive capacitors are normally encased in metal. The grounded case provides a connection to one set of conductor plates, and an insulated lead connects to the other set.

Wiring

Electrical wires are used to conduct electricity to operate the electrical and electronic devices in a vehicle. There are two basic types of wires used: solid and stranded. **Solid wires** are single-strand conductors. **Stranded wires** are made up of a number of small solid wires twisted together to form a single conductor. Stranded wires are the most commonly used type of wire in an automobile. Electronic units, such as computers, use specially shielded, twisted cable for protection from unwanted induced voltages that can interfere with computer functions **(Figure 15–55)**. In addition, some solid state components use printed circuits.

The current-carrying capacity and the amount of voltage drop in an electrical wire are determined by its length and gauge (size). The wire sizes are established by the Society of Automotive Engineers (SAE), which is the **American wire gauge (AWG)** system. Sizes are identified by a numbering system ranging from number 0 to 20, with number 0 being the largest and number 20 the smallest in a cross-sectional area. In the metric system of wire sizing, now used by manufacturers worldwide, the size is based on the cross-sectional area of the conductor in square millimetres (mm^2). Small current wires such as those used on engine sensors may use a 0.5 mm^2 wire, whereas battery cables may be 19 mm^2 or larger. A comparison of the AWG and metric wire gauge sizes is shown in **Table 15–2**.

TABLE 15–2 **WIRE GAUGE SIZES**

METRIC SIZE (MM^2)	AWG WIRE SIZE	AMPERE CAPACITY
0.5	20	4
0.8	18	6
1.0	16	8
2.0	14	15
3.0	12	20
5.0	10	30
8.0	8	40
13.0	6	50
19.0	4	60

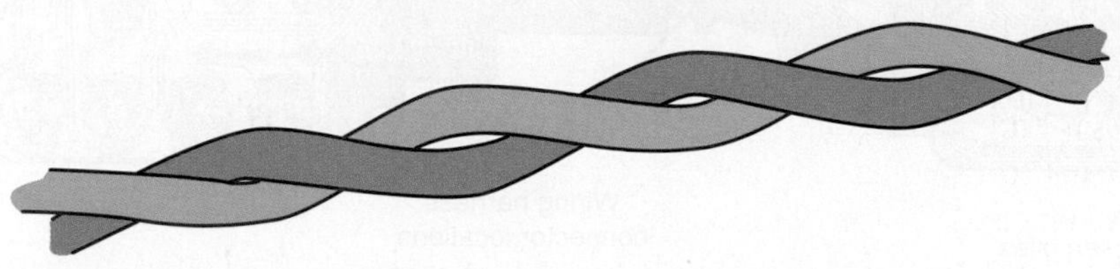

FIGURE 15–55
Electronic units use specially shielded, twisted cable for protection from unwanted induced voltages that can interfere with computer functions.

Automotive wiring can also be classified as primary or secondary. Primary wiring carries low voltage to all the electrical systems of the vehicle except to secondary circuits of the ignition system. Secondary wire, also called **high-tension cable**, has extra-thick insulation to contain high voltage from the ignition coil to the spark plugs. The conductor itself is designed for low currents.

Wires are commonly grouped together in harnesses. A single-plug harness connector may form the connections for four, six, or more circuits. Harnesses and harness connectors help organize the vehicle's electrical system and provide a convenient starting point for tracking and testing many circuits. Most major wiring harness connectors are located in a vehicle's dash or firewall area **(Figure 15–56)**.

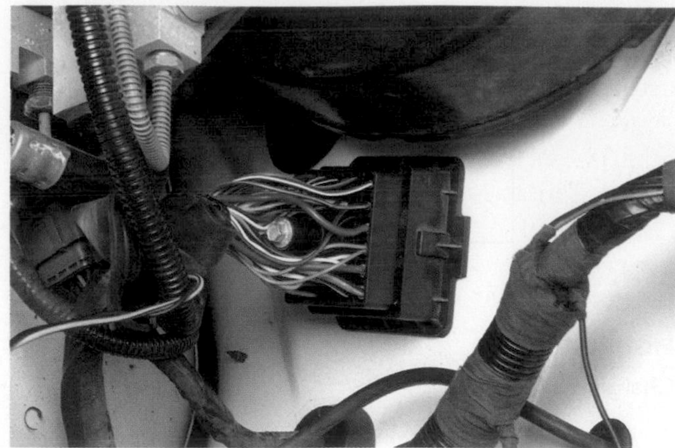

FIGURE 15–56

A typical front wiring harness bulkhead connector.

FLAT WIRING As the number of electrical and electronic devices installed in vehicles increases, so does the need for more wiring. More wiring leads to more weight and more potential problems. Also, the size and number of wiring harnesses increases, and spots to carefully tuck them away are limited. For example, a large wiring harness has a difficult time fitting between the roof and the head liner of the vehicle. To run wiring from the front of the vehicle to the rear via the roof, the harnesses are made of small groups of wires, which means there are more harnesses travelling along the roof. Another solution is to make the wiring harnesses flat.

Flat wiring reduces the bulge or thickness of a harness. The copper conductors inside these wires are flattened and no longer round in appearance. In a wiring harness, several flat wires are laid out next to each other and are covered with a plastic insulating material. The plastic offers protection and isolation to the conductors and keeps the harness flat and flexible. In the future, flat wiring may also have electronic components embedded in it. With this design, the wiring harness is not only easier to hide in body panels but also serves as a flexible printed circuit able to be located nearly anywhere in the vehicle.

PRINTED CIRCUITS Many late-model vehicles use flexible printed circuits **(Figure 15–57)** and printed circuit boards. Both types of printed circuits allow

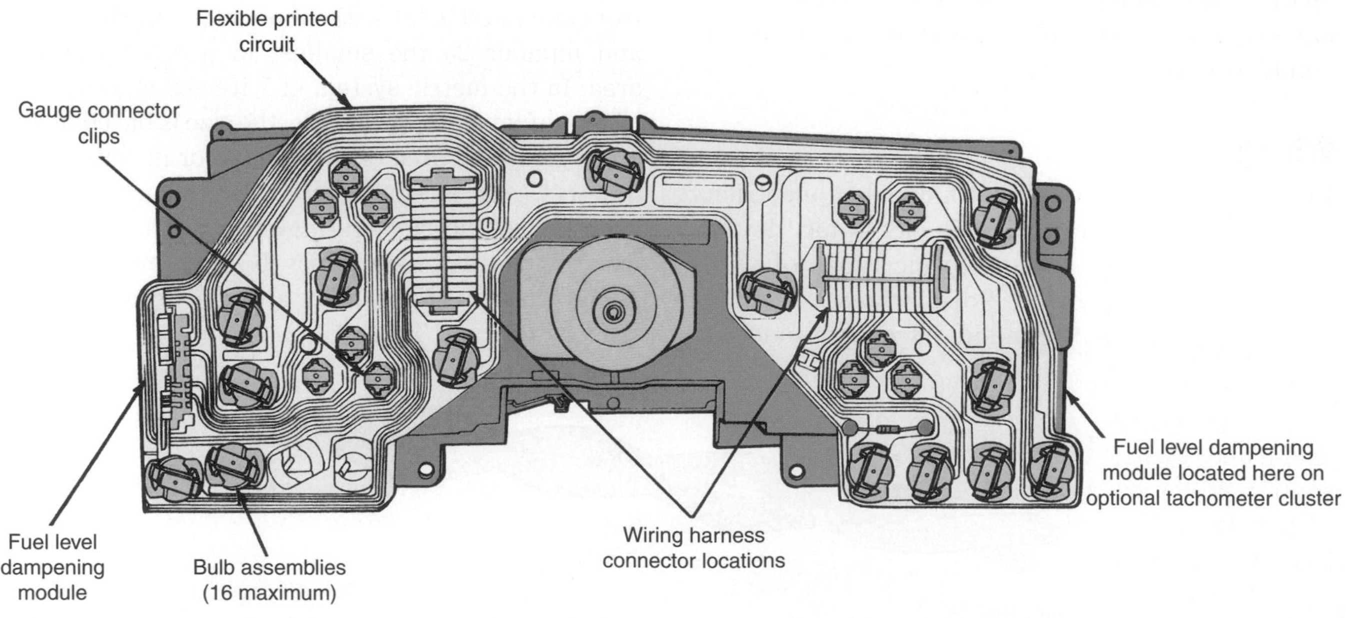

FIGURE 15–57

A typical printed circuit board.

for complete circuits to many components without having to run dozens of wires. Printed circuit boards are typically contained in a housing, such as the engine control module. These boards are not serviceable and in some cases not visible. When these boards fail, the entire unit is replaced.

A flexible printed circuit saves weight and space. It is made of thin sheets of nonconductive plastic onto which conductive metal, such as copper, has been deposited. Parts of the metal are then etched or eaten away by acid. The remaining metal lines form the conductors for the various circuits on the board. The printed circuit is normally connected to the power supply or ground wiring through the use of plug-in connectors mounted on the circuit sheet.

ELECTROMAGNETISM BASICS

Electricity and magnetism are related. One can be used to create the other. Current flowing through a wire creates a magnetic field around the wire. Moving a wire through a magnetic field creates current flow in the wire.

Many automotive components, such as alternators, ignition coils, starter solenoids, and magnetic pulse generators, operate using these principles of **electromagnetism**.

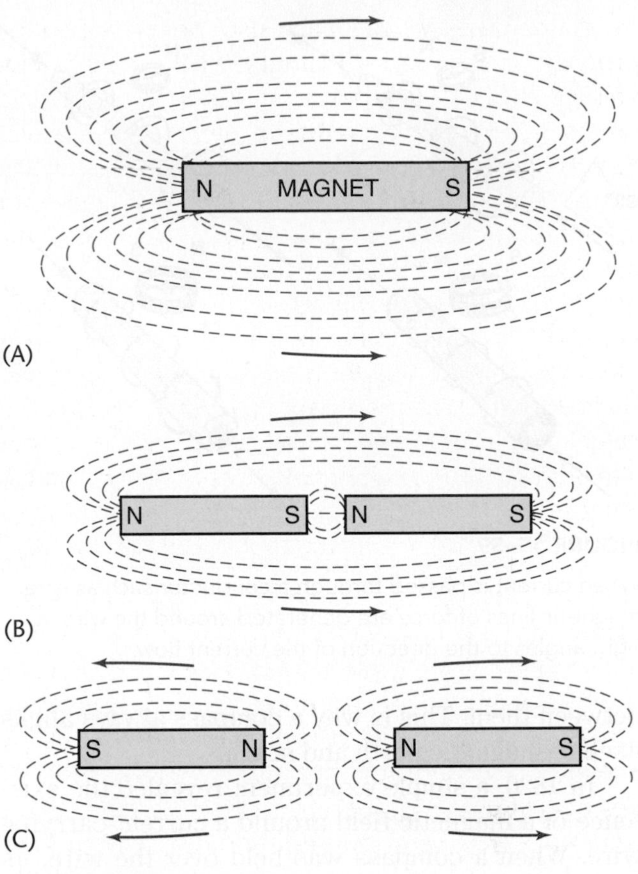

FIGURE 15–58

(A) In a magnet, lines of force emerge from the north pole and travel to the south pole before passing through the magnet back to the north pole. (B) Unlike poles attract, whereas (C) similar poles repel each other.

Fundamentals of Magnetism

A substance is said to be a **magnet** if it has the property of magnetism—the ability to attract such substances as iron, steel, nickel, or cobalt. These are called magnetic materials.

A magnet has two points of maximum attraction, one at each end of the magnet. These points are called poles, with one being designated the north pole and the other the south pole **(Figure 15–58A)**. When two magnets are brought together, opposite poles attract **(Figure 15–58B)**, whereas similar poles repel each other **(Figure 15–58C)**.

A magnetic field, called a **flux field**, exists around every magnet. The field consists of imaginary lines along which the magnetic force acts. These lines emerge from the north pole and enter the south pole, returning to the north pole through the magnet itself. None of the lines cross each other. All lines are complete.

Magnets can occur naturally in the form of a mineral called magnetite. Artificial magnets can also be made by inserting a bar of magnetic material inside a coil of insulated wire and passing direct current through the coil. This principle is very important in understanding certain automotive electrical components. Another way of creating a magnet is by stroking the magnetic material with a bar magnet. Both methods force the randomly arranged molecules of the magnetic material to align themselves along north and south poles.

Artificial magnets can be either temporary or permanent. Temporary magnets are usually made of soft iron. They are easy to magnetize but quickly lose their magnetism when the magnetizing force is removed. Permanent magnets consist of ferrites (iron) and rare earth metals, including neodymium. They are difficult to magnetize. However, once magnetized, they retain this property for very long periods.

The earth is a very large magnet, having a north and a south pole, with lines of magnetic force running

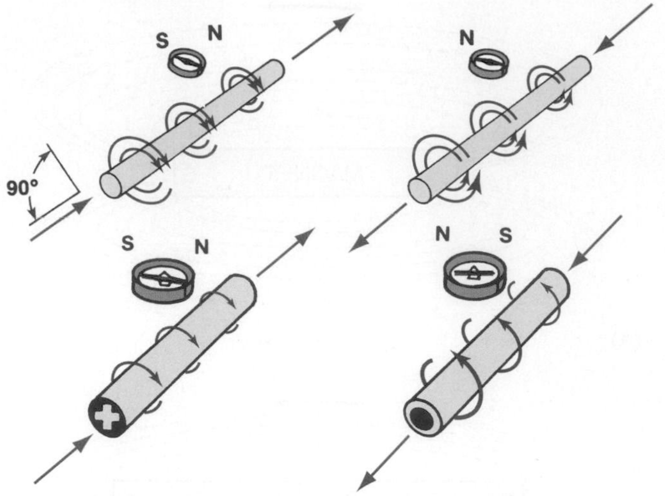

FIGURE 15–59

When current is passed through a conductor such as wire, magnetic lines of force are generated around the wire at right angles to the direction of the current flow.

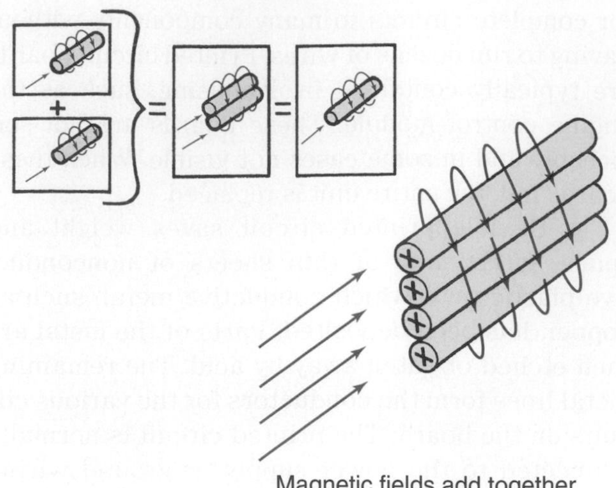

FIGURE 15–60

Increasing the number of conductors carrying current in the same direction increases the strength of the magnetic field around them.

between them. This is why a compass always aligns itself to magnetic north and south.

In 1820, a simple experiment revealed the existence of a magnetic field around a current-carrying wire. When a compass was held over the wire, its needle aligned itself at right angles to the wire **(Figure 15–59)**. The lines of magnetic force are concentric circles around the wire. The density of these circular lines of force is very heavy near the wire and decreases farther away from the wire. As is also shown in the same figure, the polarity of a current-carrying wire's magnetic field changes depending on the direction the current is flowing through the wire.

Remember, these magnetic lines of force or flux lines do not move or flow around the wire. They simply have a direction, as shown by their effect on a compass needle. These lines of force are always at right angles to the conducting wire.

FLUX DENSITY The more flux lines, the stronger the magnetic field at that point. Increasing current increases **flux density**. Also, two conducting wires lying side by side carrying equal currents in the same direction create a magnetic field equal in strength to one conductor carrying twice the current. Adding more wires also increases the magnetic field **(Figure 15–60)**.

COILS Looping a wire into a coil concentrates the lines of force inside the coil. The resulting magnetic field is the sum of all the single-loop magnetic fields added together **(Figure 15–61)**. The overall effect is the same as placing many wires side by side, each carrying current in the same direction.

Magnetic Circuits and Reluctance

Just as current can only flow through a complete circuit, the lines of flux created by a magnet can only occupy a closed magnetic circuit. The resistance that a magnetic circuit offers to a line of flux is called **reluctance**. Magnetic reluctance can be compared to electrical resistance.

Reconsider the coil of wire shown in Figure 15–61. The air inside the coil has very high reluctance and limits the strength of the magnetic field that can be produced. However, if an iron core is placed inside the coil, the strength of the magnetic

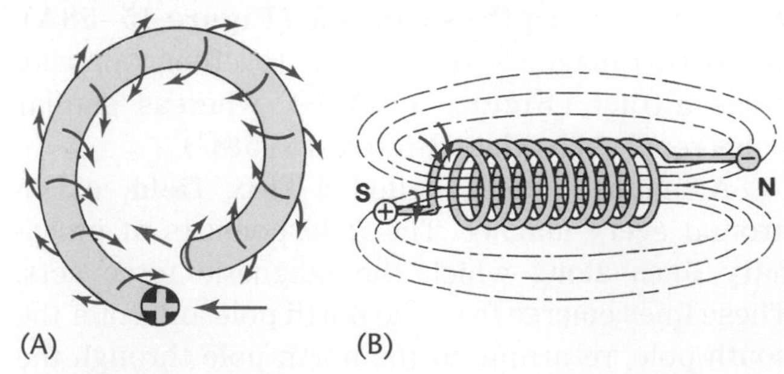

FIGURE 15–61

(A) Forming a wire loop concentrates the lines of force inside the loop. (B) The magnetic field of a wire coil is the sum of all the single-loop magnetic fields.

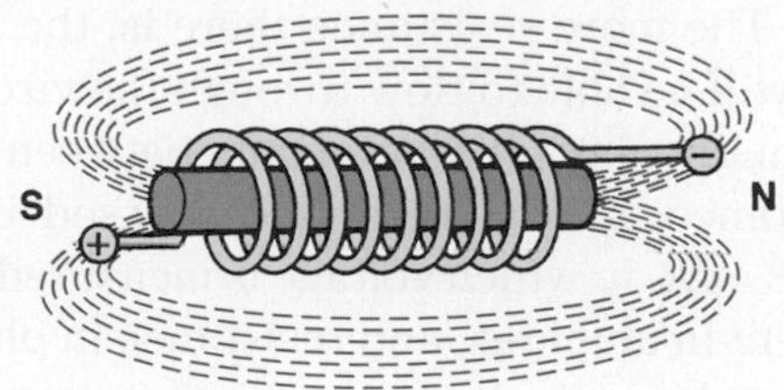

FIGURE 15–62

Placing a soft iron core inside a coil greatly reduces the reluctance of the coil and creates a usable electromagnet.

field increases tremendously because the iron core has a very low reluctance **(Figure 15–62)**.

When a coil is wound around an iron core in this manner, it becomes a usable electromagnet. The strength of the magnetic poles in an electromagnet is directly proportional to the number of turns of wire and the current flowing through them.

The equation for an electromagnetic circuit is similar to Ohm's law for electrical circuits. It states that the number of magnetic lines of flux produced is proportional to the ampere-turns divided by the reluctance. To summarize:

- Field strength increases if current through the coil increases.
- Field strength increases if the number of coil turns increases.
- If reluctance increases, field strength decreases.

Induced Voltage

Now that we have explained how current can be used to generate a magnetic field, it is time to examine the opposite effect of how magnetic fields can produce electricity.

Figure 15–63 shows a straight piece of wire with the terminals of a voltmeter attached to both ends. If the wire is moved across a magnetic field, the voltmeter registers a small voltage reading. A voltage has been induced in the wire.

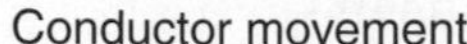

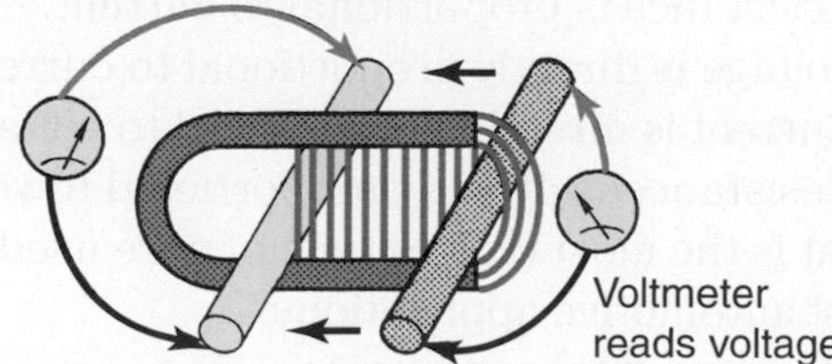

FIGURE 15–63

Moving a conductor so that it cuts across the magnetic lines of force induces a voltage in the conductor.

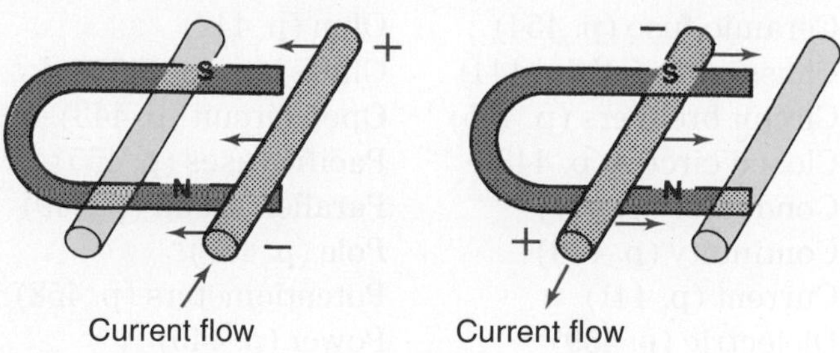

FIGURE 15–64

The polarity of the induced voltage depends on the direction in which the conductor moves as it cuts across the magnetic field.

It is important to realize that the wire must cut across the flux lines to induce a voltage. Moving the wire parallel to the lines of flux does not induce voltage.

Voltage can also be induced by holding the wire still and moving the magnetic field at right angles to the wire. This is the exact setup used in a vehicle's alternator. A magnetic field is made to cut across stationary conductors to produce voltage.

The wire becomes a source of electricity and has a polarity or distinct positive and negative end. However, this polarity can be switched depending on the relative direction of movement between the wire and magnetic field **(Figure 15–64)**. This is why charging devices produce alternating current.

The amount of voltage induced depends on four factors:

- The stronger the magnetic field, the stronger the induced voltage.
- The faster the field is being cut, the more lines of flux are cut and the stronger the voltage induced.
- The greater the number of conductors, the greater the voltage induced.
- The closer the conductor(s) and magnetic field are to right angles (perpendicular) to one another, the greater the induced voltage.

The importance of electromagnetism and induced voltage is clearly seen in chapters dealing with starting, charging, ignition, and electronic control systems.

KEY TERMS

Alternating current (AC) (p. 441)
American wire gauge (AWG) (p. 459)
Ampere (p. 441)
Blade fuses (p. 454)
Capacitor (p. 459)
Cartridge fuse (p. 454)

Ceramic fuse (p. 454)
Chassis ground (p. 444)
Circuit breakers (p. 455)
Closed circuit (p. 443)
Conductor (p. 447)
Continuity (p. 443)
Current (p. 441)
Dielectric (p. 459)
Direct current (DC) (p. 441)
Electromagnetism (p. 461)
Electromotive force (EMF) (p. 442)
Electronics (p. 439)
Fixed value resistors (p. 452)
Flux density (p. 462)
Flux field (p. 461)
Fusible links (p. 455)
Ganged (p. 457)
Ground wire (p. 451)
Grounding (p. 451)
High-tension cable (p. 460)
Induction (p. 441)
Insulators (p. 447)
Magnet (p. 461)
Magnetism (p. 440)
Maxi fuses (p. 455)
Multiple-pole multiple-throw (MPMT) (p. 457)
Normally closed (NC) (p. 457)
Normally open (NO) (p. 457)
Ohm (p. 442)
Ohm's law (p. 445)
Open circuit (p. 443)
Pacific fuses (p. 455)
Parallel circuit (p. 449)
Pole (p. 457)
Potentiometers (p. 453)
Power (p. 446)
Relay (p. 458)
Reluctance (p. 462)
Resistance (p. 442)
Resistor wire (p. 452)
Rheostats (p. 453)
Sensor (p. 458)
Series circuit (p. 448)
Series-parallel circuit (p. 450)
Sine wave (p. 442)
Single-pole double-throw (SPDT) (p. 457)
Single-pole single-throw (SPST) (p. 457)
Solenoids (p. 458)
Solid wires (p. 459)
Spade fuses (p. 454)
Stepped resistors (p. 452)
Stranded wires (p. 459)
Tapped resistors (p. 452)
Thermistor (p. 453)
Throw (p. 457)
Variable resistors (p. 452)
Voltage (p. 442)
Voltage drops (p. 448)
Voltage limiter (p. 456)
Watts (p. 446)
Watt's law (p. 447)

SUMMARY

- For electrical flow to occur, there must be an excess of electrons in one place, a lack of electrons in another, and a path between both places.
- Voltage is the force or pressure in an electrical circuit. A voltage drop across a load in a circuit indicates that work is being done.
- Current is measured in amps. This is a measurement of the actual flow rate of electrons in an electrical circuit.
- Resistance is measured in ohms. This is a measurement of the size of the restriction to current flow. The more resistance there is, the less current will be able to flow through the circuit.
- The mathematical relationship between current, resistance, and voltage is expressed in Ohm's law, $E = IR$, in which voltage is measured in volts, current in amperes, and resistance in ohms.
- The mathematical relationship between current, voltage, and power is expressed in Watt's law, $P = E \times I$. Power is measured in watts or kilowatts (1000 W).
- Three basic types of circuits are used in automobile wiring systems: series circuits, parallel circuits, and series-parallel circuits.
- Electrical schematics are diagrams with electrical symbols that show the parts and how electrical current flows through the vehicle's electrical circuits. They are used in troubleshooting.
- Fuses, fuse links, maxi fuses, and circuit breakers protect circuits against overloads. Switches control on–off and direct current flow in a circuit. A relay is an electric switch. A solenoid is an electromagnetic switch that translates current flow into mechanical movement. Resistors limit current flow.
- The strength of an electromagnet depends on the number of current-carrying conductors, the current in the coil, and the material in the core of the coil. Inducing a voltage requires a magnetic field producing lines of force, conductors that can be moved, and movement between the conductors and the magnetic field so that the lines of force are cut.

REVIEW QUESTIONS

1. How does current act in a series circuit?
   a. It varies throughout the circuit.
   b. It is constant throughout the circuit.
   c. It is calculated by multiplying volts times amps.
   d. It is calculated by dividing resistance by voltage.
2. What is stated by Ohm's law?
   a. Resistance is proportional to current.
   b. Voltage is directly proportional to current.
   c. Current is directly proportional to voltage.
   d. Resistance is directly proportional to voltage.
3. What is the most typical type of wire used in most automotive applications?
   a. stainless steel solid core
   b. solid copper core
   c. multi-strand aluminum core
   d. multi-strand copper core

4. What would have the greatest effect on resistance of a wire?
   a. location of the wire
   b. size of the wire
   c. voltage through the wire
   d. knots in the wire
5. What does "NC" indicate when shown on a switch?
   a. Current will be conducted when the switch is in the at-rest state.
   b. Current will not flow through in the at-rest state.
   c. The switch is open until current is applied.
   d. The switch is closed until current is applied.
6. The rheostat is what type of resistor?
   a. fixed value
   b. variable
   c. stepped
   d. thermistor
7. How much current will flow in a circuit that contains two 10 resistors in series with 20 volts applied?
   a. 0.5 amp b. 1.0 amp
   c. 2.0 amps d. 10.0 amps
8. What is the total resistance of a series circuit that has 12 volts (12 V) applied and has 0.5 amps of current flowing?
   a. 24 Ω b. 18 Ω
   c. 12 Ω d. 6 Ω
9. What would be the effect of adding a resistance to a series circuit?
   a. Total circuit resistance would decrease.
   b. Voltage drop across each resistor would increase.
   c. Voltage drop across each resistor would decrease.
   d. Total current flowing in the circuit would increase.
10. What would be the effect of adding a load in parallel to a parallel circuit?
    a. Total resistance would go up.
    b. Voltage drop across each branch would go up.
    c. Voltage drop across each branch would go down.
    d. Current flowing in the circuit would increase.
11. Which of the following is a characteristic of all parallel circuits?
    a. The total resistance is always higher than the highest resistance.
    b. The current through each branch is the same.
    c. The voltage drops vary for each branch of the circuit.
    d. The total resistance is always less than the smallest resistor.
12. What is the current in a 12-volt circuit with two 6-ohm resistors connected in parallel?
    a. 2 amps
    b. 6 amps
    c. 4 amps
    d. 12 amps
13. A 12-volt parallel circuit consists of a 3 Ω, a 6 Ω, and a 2 Ω resistor. What is the total resistance of this circuit?
    a. 1 Ω b. 2 Ω
    c. 11 Ω d. 12 Ω
14. A circuit protection device operates by which principle?
    a. limiting voltage
    b. limiting wattage
    c. limiting resistance
    d. limiting amperage
15. What might be considered a benefit of changing to a 42-volt electrical system?
    a. Larger capacity actuators will be required
    b. Increased starter motor current draw
    c. The use of smaller conductors due to the voltage increase
    d. Larger batteries will be required
16. How much power is consumed by a 6-ohm load with 12 volts applied?
    a. 3 watts b. 6 watts
    c. 12 watts d. 24 watts
17. What is created through induction?
    a. current
    b. resistance
    c. voltage
    d. magnetism
18. What name is given to a protection device that can be reset?
    a. circuit breaker
    b. fusible link
    c. mini fuse
    d. ATO fuse
19. What does SPST indicate when seen on a component in an electrical diagram?
    a. a solenoid that is positively switched
    b. a switch that is only on or off
    c. a switch that controls two circuits
    d. a switch that has two on positions
20. Why is wire shielding used?
    a. to protect wire from induction
    b. to protect wire from heat
    c. to protect wire from abrasion
    d. to protect wire from water

CHAPTER 16

Basics of Electronics and Computer Systems

LEARNING OUTCOMES

- Understand the purpose and operation of a capacitor.
- Describe how semiconductors, diodes, and transistors work.
- Explain the advantages of using electronic control systems.
- Explain the basic function of the central processing unit (CPU).
- List and describe the functions of the various sensors used by computers.
- Explain the principle of computer communications.
- Summarize the function of a binary code.
- Name the various memory systems used in automotive computers.
- List and describe the operation of output actuators.
- Explain the principle of multiplexing.
- Describe the precautions that must be taken when diagnosing electronic systems.
- Perform a communications check on a multiplexed system.
- Reprogram a control module in a vehicle.

Computerized engine controls and other features of today's vehicles would not be possible if it were not for electronics. For clarity, electronics is defined as the technology of controlling electricity. Capacitors, transistors, diodes, semiconductors, integrated circuits, and solid-state devices are all considered to be part of electronics rather than just electrical devices. But keep in mind that all the basic laws of electricity apply to electronic controls.

CAPACITORS

A **capacitor** is used to store and release electrical energy. Capacitors can be used to smooth out current fluctuations, store and release a high voltage, or block DC voltage. Capacitors are sometimes called condensers. Although a battery and a capacitor store electrical energy, the battery stores the energy chemically. A capacitor stores energy in an electrostatic field created between a pair of electrodes.

A capacitor can release all of its charged energy in an instant, whereas a battery slowly releases its charge. A capacitor is quick to discharge and quick to charge. A battery needs some time to discharge and charge but can provide continuous power. A capacitor provides power only in bursts.

Operation

A capacitor has a positive and a negative terminal (**Figure 16–1**). Each terminal is connected to a thin electrode or plate (usually made of metal). The plates are parallel to each other and are separated by an insulating material called a dielectric. The dielectric can be paper, plastic, glass, or anything that does not conduct electricity. Placing a dielectric between the plates allows the plates to be placed close to each other without allowing them to touch.

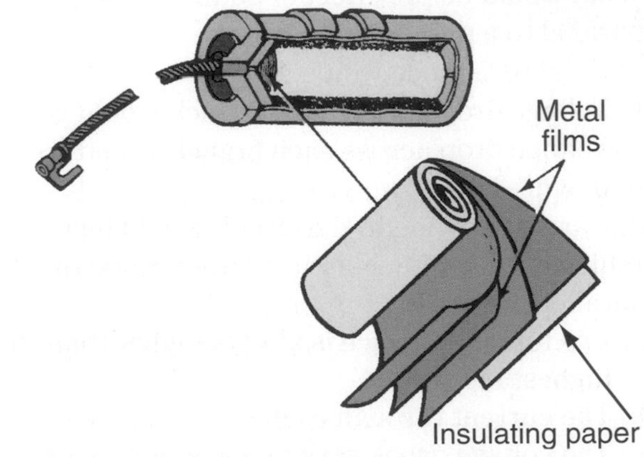

FIGURE 16–1
A construction of a condenser, which is a common name for a capacitor, that absorbs voltage spikes.

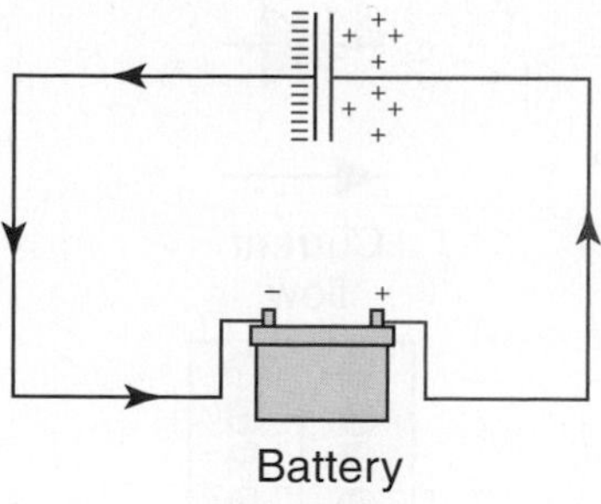

FIGURE 16–2

When voltage is applied to a capacitor, the two electrodes receive equal but opposite charges.

When voltage is applied to a capacitor, the two electrodes receive equal but opposite charges **(Figure 16–2)**. The negative plate accepts electrons and stores them on its surface. The other plate loses electrons to the power source. This action charges the capacitor. Once the capacitor is charged, it has the same voltage as the power source. This energy is stored statically until the two terminals are connected together.

The ability of a capacitor to store an electric charge is called **capacitance**. The standard measure of capacitance is the **farad (F)**. A one-farad capacitor can store one coulomb of charge at one volt. A coulomb is 6.25 *billion billion* electrons. One ampere equals the flow of one coulomb of electrons per second, so a one-farad capacitor can hold one ampere-second of electrons at one volt. A capacitor's capacitance is directly proportional to the surface areas of the plates and the nonconductiveness of the dielectric and is inversely proportional to the distance between the plates. Most capacitors have a capacitance rating of much less than a farad, and their values are expressed in one of these terms:

- Microfarads: μF (1 μF = 10^{-6} F)
- Nanofarads: nF (1 nF = 10^{-9} F)
- Picofarads: pF (1 pF = 10^{-12} F)

Capacitors oppose a change of voltage. If a battery is connected to a capacitor as shown in Figure 16–2, the capacitor will be charged when current flows from the battery to the plates. This current flow will continue until the plates have the same voltage as the battery. At this time, the current flow stops, and the capacitor is charged.

The capacitor remains charged until a circuit is completed between the plates. If the charge is routed through a voltmeter, the capacitor will discharge with the same voltage as the battery that charged it. This statement explains why capacitors are commonly used to filter or clean up voltage signals, such as sound from a stereo. Current can flow only during the period that a capacitor is either charging or discharging.

Automotive capacitors are normally encased in metal. The grounded case provides a connection to one set of conductor plates and an insulated lead connects to the other set.

Variable capacitors are called **trimmers** or tuners and are rated very low in capacity because of the reduced size of their conducting plates. For this reason, they are used only in very sensitive circuits such as radios and other electronic applications.

SEMICONDUCTORS

A semiconductor is a material or device that can serve as a conductor or an insulator. Semiconductors have no moving parts; therefore, they seldom wear out or need adjustment. Semiconductors, or solid-state devices, are also small, require little power to operate, are reliable, and generate very little heat. For all these reasons, semiconductors are being used in many applications. However, current to them must be limited, as should heat.

Because a semiconductor can function as both a conductor and an insulator, it is often used as a switching device. How it behaves depends on what it is made of and which way current flows (or tries to flow) through it. Two common semiconductor devices are diodes and transistors. Diodes are used for isolation of components or circuits, clamping, or rectification of AC to DC. Transistors are used for amplification or switching.

Semiconductor materials have less resistance than an insulator but more resistance than a conductor. They also have a crystal structure. This means their atoms do not lose and gain electrons as conductors do. Instead, the atoms in semiconductors share outer electrons with each other. In this type of atomic structure, the electrons are tightly held and the element is stable. Common semiconductor materials are silicon (Si) and germanium (Ge).

Because the electrons are not free, the crystals cannot conduct current and are called **electrically inert materials**. In order to function as semiconductors, a small amount of trace element, called **impurities**, must be added. The type of impurity determines the type of semiconductor.

N-type semiconductors have loose, or excess, electrons. They have a negative charge and can carry current. N-type semiconductors have an impurity with five electrons in the outer ring (called

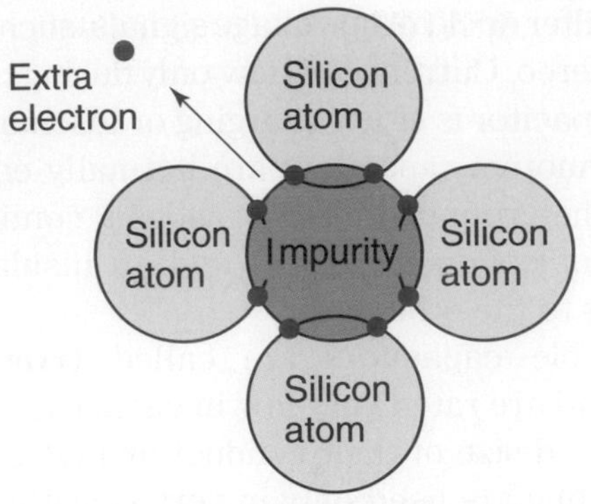

FIGURE 16–3

Atomic structure of an N-type silicon semiconductor.

pentavalent atoms). Four of these electrons fit into the crystal structure, but the fifth is free. This excess of electrons produces the negative charge. **Figure 16–3** shows an example.

P-type semiconductors are positively charged materials. They are made by adding an impurity with three electrons in the outer ring (trivalent atoms). When this element is added to silicon or germanium, the three outer electrons fit into the pattern of the crystal, leaving a hole where a fourth electron would fit. This hole is actually a positively charged empty space. This hole carries the current in the P-type semiconductor. **Figure 16–4** shows an example of a P-type semiconductor.

Hole Flow

Understanding how semiconductors carry current without losing electrons requires an understanding of the concept of hole flow. The holes in a P-type semiconductor, being positively charged, attract electrons. Although the electrons cannot be freed from their atom, they can rearrange and fill a hole in a nearby atom. Whenever this happens, the place where the electron was is now a hole. This hole is then filled by another electron, and the process continues.

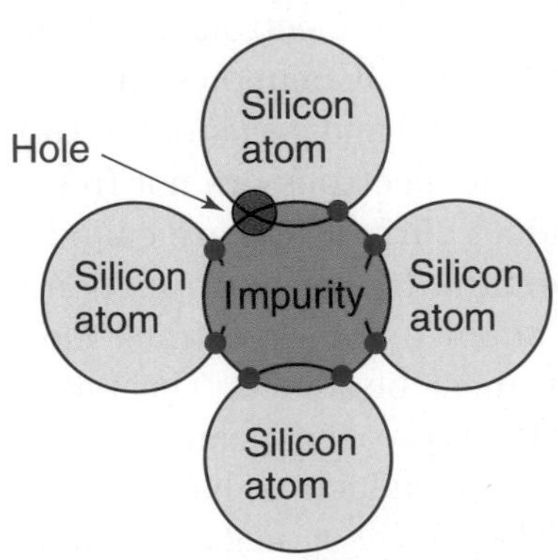

FIGURE 16–4

Atomic structure of a P-type silicon semiconductor.

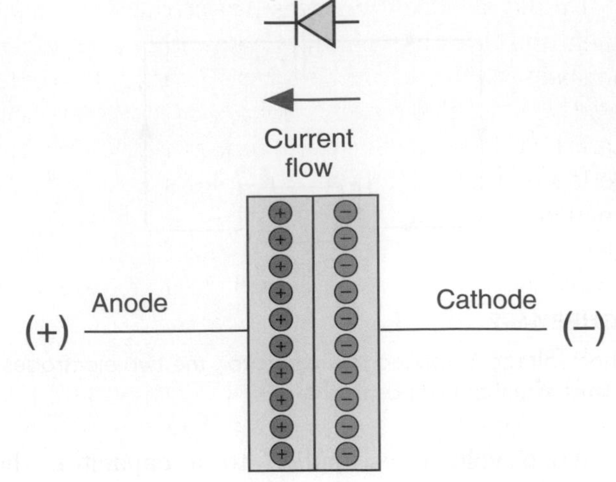

FIGURE 16–5

A diode and its schematic symbol.

The electrons move to the positive side of the structure, and the holes move to the negative side.

Diodes

Diodes are simple semiconductors. The most commonly used are regular diodes, LEDs, zener diodes, clamping diodes, and photo diodes. A **diode** allows current to flow in one direction, so it can serve as a conductor or insulator, depending on the direction of current flow **(Figure 16–5)**. In a generator, voltage is rectified by diodes. Diodes are arranged so that current can leave the AC generator in one direction only (as direct current).

Inside a diode are positive and negative areas that are separated by a boundary area. The boundary area is called the PN junction. When the positive side of a diode is connected to the positive side of the circuit, it is said to have **forward bias (Figure 16–6)**.

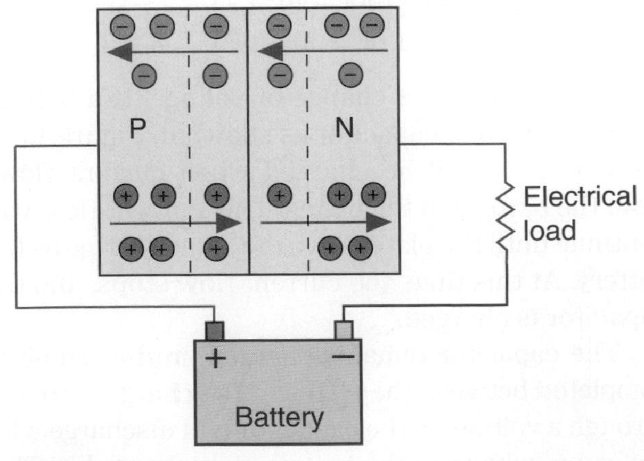

FIGURE 16–6

Forward-biased voltage allows current flow through the diode.

Unlike electrical charges are attracted to each other, and like charges repel each other. Therefore, the positive charge from the circuit is attracted to the negative sidc. The circuit's voltage is much stronger than the charges inside the diode, which causes the diode's charges to move. The diode's P material is repelled by the positive charge of the circuit and is pushed toward the N material, and the N material toward the P. This causes the PN junction to become a conductor, allowing current to flow.

When **reverse bias** is applied to the diode, the P and N areas are connected to opposite charges. Because opposites attract, the P material moves toward the negative part of the circuit, and the N material moves toward the positive part of the circuit. This empties the PN junction, and current flow stops.

A **zener diode** works like a standard diode until a certain voltage is reached. When the voltage reaches this point, the diode allows current to flow in the reverse direction. Zener diodes are often used in electronic voltage regulators **(Figure 16–7)**.

LEDs emit light as current passes through them **(Figure 16–8)**. The colour of the emitted light depends on the material used to make the LED. Typically, LEDs are made of a variety of inorganic semiconductor materials that produce different colours **(Table 16–1)**.

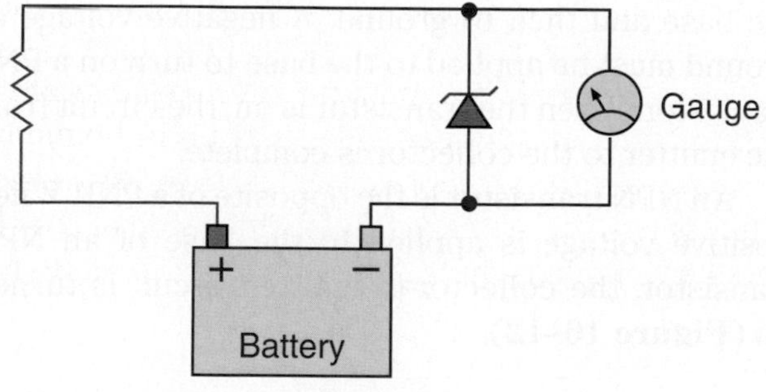

FIGURE 16–7

A simplified gauge circuit with a zener diode used to maintain a constant supply voltage to the gauge.

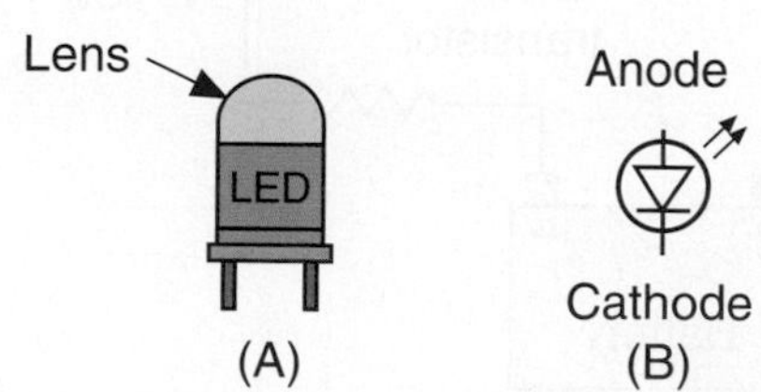

FIGURE 16–8

(A) An LED uses a lens to emit the light generated by current flow. (B) The schematic symbol for an LED.

TABLE 16–1 **MATERIAL USED FOR DIFFERENT-COLOURED LEDS**

COLOUR	SEMICONDUCTOR MATERIAL
Blue	Indium Gallium Nitride (Ingan) Sapphire (Al^2O^3) Silicon (Si) Silicon Carbide (Sic)
Bluish-green	Indium Gallium Nitride (Ingan)
Green	Aluminium Gallium Indium Phosphide (Algainp) Aluminium Gallium Phosphide (Algap) Gallium Nitride (Gan) Gallium Phosphide (Gap)
Orange	Aluminium Gallium Indium Phosphide (Algainp) Gallium Arsenide Phosphide (Gaasp)
Orange-red	Aluminium Gallium Indium Phosphide (Algainp) Gallium Arsenide Phosphide (Gaasp)
Red	Aluminium Gallium Arsenide (Algaas) Gallium Arsenide Phosphide (Gaasp) Gallium Phosphide (Gap)
White	Barrier of Aluminium Gallium Indium Nitride (Algainn)
Yellow	Aluminium Gallium Indium Phosphide (Algainp) Gallium Arsenide Phosphide (Gaasp) Gallium Phosphide (Gap)

Whenever the current flow through a coil of wire (such as that used in a solenoid or relay) stops, a voltage surge or spike is produced. This surge results from the collapsing of the magnetic field around the coil. The movement of the field across the winding induces a very high voltage spike, which can damage electronic components. In the past, a capacitor was used as a "shock absorber" to prevent component damage from this surge. On today's vehicles, a **clamping diode** is commonly used to prevent this voltage spike. By installing a clamping diode in parallel to the coil, a bypass is provided for the electrons during the time the circuit is opened **(Figure 16–9)**.

An example of the use of clamping diodes is on some air-conditioning compressor clutches. Because the clutch operates by electromagnetism, opening the clutch coil circuit produces a voltage spike. If the spike is left unchecked, it could damage the clutch coil relay contacts or the vehicle's computer. The clamping diode is connected to the circuit in reverse bias.

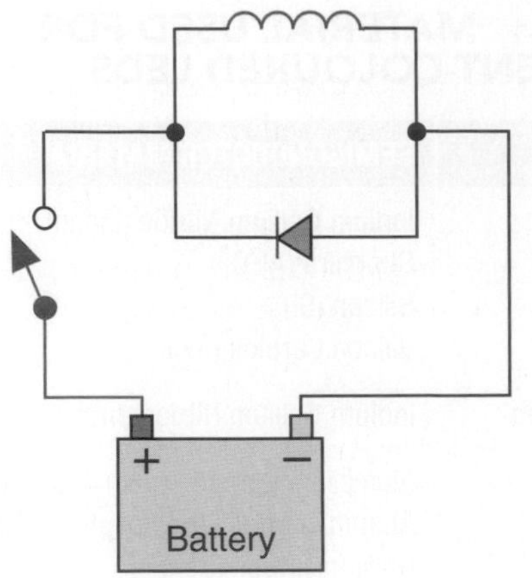

FIGURE 16–9

A clamping diode in parallel to a coil prevents voltage spikes when the switch is opened.

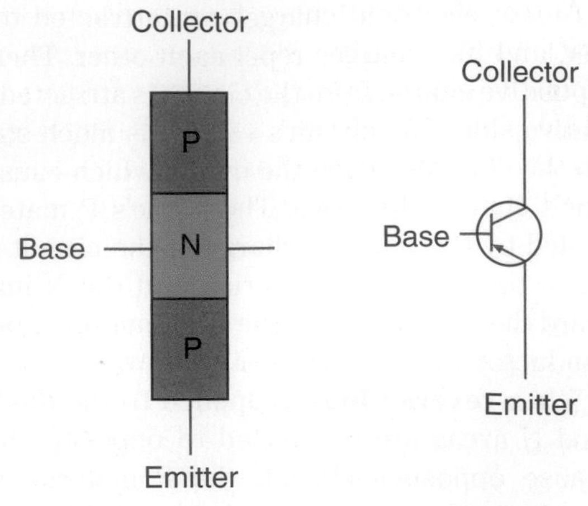

FIGURE 16–11

A PNP transistor and its schematic symbol.

Transistors

A **transistor** is produced by joining three sections of semiconductor materials. Like the diode, it is used as a switching device, functioning as either a conductor or an insulator. **Figure 16–10** shows some examples of transistors; there are many different sizes and types, depending on the application.

A transistor resembles a diode with an extra side. It can consist of two P-type materials and one N-type material or two N-type materials and one P-type material. These are called PNP and NPN types. In both types, junctions occur where the materials are joined. **Figure 16–11** shows a PNP transistor. Notice that each of the three sections has a lead connected to it. This allows any of the three sections to be connected to the circuit. The names for the legs are the **emitter**, **base**, and **collector**.

FIGURE 16–10

Typical transistors.

The centre section is called the base and is the controlling part of the circuit, or where the larger controlled part of the circuit is switched. The path to ground is through the emitter. A resistor is normally in the base circuit to keep current flow low. This prevents damage to the transistor. The emitter and collector make up the control circuit. When a transistor is drawn in an electrical schematic, the arrow on the emitter points to the direction of current flow.

The base of a PNP transistor is controlled by its ground. Current flows from the emitter through the base and then to ground. A negative voltage or ground must be applied to the base to turn on a PNP transistor. When the transistor is on, the circuit from the emitter to the collector is complete.

An NPN transistor is the opposite of a PNP. When positive voltage is applied to the base of an NPN transistor, the collector-to-emitter circuit is turned on **(Figure 16–12)**.

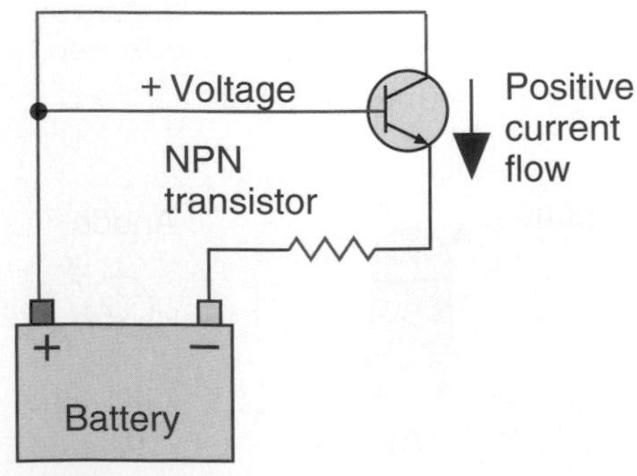

FIGURE 16–12

When positive voltage is applied to the base of an NPN transistor, current flows through the collector and the emitter.

A transistor can also function as a variable switch. By varying the voltage applied to the base, the completeness of the emitter and collector circuit will also vary. This is done simply by the presence of a variable resistor in the base circuit. This principle is used in light-dimming circuits.

Field-Effect Transistors

A field-effect transistor (FET) consists of a gate, a source, and a drain **(Figure 16–13)**, which correspond to the base, emitter, and collector of a bipolar transistor. With the exception of a junction-gate field-effect transistor (JFET), the balance of FETs have a fourth lead that is called the base, body, or, in some cases, substrate.

Each terminal name defines its function. The *gate* allows electrons to move between the *source* and the *drain* by applying a voltage, which creates a channel for movement of electrons or blocks the same channel to turn the flow off. The *body* or *substrate* is the material that contains or holds the semiconductive material for the gate, source, and drain.

The electrons, when influenced by a voltage, flow from the source to the drain. The channel of a FET is doped to create either an N-type or a P-type semiconductor, while the source and drain may be doped opposite or similarly to the channel. FETs are named for the type of insulation between the channel and the gate.

- MOSFET (metal oxide semiconductor field-effect transistor)—SiO2
- JFET (junction field-effect transistor)—reversed bias PN junction
- MESFET (metal semiconductor field-effect transistor)— Schottky barrier

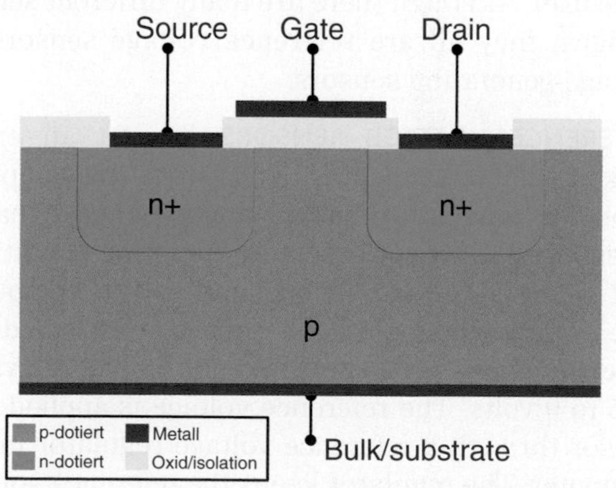

FIGURE 16–13

Example of a field-effect transistor.

Illustrated by Arne Nordmann

Integrated Circuits

The ability of one transistor or diode is limited in its ability to do complex tasks. However, when many semiconductors are combined into a circuit, they can perform complex functions.

An integrated circuit is simply a large number of diodes, transistors, and other electronic components mounted on a single piece of semiconductor material. This creates a very small package capable of performing many functions. Because of the size of an integrated circuit, many transistors, diodes, and other solid-state components can be installed in a car to make logic decisions and issue commands to other areas of the engine. This is the foundation of computerized engine control systems.

COMPUTER BASICS

Computers control nearly all of the systems in an automobile. Systems that once were controlled by vacuum, mechanical, and electromechanical devices are now controlled and operated by electronics. These are electronic control systems. They are made up of sensors, actuators, and a central processing unit, sometimes called a **microprocessor**. Electronic controls are designed to allow a system to operate in the most efficient way it can. They also provide for many driver conveniences. Although today's vehicles have several computers, they have two main computers—the powertrain control module (PCM) and body control module (BCM).

In addition to controlling various systems, the PCM and BCM continuously monitor operating conditions for possible system malfunctions. The computers compare system conditions against programmed parameters. If the conditions fall outside of these limits, the computers will detect the malfunction. A trouble code will be stored in the computers' memory to indicate the portion of the system at fault. A technician can access this code to aid in diagnosis.

The **central processing unit (CPU)** is basically thousands of transistors placed on a small chip. The CPU moves information in and out of the computer's memory **(Figure 16–14)**. Input information is processed in the CPU and checked against the programs stored in its memory. The CPU also checks for all other pertinent information held in memory. The CPU takes all of this information and uses computer logic to determine what should or should not happen. Once these decisions are made, the CPU

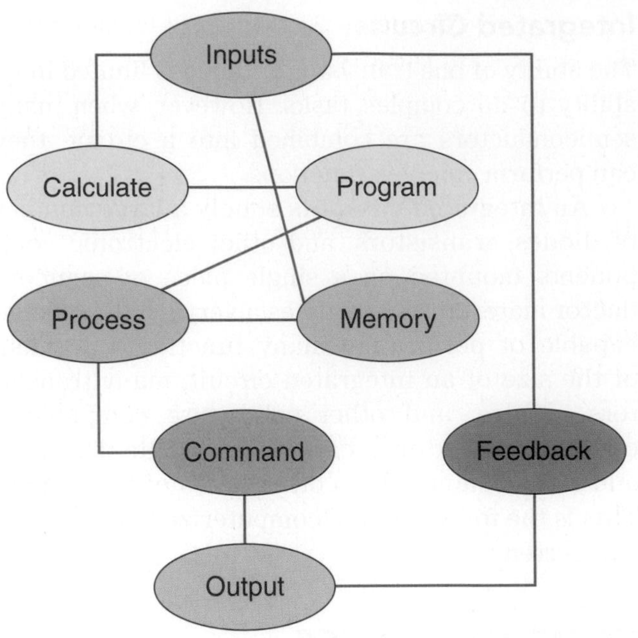

FIGURE 16–14
The basic information cycle for a computer.

sends out commands to make the required corrections or adjustments to the system.

A computer is an electronic device that stores and processes data. It relies on semiconductors and is really a group of integrated circuits. The four basic functions of a computer are as follows:

1. **Input.** A signal sent from an input device. The device can be a sensor or a switch activated by the driver, technician, or a mechanical part.
2. **Processing.** The computer uses the input information and compares it to programmed instruction. This information is processed by logic circuits in the computer.
3. **Storage.** The program instructions are stored in the computer's memory. Some of the input signals are also stored for processing later.
4. **Output.** After the computer has processed the inputs and checked its programmed instructions, it will issue commands to various output devices. These output devices may be instrument panel displays or output actuators. The output of one computer may also be an input to other computers.

Inputs

The CPU receives inputs that it checks against programmed values. Depending on the input, the computer controls the actuator(s) until the programmed results arc obtained. The inputs can come from other computers, the driver, the technician, or through a variety of sensors.

Driver input signals are usually provided by momentarily applying a ground through a switch. The computer receives this signal and performs the desired functions. For example, if the driver wishes to reset the trip odometer on a digital instrument panel, a reset button is depressed. This switch provides a momentary ground that the computer receives as an input and sets the trip odometer to zero.

Switches can be used as inputs for any operation that requires only a yes–no, or on–off, condition. Other inputs include those supplied by means of a sensor and those signals returned to the computer in the form of **feedback**. Feedback means that data concerning the effects of the computer's commands are fed back to the computer as an input signal.

If the computer sends a command signal to actuate an output device, a feedback signal may be sent back from the actuator to inform the computer that the task was performed. The feedback signal confirms both the position of the output device and the operation of the actuator. Another form of feedback is for the computer to monitor voltage when a switch, relay, or other actuator is activated. Changing positions of an actuator should result in predictable changes in the computer's voltage sensing circuit. The computer may set a diagnostic code if it does not receive the correct feedback signal.

All inputs have the same basic function. They detect a mechanical condition (movement or position), chemical state, or temperature condition and change it into an electrical signal that is used by the computer to make decisions. Each sensor has a specific job to do (for example, monitor throttle position, vehicle speed, and manifold pressure). Although there are many different sensor designs, they all are reference voltage sensors or voltage-generating sensors.

REFERENCE VOLTAGE SENSORS To get an exact look at what is happening in a system, the computer applies a constant, predetermined voltage signal to a sensor. The sensor reacts to operating conditions and sends a voltage signal back to the computer. The voltage applied by the computer is called the **reference voltage (Vref)** and normally has a value of 5 to 9 volts. The reference voltage is applied to a sensor through a reference voltage regulator in the computer. The regulator keeps the reference voltage at a predetermined value. Because the computer knows that a certain voltage value has been applied, it can indirectly interpret things such as motion,

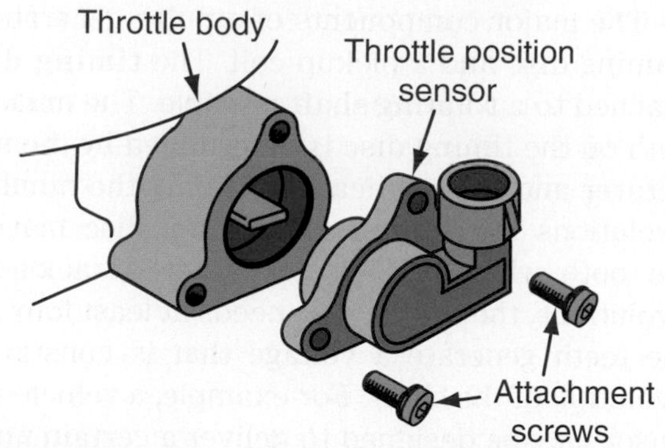

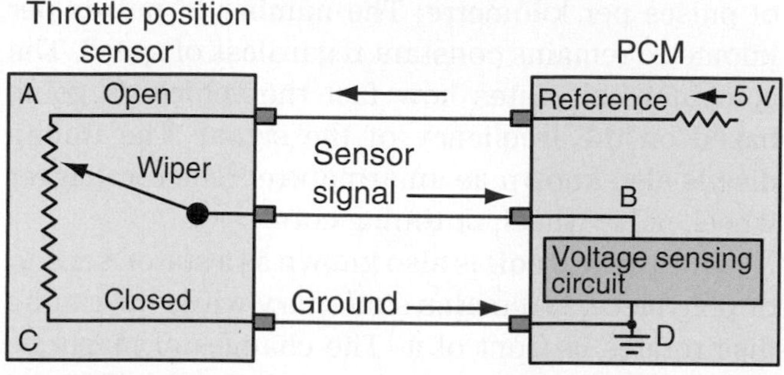

FIGURE 16–15

A TP switch is a voltage reference sensor.

temperature, and component position, based on the return signal.

Most reference voltage sensors are variable resistors or potentiometers. They modify the reference voltage, and the return voltage signal represents a condition. The computer will use this signal to calculate the condition and order changes to system operation, if necessary.

For example, consider the operation of a throttle position sensor (TP sensor). During acceleration (from idle to wide-open throttle), the computer monitors throttle plate movement based on the changing reference voltage signal returned by the TP sensor **(Figure 16–15)**. The TP sensor is a rotary potentiometer that changes circuit resistance based on throttle shaft rotation. As the sensor's resistance varies, the computer responds in a specific manner to each corresponding voltage change.

Thermistors are also variable resistors and serve as voltage reference sensors. Inputs from a thermistor allow the computer to observe small changes in temperature. Thermistors are used to monitor the temperature of engine coolant, inside and outside ambient air, intake air, and many components.

Wheatstone bridges (Figure 16–16) are also used as variable resistance sensors. These are typically constructed of four resistors connected in series-parallel between an input terminal and a ground terminal. Three of the resistors are kept at the same value. The fourth resistor is a sensing resistor. When all four of the resistors have the same value, the bridge is balanced and the voltage sensor will have a value of 0 volts. If the sensing resistor changes value, a change occurs in the circuit's balance. The sensing circuit will receive a voltage reading proportional to the amount of resistance change. If the Wheatstone bridge is used to measure temperature, temperature changes are indicated as a change in voltage by the sensing circuit. Wheatstone bridges are also used to measure pressure (**piezo-resistive**) and mechanical strain, and are also the electronic composition of a manifold absolute pressure (MAP) sensor **(Figure 16–17)**.

In addition to variable resistors, another commonly used reference voltage sensor is a switch. By opening and closing a circuit, switches provide the necessary voltage information to the computer so that vehicles can maintain the proper performance and driveability.

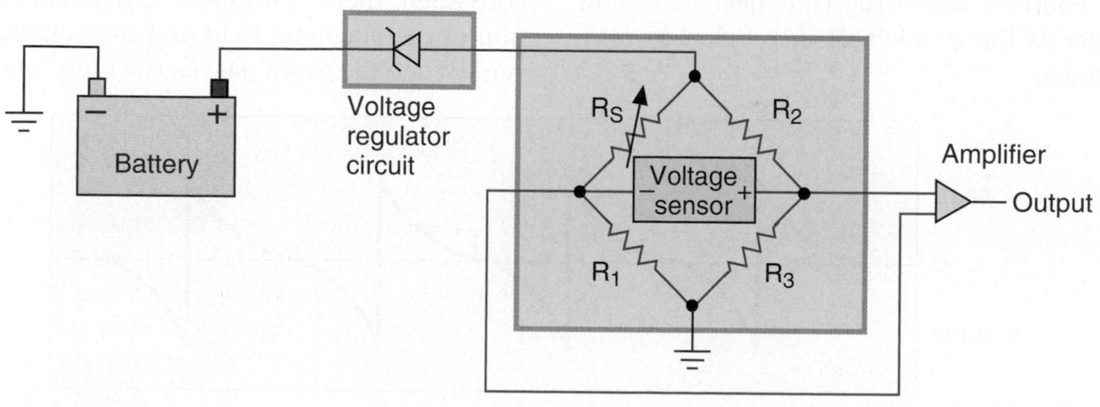

FIGURE 16–16

A wheatstone bridge.

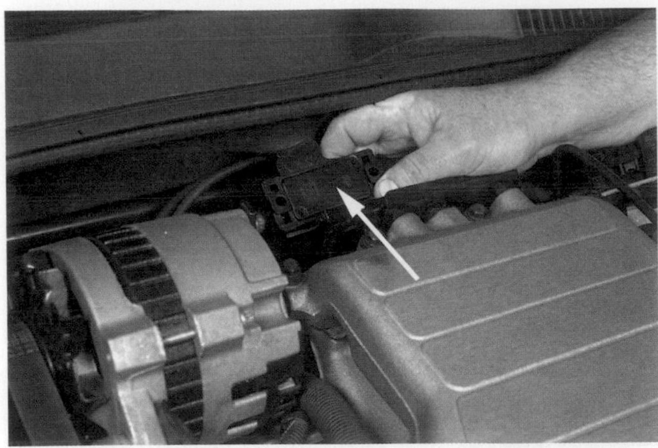

FIGURE 16–17

A MAP sensor.

VOLTAGE-GENERATING SENSORS Although many sensors are variable resistors or switches, **voltage-generating sensors** are commonly used. These sensors include speed sensors, Hall-effect switches, oxygen sensors, and knock sensors. These are capable of producing an input voltage signal for the control system. This varying voltage signal allows the computer to monitor and immediately adjust the operation of a system to meet the current needs.

Magnetic pulse generators use the principle of magnetic induction to produce a voltage signal. They are also called permanent magnet (PM) generators. These sensors are often used to send data to the computer about the speed of the monitored component. These data provide information about vehicle speed, shaft speed, and wheel speed. The signals from speed sensors are used for instrumentation, cruise control systems, antilock brake systems, ignition systems, speed-sensitive steering systems, and automatic ride-control systems. A magnetic pulse generator is also used to inform the computer about the position of a monitored device. This is common in engine controls where the CPU needs to know the position of the crankshaft in relation to rotational degrees.

The major components of a pulse generator are a timing disc and a pickup coil. The **timing disc** is attached to a rotating shaft or cable. The number of teeth on the timing disc is determined by the manufacturer and the application. If only the number of revolutions is required, the timing disc may have one tooth, whereas if it is important to track quarter revolutions, the timing disc needs at least four teeth. The teeth generate a voltage that is constant per revolution of the shaft. For example, a vehicle speed sensor may be designed to deliver a certain number of pulses per kilometre. The number of pulses per kilometre remains constant regardless of speed. The computer calculates how fast the vehicle is going based on the frequency of the signal. The timing disc is also known as an armature, reluctor, trigger wheel, pulse wheel, or timing core.

The **pickup coil** is also known as a stator, sensor, or pole piece. It remains stationary while the timing disc rotates in front of it. The changes of magnetic lines of force generate a small voltage signal in the coil. A pickup coil consists of a permanent magnet with fine wire wound around it.

An air gap is maintained between the timing disc and the pickup coil. As the timing disc rotates in front of the pickup coil, the generator sends a pulse signal **(Figure 16–18)**. As a tooth on the timing disc aligns with the core of the pickup coil, it concentrates the magnetic field. The magnetic field is forced to flow through the coil and pickup core **(Figure 16–19)**. When the tooth passes the core, the magnetic field is able to expand **(Figure 16–20)**. This action is repeated every time a tooth passes the core. The moving lines of magnetic force cut across the coil windings and induce a voltage signal.

When a tooth approaches the core, a positive current is produced as the magnetic field begins to concentrate around the coil. When the tooth and core align, there is no more expansion or contraction of the magnetic field and the voltage drops to zero. When the tooth passes the core, the magnetic

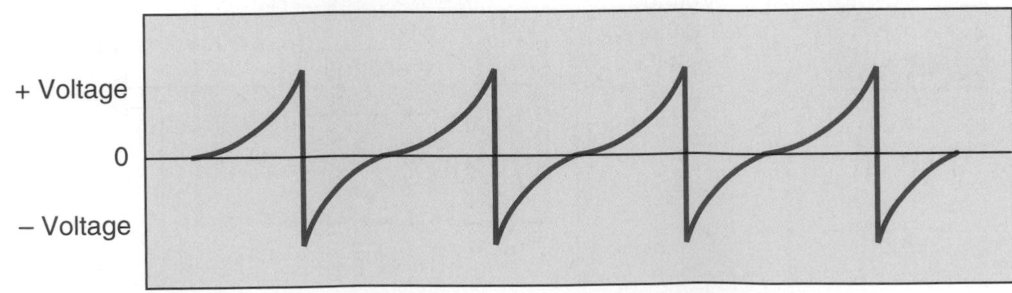

FIGURE 16–18

A pulsed voltage signal.

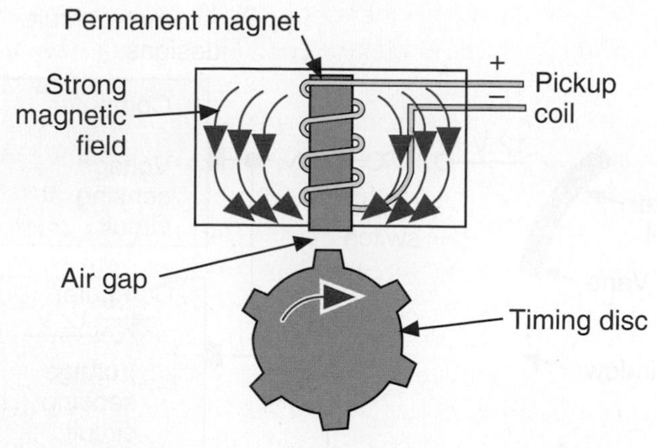

FIGURE 16–19

A strong magnetic field is produced in the pickup coil as the teeth align with the core.

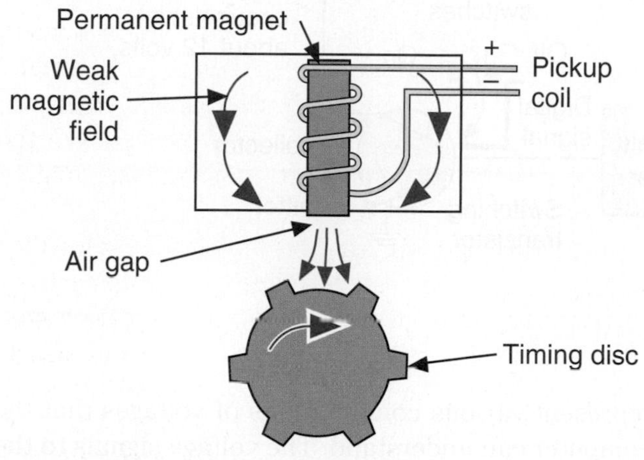

FIGURE 16–20

The magnetic field expands and weakens as the teeth pass the core.

field expands and a negative current is produced **(Figure 16–21)**. The resulting pulse signal is sent to the CPU.

The **Hall-effect switch** performs the same function as a magnetic pulse generator. It operates

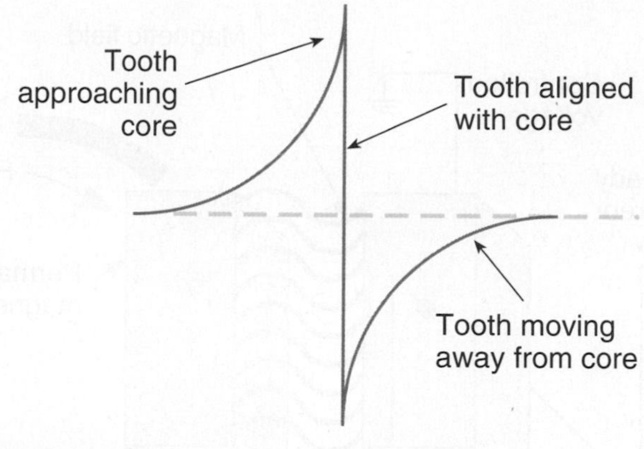

FIGURE 16–21

The waveform produced by a magnetic pulse generator.

on the principle that if a current is allowed to flow through thin conducting material exposed to a magnetic field, another voltage is produced **(Figure 16–22)**.

A Hall-effect switch contains a permanent magnet and a thin semiconductor layer made of gallium arsenate crystal (Hall layer) and a shutter wheel **(Figure 16–23)**. The Hall layer has a negative and a positive terminal connected to it. Two additional terminals located on either side of the Hall layer are used for the output circuit.

The permanent magnet is located directly across from the Hall layer, so its lines of flux bisect the layer at right angles to the current flow. The permanent magnet is stationary, and a small air gap is between it and the Hall layer.

A steady current is applied to the crystal of the Hall layer. This produces a signal voltage perpendicular to the direction of current flow and magnetic flux. The signal voltage produced is a result of the effect the magnetic field has on the electrons. When the magnetic field bisects the supply current flow, the electrons are deflected toward the Hall layer

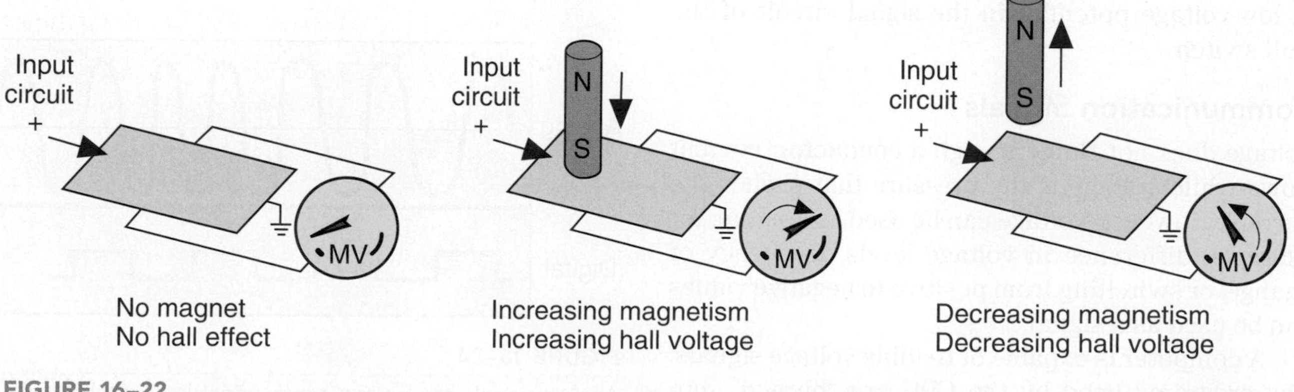

FIGURE 16–22

Hall-effect principles of voltage induction.

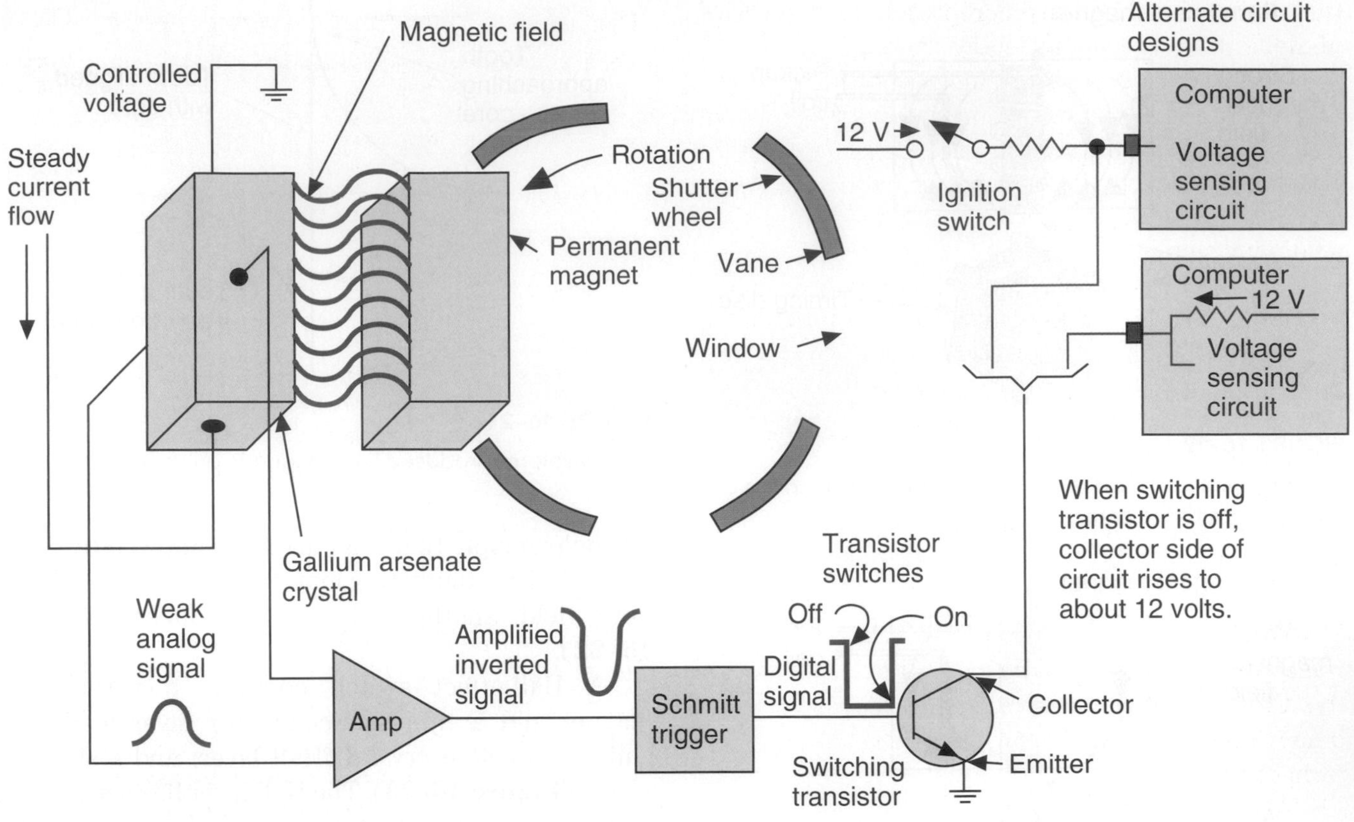

FIGURE 16–23

Typical circuit of a Hall-effect switch.

negative terminal, which results in a weak voltage potential being produced in the Hall switch.

The shutter wheel consists of a series of alternating windows and vanes. It creates a magnetic shunt that changes the strength of the magnetic field from the permanent magnet. The shutter wheel is attached to a rotating component. As the wheel rotates, the vanes pass through the air gap. When a shutter vane enters the gap, it intercepts the magnetic field and shields the Hall layer from its lines of force. The electrons in the supply current are no longer disrupted and return to a normal state. This results in low-voltage potential in the signal circuit of the Hall switch.

Communication Signals

Voltage does not flow through a conductor; current flows while voltage is the pressure that pushes the current. However, voltage can be used as a signal; for example, difference in voltage levels, frequency of change, or switching from positive to negative values can be used as a signal.

A computer is capable of reading voltage signals. The programs used by the CPU are "burned" into IC chips, using a series of numbers. These numbers represent various combinations of voltages that the computer can understand. The voltage signals to the computer can be either analog or digital. **Analog** means a voltage signal is infinitely variable, or can be changed, within a given range. **Digital** means a voltage signal that is in one of three states—either on–off, yes–no, or high–low. Digital signals produce a **square wave**. The wave represents the immediate change in the voltage signal. It is called a square wave because the digital signal creates a series of horizontal and vertical lines that connect to form a square-shaped pattern on a scope **(Figure 16–24)**.

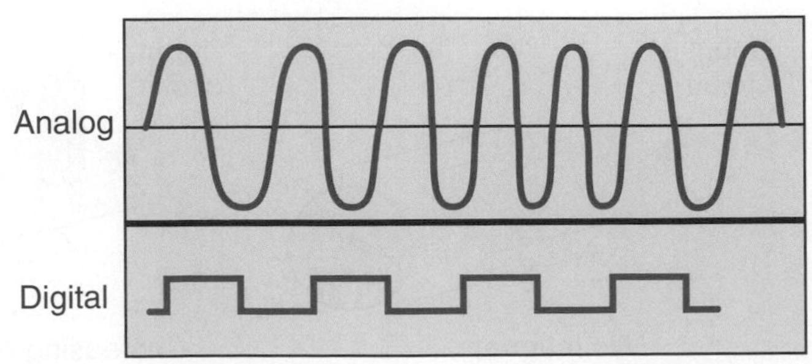

FIGURE 16–24

Analog signals can be constantly variable. Digital signals are either on–off or low–high.

In a digital signal, voltage is represented by a series of digits, which create a **binary code**.

Most input sensors produce a constantly variable voltage signal. For example, the voltage signal from an ambient temperature sensor never changes abruptly. The signal corresponds with a gradual increase or decrease in temperature and is therefore an analog signal.

A computer can only read a digital binary signal. To overcome this communication problem, all analog voltage signals are converted to a digital format by a device known as an analog-to-digital converter (**A/D converter**). The A/D converter **(Figure 16–25)** is located in the processor.

The A/D converter changes a series of signals to a binary number made up of 1s and 0s. Voltage above a given value converts to 1, and zero voltage converts to 0 **(Figure 16–26)**. Each 1 or 0 represents a **bit** of information. Eight bits equal a **byte** (sometimes referred to as a *word*). All communication between the CPU, the memories, and the interfaces is in binary code, with each information exchange in the form of a byte.

To get an idea of how binary coding works, let us see how signals from the coolant temperature sensor (CTS) are processed by the CPU. The CTS is a negative temperature coefficient (NTC) thermistor that controls a reference signal based on temperature changes. Upon receiving the CTS's analog signals, the input conditioner immediately groups each signal value into a predetermined voltage range and assigns a numeric value to each range. In

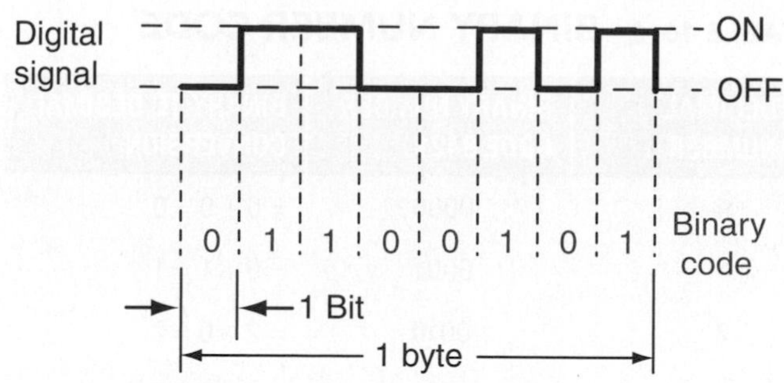

FIGURE 16–26

Each zero (0) and one (1) represents a bit of information. When eight bits are combined in specific sequence, they form a byte, or word, that makes up the basis of a computer's language.

our example, use the following ranges and numeric values: 0 to 2 volts = 1; 2 to 4 volts = 2; and 4 to 5 volts = 3 (assuming a Vref of 5 volts).

When the CTS is hot, its resistance is low, and the modified voltage signal it sends back falls into the high range (4–5 volts). Upon entering the A/D converter, the voltage value is assigned a numeric value of 3 (based on the ranges previously cited) and is ready for further translation into a binary code format. Binary numbers are represented by the numbers 0 and 1. Any number and word can be translated into a combination of binary 1s and 0s.

Without going into the finer points of binary numbering, the number 3 in binary is expressed as 11. To the thousands of tiny transistors and diodes that act as the on–off switches inside

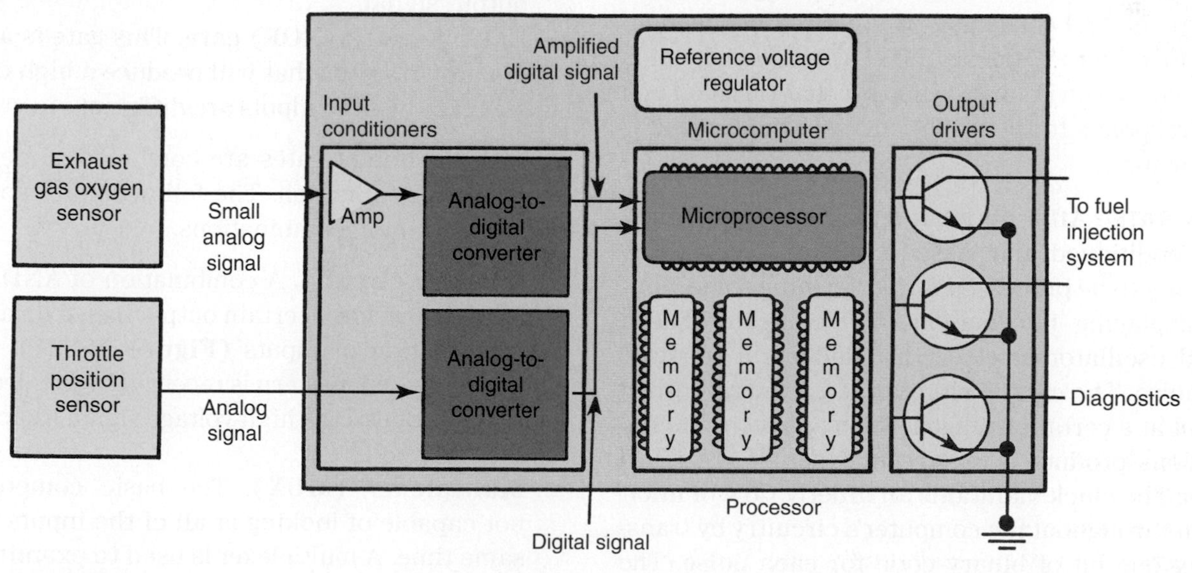

FIGURE 16–25

The A/D converter prepares input signals for the CPU.

TABLE 16–2 BINARY NUMBER CODE

DECIMAL NUMBER	BINARY NUMBER CODE 8421	BINARY TO DECIMAL CONVERSION
0	0000	= 0 + 0 = 0
1	0001	= 0 + 1 = 1
2	0010	= 2 + 0 = 2
3	0011	= 2 + 1 = 3
4	0100	= 4 + 0 = 4
5	0101	= 4 + 1 = 5
6	0110	= 4 + 2 = 6
7	0111	= 4 + 2 + 1 = 7
8	1000	= 8 + 0 = 8

microprocessors, 11 instructs the computer to turn on or apply voltage to a specific circuit for a predetermined length of time (based on its program). **Table 16–2** illustrates how binary numbers can be converted into decimal or base 10 numbers.

SCHMITT TRIGGER In addition to A/D conversion, some voltage signals require amplification before they can be processed by the computer. To do this, an input conditioner known as an amplifier is used to strengthen weak voltage signals. This is especially important for signals from Hall-effect switches. When the signal voltage leaves the sensor, it is a weak analog signal. The signal is amplified and inverted. It is then sent to a **Schmitt trigger**, which is a type of A/D converter, where it is digitized and conditioned into a clean square wave. The signal is then sent to a switching transistor that turns on and off in response to the signal and sets the frequency of the signal.

CLOCK RATE After an input signal has been generated, conditioned, and passed along to the computer, it is ready to be processed for performing some work and displaying information. The computer has a crystal oscillator or clock that delivers a constant time pulse. The crystal vibrates at a fixed rate when current at a certain voltage level is applied to it. The vibrations produce a very regular series of voltage pulses. The clock maintains an orderly flow of information throughout the computer's circuitry by transmitting one bit of binary code for each pulse. The clock enables the computer to know when one signal ends and another begins.

COMMUNICATION RATES The amount of information processed by a computer is dependent on its speed, or baud rate. **Baud rate** is the speed of communication and is roughly equal to the number of bits per second a computer can process.

Logic Gates

Logic gates are the thousands of **field-effect transistors (FETs)** incorporated into the computer's circuitry. The FETs use the incoming voltage patterns to determine the pattern of pulses leaving the gate. These circuits are called logic gates because they act as gates to output voltage signals depending on different combinations of input signals. The following are the most common logic gates and their operation. The symbols in the figures represent functions and not electronic construction.

1. **NOT gate.** A NOT gate simply reverses binary 1s and 0s and vice versa. A high input results in a low output, and a low input results in a high output.
2. **AND gate.** The AND gate has at least two inputs and one output. The operation of the AND gate is similar to two switches in series with a load. The only way the load will turn on is if both switches are closed. Before current can be present at the output of the gate, current must be present at the base of both transistors.
3. **OR gate.** The OR gate operates similarly to two switches that are wired in parallel to a light. If one switch is closed, the light will turn on. A high signal to either input will result in a high output.
4. **NAND** and **NOR gates.** A NOT gate placed behind an OR or AND gate inverts the output signal.
5. **Exclusive-OR (XOR) gate.** This gate is a combination of gates that will produce a high output signal only if the inputs are different.

These different gates are combined to perform the processing function. The following are some of the most common combinations.

1. **Decoder circuits.** A combination of AND gates used to provide a certain output based on a given combination of inputs **(Figure 16–27)**. When the correct bit pattern is received by the decoder, it will produce the high-voltage signal to activate the relay coil.
2. **Multiplexer (MUX).** The basic computer is not capable of looking at all of the inputs at the same time. A multiplexer is used to examine one of many inputs depending on a programmed priority rating **(Figure 16–28)**.

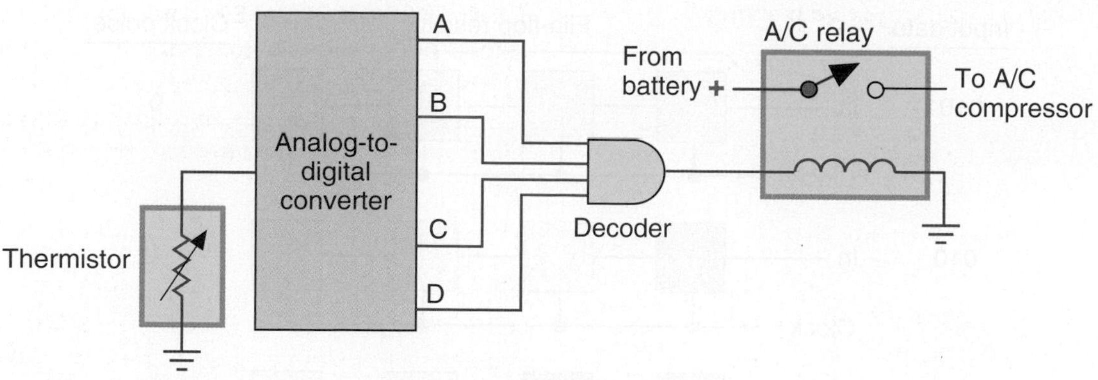

FIGURE 16–27

A simplified temperature sensing circuit that will turn on the A/C compressor when the inside temperature reaches a predetermined level.

3. **Demultiplexer (DEMUX).** Operates similarly to the MUX except that it controls the order of the outputs. The process that the MUX and DEMUX operate on is called sequential sampling. This means the computer deals with all the sensors and actuators one at a time.
4. **RS and clocked RS flip-flop circuits.** Flip-flop circuits remember previous inputs and do not change their outputs until they receive new input signals. The clocked flip-flop circuit has an inverted clock signal as an input, so circuit operations occur in the proper order. Flip-flop circuits are called sequential logic circuits because the output is determined by the sequence of inputs.
5. **Registers.** Used in the computer to temporarily store information. A register is a combination of flip-flops that transfers bits from one to another every time a clock pulse occurs **(Figure 16–29)**.
6. **Accumulators.** Registers designed to store the results of logic operations that can become inputs to other computers or modules.

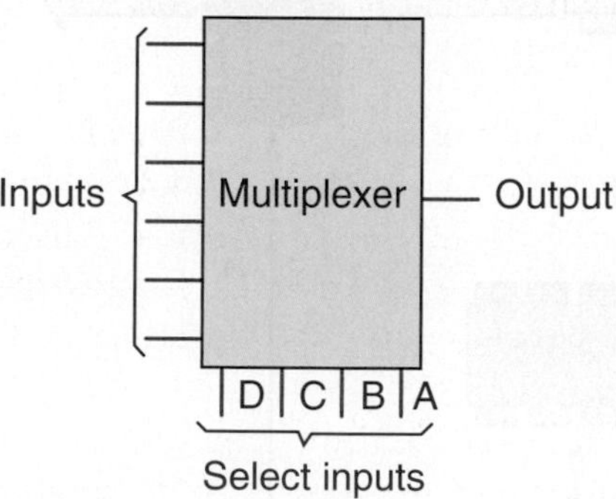

FIGURE 16–28

The selection of inputs at DCBA will determine which input data will be processed.

7. **Logic gates.** Process input information to command output devices. The order of this logic or the instructions to the computer are held in the computer's memory.

Memories

A computer's memory holds the programs and other data, such as vehicle calibrations, that the CPU refers to while making calculations. The program is a set of instructions or procedures that it must follow. Included in the program are **look-up tables** that tell the computer when to retrieve an input (based on temperature, time, etc.), how to process it, and what to do with it after it has been processed. Look-up tables are sets of instructions, and there is one for every possible condition the computer may detect.

The microprocessor works with memory in two ways: It can read information from memory or change information in memory by writing in or storing new information. To write information in memory, each memory location is assigned a number (written in binary code also) called an address. These addresses are sequentially numbered, starting with zero, and are used by the microprocessor to retrieve data and write new information into memory. During processing, the CPU often receives more data than it can immediately handle. In these instances, some information has to be temporarily stored or written into memory until the microprocessor needs it.

When ready, the microprocessor accesses the appropriate memory location (address) and is sent a copy of what is stored. By sending a copy, the memory retains the original information for future use.

Basically, three types of memory are used in automotive CPUs today **(Figure 16–30)**: read-only memory, programmable read-only memory, and random-access memory.

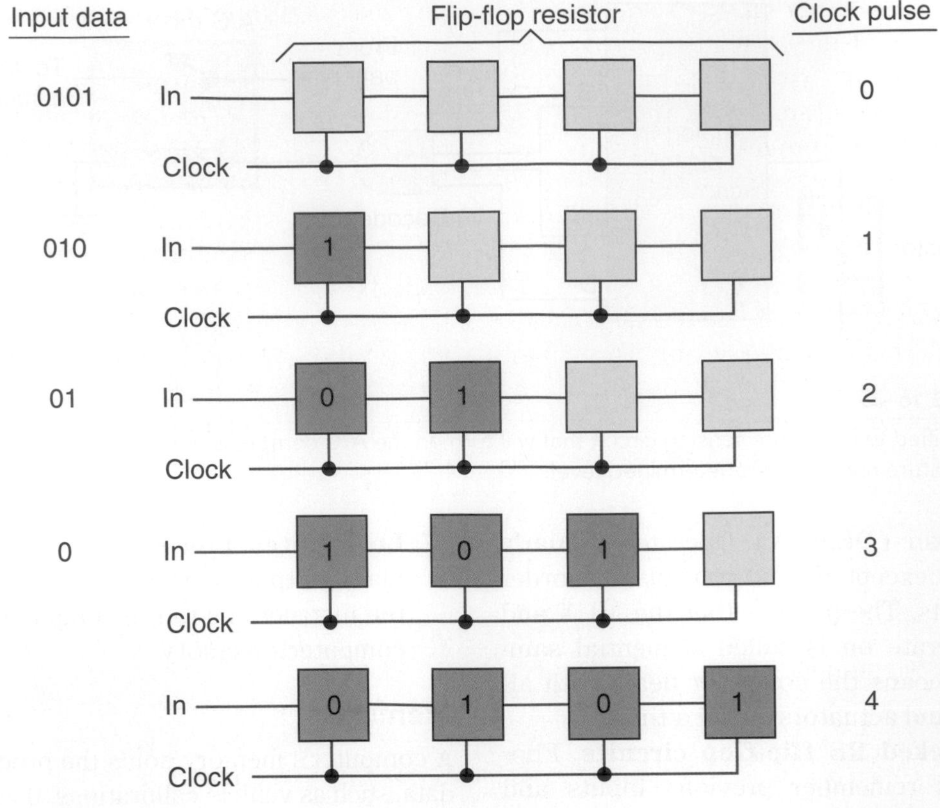

FIGURE 16–29

It takes four clock pulses to load four bits into the register.

READ-ONLY MEMORY (ROM) Permanent information is stored in **read-only memory (ROM)**. Information in ROM cannot be erased, even if the system is turned off or the CPU is disconnected from the battery. As the name implies, information can only be read from ROM.

When making decisions, the microprocessor is constantly referring to the stored information and the input from sensors. By comparing information from these sources, the CPU makes informed decisions.

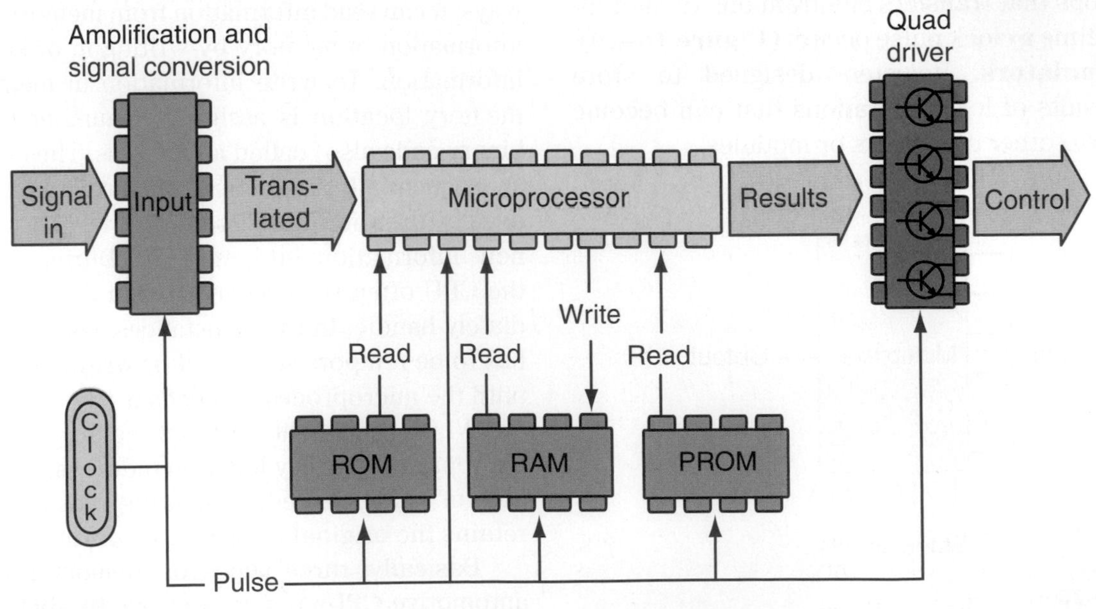

FIGURE 16–30

The three memories within a computer.

PROGRAMMABLE READ-ONLY MEMORY (PROM) The **programmable read-only memory (PROM)** differs from the ROM in that it plugs into the computer and may be reprogrammed or replaced with one containing a revised program. It contains program information specific to different vehicle model calibrations. The PROM in some computers is replaceable and can serve as a way to upgrade the system.

Erasable PROM (EPROM) is similar to PROM except that its contents can be erased to allow new data to be installed. A piece of Mylar tape covers a window. If the tape is removed, the memory circuit is exposed to ultraviolet light that erases its memory.

Electrically erasable PROM (EEPROM) allows changing the information electrically one bit at a time. Some manufacturers use this type of memory to store information concerning mileage, vehicle identification number, and options.

RANDOM-ACCESS MEMORY (RAM) The **random-access memory (RAM)** is used during computer operation to store temporary information. The CPU can write, read, and erase information from RAM in any order, which is why it is called random. One characteristic of RAM is that when the ignition key is turned off and the engine is stopped, information in RAM is erased. RAM is used to store information from the sensors, the results of calculations, and other data that are subject to constant change.

There are currently two other versions of RAM in use: volatile and nonvolatile. A volatile RAM, usually called **keep-alive memory (KAM)** has most of the features of RAM. Information can be written into KAM and can be read and erased from KAM. Unlike RAM, information in KAM is not erased when the ignition key is turned off and the engine is stopped. However, if battery power to the processor is disconnected, information in KAM is erased.

A **nonvolatile RAM (NVRAM)** does not lose its stored information if its power source is disconnected. Vehicles with digital display odometers usually store mileage information in nonvolatile RAM.

Actuators

Once the computer's programming determines that a correction or adjustment must be made in the controlled system, an output signal is sent to control devices called **actuators**. These actuators, which are solenoids, switches, relays, or motors, physically act on or carry out the command sent by the computer.

Actually, actuators are electromechanical devices that convert an electrical current into mechanical action. This mechanical action can then be used to open and close valves, control vacuum to other components, or open and close switches. When the CPU receives an input signal indicating a change in one or more of the operating conditions, the CPU determines the best strategy for handling the conditions. The CPU then controls a set of actuators to achieve a desired effect or strategy goal. For the computer to control an actuator, it must rely on a component called an **output driver**.

The output driver usually applies the ground circuit of the actuator. The ground can be applied steadily if the actuator must be activated for a selected amount of time, or the ground can be pulsed to activate the actuator in pulses.

Output drivers operate by the digital commands issued by the CPU. Basically, the output driver is nothing more than an electronic on–off switch used to control a specific actuator.

To illustrate this relationship, let us suppose the computer wants to turn on the engine's cooling fan. Once it makes a decision, it sends a signal to the output driver that controls the cooling fan relay (actuator). In supplying the relay's ground, the output driver completes the power circuit between the battery and cooling-fan motor and the fan operates. When the fan has run long enough, the computer signals the output driver to open the relay's control circuit (by removing its ground), thus opening the power circuit to the fan.

SHOP TALK

Normally, a computer will control an actuator with a low-side driver. These drivers complete the ground for the output device. Many newer systems use high-side drivers that control the outputs by varying the power to them. Most high-side drivers are metal oxide semiconductor field-effect transistors (MOSFET) controlled by another transistor. High-side drivers are used in circuits where a quick response to opens, shorts, and temperature changes is desired. Because a circuit's behaviour depends on the driver being used, it is important to check the service information before diagnosing the system.

For actuators that cannot be controlled by a digital signal, the CPU must turn its digitally coded instructions back into an analog signal. This conversion is completed by the A/D converter.

Displays can be controlled directly by the CPU. They do not require digital-to-analog conversion or output drivers because they contain circuitry that decodes the microprocessor's digital signal. The decoded information is then used to indicate such things as vehicle speed, engine rpm, fuel level, or scan tool values.

DUTY CYCLE VERSUS PULSE WIDTH Often the computer controls the results of the output by controlling the duty cycle or pulse width of the actuator. Duty cycle is a measurement of the amount of time something is on compared to the time of one cycle and is measured in a percentage. When measuring duty cycle, you are looking at the amount of time something is on during one cycle. Pulse width is similar to duty cycle except that it is the exact time something is turned on and is measured in milliseconds.

Power Supply

The CPU also contains a power supply that provides the various voltages required by the microprocessor and internal clock that provides the clock pulse, which in turn controls the rate at which sensor readings and output changes are made. Also contained are protection circuits that safeguard the microprocessor from interference caused by other systems in the vehicle and diagnostic circuits that monitor all inputs and outputs and signal a warning light if any values go outside the specified parameters. This warning light is called the malfunction indicator lamp (MIL).

Awake/Sleep Modes

The control modules are able to control or perform all of their functions in the awake mode. They enter a sleep mode when normal control or monitoring of the system functions has stopped and a time limit has passed. There is still some activity during the sleep mode; what occurs depends on the system. Basically, during the sleep mode, only enough power is used to maintain memory and for periodic monitoring of some systems. Once normal computer activity is called for, the computer wakes up and resumes its normal functions.

ON-BOARD DIAGNOSTICS

Although often thought of as something that only affects engine performance, on-board diagnostic (OBD) systems have shaped the operation of all electronic control modules. This is especially true of OBD-II.

On-board diagnostic capabilities are incorporated into a vehicle's computer to monitor virtually every component that can affect emission performance. Each component is checked by a diagnostic routine to verify that it is functioning properly.

OBD-I was the first generation of on-board diagnostic systems and was designed to monitor some of the vehicle's emission control components. Required on all 1991 and newer vehicles, OBD-I systems monitored only a few of the emission-related components and were not calibrated to a specific level of emission performance. OBD-II was developed to address these issues and to allow more accurate diagnosis by technicians.

The main goal of OBD-II systems is to detect when engine or system wear or component failure causes exhaust emissions to increase by 50 percent or more. According to the guidelines of OBD-II, all vehicles have the following:

- A universal diagnostic test connector, known as the **data link connector (DLC)**, with dedicated pin assignments
- A standard location for the DLC that must be under the dash on the driver's side of the vehicle and must be visible
- A standard list of **diagnostic trouble codes (DTCs)**
- A standard communication protocol
- The use of common scan tools on all vehicle makes and models
- Common diagnostic test modes
- Vehicle identification that must be automatically transmitted to the scan tool
- Stored trouble codes that must be able to be cleared from the computer's memory with the scan tool
- The ability to record, and store in memory, a snapshot of the operating conditions that existed when a fault occurred
- The ability to store a code whenever something goes wrong and affects exhaust quality
- A standard glossary of terms, acronyms, and definitions for all components in the electronic control systems

To meet these standards, many new technologies were incorporated into the various control systems. Because nearly anything in an automobile can affect its emissions, the technologies carried over to all electronic control systems. The result is more-efficient vehicles with more-efficient engines and accessories.

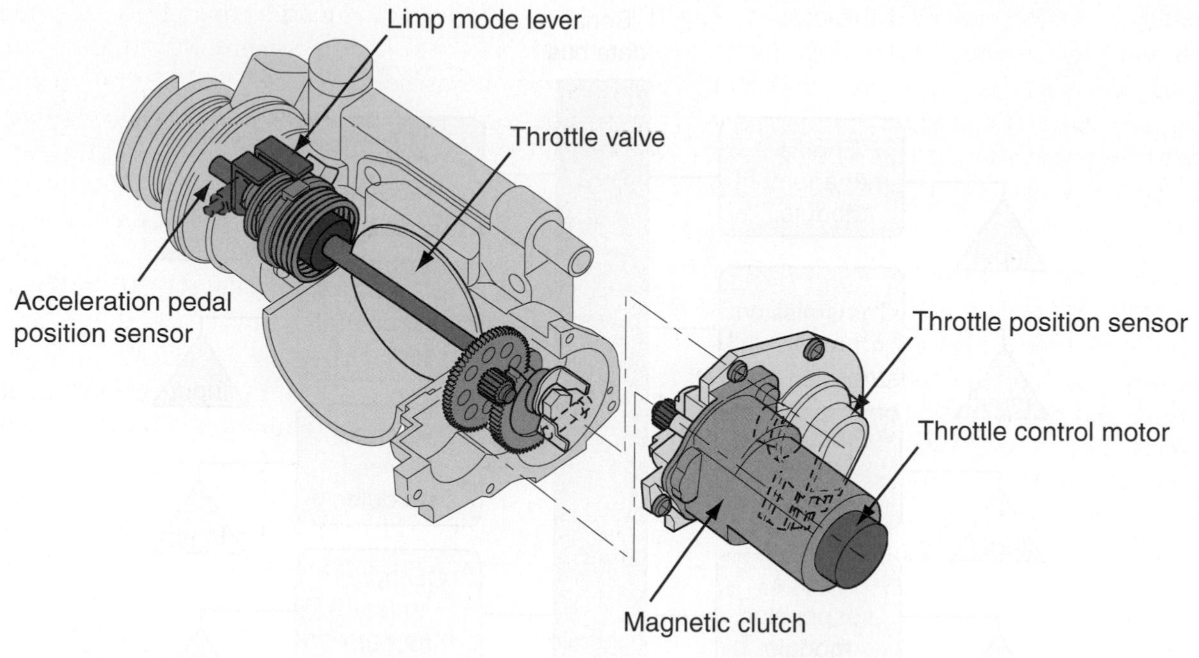

FIGURE 16–31

This electronic throttle control unit eliminates the need for a mechanical linkage between the throttle pedal and the fuel injection system.

By-Wire Technology

One of the things that has become a reality because of the high-powered computers used in today's vehicles is by-wire technology. Currently, this technology has eliminated the mechanical connection from the throttle pedal and the fuel injection system **(Figure 16–31)**. This unit is called the electronic throttle control unit **(Figure 16–32)**. Shift-by-wire technology is also used. In addition, it is being used by a few manufacturers in parking brake systems and adaptive cruise control. Soon it will be used in many other systems, such as the brake and steering systems.

FIGURE 16–32

An electronic throttle control assembly at the throttle body.

Drive-by-wire systems use sensors to translate the movement of pedals, the steering wheel, and other parts into electronic signals. The vehicle's computer receives these signals and commands electric motors to perform the function ordered by the driver. These systems respond much more quickly than mechanical linkages and can send feedback to the computer as they operate.

MULTIPLEXING

Today's vehicles have hundreds of circuits, sensors, and other electrical parts. In order for the control systems to operate correctly, there must be some communication between them. Communication can take place through wires connecting each sensor and circuit to the appropriate control module. If more than one control module is involved, additional pairs of wires must connect the sensor or circuit to the other modules. The result of this communication network is miles of wires and hundreds of connectors. To eliminate the need for all of these wires, manufacturers are using multiplexing **(Figure 16–33)**.

Multiplexing, also called in-vehicle networking, provides efficient communications

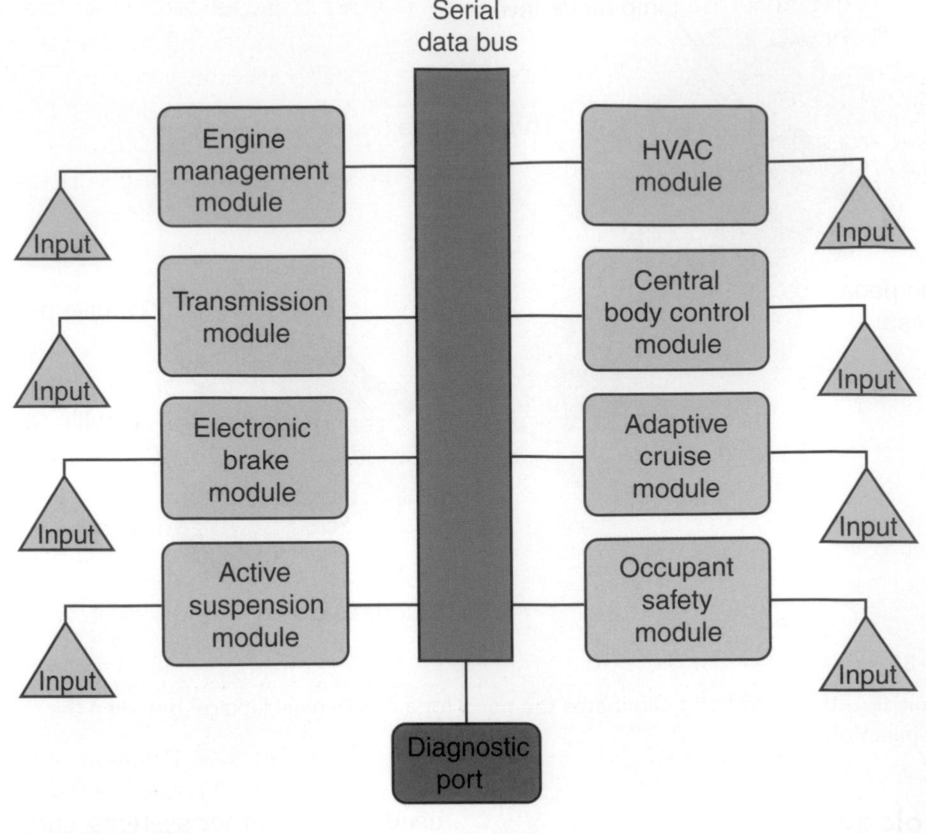

FIGURE 16–33

A multiplexed system uses a serial data bus to allow communications between the various control modules.

between vehicle systems. Multiplexing relies on one wire that allows many systems to communicate instead of many wires. A multiplex wiring system uses a **serial data bus** that connects different computers or control modules together. Each module can transmit and receive digital codes over the serial data bus, allowing one module to share information with other modules. For example, the signal relating to engine speed may be required by the engine control, transmission control, electronic brake control, and suspension control modules. Rather than have separate engine speed inputs for each module, the serial data bus carries the information, as well as other information, to all of the control modules.

Each sensor is wired directly to the control module that relies heavily on the sensor's signal. That control module sends the information, in binary code, to the serial data bus. Each control module has a code-reading device, or chip, that reads and sends messages on the serial data bus. Some chips can only send or only receive, depending on their purpose. All information on the serial data bus is available for all control modules. However, the chip of each device compares the coded message to its memory list to see whether the information is relevant to its own operation.

The chip is also used to prevent the signals from overlapping by allowing only one signal to be transmitted at a time. Each digital signal is preceded by an identification code that establishes its priority. If two modules attempt to send a message at the same time, the signal with the higher priority code is transmitted first. Because a control module processes only one input at a time, it orders the signals as it needs them. When one input is being received, the others are disregarded. Although it may appear to cause a time lapse, realize that the communication rate for most computers is between 10 000 and 1 000 000 bits per second.

Keep in mind that data are conveyed in binary numbers and therefore must be interpreted by some type of data processing before they become information. The stream of data across the bus is called serial data. It is essentially data that are transferred to and from a computer, one bit at a time. On many vehicles, serial data can be monitored with a scan tool connected to the vehicle's DLC. Monitoring serial data allows technicians to diagnose the various control modules and to check for DTCs.

The serial data bus is typically made of two wires: a ground wire and a serial bus transmission wire. These wires are twisted together to reduce magnetic interference, which can cause false information.

Advantages

Multiplexing offers many advantages over traditional wiring:

- The need for redundant sensors is eliminated because sensor data, such as vehicle speed and engine temperature, are available on the serial data bus, where they can be used by several control modules.
- Accessories and vehicle features can be easily added to the vehicle through software changes.
- Fewer wires are required for the operation of each system, which means smaller wiring harnesses; lower cost and weight; and improved serviceability, reliability, and installation. Without multiplexing, it is necessary to add a ground, a power source, and control wires whenever an electronic component is added to the vehicle.
- Improved communications between control modules allows for more accurate recording and reporting of faults, which helps in locating and solving problems.

As the electrical content of today's vehicles continues to increase, the need for networking is even more evident.

Types of Multiplexing

There are four basic techniques used for multiplexing:

- *Frequency division multiplexing (FDM).* This is an analog technique in which each communications channel is assigned a frequency. The frequencies are stacked on top of each other and many frequencies can be sent at once. To prevent interference from other channels, each is separated by a small frequency.
- *Time division multiplexing (TDM).* This is a digital technique in which a sample of each channel is inserted into the data stream. The sample period is fast enough to sample all channels within a fixed period. It works like a very fast mechanical switch.
- *Statistical time division multiplexing (STDM).* This uses chips to allocate time only to channels when it is needed. This means more channels can be connected to the bus because the chips statistically compensate for times when a particular system is not being used.
- *Wavelength division multiplexing (WDM).* This is used in fibre-optic networks where multiple signals are transmitted as light is split into different wavelengths.

Communication Protocols

A protocol is the name for the language that computers speak when they are talking to each other. The differences in protocol are based on the speed and the technique used. The Society of Automotive Engineers (SAE) has classified the different protocols by their speed and operation.

- *Class A (low-speed communication).* This is used for convenience systems, such as entertainment systems, audio, trip computer, seat controls, windows, and lighting. Most Class A functions require inexpensive, low-speed communication and use a generic universal asynchronous receiver/transmitter (UART). These functions are proprietary and have not been standardized by the industry.
- *Class B (medium-speed communication).* Class B multiplexing is used primarily with the instrument cluster, vehicle speed, and emissions data recording. Contained within this classification are different standards, designated by a number. The most commonly used is the SAE J1850 standard. Further, these standards are divided by their operation. One is a variable pulse width (VPW) type that uses a single bus wire. Another is a pulse-width-modulation (PWM) type that uses a two-wire differential bus.
- *Class C (high-speed communication).* This protocol is for real-time control of the powertrain, vehicle dynamics, and brake-by-wire. This protocol can use a twisted pair, but shielded coaxial cable or fibre optics may be used for less noise interference. The predominant Class C protocol is CAN 2.0 (controller area network version 2.0). The CAN serial data line is a high-speed serial data bus that ensures that the required real-time response is maintained. CAN assigns a unique identifier to every message. The identifier classifies the content of the message and the priority of the message being sent. Each module processes only those messages whose identifiers are stored in the module's acceptance list.

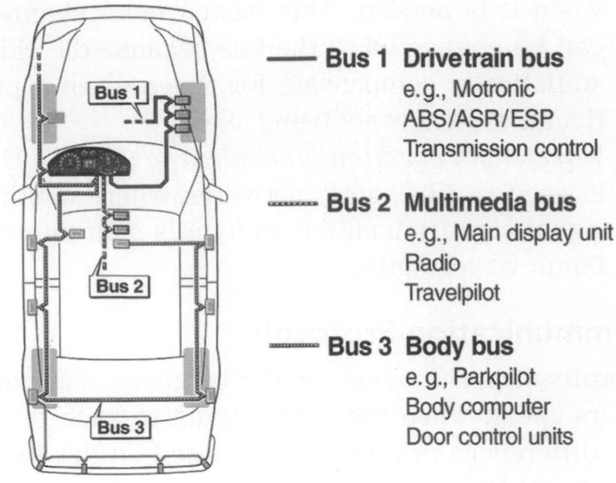

FIGURE 16–34

It is common to find a variety of protocols in a single vehicle.

Robert Bosch Corp.

It is common to find a variety of the different classes of multiplexing in a single vehicle **(Figure 16–34)**. Some systems, such as powertrain control and vehicle dynamics, require high-speed communications, whereas other systems do not.

Early multiplexed systems were often based on proprietary serial buses using generic UART or custom devices. This called for dedicated and specific scan tools, each with the ability to work only on specific systems. OBD-II called for standardized diagnostic tools, which meant standard protocols had to be implemented. Currently, Class B communications is the standard protocol; however, many systems use Class C. Since 2008, Class C communications has been the mandated protocol for diagnostics. CAN buses are found in nearly all late-model vehicles.

CAN Buses

The total network in most vehicles is comprised of two or three CAN buses. Each of these networks operates at different speeds. The different CAN buses are identified by a prefix or suffix. For example, a medium-speed bus may be called CAN B or MS-CAN. Likewise, a high-speed bus can be called CAN C or HS-CAN. Manufacturers are not consistent with these labels, so there are a variety of them.

Low- or medium-speed CANs are typically used for body functions, such as the following:

- Interior and exterior lights
- Horn
- Locks
- Windshield wipers
- Seats
- Window
- Sound systems

A high-speed bus is used for real-time functions such as the following:

- Engine management
- Antilock brake systems
- Transmission control
- Tire pressure monitoring systems
- Vehicle stability systems

These networks are integrated through the use of a gateway. A **gateway** module allows for data exchange between the different buses. It translates a message on one bus and transfers that message to another bus without changing the message **(Figure 16–35)**. The gateway interacts with each bus according to that bus's protocol. This is an important function; some information must be shared. In many vehicles, the BCM serves as the gateway for the different buses.

SHOP TALK

Although the BCM is designated as a body system, a vehicle may not start if the BCM is not operating correctly. This is due to its role as the gateway. If there is no communication between the security or antitheft system and the engine control system, the engine will not start. The PCM needs to know that it is okay to run the engine.

Each twisted wire in the CAN bus carries a different voltage **(Figure 16–36)**. Even the slightest change in voltage can affect the operation of one or more systems; therefore, all potential for voltage spikes, electrical noise, or induction must be eliminated. Twisting the wires eliminates the possibility of voltage being induced in one wire as current flows through the other. To eliminate other potential spikes and noise, two 120 Ω resistors are connected in parallel across the ends of the main CAN bus wires. These are called terminating resistors **(Figure 16–37)**. The location of the resistors varies. One may be located in the fuse block, and others can be internal to the ECM or PCM and the BCM.

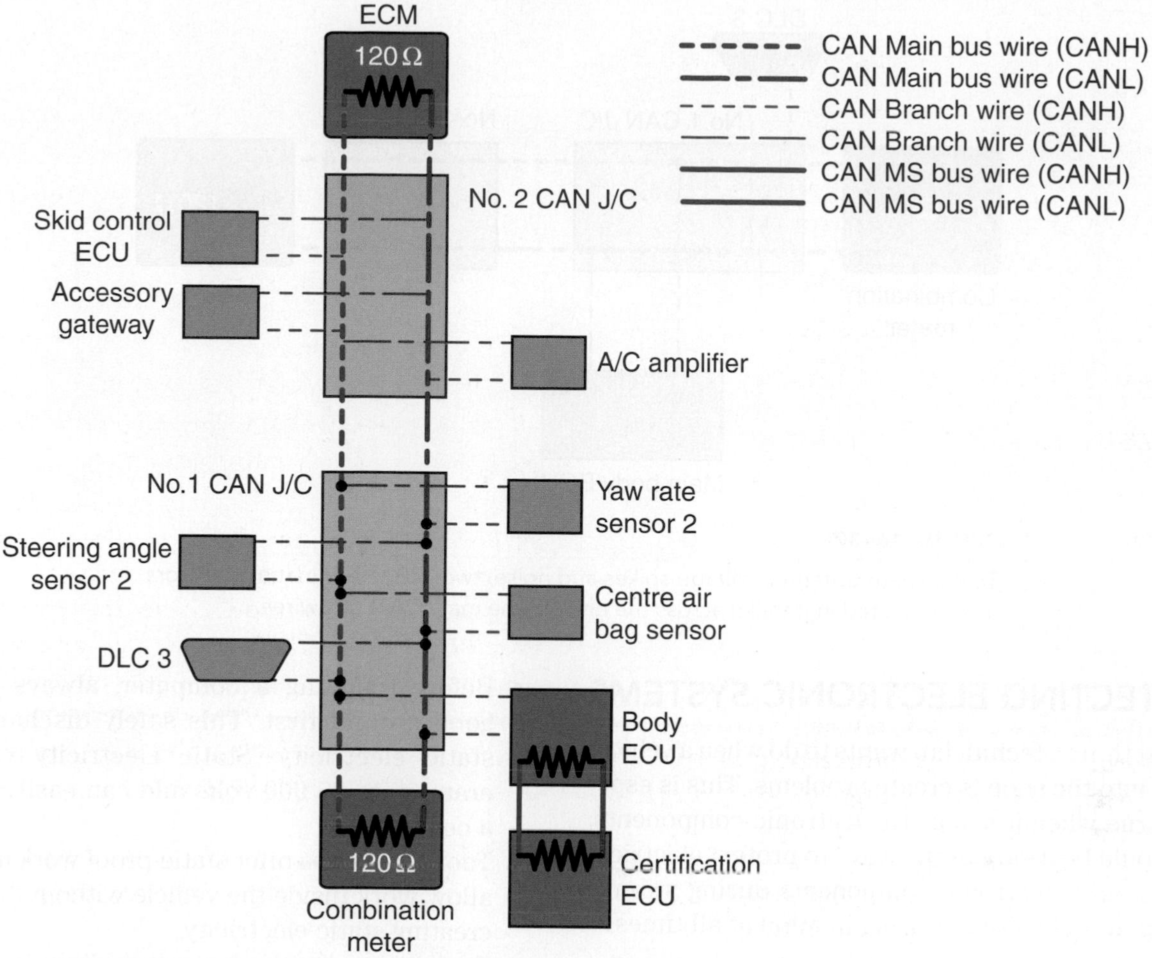

ECM
120 Ω
CAN Main bus wire (CANH)
CAN Main bus wire (CANL)
CAN Branch wire (CANH)
CAN Branch wire (CANL)
CAN MS bus wire (CANH)
CAN MS bus wire (CANL)
No. 2 CAN J/C
Skid control ECU
Accessory gateway
A/C amplifier
No.1 CAN J/C
Yaw rate sensor 2
Steering angle sensor 2
Centre air bag sensor
DLC 3
Body ECU
120 Ω
Certification ECU
Combination meter

FIGURE 16–35

A basic look at a CAN communication system.

Current drivers
OFF
ON
OFF
ON
2.55 V
2.55 V
2.51 V
2.51 V
Bus (+)
Voltage difference
0.02 V
0.100 V
0.02 V
0.100 V
Bus (–)
2.49 V
2.49 V
2.45 V
2.45 V
Binary
1
0
1
0
Message begins

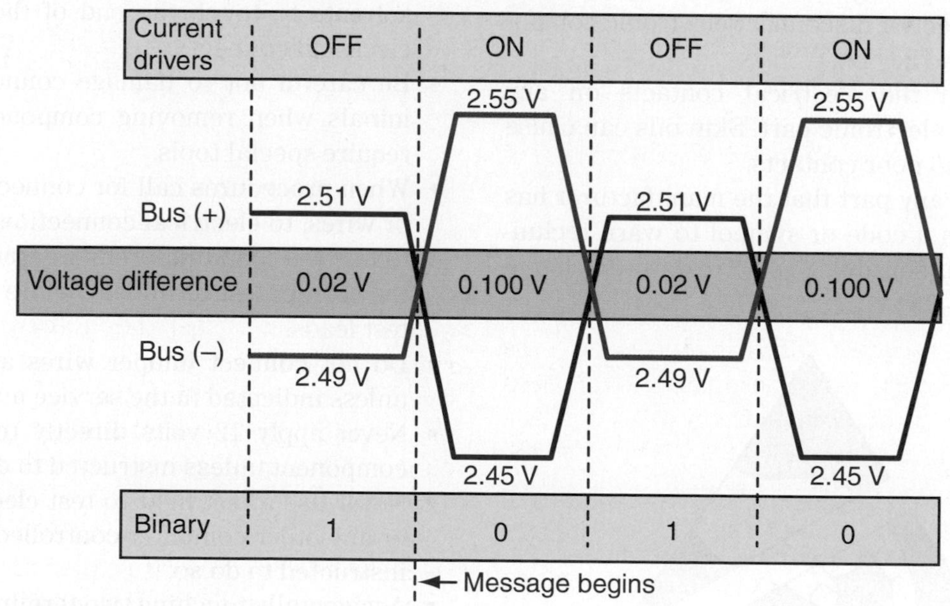

FIGURE 16–36

In order for a message to be transmitted, the drivers are energized to pull up the bias on the 1 bus and pull down on the 2 bus. There must be a voltage differential between the wires.

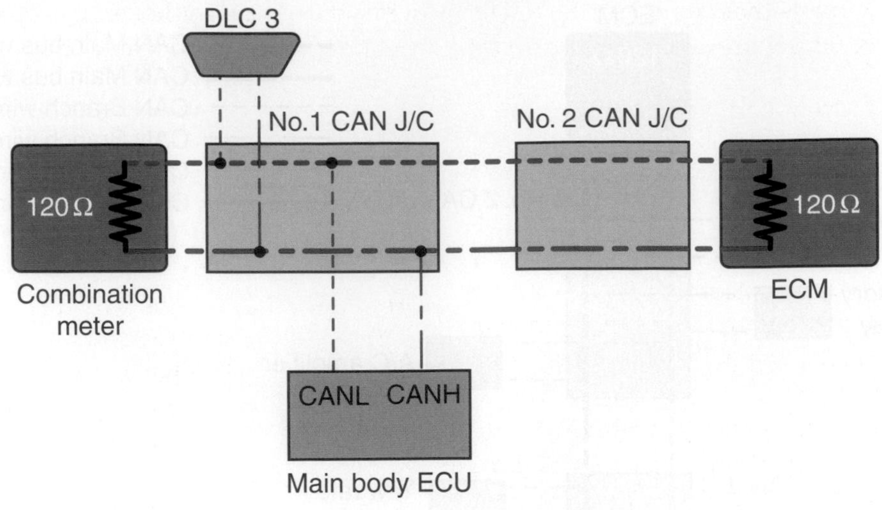

FIGURE 16–37

To eliminate potential voltage spikes and noise, two 120 V terminating resistors are connected in parallel across the ends of the main CAN bus wires.

PROTECTING ELECTRONIC SYSTEMS

The last thing a technician wants to do when a vehicle comes into the shop is create problems. This is especially true when it comes to electronic components. You should be aware of the ways to protect electrical systems and electronic components during storage and repair. Keep the following in mind at all times:

- Vehicle computer-controlled systems should avoid giving and receiving jump-starts due to the possibility of damage caused by voltage spikes.
- Do not connect or disconnect electronic components with the key on.
- Never touch the electrical contacts on any electrical or electronic part. Skin oils can cause corrosion and poor contacts.
- Be aware of any part that the manufacturer has marked with a code or symbol to warn technicians that it is sensitive to electrostatic discharge **(Figure 16–38)**.

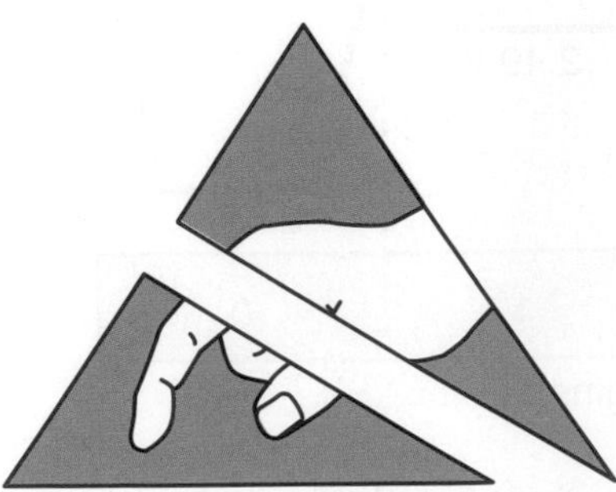

FIGURE 16–38

GM's electrostatic discharge (ESD) symbol warns technicians that a part or circuit is sensitive to static electricity.

- Before touching a computer, always touch a good ground first. This safely discharges any static electricity. Static electricity can generate up to 25 000 volts and can easily damage a computer.
- Tool companies offer static-proof work mats that allow work inside the vehicle without the fear of creating static electricity.
- Tool companies also have grounding wrist straps. A wire connects the wrist strap to a good ground.
- Never allow grease, lubricants, or cleaning solvents to touch the end of the sensor or its electrical connector.
- Be careful not to damage connectors and terminals when removing components. This may require special tools.
- When procedures call for connecting test leads, or wires, to electrical connections, use care and follow the manufacturer's instructions. Identify the correct test terminals before connecting the test leads.
- Do not connect jumper wires across a sensor unless indicated in the service manual to do so.
- Never apply 12 volts directly to an electronic component unless instructed to do so.
- Never use a test light to test electronic ignition or any other computer-controlled system unless instructed to do so.
- Accidentally touching two terminals at the same time with a test probe can cause a short circuit.
- The sensor wires should never be rerouted. When replacing wiring, always check the service manual and follow the routing instructions.

- Disconnect any module that could be affected by welding, hammering, grinding, sanding, or metal straightening.

DIAGNOSING BCMs

Before troubleshooting a system operated by a control module, check the service manual to identify any special procedures or precautions. When testing a system, you are basically trying to isolate a problem in one of the basic functions of the computer and its system.

The BCM continuously monitors operating conditions for possible system malfunctions. It compares system conditions against programmed parameters. If the conditions fall outside of these limits, it detects a malfunction and sets a trouble code that indicates the portion of the system that has the fault.

If the malfunction causes improper system operation, the computer may minimize the effects of the malfunction by using fail-safe action. During this action, the computer will control a system based on programmed values instead of the input signals. This allows the system to operate on a limited basis instead of shutting down completely.

Trouble Codes

There are as many ways to perform BCM diagnostics as there are automobile manufacturers. Nearly all vehicles require a scan tool to retrieve DTCs. The scan tool is plugged into the diagnostic connector for the system being tested. The technician chooses the system to be tested through the scan tool. Once the DTCs are retrieved, follow the appropriate diagnostic chart to isolate the fault. It is important to check the codes in the order recommended by the manufacturer.

On systems that do not retain codes after the ignition is switched off, operate the vehicle until the problem occurs again. Then retrieve the fault code before switching the ignition off. Remember that the trouble code does not necessarily indicate the faulty component; it only indicates the circuit of the system that is not operating properly. To locate the problem, follow the diagnostic procedure in the service manual for the code received.

Diagnosis should continue with a good visual inspection of the circuit involved with the code. Check all sensors and actuators for physical damage. Check all connections to sensors, actuators, control modules, and ground points. Check wiring for signs of burned or chafed spots, pinched wires, or contact with sharp edges or hot exhaust parts. Also check all vacuum hoses for pinches, cuts, or disconnects.

Communication Checks

Performing diagnostic checks on vehicles with a multiplex system should begin with a communications check. If the different control modules are not communicating with each other, there is no way to properly diagnose the systems. When a scan tool is installed, it will try to communicate with every module that could be in the vehicle. If an option is not there, the scan tool will display "No Comm" for that control module. That same message will appear if the module is present, but not communicating. Therefore, always refer to the service information to identify what modules should be present before coming to any conclusions.

The system periodically checks itself for communication errors. The different buses send messages to each other immediately after it sends a message. The message between messages checks the integrity of the communication network. All of the modules in the network also receive a message within a specific time. If the message is not received, the control module will set a DTC stating that it did not receive the message.

There are three types of DTCs used by CAN buses:

- *Loss of communication.* Loss of communication (and bus-off) DTCs are set when there is a problem with the communication between modules. This could be caused by bad connections, wiring, or the module. *Note*: In most cases, a lost communication DTC is set in modules other than the module with the communication problem.
- *Signal error.* The control modules can run diagnostics on some input circuits to determine if they are operating normally. If a circuit fails the test, a DTC will set.
- *Internal error.* The modules also run internal checks. If there is a problem, it will set an internal error DTC.

Bus Wire Service

If the bus wire needs repair due to an open, short, or high resistance, it must not be relocated or untwisted. The twisting serves an extremely important purpose. After a bus wire has been repaired by soldering, wrap that part of the wire with vinyl tape. Never run the repair wire in such a way that it bypasses the twisted sections. CAN bus wires are likely to be influenced by noise if you bypass the twisted wires.

Reprogramming Control Modules

Before going deeply into diagnosing an electronic control circuit, it is wise to check all TSBs that may relate to the problem. Often there will be one that recommends reprogramming of the computer. This is typically called **flashing** the computer. When a computer is flashed, the old program is erased and a new one written in. Reprogramming is often necessary when the manufacturer discovers a common concern that can be solved through changing the system's software. New programs are downloaded into the scan tool and then downloaded into the computer through a dedicated circuit.

Each type of scan tool has a different procedure for flashing. Always follow the manufacturer's instructions. Some scan tools are connected to a PC, and the software is transferred from a CD or a website.

TESTING ELECTRONIC CIRCUITS AND COMPONENTS

When testing or servicing electronic circuits and components, it is important to follow the manufacturer's recommended procedures and be aware of the precautions associated when doing so. Electronic circuits and components are susceptible to static charges (ESD) and voltage spikes, so it is imperative that the proper test equipment be used and proper procedures followed. Use of scan tools, DVOMs, and even lab scopes may be required to accurately pinpoint the concerns. The bottom line is this: Make sure you are using the right piece of diagnostic equipment for the job. Your diagnosis depends on it.

KEY TERMS

A/D converter (p. 477)
Actuators (p. 481)
Analog (p. 476)
Base (p. 470)
Baud rate (p. 478)
Binary code (p. 477)
Bit (p. 477)
Byte (p. 477)
Capacitance (p. 467)
Capacitor (p. 466)
Central processing unit (CPU) (p. 471)
Clamping diode (p. 469)
Collector (p. 470)
Data link connector (DLC) (p. 482)
Diagnostic trouble codes (DTCs) (p. 482)
Digital (p. 476)
Diode (p. 468)
Electrically erasable PROM (EEPROM) (p. 481)
Electrically inert materials (p. 467)
Emitter (p. 470)
Erasable PROM (EPROM) (p. 481)
Farad (F) (p. 467)
Feedback (p. 472)
Field-effect transistors (FETs) (p. 478)
Flashing (p. 490)
Forward bias (p. 468)
Gateway (p. 486)
Hall-effect switch (p. 475)
Impurities (p. 467)
Keep-alive memory (KAM) (p. 481)
Logic gate (p. 478)
Look-up tables (p. 479)
Magnetic pulse generators (p. 474)
Microprocessor (p. 471)
Multiplexing (p. 483)
Nonvolatile RAM (NVRAM) (p. 481)
Output driver (p. 481)
Pickup coil (p. 474)
Piezo-resistive (p. 473)
Programmable read-only memory (PROM) (p. 481)
Random-access memory (RAM) (p. 481)
Read-only memory (ROM) (p. 480)
Reference voltage (Vref) (p. 472)
Reverse bias (p. 469)
Schmitt trigger (p. 478)
Serial data bus (p. 484)
Square wave (p. 476)
Timing disc (p. 474)
Transistor (p. 470)
Trimmers (p. 467)
Voltage-generating sensors (p. 474)
Wheatstone bridges (p. 473)
Zener diode (p. 469)

SUMMARY

- All basic laws of electricity apply to electronic controls.
- A capacitor is used to store and release electrical energy.
- A diode allows current to flow in one direction but not in the opposite direction. It is formed by joining P-type semiconductor material with N-type semiconductor material.
- A transistor resembles a diode with an extra side. There are PNP and NPN transistors. They are used as switching devices. A very small current applied to the base of the transistor controls a much larger current flowing through the entire transistor.
- Computers are electronic decision-making centres. Input devices called sensors feed information to the computer. The computer processes this information and sends signals to controlling devices.
- Most input sensors are reference voltage sensors or voltage-generating sensors.

- Computers work digitally; therefore, they must receive digital signals or convert analog signals to digital signals before processing them.
- A typical electronic control system is made up of sensors, actuators, a microcomputer, and related wiring.
- The microcomputer and its processors are the heart of the computerized engine controls.
- There are three types of computer memory used: ROM, PROM, and RAM.
- Output sensors or actuators are electromechanical devices that convert current into mechanical action.
- On-board diagnostic capabilities are incorporated into a vehicle's computer to monitor virtually every component that can affect emission performance. Each component is checked by a diagnostic routine to verify that it is functioning properly.
- Multiplexing provides communications between vehicle systems. It uses a serial data bus that connects different computers or control modules together.
- Controller area network (CAN) is the most commonly used network protocol in today's vehicles.
- Static electricity can generate up to 25 000 volts and do damage to components. Precautions for static discharge must be taken when handling electronic components.
- Diagnosis of a computer-controlled system includes an early check for diagnostic trouble codes.
- Communication between the various control modules in a vehicle is critical to the vehicle's overall operation.
- There are three types of DTCs used by CAN buses: loss of communication, signal error, and internal error.
- At times, the manufacturer will recommend that a control module be reprogrammed; this is often called flashing the computer.
- Most electronic circuits can be checked in the same way as other electrical circuits. However, only high-impedance meters should be used.

REVIEW QUESTIONS

1. Where are the basic instructions of the computer stored?
   a. Prom **b.** RAM
   c. ROM **d.** CPU
2. The computer can read from and write to which memory?
   a. Prom **b.** RAM
   c. TPS **d.** CPU
3. What is the most common semiconductor material?
   a. copper **b.** zirconium
   c. silicon **d.** calcium
4. How many wires are commonly used for multiplexing?
   a. 1 wire **b.** 2 wires
   c. 3 wires **d.** 4 wires
5. The binary system, used by computers, has how many numbers?
   a. 6 **b.** 4
   c. 2 **d.** 1
6. To "turn on" an NPN transistor, there must be flow through which parts?
   a. collector/emitter
   b. base/collector
   c. base/emitter
   d. emitter/collector
7. Which of the following components could possibly be replaced by a transistor?
   a. a switch
   b. a variable resistor
   c. a vehicle speed sensor
   d. a circuit breaker
8. What circuit/component allows the computer to examine one of many inputs depending upon a programmed priority rating?
   a. pulse-width modulator
   b. multiplexer
   c. logic gate
   d. demultiplexer
9. As speed increases, which part of a digital signal increases?
   a. amplitude
   b. frequency
   c. negative voltage
   d. positive voltage
10. Which is the most common type of coolant or air temperature sensors?
    a. PTC resistor
    b. PTC capacitor
    c. NTC capacitor
    d. NTC resistor
11. Which sensor commonly generates a voltage?
    a. wheel speed sensor
    b. throttle position sensor
    c. MAP sensor
    d. MAF sensor

12. As speed increases, which component of a magnetic pulse generator signal increases?
   a. amplitude
   b. frequency
   c. negative voltage
   d. all of the above
13. Which sensor modifies the reference voltage signal?
   a. vehicle speed sensor
   b. throttle position sensor
   c. wheel speed sensor
   d. oxygen sensor (ZrO_2)
14. What are the requirements to forward bias a diode?
   a. The anode must be positive, and the cathode must be negative.
   b. The anode must be negative, and the cathode must be positive.
   c. The anode must be north, and the cathode must be south.
   d. The anode must be south, and the cathode must be north.
15. Which of the following diodes allow current flow in the reverse direction once a specified voltage is obtained?
   a. clamping diode
   b. light emitting diode
   c. photo diode
   d. zener diode
16. A strain gauge MAP sensor is an example of which type of sensor?
   a. Wheatstone bridge
   b. permanent magnet pulse generator
   c. frequency generator
   d. galvanic battery
17. Pulse width is measured in which units?
   a. Hz – frequency
   b. ms – time
   c. percentage – on time
   d. percentage – off time
18. A knock sensor is which kind of device?
   a. Wheatstone bridge
   b. piezo-resistive
   c. piezo-electric
   d. strain gauge
19. Which type of computer circuit controls output based on input signals?
   a. gate
   b. door
   c. window
   d. sleeve
20. What turns on and off a Hall-effect switch?
   a. heat
   b. time
   c. magnetism
   d. pressure

CHAPTER 17

General Electrical System Diagnostics and Service

LEARNING OUTCOMES

- Describe the different possible types of electrical problems.
- Read electrical automotive diagrams.
- Perform troubleshooting procedures using meters, test lights, and jumper wires.
- Describe how each of the major types of electrical test equipment are connected and interpreted.
- Explain how to use a digital multimeter (DMM) for diagnosing electrical and electronic systems.
- Explain how to use an oscilloscope for diagnosing electrical and electronic systems.
- Test common electrical components.
- Use wiring diagrams to identify circuits and circuit problems.
- Diagnose common electrical problems.
- Properly repair wiring and connectors.

Diagnosing nearly every system of a vehicle involves electrical and electronic systems. An understanding of how electrical/electronic systems work **(Figure 17–1)** and the knowledge of how to use the various types of test equipment are the keys to efficient diagnosis.

ELECTRICAL PROBLEMS

All electrical problems can be classified into one of three categories: opens, shorts, or high-resistance problems. Identifying the type of problem allows technicians to identify the correct tests to perform when diagnosing an electrical problem. An explanation of the different types of electrical problems follows.

Open Circuits

An **open** occurs when a circuit is incomplete. Without a completed path, current cannot flow **(Figure 17–2)**, and the load or component will not work. An open circuit can be caused by a disconnected wire or connector, a broken wire, or a switch in the OFF position. When a circuit is off, it is open. When the circuit is on, it is closed. Switches open and close circuits, but at times a fault will cause an open. Opening a circuit stops current flow through the circuit. Voltage is still applied up to the open point, but there is no current flow. Without current flow, there are no voltage drops across the various loads.

Shorted Circuits

When a circuit has an unwanted path for current to follow, it has a **short**. When an energized wire accidentally contacts the frame or body of the car or another wire, circuit current can travel in unintended directions through the wires. This can cause uncontrollable circuits and high current through the circuits. Shorts are caused by damaged wire insulation, loose wires or connections, improper wiring, or careless installation of accessories.

A short creates an unwanted parallel leg or path in a circuit. As a result, circuit resistance decreases and current increases. The amount that the current will increase depends on the resistance of the short. The increased current flow can burn wires or components. Preventing this is the purpose of circuit protection devices. When a circuit protection device opens due to higher than normal current, a short is the likely cause. Also, if a connector or a group of wires shows signs of burning or insulation melting, high current is the cause, which is most likely caused by a short.

A short can be caused by a number of things and can be evident in a number of ways. It can be an unwanted path to ground. The short is often in parallel to a load and provides a low-resistance path to ground. Look at **Figure 17–3**; the short is probably caused by bad insulation that is allowing the power feed for the lamp to touch the ground for the same lamp. This problem creates a parallel

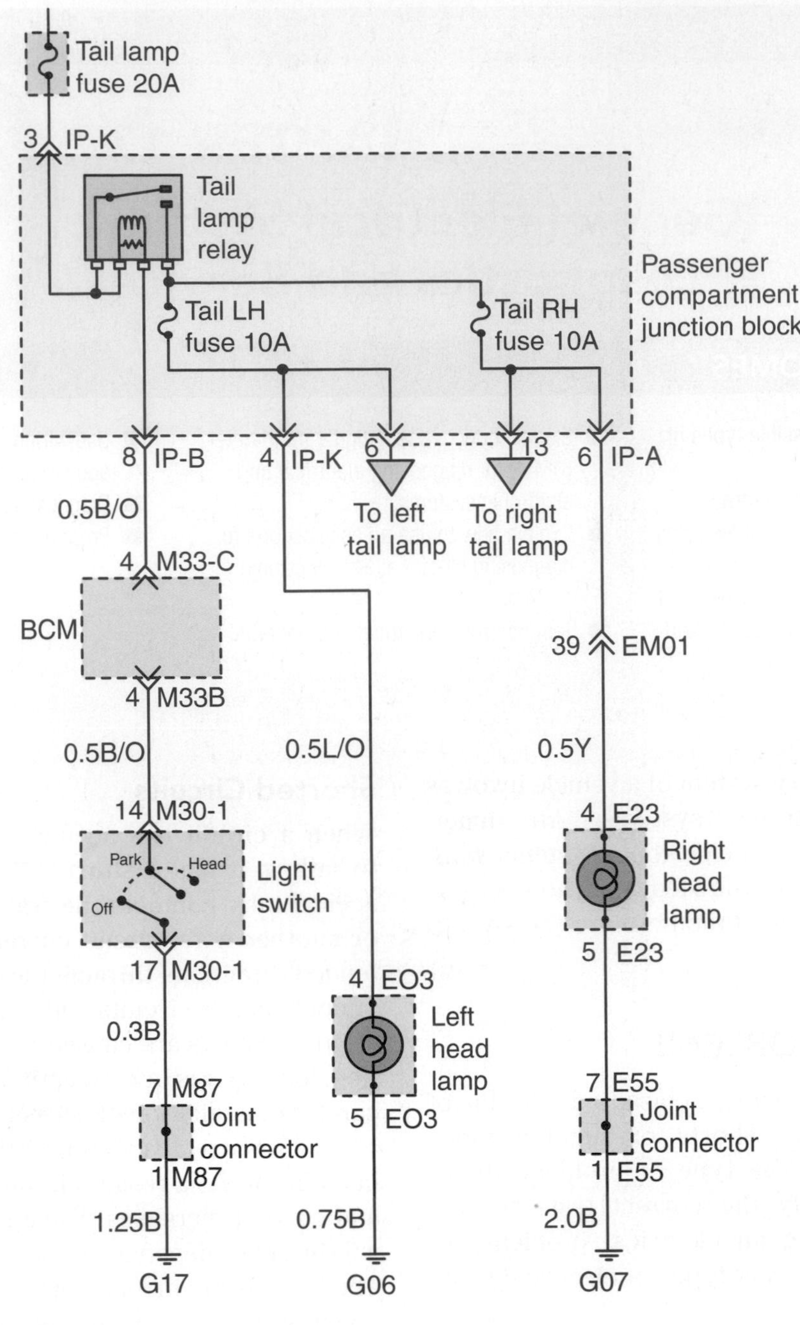

FIGURE 17–1

A wiring diagram for the tail, parking, and licence lamps on a late-model vehicle.

circuit. **Figure 17–4** represents Figure 17–3 but is drawn to show the short as a parallel leg with very low resistance. The resistance assigned to the short may be more or less than an actual short, but the value 0.001 ohm is given to illustrate what happens. With the short, the circuit has three loads in parallel: 0.001, 3, and 6 ohms. The total resistance of this parallel circuit is less than the lowest resistance, or 0.001 ohm. Using this value as the total resistance, circuit current is calculated to be more than 12 000 amps, which is much more than the fuse can handle. The high current will burn the fuse, and the circuit will not work. Some call this problem a "grounded circuit" or a "copper-to-iron" short.

Sometimes two separate circuits become shorted together. When this happens, each circuit is controlled by the other. This may result in strange happenings, such as the horn sounding every time the brake pedal is depressed **(Figure 17–5)**, or vice versa. In this case, the brake light circuit is shorted

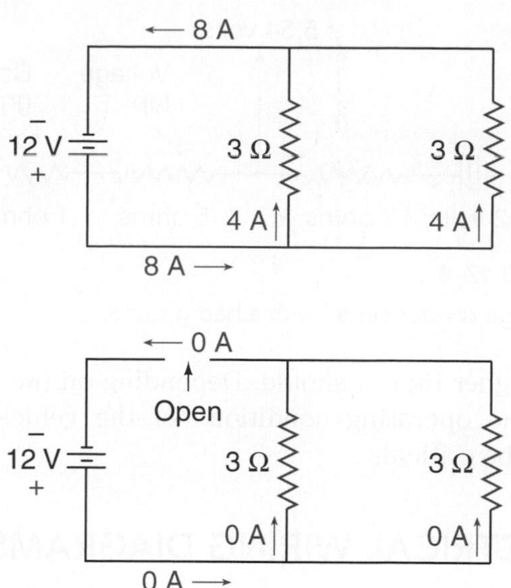

FIGURE 17–2

In an open circuit, there is no current flow. (A) is a normal circuit, and (B) is the same circuit with an open.

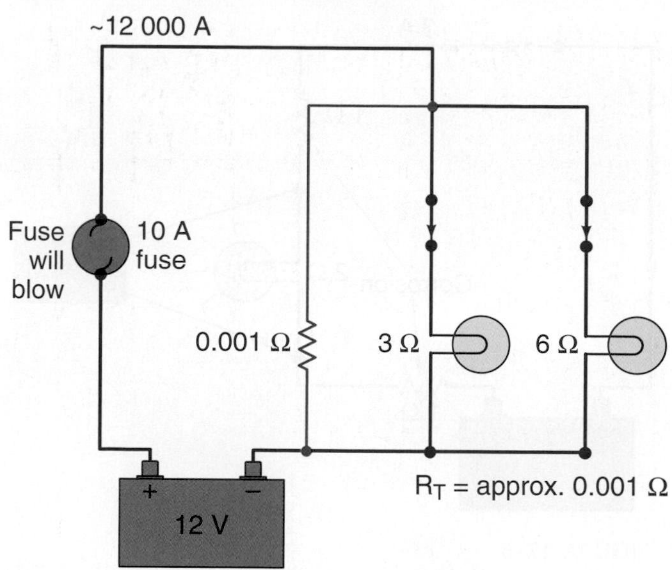

FIGURE 17–4

Ohm's law applied to Figure 17–3. Notice the rise in circuit amperage.

to the horn circuit. This is often called a "copper-to-copper" short.

High-Resistance Circuits

High-resistance problems occur when there is unwanted resistance in the circuit. The higher than normal resistance causes the current flow to be lower than normal and the components in the circuit are unable to operate properly. A common cause of this type of problem is corrosion at a connector. The corrosion becomes an additional load in the circuit **(Figure 17–6)**. This load uses some of the circuit's voltage, which prevents full voltage to the normal loads in the circuit.

Many sensors on today's vehicles are fed a 5-volt reference signal. The signal or voltage from the sensor is less than 5 volts, depending on the condition it is measuring. A poor ground in the reference voltage circuit can cause higher than normal readings back

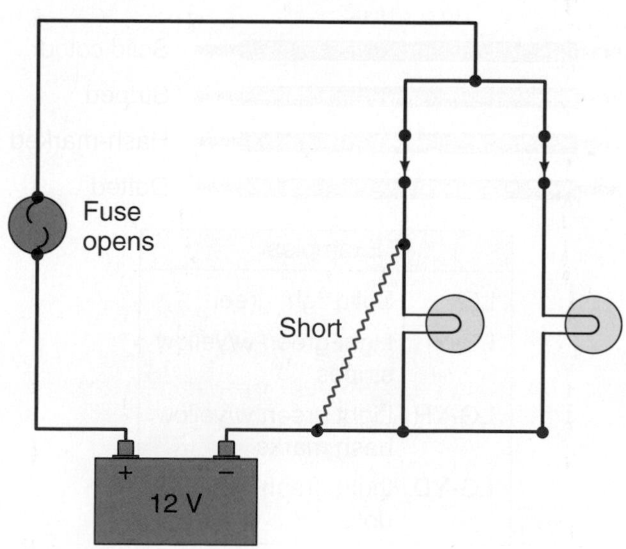

FIGURE 17–3

A short to ground.

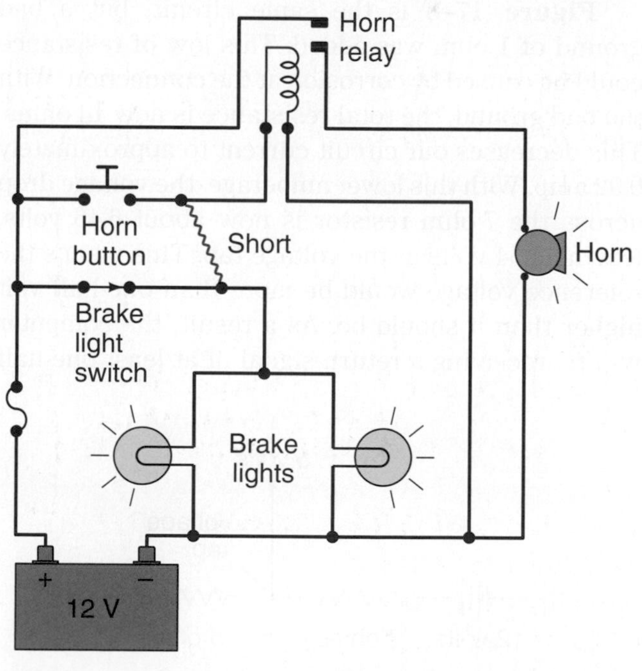

FIGURE 17–5

A wire-to-wire short.

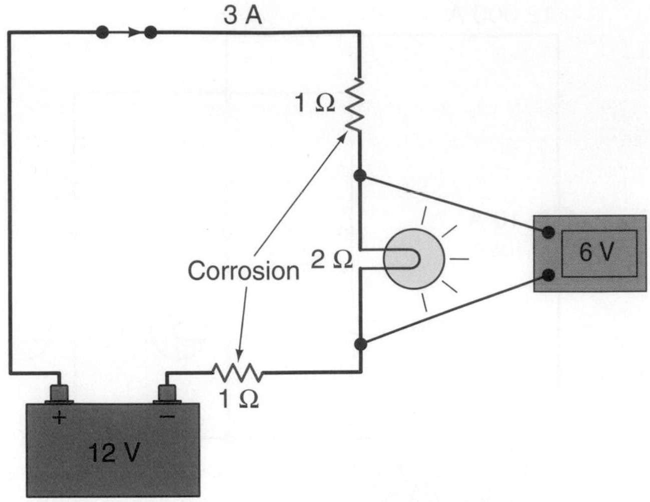

FIGURE 17–6

A simple light circuit with unwanted resistance. Notice the reduced voltage drop across the lamp and the reduced circuit current.

to the computer. This seems to be contradictory to other high-resistance problems. However, if you look at a typical voltage divider circuit used to supply the reference voltage, you will understand what is happening. Look at **Figure 17–7**. There are two resistors in series with the voltage reference tap between them. Because the total resistance in the circuit is 12 ohms, the circuit current is 1 amp. Therefore, the voltage drop across the 7-ohm resistor is 7 volts, leaving 5 volts at the tap.

Figure 17–8 is the same circuit, but a bad ground of 1 ohm was added. This low of resistance could be caused by corrosion at the connection. With the bad ground, the total resistance is now 13 ohms. This decreases our circuit current to approximately 0.92 amp. With this lower amperage, the voltage drop across the 7-ohm resistor is now about 6.46 volts, leaving 5.54 volts at the voltage tap. This means the reference voltage would be more than one-half volt higher than it should be. As a result, the computer will be receiving a return signal of at least one-half

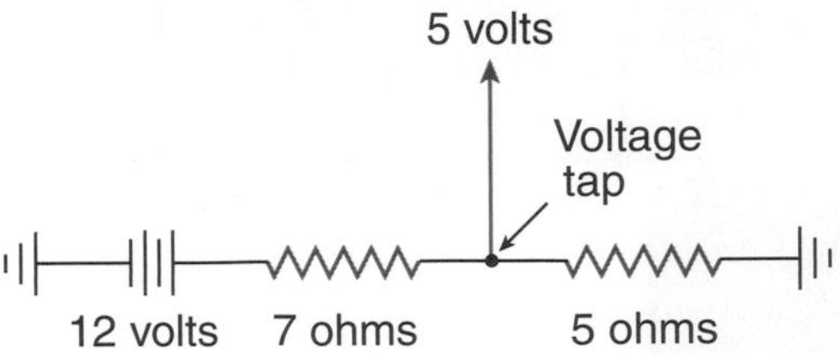

FIGURE 17–7

A voltage divider circuit.

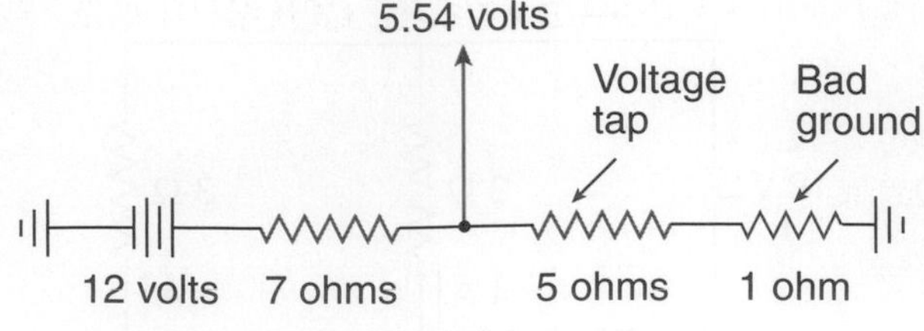

FIGURE 17–8

A voltage divider circuit with a bad ground.

volt higher than it should. Depending on the sensor and the operating conditions of the vehicle, this could be critical.

ELECTRICAL WIRING DIAGRAMS

Wiring diagrams, sometimes called **schematics**, show how circuits are constructed. A typical service manual contains dozens of wiring diagrams that can be used to diagnose and repair a vehicle's electrical system.

A wiring diagram does not show the actual location of the parts or their appearance, nor does it indicate the length of the wire that runs between components. It usually indicates the colour of the wire's insulation **(Figure 17–9)** and sometimes the wire gauge size. Typically the wire insulation is colour-coded, as shown in **Table 17–1**. The first letter or set of letters usually indicates the base colour of the insulation. The second set refers to the colour of the stripe, hash marks, or dots on the wire, if there are any. Circuits are traced through a vehicle by identifying the beginning and end of a particular coloured wire.

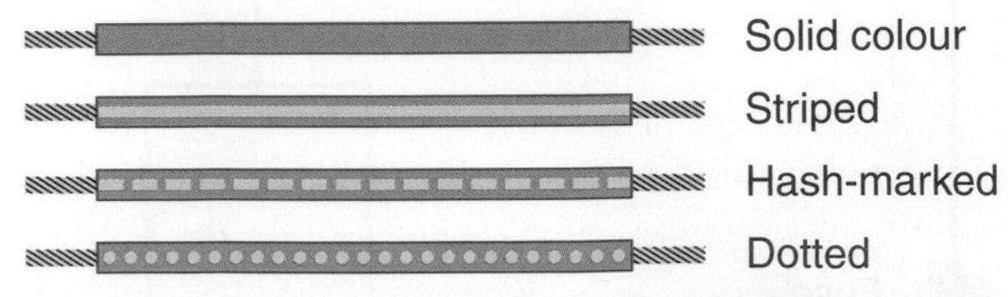

Examples	
LG	Solid light green
LG-Y	Light green w/yellow stripes
LG-YH	Light green w/yellow hash marks
LG-YD	Light green w/yellow dots

FIGURE 17–9

The different multicolour schemes of wires.

TABLE 17–1 **COMMON WIRE COLOUR CODES**

COLOUR	ABBREVIATIONS			
Aluminum	AL			
Black	BLK	BK	B	SW
Blue	BL	BLU	L	
Dark Blue	BLU DK	DB	DK BLU	
Light Blue	BLU LT	LB	LT BLU	
Brown	BRN	BR	BN	
Glazed	GLZ	GL		
Grey	GRA	GR	G	GRY
Green	GN	GRN	G	
Dark Green	GRN DK	DG	DK GRN	
Light Green	GRN LT	LG	LT GRN	
Lilac	LI			
Maroon	MAR	M		
Orange	ORN	O	ORG	
Pink	PNK	PK	P	
Purple	PPL	PR	PUR	
Red	RED	R	RD	RO
Tan	TAN	T	TN	
Violet	VLT	V		
White	WHT	W	WH	WS
Yellow	YEL	Y	YL	GE

Many different symbols are used to represent components such as resistors, batteries, switches, transistors, and many other items. Some of these symbols have already been shown in earlier discussions. Other common symbols are shown in **Figure 7–10**. Connectors, splices, grounds, and other details are also given in a wiring diagram; a sample of these is shown in **Figure 17–11**.

Wiring diagrams can become quite complex. To avoid this, most diagrams usually illustrate one distinct system, such as the backup light circuit, oil pressure indicator light circuit, or wiper motor circuit. In more complex ignition, electronic fuel injection, and computer-controlled systems, a diagram may be used to illustrate only part of the entire circuit.

USING SERVICE INFORMATION

Electrical symbols are not standardized throughout the automotive industry. Different manufacturers may have different methods of representing certain components, particularly the less common ones. Always refer to the symbol reference charts, wire colour-code charts, and abbreviation tables listed in the service information to avoid confusion when reading wiring diagrams.

ELECTRICAL TESTING TOOLS

With a basic understanding of electricity and simple circuits, it is easier to understand the operation and purpose of the various types of electrical test equipment, which are described in the following sections. Several meters are used to test and diagnose electrical systems. These are the voltmeter, ohmmeter, ammeter, and volt/amp meter. These should be used along with jumper wires, test lights, and variable resistors (piles).

Circuit Testers

Circuit testers (test lights) are used to identify shorted and open circuits in electrical circuits. Low-voltage testers are used to troubleshoot 6- to 12-volt circuits. High-voltage circuit testers diagnose higher voltage systems, such as the secondary ignition circuit. High-impedance test lights are available for diagnosing electronic systems.

A newer style of "test light" is made by several manufacturers **(Figure 17–12)**. They are connected to the battery positive and negative and can provide a safe power and ground through a thumb-activated switch. Depending on the type, they can be used on primary ignition, electrical accessories, or delicate electronic devices.

WARNING!

Do not use any type of test light or circuit tester to diagnose automotive air bag systems. Use only the vehicle manufacturer's recommended equipment on these systems.

Multimeter

A multimeter, also known as a DVOM, is one of the most important tools for troubleshooting electrical and electronic systems. A basic multimeter measures DC and AC voltage, current, and resistance. A multimeter eliminates the need for separate electrical meters. Most current multimeters also check engine

SYMBOLS USED IN WIRING DIAGRAMS			
+	Positive		Temperature switch
—	Negative		Diode
	Ground		Zener diode
	Fuse		Motor
	Circuit breaker	C101	Connector 101
	Condenser	→	Male connector
	Ohms		Female connector
	Fixed value resistor		Splice
	Variable resistor	S101	Splice number
	Series resistors		Thermal element
	Coil		Multiple connectors
	Open contacts	88:88	Digital readout
	Closed contacts		Single filament bulb
	Closed switch		Dual filament bulb
	Open switch		Light-emitting diode
	Ganged switch (N.O.)		Thermistor
	Single-pole double-throw switch		PNP bipolar transistor
	Momentary contact switch		NPN bipolar transistor
	Pressure switch		Gauge

FIGURE 17–10

Common electrical symbols used on wiring diagrams.

speed, signal frequency, duty cycle, pulse width, diodes, temperature, and pressures. The desired test is selected by turning a control knob or depressing keys on the meter. These meters can be used to test simple electrical circuits, ignition systems, input sensors, fuel injectors, batteries, and starting and charging systems.

Lab Scopes

Lab scopes are fast-reacting meters that measure and display voltages within a specific time frame. The voltage readings appear as a waveform or trace on the scope's screen. An upward movement of the trace indicates an increase in voltage, and a decrease in voltage results in a downward movement. These are especially valuable when watching the action of a circuit. Lab scopes provide the technician with "real time" data, which can be a valuable tool when trying to diagnose intermittent or hard-to-locate faults. While wiggling a harness or activating a sensor or an output device, the technician can monitor the waveform signal in an effort to locate when and possibly where the fault is occurring.

Scan Tools

When plugged into the vehicle's diagnostic connector, a scan tool can retrieve fault codes from a computer's memory and digitally display these codes **(Figure 17–13)**. A scan tool may also perform other diagnostic functions depending on the year and make of the vehicle. Many scan tools have removable modules that are updated each year. These modules are designed to test the computer systems on various makes of vehicles.

The scan tool must be programmed for the model year, make of vehicle, and type of engine. With some scan tools, this selection is made by pressing the appropriate buttons on the tester as directed by the digital tester display. On other scan tools, the appropriate memory card must be

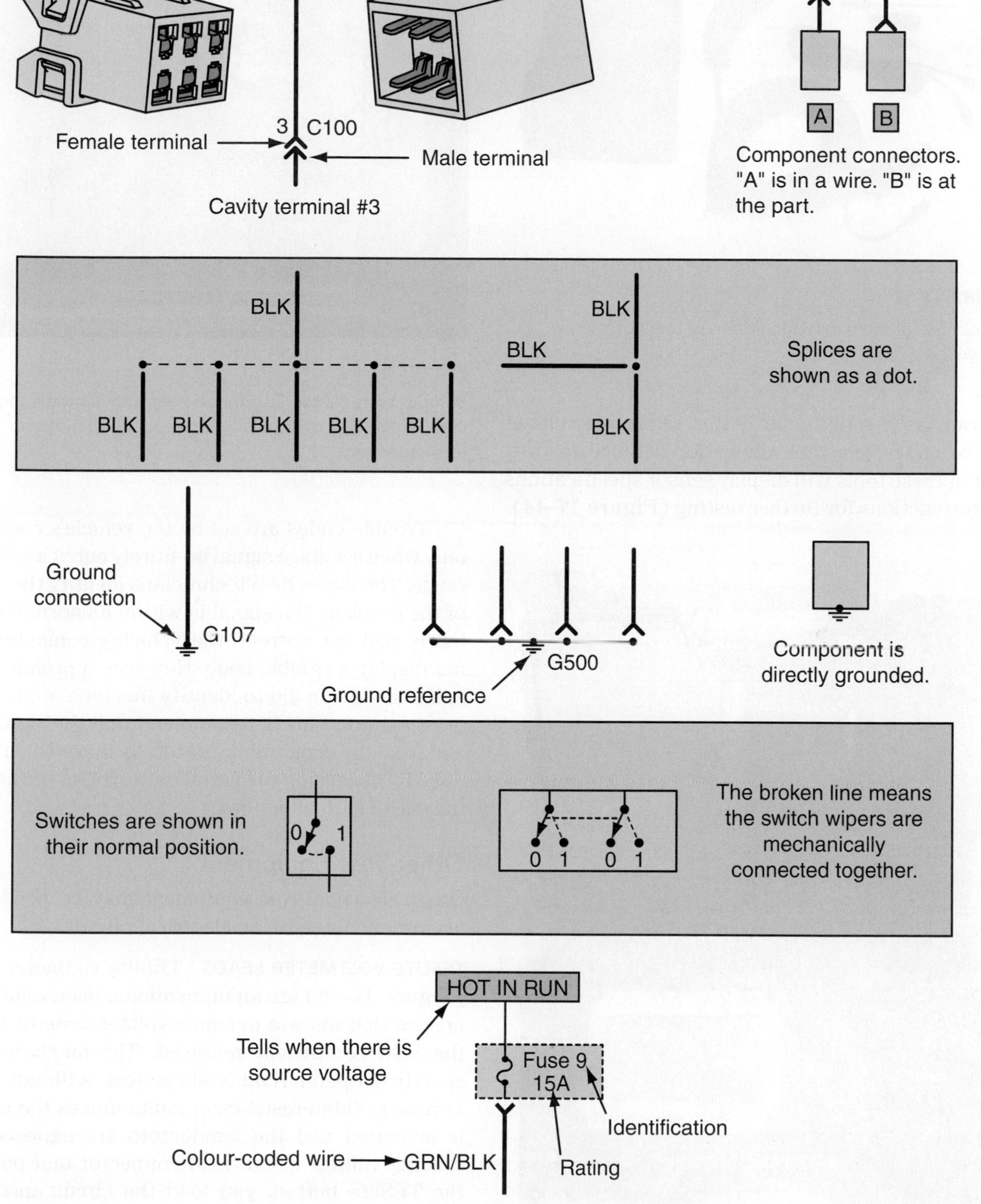

FIGURE 17–11

Examples of the information contained in a typical wiring diagram (schematic).

installed in the tester for the vehicle being tested. Some scan tools have a built-in printer to print test results, whereas others may be connected to an external printer.

Some scan tools display diagnostic information based on the trouble code. Service bulletins may also be indexed on the tool after vehicle information is entered into the tester. Other scan tools can also

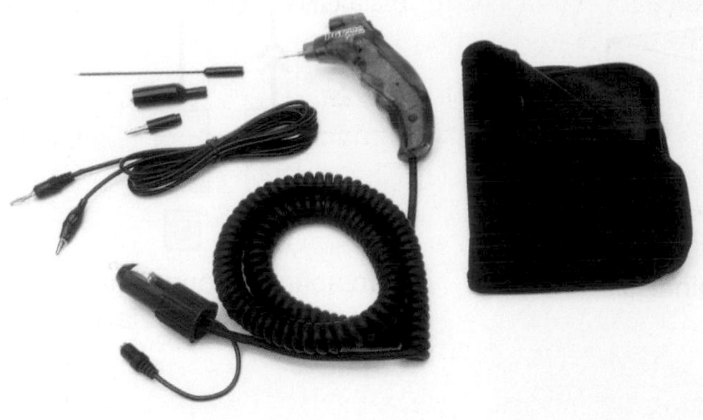

FIGURE 17–12

An example of the latest generation of test lights.

Hickok/Waekon

function as a multimeter. After retrieving related trouble codes, the tool allows for detailed testing. Most of these tools will display sensor specifications and instructions for further testing **(Figure 17–14)**.

FIGURE 17–13

A typical scan tool.

Martin Restoule

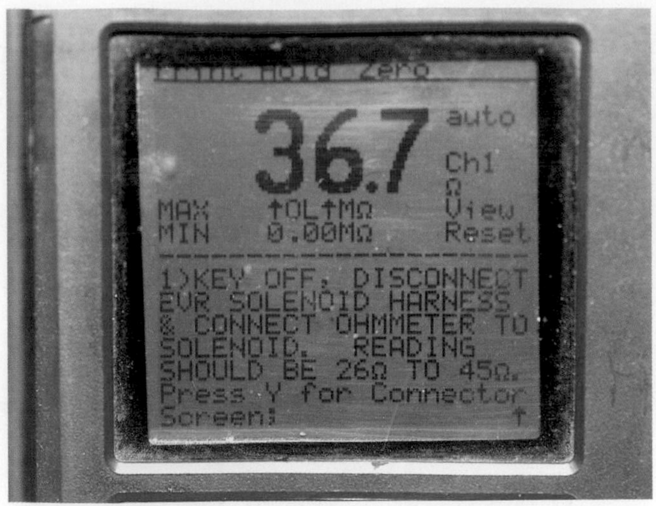

FIGURE 17–14

Some scan tools can also function as a multimeter and may display sensor specifications and instructions for further testing.

Trouble codes are set by the vehicle's computer only when a voltage signal is entirely out of its normal range. The codes help technicians identify the cause of the problem. If a signal is within its normal range but is still not correct, the vehicle's computer will not display a trouble code. However, a problem will still exist. As an aid to identify this type of problem, most manufacturers recommend that the signals to and from the computer be carefully looked at. This is done by observing the serial data stream or through the use of a breakout box.

Other Test Equipment

Other electrical test equipment may be needed to accurately diagnose an electrical circuit.

TESLITE VOLTMETER LEADS TESlite voltmeter leads **(Figure 17–15)** are an innovation in electrical diagnostics that allow a dynamic voltage drop test with the load component removed. The manufacturer-specified open-circuit voltage test will not "see" corrosion (high-resistance) faults unless the circuit is activated and the conductors are exposed. By reading voltage at the load connector and pushing the TESlite button, you load the circuit instantly, which is functionally the same as "ohmming out" the entire circuit from the load to the battery with one voltage test. TESlite leads, along with other applications of the voltmeter ("ghost" and "true zero" voltage), make it easy to now identify all possible circuit faults (opens, shorts-to-ground, and high resistance) with one voltage test.

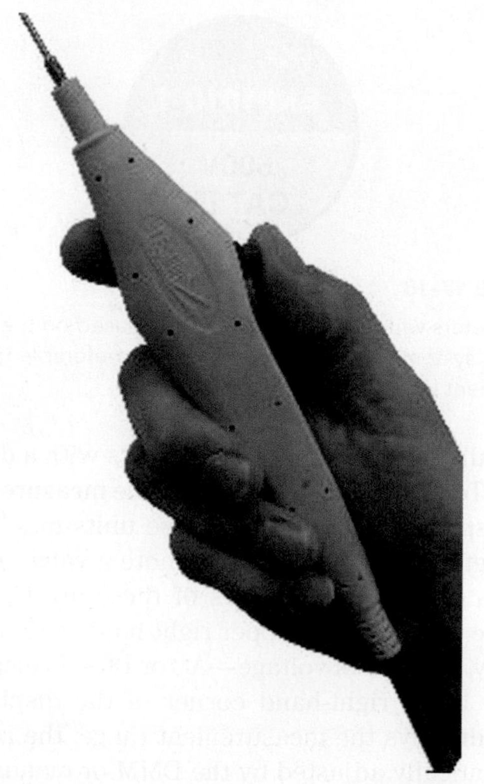

FIGURE 17–15

The TESlight voltmeter probe.

JUMPER WIRES Jumper wires are used to bypass individual wires, connectors, or components. Bypassing a component or wire helps to determine if that part is faulty **(Figure 17–16)**. If the symptom is no longer evident after the jumper wire is installed, the part bypassed is faulty. Technicians typically have jumper wires of various lengths; usually they have a fuse or circuit breaker in them to protect the circuits being tested.

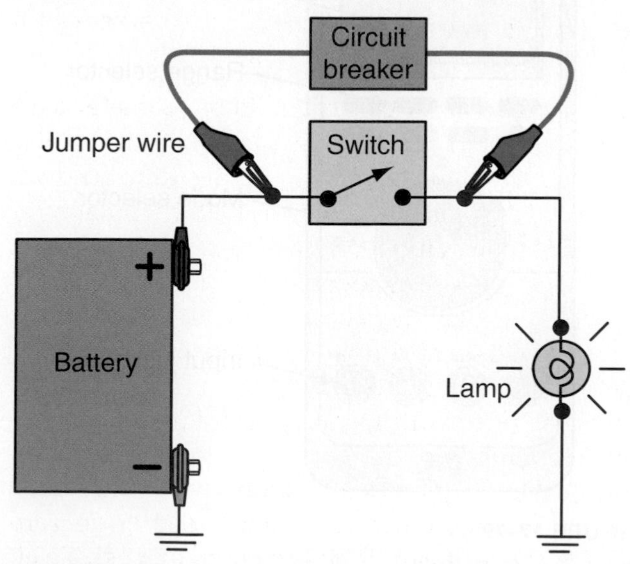

FIGURE 17–16

A jumper wire can be used to bypass a switch.

COMPUTER MEMORY SAVER Whenever a vehicle's battery needs to be disconnected, connect a memory saver to the vehicle. This saver will preserve the memory in the radio, electronic accessories, and the engine, transmission, and body computers.

Two types of memory savers are available or can be made for late-model vehicles. For a power source, use a 12-volt automotive battery or a 12-volt dry-cell lantern battery. If the vehicle's OBD II connector is continuously powered, an OBD II connector adapter plug with suitable wire leads and large alligator clips can be attached to the auxiliary battery. If the connector is controlled by the ignition switch, a set of jumper wires with alligator clips can be connected to the vehicle's electrical system.

To make either type of saver, use No. 14 or 16 wire. Install a 5-ampere in-line fuse in the positive lead, along with a diode **(Figure 17–17)**. The fuse will protect the memory saver and the electrical system from an accidental short. The diode will prevent current feedback from the vehicle's electrical system to the auxiliary battery.

If you connect the memory saver under the hood, do not connect it to the battery cable clamps. Removing and reinstalling the battery will likely dislodge the memory saver's alligator clips and make it useless. Instead, connect the saver's negative (−) lead to a good engine ground and the positive (+) lead to a point that is hot at all times, such as the battery connection at the generator or starter relay. Check a wiring diagram if you are unsure about the connection points.

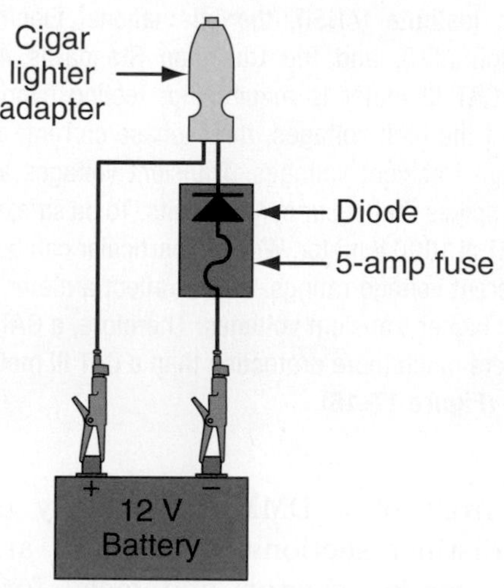

FIGURE 17–17

A homemade memory saver.

If the memory saver must stay connected while the vehicle is raised and lowered on a hoist, securely place the battery out of the way of all moving parts. If you place the battery inside the car, set it in a large plastic tray or tub to protect the vehicle from electrolyte or battery corrosion.

USING MULTIMETERS

Multimeters are available with either analog or digital displays. Analog meters use a needle to point to a value on a scale. A digital meter shows the measured value in numbers or digits. The most commonly used meter is the digital multimeter (DMM), sometimes called a digital volt/ohmmeter (DVOM). Analog meters should not be used to test electronic components. They have low internal resistance (input impedance). The low input impedance allows too much current to flow through circuits and can damage delicate electronic devices. Digital meters, on the other hand, have high input impedance, 1 mega ohm (million ohms) to 10 mega ohms. In addition, metered voltage for resistance tests is well below 5 volts, reducing the risk of damage to sensitive components and delicate computer circuits.

SHOP TALK

The DMM used on high-voltage systems in hybrids should be classified as a category III meter. There are basically four categories for electrical meters, each built for specific purposes and to meet certain standards. The categories define how safe a meter is when measuring certain circuits. The standards for the various categories are defined by the American National Standards Institute (ANSI), the International Electrotechnical Commission (IEC), and the Canadian Standards Association (CSA). A CAT III meter is required for testing hybrid vehicles because of the high voltages, three-phase current, and potential for high transient voltages. Transient voltages are voltage surges or spikes that occur in AC circuits. To be safe, you should have a CAT III 1000 V meter. Within a particular category, meters have different voltage ratings. These reflect a meter's ability to withstand higher transient voltages. Therefore, a CAT III 1000 V meter offers much more protection than a CAT III meter rated at 600 volts **(Figure 17–18)**.

The front of a DMM is normally comprised of four distinct sections: the display area, range selectors, mode selector, and jacks for the test leads **(Figure 17–19)**. In the centre of the display are large digits that represent the measured value.

FIGURE 17–18

Only meters with this symbol should be used on the high-voltage systems in a hybrid vehicle; it is preferable that the equipment is rated at 1000 volts.

Normally there are four to five digits with a decimal point. To the right of the number, the measured units are displayed (V, A, or Ω). These units may be further defined by a symbol to denote a value of more or less than one. Examples of these are shown in **Figure 17–20**. In the upper right-hand corner of the display, the type of voltage—AC or DC—is displayed. In the lower right-hand corner of the display, the DMM displays the measurement range; the range is automatically adjusted by the DMM or manually set by the technician **(Figure 17–21)**.

Setting the range on a DMM is important. If the measurement is beyond the set range, the meter will display a reading of OL, or over limit. The range on some DMMs is manually set, whereas others have an auto-range feature, in which the appropriate scale is automatically selected by the meter. Meters with auto range allow you to disable it when you want to

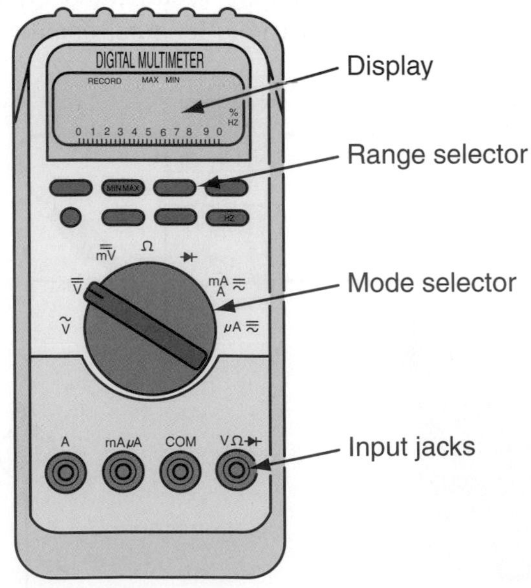

FIGURE 17–19

The front of a DVOM normally has four distinct sections: the display area, range selectors, mode selector, and jacks for the test leads.

Symbol	Name	Value
mV	millivolst	volts × 0.001
kV	kilovolts	volts × 1000
mA	milliamps	amps × 0.001
μA*	microamps	amps × 0.000001
kΩ	kilo-ohms	ohms × 1000
MΩ	megohms	ohms × 1 000 000

**Automotive technicians seldom use readings at the microamp level.*

FIGURE 17–20

Symbols used to define the value of a measurement on a DMM.

manually select ranges. Auto ranging is helpful when you do not know what value to expect. When using a meter with auto range, make sure you note the range being used by the meter. There is a big difference between 10 ohms and 10 000 000 (10 M) ohms.

WARNING!

Many DMMs with auto-range display the measurement with a decimal point. Make sure you observe the decimal and the range being used by the meter (***Figure 17–22***). A reading of 0.972 K ohm equals 972 ohms. If you ignore the decimal point you will interpret the reading as 972 000 ohms.

The mode selector defines what the meter will be measuring **(Figure 17–23)**. The number of available modes varies with meter design, but nearly all have the following:

- Volts AC
- Volts DC

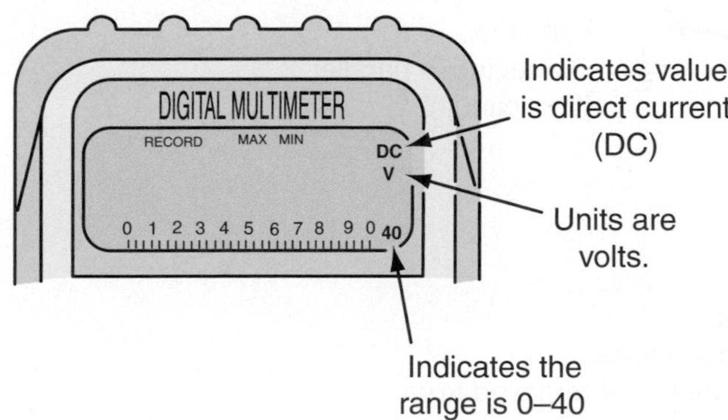

FIGURE 17–21

The measurement display also shows other important information. Some of this defines the measured value, the type of voltage—AC or DC—and the range selected on the meter.

0.345 KΩ = 345 Ω

1025 mAmps = 1.025 Amps

FIGURE 17–22

Placement of decimal point and the scale should be noticed when measuring with a meter with auto range.

- Millivolts (mV) DC
- Resistance/continuity (ohms)
- Diode check
- Amps or milliamps AC/DC

Most DMMs have two test leads and four input jacks. The black test lead always plugs into the COM input jack, and the red lead plugs into one of the other input jacks, depending on what is being measured. Often, technicians have multiple sets of test leads, each for specific purposes. For example, a lead set with small tips is ideal for probing in hard-to-reach or tight spaces. Other test leads may be fitted with clips to hold the lead at a point during testing.

Typically, the three input jacks are as follows:

- "A" for measuring up to 10 amps of current
- "A/mA" for measuring up to 400 mA of current
- "V/Ω/diode" for measuring voltage, resistance, conductance, and capacitance, and for checking diodes

Measuring Voltage

A DMM can measure source voltage, available voltage **(Figure 17–24)**, and voltage drops. Voltage is measured by placing the meter in parallel to the component or circuit being tested **(Figure 17–25)**. There are normally several voltage ranges available on a DMM. Most automotive circuits range from 50 mV to 15 volts.

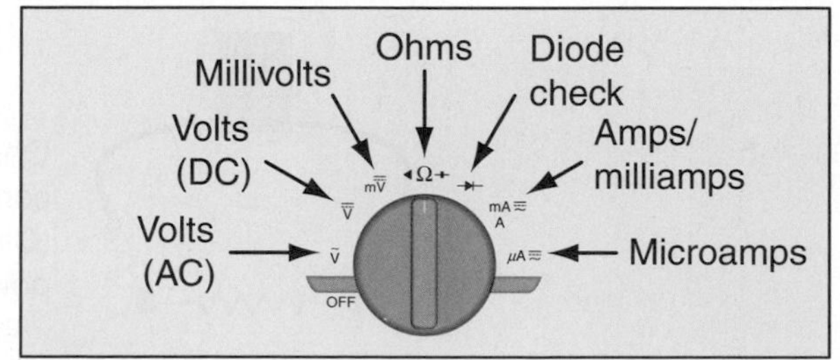

FIGURE 17–23

The mode selector defines what the meter will be measuring.

WARNING!

Hybrid vehicles have much higher voltage; always follow all safety precautions and service procedures when working with high-voltage circuits.

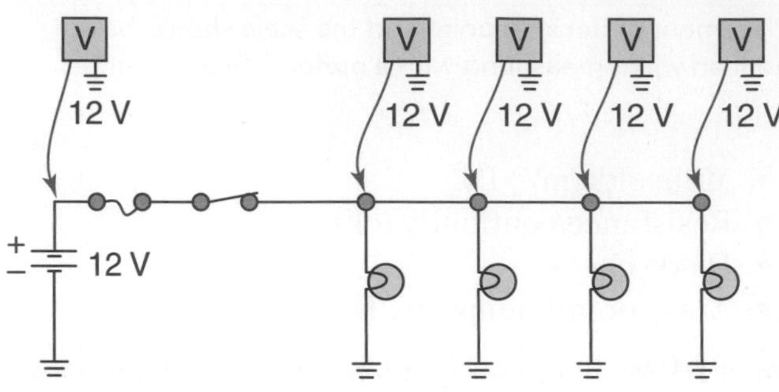

FIGURE 17–24

Checking available voltage at points within a circuit with a voltmeter.

A DMM can be used to check for proper circuit grounding. For example, if the voltmeter shows battery voltage at a lamp, but no lighting is seen, the bulb or socket could be bad or the ground connection is faulty. An easy way to check for a defective bulb is to replace it with one known to be good. You can also use an ohmmeter to check for electrical continuity through the bulb.

If the bulbs are not defective, the problem is the light sockets or ground wires. Connect the voltmeter to the ground wire and a good ground as shown in **Figure 17–26**. If 0 volts is measured, move the positive meter lead to the power feed side of the bulb. If 0 volts is measured there, the light socket or feed wire is defective. In a normal light circuit, there should be 0 volts at the negative side of the bulb. If the socket is not defective and some voltage is measured at the ground, the ground circuit is faulty. The higher the voltage, the greater the problem.

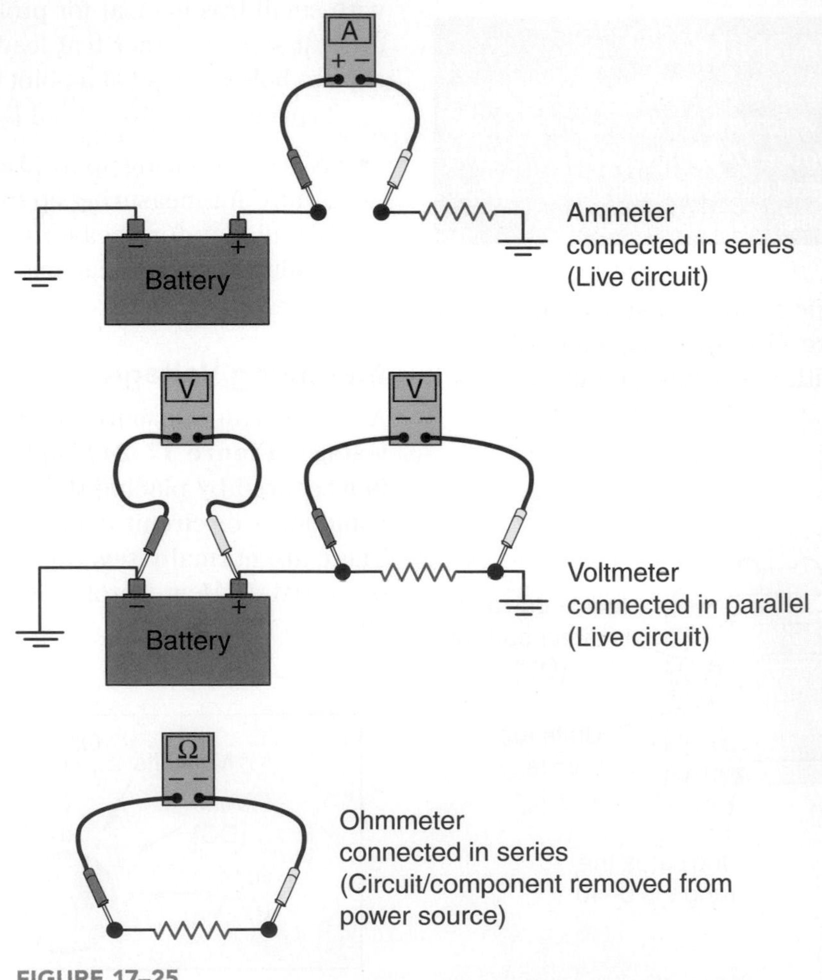

FIGURE 17–25

A voltmeter is connected in parallel or across a component or part of a circuit to measure available voltage and voltage drop.

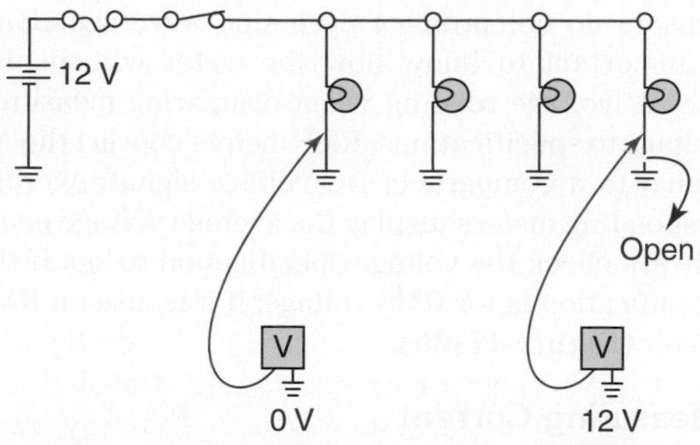

FIGURE 17–26

Using a voltmeter to check for open grounds.

PROCEDURE

To measure DC voltage:

STEP 1 Set the mode selector switch to Volts DC.

STEP 2 Select the auto-range function, or manually set the range to match the anticipated value; normally, the range is set to the closest to and higher than 12 volts.

STEP 3 Connect the test leads in parallel to the circuit or component being tested. The red lead should be connected first to the most positive side (side closest to the battery). The black lead is connected to a good ground.

STEP 4 Read the measurement on the display. If the reading is negative, it is likely that the leads are reversed.

VOLTAGE DROP TEST Measuring voltage drop is a very important test. It can identify circuits with unwanted resistances. This test is extremely valuable when working with electronic circuits; even the smallest loss of voltage will affect the performance of parts and the vehicle as a whole. The test can be performed between any two points in a circuit and across any component, such as wires, switches, relay contacts and coils, and connectors **(Figure 17–27)**. A voltage drop test can find excessive resistance in a circuit that may not be detected using an ohmmeter. An ohmmeter works by passing a small amount of current through the component being tested while it is isolated from the circuit. A voltage drop test is conducted with the circuit operating with normal amounts of current.

Consider a simple circuit. If there are 12 volts available at the battery and the switch is closed, there should also be 12 volts available at each light. If, for example, less than 12 volts are measured, that means some additional resistance is somewhere else in the circuit. The lights may light but not as brightly as they should.

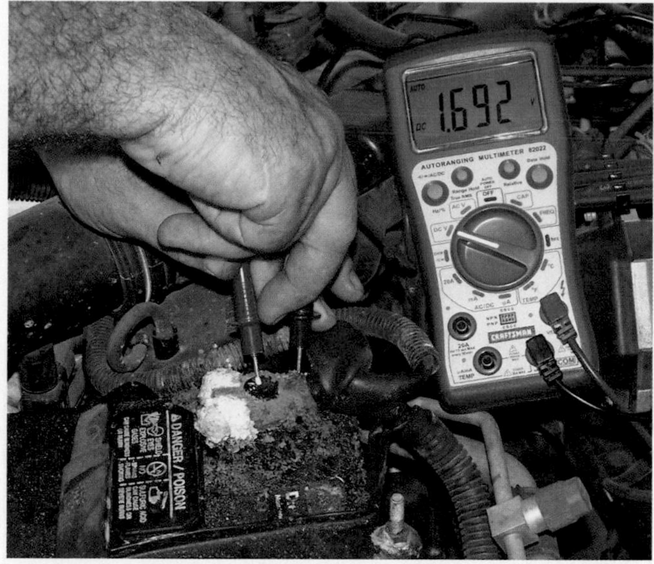

FIGURE 17–27

Measuring the voltage drop across the battery post and cable. It is easy to see why the voltage drop is high!

Figure 17–28 illustrates two headlights (2 ohms each) connected to a 12-volt battery using two wires. Each wire has an unwanted resistance of 0.05 ohm.

The two headlights are wired parallel, and total resistance of the headlights is

$$\frac{2 \text{ ohms} + 2 \text{ ohms}}{2 \text{ ohms} + 2 \text{ ohms}} = 1 \text{ ohm}$$

The total circuit resistance is

$$1 \text{ ohm} + 0.05 \text{ ohm} + 0.05 \text{ ohm} = 1.1 \text{ ohms}$$

Therefore, the current in the circuit is

$$I = E/R = 12/1.1 = 10.9 \text{ amperes}$$

The voltage drop across each wire is

$$E = I \div R$$

$$E = 10.9 \div 0.05 = 0.54 \text{ V}$$

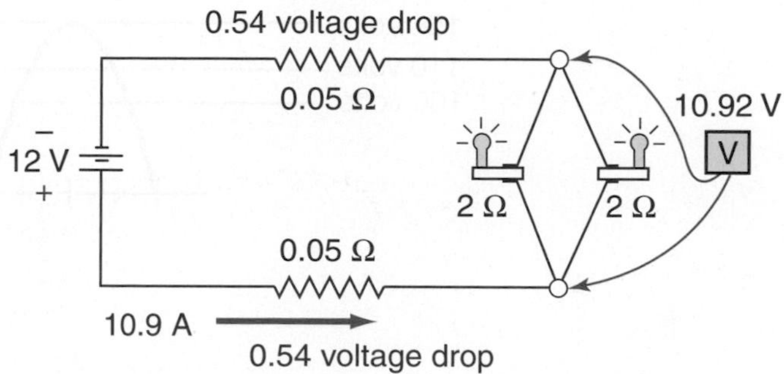

FIGURE 17–28

Wire resistance results in a slight voltage drop in the circuit.

This means there is a total of 1.08 volts dropped across the wires. When the voltage drop of the wires is subtracted from the 12-volt source voltage, 10.92 volts remain for the headlights.

Without the resistance in the wires, the headlights receive 12 amperes. With the resistance, the current flow was reduced to 10.9 amperes. The decreased current and voltage drop mean the lights will not be as bright as normal.

PROCEDURE

The procedure for measuring voltage drop is shown in Photo Sequence 11. To measure voltage drop:

STEP 1 Set the DMM to the mV or V DC mode.

STEP 2 Set the DMM to auto range or a low-voltage range.

STEP 3 Connect the positive test lead to the most positive side of the part or circuit being tested.

STEP 4 Connect the negative test lead to the least positive side of the part or circuit being tested.

STEP 5 Power the circuit. Current must be flowing in order to have a voltage drop.

STEP 6 Read the meter. Excessive voltage drops indicate excessive resistance (an unwanted load) in that portion of the circuit.

Voltage drops should not exceed the following:

- 200 mV across a wire or connector
- 300 mV across a switch or relay contacts
- 100 mV at a ground connection
- 0 mV to <50 mV across all sensor connections

MEASURING AC VOLTAGE There are two ways that multimeters display AC voltage: **root mean square (RMS)** and **average responding**. When an AC voltage signal is a true sine wave, both methods will display the same reading. Because most automotive sensors do not produce pure sine wave signals, it is important to know how the meter will display the AC voltage reading when comparing measured voltage to specifications. RMS meters convert the AC signal to a comparable DC voltage signal. Average responding meters display the average voltage peak. Always check the voltage specification to see if the specification is for RMS voltage; if it is, use an RMS meter **(Figure 17–29)**.

Measuring Current

Testing the current through a circuit gives a true picture of what is happening in the circuit. This is because the circuit is being tested under load. Low current indicates that the circuit has higher than normal resistance, and high current means the circuit has lower than normal resistance. When measuring current, you need to remember that voltage is applied and the circuit is operating. Many manufacturers do not often include current testing in their procedure due to this fact as it could result in damage to the vehicle, harness, components, and even the technician. Current flow produces heat, and caution must be taken when performing this test step. Many technicians (and manufacturers) prefer to measure the resistance of the circuit, and when the applied voltage is known, will calculate the current flow in the circuit. Very few current specifications are found in electrical manuals for individual components in automotive applications. Typically, the circuit (which may contain numerous components) will be rated for the fuse that is protecting that circuit.

SHOP TALK

If you are required to measure the current of a circuit or component, perform the test in the least intrusive manner. Always try to install the ammeter in line with the fuse or a connector, if possible. Only "tap into" a harness as a last resort or if no other access points are available.

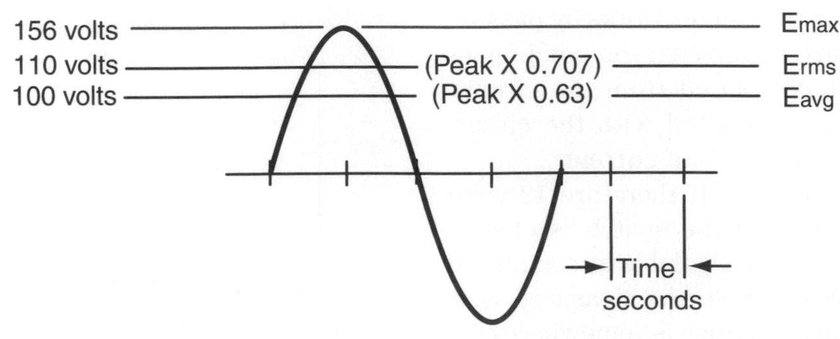

FIGURE 17–29

AC voltage: RMS.

PERFORMING A VOLTAGE DROP TEST

PHOTO SEQUENCE 11

P11–1 Tools required to perform this task: voltmeter and fender covers.

P11–2 Set the voltmeter on its lowest DC volt scale.

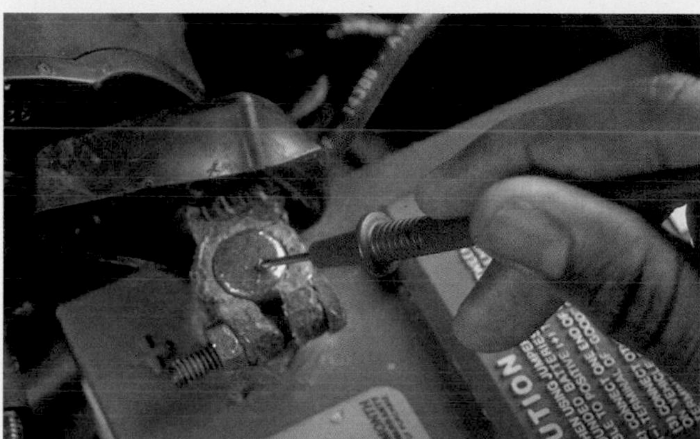

P11–3 To test the voltage drop of the entire headlamp system, connect the positive (red) lead to the battery positive terminal.

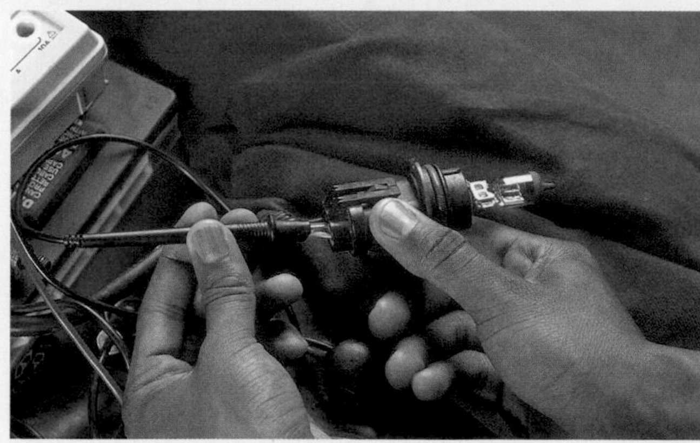

P11–4 Connect the negative (black) lead to the low beam terminal of the headlight socket. Make sure you are connected to the battery "1" or power feed wire of the headlight.

P11–5 Turn on the headlights (low beam) and look at the voltmeter reading. The voltmeter will show the amount of voltage dropped between the battery and the headlight. This reading should be very low.

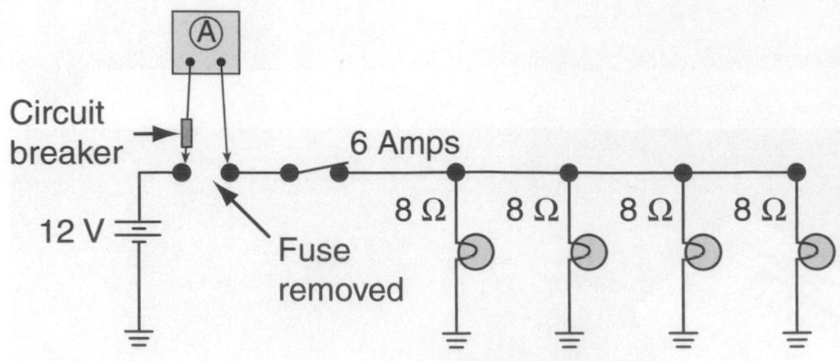

FIGURE 17–30

Checking a circuit using an ammeter.

The circuit in **Figure 17–30** normally draws 6 amps and is protected by a 10-amp fuse. If the circuit constantly blows the fuse, a short exists somewhere in the circuit. Mathematically, each light should draw 1.5 amperes (6 ÷ 4 = 1.5). To find the short, disconnect all lights by removing them from their sockets. Then close the switch and read the ammeter. With the loads disconnected, the meter should read 0 ampere. If there is any reading, the wire between the fuse and the sockets is shorted to ground.

If 0 amp was measured, reconnect each light in sequence; the reading should increase 1.5 amperes with each bulb. If, when making any connection, the reading is higher than expected, the problem is in that part of the light circuit.

WARNING!

When testing for a short, always use a fuse. Never bypass the fuse with a wire. The fuse should be rated at no more than 50 percent higher capacity than specifications. This offers circuit protection and provides enough amperage for testing. After the problem is found and corrected, be sure to install a fuse with the specified rating.

Before checking current, make sure the meter is capable of measuring the suspected amount. To check the rating of the meter, look at the rating printed next to the DMM input jacks, or check the rating of the meter's fuse (maximum current capacity is typically the same as the fuse rating). If you suspect that a measurement will have a current higher than the meter's maximum rating, use an inductive current probe.

An ammeter must be connected in series with the circuit; this allows circuit current to flow through the meter.

PROCEDURE

To measure current:

STEP 1 Turn off the circuit that will be tested.

STEP 2 Connect the test leads to the correct input jacks on the DMM.

STEP 3 Set the mode selector to the correct current setting (normally, amps or milliamps).

STEP 4 Select the auto-range function, or manually select the range for the expected current.

STEP 5 Open the circuit at a point where the meter can be inserted.

STEP 6 Connect a fused jumper wire to one of the test leads.

STEP 7 Connect the red lead to the most positive side of the circuit and the black lead to the other side.

STEP 8 Turn on the circuit.

STEP 9 Read the display on the DMM.

STEP 10 Compare the reading to specifications or your calculations.

WARNING!

Never place the leads of an ammeter across the battery or a load. This puts the meter in parallel with the circuit and will blow the fuse in the ammeter or possibly destroy the meter.

INDUCTIVE CURRENT PROBES Many DMMs have current probes (current clamps) that eliminate the need to insert the ammeter into the circuit. These probes read current by sensing the magnetic field formed in a wire by current flow **(Figure 17–31)**. Normally, to use a current probe, the DMM's mode selector is set to read millivolts (mV). The probe is then connected to the meter and turned on. Some

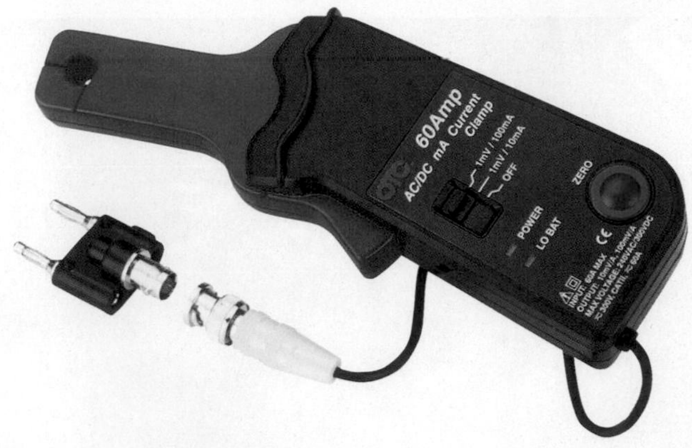

FIGURE 17–31

A low-amp current probe.

Bosch Automotive Service Solutions

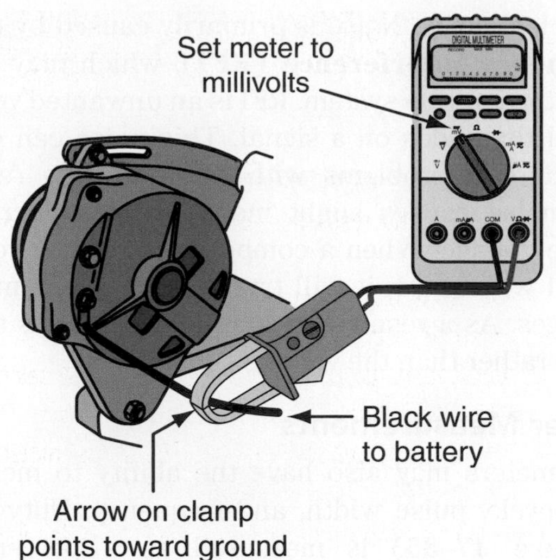

FIGURE 17–32

A current probe attached to a DVOM allows for current measurements without breaking the circuit.

probes must be zeroed prior to taking a measurement. This is done before the probe is clamped around a wire. The DMM may have a zero adjust control, which is turned until zero reads on the meter's display. The clamp is placed around a wire in the circuit being tested **(Figure 17–32)**. Make sure the arrow on the clamp is pointing in the direction of current flow. After the clamp is in place, the circuit is turned on, and the voltage read on the display. The voltage reading is then converted to an amperage reading—1 mV = 10 mA/1 mV = 100 mA.

Measuring Resistance

DMMs are used to test circuit continuity and resistance with no power applied. In other words, the circuit or component must first be disconnected from the power source **(Figure 17–33)**. Connecting an ohmmeter into a live circuit usually results in damage to the meter.

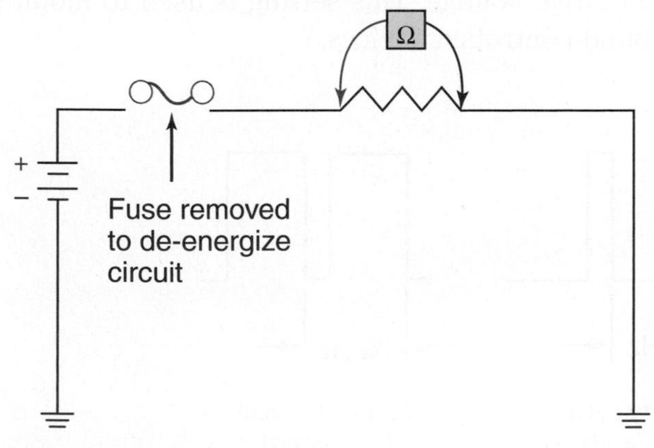

FIGURE 17–33

Measuring resistance using an ohmmeter. Note that the component has been removed from the circuit.

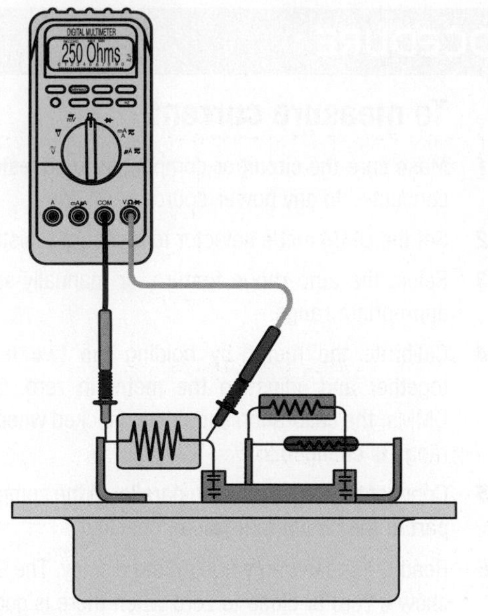

FIGURE 17–34

An ohmmeter can test the resistance of a component after it has been removed from the circuit.

Checking the resistance can be used to check the condition of a component **(Figure 17–34)** or circuit. Often, specifications list a normal range of resistance values for specific parts. If the resistance is too high, check for an open circuit or a faulty component. Excessive resistance can prevent a circuit from operating normally. Loose, damaged, or dirty connections are common causes of excessive resistance. If the resistance is too low, check for a shorted circuit or faulty component.

CAUTION!

To avoid possible damage to the meter or to the equipment under test, disconnect the circuit power and discharge all high-voltage capacitors before measuring resistance. Always follow the manufacturer's test procedures when testing air bags.

Ohmmeters also are used to trace and check wires or cables. Connect one probe of the ohmmeter to the known wire at one end of the cable, and touch the other probe to each wire at the other end of the cable. Any evidence of resistance indicates the correct wire. Using this same method, you can check for a defective wire. If low resistance is shown on the meter, the wire is sound. If no resistance is measured, the wire is open. Any indication of resistance means the wire may be shorted to another wire, and the harness is defective.

PROCEDURE

To measure current:

STEP 1 Make sure the circuit or component to be tested is not connected to any power source.

STEP 2 Set the DMM mode selector to measure resistance.

STEP 3 Select the auto-range feature, or manually select the appropriate range.

STEP 4 Calibrate the meter by holding the two test leads together and adjusting the meter to zero. On some DMMs, the calibration should be checked whenever the range is changed.

STEP 5 Connect the meter leads in parallel to the component or part of the circuit that will be checked.

STEP 6 Read the measured value on the display. The DMM will show a zero or close to zero when there is good continuity. If there is no continuity, the meter will display an infinite or over-limit reading.

CONTINUITY TESTS Many DMMs have a continuity test mode that makes a beeping noise when there is continuity through the circuit being tested. This audible sound will continue as long as there is continuity. This feature can be handy for finding the cause of an intermittent problem. By connecting the DMM across a circuit and wiggling sections of the wiring harness, a problem can be noted when the beeping stops after a particular section or wire has been moved. Please note that most meters will stop beeping at a set resistance value, usually 50 to 75 ohms.

MIN/MAX Readings

Some DMMs also feature a MIN/MAX function. This displays the maximum, minimum, and average voltage recorded during the time of the test. This is valuable when checking sensors or when looking for electrical noise. Noise is primarily caused by **radio frequency interference (RFI)**, which may come from the ignition system. RFI is an unwanted voltage signal that rides on a signal. This noise can cause intermittent problems with unpredictable results. The noise causes slight increases and decreases in the voltage. When a computer receives a voltage signal with noise, it will try to react to the minute changes. As a result, the computer responds to the noise rather than the voltage signal.

Other Measurements

Multimeters may also have the ability to measure duty cycle, pulse width, and frequency. Duty cycle **(Figure 17–35)** is measured as a percentage. A 60 percent duty cycle means that a device is on 60 percent of the time and off 40 percent of one cycle. When measuring duty cycle, you are looking at the amount of time something is on during one cycle.

Pulse width is normally measured in milliseconds. When measuring pulse width, you are looking at the amount of time something is on.

To accurately measure duty cycle, pulse width, and frequency, the meter's trigger level must be set. The trigger level tells the meter when to start counting. Trigger levels can be set at certain voltage levels or at a rise or fall in the voltage. Normally, meters have a built-in trigger level that corresponds with the voltage range setting. If the voltage does not reach the trigger level, the meter will not begin to recognize a cycle. On some meters, you can select between a rise or fall in voltage to trigger the cycle count. Most technicians refer to this as a positive or negative slope trigger. A rise in voltage is a positive increase in voltage. This setting is used to monitor the activity of devices whose power feed is controlled by a computer. A fall in voltage is negative voltage. This setting is used to monitor ground-controlled devices.

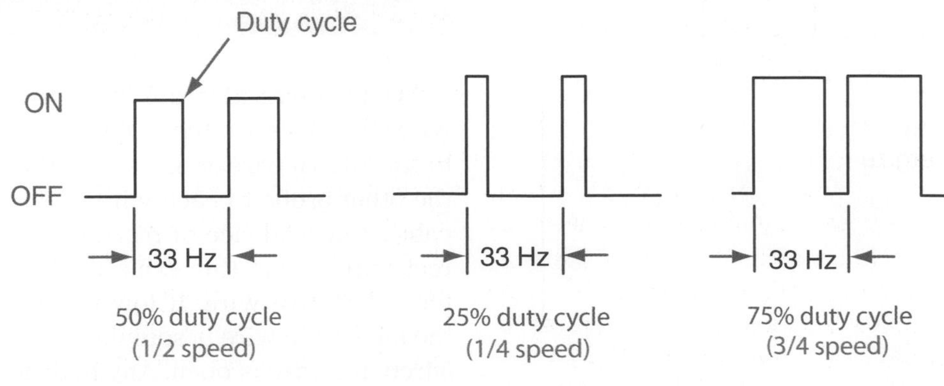

FIGURE 17–35

Duty cycle is expressed in a percentage.

Some DMMs can measure temperature. These meters are equipped with a thermocouple. Temperature readings can be made in Fahrenheit or Celsius. The thermocouple is connected to the DMM and placed on or near the object to be checked.

More elaborate DMMs have the ability to store and download data to a PC.

Safety Guidelines

To avoid possible electric shock or personal injury while using a DMM, follow these guidelines:

- Use a meter only as it was designed to be used; misuse may defeat the protection devices built into the meter.
- Never use a damaged meter.
- Make sure the battery is secure and totally enclosed with the meter.
- Inspect the test leads for damaged insulation or exposed metal. Replace damaged test leads before you use a meter.
- Never apply more than the rated voltage of the meter between the terminals or between any terminal and ground.
- Be extra careful when measuring high voltages.
- Use the proper connections, terminals, mode, and range for the measurements being taken.
- When measuring current, turn off the power to the circuit before connecting the meter in series with the circuit.
- When making electrical connections, connect the positive test lead before connecting the negative test lead; when disconnecting, disconnect the positive test lead before disconnecting the negative test lead.
- When using probes, keep your fingers behind the finger guards on the probes.
- Disconnect the power to the circuit and discharge all high-voltage capacitors before testing resistance, continuity, diodes, or capacitance.
- When testing HV components or circuits, wear high-voltage electricians gloves rated at class 0.

USING LAB SCOPES

The use of lab scopes is becoming more common with today's technicians because it offers many advantages when diagnosing intermittent and hard-to-find vehicle faults. When measuring voltage with an analog voltmeter, the meter displays only the average values at the point being probed. Digital voltmeters simply sample the voltage several times

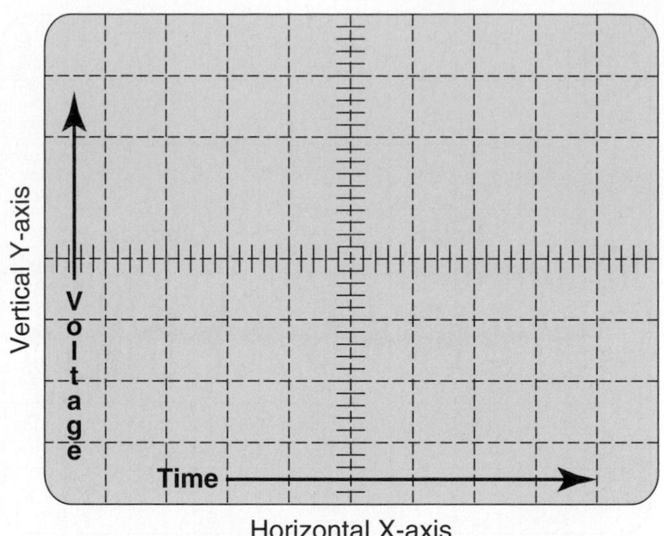

FIGURE 17–36

Grids on a scope screen that serve as a time and voltage reference.

each second and update the meter's reading at a particular rate. If the voltage is constant, good measurements can be made with both types of voltmeters. A scope, however, will display any change in voltage as it occurs. This is important for diagnosing intermittent problems.

The screen is divided into small divisions of time and voltage **(Figure 17–36)**. These divisions set up a grid pattern on the screen. Time is represented by the horizontal movement of the trace. Voltage is measured with the vertical position of the trace. Because the scope displays voltage over time, the trace moves from the left (the beginning of measured time) to the right (the end of measured time). The value of the divisions can be adjusted to improve the view of the voltage trace. For example, the vertical scale can be adjusted so that each division represents 0.5 volt, and the horizontal scale can be adjusted so that each division equals 0.005 second (5 milliseconds). This allows for viewing small changes in voltage that occur during a very short period. The grid serves as a reference for measurements.

Because a scope displays actual voltage, it will display any electrical noise or disturbances that accompany the voltage signal **(Figure 17–37)**. Electrical disturbances or **glitches** are momentary changes in the signal. These can be caused by intermittent shorts to ground, shorts to power, or opens in the circuit. These problems can occur for only a moment or may last for some time. A lab scope is handy for finding these and other causes of intermittent problems. By observing a voltage signal and

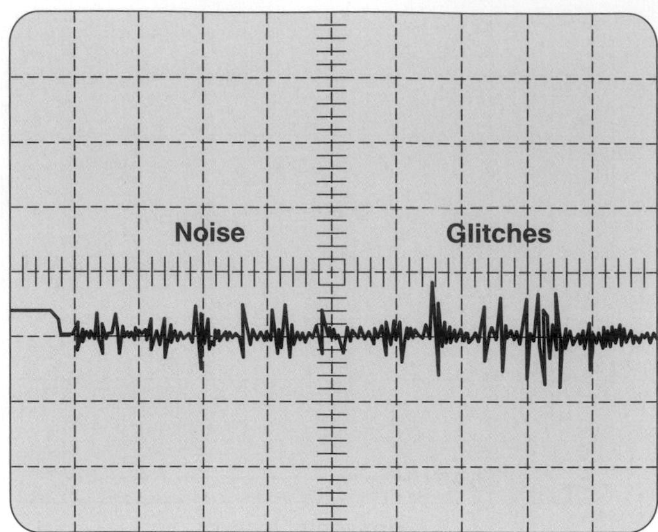

FIGURE 17–37

RFI noise and glitches may appear on a voltage signal.

wiggling or pulling a wiring harness, any looseness can be detected by a change in the voltage signal.

Analog versus Digital Scopes

Analog scopes show the actual activity of a circuit and are called real-time scopes. This simply means that what is taking place at that time is what you see on the screen. Analog scopes have a fast update rate that allows for the display of activity without delay.

A digital scope, commonly called a **digital storage oscilloscope (DSO)**, converts the voltage signal into digital information and stores it into its memory. Some DSOs send the signal directly to a computer or a printer or save it to a disk. To help in diagnostics, a technician can "freeze" the captured signal for close analysis. DSOs also have the ability to capture low-frequency signals. Low-frequency signals tend to flicker when displayed on an analog screen. To have a clean waveform on an analog scope, the signal must be repetitive and occurring in real time. The signal on a DSO is not quite real time. Rather, it displays the signal as it occurred a short time before.

This delay is actually very slight. Most DSOs have a sampling rate of one million samples per second. This is quick enough to serve as an excellent diagnostic tool. This fast sampling rate allows slight changes in voltage to be observed. Slight and quick voltage changes cannot be observed on an analog scope.

Because digital signals are based on binary numbers, the trace appears to be slightly choppy when compared to an analog trace. However, the voltage signal is sampled more often, which results

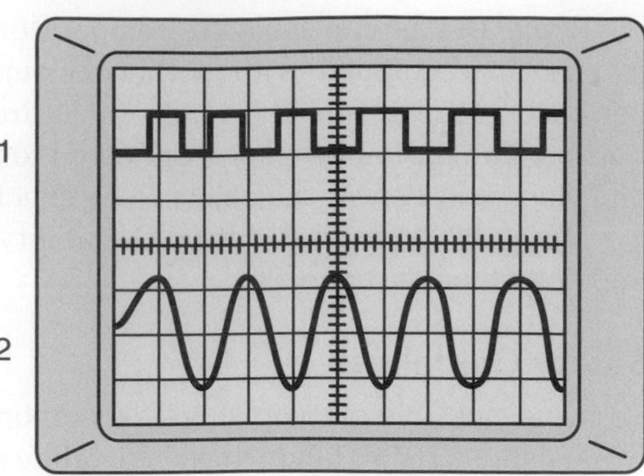

FIGURE 17–38

Two or more different signals can be observed on a multiple-channel scope. This is invaluable for diagnosis.

in a more accurate waveform. The waveform is constantly being refreshed as the signal is pulled from the scope's memory.

Both an analog and a digital scope can be dual-trace **(Figure 17–38)** or multiple-trace **(Figure 17–39)** scopes. This means they have the capability of displaying more than one trace at one time. By watching the traces simultaneously, the cause and effect of a sensor is observed, and a good or normal waveform can be compared to the one being displayed.

Waveforms

A waveform represents voltage over time. Any change in the amplitude of the trace indicates a change in the voltage. When the trace is a straight horizontal line, the voltage is constant. A diagonal line up or down represents a gradual increase or decrease in

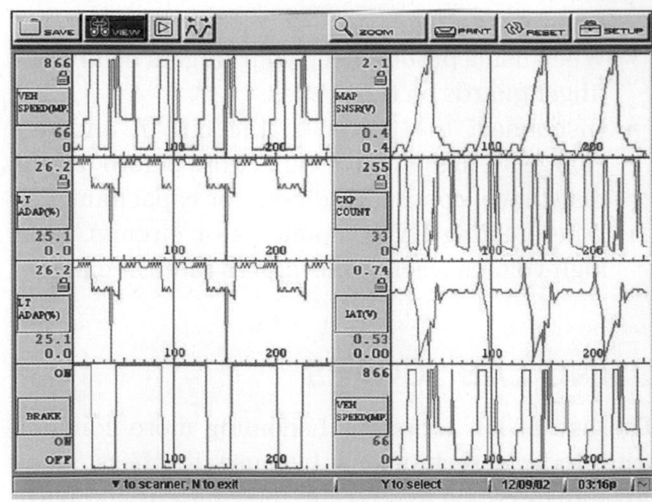

FIGURE 17–39

Some scopes and graphing scan tools can display many channels.

Snap-on Tools Company

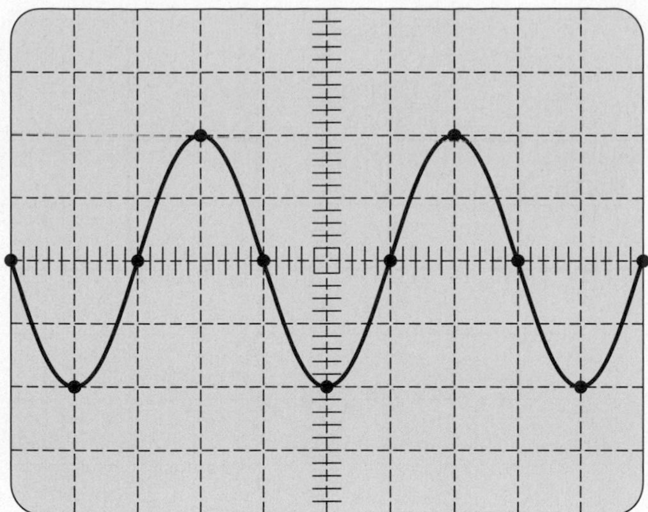

FIGURE 17–40

An AC voltage sine wave.

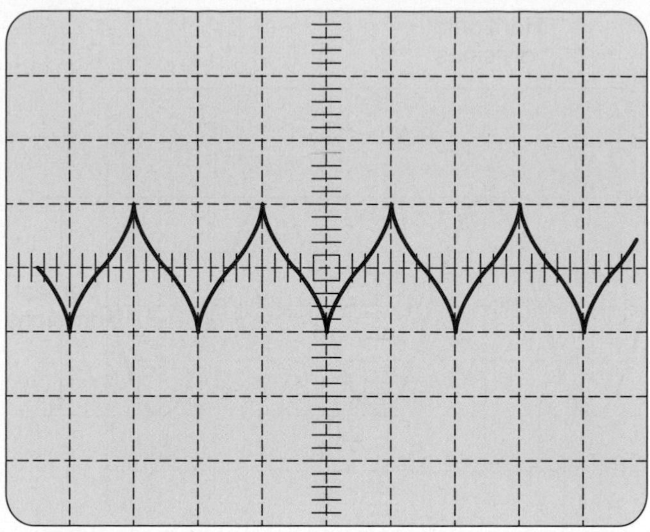

FIGURE 17–41

An AC voltage trace from a typical permanent magnet generator-type pickup or sensor.

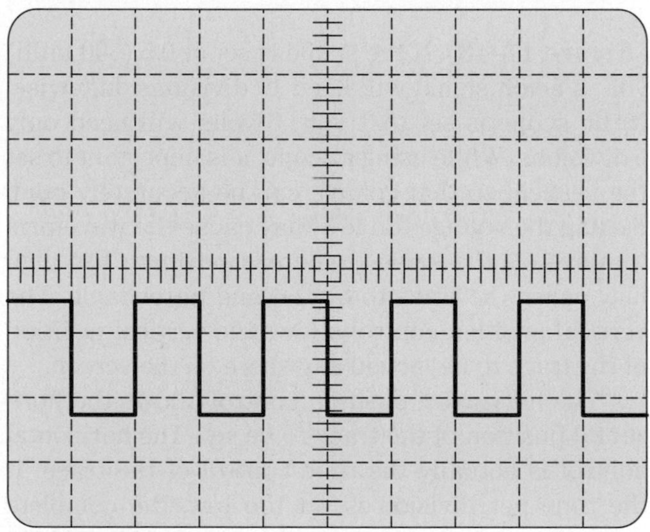

FIGURE 17–42

A typical square (on–off or high–low) wave.

voltage. A sudden rise or fall in the trace indicates a sudden change in voltage.

Scopes can display AC and DC voltage, either one at a time or both at the same time, as in the case of noise caused by RFI. Noise results from AC voltage riding on a DC voltage signal. The consistent change of polarity and amplitude of the AC signal causes slight changes in the DC voltage signal. A normal AC signal changes its polarity and amplitude over time. The waveform created by AC voltage is typically called a sine wave **(Figure 17–40)**. One complete sine wave shows the voltage moving from zero to its positive peak and then moving down through zero to its negative peak and returning to zero.

One complete sine wave is a cycle. The number of cycles that occur per second is the frequency of the signal. Frequency is measured in cycles per second (hertz or Hz). Checking frequency is one way of checking the operation of some electrical components. Input sensors are the most common components that produce AC voltage. Permanent magnet voltage generators produce an AC voltage that can be checked on a scope **(Figure 17–41)**. AC voltage waveforms should also be checked for noise and glitches. These may send false information to the computer.

DC voltage waveforms may appear as a straight line or line showing a change in voltage. Sometimes, a DC voltage waveform will appear as a square wave that shows voltage making an immediate change **(Figure 17–42)**. Square waves are identified by having straight vertical sides and a flat type. This type of wave represents voltage being applied (circuit being turned on), voltage being maintained (circuit remaining on), and no voltage applied (circuit is turned off). Of course, a DC voltage waveform may also show gradual voltage changes.

Scope Controls

Depending on manufacturer and model of the scope, the type and number of its controls will vary. However, nearly all scopes have intensity, vertical (Y-axis) adjustments, horizontal (X-axis) adjustments, and trigger adjustments. The intensity control is used to adjust the brightness of the trace. This allows for clear viewing regardless of the light around the scope screen.

The vertical adjustment actually controls the voltage that will be shown per division

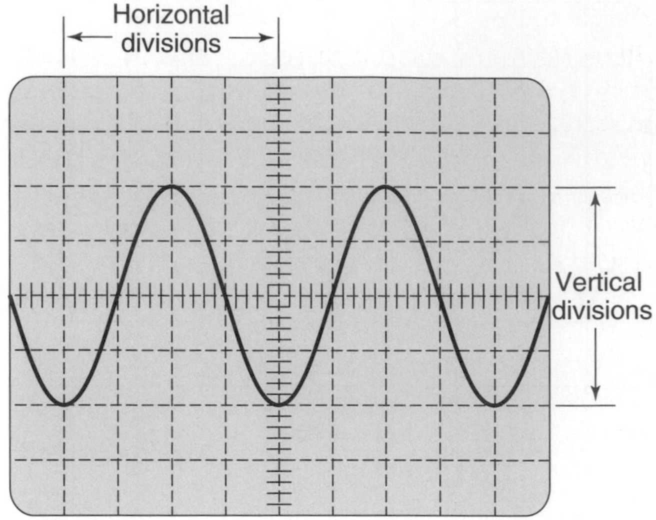

FIGURE 17–43

Vertical divisions represent voltage, and horizontal divisions represent time.

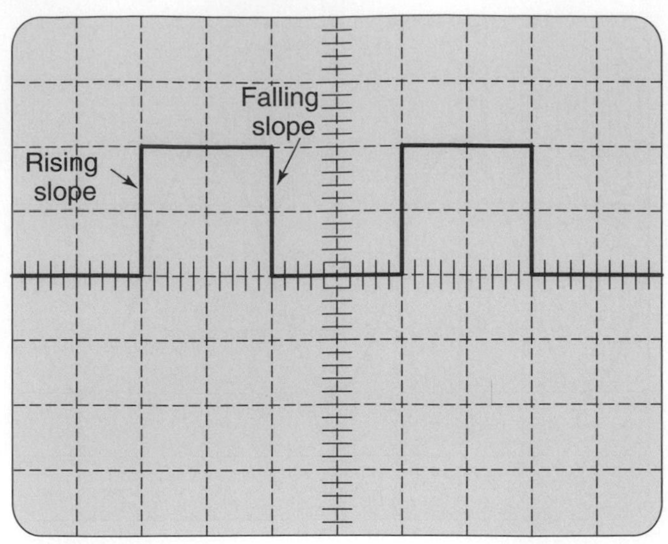

FIGURE 17–44

The trigger can be set to start the trace with a rise or fall of voltage.

(Figure 17–43). If the scope is set at 0.5 (500 milli) volt, a 5-volt signal will need 10 divisions. Likewise, if the scope is set to 1 volt, 5 volts will need only 5 divisions. While using a scope, it is important to set the vertical so that voltage can be accurately read. Setting the voltage too low may cause the waveform to move off the screen, whereas setting it too high may cause the trace to be flat and unreadable. The vertical position control allows the vertical position of the trace to be moved anywhere on the screen.

The horizontal position control allows the horizontal position of the trace to be set. The horizontal control is actually the time control of the trace. If the time per division is set too low, the complete trace may not show across the screen. Also, if the time per division is set too high, the trace may be too crowded for detailed observation. The time per division (TIME/DIV) can be set from very short periods (millionths of a second) to full seconds.

Trigger controls tell the scope when to begin a trace across the screen. Setting the trigger is important when trying to observe the timing of something. Proper triggering will allow the trace to repeatedly begin and end at the same points on the screen. There are typically numerous trigger controls on a scope. The trigger mode selector has a NORM and an AUTO position. In the NORM setting, no trace will appear on the screen until a voltage signal occurs within the set time base. The AUTO setting will display a trace regardless of the time base.

Slope and level controls are used to define the actual trigger voltage. The slope switch determines whether the trace will begin on a rising or falling of the signal **(Figure 17–44)**. The level control sets when the time base will be triggered according to a certain point on the slope.

A trigger source switch tells the scope which input signal to trigger on. This can be Channel 1, Channel 2, line voltage, or an external signal. External signal triggering is very useful when observing the trace of a component that may be affected by the operation of another component. An example of this would be observing fuel injector activity when changes in throttle position are made. The external trigger would be voltage changes at the throttle position sensor. The displayed trace would be the cycling of a fuel injector.

Graphing Multimeters

One of the latest trends in diagnostic tools is a **graphing multimeter (GMM)**. A GMM is a DMM that displays voltage, resistance, current, and frequency as a waveform. The display shows the minimum and maximum readings as a graph, as well as all current measurements **(Figure 17–45)**. By observing the graph, a technician can see any undesirable changes during the transition from a low reading to a high reading, or vice versa. These glitches are some of the more difficult problems to identify without a graphing meter or a lab scope. The waveform on a DSO may miss a change in voltage, resistance, or current that occurs too quickly for the scope to detect and display it. DSOs do not display real-time data. There is always a slight delay between what it measures and what it displays,

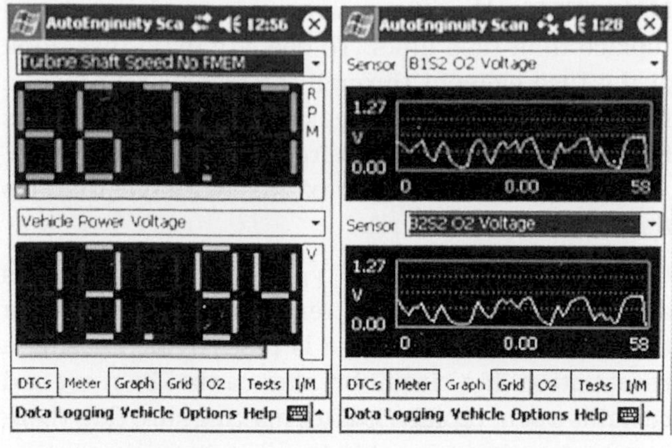

FIGURE 17–45

A sample of the information available from a GMM.

and depending on the refresh rate of the display, it may miss some changes completely. A GMM is perhaps the best tool to use when trying to find the cause of intermittent problems in low-voltage DC circuits.

The capabilities of a GMM depend on the manufacturer and the model. Some of the features found on GMMs include a signal and data recorder, individual component tests, the ability to display measurements along with a graph, glitch capture, and an audible alarm. Some also have an electronic library of known good signals for comparison. These allow for a comparison of live patterns with expected values or known good waveforms. Some even have wiring diagrams and a vehicle-specific database of diagnostic and test information.

TRANSFERRING DATA TO A PC Many DSOs and GMMs allow for the transfer of captured information to a PC through a cable or wireless interface. This feature allows for better viewing of the waveforms and other data and also allows for the creation of a personal library. The latter can be helpful in the future. There are several sources of waveforms readily available for download to a PC; these can be used as references and study guides.

TESTING BASIC ELECTRICAL COMPONENTS

All electrical components can fail. Testing them is the best way of determining if they are good or bad. For the most part, the proper way for checking electrical components is determined by what the component is supposed to do. If we think about what something is supposed to do, we can figure out how to test it. Often, removing the component and testing it on a bench is the best way to check it.

Protection Devices

When overloads or shorts in a circuit cause too much current, the wiring in the circuit heats up, the insulation melts, and a fire can result unless the circuit has some kind of protective device. Fuses, fuse links, maxi-fuses, and circuit breakers are designed to provide protection from high current. They may be used singularly or in combination.

Circuit protection devices can be checked with an ohmmeter or test light. If the fuse is good, there will be continuity through it. To test a circuit protection device with a voltmeter, check for available voltage at both terminals of the unit **(Figure 17–46)**. If the device is good, voltage will be present on both sides. A test light can be used in place of a voltmeter.

Measuring voltage drop across a fuse or other circuit protection device tells more about its condition than its continuity. If a fuse, a fuse link, or a circuit breaker is in good condition, a voltage drop of zero will be measured. If 12 volts is read, the fuse is open. Any reading between 0 and 12 volts indicates that there is unwanted resistance, and the fuse should be replaced. Make sure you check the fuse holder for resistance as well.

FUSES A cartridge fuse can be visually checked by looking at the internal metal strip. Discoloration of the glass cover or glue bubbling around the metal end caps is an indication of overheating. Pull the fuse from its holder, and look at the element through the transparent plastic housing. Look for internal breaks and discoloration **(Figure 17–47)**. A ceramic fuse is checked by looking for a break in the contact strip on the outside of the fuse.

Sometimes it is necessary to protect a device in a portion of a circuit, even though the entire circuit is already protected by a fuse in the panel. This is done

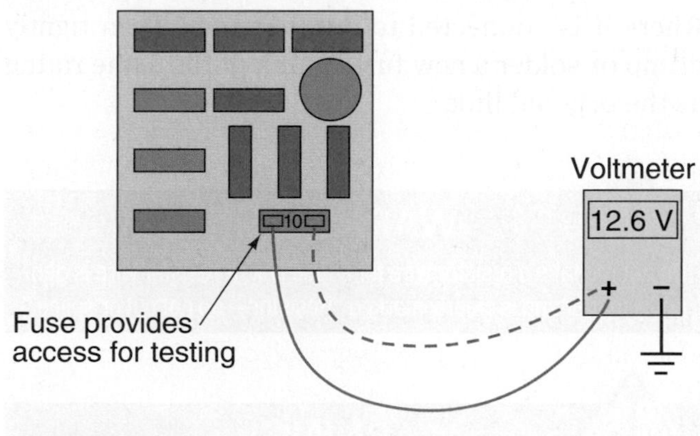

FIGURE 17–46

Circuit protection devices can be tested with a voltmeter. The DVOM is used to test for voltage on both sides of the device.

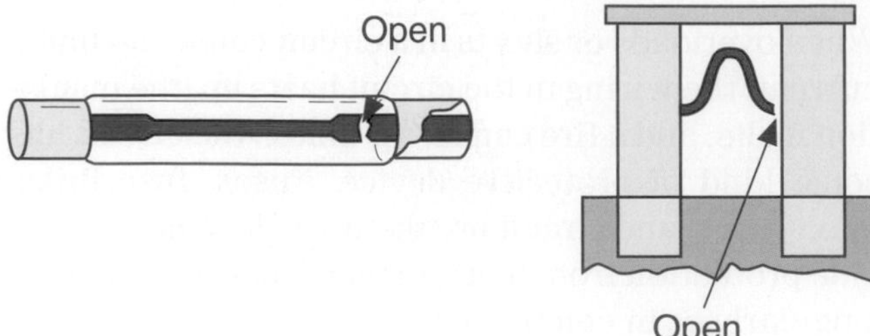

FIGURE 17–47

The condition of a fuse can often be checked visually.

by installing an **in-line fuse** in the power wire for the device. In-line fuses are used primarily on accessories that are very sensitive to power surges, such as radios and CD players. They are also used with driving lights and power antennas.

SHOP TALK

To calculate the correct fuse rating, use Watt's law: Watts divided by volts equals amperes. For example, if you are installing a 55-watt pair of fog lights, divide 55 by the battery voltage to find how much current the circuit will draw. In this case, the current is approximately 5 amperes. To allow for current surges, the correct in-line fuse should be rated slightly higher than the normal current flow. In this case, an 8- or 10-ampere fuse would do the job.

FUSE LINKS Fuse link wire is covered with a special insulation that bubbles when it overheats, indicating that the link has melted. If the insulation appears good, pull lightly on the wire. If the link stretches, the wire has melted. Of course, when it is hard to determine if the fuse link is burned out, check for continuity through the link with a test light or ohmmeter.

To replace a fuse link, cut the protected wire where it is connected to the fuse link. Then tightly crimp or solder a new fusible link of the same rating as the original link.

WARNING!

Do not mistake a resistor wire for a fuse link. A resistor wire is generally longer and is clearly marked "Resistor—do not cut or splice."

WARNING!

Always disconnect the battery ground cable prior to servicing any fuse link.

MAXI-FUSES Maxi-fuses are easier to inspect and replace than fuse links. To check a maxi-fuse, look at the fuse element through the transparent plastic housing. If there is a break in the element, it has blown. To replace it, pull it from its fuse box or panel. Always replace a blown maxi-fuse with one that has the same ampere rating.

CIRCUIT BREAKERS Two types of circuit breakers are used. One is reset by removing the power from the circuit. The other type is reset by depressing a reset button. If a circuit breaker cannot be reset and remains open, replace it after making sure that there is not excessive current in the circuit.

THERMISTORS Some systems use a positive temperature coefficient (PTC) thermistor as a protection device. When there is high current, the resistance of the thermistor increases and causes a decrease in current flow. These can be checked with an ohmmeter **(Figure 17–48)**. If an infinite reading is displayed, the thermistor is open. Another way of checking a thermistor is to change its temperature and see if its resistance changes. A negative temperature coefficient (NTC) thermistor works in the opposite way; that is, its resistance decreases as the temperature increases.

Switches

To check a switch, disconnect the connector at the switch. Check for continuity between the terminals of the switch **(Figure 17–49)** with the switch in the ON and OFF positions. While in the OFF position, there should be no continuity between the terminals. With the switch on, there should be good continuity between the terminals. If the switch is activated by something mechanical and does not

FIGURE 17–48

A temperature sensor (thermistor) can be checked with an ohmmeter.

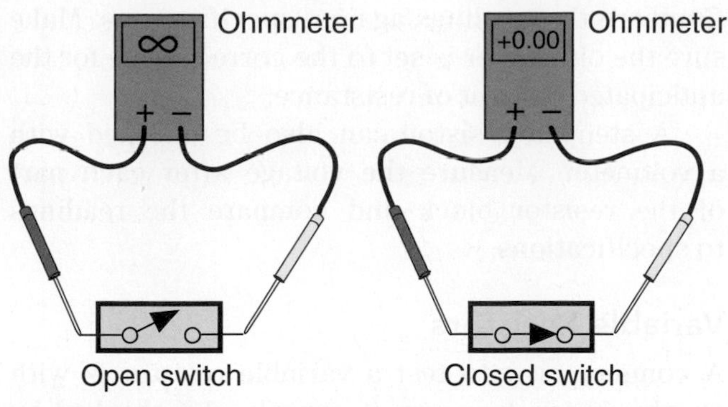

FIGURE 17–49

Checking a switch with an ohmmeter.

complete the circuit when it should, check the adjustment of the switch (some switches are not adjustable). If the adjustment is correct, replace the switch. Another way to check a switch is to bypass it with a jumper wire. If the component works when the switch is jumped, the switch is bad. A multiple-pole multiple-throw (MPMT) switch should be checked in each of its possible positions **(Figure 17–50)**. Use a wiring diagram to identify which terminals of the switch should have continuity during each switch position.

Voltage drop across switches should also be checked. Ideally, when the switch is closed, there should be no voltage drop. Excessive voltage drop indicates resistance, and the switch should be replaced.

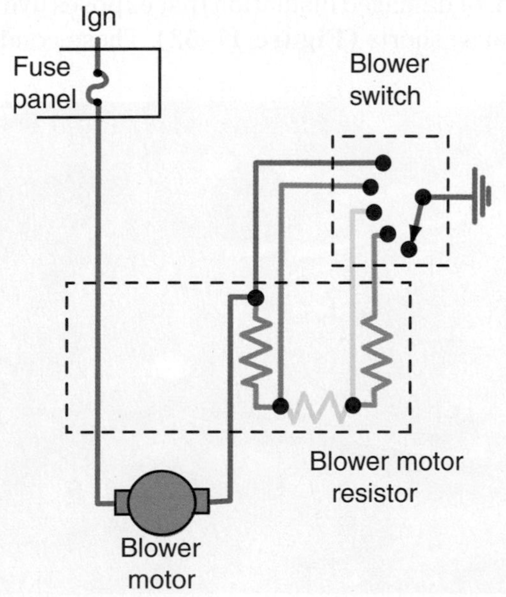

FIGURE 17–50

An MPMT switch should be checked in all of its possible positions. Use a wiring diagram to guide your tests.

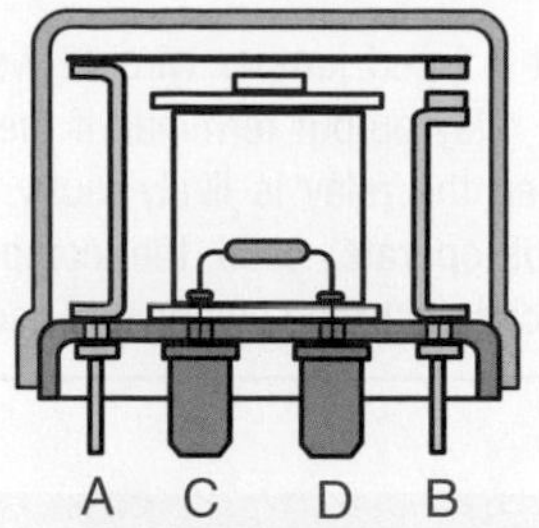

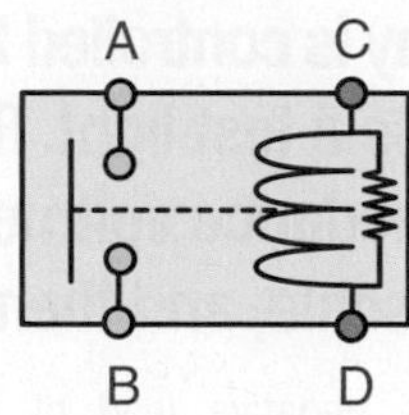

FIGURE 17–51

Use a wiring diagram to identify the terminals of a relay so that they can be tested properly.

Relays

A relay can be checked with a jumper wire, a voltmeter, an ohmmeter, or a test light. If the terminals are accessible and the relay is *not* controlled by a computer, a jumper wire and test light will be the quickest method to use. The schematic for a relay is typically shown on the outside of the relay. If not, check a wiring diagram to identify the terminals of the relay **(Figure 17–51)**. Also check the wiring diagram to determine if the relay is controlled by a power or ground switch.

PROCEDURE

If the relay is controlled on the ground side, follow this procedure to test the relay:

STEP 1 Use a DVOM to check for voltage at the relay feed circuit. If no voltage is present, then the fault is in the feed circuit. If voltage is present, continue testing.

STEP 2 Check for voltage at the relay control side (coil). If no voltage is present, then the fault is in the feed to the control side of the relay. Also check the ground side of the control circuit. This can be done using a DVOM to measure the actual resistance in the ground side circuit. Alternatively, a fused jumper wire can be connected between the ground side of the relay coil and a known good ground. If the relay works, then the fault is in the ground connection of the relay coil. If the relay doesn't work, continue testing.

(continued)

STEP 3	Connect a fused jumper wire between battery power and the relay output terminal. If the component operates, then the relay is likely faulty. If the component does not operate, then the component power and ground circuits or the component itself may be at fault.

PROCEDURE

If the relay is controlled by a computer, do not use a test light. Rather, use a high-impedance voltmeter set to the 20 V DC scale, and then:

STEP 1	Connect the negative lead of the voltmeter to a good ground.
STEP 2	Connect the positive lead to the output wire. If no voltage is present, continue testing. If there is voltage, disconnect the relay's ground circuit. The voltmeter should now read 0 volts. If it does, the relay is good. If voltage is still present, the relay is faulty.
STEP 3	Connect the positive voltmeter lead to the power input terminal. If near battery voltage is not measured there, the relay is faulty. If it is, continue testing.
STEP 4	Connect the positive lead to the control terminal. If near battery voltage is not measured there, check the circuit from the battery to the relay. If it is, continue testing.
STEP 5	Connect the positive lead to the relay ground terminal. If more than 1 volt is present, the circuit has a poor ground.

If the relay terminals are not accessible, remove the relay from its mounting. Use an ohmmeter to test for continuity between the relay coil terminals. If the meter indicates an infinite reading, replace the relay. If there is continuity, use a pair of jumper wires to energize the coil. Check for continuity through the relay contacts. If there is an infinite reading, the relay is faulty. If there is continuity, the relay is good, and the circuits need to be tested.

Check the service manual for resistance specifications for the relay's coil, and compare your readings to them. Low resistance indicates that the coil is shorted. If the coil is shorted, the transistors and/or driver circuits in the computer could be damaged due to excessive current flow.

Stepped Resistors

The best way to test a stepped resistor is with an ohmmeter. Remove the resistor, and connect the ohmmeter leads to the two ends of the resistor. Compare the readings against specifications. Make sure the ohmmeter is set to the correct scale for the anticipated amount of resistance.

A stepped resistor can also be checked with a voltmeter. Measure the voltage after each part of the resistor block and compare the readings to specifications.

Variable Resistors

A common way to test a variable resistor is with an ohmmeter; however, it can also be checked by observing the output voltage. To test a rheostat, identify the input and output terminals, and connect an ohmmeter across them. Rotate the control while observing the meter. If the resistance values do not match specifications, or if there is a sudden change in resistance as the control is moved, the unit is faulty. When using a voltmeter, the readings should be smooth and consistent as the control is moved.

To test a potentiometer, connect an ohmmeter across the resistor. The readings should be within the range listed in the specifications. Then move the leads to the input and output of the resistor. Manually change the resistance. The readings should sweep evenly and consistently within the specified resistance values. The condition of a potentiometer can also be checked with a voltmeter or a lab scope. Compare the voltage changes to the specifications.

Wiring

Wire insulation should be in good condition. Broken, frayed, or damaged insulation that exposes live wires can cause shorts **(Figure 17–52)**. These conditions

FIGURE 17–52

Broken, frayed, or damaged insulation that exposes live wires can cause shorts.

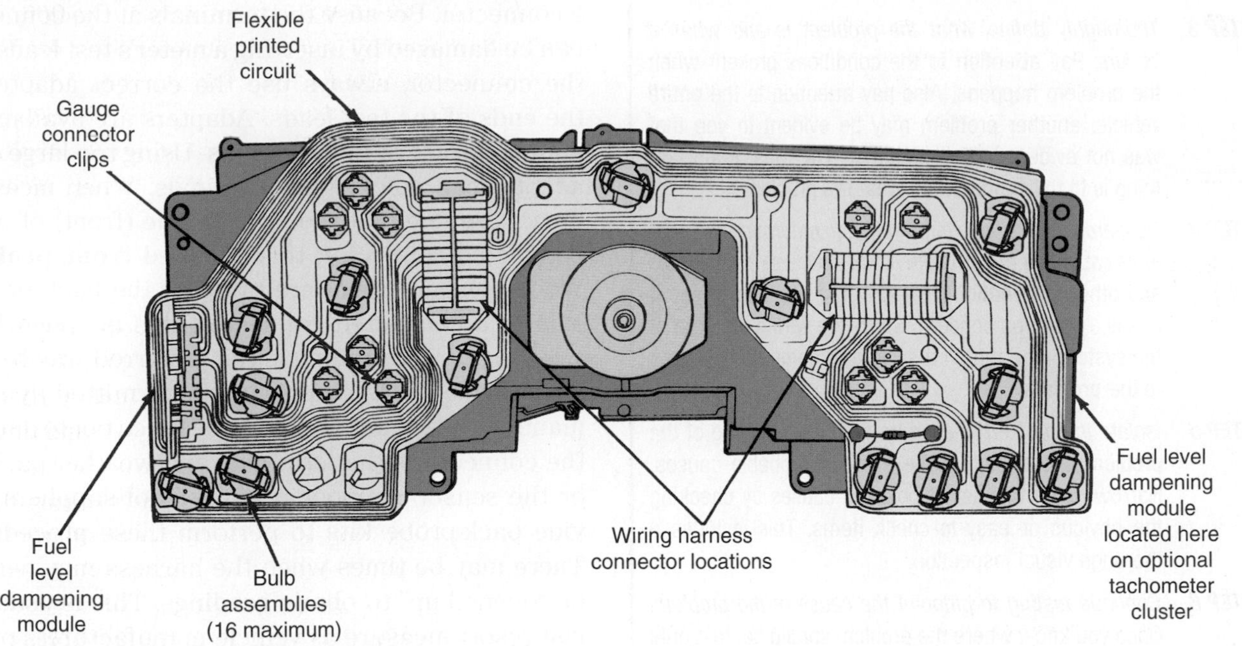

FIGURE 17–53
A typical printed circuit board.

can also create a safety hazard. Replace all wires that have damaged insulation.

When checking a circuit, make sure to check the ground connections, including the ground strap from the engine or other component to the chassis. An engine ground is typically a braided, flat cable. A bad ground cable can cause problems in many different circuits.

The best way to check a wire is to check the voltage drop across it. If the wire is in good shape, there should be very little or no voltage drop.

Wires are commonly grouped into a harness. A single-plug harness connector may form the connecting point for many circuits. Harnesses and harness connectors help organize the vehicle's electrical system and provide a convenient starting point for testing many circuits. Most major wiring harness connectors are located in the vehicle's dash or firewall area.

Printed Circuits

Late-model vehicles use flexible printed circuits boards **(Figure 17–53)**. Printed circuit boards are not serviceable and, in some cases, not visible. When these boards fail, the entire unit is replaced.

The following precautions should be observed when working with a printed circuit:

- Never touch the surface of the board. Dirt, salts, and acids on your fingers can etch the surface and set up a resistive condition.
- The copper conductors can be cleaned with a commercial cleaner or by lightly rubbing a pencil eraser across the surface.
- A printed circuit board is easily damaged because it is very thin. Be careful not to tear the surface, especially when plugging in connectors or bulbs.

TROUBLESHOOTING CIRCUITS

When troubleshooting an electrical problem in any system, it is very important that a logical approach is taken. Making assumptions or jumping to conclusions can be very expensive and a total waste of time. The basic steps for diagnosis given in Chapter 2 should be followed. Here they are again, modified to fit electrical problems:

PROCEDURE

To troubleshoot an electrical problem, follow these steps:

STEP 1 *Gather information about the problem.* From the customer, find out when and where the problem happens and what exactly happens.

STEP 2 *Verify that the problem exists.* Take the vehicle for a road test, or check the components of the customer's concerns and try to duplicate the problem, if possible.

(continued)

STEP 3 *Thoroughly define what the problem is and when it occurs.* Pay attention to the conditions present when the problem happens. Also pay attention to the entire vehicle; another problem may be evident to you that was not evident to the customer. The most important thing is to fully understand what the problem is.

STEP 4 *Research all available information to determine the possible causes of the problem.* Look at all service bulletins and other information related to the problem to see if this is a common concern. Study the wiring diagram of the system, and match a system or some components to the problem.

STEP 5 *Isolate the problem.* Based on an understanding of the problem and circuit, make a list of probable causes. Narrow down this list of possible causes by checking the obvious or easy-to-check items. This includes a thorough visual inspection.

STEP 6 *Continue testing to pinpoint the cause of the problem.* Once you know where the problem should be, test until you find it! Begin testing to determine whether the most probable cause is the problem. If this is not the cause, move to the next most probable cause. Continue this until the problem is solved.

STEP 7 *Locate and repair the problem, then verify the repair.* Once you have determined the cause, make the necessary repairs. Never assume that your work solved the original problem. Operate all features of the circuit to be sure that the original problem has been corrected and that there are no other faults in the circuit.

Troubleshooting Logic

Remember that there are three basic types of electrical problems. Knowing the type of problem that is causing the customer's concerns will dictate what tests should be conducted. If something does not work, the problem is most likely caused by a short or an open. If the fuse for that circuit is blown, the problem is a short. If the fuse is good, the problem is an open. If a part does not work correctly, such as a dim light bulb, the problem is high resistance.

Quick voltage checks will also help define the problem. Check for voltage at the part that is not working correctly. If source voltage is present at the part, the part is bad or the ground circuit is faulty. If less than source voltage is measured at the part, there is a fault in the power feed to the part. Also measure the voltage drop across the part; this can indicate a problem with the part. If a faulty part is suspected, it should be checked or replaced.

When making any checks with a meter, follow all safety precautions. Try to take all measurements at a connector. Because the terminals at the connector can be damaged by inserting a meter's test leads into the connector, always use the correct adapter on the ends of the test leads. Adapters are available to match the size of the terminals. Using too large of an adapter can deform the terminals. When measurements are taken at the mating side (front) of a disconnected connector, this is called **front probing**. When measurements are taken at the back or wire side of a connected connector, this is called **back probing**. Front probing is the preferred way to take measurements. Back probing is permitted by some manufacturers, but care must be taken not to damage the connectors, the terminals and weather packing, or the sensor/component itself. Tool suppliers provide back-probe kits to perform these procedures. There may be times when the harness may need to be "opened up" to obtain readings. This is usually a last-resort measure as vehicle manufacturers prefer the procedures be the least intrusive when dealing with harness testing and repairs. Piercing a wire's insulation could result in damage to the copper strands, so peeling back a small section of the insulation would be the preferred method. Remember to cover the exposed area with adequate insulation (shrink tube, tape, etc.) to prevent the copper core from corroding.

The key to identifying the exact cause of the problem is limiting all testing to the components and circuits that could be causing the problem. An understanding of the problem, coupled with an understanding of the circuit, will lead to the fault. A wiring diagram will serve as the map to the problem. Your understanding and knowledge will tell you where you want to go, and the wiring diagram will tell you how to get there.

Using Wiring Diagrams

During diagnosis, one of the most important sources of information is a wiring diagram. A wiring diagram shows the relationships of one circuit to the others. Based on an understanding of the diagram, electricity, and how a particular system is designed to work, testing points can be identified. Wiring diagrams are included in the electrical section of most service manuals.

Wiring diagrams contain much information about an electrical circuit **(Figure 17–54)**. They contain a comprehensive look at each circuit with all the connectors, wiring, signal connections (buses) between the devices, and electrical or electronic components of the circuit. The diagrams are

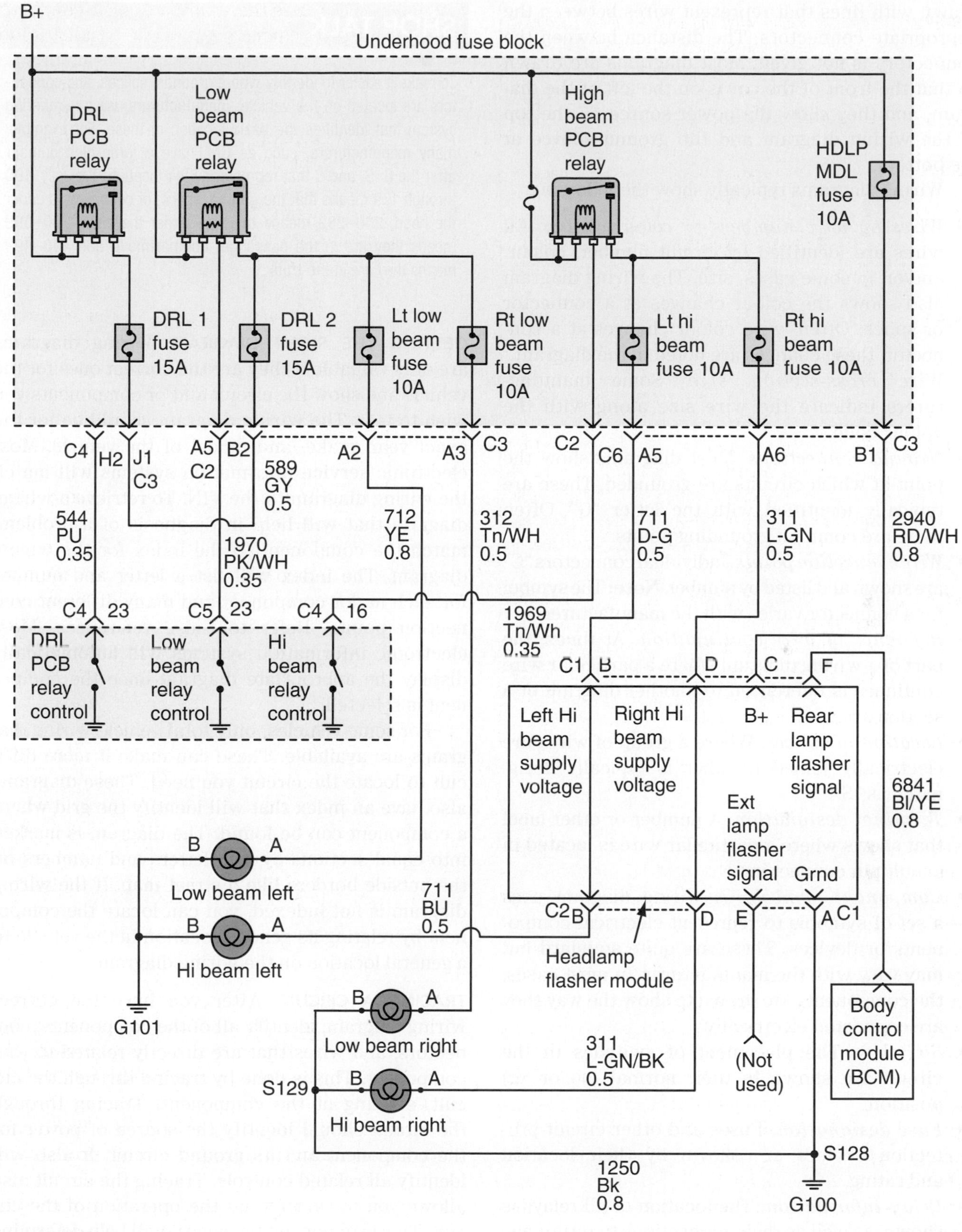

FIGURE 17–54

A wiring diagram of a headlight system.

drawn with lines that represent wires between the appropriate connectors. The distance between the connectors is not given. Most diagrams are drawn so that the front of the car is on the left of the diagram, and they show the power source on the top of the wiring diagram and the ground source at the bottom.

Wiring diagrams typically show the following:

- *Wires by wire numbers or colour-coding.* All wires are identified by circuit number, colour, and/or, in some cases, size. The wiring diagram also shows the colour changes at a connector or splice. Often, wire colour changes at a connector; these changes are noted in the diagram.
- *Wire cross-section size.* Some manufacturers indicate the wire size along with the colour code.
- *Ground connections.* Most diagrams show the point at which circuits are grounded. These are typically identified with the letter "G." Often there are common grounding points.
- *Wire connection points.* Individual connectors "C" are shown and listed by number. **Note:** The symbol for a connector varies with the manufacturer.
- *Reference of wire continuation.* At times, the part of a wiring diagram where a particular wire continues is referenced to another diagram or a section area.
- *Location of splices.* Where a group of wires are electrically joined together is typically designated as "S."
- *Terminal designation.* A number or other label that shows where a particular wire is located in a multipin connector.
- *Component symbols.* A wiring diagram uses a set of symbols to represent electrical components or devices. These are quite standard but may vary with the manufacturer. In most cases, the components are drawn to show the way they are connected electrically.
- *Switches.* The placement of switches in the circuit are shown in their normal (NO or NC) position.
- *Fuse designation.* Fuses and other circuit protection devices are shown by their location and rating.
- *Relay information.* The location of all relays is shown, as well as their connections to terminals.
- *Continuation of circuit.* This may be noted by an arrow that notes where a wire or circuit is continued. The reference is normally to a seemingly unrelated circuit and is noted by an arrow.

SHOP TALK

To make it easier to identify where grounds, splices, and connectors are located on the vehicle, manufacturers use a numbering system that identifies the basic location of these. For example, many manufacturers, such as GM, have a three-digit number after the G, S, and C that represents their location. Typically, 100 through 199 means that the ground, splice, or connector is under the hood, 200–299 means they are under the dash, 300–399 means they are in the passenger compartment, and 400–499 means they are in the trunk.

GETTING THE RIGHT DIAGRAM Wiring diagrams are only valuable if they are the correct ones for the vehicle and show the circuit and/or components you wish to test. The wiring diagram should be for the exact year, make, and model of the vehicle. Most electronic service information systems will match the wiring diagram to the VIN. To retrieve a wiring diagram that will help in diagnosis of a problem, match the component to the index for the wiring diagram. The index will list a letter and number for each major component and many different connection points. Refer to those references. Most electronic information systems will automatically display the appropriate diagram once the component is selected.

For some vehicles, only total vehicle wiring diagrams are available. These can make it more difficult to locate the circuit you need. These diagrams also have an index that will identify the grid where a component can be found. The diagram is marked into equal sections by grid letters and numbers on the outside borders like a street map. If the wiring diagram is not indexed, you can locate the component by relating its general location in the vehicle to a general location on the wiring diagram.

TRACING A CIRCUIT After you have the correct wiring diagram, identify all of the components, connectors, and wires that are directly related to that component. This is done by tracing through the circuit, starting at the component. Tracing through the circuit should identify the source of power for the component and its ground circuit. It also will identify all related controls. Tracing the circuit also allows you to understand the operation of the circuit. This is important because it will help determine where to test the circuit and what to expect at those test points. Remember, all electrical circuits have a power source, a load, and a path to ground. Make sure these are identified.

Tracing the circuit will also simplify complex wiring diagrams. Complex wiring diagrams are made up of many individual circuits. Some are directly interrelated and others are not. When the circuit of concern is pulled out of the wiring diagram, it is much easier to identify its wires and components. It also reduces the chances of being distracted by wires that probably are not the cause of the problem.

After you have traced the circuit, study it and make sure you know how the circuit is supposed to work. Then describe the problem and ask yourself what could cause this. Limit your answers to what is included in your traced wiring diagram. Also limit your answers to the description of the problem. It is wise to make a list of all probable causes of the problem and then number them according to probability. For example, if all of the lights in the instrument panel do not work, it is most probable that the cause is not all bad bulbs. Rather, it is more likely that the fuse or power feed is bad. After you have listed the probable causes, in order of probability, then look at the wiring diagram to identify how you can quickly test to find out if each is the cause.

Tracing the circuit can be done in one of two ways, neither of which involves marking a wiring diagram in a manual. Tracing can be done by drawing out the circuit on a piece of paper. It does not need to be pretty; it just needs to be accurate. Draw the component (can be a simple box) and draw all wires and controls that supply the power and ground; include all controls. Mark each wire with its colour, and make sure to note any change in colour. Also label all controls and components that are included in the circuit.

Tracing can also be done by using highlighters or markers to colour a copy of the wiring diagram. There are many different ways to do this; the one given here is just a suggestion. Developing your own method will work as long as you remain consistent. Find the component of concern in the wiring diagram and outline it in yellow. Follow the power wires toward the power source. The source can also supply power to other components, but ignore them for the time being. Trace the power wire leading to the component to the point where it connects to a control or load. Colour that wire red. That wire should have source voltage. If the control is an on–off switch, source voltage will be present when the switch is closed; colour the wire leading from the control to the component orange. If the control will pass a variable amount of voltage to the component, colour the output wire green. Now look at the ground side of the component. If the path to ground is direct with no control, colour that wire black. If the ground path has a control, colour the wire to the control blue and the wire from the output of the control to ground black.

Colouring the circuit as described allows for a simple reference as to what to expect at each wire in the circuit. The red wire(s) should always have source voltage. The orange wire will have source voltage only when the control is closed. The voltage on the green wire will vary with changes at the control. The black wire should have 0 volts at all times. However, the blue wire should have source voltage when the control is open and 0 volts when it is closed. Your testing should be based on this logic. Use the wiring diagram to identify the test points. **Figure 17–55** shows the circuit of the right low-beam headlight. This is the same diagram as used in Figure 17–54, but only the parts of the circuit that could cause a problem with one headlamp are noted. Note that all wires and components that would affect more than that headlamp are not traced. Doing this certainly simplifies diagnosing the circuit.

If the power source for the component feeds more than one component or the ground is shared by others, check the operation of those components. If they operate normally, you know that the common power and ground circuits are good. Therefore, the problem must be between the common points. Likewise if the other components do not work correctly, you know that the problem is within the common part of the circuit.

TESTING FOR COMMON PROBLEMS

It would take thousands of pages to describe all of the possible combinations of electrical problems. But you are in luck. All problems can be boiled down to one of three problems: testing for opens, testing for shorts, and testing for unwanted resistance. Identifying which one you are looking for will define what tests you need to conduct.

Testing for Opens

An open is evident by an inoperative component or circuit. Study the wiring diagram for the component. Begin your testing at the most accessible place in the circuit and work from there. Check for voltage at the positive side of the load. If there are 0 volts, move to the output of the control **(Figure 17–56)**. If there are at least 10.5 volts, the open is between the control and the load. If the reading is 10.5 volts or higher, check the ground side of the load. If the voltage there is 1 volt or lower and the load does not work, the load is

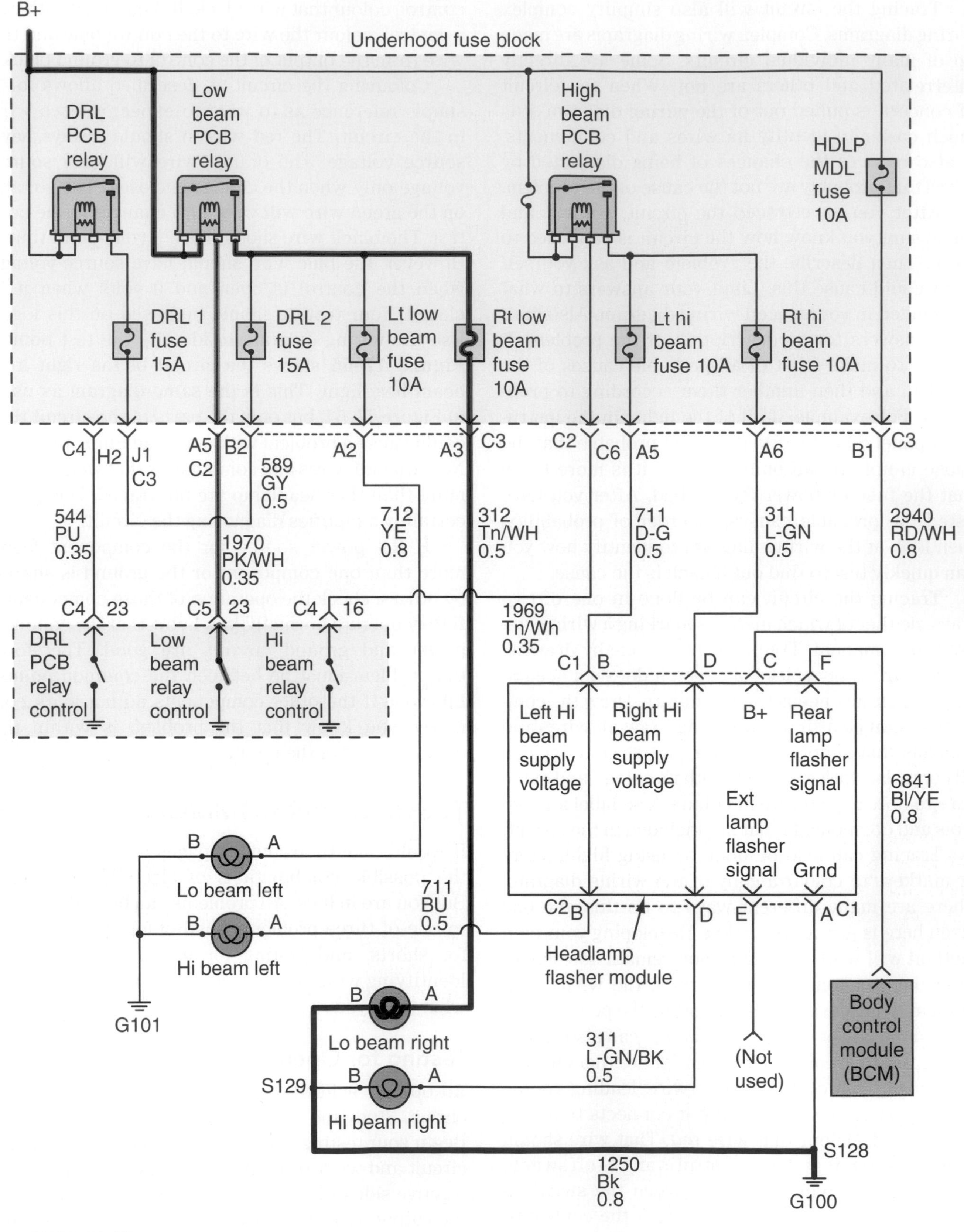

FIGURE 17–55

Figure 17–54 with the right low-beam circuit traced.

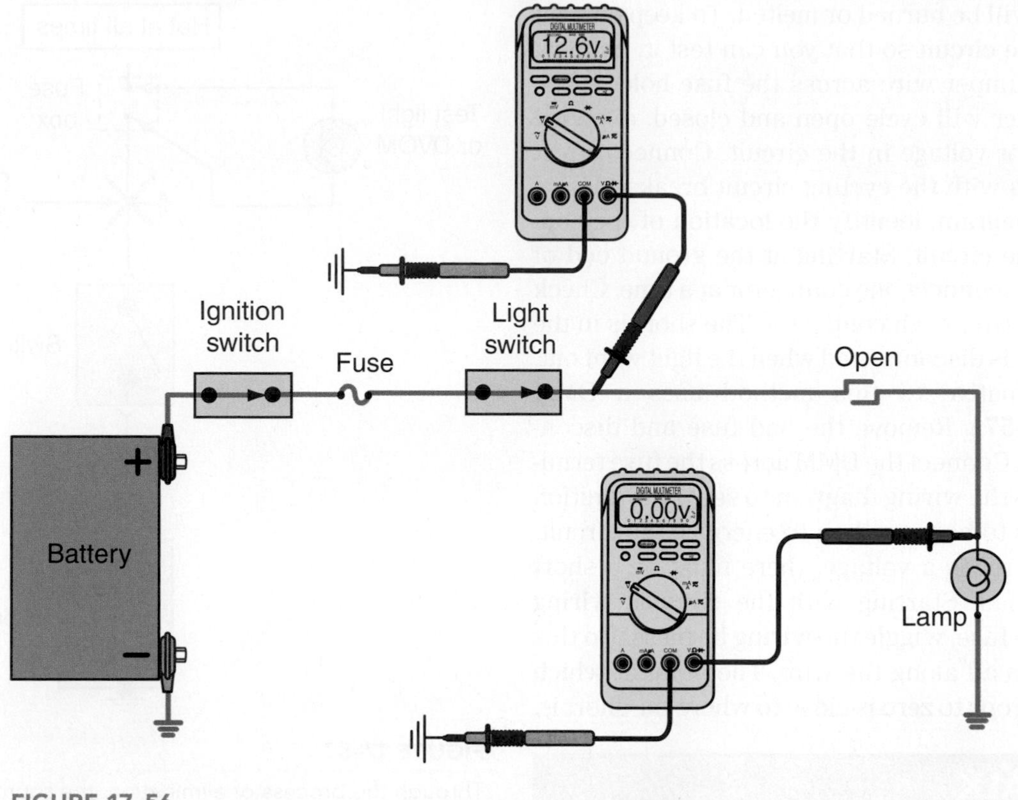

FIGURE 17–56

If you have 0 volts at the load, test the output of the switch. If there is power there, the open is between the switch and the load.

bad. If the voltage at ground is greater than 1 volt, there is excessive resistance or an open in the ground circuit. If the voltage at the positive side of the load is less than 10.5 volts but above 0 volts, move the positive lead of the voltmeter toward the battery, testing all connections along the way. If 10.5 volts or more are present at any connector, there is an open or high resistance between that point and the point previously checked. If battery voltage is present at the ground of the load, there is an open in the ground circuit. Use a jumper wire to verify the location of the open.

Testing for Shorts

Use an ohmmeter to check for an internal short in a component. If the component is good, the meter will read the specified resistance or at least some resistance. If it is shorted, it will read lower than normal or zero resistance. Also, if the component has more than two terminals or pins, check for continuity across all combinations of these. Refer to the wiring diagram to see where there should be continuity. Any abnormal readings indicate an internal short.

When a fuse is blown, this probably is due to a wire-to-wire short or a short to ground. To test these circuits, a special jumper wire with a circuit breaker should be used as a substitute for the blown fuse. The jumper wire is fit with a 10- or 20-amp self-resetting circuit breaker. This will allow for testing the circuit without causing damage to the wires and components in the circuit. Some tool companies include a buzzer in-line with the breaker.

When a wire-to-wire short is suspected, check the wiring diagram for all of the affected components. Identify all points where the affected circuits share a connector. Check the circuit protection devices for the circuits. High current due to the short will cause this. Check the wiring for signs of burned insulation and melted conductors. If a visual inspection does not identify the cause of the short, remove one of the fuses for the affected circuits. Install the special jumper wire across the fuse holder terminals. Activate that circuit and disconnect the loads that should be activated by the switch. This will create open circuits and, normally, current will not flow. If sound is coming from the buzzer in the jumper wire, current is still flowing somewhere in the circuit. Disconnect all connectors in the circuit one at a time. If the buzzer stops when a connector is disconnected, the short is in that circuit.

If the problem is a short to ground, the circuit's fuse or other protection device will be open. If the circuit is not protected, the wire, connector, or component will be burned or melted. To keep current flowing in the circuit so that you can test it, connect the special jumper wire across the fuse holder. The circuit breaker will cycle open and closed, allowing you to test for voltage in the circuit. Connect a test light in series with the cycling circuit breaker. Using the wiring diagram, identify the location of the connectors in the circuit. Starting at the ground end of the circuit, disconnect one connector at a time. Check the test light after each connector. The short is in the circuit that was disconnected when the light went out.

An alternative to this method uses a DMM **(Figure 17–57)**. Remove the bad fuse and disconnect the load. Connect the DMM across the fuse terminals. Refer to the wiring diagram to see if the ignition switch needs to be turned on to energize the circuit. If the meter reads a voltage, there must be a short before the load. Starting with the circuit's wiring closest to the fuse, wiggle the wiring harness. Do this in short steps all along the wire. The point at which the voltage drops to zero is close to where the short is.

SHOP TALK

Many manufacturers do not recommend using a circuit breaker as a substitute for the fuse during testing; rather, a sealed beam headlight (**Figure 17–58**) is connected with jumper wires across the fuse holder. The headlight serves a load and limits the current in the circuit. The headlight will light as long as current is flowing through the circuit.

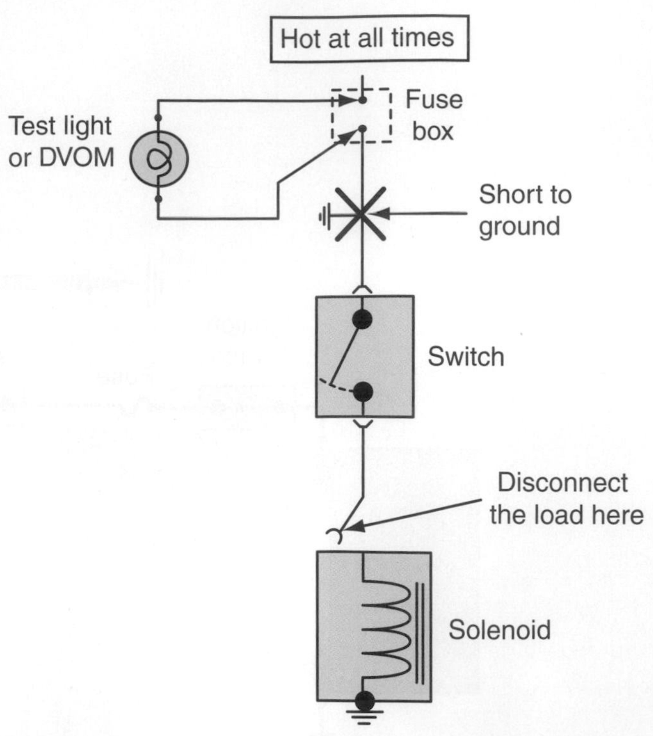

FIGURE 17–57

Through the process of elimination, the source of a short can be found.

SHORT DETECTOR Modern short detectors use a frequency generator that pulses into the system and a receiver that can pinpoint not only shorts but also open circuits **(Figure 17–59)**. Some technicians use a compass or Gauss gauge to find the location of a

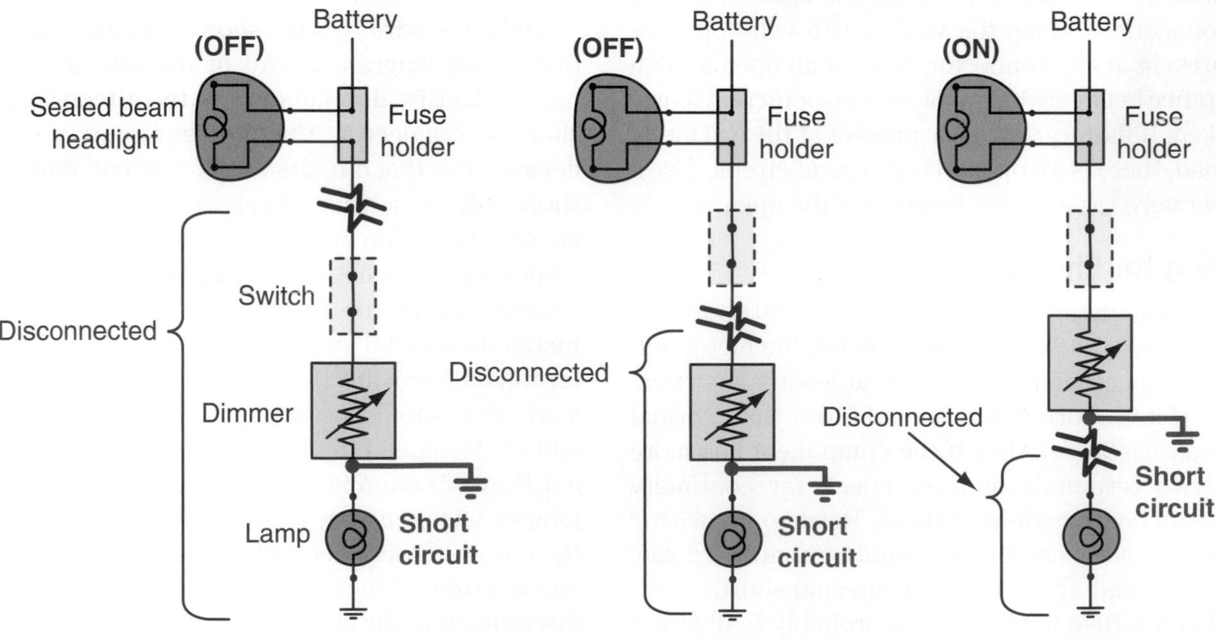

FIGURE 17–58

Many manufacturers do not recommend using a circuit breaker as a substitute for the fuse during testing; rather, a sealed beam headlight is connected with jumper wires across the fuse holder.

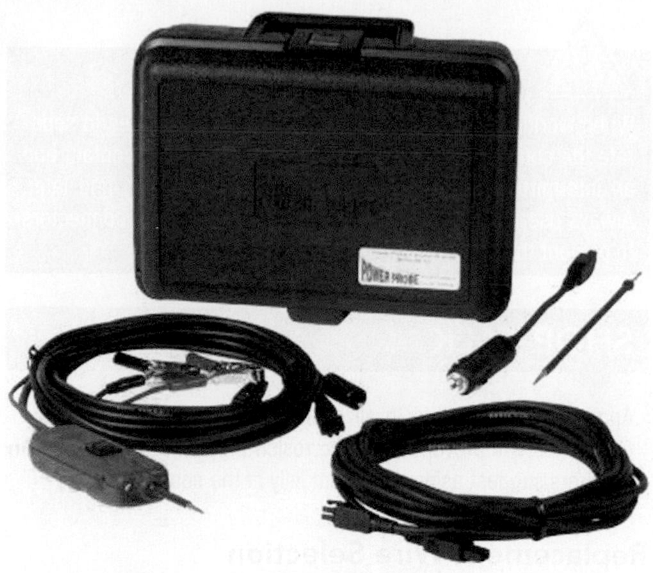

FIGURE 17–59

A frequency generator style of short detector.

Snap-on Tools Company

short **(Figure 17–60)**. A magnetic field is formed around a current-carrying conductor and a compass reacts to magnetic fields. The shorted circuit will have high current; therefore, a large magnetic field will be formed around the shorted circuit. With the wiring diagram and other service information, locate the routing of the wires in the affected circuit. Connect the jumper wire with a circuit breaker across the fuse holder for the blown fuse. Position the compass over or close to the wiring harness. The magnetic field in the wire will cause the compass's needle to move away from its north position. As the circuit breaker cycles, the needle will fluctuate. As the compass is slowly moved across the wire, it will continue to fluctuate until it passes the point where the short is. To find the exact location of the short, inspect the wire in that area. Look for signs of overheating and broken, cracked, exposed, or punctured wires.

Testing for Unwanted Resistance

High-resistance problems are typically caused by corrosion on terminal ends, loose or poor connections, or frayed and damaged wires. Carefully inspect the affected circuit.

Whenever excessive resistance is suspected, both sides of the circuit should be checked. Begin by checking the voltage drop across the load **(Figure 17–61)**. This should be close to battery voltage unless the circuit contains a resistor located before the load. If the voltage is less than desired, check the voltage drop across the circuit from the switch to the load. If the voltage drop is excessive, that part of the circuit contains the unwanted resistance. If the voltage drop was normal, the high resistance is in the switch or in the circuit feeding the switch.

Measure the voltage drop across the switch. If the voltage drop is excessive, then the switch is likely faulty. If the voltage drop is within acceptable limits, then the high resistance could be in the feed wire to the switch.

When measuring voltage before and after the load, place the red DMM lead alternately on either side of the load and the black lead to a known ground; if voltage is present on both sides of the load, then the ground circuit has excessive resistance. The ground fault could then be verified by installing a fused jumper wire to the ground side of the load.

To locate the high resistance, connect the red voltmeter lead to the ground side of the load and the black lead to the grounding point for the circuit. If the voltage drop is normal, the problem is the grounding

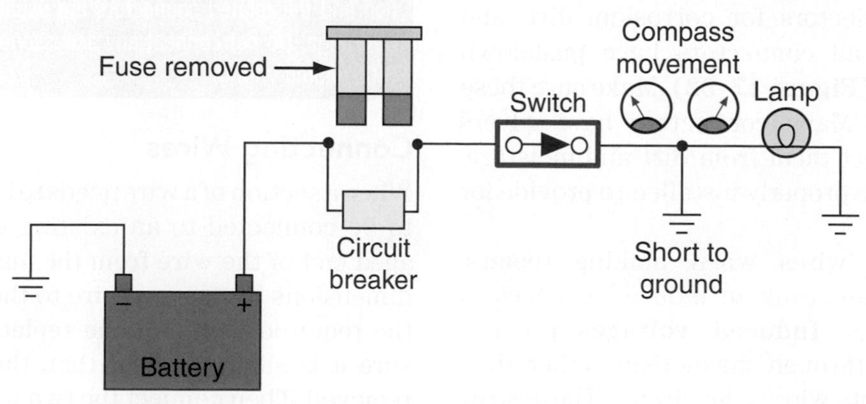

FIGURE 17–60

Use a compass to locate the cause of a short.

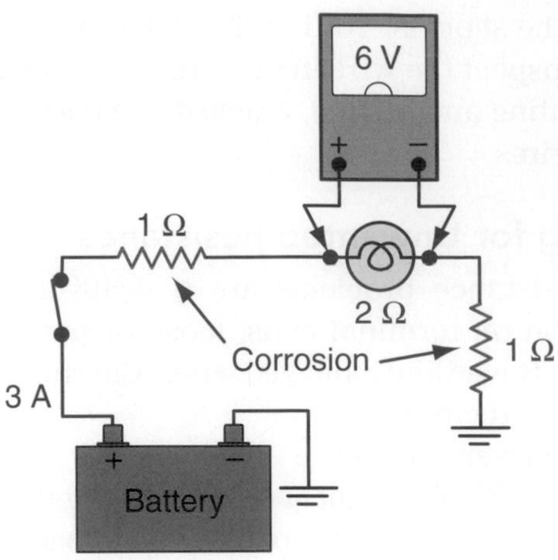

FIGURE 17–61

A simple light circuit with unwanted resistance. Notice the reduced voltage drop across the lamp and the reduced circuit current.

point. If the voltage drop is excessive, move the black meter lead toward the red. Check voltage drop at each step. Eventually, you will read a high-voltage drop at one connector and then a low-voltage drop at the next. The point of high resistance is between those two test points. If the voltage drop is normal, the high resistance is in the switch or in the circuit feeding the switch.

CONNECTOR AND WIRE REPAIRS

Many electrical problems can be traced to faulty wiring or connections. Loose or corroded terminals; frayed, broken, or oil-soaked wires; and faulty insulation are the most common causes. Wires, fuses, and connections should be checked carefully. Keep in mind that a wire's insulation does not always appear to be damaged when the wire inside is broken. Also, a terminal may be tight but still may be corroded.

Check all connectors for corrosion, dirt, and looseness. Nearly all connectors have pushdown release-type locks **(Figure 17–62)**. Make sure these are not damaged. Many connectors have covers over them to protect them from dirt and moisture. Make sure these are properly installed to provide for that protection.

Never reroute wires when making repairs. Rerouting wires can result in induced voltages in nearby components. **Induced voltages** produce unwanted signals through magnetism rather than from the components within the circuit. These stray voltages can interfere with the function of electronic circuits.

WARNING!

When working with connectors, never pull on the wires to separate the connectors. This can create an intermittent contact and an intermittent problem that can be very difficult to find later. Always use the special tools designed for separating connectors to prevent this problem.

SHOP TALK

Apply dielectric grease to all connections before you assemble them. This will prevent future corrosion problems. Some manufacturers suggest using petroleum jelly at the connectors.

Replacement Wire Selection

Often, electrical problems require the replacement of a wire or two. It is important that this is done in a way that corrects the original problem but does not create a new problem. All replacement wires should be of the same size or larger than the original. If adding an accessory, the new wire should be large enough to ensure safe and reliable performance. However, overly large wires add weight and expense, and add to the difficulty of splicing wires together. If the wire is too small, an unwanted voltage drop can occur. The two factors that should always be considered when determining the correct size of a wire are the total circuit amperage and the total length of wire (resistance increases with length) used in each circuit, including the ground. Allowance for the circuits, including grounds, has been computed in **Table 17–2**.

WARNING!

Supplemental restraint system (SRS) air bag harness insulation and the related connectors are usually colour-coded yellow or orange. Do not connect any accessories or test equipment to SRS-related wiring.

Connecting Wires

When a section of a wire needs to be replaced, it needs to be connected to an existing wire. Cut the damaged end of the wire from the main wire. Match the dimensions of the new wire to the old one. Measure the required length of the replacement wire; make sure it is slightly longer than the section that was removed. Then connect the two wires together. After the connection is made, the joint should be wrapped with tape or covered with heat shrink tubing.

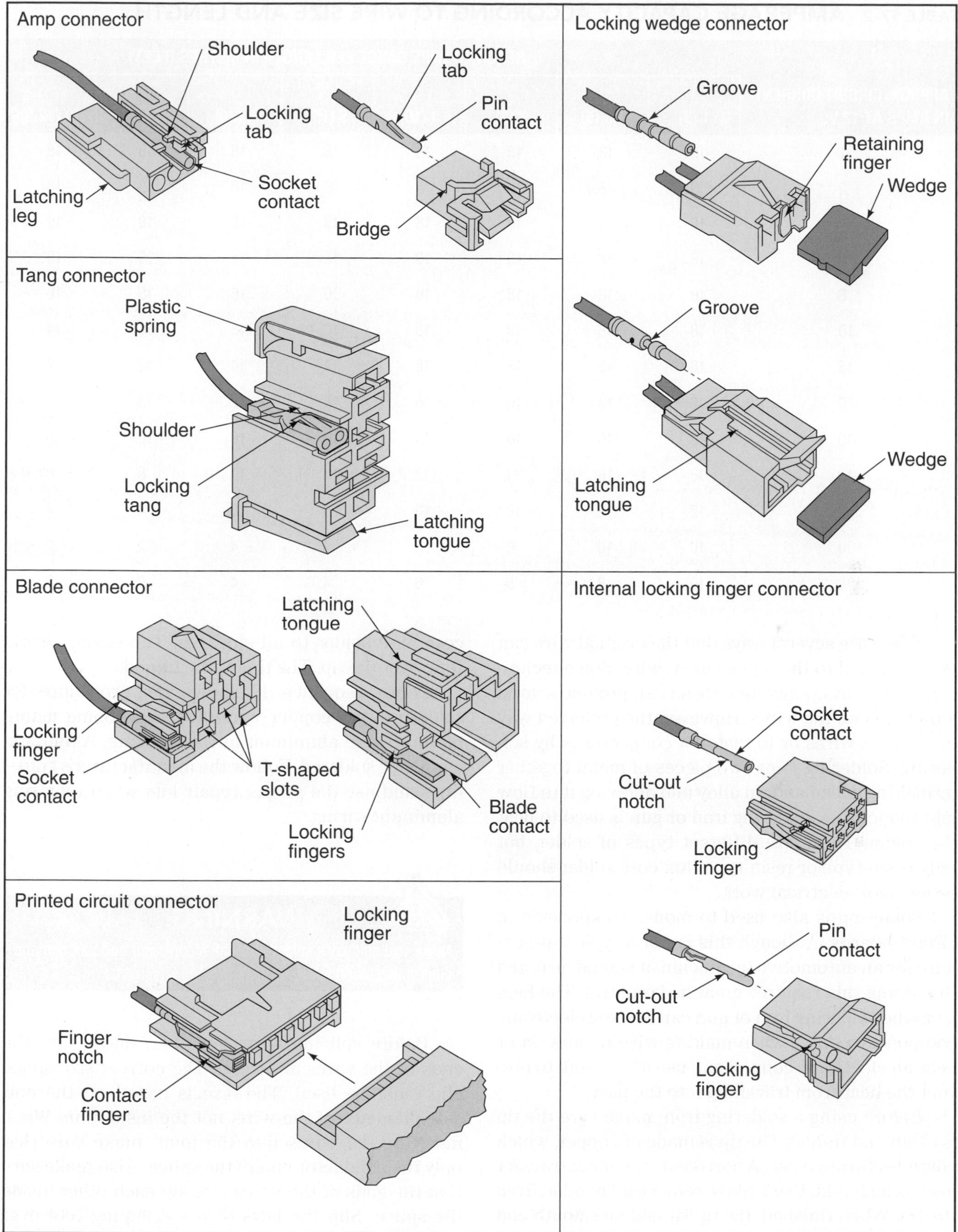

FIGURE 17–62

Different multiple-wire shell connectors and their locking mechanisms.

TABLE 17–2 AMPERAGE CAPACITY ACCORDING TO WIRE SIZE AND LENGTH

	REQUIRED WIRE GAUGE PER LENGTH IN METRES (FEET)							
APPROX. CIRCUIT CURRENT IN AMPS AT 12 V:	1 (3)	1.5 (5)	2 (7)	3 (10)	5 (15)	6 (20)	8 (25)	10 (32.8)
1	18	18	18	18	18	18	18	18
2	18	18	18	18	18	18	18	18
4	18	18	18	18	18	18	18	16
6	18	18	18	18	18	18	16	16
8	18	18	18	18	16	16	16	16
10	18	18	18	18	16	16	16	14
15	18	18	18	18	14	14	12	12
20	18	18	16	16	14	12	10	10
30	18	16	16	14	10	10	10	10
40	18	16	14	12	10	10	8	8
50	12	12	10	10	6	6	4	4
100	10	10	8	8	4	4	2	2
200	10	8	8	6	4	4	2	2

There are several ways that the original wire can be connected to the replacement wire. Some technicians use butt connectors; these can provide a good joint between the wires. However, the preferred way to connect wires or to install a connector is by soldering. Soldering joins two pieces of metal together by melting a lead and tin alloy and allowing it to flow into the joint. A soldering iron or gun is used to heat the solder. There are different types of solder, but only rosin-type or resin-type flux core solder should be used for electrical work.

Soldering is also used to mount components in circuit boards. Although this is not a typical procedure for an automotive technician, it should be noted that doing this requires great heat control. The heat from the soldering iron or gun can destroy electronic components. Also, when making wire repairs on or near an electronic component, use a heat sink to prevent the heat from travelling into the part.

Before using a soldering iron, make sure the tip is clean and tinned. The tip is made of copper, which corrodes through use. A corroded tip cannot transfer heat as it should. Use a file to remove all residue from the tip. When finished, the tip should be smooth and flat. Turn the iron on and allow it to heat. Then dip the hot tip into some soldering rosin flux. Remove the tip from the flux and immediately apply rosin core wire solder to all surfaces. The solder should flow over the tip. The tip is now tinned.

Photo Sequence 12 shows the procedure for soldering two copper wires together. Some manufacturers use aluminum in their wiring. Aluminum cannot be soldered. Follow the manufacturer's guidelines and use the proper repair kits when repairing aluminum wiring.

WARNING!

Never use acid core solder. It creates corrosion and can damage electronic components.

Before splicing wires together, make sure the ends of the wires are clean. The correct size splice clip must be used. The size is based on the outside diameter of the wire, not the insulation. When inserting the wires into the joint, make sure that only the conductor enters the splice. Also make sure that the ends of the wires overlap each other inside the splice. Slip the jaws of the crimping tool over the centre of the splice. Squeeze the crimping tool until the contact points of the crimper's jaws make contact with the splice. Then check the placement of

SOLDERING TWO COPPER WIRES TOGETHER

P12–1 Tools required to solder copper wire: 100-watt soldering iron, 60/40 rosin core solder, crimping tool, splice clip, heat shrink tube, heating gun, and safety glasses.

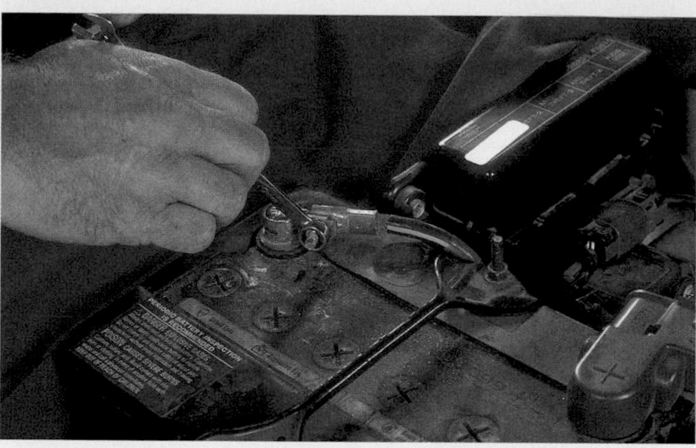

P12–2 Disconnect the fuse that powers the circuit being repaired. Note: If the circuit is not protected by a fuse, disconnect the ground lead of the battery.

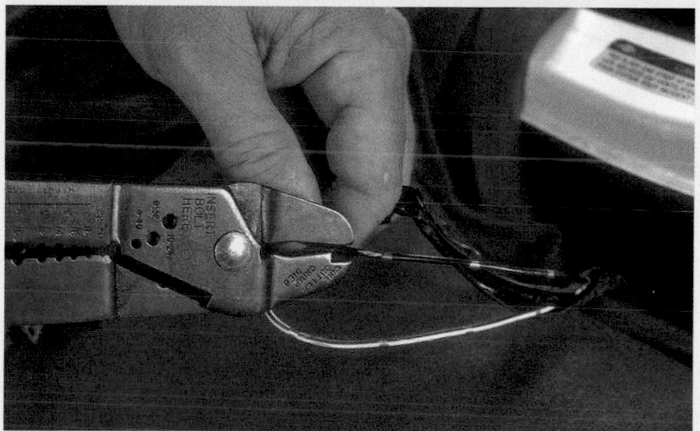

P12–3 Cut out the damaged wire.

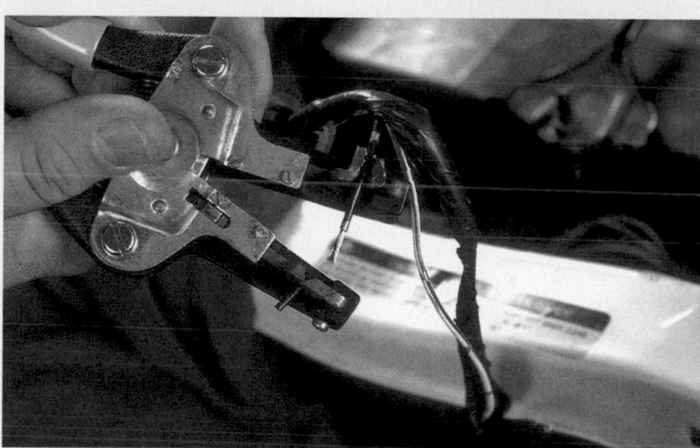

P12–4 Using the correct size stripper, remove about 1.25 cm (½ in.) of the insulation from both wires.

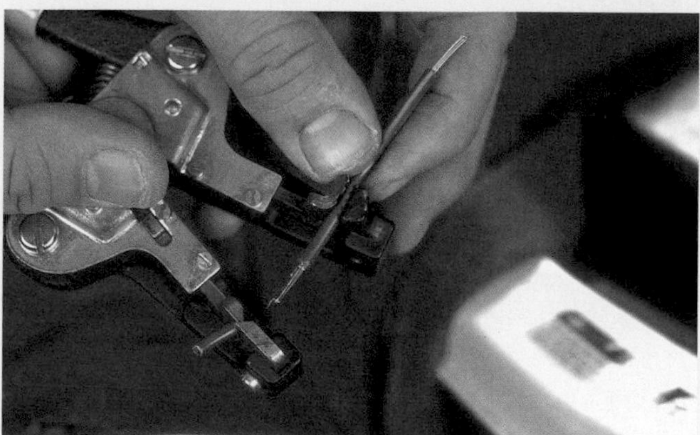

P12–5 Now remove about 1.25 cm (½ in.) of the insulation from both ends of the replacement wire. The length of the replacement wire should be slightly longer than the length of the wire removed.

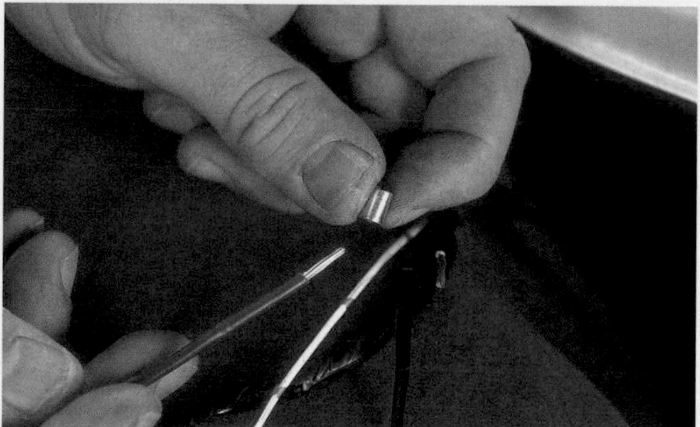

P12–6 Select the proper size splice clip to hold the splice.

SOLDERING TWO COPPER WIRES TOGETHER *(continued)*

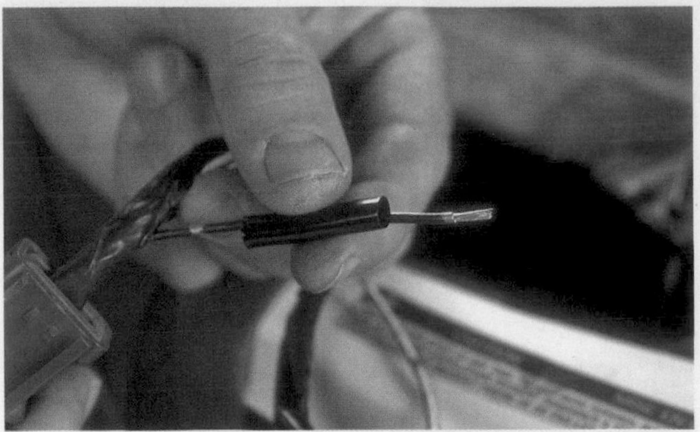

P12–7 Place the correct size and length of heat shrink tube over the two ends of the wire.

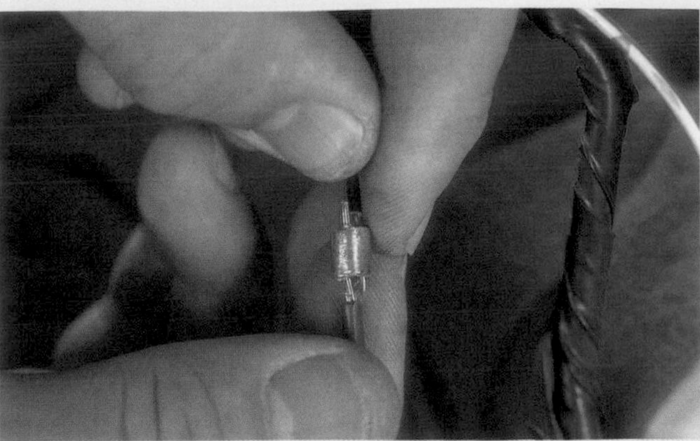

P12–8 Overlap the two splice ends, and centre the splice clip around the wires, making sure that the wires extend beyond the splice clip in both directions.

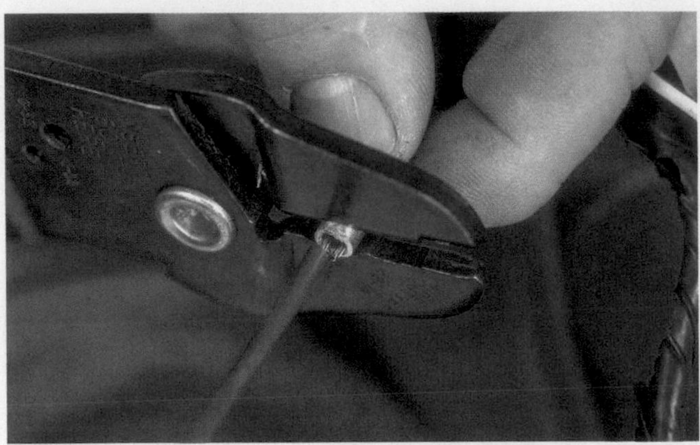

P12–9 Crimp the splice clip firmly in place.

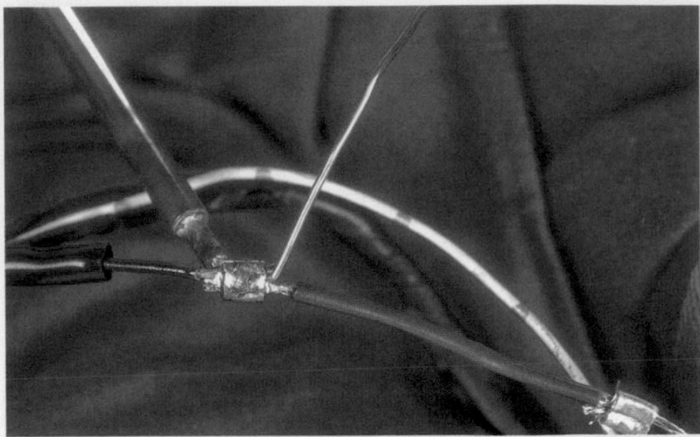

P12–10 Apply the tip flat of the soldering iron against the splice to heat it. At the same time, apply solder to the opening of the clip. Do not apply solder to the iron. The iron should be 180° away from the opening of the clip. As the splice and wires heat, the solder will flow through the splice.

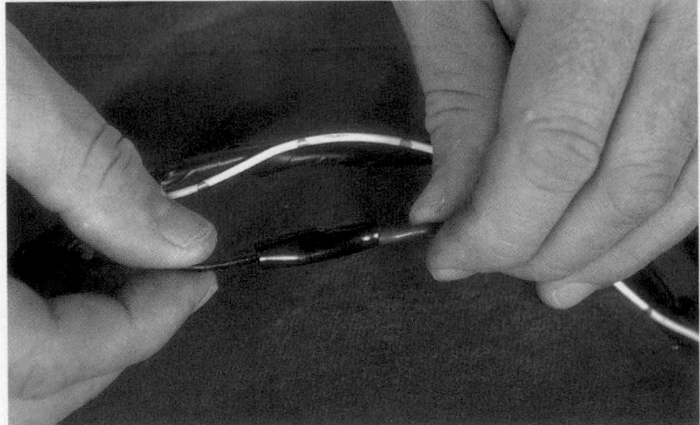

P12–11 Place the hot soldering iron in its stand and unplug it. After the solder cools, slide the heat shrink tube over the splice.

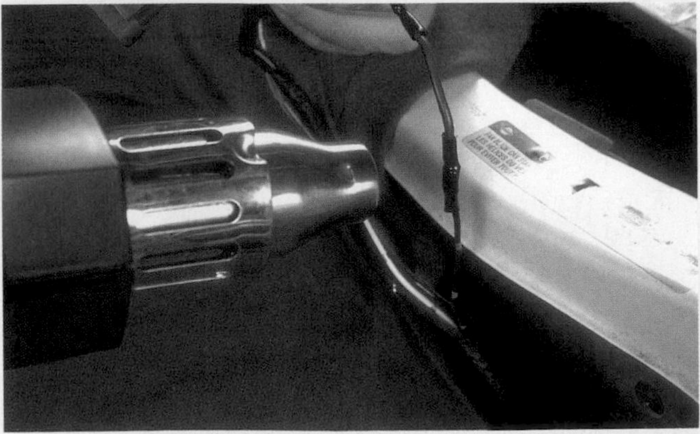

P12–12 Heat the tube with the hot air gun until it shrinks around the splice. Do not overheat the heat shrink tube.

the wires inside the splice, and apply pressure to the crimping tool to form a tight crimp.

SHOP TALK

Rather than use a splice, some technicians twist the wire ends tightly together before soldering the joint. When doing this, it is important to realize that the solder does not provide for a mechanical joint. Therefore, it is important that the twisting provides a secure joint before soldering.

After a joint has been made, it must be insulated. This can be done with heat shrink tubing or tape. When using tape, place one end on the wire about 2.5 cm (1 in.) from the joint. Tightly wrap the tape around the wire. As the tape is being wrapped, about one-half of the previous wrap should be covered by the tape as it completes one turn around the wire. Once the wrapping has reached 2.5 cm (1 in.) beyond the joint, cut the tape. Firmly press on the tape at that end to form a good seal.

When using heat shrink tubing, make sure the tubing is slightly larger than the diameter of the splice. Cut a length of the tubing so that it is longer than the splice. Before joining the wires together, slip the tubing over one of the wires. Proceed to make the joint. After the wires are connected, move the shrink tubing over the splice. Use a heat gun and heat the tubing until it shrinks tightly around the splice. The tubing will only shrink a certain amount; therefore, do not continue to heat it after it is in place. Doing this can melt the tubing and/or the insulation of the wire.

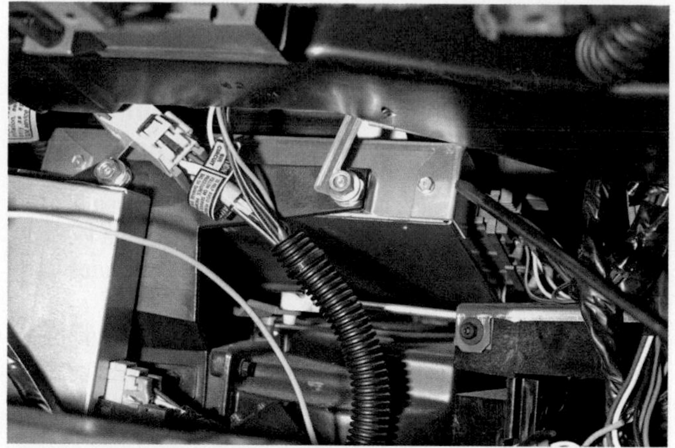

FIGURE 17–63

The wiring for SRS systems has yellow connectors, and all precautions and service procedures for dealing with these should be followed.

WARNING!

Always follow the manufacturer's wiring and terminal repair procedures. On some components and circuits, manufacturers recommend complete wiring harness replacement rather than making repairs to the wiring. For most vehicles, SRS air bag harness components, including wiring, insulation, and connectors, should not be repaired **(Figure 17–63)**. Any SRS harness damage requires replacement of the related harness.

Wire Terminals and Connectors

Many different types of connectors, terminals, and junction blocks are used on today's vehicles. In most cases, the type used in a particular application is shown in a wiring diagram. Wire end terminals are used as connecting points for wires. They are generally made of tin-plated copper and come in many shapes and sizes. They may be either soldered or crimped in place. When installing a terminal, select the appropriate size and type of terminal. Be sure it fits the unit's connecting post or prongs and it has enough current-carrying capacity for the circuit.

When a connector needs to be replaced because the original has melted or is otherwise damaged, attempt to replace it with the same type and size. Often, this is difficult because so many different types are used, and parts departments do not have all the various designs available. Normally, the available connectors are based on common shapes with a common number of terminal cavities. Therefore, it is best to use a connector that meets the need; this may mean that some of the terminal cavities are left empty. For example, if the original connector has six terminals but the available replacement has eight, arrange the wire connections to the connector as if the connector had six. This will keep the wires in order for future diagnostics and leave the end pair blank. Of course, to do this, the male and female ends of the connector must be replaced. Sometimes, the replacement connector will require different terminal ends from the original. This requires the replacement of terminal ends on the wires for the male and female connectors.

Replacing a Terminal

Terminal ends are replaced when they are damaged or to accommodate the use of a connector. The replacement process must be done to provide for good continuity and to prevent electrical problems in the future.

PROCEDURE

To replace a terminal end on a wire, follow this procedure:

STEP 1 Use the service information to identify the type of terminal that should be used, the position of the locking clips, and the terminal unlocking procedures for that connector.

STEP 2 Use a small screwdriver or terminal pick **(Figure 17–64)** to unlock the secondary locking device.

STEP 3 Gently push the terminal into the connector and hold it there.

STEP 4 Insert the terminal pick into the connector, and move the locking clip to the unlock position and hold it there.

STEP 5 Carefully pull the terminal from the connector by pulling the wire toward the rear of the connector. Do not use too much force.

STEP 6 Measure the diameter of the wire's insulation with a micrometer or vernier caliper.

STEP 7 Identify the type of terminal, and use the measurement to select the correct size for the replacement terminal.

STEP 8 Select the correct size for the replacement wire.

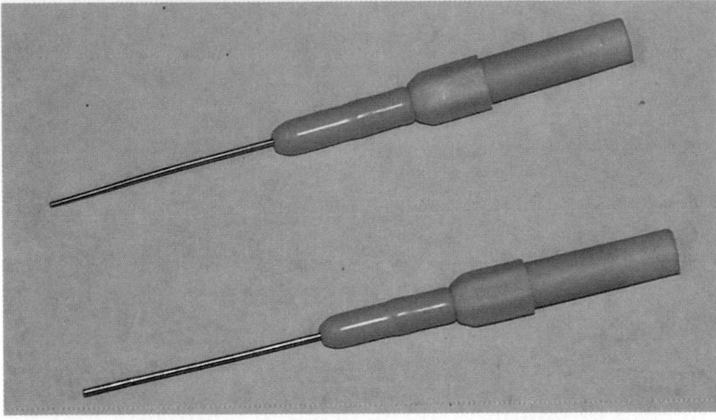

FIGURE 17–64

Terminal picks are used to unlock terminal and connector locks.

STEP 9 Cut the old terminal from the wire in the harness.

STEP 10 Use the old wire as a guide, and cut the replacement wire slightly longer. Be careful when doing this because if the wire is too short, there will be tension on the terminal, splice, or connector, which can lead to an open circuit. If the wire is too long, it may get pinched and cause a short.

STEP 11 Strip the insulation from the wire **(Figure 17–65)** in the harness and both ends of the replacement wire. Normally, 1.25 cm (½ in.) of insulation should be removed.

STEP 12 Make sure the strands of wire are not damaged while removing the insulation.

STEP 13 Place the ends of the wires into the terminal and connectors, and crimp the terminal **(Figure 17–66)**. To get a proper crimp, place the open area of the connector facing toward the anvil of the tool. Make sure the wire is compressed under the crimp.

WARNING!

Do not crimp a terminal with the cutting edge of a pair of pliers. Although this method may crimp the terminal, it also weakens it.

STEP 14 If heat shrink will be used to seal the connections, slip the appropriate length of tubing over the end of the wire that will be spliced.

STEP 15 Install the terminal into the connector. Make sure the locking clip is in the proper position. If it is not, use the terminal pick to gently bend it back to its original shape.

STEP 16 Push the terminal into the connector until a click is heard.

STEP 17 Gently pull on the wire. If the terminal is locked in the connector, it will not move.

STEP 18 Connect both sides of the connector, and secure the secondary locking device.

STEP 19 Tape the new wire to the wiring harness. If the harness is contained in conduit, make sure it is fully enclosed, and tape the outside of the conduit.

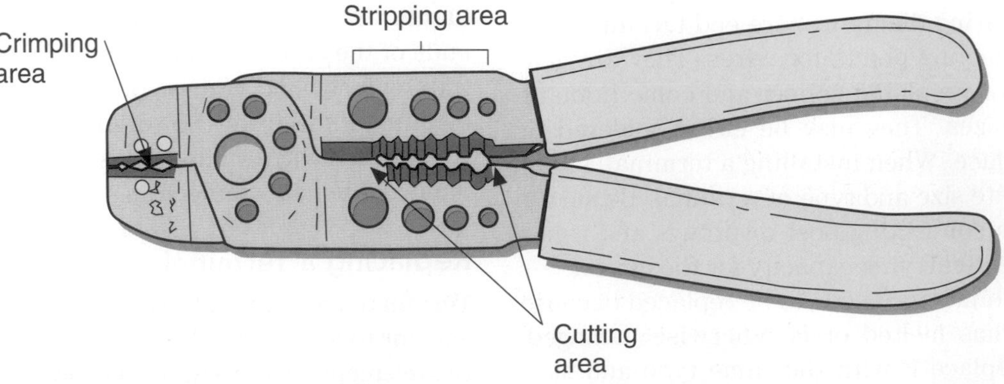

FIGURE 17–65

A typical crimping tool used for making electrical repairs.

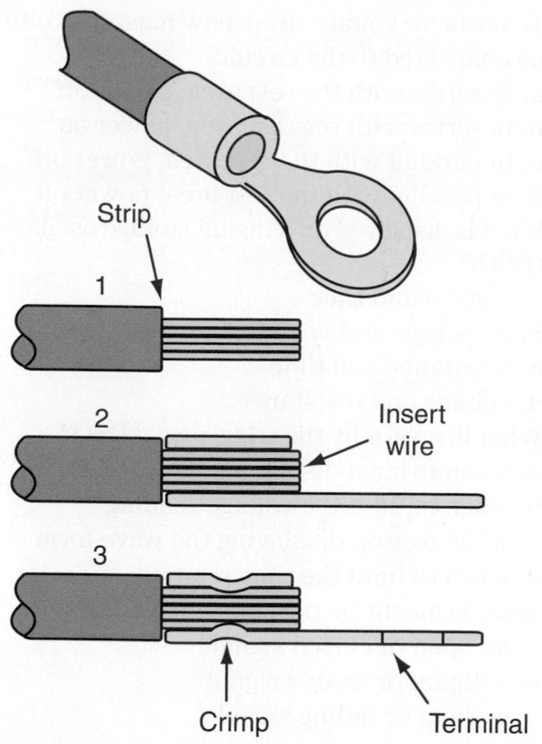

FIGURE 17–66

Placing a wire into and crimping it in a connector.

KEY TERMS

Average responding (p. 506)
Back probing (p. 520)
Digital storage oscilloscope (DSO) (p. 512)
Front probing (p. 520)
Glitches (p. 511)
Graphing multimeter (GMM) (p. 514)
Induced voltages (p. 528)
In-line fuse (p. 516)
Open (p. 493)
Radio frequency interference (RFI) (p. 510)
Root mean square (RMS) (p. 506)
Schematics (p. 496)
Short (p. 493)

SUMMARY

- All electrical problems can be classified as an open, short, or high-resistance problem. Identifying the type of problem will allow for the identification of the correct tests to conduct when diagnosing an electrical circuit.
- Wiring diagrams show where wires are connected, the circuit's components, the colour of the wires' insulation, and sometimes the wire gauge size.
- Voltmeters, ohmmeters, ammeters, and volt/amp meters are used to test and diagnose electrical systems. These are used with jumper wires, test lights, and variable resistors.
- Multimeters are multifunctional and can test DC and AC volts, ohms, and amperes. Some multimeters can also be used to measure engine rpm, duty cycle, pulse width, frequency, and temperature.
- There are two ways that DMMs display AC voltage: RMS and average responding.
- Some DMMs also feature a MIN/MAX function, which displays the maximum, minimum, and average voltage the meter recorded during the time of the test.
- On a lab scope, an upward movement of the trace indicates an increase in voltage, and a downward movement of this trace represents a decrease in voltage. As the trace moves across the screen, it represents a specific length of time.
- To troubleshoot a problem, begin by verifying the customer's complaint. Then operate the system and others to get a complete understanding of the problem. Use the correct wiring diagram, and identify testing points and probable problem areas. Test and use logic to identify the cause of the problem.
- Wiring diagrams are invaluable for diagnostics. Tracing the diagram allows you to think about how the circuit should work and where it should be tested.
- Many automotive electrical problems can be traced to faulty wiring, such as loose or corroded terminals; frayed, broken, or oil-soaked wires; and faulty insulation.
- The preferred way to connect wires or to install a connector is by soldering. Never use acid core solder. It creates corrosion and can damage electronic components.

REVIEW QUESTIONS

1. What would be the result of an open circuit?
a. high current **b.** high resistance
c. high voltage **d.** high wattage

2. What would be the result of a short to ground?
a. high current **b.** low current
c. high voltage **d.** low voltage

3. How is an ammeter connected to a circuit?
a. in parallel with the circuit turned off
b. in series with the circuit turned off
c. in parallel with the circuit turned on
d. in series with the circuit turned on

4. How is an ohmmeter connected to a circuit?
   a. with the circuit turned off
   b. with the circuit turned on
   c. in series with the circuit
   d. with both leads on the same conductor
5. What is indicated by a zero reading on an ohmmeter?
   a. too much resistance for the scale it is on
   b. too little resistance for the scale it is on
   c. the circuit is open at some point
   d. the circuit is shorted to ground
6. What would be the internal impedance on a good-quality digital multimeter?
   a. 10 ohms b. 10 K ohms
   c. 10 M ohms d. 10 G ohms
7. What does an upward movement of the trace on a lab scope indicate?
   a. increased time
   b. increased voltage
   c. increased amperage
   d. increased resistance
8. Which of the following is the most accurate statement about fuse ratings?
   a. The fuse rating should be 80 percent of the circuit current.
   b. The fuse rating should be 100 percent of the circuit current.
   c. The fuse rating should be 120 percent of the circuit current.
   d. The fuse rating should be 200 percent of the circuit current.
9. What is the preferred method to connect wires or install connectors?
   a. crimp b. twist
   c. tape d. solder
10. What would be an indication of a poor connection in a circuit?
   a. high current b. low voltage drop
   c. high voltage drop d. low resistance
11. A powered relay is being tested with a DMM. Which reading would indicate a defective relay or circuit?
   a. 80 ohms across the coil of the relay
   b. battery voltage at the control terminal, with the relay energized
   c. battery voltage at the battery side of the relay
   d. no continuity between the battery and output terminals of the relay
12. Which is the largest resistance?
   a. 10 mega ohms
   b. 10 kilo ohms
   c. 10 milli ohms
   d. 10 micro ohms
13. To measure voltage drop, how must the voltmeter be connected to the circuit?
   a. in series with the test area, power off
   b. in series with the test area, power on
   c. in parallel with the test area, power off
   d. in parallel with the test area, power on
14. What is displayed on a digital storage oscilloscope (DSO)?
   a. voltage and time
   b. amperage and voltage
   c. resistance and time
   d. voltage and resistance
15. What is meant by the trigger on a DSO?
   a. when to start displaying the wave form
   b. when to limit the voltage reading
   c. when to stop displaying the wave form
   d. when to limit the time reading
16. What is meant by the slope on a DSO?
   a. an open or closed signal
   b. a digital or analog signal
   c. a rising or falling signal
   d. a horizontal or vertical signal
17. The COM lead of a DVOM is connected to battery negative. The CH1 lead is connected to the engine block. What is indicated by a 1-volt reading while the engine is cranking?
   a. short in the ground circuit
   b. resistance in the ground circuit
   c. short in the power circuit
   d. resistance in the power circuit
18. The left rear taillight on a car does not light. Voltage before the bulb reads 12.4 volts. Voltage after the bulb reads 10.2 volts. What is the most likely cause of the problem?
   a. resistance in the power or feed circuit
   b. feedback from another circuit
   c. resistance in the ground circuit
   d. a burnt-out bulb
19. When adding electrical accessories to a car, how must the wire requirements change to meet the increased amperage?
   a. increase in size b. decrease in size
   c. increase in length d. decrease in length
20. A voltmeter's COM (black) lead is connected to the battery negative. The volt/ohm (red) lead is connected to the ground side of a light circuit. What should the meter read when the light is on?
   a. close to battery voltage
   b. close to zero voltage
   c. close to 5 volts
   d. close to 10 volts

CHAPTER 18
Batteries: Theory, Diagnosis, and Service

LEARNING OUTCOMES

- Describe how a battery works.
- List the precautions that must be adhered to when working with or around batteries.
- Describe the basic construction of an electrochemical cell.
- Explain how electrochemical cells can be connected to increase voltage and current.
- Explain the different methods used to recharge a battery.
- List and describe the various ways a battery may be rated.
- List and describe the various types of batteries according to their chemistries that may be used in automobiles.
- List the precautions that must be adhered to when working with or around high-voltage systems.
- Describe the construction and operation of a lead-acid battery.
- Describe the various types of lead-acid batteries that are available today.
- Describe the basic services and testing procedures for a lead-acid battery.

INTRODUCTION

The primary source for electrical power in all automobiles is the battery. The battery has undergone many changes through the years. Lead-acid batteries have been, and continue to be, the power source for conventional vehicles. Each of these energy-storing devices is discussed in this chapter.

BASIC BATTERY THEORY

Electrical current is caused by the movement of electrons from something negative to something positive. The strength of the attraction of the electrons (negative) to the protons (positive) determines the amount of voltage present. When a path is not present for the electrons to travel through, voltage is still present, but there is no current flow. When there is a path, the electrons move, and there is current. This is the basic operation of batteries.

Basic Construction

Batteries are devices that convert chemical energy into electrical energy. Chemical reactions that produce electrons are called **electrochemical reactions**. A battery stores DC voltage and releases it when it is connected to a circuit. Inside the battery are two **electrodes** or **plates** surrounded by an electrolyte. These three elements make up an electrochemical cell **(Figure 18–1)**. Batteries are normally made up of electrochemical cells connected together.

One of the plates has an abundance of electrons (negative plate), and the other has a lack of electrons (positive plate). The electrons want to move to the positive plate and do so when a circuit connects the two plates. Batteries have two terminals, a positive that is connected to the positive plate and a negative that is connected to the negative plate.

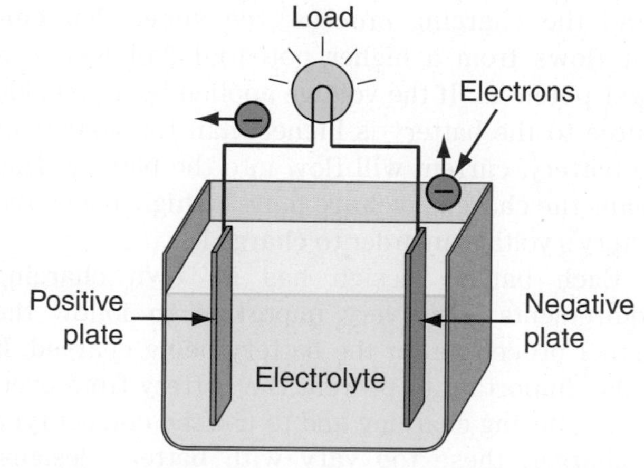

FIGURE 18–1
A simple electrochemical cell.

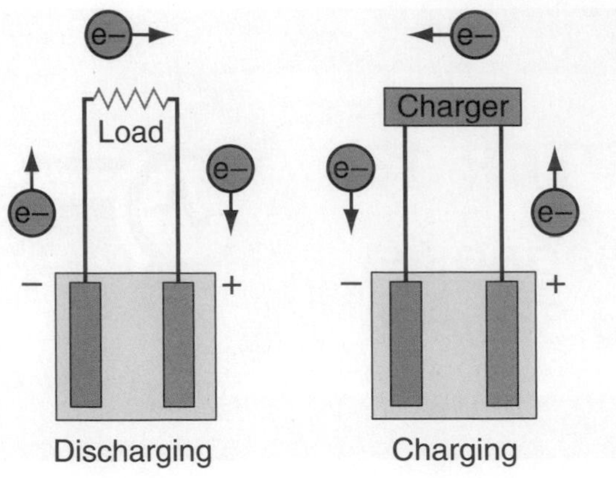

FIGURE 18–2

The flow of electrons in a battery while discharging and charging.

Electrolytes are chemical solutions that react with the metals used to construct the plates. These chemical reactions cause a lack of electrons on the positive electrode and an excess on the negative electrode. When connected into a circuit, the electrons move. The reactions continue to provide electrons for current flow until the circuit is opened or the chemicals inside the battery become weak. At that time, the battery has run out of electrons (the battery is worn out), the number of electrons on the positive and negative sides are equal, or all of the protons are matched with an electron. Recharging simply moves the electrons that moved to the positive electrode back to the negative electrode **(Figure 18–2)**.

Charging

Charging a battery restores the chemical nature of the cells. To do this, a chemical reaction takes place, causing current flow within the cells. Discharging allows for current flow outside the cell. To understand the charging process, remember that current flows from a higher potential (voltage) to a lower potential. If the voltage applied by an outside source to the battery is higher than the voltage of the battery, current will flow into the battery. This means the charging voltage must be higher than the battery's voltage in order to charge it.

Each battery design has its own charging requirements. It is very important to follow the correct procedure for the battery being charged. It is also important to prevent the battery from overheating during charging and to use the correct type of charger; these too vary with battery designs. Using the wrong charger can destroy the batteries or charger.

Cell Arrangements

The voltage produced by an individual battery cell varies with the chemicals and materials used to construct the cell. Most cells produce between 1.2 and 4 volts. To provide higher voltages, cells are connected together. In addition, there is a limited amount of current available from an individual cell, so to increase available current, cells are connected together. Cells can be connected in series or in parallel, or both.

SERIES CONNECTIONS Cells are connected in series to provide higher voltages. In this arrangement, the total voltage is the sum of the voltages in each cell. For example, a lead-acid cell, commonly used in starting batteries, produces about 2.1 volts. By connecting six together in series, the battery has a voltage of 12.6 volts. Series connections have the positive terminal of one cell connected to the negative terminal of another, the positive terminal of that cell connected to the negative of another, and so on **(Figure 18–3)**. Individual batteries can also be connected in series. Forty-two-volt systems use a 36-volt battery pack, which can be made from three 12-volt batteries **(Figure 18–4)** or eighteen 2-volt cells connected in series. Forty-two volts are provided by the charging system.

PARALLEL CONNECTIONS Cells are connected in parallel to increase the amperage of the pack of cells. The positive terminals are connected together, and all the negative terminals are connected together. The total amperage is the sum of amperages from each cell. The voltage is equal to the voltage of an individual cell.

SERIES-PARALLEL CONNECTIONS In this arrangement, groups of cells are wired in parallel, and then those groups are connected in series. This arrangement provides for increases in voltage and

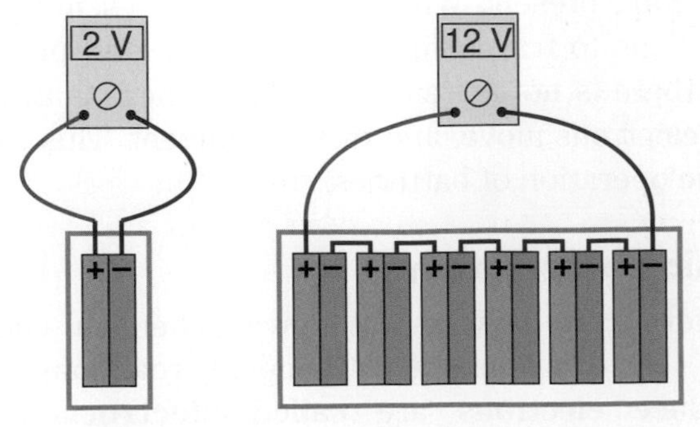

FIGURE 18–3

When individual cells are connected in series, the total voltage is equal to the sum of the cells.

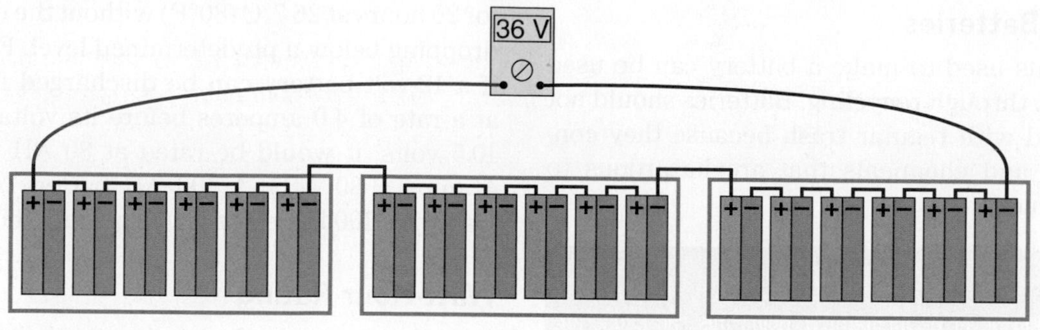

FIGURE 18–4

A 36-volt battery pack can be made from three 12-volt batteries or eighteen 2-volt cells connected in series.

amperage. Any number of cells can be connected in parallel as long as each group of parallel cells that are wired in series has the same power output.

Battery Hardware

In order to connect the battery to the vehicle's electrical system, battery cables are used. They must safely handle the voltage and current demands of the vehicle. Battery hold-downs are used to prevent damage to the battery, and heat shields are sometimes used to prevent batteries from being exposed to direct heat sources, which could result in higher battery temperatures. Most high-voltage battery packs are enclosed in a box that serves to secure the pack and to keep it within a particular temperature range.

BATTERY CABLES Battery cables must be able to carry the current required to meet all demands. Normal 12-volt cable size is 107 mm^2–170.3 mm^2 (4 or 6 gauge). Various forms of clamps and terminals are used to ensure a good electrical connection at each end of the cable **(Figure 18–5)**. Connections must be clean and tight to prevent arcing and corrosion. The positive cable is normally red, and the negative cable is black.

BATTERY HOLD-DOWNS All batteries must be held securely in the vehicle to prevent damage to the battery and to prevent the terminals from shorting to the vehicle. Battery hold-downs are made of metal or plastic **(Figure 18–6)**.

HEAT SHIELDS Some batteries may have a heat shield made of plastic or another material to protect the starting battery from high underhood temperatures. Vehicles equipped for cold climates may have a battery blanket or heater to keep the battery warm during extremely cold weather.

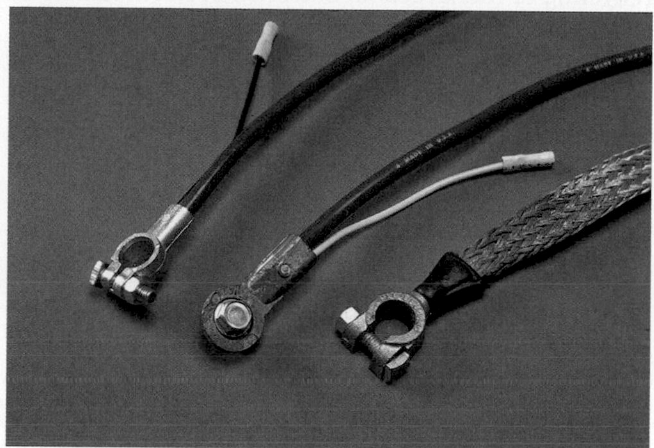

FIGURE 18–5

The battery cable is designed to carry the high current required to start the engine and supply the vehicle's electrical systems.

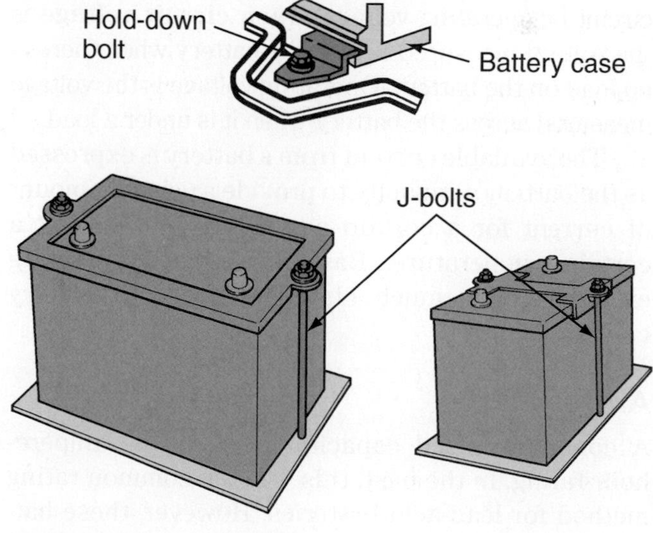

FIGURE 18–6

Examples of the different types of hold-downs used with batteries.

Chrysler Group LLC

Recycling Batteries

The materials used to make a battery can be used in the future through recycling. Batteries should not be discarded with regular trash because they contain metals and chemicals that are hazardous to the environment.

> **CAUTION!**
> A battery should never be incinerated; doing this can cause an explosion.

In 1994, the Rechargeable Battery Recycling Corporation (RBRC) was established to promote recycling of rechargeable batteries in North America. RBRC is a non-profit organization that collects batteries from consumers and businesses and sends them to recycling companies. Collected batteries are sorted by their chemical makeup. Then they are broken apart and their elements separated. The chemicals or materials are further separated and then collected.

Ninety-eight percent of all lead-acid batteries are recycled. During the recycling process, the lead, plastic, and acids are separated. The electrolyte (sulphuric acid) can be reused or is discarded after it has been neutralized. The plastic casing is cut into small pieces, scrubbed, and melted to make new battery cases and other parts. The lead is also melted and poured into ingots to be used in new batteries.

BATTERY RATINGS

The voltage rating of a battery may be expressed as open circuit or operating voltage. **Open circuit voltage** is the voltage measured across the battery when there is no load on the battery. Operating voltage is the voltage measured across the battery when it is under a load.

The available current from a battery is expressed as the battery's capacity to provide a certain amount of current for a certain amount of time and at a certain temperature. Basically, a capacity rating expresses how much electrical energy a battery can store.

Ampere-Hour

A commonly used capacity rating is the ampere-hour rating. In the past, this was the common rating method for lead-acid batteries. However, these batteries are now rated otherwise. Other battery designs are still rated in ampere-hours or milliamp-hours.

The **ampere-hour (AH) rating** is the amount of steady current that a fully charged battery can supply for 20 hours at 26.7°C (80°F) without the cell's voltage dropping below a predetermined level. For example, if a 12-volt battery can be discharged for 20 hours at a rate of 4.0 amperes before its voltage drops to 10.5 volts, it would be rated at 80 AH (20 hours × 4 amps = 80 AH). A 100 AH battery will provide 1 amp for 100 hours, or 100 amps for 1 hour.

Watt-Hour Rating

Some battery manufacturers rate their batteries in watt-hours. The **watt-hour rating** is determined at –17.8°C (0°F) because the battery's capacity changes with temperature. The rating is calculated by multiplying a battery's AH rating by the battery's voltage. The watt-hour rating of a battery may be listed in units of kilowatts. If a battery can deliver 5 AH at 200 volts, it would be rated at 1 kilowatt-hour (5 AH × 200 volts = 1000 watt-hour, or 1 kilowatt-hour).

Cold Cranking Amps

The **cold cranking amps (CCA) rating** is the common method of rating most automotive starting batteries. It is determined by the load, in amperes, that a battery is able to deliver for 30 seconds at –17.8°C (0°F) without its voltage dropping below a predetermined level. That voltage level for a 12-volt battery is 7.2 volts. The normal range for passenger car and light truck batteries is between 300 and 600 CCA; some batteries have a rating as high as 1 100 CCA.

Cranking Amps

The **cranking amps (CA) rating** is similar to CCA and is a measure of the current a battery can deliver at 0°C (32°F) for 30 seconds and maintain voltage at a predetermined level. Again, this level is 1.2 volts per cell (7.2 volts) for a 12-volt battery. This rating is more commonly used in climates that are not subject to extremely cold weather. Typically, the CCA rating of a battery is about 20 percent less than its CA rating.

Reserve Capacity

The **reserve capacity (RC) rating** is determined by the length of time, in minutes, that a fully charged starting battery at 26.7°C (80°F) can be discharged at 25 amperes before battery voltage drops below 10.5 volts. This rating gives an indication of how long the vehicle can be driven with the headlights on if the charging system fails. A battery with a reserve capacity of 120 would be able to deliver 25 amps for 120 minutes before its voltage drops below 10.5 volts.

COMMON TYPES OF BATTERIES

In addition to their use in automobiles, batteries are used in many other applications. As a result, there are many different types and designs of batteries available. Batteries differ in size, from small single cells to large battery packs comprising many cells. They also have different ratings (not always dependent on size) and service lives. The primary difference between batteries is the chemicals used in the cells.

Battery Chemistry

There are new, emerging battery technologies that are starting to appear in automotive applications that will be discussed toward the end of this chapter, so we will begin by looking at the oldest, yet still most prominent battery chemistry used in automobiles, the lead acid battery.

LEAD-ACID Lead-acid batteries are the most commonly used in automotive applications. Lead-acid batteries can provide durability (when properly maintained), provide sufficient electrical energy to power the many accessories found on today's vehicles, and have the ability to be recharged. The typical automotive battery provides in the range of 12 to 15 volts, depending on its state of charge. More than one lead-acid battery may be found in a vehicle, and depending on the electrical demands of the vehicle, they may be connected in various ways to provide different levels of electrical output. There are many variations to the basic design, but all work and are constructed in the same way. The lead-acid cell has electrodes made of lead and lead-oxide with an electrolyte that is a strong acid. The lead-acid battery is one of the oldest battery designs.

NICKEL-CADMIUM (NICAD) **Nickel-cadmium (NiCad)** batteries are mostly used in portable radios, emergency medical equipment, professional video cameras, and power tools. They provide great power and are normally the battery of choice for power tools. The electrodes in a NiCad cell are nickel hydroxide and cadmium. The electrolyte is potassium hydroxide **(Figure 18–7)**. NiCad batteries are economical and have a long service life. However, cadmium is an environmentally unfriendly metal, which is why NiCad batteries are being replaced by other designs.

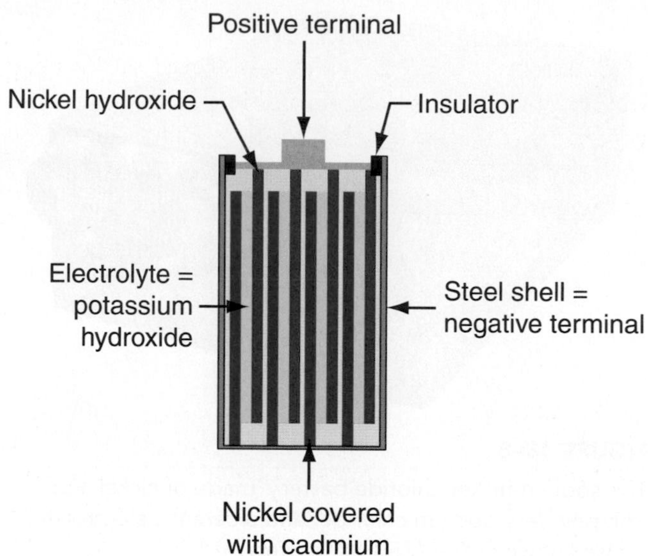

FIGURE 18–7

The basic construction of a NiCad battery.

NICKEL-METAL HYDRIDE (NIMH) **Nickel-metal hydride (niMH)** batteries are rapidly replacing NiCad batteries because they are more environmentally friendly and are more capable of receiving a full recharge. NiMH batteries also have more capacity than a NiCad but have a reduced service life and a lower current capacity under load. These batteries are commonly used in today's hybrid vehicles. The cells have electrodes made of a metal hydride and nickel hydroxide. The electrolyte is potassium hydroxide.

SODIUM-SULPHUR (NAS) The electrodes in a sodium-sulphur battery cell are made of molten sodium (negative electrode) and liquid sulphur (positive electrode). The plates are separated by a solid ceramic electrolyte made of aluminum. The battery must be kept at about 300°C (570°F) to discharge and recharge. This design of battery is very efficient and is currently being researched for possible use in vehicles.

SODIUM-NICKEL-CHLORIDE The electrodes in a sodium-nickel-chloride cell are made with nickel and iron powders and sodium chloride (table salt). The electrodes are separated by a ceramic electrolyte. Sodium-nickel-chloride batteries are also known as "ZEBRA" batteries **(Figure 18–8)**. These batteries have nearly five times the energy density as a lead-acid battery and are totally recyclable. However, they must operate at high temperatures, and the required thermal management system greatly raises their production cost. These batteries were designed to be used in automobiles and trains.

LITHIUM-ION (LI-ION) The electrodes in **lithium-ion (Li-Ion)** cells are made of a carbon compound (graphite) and a metal oxide. The electrodes are submersed in lithium salt. Overheating these cells may produce pure lithium in the cells. This metal is very reactive and can explode when hot. To prevent overheating, Li-Ion cells have built-in protective

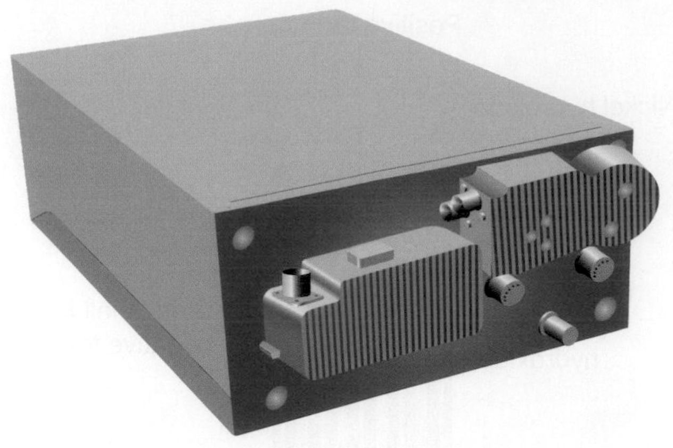

FIGURE 18–8

The sodium-nickel-chloride battery, made of nickel and iron powders, sodium chloride, and a ceramic electrolyte, is also known as the "ZEBRA" battery.

electronics and/or fuses to prevent reverse polarity and overcharging. Li-Ion batteries have very good power-to-weight ratios and are making their way into hybrid vehicles.

LITHIUM-POLYMER (LI-POLY) The **lithium-polymer (Li-Poly)** battery is nearly identical to a Li-Ion battery. Like the Li-Ion, the electrodes are made of a carbon compound (graphite) and a metal oxide. However, the lithium salt electrolyte is held in a thin, solid, plasticlike polymer rather than as a liquid. The solid polymer electrolyte is not flammable; therefore, these batteries are less hazardous if they are mistreated. These batteries can store much more energy than a lead-acid battery. They also offer many other advantages and may be used in hybrid and fuel cell vehicles in the future.

NICKEL-ZINC Nickel-zinc battery cells are also being researched and tested for possible use in vehicles. They have high specific energy and power capability, have good deep-cycle capability, can operate within a wide range of temperatures, are made of abundant low-cost materials, and are environmentally friendly. The nickel-zinc battery is an alkaline rechargeable system. These cells use a nickel/nickel-oxide electrode as the cathode and the zinc/zinc-oxide electrode as the anode. The electrolyte is normally potassium hydroxide.

LEAD-ACID BATTERIES

The most common automotive batteries are lead-acid designs. The wet cell, gel cell, absorbed glass mat (AGM), and valve regulated are versions of the lead-acid battery.

Basic Construction

A lead-acid battery consists of grids, positive plates, negative plates, separators, elements, electrolyte, a container, cell covers, vent plugs, and cell containers **(Figure 18–9)**. A **grid** is a lead alloy frame that supports the active material of each plate. Plates are typically flat, rectangular components that are either positive or negative, depending on the active material they hold.

The positive plate has a grid filled with its active material, lead peroxide. **Lead peroxide (PbO_2)** is a dark-brown, crystalline material. The material pasted onto the grids of the negative plates is **sponge lead (Pb)**. Both plates are very porous and allow the liquid electrolyte to penetrate freely.

Each battery contains a number of elements. An **element** is a group of positive and negative plates **(Figure 18–10)**. The plates are formed into a plate group, which holds a number of plates of the same polarity. The like-charged plates are welded to a lead alloy post or **plate strap**. The plate groups are placed alternately within the battery—positive, negative, positive, negative, and so on. There is usually one extra set of negative plates to balance the charge. To prevent the different plate groups from touching each other, separators are inserted between them. **Separators** are porous plastic sheets that allow for a transfer of ions between plates. When the element is placed into the battery case and immersed in electrolyte, it becomes a cell.

The electrolyte is a solution of sulphuric acid and water. The sulphuric acid supplies sulphate, which chemically reacts with both the lead and PbO_2 to release electrical energy. In addition, the sulphuric acid is the carrier for the electrons as they move inside the battery. To cause the required chemical reaction, the electrolyte must be the correct mixture of water and sulphuric acid. At 12.6 volts, the desired solution is 65 percent water and 35 percent sulphuric acid. Available voltage decreases when the percentage of acid in the solution decreases.

CASING DESIGN The container or shell of the battery is usually a one-piece, moulded assembly of polypropylene, hard rubber, or plastic. The case has a number of individual cell compartments. Cell connectors are used to join all cells of a battery in series.

The top of the battery is encased by a cell cover. The cover may be a one-piece design, or the cells might have their own individual covers. The cover must have vent holes to allow hydrogen and

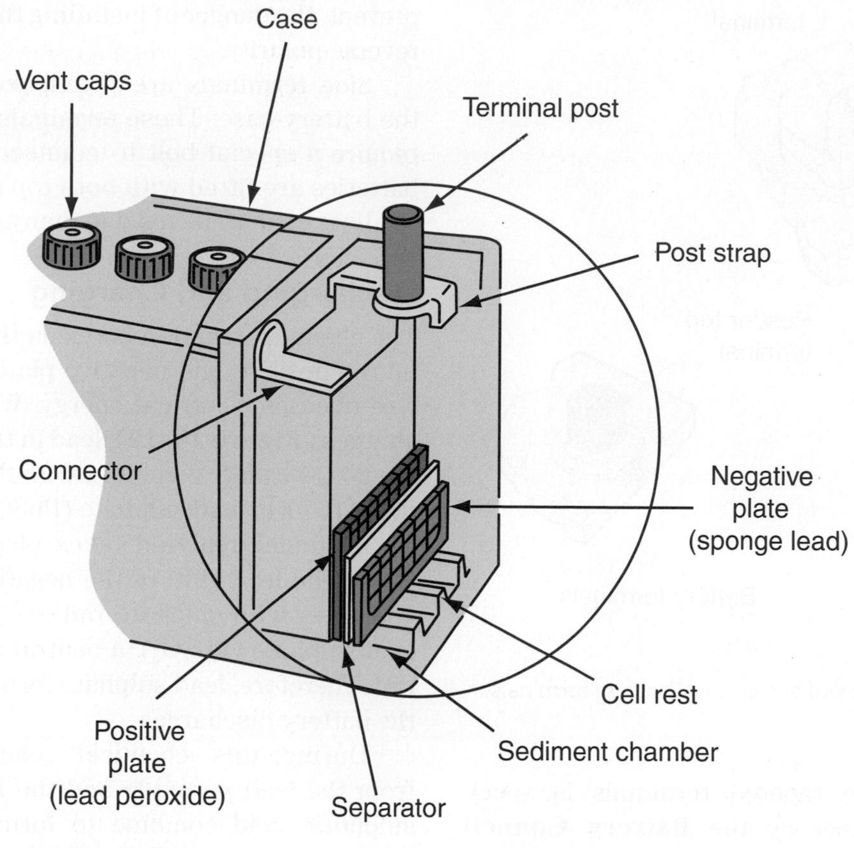

FIGURE 18–9

Components of a typical lead-acid storage battery.

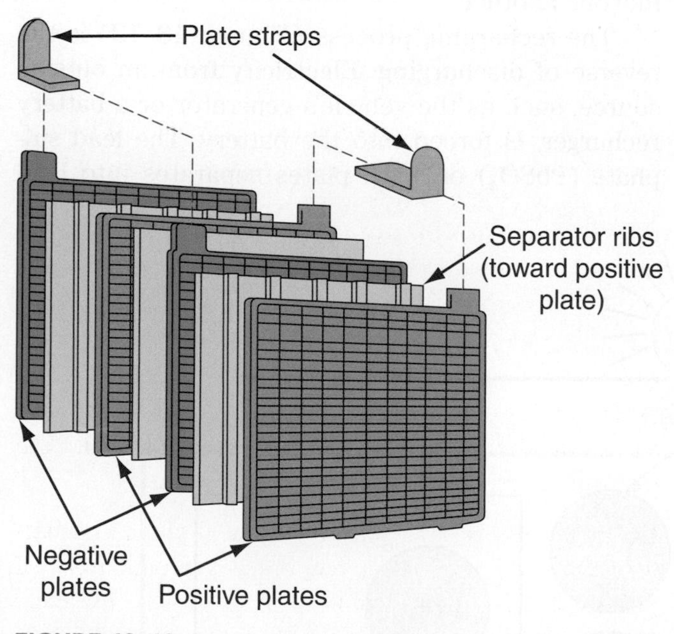

FIGURE 18–10

The parts of a typical battery element.

oxygen gases to escape. These gases are formed during charging and discharging. Battery vents can be permanently fixed to the cover or be removable, depending on the design of the battery. Vent plugs or caps are used on some batteries to close the openings in the cell cover and to allow for topping off the cells with electrolyte or water.

At the bottom of some battery casings is a sediment chamber. The chamber collects the materials that fall from the plates. If the sediments do not fall below the plates, they could cause a short between the plates. Some batteries do not have a sediment chamber; rather, the separators are used to contain all sediments and keep them from contacting the plates.

> **CAUTION!**
>
> When lifting a battery, excessive pressure on the end walls could cause acid to spew through the vent caps, resulting in personal injury. Lift with a battery carrier or with your hands on opposite corners.

TERMINALS The battery has two external terminals: a positive (+) and a negative (–). These terminals are two tapered posts, "L" terminals, threaded studs on top of the case, or two internally threaded connectors on the side **(Figure 18–11)**. The terminals have either a positive (+) or a negative (–) marking, depending on which end of the series they represent.

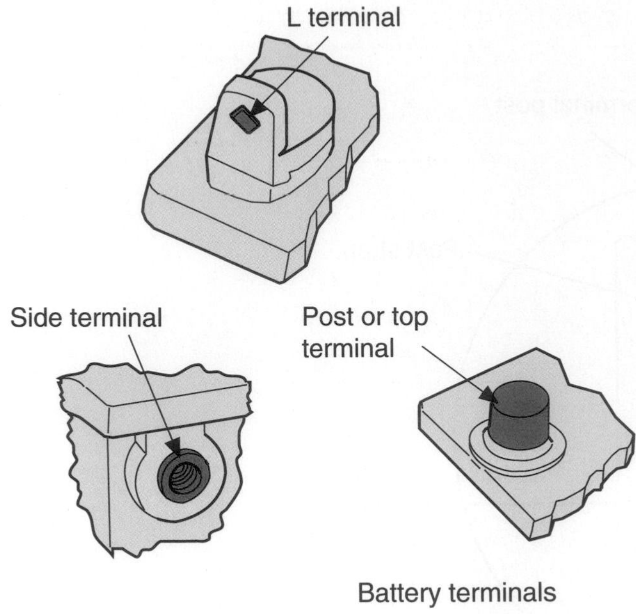

FIGURE 18–11

The most common types of automotive battery terminals.
Ford Motor Company

The size of the tapered terminals is specified by standards set by the **Battery Council International (BCI)** and Society of Automotive Engineers (SAE). This means that all positive and negative cable clamps will fit any corresponding battery terminal, regardless of the battery's manufacturer. The positive terminal is slightly larger, usually around 17.5 mm (11/16 in.) in diameter at the top, whereas the negative terminal usually has a 16 mm (5/8 in.) diameter. This minimizes, but does not prevent, the danger of installing the battery cables in reverse polarity.

Side terminals are positioned near the top of the battery case. These terminals are threaded and require a special bolt to connect the cables. Some batteries are fitted with both top and side terminals to allow them to be used in many different vehicles.

Discharging and Charging

The chemical reaction between the active materials on the positive and negative plates and the electrolyte releases electrical energy. When a battery discharges **(Figure 18–12)**, lead in the lead peroxide of the positive plate combines with the sulphate radical (SO_4) to form lead sulphate ($PbSO_4$).

A similar reaction takes place at the negative plate. The lead (Pb) of the negative active material combines with sulphate radical (SO_4) to also form lead sulphate ($PbSO_4$), a neutral and inactive material. Therefore, lead sulphate forms at both plates as the battery discharges.

During this chemical reaction, the oxygen from the lead peroxide and the hydrogen from the sulphuric acid combine to form water (H_2O). As discharging takes place, the electrolyte becomes weaker, and the positive and negative plates become like one another.

The recharging process **(Figure 18–13)** is the reverse of discharging. Electricity from an outside source, such as the vehicle's generator or a battery recharger, is forced into the battery. The lead sulphate ($PbSO_4$) on both plates separates into lead

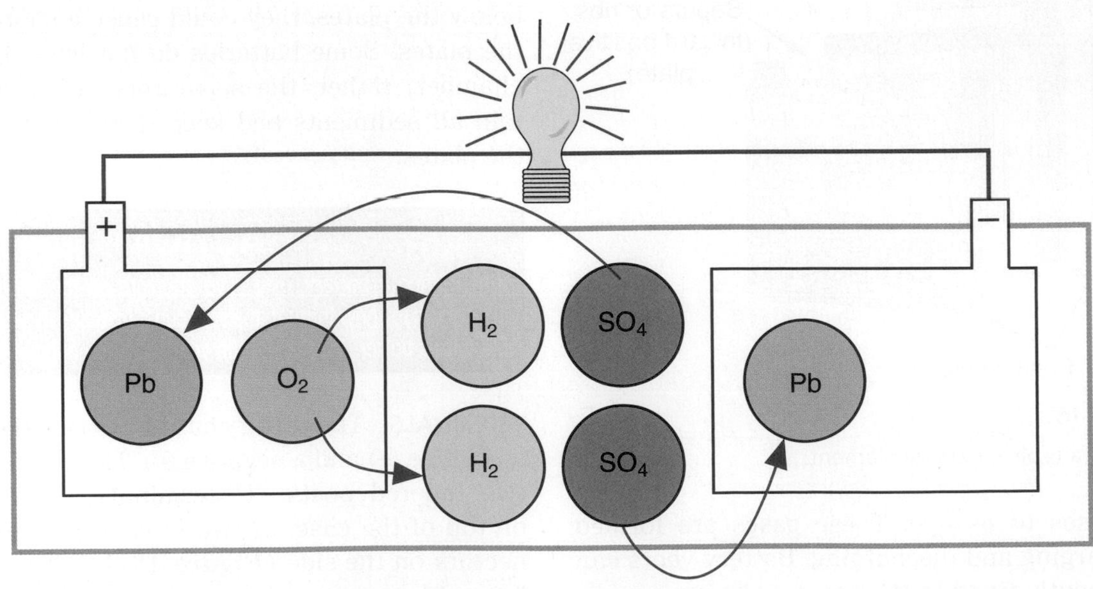

FIGURE 18–12

Chemical action that occurs inside a battery during a discharge cycle.

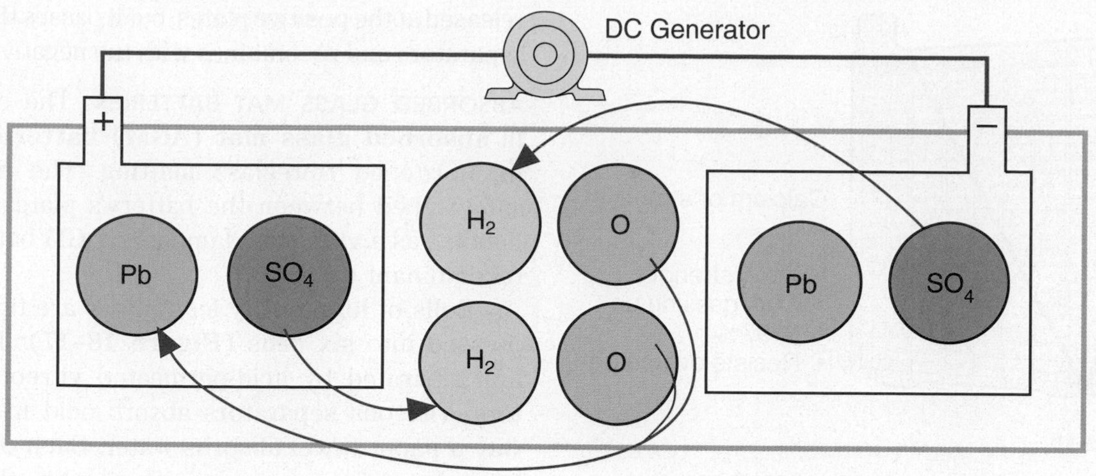

FIGURE 18–13

Chemical action that occurs inside a battery during the charge cycle.

(Pb) and sulphate (SO_4). As the sulphate (SO_4) leaves both plates, it combines with hydrogen in the electrolyte to form sulphuric acid (H_2SO_4). At the same time, the oxygen (O_2) in the electrolyte combines with the lead (Pb) at the positive plate to form lead peroxide (PbO_2). As a result, the negative plate returns to its original form of lead (Pb), and the positive plate reverts to lead peroxide (PbO_2).

An unsealed battery gradually loses water due to its conversion into hydrogen and oxygen; these gases escape the battery through the vent caps **(Figure 18–14)**. If the lost water is not replaced, the level of the electrolyte falls below the tops of the plates. This results in a high concentration of sulphuric acid in the electrolyte and permits the uncovered material of the plates to dry and harden. This will reduce the service life of a battery. This is why the electrolyte level in the battery must be frequently checked.

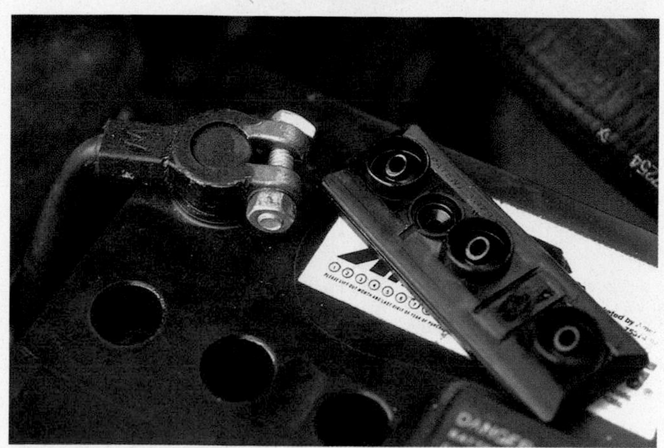

FIGURE 18–14

A battery with removable cell caps.

Designs

Lead-acid batteries can be designed as a starting battery or a deep-cycle battery. Deep-cycle batteries are designed to go through many charge and discharge cycles. They have thicker and fewer plates than a starting battery. The exact chemical composition of lead-acid batteries also depends on the designed purpose of the battery. However, all lead-acid batteries are based on the reaction of lead and acid.

MAINTENANCE-FREE AND LOW-MAINTENANCE BATTERIES The majority of batteries installed in today's vehicles are low-maintenance or maintenance-free designs. A low-maintenance battery is a heavy-duty version of a normal lead-acid battery. Many of the parts are thicker and made with different, more durable materials. Low-maintenance batteries have vent holes and caps, which allow water to be added to the cells. However, a low-maintenance battery requires additional water substantially less often than a conventional battery.

Similar in construction but made with different plate materials, a **maintenance-free battery** experiences little gassing during discharge and charge cycles **(Figure 18–15)**. Therefore, maintenance-free batteries do not have external holes or caps **(Figure 18–16)**. They are equipped with small gas vents that prevent gas pressure buildup in the case. Water is never added to maintenance-free batteries.

HYBRID BATTERIES A **hybrid battery** can withstand six deep cycles and still retain 100 percent of its original reserve capacity. A battery experiences a deep cycle when it is almost totally discharged and then recharged. The grid construction differs from other

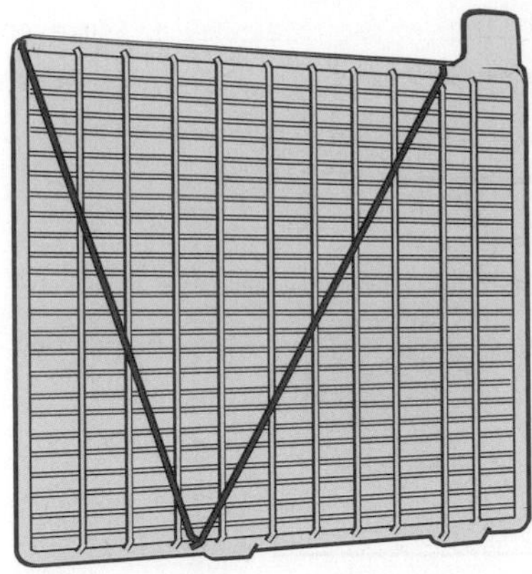

Calcium or strontium alloy. . .

- Adds strength
- Cuts gassing up to 97%
- Resists overcharge

FIGURE 18–15

Maintenance-free battery grids with support bars give increased strength and faster electrical delivery.

Ford Motor Company

batteries in that the plates have a lug located near the centre of the grid. In addition, the vertical and horizontal grid bars are arranged in a radial design. With this design, current has less resistance and a shorter path to follow. This means the battery is capable of providing more current at a faster rate.

The separators are constructed of glass covered with a resin or fibreglass. The glass separators offer low electrical resistance with high resistance to chemical contamination.

RECOMBINATION BATTERIES A **recombination (recombinant) battery** is a completely sealed maintenance-free battery that uses a gel-type electrolyte. In a gel cell battery, gassing is minimized and vents are not needed. During charging, the negative plates never reach a fully charged condition and therefore cause little or no release of hydrogen. Oxygen is released at the positive plates, but it passes through the separators and recombines with the negative plates.

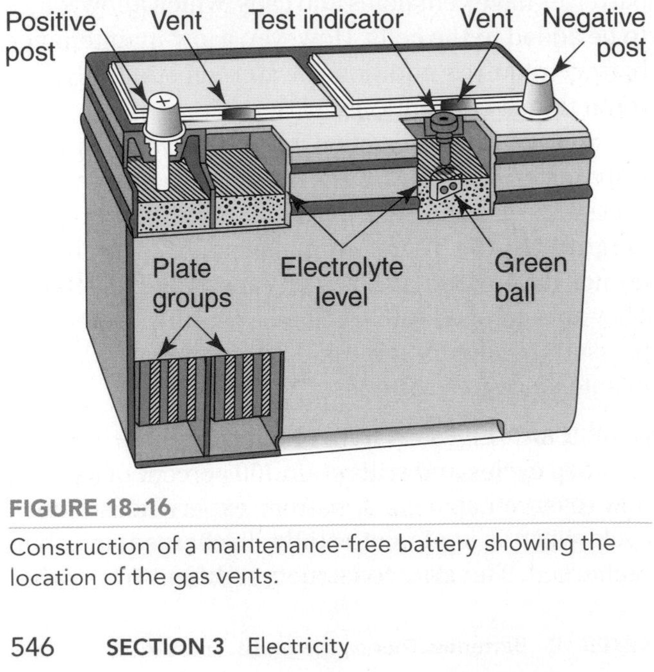

Positive post
Vent
Test indicator
Vent
Negative post
Plate groups
Electrolyte level
Green ball

FIGURE 18–16

Construction of a maintenance-free battery showing the location of the gas vents.

ABSORBED GLASS MAT BATTERIES The electrolyte in **absorbed glass mat (AGM) batteries** is held in moistened fibreglass matting. The matting is sandwiched between the battery's plates, where it doubles as a vibration dampener. AGM batteries are recombinant batteries.

Rolls of high-purity lead plates are tightly compressed into six cells **(Figure 18–17)**. The plates are separated by acid-permeated vitreous separators. Vitreous separators absorb acid in the same way a paper towel absorbs water. Each of the cells is enclosed in its own cylinder within the battery case, forming a sealed, closed system that resembles a six-pack of soda. During normal use, hydrogen and oxygen within the battery are captured and recombined to form water within the electrolyte. This eliminates the need to ever add water to the battery.

The spiral rolled plates and fibreglass mats are virtually impervious to vibration and impact. AGM batteries will never leak, have short recharging times, and have low internal resistance, which provides increased output.

VALVE-REGULATED BATTERIES **Valve-regulated lead-acid (VRLA) batteries** are similar to AGM batteries and are recombinant batteries. The oxygen

FIGURE 18–17

The construction of an AGM battery.

Exide Technologies

produced on the positive plates is absorbed by the negative plate. That, in turn, decreases the amount of hydrogen produced at the negative plate. The combination of hydrogen and oxygen produces water, which is returned to the electrolyte. Therefore, this battery never needs to have water added to its electrolyte mixture.

One plate in a VRLA is made of a lead-tin-calcium alloy with porous lead dioxide; the other is also made of a lead-tin-calcium alloy but has spongy lead as the active material. The electrolyte is sulphuric acid that is absorbed into plate separators made of a glass-fibre fabric. The battery is equipped with a valve that opens to relieve any excessive pressure that builds up in the battery. At all other times, the valve is closed, and the battery is totally sealed.

Factors Affecting Battery Life

All storage batteries have a limited service life, but many conditions can decrease it.

IMPROPER ELECTROLYTE LEVELS With non-sealed batteries, water should be the only portion of the electrolyte lost due to evaporation during hot weather and gassing during charging. Maintaining an adequate electrolyte level is the basic step in extending battery life for these designs.

TEMPERATURE Batteries do not work well when they are cold. At –17.8°C (0°F), a battery is only capable of working at 40 percent of its capacity. There is also the possibility of the battery freezing when it is very cold and its charge is low. When the battery is allowed to get too hot, the water in the electrolyte can evaporate. Batteries used in hot climates should have their electrolyte level checked very frequently, and only distilled water should be added if necessary.

CORROSION Battery corrosion is commonly caused by spilled electrolyte or electrolyte condensation from gassing. In either case, the sulphuric acid corrodes, attacks, and can destroy not only connectors and terminals but also hold-down straps and the battery tray.

Corroded connections increase resistance at the battery terminals, which reduces the applied voltage to the vehicle's electrical system. Corrosion on the battery cover can also create a path for current, which can allow the battery to slowly discharge. Finally, corrosion can destroy the hold-down straps and battery tray, which can result in physical damage to the battery and/or vehicle.

OVERCHARGING Batteries can be overcharged by either the vehicle's charging system or a battery charger. In either case, the result is a violent chemical reaction within the battery that causes a loss of water in the cells. This can permanently reduce the capacity of the battery. Overcharging can also cause excessive heat, which can oxidize the positive plate grid material and even buckle the plates, resulting in a loss of cell capacity and early battery failure.

UNDERCHARGE/SULPHATION The vehicle's charging system might not fully recharge the battery due to a fault in the charging system. This causes the battery to operate in a partially discharged condition. A battery in this condition will become sulphated when the sulphate normally formed in the plates becomes dense, hard, and chemically irreversible. This happens because the sulphate has been allowed to remain in the plates for a long period.

Sulphation of the plates causes two problems. First, it lowers the specific gravity levels of the electrolyte and increases the danger of freezing. Second, in cold weather, a sulphated battery often fails to crank the engine because of its lack of reserve power.

POOR MOUNTING Loose or missing hold-down straps allow the battery to vibrate or bounce during vehicle operation. This can shake the active materials off the grid plates, severely shortening battery life. It can also loosen the plate connections to the plate strap, loosen cable connections, or even crack the battery case. In some instances, if the battery is not properly secured, it could be launched from the vehicle during certain driving manoeuvres or even in an accident.

CYCLING Heavy and repeated cycling can cause the positive plate material to break away from its grids and fall into the sediment chambers at the base of the case. This problem reduces battery capacity and can lead to premature short circuiting between the plates. Fortunately, the new envelope design found in many batteries reduces this problem by employing the use of more robust separators, which promotes acid circulation, thus helping the battery remain cooler.

SERVICING AND TESTING BATTERIES

Testing batteries is an important part of electrical system service. Prior to conducting any tests, make sure the battery is fully charged. Also remove the surface charge of the battery by turning on the headlights with the engine off. Keep the lights on for at least three minutes. Poor and inaccurate tests can lead to serious problems and expensive and unneeded repairs. Depending on the design of the battery, state of charge and capacity can be determined in several ways: specific gravity tests, visual inspection of batteries with a built-in hydrometer, open circuit voltage tests, and the capacity test.

CAUTION!

Always wear safety glasses or goggles when working with batteries. When a battery is charging or discharging, it gives off highly explosive hydrogen gas. Some hydrogen gas is present in the battery at all times. Any flame or spark can ignite this gas, causing the battery to violently explode, propelling the vent caps at a high velocity and spraying acid in a wide area. Sulphuric acid can cause severe skin burns. If electrolyte contacts your skin or eyes, flush the area with water for several minutes. When eye contact occurs, force your eyelid open and flush your eyes with eyewash solution. Do not rub your eyes or skin, and call a doctor immediately.

Inspection

Testing a lead-acid battery should begin with a thorough inspection of the battery and its terminals **(Figure 18–18)**. The following items should be checked:

1. Check the age of the battery by looking at the date code on the battery.
2. Check the condition of the case. A damaged battery should be replaced.
3. If the battery is not sealed, check the electrolyte levels in all cells and correct them as necessary. If water is added, charge the battery before conducting any test on the battery.
4. Check the condition of the battery terminals and cables. Clean any corrosion from the cable ends and terminals. Make sure the cable ends are tightly fastened to the terminals.
5. Make sure the battery hold-downs are holding the battery securely in place.

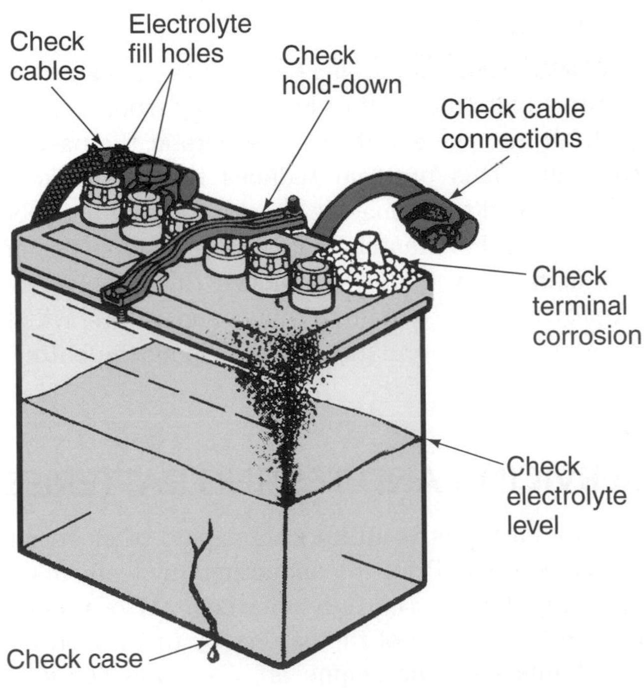

FIGURE 18–18

Batteries should be carefully checked for damage, dirt, and corrosion.

Battery Leakage Test

To perform a battery leakage test, set a voltmeter to a low DC volt range. Connect the negative lead to the battery's negative post. Then move the meter's positive lead across the top and sides of the battery case **(Figure 18–19)**. If some voltage is read on the meter, current is leaking out of the battery. The battery should be cleaned and then rechecked. If voltage is again measured, the battery should be replaced; the case is porous or cracked.

Cleaning the Battery and Terminals

Before removing the battery connectors or the battery, always neutralize any accumulated corrosion on terminals, connectors, and other metal parts. Apply a solution of baking soda and water or ammonia and water. Photo Sequence 13 shows the correct procedure for cleaning a battery, battery tray, and battery cables.

Do not splash the corrosion onto the vehicle's paint, metal, or rubber parts, or onto your hands and face. Be sure the solution cannot enter the battery cells. A stiff-bristle brush is ideal for removing heavy buildup. Dirt and accumulated grease can be removed with a detergent solution or solvent. After cleaning, rinse the battery and cable connections with clean water. Dry the components with a clean rag or low-pressure compressed air.

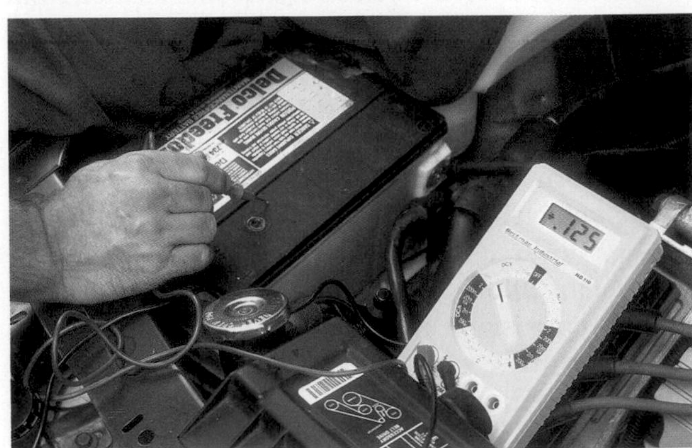

FIGURE 18–19

Performing a battery leakage test.

TYPICAL PROCEDURE FOR CLEANING A BATTERY CASE, TRAY, AND CABLES

PHOTO SEQUENCE 13

P13–1 Loosen the battery negative terminal clamp.

P13–2 Use a terminal clamp puller to remove the negative cable.

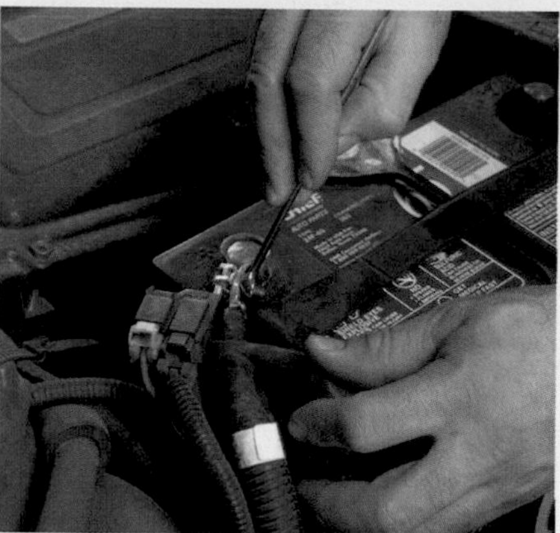

P13–3 Loosen the battery positive terminal clamp.

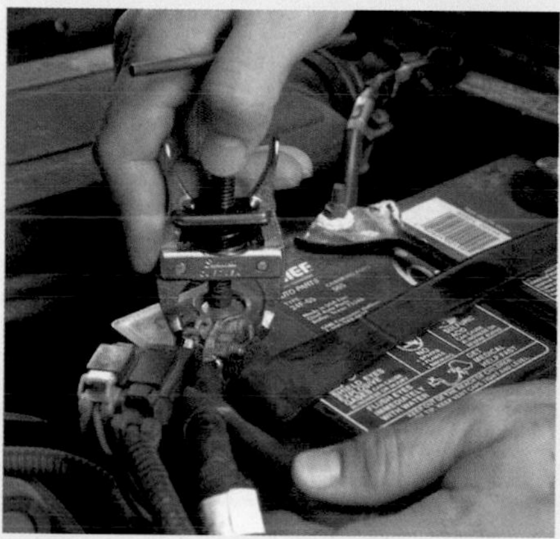

P13–4 Use a terminal clamp puller to remove the positive clamp.

P13–5 Remove the battery hold-down hardware and any heat shields.

P13–6 Remove the battery from the tray.

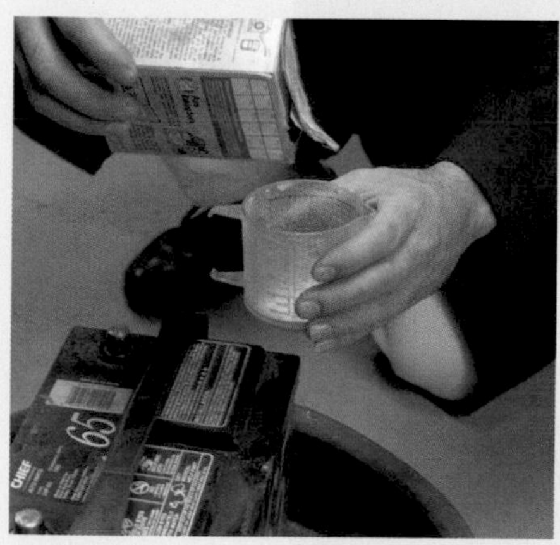

P13–7 Mix a solution of baking soda and water.

P13–8 Brush the baking soda solution over the battery case, but don't allow the solution to enter the cells of the battery.

P13–9 Flush the baking soda off with water.

PHOTO SEQUENCE 13

TYPICAL PROCEDURE FOR CLEANING A BATTERY CASE, TRAY, AND CABLES *(continued)*

P13–10 Use a scraper and wire brush to remove corrosion from the hold-down hardware.

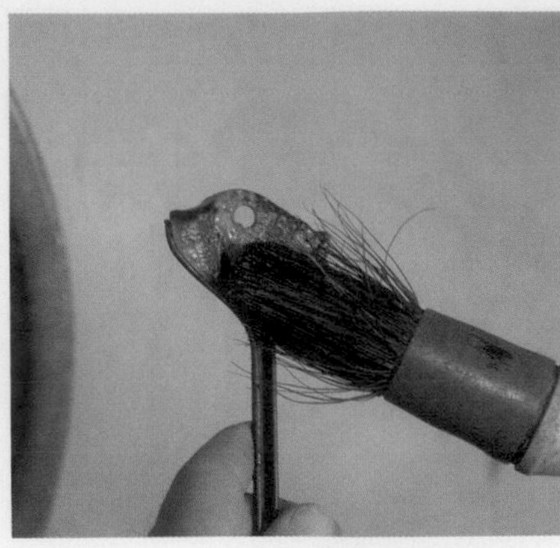

P13–11 Brush the baking soda solution over the hold-down hardware and then flush with water.

P13–12 Allow the hardware to dry and then paint it with corrosion-proof paint.

P13–13 Use a terminal cleaner brush to clean the battery cables.

P13–14 Use a terminal cleaner brush to clean the battery posts.

P13–15 Install the battery back into the tray. Also install the hold-down hardware.

P13–16 Install the positive battery cable. Then install the negative cable.

WARNING!

Nickel-metal hydride (NiMH) batteries should never be cleaned with baking soda because a rapid heating and possible fire or explosion could result.

SHOP TALK

Remember to connect a memory saver and obtain security codes for the vehicle before disconnecting the battery.

To clean the inside surfaces of the connectors and the battery terminals, remove the cables. Always begin with the ground cable. Spring-type cable connectors are removed by squeezing the ends of their prongs together with wide-jaw, vise-gripping, channel lock, or battery pliers. This pressure expands the connector so that it can be lifted off the terminal post.

For connectors tightened with nuts and bolts, loosen the nut, using a box-end wrench and/or cable-clamp pliers **(Figure 18–20)**. Using ordinary pliers or an open end wrench can cause problems. These tools might slip off under pressure with enough force to break the cell cover or damage the casing.

Always grip the cable while loosening the nut. This reduces the pressure on the terminal post that could break it or loosen its mounting in the battery. If the connector does not lift easily off the terminal when loosened, use a clamp puller. Prying with a screwdriver or bar strains the terminal post and the plates attached to it. This can break the cell cover or pop the plates loose from the terminal post.

Once the connectors have been removed, open the connector using a connector-spreading tool. Neutralize any remaining corrosion by dipping it in a baking soda or ammonia solution. Next clean the inside of the connectors and the posts, using a wire brush with external and internal bristles **(Figure 18–21)**.

Begin reinstallation by expanding the opening of the clamp **(Figure 18–22)** so that force is not needed to place it over the post. Then position and tighten the terminal on its post. Do not overtighten any nuts or bolts; this could damage the post or terminal.

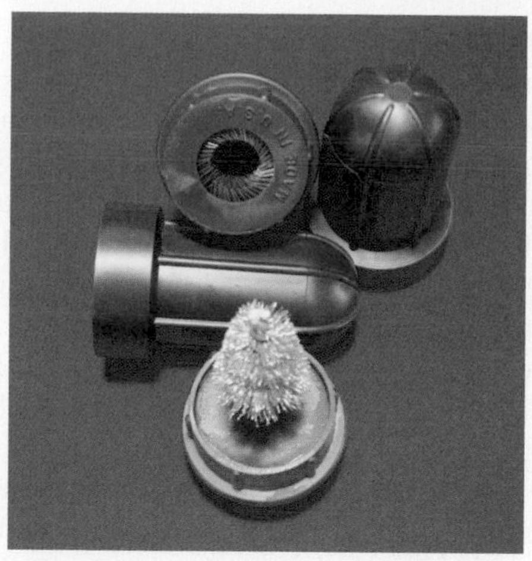

FIGURE 18–21

Combination external/internal wire brushes clean both terminals and inside cable connector surfaces.

FIGURE 18–20

It is best to loosen a battery clamp with a box-end wrench and/or cable-clamp pliers.

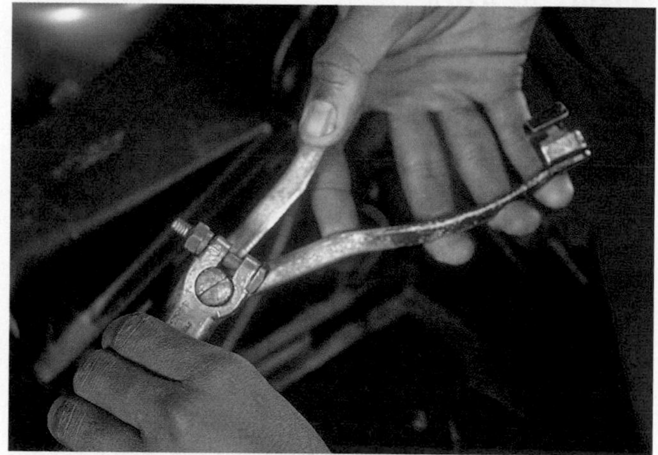

FIGURE 18–22

Before reinstalling the cable end onto the battery, use terminal end expanders to make sure the end does not need to be forced onto the battery post.

Battery Hydrometer

The electrolyte of a fully charged battery is usually about 64 percent water and 36 percent sulphuric acid. This corresponds to a specific gravity of 1.270. **Specific gravity** is the weight of a given volume of any liquid divided by the weight of an equal volume of water. Pure water has a specific gravity of 1.000, whereas battery electrolyte should have a specific gravity of 1.260 to 1.280 at 26.7°C (80°F). In other words, the electrolyte should be 1.260 to 1.280 times heavier than water.

The specific gravity of the electrolyte decreases as the battery discharges. This is why measuring the specific gravity of the electrolyte with a hydrometer can be a good indicator of how much charge a battery has lost **(Figure 18–23)**. Hydrometer readings should not vary more than 0.05 difference between cells. More variance is an indication that the battery is bad.

The electrolyte's specific gravity can be measured on unsealed batteries to give a fairly good indication of the battery's state of charge (SOC). A basic battery hydrometer uses a glass float or hydrometer in a glass tube to measure the electrolyte's specific gravity. Squeezing the hydrometer's bulb pulls electrolyte into the tube. When filled with test electrolyte, the hydrometer float bobs in the electrolyte. The depth at which the float sinks in the electrolyte indicates its relative weight compared to water. The reading is taken off the scale by sighting along the level of the electrolyte. If the hydrometer floats deeply in the electrolyte, the specific gravity is low **(Figure 18–24A)**. If the hydrometer floats high in the electrolyte, the specific gravity is high **(Figure 18–24B)**.

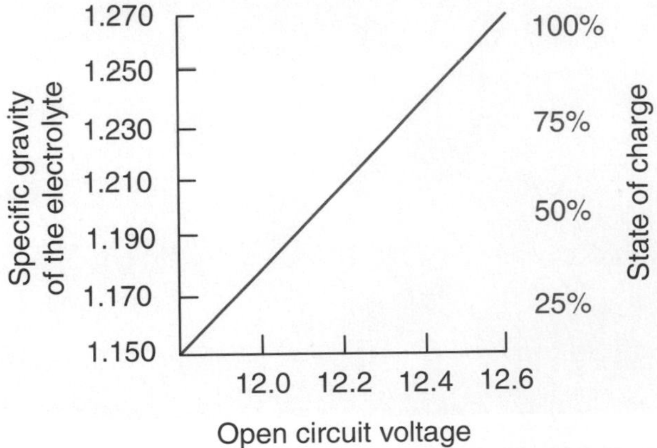

FIGURE 18–23

A graph showing the relationship between specific gravity, open circuit voltage, and the state of charge of a battery cell.

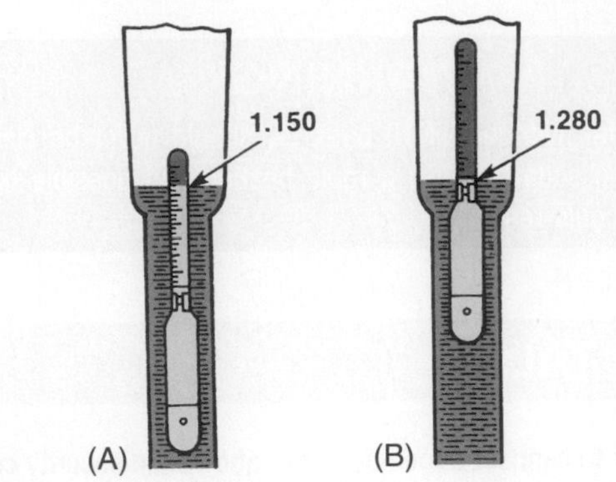

FIGURE 18–24

(A) When the scale sinks in the electrolyte, the specific gravity is low; (B) when it floats high, the specific gravity is high.

Because temperature affects the specific gravity of a substance, the specific gravity reading should be corrected by adding or subtracting 4 points (0.004) for each 5.6°C (10°F) above or below the standard of 26.7°C (80°F). Most hydrometers have a built-in thermometer to measure the temperature of the electrolyte **(Figure 18–25)**. The hydrometer reading can be misleading if it is not adjusted. For example, a reading of 1.260 taken at –6.6°C (20°F) would be 1.260–(6 × 0.004, or 0.024) = 1.236.

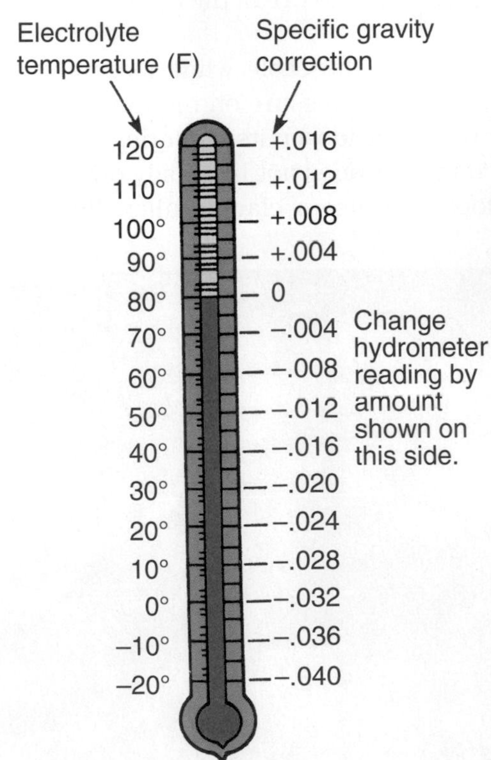

FIGURE 18–25

Hydrometers with thermometer correction scales make adjusting for electrolyte temperature easy.

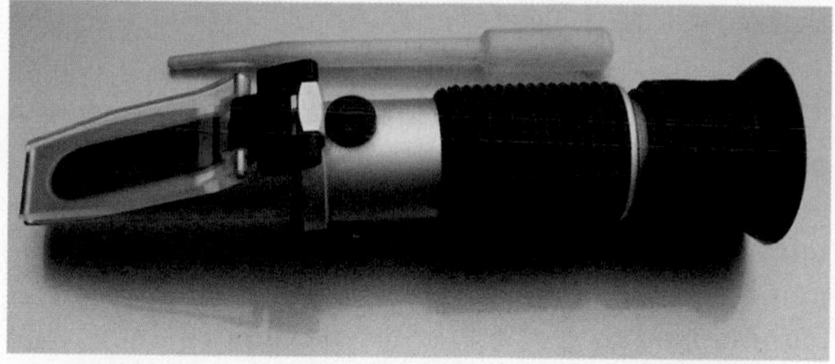

FIGURE 18–26

Electrolyte and coolant testing refractometer.

Martin Restoule

This lower reading means the cell has less charge than indicated. The specific gravity of the cells of a fully charged battery should be near 1.265 when corrected for electrolyte temperature. Recharge any battery if the specific gravity drops below an average of 1.230.

Many shops use a **refractometer (Figure 18–26)** to test not only the specific gravity of the battery but also engine coolant. A single drop of battery electrolyte is placed on a flat plate at the front of the tester. The unit is then held up to the light, and the specific gravity can be seen as the darker area on the screen **(Figure 18–27)**. Care must be taken when handling the electrolyte, and all surfaces and droppers must be cleaned after use.

Built-In Hydrometers

On some maintenance-free batteries, a special temperature-compensating hydrometer is built into the battery case **(Figure 18–28)**. A quick visual check indicates the battery's SOC **(Figure 18–29)**. It is important when observing the hydrometer that the battery has a clean top to see the correct indication. A flashlight may be required in poorly lit areas. Always look straight down when viewing the hydrometer.

A few battery designs have a charge indicator on the top of the case **(Figure 18–30)**. These batteries use a colour display to note the SOC. Typically, green means "okay," grey means "Check for recharge," and white means "Change or replace."

Many batteries do not have a built-in hydrometer. A voltage check is the only way to check the battery's SOC. The specific gravity of these batteries cannot

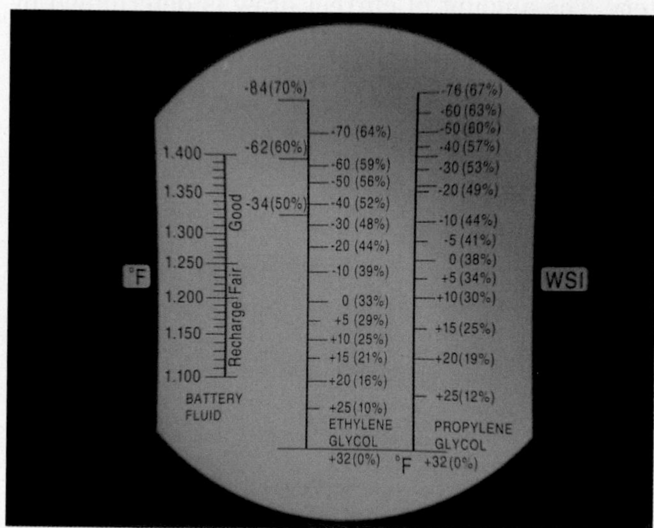

FIGURE 18–27

Typical viewing screen through a refractometer.

Martin Restoule

FIGURE 18–28

Sight glass in a maintenance-free battery.

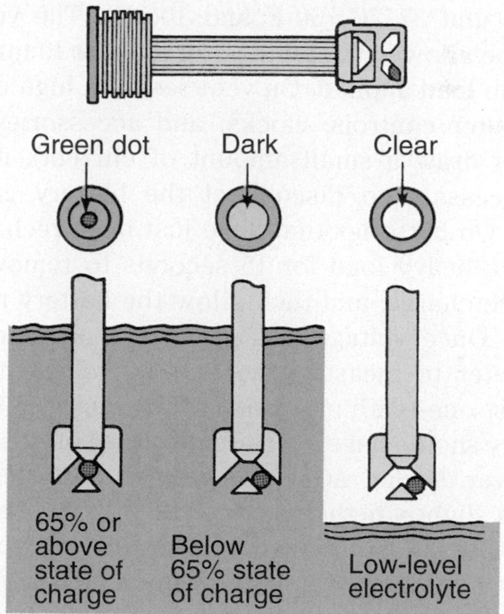

FIGURE 18–29

Design and operation of built-in hydrometers on maintenance-free sealed batteries.

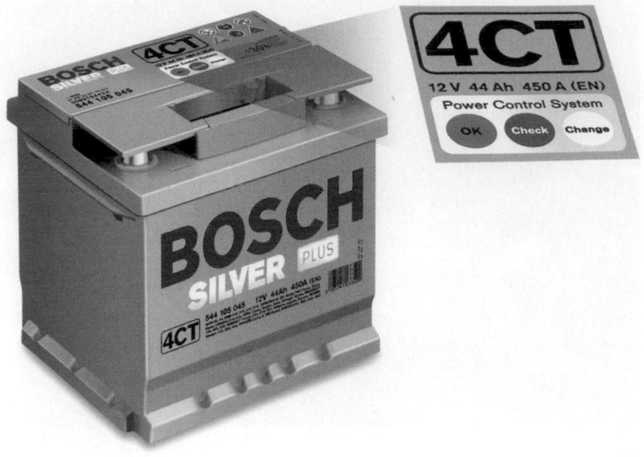

FIGURE 18–30

A battery with a built-in charge indicator.

Robert Bosch Corp.

be checked because the batteries are sealed. Never pry the cell caps off to check the electrolyte; leave the battery sealed.

Open Circuit Voltage Test

An open circuit voltage check can be used as a substitute for the hydrometer specific gravity test on maintenance-free sealed batteries with no built-in hydrometer. As the battery is charged or discharged, slight changes occur in the battery's voltage. Therefore, battery voltage with no load applied can give some indication of the SOC.

The battery's temperature should be between 15.5°C and 37.7°C (60°F and 100°F). The voltage must be allowed to stabilize for at least 10 minutes with no load applied. On vehicles with high drains (computer controls, clocks, and accessories that always draw a small amount of current), it may be necessary to disconnect the battery ground cable. On batteries that have just been recharged, apply a heavy load for 15 seconds to remove the surface charge, and then allow the battery to stabilize. Once voltage has stabilized, use a digital voltmeter to measure the battery voltage to the nearest one-tenth of a volt. A fully charged 12-volt battery should have a terminal voltage of 12.6 volts. However, sealed AGM and gel cell batteries may have a slightly higher voltage (12.8–12.9 volts). Use Figure 18–24 to interpret the results. As you can see, minor changes in battery open circuit voltage can indicate major changes in SOC. If the test indicates an SOC below 75 percent, recharge the battery, and perform the capacity test to determine battery condition.

Battery Load Test

The **load** or **capacity test** determines how well a battery performs under a load. It determines the battery's ability to furnish starting current and still maintain sufficient voltage to operate other systems. A battery load tester **(Figure 18–31)** or a volt/ampere tester (VAT) can be used for this test. A VAT can be used to test batteries, starting systems, and charging systems. Both testers have a voltmeter, an ammeter, and a carbon pile. The carbon pile is a variable resistor. When the tester is attached to the battery and operated, the carbon pile will draw current from the battery. The ammeter will read the amount of current draw. The maximum current draw from the battery, with acceptable voltage, is compared to the rating of the battery to see if the battery is okay.

Some battery load testers and VATs automatically adjust the load or carbon pile. Battery information is inputted into these machines, and they do the rest **(Figure 18–32)**.

The load or capacity test can be performed with the battery either in or out of the vehicle. The battery must be at or very near a full SOC. Use the specific gravity test or open circuit voltage test to determine the SOC, and recharge the battery if needed. For best results, the electrolyte should be as close to 26.7°C (80°F) as possible. Cold batteries show considerably lower capacity. Never load test a sealed battery if its temperature is below 15.5°C (60°F). Photo Sequence 14 shows the correct way to use a VAT-40 to conduct a load test.

During a load test, a load that simulates the current draw of a starting motor is put on the battery. The amount of current draw is determined by

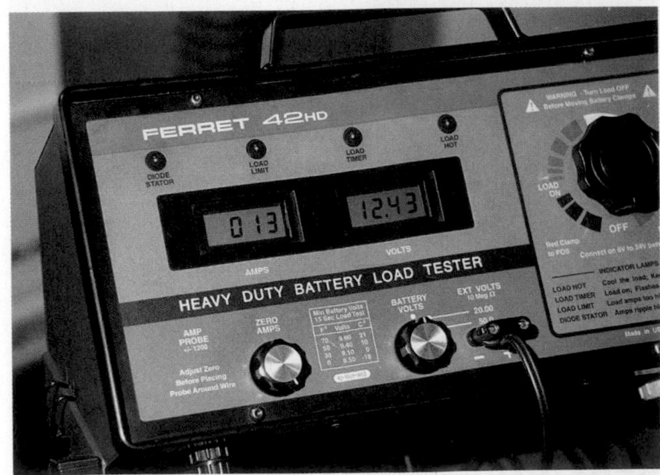

FIGURE 18–31

A battery load tester.

FIGURE 18–32

A typical automatic VAT.

SPX Corporation, Aftermarket Tool and Equipment Group

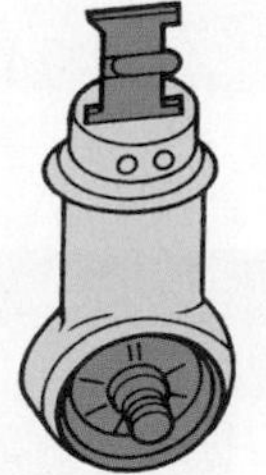
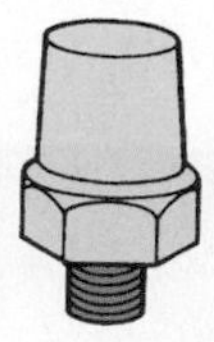
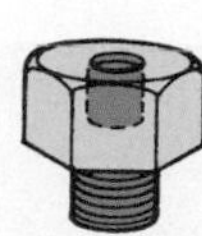

FIGURE 18–33

Adapters may be needed to test and change batteries with side-mount terminals.

the rating of the battery, and the battery's voltage is observed for 15 seconds.

When performing a battery load test, follow these guidelines:

- Make sure that the inductive pickup surrounds all of the wires from the battery's negative terminal.
- Observe the correct polarity and make sure that the test leads are making good contact with the battery posts.
- If the tester is equipped with an adjustment for battery temperature, set it to the proper setting.
- To test the battery, turn the load control knob to draw current at the rate of three times the battery's ampere-hour rating or one-half of its CCA rating.
- Discontinue the load after 15 seconds.

SHOP TALK

On some batteries with side terminals, obtaining a sound connection for the tester can be a problem. The best solution is to use the appropriate adapter **(Figure 18–33)**. If an adapter is not available, use a 3/8-inch coarse bolt with a nut on it. Bottom out the bolt. Back it off a turn. Then tighten the nut against the contact. Now attach the lead to the nut.

INTERPRETING RESULTS If the voltage reading exceeds 10.5 volts, the battery is supplying sufficient current with a good margin to safety. If the reading is right on the spec, the battery might not have the reserve necessary to handle cranking during low temperatures.

After the load is turned off, if the voltage reads below the minimum, watch the voltmeter. If it rises above 12.4, the battery is bad—it can hold a charge but cannot supply the required current. The battery can be recharged and retested, but the results are likely to be the same.

If the voltage tests below the minimum and the voltmeter does not rise above 12.4 when the load is removed, the problem may only be a low SOC. Recharge the battery and load test again.

Battery Capacitance Test

Many manufacturers recommend that a **capacitance** or **conductance test** be performed on batteries. Conductance describes a battery's ability to conduct current. It is a measurement of the plate surface available in a battery for chemical reaction. Measuring conductance provides a reliable indication of a battery's condition and is correlated to battery capacity. Conductance can be used to detect cell defects, shorts, normal aging, and open circuits, which can cause the battery to fail.

A fully charged new battery will have a high conductance reading of anywhere from 110 percent to 140 percent of its CCA rating. As a battery ages, the plate surface can sulphate or shed active material, which will lower its capacity and conductance. A conductance test is the only test that will yield accurate measurements of a battery with low SOC.

When a battery has lost a significant percentage of its cranking ability, the conductance reading will fall well below its rating, and the test decision will be to replace the battery. Because conductance measurements can track the life of the battery, they are also effective for predicting end of life before the battery fails.

PHOTO SEQUENCE 14

CONDUCTING A BATTERY LOAD TEST

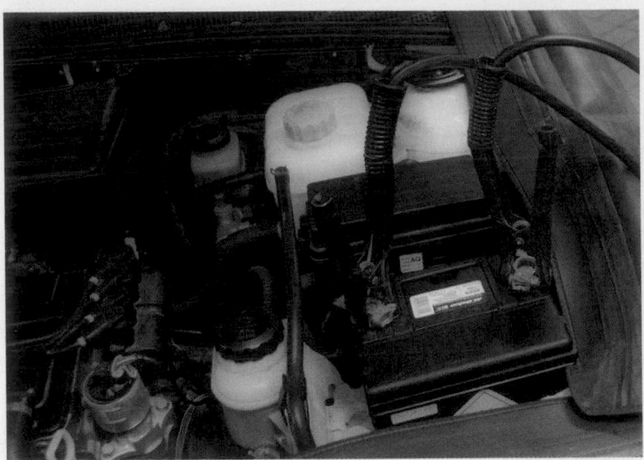

P14–1 To conduct this test, you need a VAT-40 or similar equipment. Before conducting the test, make sure the battery is fully charged and identify its CCA rating. This battery has a rating of 630 CCA.

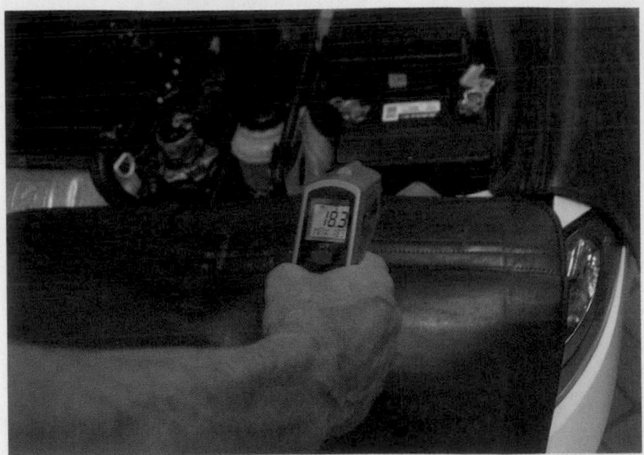

P14–2 Check the temperature of the battery; it should be around 21°C (70°F).

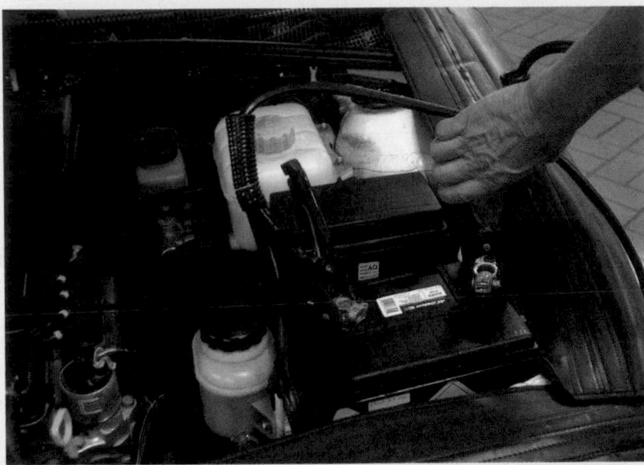

P14–3 Attach the tester's battery leads to the correct terminals on the battery.

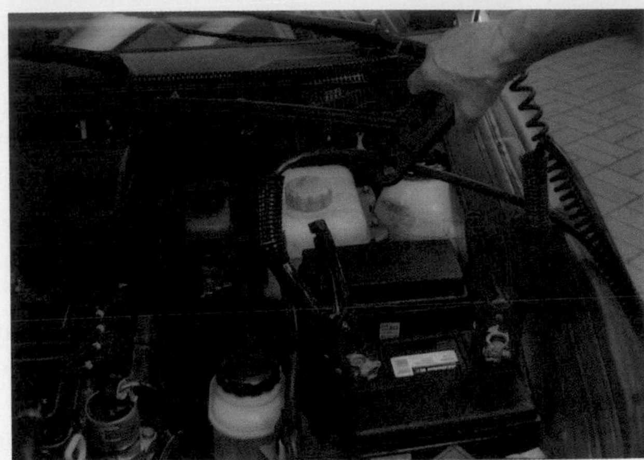

P14–4 Clamp the tester's inductive lead around the tester's negative cable. Normally, the arrow on the clamp should face away from the negative post of the battery.

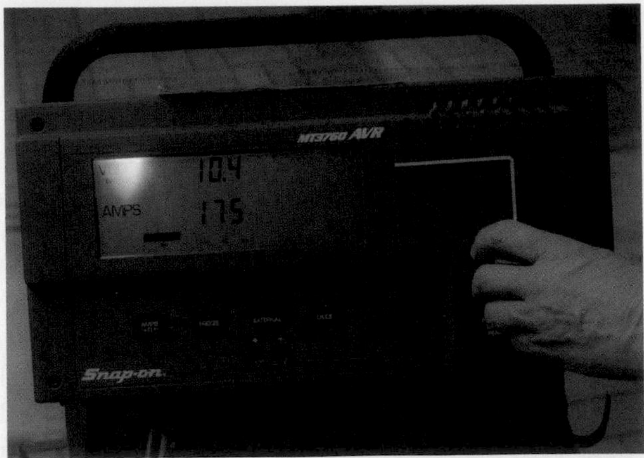

P14–5 Zero the ammeter then rotate the load control until the ammeter reads one-half of the battery's CCA rating, which in this case is 175 amps.

P14–6 After 15 seconds, read the voltage of the battery and turn the load control off. Then disconnect the tester's leads.

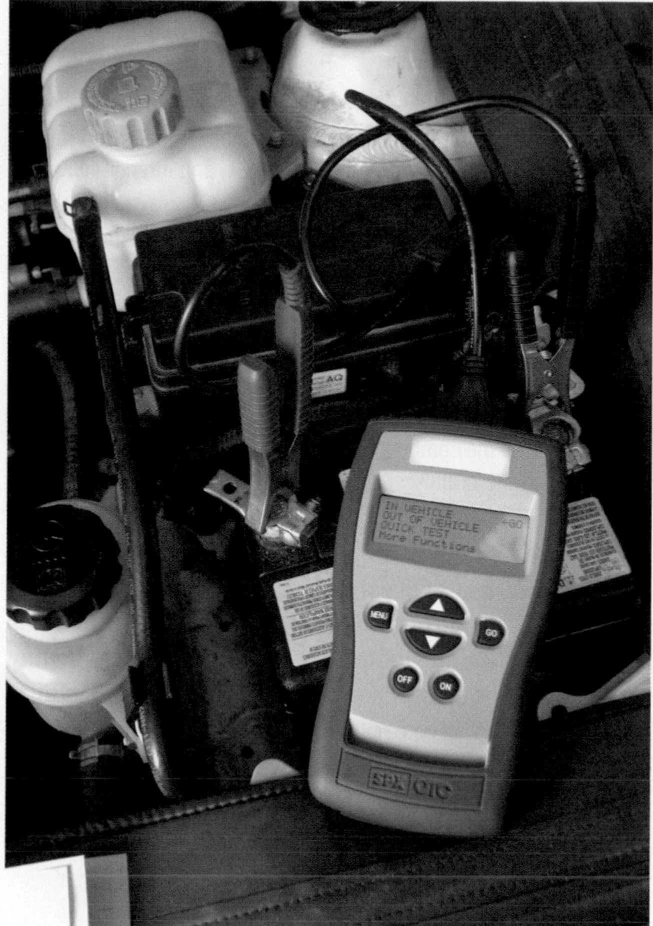

FIGURE 18–34

A conductance (capacitance) battery tester.

Sandy Matos

To measure conductance, the tester **(Figure 18–34)** creates a small signal that is sent through the battery and then measures a portion of the AC current response. The tester displays the service condition of the battery. The tester will indicate that the battery is good, needs to be recharged and tested again, has failed, or will fail shortly.

Many conductance testers will display a code at the end of the test. Be sure to look up this code in the operating manual before taking any action in response to the test results.

SHOP TALK

When testing an AGM battery with a typical capacitance tester, add 100 to the CCA rating of the battery when asked to enter the rating by the tester. If this is not done, all test results will be wrong.

Battery Drains

Many of today's vehicles have **parasitic loads**. Parasitic loads are current drains that exist when the key is off. This drain is caused by systems that operate when the engine is not running. A parasitic load is normal unless it exceeds specifications. These drains can cause a battery to lose its charge overnight or after a few days **(Table 18–1)**. The drains can also deplete the battery and cause various driveability problems. The computer may go into its backup mode, set false codes, or raise idle speeds to compensate for the low battery voltage. A constantly low battery will also have a shortened service life.

Most manufacturers have a specification for the maximum allowable amount of parasitic drain. All vehicles will have some because small amounts of current are needed to maintain the memory in various systems. A drain of 30 mA is normal for most. This will not cause a battery to lose its charge quickly. Excessive current drains are caused by problems and they will run down a battery. The most common cause is a light that is not turning off—such as the glove box, trunk, or engine compartment light. The problem can also be in an electronic system. Many of these systems are designed to periodically monitor conditions; these episodes are called wake-up times. If an electronic unit wakes up but does not shut down soon afterward, it is malfunctioning and will drain the battery.

When a battery quickly loses its charge, a battery drain test should be conducted. The current

TABLE 18–1 HOW LONG IT TAKES DIFFERENT CURRENT DRAINS TO DROP A TYPICAL BATTERY'S STATE OF CHARGE (SOC) TO 50 PERCENT

CONSTANT CURRENT DRAIN	50% SOC IN:
25 mA	30 ½ days
50 mA	16 ½ days
100 mA	8 ¼ days
250 mA	3 ¼ days
500 mA	1 ½ days
750 mA	1 day
1 A	19 hours
2 A	12 hours

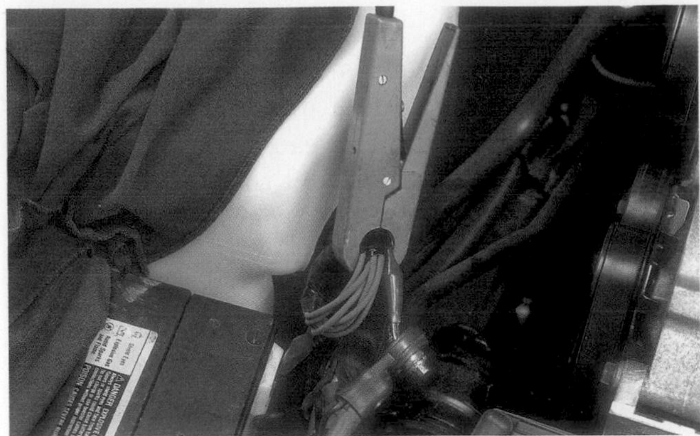

FIGURE 18–35

Using a multiplying coil to obtain accurate readings when measuring parasitic drains with a VAT-40 or similar tester.

drain can be measured with a digital multimeter (DMM) connected in series with the negative battery cable or by placing a low current probe around the cable. Current drain can also be measured with a high-current tester, such as a VAT. This requires the installation of a multiplying coil between the negative battery cable and the battery terminal post **(Figure 18–35)**. The tester's inductive probe is placed around the multiplying coil.

SHOP TALK

Remember to connect a memory saver before disconnecting the battery.

PROCEDURE

To measure the parasitic drains on a battery:

STEP 1 Make sure the ignition is off and the key is out of the switch.

STEP 2 Turn off all accessories.

STEP 3 Place the ignition switch and all vehicle accessories in the off position. Some manufacturers recommend to wait 20 minutes after the ignition is off before testing.

STEP 4 Disconnect the underhood lamp, if equipped.

STEP 5 Set the DMM to read DC amps.

STEP 6 Zero the ammeter.

STEP 7 Place the probe around the negative battery cable or insert the meter in series with the cable.

STEP 8 Read the ammeter.

STEP 9 If the parasitic load is under 2 amps, change the DMM to read mAmps.

STEP 10 Read the ammeter.

STEP 11 If the parasitic drain is more than the specified maximum, check the trunk, glove box, and interior lights to see if they are on.

STEP 12 If a light was on, remove the bulb and watch the battery drain. If the drain is now okay, check that circuit.

STEP 13 If the cause of the drain is not the lights, find the problem circuit by removing one fuse at a time while watching the ammeter. The reading will drop when the fuse on the bad circuit is removed.

STEP 14 Test the components of that circuit to identify the cause of the drain.

Sometimes, excessive battery drain will not be measured at the time of the test. If the battery runs down quickly and has passed all other tests, it may be necessary to monitor the drain overnight. Doing this will verify any problems related to the wake-up modes in the various systems. To do this, use the MIN/MAX feature of the DMM, and monitor the parasitic drain overnight.

Battery Cables

Sometimes a battery's performance is affected by its cables. Poor connections and corrosion at the terminals and cable ends will cause voltage drops, which will reduce the available voltage at all of the vehicle's systems. A visual inspection may locate these problems, but a voltage drop test across the connections is the best way to identify this problem. Heavily corroded terminals **(Figure 18–36)** are obvious, but

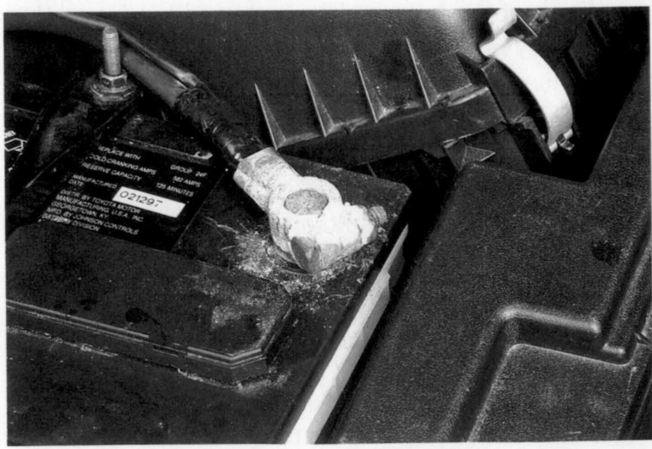

FIGURE 18–36

Corroded battery terminals reduce the efficiency of the battery and cause voltage drops.

sometimes corrosion forms a nearly invisible barrier between the battery terminals and cables. This can be unwanted resistance and will prevent the proper amount of current and voltage from going where they should.

The terminals may also be out of shape, loose, or damaged; these will cause excessive resistance. All battery cables should be inspected and tested for excessive voltage drops. All voltage drop tests should be done under a load. Operate a high-current system while measuring the voltage drop. In most cases, running the starter motor is the best load. Allow a drop of 0.1 volt for each cable connection; a total voltage drop across all battery connections should not exceed 0.2 volt.

Battery Chargers

Battery chargers are designed to supply a constant voltage or a constant current, or a mixture of the two. Constant voltage chargers provide a specific amount of voltage to the battery. The current varies with voltage of the battery. When the potential difference between the charger's voltage and the battery's voltage is great, the current is high. As the battery charges, its voltage increases, and the charging current drops off. A constant current charger varies the voltage applied to the battery in order to maintain a constant current.

Both of these techniques work fine as long as the temperature of the battery is maintained. Some chargers have a thermometer to monitor battery temperature. These chargers reduce the charging voltage and/or current in response to rising temperatures.

There are many battery chargers that are "smart" or "intelligent." These chargers are designed to charge a battery in three basic steps: bulk, absorption, and float. During bulk charging, current is sent to the battery at a maximum safe rate until the voltage reaches approximately 80 percent of its capacity. Once the battery reaches that voltage level, the charger begins the absorption step. During this time, charging voltage is held constant while the current changes according to the battery's voltage. Once the battery is fully charged, the charger switches to the float step. During this step, the charger supplies a constant voltage equal to slightly more than the voltage of the battery. The current flow is very low. This step is a maintenance charge and is intended to keep a battery charged while it is not being used.

Chargers can also be designed to supply voltage and current to a battery according to the needs of the vehicle and, often, the needs of the customer. **Fast charging** quickly charges a battery. This supplies large amounts of voltage and current. Although this charges the battery quickly, it also can overheat it if it is not closely monitored. This technique is best used when a battery is low on charge and will be installed into the vehicle in a short time. Batteries must be in good condition to accept a fast charge. Sulphation on the plates of the battery can lead to excessive gassing, boiling, and heat buildup during fast charging. Never fast-charge a battery that shows evidence of sulphation buildup or separator damage.

Slow charging or **trickle charging** applies low current to the battery and takes quite some time to fully charge it. However, it is unlikely that the battery will overheat and the battery has a good chance to be completely charged. Slow charging is the only safe way of charging sulphated batteries. The chemicals used in the construction of the battery should always be considered before fast charging or slow charging a battery.

Always charge a battery in a well-ventilated area away from sparks and open flames. The charger should be off before connecting or disconnecting the leads to the battery **(Figure 18–37)**. Remember to wear eye protection, and never attempt to charge a frozen battery.

All battery chargers have manufacturer-specific characteristics and operating instructions that must be followed. When charging a battery in the vehicle, always disconnect the battery cables to avoid damaging the generator or other electrical parts.

FIGURE 18–37

Careless use of a battery charger caused this battery to explode.

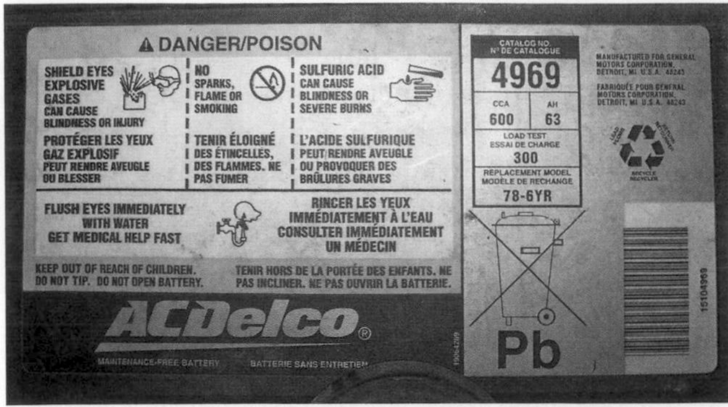

FIGURE 18–38
Battery sticker with identification and warnings.

> **CAUTION!**
>
> Do not exceed the manufacturer's battery-charging limits. Also, never charge the battery if the built-in hydrometer is clear or light yellow. Replace the battery.

Replacing a Battery

When a battery needs to be replaced, several things need to be considered when selecting the new battery: its capacity, ratings, and size. Make sure the new battery meets or exceeds the power requirement of the vehicle. The battery needs to fit the battery holding fixtures in the vehicle. The height of the battery is also important. The top of the battery and its terminals must fit safely under the hood without the possibility of shorting across the terminals. BCI group numbers are normally listed on the battery **(Figure 18–38)** and are used to state its physical size and rating.

JUMP-STARTING

When it is necessary to jump-start a car with a discharged battery, using a booster battery and jumper cables, follow the instructions shown in **Figure 18–39** to avoid damaging the charging system or creating a hazardous situation. Pay attention to the following precautions:

- Always wear eye protection when making or breaking jumper cable connections.
- Be sure the connections are done correctly and tightly.
- Make sure all electrical accessories of both vehicles are off.
- When making the connections, do not lean over the battery or accidentally let the jumper cables or clamps touch anything except the correct battery terminals.
- Use only a 12-volt supply as the booster battery.
- The gases around the battery can explode if exposed to flames, sparks, or lit cigarettes.

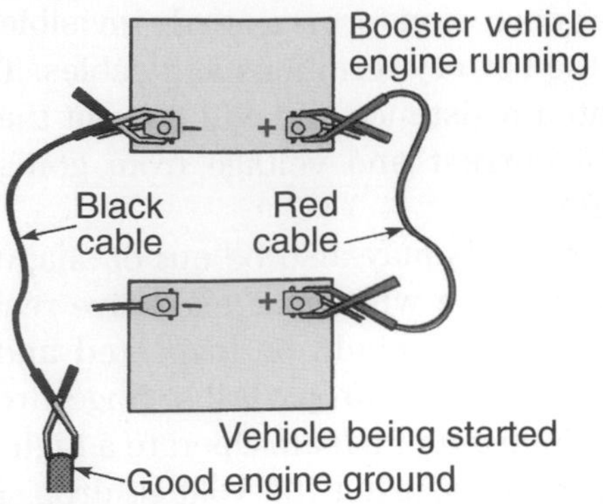

FIGURE 18–39
Proper setup and connections for jump-starting a vehicle with a low battery.

KEY TERMS

Absorbed glass mat (AGM) batteries (p. 546)
Ampere-hour (AH) rating (p. 540)
Battery Council International (BCI) (p. 544)
Capacitance test (p. 555)
Capacity test (p. 554)
Cold cranking amps (CCA) rating (p. 540)
Conductance test (p. 555)
Cranking amps (CA) rating (p. 540)
Electrochemical reactions (p. 537)
Electrodes (p. 537)
Element (p. 542)
Fast charging (p. 559)
Grid (p. 542)
Hybrid battery (p. 545)
Lead peroxide (PbO_2) (p. 542)
Lithium-ion (Li-Ion) (p. 541)
Lithium-polymer (Li-Poly) (p. 542)
Load test (p. 554)
Maintenance-free battery (p. 545)
Nickel-cadmium (NiCad) (p. 541)
Nickel-metal hydride (NiMH) (p. 541)
Open circuit voltage (p. 540)
Parasitic loads (p. 557)
Plates (p. 537)
Plate strap (p. 542)
Recombination (or recombinant) battery (p. 546)
Refractometer (p. 553)
Reserve capacity (RC) rating (p. 540)
Separators (p. 542)
Slow charging (p. 559)
Specific gravity (p. 552)
Sponge lead (Pb) (p. 542)
Sulphation (p. 547)
Trickle charging (p. 559)
Valve-regulated lead-acid (VRLA) batteries (p. 546)
Watt-hour rating (p. 540)

SUMMARY

- The primary source for electrical power in all automobiles is the battery.
- Batteries are devices that convert chemical energy into electrical energy through electrochemical reactions.
- Batteries are normally made up of electrochemical cells connected together. Each cell has three major parts: an anode (negative plate or electrode), a cathode (positive plate or electrode), and electrolyte.
- Open circuit voltage is the voltage measured across the battery when there is no load on the battery.
- Operating voltage is the voltage measured across the battery when it is under a load.
- The ampere-hour (AH) rating is the amount of steady current that a fully charged battery can supply for 20 hours at 26.7°C (80°F) without the cell's voltage dropping below a predetermined level.
- A battery's watt-hour rating is determined at –17.8°C (0°F) and is calculated by multiplying a battery's amp-hour rating by the battery's voltage.
- The cold cranking amps (CCA) rating is determined by the load, in amperes, that a battery is able to deliver for 30 seconds at –17.8°C (0°F) without its voltage dropping below a predetermined level.
- The cranking amps (CA) rating is a measure of the current a battery can deliver at 0°C (32°F) for 30 seconds and maintain voltage at a predetermined level.
- The reserve capacity (RC) rating is determined by the length of time, in minutes, that a fully charged starting battery at 26.7°C (80°F) can be discharged at 25 amperes before battery voltage drops below 10.5 volts.
- There are many different types and designs of batteries available; the primary difference is the chemicals used in the cells, such as lead-acid, nickel-cadmium (NiCad), nickel-metal hydride (NiMH), sodium-sulphur (NaS), sodium-nickel-chloride, lithium-ion (Li-Ion), lithium-polymer (Li-Poly), and nickel-zinc.
- Nickel-metal hydride (NiMH) cells are available in the cylindrical and prismatic designs and both are used in hybrid vehicles.
- Lithium-based batteries are similar in construction to nickel-based batteries but have higher energy density, suffer less from the memory effect, and are environmentally friendlier.
- There are two major types of lithium-based cells: Li-Ion and Li-Poly.
- The most common automotive batteries are lead-acid designs. The wet cell, gel cell, absorbed glass mat (AGM), and valve regulated are versions of the lead-acid battery.
- A lead-acid battery consists of grids, positive plates, negative plates, separators, elements, electrolyte, a container, cell covers, vent plugs, and cell containers. Maintenance-free batteries have no holes or caps, but they do have gas vents. Sealed maintenance-free batteries do not require the gas vents used on other maintenance-free designs.
- Improper electrolyte levels, temperature, corrosion, overcharging, undercharging/sulphation, poor mounting, and cycling affect the service life of a battery.
- Depending on the design of the battery, its condition, SOC, and capacity can be measured by different tests: battery leakage test, specific gravity tests, built-in hydrometers, open circuit voltage test, capacity test, and capacitance or conductance test.
- Parasitic loads are current drains that exist when the key is off and may be normal or caused by a problem.
- To charge a battery, a given charging current is passed through the battery for a period of time. Fast chargers are more popular, but slow charging is the only safe way to charge a sulphated battery.
- Always follow the correct procedure when jump-starting a vehicle.

REVIEW QUESTIONS

1. What would be the voltage reading of a fully charged, 12-volt, automotive, lead-acid battery?
   a. 12 volts **b.** 12.4 volts
   c. 12.6 volts **d.** 12.8 volts
2. What would be the voltage of a lead-acid battery that has one shorted cell?
   a. 12.5 volts **b.** 11.5 volts
   c. 10.5 volts **d.** 9.5 volts
3. What is the minimum testing voltage required to perform a load test?
   a. 12 volts **b.** 12.4 volts
   c. 12.6 volts **d.** 12.8 volts

4. When a load test is performed at 20°C (70°F), what is the minimum acceptable battery voltage?
   a. 12.6 volts b. 11.6 volts
   c. 10.6 volts d. 9.6 volts
5. What is mixed with water to make up electrolyte in a battery?
   a. citric acid
   b. sulphuric acid
   c. hydrochloric acid
   d. nitric acid
6. What is the preferred testing tool to determine state of charge in a lead-acid battery with removable vent covers?
   a. voltmeter
   b. thermometer
   c. hydrometer
   d. ammeter
7. What material is used in the positive plates of a lead-acid battery?
   a. sponge lead
   b. aluminium dioxide
   c. lead peroxide
   d. hydrogen peroxide
8. What is the purpose of the separators?
   a. separate the cells from each other
   b. separate the positive plates from each other
   c. separate the positive and negative plates
   d. separate the negative plates from each other
9. What is the specific gravity of a fully charged, lead-acid battery, at 26.7°C (80°F)?
   a. 1.260–1.280 b. 1.160–1.180
   c. 1.060–1.080 d. 1.006–1.008
10. Why is compensation for specific gravity reading required?
   a. changes in temperature
   b. changes in volume
   c. changes in battery size
   d. changes in humidity
11. Which of the following would describe the specific gravity the electrolyte refers to?
   a. the amount of electrolyte in the battery
   b. the weight of the electrolyte/water mix
   c. the weight of the acid/electrolyte mix
   d. the weight of the acid/water mix
12. What is indicated by the "eye," or battery indicator, in a sealed battery?
   a. the state of charge
   b. an electrolyte level that is too high
   c. the battery polarity
   d. battery plate sulphation
13. Between which points is a parasitic drain measured?
   a. the positive and negative terminals
   b. the positive post and the positive cable in series
   c. the negative post and the positive cable in series
   d. the negative post and the negative cable in series
14. Which is considered a normal range of parasitic drain?
   a. 25 to 50 amps
   b. 2.5 to 5.0 amps
   c. 0.25 to 0.50 amps
   d. 0.025 to 0.050 amps
15. CCA refers to the battery's ability to deliver current at what rate and temperature?
   a. for 30 seconds at 17.8°C (0°F)
   b. for 15 seconds at 17.8°C (0°F)
   c. for 120 minutes at 0°C (32°F)
   d. for 15 seconds at 0°C (32°F)
16. What battery rating refers to its ability to deliver 25 amps and maintain 1.75 volts per cell?
   a. cranking amperes
   b. amp-hour rating
   c. reserve capacity
   d. open circuit voltage
17. What test is being performed by placing a voltmeter's positive lead on the positive terminal and the negative lead on the case of the battery?
   a. draw test b. leakage test
   c. load test d. capacitance test
18. What is another name for an AGM battery?
   a. spiral cell battery
   b. lead-acid battery
   c. gel cell battery
   d. valve-regulated lead-acid battery
19. Which of the following statements best characterizes deep-cycle batteries?
   a. They are used in underground applications.
   b. They are used to provide maximum amperage output.
   c. They can be continually discharged and recharged.
   d. They can be continually used in an overcharged state.
20. What problem may occur in a discharged battery in cold weather?
   a. freezing b. boiling
   c. overcharging d. undercharging

CHAPTER 19
Starting Systems

LEARNING OUTCOMES

- Explain the purpose of the starting system.
- List the components of the starting system, starter circuit, and control circuit.
- Explain the different types of magnetic switches and starter drive mechanisms.
- Explain how a starter motor operates.
- Describe the operation of the different types of starter motors.
- Perform basic tests to determine the problem areas in a starting system.
- Perform and accurately interpret the results of a current draw test.
- Disassemble, clean, inspect, repair, and reassemble a starter motor.

The vehicle's starting system is designed to turn or crank the engine over until it can operate under its own power. To do this, the starter motor receives electrical power from the battery. The starter motor then converts this energy into mechanical energy, which it transmits through the drive mechanism to the engine's flywheel **(Figure 19–1)**.

The only function of the starting system is to crank the engine fast enough to run. The vehicle's ignition and fuel systems provide the spark and fuel for engine operation, but they are not considered components of the basic starting system.

FIGURE 19–1

A starter motor meshed with the engine's flex plate (or flywheel) ring gear.

STARTING SYSTEM—DESIGN AND COMPONENTS

A typical starting system has six basic components and two distinct electrical circuits. The components are the battery, ignition switch, battery cables, magnetic switch (either an electrical relay or a solenoid), starter motor, and the starter safety switch.

The starter motor **(Figure 19–2)** draws a great deal of current from the battery. A large starter motor might require 250 or more amperes of current. This current flows through the large cables that connect the battery to the starter and ground.

The driver controls the flow of this current, using the ignition switch normally mounted on

FIGURE 19–2

A cutaway of a heavy-duty starter motor.

Robert Bosch Corp.

the steering column. The battery cables are not connected to the switch. Rather, the system has two separate circuits: the starter circuit and the control circuit **(Figure 19–3)**. The starter circuit carries the heavy current from the battery to the starter motor through a magnetic switch in a relay or solenoid. The control circuit connects battery power at the ignition switch to the magnetic switch, which controls the high current to the starter motor.

Starter Circuit

The starter circuit carries the high current flow within the system and supplies power for the actual engine cranking. Components of the starter circuit are the battery, battery cables, magnetic switch or solenoid, and the starter motor.

BATTERY AND CABLES Many of the problems associated with the starting system can be solved by troubleshooting the battery and its related components.

The starting circuit requires two or more heavy-gauge cables. One of these cables connects between the battery's negative terminal and the engine block or transmission case. The other cable connects the battery's positive terminal with the solenoid. On vehicles equipped with a **starter relay**, two positive cables are needed. One runs from the positive battery terminal to the relay, and the second from the relay to the starter motor terminal. In any case, these cables carry the required heavy current from the battery to the starter and from the starter back to the battery.

Cables must be heavy enough to comfortably carry the required current load. Cranking

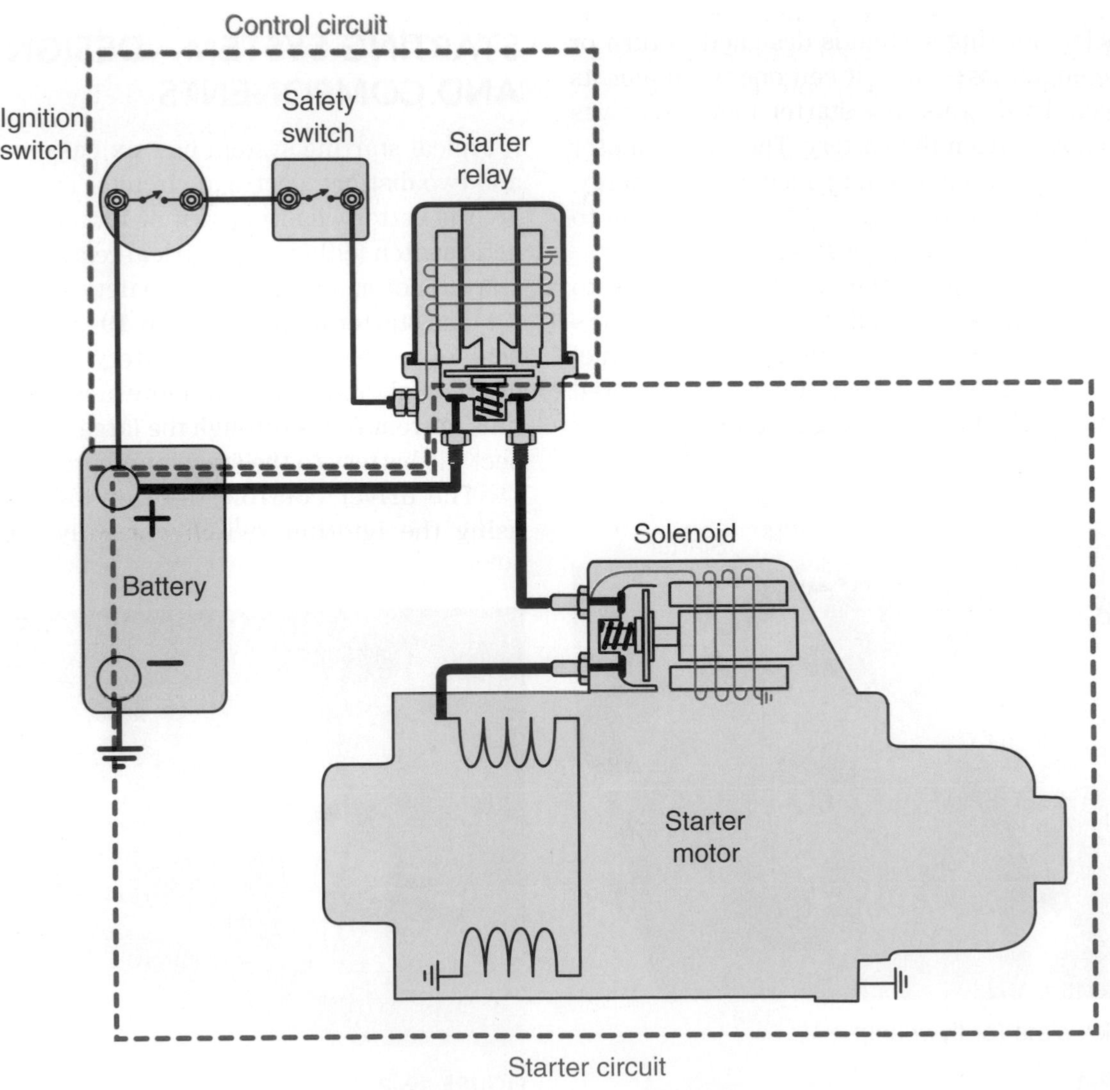

FIGURE 19–3

A simple diagram showing the starter and starter control circuits.

problems can be created when undersized cables are installed. With undersized cables, the starter motor does not develop its greatest turning effort and even a fully charged battery might be unable to start the engine.

MAGNETIC SWITCHES Every starting system contains some type of magnetic switch that enables the control circuit to open and close the starter circuit. This magnetic switch can be one of several designs.

Solenoid. The solenoid-actuated starter is by far the most common starter system used. A **solenoid** is an electromechanical device that uses the movement of a plunger to exert a pulling or holding force. As shown in **Figure 19–4**, the solenoid mounts directly on top of the starter motor.

In this type of starting system, the solenoid uses the electromagnetic field generated by its coil to perform two distinct jobs.

The first is to push the drive pinion of the starter motor into mesh with the engine's flywheel. This is the solenoid's mechanical function. The second job is to act as an electrical relay switch to energize the motor once the drive pinion is engaged. Once the contact points of the solenoid are closed, full battery current flows to the starter motor.

The solenoid assembly has two separate windings: a **pull-in winding** and a **hold-in winding**. The two windings have approximately the same number of turns but are wound from different size wire. Together these windings produce the electromagnetic force needed to pull the plunger into the solenoid coil.

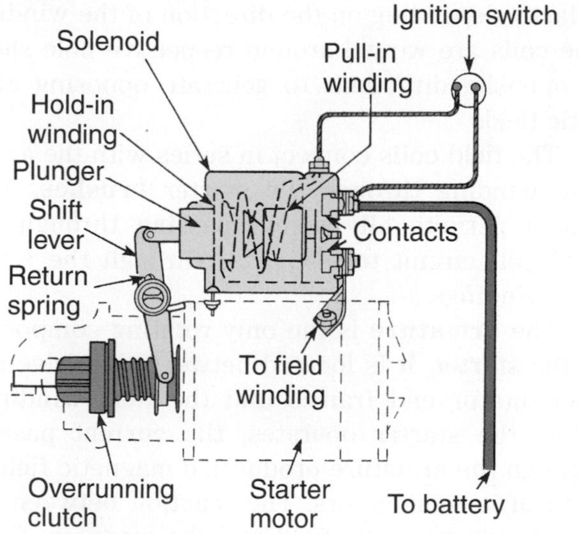

FIGURE 19–4

An example of a solenoid-actuated starter where the solenoid mounts directly to the starter motor.

The heavier pull-in windings draw the plunger into the solenoid, and the lighter-gauge hold-in windings produce enough magnetic force to hold the plunger in this position.

Both windings are energized when the ignition switch is turned to the start position. When the plunger disc makes contact with this solenoid terminal, the pull-in winding is deactivated. At the same time, the plunger contact disc makes the motor feed connection between the battery and the starting motor, directing current to the field coils and starter motor armature for cranking power.

As the solenoid plunger moves, the shift fork also pivots on the pivot pin and pushes the starter drive pinion into mesh with the flywheel ring gear. When the starter motor receives current, its armature starts to turn. This motion is transferred through an overrunning clutch and pinion gear to the engine flywheel, and the engine is cranked.

With this type of solenoid-actuated direct drive starting system, teeth on the **pinion gear** may not immediately mesh with the flywheel ring gear. If this occurs, a spring located behind the pinion compresses so that the solenoid plunger can complete its stroke. When the starter motor armature begins to turn, the pinion teeth quickly line up with the flywheel teeth, and the spring pressure forces them to mesh.

Starter Relay. Relays are the second major type of magnetic switch used. All **positive engagement starters** use a relay in series with the battery cables to deliver the high current necessary through the shortest possible battery cables. **Figure 19–5** shows a typical starter relay. It is very similar to the solenoid. However, it is not used to move the drive pinion

FIGURE 19–5

A starter relay/solenoid mounted on a vehicle.

into mesh. It is strictly an electrical relay or switch. When current from the ignition switch arrives at the ignition switch terminal of the relay, a strong magnetic field is generated in the coil of the relay. This magnetic force pulls the plunger contact disc up against the battery terminal and the starter terminal of the relay, allowing full current flow to the starter motor.

A secondary function of the starter relay is to provide an alternate electrical path to the ignition coil during cranking. This current flow bypasses the resistance wire (or ballast resistor) in the ignition primary circuit. This is done when the plunger disc contacts the ignition bypass terminal on the relay. Not all systems have an ignition bypass setup.

Some vehicles use both a starter relay and a starter motor–mounted solenoid. The relay controls current flow to the solenoid, which in turn controls current flow to the starter motor. This reduces the amount of current flowing through the ignition switch.

Most starting systems seen today are either solenoid shift or solenoid shift with relay. Simple diagrams for these systems are shown in **Figure 19–6**. The positive engagement starting motor used frequently by Ford until the early 1990s is not often encountered by technicians today. A typical starting system diagram, as illustrated in **Figure 19–7**, shows the starting motor with related circuit protection and control devices.

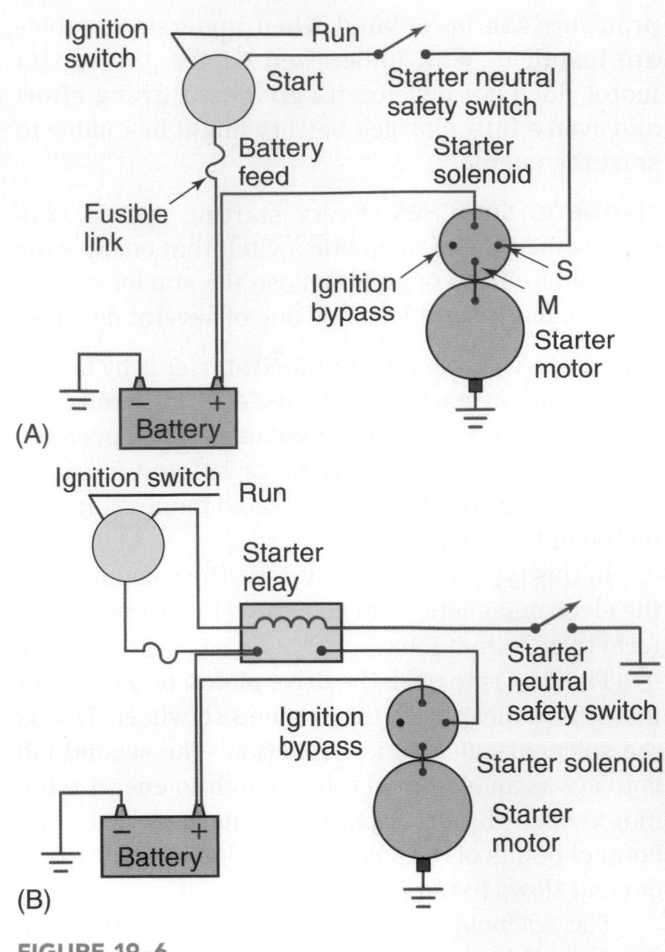

FIGURE 19–6

(A) A solenoid shift and (B) a solenoid shift with starter relay starting systems.

STARTER MOTOR The starter motor **(Figure 19–8)** converts the electrical energy from the battery into mechanical energy for cranking the engine. The starter is a special type of electric motor designed to operate under great electrical overloads and to produce very high horsepower.

All starting motors are generally the same in design and operation. Basically, the starter motor consists of a housing, field coils, an armature, a commutator and brushes, and end frames.

The **starter housing** or **starter frame** encloses the internal starter components and protects them from damage, moisture, and foreign materials. The housing supports the field coils.

The **field coils** and their **pole shoes (Figure 19–9)** are securely attached to the inside of the iron housing. The field coils are insulated from the housing but are connected to a terminal that protrudes through the outer surface of the housing.

The field coils and pole shoes are designed to produce strong stationary electromagnetic fields within the starter body as current is passed through the starter. These magnetic fields are concentrated at the pole shoe. Fields have an N or S magnetic polarity depending on the direction of the winding. The coils are wound around respective pole shoes in opposite directions to generate opposing magnetic fields.

The field coils connect in series with the armature winding through the starter **brushes**. This design permits all current passing through the field coil circuit to also pass through the armature windings.

The **armature** is the only rotating component of the starter. It is located between the drive and commutator end frames and the field windings. When the starter operates, the current passing through the armature produces a magnetic field in each of its conductors. The reaction between the armature's magnetic field and the magnetic fields produced by the field coils causes the armature to rotate. This mechanical energy is then used to crank the engine.

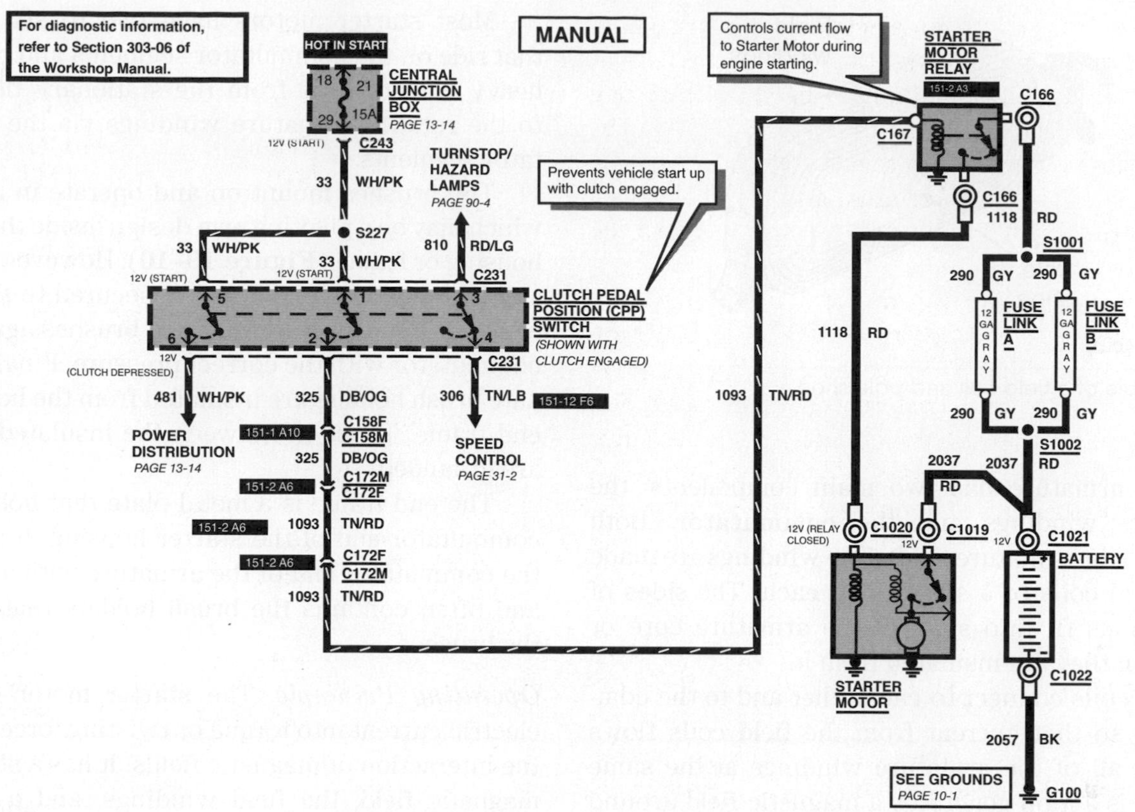

FIGURE 19–7

A schematic of a typical starting system.

Ford Motor Company

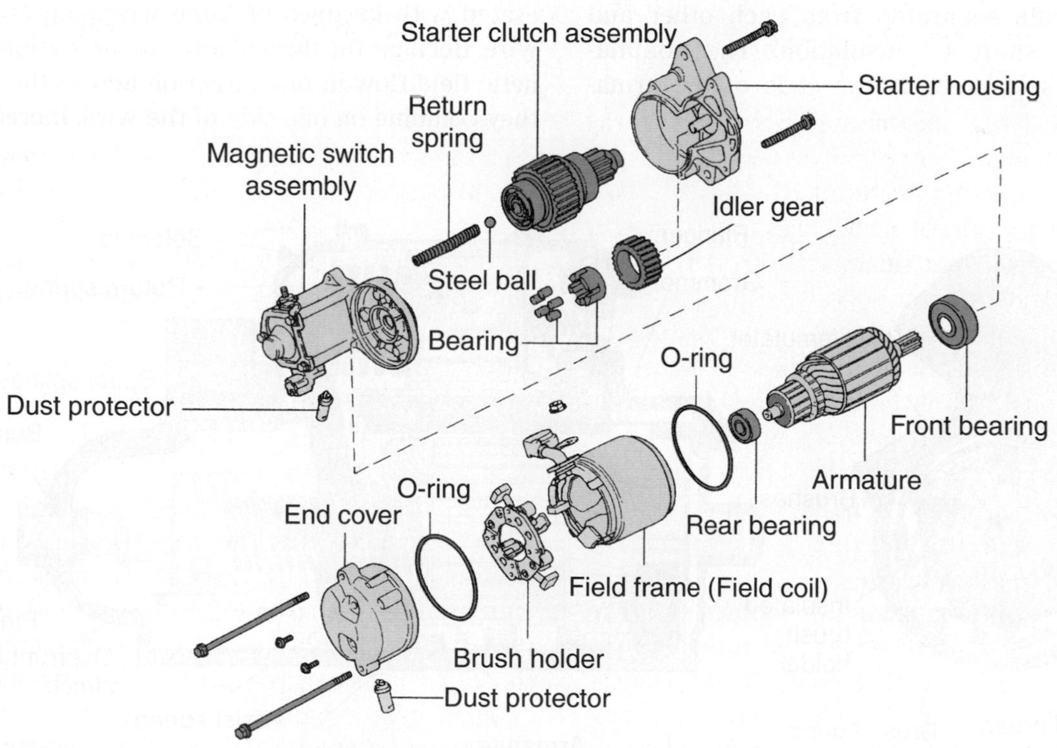

FIGURE 19–8

A typical gear reduction starter motor assembly.

Reprinted with permission.

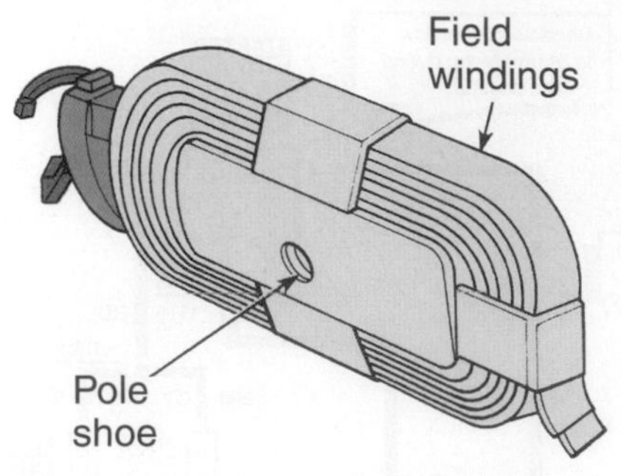

FIGURE 19–9

An example of a field coil and pole shoe.

The armature has two main components: the armature windings and the **commutator**. Both mount to the armature shaft. The windings are made of several coils of a single loop each. The sides of these loops fit into slots in the armature core or shaft, but they are insulated from it.

The coils connect to each other and to the commutator so that current from the field coils flows through all of the armature windings at the same time. This action generates a magnetic field around each armature winding, resulting in a repulsion force all around the conductor. This repulsion force causes the armature to turn.

The commutator assembly is made up of heavy copper segments separated from each other and the armature shaft by insulation. The commutator segments connect to the ends of the armature windings.

Most starter motors have two to six brushes that ride on the commutator segments and carry the heavy current flow from the stationary field coils to the rotating armature windings via the commutator segments.

The brushes mount on and operate in a holder, which may be a pivoting arm design inside the starter housing or frame **(Figure 19–10)**. However, in many starters, the brush holders are secured to the starter's end frame. Springs hold the brushes against the commutator with the correct pressure. Finally, alternate brush holders are insulated from the housing or end frame. Those in between the insulated holders are grounded.

The end frame is a metal plate that bolts to the commutator end of the starter housing. It supports the commutator end of the armature with a bushing and often contains the brush holders that support the brushes.

Operating Principle. The starter motor converts electric current into torque or twisting force through the interaction of magnetic fields. It has a stationary magnetic field, the field windings, and a current-carrying conductor, the armature windings **(Figure 19–11)**. When the armature windings are placed in this stationary magnetic field and current is passed through the windings, a second magnetic field is generated with its lines of force wrapping around the wire. Because the lines of force in the stationary magnetic field flow in one direction across the winding, they combine on one side of the wire, increasing the

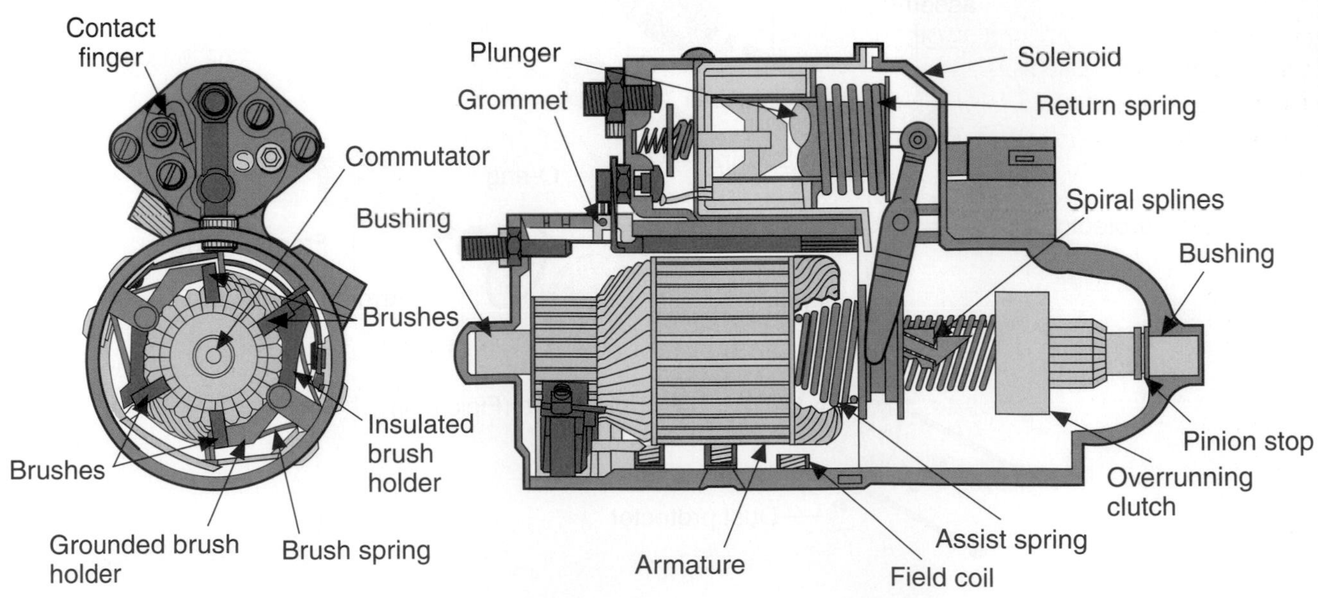

FIGURE 19–10

The location of the starter motor brushes and commutator.

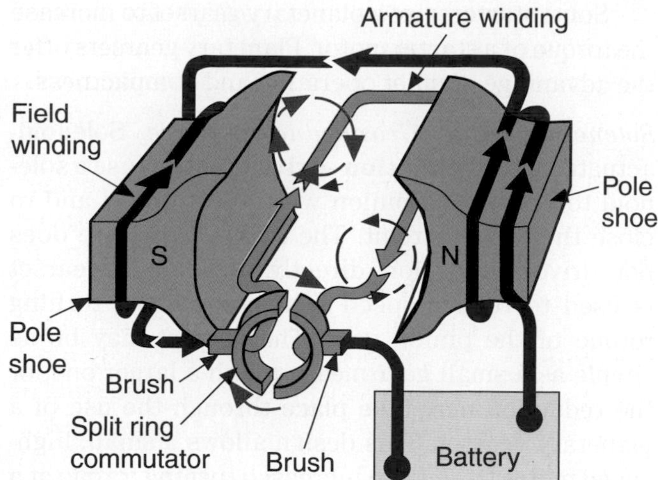

FIGURE 19–11

A simple DC series motor.

field strength, but are opposed on the other side, weakening the field strength. This creates an unbalanced magnetic force, pushing the wire in the direction of the weaker field **(Figure 19–12)**.

The armature windings are formed in loops or coils, so current flows outward in one direction and returns in the opposite direction. Because of this, the magnetic lines of force are oriented in opposite directions in each of the two segments of the loop. When placed in the stationary magnetic field of the field coils, one part of the armature coil is pushed in one direction. The other part is pushed in the opposite direction, causing the coil and the shaft to which it is mounted to rotate.

Each end of the armature winding is connected to one segment of the commutator **(Figure 19–13)**. Carbon brushes are connected to one terminal of the power supply. The brushes contact the commutator segments conducting current to and from the armature coils.

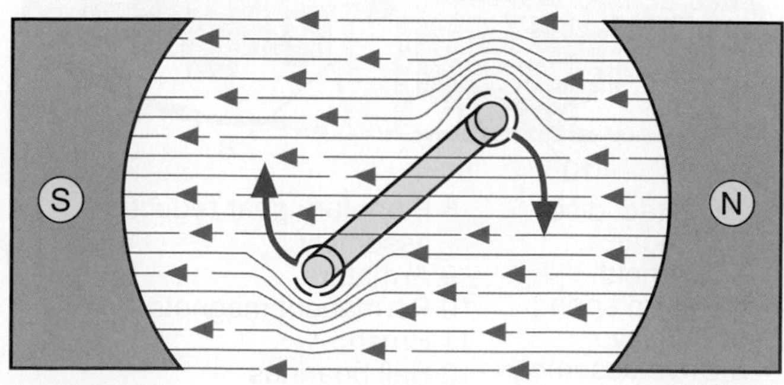

FIGURE 19–12

Rotation of the conductor is in the direction of the weaker magnetic field.

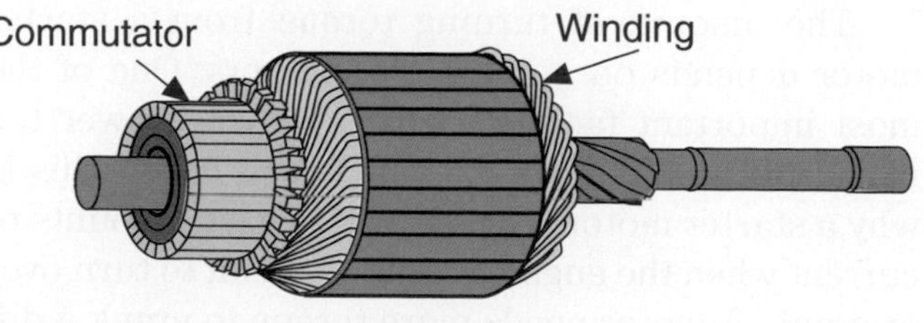

FIGURE 19–13

The armature of a starter motor.

As the armature coil turns through a half-revolution, the contact of the brushes on the commutator causes the current flow to reverse in the coil. The commutator segment attached to each coil end has travelled past one brush and is now in contact with the other. In this way, current flow is maintained constantly in one direction, while allowing the segment of the rotating armature coils to reverse polarity as they rotate.

In a starter motor, many armature segments must be used. As one segment rotates past the secondary magnetic field pole, another segment immediately takes its place. The turning motion is made uniform, and the torque needed to turn the flywheel is constant rather than fluctuating, as it would be if only a few armature coils were used.

The number of coils and brushes may differ between starter motor models. The armature may be wired in series with the field coils (**series motor**); the field coils may be wired parallel or shunted across the armature (**shunt motors**); or a combination of series and shunt wiring may be used (**compound motors**) **(Figure 19–14)**.

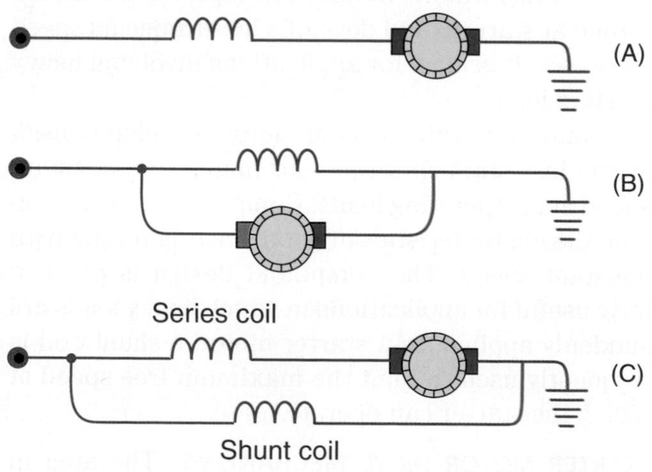

FIGURE 19–14

Starter motors are grouped according to how they are wired: (A) in series, (B) in parallel (shunt), or (C) as a compound motor using both series and shunt coils.

The amount of turning torque from a starter motor depends on a number of factors. One of the most important factors is current. The slower the motor turns, the more current it will draw. This is why a starter motor will draw excessive amounts of current when the engine is very difficult to turn over or crank. A motor needs more torque to crank a difficult-to-turn engine. The relationship between current and motor speed is explained by the principles of **counter EMF (CEMF)**.

When the armature rotates within the field windings of a motor, conditions exist to induce a voltage in the armature. Voltage is induced any time a wire is passed through a magnetic field. When the armature, which is a structure with many loops of wire, rotates past the field windings, a small amount of voltage is induced. This voltage opposes the voltage supplied by the battery to energize the armature. As a result, less current is able to flow through the armature.

The faster the armature spins, the more induced voltage is present in the armature. The more voltage in the armature, the more opposition there is to normal current flow to the armature. The induced voltage in the armature opposes or is counter to the battery's voltage. This is why the induced voltage is called CEMF.

The effects of CEMF are quite predictable. When the armature of the motor turns slowly, low amounts of voltage are induced and, therefore, low amounts of CEMF are present. The low amount of CEMF allows a high amount of current. In fact, the only time a starter motor draws its maximum amount of current is when the armature is not rotating.

A series-wound motor develops its maximum torque at start-up and develops less torque as speed increases. It is ideal for applications involving heavy starting loads.

Shunt or parallel-wound motors develop considerably less start-up torque but maintain a constant speed at all operating loads. Compound motors combine the characteristics of good starting torque with constant speed. The compound design is particularly useful for applications in which heavy loads are suddenly applied. In a starter motor, a shunt coil is frequently used to limit the maximum free speed at which the starter can operate.

STARTER MOTOR DRIVE MECHANISMS The area in which starters differ most is in their drive mechanisms used to crank the engine. The solenoid-actuated direct drive system was explained earlier in this chapter.

Some starters use a planetary gearset to increase the torque of a starter motor. Planetary gearsets offer the advantage of quiet operation and compactness.

Solenoid-Actuated Gear Reduction Drive. Solenoid-actuated **gear reduction–drive starters** use a solenoid to engage the pinion with the flywheel and to close the motor circuit. The starter armature does not drive the pinion directly. Instead, a gearset is used to reduce speed and increase the turning torque of the pinion gear. The gearset may be as simple as a small gear meshed with a larger one, or the reduction may take place through the use of a planetary gearset. This design allows a small, high-speed motor to develop increased turning torque at a satisfactory cranking rpm. The solenoid and starter drive operation is basically the same as in solenoid-actuated direct-drive systems.

PERMANENT MAGNET STARTER MOTORS The most recent change in starter motors has been in the use of permanent magnets rather than electromagnets as field coils. Electrically, this starter motor is simpler. It does not require current for field coils. Current is delivered directly to the armature through the commutator and brushes. **Figure 19–15** shows this type of starter motor. This unit functions exactly as the other styles considered. Increased use of this style is expected in the future, as production costs are greatly reduced. Maintenance and testing procedures are the same as for other designs. Notice the use of a

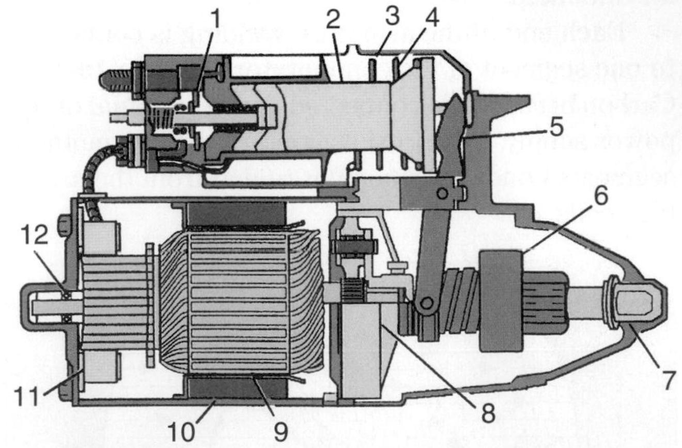

FIGURE 19–15

A permanent magnet–type starter assembly.

planetary gear reduction assembly on the front of the armature. This assembly allows the armature to spin with increased torque, resulting in improved starter cold-cranking performance.

WARNING!

Permanent magnet starters require special handling because the permanent magnet material is quite brittle and can be destroyed with a sharp blow or if the starter is dropped.

STARTER DRIVE A **starter drive** includes a pinion gear that meshes with the ring gear on the flexplate or flywheel, which mounts to the engine's crankshaft **(Figure 19–16)**. The starter drive pinion gear operates through a 15:1 to 20:1 gear ratio with the ring gear. This ratio allows the starter to rotate fast enough to develop enough power to rotate the engine efficiently. To prevent damage to the pinion gear or the flywheel's ring gear, the pinion gear must mesh with the ring gear before the starter motor rotates. To help ensure smooth engagement, the end of the pinion gear is tapered **(Figure 19–17)**. Also, the movement of the armature must always be caused by the action of the motor, not the engine. For this reason, starter drive assemblies include an overrunning clutch.

OVERRUNNING CLUTCH The **overrunning clutch** performs a very important job in protecting the starter motor. When the engine starts and runs, its speed increases. If the starter motor remained connected to the engine through the flywheel, the starter motor would spin at very high speeds, destroying the armature winding. With the engine running at 1000 rpm and a 20:1 ratio between the starter and engine, the starter would be forced to rotate at 20 000 rpm.

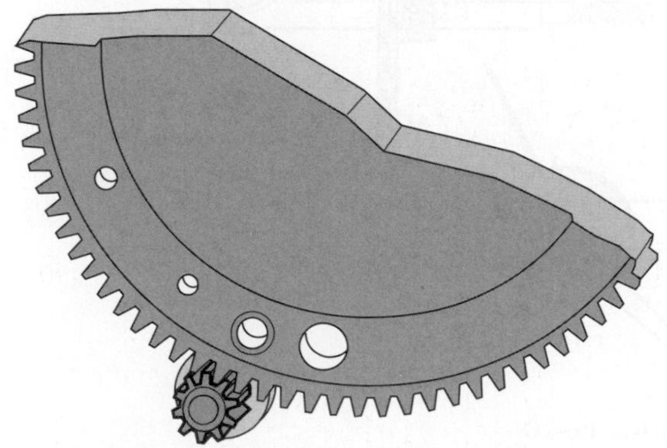

FIGURE 19–16

A starter drive pinion gear is used to turn the engine's flywheel.

FIGURE 19–17

The pinion gear teeth are tapered to allow for smooth engagement.

To prevent this, the armature must disengage from the engine as soon as the engine turns more rapidly than the starter has cranked it. However, with most starter designs, the pinion remains engaged until electricity stops flowing to the starter. In these cases, an overrunning clutch is used to disengage the starter.

The clutch housing is internally splined to the starting motor armature shaft. The drive pinion turns freely on the armature shaft within the clutch housing. When the clutch housing is driven by the armature, the spring-loaded rollers are forced into the small ends of their tapered slots and wedged tightly against the pinion barrel. This locks the pinion and clutch housing solidly together, permitting the pinion to turn the flywheel and thus crank the engine.

When the engine starts **(Figure 19–18)**, the flywheel spins the pinion faster than the armature. This releases the rollers, unlocking the pinion gear

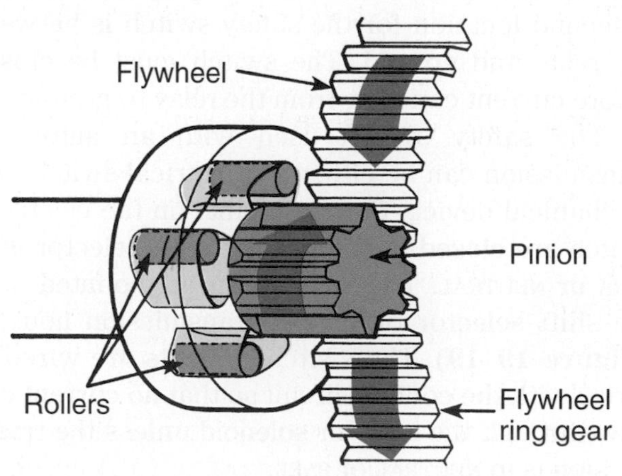

FIGURE 19–18

When the engine starts, the flywheel spins the pinion gear faster, which releases the rollers from the wedge.

from the armature shaft. The pinion then overruns the armature shaft freely until being pulled out of the mesh without stressing the starter motor. The overrunning clutch is moved in and out of mesh by the starter drive linkage.

CONTROL CIRCUIT

The control circuit allows the driver to use a small amount of battery current to control the flow of a large amount of current in the starting circuit.

The entire circuit usually consists of an ignition switch connected through normal-gauge wire to the battery and the magnetic switch (solenoid or relay). When the ignition switch is turned to the start position, a small amount of current flows through the coil of the magnetic switch, closing it and allowing full current to flow directly to the starter motor. The ignition switch performs other jobs besides controlling the starting circuit. It normally has at least four separate positions: ACCESSORY, OFF, ON (RUN), and START.

Starting Safety Switch

The **starting safety switch**, often called the **neutral safety switch**, is a normally open switch that prevents the starting system from operating when the transmission is in gear. This eliminates the possibility of a situation that could make the vehicle lurch unexpectedly forward or backward. Safety switches are used with automatic transmissions and on manual transmissions.

Starting safety switches can be located in either of two places within the control circuit. One location is between the ignition switch and the relay or solenoid. In this position, the safety switch must be closed before current can flow to the relay or solenoid. A second location for the safety switch is between the relay and ground. The switch must be closed before current can flow from the relay to ground.

The safety switch used with an automatic transmission can be either an electrical switch or a mechanical device. Contact points on the electrical switch are closed only when the shift selector is in PARK or NEUTRAL. The switch can be mounted near the shift selector or on the transmission housing **(Figure 19–19)**. The switch contacts are wired in series with the control circuit so that no current can flow through the relay or solenoid unless the transmission is in NEUTRAL or PARK.

Mechanical safety switches for automatic transmissions are sometimes simple devices that physically block the movement of the ignition key when the transmission is in a gear **(Figure 19–20)**. The ignition key can only be turned when the shift selector is in PARK or NEUTRAL.

The safety switches used with manual transmissions are usually electrical switches mounted

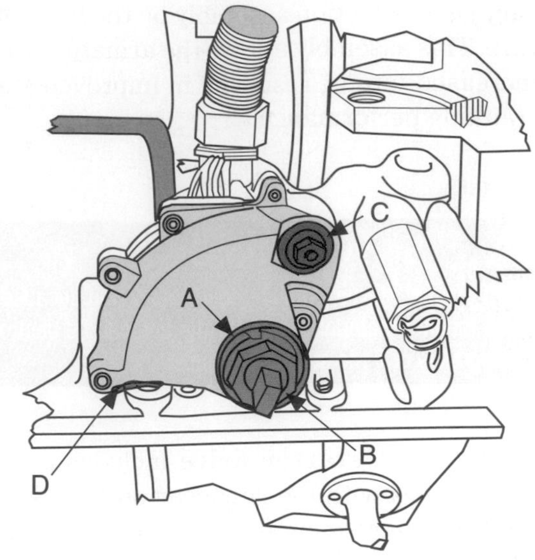

A—Locking washer

B—Switch-attaching nut

C—Switch-adjusting bolt

D—Neutral safety switch

FIGURE 19–19

A neutral safety switch attached to a transmission.

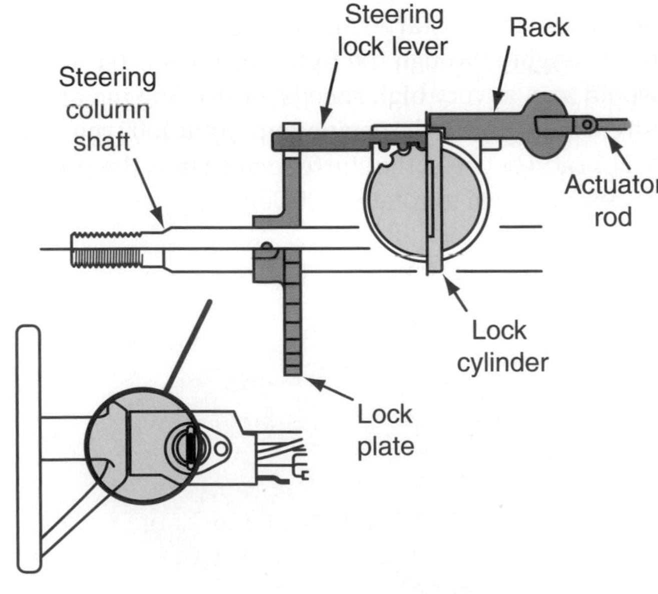

FIGURE 19–20

A mechanical linkage used to prevent starting the engine while the transmission is in gear.

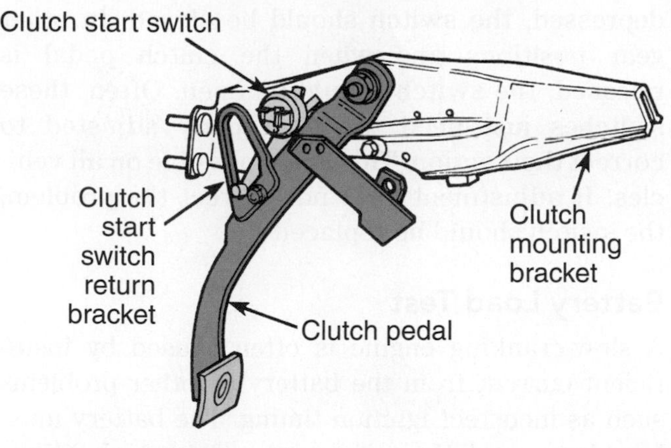

FIGURE 19–21

The clutch pedal must be fully depressed to close the clutch switch and complete the control circuit.

near the gear-shift lever or on the transmission housing. A clutch switch is a second type of safety switch used with manual transmissions. This electrical switch mounts on the floor or fire wall. Its contacts are closed only when the clutch pedal is fully depressed **(Figure 19–21)**.

STARTING SYSTEM TESTING

As mentioned earlier, the starter motor is a special type of electrical motor designed for intermittent use only. During testing, it should never be operated for more than 15 seconds without resting for 2 minutes in between operation cycles to allow it to cool.

Preliminary Checks

The cranking output obtained from the motor is affected by the condition and charge of the battery, the circuit's wiring, and the engine's cranking requirement.

The battery should be checked and charged as needed before testing the starting system.

Check the wiring and cables for clean, tight connections. Loose or dirty connections will cause excessive voltage drops. Cables can be corroded by battery acid, and contact with engine parts and other metal surfaces can fray the cable insulation. Frayed insulation can cause a short to ground that can seriously damage some of the electrical units of the vehicle.

Cables should also be checked to make sure they are not undersized (too small a gauge) or too long. Both conditions can limit the amount of current delivered to the starter motor.

When checking cables and wiring, always check any fusible links in the wiring. Most late-model vehicles are equipped with maxi fuses in place of the fusible links. Both should be checked during any routine starting system inspection. When a maxi fuse or fusible link has failed, always troubleshoot the system and locate the cause before replacing the fuse or link.

Make certain the engine is filled with proper weight oil as recommended by the vehicle manufacturer. Heavier than specified oil, when coupled with low ambient temperatures, can drastically lower cranking speed to the point where the engine does not start and excessively high current is drawn by the starter.

Check the ignition switch for loose mounting, damaged wiring, sticking contacts, and loose connections. Check the wiring and mounting of the safety switch, if so equipped, and make certain the switch is properly adjusted. Check the mounting, wiring, and connections of the magnetic switch and starter motor. Also, be sure the starter pinion is properly adjusted.

Safety Precautions

Almost all starting system tests must be performed while the starter motor is cranking the engine. However, the engine must not start and run during the test, or the readings will be inaccurate.

To prevent the engine from starting, the ignition switch can be bypassed with a remote starter switch that allows current to flow to the starting system, but not to the ignition fuel systems.

During testing, be sure the transmission is out of gear during cranking and the parking brake is set. When servicing the battery, always follow safety precautions. Always disconnect the battery ground cable before making or breaking connections at the system's relay, solenoid, or starter motor.

Troubleshooting Procedures

A systematic troubleshooting procedure is essential when servicing the starting system. Consider the fact that nearly 80 percent of starters returned as defective on warranty claims work perfectly when tested. This is often the result of poor or incomplete diagnosis of the starting and related charging systems. Testing the starting system can be divided into area tests, which check voltage and current in the entire system, and more detailed pinpoint tests, which target one particular component or segment of the wiring circuit.

Starter Solenoid Problems

A typical symptom of solenoid problems is the presence of a clicking noise when the ignition switch is turned to the start position. The clicking noise is caused by the solenoid's plunger moving back and forth. Normally, the plunger moves to the battery contacts and is held there by a magnetic field until the ignition switch is moved from the start position.

In order for the solenoid's plunger to move enough to complete the starter motor circuit and remain in that position, a strong magnetic field must be present around the solenoid's windings. The strength of the magnetic field depends on the current flowing through the windings. Therefore, anything that would reduce current flow would affect the operation of the solenoid. Common causes of the clicking are low battery voltage, low voltage available to the solenoid, or an open in the hold-in winding.

Checking voltage at the battery and at the solenoid will help you identify the cause of the problem. If the solenoid is bad, it can be replaced as a separate unit on some starter motors or replaced with the starter motor on other designs.

Starting Safety Switches

Safety switches can be checked with a voltmeter or an ohmmeter. When the transmission is placed in PARK or NEUTRAL or when the clutch pedal is depressed, the switch should be closed. In other gear positions and when the clutch pedal is released, the switch should be open. Often, these switches just need to be properly adjusted to correct their action. This is not possible on all vehicles. If adjustment does not correct the problem, the switch should be replaced.

Battery Load Test

A slow-cranking engine is often caused by insufficient current from the battery or other problems such as incorrect ignition timing. The battery must be able to crank the engine under all load conditions while maintaining enough voltage to supply ignition current for starting. Perform a battery load test before checking the starting systems.

Cranking Voltage Test

The **cranking voltage test** measures the available voltage to the starter during cranking. To perform the test, disable the fuel or use a remote starter switch to bypass the ignition switch. Normally, the remote starter switch leads are connected to the positive terminal of the battery and the starter terminal of the solenoid or relay **(Figure 19–22)**. Refer to the service manual for

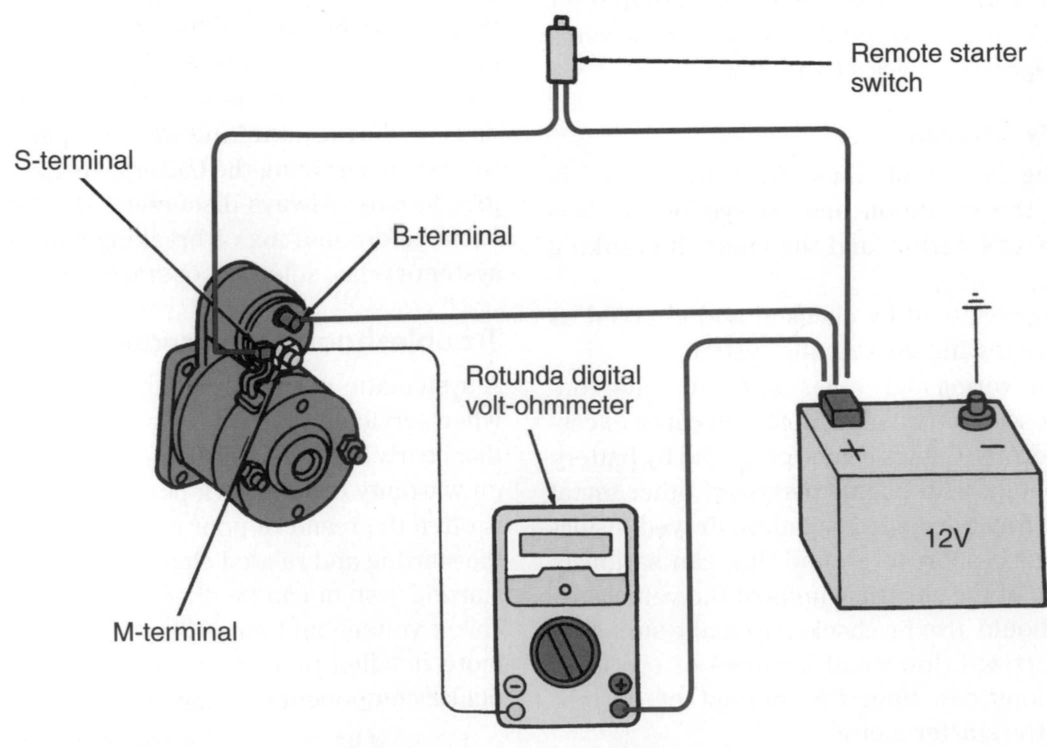

FIGURE 19–22

Using a remote starter switch to bypass the control circuit and ignition system.

Ford Motor Company

specific instructions on the model car being tested. Connect the voltmeter's negative lead to a good chassis ground. Connect the positive lead to the starter motor feed at the starter relay or solenoid. Activate the starter motor and observe the voltage reading. Compare the reading to the specifications given in the service manual. Normally, 9.6 volts is the minimum required.

TEST CONCLUSIONS If the reading is above specifications but the starter motor still cranks poorly, the starter motor is faulty. If the voltage reading is lower than specifications, a cranking current test and circuit resistance test should be performed to determine if the problem is caused by high resistance in the starter circuit or an engine problem.

Cranking Current Test

The **cranking current test** measures the amount of current the starter circuit draws to crank the engine. Knowing the amount of current draw helps to identify the cause of starter system problems.

Most starter current testers use an inductive pickup **(Figure 19–23)** to measure the current draw.

To conduct the cranking current test, connect a remote starter switch or disable the fuel prior to testing. Follow the instructions given with the tester when connecting the test leads. Crank the engine for no more than 15 seconds. Observe the voltmeter. If the voltage drops below 9.6 volts, a problem is indicated. Also, watch the ammeter and compare the reading to specifications.

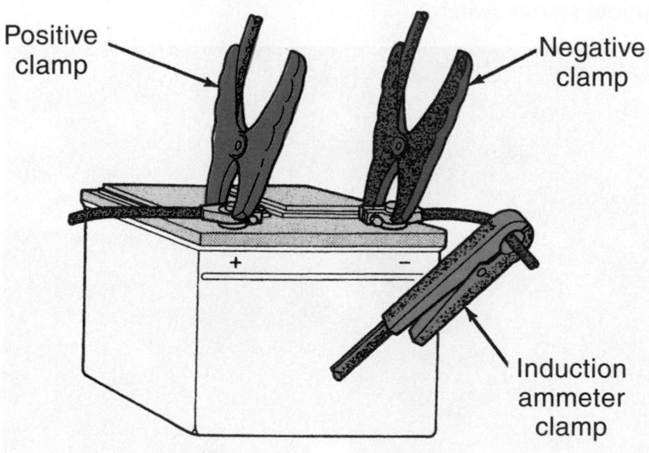

FIGURE 19–23

Connecting the test leads of a typical charging/starting/battery tester.

Chrysler Group LLC

TABLE 19–1 **RESULTS OF CRANKING CURRENT TESTING**

PROBLEM	POSSIBLE CAUSE
Low current draw	Undercharged or defective battery. Excessive resistance in circuit due to faulty components or connections.
High current draw	Short in starter motor. Mechanical resistance due to binding engine or starter system component failure or misalignment.

Table 19–1 summarizes the most probable causes of too low or high starter motor current draw. If the problem appears to be caused by excessive resistance in the system, conduct an insulated circuit resistance test.

Insulated Circuit Resistance Test

The complete starter circuit is made up of the insulated circuit and the ground circuit. The insulated circuit includes all of the high current cables and connections from the battery to the starter motor.

To test the insulated circuit for high resistance, disable the fuel or bypass the ignition switch with a remote starter switch. Connect the positive (+) lead of the voltmeter to the battery's positive (+) terminal post or nut. By connecting the lead to the cable, the point of high resistance (cable-to-post connection) may be bypassed. Connect the negative (–) lead of the voltmeter to the starter terminal at the solenoid or relay. Crank the engine and record the voltmeter reading. If the reading is within specifications (usually 0.1 to 0.5 voltage drop), the insulated circuit does not have excessive resistance. Proceed to the ground circuit resistance test outlined in the next section. If the reading indicates a voltage loss above specifications, move the negative lead of the tester progressively closer to the battery, cranking the engine at each test point. Normally, a voltage drop of 0.1 volts is the maximum allowed across a length of cable.

Photo Sequence 15 goes through the correct procedure for conducting a voltage drop test on a typical starter circuit.

TEST CONCLUSIONS When excessive voltage drop is observed, the trouble is located between that point and the preceding point tested. It is either a damaged cable or poor connection, an undersized wire, or possibly a bad contact assembly within the solenoid.

PHOTO SEQUENCE 15

VOLTAGE DROP TESTING OF A STARTER CIRCUIT

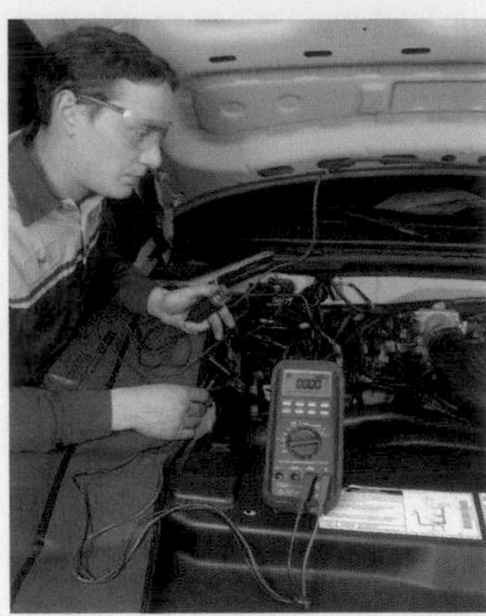

P15–1 The tools required to measure the voltage drop at various points within the starter circuit are fender covers, a digital multimeter (DMM), and a remote starter switch.

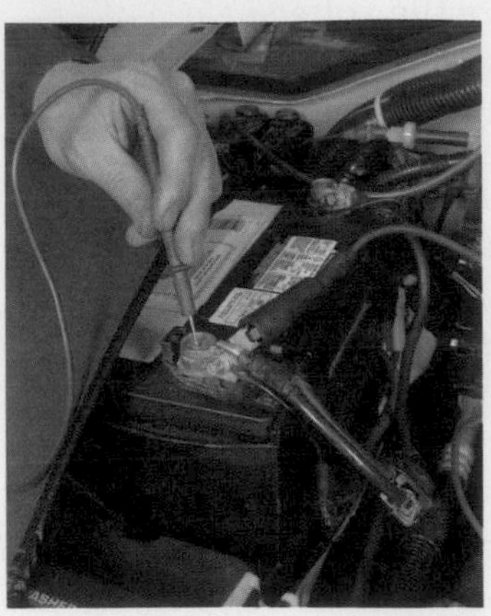

P15–2 Connect the positive lead of the meter to the positive battery post. If at all possible, do not connect it to the battery clamp.

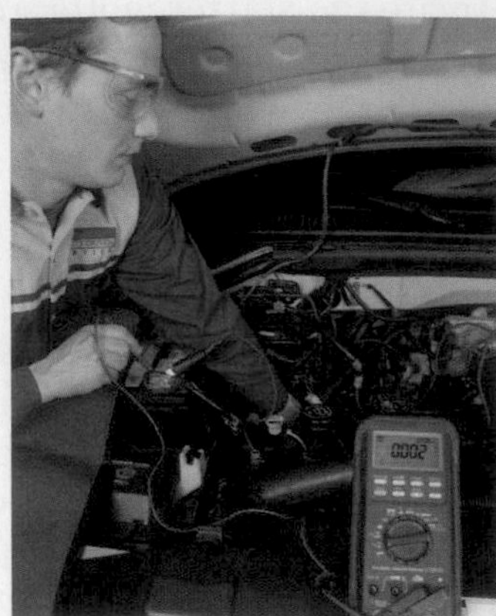

P15–3 Connect the negative lead to the main battery connection at the starter.

P15–4 Set the voltmeter to the scale that is close to, but greater than, the battery voltage.

P15–5 Disable the ignition and/or fuel and/or connect a remote starter switch.

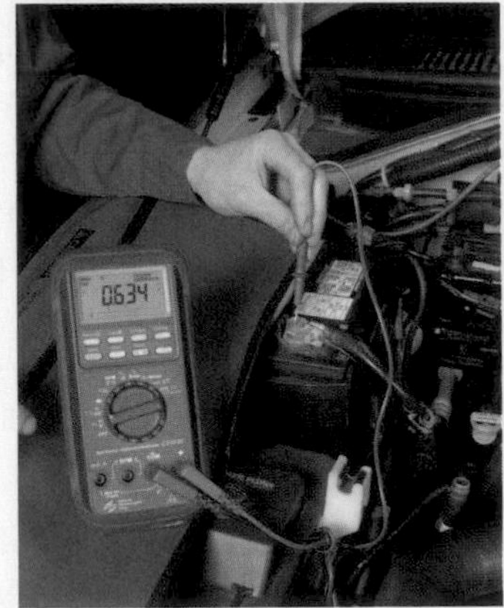

P15–6 Crank the engine and read the voltmeter. This reading shows the voltage drop on the positive side of the starter circuit.

P15–7 This reading showed excessive resistance in the circuit. To locate the resistance, move the meter's negative lead to the next location toward the battery. In this case, it is the starter side of the starter relay.

P15–8 Crank the engine and observe the reading on the meter. This is the voltage drop across the positive circuit from the battery to the output of the relay.

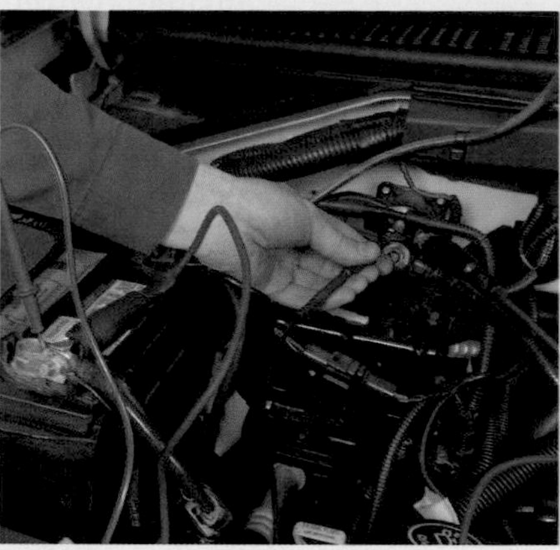

P15–9 There is still too much voltage drop, so we continue our test by moving the negative lead to the battery side of the relay.

P15–10 Crank the engine and observe the reading on the meter. This is the voltage drop across the cable from the battery to the relay. Notice that hardly any voltage was dropped. This cable is okay.

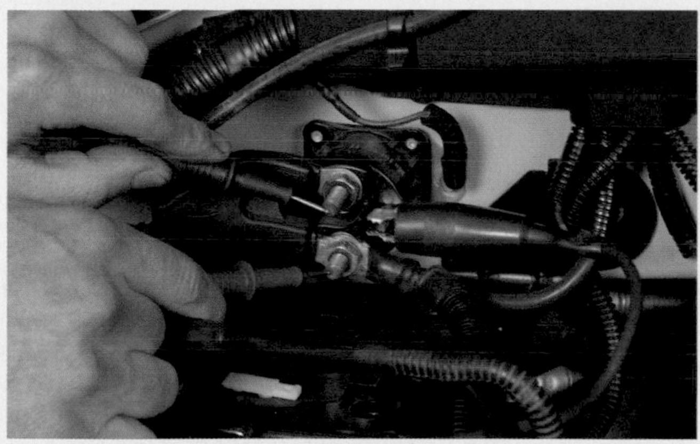

P15–11 Now connect the meter across the relay with the red lead on the battery side and black lead on the starter side.

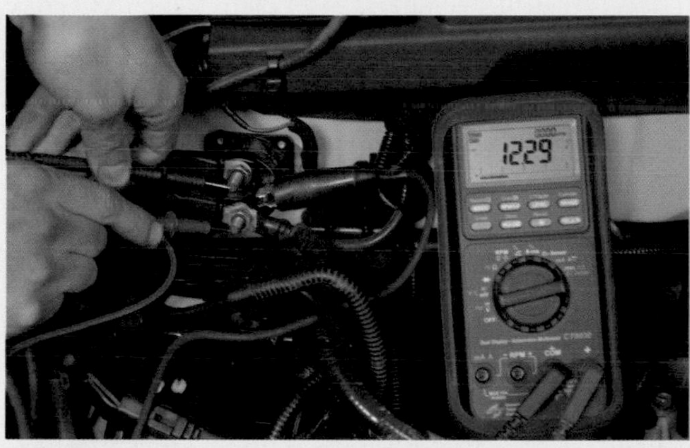

P15–12 Ignore any voltage reading you may have at this point.

P15–13 Crank the engine and observe the reading on the meter. This is the voltage drop across the contacts inside the relay.

P15–14 The reading was higher than normal, so the starter relay has high resistance and needs to be replaced.

TABLE 19–2 MAXIMUM VOLTAGE DROPS

Each large cable	0.1 V
Each connection	0.1 V
Each small wire	0.2 V
Starter relay	0.3 V

Repair or replace any damaged wiring or faulty connections. Refer to **Table 19–2** to find the maximum allowable voltage drops for the starter circuit.

Ground Circuit Resistance Test

The ground circuit provides the return path to the battery for the current supplied to the starter by the insulated circuit. This circuit includes the starter-to-engine, engine-to-chassis, and chassis-to-battery ground terminal connections.

To test the ground circuit for high resistance, disable the ignition, or bypass the ignition switch with a remote starter switch. Refer to **Figure 19–24** for the proper test connection. Crank the engine and record the voltmeter reading.

TEST CONCLUSIONS Good results would be less than a 0.2 voltage drop for a 12-volt system. A voltage drop in excess of this indicates the presence of a poor ground circuit connection, resulting from a loose starter motor bolt, a poor battery ground terminal post connector, or a damaged or undersized ground system wire from the battery to the engine block. Isolate the cause of excessive voltage drop in the same manner as recommended in the insulated circuit resistance test by moving the positive (+) voltmeter lead progressively closer to the battery. If the ground circuit tests out satisfactorily and a starter problem exists, move on to the control circuit test.

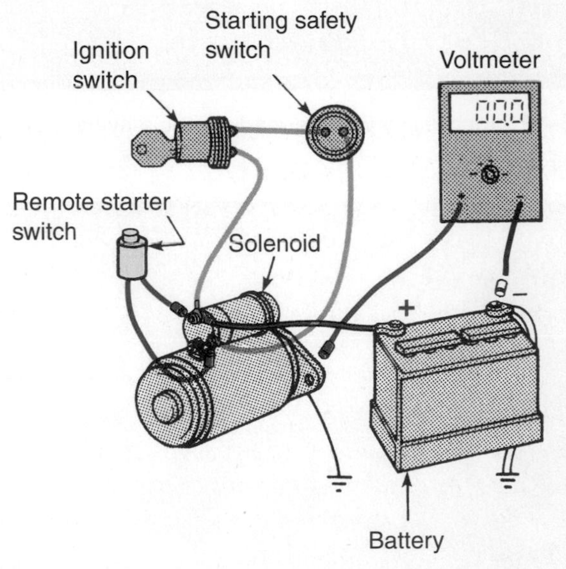

FIGURE 19–24

The setup for checking voltage drop across the ground circuit.

Voltage Drop Test of the Control Circuit

The control circuit test examines all the wiring and components used to control the magnetic switch, whether it is a relay, a solenoid acting as a relay, or a starter motor–mounted solenoid.

High resistance in the solenoid switch circuit reduces current flow through the solenoid windings, which can cause improper functioning of the solenoid. In some cases of high resistance, it may not function at all. Improper functioning of the solenoid switch generally results in the burning of the solenoid switch contacts, causing high resistance in the starter motor circuit.

Check the vehicle wiring diagram, if possible, to identify all control circuit components. These normally include the ignition switch, safety switch, the starter solenoid winding, or a separate relay.

To perform the test, disable the fuel system. Connect the positive meter lead to the battery's positive terminal and the negative meter lead to the starter switch terminal on the solenoid or relay. Crank the engine and record the voltmeter reading.

TEST CONCLUSIONS Generally, good results would be less than 0.5 volt, indicating that the circuit condition is good. If the voltage reading exceeds 0.5 volt, it is usually an indication of excessive resistance. However, on certain vehicles, a slightly higher voltage loss may be normal.

Identify the point of high resistance by moving the negative test lead back toward the battery's positive terminal, eliminating one wire or component at a time.

A reading of more than 0.1 volt across any one wire or switch is usually an indication of trouble. If a high reading is obtained across the safety switch used on an automatic transmission, check the adjustment of the switch according to the manufacturer's service manual. Clutch-operated safety switches cannot be adjusted. They must be replaced.

Test Starter Drive Components

This test detects a slipping starter drive without removing the starter from the vehicle. First, disable the fuel system or bypass the ignition switch with a remote starter switch. Turn the ignition switch to start or activate the remote starter and hold it in this position for several seconds. Repeat the procedure at least three times to detect an intermittent condition.

TEST CONCLUSIONS If the starter cranks the engine smoothly, that is an indication that the starter drive is functioning properly. If the engine stops cranking and the starter spins noisily at high speed, the drive is slipping and should be replaced.

If the drive is not slipping, but the engine is not being cranked, inspect the flywheel for missing or damaged teeth. Remove the starter from the vehicle and check its drive components. Inspect the pinion gear teeth for wear and damage. Test the overrunning clutch mechanism. If good, the overrunning clutch should turn freely in one direction, but not in the other. A bad clutch will turn freely in the overrun direction or not at all. If a drive locks up, it can destroy the starter by allowing the starter to spin at more than 15 times engine speed.

If the starter does not spin and the drive does not engage, the problem may be in the solenoid. If the solenoid is too weak to overcome the force of the return springs, the starter does not operate.

Removing the Starter Motor

If your testing indicates that the starter must be removed, the first step is to disconnect the negative cable at the battery and wrap the clamp with electrical tape. It may be necessary to place the vehicle on a hoist to gain access to the starter. Before lifting the vehicle, disconnect all wires, fasteners, and so on that can be reached from under the hood.

Disconnect the wires leading to the solenoid terminals. To avoid confusion when reinstalling the starter, it is wise to mark the wires so that they can be reinstalled on their correct terminals.

On some vehicles, you may need to disconnect the exhaust system to be able to remove the starter. Loosen the starter mounting bolts and remove all but one. Support the starter while removing the remaining bolt. Then pull the starter out and away from the flywheel. Once the starter is free, remove the last bolt and the starter.

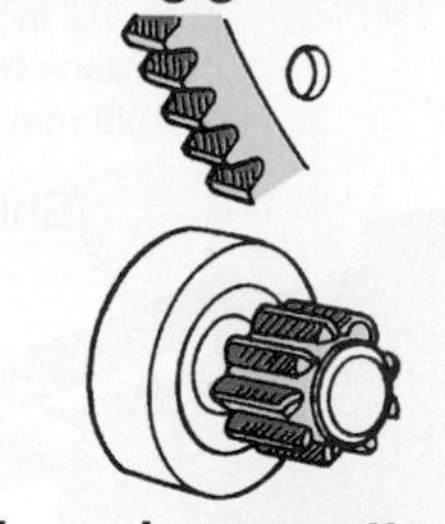

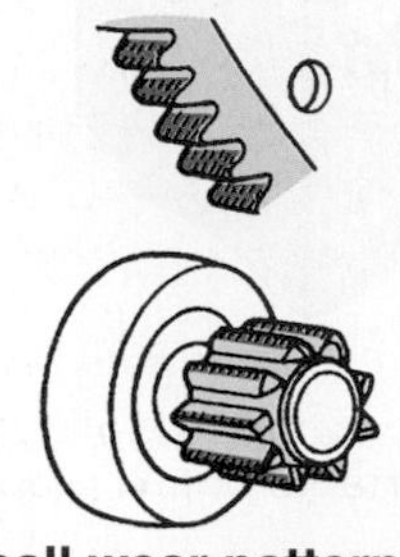

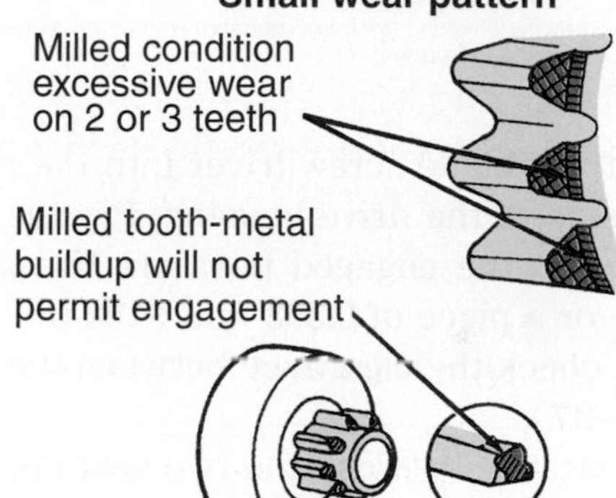

FIGURE 19–25

Starter drive and flywheel ring gear wear patterns.

Ford Motor Company

Once the starter is out, inspect the starter drive pinion gear and the flywheel ring gear **(Figure 19–25)**. When the teeth of the starter drive are abnormally worn, make sure you inspect the entire circumference of the flywheel. If the starter drive or the flywheel ring gear show signs of wear or damage, they must be replaced.

Reverse the procedure to install the starter. Make sure all electrical connections are tight. If you are installing a new or remanufactured starter, sand away the paint at the mounting point before installing it. Also, make sure you have a good hold on the starter while installing it.

Many General Motors starters use shims between the starter and the mounting pad **(Figure 19–26)**. To check this clearance, install the starter

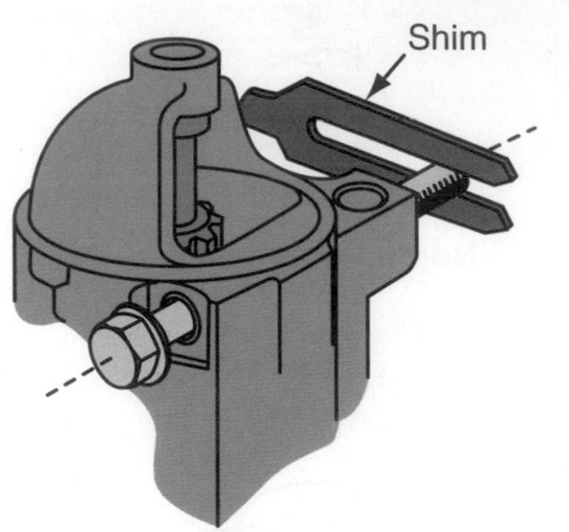

FIGURE 19–26

Shimming the starter to obtain proper pinion-to-ring gear clearance.

and insert a flat blade screwdriver into the access slot on the side of the drive housing. Pry the drive pinion gear into the engaged position. Use a wire feeler gauge or a piece of 0.508 mm (0.02 in.) diameter wire to check the clearance between the gears **(Figure 19–27)**.

If the clearance between the two gears is incorrect, shims will need to be added or subtracted to bring the clearance within specs. If the clearance is excessive, the starter will produce a high-pitched whine while it is cranking the engine. If the clearance is too small, the starter will make a high-pitched whine after the engine starts and the ignition switch is returned to the RUN position.

Free Speed (No-Load) Test

Every starter should be bench tested after it is removed and before it is installed. To conduct a free speed or no-load test on a starter, follow these steps:

PROCEDURE

Free Speed or No-Load Test

STEP 1 Clamp the starter firmly in a bench vise.

STEP 2 Connect an ammeter to the battery cable and the starter to a battery. This should cause the motor to run.

STEP 3 Check current draw and motor speed and compare them to specifications. If they meet specs when the battery has at least 11.5 volts, the starter is working properly.

If the current draw was excessive or the motor speed too low, there is excessive physical resistance, which can be caused by worn bushings or bearings, a shorted armature, shorted field windings, or a bent armature.

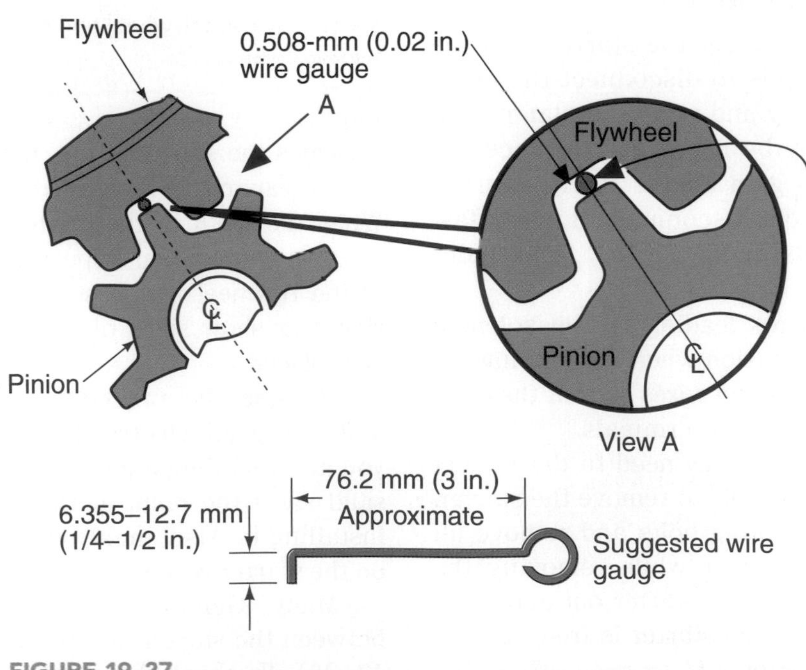

FIGURE 19–27

Checking the clearance between the pinion gear and the ring gear.

If there was no current draw and the starter did not rotate, the problem could be caused by open field windings, open armature coils, broken brushes, or broken brush springs.

Low armature speed with low current draw indicates excessive resistance. There may be a poor connection between the commutator and the brushes, or the connections to the starter are bad. If the speed and current draw are both high, check for a shorted field winding.

STARTER MOTOR SERVICE

If the starter is defective, it should be replaced or rebuilt. Often, technicians opt for replacing it rather than spending the time to repair or rebuild it. This decision, however, depends on a number of things, including the customer's desire, the shop's policies, availability of repair parts, cost, and time.

WARNING!

Do not clean the starter motor in solvent. The residue left on the parts can ignite and cause a fire and/or destroy the starter. Use denatured alcohol, compressed air regulated to 172.3 kPa (25 psi), and/or clean rags to clean the unit and its parts.

Always refer to the manufacturer's procedures when repairing a starter.

The starter should be cleaned and inspected as it is disassembled. Inspect the end frame and drive housing for cracks or broken ends. Check the frame assembly for loose pole shoes and broken or frayed wires. Inspect the drive gear for worn teeth and proper overrunning clutch operation. The commutator should be free of flat spots and should not be excessively burned. Check the brushes for wear. Replace them if worn past specifications.

Starter Motor Component Tests

With the starter motor disassembled, tests can be conducted to determine the reason for failure. The armature and field coils should be checked for shorts and opens first. Normally, if the armature or coils are bad, the entire starter is replaced.

FIELD COIL TESTS The field coil and frame assembly can be wired in a number of different ways. Accurate testing of the coils can be done only if you follow the specific guidelines of the manufacturer or if you know how the coils are wired. To do this, look at the wiring diagram and determine where the coils get their power and where they ground. When you have this information, you will know if the coils are wired in series or parallel.

The usual way to check the field coils for opens is to connect an ohmmeter between the coils' power feed wire and the field coil brush lead **(Figure 19–28)**. If there is no continuity, the field is open. To check the field coil for a short to ground, connect the ohmmeter from the field coil brush lead and the starter (field frame) housing. If there is continuity, the field coil is shorted to the housing.

ARMATURE TESTS The armature should be inspected for wear or damage caused by contact with the permanent magnets or field windings. If there is wear or damage, check the pole shoes for looseness and repair as necessary. A damaged armature must be replaced.

Next, check the commutator of the armature. If the surface is dirty or burnt, clean it with emery cloth or cut it down with an armature lathe. Measure the diameter of the commutator with an outside micrometer or vernier caliper. If the diameter is less than specifications require, replace the armature.

Measure commutator runout by mounting the armature in V-blocks. Position a dial indicator so that it rides on the centre of the commutator. If the runout is within specs, check the commutator for carbon dust or brass chips between the segments. If the commutator runout is beyond specs, replace the armature. Some commutators use material between segments that should not be undercut.

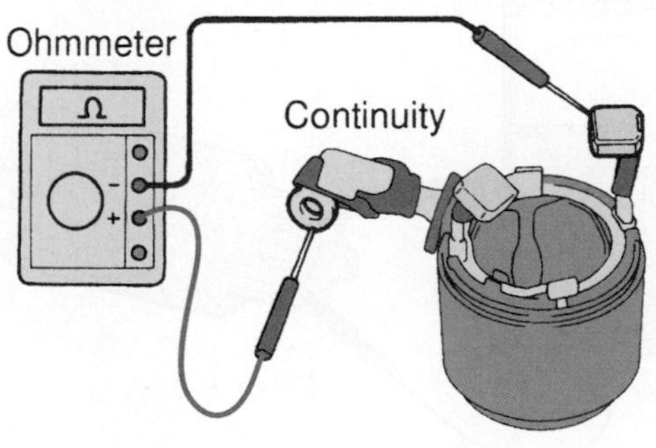

FIGURE 19–28

Checking a field coil for an open.

Reprinted with permission.

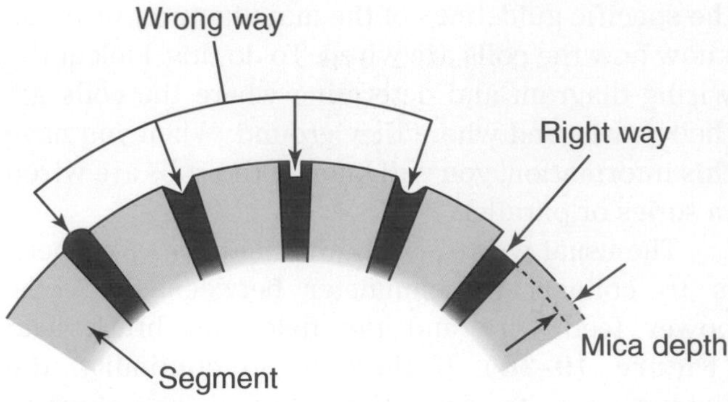

FIGURE 19–29

Check the depth of the mica between the commutator segments.

Check the depth of the insulating material (mica) between the commutator segments **(Figure 19–29)**. Check each one and compare the depth with specifications. If necessary, undercut the mica with the proper tool or a hacksaw to achieve the proper depth. If the proper depth cannot be achieved, or the insulating material is plastic, replace the armature.

Check for continuity between the segments of the commutator **(Figure 19–30)**. If an open circuit exists between any segments, replace the armature.

Place the armature in an armature tester, commonly called a **growler**. Hold a hacksaw blade on the armature core **(Figure 19–31)**. If the blade is attracted to the armature's core or vibrates while the core is turned, the armature is shorted and must be replaced.

With an ohmmeter, check the armature windings for a short to ground. Hold one meter lead to a commutator segment and the other on the armature core. Also check between the armature shaft and the commutator. If there is continuity at either of these two test points, the armature needs to be replaced.

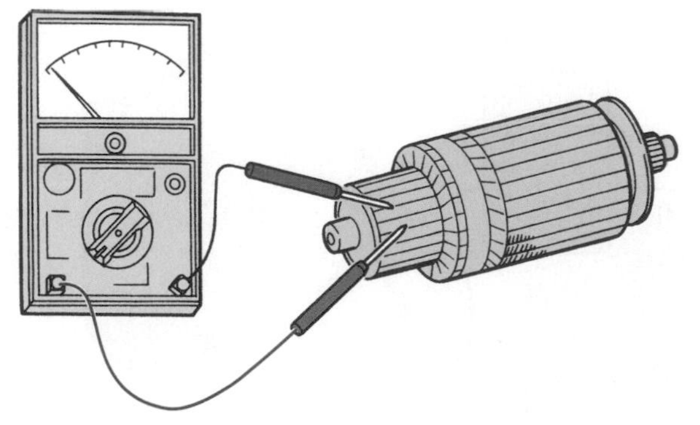

FIGURE 19–30

Checking the armature for an open circuit.

American Isuzu Motors Inc.

FIGURE 19–31

Testing an armature on a growler.

Martin Restoule

BRUSH INSPECTION Brush inspection begins with an ohmmeter check of the brush holder. Connect one meter lead to a positive brush and the other lead to a negative brush. There should be no continuity between them. If there is, replace the brush holder. Install the brushes into the brush holder and slip the unit over the commutator. Using a spring scale, measure the spring tension of the holders at the moment the spring lifts off the brush. Compare the tension with specs. If the tension is incorrect, replace the spring or the brush holder assembly.

Measure the length of the brushes **(Figure 19–32)**. If the brushes are not within specs, replace the brush or the brush holder assembly. To seat new brushes after installing them in the brush holder, slip a piece of fine sandpaper between the brush and the commutator. Then rotate the armature. This will put the contour of the commutator on the face of the brushes.

BEARINGS AND BUSHINGS Check each bearing and bushing by placing the armature into the bushing and paying attention to the fit and feel as the armature is rotated in the bushing. If the bushing or bearing feels too loose, tight, or rough, it should be replaced. Bushings can often be visually inspected for uneven and excessive wear. If the bushing is bad, replace it. Many bearings are held in the case by a retainer, while bushings are typically pressed out and into their bore.

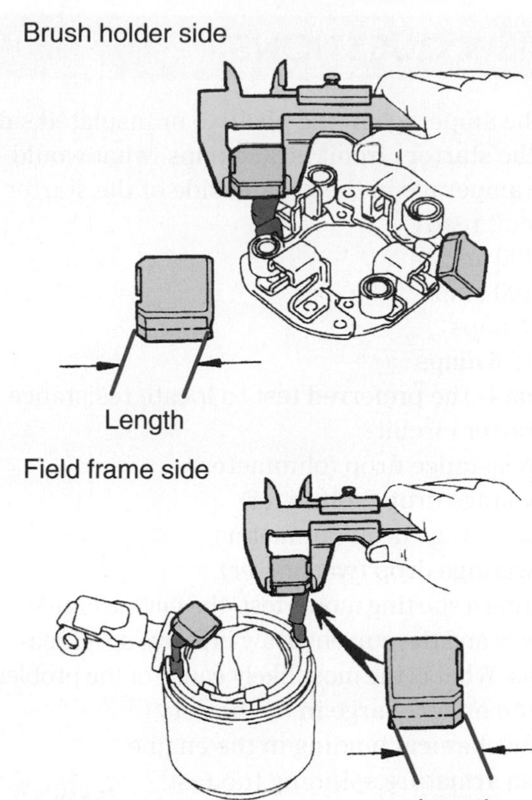

FIGURE 19–32

Measure the length of the brushes.

Reprinted with permission from Toyota.

STARTER DRIVES AND CLUTCHES Carefully inspect the teeth on the starter drive. If the teeth are chipped, excessively worn, or damaged in any way, replace the drive assembly. Also check the teeth on the starter ring gear on the engine's flywheel. Often, the same thing that caused damage to the starter drive will damage the teeth on the flywheel. If either or both are damaged, they should be replaced.

To check the operation of the overrunning clutch, slide the drive and clutch assembly onto the armature shaft. Rotate the clutch in both directions. Check its movement. It should rotate smoothly in one direction and lock in the other **(Figure 19–33)**. If it does not lock in either direction or if it locks or barely moves in both directions, the assembly must be replaced.

Starter Motor Reassembly

To reassemble the starter, basically reverse the disassembly procedures. Additional guidelines for reassembly follow:

- Lubricate the splines on the armature shaft that the drive gear rides on with a high-temperature grease.

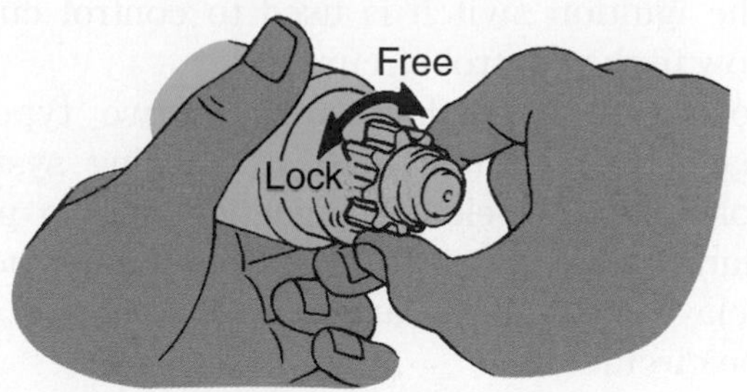

FIGURE 19–33

Check the overrunning clutch by attempting to rotate it in both directions.

- Lubricate the bearings and/or bushings with a high-temperature grease.
- Apply sealing compound to the solenoid flange before installing the solenoid to the starter motor.
- Check the pinion depth clearance.
- Perform a no-load test on the starter before installing it.

KEY TERMS

Armature (p. 566)
Brushes (p. 566)
Commutator (p. 568)
Compound motors (p. 569)
Counter EMF (CEMF) (p. 570)
Cranking current test (p. 575)
Cranking voltage test (p. 574)
Field coils (p. 566)
Gear reduction–drive starters (p. 570)
Growler (p. 582)
Hold-in winding (p. 565)
Neutral safety switch (p. 572)
Overrunning clutch (p. 571)
Pinion gear (p. 565)
Pole shoes (p. 566)
Positive engagement starters (p. 565)
Pull-in winding (p. 565)
Series motor (p. 569)
Shunt motors (p. 569)
Solenoid (p. 565)
Starter drive (p. 571)
Starter frame (p. 566)
Starter housing (p. 566)
Starter relay (p. 564)
Starting safety switch (p. 572)

SUMMARY

- The starting system has two distinct electrical circuits: the starter circuit and the control circuit.
- The starter circuit carries high current from the battery, through heavy cables, to the starter motor.
- The control circuit uses a small amount of current to operate a magnetic switch that opens and closes the starter circuit.

- The ignition switch is used to control current flow in the control circuit.
- Solenoids and relays are the two types of magnetic switches used in starting systems. Solenoids use electromagnetic force to pull a plunger into a coil to close the contact points. Relays use a hinged armature to open and close the circuit.
- The starter motor is an electric motor capable of producing very high horsepower for very short periods.
- The drive mechanism of the starter motor engages and turns the flywheel to crank the engine for starting.
- An override clutch protects the starter motor from spinning too fast once the vehicle engine starts.
- Starting safety switches prevent the starting system from operating when the transmission is engaged.
- During starter system testing, the ignition or fuel system must be bypassed or disabled so that the engine cannot start.
- Battery load, cranking voltage, cranking current, insulated circuit resistance, starter relay bypass, ground circuit resistance, control circuit, and drive component tests are all used to troubleshoot the starting system.
- With the starter removed, inspect the starter drive pinion gear and the flywheel ring gear.
- If the starter is defective, it should be replaced or rebuilt.
- To check the field coils for opens, connect an ohmmeter between the coils' power feed wire and the field coil brush lead. If there is no continuity, the field is open.
- To check the field coil for a short to ground, connect the ohmmeter from the field coil brush lead and the starter (field frame) housing. If there is continuity, the field coil is shorted to the housing.
- The armature should be inspected for wear or damage caused by contact with the permanent magnets or field windings.
- With an ohmmeter, check across the brushes to see if they are shorted. If there is a short, replace the brush holder.
- Measure the length of the brushes. If the brushes are not within specs, replace the brush or the brush holder assembly.
- Inspect the overrunning clutch by sliding it over the armature shaft and checking its movement.

REVIEW QUESTIONS

1. If the amperage in the positive or insulated side of the starter circuit is 200 amps, what would the amperage in the ground side of the starter circuit read?
   a. 100 amps
   b. 200 amps
   c. 0 amps
   d. 12.6 amps
2. What is the preferred test to locate resistance in a starter circuit?
   a. resistance drop (ohmmeter)
   b. voltage drop (voltmeter)
   c. amperage drop (ammeter)
   d. wattage drop (wattmeter)
3. During a starting motor test, the engine cranks slowly and the current draw is above specifications. What is the most likely cause of the problem?
   a. too much charge in the battery
   b. mechanical binding in the engine
   c. an armature spinning too fast
   d. open field windings
4. Which statement regarding permanent magnet gear reduction starter motors is true?
   a. They use gears to reduce maximum armature speed.
   b. They do not require an overrunning clutch.
   c. They generally have greater torque output.
   d. They generate more torque at higher rpm.
5. The starter solenoid has two windings to control pinion operation. During cranking, which of these windings is (are) energized?
   a. both the pull-in and hold-in
   b. neither the pull-in nor hold-in
   c. pull-in only
   d. hold-in only
6. What is the rotating part of the starter called?
   a. armature
   b. stator
   c. field
   d. commutator
7. What is the approximate gear ratio between the starter drive pinion and the ring gear?
   a. 1:1 **b.** 10:1
   c. 20:1 **d.** 40:1
8. If the engine requires 2000 W to crank, how much current will the system draw if the voltage drops to 10 volts?
   a. 10 volts **b.** 20 amps
   c. 200 amps **d.** 200 volts

9. Which type of starter motor generates the highest torque at low rpm?
   a. compound   b. shunt
   c. parallel   d. series
10. Which of the following best describes a solenoid?
   a. changes electrical energy into mechanical energy
   b. changes mechanical energy into movement
   c. changes chemical energy into mechanical energy
   d. changes chemical energy into electrical energy
11. What would result from an open hold-in winding in a solenoid?
   a. rapid engagement and disengagement of the solenoid
   b. high current flow
   c. no engagement of the pinion gear
   d. starter motor would not turn
12. What is the allowable voltage drop of the insulated side of the starter circuit?
   a. 9.0 volts   b. 6.0 volts
   c. 0.9 volts   d. 0.3 volts
13. What is indicated by high voltage drops in a starter system?
   a. a high current
   b. a high resistance
   c. a high voltage
   d. a closed circuit
14. What would be considered the minimum acceptable voltage available to the starting motor, during a cranking voltage test?
   a. 12.6 volts   b. 10.5 volts
   c. 9.6 volts   d. 0.3 volts
15. What is the first event to occur when the ignition switch is turned to the start position?
   a. The hold-in winding only is energized.
   b. The pull-in winding only is energized.
   c. The starter begins to rotate.
   d. The pull-in and hold-in windings are energized.
16. What could cause a starter to turn slowly and also have a low current draw?
   a. a shorted field winding
   b. a shorted armature
   c. a low charge in the battery
   d. a binding engine part
17. Which of the following could result in slow cranking?
   a. large diameter battery cables
   b. low voltage drop at the connectors
   c. a high viscosity oil in the crankcase
   d. a large (high CCA) battery
18. What would cause a whine while the engine is cranking, on a starting motor with adjustable clearance?
   a. too little pinion/ring gear clearance
   b. too much pinion/ring gear clearance
   c. too much field current
   d. too little field current
19. What would cause a whine after the engine starts, on a starting motor with adjustable clearance?
   a. too little pinion/ring gear clearance
   b. too much pinion/ring gear clearance
   c. too much field current
   d. too little field current
20. What factor limits the maximum current being drawn from the battery as the engine is cranked?
   a. EMF of the field
   b. CEMF of the armature
   c. commutator segment resistance
   d. ground loop resistance

CHAPTER 20
Charging Systems

LEARNING OUTCOMES

- Explain the purpose of the charging system.
- Identify the major components of the charging system.
- Explain the purposes of the major parts of an AC generator.
- Explain half- and full-wave rectification and how they relate to AC generator operation.
- Identify the different types of AC voltage regulators.
- Describe the two types of stator windings.
- Explain the features enabled by the use of a starter/generator unit.
- Perform charging system inspection and testing procedures using electrical test equipment.

The primary purpose of a charging system is to recharge the battery after starting the engine. After the battery has supplied the high current needed to start the engine, the battery, even a good battery, has a low charge. The charging system recharges the battery by supplying a constant and relatively low charge to the battery. The charging system must also react to supply current to the changing demands of the electrical accessories. Charging systems work on the principles of electromagnetism to change mechanical energy into electrical energy. This is done by inducing voltage.

FIGURE 20–1

An AC generator.

Robert Bosch Corp.

Voltage is induced in a wire when it moves through a magnetic field. The wire or conductor becomes a source of electricity and has a polarity or distinct positive and negative ends. However, this polarity can be switched depending on the relative direction of movement between the wire and magnetic field. This is why an **AC generator** produces alternating current **(Figure 20–1)**.

ALTERNATING CURRENT CHARGING SYSTEMS

During cranking, the battery supplies all of the vehicle's electrical power. However, once the engine is running, the charging system is responsible for producing enough energy to meet the demands of all of the loads in the electrical system, while also recharging the battery. With all of the electrical and electronic devices found on today's vehicles, the charging system has a difficult job.

Several decades ago, the charging system depended on a **DC generator**. The DC generator provided direct current (DC) and was similar to an electric motor in construction. The biggest difference between a generator and a motor is the wiring to the armature. In a motor, the armature receives current from the battery. This creates the magnetic field that opposes the magnetic fields in the motor's coils, which causes the armature to rotate. The armature in a DC generator is driven by the engine. It is not magnetized, and the windings simply rotate through the stationary magnetic field of the field windings, inducing a voltage in the conductors inside the armature. A motor can become a generator by allowing current to flow from the armature instead of to it. In a DC generator, the placement of the brushes on the commutator changes the induced AC voltage to a DC voltage output.

DC generators had a very limited current output, especially at low speeds. They could not keep up with the demands of the modern automobile and were replaced by AC generators. AC generators are capable of providing high current output even at low engine speeds. Although the DC generator has not been used for decades, the principles on which these systems operate can be found in today's hybrid vehicles and dual-purpose starting/charging systems.

SHOP TALK

With the implementation of OBD-II, new terminology was given to many parts of an automobile. Prior to then, an AC generator was called an **alternator**. In fact, in many cases and by many manufacturers, an AC generator is still referred to as an alternator. To avoid confusion, just remember that an alternator is an AC generator and vice versa.

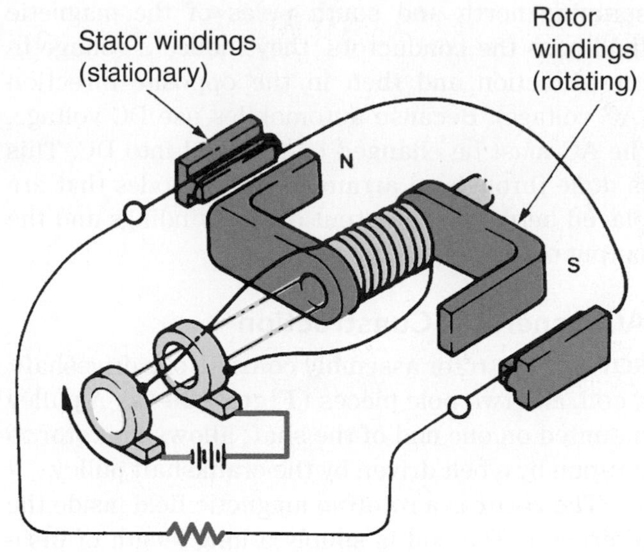

FIGURE 20–3

A simplified AC generator.

AC generators **(Figure 20–2)** use a design that is basically the reverse of a DC generator. In an AC generator **(Figure 20–3)**, a spinning magnetic field (called the rotor) rotates inside an assembly of stationary conductors (called the stator). As the

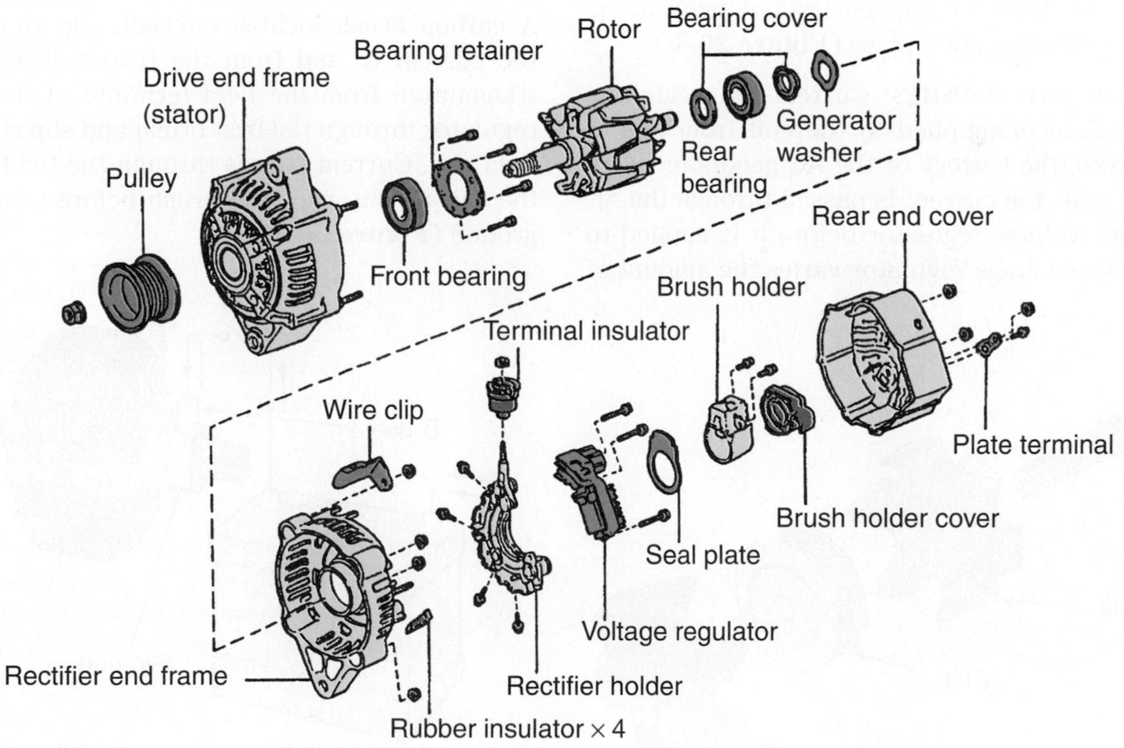

FIGURE 20–2

An exploded view of an AC generator.

Reprinted with permission

spinning north and south poles of the magnetic field pass the conductors, they induce a voltage in one direction and then in the opposite direction (AC voltage). Because automobiles use DC voltage, the AC must be changed or rectified into DC. This is done through an arrangement of diodes that are placed between the output of the windings and the output of the AC generator.

AC Generator Construction

ROTOR The rotor assembly consists of a driveshaft, a coil, and two pole pieces **(Figure 20–4)**. A pulley mounted on one end of the shaft allows the rotor to be spun by a belt driven by the crankshaft pulley.

The **rotor** is a rotating magnetic field inside the alternator. The coil is simply a long length of insulated wire wrapped around an iron core. The core is located between the two sets of **pole pieces**. A magnetic field is formed by a small amount (4.0 to 6.5 A) of current passing through the coil winding. As current flows through the coil, the core is magnetized, and the pole pieces assume the magnetic polarity of the end of the core that they touch. Thus, one pole piece has a north polarity and the other has a south polarity. The extensions of the pole pieces, known as **fingers**, form the actual magnetic poles. A typical rotor has 14 poles, 7 north and 7 south, with the magnetic field between the pole pieces moving from the N poles to the adjacent S poles **(Figure 20–5)**.

SLIP RINGS AND BRUSHES Current to create the magnetic field is supplied to the coil from one of two sources, the battery or the AC generator itself. In either case, the current is passed through the AC generator's voltage regulator before it is applied to the coil. The voltage regulator varies the amount of current supplied. Increasing field current through the coil increases the strength of the magnetic field. This, in turn, increases AC generator voltage output. Decreasing the field voltage to the coil has the opposite effect. Output voltage decreases.

Slip rings and brushes conduct current to the spinning rotor. Most AC generators have two slip rings mounted directly on the rotor shaft. They are insulated from the shaft and each other. Each end of the **field coil** connects to one of the slip rings. A carbon brush located on each slip ring carries the current to and from the field coil. Current is transmitted from the field terminal of the voltage regulator through the first brush and slip ring to the field coil. Current passes through the field coil and the second slip ring and brush before returning to ground **(Figure 20–6)**.

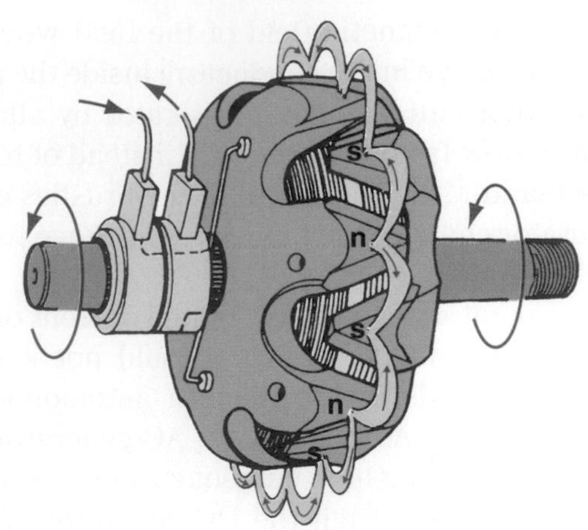

FIGURE 20–5

The magnetic field moves from the N poles, or fingers, to the S poles.

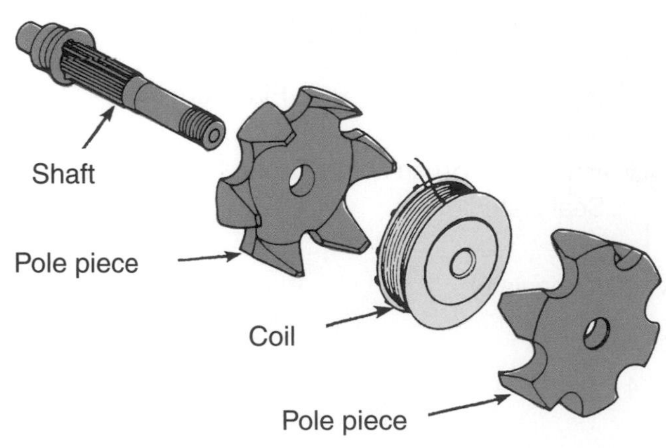

FIGURE 20–4

The rotor is made up of a coil, pole pieces, and a shaft.

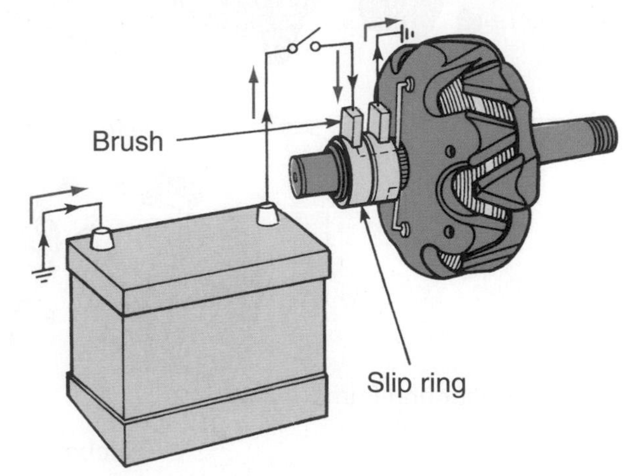

FIGURE 20–6

Current is carried by the brushes to the rotor windings via the slip rings.

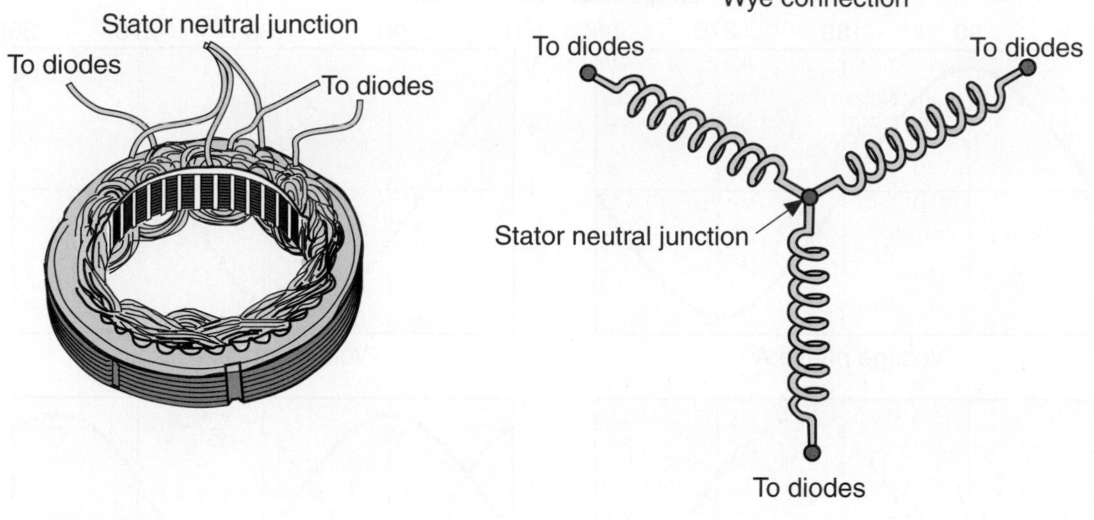

FIGURE 20–7

A wye-connected stator winding.

STATOR The **stator** is the stationary member of the alternator. It is made up of a number of conductors, or wires, into which the voltage is induced by the rotating magnetic field. Most AC generators use three windings to generate the required amperage output. They can be arranged in either a **delta** configuration or a **wye** configuration **(Figure 20–7)**. The delta winding **(Figure 20–8)** received its name because its shape resembles the Greek letter delta, D. The wye winding resembles the letter *Y*. Alternators use one or the other. Usually, a wye winding is used in applications in which high charging voltage at low engine speeds is required. AC generators with delta windings are capable of putting out higher amperages at high speeds, but low engine speed output is poor.

The rotor rotates inside the stator. A small air gap between the two allows the rotor to turn without making contact with the stator. The magnetic field of the rotor is able to energize all three of the stator windings at the same time. Therefore, the generation of AC can be quite high if needed.

Alternating current produces a positive pulse and then a negative pulse. The resultant waveform is known as a sine wave. This **sine wave** can be seen on an oscilloscope. The complete waveform starts at zero, goes positive, and then drops back to zero before turning negative. The angle and polarity of the field coil fingers are what cause this sine wave in the stator. When the north pole magnetic field cuts across the stator wire, it generates a positive voltage within the wire. When the south polarity magnetic field cuts across the stator wire, a negative voltage is induced in the wire. A single loop of wire energized by a single north and then a south results in a single-phase voltage. Remember that there are three overlapping stator windings. This produces three

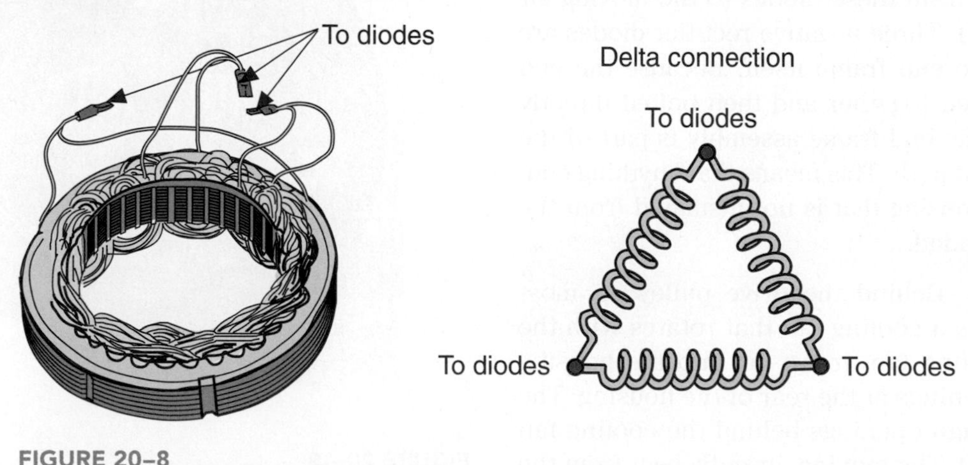

FIGURE 20–8

A delta-connected stator winding.

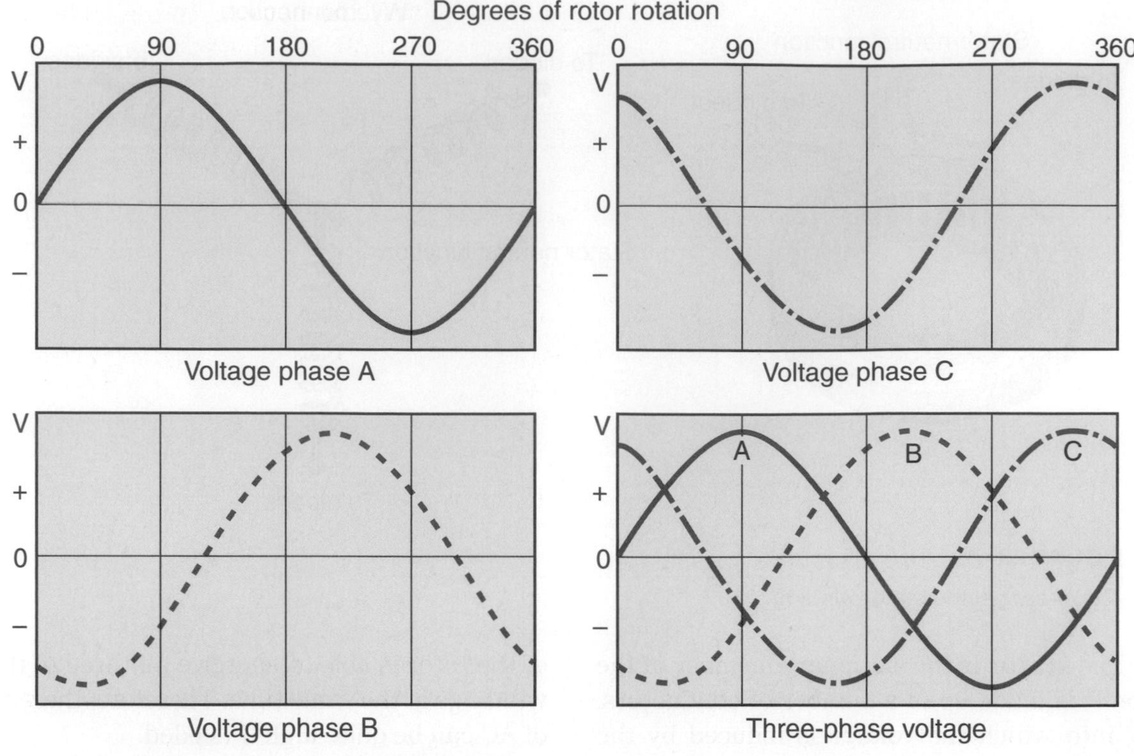

FIGURE 20–9

The voltage produced in each stator winding is added together to create a three-phase voltage.

overlapping sine waves **(Figure 20–9)**. This voltage, because it was produced by three windings, is called **three-phase voltage**.

END FRAME ASSEMBLY The end frame assembly, or housing, is made of two pieces of cast aluminum. It contains the bearings for the end of the rotor shaft where the drive pulley is mounted. Each end frame also has built-in ducts so that the air from the rotor shaft fan can pass through the AC generator. Normally, a heat sink containing three positive rectifier diodes is attached to the rear end frame. Heat can pass easily from these diodes to the moving air **(Figure 20–10)**. Three negative rectifier diodes are contained in the end frame itself. Because the end frames are bolted together and then bolted directly to the engine, the end frame assembly is part of the electrical ground path. This means that anything connected to the housing that is not insulated from the housing is grounded.

COOLING FANS Behind the drive pulley on most AC generators is a cooling fan that rotates with the rotor. This cooling fan draws air into the housing through the openings at the rear of the housing. The air leaves through openings behind the cooling fan **(Figure 20–11)**. The moving air pulls heat from the diodes, and their heat decreases.

Cooling the diodes is important for the efficiency and durability of an AC generator. Several different generator designs have been introduced recently that increase the cooling efficiency of a generator. One of these is the AD-series generator from General Motors. The "A" stands for air-cooled and the "D"

FIGURE 20–10

End frame assembly.

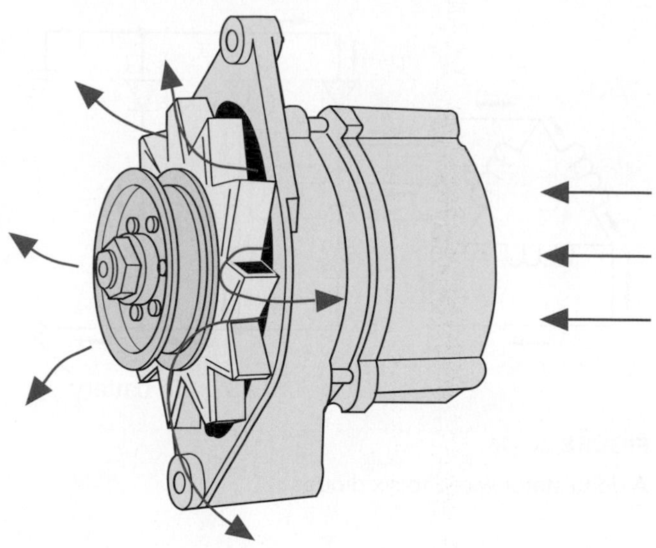

FIGURE 20–11

The cooling fan draws air in from the rear of the generator to keep the diodes cool.

Robert Bosch Corp.

means dual fans. This series is lighter than most other generators but capable of very high outputs. This type of generator does not have an external fan; instead, it has two internal fans.

LIQUID-COOLED GENERATORS Another recent design uses liquid cooling **(Figure 20–12)**. Using water or coolant to cool a generator is a very efficient way to keep diode temperatures down. But the real reason for eliminating the fan and using liquid to cool the generator is to reduce noise. The rotating fan is a source of underhood noise that some automobile manufacturers want to eliminate. These new generators have water jackets cast into the housing.

FIGURE 20–12

A water-cooled AC generator.

BMW of North America, Inc.

Hoses connect the housing to the engine's cooling system. Not only do these generators make less noise, they also have higher power output and should last longer in the high-temperature environment of the engine compartment.

AC GENERATOR OPERATION

As mentioned earlier, AC generators produce alternating current that must be converted, or rectified, to DC. This is accomplished by passing the AC through diodes.

DC Rectification

Figure 20–13 shows that when AC passes through a diode, the negative pulses are blocked off to produce the scope pattern shown. If the diode is reversed, it blocks off current during the positive pulse and allows the negative pulse to flow **(Figure 20–14)**. Because only half of the AC current pulses (either the positive or the negative) are able to pass, this is called **half-wave rectification**.

By adding more diodes to the circuit, more of the AC is rectified. When all of the AC is rectified, **full-wave rectification** occurs.

Full-wave rectification for stator windings requires another circuit with similar characteristics. **Figure 20–15** shows a wye stator with two diodes attached to each winding. One diode is insulated, or positive, and the other is grounded, or negative. The centre of the Y contains a common point for all windings. It can have a connection attached to it. It is

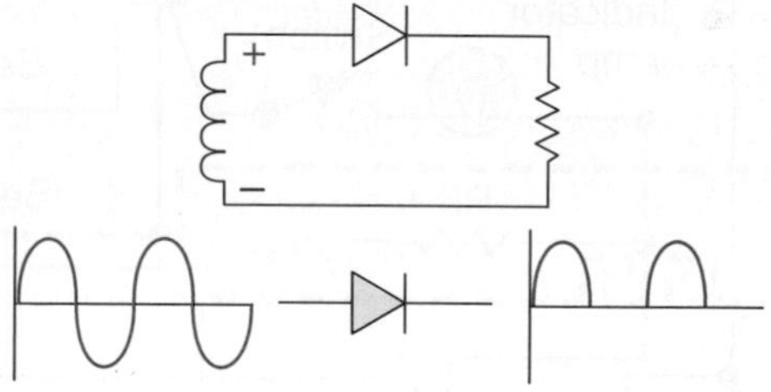

FIGURE 20–13

Half-wave rectification, diode positively biased.

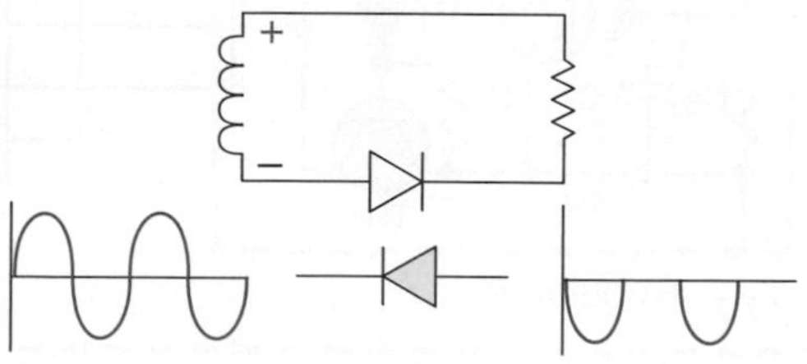

FIGURE 20–14

Half-wave rectification, diode negatively biased

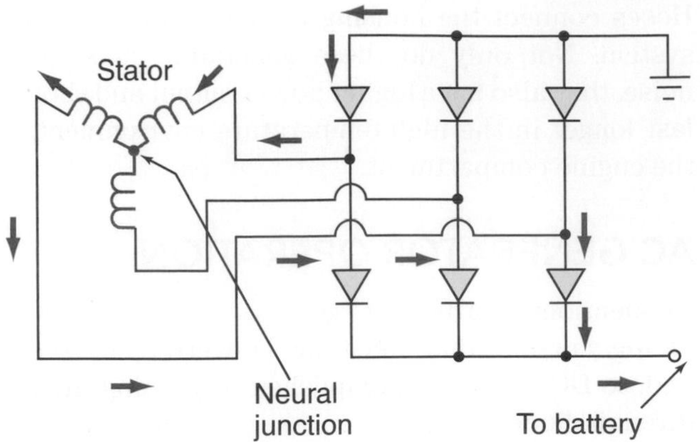

FIGURE 20–15

A wye stator wired to six diodes

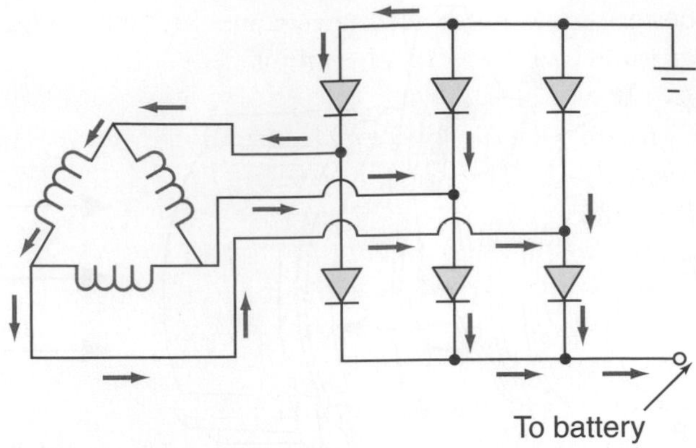

FIGURE 20–16

A delta stator wired to six diodes.

called the stator neutral junction. At any time during the rotor movement, two windings are in series, and the third coil is neutral and inactive. As the rotor revolves, it energizes the different sets of windings in different directions. However, the uniform result is that current in any direction through two windings in series produces the required DC for the battery.

The diode action does not change when the stator and diodes are wired into a delta pattern. **Figure 20–16** shows the major difference. Instead of having two windings in series, the windings are in parallel. Thus, more current is available from a delta-wound AC generator because the parallel paths allow more current to flow through the diodes. Nevertheless, the action of the diodes remains the same.

Many AC generators have an additional set of three diodes called the **diode trio**. The diode trio is used to rectify current from the stator so that it can be used to create the magnetic field in the field coil of the rotor. Using the diode trio eliminates extra wiring. To control generator output, a voltage regulator regulates the current from the diode trio and to the rotor **(Figure 20–17)**.

Voltage Regulation

Voltage output of an AC generator can reach as high as 250 volts if it is not controlled. The battery and the rest of the electrical system must be protected from this excessive voltage. Therefore, the voltage output from a charging system must be controlled. Current output

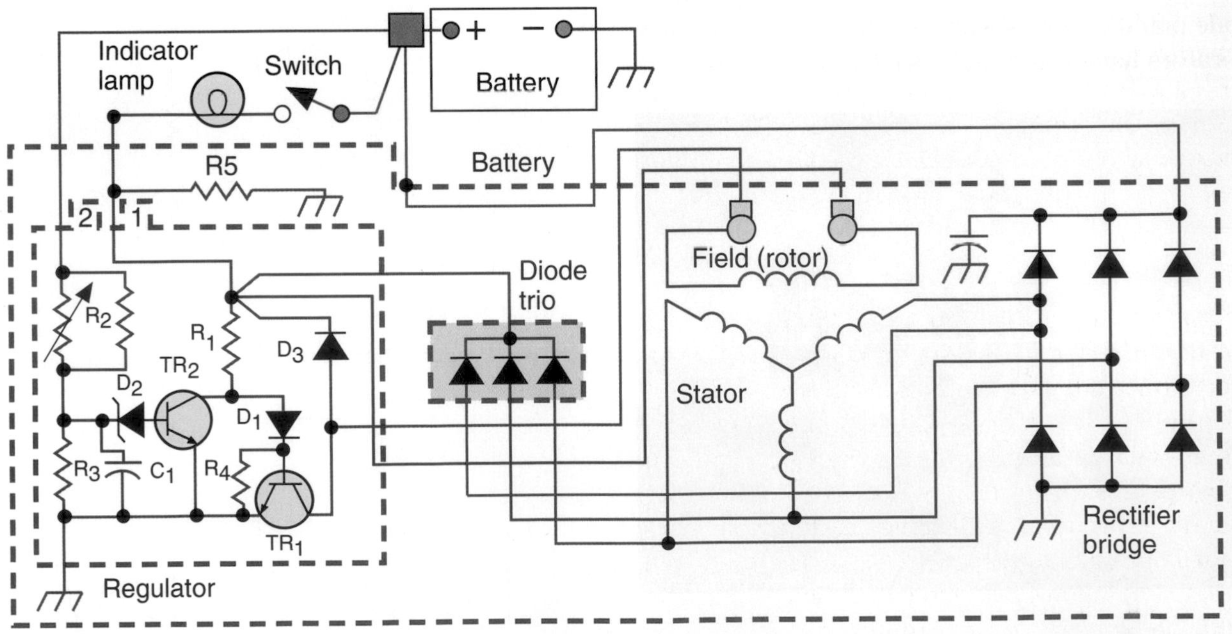

FIGURE 20–17

A wiring diagram of a charging circuit with a diode trio.

does not need to be controlled because an AC generator naturally limits the current output. The **voltage regulator** controls the voltage output of an AC generator.

Regulation of voltage is accomplished by varying the amount of field current flowing through the rotor. The higher the field current, the higher the voltage output. By controlling the amount of resistance in series with the field coil, control of the field current and voltage output is obtained. To ensure that the battery stays fully charged, most regulators are set for a system voltage between 14.5 and 15.5 volts.

The regulator must receive system voltage as an input in order to regulate the voltage output. This input voltage to an AC generator is called the **sensing voltage**. If the sensing voltage is below the regulator setting, an increase in field current is allowed, which causes an increase in charging voltage output. Higher sensing voltage will result in a decrease in field current and voltage output. The regulator will reduce the charging voltage until it is at a level to run the ignition system while putting a low charge (trickle charge) on the battery. If a heavy load is turned on, such as the headlights, the additional draw causes a decrease in battery voltage. The regulator senses the low system voltage and increases current to the rotor. This increases the strength of the magnetic field around the rotor and increases the generator's output voltage. When the load is turned off, the regulator senses the rise in system voltage and reduces the field current.

Another input that affects voltage regulation is temperature. Because ambient temperature influences the rate of charge that a battery can accept, regulators are temperature compensated. Temperature compensation is required because the battery is more reluctant to accept a charge at lower ambient temperatures. The regulator will increase the system voltage until it is at a higher level so that the battery will accept it and can become fully charged.

Field Circuits

To properly test and service a charging system, it is important to identify the type of field circuit in that system's generator. There are basically three types of field circuits. The first type is called the A-circuit. It has the regulator on the ground side of the field coil. The battery feed (B+) for the field coil is picked up inside the AC generator **(Figure 20–18)**. By placing the regulator on the ground side of the field coil, the regulator allows the control of field current by varying the current flow to ground.

The second type of field circuit is the B-circuit. In this case, the voltage regulator controls the power side

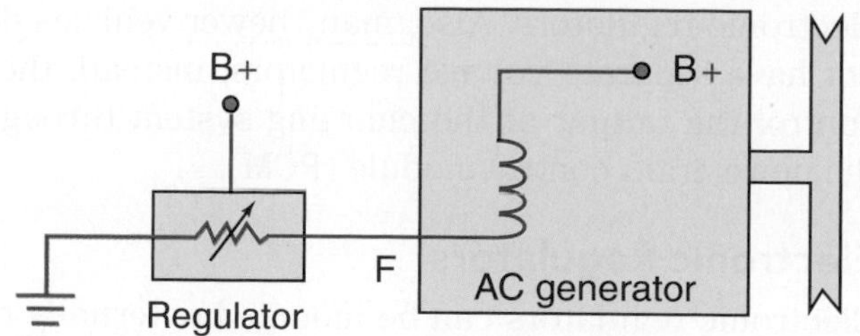

FIGURE 20–18

An A-circuit.

of the field circuit. The field coil is grounded inside the AC generator **(Figure 20–19)**. Normally the B-circuit regulator is mounted outside of the generator.

The third type of field circuit is called the isolated field. The AC generator has two field wires attached to the outside of the case. One is the ground; the other is the B+. The voltage regulator can be located on either the ground or the B+ side of the circuit **(Figure 20–20)**.

Older vehicles were equipped with electromechanical regulators, whereas newer vehicles have

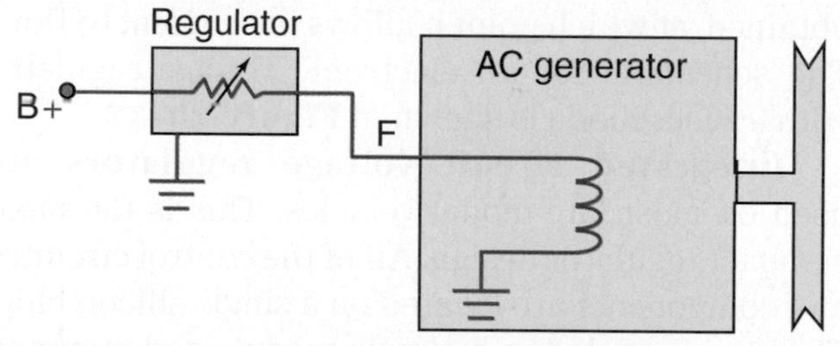

FIGURE 20–19

A B-circuit.

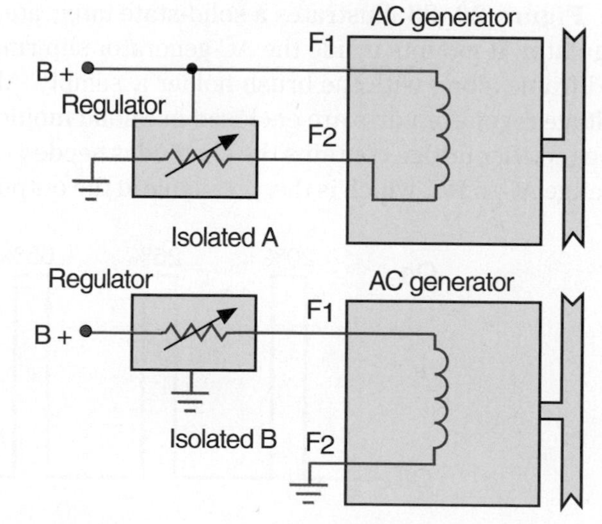

FIGURE 20–20

In the isolated field circuit AC generator, the regulator can be installed on either side of the field.

electronic regulators. Also, many newer vehicles do not have separate voltage regulators; instead, they control the output of the charging system through the powertrain control module (PCM).

Electronic Regulators

Electronic regulators can be mounted externally or internally in relation to the AC generator. The use of electronics allows for quick and accurate control of the field current. Electronic regulation is through the ground side of the field current (A-circuit control).

Pulse-width modulation controls the generator's output by varying the amount of time the field coil is energized. For example, assume that a vehicle is equipped with a 100-amp generator. If the electrical demand placed on the charging system requires 50 amps, the regulator would energize the field coil for 50 percent of the time **(Figure 20–21)**. If the electrical system's demands were increased to 75 amps, the regulator would energize the field coil 75 percent of the cycle time.

The electronic regulator uses a zener diode that blocks current flow until a specific voltage is obtained, at which point it allows the current to flow. The schematic for an electronic voltage regulator with a zener diode is shown in **Figure 20–22**.

Integrated circuit voltage regulators are used on most late-model vehicles. This is the most compact regulator design. All of the control circuitry and components are located on a single silicon chip. The chip is sealed in a plastic module and mounted either inside or on the back of the AC generator. Integrated circuit regulators are non-serviceable and must be replaced if defective.

Figure 20–23 illustrates a solid-state integrated regulator. It mounts inside the AC generator slip ring end frame along with the brush holder assembly. All voltage regulator parts are enclosed in a solid mould. The rectifier bridge contains the six diodes needed to change AC to DC, which is then available at the output battery terminal. Field current is supplied through a diode trio, which is connected to the stator windings.

FAIL-SAFE CIRCUITS To prevent simple electrical problems from causing high-voltage outputs that can damage delicate electronic components, many voltage regulators contain **fail-safe circuits**.

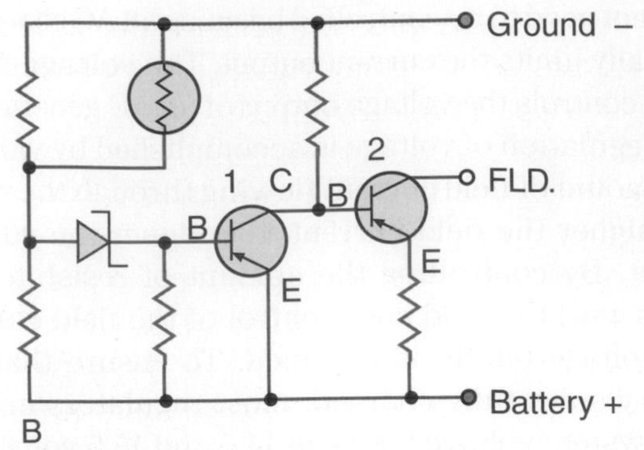

FIGURE 20–22

A simplified circuit of an electronic regulator with a zener diode.

FIGURE 20–23

Component locations of an AC generator with an internally mounted voltage regulator.

Robert Bosch Corp.

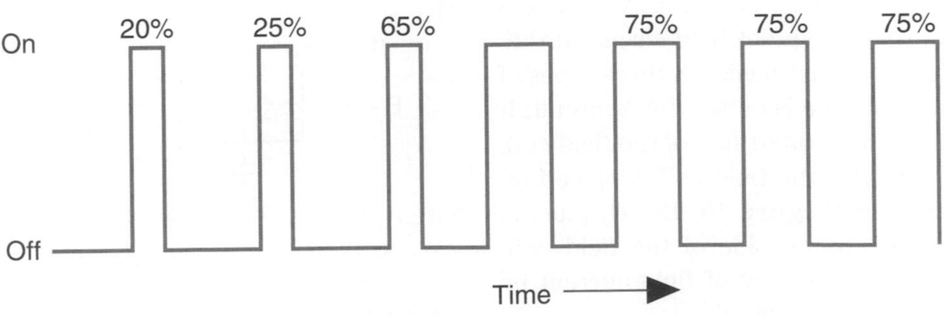

FIGURE 20–21

On-time increases with electrical demand.

A detailed explanation of how these circuits operate can be quite confusing. All you need to know is what a fail-safe circuit does, not how it does it. If wire connections to the AC generator become corroded or are accidentally disconnected, the regulator's fail-safe circuits may limit voltage output that might otherwise rise to dangerous levels. Under certain conditions, the fail-safe circuits may prevent the AC generator from charging at all. A fusible line in the fail-safe circuitry confines damage to the AC generator. Delicate electronic components in other vehicle systems are not damaged.

Computer Regulation

On a growing number of late-model vehicles, separate voltage regulators are no longer used. Instead, the voltage regulation circuitry is located in the vehicle's PCM or another control module **(Figure 20–24)**. Regardless of where the circuitry is located, it is still used to control current to the field windings in the rotor.

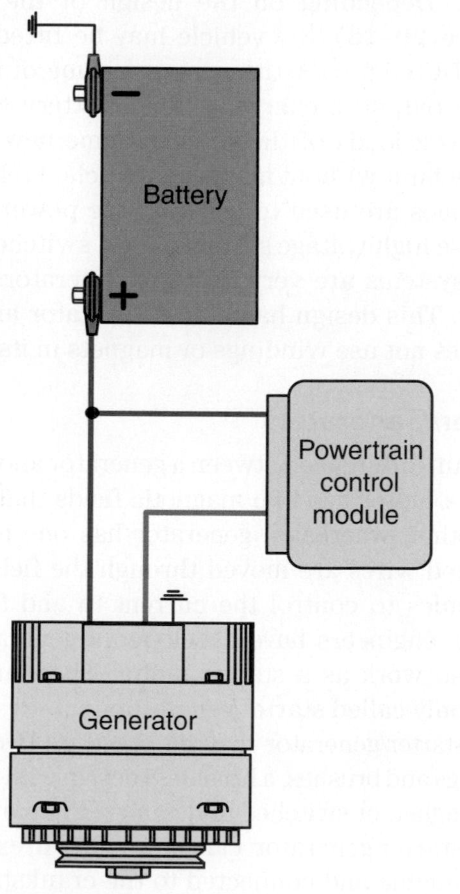

FIGURE 20–24

The basic circuit for a generator with its regulator as part of the PCM.

This type of system does not control rotor field current by acting like a variable resistor. Instead, the computer switches or pulses field current on and off at a fixed frequency of about 400 cycles per second. By varying on–off times, a correct average field current is produced to provide correct AC generator output. At high engine speeds with little electrical system load, field circuit on time may be as low as 10 percent. At low engine speeds with high loads, the computer may energize the field circuit 75 percent or more of the time, to generate the higher average field current needed to meet output demands.

A significant feature of this system is its ability to vary the amount of voltage according to vehicle requirements and ambient temperatures. This precise control allows the use of smaller, lighter storage batteries. It also reduces the magnetic drag of the AC generator, increasing engine output by several horsepower. Precise management of the charging rate can result in increased gas mileage and eliminate potential rough idle problems caused by parasitic voltage loss at low idling speeds. Most importantly, it allows the computer's diagnostic capabilities to be used in troubleshooting charging system problems, such as low- or high-voltage outputs.

Indicators

It is very important to monitor charging system performance during the course of vehicle operation. Vehicles are equipped with an ammeter, voltmeter, or indicator light. These allow the driver to monitor the charging system.

INDICATOR LIGHT The indicator light is the simplest and most common method of monitoring AC generator performance. When the charging system fails to supply sufficient current, the light turns on. However, when the ignition switch is first activated, the light also comes on because the AC generator is not providing power to the battery and other electrical circuits. Thus, the only current path is through the ignition switch, indicator light, voltage regulator, part of the AC generator, and ground, and then back through the battery **(Figure 20–25)**. Only the battery, regulator, and alternator are in the circuit. With no current flowing through the indicator light, it goes out.

With the engine running, the indicator light comes on again if the electrical load is more than the AC generator can supply, which occasionally happens when the engine is idling. If there are no problems, the light should go out as the engine speed is increased. If it does not, either the AC generator or regulator is not working properly.

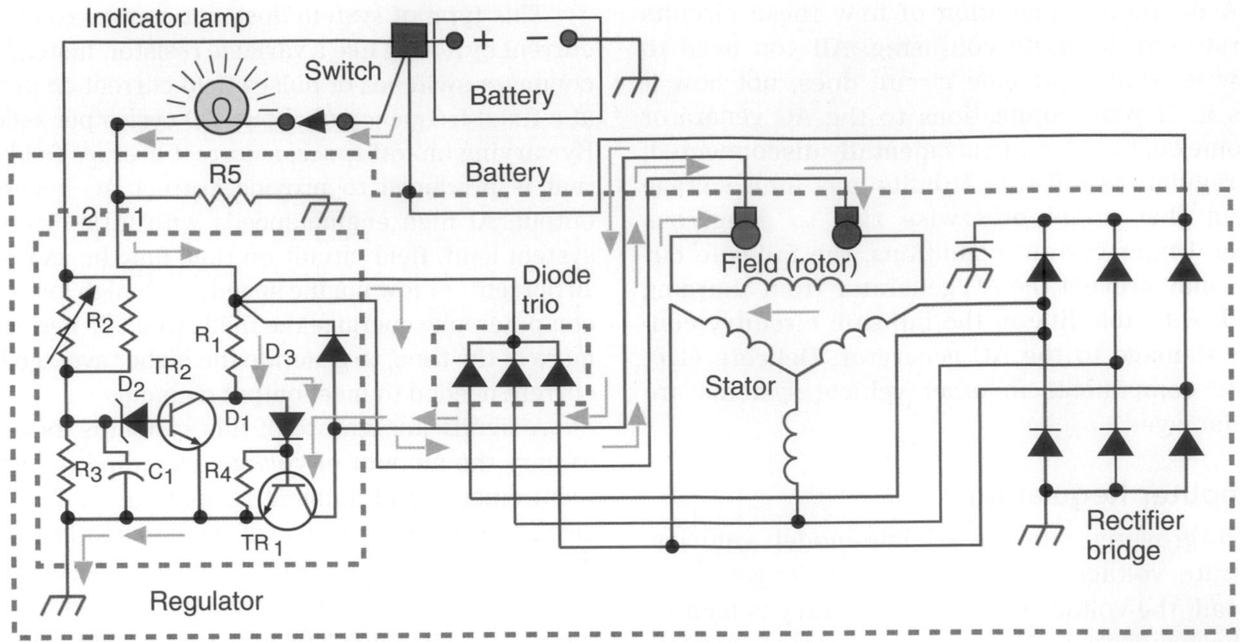

FIGURE 20–25

An electronic regulator with an indicator light on due to no AC generator output.

METERS Some vehicles have an ammeter or voltmeter in their instrument cluster. The voltmeter displays the voltage at the battery. If the charging system is working fine, the voltmeter will read more than 12.6 volts.

SHOP TALK

On late-model cars, the indicator light can be combined with the oil pressure warning light; it usually is labelled "engine." If this light turns on while the engine is running, either the AC generator is not charging or the oil pressure is low—or both.

The ammeter monitors current flow in and out of the battery. When the AC generator is delivering current to the battery, the ammeter shows a positive (+) indication. When not enough current (or none at all) is being supplied, the result is a negative (–) indication.

NEW DEVELOPMENTS

In the quest to improve fuel economy, decrease emission levels, and make vehicles more reliable, engineers have applied advanced electronics to starters and generators.

42-Volt Generators

Vehicles with a 42-volt electrical system will have an air- or liquid-cooled generator capable of producing 42 volts and 5 to 10 kW. Currently, a conventional 12-volt AC generator puts out about 1.5 kW. Depending on the design of the system **(Figure 20–26)**, the vehicle may be fitted with a DC to DC converter that changes some of the generator's output to charge a 12-volt battery or power the 12-volt loads of the vehicle. Some new generators are built without magnets. Switched reluctance techniques are used to generate the power needed for these high-voltage systems. Plus, switched reluctance systems are very efficient generators at low speeds. This design has a toothed stator and rotor and does not use windings or magnets in its rotor.

Starter/Generators

The main difference between a generator and a motor is that a motor has two magnetic fields that oppose each other, whereas a generator has one magnetic field, and wires are moved through the field. Using electronics to control the current to and from the battery, engineers have developed a generator that can also work as a starter motor. These units are commonly called starter/generators.

A starter/generator may be based on two sets of windings and brushes, a brushless design with a permanent magnet, or switched reluctance **(Figure 20–27)**.

A starter/generator can be mounted externally to the engine and connected to the crankshaft with a drive belt. Starter/generators can also be mounted directly on the crankshaft between the engine and the transmission or integrated into the flywheel

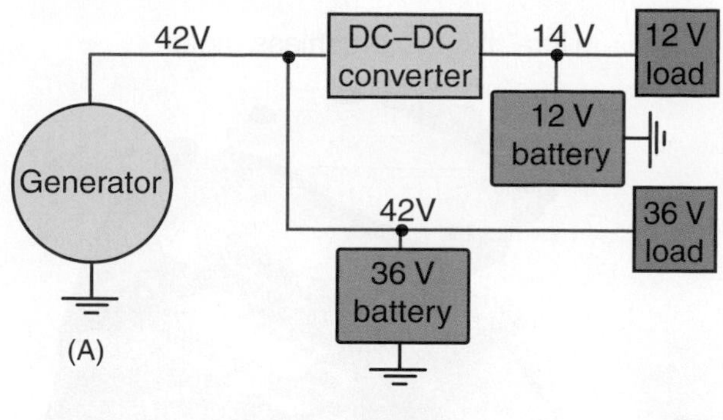

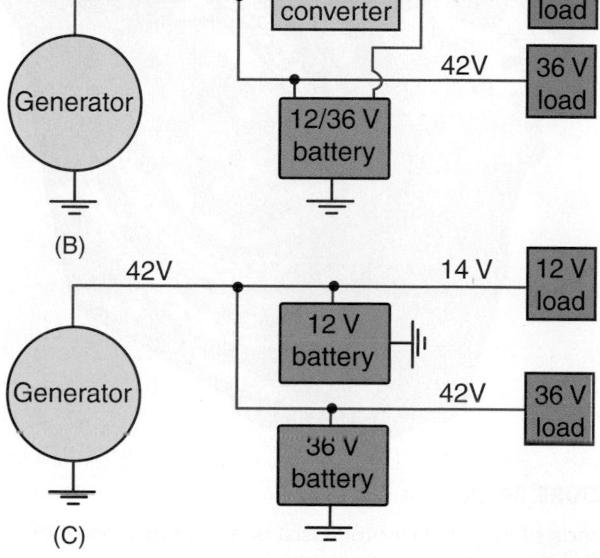

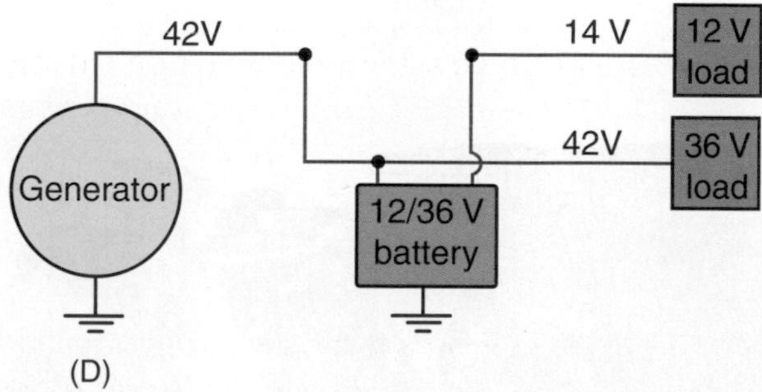

FIGURE 20–26

The different system layouts for 42-volt electrical systems: (A) a system with two batteries and a converter, (B) a system with one battery and a converter, (C) a two-battery system, and (D) a one-battery system.

(Figure 20–28). This unit works as a starter by spinning the crank at starting and serves as a generator, charging both directly from the engine and during a regenerative braking event on a hybrid vehicle (to be discussed in detail in a later chapter).

Starter/generators are capable of high charging outputs and can crank the engine at high speeds.

FIGURE 20–27

A switched reluctance starter/generator. Note the design of the rotor.

MAHLE

They also allow for other features that make the vehicle more efficient:

- *Stop–start*. When the engine is not needed, such as at a stoplight, it is automatically turned off. It restarts smoothly and instantly when any demand for power is detected by the control module.
- *Regenerative braking*. This feature collects energy created from braking and uses it to recharge the vehicle's batteries. As the vehicle decelerates and the brakes are applied, the power flow reverses, and then the wheels drive the engine.
- *Electrical assist*. The starter/generator helps the engine at start-up and during hard acceleration, providing short bursts of added power.

FIGURE 20–28

An integrated starter/generator assembly built into the flywheel.

BMW of North America, Inc.

FIGURE 20–29

A belt-drive starter/generator.

Reprinted with permission

Some starter/generators are belt driven **(Figure 20–29)** and use all of the techniques designed for regenerative braking, torque assistance, and high efficiency. The belt tensioner is mechanically or electrically controlled to allow the starter/generator to drive or to be driven by the belt. A system used by Toyota has an electromagnetic clutch fitted to the crankshaft pulley. During normal operation, the clutch is engaged and the belt is driven by the crankshaft. When the engine is stopped, the crank pulley clutch disengages and the motor/generator works as a motor to keep the accessories going. It also restarts the engine when needed.

Other Applications

With new technology in vehicle development and the demand for alternate power plants to propel vehicles, starter/generators will become more prominent. As an example: A vehicle equipped with a starter/generator can be considered a mild hybrid because it is capable of most of the functions of a hybrid vehicle. Functions such as stop–start, regenerative braking, and electrical assist are common to both a full hybrid vehicle and a mild hybrid.

There are different configurations of starter/generators, and they will be discussed in a later chapter. The locations of the starter/generator will also be unique to the vehicle application. As an example, we can look at Honda's integrated motor assist.

Honda's integrated motor assist (IMA) is a thin brushless electric motor **(Figure 20–30)** that is located between the gasoline engine and the transmission **(Figure 20–31)**. It assists the gasoline

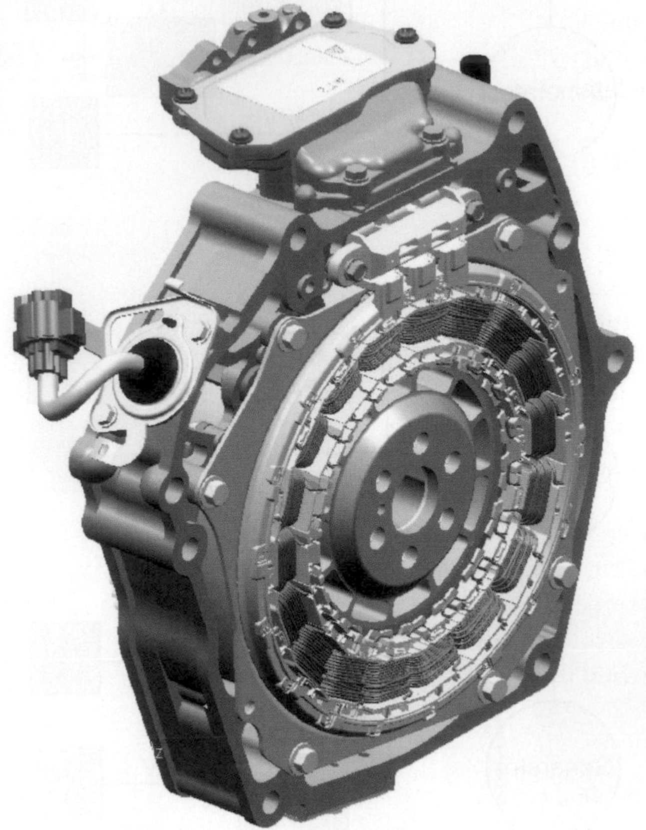

FIGURE 20–30

Honda's integrated motor assist unit is a brushless motor.

American Honda Motor Co., Inc.

FIGURE 20–31

Honda's integrated motor assist unit for hybrid vehicles fits between the engine and the transmission.

American Honda Motor Co., Inc.

engine during acceleration, functions as a generator to recharge the battery pack during deceleration, and serves as the gasoline engine's starter motor. Its modes of operation and controls will be discussed in a later chapter.

PRELIMINARY CHECKS

The key to solving charging system problems is getting to the root of the trouble the first time. Once a customer drives away with the assurance that the problem is solved, another case of a dead battery is very costly—both in terms of a free service call and a damaged reputation. Add to this the many possible hours of labour trying to figure out why the initial repair failed, and the importance of a correct initial diagnosis becomes all too clear. Some vehicles require radio/security codes for radio reprogramming. These codes should be obtained before disconnecting the battery to prevent costly retrieval procedures or a backup power supply can be used while the battery is disconnected.

Safety Precautions

- Disconnect the battery ground cable before removing any leads from the system. Do not reconnect the battery ground cable until all wiring connections have been made.
- Avoid contact with the AC generator output terminal. This terminal is hot (has voltage present) at all times when the battery cables are connected.
- The AC generator is not made to withstand a lot of force. Only the front housing is relatively strong. When adjusting belt tension, apply pressure only to the front housing to avoid damaging the stator and rectifier.
- When installing a battery, be careful to observe the correct polarity. Reversing the cables destroys the diodes. Proper polarity must also be observed when connecting a booster battery, positive to positive and negative to ground.
- Keep the tester's carbon pile off at all times, except during actual test procedures.
- Make sure all hair, clothing, and jewellery are kept away from moving parts.

Inspection

In addition to observing the ammeter, voltmeter, oscilloscope, or indicator light, there are some common warning signs of charging system trouble. For example, a low state of battery charge often signals a charging problem, as does a noisy AC generator.

FIGURE 20–32

Start your diagnosis with an inspection of the generator and its drive belt and wires.

Many charging system complaints stem from easily repairable problems that reveal themselves during a visual inspection of the system. Remember to always look for the simple solution **(Figure 20–32)** before performing more involved diagnostic procedures. Use the following inspection procedure when a problem is suspected.

PROCEDURE

Inspections

STEP 1 Before adjusting belt tension, check for proper pulley alignment, especially critical in serpentine belts.

STEP 2 Inspect the generator drive belt. Loose drive belts are a major source of charging problems.

STEP 3 Inspect the battery. It might be necessary to charge the battery to restore it to a fully charged state. If the battery cannot be charged, it must be replaced. Also make sure the posts and cable clamps are clean and tight, because a bad connection can cause reduced current flow.

STEP 4 Inspect all system wiring and connections. Many automotive electrical systems contain fusible links to protect against overloads. Fusible links can blow like a fuse without being noticed. Also look for a short circuit, an open ground, or high resistance in any of the circuits that could cause a problem that would appear to be in the charging system.

STEP 5 Inspect the AC generator and regulator mountings for loose or missing bolts. Replace or tighten as needed. Remember that the circuit completes itself through the ground of the AC generator and regulator. Most AC generators and regulators complete their ground through their mountings. If the mountings are not clean and tight, a high-resistance ground will result.

Alternator Pulley Technologies

There are three different kinds of alternator pulleys used by original equipment manufacturers (OEMs) on vehicles today: solid pulleys and two different types of clutching pulleys. Solid alternator pulleys are simply that, a solid pulley attached to the alternator shaft with a nut. They have been around for over a century but are quickly disappearing from vehicles. Clutch pulley Type I is called an overrunning alternator pulley (OAP) and is manufactured by INA. Type II is called an overrunning alternator decoupler (OAD) and is manufactured by Litens **(Figure 20–33A)**. Clutch pulleys are quickly becoming the automotive industry's standard as manufacturers begin to realize the benefits they provide.

(A)

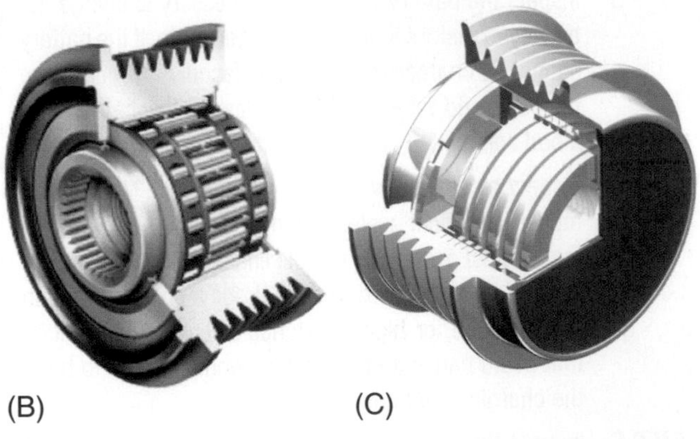

(B) (C)

FIGURE 20–33

Shown are examples of alternator pulleys: (A) a Corvette alternator with an overrunning alternator decoupler installed; (B) a cutaway view of an overrunning alternator pulley (OAP); and (C) a cutaway view of an overrunning alternator decoupler (OAD).

Al Steadman, Litens

BACKGROUND Nowadays, front-end accessory drives (FEADs) are engineered as a complete system, capable of providing years of durable, smooth, and quiet operation, as well as an increase in fuel economy. These systems are a quantum leap from the old V-belt systems of yesteryear. Before the introduction of OAPs and OADs, base engine vibrations, called in the industry "torsional vibrations," started to create serious problems for the FEAD. Issues such as noises, premature belt tensioner wear, and belt flutter/jump-off were occurring typically at idle speeds because torsional vibrations are much greater at lower rpms (start and sub-idle) and tend to diminish as rpm increases.

The alternator is typically the component with the largest inertia (mass) within the FEAD system. It tends to decelerate slower than the engine. This can cause the belt to slip at the alternator pulley during engine decelerations and generate a chirping noise. OAPs were introduced in the early 1990s to address these chirping noises. In 2000, OADs were introduced to address not only belt slip noises but also vibration and belt flutter issues that led to poor durability and noise complaints. The OAD also allows the OEM to lower the idle speed and belt tension without introducing any adverse effects such as unwanted noise, vibration, and harshness (NVH). Lowering the belt tension has many benefits. These include significant improvement in the life of all the bearings within the FEAD system (alternator, A/C compressor, coolant pump, etc.). Along with improved durability, the lower belt tension results in smaller parasitic losses throughout the FEAD system, therefore making it more efficient. The reduction in parasitic losses is also a significant contributor to the benefit of increased fuel economy and the reduction of emissions.

TYPE I – OAP (INA) The function of the OAP **(Figure 20–33B)** is achieved by introducing a one-way clutch mechanism inside the alternator pulley. This clutch mechanism allows the heavy rotor inside the alternator to overrun during an engine shutdown or a transmission shift. Allowing the rotor to overrun effectively eliminates belt chirp noises during these engine deceleration conditions.

Operation of an OAP can be checked by simply inserting the correct tool, holding the pulley and rotating the inside shaft. It must rotate smoothly in one direction and immediately lock in the other direction. During engine operation, if the clutch mechanism inside the pulley slips for whatever reason, then a no-charge situation (alt light on) may be noticed and the battery may go dead. If the clutch mechanism locks up in both directions, then belt slip noises or increased belt tensioner motion and premature wear may be noticed.

TYPE II – OAD (LITENS) The OAD **(Figure 20–33C)** has all the same features of an OAP but also utilizes an internal vibration absorbing spring. This internal spring is the difference between the OAP and the OAD. The spring effectively isolates the alternator by absorbing the torsional vibrations within the FEAD. The OAD can be thought of as a "suspension system" for the alternator, cushioning the heavy alternator rotor from those damaging vibrations that come from the firing pulses of the engine's combustion cycle. OADs are often extremely effective at absorbing the stronger vibrations of the newer, more fuel efficient engines. Smaller displacement engines, engines that include cylinder deactivation, or simply diesel engines are often very rough and therefore require that a Litens OAD be utilized in the FEAD by the original equipment manufacturer.

The operation of a Litens OAD can be verified the very same way as the OAP. Simply remove the cap, insert the correct tool, hold the pulley, and rotate the inside shaft. The OAD differs from the OAP in that the OAD must rotate smoothly in one direction and have a spring feel in the other direction. Both OADs and OAPs are considered to be wearable items and are therefore recommended to be replaced before pulley failure occurs. They should be checked and replaced (if required) whenever changing the serpentine belt and/or tensioner.

The solid pulley, OAP, and OAD are not interchangeable technology; therefore which alternator pulley the vehicle was designed with must be known first and then be replaced accordingly. Be aware that if the wrong pulley technology is used, then the FEAD components may repeatedly exhibit very early signs of failure, and the vehicle driver will notice much more vibration and strange noises, especially at idle.

GENERAL TESTING PROCEDURES

Diagnosing a charging system is a straightforward task. Tests can be conducted with a volt/ampere tester (VAT), current probe, digital multimeter (DMM), or lab scope. Charging system tests for all cars are basically the same; however, it is very important to refer to the manufacturer's specifications. Even the most accurate test results are no good if they are not matched against the correct specs.

Voltage Output Test

To check the charging system's voltage output, begin by measuring the battery's open-circuit voltage. Connect the voltmeter across the battery and note the reading on the meter. Next, start the engine and run it at the suggested rpm for this test (usually 1500 rpm). With no electrical load, the voltage reading should be about 2 volts higher than the open-circuit voltage.

A reading of less than 13.0 volts immediately after starting the engine indicates a charging problem. No increase in voltage means the system is not producing voltage. A reading of 16 or more volts indicates overcharging. A faulty voltage regulator or control voltage circuit are the most likely causes of overcharging.

If the unloaded charging system voltage is within specifications, test the output under a load. To do this, increase engine rpm to about 2000 rpm, and turn on the headlights and other high-current accessories. Under these conditions, the output should be about 0.5 volt above battery open-circuit voltage.

Current Output Test

Using a carbon pile load tester is an easy way to check the amperage output of a charging system. With the tester connected to the system, the engine is run at a moderate speed, and the carbon pile is adjusted to obtain maximum current output without voltage dropping below 12.6 volts. This reading is compared against the rated output. Normally, readings that are more than 10 percent out of specifications indicate a problem.

Field Current Check

Low generator output can be caused by worn brushes, which limit field current. To measure field current, place a current probe or the carbon pile load tester's inductive pickup over the field wire at the generator. Now load the charging system with the carbon pile to bring the generator to full output. Observe the ammeter reading on the tester. The procedure for measuring field current is different for generators with an integral regulator, and those procedures vary with the model of generator, so follow the instructions given by the manufacturer.

Diode Checks

The output of a generator is highly dependent on the condition of the diodes. Not only do the diodes rectify AC voltage to DC, they also prevent AC voltage from being present in the output. Bad diodes are indicated by the presence of more than 0.5 AC volt in the output wire. To check this, set the DMM to measure AC volts. Then connect the black meter lead to a good ground and the red lead to the generator's battery terminal.

Another check of the diodes while they are still in the generator is done with the engine off and with a low-amperage current probe. Measure the current on the generator's output wire. Any measurement greater than 0.5 milliamp indicates one or more diodes are leaking, and the generator or diodes need to be replaced.

Oscilloscope Checks

AC generator output can also be checked using an oscilloscope. **Figure 20–34** illustrates common AC generator voltage patterns for good and faulty generators. The correct pattern looks like the rounded top of a picket fence. A regular dip in the pattern indicates that one or more of the coil windings is grounded or open, or that a diode in the rectifier circuit of a diode

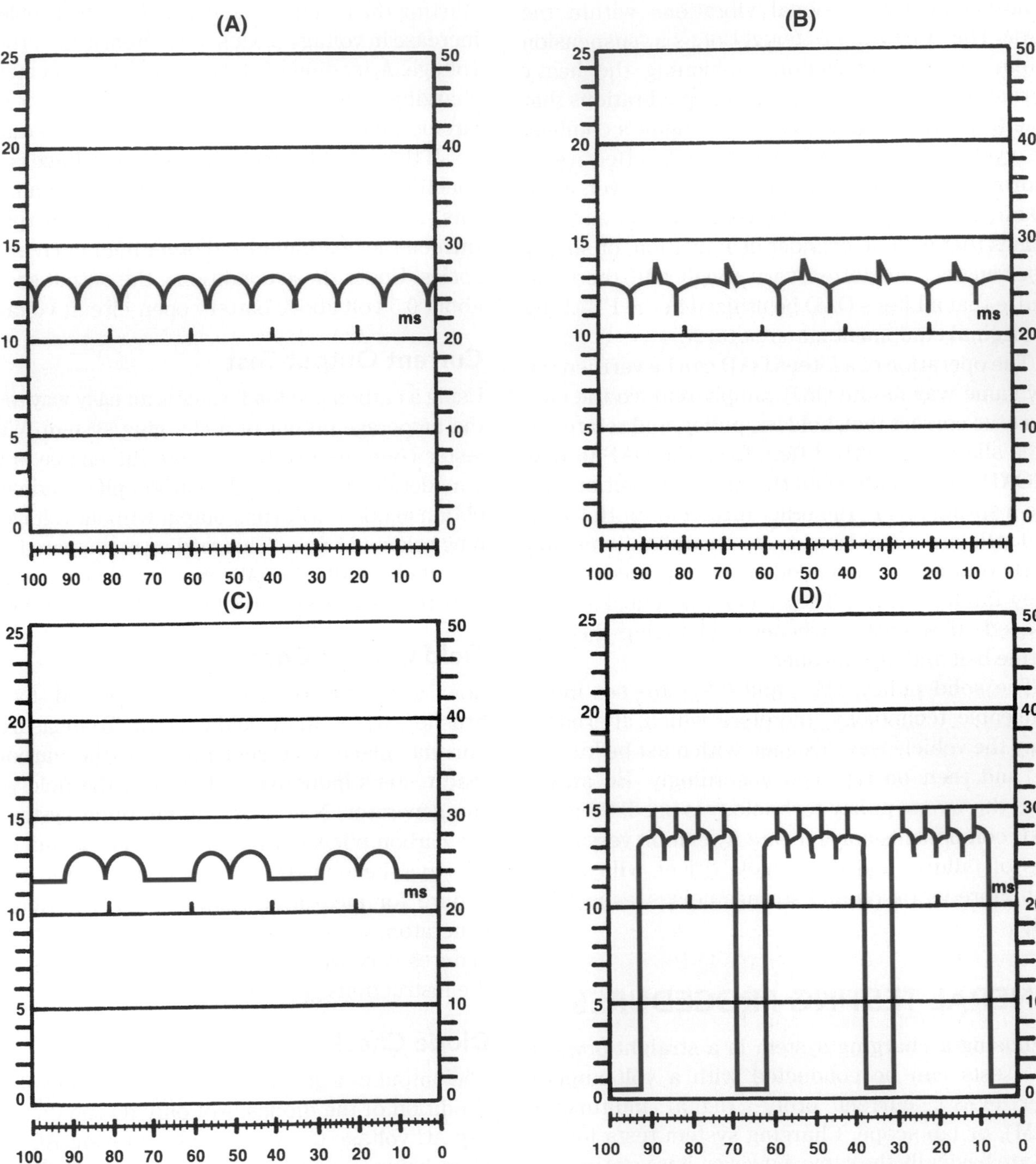

FIGURE 20–34

AC generator oscilloscope patterns: (A) a good AC generator under full load, (B) a good AC generator under no load, (C) a shorted diode and/or stator winding under full load, and (D) an open diode in diode trio.

trio circuit has failed. One or more bad or leaking diodes will decrease the output of generator.

Circuit and Ground Resistance

These tests measure voltage drop within the system wiring. They help pinpoint corroded connections or loose or damaged wiring.

Circuit resistance is checked by connecting a voltmeter to the positive battery terminal and the output, or battery terminal of the AC generator. The positive lead of the meter should be connected to the AC generator output terminal, and the negative lead to the positive battery terminal. To check the voltage drops across the ground circuit, connect the positive lead to the generator housing and the negative meter lead to the battery negative terminal. When measuring the voltage drop in these circuits, a sufficient amount of current must be flowing through the circuit. Therefore, turn on the headlights and other accessories to ensure that the AC generator is putting out at least 20 amps. If a voltage drop of more than 0.5 volt is measured in either circuit, there is a high-resistance problem in that circuit.

AC GENERATOR SERVICE

When the cause of charging system failure is the AC generator, it should be removed and replaced or rebuilt. Whether it is rebuilt or replaced depends on the type of generator it is, the time and cost required to rebuild it, your shop's policy, and your customer's desires. Many late-model AC generators are not rebuilt. They are traded in as a core toward the purchase of a new or remanufactured unit. Just in case you do need to rebuild one, make sure you follow the procedures given by the manufacturer for the generator you are working on.

To test the components of an AC generator it must be removed and disassembled.

A faulty AC generator can be the result of many different types of internal problems. Diodes **(Figure 20–35)**, stator windings **(Figure 20–36)**, and field circuits **(Figure 20–37)** may be open, shorted **(Figure 20–38** and **Figure 20–39)**, or improperly grounded. The brushes or slip rings can become worn. The rotor shaft can become bent, and the pulley can work loose or bend out of proper alignment.

Follow service manual procedures when removing and installing an AC generator. Remember, improper connections to an AC generator can destroy it.

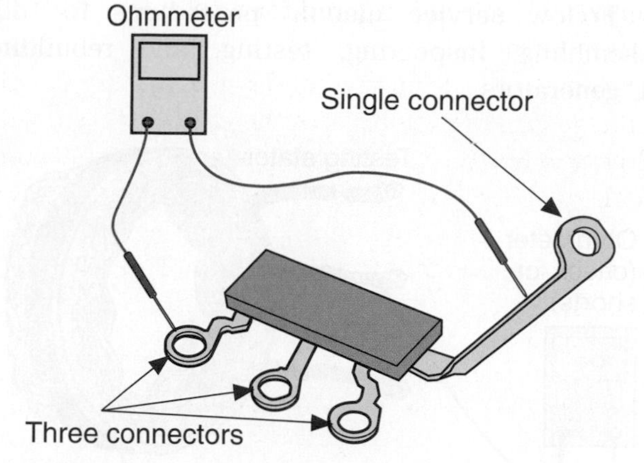

FIGURE 20–35

Using an ohmmeter to test a diode trio.

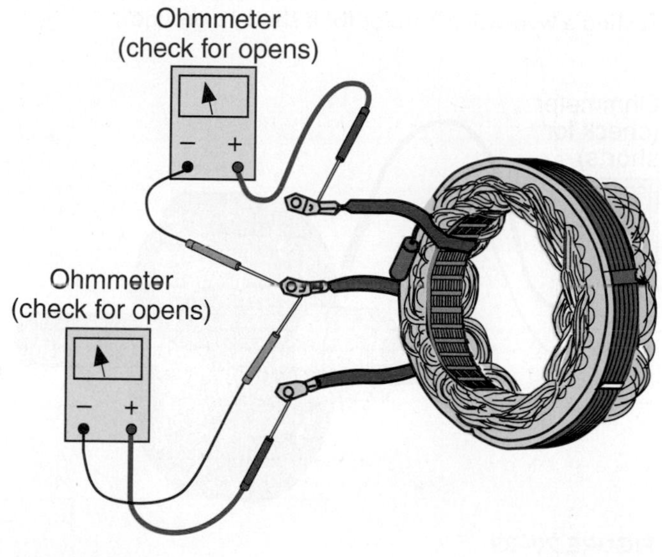

FIGURE 20–36

Testing a wye-wound stator for opens. A delta stator cannot be tested this way.

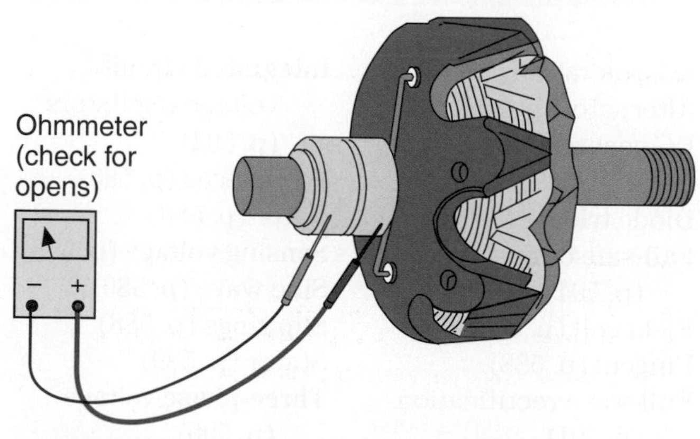

FIGURE 20–37

Testing a rotor for opens.

Follow service manual procedures for disassembling, inspecting, testing, and rebuilding AC generators.

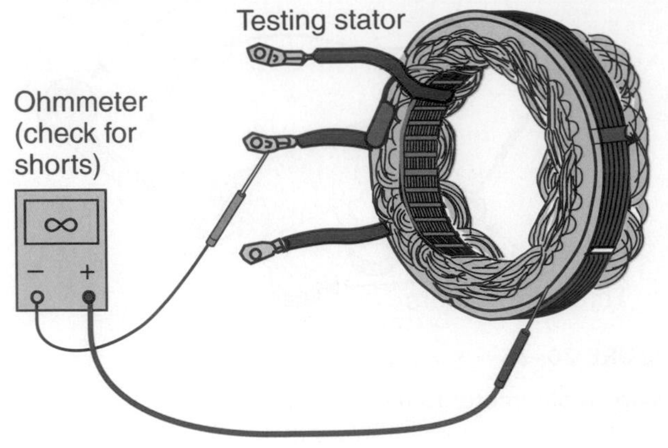

FIGURE 20–38
Testing a wye-wound stator for a short to ground.

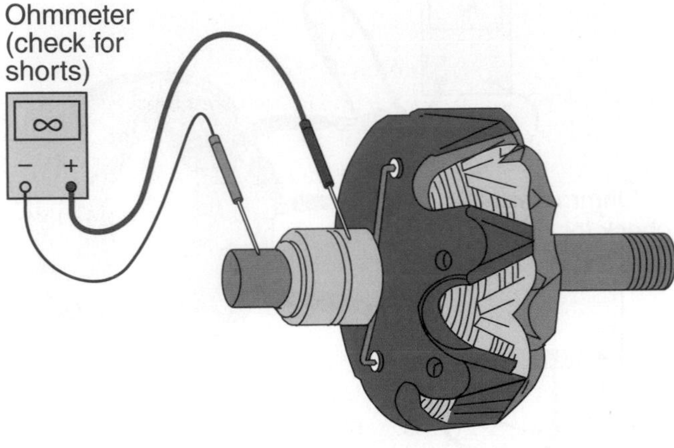

FIGURE 20–39
Testing a rotor for a short to ground.

KEY TERMS

AC generator (p. 586)
Alternator (p. 587)
DC generator (p. 586)
Delta (p. 589)
Diode trio (p. 592)
Fail-safe circuits (p. 594)
Field coil (p. 588)
Fingers (p. 588)
Full-wave rectification (p. 591)
Half-wave rectification (p. 591)
Integrated circuit voltage regulators (p. 594)
Pole pieces (p. 588)
Rotor (p. 588)
Sensing voltage (p. 593)
Sine wave (p. 589)
Slip rings (p. 588)
Stator (p. 589)
Three-phase voltage (p. 590)
Voltage regulator (p. 593)
Wye (p. 589)

SUMMARY

- Inducing a voltage requires a magnetic field, conductors, and movement between the conductors and the magnetic field so that the lines of force are cut.
- Modern vehicles use an AC generator to produce electrical current in the charging system. Diodes in the generator change or rectify the alternating current to direct current.
- A voltage regulator keeps charging system voltage above battery voltage. Keeping the AC generator charging voltage above the 12.6 volts of the battery ensures current flows into, not out of, the battery.
- Modern voltage regulators are completely solid-state devices that can be an integral part of the AC generator or mounted to the back of the generator housing. In some vehicles, voltage regulation is the job of the computer control module.
- Voltage regulators work by controlling current flow to the AC generator field circuit. This varies the strength of the magnetic field, which in turn varies current output.
- The driver can monitor charging system operation with indicator lights, a voltmeter, or an ammeter.
- Problems in the charging system can be as simple as worn or loose belts, faulty connections, or battery problems.
- Circuit resistance, current-output, voltage-output, field-current draw, and voltage regulator tests are all used to troubleshoot AC charging systems.

REVIEW QUESTIONS

1. What is the purpose of the diode trio?
   a. to prevent current flow to the field
   b. to provide current flow to the field
   c. to limit charging system output
   d. to protect the voltage regulator from voltage spikes
2. What three factors determine the amount of induced voltage produced through induction?
   a. speed, number of conductors, and strength of the magnet
   b. speed, amount of voltage, and strength of the magnet
   c. speed, amount of current, and the number of conductors
   d. speed, amount of current, and strength of the magnet

3. Voltage produced by an AC generator must be changed to DC voltage so that it can be used by the automotive system. Which device changes AC to DC?
   a. potentiometer b. diode
   c. capacitor d. transformer
4. In which part of an AC generator is the output voltage produced?
   a. stator b. rotor
   c. rectifier d. regulator
5. Current through which component is controlled by the voltage regulator?
   a. field b. stator
   c. rectifier d. diode trio
6. A method of testing an AC generator is by using a digital multimeter on the AC voltage scale. When connected between ground and the battery terminal of the generator, what is an acceptable voltage?
   a. below 0.5 volt b. above 0.5 volt
   c. below 12.6 volts d. above 14.2 volts
7. When using a battery load tester to test an AC generator, engine speed should be maintained at approximately what rpm?
   a. wide-open throttle b. 4000 rpm
   c. 2000 rpm d. idle
8. In a type-A field circuit, where is the regulator placed?
   a. in ground side of the field circuit
   b. in ground side of the stator circuit
   c. in positive side of the field circuit
   d. in positive side of the stator circuit
9. Which type of stator windings generates the maximum current output from an AC generator?
   a. delta b. wave
   c. lap d. wye
10. The output of the stator is AC voltage. What type of rectification will create DC voltage?
    a. quarter-wave rectification
    b. half-wave rectification
    c. three-quarter-wave rectification
    d. full-wave rectification
11. An automobile comes into the shop with an overcharged battery. What is the most likely cause?
    a. a loose AC generator serpentine belt
    b. high resistance in the insulated circuit
    c. a defective voltage regulator
    d. high resistance in the field circuit
12. The purpose of the brushes in AC generators is to carry current to which component?
    a. the field b. the stator
    c. the rectifier d. the regulator
13. What is the major voltage control device inside the regulator?
    a. clamping diode
    b. light-emitting diode
    c. photo diode
    d. zener diode
14. Stator windings are offset to allow a more constant voltage and current output. The arrangement of these windings produces which type of output?
    a. single-phase alternating current
    b. dual-phase alternating current
    c. three-phase alternating current
    d. quarter-phase alternating current
15. The brushes in the AC generator connect to which component?
    a. slip rings b. stator
    c. rectifier bridge d. commutator
16. How does AC generator output voltage change with outside ambient temperature?
    a. The voltage rises as temperature rises.
    b. The voltage lowers as temperature lowers.
    c. The voltage rises as temperature lowers.
    d. The voltage will remain constant.
17. What would be the duty cycle of the field current during a high-current-demand situation, in a computer-controlled AC generator?
    a. approximately 10 percent
    b. approximately 20 percent
    c. approximately 50 percent
    d. approximately 80 percent
18. What controls the amperage output of the alternator?
    a. rotor current
    b. stator current
    c. rectifier current
    d. armature current
19. What is the maximum allowable reading when testing the voltage drop on the insulated or feed circuit of an AC generator?
    a. 0.05 volt b. 0.5 volt
    c. 5 volts d. 10.5 volts
20. What is the purpose of the zener diode?
    a. to allow the voltage regulator to sense voltage
    b. to provide current flow to the field
    c. to convert AC current to DC current
    d. to convert DC current to AC current

CHAPTER 21
Lighting Systems

LEARNING OUTCOMES

- Explain the operating principles of the various lighting systems.
- Describe the different types of headlights and how they are controlled.
- Understand the functions of turn, stop, and hazard warning lights.
- Know how backup lights operate.
- Replace headlights and other burned-out bulbs.
- Explain how to aim headlights.
- Explain the purpose of auxiliary automotive lighting.
- Describe the operation and construction of the various automotive lamps.
- Diagnose lighting problems.

The lighting system provides power to both exterior and interior lights. It consists of the headlights, parking lights, marker lights, taillights, courtesy lights, dome/map lights, instrument illumination or dash lights, coach lights (if so equipped), headlight switch, and various other control switches. Other lights, such as vanity mirror lights, the underhood light, the glove box light, and the trunk compartment light, are used on some vehicles and are also part of the lighting system.

SHOP TALK

Lighting systems are largely regulated by federal laws, so the systems are similar among the various manufacturers. However, there are many variations. Before attempting to do any repairs on an unfamiliar system, you should always refer to the manufacturer's service manual or information systems.

AUTOMOTIVE LAMPS

A **lamp** is a device that provides light or illuminates something to make it visible. The term lamp is also given to light fixtures that provide light in a house, business, or outside. Today, nearly all lamps are powered by electricity; this is especially true in automobiles. The most commonly used are the incandescent lamps, halogen lamps, neon lamps, high-intensity lamps, and light-emitting diodes.

Incandescent Lamps

Perhaps the most commonly used type of lamp today is the **incandescent lamp (Figure 21–1)**. These

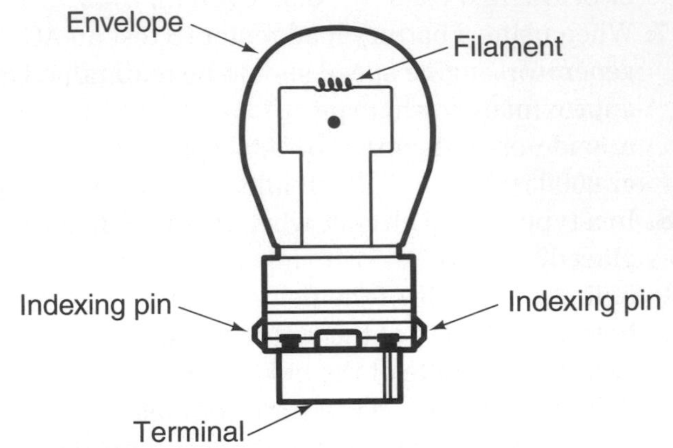

FIGURE 21–1
A single-filament bulb.

lamps produce light when current passes through a filament made of tungsten. The filament wire is heated to a high temperature, causing it to glow and give off light. Basically, electrical energy is changed to heat energy in the wire filament.

The filament is enclosed in a glass bulb that has been purged of air; therefore, the filament "burns" in a vacuum. If air enters the bulb, the oxygen will allow the filament to oxidize and burn up. As the filament burns, blackish deposits are left on the glass bulb. The presence of this coating is an indication that the bulb is bad or has little life remaining in it.

Incandescent lamps use more energy to provide light than other types of lamps. In an incandescent lamp, about 95 percent of the energy it consumes is lost through heat. Today, these lamps are being replaced by a variety of fluorescent lamps (including cold cathode fluorescent lamps [CCFL] and compact fluorescent lamps [CFL]), light-emitting diodes (LEDs), and high-intensity discharge lamps.

Halogen Lamps

Halogen lamps are actually incandescent lamps. This lamp is comprised of a tungsten filament enclosed in a halogen-filled bulb made of high-temperature resistant glass. The name **halogen** is used to identify a group of chemically related nonmetallic elements, such as iodine, chlorine, and fluorine. Most halogen lamps use iodine vapour. These lamps provide brighter light because the filament is able to burn at higher temperatures, due to the presence of halogen in the bulb. Another positive about a halogen lamp is the filament does not break down, as it does in an incandescent bulb, due to the reaction of the burning tungsten and the halogen. As a result, the metal vapour from burning is redeposited on the filament.

Neon Lamps

A **neon** lamp contains neon or similar gas (such as argon or mercury vapour) at a low pressure in a glass tube or capsule **(Figure 21–2)**. A small amount of AC or DC current flows through the tube, causing it to glow. The applied voltage must reach the striking voltage before the lamp actually lights up. At this voltage, the lamp will begin to glow. The required voltage to maintain the glow is about 30 percent less than the voltage required to start the lamp. Neon lamps have been used as the "third" brake light because of their relatively short response time and their cost.

FIGURE 21–2

Red neon light tubes.

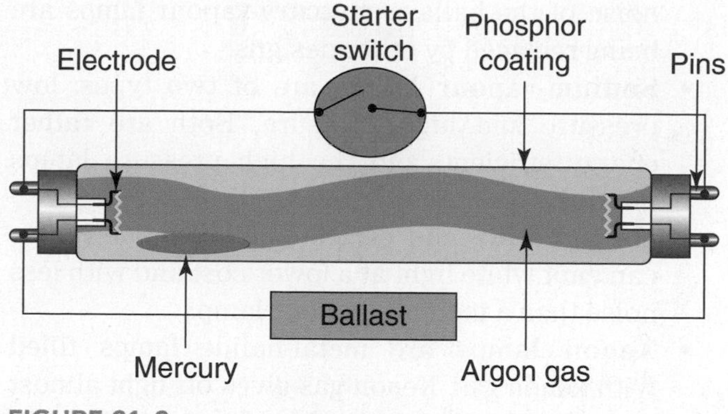

FIGURE 21–3

The construction of a fluorescent bulb.

Fluorescent Lamps

Like a neon lamp, a fluorescent lamp is a gas-discharge lamp. The excited gas in the lamp is typically mercury vapour. The enclosure of the lamp is coated with phosphor **(Figure 21–3)**. When current flows through the mercury vapour, the mercury atoms produce short-wave ultraviolet light, causing the phosphor to fluoresce and providing visible light. The advantage of using fluorescent lamps is they use less energy than a conventional incandescent lamp. However, their initial cost is higher because they need an electric ballast assembly to regulate the current through the tube.

High-Intensity Discharge Lamps

High-intensity discharge (HID) lamps provide more visible light, for the amount of consumed electrical power, compared to fluorescent and incandescent lamps. This is because less heat is lost while the lamp is lit. Light is emitted when an electric arc jumps across the tungsten electrodes inside a clear or opaque tube filled with a gas and metal salts. The gas helps the voltage jump the gap between the electrodes. Once the gap is jumped, the heat of the electrical arc reacts with the metal salts and forms a highly ionized gas that provides a great amount of light. The presence of the gas also reduces the amount of energy required to provide the light. A ballast unit is required to start and maintain the arcing in an HID lamp.

There are many types of HID lamps, including the following:

- **Mercury-vapour lamps** are commonly found in large parking lots or assembly halls. Like fluorescent lamps, they require a ballast to start. Older designs produced a bluish-green light, but the light from current versions has little colour. Due to inadequate lighting, energy costs, and the noise of the ballast, mercury-vapour lamps are being replaced by other designs.

- **Sodium-vapour lamps** are of two types: low pressure and high pressure. Both are rather energy efficient, and the high pressure lamps provide a much whiter light.
- **Metal-halide** and ceramic-metal halide lamps can emit white light at a lower cost and with less noise than a mercury-vapour lamp.
- **Xenon lamps** are metal-halide lamps filled with xenon gas. Xenon gas gives off light almost immediately after arcing begins across the electrodes of the lamp. Xenon bulbs provide a bright light that increases in intensity shortly after the arc has been established.

Light-Emitting Diodes (LED)

A **light-emitting diode (LED)** is a semiconductor that emits light when current passes through it **(Figure 21–4)**. LEDs also produce more light than heat and are much smaller than other types of lamps. The brightness of an LED depends on its temperature. More light is produced when the LED is operating with low current and is at a low temperature. Also, in order for an LED to provide a constant light output, it needs to be powered by a constant-current power supply. The power supply helps keep the LED within the desirable temperature range.

Unlike other lamps, the bulb of an LED lamp does not put off much heat. However, a great amount of heat is produced at the mounting base of the lamp assembly. That heat must be controlled to provide constant light and to protect the lamp. The need to keep LED temperatures low requires the use of heat sinks, ventilation systems, or cooling fans, which are normally quite expensive.

LED lamps are normally used in clusters because a single lamp does not provide the brightness of conventional lamps. For late-model vehicles, LEDs are commonly the bulb of choice for brake, parking, turn-signal, and daytime running lamps.

FIGURE 21–4

An LED.

HEADLIGHTS

Headlights are mounted on the front of a vehicle to light the road ahead during periods of low visibility, such as darkness or precipitation. Headlight designs and construction have been influenced by the changes in technology and safety regulations. In the past, all cars had two or four round or rectangular headlamps. Now headlights are an integral part of a vehicle's overall design **(Figure 21–5)**.

A headlamp system must offer a low and a high beam. The two beams on each side of the vehicle may have two separate lamps or a single lamp that can deliver both beams. Low beams are intended to be used whenever other vehicles are in front of a vehicle or approaching it from the opposite direction. Low beams are designed to provide adequate lighting in front of the vehicle, with some light spread to the sides of that forward beam. But they are also set to minimize the amount of light that can shine in the eyes of other drivers, which can temporarily blind them. Low beams are often called "dip" beams

FIGURE 21–5

Today's headlights are an integral part of the appearance of vehicles.

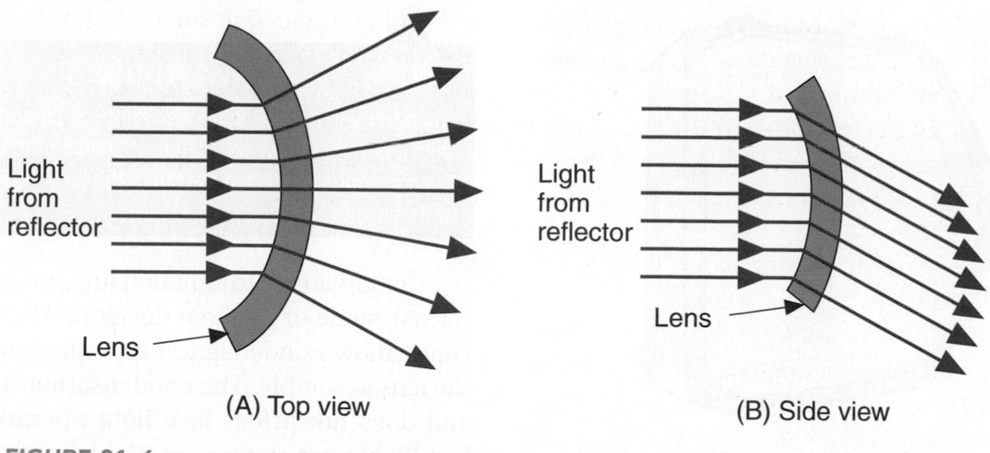

FIGURE 21–6

The reflector intensifies the light produced by the filament. (A) shows the lens top view creating a broad beam, and (B) shows the side view that is flattened and aimed toward the road, away from oncoming traffic.

because the angle of their light is lower and more to the right than high beams. Low-beam headlights are offset downward and rightward between ½° and 3° away from a straight-ahead direction.

High beams should be used when there are no other close vehicles and extra lighting is desired. High beams are designed to place the light beam parallel to the road's surface and illuminate as much of the road as possible. This means the light beams can directly hit the eyes of oncoming drivers, temporarily blinding them.

Sealed-Beam Headlights

Until recently, all vehicles had sealed-beam headlights. A **sealed-beam** headlamp is an airtight assembly with a filament, reflector, and lens. The curved reflector is sprayed with vaporized aluminum, and the inside of the lamp is normally filled with argon gas. The reflector intensifies the light produced by the filament, and the lens directs the light to form a broad flat beam **(Figure 21–6)**. To direct the light, the surface of the glass lens has concave prisms.

Low- and high-beam filaments are placed at slightly different locations within a sealed-beam bulb **(Figure 21–7)**. The filaments' location determines how light passes through the bulb's lens. This, in turn, determines the direction of the light beam. In a dual-filament lamp, the lower filament is used for the high beam, and the upper filament is used for the low beam.

There are various ways to identify sealed-beam headlights, such as #1, #2, and the "halogen" or "H" marking moulded on the front of the headlight lens. A type #1 is high beam only and has two electrical terminals. The type #2 has both low beam and high beam and three terminals. When a type #2 is switched to low beam, only one of its filaments is lit. When the high beam is selected, the second filament lights in addition to the low beam.

When there are four headlights (two on each side), the single filament of the #1 lamp lights when the high beam is selected, along with both filaments of the #2 bulb. H1 and H2 quartz halogen bulbs are the most commonly used sealed-beam headlights.

Halogen Headlamps

A halogen headlamp **(Figure 21–8)** is a glass lamp filled with halogen that encloses a tungsten filament. The tungsten-halogen combination, also called quartz-halogen, increases the light emitted by the tungsten filament. In other words, a tungsten-halogen lamp emits more lumens than other lamps

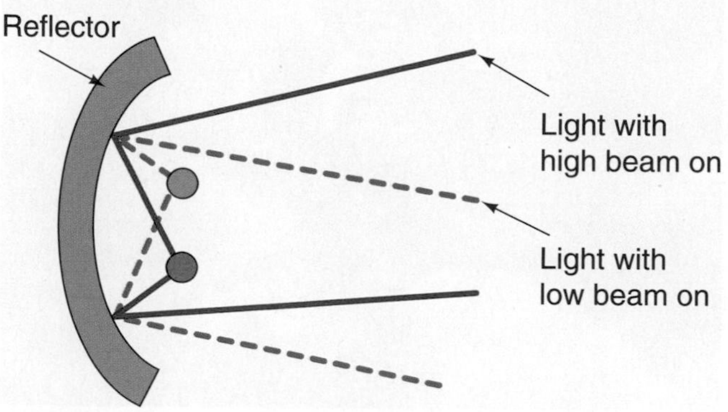

FIGURE 21–7

Filament placement controls the projection of the light beam.

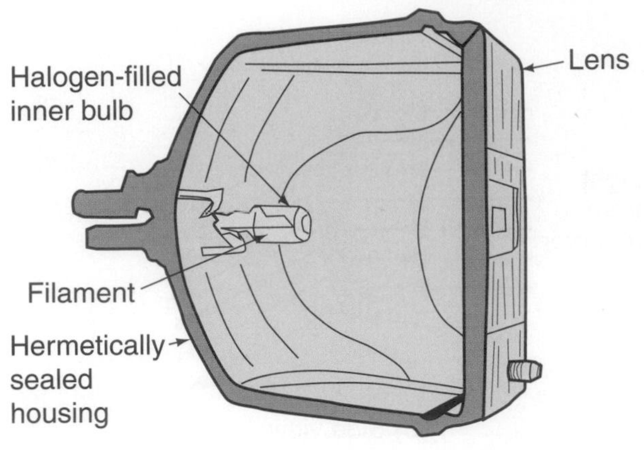

FIGURE 21–8
A halogen sealed-beam headlight with an iodine vapour bulb.

when the watts used to produce the light are considered. As you may recall, a lumen is a simple measurement of the total amount of visible light a source emits. A halogen sealed beam provides about 25 percent more lighting than conventional bulbs.

COMPOSITE HEADLIGHTS Most of today's vehicles have halogen headlight systems that use a replaceable bulb **(Figure 21–9)**. Replaceable halogen bulbs were permitted in the United States in 1983, and their popularity grew quickly. **Composite headlights** allow manufacturers to produce any style of headlight lens they desire. This improves the aerodynamics, fuel economy, and styling of the vehicle.

The bulb is inserted into the composite headlight housing. The filament is capable of withstanding high temperatures because of the presence of the halogen. Therefore, the filament can operate at higher temperatures and can burn brighter.

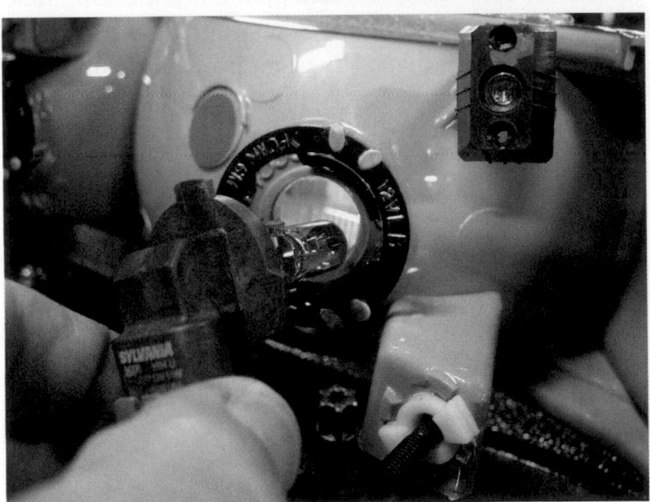

FIGURE 21–9
The mounting of a replaceable halogen bulb.
Martin Restoule

WARNING!

Whenever you replace a halogen lamp, be careful not to touch the lamp's envelope with your fingers. Staining the bulb with skin oil can substantially shorten the life of the bulb. Handle the bulb only by its base.

Composite headlight housings are often vented to release some of the heat developed by the bulbs. The vents allow condensation to collect on the inside of the lens assembly. The condensation is not a problem and does not affect headlight operation. When the headlights are turned on, the heat generated by the halogen bulb quickly dissipates the condensation. Ford uses integrated non-vented composite headlights. On these vehicles, condensation is not considered normal. The assembly should be replaced.

SHOP TALK

Because the filament is inside the inner bulb, cracking or breaking of the housing or lens does not prevent the bulb from working. As long as the filament envelope has not been broken, the filament will continue to operate. However, a broken lens will result in poor light quality, so the assembly should be replaced.

High-Intensity Discharge (HID) Headlamps

High-intensity discharge (HID), or **xenon, headlamps** create light by creating and maintaining an electrical arc across two electrodes inside a bulb. The arc excites the gas and salts inside the bulb **(Figure 21–10)**. This allows the arcing to continue

FIGURE 21–10
A xenon light bulb.
DaimlerChrysler Corporation

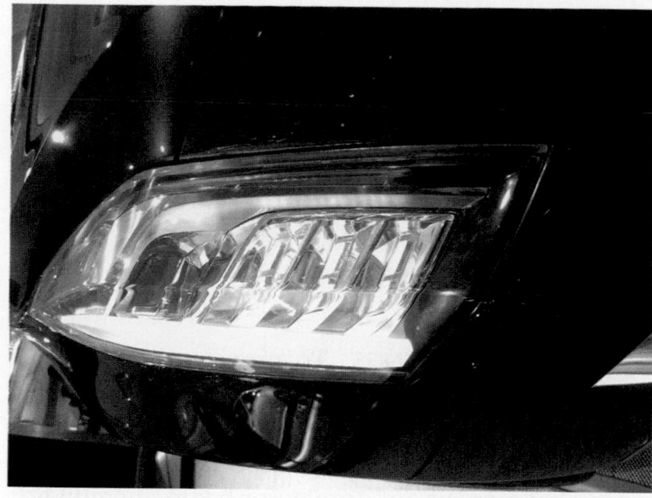

FIGURE 21–11

HID (xenon) headlights are readily identifiable by their bluish light.

to emit light. These lights are recognizable by their slightly bluish light beams **(Figure 21–11)**. They emit this coloured light because the inside of the bulb is filled with xenon gas mixed with mercury or bismuth. The result is a light that is much closer to natural daylight than that of other bulbs.

The lamps use AC voltage, and at least 30 000 volts are needed to jump the gap between the electrodes. Once the voltage bridges the gap, only about 80 volts are needed to keep current flowing across the gap.

Each HID assembly includes a lamp, ballast unit, and starter, or igniter **(Figure 21–12)**. The igniter may be part of the bulb assembly or mounted externally **(Figure 21–13)**. The ballast unit includes a DC-to-AC converter and an electronic control unit (ECU) to regulate the voltage, provide for a gradual warm-up of the lamp, and allow for instant restart when it is necessary **(Figure 21–14)**. The control unit also monitors the system. If it senses a faulty lamp, it turns off the lamp's power, to eliminate the hazardous situation of stray, high voltage.

When the headlights are switched on, it takes approximately 15 seconds for the lamps to reach maximum intensity. However, during ignition, these lamps provide more-than-adequate light for safe driving.

Xenon headlights illuminate the area to the front and sides of the vehicle with a beam that is both brighter (twice the amount) and much more consistent than the light generated by halogen

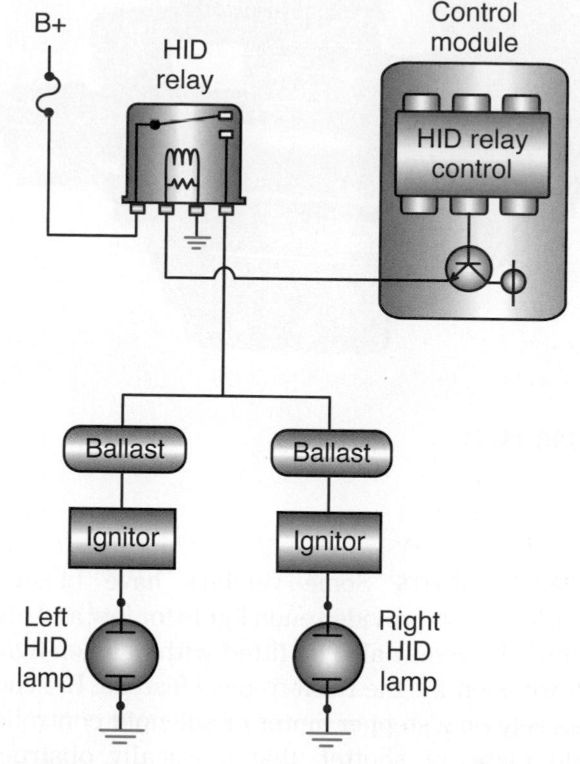

FIGURE 21–12

An HID headlamp schematic showing the lamps, ballasts, and igniters.

headlamps **(Figure 21–15)**, to make night driving safer. The light output of these lamps allows the headlamp assembly to be smaller and lighter. Xenon lights also produce significantly less heat because they use about two-thirds less power to operate, and will last two or three times longer than conventional lamps.

FIGURE 21–13

The igniter is connected to the ballast by a cable.

FIGURE 21–14

An HID ballast.

BI-XENON LIGHTS Some vehicles have bi-xenon headlamps that provide xenon lights for low and high beams. These may also be fitted with halogen lights that are used for the flash-to-pass feature. Bi-xenon lights rely on a stepper motor or solenoid-controlled shield plate, or shutter, that physically obstructs a portion of the overall light beam emitted by the arc. When the driver selects high beams, the shutter reacts and allows the headlights to project the complete, unobstructed light beam.

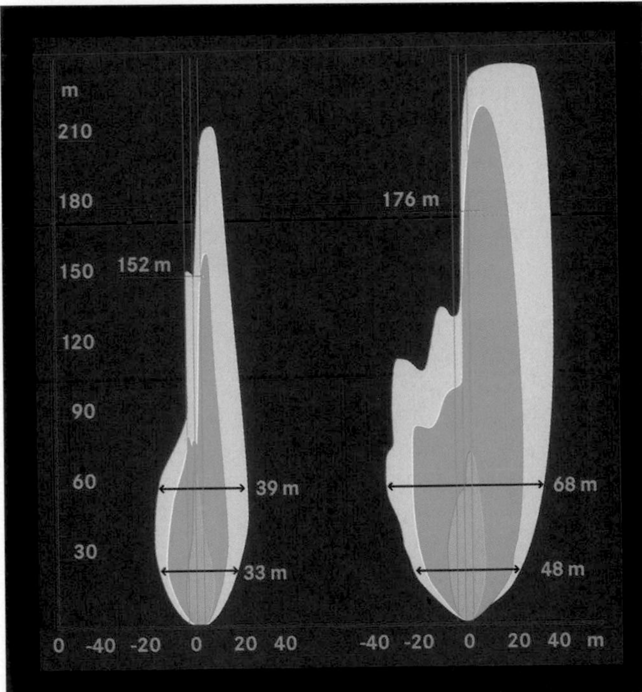

FIGURE 21–15

A comparison of the light pattern and intensity between a halogen (left) and a xenon (right) headlight.

DaimlerChrysler Corporation

SHOP TALK

HID lamps offer great visibility, but often drivers approaching a vehicle with HIDs become annoyed by the glare or brightness of the lamps. To eliminate these concerns, HID systems often have self-levelling and washer systems. Both of these are required in Europe, and most European vehicles with HIDs sold in the United States have these features.

CYLINDRICAL HOUSINGS To provide more precise light-beam patterns, many manufacturers are using headlamp assemblies that have cylindrical bulb housings **(Figure 21–16)**. These housings are the main component of projector-beam headlights, which are commonly used with HID lamps. Basically the light beam is shot out of the cylinder. The aim of the cylinder projects the beam through a lens, and light moves forward without much scattering **(Figure 21–17)**.

LED Headlights

Currently available on some vehicles are LED headlamps. An individual LED does not produce enough light to serve as a headlamp. Ten to twenty LEDs are needed for ample forward lighting **(Figure 21–18)**. There are many reasons for using LEDs in headlights:

- LEDs do not require a vacuum bulb or high voltage to work.
- LED-based lighting sources require up to 40 percent less power than traditional lighting sources. This improves a vehicle's fuel economy.
- LEDs provide a whiter light than xenon.

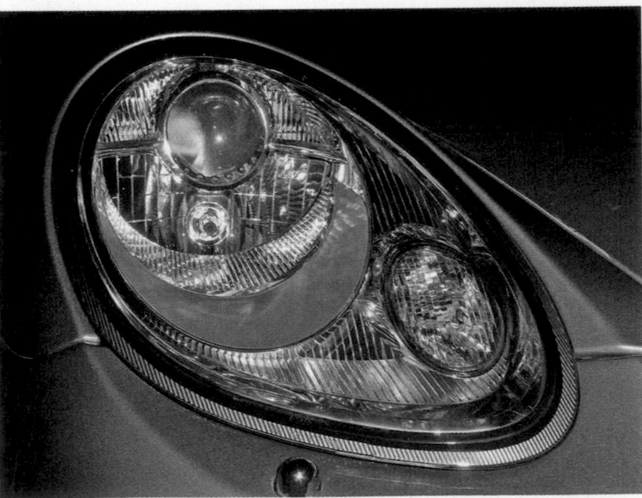

FIGURE 21–16

A headlight assembly with cylindrical bulb housings.

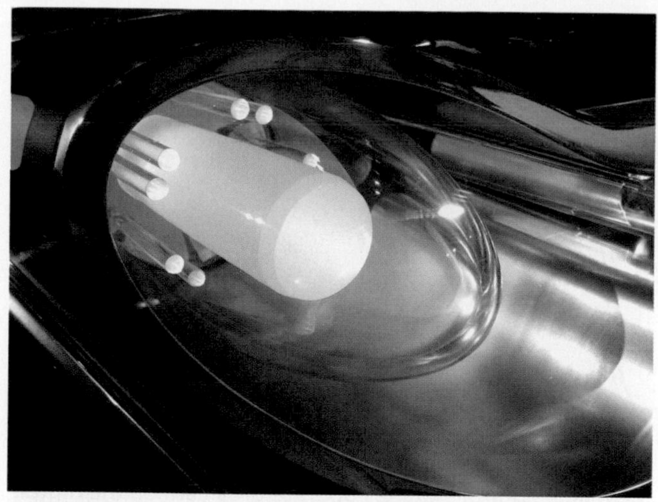

FIGURE 21–17

A cylindrical lamp assembly to focus light in a designated location.

- Prototype LED headlights have achieved a 1000-lumen output in the low-beam mode; this is the same as a xenon headlamp.
- LEDs are mercury-free, making them environmentally friendly, unlike some HID/xenon systems.
- The average operating life of an LED is twice that of the vehicle itself. This means the headlamp may never need to be replaced.
- LEDs are resistant to shock and vibration.
- LED headlamps reduce oncoming driver perception of glare.
- LED-based headlamps are up to 55 percent thinner than other designs, which give designers more flexibility and freedom.

Incorporating LED technology into headlamps presents a problem common to all semiconductors, that being heat. The heat from the engine compartment and the rear of the lamp can cause failure. Therefore, LEDs need temperature controls and must be mounted to heat-retarding materials. LED bulbs release very little heat; in fact, their lenses never get warm enough to melt snow or ice. Therefore, a lens heater fan must be incorporated in the lamp assembly.

LEDs also require precise current control, which requires complicated electronic circuitry. In spite of these obstacles, LED headlamps can open the door to other headlamp-related safety features. It is possible to have light beams that meet the current conditions automatically. This can be done by varying the number of LEDs powered or by the placement of special lenses in front of a particular group of LEDs.

Current vehicles with LED headlamps rely on several LEDs for the low beam and several more designated for the high beam **(Figure 21–19)**. The placement of the lamps is dictated by the desired light pattern and the design of the light assembly. Many assemblies use projector bulbs or lenses to channel and aim the light beams.

Daytime Running Lights

Canadian law requires that all new vehicles be equipped with **daytime running lights (DRLs)** for added safety. The control circuit is connected directly to the vehicle's ignition switch so that the lights are turned on whenever the vehicle is running.

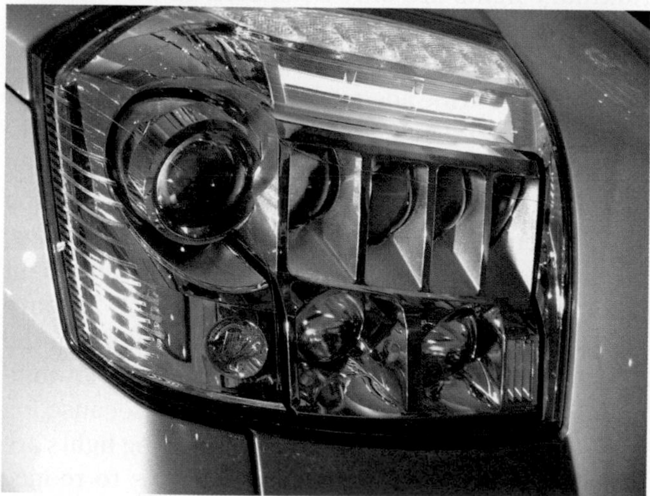

FIGURE 21–18

A headlight assembly of HID bulbs and LEDs.

FIGURE 21–19

LED lamps used for low and high beams.

Most manufacturers use the headlights, but there can be cases where only the park lights (running lights) are illuminated. The headlight circuit is equipped with a module that reduces the 12-volt battery voltage to approximately 6 volts. This voltage reduction allows the headlights to burn with less intensity and prolongs the life of the bulbs. When the headlight switch is moved to the HEADLIGHT position, the module is deactivated, and the headlights work with their normal intensity and brightness. Applying the parking brake also deactivates the DRL system, so the lights are not on when the vehicle is parked and the engine is running.

LED DRL SYSTEMS Audi introduced LED DRLs as a separate unit **(Figure 21–20)**. The lamps are on a separate circuit that does not rely on resistors to decrease normal headlamp voltage. This means less energy is wasted, and therefore the vehicle's fuel consumption is reduced. It is estimated that these lights consume more than 50 percent less power than the typical DRL. In addition, they use less space.

Concealed Headlights

Although not as common as they were a few years ago, concealed headlights are still found on some cars. Manufacturers use a concealed headlight system to improve the vehicle's aerodynamics. Today, low-profile headlight assemblies are being used instead of concealed headlights. However, there are some manufacturers that still have concealed headlights. These headlights are controlled by electric motors.

FIGURE 21–20

LED DRLs.

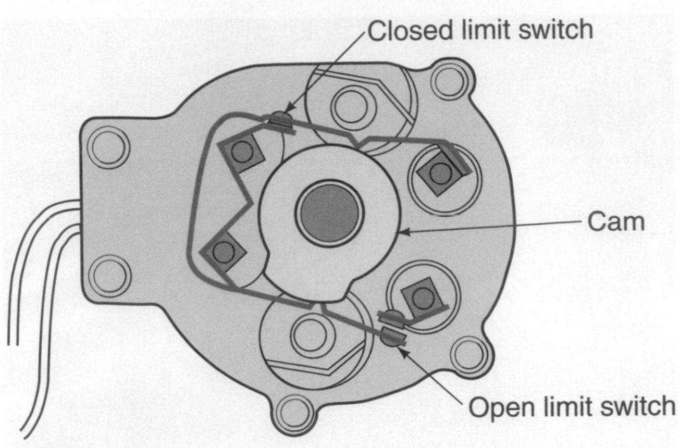

FIGURE 21–21

Most limit control switches operate off a cam on the motor.

When the headlight switch is moved to the HEADLIGHT position, the entire headlight bulb and adjuster assembly pivots upward. These headlights are controlled by electric or vacuum motors.

Typically, electrically controlled systems use a torsion bar and a single motor to open both doors or have a separate motor for each headlight door. When the headlight switch is moved to the HEADLIGHT position, current is sent to the motors. This current turns on the motors and causes the doors to open or close. Limit switches stop current flow to the motors when they are completely open or closed **(Figure 21–21)**. Electrically operated headlight doors also have a provision for manually opening the doors in case of a system failure.

Auxiliary Lights

Auxiliary lights are only a concern to a technician when the owner of the vehicle feels the need to add lighting to what he or she already has. Although headlights provide adequate illumination of the road during normal driving conditions, some want additional lighting for special conditions, such as driving in heavy fog. Normal light does not penetrate fog well. When an intense beam of light hits some fog, all the driver sees is a glare. To provide some light through the fog, fog lights **(Figure 21–22)** are designed to send a flat, wide beam of light under the blanket of fog. This is why they are mounted low and are aimed low and parallel to the road. Because fog tends to reflect light back at the driver, fog lights are often fitted with yellow or amber lenses to reduce the discomfort caused by the glare. Some vehicles have OEM fog lights that are part of the normal lighting circuit.

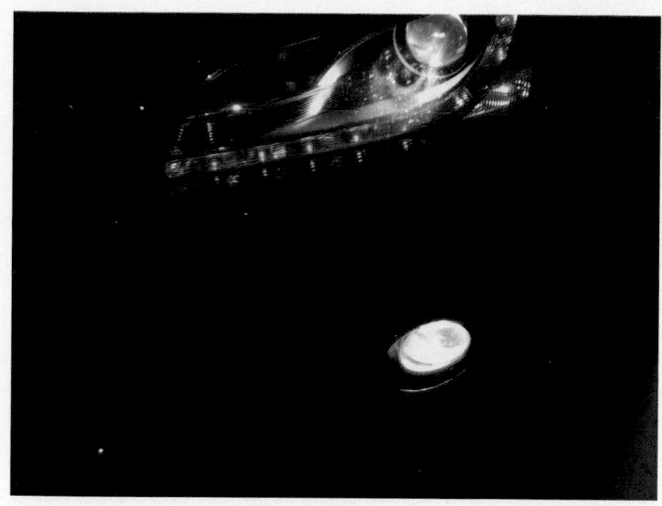

FIGURE 21–22

Fog lamps.

SHOP TALK

Auxiliary lights are easy to install when everything is carefully considered. Before beginning, become familiar with the local regulations for auxiliary lights, and adhere to them. These regulations can be obtained from your province's ministry of transportation. Also, always follow all federal, provincial, and local laws when aiming headlights, fog lights, and driving lights.

Driving lights normally use an H3 or H4 quartz halogen bulb and a high-quality reflector and lens to project an intense, pencil-thin beam of light far down the road. Proper aiming of these lights is extremely important. They are used to supplement the high beams and should be used in conjunction with them. Driving lights should be wired so that they are off when the high beams are off. One way to do this is to supply the controlling switch with current from the high-beam circuit rather than from a circuit that is live all the time.

SHOP TALK

When adding auxiliary lights, make sure that the generator and wiring are heavy enough to handle the increased wattage requirement. Installing a higher output generator may be recommended, especially if other electrical accessories are also being installed. The choice of wire size should be based on the current of the load. Most light kits include a relay as a safeguard. A relay is used because the required current for these lights, especially halogen ones, can be quite high. It is not unusual that they require as much as 25 amperes.

HEADLIGHT SWITCHES Headlight switches are either mounted on the dash panel or are part of a multifunction switch on the steering column. The headlight switch controls most of the vehicle's lighting systems. The most common style of headlight switch has three positions: OFF, PARK, and HEADLIGHT. A headlight switch normally receives direct battery voltage at two of its terminals. This allows the light circuits to be operated without having the ignition switch in the RUN or ACC (accessory) position.

When the headlight switch is in the OFF position, the open contacts prevent battery voltage from continuing on to the lights **(Figure 21–23A)**. When the switch is in the PARK position, battery voltage is applied to the parking lights, side markers, taillights, licence plate lights, and instrument panel lights **(Figure 21–23B)**. This circuit is usually protected by a 15- or 20-amp fuse that is separate from the headlight circuit.

When the switch is in the HEADLIGHT position, battery voltage is applied to the headlights. The lights lit by the PARK position remain on **(Figure 21–23C)**. Normally, a self-resetting circuit breaker is installed between the battery feed and the headlights. The circuit breaker is designed to reset itself. If a problem causes the breaker to open, the lights will go off until the breaker resets. Then the lights will come back on. If there is a serious problem in the circuit, the headlights might flash as the breaker cycles. Some vehicles have a separate fuse for the headlight on each side of the vehicle. This allows one headlight to operate if there is a problem in the circuit for one side of the vehicle.

The instrument panel lights come on whenever the headlight switch is in the PARK or HEADLIGHT position. The brightness of these lamps is adjustable. A rheostat is used to allow the driver to control the brightness of the bulbs. This control may be part of the headlight switch, in which case the driver simply rotates the headlight switch knob to adjust the panel lights. Not all headlight switches are designed to control the instrument panel lights. Many vehicles have a separate unit on the dash to control the panel lights **(Figure 21–24)**. Some modern computerized displays show a brightness scale on the instrument panel **(Figure 21–25)** to show the driver where the lights are adjusted within their possible range.

Headlight switches are basically one of three designs. A common switch setup is the rotary switch. Turning the knob of the switch to the PARK or HEADLIGHT position energizes the appropriate lights. The switch's knob also serves as the dimmer control for the instrument panel lights. Some vehicles use

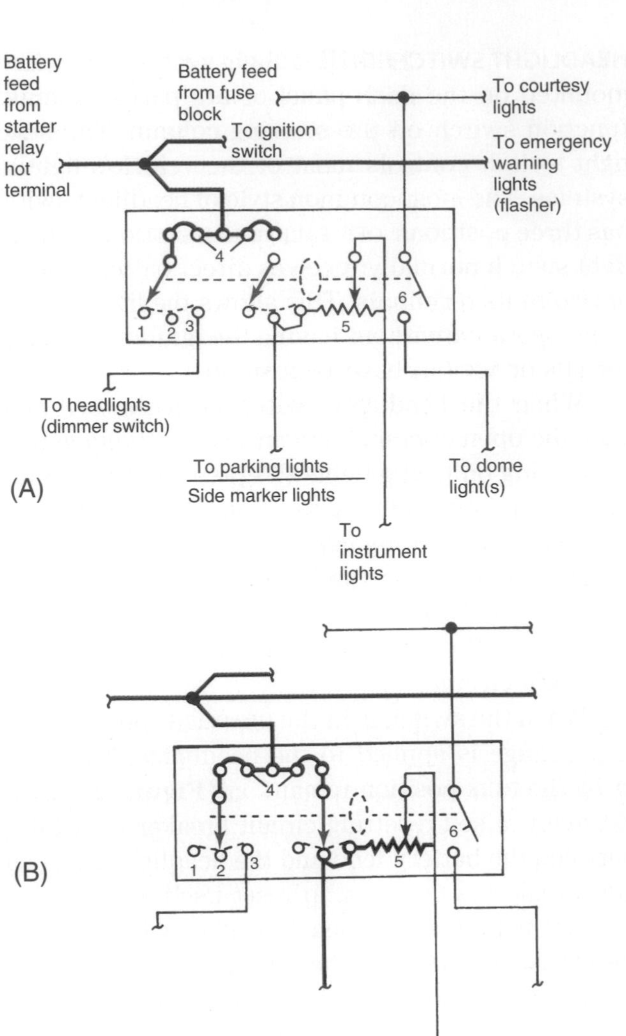

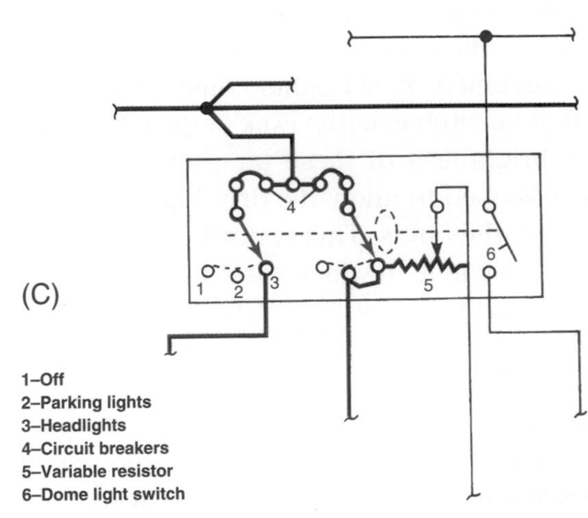

FIGURE 21–23

A headlight switch (A) in the OFF position, (B) in the PARK position, and (C) in the ON position.

Ford Motor Company

a push-button switch. The driver merely pushes a button for the desired set of lights. When this type of switch is used, there is a separate instrument panel light control. There is also a separate panel light

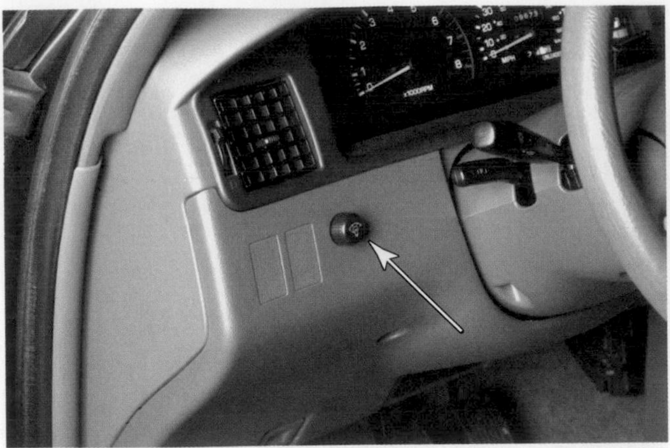

FIGURE 21–24

An instrument panel light control unit.

FIGURE 21–25

Some vehicles display a brightness scale on the instrument panel as the brightness control is moved.

control when vehicles are equipped with a steering column–mounted headlight switch **(Figure 21–26)**. To select the desired lighting mode, the driver turns the knob at the end of the switch.

Dimmer Switches

The **dimmer switch** provides a way for the driver to switch between high and low beams. A dimmer switch is connected in series with the headlight circuit and controls the current path to the headlights. The low-beam headlights are wired separately from the high-beam lights.

Many years ago, the dimmer switch was located on the floor. To switch between low and high beams, the driver used his or her foot to press the switch. This type of switch worked, but it was also subject to damage

FIGURE 21–26

A headlight switch mounted on the steering column.

because of rust and dirt. Newer vehicles have the dimmer switch on the steering column. This prevents early switch failure and increases driver accessibility.

Headlight Circuits

The complete headlight circuit consists of the headlight switch, dimmer switch, high-beam indicator, and headlights. When the headlight switch is in the HEADLIGHT position, current flows to the dimmer switch **(Figure 21–27)**. If the dimmer switch is in the LOW position, current flows through the low-beam filament of the headlights. When the dimmer switch is in the HIGH position, current flows to the high-beam headlights **(Figure 21–28)**.

The headlight circuits just discussed are designed with switches that control battery voltage and bulbs that have a fixed ground. In this system, battery voltage is present at the headlight switch. The switch must be closed to have voltage present at the headlights. Many manufacturers use a system that relies on a ground-side switch to control the headlights. In these systems, voltage is always available at the headlights. A closed headlight switch completes the circuit to ground, and the headlights turn on. In this system, the dimmer switch is also a ground control switch.

Flash to Pass

Most steering column–mounted dimmer switches have an additional feature called **flash to pass**. This circuit illuminates the high-beam headlights even with the headlight switch in the OFF or PARK position. When the driver activates the flash-to-pass feature, the contacts of the dimmer switch complete the circuit to the high-beam filaments.

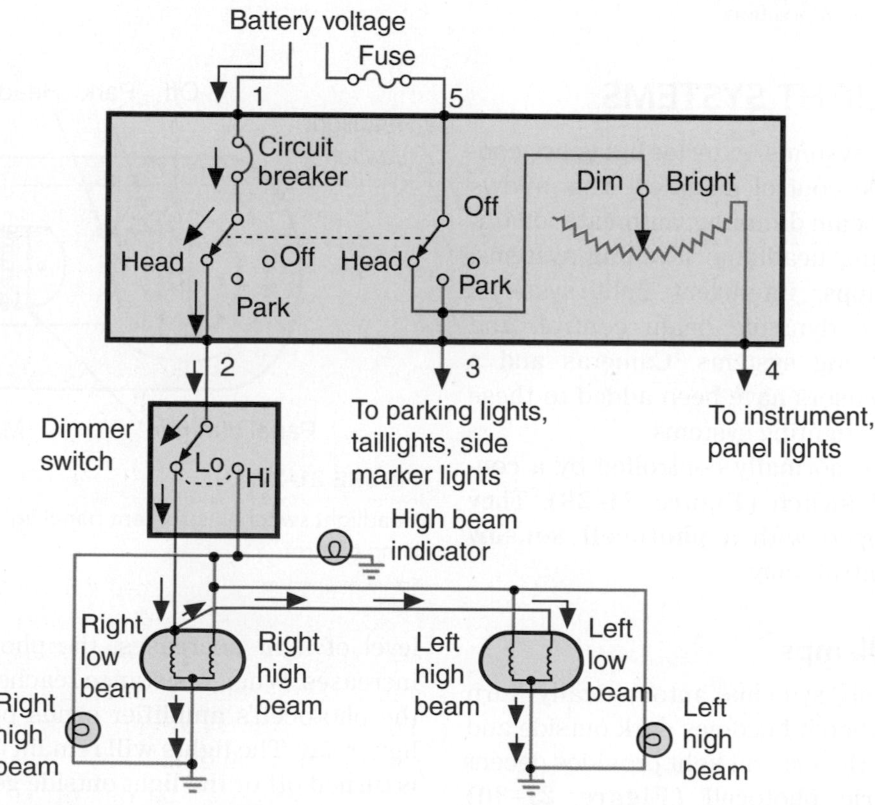

FIGURE 21–27

A headlight circuit indicating current flow with the dimmer switch in the low-beam position.

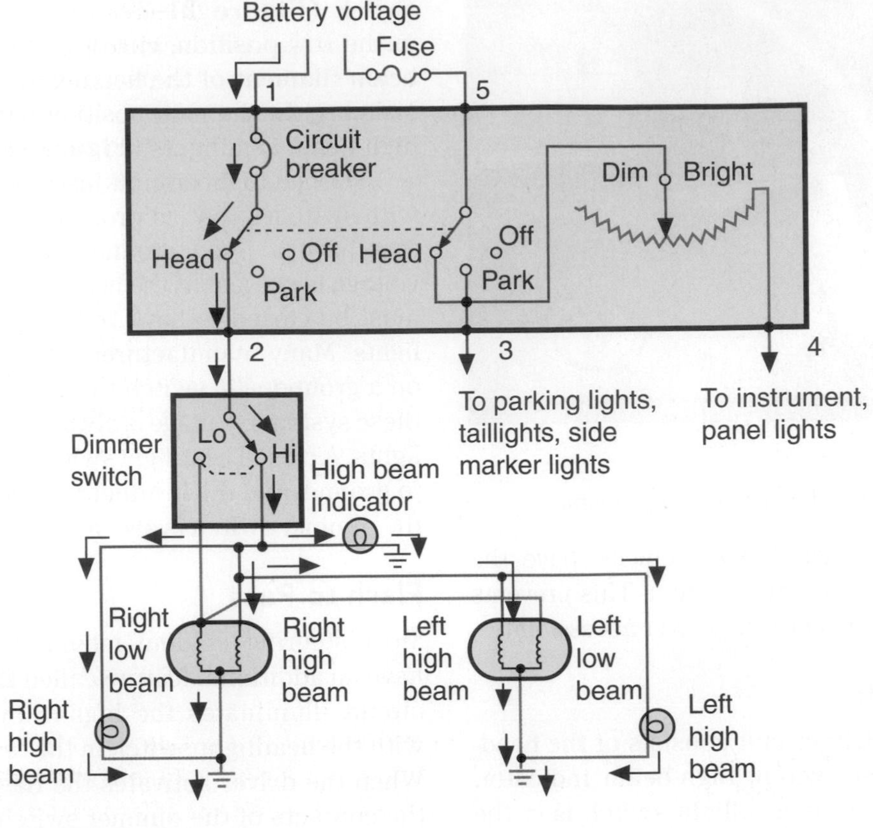

FIGURE 21–28

A headlight circuit indicating current flow with the dimmer switch in the high-beam position.

AUTOMATIC LIGHT SYSTEMS

As with many other systems, exterior lights are controlled by electronic control modules. This allows for automatic high-beam dimming, automatic on/off, glare-free high beam, headlamp levelling systems, directional headlamps, intelligent light systems, adaptive headlights, dynamic beam control, and advanced front-lighting systems. Cameras and a variety of optical sensors have been added to these computer-controlled lighting systems.

The systems are normally controlled by a conventional headlight switch **(Figure 21–29)**. They are typically equipped with a **photocell sensor/amplifier** and a control relay.

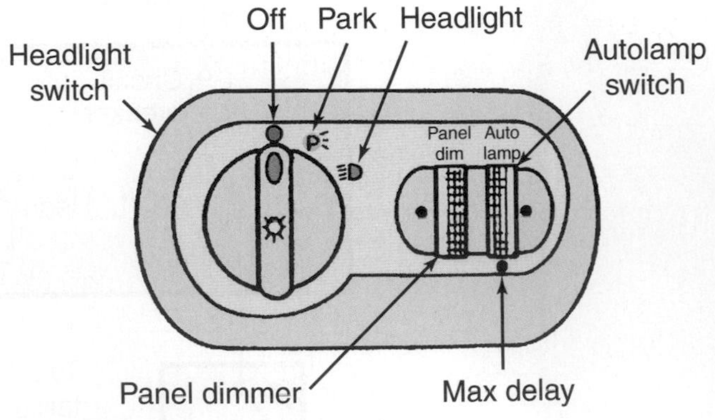

FIGURE 21–29

Headlight switch, instrument panel light control, and auto lamp control.

Ford Motor Company

Automatic Headlamps

Automatic headlamp systems automatically turn the headlights on when it becomes dark outside and turn them off when the natural light provides decent lighting. An electric photocell **(Figure 21–30)** located on the dashboard detects the current lighting conditions and activates the headlight control relay, which switches the headlamps on and off. As the level of light decreases, the photocell's resistance increases. When resistance reaches a preset amount, the photocell's amplifier sends power to the headlight relay. The lights will remain on until the system is turned off or the light outside gets bright enough.

HIGH-BEAM DETECTION This system also automatically turns the high-beam headlights on or off according to conditions. A light sensor (commonly

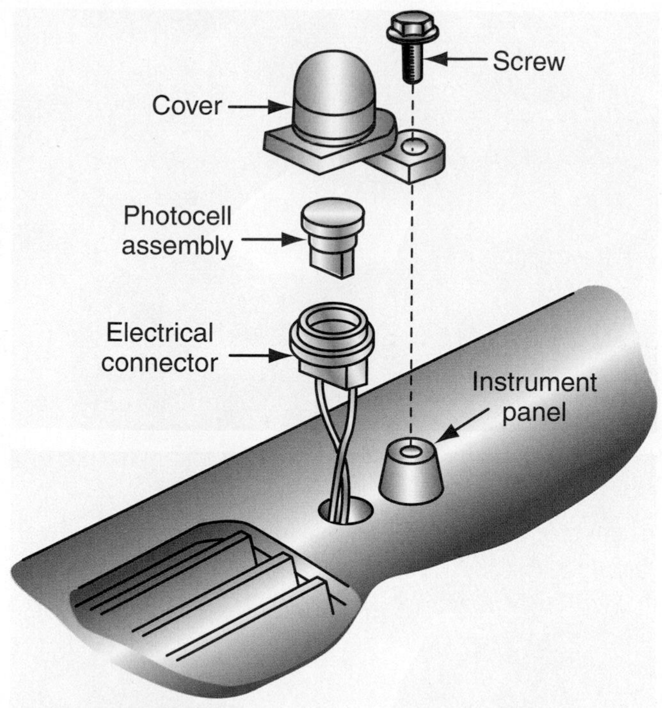

FIGURE 21–30

Most automatic headlight systems have a photocell located in the dash to sense ambient-light levels.

referred to as a camera) on the rearview mirror monitors the light in front of the vehicle **(Figure 21–31)**. When it is dark enough, the system will switch the high beams on. They will stay on until the sensor detects the headlights or taillights of another car. At that time, the high beams are switched off until the lights of the other vehicle are no longer detected. The absence of those lights causes the system to switch back to high beams. The system also fades the change from low to high to prevent abrupt changes to the light on the road ahead.

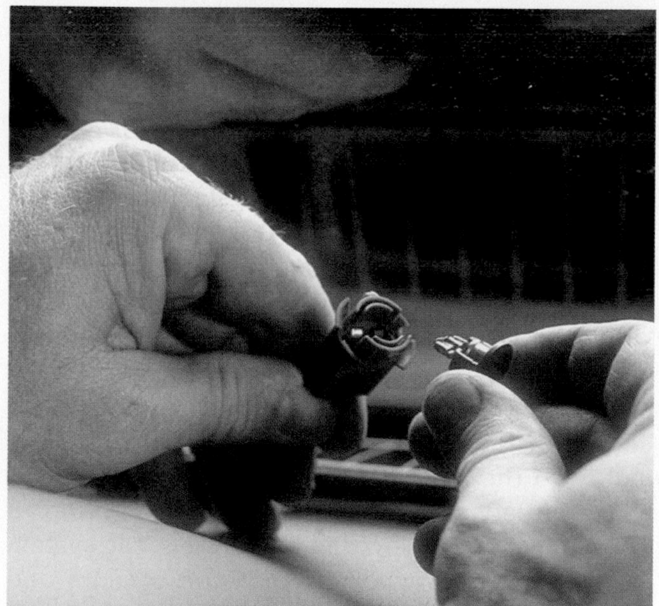

FIGURE 21–31

A photocell for an automatic lighting system.

DELAY SYSTEMS Some automatic headlamp systems allow the driver to set the length of time the vehicle's exterior lights will stay on after the passengers exit the vehicle. Normally, the variable switch can be adjusted to keep the headlights on for several minutes after the ignition is turned off. Of course, the driver can turn off the delay system, and the headlamps will shut off as soon as the ignition is turned off.

Adaptive High-Beam Assist

Adaptive high-beam assist systems continuously adjust the lighting range of the headlamps so that the beams only extend to the vehicle ahead. Doing so allows for a maximum range of lighting without causing excessive glare in the eyes of other drivers. These systems are offered by many manufacturers, and each call the system by a different name. However, most of these systems do the same basic thing. Some have additional features in addition to the basic package.

These systems provide for a continuously changing range of lighting. There is no distinct transition from low to high beams. The intensity and aim of the lamps gradually changes as conditions change. When there is no traffic close enough for glare, the system provides full high beam. When the system detects an approaching vehicle or a vehicle's tail lamps, the high beams will turn off. Most of these systems allow the headlamps to adjust about every 40 milliseconds.

A camera and/or a light sensor are located in a module at the front of the inside rearview mirror **(Figure 21–32)**. The sensor analyzes the colour of the light, its intensity, and its movement to distinguish between vehicle lights and other light sources. The systems are designed to ignore non-vehicular light sources.

Glare-Free High Beams

A glare-free high-beam system uses a camera-driven lighting control strategy. The system selectively looks at shady spots with the lighted area ahead of the vehicle and slices out segments of light from the high-beam light pattern to protect other drivers

FIGURE 21–32

A digital camera module mounted to a rearview mirror.

from glare. At the same time, the system attempts to provide the driver with maximum seeing range. The dynamic shadowing is achieved by movable shadow masks that are moved into the normal light path of the headlamp. The dynamic beam control automatically adjusts the light beams to the current traffic and ambient-lighting conditions.

When the system is activated, the camera in or around the windshield recognizes vehicles through its image processing system. The system then calculates the optimum light distribution and sends the appropriate commands to the headlights.

CUSTOMER CARE

If the customer's car is equipped with an automatic light control system, point out the location of the perforated holes or slots. Warn the customer not to place any items that could block light from the sensor/amplifier assembly. Blockage causes erratic operation of the system. The photocell must always be exposed to an outside light to function properly.

Adaptive Headlights

Adaptive headlight systems (AHSs) aim the headlight beams to follow the direction of the road directly ahead as the vehicle turns **(Figure 21–33)**. These systems are electronically controlled by a module and inputs from the steering system and vehicle speed.

Some vehicles have headlamps that are directly connected to the steering linkage to allow the lamps to follow the movement of the front wheels. As an example, the 1948 Tucker Sedan was equipped with a third headlamp, located in the centre of the front of the car, that responded to the movement of

FIGURE 21–33

A comparison of how the road is lit up with a conventional (top) headlamp system and an adaptive (bottom) system.

DaimlerChrysler Corporation

the steering system. Today, systems rely on inputs from a variety of sensors and actuators to move the headlamps.

Adaptive headlight systems are available on many vehicles, each with their own way to "bend" the light beams in anticipation of a turn. Regardless of the system's components, the system is controlled by the body control module (BCM) in response to the vehicle's current steering angle, yaw rate, and road speed.

One design swivels the entire headlamp assembly with electric motors attached at the base of the lamp assembly **(Figure 21–34)**. The motor can move the left headlamp up to 15° to the left and 5° to the right, and the right headlamp up to 5° to the left and 15° to the right. In addition to switching between low and high beams, the system can respond to suspension position sensors and operate the motor to keep the headlight beam level with the road.

Some manufacturers adjust the light beam by rotating the lamp's projector, reflector, or lens. This

FIGURE 21–34

A headlamp assembly for an adaptive headlight system.

DaimlerChrysler Corporation

system uses a motor to rotate a drum inside the light assembly. The drum changes the light pattern to provide the best illumination for the current conditions. With these systems, the aim of the beam can be adjusted in response to conditions by moving the lamp assembly or by adjusting the lamp's position within the reflector.

Another design uses extra lamps that turn on according to steering angles. For example, a newer system illuminates individual LED lamps sequentially and with various intensities. These lights may be housed in the headlamp or fog light assembly, or they can be a separate unit. These lamps normally only turn on during low-speed turning.

Adaptive headlights can also be controlled by GPS navigation and digital road maps. Information about the road ahead allows the system to anticipate curves and rotate the headlamps so that they can illuminate those curves before the driver starts to turn the steering wheel.

Headlamp Levelling Systems

Due to differences in safety regulations in Europe and North America, headlight levelling systems are much more common on European vehicles. Headlight aiming was such a concern in Europe that the 1948 Citroen 2CV had a headlamp levelling system that was manually controlled by a knob connected to the steering rod linkage. The effect of the light on the oncoming traffic was not considered with this system. However, in 1954, a system was introduced that was responsive to movements of the suspension system and kept the headlamps correctly aimed regardless of the vehicle's load and without adjustments made by the driver.

In the 1970s, other European countries required levelling systems that allowed the driver to lower the lamps' aim by a control lever or knob on the instrument panel. The rear of the vehicle sits lower if the weight of the passengers and load is high. This causes the beam from the lamps to rise and create a glare in the eyes of the drivers in front.

Headlamp levelling systems are not required for vehicles sold in North America. However, recent research strongly recommends the use of automatic levellers on all headlamps. Headlight levelling depends on many sensors and a BCM-controlled motor. In Europe, the headlights' vertical aim must be maintained regardless of vehicle load. If the vehicle is not equipped with an adaptive suspension, a headlight levelling system is needed to keep the headlamps aimed correctly.

Vehicles with xenon lamps and some high-power halogens must be equipped with headlamp self-levelling systems. These systems sense the level of the vehicle caused by load and road inclination, and then adjust the headlamps' vertical aim to keep it correctly aimed without any involvement by the driver.

HEADLIGHT SERVICE

When there is a headlight failure, it is typically caused by a burned-out bulb or light, especially if only one light fails. However, it is possible that the circuit for that one light has an open or high resistance. Check for voltage at the bulb before replacing a bulb. If there is no voltage present, the circuit needs work, and the original bulb may still be good. If more than one light (including the rear lights) is not working, carefully check the circuit. A problem there is much more likely than having a number of burned-out bulbs. Of course, if the charging system is not being regulated properly, the high voltage will cause lights to burn out prematurely. **Figure 21–35** shows common headlight problems and their probable causes.

Restoring Headlight Lenses

Many lenses are made of plastic (polycarbonate) and are affected by the environment, and deteriorate. The result is a cloudy lens that reduces the light of the lamps. The cloudiness is caused by the oxidation of the lens's protective coating installed by the lens's manufacturer. The lenses may also become pitted by road dirt and pebbles, and can crack. Cracks can allow water to enter the headlamp assembly and shorten its life.

PROBLEM AREA	SYMPTOMS	POSSIBLE CAUSES
Headlights	One low-beam headlamp does not work	Bad bulb High resistance in the power or ground circuit
	Both low beams do not work	Open in the power or ground circuit Bad (control) switch Blown fuse Bad bulbs due to overcharging
	One high-beam headlamp does not work	Bad bulb High resistance in the power or ground circuit
	Both high beams do not work	Open in the power or ground circuit Bad (control) switch Blown fuse Bad bulbs due to overcharging
	Dim headlight illumination	Poor ground connection Corroded headlight socket Poor battery cable connections Low generator output Loose or broken generator drive belt
	Lights dimmer than normal	Excessive circuit resistance on the power or ground side of the circuit Low generator output, wrong lamps, or incorrect wiring
	Lights brighter than normal	Higher than specified generator output Improper lamp application Dimmer switch stuck in the high-beam position
	Intermittent headlight operation, headlights flicker	Defective circuit breaker Overload in circuit Improper connection Defective switch Poor ground Excessive resistance
	No or improper headlight	Burned-out headlights Defective headlight switch Open circuit Defective circuit breaker Overload in circuit Improper or poor connection Poor ground Excessive resistance Defective relay Blown fuse Faulty dimmer switch Short in insulated circuit Improper bulb application Improper headlight aiming

FIGURE 21–35

Common headlight problems and their probable causes. (*continued*)

PROBLEM AREA	SYMPTOMS	POSSIBLE CAUSES
Automatic headlight system	Headlights fail to turn on	Faulty headlight switch Power feed to relays Faulty relay(s) Open, short, or high resistance in relay control circuit motor control circuits Headlamp circuit failure Burned out headlamp elements Controller power and ground circuits Faulty controller
	Headlights fail to turn on in automatic mode, headlights work in manual mode	Open input circuit from switch to control module Faulty module Faulty switch
	Headlights turn on in daytime when switch is in AUTO mode	Faulty photocell Open photocell circuit Shorted photocell circuit Faulty control module Bus communications error Immobilizer system inoperative
Automatic high–low beam	Headlights fail to automatically switch to high beam	Obstruction in front of camera Improper headlight aiming System not initialized Camera alignment System voltage to module Module ground circuit Bus network failure Faulty controller
Daytime running lights	DRLs fail to turn on	Faulty headlight switch input or circuit Power circuit to DRL relay or relay module Defective relay or relay module Parking brake switch or circuit System voltage to module Module ground circuit Bus network failure Faulty controller
Headlight leveling system	Headlights fail to level properly	Improper headlight aiming Faulty switch input or circuit Faulty headlight level sensor or sensor circuits Disconnected link to headlight level sensor(s) Circuits to motor(s) Defective motor(s) System voltage to module Module ground circuit Bus network failure System not calibrated Faulty controller

FIGURE 21–35 (*continued*)

CUSTOMER CARE

Restoring the lens is an easy procedure. There are many available products to restore the lenses. These may be special sanding pads or specially designed chemical compounds. Most damaged headlamp covers will require wet sanding to remove the worst of the damage, before a polishing compound is used on the lens. Some restoration kits include a sealer to help prevent the lens from future damage.

If the damage to the lens is minor, it may be able to be corrected by using car polish. In more severe conditions, the deterioration is deeper than the actual outer plastic material. This damage can only be corrected by replacing the entire headlamp lens.

Sanding or aggressively polishing the lenses can provide a temporary fix, but it may remove the protective coating from the lens, which can cause the lens to deteriorate more quickly and severely.

The reflector, which is a thin layer of vaporized aluminum on a metal, glass, or plastic structure, can become dirty, oxidized, or burnt. This happens when water enters the headlamp assembly, when bulbs of higher than specified wattage are used, or simply due to age and use. If degraded reflectors cannot be cleaned, they must be replaced.

LENS CLEANERS The light beams from headlights can scatter if the lens is dirty. The scattering increases the glare experienced by drivers in front of the vehicle. This is especially true if the vehicle is equipped with HID headlamps. Although lens cleaners are required in Europe on all vehicles with HID lamps, Canada has no such law. However, some vehicles do have lens cleaners. The two basic types of cleaners found on today's vehicles are a small motor-driven wiper blade or a fixed or pop-up high-pressure nozzle that cleans the lens with a steady spray of windshield washer fluid.

HEADLIGHT REPLACEMENT

There can be slight variations in procedure from one model to another when replacing headlights. For instance, on some models, the turn-signal light assembly must be removed before the headlight can be replaced. Overall, however, the procedure does not differ much from the following typical instructions.

Make sure the replacement bulb is the same type and part number as the one being replaced.

Sealed-Beam Headlight Replacement

Replacing a sealed beam headlamp is rather straightforward. The lamp is normally secured with a bezel or ring. This needs to be removed and the lamp pulled partially out of its housing. Then disconnect the electrical connector from the back of the lamp. Carefully inspect the connector for signs of damage or corrosion. Some manufacturers recommend coating the prongs and base of a new headlamp with dielectric grease for corrosion protection. Connect the new lamp to the electrical connector. Place the new lamp into the housing. Position the lamp so that the embossed number on its lens is at the top. Install the headlight bezel. Secure it with the retaining screws. Check the aim of the headlight and adjust it, if necessary.

PROCEDURE

Replacing Headlamps in a Composite Headlight Assembly

STEP 1 Place fender covers around the work area.

STEP 2 Remove the electrical connector from the back of the bulb.

STEP 3 Unlock the bulb's retaining ring by rotating it one-eighth of a turn to the left.

STEP 4 Slide the retaining ring away from the base.

STEP 5 Gently pull the bulb out of the socket.

STEP 6 Check the wire connection for corrosion.

STEP 7 Coat the terminals with dielectric grease.

STEP 8 Place the bulb in the socket **(Figure 21–36)**. *Do not* touch the bulb with your fingers.

STEP 9 Take care not to get grease on the bulb.

STEP 10 Position the mounting flange at the rear of the socket.

STEP 11 Install and lock the retaining ring.

STEP 12 Reconnect the electrical connector.

STEP 13 Adjust headlamp aim as necessary.

SHOP TALK

Because of the extremely high voltages involved, any work on xenon lighting should be done carefully and according to the manufacturer's recommendations.

HID Diagnosis and Service

Proper diagnosis of these systems depends on understanding how the circuit works. If only the low beams have an HID system, the headlights contain an arc

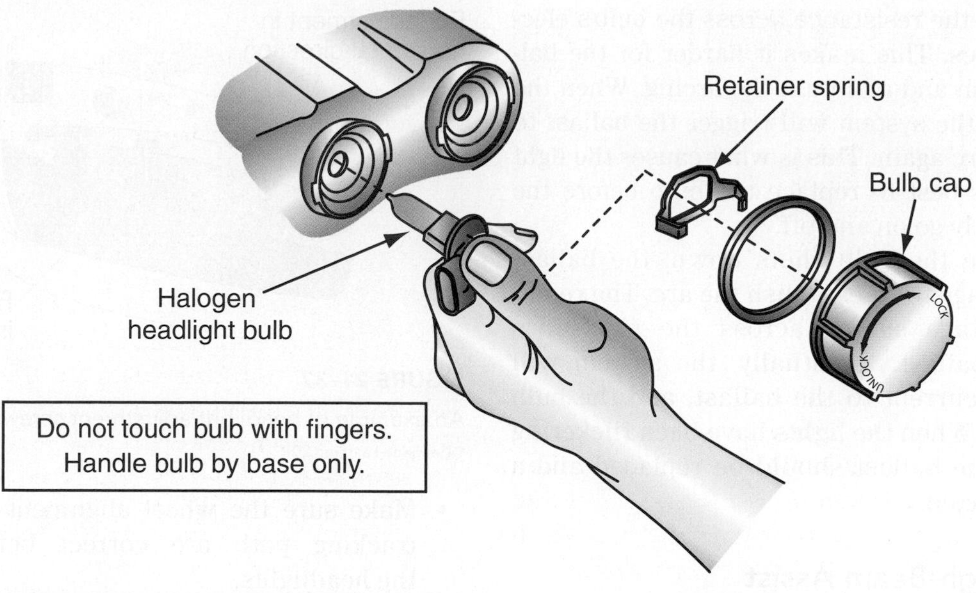

FIGURE 21–36

The correct way to install a halogen bulb into a composite headlamp.

tube and ballast. The ballast increases the voltage so that an arc can be established across the electrodes. When the headlight switch is moved to the HEADLIGHT position, the ground for the low-beam relay is completed by the control computer (BCM). This energizes the relay, and battery voltage is applied to the ballast in each headlamp assembly. The ballast increases the voltage and starts the high-voltage arcing to light the bulb. Once the arc is established, much less voltage is necessary to maintain the arc across the electrodes.

In bi-xenon systems, there is no separate high-beam bulb. Rather, a solenoid or stepper motor attached to a shutter is used to redirect the light beams from the bulb. When the headlamp dimmer switch is moved to the high-beam position, ground is applied through the dimmer switch to the BCM. In response to this signal, the BCM completes the ground for the high-beam relay, energizing the relay. Battery voltage is then applied to the left and right high-beam solenoids that move the shutter.

When an HID lamp is not working, do not automatically assume there is a faulty lamp. The ballast assembly could be bad, or there could be a fault in its electrical circuit. To make sure the circuit is operating fine, check for battery voltage at ballast, and also check its ground. Look at the lamp; a smoky or blackened lamp may indicate the bulb is burned out. If the bulb looks good but does not illuminate, it is likely that the ballast or igniter is faulty.

NORMAL DELAY Often, the wait period for xenon lights to become fully illuminated is considered a problem. This is not a problem; it is normal. The ballast needs to provide high voltage to start and keep the lamps illuminating. This takes a little time; it may take 2 seconds to establish the arc and then another 30 seconds to have full illumination. Obviously, if it takes longer than that, there is a problem.

BULB COLOUR Another normal concern is the colour of the light. HIDs produce a bluish-white light. However, some produce a light beam that appears pure white. This is not a problem, even if one side appears to be a different colour than the other side. The colour of the light depends on many normal factors. A worn bulb may have a dim pinkish glow. Often, replacement bulbs will have a yellowish-white look for the first five minutes when turned on. The lamps still provide the same amount of light as the originals and will provide a bluish light after about 100 hours of operation.

BULB REPLACEMENT HID bulbs normally do not suddenly stop working. As the bulb wears, it will shut off and then turn back on. When the bulb is in the early stages of wear, this will happen very infrequently. As time goes on, the bulb will shut down and come on again rather frequently. This problem may progress to flickering until the system totally shuts down. Each manufacturer has a procedure for monitoring and diagnosing the system; always follow those.

Over time, the resistance across the bulb's electrodes increases. This makes it harder for the ballast to establish and maintain the arcing. When the arcing is lost, the system will trigger the ballast to establish the arc again. This is what causes the light to flicker. It is best to replace the bulb before the lights constantly go on and off.

Every time the bulb shuts down, the ballast uses high voltage to re-establish the arc. The repetitive high-voltage surges across the electrodes damage the ballast. Eventually, the system will stop sending current to the ballast, and the bulb will not work. When the lights have been flickering for a while, the ballast should be replaced and a new bulb inserted.

Adaptive High-Beam Assist

Most adaptive high-beam assist systems rely on an LED to inform the driver of any system fault. The green LED is located along the bottom plane of the inside rearview mirror. If the LED is flashing slowly, the system recognizes that its camera needs to be properly aimed. Always follow the manufacturer's procedure for doing this. If the LED is continuously flashing rather quickly, a system fault has been detected, and a DTC has been set.

Headlight Adjustments

Headlights must be kept in adjustment to obtain the safest and best light beams on the road ahead. Headlights that are properly adjusted cover the correct range and afford the driver the proper nighttime view. Headlights that are out of adjustment can cause other drivers discomfort and sometimes create hazardous conditions.

Before adjusting or aiming a vehicle's headlights, make the following inspections to ensure that the vehicle is level. Any one of these conditions can result in an incorrect setting.

- If the vehicle is coated with snow, ice, or mud, clean it, especially the underside. The additional weight can alter the riding height.
- Try to make the adjustment with the fuel tank half-full; this should be the only load present on the vehicle.
- Worn or broken suspension components affect the setting, so check the springs or shock absorbers.
- Inflate all tires to the recommended air pressure levels.

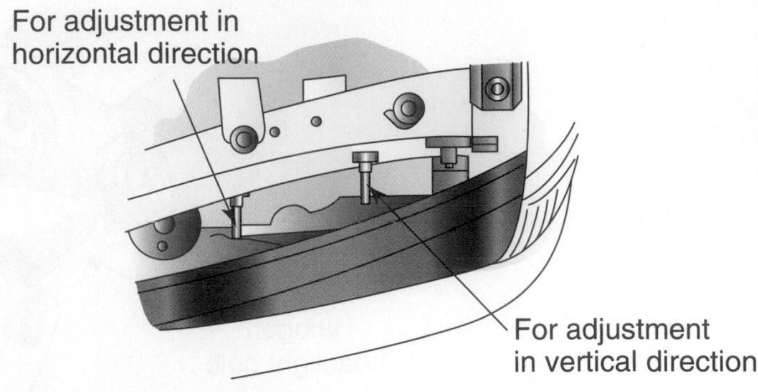

FIGURE 21–37

An example of headlight adjustment screws.

- Make sure the wheel alignment and rear axle tracking path are correct before adjusting the headlights.
- After placing the vehicle in position for the headlight test, push down on the front fender to settle the suspension.

Headlight assemblies have adjusting screws **(Figure 21–37)** to move the headlight within its assembly to obtain correct headlight aim. Lateral or side-to-side adjustment is made by turning the adjusting screw at the side of the headlight. Vertical or up-and-down adjustment is made by turning the screw at the top of the headlight.

SHOP TALK

While adjusting headlight beams, make sure they meet the standards established by your local community or province.

To properly adjust the headlights, many types of headlight-aiming equipment can be used **(Figure 21–38)**. These aimers use mirrors with split images, like split-image finders on some cameras, and spirit levels to make exact adjustments on older sealed-beam headlights. Newer optical aimers **(Figure 21–39)** are used to aim newer composite headlights. When using any mechanical aiming tool, follow the manufacturer's instructions.

When headlight-aiming equipment is not available, alignment can be checked and adjusted by projecting the beam of each light on a screen or chart placed about 8 m (25 ft.) in front of the headlight **(Figure 21–40)**. The vehicle must be exactly perpendicular (at a right angle) to the chart.

Measure the distance between the centres of the headlights. Use this measurement and draw two vertical lines on the chart that correspond with the centre

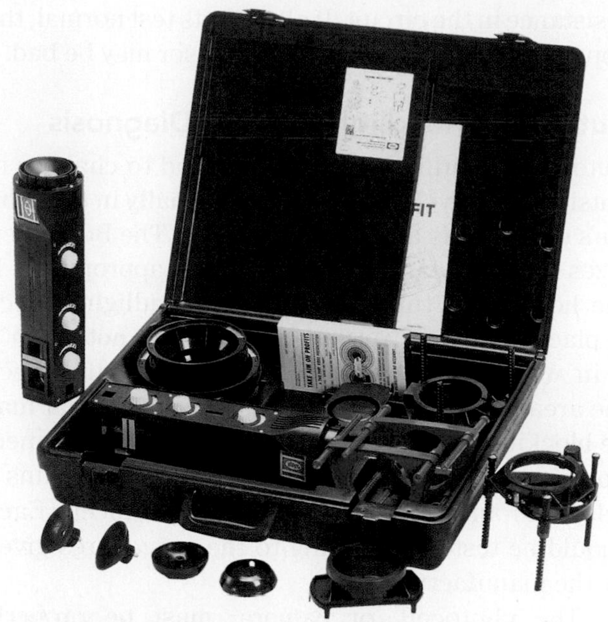

FIGURE 21–38

An old-style headlight aiming kit.

Snap-on Tools Company

FIGURE 21–39

An optical headlight aimer.

Model OPT-5405, Panther Headlight Aimer, Sayco Distributors Ltd/Canbuilt Manufacturing

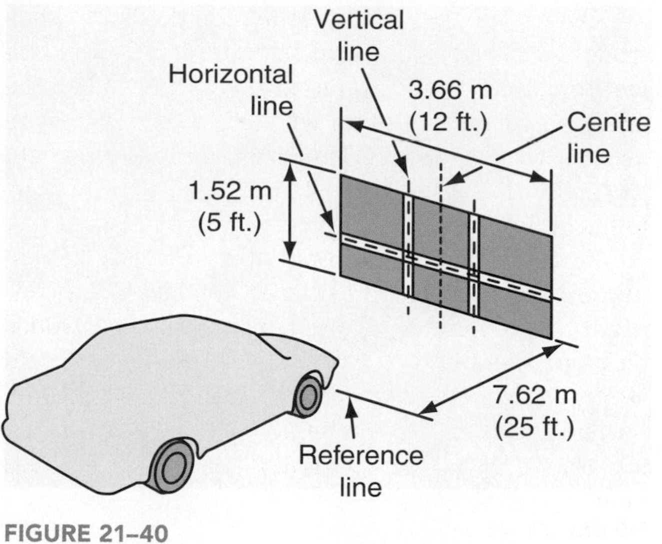

FIGURE 21–40

Acceptable beam patterns on a wall screen.

©Cengage Learning 2015

of the headlights. Then draw a vertical centreline halfway between the two vertical lines. Measure the distance from the floor to the centres of the headlights. Subtract 50 mm (2 in.) from this height and draw a horizontal line on the screen at this new height.

With the headlights on high beam, the hot spot of each projected beam should be centred at the intersection of the vertical and horizontal lines on the chart. If necessary, adjust the headlights vertically and laterally to obtain proper aim.

Some vehicles have a horizontal indicator gear at each headlamp assembly. Prior to making any adjustments, it is recommended that this indicator be at 0. A screwdriver is used to bring the gear back to 0. After this, the headlamps can be adjusted to specifications.

Also, many headlamps have a bubble level to aid alignment. The bubble level **(Figure 21–41)** is calibrated to the earth's surface; therefore, the vehicle must be on level ground when the headlights are aimed. If the headlight beam projection appears high to oncoming traffic, check the alignment on an alignment screen. If the beam pattern is above or to the left of the specified location on the screen, adjust the headlights, and then recalibrate the bubble level and magnifying window. Ideally, if the headlights are aligned, the bubble level and magnifying window will be centred. Never change the calibration of the magnifying window or bubble level if the headlights are out of alignment.

Auto-Levelling Headlamps

Some vehicles, primarily those equipped with xenon headlamps, have an automatic headlight-levelling system. With this system, there is no need to adjust

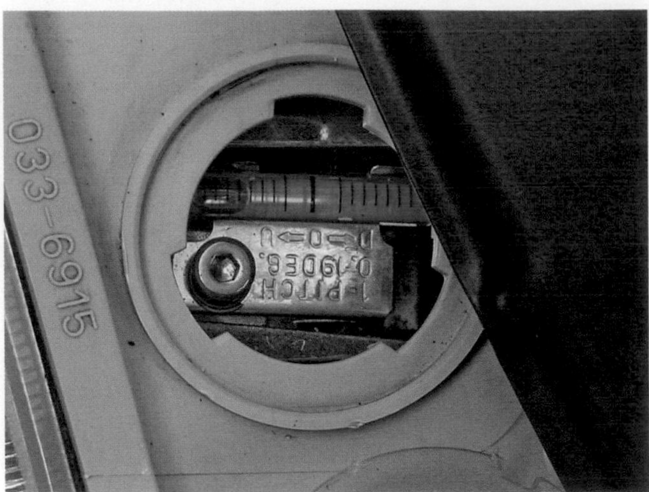

FIGURE 21–41

Some vehicles have a bubble level at the headlights to aid in alignment.

©Cengage Learning 2015

the headlamps. However, if this system fails, the headlamps will not be properly aligned, and the system must be diagnosed. Before diagnostics can take place, an understanding of how the system works is important.

DIAGNOSIS Diagnosis is done with a scan tool. The scan tool can be used to order the motors to move up and down. With the ignition on, do this and observe the headlamps. If they do not move, test the motors. If the motors move, check the voltage of the front and rear levelling-position sensors. Normally, the voltage should be between 0.5 and 4.9 volts. If the reading is not within the specified range, check the sensors.

The scan tool may also retrieve fault codes for the system. Possible fault codes can be set for a short to ground, open/high resistance, short to voltage, and abnormal signal performance for the following areas:

- Front and rear headlamp-levelling sensor
- Front and rear headlamp-levelling sensor 5-volt reference
- The headlamp-levelling motor control circuits on both sides

The motor assembly is checked with an ohmmeter. With the ignition off, disconnect the electrical connector at each headlamp-levelling motor. Take resistance readings across the various terminals of the connector, and compare your reading to the specifications. The levelling sensors are checked with a voltmeter. First, the reference voltage is checked at the specified terminals. If the voltage is higher than normal, there is a short to ground, an open, or high resistance in the circuit. If all circuits test normal, the appropriate headlamp-levelling sensor may be bad.

Automatic Headlight System Diagnosis

Automatic headlight systems respond to changes in outside light levels. A photocell, normally in the vehicle's dash, sends a signal to the BCM. The BCM energizes the low-beam circuit when it is appropriate. If the headlamps turn on when the headlight switch is placed in the HEADLIGHT position, but not when it is in AUTO, the problem is in the auto circuit. Check the area around the photocell for anything that may be blocking light from the sensor. Check all connections in the system. The circuit normally contains a relay and amplifier along with the photocell. Each should be tested according to the procedures given by the manufacturer.

The photocell, or camera, must be properly aligned in order for the system to respond to the lights of oncoming vehicles and not lights coming from the far sides of the road. This typically is a very involved process, and the procedures given by the manufacturer must be followed. This involves many measurements and conducting tests with a scan tool **(Figure 21–42)**.

Most systems also keep the lights on for a short time after the vehicle has been parked. The controlling device for this feature is a potentiometer incorporated into the headlamp switch. When the delay feature is not working properly, begin testing at the timer control.

Adaptive Headlight Diagnosis

The operation of the adaptive headlight system is controlled by the headlamp control module. Therefore,

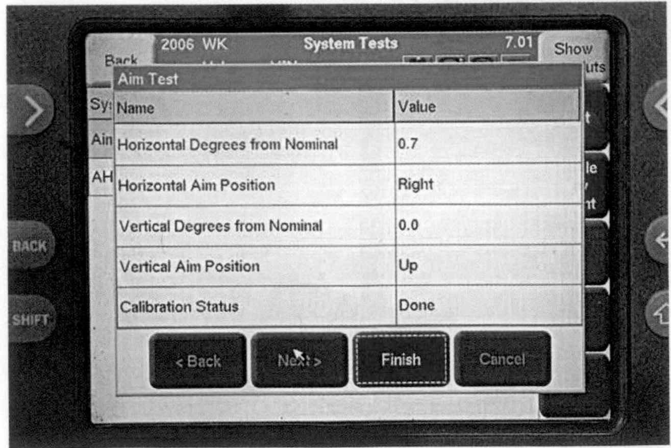

FIGURE 21–42

Calibration of the photocell includes a check with a scan tool.

©Cengage Learning 2015

diagnosis begins with a scan tool. Because the system responds to many inputs, each one should be monitored.

The headlamp control module receives serial data from the engine control module, transmission control module, electronic brake control module, and body control module. The control module calculates the desired headlamp angle and orders the actuators or motors at each headlight to move the headlamps. The control module also monitors the condition of the motor control circuits. If a problem is detected, a DTC will be set and a message displayed in the instrument panel to alert the driver of a problem.

INTERIOR LIGHT ASSEMBLIES

The types and numbers of interior light assemblies used vary significantly from one vehicle to another **(Figure 21–43)**. Following are the more common ones.

ENGINE COMPARTMENT LIGHT Operating the hood causes the engine compartment light switch to close and light the underhood area.

Some pickup trucks and SUVs are equipped with an underhood retractable magnetic base light mounted on a reel. The light can be used anywhere around the vehicle.

GLOVE BOX LIGHT Opening the glove box door closes the glove box light switch contacts, and the light comes on.

FIGURE 21–43

Full interior illumination is available with this light setup.

DaimlerChrysler Corporation

LUGGAGE COMPARTMENT LIGHT The light is mounted in the underside of the trunk deck lid in the luggage compartment.

TRUNK LID LIGHT Lifting the trunk lid causes the light switch to close, and the light comes on.

VANITY LIGHT Pivoting the sun visor downward and opening the vanity mirror cover causes the vanity light switch contacts to close, and the light to come on.

COURTESY LIGHTS There are several types of courtesy lights. Some vehicles have courtesy lights that are in the door trim panels, under each side of the instrument panel, and in the centre of the headlining. These are illuminated when one of the doors is opened, by rotating the headlight switch to the full counter-clockwise position, or by depressing the designated switch. **Figure 21–44** is a wiring diagram of a typical courtesy light circuit. The courtesy lights are also turned on by the illuminated entry or keyless entry systems, if the vehicle is equipped with one or both of these.

Front-compartment footwell courtesy lights are mounted on the lower closeout pancls at both ends of the instrument panel. The bulbs are accessible from under the instrument panel for replacement without removing other parts.

Some courtesy lights are a combination of map lights located on each side of the dome light housing. The map lights are operated independently of the dome light by two switches located at each map light housing. The dome light is actuated by turning the headlight switch control knob fully counter-clockwise.

Power is supplied from the fuse block to the courtesy or dome/map light. The ground for the light is controlled by the position of the door switch. That is, these door switches are held in open position and do not provide for a ground circuit. When the door is opened, a spring pushes the switch closed to ground the circuit, and the dome/map or courtesy lights come on.

ILLUMINATED ENTRY SYSTEM This system assists vehicle entry during the hours of darkness by illuminating the door lock cylinder so that it may be easily located for key insertion. The vehicle interior is also illuminated by the courtesy lights.

The system consists of four main components: electronic module, illuminated door lock cylinder, door handle switch, and wiring harness.

Activation of the system is accomplished by raising the outside door handle or by pressing a

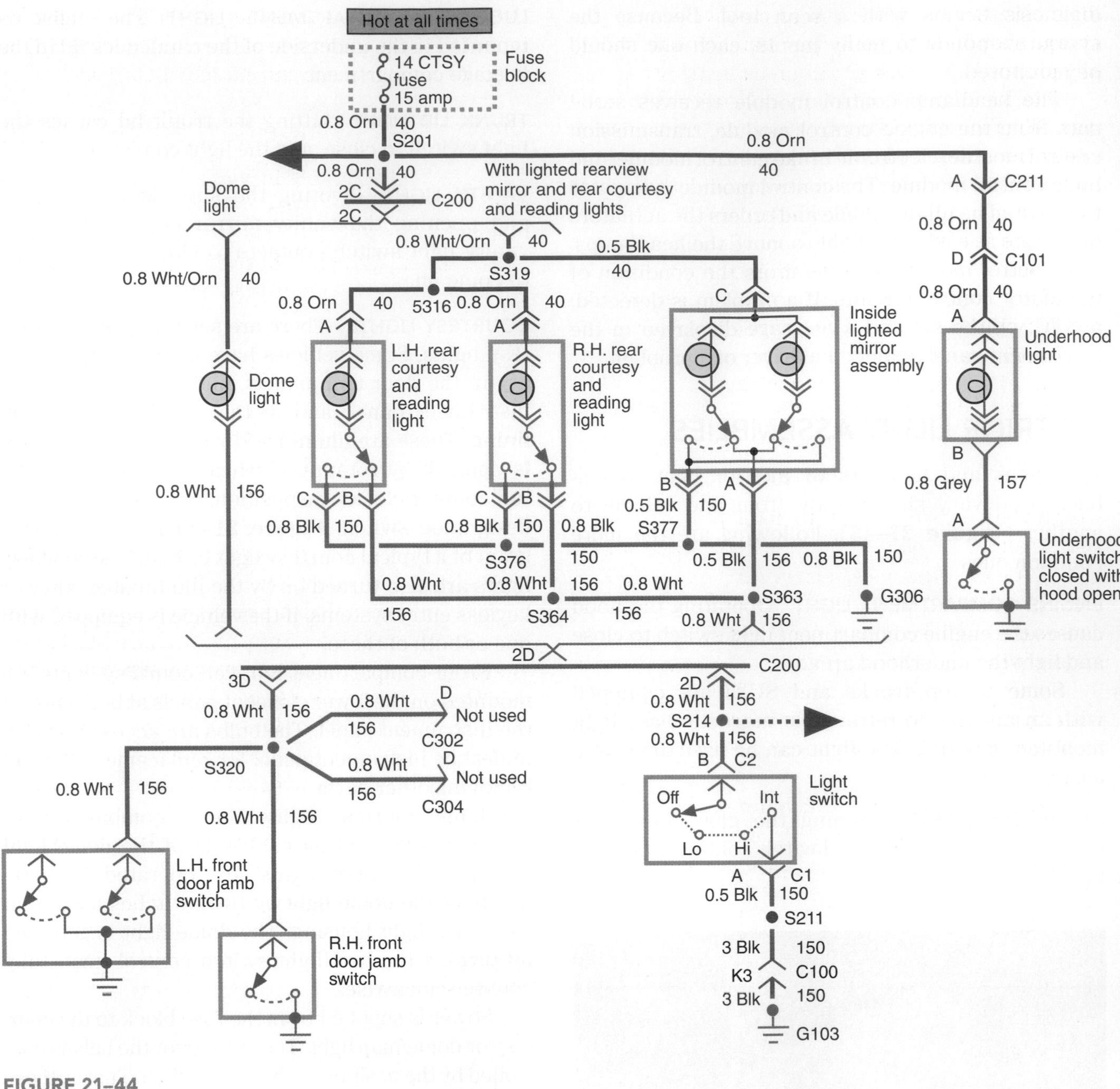

FIGURE 21–44

A typical courtesy light circuit.

code button on the keyless entry system. This action momentarily closes a switch mounted on the door handle mechanism, which completes the ground circuit of the electronic actuator module and switches the system on. The vehicle interior lights turn on, and both front door lock cylinders are illuminated by a ring of light around the area where the key enters. This illumination remains on for approximately 25 seconds and then automatically turns off. During this 25-second period, the system can be manually deactivated by turning the ignition switch to the RUN position.

The system is activated every time the outside front door handles are operated, whether the vehicle is locked or not. Opening the doors from the inside of the vehicle does not activate the system. If the outside door handle is held up indefinitely so that the handle switch is continuously closed, the system operates as normal and turns off after 25 seconds. At the completion of this cycle, if the door handle is still in the raised position, the system remains off. It is impossible to activate the system from the other front door handle until the raised handle is returned to its normal position. This function is built into the logic circuitry of the system to prevent battery discharge should the outside door handle be intentionally propped up or become jammed in any way.

Interior lights all basically operate in the same way. Whether the courtesy lights are on the door, under the seats, under the instrument panel, or on the rear interior quarter panels does not change how they are controlled. Also, whether the illumination lights are just behind the instrument panel or are also used in centre consoles or door armrests does not affect their operation. The only difference is the number of lights and variances in electrical wiring.

Interior and courtesy lights rarely give any trouble. However, if they do not operate, check the fuse, bulb, switch, and wiring.

Distributed Lighting System

Distributed lighting refers to light present at one source and transmitted through fibre optic "tubes" to one or more other locations **(Figure 21–45)**. This technology has been around for quite some time and is used to illuminate all of the instrument panel gauges with a single lamp. On today's vehicles, the light source is now most likely to be a single LED. Distributed lighting is typically used to add lighting at places where it might not be practical to have a bulb.

This technology is also favoured because fibre-optic connections won't corrode. More significantly, distributed lighting reduces noise interference on the data communications bus. As the electronic content of vehicles increases, distributed lighting will be used for more interior and exterior lighting.

REAR EXTERIOR LIGHT ASSEMBLIES

The rear light assembly includes the taillights, turn-signal/stop/hazard lights/high-mounted stoplights, rear side marker lights, backup lights, and licence plate lights. Taillights operate when the parking lights or headlights are turned on.

Turn, Stop, and Hazard Warning Light Systems

Power for the turn (directional signal), stop, and hazard warning light systems is provided by the fuse panel **(Figure 21–46)**. Each system has a switch that must close to turn on the lights in the circuit. Hazard lights are commonly referred to as four-way flashers because the lights at all four corners of the vehicle will flash when the circuit is turned on.

The turn-signal and hazard light switches on many current vehicles are part of a **multifunction switch**. When the turn or directional signal switch is activated, only one set of the switch contacts is closed—left or right. However, when the hazard switch is activated, all contacts are closed and all turn-signal lights and indicators flash together and at the same time.

The power for the turn signals is provided through the fuse panel, but only when the ignition

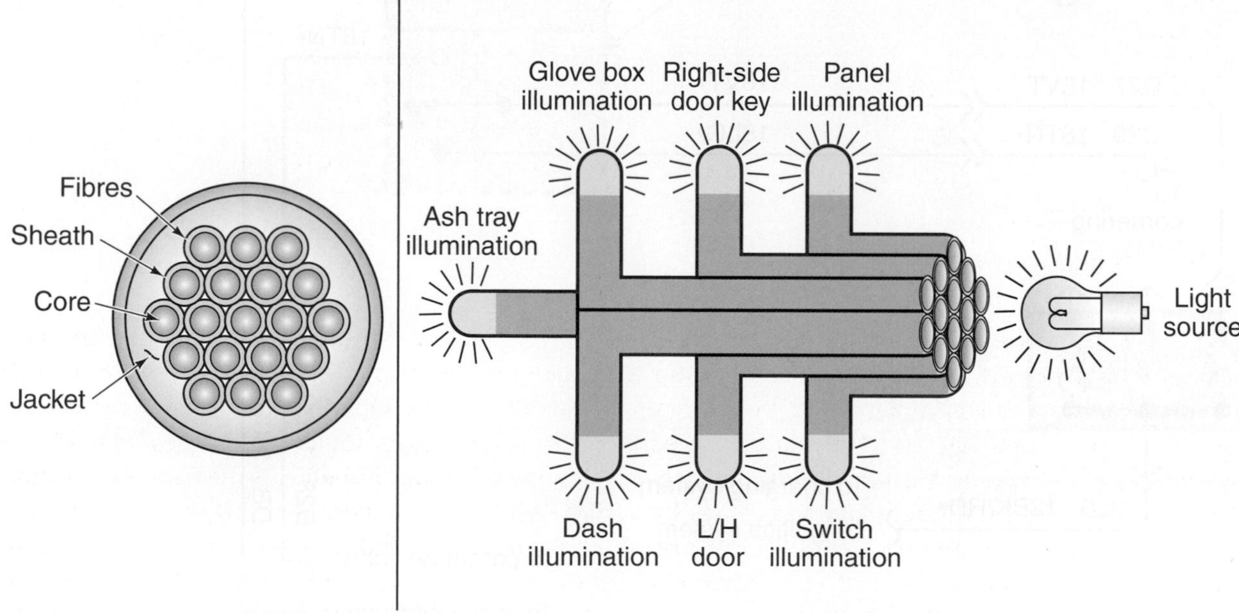

FIGURE 21–45

One light source can illuminate several locations by using fibre optics.

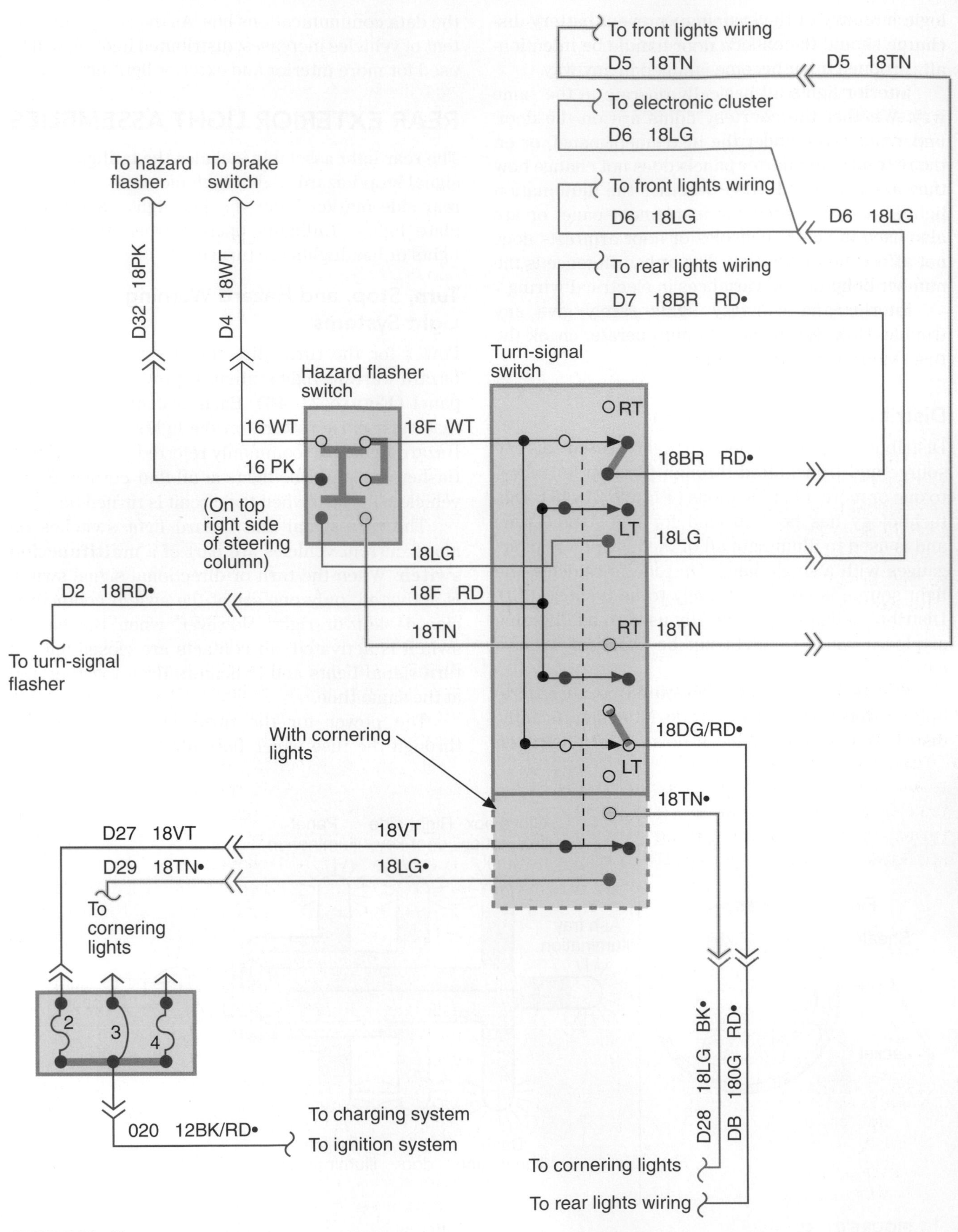

FIGURE 21–46

The turn-signal circuit for a two-bulb system.

Chrysler Group LLC

switch is on. The hazard lights are also powered through the fuse panel; however, they have power at all times regardless of ignition switch position.

Some cars are equipped with cornering side lights, which are generally fed from the multifunction main switch. When the turn-signal switch is activated, the cornering light on the appropriate side burns with a steady glow.

Side markers are connected in parallel with the feed circuit (from the headlight switch) that feeds the minor filaments of the front parking lights and rear taillights.

What a multifunction switch controls depends on the make, model, and year of the vehicle. Some control the directional signals and serve as the dimmer switch. Others control the turn and hazard signals and serve as the headlight, dimmer, windshield wiper, and windshield washer switch **(Figure 21–47)**.

This switch is not repairable and must be replaced if defective. Photo Sequence 16 outlines the typical procedure for removing a multifunction switch. Some of the steps shown in this procedure may not apply to all types of vehicles; always refer to the service manual before removing this switch. Also carefully study the procedures beforehand and identify any special warnings that should be adhered to, especially those concerning the air bag.

FLASHERS Flashers are components of both turn and hazard systems. They contain a temperature-sensitive bimetallic strip and a heating element **(Figure 21–48)**. The bimetallic strip is connected to one side of a set of contacts. Voltage from the fuse panel is connected to the other side. When the left turn-signal switch is activated, current flows through the flasher unit to the turn-signal bulbs. This current

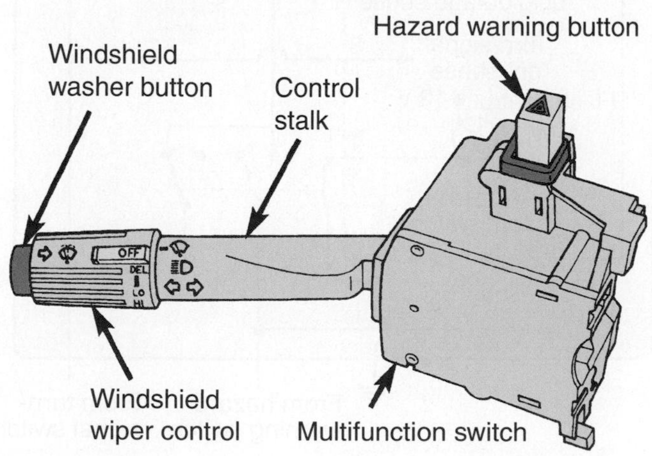

FIGURE 21–47

A typical multifunction switch.

Chrysler Group LLC

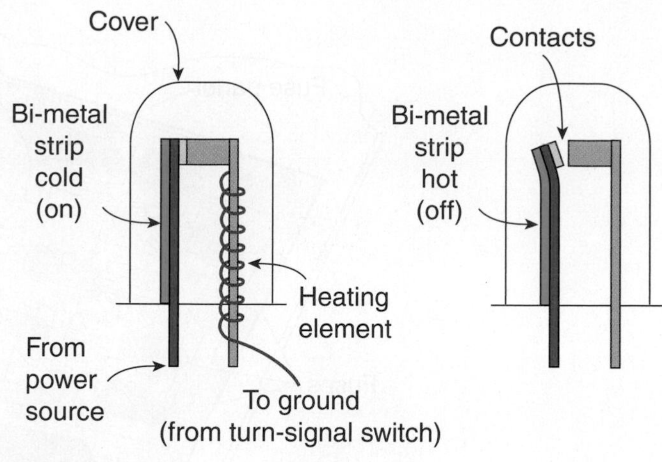

FIGURE 21–48

An inside look at a typical turn-signal flasher.

causes the heating element to emit heat, which in turn causes the bimetallic strip to bend and open the circuit. The absence of current flow allows the strip to cool and again close the circuit. This intermittent on–off interruption of current flow makes all left turn-signal lights flash. Operation of the right turn is the same as the operation of the left turn signals.

Turn-signal flashers are installed on the fuse panel on current car models and most current truck models **(Figure 21–49)**. However, on earlier models this is not true. Hazard flashers are also mounted in various locations. Refer to the service manual for locations on the model being serviced.

A test light can be used to determine which flasher is used for the turn signals and which is used for the hazard warning light. An easier way is to turn on both the directionals and the hazards. This activates both flasher units. By removing one of the flashers, the affected circuit no longer flashes. Therefore, that flash unit controls that particular circuit. If the turn signals fail to operate and the fuse is good, the flasher has probably failed.

Occasionally, the flasher does not flash as fast as it once did, or it flashes faster. This is also cause for replacement. If it flashes too slowly or not at all, check for a burned-out bulb first.

A flasher features two or three prongs that plug into a socket. Just pull the flasher out of the socket, and replace it with a new one.

Flashers are designed to operate a specific number of bulbs to give a specific **candlepower** (brightness). If the candlepower on the turn-signal bulbs is changed, or if additional bulbs are used (if a vehicle is hooked up to a trailer, for instance), a heavy-duty flasher must be used. This usually fits the socket

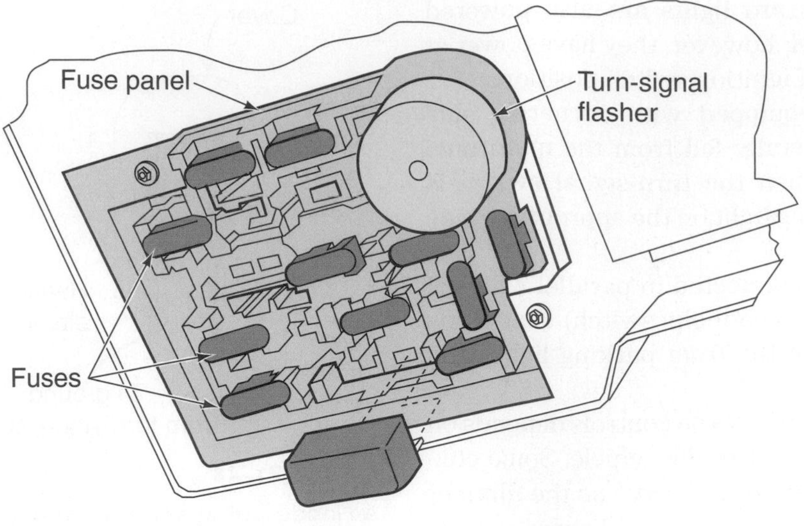

FIGURE 21–49

A common location for the flashers in the fuse panel.

Ford Motor Company

without modifications. Although heavy-duty flashers will operate additional bulbs, they have one big disadvantage and should not be used unless it is necessary. These flashers will not cause the turn signals to flash more slowly if a bulb burns out. When a turn-signal bulb fails, the driver has no idea that it did.

WARNING!

The flasher unit for turn signals should not be switched with a flasher unit for the hazard lights.

Some newer vehicles have a combination flasher unit that controls the flash rate of both the turn signals and the hazard lights. These combination flashers are electronic units **(Figure 21–50)**. The actual turning off and on of the lights is caused by the cycling of a transistor. This type flasher also senses when a bulb is burned out and causes the remaining bulbs on that side to flash faster. Because this flasher is an electronic device, it cannot be tested with normal test equipment. The only test of the flasher is to substitute it with a known good one. If the lights flash normally, the original flasher unit was bad and needs to be replaced.

Brake Lights

The brake (stop) lights are usually controlled by a stoplight switch that is normally mounted on the brake pedal arm. Some cars are equipped with a brake or stoplight switch mounted on the master cylinder, which closes when hydraulic pressure increases as the brake pedal is depressed. In either case, voltage is present at the stoplight switch at all times. Depressing the brake pedal causes the stoplight switch contacts to close. Current can then flow to the stoplight filament of the rear light assembly. These stay illuminated until the brake pedal is released.

In addition to the stoplights at the rear of the vehicle, all late-model vehicles have a **centre high-mounted stoplight (CHMSL)** that provides an

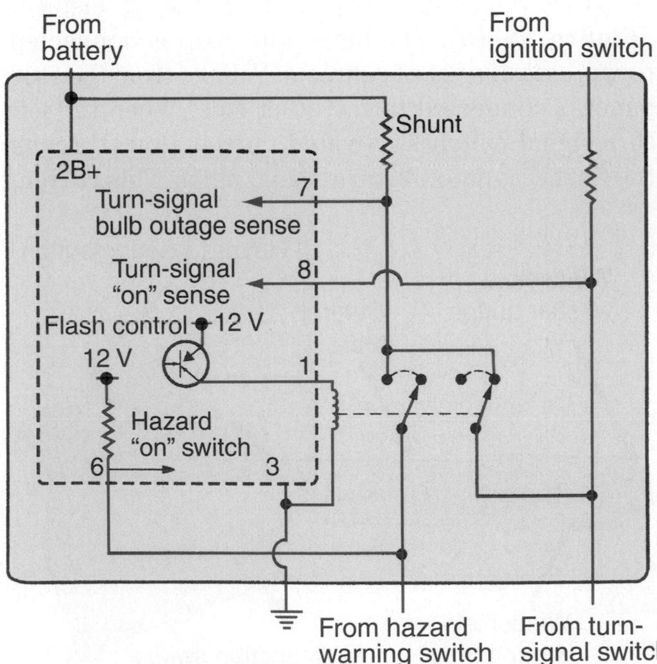

FIGURE 21–50

A typical electronic combination flasher unit.

Chrysler Group LLC

REMOVING A MULTIFUNCTION SWITCH

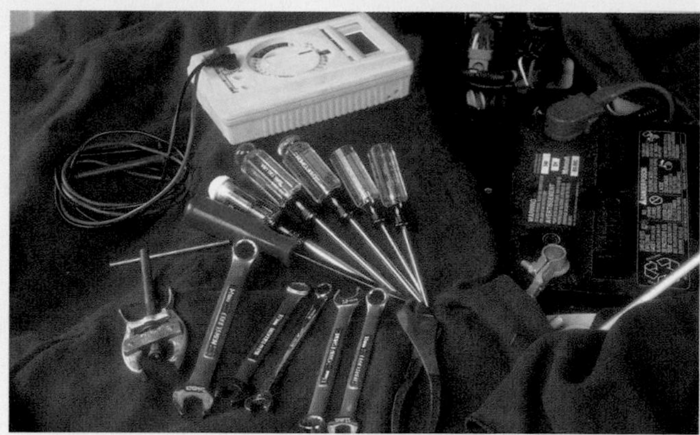

P16–1 The tools required to test and remove a multifunction switch are fender covers, battery terminal pliers and pullers, assorted wrenches, a Torx driver set, and an ohmmeter.

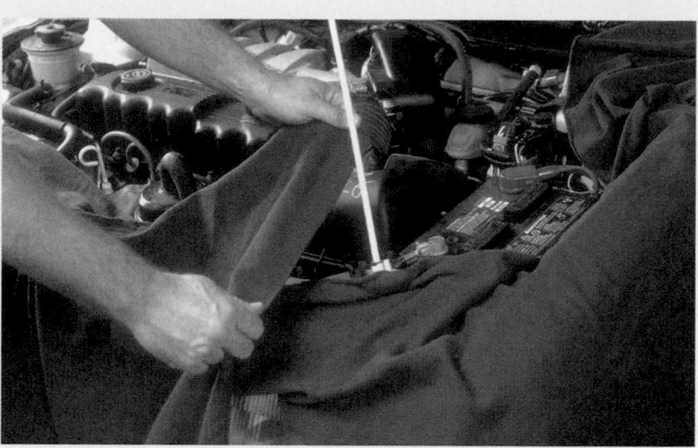

P16–2 Place the fender covers over the fenders of the vehicle. Caution: Always observe all air bag service precautions before proceeding with any steering column service.

P16–3 Loosen the negative battery clamp bolt and remove the battery clamp. Place the cable where it cannot contact the battery.

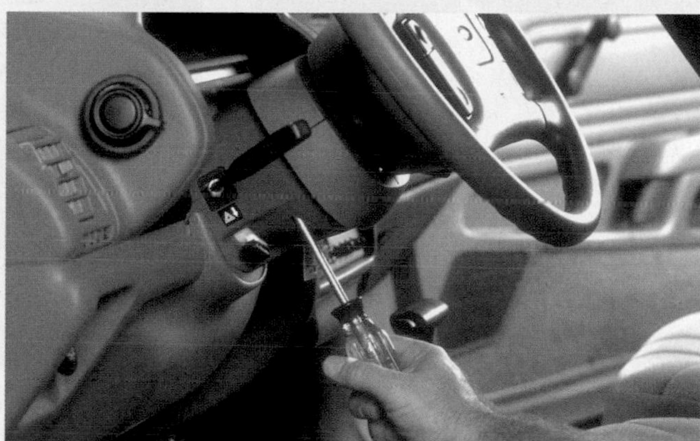

P16–4 Remove the shroud retaining screws, and remove the lower shroud from the steering column.

P16–5 Loosen the steering column attaching nuts. Do not remove the nuts.

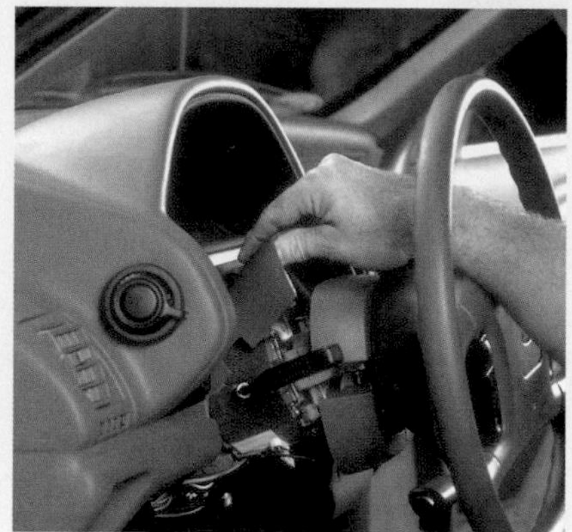

P16–6 Lower the steering column just enough to remove the upper shroud.

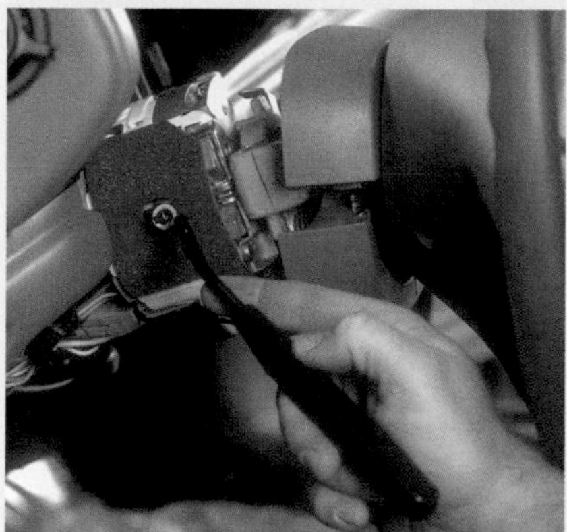

P16–7 Remove the turn-signal lever by simply rotating the outer end of the lever. Then pull it straight out.

REMOVING A MULTIFUNCTION SWITCH *(continued)*

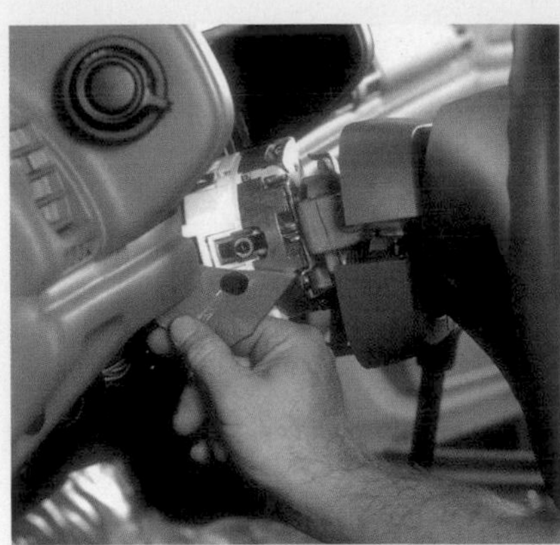

P16–8 Peel back the foam shield from the turn-signal switch.

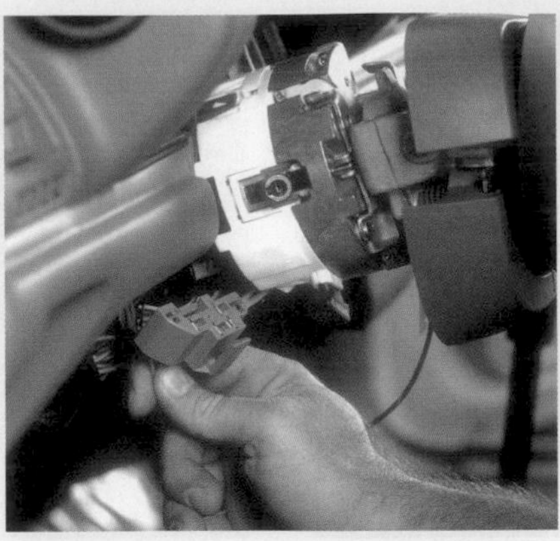

P16–9 Disconnect the turn-signal switch's electrical connectors.

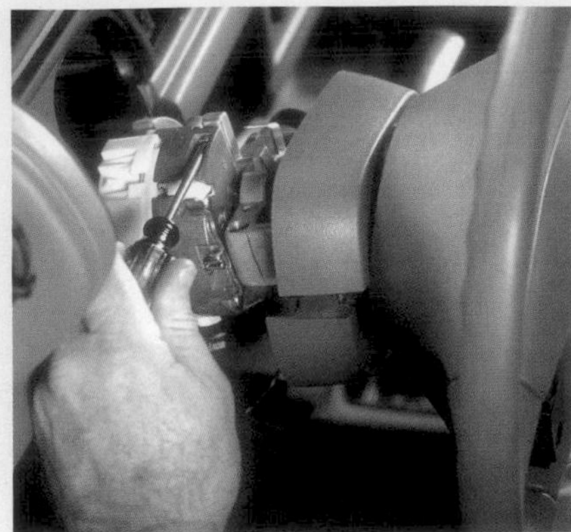

P16–10 Remove the screws that attach the switch to the lock cylinder assembly.

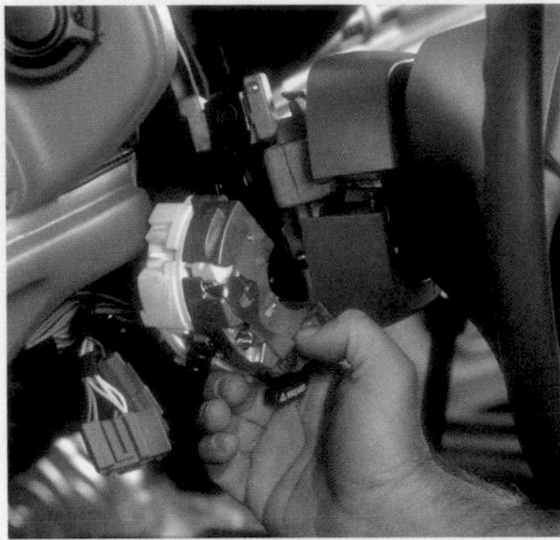

P16–11 Disengage the switch from the lock assembly.

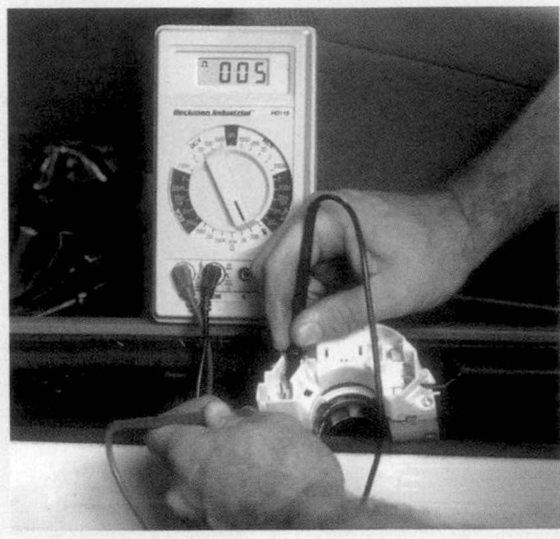

P16–12 Use an ohmmeter to test the switch. Check for continuity when the dimmer switch is in the low-beam position.

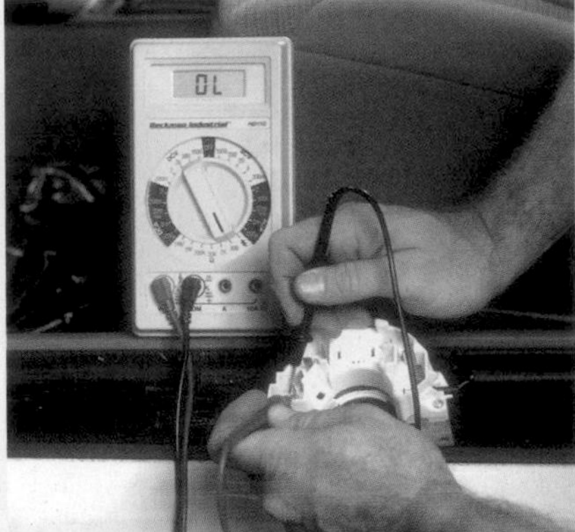

P16–13 When the switch is in the low-beam position, the circuit should be open between the high-beam terminals.

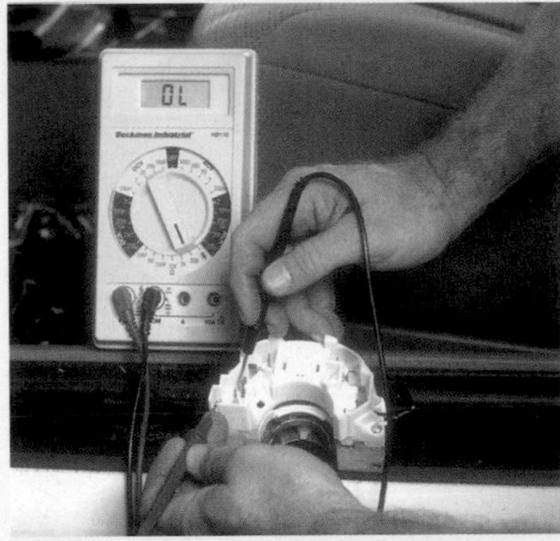

P16–14 Also check the other terminals and circuits that should be open when the dimmer switch is in the low-beam position.

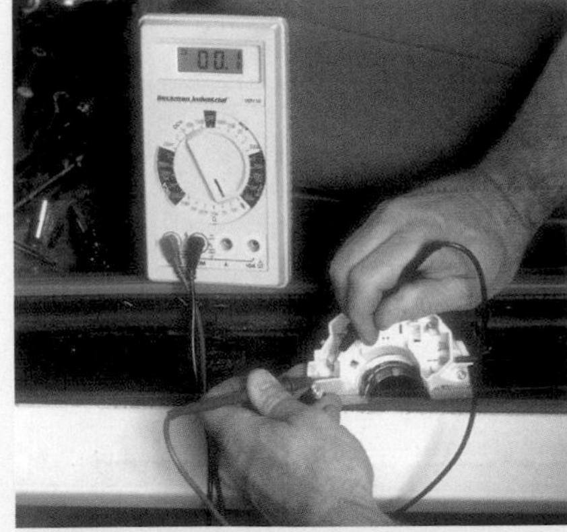

P16–15 With the switch in the high-beam position, there should be continuity across the high-beam circuit. Also check for continuity across the other circuits that should be open when the switch is in the high-beam position.

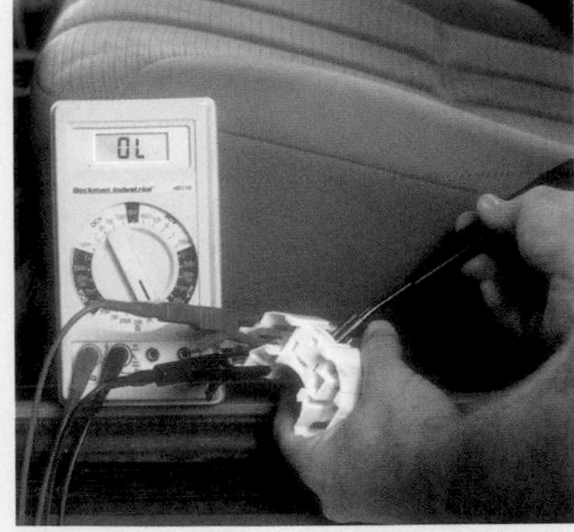

P16–16 When the dimmer switch is placed in the flash-to-pass position, there should be continuity across those designated terminals and an open across the others.

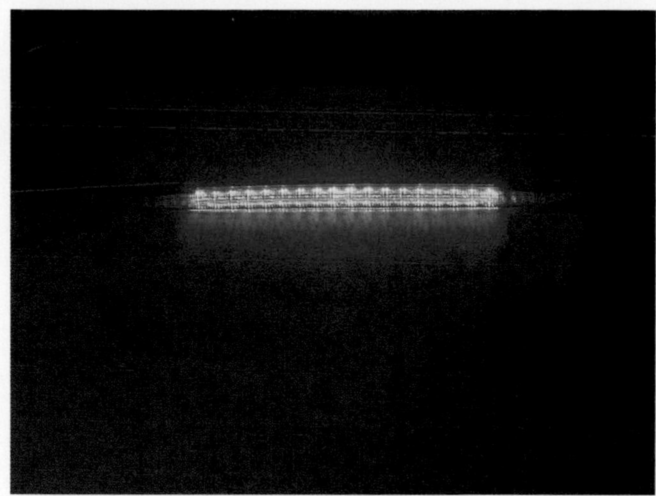

FIGURE 21–51
A CHMSL with LEDs.
©Cengage Learning 2015

FIGURE 21–52
A neon lamp used for the third brake light.
BMW of North America, Inc.

additional clear warning signal that the vehicle is braking **(Figure 21–51)**. Studies have shown the additional stoplights to be effective in reducing the number and severity of rear collisions. The high-mounted stoplight is activated when current is applied to it from the stoplight switch. It stays illuminated until the brake pedal is released. When its contacts are closed, the stoplight switch can also provide current to the speed control amplifier, antilock brake control module, and electric brake controller connector.

LED Lights

Some vehicles use neon lamps and/or LEDs for tail, brake, and turn-signal lights. Neon lights are more energy efficient and turn on more quickly than regular lights **(Figure 21–52)**. Because neon bulbs have no filament, the neon bulb will last longer than a conventional light bulb.

Whereas conventional bulbs take around 200 milliseconds (ms) to reach their full brightness, neon bulbs turn on within 3 ms. The importance of this time difference is that it gives the driver behind the vehicle an earlier warning to stop. This early warning can give the approaching driver 5.5 more metres for stopping when driving at 100 km/h.

LEDs offer the same advantages as neon bulbs and turn on even more quickly because they do not need to heat up to illuminate **(Figure 21–53)**. LEDs achieve their full output in less than a millisecond. Several LEDs are placed behind the lens and are activated at the same time to give a bright illumination of the light assembly. LEDs also require a much smaller space, so they are much less intrusive in the trunk. LEDs have a long operating life and provide a more precise contrast and signal pattern, thus attracting attention much more effectively.

FIGURE 21–53
LEDs are used in this taillight assembly.
Nissan Canada

Using the same basic technology as LEDs, laser-lit taillights consume seven times less power than incandescent sources. These savings are extremely important for electric vehicles. The light waves of a laser light beam move in the same direction, and the light is all the same colour. When used with rear exterior lights, fibre optics carry red light from a diode laser to a series of mirrors, which send the beam across a thin sheet of acrylic material.

ADAPTIVE BRAKE LIGHTS This system can select one of two available brake light areas for illumination: moderate braking activates the standard brake lights incorporated within the taillight assemblies, as well as the centre high-mounted stoplight. Under intense braking and during all braking manoeuvres with active ABS intervention, additional lamps are lit, thereby changing the size of the brake lights and their intensity **(Figure 21–54)**. By increasing the brake lights' illuminated surface area, the system alerts drivers of following vehicles that the vehicle in front has started braking and decelerating at a rapid rate. This warning allows the driver of the following vehicle to react more quickly and reduces the danger of a rear impact.

An electronic control unit processes signals supplied by the speed sensor and the antilock brake system. It then uses these data to calculate the intensity of the braking as reflected by the vehicle's rate of deceleration.

USING SERVICE INFORMATION

In addition to the taillight system, the rears of vehicles have many other lighting circuits. Most cars have brake lights, run lights, turn signals, and backup lights. Let us look at these circuits **(Figure 21–55)** and see that their diagnosis is very simple once you are aware of how they appear in the service information system. Start with the brake lights. The easiest circuit to look at first is the three-bulb circuit found on many vehicles. The drawing shows a typical taillight circuit, which contains three separate filaments for each side of the rear of the vehicle. There is a separate filament for each function: brake, turn, and run. A constant source of fused B+ is made available to the brake switch. The brake switch is usually located on the brake pedal and is closed by pushing down on the brake pedal. B+ is now available to the bulbs, wired in parallel, at the rear of the vehicle. Releasing the brake pedal allows the spring-loaded normally open (NO) switch to open and turn the brake lights off. This is a simple circuit that requires only a 12-volt test light or a voltmeter for diagnosis. The most common cause of failure is bulbs that burn out. Testing for B+ and ground at the bulb socket should verify the circuit. If B+ is not available at the socket, test for power at each connector, moving back toward the switch until it is found. Repair the open. Do not forget that the circuit is only hot if the brake pedal is depressed.

FIGURE 21–54

(Left) Normal illumination of the brake lights; (right) illumination during hard stops.

BMW of North America, Inc.

BACKUP LIGHTS When the transmission is placed in reverse gear, backup lights are turned on to illuminate the area behind the vehicle and to let other drivers know that the vehicle is in reverse. The major components in the system are the backup light switch and the lights.

Power for the backup light system is provided by the fuse panel. When the transmission is shifted to reverse, the backup light switch closes and power flows to the backup lights. That is, any time the transmission is in reverse, current flows from the fuse panel through the backup light switch to the backup lights. On many vehicles, the fuse that protects the backup light system also protects the turn-signal system.

In general, vehicles with a manual transmission have a separate switch. Those with an automatic transmission use a combination neutral start/backup light switch. The combination neutral start/backup light switch used with automatic transmissions is actually two switches combined in one housing. In PARK or NEUTRAL, current from the ignition switch is applied through the neutral start switch to the

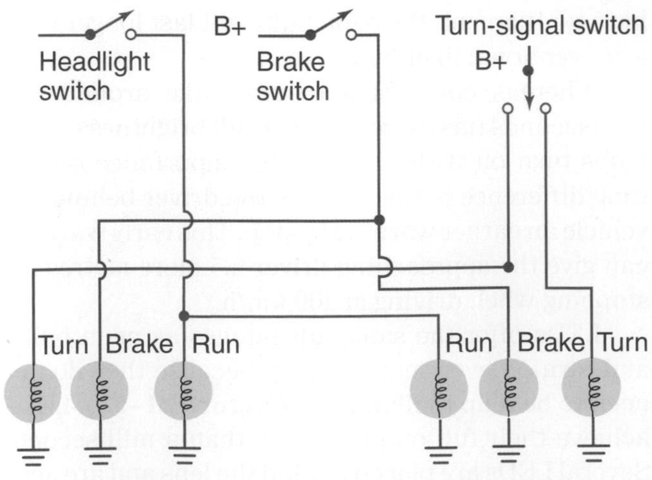

FIGURE 21–55

A typical three-bulb taillight circuit.

starting system. In reverse, current from the fuse panel is applied through the backup light switch to the backup lights.

The backup light system is relatively easy to troubleshoot. On vehicles that use one fuse to protect both the turn signals and the backup lights, the fuse can be checked. If the backup lights are not working, check turn-signal operation. If they work, the fuse is good. Check for power at the backup light switch input and outlet with the transmission in reverse. (Make sure the parking brake is set.)

If the switch is okay, or there is no power to the switch, check the wiring—especially the connectors. If the backup lights stay on when the transmission is not in reverse, suspect a short in the backup light switch.

LIGHT BULBS

Besides headlight bulbs, there are several different types of light bulbs used in modern vehicles.

Other Bulbs

Bulbs used in most other lighting fixtures fit into sockets and are held in place by spring tension or mechanical force **(Figure 21–56)**. Bulbs are coded with numbers for replacement purposes. Bulbs with different code numbers might appear physically similar but have different wattage ratings.

Light systems sometimes use one wire to the light, making use of the car body or frame to provide the ground back to the battery. Because many of the manufacturers have gone to plastic socket and mounting plates (as well as plastic body parts) to reduce weight, many lights must now use two wires to provide the ground connection. Some double-filament lights use two hot wires and a third ground wire. That is, double-filament bulbs have two contacts and two wire connections to them if grounded through the base. If not grounded through the base of the bulb, a two-filament bulb has three contacts and three wires connected to it. Single-filament bulbs may be single- or double-contact types. Single-contact types are grounded through the bulb base, whereas double-contact, single-filament types have two wires—one live and the other a ground.

When replacing a bulb, inspect the bulb socket. If the socket is rusty or corroded, the socket or light assembly base should be replaced. Also inspect the lens and gasket for damage while the lens is removed, and replace any damaged part.

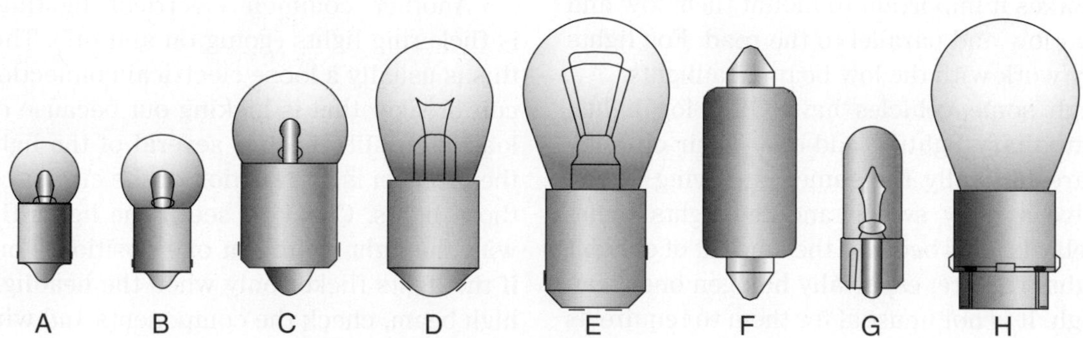

Common automotive bulbs:

A, B – Miniature bayonet for indicators and instrument lights
C – Single contact bayonet for licence and courtesy lights
D – Double contact bayonet for trunk and underhood lights
E – Double contact bayonet with staggered indexing lugs for stop, turn, and brake lights
F – Cartridge type for dome lights
G – Wedge base for instrument lights
H – Blade double contact for stop, turn, and brake lights

FIGURE 21–56

Common types of automotive bulbs.

There are two basic construction designs for exterior lights: those in which the lens is removed and then the bulb removed from the front; and those in which the light assembly must first be removed, then the socket from the back of the assembly, and finally the bulb from the socket. Removing the lens from the latter type assembly could cause serious damage to the reflector due to dust and other contaminants. Wiping the reflector surface to clean it can also seriously reduce the light's brightness. Therefore, do not remove the lens from light assemblies in which the socket and bulb are removed from the back of the assembly.

The bulbs are held in their sockets in a number of ways. Some bulbs are simply pushed into and pulled out of their sockets, and some are screwed in and out. To release a bayonet-style bulb from its socket, the bulb is pressed in and turned counter-clockwise. The blade-mount style is removed by pulling the bulb off the mounting tab, then turning the bulb and removing it from the retaining pin.

FOG LIGHTS Ordinary headlights do not penetrate fog well. Focus a powerful beam of light at the fog, and all the driver gets back is a powerful glare. To deal with that problem, fog lights use the same bulbs, but instead of trying to pierce the darkness, they produce a flat, wide beam of light under the blanket of fog. This makes it important to mount them low and to aim them low and parallel to the road. Fog lights should only work with the low-beam headlights.

Although some vehicles have OEM fog lights, most are auxiliary lighting add-ons. Their circuits, however, are basically the same as driving lights. They involve a relay switch and the lights themselves. A relay is used because the amount of current that fog lights require, especially halogen ones, can be quite high. It is not unusual for them to require as much as 25 amps.

The dash switch controls the current to one side of the relay's coil. A direct ground is supplied to the other side of the relay coil. With both battery voltage and ground applied, current flows through the coil and a magnetic field develops. The field closes the contacts in the relay. One side of the contacts is connected to a fused source of battery voltage. The other side is connected to the fog lights, which are wired in parallel. Each filament has its own remote ground connection.

Driving and fog lights have tremendous output and have correspondingly high electrical requirements. This means the car should have an efficient charging system and a heavy-duty battery.

SHOP TALK

When replacing fog light bulbs, avoid touching the glass part of the new bulb assembly. Skin oil, present on even recently washed hands, will be deposited on the glass. This oil prevents the bulb from dissipating heat. The increased heat inside the bulb causes the filament to burn out prematurely.

LIGHTING MAINTENANCE

In addition to replacing all burned-out lights and bulbs, when a vehicle comes in for servicing, periodically check to see that all wiring connections are clean and tight, that light units are tightly mounted to provide a good ground, and that headlights are properly adjusted. Loose or corroded connections can cause a discharged battery, difficult starting, dim lights, and possible damage to the AC generator and regulator. Often, moisture gets into a bulb socket and causes corrosion of the electrical contacts and the bulb. Corrosive conditions can be repaired by using emery cloth on the affected areas. For severe cases, replace the socket and bulb. After any repair, always attempt to waterproof the assembly to prevent future problems. Cracked or broken assemblies are easily replaced. They are secured by attaching hardware that is normally readily accessible to the technician.

Another common electrical lighting problem is flickering lights (going on and off). The cause of this is usually a loose electrical connection or a circuit breaker that is kicking out because of an overloaded circuit. If all or several of the lights flicker, the problem is in a section of the circuit common to those lights. Check to see if the lights flicker only with the light switch in one position. For example, if the lights flicker only when the headlights are on high beam, check the components and wiring in the high-beam section of the circuit. If only one light flickers, the problem is in that section of the circuit. Check the bulb socket for corrosion. Also make sure the bulb terminals are not worn. This could upset the electrical connection. If necessary, replace the bulb socket and bulb.

Look at the turn-signal diagram in **Figure 21–57A**. The diagram shows the inside of a turn-signal switch for a two-bulb system. The turn-signal switch determines whether one of the bulbs is used for turning or brake lighting. The rectangular bars on the diagram are stationary contacts that the circuit wires connect to. Each contact has one wire connected to it. The top connection is from the brake switch and is B+ if the brakes are applied. The

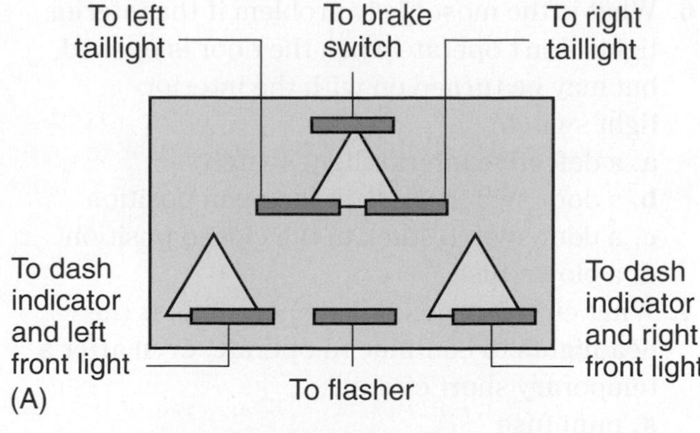

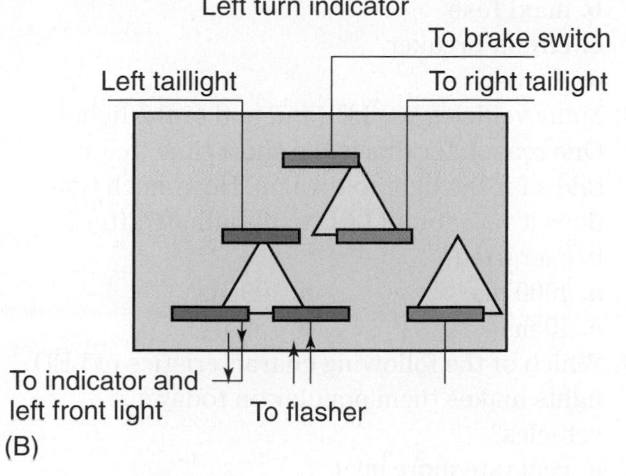

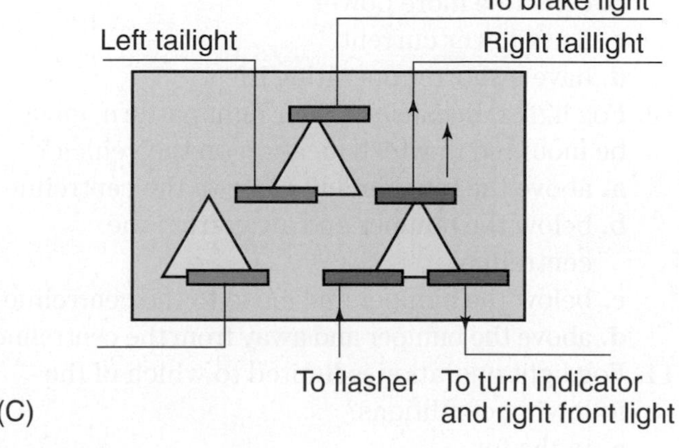

FIGURE 21–57

A turn-signal switch (A) not in use, (B) with a left turn indicated, and (C) with a right turn indicated.

Ford Motor Company

middle row connections are for rear combination lights (combination brake/turn signal). The bottom row of connections is for the front lights, including the dash indicators, and the B+ coming from the flasher. The triangles drawn over the bars are a set of three movable conductive pads that connect the different bars together depending on the position of the switch. They are drawn in the no turn or neutral position. This allows B+ from the brake switch to activate both rear lights at the same time. **Figure 21–57B** shows the same switch in a left turn. Notice that the conductive pads or triangles have moved to the right. This allows the brake switch to power only the right taillight, while the flasher connection is now in contact with the left rear taillight and the left front/indicator lights. The right taillight is being operated as a brake light, while the left one is in a turn-signal operation. **Figure 21–57C** shows the same switch in a right turn mode. Notice that the conductive pads have moved to the left and have connected the brake switch to the left taillight, while the right is now powered off the turn-signal flasher. This style of switch is very popular and normally is very durable. The most common problem encountered with the switch is usually mechanical rather than electrical. As the vehicle is driven around the corner, the cancelling mechanism must put the switch back into a neutral position so that both taillights can be used for brake warning. When this cancelling does not take place, the turn-signal switch is normally replaced to correct the problem.

KEY TERMS

Adaptive headlight systems (AHSs) (p. 620)
Candlepower (p. 633)
Centre high-mounted stoplight (CHMSL) (p. 634)
Composite headlights (p. 610)
Daytime running lights (DRLs) (p. 613)
Dimmer switch (p. 616)
Flashers (p. 633)
Flash to pass (p. 617)
Halogen (p. 607)
High-intensity discharge (HID) lamps (p. 607)
Incandescent lamp (p. 606)
Lamp (p. 606)
Light-emitting diode (LED) (p. 608)
Multifunction switch (p. 631)
Neon (p. 607)
Photocell sensor/amplifier (p. 618)
Sealed beam (p. 609)
Xenon headlamps (p. 610)

SUMMARY

- The headlight switch controls the headlights and all other light systems, with the exception of the turn signals, hazard warning, and stoplights.
- Dimmer switches permit the headlights to change from high to low beam and vice versa.
- An automatic headlight dimmer circuit switches the headlights from high to low beam in response to either light from an approaching vehicle or light from the taillights of a vehicle being overtaken.

- Depending on the make or model of the vehicle, courtesy lights can be found on the door, under the seats, under the instrument panel, on the rear interior quarter panels, or on the ceiling.
- The rear light assembly includes the taillights, turn-signal/stop/hazard lights, high-mounted stoplight, rear side marker lights, backup lights, and licence plate lights. Taillights operate when the parking lights or headlights are on.
- Headlights must be kept in adjustment to obtain maximum illumination and vehicle occupant safety.
- Flashers are used in turn-signal, hazard warning, and side marker light circuits.
- The backup light system illuminates the area behind the vehicle when the transmission is put in reverse gear.
- Headlight systems consist of two or four sealed-beam halogen headlight bulbs or separate bulb and reflector assemblies.

REVIEW QUESTIONS

1. The time it takes for a light to reach full brightness can result in a delay in warning approaching drivers. How long does it take for a conventional tungsten filament bulb to reach full brightness?

a. 2 ms **b.** 20 ms
c. 200 ms **d.** 2000 ms

2. Which of the following is a feature of composite headlights, used on modern cars?

a. removable lenses
b. sealed lenses and bulbs
c. removable bulbs
d. removable reflectors

3. What type of gas is commonly used in an HID style of headlight?

a. halogen **b.** xenon
c. argon **d.** helium

4. What voltage is required to initially start an HID headlight?

a. 12 volts
b. 80 volts
c. 1000 volts
d. 15 000 volts

5. Why do HID headlights appear blue in colour?

a. They are constructed using tinted glass.
b. They produce light closest to natural light.
c. The burning gas gives off a blue light.
d. Blue is the easiest colour to see.

6. What is the most likely problem if the interior lights don't operate when the door is opened, but may be turned on with the interior light switch?

a. a defective interior light switch
b. a door switch stuck in the open position
c. a door switch stuck in the closed position
d. a blown fuse

7. What circuit protection device enables the headlights to continue to operate, even after a temporary short circuit?

a. mini fuse
b. maxi fuse
c. circuit breaker
d. flasher

8. Many vehicles use LED tail and brake lights. One reason for this is the short time that it takes for the light to turn on. How much time does it take for an LED to illuminate after it is energized?

a. 1000 ms **b.** 100 ms
c. 10 ms **d.** 1 ms

9. Which of the following characteristics of LED lights makes them popular on today's vehicles?

a. generate more heat
b. consume more power
c. use a lower current
d. have a shorter operating life

10. Fog lights, because of their light pattern, must be mounted in which location on the vehicle?

a. above the bumper and close to the centreline
b. below the bumper and away from the centreline
c. below the bumper and close to the centreline
d. above the bumper and away from the centreline

11. Fog light operation is limited to which of the following conditions?

a. in the fog
b. with the high beams on
c. with the low beams on
d. only on the highway

12. What is the most common method of controlling daytime running lights, which are mandatory on all new Canadian automobiles?

a. by applying reduced voltage to the headlamps
b. by increasing the duty-cycled current to the headlamps
c. by applying increased voltage to the high-beam lamps
d. by applying increased voltage to the low-beam lamps

13. What is the name for the feature that illuminates the high-beam headlights even when the headlight switch is off or in low beam?
    a. adaptive switching
    b. flash to pass
    c. retractable dimming
    d. bidirectional switching
14. What type of switch is used in a simple turn-signal flasher?
    a. xenon
    b. plasma
    c. bi-metal
    d. semiconductive
15. Which of the following conditions are not recommended before headlight aiming is performed?
    a. tire inflation checked and adjusted
    b. the suspension system checked
    c. the fuel tank filled
    d. the vehicle cleaned of excess mud, snow, or ice
16. On some vehicles, headlights are constantly adjusted electronically to compensate for changing road and ride height conditions. What type of headlights use this type of compensation?
    a. halogen type **b.** HID type
    c. tungsten type **d.** composite type
17. Auxiliary lamps that are added to a lighting circuit must include what electrical device to protect the switch and wiring?
    a. a coil
    b. a diode
    c. a relay
    d. a capacitor
18. A technician tests for voltage on a light bulb that does not work. Battery voltage is recorded on both the feed and ground sides of the bulb. What is the most likely problem?
    a. blown fuse or circuit breaker
    b. bulb filament burnt open
    c. resistance on the feed side
    d. open on the ground side
19. Automatic dimming headlights use what electronic device to activate the dimming circuit?
    a. a photo diode
    b. a light-emitting diode
    c. a clamping diode
    d. a zener diode
20. What is an important precaution when replacing halogen-style headlight bulbs?
    a. Don't turn the headlamp circuit off.
    b. Don't touch the glass part of the bulb.
    c. Don't toggle the dimmer off and on.
    d. Don't turn on the circuit without the bulb installed.

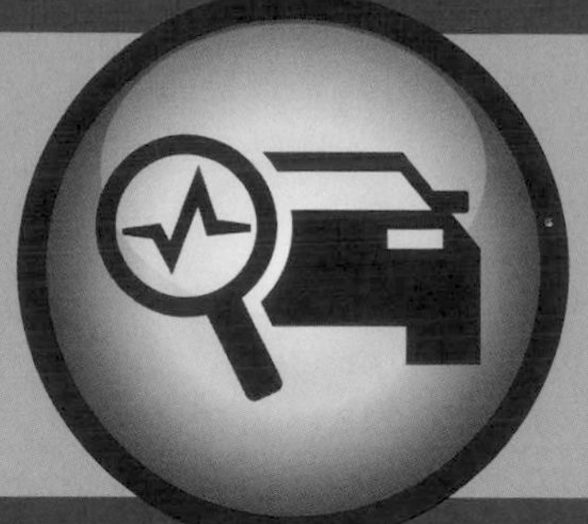

CHAPTER 22
Instrumentation and Information Displays

LEARNING OUTCOMES

- Describe the two types of instrument panel displays.
- Know the purpose of the various gauges used in today's vehicles and how they function.
- Describe the operation of the common types of gauges found in an instrument cluster.
- List and explain the function of the various indicators found on today's vehicles.
- List and explain the function of the various warning devices found on today's vehicles.
- Explain the basics for diagnosing a gauge or warning circuit.

Every automobile is equipped with a number of electrical instruments. The number and types of these systems and components vary significantly from vehicle model to vehicle model and year to year. The appearance of the gauge layout also varies from the quite simple **(Figure 22–1)** to the elaborate **(Figure 22–2)**. No matter what it looks like, a vehicle's instrumentation must be easy to read and give accurate information. Instrument gauges, lights, and warning indicators provide valuable information to the driver concerning a vehicle's various systems.

FIGURE 22–1

A simple but functional approach to the layout of the essential gauges in an instrument panel.

BMW of North America, Inc.

FIGURE 22–2

An elaborate arrangement of the essential gauges in an instrument panel.

Reprinted with permission.

INSTRUMENT PANELS

An **instrument panel** is a design element for the interior of a vehicle. It is an assembly that expands across the interior width of the vehicle and contains a glove compartment, air ducts and registers, air bags, various driver information displays, and entertainment systems. The shape, texture, and appearance of the panel are carefully considered by the manufacturer. It also contains an array of gauges, indicators, and controls connected to hidden mazes of wiring, printed circuitry, and vacuum hoses.

Displays

There are many different instrument panel designs and layouts. The two basic types of instrument panel displays are analog and digital. In a traditional

FIGURE 22–3

An analog instrument panel.

Reprinted with permission.

analog display **(Figure 22–3)**, an indicator moves in front of a fixed scale to indicate a condition. The indicator is often a needle, but it can also be a liquid crystal or graphic display. A digital display uses numbers instead of a needle or graphic symbol. Analog displays show relative change better than digital displays **(Figure 22–4)**. They are useful when the driver must see something quickly and the exact reading is not important. For example, an analog tachometer shows the rise and fall of the engine speed better than a digital display. The driver does not need to know the exact engine speed. The most important thing is how fast the engine is reaching the red line on the gauge. A digital display is better for showing exact data such as kilometres. Many speedometer-odometer combinations are examples of both analog (speed) and digital (distance).

FIGURE 22–4

A digital instrument panel.

Siemens VDO Automotive

Three types of digital electronic displays are used today.

VACUUM FLUORESCENT These displays use glass tubes filled with argon or neon gas. The segments of the display are little fluorescent lights. When current is passed through the tubes, they glow very brightly. These displays are both durable and bright.

Heads-Up Display

A **heads-up display (HUD)** projects vehicle information onto the windshield, using a vacuum fluorescent light source via a dash-mounted prismatic mirror to project the images and to complement in-dash instrumentation **(Figure 22–5)**. Because these images are projected in the driver's field of vision, the driver does not need to move his or her eyes from the road to see certain pertinent information.

HUD systems may have a central control that turns the display on and off, plus a brightness control. The brightness control is important to the comfort and sight of the driver. If the display is too bright, the road ahead can be distorted. If the image is too dim, the driver may focus too much on the windshield.

Among the images HUD may display are vehicle speed, turn-signal indicators, low-fuel warning, and a high-beam indicator. HUD systems work best with a clean windshield and dim ambient lighting.

SHOP TALK

Most warning or instrument controls use **International Standards Organization (ISO)** symbols, which were developed by this organization to provide symbols easily recognizable throughout the world.

FIGURE 22–5

The HUD system projects information on the inside of the windshield.

©Cengage Learning 2015

LIGHT-EMITTING DIODES (LEDs) These displays are used as either single-indicator lights, or they can be grouped to show a set of letters or numbers. LED displays are commonly red, yellow, or green and can be hard to see in bright light.

LIQUID-CRYSTAL DISPLAYS (LCDs) These displays are made of sandwiches of special glass and liquid. A separate light source is required to make the display work. When there is no voltage, light cannot pass through the fluid. When voltage is applied, the light passes that point of the display. The action of LCDs slows down in cold weather. These displays are also very delicate and must be handled with care. Any rough handling of the display can damage it.

An LCD display is used to show the information provided by the vehicle's systems **(Figure 22–6)**. The LCD may be part of the instrument panel or in a separate unit, called the centre stack, that extends from the instrument panel into the seating area. Today, the centre stack is mostly used to display the entertainment and climate controls. The multi-information display (MID) has been available for quite some time. In the future, there may be two or more LCDs in the centre stack. An LCD screen can also allow the driver to see navigation instructions, information about the music currently being played, high-voltage battery status, and the transactions of a cellphone.

The LCD screen may also display the view from the rearview and other cameras and may become more common as more vehicles are equipped with rearview cameras and displays on the instrument panel or centre stack. However, the display of the cameras may also be embedded in the inside rearview mirror **(Figure 22–7)**.

FIGURE 22–6

An LCD display.

©Cengage Learning 2015

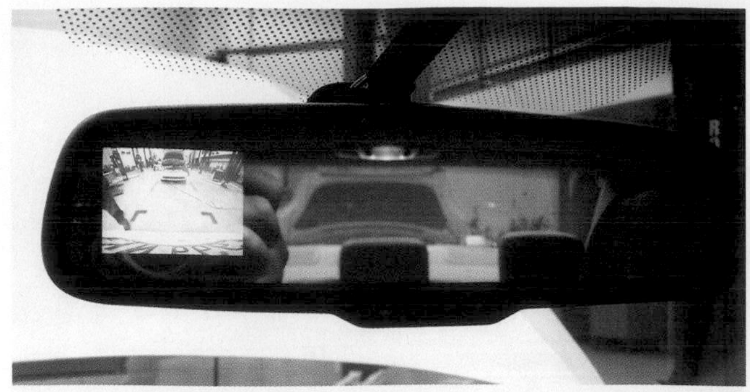

FIGURE 22–7

The backup camera's LCD display may be shown on the inside rearview mirror.

©Cengage Learning 2015

A full digital LCD instrument panel can allow the driver to switch between a digital and analog speedometer or display both at the same time. It can even allow the driver to increase the font size on the screen.

CADILLAC CUE Cadillac's User Experience (CUE) infotainment package has an 8-inch LCD touch screen **(Figure 22–8)**, integrated into the top of the instrument panel or centre console. The screen displays CUE's home page, which resembles a smartphone screen by using large, easy-to-target icons to execute commands. The touchscreen is much like a tablet and has icons that are used to execute

FIGURE 22–8

A large LCD touchscreen display in the centre of the console of this Cadillac.

©Cengage Learning 2015

FIGURE 22–9

An LCD display with an analog speedometer.

commands. Interactive motions, such as tap, flick, swipe, and spread, are functional on the screen.

FORD'S SMARTGAUGE This LCD system is primarily found in Ford's hybrids **(Figure 22–9)**. It shows the driver how well he or she is driving economically. The system has a pair of 4.5-inch LCDs on each side of the speedometer. If the driver is conserving a great amount of fuel, a collection of green leaves appears on the screen. The driver can also customize what the gauges show, including a navigation screen, phone info, infotainment information, or efficiency.

GAUGES

Gauges provide the driver with a scaled indication of the condition of a system. All gauges require an input from either a sending unit, sensor, or switch. With sending or sensor units, a change or movement made by an external component causes a change in electrical resistance. Movement may be caused by pressure against a diaphragm, by heat, or by the motion of a float. Some engines use a switch as an input for a gauge. An example is an oil pressure switch. With the engine off or operating with low oil pressure, the switch is open and the gauge reads low. Once oil pressure reaches a certain pressure, the switch closes and the gauge needle reads normal. In today's vehicles, control computers and the gauges need the same information. The information passes through the computer and then to the gauge.

Some earlier-model vehicles have an **instrument voltage regulator (IVR) (Figure 22–10)** to stabilize and limit voltage to the gauges. The use of an IVR provides more accurate gauge readings. When a vehicle has an IVR, it also has a radio choke. A radio choke is a small coil of wire installed in the battery lead to the IVR. The choke prevents radio interference caused by the pulsations of the IVR.

Air-Core Gauge

Today, the most commonly used gauge is the **air-core gauge**. This design relies on the interaction of two electromagnets and the effect of the strength of those fields on a permanent magnet **(Figure 22–11)**. The needle of the gauge is attached to the permanent magnet. The electromagnetic windings do not have a metal core; rather, the permanent magnet is placed inside the windings. One of these

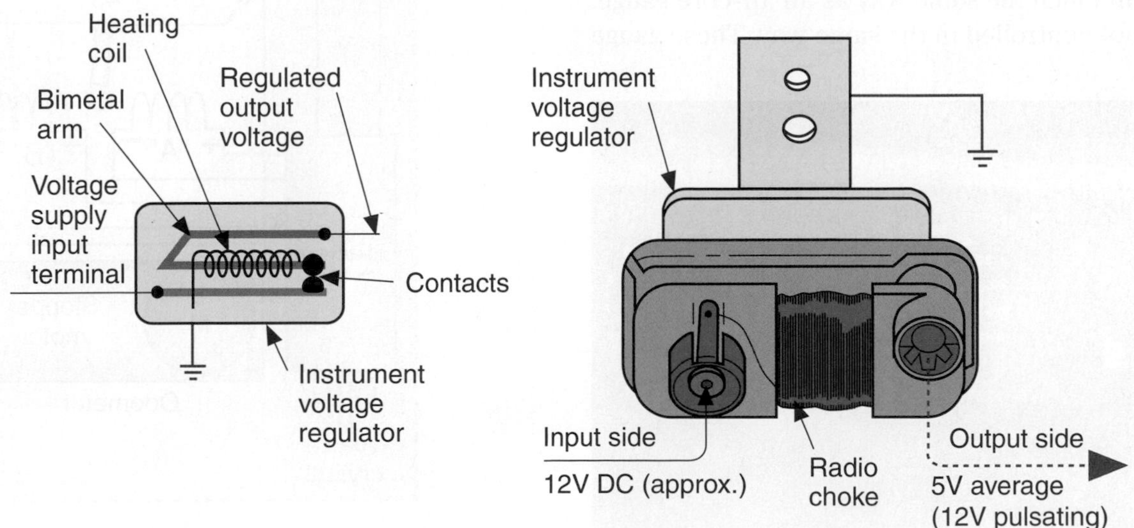

FIGURE 22–10

An instrument voltage regulator (IVR).

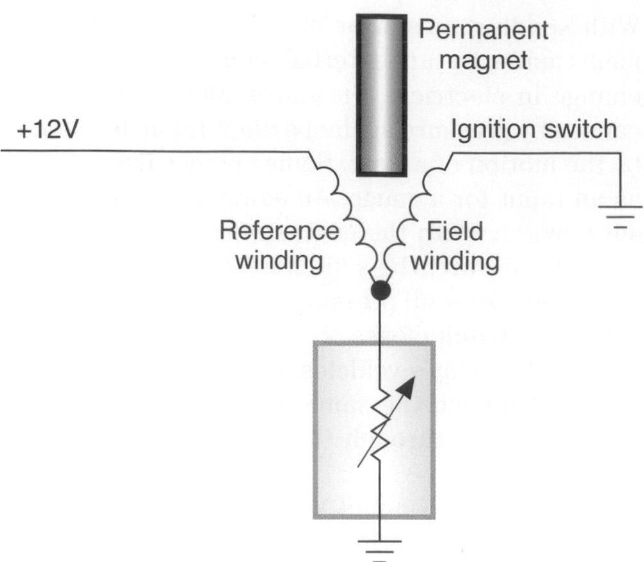

FIGURE 22–11

An air-core fuel gauge circuit.

windings is a reference winding and receives battery voltage at all times. The other winding, called the field winding, is placed at an angle to the reference winding. When **current** flows through the coils, their **magnetic fields** superimpose, and the magnet is free to align with the combined fields. The amount of movement is controlled by the resistance of the sending unit.

Quartz Analog Gauge

Computer-controlled quartz analog gauges **(Figure 22–12)** are becoming increasingly common. They operate in much the same way as an air-core gauge, but are not controlled in the same way. These gauge circuits rely on a permanent magnet generator sensor installed in the part or system that will be monitored. As the sensor rotates, a small AC signal is produced. This signal is then sent to a buffer circuit where it is changed to a digitalized signal. That signal is then sent to a quartz clock circuit and a driver circuit. The driver circuit sends voltage pulses to the windings in the gauge. The permanent magnet and the windings cause the needle to move. These gauges display very accurate readings.

The needle may be moved by a stepper motor **(Figure 22–13)**. A stepper motor is simply a motor that can rotate in small steps in response to digital signals from a computer. A stepper motor uses a

FIGURE 22–12

Quartz analog instrumentation.

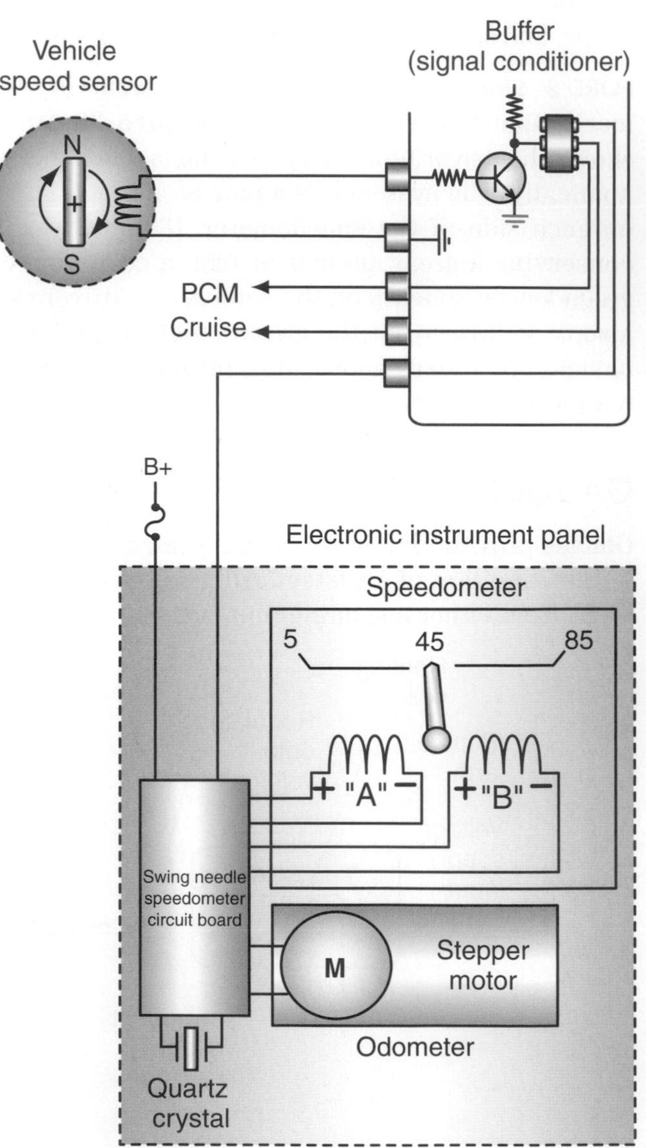

FIGURE 22–13

A schematic of a quartz analog gauge with a stepper motor controlling the position of the needle.

permanent magnet and two electromagnets. The electromagnets are controlled by the computer, which pulses the windings and changes the polarity of the windings to cause the motor's armature to rotate 90° at a time. Each signal pulse is recognized by the computer as a count or step. The computer can, therefore, know the position of the motor's armature by counting the number of steps that have been sent to the motor. These motors are widely used in GM gauges.

Magnetic Gauges

There are several types of magnetic gauges. The simplest form is the ammeter type **(Figure 22–14)**, in which a permanent magnet attracts a ferrous indicator needle connected to a pivot point and holds it centred on the gauge. An armature, or coil of wire, is wrapped around the base of the needle near the pivot point. When current flows through the armature, a magnetic field is formed. This magnetic field opposes that of the permanent magnet. Attractive or repulsive magnetic forces cause the needle to swing left or right. The direction the needle swings depends on the direction of current flow in the armature. The needle of the gauge pivots on the armature, often called a bobbin. This type of gauge is often referred to as a **D'Arsonval gauge**.

When sending-unit resistance changes, current flow through the bobbin changes, causing the strength of the magnetic field created around the bobbin to change. As the resistance in the sending unit increases, the circuit's current decreases, and a lower indicator position is shown on the gauge.

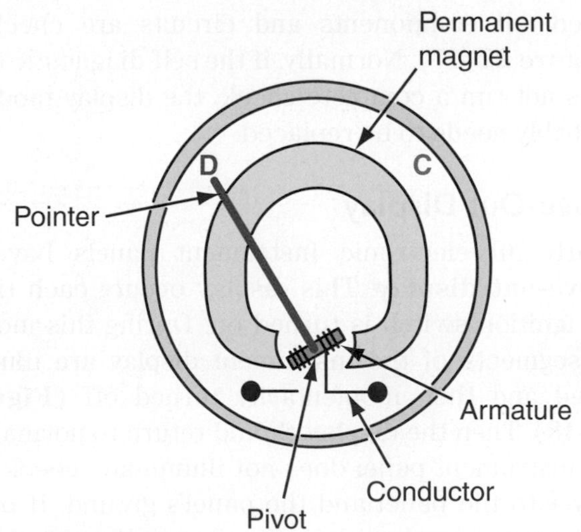

FIGURE 22–14

A simple ammeter that relies on magnetic principles.

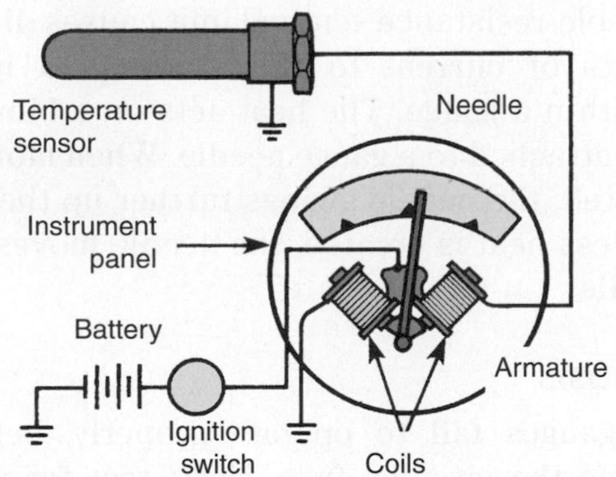

FIGURE 22–15

A balancing coil gauge.

Ford Motor Company

A **balancing coil gauge** also operates on principles of magnetism. However, a permanent magnet is not used. The base of the indicating arm is pivoted on an armature. Two or three (two plus one) coils are set on either side of the armature. **(Figure 22–15)**.

When the same amount of current flows through both coils, the needle sets between the two. When the resistance of the sending unit is low, the right-hand coil receives more current than the left-hand coil, and the needle moves to the right. When the resistance of the sending unit is high, the left-hand coil receives more current. More magnetic force is created in the left-hand coil, and the needle swings to the left.

Thermal or Bimetallic Gauge

Thermal gauges or **bimetallic gauges** are not used on modern cars but were commonly used on older vehicles. These gauges operate through heat created by current flow **(Figure 22–16)**.

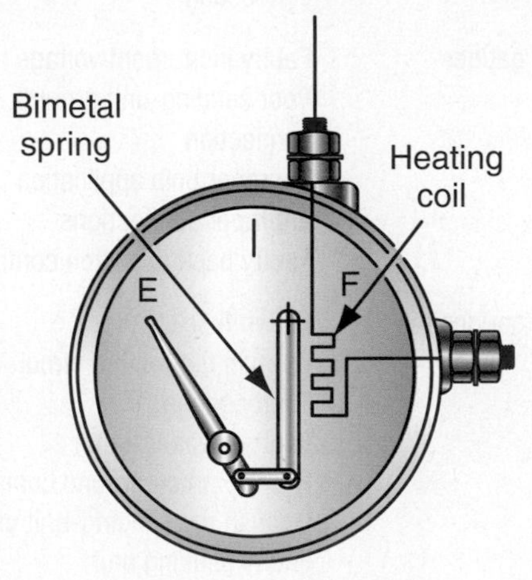

FIGURE 22–16

A bimetallic (thermal) gauge.

A variable-resistance sending unit causes different amounts of current to flow through a heating coil within a gauge. The heat acts on a bimetallic spring attached to a gauge needle. When more heat is created, the needle swings further up the scale. When less heat is created, the needle moves down the scale.

Diagnosis

If all gauges fail to operate properly, begin by checking the circuit's fuse. Next, test for voltage at the last point common to all the malfunctioning gauges. If voltage is not present, work toward the battery to find the fault. If only one gauge is not working, carefully check its circuit. **Table 22–1** lists basic gauge problems along with their possible causes.

TABLE 22–1 BASIC GAUGE PROBLEMS AND THEIR POSSIBLE CAUSES

SYMPTOMS	POSSIBLE CAUSES
One or all gauges fluctuate from low or high to normal readings.	Poor ground connection Excessive resistance Poor connections Faulty stepper motor Faulty sending unit Defective printed circuit
One or all gauges read high.	Faulty instrument voltage regulator Shorted printed circuit Faulty sending unit Short to ground in sending-unit circuit Faulty gauge Poor sending-unit ground connection
One or all gauges read low.	Faulty instrument voltage regulator Poor sending-unit ground connection Improper bulb application Improper connections Faulty backup switch contacts
One or all gauges fail to read.	Blown fuse Open in the printed circuit Faulty gauge Faulty stepper motor Poor common ground connection Open in the sending-unit circuit Faulty sending unit Faulty instrument voltage regulator Poor electrical connection to cluster Shorted sending-unit circuit

WARNING!

Many instruments and warning devices are linked to the vehicle's body control module and multiplexing network. Before troubleshooting a gauge or warning system, check the service manual or information system to identify any special procedures or precautions.

ELECTRONIC INSTRUMENT CLUSTERS

Many of the displays on an instrument panel are controlled by the same sensors used to control the vehicle's powertrain and other systems. The signals from the sensors are part the multiplex system, in which the inputs are used by many systems. The routing of the input signals is handled by the body control module (BCM).

The BCM may also be capable of running diagnostic checks on the instrument panel. If any of the input or output signals are out of the normal range, the BCM will set a diagnostic trouble code (DTC). The code is typically retrieved with a scan tool.

Most instrument-panel display modules have a self-diagnostic mode. This mode allows the module to isolate faults in the cluster. When this test is initiated, all components and circuits are checked **(Figure 22–17)**. Normally, if the self-diagnostic test does not run a complete check, the display module probably needs to be replaced.

Prove-Out Display

Nearly all electronic instrument panels have a **prove-out display**. This display occurs each time the ignition switch is turned on. During this mode, all segments of the instrument display are illuminated and then momentarily turned off **(Figure 22–18)**. Then the display should return to normal. If the instrument panel does not illuminate, check for power to the panel and the panel's ground. If both of these are good, replace the cluster. If some of the segments or parts of the display do not illuminate, replace the cluster.

Gauge	Display	Condition	Display	Condition
Oil gauge	H Norm L	Oil pressure sensor input short circuited light top 2 bars and bottom 2 bars and extinguishes oil can ISO symbol	H Norm L	Low oil pressure warning or oil pressure sensor input open circuited lights bottom bar and flashes ISO symbol
Temp gauge	H Norm C	Engine temperature sensor input short circuited lights top 2 bars and bottom 2 bars and extinguishes ISO symbol	H Norm C	Cold engine temperature indication or engine temperature sensor input open circuited lights bottom bar and ISO symbol
Fuel gauge	H Norm L	Fuel level sender input short circuited or open lights top 2 and bottom 2 bars and extinguishes ISO symbol	CO CS	Fuel level sender input short or open circuited displays "CS" (short) or "CO" (open) in driver information centre
Odometer	55 ERROR			Odometer malfunction displays ERROR in odometer display
Fuel computer	FFS			Fuel flow signal short or open circuited displays FFS in driver information centre

FIGURE 22–17

When the system is in its diagnostic mode, a gauge readout is displayed to show the condition of the gauges.

©Cengage Learning 2015

FIGURE 22–18

All segments of the electronic instrument cluster should illuminate during the prove-out display.

©Cengage Learning 2015

BASIC INFORMATION GAUGES

The following gauges are found on nearly all instrument panels. **Table 22–2** lists some of the common instrument-panel problems and their possible causes.

General Diagnosis and Testing

Diagnosis should begin with a good visual inspection of the circuit. Check all sensors and actuators for physical damage. Check all connections to sensors, actuators, control modules, and ground points. Check the wiring for signs of burned or chafed spots, pinched wires, or contact with sharp edges or hot exhaust parts. After completing the visual checks, use a scan tool to retrieve any DTCs **(Figure 22–19)**, and monitor the data stream as necessary. Many systems provide for testing of the operation of instrument panel gauges and warning lights with a scan tool. Always refer to the manufacturer's recommended procedure before beginning to diagnose a circuit.

Speedometer

In the past, the speedometer was a nonelectrical or mechanical gauge. It had a drive cable attached to a gear in the transmission that turned a magnet inside a cup-shaped metal piece at the gauge.

TABLE 22–2 COMMON INSTRUMENT-PANEL PROBLEMS AND THEIR POSSIBLE CAUSES

SYMPTOMS	POSSIBLE CAUSES
Digital display does not light.	Blown fuse Inoperative power and ground circuit Faulty instrument panel
Speedometer reads wrong speed.	Faulty speedometer Wrong gear on vehicle speed sensor (VSS) Wrong tire size
Speedometer always reads zero.	Faulty wiring Inoperative instrument panel
Odometer displays error.	Inoperative odometer memory module in instrument panel
Fuel gauge display is erratic.	Sticky or inoperative fuel gauge sender Faulty in circuit Inoperative fuel gauge
Fuel gauge will not display FULL or EMPTY.	Sticky or inoperative fuel gauge sender
Fuel gauge displays top and bottom two bars.	Open or short in circuit
Fuel computer displays CS or CO.	Open or short in fuel gauge sender Inoperative instrument panel
Fuel economy function of message centre is erratic or inoperative.	Inoperative fuel flow signal Faulty wiring Inoperative instrument panel
Extra or missing display segments.	Inoperative instrument panel
Temperature gauge displays top and bottom two bars.	Short in circuit Inoperative coolant temperature sender Inoperative instrument panel

DTC	DESCRIPTION
B1500	Open detected in the fuel level sender circuit
U0100	Lost communication with ECM/PCM
U0142	Lost communication with main BCM

FIGURE 22–19

Examples of instrument panel DTCs.

The cup was attached to a needle, which was held at zero by a hairspring (a fine wire spring). As the cable rotated faster with an increase in speed, magnetic forces acted on the cup and forced it to move. As a result, the needle moved up the speed scale.

Electric speedometers are used in nearly all late-model vehicles. Although there are several systems in use, one of the most common types receives its speed information from the transmission-mounted vehicle speed sensor (VSS) **(Figure 22–20)**. The speed sensor can be a permanent magnet (PM) generator, Hall-effect switch, or optical sensor. The VSS monitors the speed of the transmission's output shaft. The output signal from the sensor varies in AC frequency and amplitude, according to speed. The signal is sent to the computer control system where it is interpreted and used to control the reading on the speedometer. The speed signal is also used by other modules in the vehicle, including the speed control module, ride control module, engine control module, and others.

The number of pulses produced by the VSS determines the distance the vehicle has travelled. The frequency of the signal (in hertz [Hz]) determines vehicle speed in kilometres per hour (km/h). To display speed on an analog gauge, the circuit electronics are calibrated to drive the pointer to a location representing the calculated speed. As vehicle speed increases, the frequency increases. The display may be limited to 200 km/h (125 mph) or less.

Most digital speedometers have a speed limit. If vehicle speed exceeds these values, the speedometer continues to display the top of its range. Vehicle speed is displayed whether the vehicle is moving forward or backward.

Odometer

The **odometer** is a digital gauge that is usually driven by a spiral gear cut on the speedometer's magnet shaft. The odometer's numbered drums are geared so that when any one drum finishes a complete revolution, the drum to the left is turned one-tenth of a revolution.

An electronic odometer receives its information from the VSS. On some systems, a stepper motor is used to drive the odometer **(Figure 22–21)**. The motor receives signals from the powertrain control module (PCM) and causes the odometer to move in steps.

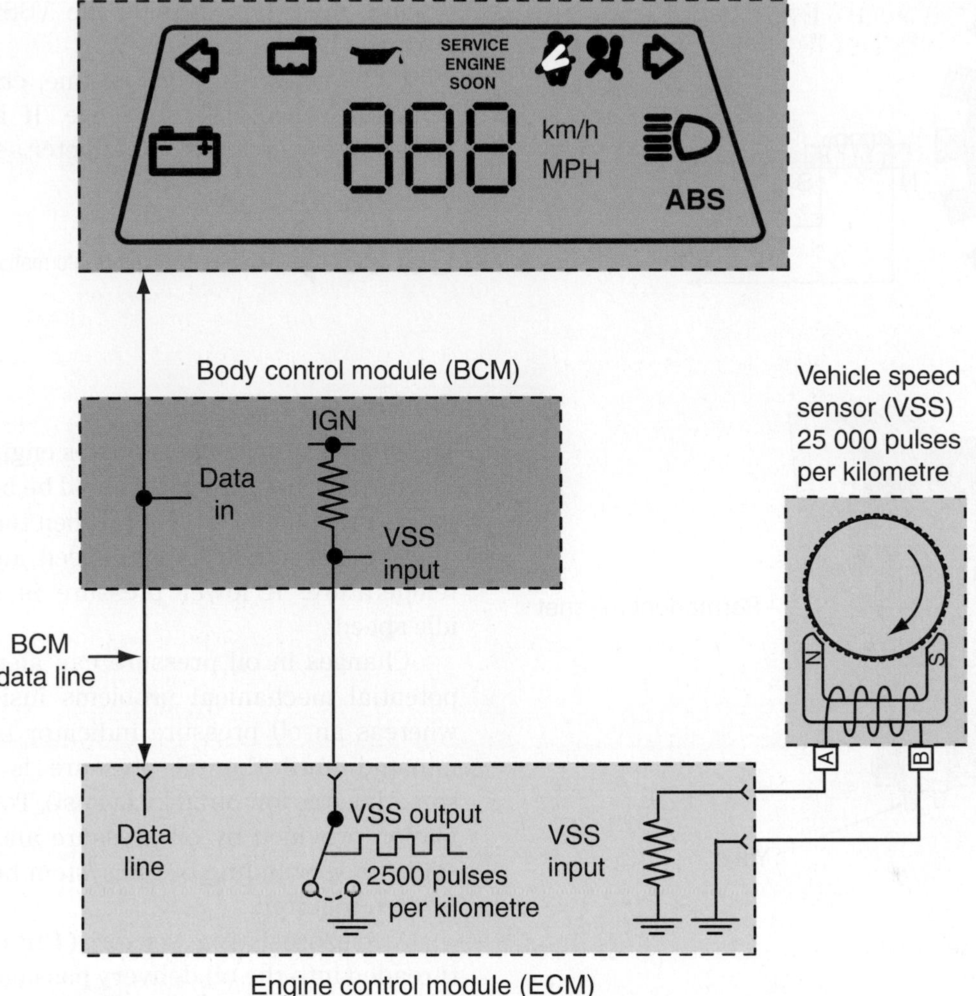

FIGURE 22–20

The signals from the VSS are shared by the instrument panel and the PCM.

Odometers display seven digits, with the last digit in tenths of a unit. The accumulated distance value of the digital display odometer is stored in a nonvolatile read-only memory (ROM) that retains the mileage value even if the battery is disconnected.

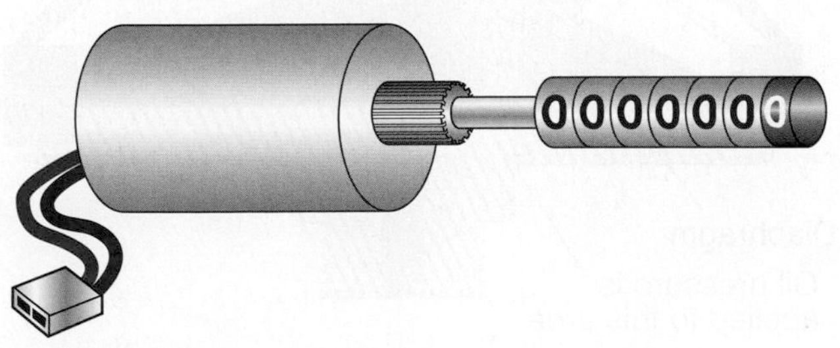

FIGURE 22–21

A stepper motor controls the movement of the odometer drum.

Because the odometer records the number of kilometres a vehicle has travelled, federal law requires that the odometer in any replacement speedometer register the same distance as that registered on the removed speedometer. Therefore, if a speedometer has been replaced, set the odometer of the new one to match the old.

With the use of electronic odometer displays, the odometer reading is often stored in more than one module. This allows for the correct distance to be displayed in the event the PCM or instrument cluster is replaced.

TRIP ODOMETER This is part of the normal odometer. It records the distance travelled in intervals.

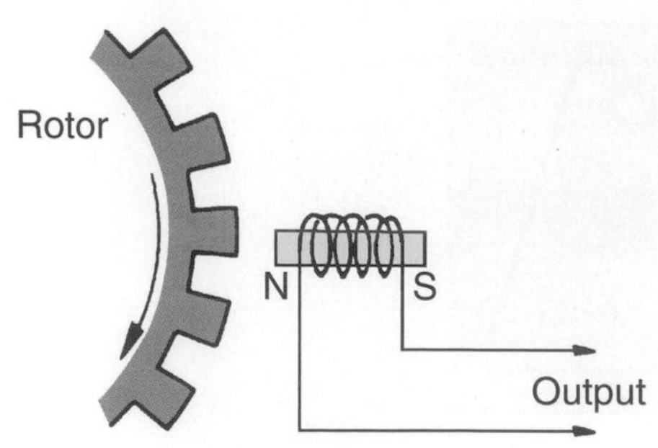

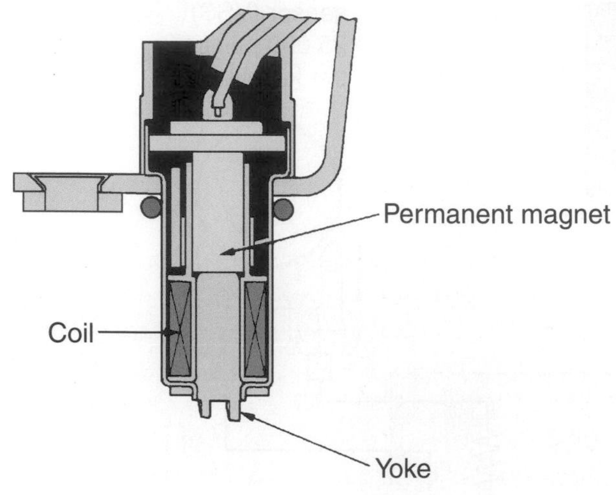

FIGURE 22–22

A vehicle speed sensor (VSS).

The driver activates this gauge, and it can be reset to zero whenever the driver desires.

DIAGNOSIS If the speedometer or odometer is not working, the operation of the VSS **(Figure 22–22)** should be checked before anything else. The VSS can be checked with a scan tool, lab scope, and DMM. A bad VSS may also cause erratic speedometer readings.

To check the VSS, remove the instrument cluster to gain access to the electrical connections. Connect a voltmeter across the appropriate terminals (refer to the service information). Turn the ignition on, and check the reading on the voltmeter. If there are zero volts, check the power source. If voltage is present, check the VSS.

The VSS can be checked in many different ways. One way is to connect a voltmeter to the circuit and rotate the VSS with a small screwdriver. If the voltmeter reading changes, there may be a problem in the wiring of the circuit. If the voltmeter reading does not change, the VSS or associated wiring is bad.

If the sensor is working fine, check the wiring from the sensor to the gauge. If all checks out, replace the speedometer/odometer assembly.

GO TO Chapter 32 for more information on checking a VSS.

Oil Pressure Gauge

The oil pressure gauge indicates engine oil pressure. The oil pressure typically should be between 310 kPa (48 psi) and 480 kPa (70 psi) when the engine is running at a specified engine speed and at operating temperature. A lower pressure is normal at low idle speed.

Changes in oil pressure can alert the driver of potential mechanical problems inside the engine, whereas an oil pressure indicator light is not illuminated until the oil pressure is critically low, sometimes as low as 30 kPa (4 psi). To interpret information provided by oil pressure and other gauges, some understanding of the system being monitored is often required.

A piezoresistive sensor **(Figure 22–23)** is threaded into the oil delivery passage of the engine. The pressure of the oil causes a flexible diaphragm to move. This movement is transferred to a contact arm that slides down the resistor. The position of

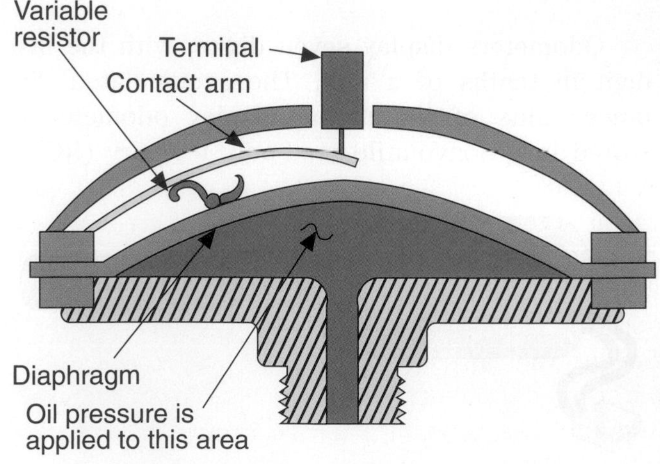

FIGURE 22–23

A piezoresistive sensor used for measuring engine oil pressure.

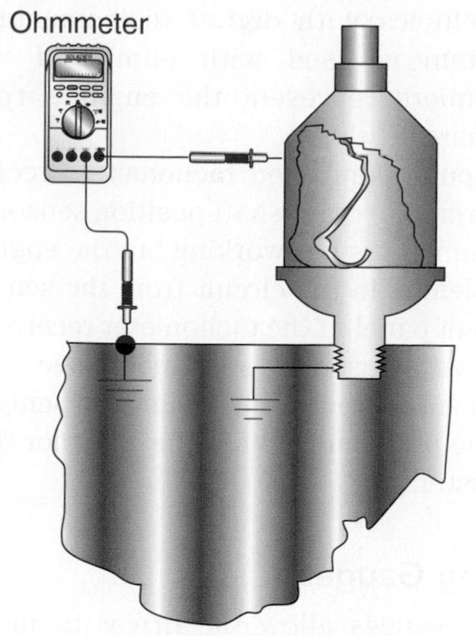

FIGURE 22–24

Using an ohmmeter to test a piezoresistive sensor.

the sliding contact arm determines the resistance value and the amount of current flow through the gauge.

DIAGNOSIS To test a piezoresistive-type sending unit, connect an ohmmeter to the sending unit's terminal and to ground **(Figure 22–24)**. Check the resistance with the engine off, and compare it to specifications. Start the engine and allow it to idle. Check the resistance value, and compare it to specifications. Before replacing the sending unit, connect a shop oil pressure gauge to confirm that the engine is producing adequate oil pressure.

Coolant Temperature Gauge

This gauge indicates engine coolant temperature (ECT). It should normally indicate between C (cold) and H (hot). The sending unit is typically a variable resistor such as a thermistor. It regulates the current flow through the temperature gauge winding **(Figure 22–25)**. With low coolant temperature, sender resistance is high, and current flow is low. The needle points to C. As coolant temperature increases, sender resistance decreases, and current flow increases. The needle moves toward H.

On a digital panel, the temperature gauge is normally a bar type with a set number of segments. With low coolant temperature, few segments are lit. As

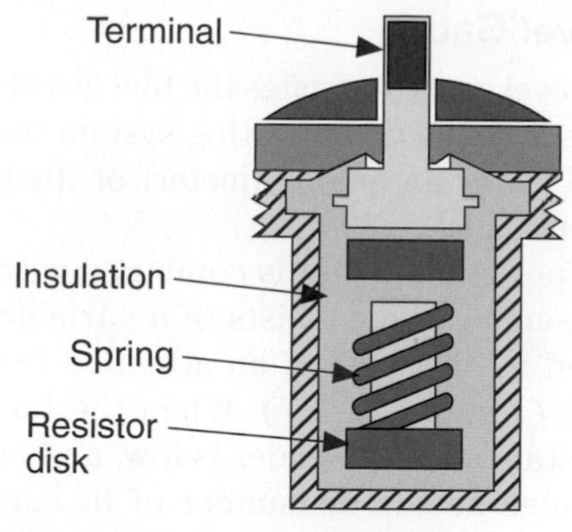

FIGURE 22–25

A typical temperature sending unit that does not use a thermistor.

coolant temperature increases, the number of illuminated segments increases.

Depending on the vehicle, there may be more than one coolant temperature sensor. One sensor may be dedicated to the coolant temperature gauge, and a second sensor is used by the PCM to monitor coolant temperature and for cooling fan operation. Before attempting to diagnose a temperature gauge concern, make sure you identify the correct sensor used by the gauge.

DIAGNOSIS To test a coolant temperature sending unit, use an ohmmeter to measure resistance between the terminal and ground **(Figure 22–26)**. The resistance value of the negative temperature coefficient (NTC) thermistor decreases with an increase in temperature. Compare your measurements with the manufacturer's specifications. On most new vehicles, the action of an ECT sensor can be monitored on a scan tool.

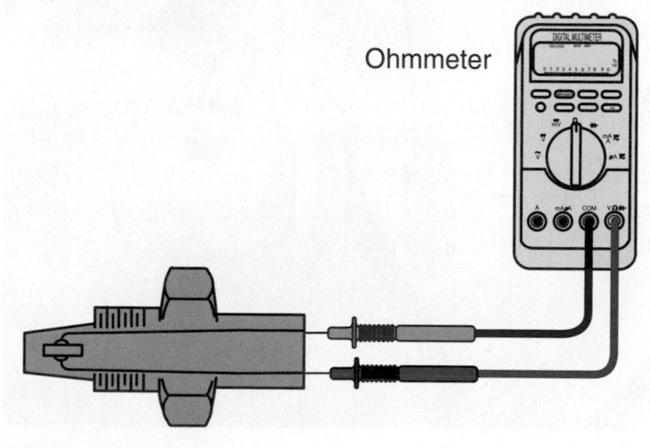

FIGURE 22–26

Testing a temperature sensor with an ohmmeter.

Fuel Level Gauge

The fuel level gauge indicates the fuel level in the fuel tank. It is a magnetic indicating system that can be found on either an analog (meter) or digital (bars) instrument panel.

The fuel sending unit is combined with the fuel pump assembly and consists of a variable resistor controlled by the level of an attached float in the fuel tank **(Figure 22–27)**. When the fuel level is low, resistance in the sender is low, and movement of the gauge needle or number of lit bars is minimal (from the empty position). When the fuel level is high, the resistance in the sender is high, and movement of the gauge indicator (from the empty position) or number of lit bars on a digital display is greater.

In some fuel gauge systems, an anti-slosh/**low-fuel warning (LFW)** module is used to reduce fuel gauge needle fluctuation caused by fuel motion in the tank and provide a low-fuel warning when the fuel tank reaches ⅛ to ¹⁄₁₆ full.

Photo Sequence 17 covers a typical procedure for bench-testing a fuel gauge sending unit.

Tachometer

The **tachometer** indicates engine rpm (engine speed). The electrical pulses to the tachometer typically come from the ignition module or PCM. The tachometer, using a balanced coil gauge, converts these impulses to rpm that can be read. The faster the engine rotates, the greater the number of impulses from the coil. Consequently, higher engine speeds are indicated.

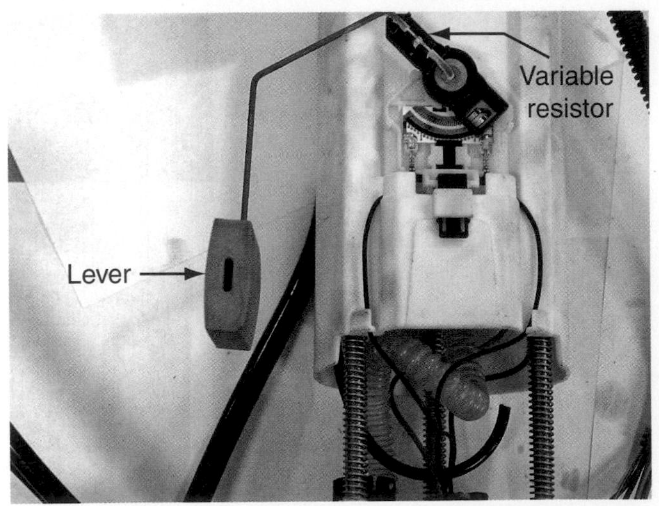

FIGURE 22–27

A fuel gauge sending unit.

In vehicles with digital instrumentation, the bar system is used with numbered segments. The numbers represent the engine's rpm times one thousand.

Computer-controlled tachometers receive their signals from the crankshaft position sensor (CKP). If the tachometer is not working but the engine starts, the problem is in the circuit from the sensor to the instrument panel. If the tachometer receives a signal from the data bus, use a scan tool to see if the signal is being sent correctly. If the signal is being sent correctly, the problem is either the gauge or the instrument cluster.

Charging Gauges

Charging gauges allow the driver to monitor the charging system. Although a few older cars use a voltmeter, most charging systems employ either an ammeter gauge or an indicator light. The ammeter gauge is placed in series with the battery and generator. When the generator is delivering current to the battery, the gauge displays a positive (+) indication. When the battery is not receiving enough current (or none at all) from the generator, a negative (–) display is obtained.

INDICATORS AND WARNING DEVICES

Light Indicators and Warnings

Indicator lights and warning devices are generally activated by the closing of a switch.

Most of these are safety or emissions related. These are controlled by sending units that act as simple on–off switches. The sending units can be normally open or normally closed switches. Warning lights notify the driver that something in the system is not functioning properly or that a situation exists that must be corrected. Some system problem may cause the system to emit an audible sound in addition to illuminating the lamp. Gauges and warning lights often work together to alert the driver to a problem.

An indicator light comes on to inform the driver that something has been turned on, such as the rear-window defogger. Warning lights notify the driver that something in the system is not functioning properly or that a situation exists that must be corrected. The function of some of the more common warning lights is described here.

BENCH-TESTING A FUEL GAUGE SENDING UNIT

PHOTO SEQUENCE 17

P17–1 The tools required to perform this task are a digital multimeter (DMM), jumper wires, and a service manual.

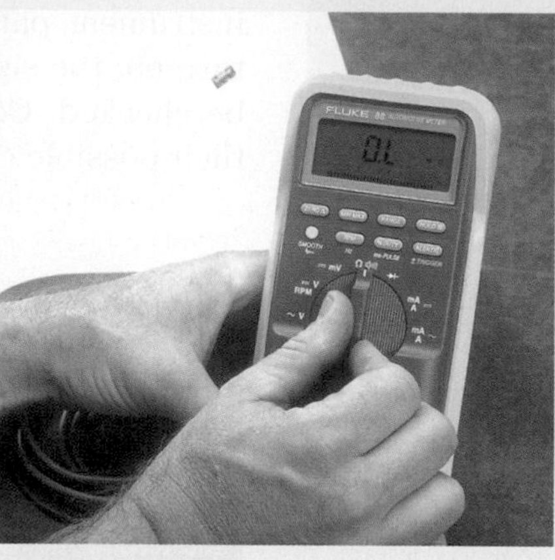

P17–2 Select the ohmmeter function of the DMM.

P17–3 Connect the DMM's negative test lead to the ground terminal of the sending unit.

P17–4 Connect the meter's positive lead to the variable resistor terminal.

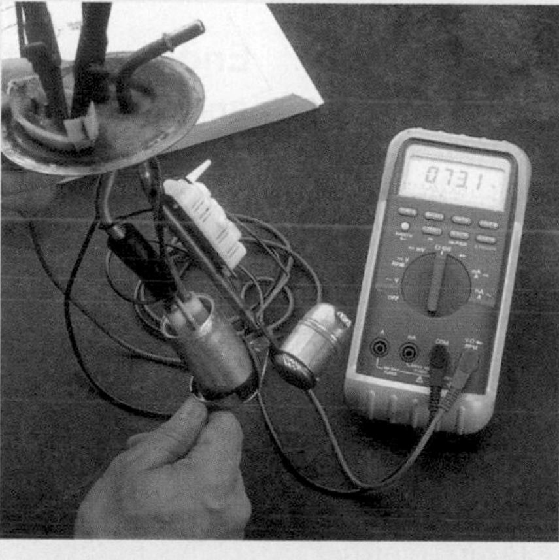

P17–5 Holding the sender unit in its normal position, place the float rod against the empty stop.

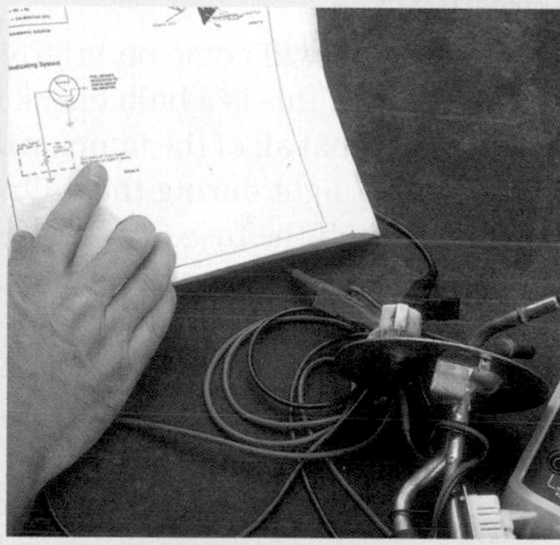

P17–6 Read the ohmmeter and check the results with the specifications.

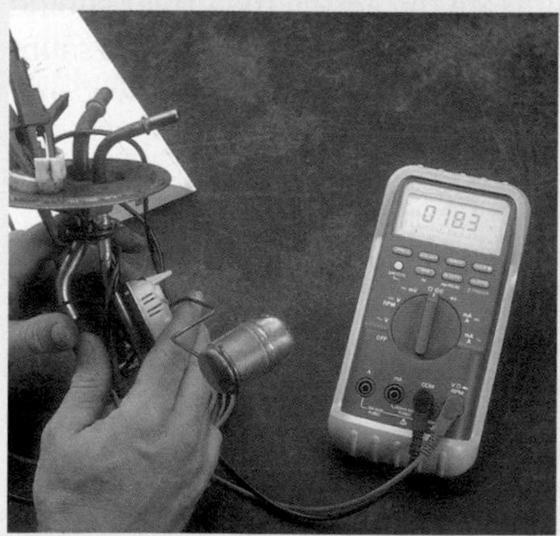

P17–7 Slowly move the float toward the full stop while observing the ohmmeter. The resistance change should be smooth and consistent.

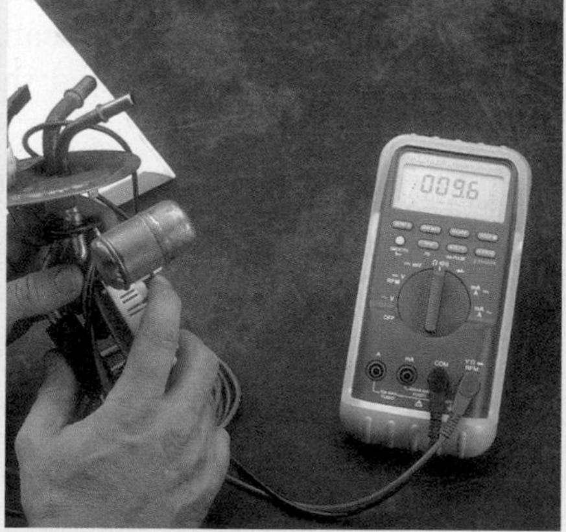

P17–8 Check the resistance value while holding the float against the full stop. Check the results with the specifications.

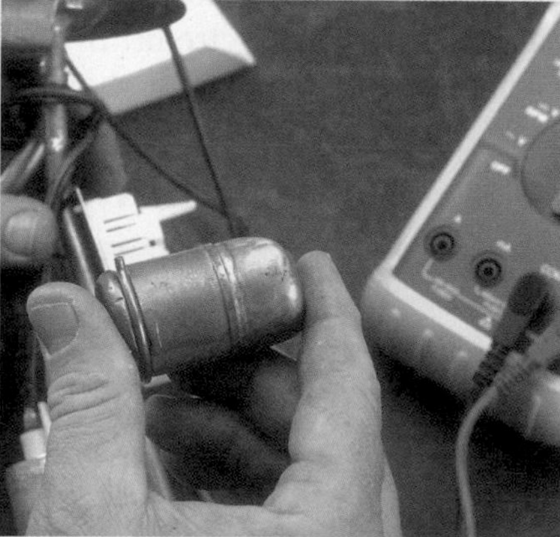

P17–9 Check the float to be sure it is not filled with fuel, distorted, or loose.

FIGURE 22–28

Most warning lights and indicators go through a bulb check when the ignition is first switched to the ON position.

Diagnosis

Warning lights will come on briefly when the engine is first started; this is a bulb check **(Figure 22–28)**. It is unlikely that all of the lamps will be bad, so if the lamps do not light during the bulb check, check the circuit's fuse. If the fuse is okay, check for voltage at the last common connection. If there is no voltage there, follow the circuit back to the battery, and check for voltage at all available points. If there was voltage at the common connection, test each circuit branch.

Unplugging the electrical connection to the sending unit should cause the warning lamp to either light or go out, depending on whether the sending unit is normally closed or open. The same is true if the disconnected wire is momentarily touched to a ground. If the lamp does not respond as it should, carefully check the circuit. If the lamp did respond normally, the sending unit should be replaced.

Computer-controlled warning lamps are diagnosed with a scan tool. The scan tool can be used to force a lamp to illuminate. If a lamp does not turn on when it is ordered to, the bulb, circuit board, or instrument panel module is bad. If the lamp does turn on, the signal to the instrument cluster should be checked. Common warning lamp problems and their possible causes are shown in **Table 22–3**.

WARNING!

Be aware that many of the warning and indicator lamps found on today's vehicles are triggered by a PCM or BCM. Often, they are part of the multiplexed system. With this in mind, always refer to the testing methods recommended by the manufacturer before testing these systems. Using conventional testing methods on a computerized system may destroy part or all of the system.

Engine-Related Warning and Indicator Lights

CHECK ENGINE LIGHT This may be also labelled as SERVICE ENGINE SOON **(Figure 22–29)**. This warning is primarily an emissions-related light. If something happens that causes the vehicle to have higher emissions, the PCM will display this warning. This warning may also be triggered by oil pressure or coolant temperature. This warning may illuminate when the computer has stored a fault or diagnostic code in its memory.

OIL PRESSURE LIGHT This lamp is operated by an oil pressure switch located in the engine's lubricating system. Some vehicles will illuminate this lamp in yellow or red to indicate the action the driver should take, red meaning the engine has an oil pressure problem and the engine should be shut down, and

TABLE 22–3 **COMMON WARNING LAMP PROBLEMS AND THEIR POSSIBLE CAUSES**

SYMPTOMS	POSSIBLE CAUSES
Warning light remains on all of the time.	Sending-unit circuit grounded Faulty sending-unit switch
Warning light fails to operate on an intermittent basis.	Loose sending-unit circuit connections Faulty sending unit
One or all warning lights fail to operate.	Blown fuse Burned-out bulb Open in the circuit Defective sending-unit switches

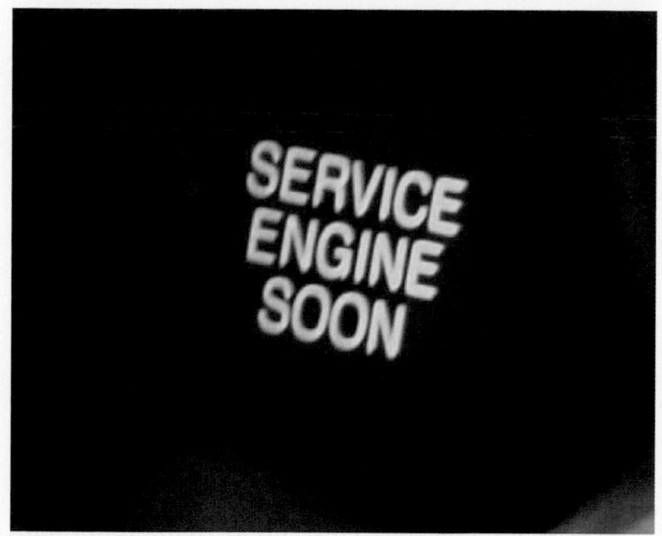

FIGURE 22–29

A typical SERVICE ENGINE SOON warning display.

yellow indicating the oil level is low and should be topped off as soon as possible.

CHARGE INDICATOR LIGHT If there is something wrong with the charging system, this light comes on while the engine is running. In some cases, this light may remain on after the engine is shut off. This also indicates a fault in the charging system.

CHECK FILLER CAP This lamp will be illuminated when the gas filler cap is not tight or off and when the engine control system senses a problem with the fuel system.

ADD COOLANT LAMP The purpose of this lamp is to inform the driver of low coolant levels in the cooling system.

Electronic Throttle Control Light

Engines equipped with electronic throttle control have a warning light that will turn on if a fault is detected in the system. If this occurs, engine and vehicle speed is typically limited as a safety precaution. This limited operation provides "limp-in" operation so that the vehicle can be driven to a repair facility.

Safety-Related Warning and Indicator Lights

SRS AIR BAG READINESS LIGHT The air bag readiness light lets the driver know the air bag system is working and ready to do its job. It lights briefly when the ignition is turned on. A malfunction in the air bag system will cause the light to stay on continuously or to flash.

Passenger or Side Air Bag Off Light

Some vehicles have an indicator light that alerts the driver when the passenger side air bag or side-impact air bags are disabled. This typically occurs when an object of insufficient weight is placed in the passenger seat. The occupant detection system in the seat recognizes that the weight is not enough to be a full-sized person and turns the air bag off.

FASTEN BELTS INDICATOR When the ignition is turned to RUN or START, the warning chime module applies voltage to illuminate the fasten belts indicator for six seconds, whether or not the driver and/or passenger belt is buckled **(Figure 22–30)**.

BRAKE WARNING LIGHT This is the traditional red brake warning light. When this light is lit, it is an indication that the parking brake is engaged. Some vehicles use the same warning light to indicate hydraulic system failure.

BRAKE PAD INDICATOR This light illuminates when the sensors at the wheel brake units see that the brake pads are worn too thin. With the indicator lit, the driver should avoid hard braking and have the brakes serviced as soon as possible.

PARKING BRAKE WARNING LIGHT Vehicles built specifically for Canada may have a separate brake warning light that indicates when the parking brake is applied.

BRAKE FLUID LEVEL WARNING LIGHT This light is connected to the brake fluid level sensor in the brake fluid reserve tank. If brake fluid decreases to less than the

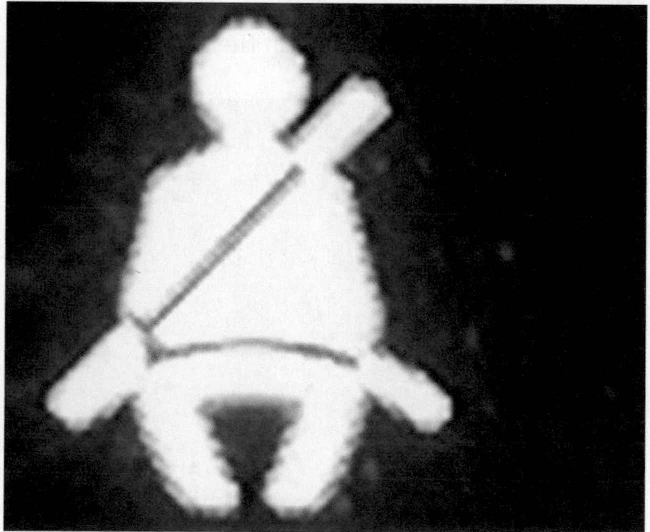

FIGURE 22–30

A typical FASTEN YOUR SEATBELT warning.

specified volume in the reservoir, the sensor is actuated and the light comes on while the engine is running.

LAMP-OUT WARNING LIGHT The lamp-out warning module is an electronic unit designed to measure small changes in voltage levels. An electronic switch in the module closes to complete a ground path for the indicator lights in the event of a bulb going out. The key to this system being able to detect one bulb out on a multibulb system is the use of the special resistance wires. With bulbs operating, the resistance wires provide 0.5-volt input to the lamp-out warning module. If one bulb in a particular system goes out, the input off the resistance wire drops to approximately 0.25 volt. The lamp-out warning module detects this difference and completes a ground path to the indicator light for the affected circuit.

STOPLIGHT WARNING LIGHT The light is controlled by the stoplight checker. This checker consists of a **reed switch** and magnetic coils. Under normal conditions, magnetic fields form around the coils by the current flowing through each light while the stoplight switch is on. These magnetic fields cancel each other because the coils are wound in opposite directions. As a result, the reed switch remains off and the warning light is off. If either the left- or right-side stoplight fails, current flows through only one coil, and the resultant magnetic field causes the reed switch to turn on. The warning light remains lit as long as the brake pedal is depressed.

ANTILOCK LIGHT If an antilock brake system (ABS) fault is present, the antilock brake module grounds the indicator circuit, and the antilock light goes on **(Figure 22–31)**. Vehicles built for the United States may have a different symbol on their warning light.

DRIVER INFORMATION WARNING AND INDICATOR LIGHTS

Blind Spot Detection, Backup, and Lane Departure Warnings

Both blind spot detection and lane departure warning systems are becoming common on modern cars and trucks. These systems may use a combination of visual alerts, such as blinking light in the mirrors, video from cameras displayed on the dash or driver information centre (DIC), audible alerts, or vibrating seats. These alerts are used to inform the driver that an object is detected within the range of view of the various sensors.

LOW-FUEL WARNING This lamp will illuminate when the fuel level is below a quarter full. An electronic switch closes, and power is applied to illuminate the lamp.

MAINTENANCE REMINDER This lamp turns on to alert the driver that maintenance needs to be performed on the vehicle **(Figure 22–32)**. The light is controlled based on calculations that take into account mileage and the driving and operating conditions since the last service. This is not a warning; it is a reminder. Often, the lamp will light for two seconds when the car is started, if the vehicle has been serviced within a specified interval. Normally, the required maintenance is an engine oil change. The name of this indicator includes "oil" on many vehicles. When the lamp is not reset during the specified distance, it will flash after the engine has started. If there has been no service for about twice the recommended distance, the lamp will stay on continuously until it is reset. The procedure for resetting this reminder varies with the manufacturer.

FIGURE 22–31

A typical ABS light illuminated.

Martin Restoule

FIGURE 22–32

The wrench on the gauge indicates that maintenance is due.

©Cengage Learning 2015

TRANSMISSION INDICATOR This is part of an automatic transmission's control system. If the system detects a fault, it may operate the transmission in the fail-safe mode and will illuminate the warning light to inform the driver of the problem and to alert him or her that the transmission may not be working normally.

DRIVE INDICATOR Some four-wheel-drive vehicles have a lamp that, when lit, indicates that the vehicle is in four-wheel-drive mode of operation.

O/D OFF INDICATOR This lamp is illuminated when the overdrive function of an automatic transmission has been switched off by the driver.

REAR DEFROST INDICATOR When this light is lit, the defroster or deicer is operating.

HIGH-BEAM INDICATOR With the headlights turned on and the main light switch dimmer switch in the high-beam position, the indicator illuminates.

LEFT AND RIGHT TURN INDICATORS With the multifunction switch in the left or right turn position, voltage is applied to the circuit to illuminate the left or right turn indicator. The turn indicator flashes in unison with the exterior turn signal bulbs.

FOG LIGHT INDICATOR This lamp is illuminated when the fog lights are turned on.

TIRE PRESSURE MONITOR (TPM) When a low-inflated or flat tire is found, this warning light turns on. Some systems will illuminate this lamp in red or yellow. Red means there is an excessively low or flat tire, and yellow means a tire has low pressure **(Figure 22–33)**.

FIGURE 22–33

The TPM display with an outline of the vehicle showing there is a tire pressure issue on the right side.

Some systems also emit a sound to alert the driver. To turn off the TPM lamp, always follow the manufacturer's procedure. Some vehicles may use two indicators to differentiate between a low tire pressure and a TPMS fault.

GO TO ▶ Chapter 46 for more information on TPMs.

TRACTION/STABILITY CONTROL LAMP This lamp is illuminated in red when there is a problem with the traction and/or stability control systems. The red lamps will also be lit when the system is turned off. Yellow lamps are lit when the system is actively regulating drive torque and braking force.

CRUISE CONTROL LIGHT This lamp is lit whenever the cruise control is turned on.

AIR SUSPENSION LIGHT Voltage is present at the air suspension indicator at all times. If an air suspension fault exists, the ground of the light circuit is closed, and the indicator illuminates.

DOOR AJAR WARNING When the ignition is turned on and if the doors are left open or are ajar, this light comes on.

ADD WASHER FLUID Obviously, the purpose of this lamp is to inform the driver of a low level in the windshield washer fluid reservoir.

FUEL CONSUMPTION GAUGE This gauge gives the driver the current fuel consumption in litres/100 km or miles per gallon (mpg). The gauge is also called an "energy control" gauge, and it informs the driver as to how the current operating conditions are affecting fuel consumption.

Sound Warning Devices

Various types of tone generators, including buzzers, chimes, and voice synthesizers, are used to remind drivers of a number of vehicle conditions. These warnings can include fasten seat belts, air bag operational, key left in ignition, door ajar, and light left on. **Figure 22–34** is a tone generator system schematic.

PARK DISTANCE CONTROL (PDC) This feature uses sensors to measure the distance the front and/or rear of the vehicle is from an object **(Figure 22–35)**. An audible signal changes in frequency as the vehicle gets closer to an object. As the distance between the vehicle and object decreases, the intervals between the tones become shorter. When the object is very

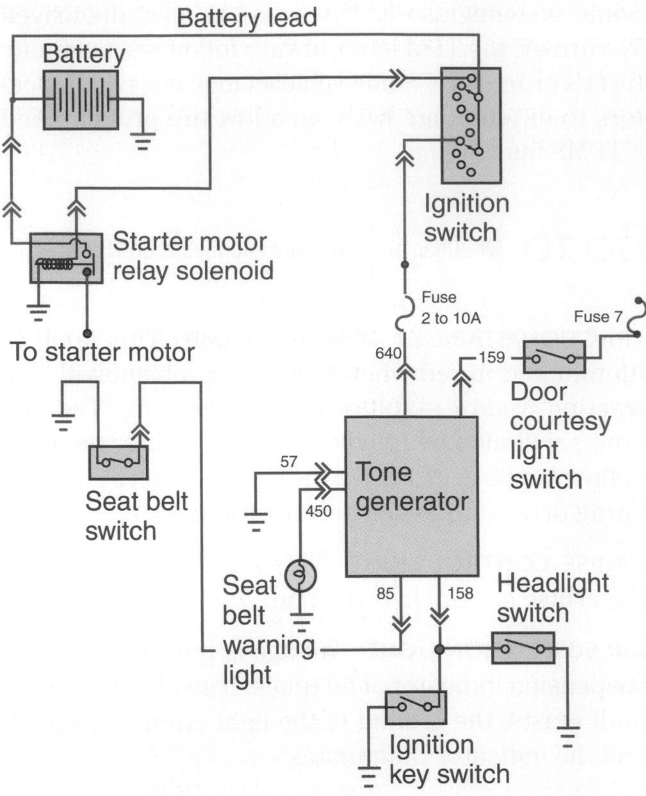

FIGURE 22–34

A tone generator warning system.

Ford Motor Company

close, the tone is emitted continuously. The system uses four ultrasonic sensors at the rear and the front of the vehicle. Some systems include a visual indication of the distances to the obstacles, in addition to the audible warning.

FIGURE 22–35

The sensors measuring the distance from the rear of the car to obstacles trigger the warning system (insert) inside the vehicle.

Robert Bosch Corp.

Some systems allow the front sensors to be manually turned off in special situations such as stop-and-go traffic. The rear sensors automatically turn on when the transmission is placed into reverse.

DRIVER INFORMATION CENTRES

The various gauges, warning devices, and comfort controls may be grouped together into a driver information centre or instrument cluster. This information centre may be simple **(Figure 22–36)**, or it may be an all-encompassing cluster of information. The purpose of this message centre is to keep the driver alert to the information provided by the system. The types and extent of information vary from one system to another.

In addition to standard warning signals, the information centre may provide such vital data as fuel range, average or instantaneous fuel economy, fuel used since reset, time, date, estimated time of arrival (ETA), distance to destination, elapsed time since rest, average car speed, percent of oil life remaining, and various engine-operating parameters.

Other electronic displays and controls can be found on today's vehicles.

Graphic Displays

Graphic displays are translucent drawings or pictures of a vehicle. These displays have lights located at various spots in the graphic. When a graphic is lit, the area by the light has a problem. These indicators can note that the trunk is open or a light bulb is not working **(Figure 22–37)**.

FIGURE 22–36

Information for the driver appears across the bottom of the instrument cluster.

Reprinted with permission.

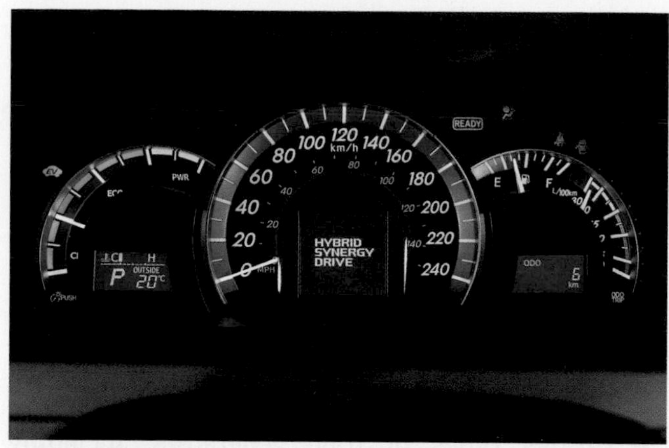

FIGURE 22–37

Notice the small graphic of the car that shows whether the trunk or any doors are open.

Toyota Canada

Hybrid Vehicles

Hybrid vehicles have some unique and interesting warnings, indicators, and displays. They all also have normal displays for the engine and other systems. The following is a brief look at the unique stuff they have.

GO TO ▶ Chapter 37 for more information on hybrid vehicles.

HONDA The instrument panel on most Honda hybrids displays the typical conditions for a gasoline engine, as well as the operation of the hybrid (IMA) system and the car's fuel efficiency. Also, on cars with a manual transmission, the panel has upshift and downshift lights that are triggered by the PCM to inform the driver when it is most economical to shift gears.

The instrument panel has a meter that displays the status of the battery and IMA system **(Figure 22–38)**. A charge/assist indicator shows when the system's electric motor is assisting the engine. The amount of assist is indicated by bars. The number of bars illuminated indicates how much assist is being provided.

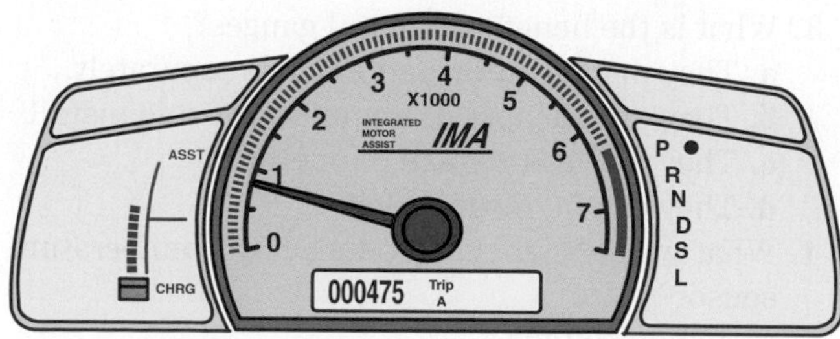

FIGURE 22–38

The instrument cluster in Honda hybrid vehicles has a meter that displays the status of the battery and IMA system.

This same display shows the amount of charge going to the batteries. When more bars are illuminated, the batteries are being recharged at a higher rate. Also on this side of the cluster is a state-of-charge indicator for the battery module. The entire cluster is designed to help the driver achieve maximum fuel economy.

TOYOTA Most Toyota hybrids have a multi-information display in the centre cluster panel. The display, a 7-inch LCD with a pressure-sensitive touch panel, serves many functions. Many of these are typical, but some are unique to hybrid technologies. One is the fuel consumption screen. This display shows average fuel consumption, current fuel consumption, and the current amount of recovered energy.

Another unique display is the energy monitor screen **(Figure 22–39)**. This shows the direction and path of energy flow through the system in real time. By observing this display, drivers can alter their driving to achieve the most efficient operation for the current conditions.

Like other vehicles, these vehicles are equipped with a variety of warning lamps and indicators. Here are some of the indicators that are unique:

- *"Ready" light.* This lamp turns on when the ignition switch is turned to START, to indicate that the car is ready to drive. The engine may or may not be running at this time but is ready to start as needed.
- *Output control warning light.* This lamp turns on when the temperature of the HV battery is too high or low. When this lamp is lit, the system's power output is limited.
- *HV battery warning light.* This lamp lights when the charge of the HV battery is too low.

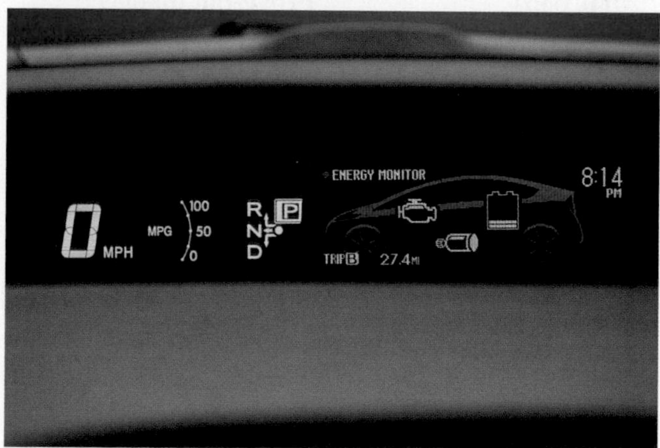

FIGURE 22–39

The energy monitor for a Toyota Prius.

Toyota Canada

- *Hybrid system warning light.* When the HV control unit detects a problem with the motor/generators, the inverter assembly, the battery pack, or the ECU itself, this lamp will be lit.
- *Malfunction indicator light.* This lamp is tied into the engine control system and will be lit when the PCM detects a fault within that control system.
- *Discharge warning light.* This lamp is tied to the 12-volt system and DC–DC converter. It will illuminate when there is am problem in that circuit.

Hybrid four-wheel-drive SUVs show more information and include lamps for the four-wheel-drive option. The latter is monitored by a four-wheel-drive warning lamp that notifies the driver of any detected fault within the MGR (motor/generator—rear) and rear transaxle. When this lamp is lit, a warning buzzer is also activated.

SHOP TALK

Starting or driving a hybrid is different. When the car is ready to be driven, the engine may not be running, but the "ready" lamp will be lit, which means the motor is ready to move the vehicle. If the "ready" lamp does not come on, the vehicle cannot be driven.

KEY TERMS

Air-core gauge (p. 647)
Balancing coil gauge (p. 649)
Bimetallic gauges (p. 649)
Current (p. 648)
D'Arsonval gauge (p. 649)
Heads-up display (HUD) (p. 645)
Instrument panel (p. 644)
Instrument voltage regulator (IVR) (p. 647)
International Standards Organization (ISO) (p. 645)
Low-fuel warning (LFW) (p. 656)
Magnetic fields (p. 648)
Odometer (p. 652)
Prove-out display (p. 650)
Reed switch (p. 660)
Tachometer (p. 656)
Thermal gauges (p. 649)

SUMMARY

- The two basic types of instrument panel displays are analog and digital. In an analog display, an indicator moves in front of a fixed scale to indicate a condition. A digital display uses numbers instead of a needle or graphic symbol.
- Three types of digital electronic displays are used today: light-emitting diode (LED), liquid-crystal display (LCD), and vacuum fluorescent.
- A heads-up display projects visual images on the windshield by a vacuum fluorescent light source to complement existing traditional in-dash instrumentation.
- A gauge circuit is often made of the gauge, a sending unit, and an instrument voltage regulator (IVR).
- Two types of electrical analog gauges—magnetic and thermal—are commonly used with sensors or sending units.
- Indicator lights and warning devices are generally activated by the closing of a switch.
- Various types of tone generators, including buzzers, chimes, and voice synthesizers, are used to remind drivers of a number of vehicle conditions.
- Park distance control uses sensors to measure the distance the front and/or rear of the vehicle is from an object and emits an audible warning as the vehicle gets closer to an object.
- The various gauges, warning devices, and comfort controls may be grouped together into a driver information centre or instrument cluster.
- Diagnosis of gauges, indicators, and warning lights should begin with a good visual inspection of the circuit. Check all sensors and actuators; connections and wires to sensors, actuators, control modules, and ground points; and all vacuum hoses.

REVIEW QUESTIONS

1. What is a benefit of analog displays?
   a. They display more-accurate readings.
   b. They show change of readings better.
   c. They are brighter at night.
   d. They have a lower failure rate.
2. What type of device is usually used to monitor oil pressure?
   a. potentiometer **b.** rheostat
   c. solenoid **d.** piezoresistive
3. What is the benefit of digital gauges?
   a. They display information more accurately.
   b. They are cheaper to manufacture and install.
   c. They use less voltage to operate.
   d. They can be reset easier.
4. What type of signal is produced by a temperature sensor?
   a. analog voltage
   b. digital voltage
   c. pulse modulated frequency
   d. sine wave

5. When in reverse, what is the most common type of driver distance warning?
   a. flashing lights
   b. vibration in the seat
   c. buzzers or chimes
   d. radio volume increase
6. What is likely the cause of all gauges reading lower than expected?
   a. a poor instrument-panel ground
   b. a high charging-system voltage
   c. a burnt-out instrument panel fuse
   d. a defective vehicle speed sensor
7. In which computer memory module does the odometer store vehicle accumulated mileage information?
   a. PROM b. ROM
   c. RAM d. EPROM
8. What would result from grounding the wire that runs from the fuel gauge to the sending unit?
   a. The fuel sender fuse would blow.
   b. The fuel sender float would rise.
   c. The fuel gauge would read empty.
   d. The fuel gauge would read full.
9. What does HUD stand for, when discussing instrumentation?
   a. hidden under dash
   b. high united discharge
   c. heads-up display
   d. head-unit diagnosis
10. Most modern vehicles often use what kind of device when displaying analog values (pressure, temperature, fuel level, etc.)?
    a. reed switches b. relays
    c. bobbin gauge d. balancing coil
11. Park distance control (PDC) uses what type of sensor to measure vehicle distance in relation to an object?
    a. ultrasonic sensor
    b. permanent magnet sensor
    c. wheel speed sensor
    d. LED referencing sensor
12. What usually triggers indicator or warning lights?
    a. a rheostat b. a PMG
    c. a switch d. a pulse generator
13. What condition will alert the driver to a problem in the turn-signal lights?
    a. The MIL is illuminated or flashes.
    b. There is a chime alert in time with the flasher.
    c. The indicator lights flash faster than usual.
    d. There is a vibration in the seat.
14. What happens to the needle of a balancing coil gauge when the power is shut off?
    a. The needle will rest at the lowest end of the scale.
    b. The needle will rest at the highest end of the scale.
    c. The needle will remain centred on the scale.
    d. The needle will remain where it was when power was removed.
15. What is the first step to diagnose any system problem?
    a. Visually inspect the system.
    b. Test all power and grounds.
    c. Verify the complaint.
    d. Check service bulletins.
16. What holds the indicator needle to zero on a speedometer?
    a. magnetic force
    b. weight of the needle
    c. hairspring
    d. buffer in the sending unit
17. An HUD projects an image by use of what kind of light source?
    a. light-emitting diode (LED)
    b. liquid-crystal display (LCD)
    c. halogen lamp array (HLA)
    d. vacuum tube fluorescent (VTF)
18. Which of the following would describe a light-emitting diode (LED) display?
    a. used to project images onto the windshield
    b. used for headlight systems
    c. used for dash illumination
    d. used for warning lamps
19. What values are most often used when testing a temperature sensor?
    a. current and resistance
    b. voltage and current
    c. resistance and voltage
    d. wattage and resistance
20. Which of the following dash warning-light statements is correct?
    a. The seat belt light will come on only if the belts are fastened.
    b. The MIL (malfunction indicator light) stays illuminated if there is a mechanical or electrical problem.
    c. The MIL (malfunction indicator light) stays illuminated if there is a computer-monitored problem.
    d. An air bag light flashes to notify the driver that the system is functional.

CHAPTER 23
Electrical Accessories

LEARNING OUTCOMES

- Explain the basic operation of electric windshield wiper and washer systems.
- Explain the operation of power door locks, power windows, and power seats.
- Understand how cruise control operates and the differences in the various systems.
- Know how to diagnose blower motor problems.
- Determine how well the defroster system performs.
- Identify the components of typical audio and video systems.
- Diagnose problems with power door and window systems.
- Understand the operation of the various security systems.

Most electrical accessories make driving safer, easier, and more pleasant for the driver and passengers. This chapter covers many of the common accessories. Other automotive electric and electronic equipment, such as passive seat belts and air bags, is described elsewhere in this book.

BODY CONTROL MODULES

Many of the accessories found on today's vehicles are controlled, operated, or monitored by a body computer or control module (BCM) system **(Figure 23–1)**. It is important to understand body computer systems and how to diagnose system problems before learning the details of these accessories.

The basic operation of a computer was explained in Chapter 16. Remember that the operation of a computer is divided into four basic functions: input, processing, storage, and output. Understanding these functions will help you organize the troubleshooting process. When a system is tested, you are basically trying to isolate a problem to one of these functions.

In the process of controlling the various systems, a BCM continuously monitors operating conditions for possible system malfunctions. The computer compares system conditions against programmed parameters. If the conditions fall outside of these limits, a BCM will detect the malfunction. A trouble code stored in a BCM memory indicates the portion of the system at fault. A technician can access this code to aid in diagnosis.

If a malfunction results in improper system operation, the computer may minimize the effects of the malfunction by using fail-safe action. During this mode of operation, the computer controls a system based on programmed values instead of the input signals, which allows the system to operate on a limited basis instead of shutting down completely.

Trouble Codes

The method used to retrieve codes from a BCM's memory varies greatly; always refer to the service manual or information system for the correct procedure. Depending on the system, the computer may store codes for long periods of time or lose the code when the ignition is turned off.

On systems that do not retain codes after the ignition is switched off, operate the vehicle until the problem occurs again. Then retrieve the fault code before switching the ignition off. Remember, the trouble code does not necessarily indicate the faulty component; it only indicates the circuit of the system that is not operating properly. To locate the problem, follow the manufacturer's diagnostic procedure for the code received.

Diagnosis should begin with a good visual inspection of the circuit involved with the code. Check all sensors and actuators for physical damage. Check all connections to sensors, actuators, control modules, and ground points. Check wiring for signs of burned or chafed spots, pinched wires, or contact with sharp edges or hot exhaust parts. Also check all vacuum hoses for pinches, cuts, or disconnects.

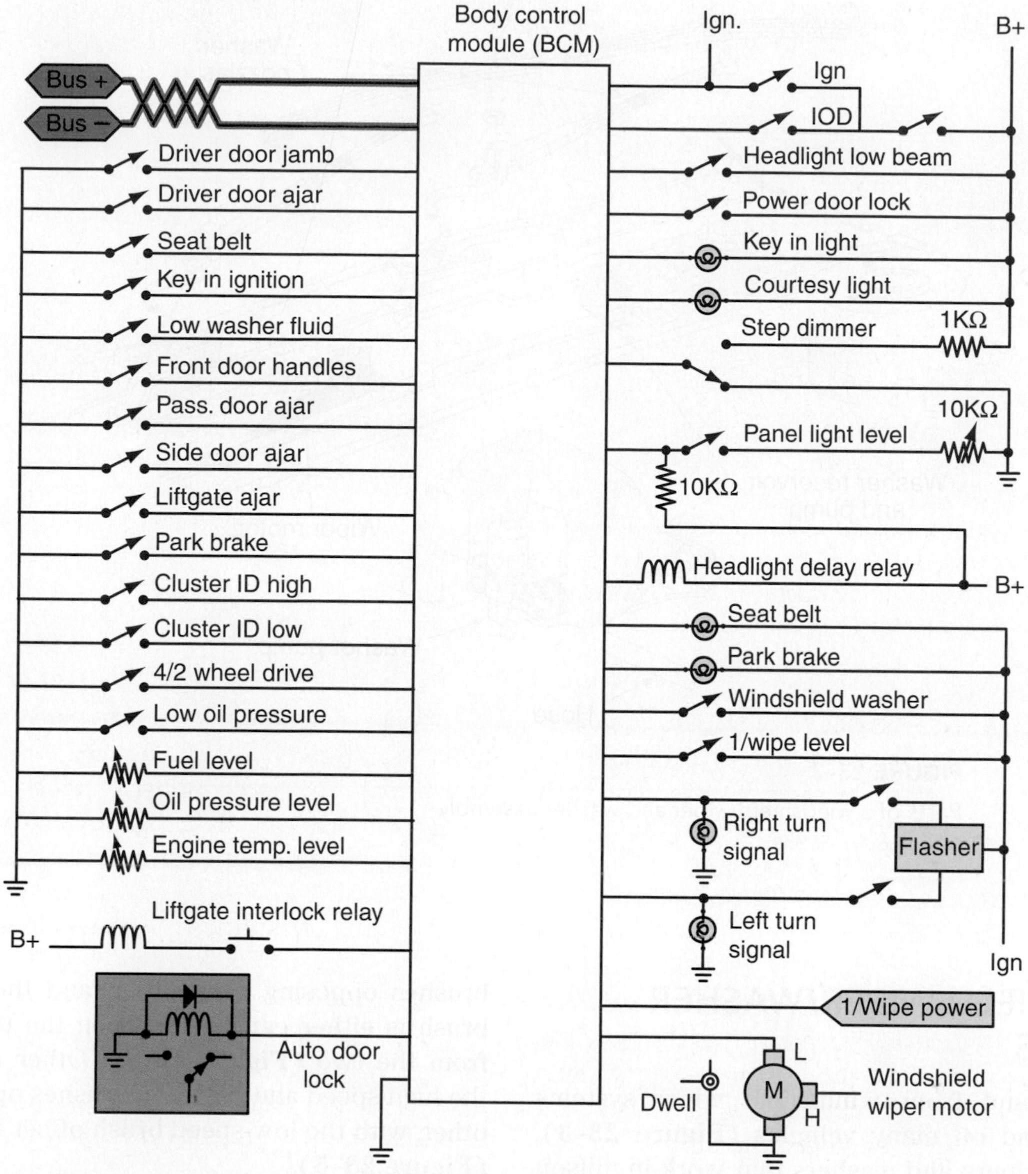

FIGURE 23–1

The body computer regulates many of the vehicle's electrical systems.

Chrysler Group LLC

Entering Diagnostics

There are as many ways to perform body control module (BCM) diagnostics as there are automobile manufacturers. Nearly all vehicles need a scan tool. The scan tool is plugged into the diagnostic connector for the system being tested. Some manufacturers provide a single diagnostic connector **(Figure 23–2)**, and the technician chooses the system to be tested through the scan tool. Use only the methods recommended by the manufacturer for retrieving diagnostic trouble codes (DTCs). Once the DTCs are retrieved, follow the appropriate diagnostic chart to isolate the fault.

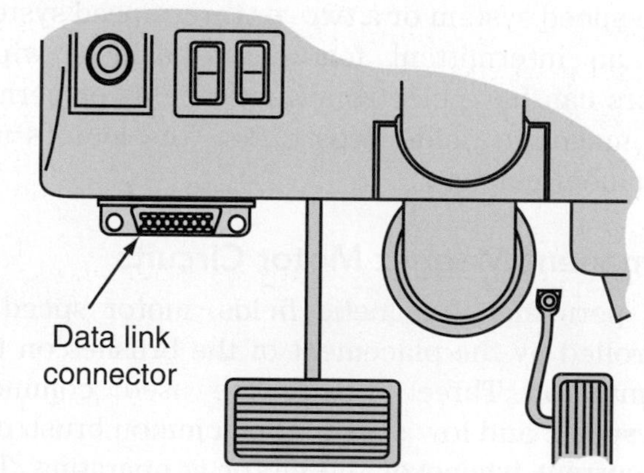

FIGURE 23–2

The data link connector (DLC) for accessing diagnostic trouble codes.

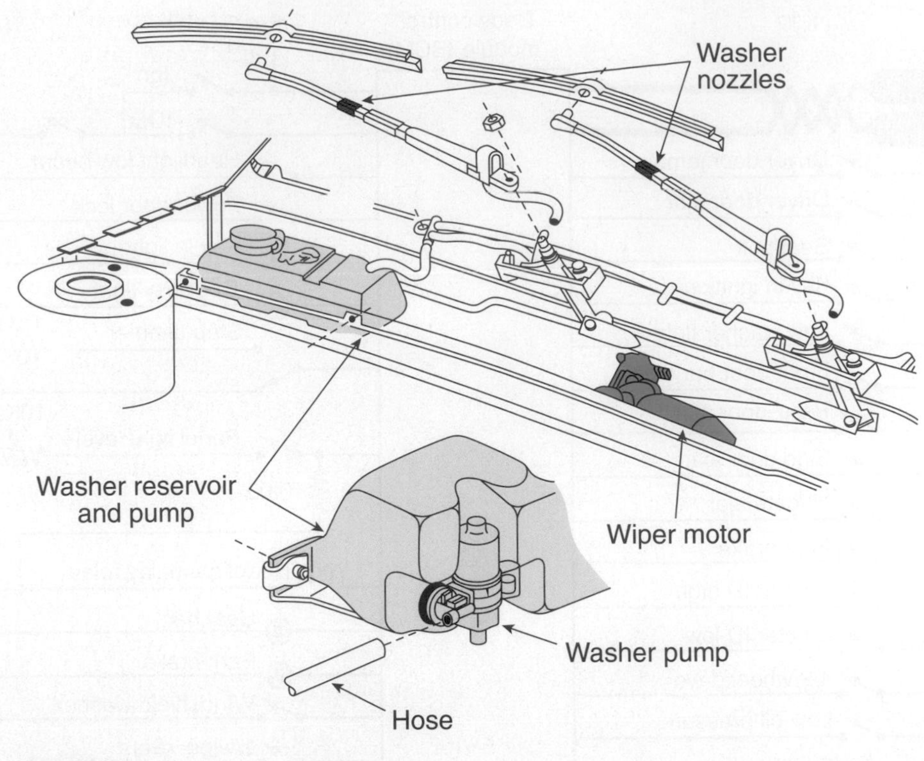

FIGURE 23–3

Parts of a windshield wiper and washer assembly.

©Cengage Learning 2015

WINDSHIELD WIPER/WASHER SYSTEMS

Both rear and front windshield wiper systems can be found on many vehicles **(Figure 23–3)**. Headlight wipers and washers that work in unison with the windshield wipers are also available. The two basic designs of windshield wiper systems used on today's vehicles are a standard two- or three-speed system or a two- or three-speed system with an intermittent feature. Windshield wiper motors can have electromagnetic fields or permanent magnetic fields. Most often, the motors use permanent magnets.

Permanent Magnet Motor Circuits

With permanent magnetic fields, motor speed is controlled by the placement of the brushes on the commutator. Three brushes are used: common, high speed, and low speed. The common brush carries current whenever the motor is operating. The low- and high-speed brushes are placed in different locations based on motor design. The most commonly used motor has the low-speed and common brushes opposing each other, and the high-speed brush is either centred between the two or offset from the two **(Figure 23–4)**. Other motors have the high-speed and common brushes opposing each other, with the low-speed brush offset from the two **(Figure 23–5)**.

The placement of the brushes determines how many armature windings are connected in the circuit. When battery voltage is applied to fewer

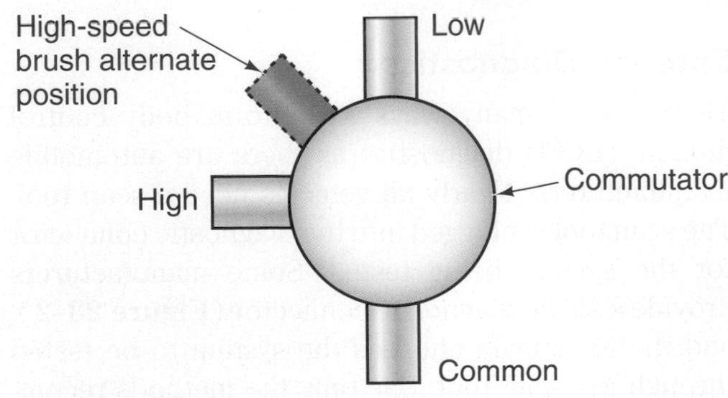

FIGURE 23–4

The most common brush arrangement has the low-speed brush opposing the common brush.

©Cengage Learning 2015

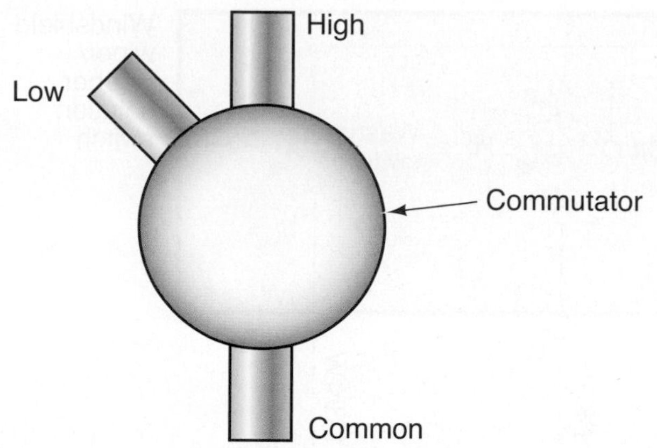

FIGURE 23–5

In this brush arrangement, the high-speed and common brushes oppose each other, and the low-speed brush is offset from the two.

windings, there is less magnetism in the armature and less counter-EMF (CEMF). With less CEMF in the armature, armature current is higher. This high current results in higher motor speeds. When more windings are energized, the magnetic field around the armature is greater, and there is more CEMF, resulting in lower current flow and slower motor speeds.

PARK SWITCH A park switch is incorporated into the motor assembly and operates off a cam or latch arm on the motor's gear **(Figure 23–6)**. The switch supplies voltage to the motor after the wiper control switch has been turned off. This allows the motor to continue running until the wipers have reached their park position.

When the wiper control is in the high-speed position, voltage is applied through the switch to the high-speed brush **(Figure 23–7)**. Wiper 2 moves with wiper 1 but does not complete any circuits. When the switch is moved to the low-speed position, voltage is applied through wiper 1 to the low-speed brush. Wiper 2 also moves but does not complete any circuits.

When the switch is returned to the OFF position, wiper 1 opens. Voltage is applied to the park switch, and wiper 2 allows current to flow to the low-speed brush. When the wiper blades are in their lowest position, the park switch is moved to the PARK position. This opens the circuit to the low-speed brush, and the motor shuts off.

DEPRESSED PARK WIPER SYSTEMS Systems with a depressed park mode have a second set of contacts with the park switch. The contacts allow current to flow to the low-speed brush rather than the common brush. The ground for the motor is established by

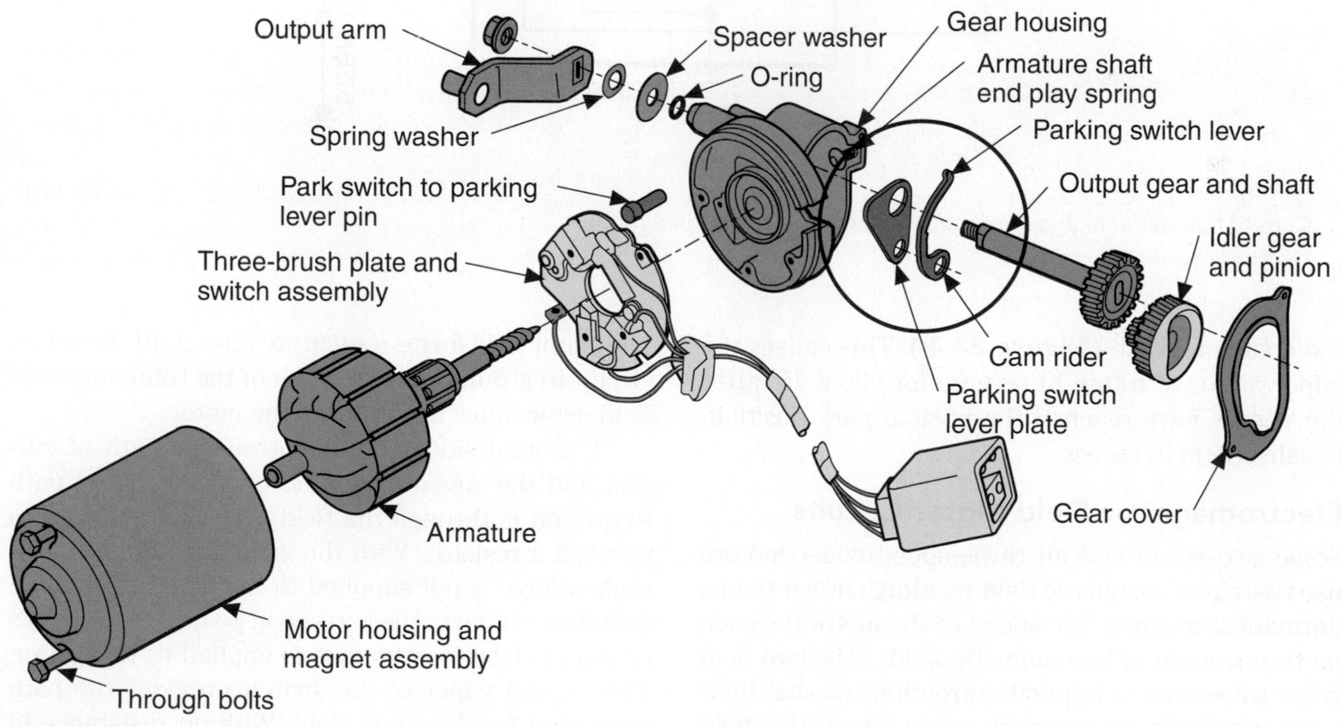

FIGURE 23–6

An exploded view of a wiper motor with a park mode.

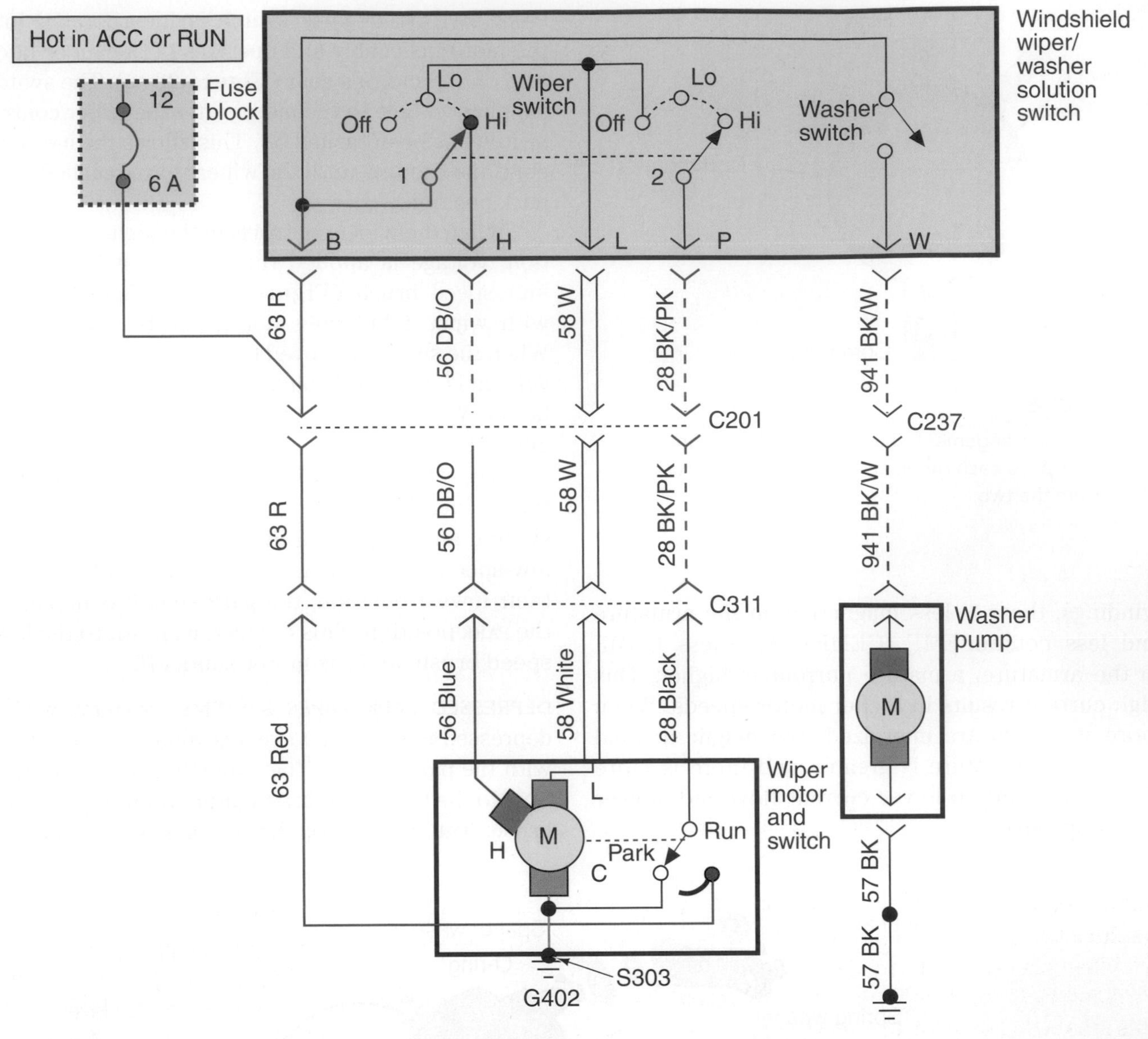

FIGURE 23–7

Current flow in the high-speed mode of a permanent magnet motor.

Ford Motor Company

the common brush **(Figure 23–8)**. This causes the wiper motor to rotate in reverse for about 15° after the wipers have reached the normal park position, causing them to recess.

Electromagnetic Field Motor Circuits

Some two-speed and all three-speed wiper motors use two electromagnetic field windings rather than a permanent magnet. The speed of the motor depends on the strength of the magnetic fields. The two field coils are wound in opposite directions so that their magnetic fields oppose each other. The series field is wired in series with the brushes and commutator. The shunt field forms a separate circuit off the series circuit to ground. The strength of the total magnetic field determines the speed of the motor.

A ground side switch controls the path of current and the speed of the motor. One current path to ground is through the field coil, and the other is through a resistor. With the switch in the OFF position, voltage is not supplied to the motor. When the switch is placed in the low-speed position, the relay's contacts close, and voltage is applied to the motor. The second wiper of the switch provides the path to ground for the shunt field. With no resistance in the shunt field coil, the shunt field is very strong

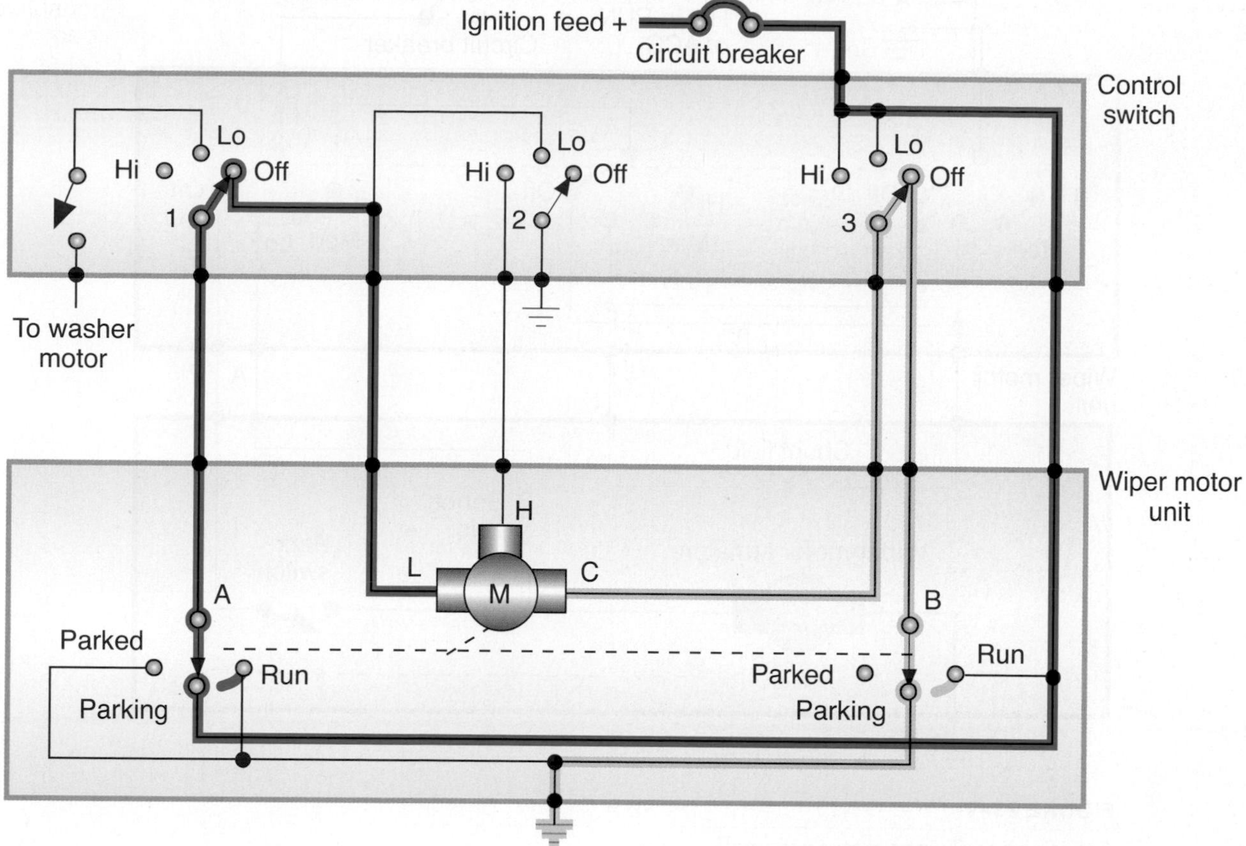

FIGURE 23–8

Circuit current flow when the wipers are parking and moving to the depressed position.

and bucks the magnetic field of the series field. This results in slow motor operation.

When the switch is in the high-speed position, the shunt field finds its ground through the resistor. This results in low current and a weak magnetic field in the shunt coil; therefore, the armature turns at a higher speed.

THREE-SPEED MOTORS The control switch for a three-speed motor determines what resistors, if any, will be connected to the circuit of one of the fields **(Figure 23–9)**. When the switch is in the low-speed position, both field coils have the same amount of current flow. Therefore, the total magnetic field is weak, and the motor runs slowly.

When the switch is in the medium-speed position, current flows through a resistor before going to the shunt field. This connection weakens the shunt coil, and the motor's speed increases.

With the switch in the high-speed position, a resistor of greater value is connected to the shunt field. This connection weakens the magnetic strength of the coil, and the motor runs faster.

WINDSHIELD WIPER LINKAGE AND BLADES Several arms and pivot shafts make up the linkage used to transmit the rotation of the motor to oscillate the windshield wipers. As the wiper motor runs, the linkage rotates the arms from left to right. The arrangement of the linkage causes the wipers' pivot points to oscillate. The wiper arms and blades are attached directly to the two pivot points.

A few wiper systems have two wiper motors that operate in opposite directions, thus creating the oscillation motion of the wipers **(Figure 23–10)**. These systems also occupy less space in the engine's cowl area.

REAR-WINDOW WIPER/WASHER SYSTEM This system is typically found on hatchbacks, vans, and SUVs and has a separate switch to control power to the wiper motor. The parking function is completed within the rear-window wiper motor and switch. Check the fuse or circuit breaker if any wiper/washer system is not working. If the wiper still does not work, trace the power flow through the system, following the electrical schematic in the service information.

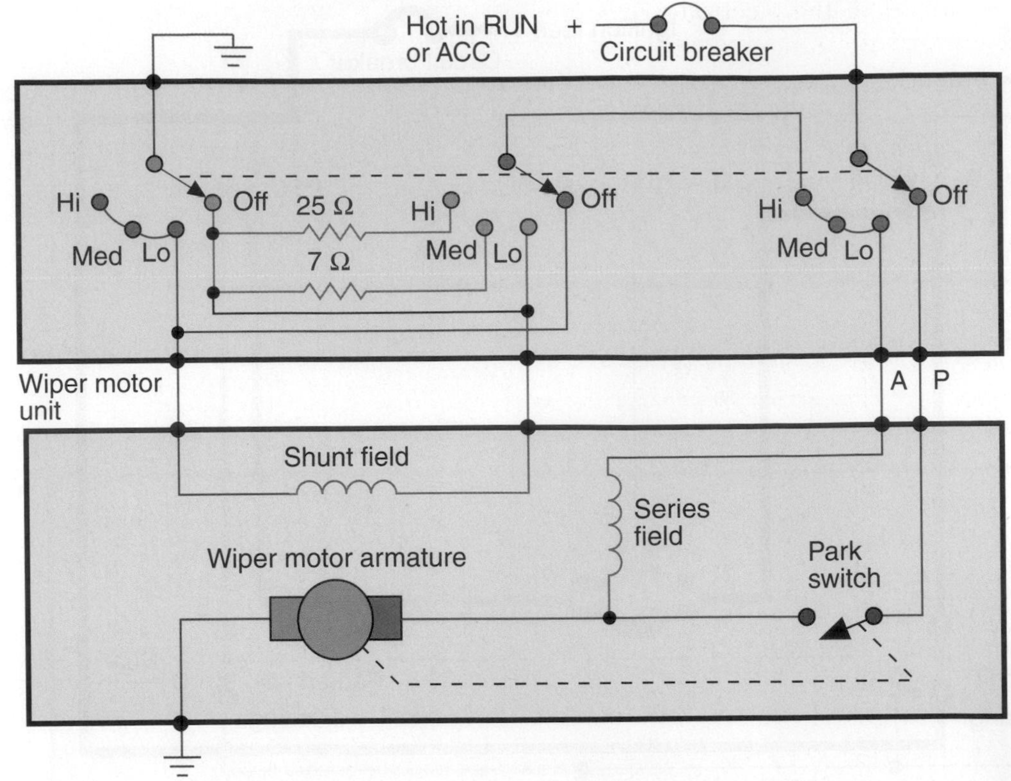

FIGURE 23–9

A three-speed wiper motor schematic.

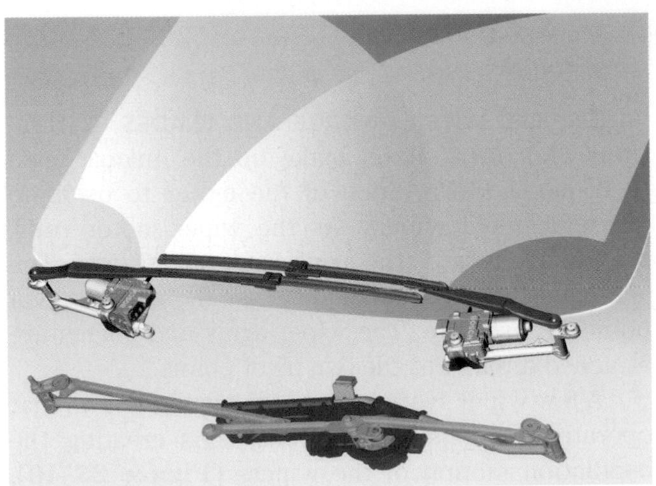

FIGURE 23–10

(Top) This system uses two motors that operate in the opposite direction. (Bottom) A typical single-motor linkage. Note the differences in required space and complexity.

Robert Bosch Corp.

Intermittent Wiper Systems

Many wiper systems offer an intermittent mode that provides a variable interval between wiper sweeps. Many of these systems use a module, or governor, mounted on or near the steering column **(Figure 23–11)**.

The delay between wiper sweeps is controlled by the driver with a potentiometer-type control. By rotating the intermittent control, the resistance value changes. The module contains a capacitor that is charged through the potentiometer. Once

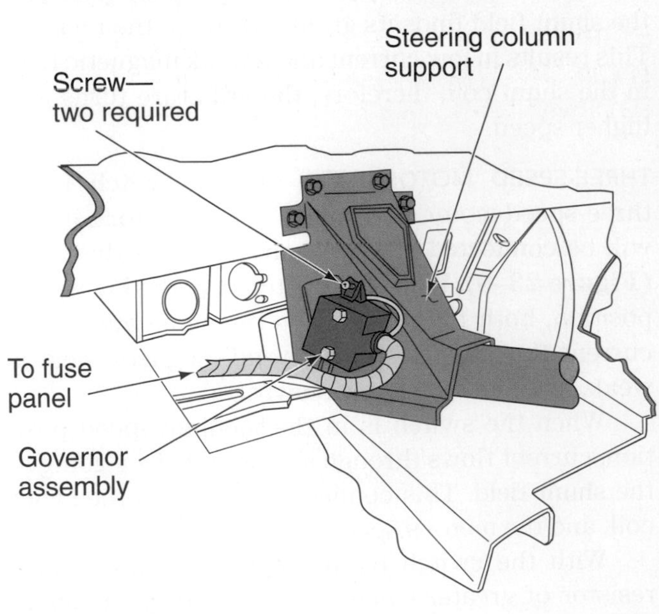

FIGURE 23–11

An intermittent wiper module.

Ford Motor Company

the capacitor is saturated, the electronic switch is triggered, and current flows to the wiper motor. The capacitor discharge is long enough to start the wiper operation, and the park switch is returned to the RUN position. The wiper will continue to run until one sweep is completed and the park switch opens. The amount of time between sweeps is based on the length of time required to saturate the capacitor. As more resistance is added to the potentiometer, it takes longer to saturate the capacitor.

RAIN-SENSING WIPERS Some vehicles have a setting for windshield wiper operation that responds to water on the windshield. The sensor for these wipers is usually in the centre and at the top of the windshield behind the rearview mirror. The sensor transmits an infrared light onto the windshield's surface through a special optical element **(Figure 23–12)**. The system has a series of LEDs that shine at an angle onto the inside of the windshield and an equal number of light collectors **(Figure 23–13)**.

When the windshield is dry, all of the light from the LEDs is reflected back to the collectors. The windshield's ability to reflect light starts to change as soon as moisture begins to accumulate on the glass. Water refracts some of the light away from the collectors. This lower level of reflected light serves as an index indicating higher levels of moisture on the windshield's surface. The rain sensor uses all changes in reflected light as the basis for determining the intensity of the rain. In response, the number of sweeps made by the windshield wipers increases or decreases. The sensitivity level of this system can be adjusted by the driver.

SPEED-SENSITIVE WIPERS Some vehicles have speed-sensitive wipers that vary the speed or intermittent intervals according to vehicle speed. This feature addresses the problem of excessive water accumulating on the windshield when the vehicle is moving fast. These systems are typically controlled by a BCM in response to inputs from the vehicle speed sensor (VSS). With some newer systems, while on the intermittent setting with the vehicle stopped, the front wipers will sweep the windshield as soon as the brakes are released. Other systems will automatically turn the rear wipers on when the front wipers are activated and the driver selects REVERSE.

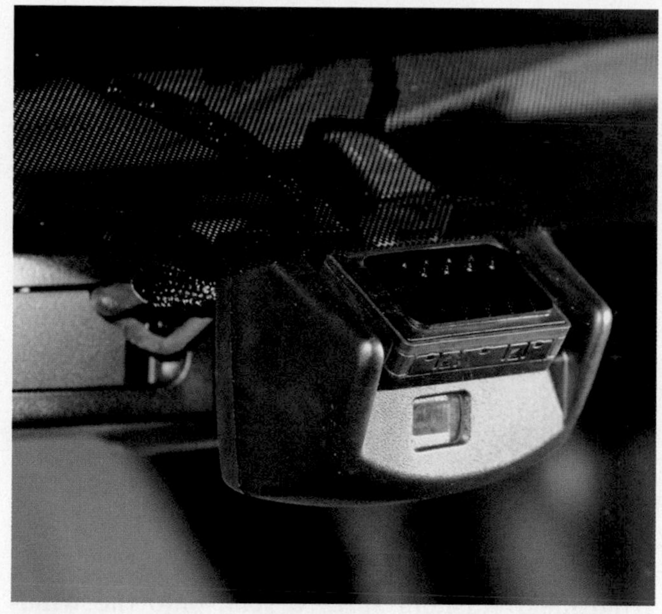

FIGURE 23–12

A typical rain-sensing module mounted on the inside of the windshield so that it can monitor the moisture on the outside of the windshield.

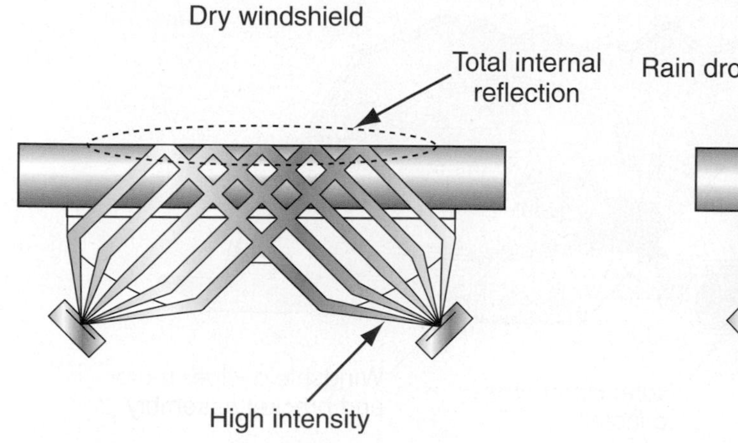

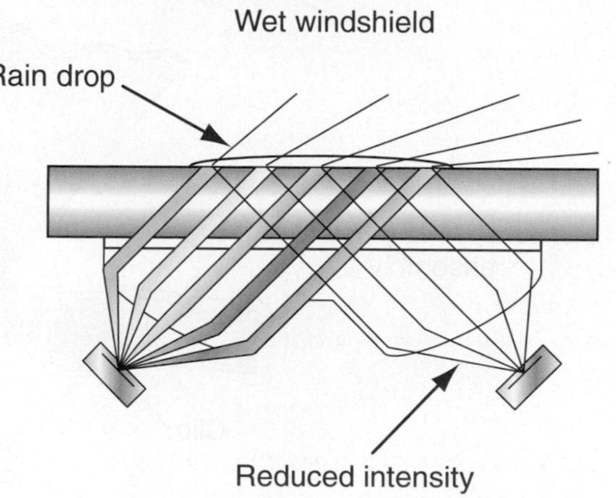

FIGURE 23–13

The basic operation of a rain sensor.

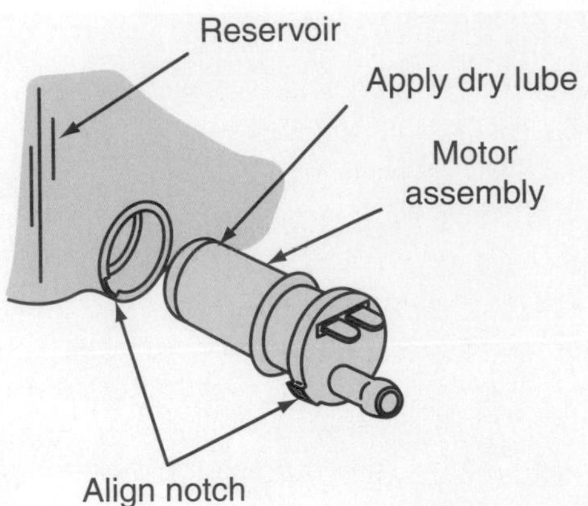

FIGURE 23–14

Installation of a washer pump and motor in a fluid reservoir.

FIGURE 23–15

Headlight wiper and washer systems may work directly with the windshield wipers or have a separate control.

Windshield Washers

Windshield washers spray a fluid onto the windshield and work in conjunction with the wiper blades to clean the windshield of dirt. Most systems have the washer pump installed in the fluid reservoir **(Figure 23–14)**. Some older vehicles use a pulse-type pump that operates off and is timed to the wiper motor.

Washer systems are activated by holding the washer switch. If the wiper/washer system also has an intermittent control module, a signal is sent to the module when the washer switch is activated. An override circuit in the module operates the wipers at low speed for a programmed length of time. The wipers either return to the parked position or operate in intermittent mode, depending on the design of the system.

Some vehicles have wipers and washers that clean the headlights **(Figure 23–15)** and fog lights for maximum visibility. Headlight washer systems may operate from their own switch and pump or work along with the windshield washer system.

Wiper System Service

Customer complaints about windshield wiper operation can include poor wiping, no operation, intermittent operation, continuous operation, and wipers that do not park. Other complaints will be related to wiper arm adjustments, such as slapping the moulding or one blade parking lower than the other.

When the wipers move as they should but do not wipe the glass surface the way they should or make noise while moving across the glass, the blades and/or arms should be replaced.

If the wipers work more slowly than expected, disconnect the wiper linkage at the motor **(Figure 23–16)**. Turn on the wiper system. If the motor runs

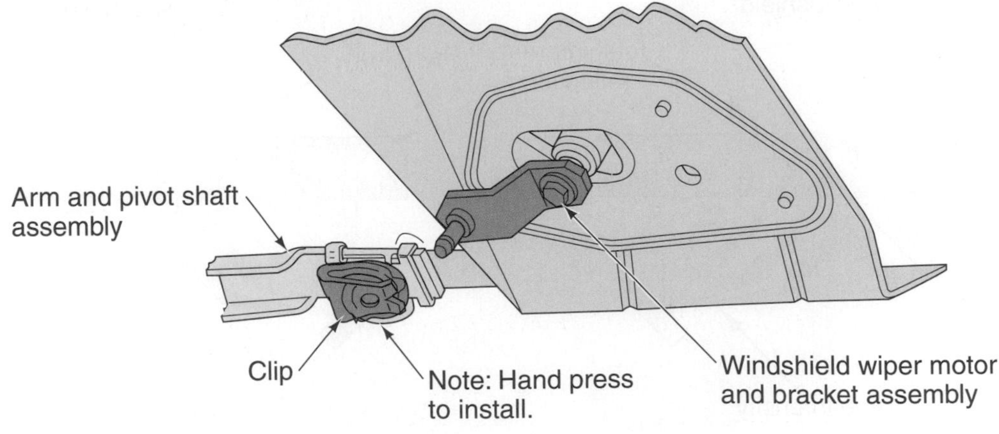

FIGURE 23–16

Disconnecting the wiper linkage arms at the wiper motor.

properly, the problem is the linkage and is not electrical. If the motor runs more slowly than normal, check for resistance in the circuit.

If the motor does not run at a particular speed or at all, the problem is electrical. Carefully inspect the motor, wires, connectors, and switch. On some late-model vehicles, wiper system inputs and operation can be monitored with a scan tool. Use the scan tool to check for DTCs, and check the operation of the wiper switch while watching the data. Pay attention to the circuits that could cause the problem, not the entire wiper circuit. Test for voltage at the motor in the various switch positions. Also check the ground circuit. If the motor is receiving the right amount of voltage at the various switch positions and the ground circuits are good, the problem must be the motor. Wiper motors are replaced, not repaired or rebuilt.

WARNING!

Most wiper motors are the permanent magnet type, which can be quite delicate. Do not throw the motor around or hammer on the case. Both of these actions can destroy the magnetic fields.

If the wiper motor runs but the blades do not move, inspect the nuts securing the wiper arms to the posts. It is possible that nuts have loosened on the posts. If the wiper arms are splined to the posts and do not move, check for stripped splines on the posts and in the wiper arms. Also check the wiper linkage. Many manufacturers use plastic ball sockets to connect components. These sockets tend to crack and break, causing the wiper motor to run but the arms to remain in place.

Washer System Service

Many washer problems are caused by restrictions in the fluid lines or nozzles. To check for restrictions, remove the hose from the pump and operate the system. If the pump ejects a stream of fluid, then the fault is in the delivery system. The exact location of the restriction can be found by reconnecting the fluid line to the pump and disconnecting the line at another location. If the fluid still streams out, the problem is after that new disconnect. If the fluid does not flow out, the problem is before where the hose was disconnected. Repeat this process until the problem is found.

If the pump does not spray out a steady stream of fluid, the problem is in the pump circuit. It should be tested in the same way as any other electrical circuit. Make sure it gets power from the switch when it should, and then check the ground. If the power to the pump is good and there is a good ground, the problem is the pump. These are not rebuilt or repaired; they must be replaced. If a vehicle has repeated washer pump failures, make sure the fluid pickup in the reservoir is clear of dirt and debris. A clogged pickup can overwork the pump, causing premature failure.

HORNS/CLOCKS/POWER OUTLET SYSTEMS

The purpose and operation of these systems are obvious, and their circuits may vary from one model and year to another. However, the overall operation remains the same.

Horns

The electrical horn in automobiles operates on basic electromagnetic principles. When turned on, a diaphragm in the horn vibrates and emits a noise **(Figure 23–17)**. The diaphragm is a thin, flexible,

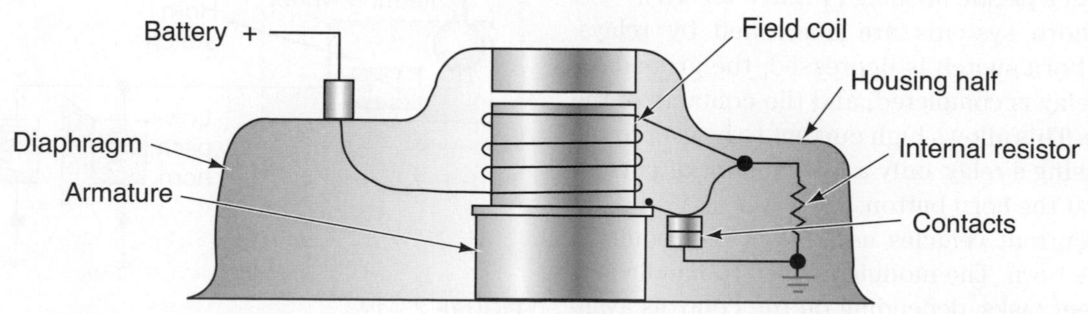

FIGURE 23–17

The basic construction of an automotive horn.

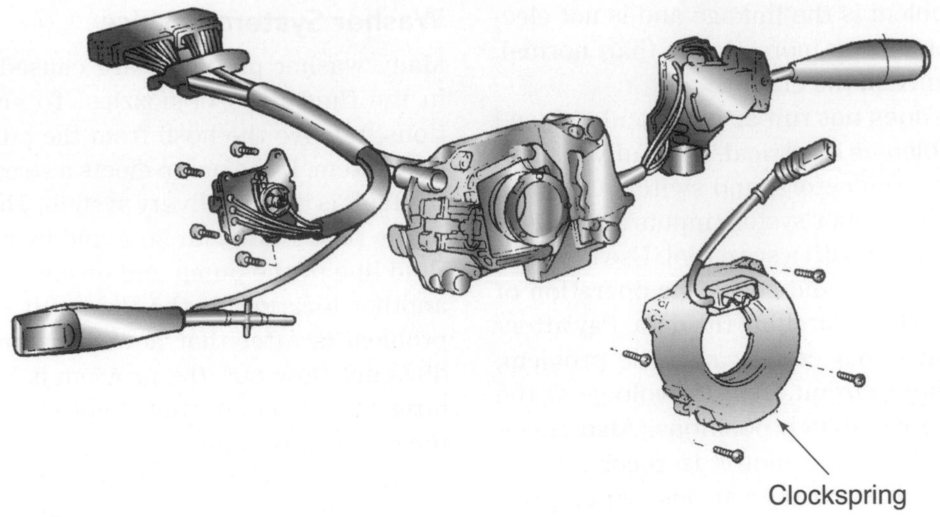

FIGURE 23–18

Nearly all vehicles use a clockspring to maintain contact between the steering wheel and the contacts for the air bag and other accessories.

©Cengage Learning 2015

circular plate mounted to an electromagnet. This diaphragm is only solidly attached to the outside of the horn assembly. This mounting allows the centre of the diaphragm to flex. The electromagnet is moved by the magnetic field set up in the field windings. The circuit to the field windings is open when the electromagnet moves. After the electromagnet moves back to its static position, current is again sent to the field winding. This process continues several times per second as long as the horn button is depressed.

A typical horn switch is either installed in the centre of the steering wheel or may be part of the multifunction switch (on some older models). With the switch in the centre of the steering wheel, a clockspring is used to provide continuity, regardless of the position of the steering wheel, between switch (or button) and the steering column wiring harness. A clockspring is a winding of conductive material contained in a plastic housing **(Figure 23–18)**.

Most horn systems are controlled by relays. When the horn switch is depressed, the ground for the horn relay is completed, and the contacts of the relay close. This allows high current to be sent to the horn. By using a relay, only a low amount of current is present at the horn button.

Many current vehicles use a control module to operate the horn. The module is used to handle several different tasks, depending on the controls available on the steering wheel. The position and status of all of those controls is sent to the steering column module and is shared with other vehicle systems.

Most vehicles have two horns wired in parallel with each other and in series with the switch. Each horn has a different shape and design to produce a different tone **(Figure 23–19)**.

The two horns provide a fuller sound than one horn can. Adjustments to either of the horns can be made by turning a screw on the outside of the horn **(Figure 23–20)**.

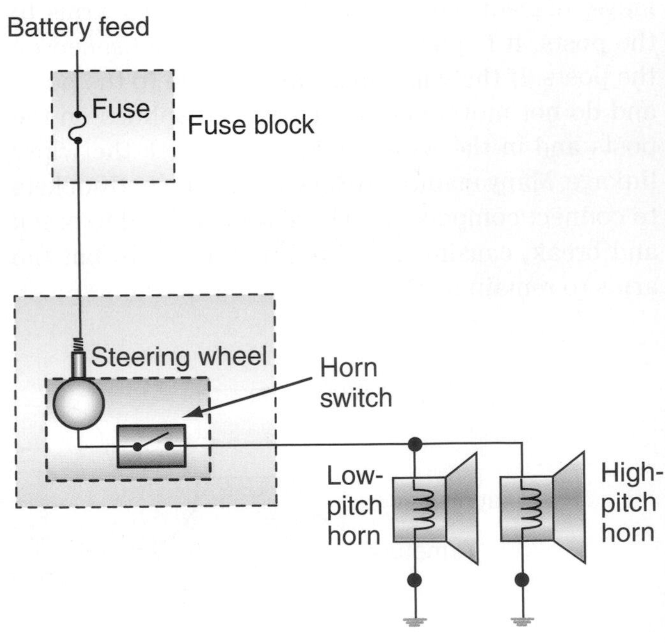

FIGURE 23–19

A basic schematic for a horn circuit that uses a power switch to control the operation of the horn.

©Cengage Learning 2015

FIGURE 23–20

The pitch of a horn can be adjusted with a screw on the horn.

HORN DIAGNOSIS If the horn does not work, check the fuse. If the fuse is bad, also check the relay. The relay is checked with a DMM. If power is not present, the problem may be that circuit or the relay. The relay can be checked for continuity with an ohmmeter.

If the relay is good, the switch and the circuit should be checked. Following the precautions for working around an air bag, disconnect the horn switch's connector. Connect the terminals of the switch together with a jumper wire. The horns should sound. If the horns work, replace the switch. If the horns do not work, check the cable reel behind the steering wheel. Jump across the cable's terminals. Again, the horns should sound. If they do not, check the wiring harness in the steering column for an open. If all power circuits and controls check out, disconnect each horn. Test each horn by connecting battery power to the terminal and grounding the bracket. The horns should sound. If they do not, replace them.

Clock

The clock receives power directly from the fuse panel. Some clocks have additional functions. These are explained in the owner's guide for the particular vehicle.

Power Outlet

Many vehicles have an additional power outlet that looks like a cigarette lighter receptacle. These can be used to power or recharge 12-volt appliances, such as cellphones. Some vehicles also have 110- to 155-volt receptacles to run normal household appliances. Of course, you must have the correct plug to put into the receptacle. These systems are protected, so the first step in troubleshooting would be to check the fuses.

BLOWER MOTORS

The blower or interior fan motor circulates air from the heating and air-conditioning (HVAC) system through ductwork and into the passenger compartment. The blower motor is normally located in the heater housing assembly. An example of a blower motor and its location is shown in **Figure 23–21**.

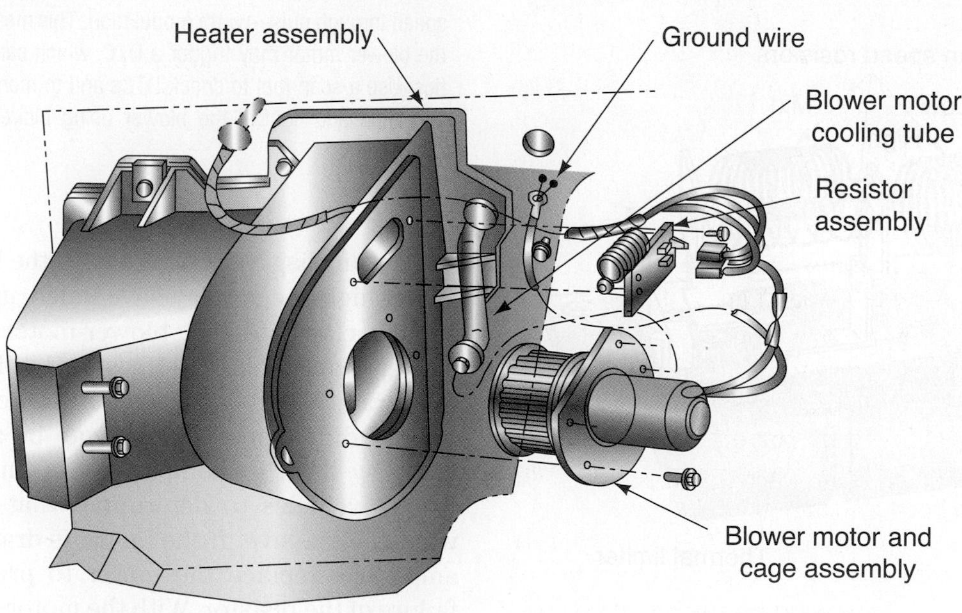

FIGURE 23–21

The location of a typical HVAC blower motor.

The speed of the fan is controlled by a multiposition switch on the HVAC control panel. The switch directs electricity to various points in a stepped resistor block that is normally located near the heater housing and connected in series to the blower motor. A typical resistor block has three or four resistors connected in series. An example of a fan circuit's resistor block is shown in **Figure 23–22**. The resistors cause a voltage drop in the fan circuit, reducing the voltage and current available for the fan. Low resistance allows for higher fan speeds. Typically, the fan receives full battery voltage for high-speed operation, and the resistor is bypassed. High resistance causes slower blower-motor speeds. On some vehicles, when the engine is running, the motor runs constantly at low speed.

Diagnosis

The first step in diagnosing the blower-motor circuit is to determine the exact problem. Blower systems typically fail in one of several ways: The blower does not work on any speed; the blower works on high speed only; the blower works on all speeds but high speed; or the blower works on one or more speeds, but one or more of the speeds do not work. Diagnosing these concerns will depend on the system, so you will need to examine a wiring diagram. Some vehicles use one fuse for the blower circuit, whereas others use two fuses and one or two relays. In addition, some vehicles supply 12 volts to the motor and control the ground circuit through the switch and resistors, and others are wired to do the opposite. In general, for systems that are not BCM-controlled, use **Table 23–1** to help diagnose the system.

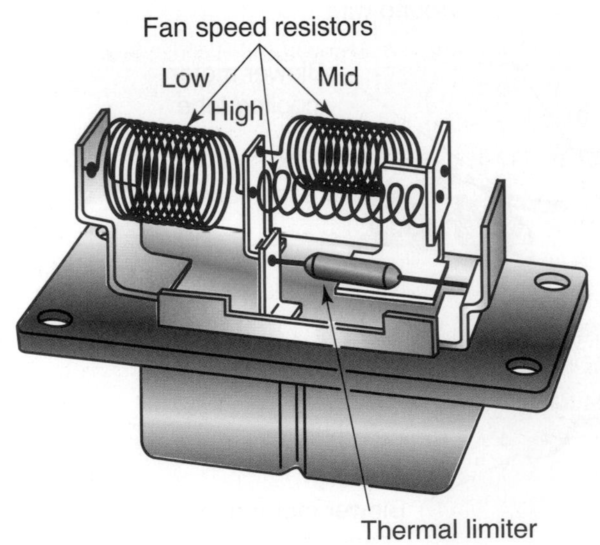

FIGURE 23–22

A typical fan motor resistor block used to control fan speed.

TABLE 23–1 **BLOWER-MOTOR DIAGNOSIS**

PROBLEM	POSSIBLE CAUSE
Blower inoperative at all speeds	Open fuse Defective relay Faulty blower motor Open ground circuit Faulty switch
Works on high speed only	Open fuse Defective relay Open blower resistor
High speed inoperative	Open high-speed fuse Defective high-speed relay Faulty blower switch Open in high-speed circuit
Inoperative at various speeds	Open in blower resistor Faulty blower switch

SHOP TALK

Many of today's vehicles use a BCM to control blower-motor speed through pulse-width modulation. This means problems with the blower motor may trigger a DTC, which can help in diagnostics. Use a scan tool to check DTCs and to monitor HVAC system operation and control the blower using bidirectional control of the system.

If your diagnosis determines the blower resistor and/or high-speed relay have failed, use an inductive current clamp to test blower-motor current draw. Slower than normal motor speed will increase current draw, which can burn out the resistor. Finding current draw specifications can be a challenge, so you may have to compare your findings to other similar vehicles to determine what is normal and what is excessive. If the motor is drawing excessive amperage, replace the motor to prevent a second failure of the resistor. With the motor removed, make sure there is not an accumulation of debris in the blower-motor housing that can obstruct airflow or interfere with blower operation.

CRUISE (SPEED) CONTROL SYSTEMS

Cruise control systems are designed to allow the driver to maintain a constant speed (usually above 50 km/h or 30 mph) without applying continual pressure on the accelerator pedal. Selected speeds are easily maintained and can be easily changed. When engaged, the system sets the throttle position to the desired speed. The speed is maintained unless heavy loads and steep hills interfere. The cruise control switch is often located near the centre or sides of the steering wheel **(Figure 23–23)**.

Electronic Cruise Control Systems

Cruise control can use electronic components rather than mechanical components. An electronic control module can be used to move a **servo unit**, or it can control an electric stepper motor. Many of today's vehicles are fitted with a stepper motor. The motor moves a strap attached to the cruise control cable, which moves the throttle linkage. The motor may be a separate unit or built into the cruise control module. Vehicles with electronic throttle control (throttle-by-wire) do not need a separate cruise control module, stepper motor, or cable to control engine speed. The powertrain control module (PCM) has full control of the throttle; therefore, the circuitry of the PCM operates the cruise control system. There are two brake-activated switches operated by the position of the brake pedal. When the pedal is depressed, the brake release switch disengages the system.

The vehicle speed sensor (VSS) is used to monitor or sense vehicle speed. The computer has several other inputs to help determine the operation of the servo **(Figure 23–24)**. A **servo** is a device that uses feedback to control parameters. These inputs include a brake release switch (clutch release switch); a speedometer, buffer amplifier, or generator speed sensor; and a lever-mounted mode switch or speed control on the steering wheel (signal to control the cruise control).

FIGURE 23–23

The cruise control switch is used to set, increase, decrease, or resume speed or turn the system on or off.

A BCM monitors the signals from the cruise control on–off, set/coast, resume/accelerate, and cancel switches to detect when the driver wants to change the current cruise control operation. It sends the status of the switches to the engine control module (ECM) as serial data. The ECM uses this information to set and maintain the selected speed by controlling the stepper motor. The ECM will disengage cruise control operation if the system is turned off or the brake and/or clutch pedal is depressed.

Cruise control can operate only if certain conditions exist, such as the vehicle's speed is within a prescribed range; a BCM does not detect a problem in the system; and system voltage is within a prescribed range.

Adaptive Cruise Control

Like other cruise control systems, adaptive cruise control automatically maintains the desired speed of the vehicle, but it also maintains a safe distance between vehicles. The desired distance between vehicles is set by the driver. The system also adjusts the speed of the vehicle to mirror that of a slower vehicle in front of it and then maintains that speed. A laser or radar sensor **(Figure 23–25)** mounted near the front bumper serves as the eyes for the system.

Other vehicles travelling within the sensor's range reflect the radar waves **(Figure 23–26)**, and the sensor picks up the returning signals. The control unit uses this information to determine the position and speed of the preceding vehicle. When the system detects a slower-moving vehicle in the same lane, it reduces the throttle or gently applies the brakes to reduce speed. The vehicle then follows behind the preceding vehicle at the speed required to maintain a predefined distance. As soon as the vehicle in front has moved or increased speed, the system will accelerate the vehicle back up to the set and desired speed. This technology is available for the full-speed range, including stop-and-go functionality.

Cruise Control System Service

Most cruise control systems are controlled by a BCM and work with the electronic throttle control system. Diagnosis of these cruise systems is the same as for

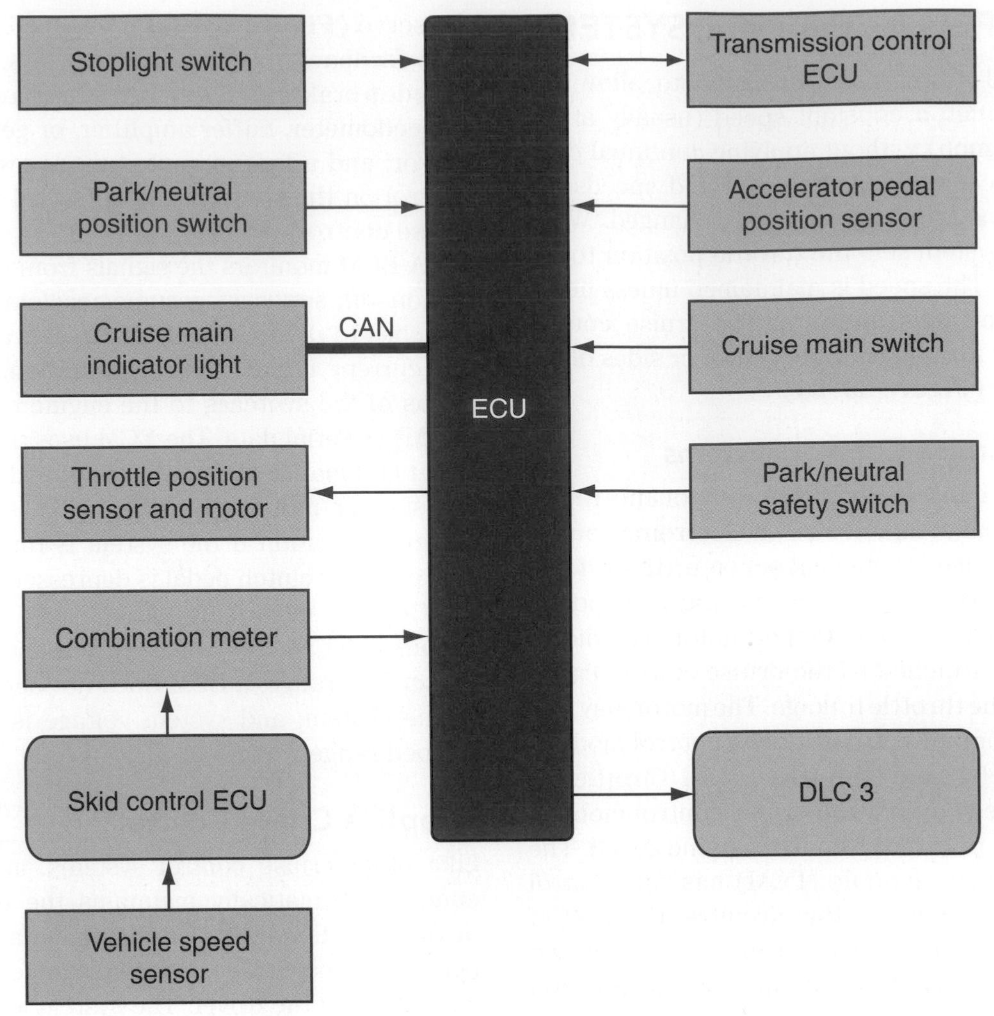

FIGURE 23–24

The main components of an electronic cruise control system.

©Cengage Learning 2015

FIGURE 23–25

The distance sensor for an intelligent or adaptive cruise control system.

DaimlerChrysler Corporation

FIGURE 23–26

The response of an adaptive cruise control system.

Delphi Corporation

any other electronic system. Diagnostic work is done with a scan tool and DMM. Typically, on most late-model vehicles, cruise control problems are caused by faulty circuits, sensors, and/or switches **(Table 23–2)**.

Before connecting the scan tool, check the operation of the cruise indicator lamp. If it does not operate properly, test and repair that circuit before proceeding. If the light is blinking, the control module detects an electrical fault in the system. Also, check all of the system's accessible components for obvious damage. Then with the scan tool, verify that communication links on the CAN bus are good. Retrieve the DTCs, and proceed with the recommended procedures for the codes.

The service information will give information on the expected voltage values at various points in the system. It will also list acceptable resistance readings through the switches. If the resistance across any switch is not within the listed values, the control assembly should be replaced. Also, the brake light circuit must be functioning correctly; if the brake lights are not working or there is a problem with the brake light switch, the cruise may be disabled.

TABLE 23–2 COMMON CRUISE CONTROL PROBLEMS AND THEIR PROBABLE CAUSES

SYMPTOM	LIKELY PROBLEM AREA
Pushing the ON–OFF button does not turn the system on.	Stop light switch circuit Clutch switch circuit (M/T) Vehicle speed sensor circuit Cruise control switch circuit Transmission range sensor circuit Control module
Vehicle speed cannot be set. (The CRUISE indicator is on.)	Cruise control switch circuit Vehicle speed sensor circuit Stop light switch Transmission range sensor circuit Clutch switch circuit Control module
Cruise control stops during operation.	Cruise control switch circuit Vehicle speed sensor circuit Stop light switch Transmission range sensor circuit Clutch switch circuit Control module
Cruise control cannot be manually cancelled.	Cruise control switch circuit Control module
Cruise control is not cancelled when vehicle speed drops below the low-speed limit.	Vehicle speed sensor circuit Control module
Depressing the brake pedal does not cancel cruise control.	Stop light switch circuit Control module
Depressing the clutch pedal does not cancel cruise control.	Clutch switch circuit Control module
Moving the shift lever does not cancel cruise control.	Transmission range sensor circuit Control module
Speed is not constant.	Vehicle speed sensor circuit Control module
CRUISE indicator is blinking.	Cruise control circuit Control module

SOUND SYSTEMS

Sound systems are available in a wide variety of models. The complexity of the system varies significantly from the basic AM radio to more complex stereo systems **(Figure 23–27)** that include an AM/FM radio receiver, a stereo amplifier, compact disc (CD) player, equalizer, several speakers, MP3 player, satellite radio, and a power antenna system.

A radio receives signals (radio waves) that are broadcast from radio station towers or antennas. Amplitude modulation (AM) waves travel far but cannot be used to broadcast in stereo. Also, AM does not have as good sound quality as frequency modulation (FM). Nearly all FM broadcasts are in stereo, but the distance range for good reception is limited.

A recent addition to traditional radio is HD Radio. HD Radio broadcasts digital information in addition to standard analog AM and FM radio signals. To receive these broadcasts, a digital HD Radio receiver is required. This improves the sound quality produced by the receiver. In addition to music, HD Radio can also transmit song titles and artist, traffic, and stock information.

Antenna

An antenna collects radio waves for both AM and FM stations. Radio stations' broadcast towers transmit electromagnetic energy through the air and induce an AC voltage in the antenna. This AC signal is then converted to an audio output by the radio receiver unit.

Some vehicles incorporate the antenna into the rear-window defogger grid. A window defogger/antenna module **(Figure 23–28)** separates the radio frequencies used by the radio from the current used by the defogger grid. When the radio is turned on, a 12-volt signal is sent to the defogger/antenna module. A coaxial cable from the module provides the radio with the radio waves.

FIGURE 23–27

A modern sound system control panel.

Martin Restoule

POWER ANTENNAS Many vehicles are equipped with electrically operated antennas that extend when the radio is turned on and lower when it is turned off. These antennas are powered by a small reversible electric motor. The motors are turned off by limit switches that open when the mast has extended or lowered to its desired height. Power antennas (even black-coloured antennas) need to be cleaned with chrome polish on a regular basis to keep them working properly. Often when there is a problem with a power antenna, it is caused by dirt or a lack of lubricant on the telescoping mast. When there is a problem with the power unit, it is normally replaced as a unit.

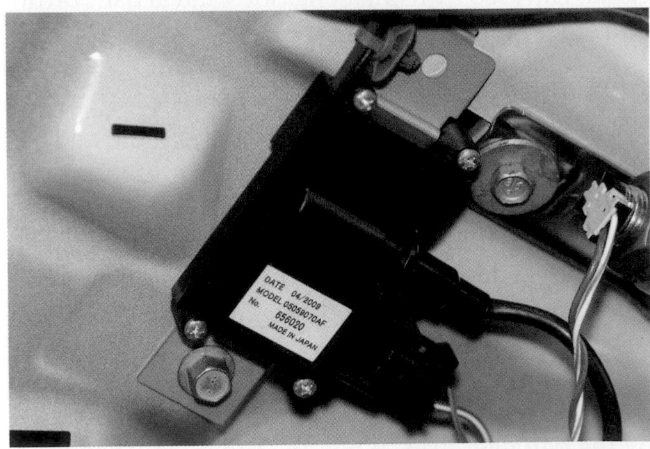

FIGURE 23–28

A rear-window defogger/antenna module.

©Cengage Learning 2015

Satellite Radio

To provide high-quality radio that is not interrupted by distance, some vehicles can be purchased with satellite radio. Satellite radios are also available as add-on items. These radios pick up electromagnetic waves from satellites many kilometres above the earth and use terrestrial repeater networks in some limited markets to support cars in areas blocked with a large number of high-rise buildings. Because the microwaves are transmitted by more than one satellite at all times, and each in their own orbit or place within the orbit, the same radio station can be heard from coast to coast **(Figure 23–29)**. Although distance does not hamper the reception, the radio waves cannot penetrate buildings, tunnels, or large groupings of trees. Select satellite radio content is also being offered through increased Internet and cellular connectivity paths such as mobile cellular and office/home Internet.

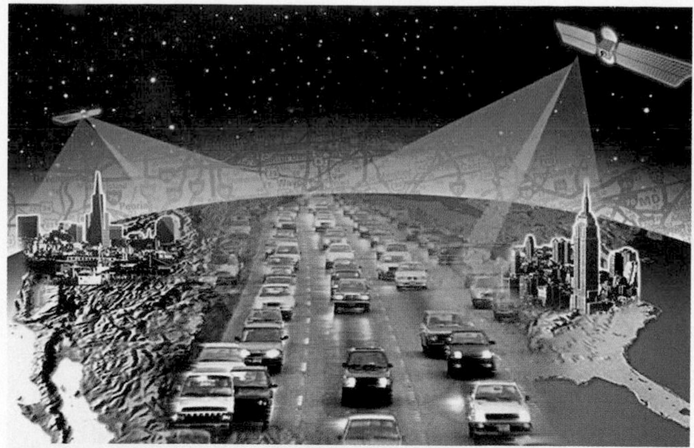

FIGURE 23–29

The reception of satellite radio systems is not affected by distance because the radio waves are transmitted by more than one satellite.

Delphi Corporation

FIGURE 23–31

An antenna for a satellite radio.

Satellite radio programs provide a wide variety of commercial-free music and talk show channels **(Figure 23–30)**. The satellite digital audio receiver (SDAR) is an audio receiver and is separate from the conventional radio receiver. A signal from a satellite is processed by the SDAR, which provides a signal to the conventional radio. The antenna for the SDAR is normally located on the centreline of the roof of the vehicle **(Figure 23–31)**.

Speakers

Sound quality depends on the basic system but especially on the quality of the speakers and their placement. Most systems are equipped with several speakers, each designed to produce a different range of sound. Matching speakers to the system is done by selecting speakers that have the same impedance as the rest of the system. The placement of the speakers is also critical to providing a good clean sound.

Sound waves from a speaker will bounce off anything they hit, including other sound waves. This bouncing of sound can cause noise or distortion. To achieve a high-quality sound system, the speakers must be placed so that all bouncing is minimized.

Because one speaker cannot reproduce the entire range of hearing frequencies (approximately 20 Hz to 20 KHz), speakers are designed to provide quality sound within a particular range.

Large speakers, called woofers, produce the low frequencies of mid-range and bass better than smaller speakers, called **tweeters**. **Woofers** are typically placed toward the rear of the vehicle **(Figure 23–32)**. Tweeters produce high frequencies of treble better than woofers and are normally placed

FIGURE 23–30

The screen of a satellite digital audio receiver.

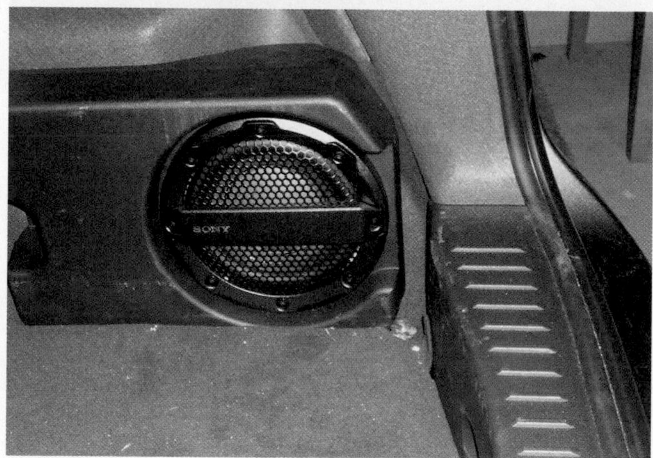

FIGURE 23–32

A woofer mounted in the rear of the vehicle.

in each front and rear door of the vehicle. High/mid-range speakers are normally placed in the centre of the front instrument panel. Additional high/mid-range speakers may be placed throughout the passenger compartment. Coaxial speakers or full-range speakers are also normally placed in the front and rear doors, have two separate speakers combined into one unit, and cover a broader frequency range. Subwoofer speakers can be coupled with a coaxial speaker, tweeter, or mid-range speaker to cover a wider range of frequency while maximizing sound quality. Surround-sound systems may have full-range speakers set inside or near the seats' headrests.

CD Players

A compact disc (CD) player uses a laser pickup to read digital signals recorded on a compact disc. By converting the digital signals to analog, it can play music and other audio formats. CD players vary from being able to insert one or more CDs into the main unit to having an auxiliary unit where many CDs can be installed. The control unit allows the operator to select the CD of choice.

Newer vehicles may have ports that allow MP3 music files to be played from an SD card or a USB flash drive **(Figure 23–33)**. Some also have connectors that allow full operation of a portable music storage device (iPod) through the sound system.

DVD Systems

Many vehicles have DVD players. Most of these display videos on a flip-down, roof-mounted monitor or on a monitor attached to back of the front seats **(Figure 23–34)**. In some cases, a DVD can be viewed on the navigation screen when the vehicle is not moving. The sound from the video is channelled to the vehicle's speakers through the radio, unless headphones are used. When the headphones are activated, the audio of the video can be heard through them, while others in the vehicle can listen to the output of the normal sound system.

FIGURE 23–33

A variety of connections for the different sound devices and storage media.

FIGURE 23–34

A DVD monitor located in the back of a front seat.

These systems are controlled by a rear-seat entertainment electronic control unit (ECU) that integrates some of the functions of the front-seat audio unit and the multi-information display.

Wireless headsets and remote controllers use infrared-emitting diodes to transmit signals. Wired headsets connect to the DVD player and allow for headset volume control. The DVD player may also have terminals that allow other audio/visual equipment, such as a video game player, to connect to the system.

SELF-DIAGNOSTICS The system performs self-diagnostics and stores DTC(s) if any faults are found. DTCs are retrieved by entering into the diagnostic mode by pressing the required series of keys on the control panel.

Amplifiers

Many optional sound systems have a very high wattage output and use several amplifiers. An amplifier increases the volume of a sound without distorting it. Amplifiers are typically rated by the maximum power (watts) they can put out. Amplifiers can be either remotely mounted or integrated in the speakers. Most remotely mounted amplifiers are connected to a data bus and receive power through an amplifier relay. When the radio is turned on, 12 volts is sent to energize the relay.

Some sound systems have an automatic sound level system that adjusts the volume to compensate for changes in ambient noise and vehicle speed. Other vehicles feature a system that allows for fine-tuning the output based on the number and position of passengers in the vehicle. To provide these features, the outputs from the sound system are sent to the data bus, where signal sensors are used to analyze the sound and adjust the volume at each speaker accordingly.

Some new vehicles are incorporating active noise cancellation functions into the audio system. Active noise cancellation uses microphones to detect engine, wind, and tire noise and then programs the audio system to play back the noise 180° out of phase. This cancels the unwanted noise and allows for a quieter passenger compartment.

In addition to power and audio signal wires, some systems have a serial data wire that allows communication between the various components of the sound system. The controls of the sound system may be housed at the radio receiver unit or may have additional remote controls at the steering wheel or another location inside the passenger compartment. The switches either can be resistive multiplexed switches or use a supplementary bus system, such as the local interconnect network (LIN), sometimes called the infotainment network. The switches send different voltage signals to the controlling module, depending on which switch is pressed. In turn, the module responds by sending a request message via the data bus to the radio.

Diagnosis

If the radio system is not working, check the fuse. If the fuses are okay, refer to the service information, and check the rest of the power and ground circuits for the radio. If you determine the radio is the problem, remove it and send it to a qualified shop or replace it. Refer to **Table 23–3** to view the possible causes of common sound system problems.

If the sound system has poor sound quality or poor radio reception, the circuit's wiring, the antenna, and the speakers should be checked. If reception is bad, use an ohmmeter to check the ground of the antenna. Also connect one lead of the ohmmeter to the antenna's mast and its case; there should be no continuity between the two.

Poor speaker or sound quality is usually caused by damaged speakers, wiring, bent sheet metal around the speaker opening, damaged or missing speaker mounting brackets, or missing or loose attaching hardware or speaker covers. Be careful not to overtighten the speaker hardware as this may bend or deform the speaker baskets, causing buzzes or distorted sound.

TABLE 23–3 COMMON SOUND SYSTEM CONCERNS AND THEIR POSSIBLE CAUSES

SYMPTOMS	POSSIBLE CAUSES
Radio will not turn on.	Open power battery feed circuit Open ignition feed circuit Poor radio ground circuit Loss of bus communications
Radio will not produce sound.	Open power battery feed circuit Open ignition feed circuit Poor ground connection Defective radio Open speaker circuit Defective amplifier Open power feed to amplifier Poor amplifier ground circuit Bus communications error Stuck MUTE button
No sound in AM or FM mode. CD audio operates normally.	Faulty antenna connection Poor antenna ground Faulty radio
Excessive noise.	Faulty antenna connection Poor antenna ground Faulty engine to chassis ground Interference from ignition system, neon lights or electrical power lines
Poor radio reception.	Faulty antenna connection Poor antenna ground Faulty radio
Poor sound quality.	Damaged speaker cones Damaged speaker mountings Damaged wiring

SHOP TALK

Antitheft audio systems have built-in devices that make the system useless if stolen. If the power source for the system is cut, even if it is later reconnected, the system will not work unless its ID number is put in. When working on vehicles equipped with this system, before disconnecting the battery terminals or removing the audio system, ask the customer for the ID number so that you can input it after service, or request that the customer input the ID after the repairs are completed. A memory saver or backup battery can be used to maintain the radio's code and settings during service.

SATELLITE RADIO RECEPTION FACTORS Several factors can affect satellite radio reception performance:

- *Antenna obstructions.* Possible obstructions include ice, snow, and items on a roof rack.
- *Reception interference.* Hills, mountains, tall buildings, bridges, tunnels, freeway overpasses, parking garages, dense tree foliage, and thunderstorms can interfere with reception.
- *Station overload.* A stronger signal may overtake a weaker one and result in an audio mute and a display of NO SIGNAL.

TELEMATICS

Telematics is a technology that allows the transfer of data over a distance and from remote sources. It is best defined as the integrated use of telecommunications and information technology (IT). IT is a branch of engineering focused on the use of computers and telecommunications equipment to retrieve, store, sort, and transmit data. Telematics is used by NASA to control the flight of unmanned spacecraft while in flight.

Telemetry is widely used by NASA to transmit data and photos from space flights. **Telemetry** relates to the transmission of measurements from a remote location to another point where the data can be studied. Telemetry relies on remote sensors that measure electrical or physical data. These inputs are converted to voltage, and a multiplexer combines the voltages and timing data to form a single data stream. The signal is sent to a receiver that translates the data stream, and the data are then displayed and processed. Telemetry commonly refers to wireless data transfer, but it also includes other ways of data transfer, such as telephone lines and computer networks. Telemetry is also used to transmit engine, chassis, and transmission data from race cars to a race team's pit area. Telemetry is used by many automotive systems, including a tire pressure monitoring system (TPMS), through LAN or radio waves.

Today in the automotive world, telematics relies on wireless communication, but wires and optical links may be used. It is used to provide displays for the following:

- Navigation assistance
- Traffic information
- Satellite radio or video
- High-speed Internet
- Automatic air bag deployment notification
- Vehicle tracking
- Vehicle conditions

Phone Systems

There are many different systems that provide for hands-free cellphone systems that allow for communication between a cellphone and a vehicle's entertainment system and on-board microphone. Most often, communication and entertainment systems are connected by Bluetooth technology or a standard USB port. Voice commands can dial or answer the phone and choose the entertainment media.

BLUETOOTH A technology available for bringing information, voice, and video into the passenger compartment is Bluetooth. Although there are other methods for doing this, Bluetooth is the most commonly used. Bluetooth is a wireless connection that uses a 2.4 GHz frequency band. When Bluetooth is built into the radio receiver, a Bluetooth-compatible cellphone can be connected to the main sound system through a wireless connection. This allows for hands-free operation of the phone without the need for a connector or cable to connect the phone **(Figure 23–35)**. Likewise, when the receiver is Bluetooth

FIGURE 23–35

Phone controls on the steering wheel.

enabled, a compatible portable audio player can be connected through a wireless connection. This allows files stored in an audio player to be heard, and the features of the audio player, such as play/stop, can be operated directly from the radio receiver assembly.

Voice Activation System

Voice activation or control systems allow for the control and operation of some accessories with voice commands. The voice commands are in addition to normal manual controls. Voice activation is commonly used for cellphone operation but can be used on other controls. Voice activation systems recognize the driver's voice and can respond with answers in response to the driver's questions. Once the system has understood the driver's request, it responds by carrying out the desired function, such as changing the stations on the radio. The system works through the microphones located near the driver. The voice activation system can recognize up to 2000 commands and numeric sequences.

The telematics transceiver sends power to the telephone microphone assembly, and the microphone sends signals to the display and navigation module display through the transceiver. The displays and microphone are connected to each other using microphone connection detection signal lines.

A microphone converts sound into an electrical signal. Most microphones use electromagnetic induction, capacitance change, piezoelectric generation, or light modulation to change mechanical vibrations into an electrical voltage signal. A microphone has a housing, a vibration sensitive transducer to send signals to other equipment, and an electronic circuit to adapt the output of the transducer to the receiver. A wireless microphone also contains a radio transmitter. In a vehicle, the microphone may be located in the instrument panel, steering wheel, or map light assembly.

USING SERVICE INFORMATION

Inspection and diagnostic procedures for the microphone circuit can be found in the service information.

NAVIGATION SYSTEMS

Navigation systems use **global positioning satellites (GPSs)** to help drivers make travel decisions. These global positioning systems set up a mathematical grid between the satellites and radio stations on the ground. The exact position of a vehicle can be plotted on the grid; therefore, the system knows exactly where the vehicle is **(Figure 23–36)**. A GPS antenna collects the signals for the control unit, which determines the vehicle's location according to latitude and longitude coordinates, and a gyroscope sensor that monitors the vehicle's direction by monitoring angular velocity. The processor compares the data from the global positioning satellites and gyro sensor to the information stored in its memory or to data found on a designated CD or DVD.

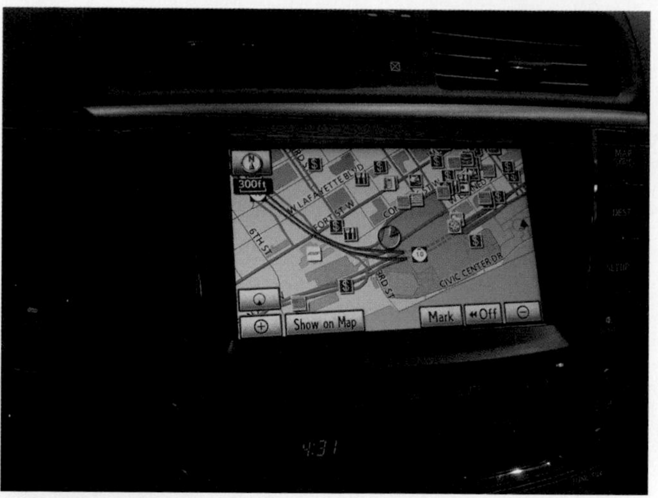

FIGURE 23–36

The screen of a typical navigational centre.

Two methods are used to determine the exact location of a vehicle, and both are used simultaneously. One is GPS navigation, which detects the location of the vehicle using radio waves from a satellite. The current position is determined by measuring the time it takes the radio waves from the satellites to reach the processor. The other method is autonomous navigation, which uses the gyro and speed sensors to monitor the vehicle's travel. The signals from these sensors are updated once per second, and if they have changed, the screen is updated.

Usually, map data provided on a DVD and navigation information are displayed on a TFT or LCD colour screen. Many systems give turn-by-turn guidance either by voice, on-screen displays, or both. Many systems feature touch-screen technology and can display and control other systems, such as air-conditioning, heating, and sound systems. Touchscreens have touch-sensitive switches. When a touch-sensitive switch is pressed, the outer glass bends to contact the inner glass at the pressed position. By doing this, the voltage ratio is measured, and the pressed position is detected. Voice prompts can be sent through the audio system. Other systems allow watching a DVD while the vehicle is parked.

Most navigation systems can display the following:

- Current traffic information regarding traffic backups, including alternative routes so that travel is not delayed
- A road map marking the exact location of the vehicle
- The best way of getting to a destination
- The number of kilometres that have been travelled and how many remain before reaching a destination, as well the expected time of arrival

Vehicle Tracking Systems

Vehicle tracking systems can identify the location of a vehicle if it has been stolen or lost. The system is based on the vehicle's navigational system and/or the driver's cellular phone. When the cellphone is the identifier, the tracking system is triggered when the thief attempts to place a phone call. If the correct code is not entered into the phone, the satellite begins to track the vehicle. This tracking signal is then monitored by an operator who can call the police in the area where the vehicle is being tracked. When the system relies on the GPS, a security or police officer can watch the movement of the vehicle on a remote computer screen.

Some systems automatically send a signal to the vehicle tracking operator if an air bag is deployed. This signal, in addition to the global positioning satellite network, allows the authorities to immediately know when and where a serious accident has occurred, and an emergency squad can respond quickly.

Systems such as GM's OnStar (a cellphone-based system) and Ford's SYNC (an internal system co-developed by Microsoft) can also provide the driver with control over various systems, such as climate control. Every system provides different conveniences to the driver, such as the following (keep in mind that not all systems provide the same features):

- Turn-by-turn navigation
- 911 emergency assistance and advice
- Roadside assistance
- Remote diagnostics (while the vehicle is driven)
- Remote door unlocking
- Email
- Full cellphone features
- Total control of the audio system

POWER LOCK SYSTEMS

Although systems for automatically locking doors vary from one vehicle to another, the overall purpose is the same—to lock all outside doors. As a safety precaution against being locked in a car due to an electrical failure, power locks can be manually operated.

When either the driver's or passenger's control switch is activated (either locked or unlocked), power from the fuse panel is applied through the switch to a reversible motor. A rod that is part of the lock assembly moves up or down to lock or unlock the door. On some models, the signal from the switch is applied to a relay that, when energized, applies an activating voltage to the door lock actuator. The door lock actuator consists of a motor and a built-in circuit breaker. Because the motors are reversible, each does not have its own ground. The ground for the lock circuits is at the master or door circuits **(Figure 23–37)**. On station wagon models with power door locks, the tailgate lock actuator is also controlled by the door lock switches. Station wagon models without power door locks can be equipped with a separate tailgate power lock system.

Today, most power door lock systems place the motor in the door latch assembly and are controlled by a BCM. This allows door lock/unlock functions from a remote control and allows for integration with the vehicle's security system. These systems depend on many different components, including a remote control door lock receiver, door modules, individual door switches and relays, individual door latch assemblies, and LAN and/or LIN serial data.

Most models use a control switch mounted in the door arm or in the door trim panel. However, some models use switches controlled by the front door push-button locks **(Figure 23–38)**.

AUTOMATIC DOOR LOCKS Some vehicles have an autolock or automatic door lock system. This provides for the automatic locking of all doors when the doors are closed, ignition is turned ON, the transmission is shifted out of PARK, and the vehicle is moving more than 7 km/h (4 mph). Depending on the vehicle, configuration of when or if automatic door locks activate may be changed using a scan tool to suit customer preferences.

Child Safety Locks

To prevent accidental door opening by a child in a rear seat, nearly all vehicles have a child safety lock. When this lock is set, the rear doors cannot be opened from the inside. These doors can be opened

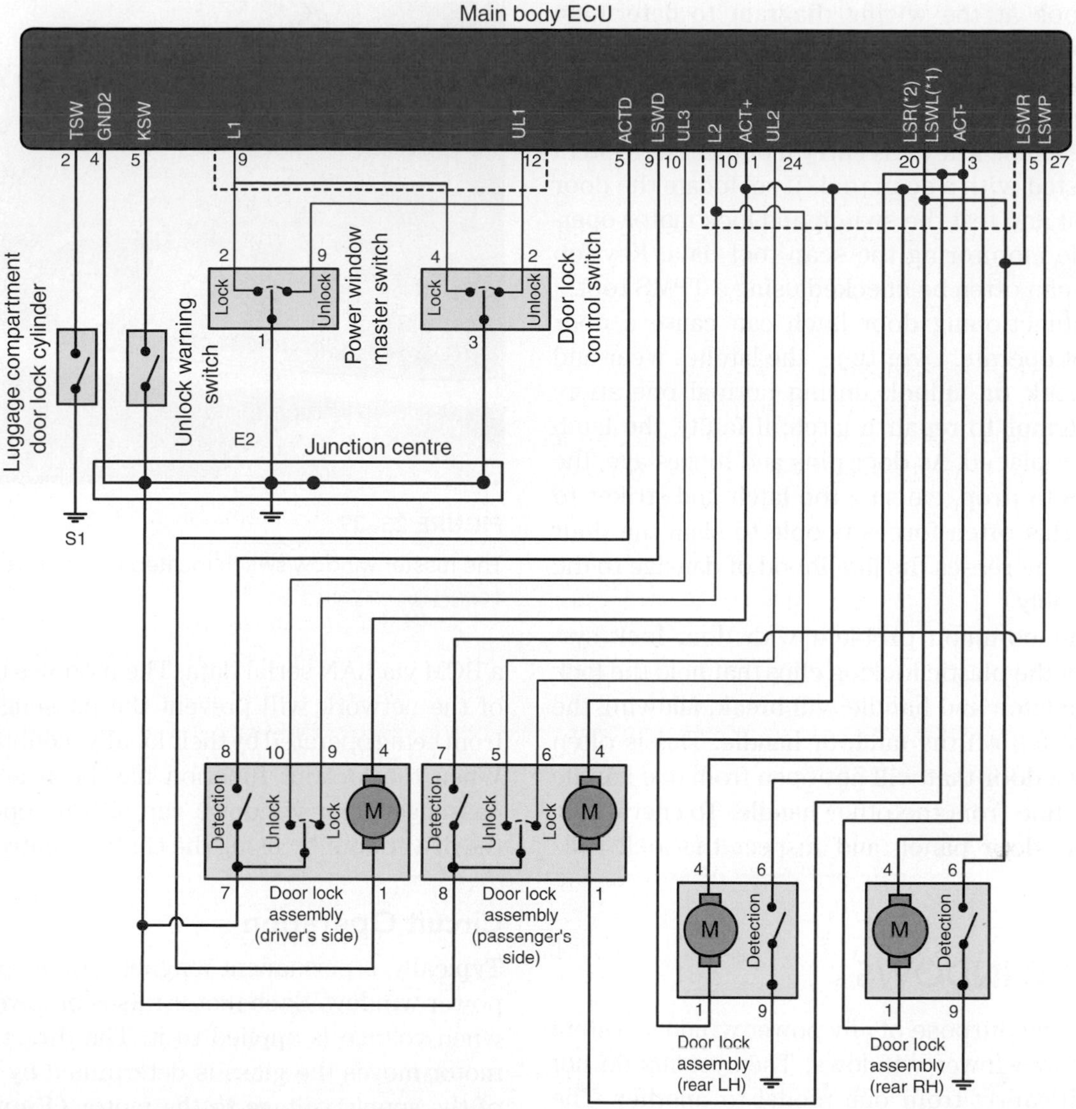

Main body ECU

TSW GND2 KSW L1 UL1 ACTD LSWD UL3 L2 ACT+ UL2 LSR(*2) LSWL(*1) ACT- LSWR LSWP

Luggage compartment door lock cylinder

Unlock warning switch

Power window master switch

Door lock control switch

Lock Unlock

E2 Junction centre

S1

Detection Unlock Lock M

Door lock assembly (driver's side)

Door lock assembly (passenger's side)

Door lock assembly (rear LH)

Door lock assembly (rear RH)

FIGURE 23–37

A wiring diagram for a power door lock system.

FIGURE 23–38

A central door lock switch.

only from the outside. The lock switch for this feature is normally located on the rear edge of the rear doors and must be set separately for each door.

Power Trunk Release

The power trunk release system is a relatively simple electrical circuit that consists of a switch and a solenoid. When the trunk release switch is pressed, voltage is applied through the switch to the solenoid. With battery voltage on one side and ground on the other, the trunk release solenoid energizes, and the trunk latch releases to open the trunk lid.

Diagnosis

Power door lock or trunk release systems rarely give trouble. However, if all of the door locks are inoperative, begin by checking the circuit fuse. If the fuse is okay, look at the wiring diagram to determine the next best location for testing. If the system uses a remote key, test the door lock operation, using the remote and the interior door controls. Vehicles with remote locks use a keyless entry module. The system may be tested with a scan tool. If so, locate the door lock circuit and test the switch and lock motor operation while monitoring the scan tool data. Key fob operation can often be checked using a TPMS tool.

A malfunctioning door latch can cause a door lock to not operate. Over time, the latches wear and may not lock or unlock during normal operation. Do not attempt to repair a latch; if faulty, the latch should be replaced. As door pins and hinges age, the door tends to drop, causing the latch and striker to misalign. This often forces people to slam the door shut, which increases the likelihood of damage to the latch assembly.

Another common problem with door lock systems is that the plastic locking clips that hold the lock rods to the latch and handle will break, allowing the lock rod to fall off the latch or handle. This is often evident by a door that will not open from one handle but works fine from the other handle. To check this, remove the door panel, and inspect the lock rods and clips.

POWER WINDOWS

Obviously, the purpose of any power window system is to raise and lower windows. The systems do not vary significantly from one model to another. The major components of a typical system are the master control switch, individual window control switches, control modules, and the window drive motors.

The master control switch provides overall system control. Power for the system comes directly from the fuse panel.

Depending on the vehicle, there may be one fuse or circuit breaker for the system or a fuse for each door. Power for the individual window control switches comes through the master control switch **(Figure 23–39)**. When the master switch allows, the other windows can be controlled by the switches located on the associated doors.

Window Lockout Systems

A lock switch included on four-door master control switches is a safety device to prevent children from opening windows at the individual window switches without the driver's knowledge. When the lockout switch is pressed, it briefly closes and sends a signal to a BCM via LAN serial data. The modules that are part of the network will prevent the passenger windows from being operated by their locally mounted switches. When the lockout function has been activated, the passenger door windows can still be operated from the driver door by using the central control switch.

FIGURE 23–39

The master window switch located in the driver's door.

Circuit Operation

Typically, a permanent magnet motor operates each power window. Each motor raises or lowers the glass when voltage is applied to it. The direction that the motor moves the glass is determined by the polarity of the supply voltage to the motor **(Figure 23–40)**. Today, most driver and front passenger doors have a motor with its own module. Each of the modules is controlled by a BCM. That signal can cause the window to move up or down and convey the action across the serial data network line to control the rear windows.

Voltage is applied to a window motor when the UP switch in the master switch assembly is activated. The motor is grounded through the DOWN contact. Battery voltage is applied to the motor in the opposite direction when any DOWN switch in the master switch assembly is activated. The motor is then grounded through the master switch's UP contact.

The operation of the individual window switches is much the same. When the UP switch is activated, voltage is applied to the window's motor. The motor is grounded through the DOWN contact at the switch and the DOWN contact at the master switch. When the DOWN switch in the window switch is activated, voltage is applied to the motor in the opposite direction. The motor is grounded through the UP contact at

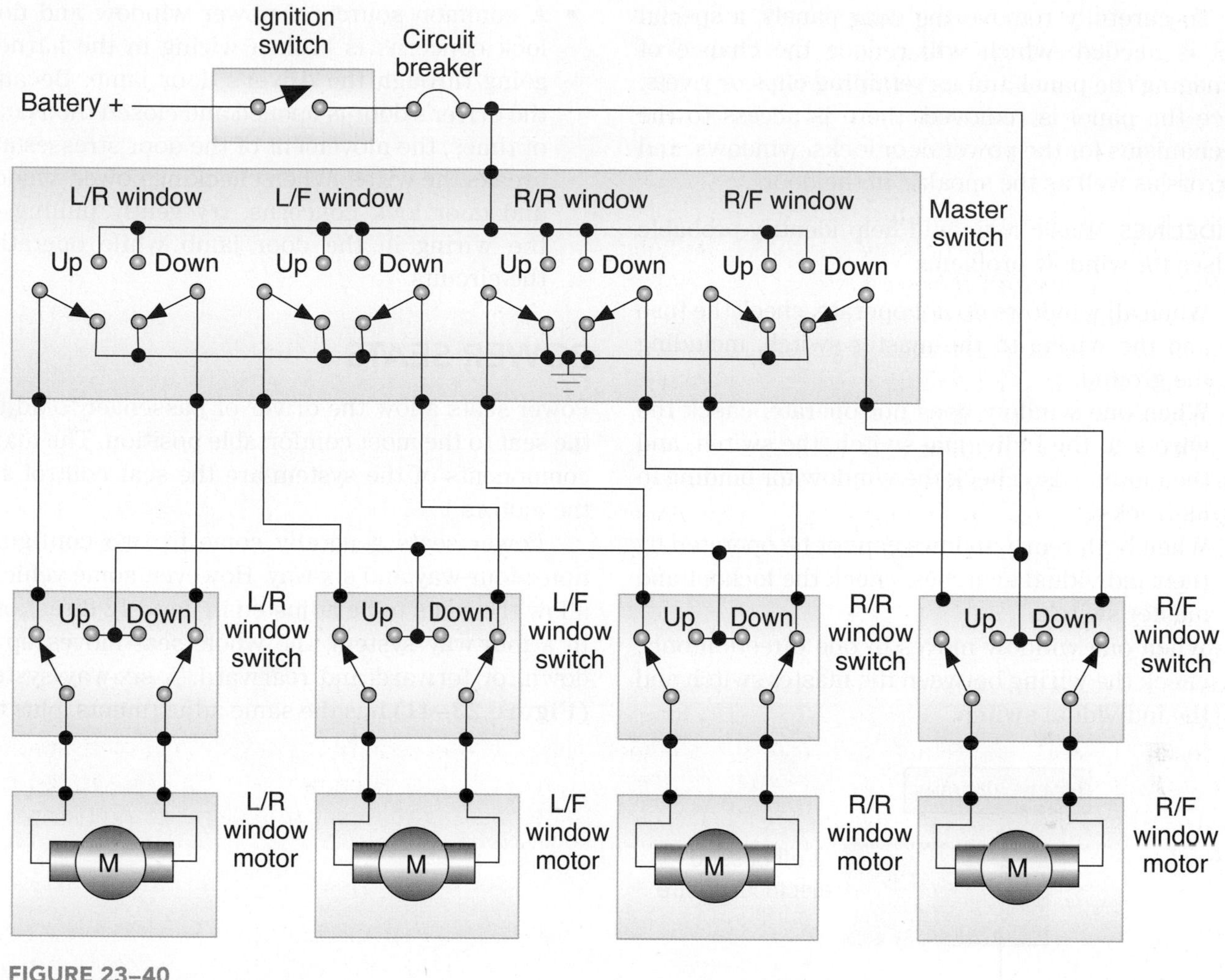

FIGURE 23–40

A typical power window circuit.

the window switch and the UP contact in the master switch. This runs the motor in the opposite direction.

Each motor is protected by an internal circuit breaker. If the window switch is held too long with the window obstructed or after the window is fully up or down, the circuit breaker opens the circuit.

EXPRESS WINDOWS Some vehicles are equipped with an express down window feature. This feature allows the windows to be fully opened or closed by momentarily pressing the window switch for more than 0.3 second and then releasing it. The window may be stopped at any time by depressing the UP or DOWN switch. The express window option relies on an electronic module and a relay. When signalled, the module energizes the relay, which completes the motor's circuit. When the window is fully down, the module opens the relay control switch, which stops power to the motor. The motor will also stop 10 to 30 seconds after the DOWN switch is depressed.

OBSTACLE-SENSING WINDOWS This feature is available on some vehicles and prevents the window from closing if the system detects something, such as fingers, that are between the window glass and window frame. This detection causes the window to reverse its direction. These systems typically rely on infrared sensors. When the light beams are broken by the presence of an obstacle, the window will reverse and go down.

Diagnosis

The first step in diagnosing a power window system is to determine whether the whole system is not working correctly or just one or two windows. If it is the whole system, the problem can be isolated to fuses, circuit breakers, or the master control switch. If only a portion of the system does not work, check the components used in the portion that is not working. Removing the door trim panels allows access to the motor and window linkage.

To carefully remove the door panels, a special tool is needed, which will reduce the chance of damaging the panel and its retaining clips or rivets. Once the panel is removed, there is access to the mechanisms for the power door locks, windows, and mirrors as well as the speaker in the door.

GUIDELINES Basic logic will help identify probable causes for window problems.

- When all windows do not operate, check the fuse and the wiring to the master switch, including the ground.
- When one window does not operate, check the wiring to the individual switch, the switch, and the motor. Also check the window for binding in its tracks.
- When both rear windows cannot be operated by their individual switches, check the lockout and master switch.
- When one window moves in one direction only, check the wiring between the master switch and the individual switch.
- A common source of power window and door lock concerns is broken wiring in the harness going through the driver's door jamb. Because the driver's door is opened and closed thousands of times, the movement of the door stresses and breaks the wires. When checking power window and door lock concerns, try gently pulling on the wiring in the door jamb while operating the circuits.

POWER SEATS

Power seats allow the driver or passenger to adjust the seat to the most comfortable position. The major components of the system are the seat control and the motors.

Power seats generally come in two configurations: four-way and six-way. However, some vehicles allow the seats to be adjusted in up to 12 directions. In a four-way system, the whole seat moves up or down, or forward and rearward. A six-way system **(Figure 23–41)** has the same adjustments, plus the

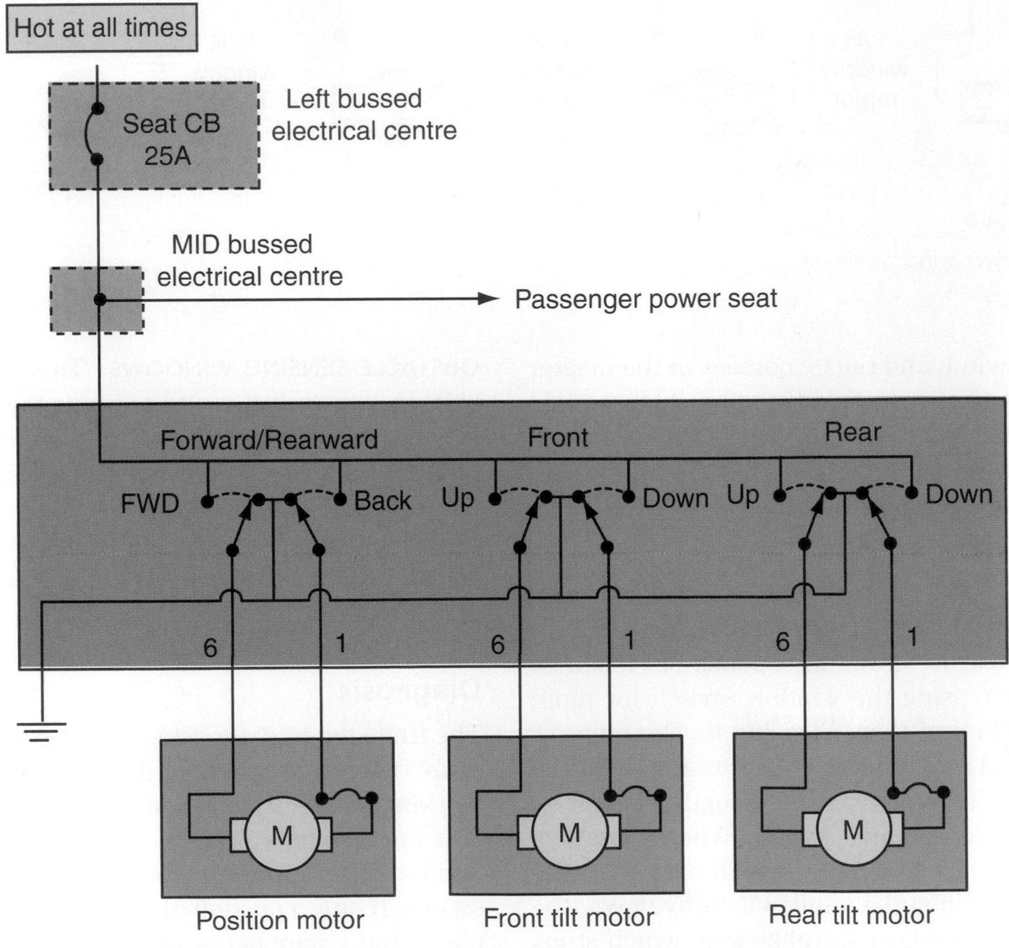

FIGURE 23–41

A power seat circuit.

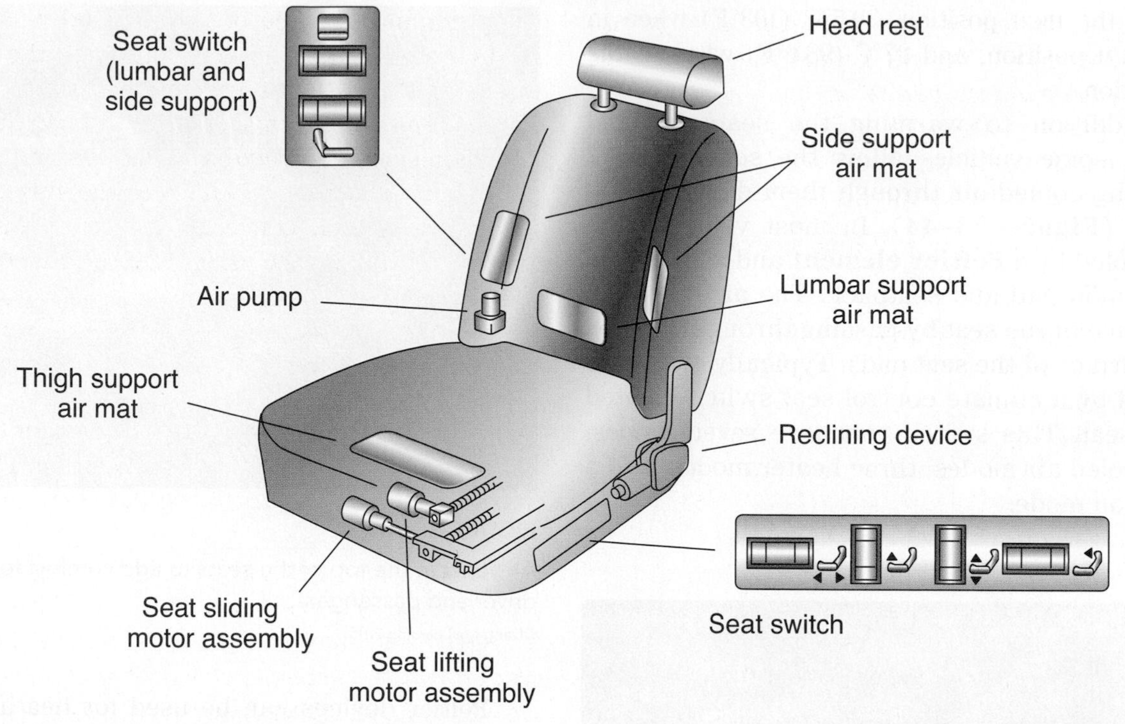

FIGURE 23–42

Seat adjustments lead to driver and passenger comfort.

capability to adjust the height of the front and rear of the seat. Generally, a four-way system is used on bench seats, and a six-way system is used on split-bench and bucket seats. Some units also control the tilt, rear/forward movement, height, and angle of the seatback. The adjuster for the seatback may also control the height of the headrest or restraint.

Two motors are typically used on four-way systems, whereas three are used on six-way systems. Many newer systems rely on a horizontal motor, front vertical motor, rear vertical motor, and recline motor. The names of the motors identify their function. To raise or lower the entire seat on a six-way system, both the front vertical height and the rear vertical height motors are operated together. The motors are generally two-directional motor assemblies that include a circuit breaker to protect against circuit overload if the control switch is held in the actuate position for long periods **(Figure 23–42)**.

Climate Control Seats

Many vehicles have an option that warms up the seats, an especially nice feature in cold climates. The system relies on heating coils in the seat cushion and back controlled by relays and switches **(Figure 23–43)**. Some systems offer a heat function at the front seats, and others also offer heated seats in the rear of the vehicle. Each seat is controlled by a separate switch. Many of these circuits are controlled by a BCM that receives inputs from various temperature sensors. The heated seat system is designed to warm the seat cushion and seatback to approximately 42°C (107.6°F)

FIGURE 23–43

This heated seat became very hot because of a shorted heating coil.

when in the HIGH position, 39.5°C (103°F) when in the MEDIUM position, and 37°C (98.6°F) when in the LOW position.

In addition to warming the seats in cold weather, some vehicles allow the seats to cool by passing cooled air through them during warm weather **(Figure 23–44)**. In most vehicles, the air is cooled by a **Peltier element** and a fan in the seat-cushion pad and seatback. The air moves to the surface of the seat by passing through grooves in the surface of the seat pads. Typically, the fan is operated by a climate control seat switch located on the seat. This switch can have seven modes: three cooled air modes, three heater modes, and a ventilation mode.

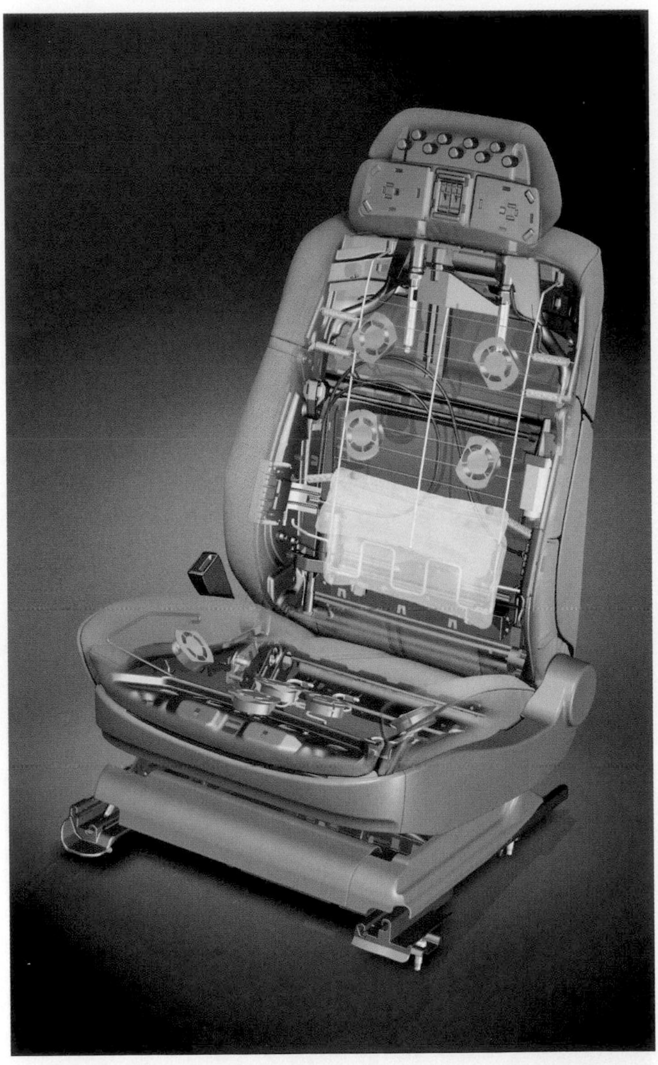

FIGURE 23–44

This seat features ventilation, heating, and multiple position adjustments, including the headrest and lumbar support; the small round objects are fans.

BMW Group

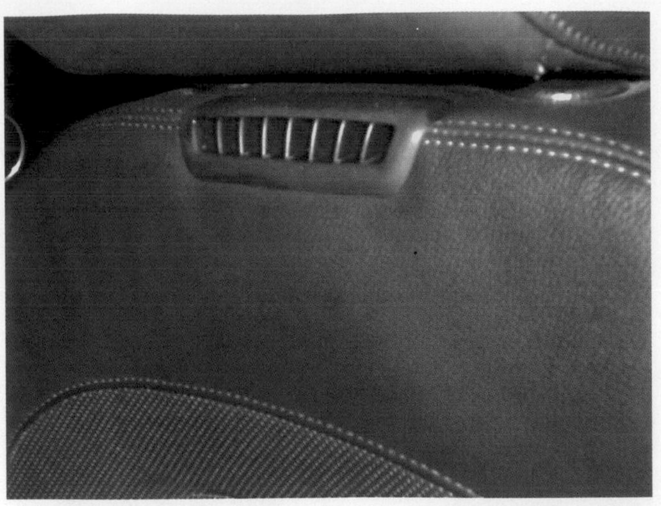

FIGURE 23–45

Air vents at the top of the seats to add comfort for the driver and passengers.

©Cengage Learning 2015

Peltier devices can be used for heating or for cooling, although they are most commonly used for cooling. When operated as a cooler, voltage is applied across the device, which creates a difference in temperature between the two sides of the device. The difference in heat creates a cool output at the seats when ambient temperatures are high. Normally, the maximum temperature difference between the hot side and cold side of the device is about 70°C (160°F).

Other systems offer air vents positioned at the top of the seats to provide warm or cool air **(Figure 23–45)**. These are especially nice when a driver of a convertible is out with the top down when the weather is either cool or hot.

Other Seat Options

Many different options are available for the seats of a vehicle. Some of the following features are available on vehicles with manual seats; others are available only with the power seat option.

POWER LUMBAR SUPPORTS A power lumbar support allows the driver to inflate or deflate a bladder located in the lower seatback. Adjusting this support improves the driver's comfort and gives support at the lower lumbar region of the spinal column.

MEMORY SEATS The memory seat option allows for automatic positioning of the driver's seat to different, often up to three, programmable positions **(Figure 23–46)**. This feature allows different

FIGURE 23–46

Seat adjustment controls and seat memory buttons.

DaimlerChrysler Corporation

drivers to have their desired seating position automatically adjusted by the system. It also allows an individual driver to set different positions for different driving situations.

Some systems with a remote key fob can be programmed to move the seat to its memory position whenever the unlock button of the key fob is depressed. Each driver can have his or her own key fob, and the desired seat position will be selected when the door is unlocked. Also, other systems automatically adjust the power mirrors to a setting for each driver.

ADAPTIVE AND ACTIVE SEATING Adaptive seating, a feature that moves the seat slightly as the driver shifts positions, is offered in some luxury vehicles. Moving the seat improves the driver's comfort and support while driving for a long time. Active seating stimulates the spine and surrounding muscles with continuous yet virtually imperceptible motion. This type of seating is designed to prevent the driver from getting saddle sore while sitting without moving for a long time. The right and left halves of the seat cushion move up and down at cyclical intervals. To do this, two pillows are integrated within the seat's upholstery. A hydraulic pump alternately inflates the two cavities with a mixture of water and glysantine.

MASSAGING SEATS To help reduce driver fatigue, groups of air bags or bladders fill with air and then release the air. In many cases, this sensation moves the seatback up and down when activated. The air bags alternately inflate and deflate to create a "kneading" sensation on the backs of the driver and occupants. A seat massaging or vibration system normally uses an air pump (located in the seatback or trunk), solenoid valves to control the airflow in and out of the air bags, and a control switch. Depending on the system, the control may adjust the intensity, speed, and placement of the air bags' activity.

The actual system and its operation varies widely. The number of air bags and their placement in the seatback greatly differs with each design. Simpler systems have a simple on–off switch, whereas others offer as many as 25 different control settings. In the very controllable systems, there may be up to 10 bladders placed strategically in the seatback. Some simpler systems have five rows of two bladders, and the placement of those bladders work the shoulders and back of the occupant; however, there is only one control setting.

Diagnosis

Before testing a power seat system, determine if the seats are BCM controlled or directly controlled. If the seats are controlled by a BCM, use a scan tool to check the system. In both cases, do a visual inspection of the wires and connectors in the system. Two types of problems affect power seats: One is a tripped or constantly tripping circuit breaker, and the other is the inability of the seat to move in a direction.

The resettable circuit breaker should protect the system from a short circuit or from high current due to an obstructed or stuck seat adjuster. The circuit breaker must be replaced if it is faulty. Before testing the circuit breaker, make sure the seat tracks are not damaged and that nothing is physically preventing the seat from moving.

When a seat does not move in a particular direction, turn on the dome light, and then move the power seat switch in the problem direction. If the dome light dims, the seat may be binding or have some physical resistance on it. Check under the seat for binding or obstructions. If the dome light does not dim, test the system.

Disconnect the negative terminal at the battery. Remove the power seat switch from the seat or door armrest. Check for battery voltage to the switch. If there is no voltage and the circuit breaker is okay, check for an open in the power feed circuit. If voltage is present, check for continuity between the ground connection at the switch and a good ground. If there is no continuity, repair the ground circuit. If there is continuity, the switch must be tested. Use an ohmmeter to test the continuity of the switch in each position. Check the service

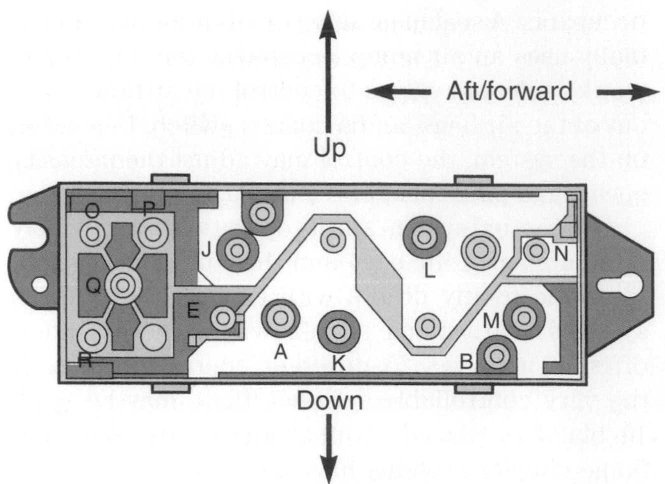

Terminals shown as viewed from rear of switch

Power seat switch	
Switch position	Continuity between
Off	B-N, B-J, B-M, B, E, B-L, B-K
Vertical up	A-E, A-M, B-N, B-J
Vertical down	A-J, A-N, B-M, B-E
Horizontal forward	A-L, B-K
Horizontal aft	A-K, B-L
Front tilt up	A-M, B-N
Front tilt down	A-N, B-M
Rear tilt up	A-E, B-J
Rear tilt down	A-J, B-E
Lumbar off	O-P, P-R
Lumbar up(inflate)	O-P, Q-R
Lumbar down(deflate)	O-R, P-Q

FIGURE 23–47

The terminals of a power seat switch identified for continuity checks.

Chrysler Group LLC

information for the expected continuity between the various terminals of the switch **(Figure 23–47)**. If the switch checks out, test the motor. If the switch is bad, replace it.

Test each motor by connecting it to a power source and a good ground. If while doing this, the motor stops running, immediately disconnect the power source. If the front up–down motor, rear up–down motor, or slide motor does not run or does not run smoothly, replace the motor.

SHOP TALK

The slide motor is normally not available as a separate unit because it is assembled into the seat track.

If the massaging seats do not work properly, the controls, air hoses, electrical wiring, sensors, and pump should be checked.

POWER MIRROR SYSTEM

The power mirror system consists of a joystick-type control switch and a dual motor drive assembly located in each mirror assembly. The driver's door switch assembly has two switches assembled as a single unit. One switch selects which side mirror, left or right, is to be adjusted. The other switch adjusts that mirror.

Rotating the power mirror switch to the left or right position selects one of the mirrors for adjustment. Moving the joystick control up and down or right and left moves the mirror to the desired position. The dual motor drive assembly is located behind the mirror glass **(Figure 23–48)**. The position of the mirrors may be tied to the memory power seats and will automatically adjust when a seating position is selected.

A few vehicles automatically tilt the passenger-side outside mirror downward whenever the transmission is placed into REVERSE. This allows the driver to see the area directly next to the vehicle during parking.

ELECTROCHROMIC MIRRORS Many new vehicles are being equipped with electrochromic side-view mirrors. Requiring no electrical connections, these mirrors operate on the intensity of the glare in much the same manner as a pair of eyeglasses with photochromatic lenses. When glare is heavy, the mirrors

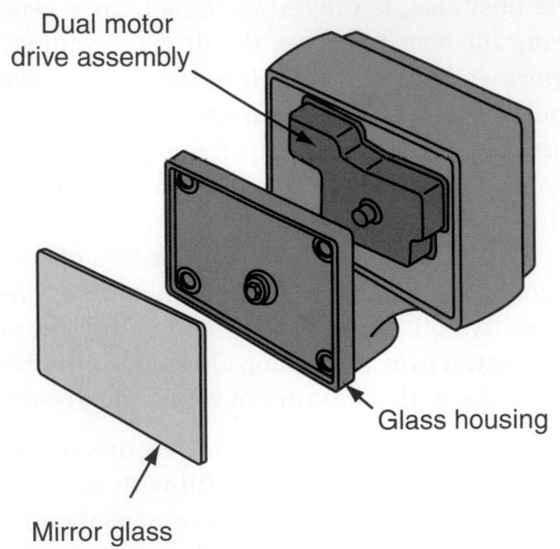

FIGURE 23–48

A power mirror motor assembly.

darken fully (down to 6 percent reflectivity). When glare is mild, the mirrors provide 20 to 30 percent reflectivity. When glare subsides, the mirrors change to the clear daytime state.

Inside Mirrors

Some automatic day/night inside rearview mirrors use a thin layer of electrochromic material between two pieces of conductive glass. A switch located on the mirror allows the driver to turn the feature on or off. When it is turned on, the mirror switch is lit by an LED. The self-dimming feature is disabled whenever the transmission is placed into REVERSE.

When the mirror is turned on, two photocell sensors monitor external light levels and adjust the reflection of the mirror **(Figure 23–49)**. The ambient photocell sensor detects the light levels outside and in front of the vehicle. The headlight photocell faces rearward to detect the level of light coming in from the rear of the vehicle. When there is a difference in light levels between the two photocells, the mirror begins to darken.

On some vehicles, the electrochromic feature of the driver's outside rearview mirror is controlled by the inside rearview mirror. The inside mirror supplies a signal to the outside mirror and causes it to darken along with the inside mirror.

The automatic day/night mirror cannot be repaired. If it is faulty, it must be replaced.

Some rearview mirrors are fitted with a directional compass display. The readings on the compass appear on the mirror's reflective surface, normally in the lower left corner.

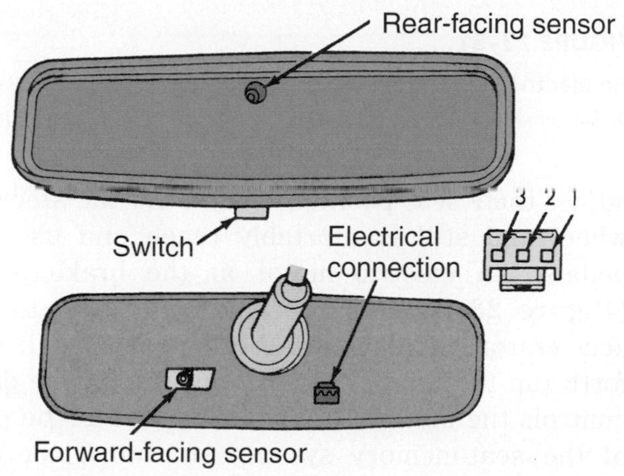

FIGURE 23–49

An automatic day/night mirror.

Chrysler Group LLC

Power Folding Mirrors

Some vehicles have power folding mirrors. This feature allows the driver to retract the outside mirrors to a fully folded position where they fold in toward the door's windows. The mirrors can also be extended to the fully unfolded position for normal use. The operation of the folding mirrors is accomplished by a switch and an additional power folding mirror motor.

REAR-WINDOW DEFROSTER AND HEATED MIRROR SYSTEMS

The rear-window defroster (also called a **defogger** or **deicer**) heats the rear-window surface to remove moisture and ice from the window. On some vehicles, the same control heats the outside mirrors. The major components of a rear-window defroster include a switch, a relay assembly, and the heating elements on the glass surface **(Figure 23–50)**.

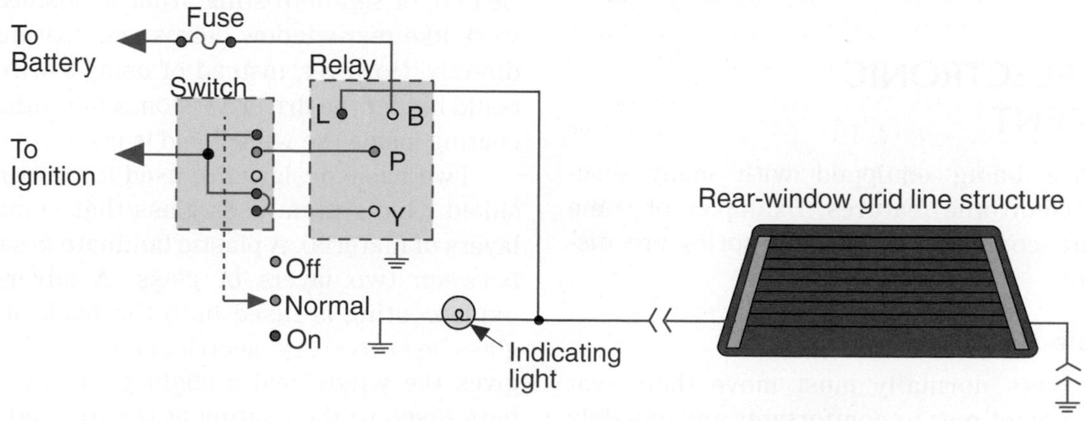

FIGURE 23–50

A rear defogger circuit schematic.

Pressing the rear-window defroster switch momentarily energizes the relay. Battery voltage is then applied through closed contacts of the relay to the rear-window defroster grid. On models with a heated mirror, current also flows through a separate fuse to the mirror's heated grid.

After about 10 minutes, a time-delay circuit opens the ground path to the relay's coil, and the coil de-energizes, shutting off power to the grids. The time-delay circuit prevents the system from remaining on during periods of extended driving. The system can also be manually turned off.

Diagnosis

One of the most common problems with rear-window defrosters is damage to the grids on the window. Damage can be caused by hard objects rubbing across the inside surface of the glass or by using harsh chemicals to clean the window. When a segment of the grid breaks, it opens the circuit. Often, the customer's complaint is that the unit does not defrost the entire window. Normally, one or two lines of the grid are open. The open can be found by using a test light. With the defroster turned on, voltage should be present at all points of the grid. If part of the grid does not have voltage, move the probe toward the positive side of the grid. Once voltage is present, you know the open is between those two points. Opens can be repaired by painting a special compound over the open. The correct procedure for doing this is shown in Photo Sequence 18.

If none of the grids heat up, check for voltage at the connection to the grids, and then check the ground circuit. If no voltage is present at the grids, examine a wiring diagram to determine where to check the circuit. If the switch and relay work correctly, there is likely an open in the wiring to the grid.

OTHER ELECTRONIC EQUIPMENT

Vehicles are being equipped with many electrical and electronic features. Examples of some of the more common newer accessories are discussed here.

Adjustable Pedals

Shorter drivers normally must move their seat very close, sometimes uncomfortably and unsafely close, to the steering wheel. By moving the pedals toward the driver, drivers may be able to adjust their seat position away from the steering wheel and still comfortably reach and use the pedals. An electric motor at the brake pedal **(Figure 23–51)**, with a cable connection to the accelerator pedal, moves both pedals back and forth (up to 76 mm [3 in.]). A switch on the dash controls the motor. This feature may also be part of the seat-memory system, so the driver can quickly bring both the seat and pedals to the most comfortable position.

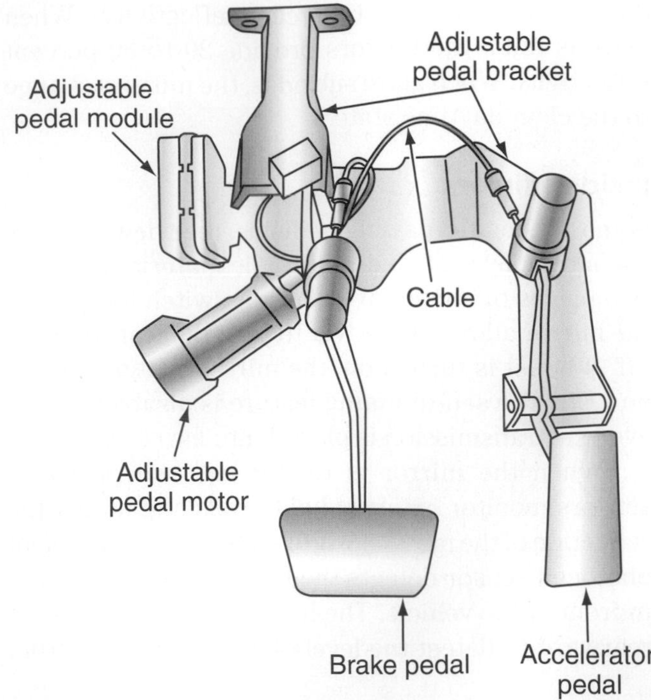

FIGURE 23–51

An electrically controlled pedal assembly.

Heated Windshields

Heated, or self-defrosting, front windshield systems work like rear-window defrosters, heating the glass directly. However, instead of using a wire grid that could hinder the driver's vision, a microthin metallic coating inside the windshield is used.

Two basic designs are used for heating a windshield. One system uses glass that contains three layers of material. A plastic laminate is sandwiched between two layers of glass. A silver and zinc oxide coating is fused onto the back of the outer glass layer to carry electrical current. This coating gives the windshield a slight gold tint. Silver bus bars fused to the coating at the top and bottom of the windshield connect the coating to power and ground circuits.

TYPICAL PROCEDURE FOR GRID WIRE REPAIR

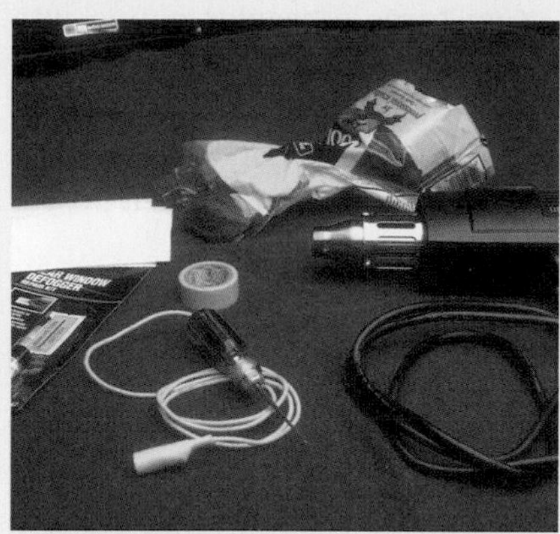
P18–1 The tools required to perform this task include masking tape, repair kit, a 260°C (500°F) heat gun, a test light, steel wool, alcohol, and a clean cloth.

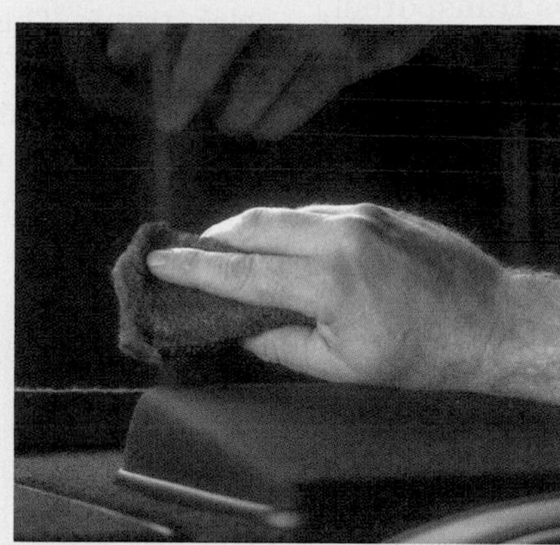
P18–2 Clean the grid line area to be repaired. Buff with fine steel wool. Wipe clean with a cloth dampened with alcohol. Clean an area about 6 mm (1/4 in.) on each side of the break.

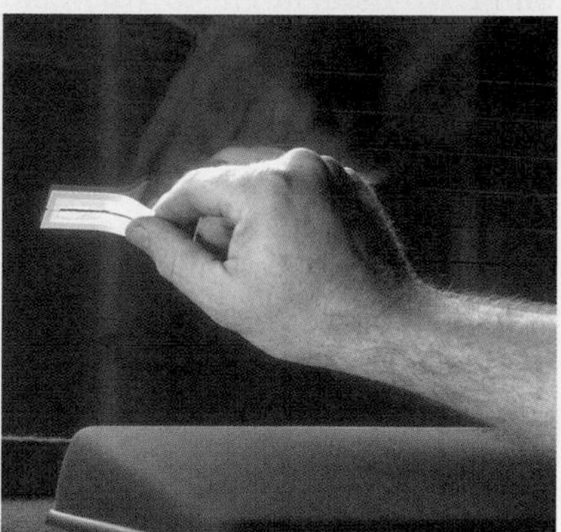
P18–3 Position a piece of tape above and below the grid. The tape is used to control the width of the repair, so try to match the width with the original grid.

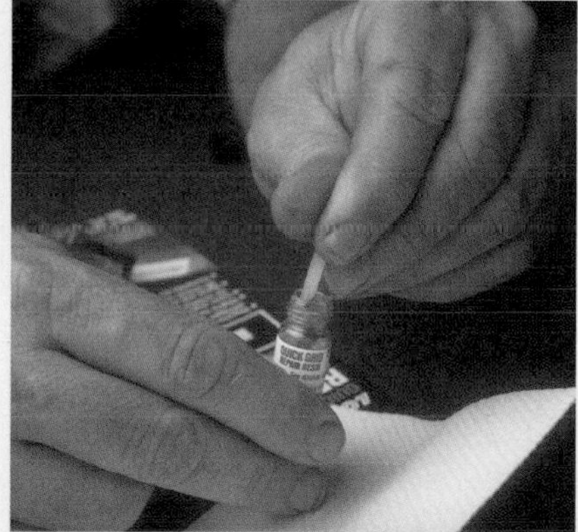
P18–4 Mix the hardener and silver plastic thoroughly. If the hardener has crystallized, immerse the packet in hot water.

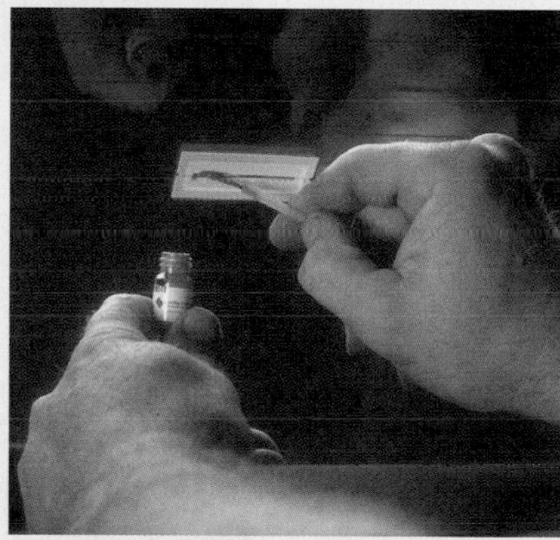
P18–5 Apply the grid repair material to the repair area, using a small stick.

P18–6 Carefully remove the tape.

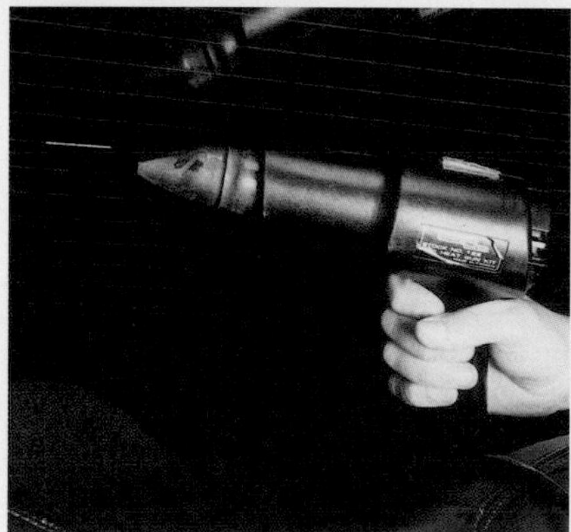
P18–7 Apply heat to the repair area for two minutes. Hold the heat gun 25 mm (1 in.) away from the repair.

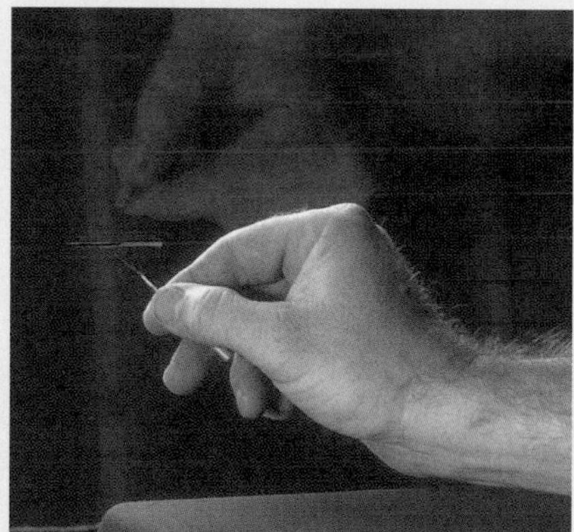
P18–8 Inspect the repair. If it is discoloured, apply a coat of tincture of iodine to the repair. Allow to dry for 30 seconds, and then wipe off the excess with a cloth.

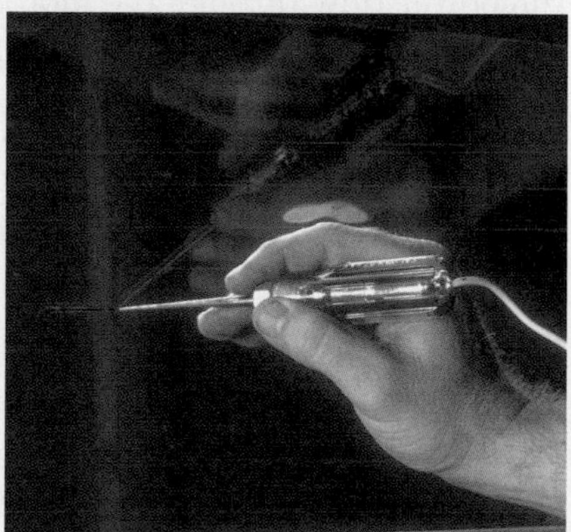
P18–9 Test the repair with a test light. Note: It takes 24 hours for the repair to fully cure.

The other design also has three layers. However, the inner layer is not plastic; rather, it is a thin film of resistive coating sprayed between the inner and outer windshield glass. This coating is transparent, so the windshield does not appear tinted.

In a typical windshield defroster system, the generator's output is redirected from the normal electrical system to the windshield circuit. This leaves all other electrical circuits to operate on power from the battery. When the voltage regulator senses a drop in battery voltage, it full-fields the generator so that it can put out between 30 and 90 volts. If battery voltage drops below 11 volts, the system is turned off, and the AC generator's output is again directed to the battery. Check the service manual for specific voltage details.

The defrost cycle generally lasts for about four minutes. After that, the generator is switched to the normal charging operation controlled by the voltage regulator. If the windshield is not clear, the system can be selected again.

Power Roof Systems

A **power roof** (moonroof, sunroof, panoramic roof) can slide the roof panel open or closed. It can also tilt the panel up in the back to allow fresh air and natural light into the passenger compartment. The major components of any power roof panel system are a relay, control switch, sliding roof panel, and motor. This circuit is normally protected by the in-line circuit breaker.

When the two-position switch is moved to the open position, the roof panel moves into a storage area between the headliner and the roof. The panel stops moving any time the switch is released. Moving the switch to the closed position reverses the power flow through the motor.

If the system is not operating, check the fuse or circuit breaker. If these are okay, check for power at the switch with the ignition switch. If the voltage is present, the relay is okay. Check for power to the motor with the switch held in the open position. Refer to the service information for additional diagnosis and testing information.

SOLAR SUNROOF This feature uses light-sensitive elements under the glass sunroof to produce electricity to power a ventilator inside the vehicle. Even with the ignition off, the interior is supplied with a continuous flow of fresh filtered air. The solar sunroof puts no additional demands on the battery or charging system and reduces the energy used by the climate control system.

Retractable Hardtops

A new trend in convertibles is retractable hardtops. The hard roof lowers and conceals itself in the vehicle's trunk. This system turns a regular coupe into a convertible. Most systems use electrohydraulic cylinders to move the hardtop **(Figure 23–52)**. In most cases, it takes about 30 seconds to stow or close the roof. Some of these hardtops have a power

FIGURE 23–52

A retractable hardtop.

BMW Group

sliding glass sunroof. Diagnosis of these systems is conducted through a BCM.

Forward and Reverse Sensing Systems

Some vehicles are equipped with forward and/or a reverse-sensing system that warns the driver when obstacles are within a designated range of the front and rear bumpers **(Figure 23–53)**. These systems have ultrasonic sensors located in the front and rear bumpers to detect any obstacles at the front and rear of the vehicle. The rear sensors are active only when the vehicle is in reverse. The reverse sensor detects obstacles up to 2 m (6 ft.) from the rear bumper.

The front sensors are active when the vehicle is in any gear but PARK or NEUTRAL, and its speed is below 10 km/h (6 mph). The front sensors can detect objects within an area about 70 cm (28 in.) from the front of the vehicle and about 25 cm (10 in.) to the front side of the vehicle.

If the sensors detect something, the distance between the vehicle and the obstacle is broadcast through the vehicle's speakers by a series of chimes. As the vehicle moves closer, the tone's rate increases. If the system detects an object more than 25 cm (10 in.) from the side of the vehicle, the tone will be emitted for only three seconds. When the obstacle is less than 25 cm (10 in.) away, the tone is continuous. Once the system detects an object approaching, the tone will sound again. While emitting the warning, radio volume is reduced to a predetermined level. Once the warning is turned off, the radio will return to its previous volume.

FIGURE 23–53

Radar sensor and a rearview camera located at the rear of this car.

Rearview Cameras

To assist the driver in safely backing the vehicle, some vehicles have a "television" camera that sends signals that display the behind-the-vehicle view on the instrument panel, rearview mirror, or the navigation screen. The display allows the driver to see what is behind the vehicle when moving in reverse. Normally, these systems are comprised of the following:

- *Display and navigation module.* This unit receives video signals that display an image of the area behind the vehicle.
- *Camera assembly.* A colour video camera, with a wide-angle lens, mounted on the rear of the vehicle, transmits images of the area behind the vehicle to the display and navigation module display.
- *Park/neutral position switch.* This sends a signal to the ECU to display the views from the camera when REVERSE gear is selected, plus it triggers other views from the rear of the vehicle.

In most cases, the display shows a variety of paths, based on colour and the current conditions **(Figure 23–54)**. These paths depend on what the system perceives is the driver's intended path. Two primary paths are displayed on the screen: active and fixed. These two displays will fade in and out on the display, depending on the position of the steering wheel. Active guidelines show the intended path of the vehicle, based on the system's inputs. Fixed guidelines show the actual direction while the vehicle is moving.

FIGURE 23–54

The blue area shows what is directly behind the vehicle and the orange lines show where the vehicle is headed according to the steering wheel.

The visual alerts are red, yellow, or green highlights, which appear on top of the video image when an object is detected by the reverse sensing system. Basically, all objects that are the closest to the vehicle are shown in the red zone. Objects that are further away are shown in the green zone. The yellow areas merely show the transition from the green to the red zone. The centreline shows where the vehicle is headed based on the position of the steering wheel.

NIGHTTIME AND DARK AREAS At nighttime or in dark areas, the camera system relies on the reverse lamp to provide the necessary light to produce an image. Both reverse lamps must operate correctly to provide a clear image in the dark on the display.

SHOP TALK

If either of the reverse lamps are not working properly, replace the lamps before allowing the customer to depend on the rear-view cameras.

Parking Assist

Using forward and rear sensing monitors and rear-view cameras allows drivers to park their vehicles easier. The parking assist control unit calculates the vehicle's location in relation to other vehicles and obstacles. The control unit receives signals from the front and rear sensing monitors, ECM, steering angle sensor, power-steering control unit, and ABS control unit. Tones from the monitoring system become more rapid as the vehicle nears an obstacle.

Normally, the view from the camera is displayed on the navigation screen or a information display in the instrument panel or centre stack. This view is displayed along with reference lines that mark off the area directly behind the vehicle and an additional area showing where the vehicle is headed.

Blind Spot Detection

A few vehicles offer an option that allows drivers to view what is on the right and/or left side of their vehicle as they drive along the road **(Figure 23–55)**. This feature allows for safe lane changing. From personal experience, this feature has given me safe and confident lane changing, especially since my car has huge blind spots. Basically, a blind spot is an area that cannot be clearly seen while operating the vehicle. A blind spot can be at the front, sides, and rear of a vehicle.

Radar sensors observe the rear blind spots on both sides of the vehicle and notify the driver of any oncoming or stationary vehicles on either side of the vehicle. When the vehicle enters the blind spot and there is a vehicle in one or both of them, a yellow light on the corresponding outside mirror lights up. In most cases, if the driver puts on the turn signal for a lane change, the yellow light starts flashing quickly to warn that a vehicle is in that lane.

FIGURE 23–55

The blind spot indicator will light up when the camera detects a vehicle entering the blind spot.

Lane-Departure Warning Systems

Lane-departure warning systems are designed to warn a driver when the vehicle begins to move out of its lane, unless a turn signal is on in that direction, when travelling on open roads.

There are two basic lane-departure systems: one type emits a warning sound or vibration at the steering wheel when the vehicle is leaving its lane; the other warns the driver and is capable of assuming control of the vehicle if the driver does not respond to the warnings.

These systems rely on inputs from a number of sensors and a camera that identifies the lane markings as the vehicle moves forward. The camera monitors the lane markings and sends data to the system's control unit. Based on all of the inputs, the control unit is able to detect when a vehicle drifts out of the lane.

Pre-Collision Systems

A recent accident prevention system is called the pre-collision system and is offered on many vehicles and will be offered on more in the future. The pre-collision system detects other vehicles and obstacles in front of a vehicle. In response, it controls all systems within the basic pre-collision system to prevent an accident, and if one occurs, the system will do what it can to prevent injuries to the driver and passengers. Basically, the system relies on a front radar sensor, an object recognition camera sensor, and a driver monitor camera, which monitors the driver's facial direction.

When the system determines there is a risk of colliding with an obstacle at the front of the vehicle, the driver will be warned of the potential. In addition, if the system determines that a crash is unavoidable, all slack in the seat belts is instantly removed. The system also increases the amount of hydraulic pressure that is applied by the brake pedal and controls the suspension system to minimize the amount of nose dive while the brakes are strongly applied.

Self-Parking

A few vehicles offer an option for self-parking, at times referred to as active park assist. This feature parallel parks a vehicle, without driver control, after the space is picked out by the driver. A variety of systems are used, but most calculate the size of the parking space and let the driver know if the vehicle will fit into the space. This technology is possible only with vehicles that have electric steering. The system can detect an available parallel parking space and automatically steer the vehicle into the space while the driver controls the accelerator, gearshift, and brakes.

The system uses ultrasonic sensors on either side of the front bumper, linked to a display in the instrument panel **(Figure 23–56)**. It also relies on wheel-speed and steering-angle sensors to monitor the vehicle's movement and alert the driver of the remaining available space. Some systems rely on global positioning systems, a steering wheel sensor, and a rearview camera.

When the system is activated, it knows what side of the street the driver wants to park the vehicle by the selection of the turn signal. Once it finds a suitable parking space, the space is displayed on the information screen, along with instructions. To begin the process, the driver is told to drive forward until a message says to stop. At that point, the transmission is placed into reverse, and the driver takes his or her hands off the steering wheel. From this point on, the steering is controlled by the system and not the driver. By applying the brakes when the driver determines the vehicle has backed up enough, the vehicle can be placed into a forward gear. This process is repeated once the vehicle has moved forward enough. The instructions will tell the driver to make backward and forward manoeuvres until the vehicle is safely parked. At that point, the system is shut down by the driver turning the steering wheel. If the vehicle is not properly placed within the parking spot, the driver should correct it.

Night Vision

A few vehicles have a feature that uses military-style thermal imaging to allow drivers to see things they normally may not see until it is too late **(Figure 23–57)**. The thermal-imaging system uses a camera with a fixed lens mounted behind the front grille and projects the image onto the bottom of the

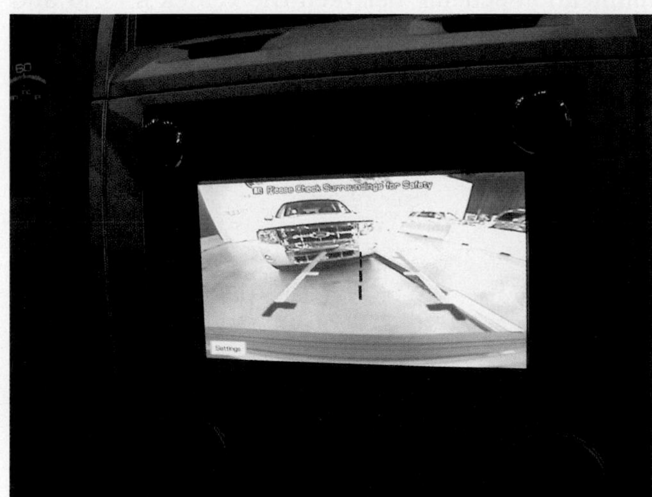

FIGURE 23–56

The display for the self-park system shows where the vehicle is headed.

FIGURE 23–57

Night vision uses an infrared camera to project unseen objects on the screen.

driver's side of the windshield by an HUD. The lens, which is an infrared sensor, is designed to operate at room temperature; therefore, the lens has its own heating and cooling system to keep it at the desired temperature.

The images seen on the HUD display are from the area in front of the light beams from the car's normal high-beam headlights. The system allows the driver to see up to five times more of the road than with just headlights.

The system works by registering small differences in temperature and displaying them in 16 different shades of grey on the HUD screen. It has the ability to display animals, or people, behind bushes or trees, and can see through rain, fog, and smoke. Cold objects appear as dark images, whereas warm objects are white or a light colour.

GARAGE DOOR OPENER SYSTEM

Most late-model vehicles have a programmable garage door opener built into the area around the inside rearview mirror. This system is programmed with the garage door's transmitter code and can also be used to control electric gates, entry gates, door locks, home lighting systems, security systems, or other transmitter code–based systems.

SECURITY AND ANTITHEFT DEVICES

Three basic types of antitheft devices are available: locking devices, disabling devices, and alarm systems. Many of the devices are available as standard equipment or as options from the manufacturers; others are aftermarket installed.

Locks and Keys

Locks are designed to deny entry to the engine, passenger, and trunk compartments of the car, as well as to prevent a thief from driving the car away. Most locks deny entry by moving a mechanical block between the vehicle's body and the door. Latches and keys simply move those blocks.

To prevent theft, manufacturers use specially cut keys that cannot be easily duplicated, and the lock mechanisms for these keys are extremely difficult to pick. The master key is often built into a remote-control key handle. The master key can lock and unlock the doors, trunk, fuel filler door, and glove compartment, all at the same time. These locks can be operated with the key or with the remote. These systems also automatically unlock the doors in case of an accident and turn on the hazard and interior lights.

A special key, often called a valet key, works only in the doors and ignition, thereby preventing entry to the trunk and glove compartment.

Both internal and external hood locks are available to prevent theft of anything under the hood. These locks are especially useful on a vehicle with a battery-powered alarm system because they prevent a thief from disconnecting the power source or disabling the alarm.

Many cars are equipped with special fuel filler doors that help to prevent the theft of gas from the fuel tank. Voltage is present at the fuel filler door release switch at all times. When the switch is closed, the door release solenoid is energized, and the fuel door opens.

Older vehicles may use a **resistance key**, which is a normally cut key with a small resistor bonded to it. When the key is inserted into the ignition switch, the circuit must recognize that that resistance is the correct amount for the vehicle before the engine will start.

Passkeys

The passkey is a specially designed key, or transponder, that is selected and programmed just for the vehicle for which it was intended. Although another key may fit into the ignition switch or door lock, the system does not allow the engine to start without the correct electrical signal from the key.

Transponder key systems are based on a communication scheme between the vehicle's PCM and the transponder in the key **(Figure 23–58)**. Each

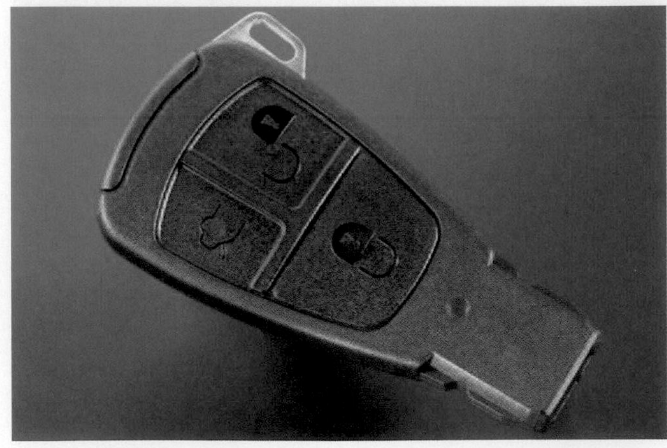

FIGURE 23–58

A totally electronic ignition key.

DaimlerChrysler Corporation

time the key is inserted into the ignition switch, the PCM sends out a different radio signal. If the key's transponder is not capable of returning the same signal, the engine will not start.

Smart Keys

Newer vehicles are equipped with a specially designed key, or fob. They are commonly used with push-button start systems **(Figure 23–59)**. To start the engine, the system must receive the correct signal from the key. These keys also control the door and trunk locks. These keys can also be programmed for specific user preferences that are adjusted when the key is inserted. Some of these preferences can include preset radio station; HVAC settings; and steering wheel, pedal, mirror, and seat positions.

Transponder key systems set up communications between the vehicle's PCM and the transponder in the key. Each time the key is inserted into the ignition switch, the PCM sends out a different radio signal. If the key's transponder is not capable of returning the same signal, the engine will not start. The battery for the remote is charged each time the key is inserted into the ignition.

On some systems, the transponder is inserted into a slot in the instrument panel. To start the engine, the transponder is pressed. On others, after the transponder is inserted into its slot, a start button is pressed. If there is a communication link between the transponder and the vehicle, and the codes both match, the engine will start.

FIGURE 23–59

A typical start–stop button in a late-model vehicle.

©Cengage Learning 2015

FIGURE 23–60

Examples of transponder keys.

©Cengage Learning 2015

On many newer vehicles, the transponder does not need to be inserted. It merely needs to be close to the vehicle **(Figure 23–60)**. This system is normally called a smart access system. The system uses a radio frequency (RF) signal to communicate with your vehicle and authorize your vehicle to respond to commands. The system can perform many functions without inserting a key or pressing a button. It can lock and unlock the doors, allow the engine to start by pressing the engine switch while depressing the brake pedal, and open the trunk. Doors can be unlocked either by touching the inside of any exterior door handle, the luggage compartment handle, or pressing a button on the transmitter. To lock the doors, press or pass by the black spot (lock area) on either front exterior door handle **(Figure 23–61)**.

When the electronic key enters into zones around the vehicle, a certification control module certifies the ID code from the key. Once the signal is certified, the control module transmits an engine immobilizer deactivation signal to the ID code box and a steering unlock signal to the steering lock ECU. A BCM also receives a certification signal and actuates the door lock motor to unlock or lock the door.

The actuation zones are set up by several oscillators. Each oscillator transmits a signal every quarter of a second when the engine is off and the doors are

FIGURE 23–61

The black area next to the key lock serves as the sensor for unlocking and locking the doors.

locked. When a signal detects the electronic key, certification begins. The actuation zone is normally about 1 m (3 ft.).

The system can also trigger a BCM to restore the position of the driver's seat (driving position memory system), shoulder belt anchor, steering wheel, and outside rearview mirror.

Keyless Entry Systems

A keyless entry system allows the driver to unlock the doors or trunk lid from outside of the vehicle without using a key. It has two main components: an electronic control module and a coded-button keypad on the driver's door or a key fob. Some keyless systems also have an illuminated entry system.

The electronic control module typically can unlock all doors, unlock the trunk, lock all doors, lock the trunk, turn on courtesy lamps, and illuminate the keypad or keyhole after any button on the keypad is pushed or either front door handle is pulled.

Remote keyless entry systems rely on a handheld transmitter, frequently part of the key fob. With a press of the unlock button on the transmitter from 7 to 14 m (25 to 50 ft.) away (depending on the type of transmitter) in any direction range, the interior lights turn on, the driver's door unlocks, and the theft security system is disarmed. The trunk can also be unlocked. Pressing the lock button locks all doors and arms the security system. For maximum security, some remote units and their receiver change access codes each time the remote is used.

Some remote units can also open and close all of the vehicle's windows, including the sunroof. They may also be capable of setting off the alarm system in the case of panic. When factory remote starter systems are installed, many remote units feature a start button as well.

AUTOMATIC LIFTGATE OPENERS Some late-model Ford SUVs and crossovers have a feature that allows the liftgate to open automatically with the swipe of a foot under the rear bumper. Opening is activated by a sensor that triggers the locking mechanism and opening motors at the liftgate.

Alarm Systems

Antitheft systems are installed by either the OEM or aftermarket companies. The basic idea of these systems is to scare away a thief and/or prevent the vehicle from starting. **Figure 23–62** shows most of the components of common antitheft systems.

The two methods for activating alarm systems are passive and active. Passive systems switch on automatically when the ignition key is removed or the doors are locked. They are often more effective than active systems. Active systems are activated manually with a key fob transmitter, keypad, key, or toggle switch.

Switches similar to those used to turn on the courtesy lights as the doors are opened are often used to arm the alarm. When a door, hood, or trunk is opened, the switch closes and the alarm sounds. It turns itself off automatically (provided the intruder has stopped trying to enter the car) to prevent the battery from being drained. It then automatically rearms itself.

For most systems, the driver's door lock is the master switch for the unit. If the alarm is activated, it can be turned off by inserting and turning the door lock or the ignition switch. The driver's door lock may be equipped with a pair of magnetic **reed-type switches**. The reed switches move with the door lock to arm and disarm the system.

Ultrasonic sensors are used to detect motion and will trigger the alarm if there is movement inside the vehicle. Current-sensitive sensors activate the alarm if there is a change in current within the electrical system, such as when a courtesy light goes on or the ignition starts. Motion detectors monitor changes in the vehicle's tilt, such as when someone is attempting to steal the tires.

Many alarm systems are designed to sound an alarm, turn on the hazard lights, and cause the high beams to flash along with the hazard lights. Indicator lights on the inside of the vehicle alert others that the alarm is set and also remind the

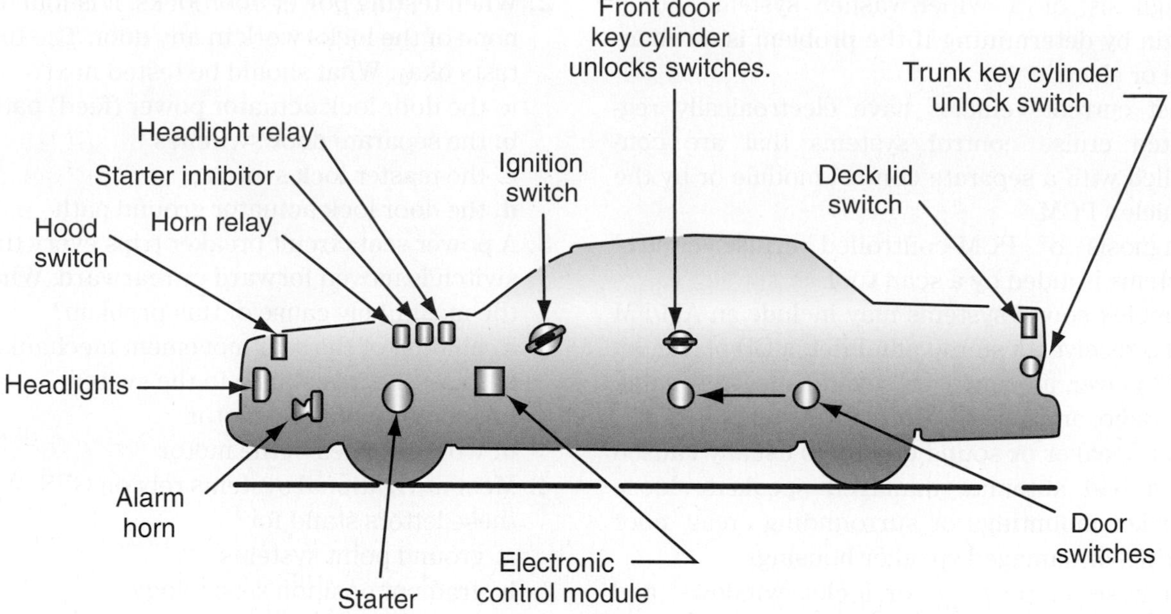

FIGURE 23–62

The basic layout of an antitheft system.

driver to turn the alarm off before entering. To avoid false alarms, some systems allow for the disabling of particular sensors, such as the motion detector inside the vehicle that could be set off by a pet inside the vehicle.

Some late-model vehicles have an immobilizer system that will prevent the engine from starting if a wrong coded key is used. When the driver (or passenger) is in the vehicle while carrying the correct key and depresses the brake pedal and pushes down on the start–stop switch, the engine will start. Starting is permitted because the key code is recognized by the control unit. In a fraction of a second, the ECU turns on all engine systems required to start the engine and to unlock the steering wheel.

DIAGNOSIS There are many different types of antitheft systems used by manufacturers; there also are many different aftermarket systems available. When diagnosing a problem, it is very important to have as much information as possible about the system. Most often, a manufacturer-specific scan tool is needed to diagnose an OEM antitheft system. In some cases, only dealership personnel have access to the necessary data for diagnosis. Before proceeding with a detailed diagnosis, inspect all fuses and relays in the system. Also check for loose wires, connectors, and components. This is especially important if the system works intermittently.

KEY TERMS

Defogger (p. 697)
Deicer (p. 697)
Global positioning satellites (GPSs) (p. 687)
Peltier element (p. 694)
Power roof (p. 700)
Resistance key (p. 704)
Servo unit (p. 679)
Telematics (p. 686)
Telemetry (p. 686)
Transponder key (p. 704)
Tweeters (p. 683)
Woofers (p. 683)

SUMMARY

- Many accessories may be controlled and operated by a body computer, and therefore diagnosis may involve retrieving trouble codes from the computer.
- The basic designs of windshield wiper systems used on today's vehicles include a standard two- or three-speed system with or without an intermittent or rain-sensing feature.
- The motors can have electromagnetic fields or permanent magnetic fields. With permanent magnetic fields, motor speed is regulated by the placement of the brushes on the commutator. The speed of electromagnetic motors is controlled by the strength of the magnetic fields.

- Diagnosis of a wiper/washer system should begin by determining if the problem is mechanical or electrical.
- Most current vehicles have electronically regulated cruise control systems that are controlled with a separate control module or by the vehicle's PCM.
- Diagnosis of PCM-controlled cruise control systems is aided by a scan tool.
- Complex sound systems may include an AM/FM radio receiver, a stereo amplifier, a CD player, an MP3 player, an equalizer, several speakers, satellite radio, and a power antenna system.
- Poor speaker or sound quality is usually caused by a bad antenna, damaged speakers, loose speaker mountings or surrounding areas, poor wiring, or damaged speaker housings.
- Diagnosis of power door locks, windows, and seats is best done by dividing the circuit into individual circuits and basing your testing on the symptoms.
- One of the most common problems with rear-window defrosters is damage to the grids on the window, which can usually be repaired.
- Navigation systems use global positioning satellites to help drivers make travel decisions while they are on the road.
- Three basic types of antitheft devices are available: locking devices, disabling devices, and alarm systems.
- Most ignition keys for late-model vehicles are either resistance or transponder passkeys that work only on the vehicle for which they were intended.
- Passive alarm systems switch on automatically when the ignition key is removed or the doors are locked. Active systems are activated manually.

REVIEW QUESTIONS

1. How are multiple speeds obtained on an electromagnetic field wiper motor?
   a. by switching current to a different brush
   b. by increasing current through the shunt field to slow the motor
   c. by decreasing current through the shunt field to speed up the motor
   d. by reversing the direction of current through the armature
2. When testing power door locks, it is found that none of the locks work in any door. The fuse tests okay. What should be tested next?
   a. the door lock actuator power (feed) path
   b. the separate door switches
   c. the master lock switch
   d. the door lock actuator ground path
3. A power seat circuit breaker trips every time the switch is moved forward or rearward. What is the most likely cause of this problem?
   a. binding of the seat movement mechanism
   b. excessive resistance in the switch
   c. bad ground at the motor
   d. worn brushes in the motor
4. Most navigational systems rely on GPS. What do these letters stand for?
   a. ground point systems
   b. gradient position seismology
   c. global positioning satellite
   d. globe potential systems
5. What power seat motor will raise and lower the front of a six-way seat?
   a. the horizontal motor
   b. the front vertical motor
   c. the rear vertical motor
   d. the recline motor
6. When testing for a defective external antenna or poor reception from an external antenna, what should an ohmmeter test show?
   a. no continuity between the mast and ground
   b. continuity between the mast and ground
   c. voltage between the mast and ground
   d. zero ohms between the mast and ground
7. A trunk release uses what device to open the latch?
   a. solenoid **b.** relay
   c. thermistor **d.** rheostat
8. With no switches operated, how are the power window motors typically connected to the circuit?
   a. with both leads connected to power
   b. with both leads connected to ground
   c. with one lead on power and the other on ground
   d. with both leads isolated from the rest of the circuit
9. If the power window switch is held down after the window has reached its lowest position, what device protects the circuit?
   a. fuse
   b. relay
   c. fusible link
   d. circuit breaker

10. Lumbar support in a seat refers to support for which body part?
    a. hips
    b. lower back
    c. upper back
    d. both sides of the back
11. Memory seats usually have which features?
    a. supplemented with movable pedals to achieve the safest seat position
    b. used only on passenger seats
    c. ability to store the driving position of several drivers
    d. require each driver to enter a personal code each trip
12. Which of the following statements describes a typical heated windshield system?
    a. A defrost cycle generally lasts one minute.
    b. In defrost mode, battery voltage can go as high as 30 volts.
    c. The voltage at the windshield may go as high as 90 volts.
    d. It uses an electric element to heat the air blowing on the glass.
13. What is the most effective type of alarm system?
    a. ultrasonic
    b. silent
    c. passive
    d. active
14. A sophisticated alarm system will use what kind of sensor to detect movement inside the vehicle?
    a. radio frequency
    b. pressure
    c. electromagnetic
    d. ultrasonic
15. What method of communication is used by a transponder style of key system?
    a. sending and receiving a signal from the PCM
    b. sending a signal to the PCM
    c. receiving a signal from the PCM
    d. receiving a signal from the radio
16. Which band has the best sound quality?
    a. FM band
    b. AM band
    c. low-pass band
    d. high-pass band
17. Adaptive cruise control systems have which of the following abilities?
    a. change speed according to road conditions
    b. allow hands-free driving
    c. decrease speed in rain and snow
    d. slow down and speed up to maintain a safe distance with the vehicle in front
18. What determines a change in speaker volume in an automatic sound level audio system?
    a. vehicle speed
    b. wind noise inside the vehicle
    c. location and number of passengers
    d. type of music being played
19. When disconnecting or changing the battery on some vehicles, what precaution must be taken with the radio?
    a. The radio must be in the OFF position.
    b. The radio security code must be known.
    c. The radio must be set to AM.
    d. The radio must be in the ON position.
20. Where is the image displayed for the advanced night vision system using thermal imaging?
    a. the rearview mirror
    b. the side-view mirrors
    c. the HUD
    d. the on-board navigation display

CHAPTER 24
Restraint Systems: Theory, Diagnosis, and Service

LEARNING OUTCOMES

- Identify and describe devices that contribute to automotive safety.
- Explain the difference between active and passive restraint systems.
- Know how to service and repair passive belt systems.
- Describe the function and operation of air bags.
- Identify the major parts of a typical air bag system.
- Safely disarm and inspect an air bag assembly.
- Know how to diagnose and service an air bag system.

Safety is foremost in the minds of automobile manufacturers. According to the Insurance Bureau of Canada (IBC), information on vehicle safety is becoming more and more important to the Canadian consumer in the market for a new or used car. Purchasing a vehicle with a higher safety rating not only provides the owner with improved protection but also is factored into the calculation of insurance premiums.

Many safety features are now available as standard equipment or as options. Some of these include side-impact barriers, crumple zone in the body, seat belts, antilock brakes, traction control, stability control, and air bags **(Figure 24–1)**. There are many safety items that have been around for many years, such as laminated and tempered glass.

Common restraint systems—seat belts and air bags—are covered in this chapter. It is important for a technician to understand how these systems work and how to diagnose and service them.

An **active restraint system** is one that a vehicle's occupant must engage manually **(Figure 24–2)**. For example, in most vehicles, the driver and passengers must fasten the seat belts for crash protection. A **passive restraint system** operates automatically **(Figure 24–3)**. No action is required of the occupant to make it functional.

Passive safety equipment includes the automatic safety belt system, air bags, rigid occupant cell, and crumple zones at the front of the vehicle. The rear and sides of the body are among the most important safety features of today's cars and are designed to dissipate most of the impact energy for the protection of vehicle occupants.

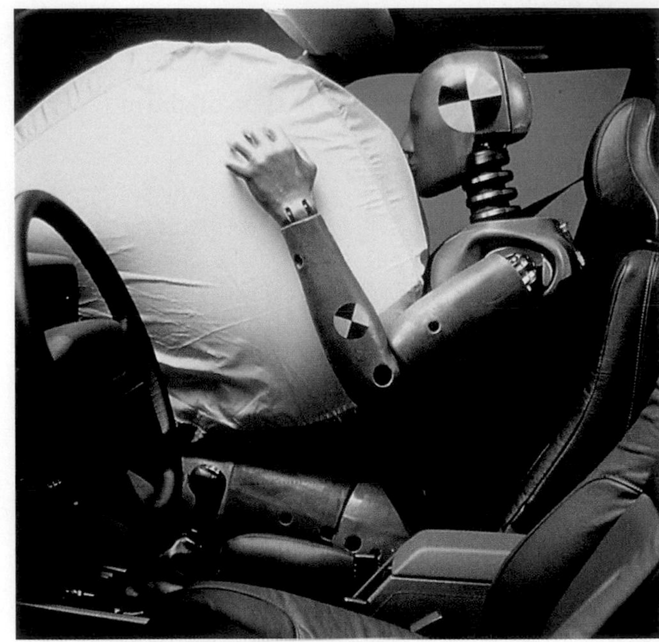

FIGURE 24–1
A passenger-side air bag.
Volvo Car Corporation

All vehicles built or sold in North America must have one or both of these passive restraints: automatic seat belts or air bags. Both have made a tremendous impact on driver and passenger safety over

Rear seat belt
Child seat anchor plate
Shoulder anchor adjuster
Front seat belt
Front-seat belt buckles
Rear centre belt and rear seat belt buckles

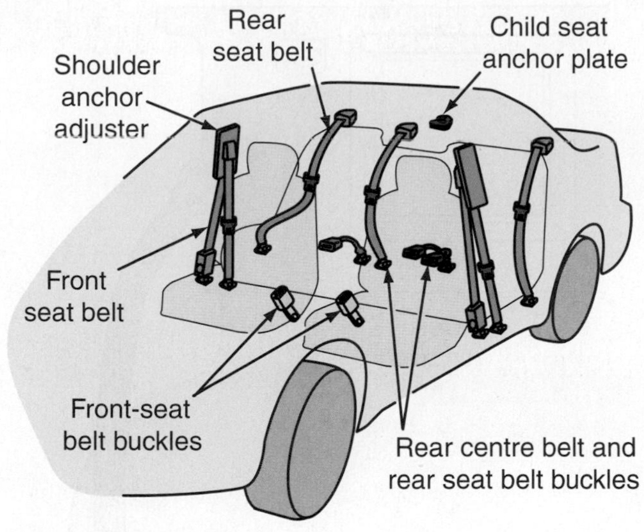

FIGURE 24–2

Active restraint systems.

the last few years. It is very important that technicians be able to service these systems.

SEAT BELTS

A passive seat belt system uses electric motors to automatically move shoulder belts across the driver and front-seat passenger. The upper end of the belt is attached to a carrier that moves in a track at the top of the door frame. The other end is secured to an **inertia lock retractor (Figure 24–4)**. The retractors are mounted in the centre console. When the door is opened, the outer end of the shoulder belt moves forward to allow for easy entry or exit. When the doors are closed and the ignition turned on, the belts move rearward and secure the occupants. The active lap belt is manually fastened by the occupant and should be worn with the passive belt.

Emergency release buckle
Shoulder belt
Shoulder anchor
Knee panel
Knee panel
Caution label
Outer belt assembly (manual lap belt)
Inner belt assembly (manual lap belt)
Belt holder
Rail and motor assembly
Rail
Shoulder belt
Locking device
Tube
Motor
Emergency locking retractor assembly (manual lap belt)
Emergency locking retractor assembly
Belt guide
Emergency locking retractor assembly (manual lap belt)

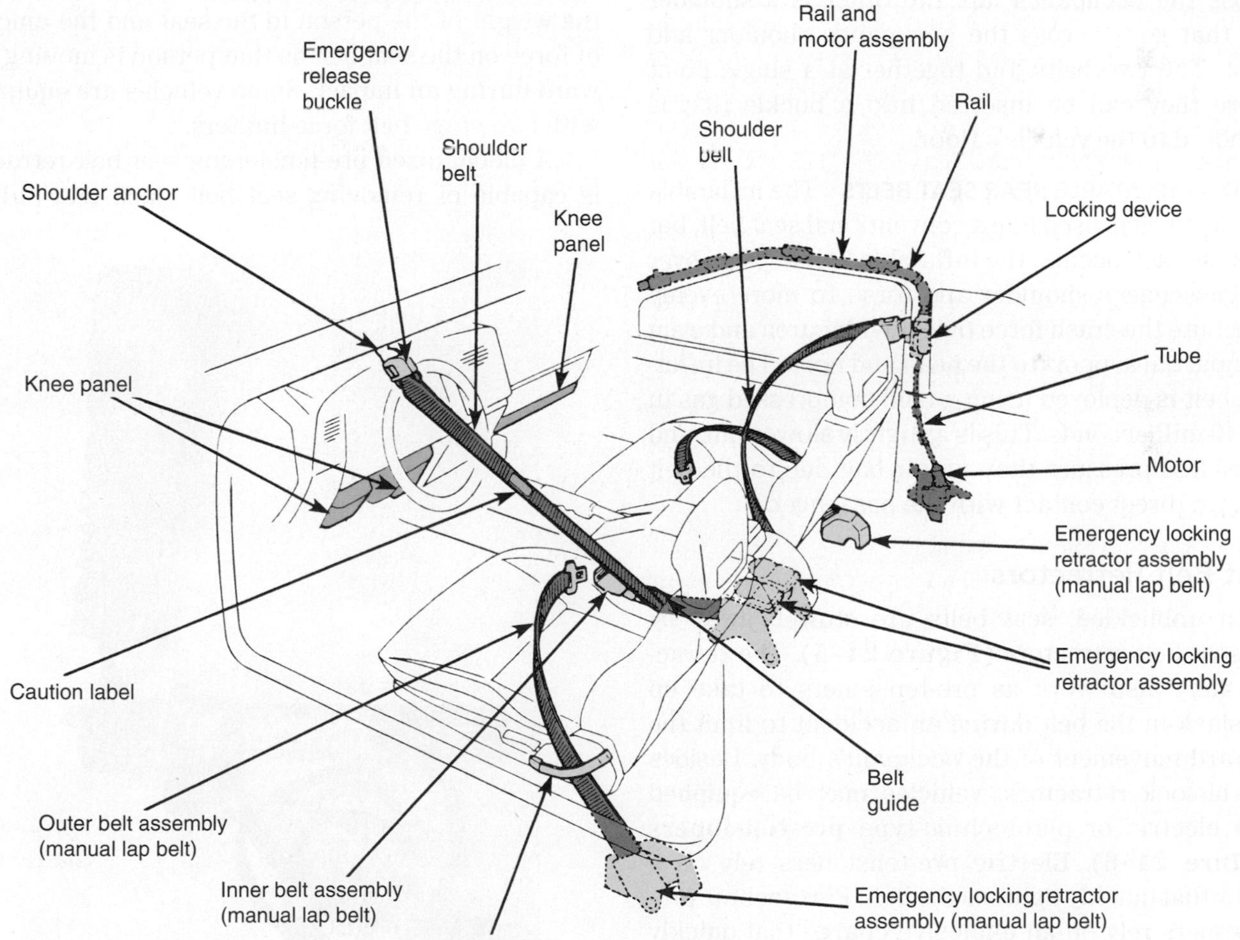

FIGURE 24–3

Passive seat belt restraint system.

Ford Motor Company

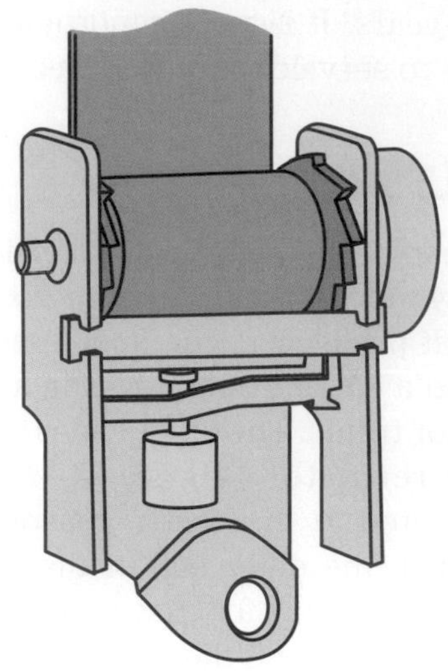

FIGURE 24–4

An inertia lock seat belt retractor.

Most vehicles, especially those with air bags, have two active belts. One is a lap belt that goes across the occupant's lap; the other is a shoulder belt that goes across the occupant's shoulder and chest. The two belts join together at a single point where they can be inserted into a buckle that is anchored to the vehicle's floor.

FORD'S INFLATABLE REAR SEAT BELTS The inflatable rear seat belt is used like a conventional seat belt, but when a crash occurs, the inflatable belt deploys over the passenger's shoulder and torso to more evenly distribute the crash force over a wider area and even offer partial support to the head and neck. The inflatable belt is deployed using a cold compressed gas in just 40 milliseconds. This is a slightly slower rate and under less pressure than an air bag due to the belt being in direct contact with the passenger.

Seat Belt Retractors

When unbuckled, seat belts are stowed away by the seat belt retractors **(Figure 24–5)**. The retractors may also work as pre-tensioners to take up the slack in the belt during an accident to limit the forward movement of the occupant's body. Besides inertia lock retractors, vehicles may be equipped with electric or pyrotechnic-type **pre-tensioners (Figure 24–6)**. Electric pre-tensioners rely on a motor that quickly tightens the belt. Pyrotechnic pre-tensioners rely on an explosive charge that quickly retracts the belt and locks it in place.

The speed and force at which the seat belt tightens as a result of a collision is variable on some vehicles. These systems rely on sensors to measure the weight of the person in the seat and the amount of force on the seat belt as that person is moving forward during an impact. Some vehicles are equipped with two-stage belt force limiters.

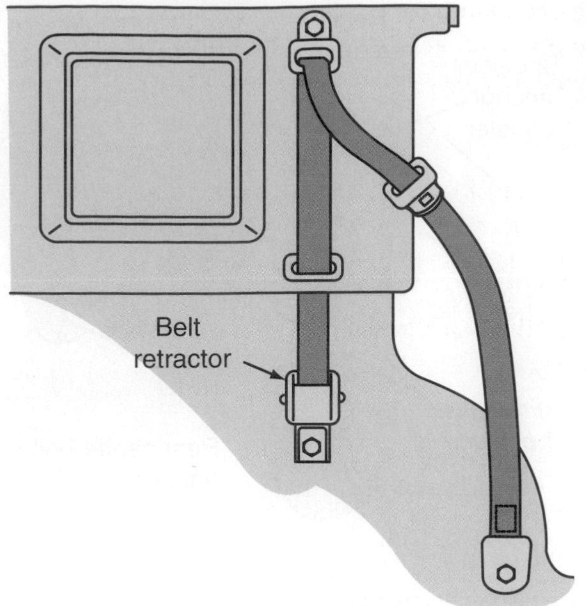

FIGURE 24–5

A typical seat belt retractor.

A mechanized pre-tensioning seat belt retractor is capable of removing seat belt slack and pulling

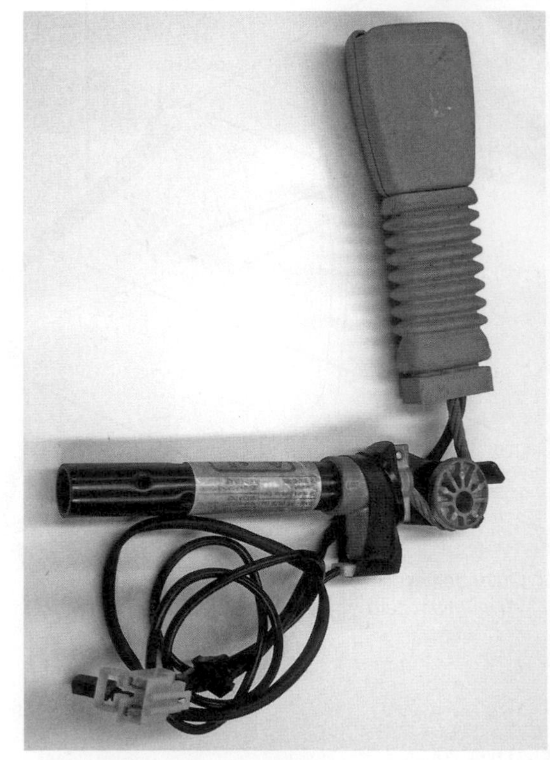

FIGURE 24–6

A buckle-mounted pre-tensioner.

the occupant into position using various inputs from the brake and chassis control systems or from a pre-crash sensing system.

Warning Lights

All modern seat belt systems have a warning light and a buzzer or chime that is turned on when the vehicle is started to remind the occupants to buckle up. Many manufacturers' vehicles will continually flash the warning lamp and sound the chime or buzzer until the seat belt is properly latched.

SEAT BELT SERVICE

Inspecting seat belt systems should follow a systematic approach. Always take as much time as necessary to do your inspection; remember that seat belts are designed to protect people.

Webbing Inspection

Pay special attention to where the webbing contacts maximum stress points, such as the buckle, D-ring, and retractor. Collision forces centre on these locations and can weaken the belt. Signs of damage at these points require belt replacement. Check for twisted webbing due to improper alignment when connecting the buckle. Fully extend the webbing from the retractor. Inspect the webbing and replace it with a new assembly if the following conditions are noted **(Figure 24–7)**: cut or damaged webbing, broken or pulled threads, cut loops at the belt edge, colour fading as a result of exposure to sun or chemical agents, or bowed webbing.

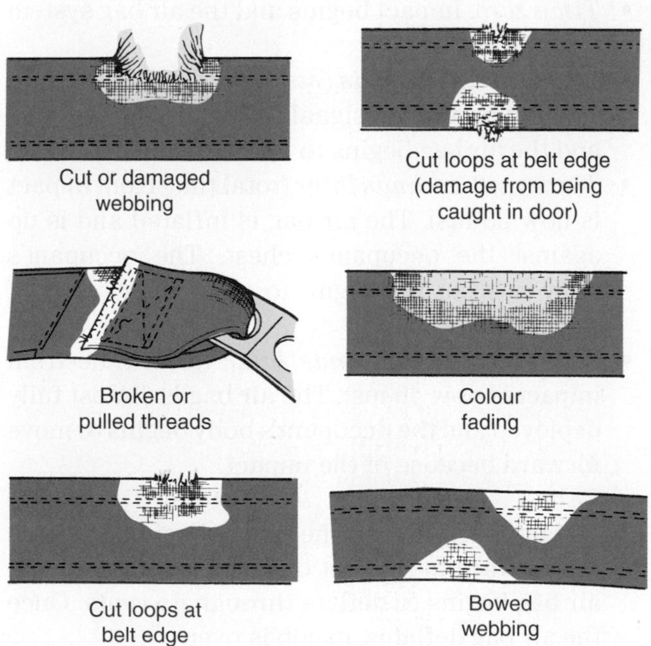

FIGURE 24–7
Examples of webbing defects.

If the webbing cannot be pulled out of the retractor or will not retract to the stowed position, check for the following conditions, and clean or correct as necessary: webbing soiled with gum, syrup, grease, or other material; twisted webbing; or the retractor or loop on the B-pillar out of position.

SHOP TALK

Never bleach or dye the belt webbing. Clean it with a mild soap solution and water.

Buckle Inspection

To determine if the buckle works or if the buckle housing has been damaged, insert the seat belt into the buckle until a click is heard. Pull back on the webbing quickly to ensure that the buckle is latched properly. Replace the seat belt assembly if the buckle does not latch. Depress the button on the buckle to release the belt. The belt should release with a pressure of approximately 2 lb. Replace the seat belt assembly if the buckle cover is cracked; the push button is loose; or the pressure required to release the button is too high. Inspect the latch assembly where the buckle inserts and look for material or debris that may prevent the latch assembly from operating correctly.

Retractor Inspection

Retractors for lap belts should lock automatically once the belt is fully out. Either webbing-sensitive or vehicle-sensitive seat belt retractors are used with passive seat belt systems. Webbing-sensitive retractors can be tested by grasping the seat belt and jerking it. The retractor should lock up; if it does not, replace the seat belt retractor.

Vehicle-sensitive belt retractors will not lock up using the same procedure. To test these belts, a braking test is required. Perform this test in a safe place. A helper is required to check the retractors on the passenger side and in the back if the vehicle is equipped with rear lap/shoulder belts.

Test each belt by driving the car at 8 to 13 km/h (5 to 8 mph) and quickly applying the brakes. If a belt does not lock up, replace the seat belt assembly. During this test, it is important for the driver and helper to brace themselves in the event the retractor does not lock up.

Most retractors are not interchangeable. That is, an R marked on the retractor tab indicates that it is for the right side only, and an L should be used on the left side only.

Drive Track Assembly and Anchor Inspection

Seat belt anchors are found where the retractors attach to the car body. High-impact forces occur here during a collision; therefore, carefully inspect the anchor areas and attaching bolts. Loose bolts should be replaced. Look for cracks and distortion in the metal where the retractors and D-rings anchor. Upper body damage must be properly repaired and brought back to exact dimensional specifications before any repairs are performed on the seat belt system. If there is damage to the metal in the mounting area, proper repairs, such as welding in reinforcement metal, must be completed before reattaching the anchor. Be sure to restore corrosion protection to the area. When spraying anticorrosion materials, make sure they do not enter the retractor. They can keep it from operating properly. Finally, look for dirt and corrosion around the anchor area.

The drive motor is usually located at the base of the track assembly behind the rear seat side trim panel. Its purpose is to pull the tape that positions the belt. If a check of the drive motor reveals that it is faulty, replace it. Like any motorized system, the seat belt parts need periodic lubrication. To service the motorized seat belt system, follow the instruction given in the service manual.

Rear Seat Restraint System

Rear seat belts are inspected in the same way as the front. However, some vehicles have a centre seat belt. These belts do not have a retractor. Check the webbing, anchors, and the adjustable locking slide for the belt. Fasten the tongue to the buckle and adjust by pulling the webbing end at a right angle to the connector and buckle. Release the webbing, and pull upward on the connector and buckle. If the slide lock does not hold, remove and replace that seat belt assembly.

Warning Light and Sound Systems

When the ignition is turned to the ON or RUN position, the FASTEN SEAT BELT light should come on. There should also be a buzzer or chime. If these warning light and sound systems do not come on, check for a blown fuse or circuit breaker. If that checks out fine, and there is sound but no light, check for a damaged or burned-out bulb. If the bulb lights but there is no sound, check for damaged or loose wiring, switches, or buzzer (voice module).

Servicing Seat Belts

If the seat belt is damaged in any way, it should be replaced. Some guidelines for servicing lap and shoulder belts follow:

- Keep sharp edges from damaging any portion of the belt buckle or latch plate.
- Avoid bending or damaging any portion of the belt buckle or latch plate.
- Do not attempt repairs on lap or shoulder belt retractor mechanisms or lap belt retractor covers. Replace them with new replacement parts.
- Torque all seat and shoulder belt anchor bolts as specified in the service manual.

AIR BAGS

An air bag is much like a nylon balloon that quickly inflates to stop the forward movement of the occupant's upper body. Air bags are passive systems designed to be used with an active system—seat belts. Air bags were not designed to replace seat belts. An air bag's job during a crash is normally over in less than one second after it begins **(Figure 24–8)**. Consider this sequence:

- *Time zero.* Impact begins and the air bag system is doing nothing.
- *Twenty milliseconds (ms) later.* The sensors are sending an impact signal to the air bag module, and the air bag begins to inflate.
- *Three milliseconds later* (total time from impact is now 23 ms). The air bag is inflated and is up against the occupant's chest. The occupant's body has not yet begun to move as a result of the impact.
- *Seventeen milliseconds later* (total time from impact is now 40 ms). The air bag is almost fully deployed and the occupant's body begins to move forward because of the impact.
- *Thirty milliseconds later* (total time from impact is just 70 ms). The air bag begins to absorb the forward movement of the occupant, and the air bag begins to deflate through its vents. Once the air bag deflates, its job is over.

The systems and parts used to deploy an air bag vary with the year and manufacturer of the vehicle,

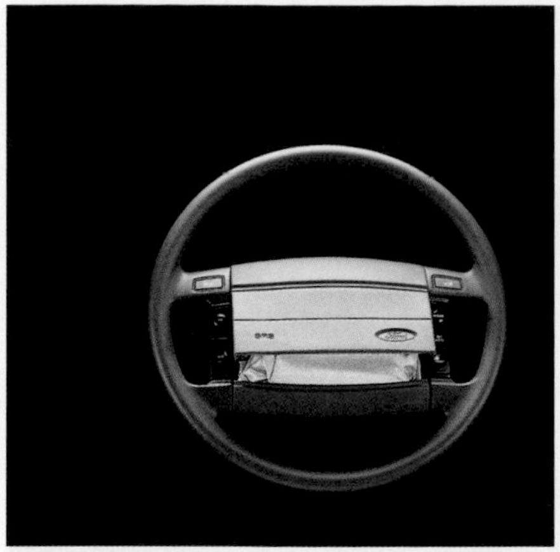
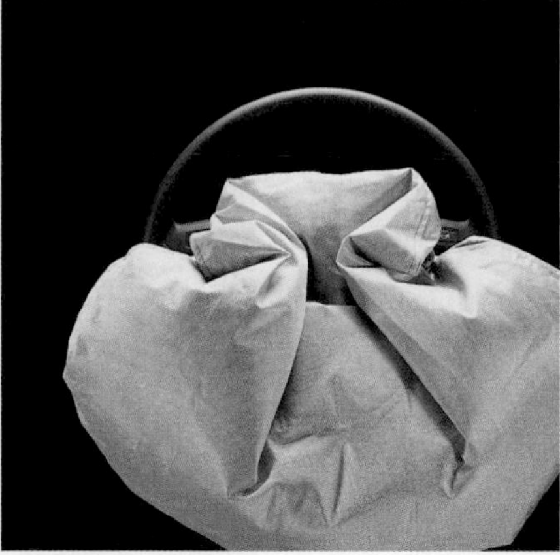
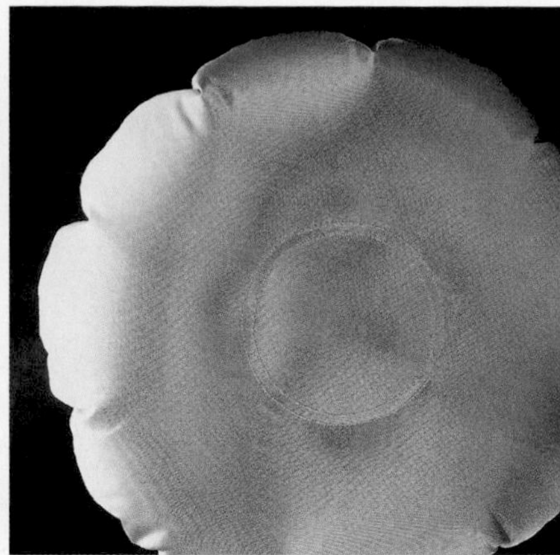

FIGURE 24–8

Various stages of air bag inflation.

TRW, Incorporated

as well as the location of the air bag. An air bag is inflated or deployed by rapid production or expansion (explosion) of a gas. The gas is fired by an igniter commonly called a **squib**.

Different manufacturers also call their air bag systems by different names, such as **supplemental inflatable restraint (SIR)** and **supplemental restraint system (SRS)**. All late-model vehicles have a driver-side and a passenger-side air bag **(Figure 24–9)**.

FIGURE 24–9

Driver-side and passenger-side air bags.

DaimlerChrysler Corporation

A driver-side air bag system may include a knee diverter, which is also called a knee bolster. This unit is designed to cushion the driver's knee from impact and help prevent the driver from sliding under the air bag during a collision. It is located underneath the steering column and behind the steering column trim.

Passenger-side air bag modules are located in the vehicle's dash. These air bags are very similar in design and operation to those on the driver's side. However, many manufacturers use a different set of sensors. The actual capacity of gas required to inflate the passenger-side air bag is much greater because the bag must span the extra distance between the occupant and the dashboard. The steering wheel and column make up this difference on the driver's side.

Because of the concern for babies and small children, pickups and other two-seat vehicles either do not have a passenger side SIR or have a switch that prevents it from deploying. The switch is typically moved with a key and allows the driver to activate or deactivate the passenger-side SIR. Deactivation is recommended whenever a child is in a rear-facing child seat. An indicator light in the instrument panel lets the driver know the current status of the passenger side SIR.

Some vehicles are equipped with more than these two air bags **(Figure 24–10)**. The occupants in the front seat may be further protected by side air bags and/or side curtain air bags. The rear passengers may be protected by air bags in the rear of the front seat backs, side air bags, and/or side curtain air bags.

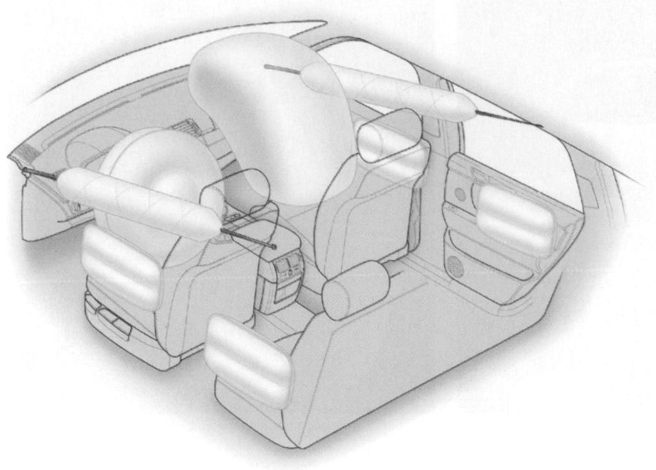

FIGURE 24–10

This vehicle has a total of eight air bags.

BMW of North America, Inc.

Side air bags can take on many different shapes and be deployed from various locations. Some manufacturers offer side air bags, called side curtains **(Figure 24–11)**, that blanket the entire side of the car. Side air bags **(Figure 24–12)** are available for the front and rear doors on some cars. These air bags are deployed from the interior trim on the door or are deployed from the outside of the seat. Side impact head protection systems inflate a long narrow air bag that extends from the windshield area to behind the front seat back **(Figure 24–13)**.

Many manufacturers now incorporate an inflatable air bag into the shoulder belts of the seat belt assembly. This also helps to limit passenger movement during a crash event.

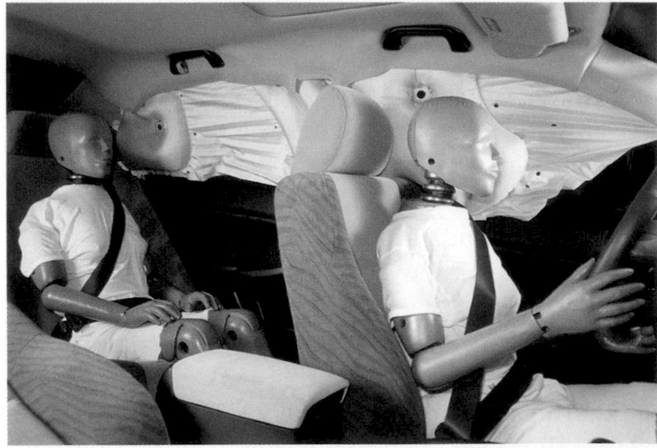

FIGURE 24–11

A side curtain.

Honda Motor Co., Inc.

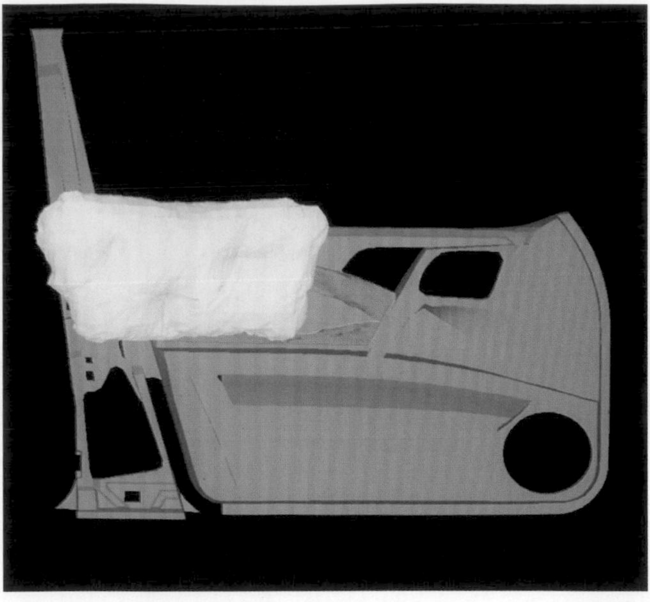

FIGURE 24–12

A side air bag.

DaimlerChrysler Corporation

Door-mounted side air bags must begin deploying in 5 to 6 milliseconds. This requirement is based on the simple fact that only a few centimetres separate the occupant from the other vehicle during a side impact. Seat-back-mounted side air bags do not need to operate at these great speeds. The head air bag, or inflatable tubular structure, is designed to stay inflated for about five seconds to offer protection against a second or third impact.

FIGURE 24–13

A side impact head air bag.

BMW of North America, Inc.

Current air bag systems work in conjunction with the seat belt pre-tensioners and retractors. When the air bag circuit is turned on, so is the pre-tensioner circuit. These actions limit the movement of the occupants.

A current trend with SIR systems is to make the impact of the bag's deployment less powerful, thereby accommodating occupants of different sizes and reducing the number of injuries caused by the air bag itself. These systems are referred to as second-generation air bags.

Second-generation air bags have been depowered to inflate with less force than earlier air bags. The air bags are depowered by reducing the peak inflation pressure and/or rise rate. Rise rate is the force and speed at which an air bag inflates. Rise rate is controlled by the type and amount of inflator gas, the size of the air bag, and the air bag's vent design.

Depending on the specific model vehicle, air bag size, and seat belt system, a second-generation air bag inflates with an average of 20 to 35 percent less energy.

Adaptive SRS Systems

All 2006 and newer vehicles must have a system that allows for air bag suppression when infants, children, or small adults are in the front passenger seat. This system uses a load sensor, a seat belt tension sensor, and an electronic control unit. The load sensor measures the weight on the seat and classifies the occupant as an adult or child and provides the classification to the air bag controller, which enables or suppresses passenger air bag deployment. A belt tension sensor identifies cinched child seats.

Some vehicles have **smart air bags** or **adaptive air bags**. Many of these systems have two possible stages of air bag deployment. The force of the expanding air bag is controlled to match the severity of the impact and/or by the size and weight of the seat's occupant. Other things considered are seat-track position and seat belt use. All of these factors require different deployment rates.

Two-stage air bags have twin chambers, comprising two air bags, two containers of gas, and two squibs. When low-pressure deployment is desired, only one squib is fired. The air bag sensor assembly calculates the extent of impact, seat position, and status of the seat belts, and controls the inflation times for the two chambers.

During a severe collision, the occupants need maximum protection, and both squibs fire. The speed of deployment can also be controlled by the firing of the squibs. For rapid deployment, both squibs fire at the same time. To phase in full deployment, one squib is fired, and then a few milliseconds later the other is. When the twin-chamber air bag is deployed, it forms the shape of two bags with a depression in the middle. The shape supports the occupant **(Figure 24–14)**.

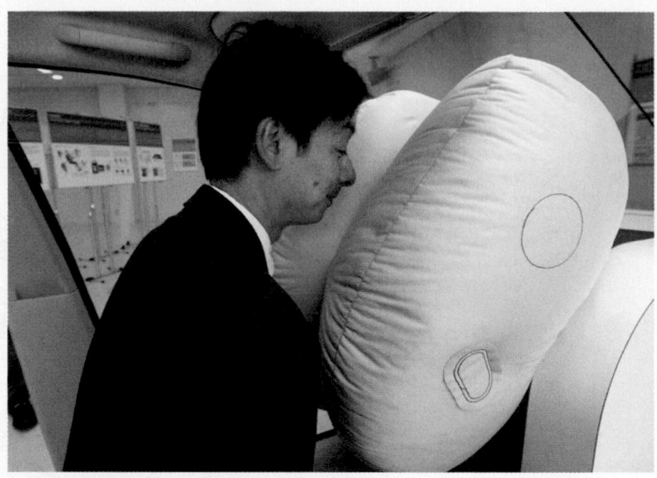

FIGURE 24–14

Full deployment of a twin-chamber air bag takes on the shape that surrounds the occupant.

AP Photo/Koji Sasahara

A few late-model systems offer rollover protection. Sensors monitor the speed of the rollover and inflate the air bags according to the sensed severity of the rollover. During a rollover, the air bags stay inflated longer than normal to keep the occupants safe until the vehicle comes to a rest. These systems also take the slack out of the seat belts, shut off the fuel pump, and disconnect the battery when a rollover is sensed.

Electrical System Components

The electrical circuit of an air bag system includes impact sensors and an electronic control module **(Figure 24–15)**. The electrical system conducts a system self-check to let the driver know that it is functioning properly, detects an impact, and sends a signal that inflates the air bag.

SENSORS To prevent accidental deployment of the air bag, most systems require that at least two sensor switches be closed to deploy the air bag **(Figure 24–16)**. The number of sensors used in a system depends on the design of the system. Some systems use only a single sensor, and others use up to five. The name used to identify the sensors also

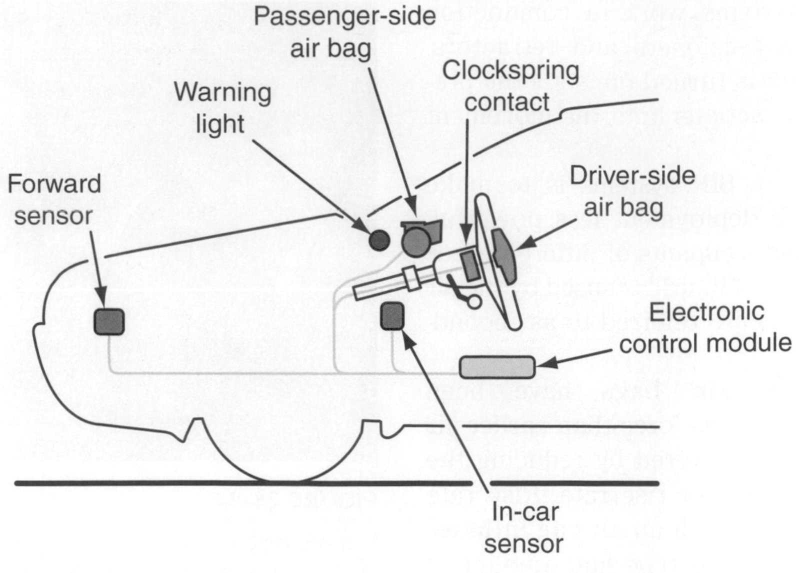

FIGURE 24–15

The location of the components for Honda's air bag system.

varies among manufacturers. Normally, sensors are located in the engine and passenger compartments, although some makes and models have the sensors located inside the restraint control module.

Typically, ignition of the air bag occurs only when an outside (impact or crash) sensor and an inside (safing or arming) sensor are closed. Once the two sensors are closed, the electrical circuit to the igniter is complete. The igniter starts the chemical chain reaction that produces heat. The heat causes the generant to produce nitrogen gas, which fills the air bag.

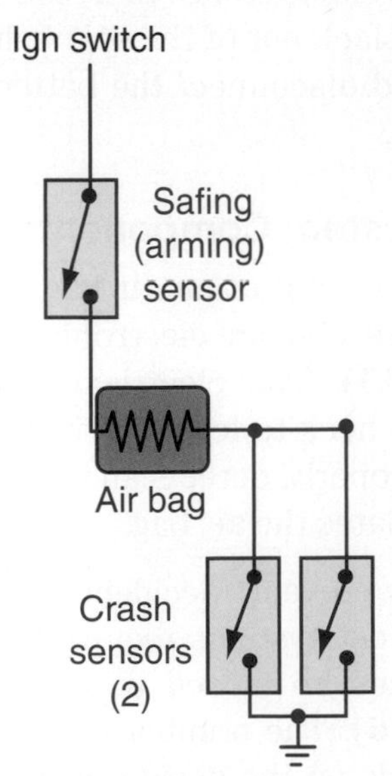

FIGURE 24–16

A simple air bag circuit with sensors.

ROLLER-TYPE SENSORS This type of sensor has a roller located on a ramp **(Figure 24–17)**. One terminal of the sensor is connected to the ramp, and the second terminal is connected to a spring contact extending through an opening in the ramp, but not touching the ramp. The roller is held against a stop by small springs. If the vehicle is involved in a collision at a high enough speed, the roller moves up the ramp

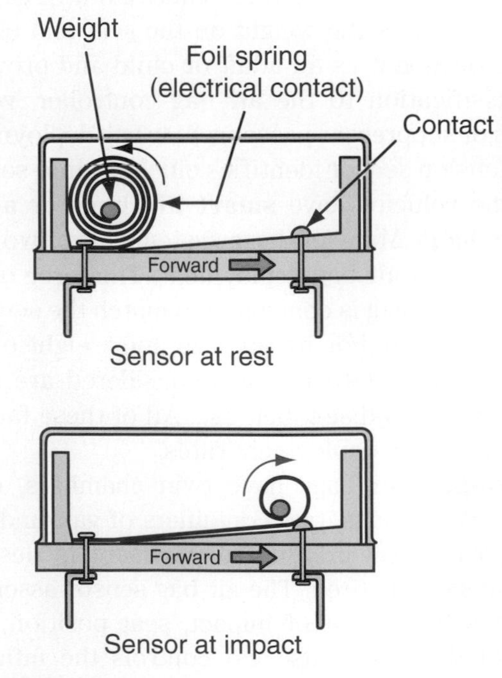

FIGURE 24–17

A roller-type air bag sensor.

Chrysler Group LLC

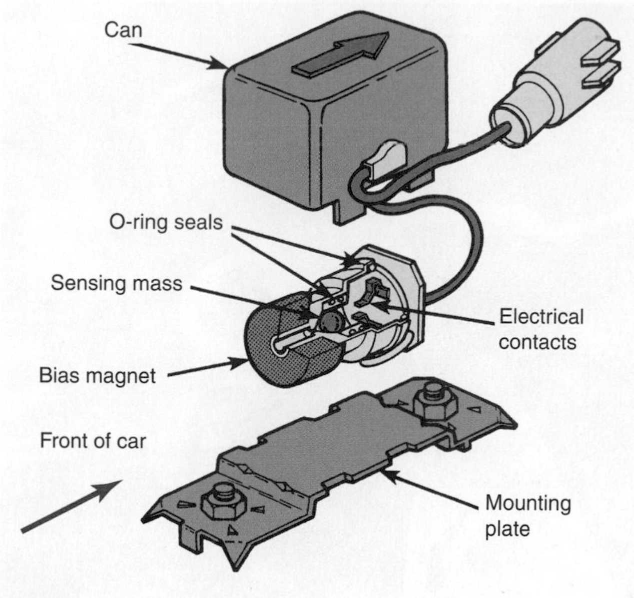

FIGURE 24–18

A typical ball and magnet sensor for an air bag system.

Chrysler Group LLC

and strikes the spring contact. This movement completes the circuit between the ramp and the spring contact through the roller, and the air bag deploys.

MASS-TYPE SENSOR This type of sensor contains a normally open set of gold-plated switch contacts and a gold-plated ball, the sensing mass, which is held in place by a magnet **(Figure 24–18)**. At the point of sufficient force, the ball will break loose from the magnet and make contact with the electrical contacts to complete the circuit.

ACCELEROMETER In the past number of years, accelerometers have been replacing the mechanical triggering sensors not only for air bag deployment but also for suspension and body control. An accelerometer **(Figure 24–19)** is a sensor built on the principle of measuring the force exerted on a test body for the measurement of acceleration (or deceleration) along a given axis. This force is recorded by stress changes on a silicon disc or wafer.

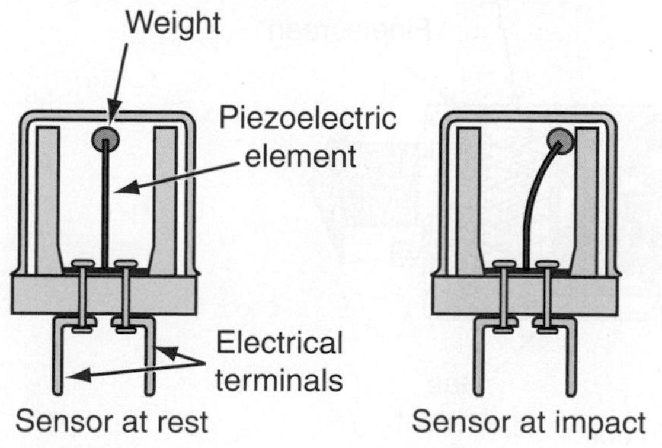

FIGURE 24–19

An accelerometer air bag sensor.

Chrysler Group LLC

DIAGNOSTIC MONITOR ASSEMBLY The **air bag sensing diagnostic monitor (ASDM)** or **restraint control module (RCM)** constantly monitors the readiness of the SIR/SRS electrical system. If the module determines there is a fault, it will illuminate the warning lamp. Depending on the fault, the SIR/SRS system may be disarmed until the fault is corrected.

The diagnostic module also supplies backup power to the air bag module in the event that the battery or cables are damaged during the accident. The control module contains a capacitor that can store a charge that can last up to 30 minutes after the battery has been disconnected.

WARNING!

The backup power supply must be depleted before any air bag service is performed. To deplete this backup power, disconnect the positive battery cable and wait at least 30 minutes. Refer to the information in the service manual or information system to determine exactly how long you should wait.

WIRING HARNESS For identification and safety purposes, the electrical harnesses of the SIR system typically have yellow connectors. Single-stage air bags have one inflator and one pair of wires that connect to the air bag module. Two-stage air bags have two inflators and two pairs of wires that connect to the air bag module.

CLOCKSPRING The **clockspring** allows for electrical contact to the air bag module at all times. Because the air bag module sits in the centre of the steering wheel, the clockspring is designed to provide voltage to the module regardless of steering wheel position. The clockspring is located between the steering wheel and the steering column **(Figure 24–20)**.

The clockspring's electrical connector contains a long conductive ribbon. The wires from the air bag's electrical system are connected from the underside of the clockspring to the conductive ribbon. The other end of the ribbon is connected to the air bag module. When the steering wheel is turned, the ribbon coils and uncoils without breaking the electrical connection.

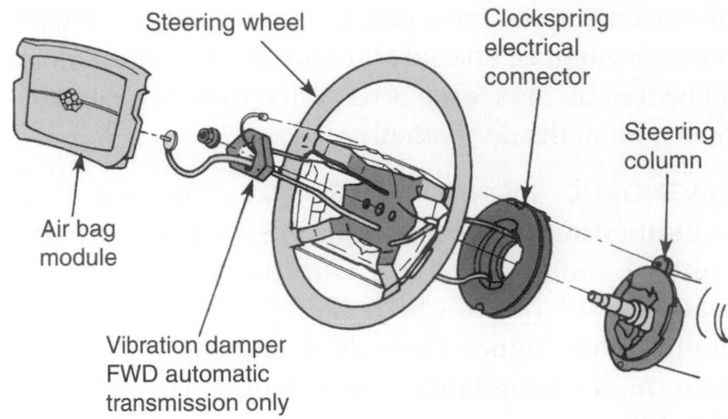

FIGURE 24–20

The clockspring is located between the steering wheel and the steering column.

Chrysler Group LLC

SIR/SRS OR AIR BAG READINESS LIGHT This light lets the driver know the air bag system is ready to do its job. The warning light is operated by the diagnostic module. The readiness light turns on briefly when the driver turns the ignition key from off to run. The light should go out once the engine is running. A malfunction in the air bag system causes the light to stay on continuously or to flash. Some systems have a tone generator that sounds if there is a problem in the system or if the readiness light is not functioning or if the bulb has been removed.

Air Bag Module

The air bag module is the air bag and inflator assembly packaged into a single unit or module. The module is located in the steering wheel for the driver and in the dash panel for the front seat passenger **(Figure 24–21)**. The various types of side

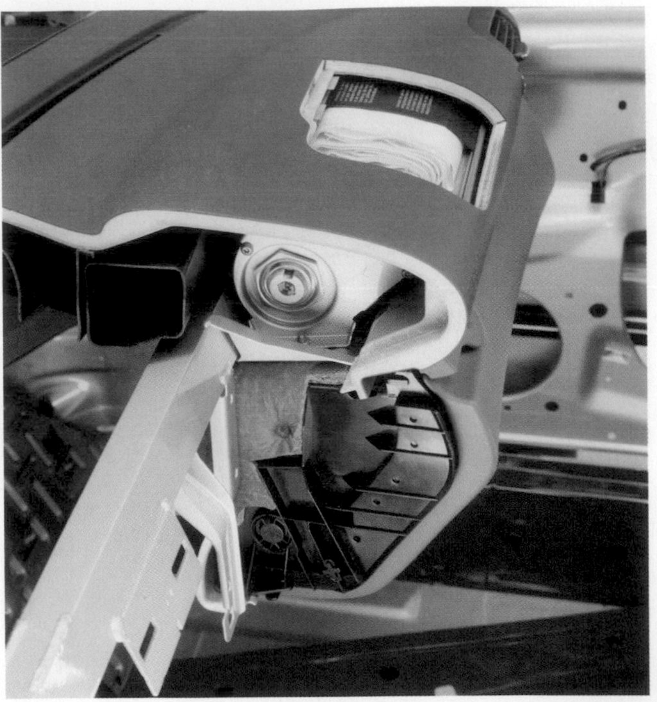

FIGURE 24–21

A cutaway view showing the complete passenger-side air bag module in the dash.

DaimlerChrysler Corporation

protection air bags have the module located at the point where the bag is deployed.

The inflation of the pyro-type air bag is accomplished by the production and rapid release of nitrogen gas. The igniter **(Figure 24–22)** is an integral part of the inflator assembly. It starts a chemical reaction to inflate the air bag. At the centre of the igniter assembly is the squib, which contains **zeronic potassium perchlorate (ZPP)**. When

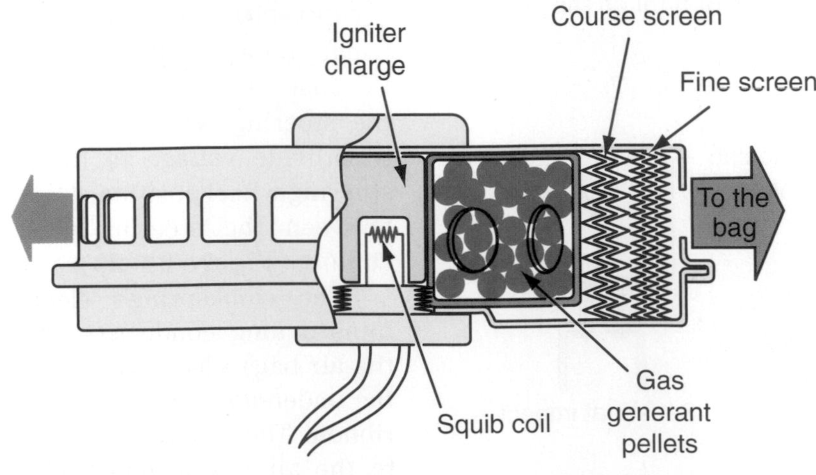

FIGURE 24–22

The action of an air bag module releasing nitrogen gas to an air bag.

voltage is supplied to the squib, an electrical arc is formed between two pins.

The inflation assembly is composed of a gas generator (called a generant) containing sodium azide and copper oxide or potassium nitrate propellant. The ZPP ignites the propellant charge. During ignition, large quantities of hot, expanding nitrogen gas are produced very quickly and rapidly inflate the air bag. As the nitrogen moves into the air bag, it is filtered to remove sodium hydroxide dust formed during the chemical reaction.

> **CAUTION!**
>
> Wear gloves and eye protection when handling a deployed air bag module. Sodium hydroxide residue may remain on the bag and cause a skin irritation.

Not all air bags use nitrogen gas to inflate the bag; some use a solid propellant and compressed argon gas **(Figure 24–23)**. Argon has a stable structure, cools more quickly, and is inert as well as nontoxic. Argon is commonly used for passenger-side and side-protection air bags.

A mounting plate and retainer ring attach the air bag assembly to the inflator.

The bag itself is made of a thin, nylon fabric that is folded into the steering wheel, dash, seat, or door. The powdery substance released from the air bag when it is deployed is regular cornstarch or talcum powder. These powders are used by the air bag manufacturers to keep air bags lubricated and

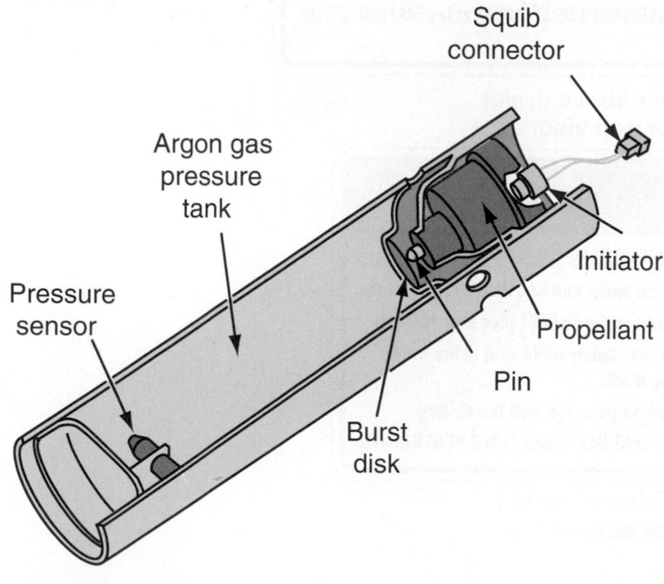

FIGURE 24–23

An inflator module that uses argon gas.

pliable while they are in storage. The entire module must be serviced as one unit when repair of the air bag system is required. It should be noted that some newer systems display an air bag readiness light on the dash to show that the system recognizes body weight in the driver and/or passenger seat.

Diagnosis

Before diagnosing the system, perform a system check by observing the air bag warning light and comparing your findings with those described in the service manual for the vehicle. Typically, an air bag system problem is indicated by any of the following warning light conditions:

- If the light does not come on when the ignition is initially turned on
- If the light does not steadily light while the engine is cranking
- If the light remains on but does flash when the ignition is turned on
- If the light flashes seven to nine times and then remains on when the ignition is turned on
- If the light comes on while the engine is running

If any of these conditions are present, the air bag system needs to be checked. A thorough visual inspection of sensor integrity is the best place to start when diagnosing a system that is disarmed because of a fault. Damage from a collision or mishandling during a non-related repair can set up a fault area, which will disarm the air bag system.

When a technician places the system into its diagnostic mode, the warning lamp may flash trouble codes or the codes can be retrieved with a scan tool.

RETRIEVING TROUBLE CODES The control module of most air bag systems will store trouble codes that can be retrieved by either a scan tool or flash codes. Since 1996, most systems require the use of a scan tool.

Scan Tool DTC Retrieval. Normally, an air bag system stores two types of faults. Active DTCs will turn the air bag warning lamp on, whereas stored codes are intermittent problems and probably will not turn on the warning lamp.

To retrieve codes, connect the scan tool to the DLC, and turn the ignition on. Follow the instructions for the scan tool to retrieve air bag information. Record all stored and active codes. Diagnose the cause of the codes in order, from the lowest number to the highest. Stored codes can be erased with the scan tool, but active codes are erased only when the problem is corrected.

Flash Codes. On vehicles that display codes with the warning light or on the digital instrument panel, make sure you follow the procedure prescribed by the manufacturer to retrieve the codes. Flash code systems do not display stored codes; therefore, the cause of any intermittent problem must be found through normal diagnostics and reasoning.

SERVICING THE AIR BAG SYSTEM

Whenever working on or around air bag systems, it is important to follow all safety warnings **(Figure 24–24)**. Examples of these warnings follow:

- Wear safety glasses when servicing the air bag system.
- Wear safety glasses when handling an air bag module.

AVERTISSEMENT

WARNING

On back of air bag

⚠WARNING

DO NOT TAMPER WITH OR DISCONNECT THE AIR BAG SYSTEM WIRING. You could inflate the bag(s) or make it inoperative which may result in injury. See Shop Manual.

⚠ AVERTISSEMENT

NE PAS MANIPULER NI DÉBRANCHER LE CÂBLAGE ÉLECTRIQUE DU DISPOSITIF D'UN SAC GONFLABLE. Cela pourrait gonfler le sac ou le mettre hors service et entraîner des blessures. Voir le manuel de réparation.

F4S8-6300014-AA

In engine compartment

Label on front of driver and passenger sun visors

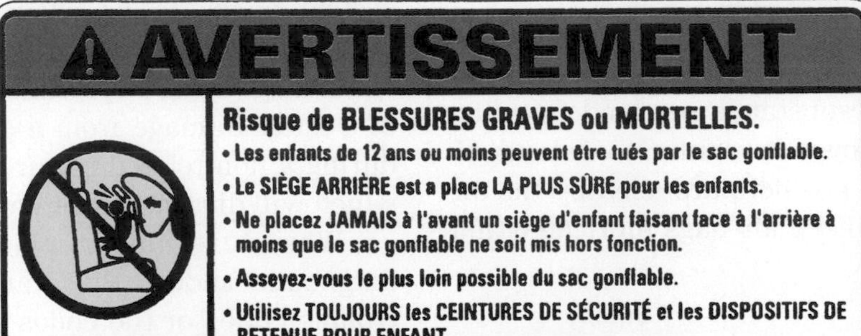

Label on headliner above driver and passenger sun visors

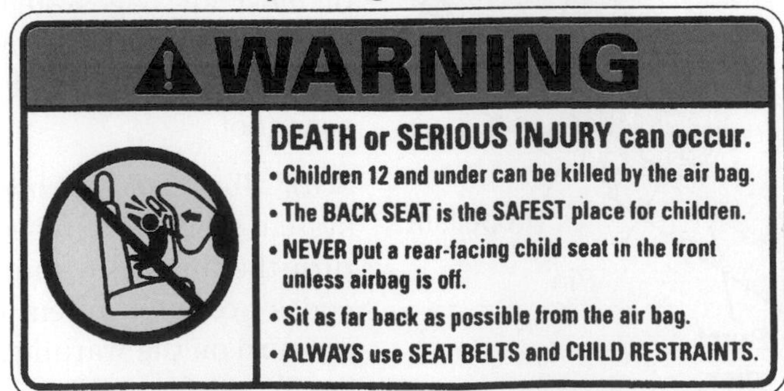

Label on back side of driver and passenger sun visors

FIGURE 24–24

Various air bag warning labels found on today's vehicles.

Ford Motor Company

- Wear a ground strap while handling any air bag system electrical connections.
- Wait at least 30 minutes after disconnecting the negative battery cable before beginning any service on or around the air bag system. The reserve energy module is capable of storing enough power to deploy the air bag for up to 30 minutes after battery voltage is lost.
- Handle all air bag sensors with care. Do not strike or jar a sensor in such a manner that deployment may occur.
- When carrying a live air bag module, face the trim and bag away from your body.
- Do not place your face or body within close proximity of the air bag module(s) when working in the interior of the vehicle.
- Do not carry the module by its wires or connector.
- When placing a live module on a bench, face the trim and air bag up.
- Deployed air bags may have a powdery residue on them. Sodium hydroxide is produced by the deployment reaction and is converted to sodium bicarbonate carbonate when it comes into contact with atmospheric moisture. It is unlikely that sodium hydroxide will still be present. However, wear safety glasses and gloves when handling a deployed air bag. Wash your hands immediately after handling the bag.
- A live air bag module must be deployed before disposal. Because the deployment of an air bag is through an explosive process, improper disposal may result in injury and fines. A deployed air bag should be disposed of according to Environment Canada and manufacturer procedures.
- Do not use a battery- or AC-powered voltmeter, ohmmeter, or any other type of test equipment not specified in the service manual. Never use a test light to probe for voltage.

CAUTION!

A two-stage air bag may appear to be fully deployed when only its first stage has deployed. Care must be taken to make sure that two-stage air bags have been fully deployed before handling them. Always assume that any deployed two-stage air bag has an active stage 2. Improper handling or servicing can activate the inflator module and cause personal injury. Always follow the manufacturer's recommended handling procedures.

Service Guidelines

Keep in mind that wiring difficulties of any kind call for careful removal of the air bag module and putting a simulator in its place. The simulator **(Figure 24–25)** replaces the air bag to eliminate any chance of air bag deployment during diagnosis. This tool allows for a safe test sequence that should identify any continuity problems or short circuits. If a condition is present in the system that would cause deployment of the air bag, the simulator will indicate it and will remain in that mode until the problem is fixed.

An air bag module is serviced as a complete assembly. Technicians repairing these systems are also advised to service crash sensors and any other related components in assembly groupings. A damaged crash sensor should be replaced. Some manufacturers may recommend replacement of the entire set if a failure or degradation of any single sensor is found.

Photo Sequence 19 covers a typical procedure for removing an air bag module.

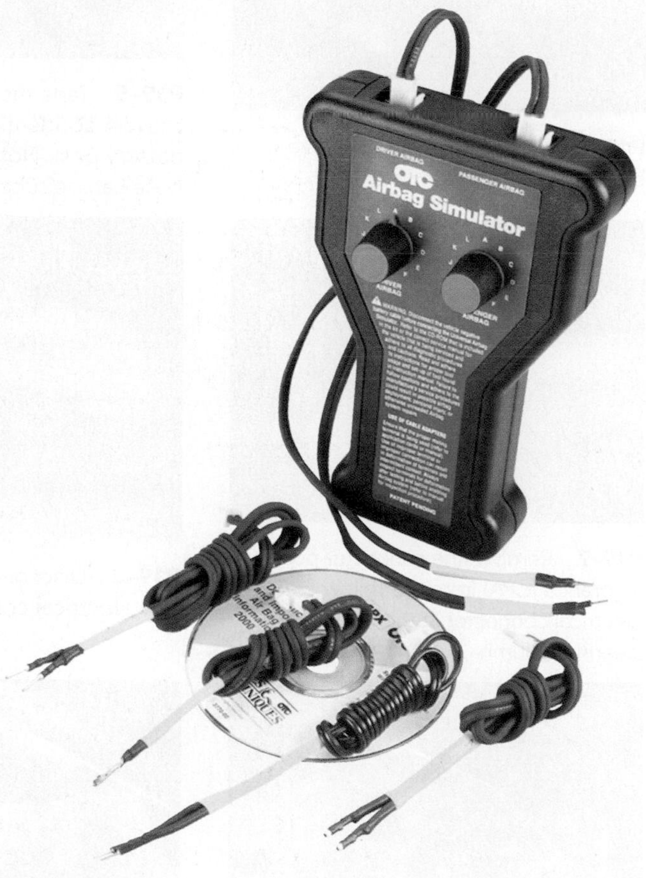

FIGURE 24–25

An air bag simulator used to replace the air bag module during testing.

SPX Corporation, Aftermarket Tool and Equipment Group

PHOTO SEQUENCE 19

REMOVING AN AIR BAG MODULE

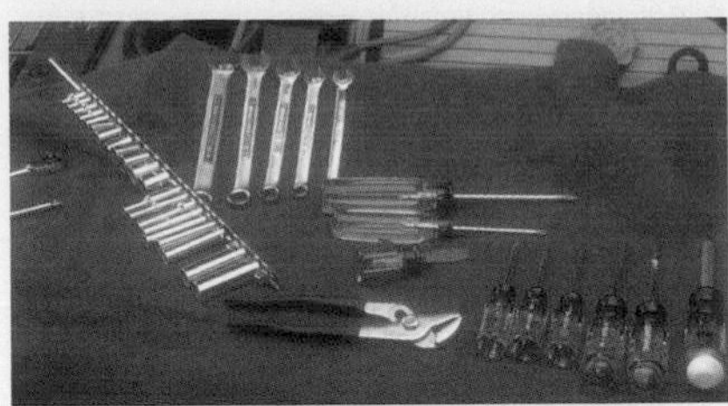

P19–1 Tools required to remove the air bag module are safety glasses, seat covers, a screwdriver set, a Torx driver set, battery terminal pullers, battery pliers, assorted wrenches, a ratchet and socket set, and the service manual.

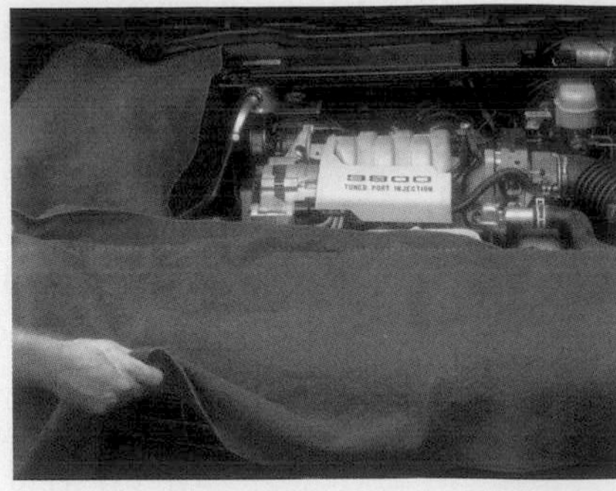

P19–2 Place the seat and fender covers on the vehicle.

P19–3 Place the front wheels in the straight-ahead position, and turn the ignition switch to the LOCK position.

P19–4 Disconnect the negative battery cable.

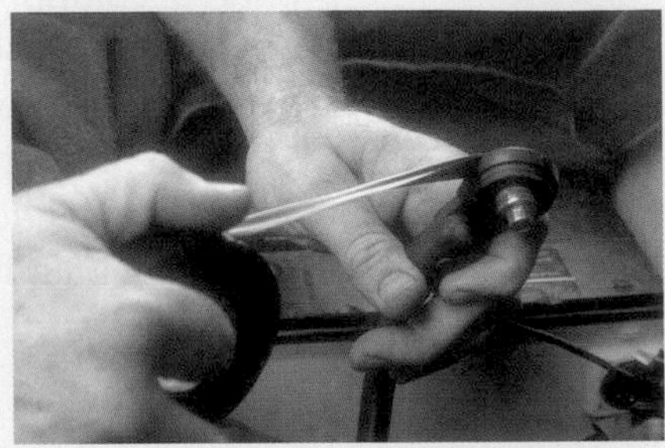

P19–5 Tape the cable terminal to prevent accidental connection with the battery post. Note: A piece of rubber hose can be substituted for the tape.

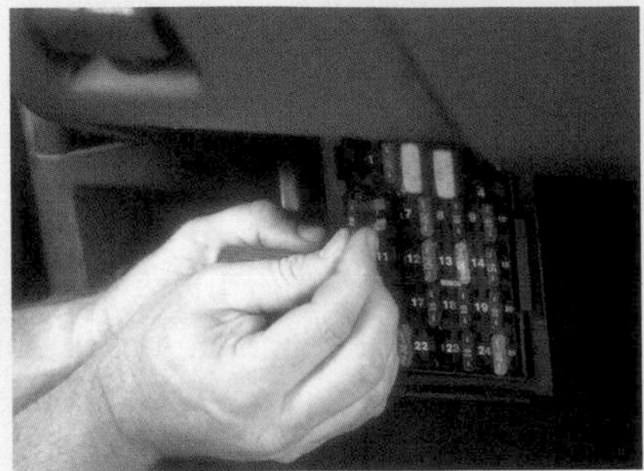

P19–6 Remove the SIR fuse from the fuse box. Wait 30 minutes to allow the reserve energy to dissipate.

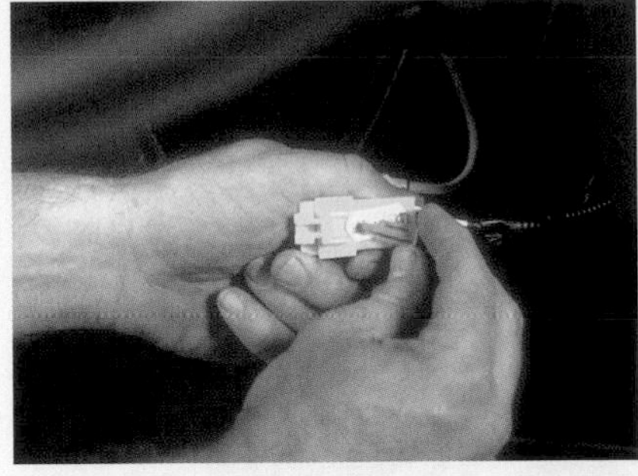

P19–7 Remove the connector position assurance (CPA) from the yellow electrical connector at the base of the steering column.

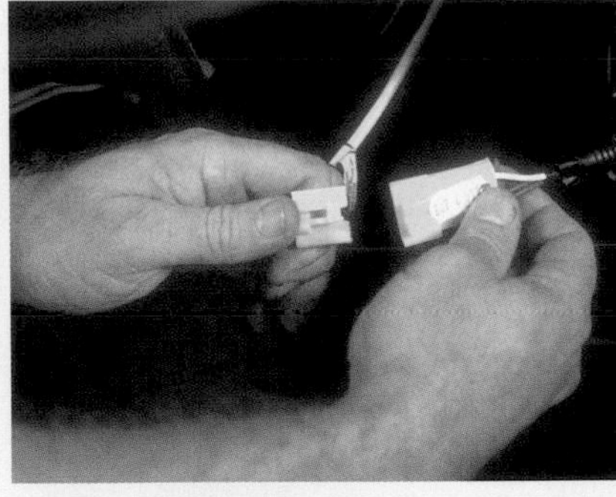

P19–8 Disconnect the yellow two-way electrical connector.

P19–9 Remove the four bolts that secure the module from the rear of the steering wheel.

P19–10 Rotate the horn lead one-quarter turn and disconnect.

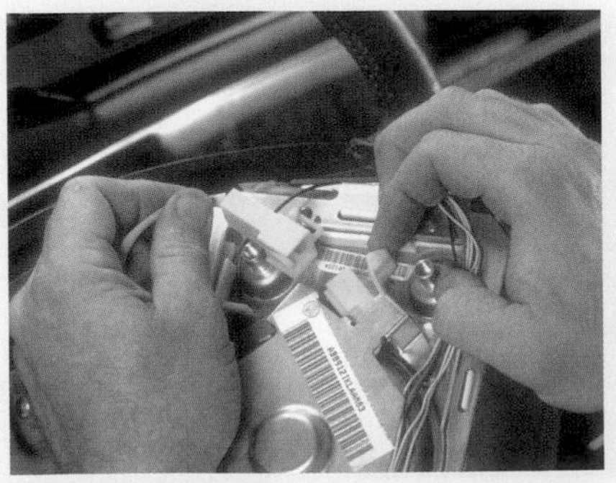

P19–11 Disconnect the electrical connectors.

P19–12 Remove the module.

The steering column clockspring should be maintained in its correct index position at all times. Failure to do so can cause damage to the enclosure, wiring, or module. Any of these situations can cause the air bag system to default into a non-operative mode. The clockspring should be replaced any time the air bag has been deployed.

Before returning the vehicle to the customer after service, make sure the sensors are firmly fastened to their mounting fixtures, with their arrows facing forward. Be certain all the fuses are correctly rated and replaced. Make sure a final check is made for codes using the approved scan tool. Carefully recheck the wire and harness routing before releasing the car.

OTHER PROTECTION SYSTEMS

To make vehicles safe and to protect the occupants inside the vehicle, manufacturers include many different systems and options. What follows is a quick look at a few of these. By no means is this discussion conclusive; there are many things about how a vehicle is made that influence the protection and safety it offers. Basically, cars that offer good protection are those that are constructed to maintain integrity when impacted on. This construction includes side door beams, crumple zones, and reinforced areas of the frame.

Crumple zones are areas of the body that will bend or break away to defuse or dissipate impact energy in order to protect the passengers inside the vehicle. These crumple zones absorb or take on the impact while keeping the passenger compartment undisturbed.

CUSTOMER CARE

Let your customers know that a well-maintained vehicle is a prerequisite for a safe vehicle. Some problems will lead to accidents and/or injury.

Headrests

Nearly two-thirds of all injuries in collisions are soft-tissue related, commonly referred to as whiplash. Good head restraint (headrest) design and proper adjustment may prevent these injuries. A properly adjusted headrest prevents the head and neck from extending backward on impact. Nine out of ten people do not adjust the headrest for their height and comfort.

CUSTOMER CARE

If a customer's vehicle has adjustable headrests, do the customer a big favour by telling him or her how to adjust them. A headrest should be at least to the top of the driver's ear and less than 10 cm (4 in.) from the back of his or her head.

In high-speed rear crashes, the backs of seats must be strong enough to transfer the energy of the impact from the occupant while keeping him or her in the seat. In low- to high-speed crashes, the headrest and upper part of the seat must reduce the relative motion between the head and neck to prevent whiplash. The reclining function of the seats must also be strong enough to withstand the force of the occupant being thrown back into the seat.

New systems have been developed that automatically adjust the headrest. This self-aligning headrest system moves the headrest up and forward when the vehicle is hit from behind.

Rollover Protection

Some convertibles have a built-in rollbar to protect the passengers in the vehicle in the case of a rollover. These units are permanent structures of the vehicle. Others have automatic systems that provide for this. These systems deploy a roll bar from behind the headrests when the vehicle experiences extreme tilting; the wheels lose contact with the ground; or the vehicle is involved in a serious accident **(Figure 24–26)**.

FIGURE 24–26

This rollbar pops out when the system anticipates a potential rollover.

DaimlerChrysler Corporation

KEY TERMS

Active restraint system (p. 710)
Adaptive air bags (p. 717)
Air bag sensing diagnostic monitor (ASDM) (p. 719)
Clockspring (p. 719)
Inertia lock retractor (p. 711)
Passive restraint system (p. 710)
Pre-tensioners (p. 712)
Restraint control module (RCM) (p. 719)
Second-generation air bags (p. 717)
Smart air bags (p. 717)
Squib (p. 715)
Supplemental inflatable restraint (SIR) (p. 715)
Supplemental restraint system (SRS) (p. 715)
Zeronic potassium perchlorate (ZPP) (p. 720)

SUMMARY

- All new vehicles built or sold in North America must have one or both types of passive restraints: seat belts or air bags.
- Restraint systems are either active or passive.
- When servicing seat belts, inspect the webbing, buckles, retractors, and anchorage.
- An air bag is inflated or deployed by rapid expansion of a gas fired by an igniter or squib.
- Second-generation air bags have been depowered to inflate with less force than earlier air bags.
- Smart air bags have two possible stages of deployment in an attempt to match the severity of the impact and/or by the size and weight of the seat's occupant.
- The electrical circuit of an air bag system includes impact sensors and an electronic control module.
- The air bag module is the air bag and inflator assembly packaged into a single unit.
- A system check of an air bag system consists of observing the air bag warning light and the use of a scan tool.
- The control module will store trouble codes that can be retrieved by a scan tool.
- Before doing any work on an air bag system, disconnect the negative battery cable.
- Care must be taken when removing a live (not deployed) air bag. Be sure the bag and trim cover are pointed away from you.

REVIEW QUESTIONS

1. What action must be taken by the driver or passenger to use an active restraint system?
   a. manually connect a seat belt
   b. close a door to activate a seat belt motor
   c. send a signal to the seat belt monitor
   d. switch on an air bag system
2. Most seat belts in modern cars have retractors, except for which occupant?
   a. front driver and passenger seats in pickups
   b. front passenger seats
   c. rear passenger seats at the right and left
   d. rear passenger seats centre
3. Most seat belts, with passive pre-tensioners, use what kind of device to remove belt slack and position the occupants?
   a. moving weight system
   b. pyrotechnic device
   c. photosensitive device
   d. coil spring system
4. Which of the following methods is used to limit the severity of air bag deployment?
   a. adding foam covering to the bag for extra padding
   b. detuning the crash sensors to slow down reaction time
   c. auto-applying brakes before the collision
   d. using a two-stage deployment
5. How long does it take for the air bag to fully inflate?
   a. 0.23 ms
   b. 2.3 ms
   c. 23 ms
   d. 230 ms
6. The driver-side air bag on most vehicles uses what gas to deploy the bag?
   a. argon
   b. nitrogen
   c. helium
   d. oxygen
7. Some passenger-side air bags use what gas to inflate the bag?
   a. argon
   b. nitrogen
   c. helium
   d. oxygen
8. Some air bags rely on two sensors to deploy an air bag. What are these two sensors called?
   a. decel sensor and crash sensor
   b. crash sensor and impact sensor
   c. vehicle speed sensor and safing sensor
   d. safing sensor and crash sensor

9. What colour is used to identify SIR wiring?
   a. yellow b. black
   c. orange d. brown
10. How long should the system be powered down before servicing an air bag?
    a. 3 seconds
    b. 3 minutes
    c. 30 minutes
    d. 3 hours
11. What is the purpose of the clockspring?
    a. to act as a timing device for air bag deployment
    b. to act as return mechanism for the air bag
    c. to provide a direct electrical connection to the air bag
    d. to gauge the severity of the crash
12. What is the most common colour used to identify air bag electrical connectors?
    a. red
    b. yellow
    c. blue
    d. green
13. When servicing or storing a live air bag module, how should the trim face be oriented?
    a. The trim cover should face down on a flat surface.
    b. The trim cover should face down on a curved surface.
    c. The trim cover should face up on a curved surface.
    d. The trim cover should face up on a flat surface.
14. What is one of the powders released from a deployed air bag?
    a. talcum powder
    b. hydrogen sulphide powder
    c. sodium azide powder
    d. calcium hydroxide powder
15. Inflation of a pyrotechnic air bag is caused by the rapid burning of what substance?
    a. nitrogen phosphate
    b. potassium percolates
    c. sodium azide
    d. copper oxide
16. What is the safest method of testing and diagnosing an air bag system connected to the system?
    a. a voltmeter
    b. a scan tool
    c. an ohmmeter
    d. flash codes
17. What should be done to the air bags on a vehicle that is being scrapped?
    a. returned to the manufacturer
    b. disassembled according to instructions
    c. removed and tagged
    d. deployed according to instructions
18. What is the purpose of the reserve energy supply?
    a. to provide power to keep the air bag module memory alive
    b. to provide a backup power supply to deploy the air bags
    c. to provide a power supply to turn on the air bag light
    d. to provide a backup power supply to record codes
19. What is the most effective location for air bag sensor placement?
    a. facing rearward in the trunk area
    b. facing forward in the front radiator support area
    c. under the driver's seat
    d. under the passenger seat
20. Modern crash sensors are usually which type?
    a. mechanical sensors
    b. accelerometers
    c. ride height sensors
    d. electromagnetic sensors

CHAPTER 25
Ignition Systems

LEARNING OUTCOMES

- Name and describe the three basic types of ignition systems.
- Name the two major electrical circuits used in all ignition systems and their common components.
- Describe the operation of ignition coils, spark plugs, and ignition cables.
- Explain how high voltage is induced in the coil secondary winding.
- Describe the various types of spark timing systems, including electronic switching systems and their related engine-position sensors.
- Explain the basic operation of a computer-controlled ignition system.
- Describe the operation of a distributor-based ignition system.
- Describe the operation of a distributorless ignition system.

WE SUPPORT THE CANADIAN INTERPROVINCIAL STANDARDS PROGRAM

The ignition system in today's vehicles is an integral part of the electronic engine control system. The engine control module (ECM) or powertrain control module (PCM) controls all functions of the ignition system and constantly corrects the spark timing. The desired ignition timing is calculated by the PCM according to inputs from a variety of sensors. These inputs allow the PCM to know the current operating conditions. The PCM matches those conditions to its programming and controls ignition timing accordingly. It is important to remember that there has always been a need for engine speed– and load-based timing adjustments. Electronic systems are very efficient at making these adjustments. Many of the inputs used for ignition system control are also used to control other systems, such as fuel injection. These inputs are available on the CAN buses **(Figure 25–1)**.

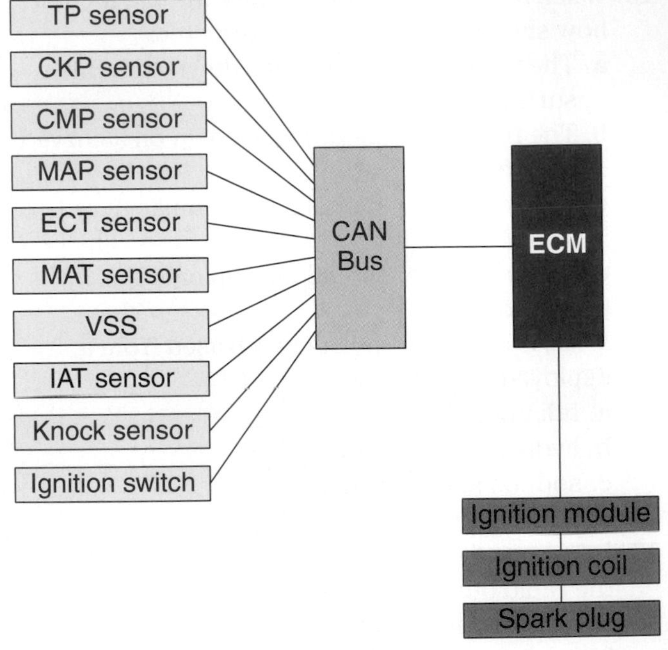

FIGURE 25–1

Many of the inputs used for ignition system control are also used to control other systems and are available on the CAN buses.

There are three basic ignition system designs: **distributor-based ignition (DI) systems**, **distributorless ignition systems**, and **direct ignition systems (DIS)**. The latter two designs are designated as electronic ignition (EI) systems by the Society of Automotive Engineers (SAE). DIS is the commonly used design on today's engines. Regardless of the type, all modern ignition systems are designed to perform the same functions: to generate sufficient voltage to force a spark across the spark plug gap, to time the arrival of the spark to coincide with the movement of the engine's pistons, and to vary the spark arrival time based on varying operating conditions.

BASIC CIRCUITRY

All ignition systems consist of two interconnected electrical circuits: a **primary** (low-voltage) **circuit** and a **secondary** (high-voltage) **circuit (Figure 25–2).**

Depending on the exact type of ignition system, components in the primary circuit include the following:

- Battery
- Ignition switch
- Ballast resistor or resistance wire (some systems)
- Starting bypass (some systems)
- Ignition coil primary winding
- Triggering device
- Switching device or control module (igniter)

The secondary circuit includes these components:

- Ignition coil secondary winding
- Distributor cap and rotor (some systems)
- High-voltage cables (some systems)
- Spark plugs

Primary Circuit Operation

When the ignition switch is on, voltage from the battery is supplied to the positive connection of the primary winding of the ignition coil. This path to the coil may be through the ignition switch, ignition control

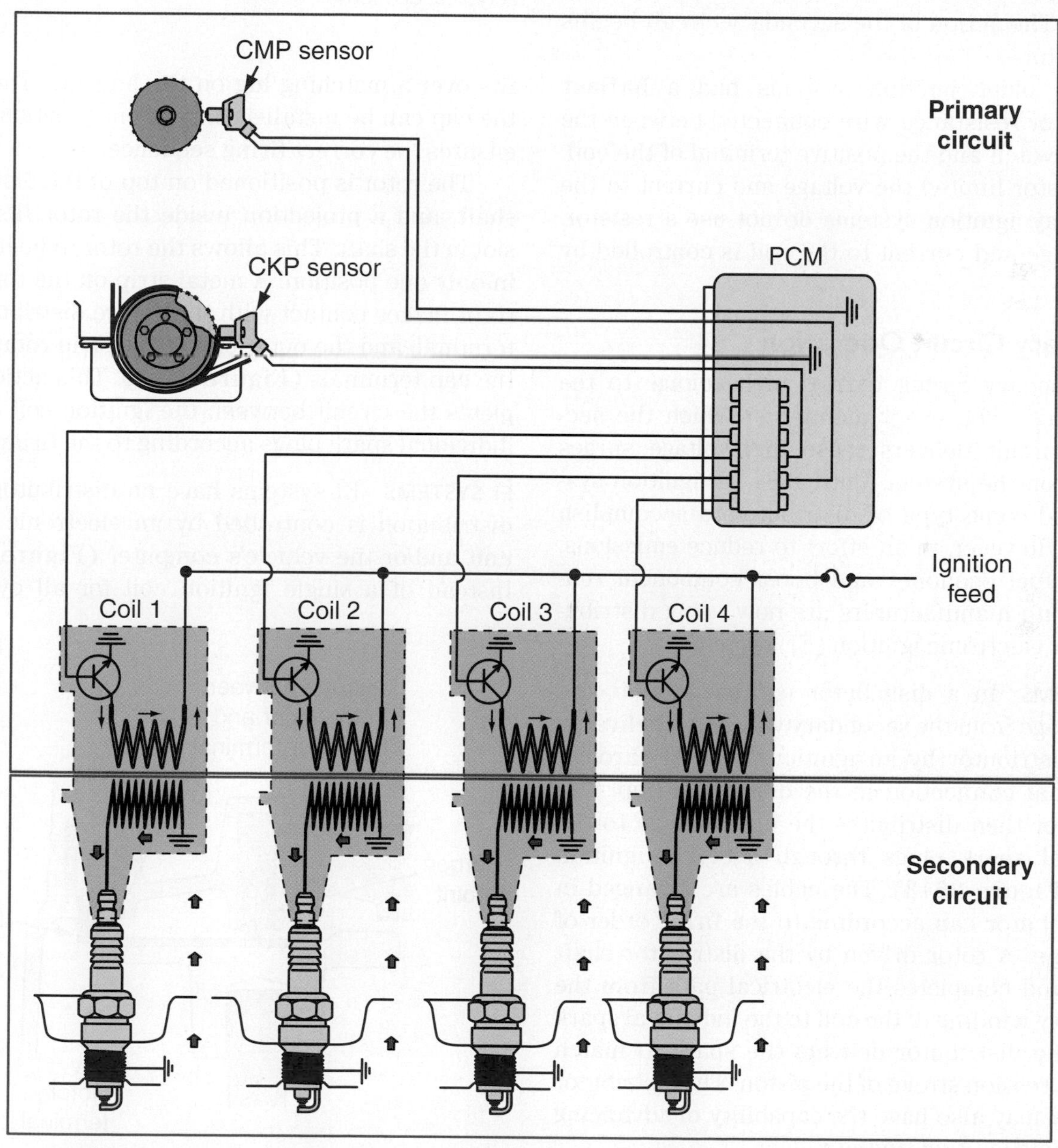

FIGURE 25–2

Ignition systems have a primary (low-voltage) and a secondary (high-voltage) circuit.

module, or PCM, depending on the system. The negative connection of the primary winding is connected to a switching device. This can be the ignition control module or the PCM. When the switch is closed, current flows through the circuit. The current flow through the ignition coil's primary winding creates a magnetic field. As the current continues to flow, the magnetic field gets stronger. When the triggering device signals to the switching unit that the piston is approaching top dead centre (TDC) on the compression stroke, the circuit opens, and current flow is stopped. This causes the magnetic field around the primary winding to collapse across the secondary winding. The movement of the magnetic field across the winding induces a high voltage in the secondary winding. The action of the secondary circuit begins at this point.

Some older ignition systems had a **ballast resistor** or resistance wire connected between the ignition switch and the positive terminal of the coil. This resistor limited the voltage and current to the coil. Today, ignition systems do not use a resistor, and voltage and current to the coil is controlled by the PCM.

Secondary Circuit Operation

The secondary circuit carries high voltage to the spark plugs. The exact manner in which the secondary circuit delivers these high-voltage surges depends on the system. Until 1984, all ignition systems used some type of distributor to accomplish this job. However, in an effort to reduce emissions, improve fuel economy, and boost component reliability, auto manufacturers are now using distributorless or electronic ignition (EI) systems.

DI SYSTEMS In a distributor ignition system, the high voltage from the secondary winding is delivered to the distributor by an ignition cable or through an internal connection in the distributor cap. The distributor then distributes the high voltage to the individual spark plugs through a set of ignition cables **(Figure 25–3)**. The cables are arranged in the distributor cap according to the firing order of the engine. A rotor driven by the distributor shaft rotates and completes the electrical path from the secondary winding of the coil to the individual spark plugs. The distributor delivers the spark to match the compression stroke of the piston. The distributor assembly may also have the capability of advancing or retarding ignition timing.

The distributor cap is mounted on top of the distributor assembly, and an alignment notch in the cap fits over a matching lug on the housing. Therefore, the cap can be installed only in one position, which ensures the correct firing sequence.

FIGURE 25–3
A typical distributor.

The rotor is positioned on top of the distributor shaft, and a projection inside the rotor fits into a slot in the shaft. This allows the rotor to be installed in only one position. A metal strip on the top of the rotor makes contact with the centre distributor cap terminal, and the outer end of the strip rotates past the cap terminals **(Figure 25–4)**. This action completes the circuit between the ignition coil and the individual spark plugs according to the firing order.

EI SYSTEMS EI systems have no distributor; spark distribution is controlled by an electronic control unit and/or the vehicle's computer **(Figure 25–5)**. Instead of a single ignition coil for all cylinders,

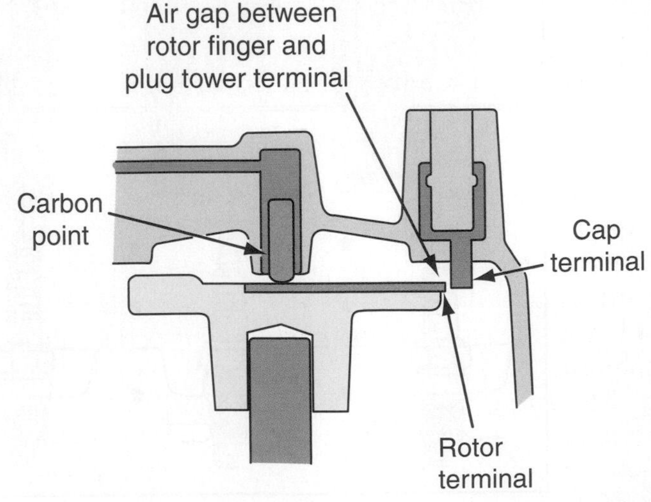

FIGURE 25–4
The relationship of a rotor and distributor cap.

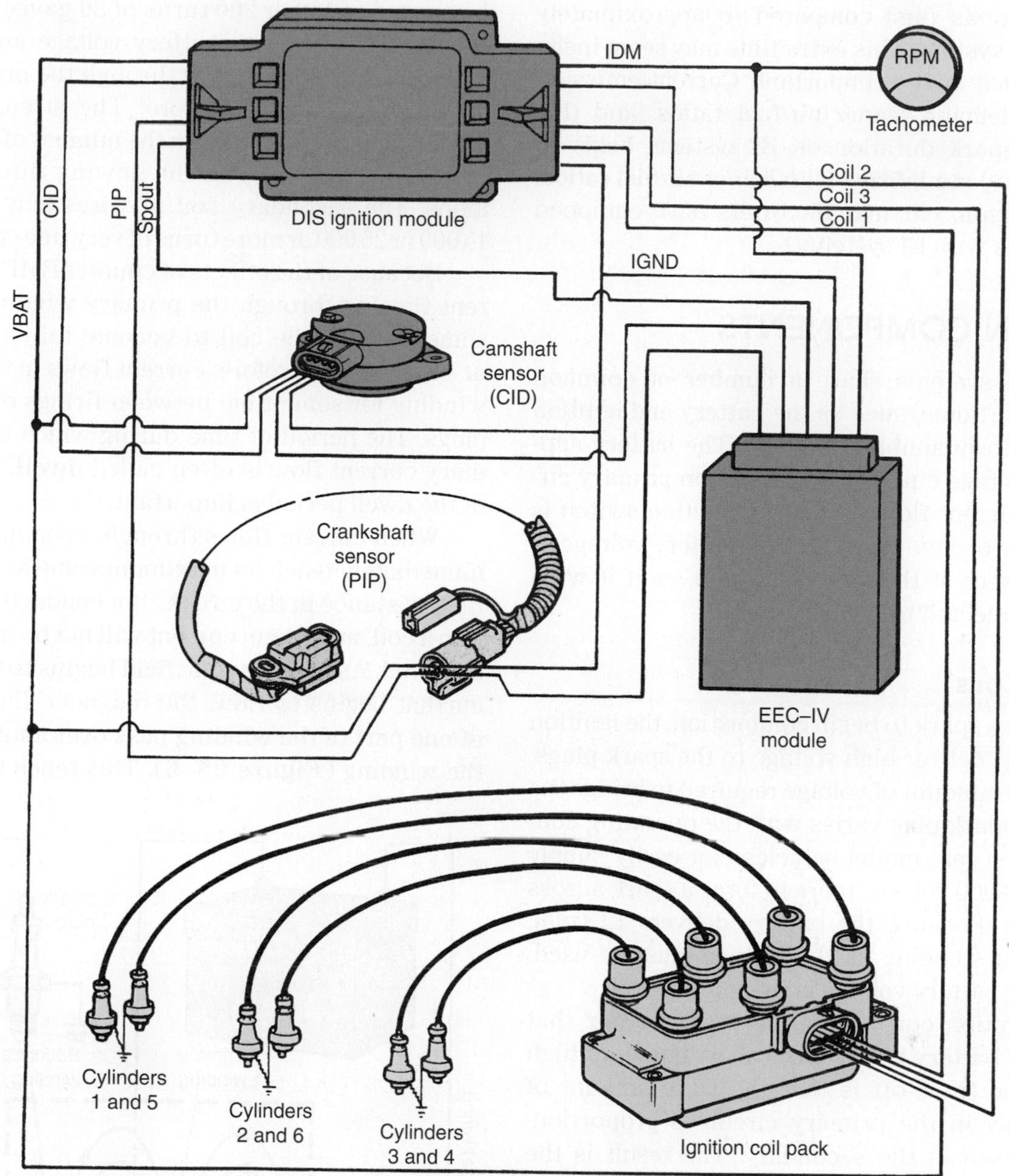

FIGURE 25–5

An electronic ignition system for a six-cylinder engine.

Ford Motor Company

each cylinder may have its own ignition coil, or two cylinders may share one coil. The coils are wired directly to the spark plug they control. An ignition control module, tied into the vehicle's computer control system, controls the firing order and the spark timing and advance.

The energy produced by the secondary winding is voltage. This voltage is used to establish a complete circuit so that current can flow. The excess energy is used to maintain the current flow across the spark plug's gap. Distributorless ignition systems are capable of producing much higher energy than conventional ignition systems. This is because having multiple coils allows for increased current flow and coil charge time.

Because DI and EI systems are firing spark plugs with approximately the same air gap across the electrodes, the voltage required to start firing the spark plugs in both systems is similar. If the additional energy in the EI systems is not released in the form of voltage, it will be released in the form of current flow. This results in higher firing current and longer spark plug firing times. The average firing time across the spark plug electrodes in an EI system is

1.5 milliseconds (ms) compared to approximately 1 ms in a DI system. This extra time may seem insignificant, but it is very important. Current emission standards demand leaner air/fuel ratios, and this additional spark duration on EI systems helps to prevent cylinder misfiring with leaner air/fuel ratios. For this reason, car manufacturers have equipped their engines with EI systems.

IGNITION COMPONENTS

All ignition systems share a number of common components. Some, such as the battery and ignition switch, perform simple functions. The battery supplies low-voltage current to the ignition primary circuit. The current flows when the ignition switch is in the START or RUN position. Full-battery voltage is always present at the ignition switch, as if it were directly connected to the battery.

Ignition Coils

To generate a spark to begin combustion, the ignition system must deliver high voltage to the spark plugs. Because the amount of voltage required to bridge the gap of the spark plug varies with the operating conditions, most late-model vehicles can easily supply 30 000 to 60 000 volts or more to force a spark across the air gap. Because the battery delivers 12 volts, a method of stepping up the voltage must be used. Multiplying battery voltage is the job of a coil.

The ignition coil is a **pulse transformer** that transforms battery voltage into short bursts of high voltage. The trade-off is that the large amount of current flow in the primary circuit is proportionally decreased in the secondary. The result is the low-voltage, high-amperage current of the primary is changed into a high-voltage, low-amperage spark from the secondary.

As explained previously, when a magnetic field moves across a wire, voltage is induced in the wire.

If a wire is bent into loops forming a coil and a magnetic field is passed through the coil, an equal amount of voltage is generated in each loop of wire. The more loops of wire in the coil, the greater the total voltage induced. If the strength of the magnetic field is doubled, the voltage output doubles.

An ignition coil uses these principles and has two coils of wire wrapped around an iron or steel core. An iron or steel core is used because it has low **inductive reluctance**. In other words, iron freely expands or strengthens the magnetic field around the windings. The first, or primary, coil is normally composed of 100 or 200 turns of 20-gauge wire. This coil of wire conducts battery voltage and current. When a current is passing through the primary coil, it magnetizes the iron core. The strength of the magnet depends directly on the number of wire loops and the amount of current flowing through those loops. The secondary coil of wires may consist of 15 000 or 25 000 or more turns of very fine copper wire.

Because of the effects of counter EMF on the current flowing through the primary winding, it takes some time for the coil to become fully magnetized or saturated. Therefore, current flows in the primary winding for some time between firings of the spark plugs. The period of time during which there is primary current flow is often called **dwell**. The length of the dwell period is important.

When current flows through a conductor, it will immediately reach its maximum value as allowed by the resistance in the circuit. If a conductor is wound into a coil, maximum current will not be immediately achieved. As the magnetic field begins to form as the current begins to flow, the magnetic lines of force of one part of the winding pass over another part of the winding **(Figure 25–6)**. This tends to cause an

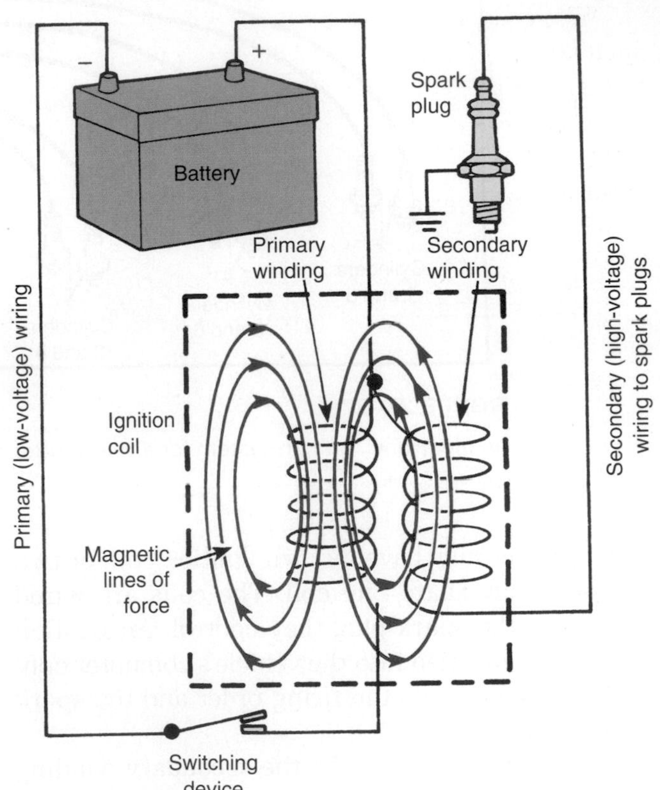

FIGURE 25–6

Current passing through the coil's primary winding creates magnetic lines of force that cut across and induce voltage in the secondary windings.

opposition to current flow. This occurrence is called **reactance**. Reactance causes a temporary resistance to current flow and delays the flow of current from reaching its maximum value. When maximum current flow is present in a winding, the winding is said to be saturated, and the strength of its magnetic field will also be at a maximum.

Saturation can occur only if the dwell period is long enough to allow for maximum current flow through the primary windings. A less than saturated coil will not be able to produce the voltage it was designed to produce. If the energy from the coil is too low, the spark plugs may not fire long enough or may not fire at all. If the current is applied longer than needed to fully saturate the winding, the coil will overheat.

A typical coil requires 2 to 6 ms to become saturated. The actual required time depends on the resistance of the coil's primary winding and the voltage applied to it. Some systems electronically limit the primary current flow at low speeds to prevent the coil from overheating. When the engine reaches higher speeds, the current limitation feature is disabled. An example of a waveform representing current flow through the coil primary is shown in **(Figure 25–7)**.

When the primary coil circuit is suddenly opened, the magnetic field instantly collapses. The sudden collapsing of the magnetic field produces a very high voltage in the secondary windings. This high voltage is used to push current across the gap of the spark plug.

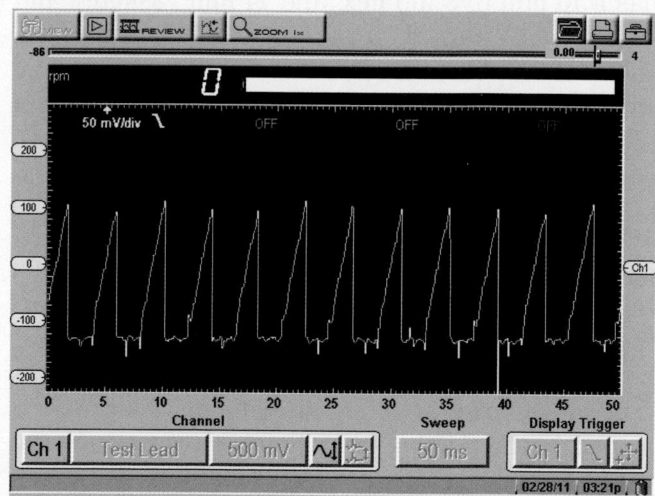

FIGURE 25–7

Waveforms showing the increasing (ramping up) amount of current in an ignition coil's primary winding before the field collapses and causes a sudden decrease in current.

IGNITION COIL CONSTRUCTION Older engines were equipped with ignition coils that were contained in a metal housing filled with oil to help cool the windings. Today's coils are air cooled. This is now possible because an individual coil is not responsible for providing the firing voltage for all spark plugs. Today's coils fire just one or two plugs. Many different coil designs are found on today's vehicles **(Figure 25–8)**. The actual design depends on the ignition system and the application.

A laminated soft iron core is positioned in the centre of each coil. The secondary winding is wound around the core, and the primary winding is wound around the secondary winding. The two ends of the primary winding are on the outside of the coil housing and are labelled as positive and negative. One end of the secondary winding is internally connected to the positive terminal of the primary winding; the other end is connected to the spark plug circuit. The wires used to make up the windings are covered with insulation to prevent the wires from shorting to each other.

SECONDARY VOLTAGE The typical amount of secondary coil voltage required to jump the spark plug gap is 10 000 volts. Most coils have at least 25 000 volts available from the secondary. The difference between the required voltage and the maximum available voltage is referred to as secondary reserve voltage. This reserve voltage is necessary to compensate for high voltages that occur due to high cylinder pressures during heavy load situations, such as wide-open throttle acceleration, and increased secondary resistances as the spark plug gap increases through use. The maximum available voltage must

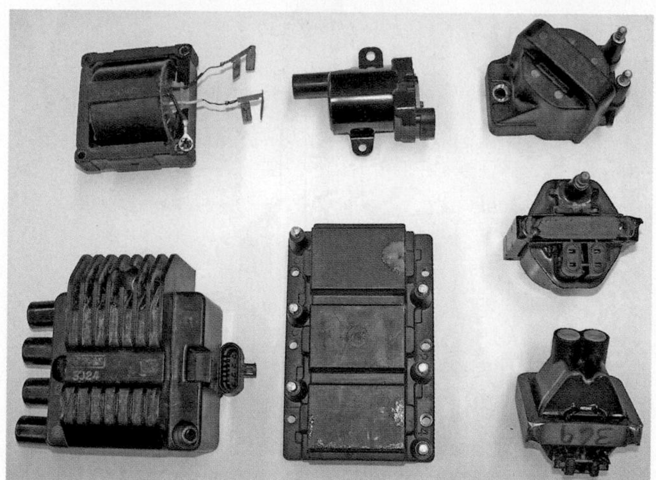

FIGURE 25–8

Many different ignition coil designs can be found on today's vehicles.

always exceed the required firing voltage, or an ignition misfire will occur. If there is an insufficient amount of voltage available to push current across the gap, the spark plug will not fire.

In most ignition systems with a distributor, only one ignition coil is used. The high voltage of the secondary winding is directed by the distributor to the various spark plugs in the system. Therefore, there is one secondary circuit with a continually changing path.

Although distributor systems have a single secondary circuit with a continually changing path, distributorless systems have several secondary circuits, each with an unchanging path.

SPARK PLUGS

Spark plugs provide the crucial **air gap** across which the high voltage from the coil causes an arc or spark. The main parts of a spark plug are a steel shell; a ceramic core or insulator, which acts as a heat conductor; and a pair of electrodes, one insulated in the core and the other grounded on the shell. The shell holds the ceramic core and electrodes in a gas-tight assembly and has threads for plug installation in the engine **(Figure 25–9)**. The insulator material may be alumina silicate or a black-glazed, zirconia-enhanced ceramic insulator to provide for increased durability and strength. The shell may be coated with corrosion resistance material and/or materials that prevent the threads from seizing to the cylinder head.

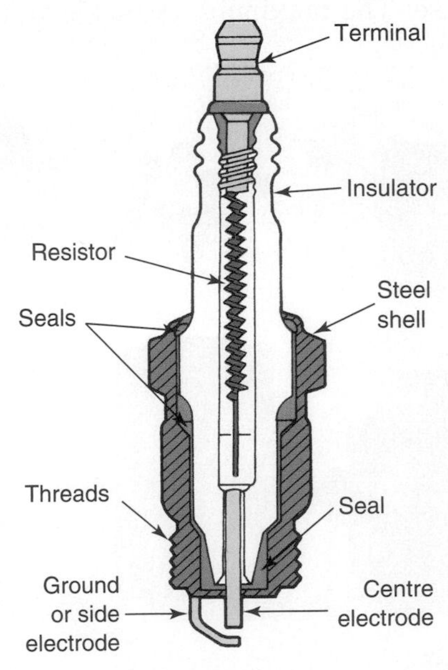

FIGURE 25–9
Components of a typical spark plug.

A terminal post on top of the centre electrode is the connecting point for the spark plug cable. Current flows through the centre of the plug, through a resistor, and arcs from the tip of the centre electrode to the ground electrode. The centre electrode is surrounded by the ceramic insulator and is sealed to the insulator with copper and glass seals. These seals prevent combustion gases from leaking out of the cylinder. Ribs on the insulator increase the distance between the terminal and the shell to help prevent electric arcing on the outside of the insulator. The steel spark plug shell is crimped over the insulation, and a ground electrode, on the lower end of the shell, is positioned directly below the centre electrode. There is an air gap between these two electrodes.

Spark plugs come in many different sizes and designs to accommodate different engines **(Figure 25–10)**.

SIZE Automotive spark plugs are available with a thread diameter of 12 mm, 14 mm, 16 mm, and 18 mm. The 18 mm spark plugs are mostly found on older engines and have a tapered seat that seals, when tightened properly, into a tapered seat in the cylinder head. The 12 mm, 14 mm, and 16 mm plugs can have a tapered seat or a flat seat that relies on a thin steel gasket to seal in its bore in the cylinder head. All spark plugs have a hex-shaped outer shell that accommodates a socket wrench for installation and removal. A 12 mm plug has a 16 or 18 mm (5/8 or 11/16 in.) hex' a 14 mm plug with a tapered seat has a 16 mm (5/8 in.) hex' and 14 mm gasketed and 18 mm plugs have a 20.6 mm (13/16 in.) hex on the shell. The 16 mm plugs, used on certain Ford V8 engines, may have a 9/16-inch or 5/8-inch hex on the shell. A tapered plug should never be used in an engine designed to use a gasketed plug, or vice versa.

REACH One important design characteristic of spark plugs is the **reach (Figure 25–11)**. This refers to the length of the shell from the contact surface at the seat to the bottom of the shell, including both threaded and nonthreaded sections. Reach is crucial because the plug's air gap must be properly placed in the combustion chamber to produce the correct amount of heat.

When a plug's reach is too short, its electrodes are in a pocket, and the arc is not able to adequately ignite the mixture. If the reach is too long, the exposed plug threads can get so hot they will ignite the air/fuel mixture at the wrong time and cause preignition. **Preignition** is a term used to describe

THREAD SIZE AND HEX SIZE							
Letter	Thread Size	Hex Size	Description	Letter	Thread Size	Hex Size	Description
L	18 mm	22.0 mm		SK	14 mm	16.0 mm	Iridium-tipped electrode
M	18 mm	25.4 mm		S	14 mm	20.6 mm	Special surface gap for Mazda
MA	18 mm	20.6 mm	Taper seat				R.E./thread reach 21.5 mm
J	14 mm	20.6 mm	Extended electrodes	SF	14 mm	20.6 mm	Surface gap
P	14 mm	20.6 mm	Platinum-tipped electrodes	T	14 mm	20.6 mm	Taper seat
PQ	14 mm	16.0 mm	Platinum-tipped electrodes	W	14 mm	20.6 mm	
Q	14 mm	16.0 mm		X	12 mm	18.0 mm	
QJ	14 mm	16.0 mm	Extended electrode	XU	12 mm	16.0 mm	
K	14 mm	16.0 mm	ISO	U	10 mm	16.0 mm	
KJ	14 mm	16.0 mm	ISO	Y	8 mm	13.0 mm	
PK	14 mm	16.0 mm	Platinum-tipped electrodes				

THREAD REACH		
Letter	Reach	Description
E (flat seat)	19.0 (3/40) or 20.0 mm	W16EXR-U W25EBR
E (taper seat)	0.7080	T16EPR-U
F	12.7 mm (1/20)	W20FP-U
FE	19.0 mm (3/40) Half thread	U24FER-9
G	21.8 mm	X27GPR-U
L	11.2 mm (7/160)	W14L
(None):		
18 mm thread (flat seat)	12.0 mm	M24S, L14-U
14 mm thread (flat seat)	9.5 mm (3/80)	W20S-U, W9PR-U
18 mm thread (taper seat)	0.4800	MA16PR-U
14 mm thread (taper seat)	0.4600 or 0.3250	T16PR-U T20M-U

SPECIAL DESIGN		
Letter	Description	Example
A	Dual ground electrodes for Mazda R.E.	W22EA
A	Electrode projection (7.0 mm)	QJ16AR-U
B	Triple ground electrodes	W20EPB
B	Electrode projection (9.5 mm)	J16BR-U
C	Electrode projection (5.0 mm)	QJ20CR11
D	4-ground electrodes for Mazda R.E.	W27EDR14
H	Electrode projection (8.5 mm)	QJ16HR-U
K	Special type for Honda CVCC	W16EKR-S11
LM	Special type for lawnmowers	W14LM-U
M	Compact type	W20M-U
N	Racing type (nickel ground electrode)	W27EN
Pt	Racing type (platinum ground electrode)	W27Ept
P	Projected insulator nose	W16EP-U
S	Regular type	W24ES-U
T	Dual ground electrodes for Toyota T.G.P.	W20ET-S
X	Extra projected insulator nose	W16EX-U

SPECIAL GAP CONFIGURATION		
Letter	Description	Example
GL	Platinum centre electrode	X22EPR-GL
L	Special type for Honda CVCC and extra project type for moped	W20ESR-L11 W14FP-UL
S	Semi-surface gap	W20EPR-S11
U	U-grooved ground electrode	W16EX-U
US	Star centre electrode with U-groove	W14-US
V	Thin centre electrode	W24ES-V
Z	Thin platinum centre electrode with tapered ground electrode	W24ES-ZU
C	Cut-back ground electrode	W27EMR-C
P	Platinum-tipped plug for DIS	PQ20R-PB

FIGURE 25–10

The different designs of spark plugs have unique part numbers that can be interpreted by charts given by the manufacturer.

DENSO Corporation

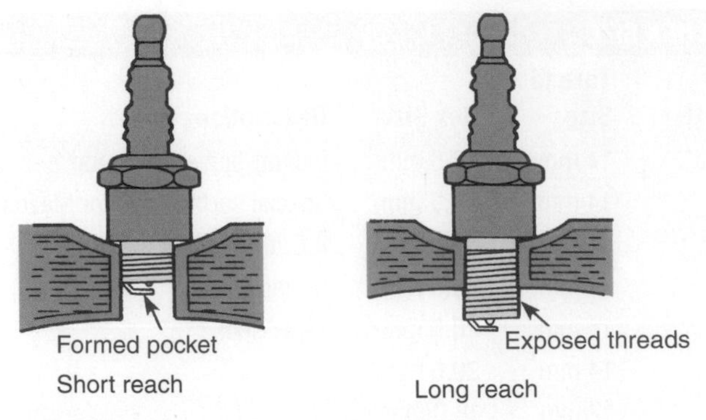

FIGURE 25–11

Spark plug reach: long versus short.

abnormal combustion, which is caused by something other than the heat of the spark.

HEAT RANGE When the engine is running, most of the plug's heat is concentrated on the centre electrode. Heat is quickly dissipated from the ground electrode because it is attached to the shell, which is threaded into the cylinder head. Coolant circulating in the head absorbs the heat and moves it through the cooling system. The heat path for the centre electrode is through the insulator into the shell and then to the cylinder head. The **heat range** of a spark plug is determined by the length of the insulator before it contacts the shell. In a cold-range spark plug, there is a short distance for the heat to travel up the insulator to the shell. The short heat path means the electrode and insulator will maintain little heat between firings **(Figure 25–12)**.

In a hot spark plug, the heat travels farther up the insulator before it reaches the shell. This provides a longer heat path, and the plug retains more heat. A spark plug needs to retain enough heat to clean itself between firings, but not so much that it damages itself or causes preignition of the air/fuel mixture in the cylinder.

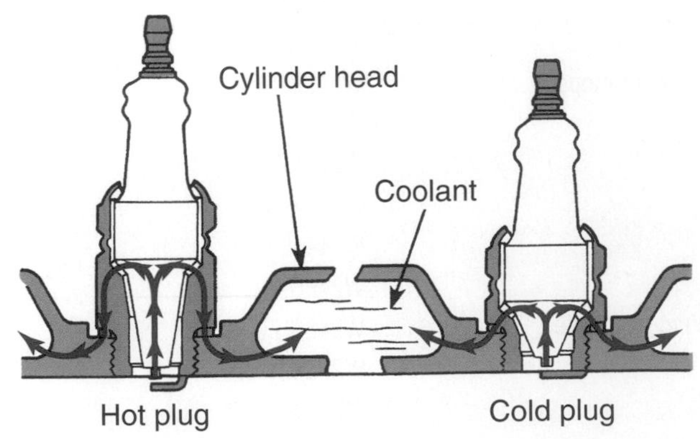

FIGURE 25–12

Spark plug heat range: hot versus cold.

The heat range is indicated by a code within the plug number imprinted on the side of the spark plug, usually on the porcelain insulator.

RESISTOR PLUGS Most automotive spark plugs have a resistor (normally about 5 K ohms) between the top terminal and the centre electrode. The resistance increases firing voltage. Some spark plugs use a semiconductor material to provide for this resistance. The resistor also reduces RFI, which can interfere with, or damage, radios, computers, and other electronic accessories, such as GPS systems. If an engine was originally equipped with resistor plugs, resistor plugs should be installed when the originals are replaced.

WARNING!

Using a nonresistor plug on some engines may cause erratic idle, high-speed misfire, engine run-on, power loss, and abnormal combustion.

SPARK PLUG GAPS The correct spark plug air gap is essential for achieving optimum engine performance and long plug life. A gap that is too wide requires higher voltage to jump the gap. If the required voltage is greater than what is available, the result is **misfiring**. Misfiring results from the inability of the ignition to jump the gap or maintain the spark. A gap that is too narrow requires lower voltages and can lead to rough idle and prematurely burned electrodes due to higher current flow.

ELECTRODES The materials used in the construction of a spark plug's electrodes determine the longevity, power, and efficiency of the plug. The construction and shape of the tips of the electrodes are also important.

The electrodes of a standard spark plug are made with copper, and some use a copper-nickel alloy. Copper is a good electrical conductor and offers some resistance to corrosion. Copper melts at 10 838°C (19 818°F), so it is more than suitable for use in an internal combustion engine.

Platinum electrodes are used to extend the life of a plug **(Figure 25–13)**. Platinum melts at 17 608°C (32 008°F) and is highly resistant to corrosion. Although platinum is an extremely durable material, it is an expensive precious metal; therefore, platinum spark plugs cost more than copper plugs.

FIGURE 25–13

A platinum-tipped spark plug.

Bob Freudenberger

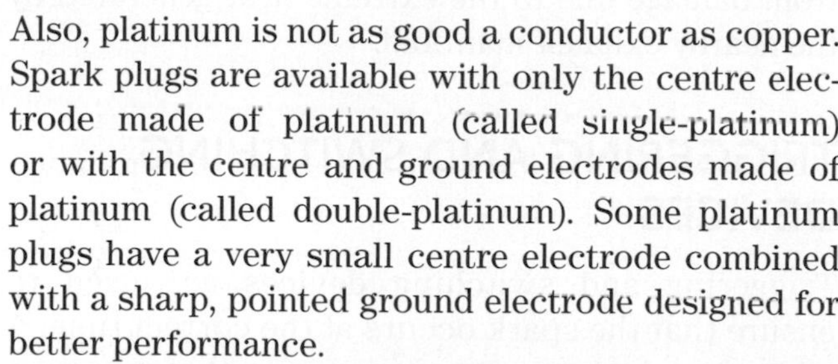

Also, platinum is not as good a conductor as copper. Spark plugs are available with only the centre electrode made of platinum (called single-platinum) or with the centre and ground electrodes made of platinum (called double-platinum). Some platinum plugs have a very small centre electrode combined with a sharp, pointed ground electrode designed for better performance.

Until recently, platinum was considered the best material to use for electrodes because of its durability. However, iridium is six times harder, eight times stronger, and has a melting point 6858°C (12 008°F) higher than platinum. Iridium is a precious, silver-white metal and one of the densest materials found on earth. A few spark plugs use an iridium alloy as the primary metal, complemented by rhodium to increase oxidation wear resistance. This iridium alloy is so durable that it allows for an extremely small centre electrode. A typical copper/nickel plug has a 2.5 mm diameter centre electrode, and a platinum plug has a 1.1 mm diameter. An iridium plug can have a diameter as small as 0.4 mm **(Figure 25–14)**, which means the firing voltage requirements are decreased. Iridium is also used as an alloying material for platinum.

Another rare and hard material used to make electrodes is yttrium. Yttrium has a silvery-metallic lustre and has a melting point of 15 238°C (27 738°F). Yttrium is fairly stable in air but oxidizes readily when heated. (Moon rocks contain yttrium.) Yttrium produces a highly adhesive oxide layer that makes the spark plug very durable and reliable, thereby extending its service life.

FIGURE 25–14

The spark plug has a small diameter iridium centre electrode and a grooved ground electrode.

DENSO Corporation

ELECTRODE DESIGNS Spark plugs are available with many different shapes and numbers of electrodes. When trying to ascertain the advantages of each design, remember that the spark is caused by electrons moving across an air gap. The electrons will always jump in the direction of the least electrical resistance. Therefore, if there are four ground electrodes to choose from, the electrons will jump to the closest. Also, keep in mind that the contents and pressure of the air in the air gap influences the resistance of the air gap. Again, the electrons will jump across the path of least resistance. Therefore, spark plugs with four ground electrodes do not typically supply a spark to all four electrodes **(Figure 25–15)**.

FIGURE 25–15

A spark plug with four ground electrodes.

Robert Bosch Corp.

The shape of the ground electrode may also be altered. A flat, conventional electrode tends to crush the spark, and the overall volume of the flame front is smaller. A tapered ground electrode increases flame front expansion and reduces the heat lost to the electrode.

Some ground electrodes have a U-groove machined into the side that faces the centre electrode. The U-groove allows the flame front to fill the gap formed by the U. This ball of fire develops a larger and hotter flame front, leading to a more complete combustion.

One brand of spark plug has a V-shaped ground electrode. This style of electrode does not block the flame front and allows it to travel upward through the V-notch into the combustion chamber. These spark plugs may be equipped with three separate points of platinum, one at each end of the V and the other at the centre electrode.

There are also different centre electrode designs. These variations are based on the diameter and shape of the electrode. A small diameter centre electrode requires less firing voltage and tends to have a longer service life. Some centre electrodes are tapered.

Some centre electrodes have a V-groove machined in them to force the spark to the outer edge of the ground electrode, placing it closer to the air/fuel mixture. This allows for quicker ignition of the air/fuel mixture. V-grooved centre electrodes also require lower firing voltages.

On some spark plugs, the centre electrode does not extend from the insulator, and the spark is generated across the end of the plug. With this design, the ground electrode does not block the flame front. This arrangement is called a surface gap and is intended to prevent carbon fouling, timing drift, and misfiring.

Ignition Cables

Spark plug cables, or ignition cables, make up the secondary wiring. These cables carry the high voltage from the distributor or the multiple coils to the spark plugs. The cables are not solid wire; instead, they contain carbon fibre cores that act as resistors in the secondary circuit **(Figure 25–16)**. They cut down on radio and television interference, increase firing voltages, and reduce spark plug wear by decreasing current. Insulated boots on the ends of the cables strengthen the connections as well as prevent dust and water infiltration and voltage loss.

Some ignition cables are called *variable pitch* resistor cables. These cables rely on tightly wound and loosely wound copper wire around a layer of ferrite magnetic material wrapped over a fibreglass strand core. This construction creates the necessary resistance, with a fraction of the impedance found in solid carbon core-type wire sets.

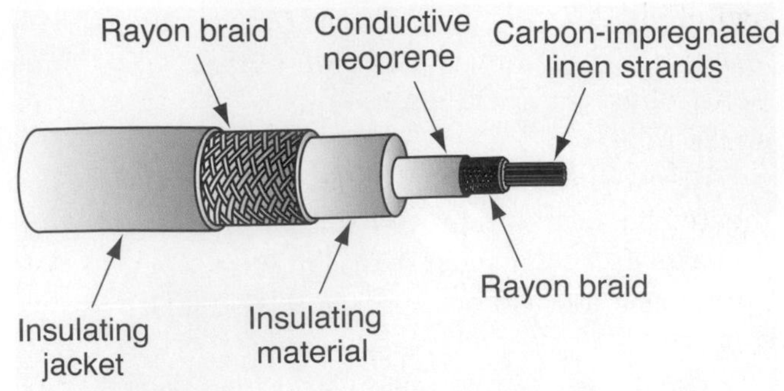

FIGURE 25–16

Spark plug cable construction.

©Cengage Learning 2015

Some engines have spark plug cable heat shields **(Figure 25–17)** fitted onto the boots at the cylinder head. These shields surround each spark plug boot and spark plug. They protect the spark plug boot from damage due to the extreme heat generated by the nearby exhaust manifold.

TRIGGERING AND SWITCHING DEVICES

Triggering and switching devices are used to ensure that the spark occurs at the correct time. A triggering device is simply a device that monitors the position of the crankshaft. A switching device is what controls current flow through the primary winding. When the triggering device sends a signal to the switching device that the piston of a particular cylinder is on the compression stroke, the

FIGURE 25–17

Spark plug boot heat shields.

Chrysler Group LLC

switching device stops current flow to the primary winding. This interruption of current flow happens when the PCM decides it is best to fire the spark plug.

Electronic switching components are normally located in an ignition control module, which may be part of the vehicle's PCM. On older vehicles, the ignition module may be built into the distributor or mounted in the engine compartment.

The ignition module advances or retards the ignition timing in response to engine conditions. Early systems had little control of timing and used mechanical or vacuum devices to alter timing. Today's computer-controlled systems have full control and can adjust ignition timing in response to the input signals from a variety of sensors and the programs in the computer.

Most electronically controlled systems use an NPN transistor to control the primary ignition circuit, which ultimately controls the firing of the spark plugs. The transistor's emitter is connected to ground. The collector is connected to the negative terminal of the coil. When the triggering device supplies a small current to the base of the transistor, current flows through the primary winding of the coil. When the current to the base is interrupted, the current to the coil is also interrupted. An example of how this works is shown in **Figure 25–18**, which is a simplified diagram of an electronic ignition system.

ENGINE POSITION SENSORS

The time when the primary circuit must be opened and closed is related to the position of the pistons and the crankshaft. Therefore, the position of the crankshaft is used to control the flow of current to the base of the switching transistor.

A number of different types of sensors are used to monitor the position of the crankshaft and control the flow of current to the base of the transistor. These engine position sensors and generators serve as triggering devices and include magnetic pulse generators, metal detection sensors, Hall-effect sensors, magneto resistive sensors, and photoelectric (optical) sensors.

These sensors can be located inside the distributor or mounted on the outside of the engine to monitor crankshaft position (CKP). In many cases, the input from a CKP is supplemented by inputs from camshaft position (CMP) sensors. On nearly all late-model engines, the CKP and CMP are magnetic pulse generators or Hall-effect switches.

Magnetic Pulse Generator

Basically, a magnetic pulse generator or inductance sensor consists of two parts: a trigger wheel and a pickup coil. The trigger wheel may also be called a reluctor, pulse ring, armature, or timing core. The pickup coil, which consists of a length of wire wound around permanent magnet, may also be called a stator, sensor, or pole piece. Depending on the type of ignition system used, the timing disc may be mounted on the distributor shaft, at the rear of the crankshaft, or behind the crankshaft vibration damper **(Figure 25–19)**.

The magnetic pulse or PM generator operates on the principles of electromagnetism. A voltage is induced in a conductor when a magnetic field passes over the conductor or when the conductor moves over a magnetic field. The magnetic field is provided by a magnet in the pickup unit, and the rotating trigger wheel provides the required movement through the magnetic field to induce voltage.

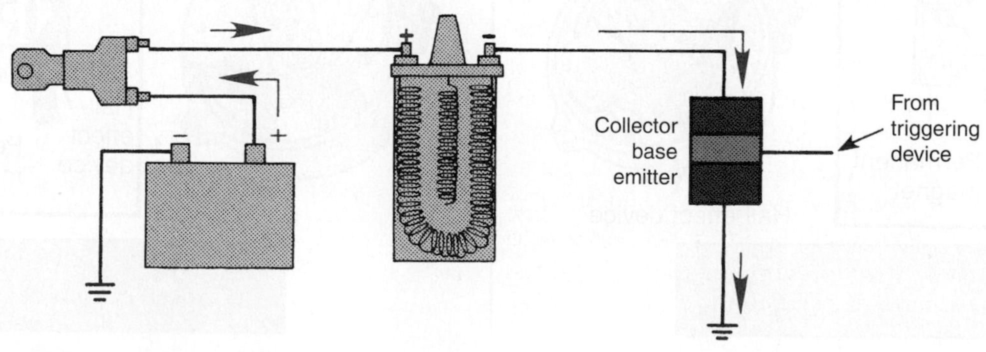

FIGURE 25–18

When the triggering device supplies a small amount of current to the transistor's base, the primary coil circuit is closed, and current flows.

FIGURE 25–19

Location of a typical crankshaft position (CKP) sensor.

As the trigger wheel rotates past the pickup coil, a weak AC voltage signal is induced in the pickup coil. This signal is sent to the ignition module. In early ignition systems, the change in polarity was used as a signal to prepare the ignition coil for another spark plug firing.

When a tooth is aligned to the pickup coil, the magnetic field is not expanding or contracting. There is no change in the magnetic field, and at that point zero voltage is induced in the pickup coil. A gap anomaly creates a voltage signal called the timing or "sync" pulse and is used by the PCM as the basis for timing the events in the ignition system. The timing pulses correspond with the position of each piston within its cylinder.

GO TO ▶ Chapter 16 for a detailed discussion of magnetic pulse generators and Hall-effect sensors.

Hall-Effect Sensor

The Hall-effect sensor or switch is the most commonly used engine position (CKP) sensor. A Hall-effect sensor produces an accurate voltage signal throughout the entire speed range of an engine. It also produces a square wave signal that is more compatible with computers. In an ignition system, the shutter blades are mounted on the distributor shaft **(Figure 25–20)**, flywheel, crankshaft pulley, or cam gear so that the sensor can generate a position signal as the crankshaft rotates. A Hall-effect sensor may be normally on or off, depending on the system and its circuitry. When a normally off sensor is used, there is maximum voltage output from the sensor when the magnetic field is blocked by the shutter. The opposite is true for normally on sensors. They have a voltage output when the magnetic field is not blocked.

A typical Hall-effect sensor has three wires connected to it. One wire is the reference voltage wire. The PCM supplies a reference voltage of 5 to 12 volts, depending on the system. The second wire delivers the output signal from the sensor to the PCM, and the third wire provides a ground for the sensor.

Hall-effect switches are also used as camshaft position (CMP) sensors. When the engine is being started, the PCM receives a signal from the CKP, but

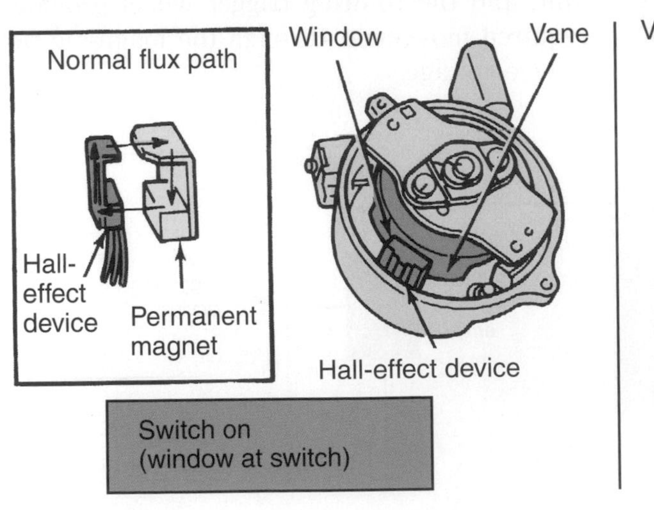

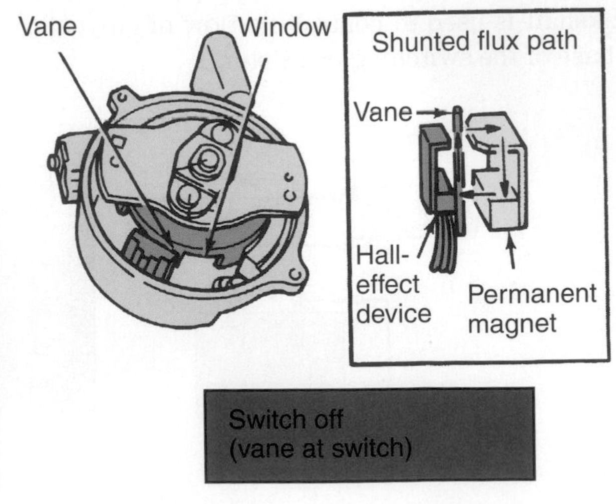

FIGURE 25–20

Operation of a Hall-effect switch.

Ford Motor Company

the spark plugs will not fire until the PCM receives a reference pulse from the CMP. That combined signal from the CKP and CMP sensors is also used to match fuel injector timing with the engine's firing order on engines equipped with sequential fuel injection. After the engine starts, the PCM no longer relies on the CMP for ignition sequencing. However, if the CMP is bad, the engine will not restart. If the CKP goes bad, the engine will not start or run.

Magneto Resistive Sensor

Magneto resistive (MR) sensors look similar to magnetic pulse generators but operate like Hall-effect and produce a digital square wave signal. An MR sensor uses a permanent magnet and two magnetic reluctance pickups spaced on either side of the magnet. As a reluctor wheel passes the sensor, the magnetic field changes and is sensed by the two pickups. The detection of the change in the field takes place at slightly different times by the two sensors and results in the output of a signal.

Like a Hall-effect, the MR sensor has three wires, but instead of requiring a voltage source to operate, the MR sensor generates a switching 5-volt signal.

Photoelectric Sensor

Some distributor ignition systems relied on photoelectric sensors **(Figure 25–21)** to monitor engine position. These sensors are also called optical sensors. They consisted of an LED, a light-sensitive phototransistor (photo cell), and a slotted disc called an interrupter. As the interrupter rotated between the LED and the photo cell, a pulsating square wave voltage signal was generated in the photo cell.

Photoelectric sensors may combine both the CKP and CMP sensors together, using one interrupter ring and two sets of LEDs and photo cells. The CKP uses 360 slots in the interrupter, one for each degree of crankshaft rotation. The CMP uses the number of slots equal to the number of cylinders of the engine. The slot for cylinder number 1 is larger, to differentiate it from the others.

FIGURE 25–21

A distributor with an optical-type pickup.

Chrysler Group LLC

Metal Detection Sensors

Metal detection sensors are found on early electronic ignition systems. They work much like a magnetic pulse generator with one major difference.

A trigger wheel is pressed over the distributor shaft, and a pickup coil detects the passing of the trigger teeth as the distributor shaft rotates. However, unlike a magnetic pulse generator, the pickup coil of a metal detection sensor does not have a permanent magnet. Instead, the pickup coil is an electromagnet. A low level of current is supplied to the coil by an electronic control unit, inducing a weak magnetic field around the coil. As the reluctor on the distributor shaft rotates, the trigger teeth pass very close to the coil **(Figure 25–22)**. As the teeth pass in and out of the coil's magnetic field, the magnetic field builds and collapses, producing a corresponding change in the coil's voltage. The voltage changes are monitored by the control unit to determine crankshaft position.

Timing Retard and Advance

One of the most important duties of an ignition system is to provide the spark at the correct time. On late-model engines, this is the job of the PCM.

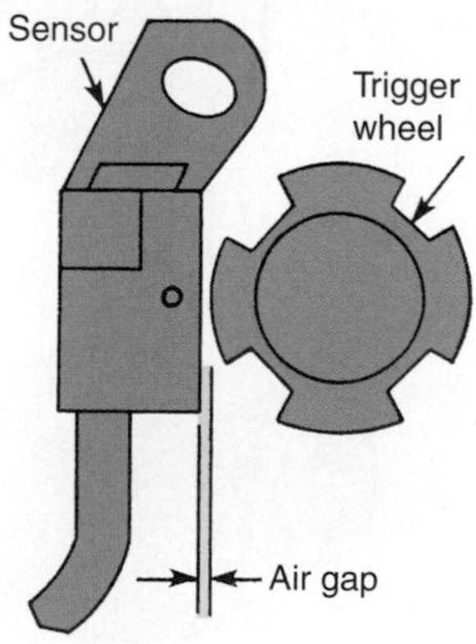

FIGURE 25–22

In a metal detection sensor, the revolving trigger wheel teeth alter the magnetic field produced by the electromagnet in the pickup coil.

The PCM uses data from input sensors such as the CKP, engine coolant sensor, mass airflow sensor, and others, to determine ignition timing. On earlier ignition systems, this was accomplished at the distributor through mechanical and vacuum-responsive devices. Mechanical advance responded to engine speed, and vacuum advance responded to engine load.

DISTRIBUTOR IGNITION SYSTEM OPERATION

The primary circuit of a DI system is controlled by a triggering device and a switching device located inside the distributor or external to it. Although these systems are no longer used by automakers, there are many of them still on the road, and they need service.

Distributor

The reluctor, or trigger wheel, and distributor shaft assembly rotate on bushings in the aluminum distributor housing. A roll pin extends through a retainer and the distributor shaft to hold the shaft in place in the distributor. Another roll pin is used to fasten the drive gear to the lower end of the shaft. This drive gear typically meshes with a drive gear on the engine's camshaft. The gear size is designed to drive the distributor shaft at the same speed as the camshaft, which rotates at one-half the speed of the crankshaft.

Through the years, there have been many different designs of DI systems. All operate in basically the same way but are configured differently. The systems described in this section represent the different designs used by manufacturers. These designs are based on the location of the electronic control module (unit) (ECU) and/or the type of triggering device used:

- DI systems with internal ignition module
- DI systems with external and remote ignition module
- DI systems with the ignition modules mounted on the distributor **(Figure 25–23)**

COMPUTER-CONTROLLED DI SYSTEMS After the manufacturers eliminated the mechanical and vacuum advance mechanisms on their distributors, the ECM

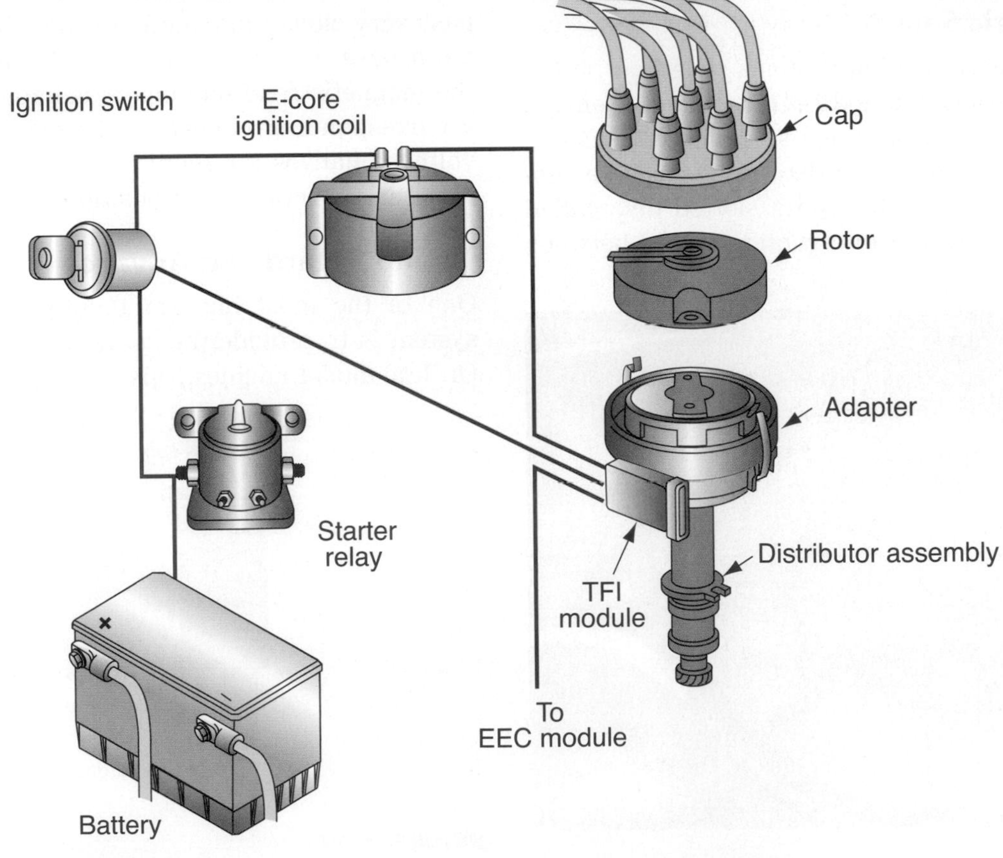

FIGURE 25–23

A distributor assembly with the ignition module mounted externally on the housing.

or PCM controlled ignition timing. This allowed for more precise control of ignition timing and provided improved combustion. The PCM adjusted the ignition timing according to engine speed, engine load, coolant temperature, throttle position, and intake manifold pressure. These systems varied with application and used a variety of triggering devices.

ELECTRONIC IGNITION SYSTEMS

Modern engines are not equipped with a distributor; rather, they have electronic ignitions. In the past, the term *electronic ignition* was designated to those ignition systems that used electronic controls. Today, electronic ignitions are those that do not use a distributor. There are two types of electronic ignition (EI) systems used on today's engines: waste spark **(Figure 25–24)** and coil-per-cylinder **(Figure 25–25)** systems. Most waste spark systems use an ignition module to control firing order while the PCM determines spark timing. Coil-per-cylinder systems typically use the PCM as the ignition module, which controls firing order and timing. A crank sensor is used to trigger the ignition system.

There are many advantages of a distributorless ignition system over one that uses a distributor. Here are some of the more important ones:

- Elimination of the distributor cap and rotor and the subsequent resistance
- No moving parts and therefore requires little maintenance
- Possible to control the ignition of individual cylinders to meet specific needs

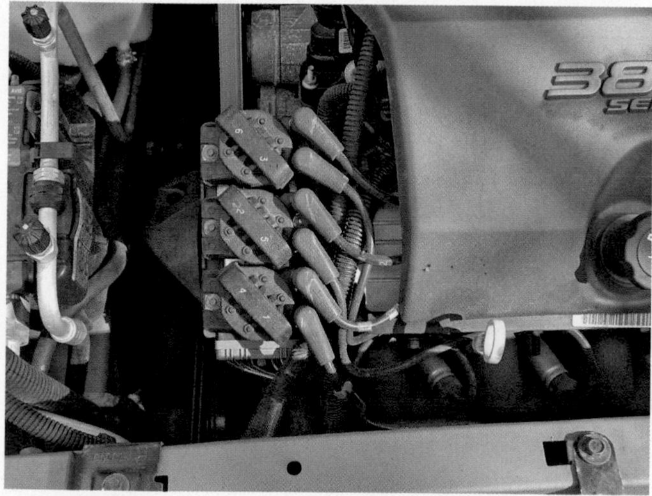

FIGURE 25–24

A coil pack for a waste spark ignition system; also called a double-ended ignition coil.

FIGURE 25–25

A coil-over-plug ignition system.

- Flexibility in mounting location—important because of today's smaller engine compartments
- Reduced radio frequency interference because there is no rotor to cap gap
- Elimination of a common cause of ignition misfire, the buildup of water and ozone/nitric acid in the distributor cap
- Elimination of mechanical timing adjustments
- Places no mechanical load on the engine in order to operate
- Increased available time for coil saturation
- Increased time between firings, which allows the coil to cool more

Double-Ended Coil or Waste Spark Systems

Double-ended or waste spark ignition systems use one ignition coil for two spark plugs **(Figure 25–26)**. Both ends of the coil's secondary side are directly connected to a spark plug, which means that two plugs are ignited at the same time; one is fired on the compression stroke of one cylinder, and the other is fired on the exhaust stroke of the companion cylinder.

A four-cylinder engine has two ignition coils; a six-cylinder has three; and an eight-cylinder has four. The computer, ignition module, and various sensors combine to control spark timing.

The computer collects and processes information to determine the ideal amount of spark advance for the operating conditions. The ignition module uses crank/cam sensor data to control the timing of the primary circuit in the coils **(Figure 25–27)**. Remember that there is more than one coil in a distributorless ignition system. The ignition module synchronizes the coils' firing sequence in relation to

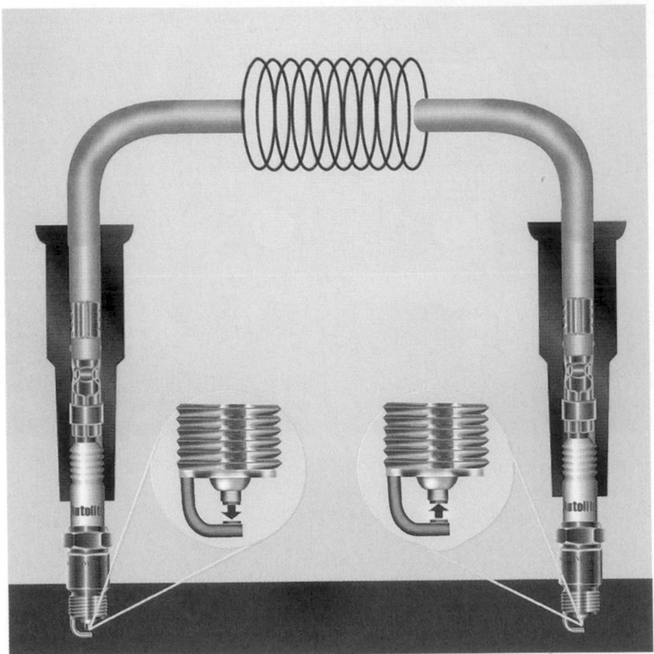

FIGURE 25–26

An EI system with a double-ended coil.

Honeywell

crankshaft position and firing order of the engine. Therefore, the ignition module takes the place of the distributor.

Primary current is controlled by transistors in the control module. There is one switching transistor for each ignition coil in the system. The transistors complete the ground circuit for the primary, thereby allowing for a dwell period. When primary current flow is interrupted, secondary voltage is induced in the coil, and the coil's spark plug(s) fire. The timing and sequencing of ignition coil action is determined by the control module and input from a triggering device.

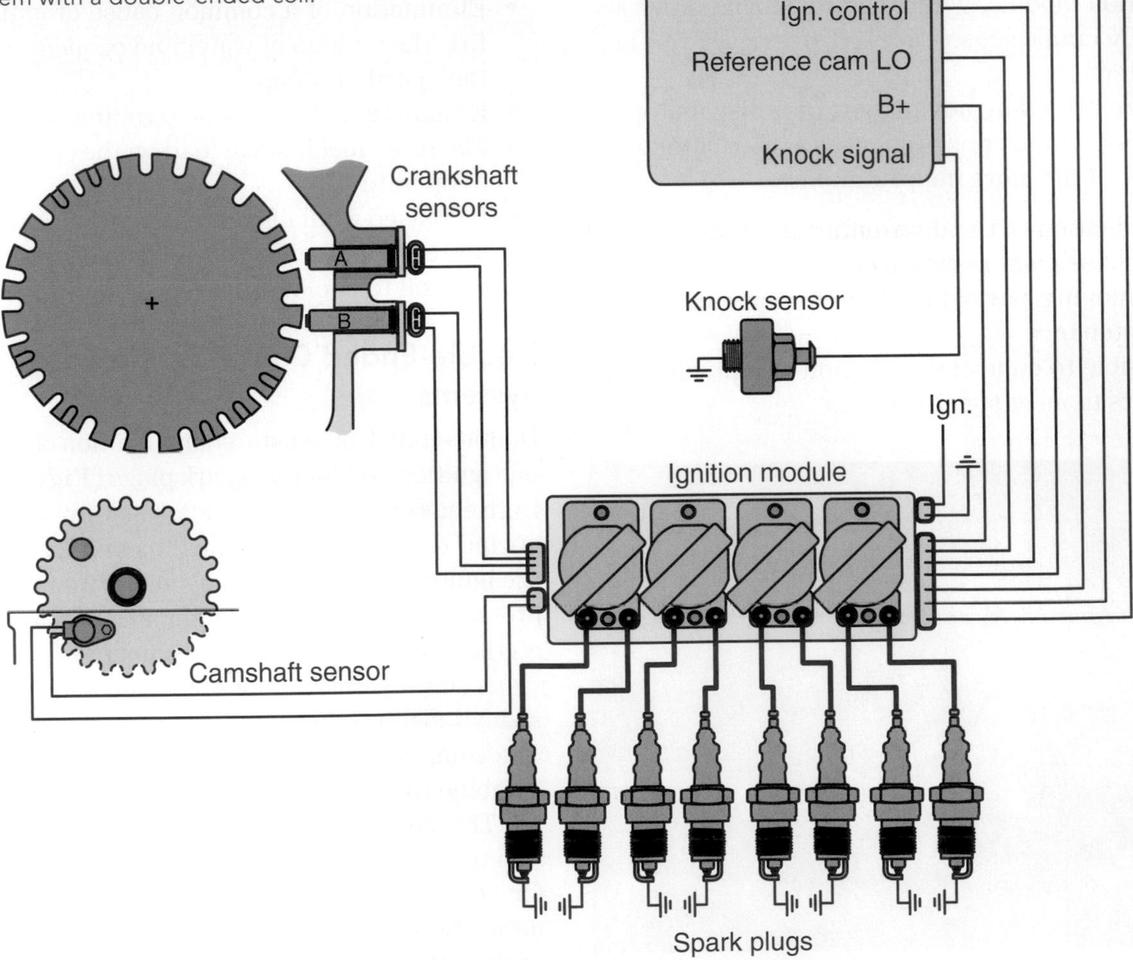

FIGURE 25–27

An EI system with two crankshaft position sensors and one camshaft position sensor.

Ford Motor Company

The control module is also responsible for limiting the dwell time. In EI systems, there is time between plug firings to fully saturate the coil. Achieving maximum current flow through the coil is great if the system needs the high voltage that may be available. However, if the high voltage is not needed, the high current is not needed, and the heat it produces is not desired. Therefore, the control module is programmed to allow total coil saturation only when the very high voltage is needed or the need for it is anticipated.

The ignition module also adjusts spark timing below 400 rpm (for starting) and when the vehicle's control computer bypass circuit becomes open or grounded. Depending on the exact EI system, the ignition coils can be serviced as a complete unit or separately. The coil assembly is typically called a **coil pack** and is comprised of two or more individual coils.

WASTE SPARK Double-ended coil systems are based on the **waste spark** method of spark distribution. Both ends of the ignition coil's secondary winding are connected to a spark plug. Therefore, one coil is connected in series with two spark plugs. The two spark plugs belong to cylinders whose pistons rise and fall together, called companion cylinders. With this arrangement, one cylinder of each pair is on its compression stroke and the other is on its exhaust stroke when the spark plugs are fired. Typically, cylinder pairings are as follows:

- Four-cylinder engines: 1 & 4 and 2 & 3
- V6 engines: 1 & 4, 2 & 5, and 3 & 6
- Inline six cylinders: 1 & 6, 2 & 5, and 4 & 3
- V8 engines: 1 & 4, 3 & 8, 6 & 7, and 2 & 5 or 1 & 6, 3 & 5, 4 & 7, and 2 & 8

(The pairings on V6s and 8s will vary as manufacturers vary how they number the cylinders.)

Due to the way the secondary coils are wired, when the induced voltage cuts across the primary and secondary windings of the coil, one plug fires in the normal direction—positive centre electrode to negative side electrode—and the other plug fires just the reverse side to centre electrode **(Figure 25–28)**. Both plugs fire simultaneously, completing the series circuit. Each plug always fires the same way on both the exhaust and compression strokes.

The coil is able to overcome the increased voltage requirements caused by reversed polarity and still fire two plugs simultaneously because each coil is capable of producing up to 100 000 volts. There is very little resistance across the plug gap on exhaust, so the plug requires very little voltage to fire, thereby providing its mate (the plug that is on compression) with plenty of available voltage. If you think about a series circuit with two unequal resistors, the majority of the voltage will be dropped by the larger value resistor, while less voltage will be used by the lower value resistor. The waste spark circuit operates in the same way; the lower resistance across the waste spark plugs gap drops less of the available voltage, leaving voltage available for the other spark plug.

Some EI systems use the waste spark method of firing but only have one secondary wire coming off

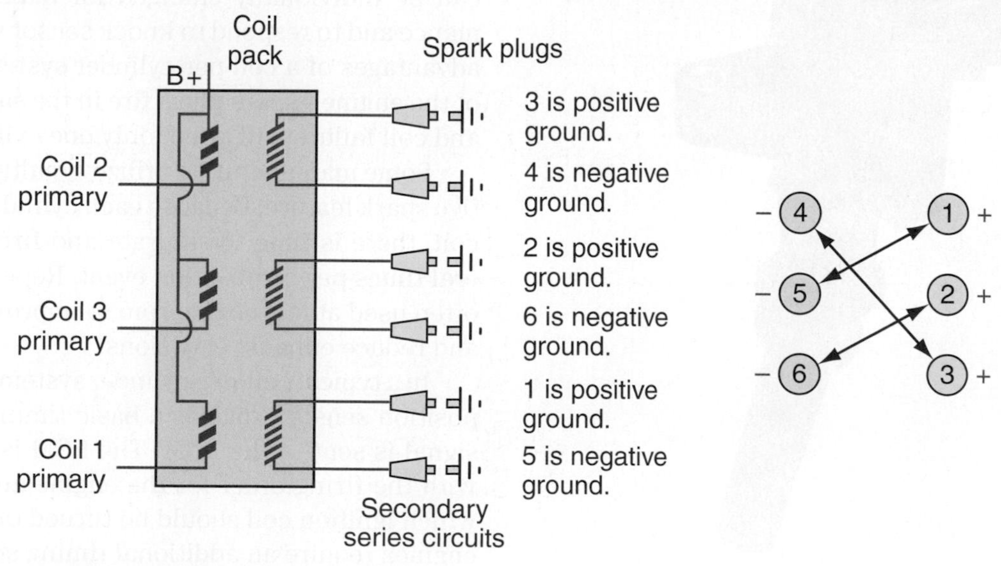

FIGURE 25–28

Polarity of spark plugs in an EI system.

Ford Motor Company

each ignition coil. In these systems, one spark plug is connected directly to the ignition coil, and the companion spark plug is connected to the coil by a high-tension cable.

Coil-per-Cylinder Ignition

The operation of a coil-per-cylinder ignition system is basically the same as any other ignition system. By definition, these systems have an individual coil for each spark plug. There are two different designs of coil-per-cylinder systems used today: the **coil-over-plug (COP)** and the separate coil. COP systems rely on a single assembly of an ignition coil and spark plug **(Figure 25–29)**. In these systems, the spark plug is directly attached to the coil and there is no spark plug wire.

The separate coil system is often called a coil-by-plug or coil-near-plug ignition system **(Figure 25–30)**. These systems have individual coils mounted near the plugs and use a short secondary plug wire to connect the coil to the plug. These systems are used when the location of the spark plug does not allow enough room to mount individual coils over the plugs or when the plugs are too close to the exhaust manifold.

Having one coil for each spark plug allows for more time between each firing, which increases the life of the coil by allowing it to cool. In addition, it also allows for more saturation time, which increases the coil's voltage output at high engine speeds. The increased output makes the coils more effective with lean fuel mixtures, which require higher firing voltages.

FIGURE 25–29

A coil-over-plug assembly.

Visteon Corporation

FIGURE 25–30

A coil-near-plug system.

Another advantage of using the coil-per-cylinder system is that the ignition timing at each cylinder can be individually changed for maximum performance and to respond to knock sensor signals. Other advantages of a coil-per-cylinder system are that all of the engine's spark plugs fire in the same direction, and coil failure will affect only one cylinder.

Some manufacturers utilize a multi-fire or repetitive spark feature. Because each cylinder has its own coil, there is time to saturate and fire the coil several times per combustion event. Repetitive spark is often used at low engine rpm to improve idle quality and reduce exhaust emissions.

In a typical coil-per-cylinder system, a crankshaft position sensor provides a basic timing signal. This signal is sent to the PCM. The PCM is programmed with the firing order for the engine and determines which ignition coil should be turned on or off. Some engines require an additional timing signal from the camshaft position sensor. On some systems, there is also a coil capacitor for each bank of coils for radio noise suppression.

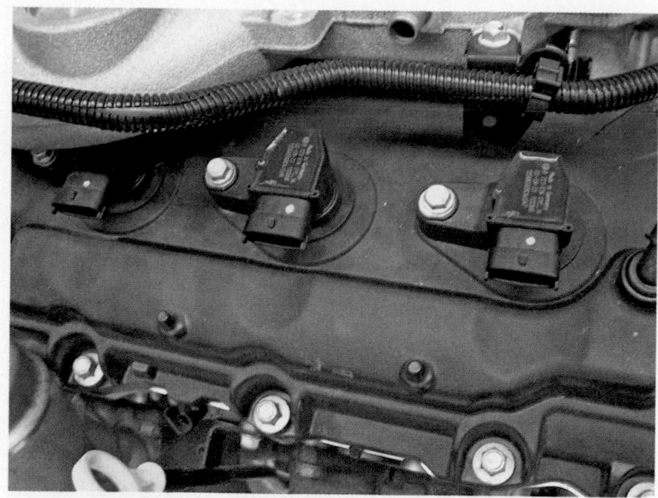

FIGURE 25–31

In a COP system, the coil is mounted directly above the spark plug.

COIL-OVER-PLUG (COP) IGNITION The true difference between COP and other ignition systems is that each coil is mounted directly atop the spark plug **(Figure 25–31)**, so the voltage from the coil goes directly to the plug's electrodes without passing through a plug wire. This means there are no secondary wires to come loose, burn, leak current, break down, or replace. Eliminating plug wires also reduces radio frequency interference (RFI) and electromagnetic interference (EMI) that can interfere with computer systems. However, the absence of plug wires also means that the coils need to be removed and reconnected with adapters or plug wires to test for spark, connect a pickup for an ignition scope, or perform a manual cylinder power balance test.

Twin Spark Plug Systems

Most engines have one spark plug per cylinder, but some have two. One spark plug is normally located on the intake side of the combustion chamber, and the other is at the exhaust side. When ignition takes place in two locations within the combustion chamber, more-efficient combustion and cleaner emissions are possible. Two coil packs are used, one for the intake side and the other for the exhaust side. These systems are called **dual plug** or **twin plug** systems **(Figure 25–32)**.

Some engines fire only one plug per cylinder during starting. The additional plug fires once the engine is running. During dual plug operation, the two coil packs are synchronized so that the two plugs of each cylinder fire at the same time. Therefore, in a waste spark system, four spark plugs are fired at a time—two during the compression stroke of a cylinder and two during the exhaust stroke of another cylinder.

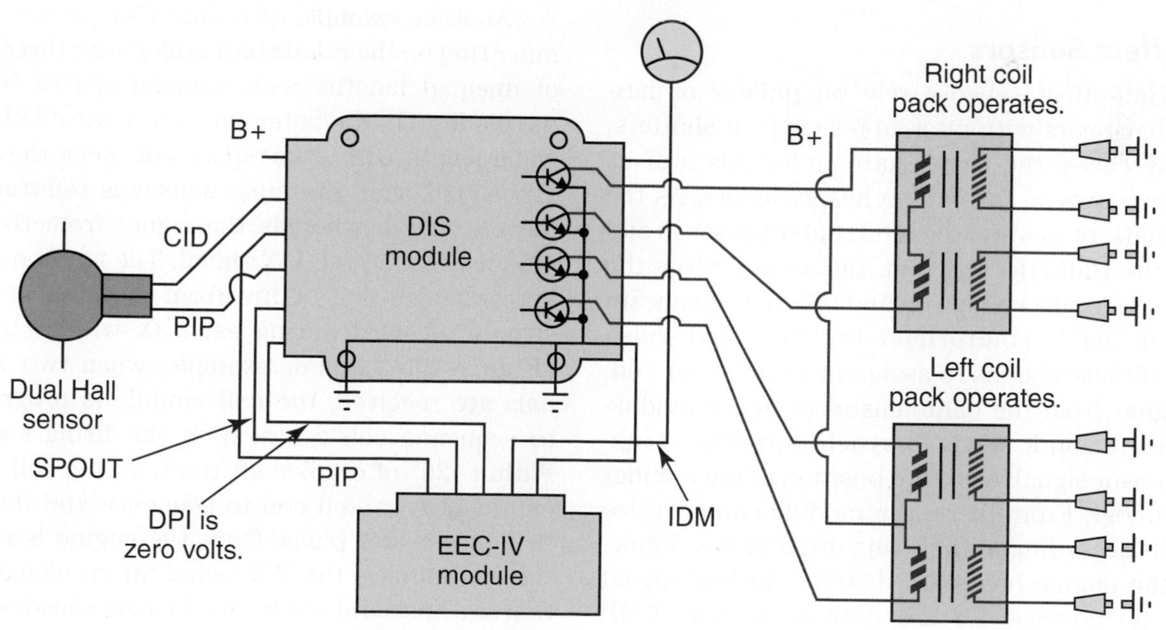

FIGURE 25–32

A dual plug system for a four-cylinder engine.

Ford Motor Company

ELECTRONIC IGNITION SYSTEM OPERATION

From a general operating standpoint, most electronic ignition systems are similar. One difference in design is the number of ignition coils. COP systems have the same number of coils as the engine has cylinders. Waste spark systems have half the number of coils as there are cylinders. Perhaps the biggest difference in system operation is based on the use of CKP and CMP sensors.

All systems have a CKP to monitor crankshaft position and engine speed. Some also monitor the relative position of each cylinder. Not all systems have a CMP; some have more than one. The signals from a CMP sensor are used for cylinder identification and for verifying the correlation between the position of the crankshaft and the camshafts. The design of the trigger wheels or rotors for these two sensors also varies. The design is primarily based on whether the sensor is a magnetic pulse (variable reluctance) or Hall-effect sensor. Both can be used for either sensor. Inputs from these sensors are critical to the operation of the fuel injection and ignition systems.

The layout and operation of these sensors are designed to provide fast engine starts and synchronization of the fuel injection and ignition systems with the position of the engine's individual pistons.

Hall-Effect Sensors

Many Hall-effect sensors rely on pulleys or harmonic balancers with interrupter rings or shutters. In many cases, the crankshaft pulley has half as many windows as the engine has cylinders. As the crankshaft rotates and the interrupter passes in and out of the Hall-effect switch, the switch turns the module reference voltage on and off. The signals are identical, and the control module cannot distinguish which of these signals to assign to a particular coil. The signal from the cam sensor gives the module the information it needs to synchronize the crankshaft sensor signals with the position of the number one cylinder. From there, the module can energize the coils according to the firing order of the engine. Once the engine has started, the camshaft signal serves no purpose for the ignition system. CMP signals are used for proper fuel injection timing on sequential injection systems.

Other systems use a dual crankshaft sensor located behind the crankshaft pulley. The pulley has two sets of two interrupter rings **(Figure 25–33)** that rotate through the Hall-effect switches at the dual crankshaft sensor. The inner ring with equally spaced blades rotates through the inner Hall-effect switch, whereas the outer ring with one opening rotates through the outer Hall effect.

FIGURE 25–33
This crankshaft pulley for a six-cylinder engine has two sets of interrupter rings.

Another example of a dual CKP sensor has an inner ring on the crankshaft pulley with three blades of unequal lengths with unequal spaces between the blades. On the outer ring, there are 18 blades of equal length with equal spaces between the blades. The signal from the inner sensor is referred to as the 3X signal, whereas the signal from the outer sensor is called the 18X signal. The ignition module knows which coil to fire from the number of 18X signals received during each 3X window rotation **(Figure 25–34)**. For example, when two 18X signals are received, the coil module is programmed to sequence coil 3-6 next in the firing sequence. Within 120° of crankshaft rotation, the coil module can identify which coil to sequence and thus start firing the spark plugs. Once the engine is running, the system uses the 18X signal for crankshaft position and speed information. The cam sensor signal is used for injector sequencing, but it is not required for coil sequencing.

Many CMP sensors are Hall-effect sensors and produce a square wave signal **(Figure 25–35)**. The sensor may respond to a single slot on a camshaft

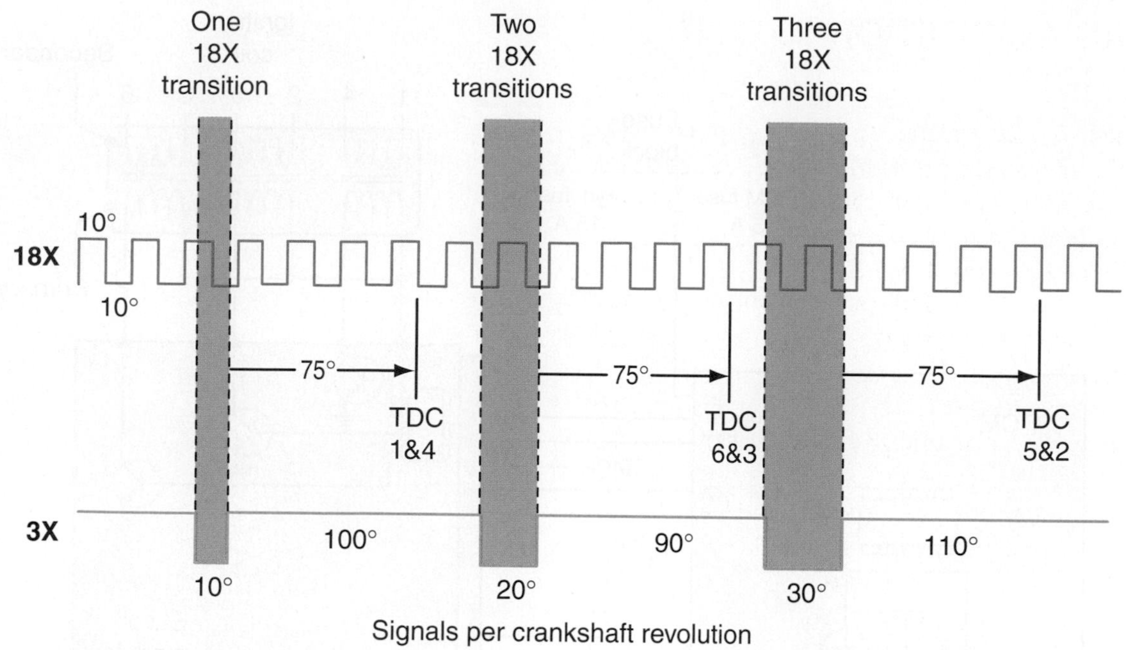

FIGURE 25–34

3X and 18X crankshaft signals.

pulley, or the pulley will have several. One design has four narrow and wide slots on the trigger wheel. The PCM uses the narrow and wide signal patterns to identify camshaft position, or which cylinder is on its compression stroke and which is on the exhaust. The PCM can then calculate the correct timing and sequencing for the spark plugs and fuel injectors.

Magnetic Pulse Generators

Many late-model engines use magnetic pulse generators as CKP sensors. The trigger wheel, also called the reluctor, can be located behind the crankshaft pulley, inside the engine in the middle of the crankshaft, or at the flywheel. Again, the design of the trigger wheel depends on the application and operation of the system.

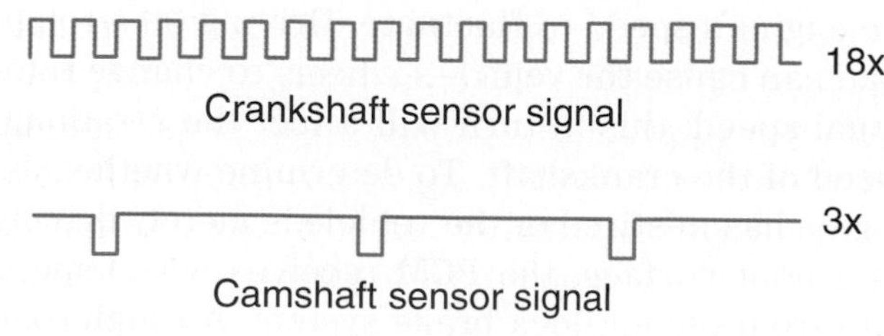

FIGURE 25–35

The relationship between crankshaft and camshaft signals on a GM 3.8L SFI engine.

In more basic systems, the trigger wheel is located behind the crankshaft pulley. If the engine is a six cylinder, there will be seven slots in the reluctor, six of which are spaced exactly 60° apart; the seventh notch is located 10° from the number six notch and is used to synchronize the coil firing sequence in relation to crankshaft position **(Figure 25–36)**. The same triggering wheel can be and is used on four-cylinder engines. The computer only needs to be programmed to interpret the signals differently than for a six-cylinder engine.

The CKP sensor generates a small AC voltage each time one of the machined slots passes by. By counting the time between pulses, the ignition module picks out the unevenly spaced seventh slot and starts the calculation of the ignition coil sequencing. Similar systems are used with more slots or teeth machined into the reluctor. There is always at least one gap for cylinder identification purposes.

In many cases, the gap is used to identify the position of the crankshaft during engine cranking. A CMP sensor is used to determine what stroke the cylinders are on.

Magneto Resistive Sensors

This type of sensor is used as a CKP sensor and is usually mounted behind the crankshaft pulley or is bolted to the block. A toothed timing wheel may be

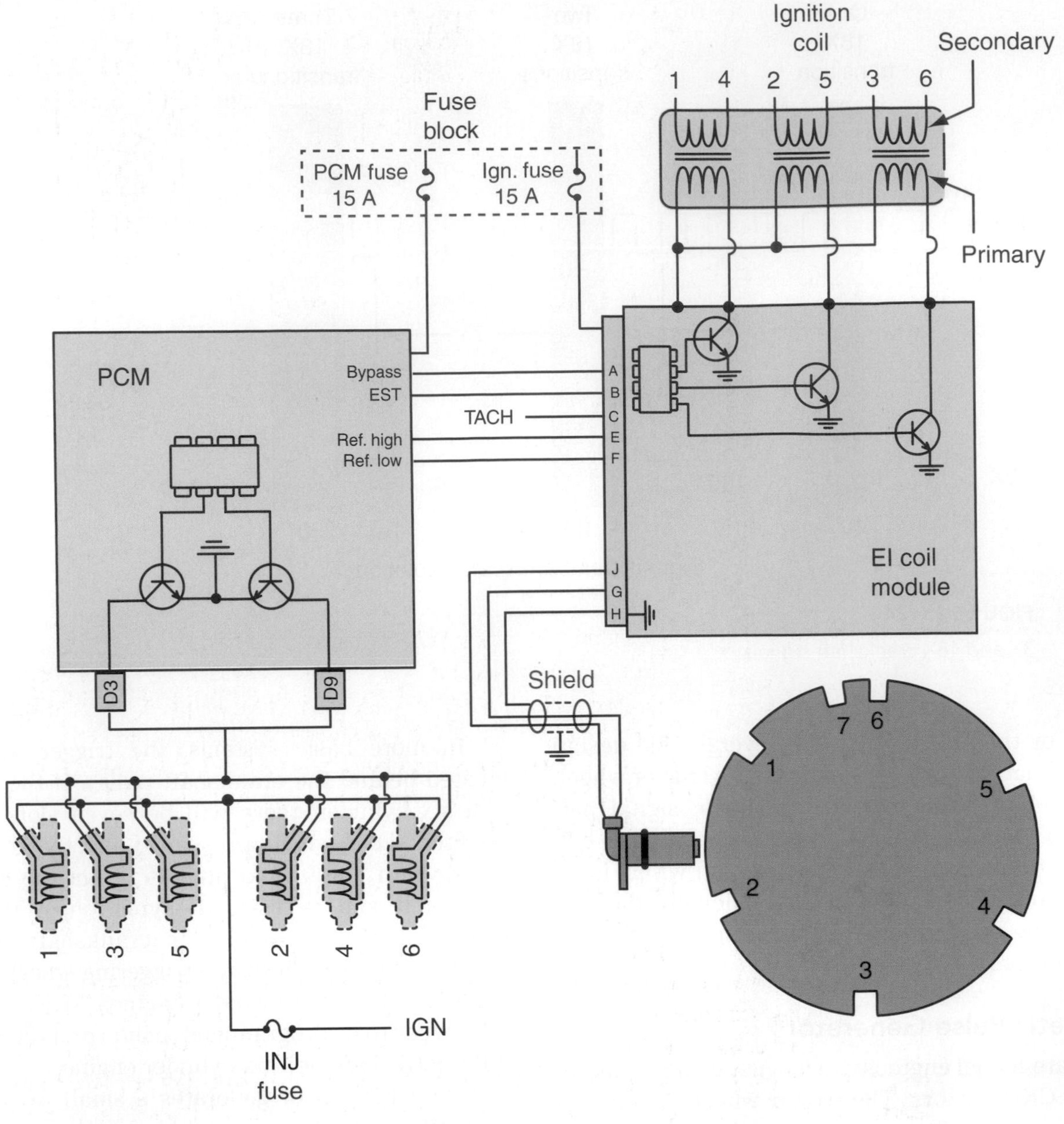

FIGURE 25–36

Schematic of an EI system with a magnetic pulse generator–type crankshaft sensor. Note the notches on the crankshaft timing wheel.

mounted on the end of the crankshaft, integrated into the crankshaft, or located on the flywheel. Magneto resistive sensors produce a digital square wave pattern that increases in frequency as engine speed increases.

Misfire Detection

A high-data-rate CKP sensor is used to detect engine misfires, and the CMP is used to identify which cylinder is misfiring. Misfires are detected by variations in crankshaft rotational speed for each cylinder. An interesting feature of most misfire monitors is the ability of the PCM to distinguish an actual misfire from other things that may cause the engine's speed to fluctuate. Driving on a rough road can cause the vehicle's wheels to change rotational speed; this in turn will affect the rotational speed of the crankshaft. To determine whether the engine has misfired or the vehicle is merely driving on a poor surface, the PCM receives wheel-speed data from the antilock brake system. A rough road will cause variances in wheel speed, and the PCM looks at that data before concluding that a misfire occurred.

Basic Timing

The PCM totally controls ignition timing with COP systems and is otherwise not adjustable. When the engine is cranked for starting, the PCM sets the timing at a fixed value. This value is used until the engine is running at a predetermined speed. Once that speed is met, the PCM looks at several inputs, including engine speed, load, throttle position, and engine coolant temperature, and makes adjustments accordingly. The PCM continues to rely on those inputs and its programmed strategy throughout operation. All PCMs have limits as to how far the timing can be retarded and advanced.

TIMING CORRECTIONS The PCM adjusts ignition timing according to its programming and sensor inputs. There are times when the timing is adjusted or corrected to compensate for slight changes in the operating conditions or abnormal occurrences:

- *Temperature.* Ignition timing is advanced when the coolant temperature is low. When the temperature is very high, the timing is retarded.
- *Engine knock.* When a knock is detected, the PCM retards the timing in fixed steps until the knock disappears. When the knocking stops, the PCM stops retarding the timing and begins to advance the timing in fixed steps unless the knocking reoccurs.
- *Stabilizing idle.* When the engine idle speed moves away from the desired idle speed, the PCM will adjust the timing to stabilize the engine speed. It is important to know that ignition timing changes are made only to correct minor idle problems. If the engine's speed is above the desired speed, the timing is retarded, and when it is too low, the timing is advanced.
- *EGR operation.* When the EGR valve opens, the timing is advanced. The amount of advance depends on intake air volume and engine speed.
- *Transition correction.* When the vehicle is accelerated immediately after deceleration, the timing is temporarily advanced or retarded to smoothen the transition.
- *Torque control.* To provide smooth shifting of an automatic transmission, the PCM will temporarily retard the ignition timing to reduce the engine's torque when the transmission is beginning to change gears.
- *Traction control correction.* When excessive wheel slippage occurs, the PCM will retard the timing to reduce the torque output from the engine. Once the slippage has been corrected, timing returns to normal.
- *E85 fuel:* Modern flex-fuel vehicles, capable of using E85 ethanol blends, are able to detect the use of E85. When E85 is detected, engine operating parameters, including ignition timing, are changed to compensate for the difference in engine output and fuel consumption.

CAUTION!

Because EI systems have considerably higher maximum secondary voltage compared to distributor-type ignition systems, greater electrical shocks are obtained from EI systems. Although such shocks may not be directly harmful to the human body, they may cause you to jump or react suddenly, which could result in personal injury. For example, when you jump suddenly as a result of an EI electrical shock, you may hit your head on the vehicle hood or push your hand into a rotating cooling fan.

KEY TERMS

Air gap (p. 734)
Ballast resistor (p. 730)
Coil-over-plug (COP) (p. 746)
Coil pack (p. 745)
Direct ignition systems (DIS) (p. 728)
Distributor-based ignition (DI) systems (p. 728)
Distributorless ignition systems (p. 728)
Dual plug (p. 747)
Dwell (p. 732)
Heat range (p. 736)
Inductive reluctance (p. 732)
Magneto resistive (MR) sensors (p. 741)
Misfiring (p. 736)
Preignition (p. 734)
Primary circuit (p. 729)
Pulse transformer (p. 732)
Reach (p. 734)
Reactance (p. 733)
Secondary circuit (p. 729)
Twin plug (p. 747)
Waste spark (p. 745)

SUMMARY

- The ignition system supplies high voltage to the spark plugs to ignite the air/fuel mixture in the combustion chambers.
- The ignition system has two interconnected electrical circuits: a primary circuit and a secondary circuit.
- The primary circuit supplies low voltage to the primary winding of the ignition coil. This creates a magnetic field in the coil.

- A switching device interrupts primary current flow, collapsing the magnetic field and creating a high-voltage surge in the ignition coil secondary winding.
- The switching device used in electronically controlled systems is an NPN transistor.
- The secondary circuit carries high-voltage surges to the spark plugs. On some systems, the circuit runs from the ignition coil, through a distributor, to the spark plugs.
- The distributor may house the switching device plus timing advance mechanisms. Some systems locate the switching device outside the distributor housing.
- Ignition timing is directly related to the position of the crankshaft. Magnetic pulse generators and Hall-effect sensors are the most widely used engine position sensors. They generate an electrical signal at certain times during crankshaft rotation. This signal triggers the electronic switching device to control ignition timing.
- Distributors are seldom found on today's engines. Nearly all of today's engines are equipped with an EI system for which there are primarily two different designs: waste spark coil and coil per cylinder.
- In computer-controlled ignitions, the computer receives input from numerous sensors. Based on these data, the computer determines the optimum firing time and signals an ignition module to activate the secondary circuit at the precise time needed.
- In some systems, the camshaft sensor signal informs the computer when to sequence the coils and fuel injectors.
- The crankshaft sensor signal provides engine speed and crankshaft position information to the computer.
- Some EI systems have a combined crankshaft and sync sensor at the front of the crankshaft.

REVIEW QUESTIONS

1. Explain how voltage is induced in the distributor pickup coil as the reluctor high point approaches alignment with the pickup coil.
2. Explain why dwell time is important to ignition system operation.
3. Name the engine operating conditions that most affect ignition timing requirements.
4. Explain how the plugs fire in a two-plug-per-coil EI system.
5. Explain the components and operation of a magnetic pulse generator.
6. What happens when the low-voltage current flow in the coil primary winding is interrupted by the switching device?
   a. The magnetic field collapses.
   b. A high-voltage is induced in the coil secondary winding.
   c. Both a and b.
   d. Neither a nor b.
7. What occurs when a spark plug with multiple ground electrodes are fired?
   a. A number of weaker sparks are sent to each ground electrode.
   b. Each ground electrode will receive a strong spark.
   c. The spark will travel to the electrode offering the path of least resistance.
   d. The spark will be directed to different electrodes, depending on engine speed.
8. Why is high voltage needed to establish a spark across the gap of a spark plug?
   a. because of the reduced flow through the ignition cables
   b. because of the high resistance of the air gap between the electrodes
   c. because of the low resistance of the air gap between the electrodes
   d. because of the materials used in both electrodes
9. What does the use of one ignition coil per cylinder produce?
   a. a weaker spark
   b. less time for the spark to be produced
   c. less coil saturation time
   d. greater coil saturation time
10. Which of the following is a function of all ignition systems?
    a. to generate sufficient voltage to force a spark across the spark plug gap
    b. to time the arrival of the spark to coincide with the movement of the engine's pistons
    c. to vary the spark arrival time based on varying operating conditions
    d. all of the above
11. Reach, heat range, and air gap are all characteristics that affect the performance of which ignition system component?
    a. ignition coils **b.** ignition cables
    c. spark plugs **d.** breaker points

12. What is the benefit of eliminating the spark plug wire in COP ignition systems?
    a. to reduce the chances of EMI and RFI
    b. to reduce manufacturing costs
    c. to provide simpler spark testing procedures
    d. to provide simpler spark plug removal
13. Why are variable pitch resistor spark plug wires used?
    a. to reduce the impedance of the wire
    b. to increase the impedance of the wire
    c. to increase the RFI around the wire
    d. to decrease the RFI around the wire
14. When does the magnetic field surrounding the coil in a magnetic pulse generator move?
    a. when the reluctor tooth approaches the coil
    b. when the reluctor tooth begins to move away from the pickup coil pole
    c. when the reluctor is aligned with the pickup coil pole
    d. both a and b
15. Which of the following electronic switching devices has a reluctor with wide shutters rather than teeth?
    a. magnetic pulse generator
    b. metal detection sensor
    c. Hall-effect sensor
    d. all of the above
16. A magnetic style crank sensor will give which kind of signal?
    a. modulated pulse-width signal
    b. variable AC signal
    c. square wave frequency signal
    d. positive-only analog signal
17. A Hall effect–style crank sensor will give which kind of signal?
    a. modulated-pulse-width signal
    b. variable AC signal
    c. square wave frequency signal
    d. positive-only analog signal
18. A misfire monitor uses which sensor to determine misfires?
    a. cam sensor
    b. crank sensor
    c. manifold absolute pressure (MAP) sensor
    d. mass airflow (MAF) sensor
19. Engine ignition timing is controlled by opening of which circuit?
    a. crank sensor circuit
    b. cam sensor circuit
    c. ignition primary circuit
    d. ignition secondary circuit
20. A knock sensor is used to detect preignition and detonation in an engine. What occurs when a signal is received?
    a. The air/fuel mixture in the cylinder is leaned out.
    b. The timing is retarded.
    c. The air/fuel mixture in the cylinder is enriched.
    d. The timing is advanced.

CHAPTER 26
Ignition System Diagnosis and Service

LEARNING OUTCOMES

- Perform a no-start diagnosis and determine the cause of the condition.
- Determine the cause of an engine misfire.
- Perform a visual inspection of ignition system components, primary wiring, and secondary wiring to locate obvious trouble areas.
- Describe what an oscilloscope is, its scales and operating modes, and how it is used in ignition system troubleshooting.
- Test the components of the primary and secondary ignition circuits.
- Test individual ignition components, using test equipment such as a voltmeter, ohmmeter, and testlight.
- Service and install spark plugs.
- Describe the effects of incorrect ignition timing.
- Check and set (when possible) ignition timing.
- Diagnose engine misfiring on electronic ignition–equipped engines.

This chapter concentrates on testing ignition systems and their individual components. It must be stressed, however, that there are many variations in the ignition systems used by auto manufacturers. The tests covered in this chapter are those generally used as basic troubleshooting procedures. Exact test procedures and the ideal troubleshooting sequence will vary among vehicle makers and individual models. Always consult the vehicle's service information when performing ignition system service.

Two important precautions should be taken during all ignition system tests:

1. Turn the ignition switch off before disconnecting any system wiring.
2. Do not touch any exposed connections while the engine is cranking or running.

MISFIRES

When something prevents complete combustion, the result is a misfire, or incomplete combustion. Misfires can cause lack of power, poor gas mileage, excessive exhaust emissions, and a rough-running engine. Misfires are not always caused by the ignition system; other systems also can cause them. A spark plug misfires when it has a weak spark, or does not fire at all. Misfires can be caused by a fouled spark plug, a bad coil, problems in the primary or secondary ignition circuit, lean fuel mixture, or an incorrect plug gap.

Abnormal Combustion

Incomplete combustion is not the only abnormal condition engines may experience; they may also experience detonation. Detonation is usually caused by engine overheating, excessively lean mixtures, or the use of low-octane gasoline. Detonation can cause physical damage to the pistons, valves, bearings, and spark plugs.

Preignition can cause pinging or spark knocking. Any hot spot within the combustion chamber can cause preignition. Common causes of preignition are incandescent carbon deposits in the combustion chamber, a faulty cooling system, too hot of a spark plug, poor engine lubrication, and cross-firing. Preignition usually leads to detonation; preignition and detonation are two separate events.

GENERAL IGNITION SYSTEM DIAGNOSIS

The ignition system should be tested whenever you know or suspect there is no spark or not enough spark, or when the spark is not being delivered at the correct time to the cylinders.

Common vs. Non-Common Problems

In most cases, ignition problems can be divided into two types: common and non-common. Common problems are those that affect all cylinders, and non-common problems are those that affect one or more cylinders, but not all. Common ignition components include the parts of the primary circuit and the secondary circuit up to the distributor's rotor in distributor-based ignition (DI) systems. Non-common parts are the individual spark plug terminals inside the distributor cap, spark plug wires, and the spark plugs. With electronic ignition (EI) systems, the individual coils are non-common parts.

The best indicator of a non-common problem is the reading on a vacuum gauge **(Figure 26–1)**. If a vacuum gauge is connected to a four-cylinder engine at idle and the needle of the gauge is within the normal vacuum range for three-fourths of the time and drops one-fourth of the time, this indicates that three of the cylinders are working normally while the fourth is not. The cause of the problem is non-common. If that cylinder is sealed and all cylinders are receiving the correct amount of air and fuel, the problem must be in the ignition system. The problem is in the non-common parts of the ignition system.

Determining if the ignition problem is common or non-common is a good way to start troubleshooting the ignition system. By dividing the ignition system into common and non-common parts, you will test only those parts that could cause the problem.

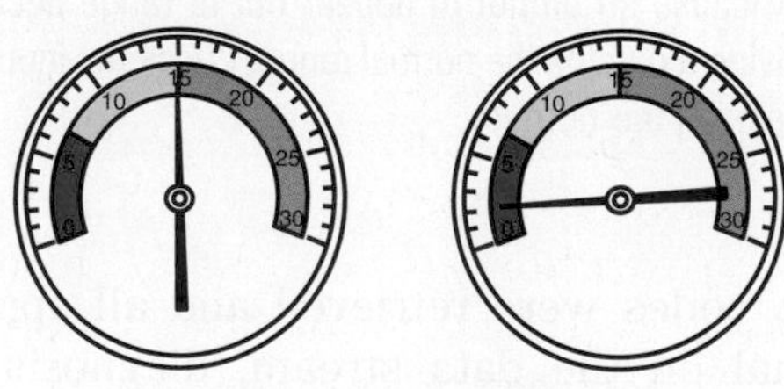

FIGURE 26–1

The reaction of a vacuum gauge when there is a common or non-common problem.

Generally, when an engine runs unevenly, the cause is a non-common problem. If the engine does not start, the problem is probably a common one. EI systems, especially coil-per-cylinder systems, make troubleshooting a little easier. The PCM or ignition module may be the only part that is common to all cylinders. The coils and all of the secondary circuit are common to only one or two cylinders. For example, if a coil in a waste spark system is bad, two cylinders will be affected and not the entire engine.

IGNITION SYSTEM INSPECTION

Begin all diagnosis by gathering as much information as possible from the customer. Then conduct a careful visual inspection. The system should be checked for obvious problems. Although no-start problems and incorrect ignition timing are caused by the primary circuit, the secondary circuit can be the cause of driveability problems and should be carefully checked. In addition to the ignition system, inspect all related electrical connectors or fuses, vacuum lines, the air intake system, and the cooling system. Also check available service information that may relate to the symptoms.

Symptoms commonly caused by ignition system problems include the following (keep in mind that the ignition system is not the only thing that can cause these):

- *Hard starting.* The engine requires an excessive amount of time to start.
- *Rough idle.* The engine idles poorly and may stall.
- *Engine stalling.* The engine quits unexpectedly. It may occur right after engine start-up, while idling, or during deceleration.
- *Hesitation.* The engine does not immediately respond to the opening of the throttle.
- *Stumble.* The engine temporarily loses power during acceleration.
- *Poor acceleration.* The vehicle accelerates slower than expected.
- *Surge.* The engine's speed fluctuates with a constant throttle during idle, steady cruise, acceleration, or deceleration.
- *Bucking.* The vehicle jerks shortly after acceleration or deceleration.
- *Knocking (pinging).* The engine makes a sharp metallic noise during acceleration.

- *Backfire and afterfire.* Backfire is a loud pop coming from the intake system, usually during rapid throttle opening. Afterfire is a popping that occurs in the exhaust system, usually during quick deceleration.

Scan Tools

Today's ignition systems are part of the engine control system. Part of the visual inspection should include a check of the malfunction indicator lamp (MIL). If it is operating correctly and a fault is emissions related, the lamp will remain on after the engine has started. If the MIL is flashing with the engine running, this indicates a catalyst-damaging misfire is occurring. Within the OBD-II diagnostic trouble code (DTC) structure, DTCs P0300 through P0399 are specific to the ignition system and misfires. DTCs related to specific components and systems are addressed in the appropriate sections in this chapter. Also check all technical service bulletins (TSBs) related to the vehicle. Look for bulletins that recommend reflashing the powertrain control module (PCM). Sometimes, the EEPROM (electrically erasable PROM) needs to be reprogrammed due to changes made in the strategy or calibrations after the vehicle was produced.

Diagnostic trouble codes and scan tool data should be retrieved during the initial diagnostic routine **(Figure 26–2)**. This includes key-on, engine-off (KOEO); key-on, engine-running (KOER); and continuous self-test DTCs. If the scan tool is unable to communicate with the system, follow the tests prescribed by the manufacturer to correct that problem. Also make sure to record any and all freeze-frame data associated with the DTCs. Review the code-setting conditions and follow the pinpoint tests to identify what set the code. Often, these will lead to the exact cause of the problem; other times, you will need to do further testing. If more than one DTC was set, refer to the vehicle's wiring diagram to identify parts and circuits that may be common to those codes. If DTCs are retrieved, you will need to determine if they are related to the concern. You may need to correct the problems that are causing the trouble codes, before moving on. Fixing the cause of the codes may correct the ignition problem.

If no DTCs were retrieved, look at the system's serial data. First identify the appropriate parameter identification (PIDs). The PID test mode allows access to PCM information, including input signals, outputs, calculated values, and the status of the system and monitors. While observing serial data, look at the inputs. Identify any signals that are outside the normal range. Do the same with the outputs. Mode $06 data can also be helpful. This mode displays the test values stored at the time a particular monitor was completed **(Figure 26–3)**. Depending on the vehicle, Mode $06 data may also provide information specific for cylinder misfires.

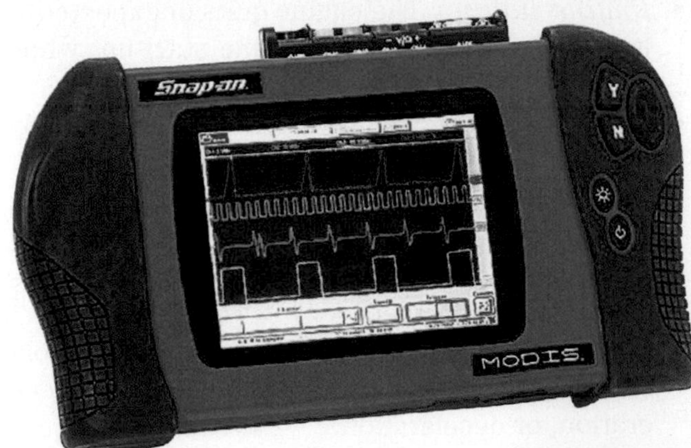

FIGURE 26–2

A diagnostic tool that performs the functions of many different testers, including a scan tool.

Snap-on Incorporated

SHOP TALK

Keep in mind that the PCM may assume that the input signals are correct while it controls an output device. This means incorrect inputs can cause an output to appear out of range because the PCM is driving it outside the normal range. Check the input signals before checking the outputs.

If no codes were retrieved and all appears to be normal on the data stream, diagnosis should be based on symptoms and detailed testing of the ignition system.

Primary Circuit

Primary ignition system wiring should be checked for tight connections. Electronic circuits operate on very low voltage. Voltage drops caused by corrosion or dirt can cause running problems. Missing or broken tab locks on wire terminals are often the cause of intermittent ignition problems due to vibration or thermal-related failure.

Test the integrity of a suspect connection by tapping, tugging, and wiggling the wires while

J1979 MISFIRE MODE $06 DATA			
Monitor ID	**Test ID**	**Description for CAN**	**Increments**
A1	$80	Total engine misfire and catalyst damage misfire rate	%
A1	$81	Total engine misfire and emission threshold misfire rate	%
A1	$82	Highest catalyst damage misfire and catalyst damage threshold misfire rate	%
A1	$83	Highest emission threshold misfire and emission threshold misfire rate	%
A1	$84	Inferred catalyst mid-bed temperature	°C
A2-AD	$0B	Misfire counts for last 10 drive cycles	events
A2-AD	$0C	Misfire counts for last/current drive cycle	events
A2-AD	$80	Cylinder "X" misfire rate and catalyst damage misfire rate	%
A2-AD	$81	Cylinder "X" misfire rate and emission threshold misfire rate	%

FIGURE 26–3

Mode $06 data from the misfire monitor.

the engine is running. Be gentle. The object is to re-create an ignition interruption, not to cause permanent circuit damage. With the engine and key off, separate the suspect connectors and check them for dirt and corrosion. Clean the connectors according to the manufacturer's recommendations.

Do not overlook the ignition switch as a source of intermittent ignition problems. A loose mounting rivet or poor connection can result in erratic spark output. To check the switch, gently wiggle the ignition key and connecting wires with the engine running. If the ignition cuts out or dies, the problem is located.

Carefully inspect the wires and belts for the charging system. Also check the charging voltage at the battery. The efficiency of an ignition system depends on the voltage it receives. If battery or charging system voltage is low, the input to the primary side of the coil will also be low.

Moisture can cause a short to ground or reduce the amount of voltage available to the spark plugs. This can cause poor performance or a no-start condition. Carefully check the ignition system for signs of moisture. **Figure 26–4** shows the common places where moisture may be present in a DI system.

Ground Circuits

Ground straps are often neglected, or worse, left disconnected after routine service. With the increased use of plastics in today's vehicles, ground straps may mistakenly be reconnected to a nonmetallic surface. The result of any of these problems is that the current that was to flow through the disconnected or improperly grounded

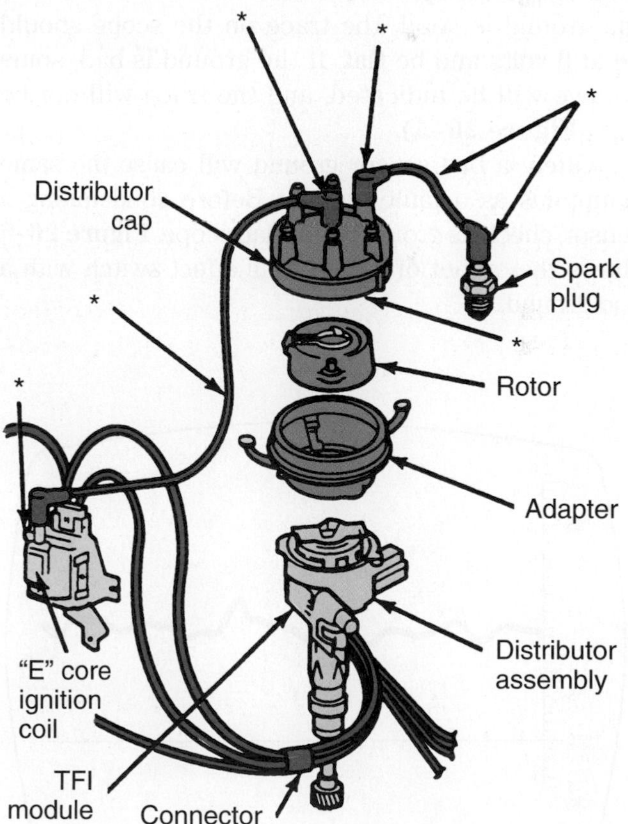

FIGURE 26–4

Places to check for moisture in an ignition system.

Ford Motor Company

strap is forced to find an alternate path to ground. Sometimes, the current attempts to back up through another circuit. This may cause the circuit to operate erratically or fail altogether. The current may also be forced through other components, such as wheel bearings or shift and clutch cables that are not meant to handle current flow, causing them to wear prematurely or become seized in their housing.

Examples of bad ground-circuit-induced ignition failures include burned ignition modules resulting from missing or loose coil ground straps and intermittent ignition operation resulting from a poor ground at the control module. Poor ground can be identified by conducting voltage drop tests and by monitoring the circuit with a lab scope.

When conducting a voltage drop test, remember that the circuit must be turned on and have current flowing through it. If the circuit is tested without current flow, the circuit will show zero voltage drop, which would indicate that it is good regardless of the amount of resistance present.

The same is also true when checking a ground with the lab scope. Make sure the circuit is on. If the ground is good, the trace on the scope should be at 0 volts and be flat. If the ground is bad, some voltage will be indicated, and the trace will not be flat **(Figure 26–5)**.

Often, a bad sensor ground will cause the same symptoms as a faulty sensor. Before condemning a sensor, check its ground with a lab scope. **Figure 26–6** shows the output of a good Hall-effect switch with a bad ground.

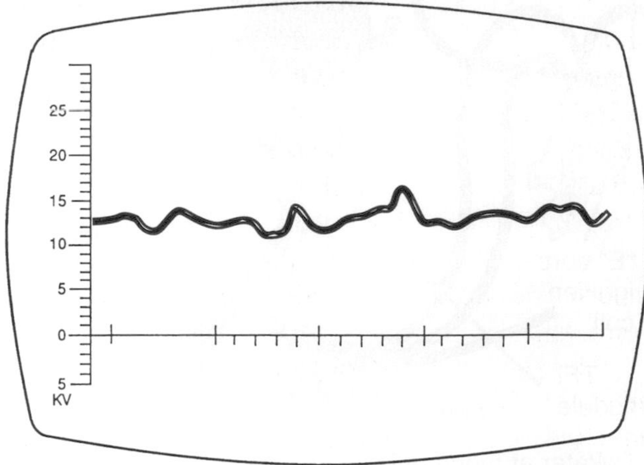

FIGURE 26–5

A voltage signal caused by a poor ignition module ground.

SPX Corporation, Aftermarket Tool and Equipment Group

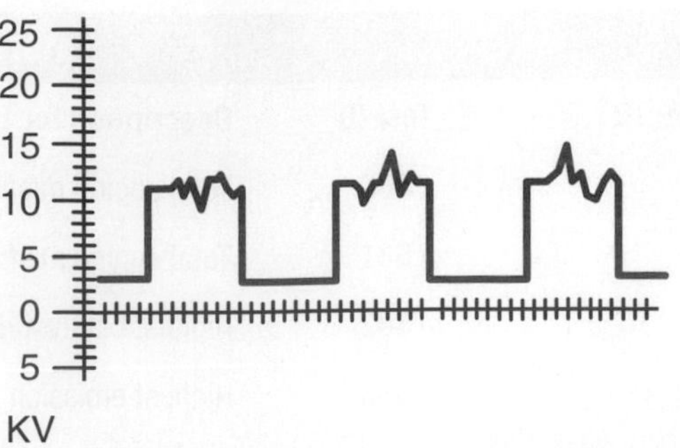

FIGURE 26–6

A voltage trace with ignition system noise due to a bad ground. Notice that the trace does not reach zero.

SHOP TALK

When checking the ignition system with a lab scope, gently tap and wiggle the components while observing the trace. This may indicate the source of an intermittent problem.

Electromagnetic Interference

Electromagnetic interference (EMI) can cause problems with the vehicle's computer. EMI is produced when electromagnetic radio waves of sufficient amplitude escape from a wire or conductor. Unfortunately, an automobile's spark plug wires, ignition coil, and AC generator coils all possess the ability to generate these radio waves. EMI can alter signals from sensors and to actuators. The result may be an intermittent driveability problem that may appear to be caused by many different systems.

To minimize the effects of EMI, check to make sure that sensor wires running to the computer are routed away from potential EMI sources. Rerouting a wire by no more than an inch or two may keep EMI from falsely triggering or interfering with computer operation.

Connecting a lab scope to voltage and ground wires can identify EMI problems. Common problems such as poor spark plug wire insulation will allow EMI.

Sensors

A voltage pulse from a crankshaft position sensor **(Figure 26–7)** activates the transistor in the control module. In most ignition systems, this sensor is either a magnetic pulse generator or Hall-effect sensor. These sensors are mounted on either the distributor shaft or the crankshaft.

Under unusual circumstances, the nonmagnetic reluctor can become magnetized and upset the

FIGURE 26–7

The wiring to the crankshaft sensor should be carefully inspected.

©Cengage Learning 2015

pickup coil's voltage signal to the control module. Use a steel feeler gauge to check for signs of magnetic attraction, and replace the reluctor if the test is positive. On some systems, the gap between the pickup and the reluctor must be checked and adjusted to manufacturer's specifications. To do this, use a properly sized nonmagnetic feeler gauge to check the air gap between the coil and reluctor. Adjust the gap if it is out of specification.

Hall-effect sensor problems are similar to those of magnetic pulse generators. This sensor produces a voltage when it is exposed to a magnetic field. The Hall-effect assembly is made up of a permanent magnet located a short distance away from the sensor. Attached to the distributor shaft or crankshaft pulley is a shutter wheel. When the shutter is between the sensor and the magnet, the magnetic field is interrupted, and voltage immediately drops to zero. This drop in voltage is the signal to the ignition module. When the shutter leaves the gap between the magnet and the sensor, the sensor produces voltage again.

Control Modules

Electronic ignitions use transistors as switches. These transistors are contained inside the control module that can be mounted to or in the distributor, remotely mounted to a surface inside the engine compartment, mounted below the ignition coil pack **(Figure 26–8)**, or integral to the PCM. Control modules should be tightly mounted to clean surfaces. A loose mount can cause intermittent misfires or no-start conditions. Often, the module is grounded through its mounting. A loose mounting

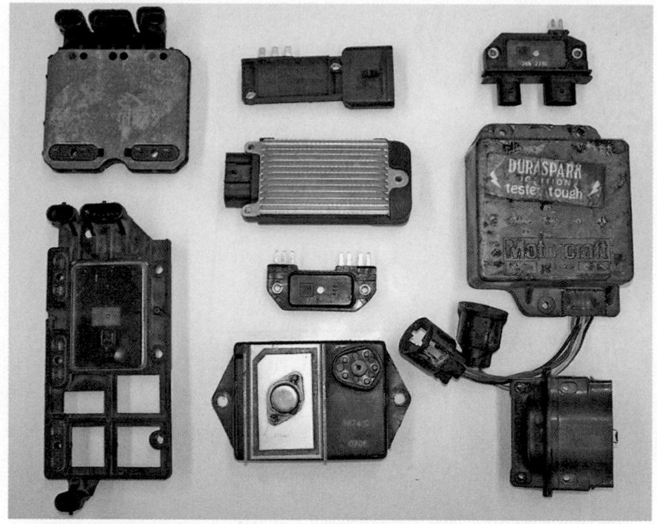

FIGURE 26–8

When the ignition module is not part of the PCM, it can have many different shapes and designs.

©Cengage Learning 2015

also can cause heat buildup that can damage and destroy the transistors and other electronic components inside the module. Some manufacturers recommend the use of special heat-conductive silicone grease between the control unit and its mounting. This helps conduct heat away from the module, reducing the chance of heat-related failure. During the visual inspection, check all electrical connections to the module. They must be clean and tight.

Secondary Circuit

Spark plug (ignition) and coil cables should be pushed tightly into the distributor cap and coil and onto spark plugs. Inspect all secondary cables for cracks and worn insulation, which cause high-voltage leaks. Inspect all of the boots on the ends of the secondary wires for cracks and hard, brittle conditions. Replace the wires and boots if they show evidence of these conditions. Inspect the terminals inside of the boots. Often, corrosion and signs of arcing are found inside the boots. This is often caused by poor contact between the terminals, which results in arcing between the terminals. Most manufacturers recommend spark plug wire replacement only in complete sets.

The secondary coil cable should also be inspected **(Figure 26–9)**. When checking this cable, check the ignition coil. The coil should be inspected for cracks or any evidence of arcing or leakage in the coil tower.

Secondary cables must be connected according to the firing order. Refer to the manufacturer's

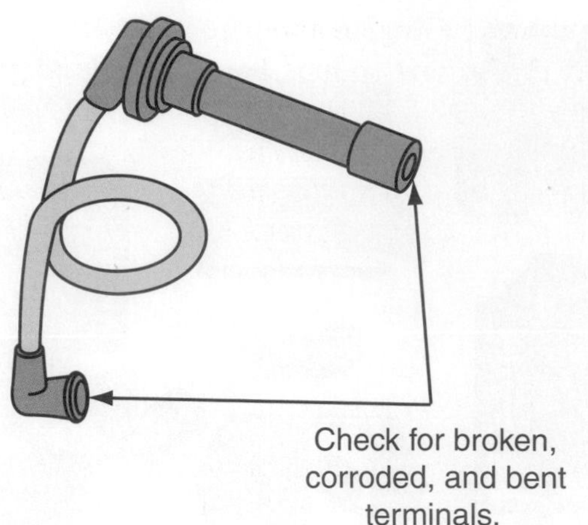
Check for broken,
corroded, and bent
terminals.

FIGURE 26–9

Carefully inspect the secondary cables.

service information to determine the correct firing order and cylinder numbering.

White or greyish powdery deposits on secondary cables at the point where they cross or near metal parts indicate that the cables' insulation is faulty. The deposits occur because the high voltage in the cable has burned the dust collected on the cable. Such faulty insulation may produce a spark that sometimes can be heard and seen in the dark. An occasional glow around the spark plug cables, known as a **corona effect**, is not harmful but indicates that the cable should be replaced.

Many OHC engines locate the spark plugs at the top of the combustion chamber. This requires the spark plug cables to pass through the valve cover to the plug. Inspect the spark plug end of the cables for oil. The seals surrounding the plug passage can leak, which allows oil to leak onto the plug wire boots. The oil then breaks down the boots, and arcing can occur.

Spark plug cables from consecutively firing cylinders should cross rather than run parallel to one another. Spark plug cables running parallel to one another can induce firing voltages in one another and cause the spark plugs to fire at the wrong time.

On distributorless or electronic ignition (EI) systems, visually inspect the secondary wiring connections at the individual coil modules. Make sure all of the spark plug wires are securely fastened to the coil and the spark plug. If a plug wire is loose, inspect the terminal for signs of burning. The coils should be inspected for cracks or any evidence of leakage in the coil tower. Check for evidence of terminal resistance. A loose or damaged wire or bad plug can lead to carbon tracking of the coil. If this condition exists, the coil must be replaced.

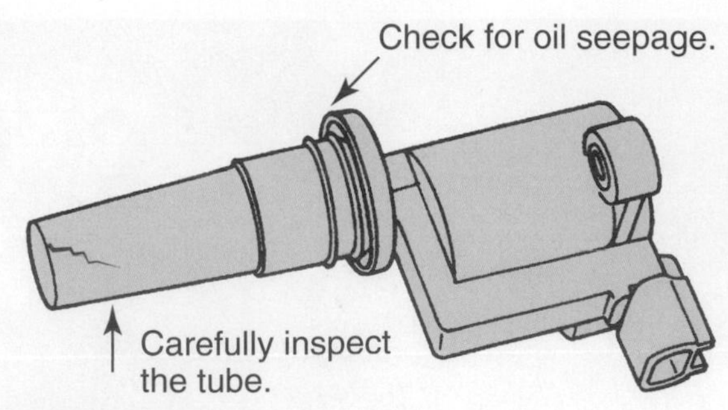
Check for oil seepage.
Carefully inspect
the tube.

FIGURE 26–10

The plastic assembly for COP assembly should be carefully inspected.

On coil-over-plug (COP) systems, carefully check the tubes that fit around the terminal of the spark plugs **(Figure 26–10)**. If the tube is cracked, voltage can leak out, jump to the cylinder head, and cause a misfire **(Figure 26–11)**. Also, make sure the coil assembly fits snugly over the spark plug and is securely mounted.

Distributor Cap and Rotor

The distributor cap should be properly seated on its base. All clips or screws should be tightened securely.

The distributor cap and rotor should be removed for visual inspection **(Figure 26–12)**. Physical or electrical damage is easily recognizable. Electrical damage from high voltage can include corroded or burned metal terminals and **carbon tracking** inside distributor caps. Carbon tracking is the formation of a line of carbonized dust between distributor cap terminals or between a terminal and the distributor housing. Carbon tracking indicates that high-voltage

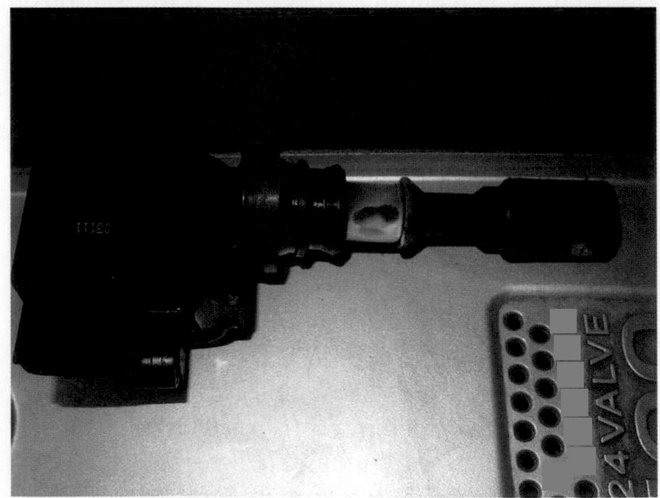
24 VALVE

FIGURE 26–11

This COP assembly was leaking voltage and arcing to the engine.

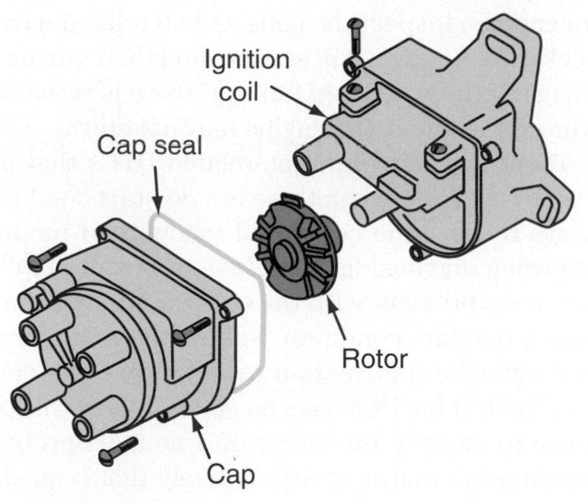

FIGURE 26–12

Inspect the distributor cap and rotor.

electricity has found a low-resistance conductive path over or through the plastic. The result is a misfire or a cylinder that fires at the wrong time. Check the outer cap towers and metal terminals for defects. Cracked plastic requires replacement of the unit. Damaged or carbon-tracked distributor caps or rotors should be replaced.

The rotor should be inspected carefully for discoloration and other damage. Inspect the top and bottom of the rotor carefully for greyish, whitish, or rainbow-hued spots. Such discoloration indicates that the rotor has lost its insulating qualities. High voltage is being conducted to ground through the plastic. This can cause a no-start condition as the spark jumps to ground inside the cap.

If the distributor cap or rotor has a mild buildup of dirt or corrosion, it should be cleaned. If it cannot be cleaned up, it should be replaced. Small round brushes are available to clean cap terminals. Wipe the cap and rotor with a clean shop towel, but avoid cleaning them in solvent or blowing them off with compressed air, which may contain moisture and may result in high-voltage leaks.

Check the distributor cap and housing vents. Make sure they are not blocked or clogged. If they are, the internal ignition module will overheat. It is good practice to check these vents whenever a module is replaced.

NO-START DIAGNOSIS

When an engine will not start, the cause is most likely a common circuit or component. If the cause is in the ignition system, simple tests can identify if the problem is in the primary or secondary circuit. Begin by cranking the engine and listening to it as it turns over. If the engine sounds like it is spinning faster than normal, the problem may be a broken or jumped timing belt or chain. If the cranking speed is normal and there is no attempt to start, the ignition may be at fault. If the engine acts as if it is trying to start, the problem may be in the fuel system.

SHOP TALK

Checking the operation of the fuel injection system when there is a no-start condition can also check the primary ignition. Connect a noid light to an injector harness **(Figure 26–13)**. If the injectors pulse while the engine is cranked, the triggering unit for the primary ignition circuit should be okay. The injection system uses the same signals to pulse the injectors.

If the problem is caused by an ignition fault, follow this procedure to determine the exact cause of the problem. Often, manufacturers include a detailed troubleshooting tree in their service information to help identify the cause of the no-start condition.

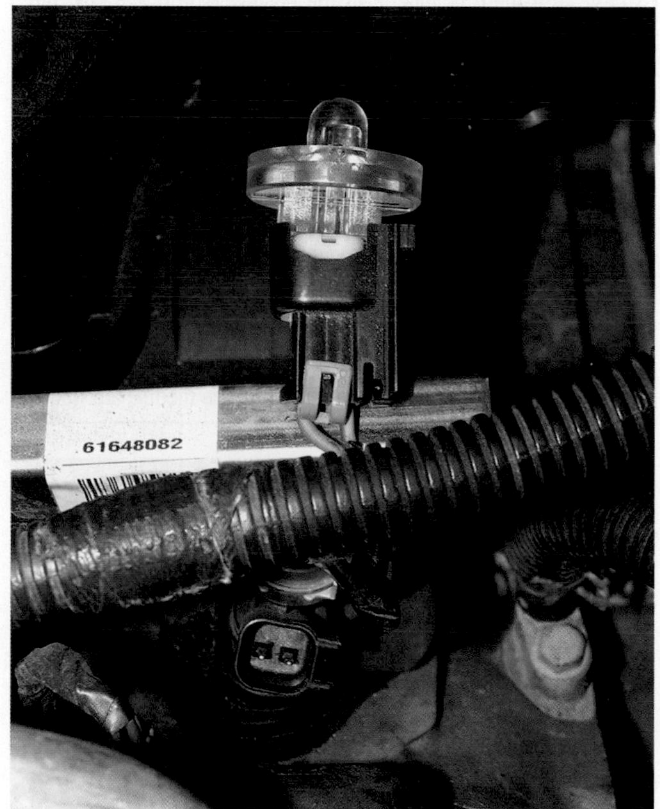

FIGURE 26–13

Use a noid light at the injectors when there is a no-start condition. If the injectors pulse while the engine is cranked, the triggering unit for the primary ignition circuit should be okay.

PROCEDURE

Basic No-Start Diagnosis

STEP 1 Connect a test spark plug to the spark plug wire and ground the spark plug case.

STEP 2 Crank the engine and observe the spark plug. If there is a bright, snapping, blue spark, the ignition is working properly.

STEP 3 If the test spark plug does not fire, check for coil output at the coil terminal.

STEP 4 If there is no spark, turn on the ignition. Connect a digital voltmeter to the positive side of the coil and ground. Read the voltage. Connect the voltmeter to the negative side of the coil and ground, and in most cases, the voltmeter should read the applied voltage. If there is no voltage on the negative side, the coil is open. If there is no voltage on the positive side, the supply is open.

STEP 5 With the testlight or DMM still connected, crank the engine. If the light flashes or a pulsing voltage is shown, the primary circuit is okay, and the problem is a bad coil.

STEP 6 If the light does not flash, check the voltage from the ignition switch to the positive side of the coil. If there is no voltage, the problem is in that circuit or the switch. If there is voltage at the positive side of the coil, the problem is the pickup unit or the control module.

STEP 7 Keep in mind that on some vehicles, the PCM will not send power to the coil until it receives a CKP signal. A magnetic pulse generator can be checked with an ohmmeter, DMM, or scope. A Hall-effect sensor should be checked with a DMM or scope. Compare your findings to specifications. A quick check for CKP operation is to connect a scan tool, crank the engine, and observe the engine data. If engine rpm PID shows cranking speed, the CKP is typically working.

STEP 8 If the pickup unit is good, suspect the ignition module. Make sure all wiring to and from the module are good.

SHOP TALK

When using a test spark plug, make sure you use the correct one for the system. There are two different types: low-voltage and high-voltage test plugs. Low-voltage plugs will fire if around 25 kV are applied to them. High-voltage plugs need 35 kV. If you use a high-voltage plug on a low-voltage system, it may not spark, leading you to believe that there is an ignition problem when there may not be. To determine which test plug to use, look at the specifications of the ignition system.

No-Start Diagnosis of EI Systems

When an engine with an EI system has a no-start problem, begin diagnosis with a visual inspection of the ignition system. Check for good primary connections. Inspect the coils and all related wiring. Check the CKP and CMP sensors and their wiring for damage. If these sensors fail or if there is resistance in the connections, the engine may not start.

There are many ignition-related DTCs that may be set by the PCM when there is a no-start condition. Always retrieve the codes and follow their pinpoint tests when diagnosing an EI system. A stored P0335 indicates a problem with the CKP circuit, which can cause a no-start condition. Follow the manufacturer's diagnostic information to determine the cause of the fault. If the PCM sets no codes, follow this procedure to identify the cause of a no-start problem. Keep in mind that it is very unlikely that a no-start

PROCEDURE

No-Start Diagnosis for EI Systems

STEP 1 Connect a test spark plug **(Figure 26–14)** to the spark plug wire and ground the spark plug case.

STEP 2 Crank the engine and observe the spark plug. If there is a bright, snapping, blue spark, the ignition is working properly.

STEP 3 If the spark is weak, check the power at each coil. With the ignition switch on, the voltmeter should read battery voltage. If the voltage is less than that, check the system's wiring diagram to determine what is included in the coil's power feed circuit. If there is no spark, check the CKP sensor input to the PCM.

STEP 4 If the CKP signal is good, check the power and ground circuits for the PCM.

STEP 5 If battery voltage is present, check the voltage drop across each of the components and wires to identify the location of an open or high resistance.

STEP 6 If none is found, check the crankshaft and camshaft position sensors. Both of these sensor circuits can be checked with a voltmeter, ohmmeter, or digital storage oscilloscope (DSO). If the sensors are receiving the correct amount of voltage and have good low-resistance ground circuits, their output should be a digital or a pulsing voltage signal while the engine is cranking. Compare their resistance readings to specifications. If any readings are abnormal, the circuit needs to be repaired, or the sensor needs to be replaced.

SHOP TALK

When using a DMM to check a digital crankshaft or camshaft sensor, crank the engine for a very short time and observe the meter. The reading should cycle from around 0 volts to 9 to 12 volts. Because digital meters do not react instantly, it is difficult to see the changes if the engine is cranked continually.

FIGURE 26–14

A test spark plug for high-voltage ignition systems.

Snap-on Tools Company

concern is caused by a non-common circuit or component, such as an ignition coil.

DIAGNOSING WITH AN ENGINE ANALYZER

It is impossible to accurately troubleshoot any ignition system without performing various electrical tests. An engine analyzer houses most of the necessary test equipment to do a complete engine performance analysis. Using an engine analyzer is a good way to determine if the driveability problem is caused by the ignition system. Also, the analyzer can be used to check individual ignition parts and circuits.

Cylinder Performance Test

During the cylinder performance test (also called the power balance test), the analyzer momentarily stops the ignition system from firing one cylinder at a time. To perform this test, the analyzer must connect to the primary ignition at the coil negative terminal. During this brief time, the rpm drop is recorded. When a cylinder is not contributing to engine power due to low compression or some other problem, there will be very little rpm drop when that cylinder stops firing. During the test, some analyzers record the actual rpm drop, whereas others show the percentage of rpm drop. Many analyzers also record the amount of hydrocarbons (HC) emitted when each cylinder stops firing. If a cylinder was misfiring prior to the test, the cylinder will have high HC emissions. Therefore, when this cylinder stops firing during the test, there will not be much of a change in HC emissions. A cylinder

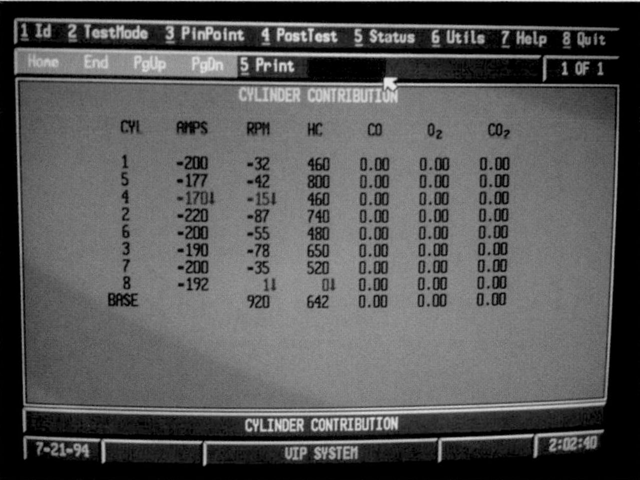

FIGURE 26–15

Cylinder performance test results showing one bad cylinder.

with low compression or a problem that causes incomplete combustion will not have much rpm drop or HC change during the cylinder performance test **(Figure 26–15)**.

Ignition Performance Tests

Ignition performance tests on most engine analyzers include primary circuit tests, secondary **kilovolt (kV)** tests, an acceleration test, scope patterns, and a cylinder miss recall. Some secondary kV tests include a snap kV test in which the analyzer directs the technician to accelerate the engine suddenly. When this action is taken, the firing kV should increase evenly on each cylinder. Some analyzers display the burn time for each cylinder with the secondary kV tests. The burn time is measured in milliseconds (ms).

The secondary kV display from an EI system includes average kV for each cylinder on the compression stroke and average kV for the matching cylinder that fires at the same time on the exhaust stroke. The burn time is also included on the secondary kV display from an EI system **(Figure 26–16)**.

Some analyzers are capable of freezing scope patterns and storing them in memory. These can be later recalled and reviewed to identify intermittent cylinder misfires.

Scope Patterns

An oscilloscope, or "scope," converts the electrical activity of the ignition system into a visual image showing voltage changes over a given period. This

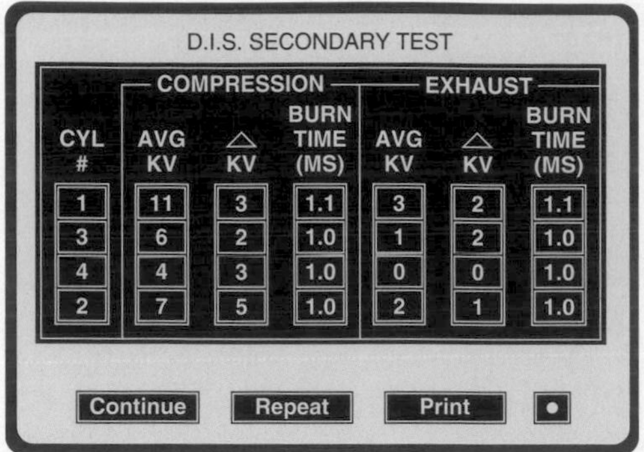

FIGURE 26–16

Secondary kV display on an EI system.

SPX Corporation, Aftermarket Tool and Equipment Group

information is displayed on a screen in the form of a continuous voltage line called a pattern or trace **(Figure 26–17)**. By studying the pattern, a technician can see what the ignition system is doing.

Always follow the instructions for the specific analyzer when connecting the test leads and operating the scope. Scopes typically have at least four leads for distributor ignitions: a primary pickup that connects to the negative terminal of the ignition coil, a ground lead that connects to a good ground, a secondary pickup that clamps around the coil's high-tension wire, and a trigger pickup that clamps around the spark plug wire of the number 1 cylinder.

Connecting the scope to a DIS system requires adapters or additional test leads. The leads are connected to the individual spark plug wires **(Figure 26–18)**. On some scopes, the companion cylinders of a waste spark system are viewed at the same time.

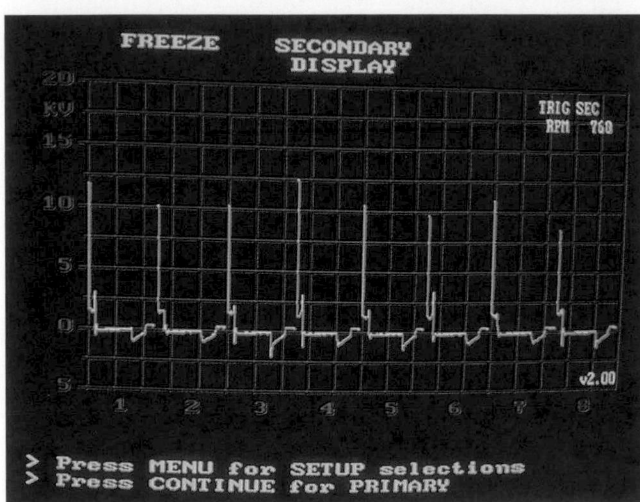

FIGURE 26–17

The secondary pattern for an eight-cylinder engine.

FIGURE 26–18

The tester's leads are connected to the individual spark plug wires.

Snap-on Tools Company

Other scopes display all of the cylinders, allowing technicians to compare the activity of the cylinders.

Adapters must also be used to monitor the secondary circuit on COP systems **(Figure 26–19)**. These adapters also allow for cylinder-to-cylinder comparisons. Most COP systems can be tested with a low-amperage probe through the ignition primary circuit. This allows you to check primary current ramp for each coil, which can help in diagnosing a defective coil.

SCALES A typical scope screen has two vertical voltage scales: one on the left and one on the right. Typically, the scale on the left is divided into increments of 1 kV (1000 volts) and ranges from 0 to 25 kV.

FIGURE 26–19

A COP test adapter.

Snap-on Tools Company

This scale is useful for testing secondary voltage. It can also be used to measure primary voltage by interpreting the scale in volts rather than kilovolts. The scale on the right side is divided into increments of 2 kV and has a range of 0 to 50 kV for testing secondary voltage. This scale can also be used to measure primary voltage in the 0 to 500 volt range.

The screen also has a horizontal time scale located at the bottom. The time may be expressed as percent of dwell or in milliseconds. The percent of dwell scale is divided into increments of 2 percentage points and ranges from 0 to 100 percent. This represents one complete ignition cycle. The millisecond scale is typically broken down into units of 0 to 5 ms or 0 to 25 ms. The 5 ms scale is often used to measure the duration of the spark **(Figure 26–20)**. The complete firing pattern can normally be displayed in the 25 ms mode. A scope displays changes in voltage over time from left to right, similar to reading a book.

UNDERSTANDING SINGLE CYLINDER PATTERNS A typical ignition waveform can represent the secondary or primary circuit. A typical secondary pattern is shown in **Figure 26–21**. A typical primary circuit pattern is shown in **Figure 26–22**. The primary pattern is used when secondary circuit connections are

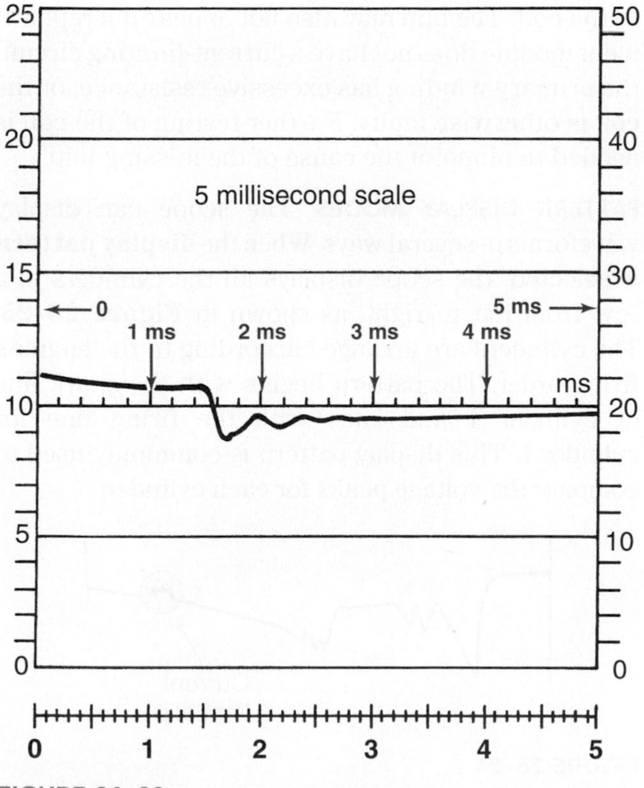

FIGURE 26–20

A 5 ms pattern showing spark duration.

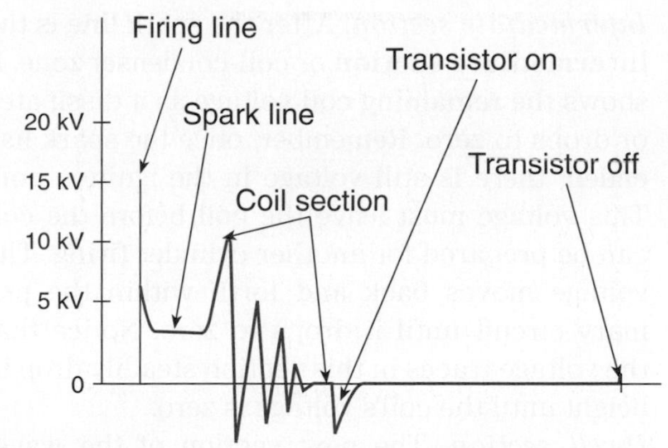

FIGURE 26–21

A typical secondary pattern.

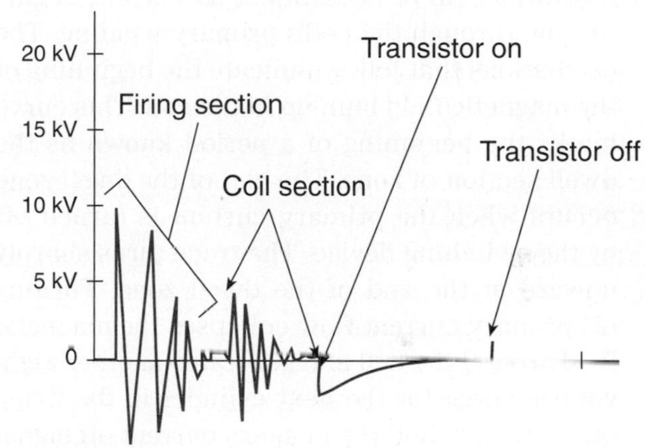

FIGURE 26–22

A typical primary pattern.

not possible or to observe cylinder timing problems. The main sections of a secondary waveform include the following:

- *Firing line.* The **firing line** appears on the left side of the screen. The height of the firing line represents the voltage needed to overcome the resistance in that secondary circuit and to initiate a spark across the gap of the spark plug. Typically, around 10 000 volts are required. Keep in mind that cylinder conditions have an effect on this resistance. Leaner air/fuel mixtures increase the resistance and increase the required **firing voltage**.
- *Spark line.* Once the resistance in the secondary is overcome, the spark jumps the plug gap, establishing current flow. The time the spark actually lasts is represented by the **spark line**. The spark line begins at the firing line and continues until the voltage from the coil drops below the level needed to keep current flowing across the gap.

- *Intermediate section.* After the spark line is the **intermediate section** or coil-condenser zone. It shows the remaining coil voltage as it dissipates or drops to zero. Remember, once the spark has ended, there is still voltage in the ignition coil. This voltage must leave the coil before the coil can be prepared for another cylinder firing. The voltage moves back and forth within the primary circuit until it drops to zero. Notice that the voltage traces in this section steadily drop in height until the coil's voltage is zero.
- *Dwell section.* The next section of the waveform begins with the primary circuit current ON signal. It appears as a slight downward turn followed by several small oscillations. The slight downward curve occurs just as current begins to flow through the coil's primary winding. The oscillations that follow indicate the beginning of the magnetic field buildup in the coil. This curve marks the beginning of a period known as the dwell section of zone. The end of the dwell zone occurs when the primary current is turned off by the switching device. The trace turns sharply upward at the end of the dwell zone. Turning off primary current flow collapses the magnetic field around the coil and generates another high-voltage surge for the next cylinder in the firing order. Remember, the primary current off signal is the same as the firing line for the next cylinder. The length of the dwell section represents the amount of time that current is flowing through the primary.

Most scope patterns look more or less like the one just described. The patterns produced by some systems may have fewer oscillations in the intermediate section. Patterns may also vary slightly in the dwell section. The length of this section depends on when the control module turns the transistor on and off.

Older DI systems used a fixed dwell period. The number of dwell degrees remained the same during all engine speeds. So if the engine has 30° of dwell at idle, it should have 30° of dwell at 2000 rpm. This is not saying that the actual amount of time has remained the same. A fixed dwell of 30° at 2000 rpm gives the ignition coil only one-quarter the saturation time, in milliseconds, that it has at 500 rpm.

Most control modules provide a variable dwell; dwell time changes with engine speed. At idle and low rpm speeds, a short dwell provides enough time for complete ignition coil saturation **(Figure 26–23A)**. The current ON and current OFF signal appear very close to each other, usually less than 20°. As engine speed increases, the control module lengthens the dwell degrees **(Figure 26–23B)**. This, of course, increases the available time for coil saturation.

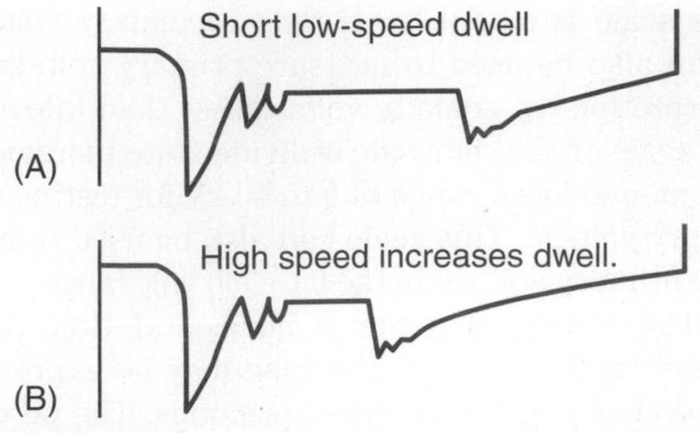

FIGURE 26–23
An example of a variable dwell ignition system.

Many late-model systems are **current limiting**. These systems saturate the ignition coil quickly by passing high current through the primary winding for a fraction of a second. Once the coil is saturated, the need for high current is eliminated, and a small amount of current flows to keep the coil saturated.

The point at which the control module cuts back from high to low current appears as a small blip or oscillation during the dwell section of the pattern **(Figure 26–24)**. At high engine speeds, this blip may be missing. In an attempt to keep the coil saturated, the module may not stop sending high current to the coil. The blip may also not appear if a replacement module does not have a current-limiting circuit; the primary winding has excessive resistance; or the coil is otherwise faulty. Further testing of the coil is needed to pinpoint the cause of the missing blip.

PATTERN DISPLAY MODES The scope can display waveforms in several ways. When the **display pattern** is selected, the scope displays all the cylinders in a row from left to right, as shown in **Figure 26–25**. The cylinders are arranged according to the engine's firing order. The pattern begins with the spark line of cylinder 1 and ends with the firing line for cylinder 1. This display pattern is commonly used to compare the voltage peaks for each cylinder.

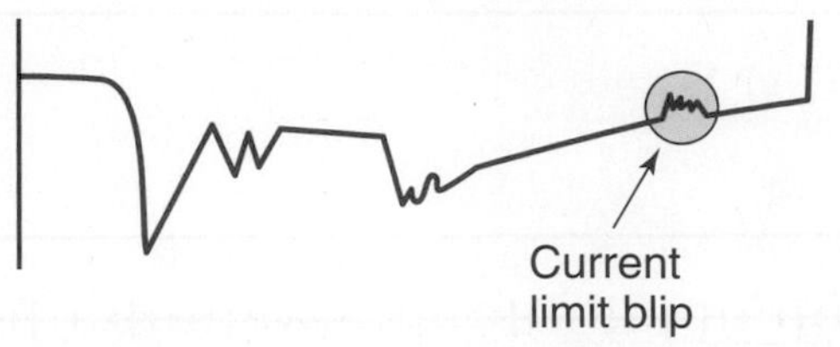

FIGURE 26–24
A pattern showing an ignition system limiting current during dwell.

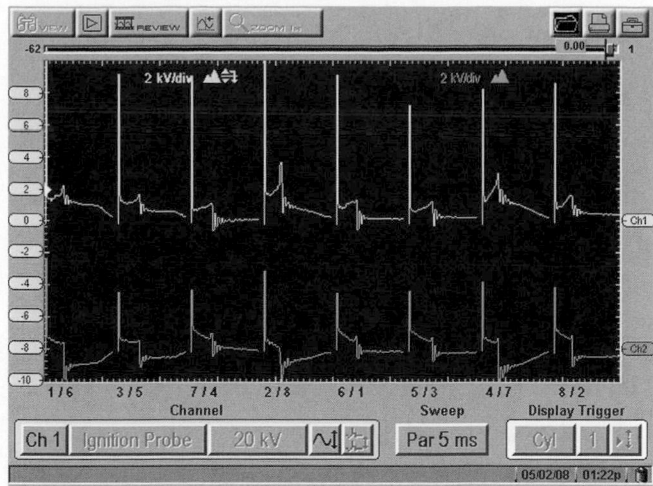

FIGURE 26–25

Typical parade (display) patterns for the primary and secondary circuits.

Snap-on Tools Company

Another display mode is the **raster pattern (Figure 26–26)**. A raster pattern stacks the waveforms of the cylinders one above the other. Cylinder 1 is displayed at the bottom of the screen, and the rest of the cylinders are arranged above it according to the engine's firing order. In a raster pattern, the waveform for each cylinder begins with the spark line and ends with the firing line. This allows for a much closer inspection of the voltage and time trends than is possible with the display pattern.

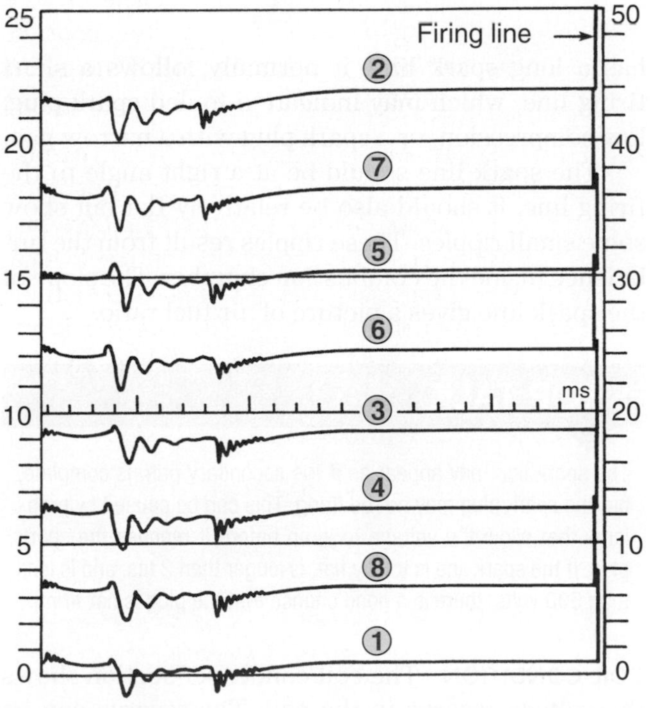

FIGURE 26–26

A secondary raster pattern.

A **superimposed pattern** displays all of the patterns one on top of the other. Like the raster pattern, the superimposed voltage patterns are displayed the full width of the screen, beginning with the spark line and ending with the firing line. A superimposed pattern is used to identify variations of one cylinder's pattern from the others.

SHOP TALK

Often, a DSO is capable of displaying only one cylinder at a time. To look at the rest of the signals, the input pickup must be moved to the other cylinders one at a time. This means it is not possible to view a parade, raster, or superimposed pattern.

SPARK PLUG FIRING VOLTAGE In a secondary pattern, the firing line is the highest line in the pattern. The firing line is affected by anything that adds resistance to the secondary circuit. This includes the condition of the spark plugs or the secondary circuit, engine temperature, fuel mixture, and compression pressures. The normal height of a firing line, with the engine at idle, should be between 7 and 13 kV with no more than a 3 kV variation between cylinders. If one or more firing lines are too low or high, the cause is something that is common to only those cylinders. If the firing lines are all too high or low, the problem is something that is common to all cylinders. **Table 26–1** covers most of the things that could cause abnormal firing lines.

High resistance in the secondary circuit also produces a spark line that is higher in voltage and has a steep slope with shorter firing durations. A good spark line should be relatively flat and measure 2 to 4 kV in height.

The voltage required to fire a spark plug increases when the engine is under load. The voltage increase is moderate and uniform if the spark plugs are in good condition and properly gapped. To test the spark plugs under load, note the height of the firing lines at idle speed. Then quickly open and release the throttle (snap accelerate) and note the rise in the firing lines while checking the voltages for uniformity. A normal rise should be between 3 and 4 kV. If the rise is not equal on all cylinders or if the rise is too low or too high, the spark plugs are probably faulty. Also, watch the decrease in their height as the throttle is closed. If the cylinders do not uniformly drop, there is high resistance in the circuits that were different.

TABLE 26–1 FIRING LINE DIAGNOSIS

CONDITION	PROBABLE CAUSE	REMEDY
Firing voltage lines the same, but abnormally high	1. Retarded ignition timing 2. Fuel mixture too lean 3. High resistance in coil wire 4. Corrosion in coil tower terminal 5. Corrosion in distributor coil terminal	1. Rest ignition timing. 2. Check fuel pressure or check for vacuum leak. 3. Replace coil wire. 4. Clean or replace coil. 5. Clean or replace distributor cap.
Firing voltage lines the same, but abnormally low	1. Fuel mixture too rich 2. Breaks in coil wire causing arcing 3. Cracked coil tower causing arcing 4. Low coil output 5. Low engine compression	1. Check for leaking injector(s) high fuel pressure. 2. Replace coil wire. 3. Replace coil. 4. Replace coil. 5. Determine cause and repair.
One or more, but not all, firing voltage lines higher than the others	1. EGR valve stuck open 2. High resistance in spark plug wire 3. Cracked or broken spark plug insulator 4. Intake vacuum leak 5. Defective spark plugs 6. Corroded spark plug terminals	1. Inspect or replace EGR valve. 2. Replace spark plug wires. 3. Replace spark plugs. 4. Repair leak. 5. Replace spark plugs. 6. Replace spark plugs.
One or more, but not all, firing voltage lines lower than the others	1. Breaks in plug wires causing arcing 2. Cracked coil tower causing arcing 3. Low compression 4. Defective or fouled spark plugs	1. Replace spark plug wires. 2. Replace coil. 3. Determine cause and repair. 4. Replace spark plugs.
Cylinders not firing	1. Cracked distributor cap terminals 2. Shorted spark plug wire 3. Mechanical problem in engine 4. Defective spark plugs 5. Spark plugs fouled	1. Replace distributor cap. 2. Determine cause of short and replace wire. 3. Determine problem and correct. 4. Replace spark plugs. 5. Replace spark plugs.

The firing lines should also be checked while the engine is cranking and the plug wires are disconnected from the plugs. Insert a test plug into the plug wire. This will give an indication of the condition of the coil(s) and the secondary insulation. Follow the specified procedures to prevent the engine from starting. Then crank the engine and observe the firing lines. They should reach at least 30 kV; less indicates a possible bad coil. If some of the firing lines are low, the secondary insulation is breaking down or the coil for those cylinders is bad.

SPARK DURATION The amount of time that the spark plug is actually firing is important. This is called spark duration. **Spark duration** is represented by the length of the spark line and is measured in milliseconds. Most engines have a spark duration of approximately 1.5 ms. Short spark durations cannot provide complete combustion and can cause increased emissions levels and a loss of power. If the spark duration is too long, the spark plug electrodes might wear prematurely. When the ignition pattern has a long spark line, it normally follows a short firing line, which may indicate a fouled spark plug, low compression, or a spark plug with a narrow gap.

The spark line should be at a right angle to the firing line. It should also be relatively flat but show some small ripples. These ripples result from the turbulence inside the combustion chamber. The slope of the spark line gives a picture of air/fuel ratio.

SHOP TALK

The spark line may appear as if the secondary path is complete, but the spark plug may not be firing. This can be caused by problems that allow the voltage to jump before it reaches the spark plug. If the spark line is totally flat, is longer than 2 ms, and is less than 500 volts, there is a good chance that the plug is not firing.

COIL CONDITION The coil/condenser section shows the voltage reserve in the coil. The reserve can be determined by looking at the height of the oscillations. The heights should uniformly decrease to a zero value. If the waveform does not have normal oscillations in its intermediate section, check for a possible short in the coil by testing the resistance of the primary and secondary windings.

The available voltage output from a coil can also be checked with the coil output test. Following are the steps to safely perform a coil output test:

1. Install a test plug in the coil wire or a plug wire if there is no coil wire.
2. Set the scope on display with a voltage range of 50 kV.
3. Crank the engine and note the height of the firing line. The firing line should exceed 35 kV. Lower than specified voltages may indicate lower than normal available voltage in the primary circuit. The control module may have developed high internal resistance. The coil or coil cable may also be faulty.

PRIMARY CIRCUIT CHECKS The primary ignition pattern shows the action of the primary circuit. To be able to spot abnormal sections of a primary waveform, you must know what causes each change of voltage and time in a normal primary waveform. Although the true cycle of the primary circuit begins and ends when the switching transistor is turned on, the displayed pattern begins right after the transistor is turned off. At this moment in time, the magnetic field around the windings collapses, and a spark plug is fired.

SHOP TALK

Always carefully check the primary circuit; a 1-volt loss in the primary circuit can reduce secondary output by up to 10 000 volts.

Looking at the primary pattern shown in **Figure 26–27**, the trace at the left represents the collapsing of the primary winding after the transistor turns off and primary current flow is interrupted. The height of these oscillations depends on the current that was flowing through the winding right before it was stopped. The amount of current flow depends on the time it was able to flow, the voltage applied to the winding, and the resistance of the winding. High primary circuit resistance will reduce the maximum amount of current that can flow through the winding. Reduced current flow through the winding will reduce the amount of voltage that can be induced when the field collapses.

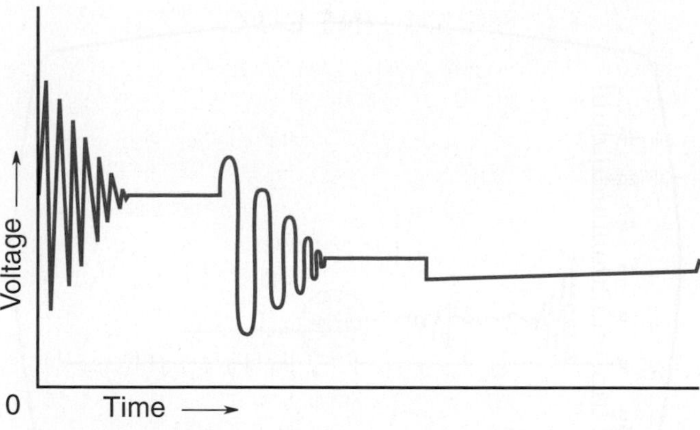

FIGURE 26–27

A typical primary pattern.

During the collapsing of the primary winding, the spark plug is firing. The primary circuit's trace shows sharp oscillations of decreasing voltages. The overall shape of this group of oscillations should be conical and should last until the spark plug stops firing.

After the firing of the plug, some electrical energy remains in the coil. This energy must be released prior to the next dwell cycle. The next set of oscillations shows the dissipation of this voltage. These oscillations should be smooth and become gradually smaller until the 0-volt line is reached. At that point, there is no voltage left and the coil is ready for the next dwell cycle.

Immediately following this dissipation of coil energy is the transistor ON signal. This is when current begins to flow through the primary circuit. It is the beginning of dwell. When the transistor turns on, there should be a clean and sharp turn in the trace. A clean change indicates that the circuit was instantly turned on. If there is any sloping or noise at this part of the signal, something is preventing the circuit from being instantly turned on. When looking at a superimposed primary pattern, any variation between cylinders will show up as a blurred or noisy transistor ON signal **(Figure 26–28)**.

If there are erratic voltage spikes at the transistor ON signal, the ignition module may be faulty; the distributor shaft bushings may be worn; or the armature is not securely fit to the distributor shaft. The problem is preventing dwell from beginning smoothly and causing the engine to have a rough idle, an intermittent miss, and/or higher than normal HC emission levels.

During dwell, the trace should be relatively flat. However, many ignition systems have features that change current flow during dwell. These features

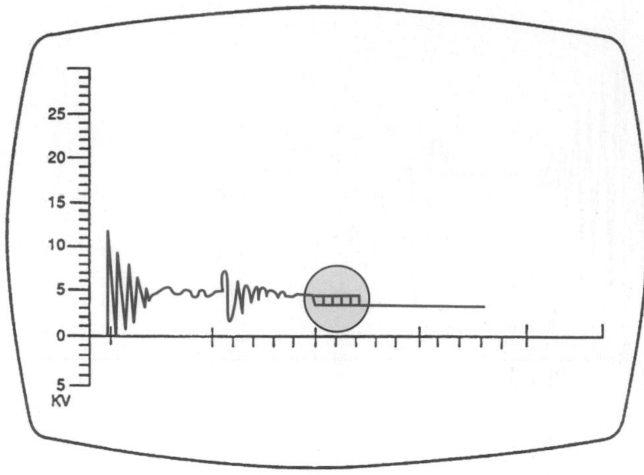

FIGURE 26–28

A superimposed primary pattern showing cylinders with different transistor ON times.

SPX Corporation, Aftermarket Tool and Equipment Group

are designed to allow complete coil saturation only when that is needed. By reducing the amount of current, the amount of voltage induced in the secondary is also reduced.

Stress Testing Components

Often, an intermittent ignition problem occurs only under certain conditions, such as extremes in heat or cold, or during rainy or humid weather. Careful questioning of the customer should lead to determining if the problem is stress-condition related. Does the problem occur on cold mornings? Does it occur when the engine is fully warmed up? Is it a rainy day problem? If the answer to any of these questions is positive, you can reproduce the same conditions in the shop during stress testing.

> **CAUTION!**
>
> When using cool-down sprays, always wear eye protection, and avoid spraying your skin or clothing. Also keep away from the belts and fans. Use extreme caution.

COLD TESTING With the scope on raster, cool major ignition components such as the control module, the pickup coil, and other major connections one at a time, using a liquid cool-down agent. After cooling a component, watch the pattern for any signs of malfunction, particularly in the dwell zone. If there is no sign of malfunction, cool down the next component after the first has warmed to normal operating temperature. Cooling (or heating) more than one component at a time provides inconclusive results.

HEAT TESTING To heat stress components, use a heat gun or hair dryer to direct hot air into the component. Heat guns intended for stripping paint and other household jobs can become extremely hot and melt plastic, wire insulation, and other materials. Use a moderate setting and proceed cautiously. Look for changes in the dwell section of the trace, particularly in the variable dwell or current-limiting areas. If connections appear to be the problem, disconnect them, clean the terminals, and coat them with dielectric compound to seal out dirt and moisture.

MOISTURE TESTING A wet stress test is performed by lightly spraying the components, coil and ignition cables, and connections with water. Do not flood the area; a light mist does the job. A scope set on raster or display helps pinpoint problems, but it is often possible to hear and feel the miss or stutter without the use of a scope. As with heat and cold testing, do not spray down more than one area at a time, or results could be misleading. If you suspect a poor connection, clean and seal it, and then retest it.

Testing with a Scan Tool

On many vehicles, cylinder power balance tests can be performed with a scan tool. Often located in the special functions or engine output controls section of the powertrain menu, this function allows you to turn off fuel injectors individually to check for each cylinder's contribution. If an engine has a misfire, this test can narrow down which cylinder is misfiring by noting how the engine responds as each injector is disabled. If the engine speed does not decrease when a cylinder's injector is turned off, the problem cylinder has been found.

DIAGNOSING WITH A DSO OR GMM

Commonly used diagnostic tools are modular units that can serve as many different tools, including a scan tool, lab scope, DMM, and GMM. They may also contain libraries of information that serve as references for diagnostics, such as the basic operation of a component or system, test procedures, identification of pins in a connector, location of the component or connector, specifications, and diagnostic tips. Some even have a database of known good waveforms in their memory, which again help in the diagnosis of a problem. All of these features make diagnosis simpler.

Each model of these tools has many unique operating procedures, and the tool's user manual must be used to have accurate results. What follows are some general guidelines for using these tools to diagnose an ignition system.

After vehicle identification information is entered, the test mode is selected. This mode allows you to test individual components. The screen will display how to properly connect the tool to the component and, in most cases, display the expected reading for those tests. There will also be the option of displaying the test results in a variety of ways: digital, digital graphs or waveforms, analog graphs or waveforms, or a combination of these.

Using the DSO or GMM

The graphing meter function is much like using a scope, except that it also displays digital MIN/MAX readings with the trace. Many have a four-channel capability, which means you can look at the signals from four different sensors and/or outputs at the same time. This is a great help during diagnosis. This mode also allows a technician to observe any glitches or noise that may affect operation.

To aid in diagnosing intermittent problems, screens can be frozen, saved, and printed for review later. The tool can typically be connected to the vehicle, and then the vehicle can be taken for a road test. Intermittent problems can be observed, and the data around those problems can be stored in the meter's memory for review after the road test.

In most cases, additional leads must be attached to the lab scope or GMM to read additional channels. These are colour-coded, and the colour of the display matches the colour of the lead. That lead is attached to the point where the desired signal can be monitored. A ground lead must also be attached. The colour-coding throughout allows for easy identification of what is being monitored **(Figure 26–29)**. The lab scope may allow a look at DC volts, low amps, ignition secondary voltages, vacuum, and pressure.

The readings or waveforms for the component or system monitored can be looked at individually or at the same time. In most cases, channel 1 will always be displayed. If the tool is capable of two or more channels, these can be selected individually. However, on many GMMs, channels 1 and 2 are always displayed, and the additional channels must be selected.

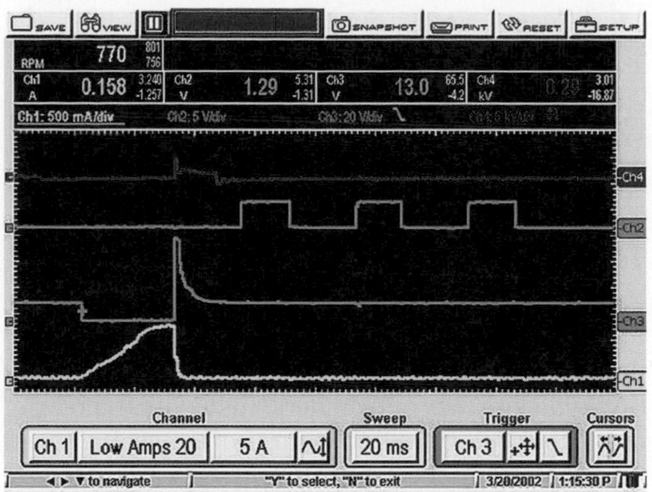

FIGURE 26–29

Waveforms on four different channels: CH1 in yellow, CH2 in blue, CH3 in green, and CH4 in red.

Snap-on Tools Company

Certain settings and controls are critical to recording a worthwhile waveform, including the following:

- *Scale.* Used to measure the events of one or more waveforms.
- *Filter.* Minimizes and cleans up waveforms from unwanted noise in order to look at voltages.
- *Threshold.* Changes the reference point for the waveform and is only used when measuring frequency, duty cycle, dwell, and pulse width. This may be manually or automatically set.
- *Peak detect.* Used to capture spikes and glitches in signals. When it is off, the tool collects just enough data to form a waveform. When it is on, the tool collects more data than are needed. This is why it allows you to capture a glitch or noise.
- *Time scale (sweep).* Sets the amount of time data will be displayed; more data can be observed with a faster sweep. Also, the usefulness of peak detect is increased with a longer sweep. However, long sweeps have a slow sample rate, and when observing ignition systems, the firing line may not appear.
- *Voltage scale (sensitivity).* Adjusts how sensitive the scope will be to voltage changes. The lower the setting, the more sensitive the ignition scope will be to detecting cylinder firing.
- *Trigger.* Sets the criteria that should be recognized to start the display of data.

- *Trigger slope.* Sets the direction that the waveform must be moving toward to start the waveform.
- *Trigger level.* Used to place the trigger point along the horizontal or vertical axis of the display.

Testing the Ignition System

Prior to testing an ignition system, the diagnostic tool must be programmed to match the type of ignition system found on the vehicle; this includes ignition type, number of cylinders, firing order, polarity of the spark plugs, and the source of the engine speed reference. Each type of ignition system also requires different adapters and different setup procedures.

SHOP TALK

It is important that the correct adapters and leads be used when using most testers. These are designed for specific purposes, and if the wrong ones are used, inaccurate measurements will result. For example, there are many different COP adapters because there are many different designs of COP systems. Also, adapters allow for connection to a waste spark system that allows a view of the total system.

All ignition scopes must have a means to monitor the secondary and a way to monitor the switching of the primary circuit. The primary is monitored with a reference or engine speed pickup. This pickup is placed around the number 1 spark plug wire on all systems. On DIS systems, the tool must be programmed with the number of cylinders, cylinder firing order, and plug polarities. On coil-per-cylinder systems, the tool must know the number of cylinders and the firing order.

It may take a few seconds for the tool to synchronize itself with the ignition system. Also, on some tools connected to waste spark systems, the firing of the plugs that are on the compression stroke will appear only on channel 1 and those that occur on the exhaust stroke will appear on channel 2.

All events of the ignition system will appear in the waveform, as they were in the waveform shown on an ignition scope. This means the interpretation of the patterns is the same regardless of what is displaying it.

The primary difference between an ignition scope and a DSO or GMM is that the latter can display the MIN/MAX values for firing kV, spark kV, and spark duration for each cylinder. On an ignition scope, those values must be visually observed.

Both the power and waste spark of an EI system can be observed and compared for each coil. Photo Sequence 20 shows the procedure for connecting a scope to a DLI and how to interpret the results. Keep in mind that if one plug in a DLI system is fouled, it will affect the other plug in the circuit. Observing the activity of both will identify the problem plug.

AN UNDETECTED CYLINDER When the firing of a cylinder is not detected by the tool, there will be a void in the ignition system's waveform. This can be caused by an ignition problem or improper setup of the tool. To determine if it is the latter, lower the sensitivity setting. If the cylinder still is not displayed, it is fair to assume that the cylinder is not firing.

IGNITION TIMING

The primary circuit controls the secondary circuit. Therefore, it controls ignition timing. Most primary circuit problems in a computer-controlled ignition system result in starting problems or poor performance due to incorrect timing.

If engine performance is poor, the cause of the problem can be many things. There can be a problem with the engine, such as poor compression, incorrect valve timing, overheating, and so on. The air/fuel mixture or the ignition timing is incorrect. When the ignition timing is not correct, many tests will point to the problem. Incorrect ignition timing will cause incomplete combustion at one or all engine speeds. Incomplete combustion will cause excessive O_2 in the exhaust. This will cause the PCM to try to correct the apparent lean mixture **(Figure 26–30)**. Incorrect

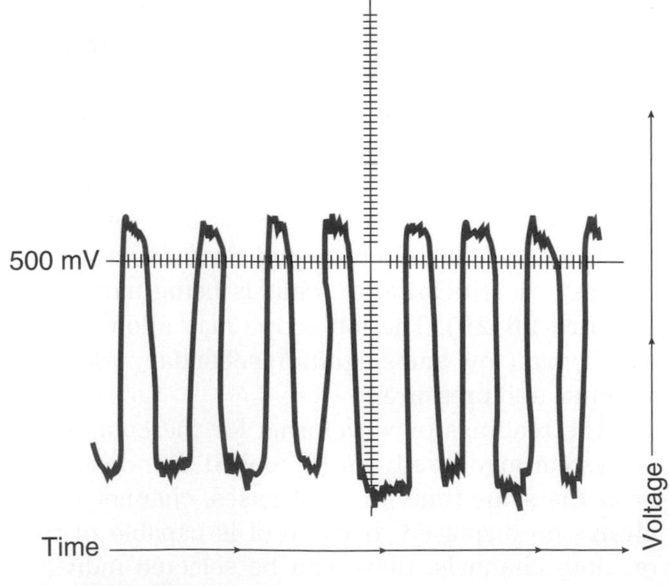

FIGURE 26–30
A lean-biased O_2 sensor.

USING A SCOPE TO TEST A DISTRIBUTORLESS IGNITION SYSTEM

PHOTO SEQUENCE 20

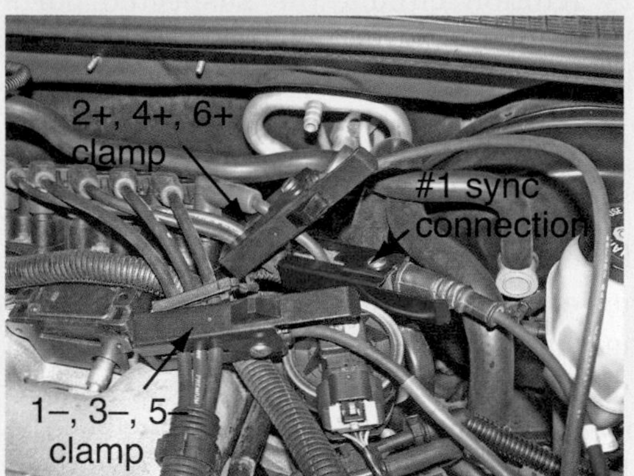

P20–1 To observe the activity of an EI system on a scope, special adapters are required.

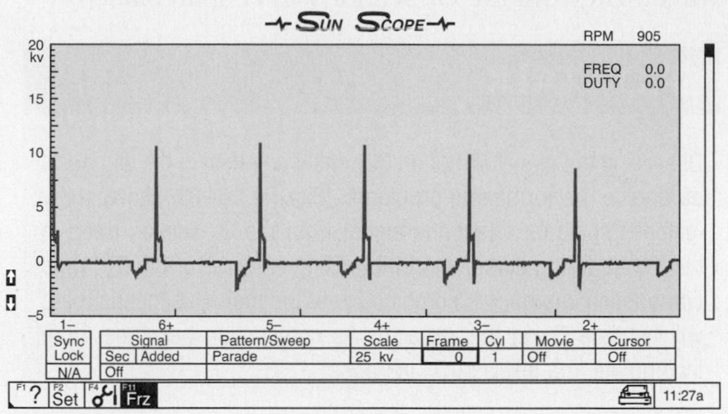

P20–2 A secondary pattern for the system. Note that the waveform for three of the cylinders (6+, 4+, and 2+) fires positively and the remaining three fire negatively (1–, 3–, and 5–).

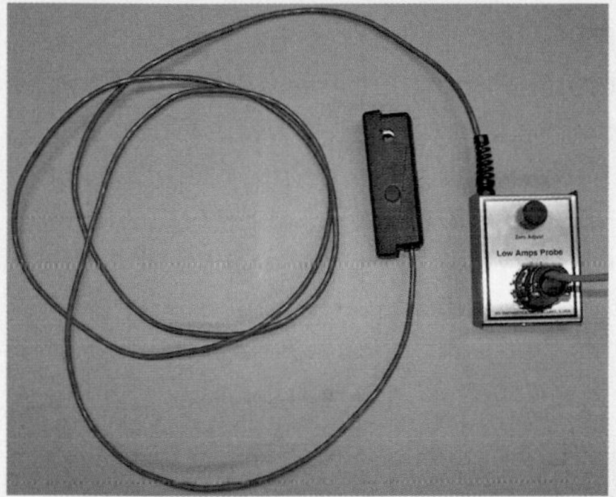

P20–3 To look at the primary circuit, connect a low-amp probe to the scope. Make sure it is compatible with the scope.

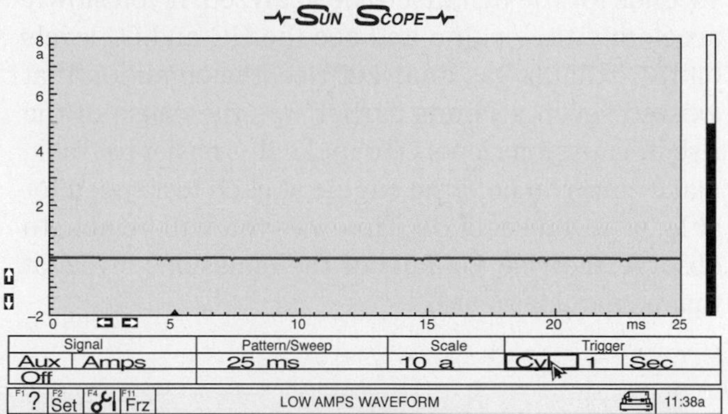

P20–4 The baseline on the scope for amperage must be set to zero before the probe is connected to the system.

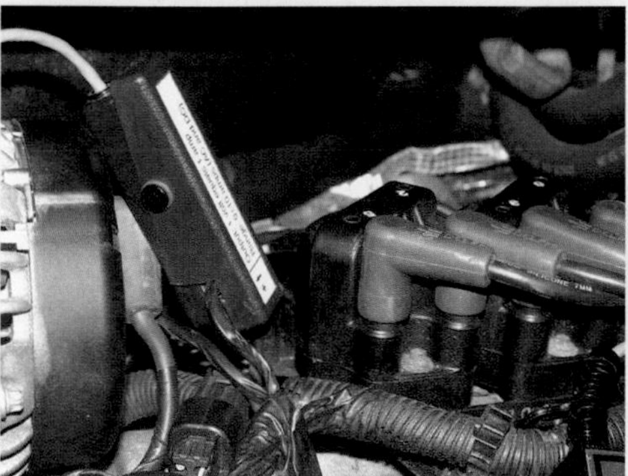

P20–5 Install the probe around the ignition feed at the ignition module.

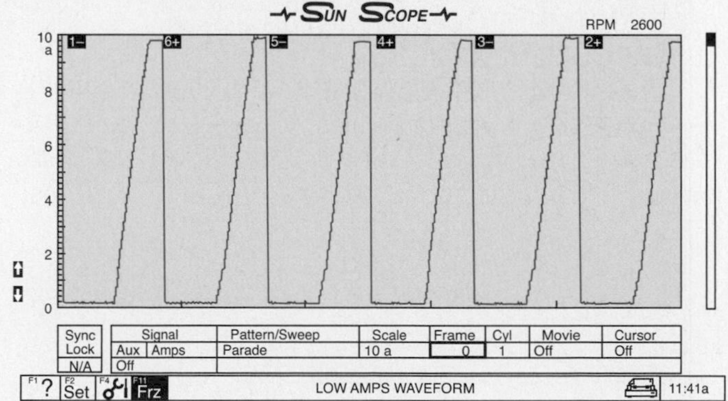

P20–6 Once the amp probe is synchronized to the firing order, primary patterns for all cylinders will appear. The slope on the front edge of the waveform shows the time it took for coil saturation. The flat tops of the waveform result from the module limiting the current to the coil.

timing is not a lean condition, but the PCM cannot tell that the timing is wrong. It only knows there is too much O_2 in the exhaust. Under this condition, the waveform from the O_2 sensor will be lean biased.

SHOP TALK

Observing the waveform of an O_2 sensor can help in the diagnosis of engine performance problems. **Figure 26–31** shows how ignition problems affect the signal from the O_2 sensor. Keep in mind that during complete combustion, nearly all of the O_2 in the combustion chamber is combined with the fuel. This means there will be little O_2 in the exhaust. As combustion becomes more incomplete, the levels of O_2 increase.

Excessive O_2 in the exhaust will also show up on an exhaust gas analyzer. With incorrect timing, you should also see higher than normal amounts of HC. Keep in mind that it takes approximately seven seconds for the exhaust to be analyzed. If you slowly accelerate the engine and see the HC and O_2 levels on the exhaust gas analyzer rise, the condition that existed seven seconds earlier was the cause of the rise in emissions levels. To make this easier to track, make sure you hold the engine at each test speed for at least seven seconds. This way you will be able to observe the rise (or fall) of the emissions levels at that particular speed.

Incorrect ignition timing will also affect manifold vacuum readings and ignition system waveforms on a scope. When anything indicates a problem with the primary ignition circuit, the suspected parts should be tested. Symptoms of overly advanced timing include pinging or engine knock. Insufficient advanced or retarded timing at higher engine speeds could cause hesitation and poor fuel economy.

Setting Ignition Timing

Only engines equipped with a distributor may need to have their ignition timing set or adjusted. On these systems, the correct base timing is critical for the proper operation of the engine. Because the computer bases its control to the base timing setting, all other ignition timing settings will also be wrong. On DI systems, the base timing is adjustable. On others, if the base timing is wrong, the ignition module, distributor, or PCM may need to be replaced.

Each ignition system has its own set of procedures to check ignition timing; always refer to the vehicle's emissions underhood label or the appropriate service information before proceeding. These give the correct procedure for disabling the computer's control of the timing, the correct timing specifications, and the conditions that must be present when checking or adjusting base ignition timing.

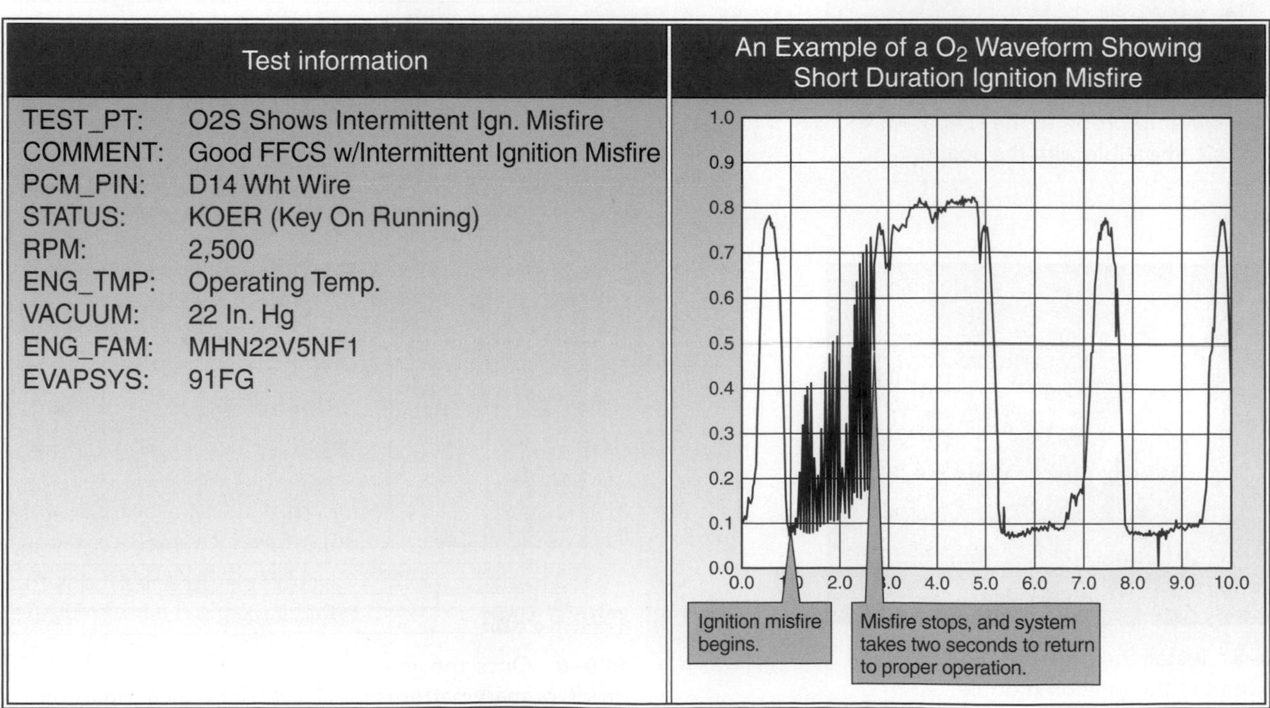

FIGURE 26–31

An O_2 sensor signal caused by an ignition problem.

To check the ignition timing, a timing light is aimed at the ignition timing marks. The timing marks are usually located on the crankshaft pulley or on the flywheel. A stationary pointer, line, or notch is positioned above the rotating timing marks. The timing marks are lines on the crankshaft pulley **(Figure 26–32)** or flywheel that represent various positions of the piston as it relates to TDC. When piston 1 is at TDC, the timing line or notch will line up with the zero reference mark on the timing plate. Usually, an engine is timed so that the number 1 spark plug fires several degrees BTDC. The timing light flashes every time the number 1 spark plug fires. When pointed at the timing marks, the strobe of the light will freeze the spinning timing marks as it passes the timing scale. The ignition timing is checked by observing the degrees of crankshaft rotation (BTDC or ATDC) when the spark plug fires.

Spark plug gap and idle speed must be correct before checking or setting ignition timing. Also, the engine must be at operating temperature. After you have a base timing reading, compare it

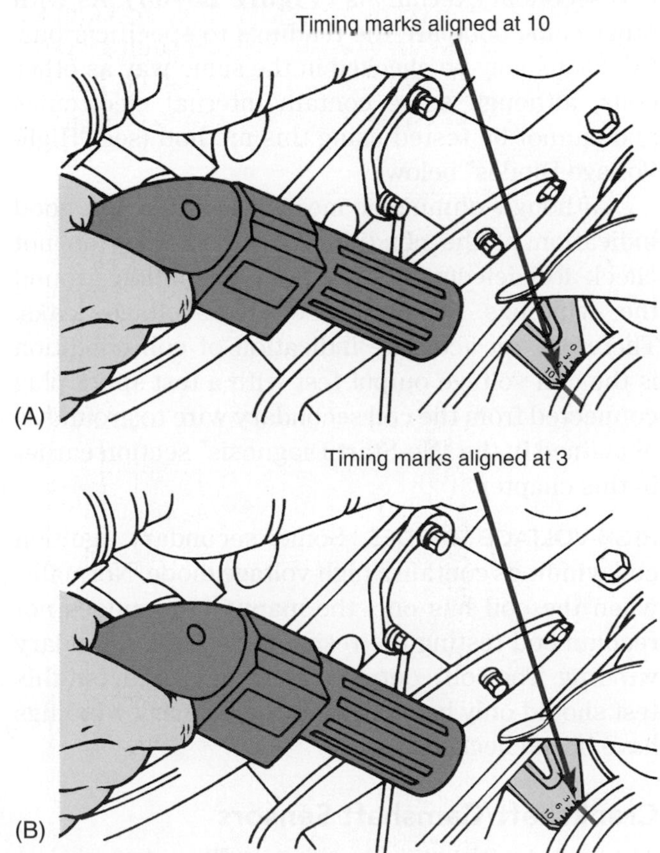

FIGURE 26–32

(A) Timing marks illuminated by a timing light at 10° BTDC and (B) timing marks at 3° BTDC.

to the specifications. As an example, if the specification calls for 10° before TDC and your reading was 3° before TDC, the timing is retarded 7°. This means the timing must be advanced by 7°. To do this, rotate the distributor until the timing marks align at 10°. Then retighten the distributor hold-down bolt.

BASIC PRIMARY CIRCUIT COMPONENTS

The primary circuit has the responsibility of controlling the action of the secondary, and in many cases, it is the cause of starting problems. The components in the primary ignition circuit vary with manufacturer and design. It is important to identify the components of the circuit and the appropriate test methods. Always work systematically through a circuit, testing each wire, connector, and component. Do not jump from one component to another. It is possible that the component inadvertently overlooked is the one causing the trouble. Always compare the readings with specifications given by the manufacturer.

Ignition Switch

On many systems, the ignition switch supplies voltage to the ignition control module and/or the ignition coil. Often, an ignition system has two wires connected to the RUN terminal of the ignition switch. One is connected to the module; the other is connected to the primary side of the coil. The START terminal of the switch is also wired to the module. On newer systems, the ECM often supplies power to the coil primary terminal once the ignition is turned on or the engine is cranked.

You can check for voltage using either a 12-volt testlight or a DMM. To use a testlight, turn the ignition key off and disconnect the wire connector at the module. Turn the key to the RUN position and probe the power wire connection to check for voltage. Also check for voltage at the battery terminal of the ignition coil using the testlight.

Next, turn the key to the START position and check for voltage at the module and the battery terminal of the ignition coil. If voltage is present, the switch and its circuit are okay.

To make the same test using a DMM, turn the ignition switch to the OFF position and back-probe, with the meter's positive lead, the power feed wire at the module. Connect the meter's negative to a good ground. Turn the ignition to the RUN or START position as needed and measure the voltage. The reading should be at least 90 percent of battery voltage.

Ignition Coil Resistance

Ignition coils, like all parts that contain electrical windings, can be checked with an ohmmeter. In an ignition coil, there are two separate windings and each has a different resistance value. This is due to the wire size and the number of windings. Always refer to the specifications prior to testing a coil. If a measurement is not within specifications, the coil or coil assembly should be replaced. It is important to remember that a resistance test such as this can confirm whether a part is faulty but should not be relied upon to prove a part is good. This is because measuring the resistance of the coil windings is a static test, without the coil being under any load or stress. The winding may test good this way but fail in operation as current flows through the winding and heat is generated.

To check the primary windings, set the ohmmeter to the auto-range mode and connect the meter across to the primary coil (BAT and TACH or + and –) terminals **(Figure 26–33)**. An infinite ohmmeter reading indicates an open winding. Higher than normal readings indicate the presence of excessive resistance. If the measurement is less than the specified resistance, the windings are shorted.

To check the secondary winding, connect the meter between the coil's secondary terminal and the positive (BAT) terminal of the coil **(Figure 26–34)**. A meter reading below the specified resistance indicates a shorted secondary winding. An infinite meter reading indicates that the winding is open. Higher than normal readings indicate the presence of excessive resistance.

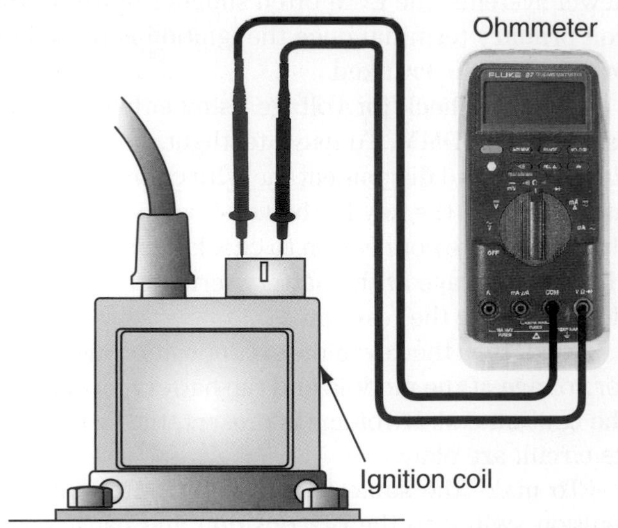

FIGURE 26–33

An ohmmeter connected to primary coil terminals.

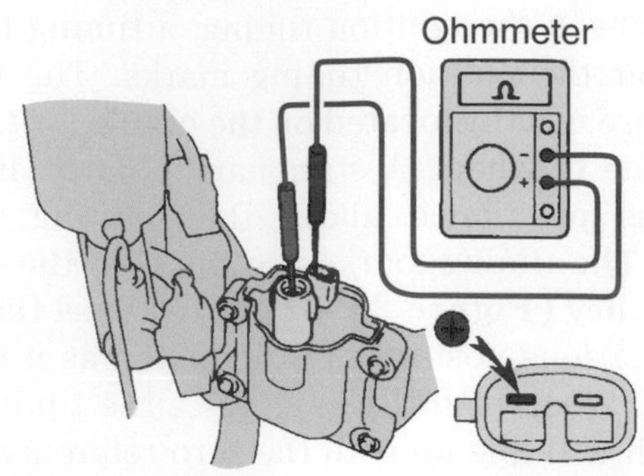

FIGURE 26–34

An ohmmeter connected from one primary terminal to the coil tower to test secondary winding.

The secondary windings of a waste spark ignition coil are not checked in the same way as other coils. Each coil has two secondary terminals. The coil is checked by connecting the meter across the two secondary terminals **(Figure 26–35)**. As with other coils, compare the readings to specifications. COP coils may be checked in the same way as other coils, although some contain internal electronics and cannot be tested using this method (see "High-Voltage Diodes" below).

Although ohmmeter measurements are a good indication of the condition of a coil, they do not check for defects such as poor insulation around the windings, which causes high-voltage leaks. Therefore, an accurate indication of coil condition is the coil voltage output test with a test spark plug connected from the coil secondary wire to ground as explained in the "No-Start Diagnosis" section earlier in this chapter.

HIGH-VOLTAGE DIODES Some secondary ignition coil windings contain a high-voltage diode. Normally, when the coil has one, the manufacturer does not recommend testing the resistance of the secondary winding. The coil's output can be measured, but this test should only be done after the primary windings have been checked.

Crankshaft/Camshaft Sensors

Although ignition timing on most EI systems is not adjustable, the air gaps at the crankshaft and camshaft sensors will affect the operation of the ignition system. On some engines, this gap is adjustable, and

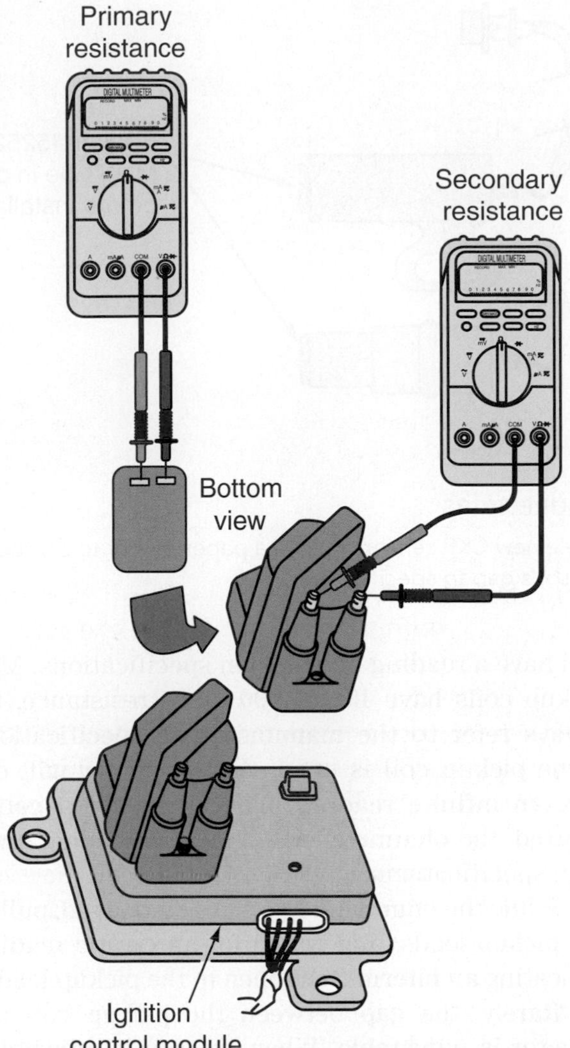

FIGURE 26–35

Meter connections for testing the resistance of a double-ended ignition coil.

on others, it is an indication that the sensor should be replaced. If there is no provision for adjusting the gap and the gap is incorrect, the sensor should be replaced.

When checking the gap, make sure there are no signs of damage to the rotating vane assembly. Measure the gap with a nonmagnetic feeler gauge. Compare your findings to specifications. If the gap is not correct, the sensor should be replaced or the gap adjusted.

The sensors can also be checked with an ohmmeter. Connect the negative lead to the ground terminal at the sensor, and measure the resistance between that terminal and the others, one at a time. The resistance should be within the specified range; if not, replace the sensor.

Also, the action of the sensors can be monitored with an AC voltmeter or a lab scope connected

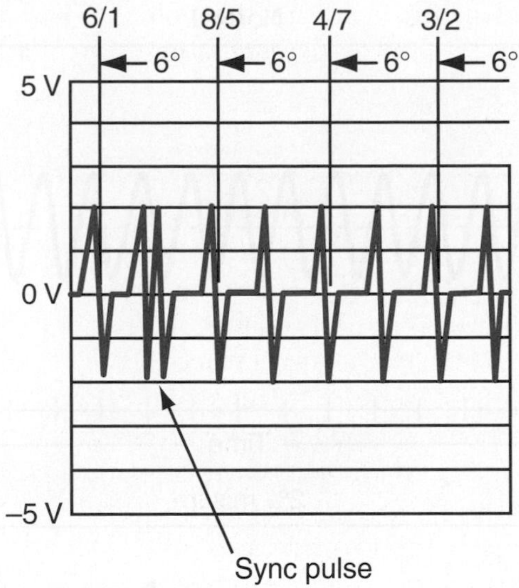

FIGURE 26–36

A waveform for a nine-slot trigger wheel for the crankshaft sensor.

across the sensor terminals to check the sensor signal while the engine is cranking. Meter readings below the specified value indicate a shorted sensor winding, whereas infinite meter readings prove that the sensor winding is open. On a lab scope, the pattern should be smooth, and all peaks, with the exception of the sync pulse, should be at the same height **(Figure 26–36)**.

SHOP TALK

While observing the wave patterns of a CKP, remember that the pattern can be altered by engine misfiring **(Figure 26–37)**.

On some engines, the crankshaft sensor is mounted inside the engine. These sensors are continuously splashed with engine oil. Often, these sensors fail because engine oil enters the sensor and shorts out the winding.

A quick check of the condition of the permanent magnet in a sensor can be made by placing a flat steel tool (such as the blade of a feeler gauge set) near the sensor. It should be attracted to the sensor if the sensor's magnet is okay.

If a crankshaft or camshaft sensor needs to be replaced, always clean the sensor tip and install a new spacer (if so equipped) on the sensor's tip. New sensors typically have a spacer already installed on the sensor **(Figure 26–38)**. Install the sensor until the spacer lightly touches the sensor ring, and tighten the sensor mounting bolt.

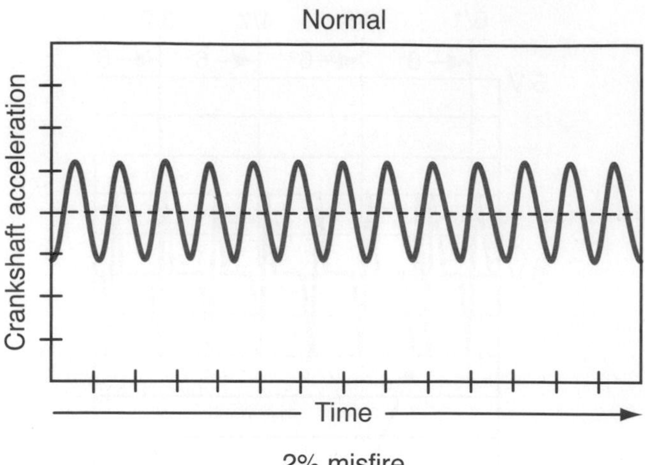

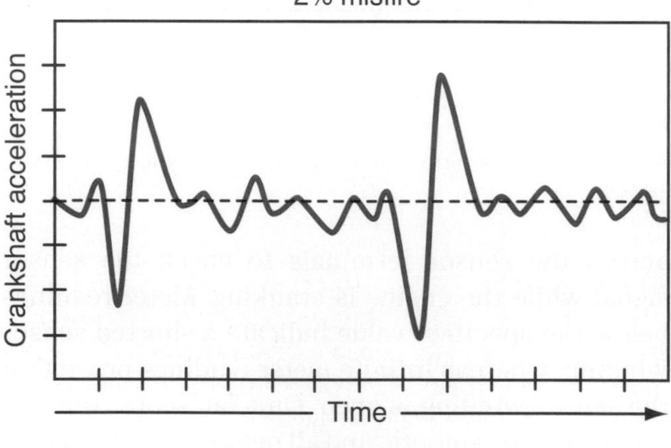

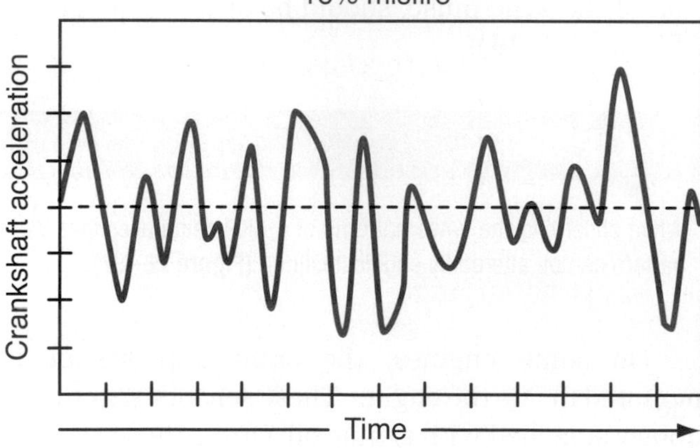

FIGURE 26–37

The reaction of a CKP's waveform to misfires.

Pickup Coils

It is important that all wires and connectors between the trigger sensor and ignition module and module to the PCM be visually checked as well as checked for excessive resistance with an ohmmeter.

To test magnetic pulse and metal detection pickup coils inside a distributor or used as a CKP or CMP with an ohmmeter, connect the ohmmeter from one of the pickup leads to ground to test for a short to ground. If there is a short to ground, the ohmmeter will have a reading of less than specifications. Most pickup coils have 150 to 900 ohms resistance, but always refer to the manufacturer's specifications. If the pickup coil is open, the ohmmeter will display an infinite reading. When the pickup coil is shorted, the ohmmeter will display a reading lower than specifications.

While the ohmmeter leads are connected, pull on the pickup leads, and watch for an erratic reading, indicating an intermittent open in the pickup leads.

Rarely, the gap between the pickup coil and reluctor is adjustable. When the gap is adjustable, it should be measured with a nonmagnetic feeler gauge placed between the reluctor high points and the pickup coil. If adjustment is required, loosen the pickup and move it until the specified air gap is obtained. Retighten the pickup coil bolts to the specified torque. Most pickup coils are riveted or integral to the pickup plate. A pickup gap adjustment is not required for these pickup coils.

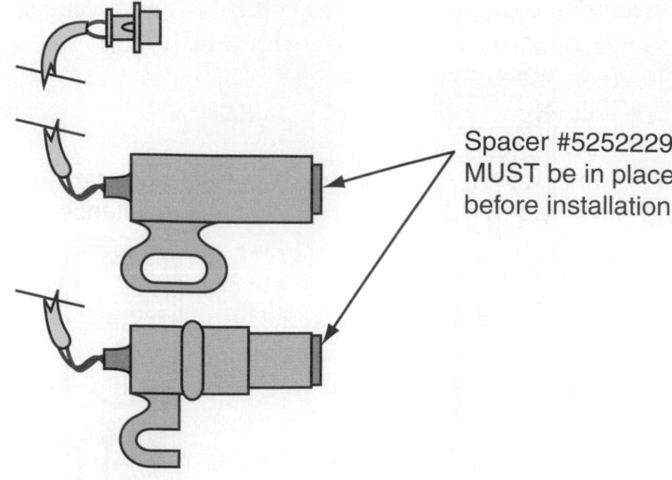

FIGURE 26–38

Some new CKP sensors rely on a paper spacer to set the sensor's gap to specifications.

VOLTMETER CHECKS If the resistance is within specifications, the circuit should be checked with a voltmeter. Disconnect the sensor or back-probe the terminals at the connector. Set the meter to measure AC volts and connect the meter leads to the sensor. Crank the engine and note the voltage. A varying AC voltage should be displayed. At cranking speeds, the sensor should produce at least 200 mVAC.

Hall-Effect Sensors

With the voltmeter hooked up, insert a steel feeler gauge or knife blade between the Hall layer and magnet. If the sensor is good, the voltmeter should read within 0.5 volt of battery voltage when the feeler gauge or knife blade is inserted and touching the magnet. When the feeler gauge or blade is removed, the voltage should read less than 0.5 volt.

In the following tests, the distributor connector is connected to the unit and the connector back-probed. With the ignition switch on, a voltmeter should be connected from the voltage input wire to ground. The specified voltage should appear on the meter. The ground wire should be tested with the ignition switch on and a voltmeter connected from the ground wire to a ground connection near the distributor. With this meter connection, the meter indicates the voltage drop across the ground wire, which should not exceed 0.2 volt.

Connect a digital voltmeter from the pickup signal wire to ground. If the voltmeter reading does not fluctuate while cranking the engine, the pickup is defective. However, if the voltmeter reading fluctuates from nearly 0 volt to between 9 and 12 volts, that indicates a satisfactory pickup. During this test, the voltmeter reading may not be accurate because of the short duration of the voltage signal. If the Hall-effect pickup signal is satisfactory and the 12-volt test lamp did not flutter during the no-start test, the ignition module is probably defective.

Using a Logic Probe

The primary can also be checked with a logic probe. There are three lights on a logic probe. The red light illuminates when the probe senses more than 10 volts. When there is a good ground, the green light turns on. The yellow light flashes whenever the voltage changes. This light is used to monitor a pulsing signal, such as one produced by a digital sensor like a Hall-effect switch.

To check the primary circuit with a logic probe, turn the ignition on. Touch the probe to both (positive and negative) primary terminals at the coil. The red light should come on at both terminals **(Figure 26–39)**, indicating that at least 10 volts are available to the coil and that there is continuity through the coil. If the red light does not come on when the positive side of the coil is probed, check the power feed circuit to the coil. If the light comes on at the positive terminal, but not at the negative, the coil has excessive resistance or is open.

Now move the probe to the negative terminal of the coil and crank the engine. The red and green lights should alternately flash, indicating that over 10 volts are available to the coil while cranking and that the circuit is switching to ground. If the lights

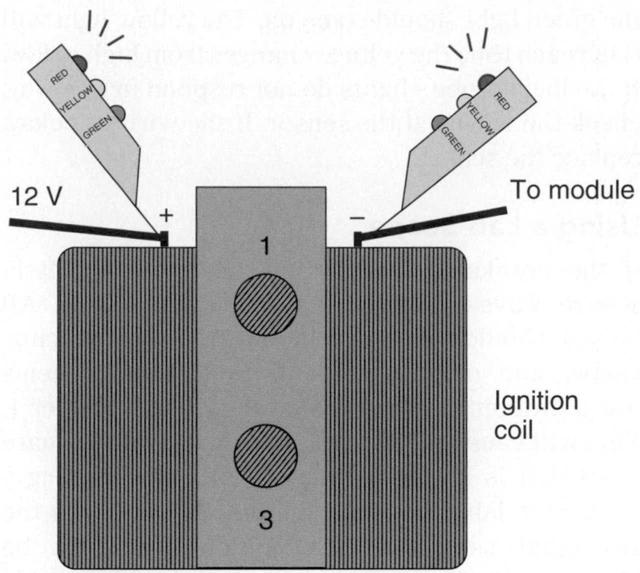

FIGURE 26–39

The red light on a logic probe should turn on when the probe is touched to both sides of the primary winding.

do not come on, check the ignition power feed circuit from the starter. If the red light comes on but the green light does not, check the crankshaft or camshaft sensor. If these are working properly, the ignition module is probably defective.

A Hall-effect switch is also easily checked with a logic probe. If the switch has three wires, probe the outer two wires with the ignition on **(Figure 26–40)**. The red light should come on when one of the wires is probed, and the green light should come on when the other wire is probed. If the red light does not turn on at either wire, check the power feed circuit to the sensor. If the green light does not come on, check the sensor's ground circuit.

Back-probe the centre wire and crank the engine. All three lights should flash as the engine is cranked. The red light will come on when the sensor's output is above 10 volts. As this signal drops below 4 volts,

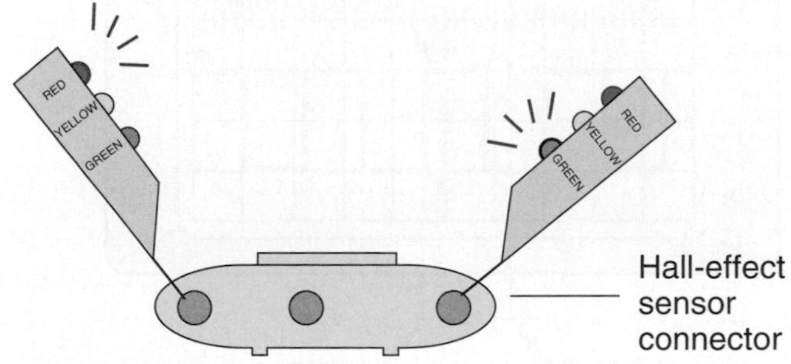

FIGURE 26–40

One end terminal of a Hall-effect sensor connector should cause the red light to come on; the other end terminal should cause the green light to come on.

the green light should come on. The yellow light will flash each time the voltage changes from high to low. If the logic probe's lights do not respond in this way, check the wiring at the sensor. If the wiring is okay, replace the sensor.

Using a Lab Scope

If the crankshaft sensor is a Hall-effect switch, square waves should be seen on the scope. All pulses should normally be identical in spacing, shape, and amplitude. Note that some systems use a different size shutter to designate cylinder 1. This will cause the waveform to display one square wave that is different than the others. By using a dual-trace lab scope, the relationship between the crankshaft sensor and the ignition module can be observed. During starting, the module will provide a fixed amount of timing advance according to its program and the cranking speed of the engine. By observing the crankshaft sensor output and the ignition module, this advance can be observed **(Figure 26–41)**. The engine will not start if the ignition module does not provide for a fixed amount of timing advance.

Knock Sensors

Most systems have knock sensors that retard timing when the engine is experiencing ping or knocking. A quick check of a knock sensor is made by watching ignition timing while the engine is running and the engine block is tapped with the handle of a screwdriver. The noise should cause a change in timing.

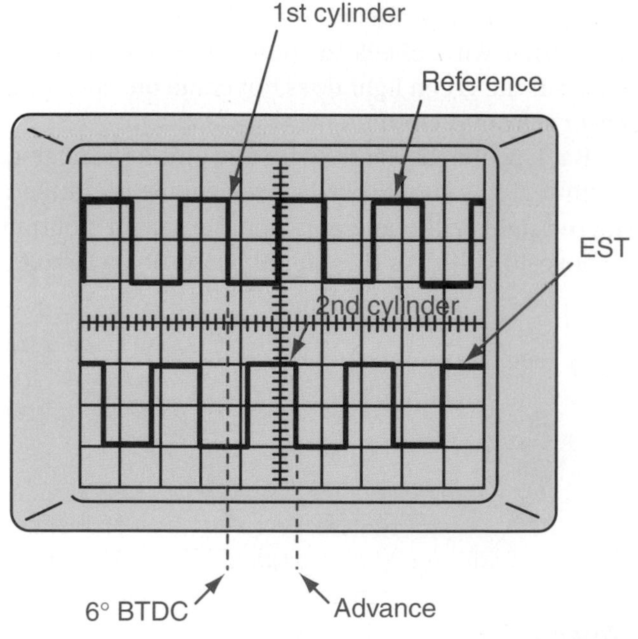

FIGURE 26–41
Electronic spark timing (EST) and crankshaft sensor signals compared on a dual-trace scope.

Control Module

The most effective way to test a control module is to use an ignition module tester. This tester evaluates and determines whether the module is operating within a given set of parameters. It does this by simulating normal operating conditions while looking for faults in the module's components. If a module tester is not available, check out all other system components before condemning the control module. Test for power and ground to the module as well as the signal from the CKP sensor. It is possible that a weak signal from the CKP may not trigger the module, resulting in the module not triggering the coil.

DISTRIBUTOR SERVICE

Typically, the distributor will need to be removed to service it. Before removing it, grasp the distributor shaft and move it from side to side. If any movement is detected, remove the distributor and check the bushing. Lightly clamp the distributor in a soft-jaw vise. Clamp a dial indicator on the top of the distributor housing. Position the plunger of the indicator so that it rests on the distributor shaft. When the shaft is pushed horizontally, watch the movement of the indicator. Compare this to the specifications given in the service information. If the movement exceeds the allowed amount, the distributor bushings and/or shaft are worn. Many manufacturers recommend complete distributor replacement rather than bushing or shaft replacement.

A typical procedure for removing a distributor follows:

PROCEDURE

To Remove a Distributor

STEP 1 Disconnect the electrical connector and the vacuum advance hose, if the distributor has them.

STEP 2 Remove the distributor cap and mark the position of the rotor. On some vehicles, it may be necessary to remove the spark plug wires from the cap prior to cap removal.

STEP 3 Make a mark to index the distributor body to another component for proper reinstallation; then remove the distributor hold-down bolt and clamp.

STEP 4 Pull the distributor from the engine. Most distributors will need to be twisted as they are pulled out of their bore. Note the direction of rotation.

STEP 5 Once the distributor is removed, install a shop towel in the distributor opening to keep foreign material out of the engine block.

SECONDARY CIRCUIT TESTS AND SERVICE

A typical procedure for installing and timing the distributor follows:

PROCEDURE

To Install the Distributor and Time It to the Engine

STEP 1 Lubricate the O-ring on the distributor shaft.

STEP 2 Position the rotor so that it is aligned with the mark made to the distributor housing prior to removal.

STEP 3 Align the distributor to the mark made on the engine block during removal.

STEP 4 Lower the distributor into the engine block; make sure the distributor drive is fully seated. Distributors equipped with a helical drive gear will rotate as the distributor is being installed, causing the distributor to move away from the reference marks. Pay attention to how much the rotor moves; then remove the distributor and move the rotor backward the same amount. This should allow the shaft to rotate while the distributor is being installed and still be aligned with the reference marks.

STEP 5 Make sure the distributor housing is fully seated against the engine block; sometimes, it may be necessary to wiggle or rock the distributor to seat it fully into the drive gear. Distributors with drive lugs must be mated with the drive grooves in the camshaft. Both are offset to eliminate the possibility of installing the distributor 180° out of time.

STEP 6 Rotate the distributor a small amount so that the timer core teeth and pickup teeth are aligned.

STEP 7 Install the distributor hold-down clamp and bolt **(Figure 26–42)**, and leave the bolt slightly loose.

STEP 8 Install the spark plug wires in the direction of distributor shaft rotation and in the correct cylinder firing order.

STEP 9 Connect the wiring for the distributor. The vacuum advance hose is usually left disconnected until the timing is set with the engine running.

FIGURE 26–42

The hold-down clamp and bolt for a distributor.

A coil secondary winding and spark plug are found in all ignition systems. DI, distributorless, and coil-near-cylinder systems also have spark plug wires. DI systems also have a distributor cap and rotor and sometimes a lead for the ignition coil.

Distributor Cap and Rotor

Because there is an air gap between the rotor and the distributor cap, electrical arcing takes place between the two. This causes a deterioration of the rotor tip and the distributor cap terminals. This added resistance could cause cylinder misfires.

When installing a new cap or rotor, it is always wise to replace both at the same time. Make sure both are fully seated before attempting to start the engine. Also, remember that the spark plug wires need to be arranged on the cap according to the firing order **(Figure 26–43)**.

Waste Spark Systems

Standard test procedures using an oscilloscope, an ohmmeter, and a timing light can be used to diagnose problems in distributorless ignition systems. Keep in mind, however, that problems involving one cylinder may also occur in its companion cylinder that fires off the same coil. Follow the testing

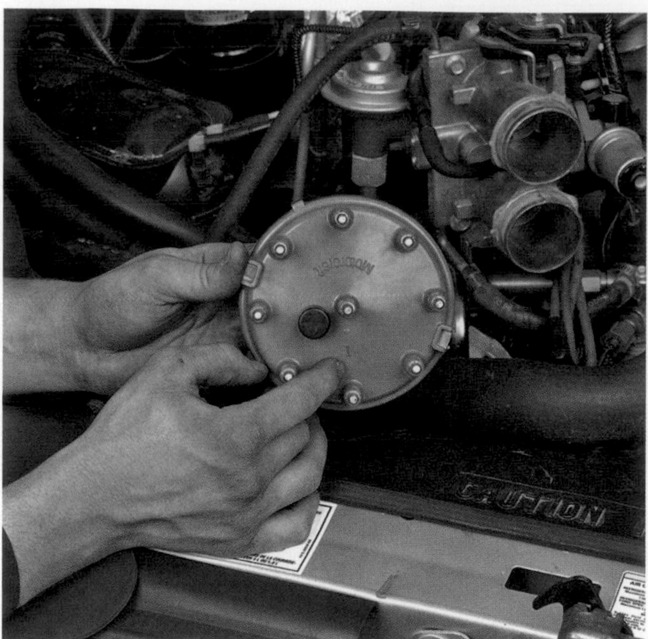

FIGURE 26–43

Before installing the spark plug wires to the distributor cap, identify the terminal for cylinder 1, and then follow the engine's firing order.

procedures given in the service information. Specific DTCs are designed to help troubleshoot ignition problems in these systems. The diagnostic procedure for EI systems varies depending on the vehicle make and model year.

There is a separate primary circuit for each coil. If one coil does not work properly, it may be caused by something common or not common to the other coils. Regardless of the system's design, there are common components in all electronic ignitions: an ignition module, crankshaft and/or camshaft sensors, ignition coils, a secondary circuit, and spark plugs. Most of these components are common to only the spark plug or spark plugs to which they are connected. The components that are common to all cylinders (such as the camshaft sensor) will, most often, be the cause of no-start problems. The other components will cause misfire problems.

Testing the secondary circuit of a DIS system is just like testing the secondary of any other type of ignition system. The spark plug wires and spark plugs should be tested to ensure that they have the appropriate amount of resistance. Because the resistance in the secondary dictates the amount of voltage the spark plug will fire with, it is important that secondary resistance be within the desired range.

A quick way to examine the secondary circuit is with an ignition scope. On the scope, the firing line and the spark plug indicate the resistance of the secondary. While observing the firing line, remember that the height of the line increases with an increase in resistance. The length of the spark line decreases as the firing line goes higher. This means that high resistance will cause excessively high firing voltages and reduced spark times.

When checking a DIS system with a scope, remember that in waste spark systems, half of the plugs fire with reverse polarity. This means half of the firing lines will be higher than the other firing lines. Normally, reverse firing requires 30 percent more voltage than normal firing.

Excessive resistance is not the only condition that will affect the firing of a spark plug. Spark plug wires and the spark plug itself can allow the high voltage to leak and establish current through another metal object instead of the electrodes of the spark plug. When this happens, the spark plug does not fire, and combustion does not take place within the cylinder.

Also, keep in mind that the secondary circuit is completed through the metal of the engine. If the spark plugs are not properly torqued into the cylinder heads, the threads of the spark plug may not make good contact, and the circuit may offer resistance. Always tighten spark plugs to their specified torque. Make sure not to cross-thread them.

Most manufacturers recommend the use of an antiseize compound on spark plug threads. This compound must be applied in the correct amounts and at the correct place. Too little compound will cause gaps in the contact between the spark plug threads and the spark plug bores. Too much may allow the spark to jump to a buildup rather than the spark plug electrode.

Coil-over-Plug Systems

Remember that an individual coil problem will cause misfiring in only one cylinder. Ignition coils are tested with an ohmmeter in much the same way as other ignition coils. If the resistance is out of specifications, the coil should be replaced. Intermittent coil problems can be caused by corrosion at the electrical connectors to the coil. The action of the secondary can also be monitored on a lab scope **(Figure 26–44)**.

Codes retrieved from the PCM will indicate whether the misfire is a general one or an individual cylinder. Because a COP ignition problem will affect only one cylinder, the general misfire code (P0300) is probably caused by a fuel delivery problem or a vacuum leak.

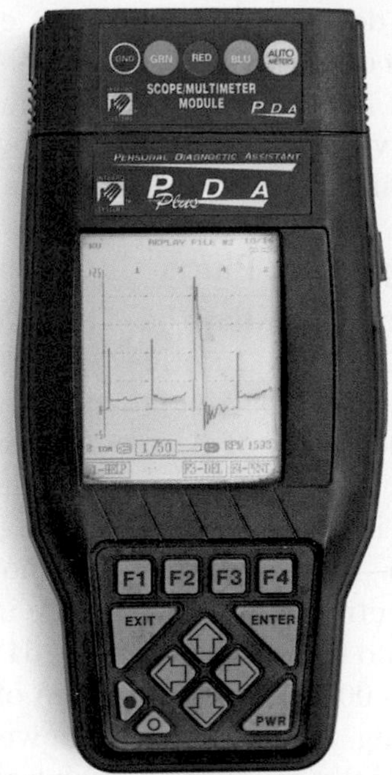

FIGURE 26–44

The scope pattern for a COP system with one faulty cylinder.

DTCs that indicate a misfire at an individual cylinder (P0301, P0302, P0303, etc.) are typically caused by a dirty or defective fuel injector, a fouled spark plug, a bad coil, or an engine mechanical problem. If the misfire is caused by a fuel injector problem, a fuel injector code (P0201, P0202, P0203, etc.) will also be retrieved; these identify the affected cylinder.

If a crankshaft position sensor is bad, there will be no timing reference, and this can prevent the engine from starting or cause it to have a hard time starting. A defective CKP may set a P0335 or P0336. Problems with the sensor connection and wiring may also cause these DTCs to set.

Each ignition coil has a driver circuit in the PCM that controls primary current flow. If there is a bad driver circuit, that spark plug will not fire. Also, keep in mind that an engine may start with a faulty camshaft sensor, but it may only run in the fail-safe or limp-in mode because the fuel injectors cannot be synchronized without the camshaft signal. Faults in the coil primary or secondary circuits or a faulty coil may cause a DTC ranging from P0351 to P0362 to set.

Coil-near-plug systems can be checked with a scope or graphing meter in the same way as other ignition systems. The pickup for secondary signals is installed over the spark plug wire. However, COP systems require special adapters to connect the scope or analyzer to the ignition system. Some low-amperage current probes can also be used to monitor the activity of an individual coil in a COP system. Also, make sure to check the spark plug wires **(Figure 26–45)**.

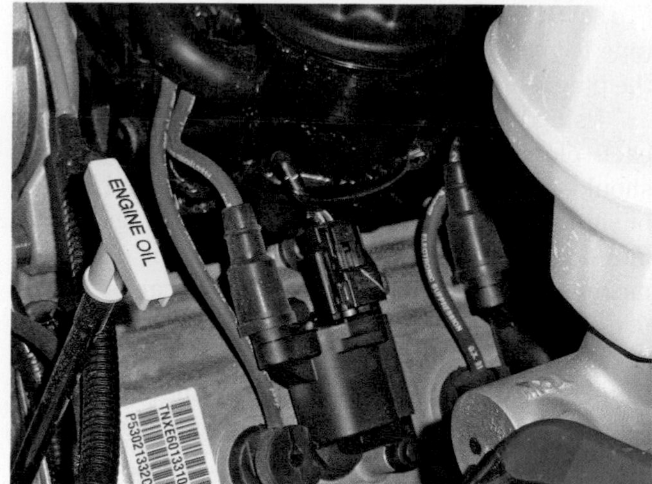

FIGURE 26–45

A coil-near-plug ignition system.

Spark Plugs

The ignition system is designed to do no more than supply the voltage necessary to cause a spark across the gap of a spark plug. This simple event is what starts the combustion process. Needless to say, a healthy spark plug is extremely important to the combustion process. Spark plug replacement is part of the preventive maintenance program for all vehicles. The recommended replacement interval depends on a number of factors but ranges from 32 000 to 160 000 km (20 000 to 100 000 miles).

Removal of an engine's spark plugs is pretty straightforward. Remove the cables from each plug, being careful not to pull on the cables. Instead, grasp the boot **(Figure 26–46)** and gently twist it off.

FIGURE 26–46

Grab hold of the boot and twist while pulling it off a spark plug.

Reprinted with permission.

(To save time and avoid confusion later, use masking tape to mark each of the cables with the number of the plug it attaches to.)

Using a spark plug socket and ratchet, loosen each plug a couple of turns. A spark plug socket should be used because it has an internal rubber bushing to prevent plug-insulator breakage. Spark plug sockets can have either a ⅜- or ½-in. drive, and most have an external hex so that they can be turned using an open-end or box wrench.

Once the plugs are loose, use compressed air to blow dirt away from the base of the plugs. Then remove the plugs, making sure their gaskets have also been removed (if applicable). When the spark plugs are removed **(Figure 26–47)**, they should be set in order so that the spark plug from each cylinder can be examined.

Check the threads in the cylinder head for damage. Normally, you can do this by feel as you remove a spark plug. If the plug does not turn out smoothly after it is loose, the threads may be damaged. Often, the threads can be cleaned up with a spark plug thread chaser. Also, check the threads on the spark plug. Look for damage or metal embedded in the threads, as these are sure signs of problems. If the cylinder head is aluminum, it may be necessary to install a threaded insert into the spark plug bore.

When working on certain Ford V8s with the extended reach spark plugs, special service procedures are often required to remove the spark plugs. Due to the design of the plug, the extended shell often breaks off down in the cylinder head. A special service tool is needed to remove the shell. Follow the Ford service information regarding servicing the spark plugs in 4.6L, 5.4L, and 6.8L 3V engines.

FIGURE 26–47

Use a spark plug socket to remove the plugs.

Inspecting Spark Plugs

Once the spark plugs have been removed, it is important to "read" them. In other words, inspect them closely, noting in particular any deposits on the plugs and the degree of electrode erosion. A normal-firing spark plug will have a minimum amount of deposits on it and will be coloured light tan or grey **(Figure 26–48)**. However, there should be no evidence of electrode burning, and the increase of the air gap should be no more than 0.0254 mm (0.001 in.) for every 16 000 km (10 000 miles) of engine operation. A plug that exceeds this wear should be replaced and the cause of excessive wear corrected **(Figure 26–49)**. Worn or dirty spark plugs may work fine at idle or low speeds, but they frequently fail during heavy loads or higher engine speeds.

It is possible to diagnose a variety of engine conditions by examining the electrodes of the spark

FIGURE 26–48

Normal spark plug.

NGK Spark Plugs Canada Limited

FIGURE 26–49

A worn spark plug.

NGK Spark Plugs Canada Limited

FIGURE 26–50

A cold- or carbon-fouled spark plug.

NGK Spark Plugs Canada Limited

plugs. Ideally, all of the plugs from an engine should look alike. Whenever plugs from different cylinders look different, a problem exists in those cylinders. The following are examples of plug problems and how they should be dealt with.

CARBON FOULING This condition is the result of an excessively rich air/fuel mixture. It is characterized by a layer of dry, fluffy, black carbon deposits on the tip of the plug **(Figure 26–50)**. **Carbon fouling**, also called cold fouling, can also be caused by an ignition fault that causes the spark plug not to fire. If only one or two of the plugs show evidence of fouling, sticking valves are the likely cause. If the plug is cleaned, it can be used again. Correct the cause of the fouling before reinstalling or replacing the plugs.

SHOP TALK

If carbon fouling is found on a vehicle that operates a great deal of the time at idle and low speeds, plug life can be lengthened by using hotter spark plugs.

WET FOULING When the tip of the plug is covered with oil, this condition is known as **wet fouling (Figure 26–51)**. The oil may be entering the combustion chamber through worn valve guides or valve guide seals. On high-mileage engines, check for worn rings or excessive cylinder wear. Before replacing the plugs, correct the mechanical concern.

FIGURE 26–51

A wet- or oil-fouled spark plug.

NGK Spark Plugs Canada Limited

GLAZING Under high-speed conditions, the combustion chamber deposits can form a shiny, yellow glaze over the insulator. When it gets hot enough, this **glazing** acts as an electrical conductor, causing the current to follow the deposits and short out the plug. Because it is virtually impossible to remove glazed deposits, glazed plugs should be replaced.

OVERHEATING This condition is characterized by white or light-grey blistering of the insulator. There may also be considerable electrode gap wear **(Figure 26–52)**. Overheating can result from using too hot a plug, over-advanced ignition timing, detonation, a malfunction in the cooling system, an overly lean air/fuel mixture, using fuel too low in octane, or an improperly installed plug. Overheated plugs must be replaced.

TURBULENCE BURNING When turbulence burning occurs, the insulator on one side of the plugs wears away as the result of normal turbulence in the combustion chamber. As long as the plug life is normal, this condition is of little consequence. However, if the spark plug shows premature wear, overheating can be the problem.

PREIGNITION DAMAGE This is caused by excessive engine temperatures and is characterized by melting of the electrodes or chipping of the electrode tips **(Figure 26–53)**. When this problem occurs, look for the general causes of engine overheating, including over-advanced ignition timing, a burned head gasket, and using fuel too low in octane. Other possibilities include loose plugs or using plugs of the improper heat range. Do not attempt to reuse plugs with preignition damage.

FIGURE 26–52

This spark plug shows signs of overheating.

NGK Spark Plugs Canada Limited

FIGURE 26–53

A spark plug with melting damage possibly caused by preignition.

NGK Spark Plugs Canada Limited

Regapping Spark Plugs

Both new and used spark plugs should have their air gaps set to manufacturer's specifications. Always use round wire gauges when checking and setting the gap **(Figure 26–54)**.

After the gap has been adjusted, make sure that the ground electrode is as horizontal as it can be.

Always check the air gap of a new spark plug before installing it. Never assume the gap is correct just because the plug is new. Do not try to reduce a plug's air gap by tapping the ground electrode on a bench. Use a spark plug gapping tool to bend the ground electrode to its correct height. When doing this, be careful not to contact or put pressure on the centre electrode. This is especially critical with fine wire platinum and iridium plugs. Also, while bending the ground electrode, try to keep it in alignment with the centre electrode.

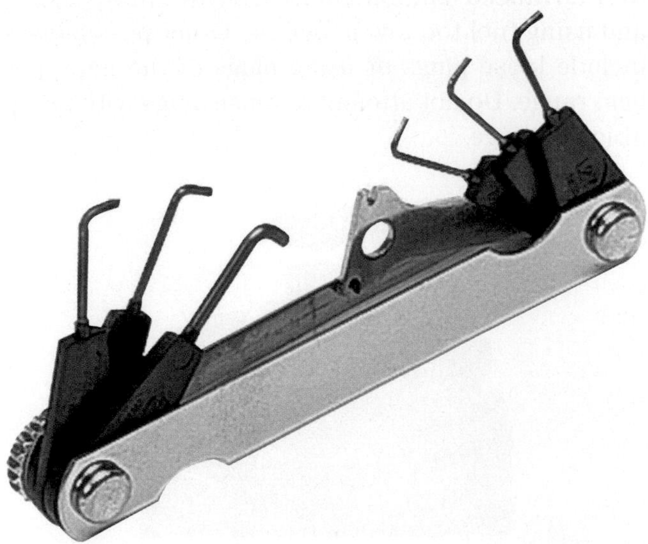

FIGURE 26–54

A round wire-type feeler gauge set.

Snap-on Tools Company

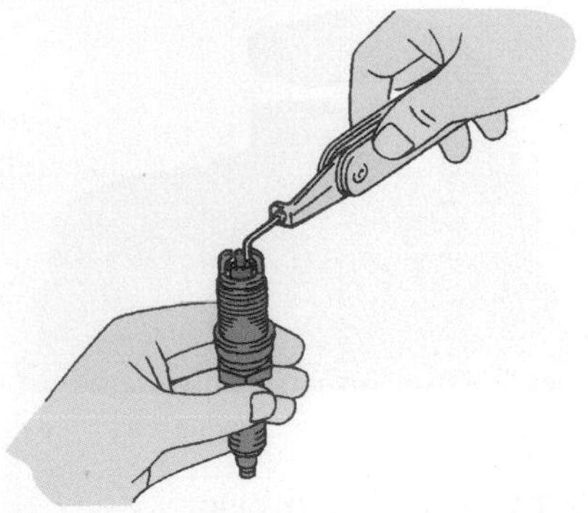

FIGURE 26–55

The gap between the centre electrode and both ground electrodes should be checked and adjusted to specifications.

Reprinted with permission.

Some engines are equipped with spark plugs that have more than one ground electrode. The gap between the centre electrode and each ground electrode should be checked **(Figure 26–55)**. If the gap between the centre electrode and one of the ground electrodes is less than that of the others, spark will occur only at the smallest gap. This is also true of V-shaped ground electrodes. If one leg of the vee is closer to the centre electrode than the other, the spark will always occur across the shortest distance.

The gap of spark plugs with a surface gap and of some with more than one ground electrode cannot be adjusted with conventional tools, and most manufacturers recommend that the gap be left alone.

PROCEDURE

Spark Plug Installation

STEP 1 Wipe dirt and grease from the plug seats with a clean cloth.

STEP 2 Verify that the replacement spark plugs are the correct ones for the engine by matching the part number to its application. Never assume that the plugs that were removed from the engine are the correct type.

STEP 3 Adjust the air gap, as needed.

STEP 4 Check the service information to see if antiseize compound should be applied to the plug's threads **(Figure 26–56)**.

STEP 5 Install the plugs and tighten them with your hand. If the plugs cannot be installed easily by hand, the threads in the cylinder head may need to be cleaned with a thread-chasing tap. Be especially careful not to cross-thread the plugs when working with aluminum heads.

STEP 6 Tighten the plugs with a torque wrench, following the vehicle manufacturer's specifications or the values shown in **Figure 26–57**.

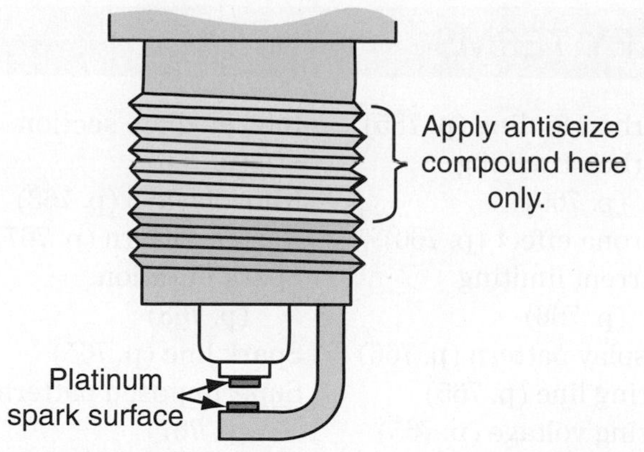
Apply antiseize
compound here
only.
Platinum
spark surface

FIGURE 26–56

Proper placement of antiseize compound on the threads of a spark plug.

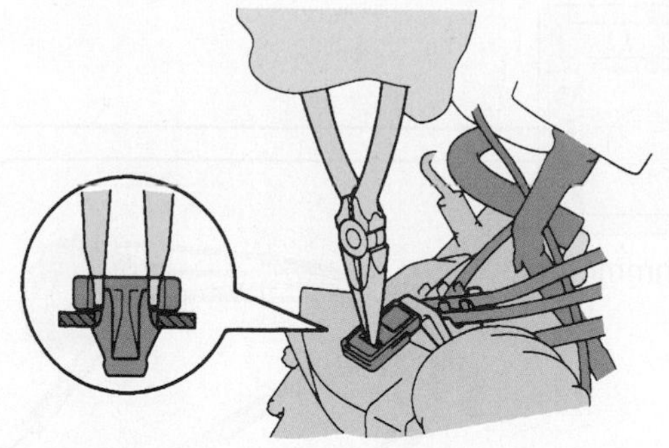

FIGURE 26–58

Use needle-nose pliers to remove the wires from the spark plug cable protector.

Secondary Ignition Wires

Inspect all the spark plug wires and the secondary coil wire for cracks and worn insulation, which cause high-voltage leaks. Inspect all the boots on the ends of the plug wires and coil secondary wire for cracks and hard, brittle conditions.

Make sure that the secondary cables are secured tightly to the spark plugs and the ignition coil(s). Also check the cables for any damage or signs of arcing. Replace the wires and boots if they show evidence of these conditions. Most manufacturers recommend that spark plug wires be replaced as a complete set.

SHOP TALK

Many ignition systems use locking tabs to secure the spark plug cable to the ignition coil. To remove the cable from the coil, squeeze the locking tabs with needle-nose pliers **(Figure 26–58)** or use a screwdriver to lift up the locking tab. When reconnecting the cable to the coil, make sure that the locking tabs are in place by pressing down on the centre of the cable terminal.

RESISTANCE CHECKS The resistance of secondary cables can be checked with an ohmmeter. Do this by removing the cable and measuring across the cable. On DI systems, the spark plug wires may be left in the distributor cap when measuring resistance **(Figure 26–59)**. This will also check the cap-to-cable connections. Set the ohmmeter to the X1,000 scale and connect the ohmmeter leads from the end of a spark plug wire to the appropriate terminal inside the distributor cap.

If the ohmmeter reading is more than specified by the manufacturer, remove the wire from the cap and check the wire alone. If the wire has more resistance than specified, replace the wire. When spark plug wire resistance is satisfactory, check the cap terminal for corrosion. Repeat the ohmmeter tests on each spark plug wire and the coil secondary wire.

Replacing Spark Plug Wires

When spark plug wires are being installed, make sure they are routed properly, as indicated in the vehicle's service information. When removing the spark plug wires from a spark plug, grasp the spark plug boot tightly, and twist while pulling the cable from the end of the plug. When installing a spark plug wire, make sure the boot is firmly seated around the top of the

SPARK PLUG TYPE	THREAD DIAMETER	CAST-IRON HEAD (N · m [lb · ft])	ALUMINUM HEAD (N · m [lb · ft])
Flat seat w/gasket	18 mm	33.9–44.7 (25–33)	33.9–44.7 (25–33)
Conical seat/no gasket	18 mm	19.0–29.8 (14–22)	19.0–29.8 (14–22)
Flat seat w/gasket	14 mm	24.4–33.9 (18–25)	24.4–29.8 (18–22)
Flat seat w/gasket	12 mm	13.6–24.4 (10–18)	13.6–20.3 (10–15)
Conical seat/no gasket	14 mm	13.6–24.4 (10–18)	9.5–20.3 (7–15)

FIGURE 26–57

Typical spark plug torque specifications.

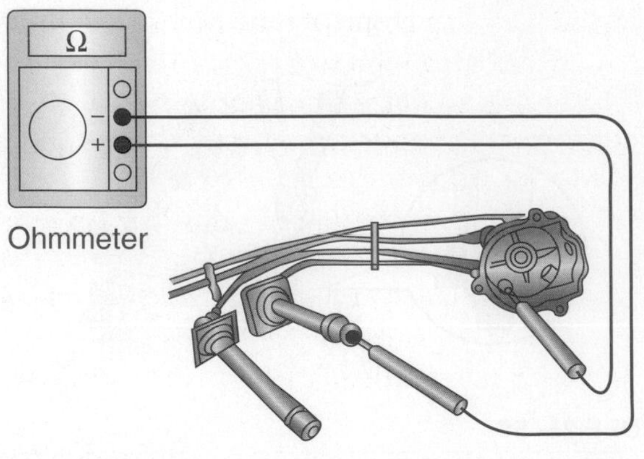

FIGURE 26–59

Check the resistance of each spark plug cable between both ends of the cable.

plug, then squeeze the boot to expel any air that may be trapped inside.

Two spark plug wires should not be placed side by side for a long span if these wires fire one after the other in the cylinder firing order. When two spark plug wires that fire one after the other are placed side by side for a long span, the magnetic field from the wire that is firing builds up and collapses across the other wire. This magnetic collapse may induce enough voltage to fire the other spark plug and wire when the piston in this cylinder is approaching TDC on the compression stroke. This action may cause detonation and reduced engine power.

Also make sure that the wires are secure in their looms and that the looms are properly placed **(Figure 26–60)**.

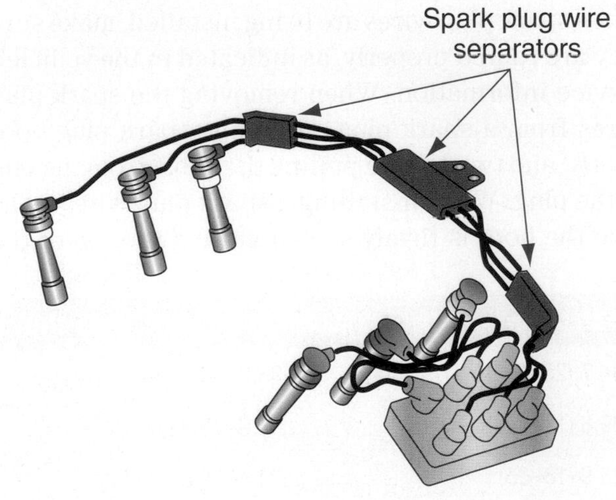

FIGURE 26–60

Make sure the wire separators are in place when reconnecting the spark plug wires.

KEY TERMS

Carbon fouling (p. 785)
Carbon tracking (p. 760)
Corona effect (p. 760)
Current limiting (p. 766)
Display pattern (p. 766)
Firing line (p. 765)
Firing voltage (p. 765)
Glazing (p. 785)
Intermediate section (p. 766)
Kilovolt (kV) (p. 763)
Raster pattern (p. 767)
Spark duration (p. 768)
Spark line (p. 765)
Superimposed pattern (p. 767)
Wet fouling (p. 785)

SUMMARY

- Secure wiring and connections are important to ignition systems. Loose connections, corrosion, and dirt can adversely affect performance.
- Wires, connections, and ignition components can be tested for intermittent failure by wiggling them or stress testing by applying heat, cold, or moisture.
- A scope provides a visual representation of voltage changes over time.
- Waveforms can be viewed in different modes and scales on a scope. Secondary and primary ignition circuits can be viewed.
- Ignition patterns can be broken down into three main sections or zones: firing section, intermediate section, and dwell section.
- The firing line and spark line display firing voltage and spark duration.
- The intermediate section shows coil voltage dissipation.
- The dwell section shows the activation of primary coil current flow and primary coil current switch off. The primary current off signal is also the firing line for the next cylinder in the firing order.
- Current-limiting ignition systems saturate the ignition coil very quickly with high current flow and then cut back or limit current flow to maintain saturation. This system extends coil life.
- Precautions must always be taken to avoid open circuits during ignition system testing. A special test plug is used to limit coil output during testing. Always use the correct test plug for the system.

- Firing voltages are normally between 7 and 13 kV with no more than 3 kV variation between cylinders.
- High secondary circuit resistance produces a higher than normal firing line and shorter spark lines.
- Individual ignition components are commonly tested for excessive internal resistance, using an ohmmeter. A voltmeter or scope can also be used to monitor their operating voltages.
- Proper spark plug gapping and installation are important to ignition system operation. Spark plug condition, such as cold fouling, wet fouling, and glazing, is often a good indication of other problems.
- Standard test procedures using an oscilloscope, GMM, and/or DMM can be used to diagnose problems in EI and DI systems.
- Often, if a crank or cam sensor fails, the engine will not start. These sensor circuits can be checked with a voltmeter. If the sensors are receiving the correct amount of voltage and have good low-resistance connections, their output should be a square wave or a pulsing analog signal while the engine is cranking.
- The resistance of some COP ignition coils can be checked in the same way as conventional coils; however, different meter connections are required to test waste spark coils.

REVIEW QUESTIONS

1. Name the three types of stress testing used to test for intermittent ignition component problems and list the procedures for conducting each type of test.
2. Why is the procedure for checking the resistance of a waste spark ignition coil different from the procedures for checking other types of ignition coils?
3. Name the three types of trace pattern display modes used on an oscilloscope and give examples of when each mode is most useful.
4. List the common types of spark plug fouling and the typical problems each type of fouling indicates.
5. List at least two methods of checking the operation of Hall-effect sensors.
6. What happens if one of the ground electrodes of a spark plug with two or more electrodes is closer to the centre electrode than the other?
   a. The spark would occur at the electrodes with the larger gap.
   b. More voltage would be required to produce a spark to the closer electrode.
   c. Less voltage would be required to produce a spark to the further electrodes.
   d. The spark would occur at the electrode with the least gap.
7. When diagnosing an ignition system problem that has not set a DTC, what should be checked first?
   a. input signals
   b. output signals
   c. mode $06 data
   d. individual component measurements
8. Which of the following would a leaner air/fuel mixture produce?
   a. Decrease the electrical resistance inside the cylinder and decrease the required firing voltage.
   b. Increase the electrical resistance inside the cylinder and increase the required firing voltage.
   c. Increase the electrical resistance inside the cylinder and decrease the required firing voltage.
   d. Have no measurable effect on cylinder resistance.
9. Which of the following would cause a "common" ignition problem?
   a. a spark plug
   b. a spark plug wire
   c. a COP coil
   d. the ignition module
10. What does the spark line represent?
    a. the amount of voltage present at the beginning of the spark
    b. the amount of voltage present at the end of the spark
    c. the time the spark actually lasts
    d. the coil saturation time required to produce the spark
11. What tool should be used to measure the air gap of the crankshaft or camshaft sensors?
    a. a nonmagnetic feeler gauge
    b. a steel go-no-go feeler gauge
    c. a regular steel feeler gauge
    d. a dial indicator

12. During the cylinder performance test (also called the power balance test), what does the analyzer measure?
    a. each cylinder's rpm drop
    b. the spark voltage at each cylinder
    c. the spark duration at each cylinder
    d. individual cylinder combustion pressures
13. While checking a pickup coil with an ohmmeter, what would a higher than normal reading indicate?
    a. shorted
    b. open
    c. has high resistance
    d. none of the above
14. What can a Hall-effect CKP sensor be checked with?
    a. ammeter
    b. voltmeter
    c. ohmmeter
    d. oscilloscope
15. Which of the following will cause one firing line to be higher than the rest?
    a. high resistance in the spark plug wire
    b. lean fuel mixture
    c. rich fuel mixture
    d. high resistance in the coil wire
16. Which of the following could cause all of the firing lines to be abnormally high?
    a. advanced ignition timing
    b. fuel mixture too rich
    c. high resistance in coil wire
    d. low engine compression
17. What problem would a P0300 series code indicate?
    a. the fuel injector system
    b. the ignition system
    c. a misfiring cylinder
    d. the fuel delivery system
18. What problem would a P0200 series code indicate?
    a. the fuel injector system
    b. the ignition system
    c. a misfiring cylinder
    d. the fuel delivery system
19. An inspection of spark plugs reveals that they are overheated. What condition would cause this?
    a. retarded ignition timing
    b. overly rich air/fuel mixture
    c. fuel of too low an octane
    d. radiator cooling fan stuck on
20. When using a logic probe, what does a red light indicate?
    a. a good ground connection
    b. less than 5 volts at the test point
    c. more than 10 volts at the test point
    d. a poor ground connection

CHAPTER 27
Gasoline, Diesel, and Other Fuels

LEARNING OUTCOMES

- Describe the basic composition of gasoline.
- Explain why materials are added to gasoline to make it more efficient.
- Name the common substances used as oxygenates in gasoline and explain what they do.
- Describe how the quality of a fuel can be tested.
- Explain the advantages and disadvantages of the various alternative fuels.
- Explain the differences between diesel fuel and gasoline.
- Describe the common types of fuel injection used on today's diesel engines.
- Describe the various techniques used to allow current diesel engines to meet emission standards.

This chapter takes a look at the fuels used to propel a vehicle. Although there are several types of fuels for automotive use, gasoline is the most commonly used and most readily available. However, there is much interest in finding suitable alternatives to gasoline, these are discussed in this chapter.

Regardless of the type of fuel used for combustion, efficiency depends on having the correct amount of air mixed with the correct amount of fuel. The ideal air/fuel or stoichiometric ratio for a gasoline engine is approximately 14.7 kilograms (kg) of air mixed with 1 kg of gasoline. This provides a ratio of 14.7:1. Different fuels have different stoichiometric ratios. Because air is so much lighter than gasoline, it takes nearly 10 000 litres of air mixed with 1 litre of gasoline to achieve this air/fuel ratio. Lean ratios of 15 to 16:1 provide the best fuel economy. Rich mixtures have a ratio below 14.7:1 and provide more power from the engine but greater fuel consumption **(Figure 27–1)**.

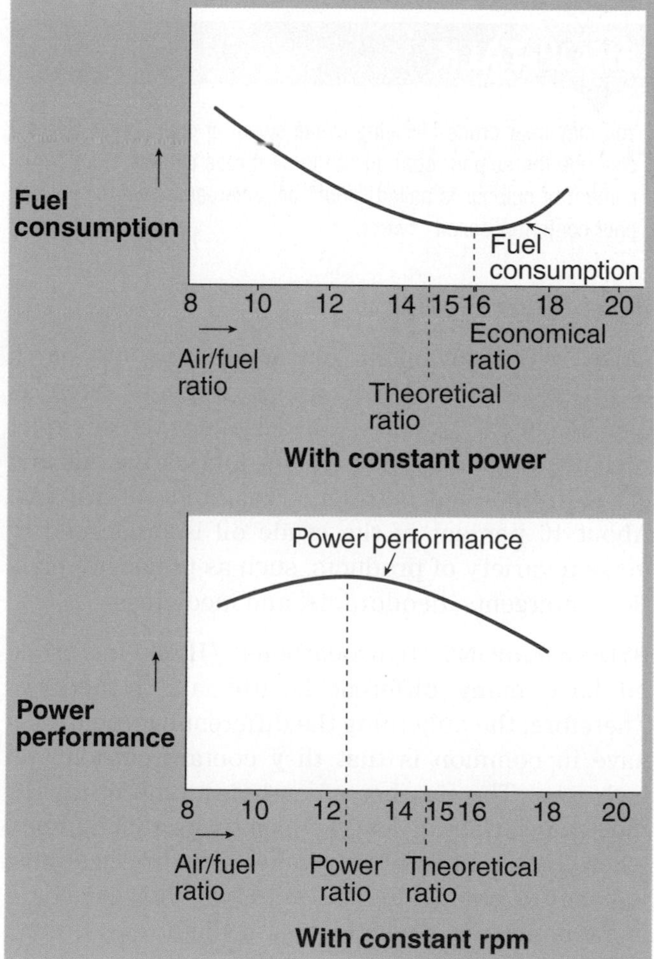

FIGURE 27–1

Fuel consumption and performance at various air/fuel ratios.

CRUDE OIL

Crude oil is also called **petroleum**, which means oil from the earth. The name fits; crude oil is drawn out of oil reservoirs and sands below the earth's surface. The oil extracted from the earth is called crude because it has yet to be processed or refined. Crude oil is commonly referred to as a fossil fuel because it is naturally produced by the decaying of plants and animals that lived a long time ago and were covered by dirt for many years. Crude oil is a liquid that varies in appearance. Normally, it has a dark brown or black colour, but it can also be yellow or greenish.

Although the composition of crude oil varies, it typically is as follows:

- 84 percent carbon
- 14 percent hydrogen
- 1 to 3 percent sulphur, in the form of hydrogen sulphide, sulphides, disulphides, and elemental sulphur
- Less than 1 percent nitrogen
- Less than 1 percent oxygen
- Less than 1 percent metals, normally nickel, iron, vanadium, copper, and arsenic
- Less than 1 percent salts, in the form of sodium chloride, magnesium chloride, and calcium chloride

The high concentration of carbon and hydrogen is why products produced from crude oil are called hydrocarbon fuels or compounds.

SHOP TALK

You may hear crude oil being called sweet or sour. These terms describe the sulphur content of the oil. Crude oil that has a high content of sulphur is called "sour" oil, whereas oil with low sulphur content is called "sweet."

Petroleum Products

Most of the petroleum extracted from the earth is processed into hydrocarbon products, such as asphalt, wax, gasoline, diesel fuel, kerosene jet fuel, heating and other fuel oils, lubricating oils and greases, liquefied petroleum gas, and natural gas. About 16 percent of the crude oil is processed to make a variety of products, such as polymers, plastics, detergents, deodorants, and medicines.

HYDROCARBONS Hydrocarbons (HCs) in crude oil have many different lengths and structures. Therefore, the only thing the different hydrocarbons have in common is that they contain carbon and hydrogen. The number of carbon atoms in an HC molecule defines its length. Sometimes, that number, when combined with the number of hydrogen atoms, is called a chain. The HC with the shortest chain is methane (CH_4), which is a very light gas. Longer chains with five or more carbons are liquids or solids. Asphalt has 35 or more carbon atoms. HCs contain a great amount of energy, which is why they have been used as a source of energy for many years.

Each of the different HCs must be separated from crude oil in order to be useful. This separation process is called refining. After refining, one barrel (159 litres [L], or 42 gallons) of crude oil will produce 75.7 L (20 gallons) of gasoline, 26.5 L (7 gallons) of diesel fuel, and smaller amounts of various other petroleum products.

Refining

A refinery is the place where the separation occurs. The easiest and most common way to separate the various HCs (called fractions) is through a process called **fractional distillation (Figure 27–2)**. The basis of this method is simply that the different HC compounds have progressively higher boiling points. Here are some examples:

- Propane will boil at less than 40°C (104°F).
- Gasoline will boil at 40°C to 205°C (104° to 401°F).
- Jet fuel will boil at 175°C to 325°C (350° to 617°F).
- Diesel fuel will boil at 250°C to 350°C (482° to 662°F).
- Lubricating oil will boil at 300°C to 370°C (572° to 700°F).
- Asphalt will boil at temperatures greater than 600°C (1112°F).

During fractional distillation, crude oil is heated with high-pressure steam to about 600°C (1112°F). This causes all of the crude oil to boil, forming vapour. The vapour moves into the fractional distillation column, which has many trays or plates. As the vapour moves up the column, it cools. The vapour condenses or becomes a liquid when it reaches the point in the column where the temperature is equal to the fractions' boiling temperature. Therefore, the fractions with the lowest boiling point will condense at the highest level within the column, and those with high boiling points will condense at lower levels. The various trays collect the condensation and pass the liquid out of the column.

Very few of the fractions that leave the column are ready to be used. They must be treated and cleaned to remove impurities. Also, refineries combine the various fractions to make a desired product. For example, different octane ratings of gasoline are possible by mixing different fractions. Some fractions are chemically altered so that they can be used for their specific application, and others are chemically processed to produce other fractions. The finished products are stored until they can be delivered to their markets.

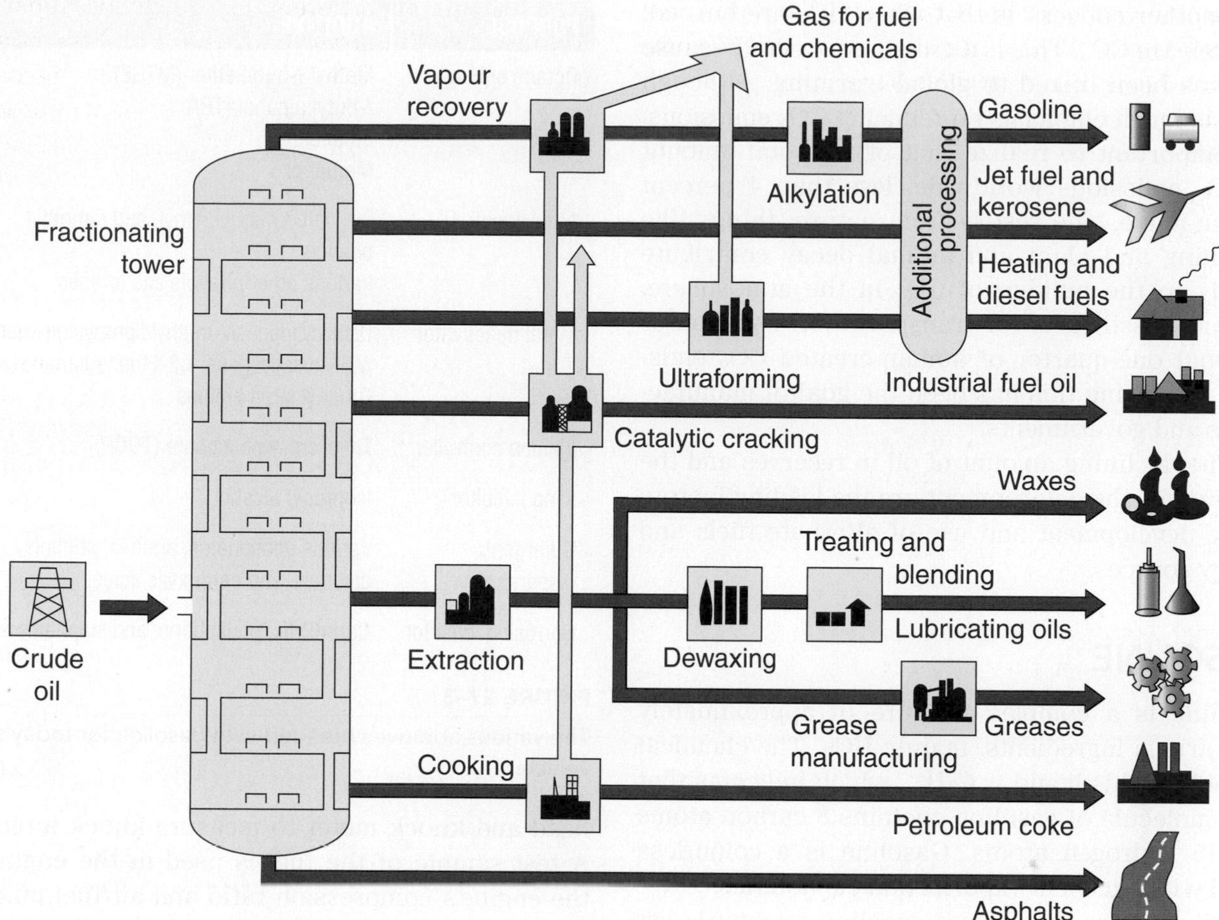

FIGURE 27–2

The refining process for crude oil.

American Petroleum Institute

CHEMICAL PROCESSING Some fractions are processed chemically to produce a different type of HC. Doing this allows the refineries to alter some HCs to meet market demands. Through processing, if the demand for gasoline is high, diesel fuel can be altered to become gasoline.

The process of breaking down HCs with a higher boiling point into an HC with a lower boiling point is called cracking. During cracking, the HCs are introduced to high-temperature and sometimes high-pressure conditions, forcing the HCs to break apart. Catalysts are often used to speed up the cracking process.

Also, the structure of HC molecules can be rearranged to provide a different HC. This is commonly used to produce octane boosters for gasoline.

CLEANING AND BLENDING The fractions captured from fractional distillation and chemical processing are treated in a variety of ways to remove all impurities. Some of the techniques used by refineries include passing the fractions through sulphuric acid to remove unsaturated HCs, a drying column to remove water, and hydrogen-sulphide scrubbers to remove sulphur.

After they are cleaned, the base fraction is blended with small amounts of other fractions to make various products, such as various grades of gasoline and lubricating oil and greases.

Concerns

Fossil fuels are the world's most important energy source. However, their use comes with costs. Although it appears that there is plenty of oil available today, we may run out in the future.

It is estimated that there are approximately 3.74 trillion barrels (440 km^3) of oil reserves, including oil sands, available. This seems like a lot; however, the current level of oil consumption is about 84 million barrels (3.6 km^3) per year. This means the oil from known oil reserves will be gone by 2039.

Another concern is that when HCs are burned, they release CO_2. This is a growing concern because CO_2 has been linked to global warming. Although there is much emphasis on reducing CO_2 emissions, it is important to realize that of the total amount of CO_2 emissions worldwide, less than 4 percent is man-made. The rest is from nature; things like breathing and plant and animal decay contribute greatly to the buildup of CO_2 in the atmosphere. Burning fossil fuels for transportation contributes to about one-quarter of human-created CO_2 emissions. Reducing this has been the goal of manufacturers and governments.

The declining amount of oil in reserves and the concern for the environment are the leading factors in the development and use of alternate fuels and energy sources.

PURPOSE	ADDITIVE
Octane enhancer	Methyl *t*-butyl ether (MTBE) *t*-butyl alcohol (TBA) Ethanol Methanol
Antioxidant	Butylated methyl, ethyl, and dimethyl phenols Various other phenols and amines
Metal deactivator	Disalicylidene-*N*-methyldipropylene-triamine *N,N*¹-disaliclidene-1,2-Ethanediamine Other related amines
Ignition controller	Tri-o-cresylphosphate (TOCP)
Icing inhibitor	Isopropyl alcohol
Detergent/ dispersant	Various phosphates, amines, phenols, alcohols, and carboxylic acids
Corrosion inhibitor	Carboxylic, phosphoric, and sulphonic acids

FIGURE 27–3

The various additives blended with gasoline for today's vehicles.

GASOLINE

Gasoline is a complex mixture of approximately 300 various ingredients, mainly HCs. The chemical symbol for this liquid is C_8H_{15}, which indicates that each molecule of gasoline contains 8 carbon atoms and 15 hydrogen atoms. Gasoline is a colourless liquid with excellent vaporization capabilities.

Oil refiners must meet gasoline standards set by the American Society for Testing and Materials (ASTM), which are adopted by Environment Canada, and any Canadian Environmental Protection Act (CEPA) regulations, and possible provincial regulations, as well as their own company standards.

Many of the performance characteristics of gasoline can be controlled during refining and blending. Many additives are blended into gasoline before it is available to the public **(Figure 27–3)**. The major factors affecting fuel performance are antiknock quality, volatility, sulphur content, and deposit control.

Antiknock Quality

An **octane** number or rating was developed by the petroleum industry so that the **antiknock** quality of a gasoline could be rated. The octane number is a measure of the fuel's tendency not to experience detonation in the engine. The higher the octane rating, the less the engine will have a tendency to knock.

Two methods are used for determining the octane number of gasoline: the **motor octane number (MON)** method and the **research octane number (RON)** method. Both use a laboratory single-cylinder engine equipped with a variable head and knock meter to measure knock intensity. A test sample of the fuel is used in the engine as the engine's compression ratio and air/fuel mixture are adjusted to develop specific knock intensity. There are two primary standard reference fuels: **isooctane** and **heptane**. Isooctane does not knock in an engine but is not used in gasoline because of its expense. Heptane knocks severely in an engine. Isooctane has an octane number of 100. Heptane has an octane number of zero.

A fuel of unknown octane value is run in the test engine equipped with a variable compression cylinder head and a knock meter. The severity of knock is measured. Various proportions of isooctane and heptane are run in the engine to duplicate the severity of the engine knock when the test fuel was run. When the knock caused by the isooctane and heptane mixture matches that caused by the fuel being tested, the octane number is established by the percentage of isooctane in the mixture. For example, if 85 percent isooctane and 15 percent heptane produced the same knock severity as the tested fuel, that fuel would be rated as having an octane rating of 85.

The octane rating required by law and the one displayed on gasoline pumps is the **Antiknock Index (AKI)**. It is the average of RON and MON. The AKI is stated as (R+M)/2.

Most modern engines are designed to operate efficiently with regular grade gasoline and do not require high-octane gasoline. One of the things to remember about high-octane fuel is that it burns slower than low-octane gasoline. This is why it is less likely to cause detonation. Most engine control systems have a sensor to detect whether a knock is occurring so that the PCM can retard the ignition timing to prevent detonation. Higher-octane gasoline is used in high-performance engines because they have high compression ratios, which provide greater power output.

Volatility

Gasoline is very volatile. It readily evaporates, so it readily mixes with air for combustion. The volatility of gasoline affects the following performance characteristics or driving conditions.

COLD STARTING AND WARM-UP A fuel can cause hard starting, hesitation, and stumbling during warm-up if it does not readily vaporize. A fuel that vaporizes too easily in hot weather can form vapour in the fuel delivery system, causing **vapour lock** or a loss of performance. If gasoline vaporizes while it is in a fuel line, it can stop the flow of gasoline. Rather than flow through the line, the pressurized fuel will compress the vapour, not move it. Vapour lock can cause a variety of driveability problems. Gasoline blended for summer (hot weather) use is less volatile (does not evaporate as easily) than winter gasoline.

ALTITUDE Gasoline vaporizes more easily at high altitudes, so volatility is controlled in blending, according to the elevation of the location where the fuel is sold.

CRANKCASE OIL DILUTION A fuel must vaporize to prevent diluting the crankcase oil with liquid fuel or break down the oil film on the cylinder walls, causing scuffing or scoring. The liquid eventually enters the oil in the crankcase, forming an accumulation of sludge, gum, and varnish, as well as affecting the lubrication properties of the oil.

There are three methods of measuring the volatility of a fuel. The most common is the **Reid vapour pressure (RVP)** test. The RVP test is performed by placing a sample of gasoline into a sealed metal container that has a pressure measuring device attached to it. The container is submerged in heated 38°C (100°F) water. As the fuel is heated, it vaporizes. Remember, the more volatile a fuel is, the easier it will vaporize. As fuel vaporizes, it creates vapour pressure within the container. Vapour pressure is measured in kilopascals (kPa) or pounds per square inch (psi).

Sulphur Content

Some of the sulphur in the original crude oil may be found in refined gasoline. Sulphur content is reduced at the refinery to limit the amount of corrosion it can cause in the engine and exhaust system. When the hydrogen in the fuel is burned, one of the by-products of combustion is water. Water leaves the combustion chamber as steam but can condense back to water when passing through a cool exhaust system. When the engine is shut off and cools, steam condenses back to a liquid and forms water droplets.

When the sulphur in the fuel is burned, it combines with oxygen to form **sulphur dioxide**. This compound can combine with water to form sulphuric acid, which is a highly corrosive compound. This acid is the leading cause of exhaust valve pitting and exhaust system deterioration. Sulphuric acid also attacks the linings of the main and rod bearings. This is one reason engine oil needs to be changed regularly. With catalytic converters, the sulphur dioxide can cause the obnoxious odour of rotten eggs during engine warm-up. To reduce corrosion caused by sulphuric acid, the sulphur content in gasoline is limited to less than 0.01 percent.

DEPOSIT CONTROL Several additives are added to gasoline to control harmful deposits; these include gum or oxidation inhibitors, detergents, metal deactivators, and rust inhibitors.

Basic Gasoline Additives

At one time, all a gasoline-producing company needed to do to provide its product was to pump the crude from the ground, run it through the refinery to separate it, dump in a couple of grams of lead per gallon, and deliver the finished product to a service station. Of course, automobiles were much simpler then, and what they burned did not seem critical. As long as the gasoline vaporized easily and did not cause engine knock, everything was fine.

Back then, lead compounds, such as **tetraethyl lead (TEL)** and **tetramethyl lead (TML)**, were added to gasoline to increase its octane rating. However, since the mid-1970s, vehicles have been designed to run on unleaded gasoline only. Leaded fuels are no longer available as automotive fuels. Because of the poisoning effect lead has on humans and on catalytic converters, today's gasoline is limited to a lead content of 0.016 gram per litre (0.06 gram per gallon). Now, to achieve the desired octane rating, methylcyclopentadienyl manganese tricarbonyl (MMT) is normally added to gasoline.

ANTI-ICING OR DEICER Isopropyl alcohol is added seasonally to gasoline as an anti-icing agent to prevent fuel line freeze-up in cold weather.

METAL DEACTIVATORS AND RUST INHIBITORS These additives are used to inhibit reactions between the fuel and the metals in the fuel system that can form abrasive and filter-plugging substances.

GUM OR OXIDATION INHIBITORS Some gasoline contains aromatic amines and phenols to prevent the formation of gum and varnish. During storage, harmful gum deposits can form due to the reaction between some gasoline molecules with each other and with oxygen. Oxidation inhibitors are added to promote gasoline stability. They help control gum, deposit formation, and staleness.

Gum content is influenced by the age of the gasoline and its exposure to oxygen and certain metals such as copper. If gasoline is allowed to evaporate, its residue can form gum and varnish.

DETERGENTS Detergent additives are designed to do only what their name implies—clean certain critical parts inside the engine. They do not affect octane.

PERFORMANCE TIP

Adding nitrous oxide to the air/fuel mixture is not something done by oil refineries. Rather, it is commonly done by those seeking more instantaneous power from their engines. Nitrous oxide is injected as a dense liquid. When nitrous oxide is heated, it breaks down into nitrogen and oxygen. This provides more oxygen inside the cylinder when the fuel ignites. Because there is more oxygen, more fuel can be injected into the cylinder. The engine therefore produces more power. Nitrous oxide also improves engine performance by cooling the gases in the cylinder, thereby making the air denser. Nitrous oxide is injected into the engine's intake when the driver pushes a button to activate the system. Nitrous kits, which include nearly all that is needed to add the system to an engine, are available for many engines. The nitrous tanks typically store enough nitrous for three to five minutes of operation.

OXYGENATES

Oxygenates are compounds, such as alcohols and ethers, that contain oxygen. By carrying oxygen, the fuel tends to lean the mixture. Oxygenates improve combustion efficiency, thereby reducing emissions. Many oxygenates also serve as excellent octane enhancers when blended in gasoline **(Figure 27–4)**. Oxygenated fuels tend to have lower CO emissions.

It should be noted that the use of oxygenated gasoline may cause a slight decrease in fuel economy in late-model vehicles. This is due to the HO_2S detecting extra oxygen and the PCM responding to this by richening the mixture.

Oxygenates added to gasoline produce what is referred to as **reformulated gasoline (RFG)**. RFG is also called "cleaner burning" gasoline and costs slightly more than normal gasoline. RFG can be used in most engines with no modifications. RFG was formulated to reduce exhaust emissions.

Ethanol

By far, the most widely used gasoline oxygenate additive is **ethanol** (ethyl alcohol), or grain alcohol. Ethanol is a noncorrosive and relatively nontoxic alcohol made from renewable biological sources. Blending 10 percent ethanol into gasoline results in an increase of 2.5 to 3 octane points. With ethanol-blended gasoline, air toxics are about 50 percent less. Ethanol decreases CO emissions due to the higher oxygen content of the fuel.

Ethanol can also loosen contaminants and residues that may have gathered in the vehicle's fuel system. All alcohols have the ability to absorb the water in the fuel system that results from condensation. This reduces the chances of fuel line freeze-up during cold weather.

Methanol

Methanol is the lightest and simplest of the alcohols and is also known as wood alcohol. It can be distilled from coal or other sources, but most of what is used today is derived from natural gas.

	ETHANOL	MTBE	ETBE	TAME
Chemical formula	CH_3CH_2OH	$CH_3OC(CH_3)_3$	$CH_3CH_2OC(CH_3)_3$	$(CH)_3CCH_2OCH_3$
Octane, (R+M)/2	115	110	111	105
Oxygen content, % by weight	34.73	18.15	15.66	15.66
Blending vapour pressure, RVP	18	8	4	1.5

FIGURE 27–4
The typical properties of the common oxygenates.

Many automakers continue to warn motorists about using a fuel that contains more than 10 percent methanol and cosolvents by volume. It is far more corrosive to fuel system components than ethanol, and it is this corrosion that has automakers concerned. Methanol is also highly toxic, and there are safety concerns with ingestion, eye or skin contact, and inhalation.

Methanol can be used directly as an automotive fuel, but the engine must be modified for its use. It can also be used in flexible fuel vehicles such as M85, which is 85 percent methanol. However, this is not very common. In the future, methanol could be the fuel of choice for providing hydrogen to power fuel-cell vehicles.

MTBE

In the past, methyl tertiary butyl ether (MTBE) was used as an octane enhancer because of its excellent compatibility with gasoline. Methanol can be used to make MTBE. However, MTBE production and its use have declined because it was found to contaminate groundwater. As of 2004, MTBE is no longer used in gasoline and has been replaced by ethanol and other oxygenates such as tertiary amyl methyl ether (TAME) and ethyl tertiary butyl ether (ETBE).

Aromatic Hydrocarbons

These are petroleum-derived compounds, including benzene, xylene, and toluene, that are being used as octane boosters.

GASOLINE QUALITY TESTING

Two tests can be done to test the quality of gasoline: the Reid vapour pressure (RVP) test and the alcohol content test.

Testing the RVP of Gasoline

RVP is a measure of the volatility of gasoline. Fuels that are more volatile vaporize more easily, creating more pressure. Increasing the RVP of a gasoline permits the engine to start more easily in cold weather. The RVP of winter blend gasoline is about 62 kPa (9.0 psi). Summer grade is typically around 48 kPa (7.0 psi).

A special fuel vapour–pressure tester is needed to test the RVP of gasoline. Make sure the gasoline that is being tested is cool. Then put a sample in the tester's container, and secure the seal in the container as soon as the gasoline is in it. Put hot water in another container, and put the container holding the fuel into it. Make sure that most of the container holding the fuel is covered by water. Connect the pressure gauge assembly to the container holding the gasoline. Put a thermometer in the water. When the water temperature is 40°C (105°F) for at least two minutes, take your pressure reading and compare it to specifications.

Alcohol in Fuel Test

Pump gasoline may contain a small amount of alcohol, normally up to 10 percent. If the amount is greater than that, problems may result, such as fuel system corrosion, fuel filter plugging, deterioration of rubber fuel system components, and a lean air/fuel ratio. These fuel system problems caused by excessive alcohol in the fuel may cause driveability complaints such as lack of power, acceleration stumbles, engine stalling, and no-start. If the correct amount of fuel is being delivered to the engine and there is evidence of a lean mixture, check for air leaks in the intake, and then check the gasoline's alcohol content.

There are many different ways to check the percentage of alcohol in gasoline. Some are more exact than others, and some require complex instruments.

PROCEDURE

To Check the Amount of Alcohol in a Sample of Gasoline

STEP 1 Obtain a 100-millilitre (ml) cylinder graduated in 1 ml divisions.

STEP 2 Fill the cylinder to the 90 ml mark with gasoline.

STEP 3 Add 10 ml of water to the cylinder so that it is filled to the 100 ml mark.

STEP 4 Install a stopper in the cylinder, and shake it vigorously for 10 to 15 seconds.

STEP 5 Carefully loosen the stopper to relieve any pressure.

STEP 6 Install the stopper and shake vigorously for another 10 to 15 seconds.

STEP 7 Carefully loosen the stopper to relieve any pressure.

STEP 8 Place the cylinder on a level surface for five minutes to allow liquid separation.

STEP 9 Observe the liquid. Any alcohol in the fuel is absorbed by the water and settles to the bottom. If the water content in the bottom of the cylinder exceeds 10 ml, there is alcohol in the fuel. For example, if the water content is now 15 ml, there was 5 percent alcohol in the fuel. *Note:* Because this procedure does not extract 100 percent of the alcohol from the fuel, the percentage of alcohol in the fuel may be higher than indicated.

ALTERNATIVES TO GASOLINE

The actual cost of using gasoline in engines is not limited to the price per gallon or litre. There are other factors, or costs, that need to be considered: our environment, our dependence on foreign oil supplies, and the depletion of future oil supplies. Any reduction in the use of fossil fuels will have benefits for generations to come. Let us look at some simple facts:

- The number of household vehicles in Canada is growing, and nearly tripled from 1969 to 2001. In 2013 in North America, 18.5 million new cars and light trucks were sold (1.74 million in Canada alone). These numbers do not include the automobiles on the road that were not bought that year. There are well over 225 million vehicles on the road.
- It is estimated that the total distance covered by those automobiles in one year is well over 3220 billion kilometres (2000 billion miles). To put this in perspective, let us assume the average fuel mileage of all those vehicles is 11.76 L/100 km (20 miles per gallon). This means over 379 billion litres (100 billion gallons) of oil are burned by automobiles each year.
- By 2020, oil consumption is expected to grow by nearly 40 percent, and our dependence on foreign oil sources is projected to rise to more than 60 percent.
- A 10 percent reduction in fossil fuel consumption by cars and light trucks would result in using 24 million fewer gallons of oil each day.
- North Americans spend close to \$100 000 per minute to buy foreign oil.
- Automobiles and gasoline contribute to environmental damage. Not only do automobiles emit pollutants **(Figure 27–5)**, but the extraction, production, and marketing of gasoline also lead to air pollution, water pollution, and oil spills.
- Because of the heavy reliance on fossil fuels, the transportation industry is one of the sources of carbon dioxide and other heat-trapping gases that cause global warming.

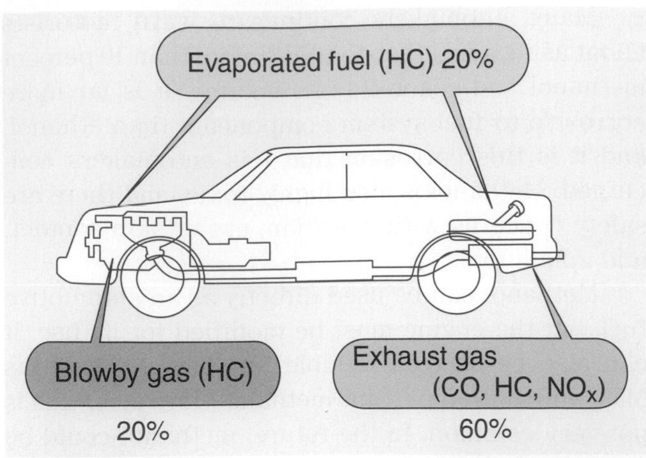

FIGURE 27–5

Gasoline-powered vehicles emit a wide assortment of pollutants.

©Cengage Learning 2015

Alternative Fuels

The concerns of burning fossil fuels and the decline of their reserves have led to a comprehensive search for alternative fuels. While looking at the viability of an alternative fuel, many things are considered, including emissions, cost, availability, energy density, safety, engine life, fuelling facilities, weight and space requirements for fuel tanks, and the range of a fully fuelled vehicle. By using alternative fuels, we not only can reduce our reliance on petroleum but also can reduce emissions and the effects an automobile's exhaust has on global warming. Many of these fuels are also being considered as the fuel of choice for fuel-cell electric vehicles.

Much attention has been paid to renewable fuel sources. **Renewable fuels** are those derived from non-fossil sources and produced from plant or animal products or wastes (biomass). Biomass fuels, such as biodiesel and ethanol, can be burned in internal combustion engines. Biomass fuels tend to be carbon neutral, which means that during combustion, they release the same amount of CO_2 that was absorbed from the atmosphere when the plant or animal was living. Combustion does not cause an increase in CO_2 emissions. Ethanol and methanol are used as oxygenates for blending with gasoline. They can also be used as the primary energy source for internal combustion engines. However, because ethanol is made from renewable sources, it is the most commonly used.

ENERGY DENSITY Each of these alternative fuels can be looked at in terms of **energy density**. This is the amount of energy provided by a standard weight of each. Energy density is typically rated as joules per kilogram. A joule is defined as the energy required to produce one watt of power for one second. Refer to **Table 27–1** to review the energy densities of common energy sources.

TABLE 27–1 THE ENERGY DENSITIES OF COMMON ENERGY SOURCES

MATERIAL	APPROXIMATE ENERGY PER KILOGRAM
Uranium 238	20 terajoules
Hydrogen	143 megajoules
Natural gas	53.6 megajoules
LPG propane	49.6 megajoules
Gasoline	47.2 megajoules
Diesel fuel	46.2 megajoules
Gasohol E10	43.54 megajoules
Biodiesel	42.20 megajoules
Gasohol E85	33.1 megajoules
Coal	32.5 megajoules
Methanol	19.7 megajoules
Supercapacitor	100 kilojoules
Lead-acid battery	100 kilojoules
Capacitor	360 joules

Ethanol

Ethanol is a high-quality, low-cost, high-octane fuel (rated at 115) that burns cleaner than gasoline. The use of ethanol as a fuel is not new. Ford's Model T was designed to run on ethyl alcohol (ethanol). Ethanol (CH_3CH_2OH), commonly called grain alcohol, is a renewable fuel made from nearly anything that contains carbon **(Figure 27–6)**. It is most commonly produced by fermenting and distilling corn, cornstalks, wheat, sugar cane, other grains, or biomass waste. Ethanol can be used as a high-octane fuel in vehicles and is often mixed with gasoline to boost its octane rating.

Because ethanol is an alcohol, it can absorb moisture that may be present in a fuel system. The absorbed water is simply passed with the fuel and burned by the engine. However, if the moisture content in the fuel becomes too high, the water will separate from the fuel and drop to the bottom of the fuel tank. If this is suspected, remove all fuel and water from the tank, and refill it with clean ethanol-blended fuel.

For automotive use, ethanol is blended with gasoline. The common blends are an E10 blend, which is 10 percent ethanol and 90 percent gasoline; E15, which is 15 percent ethanol; and E85, which is 85 percent ethanol. Most gasoline-powered vehicles in North America can run on blends of up to 10 percent ethanol, and some are equipped to run on E85.

The use of E85 has many advantages over the use of traditional gasoline:

- It is produced in the United States and can reduce North America's reliance on foreign oil.
- Vehicles do not need many modifications to use it.
- Its emissions are cleaner than those of a gasoline engine.

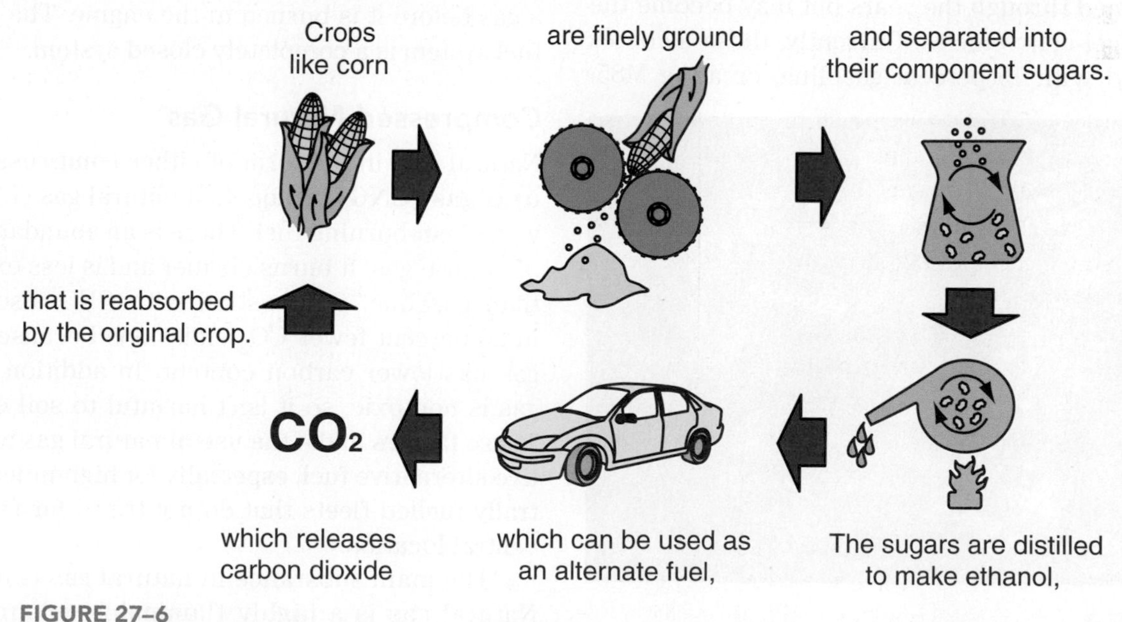

FIGURE 27–6

The carbon cycle of ethanol.

- CO_2 emissions are much lower.
- Ethanol-blended fuel keeps the fuel system clean because it does not leave varnish or gummy deposits.

However, the infrastructure for E85 is weak. There are few fuel filling stations that offer E85. The amount of energy it takes to produce E85 is more than the energy it provides. E85 also has about 25 percent less energy than gasoline; therefore, fuel economy will decrease by about that much in a typical vehicle.

SHOP TALK

After 40 years of running on methanol, a non-renewable fuel made from natural gas, the Indy Racing League (IRL) in 2007 switched to ethanol to power the engines **(Figure 27–7)** in its race series. Also, in 2011, NASCAR mandated the use of E15 in all of its racing. According to NASCAR, E15 is good for racing and good for the environment.

Methanol

Methanol (CH_3OH) is a clean-burning alcohol fuel that is often made from natural gas but can also be produced from coal and biomass. Because North America has an abundance of these materials, the use of methanol can decrease the dependence on foreign oils. Methanol is very corrosive, and an engine designed to run on it must be equipped with special plastic and rubber components, as well as a stainless-steel fuel system. Methanol use as a fuel has declined through the years but may become the fuel for fuel-cell vehicles. Currently, these alcohols are mixed with 15 percent gasoline, creating M85.

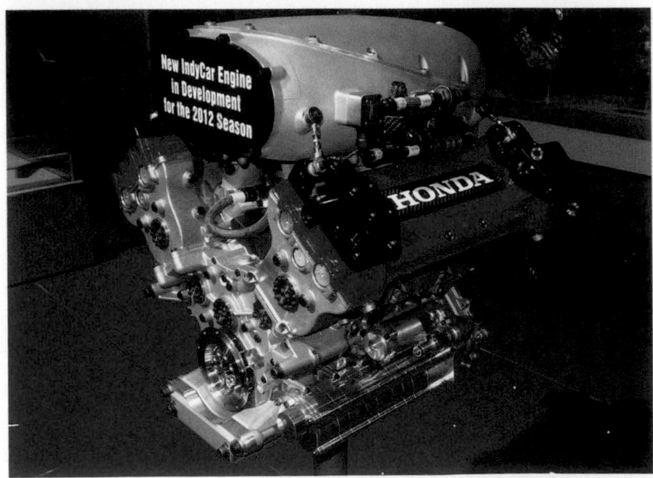

FIGURE 27–7

An Indy ethanol-fuelled race engine.

The small amount of gasoline improves the cold-starting ability of the alcohols.

Propane/LPG

Propane, also referred to as **liquefied petroleum gas or LP gas**, is used by many fleets around the world in taxis, police cars, school buses, and trucks. LP gas is similar to gasoline chemically. It is called liquid petroleum because it is stored as a liquid in a pressurized bottle. The pressure increases the boiling point of the liquid and prevents it from vaporizing. LP gas burns clean because it vaporizes at atmospheric temperatures and pressures. This means it emits fewer HC, CO_2, and CO emissions. Propane is a clean-burning fuel that provides a driving range closer to gasoline than other alternative fuels.

Propane allows for quick starting, even in the coldest of climates. It also has a higher octane rating than gasoline. However, there is a reduction of engine power output (about 5 percent) because it is difficult to fill the cylinders with the gas. Propane is a dry fuel that enters the engine as vapour. Gasoline, on the other hand, enters the engine as tiny droplets of liquid. LP gas is a good alternative to gasoline, but it is a fossil fuel and therefore is not a favoured alternative fuel for the future.

LP gas vehicles have designated engine fuel controls and special tanks or cylinders to store the gas. However, the gas is stored at about 1380 kPa (200 psi). Under this pressure, the gas turns into a liquid and is stored as a liquid. When the liquid propane is drawn from the tank, it warms and changes back to a gas before it is burned in the engine. The propane fuel system is a completely closed system.

Compressed Natural Gas

Natural gas, in the form of either **compressed natural gas (CNG)** or liquefied natural gas (LNG), is a very clean-burning fuel. There is an abundant supply of natural gas. It burns cleaner and is less expensive than gasoline. Combustion with CNG also results in 25 percent fewer CO_2 emissions because natural gas has lower carbon content. In addition, natural gas is nontoxic, so it isn't harmful to soil or water. These factors make the use of natural gas an attractive alternative fuel, especially for high-mileage, centrally fuelled fleets that do not travel far from their central location.

The main substance in natural gas is methane. Natural gas is a highly flammable colourless gas and is commonly used in homes for heaters, stoves, and water heaters. CNG and LNG are considered

FIGURE 27–8

The storage tanks for CNG.

alternative fuels under the U.S. Energy Policy Act of 1992. CNG is used in light- and medium-duty vehicles, whereas LNG is used in transit buses, train locomotives, and long-haul semi-trucks.

CNG must be safely stored in cylinders at pressures of 16 550, 20 685, or 24 820 kPa (2400, 3000, or 3600 psi) **(Figure 27–8)**. This is the biggest disadvantage of using CNG as a fuel. The space occupied by these cylinders takes away luggage and, sometimes, passenger space. As a result, CNG vehicles have a shorter driving range than comparable gasoline vehicles. Bi-fuel vehicles are equipped to store both CNG and gasoline and will run on either.

Natural gas turns into a liquid when it is cooled to –164°C (–263.2°F). Because it is a liquid, a supply of LNG takes up less room in the vehicle than does CNG. Therefore, the driving range of an LNG vehicle is longer than a comparable CNG vehicle. For vehicles needing to travel long distances, LNG is a good choice. However, the fuel must be dispensed and stored at extremely cold temperatures. This requires refrigeration units that also take up space. This is why LNG is not a practical fuel for personal use and is only used in heavy-duty applications.

The use of natural gas as a fuel has advantages due to its domestic availability, vast distribution infrastructure, low cost, and clean-burning qualities. However, the space taken by the CNG cylinders and their weight, about 136 kg (300 pounds), can be considered a disadvantage in most applications.

The basic components of a **natural gas vehicle (NGV)** are shown in **Figure 27–9**. The CNG fuel system moves high-pressure natural gas from the storage cylinder to the engine. It also reduces the pressure of the gas so that it is compatible with the engine's fuel-management system. The natural

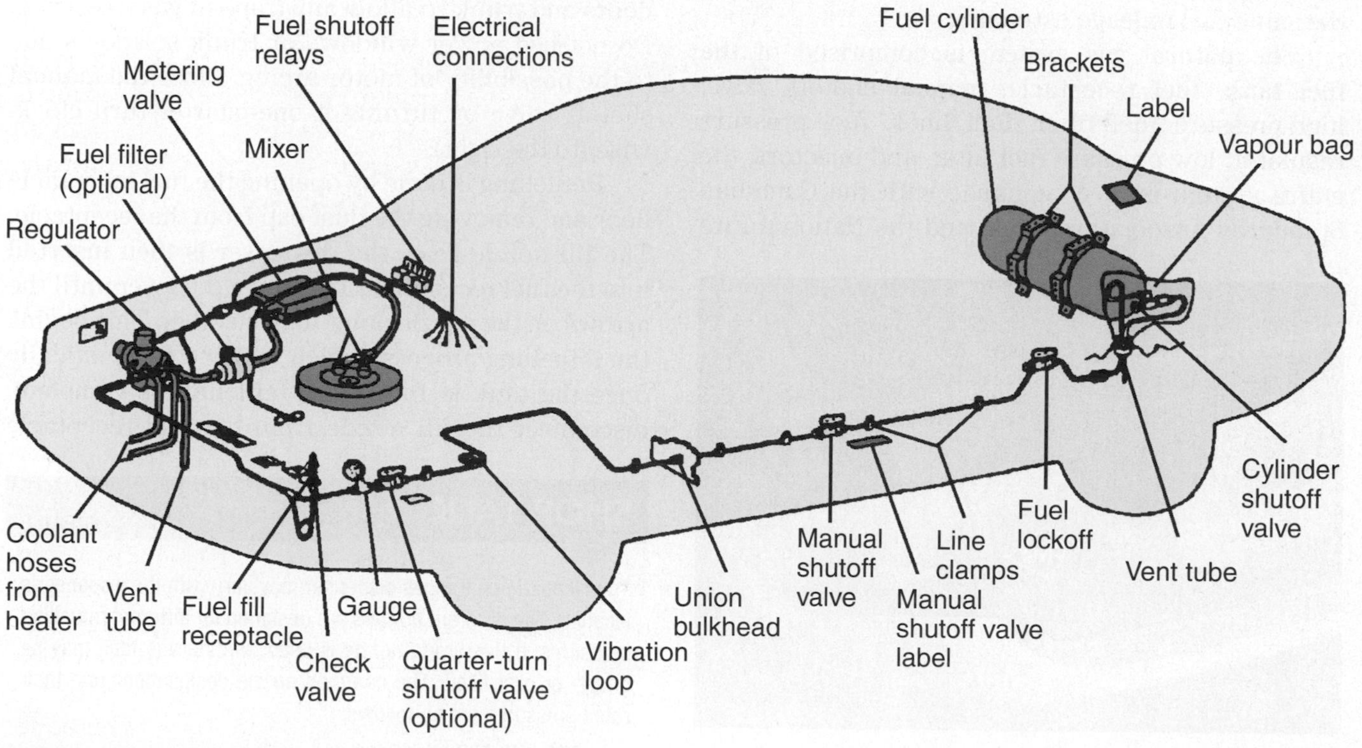

FIGURE 27–9

NCV system components.

gas is injected into the engine intake air the same way gasoline is injected into a gasoline-fuelled engine, as the high temperatures and pressures in the combustion chamber quickly ignite the gas.

There are three basic types of natural gas vehicles:

- Dedicated, which are designed to run only on natural gas. These can be light- or heavy-duty vehicles.
- Bi-fuel, which have two separate fuel systems that allow them to run on either natural gas or gasoline and they are typically light-duty vehicles.
- Dual-fuel, which are normally used only with heavy-duty applications. These have natural gas and diesel fuel systems.

HONDA CIVIC NATURAL GAS VEHICLE The Honda Civic converted to use CNG is based on a typical Civic sedan **(Figure 27–10)**: the same engine, transmission, accessories, and body. The major differences reflect the modifications to the engine. In a normal Civic, the 1.8L engine is rated at 140 hp @ 6500 rpm and 128 lb·ft @ 4300 rpm. The engine in the CNG car is rated lower, at 110 hp @ 6500 rpm and 106 lb·ft @ 4300 rpm. The CNG engine also has a higher compression ratio: 12.7:1 compared to 10.6:1. Both have the same EPA emissions rating and about the same fuel mileage estimates.

The natural gas system is comprised of the fuel tank, fuel receptacle, manual shutoff valve, high-pressure fuel filter, fuel lines, fuel pressure regulator, low-pressure fuel filter, and injectors; the entire system is in compliance with the Canadian Standards Association (CSA) and the National Fire Protection Association's Vehicular Gaseous Fuel Systems (NFPA-52) code.

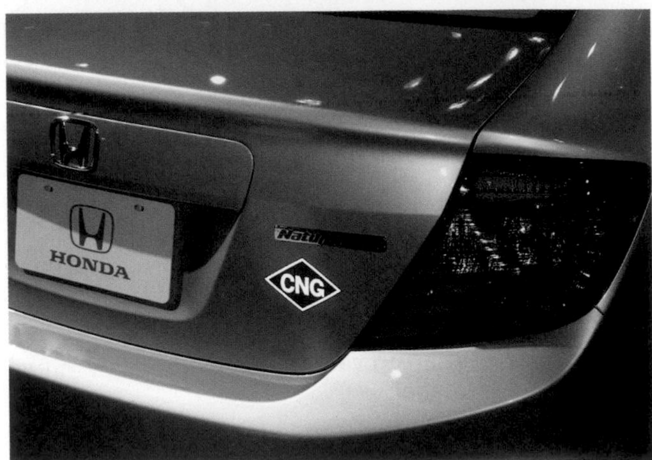

FIGURE 27–10

A CNG Honda Civic.

To make sure only quality and well-filtered gas enters the storage tank in a Civic, Honda highly recommends that the vehicle be refuelled at a public commercial-grade CNG refuelling station. Honda further recommends not installing a home refuelling station because of the wide variation of natural gas quality delivered to homes. The major worry is moisture, which can cause damage and result in expensive repairs.

WARNING!

Natural gas is a highly flammable and explosive gas. Serious injury, or death, can result from the ignition of leaking gas. If a leak is suspected, the car must be shut down and the leak identified and fixed.

The storage system is designed to hold CNG at a maximum of 24 800 kPa (3600 psi). The only time a gas smell or a hiss should be heard is during refuelling. Any other time may indicate there is gas leak. If a leak is suspected, the system should be shut down immediately and the car pushed outdoors to a well-ventilated area. Turn the ignition switch to the LOCK position. Make sure to keep the car away from heat, sparks, and flame. Then manually open the car's doors and trunk to allow any trapped gas to escape. Do not use power windows or trunk solenoids due to the possibility of motor arcing. Close the manual shutoff valve by turning it one-quarter turn clockwise (to the right).

Refuelling is done by opening the fuel receptacle door and removing the dust cap from the receptacle. The fill nozzle from the dispenser is then inserted into the fuel receptacle. Then turn the lever until the arrows on the nozzle point to each other. This begins the refuelling process, which will end automatically once the tank is full. Once refuelling is complete, disconnect the fill nozzle from the fuel receptacle

SHOP TALK

The fill nozzle on the gas dispenser may vary with the dispenser's location. The different nozzles are designed for different refuelling pressures. If the wrong nozzle is used, the storage tank may be under- or over-filled. The common nozzle designations and their rated pressures are as follows:

- P24—16 500 kPa (2400 psi)
- P30—20 700 kPa (3000 psi)
- P36—24 800 kPa (3600 psi)

by slowly turning the lever on the nozzle 180°. Then securely fasten the receptacle dust cap onto the fuel receptacle, and close the fuel receptacle door.

The fuel gauge displays the amount of gas that remains in the tank; this is determined by the system looking at the pressure and temperature of the gas in the tank. A LOW reading on the gauge means the fuel pressure has dropped to about 20 700 kPa (3000 psi), and a FULL reading indicates a tank pressure of 24 800 kPa (3600 psi). The low fuel indicator will come on whenever the fuel level is low. It may also come on during very cold weather, although there may be ample gas in the tank. This is because the pressure in the tank will decrease when it is cold.

P-SERIES FUELS P-series is a new fuel classified as an alternative fuel. It is a blend of natural gas liquids, ethanol, and biomass-derived cosolvents. **P-series fuels** are clear, colourless, 89–93 octane, liquid blends that are formulated to be used alone or freely mixed, in any proportion, with gasoline. Like gasoline, low vapour-pressure formulations are produced to prevent excessive evaporation during summer, and high vapour-pressure formulations are used for easy starting in cold weather.

Each gallon of P-series fuel emits approximately 50 percent less CO_2, 35 percent fewer HCs, and 15 percent less CO than gasoline. It also has 40 percent less ozone-forming potential.

Hydrogen

Hydrogen is cited by some as the fuel of the future because it is full of energy due to its atomic structure and abundance. It is the simplest and lightest of all elements and has one proton and one electron **(Figure 27–11)**. Hydrogen is a colourless and odourless gas. It is one of the most abundant elements on earth. The combination of hydrogen and oxygen forms water. Fossil fuels are combinations of carbon and hydrogen, or HCs.

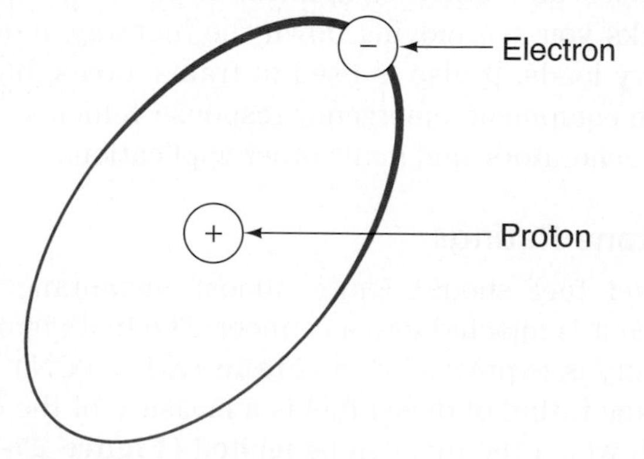

FIGURE 27–11

A hydrogen atom.

©Cengage Learning 2015

Hydrogen is extracted from various substances through a process that pulls hydrogen out of its bond with another element or elements. Hydrogen is commonly extracted from water, fossil fuels, coal, and biomass. The two most common ways that hydrogen is produced are steam reforming and electrolysis. Currently, it costs much more to produce hydrogen than it does to produce other fuels such as gasoline. This, again, is an obstacle and the focus of much research.

HYDROGEN FUEL To demonstrate the energy in hydrogen, there are hydrogen bombs. Some manufacturers are experimenting with burning hydrogen in internal combustion engines. Three major auto manufacturers have developed and tested hydrogen-fuelled internal combustion engines; actually, these vehicles have bi-fuel capabilities. BMW's bi-fuelled V12 engine uses liquefied hydrogen or gasoline as its fuel. When running on hydrogen, the engine emits zero CO_2 emissions. To store the liquefied hydrogen, the storage tank is kept at a constant temperature of –253°C (–423°F). At this temperature, the liquid hydrogen has the highest possible energy density.

Ford and Mazda have also developed internal combustion engine vehicles with hydrogen power. Mazda is using its rotary engine, which it claims is ideal for using hydrogen fuel. The concept vehicles from both manufacturers are also bi-fuel vehicles. Ford has converted a V10 and a 2.3L, I-4 engine to run on hydrogen. Engine modifications include a higher compression ratio, special fuel injectors, and a modified electronic control system. When running on hydrogen, the engine is more than 10 percent more efficient than when it runs on gasoline, and emissions levels are very close to zero. Because the fuel contains no carbon, there are no carbon-related emissions (CO, HC, or CO_2).

Typically, an engine running on hydrogen produces less power than a same-sized gasoline-powered engine. Ford added a supercharger with an intercooler to the engines to compensate for the loss of power.

INFRASTRUCTURE AND STORAGE Other than manufacturing costs, the biggest challenge for hydrogen-powered vehicles is the lack of an infrastructure. Vehicles need to be able to be refuelled quickly and conveniently.

Hydrogen is normally stored as a liquid or as a compressed gas. When stored as a liquid, it must

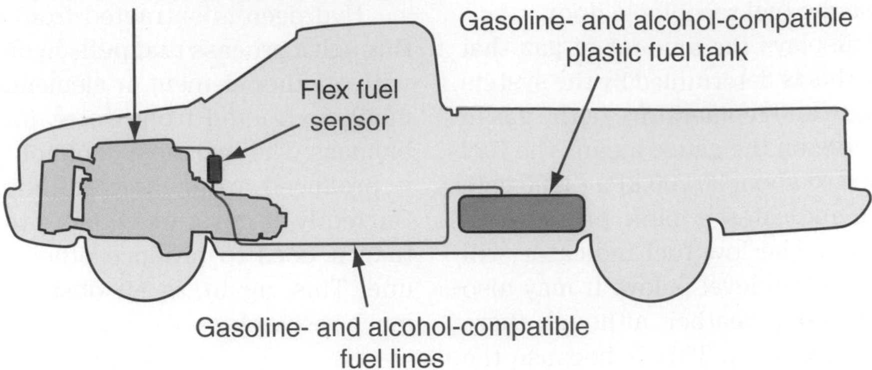

FIGURE 27–12

A flexible fuel vehicle.

be kept very cold. Keeping it that cold adds weight and complexity to the storage system. The tanks required for compressed hydrogen need to be very strong, and that translates to weight. Also, higher pressures mean more hydrogen can be packed into the tank, but the tank must be made stronger before the pressure can be increased.

Flex Fuel Vehicles

Flexible fuel vehicles (FFVs) can run on ethanol or gasoline, or a mixture of the two **(Figure 27–12)**. The alcohol fuel and gasoline are stored in the same tank. This gives the driver flexibility and convenience when refilling the fuel tank. Many vehicles are fitted with systems that allow the use of multiple fuels. These include vehicles from Chrysler, Ford Motor Co., General Motors, and Nissan. Flex fuel vehicles may have a clover leaf symbol (internally or externally) that shows they can use multiple fuels; in addition, the fuel receptacle clearly states that up to E85 or gasoline can be used **(Figure 27–13)**.

FIGURE 27–13

Common flex fuel labels at the fuel receptacle and the outside of the vehicle.

Most of these have a virtual fuel sensor that relies on inputs from the HO_2S for oxygen readings. These systems adjust the air/fuel mixture according to the oxygen readings of the different fuel compositions that may be in the fuel tank.

DIESEL FUEL

Diesel fuel is designed to be used by diesel engines and is therefore not an alternative fuel for gasoline engines. Diesel fuel is a fossil fuel, but it has different properties and characteristics than gasoline. Diesel fuel is heavier and has more carbon atoms, and it has about 15 percent more energy density. Diesel fuel also evaporates much more slowly than gasoline—its boiling point is actually higher than the boiling point of water. The downside of diesel fuel is that it costs more per litre than gasoline.

Diesel fuel is used to power a wide variety of vehicles and other equipment. It fuels the diesel trucks you see moving down the highway, moving heavy loads. It also is used in trains, boats, buses, farm equipment, emergency response vehicles, electric generators, and many other applications.

Cetane Ratings

Diesel fuel should ignite almost instantaneously when it is injected into a cylinder. The fuel's ignition quality is expressed by a **cetane rating (CN)**. The cetane rating of diesel fuel is a measure of the ease with which the fuel can be ignited **(Figure 27–14)**. Because diesel fuel ignites with compression instead of a spark, it is most efficient when it ignites as quickly as possible.

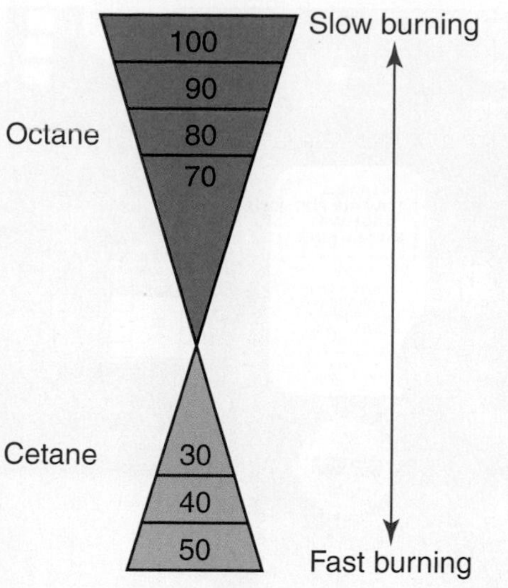

FIGURE 27–14

A comparison of cetane and octane ratings.

©Cengage Learning 2015

The cetane rating of the fuel is based on a test that uses a single-cylinder test engine with a variable compression ratio. As a result of the test, the fuel is assigned a cetane number. Cetane is a colourless, liquid hydrocarbon that ignites immediately when introduced to compression heat and pressures. Pure cetane has a numerical rating of 100, and diesel fuels are compared to cetane and rated according to its performance relative to pure cetane.

Actually, the cetane number represents the amount of time the fuel delays ignition after it has been introduced into the combustion chamber. Obviously, if the fuel ignites instantaneously (no delay), it will have a rating of 100. A short delay period allows the fuel to burn more completely, and the engine will run more smoothly with more power while producing low emissions. Fuel that takes a longer time to ignite does not allow the engine to work as efficiently. Generally speaking, the cetane number expresses the fuel's ability to start the engine at low temperatures and warm up quickly, without misfires.

Today's diesel engines run best with a fuel rated between 45 and 55 CN. The typical ignition temperature of CN 45 fuel is around 250°C (482°F). The ignition temperature for a CN 40 fuel would be higher, about 290°C (550°F). The cetane number can be improved by adding compounds such as ethyl nitrate, acetone peroxide, and amyl nitrate. The required cetane rating for an engine depends on many factors, including its engine design, size, speed of operation, and atmospheric conditions. Running a diesel engine with a fuel with a lower than recommended CN may cause abnormal noise and vibration, lower power output, excessive deposits and wear, and hard starting. The following is a list of commonly available diesel fuels with their cetane ratings:

- Regular diesel – CN 48
- Biodiesel blend (B20) – CN 50
- Premium diesel – CN 55
- Biodiesel (B100) – CN 55
- Synthetic diesel – CN 55

Grades of Diesel Fuel

The minimum quality standards for diesel fuel grades are established by the American Society for Testing Materials. They have defined two basic grades of diesel fuels: number 1 and number 2. Number 2 fuel is the most popular and most widely distributed, but it contains a significant amount of paraffin wax. However, wax does contain a great amount of energy and adds to the fuel's viscosity and ability to lubricate. Paraffin wax may cause cold weather problems. For example, as the fuel gets colder, wax crystals in the fuel lines can grow and inhibit the flow of the fuel, causing fuel starvation.

Number 1 diesel fuel is designed for extremely cold temperatures. It is less dense than number 2 and has a different boiling point and less paraffin wax. Most often, number 1 is blended with number 2 to improve cold weather starting. In moderately cold climates, the blend may be nine parts number 2 mixed with one part number 1. In very cold climates, the ratio may be as high as 50:50. Fuel economy can be expected to decrease during cold weather because of the use of number 1 fuel in the diesel fuel blend.

Biodiesel Fuels

Biodiesel is an alternative fuel for diesel engines. **Biodiesel fuel** is not made from petroleum; rather, it is made from renewable biological sources. Most commercially available biodiesel fuel is made from soybean oil. However, it is also produced from animal fats, recycled cooking oil, canola oil, corn, and sunflowers. The use of biodiesel fuel is not new; the first diesel engine ran on peanut oil. Biodiesel is considered a renewable fuel. To produce more fuel, more crops need to be planted.

Biodiesel can be used in diesel engines with little or no modifications to the engine. Pure biodiesel is biodegradable, nontoxic, and free of sulphur and aromatics. It can be used by itself or blended with petroleum-based diesel fuel. The two most common blends are B5, which is 95 percent petroleum-based fuel and 5 percent biodiesel fuel, and B20, which contains 20 percent biodiesel blended into regular diesel fuel.

An engine running on pure biodiesel (B100) emits much lower amounts of hydrocarbons, sulphur, carbon monoxide, carbon dioxide, and particulates when compared to running on petroleum-based fuel. However, U.S. Environmental Protection Agency tests have shown that the use of pure biodiesel increases an engine's NO_x emissions by 10 percent.

Other than these emission-related advantages and disadvantages, biodiesel fuel can also help reduce our dependency on imported oil. In addition, its use allows diesel engines to last longer and run more quietly and smoothly because of the lubricant qualities of the fuel. The downside is simply cost; it costs more to produce biodiesel than petroleum-based fuel.

Most late-model diesel engines can use biodiesel fuel without modifying the engine or fuel system. However, diesel engines produced prior to 1992 need modifications, which are dependent on the manufacturer of the engine. Even the manufacturers of newer "clean" diesel engines carefully define how biodiesel should be used in their engines, For example, Volkswagen (VW) recommended that only mixtures up to 5 percent biodiesel (B5) should be used in their early automotive diesel engines. To show how things change regarding this fuel, recently VW has allowed the use of fuel blends with 20 percent biodiesel and recommends the use of B5, at all times, because of its better lubricating properties. The manufacturer's recommendations are very important because vehicle warranties may become invalid if the wrong fuel is used.

Ultra-Low Sulphur Diesel Fuel

As of 2007, nearly all diesel fuel available in North America and Europe is **ultra-low sulphur diesel (ULSD) fuel**. This has become the standard for diesel fuel with low sulphur content **(Figure 27–15)**. The previous standards allowed diesel fuel to contain up to 500 ppm of sulphur (S500). S15 or ULSD is a much cleaner fuel that has a maximum sulphur content of 15 ppm. Engines running on this fuel emit less NO_x soot and other unwanted sulphur compounds. It is important to note that the use of this new fuel plus the new emissions devices for diesel engines has resulted in more than a 90 percent reduction in soot and NO_x **(Figure 27–16)**.

However, in the refining process necessary to reduce the amount of sulphur, the amount of paraffin is also reduced. As mentioned before, paraffin is vital to the lubrication of a diesel engine. Therefore, to offset the loss of paraffin and to protect the engine, injection pump, and the injectors, additives are blended into the fuel to increase lubricity. Blending ULSD fuel with a biodiesel fuel will increase its lubricity, and in the correct proportions it may allow the disadvantages of using ULSD to be dismissed.

FIGURE 27–15

A ULSD fuel dispenser.

DIESEL ENGINES

Diesel engines in cars and light trucks will become more common soon. There are many reasons for this, one of which is that low sulphur diesel fuel is now available in North America. This has eliminated some of the concerns about a diesel engine's high emission levels. Diesel engines provide more torque **(Figure 27–17)** than a gasoline engine of the same size and consume less fuel per kilometre.

The fuel injected into a diesel engine is ignited by the heat of compression. A diesel engine needs a compression ratio of at least 16:1. Intake air is

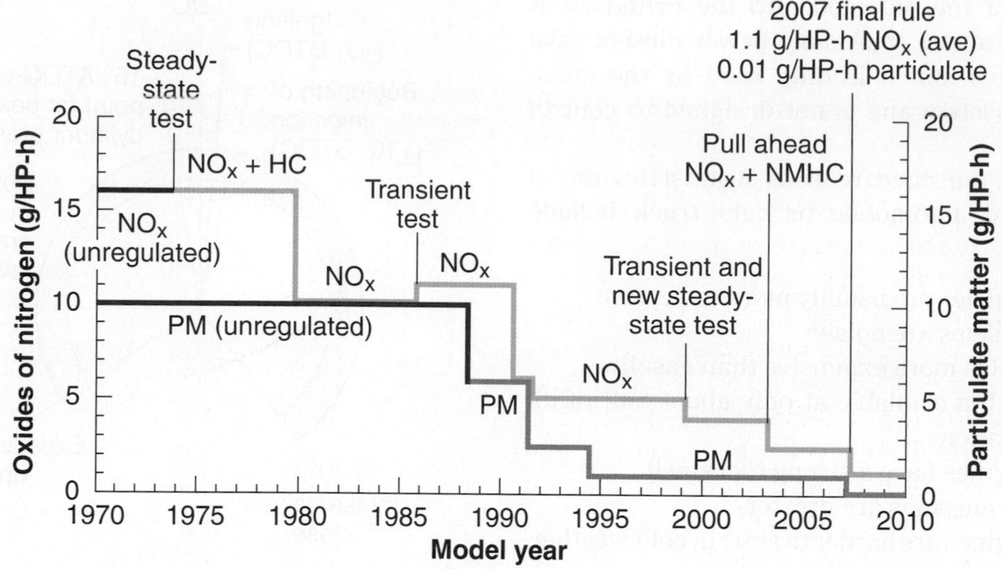

FIGURE 27–16

Current U.S. EPA NO_x emission standards for diesel engines.

compressed until its temperature reaches about 540°C (1000°F). This high compression ratio and the resulting high compression pressures make a diesel engine more fuel efficient than a gasoline engine. Because a diesel engine has very high compression pressures and produces high amounts of torque, it is made stronger and heavier than the same size gasoline engine. Because of their basic construction, diesel engines tend to last longer and therefore have higher resale values.

The fuel efficiency of a diesel also results from the fact the engines do not suffer from throttle losses because intake air is not controlled by a throttle. Engine speed and power output is controlled by

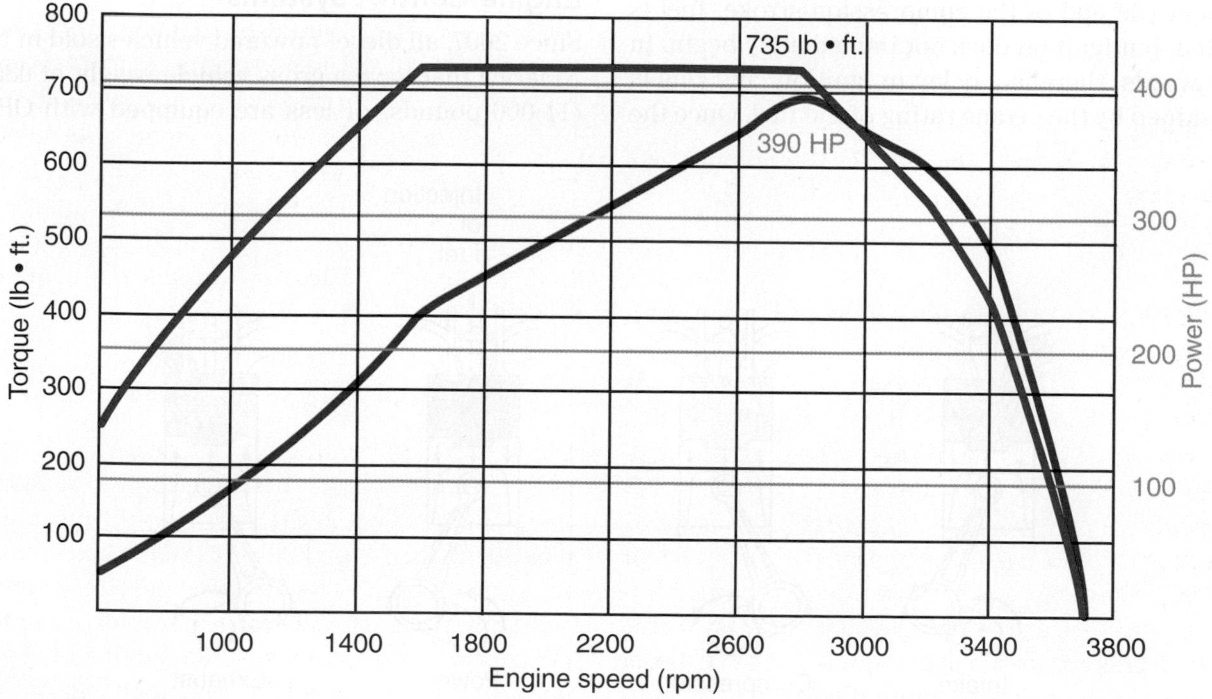

FIGURE 27–17

The power and torque curves for a late-model Ford Power Stroke diesel engine.

the amount of fuel injected into the cylinders. It is important to note that many newer diesels have a throttle valve, but it is only used by the emission control system and is not designed to control engine speed.

Typically, the cited reasons against owning a diesel-powered automobile or light truck include the following:

- Diesel vehicles are usually more expensive.
- Diesel vehicles are noisier.
- Diesel fuel is more expensive than gasoline.
- Diesel fuel is available at only about half of all service stations.
- Diesel engines have a strong fuel smell.
- Exhaust emissions are greater.
- Diesel engines are harder to start in cold weather.

Diesel Combustion

Nearly all available diesel engines for cars and light- and medium-duty trucks are based on the four-stroke cycle **(Figure 27–18)**. The primary difference between the strokes of a gasoline engine and a diesel engine takes place during the power stroke. Because the fuel in a diesel is not delivered until the piston on the compression stroke is near TDC and the heat formed during that stroke is what ignites the fuel, the actual combustion process can take three separate steps **(Figure 27–19)**.

Near the end of the compression stroke, fuel is injected, but ignition does not immediately begin. In other words, there is a delay in ignition, and this is determined by the cetane rating of the fuel. Once the

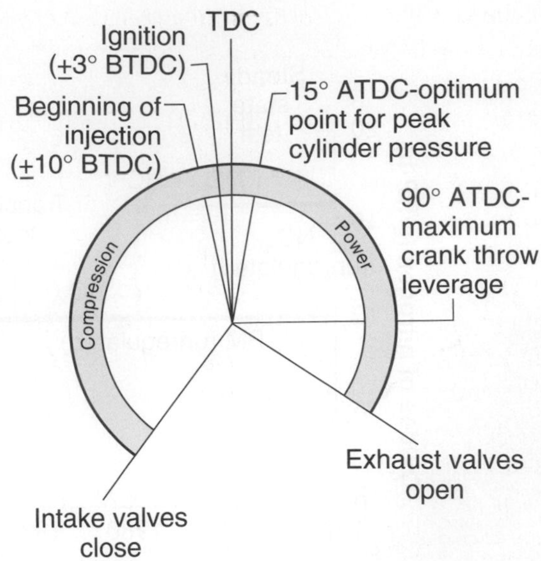

FIGURE 27–19

The events ("steps") that occur during the compression and power strokes of a diesel engine.

©Cengage Learning 2015

fuel begins to burn, pressure in the cylinder greatly increases. After this, the remaining fuel in the combustion chamber begins to burn, and this process continues until all of the fuel that is mixed with air burns.

Engine Control Systems

Since 2007, all diesel-powered vehicles sold in North America that have a gross vehicle weight of 6350 kg (14 000 pounds) or less are equipped with OBD II.

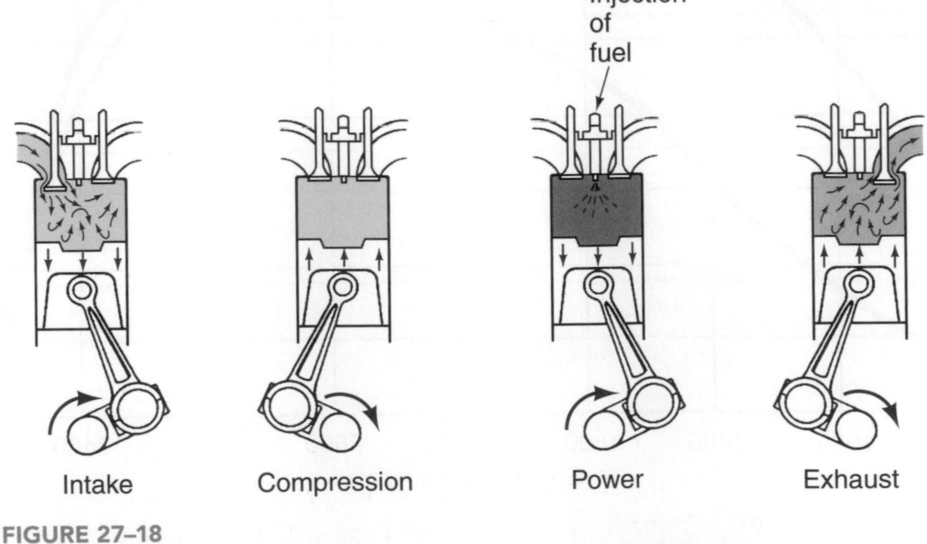

FIGURE 27–18

The four-stroke cycle of a diesel engine.

©Cengage Learning 2015

This is basically the same system as used in gasoline vehicles. Like those systems, a scan tool can be used to retrieve DTCs and other data. The malfunction indicator lamp (MIL) is illuminated when certain DTCs are set. The DTCs are set by the PCM based on input from a variety of sensors and by the various monitors in the system.

Diesel OBD-II monitors all systems that may affect the emissions from the engine. These monitors check the electrical status of the inputs and outputs of each circuit. They also look at the data being sent to determine if they make sense and are in line with other data. The monitors also watch how the actuators respond to the commands from the PCM. The mandatory monitors include the following:

- *Comprehensive component monitor.* This performs functional and rationality tests on emission-related circuits.
- *Misfire monitor:* This monitor will set the MIL when a misfire occurs, when the engine is at idle, if the vehicle has an automatic transmission. The MIL will not be set if the vehicle has a manual transmission.
- *Glow plug monitor:* If a problem with the glow plugs is detected, the MIL will be set. This occurs on vehicles that weigh less than 3856 kg (8500 pounds).

Although many of the monitors for a diesel engine are similar to those for a gasoline engine, some are unique to the diesel. Examples are the exhaust gas recirculation (EGR) cooler monitor and the **diesel oxidation catalyst (DOC)** efficiency monitor. The EGR cooler monitor checks the efficiency of the cooler by monitoring the temperature difference between the inlet and outlet sensors with the EGR valve open. The DOC monitor relies on exhaust gas temperature sensors and the monitor runs during active regeneration of the **diesel particulate filter (DPF)**. If the temperature of the exhaust gas does not increase to the desired minimum, a DTC is set and the MIL is illuminated.

The control system monitors the engine while it is running and makes adjustments to ensure overall efficiency. Inputs that are critical to efficient operation are those related to cylinder pressures. By monitoring the pressures inside the cylinders, the PCM can make accurate decisions about the combustion process.

Most engines equipped with cylinder pressure sensors are fitted with glow plugs that have an integrated pressure sensor **(Figure 27–20)**. These sensors are basically a normal glow plug and have been modified to allow the centre electrode to move up and down. The movement of the electrode is in response to cylinder pressure. The movement changes the position and pressure the electrode has on a sensing element. Depending on these factors, the sensing element sends a voltage signal to the PCM.

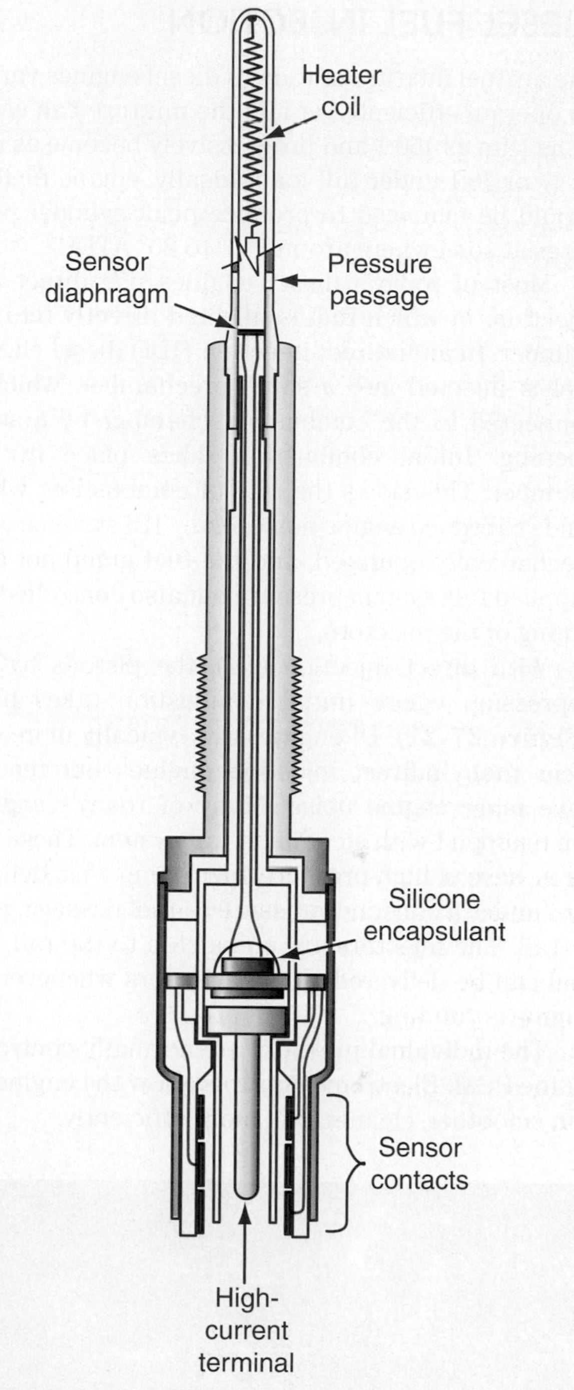

FIGURE 27–20

A piezo-resistive cylinder pressure sensor integrated into a glow plug.

DIESEL FUEL INJECTION

The air/fuel mixture of today's diesel engines varies. To operate efficiently, at idle the mixture can easily be as lean as 150:1 and progressively become as rich as 17 or 18:1 under full load. Ideally, engine fuelling should be managed to produce peak cylinder pressures at somewhere around 10° to 20° ATDC.

Most of today's diesel engines use direct fuel injection, in which fuel is injected directly into the cylinder. In an indirect injection (IDI) diesel engine, fuel is injected into a small prechamber, which is connected to the combustion chamber by a small opening. Initial combustion takes place in this chamber. This slows the rate of combustion, which tends to reduce engine noise. Many IDI systems were mechanically operated, and the fuel pump not only supplied fuel system pressure but also controlled the timing of the injectors.

With direct injection (DI), the pistons have a depression where initial combustion takes place **(Figure 27–21)**. DI engines are typically more efficient than indirect injection engines but tend to have more engine noise. Many of today's engines are equipped with a common rail system. These systems have a high-pressure fuel pump that delivers fuel under a constant pressure to fuel passage, pipe, or rail. The injectors are connected to the rail, and fuel can be delivered to the cylinders whenever the engine is running.

The individual injectors are normally controlled by the PCM. Electronic controls allow the engines to run smoother, cleaner, and more efficiently.

FIGURE 27–21

Notice the depression at the top of the pistons of this GM Duramax diesel.

Injector Nozzles

An injector nozzle is the part that delivers the fuel into the engine's cylinders. The nozzles are threaded into or clamped onto the cylinder head; there is one for each cylinder. Many nozzles are spring-loaded valves that, when the injector is activated, spray fuel into the combustion chamber or a pre-combustion chamber. Nearly all of today's diesel engines have multiple-orifice hydraulic or electrohydraulic nozzles.

The exact type of nozzle used mainly depends on the type of injection system. However, the tip of all nozzles has many orifices or holes from which atomized fuel is sprayed into the cylinders. The pressure of the fuel determines how much the fuel will be broken down into droplets. It is important to note that a change in pressure regulates the amount of fuel delivered. The time an injector is turned on also determines fuel quantity.

Electronic Unit Injection (EUI)

Electronic unit injection (EUI) systems have been used since the 1980s, but primarily in heavy-duty trucks. Volkswagen used EUIs in its TDI engines until 2010 **(Figure 27–22)**. Since then, Volkswagen has used a common rail injection system that will be discussed later.

In these systems, the injector's pressure is the result of a camshaft-driven high-pressure pump. However, the engine's PCM controls the amount of fuel sprayed into the cylinders and the timing of the spray. Although effective, EUIs did not allow for the precise control of injection needed to meet today's driving needs and emission standards.

Hydraulic Electronic Unit Injection (HEUI)

Hydraulic electronic unit injection (HEUI) relies on engine oil pressure to control the operation of the individual injectors **(Figure 27–23)**. The engine oil opens the injector by pressing down on a diaphragm inside each injector. In turn, the diaphragm pushes on the fuel inside the injector and pressurizes it from 20 700 to 144 800 kPa (3000 to 21 000 psi). Fuel under that pressure is directly injected into the cylinders. Since the injectors in this system pressurize the fuel for delivery into the cylinders, there is no need for an expensive and difficult-to-control high-pressure fuel pump.

These systems were widely used until 2010. The system as used in Ford's 6.0, 6.4, and 7.3 Power Stroke diesels is comprised of the following:

- A high-pressure engine oil pump and reservoir

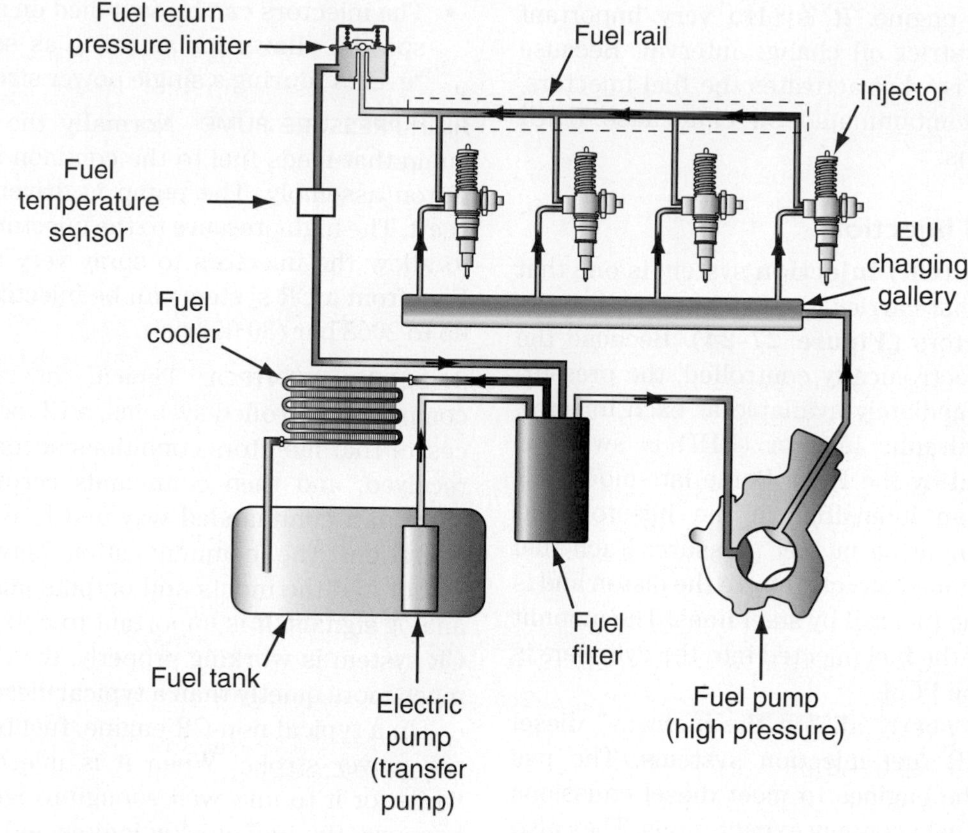

FIGURE 27–22

A pre-2010 VW TDI EUI fuel system.

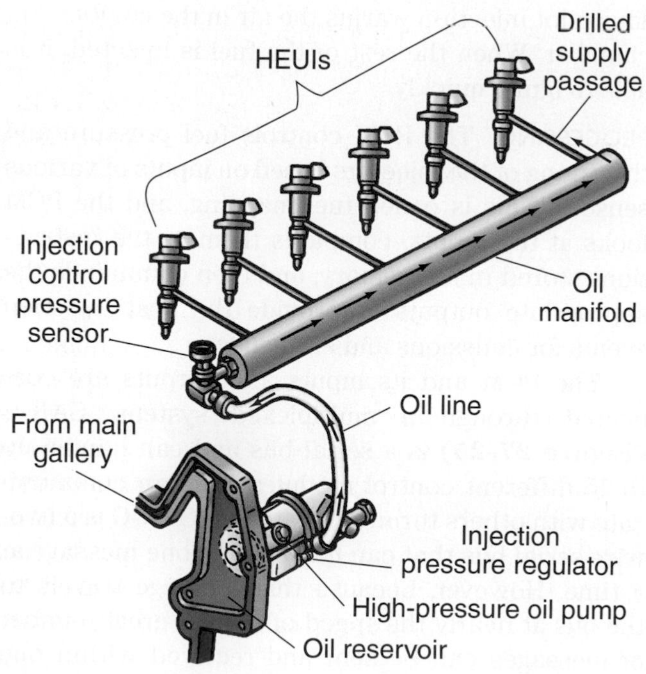

FIGURE 27–23

The oil system that forces the HEUI injector open when the PCM allows it.

- Engine oil pressure regulator
- High-pressure (stepper) pump
- Actuation oil pressure sensor
- Passages in the cylinder head for fuel flow to the injectors
- HEUI injectors
- Electronic control module (ECM)

OPERATION Fuel is drawn out of the tank by a fuel pump. The fuel, under low pressure, passes through fuel lines and the fuel filter, water separator, and fuel heater. The fuel is then delivered to the fuel passages in the cylinder head or rail. Oil from the engine's lubrication system is pumped into a high-pressure pump. This pump increases the engine oil's pressure to a precisely controlled amount. The engine oil's path includes a return to the engine's oil reservoir. The pressurized oil is then sent to the injectors. The presence of the pressurized oil actuates the injectors. The PCM controls the pressure buildup, and the actual actuation pressures are between 33 bar (485 psi) and 275 bar (4000 psi). The PCM therefore controls the amount of fuel injected into the cylinders.

IMPORTANCE OF MAINTENANCE It is very important that the correct type of oil be used in an

HEUI-equipped engine. It is also very important to adhere to a strict oil change interval. Because the engine's oil is what activates the fuel injectors, the wrong or contaminated oil can cause HEUI injector problems.

Common Rail Injection

A **common rail (CR) injection** system is one that has a fuel rail that carries high-pressure fuel to the individual injectors **(Figure 27–24)**. Because the injectors are electronically controlled, the pressurized fuel is immediately available at each injector. Each electrohydraulic injector (EHI) is switched open and closed by the PCM. Some late-model CR systems have an intensifier in the injector that allows for an increase in fuel pressure. Each fuel injector is positioned directly above the piston and is connected to the fuel rail by steel lines. The amount and pressure of the fuel injected into the cylinders is controlled by the PCM.

Since 2010, nearly all "on the highway" diesel engines use CR fuel injection systems. The use of CR allows the engines to meet diesel emissions standards and fuel economy expectations. They also provide for the following:

- Lower engine noise levels
- Balanced engine cylinder pressures
- High injection pressures independent of engine speed

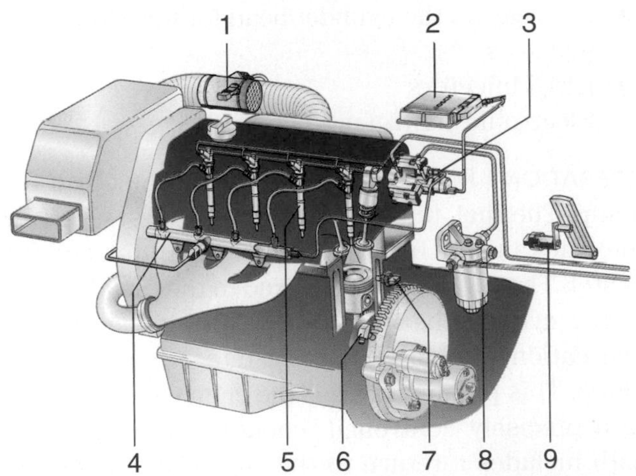

FIGURE 27–24

A common rail system for a four-cylinder engine.

- The injectors can be switched on and off at high speeds; allows for as many as seven injection "events" during a single power stroke.

HIGH-PRESSURE PUMP Normally, the high-pressure pump that feeds fuel to the common rail is a three-piston assembly. The pump is driven by the camshaft. The high pressure to the injectors is necessary to allow the injectors to spray very atomized fuel. Fuel from a CR system can be injected at pressures up to 2068 bar (30 000 psi).

INJECTOR CONTROL Typical of all late-model computer-controlled systems, a CR setup has a processor that monitors conditions according to inputs received, and then commands certain outputs to react in a commanded way and to ensure efficient operation. The communication between the processor and the inputs and outputs may be digital or analog signals. It is important to note that when the CR system is working properly, the engine will run much more quietly than a typical diesel engine.

In a typical non-CR engine, fuel is injected once per power stroke. When it is injected, it takes a while for it to mix well enough to ignite. Once this happens, the fuel quickly ignites and causes a large pressure increase in the combustion chamber. This results in the common knock of a diesel engine. In a CR system, the injection cycle begins with one or two very small sprays of fuel, called pilot injection. This pilot injection warms the air in the combustion chamber. When the rest of the fuel is injected, it is able to ignite quickly.

PROCESSING The PCM controls fuel pressure and the timing of the injectors based on inputs of various sensors. This is called fuel mapping, and the PCM looks at the inputs, compares them to the instructions stored in its memory, and then commands the appropriate outputs to provide the best injection events for emissions and efficiency.

The PCM and its inputs and outputs are connected through a multiplexed system. CAN-C **(Figure 27–25)** is a serial bus and can handle up to 15 different control modules and can communicate with others through a gateway. CAN-C is a two-wire serial bus that can handle only one message at a time. However, because that message travels to the bus at nearly the speed of light, a great number of messages can be sent and received within one second **(Figure 27–26)**.

INPUTS Many different inputs provide the current status of the engine to the PCM. Many of these are the same as used on other diesel and gasoline engines

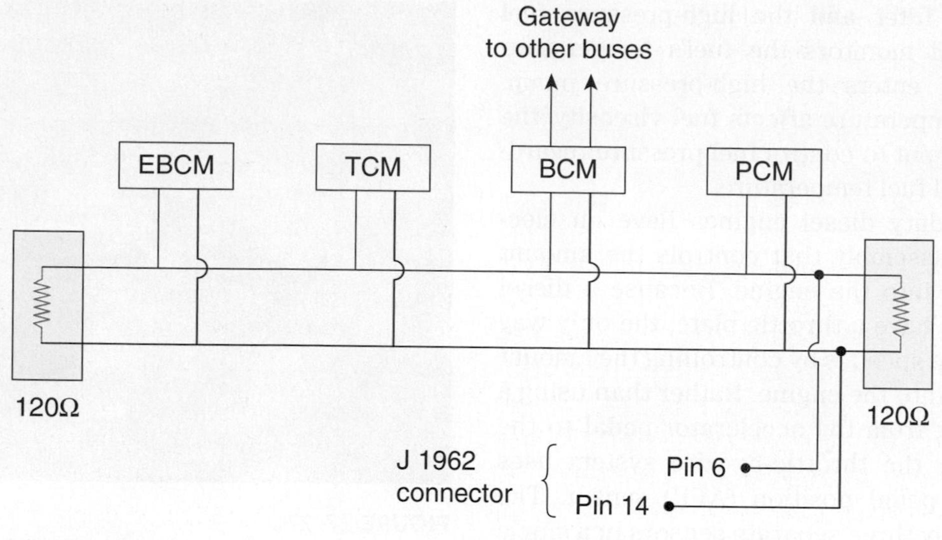

FIGURE 27–25

A typical serial data bus.

and are tied into the powertrain bus. Different from other engine control systems, the PCM in a CR system also monitors ambient air temperature and retards the engine's timing of the engine in cold weather and retards injection timing to allow for easier starting.

The system also includes a **fuel rail pressure (FRP) sensor**. The FRP sends the actual fuel pressure to the PCM. The FRP is threaded into the fuel rail. The FRP is a three-wire variable capacitance sensor. The PCM supplies a 5-volt reference signal, and the FRP sensor sends a portion of that to the PCM to indicate fuel pressure. The FRP sensor monitors fuel rail pressure continuously to provide feedback to the PCM.

There is also a **fuel rail temperature (FRT) sensor** mounted in the fuel line between the

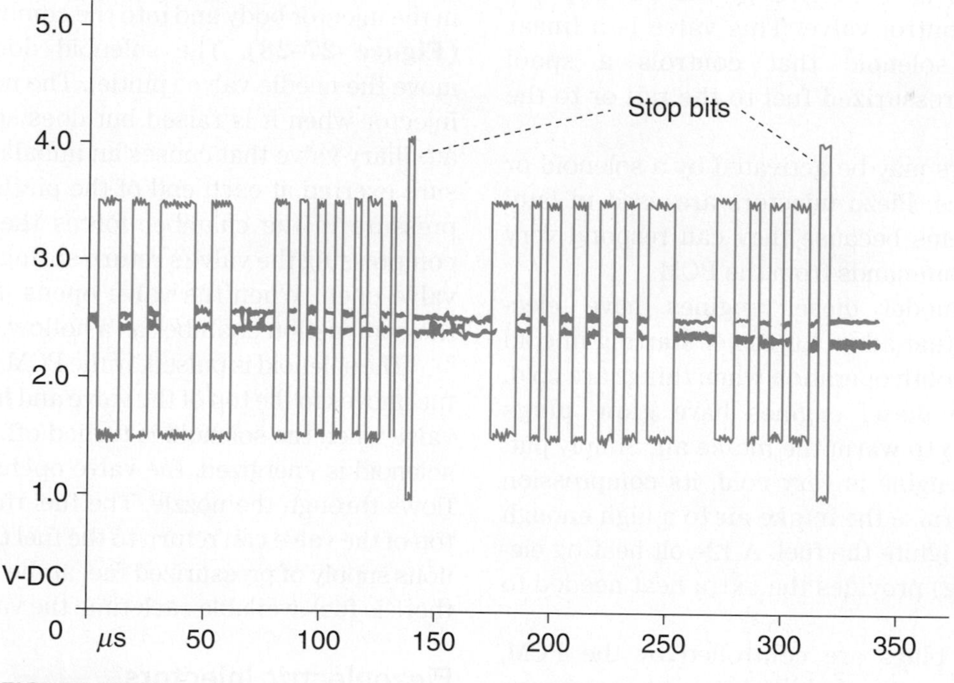

FIGURE 27–26

A scope pattern of a properly functioning CAN-C bus showing a 2.0 V differential throughout. The two 3.0 V differentials between the CAN high and low voltages indicate the end of a message.

secondary fuel filter and the high-pressure fuel pump. The PCM monitors the fuel's temperature before the fuel enters the high-pressure pump. Because fuel temperature affects fuel viscosity, the PCM uses this input to control fuel pressure regardless of the actual fuel temperature.

Some light-duty diesel engines have an electronic throttle assembly that controls the amount of fuel injected into the engine. Because a diesel engine does not have a throttle plate, the only way to control engine speed is by controlling the amount of fuel injected into the engine. Rather than using a mechanical link from the accelerator pedal to the injection pump, the throttle-by-wire system uses an accelerator pedal position (APP) sensor. The sensor is actually three separate sensors in a single assembly that changes input voltage according to the position of the accelerator pedal. The PCM looks at the voltage signals from each of the three sensors and compares them to what they should be if there are no faults. If an error is detected, engine and vehicle speed are often reduced to allow the system to operate in spite of the discrepancies from the sensors.

OUTPUTS The primary outputs to control fuel delivery in a CR system are the rail pressure control valve, injectors, and glow plugs. Fuel pressure in the rail is controlled by the PCM by the rail pressure control valve. This valve is a linear proportioning solenoid that controls a spool valve to send pressurized fuel to the rail or to the return circuit.

The injectors may be activated by a solenoid or piezo electronics. Piezo injectors are used in late-model CR systems because they can respond very quickly to the commands from the PCM.

Most late-model diesel engines have electronic controls that allow for easier starting in cold weather and smooth operation when things are cold. However, some diesel engines have **glow plugs** **(Figure 27–27)** to warm the intake air. Simply put, when a diesel engine is very cold, its compression stroke may not raise the intake air to a high enough temperature to ignite the fuel. A 12-volt heating element (glow plug) provides the extra heat needed to ignite the fuel.

Most glow plugs are controlled by the PCM, which monitors engine and intake air temperatures. Glow plugs are cycled on and off depending on the temperature of the engine. The PCM also keeps the glow plugs energized after the engine starts to improve the engine's idle quality.

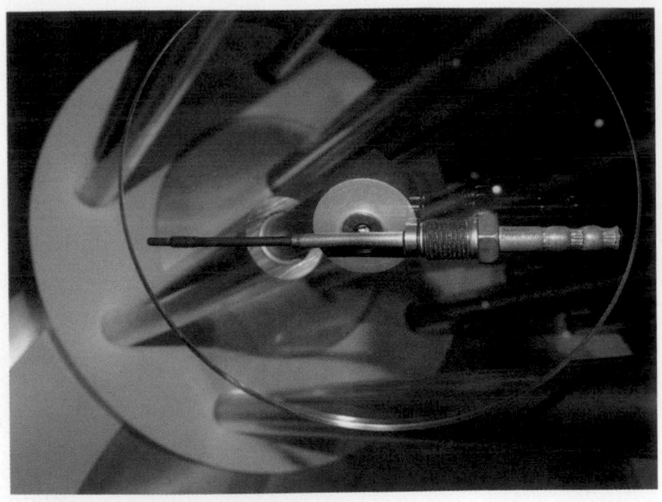

FIGURE 27–27

A glow plug.

Solenoid Injectors

In a CR system, pressurized diesel fuel is applied to the injectors, which are controlled by a PCM via a solenoid. Because the injectors are controlled by the computer, combustion can be controlled to provide maximum engine efficiency with the lowest possible noise and low exhaust emissions.

The solenoid is attached to the injector nozzle and opens to allow fuel to flow through a fuel passage in the injector body and into the combustion chamber **(Figure 27–28)**. The solenoid does not directly move the needle valve (pintle). The needle opens the injector when it is raised but does so by moving an auxiliary valve that causes an imbalance in the pressure exerted at each end of the pintle. The high fuel pressure in the chamber forces the valve upward, compressing the valve's return spring and forcing the valve open. When the valve opens, fuel sprays into the combustion chamber as a hollow cone spray.

The solenoid is pulsed by the PCM; therefore, some fuel moves to the top of the valve and helps to close the valve when the solenoid is turned off. The instant the solenoid is energized, the valve opens and fuel again flows through the nozzle. The fuel that moves to the top of the valve can return to the fuel tank. The continuous supply of pressurized fuel at the injector ensures there is fuel available each time the valve opens.

Piezoelectric Injectors

Piezoelectric injectors operate in much the same way as solenoid injectors, except they do not rely on an electrical winding or coil. The injectors have no moving parts. Rather, a piezo injector relies on many

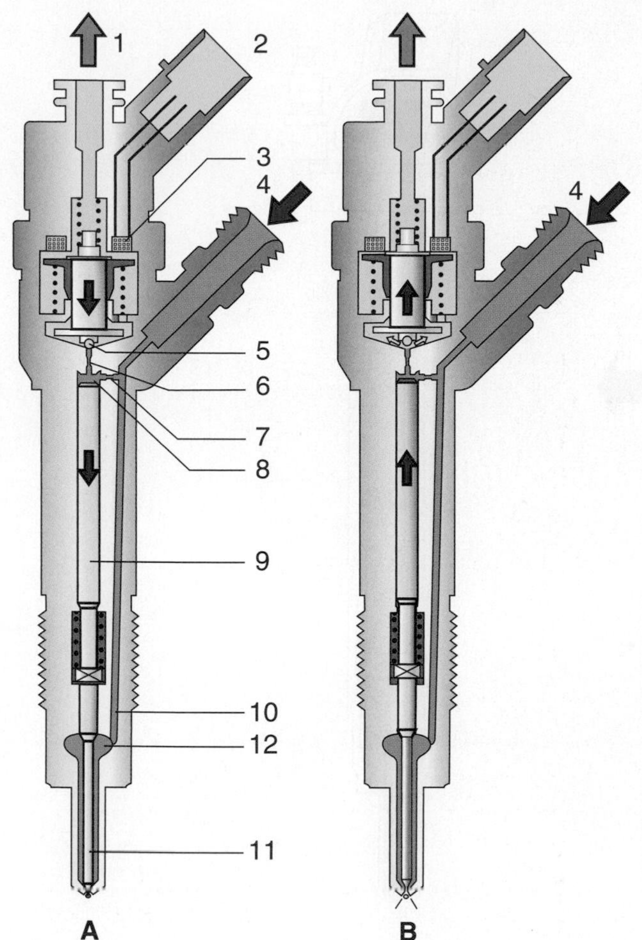
1
2
3
4
5
6
7
8
9
10
12
11
4
A
B

Injector (schematic)

A. Injector closed (at-rest status)
B. Injector opened (injection)
1. Fuel return
2. Electrical connection
3. Triggering element (solenoid valve)
4. Fuel inlet (high pressure) from the rail
5. Valve ball
6. Bleed orifice
7. Feed orifice
8. Valve control chamber upper pressure field
9. Valve control plunger
10. Feed passage to the nozzle
11. Nozzle needle
12. Lower pressure field

FIGURE 27–28

A solenoid injector.

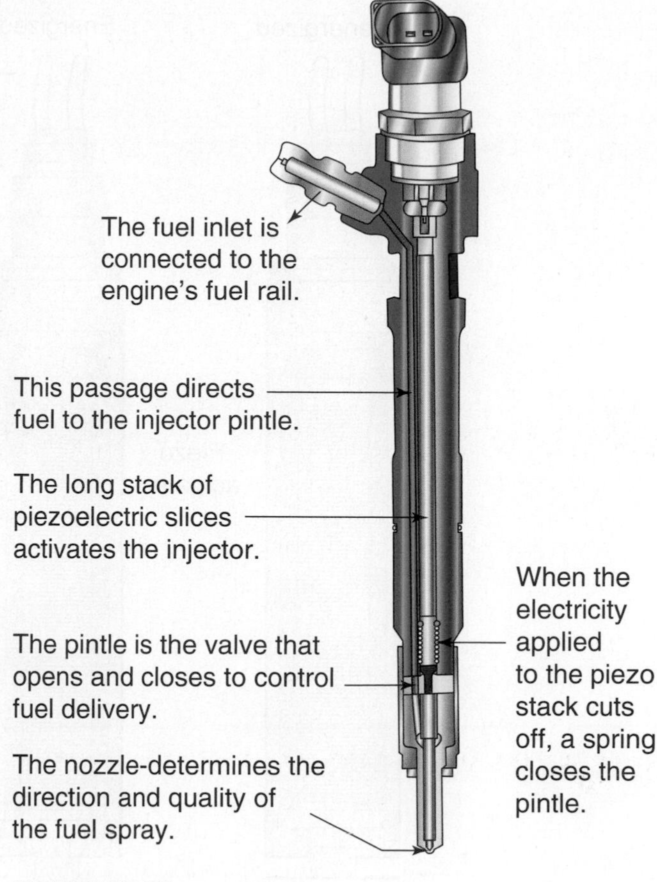
The fuel inlet is connected to the engine's fuel rail.
This passage directs fuel to the injector pintle.
The long stack of piezoelectric slices activates the injector.
When the electricity applied to the piezo stack cuts off, a spring closes the pintle.
The pintle is the valve that opens and closes to control fuel delivery.
The nozzle-determines the direction and quality of the fuel spray.

FIGURE 27–29

A piezoelectric injector.

very thin layers of piezoelectric crystal material stacked on each other at the orifice of the injector's nozzle **(Figure 27–29)**. When electricity is applied to the stack, it expands slightly (approximately 0.102 mm or 0.004 in.). The expansion, however, is enough to allow fuel to spray out of the nozzle **(Figure 27–30)**. When no electricity is applied to the stack, fuel cannot leave the nozzle because the piezo layers are tightly jammed against each other. The pintle closes by two forces, a return spring and high fuel pressure. Current applied to the injector, and the operation of the injectors is controlled by the PCM.

A crystal is a clear, transparent mineral, such as quartz. Other piezoelectric crystals include table sugar and tourmaline. Quartz is one of the most common piezoelectric minerals; it is used in many devices such as radios and watches.

Piezo material is a type of crystal that rapidly expands when it is exposed to electrical current. They also retract just as quickly when current to them is stopped. However, the amount the crystals expand and retract is very small. The slight change in shape forms small gaps between the layers of the piezo material. This allows for a precisely controlled flow of fuel out of injector.

Each injector has at least 400 separate layers of piezo material.

Piezoelectric materials can also emit a small amount of voltage when they are struck, squeezed, or exposed to vibration. Piezoelectric materials are often used in pressure sensors.

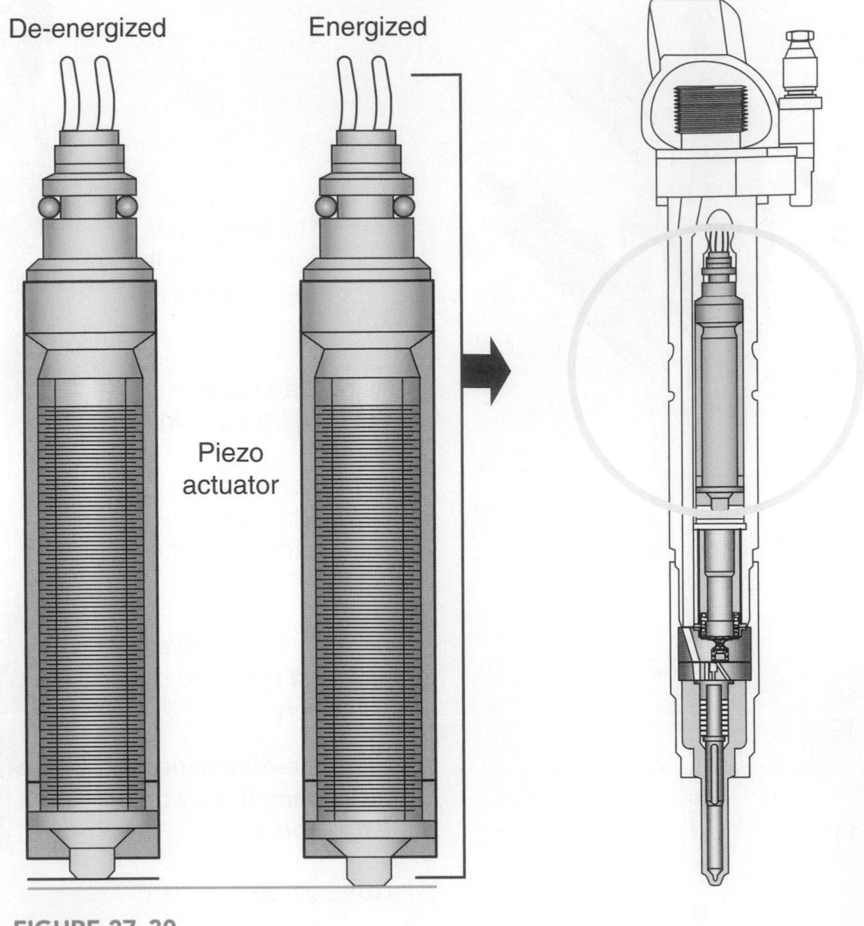

FIGURE 27–30

The action of a piezoelectric injector.

The advantages of piezo over solenoid injectors are the following:

- They offer improved fuel economy.
- Since there are no moving parts, they have a longer life.
- They reduce combustion noise.
- They allow for more precise and rapid control of injection intervals.
- They offer improved combustion.
- They reduce exhaust emissions.
- They can allow seven or more smaller and staggered sprays of fuel during a single power stroke.

Fuel Delivery

The fuel tank for a diesel engine has a larger filler neck than those found on gasoline vehicles. This is one reason some consumers make the mistake of putting gasoline in their diesel tank. Also, because diesel fuel is not as volatile as gasoline, the fuel tank is not fitted with evaporative emission control devices.

The fuel in the tank is drawn out by an in-tank transfer pump **(Figure 27–31)**. This pump is normally an electrically driven, low-pressure, high-volume pump. The pump may run continuously or be cycled by the PCM. The purpose of the pump is to supply fuel to the injection pump. The injection pump is used to increase the pressure of the fuel so that it can be injected.

Somewhere between the transfer pump and the injectors is a water/fuel separator, which may be part of the fuel filter. Because water is heavier than the fuel, it sinks to the bottom of the separator. Draining the water out of the separator is part of a normal preventive maintenance service for diesel engines. Typically, there is a float inside the separator. The float completes the circuit for a water warning lamp when the level of water in the separator reaches a point where it should be drained. It is important to drain the water because it can cause premature wear as it is not a good lubricant. It can also damage the injector nozzles because water is not easily atomized.

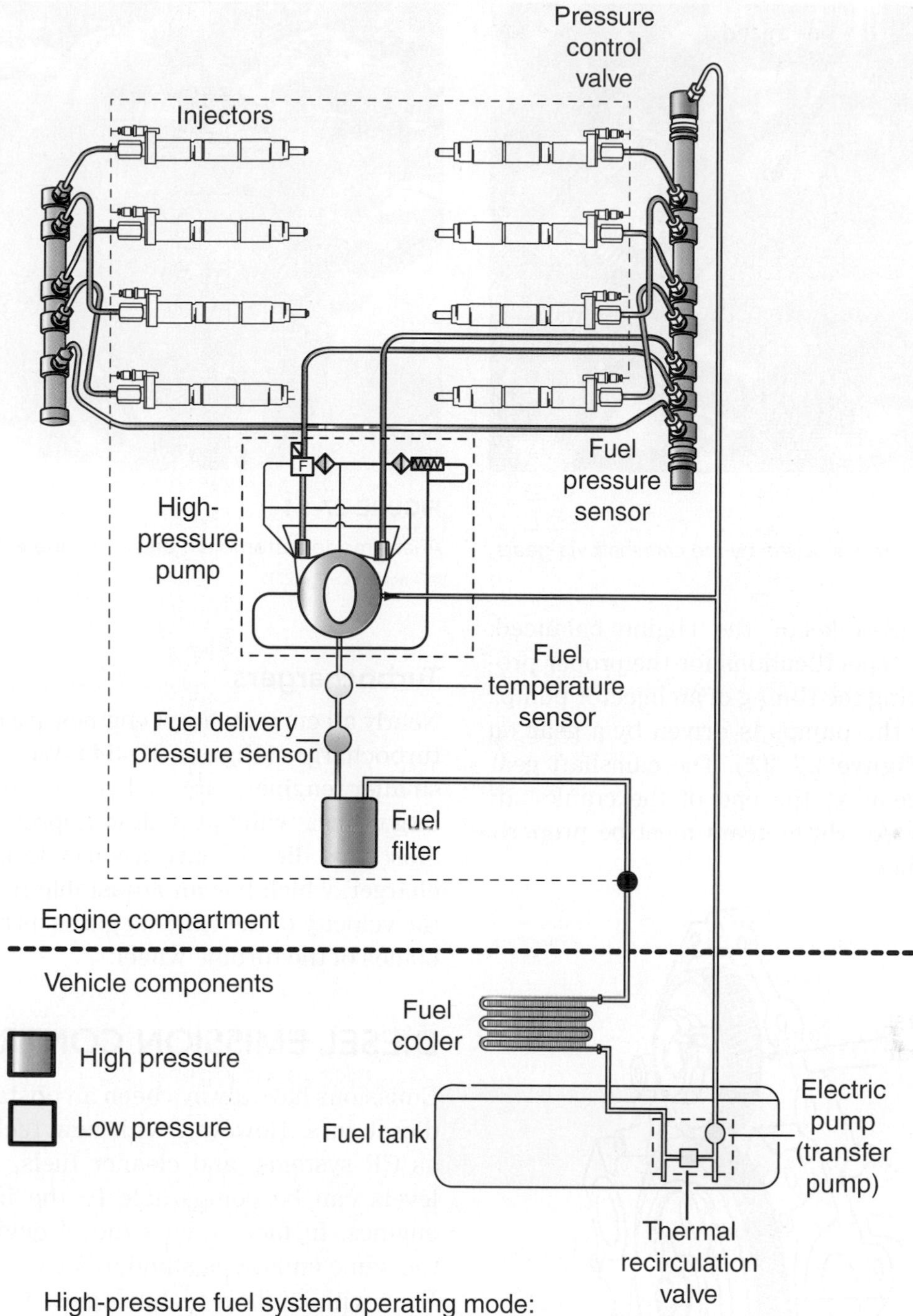

FIGURE 27–31

A fuel supply system for a V8 diesel engine with common rails.

High-pressure fuel lines connect the high-pressure fuel pump to the fuel rails and injectors. Often, the high-pressure pump is driven by a gear on the front of the camshaft.

TIMING THE PUMPS The high-pressure pumps used in common rail systems may need to be timed to the engine. Timing the pumps not only creates a starting point for any timing changes made by the engine

FIGURE 27–32

The high-pressure pump is driven by the camshaft via gears.

controls but it also keeps the engine balanced. Always check the specifications for the proper procedure for adjusting the timing of an injector pump.

The gear for the pumps is driven by a gear on the camshaft **(Figure 27–32)**. The camshaft gear is driven by a gear on the end of the crankshaft **(Figure 27–33)**. All three gears must be properly timed to each other.

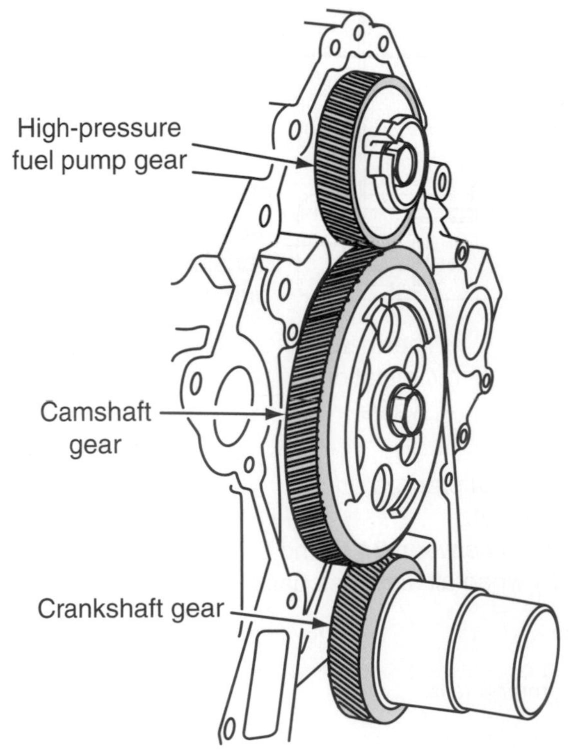

FIGURE 27–33

The crankshaft, camshaft, and fuel pump gears must be perfectly aligned.

FIGURE 27–34

A late-model automotive diesel engine with a turbocharger.

Turbochargers

Nearly all current diesel engines are equipped with a turbocharger **(Figure 27–34)**. With a turbocharger, smaller engines can deliver as much power as larger ones while providing improved fuel mileage. Most new diesels have a variable geometry turbocharger, which has an adjustable nozzle that varies the velocity of the exhaust gases as they push on the blades of the turbine wheel.

DIESEL EMISSION CONTROLS

Emissions have always been an obstacle with having diesel cars. However, with new technologies (such as CR systems) and cleaner fuels, their emissions levels can be comparable to the best of gasoline engines. In fact, today's diesel engines must meet the same emissions standards as a gasoline engine. Even without these technologies, diesel engines emit small amounts of carbon monoxide, hydrocarbons, and carbon dioxide. However, diesel engines naturally emit high amounts of particulate matter (PM), or soot, and NO_x emissions. To decrease these emissions, many new diesel vehicles have an assortment of catalysts and an EGR system to clean the exhaust before it leaves the tailpipe; others use selective catalytic reduction (SCR) systems **(Figure 27–35)**.

Diesel Oxidation Catalysts

Diesel oxidation catalysts (DOC) have been used on all light-duty diesel vehicles since 2007. They are a flow-through honeycomb-style substrate assembly

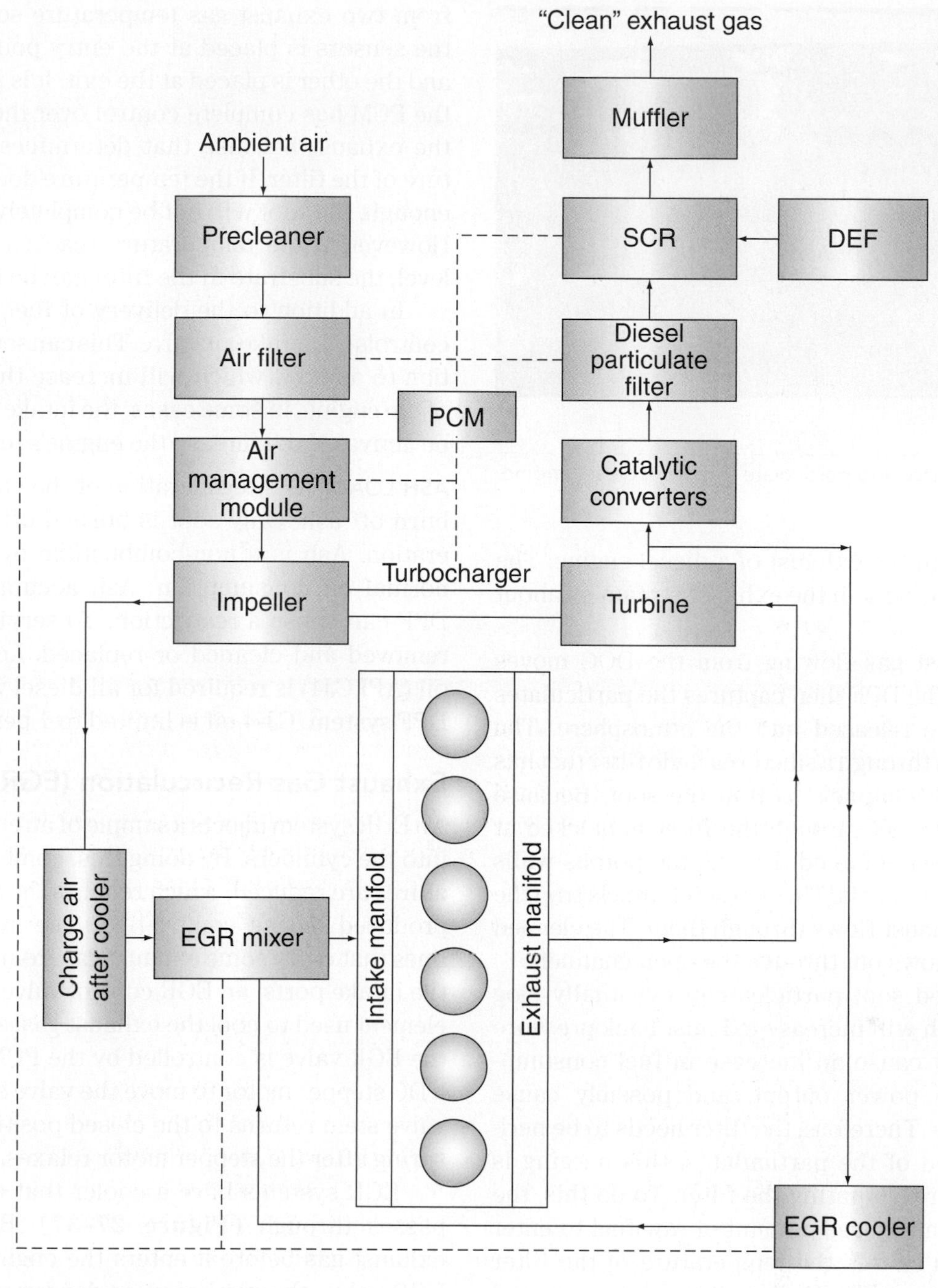

FIGURE 27–35

The flow of the intake and gases for a typical diesel engine.

placed in the exhaust stream **(Figure 27–36)**. Similar to the catalytic converters used on gasoline engines, the substrate is wash-coated with a layer of catalyst materials, such as platinum, palladium, and other base metal catalysts.

The main purpose of a DOC is to convert fuel-rich gases in the exhaust to heat, which reduces the amount of CO, HC, and other compounds that cause obnoxious exhaust odours.

Diesel Exhaust Particulate Filter

Particulate matter refers to the tiny particles of solid or semisolid material present in the exhaust of a diesel engine. Soot is a natural by-product of the combustion of diesel fuel.

Diesel particulate filters have been used in all light-duty diesel vehicles since 2007. A DPF works with the oxidation catalyst and an EGR valve to remove a majority of the NO_x, soot, and unburned

FIGURE 27–36

A catalytic converter and particulate trap for a diesel engine.

Martin Restoule

hydrocarbons in the exhaust of a diesel engine. The DPF reduces the soot in the exhaust stream by about 90 percent.

The exhaust gas flowing from the DOC moves into the DPF. The DPF then captures the particulates before they are released into the atmosphere. The exhaust passes through a silicon carbide filter that has honeycomb-cell channels to trap the soot. Because every other channel through the filter is blocked at one end, exhaust is forced through the porous walls of the blocked channels. The porous channels trap the soot as the exhaust flows through them. The cleaned exhaust then flows out through the open channels.

The trapped soot particles can eventually clog the filter, which will increase exhaust back pressure and eventually cause an increase in fuel consumption, decrease power output, and possibly cause engine damage. Therefore, the filter needs to be periodically purged of the particulates; this purging is referred to as regenerating the filter. To do this, the PCM allows a measured amount of raw fuel to enter the filter. This causes the temperature of the filter to greatly increase. The heat then burns the trapped soot, and the filter is regenerated or renewed.

Several factors can trigger the PCM to perform regeneration: distance since last regeneration, fuel used since last regeneration, engine run time since last regeneration, and exhaust differential pressure across the DPF. Differential pressure sensors (DPS) are placed at the entrance and exit points of the filter. By monitoring the exhaust pressure at two points, the PCM can determine the differential in exhaust pressure and determine how freely the exhaust is flowing through the filter.

To determine when and what amount of fuel should enter the exhaust, the PCM relies on the inputs from two exhaust gas temperature sensors. One of the sensors is placed at the entry point of the filter and the other is placed at the exit. It is important that the PCM has complete control over the fuel entering the exhaust because that determines the temperature of the filter. If the temperature does not increase enough, the soot will not be completely burned away. However, if the temperature reaches too high of a level, the substrate in the filter can be damaged.

In addition to the delivery of fuel, the PCM also controls the intake air valve. This can serve as a restriction to airflow, which will increase the temperature of the engine. In some cases, the intake air heater may be activated to increase the engine's temperature.

ASH LOADING Regeneration of the catalyst will not burn off ash. Only soot is burned off during regeneration. Ash is a non-combustible by-product from normal oil consumption. Ash accumulation in the DPF can cause a restriction. To service a DPF, it is removed and cleaned or replaced. Low ash engine oil (API CJ4) is required for all diesel vehicles with a DPF system. CJ-4 oil is limited to 1 percent ash.

Exhaust Gas Recirculation (EGR) Systems

An EGR system injects a sample of an engine's exhaust into the cylinders. By doing this, combustion temperatures are reduced, which reduces the amount of NO_x produced during combustion. The system includes lines that carry some exhaust gas from the exhaust to the intake ports, an EGR control valve, and a cooling element used to cool the exhaust gases. The action of the EGR valve is controlled by the PCM and may use a DC stepper motor to move the valve stem open. The valve stem returns to the closed position by a return spring after the stepper motor relaxes.

EGR systems have a cooler that engine coolant passes through **(Figure 27–37)**. By cooling the exhaust gas before it enters the engine through the EGR valve, the gas becomes denser and keeps combustion chamber temperatures low. The amount of EGR gases that flow through the cooler is determined by the PCM, which receives input from an EGR temperature sensor.

Selective Catalytic Reduction Systems

The **selective catalytic reduction (SCR)** unit fits in the exhaust between the DOC and DPF **(Figure 27–38)**. It reduces NO_x emissions when a reductant is injected over the SCR catalyst. The use of an SCR breaks down NO_x into water vapour and nitrogen (N_2), thereby reducing the amount of EGR needed to greatly decrease NO_x. In some cases, an

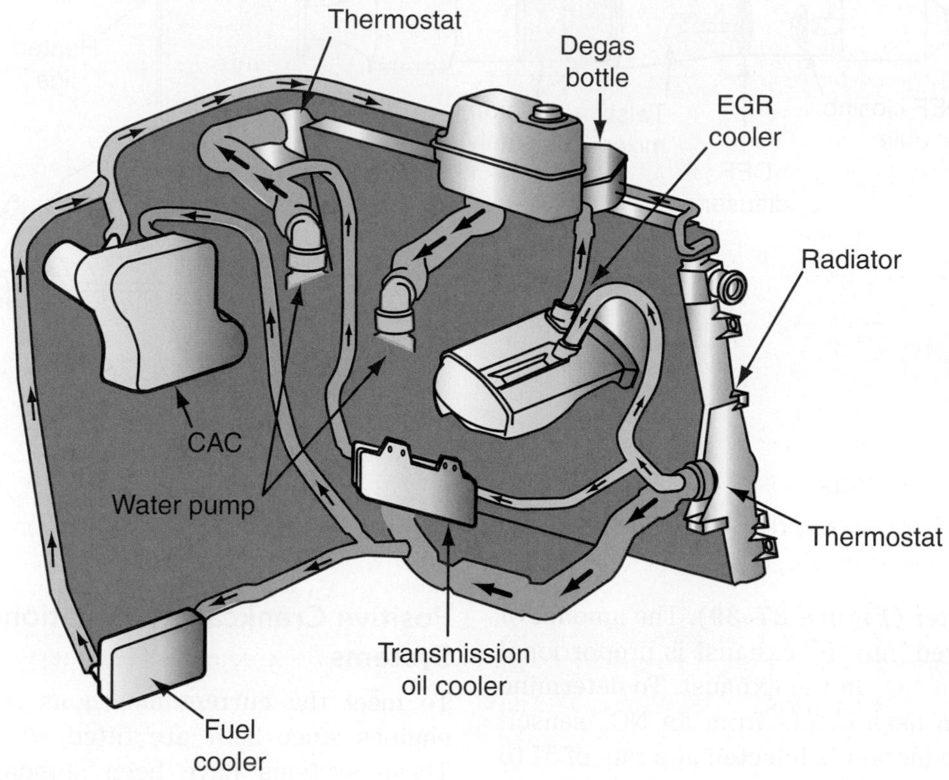

FIGURE 27–37

The various heat exchangers, including an EGR cooler, used on Ford Power Stroke engines.

SCR-equipped engine does not need an EGR valve to meet emission standards.

A **reductant**, also called a reducing agent, is a material that donates an electron to another material during the redox process. When a reductant gives up an electron, it becomes oxidized. Oxidation and reduction always occur simultaneously: One substance is oxidized by the other, which is reduced. During oxidation, a molecule provides electrons. Basically, a reductant removes oxygen from a substance and combines with the oxygen to form another compound. In the case of an SCR system, oxygen is separated from the NO_x and is combined with hydrogen to form water. It is claimed that the use of an SCR system reduces NO_x emissions by approximately 80 percent.

The common reductants used in SCR systems are ammonia and urea water solutions. Urea is commonly used as a nitrogen-rich fertilizer. In North America, urea is referred to as diesel exhaust fluid (DEF) and is called AdBlue in Europe. It should be noted that the catalytic converters are only effective when the reductant is injected into them when they are within a particular temperature range. The vehicle's PCM is programmed to keep the temperature of the exhaust within that range.

The PCM also controls when and how much DEF is injected into the exhaust stream just ahead of

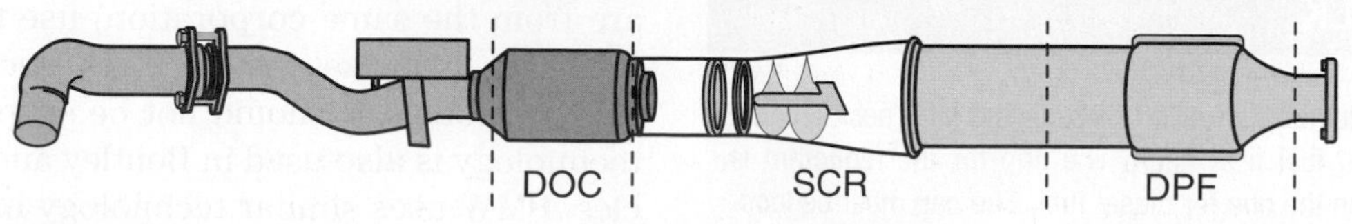

FIGURE 27–38

The main parts of an exhaust after treatment system on a diesel engine.

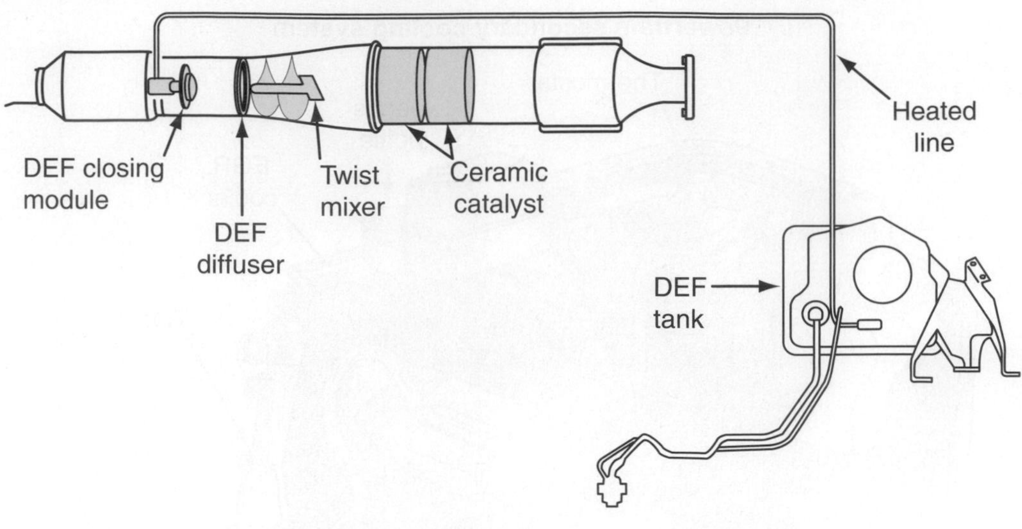

FIGURE 27–39

An SCR system and related parts.

the SCR converter **(Figure 27–39)**. The amount of reductant sprayed into the exhaust is proportional to the amount of NO_x in the exhaust. To determine this, the system uses inputs from an NO_x sensor. Normally, the reductant is injected at a rate of 57 to 113 g to 3.785 L (2 to 4 ounces to a gallon) of ultra-low sulphur diesel fuel (ULSD).

In the system, there is a mixer that mixes the reductant with the exhaust. The mixer has an atomizer and a twist mixer. The reductant is evenly distributed in the exhaust stream by the twist mixer. The atomizer breaks down the reductant into fine droplets so that it can easily mix with the exhaust.

The reductant is stored in a separate designated tank. The size of the tank varies with the manufacturer but represents the amount of urea used during a typical maintenance or required oil change cycle. Light-duty trucks typically have about a 19 L (5-gallon) DEF tank. SCR systems are equipped with a warning lamp that signals to the driver that the reductant tank needs to be refilled. If the warning lamp is ignored and the tank is not refilled, engine operation will be affected, and the engine may not start until the tank is refilled.

SHOP TALK

Normally, the reductant is refilled through the filler neck located next to the diesel fuel filler neck. The cap for the reductant is much smaller than the one for diesel fuel. The cap must be loosened and tightened with a wrench. The reductant's container is screwed into the filler neck and once it is tightened, the reductant will begin to flow into the reservoir.

Positive Crankcase Ventilation (PCV) Systems

To meet the current emissions standards, diesel engines since 2007 are fitted with PCV systems. These systems have been standard on gasoline engines for 60 years. The PCV systems in a diesel may be referred to as closed crankcase ventilation (CCV) systems. The purpose of these systems is the same as for a gasoline engine, that being preventing crankcase gases from entering the atmosphere. Crankcase gases are composed of blowby gases from the engine's cylinders and boil-off gases from the engine's oil. The later results from the heating of the engine oil as it evaporates. Engine blowby gases are those gases that escape past the piston rings and valves and end up in the engine's crankcase. Both of these gases need to be removed or limited, as they can affect the operation and endurance of an engine.

TDI Engines

Perhaps the most common "clean" diesels are the TDI engines made by Audi and Volkswagen. There are certainly other manufacturers, but Audi and VW seem to be the current dominant ones in North America. Both manufacturers (although they really are from the same corporation) use the same technologies. Because their efforts encompass many vehicle models, it should not be surprising that the technology is also used in Bentley and Porsche vehicles. BMW uses similar technology in their vehicles for North America and Europe.

Volkswagen has sold more diesel cars in North America than any other manufacturer.

FIGURE 27–40

A VW TDI engine.

TDI (turbocharged direct injection) is a design of diesel engines, which features turbocharging and direct fuel injection **(Figure 27–40)**. Currently, there are several models in the VW TDI family: Passat, Golf, Jetta, Beetle, and Toureg. All but one of these are fitted with the same four-cylinder 2.0L engine that produces 140 hp at 4000 rpm and 236 lb·ft of torque at only 1750 rpm. The only model that does not use this engine is the Toureg, which uses a 3.0L V-6 rated at 225 hp and 406 lb·ft of torque.

Engines fitted with piezoelectric injectors on a common rail **(Figure 27–41)** typically use a variable geometry turbocharger. This turbocharger has an electric motor **(Figure 27–42)** that controls the vanes inside the unit to change the speed and angle

FIGURE 27–41

The common rail system on a VW TDI engine.

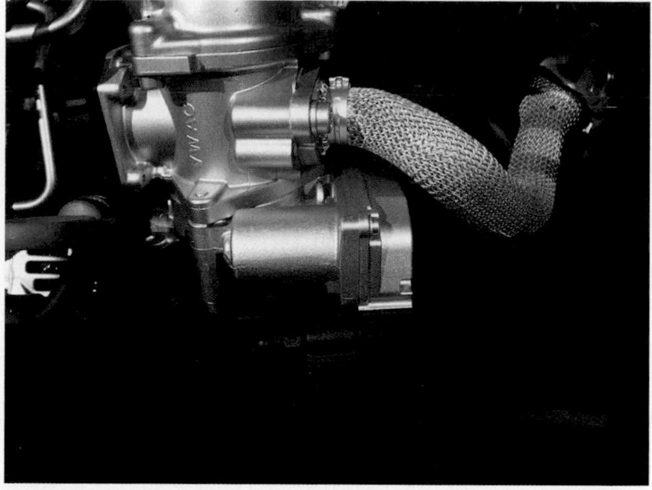

FIGURE 27–42

An electric motor controls the geometry of the vanes inside the turbocharger.

of the exhaust gases moving against the turbocharger's turbine. This not only reduces the amount of lag when the accelerator pedal is depressed but also maximizes engine output.

These are called clean diesels because they do not emit the soot or high NO_x that characterizes many diesels. Some VW diesels rely on the urea technology to treat the exhaust. Others have a particulate filter in the exhaust system.

Most TDI engines also have an intercooler to lower the temperature (and therefore increase the density) of the "charged," or compressed, air from the turbo, thereby increasing the amount of fuel that can be injected and combusted. The use of the intercooler, turbocharger, and common rail direct fuel injection system allow for greater engine efficiency. The result is greater power output, decreasing emissions, and more torque than the non-turbo and non-direct injection gasoline engines of the same size. The fuels recommended for TDI engines include pure diesel fuel, or B5, B20, or B90 biodiesel.

The engines have relatively low displacement and therefore can be quite compact. This means they have a low surface area. Because of this, they experience reduced heat loss during the combustion process. This is one of the reasons they are so efficient; however, they also do not provide as much heat as a gasoline engine to warm the passenger compartment, so there is a penalty there.

AUDI TDI R10/15 VEHICLES Audi is one of the manufacturers under the VW umbrella, so they offer many competing products with VW, as do Bentley and

FIGURE 27–43

An Audi race car equipped with a TDI engine.

Porsche. However, Audi has done more to market its TDI engines; some of its showcase efforts have been the R10 and R15 race cars **(Figure 27–43)**. These cars are diesel powered and have won many notable races, including the 24 Hours at LeMans and other endurance races. The race cars are fitted with V-12 diesel engines, and some of the later competition models have hybrid drivetrains. The technology has shown that diesel engines can more than compete with the world's most powerful cars and engines.

DIAGNOSTICS

In most cases, diesel engines can be diagnosed with a scan tool as well as with the same logic used to diagnose a gasoline engine. This is because the two engines have many of the same inputs and outputs; therefore, the data a scan tool retrieves are similar for both types of engines. The scan tool will also display the DTCs recorded by the system **(Figure 27–44)**. Diagnostics for diesel control systems varies with the manufacturer. However, certain checks are universal regardless of the manufacturer. These include checks of the following:

- The action of the solenoid and injectors
- The injection actuation pressure
- Active faults in the system
- Previous faults held in memory

Common faults include hard starting, no-start, extended cranking before starting, and low power.

FAULT CODE	CONDITION DESCRIPTION	PROBABLE CAUSES
P0122	Accelerator pedal sensor circuit low input	Grounded circuit, biased sensor, PCM
P0123	Accelerator pedal sensor circuit high input	Open circuit, biased sensor, PCM, short to 5v
P0220	Throttle switch B circuit malfunction	Short/open circuit, switch failure, operator, PCM
P0221	Throttle switch B circuit performance	Failed pedal assembly
P0230	Fuel pump relay driver failure	Open FP relay, blown fuse, open/grounded circuit
P0231	Fuel pump circuit failure	Fuse, relay, inertia switch, fuel pump, open/short circuit
P0232	Fuel pump circuit failure	Relay failure, short circuit, pump failure
P0236	Turbo boost sensor A circuit performance	Restricted inlet/exhaust/supply hose, missing hose
P0237	Turbo boost sensor A circuit low input	Circuit open, short to ground, MAP sensor
P0238	Turbo boost sensor A circuit low high	Circuit short to power, MAP sensor
P026_	Injector circuit low cylinder X	Harness short to ground
P026_	Injector circuit high cylinder X	High resistant connector or harness
P026_	Cylinder X contribution/balance fault	Power cylinder, valve train or injector problem, circuit
P030_	Fault cylinder X misfire	Mechanical engine failure
P0380	Glow plug circuit malfunction	Open/grounded circuit., solenoid open/shorted, failed PCM
P0381	Glow plug indicator circuit malfunction	Open/grounded circuit, lamp open, failed PCM

FIGURE 27–44

Examples of diesel engine–related DTCs.

Using a scan tool, check the sensor values to help pin down the source of the problem. To do so, a diagnostic tool must be connected to the computer system and/or a computer-based, diagnostic machine must be connected to the main data bus that controls the systems.

Diesel Exhaust Smoke Diagnosis

Although some exhaust smoke is considered normal operation for most diesel engines, especially older ones, the cause of high emissions, visible emissions, and excessive smoke should be diagnosed and repaired. Basically, the exhaust emissions from a diesel engine can be described as a gas, liquid, or solid. Keep in mind that not all exhaust emissions can be seen by your eye.

GASEOUS EMISSIONS From the outside, a gaseous emission cannot be seen. However, it may contain many different undesirable elements, such as NO_x, CO, and HC.

LIQUID EMISSIONS Again, according to the eye, liquid emissions may only appear as a white to grey smoke. White smoke normally occurs during cold engine starts and as the result of condensed fuel droplets. White smoke is also indicative of a cylinder misfire in a warm engine, the most common causes of white exhaust smoke are inoperative glow plugs, low engine compression, incorrect injector spray patterns, and coolant leaks into the combustion chamber.

Grey or blue smoke is normally due to oil burning caused by worn piston rings, scored cylinder walls, or bad valve stem seals. Grey or blue smoke can also be caused by defective injectors or injector O-rings.

SOLID EMISSIONS Normally, solid emissions appear as black soot, which is caused by incomplete combustion due to a lack of air or a fault in the injection system. When there is an excessive amount of soot, check the specific gravity of the fuel, the balance of the injectors, operation of the ECT and/or FRP sensor, or restrictions in the intake or turbocharger. Begin your checks by looking for restrictions in the air intake. To do so, connect a vacuum/pressure gauge to the intake air track.

Compression Testing

To test the compression on a diesel engine, remove the glow plugs or injector for the cylinder being tested. Then use a diesel compression gauge and crank the engine. A diesel engine should produce at least 2068 kPa (300 psi) at cylinders, which should be within 345 kPa (50 psi) of each other.

Cylinder Balance Test

One way to conduct a cylinder balance test on early (non-computer-controlled) diesel engines is to measure the resistance of the glow plugs after the engine has run for a while. Remember that the heating element of the glow plugs increases in resistance as their temperature increases. Therefore, all glow plugs should have about the same amount of resistance when checked with an ohmmeter before the engine has been started and after the engine has been started and has run. By looking at the resistance of the glow plugs, you may be able to identify a weak cylinder.

To test for even cylinder balance using glow plug resistance, do the following on a warm engine:

- Unplug, remove, measure, and record the resistance of the glow plug in each cylinder.
- With the electrical connectors removed from the glow plugs, start the engine.
- Allow the engine to run for several minutes to let the engine to warm the glow plugs.
- Turn the engine off, and measure the resistance of the glow plugs.
- The resistance of the glow plugs should be higher than the resistance measured at the beginning of the test. A glow plug with a different resistance is not firing correctly, and that cylinder should be checked for a loss of compression.

Another way to check cylinder balance includes the use of an infrared thermometer or pyrometer to measure exhaust temperature at each exhaust port. Misfiring cylinders will run colder than the others.

Injector Opening Testing

An injector opening (pop) tester is used to check the spray pattern of an injector nozzle. The handle of the tester is depressed, and the pop-off pressure is displayed on the gauge. The spray pattern should be a hollow cone but may vary on design. The nozzle should also be tested for leakage (dripping of the nozzle) while under pressure. If the spray pattern is not correct, cleaning, repairing, or replacing the injector nozzle will correct the problem.

KEY TERMS

Antiknock (p. 794)
Antiknock Index (AKI) (p. 794)
Biodiesel fuel (p. 805)
Cetane rating (CN) (p. 804)
Common rail (CR) injection (p. 812)
Compressed natural gas (CNG) (p. 800)
Diesel oxidation catalyst (DOC) (p. 809)

Diesel particulate filter (DPF) (p. 809)
Energy density (p. 798)
Ethanol (p. 796)
Flexible fuel vehicles (FFVs) (p. 804)
Fractional distillation (p. 792)
Fuel rail pressure (FRP) sensor (p. 813)
Fuel rail temperature (FRT) sensor (p. 813)
Glow plugs (p. 814)
Heptane (p. 794)
Hydraulic electronic unit injection (HEUI) (p. 810)
Hydrogen (p. 803)
Isooctane (p. 794)
Liquefied petroleum gas (LP gas) (p. 800)
Methanol (p. 796)
Motor octane number (MON) (p. 794)
Natural gas vehicle (NGV) (p. 801)
Octane (p. 794)
Oxygenates (p. 796)
Petroleum (p. 791)
P-series fuels (p. 803)
Piezoelectric injector (p. 814)
Reductant (p. 821)
Reformulated gasoline (RFG) (p. 796)
Reid vapour pressure (RVP) (p. 795)
Renewable fuels (p. 798)
Research octane number (RON) (p. 794)
Selective catalytic reduction (SCR) (p. 820)
Sulphur dioxide (p. 795)
TDI (turbocharged direct injection) (p. 823)
Tetraethyl lead (TEL) (p. 795)
Tetramethyl lead (TML) (p. 795)
Ultra-low sulphur diesel (ULSD) fuel (p. 806)
Vapour lock (p. 795)

SUMMARY

- Crude oil is also called petroleum. It is drawn out of oil reservoirs and sands below the earth's surface.
- The easiest and most common way to separate the various hydrocarbons (called fractions) in crude oil is through a process called fractional distillation.
- Gasoline is a complex mixture of approximately 300 various ingredients, mainly hydrocarbons.
- Gasoline's octane number or rating gives the antiknock quality of a gasoline.
- Octane ratings are measured by the motor octane number (MON) method and the research octane number (RON) method. The typical octane rating is the Antiknock Index (AKI). It is stated as (R+M)/2.
- The most common method for measuring the volatility of a fuel is the Reid vapour pressure (RVP) test.
- Several additives are put into gasoline to control harmful deposits, including gum or oxidation inhibitors, detergents, metal deactivators, and rust inhibitors.
- Oxygenates are compounds such as alcohols and ethers that are added to fuel to improve combustion efficiency and enhance octane ratings.
- The most widely used gasoline oxygenate additive is ethanol.
- Renewable fuels are those derived from nonfossil sources and produced from plant or animal products or wastes (biomass).
- Flexible fuel vehicles (FFV) can run on ethanol or gasoline or a mixture of the two. The alcohol fuel and gasoline are stored in the same tank.
- Diesel fuel is heavier and has more carbon atoms; it has about 15 percent more energy density than gasoline.
- Diesel fuel's ignition quality is measured by a cetane rating.
- Animal fats, recycled restaurant grease, and vegetable oils derived from crops such as soybeans, canola, corn, and sunflowers are used in the production of biodiesel fuel.
- Ultra-low sulphur diesel (ULSD) fuel has a maximum sulphur content of 15 ppm. Engines running on this fuel emit less NO_x, soot, and other unwanted sulphur compounds.
- Diesel engines provide more torque than a gasoline engine of the same size and consume less fuel per kilometre.
- Hydraulic electronic unit injection (HEUI) relies on engine oil pressure to control the operation of the individual fuel injectors.
- A common rail (CR) fuel system has a fuel rail that carries high-pressure fuel to electronically controlled injectors.
- Piezoelectric injectors use many, very thin layers of piezoelectric crystal material stacked on top of each other at the orifice of the injector's nozzle. When electricity is applied to the stack, it expands slightly, and fuel is sprayed into the cylinder.
- Diesel oxidation catalysts (DOC) are flow-through honeycomb-style substrate assemblies placed in the exhaust stream to reduce the amount of CO, HC, and other compounds in the exhaust.
- Diesel particulate filters (DPFs) work with the oxidation catalyst and an EGR valve to remove a majority of the NO_x, soot, and unburned hydrocarbons in the exhaust of a diesel engine. Selective catalytic reduction (SCR) reduces NO_x emissions by injecting a reductant into the exhaust stream over the SCR catalyst.

- A reductant, also called a reducing agent, is a material that donates an electron to another material during the redox process.
- Diesel engines can be diagnosed with a scan tool and with the same logic used to diagnose a gasoline engine.

REVIEW QUESTIONS

1. Pump gasoline may contain a small amount of alcohol. Name three problems that can occur if there is an excessive amount of alcohol in gasoline.
2. *True or False?* The diesel fuel's antiknock quality is measured by the cetane rating.
3. What percentage of ethanol does E85 contain?
4. List five advantages the use of piezoelectric injectors has over solenoid-based injectors.
5. What does the Reid vapour pressure (RVP) test measure?
6. Which of the following chemicals is commonly added to gasoline to increase its octane rating?
   a. isooctane **b.** heptane
   c. sulphur **d.** ethanol
7. Which of the following is the least likely to affect the performance of a fuel?
   a. antiknock quality
   b. volatility
   c. conductivity
   d. deposit control
8. Which of the following is a stoichiometric ratio?
   a. 13.5 to 1 **b.** 14.1 to 1
   c. 14.7 to 1 **d.** 15.3 to 1
9. What is the purpose of adding isopropyl alcohol to gasoline in cold weather?
   a. to increase combustion temperature
   b. to prevent fuel line freeze-up
   c. to decrease combustion temperature
   d. to keep fuel injectors clean
10. Which of the following will reduce the chances of detonation?
    a. using a hemispherical combustion chamber
    b. using low octane rated gasoline
    c. advancing the ignition timing
    d. increasing the compression ratio
11. Which of the following is a commonly used oxygenate?
    a. phenols **b.** nitrous oxide
    c. methanol **d.** tetraethyl lead
12. Which of the following does not affect engine knock?
    a. fuel detergents
    b. a lean fuel mixture
    c. over-advanced ignition timing
    d. the octane number
13. What two types of molecules are contained in gasoline?
    a. carbon and hydrogen
    b. carbon and octane
    c. hydrogen and octane
    d. heptane and octane
14. What type of fuel derives almost exclusively from corn?
    a. methanol
    b. ethanol
    c. liquid petroleum gas
    d. toluene
15. Which of the following chemicals is commonly added to gasoline to reduce gum and varnish formation?
    a. isopropyl alcohol **b.** carbon
    c. hydrogen **d.** aromatic amines
16. What is currently added to fuel to increase its octane rating?
    a. nitrous oxide
    b. isopropyl alcohol
    c. tetraethyl lead
    d. methylcyclopentadienyl manganese tricarbonyl (MMT)
17. At what temperature is a Reid vapour pressure (RVP) test performed?
    a. 10°C (50°F) **b.** 38°C (100°F)
    c. 66°C (150°F) **d.** 93°C (200°F)
18. What is the common reductant used in the selective catalytic reduction (SCR) system?
    a. ammonia **b.** urea
    c. nitrogen **d.** sulphur
19. What is a typical cetane number for diesel fuel?
    a. 30 **b.** 70
    c. 50 **d.** 90
20. In the selective catalytic reduction (SCR) system, what does the reductant do?
    a. adds oxygen to nitrogen to form NO_x
    b. removes O_2 from NO_x to form N and H_2O
    c. breaks down NO_x to N_2 and O_2
    d. adds carbon to NO_x to form N and CO_2

CHAPTER 28
Fuel Delivery Systems

LEARNING OUTCOMES

- Describe the components of a fuel delivery system and the purpose of each.
- Conduct a visual inspection of a fuel system.
- Relieve fuel system pressure.
- Inspect and service fuel tanks.
- Inspect and service fuel lines and tubing.
- Describe the different fuel filter designs and mountings.
- Remove and replace fuel filters.
- Explain how common electric fuel pump circuits work.
- Conduct a pressure and volume output test on an electric fuel pump.
- Service and test electric fuel pumps.

To have an efficient-running engine, there must be the correct amount of fuel. To provide this, fuel must be stored, pumped out of storage, piped to the engine, filtered, and delivered to the fuel injectors. In addition, the fuel system is designed to prevent fuel vapours from entering the atmosphere.

Many modern systems are returnless on-demand systems. In older systems, a fuel pump delivered fuel under pressure to the fuel injectors. A pressure regulator at the injectors controlled the fuel pressure by sending excess fuel back to the fuel tank. This is a return fuel system **(Figure 28–1)**.

In a mechanical returnless system **(Figure 28–2)**, the pressure regulator is in the fuel tank and excess

Bosch Common Rail System for Passenger Cars

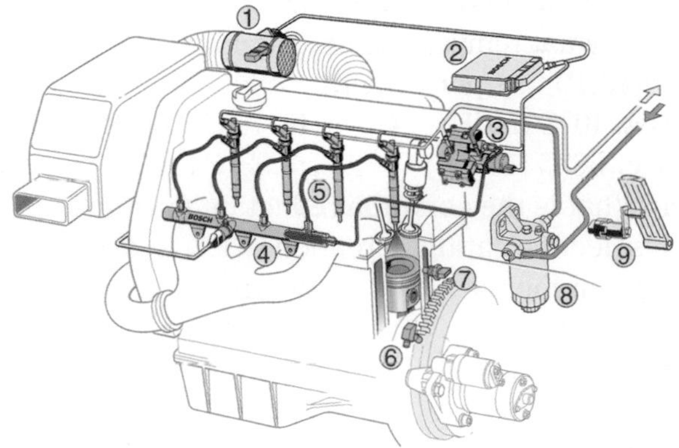

FIGURE 28–1

A return fuel delivery system.

Robert Bosch Corp.

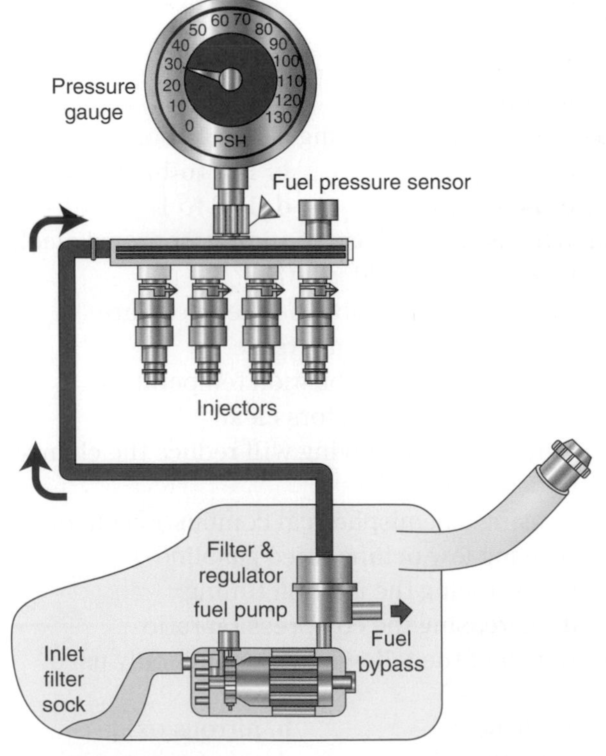

FIGURE 28–2

A returnless fuel delivery system.

© AA1 Car Auto Diagnosis & Repair Help, www.AA1Car.com

fuel is released to the tank. There is no need for a return line. Most newer systems are electronically controlled, and fuel pressure and volume are controlled by the PCM according to the existing operating conditions. In a return system, the fuel sent back to the tank has been heated by underhood temperatures. The introduction of the warm fuel to the tank causes the fuel to evaporate.

BASIC FUEL SYSTEM DIAGNOSIS

The fuel system should be checked whenever there is evidence of a fuel leak, fuel smell, or inadequate fuel supply. The fuel system should be checked whenever basic tests suggest that there is too little or too much fuel being delivered to the cylinders. Lean mixtures are often caused by insufficient amounts of fuel being drawn out of the fuel tank. Lean mixtures can cause bad results in many different diagnostic tests, including high HC, O_2, and NO_x readings on an exhaust analyzer and high firing lines on a scope.

When no fuel is delivered to the engine, the engine will not start. On carburetor and throttle-body-injected engines, it is easy to determine if fuel is being delivered. Simply look down the throttle body. If the surfaces are wet or you see fuel being sprayed while cranking the engine, fuel is there. With port injection, it is a little more difficult. Connect a fuel pressure gauge to the fuel line or rail and observe the fuel pressure while cranking the engine. Testing fuel pressure is described later in this chapter. However, if there is no fuel pressure while cranking, there is no fuel being delivered to the engine.

There are many components in the fuel system. These can be grouped into two categories: fuel delivery and the fuel injection system. Diagnosis and basic service to the fuel delivery system are covered in this chapter. All tests given in this chapter assume that the fuel is good and not severely contaminated.

CAUTION!

Gasoline is very volatile and flammable. Never expose it to open flame or extreme heat. Disconnect the negative battery cable before doing anything that may release gas vapours. Use containers to catch gasoline and rags to wipe up any spills. Use a flashlight or an enclosed fluorescent tube or LED lamp designed for safe use around fuels. When working with a fuel system, always have a Class B fire extinguisher nearby.

Contaminated Fuel

Obviously, water does not burn. Therefore, water in the fuel tank can cause a driveability problem. As mentioned in an earlier chapter, to test for contaminated fuel, place a sample of fuel in a graduated container. Special fuel test beakers are available to make this test easier. Place a measured quantity of fuel into a container, and mark the level. Next, add about 10 percent water to the container, and mark the level. Seal the container and shake for several seconds, and then let the fuel and water sit for up to five minutes. As the fuel and water separate, you will be able to tell by the marks you made if the amount of water is the same or has increased. This test can also help you determine if a customer has filled his or her non-flex fuel vehicle with E85 as the water and ethanol will show as a significant percentage of the volume.

If there is water in the fuel, drain the fuel tank, replace the fuel filter, and refill the tank with fresh gasoline. Also, as fuel sits for a while, it becomes less volatile or stale. This is because some of the lighter HCs evaporate over time. When fuel is stored a long time and exposed to air and heat, the fuel begins to break down and evaporate, leaving behind large molecules of carbon and gum. The separation of these materials from the fuel lowers its volatility. Also, the molecules can collect in and restrict the fuel lines and injectors. If the molecules are injected into the cylinders, they will not burn, and they can cause abrasion in the cylinders. If a fuel smells sour and has been stored for quite some time, the fuel is stale and may be unusable. However, it may be still usable if it is mixed with as much fresh fuel as possible.

There are many products available to partially revitalize the fuel; however, if the fuel is so stale that an engine will not run on it, drain it and refill the tank with fresh gasoline. If fuel will be stored for a while, a fuel stabilizer should be added to it before it is stored.

GUIDELINES FOR SAFELY WORKING ON FUEL SYSTEMS

Many things need to be considered before working on a fuel system. Fuel in vapour and liquid form presents many potential hazards. Fuel plus heat presents even more! Also, dispose of all drained fuel according to local regulations.

Before loosening or disconnecting fuel lines, all pressure in the system must be released. Fuel injection systems operate at high fuel pressures and are designed to hold most of that pressure when the engine is not running. This residual pressure allows for fast engine starting. When a fuel line that has pressurized fuel in it is loosened, the fuel will spray uncontrollably as soon as it can. The fuel can spray on something hot and cause a fire or spray into your eyes and cause a serious injury.

Most fuel injection fuel rails have a fuel pressure test port (often referred to as the Schrader valve) on the fuel rail **(Figure 28–3)**. The pressure can be relieved at this port. Begin by disconnecting the negative battery cable. Then loosen the fuel tank filler cap to relieve any vapour pressure built up in the tank. Wrap a shop towel around the fuel pressure test port on the fuel rail and remove the dust cap from this valve. Connect a fuel pressure gauge to the fuel pressure test port on the fuel rail. Install a bleed hose onto the gauge and put the free end into an approved gasoline container. Then open the gauge bleed valve to relieve fuel pressure from the system into the gasoline container **(Figure 28–4)**. Be sure that all the fuel in the bleed hose is drained into the gasoline container.

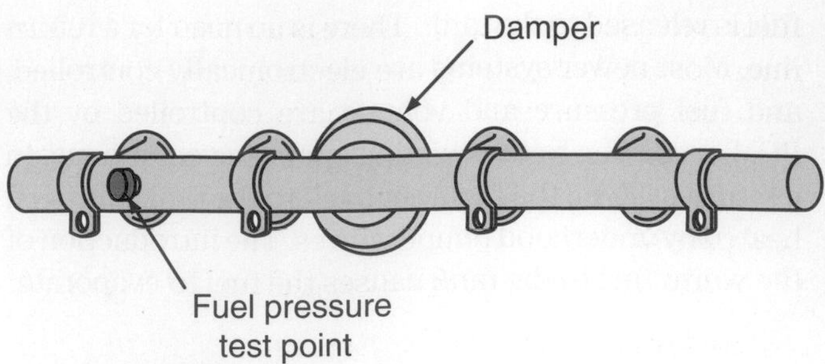

FIGURE 28–3

The typical location of the fuel pressure test port.

If the system does not have a test port, the pressure can be relieved by loosening the fuel tank

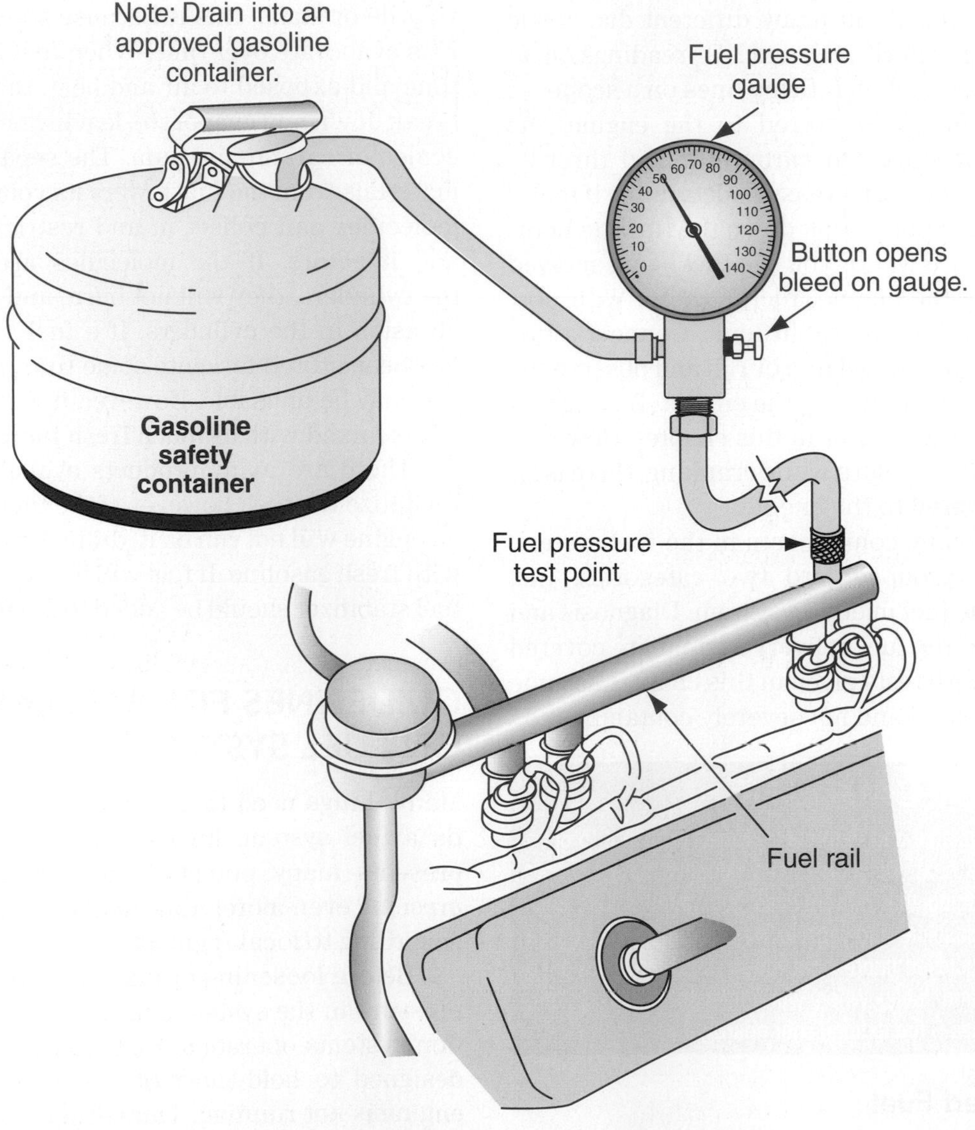

FIGURE 28–4

Relieving fuel pressure from the system.

filler cap to relieve any tank vapour pressure. Then remove the fuel pump fuse or relay. Start and run the engine until the fuel in the lines is used up and the engine stops. Crank the engine with the starter for about three seconds to relieve any remaining fuel pressure.

Additional safety guidelines include the following:

- Always wear eye protection, and follow all other safety rules to prevent injury to yourself or others when servicing fuel systems.
- When working on a fuel system in the engine compartment, disconnect the negative cable of the battery. An electrical spark may cause a fire or explosion.
- Do not smoke when working on or near any fuel-related component.
- Do not allow heat or flames to be near while working on or near the fuel system.
- Remove all electronic devices, such as cellphones and audio equipment, from your clothing when working on or near the fuel system.
- Handle and store all fuels with the utmost caution.
- Clean all fuel spills immediately; spilled fuel may be ignited by hot components.
- If a fuel line or hose is damaged in any way, replace it.
- When disconnecting or reconnecting a fuel line or hose, make sure that the mating parts are totally clean.
- After disconnecting a fuel line or hose, plug both ends to prevent dirt from entering.
- When disconnecting and reconnecting a fuel line or hose, only use the tools designed for that connection. Using the wrong tool can cause a poor connection that can result in a fuel leak.

FUEL TANKS

Fuel tanks include devices that prevent vapours from leaving the tank. For example, to contain vapours and allow for expansion, contraction, and overflow that result from changes in the temperature, the fuel tank has a separate air chamber dome at the top. All fuel tank designs provide some control of fuel height when the tank is filled. Frequently, this control is achieved by using vent lines with the filler tube or tank **(Figure 28–5)**. These fuel height controls allow only 90 percent of the tank to be filled. The remaining 10 percent is for expansion during hot weather. Some fuel tanks have an overfill limiting valve to prevent overfilling of the tank.

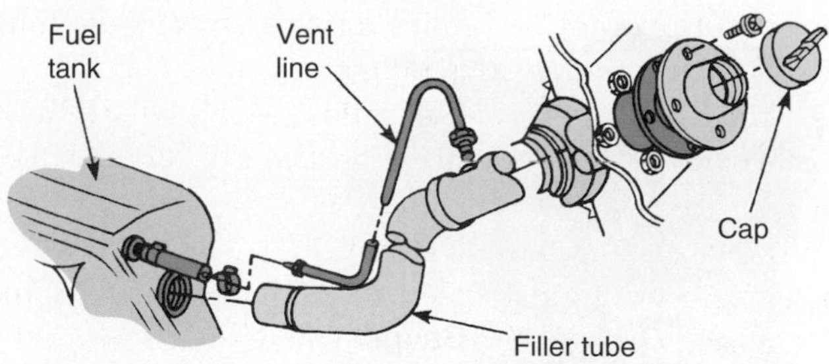

FIGURE 28–5

Vent lines within the fuel tank filler tube control the fuel level.

Fuel tanks are constructed of pressed corrosion-resistant steel, aluminum, or moulded polyethylene plastic. Aluminum and moulded plastic tanks are the most commonly used.

Most tanks have slosh baffles or surge plates to prevent fuel from splashing around inside the tank. In addition to slowing down fuel movement, the plates tend to keep the fuel pickup and sending unit for the fuel gauge immersed in the fuel during hard braking and acceleration. The plates or baffles also have holes or slots in them to permit the fuel to move from one end of the tank to the other. With few exceptions, the fuel tank is located in the rear of the vehicle.

A fuel tank has an inlet filler tube and cap. The location of the filler tube depends on the tank design. All current filler tubes have a built-in restrictor that prevents the entry of the larger leaded-fuel delivery nozzle at gas pumps. The filler tube can be a rigid one-piece tube soldered to the tank or can be made of multiple pieces.

Some form of liquid vapour separator is incorporated into nearly every fuel tank. This separator stops liquid fuel or bubbles from reaching the vapour storage canister. It can be located inside the tank, on the tank, in the fuel vent lines, or near the fuel pump. Check the service information for the exact location of the liquid vapour separator and the routing of the hoses to it.

Inside the fuel tank, there is also a sending unit that includes a pickup tube and float-operated fuel gauge sender unit. Most current fuel pumps are installed inside the tank, and the pickup and fuel gauge sensor are part of that assembly **(Figure 28–6)**. A fuel strainer attaches to the pickup tube. The fuel strainer, sometimes referred to as a sock, is made of woven plastic. The strainer serves as a filter, stopping any rust or dirt that may be in the fuel from entering into the fuel pump. The fuel tank also has vent valves that are connected via hoses to a charcoal canister that collects HC emissions when the engine is running.

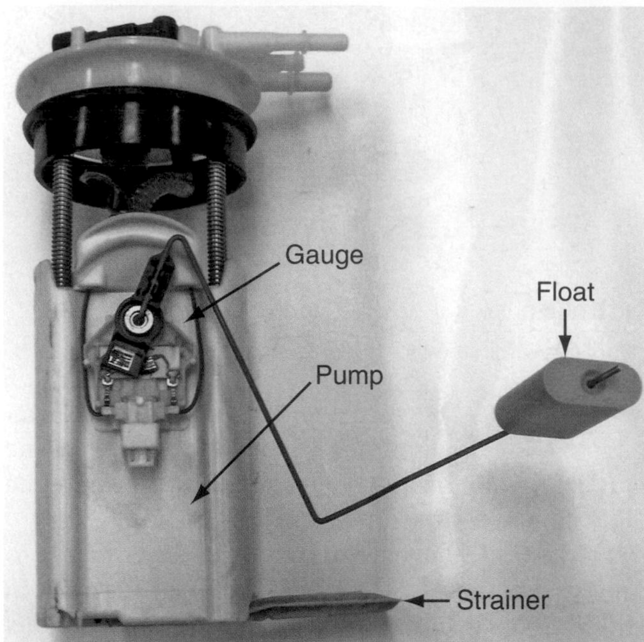

FIGURE 28–6

A combination electric fuel pump and sending unit.

Inspection

Fuel tanks should be inspected for leaks; road damage; corrosion and rust on metal tanks; loose, damaged, or defective seams; loose mounting bolts; and damaged mounting straps. Leaks in the fuel tank, lines, or filter may cause a gasoline odour in and around the vehicle, especially during low-speed driving and idling.

A weak seam, rust, or road damage can cause leaks in the metal fuel tank. The best method of permanently solving this problem is to replace the tank. Another method is to remove the tank and steam clean or boil it in a caustic solution to remove the gasoline residue. After this has been done, the leak can be soldered or brazed by an appropriately equipped specialty shop.

Holes in a plastic tank can sometimes be repaired by using a special tank repair kit. Be sure to follow manufacturer's instructions when doing the repair.

When a fuel tank is leaking or has water in it, the tank must be cleaned, repaired, or replaced.

SHOP TALK

When a fuel tank must be removed, if possible, ask the customer to bring the vehicle to the shop with a minimal amount of fuel in the tank.

Fuel Tank Draining

WARNING!

Always drain gasoline into an approved container, and use a funnel to avoid gasoline spills.

The fuel tank must be drained prior to tank removal. Begin by removing the negative cable from the battery. Then raise the vehicle on a hoist. Make sure you have an approved gasoline container and are prepared to catch all of the fuel before proceeding. If the tank has a drain bolt, remove it to drain the fuel. If the fuel tank does not have a drain bolt, locate the fuel tank drainpipe or filler pipe. Using the proper adapter, connect the intake hose from a hand-operated or air-operated pump to the pipe. Insert the discharge hose from the hand-operated or air-operated pump into an approved gasoline container, and operate the pump until all the fuel is removed from the tank.

CAUTION!

Abide by federal and provincial laws for the disposal of contaminated fuels. Be sure to wear eye protection when working under the vehicle.

Fuel Tank Service

In most cases, the fuel tank must be removed for servicing. The procedure for removing a fuel tank varies depending on the vehicle make and year. Always follow the procedure in the vehicle manufacturer's service information. What follows is a typical procedure:

PROCEDURE

Removing a Fuel Tank

STEP 1 Disconnect the negative terminal from the battery.

STEP 2 Relieve the fuel system pressure and drain the fuel tank.

STEP 3 Raise the vehicle on a hoist or lift the vehicle with a floor jack and lower the chassis onto jack stands.

STEP 4 Use compressed air to blow dirt from the fuel line fittings and wiring connectors.

STEP 5 Remove the fuel tank wiring harness connector from the body harness connector.

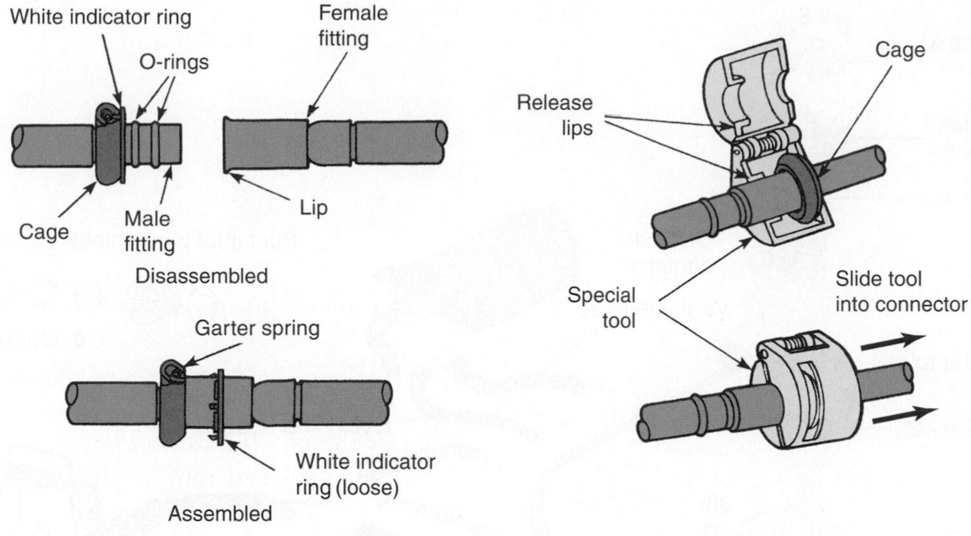

FIGURE 28–7

Some quick-connect fittings require the use of a special tool to separate them.

STEP 6 Remove the ground wire retaining screw from the chassis, if used.

STEP 7 Disconnect the fuel lines from the fuel tank. If these lines have quick-disconnect fittings, follow the manufacturer's recommended removal procedure in the service information. Some quick-disconnect fittings are hand releasable, and others require the use of a special tool **(Figure 28–7)**.

STEP 8 Wipe the filler pipe and vent pipe hose connections with a shop towel, and then disconnect the hoses from the filler pipe and vent pipe to the fuel tank.

STEP 9 Unfasten the filler from the tank. If it is a rigid one-piece tube, remove the screws around the outside of the filler neck near the filler cap. If it is a three-piece unit, remove the neoprene hoses after the clamp has been loosened.

STEP 10 Loosen the bolts holding the fuel tank straps to the vehicle **(Figure 28–8)** until they are about two threads from the end.

STEP 11 Holding the tank securely against the underchassis with a transmission jack or by hand, remove the strap bolts and lower the tank to the ground. When lowering the tank, make sure all wires and tubes are unhooked. Be careful as small amounts of fuel might still be in the tank.

WARNING!

Do not heat the bolts on the fuel tank straps in order to loosen them. The heat could ignite the fuel fumes.

To reinstall a repaired or new fuel tank, reverse the removal procedure. Be sure that all the rubber or felt tank insulators are in place. Then, with the tank straps in place, position the tank. Loosely fit the tank straps around the tank, but do not tighten them. Make sure that the hoses, wires, and vent tubes are connected properly **(Figure 28–9)**. Check the filler neck for alignment and for insertion into the tank. Tighten the strap bolts and secure the tank to the car. Install all of the tank accessories (vent line, sending unit wires, ground wire, and filler tube). Fill the tank with fuel and check it for leaks, especially around the filler neck and the pickup assembly. Reconnect the battery and check the fuel gauge for proper operation.

FILLER CAPS

Filler tube caps (commonly called gas or fuel caps) seal the fuel tank while allowing refilling of the tank. Filler caps are non-venting and have some type of pressure-vacuum relief valve arrangement

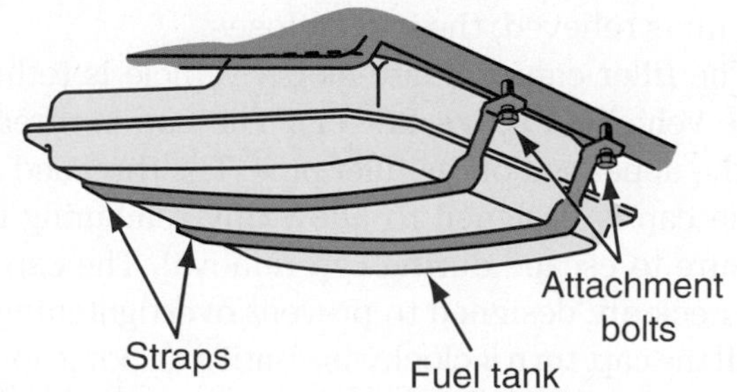

FIGURE 28–8

Front and rear fuel tank strap mounting bolts.

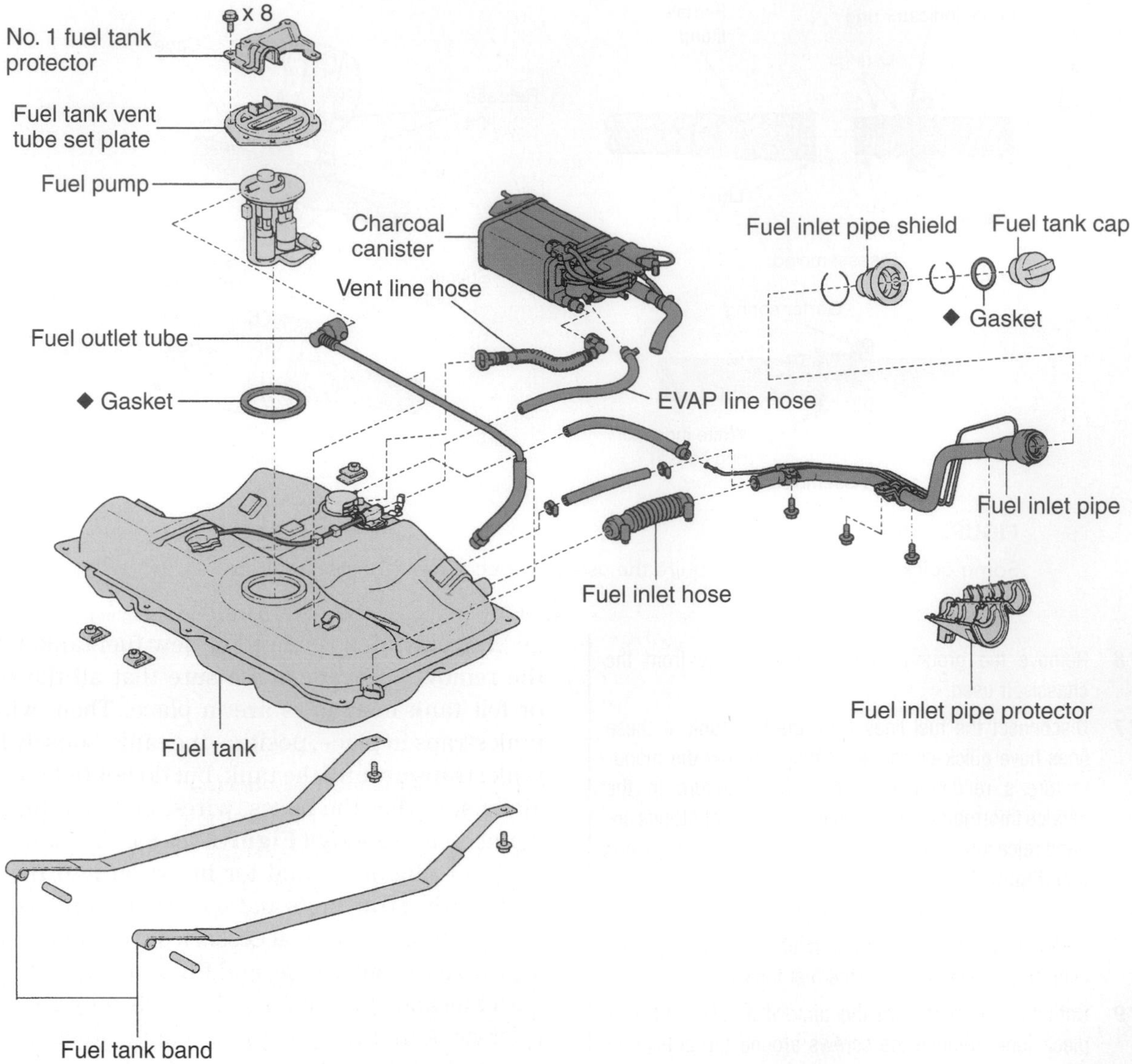

FIGURE 28–9

The hoses, wires, and tubes normally connected to a fuel tank.

Reprinted with permission.

(Figure 28–10). Under normal conditions, the valve is closed. When extreme pressure or vacuum is present, the relief valve opens to prevent the tank from ballooning or collapsing. Once the pressure or vacuum is relieved, the valve closes.

The filler cap of a late-model vehicle is tethered to the vehicle **(Figure 28–11)**. The cap is threaded into the upper end of the filler pipe. The threaded area on the cap is designed to allow any remaining tank pressure to escape during cap removal. The cap and filler neck are designed to prevent overtightening. To install the cap, turn it clockwise until a clicking noise is heard. This indicates that the cap is properly tightened and fully seated. A fuel filler cap that is not fully seated may cause a malfunction in the emission system.

FIGURE 28–10

A cutaway of a pressure-vacuum gasoline filler cap.

Stant Manufacturing Inc.

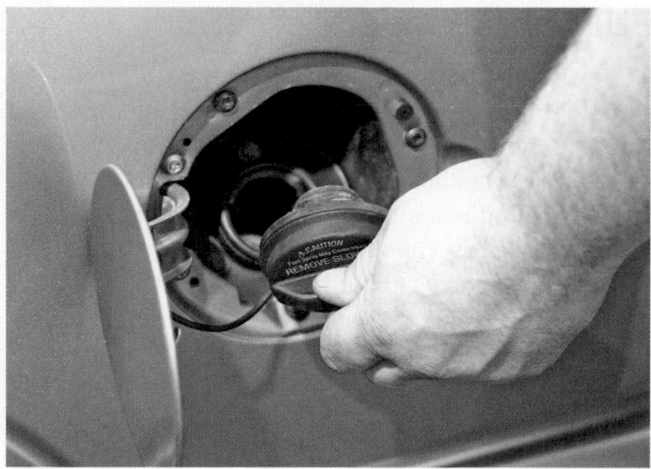

FIGURE 28–11

The gas cap for current vehicles is tethered to the vehicle and threaded into the filler tube.

WARNING!

When a tank filler cap is replaced, the replacement cap must be exactly the same as the original cap, or a malfunction may occur in the filling and venting system, resulting in higher emission levels and the escape of dangerous HCs.

OBD-II Monitor

Late-model vehicles that meet enhanced evaporative requirements have a vacuum-based evaporative system integrity check. If the gas cap is loose or missing, the ECM/PCM will detect an evaporative system leak and will illuminate a warning message. On some vehicles, the CHECK FUEL CAP message will appear each time the engine is started until the system turns the message off. The message may turn off after the cap is replaced or tightened until at least one click is heard. If the message does not turn off, there could be a leak in the system, or the circuit for the message is faulty.

A fuel tank pressure (FTP) sensor is a transducer that converts the absolute pressure in the fuel tank into an input for the PCM **(Figure 28–12)**. The integrity check is done by creating a vacuum in the tank and measuring how well it holds the vacuum. If the gas cap is off or loose, the tank will not hold a vacuum. Before a vacuum is formed in the tank, the canister vent solenoid is closed to seal the entire evaporative system. Then the vapour management valve creates a slight negative pressure in the tank. If the desired amount of vacuum cannot be established, a system leak is indicated, and the PCM will store a P0455 DTC and illuminate the warning message.

FIGURE 28–12

A fuel tank pressure (FTP) sensor.

Other possible causes for this code are disconnected or kinked vapour lines, an open canister vent solenoid, or a closed vapour management valve.

Fuel Cap Testing

A gas cap should be checked whenever the PCM detects a leak in the evaporative system and the cap is securely fastened. Also, Ontario and British Columbia mandate a gas cap check as part of its emissions tests. A gas cap is checked with a special tester. The cap is connected to the tester by an adapter specifically designed for the cap **(Figure 28–13)**. The tester then applies pressure to the cap and monitors its ability to hold the pressure. The readout on the tester simply says PASS or FAIL. A cap that has failed should be replaced. However, it is important that the correct adapter has been used for the cap. If the wrong adapter was used, the cap will fail the test even if it is good.

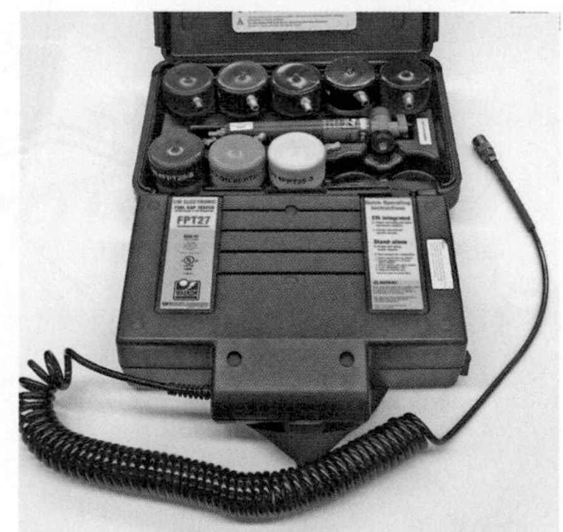

FIGURE 28–13

A fuel cap tester with various adapters.

Capless Fuel System

Ford Motor Company's Ford GT was the first modern production car to meet all emissions standards without a gas cap **(Figure 28–14)**. This technology is used on many 2008 and newer Ford vehicles. It is a very simple design. A spring-loaded flapper valve is positioned at the opening for the filler neck. This valve tightly seals the tank until a fuel nozzle is inserted into the opening. The nozzle opens the valve and allows for the refilling of fuel. As soon as the nozzle is removed, the valve is shut by the springs. A capless fuel system reduces the time the fuel vapours can escape during refuelling. It also makes it more convenient for the consumer because there is no cap to tighten. The filler door, which is outside the filler tube, helps to seal in the fuel and fuel vapours.

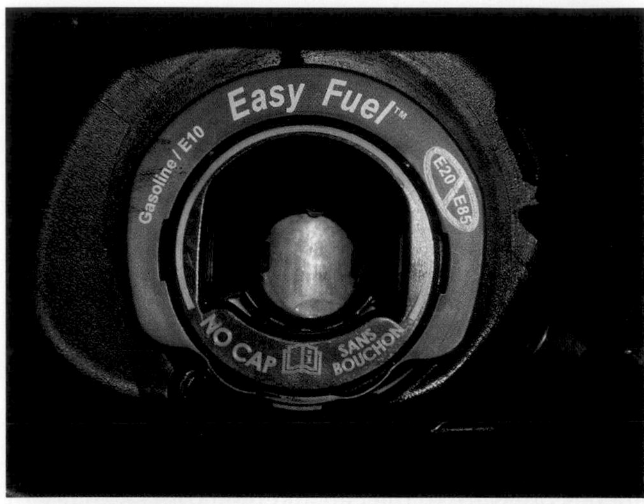

FIGURE 28–14

A capless fuel filler tube with its flapper valve.

FUEL LINES AND FITTINGS

The fuel lines **(Figure 28–15)** carry fuel from the tank to the fuel filter and fuel injection assembly. Fuel lines can be made of metal tubing or flexible nylon or synthetic rubber hose. The latter must be able to resist gasoline. The hoses must also be non-permeable so that gas and gas vapours cannot evaporate through the hose. Ordinary rubber hose, such as that used for vacuum lines, deteriorates when exposed to gasoline. Only hoses made for fuel systems should be used. Similarly, vapour vent lines must be made of material that resists attack by fuel vapours.

Fuel supply lines from the tank to the injectors are routed to follow the frame along the underchassis of the vehicle. Generally, rigid lines are used extending from near the tank to a point near the fuel pump and fuel filter. To prevent ruptures during a rear impact, the gaps between the frame and tank or fuel pump are joined by short lengths of flexible hose.

Late-model fuel tanks have hoses to allow air in the fuel tank to vent vapours into the charcoal canister when the tank is being filled with fuel. Vent hoses are usually installed alongside the filler neck. Replacement vent hoses are usually marked with the designation **EVAP** to indicate their intended use. The inside diameter of a fuel delivery hose is generally larger (7.94 to 9.35 mm [5/16 to 3/8 in.]) than that of a fuel return hose (6.35 mm [1/4 in.]). The EVAP line to the tank is also often a different size or connection type than the fuel supply and return lines. This helps to prevent the lines from incorrect installation.

To control the rate of vapour flow from the fuel tank to the vapour storage tank, a plastic or metal

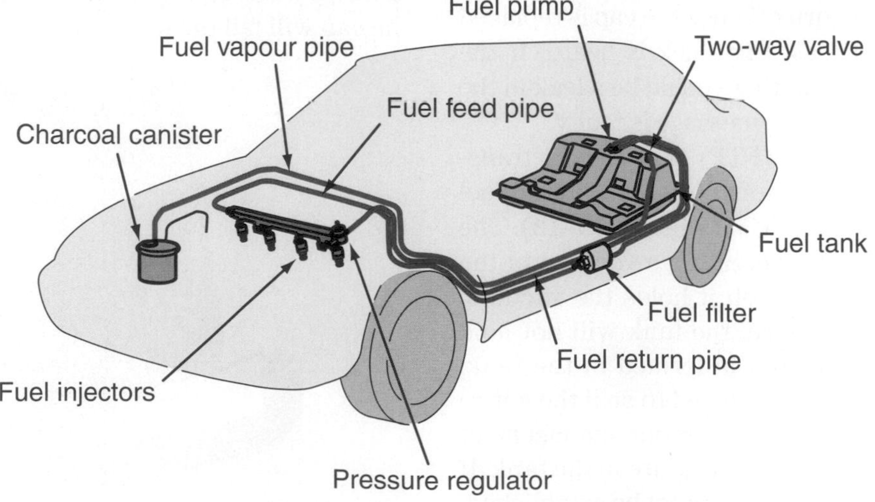

FIGURE 28–15

A typical layout of the fuel lines on a late-model car.

restrictor may be placed either in the end of the vent pipe or in the vapour-vent hose itself. When the latter hose must be replaced, the restrictor must be removed from the old vent hose and installed in the new one.

Fittings

Sections of fuel line are assembled together by fittings. Some of these fittings are a threaded-type fitting, whereas most are a quick-release design. Many fuel lines have quick-disconnect fittings with a unique female socket and a compatible male connector. These quick-disconnect fittings are sealed by an O-ring inside the female connector. Some of these quick-disconnect fittings have hand-releasable locking tabs **(Figure 28–16)**, and others require a special tool to release the fitting **(Figure 28–17)**.

WARNING!
Other types of O-rings should not be substituted for a Viton O-ring.

The interior components, such as the O-rings and spacers, of quick-connect fittings are not serviceable. If the fitting is damaged, the complete fuel tube or line must be replaced.

Some fuel lines have threaded fittings with an O-ring seal to prevent fuel leaks. These O-ring seals are usually made from Viton, which resists deterioration from gasoline. On some other fuel lines, the fuel hose is clamped to the steel line, and the hose and clamp must be properly positioned on the steel line **(Figure 28–18)**.

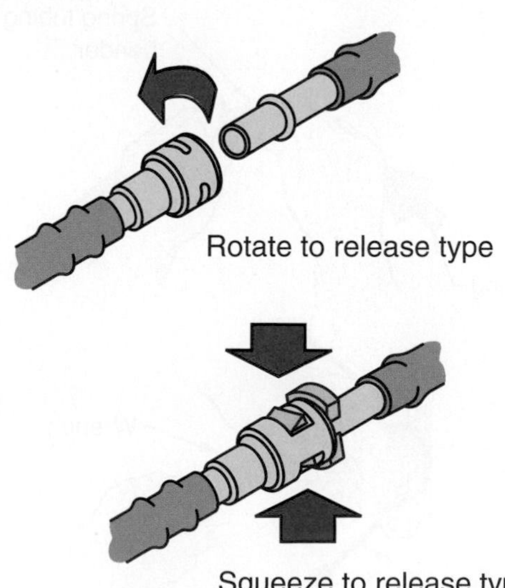

FIGURE 28–16

Quick-disconnect hand releasable fuel line fittings.

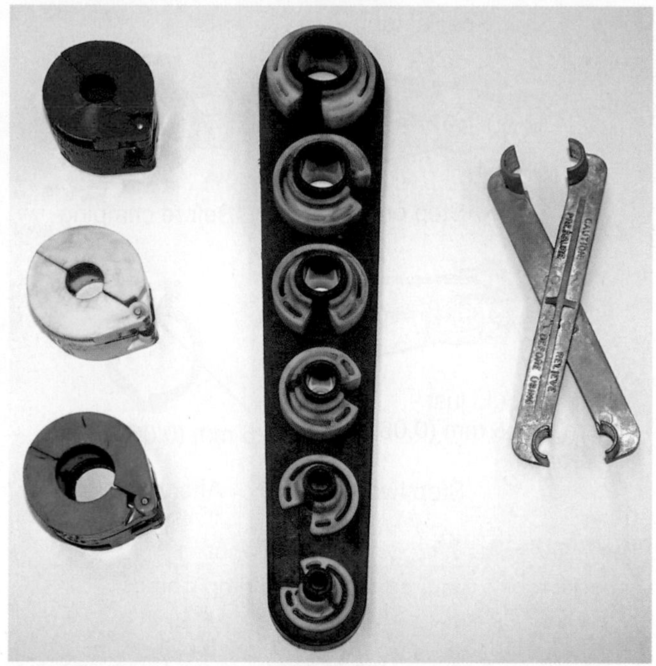

FIGURE 28–17

An assortment of quick-disconnect tools.

A variety of clamps are used on fuel system lines, including the spring and screw types. Crimp, or Oetiker, ear-type clamps **(Figure 28–19)** are the most commonly used. These clamps are made from a single, spring strap. They are available in many different sizes and designs, each made for a particular connection. They are tightened with a special crimping tool.

Inspection

All fuel lines should occasionally be inspected for holes, cracks, leaks, kinks, or dents. Because the fuel is under pressure, leaks in the line between the pump and injection assembly are relatively easy to recognize.

Rubber fuel hose should be inspected for leaks, cracks, cuts, kinks, oil soaking, and soft spots or

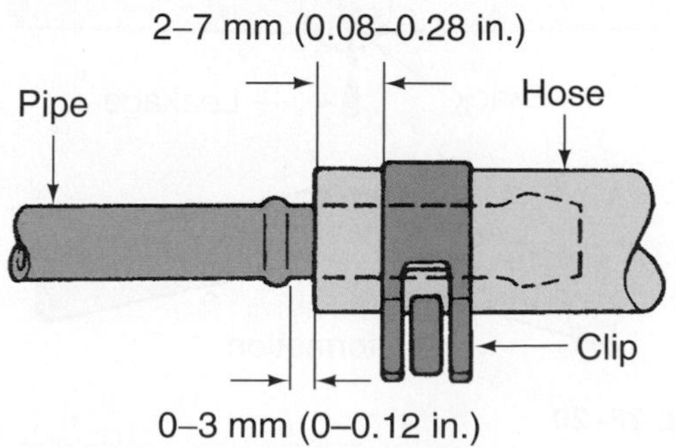

FIGURE 28–18

A fuel hose clamped to steel tubing.

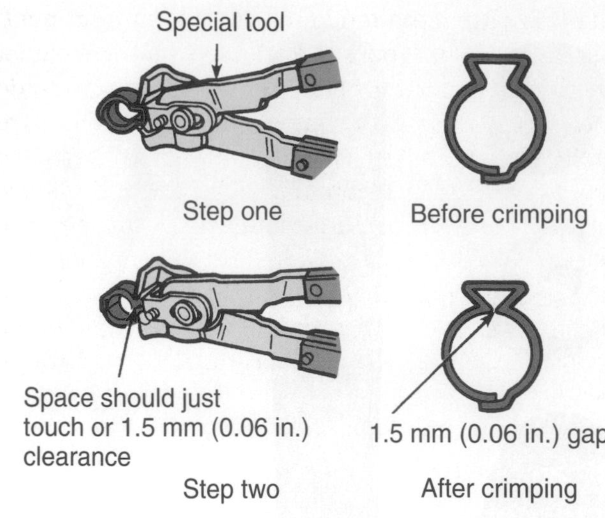

FIGURE 28–19

A special tool is required to tighten crimp clamps.

deterioration. If any of these conditions is found, the fuel hose should be replaced. When rubber fuel hose is installed, the hose should be installed to the proper depth on the metal fitting or line.

Steel tubing should be inspected for leaks, kinks, and deformation **(Figure 28–20)**. Tubing should also be checked for loose connections and proper clamping to the chassis. If the tubing's threaded connections are loose, they must be tightened to the specified torque. Some threaded fuel line fittings contain an O-ring. If the fitting is removed, the O-ring should be replaced.

Nylon fuel pipes should be inspected for leaks, nicks, scratches and cuts, kinks, melting, and loose fittings. If these fuel pipes are damaged in any way, they must be replaced. Nylon fuel pipes must be secured to the chassis at regular intervals to prevent fuel pipe wear and vibration.

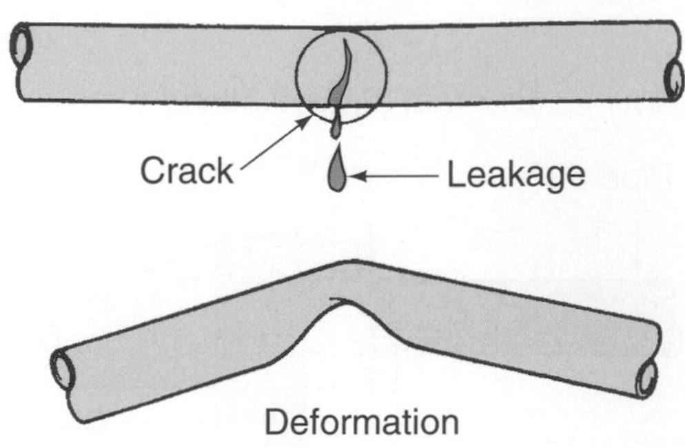

FIGURE 28–20

Steel tubing should be inspected for leaks, kinks, and deformation.

Reprinted with permission.

WARNING!

Always cover a nylon fuel pipe with a wet shop towel before using a torch or other source of heat near the line. Failure to observe this precaution may result in fuel leaks, personal injury, and property damage.

WARNING!

If a vehicle has nylon fuel pipes, do not expose the vehicle to temperatures above 90°C (194°F) for any extended period to avoid damage to the pipes.

Line Replacement

When a damaged fuel line is found, replace it with one of similar construction—steel tubing with steel, and flexible tubing with nylon or synthetic rubber. When installing flexible tubing, always use new clamps. The old ones lose some of their tension when they are removed and do not provide an effective seal when used on the new line.

Any damaged or leaking fuel line must be replaced. To fabricate a new fuel line, select the correct tube and fitting dimension and start with a length that is slightly longer than the old line. With the old line as a reference, use a tubing bender to form the same bends in the new line as those that exist in the old. Although steel tubing can be bent by hand to obtain a gentle curve, any attempt to bend a tight curve by hand usually kinks the tubing. To avoid kinking, always use a bending tool like those shown in **Figure 28–21**.

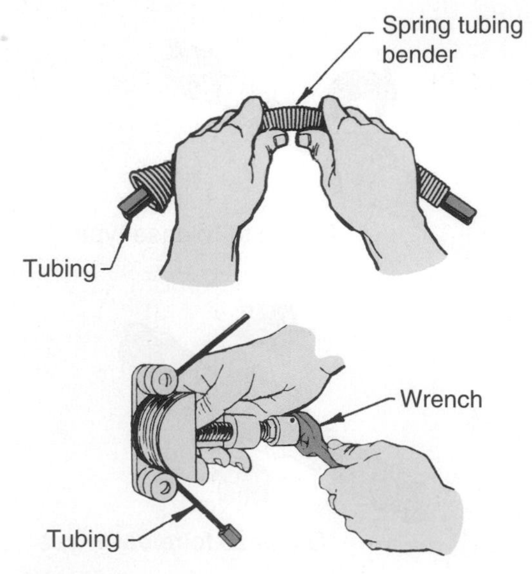

FIGURE 28–21

Two types of bending tools for steel tubing.

CAUTION!

Do not substitute aluminum or copper tubing for steel tubing. Never use a hose within 10 cm (4 in.) of any hot engine or exhaust system component.

Nylon fuel pipes provide a certain amount of flexibility and can be formed around gradual curves under the vehicle. Do not force a nylon fuel pipe into a sharp bend, because doing so may kink the pipe and restrict the flow of fuel. When nylon fuel pipes are exposed to gasoline, they may become stiffer, making them more susceptible to kinking. Be careful not to nick or scratch nylon fuel pipes.

WARNING!

When connecting threaded fittings, make sure that the threads are aligned before tightening the connection. If the fittings are cross-threaded, fuel leaks will inevitably result. Also, always tighten fuel line fittings to the specified torque. If a fitting leaks, check the O-rings and seals.

FUEL FILTERS

Automobiles and light trucks usually have an in-tank strainer and a gasoline filter. The strainer, located in the gasoline tank, is made of a finely woven fabric. The purpose of this strainer is to prevent large contaminant particles from entering the fuel system where they could cause excessive fuel pump wear or plug fuel metering devices. It also helps to prevent passage of any water that might be present in the tank. Servicing of the fuel tank strainer is seldom required.

A fuel filter is connected in the fuel line between the fuel tank and the engine. Many of these filters are mounted under the vehicle **(Figure 28–22)**, and others are mounted in the engine compartment. Most fuel filters contain a pleated paper element mounted in the filter housing, which may be made from metal or plastic. Paper filter elements are efficient at removing and trapping small particles as well as large-size contaminants. Fuel filters are typically contained in a metal case, but some have a plastic housing. On many fuel filters, the inlet and outlet fittings are identified, and the filter must be installed properly. An arrow on some filter housings indicates the direction of fuel flow through the filter.

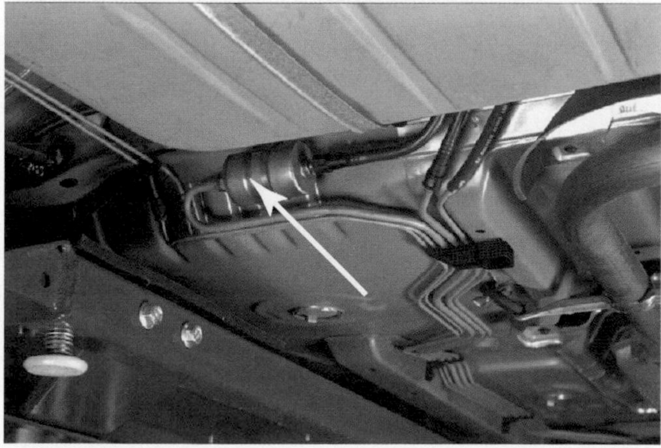

FIGURE 28–22

An inline fuel filter mounted under a vehicle.

Servicing Filters

Fuel filters **(Figure 28–23)** and elements are serviced by replacement only. Some vehicle manufacturers recommend fuel filter replacement at 48 000 km (30 000 miles). Always replace the fuel filter at the vehicle manufacturer's recommended mileage. If dirty or contaminated fuel is placed in the fuel tank, the filter may require replacing before the recommended mileage. A plugged fuel filter may cause the engine to surge and cut out at high speed or hesitate on acceleration. A restricted fuel filter causes low fuel-pump volume.

The fuel filter replacement procedure varies depending on the make and year of the vehicle and the type of fuel system. Some vehicles do not have serviceable fuel filters. Instead, the filter is located in the tank and is serviced as part of the fuel pump module. Always follow the filter replacement procedure in the appropriate service information.

FIGURE 28–23

An assortment of fuel filters.

Robert Bosch Corp.

To install a new filter, begin by wiping the male tube ends of the new filter with a clean shop towel. Apply a few drops of clean engine oil to the male tube ends on the filter. Check the quick connectors to be sure the large collar on each connector has rotated back to the original position. The springs must be visible on the inside diameter of each quick connector. Then install the filter, in the proper direction, and leave the mounting bolt slightly loose. Install the outlet connector onto the filter outlet tube and press the connector firmly in place until the spring snaps into position. Grasp the fuel line and try to pull this line from the filter to be sure the quick connector is locked in place. Then do the same with the inlet connector. Now tighten the filter-retaining bolt to the specified torque. Once everything is connected, lower the vehicle, start the engine, and check for fuel leaks at the filter.

FUEL PUMPS

A fuel pump draws fuel from the fuel tank and pushes it through fuel lines to the engine's injection system. All current vehicles use an electric fuel pump. Older vehicles with carburetors had mechanical pumps; these will not be discussed.

An electric fuel pump can be located inside or outside the fuel tank. Diaphragm, plunger, or bellows types are found on non-EFI systems and are referred to as demand pumps. When the ignition is turned on, the pump starts to run and shuts off automatically when the fuel line is pressurized. When there is a demand for more fuel, the pump turns on again. **Figure 28–24** shows a typical wiring diagram for an electric fuel pump.

An in-tank electric pump is usually the rotary type. Some vehicles have an in-tank pump and a second pump mounted under the vehicle. An in-tank fuel pump has a small DC electric motor with an impeller mounted on the end of the motor's shaft. A pump cover, with inlet and discharge ports, is mounted over the impeller. When the armature and impeller rotate, fuel is moved from the tank to the inlet port, and the impeller forces the fuel around the impeller cover and out the discharge port **(Figure 28–25)**.

Fuel moves from the discharge port through the inside of the motor and out the check valve and outlet connection, which is connected via the fuel line to the fuel filter and underhood fuel system components. A pressure relief valve near the check valve opens if the fuel supply line is restricted and pump pressure becomes very high. When the relief valve opens, fuel is returned through this valve to the pump inlet. Each time the engine is shut off, the check valve prevents fuel from draining out of the fuel system and back into the fuel tank.

Fuel pumps are mounted inside the tank to reduce noise, keep them cool, and keep the entire fuel line pressurized to prevent premature fuel

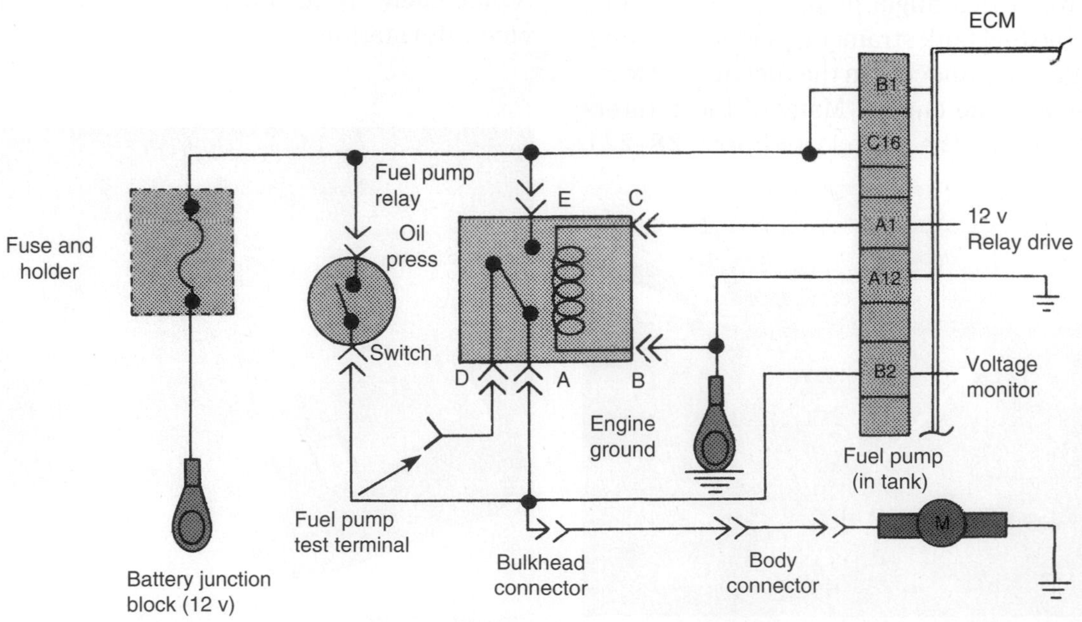

FIGURE 28–24

A typical wiring diagram for an electric fuel pump.

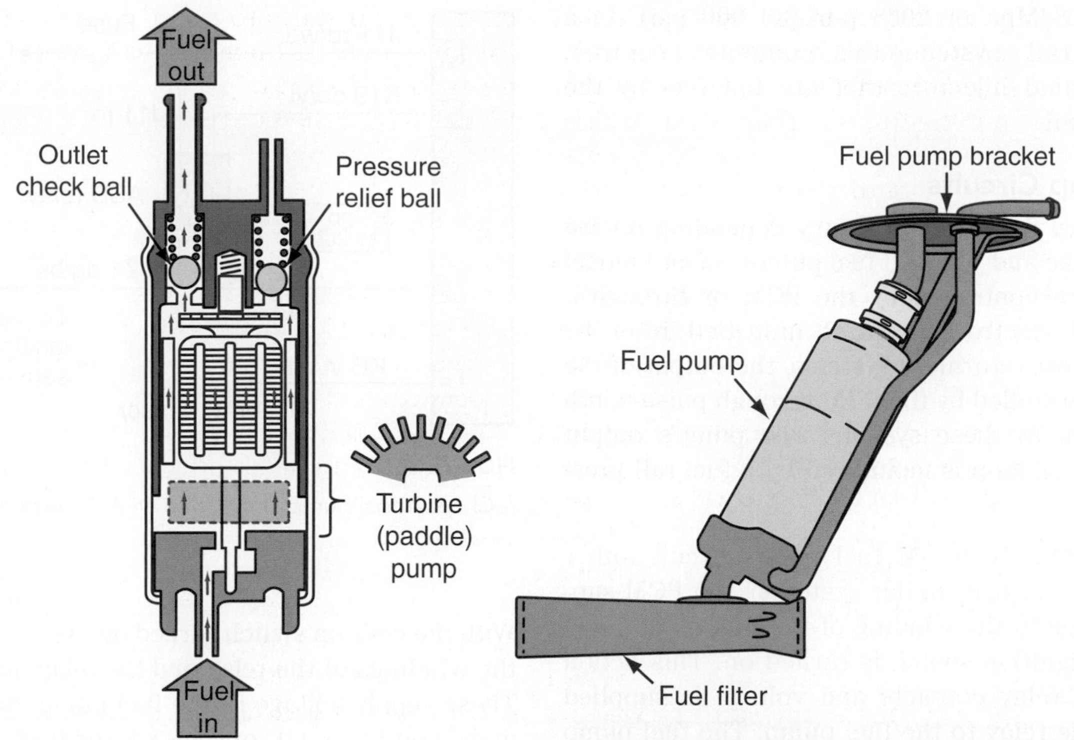

FIGURE 28–25

An electric fuel pump.

evaporation. Although it is dangerous to have a spark near gasoline, and there is a great potential for sparks between an electric motor's armature and brushes, the in-tank fuel pump is safe because there is no oxygen to support combustion in the tank. The mechanical fuel pumps on direct injection systems are driven off a dedicated lobe on the engine's camshaft **(Figure 28–26)**. The lobe pushes against a spring-loaded piston in the pump. Fuel enters the high-pressure pump from the electric supply pump and is then further pressurized by the movement of the piston. A solenoid is used to return excess fuel to the tank. The solenoid is controlled by the PCM and is used to maintain the correct amount of pressure in the fuel rail and injectors. The outlet of the pump is connected to a fuel rail.

DIESEL ENGINES Fuel injection is used on all diesel engines. Older diesel engines had a distributor-type injection pump driven and regulated by the engine. The pump supplied fuel to injectors according to the engine's firing order. Newer light-duty diesel engines are equipped with common rail or direct injection (DI) systems. In these systems, an engine-driven fuel pump delivers fuel to the injectors at a very high pressure,

FIGURE 28–26

In a GDI system, the fuel pump is driven by a dedicated lobe on the camshaft.

about 206.8 Mpa or 2068 bar (30 000 psi). In a common rail system, the computer controls the individual injectors that are fed fuel by the common rail.

Fuel Pump Circuits

Electric fuel pump circuits vary depending on the vehicle make and year. All fuel pumps on late-model vehicles are controlled by the PCM or through a designated electronic control unit tied into the PCM. In most returnless systems, the output of the pump is controlled by the PCM through pulse-width modulation. In these systems, the pump's output during closed loop is monitored by a fuel rail pressure sensor.

RETURN TYPE In a GM fuel pump circuit (other manufacturers use similar systems), the PCM supplies voltage to the winding of the fuel pump relay when the ignition switch is turned on. This action closes the relay contacts and voltage is supplied through the relay to the fuel pump. The fuel pump remains on while the engine is cranking or running. If the ignition switch is on for two seconds and the engine is not cranked, the PCM turns off the voltage to the fuel pump relay to stop the pump.

The PCM may also shut off the fuel pump when the following occur (fuel injectors may be shut off under some of the circumstances):

- The vehicle experiences long, high-speed, closed throttle coast down. Fuel is shut off to prevent damage to the catalytic converters, reduce emissions, and increase the effects of engine braking.
- The engine speed exceeds a predetermined limit.
- The speed of the vehicle exceeds the speed rating of the tires.
- The vehicle has been in a collision.
- The air bag has deployed.
- A fuel line has ruptured.
- The vehicle's antitheft system is activated.

On some systems, an oil pressure switch is connected parallel to the fuel pump relay points. If the relay becomes defective, voltage is supplied through the oil-pressure switch points to the fuel pump. This action keeps the fuel pump operating and the engine running, even though the fuel pump relay is defective. When the engine is cold, oil pressure is not available immediately, and the engine will have long crank times and be slow to start if the fuel pump relay is defective.

The fuel pump relay in Chrysler EFI systems is referred to as an automatic shutdown (ASD) relay.

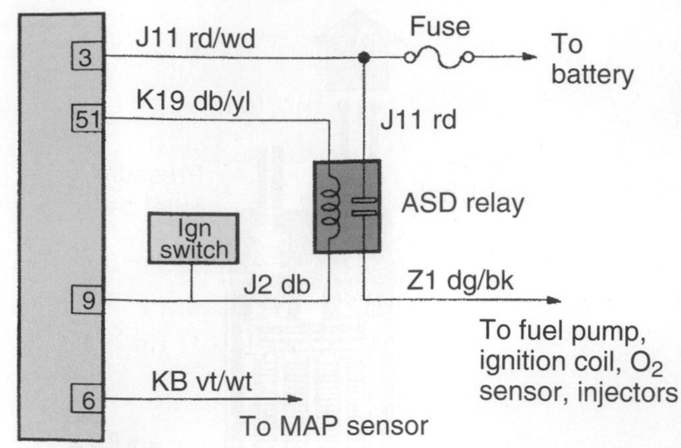

FIGURE 28–27

A Chrysler fuel pump circuit with an ASD relay.

Chrysler Group LLC

With the ignition switch turned on, the PCM grounds the windings of the relay and the relay points close. These supply voltage to the fuel pump, positive primary coil terminal, oxygen sensor heater, and the fuel injectors in some systems **(Figure 28–27)**.

Later-model Chrysler fuel pump circuits have a separate ASD relay and a fuel pump relay. In these circuits, the fuel pump relay supplies voltage to the fuel pump, and the ASD relay powers the positive primary coil terminal, injectors, and oxygen-sensor heater. The ASD relay and the fuel pump relay operate the same as the previous ASD relay. The PCM grounds both relay windings through the same wire.

Rollover Protection

Electric fuel pump circuits include some sort of rollover protection. On Ford vehicles, this includes the installation of an **inertia switch** that shuts off the fuel pump if the vehicle is involved in a collision or rolls over. A typical inertia switch **(Figure 28–28)** consists of a permanent magnet, a steel ball inside a conical ramp, a target plate, and a set of electrical contacts. The magnet holds the steel ball at the bottom of the ramp. In the event of a collision, the inertia of the ball causes it to break away from the magnetic field and roll up the ramp. When it strikes the target plate, the electrical contacts open and the circuit between the PCM and fuel pump control unit opens, causing the fuel pump to turn off. The switch has a reset button that must be depressed before the pump will operate again.

PASSIVE RESTRAINT SYSTEMS Most passive restraint systems will send a signal to the PCM when an air

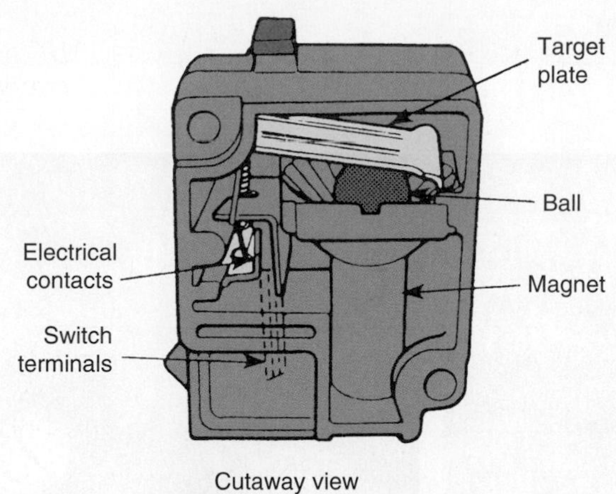

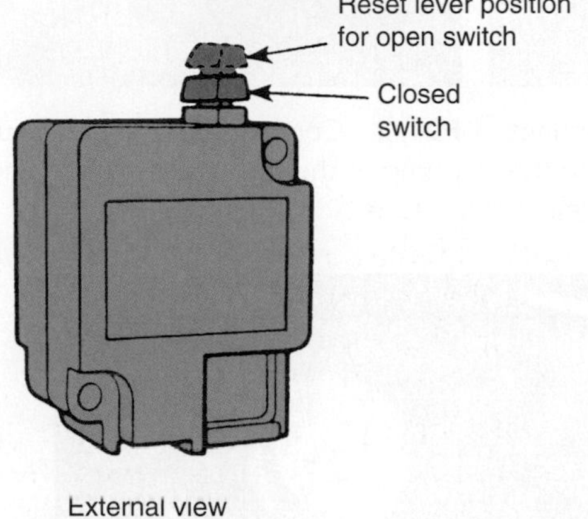

FIGURE 28–28

Details of a Ford inertia switch.

bag is deployed. The PCM, in turn, shuts down the power to the fuel pump. In most cases, it is the centre air bag sensor assembly that sends the deployment signal to the ECM through the CAN. This air bag sensor assembly contains a deceleration sensor, a safing sensor, a drive circuit, a diagnosis circuit, and an ignition control circuit. The main sensor for the deployment of an air bag and the depowering of the fuel pump is the deceleration sensor.

The circuit constantly monitors its own operation and readiness; if it detects a malfunction, the SRS warning light will illuminate and a DTC will be stored.

Returnless Systems

In a mechanical returnless system, fuel pressure is maintained by a fuel pressure regulator located in the fuel tank. Electronic returnless systems control fuel pressure by using pulse-width modulation (PWM) of the fuel pump's power circuit. These systems may use the PCM to directly control the fuel pump, or a separate fuel pump driver module may be used. In either system, the fuel pump speed is controlled by rapidly turning the fuel pump circuit on and off. The longer the time it is on, the faster the pump spins, and more fuel is delivered to the engine. As demand decreases, the pump is left off longer and the fuel supply is decreased.

Electronic returnless systems use a fuel rail pressure (FRP) sensor, which may contain a fuel temperature sensor as well. The FRP sensor data are used by the PCM to control fuel pump operation and to adjust injector pulse width.

Troubleshooting

The fuel pressure test is the commonly used test for the fuel pump and related parts. Before conducting this test, carefully inspect the system for leaks and repair them before continuing. Then relieve the pressure in the system. When doing this, make sure to collect all spilled fuel. Scan tools can be used to shut down the fuel pump on many systems. Also, the PCM may set a DTC as a result of shutting off the fuel pump, make sure to clear that code after testing.

Fuel pressure is read with a fuel pressure gauge or by using a pressure transducer and a scope. The procedure for testing fuel pump pressure with a gauge is shown in Photo Sequence 21. These photos outline the steps to follow while performing the test on an engine with port fuel injection (PFI). To conduct this test on specific fuel injection systems, refer to the service information for instructions. Most domestic systems have a **Schrader valve** on the fuel rail, which can be used to connect the fuel pressure gauge. If the system does not have a Schrader valve, a tee should be installed in the fuel supply line to connect the gauge **(Figure 28–29)**.

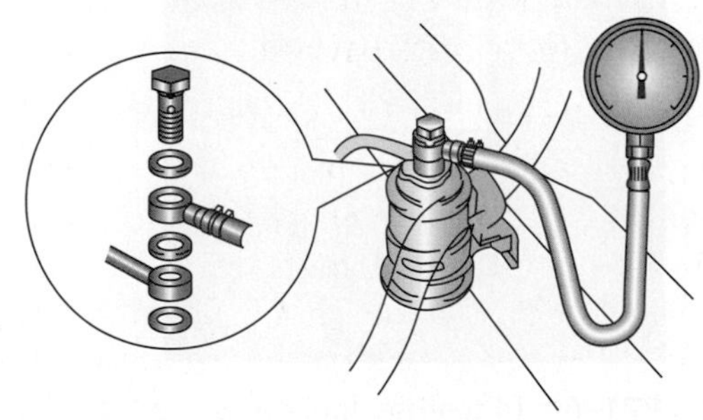

FIGURE 28–29

On fuel systems that do not have a Schrader valve, it may be necessary to fit a tee in the fuel line to connect the fuel pressure gauge.

PHOTO SEQUENCE 21

CHECKING FUEL PRESSURE ON A PFI SYSTEM

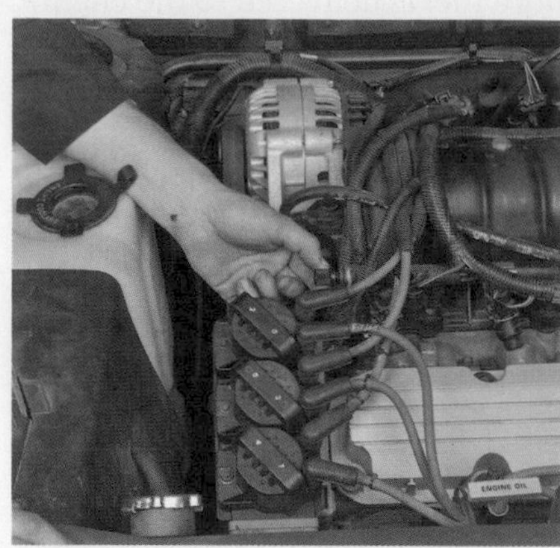

P21–1 Most fuel rails are equipped with a test fitting that can be used to relieve pressure and to test pressure.

P21–2 To test fuel pressure, connect the appropriate pressure gauge to the fuel rail test fitting (Schrader valve).

P21–3 Connect a handheld vacuum pump to the fuel pressure regulator.

P21–4 Turn the ignition switch to the RUN position, and compare the gauge reading to specifications. A reading lower than normal may indicate a faulty fuel pump or restricted fuel filter or supply lines.

P21–5 To test the fuel pressure regulator, create a vacuum at the regulator with the vacuum pump. Fuel pressure should decrease as vacuum increases. If pressure remains the same, the regulator is faulty.

P21–6 To test the fuel pressure on an engine with gasoline direct injection (GDI), you will need to use a scan tool. This is because pressure in the GDI system can exceed 17 237 kPa (2500 psi).

P21–7 With the engine running, locate the fuel pressure PID (parameter identification) and compare the reading to the fuel pressure specification.

On some engines, the fuel rail is fitted with a fuel pulsation damper. The point where the damper attaches to the fuel rail is the recommended place for connecting the pressure gauge. To connect the gauge, place a rag over the damper unit and loosen it one turn with a wrench. After all pressure is released, remove the damper unit, and connect the gauge into the damper's fitting **(Figure 28–30)**.

Often, the specifications for fuel pressure are for key-on, engine-off (KOEO) conditions. This means there will be no signal from the CKP, which means the fuel pump will not run for very long with the key on. Many systems will energize the fuel pump for only a few seconds prior to cranking. With the gauge connected, turn the key on and note the fuel pressure. You may have to cycle the key on and off several times if fuel pressure was completely relieved to install the gauge. Fuel pressure should reach the KOEO specification and should not drop once the key is turned off. The action of the fuel pump can also be checked by controlling the vacuum to the fuel pressure regulator. With 0.677 bar, 67.7 kPa (20 in. Hg), applied to the regulator, the fuel pressure should drop. When there is no vacuum to the regulator, the fuel pressure should rise. This check also verifies that the regulator is working properly. A quick check of fuel volume can be performed by watching fuel pressure with the engine running and the pressure regulator vacuum hose disconnected. The pressure should increase several kPa (psi) with the hose disconnected. Quickly snap the throttle open and watch fuel pressure. If the fuel pump maintains sufficient volume, the pressure should only drop slightly, a couple of kPa (psi), as the injectors supply more fuel to the engine.

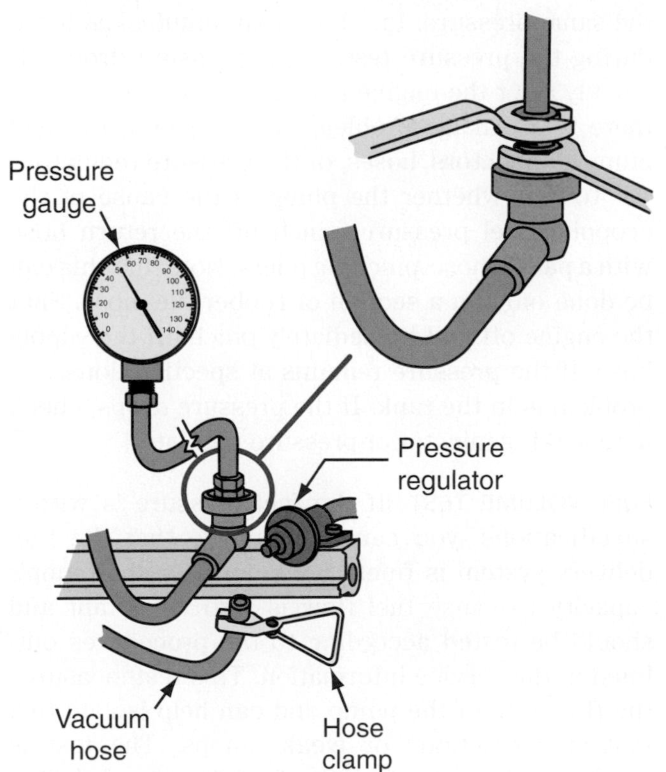

FIGURE 28–30

Connecting a fuel pressure gauge to the fuel pulsation damper fitting.

WARNING!

When testing a fuel system, ensure the following:

1. Do not let fuel contact any electrical wiring. Even the smallest spark could ignite the fuel.
2. To prevent personal injury, fire, or damage to the vehicle or equipment, only use a scan tool to check the fuel pressure on a GDI vehicle.

Some manufacturers recommend that the pressure be measured while the engine is idling. Always make sure that you are using the correct conditions and specifications. Pressure will be slightly lower when the engine is running. Typically, the pressure will rise and fall because the injectors are opening and closing, causing the gauge to fluctuate slightly. Major fluctuations, several kPa (psi), however, indicate air in the system.

Some technicians perform what is called a fuel pump deadhead pressure test. This test momentarily restricts the fuel return line and forces the pump to produce its maximum pressure. With a fuel pressure gauge installed and the engine running at idle, briefly clamp the fuel return line, and note fuel pressure. Do not leave the return clamped longer than it takes the fuel pressure to maximize. In most cases, a good pump will nearly double its engine running pressure. If the pressure does not increase significantly, suspect a weak fuel pump or clogged filter.

Remember, if the fuel pressure is outside specifications, driveability problems can result. Excessive pressure causes a rich air/fuel mixture, and insufficient pressure results in a leaner than normal mixture.

Testing returnless systems is similar to systems with return lines except that fuel pressure can be monitored using a scan tool. To check whether the fuel rail pressure (FRP) sensor is working, watch the fuel pressure PID while changing engine speed and load. If the pressure does not change with engine changes, suspect the pressure sensor is faulty. To double-check the accuracy of the fuel pressure sensor, connect a fuel pressure gauge to the service port and compare the readings. If the FRP sensor is faulty, such as reading pressure as higher than it actually is, the PCM will command decreased fuel pressure. This can cause a lean condition as there is not enough fuel available at the injectors. It is important to note that the readings between the mechanical gauge and FRP sensor on a running engine likely will not match. This is because the pressure reading from the FRP sensor is referenced to manifold pressure and not atmospheric pressure. Refer to the manufacturer's service information to determine the variation in pressure readings.

INTERPRETING THE RESULTS High fuel pressure readings usually indicate a faulty pressure regulator or an obstructed return line. To identify the cause, disconnect the fuel return line at the tank. Use a length of hose to route the returning fuel into a container. Start the engine, and note the pressure reading at the engine. If fuel pressure is now within specifications, check for an obstruction in the return system at the tank. The fuel reservoir check valve or aspirator jet might be clogged.

If the fuel pressure still reads high with the return line disconnected, note the volume of fuel flowing through the line. Little or no fuel flow can indicate a plugged return line. Shut off the engine and connect a length of hose directly to the fuel pressure regulator return port to bypass the return hose. Restart the engine and again check the pressure. If bypassing the return line brings the readings back within specifications, a plugged return line is the problem.

If pressure is still high, apply vacuum to the pressure regulator. If there is still no change, replace the pressure regulator. If applying vacuum to the regulator lowers fuel pressure, the vacuum hose to the regulator might be plugged, leaking, or misrouted.

On **mechanical returnless fuel systems**, the fuel pressure regulator is mounted in the tank with the fuel pump. To service the regulator, the fuel pump module must be removed from the tank. With PWM returnless systems, a faulty fuel rail pressure sensor can cause the PCM to increase fuel pressure if the sensor is reading too low.

Low fuel pressure, on the other hand, can be due to a clogged fuel filter, a restricted fuel line, a weak pump, a leaky pump check valve, a defective fuel pressure regulator, a leak in the supply line in the tank, excessive resistance in the fuel pump power or ground circuits, or a dirty filter sock in the tank. It is possible to rule out filter and line restrictions by checking the pressure at the pump outlet. A higher reading at the pump outlet, of at least 34.4 kPa (5 psi), means there is a restriction in the filter or line. If the reading at the pump outlet is unchanged, then the pump either is weak or is having trouble picking up fuel (clogged filter sock in the tank). Either way, it is necessary to get inside the fuel tank. If the filter sock is gummed up with dirt or debris, it is also wise to clean out the tank when the filter sock is cleaned or replaced.

Another possible source of trouble is the pump's check valve. Some pumps have one, whereas others have two (positive displacement roller vane pumps). The check valve prevents fuel movement through the pump when the pump is off so that residual pressure remains at the injectors. This can be checked by watching the fuel pressure gauge after the engine is shut off.

RESIDUAL PRESSURE Often, the cause of poor starting is the lack of residual pressure in the system. This can be checked by looking at the pressure after the engine has been run and then turned off. Using a pressure transducer and GMM or DSO allows you to monitor system pressure over a period of time without having to actually wait and watch a fuel pressure gauge. The system should hold about the same pressure, for about five minutes, as it did during the pressure test. If the pressure drops off quickly after the engine and ignition are turned off, there is a leakage problem in an injector, the fuel pump, connectors, hoses, or the pressure regulator.

To test whether the pump is the cause of the dropping fuel pressure, pinch off the return hose with a pair of hose-pinching pliers. Note that this can be done only on a section of rubber fuel hose. Shut the engine off and immediately pinch off the supply hose. If the pressure remains at specifications, the problem is in the tank. If the pressure drops, check for a leaking injector or pressure regulator.

FUEL VOLUME TEST If the fuel pressure is within specifications, you cannot conclude that the fuel delivery system is fine. Fuel volume or the pump's capacity to cause fuel flow is also important and should be tested according to the procedures outlined in the service information. This test measures the flow rate of the pump and can help isolate fuel system restrictions or weak pumps. The test is conducted by collecting the fuel dispensed during a period of time, which is normally 30 seconds. Disconnect the return line on returnable systems or the supply line on returnless systems, and connect a

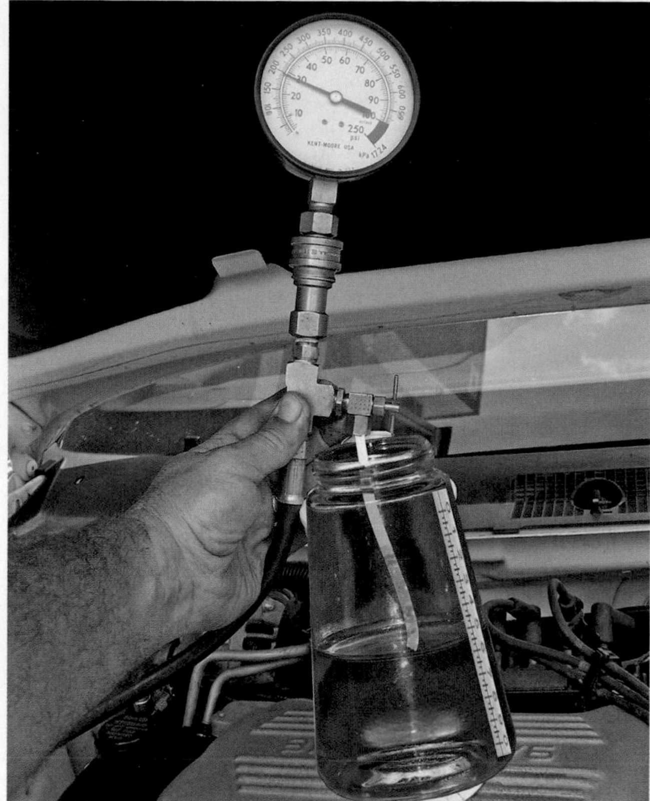

FIGURE 28–31

The setup for checking the fuel pump volume.

hose to end of the line. Put the other end of the hose into a graduated container **(Figure 28–31)**. Turn on the fuel pump for about 30 seconds and measure the amount of fuel in the container. Normally the desired amount is 0.47 L (1 pint). The flow of fuel into the container should be smooth and continuous with no signs of air bubbles. Results other than these indicate a bad pump or restrictions in the delivery system.

After checking the fuel pressure and volume, remove the pressure gauge and all adapters and hoses installed for the tests. Reinstall and tighten the fuel filler cap. Then turn the ignition on and check for fuel leaks.

USING A LAB SCOPE Monitoring the voltage to the fuel pump with a lab scope or GMM can give an idea of how well the pump is working. The voltage traces should be consistent throughout the time period set on the scope. The pattern is made up of a series of small humps. Each of the humps represents one commutator segment in the motor. Each of the humps should look the same. If there are variances in the pattern, a worn pump is indicated **(Figure 28–32)**. The voltage traces show the voltage needed to establish current flow in the pump's windings.

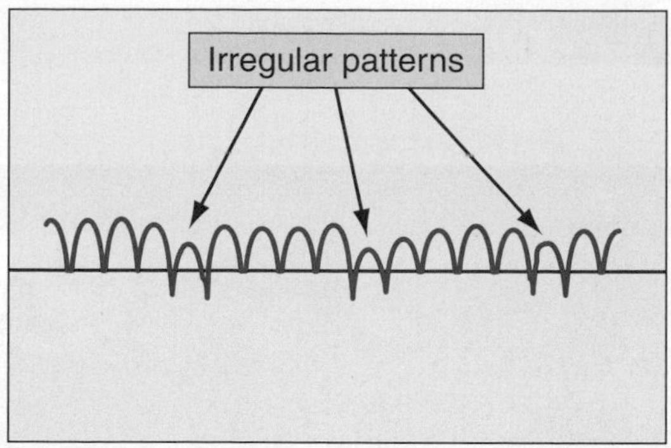

FIGURE 28–32

Monitoring a fuel pump with a lab scope or GMM can show a worn pump, as evidenced by irregular voltage spikes.

USING A DMM A pressure transducer can be connected to a lab scope and to the fuel system to measure fuel pressure. This is a safe way to monitor fuel pressure during a road test. The transducer is connected in the same way as a pressure gauge. Check the scope's information to determine how to interpret the readings.

A DMM with a current probe can be used to monitor the current flow to the fuel pump. Looking at the current can give a good picture of the condition of the pump and the overall fuel delivery system. The normal amount of current draw for a fuel pump may be listed in the vehicle's specifications. If it is not, use the values in **Figure 28–33** as a guideline. Photo Sequence 22 shows how to look at the current to the pump. This is very valuable for diagnosing a fuel pump and its circuit.

In motor circuits, excessive current means the armature is rotating slower than normal. Therefore, if the pump is drawing too much current, it is working harder than it should. This can be caused by a restricted fuel filter or return line.

The opposite is true; if the current levels are low, the motor is spinning too fast. This can be caused by very low fuel levels in the tank. Current can also be too low due to high resistance in the circuit. Check the connectors from the relay and to the pump. Also, check the ground circuit for the pump. If the circuit

MIN/MAX FUEL PRESSURE	NORMAL DRAW
70–105 kPa (10–15 psi)	2–4 amps
240–415 kPa (35–60 psi)	4–9 amps

FIGURE 28–33

A chart showing the typical current draw of low- and high-pressure fuel pumps.

PHOTO SEQUENCE 22

CHECKING CURRENT RAMPING TO THE FUEL PUMP

P22–1 To test the fuel pump, you will need a lab scope and current clamp. Begin by installing a fender cover if working at an underhood fuse box.

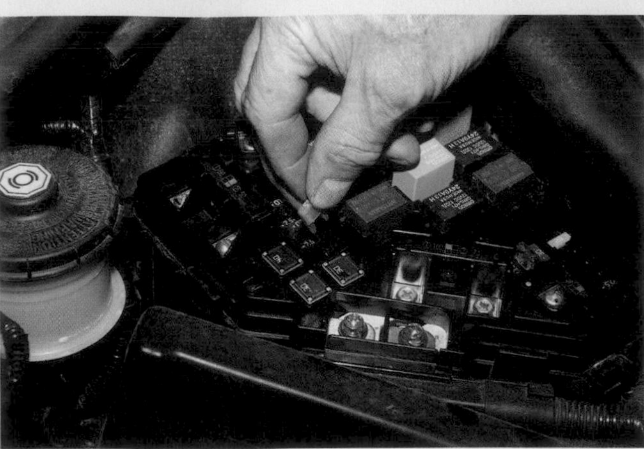

P22–2 Locate the fuel pump fuse and remove it from the fuse box.

P22–3 Install a fused jumper wire in place of the fuel pump fuse. Make sure the fused jumper wire has the same amperage rating as the fuel pump circuit.

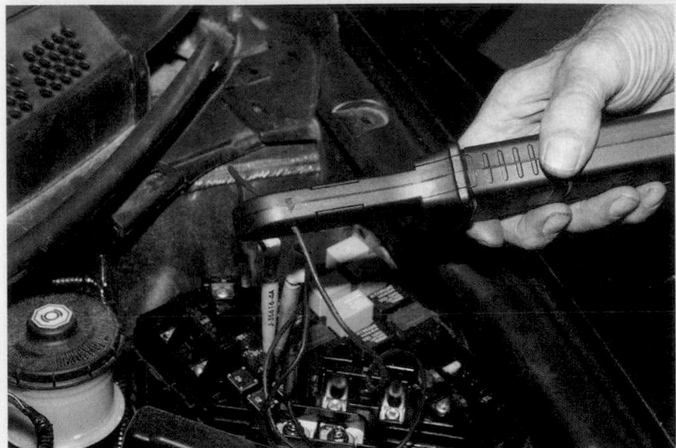

P22–4 Place the current clamp over the jumper wire with the arrow on the clamp pointing in the direction of current flow.

P22–5 Set the scope to display the signal from the clamp. In this example, the scope is set to measure millivolts (mV) and the clamp is set to 20 amps. This setting equals 10 mV = 1 amp.

P22–6 Turn the ignition on and capture the fuel pump current draw waveform.

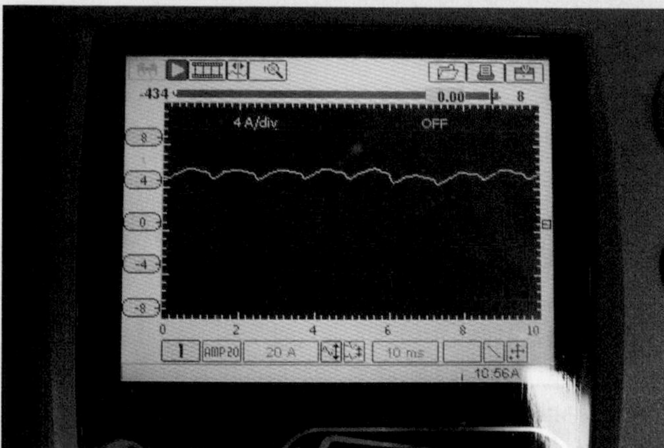

P22–7 This pattern shows a good fuel pump. Note the consistent humps of current flow through each commutator and winding. Although pump current draw will vary depending on vehicle manufacturer and fuel system pressure, typical current draw is between 4 and 8 amps.

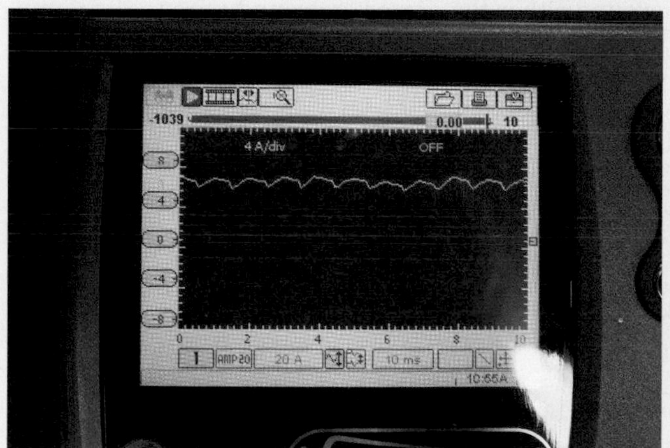

P22–8 This image shows a waveform from a vehicle with a clogged fuel filter, which is slowing the pump and increasing current draw.

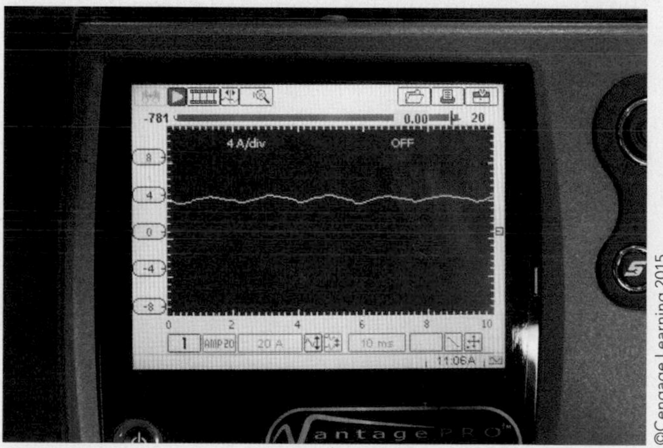

P22–9 This image shows a waveform from a vehicle with a corroded connection at the fuel pump. Note the lower than normal amperage and pump speed.

is fine, check the pump. You should not assume that the pump is the cause of a low-pressure problem until you have ruled out that the power and ground circuits for the pump are in good condition and there is not excessive voltage drop in the circuits.

SHOP TALK

All electrical tests assume that the battery is fully charged. Often, when there is a fuel pump problem, the engine has been cranked often and long. This lowers the battery's voltage and will affect all tests on the pump and its circuit.

No-Start Diagnosis

When an engine fails to start because there is no fuel delivery, the first check is the fuel gauge. A gauge that reads higher than a half tank probably means there is fuel in the tank, but not always. A defective sending unit or miscalibrated gauge might be giving a false indication. Sticking a wire or dowel rod down the fuel tank filler pipe tells whether there really is fuel in the tank. If the gauge is faulty, repair or replace it.

Listen for pump noise. When the key is turned on, the pump should hum for a couple of seconds to build system pressure. On Ford vehicles, make sure the inertia switch has not been tripped. Hitting a large enough pothole can cause the inertia switch to open, shutting off the fuel pump. The pump may be energized through an oil-pressure switch (the purpose of which is to shut off the flow of fuel in case of an accident that stalls the engine). On most late-model cars with computerized engine controls, the computer energizes a pump relay **(Figure 28–34)** when it receives a cranking signal from the crankshaft sensor. An oil-pressure switch might still be included in the circuitry for safety purposes and to serve as a backup in case the relay or computer signal fails. Failure of the pump relay or computer driver signal can cause slow starting because the fuel pump does not come on until the engine cranks long enough to build up sufficient oil pressure to trip the oil-pressure switch.

The pump might be good, but if it does not receive voltage and have a good ground, it does not run. To check the ground, use a voltmeter to check for voltage drop from the end of the circuit to a good ground. Low voltage can reduce pump operating speed and output. If the electrical circuit checks out but the pump does not run, the pump is probably bad and should be replaced.

You can also test pump current draw to determine if the pump is working. Place a current probe around either the power or ground wire to the pump. Set the probe and DMM to measure pump current, typically less than 10 amps. Turn the key on or

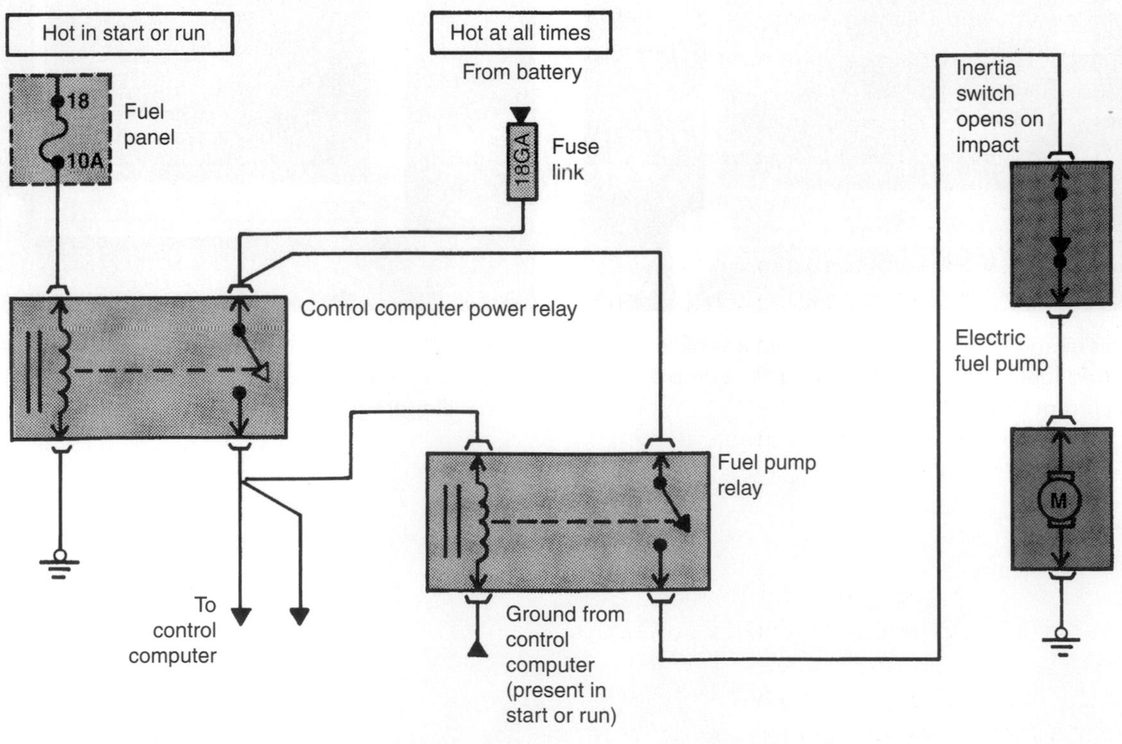

FIGURE 28–34

An electrical fuel delivery system wiring diagram.

command the pump on with a scan tool, and note the reading. Typical pump current draw is 6 to 8 amps. If no reading is shown and you have checked the power and ground circuits, the pump is faulty or the electrical connections inside the tank are open. Lower than specified current can indicate excessive voltage drop in the pump circuit or a pump that is spinning faster than normal. This can be caused by a leak in the supply line inside of the tank. Higher than normal current draw indicates the pump is spinning too slowly or there is a restriction in the supply system.

No voltage at the pump terminal when the key is on and the engine is cranking indicates a faulty oil-pressure switch, pump relay, or relay drive circuit in the computer; an open inertia switch; or a wiring problem. Check the pump fuse to see if it is blown. Replacing the fuse might restore power to the pump, but until you have found out what caused the fuse to blow, the problem is not solved. The most likely cause of a blown fuse would be a short in the wiring between the relay and pump, a short inside the oil-pressure switch or relay, or a bad fuel pump.

A faulty oil-pressure switch can be checked by bypassing it with a fused jumper wire. If doing this restores power to the pump and the engine starts, replace the switch. If an oil-pressure switch or relay sticks in the closed position, the pump can run continuously whether the key is on or off, depending on how the circuit is wired.

To check a pump relay, use a voltmeter to check across the relays and ground terminals to tell if the relay is getting battery voltage and ground. Next, turn off the ignition, wait about 10 seconds, and then turn it on. The relay should click and you should see battery voltage at the relay's pump terminal. If nothing happens, repeat the test, checking for control of the relay by the computer. Depending on the vehicle, the PCM may supply power or ground to control the relay. The presence of a control signal here means the computer is doing its job, but the relay is failing to close and should be replaced. No control signal from the computer indicates an open in that wiring circuit or a fault in the computer itself.

Replacement

When replacing an electric pump, be sure that the new or rebuilt replacement unit meets the minimum requirements of pressure and volume for that particular vehicle. This information can be found in the service information. If the fuel pump is mounted in the fuel tank, the procedure for replacement is different than if the unit is external to the tank.

CAUTION!

Avoid the temptation to test the new pump before reinstalling the fuel tank by energizing it with a couple of jumper wires. Running the pump dry can damage it because the pump relies on fuel for lubrication and cooling.

INTERNAL FUEL PUMP On many vehicles, the fuel tank must be removed to replace the fuel pump and/or fuel gauge sending unit. On other vehicles, the unit can be serviced through an opening in the vehicle's trunk **(Figure 28–35)**. Some vehicles have a separate fuel pump and gauge sending unit, whereas others have both contained in a single unit. Once the

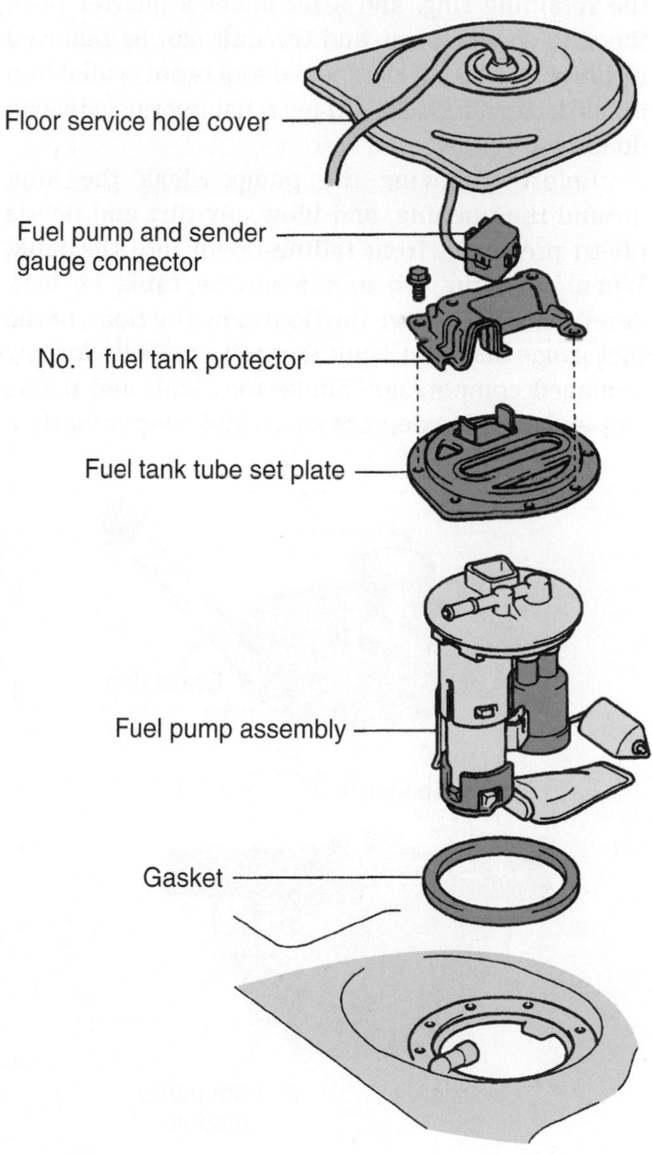

FIGURE 28–35

Some fuel pumps can be serviced through an access hole in the car's floor.

Reprinted with permission.

fuel tank is empty and out of the vehicle, if necessary, remove the unit from the tank.

CAUTION!

Never turn on the ignition switch or crank the engine with a fuel line disconnected. This action will result in gasoline discharge from the disconnected line, which may result in a fire, causing personal injury and/or property damage.

These units are often held in the tank by either a retaining ring or screws. The easiest way to remove a retaining ring is to use a special tool designed for this purpose. This tool fits over the metal tabs on the retaining ring, and after about a quarter turn, the ring comes loose, and the unit can be removed **(Figure 28–36)**. If the special tool is not available, a brass drift punch and ball-peen hammer usually can do the job.

Before removing the pump, clean the tank around the opening, and blow any dirt and debris off to prevent it from falling down into the tank. When removing the unit from the tank, be very careful not to damage the float arm, the float, or the fuel gauge sender. Check the unit carefully for any damaged components. Shake the float, and if fuel can be heard inside, replace it. Make sure the float arm is not bent. It is usually wise to replace the sock and O-ring before replacing the unit. Check the fuel gauge and sender unit, as described in the service information. When reinstalling the pickup pipe-sending unit, be very careful not to damage any of the components.

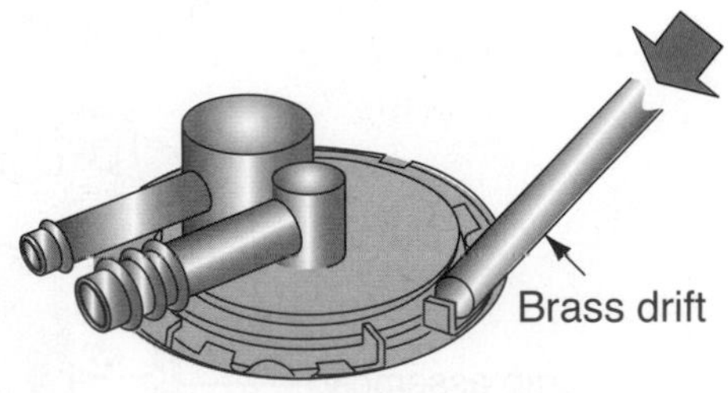

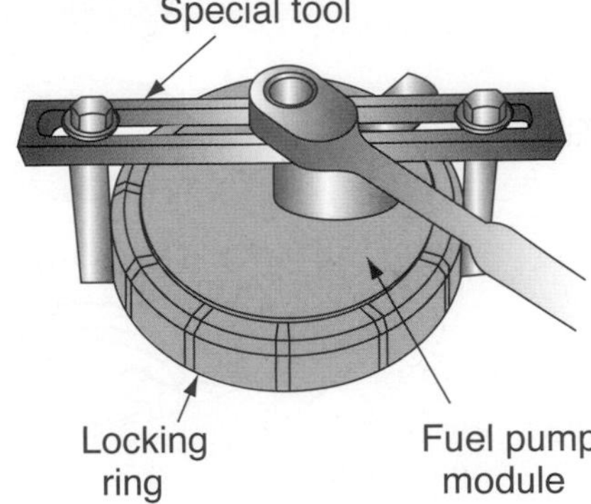

FIGURE 28–36

The locking ring for the fuel pump can be loosened and tightened with a brass drift or a tool designed for doing this.

Once the unit is removed, check the filter on the fuel pump inlet. If the filter is contaminated or damaged, replace the filter. Many technicians replace the inlet filter as part of replacing the fuel pump. Inspect the fuel pump inlet for dirt and debris. Replace the fuel pump if these foreign particles are found in the pump inlet.

If the pump inlet filter is contaminated, flush the tank with hot water for at least five minutes. Dump all the water out of the tank through the pump opening in the tank. Shake the tank to be sure all the water is removed. Allow the tank to sit and air dry before reinstalling it or adding fuel to it. Remember, gasoline fumes are extremely ignitable, so keep all open flames and sparks away from the tank while it is drying.

Check all fuel hoses and tubing on the fuel pump assembly. Replace fuel hoses that are cracked, deteriorated, or kinked. If replacing a piece of fuel hose within the tank, make sure the hose is rated for immersion in fuel. Using inferior fuel line will cause the line to fail and fall apart within the tank. When fuel tubing on the pump assembly is damaged, replace the tubing or the pump.

Make sure the sound insulator sleeve is in place on the electric fuel pump, and check the position of the sound insulator on the bottom of the pump.

Clean the pump and sending unit mounting area in the fuel tank with a shop towel, and install a new gasket or O-ring on the pump and sending unit. Install the fuel pump and gauge sending unit assembly in the fuel tank and secure this assembly in the tank using the vehicle manufacturer's recommended procedure. Many late-model fuel pump modules are spring-loaded and require you to position the unit and keep it pressed down during reassembly. Make sure the unit is properly positioned inside of the tank to prevent damage to the fuel sending unit and inlet filter.

KEY TERMS

EVAP (p. 836)
Inertia switch (p. 842)
Mechanical returnless fuel systems (p. 846)
Schrader valve (p. 843)

SUMMARY

- A typical fuel delivery system includes a fuel tank, fuel lines, a fuel filter, a fuel pressure regulator, and a fuel pump.
- The fuel system should be checked whenever there is evidence of a fuel leak or smell.
- The fuel system should also be checked whenever basic tests suggest that there is too little or too much fuel being delivered to the cylinders.
- Fuel delivery problems typically cause no-start or loss-of-power problems.
- Because electronic fuel injection systems have a residual fuel pressure, this pressure must be relieved before disconnecting any fuel system component.
- Fuel tanks have devices that prevent fuel vapours from leaving the tank. All fuel tanks have a filler tube and a non-vented cap.
- The fuel tank should be inspected for leaks; road damage; corrosion; rust; loose, damaged, or defective seams; loose mounting bolts; and damaged mounting straps.
- Leaks in the metal fuel tank can be caused by a weak seam, rust, or road damage. The best way to permanently solve these problems is to replace the tank.
- In-tank fuel pumps and fuel level gauge sending units are held in the tank by either a retaining ring or screws.
- Fuel lines are made of seamless, double-wall metal tubing, flexible nylon, or synthetic rubber hose. The latter must be able to resist gasoline and be non-permeable to fuel and fuel vapours. Rubber hose, such as that used for vacuum lines, deteriorates when exposed to gasoline. Only hoses made for fuel systems should be used for replacement.
- Vapour vent lines must be made of material that resists attack by fuel vapours. Replacement vent hoses are usually marked with the designation **EVAP** to indicate their intended use.
- All fuel lines should occasionally be inspected for holes, cracks, leaks, kinks, or dents.
- Automobiles and light trucks have an in-tank strainer and a fuel filter. The strainer, located in the gasoline tank, is made of a fine woven fabric. The strainer prevents large contaminant particles from entering the fuel system where they could cause excessive fuel pump wear.
- To determine if the fuel pump is in satisfactory operating condition, tests for both fuel pump pressure and fuel pump capacity should be performed.
- High fuel pressure readings normally indicate a faulty pressure regulator or an obstructed return line.
- Low pressure can be caused by a clogged fuel filter, restricted fuel line, weak pump, leaky pump check valve, defective fuel pressure regulator, or dirty filter sock in the tank.
- An inertia switch in the fuel pump circuit opens the fuel pump circuit immediately if the vehicle is involved in a collision.
- SRSs automatically shut off the fuel pump when an air bag is deployed.

REVIEW QUESTIONS

1. Fuel pump ________ is a statement of the amount of flow of the pump.
2. Explain the purpose of the relief valve and one-way check valve in an electric fuel pump.
3. What type of fire extinguisher should you have close by when you are working on fuel system components?
4. What is the first thing that should be disconnected when removing a fuel tank?
5. Why is a plastic or metal restrictor placed in either the end of the vent pipe or in the vapour-vent hose on some vehicles?
6. Which of the following could cause a high fuel pressure reading?
   a. a clogged fuel filter
   b. a leaking fuel line
   c. a faulty fuel pump
   d. a faulty fuel pressure regulator
7. On a vehicle equipped with port fuel injection, where is the test port usually located?
   a. on the fuel pump
   b. on the fuel rail
   c. on the fuel tank
   d. on the fuel filter
8. Which of the following gasoline tank filler cap statements is correct?
   a. They vent gasoline vapours to the atmosphere.
   b. They will not release extreme pressures.
   c. They will allow air in to relieve high tank vacuum.
   d. They will not allow air in to relieve high tank vacuum.

9. Which of the following fuel line fittings require the use of O-rings?
   a. quick-connect fittings
   b. double flare fittings
   c. compression fittings
   d. clamp fittings
10. What is a typical fuel filter replacement interval?
   a. 24 000 km (15 000 miles)
   b. 48 000 km (30 000 miles)
   c. 72 000 km (45 000 miles)
   d. 96 000 km (60 000 miles)
11. What could cause a rich air/fuel mixture?
   a. a clogged fuel return line
   b. a faulty fuel pump
   c. a clogged fuel filter
   d. a leaking fuel line
12. What could cause no voltage at the fuel pump terminal with the key on and engine cranking?
   a. a poor pump ground
   b. a faulty fuel pressure regulator
   c. a faulty fuel pump relay
   d. an oil-pressure switch stuck in the closed position
13. What is the purpose of the inertia switch?
   a. to open the fuel pump circuit in the case of a collision
   b. to close the fuel pump circuit in the case of a collision
   c. to restrict fuel pressure at wide-open throttle
   d. to cycle the fuel pump at initial start-up
14. What could a loss of vacuum at the fuel pressure regulator cause?
   a. a gasoline leak
   b. low fuel pressure
   c. high fuel pressure
   d. a lean-running engine
15. What is the first step in removing a fuel tank?
   a. Disconnect the negative battery cable.
   b. Raise the vehicle on the hoist.
   c. Disconnect the fuel lines from the tank.
   d. Loosen the fuel tank strap mounting bolts.
16. What material are most fuel line O-rings made of?
   a. rubber
   b. silicone
   c. neoprene
   d. Viton
17. When a steel fuel line is damaged and requires replacement, what type of line should be used?
   a. steel
   b. flexible tubing
   c. nylon
   d. synthetic rubber
18. What does a buzzing sound coming from the fuel tank indicate?
   a. a poor fuel pump tank ground
   b. faulty fuel pump contacts
   c. an operating fuel pump
   d. a clogged fuel pump screen
19. What does the PCM ground for the fuel pump to operate on most Chrysler cars with electronic fuel injection?
   a. the fuel pump relay
   b. the automatic shutdown relay
   c. the fuel pump
   d. the ignition switch
20. What component, located in the fuel tank, prevents large contaminants from entering the fuel system?
   a. the fuel filter
   b. liquid separator
   c. vapour separator
   d. strainer

CHAPTER 29
Electronic Fuel Injection

LEARNING OUTCOMES

- Explain the differences in the point of injection in throttle body or port injection systems.
- Describe the difference between a sequential fuel injection (SFI) system and a multiport fuel injection (MPFI) system.
- Explain the design and function of major EFI components.
- Describe the inputs used by the computer to control the idle air control (IAC) and IAC bypass air motors.
- Describe how the computer supplies the correct air/fuel ratio on a throttle body injection (TBI) system.
- Explain how the clear flood mode operates on a TBI system.
- Explain why manifold vacuum is connected to the pressure regulator in an MPFI system.
- Describe the operation of the pressure regulator in a returnless EFI system.
- Describe the operation of the central injector and poppet nozzles in a central multiport fuel injection (CMFI) system.
- Describe the operation of direct gasoline injection systems.
- Describe the operation of the injection systems used in light- and medium-duty diesel engines.

This chapter discusses the components of electronic fuel injection (EFI) systems and explains how the various designs of EFI work. EFI systems are computer controlled and designed to provide the correct air/fuel ratio for all engine loads, speeds, and temperature conditions. The computer monitors the operating conditions and attempts to provide the engine with the ideal air/fuel ratio. The ideal fuel ratio is often called the **stoichiometric ratio.** A stoichiometric mixture is one that has the air-to-fuel ratio necessary for complete combustion of the fuel **(Figure 29–1)**. This means all of the fuel and the oxygen in the air are completely consumed during combustion. Different fuels have a different stoichiometric ratio **(Figure 29–2)**. The stoichiometric mixture for gasoline is 14.7:1.

A stoichiometric mixture is theoretically the best combination of fuel and air to provide for total combustion. The ratio, however, is based on an ideal environment for combustion. This environment rarely exists, and the injection system's controls vary the ratio in response to the inefficiencies. It also responds to changes in conditions that affect the combustion process. The air/fuel ratio also changes for starting, maximum power, and maximum fuel economy. The stoichiometric ratio allows the catalytic converters to work more efficiently.

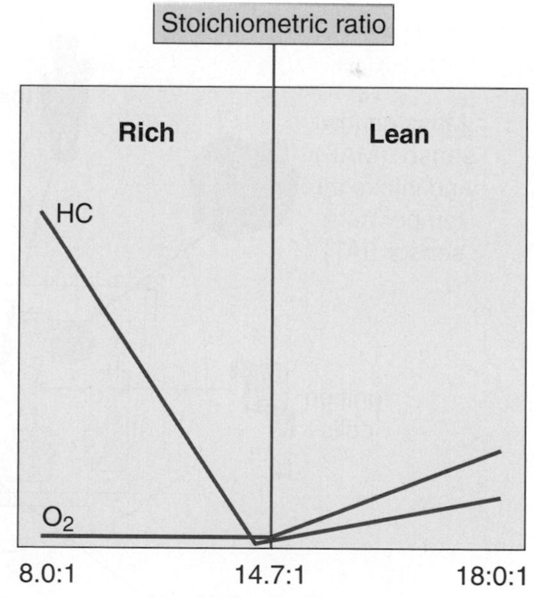

FIGURE 29–1

A graph showing how the amount of raw gasoline (HC) and air (O_2) released in the exhaust changes with a change in the air/fuel ratio.

FUEL TYPE	STOICHIOMETRIC A/F RATIO
Natural gas	17.2:1
Typical gasoline*	14.7:1
Diesel	14.6:1
Ethanol	9.0:1
Methanol	6.4:1

* Gasoline types vary according to the percentage of alcohol added.

FIGURE 29–2

Different fuels have different stoichiometric air/fuel ratios.

BASIC EFI

In an EFI system **(Figure 29–3)**, the computer must know the amount of air entering the engine so that it can supply the correct amount of fuel for that amount of air. In systems with a manifold absolute pressure (MAP) sensor, the computer calculates the amount of intake air based on MAP and rpm input signals. This type of EFI system is referred to as a **speed density** system, because the computer calculates the air intake flow according to engine speed and intake manifold vacuum. Because air density changes with air temperature, an intake air temperature sensor must also be used.

Today, the most commonly used EFI system is the mass airflow (MAF) system. This system relies on an MAF sensor that directly measures the amount of intake air. The most common type of MAF sensor is the hot wire design. MAF systems are very responsive to changes in operating conditions because they actually measure, rather than compute, airflow.

During closed loop, EFI systems rely on the input from a variety of sensors before adjusting the air/fuel ratio. Based on all of the inputs, the PCM is able to determine the current operating conditions of the engine, such as starting, idle, acceleration, cruise, deceleration, and operating temperature. The PCM gathers the inputs and refers to the look-up tables in its memory to determine the ideal air/fuel ratio for the current conditions.

When conditions, such as starting or wide-open throttle, demand that the signals from the oxygen sensor be ignored, the system operates in open loop. During open loop, fuel is delivered according to predetermined parameters held in the PCM's memory. Pre-OBD-II systems may also go into the open-loop mode while idling, or at any time that the oxygen sensor cools off enough to stop sending a good signal.

Fuel Injectors

Fuel injectors are electromechanical devices that meter and atomize fuel so that it can be sprayed into the intake manifold. Injectors used on gasoline

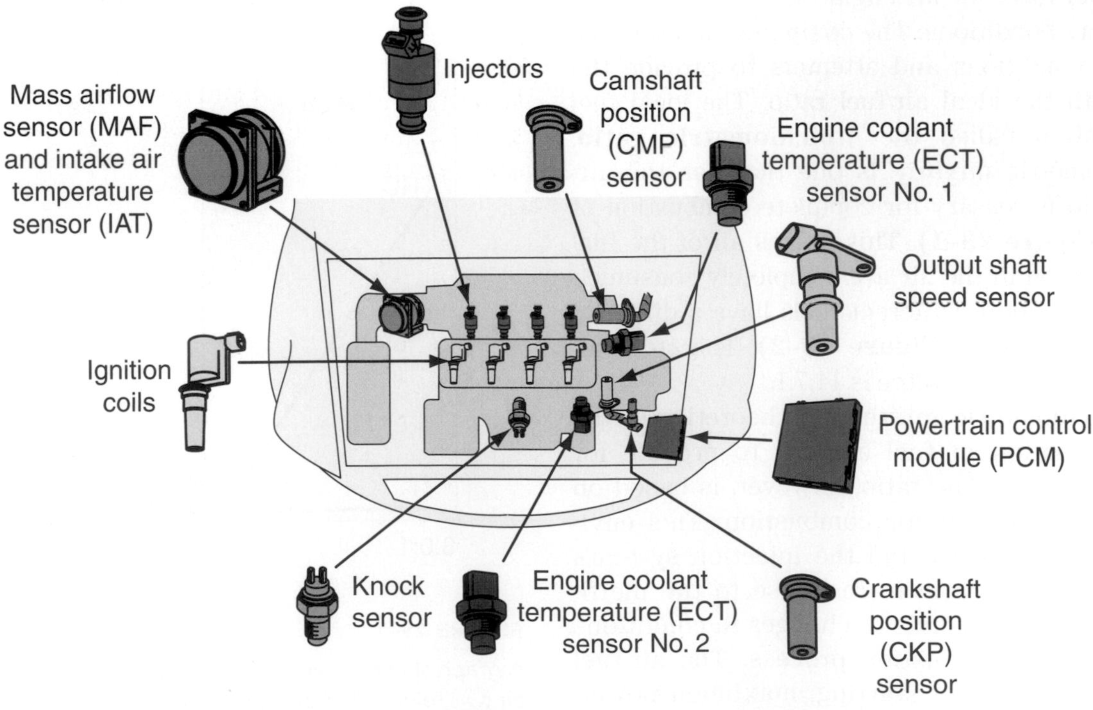

FIGURE 29–3

A typical electronic fuel injection system.

direct injection systems are bolted to the cylinder head with the fuel rail and spray fuel directly into the combustion chamber. Fuel injectors resemble a spark plug in size and shape. O-rings are used to seal the injector at the intake manifold, throttle body, and/or fuel rail mounting positions. These O-rings provide thermal insulation to prevent the formation of vapour bubbles and promote good hot-start characteristics. They also dampen potentially damaging vibration. When the injector is electrically energized, a fine mist of fuel sprays from the injector tip.

Most injectors consist of a solenoid, a needle valve, and a nozzle **(Figure 29–4)**. The solenoid is attached to the nozzle valve. The PCM controls the injector by controlling its ground circuit through a driver circuit. When the solenoid winding is energized, it creates a magnetic field that draws the armature back and pulls the needle valve from its seat. Fuel then sprays out of the nozzle. When the solenoid is de-energized, the magnetic field collapses, and a helical spring forces the needle valve back on its seat, shutting off fuel flow.

Another injector design uses a ball valve and valve seat. In this case, the magnetic field created by the solenoid coil pulls a plunger upward, lifting the ball valve from its seat. A spring is used to return the valve to its seated or closed position.

Each fuel injector has a two-wire connector. One wire supplies voltage to the injector. This wire may connect directly to the fuse panel or to the PCM, which, in turn, connects to the fuse panel. In a few older systems, a resistor under the hood or in the PCM is used to reduce the current flow through the low-resistance injectors. The second wire is a ground wire. This ground wire is connected to the driver circuit inside the PCM.

The amount of fuel released by an injector depends on fuel pressure, the size of the injector's nozzles, and the length of time the injector is energized. Fuel pressure is mainly controlled by a pressure regulator, and the injector's pulse width is controlled by the PCM. Typical pulse widths range from 1 millisecond (ms) at idle to 10 ms at full load. The PCM controls the pulse width according to various input sensor signals, operating conditions, and its programming. The primary inputs are related to engine load and engine coolant temperature. Different engines require different injectors. Injectors are designed to pass a specified amount of fuel when opened. In addition, the number of holes at the tip of the injector varies with engines and model years. Fuel injectors can be top fuel feeding, side feeding, or bottom feeding **(Figure 29–5)**. Top- and side-feed injectors are primarily used in port injection systems that operate using high fuel system

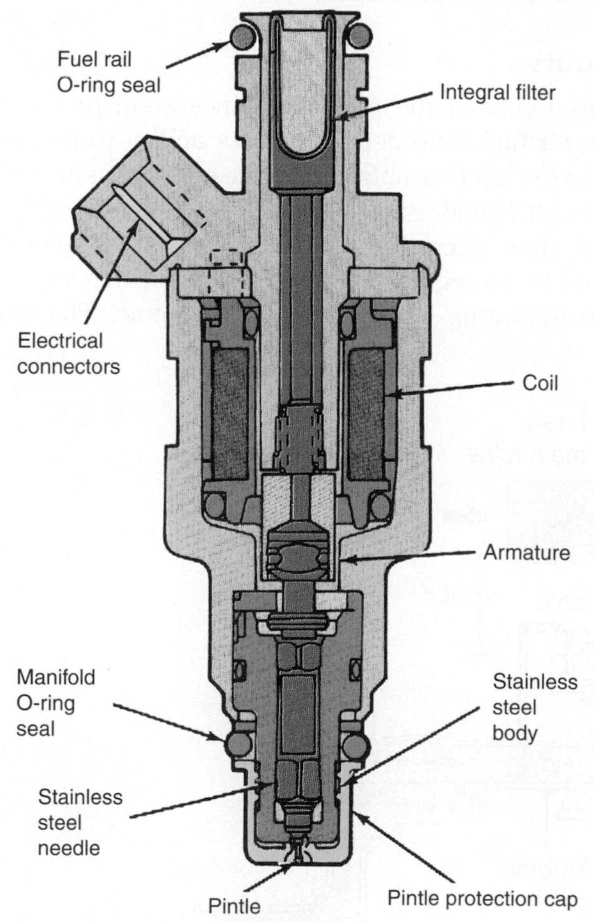

FIGURE 29–4

A typical fuel injector used in multiport fuel injection systems.

Ford Motor Company

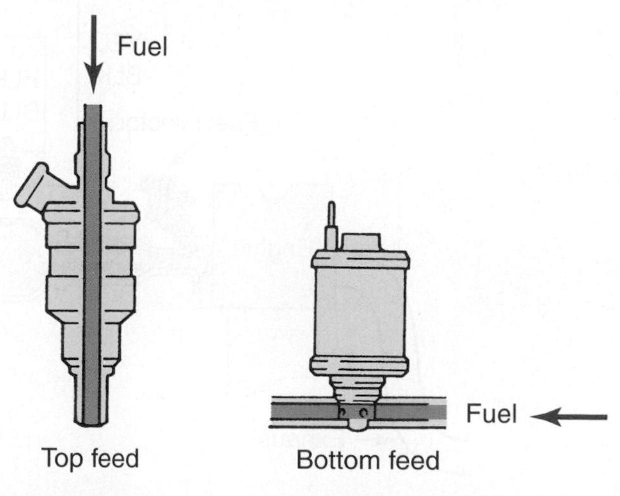

FIGURE 29–5

Examples of top-feed and bottom-feed injectors.

pressures. Bottom-feed injectors are used in throttle body systems. Bottom-feed injectors are able to use fuel pressures as low as 70 kPa (10 psi).

There have been some problems with deposits on injector tips. Because small quantities of gum are present in gasoline, injector deposits usually occur when this gum bakes onto the injector tips after a hot engine is shut off. Most manufacturers use fuel injectors designed to reduce the chance of deposit buildup at the tips. Also, oil companies have added a detergent to their gasoline to help prevent injector tip deposits.

Idle Speed Control

Idle speed control is a function of the PCM. Based on operating conditions and inputs from various sensors, the PCM regulates the idle speed. In throttle body and port EFI systems, engine idle speed is controlled by bypassing a certain amount of airflow past the throttle valve in the throttle body housing. Two types of air bypass systems are used: auxiliary air valves and **idle air control (IAC)** valves. IAC valve systems are more common **(Figure 29–6)**. Engines that use electronic throttle control do not use an IAC valve. Instead, the PCM commands the throttle plates open a small amount, approximately 10 percent, to allow airflow into the engine. The IAC system is a stepper motor or actuator that positions the IAC valve in the air bypass channel around the throttle valve. The IAC valve is part of the throttle body casting. The PCM calculates the amount of air needed for smooth idling based on input data, such as coolant temperature, engine load, engine speed, and battery voltage.

If engine speed is lower than desired, the PCM activates the motor to retract the IAC valve. This opens the channel and diverts more air around the throttle valve. If engine speed is higher than desired, the valve is extended and the bypass channel is made smaller, which reduces airflow and engine speed. The PCM typically increases engine speed when generator output is high, such as when the A/C compressor is engaged and the engine cooling fans are operating.

During cold starts, idle speed can be as high as 2100 rpm to quickly raise the temperature of the catalytic converter. The PCM maintains cold idle speed for approximately 40 to 50 seconds.

Some older engines are equipped with an auxiliary air valve to aid in the control of idle speed. Unlike the IAC valve, the auxiliary air valve is not controlled by the PCM. Like the IAC system, however, the auxiliary air valve provides additional air during cold engine idling.

Inputs

The ability of the fuel injection system to control the air/fuel ratio depends on its ability to properly time the injector pulses with the compression stroke of each cylinder and its ability to vary the injector "on" time, according to changing engine demands. Both tasks require the use of sensors that monitor the operating conditions of the engine. The PCM

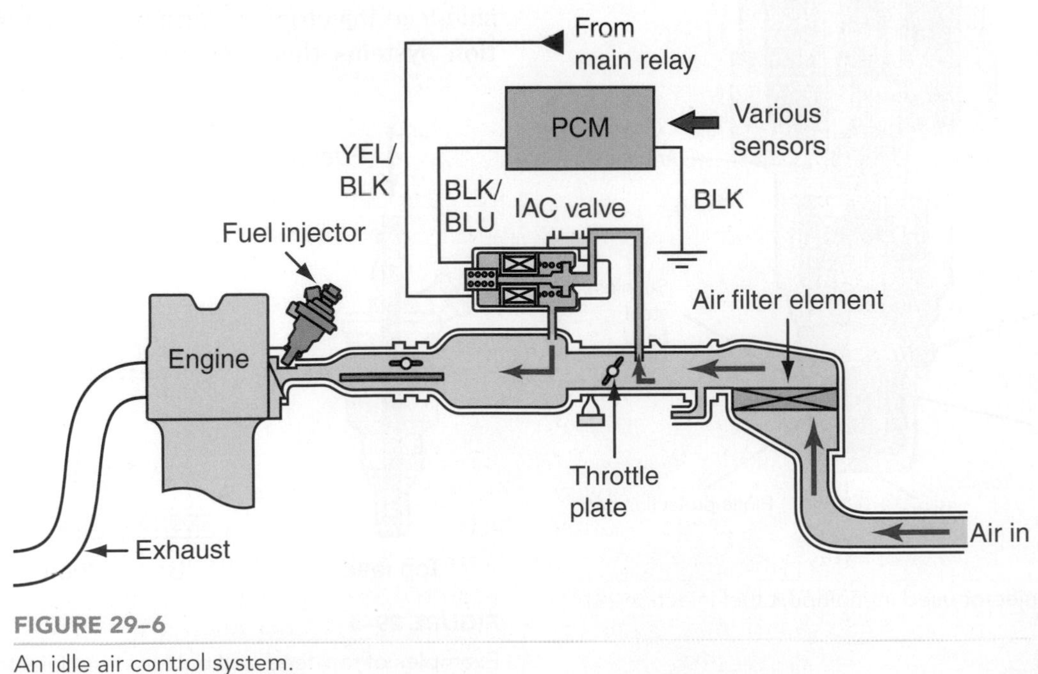

FIGURE 29–6

An idle air control system.

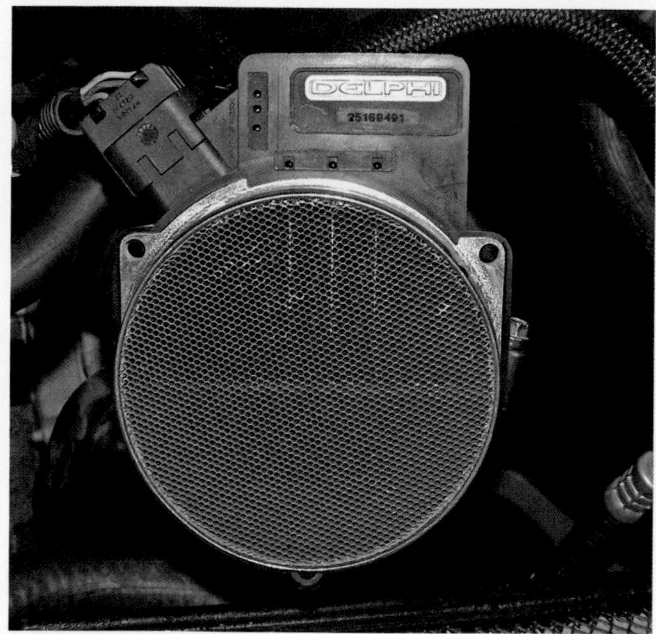

FIGURE 29–7

A mass airflow (MAF) sensor.

receives these signals from the CAN bus and inputs sent directly to the computer.

MAF SENSOR The mass airflow (MAF) sensor converts air flowing past a heated sensing element into an electronic signal **(Figure 29–7)**. The strength of this signal is determined by the energy needed to keep the element at a constant temperature above the incoming ambient air temperature. As the volume and density (mass) of airflow across the heated element changes, the temperature of the element is affected and the current flow to the element is adjusted to maintain the desired temperature. The varying current parallels the particular characteristics of the incoming air (hot, dry, cold, humid, high/low pressure). The PCM monitors the changes in current to determine air mass and to calculate fuel requirements. An airflow sensor known as a Karman Vortex is also being used in a number of modern engines.

MAP SENSOR The manifold absolute pressure (MAP) sensor measures changes in the intake manifold pressure that result from changes in engine load and speed. At closed throttle, the engine produces a low MAP value. A wide-open throttle produces a high value. MAP output is the opposite of what is measured on a vacuum gauge. Low MAP output means low absolute pressure or a large vacuum. The use of this sensor also allows the control computer to adjust automatically for different altitudes. The MAP signal may also be used to inform the PCM when the EGR valve is open during the EGR monitor test.

FIGURE 29–8

An oxygen sensor.

OXYGEN SENSORS (O_2S) The signals from the exhaust gas oxygen sensor (O_2S) (or lambda sensor) are used to monitor the air/fuel mixture **(Figure 29–8)**. When the sensor's signal indicates a lean mixture, the computer enriches the air/fuel mixture to the engine. When the sensor reading is rich, the computer leans the air/fuel mixture. Because an O_2S must be hot to operate properly, late-model engines use heated oxygen sensors (HO_2S). These sensors have an internal heating element that allows the sensor to reach operating temperature quickly and to maintain its temperature during periods of idling or low engine load.

OBD-II systems have two or more O_2Ss in each exhaust system, one before the catalytic converter (upstream) and one after it (downstream). The signals from the upstream sensor readings are used to monitor the air/fuel ratio, and the downstream sensors are used to monitor the effectiveness of the catalytic converter.

Many engines are fitted with air/fuel (A/F) ratio sensors in the upstream position. These sensors can react to very minor changes in the air/fuel ratio and over a wider range of air/fuel ratios compared to a traditional O_2S. This allows the PCM to have precise control of the fuel injection system.

IAT SENSOR Cold air is denser than warm air. Cold, dense air can burn more fuel than the same volume of warm air because it contains more oxygen. The intake air temperature (IAT) sensor measures air temperature and sends an electronic signal to the PCM. The computer uses this input along with the air volume input in determining the amount of oxygen entering the engine. This sensor is often combined with the MAF sensor.

ECT SENSOR The engine coolant temperature (ECT) sensor signals the PCM when the engine needs cold enrichment, as it does during warm-up. This adds to the base pulse width but decreases to zero as the engine warms up. The PCM may also order a richer mixture when the engine is overheating. A rich mixture burns cooler.

APP/TP SENSORS The accelerator pedal position (APP) and throttle position (TP) sensors allow the PCM to monitor driver intent and throttle position. The signals from these sensors are used to clarify load and operating conditions. Both sensors use potentiometers that react directly to the movement of the accelerator pedal/throttle plate. A sudden increase in APP voltage tells the PCM to momentarily enrich the mixture to prevent hesitation and stumbling during acceleration.

CKP SENSOR The crankshaft position (CKP) sensor is used to monitor engine speed. This signal advises the PCM to adjust the pulse width and frequency of the injectors to match the engine speed. This input is the most important input in the engine management system. It is used to synchronize the injectors with events in the cylinders. The signals from the CKP are often used with the signals from the CMP to determine which cylinders are on the compression stroke.

CMP SENSOR The camshaft position (CMP) sensor is used to synchronize the firing of the injectors with the individual cylinders in the engine. By using the signals from the CMP, the PCM can determine that piston number 1 is on the compression stroke. This is used for fuel injection timing.

ADDITIONAL INPUT INFORMATION SENSORS Additional sensors are also used to provide the following information on engine conditions:

- Vehicle speed
- Air-conditioner operation
- Gearshift lever position
- Battery voltage and generator load
- EGR valve position
- Power-steering pressure switch
- Knock sensor

Operational Modes

All fuel injection systems operate in response to inputs. However, the PCM's programming allows it to define the conditions and establish a summary of those conditions. The PCM then controls the delivery of fuel according to that mode of operation. Different EFI systems have different operational modes, but most have starting, run, clear flood, acceleration, and deceleration.

STARTING MODE When the ignition switch is initially moved to the START position, the PCM turns on the fuel pump for about two seconds. When the PCM receives a good signal from the CKP sensor, it energizes the fuel pump to allow for starting. If a CKP signal is not present, the fuel pump is shut off. With a CKP signal, the PCM controls injector timing and bases the pulse width of the injectors entirely on the engine's coolant temperature and load. Once the engine is cranking, the PCM sets the injectors' pulse width according to inputs from the MAF, IAT, ECT, and TP sensors. In some cases, as the engine is cranking, the injectors may prime the cylinders with a spray of fuel to help get the engine started. The system stays in starting mode until the engine is rotating at a predetermined speed.

RUN MODE Once the engine has started and is running above a predetermined speed, the system will operate in open loop. In open loop, the PCM sets injector pulse width according to MAP or MAF, IAT, ECT, and TP sensor signals. The system stays in open loop until the PCM receives good signals from the (HO_2S) and a predetermined engine temperature has been reached. Once these conditions have been met, the system moves into closed loop. In closed loop, the PCM adjusts the pulse width according to inputs from a variety of sensors, but primarily the (HO_2S) or A/F ratio sensors **(Figure 29–9)**.

CLEAR FLOOD MODE At times, the engine will not start because it has received too much fuel; this is called flooding. When an engine floods, the excess fuel must be pumped out of the cylinders. This is done by fully depressing the accelerator pedal and cranking the engine. The **clear flood mode** is not an automatic process; it is initiated by depressing

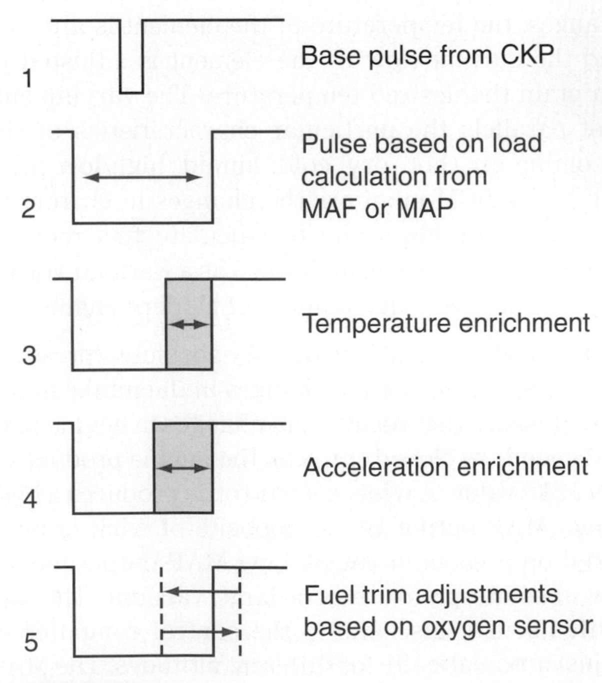

FIGURE 29–9
A graphical representation of how pulse width is calculated.

and holding the accelerator pedal down. When the PCM detects a wide-open throttle, it will go into the acceleration enrichment mode for three seconds. If the throttle is held open and the engine's speed is below a predetermined rpm, the system will return to the start mode. In some cases, the PCM will completely turn off the injectors if engine cranking continues for a long period.

ACCELERATION MODE Based on signals from the TP and MAP or MAF sensors, the PCM can tell when the vehicle is being accelerated. To compensate for the sudden rush of intake air as the throttle is opened, the PCM increases the injectors' pulse width. The pulse-width change is calculated by the PCM according to inputs from the CKP, MAP, ECT, MAF, and TP sensors. Once the PCM determines that the vehicle is no longer accelerating, the EFI system is returned to the run mode.

DECELERATION MODE Inputs from the MAP or MAF and TP sensors are also used by the PCM to detect deceleration. During deceleration, the PCM reduces injector pulse width. Some systems will shut off the fuel injectors when the vehicle is rapidly decelerating, although a few vehicles will totally shut down the fuel system for a brief period during deceleration.

Fuel Trim

OBD-II systems constantly monitor the signals from the oxygen or air/fuel ratio sensors while operating in closed loop. Based on these inputs and programming instructions, the PCM alters the injectors' pulse width to provide the best possible combination of driveability, fuel economy, and emission control. The adjustment made to the base (programmed) pulse width for operating conditions is called fuel trim. Fuel trim can be monitored for diagnostic purposes. The base pulse width is given a value of 0 percent, and all changes are expressed as a negative or positive value. A positive fuel trim value means the PCM detects a lean mixture and is increasing the pulse width to add more fuel to the mixture. A negative fuel trim value means the PCM is reducing the amount of fuel by decreasing the pulse width to compensate for a rich condition. Some vehicles use equivalence ratio (EQ RAT) instead of fuel trim. If no changes are required to the air/fuel ratio, the EQ ratio is 1.0. If the commanded EQ ratio is 0.95, the air/fuel ratio is commanded rich. Numbers greater than 1.0 indicate a lean command. To determine the air/fuel ratio, multiply the stoichiometric ratio of 14.64:1 by the EQ ratio. For example, an EQ ratio of 0.95 equals an air/fuel ratio command of 13.9:1 ($14.64 \times 0.95 = 13.908$).

The system allows for short-term and long-term fuel trim. Short-term fuel trim (STFT) represents changes made immediately in response to (HO_2S) signals. Long-term fuel trim (LTFT) represents changes made to set a new base pulse width. LTFT responds to the trends of the STFT. For example, if the STFT is commanded rich and remains at +12 percent for a period of time, the LTFT will increase until STFT returns close to 0 percent correction. The LTFT will then show as a positive number in the scan data. To determine the total fuel correction by the PCM, add the STFT and LTFT numbers together. If STFT is 4 percent and the LTFT is 8 percent, total fuel correction is 12 percent rich. STFT is erased when the ignition is turned off. LTFT remains in the PCM's memory.

THROTTLE BODY INJECTION (TBI)

For many auto manufacturers, **throttle body injection (TBI)** served as a stepping stone from carburetors to more-advanced fuel injection systems. TBI units were used on many engines during the 1980s and 1990s. The throttle body unit is mounted directly to the intake manifold. The injector(s) spray fuel down into a throttle body chamber leading to the intake manifold. The intake manifold feeds the air/fuel mixture to all cylinders.

Four-cylinder engines have a single throttle body assembly with one injector and throttle plate, whereas V6 and V8 engines are usually equipped with dual injectors and two throttle plates on a common throttle shaft **(Figure 29–10)**.

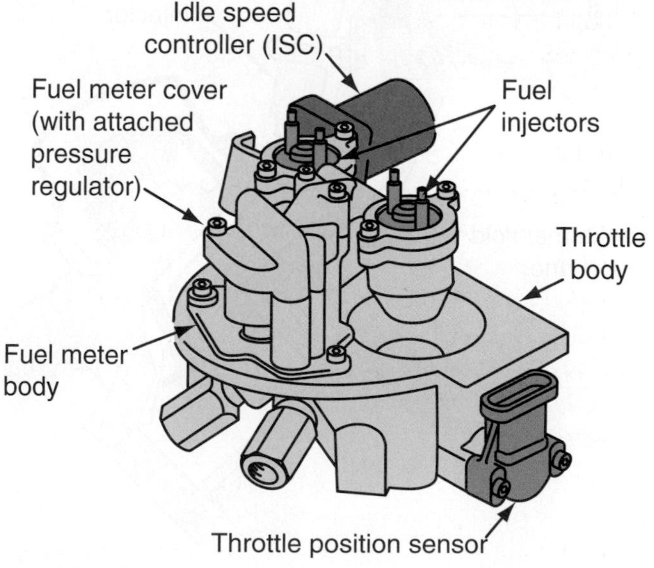

FIGURE 29–10

Dual throttle body assembly used on a six- or an eight-cylinder engine.

Throttle body systems are not as efficient as port injection systems: Fuel is not distributed equally to all cylinders, the air and fuel from the TBI unit pass through the intake manifold, and the length and shape of the manifold's runners affect distribution. There is also a potential problem of fuel condensing and forming puddles in the manifold when the manifold is cold.

PORT FUEL INJECTION (PFI)

Port fuel injection (PFI) systems use at least one injector at each cylinder. They are mounted in the intake manifold near the cylinder head where they can inject a fine, atomized fuel mist as close as possible to the intake valve **(Figure 29–11)**. Delivering the fuel mist right outside the combustion chamber allows the fuel to break down and vaporize a little more before it enters the cylinder. Some port injection systems have a director plate right under the nozzle of the injector. This plate has several small holes that break up the fuel as it is sprayed through the plate.

Through the years, many different PFI systems have been used, and although they have things in common, they do not fire the injectors in the same way. PFI systems can be divided into two basic categories: **multiport fuel injection (MPFI)** systems and **sequential fuel injection (SFI)** systems; each is defined by injector control.

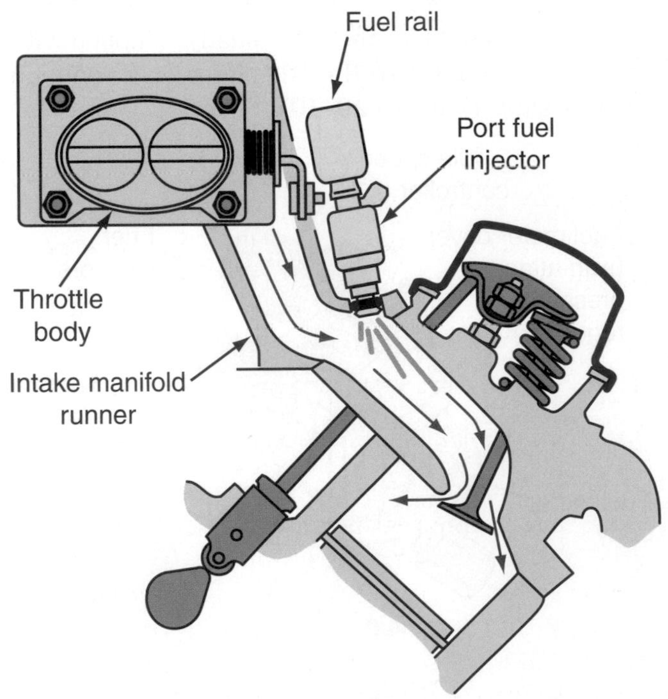

FIGURE 29–11

Port injection sprays fuel into the intake port and fills the port with fuel vapour before the valve opens.

Multiport Fuel Injection Systems (MPFI)

Due to OBD-II regulations, MPFI systems are no longer used. In MPFI systems, the injectors were grouped together and the injectors in each group or bank fired at the same time. Some MPFI systems fired all of the injectors simultaneously. This system offered easy programming and relatively fast adjustments to the air/fuel mixture. The injectors were connected in parallel, so they shared one driver circuit, and the PCM sent out one signal for all injectors. The amount of fuel required for each four-stroke cycle was divided in half and delivered in two injections, one for every 360° of crankshaft rotation. The fact that the intake charge had to wait in the manifold for varying periods was the system's major drawback.

In grouped systems, there is a power and ground circuit for each group. When the injectors are split into two equal groups, the groups are fired alternatively, with one group firing each engine revolution.

Because only two injectors can be fired relatively close to the time when the intake valve is about to open, the fuel charge for the remaining cylinders must stand in the intake manifold for varying periods. These periods are very short; therefore, the standing of fuel in the intake manifold was not that great a disadvantage of MPFI systems. At idle speeds, this wait was about 150 ms, and at higher speeds the time was much less.

Sequential Fuel Injection Systems

SFI systems use an MAF or MAP sensor as well as a variety of other sensors. These systems control each injector separately so that it fires just before the intake valve opens. This means that the mixture is never static in the intake manifold, and adjustments to the mixture can be made instantaneously between the firing of one injector and the next. Sequential firing is the most accurate and desirable method of regulating port injection.

To meet OBD-II regulations, SFI systems are capable of turning off the injector at a misfiring cylinder.

In SFI systems, each injector is connected individually to the computer, and the computer completes the ground for each injector one at a time. A sequential system has one injector per injector driver.

When the injection system fires according to crankshaft speed, it is called a synchronous system. The action of the injectors is timed to the crankshaft.

During starting and acceleration, the PCM may inject extra fuel into all of the cylinders without referring to inputs from the CKP. When the system does this, it is called asynchronous injection.

Throttle Body

The throttle body in a PFI system controls the amount of air that enters the engine as well as the amount of vacuum in the intake manifold. On non-electronic throttle control systems, the throttle body assembly is comprised of an IAC valve assembly, an idle air orifice, single or double bores with throttle plates, and a TP sensor **(Figure 29–12)**.

The throttle body housing assembly is a single-piece aluminum or plastic casting. The throttle bore and throttle plates control the amount of intake air that enters the engine. The throttle shaft and plate(s) are controlled by the accelerator pedal, via a linkage comprised of a cam and cable. The TP sensor monitors the movement and position of the plates and sends a signal to the PCM. On some systems, a small amount of coolant is routed through a passage in the throttle body to prevent icing during cold weather.

The throttle bores and plates in many throttle bodies are coated with a special sealant. This sealant has two purposes: It helps seal the plates to the bores when the throttle is closed, and it protects the throttle body from damage caused by contaminants, such as carbon sludge, that may build up in the intake manifold.

FIGURE 29–12

A TP sensor mounted to a throttle body.

SHOP TALK

Throttle body assemblies that have this protective coating should not be cleaned, because any cleaning may remove the sealant. Most coated throttle bodies have a warning sticker affixed to them for identification.

The MAF sensor measures the amount of intake air as well as the air used by the idle air orifice and the positive crankcase ventilation (PCV) system. This allows the PCM to know how much air is entering the intake manifold. Some throttle bodies also have a fresh air outlet above the throttle plates for the PCV and IAC systems.

When the engine is at idle and the throttle plates are closed, a small amount of air enters the engine through the idle air orifice. The IAC system allows additional air to enter if the PCM commands it to do so. Most throttle bodies have a base-idle screw, which is adjusted rarely and only as part of an idle speed adjustment procedure. Idle speed is totally controlled by the PCM, and there are typically no provisions for adjusting it under normal operating conditions. Some throttle bodies have a throttle stop screw that prevents the throttle plate from seizing in its bore when it is quickly closed.

Throttle bodies also have several vacuum taps or ports located below the throttle plates. These are used to monitor engine vacuum and provide vacuum to the PCV valve, exhaust gas recirculation (EGR) valve, evaporative emission (EVAP) system, and A/C controls.

Fuel Delivery

The fuel injectors are separately controlled and are mounted to the intake manifold and supplied pressurized fuel through fuel lines. These fuel lines run to each cylinder from a fuel manifold, usually referred to as a **fuel rail (Figure 29–13)**. The fuel rail also has a fuel pressure test port, and on returnless fuel systems, a fuel rail pressure (FRP) sensor. The rail is supplied fuel by the fuel pump and distributes the same amount of fuel at the same pressure to each cylinder. Because the fuel is dispensed at each cylinder and not the intake manifold, there is little or no fuel to wet the manifold walls; therefore, there is no need for manifold heat or early fuel evaporation system. Many PFI engines use plastic intake manifolds because the runners are no longer carrying fuel to the cylinders.

The fuel rail assembly on a PFI system of V6 and V8 engines usually consists of a left- and right-hand

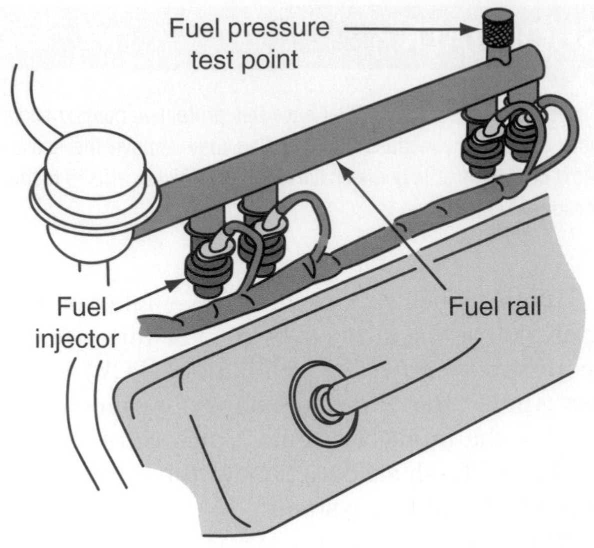

FIGURE 29–13

A typical fuel rail.

rail assembly. The two rails can be connected by crossover and return fuel tubes. A typical fuel rail for a V-type engine is shown in **Figure 29–14**.

Some engines are equipped with variable induction intake manifolds that have separate runners for low and high speeds. This technology is only possible with port injection.

Engines that use returnless fuel systems incorporate a fuel pressure sensor that monitors the pressure at the fuel injectors. This sensor may also include a fuel temperature sensor as well. The PCM controls fuel pressure based on this input. The sensor receives a reference voltage of 5 volts, and changes in fuel rail pressure are reflected by the signal back to the PCM. Basically, when fuel pressure is high, the signal has high voltage, and when the pressure is low, the signal is low.

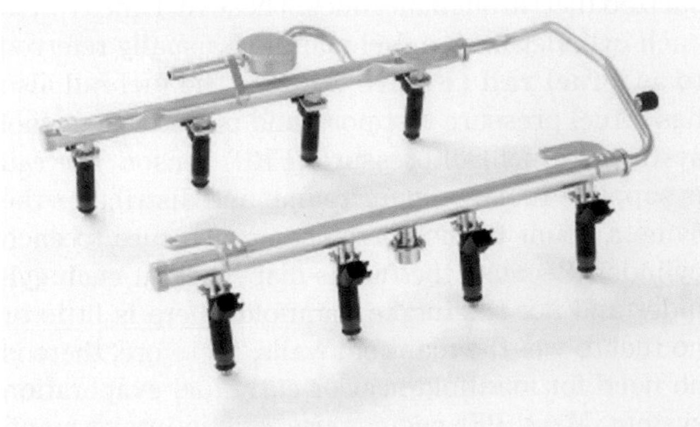

FIGURE 29–14

A one-piece fuel rail for a V8 engine.

PULSATION DAMPER Some engines have a pulsation damper on their fuel rails. These dampers reduce the pressure pulsations caused by the rapid opening and closing of the fuel injectors. Without control over these fluctuations, the fuel pressure at each injector can be affected. The damper works to control the volume of fuel in the fuel rail. When the pressure in the rail quickly drops, the damper temporarily reduces the volume of fuel in the rail to prevent the fuel pressure from becoming too low. The opposite occurs when the pressure quickly increases. The damper also reduces fuel noise and maintains pressure during engine cooldown.

Injector Control

The computer is programmed to ground the injectors slightly ahead of the actual intake valve openings so that the intake ports are filled with fuel vapour just before the intake valves open. In both SFI and MPFI systems, the computer supplies the correct injector pulse width to provide a stoichiometric air/fuel ratio. The computer increases the injector pulse width to provide air/fuel ratio enrichment while starting a cold engine.

SFI requires inputs from the CKP and CMP sensors in order to determine when to fire each injector. On many systems, once the position of piston number 1 from the CKP is received by the PCM, signals from the CMP are used to synchronize the injectors to the engine's firing order.

Pressure Regulators

PFI systems require an additional control not required by TBI units. In a TBI, the injectors are mounted above the throttle plates and are not affected by fluctuations in manifold vacuum; PFI have their tips located in the manifold where constant changes in vacuum would affect the amount of fuel injected. To compensate for these fluctuations, port injection systems have a fuel pressure regulator that senses manifold vacuum and continually adjusts the fuel pressure to maintain a constant pressure drop across the injector. The pressure regulators respond to changes in manifold pressure due to engine load.

On return PFI systems, a fuel pressure regulator located on or near the fuel rail **(Figure 29–15)**. A diaphragm and valve assembly is positioned in the centre of the regulator, and a diaphragm spring seats the valve on the fuel outlet. When fuel pressure reaches the regulator setting, the diaphragm moves against the spring tension and the valve opens. This allows fuel

FIGURE 29–15

A typical fuel pressure regulator for a port injection system.

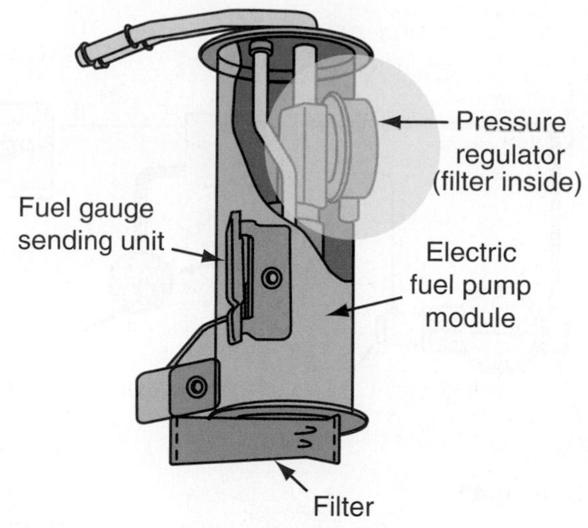

FIGURE 29–16

A returnless fuel system with the pressure regulator and filter mounted in the fuel tank together with the fuel pump and fuel gauge sending unit.

to flow through the return line to the fuel tank. The fuel pressure drops slightly when the regulator valve opens and builds when the spring closes the valve.

A vacuum hose is connected from the intake manifold to the vacuum inlet on the pressure regulator. This hose supplies vacuum to the diaphragm. The vacuum works against fuel pressure to move the diaphragm and open the valve. When the engine is running at idle speed, high manifold vacuum is applied to the pressure regulator. Under this condition, fuel pressure opens the regulator valve, reducing the pressure.

If the engine is operating at wide-open throttle (WOT), no vacuum (high manifold pressure) is applied to the pressure regulator. This high pressure increases the fuel pressure. This is good because the injectors are discharging fuel into a higher pressure when compared to idle speed conditions. If the fuel pressure remained constant at idle and WOT conditions, the injectors would discharge less fuel into the higher pressure in the intake manifold at WOT. The increase in fuel pressure supplied by the pressure regulator at WOT maintains the same pressure drop across the injectors at idle speed and WOT. When this same pressure drop is maintained, the change in pressure at the injector does not affect the amount of fuel discharged by the injectors.

RETURNLESS SYSTEMS Nearly all late-model injection systems are returnless systems. Those systems that have a fuel pressure regulator as part of the fuel sender and pump assembly in the fuel tank **(Figure 29–16)** are called mechanical returnless systems. The term is applied because the regulator sits inside the fuel tank with no designated return fuel line, and its action is not related to engine vacuum. With a returnless system, only the fuel needed by the engine is filtered, thus allowing the use of a smaller fuel filter. When the engine is at idle, the fuel pressure should be between 80 and 410 kPa (55 and 60 psi). If the pressure regulator supplies a fuel pressure that is too low or too high, a driveability problem can result. The system also has a switch that shuts off the fuel when a collision has occurred.

Many newer returnless systems are called electronic returnless systems. This is because the PCM controls the speed of the fuel pump to control fuel pressure. These systems may use the PCM to directly control the fuel pump, or the PCM may control a second module that actually drives the pump. Regardless of which system is used, the fuel pump speed is controlled using a pulse-width-modulated (PWM) signal. As fuel demand increases, the pulse width is increased, and more fuel is supplied.

Throttle-by-Wire Systems

All new vehicles now use throttle-by-wire systems. These systems eliminate the need for a throttle cable and linkage. They also provide many other benefits over traditional throttle controls, such as improved driveability, increased fuel efficiency, and decreased emissions. Throttle-by-wire systems used by manufacturers are known by different names, the most common being electronic throttle body (ETB) and electronic throttle control (ETC). There are differences in the programming and hardware used.

Based on driver inputs from the APP (accelerator pedal position) sensors and its programming, the PCM electronically controls the position of the throttle plate **(Figure 29–17)**. A DC motor that is

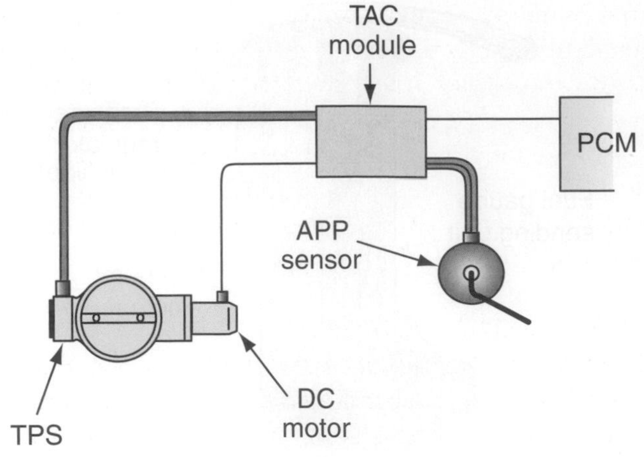

FIGURE 29–17

A simplified diagram for a throttle actuator control (TAC) system.

controlled by the PCM is used to move the plate via gears **(Figure 29–18)**. The motor may be an integral part of the throttle body or a separate unit mounted to the throttle body.

FIGURE 29–18

An electric motor, in response to commands from the PCM, moves the throttle plate.

As a safety feature, two springs are attached to the throttle plate shaft. One, which is the stronger of the two, is used to open the throttle plate slightly if the PCM loses control of engine speed. This allows for limp-in operation. The other spring is used to close the throttle. To prevent the throttle plate from closing too much and possibly binding its bore, a hard stop is on the throttle body assembly to limit the closure of the throttle plate.

Driver input to the PCM is delivered by accelerator pedal sensors. A system may have more than one pedal sensor or may use a sensor that sends out more than one signal. In either case, the PCM will receive multiple signals regarding accelerator pedal position. This redundancy is important because it allows the PCM to closely monitor the action of the system. The multiple signals ensure that the PCM is receiving correct information.

The TP sensor also provides redundant signals. The multiple signals ensure that the PCM knows where the throttle plate is positioned at all times. These TP sensors are recognizable because they have four electrical leads.

Redundancy also occurs in the processing of the inputs and the controlling of the throttle motor. The throttle system is monitored by two separate processors inside the PCM. Both are looking at the same things, and if one determines that something is wrong with the other, it will override the commands of that processor.

In addition to eliminating the throttle linkage, electronic throttle systems also eliminate the need for an IAC valve and idle air orifice. These are not needed because the PCM can control throttle plate opening to meet the air needs of the engine when it is idling.

ETCs are also used with variable valve timing. Because the engine's power changes with a change in valve timing, the desired position of the throttle for those conditions also changes. The PCM can instantly change the throttle position to eliminate any noticeable change in engine power.

ETC works with electronically controlled automatic transmissions. The PCM can alter throttle position when the transmission is shifting gears, to improve shift quality, allowing the transmission to make quicker, smoother gear changes.

Other advantages include the elimination of cruise control actuators while providing improved speed control. ETC also allows the PCM to be more effective when limiting engine rpm, vehicle speeds, and engine output torque if the automatic traction control systems detect wheel slip.

CENTRAL PORT INJECTION (CPI)

In a **central port injection (CPI)** or **central multiport fuel injection (CMFI)** system, a central injector assembly is mounted in the lower half of the intake manifold. The CPI system uses one injector to control the fuel flow to individual **poppet nozzles (Figure 29–19)**. The CPI injector assembly consists of a fuel metering body, a pressure regulator, one fuel injector, poppet nozzles with nylon fuel tubes, and a gasket seal. The injector distributes metered fuel through the lines connected to the nozzles. To meet OBD-II regulations, an updated version of CPI uses individual injectors for each poppet nozzle. This system, called **central sequential fuel injection (CSFI)**, allows each injector to be shut off in the event of a misfire.

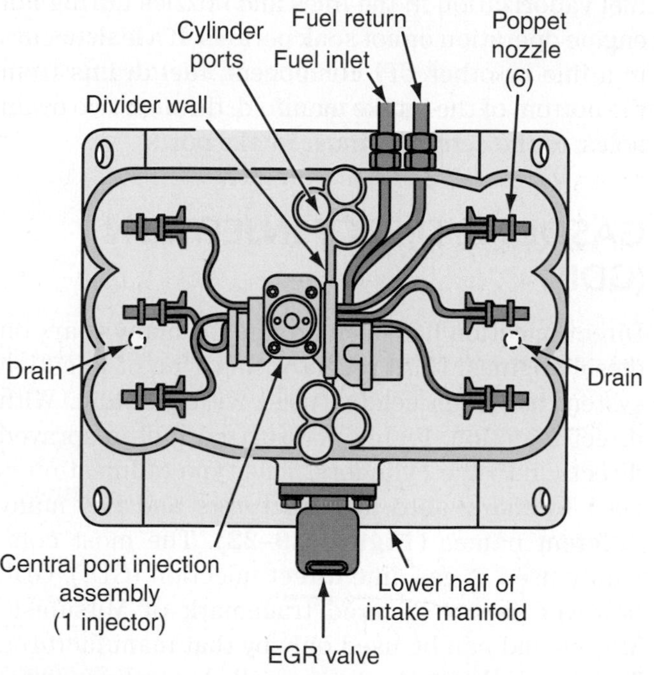

FIGURE 29–19

Central multiport fuel injection components in the lower half of the intake manifold.

Pressure Regulator

The pressure regulator is mounted with the central injector. The regulator is inside the intake manifold, and intake vacuum is supplied through an opening in the regulator cover. Normally, a regulator spring pushes downward on a diaphragm and closes the valve. Fuel pressure from the fuel pump pushes the diaphragm upward. When the pressure exceeds that of the spring, the valve opens. Pressure is decreased as fuel flows through the valve and a return line to the fuel tank **(Figure 29–20)**.

The pressure regulator maintains fuel pressure at 370 to 440 kPa (54 to 64 psi), which is higher than many PFI systems. Higher pressure is required in the CSFI system to prevent fuel vaporization from the extra heat encountered with the CSFI assembly, poppet nozzles, and lines mounted inside the intake manifold. Fuel pressure on the CSFI system is maintained between 415 to 455 kPa (60 to 66 psi).

Injector Design and Operation

CPI systems use one maxi injector, which supplies fuel to each of the poppet nozzles. A failure of the injector or of the supply tubes or nozzles requires

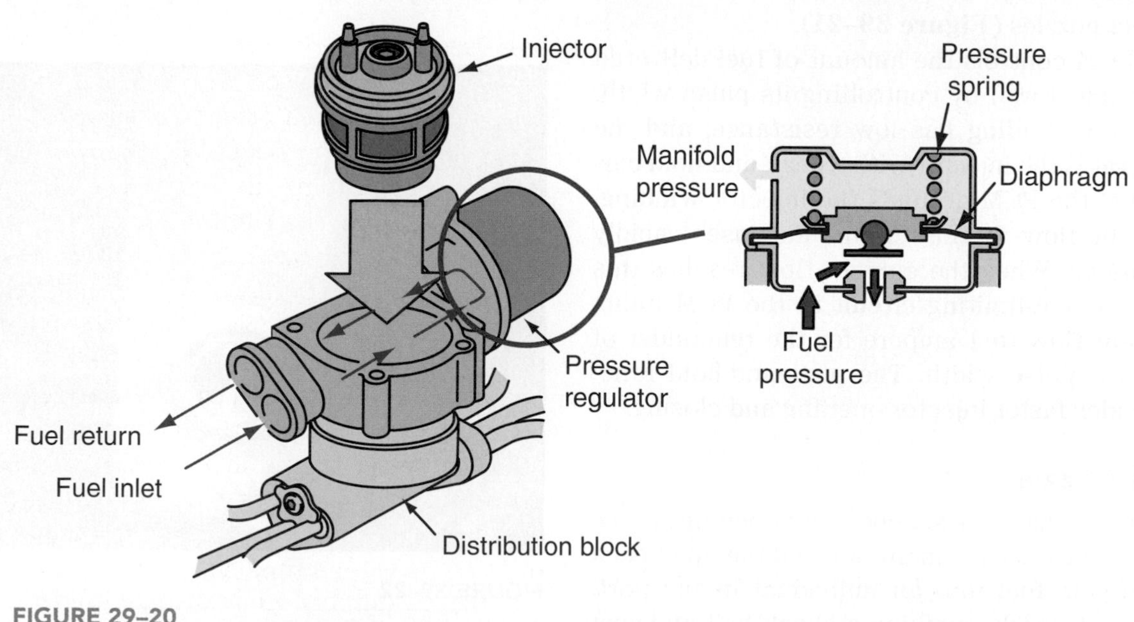

FIGURE 29–20

A pressure regulator for a CMFI system.

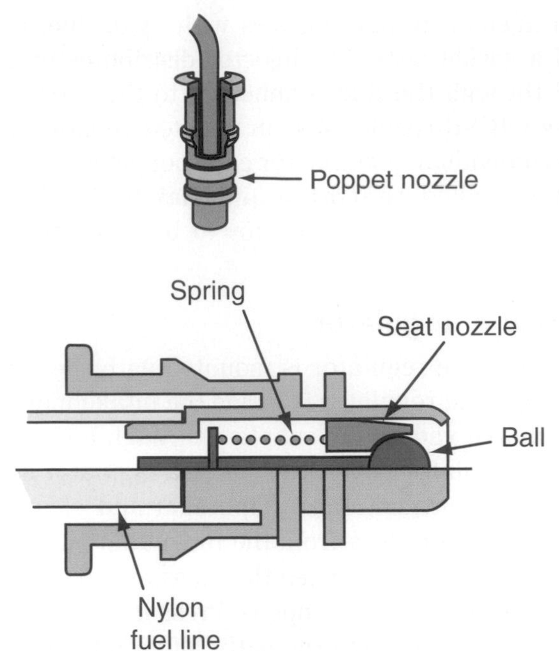

FIGURE 29–21

The internal design of a poppet nozzle.

replacement of the entire CPI unit. Because the CSFI uses individual injectors for each cylinder, each injector and nozzle can be serviced separately.

In both systems, the armature is placed under the injector winding in the injector. The lower side of this armature acts as a valve that covers the outlet ports to the nylon tubes and poppet nozzles. A supply of fuel at a constant pressure surrounds the armature while the ignition switch is on. Each time the PCM grounds the injector, the armature is lifted up, which opens the injector ports. Fuel is then forced to the poppet nozzles **(Figure 29–21)**.

The PCM controls the amount of fuel delivered from the injector(s) by controlling its pulse width. The injector winding has low resistance, and the PCM operates the injector with a peak-and-hold current. When the PCM grounds the injector winding, the current flow in this circuit increases rapidly to 4 amperes. When the current flow reaches this value, a current-limiting circuit in the PCM limits the current flow to 1 ampere for the remainder of the injector pulse width. The peak-and-hold function provides faster injector opening and closing.

Poppet Nozzles

The poppet nozzles are snapped into openings in the lower half of the intake manifold, and the tip of each nozzle directs fuel into an individual intake port. Each poppet nozzle contains a check ball and seat at the tip of the nozzle. A spring holds the check ball in the closed position. When fuel pressure is applied to the nozzles, the pressure forces the check ball to open against spring pressure. The poppet nozzles open when the fuel pressure exceeds 345 kPa (50 psi), and the fuel sprays from these nozzles into the intake ports.

When fuel pressure drops below this value, the nozzles close. Under this condition, approximately 276 kPa (40 psi) fuel pressure remains in the nylon lines and poppet nozzles. This pressure prevents fuel vaporization in the lines and nozzles during hot engine operation or hot soak periods. If a leak occurs in a line or other CPI component, fuel drains from the bottom of the intake manifold, through two drain holes, to the centre cylinder intake ports.

GASOLINE DIRECT INJECTION (GDI)

Direct injection has been around for many years on diesel engines. Until recently, this type of injection system has been seldom used with gasoline. With direct injection, highly pressurized fuel is sprayed directly into the cylinders. This type of injection is used by many auto manufacturers and has many different names **(Figure 29–22)**. The most commonly used is gasoline direct injection (GDI); GDI, however, is a registered trademark of Mitsubishi Motors and can be used only by that manufacturer. Therefore, the systems are called gasoline direct injection (GDI), fuel stratified injection (FSI), high precision injection, direct injection (DI), or direct injection spark ignition (DISI). Because the fuel is

FIGURE 29–22

A turbocharged direct injection–equipped engine from Volkswagen.

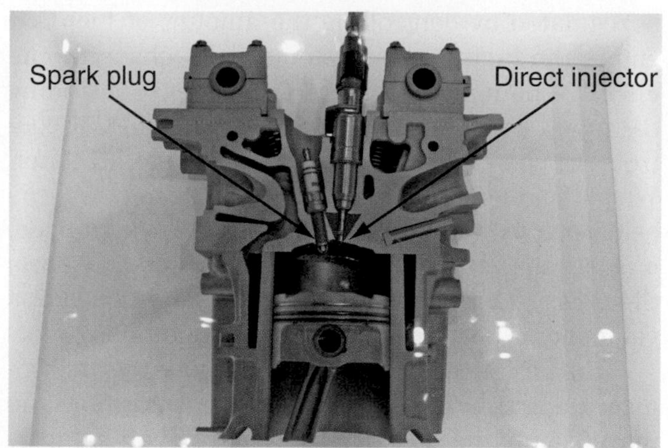

FIGURE 29–23

The direct injector is placed in the combustion chamber so that its spray of fuel is aimed at the spark plug.

FIGURE 29–24

An injector for gasoline direct injection.

under very high pressures, it is injected into the cylinders as very fine droplets.

Injectors

With gasoline direct injection, specially designed injectors deliver the fuel into the high pressures and temperatures in the cylinders **(Figure 29–23)**. To prevent the heat from igniting the fuel in the injector, the injectors are designed to completely seal after the fuel is sprayed. The injectors must also be able to spray the fuel at a much higher pressure than what is in the cylinder. If this did not happen, the fuel would not enter the cylinder; instead, the cylinder's pressure would enter the injector. Remember, a high pressure always moves to a point of lower pressure. The injectors need more electrical power to work with the high pressure; therefore, the PCM has separate high-voltage and driver circuits for each injector. Voltages for direct injection systems are initially boosted, up to 75 volts, depending on the system. The boosted voltage is to ensure very fast response times and precise control of the injector. The injectors must also be resistant to deposit formations and provide a highly atomized and directed spray of fuel. The fuel injectors are normally made with a small extended tip **(Figure 29–24)**. This allows the injector to quickly move the heat, which is absorbed during combustion, to the cooling jackets in the cylinder head.

The PCM controls the pulse width and timing of each injector and allows the system to operate in very distinctly different modes. Fuel is injected before or after the intake valve is closed, depending on the operational mode. The pulse width also changes with the operational mode, and adjustments are made according to inputs from the MAF and IAT sensors.

Solenoid injectors are used in most direct-injection systems; however, some are piezoelectrically (piezo)-actuated injectors. Piezo injectors rely on stacked piezoelectric crystals. When voltage is applied to the crystals, they change size to operate the injector. Piezoelectric injectors have a much faster response time than solenoid injectors.

When the injector is at rest, high-pressure fuel is present at both the top of the injector pintle holding the pintle against its seat and also at the nozzle end of the pintle trying to force it upward off its seat. When energized, the expansion of the stack opens a valve above the injector pintle, allowing the upper pressurized fuel to return to the fuel tank. This reduced upper pressure allows the high-pressure fuel at the nozzle end of the pintle to force the pintle upward off its seat, allowing injection.

With GDI, fuel can be injected at any time, not only when the intake valve is open. Also, the injectors can pulse twice during the transition from the compression stroke to combustion. The two pulses promote complete combustion when the PCM senses that operating conditions may prevent a complete burning of the fuel.

High-Pressure Fuel Pump

Gasoline is moved from the fuel tank to the engine in a conventional way, which is by an in-tank electric pump. The fuel is delivered to a mechanical, high-pressure pump driven by an eccentric on the end of

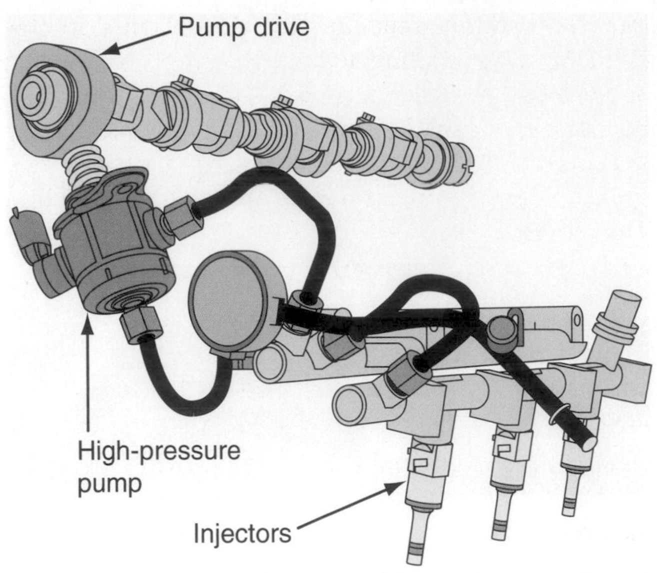

FIGURE 29–25

The high-pressure fuel pump for a GDI system is driven by a camshaft.

a camshaft **(Figure 29–25)**. This pump supplies fuel to a variable-pressure fuel rail. The individual injectors are attached to the rail. A GDI system can operate with pressures of 33 to 130 bar (435 to 1885 psi). The pump is a volume-controlled, high-pressure pump. It moves only the required amount of fuel to the fuel rail.

The PCM regulates the pressure in the fuel rail, based on signals from the inputs from the fuel rail pressure regulator that is located on the fuel rail or is part of the high-pressure pump. The pressure is regulated by controlling the amount of fuel that enters the high-pressure pump or by changing the effective pumping stroke of the pump. Controlling the pump inlet is the most common. The PCM controls the power and ground circuits of the regulator. When the regulator is de-energized, the inlet valve is held open by spring pressure. When it is energized, it closes the valve. Through pulse-width modulation, the pressure from the pump can be maintained at a value that is best for the operating conditions.

The PCM uses CKP and CMP signals to synchronize the action of the regulator with position and movement of the eccentric on the camshaft.

To protect the system from excessive pressures, there are pressure relief valves in the pump or fuel rail. When pressure reaches a predetermined value, some fuel leaves the relief valve and returns to the fuel tank.

Operational Modes

Most direct-injection systems can operate in three distinct modes: lean burn, stoichiometric, and full power. Each of these modes has different air/fuel ratios, injection timing, and pump pressures. The PCM chooses the mode based on operating conditions. The lean mode relies on a stratified charge for combustion, and the stoichiometric and full-power modes rely on a homogeneous mixture, which means that the air and fuel are well mixed **(Figure 29–26)**. The PCM must be able to smoothly transition the move from one to another. The systems also have a limp-in mode if the PCM detects a problem. During this time, the PCM

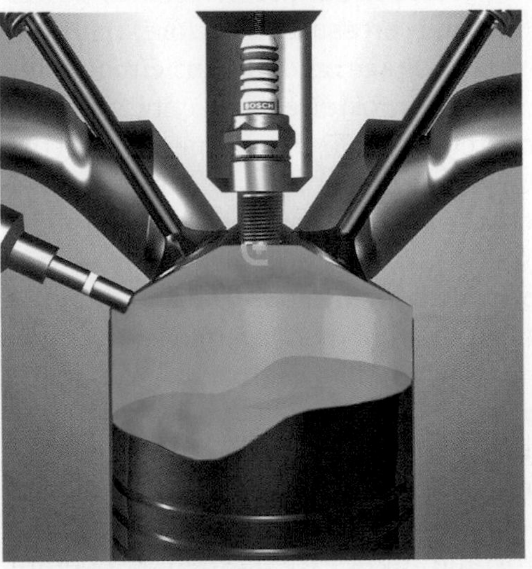

FIGURE 29–26

Two of the basic operation modes of a GDI system.

Robert Bosch Corp.

will use a fuel strategy that allows the engine to run well enough to drive the vehicle in for service.

It is important to note that not all direct-injection systems have a lean burn mode. These systems use this technology for power gains and emission reduction, not to minimize fuel consumption.

LEAN BURN MODE Direct-injected engines are able to run at very lean mixtures, with air/fuel ratios as high as 40:1. This benefit is why GDI engines are capable of drastically reducing fuel consumption. The engines run in the lean mode when the vehicle is cruising with a very light load. In addition to a reduction in fuel consumption, exhaust emissions are also lowered. The lean mixtures are possible because the system allows for a stratified charge.

The placement of the injector tip in the combustion chamber is an important feature of a direct-injection system, especially when it is operating in the lean mode. A small amount of fuel is sprayed near the spark plug when the piston has nearly completed its compression stroke, but before ignition occurs. There is enough fuel around the spark plug to cause combustion. This local area of combustion is the **stratified charge**. The area surrounding that small area has little or no fuel and is mostly air or recirculated exhaust gas. The air surrounding the stratified charge forms an insulating cushion to keep fuel away from the cylinder walls.

STOICHIOMETRIC MODE During medium-load operation, the system operates in the stoichiometric mode. The air/fuel ratio is near stoichiometric, and the fuel is injected during the intake stroke.

FULL-POWER MODE The full-power mode is used during heavy loads and hard acceleration. The air/fuel mixture is slightly richer than stoichiometric, and the fuel is again injected during the intake stroke.

Compression Ratios

The ability of a direct-injection system to change air/fuel ratios and injection timing over a wide range also allows it to eliminate most conditions that would cause engine knock. This means GDI engines can operate at higher compression ratios without requiring the use of high-octane gasoline. The benefit of high compression is simply that higher compression extracts more energy from each droplet of fuel. Therefore, running higher compression ratios provides increased engine horsepower and torque without consuming more fuel.

Higher compression ratios are also possible because the small droplets of fuel injected directly into the cylinder tend to cool the mixture in the cylinder. This cooling makes the mixture denser, meaning more power can be produced, and makes the mixture less likely to detonate. Regardless of the operational mode, fuel is sprayed around the spark plug allowing the mixture to burn quickly. This means there is less need for spark advance decreasing the chances of detonation.

Advantages of GDI

When compared to other injection systems, direct injection has the following advantages:

- Increases fuel efficiency
- Provides higher power output
- Increases the engine's volumetric efficiency
- Lowers engine thermal losses
- Decreases emissions (NO_x increase if combustion temperatures go too high)
- Allows the engine to have high compression without the need to use high-octane fuel
- Reduces most of the turbo lag when used with a turbocharger

GDI Plus SFI

Some engines from Toyota and Lexus use a combination of direct and indirect injection. Each cylinder has two injectors: one at the intake port and the other directly in the cylinder **(Figure 29–27)**. Both sets of injectors work in the same way as they would in a normal setup. However, the PCM shuts down the port injection when it is not needed. When the engine is cold or running at low to middle speeds with a light load, both injection systems operate. During high engine speeds and heavy loads, only direct injection is used. During cold starts, the port and direct injectors work together to create a lean stratified charge. Fuel is initially injected into the intake port during the intake stroke. Fuel is then injected from the direct injector near the end of the piston's compression stroke. This creates a stratified charge. This charge burns rapidly and therefore quickly heats the chamber; this in turn quickly warms up the catalytic converter.

The PCM controls the injection volume and timing of each injector according to engine load, intake airflow, temperature, and other inputs. This system takes advantage of the benefits of both types of injection systems. A port injection system is more efficient at slight throttle openings, whereas a direct-injection system is best with engine speed and load. The goal of the system is improved

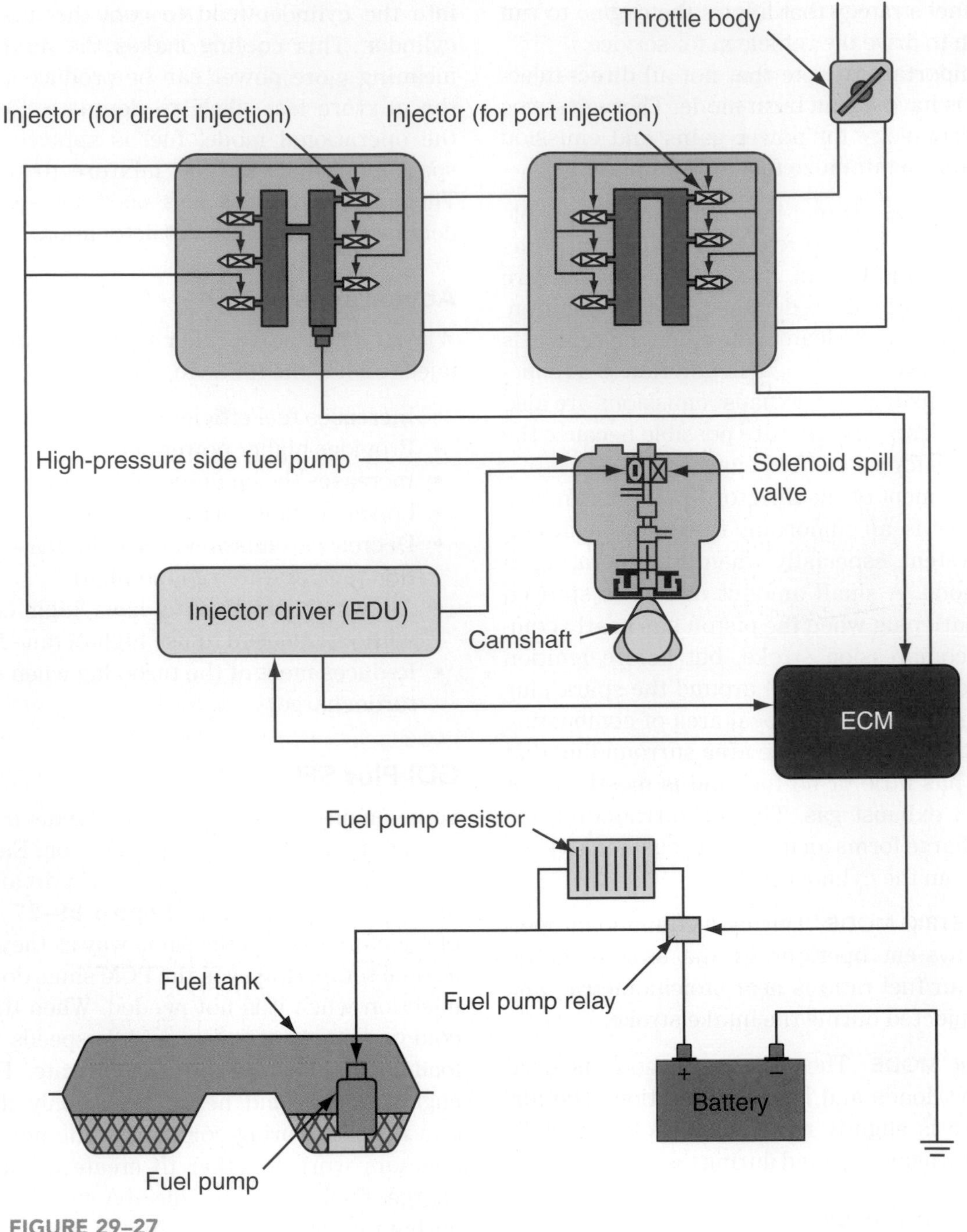

FIGURE 29–27

Toyota's dual system.

performance and decreased fuel consumption and emissions during all operating conditions.

Fuel is delivered to both injection systems by the low-pressure fuel pump in the tank. The fuel for the direct-injection system is sent to a high-pressure pump driven by the exhaust camshaft. At the end of the camshaft, there is a three-sided lobe that rides on the plunger of the high-pressure pump. The plunger is moved up and down, or stroked, three times for every one complete revolution of the lobe.

The injectors in the direct system have a two-slit orifice at the nozzle. This nozzle is designed to provide a fuel spray that fans out toward the spark plug. The injectors are pulsed by an electronic driver unit (EDU) that sends a high-voltage signal to the injectors. The EDU is controlled by the PCM using pulse-width modulation.

A fuel pressure sensor on the fuel rail monitors the pressure and sends a signal to the PCM. The PCM calculates the required pressure for the current conditions and orders the EDU to alter the action of a spill control valve, if necessary.

The spill valve is located at the inlet passage of the high-pressure pump and is used to control the pump discharge pressure. It is electrically opened and closed by the EDU. The pressure is regulated by controlling the amount of time the valve is closed.

Fuel is drawn in by the pump when the valve is open, and the pump's plunger is moved downward by its spring. The EDU will then close the spill valve, and the pump's plunger will move up. The force of the lobe on the plunger will put pressure on the fuel inside the pump. Once the fuel pressure is strong enough to open a check valve at the outlet of the pump, fuel will flow through the fuel rail to the injectors.

Therefore, because the PCM is able to control the timing of the injectors as well as the high-pressure fuel system, the PCM can accurately inject the precise amount of fuel to maximize engine output and minimize exhaust emissions.

KEY TERMS

Central multiport fuel injection (CMFI) (p. 867)
Central port injection (CPI) (p. 867)
Central sequential fuel injection (CSFI) (p. 867)
Clear flood mode (p. 860)
Fuel rail (p. 863)
Idle air control (IAC) (p. 858)
Multiport fuel injection (MPFI) (p. 862)
Poppet nozzles (p. 867)
Port fuel injection (PFI) (p. 862)
Sequential fuel injection (SFI) (p. 862)
Speed density (p. 856)
Stoichiometric ratio (p. 855)
Stratified charge (p. 871)

SUMMARY

- EFI systems are computer controlled and designed to provide the correct air/fuel ratio for all engine loads, speeds, and temperature conditions. The ideal fuel ratio is called the stoichiometric ratio, which means the air-to-fuel ratio can allow for complete combustion.
- EFI systems rely on inputs from various sensors; these include airflow, air temperature, mass airflow, manifold absolute pressure, exhaust oxygen content, coolant temperature, and throttle position sensors.
- The volume airflow sensor and mass airflow sensors determine the amount of air entering the engine. The MAP sensor measures changes in the intake manifold pressure that result from changes in engine load and speed.
- The heart of the fuel injection system is the electronic control unit. The PCM receives signals from all the system sensors, processes them, and controls the fuel injectors.
- In an EFI system, the computer supplies the proper air/fuel ratio by controlling injector pulse width.
- Two types of fuel injectors are commonly used: top feed and bottom feed. Top-feed injectors are used in port injection systems. Bottom-feed injectors are used in throttle body injection systems.
- In any EFI system, the fuel pressure must be high enough to prevent fuel boiling.
- Most computers provide a clear flood mode if a cold engine becomes flooded. Pressing the gas pedal to the floor while cranking the engine activates this mode.
- In an SFI system, each injector has an individual ground wire connected to the computer.
- The pressure regulator maintains the specified fuel system pressure and returns excess fuel to the fuel tank.
- In a returnless fuel system, the pressure regulator and filter assembly is mounted with the fuel pump and gauge sending unit assembly on top of the fuel tank. This pressure regulator returns fuel directly into the fuel tank.
- A central multiport injection system has one central injector and a poppet nozzle in each intake port. The central injector is operated by the PCM, and the poppet nozzles are operated by fuel pressure.
- GDI systems inject gasoline directly into the combustion chamber and produces a stratified air/fuel charge that allows for complete combustion with lean air/fuel ratios.
- GDI systems use special injectors that spray the gasoline at very high pressures and seal extremely well when they are not open.

REVIEW QUESTIONS

1. Explain the major differences between throttle-body fuel injection and port fuel injection systems.
2. What is meant by sequential firing of the fuel injectors?

3. Explain the purpose of the TP sensor input in a speed-density fuel injection system.
4. Describe the purpose of an ECT signal on an EFI system.
5. Explain how the computer controls the air/fuel ratio in an EFI system.
6. What does the O-ring at the base of the fuel injector prevent?
   a. a vacuum leak at the injector
   b. a fuel leak at the injector and fuel rail
   c. a fuel leak between the intake manifold and the injector
   d. air from entering the fuel charge
7. What is the percentage of time a solenoid is energized relative to the total cycle time called?
   a. pulse-width modulation
   b. frequency
   c. duty cycle
   d. amplitude
8. Which of the following describes gasoline direct injection (GDI)?
   a. one or two injectors centrally located in a throttle body housing
   b. injectors located in the intake manifold at each intake port
   c. fuel injected on the intake manifold side of the intake valve
   d. fuel injected into the cylinder
9. Which of the following is a condition that would allow closed loop?
   a. wide-open throttle operation
   b. during starting
   c. a warm engine operating at cruising speed
   d. a cold engine operating at cruising speed
10. Which electronic fuel injection system fires each injector individually?
    a. multiport fuel injection
    b. sequential fuel injection
    c. throttle body injection
    d. port fuel injection
11. When does the fuel pressure regulator have the most vacuum controlling it?
    a. at idle
    b. at cruising speed
    c. during moderate acceleration
    d. at wide-open throttle
12. What is a typical fuel injector pulse-width range?
    a. from 0.5 ms to 5 ms
    b. from 1 ms to 10 ms
    c. from 5 ms to 15 ms
    d. from 10 ms to 20 ms
13. What does an MAP sensor measure?
    a. manifold air temperature
    b. multiple air-valve positions
    c. engine coolant temperature
    d. engine load
14. How does the IAC control idle speed?
    a. by changing the amount of fuel injected
    b. by changing the amount of air bypassing the throttle plate
    c. by opening and closing the throttle
    d. by changing the ignition advance
15. What is injector pulse width?
    a. the size of the injector pintle
    b. the amount of current flow through the injector
    c. the discharge pressure of the injector
    d. the injector-on time
16. What is the typical air/fuel ratio during clear flood mode?
    a. 14:1 b. 16:1
    c. 18:1 d. 20:1
17. What does a mass airflow sensor measure?
    a. airflow into the intake manifold
    b. the density of the air entering the intake manifold
    c. the moisture in the air entering the intake manifold
    d. the pressure inside the intake manifold
18. What is another name for a lambda sensor?
    a. exhaust gas oxygen sensor
    b. mass airflow sensor
    c. MAP sensor
    d. coolant temperature sensor
19. What component is responsible for adjusting injection duration?
    a. the MAP sensor b. the IAC motor
    c. the O_2S d. the PCM
20. What is the main difference between port fuel injectors and direct injection injectors?
    a. Direct injectors are capable of higher fuel volumes.
    b. Port fuel injectors are capable of higher fuel volumes.
    c. Direct injectors spray fuel at higher pressure.
    d. Port fuel injectors spray fuel at higher pressure.

CHAPTER 30
Fuel Injection System Diagnosis and Service

LEARNING OUTCOMES

- Perform a preliminary diagnostic procedure on a fuel injection system.
- Remove, clean, inspect, and install throttle body assemblies.
- Explain the results of incorrect fuel pressure in a TBI, MPFI, or SFI system.
- Perform an injector balance test and determine the injector condition.
- Clean injectors on an MPFI or SFI system.
- Perform an injector sound, ohmmeter, noid light, and scope test.
- Perform an injector flow test and determine injector condition.
- Perform an injector leakage test.
- Remove and replace the fuel rail, injectors, and pressure regulator.
- Diagnose causes of improper idle speed on vehicles with fuel injection.

This chapter contains diagnosis and service information for a number of current and some older technologies. Due to the significant number of older vehicles still on Canadian roads, older systems such as TBI were included. These systems will probably disappear in the not-so-distant future.

Troubleshooting fuel injection systems requires systematic step-by-step test procedures. With so many interrelated components and sensors controlling fuel injection performance **(Figure 30–1)**, a hit-or-miss approach to diagnosing problems can quickly become frustrating, time-consuming, and costly.

Most fuel injection systems are integrated into engine control systems **(Figure 30–2)**. The self-test modes of these systems are designed to help in engine diagnosis. Unfortunately, when a problem upsets the smooth operation of the engine, many service technicians automatically assume that the computer (PCM) is at fault. But in the vast majority of cases, complaints about driveability, performance, fuel mileage, roughness, or hard starting or no-starting are due to something other than the computer itself (although many problems are caused by sensor malfunctions that can be traced using the self-test mode).

Before condemning sensors as defective, remember that weak or poorly operating engine components can often affect sensor readings and result in poor performance. For example, a sloppy timing chain or bad rings or valves reduce vacuum and cylinder pressure, resulting in a lower exhaust temperature. This can affect the operation of a perfectly good oxygen or lambda sensor, which must heat up to approximately 315°C (600°F) before functioning in its closed loop mode.

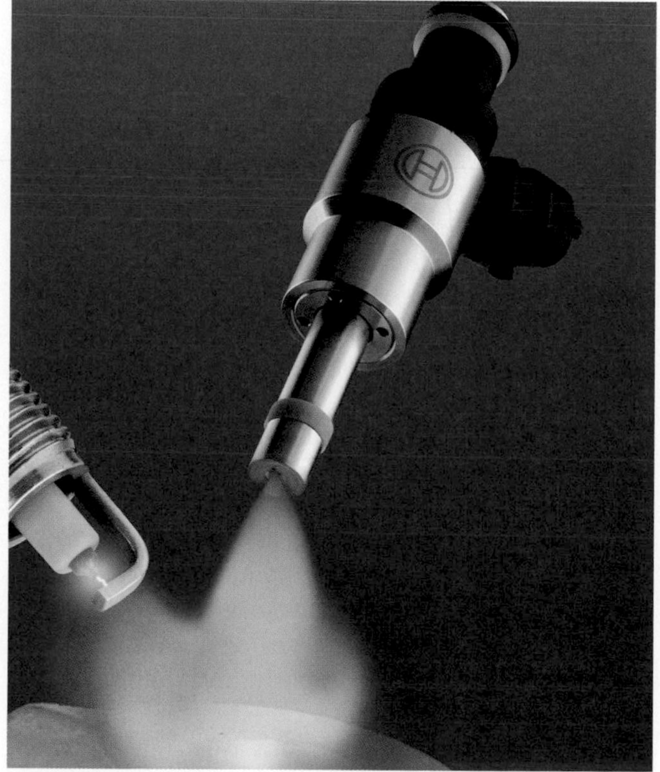

FIGURE 30–1

The action of a fuel injector and spark plug to start combustion.

Robert Bosch Corp.

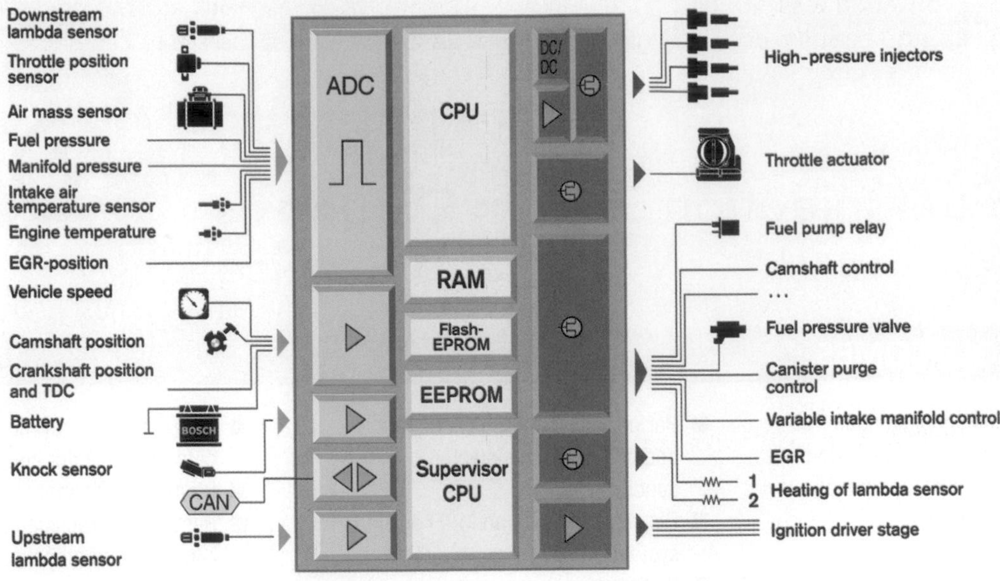

FIGURE 30–2

The layout of a late-model EFI system.

Robert Bosch Corp.

A problem such as an intake manifold leak can cause a sensor, the MAP sensor in this case, to adjust engine operation to less than ideal conditions.

PRELIMINARY CHECKS

The best way to approach a problem on a vehicle with electronic fuel injection is to treat it as though it had no electronic controls at all.

Before proceeding with specific fuel injection checks and electronic control testing, be certain of the following:

- The battery is in good condition, fully charged, with clean terminals and connections.
- The charging and starting systems are operating properly.
- All fuses and fusible links are intact.
- All wiring harnesses are properly routed with connections free of corrosion and tightly attached **(Figure 30–3)**.
- All vacuum lines are in sound condition, properly routed, and tightly attached.
- The positive crankcase ventilation (PCV) system is working properly and maintaining a sealed crankcase.
- All emission control systems are in place, hooked up, and operating properly.
- The level and condition of the coolant/antifreeze is good, and the thermostat is opening at the proper temperature.
- The secondary spark delivery components are in good shape with no signs of cross-firing, carbon tracking, corrosion, or wear.
- The engine is in good mechanical condition.

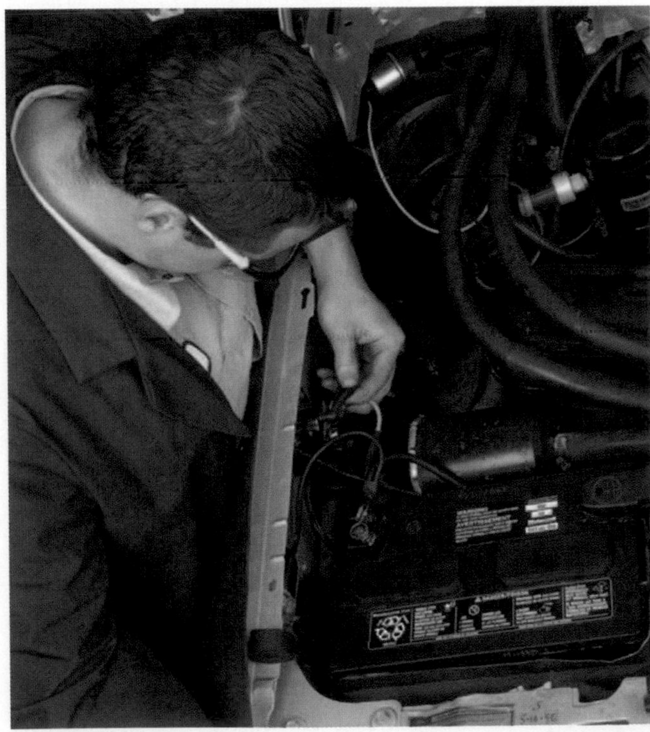

FIGURE 30–3

All underhood wiring should be carefully inspected.

- The gasoline in the tank is of good quality and has not been substantially cut with alcohol or contaminated with water.

EFI System Component Checks

In any electronic throttle body or port injection system, three things must occur for the system to operate:

1. An adequate amount of air must be supplied for the air/fuel mixture.
2. A pressurized fuel supply must be delivered to properly operating injectors.
3. The injectors must receive a trigger signal from the control computer, which must first receive an rpm signal from the ignition.

SERVICE PRECAUTIONS

These precautions must be observed when electronic fuel injection systems are diagnosed and serviced:

- Always relieve the fuel pressure before disconnecting any component in the fuel system.
- Never turn on the ignition switch when any fuel system component is disconnected.
- Use only the test equipment recommended by the vehicle manufacturer.
- Always turn off the ignition switch before connecting or disconnecting any system component or test equipment.
- When arc welding is necessary on a computer-equipped vehicle, disconnect both battery cables before welding is started. Always disconnect the negative cable first.
- Never allow electrical system voltage to exceed 16 volts. This could be done by disconnecting the circuit between the alternator and the battery, with the engine running.
- Avoid static electricity discharges when handling computers, modules, and computer chips.

BASIC EFI SYSTEM CHECKS

If all of the preliminary checks do not reveal a problem, proceed to test the electronic control system and fuel injection components. Some older control systems require involved test procedures and special test equipment, but most newer designs have a self-test program designed to help diagnose the problem. These self-tests perform a number of checks on components within the system. Input sensors, output devices, wiring harnesses, and even the computer itself may be among the items tested.

The results of the testing are converted into trouble codes that the technician may read using a scan tool or the **malfunction indicator lamp (MIL)**. On pre-OBD-II vehicles, the meaning of trouble codes vary from manufacturer to manufacturer, year to year, and model to model, so it is important to consult the appropriate service information. Although SAE standard J 2012 calls for common diagnostic trouble codes, a variety of service literature is also required to service OBD-II vehicle diagnostic trouble codes (DTCs). Many DTCs encountered by the technician may be manufacturer specific, or the testing procedures may be unique to a particular make of vehicle.

Always remember that trouble codes only indicate the particular circuit in which a problem has been detected. They do not pinpoint individual components. So if a code indicates a defective lambda or oxygen sensor, the problem could be the sensor itself, the wiring to it, or its connector. Trouble codes are not a signal to replace components. They indicate that a more thorough diagnosis is needed in that area.

The following sections outline general troubleshooting procedures for the most popular electronic fuel injection (EFI) designs in use today.

Oxygen Sensor Diagnosis

Oxygen sensors **(Figure 30–4)** produce a voltage based on the amount of oxygen in the exhaust. Large amounts of oxygen result from lean mixtures and result in low voltage output from the O_2 sensor. Rich mixtures release lower amounts of oxygen into the exhaust; therefore, the O_2 sensor voltage is high.

Carefully check the wiring and connectors to the oxygen sensors for damage and evidence of unwanted resistance. Also check for intake and exhaust system leaks. These conditions would tend to create more oxygen in the exhaust, and the PCM will respond by adding fuel to the mixture. A misfiring spark plug allows unburned fuel and oxygen in the exhaust, which also causes the sensor to give a false lean reading.

Before testing an O_2 sensor, refer to the correct wiring diagram to identify the terminals at the sensor. Most late-model engines use heated oxygen sensors (HO_2Ss). These sensors have an internal heater that helps to stabilize the output signals. Most

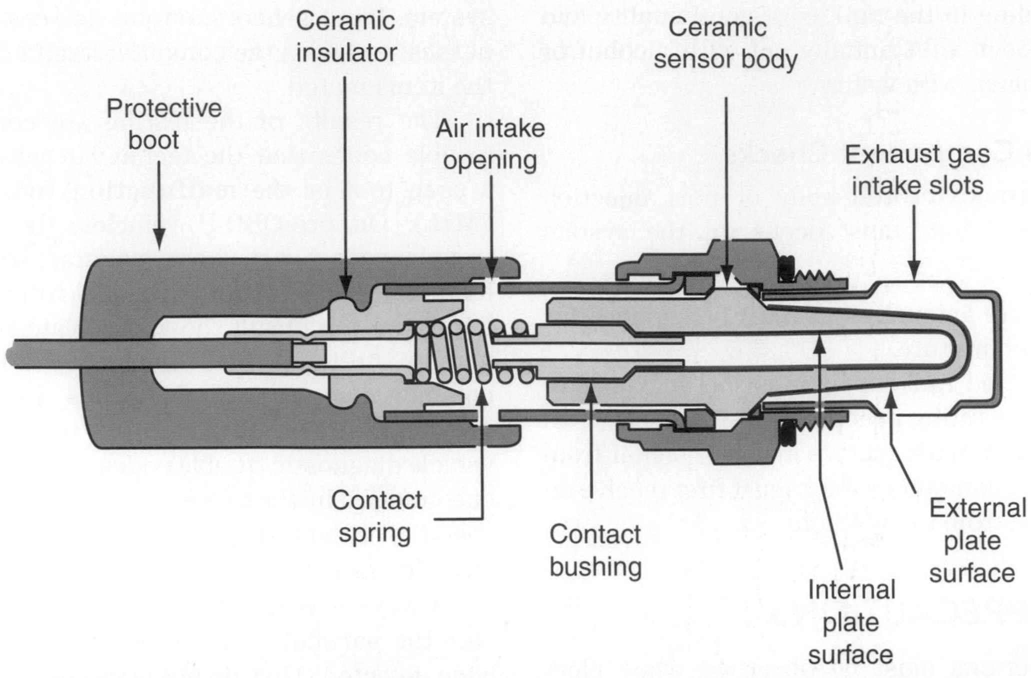

FIGURE 30–4

The components of an oxygen sensor.

heated oxygen sensors have four wires connected to them. Two are for the heater, and the other two are for the sensor **(Figure 30–5)**.

SHOP TALK

On some engines, O_2S heater problems (and faulty O_2S signals) can result from a poorly grounded exhaust system. A poor ground adds resistance to the circuit, which prevents the heater from heating the sensor sufficiently. This can cause faulty sensor output signals. Often, the problem can be corrected by tightening the exhaust-manifold-to-engine bolts.

An O_2 sensor can be checked with a voltmeter. Connect it between the O_2 sensor wire and ground. The sensor's voltage should be cycling from low voltage to high voltage. The signal from most O_2 sensors varies between 0 and 1 volt. If the voltage is continually high, the air/fuel ratio may be rich or the sensor may be contaminated. When the O_2 sensor voltage is continually low, the air/fuel ratio may be lean; the sensor may be defective; or the wire between the sensor and the computer may have a high-resistance problem. If the O_2 sensor voltage signal remains in a mid-range position, the computer may be in open loop, or the sensor may be defective.

If the O_2S voltage signal sits at, or close to, 0, unplug the sensor. If the voltage signal increases while the sensor is being unplugged, the sensor is probably shorted to ground. If the O_2S voltage

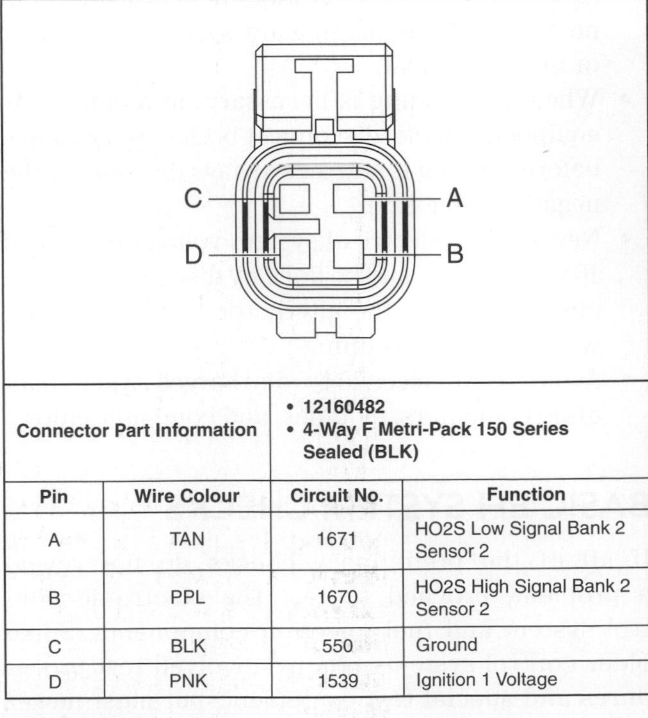

Connector Part Information	• 12160482 • 4-Way F Metri-Pack 150 Series Sealed (BLK)		
Pin	**Wire Colour**	**Circuit No.**	**Function**
A	TAN	1671	HO2S Low Signal Bank 2 Sensor 2
B	PPL	1670	HO2S High Signal Bank 2 Sensor 2
C	BLK	550	Ground
D	PNK	1539	Ignition 1 Voltage

FIGURE 30–5

Use service information to identify the terminals on a heated oxygen sensor.

signal sits at, or close to, 1 volt, check the wiring at the sensor to see if the heater power feed wire or connector is shorted to the sensor's output signal wire.

Some engines are equipped with oxygen sensors that use a 1-, 3-, or 5-volt reference signal. These systems will not show the usual voltage of 0 to 1 volt when monitored on a scan tool, voltmeter, or lab scope. This means the reading on a scan tool, voltmeter, or lab scope is 5 to 6 volts instead of the typical 0 to 1 volt.

The activity of an O_2 sensor is best monitored with a lab scope. The scope is connected to the sensor in the same way as a voltmeter **(Figure 30–6)**. The switching of the sensor should be seen as the sensor signal goes from lean to rich to lean continuously **(Figure 30–7)**.

The activity of the sensor can also be monitored on a scanner. While the engine is running, the scanner should show that the O_2 sensor voltage moves to nearly 1 volt and then drops back to close to 0 volts. Immediately after it drops, the voltage signal should move back up. This immediate cycling is an important function of an O_2 sensor. If the response is slow, the sensor is lazy and should be replaced. With the engine running at about 2500 rpm, a zirconia-style sensor should cycle between 200 mV and 800 mV for voltage and take approximately 100 mS per change.

A scan tool is also an excellent way to monitor the fuel control of a fuel-injected engine. Many factors determine the pulse width of the injectors, but it should always respond to the O_2 readings.

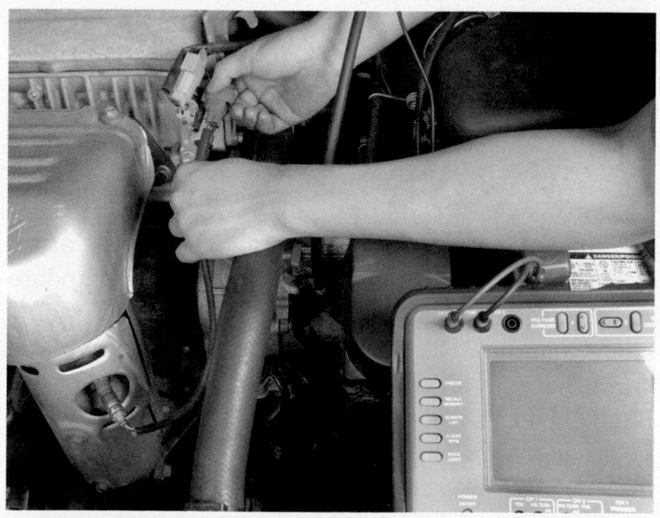

FIGURE 30–6

The correct way to connect a lab scope to an oxygen sensor.

On pre-OBD-II GM vehicles, fuel correction was shown on a scan tool as Integrator or Block Learn. **Integrator** represents a short-term correction to the amount of fuel delivered during closed loop. **Block Learn** makes long-term corrections. Injector pulse width is adjusted according to both Integrator and Block Learn.

A scale of 0 to 255 is used for Integrator and Block Learn, and a mid-range reading of 128 is preferred. The oxygen (O_2) sensor signal is sent to the Integrator chip and then to the pulse-width calculation chip in the PCM, and the Block Learn chip is connected parallel to the Integrator chip. If the O_2 sensor voltage changes once, the Integrator chip and the pulse-width calculation chip change the injector pulse width. If the O_2 sensor provides four continually high- or low-voltage signals, the Block Learn chip makes a further injector pulse-width change. When the Integrator, or Block Learn, numbers are considerably above 128, the PCM is continually attempting to increase fuel; therefore, the O_2 sensor voltage signal must be continually low, or lean. If the Integrator, or Block Learn, numbers are considerably below 128, the PCM is continually decreasing fuel, which indicates that the O_2 sensor voltage must be always high, or rich.

OBD-II Adaptive Fuel Control Strategy

OBD-II systems continuously check the fuel system with its comprehensive and misfire monitors. OBD-II's **short-term fuel trim (STFT)** and **long-term fuel trim (LTFT)** strategies monitor the oxygen sensor signals. The information gathered by the PCM is used to make adjustments to the fuel control calculations. The adaptive fuel control strategy allows for changes in the amount of fuel delivered to the cylinders according to operating conditions. The STFT and LTFT are similar in operation to the Block Learn and Integrator.

During open loop, the PCM changes pulse width without any feedback from the O_2S, and no adjustments to the short-term fuel trim are made, so the scan tool will report a value of 0 percent. Once the sensor warms up, the PCM moves into closed loop and begins to recognize the signals from the O_2S. The system remains in closed loop until the engine stops, unless the throttle is fully opened or the engine's temperature drops below a specified limit, then it goes into open loop.

When the system is in open loop, the injectors operate at a fixed base pulse width. During closed loop operation, the pulse width is either lengthened

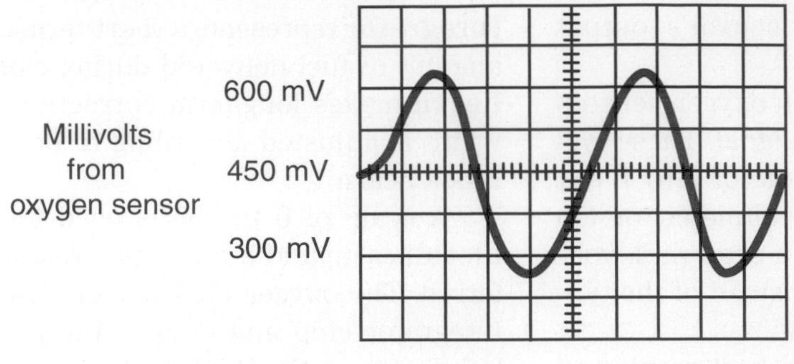

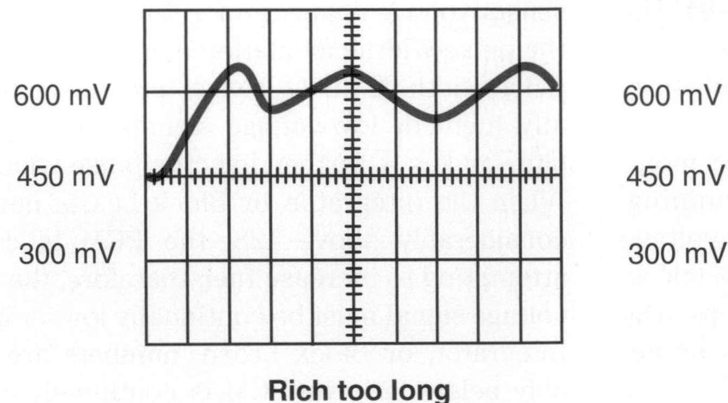

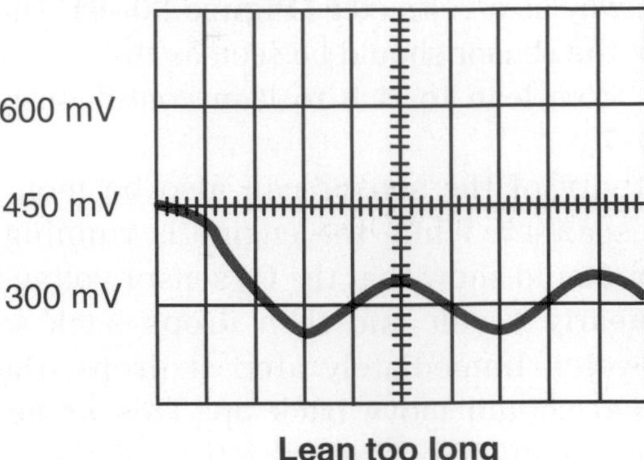

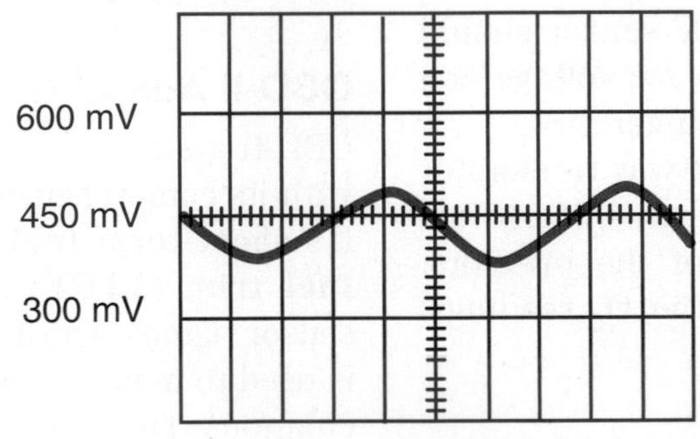

FIGURE 30–7

Normal and abnormal O_2 sensor waveforms.

or shortened. As the voltage from the oxygen sensor increases in response to a rich mixture, the short-term fuel trim decreases, which means the base pulse width is shortened. Decreases in STFT are indicated on a scan tool as a negative number. For example, –%25 showing on a scan tool means that the pulse width is being shortened by 25 percent. An STFT value of 25 or %25 would indicate that a lean mixture is being reported by the O_2S, and the fuel injector pulse width is being lengthened by 25 percent.

Once the engine reaches a specified temperature (normally 82°C [180°F]), the PCM begins to update the LTFT. The adaptive setting is based on engine speed and the STFT. If the STFT moves 3 percent and stays there for a period of time, the PCM adjusts the LTFT. The LTFT becomes a new but temporary base. In other words, the LTFT changes the length of the pulse width that is being changed by the STFT. STFT works to bring LTFT close to 0 percent correction.

If a lean condition exists because of a vacuum leak or restricted fuel injectors, the LTFT will have a plus number on the scan tool. If the injectors leak or the fuel pressure regulator is blocking the return line, there will be a rich condition. The rich condition will be evident by minus LTFT numbers on the scan tool. If the engine's condition is too far toward either the lean or rich side, the LTFT will not compensate, and a DTC will be set.

High LTFT readings indicate there may be a dirty mass airflow (MAF) sensor, clogged injectors, a clogged fuel filter, or a bad fuel pump.

On an OBD-II vehicle, zero is the midpoint of the fuel strategy during closed loop. The correction to fuel delivery is illustrated as a percentage; numbers without a minus sign indicate fuel is being added, and numbers with a minus sign indicate fuel is being subtracted. The constant change or crossing above and below the zero line indicates proper system operation. If the STFT readings are constantly on either side of the zero line, the engine is not operating efficiently.

Some intermittent fuel-related driveability problems may be diagnosed by recording the computer's data stream. Before you jump into the data stream, look at the STFT and LTFT.

To diagnose the fuel control system, connect the scan tool and start the engine. Then pull up the PCM's data stream display on the scan tool. Observe the STFT and note the value. With the engine running, pull off a large vacuum hose, and watch for the STFT value to rise above zero to as high as 33 percent. Now look at the LTFT value; it should have risen above zero as it learned the engine's condition. A smaller vacuum leak must be made to cause the STFT to return to zero as the result of LTFT compensation.

Because V-type engines have two sets of oxygen sensors, a pair of STFT and LTFT will appear on the scan tool. Use them to isolate the side of the engine that is running rich or lean by simply observing the codes. A high LTFT on one bank indicates there is a problem in that bank, possibly a vacuum leak or EGR problem.

Air Induction System Checks

In a fuel injection system (particularly designs that rely on airflow meters or mass airflow sensors), all the air entering the engine must be accounted for by the air measuring device. If it is not, the air/fuel ratio becomes overly lean. For this reason, cracks or tears in the plumbing between the airflow sensor and throttle body are potential air leak sources that can affect the air/fuel ratio.

FIGURE 30–8

Carefully inspect the ductwork and hoses of the air induction system.

During a visual inspection of the air control system, pay close attention to these areas, looking for cracked or deteriorated ductwork **(Figure 30–8)**. Also make sure all induction hose clamps are tight and properly sealed. Look for possible air leaks in the crankcase, for example, around the dipstick tube and oil filter cap. Any extra air entering the intake manifold through the PCV system is not measured either and can upset the delicately balanced air/fuel mixture at idle.

It is important to note that vacuum leaks may not affect the operation of engines fitted with a speed-density fuel injection system. This does not mean that vacuum leaks are okay; it just means that the operating system may be capable of adjustments that allow the engine to run well in spite of the vacuum leak. This is true for vacuum leaks that are common to all cylinders. If the vacuum leak affects one or two cylinders, the computer cannot compensate for the unmetered air, and those two cylinders will not operate efficiently.

Airflow Sensors

When looking for the cause of a performance complaint that relates to poor fuel economy, erratic performance, hesitation, or hard starting, make the following checks to determine if the airflow sensor is at fault. Detailed testing of these and other sensors is included in Chapter 31.

MASS AIRFLOW SENSORS Both mass airflow (MAF) and volume airflow (VAF) sensors measure intake air before it reaches the throttle plate. If any air bypasses these sensors and enters the combustion chambers without being measured, the engine will run lean. An intake manifold vacuum leak or a leak between the sensors and the throttle plate also reduces fuel delivery. The resulting lean mixture causes severe driveability problems, and the PCM will store DTCs, indicating excessive fuel corrections and/or lean misfires.

If a bad MAF is suspected as causing a lean condition, remove the MAF, and inspect the hot wire for signs of debris. This wire can be cleaned by gently wiping it with a dry cotton swab. Never soak or immerse the MAF sensor in any cleaner. Some manufacturers recommend that the sensor be replaced if the wire is dirty.

VOLUME AIRFLOW SENSORS To check a volume airflow (VAF) sensor, remove the air intake duct from the sensor to gain access to the sensor's flap. Check it for binding, sticking, or scraping by rotating the sensor flap (evenly and carefully) through its operating range. It should move freely, make no noise, and feel smooth. If it does not, a quick spray of throttle body cleaner may loosen it up. If that does not help, replace the sensor.

On some systems, the movement of the sensor flap turns on the fuel pump. If the system is so equipped, turn the ignition on. Do not start the engine. Move the flap toward the open position. The electric fuel pump should come on as the flap is opened. If it does not, turn off the ignition, remove the sensor harness, and check for specific resistance values with an ohmmeter at each of the sensor's terminals.

To check a VAF, connect a voltmeter or scope across its terminals. Watch the meter as you move the flap to its full open position, and then slowly release it. A good VAF should produce a smooth change on the meter or scope screen. If the voltage sweeps are not smooth, the VAF should be replaced.

On most VAFs, it is possible to check the resistance values of the potentiometer by moving the air flap, but in either case, service information is needed to identify the various terminals and to look up resistance specifications. Readings that are out of range indicate the VAF should be replaced.

KARMAN VORTEX SENSORS Karman Vortex sensor problems typically result in surging, hesitation, stalling, and increased emission levels. These sensors should be carefully inspected, as most problems are caused by loose or corroded electrical connectors.

The best way to check a vortex-type airflow sensor is by observing its frequency signal on a scope or a digital meter. The output of Karman Vortex sensors is typically two separate signals: an airflow frequency signal and an air temperature or barometric pressure voltage signal. The airflow signal should be a digital signal that cycles between 0 and 5 volts. The frequency of the signal should increase as airflow increases. Its frequency should increase smoothly and steadily with an increase in engine speed.

SPEED-DENSITY (MAP) SYSTEMS Speed-density systems rely heavily on manifold pressure readings when they are making their fuel calculations. Increased manifold pressure increases pulse width, regardless of the cause of the pressure increase (open throttle, vacuum leak, recirculated exhaust gas, or exhaust pressure).

MAP systems measure manifold pressure after the throttle plate; therefore, their response is opposite to that of MAFs and VAFs. Air leaks in speed-density systems increase injector pulse width if there is a vacuum leak. This can be verified by watching the idle air control (IAC) counts on a scan tool. Low IAC counts typically indicate a vacuum leak in MAP systems because the vacuum leak is causing a lower pressure in the manifold. The PCM, in turn, increases the amount of fuel to the air, which causes the idle speed to increase. Observing this, the PCM attempts to reduce the idle speed by closing the IAC.

Throttle Body

The throttle body **(Figure 30–9)** allows the driver to control the amount of air that enters the engine, thereby controlling the speed of the engine. Each type of throttle body assembly is designed to allow a certain amount of air to pass through it at a particular amount of throttle opening. If anything accumulates on the throttle plates or in the throttle bore, the amount of air that can pass through is reduced. This normally causes an idle problem.

These deposits can be cleaned off the throttle assembly, and the airflow through them restored. Begin by removing the air duct from the throttle assembly; that will give access to the plate and bore. The deposits can be cleaned with a spray cleaner or wiped off with a cloth. If either of these cleaning methods does not remove the deposits, the throttle

FIGURE 30–9

A throttle body positioned between the banks of a V8 engine.

PROCEDURE

Throttle Body Inspection

STEP 1 Check for smooth movement of the throttle linkage from fully closed to wide-open position.

STEP 2 Check the throttle linkage and cables for wear or looseness.

STEP 3 Check all vacuum lines, coolant tubes, and wiring for damage and secure connections.

STEP 4 Connect a smoke machine to a manifold vacuum port and inject smoke into the engine. Look for signs of smoke around the throttle body and vacuum hoses.

STEP 5 Operate the engine until it reaches normal operating temperature, and check the idle speed on a tachometer. The idle speed should be 700 to 800 rpm. Refer to the manufacture's idle speed specifications.

body should be removed, disassembled, and placed in an approved cleaning solution.

A pressurized can of throttle body cleaner may be used to spray around the throttle area without removing and disassembling the throttle body. The throttle assembly can also be cleaned by soaking a cloth in carburetor solvent and wiping the bore and throttle plate to remove light to moderate amounts of carbon residue. Also clean the backside of the throttle plate. Then remove the IAC valve from the throttle body (if so equipped), and clean any carbon deposits from the pintle tip and the IAC air passage.

If on the scan tool you observed IAC counts around 40 or greater, these should drop down to about 15 after cleaning. If adjustments to the minimum idle air settings are required, it is critical that the specified procedures be followed.

If the intake manifold setup has diaphragms or solenoids that control the selection of manifold runners, make sure you allow some of the cleaning solution to be drawn in by the engine while it is running in order to clean the valves, sometimes called *butterflies*, controlled by the solenoids or diaphragms. Dirty switchover valves can cause hard starting, poor performance, increased oil consumption, and DTCs.

THROTTLE BODY INSPECTION Throttle body inspection and service procedures vary widely depending on the year and make of the vehicle. However, some components such as the throttle position sensor (TPS) are found on nearly all throttle bodies. Because throttle bodies have some common components, inspection procedures often involve checking common components.

THROTTLE BODY REMOVAL AND CLEANING Whenever it is necessary to remove the throttle body assembly for replacement or cleaning, make sure you follow the procedures outlined by the manufacturer. Throttle bodies often have coolant circulating through them, so it may be necessary to drain coolant from the radiator before removing the throttle body. Once the assembly has been removed, remove all nonmetallic parts, such as the TPS, IAC valve, and the throttle body gasket, from the throttle body. Now it is safe to clean the throttle body assembly in the recommended throttle body cleaner and blow dry with compressed air. Blow out all passages in the throttle body assembly.

Before reinstalling the throttle body assembly, make sure all metal mating surfaces are clean and free from metal burrs and scratches. Make sure you have new gaskets and seals for all sealing surfaces before you begin to reinstall the assembly. After everything that was disconnected is reconnected, start the engine and verify there are no leaks and that the engine idle speed is correct.

CAUTION!

Not all manufacturers endorse throttle bore cleaning. Also, never immerse an electronically controlled throttle body in a cleaning solution.

Fuel System Checks

If the air control system is in working order, move on to the fuel delivery system. It is important to always remember that fuel injection systems operate at high fuel-pressure levels. This pressure must be relieved before any fuel line connections can be broken. Spraying gasoline (under a pressure of 240 kPa [35 psi] or more) on a hot engine creates a real hazard when dealing with a liquid that has a flash point of 43°C (–45°F).

Follow the specific procedures given in the service information when relieving the pressure in the fuel lines.

CAUTION!

Dispose of the fuel-soaked towel or rag by placing it in a fireproof container.

FUEL DELIVERY When dealing with an alleged fuel complaint that is preventing the vehicle from starting, the first step (after spark, compression, and so forth have been verified) is to determine if fuel is reaching the cylinders (assuming there is gasoline in the tank). Checking for fuel delivery is a simple operation on throttle body systems. Remove the air cleaner, crank the engine, and watch the injector for signs of a spray pattern. If a better view of the injector's operation is required, an ordinary strobe light does a great job of highlighting the spray pattern.

It is impossible to visually inspect the spray pattern and volume of port system injectors. However, an accurate indication of their performance can be obtained by performing simple fuel pressure and fuel volume delivery tests. Keep in mind that fuel pressure affects the output of a fuel injector. If an injector has the same pulse rate but receives low pressure, there is less fuel; if the pressure is high, the amount of fuel is increased.

Low fuel pressure can cause a no-start or poor-running condition. It can be caused by a faulty fuel pump, a fault in the fuel pump electrical circuit, a clogged fuel filter, a faulty pressure regulator, or a restricted fuel line anywhere from the fuel tank to the fuel filter connection.

If a fuel volume test shows low fuel volume, it can indicate a bad fuel pump or a blocked or restricted fuel line. When performing the test, visually inspect the fuel for signs of dirt or moisture. Excessive dirt or moisture indicate a contaminated fuel system and may require major fuel system service, including removal of the fuel tank for cleaning.

High fuel pressure readings will result in a rich-running engine. A restricted fuel return line to the tank or a bad fuel regulator may be the problem. To isolate the cause of high pressure, relieve system pressure and connect a tap hose to the fuel return line. Direct the hose into a container and energize the fuel pump. If fuel pressure is now within specifications, the fuel return line is blocked. If pressure is still high, the pressure regulator is faulty.

Ford and other manufacturers are using returnless fuel delivery systems. Rather than a pressure regulator, they use a pressure sensor on the fuel rail **(Figure 30–10)** and modulate the fuel pump's pulse width to control pressure. In these systems, overly high pressures are caused by the pressure sensor, fuel pump, PCM, or the wiring of the circuit.

If the first fuel pressure reading is within specs but the pressure slowly bleeds down, there may be a leak in the fuel pressure regulator, the fuel pump check valve, or the injectors themselves. Remember, hard starting is a common symptom of system leaks. Also, fuel starvation and lean conditions can occur when the injectors drain the fuel rail faster than the fuel pump can fill it. This could be caused by low fuel pressure or delivery volume.

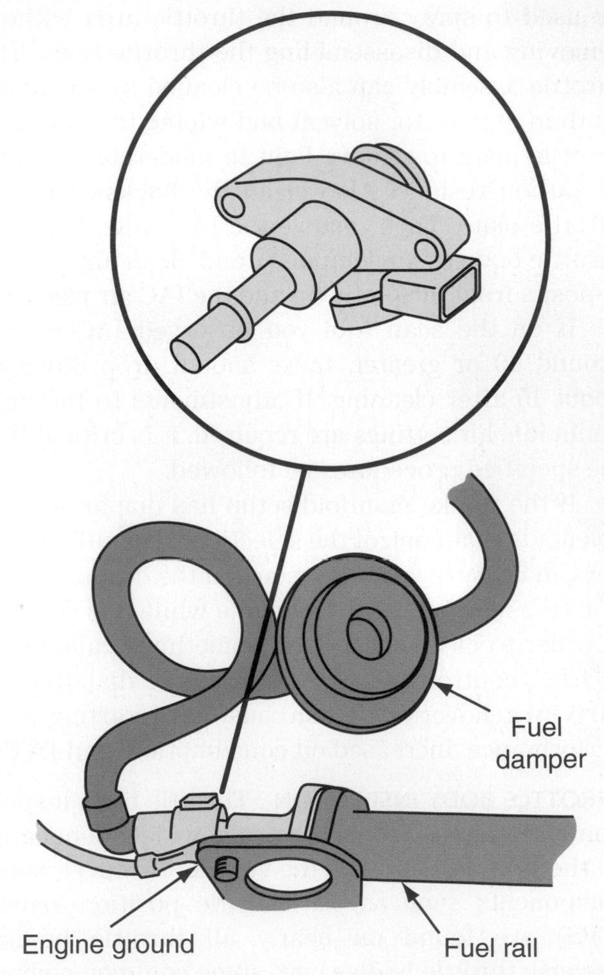

FIGURE 30–10

The location of a fuel pressure sensor in a fuel rail.

One of the best ways to check for injector leaks is to conduct the injector rest test. Connect the fuel pressure gauge, and run the engine until it reaches normal operating temperature. Then turn off the engine and look at the pressure gauge. The fuel pressure may increase initially after the engine is shut down because of the expansion of the fuel. If the pressure drops faster than specifications, there is a leak in the fuel system.

To find the leak, operate the fuel pump just long enough to restore normal pressure; then stop it and immediately restrict the fuel return line between the fuel regulator and the tank. If pressure holds with the return line restricted (or plugged, if recommended by the manufacturer), the leak is in the regulator or return line. If the pressure still drops quickly, remove the restriction from the line. Run the fuel pump to restore normal pressure, and then immediately restrict the supply hose between the fuel pump and the inlet on the fuel rail. If the system now maintains pressure, the fuel pump check valve is probably leaking. If the pressure still drops, there is an external leak in the rail or a rupture in the fuel pressure regulator, or the injectors are leaking.

Injector Checks

A fuel injector is nothing more than a solenoid-actuated fuel valve. Its operation is quite basic in that as long as it is held open and the fuel pressure remains steady, it delivers fuel until it is told to stop.

Because all fuel injectors, except piezo injectors, operate in a similar manner, fuel injector problems tend to exhibit the same failure characteristics. The main difference is that, in a throttle body injection (TBI) design, generally all cylinders will suffer if an injector malfunctions, whereas in port systems, the loss of one injector will affect only one cylinder.

An injector that does not open causes hard starts on port-type systems and an obvious no-start on single-point TBI designs. An injector that is stuck partially open causes loss of fuel pressure (most noticeably after the engine is stopped and restarted within a short time period) and flooding due to raw fuel dribbling into the engine. In addition to a rich-running engine, a leaking injector also causes the engine to diesel or run on when the ignition is turned off. Buildups of gum and other deposits on the tip of an injector can reduce the amount of fuel sprayed by the injector, or they can prevent the injector from totally sealing, allowing it to leak. Because injectors on MPFI and SFI systems are subject to more heat than TBI injectors, port injectors have more problems with tip deposits.

Because an injector adds the fuel part to the air/fuel mixture, any defect in the fuel injection system will cause the mixture to go rich or lean. If the mixture is too rich and the PCM is in control of the air/fuel ratio, a common cause is that one or more injectors are leaking. An easy way to verify this on port-injected engines is to use an exhaust gas analyzer.

With the engine warmed up, but not running, remove the air duct from the airflow sensor. Then insert the gas analyzer's probe into the intake plenum area. Be careful not to damage the airflow sensor or throttle plates while doing this. Look at the hydrocarbon (HC) readings on the analyzer. They should be low and drop as time passes. If an injector is leaking, the HC reading will be high and will not drop. This test does not locate the bad injector, but does verify that one or more are leaking.

Another cause of a rich mixture is a leaking fuel pressure regulator. If the diaphragm of the regulator is ruptured, fuel will move into the intake manifold through the diaphragm, causing a rich mixture. The regulator can be checked by using two simple tests. After the engine has been run, disconnect the vacuum line to the fuel pressure regulator **(Figure 30–11)**. If there are signs of fuel inside the hose or if fuel comes out of the hose, the regulator's diaphragm is leaking. The regulator can also be tested with a hand-operated vacuum pump. Apply 127 mm Hg (5 in. Hg) to the regulator. A good regulator diaphragm will hold that vacuum.

CHECKING VOLTAGE SIGNALS When an injector is suspected as the cause of a lean problem, the first step is to determine if the injector is receiving a signal (from the PCM) to fire. Determining if the

FIGURE 30–11

A fuel pressure regulator.

injector is receiving a signal is best done with a lab scope. In some cases, the location of the harness makes connection difficult.

WARNING!

When performing this test, make sure to keep off the accelerator pedal. On some models, fully depressing the accelerator pedal activates the clear flood mode, in which the voltage signal to the injectors is automatically cut off. Technicians unaware of this waste time tracing a phantom problem.

In the past, a **noid light** was used to check the ability of the injector to be turned on and off. The current test method is to check the voltage and the current patterns by using an oscilloscope. These are dynamic tests that will show problems that can be masked by a noid or test light.

An ohmmeter can be used to test the electrical soundness of an injector. Connect the ohmmeter across the injector terminals **(Figure 30–12)** after the wires to the injector have been disconnected. If the meter reading is infinite, the injector winding is open. More resistance than the specifications means there is high resistance in the winding. A reading that is lower than the specifications indicates that the winding is shorted. If the injector is out of specifications, it must be replaced. Be sure to test the resistance when both hot and cold. An injector winding may test within specifications at room temperature but fail when hot. It is also important to remember that a resistance test does not prove a component is good as the part being tested is not under load at the time of the test.

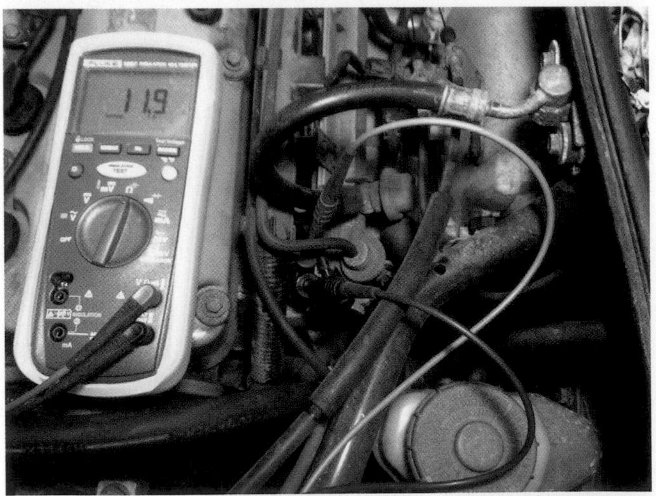

FIGURE 30–12

Checking an injector with an ohmmeter.

INJECTOR BALANCE TEST If the injectors are electrically sound, perform an injector pressure balance test. This test will help isolate a clogged or dirty injector. Depending on the vehicle, the on-board computer system may have an injector test mode accessible through a scan tool. If the vehicle does not have an injector test mode, an electronic injector pulse tester is used for this test. As each injector is energized, a fuel pressure gauge is observed to monitor the drop in fuel pressure. The tester is designed to safely pulse each injector for a controlled length of time. The tester is connected to one injector at a time. The ignition is turned on until a maximum reading is on the pressure gauge. That reading is recorded and the ignition turned off. With the tester, activate the injector and record the pressure reading after the needle has stopped pulsing. This same test is performed on each injector. Start the engine, and let it run briefly between each injector test to allow the fuel to burn out of the cylinders. Always follow the recommended test procedures outlined by the manufacturer to ensure proper safety considerations and equipment used, and that cranking or starting is involved during the test.

The difference between the maximum and minimum reading is the pressure drop. Ideally, each injector should drop the same amount when opened. Typically, a variation of 20 kPa (1.5 to 3 psi or more) after each injector is energized is a cause for concern. If there is no pressure drop or a low pressure drop, suspect a restricted injector orifice or tip. A higher than average pressure drop indicates a rich condition. When an injector plunger is sticking in the open position, the fuel pressure drop is excessive. If there are inconsistent readings, the nonconforming injectors either have to be cleaned or replaced.

If the injector's orifice is dirty or otherwise restricted, there will not be much pressure decrease when the injector is energized. Stumbles during acceleration, engine stalling, and erratic idle are all caused by restricted injector orifices.

If an excessive amount of pressure drop is observed, it is likely that an injector's plunger is sticking open. A sticking injector may result in a rich air/fuel mixture.

INJECTOR SOUND TEST If the injector's electrical leads are difficult to access, an injector power balance test may be hard to perform. As an alternative, start the engine and use a technician's stethoscope to listen for correct injector operation. A good injector makes a rhythmic clicking sound as the solenoid is energized and de-energized several times each second. If a clunk-clunk instead of a steady click-click is heard, chances are the problem injector has been found. Cleaning or replacement is in order. If an injector does not produce any clicking noise, the injector, connecting wires, or PCM may be defective. When the injector clicking noise is erratic, the injector plunger may be sticking. If there is no injector clicking noise, proceed with the injector resistance test and noid light test to locate the cause of the problem. If a stethoscope is not handy, use a thin steel rod, wooden dowel, or fingers to feel for a steady on–off pulsing of the injector solenoid.

INJECTOR FLOW TESTING Some vehicle manufacturers recommend an injector flow test rather than the balance test. To conduct this test, remove the injectors and fuel rail from the engine, and place the tip of the injector to be tested in a calibrated container. Leave all of the injectors in the fuel rail. Then connect a jumper wire across the specified terminals in the data link connector (DLC) for fuel pump testing. Turn on the ignition switch, and connect a jumper wire from the terminals of the injector to the battery terminals **(Figure 30–13)**. Disconnect the jumper wire from the negative battery cable after 15 seconds. Record the amount of fuel in the calibrated container.

Repeat the procedure on each injector. If the volume of fuel discharged from any injector varies more than 5 cc (0.3 cu. in.) from the specifications or the others, the injector should be replaced. When you have completed your testing, reconnect the negative battery cable, and disconnect the 12-volt power supply. This test should only be performed if suggested by the manufacturer to do so. On many low resistance injectors, current is limited by the PCM. Applying direct battery power to the injector for any duration could overheat and permanently damage the injector windings.

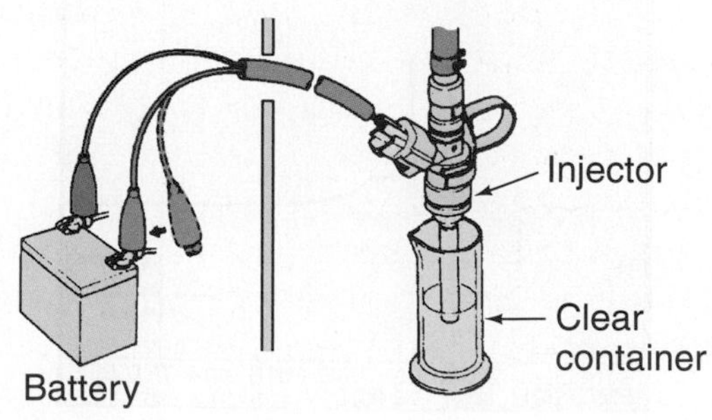

FIGURE 30–13

The setup for testing the flow of an injector.

Reprinted with permission.

Before reinstalling the fuel rail and injectors into the engine, connect a fuel pressure gauge to the fuel system. While the fuel system is pressurized, observe each exposed injector for leakage from the injector tip. Injector leakage must not exceed the manufacturer's specifications. If the injectors leak into the intake ports on a hot engine, the air/fuel ratio may be too rich when a restart is attempted a short time after the engine is shut off. When the injectors leak, they drain all the fuel out of the rail after the engine is shut off for several hours. This may result in slow starting after the engine has been shut off for a longer period of time.

While checking leakage at the injector tips, observe the fuel pressure in the pressure gauge. If the fuel pressure drops off and the injectors are not leaking, the fuel may be leaking back through the check valve in the fuel pump. Repeat the test with the fuel line plugged. If the fuel pressure no longer drops, the fuel pump check valve is leaking. If the fuel pressure drops off and the injectors are not leaking, the fuel pressure may be leaking through the pressure regulator and the return fuel line. Repeat the test with the return line plugged. If the fuel pressure no longer drops off, the pressure regulator valve is leaking.

OSCILLOSCOPE CHECKS An oscilloscope can be used to monitor the injector's pulse width and duty cycle when an injector-related problem is suspected. The **pulse width** is the time in milliseconds that the injector is energized. The duty cycle is the percentage of on-time to total cycle time.

To check the injector's firing voltage on the scope, a typical hookup involves connecting the scope's positive lead to the negative or the PCM side of the circuit and the negative lead to a good ground.

Fuel injection signals vary in frequency and pulse width. The pulse width is controlled by the PCM, which varies it to control the air/fuel ratio. The frequency varies depending on engine speed. The higher the speed, the more pulses per second there are. Most often, the injector's ground circuit is completed by a driver circuit in the PCM. All of these factors are important to remember when setting a lab scope to look at fuel injector activity. Set the scope to read 12 volts per division; then set the sweep and trigger to allow you to clearly see the ON signal on the left and the OFF signal on the right. Make sure the entire waveform is clearly seen. Also remember that the setting may need to be changed as engine speed increases or decreases.

Fuel injectors are either fired individually or in groups. When the injectors are fired in groups, a driver circuit controls two or more injectors. On some V-type engines, one driver fires the injectors on one side of the engine, while another fires the other side. Each fuel injector has its own driver transistor in sequential and TBI. It is extremely important while troubleshooting that you recognize how the injectors are fired. When the injectors are fired in groups, there can be a common or noncommon cause of the problem. For example, a defective driver circuit in the PCM would affect all of the injectors in a group, not just one. Conversely, if one injector in the group is not firing, the problem cannot be the driver.

To read the injector waveform on group fuel injection systems, the scope must be connected to one injector harness for each group. Because all of the injectors in the group share the same circuit, a problem in one will affect the entire waveform for the group. The only way to isolate an injector electrical problem is to disconnect the injectors, one at a time. If the waveform improves when an injector is disconnected, that injector has a problem. If the waveform never cleans up, the problem is in the driver circuit or the wiring harness.

In sequential fuel injection systems, each injector has its own driver circuit and wiring. To check an individual injector, the scope must be connected to that injector. This is great for locating a faulty injector. If the scope has a memory feature, a good injector waveform can be stored and recalled for comparison to the suspected bad fuel injector pattern. To determine if a problem is the injector itself or the PCM and/or wiring, simply swap the injector wires from an injector that had a good waveform to the suspect injector. If the waveform cleans up, the wiring harness or the PCM is the cause of the problem. If the waveform is still not normal, the injector is to blame.

There are three types of fuel injector driver circuits. In a saturated switch driver circuit, voltage is applied to the injector through a fuse. The PCM provides a ground to the circuit to energize the injector. Current flow in the circuit is limited by the resistance of the injector winding. The waveform for this type of injector circuit is shown in **Figure 30–14**. Notice that there is a single voltage spike at the point where the injector is turned off. The total on-time of

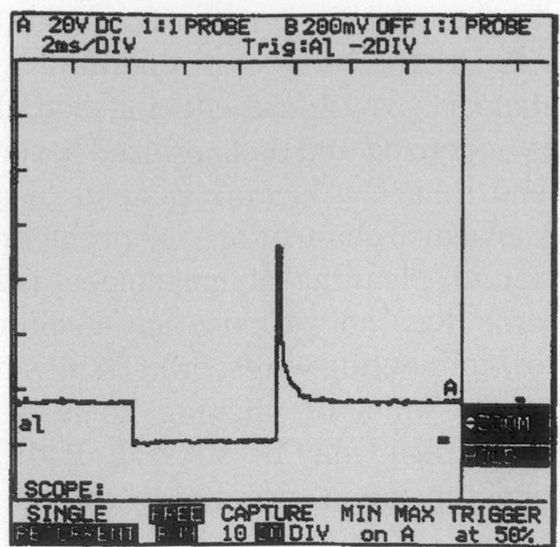

FIGURE 30–14

The waveform for a conventional fuel injector driver circuit.

Fluke Corporation

the injector is measured from the point where the trace drops (on the left) to the point where it rises up (next to the voltage spike).

Peak and hold injector circuits use two driver circuits to control injector action. Both driver circuits complete the circuit to open the injector. This allows for high current at the injector, which forces the injector to open quickly. After the injector is open, one of the circuits turns off. The second circuit remains on to hold the injector open. This is the circuit that controls the pulse width of the injector and limits current flow during on-time. When this circuit turns off, the injector closes. When looking at the waveform for this type of circuit **(Figure 30–15)**, there will be two volt spikes. One is produced when each circuit opens. To measure the on-time of this type injector, measure from the drop on the left to

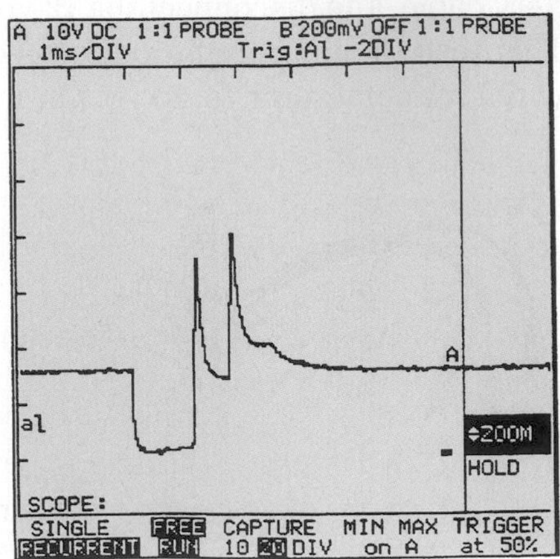

FIGURE 30–15

The waveform for a peak and hold fuel injector driver circuit.

Fluke Corporation

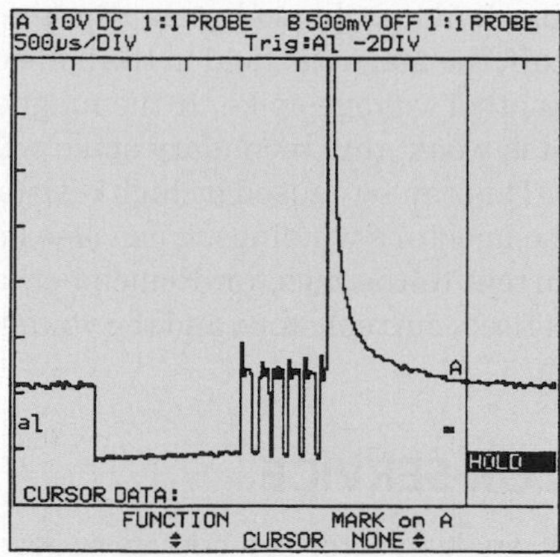

FIGURE 30–16

The waveform for a pulse-width-modulated fuel injector driver circuit.

Fluke Corporation

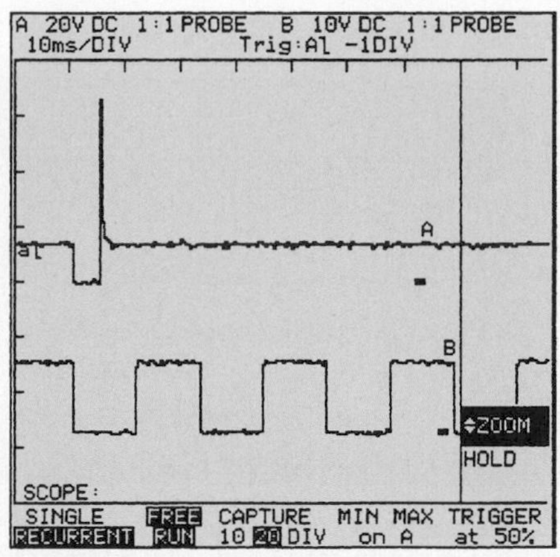

FIGURE 30–17

On a dual trace scope, you can compare the ignition reference signal to the injector's own signal.

Fluke Corporation

the point where the second voltage spike is starting to move upward.

A **pulse-width-modulated injector** driver circuit uses high current to open the injector. Again, this allows for quick injector firing. Once the injector is open, the circuit ground is pulsed on and off to allow for a long on-time without allowing high current flow through the circuit. To measure the pulse width of this type of injector **(Figure 30–16)**, measure from the drop on the left to the beginning of the large voltage spike, which should be at the end of the pulses.

For all types of injectors, the waveform should have a clean, sudden drop in voltage when it is turned on. This drop should be close to 0 volts. Typically, the maximum allowable voltage during the injector's on-time is 600 mV. If the drop is not perfectly vertical, either the injector is shorted or the driver circuit in the PCM is bad. If the voltage does not drop to below 600 mV, there is resistance in the ground circuit or the injector is shorted. When comparing one injector's waveform to another, check the height of the voltage spikes. The voltage spike of all injectors in the same engine should have approximately the same height. If there is a variance, the injector is either shorted or the winding has high resistance. Other causes include the power feed wire to the injector with the variance or the PCM's driver for that injector is faulty.

While checking the injectors with a lab scope, make sure the injectors are firing at the correct time. To do this, use a dual trace scope, and monitor the ignition reference signal and a fuel injector signal at the same time. The two signals should have some sort of rhythm between them. For example, there can be one injector firing for every four ignition reference signals **(Figure 30–17)**. This rhythm is dependent upon the synchronization of the CKP and CMP sensors. Most sequential systems use the CMP signal for injector timing. If the injector's waveform is fine but the timing is wrong, the ignition reference circuit is not allowing the injector to fire at the correct time. This is usually caused by incorrect camshaft timing. If the ignition signal is lost because of a faulty sensor, the injection system will also shut down. If the injector circuit and the ignition reference circuit shuts down at the same time, the cause of the problem is probably the ignition reference sensor. If the injector circuit shuts off before the ignition circuit, the problem is the injector circuit or the PCM.

CURRENT RAMPING When using a DSO or GMM, the ramping current should be compared to the voltage **(Figure 30–18)**. When the injector is grounded, current flows to it. The current trace should show a steady build until the current limiter injector's driver circuit activates. At that point, the current should level out until the injector is turned off. When the circuit is turned off, current should drop sharply. When one looks at the build of current, any unwanted resistance in the injector's coil will show up as a decrease in current flow during its ramp. Although the injector may be firing in spite of the resistance, the injector is bad or will be soon. If the current flow rises too quickly, this indicates a

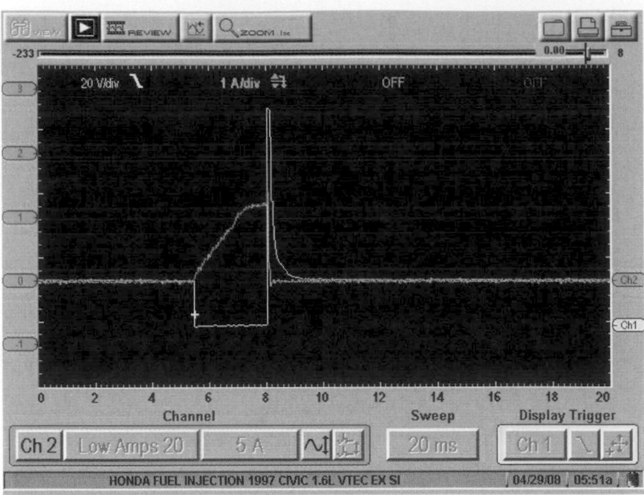

FIGURE 30–18

Comparing injector current ramping (green) with the voltage signals (yellow).

Snap-on Incorporated

shorted injector that is drawing too much current **(Figure 30–19)**.

It is important to understand the stages of voltage at the injector to understand what the current should be doing. Before the injector fires, there is system voltage to it, but no current flow due to the open ground circuit. Once the PCM grounds the injector, current begins to flow. When there is enough current to turn on the injector, the injector opens. At this point, if the voltage on the ground side does not drop close to zero, there could be ground problems or a weak driver circuit. Also, the current may have a rough time building. Immediately after the injector is turned off, the magnetic field in the coil collapses, causing another voltage spike. If the magnetic field in the coil is weak, this secondary spike will not be very high. This can be caused by high resistance or a short in the injector's windings. It can also be caused by low current to the injector. Remember, after the injector is fired, current stops and the voltage should move high.

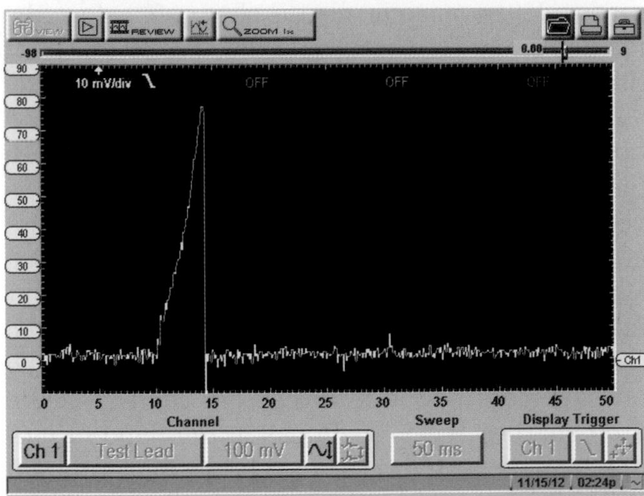

FIGURE 30–19

This injector is shorted; it is drawing too much current.

Snap-on Incorporated

INJECTOR SERVICE

Because a single injector can cost up to several hundred dollars, arbitrarily replacing injectors when they are not functioning properly, especially on multiport systems, can be an expensive proposition. If injectors are electrically defective, replacement is the only alternative. However, if the injector balance test indicates that some injectors are restricted or if the vehicle is exhibiting rough idle, stalling, or slow or uneven acceleration, the injectors may just be dirty and require a good cleaning.

Injector Cleaning

Before discussing the typical cleaning systems available and how they are used, several cleaning precautions are in order. First, never soak an injector in cleaning solvent. Not only is this an ineffective way to clean injectors, but it most likely will destroy the injector in the process. Also, never use a wire brush, pipe cleaner, toothpick, or other cleaning utensil to unblock a plugged injector. The metering holes in injectors are drilled to precise tolerances. Scraping or reaming the opening may result in a clean injector, but it may also be one that is no longer an accurate fuel-metering device.

The basic premise of all injection cleaning systems is similar in that some type of cleaning chemical is run through the injector in an attempt to dissolve deposits that have formed on the injector's tip. The methods of applying the cleaner can range from single-shot, premixed, pressurized spray cans to self-mix, self-pressurized chemical tanks resembling bug sprayers. The premixed, pressurized spray-can systems are fairly simple and straightforward to use as the technician does not need to mix, measure, or otherwise handle the cleaning agent.

Automotive parts stores usually sell pressurized containers of injector cleaner with a hose for Schrader valve attachment. During the cleaning process, the engine runs on the pressurized container of unleaded fuel and injector cleaner. Fuel pump operation must be stopped to prevent the pump from forcing fuel up to the fuel rail. The fuel return line must be plugged to prevent the solution in the cleaning container from flowing through the return line into the fuel tank. Disconnect the wires from the in-tank fuel pump or the fuel pump relay to disable the fuel pump. If you disconnect the fuel pump relay on some General Motors products, the oil-pressure switch in the fuel pump circuit must also be disconnected to prevent current flow through this switch to the fuel pump. Plug the fuel return line from the fuel rail to the tank. Connect a can of injector cleaner to the Schrader valve on the fuel rail, and run the engine for about 20 minutes on the injector solution.

Other systems require the technician to assume the role of chemist and mix up a desired batch of cleaning solution for each application. The chemical solution then is placed in a holding container and pressurized by hand pump or shop air to a specified operating pressure. The injector cleaning solution is poured into a canister on some injector cleaners, and shop air supply is used to pressurize the canister to the specified pressure. The injector cleaning solution contains unleaded fuel mixed with injector cleaner.

The container hose is connected to the Schrader valve on the fuel rail **(Figure 30–20)**. Disable the fuel pump according to the car manufacturer's instructions (for example, pull the fuel pump fuse or disconnect a lead at pump). Clamp off the fuel pump return line at the flex connection to prevent the cleaner from seeping into the fuel tank. Set and connect the cleaning system so that it can circulate the cleaning solution through the fuel rail with the engine off. To do this, adjust the machine's delivery pressure to a pressure higher than normal delivery pressure. To flush the entire fuel rail, including the injector inlet screens and regulator, adjust the machine's delivery pressure higher than the fuel pressure regulator's normal pressure setting.

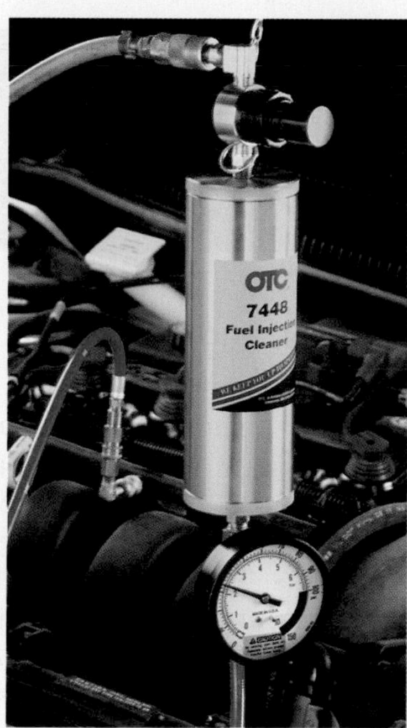

FIGURE 30–20

An injector cleaner connected to a fuel rail.

SPX Corporation, Aftermarket Tool and Equipment Group

Readjust the machine to a pressure slightly lower than the normal regulated pressure, and then open the cleaner's control valve one-half turn or so to prime the injectors, and then start the engine. If available, set and adjust the cleaner's pressure gauge to approximately 35 kPa (5 psi) below the operating pressure of the injection system, and let the engine run at 1000 rpm for 10 to 15 minutes or until the cleaning mix has run out. If the engine stalls during cleaning, simply restart it. Run the engine until the recommended amount of fluid is exhausted and the engine stalls. Shut off the ignition, remove the cleaning setup, and reconnect the fuel pump.

After removing the clamping devices from around the fuel lines, start the car. Let it idle for five minutes or so to remove any leftover cleaner from the fuel lines. In the more severely clogged cases, the idle improvement should be noticeable almost immediately. With more subtle performance improvements, an injector balance test verifies the cleaning results. Once the injectors are clean, recommend the use of an in-tank cleaning additive or a detergent-laced fuel.

The more advanced units feature electrically operated pumps neatly packaged in roll-around cabinets that are quite similar in design to an A/C charging station **(Figure 30–21)**.

After the injectors are cleaned or replaced, rough engine idle may still be present. This problem occurs because the adaptive memory in the computer has learned previously about the restricted injectors. If the injectors were supplying a lean air/fuel ratio, the computer increased the pulse width to try to bring the air/fuel ratio back to stoichiometric. With the cleaned or replaced injectors, the adaptive computer memory is still supplying the increased pulse width. This action makes the air/fuel ratio too rich now that the restricted injector problem does not exist. With the engine at normal operating temperature, drive the vehicle for at least five minutes to allow the adaptive computer memory to learn

FIGURE 30–21

A fuel injector cleaner cart.

about the cleaned or replaced injectors. Afterward, the computer should supply the correct injector pulse width, and the engine should run smoothly. This same problem may occur when any defective computer system component is replaced.

If the fuel delivery system is equipped with a cold-start injector, make sure it gets cleaned along with the primary injectors. This step is especially critical if the owner complains of cold-starting problems.

To effectively clean the cold-start injector, hook up the cleaning system, and remove the cold-start injector from the engine. Open the control valve on the solvent containers, and use an electronic triggering device to manually pulse the injector. Direct the spray into a suitable container. Pulse the injector until the spray pattern looks healthy.

On models where the cold-start injector is not readily accessible, check the fuel flow through the injector before and after the cleaning procedure. Pulse as necessary until the maximum flow through the injector is obtained.

Ultrasonic Cleaning

Another method of injector cleaning is ultrasound. The injectors are removed from the engine and placed in a liquid bath that contains a mild detergent. The liquid is exposed to a high-frequency sound (30 kHz) that creates alternating high- and low-pressure waves. This in turn causes bubbles, which separate the dirt from the surfaces of the injector in a process known as cavitation.

FUEL RAIL, INJECTOR, AND REGULATOR SERVICE

There are service operations that will require removing the fuel injection fuel rail, pressure regulator, and/or injectors. Most of these are not related to fuel system repair. However, when it is necessary to remove and refit them, it is important that it be done carefully and according to the manufacturer's recommended procedures.

Injector Replacement

Photo Sequence 23 outlines a typical procedure for removing and installing an injector. Consult the vehicle's service information for instructions on removing and installing injectors. Before installing the new one, always check to make sure the sealing O-ring is in place. Also, prior to installation, lightly lubricate the sealing ring with engine oil or automatic transmission fluid (avoid using silicone grease, which tends to clog the injectors) to prevent seal distortion or damage.

WARNING!

Cap injector openings in the intake manifold to prevent the entry of dirt and other particles. Also, after the injectors and pressure regulator are removed from the fuel rail, cap all fuel rail openings to keep dirt out of the fuel rail.

Fuel Rail, Injector, and Pressure Regulator Removal

The procedure for removing and replacing the fuel rail, injectors, and pressure regulator varies depending on the vehicle. On some applications, certain components must be removed to gain access to these components. The system must be relieved of any and all pressure before the fuel lines are opened to remove any of the components. Before removing the fuel rail, use compressed air to blow dirt and debris from around where the injectors pass through the intake, and wipe off any dirt from the fuel rail with a shop towel. Then loosen the fuel line clamps on the fuel rail, if so equipped. If these

REMOVING AND REPLACING A FUEL INJECTOR ON A PFI SYSTEM

PHOTO SEQUENCE 23

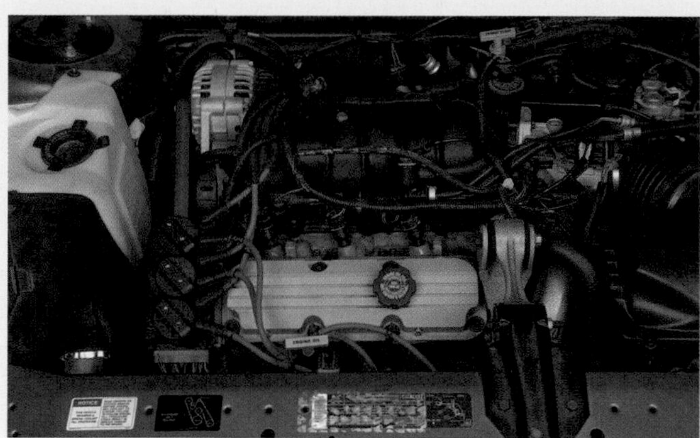

P23–1 Often, an individual injector needs to be replaced. Random disassembly of the components and improper procedures can result in damage to one of the various systems located near the injectors.

P23–2 The injectors are normally attached directly to a fuel rail and inserted into the intake manifold or cylinder head. They must be positively sealed because high-pressure fuel leaks can cause a serious safety hazard.

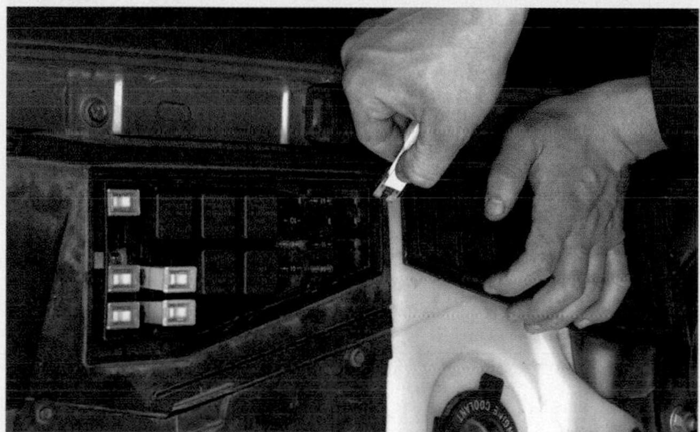

P23–3 Prior to loosening any fitting in the fuel system, the fuel pump fuse should be removed.

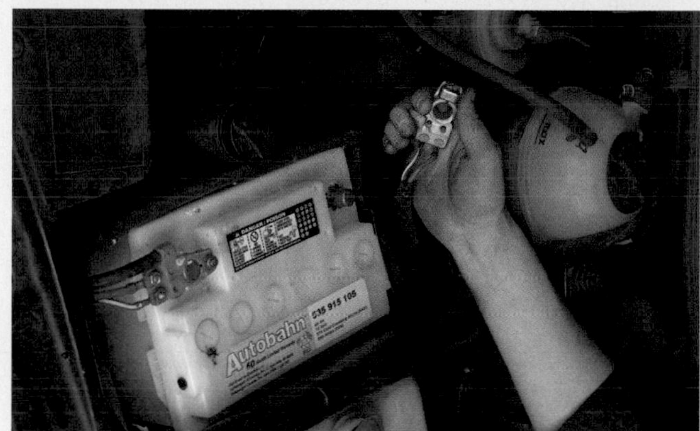

P23–4 As an extra precaution, many technicians disconnect the negative cable at the battery.

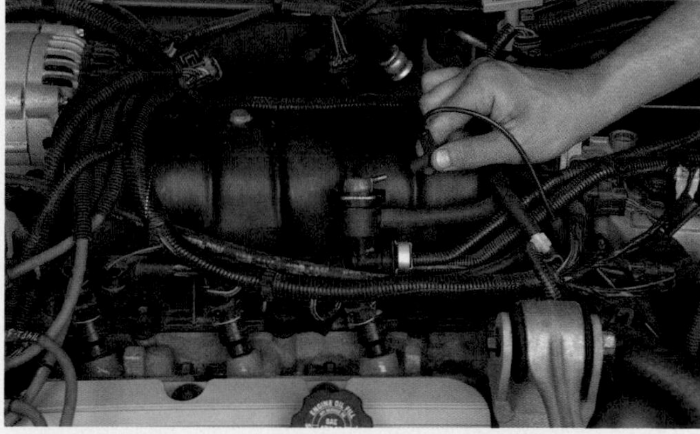

P23–5 To remove an injector, the fuel rail must be able to move away from the engine. The rail-holding brackets should be unbolted and the vacuum line to the pressure regulator disconnected.

P23–6 Disconnect the wiring harness to the injectors by depressing the centre of the attaching wire clip.

REMOVING AND REPLACING A FUEL INJECTOR ON A PFI SYSTEM *(continued)*

P23–7 The injectors are held to the fuel rail by a clip that fits over the top of the injector. O-rings at the top and at the bottom of the injector seal the injector.

P23–8 Pull up on the fuel rail assembly. The bottoms of the injectors will pull out of the manifold while the tops are secured to the rail by clips.

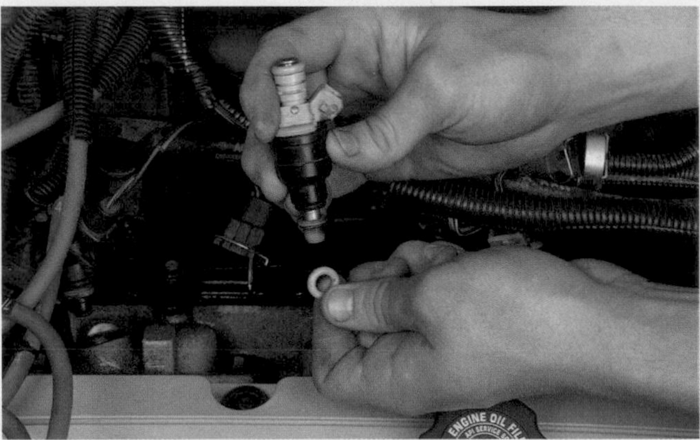

P23–9 Remove the clip from the top of the injector, and remove the injector unit. Install new O-rings onto the new injector. Be careful not to damage the seals while installing them, and make sure they are in their proper locations.

P23–10 Install the injector into the fuel rail, and set the rail assembly into place.

P23–11 Tighten the fuel rail hold-down bolts according to the manufacturer's specifications.

P23–12 Reconnect all parts that were disconnected. Install the fuel pump fuse, and reconnect the battery. Turn the ignition switch to the RUN position, and check the entire system for leaks. After a visual inspection has been completed, conduct a fuel pressure test on the system.

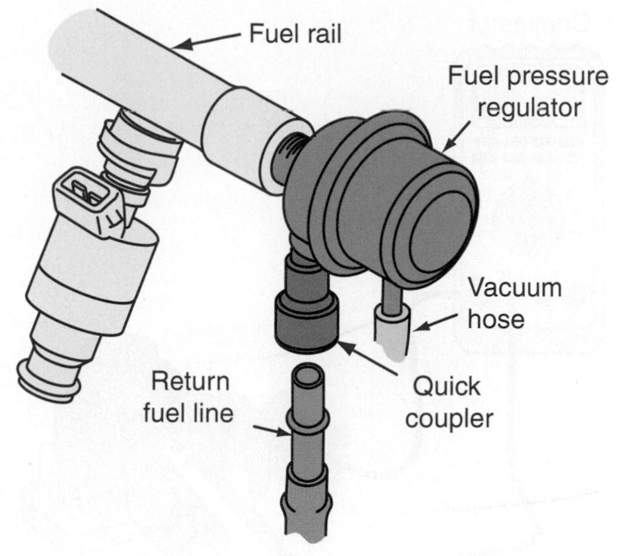

FIGURE 30–22

Fuel supply and return lines connected to the fuel rail.

lines have quick-disconnect fittings, grasp the larger collar on the connector, and twist in either direction while pulling on the line to remove the fuel supply and return lines **(Figure 30–22)**. Now remove the vacuum line from the pressure regulator, and disconnect the electrical connectors from the injectors. The fuel rail is now ready to be removed. On some engines, the fuel rail is held in place by bolts; they need to be removed before pulling the fuel rail free. When pulling the fuel rail away from the engine, pull with equal force on each side of the fuel rail to remove the rail and injectors.

WARNING!

Do not use compressed air to flush or clean the fuel rail. Compressed air contains water, which may contaminate the fuel rail.

Prior to removing the injectors and pressure regulator, the fuel rail should be cleaned with a spray-type engine cleaner. Normally, the approved cleaners are listed in the service information. After the rail is cleaned, the injectors can be pulled from the fuel rail **(Figure 30–23)**. Use snap-ring pliers to remove the snap ring from the pressure regulator cavity. Note the original direction of the vacuum fitting on the pressure regulator, and pull the pressure regulator from the fuel rail. Clean all components with a clean shop towel. Be careful not to damage fuel rail openings and injector tips. Check all injector and

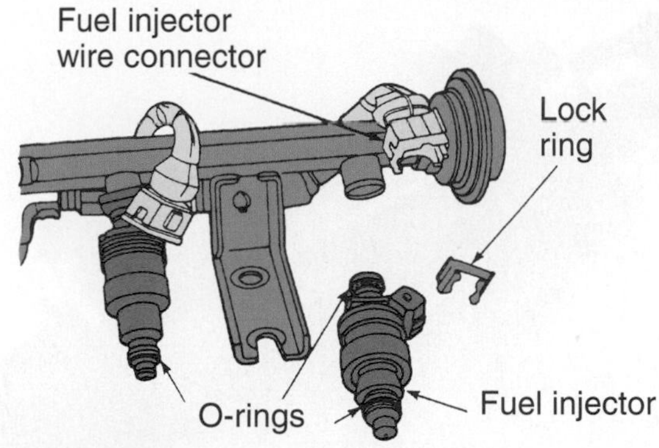

FIGURE 30–23

To remove the injectors from the fuel rail, remove the lock ring and pull the injector out.

pressure regulator openings in the fuel rail for metal burrs and damage.

CAUTION!

Do not immerse the fuel rail, injectors, or pressure regulator in any type of cleaning solvent. This action may damage and contaminate these components.

When reassembling the fuel rail with the injectors and pressure regulator, make sure all O-rings are replaced and lightly coated with engine oil. Assemble the fuel rail in the reverse order as that used for disassembly. After the rail and injectors are in place and everything connected to them, reconnect the negative battery terminal, and disconnect the 12-volt power supply from the cigarette lighter. Then start the engine, check for fuel leaks at the rail, and be sure the engine operation is normal.

Special GDI Checks

The injectors in a GDI system **(Figure 30–24)** are best checked with an ohmmeter. Resistance checks can identify whether the injector has an open or a short. If an injector does not have the specified resistance, it should be replaced. Of course, there are also designated DTCs that may lead you to suspect a problem with an injector. Those DTCs are related to particular injectors, unless the malfunction is something that affects many of them.

Because GDI operates under very high pressures, a typical fuel volume check on the high-pressure pump should not be done. However, fuel pressure and volume tests can be performed on the supply pump. The fuel pressure of the high-pressure pump is tested with a scan tool. If the pressure

FIGURE 30–24

A GDI injector.

displayed on the scan tool is not within the specified range, first confirm the fuel pressure sensor is operating correctly. If the fuel pressure sensor is reading correctly, check the high-pressure pump. Some engines have experienced increased wear on the pump drive, which reduces fuel pressure. The fuel injector **(Figure 30–25)** and pump control can be checked with an ohmmeter **(Figure 30–26)**. Connect the meter across the terminals at the connector. Compare your readings to specifications. If the reading does not match specs, replace the injector or pump. The fuel pressure sensor can also be checked with a voltmeter. With the engine running

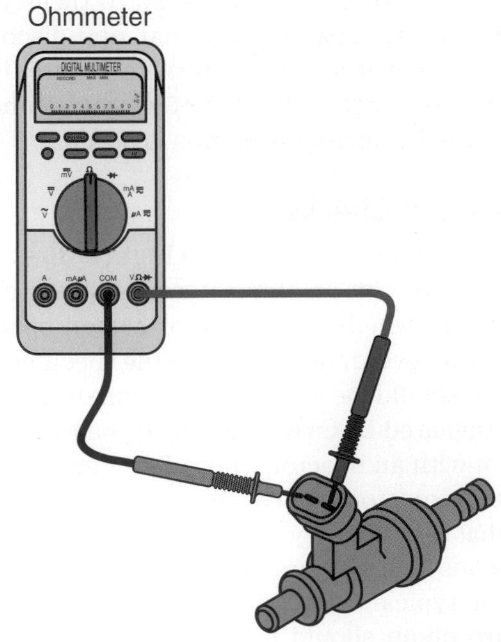

FIGURE 30–25

A GDI injector can be checked with an ohmmeter.

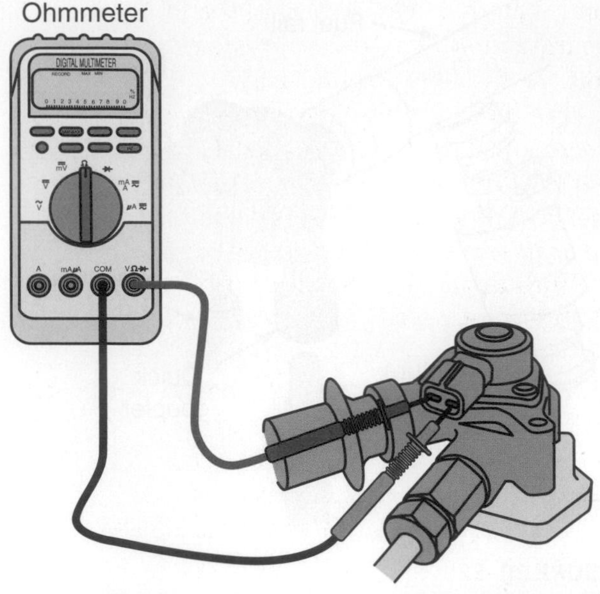

FIGURE 30–26

The control unit for the high-pressure fuel pump in a GDI system can be checked with an ohmmeter.

and a fuel pressure gauge connected, back probe the sensor's output signal, and compare the pressure gauge reading and voltage to specifications.

ELECTRONIC THROTTLE CONTROLS

As explained in Chapter 29, an electronic throttle control (ETC) system **(Figure 30–27)** normally includes a throttle actuator or motor, TP sensors, accelerator pedal position (APP) sensors, an electronic throttle control module, and a relay. The

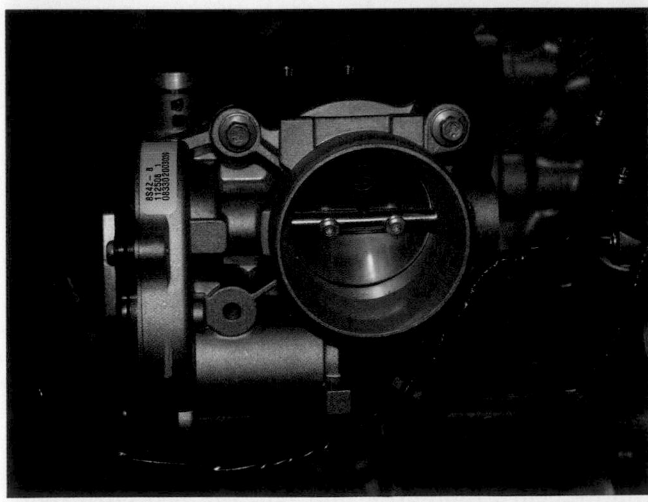

FIGURE 30–27

An electronically controlled throttle assembly.

control module may be part of the PCM or may be a separate unit. The actuator responds to commands from the PCM **(Figure 30–28)**.

The ETC system relies on redundant inputs and processors. Two accelerator pedal position signals are sent to the PCM, each having a different voltage range. The PCM processes these signals along with other inputs to set the throttle plate at the required position. The throttle plate is controlled by a DC motor that is controlled by the PCM. The PCM has two driver circuits for the motor; one causes the plate to open, and the other closes it. The exact position of the throttle plate is monitored by two TP sensors; the signal range of these also differ. It also has redundant monitors that track the system's effectiveness. When the PCM detects a fault, the monitor will set a DTC and/or put the system into a limp-in mode **(Figure 30–29)**.

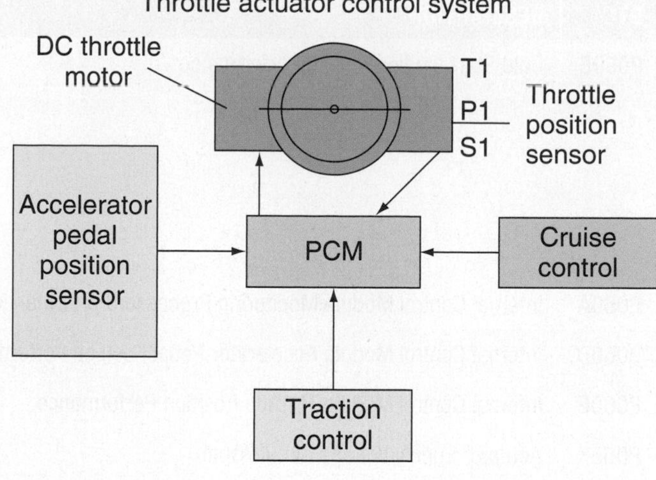

FIGURE 30–28

A simplified look at an electronic throttle control system.

DTC	DESCRIPTION	PROBABLE CAUSES	MIL
P0120	Throttle/Pedal Position Sensor/Switch "A" Circuit Malfunction	1. TP Sensor 2. ECM	ON
P0121	Throttle/Pedal Position Sensor/Switch "A" Circuit Range/Performance Problem	1. TP Sensor	ON
P0122	Throttle/Pedal Position Sensor/Switch "A" Circuit Low Input	1. TP Sensor 2. Short in APP circuit 3. Open in TP circuit 4. ECM	ON
P0123	Throttle/Pedal Position Sensor/Switch "A" Circuit High Input	1. TP Sensor 2. Open in APP circuit 3. Open in ETC circuit 4. Short between TP and APP circuits 5. ECM	ON
P0220	Throttle/Pedal Position Sensor/Switch "B" Circuit	1. TP Sensor 2. ECM	ON
P0222	Throttle/Pedal Position Sensor/Switch "B" Circuit Low Input	1. TP Sensor	ON
P0223	Throttle/Pedal Position Sensor/Switch "B" Circuit High Input	1. TP Sensor	ON
P0505	Idle Control System Malfunction	1. ETC 2. Air induction system 3. PCV hose connection 4. ECM	ON
P050A	Cold Start Idle Air Control System Performance	1. Throttle body assembly 2. MAF sensor 3. Air induction system 4. PCV hose connections 5. VVT system 6. Air cleaner element 7. ECM	ON

FIGURE 30–29

Some examples of generic DTCs related to the electronic throttle control system. *(continued)*

DTC	DESCRIPTION	PROBABLE CAUSES	MIL
P050B	Cold Start Ignition Timing Performance	1. Throttle body assembly 2. MAF sensor 3. Intake system 4. PCV hose connections 5. VVT system 6. Air filter element 7. ECM	ON
P060A	Internal Control Module Monitoring Processor Performance	ECM	ON
P060D	Internal Control Module Accelerator Pedal Position Performance	ECM	ON
P060E	Internal Control Module Throttle Position Performance	ECM	ON
P0657	Actuator Supply Voltage Circuit/Open	ECM	ON

FIGURE 30–29 ***(continued)***

Diagnostic Monitors

The PCM monitors the voltage levels of the APP sensors, TP sensors, and throttle motor circuit. It also monitors the return rate of the two return springs. The APP sensors are potentiometers mounted at the accelerator pedal. Depending on the vehicle, there can be two separate APP sensors or one sensor capable of sending two separate signals. One signal is used to define the position of the pedal. The signal varies between 0 and 5 volts, according to the pedal's position. The other signal is used to monitor the sensor itself. The PCM compares the two signals, and if there is a difference in the reported pedal position, the PCM determines that the sensor is bad and will illuminate the MIL and set a DTC.

The TP sensor (there may be two separate sensors or one capable of providing two separate signals) is mounted to the throttle body. This sensor sends a signal noting the position of the throttle. The PCM uses one of the signals as feedback. This is done by adding the TP voltage to the APP voltage. This sum must equal 5 volts. If the sum is greater or less than this, the PCM will illuminate the MIL and set a DTC.

The PCM also looks at the voltages to the control unit and the actuator assembly, in addition to the amount of current required to move the throttle plate. If it detects lower than normal voltage to the control unit or actuator, it will disable the throttle control system and set it into the limp-in mode. It will also illuminate the MIL and set the DTC. The PCM continuously monitors the current flow through the actuator. If the current is too high or low, the PCM recognizes this as a problem. The PCM also looks at the position of the throttle and the current flowing to it. If the position does not change when it is commanded to do so, the MIL will be illuminated and a DTC set.

FAIL-SAFE MODE When DTCs relating to the ETC system are set, the PCM will enter into a fail-safe or limp-in mode. During this time, the actuator is no longer controlled by the PCM. However, the throttle plate is held slightly open by a spring. This limited throttle opening restricts the ability of the engine to provide power. If the accelerator pedal is depressed, the PCM will control engine output by controlling the fuel injection and ignition systems. This will allow the vehicle to be driven slowly.

Idle Speed

Although there is no adjustment for idle speed, idle speed checks can give an indication of the condition of the electronic throttle system. The best way to monitor idle speed is with a scan tool. Before checking the speed, make sure that the MIL is not illuminated and that there are no DTCs set. Also make sure that the ignition, air induction, and PCV systems are okay.

Set the parking brake, and disconnect the connector to the evaporative emission (EVAP) canister purge valve. Connect the scan tool to the DLC, and make sure there is communication between the tool and the vehicle. Start the engine and make sure all accessories are turned off. Increase the speed to about 3000 rpm, and hold it there until the cooling fan turns on. Allow the engine to idle; check the idle speed and compare it to specs. Then turn on some heavy load accessories, such as the high-beam headlights and max air conditioning with the blower on its HIGH setting. Check the idle speed. If the idle speed does not match specifications, conduct the idle learn procedure (see below). If the speed is still wrong, further diagnostics of the system and fuel injection system are necessary. Once the idle speed matches the specifications, reconnect the EVAP canister purge valve.

IDLE LEARN All throttle positions reported to the PCM reflect a change from a base reading. It is extremely important that the PCM knows that base. Manufacturers have prescribed procedures for teaching the PCM this base. Those procedures must be followed exactly as stated. The learn or relearn procedure should be completed any time the PCM has been replaced or updated, and after the throttle body has been cleaned or replaced. Often, no-DTC idle speed problems are solved by completing the idle learn or relearn procedure. It is important to note that the system does not need to relearn if the battery has been disconnected. The system will automatically go through the procedure when the engine is restarted.

Idle learn is normally done by running the engine at idle for a prescribed time after it has reached normal operating temperature and the PCM has been reset by a scan tool. The idle learn process can be monitored and verified with a scan tool.

General Diagnostics

There are several DTCs assigned to problems with the ETC system. All of the major components have several DTCs assigned to them. There are also pinpoint tests for each component. Most of the components can be checked with a voltmeter, ohmmeter, or scope. The parts should be inspected prior to testing. This is especially true of the throttle body assembly. It should be checked for any buildup of dirt on the plate and in the bores. If needed, clean the assembly. Also make sure that the plate moves freely in the bore.

Ohmmeter checks are made across specific terminals at the throttle body. On most units, the connection is made across the positive (M+) feed to the motor and the negative (M–) terminals. This setup measures the resistance of the windings in the actuator's motor. Some throttle bodies have the TP sensor built into the assembly. In these cases, the TP sensor is checked at different terminals in the same connector. Always refer to the service information to identify the test terminals and resistance specifications.

A voltmeter can be used to check the operation of the TP and APP sensors. This is done in the same way as when checking typical sensors, except that with ETC systems there are two output signals to monitor. The sensors can also be checked with a lab scope.

IDLE SPEED CHECKS

The idle speed of fuel-injected engines without an electronic throttle is regulated by controlling the amount of air that bypasses the airflow sensor or throttle plates. When one of these vehicles has an idling problem, check the linkage and vacuum lines before going any further with your diagnosis. Although idle speed is not typically adjustable, some engines do have provisions for setting the speed. Always refer to the decal in the engine compartment to identify the conditions that must be present before adjusting the speed. Also refer to the service information before making any adjustments. The idle speed and quality for most engines are controlled by the PCM through the IAC.

IAC Checks

If the idle speed is not within specifications, the system should be looked at with a scan tool. Make sure there is proper communications between the scan tool and the vehicle. Also make sure that all CAN communications are good.

Many different inputs will affect the performance of the IAC as well as cause other driveability problems. Following are examples:

- With higher than normal TP signals at idle, the PCM interprets this as the throttle being opened and will close the IAC to decrease idle speed.
- If the resistance of the ECT is too high, a higher than normal voltage signal is sent to the PCM, which interprets this as a cold engine. The PCM will then open the IAC to increase speed.
- If the resistance of the ECT is too low, a lower than normal voltage signal is sent to the PCM, which interprets this as a hot engine. The PCM will then open the IAC to increase speed.
- A stuck closed A/C switch will signal to the PCM that the A/C is always on, and the PCM will always order an increased idle speed.
- If battery voltage is low, the PCM may command higher than normal idle speed to increase charging system output.

SHOP TALK

On most vehicles, the scan tool will indicate the status of input switches as closed or open or high or low. Most input switches provide a high-voltage signal to the PCM when they are open and a low-voltage signal if they are closed.

If the sensors and circuits seem to be fine, check the IAC. The IAC can be monitored with a scan tool. Most will have a test mode that allows for manual control of the IAC. To test the IAC motor, use the scan tool to command the IAC motor to open and close while watching engine speed. As the IAC opens, engine speed should increase, and as the IAC closes engine speed should decrease. If the scan tool data shows the IAC commanded to move, but the engine speed does not change, suspect a faulty IAC motor. Remember that the IAC counts PID is a command by the PCM. Just because the IAC is commanded does not mean the motor actually performed the command.

Some vehicles have an active IAC test mode. When the IAC is operated in its low range, the PCM will move the IAC in approximately 16 steps **(Figure 30–30)**. When operated in its high range, there should be approximately 112 steps or counts **(Figure 30–31)**. The normal range of counts is given in the appropriate service information.

Most technicians look for consistent idle counts on a warm engine to verify that the IAC is not sticking or malfunctioning. During the check, devices such as the A/C system can be operated and the scan tool watched. If the counts and idle speed change when the A/C is turned on and off, the IAC,

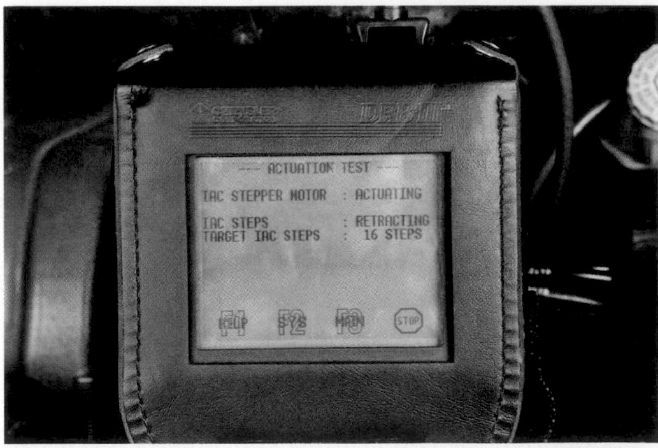

FIGURE 30–30

When the IAC is operated in its low range, the PCM will move the IAC in approximately 16 steps.

FIGURE 30–31

When operated in its high range, the IAC should move approximately 112 steps or counts.

connecting wires, and PCM are working fine. If the counts change but engine speed does not, or if the counts do not change, further diagnosis is required. Always follow the diagnostic procedures given by the manufacturer.

SHOP TALK

If the IAC motor counts are below 12 on the scan tool, look for a vacuum leak. The PCM will run the valve all the way down to compensate for the leak. If no leak is found, then check the circuit between the PCM and the motor. The circuit is likely open between the PCM and the motor. Wiggle the wires on the IAC motor, and observe the reading on the scan tool. If the count reading changes while wiggling the wires, you have found the problem.

IAC problems on OBD-II vehicles will set one or more DTCs. Although there are DTCs designated for the IAC system, keep in mind that the operation of the entire engine performance system can affect idle speed and quality. Therefore, all DTCs could be the cause of an idle problem.

It is important to note that the PCM cannot effectively control idle if it does not know when the throttle is closed. This is also important for all other engine speeds. The PCM sets the closed throttle reference according to the lowest TP voltage signal since the engine was started. The PCM does this each time the engine is started. This reference voltage is called the TPREL PID on most systems. When looking at this PID with the throttle closed, the value should be C/T (closed throttle). If any other value is noted, the engine will have a higher than normal idle speed. Abnormal readings are typically caused by a bad TP sensor, loose or worn throttle plates, or excessive noise in the TP or related circuits.

Most late-model vehicles have a feature that adjusts the calibration of the IAC according to the wear of system components; this is called idle air trim. The system constantly monitors the engine and its systems and determines the ideal idle speed. The idle speed is based on look-up tables. If the corrections exceed predetermined levels, the PCM will set a DTC. Once the problem is solved and corrected, the system must relearn its base idle trim values. This process is completed with a scan tool and should be done whenever a part of the IAC system is replaced or a repair is made to something that affects idle speed.

Servicing the IAC Motor

On some vehicles, there is a provision to manually move the IAC plunger through the scan tool. When this test mode is selected, the PCM is ordered to extend and retract the IAC motor plunger every 2.8 seconds. When this plunger extends and retracts properly, the motor, connecting wires, and PCM are in normal condition. If the plunger does not extend and retract, further diagnosis is necessary to locate the cause of the problem. On other vehicles, this same test can be conducted with a jumper wire connecting two terminals of the DLC together.

Carbon deposits in the IAC motor air passage in the throttle body or on the IAC motor's pintle can cause erratic idling and engine stalling. Remove the motor from the throttle body, and inspect the throttle body air passage for carbon deposits. If there are heavy carbon deposits, remove the complete throttle body for cleaning. Clean the IAC air passage, motor sealing surface, pintle valve, and pintle seat with throttle body cleaner.

WARNING!

Be careful while cleaning the assembly; the IAC motor can be damaged if throttle body cleaner is allowed to get inside the motor.

The motor can be checked with an ohmmeter. If the ohmmeter readings are not within specifications, replace the motor. If a new IAC motor is installed, make sure that the part number, pintle shape, and diameter are the same as those on the original motor. Measure the distance from the end of the pintle to the shoulder of the motor casting **(Figure 30–32)**. Move the pintle until it is at the specified distance. If the pintle is extended too far, the motor can be damaged during installation.

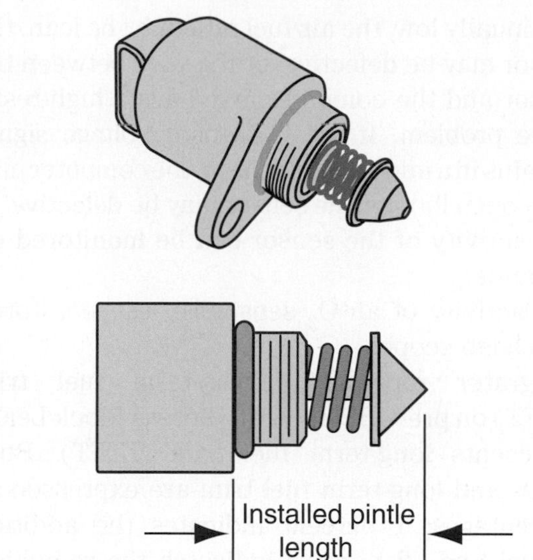

FIGURE 30–32

Measure the distance the pintle of an IAC motor that is extended before installing the motor.

Install a new gasket or O-ring on the motor. If the motor is sealed with an O-ring, lubricate the ring with transmission fluid. If the motor is threaded into the throttle body, tighten the motor to the specified torque. When the motor is bolted to the throttle body, tighten the mounting bolts to the specified torque.

KEY TERMS

Block Learn (p. 879)
Integrator (p. 879)
Long-term fuel trim (LTFT) (p. 879)
Malfunction indicator lamp (MIL) (p. 877)
Noid light (p. 886)
Peak and hold injector (p. 888)
Pulse width (p. 887)
Pulse-width-modulated injector (p. 889)
Short-term fuel trim (STFT) (p. 879)

SUMMARY

- Always relieve the fuel pressure before disconnecting any component in the fuel system.
- Always turn off the ignition switch before connecting or disconnecting any system component or test equipment.
- The signal from most O_2 sensors varies between 0 and 1 volt. If the voltage is continually high, the air/fuel ratio may be rich or the sensor may be contaminated. When the O_2 sensor voltage is

continually low, the air/fuel ratio may be lean, the sensor may be defective, or the wire between the sensor and the computer may have a high-resistance problem. If the O_2 sensor voltage signal remains in a mid-range position, the computer may be in open loop or the sensor may be defective.

- The activity of the sensor can be monitored on a scanner.
- The activity of an O_2 sensor is best monitored with a lab scope.
- Integrator represents short-term fuel trim (STFT) on pre-OBD-II cars, whereas Block Learn represents long-term fuel trim (LTFT). Both short- and long-term fuel trim are expressed as percentages; 1 percent indicates the addition of fuel and –2 percent indicates the reduction of fuel.
- An injector that does not open causes hard starts on port-type systems and an obvious no-start on single-point TBI designs.
- An injector that is stuck partially open causes loss of fuel pressure and flooding due to raw fuel dribbling into the engine. In addition to a rich-running engine, a leaking injector also causes the engine to diesel or run on when the ignition is turned off.
- Buildups of gum and other deposits on the tip of an injector can reduce the amount of fuel sprayed by the injector or they can prevent the injector from totally sealing, allowing it to leak.
- Another cause of a rich mixture is a leaking fuel pressure regulator. If the diaphragm of the regulator is ruptured, fuel will move into the intake manifold through the diaphragm, causing a rich mixture.
- When an injector is suspected as the cause of a lean problem, the first step is to determine if the injector is receiving a signal to fire. Check for voltage at the injector using a high impedance test light or a convenient noid light that plugs into the connector.
- An injector pressure balance test will help isolate a clogged or dirty injector.
- An oscilloscope can be used to monitor the injector's pulse width and duty cycle when an injector-related problem is suspected.
- The pulse width is the time in milliseconds that the injector is energized. The duty cycle is the percentage of on-time to total cycle time.
- For all types of injectors, the waveform on the scope should have a clean, sudden drop in voltage when it is turned on.
- Never soak an injector in cleaning solvent or use a wire brush, pipe cleaner, toothpick, or other cleaning utensil to unblock a plugged injector.
- In a fuel injection system, idle speed is regulated by controlling the amount of air that is allowed to bypass the airflow sensor or throttle plates. When a car tends to stall or idles too fast, look for obvious problems such as binding linkage and vacuum leaks first. If no problems are found, go through the minimum idle checking/setting procedure described on the underhood decal.
- If the idle speed is not within specifications, the input sensors and switches should be checked carefully with the scan tester.
- If the engine coolant temperature sensor's resistance is higher than normal, it sends a higher than normal voltage signal to the PCM. The PCM thinks the coolant is colder than it actually is, and under this condition, the PCM operates the IAC or ISC motor to increase idle speed.

REVIEW QUESTIONS

1. List the three things that must occur for an EFI to operate properly.
2. What is indicated by trouble codes?
3. What is the correct procedure for checking an oxygen sensor with a DMM?
4. What is the difference between STFT and LTFT?
5. What is indicated by a Block Learn fuel trim number that is consistently below 100?
6. Which of the following is a likely cause of a no-start condition, if the engine starts when the electrical connector to the MAF is disconnected?
   a. a defective oxygen sensor
   b. a defective PCM
   c. a defective MAP sensor
   d. a defective MAF sensor
7. Which of the following could set off an O_2 sensor trouble code?
   a. a faulty O_2 sensor
   b. damaged O_2 sensor wires
   c. a melted O_2 sensor connector
   d. any item in the O_2 sensor circuit
8. Which of the following would likely cause a rough engine idle after injectors are cleaned or replaced?
   a. a faulty O_2 sensor
   b. a faulty cold start injector
   c. the computer's adaptive memory
   d. a faulty throttle position sensor

9. How should an ohmmeter be connected to an injector for testing?
   a. positive lead to the positive injector terminal and the negative lead to an engine ground
   b. positive lead to the negative injector terminal and the negative lead to an engine ground
   c. connecting the meter leads to both injector terminals
   d. positive lead to an engine ground and the negative lead to the negative injector terminal
10. Which of the following injector balance test statements is correct?
    a. A low-pressure drop on one injector indicates a restricted injector.
    b. A high-pressure drop on one injector indicates a restricted injector.
    c. The vehicle should be running.
    d. The fuel pump must run continuously.
11. What is the duty cycle of an injector?
    a. the percentage of on-time to total cycle time
    b. the percentage of off-time to total cycle time
    c. the time in milliseconds that the injector is on
    d. the time in milliseconds that the injector is off
12. What should be used on the injector sealing O-ring when installing an injector?
    a. silicone grease
    b. engine oil
    c. silicone sealant
    d. gasket cement
13. How can a zirconium-dioxide oxygen sensor be checked with a multimeter?
    a. Use the ohmmeter between the sensor lead and ground.
    b. Use the voltmeter between the sensor lead and ground.
    c. Use the ammeter between the sensor lead and ground.
    d. Use the ohmmeter between the sensor lead and sensor body.
14. What temperature must an O_2 sensor reach to function in closed loop mode?
    a. 65.6°C (150°F) b. 232.2°C (450°)
    c. 148.9°C (300°F) d. 315.6°C (600°F)
15. Why is an injector balance test performed?
    a. to verify injector delivery volume
    b. to check injector winding resistance
    c. to verify injector source voltage
    d. to check PCM driver circuits
16. What is the correct method of cleaning an injector?
    a. Soak the injector in a cleaning solution.
    b. Run a cleaning chemical through the injector.
    c. Use a pipe cleaner in the nozzle orifices.
    d. Use a wire brush to clean the nozzle tip.
17. Which of the following zirconium-dioxide oxygen sensor output statements is correct?
    a. A steady 0.43 volt indicates closed loop operation.
    b. A steady 0.42 volt indicates a lean mixture.
    c. A steady 0.42 volt indicates a rich mixture.
    d. A varying voltage of 0.2 to 0.8 volt indicates normal operation.
18. How many wires does a heated oxygen sensor usually require?
    a. one b. three
    c. two d. four
19. Which short- and long-term fuel trim reading indicates a correction for a lean condition?
    a. 0 (zero) b. –5%
    c. –10% d. +10%
20. Which short- and long-term fuel trim reading indicates a correction for a rich condition?
    a. 0 (zero) b. –5%
    c. –10% d. +10%

CHAPTER 31
Engine Performance Systems

LEARNING OUTCOMES

- State the purpose of the major engine performance systems/components.
- Explain what is meant by open loop and closed loop.
- Explain the reasons for OBD-II.
- Explain the requirements to illuminate the malfunction indicator light in an OBD-II system.
- Briefly describe the monitored systems in an OBD-II system.
- Describe an OBD-II warm-up cycle.
- Explain trip and drive cycle in an OBD-II system.
- Describe how engine misfire is detected in an OBD-II system.
- Describe the differences between an A misfire and a B misfire.
- Describe the purpose of having two oxygen sensors in an exhaust system.
- Briefly describe what the comprehensive component monitor looks at.
- Retrieve and record stored diagnostic trouble codes, and clear codes.
- Diagnose the causes of emissions or driveability concerns resulting from malfunctions in the computerized engine control system with stored diagnostic trouble codes.
- Diagnose emissions or driveability concerns resulting from malfunctions in the computerized engine control system with no stored diagnostic trouble codes, and determine necessary action.
- Obtain and interpret scan tool data.

How well an engine runs depends on the combustion process. To ensure that combustion takes place efficiently, various engine performance systems are used. Today's systems are designed to achieve as close to complete combustion as possible. Basically, if all of the fuel that enters an engine's cylinder is burned, combustion is complete.

The requirements for complete combustion are simple. However, achieving these requirements is not. Complete combustion will occur when the correct amount of air is mixed, in a sealed container, with the correct amount of fuel. The mixture is compressed and shocked by the correct amount of heat at the correct time. Air entering the cylinder via the intake valve is mixed with fuel. This mixture is compressed, which greatly increases the amount of energy released during combustion. Once compressed, the mixture is ignited and burns. The spent exhaust gases then flow from the cylinder through the exhaust valve. The amount of air and fuel must be precisely controlled, as should the spark for ignition. Because the engine runs at different speeds, loads, and temperatures, these requirements are very difficult to meet.

Emission control devices are added to the vehicle because complete combustion at all times has not been achieved. These devices reduce the amount of unwanted vehicle emissions. They also affect the operation of the engine and are therefore an engine performance system.

IGNITION SYSTEMS

For complete combustion, the ignition system must supply properly timed, high-voltage surges across the spark plugs' electrodes **(Figure 31–1)** at the proper time under all engine operating conditions. This is quite a task: Consider a six-cylinder engine running

FIGURE 31–1

An ignition system has the sole purpose of providing the spark to start combustion.

Ted Kinsman/Science Source

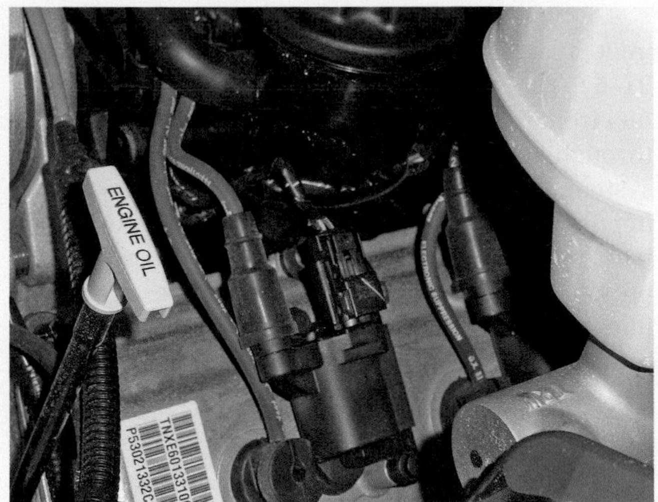

FIGURE 31–2

An ignition coil for an electronic (distributorless) ignition system.

at 4000 rpm; the ignition system must supply 12 000 sparks per minute because it must fire three spark plugs per revolution. These plug firings must also occur at the correct time and generate the correct amount of heat. If the ignition system fails to do these things, fuel economy, engine performance, and emission levels will be adversely affected. There are basically two types of ignition systems: **distributor ignition (DI)** and **electronic ignition (EI) (Figure 31–2)** or distributorless ignition systems (DIS).

Purpose of the Ignition System

For each cylinder, the ignition system has three primary jobs:

- It must generate an electrical spark with enough heat to ignite the air/fuel mixture in the combustion chamber.
- It must maintain that spark long enough to allow total combustion of the fuel in the chamber.
- It must deliver the spark to each cylinder to allow combustion to begin at the right time during the compression stroke.

Ignition Timing

Ignition timing refers to the precise time that spark occurs and is specified by referring to the position of the number 1 piston in relation to crankshaft rotation. Engines with distributor ignition systems have ignition timing reference marks located on a pulley or flywheel to indicate the position of the number 1 piston at TDC on the compression stroke. This reference is used to set initial ignition timing. For those engines, the manufacturers typically list an initial or **base ignition timing** specification. The specification, such as 10 BTDC, indicates that the spark occurs at 10° before top dead centre of the compression stroke.

Timing may be advanced, meaning that the process to generate the spark occurs sooner as engine speed increases **(Figure 31–3)**. This is so there is

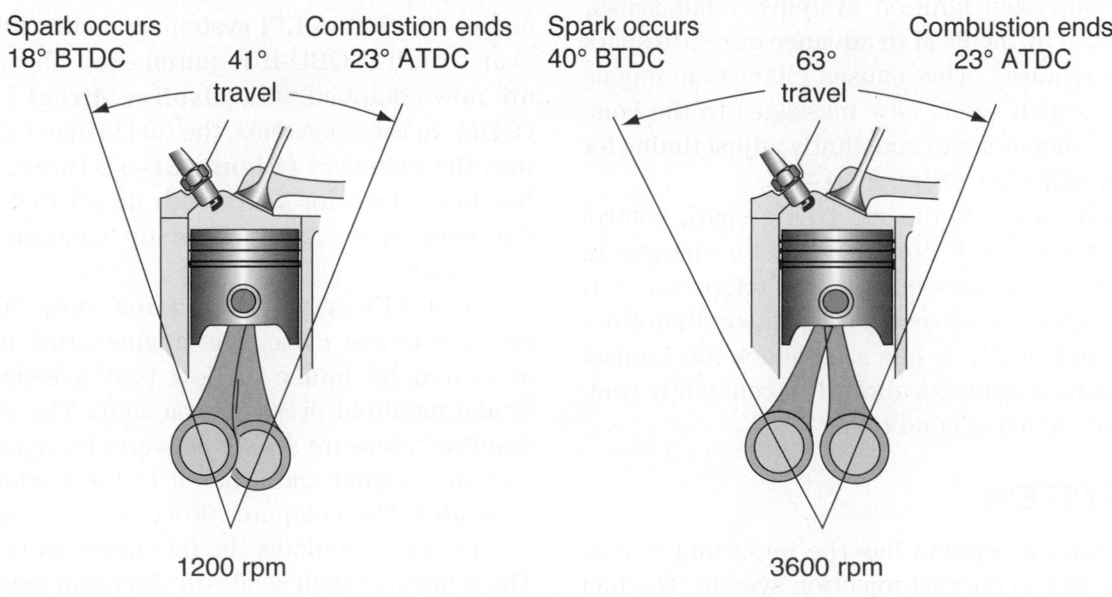

FIGURE 31–3

With an increase in speed, ignition must begin earlier to end by 23° ATDC.

sufficient time for the spark to be produced in the coil and delivered to the combustion chamber. Timing also may be retarded, meaning the spark is delivered later.

Firing Order

Each cylinder must produce power once every 720° of crankshaft rotation. Therefore, the ignition system must provide a spark at the right time so that each cylinder can have a power stroke at its own appropriate time. To do this, the ignition system must monitor the rotation of the crankshaft and the relative position of each piston to determine which piston should be delivered the spark. The spark for all cylinders must be delivered at the right time. How the ignition system does this depends on the design of the system.

The firing order is the sequence in which each spark plug ignites the air/fuel mixture in the cylinders. This order cannot be altered without adversely affecting engine performance. Each engine has a mechanical sequence of valves opening and closing as the pistons move up and down. The sequence defines when each cylinder produces power. The firing order corresponds to each cylinder as it moves through the compression stroke. If the spark occurs at any time other than during the end of the compression stroke, performance will be affected.

With distributor system, the firing order follows the placement of the ignition wires along the distributor cap. EI systems trigger the ignition coils to fire in order.

Computer-Controlled Systems

With computerized ignition systems, input sensor data are used by the PCM to advance or retard spark timing as required. This causes changes in engine operation, which sends new messages to the computer. The computer can constantly adjust timing for maximum efficiency **(Figure 31–4)**.

The advantage of the electronic spark control system is threefold. It compensates for changes in engine (and sometimes outside air) temperature. It makes changes at a rate many times faster than older systems. And finally, it has a feedback mechanism in which sensor readings allow it to constantly compensate for changing conditions.

FUEL SYSTEM

The fuel delivery system has the important role of delivering fuel to the fuel injection system. The fuel must also be delivered in the right quantities and at the right pressure. The fuel must also be clean when it is delivered.

A typical fuel delivery system includes a fuel tank, fuel lines, fuel filters, and a pump **(Figure 31–5)**. The system works by using a pump to draw fuel from the fuel tank and passing it under pressure through fuel lines and filters to the fuel injection system. The filter removes dirt and other harmful impurities from the fuel. A fuel line pressure regulator maintains a constant high fuel pressure. This pressure generates the spraying force needed to inject the fuel. Excess fuel not required by the engine returns to the fuel tank through a fuel return line.

Fuel Injection

Electronic fuel injection (EFI) has proven to be the most precise, reliable, and cost-effective method of delivering fuel to the combustion chambers of today's engines. EFI systems are computer controlled and designed to provide the correct air/fuel ratio for all engine loads, speeds, and temperature conditions.

Although fuel injection technology has been around since the 1920s, it was not until the 1980s that manufacturers began to replace carburetors with **electronic fuel injection (EFI)** systems. Many of the early EFI systems were **throttle body injection (TBI)** systems in which the fuel was injected above the throttle plates. A similar system, **central port injection (CPI)**, has the injector assembly located in the lower half of the intake manifold. TBI systems have been replaced by **port fuel injection (PFI)**, which has injectors located in the intake ports of the cylinders. All new cars have been equipped with an EFI system since the 1995 model year to fulfill OBD-II requirements. Many engines are now equipped with **gasoline direct injection (GDI)**. In these systems, the fuel is injected directly into the cylinders **(Figure 31–6)**. Direct injection has been used for years with diesel fuels but has not been successfully used on gasoline engines until lately.

Most EFI systems inject fuel only during the engine's intake cycle. The engine's fuel needs are measured by intake airflow past a sensor or by intake manifold pressure (vacuum). The airflow or manifold pressure sensor converts its reading to an electrical signal and sends it to the engine control computer. The computer processes this signal (and others) and calculates the fuel needs of the engine. The computer then sends an electrical signal to the fuel injector or injectors. This signal determines the amount of time the injector opens and sprays fuel. This interval is known as the injector pulse width.

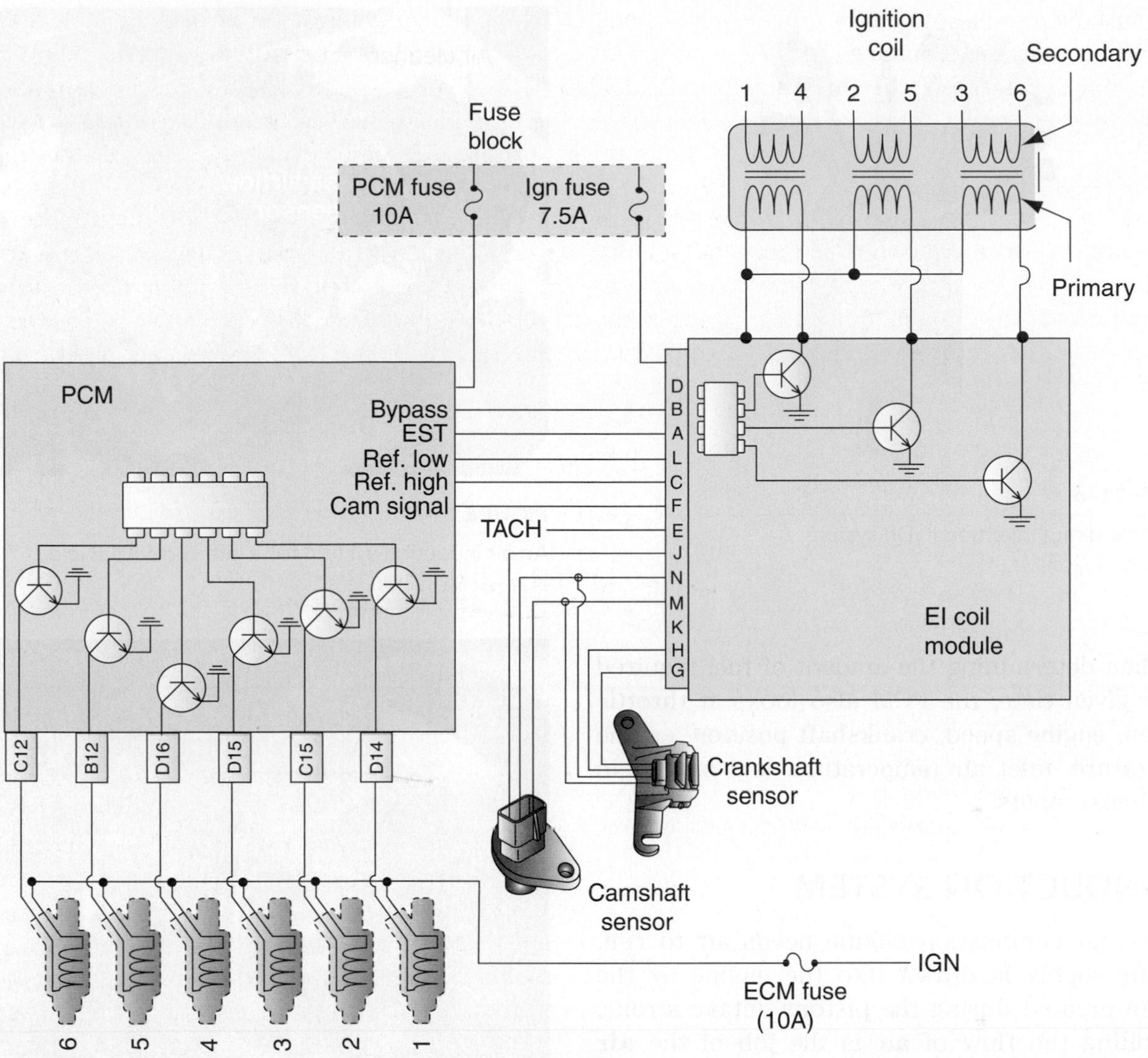

FIGURE 31–4

Basic layout for a computer-controlled ignition system.

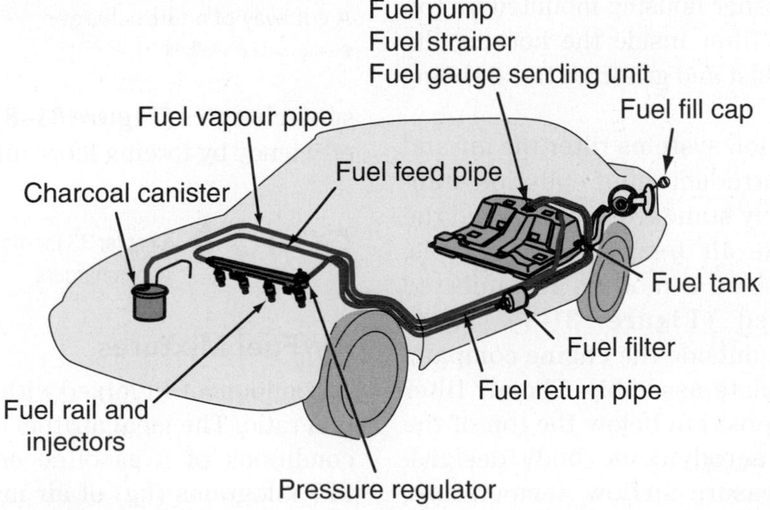

FIGURE 31–5

The fuel delivery system for a late-model car.

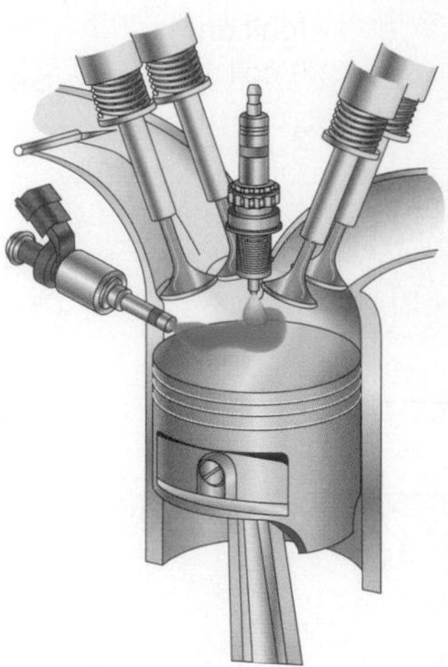

FIGURE 31–6

A gasoline direct injection (GDI) system.

When determining the amount of fuel required at any given time, the PCM also looks at throttle position, engine speed, crankshaft position, engine temperature, inlet air temperature, and oxygen in the exhaust inputs.

AIR INDUCTION SYSTEM

An internal combustion engine needs air to run. This air supply is drawn into the engine by the vacuum created during the pistons' intake stroke. Controlling the flow of air is the job of the **air induction system**.

Prior to the introduction of emission control devices, the induction system was quite simple. It consisted of an air cleaner housing mounted on top of the engine, with a filter inside the housing. Its function was to filter dust and grit from the air being drawn into the engine.

Modern air induction systems filter the air and do much more. The introduction of emission standards and fuel economy standards encouraged the development of intake air temperature controls. The air intake system on a modern fuel-injected engine is complicated **(Figure 31–7)**. Ducts channel cool air from outside the engine compartment to the throttle plate assembly. The air filter has been moved to a position below the top of the engine to allow for aerodynamic body designs. Electronic meters measure airflow, temperature, and density.

To improve fuel economy, manufacturers are using smaller engines equipped with turbo- or superchargers **(Figure 31–8)**. These increase engine efficiency by forcing more air into the cylinders.

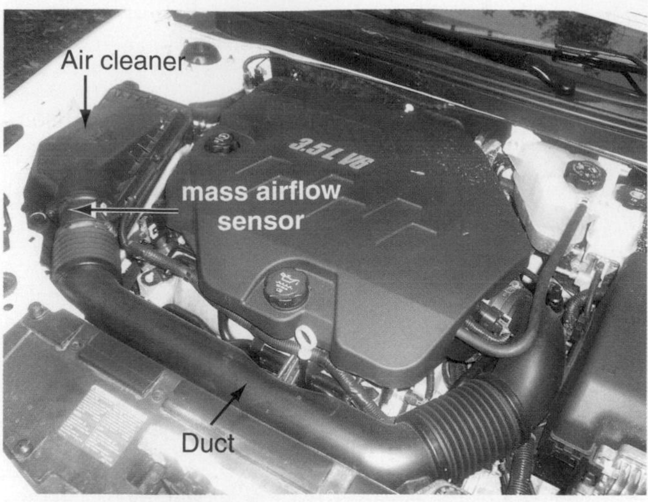

FIGURE 31–7

An air induction system for a late-model vehicle.

FIGURE 31–8

A cutaway of a turbocharger.

GO TO Chapter 33 for more information about turbo- and superchargers.

Air/Fuel Mixtures

The amount of air mixed with the fuel is called the air/fuel ratio. The ideal air/fuel ratio for most operating conditions of a gasoline engine is approximately 14.7 kilograms (kg) of air mixed with 1 kg of gasoline. This provides an ideal ratio of 14.7:1. Because air is so much lighter than gasoline, it takes nearly 10 000 litres (L) of air mixed with 1 L of gasoline to achieve an air/fuel ratio of 14.7:1. This is why proper air delivery is as important as fuel delivery.

When the mixture has more air than the ideal ratio calls for, the mixture is said to be lean. Ratios of 15 to 16:1 provide the best fuel economy from gasoline engines. Mixtures that have a ratio below 14.7:1 are considered rich mixtures. Rich mixtures (12 to 13:1) provide more power production from the engine but increase fuel consumption. An advantage of GDI systems is the ability to provide very lean mixtures under certain operating conditions, which results in increased fuel economy and reduced exhaust emissions.

EMISSION CONTROL SYSTEMS

Emission controls have one purpose and that is to reduce the amount of pollutants and environmentally damaging substances released by vehicles. The consequences of the pollutants are grievous **(Figure 31–9)**. The air we breathe and water we drink have become contaminated with chemicals that adversely affect our health. It took many years for the public and industry to address the problem of these pollutants. Not until smog became an issue did anyone in power really care and do something about these pollutants.

Smog not only appears as dirty air, it is also an irritant to the eyes, nose, and throat. The things necessary to form photochemical smog are hydrocarbons (HC) and oxides of nitrogen (NO_x) exposed to sunlight in stagnant air. HC in the air that reacts with the NO_x causes these two chemicals to react and form photochemical smog.

There are three main automotive pollutants: HC, carbon monoxide (CO), and NO_x. Particulate (soot) emissions are also present in diesel engine exhaust. HC emissions are caused largely by unburned fuel from the combustion chambers. HC emissions can also originate from evaporative sources such as the gasoline tank. CO emissions are a by-product of the combustion process, resulting from incorrect air/fuel mixtures. NO_x emissions are caused by nitrogen and oxygen uniting at cylinder temperatures above 1261°C (2300°F). Current concerns also include carbon dioxide emissions. CO_2 is a by-product of complete combustion but is said to contribute to global warming.

FIGURE 31–9
Dirty exhaust is bad for everyone.
Martin Restoule

Computer-Controlled Systems

The EGR valve, air pump, and evaporative emissions canister are controlled by the PCM. The computer keeps the level of the three major pollutants (CO, HC, and NO_x) at acceptably low levels. Other emission control devices may also be wholly or partly controlled by the computer. The control of the air, fuel, and ignition systems also contributes significantly to the control of emissions.

ENGINE CONTROL SYSTEMS

As manufacturers come closer to achieving complete combustion, engines are able to produce more power, use less fuel, and emit fewer pollutants. This has been made possible by technological advances, primarily in electronics.

GO TO ▶ Chapter 16 for a detailed discussion of computers and control systems.

The computer is an engine control system that functions like other computers. It receives inputs, processes information, and commands an output. The primary computer in an engine control system is the **engine control module (ECM)** or the **powertrain control module (PCM)**.

SHOP TALK

A PCM controls more than the engine, and it often performs the functions of the ECM as well. In this text, the term *PCM* is used unless something is specifically part of an ECM.

The engine control system relies on sensors that convert engine operating conditions such as temperature, engine and vehicle speeds, throttle position, and other conditions into electrical signals that are constantly monitored by the PCM **(Figure 31–10)**.

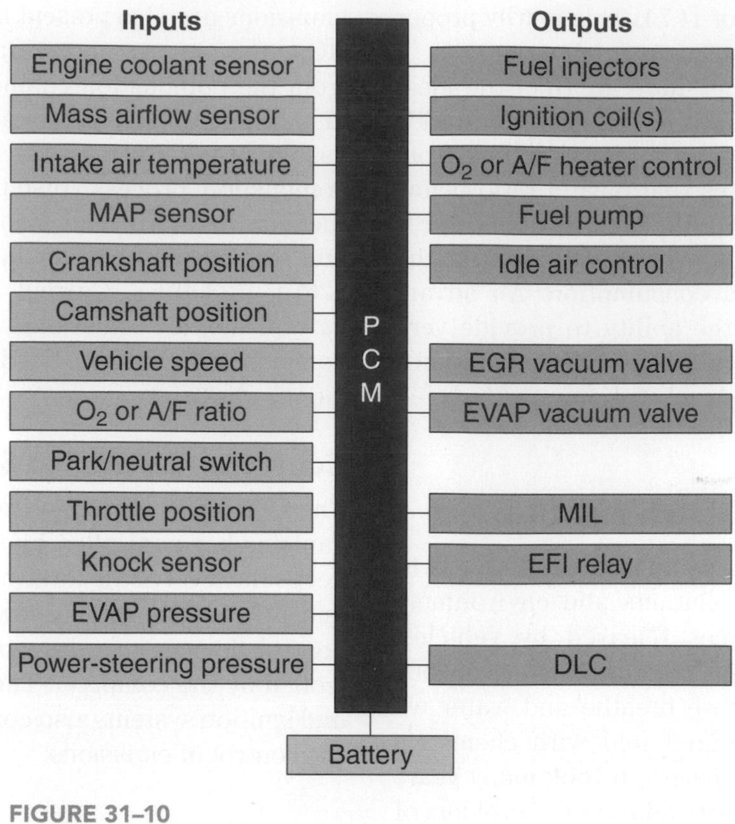

FIGURE 31–10

A basic look at an engine control system.

The PCM also senses some conditions through electrical connections. These include voltage changes at various components.

The PCM sorts the input signals and compares them to parameters programmed in it. This is the processing role of the computer. Based on the comparison, the computer may command a change in the operation of a component or system. The PCM also monitors the activity of the system and can detect any problems that occur. At that point, it will set a DTC and store other diagnostic information. The PCM may also store vehicle information, such as the vehicle identification number (VIN), calibration identification (CAL ID), and calibration verification. These are used to ensure that all calibration settings match the vehicle.

Based on the input information and the programs, the PCM decides the best operating parameters and sends out commands to various outputs or actuators. These commands are first sent to output drivers that cause an output device to turn on or off. These outputs include solenoids, relays, lights, motors, clutches, and heaters.

The PCM is linked with several other control modules. They share information, and in some cases, one control module controls another. The shared information is present on the CAN data bus in most vehicles **(Figure 31–11)**.

GO TO Chapter 16 for a detailed discussion of CAN and other multiplexing systems.

FIGURE 31–11

The twisted pairs of wires that serve as data buses in a multiplexed system.

System Components

The sensors, actuators, and computer communicate through the use of electronic and multiplexed circuits. For example, when the incoming voltage signal from the coolant sensor tells the PCM that the engine is getting hot, the PCM sends out a command to turn on the electric cooling fan. The PCM does this by grounding the relay circuit that controls the electric cooling fan. When the relay clicks on, the electric cooling fan starts to spin and cools the engine. The information may also be used to alter the air/fuel ratio and ignition timing.

In a PCM **(Figure 31–12)**, RAM is used to store data collected by the sensors, the results of calculations, and other information that is constantly changing during engine operation. Information in volatile RAM is erased when the ignition is turned off or when the power is disconnected. Nonvolatile RAM does not lose its data if its power source is disconnected.

The computer's permanent memory is stored in ROM or PROM and is not erased when the power source is disconnected. ROM and PROM are used to store computer-controlled system strategy and look-up tables. PROM normally contains the specific information about the vehicle.

The look-up tables (sometimes called maps) contain calibrations and specifications. Look-up tables indicate how an engine should perform. For example, information indicating a vacuum reading of 20 in. Hg is received from the manifold absolute pressure (MAP) sensor. This information and the information from the engine speed sensor are compared to a table for spark advance. This table tells the computer what the spark advance should be for that throttle position and engine speed **(Figure 31–13)**. The computer then modifies the spark advance.

FIGURE 31–12

A cutaway of a current "advanced" powertrain control module.

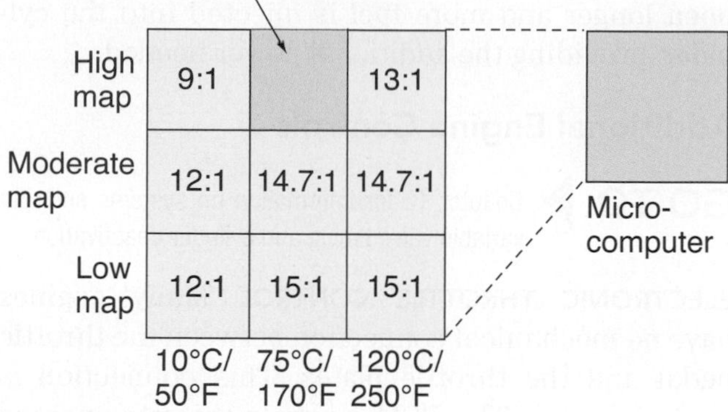

FIGURE 31–13

Example of a base look-up table.

When making decisions, the PCM is constantly referring to three sources of information: the look-up tables, system strategy, and the input from sensors. The computer makes informed decisions by comparing information from these sources.

Computer Logic

In order to control an engine system, the computer makes a series of decisions. Decisions are made in a step-by-step fashion until a conclusion is reached. Generally, the first decision is to determine the engine mode. For example, to control air/fuel mixture, the computer first determines whether the engine is cranking, idling, cruising, accelerating, or decelerating. Then the computer can choose the best system strategy for the present engine mode. In a typical example, sensor input indicates that the engine is warm, rpm is high, manifold absolute pressure is high, and the throttle plate is wide open. The computer determines that the vehicle is under heavy acceleration or has a wide-open throttle. Next, the computer determines the goal to be reached. For example, with heavy acceleration, the goal is to create a rich air/fuel mixture. At wide-open throttle, with high manifold absolute pressure and coolant temperature of 77°C, the table indicates that the air/fuel ratio should be 13:1—that is, 13 kg of air for every 1 kg of fuel. An air/fuel ratio of 13:1 creates the rich air/fuel mixture needed for heavy acceleration.

In a final series of decisions, the computer determines how the goal can be achieved. In our example, a rich air/fuel mixture is achieved by increasing fuel injector pulse width. The injector nozzle remains open longer and more fuel is injected into the cylinder, providing the additional power needed.

Additional Engine Controls

GO TO Chapter 12 for information on systems such as variable valve timing and cylinder deactivation.

ELECTRONIC THROTTLE CONTROL Many engines have no mechanical connection between the throttle pedal and the throttle plates. This connection is made by wire. The PCM controls throttle opening based on several inputs, including a throttle pedal sensor **(Figure 31–14)**. Electronic throttle control is used because it improves driveability and idle speed stability, and eliminates cruise control actuators.

VARIABLE INTAKE MANIFOLDS In an attempt to match airflow into the engine with engine speed and other variables, some engines have intake manifolds with two or more runners for each cylinder. The PCM opens and closes runners of different lengths to match the engine's needs. The length of the runner determines the amount of air that can be moved into the cylinders. It also affects the turbulence of that air.

Control of Non-Engine Functions

Some devices that are not directly connected to the engine are also controlled by the PCM to ensure maximum efficiency. For example, air-conditioner compressor clutches can be turned on or off, depending on various conditions. One common control procedure turns off the compressor when the throttle is fully opened. This allows maximum engine acceleration by eliminating the load of the compressor.

On some vehicles, the torque converter lockup clutch is applied and released by a signal from the computer. The clutch is applied by transmission hydraulic pressure, which is controlled by electrical solenoids that are in turn controlled by the computer.

In most cases, the PCM works with other control modules to control a system. Examples of these are antilock brake systems, traction and stability control systems, and other accessories.

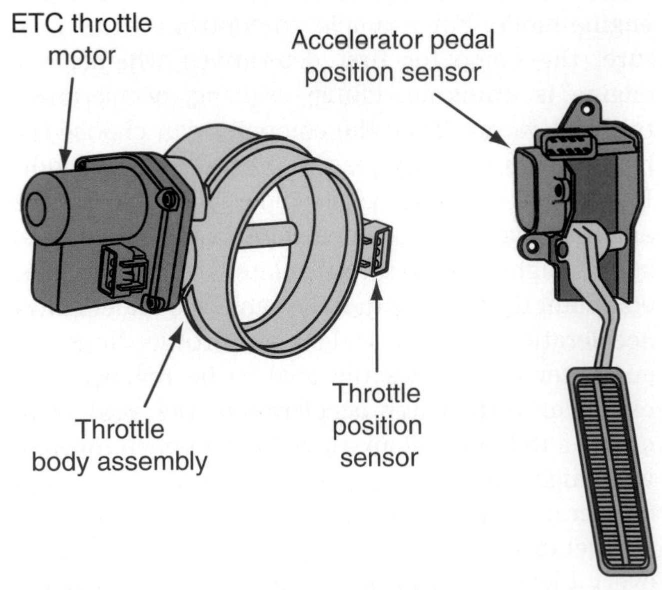

FIGURE 31–14

An electronic throttle control system.

ON-BOARD DIAGNOSTIC SYSTEMS

Because all manufacturers have continually updated, expanded, and improved their computerized control systems, there are hundreds of different domestic and import systems on the road. Fortunately for technicians, OBD-II called for all vehicles to use the same terms, acronyms, and definitions to describe their components. They also have the same type of diagnostic connector and basic test sequences, and display the same trouble codes. OBD-II began in 1996 and has been on all vehicles sold in North America since 1997.

The primary goal of OBD systems is to reduce vehicle emissions and reduce the possibility of future emission increases by detecting and reporting system malfunctions.

Vehicle Emission Control Information (VECI) Decal

In the mid-1990s, OBD-I systems were being phased out and OBD-II was being phased in. This meant that, depending on model year, make, and model, a vehicle could be OBD-I and have an OBD-II diagnostic connector or have OBD-I and OBD-II diagnostic connectors. To determine which system is actually used, refer to the VECI decal **(Figure 31–15)** located under the hood. This decal provides information about which OBD system is used as well as listing the installed emission control devices and what emission year the vehicle conforms with. Other information commonly found on the VECI includes a vacuum diagram for the emission control system, engine size, spark plug gap, and valve lash specifications.

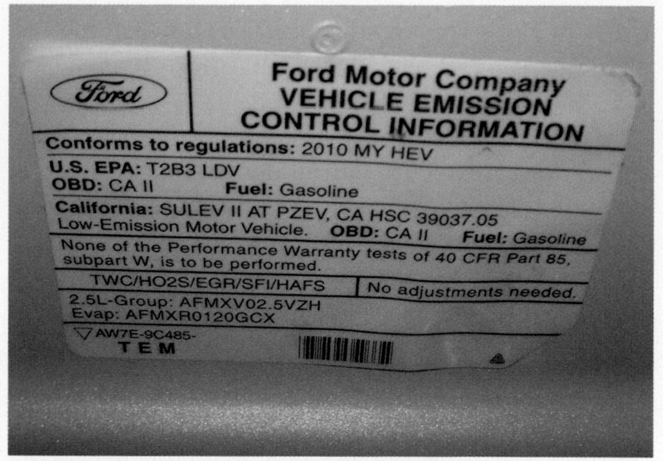

FIGURE 31–15

The VECI from a late-model car.

OBD-I (On-Board Diagnostic System, Generation 1)

OBD-I systems were first used in 1988. The ECM was capable of monitoring critical emission-related parts and systems and illuminating a malfunction indicator if a defect was found. The **malfunction indicator lamp (MIL)** was in the instrument panel. Most OBD-I systems used flash codes to display DTCs. The codes were displayed with the MIL. Often, the codes were displayed by jumping across terminals at a diagnostic **data link connector (DLC)**. Manufacturers provided lists of what the codes represented, along with step-by-step diagnostic procedures for identifying the exact fault.

Typically, the DTCs represented problems with the sensors, the fuel metering system, and the operation of the EGR valve. If any of these were open or shorted, had high resistance, or were operating outside a normal range, a code was set. The MIL not only helped with diagnostics but also alerted the driver that there was a problem. The MIL would turn off when the condition returned to normal; however, the DTC remained in memory until it was erased by a technician.

OBD-I was a step in the right direction, but it had several faults. It monitored few systems, had a limited number of DTCs (these were not standardized, so each manufacturer had its own), and allowed a limited use of serial data; most manufacturers required a specific scan tool and procedure, and the names used to describe a component varied across the manufacturers and their model vehicles.

OBD-II (On-Board Diagnostic System, Generation 2)

OBD-II was established to overcome some of the weaknesses of OBD-I. This was possible because of the advances made in computer technology and was necessary because of stricter emissions standards. OBD-I systems monitored only a few emission-related parts and were not set to maintain a specific level of emissions. OBD-II was developed to be a more comprehensive monitoring system and to allow more accurate diagnosis by technicians.

Studies estimate that approximately 50 percent of the total emissions from late-model vehicles are the result of emission-related problems. OBD-II systems are designed to ensure that vehicles remain as clean as possible over their entire life. During an emissions or “smog” check, an inspection computer can be plugged into the DLC of the vehicle and read the data from the vehicle’s computer. If emission-related DTCs are present, the vehicle will fail the test.

OBD-II added monitor functions for such things as catalyst efficiency, engine misfire detection, evaporative system, secondary air system, and EGR system flow rate. These monitors detect problems that would affect emissions levels. Also, a serial data stream of 20 basic data parameters and common DTCs was adopted.

OBD-II systems monitor the effectiveness of the major emission control systems and anything else that may affect emissions, and will illuminate the MIL when a problem is detected. During the monitoring functions, every part that can affect emission performance is checked by a diagnostic routine to verify that it is functioning properly. OBD-II systems must illuminate the MIL **(Figure 31–16)** if the vehicle’s conditions allow emissions to exceed 1.5 times the allowable standard for that model year based on a federal test procedure (FTP). When a component or strategy failure would allow emissions to exceed this level and the fault was detected during two consecutive trips, the MIL would illuminate to inform the driver of a problem, and a DTC would be stored in the PCM.

Besides increasing the capability of the PCM, additional hardware is required to monitor and maintain emissions performance. Examples of these are the addition of a heated oxygen sensor down the exhaust stream from the catalytic converter, the upgrade of specific connectors, components designed to last the mandated 128 748 km or 8 years, more precise crankshaft or camshaft position sensors, and a new standardized 16-pin DLC.

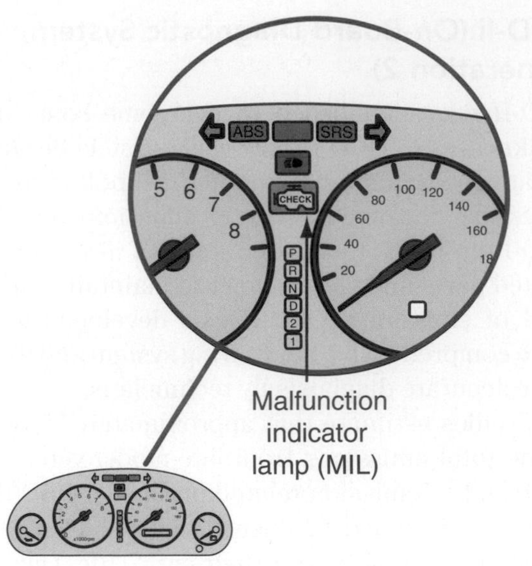

FIGURE 31–16

A typical MIL.

Rather than use a fixed, unalterable PROM, OBD-II PCMs have an EEPROM to store a large amount of information. The EEPROM stores data without the need for a continuing source of electrical power. It is an integrated circuit that contains the program used by the PCM to provide powertrain control. It is possible to erase and reprogram the EEPROM without removing it from the computer. When a modification to the PCM operating strategy is required, the EEPROM may be reprogrammed through the DLC using a scan tool or pass-through device and a computer.

GO TO Chapter 16 for more information on computer memory and for instructions on flashing a computer.

For example, if the vehicle calibrations are updated for a specific car model sold in California, a computer may be used to erase the EEPROM. After the erasing procedure, the EEPROM is reprogrammed with the updated information. The new program may be accessed via the vehicle manufacturer's service information website or from a disc. Manufacturers periodically send authorized service facilities the disks required for current updating of the EEPROMs. PCM recalibrations must be directed by a service bulletin or recall letter.

DATA LINK CONNECTOR OBD-II standards require the DLC to be easily accessible while sitting in the driver's seat **(Figure 31–17)**. The DLC cannot be hidden behind panels and must be accessible without tools. The connector pins are arranged in two rows and are numbered consecutively. Seven of the sixteen pins have been assigned by the OBD-II standard. They are used for the same information, regardless of the vehicle's make, model, and year. The remaining nine pins can be used by the individual manufacturers to meet their needs and desires.

The connector is D-shaped and has guide keys that allow the scan tool to be installed only one way. Using a standard connector and designated pins allows data retrieval with any scan tool designed for OBD-II. Some vehicles meet OBD-II standards by

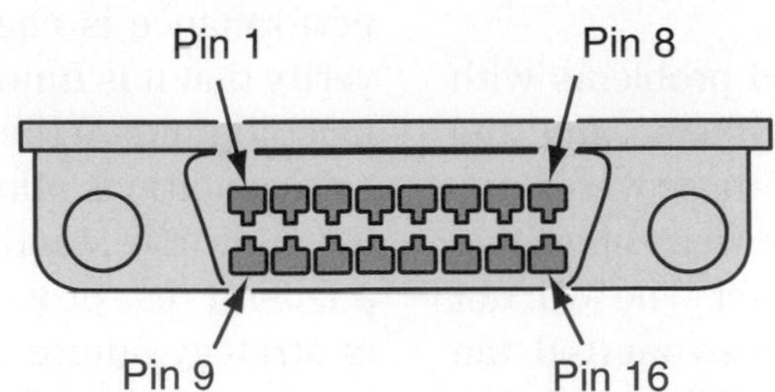

Pin 1: Manufacturer discretionary
Pin 2: J1850 bus positive
Pin 3: Manufacturer discretionary
Pin 4: Chassis ground
Pin 5: Signal ground
Pin 6: Manufacturer discretionary
Pin 7: ISO 1941-2 "K" line
Pin 8: Manufacturer discretionary

Pin 9: Manufacturer discretionary
Pin 10: J1850 bus negative
Pin 11: Manufacturer discretionary
Pin 12: Manufacturer discretionary
Pin 13: Manufacturer discretionary
Pin 14: Manufacturer discretionary
Pin 15: ISO 9141-2 "L" line
Pin 16: Battery power

FIGURE 31–17

A standard OBD-II DLC with pin designations.

providing the designated DLC along with their own connector for their own scan tool. Often, a vehicle will have more than one DLC, each with its own purpose. Due to OBD standards, the OBD DLC will always be located within a foot, to the right or left, of the steering column.

SHOP TALK

When a vehicle has a 16-pin DLC, this does not necessarily mean that the vehicle is equipped with OBD-II.

OBD-II TERMS All vehicle manufacturers must use the same names and acronyms for all electric and electronic systems related to the engine and emission control systems. Previously, there were many names for the same component. Now all similar components will be referred to with the same name. Beginning with the 1993 model year, all service information has been required to use the new terms. This new terminology is commonly called J1930 terminology because it conforms to the SAE standard J1930.

OBD-II FOR LIGHT-DUTY DIESELS The OBD-II systems are also mandated for all diesel engine vehicles that weigh 6350 kg (14 000 pounds) or less. These systems are very similar to those found in gasoline engines. The exceptions to this are the exclusion of the systems that are unique to gasoline engines and the inclusion of those systems unique to a diesel engine.

OBD-III (On-Board Diagnostic System, Generation 3)

Although not implemented at the printing of this book, the basic functions and operation of OBD-III will likely remain very similar to OBD-II. The exception will be for the reporting of data collected of emission-related faults. How the data collection and reporting will be performed has not been determined. One likely possibility is using a built-in communication system, such as OnStar.

The main goal of OBD-III is to minimize the delay between the detection of an emissions failure by the OBD-II system and the actual repair of the vehicle. It has been said that the check engine light is a poor motivator for prompt repair, and many repairs to the emissions-related parts of vehicles are being delayed until the mandatory emissions inspection is approaching. In other words, vehicles are running around with problems that increase emissions, and some owners are doing nothing about it.

If OBD-III is adopted with remote monitoring and reporting as a function, there will probably be one centralized data collection agency for each province. Once the vehicle's emission data are collected, the vehicle's owner will be notified about the results. If a fault is present that increases emission, the owner will be given a certain amount of time to have the problem corrected and the vehicle retested.

This type of vehicle monitoring has raised some fear in car owners. They feel that the government will be able to know too much about their driving habits and driving routes, and they want their privacy protected. For this reason, the final design and method for OBD-III have not been decided.

SYSTEM OPERATION

The PCM will operate in different modes based on the conditions. These modes are often referred to as control loops. The PCM does not always process all the information it receives. It is programmed to ignore or modify some inputs according to the current operating conditions.

Closed-Loop Mode

During the **closed-loop** mode, the PCM receives and processes all information available. Sensor inputs are sent to the PCM; the PCM compares those values to its programs, and it then sends commands to the output devices. The output devices adjust ignition timing, air/fuel ratio, and emission control operation. The resulting engine operation will result from the new inputs from the sensors. This continuous cycle of information is called a closed loop.

Closed control loops are often referred to as feedback systems. This means that the sensors provide constant information, or feedback, on what is taking place in the engine. This allows the PCM to constantly monitor, process, and send out new output commands.

Open-Loop Mode

When the engine is cold, most electronic engine controls go into **open-loop** mode. In this mode, the control loop is not a complete cycle and the computer does not react to feedback information from the oxygen sensors. Instead, the computer makes decisions based on preprogrammed information that allows it to make basic ignition or air/fuel settings based on coolant temperature, throttle position, and other inputs. The open-loop mode is activated when a signal from the temperature sensor indicates that the engine temperature is too low for the fuel to properly vaporize and burn in the cylinders. Systems with unheated oxygen sensors may also go into the open-loop mode while idling, or at any time that the oxygen sensor cools off enough to stop sending a good signal, and at wide-open throttle.

SHOP TALK

Most late-model engines have a heated oxygen sensor that reduces the time a PCM will be in open loop. If the system can go to closed loop just one minute sooner, the amount of pollutants released will be cut nearly in half.

Fail-Safe or Limp-In Mode

Most computer systems also have what is known as the fail-safe or limp-in mode. The limp-in mode is nothing more than the computer's attempt to take control of vehicle operation when input from one of its critical sensors has been lost or is well out of its normal range. To be more specific, if the computer sees a problem with the signal from a sensor, it either works with fixed values in place of the failed sensor input, or, depending on which input was lost, it can also generate a modified value by combining two or more related sensor inputs.

To illustrate this, if a fault occurs in the electronic throttle control system, such as readings from the accelerator position sensors that are out of range, the PCM will disable throttle control and run the engine at a predetermined rpm. This limited function limits engine speed, often to only 1200 rpm but does allow the engine to run until the driver can reach a service location.

Adaptive Strategy

A system's adaptive strategy is based on a plan for the timing and control of systems during different operating conditions. If a computer has adaptive strategy capabilities, it can actually learn from past experiences. For example, the normal voltage signals from the TP sensor to the PCM range from 0.6 to 4.5 volts. If a 0.2-volt signal is received, the PCM may regard this signal as the result of a worn TP sensor and assign this lower voltage to the normal low-voltage signal. The PCM will add 0.4 volt to the 0.2 volt it received. All future signals from the various throttle positions will also have 0.4 volt added to the signal. Doing this calculation adjusts for the worn TP sensor and ensures that the engine will operate normally. If the input from a sensor is erratic or considerably out of range, the PCM may totally ignore the input.

Most adaptive strategies have two parts: short term and long term. Short-term strategies are those immediately enacted by the computer to overcome a change in operation. These changes are temporary. Long-term strategies are based on the feedback about the short-term strategies. These changes are more permanent.

OBD-II MONITORING CAPABILITIES

OBD-II monitors the performance of emission and other related systems. The purpose of these monitors **(Figure 31–18)** is to detect failing systems and not wait until they fail before illuminating the MIL. OBD-II systems will perform certain tests on various subsystems of the engine management system. If one or more monitored systems are found to have a malfunction, the MIL will illuminate to alert the driver of a problem. Some monitors run continuously, whereas others will run only when certain operating conditions are present during the drive cycle.

These conditions are called the **enable criteria**. The following are examples of enable criteria for a particular monitor:

- Time since engine start greater than 300 seconds
- Engine coolant temperature between 77°C and 105°C (170°F and 220°F)
- Throttle position between 1.5 and 3 volts
- Vehicle speed between 16 and 100 km/h (10 and 60 mph)
- Fuel level between 20 and 80 percent

Readiness monitors	
ID: $	
Misfire	Test complete
Fuel system	Test complete
Components	Test complete
Heated catalyst	Test complete
Evaporative system	Test complete
Sec. air system	Test complete
A/C system refrig.	Test complete
O_2 sensor	Test complete

FIGURE 31–18

A report from a scan tool noting which monitor has been completed.

If the enable criteria for a specific monitor are not met during routine driving, the monitor will not run. In addition, a fault in one system can prevent the monitor for another system to run. For example, a thermostat that is stuck open can cause engine coolant temperature to remain below the value needed for a monitor to begin. This could keep many of the monitors from running and completing.

DRIVE CYCLE The OBD-II **drive cycle** is a defined set of operating conditions that must take place for all monitors to run and complete. If the monitor does not complete, some aspects of self-diagnosis cannot take place. A drive cycle **(Figure 31–19)** includes an engine start and operation that brings the vehicle into closed loop and includes whatever specific operating conditions are necessary either to initiate and complete a specific monitoring sequence or to verify a symptom or repair. Each manufacturer has guidelines that define how the vehicle is to be driven to complete a drive cycle.

OBD-II TRIP A **trip** is a partial drive cycle that includes all of the conditions (enable criteria) required for a particular monitor to run. To run a monitor, the vehicle must be driven at different speeds and conditions, similar to when performing a drive cycle.

During diagnosis, it may be necessary to complete a trip for a monitor to verify the problem or the repair. Depending on the monitor, the system tests the component or system once per trip. It is important to note that once a repair has been made, the vehicle will need to complete a trip or the drive cycle so that affected monitors can run and pass. If a monitor will not pass, ensure that the enable criteria are met and the drive cycle is correct. Failure of a monitor to complete can also be caused by a fault in another circuit or system that prevents the monitor from running to completion.

WARM-UP CYCLE OBD-II standards define a warm-up cycle as the period from when the engine is started until the engine temperature has increased by at least 16°C (60°F) and has reached at least 88°C (160°F).

Catalyst Efficiency Monitor

OBD-II vehicles use a minimum of two oxygen sensors. One of these is used for feedback to the PCM for fuel control, and the other, located at the rear of the catalytic converter, gives an indication of the efficiency of the converter and may also be used for fuel control. The downstream O_2 sensor is sometimes called the **catalyst monitor sensor (CMS)**. The catalyst efficiency monitor compares the signals between the two O_2 sensors to determine how well the catalyst is working.

One heated O_2 sensor (HO_2S) is mounted near the exhaust manifold and the additional HO_2S is mounted downstream from the catalytic converter **(Figure 31–20)**. The HO_2S are identified by their

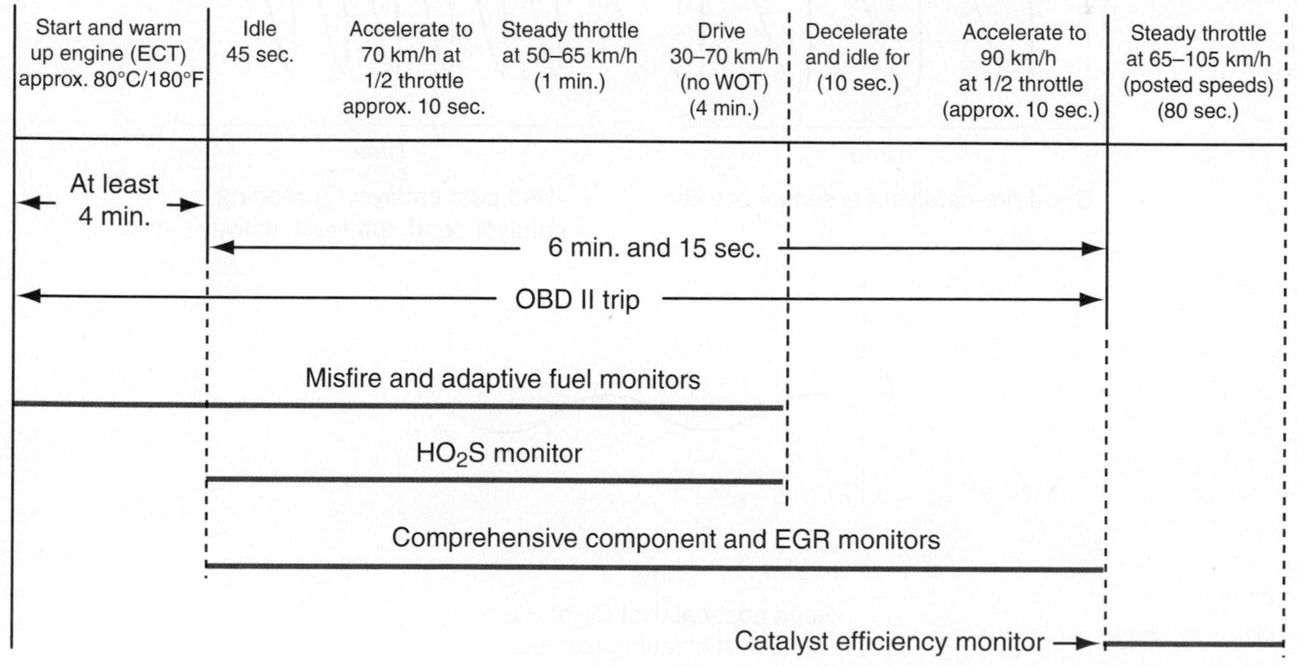

FIGURE 31–19

OBD trip cycle.

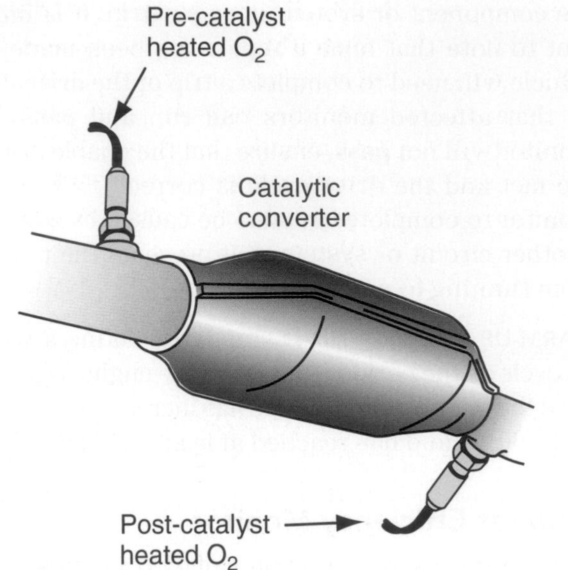

FIGURE 31–20

Pre-catalyst and post-catalyst heated O_2 sensors (HO_2S).

position and location relative to the converters. S1 means the O_2 sensor is upstream or before the catalytic converter, and S2 is downstream or after the catalytic converter. On V-type engines, the additional designation B1 means the sensor is installed in the bank for cylinder number 1. B2 indicates it is for the other bank of cylinders.

The downstream HO_2S are designed to prevent the collection of condensation on their ceramic material. The internal heater is not turned on until the ECT sensor indicates a warmed-up engine. This action prevents cracking of the ceramic. Gold-plated pins and sockets are used in the HO_2S, and the downstream and upstream sensors have different wiring harness connectors.

A catalytic converter stores oxygen during lean engine operation and gives up this stored oxygen during rich operation to burn up excessive HCs. Catalytic converter efficiency is measured by monitoring the O_2 storage capacity of the converter during closed-loop operation.

When the catalytic converter is storing oxygen properly, the downstream HO_2S provides fewer cross counts (low-frequency) voltage signals. If the catalytic converter is not storing O_2 properly, the cross counts of the voltage signal increase on the downstream HO_2S. When the signals from the downstream HO_2S approach the cross counts of the upstream sensors **(Figure 31–21)**, a DTC is set in the PCM

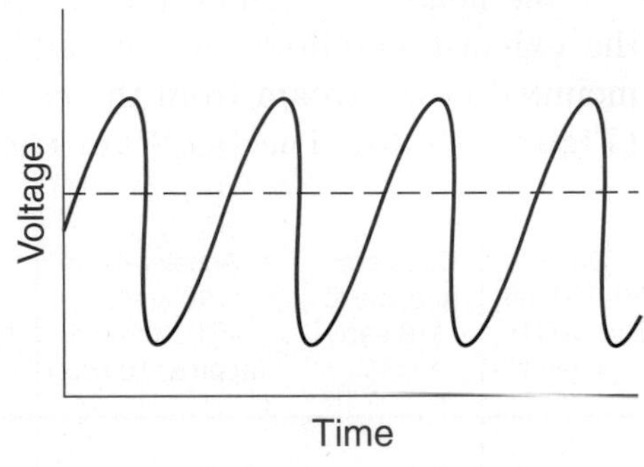

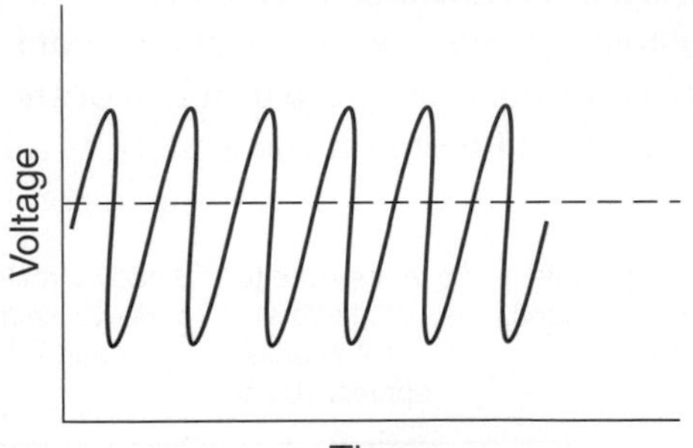

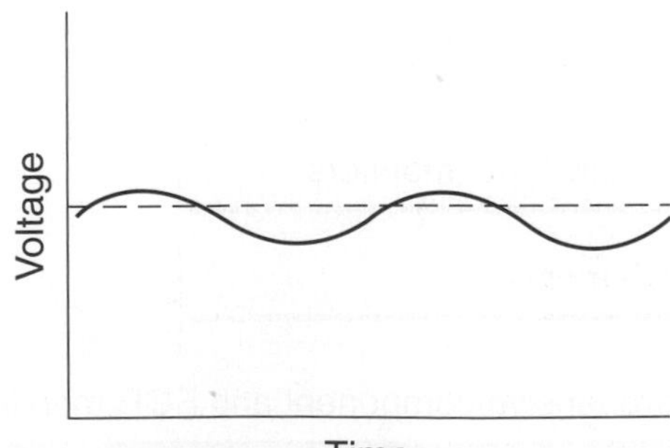

FIGURE 31–21

Oxygen sensor signal for a good and bad catalytic converter.

memory. If the fault occurs on three drive cycles, the MIL light is illuminated.

Misfire Monitor

If a cylinder misfires, HCs are exhausted from the cylinder and enter the catalytic converter. A misfire means a lack of combustion in at least one cylinder for at least one combustion event. This often allows unburned fuel to pass through the cylinder and into the exhaust. Although the converter can handle an occasional sample of raw fuel, too much fuel to the converter can overheat and destroy it. The honeycomb material in the converter may melt into a solid mass. If this happens, the converter is no longer efficient in reducing emissions.

Cylinder misfire monitoring requires measuring the contribution of each cylinder to total engine power. The misfire monitoring system uses a highly accurate crankshaft angle measurement to measure the crankshaft acceleration each time a cylinder fires **(Figure 31–22)**. If a cylinder is contributing normal power, a specific crankshaft acceleration time occurs. When a cylinder misfires, the cylinder does not contribute to engine power, and crankshaft acceleration for that cylinder is slowed. With a few exceptions, this monitor runs continuously while the engine is running. An example of an exception is during closed-throttle deceleration.

Most OBD-II systems allow a random misfire rate of about 2 percent before a misfire is flagged as a fault. It is important to note that this monitor only looks at the crankshaft's speed during a cylinder's firing stroke. It cannot determine if the problem is fuel-, ignition-, or mechanically related. Misfires are categorized as type A, B, or C. Type A could cause immediate catalyst damage. Type B could cause emissions of 1.5 times the design standard, and type C could cause an I/M failure. When there is a type A misfire, the MIL will flash. If there is a type B misfire, the MIL will turn on but will not flash.

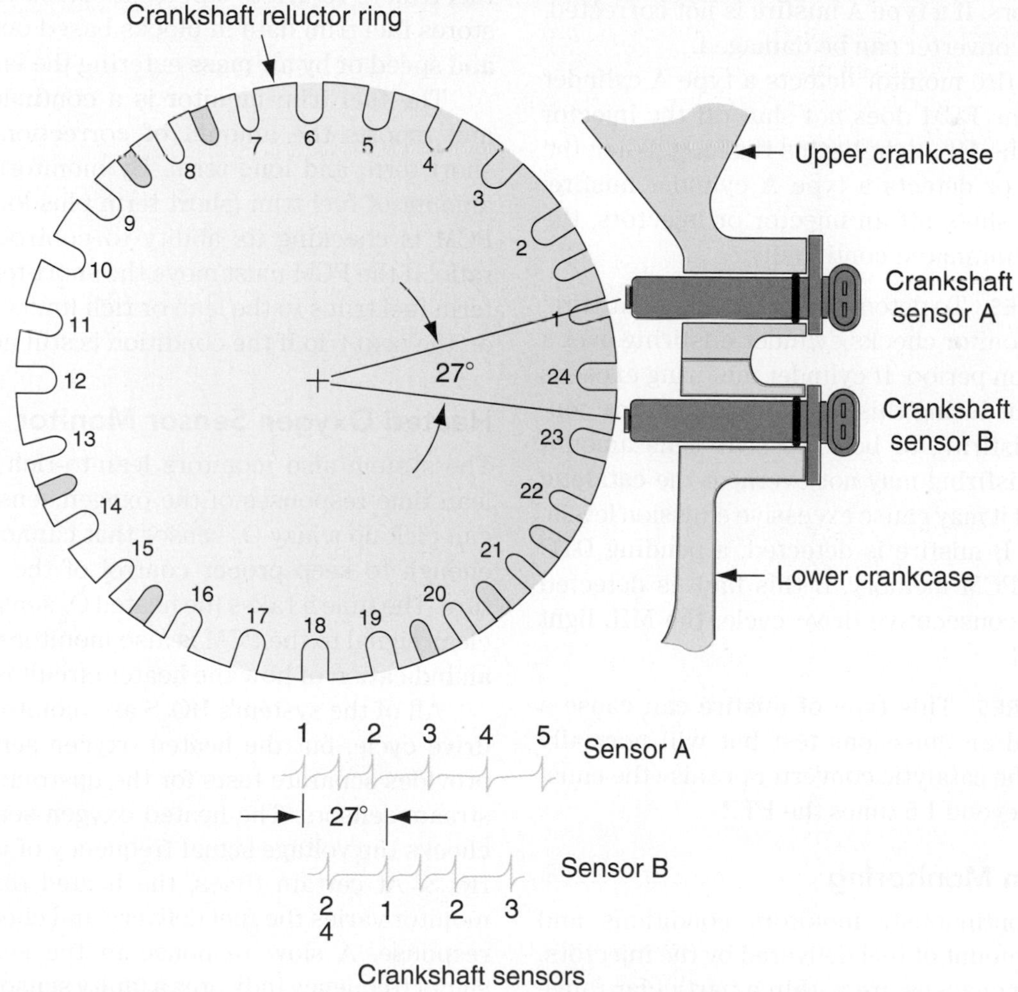

FIGURE 31–22

An example of using two crankshaft sensors to monitor the engine and detect misfire.

Type C misfires will typically not cause the MIL to light or flash.

The misfire monitoring sequence includes an adaptive feature that compensates for variations in engine operation caused by manufacturing tolerances and component wear. It also has the capability to allow vibration at different engine speeds and loads. When an individual cylinder's contribution to engine speed falls below a certain threshold, the misfire monitoring sequence calculates the vibration, tolerance, and load factors before setting a misfire code.

TYPE A MISFIRES The monitor checks for type A misfires in 200 revolution increments. If a cylinder misfires between 2 and 20 percent of the time, the monitor considers the misfiring to be excessive. This condition may cause the PCM to shut off the fuel injectors at the misfiring cylinder or cylinders to limit catalytic converter heat. When the engine is operating under heavy load, the PCM will not turn off the injectors. If a type A misfire is not corrected, the catalytic converter can be damaged.

If the misfire monitor detects a type A cylinder misfire and the PCM does not shut off the injector or injectors, the MIL light begins flashing. When the misfire monitor detects a type A cylinder misfire, and the PCM shuts off an injector or injectors, the MIL light is illuminated continually.

TYPE B MISFIRES To detect a type B cylinder misfire, the misfire monitor checks cylinder misfiring over a 1000-revolution period. If cylinder misfiring exceeds 2 to 3 percent during this period, the monitor considers the misfiring to be excessive. This amount of cylinder misfiring may not overheat the catalytic converter, but it may cause excessive emission levels. When a type B misfire is detected, a pending DTC is set in the PCM memory. If this fault is detected on a second consecutive drive cycle, the MIL light is illuminated.

TYPE C MISFIRES This type of misfire can cause a vehicle to fail an emissions test but will normally not damage the catalytic converter or raise the emissions levels beyond 1.5 times the FTP.

Fuel System Monitoring

The PCM continuously monitors conditions and adjusts the amount of fuel delivered by the injectors. Most of these changes are within a particular range around a base setting. The PCM checks the effectiveness of these changes through feedback from the oxygen sensor. The PCM then makes corrections as necessary. When the PCM needs to make a change outside that range, it will make a fuel trim adjustment. Fuel trim is the required new setting compared to the basic injector pulse width. Remember, the amount of fuel injected into the engine is controlled by the amount of time the injector is turned on.

The system allows for short-term and long-term fuel trim. **Short-term fuel trim (STFT)** makes minor adjustments to the pulse width. These adjustments are temporary and not held in memory when the ignition is switched off. **Long-term fuel trim (LTFT)** is set by the effectiveness of the short-term trim. If the short-term trim meets the requirements of the engine during different operating conditions, the PCM may use those adjustments as the new base for injection timing. Fuel requirements will change as a vehicle ages. As short-term fuel trim averages increase or decrease, a new long-term fuel trim value will be learned and remain in memory. This new long-term value will be such to allow the short-term fuel trim to return to 0 percent correction. The PCM stores fuel trim data in blocks based on engine load and speed or by air mass entering the engine.

The fuel trim monitor is a continuous monitor and reports the amount of correction made with short term and long term. By monitoring the total amount of fuel trim (short term plus long term), the PCM is checking its ability to control the air/fuel ratio. If the PCM must move the short-term and long-term fuel trims to the lean or rich limits, a DTC is set on the next trip if the condition is still necessary.

Heated Oxygen Sensor Monitor

The system also monitors lean-to-rich and rich-to-lean time responses of the oxygen sensor. This test can pick up a lazy O_2 sensor that cannot switch fast enough to keep proper control of the air/fuel mixture. The time it takes the heated O_2 sensors to send a clean signal to the PCM is also monitored. This gives an indication of how the heater circuit is working.

All of the system's HO_2S are monitored once per drive cycle, but the heated oxygen sensor monitor provides separate tests for the upstream and downstream sensors. The heated oxygen sensor monitor checks the voltage signal frequency of the upstream HO_2S. At certain times, the heated oxygen sensor monitor varies the fuel delivery and checks for HO_2S response. A slow response in the sensor voltage signal frequency indicates a faulty sensor. The sensor signal is also monitored for excessive voltage.

The heated oxygen sensor monitor also checks the frequency of the rear HO_2S signals and checks these signals for excessively high voltage. If the monitor does not detect signal voltage frequency within a specific range, the rear HO_2S are considered faulty. The heated oxygen sensor monitor will command the PCM to vary the air/fuel ratio to check the response of the rear HO_2S.

EGR System Monitoring

The EGR monitors use different strategies to determine if the system is operating properly. Some monitor the temperature within the EGR passages. A high temperature indicates that the valve is open and exhaust gases are moving through the passages. Other systems look at the MAP signal, energize the EGR valve, and look for corresponding change in vacuum levels. As the valve is opening, there should be a drop in vacuum. Some systems open the EGR during coast down and monitor the change in STFT.

The EGR monitor looks at the operation of the EGR valve and the flow rates of the system. It also looks for shorts or opens in the circuit. If a fault is detected in any of the EGR monitor tests, a DTC is set in the PCM memory. If the fault occurs during two drive cycles, the MIL light is illuminated. The EGR monitor operates once per OBD-II trip.

There are many different EGR systems used today. The pressure feedback EGR, linear EGR, and delta pressure feedback EGR systems are the most common. If the system uses a delta pressure feedback EGR (DPFE), there is an orifice located under the EGR valve. Small exhaust pressure hoses are connected from each side of this orifice to the DPFE sensor. During the EGR monitor, the PCM first checks the DPFE signal. If this sensor signal is within the normal range, the monitor proceeds with the tests.

The PCM checks EGR flow by checking the DPFE signal against an expected DPFE value for the operating conditions at steady throttle within a specific rpm range.

With the EGR valve closed, the PCM checks for pressure difference at the two pressure hoses connected to the DPFE sensor. When the EGR valve is closed and there is no EGR flow, the pressure should be the same at both pipes. If the pressure is different at these two hoses, the EGR valve is stuck open.

The PCM commands the EGR valve to open and then checks the pressure at the two exhaust hoses connected to the DPFE sensor. With the EGR valve open and EGR flow through the orifice, there should be higher pressure at the upstream hose than at the downstream hose **(Figure 31–23)**.

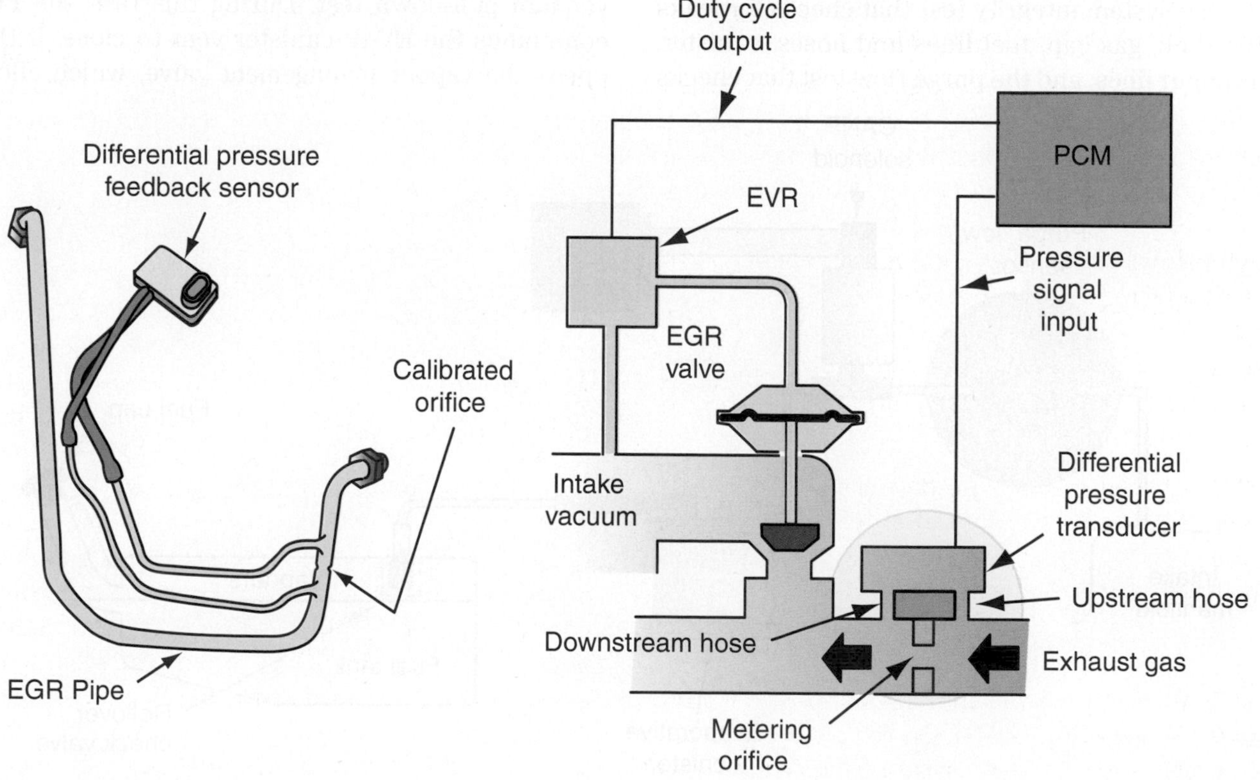

FIGURE 31–23

An EGR system with a delta pressure feedback sensor.

Evaporative (EVAP) Emission System Monitor

In addition to the various components and failures that could affect the tailpipe emissions on a vehicle, OBD-II monitors the fuel evaporative systems. The system tests the ability of the fuel tank to hold pressure and the purge system's ability to vent the gas fumes from the charcoal canister when commanded to do so by the PCM.

Chrysler and others often use a leak detection pump (LDP) to detect leaks in the EVAP system. When specific operating conditions are met, the PCM powers the pump to test the EVAP system. The pump pressurizes the system. As pressure builds, the cycling rate of the pump decreases. If the system does not leak, pressure will continue to build until the pump shuts off. If there is a leak, pressure will not build up, and the pump will continue to run. The pump continues to run until the PCM determines it has run a complete test cycle and then sets a DTC. If there are no leaks, the PCM will run the purge monitor. Because of the pressure in the system from the pump, the cycle rate should be high. If no leaks are present and the purge cycle is high, the system passed the test.

In most systems, the EVAP monitor has two parts: the system integrity test that checks for leaks in the tank, gas cap, fuel lines and hoses, canister, and vapour lines, and the purge flow test that checks for blockage in the vapour lines and purge solenoid. During the integrity test, the PCM closes the canister vent solenoid and pulses the purge solenoid until the tank sensor reads a negative pressure of 20 inches of water. The PCM then closes the purge solenoid and measures the time it takes for the vacuum in the tank to decay. The EVAP purge volume test opens the canister vent solenoid and then varies the duty cycle of the purge solenoid until a measured change in STFT occurs. The EVAP monitor will be suspended whenever the fuel tank is less than one-quarter full or more than three-quarters full.

A few EVAP systems have a purge flow sensor (PFS) connected to the vacuum hose between the canister purge solenoid and the intake manifold **(Figure 31–24)**. The PCM monitors the PFS signal once per drive cycle to determine if there is vapour flow or no vapour flow through the solenoid to the intake manifold.

ENHANCED EVAP SYSTEMS All new vehicles since 2003 have an enhanced evaporative system monitor. This system detects leaks and restrictions in the EVAP system **(Figure 31–25)**. This monitor first checks the integrity of the EVAP system. This test is run only when certain enable criteria are met. Once the integrity test is run, the monitor will conduct a vacuum pull-down test. During this test, the PCM commands the EVAP canister vent to close. It then opens the vapour management valve, which allows

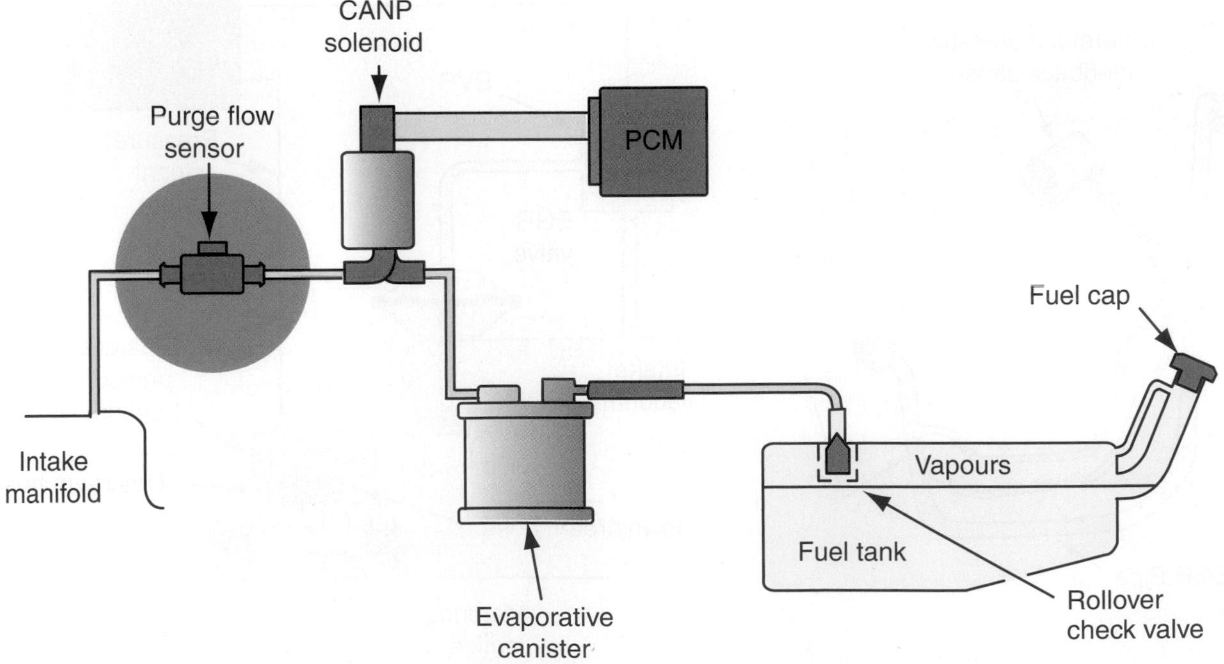

FIGURE 31–24

An EVAP system with a purge flow sensor.

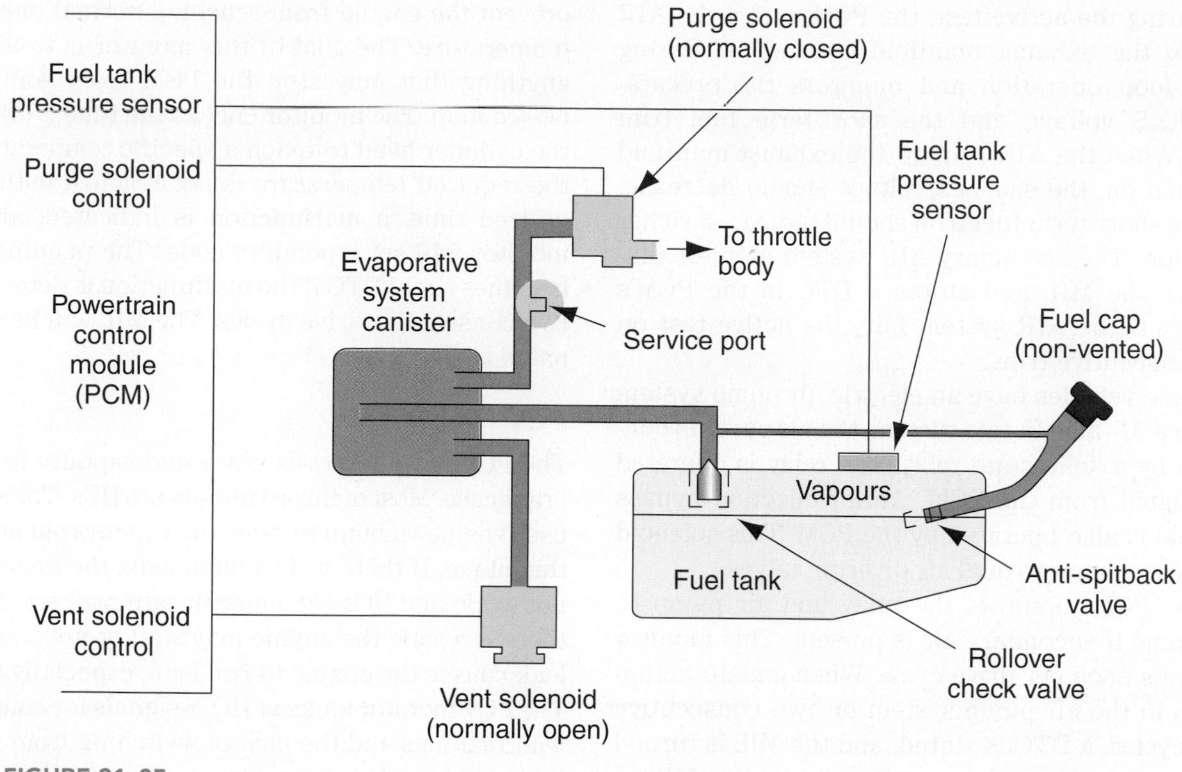

FIGURE 31–25

An enhanced EVAP system.

intake vacuum to draw a small amount of vacuum on the system. If the system has a leak, vacuum will not be held in the system, and the monitor will set a DTC. If there are no leaks, the monitor will close the vapour management valve and monitor the system's ability to hold a vacuum. The final check is the vapour generation test. The PCM releases the vacuum in the system and then closes the EVAP system. The monitor looks for pressure changes in the system. An increase in pressure indicates excessive vapour generation.

A specially designed fuel tank filler cap is used on these systems. In these systems, an evaporative system leak or a missing fuel tank cap will cause the MIL to turn on. Also, if the fuel filler cap is not on tight enough, the OBD system can detect leaking vapour and the MIL will light up. If the filler cap is then tightened, the indicator will generally go out after a short period. Some vehicles illuminate a CHECK GAS CAP light instead of the MIL.

Secondary Air Injection (AIR) System Monitor

The AIR system operation can be verified by turning the system on to inject air upstream of the O_2 sensor while monitoring its signal. Many older designs inject air into the exhaust manifold when the engine is in open loop and switch the air to the converter when it is in closed loop. If the air is diverted to the exhaust manifold during closed loop, the O_2 sensor thinks the mixture is lean, and the signal should drop.

SHOP TALK

Many engines do not need to have a secondary air injection system; therefore, this monitor does not run.

On some vehicles, the AIR system is monitored with passive and active tests. During the passive test, the voltage of the precatalyst HO_2S is monitored from start-up to closed-loop operation. The AIR pump is normally on during this time. Once the HO_2S sensor is warm enough to produce a voltage signal, the voltage should be low if the AIR pump is delivering air to the exhaust manifold. The secondary AIR monitor will indicate a pass if the HO_2S voltage is low at this time. The passive test also looks for a higher HO_2S voltage when the AIR flow to the exhaust manifold is turned off by the PCM. When the AIR system passes the passive test, no further testing is done. If the AIR system fails the passive test or if the test is inconclusive, the AIR monitor in the PCM proceeds with the active test.

During the active test, the PCM cycles the AIR flow to the exhaust manifold on and off during closed-loop operation and monitors the precatalyst HO_2S voltage and the short-term fuel trim value. When the AIR flow to the exhaust manifold is turned on, the sensor's voltage should decrease, and the short-term fuel trim should indicate a richer condition. The secondary AIR system monitor illuminates the MIL and stores a DTC in the PCM's memory if the AIR system fails the active test on two consecutive trips.

Some vehicles have an electric air pump system **(Figure 31–26)**. In this system, the air pump is controlled by a solid-state relay. The relay is operated by a signal from the PCM. An air-injection bypass solenoid is also operated by the PCM. This solenoid supplies vacuum to dual air diverter valves.

The PCM monitors the relay and air pump to determine if secondary air is present. This monitor functions once per drive cycle. When a malfunction occurs in the air pump system on two consecutive drive cycles, a DTC is stored, and the MIL is turned on. If the malfunction corrects itself, the MIL is turned off after three consecutive drive cycles in which the fault is not present.

Thermostat Monitor

Present on all 2002 and newer vehicles, the thermostat monitor checks the engine and its cooling system for defects that would affect engine temperature and prevent the engine from reaching normal operating temperature. The goal of this monitor is to identify anything that may stop the PCM from going into closed loop. The monitor checks the time it takes for the cylinder head to reach a specific temperature. If the required temperature is not reached within the desired time, a malfunction is indicated, and the monitor will set a pending code. The pending code becomes a real DTC if the malfunction is detected on two consecutive drive cycles. The MIL will be illuminated at that time.

PCV Monitor

The PCV valve removes unwanted vapours from the crankcase. Most of these vapours are HCs. The system uses engine vacuum to draw the vapours out and into the intake. If there is a vacuum leak, the system will not work, and HCs can enter the atmosphere. Also, if there is a leak, the engine may stall or not start. The leak causes the engine to run lean, especially at idle. The PCV monitor looks at HO_2S signals for consistent lean readings and the lack of switching from rich to lean. Abnormal signals will cause the monitor to illuminate the MIL after two consecutive driving cycles and will store one or more of the DTCs.

Variable Cam Timing System Monitor

Late-model vehicles equipped with variable valve timing (VVT) use a monitor that checks system operation. Because VVT systems are often used

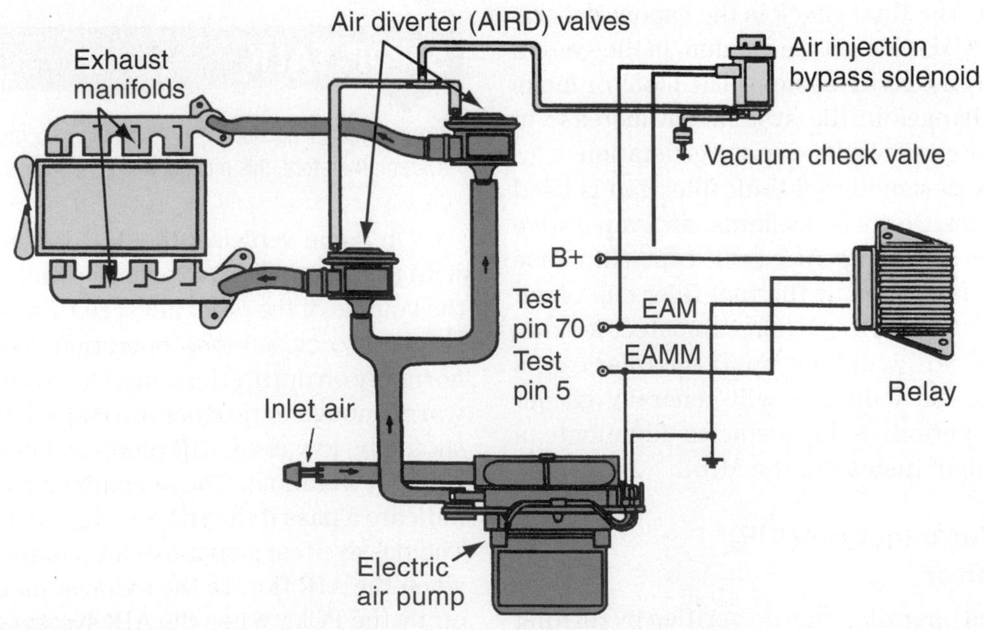

FIGURE 31–26

An electric air pump system.

Ford Motor Company

FIGURE 31–27

Location of camshaft sensors.

for EGR purposes, the operation of the VVT can directly affect exhaust emissions. The monitor examines the response time for the camshaft position to change based on VVT command and camshaft position measured by the CMP sensor **(Figure 31–27)**. If the response rate is slow, a DTC will set for VVT performance.

Electronic Throttle Control System Monitor

Engines equipped with ETC use a separate monitor to evaluate system operation. This is because of the safety concerns related to ETC operation. Instead of using one computer processor for ETC operation, two separate processors and monitoring systems are used. This provides redundant control and monitoring of the system.

If a fault is detected, the monitor may limit rpm, force rpm to remain at idle speed, or even force engine shutdown by turning off the fuel injectors.

Comprehensive Component Monitor

The comprehensive component monitor (CCM) is a continuous monitor. It looks at the inputs and outputs that would affect emission levels. The system looks at any electronic input that could affect emissions. The strategy is to look for opens and shorts or input signal values that are out of the normal range. It also looks to see if the actuators have their intended effect on the system and to monitor other abnormalities.

The CCM uses several strategies to monitor inputs and outputs. One strategy for monitoring inputs involves checking inputs or electrical defects and out-of-range values by checking the input signals at the analog digital converter. This monitor also checks the circuits for the various outputs. These circuit checks look for continuity and out-of-range values. If an open circuit is detected, a DTC will be set.

The CCM also checks frequency signal inputs by performing rationality checks. During a rationality check, the monitor uses other sensor readings and calculations to determine if a sensor reading is proper for the present conditions. The following get rationality checks:

- Crankshaft position sensor (CKP)
- Output shaft speed sensor (OSS)
- Camshaft position sensor (CMP)
- Vehicle speed sensor (VSS)

The functional test of the CCM checks most of the outputs by monitoring the voltage of each output solenoid, relay, or actuator at the output driver in the PCM. If the output device is off, this voltage should be high. This voltage is pulled low when the output is turned on.

OBD-II SELF-DIAGNOSTICS

It is important to remember that although the diagnostic capabilities of OBD-II are great, the system does not check everything and is not capable of finding the cause of all driveability problems. Not all problems will activate the MIL or store a DTC. A DTC only indicates that a problem exists somewhere in a system, in the circuit of the sensor or an output. A technician's job is to find the problem. Often, DTCs are set that indicate an out-of-range reading by a sensor. This does not mean the sensor is bad. The abnormal signals can be caused by engine mechanical problems or faults in the air, fuel, ignition, emission control, and other systems. Retrieving the information stored in the PCM is a starting point for diagnosis.

When no DTCs are present but there is a driveability problem, or when determining the true cause of the DTC, the basic engine should be checked before moving to the support systems. Many engine control systems offer a running compression test. This test runs much like the misfire monitor. The speed of the crankshaft between cylinder firings is measured and compared. If one or more cylinders accelerate slower on the power stroke, there may be a compression problem. This test should be followed up with the other standard engine mechanical tests.

GO TO Chapter 9 for the procedures for conducting the various engine mechanical tests.

MIL

According to OBD-II regulations, the PCM must illuminate the MIL when it detects a problem that affects emissions. Also, when a malfunction is detected, a DTC must be set. Depending on the problem, the MIL will either light and stay on, or it will blink. The action of the MIL depends on the monitor and the problem. For example, a misfire that would allow emissions to exceed regulations, but not damage the catalyst, will light the MIL. If the misfire will raise the catalytic converter's temperature enough to destroy it, the MIL will blink.

The MIL will be turned off if the misfire did not occur during three consecutive drive cycles. This requirement is true for most monitors. But an appropriate DTC will be recorded, as will freeze-frame data. If the same fault is not detected during 40 warm-up cycles, the DTC and freeze-frame data are erased from active memory. However, the DTC and freeze-frame data will be stored as a history code with freeze-frame data until they are cleared.

Diagnostic Trouble Codes

OBD-II codes are standardized, which means that most DTCs mean the same thing regardless of the vehicle. However, vehicle and scan tool manufacturers can have additional DTCs and add more data streams, report modes, and diagnostic tests. DTCs are designed to indicate the circuit and the system where a fault has been detected.

An OBD-II DTC is a five-character code with both letters and numbers **(Figure 31–28)**. This is called the alphanumeric system.

The first character of the code is a letter. This defines the system where the code was set. Currently, there are four possible first character codes: "B" for body, "C" for chassis, "P" for powertrain, and "U" for undefined. The U-codes are designated for future use.

The second character is a number. This defines the code as being a mandated code or a special manufacturer code. A "0" code means that the fault is defined or mandated by OBD-II. A "1" code means the code is manufacturer specific. As an example, all OBD-II vehicles can set a P0300 random cylinder misfire code. The P0 indicates this is an OBD-II code common to all makes, models, and engines. A manufacturer code, such as P1259, is specific to a particular system or component that is not common across all engines. In this case, a P1259 refers to a problem with the V-TEC system used by Honda and Acura vehicles. Codes of "2" or "3" are designated for future use.

The third through fifth characters can be letters or numbers. These describe the fault. The third character indicates where the fault occurred. The remaining two characters describe the exact condition that set the code. The digits are organized so that the various codes related to a particular sensor or system are grouped together.

Not all DTCs will cause the MIL to light; this depends on the monitor and the problem. DTCs that will not affect emissions will never illuminate the MIL. Basically, there are three types of DTCs. An active or current DTC represents a fault that was detected and occurred during two trips. When a two-trip fault is detected for the first time, a DTC is stored as a **pending code**. A pending DTC is a code representing a fault that has occurred but that has not occurred enough times to illuminate the MIL. There are some DTCs that will set in one trip; again, this depends on the monitor.

First digit – Letter indicates component group area
- P = Powertrain
- B = Body
- C = Chassis
- U = Network communications

Second digit
- 0 = SAE or OBD mandated
- 1 = Manufacturer specific

Third digit – Subgroup, powertrain subgroups:
- 0 = Total system
- 1 = Fuel and air metering
- 2 = Fuel and air metering
- 3 = Ignition system or misfire
- 4 = Auxiliary emission controls
- 5 = Idle speed control
- 6 = PCM and auxiliary inputs
- 7 = Transmission
- 8 = Transmission

Fourth and Fifth digits – Defines the area or component and basic problem

FIGURE 31–28
Interpreting OBD-II DTCs.

Freeze-Frame Data

One of the mandated capabilities of OBD-II is the **freeze frame**, or snapshot, feature. Although the regulations mandate just emission-related DTCs, manufacturers can choose to include this feature for other systems. With this feature, the PCM takes a snapshot of the activity of the various inputs and outputs at the time the PCM illuminated the MIL. The PCM uses these data for identification and comparison of similar operating conditions if the same problem occurs in the future. This feature is also valuable to technicians, especially when trying to identify the cause of an intermittent problem. The action of sensors and actuators when the code was set can be reviewed. This can be a great help in identifying the cause of a problem. The information held in freeze frame is actual values; they have not been altered by the adaptive strategy of the PCM **(Figure 31–29)**.

Once a DTC and the related freeze-frame data are stored in memory, they will stay there, even if other emission-related DTCs are set. However, if a fault occurs that has a greater effect on emission, the freeze frame of a lower priority fault may be overwritten by the data of the more serious fault. For example, if a freeze frame is stored for a small EVAP leak, these data may be overwritten if a catalyst damaging misfire occurs. The data are stored by priority; information related to misfire and fuel control have priority over other DTCs. These data are lost if the vehicle's battery is disconnected, or they can be erased with a scan tool. When a scan tool is used to erase a DTC, it automatically erases all associated freeze-frame data.

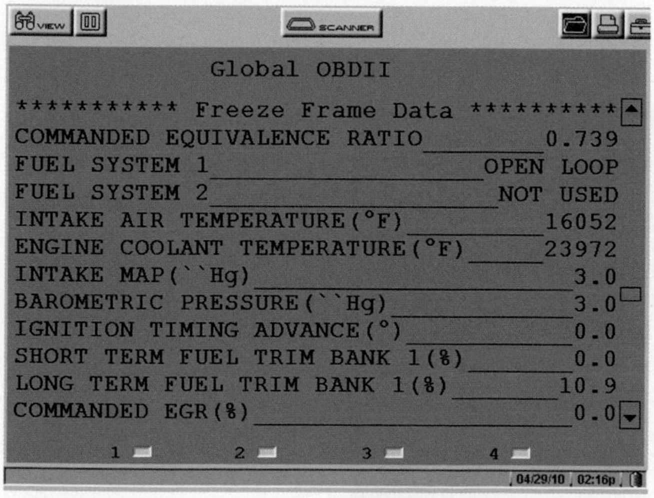

```
VIEW                 SCANNER
              Global OBDII
*********** Freeze Frame Data **********
COMMANDED EQUIVALENCE RATIO________0.739
FUEL SYSTEM 1__________________OPEN LOOP
FUEL SYSTEM 2___________________NOT USED
INTAKE AIR TEMPERATURE(°F)_______16052
ENGINE COOLANT TEMPERATURE(°F)______23972
INTAKE MAP(``Hg)_____________________3.0
BAROMETRIC PRESSURE(``Hg)____________3.0
IGNITION TIMING ADVANCE(°)___________0.0
SHORT TERM FUEL TRIM BANK 1(%)_______0.0
LONG TERM FUEL TRIM BANK 1(%)_______10.9
COMMANDED EGR(%)_____________________0.0
    1        2        3        4
                        04/29/10  02:16p
```

FIGURE 31–29

An example of the data captured in a freeze frame for a DTC.

Generic and Enhanced Data

Within the OBD-II computer, there are two different types of access: generic, also called Global OBD-II, and enhanced. Enhanced is sometimes called OE or the manufacturer side of the system. These two interconnected systems both provide access to codes and data, but do so differently.

Generic OBD-II provides access to the test modes discussed in the next section. This access can be used by tools such as code readers and aftermarket scan tools, such as those made by Snap-On, Bosch, OTC, and others. The data available are often limited compared to that available when using the OE scan tool or one that can access enhanced data. To obtain data using the generic mode, the tool does not need to be programmed to the vehicle. This means that entering VIN information is not necessary to access the serial data stream. It is important to remember that generic access is for emission control–system purposes only. You will not be able to see other systems, such as ABS or body systems, when using generic mode.

Enhanced data often provide access to a greater number of PIDs and to more systems for bidirectional control. To utilize enhanced functions, the scan tool must be programmed to the vehicle, requiring entering VIN information into the scan tool. Late-model systems often have automatic VIN identification by the scan tool, so manual VIN entry is not necessary. When using enhanced data, especially with the OE scan tool, you will find you have access to systems other than powertrain and emissions.

Many technicians begin their diagnosis by using generic mode to check for DTCs and freeze-frame data. In many cases, the generic data are sufficient to diagnose and repair a fault.

Generic Test Modes

All OBD-II systems have the same basic test modes and all are accessible with a generic OBD-II scan tool. Always refer to the manufacturer's information when using these test modes for diagnosis. Photo Sequence 24 shows the procedure for preparing a modern scan tool to read OBD-II data.

Mode 1 is the **parameter identification (PID)** mode. This mode allows access to current emission-related data values of inputs and outputs, calculated values, and system-status information. Some PID values are manufacturer specific; others

PHOTO SEQUENCE 24

PREPARING A MODERN SCAN TOOL TO READ OBD-II DATA

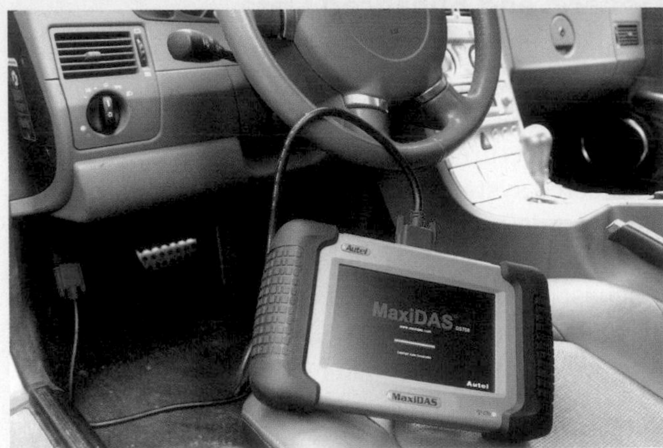

P24–1 Connect the scan tool to the vehicle.

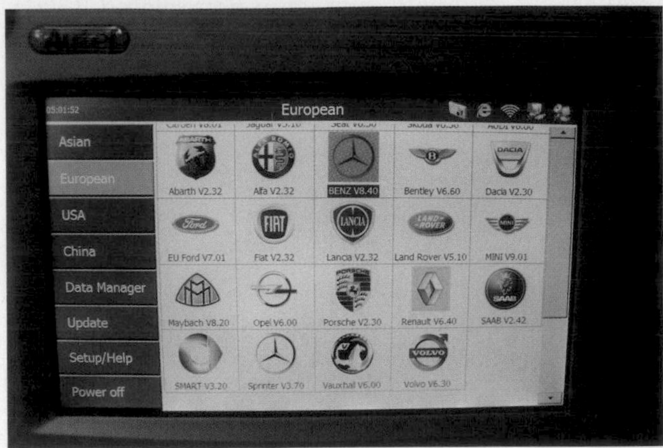

P24–2 Select the make of the vehicle.

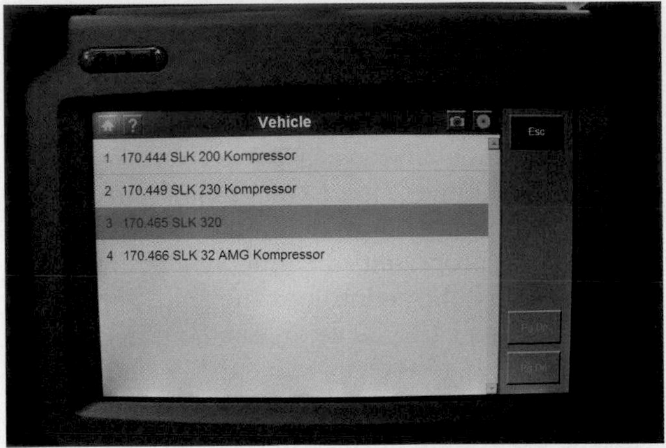

P24–3 Then select the model.

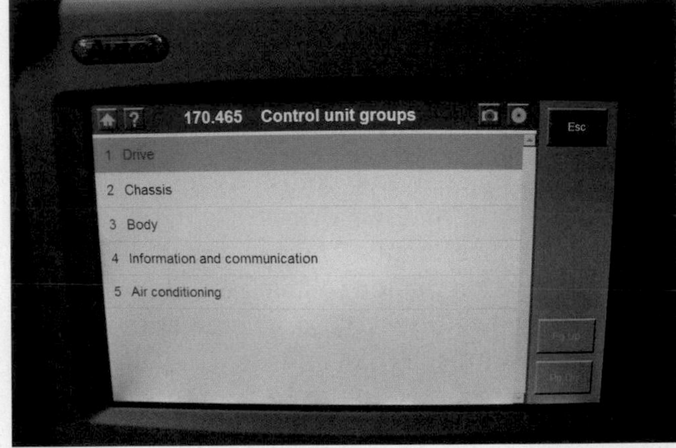

P24–4 The tool will display the basic control systems.

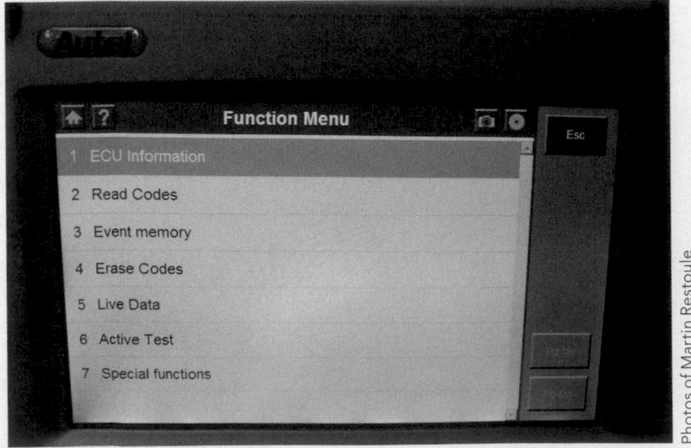

Photos of Martin Restoule

P24–5 The display will ask what you want to do.

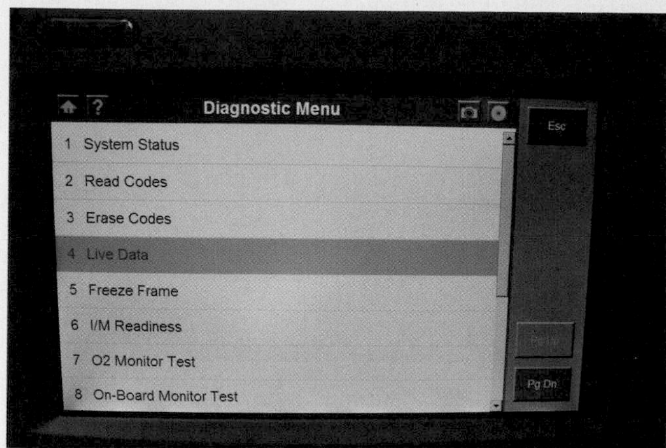

P24–6 A diagnostic menu will allow you to select various diagnostic procedures and tests.

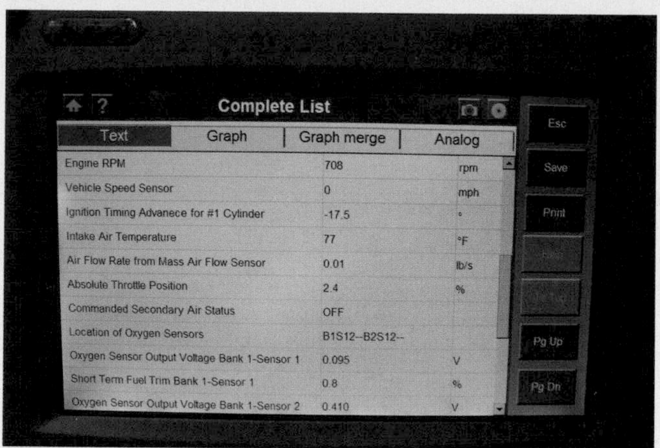

P24–7 Data lists or test results from a selected test will appear on the tool.

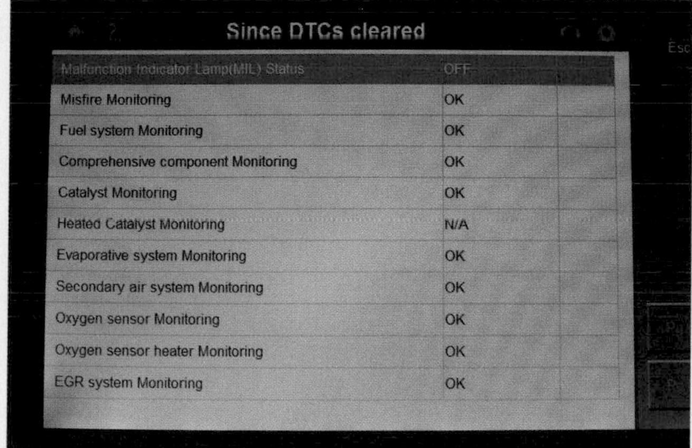

P24–8 The various monitor systems status can also be displayed.

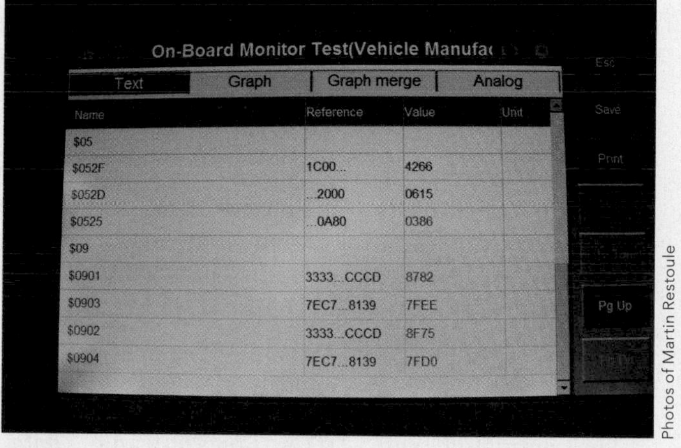

P24–9 Mode $06 data can be retrieved by entering specific vehicle information, such as the VIN.

Photos of Martin Restoule

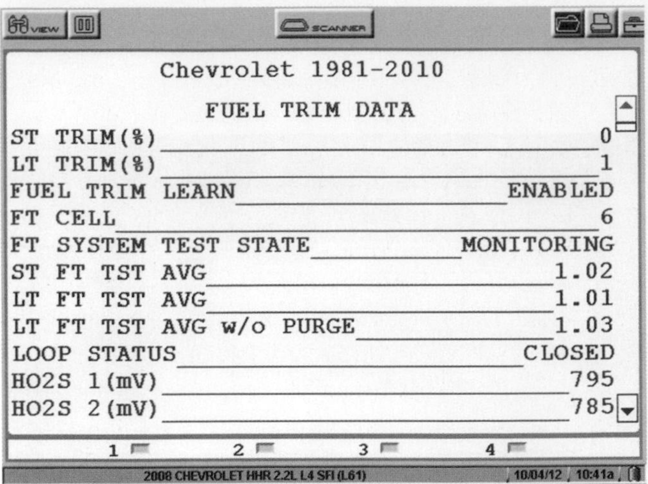

FIGURE 31–30

An example of what is available when a technician looks at the serial data on a scan tool.

Snap-on Incorporated

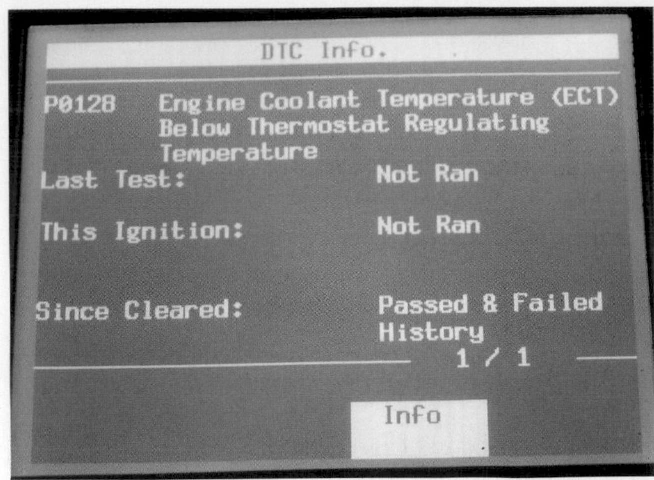

FIGURE 31–31

In test Mode 7, the DTCs reported in Continuous Tests and Pending Codes are pending DTCs. If the conditions are detected again, the DTCs will be stored as active DTCs.

Snap-on Incorporated

are common to all vehicles. This information is referred to as **serial data (Figure 31–30)**.

Mode 2 is the freeze frame–data access mode. This mode permits access to emission-related data values from specific generic PIDs. The number of freeze frames that can be stored is limited. The freeze-frame information updates if the condition recurs.

Mode 3 permits scan tools to obtain stored DTCs. The information is transmitted from the PCM to the scan tool following a Mode 3 request. The DTC or its descriptive text, or both, can be displayed on the scan tool.

Mode 4 is the PCM reset mode. It allows the scan tool to clear all emission-related diagnostic information from its memory. When this mode is activated, all DTCs, freeze-frame data, DTC history, monitoring test results, status of monitoring test results, and on-board test results are cleared and reset.

Mode 5 is the oxygen sensor monitoring test. This mode gives the actual oxygen sensor outputs during the test cycle. These are stored values, not current values that are retrieved in Mode 1. This information is used to determine the effectiveness of the catalytic converter.

Mode 6 is the output state mode (OTM) and can be used to identify potential problems in the non-continuous monitored systems.

Mode 7 reports the test results for the continuous monitoring systems **(Figure 31–31)**.

Mode 8 is the request for control of an on-board system test or component mode. It allows the technician to control the PCM, through the scan tool, to test a system. In some cases, the scan tool will only set the conditions for conducting a test and not actually conduct the test. An example of this is the EVAP leak test, which is done with other test equipment.

Mode 9 is the request for vehicle information mode. This mode reports the vehicle's identification number, calibration identification, and calibration verification. This information can be used to see if the most recent calibrations have been programmed into the PCM.

BASIC DIAGNOSIS OF ELECTRONIC ENGINE CONTROL SYSTEMS

Diagnosing a computer-controlled system is much more than accessing DTCs. You need to know what to test, when to test it, and how to test it. Because the capabilities of the engine control computer have evolved from simple to complex, it is important to know the capabilities of the system you are working with before attempting to diagnose a problem. Refer to the service information for this. After you understand the system and its capabilities, begin your diagnosis using your knowledge and logic.

The importance of logical troubleshooting cannot be overemphasized. The ability to diagnose a problem (to find its cause and its solution) is what separates an automotive technician from a parts changer.

Logical Diagnosis

When faced with an abnormal engine condition, the best automotive technicians compare clues (such as scan tool data, meter readings, oscilloscope readings, and visible problems) with their knowledge of proper conditions and discover a logical reason for the way the engine is performing. Logical diagnosis means following a simple basic procedure. Start with the most likely cause and work to the most unlikely. In other words, check out the easiest, more obvious solutions first before proceeding to the less likely and more difficult solutions. Do not guess at the problem or jump to a conclusion before considering all of the factors.

The logical approach has a special application to troubleshooting electronic engine controls. Always check all traditional non-electronic engine control possibilities before attempting to diagnose the electronic engine control system. For example, low battery voltage might result in faulty sensor readings. Remember, even the most advanced on-board computer system is not capable of correcting for mechanical engine problems. Incorrect valve timing, sticking valves, low compression, and vacuum leaks may present problems that appear to be computer or sensor faults.

Repair Information

All late-model engine controls have self-diagnosis capabilities. A malfunction in any sensor, output device, or in the computer itself is stored as a DTC. DTCs are retrieved, and the indicated problem areas checked further. Correct diagnosis depends on correctly interpreting all data collected and performing all subsequent tests properly. The following service information will help:

- Instructions for retrieving DTCs and obtaining freeze frames
- Instructions for diagnosing a no-communication problem between the system and the scan tool
- Current technical bulletins for the vehicle
- A fail-safe chart that shows what strategy is taken when certain DTCs are set
- The operating manual for the scan tool
- A DTC chart with the codes and possible problem areas
- A parts locator for the system
- An electrical wiring diagram for the system
- Identification of the various terminals of the PCM
- A DTC troubleshooting guide
- Component testing sequences

DIAGNOSING OBD-I SYSTEMS

OBD-I systems have limited self-diagnostic capabilities. By entering into a self-test mode, the system is able to evaluate its condition. If problems are found, they may be identified as either hard faults (on-demand) or intermittent failures. Each type of fault or failure is assigned a numerical trouble code that is stored in memory.

A hard fault is a problem found in the system at the time of the self-test. An intermittent fault, on the other hand, indicates that a malfunction occurred (for example, a poor connection causing an intermittent open or short) but was not present during the self-test. Nonvolatile RAM allows intermittent faults to be stored for up to a specific number of ignition key on–off cycles. If the trouble does not reoccur during that period, it is erased from the memory.

There are various ways to access the trouble codes stored in the computer. Most manufacturers have specific equipment designed to monitor and test the system's components. Aftermarket companies also manufacture scan tools that have the capability to read and record the system's input and output signals.

Visual Inspection

Begin diagnostics with a visual check of the engine and its systems. Include the following:

1. Inspect the air filter and related hardware around the filter.
2. Inspect the entire PCV system.
3. Check to make sure the EVAP canister is not saturated or flooded.
4. Check the charging system for loose, damaged, and corroded wires or connections.
5. Check the condition of the battery and its terminals and cables.
6. Make sure all vacuum hoses are connected and are not pinched or cut.
7. Check all sensors and actuators for signs of physical damage.

Unlocking Trouble Codes

Although the parts in any computerized system are amazingly reliable, they do occasionally fail. Diagnostic charts in service information can help you through troubleshooting procedures. Start at the top and follow the sequence down.

The MIL should not be illuminated after the engine has started. If it is, the computer has found a problem. OBD-I diagnostic procedures and DTCs differ with vehicle make and year. Each time the key is turned to the ON position, the system does a self-check. The self-check makes sure that all of the bulbs, fuses, and electronic modules are working. If the self-test finds a problem, it might store a code for later servicing. It may also instruct the computer to turn on the MIL to show that service is needed.

Retrieving Trouble Codes

Prior to OBD-II, each car manufacturer required a different method for retrieving DTCs. In fact, there were different procedures for different models and engines made by the same manufacturer. The diagnostic connectors looked different and were located in different places. Also, the diagnostic codes represented different things. Although all of the computer systems did basically the same thing, diagnostic methods were as different as day and night. Always check the service information for the vehicle being serviced before attempting to retrieve DTCs.

To retrieve DTCs, some manufacturers require that certain pins at the DLC be connected, particular connectors be disconnected, or a specific key-on and key-off sequence be followed. In these cases, the DTCs appear as flash codes displayed with the MIL or on a voltmeter. In most cases, a scan tool can be used to view the DTCs.

Using a Scan Tool

Scan tools are available to diagnose nearly all engine control systems. When using a scan tool on these early systems, make sure it is compatible with the system. Often, a manufacturer-specific scan tool is required.

Photo Sequence 25 shows a typical procedure for using a scan tool on an OBD system. The use of a scan tool varies with the make of the tool, but most require an initial entry of vehicle information, including the VIN.

After the tool has been programmed, the technician selects the desired test sequence on the scan tool. These selections also vary with the type of scan tool and the vehicle being tested.

DIAGNOSING OBD-II SYSTEMS

At least one drive cycle is required for the monitors to run. All OBD-II scan tools include a readiness function that shows all of the monitoring sequences and the status of each—complete or incomplete. Some systems may show the monitors as ready or not ready. Incomplete can mean the monitor did not complete; judgment is withheld pending further testing; the monitor did not operate; or the monitor operated and recorded a failure. The "Readiness Test" and "Monitor Status" screens on most scan tools contain identical information.

Readiness monitor tools are available. These plug into the DLC and alert the technician when a drive cycle is completed. These can save much time while trying to gather data before and after a repair. They are especially handy after a repair has been made. The vehicle is taken for a test drive, and when the readiness monitor flashes and/or beeps, the vehicle can be brought back into the shop and checked for DTCs.

Troubleshooting OBD-II Systems

The following steps provide a general outline for troubleshooting OBD-II systems. There are slight variations in different years and with different models. Always refer to the manufacturer's information before beginning your diagnosis. Troubleshooting OBD-II systems involves a series of steps, as listed in **Figure 31–32**. Here is a brief discussion of each step:

1. **Interview the customer.** Gather as much information as possible from the customer. Ask the customer to describe the driving conditions present when the problem appears. This should include weather, traffic, and speed.
2. **Check the MIL.** The MIL should turn on when the ignition is turned on and the engine is not running. When the engine is started, the MIL should go off. If either of these does not occur, troubleshoot the lamp system before continuing.
3. **Connect the scan tool.** Make sure the tool is OBD-II compliant.
4. **Check DTC(s) and freeze-frame data.** When using a scan tool, an asterisk (*) next to the DTC often indicates there is stored freeze-frame data associated with that DTC. During diagnostics, it is helpful to know the conditions present when the DTC was set, such as whether the vehicle was running or stopped, what the engine's temperature was, and whether the air/fuel ratio was rich or lean. Print or record all DTC and related information. If there was a no- or poor-communication DTC, solve that problem before continuing.
5. **Check service history and service publications.** There may be a TSB or other service alert that may have the necessary repair information. Check these, and follow the procedures outlined in them before continuing. This is especially important to note whether an updated

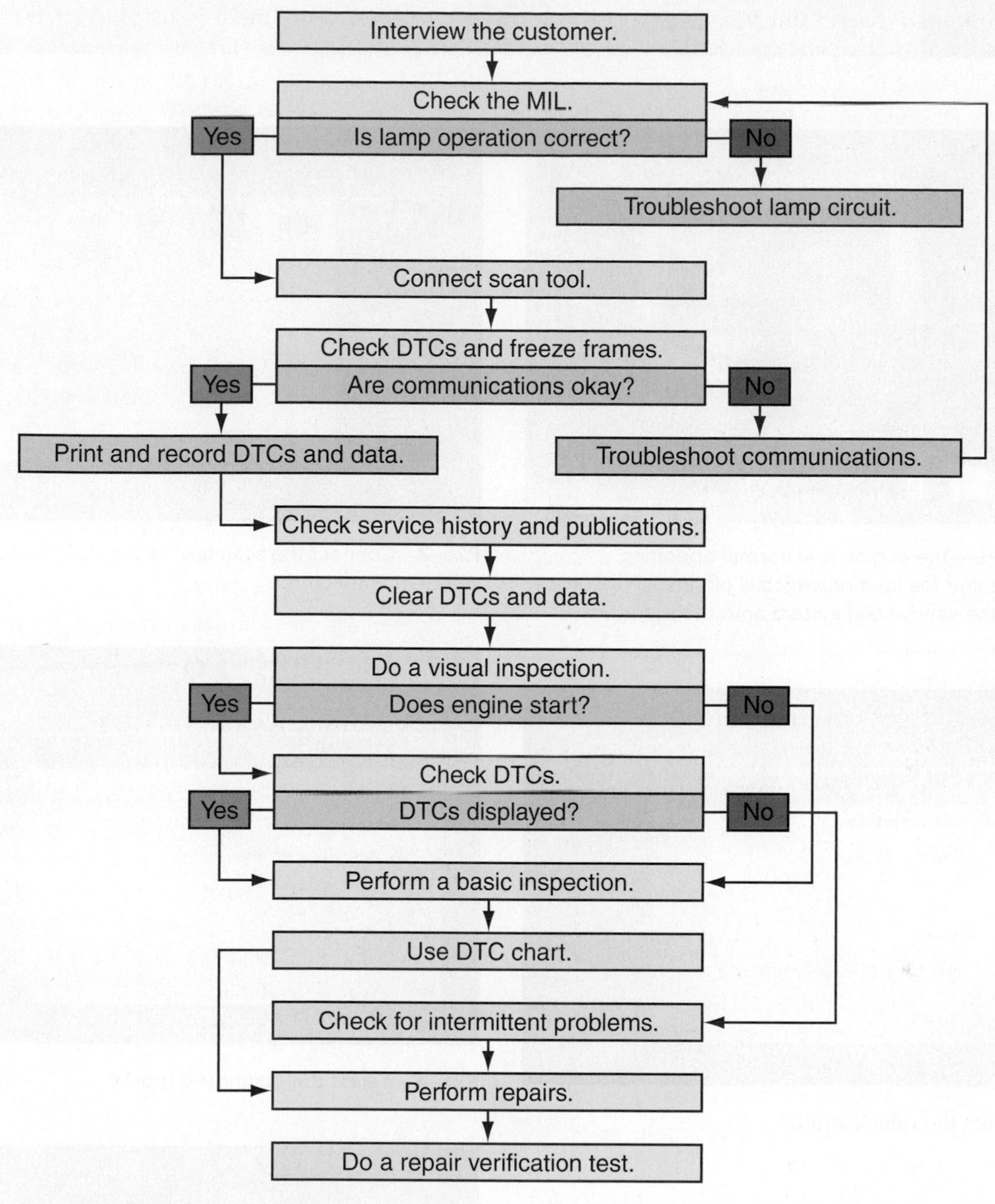

FIGURE 31–32

The steps that should be followed when troubleshooting OBD-II systems.

calibration or flash is available. Hours of diagnostic time can be wasted trying to diagnose and solve a problem addressed by a new software release. Service history can give clues about the cause of the problem because the problem may be related to a recent repair.

6. **Visual inspection.** Take a quick look at the very basics. Check all wires to make sure they are firmly connected and not damaged. Try not to wiggle the wires while doing this; a wiggle may correct an intermittent problem, and it may be hard to find later. Conduct a visual inspection on the battery and fuel level. Correct any problems if necessary. If the engine does not start, move down to step 8.
7. **Check DTCs.** Repeat step 4. If there are no DTCs, check the status of the monitors' readiness and the pending codes on the scan tool. Do what is necessary to complete the necessary drive cycles before continuing. For many DTCs, the PCM will enter into the fail-safe mode. This means the PCM has substituted a value to allow

PHOTO SEQUENCE 25

DIAGNOSIS WITH A SCAN TOOL

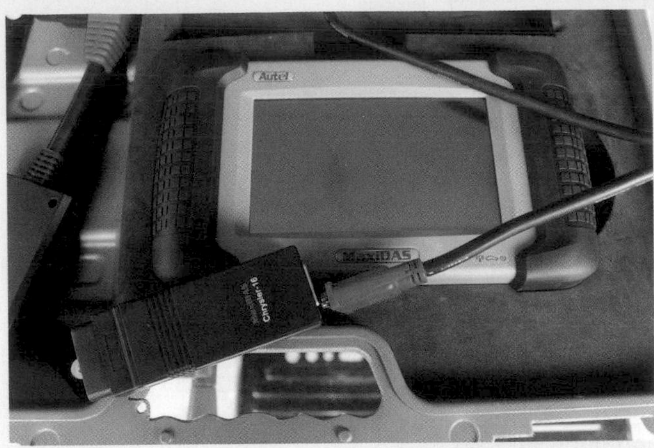

P25–1 Be sure the engine is at normal operating temperature and the ignition switch is off. Install the proper adapter for the vehicle and system onto the scan tool.

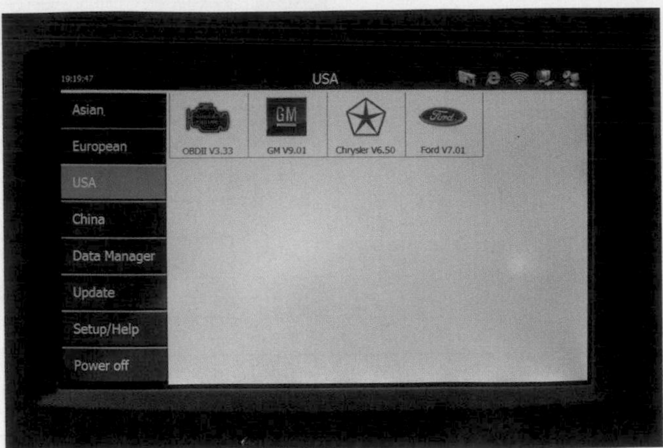

P25–2 Connect the scan tool to the DLC. Select the vehicle manufacturer.

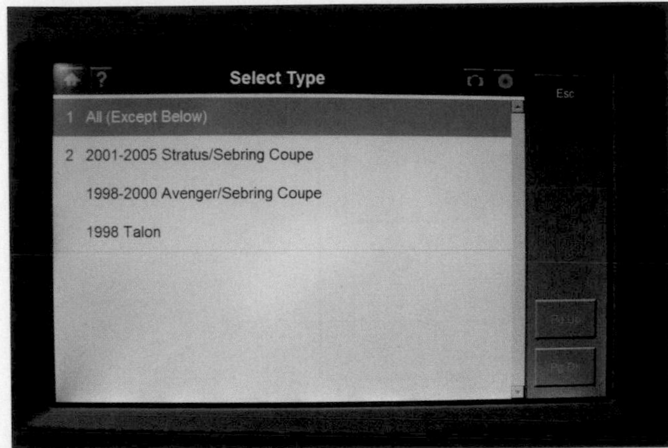

P25–3 Select the vehicle model.

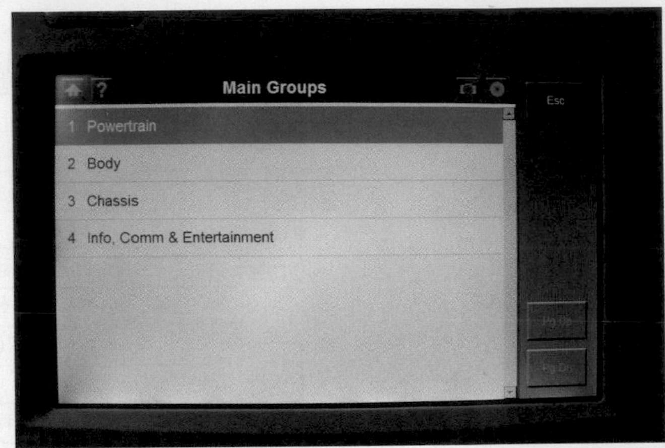

P25–4 Select the diagnostic group.

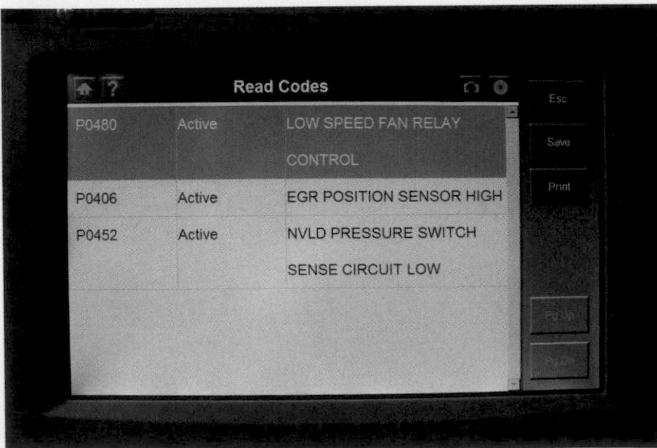

P25–5 Retrieve the DTCs with the scan tool, and then interpret the codes by using the service information.

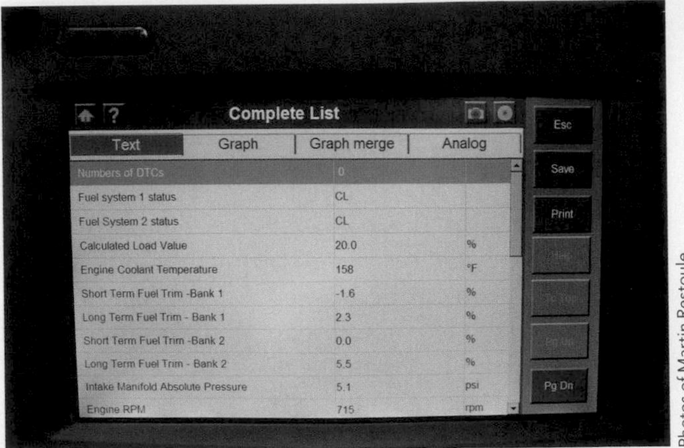

P25–6 Start the engine and obtain the input sensor and output actuator data on the scan tool. If a printer for the tool is available, print out the data report. Compare data to specifications.

the engine to run. Refer to the service information to determine if any DTCs indicate the fail-safe mode; if so, follow the appropriate diagnostic steps. If there is a DTC, proceed to step 9. If no DTCs are present, go to step 10.

8. **Basic inspection.** When the DTC is not confirmed in the previous check, diagnosis of the engine's support system should be done. Use the problem symptoms chart in the service information to diagnose when no code is displayed but the problem is still occurring.
9. **DTC chart.** Use the DTC chart to determine what was detected, the probable problem areas, and how to diagnose that DTC **(Figure 31–33)**.
10. **Check for intermittent problems.** If the cause of the problem has not yet been determined, proceed to check for an intermittent problem.
11. **Perform repairs.** Once the cause of the concern has been identified, perform all required services.
12. **Repair verification test.** After making a repair, check your work by repeating steps 2, 3, and 4.

Intermittent Faults

An intermittent fault is a fault that is not always present. It may not activate the MIL or cause a DTC to be set. Therefore, intermittent problems can be difficult to diagnose. By studying the system and the relationship of each component to another, you should be able to create a list of possible causes for the intermittent problem. To help identify the cause, follow these steps:

1. Observe the history DTCs, DTC modes, and freeze-frame data.
2. Dealerships can access technical assistance centres for possible solutions, and independent shops can use service sites and networks such as iATN and Identifix. Combine your knowledge of the system with the service information that is available.
3. Evaluate the symptoms and conditions described by the customer.
4. Use a check sheet to identify the circuit or electrical system component that may have the problem.
5. Follow the suggestions for intermittent diagnosis found in service material.
6. Visually inspect the suspected circuit or system.
7. Use the data-capturing capabilities of the scan tool.
8. Test the circuit's wiring for shorts, opens, and high resistance. This should be done with a DMM and in the typical manner, unless instructed differently in the service information.

Most intermittent problems are caused by faulty electrical connections or wiring. Refer to a wiring diagram for each of the suspected circuits or components. This will help identify all of the connections and components in that circuit. The entire electrical system of the suspected circuit should be carefully and thoroughly inspected. Check for burnt or damaged wire insulation, damaged terminals at the connectors, corrosion at the connectors, loose connectors, wire terminals loose in the connector, and disconnected or loose ground wire or straps. The absence of a visual problem with a connector or terminal does not mean that a problem does not exist. Perform a pin drag test to check the connection

DTC #	DETECTION ITEM = THE SYSTEM THAT HAS THE PROBLEM	TROUBLE AREA = THE SUSPECTED PROBLEM AREA OF THE SYSTEM	THE MIL IS LIT (Y/N)	INFORMATION IS HELD IN MEMORY (Y/N)
P0100	Mass airflow circuit malfunction	• Open or short in MAF circuit • Faulty MAF sensor • Faulty PCM	N	Y
P0101	MAF circuit – range/performance problem	Faulty MAF sensor	Y	Y
P0110	Faulty IAT circuit	• Open or short in IAT circuit • Faulty IAT • Faulty PCM	Y	Y
P0115	Faulty ECT circuit	• Open or short in ECT circuit • Faulty ECT • Faulty PCM	Y	Y

FIGURE 31–33

An explanation of a typical DTC chart.

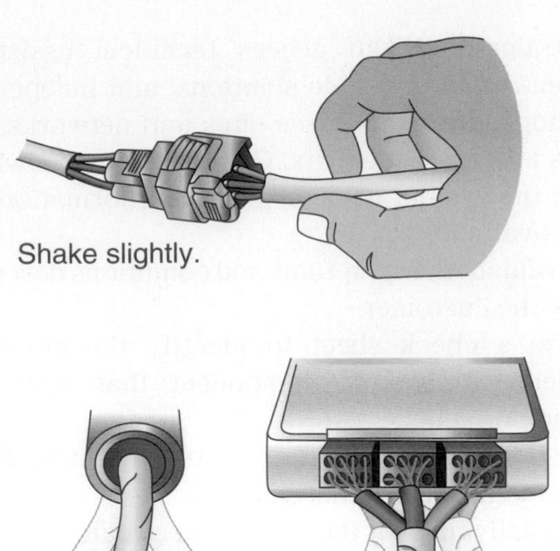

FIGURE 31–34

The wiggle test can be used to locate intermittent problems.

between the male and female terminals in a suspect connector. This requires inserting the correct male terminal into the female terminal and noting the amount of drag or tension between the two. If the male terminal easily slides or falls out of the female terminal, examine the cavity of the female terminal for damage.

To locate the source of the problem, a voltmeter can be connected to the suspected circuit and the wiring harness wiggled **(Figure 31–34)**. As a guideline for what voltage should be expected in the circuit, refer to the reference value table in the service information. If the voltage reading changes with the wiggles, the problem is in that circuit. The vehicle can also be taken for a test drive with the voltmeter connected. If the voltmeter readings become abnormal with changing operating conditions, the circuit being observed probably has the problem.

The vehicle can also be taken for a test drive with the scan tool connected. The scan tool can be used to monitor the activity of a circuit while the vehicle is being driven. This allows you to look at a circuit's response to changing conditions. The snapshot or freeze-frame feature stores engine conditions and operating parameters at command or when the PCM set a DTC. If the snapshot can be taken when the intermittent problem is occurring, the problem will be easier to diagnose.

With an OBD-II scan tool, actuators can be activated and their functionality tested. The results of the change in operation can be monitored. Also, the outputs can be monitored as they respond to input changes. When an actuator is activated, watch the response on the scan tool. Also listen for the clicking of the relay that controls that output. If no clicking is heard, measure the voltage at the relay's control circuit; there should be a change of more than 4 volts when it is activated.

To monitor how the PCM and an output respond to a change in sensor signals, use the scan tool. Select the mode that relates to the suspected circuit and view, then record, the scan data for that circuit. Compare the reading to specifications. Then create a condition that would cause the related inputs to change. Observe the data to see if the change was appropriate.

Serial Data

PIDs are codes used to request data from a PCM. A scan tool is used for the request and receipt of the serial data. To do this, the desired PID **(Figure 31–35)** is entered into the scan tool. The device connected to the CAN bus that is responsible for this PID reports the value for that PID back to the bus and is read on the scan tool. Many PIDs are standard for all OBD-II systems; however, not all vehicles will support all PIDs, and there are manufacturer-defined PIDs that are not part of the OBD-II standard.

Manufacturers list and define the various applicable PIDs in their service information. This information should be compared to the observed data. If an item is not within the normal values, record the difference, and diagnose that particular item.

Using Mode $06

Mode $06 allows access to the results of various monitor diagnostic tests. The test values are stored at the time each monitor is completed. The information available in Mode $06 can be extremely helpful, but it also can be very difficult to decipher. Mode $06 data are given with $ signs, test IDs (TIDs), content IDs (CIDs), and a mixture of letters and numbers. Most of this has no face value because it is given in "hexadecimals," which is like a foreign language. For example, the hex code for the number 10 is $0A. This is why it is recommended that only scan tools capable of interpreting these data be used. The way to observe Mode $06 data on a scan tool varies with the make and model of the tool. Always refer to the tool's instruction manual.

FREEZE FRAME	ACRONYM	DESCRIPTION	MEASUREMENT UNITS
X	AAT	Ambient Air Temperature	Degrees
X	AIR	Secondary Air Status	On–Off
X	APP_D	Accelerator Pedal Position D	%
X	APP_E	Accelerator Pedal Position E	%
X	APP_F	Accelerator Pedal Position F	%
X	CATEMP11	Catalyst Temperature Bank 1, Sensor 1	Degrees
X	CATEMP12	Catalyst Temperature Bank 1, Sensor 2	Degrees
X	CATEMP21	Catalyst Temperature Bank 2, Sensor 1	Degrees
X	CATEMP22	Catalyst Temperature Bank 2, Sensor 2	Degrees
	IAT	Intake Air Temperature	Degrees
X	LOAD	Calculated Engine Load	%
X	LOAD_ABS	Absolute Load Value	%
X	$LONGFT_1$	Current Bank 1 Fuel Trim Adjustment (kamref1) from Stoichiometry, Which is Considered Long Term	%
X	$LONGFT_2$	Current Bank 2 Fuel Trim Adjustment (kamref2) from Stoichiometry, Which is Considered Long Term	%
X	MAF	Mass Airflow Rate	gm/s-lb/min
	MIL_DIST	Distance Travelled with MIL on	kilometre
X	O_2S11	Bank 1 Upstream Oxygen Sensor (11)	Volts
X	O_2S12	Bank 1 Downstream Oxygen Sensor (12)	Volts
X	O_2S13	Bank 1 Downstream Oxygen Sensor (13)	Volts
X	O_2S21	Bank 2 Upstream Oxygen Sensor (21)	Volts
X	O_2S22	Bank 2 Downstream Oxygen Sensor (22)	Volts
X	O_2S23	Bank 2 Downstream Oxygen Sensor (23)	Volts
X	$SHRTFT_1$	Current Bank Fuel Trim Adjustment (lambse1) from Stoichiometry, Which is Considered Short Term	%
X	$SHRTFT_2$	Current Bank 2 Fuel Trim Adjustment (lambse2) from Stoichiometry, Which is Considered Short Term	%
X	SPARKADV	Spark Advance Request	Degrees
X	SPARK_ACT	Spark Advance Actual	Degrees
	TAC_PCT	Commanded Throttle Actuator	%
X	TP	Throttle Position	%
X	VSS	Vehicle Speed Sensor	km/h-mph

FIGURE 31–35

A sample list of generic PIDs.

If the scan tool you are using reads Mode $06 data but does not translate it into usable data, you can still use the data supplied by the tool. You will need to access the Mode $06 data supplied on the websites of many auto manufacturers. Using the information from the scan tool and the information from the manufacturer, you should be able to determine what each TID means and what the reported values represent.

Mode $06 data can help identify the cause of problems when a DTC has not been retrieved. Manufacturers normally list the various normal Mode $06 values. To effectively use Mode $06 data, compare the captured reading with the normal values. Then use logic and your knowledge of the system, part, and electricity to determine what is abnormal and what could be causing that. Some manufacturers recommend using Mode $06 to diagnose specific systems. Using these data is also very handy in diagnosing any system of the control circuit.

Repairing the System

After identifying the cause of the problem, repairs should be made. When servicing or repairing OBD-II circuits, the following guidelines are important:

- Do not connect aftermarket accessories into an OBD-II circuit.
- Do not move or alter grounds from their original locations.
- Always replace a relay with an exact replacement. Damaged relays should be thrown away, not repaired.
- Make sure all connector locks are in good condition and are in place.
- After repairing connectors or connector terminals, make sure the terminals are properly retained and the connector is sealed.
- When installing a fastener for an electrical ground, be sure to tighten it to the specified torque.

After repairs, the system should be rechecked to verify that the repair took care of the problem. This may involve road-testing the vehicle in order to verify that the complaint has been resolved.

Using a Wireless Interface

Wireless OBD-II diagnostic scanners and code-reader interfaces are now available. These usually offer both Wi-Fi and Bluetooth technology that can allow a laptop, tablet, or smart/super-phone to access the vehicle information **(Figure 31–36)**. The receiving device, whether a laptop or smartphone, must have a program loaded on it that allows communication with the interface and the transfer of information to occur. These devices do not normally have all the options that a proper scan tool offers, but they do allow for clearing engine codes, resetting the check engine light, and viewing real-time vehicle data during operation.

FIGURE 31–36

A typical Bluetooth interface for an Android-type smart/super-phone.

Martin Restoule

KEY TERMS

Air induction system (p. 908)
Base ignition timing (p. 905)
Catalyst monitor sensor (CMS) (p. 917)
Central port injection (CPI) (p. 906)
Closed-loop (p. 915)
Data link connector (DLC) (p. 913)
Distributor ignition (DI) (p. 905)
Drive cycle (p. 917)
Electronic fuel injection (EFI) (p. 906)
Electronic ignition (EI) (p. 905)
Enable criteria (p. 916)
Engine control module (ECM) (p. 909)
Freeze frame (p. 927)
Gasoline direct injection (GDI) (p. 906)
Ignition timing (p. 905)
Long-term fuel trim (LTFT) (p. 920)
Malfunction indicator lamp (MIL) (p. 913)
Open-loop (p. 915)
Parameter identification (PID) (p. 927)
Pending code (p. 926)
Port fuel injection (PFI) (p. 906)
Powertrain control module (PCM) (p. 909)
Serial data (p. 930)
Short-term fuel trim (STFT) (p. 920)
Throttle body injection (TBI) (p. 906)
Trip (p. 917)

SUMMARY

- To have complete combustion, there must be the correct amount of fuel mixed, in a sealed container, with the correct amount of air, and this must be shocked by the correct amount of heat at the right time.
- The ignition system is responsible for delivering the spark that causes combustion.
- The fuel system must deliver the fuel from the fuel tank to the fuel injectors, which spray fuel into the cylinders.
- The air induction system delivers the air to the cylinders.
- Emission control devices are added to engines because an engine does not experience complete combustion during all operating conditions.
- Engine control systems operate in a loop—input, process, and control.
- An engine control module will operate in open or closed loop. In closed loop, the computer processes all inputs.
- Most engine control systems have self-diagnostic capabilities. By entering this mode, the computer is able to evaluate the entire control system, including itself.
- According to OBD-II standards, all vehicles have a universal DLC with a standard location, a standard list of DTCs, a standard communication protocol, common use of scan tools on all vehicle makes and models, common diagnostic test modes, the ability to record and store a snapshot of the operating conditions that existed when a fault occurred, and a standard glossary of terms, acronyms, and definitions that must be used for all components in electronic control systems.
- An OBD-II system has many monitors to check system operation, and the MIL light is illuminated if vehicle emissions exceed 1.5 times the allowable standard for that model year.
- Monitors included in OBD-II are the following: catalyst efficiency, engine misfire, fuel system, heated exhaust gas oxygen sensor, EGR, EVAP, secondary air injection, thermostat, and comprehensive component.
- OBD-II vehicles use a minimum of two oxygen sensors. One of these is used for feedback to the PCM for fuel control, and the other gives an indication of the efficiency of the converter.
- An OBD-II drive cycle includes whatever specific operating conditions are necessary either to initiate and complete a specific monitoring sequence or to verify a symptom or a repair.
- OBD-II's short-term fuel trim and long-term fuel trim strategies monitor the oxygen sensor signals and use this information to make adjustments to the fuel control calculations. The adaptive fuel control strategy allows for changes in the amount of fuel delivered to the cylinders according to operating conditions.
- An OBD-II system monitors the entire emissions system, switches on a MIL if something goes wrong, and stores a fault code in the PCM when it detects a problem.
- OBD-II regulations require that the PCM monitor and perform some continuous tests on the engine control system and components. Some OBD-II tests are completed at random, at specific intervals, or in response to a detected fault.
- OBD-II systems note the deterioration of certain components before they fail, which allows owners to bring in their vehicles at their convenience and before it is too late.
- The MIL informs the driver that a fault that affects the vehicle's emission levels has occurred. After making the repair, technicians may need to take the vehicle for three trips to ensure that the MIL does not illuminate again.
- Most intermittent problems are caused by faulty electrical connections or wiring.
- OBD-I systems have limited and unique self-diagnostic routines.

REVIEW QUESTIONS

1. Describe the difference between an open-loop and a closed-loop operation.
2. Explain the use and importance of system strategy and look-up tables in the computerized control system.
3. Describe an OBD-II warm-up cycle.
4. Explain the trip and drive cycle in an OBD-II system.
5. Describe how engine misfire is detected in an OBD-II system.
6. When type B cylinder misfires occur, how many consecutive drive cycles must they occur to illuminate the MIL?
   - **a.** 1 drive cycle
   - **b.** 2 drive cycles
   - **c.** 4 drive cycles
   - **d.** 8 drive cycles

7. Which sensor is used for misfire monitoring on OBD-II systems?
   a. oxygen sensor
   b. camshaft position sensor
   c. crankshaft position sensor
   d. DPFE sensor
8. Which of the following statements is *not* true?
   a. DTCs that will not affect emissions will never illuminate the MIL.
   b. An active or current DTC represents a fault that was detected and occurred during two trips.
   c. When a two-trip fault is detected for the second time, a DTC is stored as a pending code.
   d. There are some DTCs that will set in one trip.
9. Which of the following is a computer not able to do?
   a. receive input data
   b. process input data according to a program and monitor output action
   c. control the vehicle's operating conditions
   d. store data and information
10. Which of the following memory circuits is used to store trouble codes and other temporary information?
   a. read-only memory
   b. programmed read-only memory
   c. random access memory
   d. all of the above
11. Which of the following is not part of the fuel delivery system?
   a. fuel lines **b.** fuel injector
   c. fuel filter **d.** fuel pump
12. How many pins in the data link connector are assigned by the OBD-II standard?
   a. 6 of the 12 pins
   b. 6 of the 14 pins
   c. 7 of the 16 pins
   d. 7 of the 18 pins
13. Which of the following faults may not be picked up by the on-board computer?
   a. a slow-responding O_2 sensor
   b. a vacuum leak
   c. a cylinder misfire
   d. a faulty EGR solenoid
14. What system uses a leak detection pump (LDP)?
   a. evaporative emission (EVAP) system
   b. exhaust gas recirculation (EGR) system
   c. engine vacuum system (EVS)
   d. positive crankcase ventilation (PCV) system
15. What is the purpose of the S2 or downstream HO_2S?
   a. to slow down the operation of the S1 HO_2S
   b. to speed up the operation of the S1 HO_2S
   c. to monitor the condition of the catalytic converter
   d. to verify the temperature of the exhaust gases
16. What is the purpose of using Mode $06 data?
   a. to allow access to the results of various monitor diagnostic tests
   b. to allow the scan tool to reprogram the computer
   c. to allow the scan tool to locate intermittent faults
   d. to allow the scan tool to log the various DTCs
17. What would a diagnostic trouble code starting with CO identify?
   a. a catalytic OBD-II standard code
   b. a chassis OBD-II standard code
   c. a catalytic manufacturers code
   d. a chassis manufacturers code
18. While detecting type A misfires, what is the revolution period that the monitor examines?
   a. 200 revolutions **b.** 500 revolutions
   c. 800 revolutions **d.** 1000 revolutions
19. Under what operating condition does the computer not react to oxygen sensor feedback information?
   a. open loop **b.** closed loop
   c. high engine loads **d.** idle
20. What will a catalytic converter that is not reducing emissions properly cause?
   a. a voltage frequency increase on the downstream HO_2S
   b. a voltage frequency decrease on the downstream HO_2S
   c. a voltage frequency increase on the upstream HO_2S
   d. a voltage frequency decrease on the upstream HO_2S

CHAPTER 32
Detailed Diagnosis and Sensors

LEARNING OUTCOMES

- Perform a scan tester diagnosis on various vehicles.
- Conduct preliminary checks on an OBD-II system.
- Use a symptom chart to set up a strategic approach to troubleshooting a problem.
- Monitor the activity of OBD-II system components.
- Diagnose computer voltage supply and ground wires.
- Test and diagnose switch-type input sensors.
- Test and diagnose variable resistance-type input sensors.
- Test and diagnose generating-type input sensors.
- Test and diagnose output devices (actuators).
- Perform active tests of actuators using a scan tool.

On-board diagnostic systems will lead a technician to the area of a driveability or emissions problem. Many different input sensors are involved in the overall driveability of a vehicle. This includes diesel- and gasoline-powered engines. Because of the use of computer networks, the inputs from the various sensors play an important part in the operation of all engine performance systems. Because the signals from these sensors are shared by many control modules, one sensor cannot be designated as affecting only one system. Often, a diagnostic trouble code (DTC) will be set that reflects a problem affecting more than one system. These problems are typically caused by a faulty sensor or sensor circuit. This chapter looks at the most common sensors and how to test each.

USING SCAN TOOL DATA

The engine control module (ECM), or powertrain control module (PCM), constantly monitors information from various switches, sensors, and other control modules. It controls the operation of systems that affect vehicle performance and emission levels, and monitors emission-related systems for deterioration. OBD-II monitors set a DTC when the performance of a system can cause elevated emissions. The PCM also alerts the driver of emission-related concerns by illuminating the malfunction indicator lamp (MIL). At the same time, a DTC is set, defining the area that caused the MIL to be lit **(Figure 32–1)**.

Diagnostic System Checks

Before starting to diagnose a concern, make sure you have covered the basics. At the beginning of most diagnostic routines, you are asked if you have performed a diagnostic system check. This

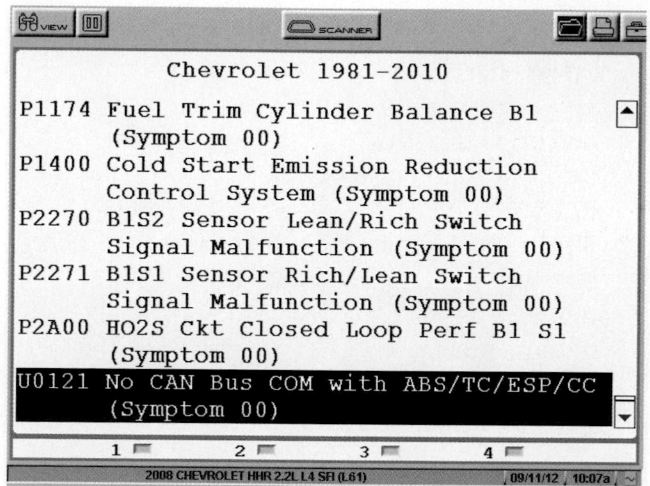

FIGURE 32–1

The DTCs that caused the MIL to light will be listed on the scan tool.

Snap-on Incorporated

usually means making basic checks of items such as the following:

- Battery voltage and charging system operation
- Any open fuses
- Visual inspection of obvious components of the system in question
- Check for the installation of any aftermarket devices

Connecting the Scan Tool

1. Be sure the engine is at normal operating temperature and the ignition switch is OFF.
2. Install the proper module for the vehicle and the system into the scan tool.
3. Connect the scan tool power leads to a power source.
4. Connect the scan tool to the DLC.

SHOP TALK

Rather than rely on exchanging specific modules and manually inputting information into the scan tool's interface with the data link connector (DLC) to program it for a particular vehicle and a system of the vehicle, Snap-On uses personality keys **(Figure 32–2)**. These keys tell the scan tool all it needs to know. But a GM vehicle may require a Key 20 to talk to powertrain and a Key 17 to talk to body. Different cars and different communication protocols use different keys to program the tool to the car. As scan tool software has been updated, some cars, particularly GMs, don't need a key to talk to the powertrain but may need a key to talk to the ABS or chassis.

FIGURE 32–2

A personality key position between the scan tool's connector and the interface for the DLC.

Establish communication with the computer system. If the scan tool does not power up, check the fuses and wiring for the DLC. If the scan tool powers up but does not communicate, you will need to diagnose this problem before continuing.

GO TO ▶ Chapter 31 for diagnosing computer communication system faults.

Quick Tests

Used primarily on Ford vehicles, quick tests provide three methods of obtaining diagnostic information from the on-board computer system. The three quick tests are as follows:

- Key-on engine-off (KOEO) self-test
- Key-on engine-running (KOER) self-test
- Continuous memory self-test

The KOEO self-test performs a functional test of certain input sensors and output actuators. If the concern is present during the KOEO self-test, a DTC will be generated and stored in memory. However, if the problem is intermittent or not occurring during the test, this test will not detect the problem, and no codes will be set.

The KOER self-test is similar to the KOEO test, except the engine is running during the test. For the test to complete correctly, the engine should be at normal operating temperature. In addition, action is often need by the technician during this test to press the brake pedal, turn the steering wheel, and cycle overdrive switches. Like the KOEO test, the fault must be present during the KOER test to be detected and for a code to be set.

The continuous memory self-test allows you to retrieve both emission- and non-emission-related DTCs. All stored DTCs are read during this test.

Once you have confirmed the customer complaint and have determined that a problem does exist, you will need to use the scan tool to access other information. The information provided by the scan tool varies depending on the scan tool and the vehicle tested. Because the standardization provided by OBD-II, scan tools are able to display the same basic type of information regardless of the vehicle.

Parameter Identifications (PIDs)

Parameter identification codes, usually called PIDs, identify which pieces of data the scan tool is requesting. When a scan tool is connected to the DLC and serial data are displayed, the scan tool is asking the computer to see sensor data that are categorized by the parameter ID. For example, when you use the scan tool to observe engine coolant temperature sensor data, the scan tool requests data from Mode $01 PID 5. Regardless of what scan tool is used on an OBD-II-equipped vehicle, requesting generic data from Mode $01 PID 5 will show electronic throttle control (ETC) data. The list of PIDs is standardized and has been updated since the introduction of OBD-II.

As discussed in Chapter 31, there are two halves to the OBD-II computer system, generic and enhanced, sometimes called the manufacturer sides. Accessing generic PIDs can be done by any type of scan tool with generic functions. To access enhanced or manufacturer PIDs requires using either an enhanced or manufacturer's scan tool. Because of these two separate sets of PIDs, not all data are available on a generic scan tool using the Mode $01. The standardized data enable the use of generic scan tools and code readers to access OBD-II data under the SAE J1979 standards.

The scan tool often has the ability to show or hide PIDs based on user preferences. Reducing the number of PIDs shown can speed up the transfer of data between the vehicle and the scan tool, which allows the data displayed to update more quickly. The scan tool and/or service information may also have details about typical values, minimum and maximum values, and complete descriptions for each PID.

DTCs and Service Information

To access stored DTCs, the scan tool requests data using Mode $03. Depending on the scan tool, a description of the DTC may be provided on the scan tool. If the scan tool does not provide this information, you will need to find the details using the service information. Check for current, pending, and history DTCs stored in memory.

After retrieving the DTCs, find their descriptions in the service information. Normally, the descriptions are followed by additional information to help diagnosis. As can be seen in **Figure 32–3**, there is more than one possible cause of the problem. One is the sensor itself and the other two concern the sensor's circuit. Detailed testing will identify the exact cause. It is important to understand what conditions cause a DTC to set for a particular vehicle as the conditions are often slightly different for different vehicle manufacturers or even different engines from the same manufacturer.

Notice that the description also leads to pinpoint tests. These are designed to guide the technician through a step-by-step procedure. To be effective, each step should be performed, in the order given, until the problem has been identified.

Make sure to check all available service information related to the DTCs. There may be a technical service bulletin (TSB) related to the code, and following those procedures may fix the problem. Also, make sure the ECM/PCM is programmed with the most current software.

Monitor Failures

When checking for DTCs, check the status of the OBD-II monitors. All OBD-II scan tools have a readiness function showing all of the monitoring sequences and the status of each: complete or incomplete. If vehicle travel time, operating conditions, or other parameters were insufficient for a monitoring sequence to complete a test, the scanner will indicate which monitoring sequence is not yet complete.

The specific set of driving conditions that will set the requirements for OBD-II monitoring sequences is described in Chapter 31.

GO TO ▶ Chapter 31 for more information about monitors and drive cycles.

	P0117 – ENGINE COOLANT TEMPERATURE (ECT) SENSOR 1 CIRCUIT LOW
Description:	Indicates the sensor signal is less than the self-test minimum. The ECT sensor minimum is 0.2 volt or 121°C (250°F).
Possible Causes:	• Grounded circuit in the harness • Damaged sensor • Incorrect harness connection
Diagnostic Aids:	A concern is present if an ECT PID reading is less than 0.2 volt with the key on and engine off, or during any engine operating mode.

FIGURE 32–3

A description of a DTC, as given in typical service information, with additional information to help diagnosis.

When most monitor tests are run and a system or component fails a test, a pending code is set. When the fault is detected a second time, a DTC is set and the MIL is lit **(Figure 32–4)**. It is possible that a DTC for a monitored circuit may not be entered into memory even though a malfunction has occurred. This may happen when the monitoring criteria have not been met.

MONITOR	FAILURE CAN BE CAUSED BY:
Catalyst monitor	Fuel contaminants Leaking exhaust Engine mechanical problems Defective upstream or downstream oxygen sensor circuits Defective PCM
Fuel system monitor	Defective fuel pump Abnormal signal from the upstream HO_2S sensors Engine temperature sensor faults Malfunctioning catalytic converter MAP- or MAF-related faults Cooling system faults EGR system faults Fuel injection system faults Ignition system faults Vacuum leaks Worn engine parts
EGR monitor	Faulty EGR valve Faulty EGR passages or tubes Loose or damaged EGR solenoid wiring and/or connectors Damaged DPFE or EGR VP sensor Disconnected or loose electrical connectors to the DPFE or EGR VP sensors Disconnected, damaged, or misrouted EGR vacuum hoses
EVAP monitor	Disconnected, damaged, or loose purge solenoid connectors and/or wiring Leaking hoses, tubes, or connectors in the EVAP system Vacuum and/or vent hoses to the solenoid and charcoal canister are misrouted Plugged hoses from the purge solenoid to the charcoal canister Loose or damaged connectors at the purge solenoid Fuel tank cap not tightened properly or it is missing
Misfire monitor	Fuel level too low during drive cycle Dirty or defective fuel injectors Contaminated fuel Defective fuel pump Restricted fuel filter EGR system faults EVAP system faults Restricted exhaust system Fault secondary ignition circuit Damaged, loose, or resistant PCM power and/or ground circuits
Oxygen sensor monitor	Malfunctioning upstream and/or downstream oxygen sensor Malfunctioning heater for the upstream or downstream oxygen sensor Faulty PCM Defective wiring to and/or from the sensors

FIGURE 32–4

Possible causes of OBD-II monitor failures.

Freeze-Frame Data

Freeze-frame data associated with a DTC should also be retrieved. These data contain values from specific generic PIDs that provide information about the operating conditions present when the code was set. It can also be used to identify components that should be tested.

Misfire freeze-frame (MFF) data contain some unique PIDs. MFF data are not part of the freeze-frame data that are stored with a DTC. They are used only for identifying the cause of a misfire. Generic freeze-frame data are also captured when a misfire occurs. However, those data will not represent what happened at the time of the misfire; rather, they will show what happened after the misfire. MFF data are captured at the time of the highest rate of misfire and not when the DTC is set.

Recording the freeze-frame information is helpful so that once the vehicle is repaired, it can be driven under conditions similar to those in the freeze frame. Some technicians call this "driving the freeze frame." This allows you to recreate the operating conditions that were present with the DTC set.

SHOP TALK

Many electronic control systems have adaptive strategies that allow the engine to run when one or more inputs fail. The ECM/PCM relies on a predetermined value when it senses an out-of-limit value from a sensor. When looking at enhanced-mode data, the input signal may not change with changes in conditions, and this can be misleading. The serial data may show a substitute value when using enhanced data. When using generic or global OBD-II data, the PIDs will show actual data as substitute values that cannot be displayed as part of generic data.

Mode $06 Data

The data available in Mode $06 can also be used to identify the cause of a concern. Mode $06 allows access to the results of the various monitor diagnostic tests **(Figure 32–5)**. These values are stored when a specific monitor has completed a test. In addition to providing specific test results, Mode $06 data can be used to determine component condition and misfire rates. In some cases, Mode $06 data can also help determine that a component is near failure.

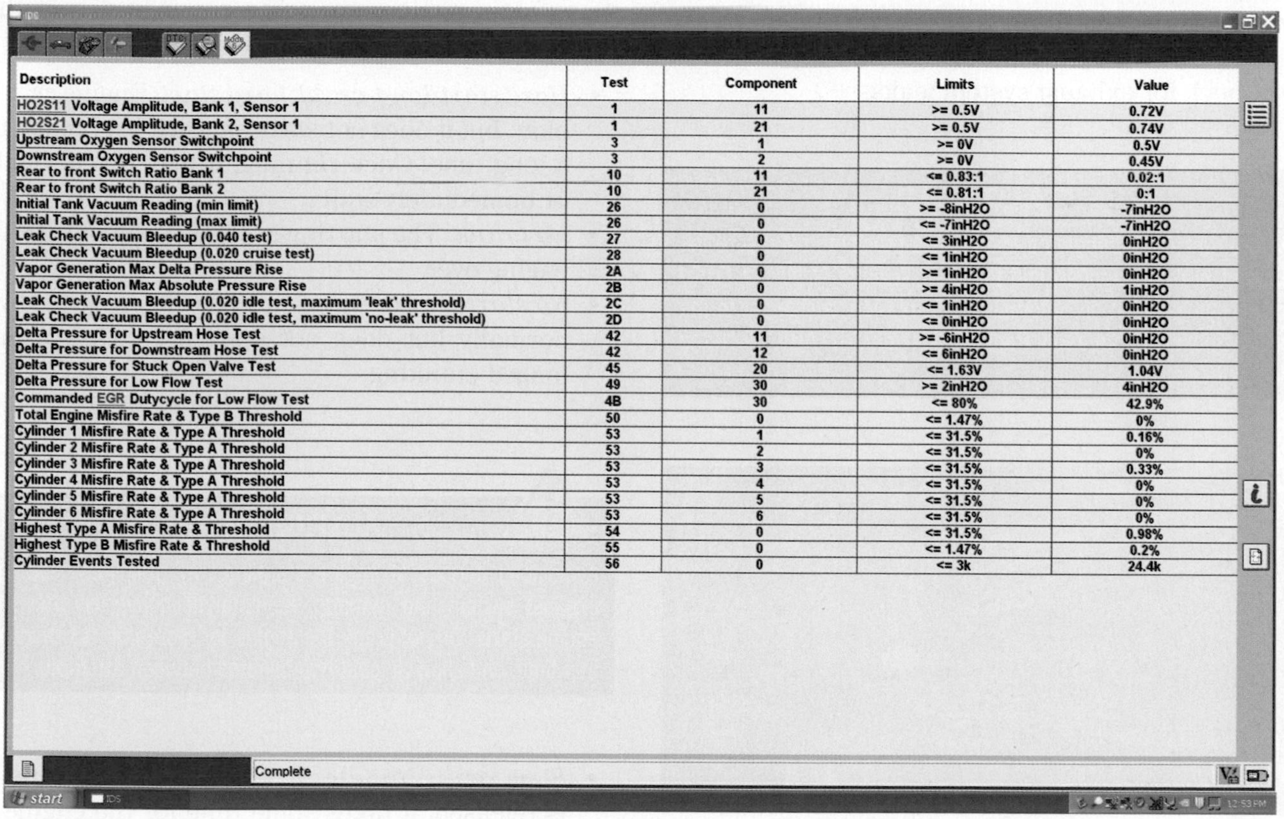

Description	Test	Component	Limits	Value
HO2S11 Voltage Amplitude, Bank 1, Sensor 1	1	11	>= 0.5V	0.72V
HO2S21 Voltage Amplitude, Bank 2, Sensor 1	1	21	>= 0.5V	0.74V
Upstream Oxygen Sensor Switchpoint	3	1	>= 0V	0.5V
Downstream Oxygen Sensor Switchpoint	3	2	>= 0V	0.45V
Rear to front Switch Ratio Bank 1	10	11	<= 0.83:1	0.02:1
Rear to front Switch Ratio Bank 2	10	21	<= 0.81:1	0:1
Initial Tank Vacuum Reading (min limit)	26	0	>= -8inH2O	-7inH2O
Initial Tank Vacuum Reading (max limit)	26	0	<= -7inH2O	-7inH2O
Leak Check Vacuum Bleedup (0.040 test)	27	0	<= 3inH2O	0inH2O
Leak Check Vacuum Bleedup (0.020 cruise test)	28	0	<= 1inH2O	0inH2O
Vapor Generation Max Delta Pressure Rise	2A	0	>= 1inH2O	0inH2O
Vapor Generation Max Absolute Pressure Rise	2B	0	>= 4inH2O	1inH2O
Leak Check Vacuum Bleedup (0.020 idle test, maximum 'leak' threshold)	2C	0	<= 1inH2O	0inH2O
Leak Check Vacuum Bleedup (0.020 idle test, maximum 'no-leak' threshold)	2D	0	<= 0inH2O	0inH2O
Delta Pressure for Upstream Hose Test	42	11	>= -6inH2O	0inH2O
Delta Pressure for Downstream Hose Test	42	12	<= 6inH2O	0inH2O
Delta Pressure for Stuck Open Valve Test	45	20	<= 1.63V	1.04V
Delta Pressure for Low Flow Test	49	30	>= 2inH2O	4inH2O
Commanded EGR Dutycycle for Low Flow Test	4B	30	<= 80%	42.9%
Total Engine Misfire Rate & Type B Threshold	50	0	<= 1.47%	0%
Cylinder 1 Misfire Rate & Type A Threshold	53	1	<= 31.5%	0.16%
Cylinder 2 Misfire Rate & Type A Threshold	53	2	<= 31.5%	0%
Cylinder 3 Misfire Rate & Type A Threshold	53	3	<= 31.5%	0.33%
Cylinder 4 Misfire Rate & Type A Threshold	53	4	<= 31.5%	0%
Cylinder 5 Misfire Rate & Type A Threshold	53	5	<= 31.5%	0%
Cylinder 6 Misfire Rate & Type A Threshold	53	6	<= 31.5%	0%
Highest Type A Misfire Rate & Threshold	54	0	<= 31.5%	0.98%
Highest Type B Misfire Rate & Threshold	55	0	<= 1.47%	0.2%
Cylinder Events Tested	56	0	<= 3k	24.4k

Complete

FIGURE 32–5

Mode $06 allows access to the results of the various monitor diagnostic tests.

For example, as catalyst efficiency decreases, the rear O_2 sensor switch rate will increase. By examining the Mode $06 data for the rear O_2 sensor switch rate, catalyst degradation can be monitored. Once the test values reach the programmed limits, a DTC will set.

Visual Inspection

Once you have checked for DTCs, perform a visual inspection. Pay close attention to wiring and components that are related to the DTCs. Often, the cause of a driveability problem can be discovered by doing the following:

- Check all vacuum hoses. Make sure they are connected and are not pinched, cut, or cracked.
- Check the condition and tension of the generator drive belt.
- Check the battery and battery connections. Look for loose and corroded connections and damaged **(Figure 32–6)** or burned wires under the hood.
- Inspect the wiring for sensors and outputs.
- Check the level and condition of the engine's coolant.
- Check the air filter. Also check the air intake system for restrictions and leaks.
- Check for exhaust system leaks.

CAUTION!

Before running the engine to diagnose a problem, make sure the parking brake is applied and the gear selector is placed firmly in the PARK position on automatic transmission vehicles or in the NEUTRAL position on manual transmission vehicles. Also, block the drive wheels.

FIGURE 32–6

Check all of the wires in the engine compartment. Look for damaged or burned wires.

SYMPTOM-BASED DIAGNOSIS

At times, no DTCs are set by the computer, but a driveability problem exists. This is when a technician must look at the various engine systems to discover the cause of the concern. What system or part to test is dictated by a description of the problem or its symptoms. Before diagnosing a problem based on its symptoms, perform a visual inspection as listed above. In addition, make sure to confirm the following:

- All data observed on the scan tool are within normal ranges.
- The PCM power and ground circuits are intact and in good condition.
- All vehicle modifications are identified.
- The vehicle's tires are properly inflated and are the correct size.

Common Symptoms

Service information typically has a section dedicated to symptom-based diagnosis. Although a customer may describe a problem in nontechnical terms, you should summarize the concern to match one or more of the various symptoms listed by the manufacturer. Here are some common driveability symptoms and a brief description of each:

- *Hard start/long-crank hard start.* Engine cranks okay, but it does not start unless it is cranked for a long time. Once running, it may run normally or immediately stall.
- *No crank.* The starting system does not turn the engine over.
- *No start (engine cranks).* The engine turns over normally but does not start even after a prolonged cranking.

WARNING!

Avoid long periods of engine cranking; raw fuel may load the exhaust system and damage the catalytic converter after the engine starts. Do not crank the engine for more than 15 seconds at a time, and allow at least two minutes between attempts.

- *Slow return to idle.* When the accelerator pedal is released, it takes some time for the engine to return to its normal idle speed.
- *Fast idle.* The engine idles at a higher than normal speed or does not return to the normal idle speed when the throttle is released.

- *Runs on (diesels).* A fast idle can cause the engine to attempt to run after the ignition is turned off. This is called dieseling because combustion is caused by the heat inside the combustion chamber.
- *Rough or unstable idle and stalling.* The engine shakes while idling or changes idle speed constantly. This concern may cause the engine to stall.
- *Low/slow idle or stalls/quits during deceleration.* Engine speed drops below its normal idle speed. This can cause stalling when the accelerator pedal is released.
- *Backfire.* Fuel ignites in the intake manifold or exhaust system. This causes a loud popping noise.
- *Lack of or loss of power.* The engine is sluggish and provides less power than is normally expected. The engine seems not to increase in speed when the accelerator pedal is depressed.
- *Cuts out/misses.* A steady pulsation or jerking at low engine speeds, especially during heavy engine loads. The exhaust may have a spitting sound at idle or low speeds.
- *Hesitation or stumble.* A momentary lack of response to the accelerator pedal. This concern can occur at any vehicle speed but is more noticeable during acceleration from a stop.
- *Surges.* The power output from the engine seems to change while operating with a steady throttle or while cruising.
- *Detonation/spark knock.* The engine makes sharp metallic knocking sounds that are usually worse during acceleration.
- *Poor fuel economy.* Fuel economy is noticeably lower than expected or lower than it was before.

Figure 32–7 is a symptom chart for common concerns that have not set a DTC and the component or system that could cause the problem. Each potential problem area should be checked. It is important to realize that some problems may cause more than one symptom. In addition, remember to check if new software calibrations are available before attempting to solve a problem based on symptoms only.

BASIC TESTING

Diagnosis of electronic engine control systems includes much more than retrieving DTCs. Individual components and their circuits must be inspected and tested. The operation of some components can be monitored with the scan tool; however, additional tests are normally necessary. These tests include the following:

- *Ohmmeter checks.* Most sensors and output devices can be checked with an ohmmeter. For example, an ohmmeter can be used to check a temperature sensor. Normally, the ohmmeter reading is low on a cold engine and high or infinity on a hot engine if the sensor is a positive temperature coefficient (PTC). If the sensor is a negative temperature coefficient (NTC), the opposite readings would be expected. Output devices such as coils or motors can also be checked with an ohmmeter.
- *Voltmeter checks.* Many sensors, output devices, and their wiring can be diagnosed by checking the voltage to them and, in some cases, from them. Even some oxygen sensors can be checked in this manner.

CONCERN	COMPONENT/SYSTEM	POSSIBLE CAUSE
Hard start/long crank	• Starting system	• Weak battery, poor battery connections, defective starter
	• Fuel/ignition/computer	• Low fuel pressure or volume, weak spark, incorrect spark timing, TP, ETC, MAF/MAP sensor faults
	• Intake air system	• Restriction in intake air system, plugged air filter, induction air leak
	• MAF sensor	• Dirty, misreading, or defective MAF sensor
	• Exhaust system	• Restricted exhaust system
	• PCV system	• Damaged or disconnected PCV hose
	• EVAP system	• Fuel vapour purge during cranking
No crank	• Antitheft devices	• Fault with key, encoder, remote, or lack of network communication
	• Base engine	• Seized or hydraulically locked engine
	• Starting system	• Defective starter, starter relay, battery, or ignition/safety switch; poor battery cable connections

FIGURE 32–7

A no-DTCs-present symptom chart.

- *Lab scope checks.* The activity of sensors and actuators can be monitored with a lab scope or a graphing multimeter (GMM). By watching their activity, the technician is doing more than testing them. Often, problems elsewhere in the system will cause a device to behave abnormally. These situations are identified by the trace on a scope and by the technician's understanding of a scope and the device being monitored.

In some cases, a final check can be made only by substitution. Substitution is not an allowable diagnostic method under the mandates of OBD-II, nor is it the most desirable way to diagnose problems. However, sometimes it is the only way to verify the cause of a problem. When substituting, replace the suspected part with a known good unit and recheck the system. If the system now operates normally, the original part is defective.

Testing Sensors

To monitor engine conditions, the computer uses a variety of sensors. All sensors perform the same basic function. They detect a mechanical condition (movement or position), chemical state, or temperature condition and change it into an electrical signal that can be used by the PCM to make decisions.

If a DTC directs you to a faulty sensor or sensor circuit, or if you suspect a faulty sensor, it should be tested. Always follow the procedures outlined by the manufacturer when testing sensors and other electronic components. Also, make sure you have the correct specifications for each part tested. Sensors are tested with a digital multimeter (DMM), scan tool, and/or lab scope or GMM.

Because the controls are different on the various types of automotive lab scopes **(Figure 32–8)** and GMMs, make sure you follow the instructions of the scope's manufacturer. If the scope is set wrong, the scope will not break. It just will not show you what you want to be shown. To help with understanding how to set the controls on a scope, keep the following things in mind. The vertical voltage scale must be adjusted in relation to the expected voltage signal. The horizontal time base or milliseconds per division must be adjusted so that the waveform appears properly on the screen. Many waveforms can be clearly displayed when the horizontal time base is adjusted correctly.

The trigger is the signal that tells the lab scope to start drawing a waveform. A marker indicates the trigger line on the screen, and minor adjustments of the trigger line may be necessary to position the waveform in the desired vertical position. Trigger slope indicates the direction in which the voltage signal is moving when it crosses the trigger line. A positive trigger slope means the voltage signal is moving upward as it crosses the trigger line, whereas a negative trigger slope indicates that the voltage signal is moving downward when it crosses the trigger line.

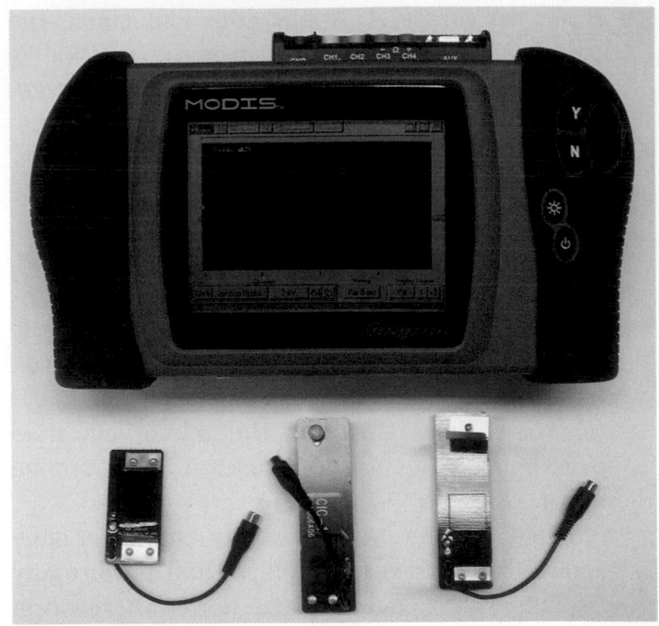

FIGURE 32–8

Many different types of lab scopes are available to diagnose electronic control systems; make sure to follow the operating procedures given by the scope's manufacturer.

Snap-on Incorporated

Software packages, often programmed in a lab scope or GMM, are available to help you properly interpret scope patterns and set up a lab scope. These also contain an extensive waveform library that you can refer to in order to find what the normal waveform of a particular device should look like. The library also contains the waveforms caused by common problems. You can also add to the library by transferring waveforms to a PC from the lab scope. Another source of waveforms, both good and bad, is iATN, the International Automotive Technicians Network. The library of good and bad waveforms allows you to compare your findings with those from similar vehicles and components. After the waveforms have been transferred, notes can be added to the file. The software may also include the theory of operation, scope setup information, and diagnostic procedures for common inputs and outputs.

There are many different types of sensors; their design depends on what they are monitoring. Some sensors are simple on–off switches. Others are some form of variable resistor that changes resistance according to temperature changes. Some sensors are voltage or frequency generators, whereas others send varying signals according to the rotational speed of a device. Knowing what they are measuring and how they respond to changes are the keys to being able to accurately test an input sensor.

Some inputs to the PCM come from another control module or are simply a connection from a device. Examples of this are the battery voltage input and the heated windshield module. The battery's voltage is available on the data bus, and many control modules need this information. There is no sensor involved, just a connection from the battery to the bus. The heated windshield module tells the computer when the heated windshield system is operating. This helps the PCM to accurately determine engine load and control idle speed.

DIAGNOSIS OF COMPUTER VOLTAGE SUPPLY AND GROUND WIRES

SHOP TALK

Never replace a computer unless the ground and voltage supply circuits are proven to be in satisfactory condition.

No PCM (OBD-II and earlier designs) can operate properly unless it has good ground connections and the correct voltage at the required terminals. A wiring diagram for the vehicle being tested must be used for these tests. Back-probe the battery terminal at the PCM, and connect a digital voltmeter from this terminal to ground **(Figure 32–9)**. Always use a good engine ground.

The voltage at this terminal should be 12 volts with the ignition switch in the OFF position. If 12 volts are not available at this terminal, check the computer fuse and related circuit. Turn on the ignition switch, and connect the red voltmeter lead to the other battery terminals at the PCM with the black lead still grounded. The voltage measured at these terminals should also be 12 volts with the ignition switch in the ON position. When the specified voltage is not available, test the voltage supply wires to these terminals. These terminals may be connected through fuses, fuse links, or relays.

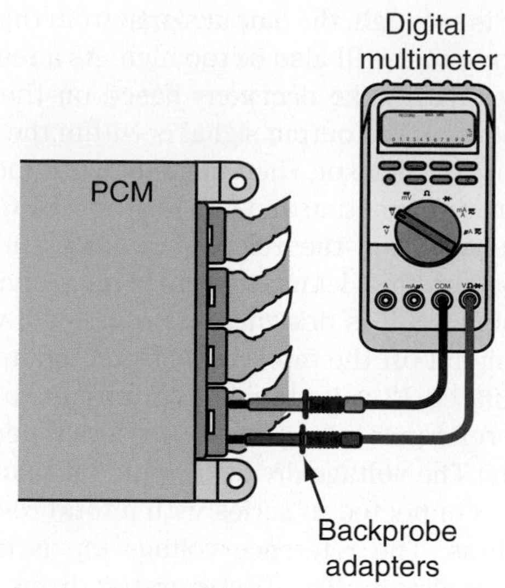

FIGURE 32–9

Using a digital voltmeter to check the PCM's circuit.

Ground Circuits

Ground wires usually extend from the computer to a ground connection on the engine or battery. With the ignition switch in the ON position, connect a digital voltmeter from the battery ground to the computer ground. The voltage drop across the ground wires should be 30 millivolts (mV) or less. If the voltage reading is greater than that or more than that specified by the manufacturer, repair the ground wires or connection.

Not only should the computer ground be checked, but so should the ground (and positive) connection at the battery. Checking the condition of the battery and its cables should always be part of the initial visual inspection before beginning diagnosis of an engine control system.

A voltage drop test is a quick way of checking the condition of any wire. To do this, connect a voltmeter across the wire or device being tested. Place the positive lead on the most positive side of the circuit. Then turn on the circuit. Ideally, there should be a 0-volt reading across any wire unless it is a resistance wire, which is designed to drop voltage; even then, check the drop against specifications to see if it is dropping too much.

A good ground is especially critical for all reference voltage sensors. The problem here is not obvious until it is thought about. A bad ground will cause the reference voltage (normally 5 volts) to be higher than normal. Normally in a circuit, the added resistance of a bad ground would cause less voltage at a load. Because of the way reference voltage sensors are wired, the opposite is true. If the reference voltage to a sensor is too high, the output signal from the sensor to the computer will also be too high. As a result, the computer will make decisions based on the wrong information. If the output signal is within the normal range for that sensor, the computer will not notice the wrong information and will not set a DTC.

To show how the reference voltage increases with a bad ground, let us look at a voltage divider circuit. This circuit is designed to provide a 5-volt reference signal off the tap. A vehicle's computer feeds the regulated 12 volts to a similar circuit to ensure that the reference voltage to the sensors is very close to 5 volts. The voltage divider circuit consists of two resistors connected in series with a total resistance of 12 ohms. The reference voltage tap is between the two resistors. The first resistor drops 7 volts **(Figure 32–10)**, which leaves 5 volts for the second resistor and for the reference voltage tap. This 5-volt reference signal will be always available at the tap, as long as 12 volts are available for the circuit.

If the circuit has a poor ground, one that has resistance, the voltage drop across the first resistor will be decreased. This will cause the reference voltage to increase. In **Figure 32–11**, to simulate a bad ground, a 4-ohm resistor was added into the circuit at the ground connection at the battery. This increases the total resistance of the circuit to 16 ohms and decreases the current flowing throughout the circuit. With less current flow through the circuit, the voltage drop across the first resistor decreases to 5.25 volts **(Figure 32–12)**. This means the voltage available at the tap will be higher than 5 volts; it will be 6.75 volts.

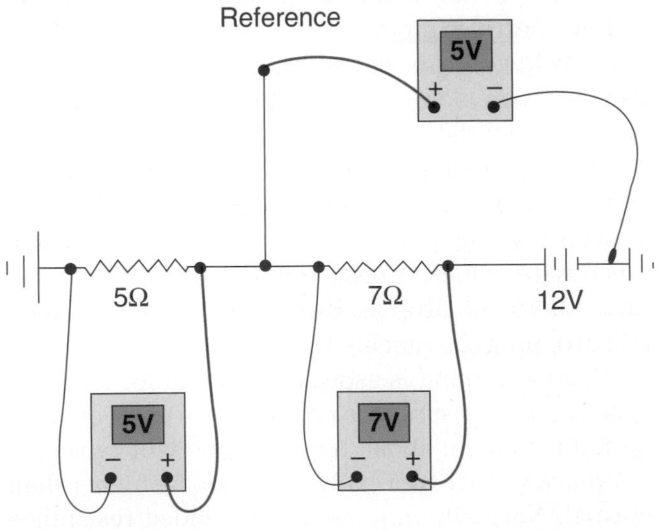

FIGURE 32–10

A voltage divider circuit with voltage values.

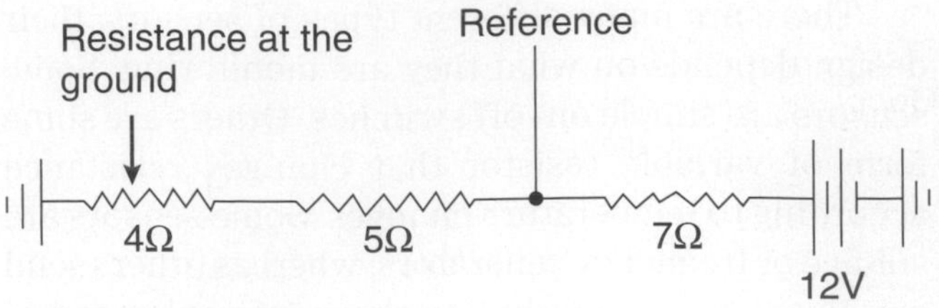

FIGURE 32–11

Voltage divider circuit with a bad ground.

ELECTRICAL NOISE Poor grounds can also allow electromagnetic interference (EMI), or noise, to be present on the reference voltage signal. This noise causes small changes in the voltage going to the sensor. Therefore, the output signal from the sensor will also have these voltage changes. The computer will try to respond to these small rapid changes, which can cause a driveability problem. The best way to check for noise is to use a lab scope.

Connect the lab scope between the 5-volt reference signal into the sensor and the ground. The trace on the scope should be flat **(Figure 32–13)**. If noise is present, move the scope's negative probe to a known good ground. If the noise disappears, the sensor's ground circuit is bad or has resistance. If the noise is still present, the voltage feed circuit is bad or there is EMI in the circuit from another source, such as the AC generator. Find and repair the cause of the noise.

Circuit noise may be present at the positive side or negative side of a circuit. It may also be evident by a flickering MIL, a popping noise on the radio, or an intermittent engine miss. Noise can cause a variety of problems in any electrical circuit. The most common

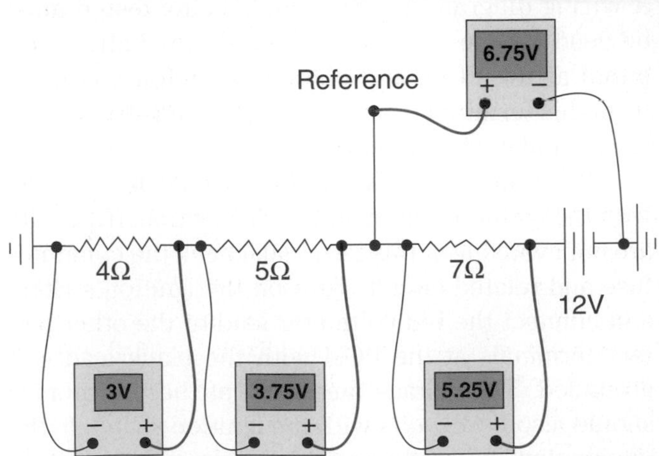

FIGURE 32–12

Figure 32–11 with voltage readings.

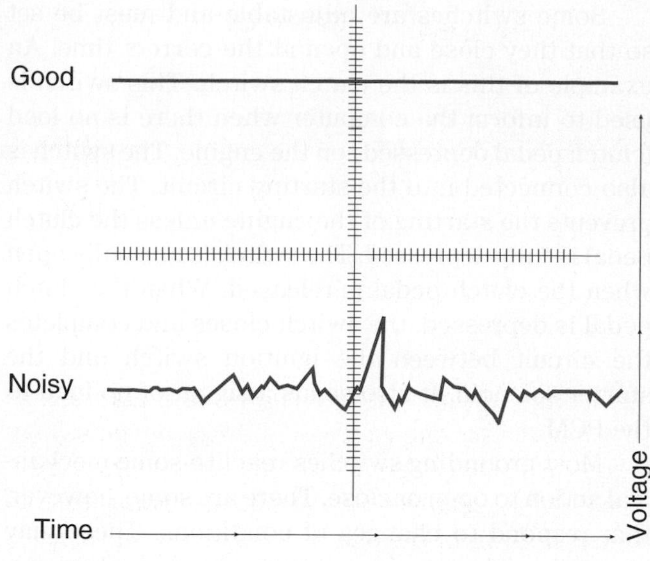

FIGURE 32–13

(Top) A good voltage signal. (Bottom) A voltage signal with noise.

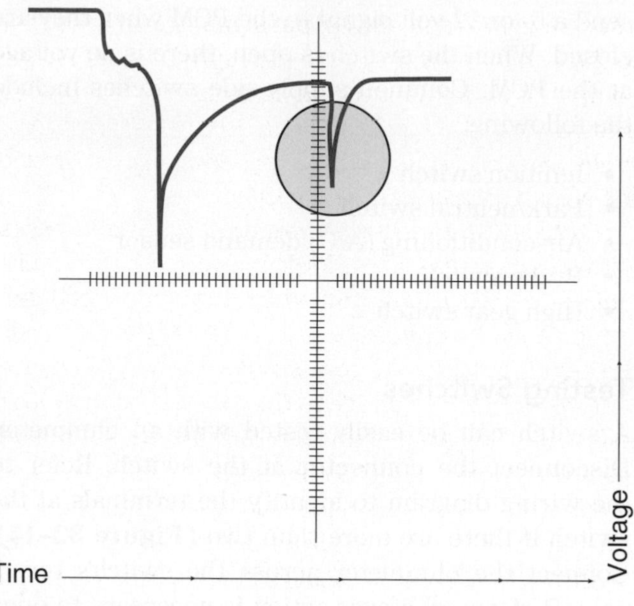

FIGURE 32–14

A trace of an electromagnetic clutch with a bad clamping diode.

sources of noise are electric motors, relays and solenoids, AC generators, ignition systems, switches, and A/C compressor clutches. Typically, noise is the result of an electrical device being turned on and off. Sometimes, the source of the noise is a defective suppression device. Manufacturers include these devices to minimize or eliminate electrical noise. Some of the commonly used noise suppression devices are resistor-type secondary cables and spark plugs, shielded cables, capacitors, diodes, and resistors. Capacitors or chokes are used to control noise from a motor or generator. If the source of the noise is not a poor ground or a defective component, check the suppression devices.

CLAMPING DIODES Clamping diodes are placed in parallel to coil windings to limit high-voltage spikes. These voltage spikes are induced by the collapsing of the magnetic field around a winding in a solenoid, relay, or electromagnetic clutch. The field collapses when current flow to the winding is stopped. The diode prevents the voltage from reaching the computer and other sensitive electronic parts. When the diode fails to suppress the voltage spikes, the transistors inside the computer can be destroyed. If the diode is bad, a negative spike will appear in a voltage trace **(Figure 32–14)**. Resistors are also used to suppress voltage spikes. They do not eliminate the spikes; rather, they limit the intensity of the spikes. If a voltage trace has a large spike and the circuit is fitted with a resistor to limit noise, the resistor may be bad.

SWITCHES

Switches are turned on and off through an action of a device or by the actions of the driver. Switches are either normally open or normally closed. Switches send a digital signal to the PCM; they are either open or closed. Some switches are provided with a reference voltage of 5 or 12 volts by the PCM. An example of this is a neutral drive/neutral gear switch (NDS). This switch lets the PCM know when the transmission has been shifted into a gear. If the transmission is in park or neutral, the switch is closed. It sends a voltage signal of 1 volt or less to the PCM. When the transmission is placed into a gear, the switch opens and sends a signal above 5 volts to the PCM.

Some switches control the ground side of the circuit. These circuits contain a fixed resistor connected in series with the switch. When the switch is closed, the voltage signal to the PCM is low or zero. When the switch is open, there is a high-voltage signal. Common grounding switches include the following:

- Idle tracking switch
- Power-steering pressure switch
- Overdrive switch
- Clutch pedal position switch

Supply- or power-side switches are the most commonly used and work in the opposite way. They send a 5- or 12-volt signal to the PCM when they are closed. When the switch is open, there is no voltage at the PCM. Common supply-side switches include the following:

- Ignition switch
- Park/neutral switch
- Air-conditioning (A/C) demand sensor
- Brake switch
- High gear switch

Testing Switches

A switch can be easily tested with an ohmmeter. Disconnect the connector at the switch. Refer to the wiring diagram to identify the terminals at the switch if there are more than two **(Figure 32–15)**. Connect the ohmmeter across the switch's terminals. Perform whatever action is necessary to open and close the switch. When the switch is open, the ohmmeter should have an infinite reading. When the switch is closed, there should zero resistance. If the switch reacts in any other way, the switch is bad and should be replaced.

Switches can also be checked with a voltmeter. The signal to the PCM from the supply-side switches should be 0 volts with the switch open and should supply voltage when the switch is closed. This indicates to the ECM that a change has taken place. Using a voltmeter is preferred because it tests the circuit as well as the switch. If less than supply voltage is present with the switch closed, there is unwanted resistance in the circuit. Again, expect the opposite readings on a ground-side switch.

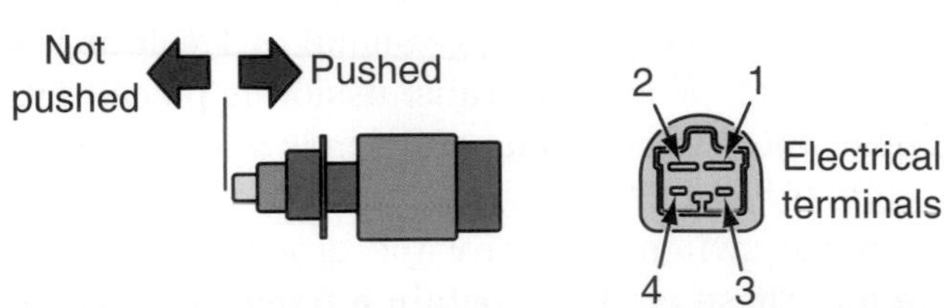

TESTER CONNECTION	SWITCH CONDITION	SPECIFIED VALUE
1–2	Switch pin released Switch pin pushed in	Below 1Ω 10kΩ or higher
3–4	Switch pin released Switch pin pushed in	10kΩ or higher Below 1Ω

FIGURE 32–15

Check the service information for the proper testing points for a switch.

Some switches are adjustable and must be set so that they close and open at the correct time. An example of this is the clutch switch. This switch is used to inform the computer when there is no load (clutch pedal depressed) on the engine. The switch is also connected into the starting circuit. The switch prevents the starting of the engine unless the clutch pedal is fully depressed. The switch is normally open when the clutch pedal is released. When the clutch pedal is depressed, the switch closes and completes the circuit between the ignition switch and the starter solenoid. It also sends a signal of no-load to the PCM.

Most grounding switches react to some mechanical action to open or close. There are some, however, that respond to changes of conditions. These may respond to changes in pressure or temperature. An example of this type of switch is the power-steering pressure switch. This switch informs the PCM when power-steering pressures reach a particular point. When the power-steering pressure exceeds that point, the PCM knows there is an additional load on the engine and will increase idle speed.

To test this type of switch, monitor its activity with the scan tool, a DMM, or lab scope. With the engine running at idle speed, turn the steering wheel to its maximum position on one side. The data PID should change from LOW to HIGH or NO to YES. If testing with a DMM or scope, the voltage signal should drop as soon as the pressure in the power-steering unit has reached a high level. If the voltage does not drop, either the power-steering assembly is incapable of producing high pressures or the switch is bad.

Temperature-responding switches operate in the same way. When a particular temperature is reached, the switch opens. This type of switch is best measured by removing it and submerging it in heated water. Watching the ohmmeter as the temperature increases, a good temperature-responding switch will open (have an infinite reading) when the water temperature reaches the specified amount. If the switch fails this test, it should be replaced.

TEMPERATURE SENSORS

The PCM changes the operation of many components and systems based on temperature. Nearly all temperature sensors are NTC thermistors and operate in the same way. Their resistance changes with a change in temperature. The PCM supplies a reference voltage of 5 volts to the sensor. That voltage is changed by the change of the resistor's resistance and is fed back through a ground wire to the PCM. Based on the return voltage, the PCM calculates the exact temperature. When the sensor is cold, its resistance is high, and the voltage signal is also high. As the sensor warms up, its resistance drops, and so does the voltage signal.

Engine Coolant Temperature (ECT) Sensor

The **engine coolant temperature (ECT) sensor** is a thermistor. By measuring ECT, the PCM knows the average temperature of the engine. Temperature is used to regulate many engine functions, such as the fuel injection system, ignition timing, variable valve timing, transmission shifting, EGR, and canister purge, as well as controlling the open- and closed-loop operational modes of the system. The ECT sensor is normally located in an engine coolant passage just before the thermostat. On cars built prior to OBD-II, a coolant switch may be used. This type of sensor may be designed to remain closed within a certain temperature range or to open only when the engine is warm.

A faulty ECT sensor or sensor circuit can cause a variety of problems. The most common is the failure to switch to the closed-loop mode once the engine is warm. ECT sensor problems are often caused by wiring faults or loose or corroded connections rather than the sensor itself. Many testers are able to show where to place the probes of the tester to check things like the ECT **(Figure 32–16).** A defective ECT sensor or circuit may cause the following problems:

- Hard engine starting
- A no-start cold condition
- Rich or lean air/fuel ratio
- Improper operation of emission devices
- Reduced fuel economy
- Improper converter clutch lock-up
- Hesitation on acceleration
- Engine stalling
- Transmission not shifting into high gear or shifting late

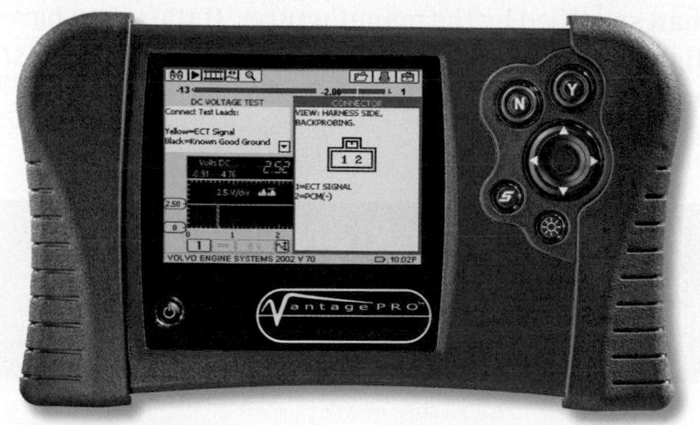

FIGURE 32–16

Engine coolant temperature (ECT) sensor information shown on a test meter.

Snap-on Incorporated

Intake Air Temperature (IAT) Sensor

The **intake air temperature (IAT) sensor** is also called an air charge temperature sensor. Its resistance decreases as the incoming air temperature increases, and increases as the incoming air temperature decreases **(Figure 32–17)**. The PCM uses the air temperature information as a correction factor in the calculation of fuel, spark, and airflow. For example, the PCM uses this input to help calculate fuel delivery. Because cold intake air is denser, a richer air/fuel ratio is required.

On engines equipped with an MAP sensor, the IAT is installed in an intake air passage. On other engines, the IAT **(Figure 32–18)** is normally an integral part of the mass airflow (MAF) sensor. Most control systems compare the inputs from the IAT and the ECT to determine if the engine is attempting a cold start.

A defective IAT sensor may cause the following problems:

- Rich or lean air/fuel ratio
- Hard engine starting
- Engine stalling or surging
- Acceleration stumbles
- Excessive fuel consumption

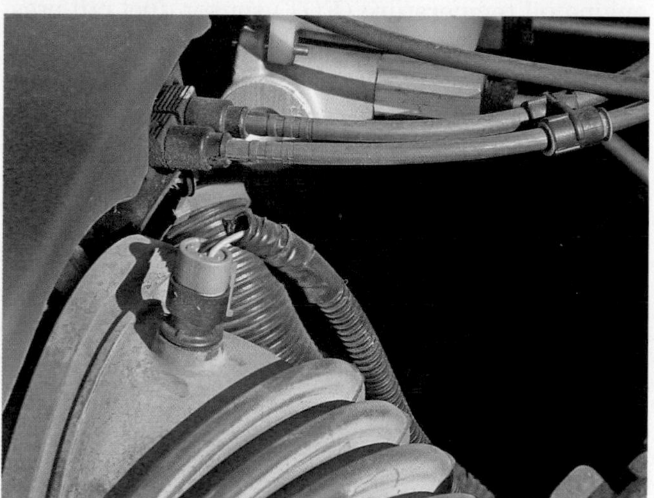

FIGURE 32–17

An air temperature sensor.

©Cengage Learning 2015

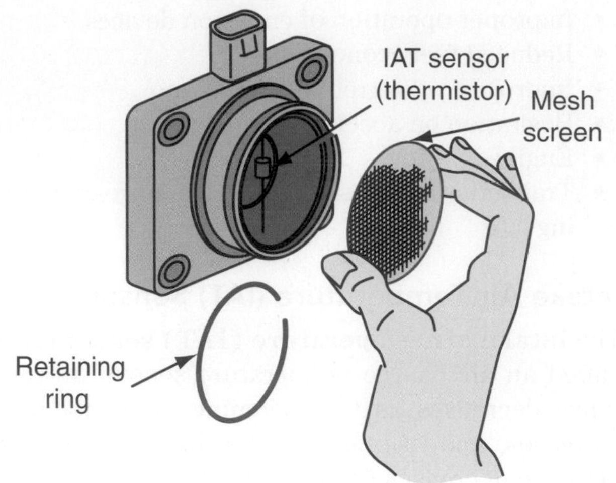

FIGURE 32–18

A typical intake air temperature (IAT) sensor incorporated into an MAF sensor.

Other Temperature Sensors

Many other temperature sensors may be used on engines. Their application depends on the engine and the control system. Some turbo- or supercharged engines have two IAT sensors: one before the charger and one after. This is done to monitor the change in temperature as the air is forced into the cylinders.

Some engines have a cylinder head temperature (CHT) sensor installed in the cylinder head to measure its temperature. The primary function of this is to detect engine overheating. When high metal temperatures are reported to the PCM, it will enter into its fail-safe cooling strategy mode.

Other common temperature sensors include an engine oil temperature (EOT) sensor, fuel rail pressure temperature (FRPT) sensor, and EGR temperature sensor. Also, many vehicles built prior to OBD-II use a temperature sensor to directly control the electric radiator fans.

Testing

Temperature sensor circuits should be tested for opens, shorts, and high resistance. Often, if one of these problems exist, a DTC will be set. Scan tool data can also give an indication of the condition of the sensor and related circuits. If the observed temperature is the coldest possible value, the circuit is open. If the temperature is the highest possible, the circuit has a short. Most service information systems give a procedure for checking for opens and shorts. A jumper wire is inserted across specific terminals of the circuit, and the data are observed. This should cause the readings to go high or hot. If the connector to the sensor is disconnected, the readings should drop to cold. On some vehicles, this test will not work because the PCM will react by using a PID value. High-resistance problems will cause the PCM to respond to a lower temperature than the actual temperature. This can be verified by using a good thermometer (infrared is best) to measure the temperature and compare it to the readings on the scan tool. There will be some difference, but it should be minor if the sensor circuit is working properly. Unwanted resistance in the circuit can cause poor engine performance, poor fuel economy, and engine overheating.

Temperature sensors can be tested by removing them and placing them in a container of water with an ohmmeter connected across the sensor terminals. A thermometer is also placed in the water. When the water is heated, the sensor should have the specified resistance at any temperature **(Figure 32–19)**. Replace the sensor if it does not have the specified resistance. Manufacturers give a temperature and resistance chart for each of the temperature sensors.

WARNING!

Never apply an open flame or a heat gun to an ECT or IAT sensor for test purposes. This action will damage the sensor.

The wiring to the sensor can also be checked with an ohmmeter. With the wiring connectors disconnected from the sensor and the computer, connect an ohmmeter from each sensor terminal to the computer terminal to which the wire is connected. Both sensor wires should indicate less resistance than specified by the manufacturer. If the wires have high resistance, the wires or wiring connectors must be repaired.

WARNING!

Before disconnecting any computer system component, be sure that the ignition switch is turned off. Disconnecting components may cause high induced voltages and computer damage.

With the sensor installed in the engine, the sensor terminals may be back-probed to connect a digital voltmeter to them. The sensor should provide the specified voltage drop at any coolant temperature

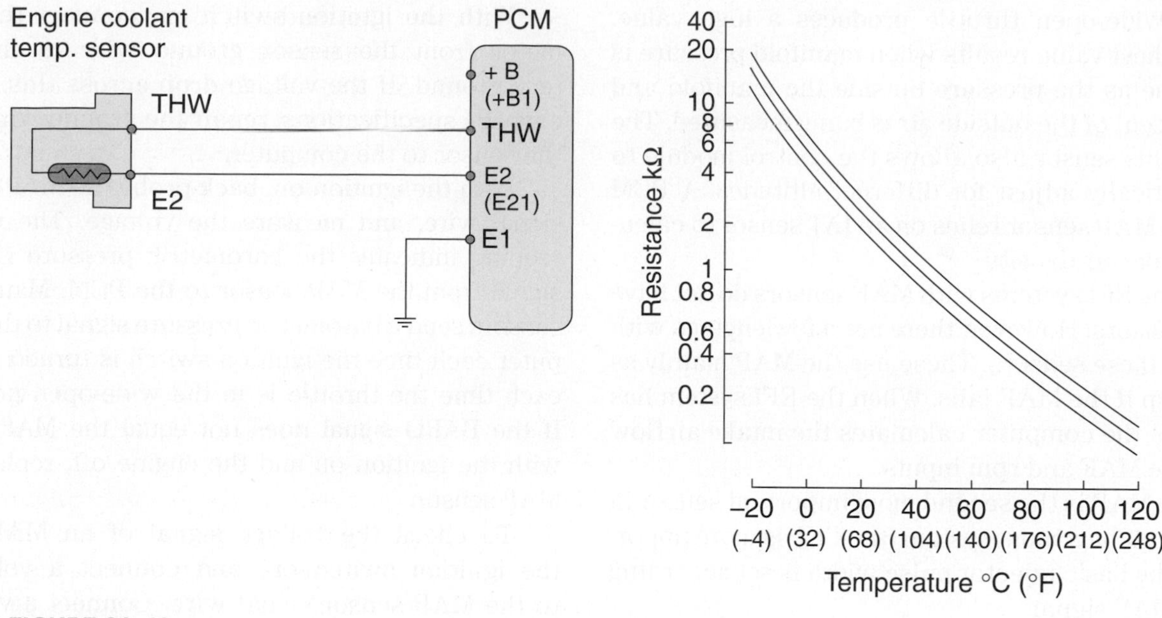

FIGURE 32–19

Specifications for an ECT sensor.

(Figure 32–20). To record accurate readings, make sure the meter leads have a good connection.

PRESSURE SENSORS

Most pressure sensors are piezoresistive sensors. A silicon chip in the sensor flexes with changes in pressures. The amount of flex dictates the voltage signal sent out from the sensor. One side of the chip is exposed to a reference pressure, which is either a perfect vacuum or a calibrated pressure. The other side is the pressure that will be measured. As the chip flexes in response to pressure, its resistance changes. This changes the voltage signal sent to the PCM. The PCM looks at the change and calculates the pressure change.

COLD—10 000-OHM RESISTOR	HOT—090-OHM RESISTOR
−20°F 4.7V	110°F 4.2V
0°F 4.4V	130°F 3.7V
20°F 4.1V	150°F 3.4V
40°F 3.6V	170°F 3.0V
60°F 3.0V	180°F 2.8V
80°F 2.4V	200°F 2.4V
100°F 1.8V	220°F 2.0V
120°F 1.25V	240°F 1.62V

FIGURE 32–20

Voltage drop specifications for an ECT sensor.

Manifold Absolute Pressure (MAP) Sensor

A **manifold absolute pressure (MAP) sensor** senses air pressure or vacuum in the intake manifold. The sensor measures manifold air pressure against an absolute pressure. The MAP sensor uses a perfect vacuum as a reference pressure. The MAP sensor measures changes in the intake manifold pressure that result from changes in engine load and speed. The PCM sends a voltage reference signal to the MAP sensor. As the pressure changes, the sensor's resistance also changes. The control module determines manifold pressure by monitoring the sensor output voltage.

The PCM uses the MAP signals to calculate how much fuel to inject in the cylinders and when to ignite the cylinders. A defective MAP sensor may cause a rich or lean air/fuel ratio, excessive fuel consumption, hard engine starting, a no-start condition, engine surging, and improper transmission operation. The MAP signal may also be used to regulate the EGR or to monitor EGR operation.

High manifold pressure (low vacuum) resulting from full throttle operation requires more fuel. Low pressure or high vacuum requires less fuel. At closed throttle, the engine produces a low MAP value. Wide-open throttle produces a high value. The highest value results when manifold pressure is the same as the pressure outside the manifold and 100 percent of the outside air is being measured. The use of this sensor also allows the control module to automatically adjust for different altitudes. A PCM with an MAP sensor relies on an IAT sensor to calculate intake air density.

Many EFI systems with MAF sensors do not have MAP sensors. However, there are a few engines with both of these sensors. These use the MAP mainly as a backup if the MAF fails. When the EFI system has an MAF, the computer calculates the intake airflow from the MAF and rpm inputs.

The MAP is the second most important sensor in the fuel management system (the CKP is more important). The basic injector pulse width is set according to the MAP signal.

Testing an MAP

The sensor is mounted on the intake manifold or someplace high in the engine compartment. A hose supplies the sensor with engine vacuum. Inspect the sensor, its electrical connectors, and the vacuum hose. The hose should be checked for cracks, kinks, and proper fit.

The PCM supplies a 5-volt reference signal to the sensor. Begin your diagnosis of the MAP circuit by measuring that voltage. With the ignition switch on, back-probe the reference wire, and measure the voltage. If the reference wire does not have the specified voltage, check the reference voltage at the PCM. If the voltage is within specifications at the PCM but low at the sensor, repair the wire. When this voltage is low at the PCM, check the voltage supply wires and ground wires for the PCM. If the wires are good, replace the computer.

An MAP sensor can be monitored with a scan tool through specific PIDs **(Figure 32–21)**. When using a scan tool, make sure to use the correct specifications and follow the subsequent tests given in the service information.

With the ignition switch on, connect the voltmeter from the sensor ground wire to the battery ground. If the voltage drop across this circuit exceeds specifications, repair the ground wire from the sensor to the computer.

With the ignition on, back-probe the MAP sensor signal wire, and measure the voltage. The voltage reading indicates the barometric pressure (BARO) signal from the MAP sensor to the PCM. Many MAP sensors send a barometric pressure signal to the computer each time the ignition switch is turned on and each time the throttle is in the wide-open position. If the BARO signal does not equal the MAP signal with the ignition on and the engine off, replace the MAP sensor.

To check the voltage signal of an MAP, turn the ignition switch on, and connect a voltmeter to the MAP sensor signal wire. Connect a vacuum pump to the MAP sensor vacuum connection, and apply vacuum to the sensor. Manufacturers list the expected voltage drop at different vacuum levels **(Figure 32–22)**. If the MAP sensor voltage is not within specifications at any vacuum, replace the sensor.

SHOP TALK

MAP sensors have a much different calibration on turbocharged engines than on non-turbocharged engines. Typically, the MAP sensor output voltage with the KOEO is half of the normal expected reading when compared to a non-turbocharged engine. Be sure you are using the proper specifications for the sensor being tested.

To check an MAP sensor with a lab scope, connect the scope to the MAP output and a good ground. When the engine is accelerated and returned to idle, the output voltage should increase and decrease **(Figure 32–23)**. If the engine is accelerated and the MAP sensor voltage does not rise and fall, or if the signal is erratic, the sensor or sensor wires are defective.

SENSORS/ INPUTS	PCM PIN	MEASURED/PID VALUES				UNITS MEASURED/PID
		KOEO	HOT IDLE	48 KM/H (30 MPH)	89 KM/H (55 MPH)	
MAP	E62	4*	1–1.4*	1.6–2.1*	1.9–2.3*	DCV

*Value may vary 20% depending on altitude, operating conditions, weather, and other factors.

FIGURE 32–21

The PIDs for a typical MAP sensor.

APPLIED VACUUM					
In. Hg	4	8	12	16	20
kPa	14	27	40	54	68
Voltage Drop	0.3–0.5	0.7–0.9	1.1–1.3	1.5–1.7	1.9–2.1

FIGURE 32–22

An example of how the voltage drop across an MAP changes with the amount of vacuum applied to the sensor.

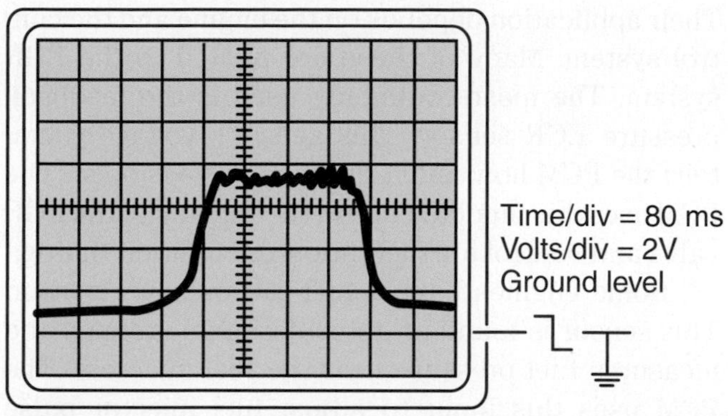

FIGURE 32–23

Trace of a normal MAP sensor.

Some MAP sensors produce a digital voltage signal of varying frequency; begin diagnosis by checking the voltage reference wire and the ground wire. Continue testing by following these steps:

PROCEDURE

Testing an MAP Sensor

STEP 1 Turn off the ignition switch, and disconnect the wiring connector from the MAP sensor.

STEP 2 Connect the connector on the MAP sensor tester to the MAP sensor.

STEP 3 Connect the MAP sensor tester battery leads to a 12-volt battery.

STEP 4 Connect a pair of digital voltmeter leads to the MAP tester signal wire and ground.

STEP 5 Turn on the ignition switch, and observe the barometric pressure voltage signal on the meter. If this voltage signal does not equal specifications, replace the sensor.

STEP 6 Supply the specified vacuum to the MAP sensor with a vacuum pump.

STEP 7 Observe the voltmeter reading at each specified vacuum. If the MAP sensor voltage signal does not equal the specifications at any vacuum, replace the sensor.

On vehicles that use a frequency-generating MAP sensor, the sensor can be checked with a DMM that measures frequency. Set the meter to read 100 to 200 hertz, and connect it to the MAP sensor. Measure the voltage, duty cycle, and frequency at the sensor with no vacuum applied. Then apply about 457 mm Hg (18 in. Hg) of vacuum to the MAP. Observe and record the same readings. A good MAP will have about the same amount of voltage and duty cycle with or without the vacuum. However, the frequency should decrease. Normally, a frequency of about 155 hertz is expected at sea level with no vacuum applied to the MAP. When vacuum is applied, the frequency should decrease to around 95 hertz.

A lab scope can be used to check an MAP sensor. The upper horizontal line of the trace should be at 5 volts, and the lower horizontal line should be close to zero **(Figure 32–24)**. Check the waveform for unusual movements of the trace. If the waveform is anything but normal, replace the sensor.

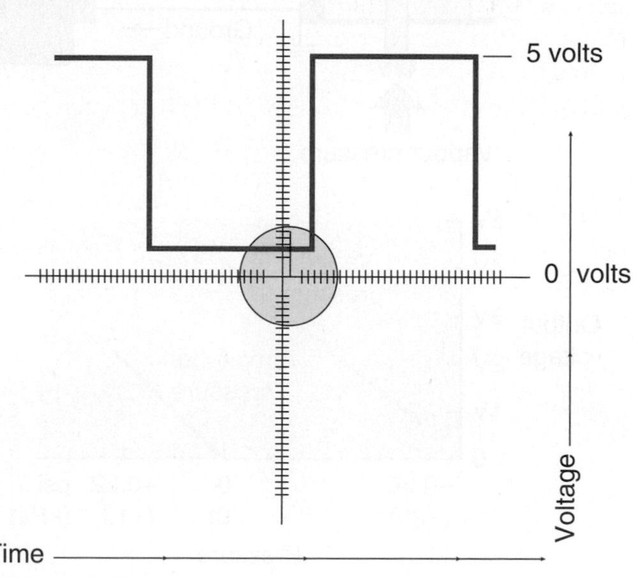

FIGURE 32–24

A good Ford MAP sensor signal.

Vapour Pressure Sensor (VPS)

The **vapour pressure sensor (VPS)** measures the vapour pressure in the evaporative emission control system. This sensor is capable of responding to slight pressure changes. The sensor has two chambers divided by a silicon chip. One chamber, called the reference chamber, is exposed to atmospheric pressure. The other chamber is exposed to vapour pressure. Changes in vapour pressure cause the chip to flex, which causes the signal to the PCM to change **(Figure 32–25)**. In most cases, the sensor receives a 5-volt reference voltage. The voltage signal represents the difference between vapour pressure and atmospheric pressure. The return or signal voltage will be high when the vapour pressure is high.

The sensor may be on the fuel pump or in a remote location. When the VPS is remotely mounted, a hose connects the sensor to a vapour pressure port. In some cases, the sensor has an additional hose that supplies atmospheric pressure to the VPS. If these hoses are reversed, the PCM will see high vapour pressure and set a DTC. All hoses should be checked for leaks, kinks, and secure connections. The reference voltage should also be checked. The operation of the sensor is checked in the same way as an MAP, except pressure, not vacuum, is applied to the sensor. Refer to the specifications for the amount of testing pressure to apply and the subsequent voltage signal from the sensor.

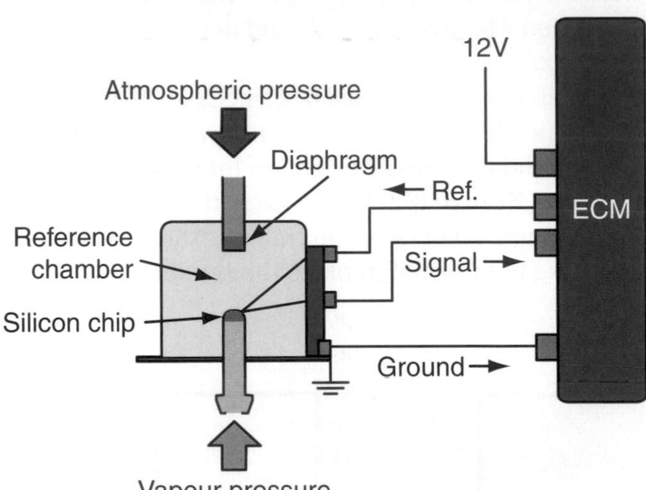

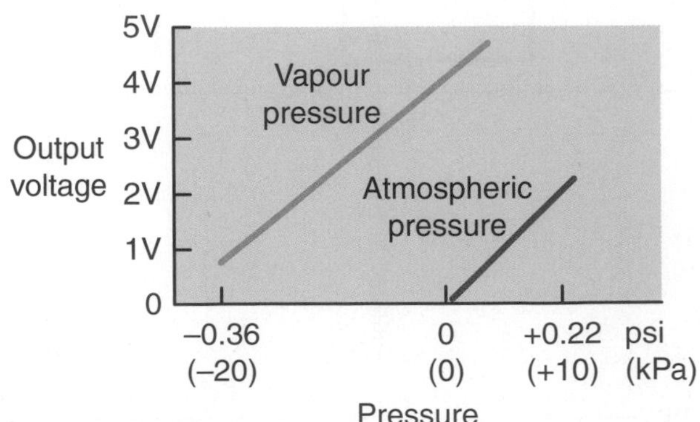

FIGURE 32–25

The operation of a vapour pressure sensor.

Other Pressure Sensors

Other pressure sensors are used on some engines. Their application depends on the engine and the control system. Many of these are related to the EGR system. The most commonly used is the feedback pressure EGR sensor. This sensor's voltage signal tells the PCM how much the EGR valve is open. The PCM uses this input to control the vacuum to the EGR valve and control air/fuel ratios and ignition timing.

Some engines have a fuel rail pressure sensor. This sensor is found on returnless fuel systems and measures fuel pressure near the fuel injectors. The PCM uses this input to adjust fuel injector pulse width. Turbo- or supercharged engines have a separate pressure sensor that monitors the amount of boost.

MASS AIRFLOW (MAF) SENSORS

The **mass airflow (MAF) sensor** measures the flow of air entering the engine **(Figure 32–26)**. This measurement of intake air volume is used to calculate engine load (throttle opening and air volume). It is similar to the relationship of engine load to MAP or vacuum sensor signal. Engine load inputs are used to control the fuel injection and ignition systems, as well as shift timing in automatic

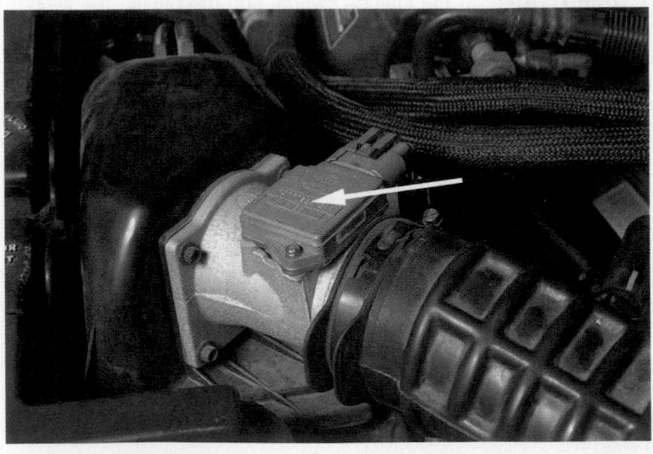

FIGURE 32–26

A mass airflow sensor.

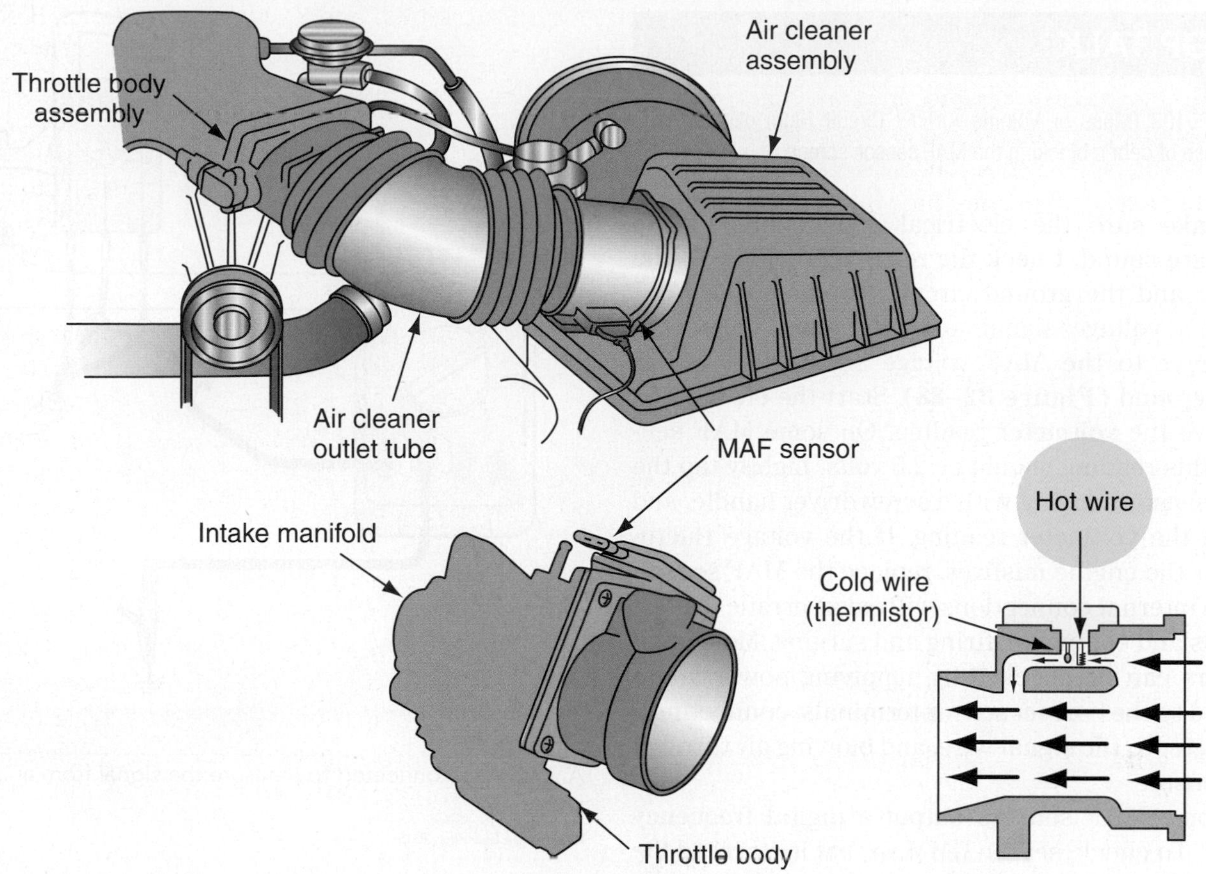

FIGURE 32–27

A hot-wire MAF.

transmissions. The airflow sensor is placed between the air cleaner and throttle plate assembly or inside the air cleaner assembly.

There are different types of MAF sensors. The most commonly used design is the hot-wire MAF. In a hot-wire-type MAF **(Figure 32–27)**, a wire, called the hot wire, is positioned so that the intake air flows over it. The sensor also has a thermistor, sometimes referred to as the cold wire, located beside the hot wire that measures intake air temperature. The sensor also contains a control module. Current from the PCM keeps the hot wire at a constant temperature above ambient temperature, normally 200°C (392°F). Airflow past the hot wire causes it to lose heat, and the PCM responds by sending more current to the wire. The increased current flow keeps the hot wire at its desired temperature. The current required to maintain the hot wire's temperature is proportional to mass airflow. The sensor measures the current and sends a voltage signal to the PCM. The PCM interprets the signal to determine mass airflow. Most MAF sensors have an integrated IAT sensor.

Testing an MAF Sensor

The test procedure for hot-wire MAF sensors varies with the vehicle make and year. Always follow the test procedure in the appropriate service information. Most often, diagnosis of an MAF sensor involves visual, circuit, and component checks. The MAF sensor passage must be free of debris to operate properly. If the passage is plugged, the engine will usually start but will run poorly or stall and may not set a DTC.

Check the air inlet system (air filter, housing, and ductwork) for obstructions, blockage, proper installation, and sealing. Closely inspect the air filter element. Poor quality filters can shed fibres, which can collect on the sensing wire in the MAF. Also, if an aftermarket performance air filter is installed that requires cleaning and oiling, ensure that there is not too much oil in the filter as the oil can also travel downstream and contaminate the MAF. Check the screen of the MAF sensor for dirt and other contaminants. Check the throttle plate bore for dirt buildup.

SHOP TALK

DTC P0103 (Mass or Volume Airflow Circuit High) can be set because of debris blocking the MAF sensor screen.

Make sure the electrical connections to the MAF are sound. Check the reference voltage to the sensor and the ground circuit. To check the MAF sensor's voltage signal and frequency, connect a voltmeter to the MAF voltage signal wire and a good ground **(Figure 32–28)**. Start the engine and observe the voltmeter reading. On some MAF sensors, this reading should be 2.5 volts. Lightly tap the MAF sensor housing with a screwdriver handle, and watch the voltmeter reading. If the voltage fluctuates or the engine misfires, replace the MAF sensor. Loose internal connections will cause erratic voltage signals and engine misfiring and surging. Most MAF sensors can be checked by supplying power and a ground to the correct sensor terminals, connecting a voltmeter to the signal wire, and blowing air through the sensor.

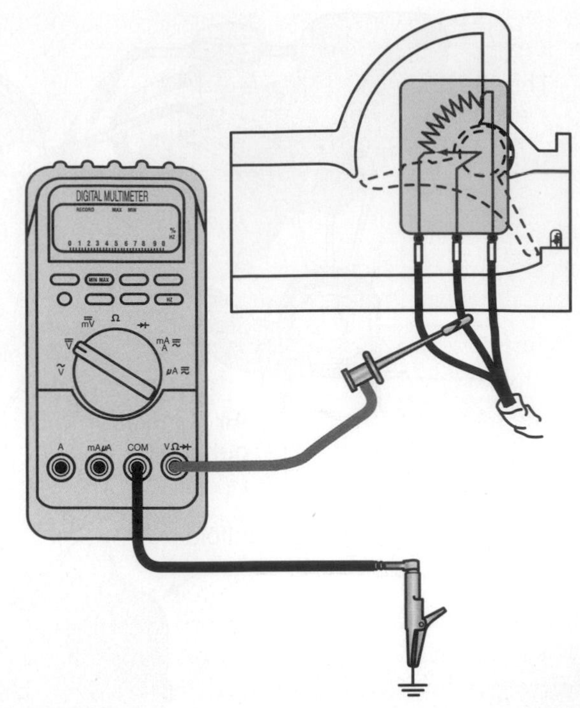

FIGURE 32–28

A voltmeter connected to measure the signal from an MAF sensor.

Some MAF sensors output a digital frequency signal. To check, set the DMM so that it can read the frequency of DC voltage. With it still connected to the signal wire and ground, the meter should read about 30 hertz (Hz) with the engine idling. Now increase the engine speed, and record the meter reading at various speeds. Graph the frequency readings. The MAF sensor frequency should increase smoothly and gradually in relation to engine speed. If the MAF sensor frequency reading is erratic, replace the sensor **(Figure 32–29)**.

An MAF sensor can be monitored with a scan tool through specific PIDs. When using a scan tool, make sure to use the correct specifications and follow the subsequent tests given in the service information. Normally, the engine is run at 1500 rpm for five seconds and then allowed to idle. The MAF return signal and operation of the sensor can be observed **(Figure 32–30)**. Often, the manufacturer recommends that the sensor be observed at different speeds.

Although diagnosing some MAF sensors with a scan tool, the test may display grams per second. This mode provides an accurate test of the MAF sensor.

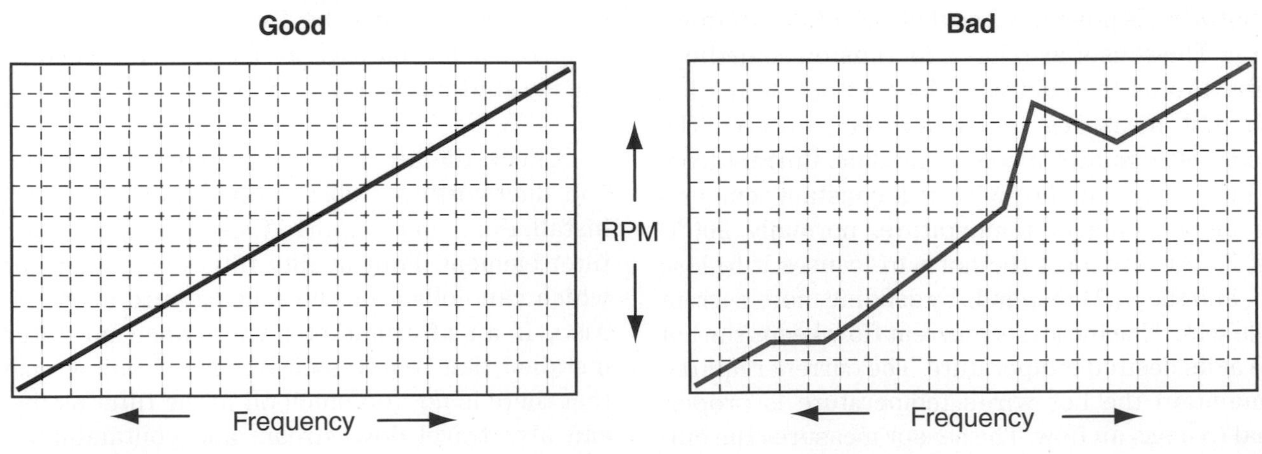

FIGURE 32–29

Satisfactory and unsatisfactory MAF sensor frequency readings.

SENSORS/ INPUTS	PCM PIN	MEASURED/PID VALUES				UNITS MEASURED/PID
		KOEO	HOT IDLE	48 KM/H (30 MPH)	89 KM/H (55 MPH)	
MAP	E25	0	0.6	1.1	1.3	DCV

FIGURE 32–30

The PIDs for a typical MAF sensor.

The grams-per-second reading should normally be 4 to 7 with the engine idling. This reading should gradually increase as the engine speed increases. When the engine speed is constant, the grams-per-second reading should remain constant. If the grams-per-second reading is erratic at a constant engine speed, or if this reading varies when the sensor is tapped lightly, the sensor is defective. An MAF sensor DTC may not be present with an erratic grams-per-second reading, but the erratic reading indicates a defective sensor.

Driveability concerns such as a hesitation, along with DTCs for a lean exhaust condition, are often caused by misreporting MAF sensors. Set the scan tool to record MAF, rpm, throttle, engine load, and O_2 activity, and perform a test drive. Under wide-open throttle (WOT) acceleration, the MAF should reach at least 100 grams/second as the transmission shifts from first to second on a four-cylinder engine and even more with a six- or eight-cylinder. If the grams/second reading is low and the O_2 readings are low, suspect a dirty or defective MAF. Look for about 3.78 grams/second (0.5 lbs./min.) at idle up to 98.28 grams/second (13 lbs./min.) or more under WOT acceleration.

Frequency-varying types of MAF sensors can be tested with a lab scope. The waveform should appear as a series of square waves. When the engine speed and intake airflow increase, the frequency of the MAF sensor signals should increase smoothly and proportionately to the change in engine speed. If the MAF or connecting wires are defective, the trace will show an erratic change in frequency **(Figure 32–31)**.

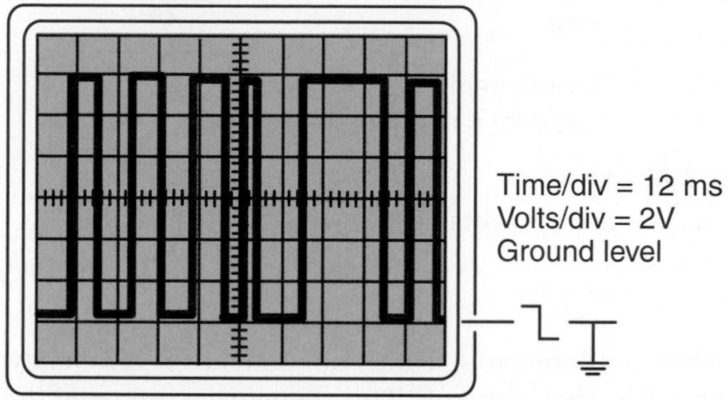

FIGURE 32–31

The trace of a defective frequency-varying MAF sensor.

SHOP TALK

Often, O_2S are called lambda sensors. The term *lambda* is used to refer to air/fuel ratio and normal. Technically, it refers to normal and is represented by the Greek letter λ. It is best to think of lambda as meaning a reference point for normal or ideal air/fuel mixture. This mixture is typically called stoichiometric. A lambda sensor measures the variance from stoichiometric, which is about the same thing an O_2S does.

OXYGEN SENSORS (O_2S)

The exhaust gas oxygen sensor (O_2S) **(Figure 32–32)**, or air/fuel ratio sensor, is the key sensor in the closed-loop mode. The O_2S is threaded into the exhaust manifold or into the exhaust pipe near the engine. The PCM uses an O_2S to ensure that the air/fuel ratio is correct for the catalytic converter. The PCM adjusts the amount of fuel injected into the cylinders based on the O_2S signal.

OBD-II standards require an O_2S before and after the catalytic converter **(Figure 32–33)**. The O_2S before the converter is used for short-term air/fuel ratio adjustments. This sensor is referred to as Sensor 1. On V-type engines, one sensor is referred to as Bank 1 Sensor 1, based on the sensor being on the same bank as cylinder 1, and the other as Bank 2 Sensor 1. The O_2S after the catalytic converter is

FIGURE 32–32

A typical oxygen sensor.

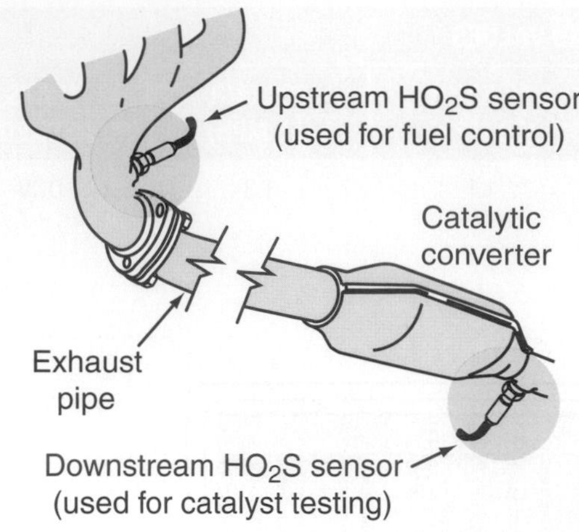

FIGURE 32–33

The pre- and post-catalytic oxygen sensors in an OBD-II system.

used to determine catalytic converter efficiency and for fuel control. This sensor is referred to as Sensor 2. With two catalytic converters, one sensor is Bank 1 Sensor 2, and the other is Bank 2 Sensor 2. Often, Sensor 2 is called the catalyst monitor sensor. Some engines have more than two O_2S in an exhaust bank. The sensor with the highest number designation is the catalyst monitor sensor.

O_2S generate a voltage signal based on the amount of oxygen and other gases present in the exhaust gas. A perfectly balanced, stoichiometric, air/fuel mixture of 14.7:1 produces an output of around 0.5 volt. This equals a lambda of 1.0. When the sensor detects an excess in oxygen, the reading is lean, and the computer enriches the air/fuel mixture to the engine. When the sensor reading is rich, the computer leans the air/fuel mixture.

Heated Oxygen Sensors

To generate an accurate signal, an O_2S must operate at a minimum temperature of 400°C (750°F). Current O_2S have a built-in heating element to quickly heat them and keep them hot at idle and light-load conditions. The heater is controlled by the PCM. Early O_2S did not have a heater and required some time for the exhaust to warm them. This resulted in extended periods of open-loop operation.

Heated oxygen sensors (HO_2Ss) have three or four wires connected to them. The additional wires provide voltage for the internal heater in the sensors. HO_2S are sometimes referred to as heated exhaust gas sensors (HEGOs). The heater is not on all of the time. The PCM opens and closes, duty cycles, the ground for the heater circuit as needed. The cycling of current to the heater protects the ceramic material of the heater from being overheated, which would cause it to break.

ZIRCONIUM DIOXIDE OXYGEN (ZrO_2) SENSORS Zirconium dioxide oxygen (ZrO_2) sensors are the most commonly used O_2S, although they are being replaced with air/fuel ratio sensors on current vehicles. These have a zirconia (zirconium dioxide) element, platinum electrodes **(Figure 32–34)**, and a heater. The zirconia element has one side exposed to the exhaust stream and the other side open to the atmosphere through the sensor's wires. Each side has a platinum electrode attached to the zirconium dioxide element. The platinum electrodes conduct the voltage generated. Contamination or corrosion of the platinum electrodes or zirconia elements will reduce the voltage signal output. This type of O_2S is sometimes referred to as a narrow range sensor because it is effective only with air/fuel ratios around 14.7:1. This

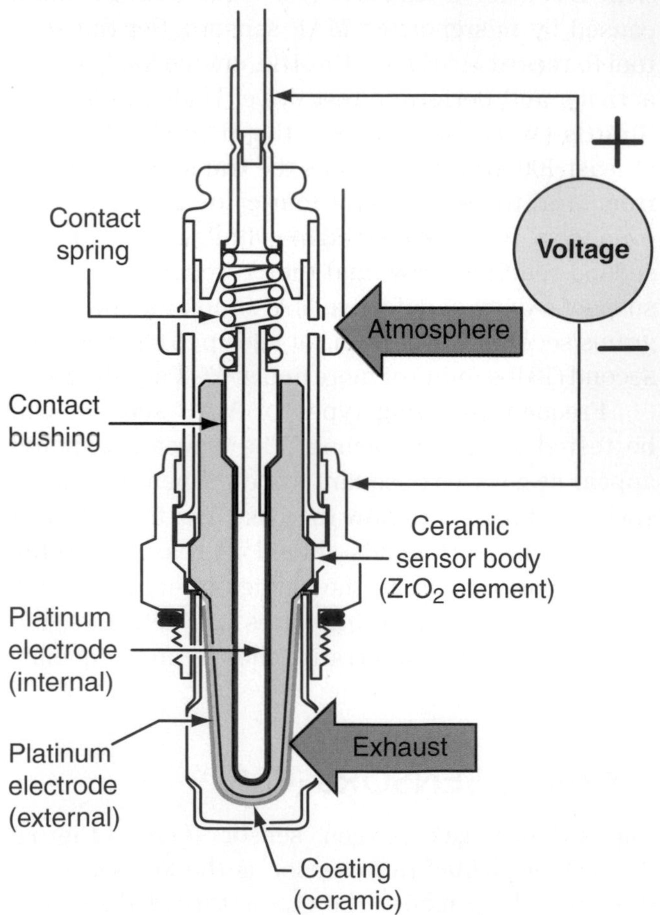

FIGURE 32–34

The voltage signal from a ZrO_2 sensor results from the difference in voltage at the two platinum plates inside the sensor.

type of sensor acts more like a lambda switch; the output varies high or low, depending on whether the exhaust gases are rich or lean compared to lambda.

TITANIUM DIOXIDE (TIO_2) SENSORS Titanium dioxide oxygen (TiO_2) sensors are found on a few vehicles. These sensors do not generate a voltage signal. Instead, they act like a variable resistor, altering a 5-volt reference signal supplied by the control module. Titanium sensors send a low-voltage signal (below 2.5 volts) with low oxygen content and a high-voltage signal (above 2.5 volts) with high oxygen content. Variable-resistance O_2S do not need an outside air reference. This eliminates the need for internal venting to the outside.

Air/Fuel Ratio (A/F) Sensor

Although an **air/fuel ratio (A/F) sensor** looks like a conventional O_2S, internally it is different. It also operates differently; its voltage output increases as the mixture becomes leaner. The sensor does not directly produce a voltage signal; rather, it changes current. A detection circuit in the PCM monitors the change and strength of the current flow from the A/F sensor and generates a voltage signal proportional to the exhaust oxygen content **(Figure 32–35)**. Based on the voltage signal, the PCM is able to calculate the air/fuel ratio over a wide range of conditions and quickly adjust the amount of fuel required to maintain a stoichiometric ratio. The wide range of operation allows the sensor to measure very lean conditions. When the engine is running in lean, the oxygen content of the exhaust is higher than a normal O_2S is capable of measuring. An A/F sensor also operates at a much higher temperature (650°C [1200°F]) than a conventional O_2S.

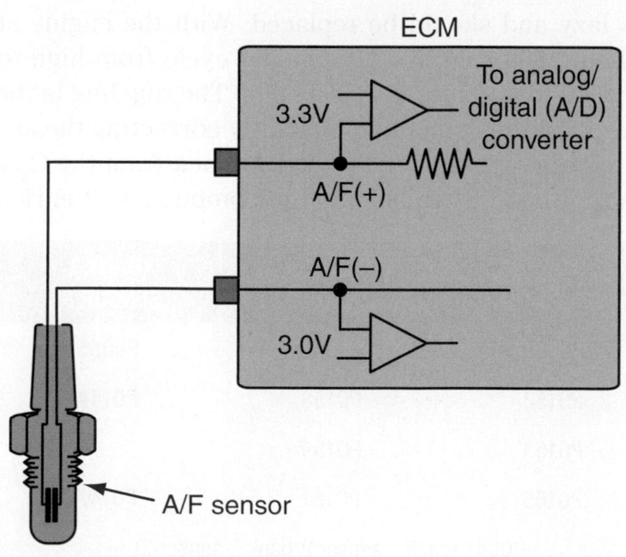

FIGURE 32–35

The detection circuit for an A/F sensor.

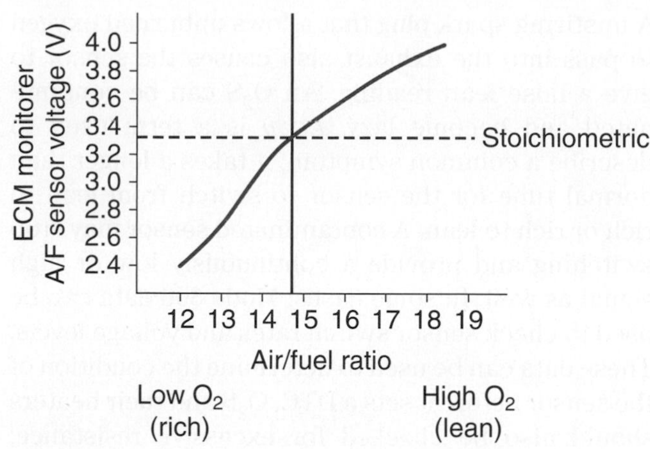

FIGURE 32–36

A chart showing the voltage outputs from an A/F sensor at different air/fuel ratios.

The detection circuit is always measuring the direction and how much current is being produced. When the mixture is at a stoichiometric ratio, no current is generated by the sensor. The voltage signal from the detection circuit is 3.3 volts **(Figure 32–36)**. When the fuel mixture is rich (low exhaust oxygen content), the A/F produces a negative current flow and the detection circuit will produce a voltage below 3.3 volts. A lean mixture, which has more oxygen in the exhaust, produces a positive current flow, and the detection circuit will produce a voltage signal above 3.3 volts.

Checking Oxygen Sensors and Circuits

SHOP TALK

The engine must be at normal operating temperature before the O_2S is tested. Always refer to the specifications and testing procedures supplied by the manufacturer.

Keep in mind that several things can cause an O_2S to appear bad other than a faulty sensor. Common causes of abnormal O_2S operation include incorrect fuel pressure, a malfunctioning AIR system, an EGR leak, a leaking injector, a vacuum leak, an exhaust leak and a contaminated MAF sensor. It is important to determine if it is the O_2S itself or some other factor that is causing the O_2S to behave abnormally.

The accuracy of the O_2S reading can be affected by air leaks in the intake or exhaust manifold.

A misfiring spark plug that allows unburned oxygen to pass into the exhaust also causes the sensor to give a false lean reading. An O_2S can be contaminated and become lazy. *Lazy* is a term used to describe a common symptom; it takes a longer than normal time for the sensor to switch from lean to rich or rich to lean. A contaminated sensor may stop switching and provide a continuously low or high signal as well. In some cases, Mode $06 data can be used to check sensor switch rates and voltage levels. These data can be used to determine the condition of the sensor before it sets a DTC. O_2S and their heaters should also be checked for excessive resistance, opens, and shorts to ground.

IDENTIFYING THE CAUSE OF O_2S CONTAMINATION Many things can cause an O_2S to become contaminated. Before simply replacing a contaminated sensor, find out why and how it was contaminated. Begin by examining the engine for leaks; oil, coolant, and other liquids can plug the pores of the sensor and cause it to respond slowly and inaccurately to the amount of oxygen in the outside air or in the exhaust. If no leaks are evident, check the vehicle's service history. It is possible that recent problems that may or may not have been corrected are the cause of the contamination. For example, if the engine had some service done to it, RTV (room temperature vulcanizing) sealant that was not designed for use around O_2S may have been used.

Using incorrect engine oil, particularly oils with high levels of sulphated ash, phosphorous, and sulphur (SAPS), has been found to contaminate both catalytic converters and oxygen sensors. This has led to the reduction in SAPS in newer engine oil specifications.

You may also discover the cause by removing the sensor. The colour and smell of the sensor may indicate the problem. If the sensor has a sweet smell, it is undoubtedly contaminated by engine coolant. If it smells burnt, there is a good chance that oil has melted onto the sensor. Silicone and engine coolant will leave white deposits on the sensor. Brown colouring may indicate oil contamination, and black means it was contaminated by a rich air/fuel mixture.

TESTING WITH A SCAN TOOL The OBD-II O_2S monitor checks for sensor circuit faults, slow response rate, switch-point voltages, and malfunctions in the sensor's heater circuit. There is a separate DTC for each condition for each sensor **(Figure 32–37)**. The catalyst monitor sensor (Sensor 3) is not monitored for response rate; however, its peak rich and lean voltage values are. In most cases, the O_2S is monitored by the PCM once per trip. The PCM tests the sensor by looking at the return signal after the air/fuel ratio has been changed. The faster the sensor responds, the better the sensor. Diagnostic Mode 5 and Mode 6 report the results of this monitor test.

The PCM will store a code when the sensor's output is not within the desired range. The normal range is between 0 and 1 volt, and the sensor should constantly toggle between about 0.2 to 0.8 volt. If the range that the sensor toggles in is within the specifications, the computer will think everything is normal and respond accordingly. This, however, does not mean the sensor is working properly.

Watching the scan tool while the engine is running, the O_2S voltage should move to nearly 1 volt and then drop back to close to 0 volts. Immediately after it drops, the voltage signal should move back up. This immediate cycling is an important function of an O_2S. If the response is slow, the sensor is lazy and should be replaced. With the engine at about 2500 rpm, the O_2S should cycle from high to low 10 to 40 times in 10 seconds. The toggling is the result of the computer constantly correcting the air/fuel ratio in response to the feedback from the O_2S. When the O_2S reads lean, the computer will enrich

DEFINITION	(BANK 1 SENSOR 1)	(BANK 1 SENSOR 2)	(BANK 2 SENSOR 1)	(BANK 2 SENSOR 2)	(BANK 1 SENSOR 3)
Heater resistance	P0053	P0054	P0059	P0060	P0055
High voltage	P0132	P0138	P0152	P0158	P0144
Slow response	P0133	P0139	P0153	P0159	
Heater circuit	P0135	P0141	P0155	P0161	P0147
Signals swapped	P0040 (bank 1 sensor 1/bank 2 sensor 1)		P0041 (bank 1 sensor 2/bank 2 sensor 2)		

FIGURE 32–37

A bad oxygen sensor can cause a number of problems and set a variety of DTCs.

the mixture. When the O_2S reads rich, the computer will lean the mixture. When the computer does this, it is in control of the air/fuel mixture.

If the voltage is continually high, the air/fuel ratio may be rich, or the sensor may be contaminated by RTV sealant, antifreeze, or lead from leaded gasoline. If the O_2S voltage is continually low, the air/fuel ratio may be lean, the sensor may be defective, or the wire between the sensor and the computer may have a high-resistance problem. If the O_2S voltage signal remains in a mid-range position, the computer may be in open loop, or the sensor may be defective.

If the scan tool you are using has the ability to graph data, use this function to examine sensor activity. Using the graphing function may allow you to look at several pieces of information at once, such as O_2 sensor activity and fuel trim.

TESTING WITH A DMM 24 If a defect in the O_2S signal wire is suspected, back-probe it at the computer and connect a digital voltmeter from the signal wire to ground with the engine idling. The difference between the voltage readings at the sensor and at the computer should not exceed the vehicle manufacturer's specifications. A typical specification for voltage drop across the average sensor wire is 0.02 volt. Photo Sequence 26 shows how to test an oxygen sensor.

Now check the sensor's ground. With the engine idling, connect the voltmeter from the sensor case to the sensor ground wire on the computer. Typically, the maximum allowable voltage drop across the sensor ground circuit is 0.02 volt. If the voltage drop across the sensor ground exceeds specifications, repair the ground wire or the sensor ground in the exhaust manifold.

With the O_2S wire disconnected, connect an ohmmeter across the heater terminals in the sensor connector. If the heater does not have the specified resistance, replace the sensor.

Most engines are fitted with heated O_2S A PTC thermistor inside the O_2S heats up as current passes through it. The PCM turns on the circuit based on ECT and engine load (determined from the MAF or MAP sensor signal). The higher the temperature of the heater, the greater its resistance. If the heater is not working, the sensor warm-up time is extended and the computer stays in open loop longer. In this mode, the computer supplies a richer air/fuel ratio. As a result, the engine's emissions are high and its fuel economy is reduced.

To test the heater circuit, disconnect the O_2S connector and connect a voltmeter between the heater voltage supply wire and ground **(Figure 32–38)**. With the ignition switch on, 12 volts should

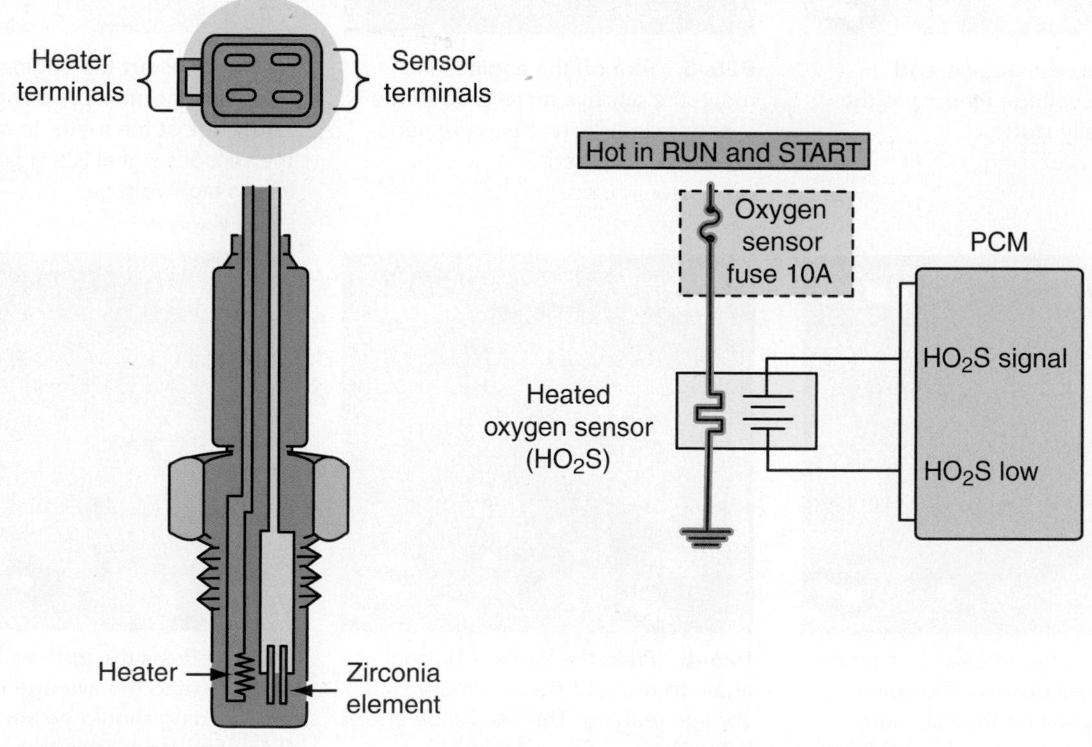

FIGURE 32–38

Using a wiring diagram to identify the terminals on a heated oxygen sensor.

PHOTO SEQUENCE 26

TESTING AN OXYGEN SENSOR

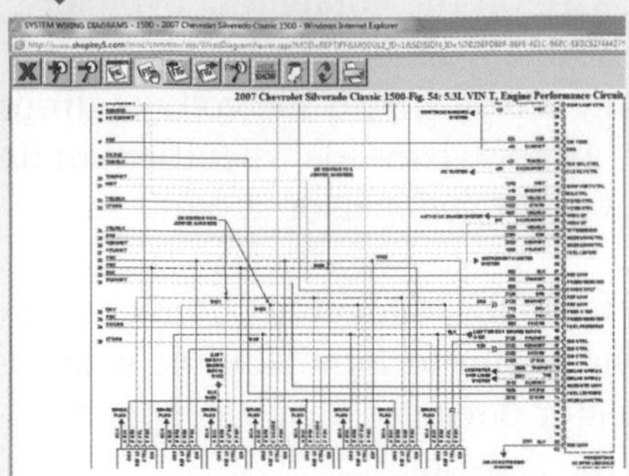

P26–1 Locate oxygen sensor in a wiring diagram for the vehicle, and identify what part of the sensor each wire is connected to.

Martin Restoule

P26–2 Connect the positive lead of the meter to the power wire for the sensor's heater. Connect the meter's negative lead to a good ground.

Snap-on Incorporated

P26–3 Place the meter where you can see it from the driver's seat.

©Cengage Learning 2015

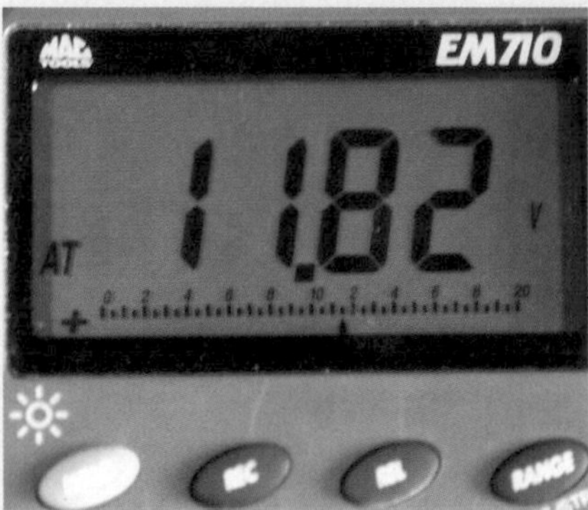

P26–4 Start the engine, and observe the voltage reading as the engine initially starts.

Snap-on Incorporated

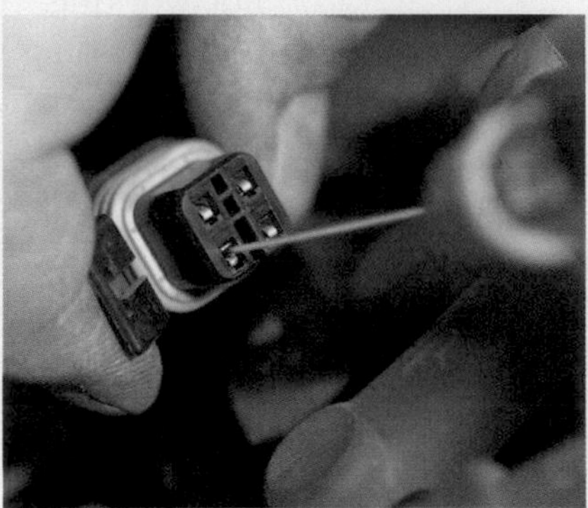

P26–5 Turn off the engine, and move the positive meter lead to the sensor's signal wire. Keep the negative lead grounded.

©Cengage Learning 2015

P26–6 Restart the engine, and allow it to reach normal operating temperature. Look at the meter to make sure the sensor's signal is toggling from low to high voltage.

Snap-on Incorporated

P26–7 Press the MIN/MAX button on the meter, and observe the voltage. This reading will be the minimum voltage and should be about 0.1 volt.

©Cengage Learning 2015

P26–8 Press the MIN/MAX button again to observe the maximum voltage reading. This should be about 0.9 volt.

©Cengage Learning 2015

P26–9 Press the MIN/MAX button again to read the average voltage. This reading should be about 0.45 volt. Repeat this test at different speeds to get a good look at how well the O_2 sensor responds.

©Cengage Learning 2015

be supplied on this wire. If the voltage is less than 12 volts, repair the fuse in this voltage supply wire or the wire itself.

TESTING WITH A LAB SCOPE A lab scope or GMM allows a look at the operation of the O_2S while it responds to changes in the air/fuel mixture. Connect the lab scope to the sensor's signal wire and a good ground. Set the scope to display the trace at 200 mV per division and 500 ms per division.

SHOP TALK

ZrO_2 and TiO_2 sensors produce an analog signal. A/F ratio sensors produce a signal only after their current output has been sent to the PCM; make sure you refer to the manufacturer's information before making judgment on an A/F sensor.

The O_2S can be biased rich or lean, or not work at all, or work too slowly. Begin your testing by allowing the engine and O_2S to warm up. Watch the waveforms **(Figure 32–39)**. If the sensor's voltage toggles between 0 and 500 mV, it is toggling below its normal range and is not operating normally. It is biased low or lean. As a result, the computer will constantly add fuel to try to reach the upper limit of the sensor. Something is causing the sensor to be biased lean. If the toggling occurs only at the higher limits of the voltage range, the sensor is biased rich.

SHOP TALK

Some PCMs are programmed with a bias for the O_2S. This means the voltage output of the sensor does not exactly reflect how the PCM interprets the signals from the sensor. Normally, the bias is added to the actual reading. Make sure to refer to the service information to identify any voltage biases that may occur. If you do not do this, you will end up with a faulty diagnosis.

When the mixture is rich, combustion has a better chance of being complete. Therefore, the oxygen levels in the exhaust decrease. The O_2S

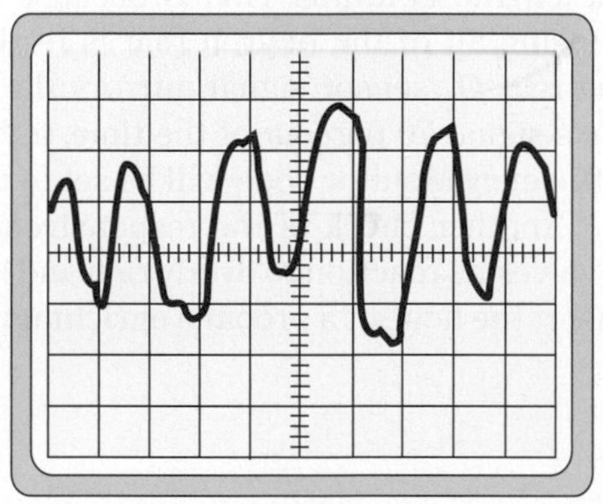

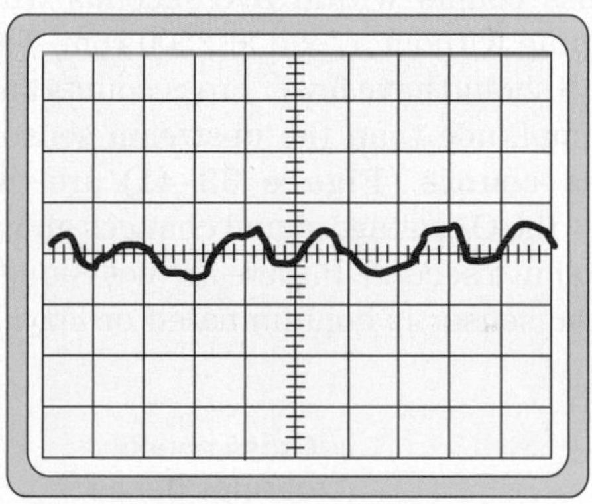

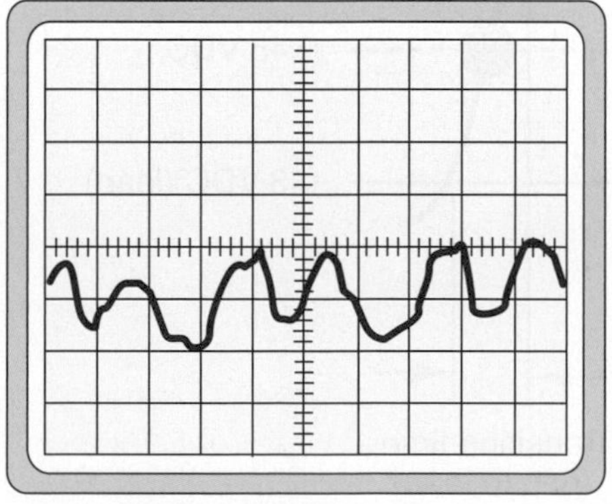

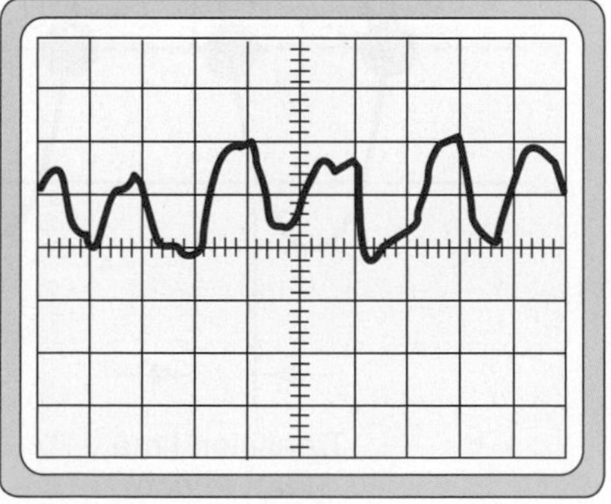

FIGURE 32–39

Normal and abnormal O_2S waveforms.

output will respond to the low O_2 with a high-voltage signal. Remember that the PCM will always try to do the opposite of what it receives from the O_2S. When the O_2 shows lean, the PCM goes rich, and vice versa. When a lean exhaust signal is not caused by an air/fuel problem, the PCM does not know what the true cause is and will enrich the mixture in response to the signal. This may make the engine run worse than it did.

The signals from the front (upstream) and rear (downstream) O_2S should be compared. This will not only help in determining the effectiveness of the catalytic converter, but it will also help determine the condition of each sensor. Use a two-channel scope or GMM. Connect the meter to the signal wire in both harnesses. Start the engine and allow the sensors to warm up. Then raise the engine's speed to 2000 to 2500 rpm and observe the waveforms **(Figure 32–40)**. In the waveform, it is evident that the upstream sensor is toggling correctly, and the upstream sensor shows the catalyst is working properly.

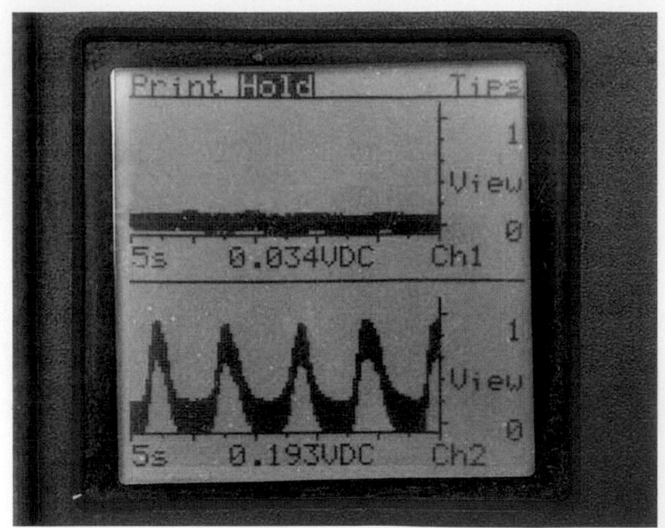

FIGURE 32–40

A comparison of the upstream (Ch 2) and the downstream (Ch 1) O_2 sensors.

The voltage signal from an upstream O_2S should have seven cross counts within five seconds with the engine running without a load at 2500 rpm. The downstream O_2S should have fewer cross counts and have a lower amplitude than the upstream sensor. O_2 signal **cross counts (Figure 32–41)** are the number of times the O_2 voltage signal changes above or below 0.45 volt in a second. If there are not enough cross counts, the sensor is contaminated or lazy. It should be replaced. The downstream O_2S signal does not toggle when the converter is warmed up and the engine is idling. This is because the converter is using all of the oxygen that is in the exhaust. If the rear O_2 sensor signal mirrors the front O_2 sensor's signal 70 percent of the time, a P0420 Catalyst Efficiency Monitor code will be set in the PCM.

Another check of the responsiveness of an O_2S involves its reaction to overly rich and lean mixtures. Insert the hose of a propane enrichment tool into the

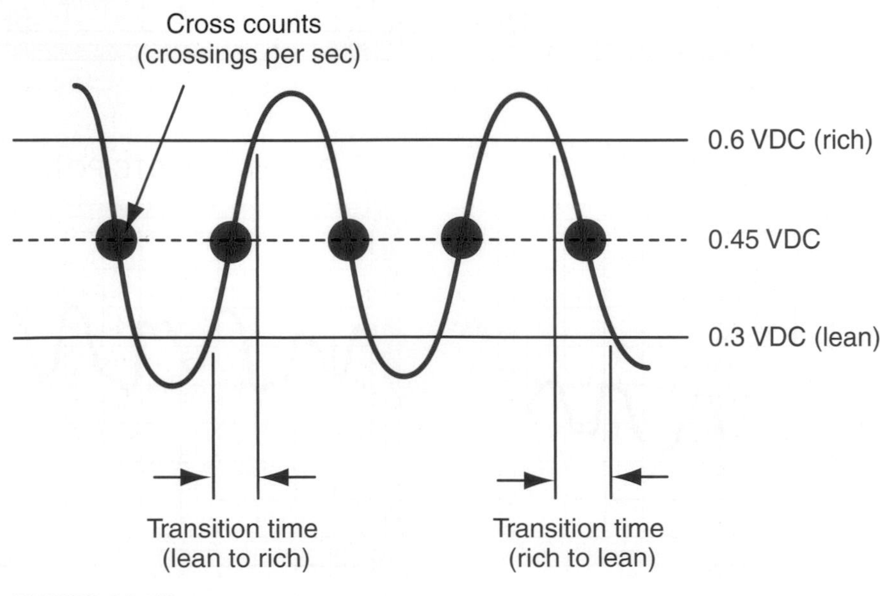

FIGURE 32–41

O_2 sensor signal cross counts.

power-brake booster vacuum hose, or simply install it into the nozzle of the air cleaner assembly. This will drive the mixture rich. Most good O_2S will produce almost 1 volt when driven full rich. The typical specification is at least 800 mV. If the voltage does not go high, the O_2S is bad and should be replaced. Now remove the propane bottle and cause a vacuum leak by pulling off an intake vacuum hose. Watch the scope to see how the O_2S reacts. It should drop to under 175 mV. If it does not, replace the sensor. These tests check the O_2S, not the system; therefore, they are reliable O_2S checks.

Testing Air/Fuel Ratio (A/F) Sensors

A/F sensors cannot be tested in the same way as an O_2S. The A/F sensor has two signal wires. The PCM supplies 3 volts to one wire and 3.3 volts to the other. If a voltmeter is connected across these two wires, the difference in potential will be measured. This reading has no meaning because the 0.3-volt difference will always be there regardless of the oxygen content. The heater circuit can be checked with a voltmeter, and the heater can be checked with an ohmmeter.

To observe the changes in voltage in relationship to changes in oxygen content, use a scan tool that has the correct software. Some scan tools are not able to read the data from the sensor detection circuit, and a PID for the A/F sensor will not be available. Some scan tools will convert the range of the voltage signals to 0 to 1 volt. This is done by dividing the output from the detection circuit by five. To calculate the actual voltage signal, multiply the measured voltage by five. A reading of 0.66 volt equals 3.3 volts, which is the amount that should be present when the mixture is stoichiometric. The voltage from an A/F sensor will increase as the mixture goes lean and will decrease with a rich mixture.

Unless there is a large change in the air/fuel ratio, the voltage readings will toggle very little. The air/fuel ratio is tightly controlled by the PCM, and only minor adjustments are normally made. However, if the output voltage of the A/F sensor stays at 3.30 V regardless of conditions, the sensor circuit may be open. If voltage from the sensor stays at 2.8 V or less or 3.8 V or more, the sensor circuit may be shorted.

The action of the sensor can also be observed with an ammeter. Place the ammeter in series with the 3.3-volt signal wire. Separate the connector to the sensor. Use jumper wires to connect the terminals for the heater and the 3-volt signal wire. Then connect the positive lead of the meter to the 3.3-volt signal wire terminal on the PCM side of the connector and the negative lead to the terminal in the other half of the connector. Run the engine and observe the meter. When the mixture is at stoichiometric, there should be 0 amps. When the mixture is rich, there should be negative current flow. And when there is a lean mixture, the current should move to positive.

The A/F sensor monitor is similar to the O_2S monitor but has different operating parameters. The monitor checks for sensor circuit malfunctions and slow response rate as well as for problems in the sensor's heater circuit. If a fault is found, a DTC that identifies the sensor and type of fault will be set. The PCM tests the performance of A/F sensors by measuring the signal response as the amount of fuel injected into the cylinders is changed. A good sensor will respond very quickly. The results of the monitor test are not reported in Mode 5. Mode 6 is used to determine if the A/F sensors passed or failed the test.

HO_2S and A/F Sensor Repair

If the HO_2S wiring, connector, or terminal is damaged, the entire O_2S assembly should be replaced. Do not attempt to repair the assembly. In order for this sensor to work properly, it must have a clean air reference. The sensor receives this reference from the air that is present around the sensor's signal and heater wires. Any attempt to repair the wires, connectors, or terminals could result in the obstruction of the air reference and degraded O_2S performance.

Additional guidelines for servicing an HO_2S follow:

- Do not apply contact cleaner or other materials to the sensor or wiring harness connectors. These materials may get into the sensor, causing poor performance.
- Ensure that the sensor pigtail and harness wires are not damaged in such a way that the wires inside are exposed. This could provide a path for foreign materials to enter the sensor and cause performance problems.
- Ensure that neither the sensor nor the wires are bent sharply or kinked. Sharp bends, kinks, and so on could block the reference air path through the lead wire.
- Do not remove or defeat the O_2S ground wire. Vehicles that utilize the ground wired sensor may rely on this ground as the only ground contact to the sensor. Removal of the ground wire will cause poor engine performance.

- To prevent damage due to water intrusion, be sure that the sensor's seal remains intact on the wiring harness.
- Use a socket or tool designed to remove the sensor; failure to do this may result in damage to the sensor and/or wiring.
- If suggested by the manufacturer, apply a light coat of antiseize lubricant to the threads of the sensor before installing it.

POSITION SENSORS

Position sensors are used to monitor the position of something from totally closed to totally open positions. Basically, they are potentiometers. A wiper arm is connected to a moving part on one end, and the other end is in contact with a resistor; the resistor is supplied with a reference voltage. As the part moves, so does the wiper arm. As the wiper arm moves on the resistor, the available voltage at the point where the wiper arm contacts the resistor is the signal voltage sent to the PCM. The PCM then interprets the part's position according to the voltage.

Throttle Position (TP) Sensor

Throttle position (TP) sensors (Figure 32–42) send a signal to the PCM regarding the rate of throttle opening and the relative throttle position. The wiper arm in the sensor is rotated by the throttle shaft **(Figure 32–43)**. As the throttle shaft moves, the wiper arm moves to a new location on the resistor. The return voltage signal tells the PCM how much the throttle plates are open. As the signal tells the PCM that the throttle is opening, the PCM enriches the air/fuel mixture to maintain the proper air/fuel ratio. The TP sensor is mounted on the throttle body. A separate idle contact switch or wide-open throttle (WOT) switch may also be used to signal when the throttle is in those positions.

FIGURE 32–42

A throttle position (TP) sensor.

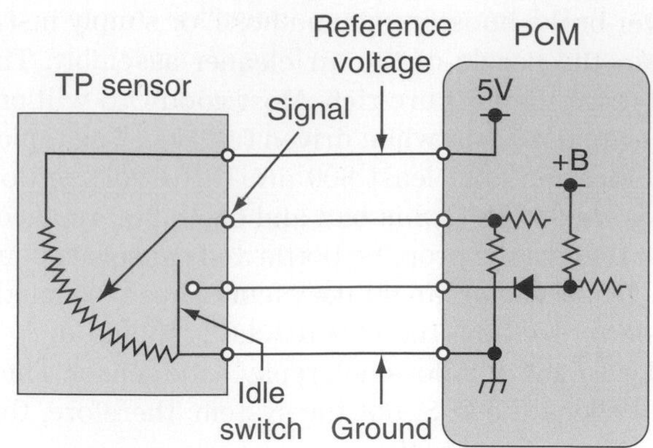

FIGURE 32–43

Basic circuit for a TP sensor.

A basic TP sensor has three wires. One wire carries the 5-volt reference signal, another serves as the ground for the resistor, and the third is the signal wire. When the throttle plates are closed, the signal voltage will be around 0.5 to 0.9 volt. As the throttle opens, there is less resistance between the beginning of the resistor and the place of wiper arm contact. Therefore, the voltage signal increases. At WOT, the signal will be approximately 3.5 to 4.7 volts. Often, the terminals in the connector for the sensor are gold-plated. The plating makes the connector more durable and corrosion resistant.

TP SENSORS FOR ELECTRONIC THROTTLE CONTROL TP sensors for electronic throttle systems have two wiper arms and two resistors in a single housing **(Figure 32–44)**. Therefore, they have two signal

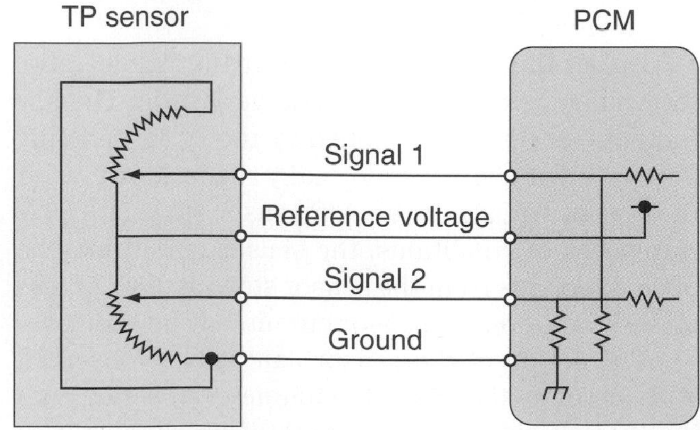

FIGURE 32–44

Basic circuit of a TP sensor for an electronic throttle system.

wires. This is done to ensure accurate throttle plate position in case one sensor fails. Some of these TP sensors have different voltage signals on the signal wire. However, they work in the same way. One of the signals starts at a higher voltage and has a different change rate. Other designs of this sensor have a signal that decreases with throttle opening, and the other increases with the opening. In either case, the PCM uses both signals to determine throttle opening and depends on one if the other sends an out-of-range signal.

Testing a TP Sensor

The most common symptoms of a bad or misadjusted TP sensor are engine stalling, improper idle speed, and hesitation or stumble during acceleration. The fuel mixture leans out because the computer does not receive the right signal telling it to add fuel as the throttle opens. Eventually, the O_2S senses the problem and adjusts the mixture, but not before the engine stumbles.

The initial setting of the sensor is critical. The voltage signal that the computer receives is referenced to this setting. Most service information systems list the initial TP sensor setting to the nearest 0.01 volt, a clear indication of the importance of this setting.

With the ignition switch on, connect a voltmeter between the reference wire to ground. Normally, the voltage reading should be 5 volts. If the reference wire is not supplying the specified voltage, check the voltage on this wire at the computer terminal. If the voltage is within specifications at the computer but low at the sensor, repair the reference wire. When this voltage is low at the computer, check the voltage supply wires and ground wires on the computer. If these wires are satisfactory, replace the computer.

SHOP TALK

When the throttle is opened gradually to check the TP sensor voltage signal, tap the sensor lightly, and watch for fluctuations on the voltmeter pointer, indicating a defective sensor.

With the ignition switch on, connect a voltmeter from the sensor signal wire to ground. Slowly open the throttle, and observe the voltmeter. The voltmeter reading should increase smoothly and gradually. If the TP sensor does not have the specified voltage or if the voltage signal is erratic, replace the sensor.

A TP sensor can also be checked with an ohmmeter. Most often, the total resistance of the sensor is given in the specifications. If the sensor does not meet these, it should be replaced.

Adjustment of the TP sensor can be made on some engines. Incorrect TP sensor adjustment may cause inaccurate idle speed, engine stalling, and acceleration stumbles. Follow these steps to adjust a typical TP sensor:

1. Back-probe the TP sensor signal wire, and connect a voltmeter from this wire to ground.
2. Turn on the ignition switch, and observe the voltmeter reading with the throttle in the idle position.
3. If the TP sensor does not provide the specified voltage, loosen the TP sensor mounting bolts, and move the sensor housing until the specified voltage is indicated on the voltmeter **(Figure 32–45)**.
4. Hold the sensor in this position, and tighten the mounting bolts to the specified torque.

TP sensors can also be tested with a lab scope. Connect the scope to the sensor's output and a good ground, and watch the trace as the throttle is opened and closed. The resulting trace should look smooth and clean without any sharp breaks or spikes in the signal **(Figure 32–46)**. A bad sensor will typically have a glitch (a downward spike) somewhere in the trace **(Figure 32–47)** or will not have a smooth transition from high to low. These glitches indicate an open or short in the sensor.

The action of a TP sensor can also be monitored on a scan tool. Compare the position, expressed in a percentage, to the voltage specifications for that throttle position.

Be careful: Some TP sensors have four wires. The additional wire is connected to an idle switch.

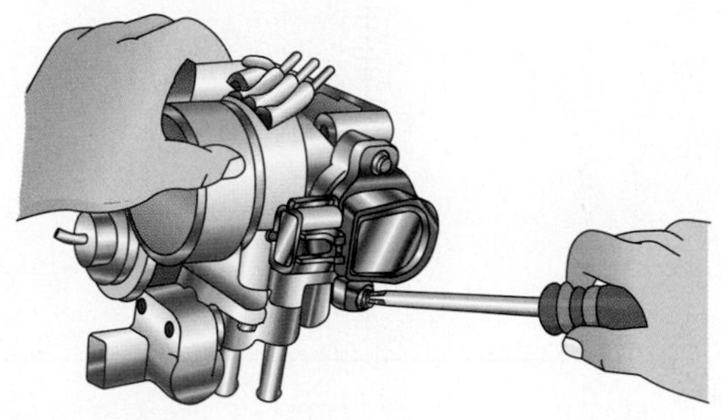

FIGURE 32–45

Loosen the TP sensor's mounting screws to adjust the sensor.

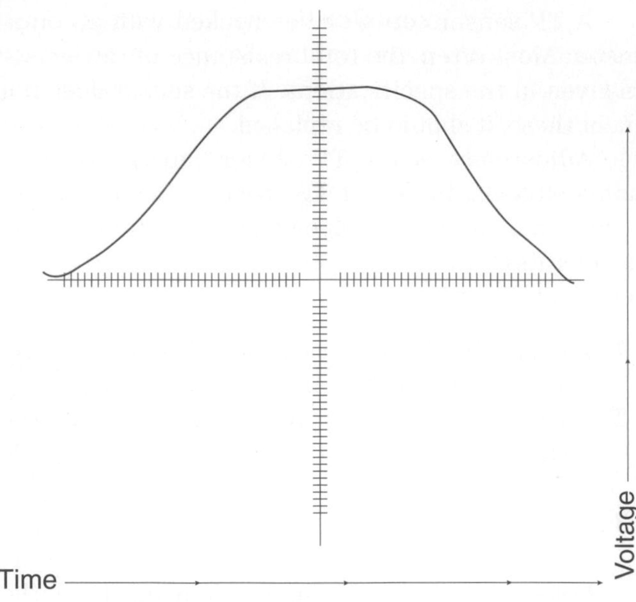

FIGURE 32–46

The waveform of a normal TP sensor as the throttle opens and closes.

Normally, when the switch is closed, there will be 0 volts, and battery voltage when the switch is open. Check the wiring diagram before measuring voltage here and before deciding if the switch and circuit are good.

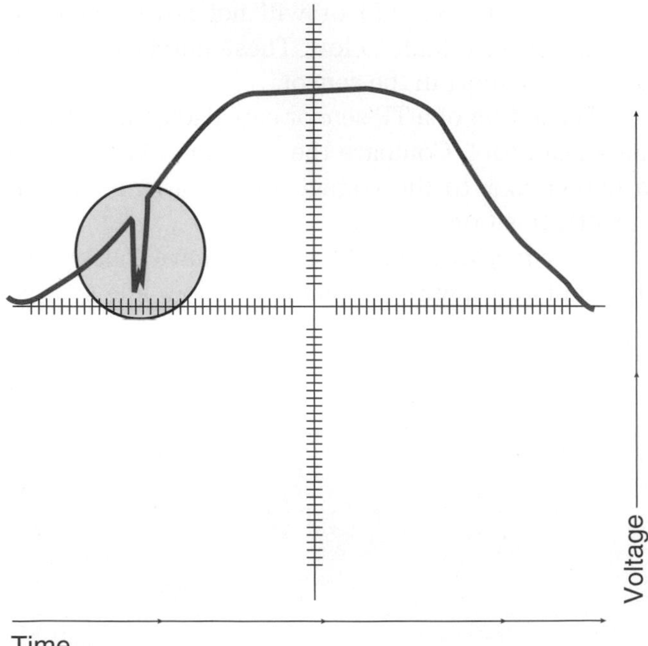

FIGURE 32–47

The waveform of a defective TP sensor. Notice the glitch while the throttle opens.

EGR Valve Position Sensor

Manufacturers use a variety of sensors or switches to determine when and how far the EGR valve is open. This information is used to adjust the air/fuel mixture and EGR flow rates. The exhaust gases introduced by the EGR valve into the intake manifold reduce the available oxygen, and thus less fuel is needed in order to maintain low HC levels in the exhaust. Most EGR valve position sensors are linear potentiometers mounted on top of the EGR valve and detect the height of the EGR valve. When the EGR valve opens, the potentiometer stem moves upward, and a higher voltage signal is sent to the PCM. These sensors work in the same way as a TP sensor.

Most EGR valve position sensors have a 5-volt reference wire, voltage signal wire, and ground wire. To test one, measure the voltage at the signal wire with the ignition on and the engine not running. The meter should read about 0.8 volt. Then connect a vacuum pump to the EGR valve and slowly increase the vacuum to about 50.8 cm Hg (20 in. Hg). The voltage signal should smoothly increase to 4.5 V at 50.8 cm Hg (20 in. Hg). If the signal voltage does not reach the specified voltage, replace the sensor.

These sensors can also be checked with a lab scope or GMM. Watch the rise of the waveform as vacuum is applied. Also look for any glitches in the waveform. These are hard to see on a voltmeter unless they are very severe. The trace should be clean and smooth.

Accelerator Pedal Position (APP) Sensor

The APP sensor is used with electronic throttle control systems. It converts accelerator pedal movement into electrical signals. Like the TP sensor for the electronic throttle system, the APP is based on two potentiometers. The PCM uses the signals from this sensor to determine power or torque demand. In turn, the PCM opens or closes the throttle plate and adjusts the amount of fuel that is injected into the cylinders.

An APP is tested in the same way as other variable resistor sensors.

SPEED SENSORS

Speed sensors measure the rotational speed of something. The PCM uses these signals in a number of ways, depending on the system. Speed sensors are either Hall-effect switches or magnetic pulse generators. Identifying the type of sensor used in a particular application dictates how the sensor should be tested.

Vehicle Speed Sensor (VSS)

The most common **vehicle speed sensor (VSS)** is a magnetic pulse generator (variable reluctance) sensor **(Figure 32–48)**. However, some use a Hall-effect switch. A VSS generates a waveform at a frequency that is proportional to the vehicle's speed. When the vehicle is moving at a low speed, the sensor produces a low-frequency signal. As vehicle speed increases, so does the frequency of the signal. The PCM uses the VSS signal to help control the fuel injection system, ignition system, cruise control, EGR flow, canister purge, transmission shift timing, variable steering, and torque converter clutch lock-up timing. The signal is also used to initiate diagnostic routines. The VSS is also used on some vehicles to limit the vehicle's speed. When a predetermined speed is reached, the PCM limits fuel delivery.

GO TO ▶ Chapter 16 for more details on magnetic pulse generators.

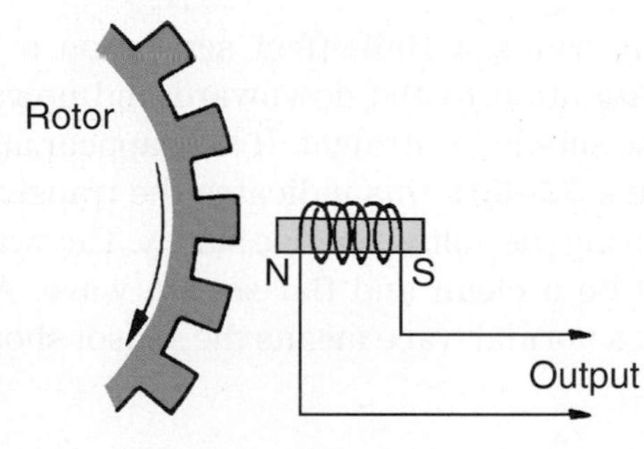

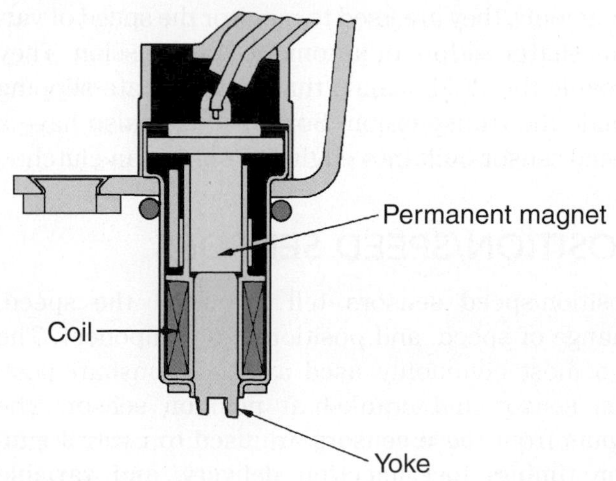

FIGURE 32–48

A vehicle speed sensor (VSS).

The VSS is mounted on the transmission or transaxle case in the speedometer cable opening, where it can measure the rotational speed of the output shaft. On earlier models, the VSS was connected to the speedometer cable. The rotation of a trigger wheel on the output shaft creates a pulsating voltage in the sensor. The frequency of the voltage increases with an increase in speed.

Troubleshooting a VSS

A defective VSS may cause different problems depending on the control systems. If the PCM does not receive a VSS signal, it will set a DTC; it may also set a code if the signal does not correlate with other inputs. A defective VSS circuit can cause increased fuel consumption, poor idle, improper converter clutch lock-up, improper cruise control operation, and inaccurate speedometer operation.

A VSS should be carefully inspected before it is tested. Check the wiring harness and connectors at the sensor and the control modules that rely on the VSS signal. Make sure the connections are tight and not damaged.

The voltage in the VSS circuit should be checked for evidence of an open, a short, or high resistance. Most of these measurements can be taken at the PCM connector; refer to the service information to identify the exact measuring points.

The VSS can be tested with a scan tool or lab scope **(Figure 32–49)**; however, make sure to

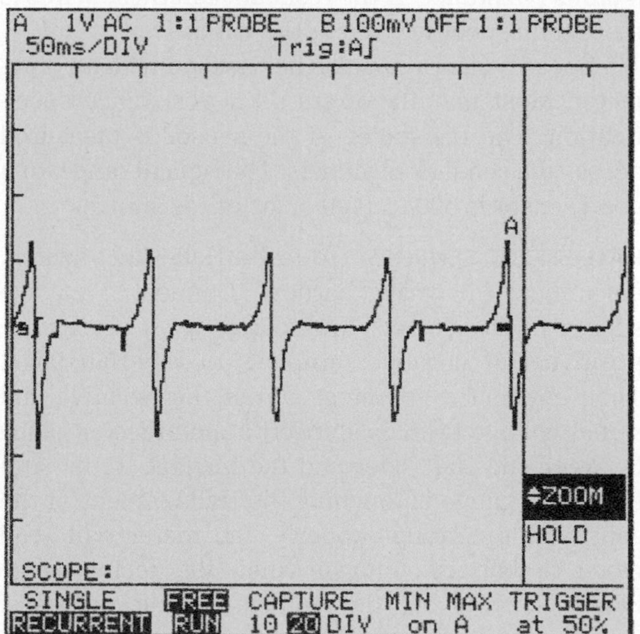

FIGURE 32–49

The waveform for a good VSS (PM generator).

follow the manufacturers' test procedures when doing this. The action of a VSS is best observed while the vehicle is moving. Connect a lab scope, GMM, or scan tool and operate the vehicle at the reference speeds given in the service information. Compare the measurements to specifications. If the measurements meet the specifications, the VSS is working properly, and any VSS-related problem is probably caused by the PCM. If the measurements are outside the specifications and the wiring is sound, the sensor is bad.

A VSS can also be checked with the vehicle on a hoist. The vehicle should be positioned so that the drive wheels are free to rotate. Back-probe the VSS output wire, and connect the voltmeter leads from this wire to ground. Select the 20 V AC scale on the voltmeter and then start the engine.

Place the transmission into a forward gear. If the VSS voltage signal is not 0.5 volt or more, replace the sensor. If the VSS signal is correct, back-probe the VSS terminal at the PCM and measure the voltage with the drive wheels rotating. If 0.5 volt is available at this terminal, the trouble may be in the PCM.

When 0.5 volt is not available at this terminal, turn the ignition switch off and disconnect the wire from the VSS to the PCM. Connect the ohmmeter leads across the wire. The meter should read near 0 ohms. Repeat the test with the ohmmeter leads connected to the VSS ground terminal and the PCM ground terminal. This wire should also have near 0 ohms resistance. If the resistance in these wires is more than specified, repair the wires.

Speed sensors can also be checked with an ohmmeter. Most manufacturers list a resistance specification. The resistance of the sensor is measured across the sensor's terminals. The typical range for a good sensor is 800 to 1400 ohms of resistance.

HALL-EFFECT SENSORS To test a Hall-effect switch, disconnect its wiring harness. Connect a voltage source of the correct low-voltage level across the positive and negative terminals of the Hall layer. Then connect a voltmeter across the negative and signal voltage terminals. Insert a metal feeler gauge between the Hall layer and the magnet. Make sure the feeler gauge is touching the Hall element. If the sensor is operating properly, the meter will read close to battery voltage. When the feeler gauge blade is removed, the voltage should decrease. On some units, the voltage will drop to near zero. Check the service information to see what voltage you should observe when inserting and removing the feeler gauge.

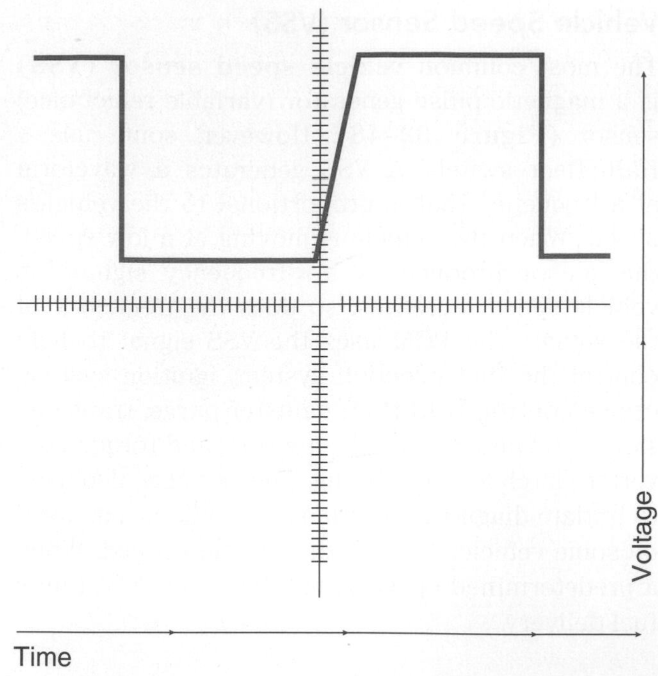

FIGURE 32–50

A Hall-effect switch with a bad transistor.

When observing a Hall-effect sensor on a lab scope, pay attention to the downward and upward pulses. These should be straight. If they appear at an angle **(Figure 32–50)**, this indicates the transistor is faulty, causing the voltage to rise slowly. The waveform should be a clean and flat square wave. Any change from a normal trace means the sensor should be replaced.

Other Speed Sensors

Speed sensors are used in several other places. Most commonly, they are used to monitor the speed of various shafts within an automatic transmission. They provide the PCM signals that may indicate slipping inside the transmission. Some vehicles also have a speed sensor built into engine's cooling fan clutch.

POSITION/SPEED SENSORS

Position/speed sensors tell the PCM the speed, change of speed, and position of a component. The two most commonly used are the camshaft position sensor and crankshaft position sensor. The inputs from these sensors are used to control ignition timing, fuel injection delivery, and variable valve timing. The sensor does this using a Hall-effect switch or magnetic pulse generator; both designs generate a voltage signal.

These sensors are most often magnetic pulse generators. The sensor is mounted close to a pulse wheel or rotor that has notches or teeth. As each tooth moves past the sensor, an AC voltage pulse is induced in the sensor's coil. As the rotor moves faster, more pulses are produced. Speed is calculated by the frequency of the signal, which is the number of pulses in a second. To provide a clean strong signal, the distance between the sensor and the rotor has a specification. If the sensor is too far away from the rotor, the signal will be weak. The ECM determines the speed that the component is revolving based on the number of pulses.

Magnetic pulse generators create an AC voltage signal, and their wiring harness is normally a twisted and/or shielded pair of wires.

Crankshaft Position (CKP) Sensor

The rotor for a **crankshaft position (CKP) sensor** has several teeth (the number varies with application) equally spaced around the outside of the rotor. One or more teeth are missing at fixed locations. These missing teeth provide a reference point for the PCM to determine crankshaft position **(Figure 32–51)**. For example, the pulse wheel may have a total of 35 teeth spaced 10° apart and an empty space where the 36th tooth would have been. The 35 teeth are used to monitor crankshaft speed; the gap is used to identify which pair of cylinders is approaching TDC. The input from the camshaft position sensor signals is used in order to determine which of these two cylinders is on its firing stroke and which is on the exhaust stroke.

Input from the CKP **(Figure 32–52)** is critical to the operation of the ignition system. This input is also used by the PCM to determine if a misfire has occurred. This is done by looking at the time intervals between the teeth. If a misfire occurs, the transition from one tooth to another will be slower than normal.

CHECKING A CKP Like all electrical devices, the sensor and its wiring should be checked for corrosion and damage. If the PCM detects abnormal signals, it will set a DTC. The CKP is checked during

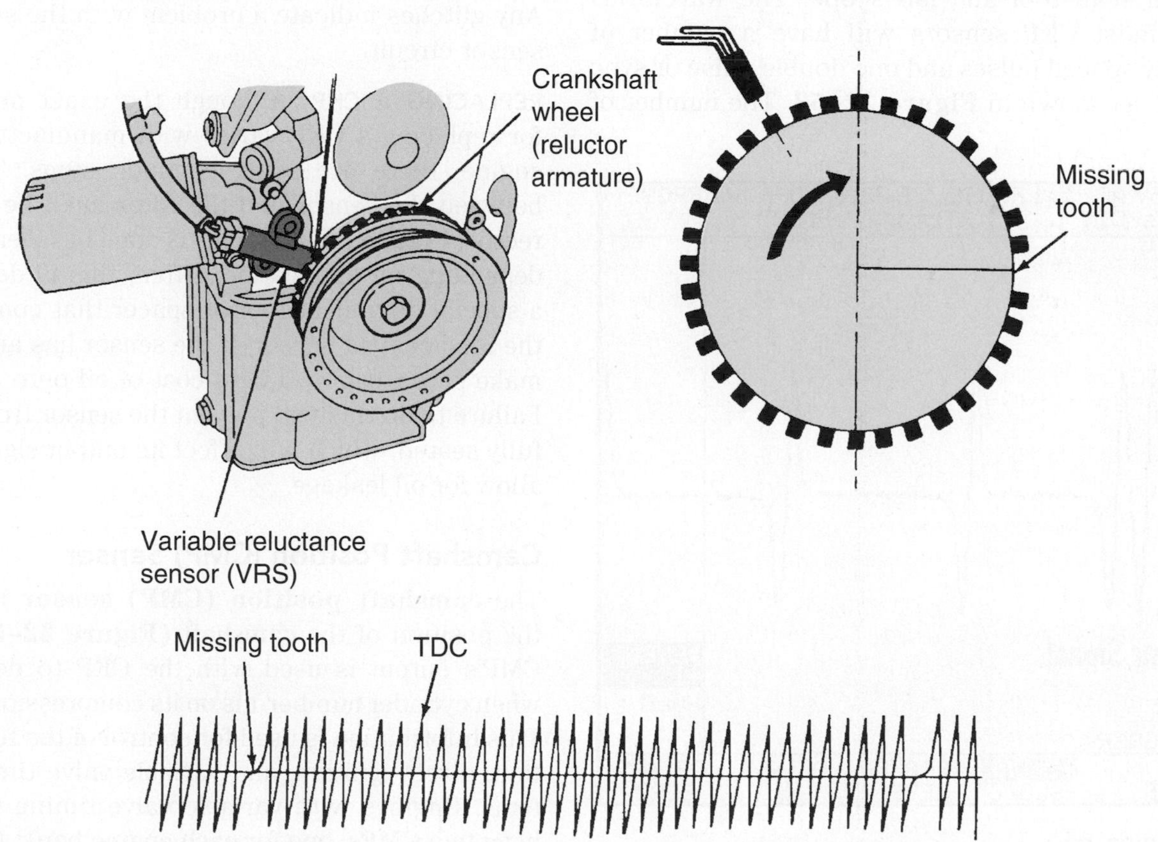

FIGURE 32–51

Sensor activity to monitor crankshaft speed and position, as well as the location of the number 1 piston.

Ford Motor Company

FIGURE 32–52

A crankshaft position (CKP) sensor.

several monitor tests. On engines with both a CKP and camshaft position sensor, the PCM will compare those two inputs and set a code if they do correlate with each other.

The operation of the sensor can be monitored with a scan tool and lab scope. The waveforms from most CKP sensors will have a number of equally spaced pulses and one double pulse or sync signal, as shown in **Figure 32–53**. The number of evenly spaced pulses equals the number of cylinders that the engine has. A Hall-effect sensor generates a digital signal. Carefully examine the trace. Any glitches indicate a problem with the sensor or sensor circuit.

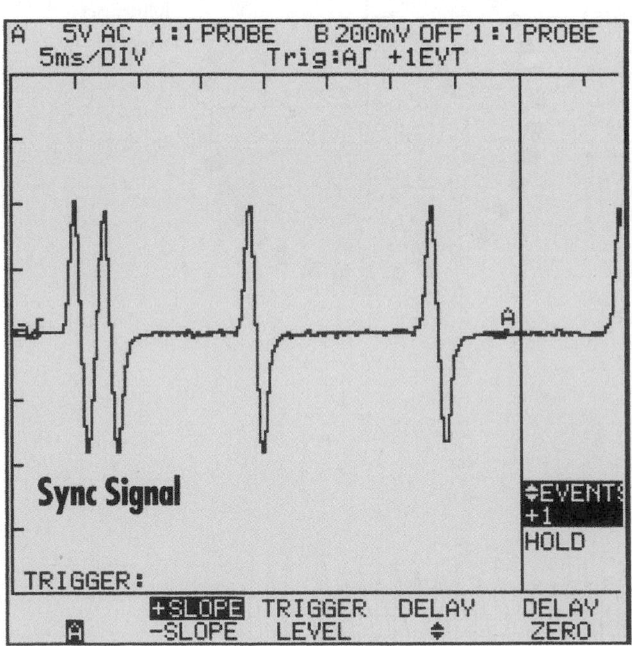

FIGURE 32–53

The trace of a good crankshaft position sensor.

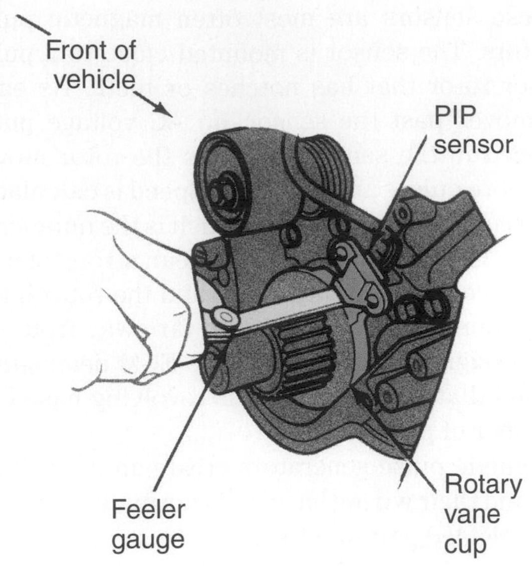

FIGURE 32–54

The gap between the CKP (PIP) sensor and the rotor is critical.

REPLACING A CKP Although the exact procedure for replacing a CKP varies with manufacturer and engine, there are some common steps. The gap between the sensor and the rotor must be set correctly **(Figure 32–54)**. This is done in several ways, depending on application. Often, this is done with a special alignment tool or spacer that comes with the replacement sensor. If the sensor has an O-ring, make sure to apply a light coat of oil onto the seal. Failure to do this will prevent the sensor from being fully seated, which will affect its output signals and allow for oil leakage.

Camshaft Position (CMP) Sensor

The **camshaft position (CMP) sensor** monitors the position of the camshaft **(Figure 32–55)**. The CMP's output is used with the CKP to determine when cylinder number 1 is on its compression stroke. This information is used for control of the fuel injection, direct ignition, and variable valve timing systems. Engines with variable valve timing typically have two CMPs, one for each engine bank. On some engines, a bad CMP will prevent the misfire monitor from completing.

FIGURE 32–55

Camshaft sensors mounted to the ends of the intake and exhaust camshafts.

©Cengage Learning 2015

CMP sensors can be magnetic pulse generators or Hall-effect sensors **(Figure 32–56)**. The type of sensor is identified by the number of wires connected to the sensor: A magnetic pulse sensor has two wires, and the Hall-effect sensor has three.

CMP SENSOR SERVICE Most diagnostic procedures for the CKP apply to the CMP sensor. There are many DTCs related to the CMP. If one of these

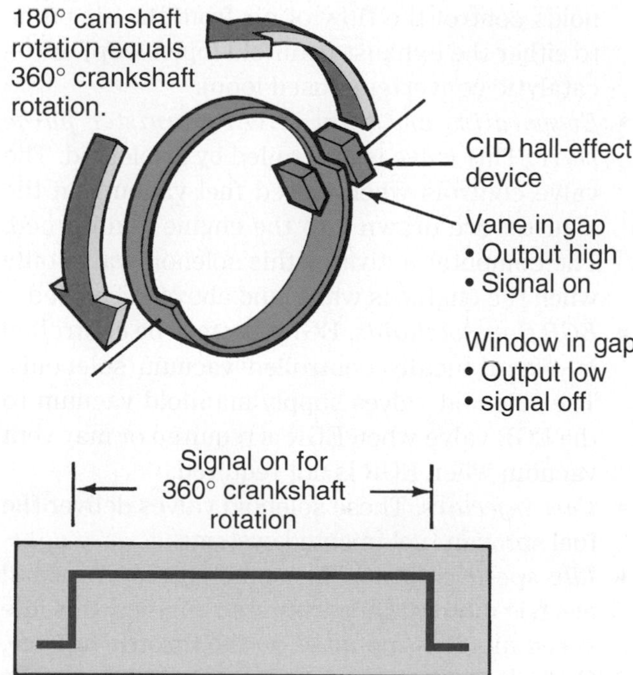

FIGURE 32–56

A Hall-effect camshaft sensor and its resultant signal.

©Cengage Learning 2015

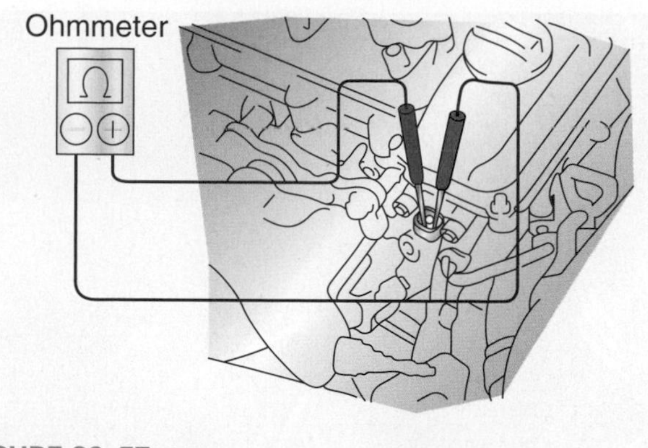

FIGURE 32–57

Checking a camshaft sensor with an ohmmeter.

©Cengage Learning 2015

is set, follow the testing procedures given by the manufacturer. These include a thorough inspection of the sensor and its wiring. When observing a CMP on a lab scope, make sure you know what type the sensor is. Magnetic pulse generators create an analog signal, whereas Hall-effect sensors provide a digital signal. Magnetic pulse generators can also be checked with an ohmmeter **(Figure 32–57)**. The procedures may also include a verification of camshaft timing. Make sure to lubricate the O-ring when installing a new sensor, and tighten the retaining bolt to specifications.

GO TO Chapter 11 for the procedure for verifying and adjusting camshaft timing.

KNOCK SENSOR (KS)

The **knock sensor (KS)** tells the PCM that detonation is occurring in the cylinders. In turn, the computer retards the timing **(Figure 32–58)**. The KS is a piezoelectric device that works like a microphone and converts engine knock vibrations into a voltage signal. Piezoelectric devices generate a voltage when pressure or a vibration is applied to them. Engine knock typically is within a specific frequency range, and a KS is set to detect vibrations within that range. The KS is located in the engine block, on the cylinder head, or below the intake manifold.

Testing a Knock Sensor

A defective KS may cause engine detonation or reduced spark advance and fuel economy. When a KS is removed and replaced, the sensor must be tightened to its specified torque. The procedure for

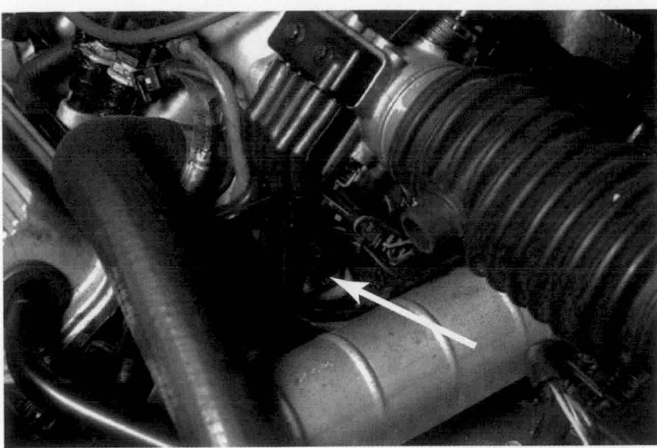

FIGURE 32–58

A knock sensor (KS).

checking a KS varies, depending on the vehicle make and year. The KS is checked with several monitors by the PCM. The PCM checks the input from the sensor for abnormal readings that may be caused by an open, a short, or high resistance. Always follow the vehicle manufacturer's recommended test procedure and specifications.

A KS can be checked by observing it on a scope while tapping the engine at a point close to the sensor. The waveform should react to the tapping. This test may not work on all KS. The sensor will not respond to the tapping if it is not synchronized to the CKP signal, which is the normal situation on some engines.

A KS and its circuit can also be checked with a voltmeter. Check the voltage input at the sensor and the power feed from the PCM. If the specified voltage is not available to the sensor, back-probe the KS wire at the PCM, and measure the voltage. If the voltage is satisfactory at this terminal, repair the KS wire. If the voltage is not within specifications at the PCM, replace the PCM.

A scan tool can also be used to diagnose a KS. During a road test, open the throttle quickly, and observe the scan tool. There should be at least one count when the throttle is opened.

COMPUTER OUTPUTS AND ACTUATORS

Once the PCM's programming instructs that a correction or adjustment must be made in the controlled system, an output signal is sent to a control device or actuator. These actuators—solenoids, switches, relays, or motors—physically act or carry out the command sent by the PCM.

Actuators are electromechanical devices that convert an electrical current into mechanical action. This mechanical action can then be used to open and close valves, control vacuum to other components, or open and close switches. When the PCM receives an input signal indicating a change in one or more of the operating conditions, the PCM determines the best strategy for handling the conditions. The PCM then controls a set of actuators to achieve a desired effect or strategy goal. In order for the computer to control an actuator, it must rely on a component called an output driver.

The circuit driver usually applies the ground circuit of the actuator **(Figure 32–59)**. The ground can be applied steadily if the actuator must be activated for a selected amount of time, or the ground can be pulsed to activate the actuator in pulses. Output drivers are transistors or groups of transistors that control the actuators. These drivers operate by the digital commands from the PCM. If an actuator cannot be controlled digitally, the output signal must pass through an A/D converter before flowing to the actuator. The major actuators in a computer-controlled engine include the following components:

- *Air management solenoids.* Found in older systems, a secondary air bypass and diverter solenoids control the flow of air from the air pump to either the exhaust manifold (open loop) or the catalytic converter (closed loop).
- *Evaporative emission (EVAP) canister purge valve.* This valve is controlled by a solenoid. The valve controls when stored fuel vapours in the canister are drawn into the engine and burned. The computer activates this solenoid valve only when the engine is warm and above idle speed.
- *EGR flow solenoids.* EGR flow may be controlled by electronically controlled vacuum solenoids. The solenoid valves supply manifold vacuum to the EGR valve when EGR is required or may vent vacuum when EGR is not required.
- *Fuel injectors.* These solenoid valves deliver the fuel spray in fuel-injected systems.
- *Idle speed controls.* These actuators are small electric motors. On carbureted engines, this idle speed motor is mounted on the throttle linkage. On fuel-injected systems, a stepper motor may be used to control the amount of air bypassing the throttle plate.
- *Ignition module.* This is actually an electronic switching device triggered by a signal from the

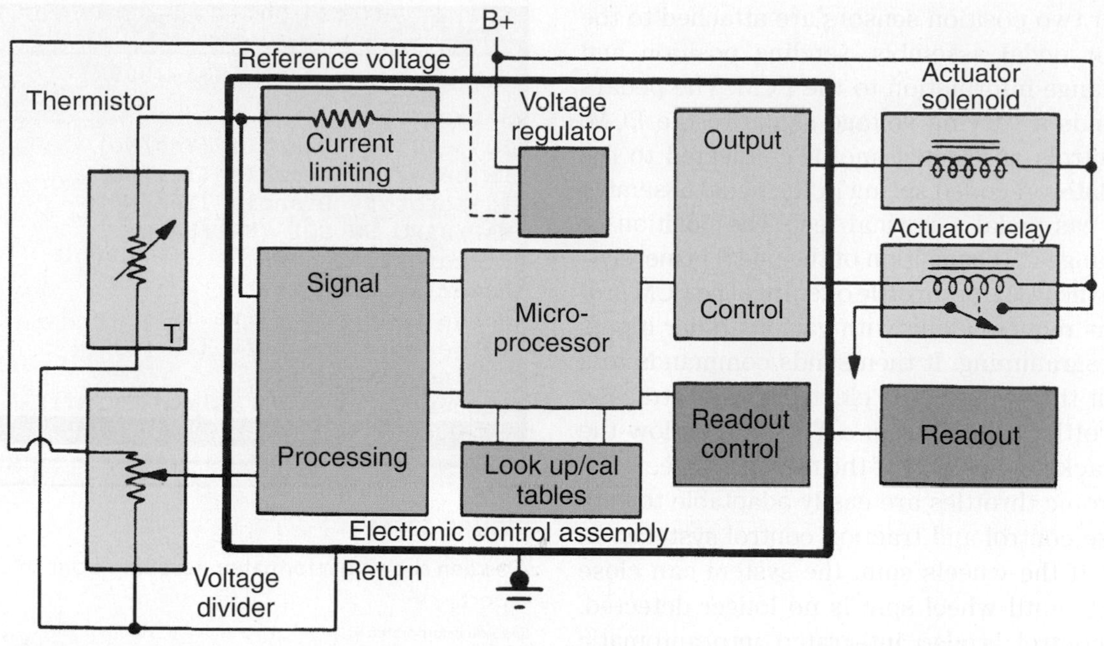

FIGURE 32–59

Output drivers in the computer usually supply a ground for the actuator solenoids and relays.

Ford Motor Company

control computer. The ignition module may be a separate unit or may be part of the PCM.

- *Motors and lights.* Using electrical relays, the computer is used to trigger the operation of electric motors, such as the fuel pump, or various warning light or display circuits.
- *Other solenoids.* Computer-controlled solenoids may also be used in the operation of cruise control systems, torque converter lock-up clutches, automatic transmission shift mechanisms, and many other systems where mechanical action is needed.

Electronic Throttle Control

Like modern aircraft, the acceleration on some late-model vehicles works on the *drive-by-wire* principle, which is typically called **electronic throttle control (ETC)**. ETC interprets gas pedal movement by the driver and allows for precise throttle control, which helps to improve fuel economy and performance while reducing emissions.

Instead of a throttle cable and mechanical linkage to the throttle body, the connection is made through wires **(Figure 32–60)**. Although these systems are electronically controlled and operated, some still have a mechanical backup system, or they resort to partial throttle if something goes wrong with the electronic system.

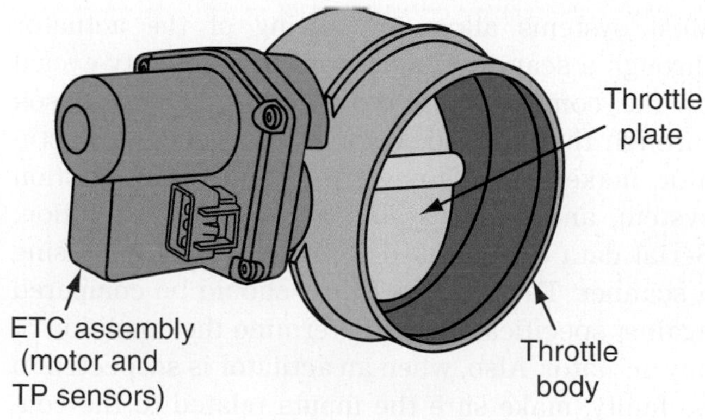

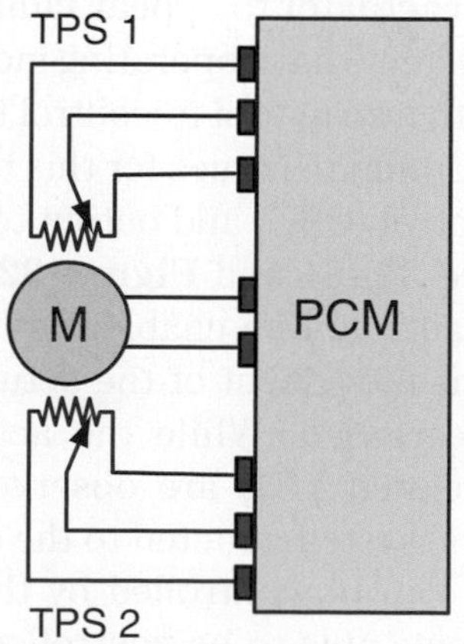

FIGURE 32–60

A cutaway of an electronic throttle control assembly.

©Cengage Learning 2015

One or two position sensors are attached to the accelerator pedal assembly, sending position and rate-of-change information to the PCM. The pedal's sensor sends a varying voltage signal to the PCM, which controls an electric motor connected to the throttle plates. A coiled spring in the pedal assembly gives the gas pedal a normal feel. The position or rate of change in the position of the pedal is merely a request to the PCM for throttle opening. The PCM processes this request along with various other inputs and its programming. It then sends commands to a driver unit that powers the electric motor attached to the throttle. Signals from a TP sensor allow the PCM to track the position of the throttle plate.

Electronic throttles are easily adaptable to support cruise control and traction control systems. In the latter, if the wheels spin, the system can close the throttle until wheel spin is no longer detected. Throttle control is also integrated into automatic shifting. With electronic control, the throttle can be closed slightly to reduce engine output during a shift, providing smoother gear changes.

Testing Actuators

Most systems allow for testing of the actuator through a scan tool. Actuators that are duty-cycled by the computer are more accurately diagnosed through this method. Prior to diagnosing an actuator, make sure the engine's compression, ignition system, and intake system are in good condition. Serial data can be used to diagnose outputs using a scanner. The displayed data should be compared against specifications to determine the condition of any actuator. Also, when an actuator is suspected to be faulty, make sure the inputs related to the control of that actuator are within normal range. Faulty inputs will cause an actuator to appear faulty.

Many control systems have operating modes that can be accessed with a scan tool to control the operation of an output. Common names for this mode are the output state control (OSC) and output test mode (OTM). (See **Figure 32–61** and **Figure 32–62**.) In this mode, an actuator can be enabled or disabled; the duty cycle or the movement of the actuator can be increased or decreased. While the actuator is being controlled, related PIDs are observed as an indication of how the system reacted to the changes. The actuators that can be controlled by this mode vary. Common outputs able to be controlled via the scan tool include fuel injectors, the fuel pump, idle speed or throttle control motors, cooling fans, EGR solenoids and valves, and EVAP solenoids. Always refer to the service information to determine what can be checked and how it should be checked.

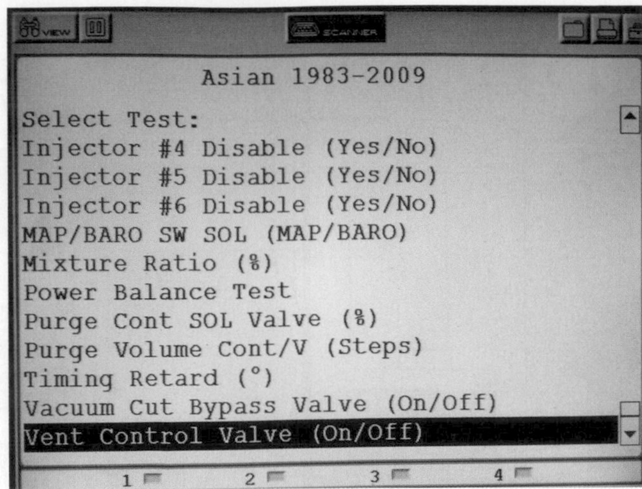

FIGURE 32–61

A screen showing information resulting from an OSC report.

Snap-on Incorporated

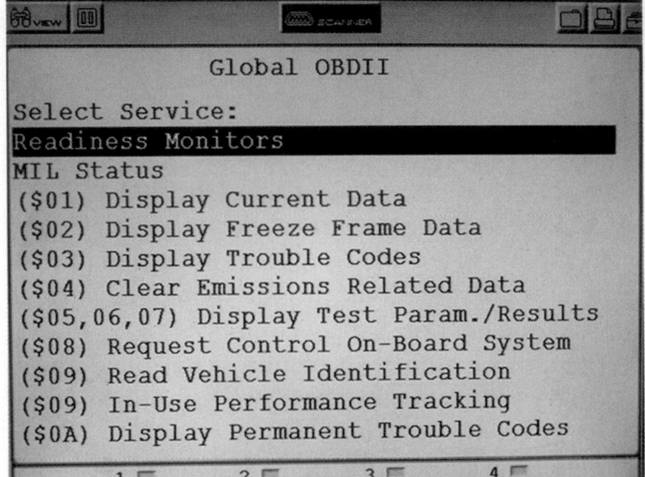

FIGURE 32–62

A screen showing information resulting from a global OTM report.

Snap-on Incorporated

Testing with a DMM

If the actuator is tested by means other than a scanner, always follow the manufacturer's recommended procedures. Because many actuators operate with 5 to 7 volts, never connect a jumper wire from a 12-volt source unless directed to do so by the appropriate service procedure. Some actuators are easily tested with a voltmeter by testing for input voltage to the actuator. If there is the correct amount of input voltage, check the condition of the ground. If both of these are good, then the actuator is faulty. If an ohmmeter needs to be used to measure the resistance of an actuator, disconnect it from the circuit first.

When checking anything with an ohmmeter, logic can dictate good and bad readings. If the meter reads infinite, this means there is an open. Based on what you are measuring across, an open could be good or bad. The same is true for very low resistance readings. Across some things, this would indicate a short. For example, you do not want an infinite reading or very low resistance across the windings of a solenoid. You want some resistance, typically between 50 and 120 ohms. However, you want an infinite reading from one winding terminal to the case of the solenoid. If you have low resistance, the winding is shorted to the case.

Testing Actuators with a Lab Scope

Most computer-controlled circuits are ground-controlled circuits. The PCM energizes the actuator by providing the ground. On a scope trace, the on-time pulse is the downward pulse. On positive-feed circuits, where the computer is supplying the voltage to turn a circuit on, the on-time pulse is the upward pulse. One complete cycle is measured from one on-time pulse to the beginning of the next on-time pulse.

Actuators are electromechanical devices, meaning they are electrical devices that cause some mechanical action. When actuators are faulty, it is because they are electrically faulty or mechanically faulty. By observing the action of an actuator on a lab scope, you will be able to watch its electrical activity. Normally, if there is a mechanical fault, this will affect its electrical activity as well. Therefore, you get a good sense of the actuator's condition by watching it on a lab scope.

To test an actuator, you need to know what it basically is. Most actuators are solenoids. The computer controls the action of the solenoid by controlling the pulse width of the control signal. You can see the turning on and off of the solenoid **(Figure 32–63)** by watching the control signal. The voltage spikes are caused by the discharge of the coil in the solenoid.

Some actuators are controlled pulse-width-modulated signals **(Figure 32–64)**. These signals show a changing pulse width. These devices are controlled by varying the pulse width, signal frequency, and voltage levels.

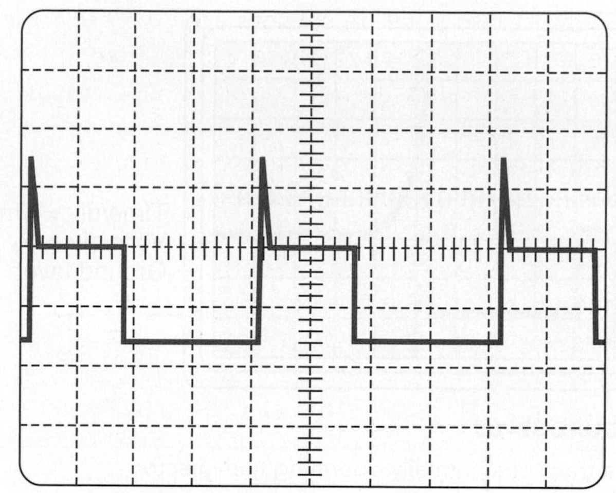

FIGURE 32–63

A typical solenoid control signal.

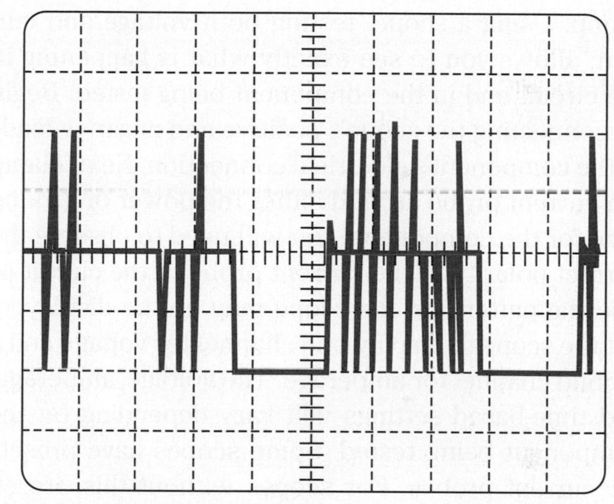

FIGURE 32–64

A typical pulse-width-modulated solenoid control signal.

Both waveforms should be checked for amplitude, time, and shape. You should also observe changes to the pulse width as operating conditions change. A bad waveform will have noise, glitches, or rounded corners. You should be able to see evidence that the actuator immediately turns off and on according to the commands of the computer.

A fuel injector is actually a solenoid. The PCM's signals to an injector vary in frequency and pulse width. Frequency varies with engine speed, and the pulse width varies with fuel control. Increasing an injector's on-time increases the amount of fuel delivered to the cylinders. The trace of a normally operating fuel injector is shown in **Figure 32–65**.

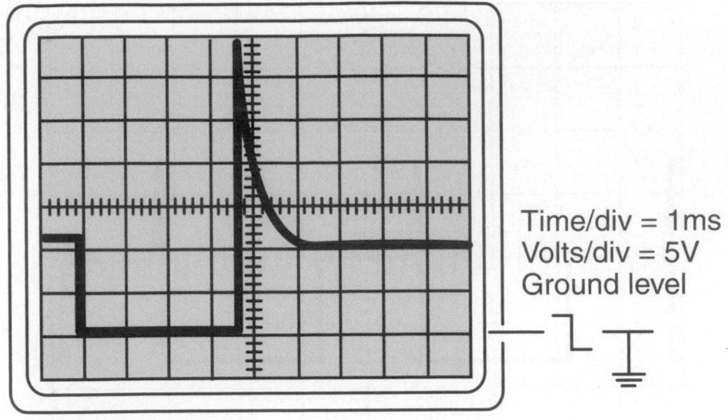

FIGURE 32–65

The trace of a normally operating fuel injector.

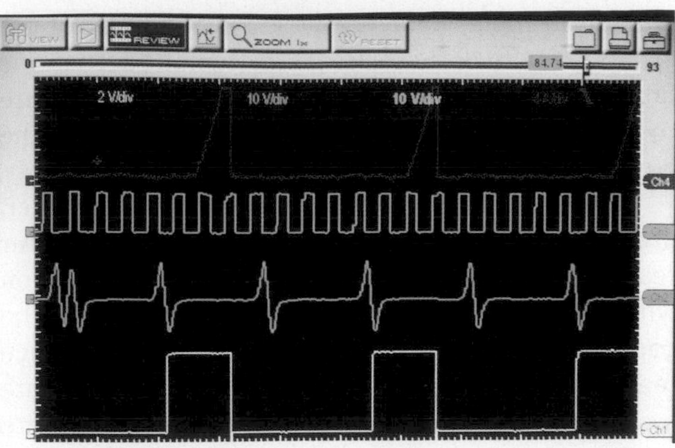

FIGURE 32–66

The waveform for a normally operating fuel injector shown with the waveforms of other sensors and the ignition system.

Testing with a Current Probe

Some actuators, such as fuel injectors and fuel pumps, are commonly tested with a current probe or current clamp. Using a scope, testing both voltage and current, allows you to see exactly what is happening in the circuit and in the component being tested. Begin by connecting the scope's positive and negative leads to the component's electrical connection. Next, clamp the current probe around either the power or ground wire for the component. You will need to observe the correct polarity of the current probe to the circuit or else the pattern on the scope may not be displayed. Set the scope to display one channel for voltage and a second channel for amperage. The voltage, amperage, and time-based settings will vary depending on the component being tested. Some scopes have presets for current probes. For scopes without this preset, you will need to select a voltage scale based on the current probe used.

Once the scope is configured, operate the circuit, and obtain the voltage and amperage waveforms **(Figure 32–66)**. In this case, the fuel injector signal from the PCM is shown with the current draw by the injector in operation. The slight dip in the leading edge of the current waveform shows where the injector pintle opened. Based on the information in the image, you can conclude that the PCM, wiring, and injector are operating correctly. The unknown in this example is how well the fuel sprays from the injector nozzle.

Repairing the System

After isolating the source of the problem, the repairs should be made. The system should then be rechecked to verify that the repair took care of the problem. This may involve road-testing the vehicle in order to verify that the complaint has been resolved.

When servicing or repairing OBD-II circuits, the following guidelines are important:

- Do not connect aftermarket accessories into an OBD-II circuit.
- Do not move or alter grounds from their original locations.
- Always replace a relay in an OBD-II circuit with an exact replacement. Damaged relays should be thrown away, not repaired.
- Make sure all connector locks are in good condition and are in place.
- After repairing connectors or connector terminals, make sure the terminals are properly retained and the connector is sealed.
- When installing a fastener for an electrical ground, be sure to tighten it to the specified torque.

Verification of repair is more comprehensive for vehicles with OBD-II system diagnostics than earlier vehicles. Following a repair, the technician should perform the following steps:

1. Review the fail records and the freeze-frame data for the DTC that was diagnosed. Record the fail records or freeze-frame data.
2. Use the scan tool's clear DTCs or clear information functions to erase the DTCs.
3. Operate the vehicle within the conditions noted in the fail records or the freeze-frame data.
4. Monitor the status information for the specific DTC until the diagnostic test associated with that DTC runs.

KEY TERMS

Air/fuel ratio (A/F) sensor (p. 963)
Camshaft position (CMP) sensor (p. 976)
Crankshaft position (CKP) sensor (p. 975)
Cross counts (p. 968)
Electronic throttle control (ETC) (p. 979)
Engine coolant temperature (ECT) sensor (p. 953)
Heated oxygen sensors (HO_2Ss) (p. 962)
Intake air temperature (IAT) sensor (p. 953)
Knock sensor (KS) (p. 977)
Manifold absolute pressure (MAP) sensor (p. 955)
Mass airflow (MAF) sensor (p. 958)
Throttle position (TP) sensors (p. 970)
Vapour pressure sensor (VPS) (p. 958)
Vehicle speed sensor (VSS) (p. 973)

SUMMARY

- Troubleshooting electronic engine control systems involves much more than retrieving trouble codes.
- The following can be used to check individual system components: visual checks, ohmmeter checks, voltmeter checks, and lab scope checks.
- Service bulletin information is absolutely essential when diagnosing engine control system problems.
- No PCM (OBD-II and earlier designs) can operate properly unless it has good ground connections and the correct voltage at the required terminals.
- A voltage drop test is a quick way of checking the condition of any electrical conductor or terminal connector.
- Poor grounds can also allow EMI or noise to be present on the reference voltage signal. The best way to check for noise is to use a lab scope.
- It is important to understand how a sensor works and what it measures before testing it.
- Sensors measure temperature, chemical characteristics, pressure, speed, position, and sound.
- Most sensors can be checked with a voltmeter, ohmmeter, scan tool, lab scope, and GMM.
- Most computer-controlled actuators are electromechanical devices that convert the output commands from the computer into mechanical action. These actuators are used to open and close switches, control vacuum flow to other components, and operate valves depending on the system's requirements.
- Most systems allow for testing of the actuator through a scan tool.
- When checking anything with an ohmmeter, logic can dictate good and bad readings. If the meter reads infinite, this means there is an open. Based on what you are measuring across, an open could be good or bad. The same is true for very low-resistance readings. Across some things, this would indicate a short.
- Actuators can be accurately tested with a lab scope.

REVIEW QUESTIONS

1. OBD-II systems use several modes of operation. List three of them.
2. List the four ways that individual components can be checked.
3. *True or False?* A bad ground can cause an increase in the reference voltage to a sensor.
4. A typical normal oxygen sensor signal will toggle between _____ and _____ volts.
5. Describe the typical procedure for adjusting a TP sensor.
6. Which of the following is the least likely cause of a no-start condition?
   a. faulty CKP sensor
   b. faulty KS
   c. fuel or ignition system faults
   d. faulty PCM wiring
7. What is the first step in the diagnostic procedure for fault finding?
   a. List and erase the codes.
   b. Road-test the car on the highway.
   c. Verify the customer complaint.
   d. Disconnect the battery to erase memory.
8. What is the effect on the signal voltage of an engine coolant temperature sensor with a poor ground?
   a. higher than normal current
   b. higher than normal voltage
   c. lower than normal voltage
   d. lower than normal resistance
9. What is the most effective device for testing computer inputs?
   a. oscilloscope
   b. scan tool
   c. voltmeter
   d. ohmmeter

10. What signal would be generated by a magnetic pulse generator, running at a constant rpm?
    a. a variable pulse width
    b. a uniform AC signal
    c. a variable frequency
    d. a digital signal
11. Which component of the signal produced by a Hall-effect sensor changes with rpm?
    a. amplitude
    b. voltage
    c. duty cycle
    d. frequency
12. Which device should be used to extract codes from an OBD-II system?
    a. scan tool
    b. analog multimeter
    c. test light
    d. flash codes
13. How should the waveform produced on an oscilloscope change when testing a throttle position sensor?
    a. rise smoothly on deceleration
    b. increase frequency on acceleration
    c. rise smoothly on acceleration
    d. increase frequency on deceleration
14. What is a common throttle position sensor failure?
    a. reference voltage too high
    b. drop out on signal voltage
    c. frequency too high at idle
    d. amplitude increase on acceleration
15. What is the ideal number of cross counts an HO_2S oxygen sensor should have in a five-second period at a constant engine speed of 2500 rpm?
    a. three
    b. five
    c. seven
    d. nine
16. What is meant by the transition time of an HO_2S oxygen sensor?
    a. the time required to reach operating temperature
    b. the time required to sample the air/fuel mixture
    c. the time required to switch from lean to rich or rich to lean
    d. the time required for the signal to reach the computer
17. An MAP sensor should be checked by attaching a vacuum pump to the sensor and monitoring the output. By varying the vacuum, the technician should look for what change in the signal produced?
    a. a change in current
    b. a change in voltage
    c. a change in frequency
    d. a change in resistance
18. Actuators are devices that the computer uses to perform what function?
    a. do mechanical work in response to a signal
    b. vary pulse width as conditions change
    c. sense temperature changes
    d. create a signal as engine conditions change
19. What type of actuators are most commonly found on computer-controlled engines?
    a. piezoelectric devices
    b. thermistors
    c. permanent magnet generators
    d. solenoids
20. What kind of signal is produced with the opening of a solenoid circuit such as a fuel injector?
    a. voltage spike
    b. varying frequency
    c. high-frequency noise
    d. current spike

CHAPTER 33
Intake and Exhaust Systems

LEARNING OUTCOMES

- Describe how the engine creates vacuum and how vacuum is used to operate and control many automotive devices.
- Inspect and troubleshoot vacuum and air induction systems.
- Explain the operation of the components in the air induction system, including ductwork, air cleaners/filters, and intake manifolds.
- Explain the purpose and operation of a turbocharger.
- Inspect a turbocharger, and describe some common turbocharger problems.
- Explain supercharger operation, and identify common supercharger problems.
- Explain the operation of exhaust system components, including exhaust manifold; gaskets; exhaust pipe and seal; catalytic converter; muffler; resonator; tailpipe; and clamps, brackets, and hangers.
- Properly perform an exhaust system inspection, and service and replace exhaust system components.

An internal combustion engine needs air for combustion. It also needs to allow spent gases to leave the cylinder after combustion. The focus of this chapter is on the intake and exhaust systems. Both of these are often overlooked but are very important.

The reason air enters a cylinder is simply a basic law of physics—high pressure always moves toward an area of low pressure. Therefore, outside air moves into the cylinders because of the low pressure formed on the intake stroke.

VACUUM SYSTEM

The vacuum in the intake manifold not only draws air into the cylinders but it also is used to operate many systems, such as emission controls, brake boosters, parking brake releases, ventilation system components, and cruise controls on older vehicles **(Figure 33–1)**. Vacuum is applied to these systems through a system of hoses and tubes that can become quite elaborate.

GO TO Chapter 7 for a detailed explanation of atmospheric pressure and vacuum.

Vacuum Basics

Vacuum refers to any pressure that is lower than the earth's atmospheric pressure at any given altitude. The higher the altitude, the lower the atmospheric pressure.

Vacuum is measured in relation to atmospheric pressure. Atmospheric pressure is the pressure exerted on every object on earth and is caused by the weight of the surrounding air. At sea level, the pressure exerted by the atmosphere is 101.3 kPa (14.7 psi). Atmospheric pressure appears as zero on most pressure gauges. This does not mean there is no pressure; rather, it means the gauge is designed to read pressures greater than atmospheric pressure. All measurements taken on this type of gauge are given in kilopascals (pounds per square inch) and should be referred to as kPag (kilopascals gauge) and psig (pounds per square inch gauge). Gauges and other measuring devices that include atmospheric pressure in their readings also display their measurements in kPa or psi. However, these should be referred to as kPaa (kilopascals absolute) and psia (pounds per square inch absolute). There is a big difference between 70 kPaa and 70 kPag (10 psia and 10 psig). A reading of 70 kPaa (10 psia) is less than atmospheric pressure at sea level and therefore would represent a vacuum, whereas 70 kPag (10 psig) would be approximately 170 kPaa (26.7 psia). Because vacuum is defined as any pressure less than atmospheric, vacuum is any pressure less than 0 kPag or 100 kPaa (0 psig or 14.7 psia) at sea level. The normal measure of vacuum is in millimetres of mercury (mm Hg) or inches of mercury (in. Hg) instead of kPa or

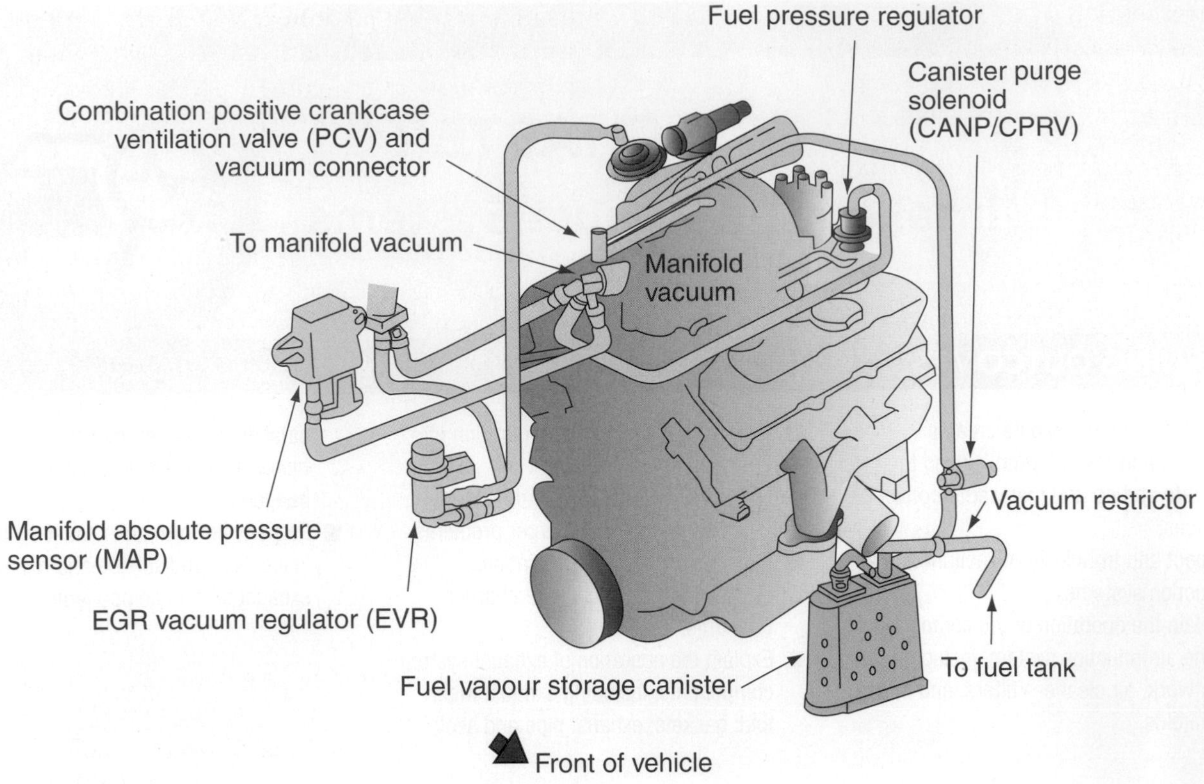

FIGURE 33–1

Typical vacuum-controlled emission systems components.

psi. Other units of measurement for vacuum are kilopascals and bars. Normal atmospheric pressure at sea level is about 1 bar, or 100 kPa.

Engine vacuum is created by the downward movement of the piston during its intake stroke. With the intake valve open and the piston moving downward, a low pressure or vacuum is created within the cylinder and intake manifold. The air passing the intake valve does not move fast enough to fill the cylinder, thereby causing the lower pressure. This vacuum is continuous in a multicylinder engine, as at least one cylinder is always at some stage of its intake stroke. The amount of low pressure produced by the piston during its intake stroke depends on a number of things. Basically, it depends on the cylinder's ability to form a vacuum and the intake system's ability to fill the cylinder. When there is high vacuum (380 to 560 mm Hg [15 to 22 in. Hg]), we know the cylinder is well-sealed, and not enough air is entering the cylinder to fill it. At idle, the throttle plate is almost closed, and nearly all airflow to the cylinders is stopped. This is why vacuum is high during idle. Because there is a correlation between throttle position and engine load, it can be said that load directly affects engine manifold vacuum. Therefore, vacuum will be high whenever there is no, or low, load on the engine.

Vacuum Controls

Engine manifold vacuum has been used to control or operate several devices and systems on vehicles, including power assist brakes, emission controls, automatic transmission modules, air-conditioning, and speed control components.

Most of these functions are now carried out by electronic controls, although vacuum is still used on some emission systems (solenoid controlled) and non-ABS braking systems.

Diagnosis and Troubleshooting

Vacuum system problems can produce or contribute to the following driveability symptoms:

- Stalls
- No-start (cold)
- Hard start (hot soak)
- Backfire (deceleration)
- Rough idle
- Poor acceleration
- Rich or lean stumble

- Overheating
- Detonation, or knock or pinging
- Rotten eggs exhaust odour
- Poor fuel economy

As a routine part of problem diagnosis, a technician who suspects a vacuum problem should first do the following:

- Inspect vacuum hoses for improper routing or disconnections (engine decal identifies hose routing).
- Look for kinks, tears, or cuts in vacuum lines.
- Check for vacuum hose routing and wear near hot spots, such as the exhaust manifold or the EGR tubes.
- Make sure there is no evidence of oil or transmission fluid in vacuum hose connections. (Valves can become contaminated by oil getting inside.)
- Inspect vacuum system devices for damage (dents in cans, bypass valves, broken nipples on vacuum control valves, broken "tees" in vacuum lines, and so on).

Broken or disconnected hoses allow vacuum leaks that admit more air into the intake manifold than the engine is calibrated for. The most common result is a rough-running engine due to the leaner air/fuel mixture created by the excess air.

Kinked hoses can cut off vacuum to a component, thereby disabling it. For example, if the vacuum hose to the EGR valve is kinked, vacuum cannot be used to move the diaphragm. Therefore, the valve will not open.

To check vacuum controls, refer to service information for the correct location and identification of the components.

Tears and kinks in any vacuum line can affect engine operation. Any defective hoses should be replaced one at a time to avoid misrouting. OEM vacuum lines are installed in a harness consisting of 3.18 mm (⅛ in.) or larger outer diameter and 1.59 mm (1⁄16 in.) inner diameter nylon hose with bonded nylon or rubber connectors. Occasionally, a rubber hose might be connected to the harness. The nylon connectors have rubber inserts to provide a seal between the nylon connector and the component connection (nipple). In recent years, many domestic car manufacturers have been using ganged steel vacuum lines.

Vacuum Test Equipment

The vacuum gauge is one of the most important engine diagnostic tools used by technicians. With the gauge connected to the intake manifold and the engine warm and idling, watch the action of the gauge's needle. A healthy engine will give a steady, constant vacuum reading between 430 and 560 mm Hg (17 and 22 in. Hg). On some four- and six-cylinder engines, however, a reading of 380 mm Hg (15 in. Hg) is considered acceptable. With high-performance engines, a slight flicker of the needle can also be expected. Keep in mind that the gauge reading will drop about 25.4 mm (1 in.) for each 305 m (1000 ft.) above sea level.

If the amount of vacuum produced by each cylinder is the same, the vacuum gauge will show a steady reading. If one or more cylinders are producing different amounts of vacuum, the gauge will show a fluctuating reading. The amount of vacuum read on the gauge, as well as the movement of the gauge's needle, can tell you quite a bit about the engine. If the gauge reading is low but steady, there is a problem that is common to all cylinders. The severity of the problem is indicated by how low it is. For example, if a vacuum reading is a steady 254 mm Hg (10 in. Hg), the problem is something common to all cylinders, such as a fairly good-sized intake manifold vacuum leak. If the gauge reads a steady 380 mm Hg (15 in. Hg), the problem is less severe but still common to all cylinders, such as retarded ignition timing.

A fluctuating needle indicates a problem that is not common to all cylinders. If the gauge is connected to a four-cylinder engine and the gauge's needle bounces at an even pace between 254 and 432 mm Hg (10 and 17 in. Hg), we can assume that two of the four cylinders are producing less vacuum than the other two. If the needle spends more time at the 432 mm Hg (17 in. Hg) reading, we can assume one cylinder is producing less vacuum. Again, how low the needle dips is an indication of the severity of the problem. If the needle dips to zero, we could suspect a hole in the piston or a severely damaged valve in one cylinder. If the needle dips to 381 mm Hg (15 in. Hg), the problem might be worn piston rings on one cylinder.

As shown in **Figure 33–2**, a hand-held vacuum pump/gauge is used to test vacuum-actuated valves and motors. If the component does not operate when the proper amount of vacuum is applied, it should be serviced or replaced.

An emissions vacuum schematic is given on an underhood decal. This shows the vacuum hose routings and vacuum source for all emissions-related equipment. The vacuum schematic in **Figure 33–3** shows the relationship and position of components as they are mounted on the engine. It is important to

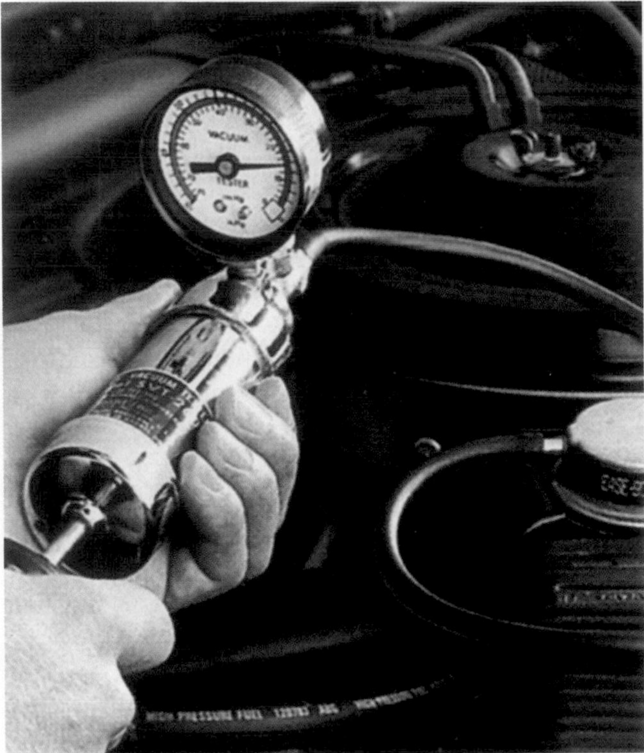

FIGURE 33–2

A hand-operated vacuum pump being used to test an air-cleaner vacuum motor.

Stant Manufacturing Inc.

remember that these schematics show only the vacuum-controlled parts of the emission system. The location and hose routing for other vacuum devices can be found in the service information.

Many technicians use smoke machines to locate vacuum leaks. Smoke machines inject smoke into the vacuum system or into the engine. If there are no leaks, smoke will not escape from the engine. However, if a vacuum leak is present, the smoke will show the location of the leak **(Figure 33–4)**.

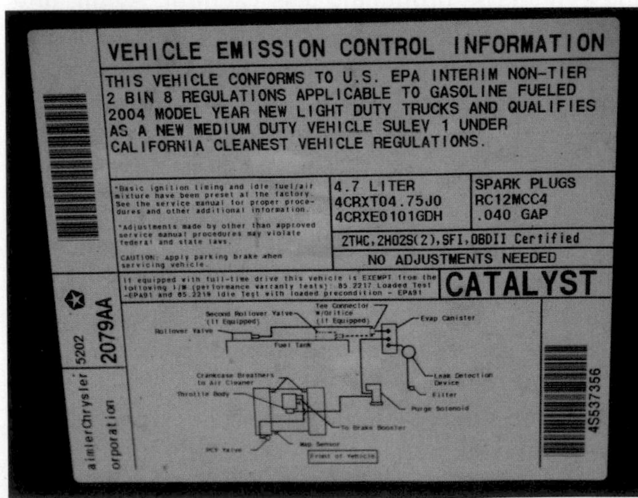

FIGURE 33–3

An underhood vacuum-hose routing diagram.

Martin Restoule

FIGURE 33–4

A smoke machine used to find a vacuum leak.

Martin Restoule

SHOP TALK

A smoke machine can be used for locating vacuum, induction, and exhaust leaks.

THE AIR INDUCTION SYSTEM

The air induction system directs outside air to the engine's cylinders. The induction system is comprised of ductwork that channels outside air to an air cleaner that removes dirt from the air, ductwork that connects the filter to the throttle body, and an intake manifold that distributes the air to the

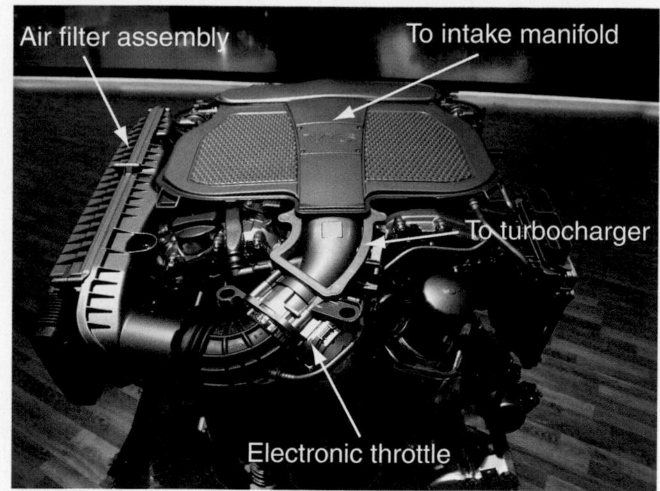

FIGURE 33–5

The air induction system for a late-model turbocharged engine.

FIGURE 33–6

An example of an air filter used on current engines.

engine's cylinders **(Figure 33–5)**. Within the induction system are sensors that measure intake air temperature and airflow.

An inspection of the air induction system should be part of diagnosing a driveability problem. Make sure that the intake ductwork is properly installed and that all connections are airtight—especially those between an airflow sensor or remote air cleaner and the throttle body.

Air Cleaner/Filter

The primary purpose of the air filter is to prevent airborne contaminants and abrasives from entering the cylinders. These contaminants can cause serious damage and appreciably shorten engine life. Therefore, all intake air should pass through the filter before entering the engine.

The air filter is inside a sealed air cleaner assembly. This assembly is also used to direct the airflow and reduce the noise caused by the movement of intake air **(Figure 33–6)**. The air cleaner also provides filtered air to the positive crankcase ventilation (PCV) system and provides engine-compartment fire protection in the event of backfire.

If the air filter becomes very dirty, the dirt can block the flow of air into the engine. Restricted airflow to the engine can cause poor fuel economy, poor performance, and high emissions.

GO TO Chapter 8 for information on servicing and replacing an air filter.

INDUCTION HOSES

To route air into the engine, a network of induction hoses and pipes are used. Air enters through the fresh air inlet and passes through induction hoses and pipes. Induction hoses are typically flexible rubber hoses to allow for movement between the engine and other induction components. Some vehicles use rigid plastic pipes to move air; these pipes may also have a Helmholtz chamber or resonator. A Helmholtz resonator looks like an additional chamber sticking out of a section of induction pipe. The purpose of the chamber is to reduce noises generated by air moving through the induction system and intake manifold.

Intake Manifold

The intake manifold distributes the clean air or air/fuel mixture as evenly as possible to each cylinder of the engine.

Older engines had cast-iron intake manifolds. The manifold delivered air and fuel to the cylinders and had short runners. These manifolds were either wet or dry. Wet manifolds had coolant passages cast directly in them. Dry manifolds did not have these coolant passages, but some had exhaust passages. Exhaust gases or coolant was used to heat up the floor of the manifold. This helped to vaporize the fuel before it arrived in the cylinders. Other dry manifold designs used some sort of electric heater unit or grid to warm up the bottom of the manifold. Heating the floor of the manifold also stopped the fuel from condensing in the manifold's plenum area. Good fuel vaporization and the prevention of condensation

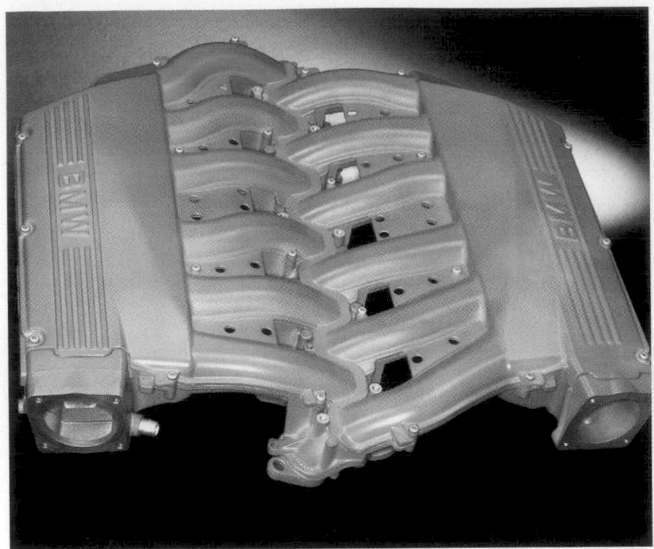

FIGURE 33–7

A die-cast aluminum intake manifold.

BMW of North America, Inc.

allowed for delivery of a more uniform air/fuel mixture to the individual cylinders.

Modern intake manifolds for engines with port fuel injection are typically made of die-cast aluminum **(Figure 33–7)** or plastic **(Figure 33–8)**. These materials are used to reduce engine weight.

FIGURE 33–8

A plastic intake manifold for an in-line five-cylinder engine.

Because intake manifolds for port-injected engines only deliver air to the cylinders, fuel vaporization and condensation are not design considerations. The primary consideration of these manifolds is the delivery of equal amounts of air to each cylinder. This style manifold is often called a *tuned* intake manifold.

Intake manifolds also serve as the mounting point for many intake-related accessories and sensors **(Figure 33–9)**. Some include a provision for

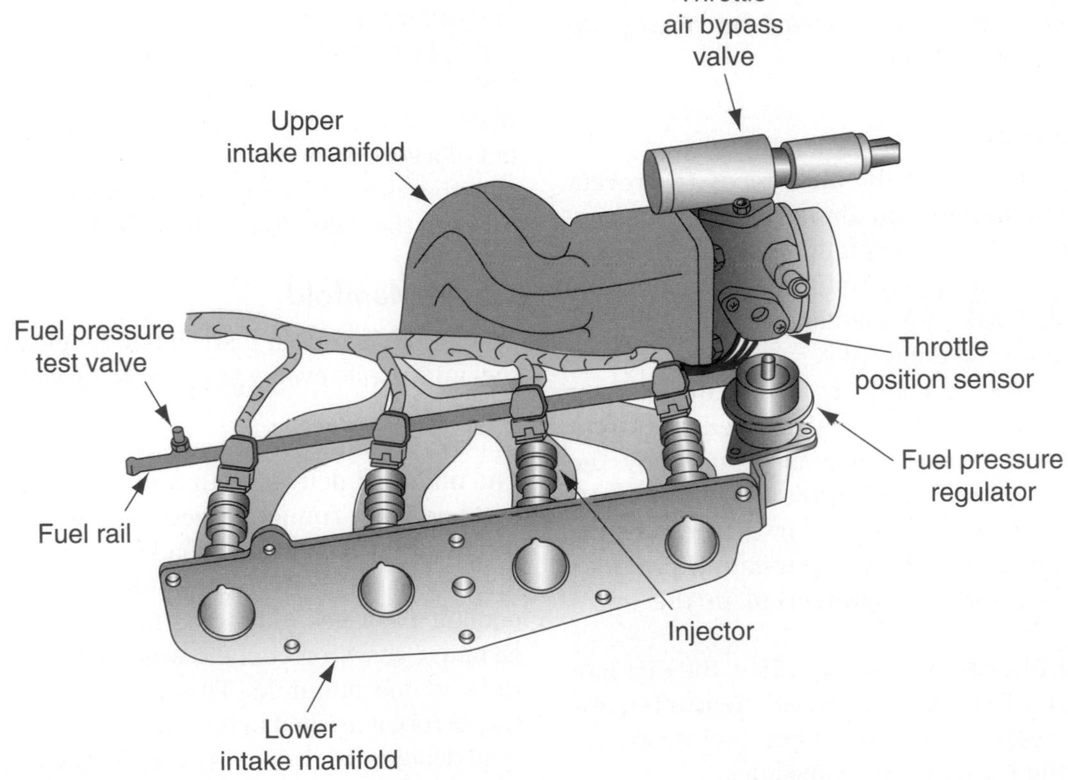

FIGURE 33–9

A late-model intake manifold for an in-line four-cylinder engine.

©Cengage Learning 2015

mounting the thermostat and thermostat housing. In addition, connections to the intake manifold provide a vacuum source for the exhaust gas recirculation (EGR) system, automatic transmission vacuum modulators, power brakes, and heater and/or air-conditioning airflow control doors. Other devices located on or connected to the intake manifold include the manifold absolute pressure (MAP) sensor, the knock sensor, various temperature sensors, and EGR passages.

Attached to the inlet of the intake manifold is the throttle body. The throttle body **(Figure 33–10)** controls the amount of air entering the engine. Inside of the throttle body is the throttle plate. At idle, the throttle plate is closed or very nearly closed, and little air enters the engine. At wide-open throttle (WOT), the throttle plate is horizontal to allow maximum airflow into the engine. On older vehicles, the throttle plate is controlled by the throttle cable, which is attached to the accelerator or gas pedal. Modern vehicles use electronic throttle control (ETC) systems.

Design Variations

Basic manifold design varies with the different types of engine. For example, the intake manifold for a four-cylinder engine has either four runners or two runners that break into four near the cylinder head. Inline six-cylinder engines have six runners or three that branch off into six near the cylinder head. On V-type engines, there are individual runners for each cylinder.

FIGURE 33–10

A throttle body assembly.

An intake manifold has two basic components: a plenum area and runners. As air first enters the intake manifold, it moves into the plenum. The air then moves from the plenum, through the runners, to the cylinders. The size and shape of the plenum and runners are designed for a specific engine and application.

The plenum serves as a reservoir for the air and is used to distribute the intake charge evenly and to enhance engine breathing. The shape of the runners is different for port fuel injection (PFI) and gasoline direct injection (GDI) engines than other engines. An intake manifold that delivers both air and fuel is designed to cause turbulence so that the air and fuel mix while they are being delivered to the cylinders. When the intake manifold delivers only air, there is no need for turbulence, and the runners provide a smooth direct flow of the air. Air-only runners have smooth finishes and a minimum number of bends. The length of the runners is designed to achieve the best performance during a particular range of engine speeds.

An engine is most volumetrically efficient when a maximum amount of air enters the cylinders. Peak engine torque occurs when the engine is most efficient. Generally, an engine designed for maximum torque and horsepower at high speeds has shorter runners than one that provides high torque at lower speeds. The length of the runners changes when the engine develops its peak torque.

One of the things that most do not think about is the behaviour of air when it enters a runner. A simple thought is the air arrives and stays there until the intake valve opens. It then flows past the valve into the cylinder. Actually, the air is moving at a pretty good speed when it is pulled into the cylinder, and it must come to a halt when the valve closes. The air does not sit there until the valve opens again; rather, it bounces off the closed valve and heads toward the plenum area. When the air reaches the plenum, it meets a rush of incoming air and bounces back toward the intake valve. The air, however, bounces back more quickly than it did when it left the intake valve. This is due to a push given by the intake air.

A runner is designed to take this bouncing air and send it back to the intake valve in time for the next opening. This timing determines the length of the runner and results in a stronger intake charge because the air is under pressure. In most manifolds, this air wave bounces several times before the intake valve opens again. The bouncing effect and resultant pressure of the air is called acoustic supercharging.

The inside diameter of the runners also affects the delivery of air. When a small diameter runner is used, the air will move into the cylinders faster. This increases volumetric efficiency at low engine speeds. When the engine is running at higher speeds, it needs a lot of air. Small diameter runners would restrict the airflow and hurt the engine's efficiency. Therefore, larger diameter runners are needed at high engine speeds.

Variable Intake Manifolds

Many engines have variable intake manifolds, which are controlled by the PCM. These manifolds change the size of the plenum area and/or the length and effective diameter of the runners according to engine speed and load. The use of these manifolds allows the engine to experience high volumetric efficiency with more than one range of engine speeds.

The operation and design of these manifolds vary with manufacturer and engine. Systems that alter the plenum area have two small plenums. Depending on the system, only one of the plenums is used during low speeds. In other systems, the plenums are divided and are used for specific cylinders. In both cases, when the engine reaches a particular speed, the plenums are opened and work together to create a larger plenum area. These systems are commonly referred to as **intake manifold tuning (IMT)** systems.

IMT systems have a motor connected to a butterfly valve in the centre of the manifold **(Figure 33–11)**. The valve is closed during low speeds, keeping the two plenum areas separated. When commanded by the PCM, the valve opens and allows the two plenums to become one large plenum. The PCM receives feedback on the position of the valve through a position sensor on the motor.

WARNING!

The butterfly valve is moved with great force. Always keep your fingers away from the valve when the system is energized. Failure to do this may result in a serious injury.

The most common variable intake manifold designs change the path of air between long and short runners or between small-diameter and large-diameter runners according to engine speed. These systems are typically called **intake manifold runner control (IMRC)** systems. Intake air passes through long- or small-diameter runners at low speeds and is

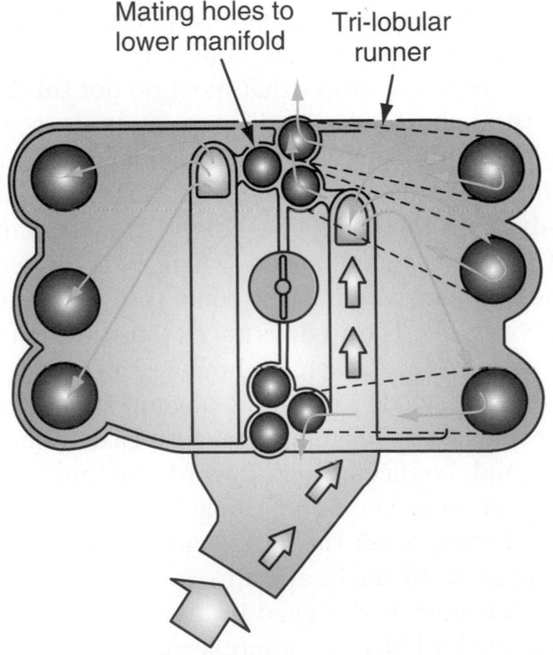

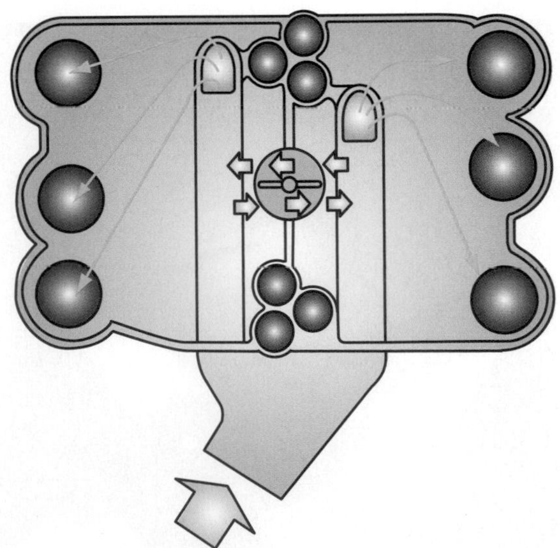

FIGURE 33–11

The action of an intake manifold tuning control valve.

routed through short- or large-diameter runners at high speeds. The switching of runners is mostly done by controlling a butterfly valve that opens and closes the short- or large-diameter runners. Ultimately, the overall volume of air is controlled by the throttle plate; the butterfly valve in the manifold merely controls the routing of the air.

Changing the runners for different speeds allows for the benefits of acoustic supercharging as well as providing increased airflow at high speeds. It is important to realize that too much airflow at low speeds can actually hurt engine performance because the engine does not need it, and the resultant air waves are hard to time to intake valve opening. All IMRC systems must have a feedback system, according to OBD-II standards. If the IMRC system is not working correctly, a DTC is set.

The butterfly valve in IMRC systems is vacuum or electrically controlled. In vacuum systems, a vacuum actuator is mounted on the manifold **(Figure 33–12)**. Vacuum to the actuator is controlled by a PCM-regulated solenoid. Linkage connects the actuator to the butterfly valve. The PCM relies mainly on inputs from the TP, ECT, and CKP sensors to determine when to open or close the butterfly valve. At low speeds, the solenoid is energized. This allows manifold vacuum to hold the valve closed. Once engine speed and other conditions are met, the solenoid is turned off and springs on the butterfly valve force the valve open.

In electrical systems, a motor is used to move the butterfly valve. There can be one valve per cylinder or one valve per bank of runners **(Figure 33–13)**. The valve can be normally open and remain open, or it is closed by the solenoid when the valve for the other runner is opened. Often, there is a butterfly valve common to the high-speed runners. This valve is located in the plenum area and switches the routing of the air inside the plenum. The action of the motor is controlled by the PCM and can be duty-cycled to be able to respond to current engine speeds.

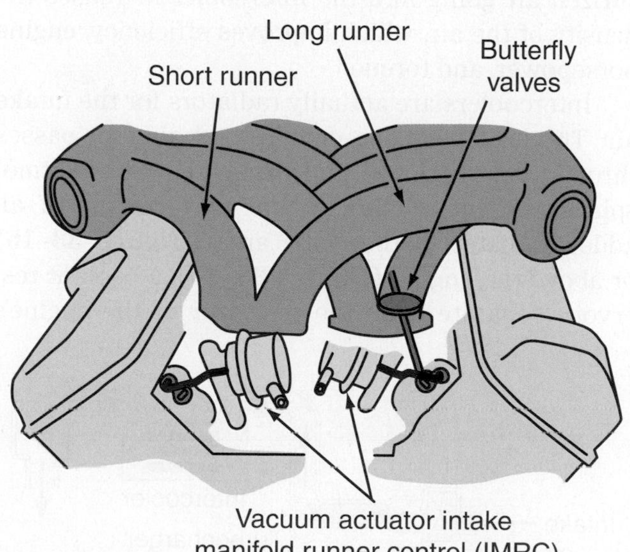

FIGURE 33–12

The vacuum controls and butterflies used to switch between the long and short intake runners.

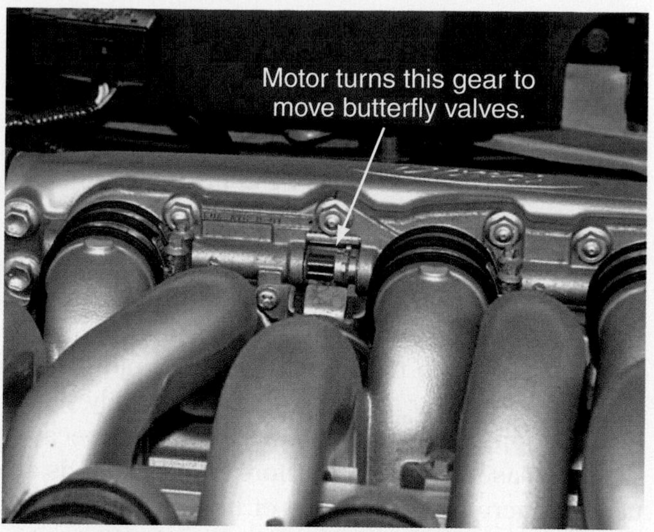

FIGURE 33–13

An electric motor meshes with this gear to rotate the shaft for the butterfly valves in this variable intake manifold.

SERVICING AN INTAKE MANIFOLD There are few reasons why an intake manifold would need to be replaced or repaired. If the manifold is cracked or the sealing surfaces are severely damaged, it should be replaced. When the manifold is removed or there is evidence of a leak between the manifold and the cylinder head, the sealing surfaces should be checked for flatness. This check should include coolant and oil passages. Minor imperfections on the surface can be filed smooth; however, never try to repair any serious damage.

A leaking manifold gasket is one of the most common causes of internal coolant leaks. A bad gasket can also cause vacuum leaks. When installing an intake manifold, use new gaskets and seals. Also, make sure you arc installing the most recent gasket designs. Make surc that the manifold and gasket are aligned properly. On V-type engines, the use of guide bolts helps to ensure proper alignment. Make sure that all of the attaching bolts are tightened to specifications and in the correct order.

Some engines have two-piece intake manifolds. The two parts are sealed together with gaskets and seals. These should be replaced when the manifold is separated, and bolts should be tightened to specifications.

FORCED INDUCTION SYSTEMS

Engines cannot produce the amount of power they are capable of at high speeds because they do not receive enough air. This is the reason why many race cars have hood scoops. Hood scoops deliver cool air under pressure to the intake manifold and provide an open source for the air. With today's body styles, hood scoops are not particularly desired because they increase air drag. Therefore, other methods are used to increase the volume of intake air and compression. Variable intake manifolds and valve timing certainly help with this. Some manufacturers are exploring variable compression engines, but none are yet in production. Performance gains have been made through air filter designs **(Figure 33–14)**. The shape of the filter and its direct exposure to the air allow for increased intake flow. Although these filters can increase horsepower and decrease fuel consumption, manufacturers do not use these filters on normal cars. With the filter exposed, there is nothing to reduce intake noise, and owners may find the noise offensive.

Keep in mind that the power generated by the internal combustion engine is directly related to the amount of air that is compressed in the cylinders. In other words, the greater the compression (within reason), the greater the output of the engine. Two approaches can be used to increase an engine's effective compression. One is to modify the engine to increase its compression ratio. This can be done in many ways, including the use of domes or high-top pistons, altered crankshaft strokes, or changes in the shape and structure of the combustion chamber.

Another, less expensive way to increase compression (and engine power) without physically changing the shape of the combustion chamber is to simply increase the intake charge. By pressurizing the intake mixture before it enters the cylinder, more air and fuel molecules can be packed into the combustion chamber. The two ways to artificially increase the amount of airflow into the engine are known as turbocharging and supercharging.

Both of these systems force more air into the intake manifold by compressing the air before it reaches the manifold. Turbocharging uses exhaust gases, and supercharging relies on the rotation of the engine to compress the air. Both systems offer benefits but have some limitations. The biggest disadvantage of using either system is related to the compression of air. When air is compressed, its heat increases. High-temperature air is less dense, meaning there is less air or oxygen in the air. Most turbocharged or supercharged systems use an intercooler to increase the density of the air.

FIGURE 33–14

A specialty cone air filter can increase the power output of an engine.

Intercoolers

The **intercooler** cools the turbocharged or supercharged air before it reaches the combustion chamber **(Figure 33–15)**. The removal of heat from the pressurized air going into the intercooler increases the density of the air, which improves efficiency, engine horsepower, and torque.

Intercoolers are actually radiators for the intake air. The heat from the compressed air that passes through it is removed and dissipated to the atmosphere. An intercooler system may consist of an added radiator in the grille area **(Figure 33–16)** or above the engine **(Figure 33–17)**, a coolant reservoir (separate from the reservoir of the engine's

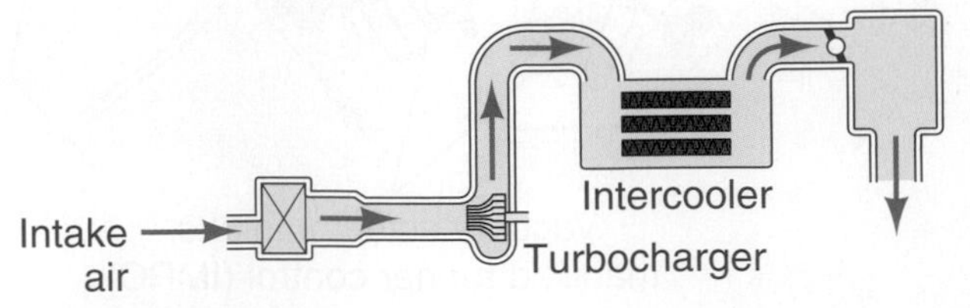

FIGURE 33–15

Routing of the boosted air in and out of an intercooler.

FIGURE 33–16

A twin turbocharged engine with an intercooler located in front of the engine.

©Cengage Learning 2015

FIGURE 33–17

The intercooler for this supercharged engine is located on top of the engine.

©Cengage Learning 2015

cooling system), a pump, and hoses and tubes to connect the components **(Figure 33–18)**. An intercooler is always located after the turbocharger or supercharger and before the intake manifold. As the heated air flows through the intercooler, heat is transferred to the coolant circulating through the intercooler. The coolant is cooled by the air passing through the intercooler. The amount of coolant moving through the intercooler is normally controlled by the PCM. Therefore, the PCM controls the temperature of the incoming air.

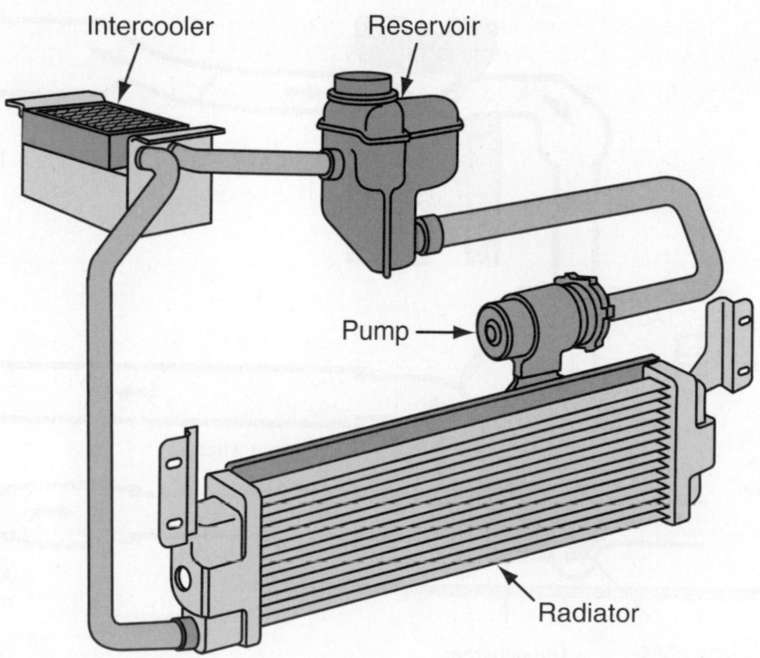

FIGURE 33–18

The main parts of a water-to-air intercooler.

©Cengage Learning 2015

TURBOCHARGERS

Turbochargers are used to increase engine power by compressing intake air before it enters the engine. Turbochargers are air pumps driven by the engine's rapidly expanding exhaust stream. The heat and pressure of the exhaust gases spin the turbine blades (hence the name turbocharger) of the pump. The turbine wheel is connected to a compressor wheel. As the turbine spins, so does the compressor wheel. The compressor wheel spins at very high speeds and compresses the intake air. The compressed air is then sent to the cylinders. Because exhaust gas is a waste product, the energy developed by the turbine is said to be free because it theoretically does not use any of the engine's power it helps to produce.

Turbochargers are used on both diesel and gasoline engines. The main advantage of their use is that they allow for an increase of power without a substantial decrease in fuel economy. This is because they boost the engine's power output only when extra power is needed. A small engine, in most cases, can be used to provide low fuel consumption and emissions levels. When increased power is needed, the turbocharger is activated.

Construction

A turbocharger is normally located close to the exhaust manifold. An exhaust pipe runs between the exhaust manifold and the turbine housing to carry the exhaust flow to the turbine wheel **(Figure 33–19)**. Another pipe connects the compressor housing intake to the throttle plate assembly or intake manifold.

A typical turbocharger, usually called a turbo, has the following components **(Figure 33–20)**:

- Turbine wheel
- Shaft
- Compressor wheel
- Centre housing and rotating assembly (CHRA) or cartridge
- Wastegate valve

Inside the turbocharger, the turbine wheel (hot wheel) and the compressor wheel (cold wheel) are mounted on the same shaft. The CHRA or cartridge houses the shaft, shaft bearings, turbine seal assembly, and compressor seal assembly. Each wheel is encased in its own spiral-shaped enclosure in the housing that serves to control and direct the flow of exhaust and intake air. Because the turbine wheel is in the exhaust path, it gets very hot. It also spins at very high speeds; therefore, it is normally made of a heat-resistant cast iron.

A conventional turbocharger normally begins to compress intake air when the engine's speed is above 2000 rpm. Some late-model engines have low inertia turbochargers that begin to compress intake air at lower engine speeds. The force of the exhaust flow is directed against the side of the turbine wheel. As

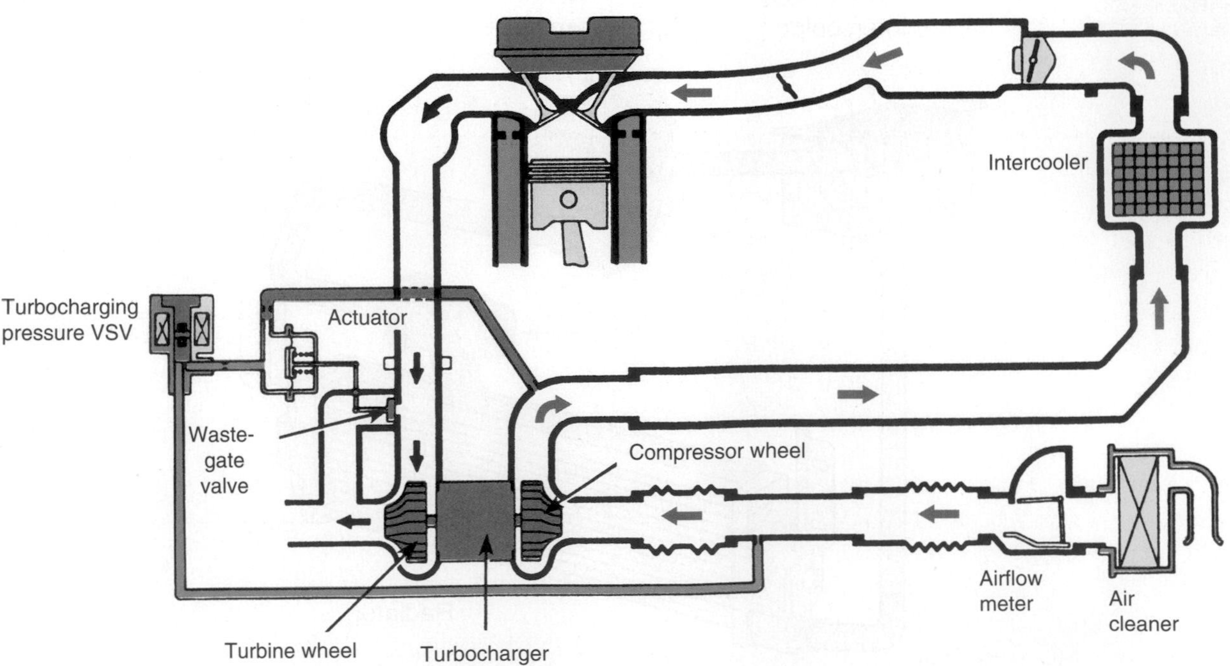

FIGURE 33–19

Exhaust gas and airflow in a typical turbocharger system.

Reprinted with permission.

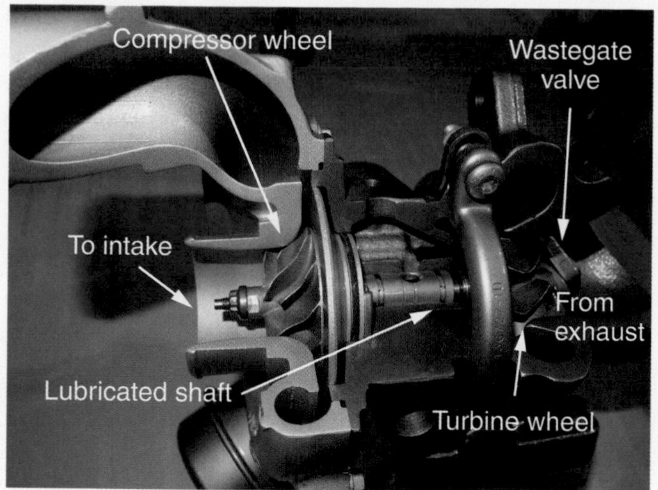

FIGURE 33–20

A cross-section of a turbocharger shows the turbine wheel, the compressor wheel, and their connecting shaft.

the hot gases hit the turbine wheel, causing it to spin, the turbine fins direct the exhaust gases toward the centre of the housing, where they exit. This action creates a flow called a vortex. The compressor wheel (shaped like a turbine wheel in reverse) spins with the turbine. Intake air is drawn into the housing and is caught by the whirling blades of the compressor and thrown outward by centrifugal force. From there, the air exits under pressure to the intake manifold and the individual cylinders.

Air is typically drawn into the cylinders by the difference in pressure between the atmosphere and engine vacuum. A turbocharger, however, is capable of pressurizing the intake charge above normal atmospheric pressure. *Turbo boost* is the term used to describe the positive pressure increase created by a turbocharger. For example, 10 psi of boost means the air is being fed into the engine at 170 kPa (101 kPa atmospheric plus 69 kPa boost) or 24.7 psi (14.7 psi atmospheric plus 10 pounds of boost). A turbocharger is equipped with a wastegate valve to control the pressure of the air delivered to the cylinders.

Turbo Lag

Increases in horsepower are normally evidenced by an engine's response to a quick opening of the throttle. The lack of immediate throttle response is felt with some turbocharged engines. This delay, or **turbo lag**, occurs because exhaust gas requires a little time to build enough energy to spin the wheels fast enough to respond to the engine's speed. This causes the power from the engine to temporarily lag behind what is needed for the conditions.

Wastegate Valve

If the pressure of the air from a turbocharger becomes too high, knocking occurs; engine output decreases; and the pressure created by the combustion of the air/fuel mixture can become so great that the engine may self-destruct. To prevent this, turbochargers have a wastegate valve. The **wastegate** valve is part of the turbine housing. It allows a certain amount of exhaust gas to bypass the turbine when the boost pressure exceeds a certain value. This action reduces the pressure.

The action of the wastegate can be controlled directly by manifold pressure **(Figure 33–21)** or by the PCM according to manifold pressure **(Figure 33–22)**. Most late-model systems have

FIGURE 33–21

The output of this turbocharger is controlled directly by manifold pressure.

FIGURE 33–22

The output of this turbocharger is controlled electronically.

PCM-controlled wastegates. In non-PCM systems, an actuator that senses the air pressure in the induction system opens the wastegate when the pressure becomes too high. This action decreases the amount of exhaust that reaches the turbine. This, in turn, reduces turbine and compressor wheel speed and decreases the output pressure from the turbocharger. When the pressure in the intake manifold is not great enough to override the spring in the actuator, the wastegate is closed, and all of the exhaust flows past the turbine; therefore, the turbine spins accordingly. When the boost pressure overcomes the tension of the actuator's spring, the actuator opens the wastegate valve and some exhaust gas is diverted around the turbine wheel. As a result, turbine speed is controlled, as is boost pressure.

On late-model engines, the wastegate is controlled by the PCM, which directly controls a solenoid that controls vacuum to the wastegate. When vacuum is introduced to the wastegate, it opens to allow exhaust gases to bypass the turbine. The action of the solenoid is controlled by the PCM according to various inputs. The PCM also adjusts ignition timing and air/fuel mixtures according to turbocharger output. Retarding spark timing is an often used method of controlling detonation on turbocharged engines. Unfortunately, any time the ignition is retarded, power is lost, fuel economy suffers, and the engine tends to run hotter. Most systems use knock-sensor signals to retard timing only when detonation is detected. These sensors are also used to limit boost pressure according to the octane rating of the fuel being used. This maximizes engine performance and reduces the chances of engine knocking during all conditions, regardless of the fuel's octane rating.

When the PCM detects excessive manifold pressure, it opens the wastegate and also enriches the mixture. This rich mixture reduces combustion temperature, which helps to cool the turbocharger and combustion chamber.

Turbo Lubrication and Cooling

Keeping a turbocharger cool and well-lubricated is essential to the unit's durability. The turbine wheel faces very high temperatures. The temperature of the exhaust from a gasoline engine can surpass 982°C (1800°F). The exhaust from a diesel engine is cooler and ranges from 768°C to 816°C (1400°F to 1500°F). This heat can destroy the turbocharger if it is not controlled.

The turbine and compressor wheels rotate at very high speeds (over 200 000 rpm). Their shaft is mounted on full-floating bearings designed to keep the shaft secure while it is rotating. The bearings are lubricated by engine oil and rotate freely on the shaft and in the housing. Engine oil is delivered to the turbocharger through an oil inlet pipe and then circulated to the bearings. The bearings have outer seals to prevent oil leakage. The oil passes over the bearings and returns to the engine's oil pan through an oil outlet pipe. The circulation of the oil cools the turbocharger.

A turbocharger may be cooled by engine coolant that circulates through coolant channels built into the housing. Coolant is sent from the engine's thermostat housing, through a coolant inlet pipe, to the turbo housing. After the coolant has circulated through the housing, it is returned to the water pump through the coolant outlet pipe.

Various Turbocharger Designs

In an effort to increase the efficiency of turbocharged engines, manufacturers have developed various designs of turbochargers and their control systems. Common alternative designs are called variable nozzle turbine and variable geometry turbochargers. In these, the cross-sectional area through which the exhaust flows is variable. This area is adjusted via movable vanes that change their angles according to turbine speed. At lower engine speeds, the vanes restrict exhaust flow, thereby increasing boost pressure. At higher engine speeds, the vanes open wider, and exhaust back pressure decreases. This allows the turbocharger to provide more boost at lower engine speeds without producing too much boost at higher speeds. It is claimed that the use of a variable turbocharger can reduce a gasoline engine's fuel consumption by 20 percent.

Because variable geometry allows for more precise control of the turbo, some variable turbos do not have a wastegate. They provide a higher boost at lower engine speeds and are more responsive to changes in engine load. They also help reduce the effects of turbo lag.

Twin Turbochargers

Some engines have two or more turbochargers. Their action depends on the application. Some engines have a turbo for half of the cylinders and another for the other half **(Figure 33–23)**. Some V-type engines have a turbocharger for each bank of cylinders. The turbos use the exhaust from specific cylinders to compress the air for those same cylinders.

Some late-model engines have dual turbochargers that use a common housing and exhaust

FIGURE 33–23

A twin turbocharger setup on an inline six-cylinder engine.

stream to send compressed air out to two separate paths **(Figure 33–24)**.

Other engines have two different-sized turbos. Each turbo is designed for specific conditions. Normally, the smaller of the two spools (spins) up to speed very quickly. This reduces turbo lag. The larger one is slower to get up to speed but adds the boost at higher engine speeds. This is a two-stage design: one for lower engine speeds and immediate increase of speed, and one for sustained power.

A few engines place the twin turbochargers within the V between the cylinder banks. This reverses the airflow through the heads and places the intake manifolds on the outside of the engine instead of between the heads. This design increases exhaust flow into the turbochargers **(Figure 33–25)**.

FIGURE 33–24

A dual turbocharger assembly.

FIGURE 33–25

This V8 engine has twin turbochargers within the V between the cylinder banks.

In a typical twin turbocharger system, the function of the two turbochargers is controlled by the operating mode of the larger turbocharger. Its operation is controlled by control valves that regulate exhaust gases to it and the amount of air from it.

During low engine speeds and loads, boost is provided only by the smaller unit. The control valves of the other turbo have it disabled. When boost pressure from the first turbo reaches a predetermined level, exhaust gas is allowed to flow to the second turbo. At this time, it is spinning, but not providing any boost. This is a preparation step; the second turbine is spinning before it is needed. Once the load or speed conditions demand more power, a valve opens and allows the boost pressure from the second turbo to enter the intake manifold. Now, boost from both turbos is sent to the intake manifold. In some systems, the higher pressure from the larger turbo causes the wastegate valve, which is an integral part of the smaller turbo, to open and control the maximum boost.

When the engine moves from high to low speed, a control valve stops the flow from the second turbo to the intake. Another control valve then blocks off exhaust flow to the turbo. These actions prevent high boost during deceleration.

DualBoost Single Sequential Turbocharger

In 2011, Ford installed a DualBoost turbocharger on its 6.7L PowerStroke diesel engines. This turbocharger has two compressor wheels and one turbine wheel. The compressor wheels have different shapes

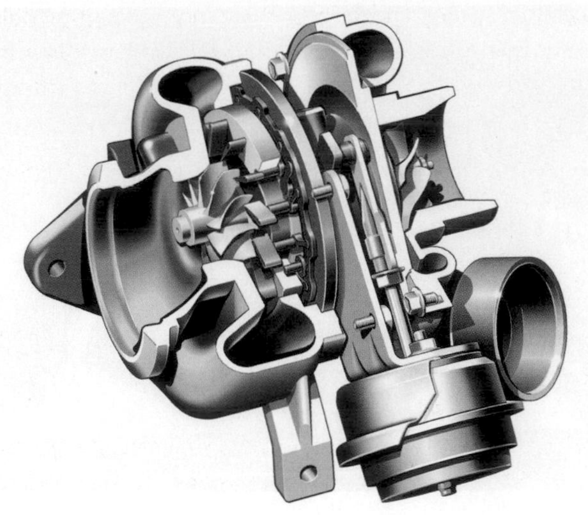

FIGURE 33–26

A typical variable geometry turbine vane system.

DaimlerChrysler Corporation

and sizes, but they are mounted onto the same shaft. The two compressor wheel housings are also different sizes. This creates a sequential turbo by allowing fast spooling-up of the small compressor, followed by increased air volume from the larger wheel. The volume of the single exhaust inlet housing is controlled by the vanes of the variable geometry system **(Figure 33–26)**. A wastegate is used on the variable turbine to bypass exhaust gas as required. Because of its small size, this turbo can be mounted inside of the V of the engine. The airflow through the cylinder heads is reversed, and exhaust gas is supplied directly from the heads into the turbo unit.

Turbocharger Inspection

To inspect a turbocharger, start the engine, and listen to the sound that the turbo system makes. As you become more familiar with this characteristic sound, it will be easier to identify an air leak between the compressor outlet and engine or an exhaust leak between engine and turbo by the presence of a higher-pitched sound. If the turbo sound cycles or changes in intensity, the likely causes are a plugged air cleaner or loose material in the compressor inlet ducts or dirt buildup on the compressor wheel and housing.

USING SERVICE INFORMATION

General service procedures for turbocharger systems are normally in a separate section of the manufacturers' service information. Individual inputs and outputs that are part of the electronic control system are covered in the engine control or performance sections.

After listening, check the air cleaner, and remove the ducting from the air cleaner to the turbo; then look for dirt buildup or damage from foreign objects. Check for loose clamps on the compressor outlet connections, and check the engine intake system for loose bolts or leaking gaskets. Then disconnect the exhaust pipe, and look for restrictions or loose material. Examine the exhaust system for cracks, loose nuts, or blown gaskets. Rotate the turbo shaft assembly. Does it rotate freely? Are there signs of rubbing or wheel impact damage?

Visually inspect all hoses, gaskets, and tubing for proper fit, damage, and wear. Check the low-pressure, or air cleaner, side of the intake system for vacuum leaks.

PRESSURE TESTING The performance of a turbocharger can be checked with a pressure gauge. The gauge is connected to the intake manifold and the amount of boost observed. This is best done on a road test. This check can also verify the action of the wastegate and its controls. Manifold pressure can be monitored with a scan tool as well.

SHOP TALK

When oil leakage is noted at the turbine end of the turbocharger, check the turbocharger oil drain tube and the engine crankcase breathers for restrictions. When sludged engine oil is found, the engine's oil and oil filter must be changed.

On the pressure side of the system, you can check for air leaks with soapy water. After applying the soap mixture, look for bubbles to pinpoint the source of the leak.

Exhaust leaks to the turbocharger housing will also affect turbo operation. If exhaust gases escape prior to entering the housing, the reduced temperature and pressure will cause a proportionate reduction in boost and a loss of power.

SHOP TALK

Because of the needed balance and sealing of the units, all repairs to turbochargers are done at specialty shops.

WASTEGATES Wastegate problems can usually be traced to carbon buildup, which keeps the unit from closing or causes it to bind. A defective diaphragm or leaking vacuum hose can result in an inoperative wastegate. Before condemning the wastegate, check the ignition timing, the spark-retard system, vacuum hoses, knock sensor, oxygen sensor, and computer. If the wastegate appears not to be operating properly (too much or too little boost), check to make sure that the connecting linkage is operating smoothly and is not binding. Also make sure that the pressure-sensing hose is clear and properly connected.

CAUTION!

When removing carbon deposits from turbine and wastegate parts, never use a hard metal tool or sandpaper. Remember that any gouges or scratches on these parts can cause severe vibration or damage to the turbocharger. To clean these parts, use a soft brush and a solvent.

COMMON TURBOCHARGER PROBLEMS With proper care and servicing, a turbocharger will provide years of reliable service. Most turbocharger failures are caused by lack of lubrication, ingestion of foreign objects, or contamination of the lubricant **(Figure 33–27)**.

Replacing a Turbocharger

If the turbocharger is faulty, it should be replaced with a new or rebuilt unit. Always follow the procedure given in the service information. Before removing the turbocharger, plug the intake and exhaust ports and the oil inlet to prevent the entry of dirt or other foreign material. While replacing it, check for an accumulation of sludge in the oil pipes and, if necessary, clean or replace the oil pipes. Do not drop or bang the new unit against anything or grasp it by easily deformed parts.

When installing a new turbocharger, put 20 cc (0.68 oz.) of oil into the turbocharger oil inlet and turn the compressor wheel by hand several times to spread oil to the bearings.

After installing the new turbocharger, or after starting an engine that has been sitting for a long period of time, there can be a considerable delay after engine start-up before the oil pressure is sufficient to deliver oil to the turbo's bearings. To prevent this problem, follow these simple steps:

1. Make certain that the oil inlet and drain lines are clean before connecting them.
2. Be sure the engine oil is clean and at the proper level.
3. Fill the oil filter with clean oil.
4. Leave the oil drain line disconnected at the turbo, and crank the engine without starting it, until oil flows out of the turbo drain port.
5. Disconnect the fuel pump fuse or relay, and crank the engine for 30 seconds to distribute oil throughout the engine.
6. Allow the engine to idle for 60 seconds.
7. Connect the drain line, start the engine, and operate it at low idle for a few minutes before running it at higher speeds.

FIGURE 33–27

Damaged turbocharger wheels.

Maintenance

Turbochargers require no maintenance; however, the engine's oil and filter must be changed on a regular basis. The units operate at high speeds and high temperatures. Poor lubrication will cause the turbo to self-destruct. Also, the high heat breaks down the oil, and therefore its service life is shorter than it would be in a conventional engine. Manufacturers typically recommend that a specific type of oil be used in these engines. After the oil and filter have been changed, the engine should run at idle for at least 30 seconds. Doing this allows oil to circulate through the turbo.

In addition to frequent oil changes, the air cleaner and filter assembly needs to be maintained. This includes making sure that there are no air leaks in the system. The slightest amount of dirt entering the turbo can damage its turbine and compressor wheels. PCV valves and filters also need to be maintained on a regular basis.

Turbo Start-up and Shutdown

The number-one cause of turbo failure is poor lubrication. Because its bearings are not well-lubricated immediately after the engine is started, allow the engine to idle for some time before putting it under a load. If the engine has not been run for a day or more, start the engine and allow it to idle for three to five minutes to prevent oil starvation to the turbo. Engine lube systems have a tendency to bleed down. When the engine has been sitting for a long time, it is wise to crank the engine without starting it, until a steady oil pressure reading is observed. This is called priming the lubricating system. A turbocharged engine should never be operated under load if the engine has less than 207 kPa (30 psi) of oil pressure. The same starting procedure should be followed in cold weather. The thick engine oil will take a longer period to flow. Low oil pressure and slow oil delivery during engine starting can destroy the bearings in a turbocharger.

When the engine has been operated at high speeds or heavy loads, the turbine wheel has been exposed to very hot exhaust gases. This heat is transferred to the turbine's shaft and the compressor wheels. If the engine is turned off immediately after high-speed driving, oil and coolant to the turbo unit will immediately stop. The shaft, however, will still spin due to its inertia. Poor lubrication at this point will destroy the bearings and shaft. Therefore, the engine should idle for 20 to 120 seconds after a hard run to allow the shaft to cool. When the engine is idling, exhaust temperatures drop to 300°C to 400°C (573°F to 752°F). The cooler exhaust gases help cool the shaft and its lubricating oil.

SUPERCHARGERS

Today, some engines have a supercharger as an alternative to the turbocharger **(Figure 33–28)**. **Superchargers** are positive displacement air pumps driven directly by the engine's crankshaft via a V-ribbed belt **(Figure 33–29)**. They improve horsepower and torque by increasing air pressure and density in the intake manifold. The pressure boost is proportional to engine speed.

FIGURE 33–28

A supercharger on a late-model high-performance engine.

FIGURE 33–29

Superchargers are driven by a drive belt.

A typical supercharger is normally made up of a magnetic clutch, two rotors, two shafts, two rotor gears, housing, and a rear plate and cover **(Figure 33–30)**. The drive belt connects the engine to a pulley connected to one rotor shaft **(Figure 33–31)**. Gears connect the two rotor shafts and drive them in opposite directions **(Figure 33–32)**. The rotors have three helical lobes. The lobes are press-fit onto the rotor shafts and then held in position by pins and serrations. The rotors turn within a sealed housing and pressurize the air as they rotate. The rotors' shafts are supported by ball or needle bearings in the rear plate and are lubricated by oil in the supercharger unit.

FIGURE 33–30

The rotors, front plate, and drive gears for a typical supercharger.

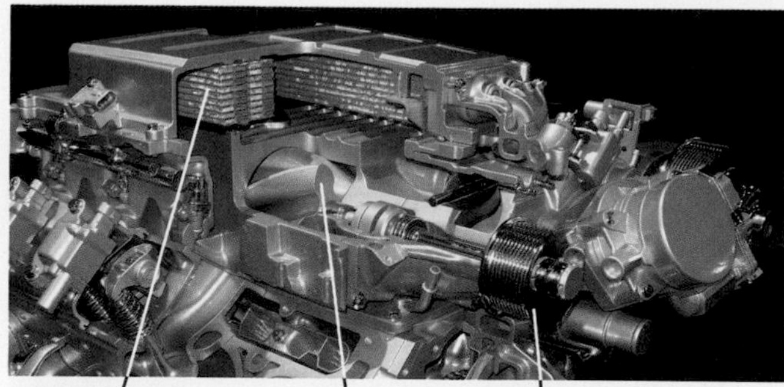

FIGURE 33–31

The drive pulley is attached to a single driveshaft.

To handle the higher operating temperatures imposed by supercharging, the engine is typically fitted with an engine oil cooler. This water-to-oil cooler is generally mounted between the engine front cover and oil filter.

Supercharger Operation

Figure 33–33 illustrates the flow of the air through a supercharger system. The air comes in through the remote-mounted air cleaner and the mass airflow meter. It then moves through the throttle plate assembly and passes through the supercharger inlet plenum assembly, which is bolted to the back of the supercharger.

FIGURE 33–32

Gears transfer the motion of the driveshaft to both rotors.

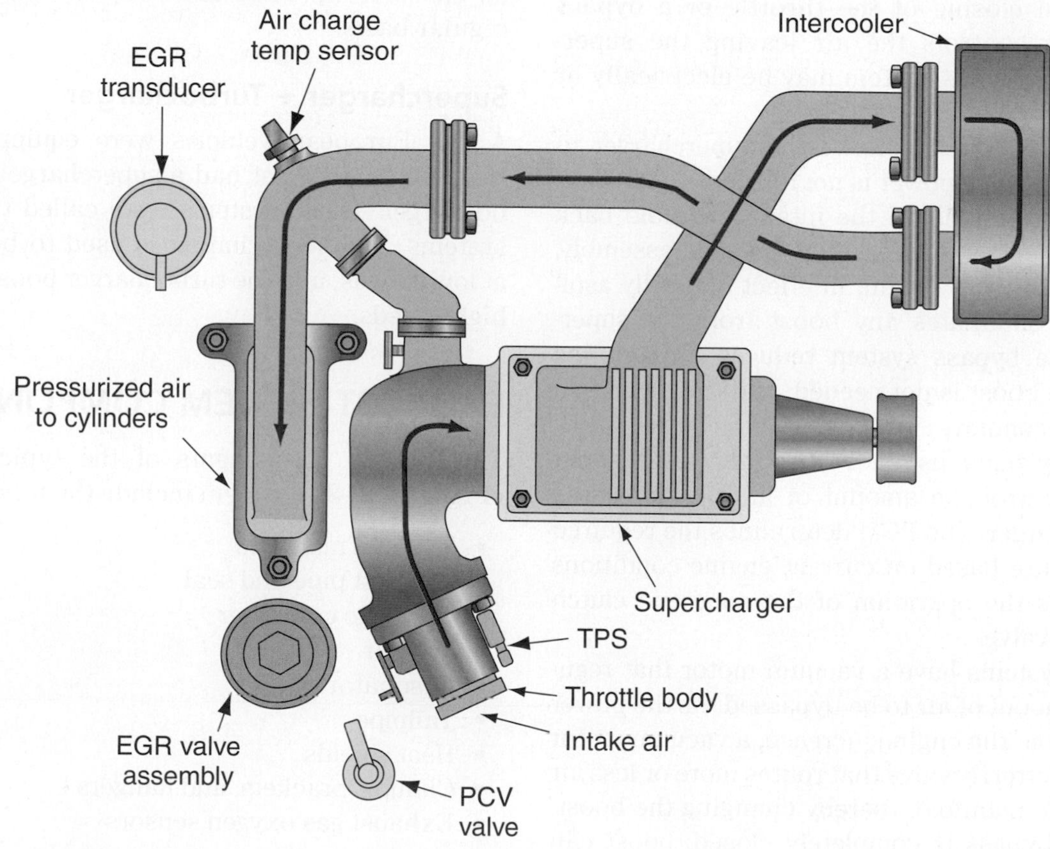

FIGURE 33–33

Airflow through a supercharger assembly into the engine.

The air enters the supercharger and is pressurized by the spinning rotors. It then exits through the top of the supercharger by way of the air outlet adapter. As the air is compressed, its temperature increases. Because cooler, denser air is desired for increased power, the heated air is routed through an intercooler. An intercooler can decrease the temperature of the air by 66°C (150°F).

This cooled air then passes through to the intake manifold adapter assembly, which is bolted to the rear of the intake manifold. When the intake valves open, the air is forced into the combustion chambers, where it is mixed with fuel delivered by the fuel injectors.

Supercharger Bypass System

Unlike a turbocharger, the supercharger does not require a wastegate to limit boost and prevent a potentially damaging overboost condition. Because the speed of the supercharger is directly linked to the engine speed, its pumping power is limited by the rpm of the engine itself rather than revolutions produced by exhaust gases. Supercharger boost is therefore directly controlled through the opening and closing of the throttle or a bypass system that controls the air leaving the supercharger. The bypass system may be electrically or vacuum controlled.

The bypass circuit allows the supercharger to idle when the extra power is not needed. The bypass routes any excess air in the intake manifold back through the supercharger inlet plenum assembly, allowing the engine to run, in effect naturally aspirated. This eliminates any boost from the supercharger. The bypass system reduces air-handling losses when boost is not needed, and this results in better fuel economy.

Some systems use a PCM-controlled stepper motor to control the amount of air that bypasses the supercharger. The PCM determines the required boost pressure based on current engine conditions and controls the operation of the magnetic clutch and bypass valve.

Other systems have a vacuum motor that regulates the amount of air to be bypassed. As the power demands from the engine increase, a vacuum motor controls a butterfly valve that routes more or less air to the intake manifold, thereby changing the boost. When this bypass is completely closed, boost can reach about 83 kPa (12 psi). When the actuator is open, during high-vacuum engine conditions, the air bypasses the supercharger. As the throttle is opened and engine vacuum decreases, the actuator closes and allows more air into the supercharger.

Supercharger Problems

Many of the problems and their remedies given for turbochargers hold good for superchargers. There are also problems associated specifically with the supercharger. Refer to the service information for the symptoms of supercharger failure and a summary of the causes and the recommended repairs.

USING SERVICE INFORMATION

Procedures for servicing supercharger systems are normally in a separate section in the service information.

Maintenance

A supercharger normally has its own oil supply and requires special oil for lubrication. Its oil level must be checked and corrected periodically. Also, the unit ought to be inspected regularly for oil leaks. Like a turbocharger, any dirt in the intake air can destroy the unit. The induction system must be checked for leaks and the filter changed on a regular basis.

Supercharger + Turbocharger

A few European vehicles were equipped with a gasoline engine that had a supercharger and a turbocharger. These systems were called twincharger systems. The supercharger is used to boost power at low speeds, and the turbocharger boosts power at high speeds.

EXHAUST SYSTEM COMPONENTS

The various components of the typical exhaust **(Figure 33–34)** system include the following:

- Exhaust manifold
- Exhaust pipe and seal
- Catalytic converter
- Muffler
- Resonator
- Tailpipe
- Heat shields
- Clamps, brackets, and hangers
- Exhaust gas oxygen sensors

All the parts of the system are designed to conform to the available space of the vehicle's undercarriage and yet be a safe distance above the road.

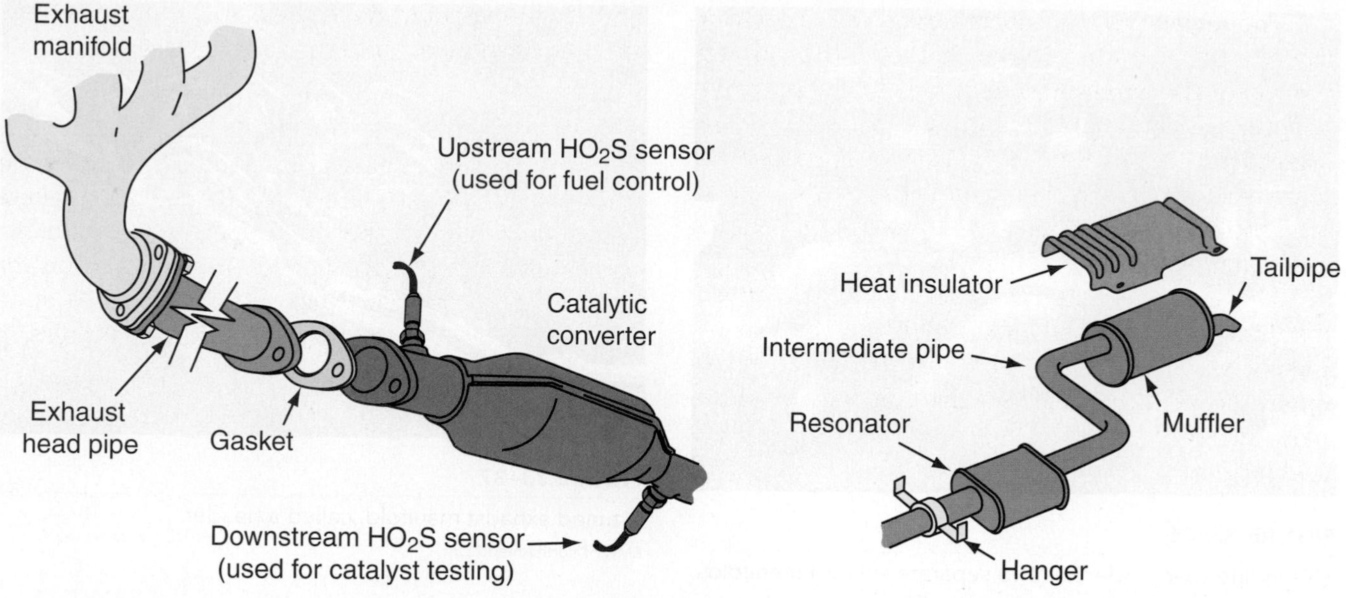

FIGURE 33–34

An exhaust system for a late-model vehicle.

CAUTION!

When inspecting or working on the exhaust system, remember that its components get very hot when the engine is running. Contact with them could cause a severe burn. Also, always wear safety glasses or goggles when working under a vehicle.

Exhaust Manifold

The exhaust manifold **(Figure 33–35)** collects the burnt gases as they are expelled from the cylinders and directs them to the exhaust pipe. Exhaust manifolds for most vehicles are made of cast or nodular iron. Many newer vehicles have stamped,

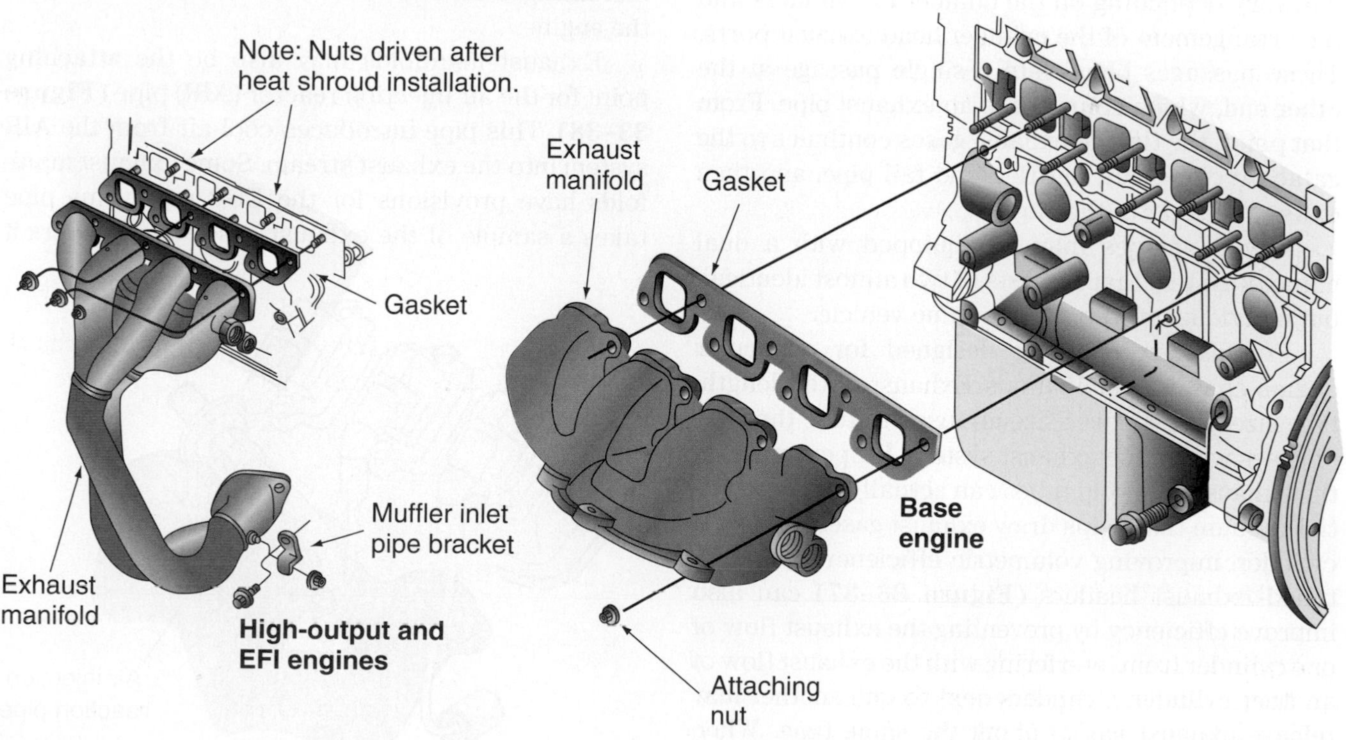

FIGURE 33–35

Two examples of exhaust manifolds.

FIGURE 33–36

This in-line six-cylinder has two separate exhaust manifolds.

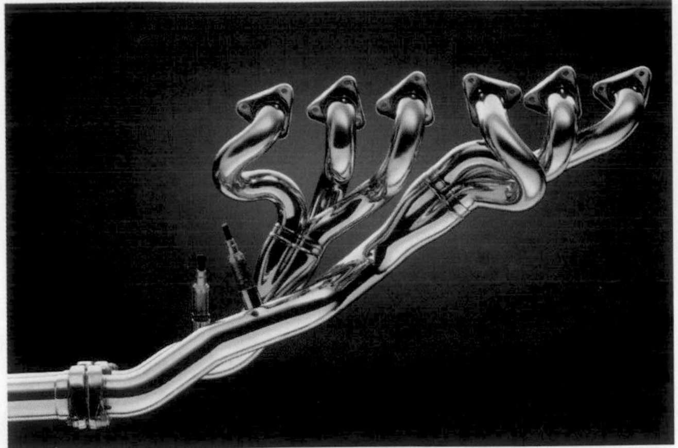

FIGURE 33–37

A tuned exhaust manifold, called a header.

BMW of North America, Inc.

heavy-gauge sheet metal or stainless steel units. A few engines incorporate the exhaust manifold into the design of the cylinder head, such as the Pentastar V6 from Chrysler.

Most in-line engines have one exhaust manifold, though two manifolds are common on BMW inline six-cylinder engines **(Figure 33–36)**. V-type engines have an exhaust manifold on each side of the engine. The numbers of passages into the exhaust manifold will vary depending on the number of cylinders and the arrangement of the cylinder head exhaust ports. These passages blend into a single passage at the other end, which connects to an exhaust pipe. From that point, the flow of exhaust gases continues to the catalytic converter, muffler, and tail pipe, and then exits at the rear of the car.

V-type engines may be equipped with a dual exhaust system that consists of two almost identical, but individual systems in the same vehicle.

Exhaust systems are designed for particular engine-chassis combinations. Exhaust system length, pipe size, and silencer size are used to tune the flow of gases within the exhaust system. Proper tuning of the exhaust manifold tubes can actually create a partial vacuum that helps draw exhaust gases out of the cylinder, improving volumetric efficiency. Separate, tuned exhaust headers **(Figure 33–37)** can also improve efficiency by preventing the exhaust flow of one cylinder from interfering with the exhaust flow of another cylinder. Cylinders next to one another may release exhaust gas at about the same time. When this happens, the pressure of the exhaust gas from one cylinder can interfere with the flow from the other cylinder. With separate headers, the cylinders are isolated from one another; interference is eliminated; and the engine breathes better.

Perhaps the largest performance gain from using an exhaust header is that it increases the engine's volumetric efficiency. A low pressure is present in the exhaust each time a pulse of exhaust ends. A header uses this low pressure to pull exhaust gases out of the cylinder when the exhaust valve opens. The low pressure also helps draw more air into the cylinder during valve overlap. Enhancing exhaust flow and increasing intake flow increases the efficiency of the engine.

Exhaust manifolds may also be the attaching point for the air injection reactor (AIR) pipe **(Figure 33–38)**. This pipe introduces cool air from the AIR system into the exhaust stream. Some exhaust manifolds have provisions for the EGR pipe. This pipe takes a sample of the exhaust gases and delivers it

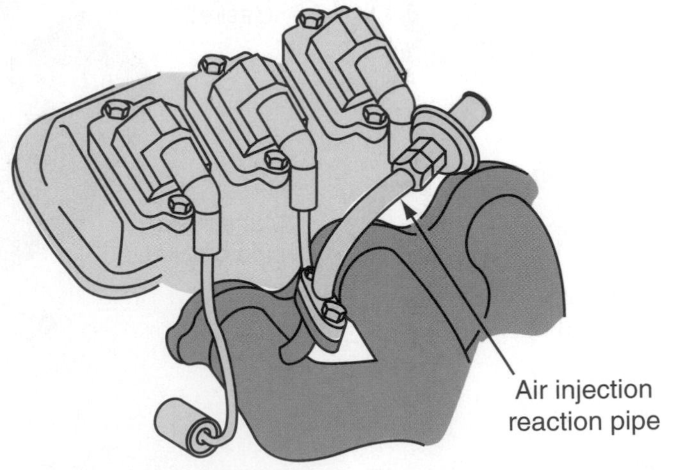

FIGURE 33–38

An AIR pipe mounting on an exhaust manifold.

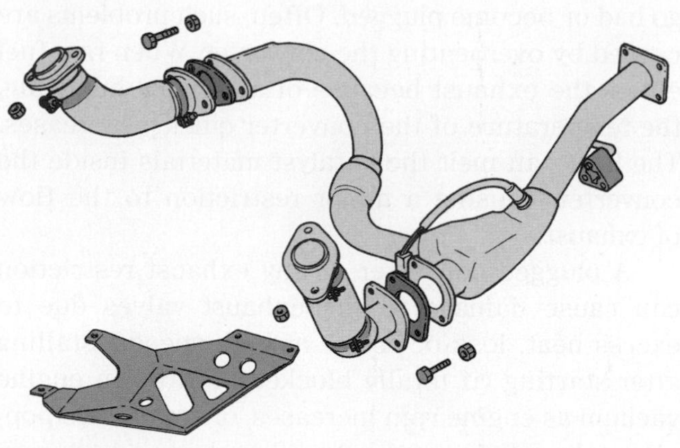

FIGURE 33–39

The front exhaust pipe assembly for a V6 engine.

American Isuzu Motors Inc.

to the EGR valve. Also, exhaust manifolds have a tapped bore that retains the oxygen sensor.

Exhaust Pipe and Seal

The exhaust pipe is a metal pipe—either aluminized steel, stainless steel, or zinc-plated heavy-gauge steel—that runs under the vehicle between the exhaust manifold and the catalytic converter **(Figure 33–39)**.

SHOP TALK

The exhaust manifold gasket seals the joint between the head and exhaust manifold. Even engines that are assembled without exhaust manifold gaskets may require one once the manifold is removed and replaced. Exhaust manifolds go through many heating/cooling cycles. This causes stress and some corrosion in the exhaust manifold. Removing the manifold usually distorts the manifold slightly so that it is no longer flat enough to seal without a gasket. Exhaust manifold gaskets are normally used to eliminate leaks when exhaust manifolds are reinstalled.

Catalytic Converters

A **catalytic converter** is part of the exhaust system and a very important part of the emission control system. Because it is part of both systems, it has a role in both. As an emission control device, it is responsible for converting undesirable exhaust gases into harmless gases. As part of the exhaust system, it helps reduce the noise level of the exhaust. A catalytic converter contains a ceramic element coated with a catalyst. A catalyst is a substance that causes a chemical reaction in other elements without actually becoming part of the chemical change and without being used up or consumed in the process.

Earlier catalytic converters were pellet type, whereas current ones are monolithic. A pellet-type converter contains a bed made from hundreds of small beads. Exhaust gases pass over this bed. In a monolithic-type converter, the exhaust gases pass through a honeycomb ceramic block. The converter beads or ceramic block are coated with a thin coating of cerium, platinum, palladium, and/or rhodium, and are held in a stainless steel container. These elements are used alone or in combination with each other to change the undesirable emissions into harmless compounds.

Since the late 1980s, vehicles have had a three-way converter (TWC) that treats all three controlled-emission gases. It oxidizes HC and CO by adding oxygen and reduces NO_x by removing oxygen from the nitrogen oxides. Diesel engines have a particulate oxidizer catalytic converter **(Figure 33–40)** that collects and cleans the particulates from diesel fuel that would normally be emitted from a diesel engine as black smoke.

Some vehicles are equipped with a minicatalytic converter that is either built into the exhaust manifold or is located next to it **(Figure 33–41)**. These converters are used to clean the exhaust during engine warm-up and are commonly called warm-up converters.

Catalytic converters may have an air tube connected from the AIR system to the oxidizing catalyst. This air helps the converter work by making extra oxygen available. The air from the AIR system is not always forced into the converter; rather, it is controlled by the vehicle's PCM. Fresh air added to the exhaust at the wrong time could overheat the converter and produce NO_x, something the converter is trying to destroy.

FIGURE 33–40

A particulate oxidizer catalytic converter.

©Cengage Learning 2015

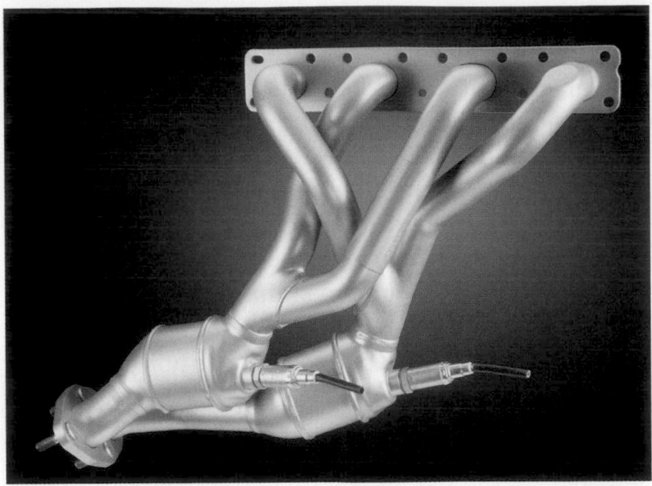

FIGURE 33–41

This exhaust manifold has two separate mini-, or warm-up, catalytic converters.

BMW of North America, Inc.

OBD-II regulations call for a way to inform the driver that the vehicle's converter has a problem and may be ineffective. The PCM monitors the activity of the converter by comparing the signals of an HO_2S located at the front of the converter with the signals from an HO_2S located at the rear **(Figure 33–42)**. If the sensors' outputs are the same, the converter is not working properly, and the MIL on the dash will light.

CONVERTER PROBLEMS The converter is normally a trouble-free emission control device; however, it can go bad or become plugged. Often, such problems are caused by overheating the converter. When raw fuel enters the exhaust because of an engine misfiring, the temperature of the converter quickly increases. The heat can melt the catalyst materials inside the converter, causing a major restriction to the flow of exhaust.

A plugged converter or any exhaust restriction can cause damage to the exhaust valves due to excess heat, loss of power at high speeds, stalling after starting (if totally blocked), a drop in engine vacuum as engine rpm increases, or sometimes popping or backfiring at the throttle valve.

The best way to determine if a catalytic converter is working is to check the quality of the exhaust. This is done with a five-gas exhaust analyzer. The results of this test should show low emission levels if the converter is working properly.

Another way to test a converter is to use a handheld digital **pyrometer**, an electronic device that measures heat. By aiming the infrared laser pointer to the exhaust pipe just ahead of and just behind the converter, it is possible to read an increase of at least 37.7°C (100°F) as the exhaust gases pass through the converter. If the outlet temperature is the same or lower, nothing is happening inside the converter. This means the converter should be replaced.

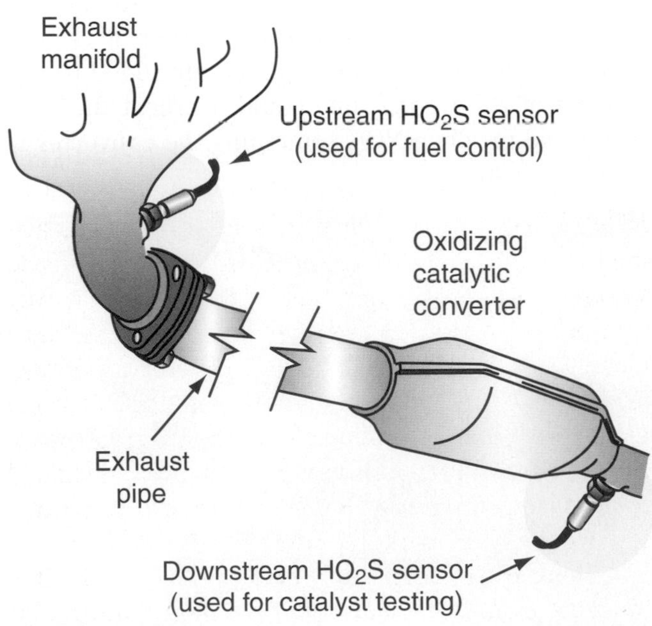

FIGURE 33–42

The basic configuration of an OBD-II exhaust manifold and pipe with its oxygen sensors and catalytic converter.

CAUTION!

To properly perform this test, one spark plug lead should be disconnected and the engine run for a very short period of time. This will ensure that the catalytic convertor is "working" and will not be replaced unnecessarily. Codes will be set and must be erased.

If there is only a slight difference in temperature, check the activity of the oxygen sensor before condemning the converter. The efficiency of today's converters depends on the normal swings of rich and lean mixtures. A biased O_2S can affect converter activity. If the O_2S is working fine, the converter should be replaced. Information about further testing of a catalytic converter is included in Chapter 34.

Mufflers

The **muffler** is a cylindrical or oval-shaped component, generally about 60 cm (2 ft.) long, mounted in the exhaust system at about the midway point of the car or toward the rear of the car. Inside the muffler is a series of baffles, chambers, tubes, and holes to break up, cancel out, or silence the pressure pulsations that occur each time an exhaust valve opens.

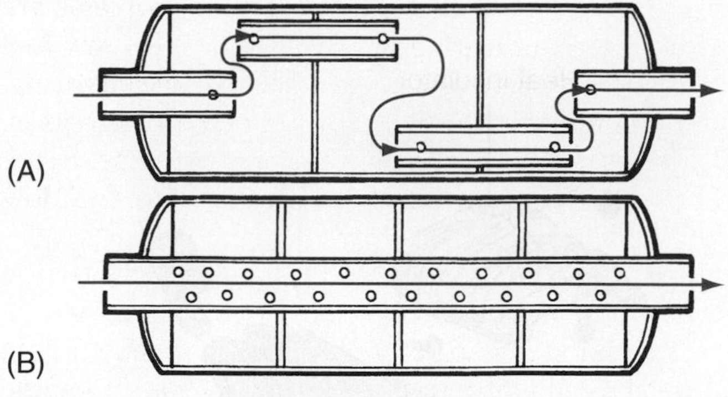

FIGURE 33–43

(A) A reverse-flow muffler, and (B) a straight-through muffler.

Two types of mufflers are frequently used on passenger vehicles **(Figure 33–43)**. Reverse-flow mufflers change the direction of the exhaust gas flow through the inside of the unit. This is the most common type of automotive muffler. Straight-through mufflers permit exhaust gases to pass through a single tube. The tube has perforations that tend to break up pressure pulsations. They are not as quiet as the reverse-flow type.

In recent years, there have been several important changes in the design of mufflers. Most of these changes have been centred at reducing weight and emissions, improving fuel economy, and simplifying assembly. These changes include the following:

- *New materials.* More and more mufflers are being made of aluminized and stainless steel. Using these materials reduces the weight of the units and extends their lives.
- *Double-wall design.* This design is primarily used to retain heat, which can minimize the amount of water that is in the system. Also, many cars use a double-wall exhaust pipe to better contain the sound and reduce pipe ring.
- *Rear-mounted muffler.* More and more often, the only space left under the car for the muffler is at the very rear. This means the muffler runs cooler than before and is more easily damaged by condensation in the exhaust system. This moisture, combined with nitrogen and sulphur oxides in the exhaust gas, forms acids that rot the muffler from the inside out. Many mufflers are being produced with drain holes drilled into them.
- *Back pressure.* Even a well-designed muffler produces some **back pressure** in the system. Back pressure reduces an engine's volumetric efficiency, or ability to "breathe." Excessive back pressure caused by defects in a muffler or other exhaust system part can slow or stop the engine. However, a small amount of back pressure can be used intentionally to allow a slower passage of exhaust gases through the catalytic converter. This slower passage results in more-complete conversion to less harmful gases. Also, no back pressure may allow intake gases to enter the exhaust.

Resonator

On some vehicles, there is an additional muffler, known as a **resonator** or silencer. This unit is designed to further reduce or change the sound level of the exhaust. It is located toward the end of the system and generally looks like a smaller, rounder version of a muffler. These are also used on some vehicles that employ cylinder deactivation to reduce exhaust noise.

Tailpipe

The **tailpipe** is the last pipe in the exhaust system. It releases the exhaust fumes into the atmosphere beyond the back end of the car.

Heat Shields

Heat shields are used to protect other parts from the heat of the exhaust system and the catalytic converter **(Figure 33–44)**. They are usually made of pressed or perforated sheet metal. Heat shields trap the heat in the exhaust system, which has a direct effect on maintaining exhaust gas velocity.

Clamps, Brackets, and Hangers

Clamps, brackets, and hangers are used to properly join and support the various parts of the exhaust system. These parts also help to isolate exhaust noise by preventing its transfer through the frame **(Figure 33–45)** or body to the passenger compartment. Clamps help to secure exhaust system parts to one another. The pipes are formed in such a way that one slips inside the other. This design makes a close fit. A U-type clamp usually holds this connection tight **(Figure 33–46)**. Another important job of clamps and brackets is to hold pipes to the bottom of the vehicle. Clamps and brackets must be designed to allow the exhaust system to vibrate without transferring the vibrations through the car.

There are many different types of flexible hangers available, each designed for a particular application. Some exhaust systems are supported by doughnut-shaped rubber rings between hooks on the exhaust component and on the frame or car body. Others are supported at the exhaust pipe and

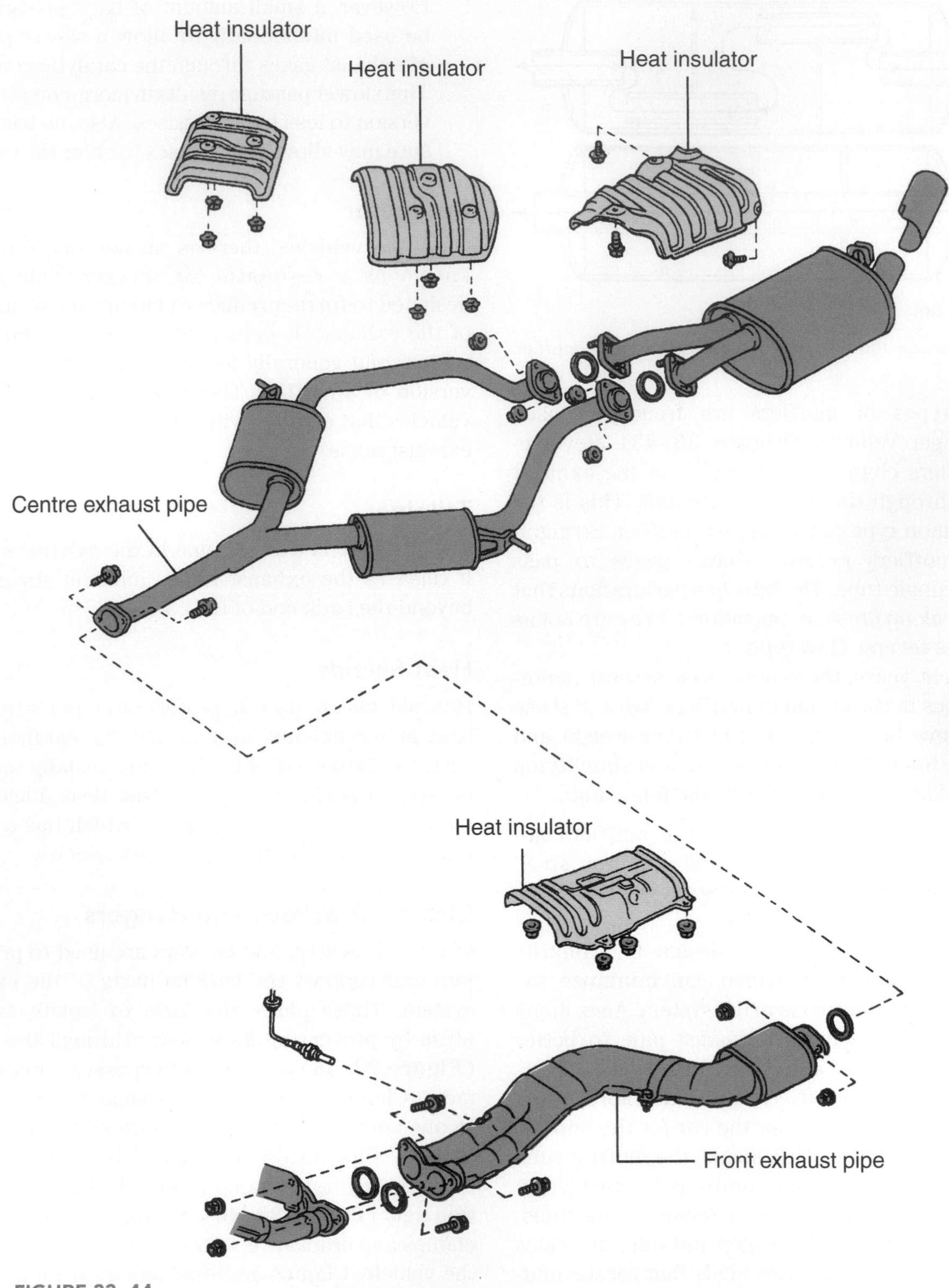

FIGURE 33–44

The typical location of heat shields in an exhaust system.

Reprinted with permission.

tailpipe connections by a combination of metal and reinforced fabric hanger. Both the doughnuts and the reinforced fabric allow the exhaust system to vibrate without breakage that could be caused by direct physical connection to the vehicle's frame.

Some exhaust systems are single units in which the pieces are welded together by the factory. By welding instead of clamping the assembly together, car makers save the weight of overlapping joints as well as that of clamps.

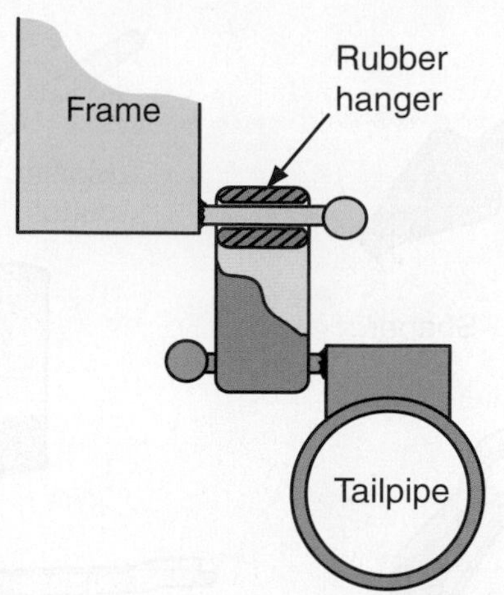

FIGURE 33–45

Rubber hangers are used to keep the exhaust system in place without allowing it to contact this pickup's frame.

EXHAUST SYSTEM SERVICE

Exhaust system components are subject to both physical and chemical damage. Any physical damage to an exhaust system part that causes a partially restricted or blocked exhaust system usually results in loss of power or creates backfiring up through the intake manifold. In addition to improper engine operation, a blocked or restricted exhaust system may cause increased air pollution. Leaks in the exhaust system caused by either physical or chemical (rust) damage could result in illness, asphyxiation, or even death. Remember that vehicle exhaust fumes can be very dangerous to one's health.

Exhaust System Inspection

Most parts of the exhaust system, particularly the exhaust pipe, muffler, and tailpipe, are subject to rust, corrosion, and cracking. Broken or loose clamps and hangers can allow parts to separate or hit the road as the car moves.

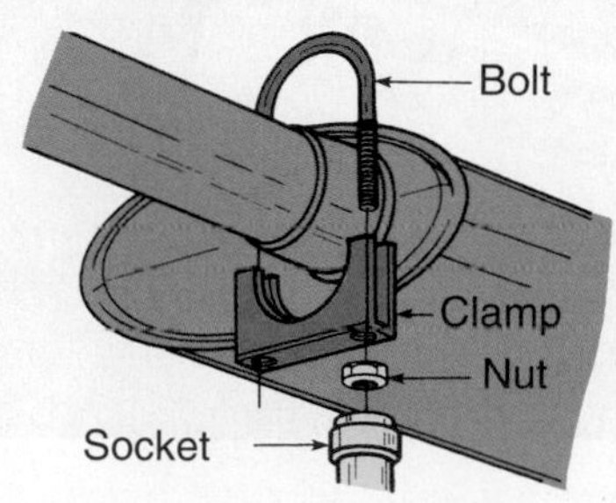

FIGURE 33–46

A U-type clamp is often used to secure two pipes that slip together.

CAUTION!

During all exhaust inspection and repair work, wear safety glasses or equivalent eye protection.

Complete exhaust system inspection should include listening for hissing or rumbling that would result from a leak in the system. An on-lift inspection should pinpoint any of the following types of damage:

- Holes, road damage, separated connections, and bulging muffler seams
- Kinks and dents
- Discoloration, rust, soft corroded metal, and so forth
- Torn, broken, or missing hangers and clamps
- Loose tailpipes or other components
- Bluish or brownish catalytic converter shell, which indicates overheating

EXHAUST RESTRICTION TEST Often, leaks and rattles are the only things looked for in an exhaust system. The exhaust system should also be tested for blockage and restrictions. Collapsed pipes or clogged converters and/or mufflers can cause these blockages.

There are many ways to check for a restricted exhaust, the most common of which is the use of a vacuum gauge. Connect a vacuum gauge to an intake manifold vacuum source. Bring the engine to a moderate speed, and hold it there. Watch the vacuum gauge. If everything is right, the vacuum reading will be high and will either stay at that reading or increase slightly as the engine runs at this speed. If the exhaust is restricted, the vacuum will begin to decrease after a period of time. This is caused by the cylinder's inability to purge itself of all of its exhaust gases during the exhaust stroke. The presence of exhaust in the cylinder when the intake stroke begins will decrease the amount of vacuum that can be formed on that stroke.

Replacing Exhaust System Components

Before beginning work on an exhaust system, make sure it is cool to the touch. In Canada, the under-car environment is harsh on exhaust systems. The hot, oxidized components are susceptible to accelerated rust buildup internally and externally. The road salt used on many Canadian roads eats away at the exhaust components from the outside, and condensation forming internally in cold winter weather increases the deterioration rate of most exhaust systems.

The usual selection of exhaust tools are not commonly used for removal. Most exhaust components

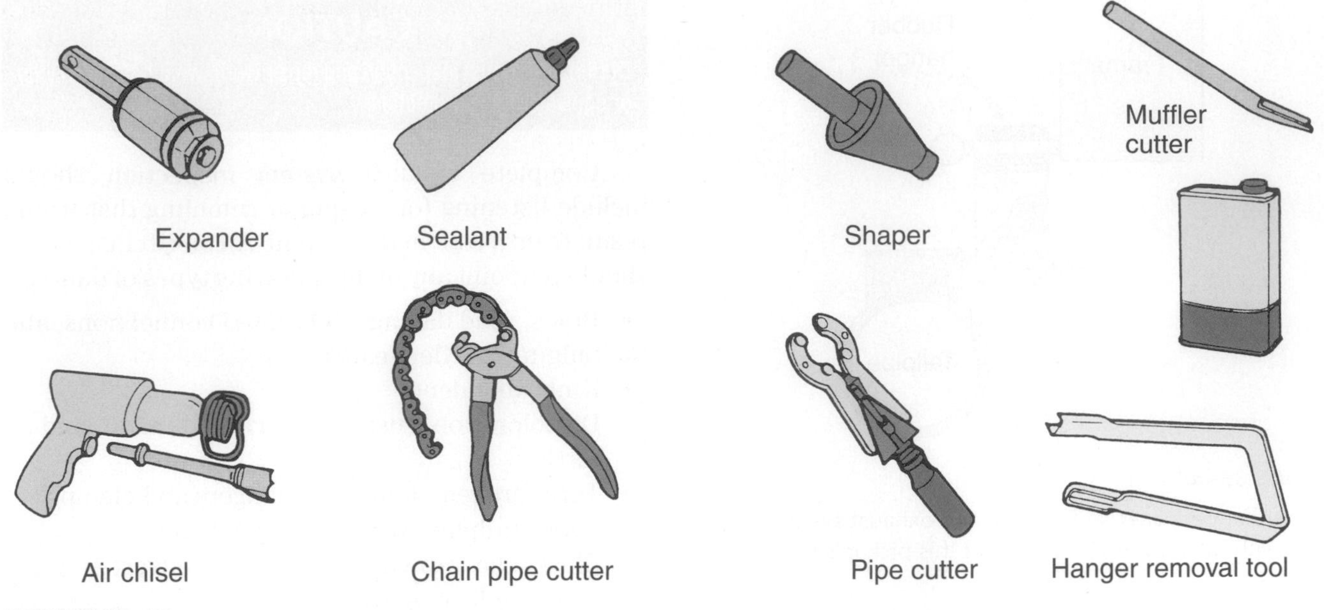

FIGURE 33–47

Special tools required for exhaust work.

are removed with the use of air-powered cutting tools and heating and cutting torches due to rust seizure and badly rusted fasteners. If the exhaust system is not very old and removal is necessary for access to other components, soak all rusted bolts, nuts, and other removable parts with a good penetrating oil. Finally, check the system for critical clearance points so that they can be maintained when new components are installed.

Most exhaust work involves the replacement of parts. When replacing exhaust parts, make sure the new parts are exact replacements for the original parts. Doing this will ensure proper fit and alignment, as well as ensure acceptable noise levels.

Exhaust system component replacement might require the use of special tools **(Figure 33–47)** and welding equipment. If it is necessary to weld exhaust components together, disconnect the battery negative cable to prevent damage to electronic components.

A new tool that often makes exhaust work easier is the induction heater. This tool uses high-frequency magnetic fields to excite and heat ferrous metal parts quickly and safely without a flame **(Figure 33–48)**.

EXHAUST MANIFOLD AND EXHAUST PIPE SERVICING As mentioned, the manifold itself rarely causes any problems. On occasion, an exhaust manifold will warp because of excess heat. A straightedge and feeler gauge can be used to check the machined surface of the manifold.

Another problem—also the result of high temperatures generated by the engine—is a cracked manifold. This usually occurs after the car passes through a large puddle and cold water splashes on the manifold's hot surface. If the manifold is warped beyond manufacturer's specifications or is cracked, it must be replaced. Also, check the exhaust pipe for signs of collapse. If there is damage, repair it. These repairs should be done as directed in the vehicle's service information.

REPLACING LEAKING GASKETS AND SEALS The most likely spot to find leaking gaskets and seals is between the exhaust manifold and the exhaust pipe **(Figure 33–49)**.

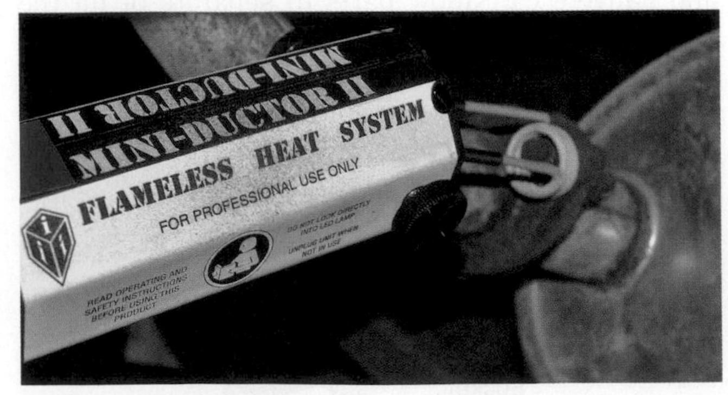

FIGURE 33–48

An inductive heater can be used to heat bolts for easier removal.

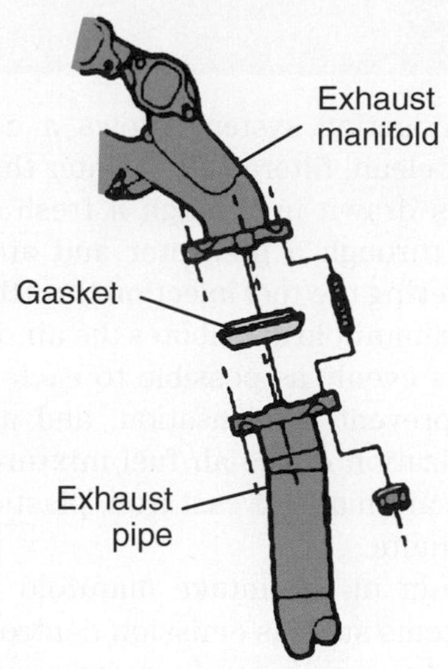

FIGURE 33–49

Leaking gaskets and seals are often found between the exhaust manifold and pipe.

When installing exhaust gaskets, carefully follow the recommendations on the gasket package label and instruction forms. Read through all installation steps before beginning. Take note of any of the original equipment manufacturer's recommendations in service information that could affect engine sealing. Manifolds warp more easily if an attempt is made to remove them while they are still hot. Remember, heat expands metal, making assembly bolts more difficult to remove and easier to break.

To replace an exhaust manifold gasket, follow the torque sequence in reverse to loosen each bolt. Repeat the process to remove the bolts. Doing this minimizes the chance that components will warp.

Any debris left on the sealing surfaces increases the chance of leaks. A good gasket remover will quickly soften the old gasket debris and adhesive for quick removal. Carefully remove the softened pieces with a scraper and a wire brush. Be sure to use a nonmetallic scraper when attempting to remove gasket material from aluminum surfaces.

Inspect the manifold for irregularities that might cause leaks, such as gouges, scratches, or cracks. Replace it if it is cracked or badly warped. File down any imperfections to ensure proper sealing of the manifold.

Due to high heat conditions, it is important to retap and redie all threaded bolt holes, studs, and mounting bolts. This procedure ensures tight, balanced clamping forces on the gasket. Lubricate the threads with a good high-temperature antiseize lubricant. Use a small amount of contact adhesive to hold the gasket in place. Align the gasket properly before the adhesive dries. Allow the adhesive to dry completely before proceeding with manifold installation.

Install the bolts finger-tight. Tighten the bolts in three steps—one-half, three-quarters, and full torque—following the torque tables in the service information or gasket manufacturer's instructions. Torquing is usually begun in the centre of the manifold, working outward in an X pattern.

REPLACING EXHAUST PIPES To replace a damaged exhaust pipe, begin by supporting the converter to keep it from falling. Carefully remove the oxygen sensor if there is one. Remove any hangers or clamps holding the exhaust pipe to the frame. Unbolt the flange holding the exhaust pipe to the exhaust manifold. When removing the exhaust pipe, check to see if there is a gasket. If so, discard it and replace it with a new one. Once the joint has been taken apart, the gasket loses its effectiveness. Disconnect the pipe from the converter, and pull the front exhaust pipe loose and remove it.

SHOP TALK

An easy way to break off rusted nuts is to tighten them instead of loosening them. Sometimes, a badly rusted clamp or hanger strap will snap off with ease. Sometimes, the old exhaust system will not drop free of the body because a large part is in the way, such as the rear drive axle or the transmission support. Use a large cold chisel, pipe cutter, hacksaw, muffler cutter, or chain cutter to cut the old system at convenient points to make the exhaust assembly smaller.

Although most exhaust systems use flanges or a slip joint and clamps to fasten the pipe to the muffler, a few use a welded connection. If the vehicle's system is welded, cut the pipe at the joint with a hacksaw or pipe cutter. The new pipe need not be welded to the muffler. An adapter, available with the pipe, can be used instead. When measuring the length for the new pipe, allow at least 50.8 mm (2 in.) for the adapter to enter the muffler.

CAUTION!

Be sure no exhaust part comes into direct contact with any section of the body, fuel lines, fuel tank, or brake lines.

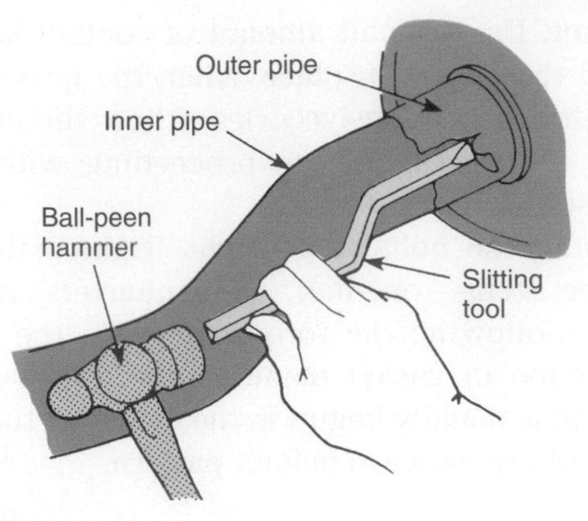

FIGURE 33–50

Removing a rusted-on muffler.

When trying to replace a part in the exhaust system, you may run into parts that are rusted together. This is especially a problem when a pipe slips into another pipe or the muffler. If you are trying to reuse one of the parts, you should carefully use a cold chisel or slitting tool **(Figure 33–50)** on the outer pipe of the rusted union. You must be careful when doing this because you can easily damage the inner pipe. It must be perfectly round to form a seal with a new pipe.

Slide the new pipe over the old. Position the rest of the exhaust system so that all clearances are evident and the parts aligned; then put a U-clamp over the new outer pipe to secure the connection.

CAUTION!

Be sure to wear safety goggles to protect your eyes and work gloves to prevent cutting your hands on the rusted parts.

KEY TERMS

Back pressure (p. 1009)
Catalytic converter (p. 1007)
Intake manifold runner control (IMRC) (p. 992)
Intake manifold tuning (IMT) (p. 992)
Intercooler (p. 994)
Muffler (p. 1008)
Pyrometer (p. 1008)
Resonator (p. 1009)
Superchargers (p. 1002)
Tailpipe (p. 1009)
Turbochargers (p. 996)
Turbo lag (p. 997)
Wastegate (p. 997)

SUMMARY

- The air induction system allows a controlled amount of clean, filtered air to enter the engine. Cool air is drawn in through a fresh air tube. It passes through a preheater and air cleaner before entering the fuel injection throttle body.
- The intake manifold distributes the air or air/fuel mixture as evenly as possible to each cylinder, helps to prevent condensation, and assists in the vaporization of the air/fuel mixture. Intake manifolds are made of cast iron, plastic, or die-cast aluminum.
- The vacuum in the intake manifold operates many systems such as emission controls, brake boosters, heater/air conditioners, cruise controls, and more. Vacuum is applied through an elaborate system of hoses, tubes, and relays. A diagram of emission system vacuum hose routing is located on the underhood decal. Loss of vacuum can create many driveability problems.
- A vehicle's exhaust system carries away gases from the passenger compartment, cleans the exhaust emissions, and muffles the sound of the engine. Its components include the exhaust manifold, exhaust pipe, catalytic converter, muffler, resonator, tailpipe, heat shields, clamps, brackets, and hangers.
- The exhaust manifold is a bank of pipes that collects the burned gases as they are expelled from the cylinders and directs them to the exhaust pipe. Engines with all the cylinders in a row have one exhaust manifold. V-type engines have an exhaust manifold on each side of the engine. The exhaust pipe runs between the exhaust manifold and the catalytic converter.
- The muffler consists of a series of baffles, chambers, tubes, and holes to break up, cancel out, and silence pressure pulsations. Two types commonly used are the reverse-flow and the straight-through mufflers.
- The tailpipe is the end of the pipeline carrying exhaust fumes to the atmosphere beyond the back end of the car. Heat shields protect vehicle parts from exhaust system heat. Clamps, brackets, and hangers join and support exhaust system components.
- Exhaust system components are subject to both physical and chemical damage. The exhaust can

be checked by listening for leaks and by visual inspection. Most exhaust system servicing involves the replacement of parts.

- The turbocharger relies on the rapid expansion of hot exhaust gases exiting the cylinder to spin turbine blades, which compresses the intake air.
- A typical turbocharger consists of a turbine (or hot wheel), shaft, compressor (or cold wheel), turbine housing, compressor housing, and centre housing and rotating assembly. A wastegate manages turbo output by controlling the amount of exhaust gas that is allowed to enter the turbine housing. Turbo boost is the positive pressure increase created by a turbocharger.
- Most turbochargers are lubricated by pressurized and filtered engine oil that is line-fed to the unit's oil inlet. Some turbocharged engines are equipped with an intercooler, which is designed to cool the compressed air from the turbocharger.
- To control detonation, most turbocharger systems use knock-sensing devices and an electronic control unit to operate the wastegate control valve through sensor signals.
- If the turbo sound cycles or changes in intensity, the likely causes are a plugged air cleaner or loose material in the compressor inlet ducts, or dirt buildup on the compressor wheel and housing. Most turbocharger failures are caused by one of the following reasons: lack of lubricant, ingestion of foreign objects, or contamination of lubricant. Turbo lag occurs when the turbocharger is unable to meet the immediate demands of the engine.
- Superchargers are air pumps connected directly to the crankshaft by a belt. The positive connection yields instant response and pumps air into the engine in direct relationship to crankshaft speed.

REVIEW QUESTIONS

1. What can be used to check for leaks on the pressure side of a turbocharger system?
2. What advantages are there to preheating intake air?
3. Name three purposes of the intake manifold.
4. How can the effectiveness of a catalytic converter be checked?
5. A late-model vehicle has at least _____ catalytic converters in its exhaust system.
6. When is a miniconverter used?
   a. on small engines where a normal converter will not fit properly
   b. on engines that used leaded fuels
   c. in conjunction with EGR systems to supply clean exhaust for the cylinders
   d. to reduce emissions during engine warm-up
7. What controls the movement of the turbocharger wastegate in a non-computer-controlled system?
   a. manifold vacuum
   b. manifold pressure
   c. exhaust back pressure
   d. reversing DC motors
8. Which of the following is a catalyst in a catalytic converter?
   a. palladium
   b. vanadium
   c. aluminum
   d. chromium
9. Which of the following components is not found in the exhaust system?
   a. the muffler
   b. the catalytic converter
   c. the turbocharger
   d. the supercharger
10. In Canada, what is a common tool for exhaust system component removal?
    a. an air ratchet
    b. hammer and chisel
    c. acetylene cutting torch
    d. impact gun
11. At what engine rpm will a typical turbocharger begin to create boost?
    a. 1000 rpm
    b. 3000 rpm
    c. 2000 rpm
    d. 4000 rpm
12. What is the purpose of using an intercooler?
    a. to cool the exhaust gases
    b. to cool the turbocharger housing
    c. to cool the intake air to make it less dense
    d. to cool the intake air to make it more dense
13. Which of the following does not require engine horsepower to produce a boost?
    a. a turbocharger
    b. a straight rotor supercharger
    c. a spiral rotor supercharger
    d. a G-Lader supercharger

14. Which turbocharger component is responsible for forcing air into the intake system?
    a. the compressor wheel
    b. the wastegate
    c. the turbocharger centre housing
    d. the turbine wheel
15. What are modern port fuel injection intake manifolds made of?
    a. die-cast aluminum
    b. cast iron
    c. stainless steel
    d. stamped steel
16. What manages turbo output by controlling the amount of exhaust gas entering the turbine housing?
    a. wastegate
    b. turbine wheel
    c. turbine seal assembly
    d. compressor wheel
17. What is the suggested minimum allowable oil pressure for safe turbocharger operation?
    a. 100 kPa (15 psi)
    b. 310 kPa (45 psi)
    c. 205 kPa (30 psi)
    d. 410 kPa (60 psi)
18. What is the first step in turbocharger inspection?
    a. Check the air cleaner for a dirty element.
    b. Open the turbine housing at both ends.
    c. Start the engine and listen to the system.
    d. Remove the ducting from the air cleaner to turbo and examine the area.
19. Which of the following statements concerning superchargers is correct?
    a. Superchargers must overcome inertia and spin up to speed as the flow of exhaust gas increases.
    b. Superchargers require a wastegate to limit boost.
    c. Superchargers do not use engine power to produce boost.
    d. Superchargers improve horsepower and torque.
20. What is an advantage of tuned exhaust headers?
    a. increases exhaust system back pressure
    b. prevents one cylinder's exhaust flow from affecting others
    c. reduces volumetric efficiency
    d. increases the exhaust system's noise level

CHAPTER 34
Emission Control Systems

LEARNING OUTCOMES

- Explain how hydrocarbon (HC) emissions are released from an engine's exhaust.
- Explain how carbon monoxide (CO) emissions are formed in the combustion chamber.
- Describe how oxides of nitrogen (NO_x) are formed in the combustion chamber.
- Describe how carbon dioxide (CO_2) is formed in the combustion chamber.
- Describe oxygen (O_2) emissions in relation to air/fuel ratio.
- Describe the operation of an evaporative control system during the canister purge and non-purge modes.
- Explain the purpose of the positive crankcase ventilation (PCV) system.
- Describe the operation of the knock sensor and electronic spark control module.
- Describe the purpose and operation of an exhaust gas recirculation (EGR) valve.
- Define the purpose of a catalytic converter.
- Describe the operation of a secondary air injection system.
- Describe the emission controls commonly found on today's light-duty diesel engines.

The emission controls on today's cars and trucks are an integral part of the engine and electronic engine control system. Perhaps it is better to say that the electronic control systems are really emission control systems. The goal by manufacturers to produce cleaner and more fuel-efficient vehicles has led to many of the control systems now in place. These systems have also contributed to significant increases in power and reliability and improved driveability.

POLLUTANTS

Automotive emissions that are of the most concern to environmentalists, engineers, and technicians are hydrocarbon (HC), carbon monoxide (CO), oxides of nitrogen (NO_x), carbon dioxide (CO_2), and oxygen (O_2). These gases contribute to air and water pollution and climate change. The exception to this is O_2, which is not a pollutant. CO_2 emissions are, however, an indication of the efficiency of the combustion process. The O_2 content in an engine's exhaust also is monitored during emissions inspections. Excessive amounts of O_2 in tailpipe emissions may indicate an engine misfire condition or perhaps even a leak, crack, or hole in exhaust pipes, which could dilute the exhaust sample.

An exhaust gas that is not monitored in most emissions testing is sulphur dioxide (SO_2). This is a colourless gas that has a rotten egg smell. It is caused by large amounts of sulphur content in gasoline and is produced by the action of the catalytic converter. SO_2 can cause heart problems, asthma, and other respiratory problems.

HC, NO_x, and CO emissions are caused by many different engine operating conditions. Two of the most important factors are combustion chamber temperatures and air/fuel ratios. It is interesting to note that the conditions that may cause a reduction in one tailpipe emission could cause an increase in another. That is why it is important to understand the relationships, causes, and effects of these various emissions **(Figure 34–1)**.

The allowable amounts of automobile emissions are regulated by the government. Through the years, the maximum allowable amount emitted by an automobile has decreased **(Figure 34–2)**. Before you can understand the purpose of each emission control device and how to diagnose and service them, you must have an understanding of why these gases are emitted by an automobile.

All vehicles for the past 40 or more years have been equipped with devices that reduce the levels of emissions released by the exhaust system. These are commonly referred to as tailpipe emissions. The following discussion of the various gases addresses their formation before they are reduced by the various emission control devices.

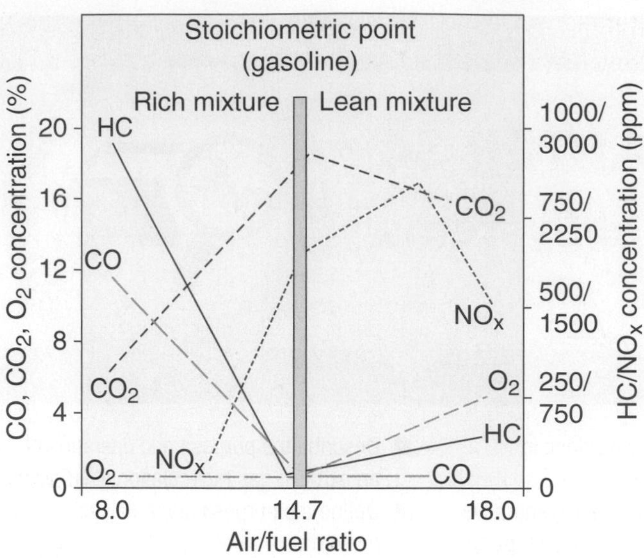

FIGURE 34–1

This chart shows how difficult it is to control exhaust emissions. Notice that NO_x and CO_2 are high when the HCs are low.

Hydrocarbons

Hydrocarbon (HC) exhaust emissions may be caused by incomplete combustion, which may lead to raw, unburned fuel exiting the tailpipe. They can also be caused by excessive amounts of oil entering the combustion chamber due to poor internal engine sealing of the cylinder. Even an engine in good condition with satisfactory ignition and fuel systems will release some HC. Evaporative emissions from the fuel storage and delivery system are also a source of HC emissions.

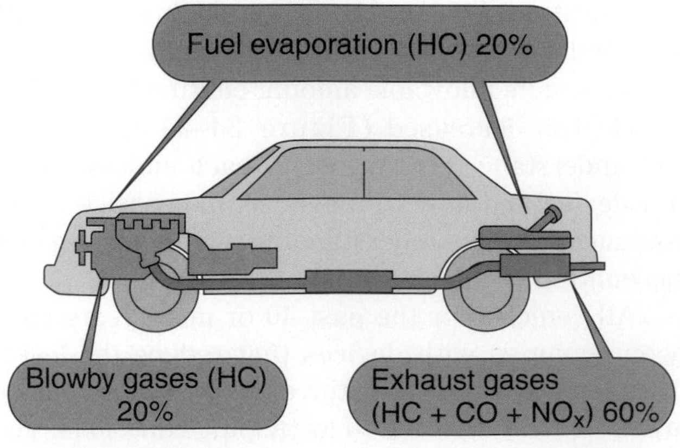

FIGURE 34–2

The main sources of automotive emissions.

HC emissions from the combustion process result when the following occur:

- The air/fuel mixture is compressed into the small sheltered areas of the combustion chamber, where the flame front cannot ignite the fuel. These include the areas formed by the top piston ring and the cylinder wall, any crevices formed by the head gasket, spark plug threads, and the valve seat.
- Fuel is absorbed by the oil on the walls of the cylinder.
- Fuel is absorbed by and/or is contained within carbon deposits in the combustion chamber.
- When the flame front approaches the cooler cylinder wall, the flame front quenches, leaving some unburned HCs.
- Some of the air/fuel mixture is left unburned because the flame front stops before igniting all of the mixture (misfire).
- The fuel does not adequately mix with the air and ignite prior to the end of combustion.
- There is a misfire caused by an ignition system fault.

An excessively lean air/fuel ratio results in cylinder misfiring and high HC emissions. A very rich air/fuel ratio also causes higher than normal HC emissions. At the stoichiometric air/fuel ratio, HC emissions are low.

HC tailpipe emissions can also increase due to engine mechanical faults that may cause excess amounts of engine oil to enter the combustion chamber. Some examples would be as follows:

- Worn valve seals and guides that allow oil to leak past into the combustion chamber
- Worn piston oil control rings that allow engine oil to bypass the rings into the combustion chamber
- Worn engine bearings that allow excess oil throw-off onto the cylinder walls.

Carbon Monoxide

Carbon monoxide (CO) is a by-product of combustion. CO is a poisonous chemical compound of carbon and oxygen. CO is a colourless, odourless, and highly poisonous gas that can cause dizziness, headaches, impaired thinking, and death by O_2 starvation. It forms in the engine when there is not enough O_2 to combine with the carbon during combustion. When there is enough O_2 in the mixture, carbon dioxide (CO_2) is formed. CO_2 is used by plants to manufacture oxygen. CO is primarily found in the exhaust but can also be in the crankcase.

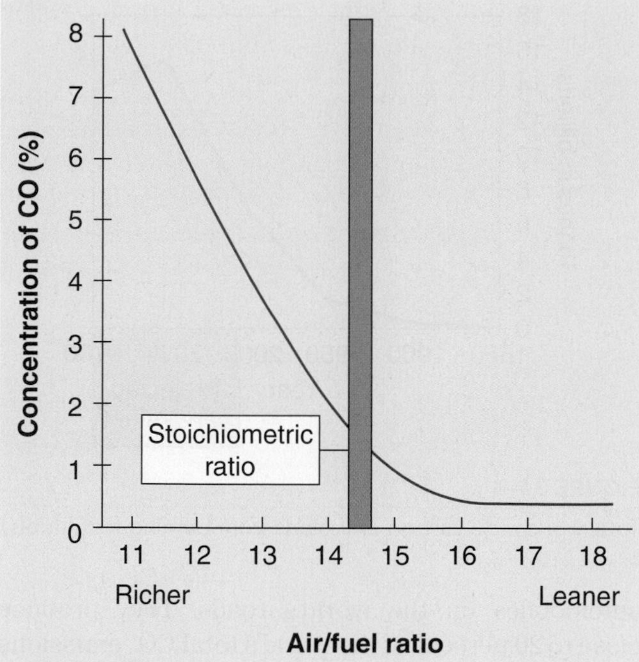

FIGURE 34–3

The amount of CO in the exhaust varies with the air/fuel mixture.

CO emissions are caused by a lack of air or too much fuel in the air/fuel mixture. CO will not occur if combustion does not take place in the cylinders; therefore, the presence of CO means combustion is taking place. As the air/fuel ratio becomes richer, the CO levels increase **(Figure 34–3)**. At the stoichiometric air/fuel ratio, the CO emissions are very low. If the air/fuel ratio is leaner than stoichiometric, the CO emissions remain very low. Therefore, CO emissions are a good indicator of a rich air/fuel ratio, but they are not an accurate indication of a lean air/fuel ratio.

Nitrogen Oxides

Nitrogen oxides (NO_x) are the various compounds of nitrogen and O_2 formed during the combustion process. Both nitrogen and O_2 are present in the air used for combustion. Exposure to NO_x can cause respiratory problems, such as lung irritation, bronchitis, or pneumonia. Photo-chemical smog results when HCs and NO_x are combined with sunlight. Smog appears as brownish ground-level haze. Smog can cause many health problems, including chest pains, shortness of breath, coughing and wheezing, and eye irritation. When NO_x in the atmosphere mixes with rain (H_2O), nitric acid (HNO_3) or acid rain is created.

The formation of NO_x is the result of high combustion temperatures and pressures. When combustion temperatures reach more than 1261°C (2300°F), the N and the O_2 in the air begin to combine and form NO_x.

SHOP TALK

It is often said that NO_x forms at 1371°C (2500°F); studies have shown that the formation of NO_x actually begins at around 1261°C (2300°F). The actual temperature is not really important; what is important is the thought that keeping combustion temperatures low keeps NO_x low.

Because outside air is 78 percent N_2, the gas cannot be prevented from entering the engine. Therefore, the only way to control NO_x is to prevent N_2 from joining with oxygen during the combustion process. This is done by controlling the temperature of the combustion process. This, however, can be very tricky. When the mixture is slightly rich, the combustion temperature drops, and there is less chance for the production of NO_x. However, a rich mixture will also lead to an increase in CO and HC emissions. Likewise, when the mixture is slightly lean, there is less of a chance for CO and HC emissions, but there is a greater chance for the production of NO_x because combustion temperature increases.

The "x" in NO_x stands for the proportion of oxygen mixed with a nitrogen atom. The "x" is a variable, which means it could be the number 1, 2, 3, and so on; therefore, the term *NO_x* refers to many different oxides of nitrogen (NO, NO_2, NO_3, etc.). NO_x emissions from an engine are mostly nitric oxide (NO), with less than 1 percent of the total NO_x being nitrogen dioxide (NO_2). NO is unhealthy and contributes to the greenhouse effect. NO_2 is a very toxic gas and contributes to the formation of smog, ozone, and acid rain.

It is important to note that diesel engines (because combustion temperatures of the fuel are low) produce more NO_2 than gasoline engines. About one-third of the nitrogen converted in a diesel engine becomes NO_2.

Carbon Dioxide

Carbon dioxide (CO_2) is not a pollutant; however, it has been linked to another environmental concern—the greenhouse effect. CO_2 is a **greenhouse gas** and may be the major cause of global warming. It is claimed that 14 percent of all CO_2 emissions in North America are from the exhaust of automobiles, and 27 percent of the CO_2 is caused by all transportation.

From an efficiency standpoint, CO_2 is an ideal by-product of combustion. Therefore, large amounts of

CO_2 in the exhaust are desired. As the air/fuel ratio goes from 9:1 to 14.7:1, the CO_2 levels gradually increase from approximately 6 percent to 13.5 percent. CO_2 levels are highest when the air/fuel ratio is slightly leaner than stoichiometric. The production of CO_2 is directly related to the amount of fuel consumed; therefore, more fuel-efficient engines produce less. Each gallon of gasoline produces about 8.8 kg (19.4 pounds) of CO_2.

To put this in perspective, consider the following:

- A vehicle that achieves 23.5 L/100km (10 mpg) will produce 0.88 kg (1.94 pounds) of CO_2 per 1.6 km (1 mile), which equates to 10 523 kg (11.6 tonnes) a year if operated for a distance of 19 300 km (12 000 miles) per year.
- A vehicle that achieves 11.76 L/100km (20 mpg) will produce 0.44 kg (0.97 pound) of CO_2 per 1.6 km (1 mile), which equates to 5443 kg (6 tonnes) a year if operated for a distance of 19 300 km (12 000 miles) per year.
- A vehicle that achieves 7.84 L/100 km (30 mpg) will produce 0.29 kg (0.65 pound) of CO_2 per 1.6 km (1 mile), which equates to 3538 kg (3.9 tonnes) a year if operated for a distance of 19 300 km (12 000 miles) per year.

To reduce CO_2 levels in the exhaust, engineers are working hard to decrease fuel consumption. This, however, is difficult, because many of the methods used to increase fuel efficiency, such as lean mixtures, increase CO_2 and other emission levels. The concern for CO_2 emissions is one of the primary reasons for the continued exploration of alternative fuel and power sources for automobiles. It is also one of the main reasons the government has raised the CAFC/CAFE mark to 6.72 L/100km (35 mpg). In 2012, the Canadian and U.S. governments passed legislation raising the CAFC/CAFE standard to 4.32 L/100km (54.5 mpg) by 2025.

Other than being emitted from an engine's exhaust or the burning of fossil fuels, CO_2 is also emitted by nature. There are many natural sources of CO_2 such as the oceans and decaying plant life. Plants also use CO_2 in their photosynthesis process, which results in a release of oxygen to the atmosphere. The current concern on global warming suggests that there is more CO_2 in the atmosphere than can be used by plant life. In fact, the more CO_2 molecules there are in the atmosphere, the warmer the earth gets.

To bring the amount of CO_2 in the atmosphere to an acceptable level, we need more plant life, and we need to reduce the amount that is released to the atmosphere. A target for reducing CO_2 is the automobile. There are approximately one billion automobiles on the world's roads. They produce close to 20 percent of the world's total CO_2 emissions from burning fossil fuels **(Figure 34–4)**. The quantity of CO_2 emissions from vehicles is measured in grams per mile (g/mi) or grams per kilometre (g/km).

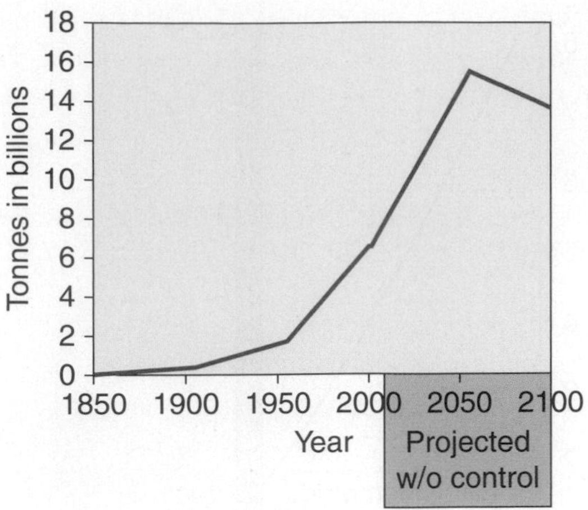

FIGURE 34–4

Total worldwide carbon emissions from burning fossil fuels.

Other substances are also considered greenhouse gases. These include water vapour, methane, and nitrous oxide. Greenhouse gases allow the light from the sun to freely enter the atmosphere. When the sunlight reaches the earth's surface, some of it is absorbed and warms the earth. The rest of the sunlight is reflected back to the atmosphere as heat. Greenhouse gases absorb and trap the heat. This process is considered to be the cause of global warming.

The European Union has standards that all new vehicles must meet. The U.S. Environmental Protection Agency (EPA) and the National Highway Traffic Safety Administration (NHTSA) has had a National Program for the reduction of automobile fuel consumption (therefore CO_2 reduction) since 2012. By 2017, CO_2 emissions are projected to be down to 131.8 grams/kilometre (g/km) and even further reduced to 88.9 g/km by 2025. In Canada, fuel consumption and emission level legislation mirrors that of the United States.

CO_2 REDUCTION AT FACTORIES Not only are manufacturers trying to reduce CO_2 levels emitted by the exhaust, they are also working to reduce the amount of greenhouse gases from their factories. GM, for example, has continued to reduce CO_2 emissions from their facilities over the years. A number of their buildings generate 35 percent fewer greenhouse gas emissions and consume 35 percent less energy than comparable size and use buildings. The strategies to this included tight monitoring and stringent energy control, reducing waste, increasing the use of renewable resources, and increasing the efficiency of the entire manufacturing process.

Oxygen

Oxygen (O_2) is not a pollutant; therefore, its presence in the exhaust does not pose any threat to our environment. However, too much oxygen in the exhaust does indicate that an improper air/fuel mixture or poor combustion has occurred in the engine.

If the air/fuel ratio is rich, all the oxygen in the air is mixed with fuel, and the O_2 levels in the exhaust are very low. When the air/fuel ratio is lean, there is not enough fuel to mix with all the air entering the engine, and O_2 levels in the exhaust are higher. Therefore, O_2 levels are a good indicator of a lean air/fuel ratio, and they are not affected by catalytic converter operation.

Water (H_2O)

H_2O should be emitted from a vehicle's exhaust. It is the result of oxidation, whereby HC is oxidized by the converter to form CO_2 and H_2O. The amount of water emitted depends on a number of things, including the effectiveness of the converter and the amount of HC and O_2 in the exhaust before it passes through the converter. Normally, during cold engine start-up, steam is emitted from the exhaust. The steam results from the heating of the water present in the converter and exhaust system. Much of this water is formed by condensation due to the cooling of the exhaust system.

Diesel Emissions

Diesel engines are the most efficient of all internal combustion engines. They have low fuel-consumption rates and produce low amounts of greenhouse gases. However, there are exhaust emissions that are of concern. Some of these are currently regulated, and others will be in the future. A typical diesel engine emits the following:

- Carbon (soot) and various carbon-based compounds
- NO_x water
- Carbon monoxide
- Sulphur dioxide
- Various hydrocarbon-based compounds

Soot is the most obvious emission and is often referred to as diesel particulates. These particulates are mostly comprised of carbon-based substances that tend to absorb the other contaminants in a diesel engine's exhaust. Many factors influence the amount of soot released, including the fuel used and the engine design. Particulate emissions are the main obstacle for using diesel engines in passenger cars and light- and medium-duty trucks. California and other states have set a standard for particulate emissions. These standards will become more stringent as diesel technology advances. These emissions are measured in grams per kilometre or mile.

EMISSION CONTROL DEVICES

According to the Ontario Ministry of the Environment, 39.5 percent of the NO_x emissions and 26.5 percent of the HC emissions produced in Ontario are from on-road vehicles. According to Environment Canada, passenger vehicles account for 21 percent of the NO_x emissions and 51 percent of the HC emissions produced by the transportation sector. After more than 40 years of emission regulations, these figures remain staggering! Imagine what these figures would be if automotive and industrial emissions had remained unregulated during the past 40 years!

Emission standards have been one of the driving forces behind many of the technological changes in the automotive industry. Catalytic converters and other emission systems were installed to meet emission standards. These standards have become progressively more stringent through the years, and the engine designs and devices required to meet the standards have become very complex.

Legislative History

Canadian vehicle emission regulations have been in effect through various federal agencies since 1971, including Transport Canada and Environment Canada. These regulations have been enforced through Memorandums of Understanding (MOU) with the various automobile manufacturers' associations (both national and international). The Canadian standards have mirrored those of the United States to produce harmonized standards that update with the U.S. regulations.

SHOP TALK

Through the years, vehicles have been built specifically for the State of California. The California Air Resources Board (CARB) has led the way for many emissions laws and standards. Many are specific to the state, although most of these have eventually been implemented by other states, provinces, and countries. Check the labelling on the vehicle to see if the vehicle has special equipment or calibrations designed for use in California and not found in vehicles for other provinces or states.

These vehicle emission regulations have continued to progressively restrict the amount of emissions produced. The new On-Road Vehicle and Engine Emission Regulations came into effect on January 1, 2004. These regulations are under the Canadian Environmental Protection Act (CEPA) created in 1999. The new regulations phased in more stringent standards between 2004 and 2009.

The driving force behind the development of emission control devices has been the various federal clean air acts put into effect over the years. The following are some key acts that have shaped the design of today's vehicles:

- The Clean Air Act of 1963 (U.S.) identified the automobile as a major contributor to air pollution, responsible for as much as 40 percent of all emissions.
- The Clean Air Act Amendments of 1965 (U.S.) required auto manufacturers to install emissions control devices on all passenger cars and light trucks by the 1968 model year.
- The Clean Air Act Amendments of 1970 (U.S.) established nationwide air-quality standards and linked federal building and highway funds to meeting those standards. Areas that did not meet the standards were required to institute a plan to correct the problem or lose the funds.
- The Clean Air Act Amendments of 1977 (U.S.) mandated that areas that did not meet air-quality standards must establish and enforce basic inspection and maintenance (basic I/M) programs for all passenger cars and light trucks. The purpose of the I/M programs was to test and repair the effectiveness of all systems that affected vehicle emissions.
- The Clean Air Act Amendments of 1990 (U.S.) again reduced the allowable amount of exhaust emissions. However, the key part of this act was the required establishment of enhanced inspection and maintenance (enhanced I/M) programs in areas that did not meet air-quality standards.
- Canadian laws and regulations proposed changes to vehicle fuel economy and reducing CO_2 emissions as was done in the United States. There, the Energy Independence and Security Act of 2007 (U.S.) required a 40 percent increase in vehicle fuel economy by the year 2025. It also mandates for increased use of renewable fuels. This act also focused on the reduction of CO_2 emissions. Current Canadian laws mirror those of the United States.

Inspection and Maintenance Programs

Emission standards set the maximum allowable amounts of emissions from a new automobile. These standards have become stricter through time. Amendments to the acts have led some provinces to mandate emission testing of vehicles on the road. These periodic tests are part of an inspection and maintenance (I/M) program. The purpose of these inspections is simply to identify those vehicles that have been tampered with or have not received good maintenance. Studies have shown that 20 percent of the vehicles on the road are not being properly maintained, and those vehicles account for more than 90 percent of the emissions from automobiles. I/M programs are designed to identify these vehicles and make the necessary repairs to allow those vehicles to have acceptable amounts of emissions.

The first U.S. Clean Air Act set emission standards for new cars. To make sure the new vehicles met these standards, the federal test procedure (FTP) was instituted. The test is performed on a random sample of preproduction vehicles. These vehicles are used to represent the vehicles for the next model year. Their emissions are carefully checked and compared to the standards established for that model year. If the emission levels meet or are lower than the standards, the vehicle is then certified.

The FTP uses an inertia weight dynamometer (dyno) that allows the vehicle to be driven under varying loads. The dyno is capable of simulating actual driving conditions that the test vehicle would encounter by changing the load applied to the drive wheels. Emission levels are measured with a constant volume sampling (CVS) system that measures the mass of HC, CO, CO_2, and NO_x emitted from the vehicle in grams per mile or grams per kilometre. These exhaust analyzers are much more precise, complex, and expensive than the exhaust analyzer found in most shops.

The act also prompted Californians to create the California Air Resources Board (CARB). CARB's purpose was to implement strict air standards, which later became U.S. federal standards. One of the approaches made by CARB to clean the air was to start periodic motor vehicle inspection (PMVI) programs. This inspection included a tailpipe emissions test and an underhood inspection. The tailpipe test certifies that the vehicle's exhaust emissions (HC and CO) are within the limits of the law. The emissions were measured in parts per million (ppm) and were taken with the engine at idle. The allowable emissions are three to four times higher than those required for a new vehicle. This provides some tolerance for engine and system wear. The underhood and/or vehicle inspection verifies that the emission control systems have not been tampered with or disconnected. During the 1980s, this basic I/M program was changed to include tests conducted under a load with a dyno. This test was called the accelerated simulated mode (ASM). This test measured CO, HC, and NO_x emissions.

After the implementation of the Clean Air Act of 1990, more precise testing was instituted. The result is called the I/M 240 test. The I/M 240 (or enhanced I/M) tests vehicle emissions while the vehicle is operating under a variety of load conditions and speeds. While on the dyno, the vehicle is operated for up to 240 seconds and under different load conditions **(Figure 34–5)**. The test drive on the dyno simulates both in-traffic and highway driving and stopping, and includes the same conditions as the FTP. However, the I/M 240 test is only a small portion of a complete FTP.

The I/M 240 consists of three separate tests: (1) a transient, mass emission tailpipe test; (2) an evaporative system purge flow test; and (3) an evaporative system pressure test. The test results are given in the same increments as the FTP (grams per kilometre) and therefore can be directly related to the FTP standards. The test also measures HC, CO, CO_2, and NO_x emissions during the I/M 240 drive cycle.

Basically, the same equipment used for the FTP is used for an I/M 240 test. A variable inertia weight dynamometer is used because it can be adjusted to match the weight of the vehicle and allows the vehicle to be driven under a variety of loads. Emission levels are determined by collecting the exhaust in a CVS and analyzing its contents through the use of very sophisticated equipment. This equipment is also similar to those used during the FTP; they include the following:

- Nondispersive infrared (NDIR) tester for measuring CO and CO_2
- Flame ionization detector (FID) for measuring HC
- Chemiluminescence detector for measuring NO_x

HC emissions from a vehicle's EVAP can be higher than the HC emissions in an engine's exhaust. Therefore, monitoring this system is an important part of the I/M 240 test. The I/M 240 test includes a visual inspection of the EVAP system and a purge volume and fuel tank pressure test. The purge test measures the flow of fuel vapours into the engine's intake during the test's drive cycle. The pressure test is used to check for leaks that would allow vapours to be released into the atmosphere.

Another available test is the acceleration simulation mode (ASM) or ASM 2 test. In this test, a dynamometer is used to load the vehicle while an exhaust gas analyzer measures hydrocarbon, carbon monoxide, and nitrogen oxide levels. The ASM 2 test accelerates the vehicle to 25 km/h, using 50 percent of the vehicle's horsepower, and the second test accelerates to 40 km/h, using 25 percent of the vehicle's horsepower.

The two-speed idle (TSI) test is performed in some areas of the country. This test samples exhaust emissions at 2500 rpm and at idle speed.

In addition to emission testing, visual inspections are also performed. The inspections typically include checking for any type of fuel leak, verifying the existence of the vehicle emission control information decal, checking to see that all vacuum hoses are attached, and verifying that the proper emission control devices are installed.

Today, British Columbia and Ontario are the only provinces that have mandatory emission testing. British Columbia's AirCare and Ontario's Drive Clean programs are not province-wide, but apply to areas of high population and vehicle density. These programs use a number of different tests, depending on the type of vehicle and the type of engine used in the vehicle being tested.

Both provinces' programs have used one or a combination of these existing test methods, including the I/M 240, ASM, two-speed idle, or

FIGURE 34–5

Emission levels are commonly checked with the vehicle running on a dyno.

Robert Bosch Corp.

opacity tests. Recently, the AirCare and Drive Clean programs have included an on-board diagnostic system test (OBD Test). This test involves using the vehicle's on-board computer to monitor the emission system and determining if the vehicle is emitting excessive amounts of tailpipe emissions. "Monitors" inside the vehicle PCM must be met under specific driving conditions in order for the PCM to determine that the vehicle's emission components are operating correctly.

Vehicle Emission Control Information (VECI)

All vehicles have a **vehicle emission control information (VECI)** decal that gives specific emission control information for that vehicle and engine (**Figure 34–6**). These decals are normally located on the underside of the hood or on the radiator support frame. The information contained on the VECI is important when conducting an I/M test and when diagnosing or repairing an emissions-related problem. Most of the information contained on the decal is expressed by acronyms; **Figure 34–7** gives some examples of these.

ENGINE/EVAPORATIVE EMISSION SYSTEM INFORMATION Since 1994, all manufacturers must use a standardized system to identify their individual engine and EVAP system families. These names must be 12 characters long and are shown in a box on the VECI. The first 12-character ID contains the size of the engine and its family group. On the second line is another 12-character ID. This identifies the family name of the EVAP system. Both of these names are specific to that vehicle.

BASE ENGINE CALIBRATION INFORMATION Important engine (powertrain) calibration information is normally given in the lower right-hand corner of the vehicle's certification label. The vehicle certification label is typically affixed on the left front door or door post. Base engine calibration information is limited to a maximum of five characters per line and no more than two lines. This coding is used during diagnostics and service. The certification label also contains a coded description of the vehicle.

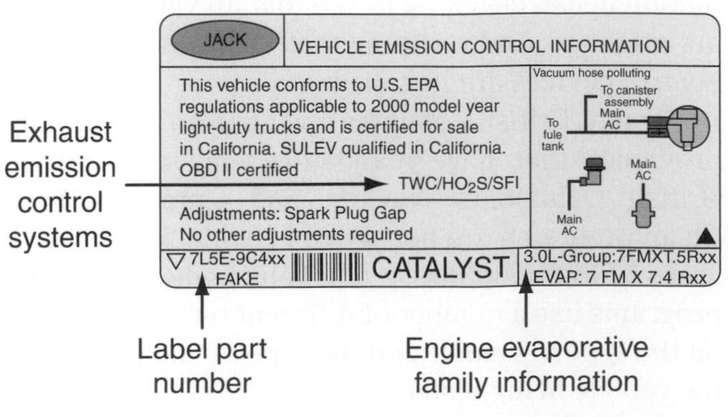

FIGURE 34–6

An example of a VECI.

ACRONYM	DEFINITION
CARB	California Air Resource Board
CI	Cylinder injection
EPA	Environmental Protection Agency
EVAP	Evaporative emissions
GVW	Gross vehicle weight
GVWR	Gross vehicle weight rating, curb weight plus payload
HO_2S	Heated oxygen sensor
ILEV	Inherently low-emission vehicle
LDDT	Light-duty diesel truck categories
LEV	Low-emission vehicle
LVW	Loaded vehicle weight, curb weight plus 136 kg (300 pounds)
MDV	Medium-duty vehicle
MHDDE	Medium heavy-duty diesel engine
MPI	Multiport injection
MY	Model year
NCP	Noncompliance penalty
OBD	On-board diagnostic
ORVR	On-board refuelling vapour recovery
PC	Passenger car
PZEV	Partially zero-emission vehicle
SFI	Sequential fuel injection
SULEV	Super ultra-low-emission vehicle
TWC	Three-way catalyst
ULEV	Ultra-low emission vehicle
ZEV	Zero-emission vehicle

FIGURE 34–7

Vehicle emission control information (VECI) acronym definitions.

Classifications of Emission Control Devices

All emission control systems fall into one of three classifications: evaporative control, pre-combustion control, and post-combustion control.

The **evaporative control (EVAP)** system is a sealed system. It traps the fuel vapours (HC) that would normally escape from the fuel delivery system into the air.

Most pollution control systems used today prevent emissions from being created in the engine, either during or before the combustion cycle. The common **pre-combustion control** systems are the PCV, engine modifications, spark control, and exhaust gas recirculation (EGR) systems.

Post-combustion control systems clean up the exhaust gases after the fuel has been burned. Secondary air or air injector systems put fresh air into the exhaust to reduce HC and CO to harmless water vapour and carbon dioxide by a chemical (thermal) reaction with oxygen in the air. Catalytic converters are the most effective post-combustion emission control; they reduce NO_x, HC, and CO emissions.

EVAPORATIVE EMISSION CONTROL SYSTEMS

Fuel vapours from the gasoline tank and the carburetor float bowl were brought under control with the introduction of EVAP systems. These systems were first installed in 1970 model cars sold in California and in most domestic-made cars beginning with the 1971 models. Through the years, EVAP emissions have been closely monitored and the control systems modified to minimize the chances of vapours entering the atmosphere. EVAP emissions are limited by law, and the current limit in Canada is 2 grams of HC per hour. EVAP emissions are even lower on vehicles classified as CARB LEV II vehicles.

Current systems are computer controlled and are monitored by the OBD-II system. Most current EVAP systems include the following components:

- Domed fuel tank in which its upper portion is raised; fuel vapours rise to this upper portion and collect
- Canister vent solenoid valve
- Special filler design to limit the amount of fuel that can be put in the tank
- Fuel tank vacuum or pressure sensor
- Fuel lines
- Vapour lines
- Fuel tank cap
- Vapour separator in the top of the fuel tank **(Figure 34–8)**
- Charcoal (EVAP) canister
- Purge lines
- Purge solenoid valve (vapour management valve)

GO TO Chapter 27 for a detailed discussion on engine control systems.

Fuel tanks are sealed units and are designed to prevent vapours that result from the evaporation of the gasoline from entering the atmosphere. Also, special devices are used to reduce the amount of gas vapours that escape from the fuel tank while the vehicle is being refuelled. When the fuel cap is removed on all late-model vehicles, the filler neck is sealed with a hinged and spring-loaded flap. The

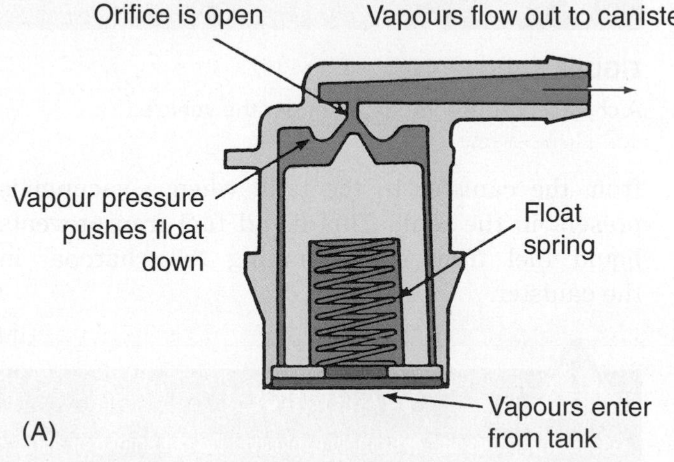

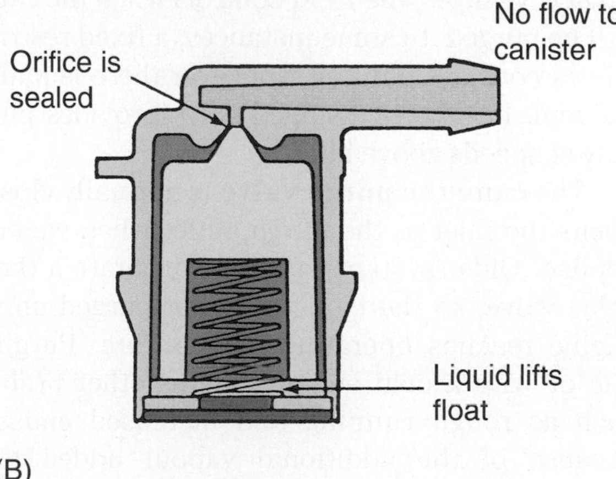

FIGURE 34–8

(A) Normal operation of a vapour separator, (B) with liquid in the separator.

size of the flap is large enough to allow an unleaded fuel nozzle to enter, but too small for a leaded nozzle. Gasoline pump nozzles have also been modified to prevent fuel vapours from entering the atmosphere during refuelling. Some nozzles are equipped with a rubber boot that seals the nozzle to the filler neck. Other nozzles are designed to create a vacuum that draws in vapours as the liquid fuel is being pumped into the tank. Late-model vehicles have an on-board refuelling vapour recovery (ORVR) system, and the special nozzles are redundant and not really needed. ORVR systems seal the filler pipe during refuelling, and fuel vapours are sent to the EVAP canister.

The EVAP system moves the built-up vapours to a canister, where they are stored until the system purges the canister and allows the vapours to enter the intake manifold. The EVAP system also allows some atmospheric pressure to enter the tank. This prevents the buildup of vacuum in the tank that could cause the tank to collapse.

Fuel vapours inside the fuel tank are vented at the top of the tank through the vapour separator. The separator collects droplets of liquid fuel and directs them back into the tank. The vapours leave the separator and move to the canister through the vapour line.

The **charcoal (carbon) canister (Figure 34–9)** is normally located in the engine compartment, but may be under the vehicle close to the fuel tank **(Figure 34–10)**. Fuel vapours from the gas tank are routed to and absorbed onto the surfaces of the canister's charcoal granules. When the vehicle is restarted, vapours are drawn by the vacuum into the intake manifold to be burned by the engine. Canister purging varies widely with vehicle make and model. On new vehicles, the PCM controls when the canister will be purged. In some instances, a fixed restriction allows constant purging whenever there is manifold vacuum. In others, a staged valve provides purging only at speeds above idle.

The **canister purge valve** is normally closed. It opens the inlet to the purge outlet when vacuum is applied. Older systems often incorporate a thermal delay valve, so the canister is not purged until the engine reaches operating temperature. Purging at idle or with a cold engine creates other problems, such as rough running and increased emissions, because of the additional vapour added to the intake manifold.

The canister contains a liquid fuel trap that collects any liquid fuel entering the canister. Condensed fuel vapour forms liquid fuel. This liquid is returned from the canister to the tank when a vacuum is present in the tank. This liquid fuel trap prevents liquid fuel from contaminating the charcoal in the canister.

FIGURE 34–9

A charcoal canister.

FIGURE 34–10

A charcoal canister mounted under the vehicle.

WARNING!

Gasoline vapours are extremely explosive! Do not smoke or allow sources of ignition near any component of the EVAP system. Explosion of gasoline vapours may result in property damage or personal injury.

Early EVAP systems were not controlled by the PCM, but instead relied on a ported vacuum purge port, a vacuum check valve, and a thermo vacuum valve (TVV). The latter was used to prevent purging when the engine was cold. The canister was purged whenever the throttle plate was open enough to expose the vacuum port to engine vacuum. The vacuum then opened the check valve to allow vapours to move to the intake manifold. The check valve also served to keep the system sealed when the engine was not running. The amount of purged vapours was controlled by a fixed orifice. This meant that any time the controlling port had vacuum, the same amount of fuel vapours was sent to the engine, regardless of engine load or speed. This resulted in driveability problems during some conditions because the air/fuel mixture became too rich.

To gain more control of canister purging, the EVAP operation is now controlled by the PCM. These systems use a purge solenoid valve that is duty-cycled by the PCM. The solenoid controls the vacuum to the canister, therefore controlling the amount of vapours purged. The purge valve is open and controlled only when the system is in closed loop, and then it opens only when the engine and conditions can respond to the extra enrichment of the vapours. Factors that can determine when purging takes place include engine coolant temperature and vehicle speed.

PCM-controlled EVAP systems allow for precise control of purge flow and vapour volume. Because the system responds to current engine and operating conditions, the purging of the vapours does not affect driveability.

Enhanced Evaporative Emission (EVAP) Systems

To meet OBD-II regulations, late-model vehicles have an enhanced EVAP system. Enhanced EVAP systems operate in much the same way as previous PCM-controlled EVAP systems, but they also conduct tests that can detect small, 0.5 mm (0.020 in.) system leaks and monitor canister purge flow. The tests are run only when certain conditions are present. These conditions vary with make, model, and engine type.

These systems have a fuel tank pressure (FTP) sensor and EVAP canister vent (CV) solenoid in addition to typical EVAP components **(Figure 34–11)**.

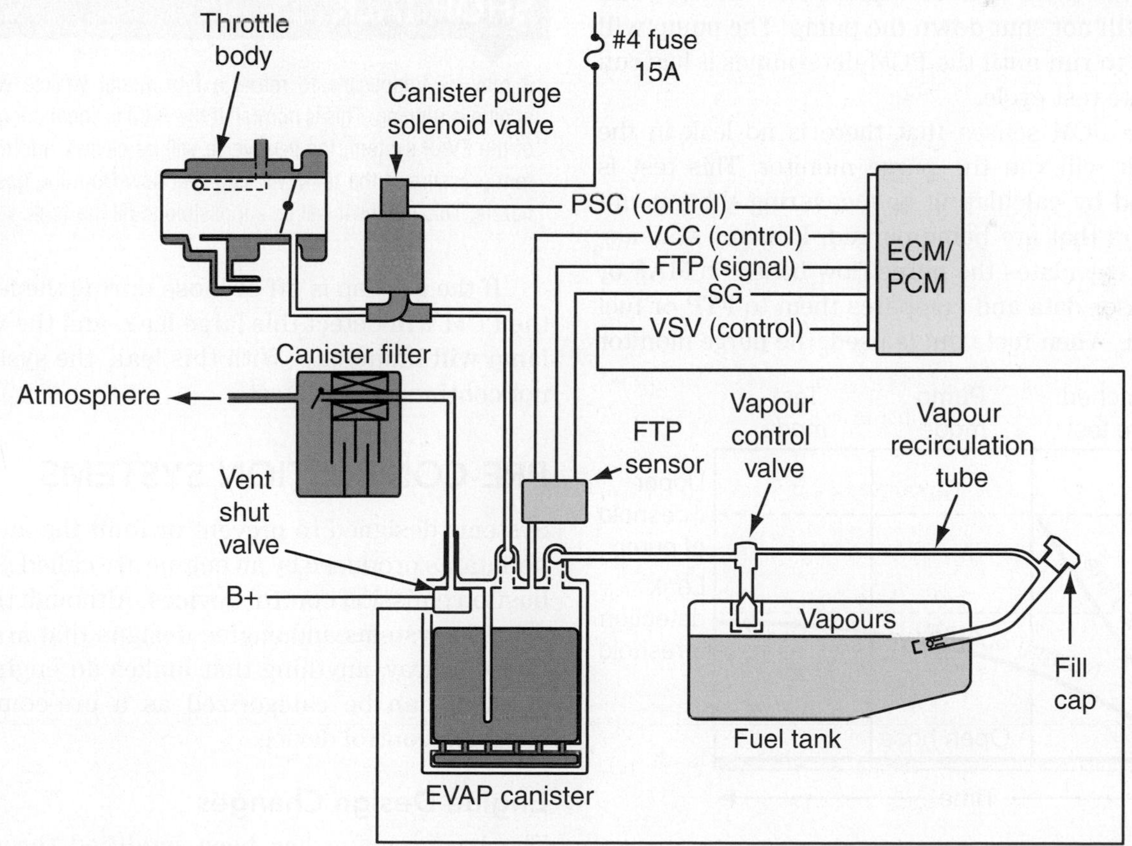

FIGURE 34–11

A diagram of an enhanced EVAP system.

Some systems are also equipped with a pump often called the leak detection pump (LDP). The CV solenoid seals the charcoal canister from the atmosphere when the system is conducting the leak check monitor. The FTP sensor measures pressure or vacuum in the fuel tank and compares it to atmospheric pressure. This input is also used to check for pinched or restricted vapour lines **(Figure 34–12)**. A vacuum is created in the system during the leak test; the FTP sensor measures the vacuum at the beginning of the test and again after a fixed period. If vacuum cannot be built into the system, or if the system will not hold a vacuum, a leak is evident. A signal fuel level input (FLI) sensor is used during the leak check to determine how much fuel is in the tank. This determines how long it should take to build a vacuum in the tank. The FLI input is also used to determine if the EVAP and other monitors can run. Too low or too high of a fuel level will not allow the monitors to run.

On systems with an LDP, the PCM turns on the pump when it is checking the system. The pump pressurizes the system. As pressure builds, the cycling rate of the pump decreases. If there is no leak in the system, the pressure will build until the pump shuts off. If there is a leak, pressure will not build up and will not shut down the pump. The pump will continue to run until the PCM determines it has run a complete test cycle.

If the PCM senses that there is no leak in the system, it will run the purge monitor. This test is completed by calculating or measuring the amount of vapours that are being purged. In many systems, the PCM calculates the purge flow based on MAF or MAP sensor data and compares them to FTP or fuel trim data. When fuel trim is used, the purge monitor

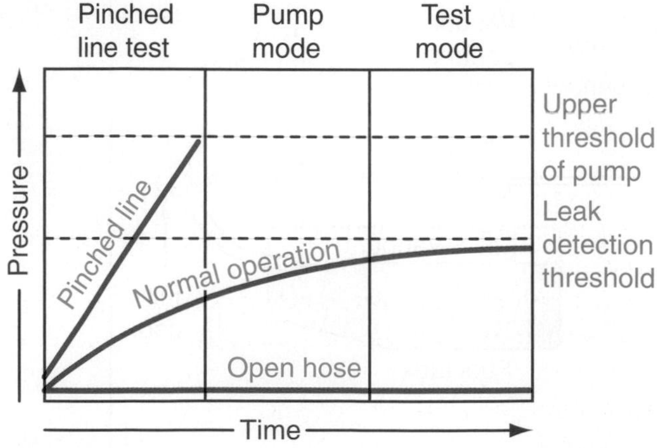

FIGURE 34–12

The rate of pressure buildup determines if the system has pinched lines or a leak, or if it is operating normally.

determines if the purge system is functioning properly by applying a long duty cycle to the purge solenoid. The subsequent change in short-term fuel trim (STFT) is monitored. If the system is working correctly as the duty cycle of the solenoid is increased, the STFT should change proportionately. This is the most commonly used method for checking the operation of the purge system.

Other systems use a purge flow sensor connected to a vacuum hose between the purge solenoid and the intake manifold. The PCM monitors the signal from the sensor once per drive cycle to determine if there is vapour flow or no vapour flow through the solenoid to the intake manifold. Other EVAP systems have a vapour management valve connected in the vacuum hose between the canister and the intake manifold. The vapour management valve is a normally closed valve. The PCM operates the valve to control vapour flow from the canister to the intake manifold. The PCM also monitors the valve's operation to determine if the EVAP system is purging vapours properly. If no leaks are present and the purge cycle is correct, the system has passed the test.

SHOP TALK

It may be impossible to refuel a late-model vehicle when the engine is running. This is normal! If the PCM is conducting a check of the EVAP system, the vent valve will be closed, and the resultant pressure in the tank will stop fuel flow from the gas pump's nozzle. This means it will be impossible to fill the tank.

If the gas cap is off or loose during the leak test, the PCM will detect this large leak, and the warning lamp will illuminate. With this leak, the system will not continue its leak test.

PRE-COMBUSTION SYSTEMS

Systems designed to prevent or limit the amount of pollutants produced by an engine are called pre-combustion emission control devices. Although there are specific systems and engine designs that are classified this way, anything that makes an engine more efficient can be categorized as a pre-combustion emission control device.

Engine Design Changes

The basic engine has been modified through the years to increase its overall efficiency. Many of the changes have occurred inside the engine. Others involve the fuel and ignition systems. The result of these changes is improved performance and driveability and a decrease in exhaust emissions. Following is a summary of some of those changes:

- *Better sealing pistons.* Blowby gases are reduced through the use of better sealing piston rings and improved cylinder wall surfaces. Many engines are also fitted with hypereutectic pistons and low-friction piston rings. This increases fuel economy and engine power. In addition, the piston rings are located higher up on the piston, reducing the space for the mixture to get trapped between the piston and cylinder wall.
- *Combustion chamber designs.* The primary goal in designing combustion chambers is the reduction or elimination of the quench area. Another trend in combustion chamber design is locating the spark plug closer to the centre of the chamber. Manufacturers have also worked with designs that cause controlled turbulence in the chamber. This turbulence improves the mixing of the fuel with the air, which improves combustion.
- *Lower compression.* For some engine designs, by keeping the compression ratio low, combustion temperatures can be kept below the point where NO_x is formed. However, new developments have allowed the use of higher compression ratios on some high-performance engines.
- *Higher compression.* Many newer engines are using high compression to extract more power out of combustion. This is possible with improved designs for pistons and exhaust systems, and by using gasoline direct injection to create a stratified air/fuel charge near the spark plug.
- *Decreased friction.* Overcoming the friction of the engine's moving parts reduces the power and energy lost during engine operation. Improved engine oils, new component materials, and weight reductions have had the biggest impact on reducing friction.
- *Intake manifold designs.* Thanks to the wide and successful use of port fuel injection, intake manifolds are designed to distribute equal amounts of air to each cylinder. The use of plastic intake manifolds has allowed for smoother runners and better heat control of the air.
- *Improved cooling systems.* High engine temperatures reduce HC and CO emissions. However, they also make the formation of NO_x harder to control. Engine cooling systems are designed to run at high temperatures but are prevented from getting too hot, thereby limiting the production of NO_x. Today's engine control systems incorporate many features that change the air/fuel mixture, ignition timing, and idle speed to control the engine's temperature.
- *Spark control systems.* Spark control systems have been in use since the earliest gasoline engines. It was discovered that the proper timing of the ignition spark helped to reduce exhaust emissions and develop more power output. Incorrect timing affects the combustion process. Incomplete combustion results in HC emissions. High CO emissions are also a by-product of incomplete combustion and can also result from incorrect ignition timing. Advanced timing can also increase the production of NO_x. When timing is too far advanced, combustion temperatures rise. For every 1° of over-advance, the temperature increases by 51.57°C (125°F). Spark control on today's engines is handled by the PCM. Through input signals from various sensors, the PCM adjusts ignition timing for optimal performance with minimal emissions levels.

PCV Systems

In late 1959, California established the first standards for automotive emissions. In 1967, the U.S. federal Clean Air Act was amended to provide standards that applied to automobiles. The first controlled automotive emission was crankcase vapours. During combustion, some unburned fuel and other products of combustion leak past the piston rings and move into the crankcase. This leakage is called blowby. Blowby gases are largely HC gases. The PCV systems route the gases, which are mixed with outside air, into the engine's intake. From there, the gases are drawn into the cylinders and burned. PCV systems were installed on all cars, beginning with the 1963 models.

Blowby must be removed from the engine before it condenses in the crankcase and reacts with the oil to form sludge. Sludge, if allowed to circulate with engine oil, corrodes and accelerates the wear of pistons, piston rings, valves, bearings, and other internal parts of the engine. Blowby also carries some unburned fuel into the crankcase. If not removed, the unburned fuel dilutes the engine's oil. When oil is diluted, it does not lubricate the engine properly, which causes excessive wear.

Blowby gases must also be removed from the crankcase to prevent premature oil leaks. Because these gases enter the crankcase by the pressure created during combustion, they pressurize the crankcase. The gases exert pressure on the oil pan gasket and crankshaft seals. If the pressure is not relieved, oil is eventually forced out of these seals.

OPERATION The PCV system uses engine vacuum to draw fresh air through the crankcase. This fresh air enters through the air filter or through a separate PCV breather filter.

When the engine is running, intake manifold vacuum is supplied to the PCV valve. This vacuum moves air through the clean air hose into the rocker arm or camshaft cover. From there, the air flows through openings in the cylinder head into the crankcase, where it mixes with blowby gases. The mixture of blowby gases and fresh air flows up through the cylinder head to the PCV valve. Vacuum draws the blowby gases through the PCV valve into the intake manifold **(Figure 34–13)**. The blowby gases mix with the intake air and enter the combustion chambers, where they are burned.

PCV Valve

The PCV valve **(Figure 34–14)** is usually mounted in a rubber grommet. A hose connects the valve to the intake manifold **(Figure 34–15)**. A clean air hose is connected from the air filter to the opposite rocker arm cover. On some systems, the PCV valve is mounted in a vent module, and the clean air filter is located in this module.

A PCV valve contains a tapered valve. When the engine is not running, a spring keeps the valve seated against the valve housing **(Figure 34–16)**. During idle or deceleration, high intake manifold vacuum moves the valve upward against the spring tension. Under this condition, the blowby gases

FIGURE 34–14

A PCV valve.

©Cengage Learning 2015

FIGURE 34–15

Routing of the vacuum line to a PCV valve.

©Cengage Learning 2015

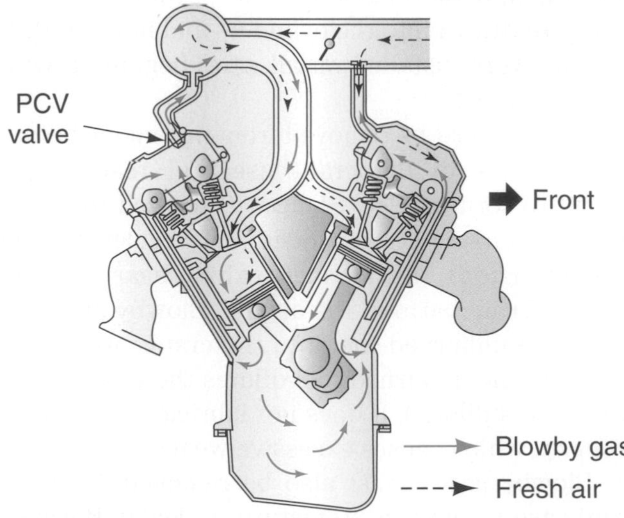

FIGURE 34–13

A typical PCV system.

©Cengage Learning 2015

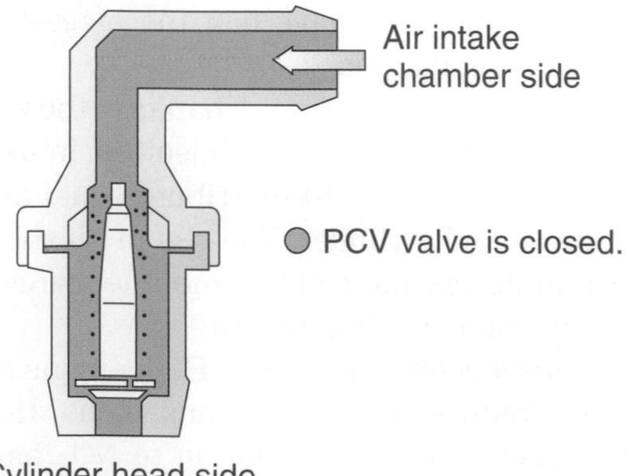

FIGURE 34–16

The PCV valve position with the engine not running.

©Cengage Learning 2015

flow through a small opening in the valve. Because the engine is not under heavy load, the amount of blowby gas is minimal, and the small valve opening is all that is needed to move the blowby gases out of the crankcase.

Manifold vacuum drops off during part-throttle operation. As the vacuum signal to the PCV valve decreases, a spring moves the tapered valve downward to increase the opening **(Figure 34–17)**. Because engine load is higher at part-throttle operation than at idle, blowby gases are increased. The larger opening of the valve allows all the blowby gases to be drawn into the intake manifold.

When the engine is operating under heavy load and at wide-open throttle, the decrease in intake manifold vacuum allows the spring to move the tapered valve further down in the PCV valve. This provides a larger opening through the valve. Because higher engine load results in more blowby gases, the larger PCV valve opening is necessary to allow these gases to flow through the valve into the intake manifold.

When worn rings or scored cylinders allow excessive blowby gases into the crankcase, the PCV valve opening may not be large enough to allow these gases to flow into the intake manifold. Under this condition, the blowby gases create a pressure in the crankcase, and some of these gases are forced through the clean air hose and filter into the air cleaner. When this action occurs, there is oil in the PCV filter and air cleaner. This same action occurs if the PCV valve is restricted or plugged.

When there is high crankcase pressure and the engine is under heavy load, the PCV gases can experience reverse flow. The high pressure and high concentration of gases, accompanied with very low engine vacuum, allow the gases to move out of the intake manifold. The vacuum is too weak to draw in the gases. This results in an accumulation of oil in the throttle body. Therefore, oil buildup inside the throttle body can be an indication of high crankcase pressure.

If the PCV valve sticks in the wide-open position, excessive airflow through the valve causes rough idle operation. If a backfire occurs in the intake manifold, the tapered valve is seated in the PCV valve as if the engine was not running. This action prevents the backfire from entering the engine, where it could cause an explosion.

HEATED PCV SYSTEMS Crankcase vapours contain some moisture, which means the water in the vapours can freeze when the engine has been sitting in cold weather. This can cause the PCV system not to work until the ice melts, which may be awhile. When the PCV system is not working, excess pressures can build up in the crankcase and force blowby gases out. To prevent this from occurring, some engines are equipped with heated PCV systems.

Heated systems have a heated PCV valve or heated PCV tube. The valves can be coolant heated **(Figure 34–18)** or electrically heated. Coolant-heated valves have passages that allow the coolant to flow around the valve. Some electrically heated valves are controlled by the PCM, whereas others use a thermistor in the heater's wiring harness. Heated tubes rely on electrical heaters that are either PCM controlled or use a thermistor to control the heat.

Electrically heated tubes or valves have a heating element as part of the valve, the connection between the valve and the PCV tube, or in the tube **(Figure 34–19)**. When the heating element is controlled by a thermistor, voltage is applied to the

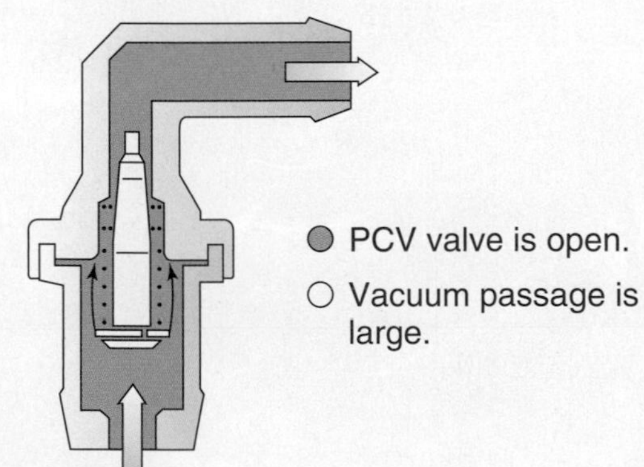

FIGURE 34–17

The PCV valve position during part-throttle operation.

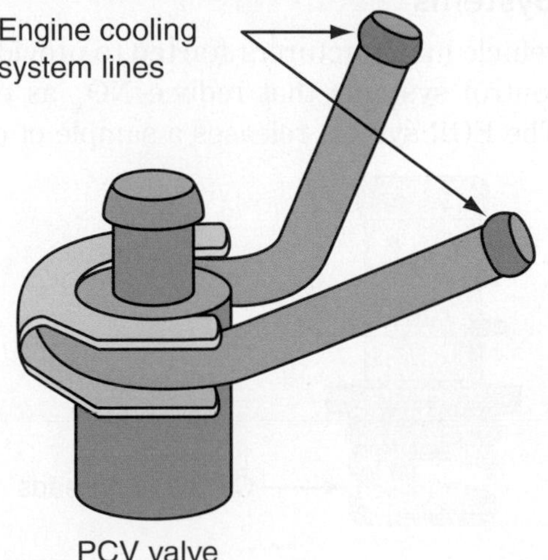

FIGURE 34–18

A coolant-heated PCV valve.

FIGURE 34–19

An electrically heated PCV valve.

©Cengage Learning 2015

element when it is cold. Once warm, the resistance is so high that the voltage and amperage is too low to energize the element.

When the PCM controls the heating element, the heater is directly controlled by the PCM. The PCM uses IAT signals to determine when to energize the heating element.

PCV MONITOR Vehicles that have OBD-II PCV monitoring capabilities use special PCV valves. These valves are designed so that they create a total seal when installed. Most use a cam-lock thread design that requires one-quarter to one-half turn to lock them in place **(Figure 34–20)**. The locking mechanism is designed to eliminate the chance of the valve accidentally becoming loose in its grommet.

EGR Systems

Most vehicle manufacturers started to provide emission control systems that reduce NO_x as early as 1970. The EGR system releases a sample of exhaust

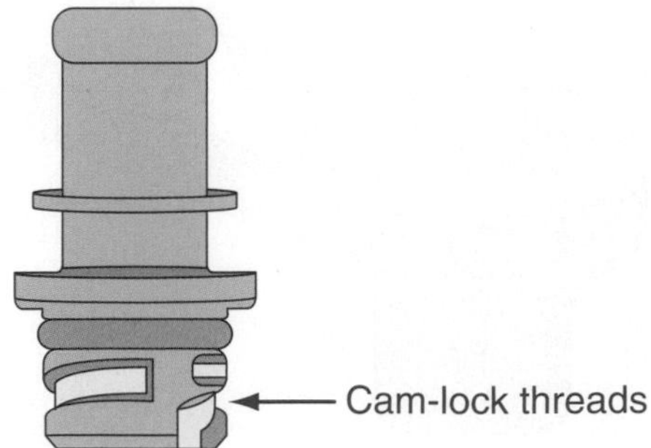

FIGURE 34–20

A PCV valve with cam-lock threads that prevent it from accidentally becoming loose.

©Cengage Learning 2015

gases into the intake's air/fuel mixture. This lowers the peak temperature of combustion and therefore reduces the chances of NO_x being formed. The recirculated exhaust gas dilutes the air/fuel mixture. Because exhaust gas does not burn, this lowers the combustion temperature and reduces NO_x emissions. At lower combustion temperatures, the nitrogen in the incoming air is simply carried out with the exhaust gases.

Driveability problems can result from having too much recirculated exhaust gas in the combustion chamber. This is especially true when there is a high demand for engine power. Also, poor control of EGR flow can cause starting and idling problems. This is why EGR flow is disabled during cold starting, at idle, and at throttle openings of more than 50 percent. There is maximum EGR flow only when the vehicle is at a cruising speed with a very light load.

Many late-model engines do not have an EGR system. Rather, they rely on variable valve timing to prevent all of the exhaust gases from leaving the cylinder during the exhaust stroke during some operating conditions. The retention of the exhaust serves the same purpose as the EGR system. Other engines without an EGR system use other technologies to reduce combustion temperatures.

OBD-II systems monitor the EGR system to determine if the system is operating properly. These monitors use a variety of sensors and methods. If a fault is detected in any of the EGR monitor tests, a DTC is set. If the fault occurs during two drive cycles, the MIL is illuminated. The EGR monitor operates once per OBD-II trip.

EGR VALVE Many older engines are equipped with a vacuum-operated EGR valve **(Figure 34–21)** to

FIGURE 34–21

An EGR valve.

©Cengage Learning 2015

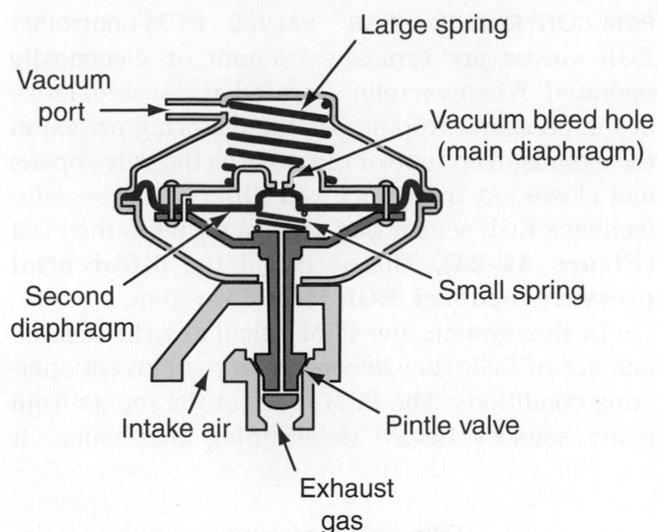

FIGURE 34–22

A typical design of a vacuum-controlled EGR valve.

regulate the flow of exhaust gas into the intake manifold. Most late-model engines have an electrically controlled EGR valve. Typically, the EGR valve is mounted to the intake manifold.

Figure 34–22 illustrates the basic design of a vacuum-controlled EGR valve. The EGR valve is a flow-control valve. A small exhaust crossover passage in the intake manifold admits exhaust gases to the inlet port of the EGR valve. Opening the EGR valve allows exhaust gases to flow through the valve **(Figure 34–23)**. Here, the exhaust gas mixes with the intake air or air/fuel mixture in the intake manifold. This dilutes the mixture so that combustion temperatures are minimized.

On some engines, the exhaust gas from the EGR system is distributed through passages in the cylinder heads and distribution plates to each intake port. The distribution plates are positioned between the cylinder heads and the intake manifold. Because the exhaust gas from the EGR system is distributed equally to each cylinder, smoother engine operation results.

To regulate the amount of EGR flow, EGR valves have a fixed orifice or a tube with a narrow inside diameter. On some engines, the gasket for the EGR valve provides the orifice.

Vacuum EGR Valve Controls. Vacuum is used to control the operation of EGR valves on older engines. Many different vacuum controls have been used. Ideally, the EGR system should operate when the engine reaches operating temperature and/or when the engine is operating under conditions other than idle or wide-open throttle (WOT). The following

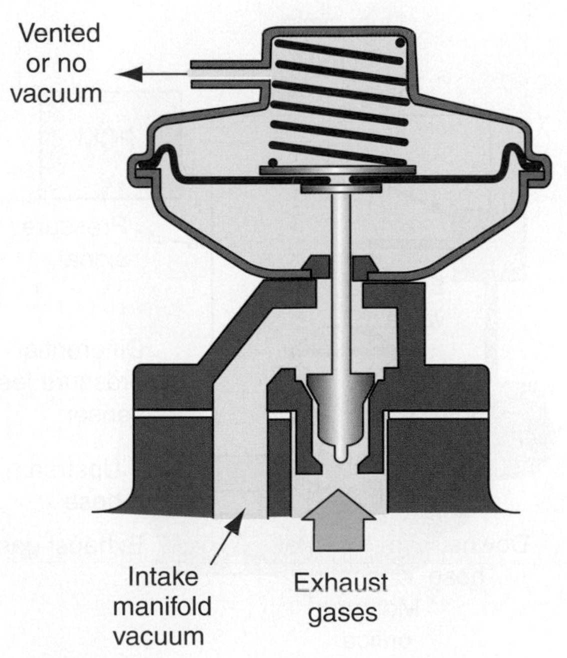

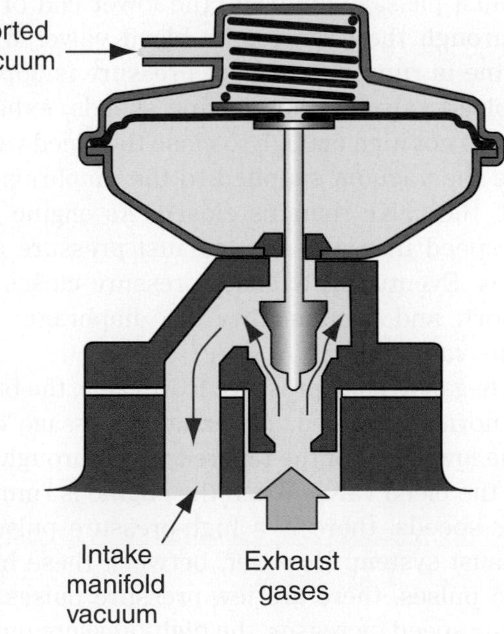

FIGURE 34–23

The operation of a normal ported EGR valve.

are various controls that relate directly to vacuum-controlled EGR systems:

- The **thermal vacuum switch (TVS)** senses the air temperature. When the engine reaches operating temperature, the TVS opens to supply vacuum to the EGR valve.
- The **ported vacuum switch (PVS)** senses coolant temperature. The PVS cuts off vacuum when the engine is cold and allows vacuum to the EGR valve when the engine is warm.
- Some engines have an **EGR delay timer control** system, which prevents EGR operation for a predetermined amount of time after warm engine start-up.
- Some applications have a WOT valve to cut off EGR flow at WOT.

BACK PRESSURE EGR Many engines have a **back pressure transducer** that modulates, or changes, the amount the EGR valve opens. It controls the amount of air bleed in the EGR vacuum line according to the level of exhaust gas pressure, which is dependent on engine speed and load. The back pressure transducer may be a separate unit or incorporated into the EGR valve. There are two basic types of back pressure EGR systems.

A positive back pressure EGR valve has a bleed port and valve positioned at the centre of a diaphragm. A spring holds the bleed valve open. An exhaust passage connects the lower end of the valve through the stem to the bleed valve. When the engine is running, exhaust pressure is applied to the bleed valve. At low engine speeds, exhaust pressure is not high enough to close the bleed valve. Because the vacuum supplied to the diaphragm is bled off, the valve remains closed. As engine and vehicle speed increase, the exhaust pressure also increases. Eventually, exhaust pressure closes the bleed port, and vacuum lifts the diaphragm and opens the valve.

In a negative back pressure EGR valve, the bleed port is normally closed. An exhaust passage connects the lower end of the tapered valve through the stem to the bleed valve. When the engine is running at lower speeds, there is a high-pressure pulse in the exhaust system. However, between these high-pressure pulses, there are low-pressure pulses. As the engine speed increases, the high-pressure pulses become closer together. The negative exhaust pressure pulses decrease, and the bleed valve closes and opens the EGR valve.

PCM-CONTROLLED EGR VALVES PCM-controlled EGR valves are typically vacuum or electrically operated. When vacuum operated, the system looks at the pressure drop across the metering orifice in the exhaust feed tube or the valve as the valve opens and closes. At the orifice is a differential pressure feedback EGR sensor that sends a signal to the PCM **(Figure 34–24)**. This is called the **differential pressure feedback EGR (DPFE)** system.

In this system, the PCM calculates the desired amount of EGR flow according to the current operating conditions. The PCM looks at the inputs from many sensors before determining this value. It

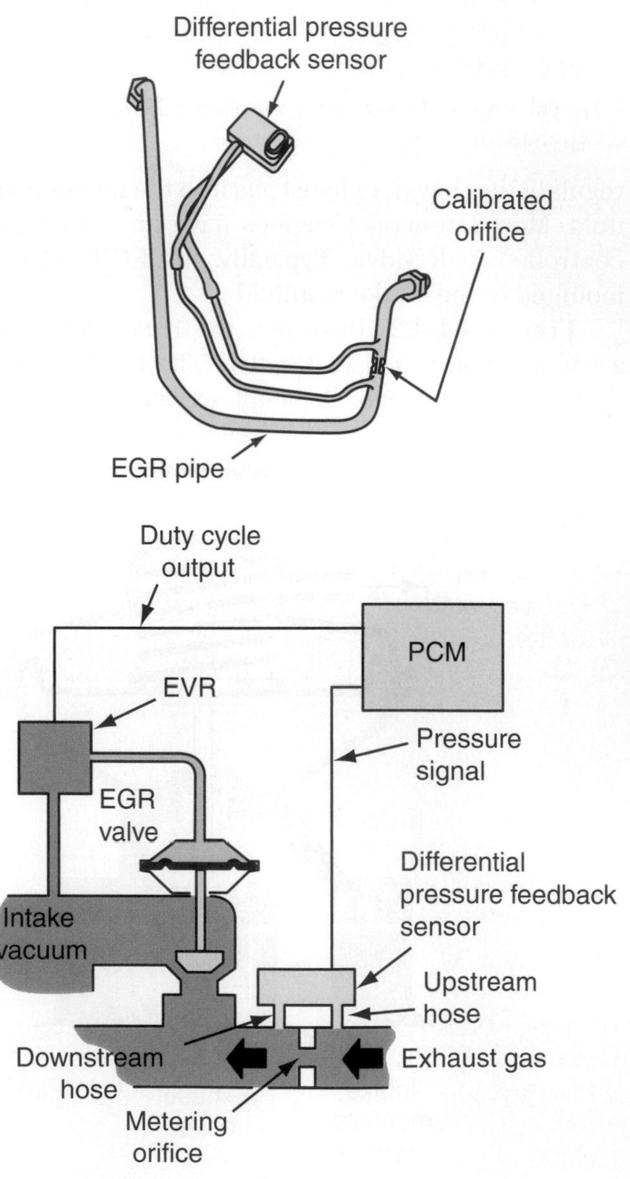

FIGURE 34–24

An EGR system with a DPFE sensor.

then calculates the necessary pressure drop across the orifice to obtain this flow. Once the value is determined, the PCM sends commands to the EGR vacuum regulator solenoid. The solenoid is duty-cycled by the PCM. As the duty cycle increases, more vacuum is sent to the valve, and it remains open for longer periods.

As exhaust gases pass through the valve, they must also pass through the orifice. The DPFE sensor measures the pressure drop across the orifice and sends a feedback signal to the PCM. Based on this signal, the PCM can make corrections to the operation of the EGR valve. Normally, the voltage signal from the DPFE sensor is 0 to 5 volts, and the voltage is directly proportional to the pressure drop.

Some EGR valves have an exhaust gas temperature sensor. This sensor contains an NTC thermistor; an increase in exhaust temperature decreases the sensor's resistance. Two wires are connected from the temperature sensor to the PCM. The PCM senses the voltage drop across this sensor. Cool exhaust temperature and higher sensor resistance cause a high-voltage signal to the PCM, whereas hot exhaust temperature and low sensor resistance result in a low-voltage signal.

ELECTRIC EXHAUST GAS RECIRCULATION (EEGR) SYSTEM The EEGR system allows for precise control of NO_x production without relying on engine vacuum. The EEGR system uses an electric motor in the EGR valve. Normally, the EEGR valve is water or air cooled. The PCM controls the stepper motor that controls the position of the EGR valve's pintle valve. The position of the valve determines the rate of EGR flow. A spring keeps the valve closed and must be overcome by the force of the motor. By using an electric motor to control the valve, the system has no need for a vacuum diaphragm, vacuum regulator solenoid, orifice or orifice tube, or DPFE sensor.

The PCM receives signals from various sensors to determine the current operating conditions. The PCM then calculates the desired amount of EGR for those conditions. Then the PCM commands the motor to move (advance or retract) a specific number of discrete steps. Normally, the stepper motor has a fixed number of possible steps, each relating to the position of the pintle valve. The position of the pintle determines the EGR flow.

Other electric EGR valves rely on solenoids that control the amount of EGR flow. A **digital EGR valve** contains up to three electric solenoids operated directly by the PCM **(Figure 34–25)**. Each solenoid contains a movable plunger with a tapered tip that seats in an orifice. When a solenoid is energized, its plunger is lifted, and exhaust gas is allowed to recirculate through the orifice into the intake manifold. Each of the solenoids and orifices has a different size. The PCM can operate one, two, or three solenoids to supply the amount of exhaust recirculation required to provide optimum control of NO_x emissions.

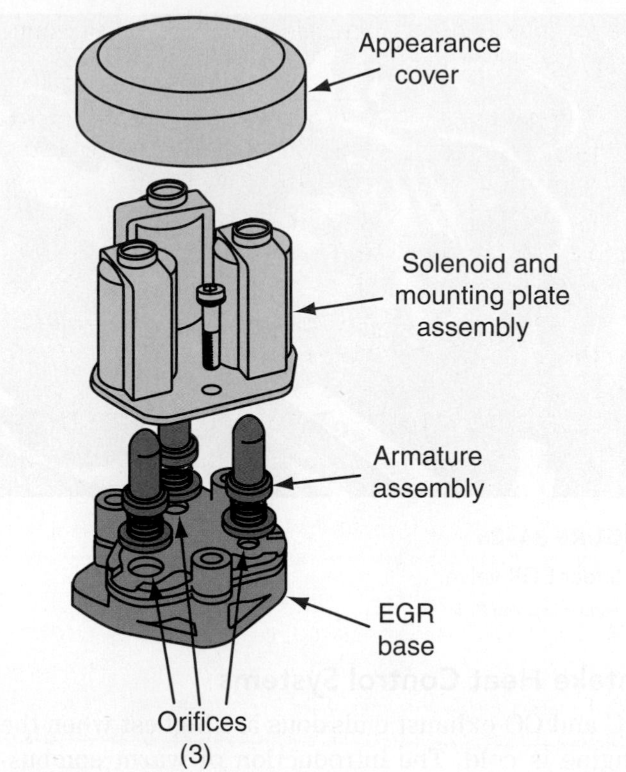

FIGURE 34–25

A digital EGR with three solenoids.

A **linear EGR valve (Figure 34–26)** has a single solenoid or stepper motor operated by the PCM. A tapered pintle is positioned on the end of the solenoid's plunger. When the solenoid is energized, the plunger and tapered valve are lifted, and exhaust gas is allowed to recirculate into the intake manifold **(Figure 34–27)**. The EGR valve contains an EGR valve position sensor, which is a linear potentiometer. The signal from this sensor varies from approximately 1 V with the EGR valve closed to 4.5 V with the valve wide open. The PCM controls the EGR solenoid winding through pulse-width modulation to provide accurate control of the plunger and EGR flow. A sensor sends feedback to the PCM to let it know that the commanded valve position was achieved.

FIGURE 34–26

A linear EGR valve.

Intake Heat Control Systems

HC and CO exhaust emissions are highest when the engine is cold. The introduction of warm combustion air improves the vaporization of the fuel in the fuel injector throttle body or intake manifold. On older engines equipped with a throttle body injection (TBI) unit, the fuel is delivered above the throttles, and the intake manifold is filled with a mixture of air and gasoline vapour. Some intake manifold heating is required to prevent fuel condensation, especially when the intake manifold is cool or cold. Therefore, these engines have intake manifold heat control devices such as heated air inlet systems, manifold heat control valves, and **early fuel evaporation (EFE)** heaters.

A heated air inlet control may be used on engines with TBI. This system controls the temperature of the air on its way to the throttle body. Another system used is an exhaust manifold heat control (similar to the old style heat riser) valve that routes exhaust gases to warm the intake manifold when the engine is cold. This heats the air/fuel mixture in the intake manifold. These control valves can be either vacuum or thermostatically operated.

Some intake heat systems are computer controlled. These systems use an EFE heater. The EFE heater is a resistance grid that heats the mixture as it passes from the throttle body to the manifold. The engine coolant temperature sensor sends a signal to the PCM in relation to coolant temperature. At a preset temperature, the PCM grounds the mixture heater relay winding, which closes the relay's contacts. When the coolant temperature reaches a preset point, the PCM opens the ground circuit, and the relay contacts open and shut off the current flow to the heater.

Port injection engines do not need heat risers or EFE heaters because the intake manifold delivers only air to the cylinders. Fuel is discharged into the intake ports near the intake valves or directly into the cylinders. Therefore, there is no need to warm the fuel.

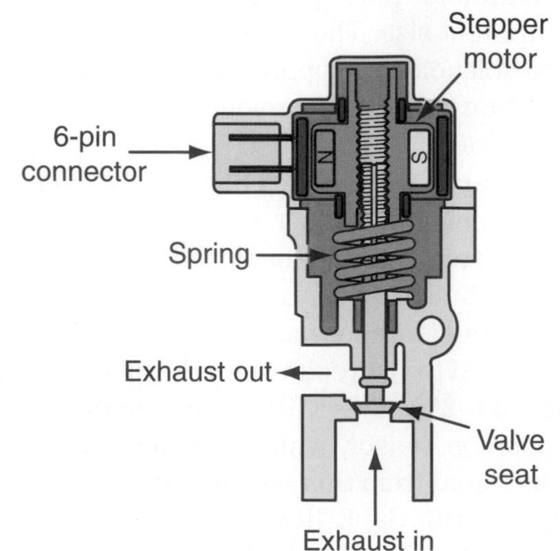

FIGURE 34–27

Basic construction of an electronically operated EGR valve with a stepper motor.

POST-COMBUSTION SYSTEMS

Post-combustion emission control devices clean up the exhaust after the fuel has been burned, but before the gases exit the vehicle's tailpipe. An excellent example of this is the catalytic converter. A converter is one of the most effective emission control devices on a vehicle for reducing HC, CO, and NO_x.

Another post-combustion system is the secondary air or air injection system. This system forces fresh air into the exhaust stream to cause a secondary combustion and reduce HC and CO emissions.

Catalytic Converters

One of the most important developments for lowering emission levels has been the availability and use of unleaded gasoline. Since 1971, engines have been designed to operate on unleaded fuels. In 1990, the use of leaded fuels in Canada was banned. Recent environmental studies since the 1990 ban have shown lead levels in the air of most Canadian cities have dropped below detectable limits.

Removing lead from gasoline eliminates lead particles in the exhaust. It also increases spark plug life, which is important for decreasing emissions. Also, the use of unleaded fuel avoids the formation of lead deposits in the combustion chambers that tend to increase HC emissions. Unleaded fuels also led to the use of catalytic converters, which provide a way to oxidize CO and HC emissions in an engine's exhaust.

Beginning with the 1975 model year, passenger cars and light trucks have been equipped with converters. A catalytic converter is positioned within the exhaust system and converts various emissions into less harmful gases. Today's catalytic converters are extremely effective in reducing the amount of HC, CO, and NO_x emitted from a vehicle's tailpipe.

Most current vehicles have two converters in each exhaust stream. If the engine has more than one exhaust manifold, there is an exhaust stream from each, and each of those streams has two converters. The first converter is located close to the exhaust manifold. Because the effectiveness of a converter depends on its temperature, placing a converter close to the manifold allows it to warm up quickly. The converters are also small, which helps them to heat up quickly. These converters are called light-up or warm-up converters, or pre-cats **(Figure 34–28)**. Their primary purpose is to reduce emissions while the main converters are warming up.

The main converter is located behind the pre-cat. On some vehicles, it may be connected to two pre-cats by a Y-pipe. Depending on the engine, a vehicle can have one or two main converters plus up to four pre-cats.

The effectiveness of the converter is measured by the catalyst monitor of OBD-II systems. The monitor relies on heated oxygen sensors to measure the converter's effectiveness. The location and number of HO_2S found in an exhaust stream vary with vehicle design and the emission certification level (LEV, ULEV, PZEV, etc.) of the vehicle. Most vehicles have two HO_2S in each exhaust stream. In each stream, there is an HO_2S in the front exhaust pipe before the catalyst. The front sensors (HO_2S11/HO_2S21) are used for fuel control. An additional sensor is located after the catalyst and is used to monitor catalyst efficiency, and is commonly referred to as a catalyst efficiency monitor, or CEM, sensor.

FIGURE 34–28

A pre-cat.

Many PZEVs have three HO_2S in each exhaust stream. The first HO_2S is located near the exhaust manifold and is used for fuel control. The second is in the centre of the converter and monitors the amount of oxygen available in the converter. The third is after the converter and is used for long-term fuel trim control and for monitoring the effectiveness of the converter.

The converter is designed to respond to ever-changing exhaust quality. The amount and type of undesired gases change with operating conditions and driving modes **(Figure 34–29)**.

A catalytic converter is basically a housing shaped like a muffler that contains two or more ceramic elements coated with a catalyst. The catalysts are responsible for the chemical changes that occur in the converter. A catalyst is something that

EMISSION	PERCENTAGE OF EXHAUST DURING			
	Idle	Acceleration	Cruise	Deceleration
CO	5.2%	5.2%	0.8%	4.2%
HC	0.08%	0.04%	0.03%	0.4%
NO_x	0.003%	0.3%	0.15%	0.006%

FIGURE 34–29

The approximate emission amounts during different driving modes.

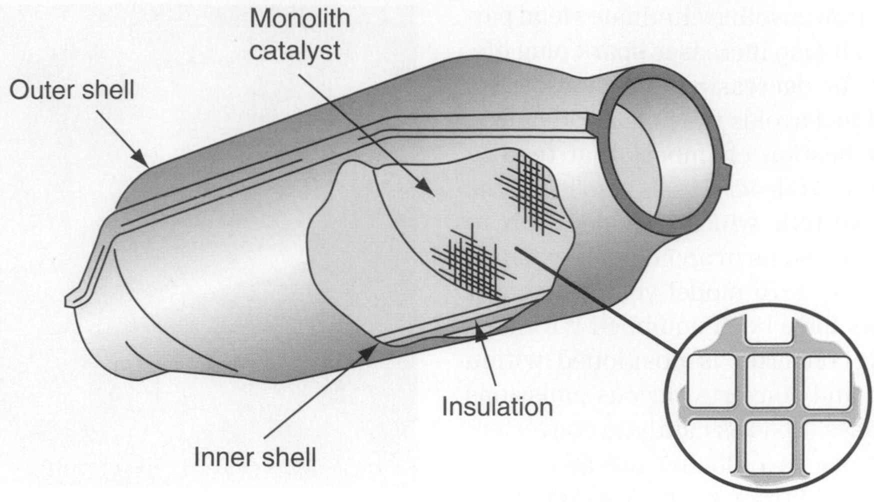

FIGURE 34–30

A honeycomb monolith-type catalytic converter.

causes a chemical reaction without being part of the reaction. As the exhaust gases pass over the catalyst, most of the harmful gases are changed to harmless gases. Internally, the ceramic elements are designed to expose the exhaust gases to as much surface area as possible. Ceramic materials are coated with the catalyst material to minimize the amount of catalyst material necessary. Most catalyst materials are precious metals that are quite expensive.

The catalyst-coated ceramic elements have a **honeycomb monolith** design or are **ceramic beads**. Nearly all converters used on today's vehicles have the honeycomb structure **(Figure 34–30)**. Early converters were made with either design. The beads, or pellets, have a porous surface and are approximately 3 mm (1/8 in.) in diameter. The honeycomb monolith design looks like a honeycomb, and each opening has 1000 to 2000 pores that are about 1 mm (0.04 in.) in size; the pores are separated by thin walls. This allows for an extremely large area for the gases to adhere and react to.

GO TO Chapter 7 for a detailed discussion of catalysts, reduction, and oxidation.

Prior to OBD-II, catalytic converters contained two different types of catalysts: a **reduction catalyst** and an **oxidation catalyst**. The two separate catalysts created a dual-bed converter. Exhaust gases passed over the first, or reduction, bed, where NO_x emissions were eliminated. Then the exhaust passed to the second bed, where they were oxidized to eliminate CO and HC emissions.

During reduction, as NO_x gases pass over the catalyst, the N atoms are pulled from the NO_x molecules and combined with other N atoms to form N_2, which passes through the converter. The released O_2 atoms react with the CO in the exhaust stream and form CO_2 or pass through to the second bed. The result of NO_x reduction is pure N_2 plus O_2 or CO_2.

During the oxidation phase inside the converter, HC and CO molecules experience a second combustion. This occurs because of the presence of O_2 and the temperature of the converter. The result of this combustion or oxidation process is water vapour (H_2O) and CO_2.

All late-model vehicles use a **three-way converter (TWC)** that decreases HC, CO, and NO_x emissions **(Figure 34–31)**. The catalyst used in

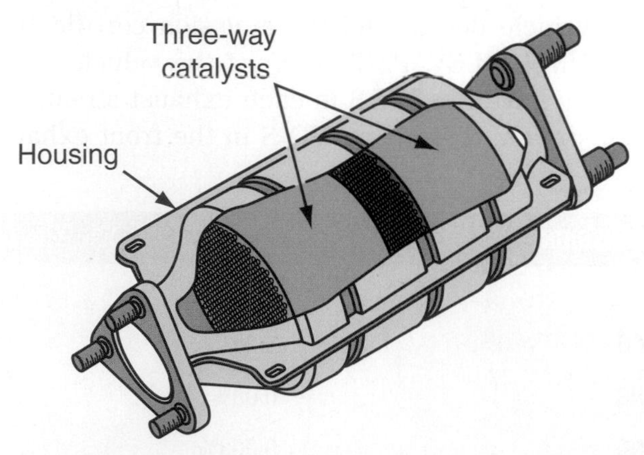

FIGURE 34–31

A typical three-way catalytic converter.

EMISSION	PROCESS	ACTION	RESULT
NO_x	Reduction	$2NO + 2CO$	$N_2 + 2CO_2$
HC	Oxidation	$HC + O_2$	$CO_2 + H_2O$
CO	Oxidation	$2CO + O_2$	$2CO_2$

FIGURE 34–32

The action of a dual-bed TWC.

the reduction bed is either platinum or rhodium. When NO_x is exposed to hot rhodium (Rh), it breaks down into O_2 and N_2 molecules. Some of the free O_2 molecules combine with CO molecules, and the resultant gases are O_2, CO_2, and N_2, which move to the oxidation catalyst. The oxidation catalyst, normally platinum (Pt) and palladium (Pd), combines CO and HC with the O_2 released by the reduction catalyst or in the exhaust to form CO_2 and H_2O **(Figure 34–32)**.

The presence of O_2 is important to the reduction and oxidizing processes. Early TWCs relied on fresh air injected by the secondary air system between the two catalysts. This air intake was controlled by the secondary AIR system. Other converters had a layer of cerium in the centre section of the converter. The element cerium has the ability to store O_2. Late-model converters rely on the O_2 content in the exhaust. The amount of O_2 in the exhaust bounces up and down as the PCM makes slight changes to the air/fuel mixture during closed-loop operation.

As the PCM adjusts the air/fuel ratio around the desired stoichiometric ratio, it constantly toggles the mixture between slightly lean to slightly rich. This action provides the necessary O_2 and CO for the TWC. High CO content is necessary for reducing NO_x emissions, whereas high O_2 is required for the oxidation of CO and HC. When the mixture is lightly rich, more CO and less O_2 are present in the exhaust. When the mixture is slightly lean, CO content decreases and O_2 content increases **(Figure 34–33)**. The converter stores some of the O_2 to allow for better oxidation during the rich cycle.

The efficiency of a catalytic converter is affected by its temperature. The temperature of the exhaust gases heat up the converter. The normal operating temperature for most converters is about 500°C (900°F). As the temperature of a converter increases, its efficiency increases. During converter warm-up, the point at which the converter is operating at more than 50 percent efficiency is called catalyst light-off.

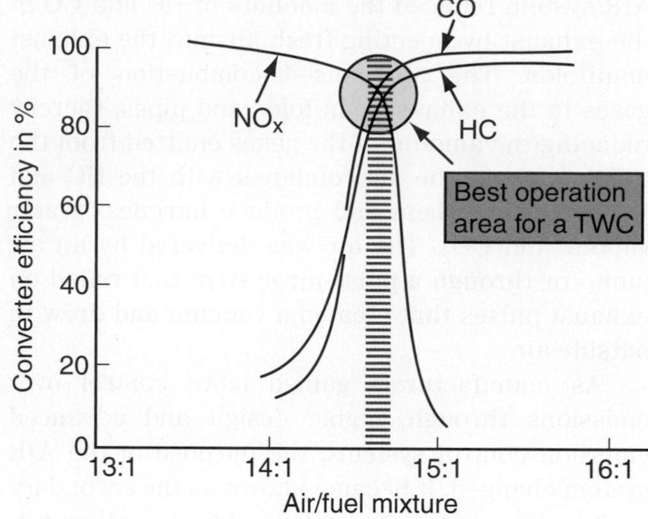

FIGURE 34–33

The efficiency of a catalytic converter is at its highest level when there is a stoichiometric mixture.

This normally occurs at 246°C to 302°C (475°F to 575°F). It takes awhile for the converter to reach this temperature, especially when it is mounted away from the engine and under the vehicle. During this warm-up time, exhaust emissions levels are high. To provide cleaner exhaust after a cold start, pre-cats are used.

OTHER CONVERTER DESIGNS Engines that run on very lean mixtures produce high amounts of NO_x. To reduce these emissions, many of these vehicles have an additional catalytic converter, called a storage or adsorber converter. After the exhaust gases leave a three-way converter, they flow through a special NO_x storage converter. This converter is coated with barium and extracts the nitrogen oxides from the exhaust and stores them until its nitrogen oxide sensor senses that the storage converter is filled. At that time, the sensor sends a signal to the PCM, and the system starts to deliver a richer air/fuel mixture. As this richer exhaust flows through the storage converter, it regenerates the converter, and the nitrogen oxides are converted into harmless nitrogen. When the converter is free of nitrogen oxide, the sensor signals the system to run lean mixtures again.

Air Injection Systems

An **air injection reactor (AIR)** system was built into cars and light trucks sold in California in 1966 and used in all automobiles for several years. The

AIR system reduced the amounts of HC and CO in the exhaust by injecting fresh air into the exhaust manifolds. The air caused combustion of the gases in the exhaust manifolds and pipes, thereby reducing the amount of the gases emitted from the tailpipe. O_2 in the air combines with the HC and CO to oxidize them and produce harmless water vapour and CO_2. The air was delivered by an air pump or through a pulse-air system that relied on exhaust pulses that created a vacuum and drew in outside air.

As manufacturers gained more control over emissions through engine design and advanced emission control systems, the purpose of the AIR system changed. It became known as the secondary air injection system and was modified to allow catalytic converters to operate more efficiently. The system injected air into the catalytic converter. This helped the converter oxidize and reduce the gases entering into it. AIR systems are not commonly used today. Improved combustion and better catalytic converters have eliminated their need.

ELECTRONIC SECONDARY AIR SYSTEMS The role and use of the secondary AIR system has decreased through the years. This is because there are fewer HC and CO emissions in the exhaust of a typical engine. When engines are fitted with a secondary AIR system, they are monitored by the OBD-II system and are solely used to supply O_2 to the catalytic converter. The air from the system not only helps clean up the emissions by causing combustion, but it also serves to heat the converter so it can work more effectively by reaching its light-off temperature more quickly.

The typical electronic secondary air system, like the conventional air injection system, consists of an air pump connected to a secondary air bypass valve, which directs the air either to the atmosphere or to the catalytic converter.

The air pump is driven by an electric motor controlled by the PCM **(Figure 34–34)**. Intake air passes through a centrifugal filter fan at the front of the pump, where foreign materials are separated from the air by centrifugal force. In some systems, air flows from the pump to an AIR bypass (AIRB) valve, which directs the air either to the atmosphere or to the AIR diverter (AIRD) valve. The AIRD valve directs the air to the catalytic converter or exhaust manifolds. The secondary air may be diverted to the exhaust manifolds upstream or to the catalytic converter downstream during different engine operating conditions.

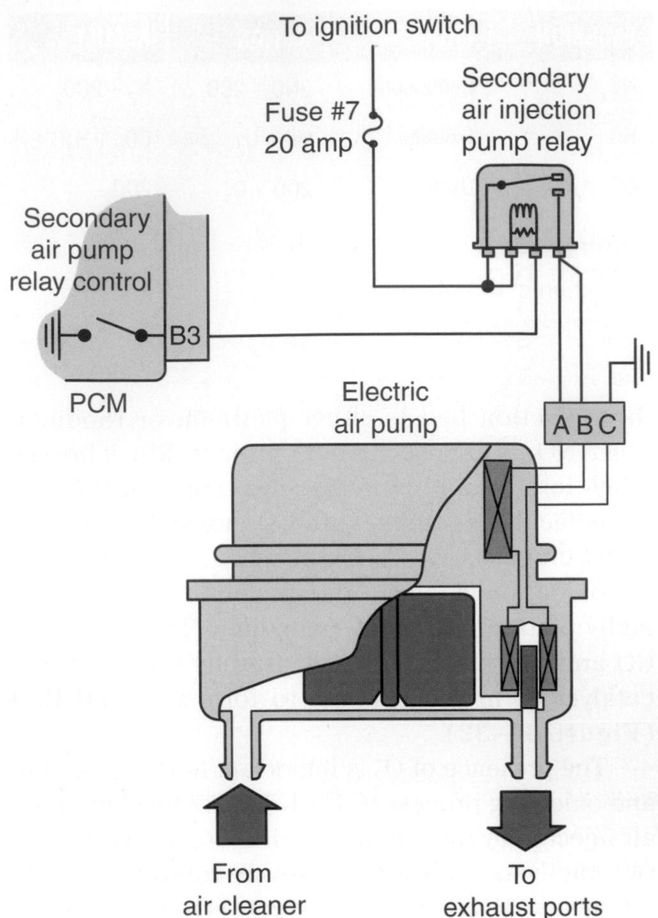

FIGURE 34–34

An electric air pump circuit.

DIESEL EMISSION CONTROLS

In the early 1980s, diesel-powered cars were offered by some manufacturers in North America, but that trend ended quickly because they were noisy, unreliable, and dirty. However, with technological advances and the availability of low-sulpur diesel fuel, diesel engines in cars and light trucks will become more common soon. Today's diesel engines are very durable. This is because they tend to be overbuilt to withstand very high-compression pressures and the shock loading from the detonation of the air/fuel mixture. The downside is that diesel engines tend to be heavier than comparably sized gasoline engines. All reciprocating parts must be built stronger and therefore are heavier. Also, to produce the horsepower needed for an application, diesel engines must have a higher displacement than would be used in a gasoline engine. In spite of the increased size, diesel engines consume less fuel. To counter the increased weight and displacement, many diesels are fitted with turbochargers.

GO TO Chapter 9 for a basic description of how diesel engines operate.

Diesel engines, especially those equipped with turbochargers, produce a substantial amount of torque at low engine speeds. Diesel engines also have a much longer stroke than a gasoline engine. The longer stroke and long burning time produce high-torque outputs. The turbocharger used on a diesel engine can provide up to 207 kPa (30 psi) of boost, which is much more than a gasoline engine can withstand. This boost significantly increases the engine's output. Diesel engines also waste less heat, which means that more of the energy of the fuel is used to power the vehicle.

The use of cleaner fuels and new technologies, such as PCM-controlled fuel injection systems, allows today's diesel engines to have emission levels that match those of gasoline engines. Clean diesel exhaust is further possible because the vehicles can be fitted with EGR and PCV valves, catalytic converters, and particulate filters. Technology has also allowed diesel engines to run quietly.

Today's diesel engines are fitted with OBD-II systems similar to those found on gasoline engines. These systems monitor the effectiveness of various emission-related devices, such as the fuel system, EGR system, catalyst, oxidation catalyst, particulate filter, and PCV system. They also have misfire and comprehensive component monitors. **Figure 34–35** shows how the emission standards for diesel engines have changed through the years.

YEAR	CO_2	NO_X	HC	PM
1988	15.5	6.0	1.3	0.60
1991	15.5	5.0	1.3	0.25
1994	15.5	5.0	1.3	0.10
1998	15.5	4.0	1.3	0.10
2004	15.5	**2.4***	–	0.10
2007#	15.5	0.2	0.14	0.01
2010	15.5	0.2	0.14	0.01

*This value is not just an NO_x standard; it is for non-methane hydrocarbons (NMHC) + NO_x, which is why there is no standard for HC.

#The PM standard went into full effect in the 2007 model year. The NO_x and HC standards have been phased in from that time and were met by 2010.

FIGURE 34–35

Emission standards for diesel engines (given in grams per brake horsepower per hour).

Low-Sulphur Fuel

Legislation has required fuel suppliers to remove nearly all sulphur from most of the diesel fuel they produce for on-highway use. Previously, up to 500 ppm were allowed. The limit for 2007 and newer diesel engines is 15 ppm. The use of this fuel not only reduces sulphur-related emissions, but it also allows diesel engines to be equipped with typically sulphur-intolerant exhaust emission controls, such as particulate filters and NO_x catalysts. This means diesel exhaust now has the capability to be as clean as that from a gasoline engine.

Diesel Fuel Injection

Most current light-duty diesel engines use a common rail injection system. These systems use high pressure, and this pressure is provided equally to all injectors. This allows the PCM to precisely control and monitor the amount of fuel injected into the cylinders. The introduction of highly pressurized fuel into the cylinders provides improved atomization of the fuel. This, along with precise timing, makes it possible for the engines to run cleaner and quieter. Common rail systems use solenoid-operated or piezoelectric injectors **(Figure 34–36)**.

GO TO Chapter 27 for a discussion and explanation of common fuel injection systems for diesel engines.

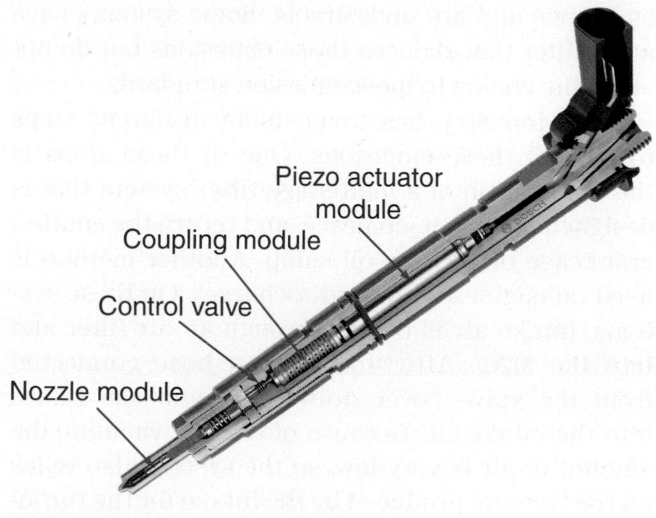

FIGURE 34–36

A typical piezoelectric injector.

Robert Bosch GmnH

The use of piezoelectric injectors increases the PCM's ability to control injection timing. New common rail designs have injectors that are activated by hundreds of thin piezo crystal wafers. Piezo crystals expand quickly when a current is applied to them. The ability of these injectors to quickly respond also allows the PCM to control several injector firings in a single combustion stroke. Some systems fire the injectors five times per stroke.

Many systems fire the injector three times. To reduce combustion noise, the PCM fires the injectors to allow a small amount of fuel to enter the cylinder a few ten-thousandths of a second before firing the injector for combustion. This small spray of fuel begins the combustion process and is called the pilot injection. The pilot injection decreases the harshness of the combustion that occurs when the main injection takes place. A third injection of fuel takes place at the end of the combustion stroke to lower the temperature inside the cylinder.

PCV System

Diesel engines emit crankcase gases just like gasoline engines. However, control of these gases is much different in diesel engines because they produce very little vacuum; therefore, conventional PCV systems do not work. Vacuum is produced in a diesel engine in the opposite way as a gasoline engine—little to no vacuum at idle with an increase as engine speed increases. Therefore, a conventional PCV system cannot work on a diesel engine. Many diesel engines release these gases to the atmosphere through a crankcase breather or downdraft tube just like early gasoline engines. Both are a source of HC and PM emissions and are undesirable. Some systems have a PM filter that reduces those emissions but do not allow the engine to meet emission standards.

The industry has taken many different steps to control these emissions. One of these steps is the installation of a multistage filter system that is designed to collect, coalesce, and return the emitted crankcase oils to the oil sump. Another method is used on engines with a turbocharger. On these systems, intake air is drawn through an air filter and into the MAF. After the MAF, a hose connected from the valve cover draws in crankcase fumes into the intake air. Because of the low vacuum, the amount of air is very low, so the system also relies on the vacuum produced by the intake for the turbocharger. The movement of air into the turbo creates a vacuum that draws the crankcase gases into the intake track.

CRANKCASE DEPRESSION REGULATOR (CDR) The **crankcase depression regulator (CDR)** valve is very similar to the PCV valve used in gasoline engines. It directs crankcase vapours back into the combustion chambers but is designed to work at very low levels of vacuum. It also maintains crankcase pressures to prevent oil consumption and oil leaks due to excessive buildup of pressure in the crankcase. A CDR contains a large silicone rubber/synthetic diaphragm and return spring. When vacuum is introduced to the valve, the valve opens against spring tension and allows crankcase gases to flow into the intake. CDR valves are used on both turbo and non-turbo engines but must be calibrated for the application.

EGR Systems

Normally, the efficiency of diesel engines results in high amounts of NO_x emissions. Therefore, EGR systems are integrated into the latest diesel engines. The EGR systems are the same as those used on gasoline engines, which means a sample of exhaust is introduced into the combustion chambers to reduce combustion temperatures. One of the main differences is that most manufacturers cool the incoming EGR gases before introducing them into the cylinders. This reduces the temperature of combustion and therefore reduces the amount of NO_x emitted by the exhaust **(Figure 34–37)**. Most systems with EGR coolers use engine coolant that passes through a separate circuit to cool the recirculated exhaust gases.

The PCM operates and monitors the EGR system. EGR flow is controlled by the PCM through a digital EGR valve. EGR flow will occur only when the engine is at a predetermined level and other conditions are met. The PCM's EGR monitor consists of a series of electrical and functional tests that monitor the operation of the EGR system.

Catalytic Converters

The most common catalytic converter for diesel engines is an oxidation catalyst. This converter uses the O_2 in the exhaust to oxidize CO to form CO_2 and oxidize HC to form H_2O and CO_2. The converter also reduces the amount of soot. Some engines are fitted with diesel particulate filters that capture the remaining soot.

To control NO_x emissions, engines may also have an NO_x adsorber catalyst built into the oxidation converter or as a separate unit. This converter is comprised of an alkaline metal (typically barium) to store NO_x and an NO_x reduction catalyst.

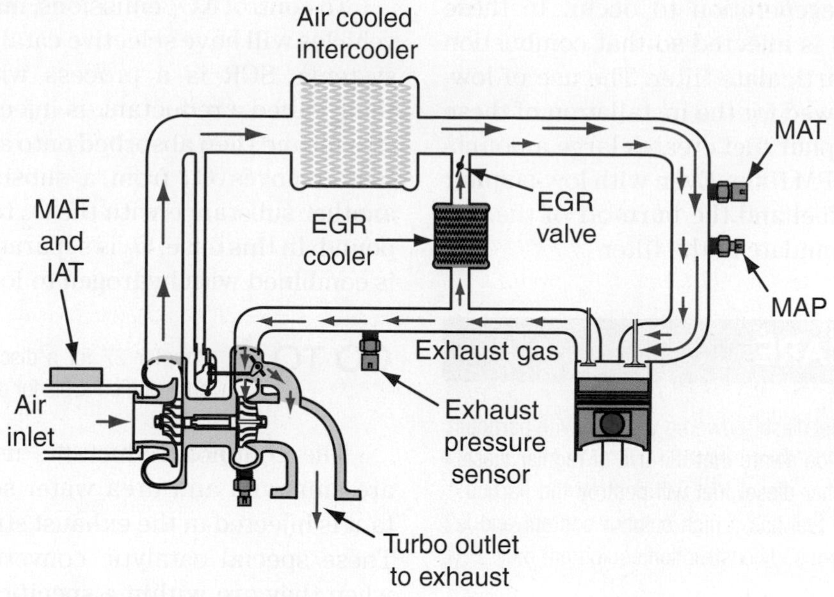

FIGURE 34–37

Intake airflow on a turbocharged diesel engine with an EGR cooler.

Particulate Filter

Limiting the amount of PM or soot from the exhaust of a diesel engine is a top priority. Research has suggested that long-term exposure to diesel soot can cause cancer. Soot can be controlled by running the engine on a lean mixture and using a high-pressure common rail injection system. This brings soot emissions down significantly, as does the oxidizing catalytic converter. However, to meet U.S. EPA emission standards for PM emissions, late-model vehicles also have a particulate filter.

Particulate filters are placed into the exhaust system after the catalytic converter. Sometimes, the PM filter is part of the converter assembly **(Figure 34–38)**. Particulate filters are designed to trap PM. Early PM filters needed to be cleaned as part of a preventive maintenance program. Newer designs periodically burn off the collected PM and are designed to last the life of the vehicle without any special maintenance. This is done by a special cleaning mode, often called regeneration, initiated by the PCM. This combustion cleans the soot from the filter. Regeneration can be passive or active. Passive regeneration occurs during normal vehicle operation, such as during highway driving. Active regeneration occurs when driving conditions do

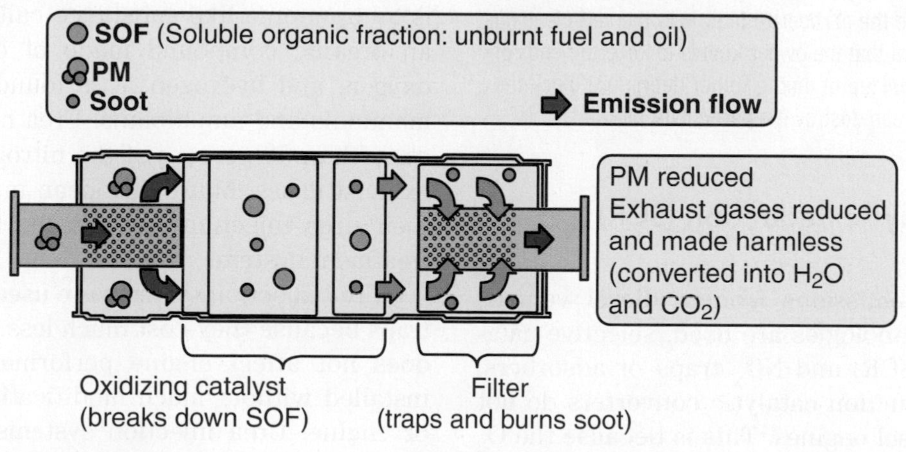

FIGURE 34–38

A continuous-regenerative diesel particulate catalyst assembly.

not allow passive regeneration to occur. In these instances, extra fuel is injected so that combustion occurs within the particulate filter. The use of low-sulphur fuel has allowed for the installation of these new filters. High-sulphur fuel creates large amounts of ash buildup in the PM filter. Even with low-sulphur fuels, ash from the fuel and the burn-off of the collected soot will accumulate in the filter.

CUSTOMER CARE

Customers with late-model diesel-powered vehicles with particulate filters need to be made aware that the use of regular diesel fuel in place of low-sulphur diesel fuel will destroy the particulate filter. Regular diesel fuel has a high sulphur content and is intended to be used in farm and construction equipment only.

The filter is fitted with a sensor that measures exhaust back pressure. A high concentration of ash buildup in the filter will cause the exhaust back pressure to increase. When the sensor informs the PCM that back pressure has reached a specified level, the PCM will order the fuel injection to spray an additional amount of fuel. The extra fuel will cause the oxidation catalytic converter to heat up, and this heat will burn off the ash in the PM filter. The period of time the PCM adds fuel to clean the filter is called the regeneration cycle.

SHOP TALK

If the owner of a diesel-powered vehicle wants to install chrome tips on the tailpipes, there may not be the correct size available. The size and shape of the pipe is designed to draw cold air in to cool the exhaust while the particulate filter is in the self-cleaning mode. Also, make sure that the owner knows to keep the ends of the pipe clean. Any buildup of mud or other debris will block the intake of air, and this can destroy the particulate filter.

Selective Catalytic Reduction (SCR) Systems

To reduce NO_x emissions from a diesel vehicle, two separate technologies are used: selective catalytic reduction (SCR) and NO_x traps or adsorbers. Conventional reduction catalytic converters do not work well on diesel engines. This is because the O_2 content in the exhaust is high, and their converters operate at a low temperature. This means a reduction converter would not be very effective.

To control NO_x emissions, many of the new diesel vehicles will have selective catalytic reduction (SCR) systems. SCR is a process where diesel exhaust fluid, called a reductant, is injected into the exhaust stream and then absorbed onto a catalyst. The reductant removes O_2 from a substance and combines another substance with the O_2 to form another compound. In this case, O_2 is separated from the NO_x and is combined with hydrogen to form water.

GO TO Chapter 27 for a discussion and explanation of emission controls for diesel engines.

The common reductants used in SCR systems are ammonia and urea water solutions. The reductant is injected in the exhaust stream over a catalyst. These special catalytic converters work well only when they are within a specific temperature range. The engine's control unit is programmed to keep the temperature of the exhaust within that range. Also, the amount of reductant sprayed into the exhaust must be proportioned to the amount of exhaust flow. This is also something controlled by the engine's control module.

When the tanks are empty, the emission levels will not be satisfactory. The tanks that hold the reductant must be capable of being refilled. This is a concern of the U.S. EPA, and different ways of enforcing drivers to refill the reductant tank are being tried. Some systems have a warning lamp that informs the driver of a low tank; other vehicles have a warning lamp, and the PCM will put the vehicle into a limp-home mode when the tank is very low or empty. Some vehicles will not start if the tank is empty.

UREA The common reductant used in SCR systems is an ammonia-like substance called urea. **Urea** is an organic compound made of carbon, nitrogen, oxygen, and hydrogen. It is found in the urine of mammals and amphibians. Urea helps to eliminate more than 90 percent of the nitrogen oxides in the exhaust gases. Many European manufacturers call their urea injection systems the Blutec emissions treatment system.

Urea injection systems are used rather than NO_x traps because they cost much less. Also, the system does not affect engine performance and can be installed without much modification to the vehicle or engine. Urea injection systems squirt the urea solution into the exhaust pipe, before the converter, where it evaporates and mixes with the exhaust gases and causes a chemical reaction that reduces NO_x.

KEY TERMS

Air injection reactor (AIR) (p. 1039)
Back pressure transducer (p. 1034)
Canister purge valve (p. 1026)
Carbon dioxide (CO_2) (p. 1019)
Ceramic beads (p. 1038)
Charcoal (carbon) canister (p. 1026)
Crankcase depression regulator (CDR) (p. 1042)
Differential pressure feedback EGR (DPFE) (p. 1034)
Digital EGR valve (p. 1035)
Early fuel evaporation (EFE) (p. 1036)
EGR delay timer control (p. 1034)
Evaporative control (EVAP) (p. 1025)
Greenhouse gas (p. 1019)
Honeycomb monolith (p. 1038)
Linear EGR valve (p. 1035)
Oxidation catalyst (p. 1038)
Ported vacuum switch (PVS) (p. 1034)
Post-combustion control (p. 1025)
Pre-combustion control (p. 1025)
Reduction catalyst (p. 1038)
Thermal vacuum switch (TVS) (p. 1034)
Three-way converter (TWC) (p. 1038)
Urea (p. 1044)
Vehicle emission control information (VECI) (p. 1024)

SUMMARY

- Unburned hydrocarbons, carbon monoxide, and oxides of nitrogen are three types of emissions being controlled in gasoline engines. HC emissions are unburned gasoline released by the engine because of incomplete combustion. CO emissions are a by-product of combustion and are caused by a rich air/fuel ratio. Oxides of nitrogen (NO_x) are formed when combustion temperatures reach more than 1261°C (2300°F).
- CO_2 is a product of combustion and not a pollutant; however, it is considered a greenhouse gas.
- An evaporative (EVAP) emission system stores vapours from the fuel tank in a charcoal canister until certain engine operating conditions are present. When those conditions are present, fuel vapours are purged from the charcoal canister into the intake manifold.
- Pre-combustion control systems prevent emissions from being created in the engine, either during or before the combustion cycle. Post-combustion control systems clean up exhaust gases after the fuel has been burned. The evaporative control system traps fuel vapours that would normally escape from the fuel system into the air.
- The PCV system removes blowby gases from the crankcase and recirculates them to the engine intake.
- With the engine running at idle speed, the high intake manifold vacuum moves the PCV valve toward the closed position.
- During part-throttle operation, the intake manifold vacuum decreases, and the PCV valve spring moves the valve toward the open position. As the throttle approaches the wide-open position, intake manifold vacuum decreases, and the spring moves the PCV valve further toward the open position.
- A digital EGR valve has up to three electric solenoids operated by the PCM.
- A linear EGR valve contains an electric solenoid that is operated by the PCM with a pulse-width-modulation (PWM) signal.
- A pressure feedback electronic (PFE) sensor sends a voltage signal to the PCM in relation to the exhaust pressure under the EGR valve.
- Early secondary air injection systems pumped air into the exhaust ports during engine warm-up and delivered air to the catalytic converters with the engine at normal operating temperatures. Newer ones move air into the catalytic converter to help in the oxidation process.
- Today's vehicles have a three-way catalytic converter that reduces NO_x and oxidizes HC and CO.
- Modern diesel engines are equipped with typical emission controls but also may have an SCR system and/or a particulate filter.

REVIEW QUESTIONS

1. Name the three emissions being monitored and controlled in gasoline engines.
2. At what temperature do nitrogen atoms combine with oxygen atoms to form NO_x?
3. What system, other than the vehicle engine, may produce excess HC emissions?
4. What is a catalyst?
5. What is the purpose of injecting secondary air into the exhaust ports during warm up?
   - **a.** to help reduce NO_x
   - **b.** to compensate for the rich mixture on cold starts
   - **c.** to promote a quicker light-off of the O_2 sensor and converter
   - **d.** to reduce CO_2 emissions

6. Which location's governing body was instrumental in developing strict emission standards?
   a. New York b. California
   c. Ontario d. British Columbia
7. The PCV system prevents which emission from escaping to the atmosphere?
   a. NO_x b. CO
   c. CO_2 d. HC
8. Which emission is the AIR system designed to reduce?
   a. HC b. CO_2
   c. NO_x d. O_2
9. What is the common name given to the fluid used in selective catalytic reduction systems?
   a. urea b. methanol
   c. ether d. iso-alcohol
10. Which emission is produced due to high combustion chamber temperatures?
    a. CO b. HC
    c. NO_x d. N_2
11. Which combination below produces photochemical smog?
    a. sunlight, NO_x, and CO
    b. sunlight, NO_x, and HC
    c. sunlight, HC, and CO
    d. sunlight, CO_2, and CO
12. Which of the following systems is designed to reduce NO_x and has little or no effect on overall engine performance?
    a. PCV b. EGR
    c. AIR d. EVAP
13. While trying to control HC and CO emissions, which process would you expect to see taking place inside the oxidizing section of a functioning catalytic converter?
    a. oxidize to H_2O and CO_2
    b. oxidize to H_2O, O_2, and CO_2
    c. oxidize to H_2O and CO_2
    d. oxidize to H_2O and CO
14. Which of the following statements about carbon dioxide is *not* true?
    a. CO_2 levels are highest when the air/fuel ratio is slightly leaner than stoichiometric.
    b. The production of CO_2 is directly related to the amount of fuel consumed; therefore, more fuel-efficient engines produce less CO_2.
    c. To reduce CO_2 levels in the exhaust, engineers are working hard to decrease fuel consumption, to keep CO_2 low.
    d. The concern for CO_2 emissions is one of the primary reasons for the delayed use of alternative fuels for automobiles.
15. What device is commonly found on late-model diesel vehicles to help reduce NO_x emissions?
    a. intercooler
    b. EGR cooler
    c. fuel heater
    d. oil cooler
16. Which two provinces require mandatory vehicle emission testing in certain areas?
    a. Manitoba and Quebec
    b. British Columbia and Alberta
    c. Alberta and Manitoba
    d. British Columbia and Ontario
17. Where does the PCV direct crankcase vapours to?
    a. the intake manifold
    b. the atmosphere
    c. the fuel tank
    d. the catalytic converter
18. What is the one criteria that most EVAP systems must meet before allowing a purge cycle?
    a. Vehicle must be operating at a predetermined speed.
    b. Vehicle must be operating under a predetermined load.
    c. Vehicle must be operating at a predetermined temperature.
    d. Vehicle must be operated in a predetermined gear range.
19. Under which driving condition would EGR flow into the engine be desirable?
    a. cold engine warm-up
    b. at idle
    c. at cruising speeds
    d. at WOT
20. Although a high CO_2 levels in the exhaust emission is a good indicator of engine operation, what does it also contribute to?
    a. greenhouse gases
    b. acid rain
    c. depleted ozone
    d. increased smog

CHAPTER 35
Emission Control Diagnosis and Service

LEARNING OUTCOMES

- Use DTCs to help diagnose emissions problems.
- Briefly describe the emissions-related monitoring capabilities of an OBD-II system.
- Describe the reasons why certain gases are formed during combustion.
- Describe the inspection and replacement of PCV system parts.
- Diagnose engine performance problems caused by improper EGR operation.
- Diagnose and service the various types of EGR valves.
- Diagnose EGR vacuum regulator (EVR) solenoids.
- Diagnose and service the various intake heat control systems.
- Check the efficiency of a catalytic converter.
- Diagnose and service secondary air injection systems.
- Diagnose and service evaporative (EVAP) systems.

The quality of an engine's exhaust depends on two things: One is the effectiveness of the emission control devices; the other is the efficiency of the engine. A totally efficient engine changes all of the energy in the fuel into heat energy. This heat energy is the power produced by the engine. To run, an engine must receive fuel, air, and heat.

In order for an engine to run efficiently, it must have fuel mixed with the correct amount of air. This mixture must be **ignited** by the correct amount of heat (spark) at the correct time. All of this must happen in a sealed container or cylinder. When these conditions are met, a great amount of heat energy is produced, and the fuel and air combine to form water and carbon dioxide.

Because it is nearly impossible for any engine to receive precisely the correct amounts of air and fuel under all operating conditions, even a properly tuned running engine will emit a certain amount of pollutants. It is the job of the emission control devices to clean them up. It is important to remember that the primary purpose of OBD-II systems is to reduce emissions; therefore, the circuits and controls in those systems must work properly in order to ensure low emissions.

The three emissions controlled in gasoline and diesel engines are unburned hydrocarbons (HC), carbon monoxide (CO), and oxides of nitrogen (NO_x). Federal laws require new cars and light trucks to meet specific emissions levels. Provincial governments have also passed laws requiring that car owners maintain their vehicles so that the emissions remain below an acceptable level. Most provinces require an annual emissions inspection to meet that goal. The most common is the OBD-II test.

OBD-II TEST

Because of the ability of OBD-II systems to monitor and report emission-related failures, many provinces now use a check of the OBD-II system as part or all of their mandatory emissions testing programs. A major advantage of using the OBD-II system is decreased costs because testing does not require the use of a dynamometer or gas analyzer.

Checking the OBD-II system usually consists of three parts: MIL operation, checking for DTCs, and a readiness monitor check. The MIL operation test verifies that the MIL illuminates with the ignition on and that it goes out once the engine is started. If the light does not turn on or if it stays on once the engine is running, the vehicle fails this test. If the MIL stays on, DTCs are checked. If the MIL is off and no DTCs are stored, the readiness monitors are checked. Depending on the province, a vehicle may have to have all monitors ready and complete; some provinces may allow a monitor to be incomplete and still pass the test.

I/M 240 TEST

An I/M 240 test checks the emission levels of a vehicle while it is operating under a variety of conditions and loads. This test gives a true look at the exhaust quality of a vehicle as it is working. The testing sequence normally begins with a leakage test **(Figure 35–1)** and a functional test **(Figure 35–2)** of the EVAP system and a complete visual inspection of the emission control system. Many local governments have plans to phase out I/M 240 testing and rely totally on OBD-II tests.

During the test, the vehicle is loaded to simulate a short drive on city streets and then a longer drive on a highway. The complete test cycle includes acceleration, deceleration, and cruising. During the test, the vehicle's exhaust is collected by a constant volume sampling (CVS) system that makes sure a constant volume of ambient air and exhaust pass through the exhaust analyzer. The CVS exhaust hose covers the entire exhaust pipe; therefore, it collects all of the exhaust. The hose contains a mixing tee that draws in outside air to maintain a constant volume to the gas analyzer. This is important for the calculation of mass exhaust emissions.

The test is conducted on a chassis dynamometer **(Figure 35–3)**, which loads the drive wheels to simulate real-world conditions. Testing a vehicle while it is driven through various operating conditions on a chassis dynamometer (dyno) is called transient testing, and testing a vehicle at a constant load on a dyno is called steady-state testing.

The end result of the I/M 240 test is the measurement of the pollutants emitted by the vehicle during normal, on-the-road driving. After the test, the customer receives an inspection report. If the car failed the emissions test, it must be fixed.

The I/M 240 test measures HC, CO, NO_x, and CO_2. The job of a technician is to analyze the emission levels in order to determine the cause of the failure

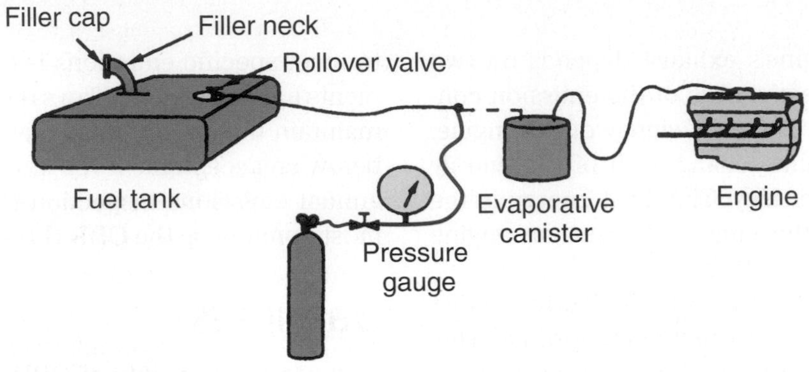

FIGURE 35–1

A typical setup for conducting a leakage test on an evaporative emission control system.

U.S. EPA

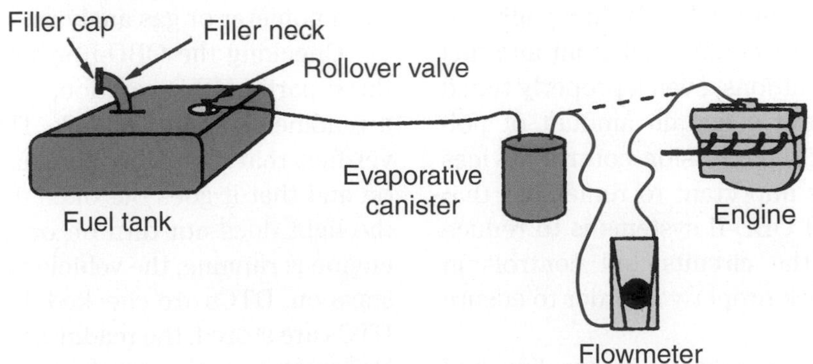

FIGURE 35–2

A typical setup for conducting a functional test of the evaporative emission control system.

U.S. EPA

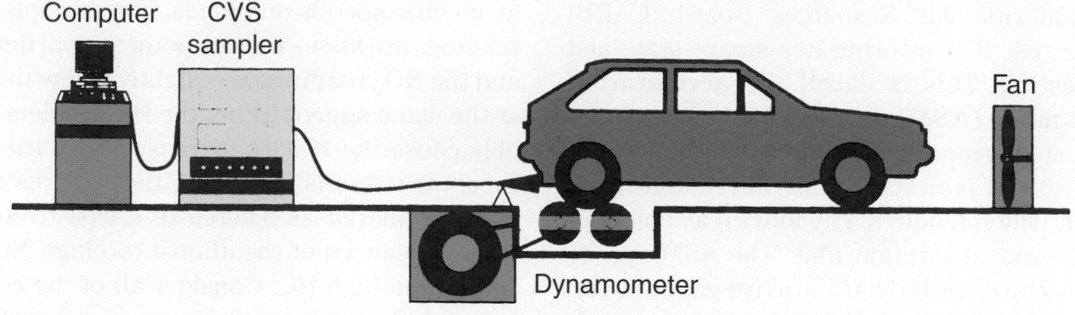

FIGURE 35–3

Components of a typical I/M 240 test station.

©Cengage Learning 2015

and under which driving conditions the vehicle failed. The drive cycle is actually six operating modes. These modes need to be duplicated to identify why the vehicle failed and if the problem has been fixed:

Mode one—idle, no load at 0 km/h
Mode two—acceleration from 0 to 55 km/h
Mode three—acceleration from 55 to 90 km/h
Mode four—a steady cruise at 55 km/h
Mode five—a steady high cruise at 90 km/h
Mode six—decelerations from 55 km/h to 0 and from 90 km/h to 0

Chassis Dynamometer

Chassis dynamometers are used during the I/M 240 test and can be valuable when diagnosing other driveability problems such as finding the cause of low power, overheating, and speedometer accuracy. A chassis **dynamometer** is designed to simulate the various road conditions in which a vehicle is driven. With a chassis dyno, a vehicle can be driven on rollers **(Figure 35–4)**. This allows the vehicle to be driven through the test conditions while it is stationary.

It is important to realize that OBD-II systems do not return all systems back to their normal state after repairs. The repaired vehicles should be run through a drive cycle to make sure that the repairs have been made correctly and that all vehicle parameters are within acceptable limits.

It should be also noted that all emission testing programs will fail a vehicle if the check engine light is illuminated or if the readiness code is not at its normal or ready state.

Other I/M Testing Programs

Many I/M programs only measure the emissions from a vehicle while it is idling. The test is conducted with a certified exhaust gas analyzer. The measurements of the exhaust sample are then compared to standards dictated by the local governments according to the production year of the vehicle.

Some test procedures also include a preconditioning mode at which the engine is run at a high idle (approximately 2500 rpm), or the vehicle is run at 50 km/h (30 mph) on a dynamometer for 20 to 30 seconds prior to taking the idle tests. This preconditioning mode heats up the catalytic converter, allowing it to work at its best. Some programs include the measurement of the exhaust gases during a low constant load on the dyno or during a constant high idle. These measurements are taken in addition to the idle tests. A visual inspection and functional test of the emission control devices is an important component to I/M programs. If the vehicle has emission control devices that are missing, non-functioning, or show signs of tampering, the vehicle will fail the inspection.

FIGURE 35–4

The wheels of this vehicle are ready to be dropped into the rollers of a chassis dyno.

©Cengage Learning 2015

The California Air Resources Board (CARB) developed a test that incorporates steady-state and transient testing. This is called the **acceleration simulation mode (ASM)** test. The ASM test includes a high-load steady-state phase and a 90-second transient test. This test is an economical alternative to the I/M 240 test, which requires a dyno with a computer-controlled power absorption unit. The ASM can be conducted with a chassis dyno and a five-gas analyzer.

Another alternative to the I/M 240 test is the **repair grade (RG-240)** test. This program uses a chassis dynamometer, constant volume sampling, and a five-gas analyzer. It is very similar to the I/M 240 test, but it is more economical. The primary difference between the I/M 240 and the RG-240 is the chassis dyno. The dyno for the RG-240 is less complicated but nearly matches the load simulation of the I/M 240 dyno.

Interpreting the Results of an I/M Test

The report from an I/M 240 test shows the amount of gases emitted during the different speeds and loads of the test. The report also gives the average output for each of the gases and shows the cut-point for the various gases. The **cut-point** is the maximum allowable limit of each gas. Nearly all vehicles will have some speeds and conditions where the emissions levels are above the cut-point. A vehicle that fails the test will have many areas of high pollutant output.

When using the report as a diagnostic tool, pay attention to all of the gases. Also attempt to identify the loads and speeds at which the vehicle went over the cut-point **(Figure 35–5)**. Think about what system or systems are responding to the load or speed. Pay attention not only to the gases that are above the cut-point but also to those that are below. Also think about the relationship of the gases at specific speeds and loads. For example, if the HC readings are above the cut-point at a particular speed and the NO_x readings are slightly below the cut-point at the same speed, fixing the HC problem will probably cause the NO_x to increase above the cut-point. As combustion is improved, the chances of forming NO_x also increases. Therefore, it wise to consider the possible causes of the almost too high NO_x in addition to the high HC. Consider all of the correlations between the gases when diagnosing a problem.

TESTING EMISSIONS

Testing the quality of the exhaust is both a procedure for testing emission levels and a diagnostic routine. An exhaust analyzer is one of the valuable diagnostic tools for diagnosing driveability problems **(Figure 35–6)**. By looking at the quality of an

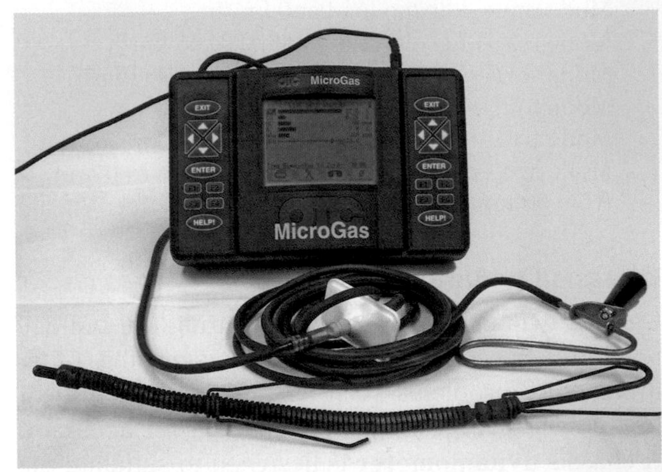

FIGURE 35–6

A handheld exhaust gas analyzer.

©Cengage Learning 2015

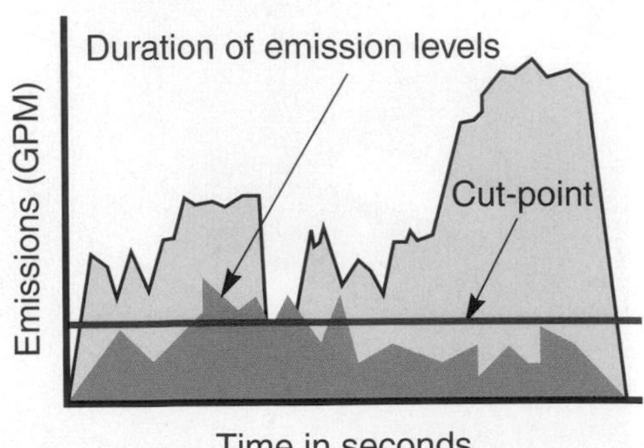

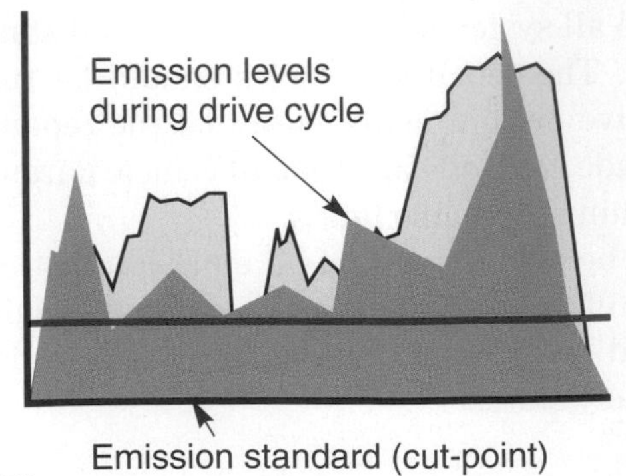

FIGURE 35–5

(Left) The emissions level is acceptable; (right) the emissions level is not acceptable.

©Cengage Learning 2015

engine's exhaust, a technician is able to look at the effects of the combustion process. Any defect can cause a change in exhaust quality. The amount and type of change serves as the basis of diagnostic work.

Exhaust Analyzer

Early emission analyzers measured the amount of HCs and COs in the exhaust. High HC emissions indicate that there are trace amounts of raw, unburned fuel and/or oil vapour in the exhaust stream. Emissions analyzers measure HC in **parts per million (ppm)** or grams per kilometre (g/km). CO is an odourless, toxic gas that is the product of combustion and is typically caused by a lack of air or excessive fuel. CO is typically measured as a percent of the total exhaust.

SHOP TALK

The exhaust analyzers used in I/M 240 and similar tests do not measure the emissions levels in ppm or a percentage. Rather, they measure the gases in grams per kilometres (g/km). This is because an I/M test is dynamic; during the test, the vehicle is operated at a variety of speeds and loads. There is no direct equivalency between the two measuring systems.

An exhaust analyzer has a long sample hose with a probe in the end of the hose. Before measuring the exhaust gases, disable the air injection system, if the engine is equipped with one. Start the engine. Once the engine is warmed up, insert the probe in the tailpipe **(Figure 35–7)**. When the analyzer is turned on, an internal pump moves an exhaust sample from the tailpipe through the sample hose and the analyzer. A water trap and filter in the hose removes moisture and carbon particles.

The pump forces the exhaust sample through a sample cell in the analyzer. The exhaust sample is then vented to the atmosphere. In the sample cell, the analyzer determines the quantities of HC and CO if the analyzer is a two-gas analyzer, or HC, CO, carbon dioxide (CO_2), and oxygen (O_2) if it is a four-gas analyzer. Most newer analyzers also measure NO_x and are called five-gas analyzers **(Figure 35–8)**.

By measuring NO_x, CO_2, and O_2, in addition to HC and CO, a technician gets a better look at the efficiency of the engine **(Figure 35–9)**. Ideally, the combustion process will combine fuel (HC) and O_2 to form water and CO_2. Although most of the air brought into the engine is nitrogen, this gas should not become part of the combustion process and should pass out of the exhaust as nitrogen.

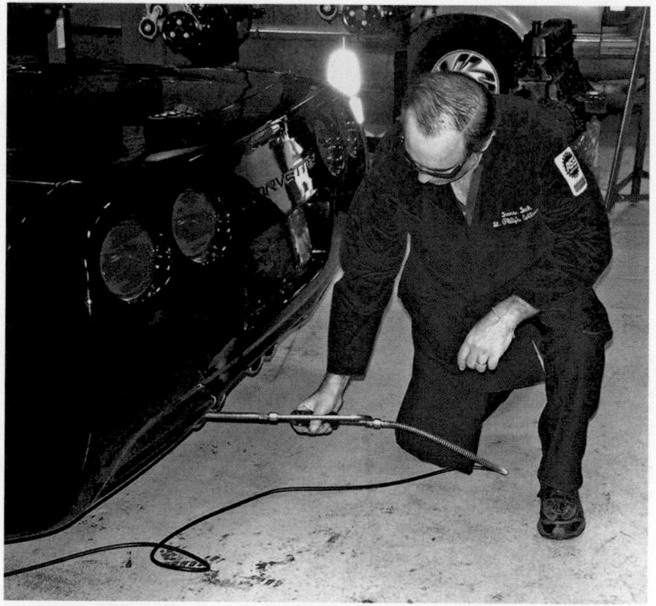

FIGURE 35–7

The analyzer's probe is inserted into the vehicle's tailpipe.

©Cengage Learning 2015

Maximum limits for the measured gases are set by regulations according to the model year of the vehicle. These limits also vary by province or locale. It is always desirable to have low amounts of four of the five measured gases at all engine speeds. The only gas that is desired in high percentages is CO_2. Normally, the desired amount of CO_2 is 13.4 percent or more. The other gases should be kept to the lowest

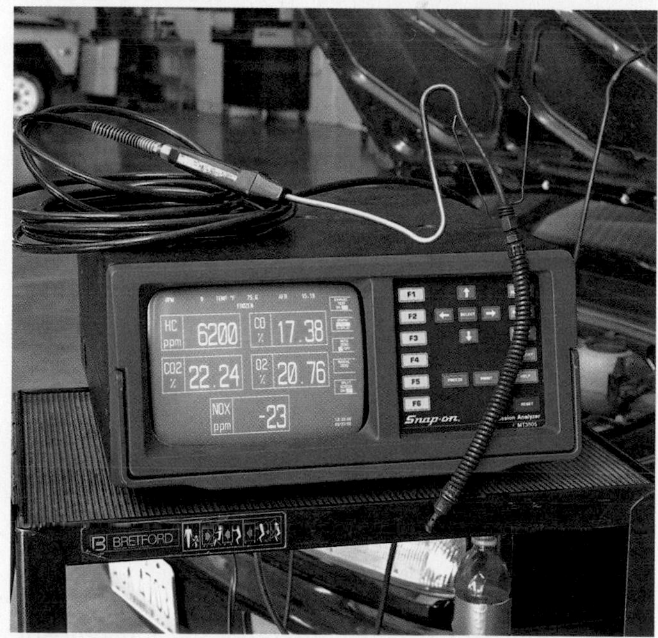

FIGURE 35–8

A five-gas exhaust analyzer.

©Cengage Learning 2015

	IDLE	2500 RPM	PROBABLE CAUSE
HC ppm CO% CO_2% O_2% NO_xppm	0–150 1–15 10–12 0.5–2.0 100–300	0–75 0–0.8 11–13 0.5–1.25 200–1000	Normal reading
HC ppm CO% CO_2% O_2% NO_x ppm	0–150 3.0+ 8–10 0–0.5 0–200	0–75 3.0+ 9–11 0–0.5 100–500	Rich mixture
HC ppm CO% CO_2% O_2% NO_x ppm	0–150 0–1.0 8–10 1.5–3.0 300–1000	0–75 0–0.25 11 1.0–2.0 1000+	Lean mixture
HC ppm CO% CO_2% O_2% NO_x ppm	50–850 0–0.3 5–9 4–9 300–1000	50–750 0–0.3 6–10 2–7 1000+	Lean misfire
HC ppm CO% CO_2% O_2% NO_x ppm	50–850 0.1–1.5 6–8 4–12 0–200	50–750 0–0.8 8–10 4–12 100–500	Misfire

FIGURE 35–9

The readings of the chemicals in the exhaust can lead the technician to the cause of a driveability problem.

levels possible. Typically, CO should be 0.5 percent or less; O_2 should be about 0.1 to 1 percent; HC should be about 100 ppm or less; and NO_x should be 200 ppm or less.

Many of the emission control devices that have been added to vehicles over the past 30 years have decreased the amount of HC and CO in the exhaust. This is especially true of catalytic converters. These devices alter the contents of the exhaust. Therefore, checking the HC and CO contents in the exhaust may not be a true indication of the operation of an engine. However, a look at the other measured gases, those not altered by the system, can give a good picture of the engine and its systems.

Interpreting the Results

When using the exhaust analyzer as a diagnostic tool, it is important to realize that the severity of the problem dictates how much higher than normal the readings will be.

The levels of HC and CO in the exhaust are a direct indication of engine performance. Unburned HCs are particles, usually vapours, of gasoline that have not been burned during combustion. They are present in the exhaust and in crankcase vapours. Of course, any raw gas that evaporates out of the fuel system is classified as HC. Today's engines are designed to be efficient and therefore release lower amounts of HCs in the exhaust. However, it is only possible to have complete combustion in a laboratory; therefore, all engines will release some HCs. This is because there is always some fuel in the combustion chamber that the flame front cannot reach.

CO forms when there is not enough O_2 to combine with the carbon during combustion. CO is a by-product of combustion. If combustion does not take place, CO will be low. When the engine receives enough O_2 in the mixture, CO_2 is formed. Lower levels of CO are to be expected when HC readings are high. When misfiring occurs, there is less total combustion in the cylinders. Therefore, CO will decrease slightly.

High NO_x readings indicate high combustion temperatures. CO_2 is a desired element in the exhaust stream and is only present when there is combustion; thus the more CO_2 in the exhaust stream, the better.

O_2 is used to oxidize CO and HC into water and CO_2. Ideally, there should be very low amounts of O_2 in the exhaust. CO and O_2 are inversely related: When one goes up, the other goes down.

SHOP TALK

Due to the inter-relationship between exhaust gases, it is quite possible to have one exhaust emission increase (or decrease) once you have corrected another.

GENERAL GUIDELINES When using gas measurements to diagnose a driveability problem, keep the following in mind:

- HC is raw unburned fuel or oil vapours.
- High CO indicates an excessively rich air/fuel mixture caused by a restriction in the air intake system or too much fuel being delivered to the cylinders.
- When CO goes up, O_2 goes down.
- O_2 is used by the catalytic converter and may be low.
- O_2 readings are higher on vehicles with properly operating air injection systems.
- If there is a high O_2 reading without a low CO reading, an air leak in the exhaust system may be indicated.
- As the mixture goes lean, O_2 levels increase.
- As the mixture goes rich, O_2 moves and stays low.
- If there is high HC and low CO in the exhaust because of a lean misfire, the amount of O_2 will be high.
- NO_x goes high when there is a lean mixture or when combustion chamber temperatures go excessively high
- NO_x is highest when CO and HC are lowest.
- Minimal NO_x is produced while an engine idles and the vehicle is not under load.
- At stoichiometric (lambda—λ), HC and CO will be at their lowest levels, and CO_2 will be at its highest.
- If there is an extremely rich air/fuel mixture, there will be high HC and CO and low NO_x and O_2 readings.
- A deficiency of O_2 reduces NO_x.

For specific causes of undesirable amounts of exhaust gases, see **Figure 35–10**.

Problems with emission control devices not only will cause an increase in emission levels but

CONDITION	POSSIBLE CONDITION/PROBLEM
Excessive HC	Ignition system misfiring Incorrect ignition timing Excessively lean air/fuel ratio Dirty fuel injector Low cylinder compression Excessive EGR dilution Defective valves, guides, or lifters Vacuum leaks Defective system input sensor
Excessive CO	Rich air/fuel mixtures Plugged PCV valve or hose Dirty air filter Leaking fuel injectors Higher than normal fuel pressures Ruptured diaphragm in the fuel pressure regulator Defective system input sensor

FIGURE 35–10

Possible conditions causing various emission levels. *(continued)*

CONDITION	POSSIBLE CONDITION/PROBLEM
Excessive HC and CO	Plugged PCV system Excessively rich air/fuel ratio AIR pump inoperative or disconnected Engine oil diluted with gasoline
Lower than normal O_2	Rich air/fuel mixtures Dirty air filter Faulty injectors Higher than normal fuel pressures Defective system input sensor Restricted PCV system Charcoal canister purging at idle and low speeds
Lower than normal CO_2	Leaking exhaust system Rich air/fuel mixture
Higher than normal O_2	Engine misfire Lean air/fuel mixtures Vacuum leaks Lower than specified fuel pressures Defective fuel injectors Defective system input sensor
Higher than normal NO_x	Overheated engine Carbon deposits on the intake valves Lean air/fuel mixtures Vacuum leaks Overadvanced ignition timing Defective EGR system Ineffective reduction catalytic converter
Higher than normal NO_x and HC	Poor flow through the radiator Partially closed thermostat Bad water pump Inactive EGR system
Higher than normal HC, O_2, and NO_x	False signal from the O_2 sensor Out-of-calibration MAP sensor Plugged injector Low fuel pressure Vacuum leak Engine knock Overadvanced ignition timing
Higher than normal HC, CO, and NO_x	Carbon buildup on top of the piston or cylinder walls Worn or slipping timing belt

FIGURE 35–10 *(continued)*

also may cause driveability problems. A no-start or hard-to-start problem can be caused by any number of faults, which in some cases may include emission control devices such as EVAP or EGR system components. If the engine runs rough or stalls, the fault may be related to the EVAP system, a disconnected or damaged vacuum hose, a stuck-open EGR valve, or a faulty PCV system. Excessive oil consumption can be caused by bad piston rings, poor valve oil seals, or a plugged PCV hose. Poor fuel economy can also be attributed to emission system faults such as a malfunctioning EGR system.

BASIC INSPECTION

Before diagnosing the cause of excessive emissions, make sure you inspect the entire vehicle for obvious problems. The results of the inspection may help pinpoint the cause of the problem.

- Test battery voltage.
- Inspect the air filter, and service as necessary. Also inspect the intake air system for debris or contamination.
- Disconnect the air inlet line hose from the charcoal canister. Then check that air can flow freely into the air inlet. If air cannot flow freely, repair or replace it.
- On older vehicles, check and adjust the idle speed.
- On older vehicles, make sure that base timing is set to specifications.
- Check the fuel pressure to ensure it is at specifications.
- Verify the engine's mechanical condition by using a vacuum gauge and compression tester, if needed.
- Check all accessible electrical connections and vacuum and air induction hoses and ducts.
- Check the condition of the PCM main grounds.
- Check all EVAP, EGR, and PCV hoses, connections, and seals for signs of leakage and damage.
- Inspect for unwanted fuel entering the intake manifold from the EVAP system, fuel pressure regulator diaphragm, or PCV system.
- Ensure there is good sealing at the fuel tank and filler pipe.

Part of the inspection should also include a look at the MIL. Make sure the MIL illuminates during key-on bulb check. If the MIL remains on once the engine is started, a DTC is stored, and this can lead to the cause of excessive emissions. Keep in mind that the primary purpose of an OBD-II system is to limit the amount of pollutants emitted by the engine. This is why retrieving DTCs from the control system is an important step when diagnosing emission problems. OBD-II systems include several system monitors that relate to exhaust emission.

OBD-II Monitors

As explained in the previous chapter, the OBD-II system monitors the emission control systems and components that can affect tailpipe or evaporative emissions. Through this monitoring process, problems are detected before emissions are 1.5 times greater than applicable emission standards. Each monitor requires one or more diagnostic tests. The monitors run continuously or non-continuously and only when certain enabling criteria have been met. If a system or component fails these tests, a DTC is stored and the MIL is illuminated within two driving cycles.

GO TO ▶ Chapter 31 for a discussion of OBD-II monitors.

A pending DTC is stored in the PCM when a concern is initially detected. Pending DTCs are displayed as long as the concern is present. Note that OBD-II regulations require a complete concern-free monitoring cycle to occur before erasing a pending DTC. This means that a pending DTC is erased on the next power-up after a concern-free monitoring cycle. However, if the concern is still present after two consecutive drive cycles, the MIL is illuminated. Once the MIL is illuminated, three consecutive drive cycles without a concern detected are required to turn off the MIL. The DTC is erased after 40 engine warm-up cycles once the MIL is extinguished.

A requirement of many mandatory I/M inspections is to look at the OBD-II system tests. This is done by observing the system status display on a scan tool. If the diagnostic tests for a particular monitor have been completed, the scan tool will indicate this with a YES or COMPLETE **(Figure 35–11)**. A system monitor is complete when either all of the tests comprising the monitor have been run and had satisfactory results, or when DTCs related to the

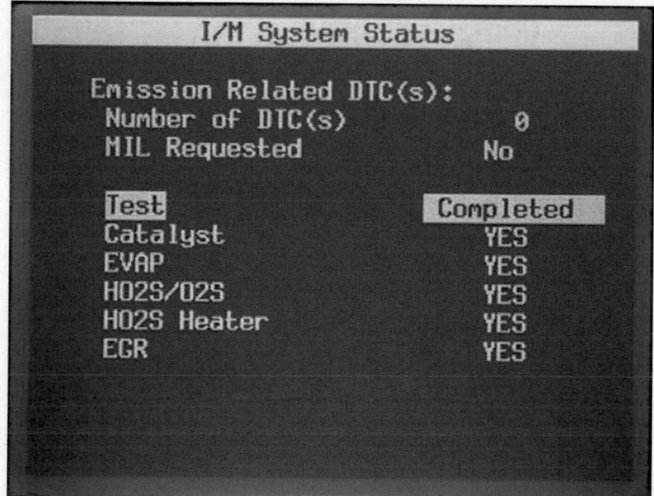

FIGURE 35–11

An I/M status screen on a scan tool, showing that the monitors have been completed.

monitor have turned on the MIL. If for any reason the tests are not completed, the system status display will indicate NO or NOT COMPLETE under the completed column. If the system has set a DTC that is associated with one of the regulated systems, the required tests may not be able to run. If there are no DTCs that would prevent the vehicle from completing a particular test, the vehicle should be operated according to the recommended drive cycle so that all enable criteria are met.

Not all vehicles have the same emission control systems. For example, a vehicle may not be equipped with an AIR or EGR. Therefore, the status of these monitors will not be shown, or the scan tool may display NOT SUPPORTED. Also, the status of some monitors, such as the misfire and comprehensive component monitors, may not be listed. These monitors run continuously and therefore their status is not required by OBD-II.

Once repairs are completed, the vehicle should be operated according to the drive cycle recommended by the manufacturer. This will allow the monitors to run and complete their tests. Keep in mind that a monitor may require very specific conditions to run. Refer to the service information for monitor-enable criteria to ensure that the vehicle and environmental conditions will allow the drive cycle to complete.

PROCEDURE

Typical Procedure for Allowing the I/M System Status Tests to Complete

STEP 1 Observe the DTCs with a scan tool. If there is a DTC that would prevent the system status tests from completing, diagnose it before continuing.

STEP 2 Check the available service information to see if there are software updates for the vehicle.

STEP 3 Check the scan tool for the current status of the monitors. If the EVAP monitor has not been completed, diagnose and repair the system before continuing.

STEP 4 Turn off all accessories.

STEP 5 Open the hood. Set the parking brake, and place the transmission in PARK for automatic transmissions or NEUTRAL for manual transmissions.

STEP 6 Start the engine and allow it to idle for about two minutes.

STEP 7 Close the hood. Release the parking brake, and take the vehicle on a test drive.

STEP 8 Lightly accelerate to 70 to 80 km/h (45 to 50 mph), and maintain this speed for about five minutes after the engine reaches normal operating temperature.

STEP 9 Lightly accelerate to 90 km/h (55 mph), and maintain this speed for about two minutes.

STEP 10 Let off the throttle and allow the vehicle to decelerate for about 10 seconds.

STEP 11 Stop the vehicle, and allow it to idle for about two minutes.

STEP 12 Turn off the ignition, and allow the vehicle to sit undisturbed for about 45 minutes.

STEP 13 Check the status of the monitors with a scan tool.

STEP 14 If all tests were completed, retrieve all hard and pending DTCs.

STEP 15 If the tests were not completed, repeat the procedure.

EVAPORATIVE EMISSION CONTROL SYSTEM DIAGNOSIS AND SERVICE

WARNING!

If gasoline odour is present in or around a vehicle, check the EVAP system for cracked or disconnected hoses, and check the fuel system for leaks. Gasoline leaks or vapours can cause an explosion, resulting in personal injury and/or property damage. The cause of fuel leaks or fuel vapour leaks should be repaired immediately.

Most I/M inspections begin with a check of the EVAP system. This includes checking the results of the EVAP monitor's diagnostic tests. The EVAP system is monitored for leaks and its ability to move fuel vapour when commanded by the PCM to do so. Problems that are detected will set a DTC. Normally, this is the only way a driver knows there is an EVAP problem. At times, the owner may complain of a gasoline odour in or around the vehicle. The common tools for troubleshooting an EVAP system are a diagnostic tester, smoke machine, and pressure kit. The exact procedures and tools will vary with each application.

EVAP Monitors

Many different EVAP systems and monitoring systems have been used by manufacturers. It is important to correctly identify the system used on a particular vehicle before doing any diagnostic procedures or service on the vehicle. A system leak can generate multiple DTCs, depending on the component and location of the leak. The monitor also checks the electrical integrity of the system and the vapour purge rate **(Figure 35–12)**. All EVAP monitor DTCs require two trips.

To monitor the vapour purge in most systems, the PCM changes the duty cycle of the purge solenoid while monitoring the short-term fuel trim (STFT). A problem is indicated by uncorrelated changes in the STFT. Other systems use a pressure sensor to monitor the pressure in the system and the purge side of the charcoal canister. Normally, the changes in the pressures between the two are very small. A difference in pressure should be evident when the PCM commands the purge solenoid to open. If there is no difference in pressure, the PCM determines that the purge system is not working properly and will set a DTC. Enhanced EVAP systems may also calculate purge flow rate by looking at the inputs from the mass airflow (MAF) sensor or by comparing manifold absolute pressure (MAP) sensor readings with the input from the tank pressure sensor.

Early EVAP systems were designed to detect leaks that were 1 mm (0.04 in.) and greater. Starting with the 2000 model year, a new EVAP monitor system has been implemented. This system is designed to detect leaks of 0.5 mm (0.02 in.). There are two different methods of leak detection used by enhanced OBD-II systems: system pressure and vacuum. Vacuum testing requires two solenoid valves and a fuel tank absolute-pressure sensor. The vacuum comes from the engine or an on-board vacuum pump (commonly referred to as the leak detection pump, or LDP). Using the pressure method of checking the system tends to be more precise but entails the use of more components. The method used by manufacturers varies with vehicle make and model year.

EVAP CODE	DESCRIPTION
P0441	Incorrect or uncommanded purge flow
P0442	Small leak
P0443	Purge solenoid electrical fault
P0446	Blocked canister vent
P0449	Canister vent solenoid electrical fruit
P0452	Tank pressure sensor voltage low
P0453	Tank pressure sensor voltage high
P0454	Tank pressure sensor voltage noisy
P0455	Large leak
P0456	Very small leak
P0457	Gross leak
P0460	Fuel level circuit
P0461	Fuel level sensor stuck/noisy
P0462	Fuel level sensor voltage low
P0463	Fuel level sensor voltage high
P0464	Fuel level sensor voltage noisy at idle
P1443 (Ford)	Gross leak, no flow
P1450 (Ford)	Excessive vacuum
P1451 (Ford)	Vent valve circuit fault
P1486 (Chrysler)	Pinched hose
P1494 (Chrysler)	LDP fault
P1495 (Chrysler)	LDP solenoid circuit

FIGURE 35–12

Common EVAP codes.

SCAN TOOL Being a system controlled and monitored by the OBD-II, the MIL will illuminate to alert the driver that there is a problem if a failure is detected. When a defect occurs in the canister purge solenoid and related circuit, a DTC is normally set. If a DTC related to the EVAP system is set, always correct the cause of the code before doing any further diagnosis of the EVAP system.

When the scan tool is set to the appropriate mode, it will indicate whether the purge solenoid is on or off. With the engine idling, the purge solenoid should be off. Be sure all the conditions required to energize the purge solenoid are present; note that this often requires the fuel level to be between 15 and 85 percent full. If enable criteria are met, leave the scan tool connected, road-test the vehicle, and observe this solenoid status on the scan tool. The tester should indicate when the purge solenoid is on. If the purge solenoid is not on under the necessary conditions, check the power supply wire to the solenoid, the solenoid winding, and the wire from the solenoid to the PCM.

Diagnosis

If the EVAP system is purging vapours from the charcoal canister when the engine is idling or operating at very low speed, rough engine operation may occur, depending on manufacturer, especially at higher ambient temperatures. Cracked hoses or a canister saturated with gasoline may allow gasoline vapours

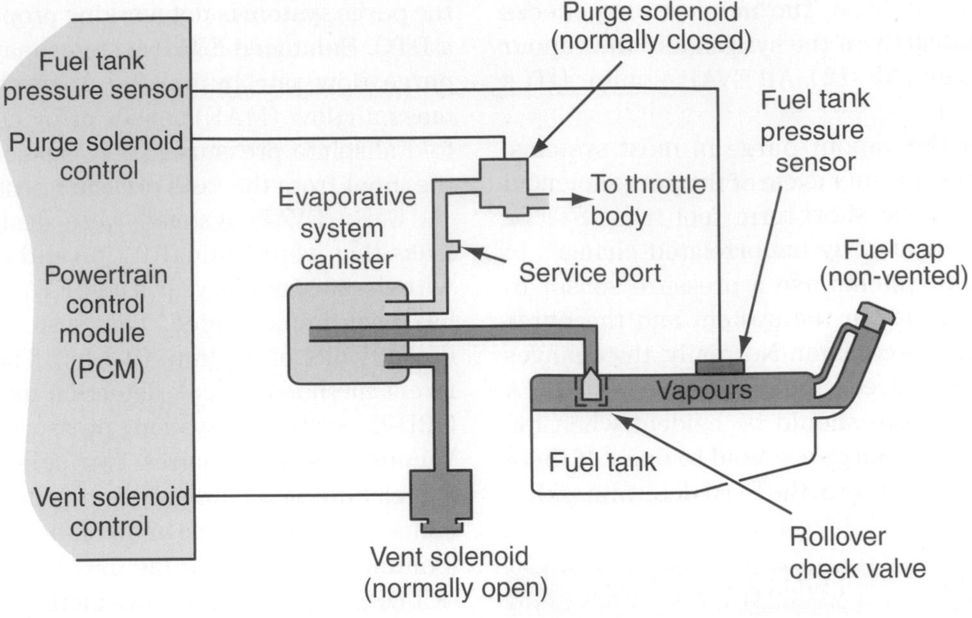

FIGURE 35–13

Check all hoses and electrical connectors and wires in the EVAP system.

to escape to the atmosphere, resulting in gasoline odour in and around the vehicle.

The entire EVAP system should be carefully inspected **(Figure 35–13)**. All of the hoses in the system should be checked for leaks, restrictions, and loose connections. Repair or replace them as necessary. Also check the canister for cracks or damage; if they are found, replace the charcoal canister assembly.

The operation of the charcoal canister **(Figure 35–14)** can be quickly checked on most vehicles. Close the purge port, gently blow into its vent port, and check that air flows from the port. If this does not happen, replace the charcoal canister assembly. Close the vent port, and gently blow into its air inlet port. Check that air flows from the purge port. If this does not happen, replace the charcoal canister assembly. With the purge port and air inlet port closed, apply low air pressure (about 19.6 kPa [2.8 psi]) to the vent port. Make sure that the canister can hold that pressure for at least one minute. If this does not happen, replace the charcoal canister assembly.

Also make certain that the canister filter is not completely saturated. Remember that a saturated charcoal filter can cause symptoms that can be mistaken for fuel system problems. Rough idle, flooding, and other conditions can indicate a canister problem. A canister filled with liquid or water

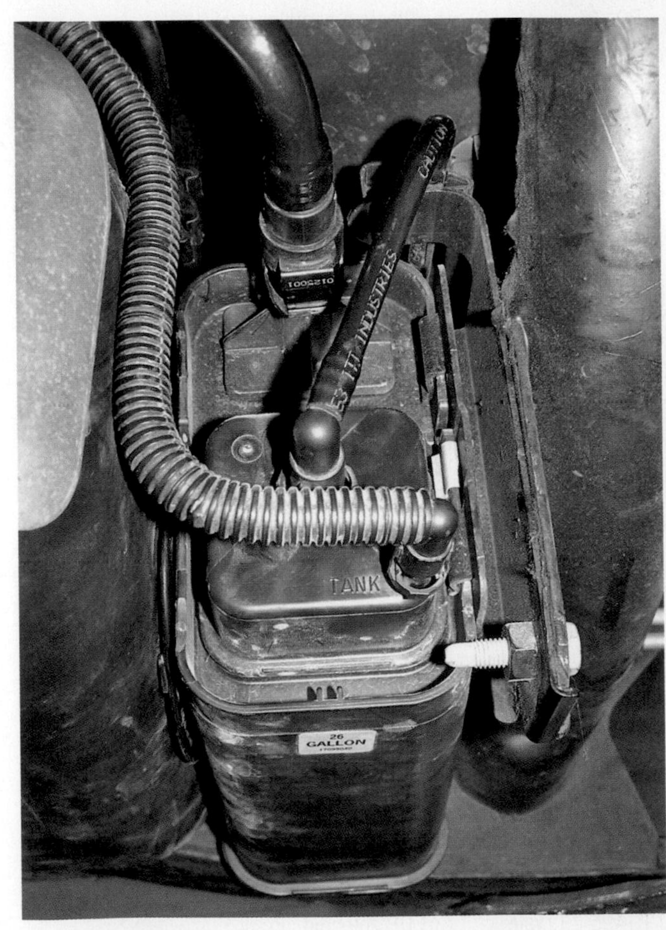

FIGURE 35–14

The charcoal canister for a typical EVAP system.

causes back pressure in the fuel tank. It can also cause richness and flooding symptoms during purge or start-up.

To test for saturation, unplug the canister momentarily during a diagnosis procedure, and observe the engine's operation. If the canister is saturated, either it or the filter must be replaced, depending on its design; that is, some models have a replaceable filter, whereas others do not.

A vacuum leak in any of the evaporative emission components or hoses can cause starting and performance problems, as can any engine vacuum leak. It can also cause complaints of fuel odour. Incorrect connection of the components can cause rich stumble or lack of purging (resulting in fuel odour).

The canister purge solenoid **(Figure 35–15)** winding may be checked with an ohmmeter. You can also test the solenoid by commanding it open and testing if there is flow through the solenoid. With the tank pressure control valve removed, try to blow air through the valve from the tank side of the valve. Use controlled shop air, equal to about the same pressure you can produce from your mouth. Some restriction to airflow should be felt until the air pressure opens the valve. Connect a vacuum hand pump to the vacuum fitting on the valve, and apply 254 mm Hg (10 in. Hg) to the valve. Now try to blow air through the valve from the tank side. Under this condition, there should be no restriction to airflow. If the tank pressure control valve does not operate properly, replace the valve.

The electrical connections in the EVAP system should be checked to ensure that they are secure, and also inspected for corroded terminals and worn insulation. Measure the voltage drop across the solenoid, and compare your readings to specifications. If the readings are outside specifications, replace the charcoal canister assembly.

SHOP TALK

It is recommended to never wiggle hoses or tighten fittings and caps during a visual inspection until the system has been pressurized for further testing.

Purge Test

Certain conditions will increase the amount of fuel vapours stored in the charcoal canister and therefore should be avoided prior to I/M tests. These include prolonged periods of idling or prolonged periods of sitting in the sun on a hot day. Most vehicles should be driven at highway speeds for about five minutes to allow them to complete their normal purge cycle before I/M testing. Excessive fuel vapours can cause an increase in CO during testing. To determine if the canister is the cause of high CO, isolate the EVAP system, and retest the emission levels. The EVAP can be isolated by disconnecting the purge hose from the throttle body or intake manifold. If the EVAP system appears to be the cause of high emissions, the system needs to be carefully checked.

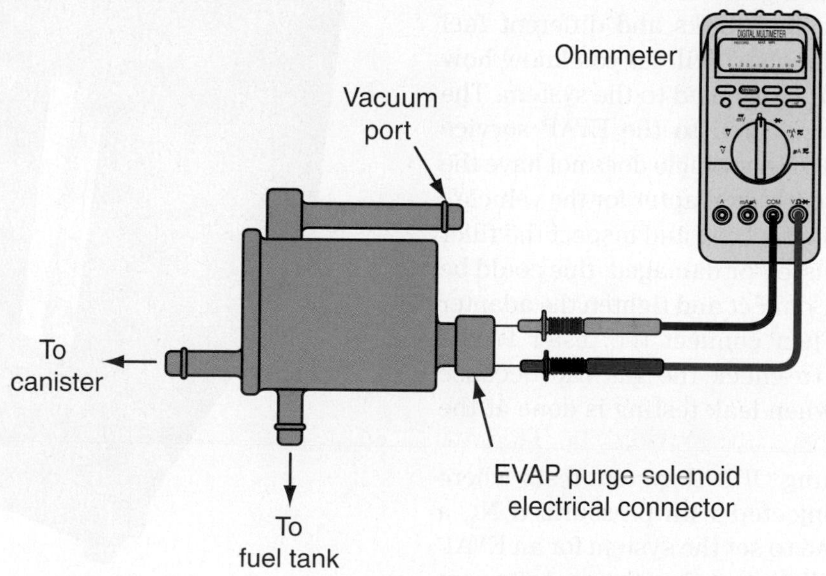

FIGURE 35–15

Checking an EVAP purge solenoid.

Before checking the purge rate of the EVAP system, allow the engine to run until it reaches normal operating temperature. Connect the purge flow tester's flow gauge in series with the engine and evaporative canister. Zero the gauge of the tester with the engine off, and then start the engine. With the engine at idle, turn on the tester, and record the purge flow rate and accumulated purge volume. Gradually increase engine speed to about 2500 rpm, and record the purge flow. A good working system will have at least 1 L of flow within a few seconds. Most vehicles will have a flow of 25 L during the time period required for an I/M 240 test. If the system does not establish at least 1 L of flow during the 240 seconds, the system needs to be carefully diagnosed, repaired, and retested.

Leak Tests

The operation of a pressure test is often called a pressure decay test. It checks for leaks that allow fuel vapours to escape into the atmosphere. EVAP leak testers pressurize the system with a very low pressure 356 mm (14 in.) of water, or about 3.5 kPa (0.5 psi). The tester typically uses pressurized nitrogen to fill the system. Nearly all OEMs recommend nitrogen (N_2) rather than compressed air. In fact, many are concerned about introducing oxygen (outside air has much oxygen in it) into the system because it can cause an explosion.

Before using an EVAP leak tester, make sure that the fuel tank is at least half full. The tester kit will include a pressure guideline chart. The chart contains various-size fuel tanks and different fuel levels. Interpreting the chart will let you know how much pressure should be applied to the system. The tester is normally connected to the EVAP service port **(Figure 35–16)**. If the vehicle does not have the service port, use the correct adapter for the vehicle's filler neck. Remove the fuel cap and inspect the filler neck. If the neck is rusted or damaged, this could be a source of leakage. Connect and tighten the adapter to the filler neck. Then connect the tester to the adapter. Remember to check the gas cap because it will be bypassed when leak testing is done at the filler neck.

Prior to performing OBD-II system tests where the system will be injected with pressurized N_2, a scan tool must be used to set the system for an EVAP test. This process will ensure that the canister vent solenoid is in the closed position. Ground the tester **(Figure 35–17)** to the vehicle before turning it on, which is typically done by connecting a jumper

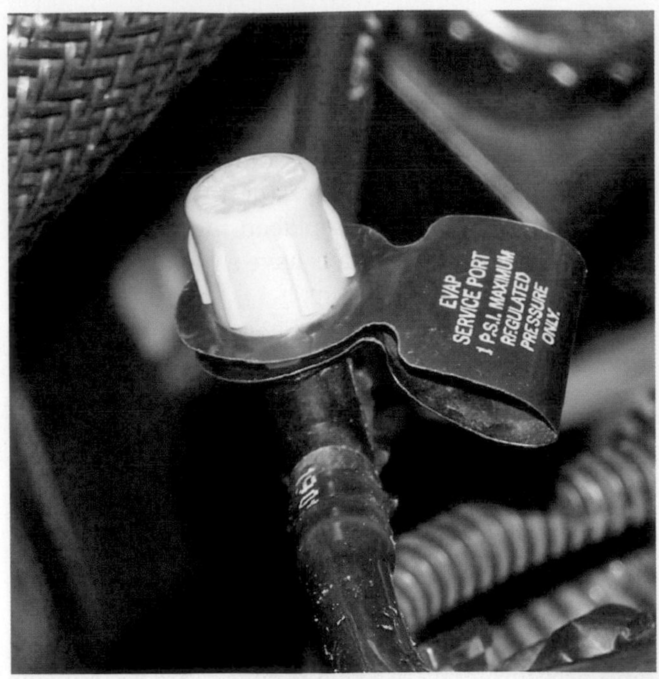

FIGURE 35–16

An EVAP service port.

©Cengage Learning 2015

FIGURE 35–17

An EVAP leak tester.

Snap-on Incorporated

wire from the ground screw on the tester to a bolt at the filler neck. Doing this eliminates the chance of **electrostatic discharge (ESD)**. ESD can ignite the fuel vapours. Then pinch off the vent hose at the carbon canister on non-OBD-II systems.

Adjust the output pressure from the tester so that it agrees with the chart mentioned previously. Once the tester applies pressure to the system, its gauge will show how much pressure the system is able to hold **(Figure 35–18)**. The pressure will initially fall and rise before it stabilizes. If the system remains above 203.2mm (8 in.) of water after two minutes, the vehicle has passed the test. Obviously, if there is a leak, the system will not be able to hold pressure. If the pressure drops dramatically, listen for leaks at the fuel cap, tank, connections, valves, and hoses.

The source of the leak may be found by spraying the suspected areas with soapy water and looking for bubbles. Leaks can also be found by using an ultrasonic leak detector that will detect the sound made by the leaking vapours. An additional way to find the leak is through the use of an exhaust analyzer or combustible gas detector. This method can be a bit frustrating because it will respond only to fuel vapours. Because the nitrogen in the system will push the vapours out, the probe of the analyzer must be placed at the suspected area prior to or immediately after the system is pressurized. Obviously, the analyzer will read high HCs at the point of the leak. Once the source of the leak is found, make the necessary repairs and retest the system.

USING DYE To help identify the source of a leak, ultraviolet dye can be added to the system. The dye will identify the source of a leak when the system is inspected with a UV lamp. The dye is approved by many OEMs and will not harm the system, catalytic converters, or H_2OS.

SMOKE TEST A very popular way to identify leaks is with a smoke machine. Nearly all OEMs recommend this method. The smoke machine vaporizes a specially formulated, highly refined mineral oil–based fluid. The machine then introduces the resultant smoky vapour into the EVAP system.

Pressurized nitrogen pushes the smoke through the system. The source of a leak is identified by the escaping smoke **(Figure 35–19)**. Similar to when pressure testing the system, the system must be sealed to get accurate results. This means the canister vent port must be blocked off. On some vehicles, the solenoid can be closed through commands input with a scan tool.

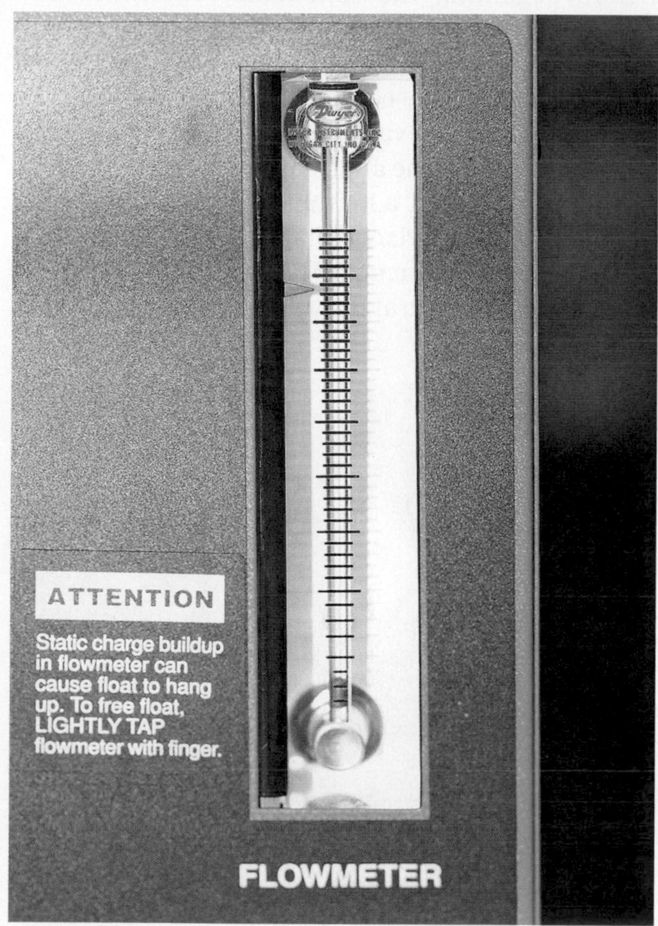

FIGURE 35–18

The flowmeter for an EVAP leak tester.

FIGURE 35–19

During a smoke test, smoke will be seen at all points of leakage. Be sure to inspect the entire system.

FUEL CAP TESTER Most fuel cap testers come with a variety of fuel cap adapters. Always use the one that is appropriate for the vehicle being tested. Tighten the cap to the adapter and connect the cap and adapter to the tester. Turn on the tester. The tester will create a pressure on the cap and monitor the cap's ability to hold the pressure. In most cases, the tester will illuminate lights, indicating that the cap is good or bad.

PCV SYSTEM DIAGNOSIS AND SERVICE

No adjustments can be made to the PCV system. Service of the system involves careful inspection, functional tests, and replacement of faulty parts. Some engines use a fixed orifice tube in place of a valve. These should be cleaned periodically with a pipe cleaner soaked in carburetor cleaner. Although there is no PCV valve, this type of system is diagnosed in the same way as those systems with a valve. When replacing a PCV valve, match the part number on the valve with the vehicle maker's specifications for the proper valve. If the valve cannot be identified, refer to the part number listed in the manufacturer's service manual. Newer PCV valves have locking devices that prevent them from becoming loose or falling out. Make sure that the lock is fully engaged when installing the valve.

Consequences of a Faulty PCV System

If the PCV valve is stuck open, excessive airflow through the valve causes a lean air/fuel ratio and possible rough idle or engine stalling. When the PCV valve or hose is restricted, excessive crankcase pressure forces blowby gases through the clean air hose and filter into the air cleaner. Worn rings or cylinders cause excessive blowby gases and increased crankcase pressure, which forces blowby gases through the hose and filter into the air cleaner. A restricted PCV valve or hose may also cause an accumulation of moisture and sludge in the engine and engine oil.

Leaks at the engine gaskets not only will cause oil leaks but also will allow blowby gases to escape into the atmosphere. The PCV system will also draw unfiltered air through these leaks. This can result in premature wear of engine parts, especially when the vehicle is operated in dusty conditions. Check the engine for signs of oil leaks. Be sure the oil filler cap and dipstick fit and seal properly. If these do not seal, they can be the source of false air and cause a change in long-term fuel trim (LTFT) on MAF systems.

Visual Inspection

The PCV valve can be located in several places. The most common location is in a rubber grommet in the valve cover. It can be installed in the middle of the hose connections, as well as directly in the intake manifold.

Once the PCV valve is located, make sure that all the PCV hoses are properly connected and have no breaks or cracks. Remove the air cleaner and inspect the air and crankcase filters. Crankcase blowby can clog these with oil. Clean or replace such filters. Oil in the air cleaner assembly indicates that the PCV valve or hoses are plugged. Make sure you check these and replace the valve and clean the hoses and air cleaner assembly. When the PCV valve and hose are in satisfactory condition and there was oil in the air cleaner assembly, perform a cylinder compression test to check for worn cylinders and piston rings.

Functional Checks of the PCV System

Before beginning the functional checks, double-check the PCV valve part number to make certain that the correct valve is installed. If the correct valve is being used, continue by disconnecting the PCV valve from the valve cover, intake manifold, or hose. Start the engine and let it run at idle. If the PCV valve is not clogged, a hissing is heard as air passes through the valve. Place a finger over the end of the valve to check for vacuum **(Figure 35–20)**. If there is little or no vacuum at the valve, check for a plugged

FIGURE 35–20

With the engine at idle, vacuum should be felt at the PCV valve.

or restricted hose. Replace any plugged or deteriorated hoses. Turn off the engine and remove the PCV valve. Shake the valve and listen for the rattle of the check needle inside the valve. If the valve does not rattle, replace it.

Some vehicle manufacturers recommend that the valve be checked by removing it from the valve cover and hose. Connect a hose to the inlet side of the PCV valve, and blow air through the valve with your mouth while holding your finger near the valve outlet **(Figure 35–21)**. Air should pass freely through the valve. If air does not pass freely through the valve, replace the valve. Move the hose to the outlet side of the PCV valve and try to blow back through the valve **(Figure 35–22)**. It should be difficult to blow air through the PCV valve in this direction. If air passes easily through the valve, replace the valve.

CAUTION!

Do not attempt to suck through a PCV valve with your mouth. Sludge and other deposits inside the valve are harmful to the human body.

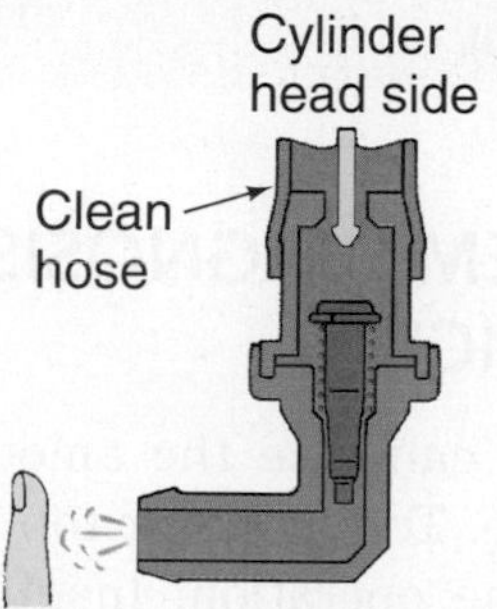

FIGURE 35–21

When air is blown through the inlet of a PCV valve, air should freely flow through.

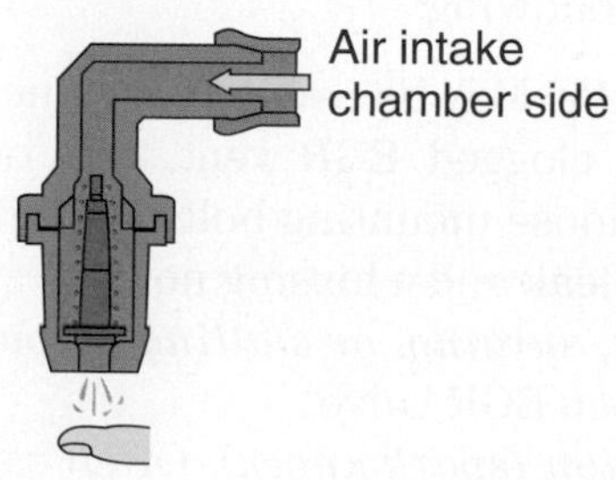

FIGURE 35–22

When air is blown through the outlet of a PCV valve, air should barely flow through.

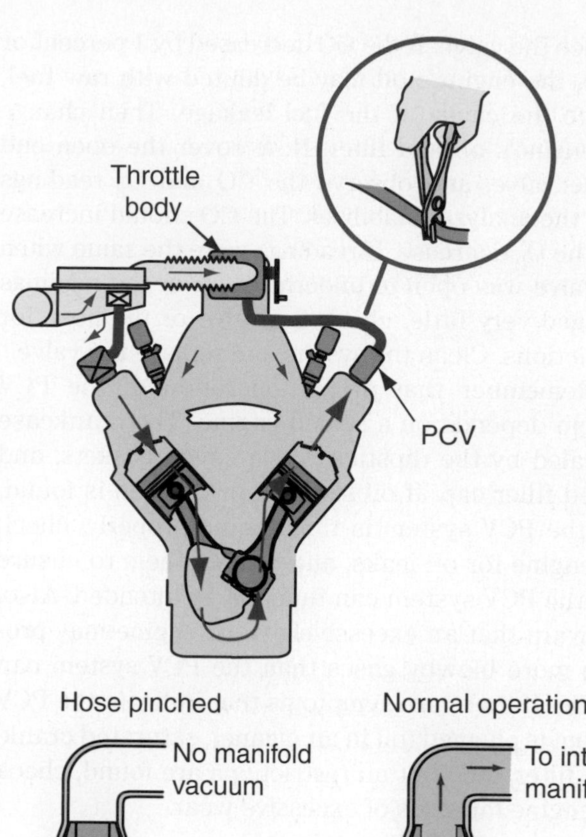

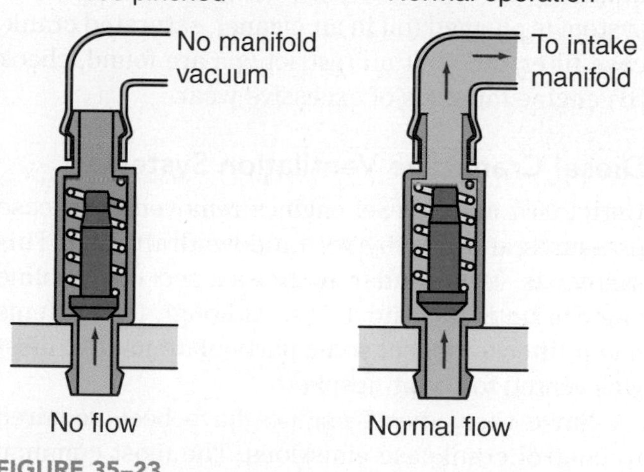

FIGURE 35–23

When the PCV hose is pinched, the valve should click.

Another simple check of the PCV valve can be made by pinching the hose between the valve and the intake manifold **(Figure 35–23)** with the engine at idle. You should hear a clicking sound from the valve when the hose is pinched and unpinched. A sound scope or stethoscope will help you hear the clicking. If no clicking sound is heard, check the PCV valve grommet for cracks or damage. If the grommet is okay, replace the PCV valve.

The PCV system can be checked with an exhaust analyzer. Check and record the CO and O_2 measurements with the engine at idle. Then pull the PCV valve out of the engine, and allow it to draw in air from under the hood. Check the CO and O_2 readings. The CO should decrease and the O_2 increase. If the readings did not change, clean the PCV system and replace the valve. If the CO decreased by 1 percent or more, the engine's oil may be diluted with raw fuel. Locate the cause of the fuel leakage. Then change the engine's oil and filter. Now cover the open end of the valve, and observe the CO and O_2 readings after the analyzer stabilizes. The CO should increase and the O_2 decrease. If readings were the same when the valve was open to underhood air or the readings changed very little, check the valve or the hose for restrictions. Clean the system and replace the valve.

Remember that proper operation of the PCV system depends on a sealed engine. The crankcase is sealed by the dipstick, valve cover, gaskets, and sealed filler cap. If oil sludging or dilution is found, and the PCV system is functioning properly, check the engine for oil leaks, and correct them to ensure that the PCV system can function as intended. Also, be aware that an excessively worn engine may produce more blowby gases than the PCV system can handle. If there are symptoms that indicate the PCV system is plugged (oil in air cleaner, saturated crankcase filter, etc.) but no restrictions are found, check the engine for signs of excessive wear.

Diesel Crankcase Ventilation Systems

Until 2007, most diesel engines removed crankcase pressures and blowby with a downdraft tube. This system is very similar to those used on gasoline engines until the late 1950s. Although the systems had a filter to collect some particulate matter, most was vented to the atmosphere.

Since then, diesel engines have been required to control crankcase emissions. The most common systems have a filter that collects the blowby gases and a system to return them to the engine's intake or crankcase. These systems are designed to self-clean, but problems in this system can occur. Therefore, it is important to inspect and clean the system as necessary, as well as check the engine for excess crankcase pressures. When pressure builds up because the valve is stuck closed, crankcase pressure can force oil past some gaskets and seals. If the valve is stuck open, oil from the crankcase will be drawn into the engine and burned as fuel; however, it is heavier and thicker than diesel fuel and will cause excessive heat in the cylinders. If oil or oil mist is present in the intake, check the valve.

Measuring Crankcase Pressure

To measure crankcase pressure, bring the engine up to normal operating temperature. Then connect a water manometer to the dipstick tube after the dipstick has been removed. With the engine at idle, observe the reading on the meter. If the reading is greater than 25.4 mm (1 in.) H_2O, check the crankcase valve. If the valve is clean and working properly, check the compression and leakage of the cylinders. If the crankcase pressure is less than 25.4 mm (1 in.) H_2O, raise the engine to above 2000 rpm. If the pressure drops to a negative value of 101.6 to 127 mm (4 to 5 in.) H_2O, the system is okay. If the drops are more than that, replace the valve. Keep in mind that a restriction in the air intake system will increase inlet vacuum in turbocharged engines and the intake vacuum on non-turbocharged engines.

SHOP TALK

If a water manometer is not available, you can make one with clear plastic tubing that will fit over the dipstick tube. Form a 75 mm (3 in.) "U" bend in the hose with at least 300 mm (12 in.) on each side of the "U." Attach the hose to a board with large staples spaced at 25.4 mm (1 in.) intervals. These intervals will be used for measurement. Fill the tube with water so that approximately one-third of the tube has water in it. On the wood, mark the level of the water. Connect one end of the tube to the dipstick tube and start the engine. Measure the amount the water moved; this is the pressure inside the crankcase as measured in mm of H_2O (inches of H_2O).

EGR SYSTEM DIAGNOSIS AND SERVICE

Manufacturers calibrate the amount of EGR flow for each engine. Too much or too little may inhibit optimum engine operation. Insufficient EGR flow may cause the engine to overheat, detonate, and emit excessive amounts of NO_x. When any of these problems exist, and it seems likely that the EGR system is at fault, check the system. Typical symptoms that are present during an EGR system malfunction include the following:

- *Rough idle.* May be caused by a stuck-open EGR valve, a clogged EGR vent, dirt on the valve seat, or loose mounting bolts (this also causes a vacuum leak and a hissing noise).
- *No-start, surging, or stalling.* Typically caused by an open EGR valve.
- *Detonation (spark knock).* Often caused by any condition that prevents proper EGR gas flow, such as a valve stuck closed, leaking valve diaphragm, restrictions in flow passages, a disconnected EGR, or a problem in the vacuum source.

- *Excessive* NO_x *emissions.* May be caused by any condition that prevents the EGR from allowing the correct amount of exhaust gases into the cylinder or anything that allows combustion temperatures.
- *Poor fuel economy.* Typically caused by the EGR system if it relates to detonation or other symptoms of restricted or zero EGR flow.

Scan Tool

On OBD-II systems, the EGR system monitor is designed to test the integrity and flow characteristics of the EGR system. The monitor is activated during EGR operation and after certain engine conditions are met. Input from the ECT, IAT, TP, and CKP sensors are required to activate the monitor. Once activated, a typical EGR monitor carries out these tests:

- The differential pressure feedback EGR (DPFE) sensor and circuit are continuously tested for opens and shorts.
- The EGR vacuum regulator solenoid is continuously tested for opens and shorts. The monitor compares circuit voltage to what should exist when the EGR valve is in its commanded state.
- Non-DPFE systems monitor STFT while opening the EGR value; the STFT should decrease. Some non-DPFE systems use the MAP sensor to determine if the EGR is functioning by monitoring pressure in the intake with the valve open.
- The EGR flow rate test is checked by comparing the actual DPFE circuit voltage to the desired EGR flow voltage for the current operating conditions to determine if the EGR flow rate is acceptable.
- The hoses connected to the DPFE sensor hoses are tested once per drive cycle. They are checked for restrictions and opens when the EGR valve is closed and during acceleration. The monitor checks the voltage from the DPFE sensor. A reading not within the normal range for a closed valve indicates a problem with the hose.
- Checking for a stuck-open EGR valve or some EGR flow at idle is continuously done at idle. The monitor compares the DPFE circuit voltage to the DPFE circuit voltage stored during key-on, engine-off operation to determine if there is EGR flow.
- The MIL is illuminated after one of these tests fails on two consecutive drive cycles.

Different methods of EGR-flow monitoring are used; these include temperature sensors, manifold pressure changes, fuel trim changes, and differential pressure measurement. Using temperature sensors, an EGR temperature sensor is installed in the EGR passageway. During normal EGR flow, the temperature of the EGR temperature sensor will rise at least 35°C (95°F) above ambient air temperature. When the EGR valve is open, the ECM compares EGR temperature to intake air temperature. If the temperature does not increase a specified amount over ambient temperature, the ECM assumes there is a problem in the system, and this information is stored in the ECM.

MAP systems base their calculations on the assumption that low intake manifold vacuum means the engine is under a heavier load and greater airflow. Too much EGR flow will increase manifold pressure, and the PCM will interpret this as an increase in airflow and will increase the amount of fuel delivered to the cylinders. The system compares MAP readings with the reading from the H_2OS. The PCM will also try to compensate for this by correcting the fuel trim. Excessive EGR flow will cause negative fuel trims, and low EGR flow will cause higher than normal fuel trim readings. If fuel trim readings seem to be out of line, EGR flow may be the cause.

When the PCM detects a problem in the system, it will set a DTC. It is important to note that the conditions required to set a DTC vary with manufacturer and model. It is also important to realize that inputs from the engine control system impact EGR operation. Therefore, all engine-related DTCs should be dealt with before moving into detailed diagnostics of the EGR system. Some DTCs pertain solely to the EGR valve and its solenoids **(Figure 35–24)**.

EGR System Troubleshooting

Before attempting to troubleshoot or repair a suspected EGR system on a vehicle, make sure that the engine is mechanically sound, that the injection system is operating properly, and that the spark control system is working properly.

Most often, an electronically controlled EGR valve functions **(Figure 35–25)** in the same way as a vacuum-operated valve. Apart from the electronic control, the system can have all of the problems of any EGR system. Those that are totally electronic and do not use a vacuum signal can have the same problems as others, with the exception of vacuum leaks and other vacuum-related problems. Sticking valves, obstructions, and loss of vacuum produce the same symptoms as on non–electronic-controlled systems. If an electronic control component is not

DTC	DESCRIPTION
P0106	MAP sensor rationality error
P0107	MAP sensor voltage low
P0108	MAP sensor voltage high
P0109	MAP sensor intermittent (non-MIL) problem
P0400	EGR system leak detected
P0401	Insufficient EGR flow
P0402	DPFE EGR stuck open
P0403	EVR circuit open or shorted
P0404	EGR control circuit range/performance problem
P0405	EGR valve position sensor circuit low
P0406	EGR valve position sensor circuit high
P1400 (Ford)	DPFE circuit low
P1401 (Ford)	DPFE circuit high
P1405 (Ford)	Upstream hose off or plugged
P1406 (Ford)	Downstream hose off or plugged
P1409 (Ford)	EVR circuit open or shorted
P2413 (Honda)	EGR system malfunction
P2457 (Ford)	Insufficient EGR cooler performance

FIGURE 35–24

Samples of EGR-related DTCs.

functioning, the condition is usually recognized by the PCM. The solenoids, or the EGR vacuum regulator (EVR), should normally cycle on and off frequently when EGR flow is being controlled (warm engine and cruise rpm). If they do not, a problem in the electronic control system or the solenoids is indicated. Generally, an electronic control failure results in low or zero EGR flow and might cause symptoms such as overheating, detonation, and power loss.

EGR solenoids are used with all types of EGR valves, especially back pressure–type valves. The PCM uses a solenoid to regulate ported or manifold vacuum to the EGR valve.

The solenoid is actually a vacuum switch. The PCM controls the switch through pulse-width modulation. No vacuum is sent to the valve unless the PCM allows it. The EGR solenoid has two or more vacuum lines and an electrical connector. The solenoid also has an air bleed and sometimes an air filter. Vacuum is bled off through the filter vent. If the filter becomes clogged, the EGR valve will open too much and cause a driveability problem.

Before attempting any testing of the EGR system, visually inspect the condition of all vacuum hoses for kinks, bends, cracks, and flexibility. Replace defective hoses as required. Check vacuum hose routing against the underhood Vehicle Emission Control Information (VECI) decal or the service manual; correct any misrouted hoses. If

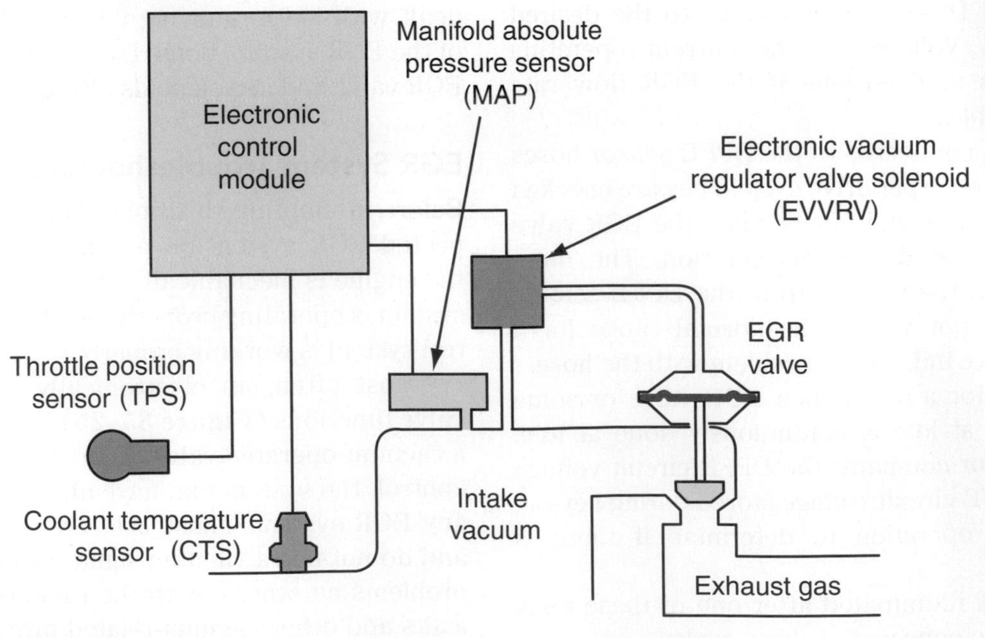

FIGURE 35–25

The circuit of a typical computer-controlled EGR system.

the system is fitted with an EVP sensor, the wires routed to it should also be checked.

If the EGR valve remains open at idle and low engine speed, the idle operation is rough, and surging occurs at low speed. When this problem is present, the engine may hesitate on low-speed acceleration or stall after deceleration or after a cold start. If the EGR valve does not open, engine detonation occurs. When a defect occurs in the EGR system, a DTC is usually set in the PCM memory.

EGR Valves and Systems Testing

On many older engines, a single diaphragm EGR valve can be checked with a hand-operated vacuum pump **(Figure 35–26)**. Before conducting this test, make sure that the engine produces enough vacuum to properly operate the valve. This is done by connecting a vacuum gauge to the engine's intake manifold. Then start the engine and gradually increase speed to 2000 rpm with the transmission in neutral. The reading should be above 406.5 mm Hg (16 in. Hg) of vacuum. If not, there could be a vacuum leak or exhaust restriction. Before continuing to test the EGR, check the MAP and/or correct the problem of low vacuum.

To check the valve with a vacuum pump, remove the vacuum supply hose from the EGR valve port. Connect the vacuum pump to the port and supply 457 mm Hg (18 in. Hg) of vacuum. Observe the EGR diaphragm movement. When the vacuum is applied, the diaphragm should move. If the valve diaphragm did not move or did not hold the vacuum, replace the valve.

FIGURE 35–26

Watch the action of the valve when vacuum is applied to it and released.

With the engine at normal operating temperature, observe the engine's idle speed. If necessary, adjust the idle speed to the emission decal specification. Slowly apply 127 to 254 mm Hg (5 to 10 in. Hg) of vacuum to the EGR valve. The idle speed should drop more than 100 rpm (the engine may stall) and then return to normal again when the vacuum is removed. If the idle speed does not respond in this manner, remove the valve and check for carbon in the passages under the valve. Clean the passages as required or replace the EGR valve. Carbon may be cleaned from the lower end of the EGR valve with a wire brush, but do not immerse the valve in solvent and do not sandblast the valve. Also, make sure that the vacuum hoses are in good condition and properly routed.

Diagnosis of a Negative Back Pressure EGR Valve

A negative back pressure EGR valve is identified by the letter N stamped on it. The valve is opened by a combination of engine vacuum, and the negative exhaust system pulses that occur as each exhaust valve closes. As soon as the valve opens, back pressure is reduced slightly, which opens a vacuum bleed and the valve quickly closes. This causes the opening of the valve to modulate according to negative exhaust system pulses.

With the engine at normal operating temperature and the ignition switch off, disconnect the vacuum hose from the EGR valve and connect a hand vacuum pump to the vacuum fitting on the valve. Supply 457 mm Hg (18 in. Hg) of vacuum to the EGR valve. The EGR valve should open and hold the vacuum for 20 seconds. If the valve does not open or cannot hold the vacuum, it must be replaced.

If the valve was okay in the first test, continue by applying 457 mm Hg (18 in. Hg) of vacuum to the valve and start the engine. The vacuum should drop to zero, and the valve should close. If the valve does not react this way, replace it.

Diagnosis of a Positive Back Pressure EGR Valve

A positive back pressure EGR valve can be identified by the letter P stamped next to the part number and date code. It has a thicker than normal pintle shaft because it is hollow. The hollow design allows exhaust gases to flow into the shaft and push up on it. With positive back pressure from the exhaust system, the shaft rises and seals the control valve. Once the control valve is closed, it allows applied vacuum to pull up on the diaphragm. With low back pressure, the valve will not hold vacuum, and the vacuum is bled to the atmosphere. As engine load increases, so does engine back pressure, which causes the control valve inside the EGR to trap vacuum and open up. To test this valve, bring the engine up to 2000 rpm to create back pressure, and then apply vacuum. EGR should open and cause a 100 rpm drop or more. Positive back pressure EGR valves are used in simple vacuum-controlled systems as well as more complex pulse-width-modulated applications.

Diagnosis of a Digital EGR Valve

Digital EGR valves are found only on GM products. They are totally electronically controlled units. They have two or three solenoids, and part of the valve is always open. Use a scan tool to check a digital EGR valve. Start the engine and allow it to run at idle speed. Select the EGR control on the scan tool, and then energize the solenoids one at a time. Engine speed should drop slightly as each EGR solenoid is energized.

If the EGR valve does not respond correctly, make sure 12 volts are applied to the EGR valve. Then check the resistance of the valve. Connect an ohmmeter across the electrical terminals on the valve **(Figure 35–27)**; the windings can be checked for opens, shorts, and excessive resistance. If any resistance reading is not within specs, the valve should be replaced. Visually check all of the wires between the EGR valve and the PCM. Also make sure that the EGR passages are not restricted or plugged. To do this, you will need to remove the valve.

Linear EGR Valve Diagnosis

The correct procedure for diagnosing a linear EGR valve **(Figure 35–28)** will vary, depending on the vehicle make and model year. Always follow the recommended procedure in the vehicle manufacturer's service manual. A scan tool may be used to diagnose a linear EGR valve. The engine should be at normal operating temperature. Because the linear EGR valve has an EVP sensor, the actual pintle position may be checked on the scan tool. The pintle position should not exceed 3 percent at idle speed. The scan tool may be operated to command a specific pintle position, such as 75 percent, and this commanded position should be achieved within two seconds. With the engine idling, select various pintle positions and check the actual pintle position. The pintle position should always be within 10 percent of the commanded position.

If a linear EGR valve does not operate properly, check the fuse in the supply wire to the EGR valve. Also check for open circuits, grounds, and shorts in the wires connected from the EGR valve to the PCM **(Figure 35–29)**. Verify that the EGR valve position (EVP) sensor is receiving a 5-volt reference signal and verify that the ground circuit is good. If these are okay, remove the valve with the wiring harness still connected to it. Then connect a digital multimeter (DVOM) across the pintle position wire at the

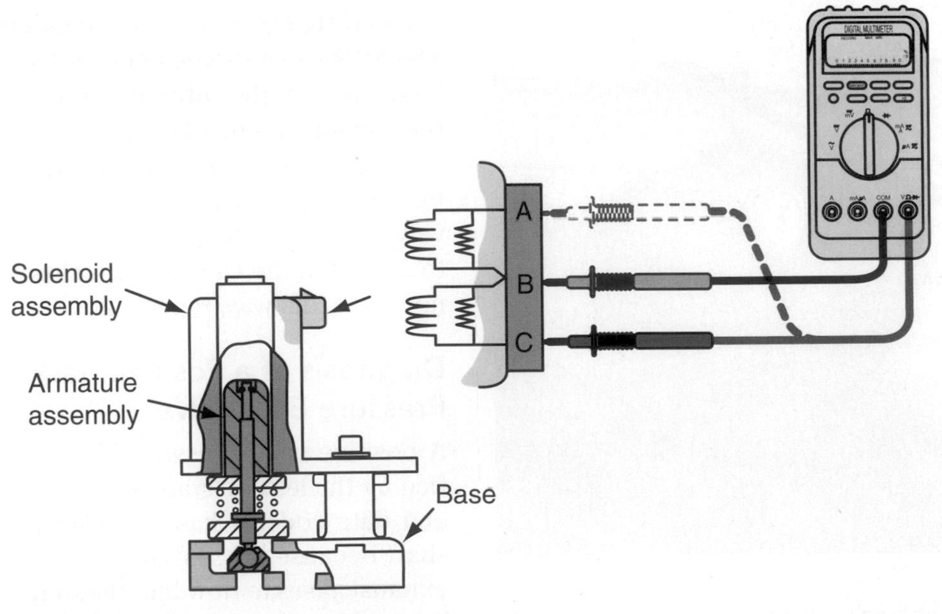

FIGURE 35–27

Ohmmeter connections for checking a digital EGR valve.

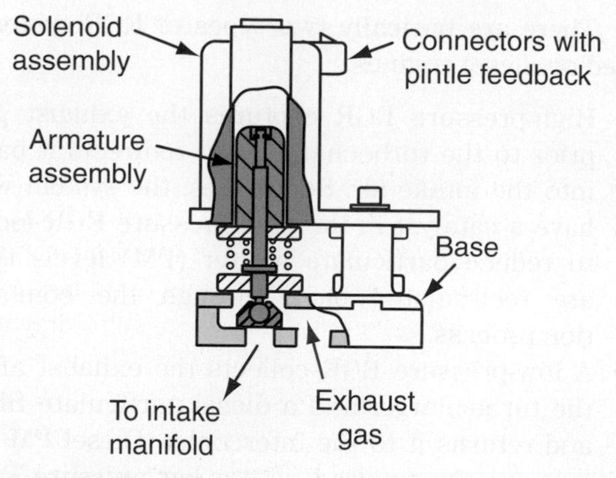

FIGURE 35–28

A linear EGR valve relies on a solenoid to move the pintle.

EGR valve to ground and manually push the pintle upward. The voltmeter reading should change from approximately 1 to 4.5 volts. If the EGR valve did not operate properly, it should be replaced.

Checking EGR Efficiency

Although most testing of EGR valves involves the valve's ability to open and close at the correct time, we are not really testing what the valve was designed to do—control NO_x emissions. EGR systems should

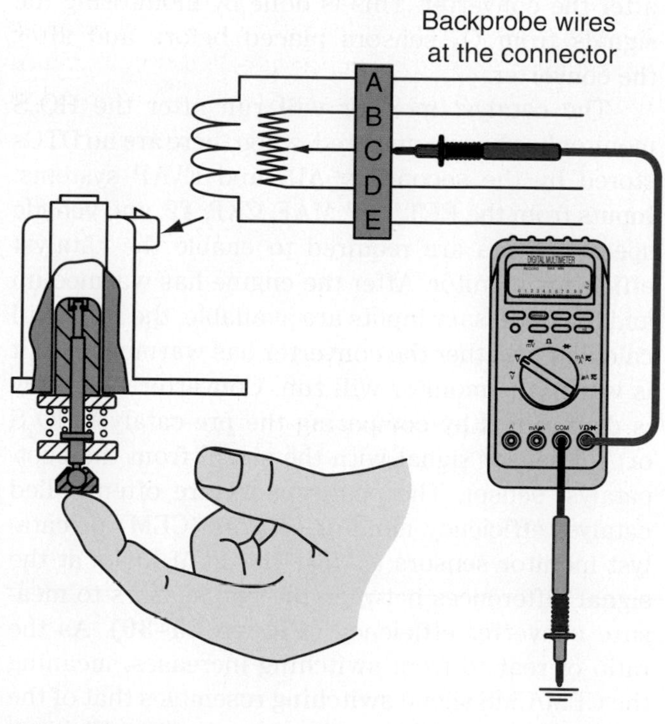

FIGURE 35–29

To check a linear EGR valve, check the voltage at the various pins of its connector.

be tested to see if they are doing what they were designed to do.

Many technicians wrongly conclude that an EGR valve is working properly if the engine stalls or idles very rough when the EGR valve is opened. Actually, this test just shows that the valve was closed and that it will open. A good EGR valve opens and closes, but it also allows the correct amount of exhaust gas to enter all of the cylinders. EGR valves are normally closed at idle and open at approximately 2000 rpm. This is where the EGR system should be checked.

To check an EGR system, use a five-gas exhaust analyzer. Allow the engine to warm up, and then raise the engine speed to around 2000 rpm. Watch the NO_x readings on the analyzer. In most cases, NO_x should be below 1000 ppm. It is normal to have some temporary increases over 1000 ppm; however, the reading should be generally less than 1000. If the NO_x is above 1000, the EGR system is not doing its job. The exhaust passage in the valve is probably clogged with carbon.

If only a small amount of exhaust gas is entering the cylinder, NO_x will still be formed. A restricted exhaust passage of only 3.175 mm (1/8 in.) will still cause the engine to run rough or stall at idle, but it is not enough to control combustion chamber temperatures at higher engine speeds. Never assume that the EGR passages are okay just because the engine stalls at idle when the EGR is fully opened. Plugged EGR passages will cause a disproportionate amount of EGR gas into the cylinders without plugged passages. This can cause the engine to run rough, but not adequately control NO_x during normal engine operation.

Electronic EGR Controls

When the EGR valve checks out and everything looks fine visually but a problem with the EGR system is evident, the EGR controls should be tested. Often, a malfunctioning electronic control will trigger a DTC. Typically, the service information gives the specific directions for testing these controls; always follow them.

Some EGR valves are electronic/mechanical EGR valves. These valves have different names depending on the application. These types of valves operate in the same way as a single diaphragm EGR valve. However, they have a position sensor above the EGR diaphragm. This tells the PCM how far the valve is open. The valve position sensor is a potentiometer and can wear. The sensor can be checked with a DVOM or lab scope. The pattern should show a clean sweep as the valve is opened and closed. The EGR monitor system watches the output from the sensor and if the sensor's voltage reading is too high or low, a DTC will be set.

EGR Vacuum Regulator (EVR) Tests

Connect a pair of ohmmeter leads to the EVR terminals to check the winding for open circuits and shorts. An infinite ohmmeter reading indicates an open circuit, whereas a lower than specified reading means the winding is shorted. Then connect the ohmmeter leads from one of the EVR solenoid terminals to the solenoid case. You should get an infinite reading; a low ohmmeter reading means the winding is shorted to the case. A scan tool can also be used to diagnose the operation of an EVR solenoid.

SHOP TALK

The same driver in a PCM may operate several outputs. On General Motors' computers, drivers sense high current flow. If a solenoid winding is shorted and the driver senses high current flow, the driver shuts down all the outputs it controls. This prevents damage caused by the high current flow. When the PCM does not operate an output or outputs, always check the resistance of the output's solenoid windings before replacing the PCM. A lower than specified resistance in a solenoid winding indicates a shorted condition, and this problem may explain why the PCM driver is not operating the outputs. Also, in some EGR systems, the PCM energizes the EVR solenoid at idle and low speeds. Under this condition, the solenoid shuts off vacuum to the EGR valve. When the proper input signals are available, the PCM de-energizes the EVR solenoid and allows vacuum to the EGR valve.

Exhaust Gas Temperature Sensor Diagnosis

To test an exhaust gas temperature sensor, remove it and place it in a container of oil. Place a thermometer in the oil and heat the container. Connect the ohmmeter leads to the exhaust gas temperature sensor terminals. The exhaust gas temperature sensor should have the specified resistance at various temperatures.

Diesel Engines

Recently, diesel engines have been equipped with EGR valves. These systems release a sample of exhaust gases into the intake of the turbocharger or the intake manifold. When the exhaust gases pass through the intercooler, the temperature is decreased, which lowers the chances of NO_x formation.

There are basically two types of EGR systems used on diesel engines:

- High-pressure EGR captures the exhaust gas prior to the turbocharger and redirects it back into the intake air. Sometimes, the system will have a catalyst in the high-pressure EGR loops to reduce particulate matter (PM) levels that are recirculated back through the combustion process.
- A low-pressure EGR collects the exhaust after the turbocharger and a diesel particulate filter and returns it to the intercooler. Diesel PM filters are always used with a low-pressure EGR system to make sure large amounts of particulate matter are not recirculated to the engine, which would result in accelerated wear in the engine and turbocharger.

CATALYTIC CONVERTER DIAGNOSIS

The catalytic converter monitor looks at a converter's ability to store O_2. O_2 storage is only one function of a converter but is a good indication of how efficient the converter is. As the catalyst efficiency declines due to thermal and chemical deterioration, its ability to store O_2 also declines. Therefore, OBD-II systems compare the O_2 content in the exhaust before and after the converter. This is done by monitoring the signals from O_2 sensors placed before and after the converter.

The catalyst monitor will run after the HO_2S monitor has been completed, when there are no DTCs stored by the secondary AIR and EVAP systems. Inputs from the ECT, IAT, MAF, CKP, TP, and vehicle speed sensors are required to enable the catalyst efficiency monitor. After the engine has warmed up and the necessary inputs are available, the PCM will calculate whether the converter has warmed up. If it is warm, the monitor will run. Converter efficiency is determined by comparing the pre-catalyst HO_2S or A/F sensor signal with the signal from the post-catalyst sensor. The post-sensors are often called catalyst efficiency monitor sensors (CEM) or catalyst monitor sensors (CMS). The PCM looks at the signal differences between the two sensors to measure converter efficiency **(Figure 35–30)**. As the ratio of rear to front switching increases, meaning the CEM/CMS signal switching resembles that of the front HO_2S, the PCM will set a catalyst efficiency DTC, either a P0420 or P0430.

During normal operation, the front HO_2S switches more often and with a greater amplitude

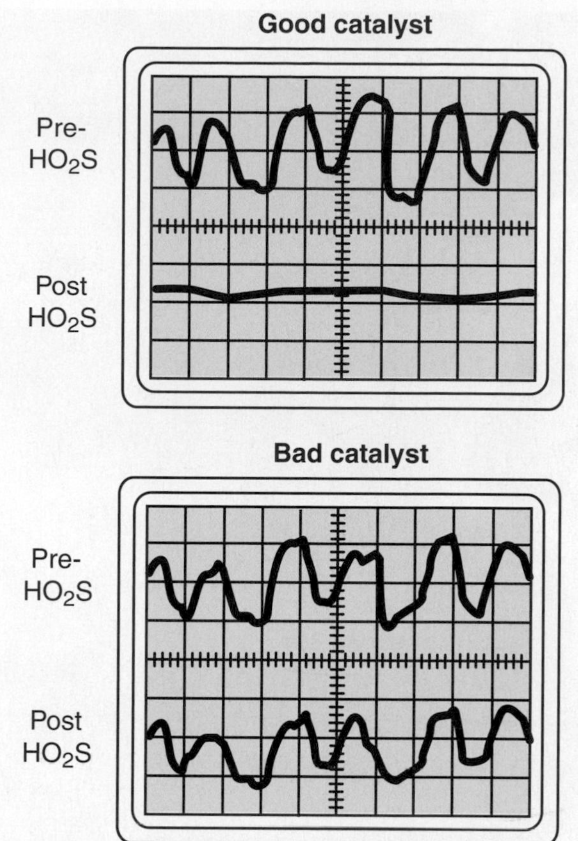

FIGURE 35–30

A comparison of the HO_2S signals for a good and a bad catalytic converter.

than the rear CEM/CMS. The catalyst efficiency monitor sensor also has a shorter signal. The monitor compares the cross counts of each sensor as well as the signal length. When the converter has lost some of its ability to store O_2, the post-catalyst or downstream signal begins to switch more rapidly with increasing amplitude and signal length. It starts to look like the signal from the pre-catalyst or upstream HO_2S.

When the signals become alike and stay that way through a number of drive cycles, the PCM will set a DTC and illuminate the MIL. The activity of the HO_2S can be monitored on a scan tool or lab scope in an effort to determine if either the sensors or the converter is bad. A converter-related DTC does not always indicate that the converter is bad. These DTCs can be set for a number of other reasons, such as the following:

- A small leak in the secondary AIR system
- A slight misfire that is causing extra O_2 to enter the exhaust stream
- An exhaust leak downstream of the front HO_2S

Converter Diagnosis

Typically, catalytic converters fail because of deterioration of the catalyst material or because of physical damage. A converter should be checked for cracks and dents. It is also possible that the internal components of the converter are damaged or broken. A quick test of internal damage is done with a rubber mallet. The converter is smacked with the mallet. If the converter rattles, it needs to be replaced and there is no need to do other testing. A rattle indicates loose catalyst substrate, which will soon rattle into small pieces. This test is not used to determine if the catalyst is good.

Converters often fail because the catalyst material becomes coated with foreign materials. This normally is the result of contaminated fuel, sealant contamination, incorrect motor oil use, or coolant entering the exhaust stream. A buildup of this material reduces the catalysts' ability to reduce NO_x and oxidize HC and CO. Current catalytic converters should be at least 90 percent efficient for HC and CO control. Earlier converters (pre-1992) could operate as low as 80 percent efficient.

An overheated converter can become plugged and restrict exhaust flow. The typical cause of an overheated converter is engine misfiring. A plugged converter or any exhaust restriction can cause loss of power at high speeds, stalling after starting (if totally blocked), or sometimes popping or backfiring at the intake manifold.

A vacuum gauge can be used to watch engine vacuum while the engine is accelerated. Another way to check for a restricted exhaust or catalyst is to insert a pressure gauge in the exhaust manifold's bore for the O_2 sensor **(Figure 35–31)**. With the gauge in place, hold the engine's speed at 2000 rpm and watch the gauge. The desired pressure reading will be less than 8.6 kPa (1.25 psi). A very bad restriction will give a reading of over 19 kPa (2.75 psi.)

You can use Mode $06 data to monitor how the PCM is measuring catalyst efficiency. As the catalyst ages and efficiency decreases, the Mode $06 data will show an increase in the catalyst monitor index ratio.

OXYGEN STORAGE TEST The O_2 storage test is based on the fact that a good converter stores O_2. Begin by warming up a four- or five-gas analyzer. Disable the air injection system **(Figure 35–32)**. Once the converter is warmed up, insert the analyzer's probe into the tailpipe, and hold the engine at 2000 rpm. Watch the O_2 readings. Once the numbers stop dropping, check the O_2 level on the gas analyzer.

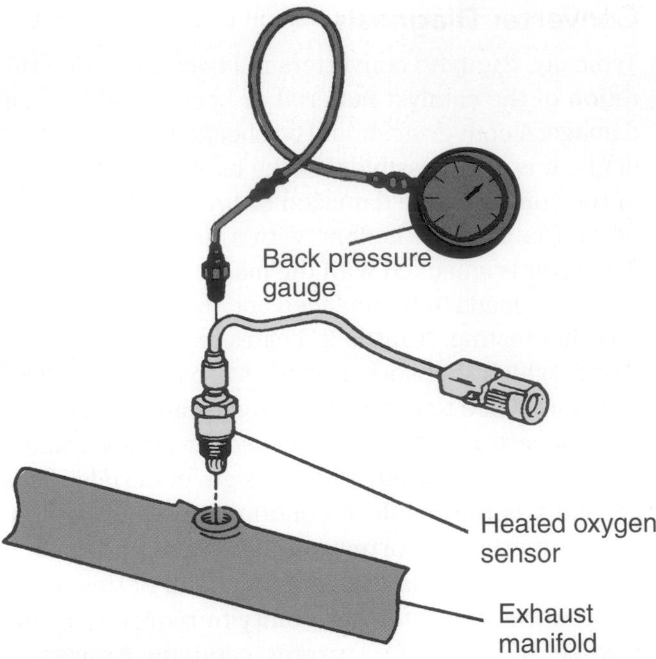

FIGURE 35–31

To measure exhaust system back pressure, insert a pressure gauge into the oxygen sensor's bore in the exhaust.

©Cengage Learning 2015

The O_2 readings should be about 0.0 percent. This shows that the converter is using the available O_2. Immediately after the O_2 drops, quickly snap the throttle, and watch the O_2 reading just as the CO begins to increase. If the O_2 now exceeds 1.2 percent, the converter is failing this test. If the O_2 readings never reached zero, the test may need to be repeated after adding some propane through the air intake until all of the O_2 stored in the converter is depleted.

CHECKING CONVERTER EFFICIENCY This converter test uses a principle that checks the converter's efficiency. Before beginning this test, make sure that the converter is warmed up and there are no ignition problems, vacuum leaks, or fuel restrictions. Disable the air injection system, and then disconnect the HO_2S. Calibrate a four- or five-gas analyzer, and insert its probe into the tailpipe. With a propane enrichment tool, enrich the air/fuel mixture until the CO reading is about 2 percent. Then reconnect the air injection system. Observe the HC, CO, and O_2 readings. If the converter is working correctly, the O_2 should increase, and HC and CO should decrease when the air injection system is reconnected. If the O_2 increased but the CO and HC did not change much, or if the O_2 is higher than the CO and the CO is greater than 0.5 percent, the converter is faulty. If the O_2 is lower than the CO, the converter is not oxidizing HC and CO.

FIGURE 35–32

Before conducting an oxygen storage test, disable the air injection system.

©Cengage Learning 2015

AIR SYSTEM DIAGNOSIS AND SERVICE

Not all engines are equipped with an air injection system; only those that need them to meet emissions standards have them. Therefore, air injection systems are vital to proper emissions on engines equipped with them. Each system has its own test procedure; always follow the manufacturer's recommendations for testing.

Most AIR systems are computer controlled and rely on solenoids to control the direction of airflow to the exhaust manifold or to the converter. When the system is in closed loop, the air from the air injection system must be directed away from the O_2 sensor. Some systems have switching valves that allow a small amount of air to flow past the O_2 sensor. The computer knows how much and adjusts the O_2 input accordingly. Sometimes, the amount of air that can move through a closed switching valve is marked on its housing. The pump has at least two hoses. The inlet hose is the larger of the two and connects to the air filter assembly or a small dedicated air filter. The output hose carries output air through the valve and into the exhaust.

Secondary AIR Monitor

The operation of the secondary air injection (AIR) system is checked by an OBD-II monitor. The monitor looks at the complete electrical circuit for the AIR system, especially the electric pump (if so equipped) and pump relay. It checks it for shorts, opens, and high resistance. It checks the ability of the system to inject air into the exhaust. It does this through input from the HO_2S. It compares the HO_2S signals when the system is off and when it is energized. The condition of the pump and hoses is also checked at this time.

The monitor runs when the AIR system is operating but only when the enable criteria are met. Most AIR systems will set DTCs in the PCM if there is a fault in the solenoids and related wiring. In some AIR systems, DTCs are set in the PCM memory if the airflow from the pump is continually upstream or downstream. Always use a scan tool to check for any DTCs related to the AIR system, and correct the causes of these codes before proceeding with further system diagnosis. When a fault is detected, a DTC is set, and the MIL will be lit if the fault is detected during two consecutive drive cycles.

Some late-model AIR systems use an electric air pump controlled by the PCM **(Figure 35–33)**. These systems have an AIR solenoid and solenoid relay. When the PCM provides a ground for the relay, battery voltage is applied to the solenoid and the pump. Typically, DTCs will be set if one of the components fails or if the hoses or check valves leak. A quick check of the system can be made with a scan tool.

Set the scan tool to watch the voltage at the O_2 sensor(s). Start the engine and allow it to idle. Once the engine has reached normal operating temperature, enable the AIR system, and check the HO_2S voltages. If the voltages are low, the AIR pump, solenoid, and shutoff valve are working properly. If the voltages are not low, each component of the system needs to be checked and tested. Most bidirectional scan tools can turn the AIR pump on for testing purposes.

Secondary AIR System Service and Diagnosis

The first step in diagnosing a secondary air injection system is to check all vacuum hoses and electrical connections in the system. Most belt-driven AIR pumps have a centrifugal filter behind the pulley to keep dirt out of the pump. Air flows through this filter into the pump. The pulley and filter are bolted to the pump shaft and are serviced separately **(Figure 35–34)**. If the pulley or filter is bent, worn, or damaged, it should be replaced. Also check the AIR pump's belt for condition and tension, and correct it as necessary. The pump assembly is usually not serviced.

In some AIR systems, pressure relief valves are mounted in the AIR bypass (AIRB) and AIR diverter (AIRD) valves. Other AIR systems have a pressure-relief valve in the pump. If the pressure-relief valve

FIGURE 35–33

An electric air pump.

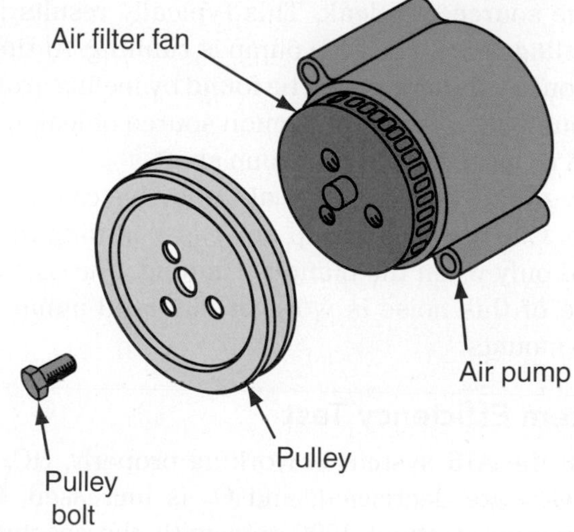

FIGURE 35–34

An AIR pump pulley and filter assembly.

is stuck open, airflow from the pump is continually exhausted through this valve, which causes high tailpipe emissions.

If the hoses in the AIR system show evidence of burning, the one-way check valves are leaking, which allows exhaust to enter the system. Leaking air manifolds and pipes result in exhaust leaks and excessive noise.

If the AIR system does not pump air into the exhaust ports during engine warm-up, HC emissions are high during this mode, and the HO_2S, or sensors, take longer to reach normal operating temperature. Under this condition, the PCM remains in open loop longer. Because the air/fuel ratio is richer in open loop, fuel economy is reduced.

When the AIR system pumps air into the exhaust ports with the engine at normal operating temperature, the additional air in the exhaust stream causes lean signals from the HO_2S or sensors. The PCM responds to these lean signals by providing a richer air/fuel ratio. This increases fuel consumption. A vehicle can definitely fail an emission test because of air flowing past the HO_2S when it should not be. If the HO_2S is always sending a lean signal back to the computer, check the air injection system.

Noise Diagnosis

Leaks in the AIR system can cause a noise. It may sound like an exhaust leak or a hissing, depending on where the leak is. To verify that the system is leaking, disconnect the pressure hose from the switching or combination valve. Plug the end of the hose, and run the pump. Normally, the sound will be amplified and can be found. The pump itself can be the source of a leak. This typically results in a whistling noise when the pump is running. At times, the source of the leak can be found by feeling around the pump as it runs. A common source of leakage is a bad or loose seal at the pump shaft.

A pump problem can also be the cause of a noise. One common pump noise is a rattling that is heard only when the pump is running. The common cause of this noise is worn or damaged pump isolator mounts.

System Efficiency Test

When the AIR system is working properly, HC, CO and CO_2 are decreased, and O_2 is increased. Run the engine at about 1500 rpm with the secondary AIR system on (enabled). Using an exhaust gas analyzer, measure and record the emission levels. Next, disable the secondary AIR system, and continue to allow the engine to idle. Again, measure and record the emission levels in the exhaust gases. The O_2 readings should be at least 4 percent less than they were when the AIR system was enabled. Less than that indicates an AIR problem.

AIR Component Diagnosis

Not all AIR systems have the same components. The following are some of the more common parts used in today's AIR systems.

AIRB SOLENOID AND VALVE When the engine is started, listen for air being exhausted from the AIRB valve for a short period. If this air is not exhausted, remove the vacuum hose from the AIRB, and start the engine. If air is now exhausted from the AIRB valve, check the AIRB solenoid and connecting wires. When air is still not exhausted from the AIRB valve, check the air supply from the pump to the valve. If the air supply is available, replace the AIRB valve.

During engine warm-up, remove the hose from the AIRD valve to the exhaust ports, and check for airflow from this hose. If airflow is present, the system is operating normally in this mode. When air is not flowing from this hose, remove the vacuum hose from the AIRD valve, and connect a vacuum gauge to this hose. If vacuum is above 300 mm Hg (12 in. Hg), replace the AIRD valve. When the vacuum is zero, check the vacuum hoses, the AIRD solenoid, and connecting wires.

With the engine at normal operating temperature, disconnect the air hose between the AIRD valve and the catalytic converters, and check for airflow from this hose. When airflow is present, system operation in the downstream mode is normal. If there is no airflow from this hose, disconnect the vacuum hose from the AIRD valve, and connect a vacuum gauge to the hose. When the vacuum gauge indicates zero, replace the AIRD valve. If some vacuum is indicated on the gauge, check the hose, the AIRD solenoid, and connecting wires.

COMBINATION VALVE This valve is typically part of the AIR monitoring system. It can be quickly checked with a handheld vacuum pump. Apply vacuum to the valve. It should open. If it does not, replace it. If it does open, check the valve's vacuum source for adequate vacuum and the solenoids that control vacuum to it.

CHECK VALVE All of the types of air injection systems have at least one thing in common—a one-way check valve. The valve opens to let air in but closes to keep exhaust from leaking out. The check valve can be checked with an exhaust gas analyzer. Start the engine, and hold the probe of the exhaust gas analyzer near the check valve port. If any amount of CO or CO_2 is read, the valve leaks. If this valve is leaking, hot exhaust is also leaking, which could ruin the other components in the air injection system.

KEY TERMS

Acceleration simulation mode (ASM) (p. 1050)
Cut-point (p. 1050)
Dynamometer (p. 1049)
Electrostatic discharge (ESD) (p. 1061)
Ignited (p. 1047)
Parts per million (ppm) (p. 1051)
Repair grade (RG-240) (p. 1050)

SUMMARY

- The quality of an engine's exhaust depends on the effectiveness of the emission control devices and the efficiency of the engine.
- The three emissions controlled in gasoline and diesel engines are unburned hydrocarbons (HC), carbon monoxide (CO), and oxides of nitrogen (NO_x).
- Most provinces require an annual emission inspection, the most common of which is the OBD-II system test.
- Chassis dynamometers are used during the I/M 240 test and can be valuable when diagnosing other driveability problems, some of which include lack of power, overheating, and speedometer accuracy.
- The report from an I/M 240 test shows the amount of gases emitted during the different speeds and loads of the test. These can be valuable when diagnosing an emission or engine problem.
- An exhaust analyzer is used to look at the quality of an engine's exhaust, which can indicate the quality of the combustion process taking place in the engine.
- Unburned hydrocarbons are particles of gasoline that have not been burned during combustion. They are present in the exhaust and in crankcase vapours.
- Carbon monoxide forms when there is not enough oxygen to combine with the carbon during combustion.
- High NO_x readings indicate high combustion temperatures.
- CO_2 is a desired element in the exhaust stream and is present only when there is combustion.
- O_2 is used to oxidize CO and HC into water and CO_2; therefore, very low amounts of O_2 in the exhaust are desirable.
- The OBD-II system monitors the emission control systems and components that can affect tailpipe or evaporative emissions.
- EVAP systems can be tested with a scan tool, DVOM, handheld vacuum pump, pressure gauge, and leak tester.
- PCV systems are most commonly checked visually or with an exhaust gas analyzer.
- EGR systems can be checked visually or with a scan tool, exhaust analyzer, handheld vacuum pump, and DVOM.
- The most common ways to check the efficiency and operation of a catalytic converter are to monitor the HO_2S and CEM/CMS sensors, retrieve DTCs, or conduct the delta temperature, oxygen storage, and efficiency tests.
- Secondary AIR systems are typically checked with a scan tool or exhaust analyzer.

REVIEW QUESTIONS

1. What effect would a rich air/fuel mixture have on HC emissions?
2. What emission will result from too little EGR flow?
3. What driveability issue can result from a charcoal canister that is filled with liquid or water?
4. What happens if a PCV valve is stuck in the open position?
5. Explain why the I/M 240 and similar tests are being replaced by the OBD-II system test.
6. What is the minimum efficiency of a functioning current-model catalytic converter?
   a. 60% **b.** 75%
   c. 80% **d.** 90%
7. How is carbon monoxide formed?
   a. when there is not enough oxygen to combine with the carbon during combustion
   b. when there is too much oxygen combining with the carbon during combustion
   c. when combustion temperatures are too high
   d. when combustion temperatures are too low

8. In readings of the chemicals in the exhaust, which is measured in parts per million (ppm)?
   a. HC **b.** CO
   c. CO_2 **d.** O_2
9. What gas is measured by a five-gas analyzer and not a four-gas analyzer?
   a. HC **b.** CO_2
   c. O_2 **d.** NO_x
10. Which of the following conditions cannot be caused by a restricted catalytic converter?
    a. stalling after the engine starts
    b. popping or backfiring through the intake manifold
    c. a drop in engine vacuum
    d. an increase power at high speeds
11. Which of the following statements about EVAP systems is *not* true?
    a. If the system is purging vapours from the charcoal canister when the engine is running at high speeds, rough engine operation will occur.
    b. Cracked hoses or a canister saturated with gasoline may allow gasoline vapours to escape to the atmosphere, resulting in gasoline odour in and around the vehicle.
    c. Rough idle, flooding, and other similar conditions can indicate a saturated canister.
    d. A vacuum leak in the system can cause starting and performance problems.
12. As a catalytic converter begins to deteriorate, how does the signal from the post-catalyst CEM/CMS sensor change as compared to the signal of the pre-catalyst HO_2S?
    a. becomes shorter than
    b. becomes more like
    c. becomes larger than
    d. becomes flatter than
13. How much pressure does a typical EVAP pressure tester apply to the system during testing?
    a. 356 mm Hg (14 in. Hg)
    b. 96.5 kPa (14 psi)
    c. 2.5 cm (1 in.) of water
    d. 3.4 kPa (0.5 psi)
14. Which of the following exhaust gases is typically not measured during an I/M 240 test?
    a. HC **b.** O_2
    c. NO_x **d.** CO
15. A vehicle is setting a lean exhaust code, and the HO_2S data show the exhaust to be very lean. Which of the following would be the likely cause of the concern?
    a. a faulty AIRD valve or solenoid
    b. poor connection at the EGRC solenoid
    c. open circuit in the EVAP canister purge solenoid
    d. unmetered air leak ahead of the MAF sensor
16. Which of the following would *not* be an indicator of a plugged or restricted catalytic converter?
    a. outlet temperature less than inlet temperature
    b. rattling at the converter when tapped with a mallet
    c. drop in engine vacuum when engine rpm increases
    d. melted PFE/DPFE hoses
17. Which tailpipe emission gases would increase during a cylinder misfire?
    a. CO and CO_2
    b. HC and O_2
    c. CO and NO_x
    d. N and CO_2
18. When performing catalytic converter diagnosis, what is the minimum temperature at which the converter can be accurately tested?
    a. 150°C (300°F)
    b. 204°C (400°F)
    c. 260°C (500°F)
    d. 316°C (600°F)
19. An EGR valve that does not open when commanded could cause which symptom?
    a. hesitation on acceleration
    b. pinging or detonation under heavy load
    c. stalling on deceleration
    d. hard start situations when hot
20. What is the preferred method of pressure testing that most OEMs recommend for EVAP systems?
    a. High-pressure shop air can be used with caution.
    b. Nitrogen is used to induce smoke into the system.
    c. A soapy solution is sprayed on the lines and seals.
    d. Pressurize the system with oxygen, and listen for leaks.

CHAPTER 36
Hybrid Vehicles

LEARNING OUTCOMES

- Explain the differences between the different platforms for hybrid vehicles.
- List and explain the purpose of the basic components used in a power-split hybrid system.
- Explain how manufacturers provide four-wheel drive in hybrid SUVs.
- Describe the purpose of an inverter.
- Explain how the stop–start feature operates.
- Explain how regenerative brakes work.
- Describe the primary advantage of plug-in hybrid vehicles.
- Describe the operation of a flywheel alternator starter hybrid system.
- Explain how a belt alternator starter system works.
- Describe the basic operation of the hybrid system used in Honda's IMA system.
- Describe the operation of the two-mode hybrid system.
- List and describe the common-sense precautions that should be adhered to while working around or on a hybrid vehicle.
- List the tools and equipment that are needed to safely diagnose, service, and repair hybrid vehicles.
- Describe how the high-voltage battery pack should be serviced.
- Describe what preventive maintenance procedures are unique to a hybrid vehicle.
- Explain the proper steps to take when diagnosing a problem in a typical hybrid vehicle.

INTRODUCTION

Hybrid electric vehicles (HEVs) are or will soon be available from all of the major automobile manufacturers. Any vehicle that combines two or more sources of power is called a hybrid. Current HEVs have an internal combustion engine and one or more electric motors. Toyota introduced the first mass-produced hybrid vehicle in 1997. That hybrid, the Prius, was available only in Japan until 2000, when it was brought to North America. Since then, many different hybrid models have been available **(Figure 36–1)**.

FIGURE 36–1

The main powertrain components for GM's hybrid pickups and SUVs.

HYBRID VEHICLES

The logic for using two power sources is simple. A vehicle with an internal combustion engine has more power available than it needs for most driving situations. Most engines can produce more than 112 kilowatts (kW) (150 horsepower [hp]); however, only 15 to 30 kW (20 to 40 hp) are normally needed to maintain a cruising speed. The rest of the power is used only for acceleration and overcoming loads, such as climbing a hill. Hybrid vehicles use a smaller engine and the output from an electric motor to provide power for acceleration and overcoming loads. Since the electric motor does not use gasoline, there is a savings in fuel costs.

The power from the electric motor supplements the engine's power. As a result, HEVs use much less fuel in city driving than comparable gasoline vehicles because the engine does not need to supply all of the power required for stop-and-go traffic. HEVs also use less gasoline when travelling on the highway. Because highway speeds can be maintained with smaller and highly efficient engines. Also, hybrids can have more than 90 percent fewer emissions than the cleanest conventional vehicles.

The overall efficiency of a hybrid can be, and in most cases is, enhanced by a number of other features. One of these is the stop–start system. When a hybrid vehicle is stopped in traffic, the engine is temporarily shut off. It restarts automatically when the driver presses the accelerator pedal, releases the brake pedal, or shifts the vehicle into a gear. In addition, to reduce the required energy to drive the generator, hybrids have regenerative braking. Rotated by the vehicle's wheels, the motor acts as a generator to charge the batteries when the vehicle is slowing down or braking. This feature recaptures part of the vehicle's kinetic energy that would otherwise be lost as heat in a conventional vehicle.

Most hybrids use transmissions specifically designed to keep the engine operating at its most efficient speed. Efficiency can also be increased by the use of low-rolling resistance (LRR) tires, which are stiff and narrow to minimize the amount of energy required to turn them. Hybrids may also be designed to minimize aerodynamic drag and made lighter.

Types

Often, hybrids are categorized as series or parallel designs. Many hybrid vehicles are parallel types and rely on power from an electric motor and the engine. When necessary, the motor and engine work together (in parallel) to drive the vehicle **(Figure 36–2)**.

In a series hybrid **(Figure 36–3)**, the engine never directly moves the vehicle. The gasoline or diesel engine drives a generator, and the generator either charges the batteries or directly powers the electric motor that drives the wheels. A computer controls the operation of the engine depending on the power needs of the battery and/or motor. When the computer senses that system voltage is low, the engine quickly starts and drives the generator. Currently there are few series hybrids available to the public. These vehicles are widely marketed as **extended-range electric vehicles (EV)**.

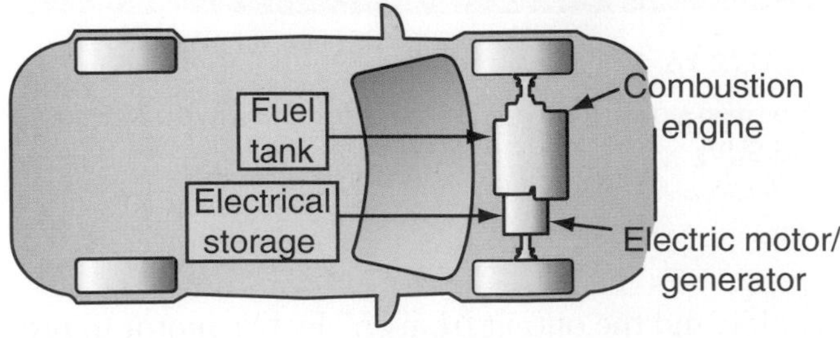

FIGURE 36–2

The basic layout for a parallel hybrid.

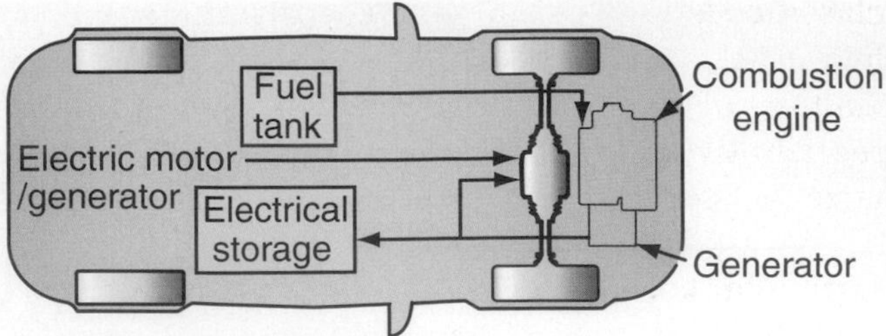

FIGURE 36–3

The basic layout for a series hybrid.

SHOP TALK

Do not get confused between the motor and the engine in a hybrid vehicle. A vehicle's engine has been, and still is, often called a motor. It is not a motor in spite of the fact that we put motor oil in it. The real name for motor oil should be engine oil. By definition, a motor is a machine that converts electrical energy into mechanical energy. An engine converts chemical energy into mechanical energy.

Most current hybrids are classified as series-parallel designs. With this design, the vehicle can be powered by the electric motor, the engine, or by both. The engine also drives the motor/generator to charge the high-voltage battery pack. During deceleration, the motor works as a generator to charge the batteries and to help slow down the vehicle.

Some hybrids are capable of shutting down the engine when the vehicle is travelling at highway speeds with light loads. The decision to power the vehicle by electricity, gasoline, or both is made by an electronic control system.

Other Hybrid Classifications

Hybrid configurations are further defined by the role of the electric motor. Keep in mind that whenever the gasoline engine does not run, there will be a savings in fuel and lower emissions will result. Although there are many variations in designs, hybrids can be classified as mild or full. Also, keep in mind that non-hybrid vehicles may be equipped with some of the features of a hybrid vehicle, such as stop–start and regenerative braking. These are designed to minimize fuel consumption.

A mild (micro) hybrid has stop–start, regenerative braking, and electric motor assist available when the engine needs added power to overcome the load. An electric motor helps or assists the engine to overcome increased load, but the vehicle is never powered by only the electric motor. A **full hybrid** can run on just the engine, just the batteries, or a combination of the two. A full hybrid has stop–start, regenerative braking, and electric motor assist, and can be driven by electricity only.

There are two additional classifications: the performance hybrid (some call this "muscle hybrid"), which is a full hybrid designed for improved acceleration without using more fuel, and a plug-in hybrid, which is a full hybrid that uses an external electrical source to charge the batteries, thereby extending the electric-only driving range by fully charging the battery pack when the vehicle is not in use.

Plug-In Hybrids

Plug-in hybrid electric vehicles (PHEVs) are full hybrids with larger batteries. The battery charger can be plugged into a normal 110-volt outlet. Charged overnight, PHEVs can drive up to 100 km without the engine ever turning on. When the batteries run low, the engine starts and powers the vehicle and the generator to charge the batteries.

The biggest advantage of plug-in hybrids is they can be driven in an electric-only mode for a much greater distance. During that time, the vehicle consumes no fuel. Under normal conditions, a plug-in hybrid can be twice as fuel efficient as a regular hybrid. A fully charged PHEV will produce half the exhaust emissions of a normal HEV. This is simply due to the fact there are no exhaust emissions when the engine is not running. **Figure 36–4** compares the various hybrid configurations and the resultant fuel economy.

The manufacturing costs of a PHEV are about 20 percent higher than a regular HEV. The increase in cost is mainly due to the price of the larger batteries. Of course, as battery technology advances and more "high-tech" batteries are produced, the cost will decrease.

HYBRID TECHNOLOGY

Hybrids are rolling examples of modern technology. The control systems attempt to precisely control the engine and electric motor. To do this, they need very complex electronics that are capable of

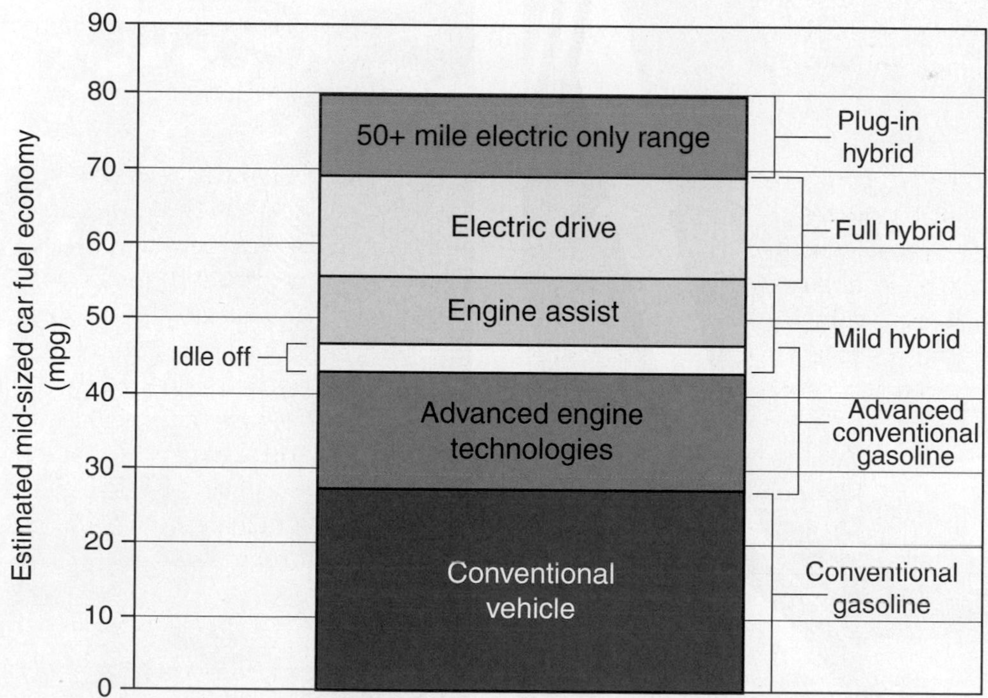

FIGURE 36–4

Estimated fuel economy potential for various hybrid classifications.

controlling and synchronizing the operation of the engine and the motors.

Batteries

The available voltage of a hybrid's battery pack **(Figure 36–5)** depends on the system and the manufacturer. The voltage range is from 42 to 360 volts. Most battery packs are basically several small batteries connected together to provide the required voltage. Most hybrids also have an additional 12-volt battery to power conventional electrical items, such as lighting, wipers, sound systems, and so on.

NICKEL-METAL HYDRIDE (NIMH) Nickel-metal hydride batteries are more environmentally friendly than some other designs and are more capable of receiving a full recharge. The cells have electrodes made of a metal hydride and nickel hydroxide. The electrolyte is potassium hydroxide.

LITHIUM-ION (LI-ION) The electrodes in lithium-ion cells are made of a carbon compound (graphite) and a metal oxide. The electrodes are submersed in lithium salt. Overheating these cells may produce pure lithium in the cells. This metal is very reactive and can explode when hot. To prevent overheating, Li-Ion cells have built-in protective electronics and/or fuses to prevent reverse polarity and overcharging. Li-Ion batteries have very good power-to-weight ratios and are making their way into hybrid vehicles.

Motor/Generators

The main difference between a generator and a motor is that a motor has two magnetic fields that oppose each other, whereas a generator has one magnetic field and wires are moved through the field. Using electronics to control the current to and from the battery, a motor that also works as a generator is used in hybrids; these are commonly referred to as motor/generators. A motor/generator may be based on two sets of windings and brushes, a brushless design with a permanent magnet, or switched reluctance. It **(Figure 36–6)** may be mounted externally to the engine and connected to the crankshaft with a drive belt. They may also be mounted directly on

FIGURE 36–5

The battery pack in a Toyota Camry hybrid.

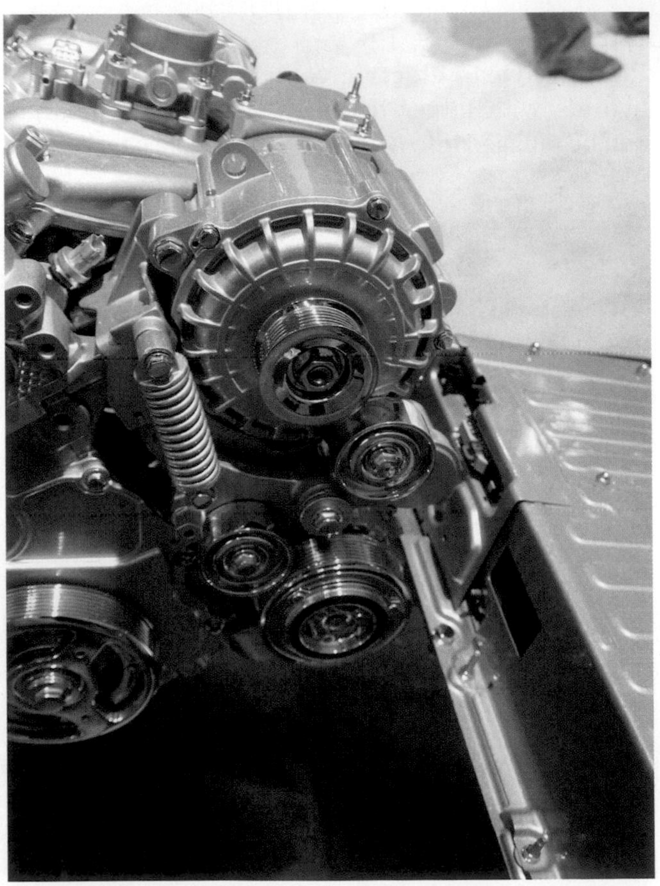

FIGURE 36–6

The unit at the top is called a belt alternator starter unit.

the crankshaft between the engine and the transmission or integrated into the flywheel. Many hybrids place the motor/generators within the transmission or transaxle assembly.

Internal Combustion Engine

The engine used in most hybrids is a four-stroke cycle engine that burns gasoline. These engines are very similar to those used in conventional vehicles. The engine relies on advanced technologies to reduce emissions and increase overall efficiency. Many of the engines are Atkinson-cycle engines.

In other countries, where diesel fuel is commonly used, diesel hybrids are being tested. Diesel engines have the highest thermal efficiency of any internal combustion engine. Because of this efficiency, diesel hybrids can achieve outstanding fuel economy.

Transmissions

The transmission used in an HEV can be a normal transmission or one especially designed for the vehicle. Often, a continuously variable transmission (CVT) is used, whose gear ratios change according to load **(Figure 36–7)**. In either case, the gear ratios are designed to allow the engine to run at its most efficient speed according to its current operating condition.

Stop–Start Feature

All hybrids have some sort of stop–start system, as do some non-hybrid vehicles. These automatically turn off the engine when the driver applies the brakes and brings the vehicle to a complete stop. This prevents wasting energy while the engine is idling and can increase fuel economy by more than 5 percent, although that varies with the vehicle.

Although the engine is off, the heating and air-conditioning systems and basic electrical systems may continue to run using battery power. The engine is restarted automatically the moment the driver releases the brake pedal, or when the control system senses the need.

Normally, stop–start systems rely on new software for the engine control system, a more powerful battery, a powerful starter, various sensors, and an electric water pump.

FIGURE 36–7

A CVT uses a belt and adjustable pulleys to change gear ratios according to conditions.

Regenerative Brakes

Regenerative braking is the process that allows a vehicle to recapture and store part of the kinetic energy that would ordinarily be lost during braking. A vehicle has more kinetic energy when it is moving fast; therefore, regenerative braking is more efficient at higher speeds **(Figure 36–8)**. When the brakes are applied in a conventional vehicle, friction at the wheel brakes converts the vehicle's kinetic energy into heat. With regenerative braking, that energy is used to recharge the batteries.

In a regenerative braking system, the rotor of the generator is turned by the wheels as the vehicle is slowing down. The activation of the generator applies resistance to the drivetrain, causing the wheels to slow down. The kinetic energy of the vehicle is changed to electrical energy until the vehicle is stopped. At that point, there is no kinetic energy. Regenerative braking can capture approximately 30 percent of the energy normally lost during braking in conventional vehicles.

In most hybrids, the control system changes the circuitry at the motor, making it act as a generator. The motor now converts motion into electricity rather than converting electricity into motion. The captured energy is sent to the batteries.

Regenerative braking is not used to completely stop the vehicle. A combination of conventional hydraulic brakes and regenerative braking is used. Hydraulic, friction-based brakes must be used when sudden and hard braking is needed.

The amount of energy captured by a regenerative braking system depends on many things, such as the state of charge of the battery, the speed at which the generator's rotor is spinning, and how many wheels are part of the regenerative braking system. Most current HEVs are front-wheel drive; therefore, energy can be reclaimed only at the front wheels. The rear brakes still produce heat that is wasted.

MG2 driven by wheels
and acts as generator
Inverter
assembly
MG2
Drive
axle
HV battery
MG1
Combustion
engine
Deceleration or braking
Speed
0
Time
0
MG1
Engine
MG2

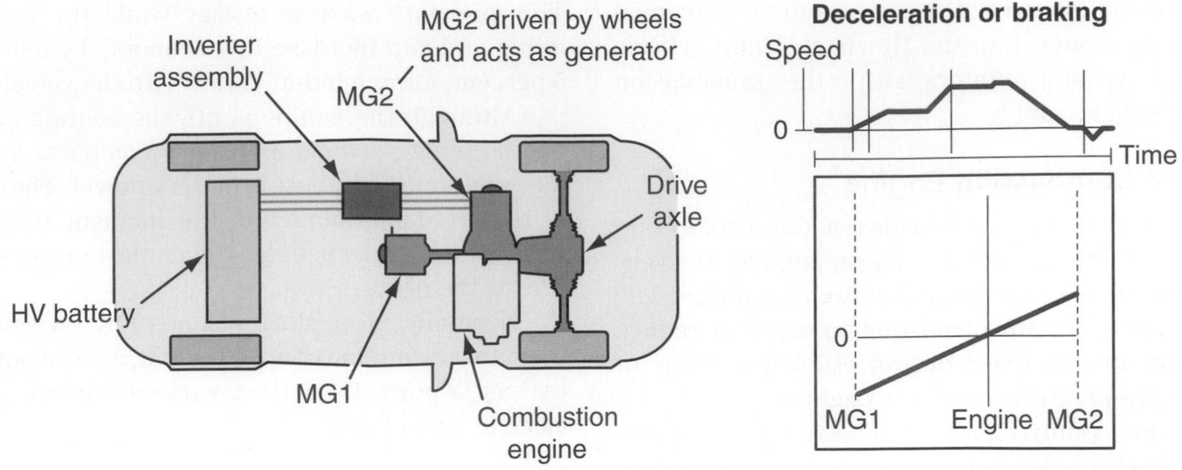

FIGURE 36–8

The power flow for a hybrid with two motor/generators during regenerative braking.

ECM
Battery
ECU
HV control
ECU
Skid control
ECU
Power-steering
ECU
Gateway
ECU
Multi-
display
CAN
BEAN
AVC-LAN
Body ECU
Air-conditioning
amplifier
Transponder
key ECU
Power source
control ECU
Entry key
ECU
Combination
meter

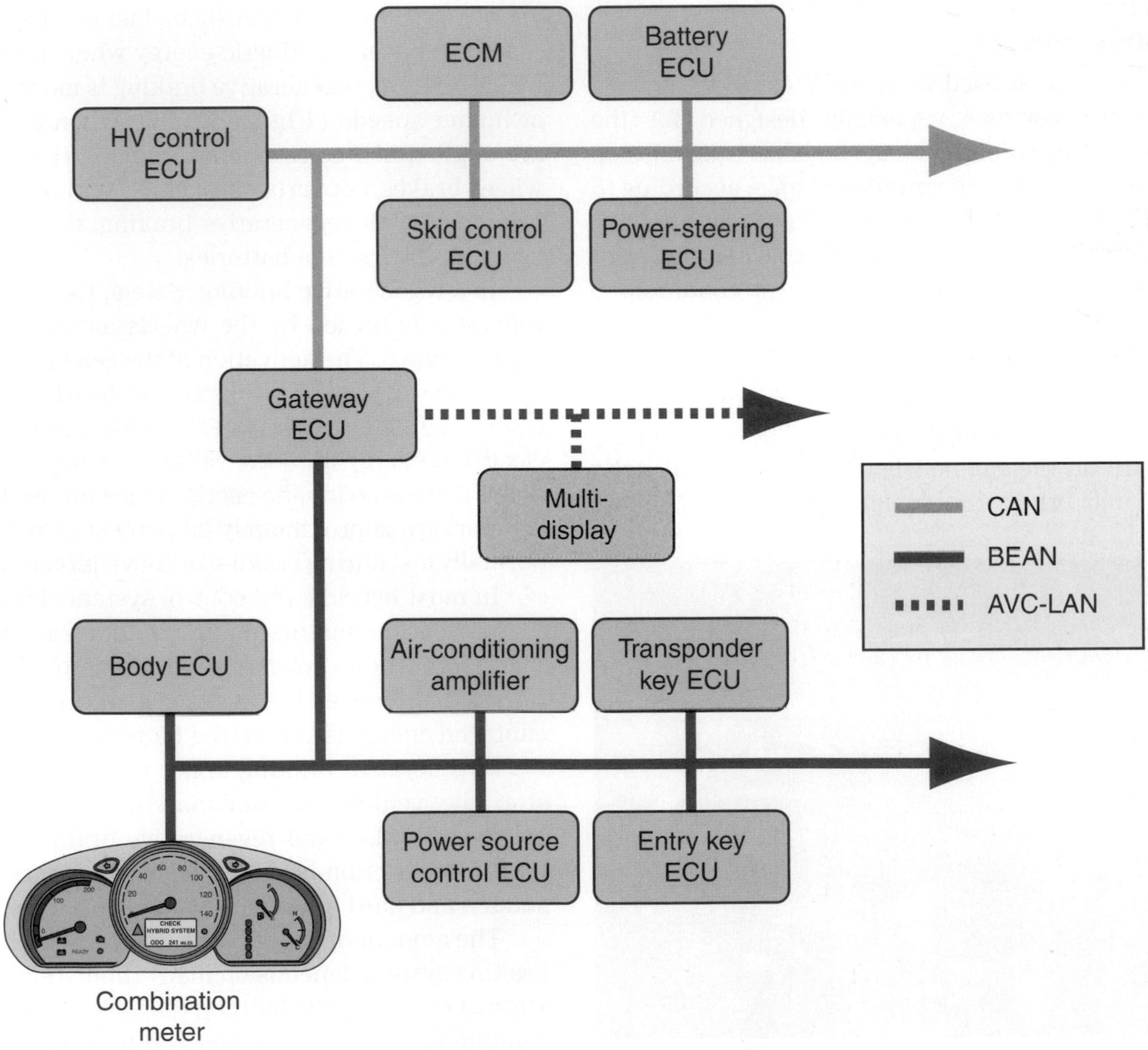

FIGURE 36–9

The individual computers in the control system are linked together and communicate with each other by high-speed communication buses, known as the controller area network (CAN).

Control System

The switching between the electric motor and gasoline engine is controlled by computers, as are other features of vehicle. The control systems are extremely complex. They have very fast processing speeds and real-time operating systems. The individual computers are linked together and communicate with each other through CAN communications **(Figure 36–9)**. The various computers include the electric motor controller, engine controller, battery management system, brake system controller, transmission controller, and electrical grid controller, and some systems also have 12- or 42-volt components.

A controller is used to manage the flow of electricity from the batteries and thereby controls the speed of the electric motor. A sensor located by or connected to the throttle pedal sends input regarding the pedal's position to the controller. The controller then sends the appropriate amount of voltage to the motor. The controller also looks at inputs from various other sensors to determine the current operating conditions of the vehicle. To provide precise control of the motor, the controller may directly or indirectly duty-cycle voltage to the motor; most controllers pulse the voltage more than 15 000 times per second.

Most of the electronics for a hybrid system are contained in a single water- or air-cooled assembly. This unit may contain an inverter, DC–DC converter, boost converter, and air-conditioning inverter. During operation, these components generate a great amount of heat. This heat must be controlled to protect the circuits, especially the transistors.

An inverter may be part of the controller assembly or be a separate unit. The inverter **(Figure 36–10)** is a power converter that changes the high DC voltage of the battery to a three-phase

FIGURE 36–10

A cutaway view of an inverter.

©Cengage Learning 2015

AC voltage for the electric motors. DC voltage from the battery is fed to the primary winding of a transformer in the inverter **(Figure 36–11)**. The direction of the current is controlled by a number of electronic switches (generally, a set of isolated gate bipolar transistors or IGBTs). Current flows through the primary winding and then is quickly

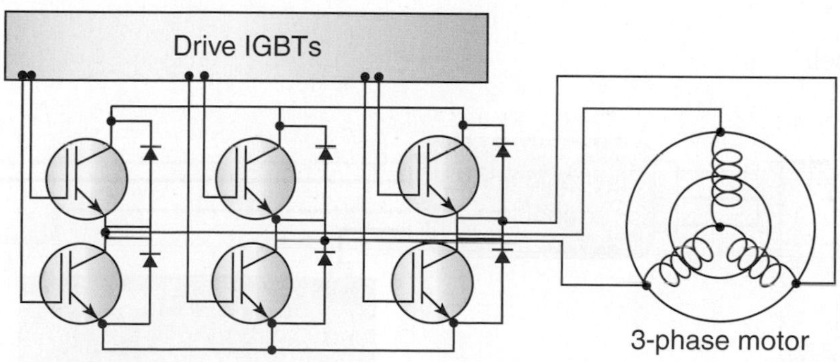

FIGURE 36–11

An electrical diagram of the connections to the motor through the IGBTs inside an inverter.

©Cengage Learning 2015

stopped and its direction reversed. This change of direction induces an AC voltage in the transformer's secondary winding. The inverter may also rectify the AC generated by the motor/generators so that it can recharge the DC battery pack.

Most housings for the electronics also contain a converter, although it may be in a separate housing. A converter changes the amount of voltage from a power source. There are two types of converters, one that increases voltage, called a step-up converter, and one that decreases the voltage, called a step-down converter. The latter is common in electric-drive vehicles to drop some of the high DC voltage to the low voltage required to power accessories such as sound systems, lights, blower fans, and the controller, and to charge the 12-volt auxiliary battery.

Also, the available voltage may be increased (boosted) before it is sent to the motor. A boost converter can supply up to 500 volts to the motor. This increased voltage increases the power output of the motor.

Basic Systems

Many different layouts and systems are used in today's hybrid vehicles. Hybrids can be further defined by the location and purpose of the electric motor(s) in the system.

BELT ALTERNATOR STARTER The least complex, but a commonly used system, is the **belt alternator starter (BAS)** system. The BAS replaces the traditional starter and generator in a conventional vehicle. The motor/generator is located where the generator would normally be and is connected to the engine's crankshaft by a drive belt. When the engine is running, the drive belt spins the motor's rotor, and the motor acts as a generator to charge the batteries. To start, or restart, the engine, the rotor spins and cranks the engine. Some BAS units also offer a small amount of engine assist.

BAS systems are typically connected to a 42-volt power source **(Figure 36–12)**, though newer systems operate at 115 volts. A belt tensioner is mechanically or electrically controlled to allow the motor/generator to drive or be driven by the belt. Some systems have an electromagnetic clutch fitted to the crankshaft pulley. When the engine is running, the clutch is engaged, and the motor acts as a generator. When the vehicle stops, the clutch disengages, and the unit is ready to act as the starting motor when the vehicle is ready to move again.

INTEGRATED STARTER ALTERNATOR DAMPER (ISAD) The **integrated starter alternator damper (ISAD)** system replaces the conventional starter, generator, and flywheel with an electronically controlled compact electric motor. Also called integrated motor assist, or IMA, the unit is typically housed in the transmission's bell housing between the engine and the transmission **(Figure 36–13)**. The electricity generated by the unit is used to recharge the 12V and high-voltage batteries; both of these are used to power the various vehicle systems.

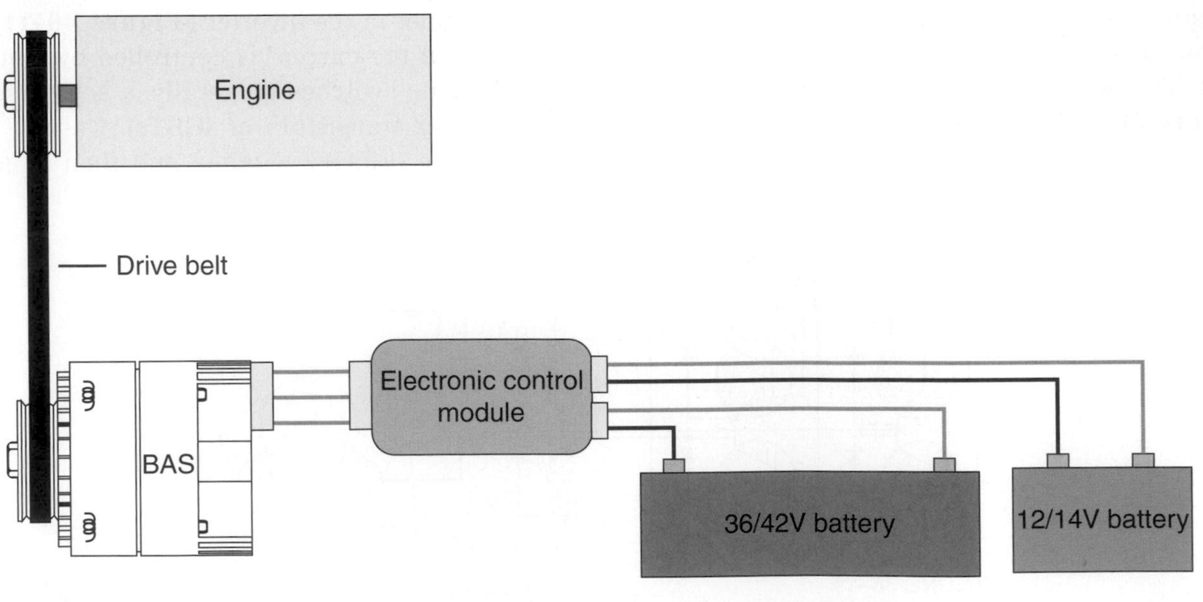

FIGURE 36–12

The layout for a typical BAS system.

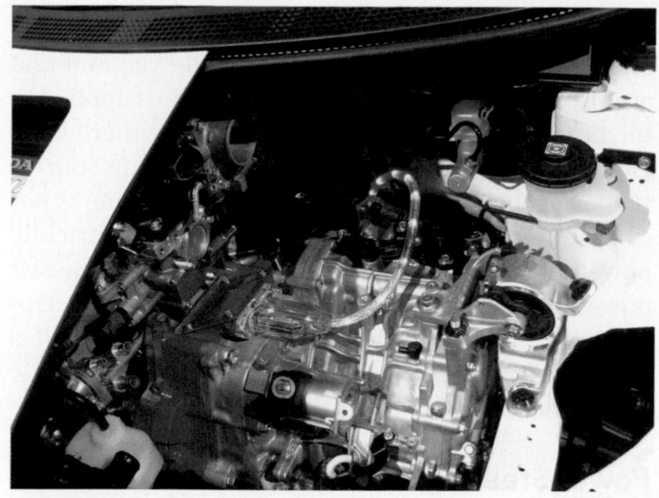

FIGURE 36–13

The red unit is an ISAD assembly sandwiched between the engine and transaxle.

POWER-SPLIT SYSTEM Currently, many full-hybrid vehicles use a power-split system **(Figure 36–14)**. They are the foundation for series-parallel hybrids and are capable of instantaneously switching from one power source to another or combining the two. The power-split device functions as a continuously variable transaxle, although it does not use the belts and pulleys normally associated with CVTs. Also, the transmission does not have a torque converter or clutch. Rather, a damper is used to cushion engine vibration and the power surges that result from the sudden engagement of power to the transaxle. The unit basically comprises a planetary gearset and two electric motors. When used with high-output engines, the power-split unit also has an additional reduction planetary gearset.

GO TO ▶ Chapter 43 for a detailed discussion of power-split units and other hybrid transmissions.

MOTORS IN TRANSMISSION This system relies on electric motors built into the transmission housing and connected to the transmission's planetary gearsets. Most of these systems are based on simple planetary gearsets coupled to two electronically controlled AC motors **(Figure 36–15)**. The gears work to increase the torque output of the motors and the engine. The result is a continuously variable transmission that responds to the needs of the vehicle. Some hybrids have a single motor within the transmission, whereas others have a motor at the transmission's input shaft and another on the output shaft.

ELECTRIC 4WD Some 4WD hybrids use an electric motor, differential, and rear transaxle housing to drive the rear wheels. This unit is not mechanically tied to the front drive axles; rather, its action is controlled by electronics. This allows the system to be capable of responding to operating conditions by varying the distribution of torque between the front and rear axles.

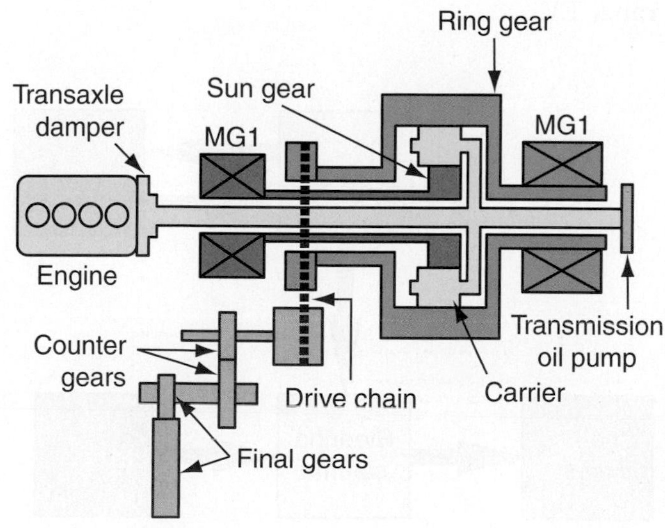

FIGURE 36–14

The layout of the main components connected to Toyota's power-splitting device.

FIGURE 36–15

An automatic transmission fitted with two electric motors. They are located inside the two large drums.

Accessories

In an HEV, the accessories are powered by either the battery or the engine, depending on the model. Some systems, such as the radio, lights, and horn, operate the same way as they do in a conventional vehicle. Other systems, such as the power-steering and power brakes, may be operated by small electric motors. It must be remembered that when working on HEVs, these auxiliaries and accessories may be powered by high voltage. Never attempt to work on these components (or the main propulsion system components) without thorough training that includes all safety procedures. Normally, most of the high-voltage components are clearly identified, and the high-voltage cables are orange.

HVAC

The engine can be used to supply the heat, so heating and defrosting systems are similar to those used in conventional vehicles. Some hybrids, however, have additional electrical heaters. These keep the passenger compartment warm when the engine is off.

HEV air-conditioning systems are identical to those used in a conventional vehicle, except a high-voltage motor may be used to rotate the compressor **(Figure 36–16)**. This increases the efficiency of the engine and allows for conditioned air when the engine is off. Compressors driven off the high-voltage (HV) system require special servicing and refrigerant oils.

GO TO ▶ Chapter 55 for a detailed discussion of high-voltage A/C systems.

FIGURE 36–16

An electric A/C compressor for a hybrid vehicle.

Power Brakes

Many power-brake systems use engine vacuum and atmospheric pressure to multiply the effort applied to the brake pedal during braking. Because there is an engine in an HEV, there is a natural vacuum source. However, there is no vacuum when the engine is not running. Therefore, some HEVs have an electrically powered vacuum pump fitted to the vacuum assist power-brake system. Other hybrids have an electro-hydraulic brake system. An electric pump provides the necessary hydraulic pressure for a hydraulic brake booster.

Power-Steering

Power-steering systems in HEVs are normally pure electrical and mechanical systems **(Figure 36–17)**. An electric motor directly moves the steering linkage. These systems are also very programmable, and the energy consumed by the motor depends on the amount the steering wheel is turned. While driving straight, the motor may not run. However, when the steering wheel is fully turned, the motor is drawing its maximum current.

GM's Series Hybrids

The Chevrolet Volt **(Figure 36–18)** and Cadillac ELR **(Figure 36–19)** are hatchbacks called extended-range EVs. According to the SAE, they should be classified as series or plug-in hybrids. Most of the time, the Volt acts as an EV or series hybrid. By combining the features of a battery-operated electric vehicle and a series hybrid, you have an extended-range EV.

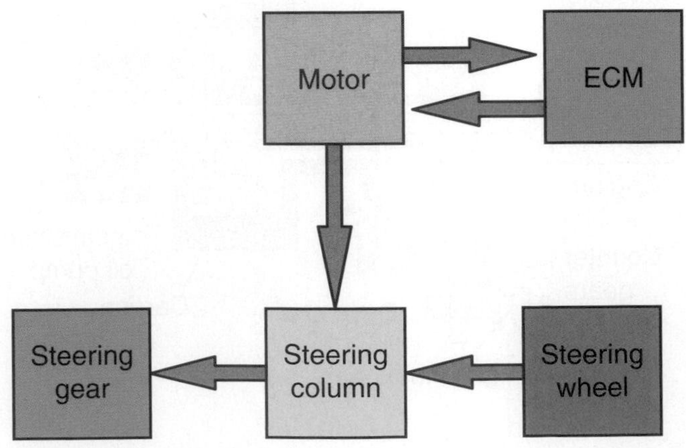

FIGURE 36–17

The command circuit for an electronically controlled power-steering system.

FIGURE 36–18

A Chevrolet Volt, an extended-range electric vehicle.

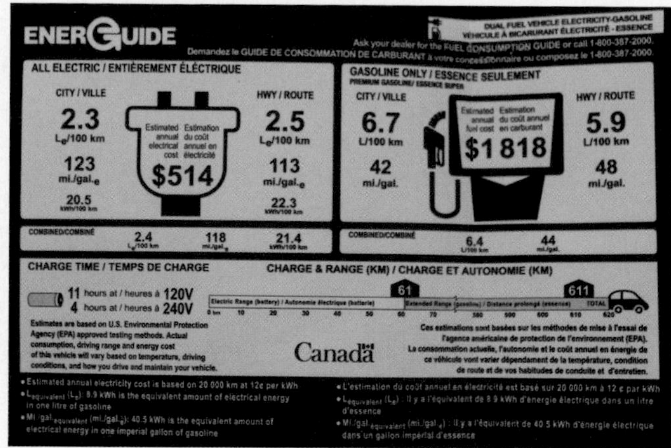

FIGURE 36–20

An EPA sticker for a Chevrolet Volt.

The Volt is powered by a Li-Ion battery and uses an engine to run a generator, when necessary. The generator's output powers the motors when the battery's charge is low. The Volt can use the energy in the battery to power the drive wheels during the first 40 to 80 km of operation. Once battery power is depleted, the engine turns on to provide the power to extend the driving range by up to 500 km. Rather than rely on a generator to recharge to battery, the Volt uses the electrical grid to serve as the primary source for recharging.

The Volt's fuel efficiency **(Figure 36–20)** is rated by the U.S. EPA and labelled in Canada by EnerGuide as 2.4 L_e/100km for all electric (city/hwy), 6.4 L/100km for gasoline only (city/hwy), and 3.92 L_e/100km for combined (electric and gasoline).

FIGURE 36–19

A Cadillac ELR.

The $L_{equivalent}$ (Le) rating refers to a comparable amount of electrical energy used over 100 km in proportion to litres of gasoline (8.9 kWh is the equivalent amount of electrical energy in 1 L of gasoline). The car is classified as an ultra-low emission vehicle (ULEV) by CARB. The only true exhaust emissions released occur when it is operating in the extended-range mode.

Powertrain

The powertrain has two AC permanent magnet electric motors—a 111 kW (149 hp) main traction motor and a 55 kW (74 hp) motor/generator—plus a 1.4L four-cylinder gasoline engine rated at 63 kW (84 hp) and 122 N·m (90 lb·ft) of torque. The motors are powered by the energy stored in the battery pack or by the energy produced by the generator. The engine is primarily used to spin the generator. A planetary gearset and three clutches manage and distribute power from the motors and engine to power the wheels **(Figure 36–21)**.

Battery

The Volt has a 1.8 m (6 ft.), 197 kg (435-pound), 16 kilowatt-hour (kWh) Li-Ion battery pack. The pack has 288 cells **(Figure 36–22)** wired in series and parallel. The rectangular cells are separated into nine modules of 32 cells. The cells and modules are arranged around aluminum cooling fins to prevent hot or cool spots on the cells.

The Volt has a thermal management system to monitor and maintain the battery's temperature. The battery can be warmed or cooled by a liquid cooling circuit that is similar to the engine's cooling system.

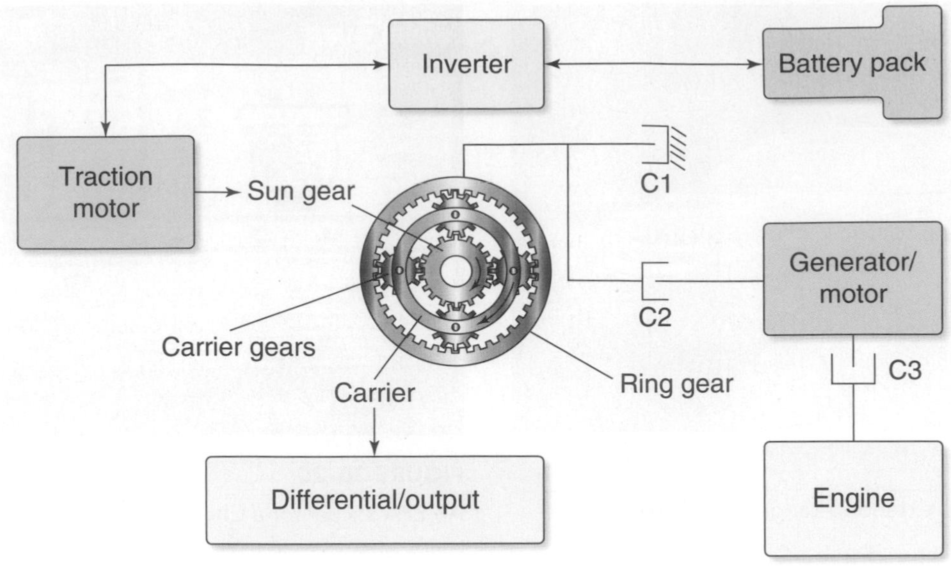

FIGURE 36–21

The main components of a Volt's powertrain.

FIGURE 36–22

The Volt's battery pack consists of 288 of these Li-Ion cells.

LG Chem Power

The system can preheat the pack during cold weather and cool it in hot weather.

Because the energy in the pack is never completely depleted and therefore never receives a full charge, the life of the pack is extended. The management system allows the battery to operate only within a predetermined state of charge (SOC) level; once that level is reached, the engine starts to maintain a charge or provide power for the motors. It is important to note that the car has a normal 12-volt battery, in addition to the high-voltage battery pack.

Basic Operation

The drivetrain (called the Voltec platform) allows the Volt to operate as a pure electric vehicle. The distance it can travel on electricity alone is affected by many things including the battery's SOC, road conditions, driving style, driver comfort settings (e.g., HVAC), and weather. Once the battery is mostly depleted, the engine starts and powers the generator, and the car operates as a series hybrid.

While operating in the series mode at higher speeds and loads (normally above 50 km/h and/or under acceleration), the generator functions as a motor to aid the traction motor (MG-B). Also, under particular conditions, the engine can be mechanically linked to the output of the gearset and assist both electric motors to drive the wheels; therefore, the Volt can operate as a series-parallel hybrid when additional power is required.

POWER MODE	CHARGE STATUS	OPERATING MODE	POWER SOURCE
All-electric	Charge depleting	Low-speed, 1 motor (MG-B)	Battery
All-electric	Charge depleting	High-speed, 2 motors (MG-B + MG-A)	Battery
Extended range	Charge sustaining	Low-speed, 1 motor (MG-B)	Battery + engine-driven generator
Extended range	Charge sustaining	High-speed, 2 motors (MG-B + MG-A)	Battery + engine-driven generator; + supplemental torque from MG-A and engine

FIGURE 36–23

An overview of the different operating modes for a Volt and ELR.

The Volt's drivetrain includes the engine, the motor/generator (MG-A), the planetary gearset, three clutches (C1, C2, and C3), the traction motor (MG-B), and the power electronics unit. This unit has three IGBT inverters, one for each motor and one for an electric oil pump. MG-A, MG-B, and the engine are connected to the planetary gearset, which allows the unit to function as a variable speed transmission.

MG-B is always connected to the sun gear, and the final drive gears are always connected to the planetary carrier. The ring gear is either held stationary by a clutch or driven by MG-A or the engine. Two of the clutches are used to lock the ring gear or connect it to MG-A. The engine and MG-A are only connected to the gearset when the appropriate clutches are applied. The third clutch connects the engine to MG-A to provide an extended driving range.

The Volt has four basic power modes: all battery-electric (charge depleting); low and high speed, in which the battery is the only source of power for the electric motors; and extended-range (charge sustaining) low and high speed, in which the battery and engine work together to power the traction motor and to improve overall efficiency **(Figure 36–23)**.

GM'S PARALLEL HYBRIDS

GM used a BAS hybrid system in 2006 through 2008 on some Saturn models, 2009 through 2010 Chevrolet Malibus, and recently in many Chevrolet and Buick models. Early systems were based on the dual voltage of a 12V/42V battery pack.

The system's electronics monitor many operating conditions, and control the operation of the motor/generator and the engine. The motor/generator can serve as the starter, assist motor, and generator. When working as a generator, it provides more than twice the output of a typical generator and is capable of providing 3000 watts of continuous power.

During operation, the generator's control module can get very hot, and excessive heat can destroy it. So there is a coolant pump to keep engine coolant circulating through the module when the engine is off during stop–start. That pump shares an electrical circuit with a second electric pump that keeps coolant circulating through the heater core when the engine is off and the vehicle is in the stop–start mode. A third pump is used to keep transmission fluid circulating in the transmission when the engine is off during stop–start.

General Motors eAssist

eAssist is available on late-model GM cars. The system is based on the previously used BAS systems. However, the new system is more powerful and provides additional torque to the driveline during heavy loads and improved regenerative braking. As a result, the new system provides close to a 25 percent increase in fuel economy over previously used systems.

The BAS unit is connected by a drive belt to the engine's crankshaft. The BAS is a three-phase AC induction motor powered by a 115-volt Li-Ion battery **(Figure 36–24)**. The air-cooled battery and its electronic controllers, along with a conventional 12-volt battery, are housed in a single unit behind the rear seat. The motor can provide an 11 kW and 107 N·m (15 hp and 79 lb·ft) boost during acceleration. It can also recover 15 kW of electricity through regenerative braking to charge the battery.

FIGURE 36–24

The main components for GM's BAS system.

FIGURE 36–25

The motors and clutches inside a two-mode transmission.

GM Two-Mode Hybrid System

GM, BMW, and DaimlerChrysler (this latter company no longer exists) co-developed a **two-mode full-hybrid system** that can be used with gasoline or diesel engines. It is claimed that the fuel consumption of a full-sized truck or SUV is decreased by at least 25 percent when it is equipped with this parallel-hybrid system.

GM offers this technology in many of its vehicles, such as the Chevrolet Silverado, Chevrolet Tahoe, GMC Yukon, and Cadillac Escalade. The system works with a 300-volt NiMH battery pack and a two-mode transmission, called an electrically variable transmission (EVT).

The transmission has four fixed gear ratios, but the motors can alter the gear ratios between each of those ratios. The system fits into a standard transmission housing and has three planetary gearsets coupled to two AC synchronous 60 kW motor/generators **(Figure 36–25)**. The gears work to increase the torque output of the motors. This combination results in a continuously variable transmission and motor/generators for hybrid operation. Multidisc clutches are used to transition the transmission from one gear ratio to another **(Figure 36–26)**.

The control of the motors' speed relies on a relatively low voltage and current, which in turn means

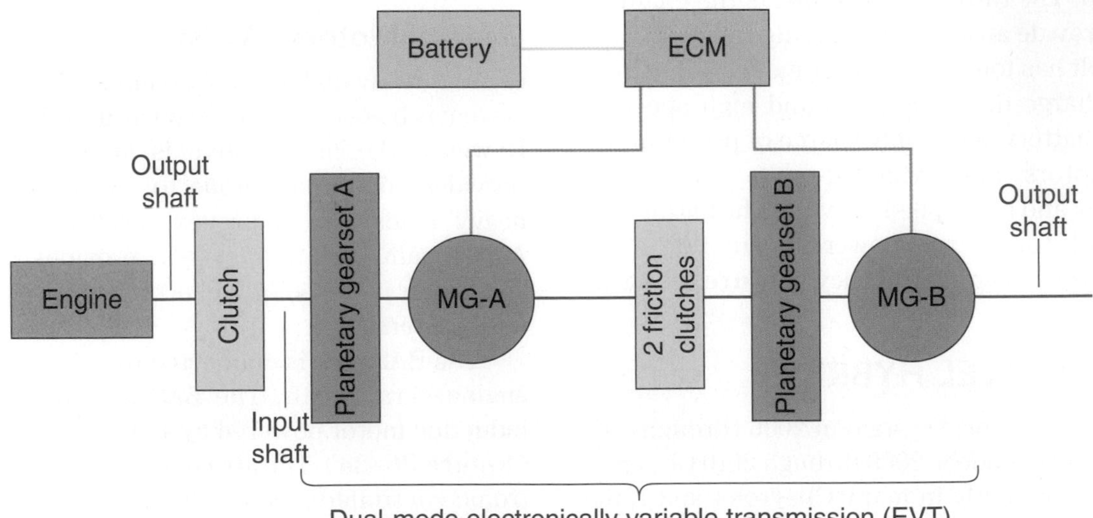

FIGURE 36–26

The two-mode system relies on two electric motors connected to planetary gearsets to move the vehicle or assist the engine during propulsion.

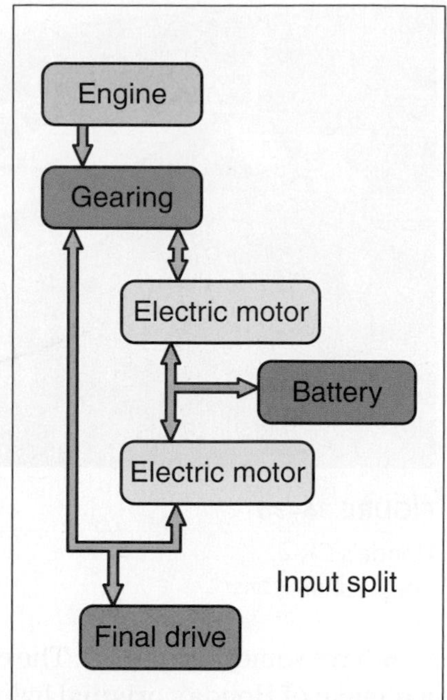

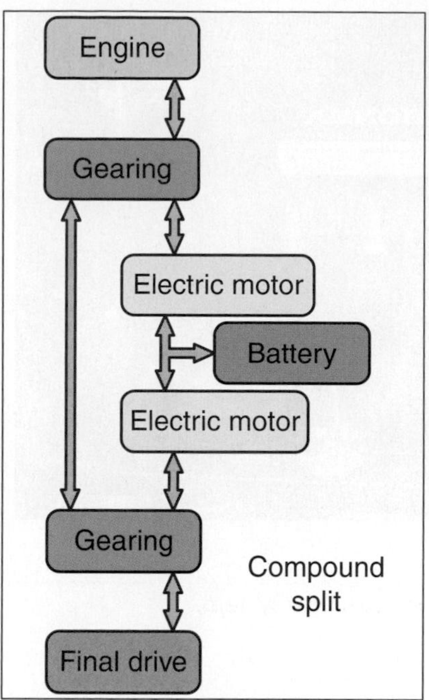

FIGURE 36–27

The flow of energy during the two operating modes.

the inverter, converter, and controller can be made lighter and smaller. The NiMH battery pack has a nominal voltage of 300 volts and is contained in a housing equipped with a cooling circuit. The system's power electronics are located under the hood and have a unique temperature control unit.

Operation

The two-mode hybrid system has two distinct modes of operation. It can operate solely on electric or engine power or by a combination of the two. Electronic controls are used to control the output of the motors and the engine. Typically, when one of the motors is not providing propulsion power, it is working as a generator driven by the engine or by the drive wheels for regenerative braking.

The first operational mode is called the input split, and the second is the compound split **(Figure 36–27)**. During input split, the vehicle can be propelled by battery power, engine power, or both. This mode of operation occurs when the vehicle is slowly accelerating and when cruising at slow speeds. When the control unit determines battery power is sufficient for the current driving conditions, the engine shuts off or some of its cylinders are deactivated. During this time, one motor is working to move the vehicle, while the other may be working as a generator to supply power for the traction motor or to recharge the battery. If the engine is commanded to start, the traction motor may shut down, and the second motor can continue to operate as a generator if needed.

Mode two comes into play when pulling heavy loads, running at highway speeds, or during heavy acceleration. The electric motor(s) assist the engine to overcome the loads, and the engine is set to provide full power. Current is no longer being generated, and the vehicle is powered by gasoline and the battery.

HONDA'S IMA SYSTEM

Most Honda hybrids use an ISAD system, called the integrated motor assist (IMA) system **(Figure 36–28)**. Honda introduced the Insight in December 1999 and became the first manufacturer to offer hybrid vehicles in North America. With this introduction came a new technology, placing an electric motor between the engine and transmission. Since then, Honda has released many variations of this design and has offered many different hybrid models.

With this design, Honda is able use a small, efficient engine in a vehicle. The power deficiencies of the engine are overcome by a small, efficient electric motor. Through the years, this platform has been improved to allow the vehicle to be powered by the engine, electric motor, or both. Early Honda hybrids were assist-only hybrids.

FIGURE 36–28

An underhood look at Honda's IMA system.

©Cengage Learning 2015

FIGURE 36–30

Honda's CR-Z.

©Cengage Learning 2015

As time passed, Honda was able use larger gasoline engines and more powerful electric motors while continuing to decrease fuel consumption and exhaust emissions. Most current Honda hybrids are rated by CARB as an AT-PZEV (advanced technology partial zero-emission vehicle, and they offer some of the best fuel economy numbers of all cars. The gains made in both areas are due to the use of improved aerodynamics, lighter construction materials, and other fuel-saving technologies.

Currently, Honda has some hybrids of particular interest: a new Insight, the CRZ, and a PHEV Accord. The new Insight **(Figure 36–29)** and the CRZ are full hybrids, although they seldom operate in an all-electric mode. They both use the same IMA system, but because they are intended for different markets, they have some differences. The current Insight is not a remake of Honda's original hybrid. Rather, it is now totally redesigned with a 1.3L four-cylinder gasoline engine and a 10kW (13hp) permanent magnet AC synchronous electric motor. These power sources are connected to a CVT.

The CR-Z (compact renaissance zero) is a two-passenger "sporty" hybrid. The CR-Z **(Figure 36–30)** offers a six-speed manual transmission in addition to a CVT. It has a 1.5L four-cylinder engine, and the Insight's IMA system delivers a combined output of 91 kW and 174 N·m (122 hp and 128 lb·ft) of torque with the manual transmission, and 167 N·m (123 lb·ft) with the CVT. This version of IMA is identical to that used in the Insight.

Recently, Honda released the Accord plug-in hybrid (PHEV). This model **(Figure 36–31)**

FIGURE 36–29

A Honda Insight.

©Cengage Learning 2015

FIGURE 36–31

A Honda plug-in hybrid Accord.

©Cengage Learning 2015

represents a significant change in Honda's hybrid approach. An Atkinson-cycle 2.0L I-4 is used with two electric motors and a Li-Ion battery pack. The latter allows for plug-in charges to enable the car to travel up to 25 km on battery power alone. This PHEV is rated at a combined rating of 2.05 L_e/100km by the EPA, and the battery can be charged in about three hours when using a 120-volt outlet or in less than one hour when using a 240-volt charger.

IMA

The IMA is positioned between the engine and the transaxle **(Figure 36–32)**. The synchronous AC motor has a three-phase stator and a permanent magnet rotor that is directly connected to the engine's crankshaft. There are three commutation sensors mounted inside the motor/generator that give the control module information about the rotor's position.

FIGURE 36–32

The IMA unit fits between the engine and the transaxle.

The IMA in late-model Hondas is used to start the engine, both during initial start-up and during stop–start. Most have an additional 12-volt starter motor and 12V battery. These are used when the SOC of the high voltage battery is low. The auxiliary starter is also used when outside temperature is extremely low or if there is a problem with the IMA system.

In the basic system, when the driver depresses the accelerator, the IMA provides power assist to the engine **(Figure 36–33)**. As the engine overcomes the load, the motor is turned off, and the car is powered only by the engine. While the car is cruising at a steady speed, the IMA can work as a generator to charge the battery and power the 12-volt system.

Some late-model hybrids have an electric motor inside the transaxle in addition to the IMA. This motor serves to provide electrical boost to the engine's output and transforms the transaxle into an electronic continuously variable transmission (E-CVT). The motor(s) are powered by a NiMH or Li-Ion battery pack.

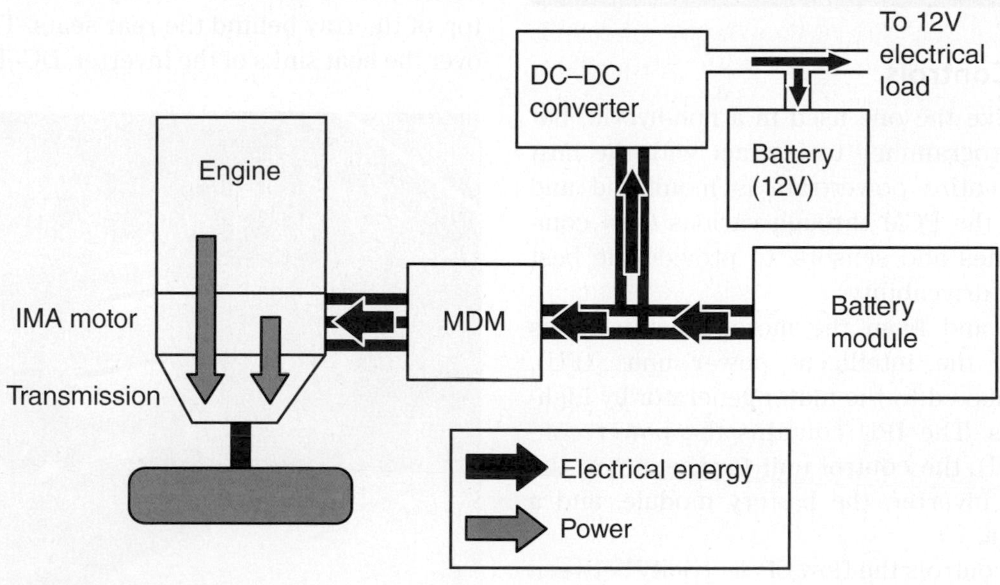

FIGURE 36–33

When the driver depresses the accelerator, the output of the motor supplements the engine's output to help acceleration.

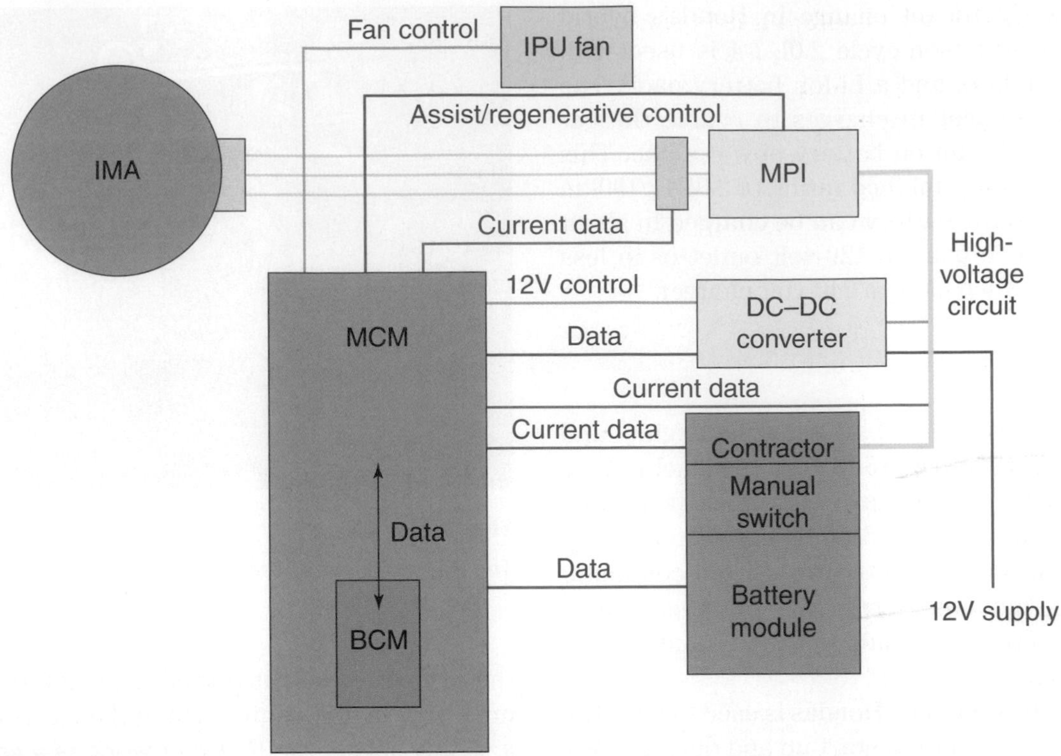

FIGURE 36–34

The basic control circuit for an IMA system.

> **CAUTION!**
>
> Because the IMA motor/generator has a permanent magnet rotor, it ***always*** generates electricity when the engine is rotating! Therefore, any time that the engine is spinning, the orange high-voltage cables could have high voltage. Keep away from them.

Electronic Controls

The PCM is like the one used in a non-hybrid, but it has been programmed to interact with the IMA system. The entire powertrain is monitored and controlled by the PCM through various CAN communication lines and sensors, to provide the best efficiency and driveability.

Power to and from the motor/generator(s) is controlled by the intelligent power unit (IPU), which is connected to the motor/generator by high-voltage cables. The IPU contains the power control unit (PCU), the control unit for the motor, the motor power inverter, the battery module, and a cooling system.

The PCU controls the flow of electricity between the IMA and battery pack. The motor control module (MCM) controls the IMA through the motor power inverter. The MCM monitors the state of charge of the battery pack and controls the IPU module fan. The MCM uses inputs of the batteries' voltage, temperature, and input and output current readings to determine the batteries' state of charge **(Figure 36–34)**.

The IPU is equipped with a cooling system mounted in the battery pack box **(Figure 36–35)**. Air is pulled into the battery module through the top of the tray behind the rear seats. The air passes over the heat sinks of the inverter, DC–DC converter,

FIGURE 36–35

The battery box of a Honda HEV.

and A/C compressor driver before it is exhausted to the outside.

A revised battery box was required with the introduction of Li-Ion batteries. The module is fitted with temperature sensors and a cooling system run by the A/C system, outside air, and a cooling fan. This system also cools the inverter, motor control module, DC–DC converter, and heat sink for the air-conditioning compressor.

The A/C compressor is powered by the high-voltage HV battery and controlled by the PCM.

Engine

The engines used with the IMA system have varied from a small three-cylinder engine to much larger V-6s. The engines have incorporated many of Honda's fuel-saving technologies, such as the variable valve timing and lift electronic control (VTEC), intelligent – dual & sequential ignition (i-DSI), variable cylinder management (VCM), and electronic throttle systems.

In addition to these, the construction of the engines includes many technologies to reduce internal friction, such as low-friction pistons and roller rocker arms.

Transmission

Most of Honda's hybrids are equipped with either a manual transmission or a continuously variable transmission (CVT). The manual transmissions are designed to be light, reduce power loss through friction, and make shifting easy. The CVT (Honda Multimatic) uses computer-controlled drive and driven pulleys and a metal "push" belt running between the variable-width pulleys. The CVT constantly adjusts to provide the most efficient drive ratio possible, depending on torque load.

Acura Hybrids

Acura, the high-end brand of vehicles from Honda, offers a line of hybrid vehicles. Most of the Acura hybrids follow the same recipe as Honda hybrids; however, some are designed primarily for increased performance as well as fuel economy. These vehicles are all-wheel-drive full hybrids. The engine in these hybrids powers the front or rear wheels, with electric motors driving the additional front or rear drive axle. In some cases, there is an additional electric motor in the transmission to assist the engine. When the rear axle is equipped with the electric motors, differential action is accomplished by electronically controlling the speed of the individual motors.

FIGURE 36–36

An Acura NSX hybrid.

The new Acura NSX Hybrid **(Figure 36–36)** uses Honda's Sport Hybrid super-handling all-wheel drive (SH-AWD) technology. This system is fitted with a gasoline engine and two electric motors at the front wheels that operate independently to provide positive or negative torque to the wheels during cornering, to improve vehicle handling. The system also has an additional motor between a dual-clutch transmission and a mid-mounted V-6 engine.

TOYOTA'S POWER-SPLIT HYBRIDS

Since its introduction of a hybrid vehicle in 2000, Toyota (and Lexus) has offered many different hybrid models.

Toyota's approach to hybridization is a combination of series- and parallel-hybrid platforms. The system relies on two electric motor/generators, a power-split transaxle, and a high-voltage battery. The power-split device mechanically blends the output from the motors and an Atkinson-cycle engine. This system **(Figure 36–37)** was called the Toyota Hybrid System (THS) when it was released, and the newer designs of the THS are called the Hybrid Synergy Drive (HSD) system. In all cases, the engine can power the vehicle or a motor/generator.

One motor/generator is primarily used to start the engine and recharge the battery pack after the engine is running. The other motor/generator assists the engine while moving the vehicle or powers the vehicle for a short distance by itself. This is the traction motor, and it can also work as a generator to provide for regenerative braking.

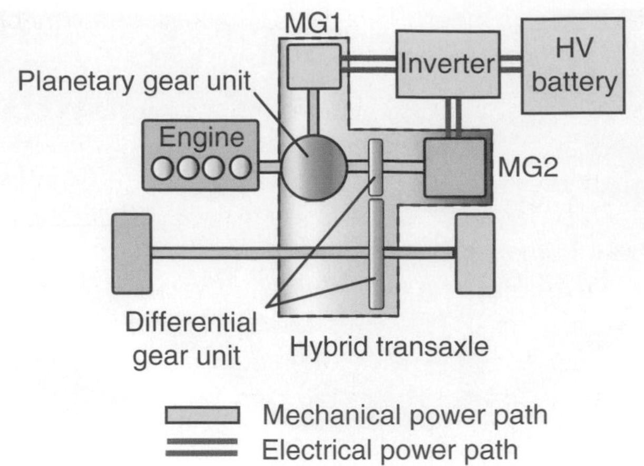

FIGURE 36–37

The THS uses a combination of two motive forces: an engine and electric motors.

©Cengage Learning 2015

Recently, Toyota has released different-size variants of the original Prius **(Figure 36–38)**. Basically, these versions are smaller or larger versions of the base model. As an example, the Prius V **(Figure 36–39)** has 50 percent more cargo space than the base model. The wheelbase of the V is longer, and the overall length has been increased; it is also taller and wider. However, the powertrain in the Prius V is identical to that found in the base model.

OTHER HYBRIDS FROM TOYOTA Several hybrid cars are available from Toyota and Lexus. Lexus has the CT200h, HS250h, GS450h, and the LS750hL. All of these are based on the same architecture as the Prius. However, the base engine, traction motor, and battery have been made more powerful to offset the increased weight of these cars over the Prius. The

FIGURE 36–38

A late-model Prius.

©Cengage Learning 2015

FIGURE 36–39

A Prius V.

©Cengage Learning 2015

total power output from these hybrid systems ranges from 56 to 438 hp. The power split devices in the cars and Lexus SUVs equipped with larger engines have an additional planetary gearset to keep the speed of the electric motors low.

Battery

Most Toyota hybrids rely on an NiMH battery pack; they also have an auxiliary 12V battery that is the power source for the ECM, lights, and other systems. The battery module **(Figure 36–40)** contains the hybrid (HV) battery pack, battery ECU, and the system main relay (SMR). The module is positioned behind the rear seat, in the trunk. A service plug and main high-voltage fuse is inserted in the high-voltage circuit. The fuse protects the circuit by opening if

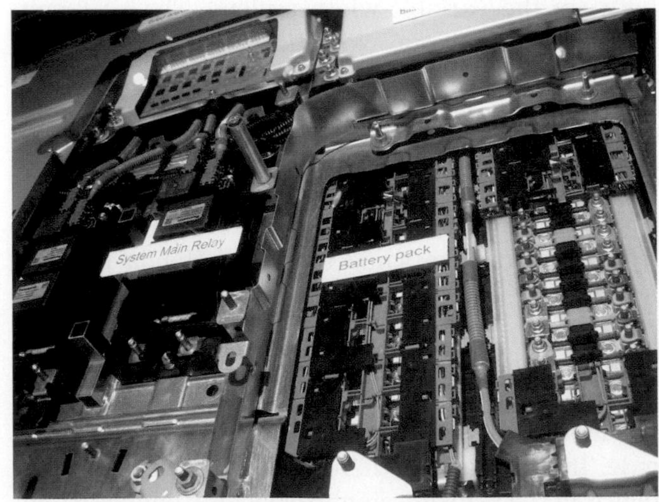

FIGURE 36–40

The main components inside a Toyota battery pack.

©Cengage Learning 2015

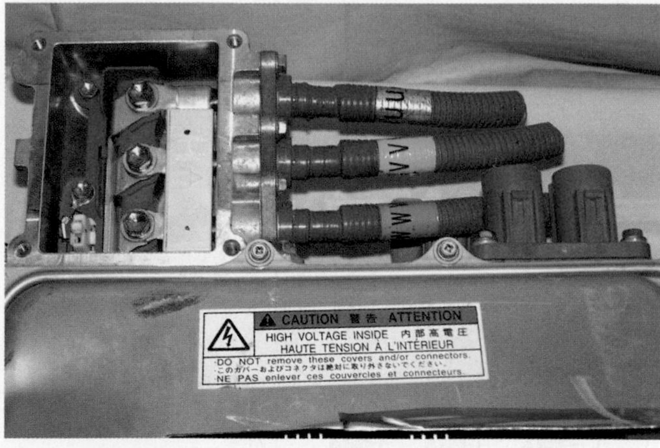

FIGURE 36–41

Individual high-voltage cables connect the inverter to the battery pack.

there is excessive current in the circuit. The service plug is used to disconnect or isolate the high-voltage circuit so that service can be performed on the circuit. The service plug is positioned in the middle of the battery modules. When removed, the circuit is open.

The nominal battery voltage varies with model and application. Late-model systems use a lower-voltage battery than earlier vehicles, but these systems can boost the voltage up to 650V. This additional voltage reduces the required amount of current to power the motor(s). If the motors' power output in watts is held constant, the amount of current drawn by the motor is inversely increased or decreased with a decrease or increase in voltage. Therefore, if the voltage is doubled, the current will be reduced by half. Also, if the current to the motor is held constant and the voltage is increased, the motor's power will be increased.

Individual high-voltage cables **(Figure 36–41)** connect the battery pack to the inverter, the inverter to the motors, and the inverter to the air-conditioning compressor.

Operation

The basic operation of all generations of the HSD system is much the same. The HSD relies on an engine, a motor/generator that serves as the starter motor and a generator (referred to as MG1, or motor generator 1), and a traction motor and generator (called the MG2, or motor generator 2). The engine, MG1, and MG2 are connected to different members of a planetary gearset inside the power-split unit. MG1, MG2, and the engine control the output of that planetary gearset. This is how those components respond to operating conditions:

- During initial acceleration, power from the battery pack to MG2 provides the energy to drive the wheels. While MG2 is powering the vehicle, the engine is off, and MG1 rotates freely and does not operate as a generator.
- This continues until the battery's voltage drops, when the driver calls for rapid acceleration, or when battery's state of charge, battery temperature, engine coolant temperature, and electrical load suggest a need for more battery energy, at which point the engine will start. To charge the battery pack, MG1 is rotated by the engine through the planetary gears.
- When the system determines the engine should power the vehicle, the engine will start if it is not already running. Engine power is split according to the needs of the system. The amount of power sent to MG1 and the drive wheels is controlled by the system.
- When the vehicle is at a cruising speed, both the engine and MG2 may power the vehicle. If engine power is not needed to maintain the speed, the engine will shut down, and the vehicle is powered by electricity alone. If the battery's SOC gets low during this time, the engine restarts to drive MG1.
- To overcome a heavy load or for hard acceleration, climbing a hill, or passing another vehicle, both the engine and MG2 power the car. MG2 receives energy from MG1 and the battery pack. This allows MG2 to work with full power. Once the vehicle returns to a normal cruising speed, battery power for MG2 is shut off, and the battery is recharged by MG1.
- When the vehicle is decelerating, the engine shuts down. MG2 then becomes a generator driven by the vehicle's drive wheels, and regenerative braking begins. The vehicle's kinetic energy is used to charge the battery pack. Most of the initial braking force is the force required to turn MG2. The hydraulic brake system supplies the rest of the braking force and brings the car to a halt.
- When reverse gear is selected, MG2 rotates in a reverse direction. The engine remains off, as does MG1.

Electronic Controls

Needless to say, the coordination of the motors and engine requires very complex electronic control systems to monitor operating and driving conditions

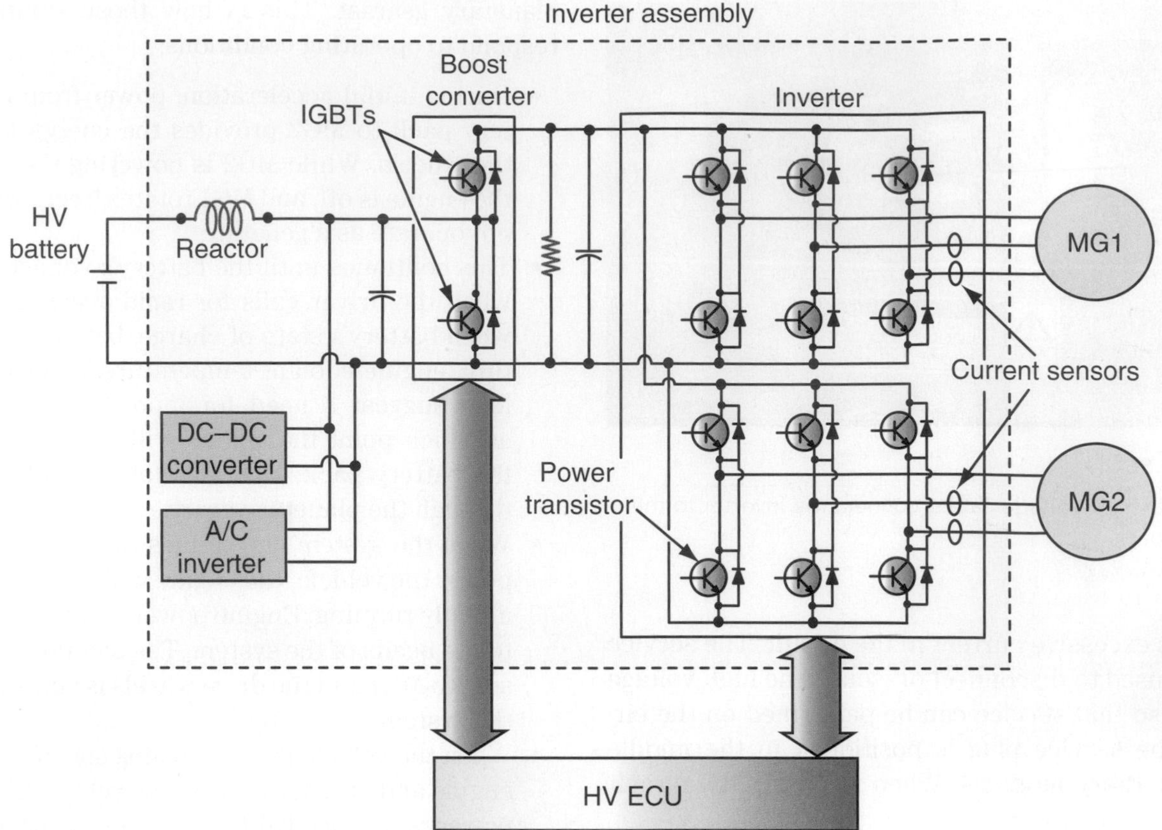

FIGURE 36–42

An electrical schematic for a THS vehicle.

and control current flow to and from the motor/generators. The ultimate control of the system is the responsibility of the hybrid vehicle control unit (HV ECU). This module receives information from sensors and other processors and in turn sends commands to a variety of actuators and controllers **(Figure 36–42)**. The ECU coordinates the engine's activity with the hybrid system. It starts and stops the engine as needed, as well as controls the operation of the engine. The HV ECU also makes sure there is proper phasing of AC to the motors. This circuit is connected between the motor/generator and the battery pack.

CAN communications are used to link various microprocessors together. The system runs continuous self-diagnostic routines. If a fault is detected, the unit stores a diagnostic code and controls the system according to data stored in its memory rather than current conditions (fail-safe mode), or it may shut down the entire system, depending on the malfunction.

The entire system is monitored by the ECU, which memorizes all conditions and operating parameters that are outside a specified range. Depending on the type and severity of the problem, the ECU will illuminate or blink the MIL, master warning light, or the high-voltage (HV) battery warning light.

The MG ECU has final control of the inverter, boost converter, and DC–DC converter based on commands from the ECU. If the ECU detects a problem in the high-voltage circuit or if the transmission is placed in neutral, the inverter is turned off to stop the operation of the motor/generators.

Depending on the model and generation of the HSD system, the electronic controls may be located in a variety of places, and the individual parts may have additional unique functions. Everything possible is done to keep the inverter assembly, MG1, and MG2 within a specified temperature by a cooling and heating system. In SUVs, the radiator for the inverter and motors is part of the engine's radiator, but it is totally isolated from it.

Most late-model hybrids have a boost converter to provide up to 650V to MG2. This converter has an integrated power module (IPM) that contains two insulated gate bipolar transistors (IGBTs), a reactor to store the energy, and a signal processor.

BATTERY ECU The battery's ECU receives information about the battery's SOC, temperature, and voltage from various sensors. This information is then sent to the HV ECU, which controls MG1 to keep the battery pack at the proper charge. The battery ECU also calculates the charging and discharging amperage required to allow MG2 to power the car. This information is also sent to the HV ECU, which sends commands to the ECM to control the engine's output. This continuous loop of information is done to maintain at least a 60 percent SOC at the battery.

The ECU also monitors the temperature of the batteries during the charge and discharge cycles, via three temperature sensors housed in the battery module and a temperature sensor in the air intake for the module. It also estimates the temperature change that will result from the cycling. Based on this information, it can adjust the battery's cooling fan, or if a malfunction is present, it can slow down or stop charging and discharging to protect the battery.

REGENERATIVE BRAKING The skid control, or "brake," ECU calculates the total amount of braking force needed to stop or slow down the car, based on the pressure exerted on the brake pedal. This in turn determines how much regenerative braking should take place and how much pressure should be sent to the brakes through the hydraulic system. This information is sent to the HV ECU, which controls the regenerative braking of MG2.

The brake ECU also controls the hydraulic brake actuator solenoids and generates pressure at the individual wheel cylinders. The total amount of force applied to the hydraulic brake system is the total required brake force minus the force supplied through regenerative braking. The skid control ECU also controls the operation of the antilock brake system.

Motor/Generators

MG1 and MG2 are permanent magnet AC synchronous motors that can also function as generators **(Figure 36–43)**. The electric 4WD-i system used in SUVs has a permanent magnet AC synchronous motor/generator, called MGR, built into the rear drive axle assembly. Unlike conventional 4WD vehicles, there is no physical connection between the front and rear axles **(Figure 36–44)**. The aluminum housing of the rear transaxle contains the MGR, a counter drive gear, counter driven gear, and a differential. The final drive ratio in the rear drive axle is very low. This provides a large amount of torque

FIGURE 36–43

MG1 and MG2 separated by a planetary gearset.

to the rear wheels. The SUVs also have a stronger MG2 that is capable of higher rotational speeds. The power-split unit in these vehicles has been modified to include an additional planetary gearset (the motor speed reduction unit). This gearset reduces the speed of MG2 and thereby increases the torque available at the front wheels.

AC synchronous motors require a sensor to monitor the position of the rotor within the stator. It is necessary to time, or phase, the three-phase AC so that it attracts the rotor's magnets and keeps it rotating and producing torque. AC creates a rotating magnetic field in the stator and the rotor chases that field. The control system monitors the position and speed of the rotor and controls the frequency of the stator's voltage, which controls the torque and speed of the motor. Toyota uses a sensor, called a **resolver**, to monitor the position of the rotor. The motors are also fitted with a temperature sensor. The ECU monitors the temperature of the motors and alters the power to them if there is evidence of overheating.

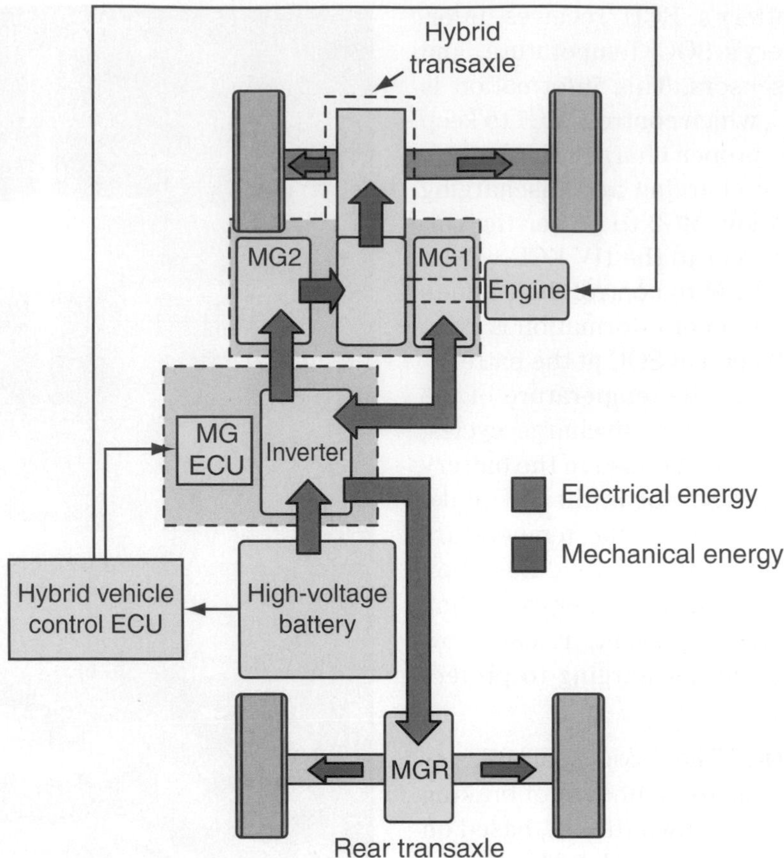

FIGURE 36–44

The rear axle is driven by an electric motor (MGR) and is not mechanically linked to the front drive axles.

Power-Split Unit

The power-split device is also called the hybrid transaxle assembly. It functions as a continuously variable transaxle, although it does not use the belts and pulleys normally associated with CVTs. The variability of this transaxle depends on the action of MG1 and the torque supplied by MG2 and/or the engine.

Conventional final drive and differential units are used to allow for good handling and ample torque to drive the wheels. The transaxle does not have a torque converter or clutch. Rather, a damper is used to cushion engine vibration and the power surges that result from the sudden engagement of power to the transaxle.

In the planetary gearset, the sun gear is attached to MG1. The ring gear is connected to MG2 and the final drive unit in the transaxle. The planetary carrier is connected to the engine's output shaft. The key to understanding how this system splits power is to realize that when there are two sources of input power, they rotate in the same direction, but not at the same speed. Therefore, one can assist the rotation of the other, slow down the rotation of the other, or they can work together. Also, keep in mind that the rotational speed of MG2 largely depends on the power generated by MG1. Therefore, MG1 basically controls the continuously variable transmission function of the transaxle.

Prius Plug-In

In 2012, Toyota released a plug-in version of the Prius **(Figure 36–45)**. The Prius PHEV is based on the base Prius but is fitted with a 4.4 kWh lithium-ion battery pack. The pack allows the Prius to operate as a pure EV for longer distances and at higher speeds. The estimated all-electric range is 21 km, which results in an expected total range of 765 km. The car is also capable of driving up to 100 km/h while in the electric mode. The estimated fuel economy while operating as a gasoline-electric hybrid is 3.27 L_e/100km.

The battery pack sits under the rear cargo floor and includes a battery charger with 7.3 m (24 ft.) long cables. The charger is designed for household current and can be plugged into any wall outlet. A full

FIGURE 36–45

Under the hood of a Prius plug-in hybrid.

charge using a 120V AC outlet takes approximately 2.5 to 3.0 hours. The charging cables connect to the charging port located behind a door on the right-rear fender **(Figure 36–46)**. The port has LED lighting to allow for safe nighttime charging. The battery pack **(Figure 36–47)** has internal and external cooling fans to control heat. The inverter has also been reworked to be compatible with the new battery. Plus, the hybrid cooling system has a larger heat exchanger and higher-capacity electric fans.

FORD HYBRIDS

In 2004, Ford released the Escape Hybrid. This was the first hybrid SUV and the first hybrid vehicle built in North America. The standard Escape Hybrid was front-wheel drive, and an intelligent 4WD system was optional. This option made the Escape the first

FIGURE 36–46

A charge nozzle inserted into a Prius PHEV.

FIGURE 36–47

The battery pack for a Prius PHEV.

4WD hybrid. In 2006, Ford released a hybrid version of the Escape's cousin, the Mercury Mariner. These SUVs were full hybrids and featured a CVT transmission and stop–start technology, as well as the ability to be powered solely by battery power.

In 2010, Ford released a hybrid edition of its mid-sized car, the Fusion. Creating a hybrid system for the Fusion led to many changes in the system used in the Escape. These hybrid systems were based on a four-cylinder engine and two electric motors. The combined power output from the engine and the traction motor is the equivalent of 142 kW (191 hp).

Since the introduction of the Fusion Hybrid, the car has been redesigned **(Figure 36–48)**, and two distinct hybrid models are available: the Fusion Hybrid and the Fusion Energi. The difference between the two models is that the Energi is a PHEV. Ford also released two hybrid versions of a new compact car, the C-Max **(Figure 36–49)**. As with the Fusion, hybrid and Energi models are available. Mechanically, Fusion and C-Max hybrids are identical. They are rated as advanced technology partial zero-emission vehicles (AT-PZEV) by CARB. They are also equipped with a high-output electric motor powered by a lithium-ion battery and an Atkinson-cycle gasoline engine. Both the Hybrid and Energi models have a projected 140 total system kW (188 hp). With the introduction of these new hybrids, Ford discontinued the Escape hybrid.

LINCOLN MKZ This Lincoln hybrid is based on the Fusion hybrid, and most of the mechanicals are the same, with some upgrades. The hybrid is capable of all-electric driving at up to 75 km for short periods.

FIGURE 36–48

A late-model Fusion hybrid.

The car is also equipped with Ford's SmartGauge with an interactive technology, called EcoGuide. This system provides real-time information to help a driver achieve maximum fuel efficiency.

Operation

The basic components and operation of Ford's hybrid system are very similar to what are found in Toyota hybrids. This has led many to conclude that Ford is simply buying the system from Toyota. This is not true. Due to the similarities and to avoid legal problems, Ford licensed some of the technology from Toyota, and Toyota licensed some technology from Ford. Toyota does not supply hybrid components to Ford. Both Ford and Toyota state that Ford received no technical assistance from Toyota during the development of the hybrid system. Very simply, Aisin supplies the transmission used in the Ford hybrids and Toyota makes its own.

FIGURE 36–49

A Ford C-Max.

FIGURE 36–50

The engine and transaxle for a late-model Ford hybrid.

These are series-parallel hybrid vehicles **(Figure 36–50)**. Ford divides the operation of the hybrid system into three different modes: positive-split, negative-split, and electric modes. During the positive-split (series) mode, the engine is running and driving the generator, to recharge the battery or directly power the traction motor. The system is in this mode whenever the battery needs to be charged or when the vehicle is operating under moderate loads and at low speeds.

During the negative-split (parallel) mode, the engine is running, as is the traction motor. The output of the traction motor tends to reduce the speed of the engine through the action of the planetary gearset. The engine's output is, however, supplemented by the power from the traction motor. During this mode, the traction motor can function as a motor or a generator, depending on the current operating conditions and the demands of the driver.

In the electric mode, the engine is off and the vehicle is propelled solely by battery power. This is the mode of operation when the battery is fully charged and during slow acceleration and low speeds, as well as when reverse gear is selected by the driver.

A feature of the Energi models is called the EV mode. There is a button that allows the driver to switch the vehicle between EV Now, Auto EV, and EV Later. Basically, this switch allows the driver to select the power source for the vehicle—electric only, gasoline only, or a combination of gas and electric.

Motors

Two separate permanent magnet AC synchronous motor/generators are used. One of these is primarily used as a generator but also serves as the starter motor for the engine and controls the activity of the transaxle. The other motor/generator is used to propel the vehicle during low-speed and low-load conditions and to assist the engine during hard acceleration, carrying heavy loads, and/or during high-speed driving.

The operation of the two motors is ultimately controlled by the primary control unit (the VSC) through inputs concerning speed and rotor position. The non-traction motor is powered by the battery pack, while the traction motor can be powered by the battery pack and/or the other motor/generator.

Controls

The vehicle system controller (VSC) is the primary control unit. Based on information from several other control units and inputs, it controls the charging, drive-assist, and engine-starting functions of the system according to current conditions. The VSC is part of the PCM.

The control system uses CAN communications and has diagnostic capabilities. The PCM monitors the activity of the system and has direct control of engine operation. The VSC communicates with the other modules and receives inputs from the gear selector sensor, the accelerator pedal position sensor, the brake pedal position sensor, and many other inputs. Based on this information, the VSC manages the charging of the battery pack, controls the stop–start function, and controls the operation of the transmission control module (TCM) that directly controls the operation of the motor/generators and, therefore, controls the operation of the transaxle. This module is housed inside the transmission case.

Battery

Both models have a Li-Ion battery pack. However, the Energi has a more powerful battery pack (4 kWh vs. 7.6 kWh). The batteries in both models are located in the rear of the car **(Figure 36–51)**. These cars can travel up to 100 km/h on electric power alone.

The high-voltage lithium-ion battery pack in the Energi provides enough power to operate in all-electric mode for short commutes. Thanks to plug-in capability, the battery can be charged overnight using a 120-volt outlet, or in less than three hours using a 240-volt outlet. The C-Max Energi has a range of more than 800 km using the battery and engine.

FIGURE 36–51

The battery pack in a Fusion hybrid.

When the connector cord is plugged into the Energi's charge port, located between the driver's door and front wheel, the port displays lights that indicate the battery's SOC. The port is surrounded by a light ring made up of four parts. When all four lights are illuminated, the vehicle has fully charged.

The hybrid and Energi's battery has a temperature management system. This system has an electric heater and a forced-air cooling system to keep the battery temperature within a specified range **(Figure 36–52)**. There are also two inertia-type switches, one in the front and the other in the rear, that can disconnect the high-voltage system if the vehicle is in an accident.

The battery energy control module (BECM) is housed in the battery pack and controls the activity of the battery. It receives commands from the VSC and sends feedback to the VSC to verify that the hybrid components are operating within the parameters set for the current condition of the battery. The

FIGURE 36–52

The cooling system for the battery in a C-Max hybrid.

battery is divided into modules. The voltage of each of the modules is constantly monitored, as is the current flow to and from them. There are eight temperature sensors in each unit to help the BECM keep the battery pack within a specified temperature range. If the temperature is outside that range or if the voltage and current flow is outside their range, the BECM will order the PCM to set a fault code and the system will move to a default setting or shut down.

There is also a lead-acid 12-volt battery, located under the hood, to provide power for the various 12-volt systems of the vehicle. This battery is recharged by the DC–DC converter, also located under the hood.

Engine/Transmission

Current Ford hybrids are equipped with a 2.0L, four-cylinder, DOHC, Atkinson-cycle engine. The early hybrids were equipped with a 2.5L, four-cylinder, DOHC, Atkinson-cycle engine. The engine is coupled to an electronically controlled continuously variable transmission (E-CVT).

The E-CVT is based on a simple planetary gearset; like the Toyota's, the overall gear ratios are determined by the motor/generator. Ford's transaxle is different in construction from that found in the Prius. In a Ford transaxle, the traction motor is not directly connected to the ring gear of the gearset. Rather, it is connected to the transfer gear assembly **(Figure 36–53)**. The transfer gear assembly is composed of three gears: one connected to the ring gear of the planetary set, a counter gear, and the drive gear of the traction motor.

The effective gear ratios are determined by the speed of the members in the planetary gearset. They are controlled by the VSC through the TCM, which calculates the required ratio according to the information it receives from a variety of inputs.

The timing of the phased AC is critical to the operation of the motors, as is the amount of voltage applied to each stator winding. Angle sensors (resolvers) at the motors' stator track the position of the rotor within the stator. The signals from the resolvers also are used for the calculation of rotor speed. These calculations are shared with other control modules. The TCM monitors the activity of the inverter and constantly checks for open circuit, excessive current, and out-of-phase cycling. The TCM also monitors the temperature of the inverter and transaxle fluid.

Cooling System

The A/C system has two parallel refrigerant loops, one for the passenger compartment and the other for the high-voltage (HV) battery. Both loops are connected to the same compressor and have their own shutoff valve. The system can cool the two zones independently.

The motor electronics (M/E) cooling system is completely separate from the engine's cooling system. The M/E cooling system cools the transaxle, motors, and the DC–DC converter. The system uses a PCM-controlled 12V coolant pump mounted near the bottom of the radiator to move the coolant through the system. The M/E system has a separate degas bottle that is part of an assembly that also includes the engine cooling system's degas bottle.

These vehicles have a PCM-controlled helper heater pump. This pump is located in series with the heater hose leading to the heater core. The pump is

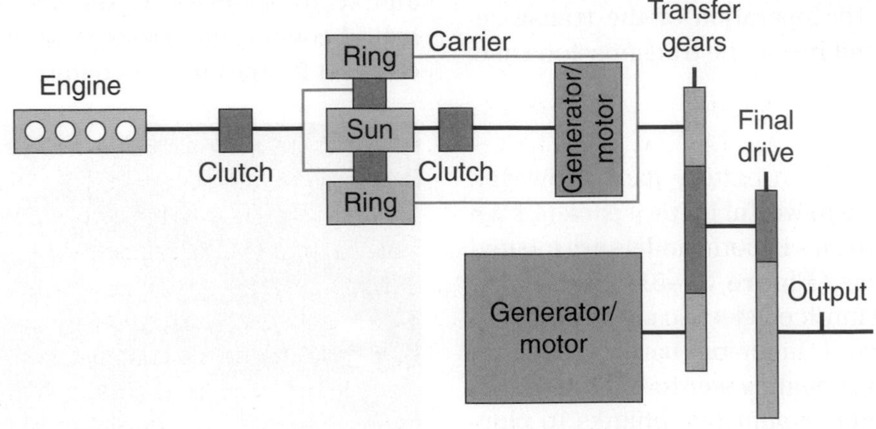

FIGURE 36–53

In a Ford hybrid transaxle, the traction motor is connected to the transfer gear assembly.

activated when the engine is off, such as during stop–start. This allows for some heat when the engine is not running.

Older models relied on an engine-driven compressor. Because cooling the battery always has precedence over passenger comfort, the engine may turn on just to run the A/C. It is important to note that the A/C unit will not run unless the engine is running. When the driver selects the MAX A/C or defrost modes, the engine will run continuously and not shut down during normal stop–start conditions. Late-model vehicles have a high-voltage electric compressor. The use of an electric compressor allows the A/C to run any time it is necessary, with the engine on or off.

4WD

Unlike the Toyotas with 4WD, the Escape and Mariner hybrids did not have a separate motor to drive the rear wheels. Rather, these wheels were driven in a conventional way with a transfer case, rear driveshaft, and a rear axle assembly. The 4WD system was fully automatic and had a computer-controlled clutch that engaged the rear axle when traction and power at the rear were needed. The system relies on inputs from sensors located at each wheel and the accelerator pedal; the system calculates how much torque should be sent to the rear wheels. By monitoring these inputs, the control unit can predict and react to wheel slippage. It can also make adjustments to torque distribution when the vehicle is making a tight turn; this eliminates any driveline shutter that can occur when a 4WD vehicle is making a turn.

PORSCHE AND VOLKSWAGEN HYBRIDS

The same basic hybrid system is used in the Porsche Cayenne, Porsche Panamera **(Figure 36–54)**, and Volkswagen Touareg. The main difference between the various models is defined by the vehicle's intended purpose and the body style. All are equipped with an Audi supercharged 3.0L V-6 engine. An electric motor is placed between the engine and an eight-speed automatic transmission. The engine and motor offer a maximum combined output of 283 kW and 580 N·m (380 hp and 428 lb·ft) of torque at 1100 rpm. They are rated as super-low ULEV-II vehicles by CARB.

If the vehicle is equipped with AWD, "VW's 4Motion," or a similar drive system that uses a Torsen centre differential with a front driveshaft and drive axle, the adaptive torque distribution system is used.

FIGURE 36–54

A Porsche Panamera hybrid.

Power for the electric motor comes from a 288V, 1.7 kWh NiMH battery pack. In electric-only mode, the vehicles can be driven up to 50 km/h and for about 2 km. An engine management system monitors the driving conditions and SOC of the battery. Based on these inputs, it controls the engine and the motor.

The vehicles have a clutch between the motor and the engine. This clutch provides a coasting feature that disengages the engine and motor from the driveline when the vehicle is operating with no load and at speeds below 160 km/h. This saves fuel by removing the engine's drag on the driveline while coasting, such as when going down hills.

The driver can manually disengage the clutch by pressing the E POWER button. This disconnects the engine from the rest of the powertrain and allows the motor to power the vehicle by itself. This feature works only when the vehicle is moving at less than 85 km/h and for only about 1.6 km. After that, the battery's charge is too low to maintain the speed, and the engine starts to propel the vehicle.

When the motor functions as a generator, it will charge the battery when the engine is in part-throttle mode and through regenerative braking. A battery management and cooling system monitors the battery's temperature to protect it from overheating. It also monitors the charge/discharge processes.

Systems that are usually driven by the engine, such as the A/C, power-steering, and power-brake systems, operate solely on electricity. Because they do not rely on the engine, they remain active when the car is running in electric mode or when it is coasting with the engine turned off.

VW Jetta Hybrid

The Jetta takes a similar approach to other VW hybrids in that it connects the engine to the transmission with an electric motor. The engine, in this case, is a turbocharged engine connected to a seven-speed, direct-shift gearbox (DSG) automatic transmission. The Jetta was the first hybrid to use a dual-clutch transmission. The DSG uses a dry clutch pack, which reduces weight and improves efficiency by eliminating the power losses that come from the constant need to lubricate a wet clutch or fill a torque converter.

The water-cooled motor is rated at 20 kW (27 hp) with 155 N·m (114 lb·ft) torque, and when combined with the engine, 127 kW (170 hp) and 250 N·m (184 lb·ft) of torque are available. Power for the electric motor is supplied by a 60-cell, 1.1 kWh Li-Ion battery pack located behind the rear seat. Like other VW Group hybrids, this system uses a clutch to decouple the engine from the drivetrain. This allows the engine to shut off and the car to coast on electricity at speeds up to 135 km/h. Also, the driver can select an electric-only mode to provide for gasoline- and emission-free operation for up to 2 km and a maximum of 70 km/h. VW is also using this system in a diesel-powered plug-in hybrid **(Figure 36–55)**.

HYUNDAI AND KIA HYBRIDS

The Hyundai Sonata hybrid **(Figure 36–56)** can go up to 120 km/h propelled only by its electric motor. It has an EPA estimated fuel economy of 6.53 L/100km in the city and 5.88 L/100km on the highway. The combined power of the engine and electric motor for this parallel hybrid is 154 kW and 264 N·m (206 hp and 195 lb·ft) of torque.

FIGURE 36–56

A Hyundai Sonata hybrid.

The electric motor **(Figure 36–57)** is sandwiched between the engine and a six-speed automatic transmission. The motor, called a transmission-mounted electrical device (TMED), replaces a conventional torque converter. The TMED is composed of two main assemblies: a 30 kW (40 hp) electric-drive motor and a solenoid-activated clutch pack. The clutch allows the power from the engine, the electric motor, or both to pass through the transmission. The drive motor also serves as a generator for regenerative braking.

FIGURE 36–55

A VW diesel-powered plug-in hybrid.

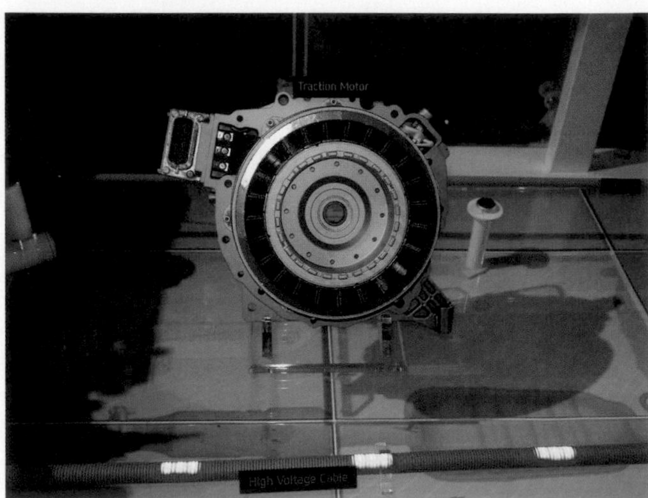

FIGURE 36–57

The ISAD assembly used in Hyundai and Kia hybrids.

The system also has an 8.5 kW BAS motor driven by the engine's crankshaft. This unit does not assist the engine; rather it only provides for the stop–start operation of the engine. The engine is an Atkinson-cycle 2.4L four-cylinder. The electric motor is powered by a 270V, 1.4 kWh, 72-cell Li-Poly battery pack. There is also a conventional 12V battery for 12V systems.

KIA OPTIMA HYBRID The Optima hybrid uses the same 2.4L, four-cylinder Atkinson-cycle engine and electric motor found in the Sonata hybrid. The ISAD-type motor is connected to the engine and the six-speed automatic transmission by a wet clutch. The motor can provide up to 30 kW (40 hp) and 205 N·m (151 lb·ft.) of torque. Electrical power is supplied by an air-cooled 270V Li-Poly battery. The car is capable of reaching about 100 km/h in its full-electric mode. The EPA has rated the Kia's fuel economy at 6.72 L/100km in the city and 5.88 L/100km on the highway.

NISSAN AND INFINITI HYBRIDS

In an Infiniti M hybrid **(Figure 36–58)**, the electric motor is sandwiched between the engine and a seven-speed automatic transmission. The engine can power the car with or without electrical assist. The 50 kW (67 hp,) 346V motor can power the car, assist the engine, start the engine, and serve as a generator. The engine and electric motor combine to produce about 268 kW (360 hp). The car is EPA rated at 8.71 L/100km in the city and 7.35 L/100km on the highway.

The hybrid system, called Infiniti Direct Response Hybrid, uses technologies developed for the Nissan Leaf electric vehicle, including the lithium-ion battery and electric motor. A 1.4 kWh Li-Ion battery pack is located under the trunk's floorboard. The battery pack is composed of laminated cells to improve battery cooling.

Infiniti uses a single motor/generator, two clutches, and a standard seven-speed automatic transmission with no torque converter. The system is also called a parallel two-clutch (P2) system. The first of the two clutches is a dry clutch located between the engine and the motor. This clutch can couple and decouple the engine from the motor. This allows the engine to shut down any time the accelerator pedal is released, such as during deceleration and coasting. The clutch eliminates the need for a torque converter and allows the full decoupling and the shutting down of the engine when there is adequate battery energy to power the vehicle solely by electricity.

The second clutch is a wet clutch at the rear of the transmission that allows the engine to turn the motor/generator to charge the batteries with the vehicle stationary.

Nissan's last hybrid was the Altima, which used the Toyota hybrid system and was available only in very limited markets. The Altima hybrid was ultimately discontinued. Future Altima hybrids will use a system similar to that used in the Infiniti M.

BMW HYBRIDS

Several different BMW models are currently available as hybrids. The technology used in the models varies. BMW's 3 and 5 Series ActiveHybrids are full hybrids **(Figure 36–59)**. A 40 kW (54 hp) electric motor is

FIGURE 36–58

An Infiniti M-series hybrid.

FIGURE 36–59

A look under the hood of a BMW ActiveHybrid.

sandwiched between the engine and an eight-speed transmission. The electric motor replaces the conventional starter and a belt-driven generator. When the motor is a generator, it recharges the battery mounted below the floor in the trunk.

The engine in the 3 and 5 Series hybrids is BMW's TwinPower Turbo, an inline six-cylinder engine. The engine's efficiency (fuel economy and performance) is enhanced by the incorporation of the twin-scroll turbocharger with Valvetronic, double-VANOS, and high-precision fuel injection. The engine and motor can provide a combined power output of 254 kW (340 hp) and 400 N·m (295 lb·ft) of torque.

A 120V Li-Ion battery is used, and the car can move in the all-electric mode for up to a maximum speed of 60 km/h. Electrical power is also available to assist the engine during heavy acceleration and heavy loads. The hybrid has both brake regeneration and auto start–stop systems.

The BMW X6 and 7 series ActiveHybrids combine an engine with two electric motors inside a transmission. This transmission was co-developed with General Motors and the old DaimlerChrysler and is commonly referred to as a two-mode transmission. The transmission used in this hybrid is a seven- or eight-speed automatic that operates as an electronic continuously variable transmission.

The transmission has two synchronous AC motors, three planetary gearsets, and four sets of multidisc clutches. The two electric motors either serve as a generator to charge the high-voltage battery pack or power the hybrid with up to 68 kW (91 hp). In all-electric driving, the hybrid can reach up to 60 km/h and travel about 2.6 km.

The transmission has two primary modes of operation: low and high speed. In each of these modes, one motor powers the hybrid while the other works as a generator. This is called the power-split drive mode, and it allows the powertrain to run at continuously variable speeds to achieve maximum efficiency regardless of load or speed. To operate the transmission, the hybrid uses a high-voltage battery pack, a power electronics unit with an integrated inverter, and high-voltage cables.

The intelligent energy management system consists of the Li-Ion battery, a 12V battery, and two on-board networks. There is a separate network for each of the power sources, but they are wired in parallel to each other. The 12V network contains all of the necessary components to operate and control the 12-volt systems in the car. The high-voltage network not only delivers to and receives high voltage from the motors but also is used to operate and control other high-voltage systems, such as the air-conditioning compressor.

A 312V NiMH battery pack with a capacity of 2.4 kWh is installed in the rear of the vehicles. The battery pack is liquid cooled and works with the A/C system or the power-steering cooling system to control the battery's temperature. If the battery's temperature rises too much, the system will automatically turn on the A/C system. A control unit is part of the battery pack and constantly monitors current battery and power levels.

The engine is a 4.4L, V-8 engine with TwinPower Turbo technology, piezo direct fuel injection, and the double-VANOS variable valve timing system. The total available power output is 358 kW (480 hp) with 780 N·m (575 lb·ft) of torque.

BMW i Concepts

BMW has developed a number of "i" concept cars. They use the prefix "i" for all vehicles that use electricity for mobility. Some of these concepts are series-type hybrids; others are pure electric vehicles.

i3 CONCEPT The BMW i3 **(Figure 36–60)** is an extended-range EV that is powered by a 125 kW (168 hp) electric motor that offers a 130 to 160 km electric driving range. The i3 body, based on the BMW 1 Series, sits on an aluminum structure, and in the centre of the structure is a liquid-cooled 32 kWh lithium-ion battery pack. The traction motor is mounted to the front of the rear axle. To serve as an extended-range EV, there is a two-cylinder 600cc

FIGURE 36–60
BMW's i3 concept hybrid.

gasoline generator that can recharge the battery pack while driving.

An intelligent liquid heating/cooling system keeps the battery at its optimal operating temperature at all times, which helps to significantly boost the performance and life expectancy of the cells. The battery can be fully recharged in four hours, using an installed outlet. If a DC-fast charger is used, an 80 percent charge can be achieved in around 30 minutes.

i8 CONCEPT The i8 Concept is a plug-in hybrid that combines an electric-drive system fitted over the front axle, with a three-cylinder engine producing about 164 kW (220 hp) and 300 N·m (221 lb·ft) of torque at the rear. The i8 is an all-wheel-drive hybrid. The electric motor powers the front drive axles, and the engine powers the rear wheels. The combined power output of 260 kW (349 hp) should move the car from 0 to 100 km/h in less than five seconds while providing at least 2.94 L_e/100km.

The body sits on an aluminum structure called the DriveCell, through the centre of which is a liquid-cooled 7.2 kWh Li-Ion battery. Having the battery in this central position allows the car to have a low centre of gravity, which enhances its handling.

MERCEDES-BENZ HYBRIDS

The Mercedes-Benz M- and S-classes have, for years, offered buyers a choice of hybrid and diesel-powered vehicles. The ML450 relies on two electric motors and a 205 kW (275 hp) 3.5L V-6 Atkinson-cycle gasoline engine to achieve an estimated 11.2 L/100km around town and 9.8 L/100km on the highway. This vehicle is classified as a full hybrid. Each of the motors is integrated into a two-mode transmission **(Figure 36–61)**. The two-mode transmission was developed with other manufacturers. Each electric motor has a specific purpose.

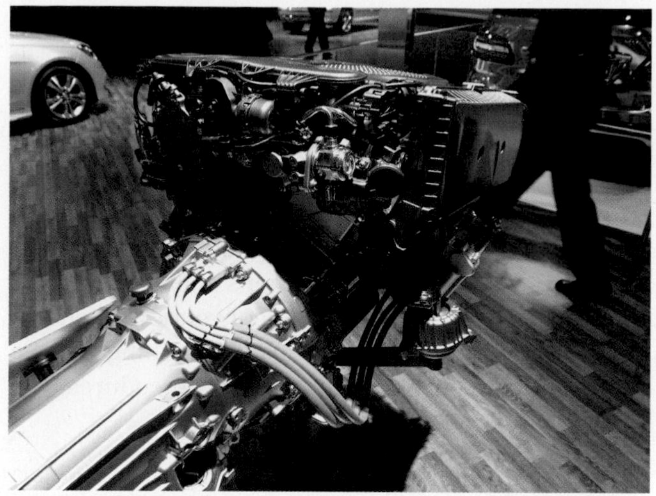

FIGURE 36–61

A two-mode transmission connected to a Mercedes-Benz V6 engine.

The motor positioned on the output shaft of the transmission has the primary purpose of moving the vehicle on electric power only. The second motor, located closer to the engine, provides the needed assist to the engine when needed. During parking and low speeds, the SUV runs on the electric drive only. The motors are powered by a liquid-cooled, 288-volt NiMH battery pack located under the rear cargo floor.

The S400 is classified as a mild hybrid. It has the Atkinson-cycle 3.5L V-6 engine used in the ML450. This was the first hybrid to use a Li-Ion battery. The battery powers a 15 kW (20 hp) electric motor. The motor/generator is contained within the torque converter housing, between the engine and transmission. Other than assisting the engine during times of need, the motor also helps to dampen drivetrain noise and vibration.

The 120V Li-Ion battery is located in the engine compartment. The battery is housed in a high-strength steel assembly with a separate cooling circuit, and its cells are separated by a gel that dampens jolts and vibrations. The 120V motor is designed to provide a large amount of torque during acceleration.

Because current flow to and from the battery can be as high as 150 amps, the battery has its own cooling circuit. In addition, a transformer in the right front wheel generates power to recharge the traditional 12V battery located in the vehicle's trunk.

It is important to note that since the oil pumps of most transmissions are engine-driven, an electric auxiliary oil pump is used to ensure the transmission is properly pressurized and lubricated when the engine is off.

E400 Hybrid

Mercedes recently released the E400 hybrid. The hybrid system used in this car is similar to the one used in the S-series. However, the E400 is considered a full hybrid and is capable of travelling over a half-mile on electricity alone. A 20 kW (27 hp) electric motor is placed between a 225 kW (302 hp), direct-injection, 3.5L V-6, and seven-speed automatic transmission. The car also has its 0.8 kWh Li-Ion battery under the hood, next to the engine.

MAINTENANCE AND SERVICE

Hybrid vehicles are maintained and serviced in much the same way as conventional vehicles, except for the hybrid components. The latter includes the high-voltage battery pack and circuits, which must be respected when doing any service on the vehicles. Manufacturers list the recommended service intervals in their service and owner's manuals. Nearly all of the items are typical of a conventional vehicle. Care needs to be taken to avoid anything orange while carrying out the maintenance procedures.

GO TO ▶ Chapter 8 for guidelines on performing preventive maintenance on a hybrid vehicle.

For the most part, actual service to the hybrid system is not something that is done by technicians unless they are certified to do so by the automobile manufacturer. Diagnosing the systems varies with the manufacturer, although certain procedures apply to all. Keep in mind that a hybrid has nearly all of the basic systems as a conventional vehicle, and they are diagnosed and serviced in the same way.

Before performing any maintenance, diagnosis, or service on a hybrid vehicle, make sure you understand the system found on the vehicle, and try to experience what it is like to drive a normal operating model. These vehicles offer a unique driving experience, and it is difficult to say what is working correctly if you do not have firsthand seat experience.

Safety Issues

Hybrid systems rely on very high voltages. Always follow the correct procedures for disarming the high-voltage system before performing any service on or near the high-voltage circuits. This is important for all services, not just electrical. Air-conditioning, engine, transmission, and body work can require services completed around and/or with high-voltage systems. If there is any doubt as to whether something has high voltage, or if the circuit is sufficiently isolated, test it before touching anything.

High-voltage circuits are identifiable by size and colour. The cables have thicker insulation and are coloured orange. The connectors are also coloured orange. On some vehicles, the high-voltage cables are enclosed in an orange shielding or casing; again, the orange indicates high voltage. In addition, the high-voltage battery pack and other high-voltage components have HIGH VOLTAGE caution labels **(Figure 36–62)**. It is important to remember that

FIGURE 36–62

A high-voltage warning label.

high voltage is also used to power some vehicle accessories. Avoid all orange-coloured cables, connectors, and wires unless you have disconnected the high-voltage power source and know what you are doing.

Precautions

- Always precisely follow the correct procedures. If a repair or service is done incorrectly, an electrical shock, fire, or explosion can result.
- Systems may have a high-voltage capacitor that must discharge after the high-voltage system has been isolated. Make sure to wait the prescribed amount of time (about 10 minutes) before working on or around the high-voltage system.
- Move the key and/or key fob a good distance away from the vehicle before starting any service.
- After removing a high-voltage cable, cover the terminal with vinyl electrical tape.
- When working on or near the high-voltage system, even when it is depowered, always use insulated tools.
- Never leave tools or loose parts under the hood or close to the battery pack. These can easily cause a short.
- Never wear anything metallic, such as rings, necklaces, watches, and earrings, when working on a hybrid vehicle.
- Alert other technicians that you are working on the high-voltage systems with a warning sign such as HIGH-VOLTAGE WORK: DO NOT TOUCH.
- Keep in mind that the engine can start and stop on its own if it is left in the IDLE STOP or READY mode.
- Make sure the READY light in the instrument panel is off.

- If the vehicle needs to be towed into the shop for repairs, make sure it is not towed on its drive wheels. Doing this will drive the generator(s) to work, which can overcharge the batteries and cause them to explode. Always tow these vehicles with the drive wheels off the ground, or move them on a flat bed.
- In case of a fire, use a Class ABC powder-type extinguisher or very large quantities of water.
- When checking for trouble codes, if DTC P3009 or P0AA6 are present, make sure you use caution when working on the high-voltage system. P0AA6 indicates there is decreased resistance in the high-voltage insulation, and DTC P3009 suggests there is a short in the high-voltage circuit.

Gloves

Always wear safety gloves during the process of depowering and powering the system back up again. These gloves must be class 0 rubber **lineman's gloves**, rated at 1000 volts. The condition of the gloves must be checked before each use. Make sure there are no tears or signs of wear. All gloves, new and old, should be checked before they are used **(Figure 36–63)**.

GO TO ▶ Chapter 6 for more information on hybrid-vehicle tools.

The insulated gloves must be sent out for testing and recertified by an accredited laboratory every six months. If recertification is not possible, new gloves should be purchased. After recertification, the laboratory marks the date of certification on each rubber glove. Used gloves should be stored in their natural shape and protected from physical damage in a cool, dark, and dry location.

Keep in mind that these insulating gloves are special gloves and not the thin surgical gloves you may be using for other repairs. You should never expose these gloves to petroleum products. Degreasers, detergents, and hand soaps may contain petroleum and should not come in contact with the gloves. Also, to protect the integrity of the insulating gloves—as well as you—while doing a service, wear leather gloves over the insulating gloves **(Figure 36–64)**. However, never use the leather gloves without the insulating gloves when working on high-voltage systems.

BUFFER ZONE When working on a high-voltage system, it is best to keep anyone who is not part of the service away from you and the car. This can be accomplished by creating a buffer zone around the car. The outside edges of the zone should be at least a metre away from the car. Orange cones should be placed to define the outer boundaries of the zone. If the vehicle is sitting unattended, it should be marked off with DO NOT ENTER tape along with the cones.

SAFETY HOOK If a "hot" high-voltage cable is loose and you cannot safely turn off the power to it, use a fibreglass reach pole and hook **(Figure 36–65)** or a dry board to move or remove the wire. The reach pole can also be used to push or pull someone away from the wire.

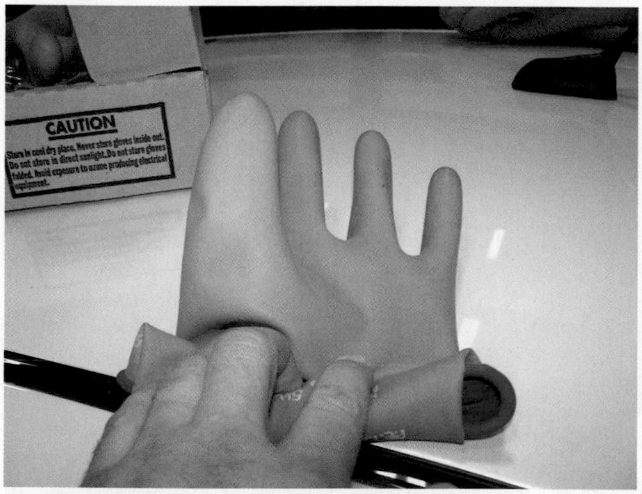

FIGURE 36–63

Blow up an insulated glove and attempt to roll it up to check its integrity.

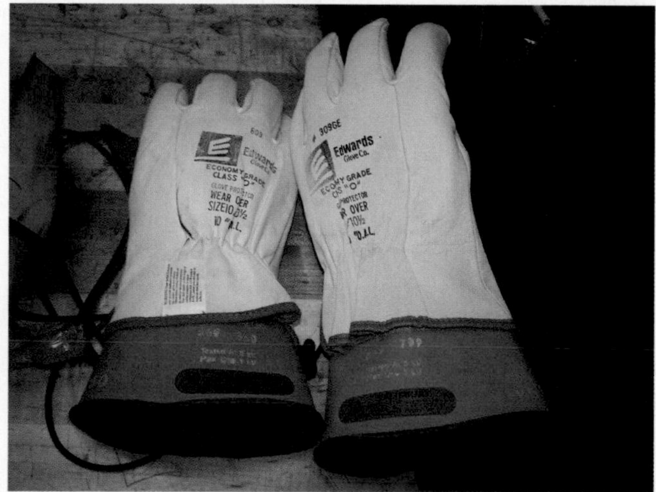

FIGURE 36–64

A pair of lineman's gloves covered with leather work gloves.

FIGURE 36–65

A fibreglass safety hook.

©Cengage Learning 2015

FIGURE 36–66

Most hybrid engines require the use of a very light engine oil.

©Cengage Learning 2015

Maintenance

The engines used in hybrids are modified versions of engines found in other models offered by the manufacturer. Other than fluid checks and changes, there is little maintenance required on these engines. However, there is less freedom in deciding the types of fluids that can be used and the parts that can replace the original equipment. Hybrids are not very forgiving. Always use the exact replacement parts and the fluids specified by the manufacturer.

Typically, the weight of the engine oil used in a hybrid is very light **(Figure 36–66)**. If the weight is increased, it is possible that the computer system will see this as a problem. This is simply caused by the extra current needed to turn over the engine. If the computer senses very high current draw while attempting to crank the engine, it will open the circuit in response.

Special coolants are required in most hybrids because the coolant not only cools the engine but also cools the inverter assembly. Cooling the inverter is important, and checking its coolant condition and level is an additional check during preventive maintenance **(Figure 36–67)**. The cooling systems used in some hybrids feature electric pumps and storage tanks. The tanks store heated coolant and can cause injury if you are not aware of how to carefully check them. The battery cooling system may need to be serviced at regular intervals. There is a filter in the ductwork from the outside of the vehicle to the battery box. This filter needs to be periodically changed. If the filter becomes plugged, the temperature of the battery will rise to dangerous levels. In fact, if the computer senses high temperatures, it may shut down the system.

A normal part of preventive maintenance is checking the power-steering and brake fluids. Some power-steering systems have a belt-driven pump, some have an electrically driven pump, and others have a pure electric and mechanical steering gear.

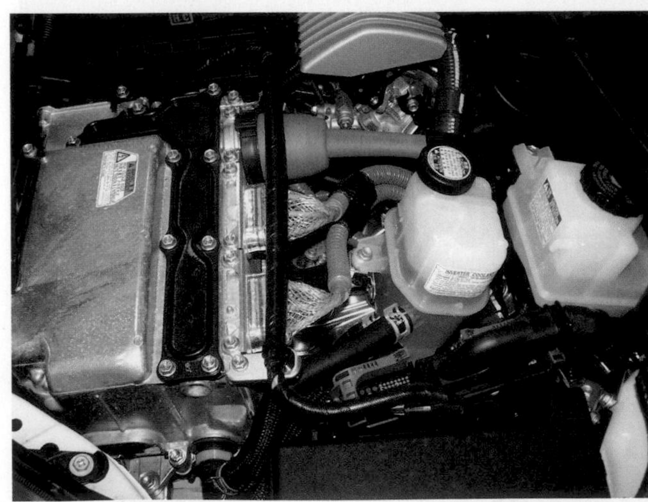

FIGURE 36–67

A coolant reservoir for an inverter.

©Cengage Learning 2015

Each variety requires different care; therefore, always check the service information for specific information before doing anything to these systems. Also, keep in mind that some hybrids use the power-steering pump as the power booster for the brake system.

Batteries

Most hybrids have two separate battery packs. One is the high-voltage pack, and the other is a 12-volt battery **(Figure 36–68)**. The high-voltage battery pack supplies the power to start the engine, assist the engine during times of heavy load, and in full hybrid, supply the energy to move the vehicle without the engine's power. The battery pack is the one that is most associated with the hybrid system. The 12-volt battery is associated with the rest of the vehicle, such as the lights, accessories, and power equipment. The 12-volt battery also supplies the power for the electronic controls that monitor and regulate the operation of the hybrid system. If this power source is not working correctly, the hybrid system will not work. Therefore, this low-voltage power source should never be ignored when working on a hybrid or a conventional system.

The procedure for depowering and isolating the high-voltage system is very important and not very difficult. However, each manufacturer has its own procedure that must be followed in the order presented. Make sure you are following the correct procedure for the specific vehicle you are working on. With the correct information and following the procedures, you can safely work on a hybrid vehicle. The following procedure has steps common to many hybrid vehicles but is not specific to any manufacturer or model and should not be used in place of the correct service procedures.

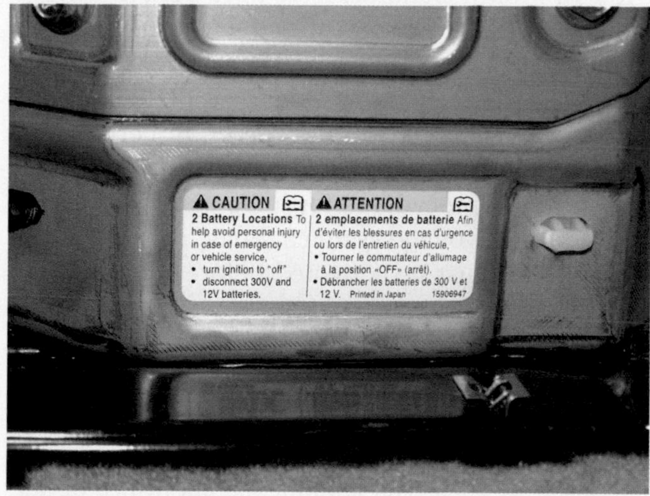

FIGURE 36–68

This warning label states there are two separate batteries.

PROCEDURE

Depowering and Isolating the High-Voltage System

STEP 1 Remove the key from the ignition switch. If the vehicle has a smart key, turn the smart key system off. This may be done by applying pressure to the brake pedal while depressing the START button for at least two seconds. If the READY lamp goes off, continue. If it does not, diagnose the problem before continuing.

STEP 2 Disconnect the negative (–) terminal cable from the auxiliary 12-volt battery. This should turn the high-voltage system off but does not complete the de-powering process.

STEP 3 Move the carpeting from the floor in the trunk or the rear of the vehicle.

STEP 4 Make sure you are wearing insulated gloves and reach in at the location of the disconnect plug at the battery box.

STEP 5 Unlatch the lever on the plug, and pull the lever down. Then remove the service plug from the battery module.

STEP 6 Put the service plug in your toolbox or elsewhere to prevent others from reinstalling it before the system is ready or while you are working on the vehicle.

STEP 7 Put electrical insulating tape over the service-plug connector.

STEP 8 Wait at least five minutes before proceeding or doing any work on or around the high-voltage system.

STEP 9 Prior to handling any high-voltage cable or part, check the voltage at the terminals. There should be less than 12 volts.

STEP 10 If a high-voltage cable must be disconnected for service, wrap its terminal with insulating tape to prevent a possible short.

STEP 11 When the service plug is reinstalled, make sure its handle is in the upright position; failure to do this may result in a loose plug that may set DTCs.

Because of the advanced electronics in hybrids, steps must be taken after reconnecting or installing a battery. The regenerative braking system needs to relearn the initial position of the brake pedal. After the battery is reconnected, slowly depress and release the brake pedal one time. The engine also needs to relearn its idle and fuel-trim strategy. If this is not done immediately after reconnecting the battery, the engine will idle and run poorly until it sets up its strategy. A typical procedure begins with turning off all accessories and starting the engine. The engine is idled until it reaches normal operating temperature, then it should be allowed to run at idle for one minute. After that time, the air-conditioning is turned on, and the engine again is allowed to idle for at least one minute. Now the vehicle should be driven for about 16 km. All manufacturers have their own sequence for doing this, so be sure to follow their procedure.

RECHARGING Recharging the high-voltage battery pack is best done by the vehicle itself; however, there are times when it may be necessary to recharge the battery in the shop. Doing this is not a typical procedure. Chances are your shop will not have the correct charger. For example, many hybrid batteries require a special charger that is not sold to its dealerships. If there is a need for one, the dealership must contact the regional office and have one delivered, and only someone from that office is allowed to operate it. This charger has the normal connections plus a cable to power the battery's cooling system. The charger is designed to bring the battery pack to a 40 to 50 percent state of charge within three hours. This is enough to start the vehicle and allow the engine to bring the battery back to full charge.

PHEV CHARGING METHODS All plug-in hybrids have a specific procedure for recharging the battery. Normally, the battery pack can be charged by a 120- or 240-volt outlet. The required charger can be built into the battery pack or connected externally. For example, a Volt is equipped with a 120V charger **(Figure 36–69)** that plugs into a household wall outlet; this allows the battery to receive a full charge in 10 to 12 hours. The charger is located in the trunk along with the tire emergency repair kit. One end of the charger's cord has a fuel nozzle–type plug that fits into a charge port located on the left side of the car.

When the charging plug is connected and all is well, the green indicators on the charger will illuminate. If the red indicators are flashing, the battery cannot be charged. A flashing red light can mean one of the following: the AC voltage is out of range; the AC outlet does not have a proper safety ground; or there is a fault in the charge cord or charger.

JUMP-STARTING If the vehicle will not start, several things can be the cause. Like conventional vehicles, it must have fuel and there must be ignition, intake air, compression, and exhaust. Before proceeding with a no-start diagnosis, make sure the immobilizing system is working properly. If the auxiliary or high-voltage battery is discharged, the engine will not start, nor will the vehicle be able to operate on electric power only. Manufacturers have built-in ways to jump-start these vehicles, if and when the batteries go dead. The basic connection from a booster battery to the dead battery is the same, but the connecting points may be different, and there are certain precautions to consider when jump-starting. There are also separate procedures for jump-starting with the low- and high-voltage systems. Some hybrids have a control that must be activated before attempting to jump-start the vehicle.

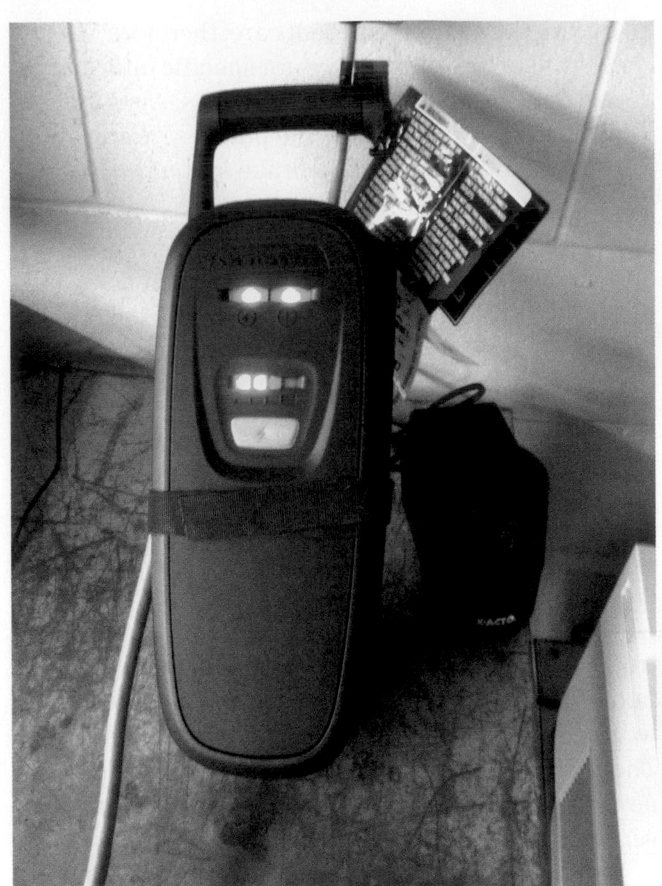

FIGURE 36–69

The Chevrolet Volt has an auxiliary 120V battery charger in its trunk.

BATTERY COOLING SYSTEM FILTER Normally, the battery has its own cooling system. The control module monitors the temperature of the cells and activates fans and/or the rear air conditioning system when the temperature rises. The cooling system draws in outside air. Within the ductwork for that vent, there is an air filter that requires periodic replacement. If the filter is dirty or restricts airflow, the battery can overheat.

Diagnostics

It is important that you have good information when diagnosing driveability problems. The problem can be caused by the hybrid system, engine, or transmission. Determining which system is at fault can be difficult. On some hybrids, it is possible to shut down the hybrid system and drive the vehicle solely by engine power. On others, such as the Toyota and Ford hybrids, this is not possible. If electric power can be shut off and the vehicle still drives poorly, the problem is the engine or transmission. If it is not possible to shut down either power source, your diagnosis must be based on the symptom and information retrieved with a scan tool.

With hybrids, it is often difficult to control the operation of the hybrid system so that certain tests can be conducted. Ford has built into its control system two scan tool–controllable modes for diagnostics. The engine-cranking mode allows the engine to crank without the engine starting. During this mode, the TCM orders the starter/generator to rotate the engine at 900 to 1200 rpm.

Ford also offers a running diagnostic mode. In this mode, the engine will run until it is ordered to stop by the scan tool or the ignition is turned off. In normal operation, the engine will not idle for very long with being shut down by the system. Therefore, when diagnostics requires that the engine be idling, the engine-running mode allows for this.

All warning lamps in the instrument panel should be checked **(Figure 36–70)**. If any of these remain on after the engine is started, the cause should be identified and corrected before continuing with diagnosis. Lastly, a scan tool should be used to retrieve any fault codes held in the computer's memory. In many cases, a manufacturer-specific scan tool is required to test hybrids. Follow the prescribed sequence for retrieving and responding to all diagnostic trouble codes (DTCs).

The scan tool also allows for some active tests that enable you to excite or disable certain outputs so that their operation can be monitored. These *inspection modes* can crank the engine to conduct a compression test, turn the traction control on and off, and turn the inverter on and off. The value of these modes is the ability to isolate systems, which will definitely help in diagnosis. The scan tool is also used to reset or calibrate the electric motor's rotor.

Test Equipment

To test high-voltage systems you need a digital multimeter (DVOM) **(Figure 36–71)** and, of course, a good pair of insulating gloves. Although the high-voltage system can be isolated from the rest of the vehicle, high voltage is still at and around the battery pack and inverter.

The DVOM used to diagnose a hybrid vehicle is not the same meter you would use on a conventional vehicle. The meter used on hybrids, electric vehicles, and fuel-cell vehicles must be classified as a **category 3 or 4 (CAT III or CAT IV)** meter. There are four categories for low-voltage electrical meters, each built for specific purposes and to meet certain standards. Low voltage, in this case, means voltages less than 1000 volts.

FIGURE 36–70

An example of some of the displays and warning lamps on a hybrid vehicle.

FIGURE 36–71

A CAT III DVOM is a must for diagnosing the high-voltage systems.

WARNING!

When using a CAT III or CAT IV meter, make sure the meter's leads are also rated for high voltage. Make sure the meter's probes have safety ridges or finger positioners. These help prevent physical contact between your fingers and the probes.

Another valued tool during diagnosis is an **insulation resistance tester**. This meter is not one commonly used by automotive technicians, but should be used by anyone who might service a damaged hybrid vehicle, such as doing body repair. These meters check for voltage leakage from the insulation of the high-voltage cables **(Figure 36–72)**. Minor leakage can cause hybrid system–related driveability problems. This meter should also be a CAT III or CAT IV meter.

To check the insulation, the approximate system voltage should be selected on the meter and the probes placed at their test position. The meter is measuring the insulation's effectiveness and not its resistance. The meter will display the voltage it detects; any voltage is not good, and the cable should be carefully examined.

Air Conditioning

Most hybrids are equipped with air-conditioning systems with either a belt-driven or an electrically powered A/C compressor. The electrical units are powered by high voltage, and all precautions should be taken to work safely with these units. Always wear lineman's gloves when inspecting or servicing high-voltage air-conditioning systems.

The refrigerant oil used in all electrically operated compressors must meet the specifications given by the manufacturer. In nearly all cases, the oil is synthetic and nonconductive and able to insulate the various electrical parts of the compressor from each other. The most common is **polyvinyl ether (PVE) oil**. The use of the regular refrigerant oil (PAG) can cause the destruction of the compressor **(Figure 36–73)**.

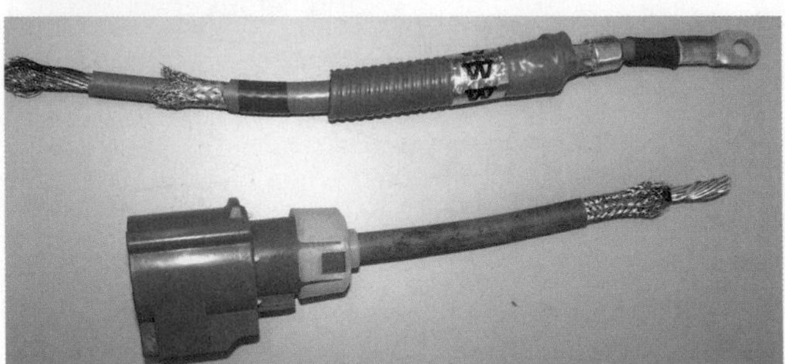

FIGURE 36–72

An insulation resistance tester (megger meter) is used to check the insulation of high-voltage cables.

FIGURE 36–73

This is what can happen to a compressor if the wrong refrigerant oil is used.

KEY TERMS

Belt alternator starter (BAS) (p. 1084)
Category 3 or 4 (CAT III or CAT IV) (p. 1115)
Extended-range electric vehicles (EV) (p. 1078)
Full hybrid (p. 1079)
Insulation resistance tester (p. 1116)
Integrated starter alternator damper (ISAD) (p. 1084)
Lineman's gloves (p. 1111)
Plug-in hybrid electric vehicles (PHEVs) (p. 1079)
Polyvinyl ether (PVE) oil (p. 1116)
Resolver (p. 1099)
Two-mode full-hybrid system (p. 1090)

SUMMARY

- Any vehicle that combines two or more sources of power is called a hybrid. Current HEVs have a gasoline engine and one or more electric motors.
- In a series hybrid, the engine never directly powers the vehicle.
- Many hybrid vehicles are parallel types and rely on power from an electric motor and the engine.
- A series-parallel hybrid can run on just the engine, just the batteries, or a combination of the two.

- Plug-in hybrid electric vehicles (PHEVs) are full hybrids with larger batteries.
- Regenerative braking is the process that allows a vehicle to recapture and store part of the kinetic energy that would ordinarily be lost during braking.
- A controller looks at inputs from various sensors to determine the operating conditions of the vehicle and manages the flow of electricity to control the speed of the motor.
- An inverter is a power converter that changes the high DC voltage of the battery to a three-phase AC voltage for the electric motors.
- The belt alternator starter (BAS) system replaces the traditional starter and generator in a conventional vehicle and is connected to the engine's crankshaft by a drive belt.
- The integrated starter alternator damper (ISAD) system replaces the conventional starter, generator, and flywheel with an electronically controlled compact electric motor housed in the transmission's bell housing between the engine and the transmission.
- A power-split unit has a planetary gearset and two electric motors and is capable of instantaneously switching from one power source to another or combining the two, while functioning as a continuously variable transaxle.
- Some hybrid systems are based on the presence of one or more motors inside a conventional transmission.
- Some 4WD hybrids use an electric motor, differential, and rear transaxle housing to drive the rear wheels.
- HEV air-conditioning systems are identical to those used in a conventional vehicle, except a high-voltage motor may be used to rotate the compressor.
- The Chevrolet Volt and Cadillac ELR are called extended-range EVs but can be also classified as series or plug-in hybrids.
- General Motors eAssist system is based on a BAS system and can provide additional torque to the driveline during heavy loads.
- A two-mode full-hybrid system fits into a standard transmission housing and has three planetary gearsets coupled to two AC synchronous 60 kW motor/generators. This results in a continuously variable transmission and motor/generators for hybrid operation.
- Most Honda hybrids use an ISAD system, called the integrated motor assist (IMA) system, in which a small, efficient engine in a vehicle and the power deficiencies of the engine are overcome by the electric motor.
- Toyota's approach to hybridization is a combination of series- and parallel-hybrid platforms. The system relies on two electric motor/generators, a power-split transaxle, and a high-voltage battery.
- The basic components and operation of Ford's hybrid system are very similar to what is found in Toyota hybrids; however, each manufacturer uses a unique design to accomplish the same thing.
- Porsche Cayenne, Porsche Panamera, Volkswagen Jetta, and Volkswagen Touareg hybrids are based on an ISAD system.
- The Hyundai Sonata and Kia Optima hybrids have an electric motor sandwiched between the engine and transmission and a BAS unit for stop–start.
- The Infiniti M hybrid is also a basic ISAD system but has two clutches; one is a dry clutch located between the engine and the motor, and the other is a wet clutch located at the rear of the transmission.
- BMW's 3 and 5 Series ActiveHybrids are based on the ISAD system.
- The BMW X6 and 7 series ActiveHybrids use the two-mode hybrid system.
- The Mercedes-Benz M hybrid is based on a two-mode transmission.
- The Mercedes-Benz S hybrid is a mild hybrid with an ISAD. The same platform is used in its E400 full hybrid.
- Hybrid systems rely on very high voltages. Always follow the correct procedures for disarming the high-voltage system, and pay strict attention to the stated precautions before performing any service on or near the high-voltage circuits.
- Always wear lineman's gloves rated at 1000V during the process of depowering and powering the system back up again.
- The condition of the gloves must be checked before each use. Make sure there are no tears or signs of wear.
- When working on a high-voltage system, keep anyone who is not part of the service away from you and the car by creating a visual buffer zone around the car.
- If a "hot" high-voltage cable is loose and you cannot safely turn off the power to it, use a fibreglass reach pole and hook or a dry board to move or remove the wire. The reach pole can also be used to push or pull someone away from the wire.
- The cooling system in many hybrids is composed of several independent cooling loops; each of these must be maintained.

- Recharging the high-voltage battery pack is best done by the vehicle itself; however, there are times when it may be necessary to recharge the battery in the shop. To do this, a special charger is required.
- To test high-voltage systems, you need a CAT III DMM.
- An insulation resistance tester should be used to check the effectiveness of all high-voltage cables.

REVIEW QUESTIONS

1. What are the basic components of a belt alternator starter hybrid system?
2. What are the main reasons that a mild hybrid consumes less fuel than a conventional vehicle?
3. List five common-sense rules that should be followed when working on a hybrid vehicle.
4. The Prius PHEV offers many advantages over the basic Prius. How is that possible?
5. What is the purpose of a typical inverter?
6. What does an insulation resistance tester measure?
   a. voltage **b.** resistance
   c. amperage **d.** current flow
7. What does the SAE classify the Chevy Volt as?
   a. an extended-range EV
   b. a series hybrid
   c. a parallel hybrid
   d. an electric vehicle
8. Which of the following is true of the Toyota Prius planetary gearset?
   a. The ring gear is connected to MG1.
   b. The sun gear is attached to MG2.
   c. The ring gear is connected to the engine's output shaft.
   d. The planetary carrier is connected to the engine's output shaft.
9. After isolating the high-voltage system, what is the minimum time you should wait before beginning to work on or around the hybrid system?
   a. 1 hour **b.** 30 minutes
   c. 15 minutes **d.** 5 minutes
10. How often must insulated lineman's gloves be tested and recertified?
    a. every 6 months **b.** every 12 months
    c. every 18 months **d.** every 2 years
11. What is the minimum safe distance for the buffer zone around you and the hybrid vehicle being worked on?
    a. 1 metre **b.** 2 metres
    c. 3 metres **d.** 4 metres
12. What does the Ford hybrid motor electronics (M/E) cooling system cool?
    a. the engine, transaxle, motors, and DC–DC converter
    b. the transaxle, motors, and DC–DC converter
    c. the motors and DC–DC converter
    d. the DC–DC converter
13. During diagnostics, the DTC P3009 is displayed. What does this indicate?
    a. There is a short in the high-voltage circuit.
    b. There is decreased resistance in the high-voltage insulation.
    c. There is an open in the high-voltage circuit.
    d. There is decreased voltage in the high-voltage circuit.
14. Which of the following is least likely to decrease fuel consumption of a hybrid vehicle?
    a. low-rolling resistance tires
    b. increased aerodynamic drag
    c. stop–start systems
    d. lighter and less-powerful engines
15. What is a step-up converter used for?
    a. to supply power to sound systems
    b. to supply power to the drive motors
    c. to power lights
    d. to charge the 12-volt auxiliary battery
16. How much of the normally lost braking energy is captured by regenerative braking?
    a. approximately 10 percent
    b. approximately 30 percent
    c. approximately 50 percent
    d. approximately 70 percent
17. What classification of meter must be used when working on the high-voltage system?
    a. an analog multimeter
    b. an auto-ranging DMM
    c. a manual-ranging multimeter
    d. a CAT III DMM
18. What component is located where the generator would normally be and can either charge the 12V battery or start the engine?
    a. the BAS **b.** the IMA
    c. the ISAD **d.** the CVT
19. What is used to start late-model Honda hybrids in extreme cold weather?
    a. the IMA
    b. a traditional 12V starter
    c. a BAS
    d. an additional electric motor inside the CVT
20. What is the voltage rating of the lineman's gloves used during powering and de-powering hybrids?
    a. 10V **b.** 100V
    c. 1000V **d.** 10000V

CHAPTER 37
Electric Vehicles

LEARNING OUTCOMES

- List some of the advantages and disadvantages of owning an electric vehicle.
- Describe the major systems that make up a BEV.
- Describe the purpose and function of a battery control system.
- Explain the differences between conductive and inductive battery charging.
- Explain how most electric vehicles' problems are diagnosed.
- Describe some precautions that should be followed when troubleshooting and repairing an electric vehicle.
- Describe the basic configurations for the powertrain in a fuel cell vehicle.
- Describe the major components of a fuel cell vehicle.
- Explain how a fuel cell works.

INTRODUCTION

Today, there are only a few pure **battery electric vehicles (BEVs)** manufactured by the major automobile companies. However, nearly all are planning to release new BEVs in the near future **(Figure 37–1)**. Market resistance has resulted from short drive ranges and high cost. However, today's technologies allow BEVs to have longer ranges than in the past; unfortunately, the cost of the vehicles is still high.

FIGURE 37–1
An Audi "e-tron," which is an electric concept car.

A LOOK AT HISTORY

Electric-drive vehicles have been around for a long time. Early automobiles were mostly electric- or steam-powered. "Steamers" were the most common until the late 1800s.

In 1900, 38 percent of the cars sold were electrically powered; the others ran on steam or gasoline. Starting the engine and changing gears were the most difficult things about driving a gasoline-powered vehicle. Electric-drive vehicles did not need to be manually cranked to get going and had no need for a transmission or change of gears. These were the primary reasons the public accepted electric drive over the gasoline vehicles. However, the internal combustion engine became popular because it allowed a vehicle to travel great distances and to achieve a decent high speed, and it was much less expensive to buy.

Let's take a quick look at some interesting developments of electric vehicles throughout history. It is said that the first practical electric vehicle was made either by Thomas Davenport in the United States or by Robert Davidson in Edinburgh, Scotland, in 1842. Both vehicles had nonrechargeable batteries and therefore had a limited driving range, and most consumers did not want them.

In 1865, the storage battery was invented and then further improved in 1881. More significantly, between the years 1890 and 1910, battery technology drastically improved with the development of the modern lead-acid battery by Henri Tudor in 1881. In 1909, Thomas Edison perfected his nickel-iron battery and marketed it to automakers, and electric vehicle popularity grew and reached its peak in 1912.

In 1904, Henry Ford overcame some of the common objections to gasoline-powered cars and, thanks to assembly line production, offered gasoline-powered vehicles at very low prices ($500 to $1000). The cost of electric vehicles was much higher and rising each year.

Further advances made to the gasoline vehicle, such as an electric starter and the availability of gasoline, provided cheap and more practical transportation. For the most part, electric-drive vehicles were a thing of the past from 1920 to 1965.

Legislative bills were introduced in the mid-1960s, recommending the use of electric vehicles as a way to reduce air pollution. Subsequent laws put mandates on auto manufacturers to clean up exhaust emissions. The initial result of these laws was altering the basic engine with a variety of emission controls, many of which adversely affected fuel economy and performance. This forced the manufacturers to look into alternative ways to provide transportation.

In 1973, the price of gasoline drastically increased as the result of an Arab oil embargo. The rising cost led to increased attention to the development of electric-drive vehicles.

The U.S. Congress passed into public law the Electric and Hybrid Vehicle Research, Development, and Demonstration Act of 1976. One objective of the law was to work with industry to improve batteries, motors, controllers, and other hybrid-electric vehicle components. The goal was to double fuel efficiency in all vehicles.

In 1990, the California Air Resources Board (CARB) adopted a requirement that 10 percent of all new cars offered for sale in California in 2003 and beyond must be zero-emission vehicles (ZEVs). But in 1998, CARB modified those requirements. The change allowed manufacturers to satisfy up to 6 percent of their ZEV requirement with automobiles that qualify as partial ZEVs. Today, only fuel cell electric vehicles (FCEVs) running on pure hydrogen and BEVs qualify as ZEVs.

In 1991, the U.S. Advanced Battery Consortium (USABC) began a project that would lead to the production of a battery that would make electric vehicles a viable option for consumers. The initial result was the development of the NiMH battery. This battery can accept three times as many charge cycles as lead acid, and can work better in cold weather.

In 1996, GM started to lease its ev1, the first modern electric car. Its range started at 110 to 160 km (70 to 100 miles). An upgraded version with a Nimh battery became available three years later with 160 to 225 km (100 to 140 miles) of range. By 1998, a few electric vehicles were made available (but very few were purchased or leased) in the United States, including Honda's EV Plus, GM's EV1 and S-10 pickup, the Ford Ranger pickup, and Toyota's RAV4 EV.

In 2003, CARB ended its initiative to require zero-emission vehicles. GM, along with Toyota and others, formally ceased production of electric vehicles. However, hybrid and electric-vehicle development did continue without the mandate. For example, note the following:

- From 2004 to present, many hybrid vehicles have been produced and are being sold at great numbers.
- In 2008, the Lotus Elise–based, Lithium-Ion Tesla Roadster went on sale.
- In 2010, Chevrolet introduced an extended-range BEV called the Volt, which was named car of the year in 2011.
- In 2010, Nissan released a BEV called the Leaf, as well as some mid-sized hybrids. Also, Mitsubishi released a BEV called the Mi-EV.
- In 2011, many new BEVs and plug-in hybrids were introduced, including the Tesla Model S; the Honda Fit; Ford's Escape, Fusion, Focus, and C-MAX; Smart ED; and a Prius Plug-in and a Tesla-powered RAV4 from Toyota. All of these will make the near future the busiest years in electric cars since early in the 20th century.

In Canada, there are no distinct laws or legislation set out specifically for electric or hybrid vehicles; rather, Canadian regulations pertain to the emissions of a vehicle. Canadian regulations harmonize vehicle emission levels to those set out by the U.S. Environmental Protection Agency (EPA).

ZERO-EMISSIONS VEHICLES

BEVs use electrical energy stored in batteries to power the traction motors **(Figure 37–2)**. BEVs have zero emissions. The only emissions related to a BEV are those released when coal, oil, or natural gas are used in power plants to generate the electrical energy required to recharge the batteries. The use of hydroelectric, wind, sunlight, or other renewable sources to generate electricity would eliminate all emissions associated with EVs. It is impossible

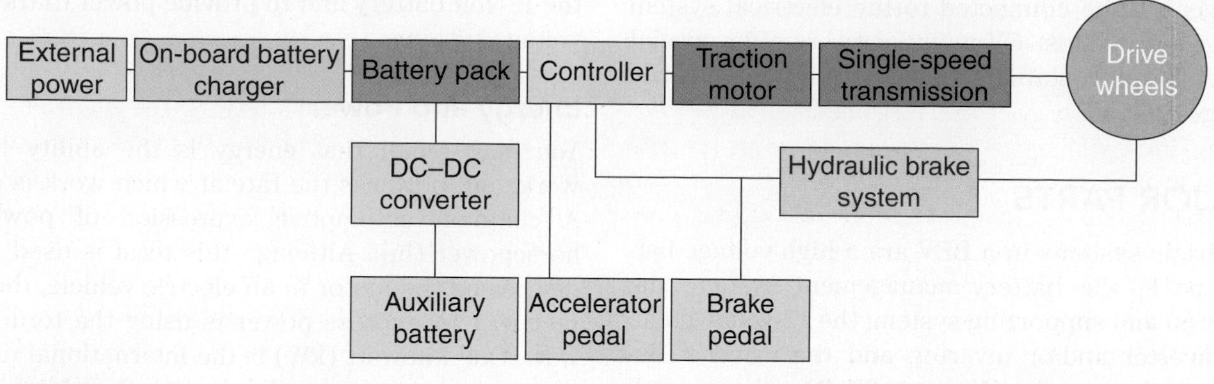

FIGURE 37–2

The major components of a BEV.

to have zero emissions from an internal combustion engine.

Fuel cell electric vehicles (FCEVs) are also electrically powered zero-emission vehicles, but they rely on hydrogen as the fuel. There is no infrastructure for dispensing hydrogen, although fuel reformers can be used to extract hydrogen from other fuels. The use of reformers does cause the release of some emissions.

Advantages

With a BEV, there is the convenience of being able to "fill up" at home, eliminating the need to go to a gasoline station. In the United States, some remote charging stations are becoming available. The cost to refuel is very low; typically, it costs less than $4.

Because of the limited range, BEVs are ideal for commuting or travelling within a limited area. Studies have shown that 80 percent of commuters travel fewer than 65 km (40 miles) per day; this is well within the range of most BEVs.

Cost

The initial cost of a BEV tends to be higher than a conventional vehicle. This is due to their limited availability and the cost of the batteries. Current estimates put the cost of a typical EV battery at around $10 000 to $15 000, with some as high as $26 000. New batteries developed to extend the range of a BEV are, unfortunately, more expensive. However, as more BEVs are produced and sold, their cost should decrease. In a number of Canadian provinces, the initial cost of an EV can be reduced by rebates or tax breaks ranging from $2000 to $8500. Some provinces also offer various green incentives that can include the use of the less-congested high-occupancy vehicle (HOV) lanes, access to future public-charging facilities, and free parking at transit lots.

The motor in an EV has few moving parts. The armature or rotor of the motor is the only moving part. An engine has hundreds of moving parts, each requiring clean lubrication and each is subject to wear. The rotor in a motor is normally mounted on sealed bearings and requires little, if any, additional lubrication throughout its life. The controller and battery charger are electronic units and require little or no maintenance. The batteries also are sealed and maintenance-free. All of these reasons explain why a BEV has very low maintenance costs.

The true cost of driving a BEV depends on the cost of electricity per kilowatt-hour (kWh) and the efficiency of the vehicle. Actual operating costs are reduced by making the cars lighter, more aerodynamic, and with less rolling resistance.

Disadvantages

Perhaps the biggest disadvantage of a BEV is the very limited driving range. The typical range between recharging the batteries is 80 to 240 km (50 to 150 miles). Although some new battery designs have extended this range, long travel in a BEV is still not practical. It is important to understand that a battery's size and the amount of power it stores do not directly determine the range of an EV. Remember, the smallest, lightest, and most aerodynamic electric vehicles will provide the longest range, with the same battery.

Long recharge times are also a problem. In addition to the recharge times, there is a problem of where they can be recharged. If the owner is at home, the charger can be connected to the electrical system of the house. Most EV manufacturers offer special home-charging stations that shorten the required charge time.

MAJOR PARTS

The basic systems in a BEV are a high-voltage battery pack, the battery management system, the motor(s) and supporting system, the 12-volt system, a converter and/or inverter, and the driver's displays and controls **(Figure 37–3)**. The propulsion system has a traction motor that provides the power to rotate the drive wheels and a controller to control the power output of the motor. The 12-volt system supplies the electrical power for the vehicle's accessories, such as the radio and lights. An inverter and converter are required to convert AC to DC and DC to AC electricity. A converter is used to reduce the system's high voltage in order to charge the 12-volt battery and to provide power to the low-voltage systems.

Energy and Power

You may recall that energy is the ability to do work, and power is the rate at which work is done. A common automotive expression of power is horsepower (hp). Although this term is used when discussing the motor in an electric vehicle, the correct way to express power is using the term kilowatt. One kilowatt (kW) is the international unit to measure power (not only electrical); a kilowatt is 1000 watts. One kW equals 1.34 horsepower, and 746 watts equals 1 horsepower. Therefore, a 149 kW motor can provide about 200 hp. An electric motor provides a maximum torque when it is spinning at 0 rpm. So it is very hard to compare the power output of an electric motor to a gasoline-powered engine that produces a maximum amount of torque at a much higher engine speed **(Figure 37–4)**.

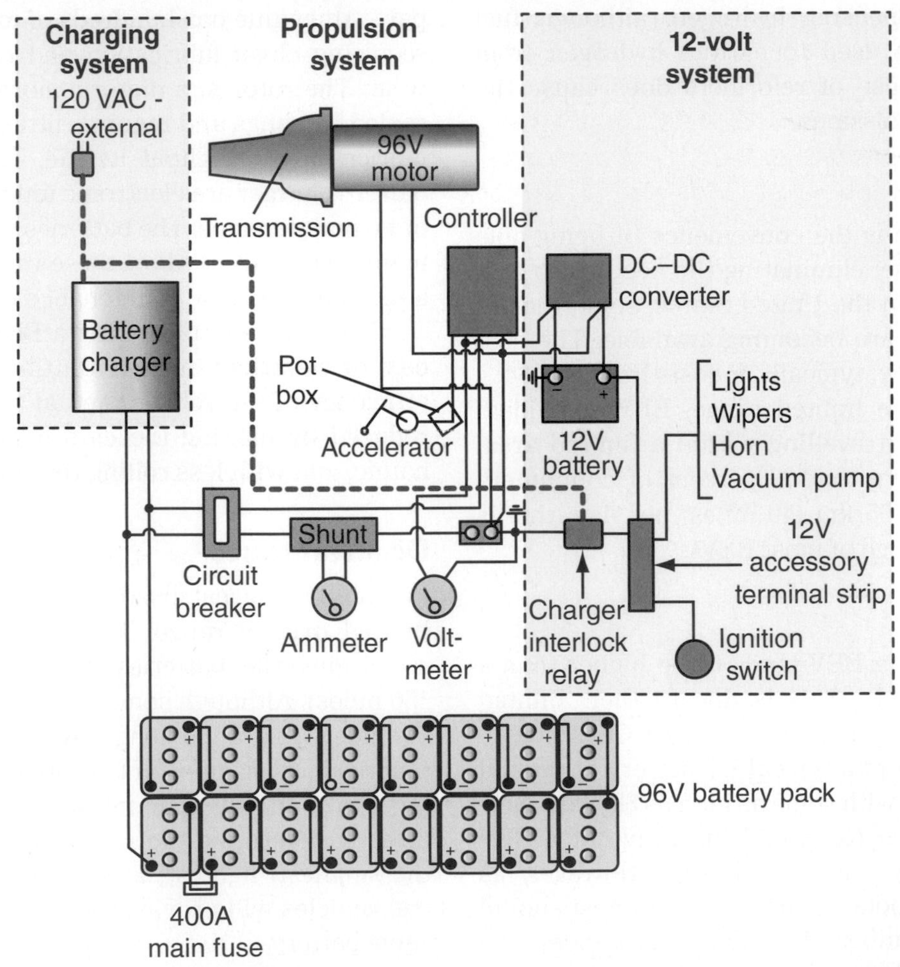

FIGURE 37–3

A basic wiring diagram of an electric vehicle.

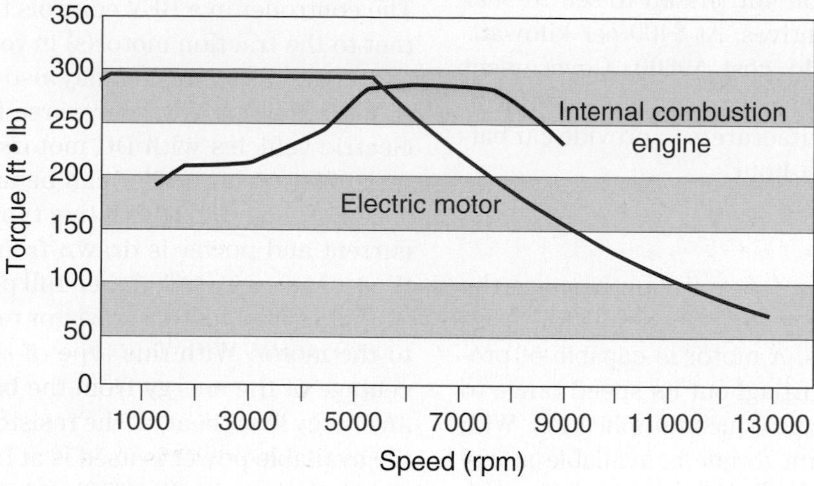

FIGURE 37–4

A comparison of the amount of torque produced by a gasoline engine and an electric motor.

©Cengage Learning 2015

The power rating of an electric motor (or gasoline engine) indicates how quickly energy can be changed into work, such as acceleration. A motor relies on the energy stored in a battery or some other source. The amount of energy available is expressed in kilowatt-hours (kWh). Kilowatt-hours express what can be accomplished by one kilowatt acting for one hour. For example, when a light bulb with a power rating of 100W has been on for one hour, it has used 100 watt-hours (0.1 kWh). This is the same amount of energy used to keep a 50W light bulb on for two hours.

When comparing systems and the available power of a battery, it is important that a battery's rating is looked at in regards to the system's voltage. If a battery has a rating of 100 amp-hours and the battery provides 12 volts, the amp-hours should be multiplied by the voltage to determine the total energy available. In this case, the energy source can provide 1200 watt hours (1.2 kW).

So if we look at a 300V battery pack rated at 24 kWh, the battery can provide 80 amps at 300 volts for one hour. Keep this in mind when looking at the ratings of batteries and battery chargers. Also, do not be fooled by the manufacturers' estimates.

Nissan says the 24 kWh pack in a Leaf provides for a 160 km (100-mile) drive range. That means 150 watts are needed to provide enough energy for one kilometre of travel. So theoretically, the battery should supply enough energy for 160 km (100 miles) of travel. This number is close to what the EPA has estimated as the driving range of the Leaf **(Figure 37–5)**.

The traction motors are either AC or DC motors and are specifically designed for this use. Most production BEVs use AC motors, and FCEVs and many conversion EVs use a DC motor. The latter is a consequence of cost. DC motors can be powered directly by the batteries, whereas AC motors require inverters to change the DC voltage stored in the batteries into the AC required by the motors. FCEVs have a DC motor because the electricity generated is not AC; therefore, there is no need for an inverter or other similar conversion equipment.

The cost of high-voltage batteries has declined through the years. For example, the Nissan Leaf has a 24 kWh battery. If batteries cost $1000 per kWh, the

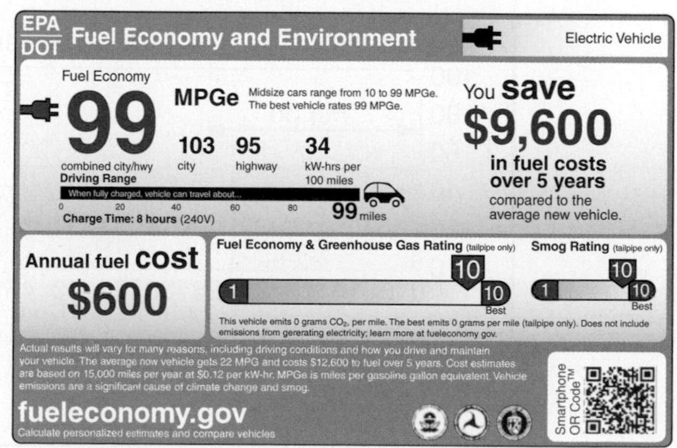

FIGURE 37–5

The new Monroney sticker for electric vehicles.

©Cengage Learning 2015

battery in a Leaf would cost $24 000, which would make it nearly impossible for Nissan to sell the car for $32 800 before incentives. At $400 per kilowatt-hour, the battery would cost $9600. Government environmental and energy departments have been pressuring battery manufacturers to provide car batteries for $250 a kilowatt-hour.

Electric Motor

In most EVs, there is no transmission because the rotary motion of the motor can be applied directly to the differential gears. A motor is capable of providing enough torque throughout its speed range to move the vehicle without torque multiplication. With an electric motor, instant torque is available at any speed. The entire rotational force of a motor is available the instant the accelerator pedal is pressed. Peak torque stays constant to nearly 6000 rpm, and then it begins to slowly decrease.

The wide torque band eliminates the need for multispeed transmissions. There is no need for a reverse gear either, because switching the polarity of the stator will cause the rotor to turn in reverse. The absence of a typical transmission saves weight and makes the powertrain much less complex.

Controller

The controller in a BEV controls the voltage and current to the traction motor(s) in response to the driver's input. The controller may also reverse the current flow to the motor when reverse gear is selected. In electric vehicles with DC motors, a simple variable resistor–type controller can be used to regulate the speed of the motor. With this type of controller, full current and power is drawn from the battery at all times. At slow speeds, when full power is not needed, a high resistance in the resistor reduces current flow to the motor. With this type of system, a large percentage of the energy from the battery is wasted as an energy loss (heat) at the resistor. The only time all the available power is used is at high speeds.

Modern controllers adjust motor speed through pulse-width modulation (PWM). Pulse width is the length of time, in milliseconds, that a component is energized. Controllers rely on transistors to rapidly interrupt the flow of electricity to the motor. High electrical power (during high speed, acceleration, and/or heavy loads) is available when the intervals during which the current is stopped are short. During slow speeds, little power is needed, and the intervals of no current flow are longer **(Figure 37–6)**.

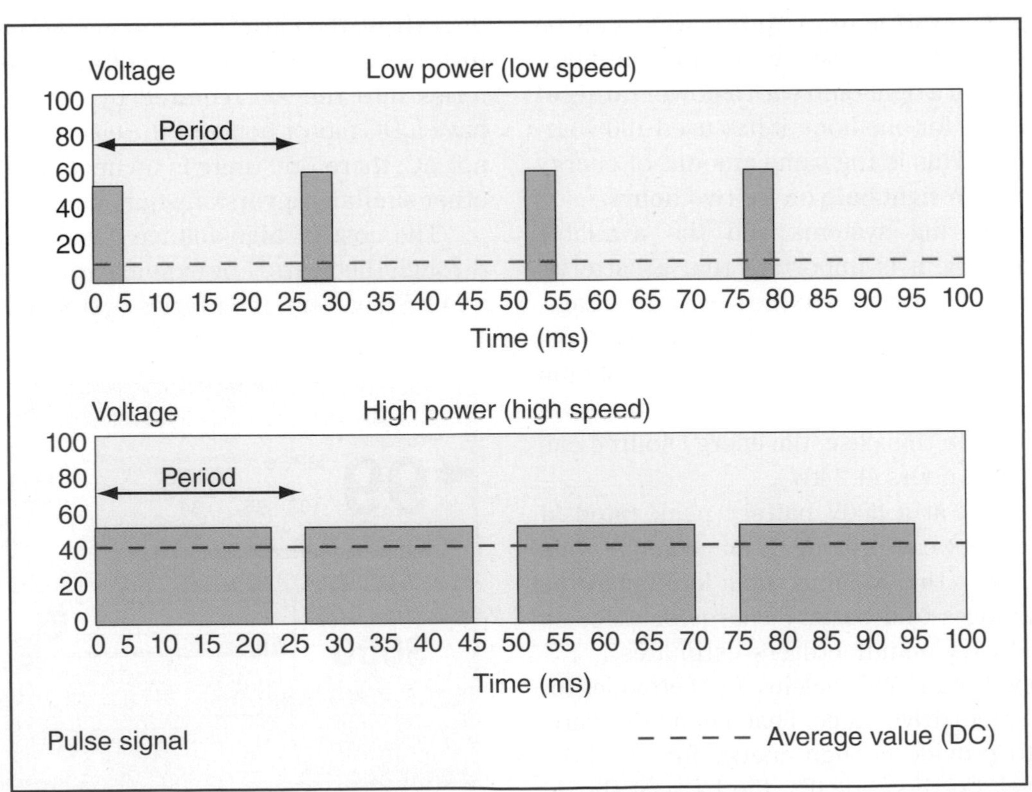

FIGURE 37–6

An explanation of pulse-width modulation at low and high speeds.

Inverter/Converter

An AC power inverter converts the battery's DC voltage into three-phase AC voltage to power the traction motor. The output voltage varies according to the demands of the driver and the vehicle. Normally, the inverter is controlled by an electronic control module. The output from a typical inverter is constantly being calculated using input signals from the accelerator pedal, the motor's shaft speed sensor, the motor's direction sensor, and the brake pedal.

The inverter is liquid cooled, and the heat from the inverter can be used to supplement the passenger compartment's heater to save energy. This is done automatically whenever the controls are set for heat.

BATTERY CHARGING

Refuelling a BEV simply means charging the batteries. Recharging involves connecting a battery charger to a source of electricity and connecting the charger to the battery pack. Battery chargers **(Figure 37–7)** may be internal (in the vehicle) or external (at a fixed location). There are advantages and disadvantages to both. An on-board charger allows the batteries to be recharged wherever there is an electrical outlet. The disadvantage of on-board chargers is their added weight and bulk. To minimize this, manufacturers normally equip the vehicles with low-power chargers that require long charge times. External chargers, however, force the driver to charge the batteries at specific locations but offer more power and decrease the time required to charge the batteries. Some BEVs with off-board chargers also have a convenience charger. These on-board chargers plug into standard 110-volt outlets and allow the driver to recharge batteries wherever electricity is available.

Most EVs have an on-board charger that uses a rectifier circuit to transform AC from the electrical grid to DC, which is necessary for recharging the battery pack. The rectifier can handle only a certain amount of power and develops a great amount of heat while it is changing AC to DC. Rectifiers can be built to handle more power and heat, but they would cost quite a lot. Based on these concerns, most conventional charging stations in North America and Japan are based on 240V/30A service. This power level seems to be a safe limit for the rectifiers. But charging at these levels takes several hours to

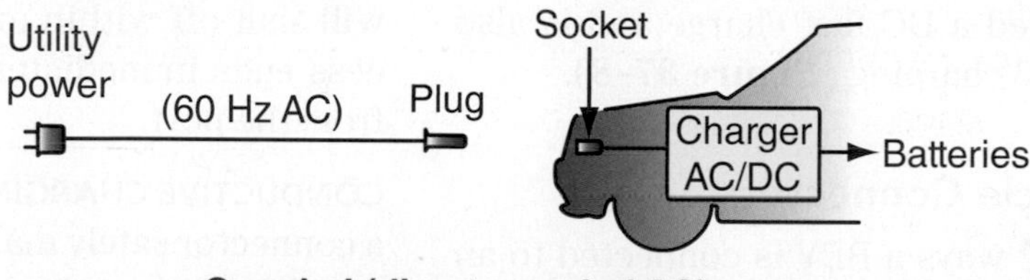

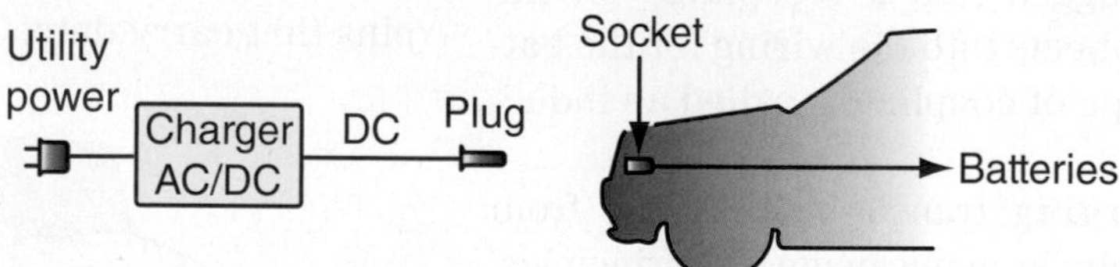

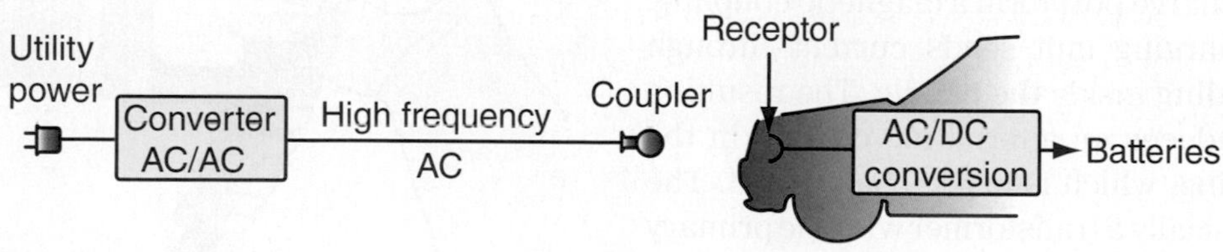

FIGURE 37–7

EV battery chargers may be internal (on-board) or external (off-board).

VOLTAGE TYPE	CHARGE LEVEL	MAX. VOLTAGE	PEAK CURRENT
AC	1	120 VAC	16A
AC	2	240 VAC	32A (2001)/ 80A (2009)
DC	1	450 VDC	80A
DC	2	450 VDC	200A
DC	3	600 VDC	400A

FIGURE 37–8

Current standards for the various charging levels.

recharge the battery pack. The required time varies with the size and type of battery and battery charger.

A solution to these problems is to use an external charging station capable of delivering DC directly to the vehicle's battery pack. Doing this would call for dedicated chargers at permanent locations. These new chargers may be able to recharge a battery in less than 20 minutes. They use sophisticated electronics to monitor the cells and regulate the charging voltage and current. Being able to quickly charge the batteries would certainly make an electric vehicle more practical. The charging setup using high voltage and high current is called a DC fast charge and is also referred to as level-3 charging **(Figure 37–8)**.

Charger to Vehicle Connectors

There are two basic ways a BEV is connected to an external source of electricity for charging. One is the traditional plug, called a conductive coupling. The coupling is plugged into a receptacle on the vehicle where it connects into the wiring for the batteries. The other type of coupling is called an inductive coupling.

Inductive charging transfers electricity from a charger to the vehicle using magnetic principles. To charge the batteries, a weatherproof paddle is inserted into the vehicle's charge port **(Figure 37–9)**. The paddle and charge port form a magnetic coupling. The external charging unit sends current through the primary winding inside the paddle. The resulting magnetic flux induces an alternating current in the secondary winding, which is in the charge port. The connection is basically a transformer with the primary winding in the paddle and the secondary winding in the vehicle. The induced AC is then converted to DC (within the vehicle) to recharge the batteries.

There is no metal-to-metal contact between the charge paddle and the charge port of the vehicle. This system provides a safe and easy-to-use way to recharge the batteries.

Inserting the paddle begins the charging process. The insertion of the paddle completes a communication link between the charger and the vehicle. The charger displays what percent of charge remains in the batteries and an estimate of the time needed to fully charge the batteries.

This link also allows the charging unit to enter into self-diagnostics and prevents the vehicle from being driven while the paddle is inserted in its port. If the charging cable becomes damaged or cut, power will shut off within milliseconds. The charging process ends immediately after the paddle is removed from the port.

CONDUCTIVE CHARGING With a conductive charger, a connector safely makes the link between the power supply and the vehicle's charge port. The connector makes a weatherproof direct electrical connection to the vehicle's charge port. The connector has multiple pins that carry data. These data are used to control

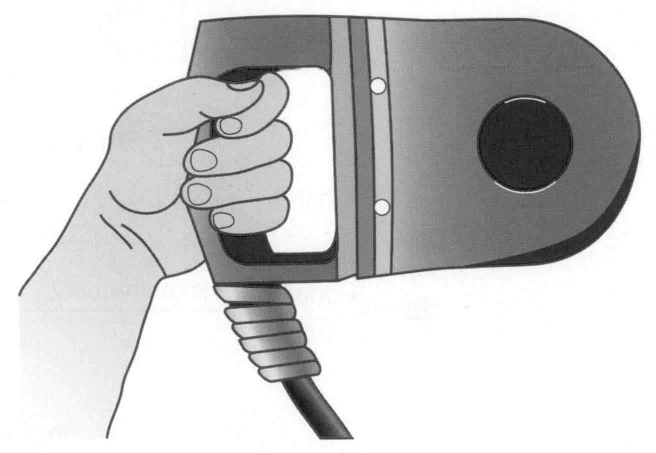

FIGURE 37–9

An inductive charging connector (paddle).

FIGURE 37–10
An on-the-wall charging unit with a special circuit box (load centre) installed to handle the high voltage.

FIGURE 37–11
An ODU connected to an electric vehicle.

the action of the charger based on the conditions of the battery pack. External chargers are available in many different sizes and can be either wall- or pedestal-mounted **(Figure 37–10)**.

Conductive charging can be accomplished with a fuel nozzle–style connector called the ODU **(Figure 37–11)**. The connector has many round male pins that mate to female ends in the vehicle. Similar to adding fuel to the vehicle, the connector is placed into an opening on the vehicle, and refuelling or recharging can take place.

Recharging Standards and Regulations

Like nearly everything designed for an automobile, there are standards and regulations that pertain to charging a high-voltage battery pack. The most recognized standard is the North American standard for electrical connectors for electric vehicles as defined by the Society of Automotive Engineers (SAE). This standard is referred to as the "SAE Surface Vehicle Recommended Practice J1772, SAE Electric Vehicle Conductive Charge Coupler." Basically, this standard covers the basic physical, electrical, communication protocol, and performance requirements for an EV's conductive charge system and coupler. The purpose of the standard is to ensure that EVs from different manufacturers will not need special or unique chargers or charging connectors.

Early electric vehicles, such as GM's EV1, used inductive charger couplers. These were replaced with conductive couplers in 2001, according to the new SAE J1772 standard initiated by the California Air Resources Board (CARB). AVCON manufactured a rectangular connector compliant with the SAE J1772 specification and was capable of delivering up to 6.6 kW of electrical power. The AVCON conductive interface was also used by the Ford Ranger EV truck and the Honda EV Plus. The connector had a rectangular charging head that plugged into an AVCON inlet mounted on the vehicle. Many public EV charging stations, funded by CARB, relied on conductive AVCON charging connections.

In 2008, CARB proposed changes to the 2001 CARB regulations that were aimed at higher charging current than the AVCON could handle. This led to the design of a new connector that would be used starting in 2010. As a result, all AVCON charging stations were converted to the new J1772 or phased out starting in 2011.

With the advent of improved chargers, and the desire to charge with AC and DC voltage, new specifications for the J1772 ODU have been released. Using AC and DC at the same time to charge a battery is faster than the previous regulated process. Therefore, the new J1772 standard allows for a single inlet in the vehicle and a single plug to be used for both AC and DC charging. The new J1772 standard incorporates AC levels 1 and 2 (up to 80 amps) and DC levels 1 and 2 (up to 200 amps).

The new "combo" connector is similar to the first-generation J1772 plug but also has pins to fit into the lower portion of the inlet. The first-generation J1772 plug fits into the upper part of the inlet, while DC charging takes place across two dedicated pins across the bottom of the connector. Above these pins is a round receptacle that has five pins **(Figure 37–12)**. These pins complete the circuit from the electric grid to the vehicle through AC power line 1 and AC power line 2 pins and a designated ground pin. The two other pins are for proximity detection and the control pilot. The proximity detection feature prevents the car from moving while it is connected to the charger. The control pilot is the communication line used to transfer information between the charger and vehicle in order to safely and efficiently charge the battery. When the male and female halves of the connector are not mated, there is no voltage at the pins, and charging does not begin until it is commanded by the vehicle.

This combination connector allows for DC fast charging and has been endorsed by Audi, BMW, Chrysler, Daimler, Ford, General Motors, Porsche, and Volkswagen. Their goal is to provide a way for consumers to charge their EVs in 15 to 20 minutes.

FIGURE 37–12

The female connections for an SAE J1772 connector.

CHADEMO PROTOCOL The new SAE standard offers a number of advantages over the competing CHAdeMO standard **(Figure 37–13)**. However, CHAdeMO (short for "charge and move") is already well established in Japan, and is also used by most existing vehicles and chargers in the North American market. In spite of CHAdeMO's current dominance, there is strong support for J1772.

The connector currently used for DC fast charging on the Nissan Leaf and Mitsubishi "i" vehicles is based on the CHAdeMO standard. Some Asian EV models have two vehicle electric inlets, one for CHAdeMO DC charging and one for J1772 AC charging **(Figure 37–14)**.

FIGURE 37–13

A CHAdeMO on-vehicle connector.

FIGURE 37–14

The Nissan Leaf has receptacles for a CHAdeMO and an SAE J1772 connector.

Charging Precautions

There are three primary things that affect the required time to recharge the batteries: the current state of charge of the battery, the chemicals used in the cells of the battery, and the type of charger used.

Each EV has a specific charging procedure. These procedures vary with the type of charger, charger coupling, and battery. Always follow the procedure for the vehicle being worked on. Here are some general guidelines to follow:

- Make sure the gear selector is in the PARK position and the parking brake is applied before charging.
- Before charging, make sure the motor switch is off and the key is removed.
- To avoid getting an electric shock, never operate the charger with wet hands.
- Avoid charging under high temperatures or direct sunlight.
- Never touch the terminals of the conductive terminals on the vehicle or coupler; you may get an electric shock.
- Do not modify the charge coupler.
- Make sure the charge coupler is firmly installed, without any tension on the cable.
- If the charge coupler is damaged, repair or replace it as soon as possible.
- Make sure water, dirt, or other foreign objects do not enter the charge port on the vehicle.
- Do not disconnect the charge coupler until the batteries are fully charged, unless it is necessary to prematurely stop charging.

ACCESSORIES

Some systems, such as the radio, lights, and horn, operate the same way as they do in a conventional vehicle. Other systems, such as the power-steering and power brakes, require additional small electric motors, which have an impact on the vehicle's driving range. Because all accessories and auxiliary systems operate on electricity, their electrical power needs reduce the capacity of the battery.

HVAC

To meet federal safety standards, all vehicles must be equipped with passenger compartment heating and windshield defrosting systems. Vehicles with an internal combustion engine use the heat of the engine's coolant to warm the passenger compartment. In a BEV, there is no engine and therefore there is no direct source for heat. The heat must be provided by an auxiliary heating system.

Some electric vehicles use an electric resistance heater with a fan. Other BEVs have liquid heaters. Water, or a mixture of water mixed with ethylene glycol, is held in a tank. The liquid in the tank is kept heated by a resistive heating element submerged in the tank. When the driver turns on the heating system, a small pump circulates the heated liquid through a heater core in the passenger compartment. A fan moves air over the core to provide heated air.

BEV air-conditioning systems also have a significant impact on the driving range. In many cases, the air-conditioning system uses a high-voltage motor to rotate the compressor. Obviously, the energy used to power the air conditioning puts a drain on the battery pack. The amount of energy consumed by the air-conditioning system depends on how often it is used, the outside temperature, and the selected temperature for the passenger compartment.

Power Brakes

Many power-brake systems use engine vacuum and atmospheric pressure to multiply the effort applied to the brake pedal during braking. Because there is no engine in a BEV, there is no direct vacuum source. However, normal vacuum-assist power-brake systems can be used if fitted with an electric vacuum pump. These pumps are similar to those used on diesel-engine vehicles. The pump may be connected to a storage tank. The tank reduces the time the pump needs to operate and therefore minimizes the effect the pump has on driving range.

Another type of power-brake system uses hydraulic pressure, from a pump, to reduce the pedal effort required to apply the brakes. Some BEVs use an electric pump to provide the necessary hydraulic pressure **(Figure 37–15)**. These systems are called electrohydraulic brake systems. Because both types of power-brake systems for BEVs operate on electrical power, brake boost is available at all times.

Power-Steering

Hydraulic pressure is often used to reduce steering effort. This pump can be driven by an electric motor, which is how some BEVs are equipped. The control for the pump can be programmed to provide more assist at lower speeds and less at higher speeds. The system can also be programmed to run only the pump when it is needed; this reduces the effect that power-steering has on the driving

FIGURE 37–15

The master cylinder and pump assembly for an electrohydraulic brake system.

range. These systems are called electrohydraulic steering systems.

Many power-steering systems are purely electrical and mechanical systems. An electric motor moves the steering linkage. These systems are programmable, and the energy consumed by the motor depends on the amount the steering wheel is turned. While driving straight, the motor may not run. However, when the steering wheel is fully turned, the motor is drawing its maximum current.

DRIVING AN EV

Driving a BEV is like driving any other vehicle, but with some notable exceptions. There is still a steering wheel, a brake pedal, and an accelerator pedal. A BEV typically has adequate acceleration and can travel at highway speeds. The biggest difference for the driver is that attention must be paid to the consumption of energy. Failure to minimize consumption and carefully plan travel routes can lead to reduced power and a need to recharge the batteries at inconvenient locations or times. If the batteries are not charged, the vehicle will not move.

Starting

The biggest adjustment a driver needs to make when preparing to drive a BEV is starting it or getting it ready for action. A BEV has no noise or vibration when it is ready to go. The driver must look at the instrument cluster to determine it is ready. Make sure the gear lever is in the PARK position and that the parking brake is applied. The accelerator should never be depressed during starting.

The ignition (motor) switch has several positions. One is LOCK, during which the traction motor is off and the steering wheel is locked. The key can be removed only at this position. The ACCESSORIES position allows some accessories to work, but the traction motor is off. The START position actually gets the traction motor ready to work, and ON is the normal position for driving. Never leave the switch in the ON position when the vehicle is not in use.

To turn on the traction motor, turn and hold the motor switch to START, with the brake pedal depressed until the READY light in the instrument cluster comes on **(Figure 37–16)**. On some vehicles, a buzzer will sound when this happens. Once the READY lamp is lit, the motor switch can be released to allow it to move to the ON position. At this point, the traction motor will run when the accelerator is depressed, and all accessories are ready to operate. If the READY light does not illuminate during the start process, there is a problem

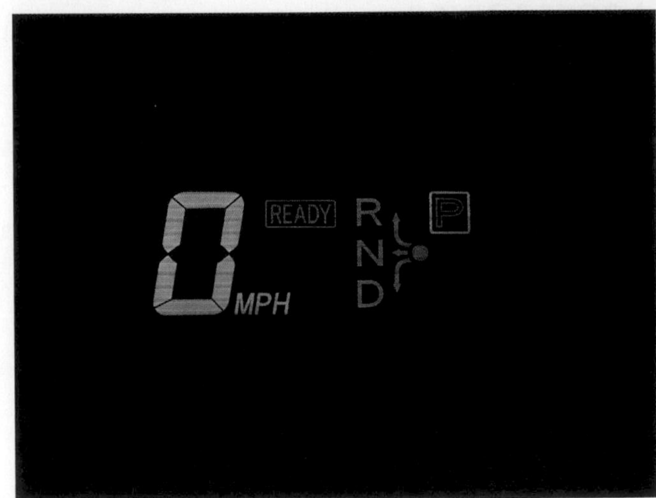

FIGURE 37–16

The instrument panel displays for a typical EV.

with the traction motor or its circuit, or the auxiliary battery is discharged.

Driving and Braking

Most BEVs have a single-speed automatic transmission, and the gearshift lever has five positions **(Figure 37–17)**. Normally, the shift lever can only be shifted out of P when the motor switch is in the ON position. When moving out of P into D or R, the brake pedal must be depressed. It is important that the accelerator is not depressed when shifting gears. Doing this can cause the vehicle to unsafely and quickly move and can cause damage to the motor. Once the shift lever has been moved and with the brake pedal still depressed, the parking brake can be released.

To begin moving, press the accelerator. Drive normally with the realization that the accelerator is the only thing that controls vehicle speed. When the accelerator is released, vehicle speed will decrease because the wheels are now turning the motor that just became a generator.

FIGURE 37–17

The gear shifter positions for an EV.

To back up, bring the vehicle to a complete stop. Then depress the brake pedal and move the shift lever into the R position. It is important to keep in mind that a BEV can accelerate just as quickly in reverse as it does in drive. However, it is more difficult to steer any vehicle in reverse; therefore, the accelerator should be gently pressed when backing up.

To park and shut down the vehicle, come to a complete stop. Then apply the parking brake. While depressing the brake pedal, move the shift lever to the P position. Now turn the motor switch to the LOCK position, and remove the key.

Maximizing Range

The driving range of a BEV is reduced by cold weather (requiring use of the heater), warm weather (requiring use of the air conditioner), and the condition and age of the battery. There are certain other things a driver can do to extend the range and the life of the batteries.

- Avoid high-speed driving. Maintain a moderate speed on highways.
- Avoid driving up inclines.
- Avoid frequent speed increases or decreases. Attempt to drive at a steady pace.
- Avoid unnecessary stopping and braking.
- Avoid full-throttle acceleration; accelerate slowly and smoothly.
- Make sure the vehicle is well maintained, including proper tire inflation pressures.
- Avoid unnecessary weight in the vehicle; it will shorten the driving range.

FORD FOCUS

The Ford Focus Electric **(Figure 37–18)** uses a 23 kWh, liquid-cooled lithium-ion battery pack that provides an all-electric range of about 160 km (100 miles). It relies on a synchronous PM electric motor rated at 107 kW (143 hp) and 250 N·m (184 lb·ft) of torque. The motor's output is transferred to the front wheels through a single-speed transmission.

The L-shaped battery pack is located under the rear seat and between the rear wheels. The battery uses a liquid cooling and heating thermal management system to precondition and regulate the temperature of the battery. The thermal management system heats or chills a coolant before passing it through the battery's cooling system.

The Focus has a 6.6 kW on-board charger. A full recharge is possible after three to four hours.

FIGURE 37–18

A Ford Focus EV.

FIGURE 37–20

The charging point on a Focus is on the left front fender.

This requires plugging in to Ford's 240-volt, 32-amp, level 2 home-recharging unit with a J1772 connector **(Figure 37–19)**. The Focus also is equipped with a 120V cord that allows the charger to be connected to a standard household outlet. At 120V, the battery pack can be recharged in close to 20 hours.

FIGURE 37–19

An at-home high-voltage charger for a Focus EV.

The instrument cluster allows the driver to monitor energy consumption. There is a smartphone application that allows the driver to remotely track the car's charging status.

Ford also provides a 2.5 kW rooftop solar-panel system through the solar system manufacturer SunPower. The solar panels can produce an average of 3000 kWh annually, theoretically enough to accommodate a customer who drives 20 000 km (12 500 miles) a year.

The charging port is in the left front fender **(Figure 37–20)**. The port illuminates when the charger's cord connector is plugged into the charge port. A blue light indicates GO and that the charger is connected and charging.

NISSAN LEAF

The Nissan Leaf **(Figure 37–21)** is a true ZEV with zero tailpipe emissions. It is rated at 2.22 L/100 km (106 mpge) for city driving, 2.56 L/100 km (92 mpge) on the highway, and 2.38 L/100km (99 mpge) for combined city and highway driving. The Leaf is equipped with an AC synchronous motor that can provide up to 80 kW (107 hp) and 280 N·m (207 lb·ft) of torque to the front drive wheels.

The EPA estimates the driving range for a Leaf with a full battery is 117 km (73 miles), although Nissan advertises a range of 160 km (100 miles). There are many things that can explain the discrepancy, but those will not be discussed here; simply

FIGURE 37–21

A Nissan Leaf.

look at the things that affect driving range, and you will find the reasons why the two disagree.

Power from the motor moves through a single-speed reducer-type transmission. But through electronic controls, there are two optional forward-drive modes: DRIVE and ECO. DRIVE provides quicker acceleration, but uses a great amount of the battery's reserve. ECO extends the driving range by limiting acceleration and reducing the power to the climate control system. It also provides additional brake regeneration, causing the car to decelerate more rapidly, but also adding electrons to the battery.

The battery is a 24 kWh lithium manganese (Li-Ion) that is capable of delivering up to 90 kW of power. The battery pack **(Figure 37–22)** is made up of 48 modules and each module contains four laminated flat cells, arranged in three stacks. The 192 stacked laminar cells have a lithium manganate cathode. The battery weighs about 272 kg (660 pounds) and is located under the floor pan directly beneath the front and rear rows of seats. The battery pack is air-cooled (and heated when necessary) to protect the cells.

Charging time varies with the type of charging used. Customers can purchase a 240V home-charging station through Nissan. Some Leaf models have a quick charge port with a 3.3 kW on-board charger. With this charger, the battery can be fully recharged by a 220/240-volt 30 amp within eight hours.

The Leaf's charging port at the front of the car has two charging inlets **(Figure 37–23)**. One is a standard J1772 connector for level 1 and 2 recharging. The other is a level 3 DC connector that uses the CHAdeMO protocol.

The Leaf also has an auxiliary 12-volt lead-acid battery that provides power for the basic systems and accessories in the car, such as the sound system, headlights, and windshield wipers. An interesting touch is that some models of the Leaf have a small solar panel on the rear spoiler **(Figure 37–24)** to help trickle-charge this auxiliary battery.

TELEMATICS The Nissan Leaf uses an advanced telematics system called Carwings. Carwings is connected any time the car is within the range of a cell tower and provides information to the driver, such as the car's position, remaining range, and the location of charging stations available within that range.

FIGURE 37–22

The battery pack in a Leaf.

FIGURE 37–23

The Leaf's charging port at the front of the car and the two charging inlets.

FIGURE 37–24

The solar panel built in to the spoiler on some models of the Leaf.

The system also monitors and compiles information about distances travelled and the amount of energy consumed **(Figure 37–25)**. It also provides daily, monthly, and annual reports of those data, and that information can be viewed on the car's digital screens. Through Carwings, cellphones can be used to remotely to turn on the air conditioner and heater and to reset all charging functions.

PEDESTRIAN SOUNDS Because BEVs emit very little noise while they are moving, the Leaf is programmed to emit digital warning sounds, one for forward motion and another for reverse, to alert pedestrians, the blind, and others that it is moving close to them. This is called the vehicle sound for pedestrians (VSP) system. This sound system moves from 2.5 kHz at the high end to a low of 600 Hz, which makes it audible for all age groups. The sound stops when the Leaf reaches 30 km/h (19 mph) and begins again when the Leaf slows to less than 25 km/h (16 mph). The VSP system is controlled by a computer and synthesizer, and the sound is emitted from a speaker in the driver's side front wheel well.

FIGURE 37–25

The Carwings display in a Leaf.

MITSUBISHI I-MIEV

The Mitsubishi i-MiEV (Mitsubishi innovative electric vehicle) is an EV hatchback **(Figure 37–26)**. The car has a PM synchronous motor mounted on the rear axle. The water-cooled 49 kW motor can provide 66 hp and 197 N·m (145 lb·ft) of torque. The motor is placed above the rear axle, and the 16 kWh lithium-ion battery pack and the motor control unit is under the floor. The motor's output is sent to the rear wheels through a single-speed, fixed-reduction transmission.

The EPA initially rated this Mitsubishi capable of providing a combined fuel mileage rating of 2.1 L/100 km (112 mpge). The estimated driving range in the city is 158 km (98 miles).

The 16 kWh SCiB battery pack uses lithium-titanate oxide in the anode, which provides increased safety and decreased charging times. The battery pack is made up of two 4-cell modules placed vertically at the centre of the pack and ten 8-cell modules placed horizontally that are connected in series. These 88 cells provide 330 volts.

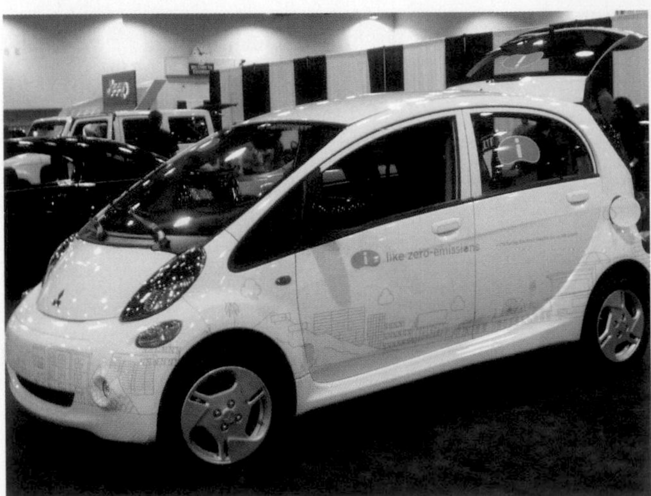

FIGURE 37–26

A Mitsubishi i-MiEV.

It is estimated that it takes 22 hours to recharge the battery with a 110-volt power supply and seven hours with 220 volts. And if the level 3 480-volt quick-charging station with CHAdeMO charging technology is available, the battery can be recharged to about 80 percent of full capacity in about 15 minutes, about 50 percent in 10 minutes, and about 25 percent in 5 minutes. This is much less than the required charge time for a typical Li-Ion battery charged under the same conditions. The SCiB also generates little heat while recharging, eliminating the need for a complex and power-robbing system to cool the battery module.

The system offers three distinct driver-selected drive modes: D, ECO, and B. Each has been designed to provide the best performance for different driving conditions. The D mode is the default position and is the best mode for driving on highways and interstates. The ECO mode limits the motor's output to increase the range by decreasing the amount of power available for acceleration. The B mode adds more regenerative braking when the car is coasting to a stop or braking on downhill stretches to more aggressively recharge the battery.

TESLA

Tesla Motors in California is an independent auto manufacturer; its focus is to manufacture high-technology EVs. The company has been strongly supported by its founder and other investors, and is dedicated to providing fun and practical EVs. This effort was further helped by the U.S. government when Tesla was granted investment dollars. Tesla's first car available to the public was its Roadster. This was actually a proof-of-concept vehicle, as it demonstrated the advantages and disadvantages of an EV.

The Tesla Roadster is a BEV produced by Tesla Motors in California **(Figure 37–27)**. It is a mid-motor rear-wheel-drive car based on the Lotus Elise. It has a 185 kW (248 hp) three-phase, four-pole AC induction motor **(Figure 37–28)** powered by a 53 kWh lithium-ion battery. The Roadster has a single-speed fixed-gear transmission. The EPA's estimated range for the Roadster is 393 km (244 miles).

Tesla calls its battery pack the energy storage system (ESS). This battery has 6831 lithium ion cells arranged into sheets and bricks. Each brick has 69 cells connected in parallel, and each sheet has nine bricks connected in series. Each of the cells is similar to those used as batteries for laptop computers.

FIGURE 37–27

The Tesla Roadster is an EV based on the gasoline-powered Lotus Elise.

The battery weighs 450 kg (990 lbs), stores 56 kWh of electric energy, and can deliver as much as 215 kW of power. Coolant is pumped continuously through the ESS when the car is operating.

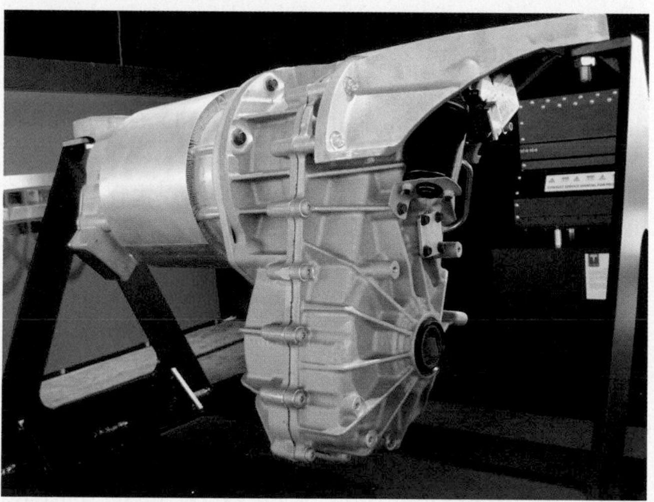

FIGURE 37–28

The motor and transmission in a Tesla Roadster.

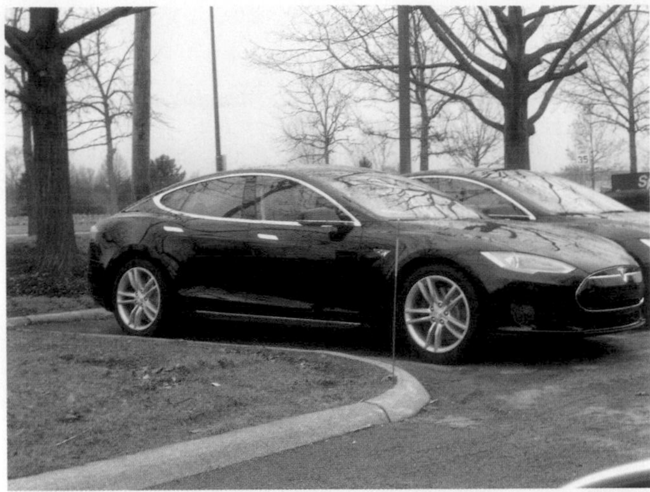

FIGURE 37–29

A Tesla Model S.

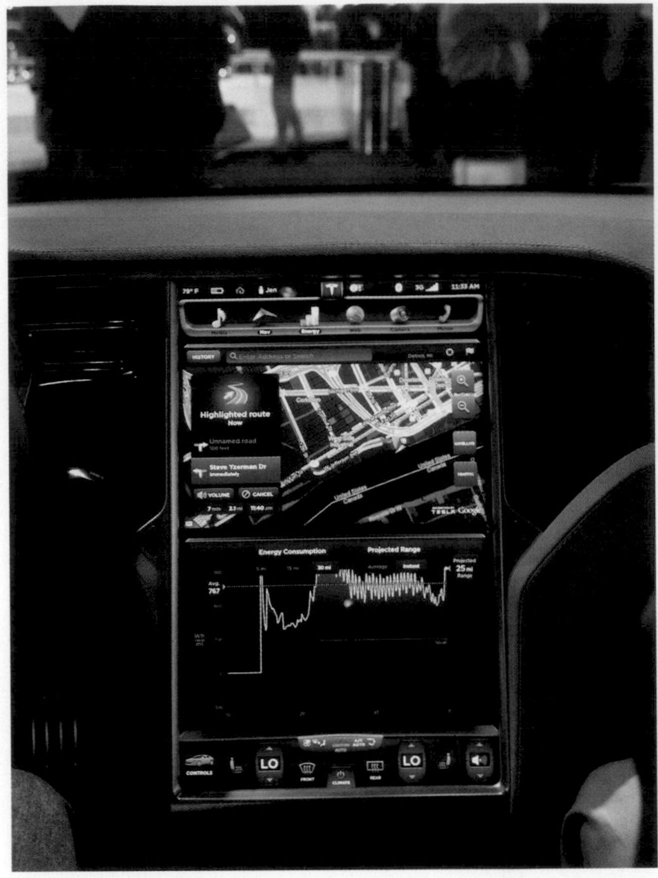

FIGURE 37–30

The display in the centre console of a Model S.

MODEL S The Tesla Model S **(Figure 37–29)** is a full-size four-door sedan engineered and produced by Tesla Motors. The Model S is a high-performance electric sedan. Much of the technology used in the Roadster has been used in the Model S. The motor is an AC induction motor that can provide up to 310 kW (416 hp) and 600 N·m (443 lb·ft) of torque.

There are various models and options, each providing various power and driving range options, up to the 85 kWh option rated at 426 km (265 miles).

Model S comes with everything needed to plug in to the most common 120V and 240V outlets. A full battery recharge requires less than four hours (or at a rate of 100 km or 62 mile range per hour) when charged with 70 amps and 240 volts. Using a 120V charger, 8 km of travel is gained for every hour of charging, and a full recharge requires about 48 hours. The Tesla uses a proprietary charging connector, but an adapter is available through Tesla that allows recharging with a J1772 connector. The system controls temperature and voltage of the battery pack by monitoring more than 100 sensors.

A 50 percent charge in 30 minutes is possible with a Tesla supercharger. These superchargers are exclusive to Tesla and will not charge other EVs. There is no cost to Tesla owners for using these chargers; Tesla pays for the electricity, which mostly justifies this proprietary arrangement. These charging stations are positioned in various locations on the West Coast of Canada and the United States as well as the U.S. Northeast.

To charge the battery, all the driver needs to do is move toward the driver's side taillight, holding a connector, and press the button on the connector, and a triangle opens to reveal the charge port. There is a light ring around the charge port. The colour of the lights signifies the status of the recharging process. In the centre console, there is a display that shows the charging status and gives a summary the driver's past performance in regards to efficiency and other information **(Figure 37–30)**.

HONDA FIT EV

This car is powered by a PM motor with 92 kW (123 hp) and 256 N·m (189 lb·ft) of torque. This motor is also used in Honda's FCEV, the Clarity. The motor, with a single-speed reduction transmission, is located at the front of the car **(Figure 37–31)**. One of the drive axles passes through the motor's rotor hollow shaft to connect to the drive wheel. The Fit's 20 kWh SCiB lithium-ion battery is located below the car's floor. This required that the car's chassis be raised nearly 50 mm (2 in).

FIGURE 37–31

An underhood look at the powerplant of a Honda Fit EV.

A J1772 charging connector is behind a flap in front of the driver's door **(Figure 37–32)**. The connector feeds power to the 6.6 kW on-board charger. This connection allows for recharging the battery in about three hours when using a 240V source. Honda has selected Leviton as the vendor to provide a home-charging station.

A unique feature of the Fit EV is its brake system. Normal braking, the feel of the brake pedal, is totally simulated. The Fit EV attempts to come to a stop using only regenerative braking. The system does provide normal friction braking to bring the car to a complete halt and during hard braking. However,

FIGURE 37–32

The charging port's door is in front of the driver's door on a Honda Fit EV.

even during those conditions, braking is computer controlled (brake by wire). In the brake system's hydraulic system, there is a fast-reacting electric motor that pressurizes the fluid at the brake calipers. This brake setup maximizes the amount of electricity generated during braking to help extend the driving range of the car.

OTHER POSSIBILITIES

Many manufacturers have developed other EVs that may or may not be available when this book is released. However, it is important to look at where the manufacturers are headed. Also, some of these may actually be common sights on our roads.

Mini E

In 2010, BMW released its first electric car, available by lease to the public. This was a limited-edition, all-electric version of the Mini Cooper. These cars were designed to be proof-of-concept test vehicles. This car had a respectable driving range (161 km/100 miles) and was able to perform almost as well as a conventional Mini.

The Mini E is equipped with a 380V (35 kWh) Li-Ion battery that can be recharged using a standard 110V electrical outlet. There is also an available 240V high-amperage, wall-mounted charger available through BWM. This charger allows the battery to receive a full charge in less than three hours.

BMW ActiveE

Continuing the experimentation with the Mini E, BMW began to lease a limited number of the four-passenger ActiveE. The cars are basically all-electric versions of the BMW 1 Series cars.

These cars also feature a Li-Ion battery pack that requires about four hours to fully charge. The battery pack is equipped with a complex battery management system that continuously monitors and controls the temperature of each cell in the pack. Perhaps the most unique thing about the battery pack is that it consists of three separate blocks. The blocks are strategically placed in the car to spread the added weight throughout the chassis. One block sits where the engine did, another is housed in the transmission tunnel, and the third sits where the fuel tank should be. The individual blocks are connected together by high-voltage cables.

Also, to maintain a characteristic of BMW vehicles, the car is rear-wheel drive. The rear axle is fitted with a synchronous motor that can provide up to 125 kW (170hp) and 250 N·m (184 lb·ft) of torque.

Toyota RAV4 EV

Toyota has built several different editions of the RAV4 EV through the years. The latest is based on the conventional RAV 4 and relies on a drivetrain developed with Tesla Motors. The Tesla-supplied PM AC motor can supply up to 112 kW (150 hp), and the motor is powered by a 50 kWh battery. Its driving range is estimated to be 129 to 193 km (80 to 120 miles).

The battery is a unique and "secret" Tesla version of a Li-Ion pack. (Tesla has not shared its battery chemistry with others, so the formula is not known at this time and therefore it is a secret.) In spite of the advancements of this breed of battery, it currently takes 12 hours at 240 volts or 28 hours at 120 volts to fully recharge the battery. The battery is very heavy—300 kg (660 lbs)—and is mounted centrally under the floor of the SUV.

SCION IQ EV This relies on the same basic electric driveline as the RAV4 and is built with the cooperation of Tesla. The Scion IQ EV is based on a gasoline car already made by Toyota.

Smart ED

The Smart fortwo electric drive (or Smart ED) is a BEV version of its normal model **(Figure 37–33)**. During the development of this EV, Smart developed and tested three different generations of this variant. The first version was powered by a rear-mounted 30 kW (41 hp) motor that drove the rear wheels. It had a sodium-nickel chloride (Zebra) battery pack with an output of 13.2 kWh. The estimated range was 109 km (68 miles) and had a top speed of 120 km/h (75 mph). Eight hours were needed to totally recharge the batteries with a 240V source.

The second generation used a lithium-ion battery supplied by Tesla Motors with a capacity of 16.5 kWh. This increased the range to 135 km (84 miles). However, a 20 kW (27 hp) motor was used, which lowered the top speed of the car to 100 km/h (62 mph). The battery pack needed three hours to charge from 20 to 80 percent of its capacity with a standard 240V outlet.

The third generation has a more powerful electric motor, 55 kW (74 hp) and 129 N·m (96 lb·ft) of torque. The car now has a top speed of 120 km/h (75 mph). It also uses a new lithium-ion battery pack that increases the range to 140 km (87 miles).

Chevy Spark EV

The Spark is a four-door, four-passenger wagon **(Figure 37–34)**. An all-electric plug-in version went on sale in select U.S. markets in 2013. The Spark is smaller than a typical subcompact and is called a minicar. It's a couple of inches shorter than a Mini Cooper. The basic design and development of the Korean-based Spark is to provide an economical commuter car that is capable of spending most of its life with electricity as its fuel.

The Spark EV relies on an A123-supplied nanophosphate lithium-ion battery pack that should provide an extended driving range for owners.

FIGURE 37–33

A Smart EV.

FIGURE 37–34

Chevrolet Spark EV.

Subaru R1e Electric Car

The Subaru R1e is a two-seater, about 50 cm (20 in.) longer than a Smart fortwo. Because it is a proof-in-concept model, it was not really developed to be practical or for sales. It only has a top speed of 105 km/h (65 mph) and a range of 80 km (50 miles). But the time to recharge the 346V Li-Ion battery pack has been reduced to about 15 minutes, using a manufacturer-specified charger. The use of this charger to obtain fast charging does not affect battery life. Also, a full charge can be completed overnight with a standard household 110V outlet. A typical charge costs about $2, the automaker says. Subaru says the life of the battery is 10 years or 160 000 km (100 000 miles).

BASIC DIAGNOSIS

Diagnosis of BEV concerns can be simpler than diagnosing concerns on a conventional vehicle because there are fewer components. However, most manufactured BEVs have complex electronics that are unique.

Manufacturer-supplied checklists are especially helpful when deciding what should be known about a particular problem and repair **(Figure 37–35)**. In the vehicle's service information, there may be symptom-based diagnostic aids. These can guide you through a systematic process. As you answer the questions given at each step, you are guided to the next step.

CUSTOMER PROBLEM ANALYSIS CHECK

EV CONTROL SYSTEM Check Sheet	Inspector's Name

Customer's Name		Model	
Driver's Name		Model Year	
Date Vehicle Brought in		Frame No.	
Licence No.		Odometer Reading	km miles

Problem Symptoms	☐ READY does not turn ON ☐ Vehicle does not move ☐ Poor acceleration ☐ Noise ☐ Vibration ☐ Harshness ☐ Smoke is rising ☐ Smell of or the likes burn ☐ Other ______

	Date Problem Occurred	
	Problem Frequency	☐ Constant ☐ Sometimes(times per day/month) ☐ Once only ☐ Other ______
Condition When Problem Occurred	Weather	☐ Fine ☐ Cloudy ☐ Rainy ☐ Snowy ☐ Various/Other ______
	Outdoor Temperature	☐ Hot ☐ Warm ☐ Cool ☐ Cold(approx. ____°F/____°C)
	Place	☐ Highway ☐ Suburbs ☐ Inner City ☐ Uphill ☐ Downhill ☐ Rough road ☐ Other ______
	Traction Motor	☐ Just after starting vehicle(min.) ☐ Standing with READY ON ☐ Driving ☐ Constant speed ☐ Acceleration ☐ Deceleration ☐ Other ______

Condition of MIL	☐ Remains on	☐ Sometimes lights up	☐ Does not light up
DTC Inspection	☐ Normal	☐ Malfunction code(s) (code) ☐ Freeze frame data ( )	

FIGURE 37–35

A checklist for inspecting and road-testing a BEV.

When these diagnostic aids are not available or prove to be ineffective, good technicians conduct a visual inspection and then take a logical approach to solving the problem.

Precautions

During diagnosis and repair of a BEV, always keep in mind that the vehicle has very high voltage. This voltage can kill you! Therefore, always adhere to the safety guidelines given by the manufacturer. Keep in mind that the same concerns stated in the chapters dealing with any high-voltage system should be adhered to now.

GO TO Chapters 8, 18, and 36 for the procedures for working with and around high-voltage systems.

Self-Diagnostics

The vehicle's control system has a built-in self-diagnostic system. When a fault is detected, the computer will store that information and may illuminate the malfunction indicator lamp (MIL) on the instrument panel. The faults held in the computer's memory can be retrieved as diagnostic trouble codes (DTCs). To retrieve these codes, follow these steps:

PROCEDURE

Retrieving DTCs

STEP 1 Measure the voltage of the auxiliary battery. If the voltage is lower than specifications, recharge it before continuing with your tests.

STEP 2 Inspect all fuses, fusible links, wiring harness, connectors, and ground in the low-voltage circuit. Repair them as necessary.

STEP 3 Connect the handheld scan tool to the data link connector (DLC) on the vehicle.

STEP 4 Turn the motor switch to the ON position and make sure the MIL is lit. If the MIL does not light, check for a burnt bulb, a bad circuit fuse, or an opening in the circuit. Again, correct the problem before proceeding. The MIL should go off when the READY lamp lights. If the MIL stays on, the computer has found a problem, and related information is stored in its memory. Turn the motor switch to the OFF position.

STEP 5 Make sure the scan tool is set up for the vehicle being tested.

STEP 6 Turn the motor switch to the ON position, and turn the scan tool on. Check for DTCs and freeze-frame data, and record all codes and data displayed on the scan tool. Refer to the manufacturer's reference to determine what the DTCs indicate.

STEP 7 Following the correct procedures, verify the concern and repair the problem. After completing any repair of the motor or related parts, erase the DTCs retained in the computer's memory with the scan tool. Then test it again to make sure the fault is no longer present.

Reduced Range

If you have a customer with an EV that is experiencing reduced driving range, do not forget that conditions that can cause reduced fuel economy in a standard car or truck can cause a reduction in range for an EV. Check the following:

- Inspect the vehicle for additional weight, such as from leaving heavy items in the trunk or cargo areas.
- Make sure the brake system is operating properly and that neither the service brakes nor parking brake are sticking on.
- Ensure the vehicle has the correct tires and that they are inflated to the proper pressure.
- Inspect the underside of the vehicle for damaged or missing covers that can affect airflow underneath the vehicle.

FUEL CELL VEHICLES

Fuel cell electric vehicles (FCEVs) are the result of many years of research and development on electric and hybrid vehicles. They share many of the same technologies but differ greatly in the source of energy used to power the electric motors that are used to move the vehicle. Although the government no longer funds fuel cell research and development, the manufacturers who have been exploring the possibilities say they will continue. But most industry analysts do not expect fuel cell cars to be available to the public until 2020 at the earliest.

The technology is not new, nor is it unproven; NASA (the U.S. National Aeronautics and Space Administration) has been using this technology in its spacecraft for years. Fuel cells provide the energy for the various electronic devices on board the spacecraft. A fuel cell vehicle is much like a battery-operated electric vehicle. It operates like one and has many of the same characteristics: Electricity powers a motor to drive the vehicle; the vehicle operates very quietly; and the output of CO_2 and other harmful emissions is zero.

FCEVs have electric motors, but the immediate energy source for those motors is not necessarily batteries. Some FCEVs use an ultra-capacitor in place of a battery pack. Regardless of where the energy is stored, all FCEVs rely on the DC voltage generated by an on-board fuel cell assembly. That energy can directly power the DC motors, or it can be sent to the storage device **(Figure 37–36)**. An external energy source is not required to refill the electrical storage unit; however, the fuel used in the fuel cell must be refilled. Pure water and heat are the only emissions from a fuel cell. Fuel cells can continue to work until the fuel supply is depleted. In other words, the driving range of a fuel cell vehicle is largely dependent on the amount of fuel it can carry.

Hydrogen

Fuel cell vehicles use hydrogen as their fuel or energy source. Hydrogen is found only in compound form, such as in water, which is a combination of hydrogen and oxygen. Fossil fuels are combinations of carbon and hydrogen, which is why they are called hydrocarbons.

Due to its atomic structure, hydrogen is full of energy and is used to make reformulated gasoline, ammonia for fertilizer, and many different food products. Hydrogen contains more energy per weight than any other fuel but contains much less energy by volume.

The auto industry has long used the energy released by separating hydrogen from a substance and recombining it with oxygen. In a gasoline-fuelled engine, gasoline is forced (by heat) to combine with oxygen. The result is combustion, which releases energy. That energy is used as mechanical energy. In a fuel cell, the same basic thing happens, but the chemical energy is released as electrical energy.

The Practicality of FCEVs

A major obstacle in the practicality of a fuel cell vehicle is the absence of an infrastructure for supplying pure hydrogen. Hydrogen production is commonly done, but it is very costly. It can be extracted

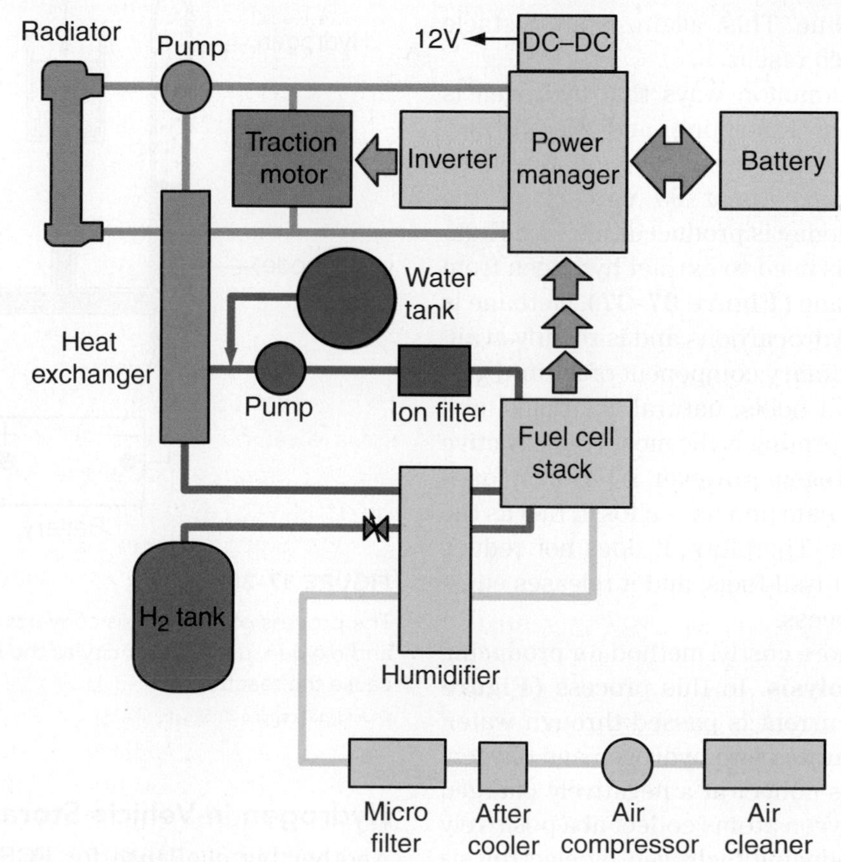

FIGURE 37–36

The basic layout for a fuel cell vehicle.

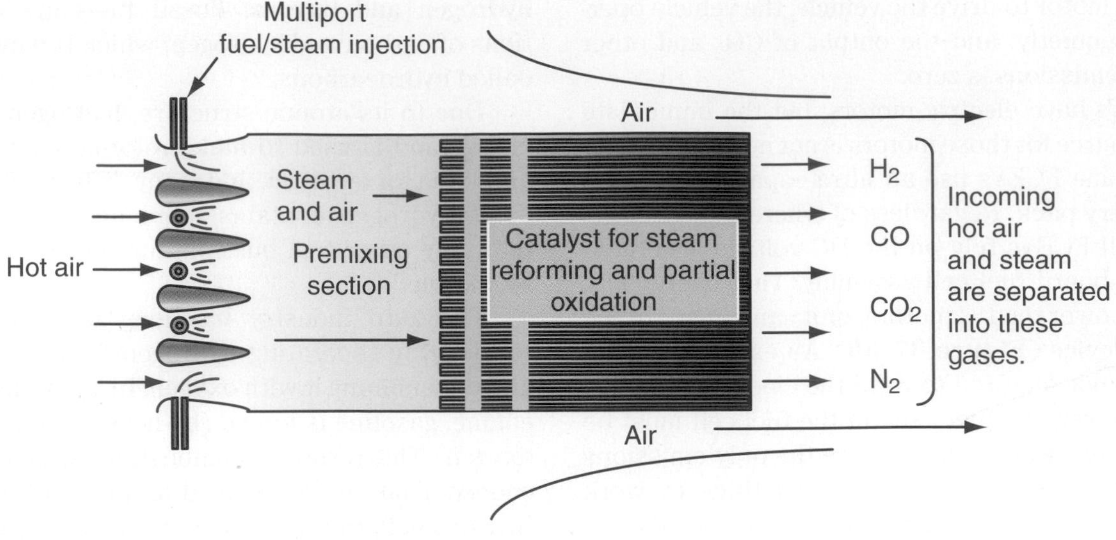

FIGURE 37–37

Basic view of how a steam reformer produces hydrogen.

from water, fossil fuels, coal, and biomass by a process that pulls hydrogen out of its bond with another element or elements. Currently, it costs much more to produce hydrogen than it does to produce other fuels, such as gasoline. This, again, is an obstacle and the focus of much research.

The two most common ways that hydrogen is produced are steam reforming and electrolysis. **Steam reforming** is the most common way used to produce hydrogen. About 95 percent of the hydrogen available today is produced this way. High-temperature steam is used to extract hydrogen from natural gas or methane **(Figure 37–37)**. Methane is the simplest of all hydrocarbons and is readily available. It is also the primary component of natural gas, which is found in oil fields, natural gas fields, and coal beds. Steam reforming is the most cost-effective way to produce hydrogen. However, it relies on fossil fuels to create the steam and uses a fossil fuel as the source for hydrogen. Therefore, it does not reduce our dependence on fossil fuels, and it releases emissions during the process.

A cleaner, but more costly, method for producing hydrogen is **electrolysis**. In this process **(Figure 37–38)**, electrical current is passed through water. The water then separates into hydrogen and oxygen. The hydrogen atoms collect at a negatively charged cathode, and the oxygen atoms collect at a positively charged anode. Producing hydrogen by electrolysis costs approximately 10 times more than using steam reforming. However, the process does result in pure hydrogen and oxygen.

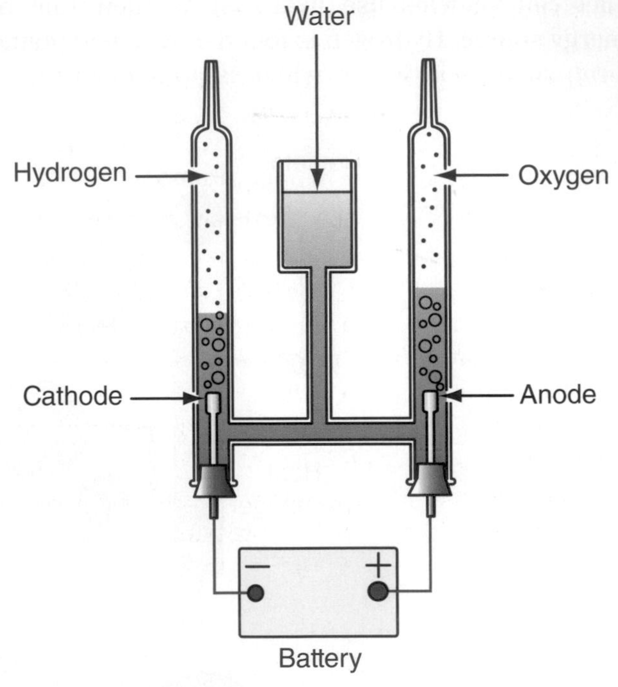

FIGURE 37–38

The process of electrolysis converts water into hydrogen and oxygen, using electricity as the source of energy to cause the reaction.

Hydrogen In-Vehicle Storage

Another big challenge for FCEVs is the storage of hydrogen. To be practical, any vehicle must have a decent driving range of at least 483 km (300 miles). Obviously, the more hydrogen that can be stored in

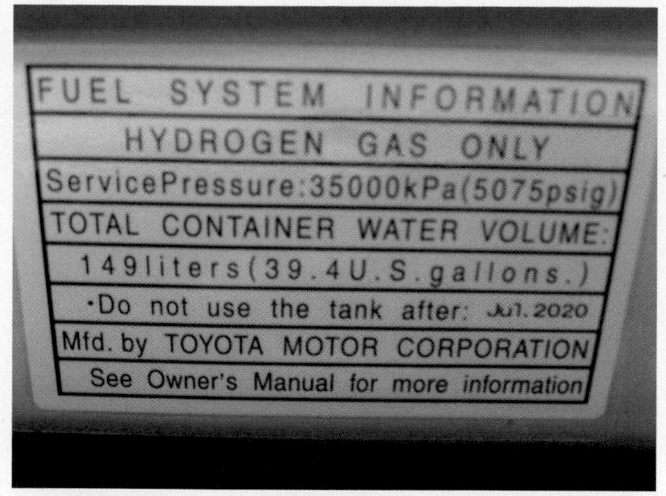

FIGURE 37–39

This sticker defines the amount and form of the stored hydrogen fuel.

©Cengage Learning 2015

the vehicle, the longer the car will be able to drive without refuelling.

Hydrogen can be stored as a liquid or as a compressed gas **(Figure 37–39)**. When stored as a liquid, hydrogen must be kept very cold. Keeping it that cold adds weight and complexity to the storage system. At cryogenic (icy cold) temperatures, more hydrogen can be stored in a given space. Cryogenic fuels are used in the rockets of NASA's space shuttle. However, liquid storage has some safety issues that are not present with compressed hydrogen storage.

The tanks required for compressed hydrogen need to be very strong, which translates to very heavy and expensive tanks. Most hydrogen tanks have an aluminum liner covered with carbon fibre and fibreglass. Also, higher pressures mean more hydrogen can be packed into a tank, but the tank must be made stronger to hold the higher pressures.

High-pressure tanks are very expensive. The typical fuel cell vehicle stores hydrogen at 352 kg/cm^2 (5000 psi) and has a driving range of about 240 km (150 miles). Doubling the pressure would nearly double the driving range. To double the pressure, stronger or more tanks are required, which adds to the cost of the vehicle.

These considerations explain why storing enough hydrogen for an acceptable driving range is very difficult. Naturally, you can store more in a larger container, but that container would consume more space and add considerable weight to the vehicle.

In-vehicle hydrogen storage is another area of much attention for researchers and engineers, and other storage technologies are being developed. Two of the technologies getting the most attention are systems based on metal hydrides and carbon nanotubes. The use of metal hydrides offers the possibility of storing three times more hydrogen in a given volume than when it is compressed. Carbon nanotubes are microscopic tubes of carbon that can store hydrogen in their pores. Because the surface of these tubes is quite irregular, the actual surface area is larger than the size of the tubes. The use of metal hybrids and carbon nanotubes may help solve the hydrogen storage problem for the future.

Reformers

The supply of hydrogen can be provided by a **reformer** that extracts hydrogen from another fuel, such as gasoline, methanol, or natural gas. Therefore, a reformer can solve the hydrogen storage problem, because storing these fuels requires less space and is much simpler than storing pure hydrogen. The objections to using a reformer are plentiful. A reformer has undesirable emissions, such as carbon dioxide. Using reformers does not reduce our dependence on fossil fuels. Reformers **(Figure 37–40)** are expensive and slow, and require long run times before they can provide enough hydrogen to move a vehicle a few feet. Plus, the cost of the reformer adds to the already high cost of a fuel cell.

Reformers make an FCEV more practical because the fuel supply is easily replenished. However, reformers have some emissions issues and consume valuable vehicle space. There is also an issue with the purity of the fuels that will be reformed. Many of these fuels have a substantial amount of sulphur. The sulphur can contaminate the catalysts used in the fuel cell and may not be totally filtered out of the hydrogen during the reforming process.

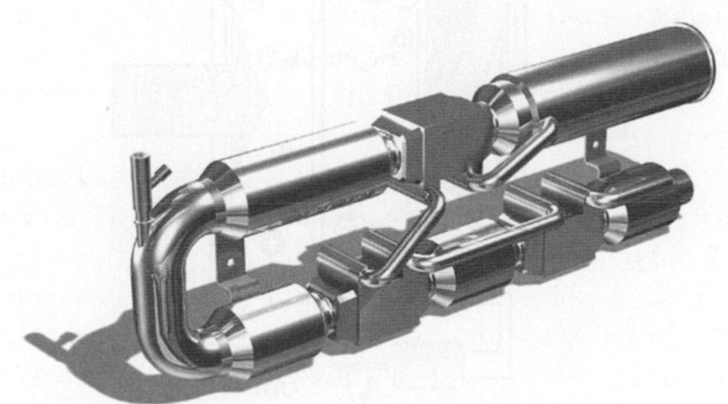

FIGURE 37–40

An example of a reformer.

PowerCell Sweden AB

FUEL CELLS

A fuel cell produces electricity through an electrochemical reaction that combines hydrogen and oxygen to form water. The basic principle of operation is the opposite of electrolysis. Electrolysis is the process of separating a water molecule into oxygen and hydrogen atoms by passing a current through an electrolyte placed between two electrodes **(Figure 37–41)**. In a fuel cell, catalysts are used to combine the fuel (hydrogen) with oxygen. The reaction releases electrons or electrical energy.

A single fuel cell produces very low voltage, normally less than one volt. To provide the amount of power needed to propel a vehicle, several hundred fuel cells are connected in series. This assembly is the **fuel cell stack (Figure 37–42)**, called this because the cells are layered or stacked next to each other.

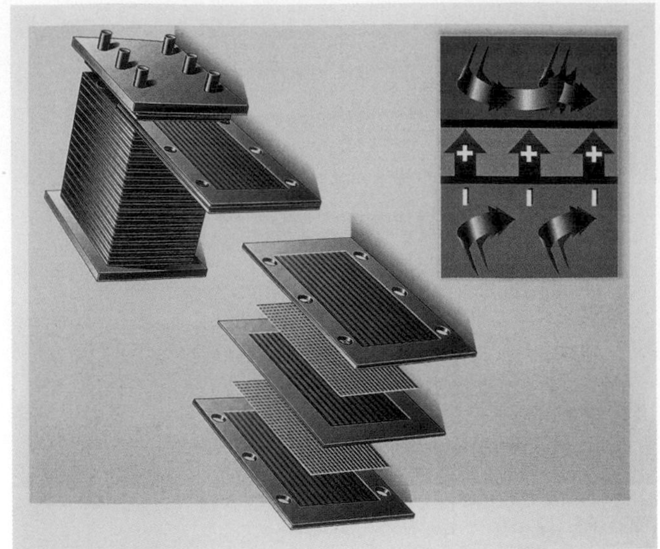

FIGURE 37–42

A fuel cell stack.

Chrysler Group LLC

FCEV Configurations

There are three basic configurations that describe the design of an FCEV powertrain **(Figure 37–43)**. When the powertrain has a direct-supply system, the energy from the fuel cell is delivered directly to the electric traction motor(s). With this configuration, the FCEV cannot have regenerative braking, and propulsion power depends entirely on the output of the fuel cell.

In a battery-hybrid powertrain system, the energy from the fuel cell is sent to the motor(s), the battery pack, or both. This configuration can use regenerative braking. The battery can also supplement the fuel cell's energy to improve performance. This system requires more electronic controls than the direct-supply system.

The third configuration uses ultra-capacitors rather than a battery. The ultra-capacitors are charged by the fuel cell and regenerative braking. Ultra-capacitors charge and discharge quickly, which allows the powertrain to respond quicker to changing conditions. Complex electronic systems are also required for this type of system.

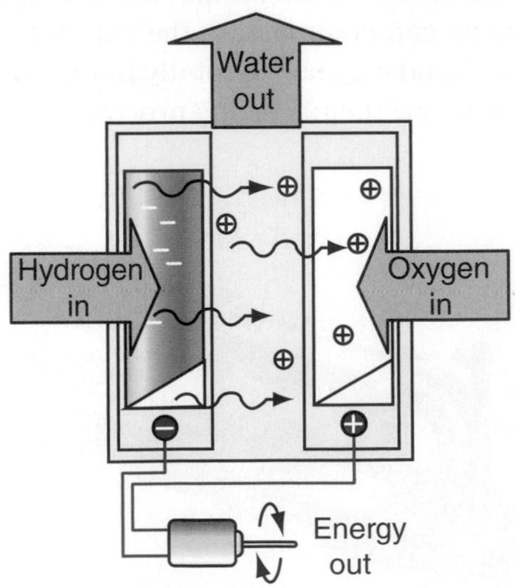

FIGURE 37–41

The basics of fuel cell operation.

©Cengage Learning 2015

Controls

To control the output of a fuel cell and therefore the speed of the vehicle, advanced electronics are necessary. Much of this technology is already used in hybrid vehicles, but the uniqueness of the fuel cell demands additional new controls. FCEVs have high- and low-voltage systems, and electronic controls are necessary to allow the fuel cell to power both. This means that all FCEVs need a DC–DC converter **(Figure 37–44)** to reduce the high-voltage from the fuel cell. These controls are in addition to the typical computer systems of other vehicle types. However, because a fuel cell generates DC voltage, an inverter is not needed unless the traction motors and the accessories require AC voltage.

Temperature Concerns

Most fuel cells take some time to start, especially when they are cold. In fact, freezing temperatures can kill a fuel cell. An exhaust system plugged with ice will also shut down a fuel cell. This is an area of much research, and some manufacturers have had

	Basic configuration	System features	Efficiency	Power performance
Fuel cell direct-supply system	Motor ← Fuel cell stack	Simple high-voltage system start-up device required	Good efficiency No regenerative braking	Responsiveness depends on the output of the fuel cell stack
Battery-hybrid system	Motor ← High-voltage control device – Fuel cell stack; Battery → High-voltage control device	High-voltage distribution system required	Heat losses affect efficiency Regenerative braking	Output assist is possible
Capacitor-assisted system	Motor ← Fuel cell stack; Ultra-capacitor	High-voltage distribution system not required	Good efficiency Regenerative braking	Instantaneous high-output assist is possible

FIGURE 37–43

The different configurations of an FCEV.

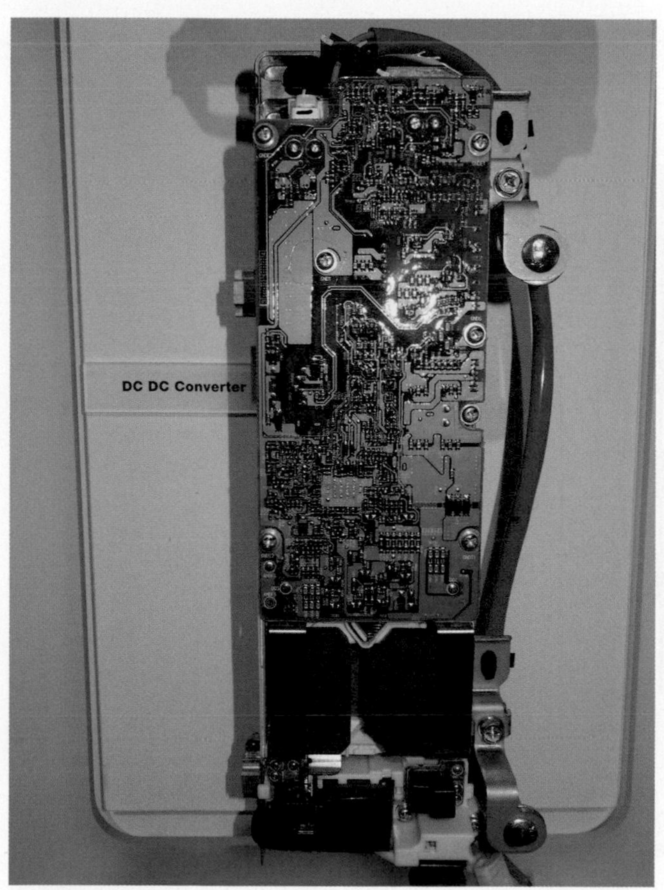

FIGURE 37–44

A DC–DC converter.

some success dealing with the problem. The basic thrust has been making sure all water is removed from the fuel cell after it has been shut down. This requires energy from a storage device. There is also research being done on mixing special coolants in the water. The fact that a fuel cell does not generate electricity until it has a temperature of 0°C (32°F) is an obstacle that needs to be overcome.

On the other side of the temperature scale, heat must be carefully controlled. Fuel cells become very hot while they operate, and they operate best within a particular temperature range. That range depends on the type of fuel cell. Some fuel cells operate best at a lower temperature than a conventional engine **(Figure 37–45)**. This presents a major challenge as it is more difficult to get rid of low heat than it is high heat. This means the cooling system may need larger and/or more radiators. This means more space is needed in the vehicle just for the cooling system. The result is less useable space for passengers and luggage.

When the space for the cooling system is added to the required space for the fuel cell stack and other components, very careful planning is necessary. This becomes more of a challenge when one considers that the electronics and traction motors must also be kept cool. The cooling of these requires an additional cooling system because they operate at a different temperature range than the fuel stack. An additional cooling problem enters when the vehicle

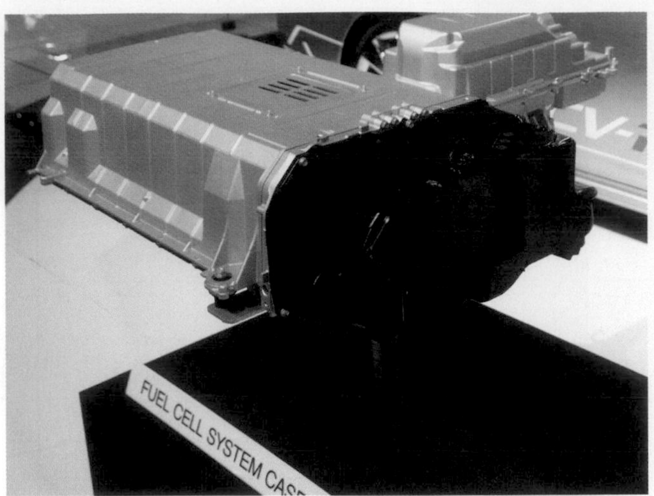

FIGURE 37–45

A fuel cell case with its cooling system.

is equipped with a high-voltage battery pack and/or ultra-capacitors.

Another heat-related problem is generated by the air compressor that feeds outside air into the fuel cell. As air is compressed, its temperature increases. Because the fuel cell works best with a specific temperature range, the compressed air can heat up the cell beyond that range. To eliminate this, intercoolers must be added to the air compressor system. These, again, occupy space. There is also the problem of filtering the incoming air. Ideally, the incoming air would be free of all dirt and other contaminants. A filtering system occupies space and has an impact on the overall layout and design of the vehicle.

Fuel Cell Types

The different types of fuels cells vary by size, weight, fuel, cost, and operating temperature. Regardless of design, all fuel cells have two electrodes coated with a catalyst. The catalyst, normally platinum, on the electrodes causes the chemical reaction in the fuel cell, but it does not materially take part in the reaction. One of the electrodes has a positive polarity, the anode, and the other is negative, the cathode. The electrodes are separated from each other by an electrolyte and separators **(Figure 37–46)**.

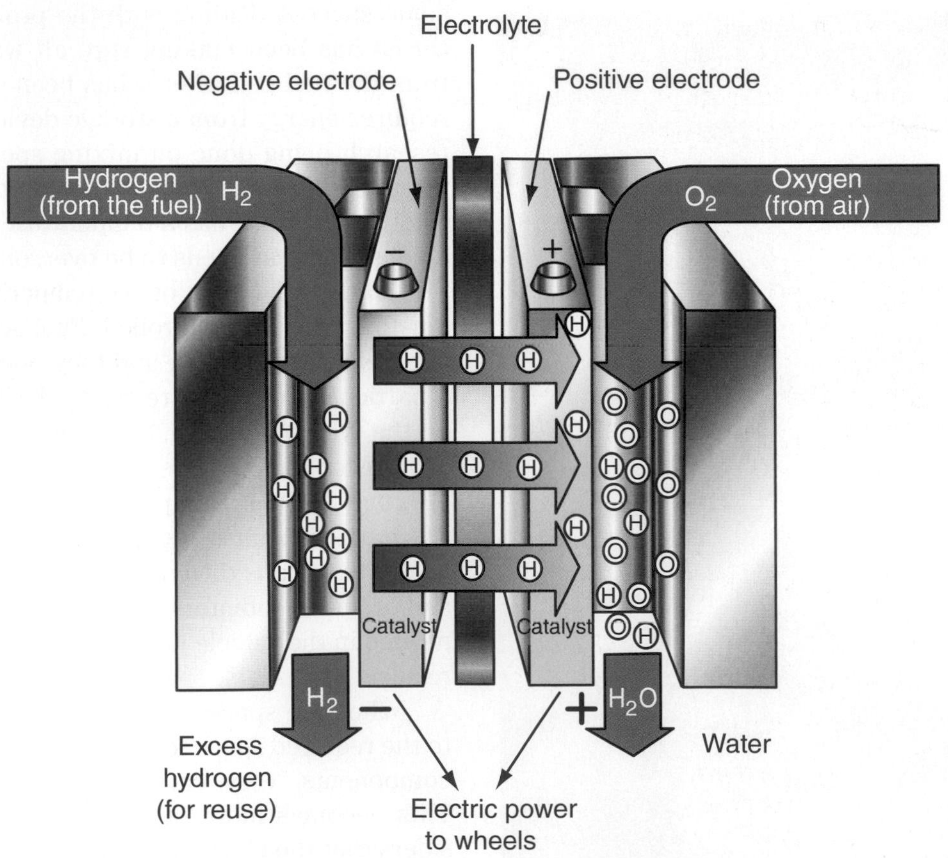

FIGURE 37–46

All fuel cells contain two electrodes—one positively and one negatively charged—and an electrolyte sandwiched between them.

Because the catalyst materials are not consumed during the operation of a fuel cell, the fuel cell consumes only hydrogen and oxygen. The oxygen is delivered to the cell by an air compressor that draws air in from outside the fuel cell. Hydrogen is fed into the fuel cell from a pressurized tank or from a reformer. The actual reactions that take place within a fuel cell depend on its design. What follows are descriptions of various fuel cells that may find their way into an automobile.

Proton Exchange Membrane Fuel Cell

The **proton exchange membrane (PEM) fuel cell (Figure 37–47)**, or derivatives of it, is a favoured design for use in vehicles because it allows for adjustable outputs, which are necessary for driving. The speed of the vehicle can be controlled by controlling the output of the fuel cell. Although it is quite compact, it is capable of providing high outputs. When compared to other fuel cell designs, it is most efficient at relatively low temperatures of 30°C to 100°C (86°F to 212°F); however, it is expensive to manufacture.

In a PEM, the electrolyte is a polymer membrane, called the proton or ion exchange membrane. Polymers can be very resistant to chemicals and can serve as an electrical insulator or separator. The polymer membrane in a fuel cell does both.

When hydrogen is delivered to the anode, the catalyst causes the hydrogen atoms to separate into electrons and protons. Oxygen enters the other side of the fuel cell and reacts to the catalyst on the cathode. This splits the oxygen molecules into oxygen ions. The protons (hydrogen ions) that were released from the hydrogen at the anode move toward the oxygen ions. Keep in mind that electrons always move to something more positive but cannot pass through the membrane. Therefore, their only path to the positive side of the fuel cell is through an external circuit. The movement of the electrons through that circuit results in direct current flow.

Because the membrane that separates the two electrodes will only allow protons to pass through, the hydrogen ions move to the cathode, where they bond with oxygen ions to form water. Some of the water produced by the fuel cell is used to humidify the incoming hydrogen and oxygen. This is important because in order for the fuel cell to work properly, the ion exchange membrane must be kept moist. The remaining amount of water produced by the fuel cell is emitted as exhaust from the fuel cell. Some heat is also emitted by the fuel cell. The heat is either released to the outside air or used to heat the fuel cell. It can also be used to heat the passenger compartment.

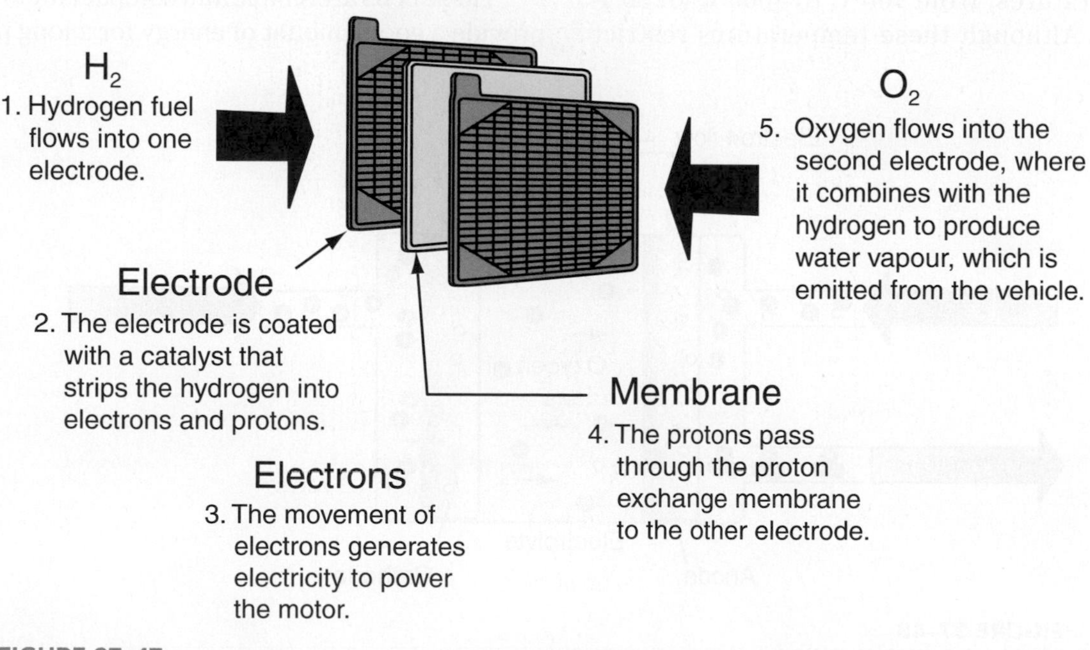
Single fuel cell
H2
1. Hydrogen fuel flows into one electrode.
O2
5. Oxygen flows into the second electrode, where it combines with the hydrogen to produce water vapour, which is emitted from the vehicle.
Electrode
2. The electrode is coated with a catalyst that strips the hydrogen into electrons and protons.
Membrane
4. The protons pass through the proton exchange membrane to the other electrode.
Electrons
3. The movement of electrons generates electricity to power the motor.

FIGURE 37–47

The basic operation of a proton exchange membrane fuel cell.

One of the biggest disadvantages of the PEM cell is the need to keep the membrane moist. In cold temperatures, the water can freeze, making the fuel cell very difficult to get started. Also, carbon monoxide (CO) can weaken the platinum catalysts. Because outside air is delivered to one side of the cell, the presence of CO in that air will reduce the output of the cell. Much research and development is taking place to alleviate these obstacles.

Solid Oxide Fuel Cell

The **solid oxide fuel cell (SOFC)** may be the first design used in a mass-produced automobile. However, it will not be used to power a traction motor. Rather, it may be used to replace the belt-driven generator (alternator). Current alternators are not very efficient, and their output is dependent on rotational speed. They also rely on engine power to operate, which means they contribute to an engine's fuel consumption. Removing the alternator and using an SOFC will increase the efficiency of the engine. The SOFC can also provide much higher power levels, which means more accessories can be electrically driven. This again will increase the efficiency of the engine. Using a SOFC will also allow accessories to operate when the engine is not running and without draining the battery.

These cells have a ceramic anode, ceramic cathode, and a solid electrolyte **(Figure 37–48)**. To be efficient, these cells must operate at very high temperatures, from 700°C to 1000°C (1290°F to 1830°F). Although these temperatures restrict the type of materials used in the cells to ceramics, they also eliminate the need for expensive catalysts. This means SOFCs have low production costs. These fuel cells can operate with a simple, single-stage, built-in reformer because of the high operating temperatures. Also, the high temperatures eliminate the chances of CO poisoning the electrodes. Efficiency estimates for this type of fuel cell vary from 40 to 45 percent, as compared to 20 to 30 percent for an internal combustion engine.

The high operating temperature is also a reason why these fuel cells may not be used to power a vehicle. When this high heat is generated, it must be released. Releasing a large quantity of high heat can cause many problems to other automotive systems.

Direct-Methanol Fuel Cell

The **direct-methanol fuel cell (DMFC)** is a type of PEM fuel cell. Liquid methanol, rather than hydrogen, is oxidized at the anode, and oxygen is reduced at the cathode. Methanol is considered an ideal hydrogen carrier because it takes little energy to release its hydrogen. The methanol is delivered directly into the cell, and therefore a reformer is not needed. Thus, the cost of these cells is lower than the PEM. Liquid methanol is also easier to store than hydrogen and has a much higher energy density than compressed hydrogen.

These cells are simple and compact units that can provide a good amount of energy for a long period of

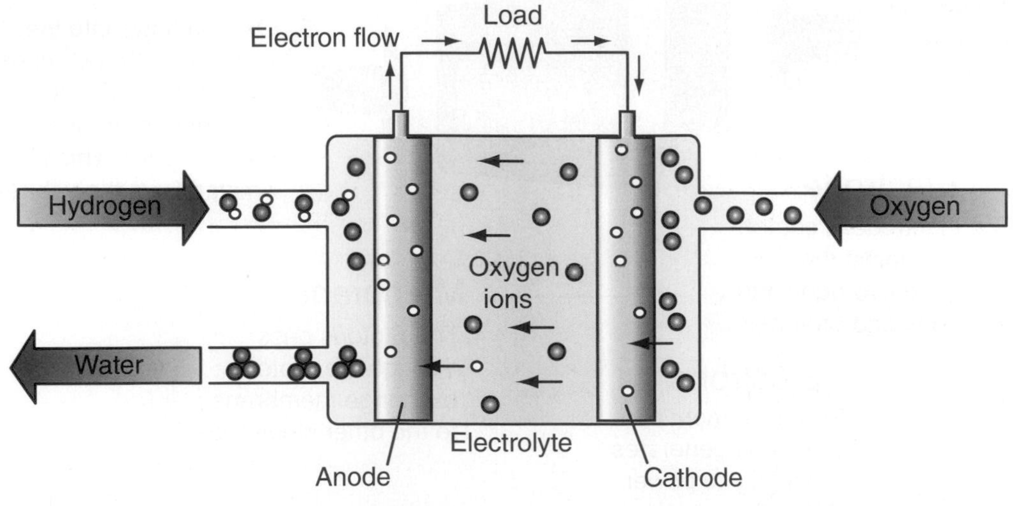

FIGURE 37–48

A solid oxide fuel cell (SOFC).

time. They operate at about the same temperature as PEMs, but the cells are not as efficient as PEMs, and their response time is slower than a PEM. Also, they have emissions that are not present with other fuel cells designs. As the hydrogen is removed from the methanol, carbon is released. The carbon and oxygen atoms combine to form CO_2 and the oxygen and hydrogen form water.

Alkaline Fuel Cell

The **alkaline fuel cell (AFC)** is the one used by NASA. It is expensive, but highly efficient. In a spacecraft, the water (its exhaust) is used as drinking water for the space travellers. This fuel cell will undoubtedly never be used in automobiles because of its cost. It is also very sensitive to carbon dioxide, which means it does best where all CO_2 can be removed from the incoming supply of air. This fuel cell operates in the same manner as a PEM.

Alkaline fuel cells use a water-based solution of potassium hydroxide (KOH) as the electrolyte. The electrodes are coated with a catalyst, although due to the operating temperature and the purity of the incoming gases, platinum is not required.

PROTOTYPE FCEVs

Fuel cell vehicles for everyday use are still years away; however, there are many fuel cell prototypes and concept vehicles on the road all over the world. All of these are part of the ongoing research that is taking place. Every major manufacturer has built at least one type of FCEV, and many have developed a new model nearly every year. These vehicles are testing different technologies in a real-world setting. It is difficult to predict exactly how a mass-produced FCEV will be equipped, but it is certain that some of the technology used in today's prototypes will be part of that final design.

The development of practical FCEVs will involve the cooperation of different auto manufacturers. Several joint ventures have already been announced. Daimler, Ford, and Nissan are working together to make a common propulsion system that can be used in a variety of vehicles. Their target date is 2017. They are also working on improving the infrastructure for hydrogen. Another joint venture has also been established. This one is with BMW and Toyota. They are working together to design a platform for a mid-size sports vehicle by 2020.

Daimler

Daimler started developing fuel cell vehicles in 1994 and has produced well over 100 vehicles for testing purposes. These vehicles include cars, buses, and vans. Their vehicles are on the roads in the United States, Europe, China, Australia, Japan, and Singapore. At each location, feasibility studies are being made while the vehicles are being used in varying driving and climate conditions.

Through continuous research, Daimler has been able to extend driving range, minimize the space required for the fuel cell components, and improve cold weather starting and operation. To minimize the space requirement, engineers have fit the entire fuel cell drive system in the floor. Doing this allowed them to convert a small car, the Mercedes-Benz A-Class, and small SUVs into FCEVs and still have room for passengers and luggage.

One prototype is a Mercedes B-Class F-Cell that has a 100 kW (136 hp) electric motor. This vehicle uses battery power to start the vehicle and assist the fuel cell in delivering power to the electric motors during acceleration. The battery pack is recharged by the fuel cell and with regenerative braking. The hydrogen is stored at 703 kg/cm^2 (10 000 psi) and has a driving range of about 450 km (280 miles).

General Motors Corporation

With several experimental FCEVs, GM has worked to minimize the required space for hydrogen storage and to increase driving range. In 2002, it introduced the "skateboard" concept **(Figure 37–49)**. In this design, all of the fuel cell–related components are packed into a carbon fibre structure that also serves as the vehicle's chassis. This propelled chassis was designed with the intent of placing any body configuration on it. The first of these vehicles was the AUTOnomy and the Hy-wire. They featured many futuristic concepts, including total drive-by-wire systems.

A recent concept GM fuel cell vehicle is the Sequel. The Sequel uses the technologies that worked well in the AUTOnomy and the Hy-wire. This battery-hybrid FCEV uses a Li-Ion battery to provide electrical energy to the three electric motors during acceleration and to capture energy during braking. A transverse-mounted, three-phase AC motor drives the front wheels, and two three-phase AC wheel hub motors drive the rear wheels.

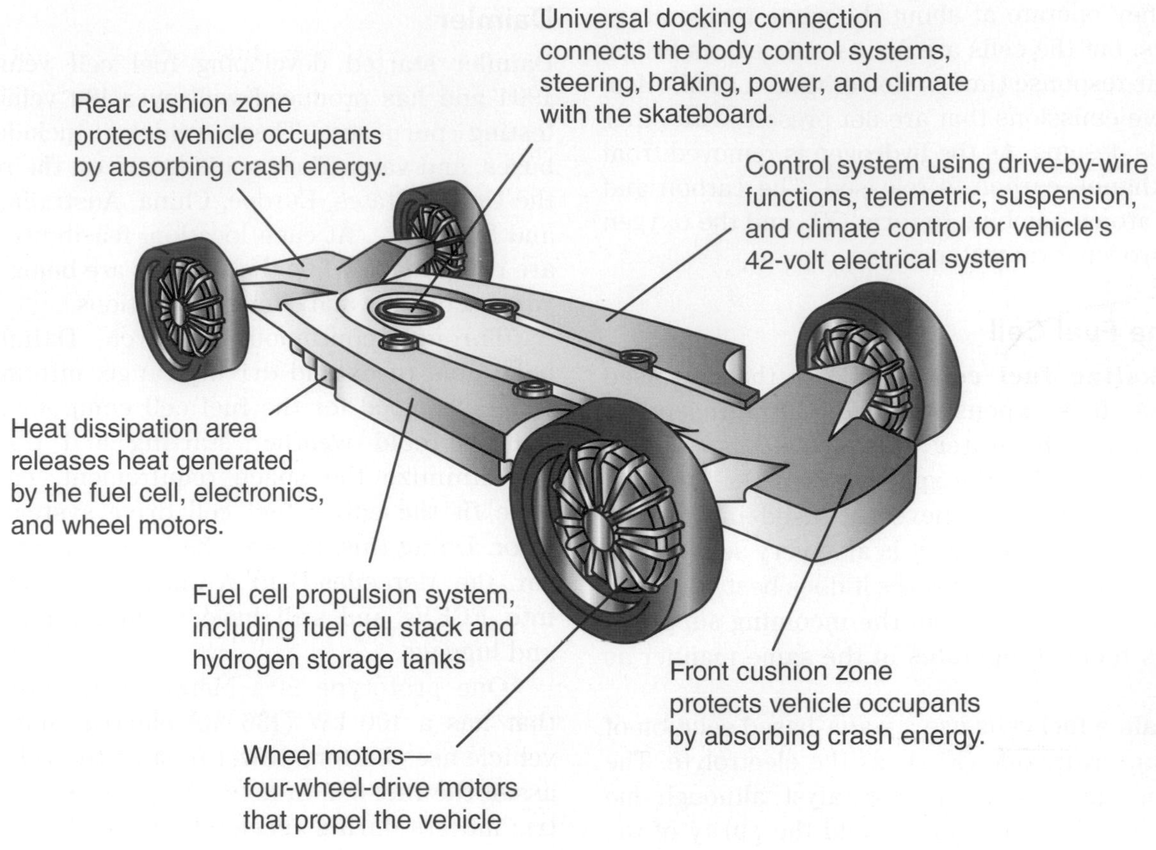
Universal docking connection
connects the body control systems,
steering, braking, power, and climate
with the skateboard.
Rear cushion zone
protects vehicle occupants
by absorbing crash energy.
Control system using drive-by-wire
functions, telemetric, suspension,
and climate control for vehicle's
42-volt electrical system
Heat dissipation area
releases heat generated
by the fuel cell, electronics,
and wheel motors.
Fuel cell propulsion system,
including fuel cell stack and
hydrogen storage tanks
Front cushion zone
protects vehicle occupants
by absorbing crash energy.
Wheel motors—
four-wheel-drive motors
that propel the vehicle

FIGURE 37–49

The basic layout of General Motors' "skateboard" chassis.

There is a separate inverter for each motor. The electrical system includes three separate systems with three different voltages. A high-voltage system provides energy for the traction motors; the 42-volt system supplies energy for the brakes, steering, air-conditioning, and other by-wire systems; and the 12-volt system is used for the conventional accessories and lights.

The fuel cell stack, hydrogen and air processing subsystems, high-voltage distribution system, and hydrogen storage tanks are housed in the skateboard. The storage tanks are designed to hold compressed hydrogen at 703 kg/cm^2 (10 000 psi).

Recently, GM unveiled a fuel cell–powered version of its Chevrolet Volt. In this concept car, the generator was replaced by a fuel cell. To accommodate the hydrogen storage tanks, the battery pack was made to half of its original size. It also uses a more advanced fuel cell design, and the vehicle is lighter. The lithium-ion battery pack can be recharged by plugging it in, from the fuel cell, or through regenerative braking.

When the car is first started, it is powered by the fuel cell. When more power is needed, the battery pack provides additional energy to the traction motor. When the battery is not powering the motor, it is being recharged.

Toyota

Much of what Toyota has learned with its hybrids is transferrable to its fuel cell vehicles. In fact, many of the same components can be transferred as well. Toyota's first FCEV, the RAV 4 FCEV, used a Toyota-developed PEM fuel cell and was configured as a battery-hybrid FCEV. There were two generations of the RAV FCEV; the first stored the hydrogen in metal hydrides, and the other had a methanol reformer and provided a range of 500 km (310 miles).

As Toyota continued to make advancements in hybrid technology, it applied the technology to new fuel cell prototypes and concept vehicles. To take a quick look at how hybrid technology and components are used in a fuel cell vehicle, consider its latest concept vehicle, the FCHR **(Figure 37–50)**.

FIGURE 37–50

A Toyota fuel cell concept car.

©Cengage Learning 2015

The vehicle is a battery-hybrid FCEV and is propelled by 80 kW (109 hp) electric motors. The compressed fuel is stored in four hydrogen fuel tanks at 700 kg/cm^2 (10 000 psi). The vehicle uses the same battery pack as the Prius, including all of the associated electronics. A power control unit is in the engine compartment and is a slightly modified version of that used in the Prius. It monitors the current operating conditions and determines when to use the battery, fuel cell, or both to propel the vehicle and to charge the battery. This is the same strategy used in hybrid vehicles. However, in the FCEV, the fuel cell and its output replace the engine **(Figure 37–51)**.

FIGURE 37–51

The fuel cell and its components fill the engine compartment of this FCEV.

©Cengage Learning 2015

Honda

Honda has also been busy with fuel cell vehicle research. Unlike Toyota, the hybrid system used in Honda's EVs does not easily adapt to an FCEV. However, many of the controls and features do transfer rather nicely. Plus, Honda also has much experience with battery electric vehicles and DC brushless motors. Honda started its venture into fuel cell vehicles in 1989 and has been road-testing vehicles since 1999, when it introduced the FCX. Through the years, Honda has had four versions of this original model, the FCX-V1, V2, V3, and V4. After the FCX-V4, it continued with new models, but those are referred to as the FCX.

In 2007, Honda announced that it will begin production of its next generation; this vehicle, called the Clarity, is not for sale but can be leased for three years to those who are interested.

The Honda FCX Clarity is powered by a 100 kW fuel cell stack, a lithium-ion battery pack, and a 95 kW (127 hp) electric motor, and it has a 352 kg/cm^2 (5000 psi) compressed hydrogen gas storage tank that yields a range of 435 km (270 miles). Honda uses its latest design of fuel cell, one that is compact but powerful and can operate in very low temperatures. The fuel cell has been developed to control water flow, because this is critical to fuel cell efficiency and start-up times. This new fuel cell is a PEFC (polymer electrolyte fuel cell). Oxygen and hydrogen flow from the top to the bottom of the fuel cell stack, and the fuel cells are arranged vertically to achieve efficient packaging. However, the biggest improvement to the fuel cell is that it is designed to allow gravity to get rid of the unwanted water in the system. By disposing of the water built up during operation, Honda's fuel is capable of starting in temperatures as low as –30°C (–22°F).

The fuel cell stack has metal separator structures that are easier to make, thereby reducing overall costs. Costs are further reduced by the use of an aromatic electrolyte membrane that also increases the fuel cell's efficiency through a broad range of temperatures. A new FCX uses Honda's ultra-capacitor and three separate electric motors. One motor drives the front wheels, and there is one motor at each rear wheel. This arrangement minimizes the space required for the drive system. The new FCX also has a special material in its hydrogen storage tanks. This material, not described at this time, doubles the storage capability of a tank and allows the new FCX to have a range of nearly 563 km (350 miles).

KEY TERMS

Alkaline fuel cell (AFC) (p. 1149)
Battery electric vehicles (BEVs) (p. 1119)
Conductive charging (p. 1127)
Direct-methanol fuel cell (DMFC) (p. 1148)
Electrolysis (p. 1142)
Fuel cell electric vehicles (FCEVs) (p. 1121)
Fuel cell stack (p. 1144)
Inductive charging (p. 1126)
Proton exchange membrane (PEM) fuel cell (p. 1147)
Reformer (p. 1143)
Solid oxide fuel cell (SOFC) (p. 1148)
Steam reforming (p. 1142)

SUMMARY

- Battery electric vehicles (BEVs) use electrical energy stored in batteries to power the traction motors and have zero emissions.
- Fuel cell electric vehicles (FCEVs) are also electrically powered zero-emission vehicles, but they rely on hydrogen as the fuel.
- The basic systems in a BEV are a high-voltage battery pack, the battery management system, the motor(s) and supporting system, the 12-volt system, a converter and/or inverter, and the driver's displays and controls.
- One kilowatt (kW) is the international unit to measure power (not only electrical); a kW is 1000 watts and equals 1.34 horsepower.
- The amount of energy available from a battery or other electrical source is expressed in kilowatt-hours (kWh); this states what can be accomplished by one kilowatt acting for one hour.
- Most production BEVs use AC motors, and FCEVs, and many conversion EVs use a DC motor.
- In most EVs, there is no transmission because a motor is capable of providing enough torque throughout its speed range to move the vehicle without torque multiplication.
- Inductive charging transfers electricity from a charger to the vehicle, using magnetic principles. There is no metal-to-metal contact between the charge paddle and the charge port of the vehicle.
- Conductive charging takes place by making a connection between the charging connector and the charge port.
- Conductive charging is normally accomplished with a fuel nozzle–style connector called the ODU.
- SAE's J1772 standard covers the basic physical, electrical, communication protocol, and performance requirements for an EV's conductive charge system and coupler.
- During diagnosis and repair of a BEV, always keep in mind that the vehicle has very high voltage. Always adhere to the safety guidelines given by the manufacturer.
- Fuel cell vehicles use hydrogen as their fuel or energy source.
- Hydrogen is full of energy. Due to its atomic structure and abundance, it may be the fuel of the future.
- The main powertrain components in a typical fuel cell vehicle include a fuel cell stack, high-pressure hydrogen supply or a reformer with a fuel tank, air supply system, humidification system, fuel cell cooling system, storage battery or ultra capacitor, traction motor and transmission, and control module and related inputs and outputs.
- A fuel cell produces electricity through a process that works in the opposite way as electrolysis, in which hydrogen and oxygen are combined to form water.

REVIEW QUESTIONS

1. Why are carbon dioxide emissions a concern?
2. List five things a driver of a BEV can do to extend the driving range of the vehicle.
3. What makes up the propulsion system in a BEV?
4. What basic factors affect the required time to recharge the battery pack in a BEV?
5. There are two basic ways a BEV is connected to an external source of electricity for charging: conductive and inductive. What is the difference between the two?
6. Which of the statements about hydrogen is true?
   a. A hydrogen atom is one proton and two electrons.
   b. Hydrogen is one of the heaviest elements known and is full of energy.
   c. Fossil fuels are combinations of carbon and hydrogen.
   d. Hydrogen is produced in a fuel cell when water is broken down into its basic elements.
7. Which of the following *cannot* be used as a source for the production of hydrogen?
   a. gasoline **b.** methanol
   c. carbon dioxide **d.** natural gas

8. Which of the following statements about fuel cells is *not* true?
   a. A single fuel cell produces very low voltage, normally less than one volt.
   b. A fuel cell produces electricity through an electrochemical reaction that combines hydrogen and oxygen to form water.
   c. A fuel cell is composed of two electrodes coated with a catalyst and separated from each other by an electrolyte and from the case by separators.
   d. In a fuel cell, catalysts are used to ignite the hydrogen; this causes a release of electrons or electrical energy.
9. What is the basic fuel for a fuel cell?
   a. hydrogen b. methanol
   c. electricity d. gasoline
10. What component changes the molecular structure of hydrocarbons into hydrogen-rich gas?
   a. pressure container b. fuel cell
   c. fuel injector d. reformer
11. Which of the following statements about hydrogen is *not* true?
   a. Hydrogen displaces air, so any release in an enclosed space could cause asphyxiation.
   b. Hydrogen must be stored as a compressed gas.
   c. Hydrogen is nontoxic.
   d. Hydrogen is highly flammable and there is risk for an explosion.
12. What BEV charge connector allows for both AC and DC charging?
   a. AVCON
   b. combo SAE J1772
   c. CHAdeMO
   d. inductive
13. What is the ability to do work called?
   a. power b. torque
   c. pressure d. energy
14. What is found in the exhaust of a fuel cell electric vehicle (FCEV)?
   a. water and heat
   b. hydrogen and heat
   c. nitrogen and hydrogen
   d. nitrogen and oxygen
15. What is the most common way to produce hydrogen?
   a. electrolysis b. hydrolysis
   c. steam reforming d. refraction
16. Which of the following can be used to create hydrogen?
   a. gasoline b. carbon dioxide
   c. carbon monoxide d. ammonia
17. What are the groupings of fuel cell called?
   a. fuel cell columns
   b. fuel cell stacks
   c. fuel cell arrangements
   d. fuel cell stand
18. What maintains the Nissan Leaf's auxiliary 12-volt lead-acid battery when not in operation?
   a. the home-charging station, through the standard J1772 connector
   b. the home-charging station, through the CHAdeMO connector
   c. power from the solar panel in the rear spoiler
   d. power from the vehicle battery
19. What is the purpose of the AC power inverter?
   a. to convert the battery's DC voltage to single-phase AC voltage to power the traction motor
   b. to convert the battery's DC voltage to three-phase AC voltage to power the traction motor
   c. to convert the battery's AC voltage to 12V DC to power the traction motor
   d. to convert the battery's AC voltage to 240V DC to power the traction motor
20. What is the purpose of the PEM in a fuel cell?
   a. to prevent the hydrogen protons from passing through to the oxygen and forming water
   b. to prevent the oxygen protons from passing through to the hydrogen and forming water
   c. to prevent the electrons from passing directly from the positive side of the fuel cell
   d. to prevent the electrons from passing directly to the positive side of the fuel cell

CHAPTER 38
Clutches

LEARNING OUTCOMES

- Describe the various clutch components and their functions.
- Name and explain the advantages of the different types of pressure plate assemblies.
- Name the different types of clutch linkages.
- List the safety precautions that should be followed during clutch servicing.
- Explain how to perform basic clutch maintenance.
- Name the most common problems that occur with clutches.
- Explain the basics of servicing a clutch assembly.

WE SUPPORT THE CANADIAN INTERPROVINCIAL STANDARDS PROGRAM

The clutch assembly is located between the transmission and engine, where it provides a mechanical coupling between the engine's flywheel and the transmission's input shaft. The driver operates the clutch through a linkage that extends from the passenger compartment to the **bell housing** (also called the clutch housing) between the engine and the transmission.

All manual transmissions require a clutch to engage or disengage engine power to the transmission smoothly and quietly. If the vehicle had no clutch and the engine was always connected to the transmission, the engine would stop every time the vehicle was brought to a stop. The clutch allows the engine to idle while the vehicle is stopped. It also allows for easy shifting between gears. (Of course, all of this applies to manual transaxles as well.)

The clutch engages the transmission gradually by allowing a certain amount of slippage between the transmission's input shaft and the flywheel. **Figure 38–1** shows the components needed to do this: the flywheel, clutch disc, pressure plate assembly, clutch release bearing (or throwout bearing), and clutch fork.

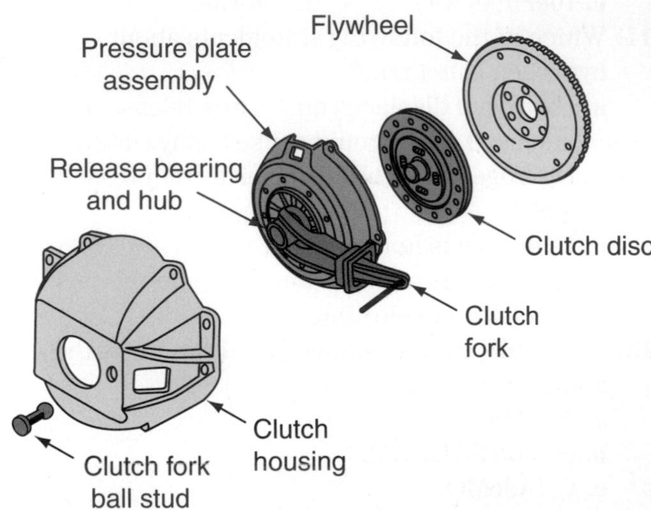

FIGURE 38–1
Major parts of a clutch assembly.

OPERATION

The basic principle of engaging a clutch is demonstrated in **Figure 38–2**. The flywheel and the pressure plate are the drive members of the clutch. The driven member connected to the transmission input shaft is the **clutch disc**, also called the **friction disc**. As long as the clutch is disengaged (clutch pedal depressed), the drive members turn independently of the driven member, and the engine is disconnected from the transmission. However, when the clutch is engaged (clutch pedal released), the pressure plate

WARNING!

Use the appropriate cleaning liquid and equipment before and during disassembling a clutch assembly. Some clutch discs may contain asbestos. Wearing a mask with HEPA filtration will prevent inhalation of asbestos and other clutch material dusts that can cause lung irritation and serious illness.

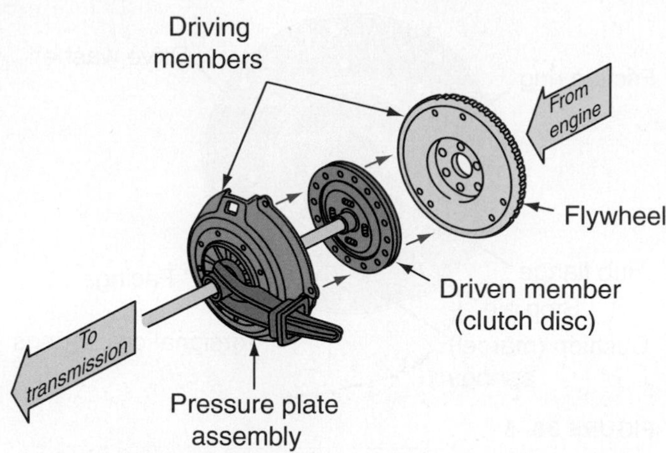

FIGURE 38–2

When the clutch is engaged, the driven member is squeezed between the two driving members. The transmission is connected to the driven member.

FIGURE 38–3

A typical flywheel mounted to the rear of an engine's crankshaft.

moves toward the flywheel, and the clutch disc is squeezed between the two revolving drive members and forced to turn at the same speed.

Torque Transfer

When the clutch is firmly clamped between the flywheel and pressure plate, engine torque moves into the transmission. A clutch assembly is designed to prevent any loss of torque when it is engaged, but it needs to allow some slippage as the disc begins to be squeezed between the pressure plate and the flywheel. This slippage prevents jarring that can result from the sudden transfer of power to the wheels.

Flywheel

The **flywheel**, an important part of the engine, is also the main driving member of the clutch **(Figure 38–3)**. It is normally made of nodular or grey cast iron, which has a high graphite content to lubricate the engagement of the clutch. Welded to or pressed onto the outside diameter of the flywheel is the starter ring gear. The starter ring gear is replaceable on most flywheels. The large diameter of the flywheel allows for an excellent gear ratio of the starter drive to ring gear, which provides for ample engine rotation during starting. The rear surface of the flywheel is a friction surface machined very flat to ensure smooth clutch engagement. The flywheel also provides some absorption of torsional vibration of the crankshaft. It further provides the inertia to rotate the crankshaft through the four strokes.

The flywheel has two sets of bolt holes drilled into it. The inner set is used to fasten the flywheel to the crankshaft, and the outer set is threaded to allow for mounting of the pressure plate assembly. A bore in the centre of the wheel and crankshaft holds the **pilot bushing** or bearing, which supports the front end of the transmission input shaft and maintains alignment with the engine's crankshaft. Sometimes, a ball or roller needle bearing is used instead of a pilot bushing. Some transaxles have a short, self-centring input shaft that does not require a pilot bushing or bearing.

DUAL-MASS FLYWHEEL A few cars and light trucks use a dual-mass flywheel. These flywheels are used to reduce vibrations transmitted through the transmission, provide for smoother shifting, and reduce gear noise. Dual-mass flywheels can reduce the oscillations of the crankshaft before they move through the transmission **(Figure 38–4)**. The flywheel consists of two rotating plates connected by a spring and damper system. The forward-most portion of the flywheel is bolted to the end of the crankshaft and smooths out the crankshaft's oscillations. The pressure plate of the clutch is bolted to the rearward portion of the flywheel. Engine torque moves from the front plate, through the damper and spring assembly, to the rear plate before it enters the transmission.

Some have a torque-limiting feature that prevents damage to the transmission during peak torque loads. The rotation of the two flywheel plates can differ by as much as 360°. This allows the forward plate to absorb torque spikes and not pass them along through the transmission.

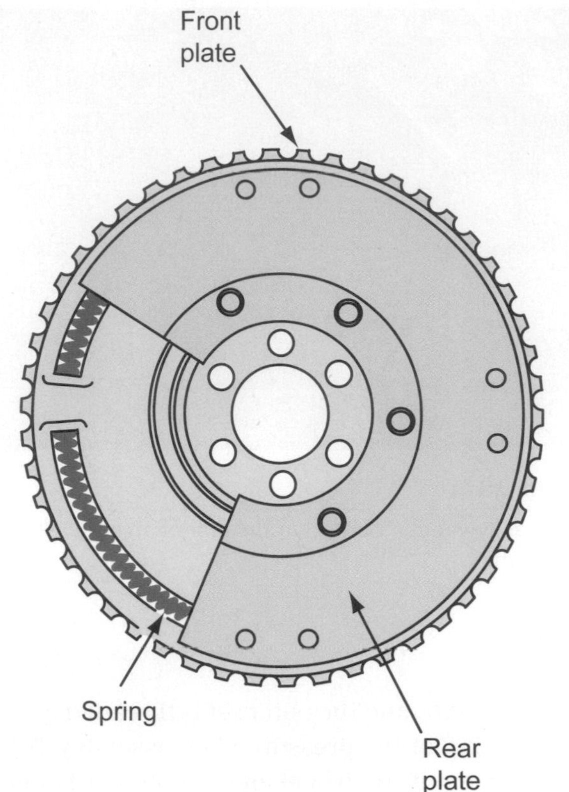

FIGURE 38–4

A dual-mass flywheel.

Clutch Disc

The clutch disc **(Figure 38–5)** receives the driving motion from the flywheel and pressure plate assembly and transmits that motion to the transmission input shaft. The parts of a clutch disc are shown in **Figure 38–6**.

FIGURE 38–5

A clutch disc.

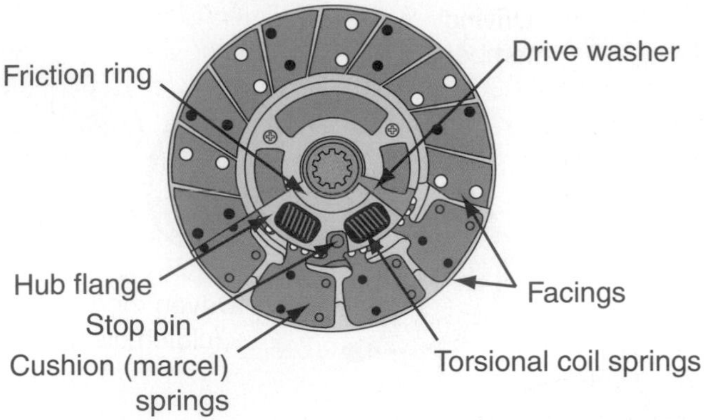

FIGURE 38–6

The major parts of a clutch disc.

The hub of the clutch plate has internal splines that fit over the external splines on the transmission's input shaft. As the clutch disc is engaged and disengaged, it slides back and forth on the splines. The clutch disc is designed to absorb such things as crankshaft vibration, abrupt clutch engagement, and driveline shock. The disc has a damper to reduce the torsional vibrations caused by the engine's power pulses. The damper is actually the disc hub and related springs. These torsional coil springs allow the disc to rotate slightly in relation to the pressure plate while they absorb the torque forces. The number and tension of these springs is determined by engine torque and vehicle weight. Stop pins limit this torsional movement to approximately 1 cm (⅜ in.).

CLUTCH FACINGS The facing of a disc is the frictional material that covers the steel clutch disc. The facing must be able to withstand the heat generated by the friction between the disc and the pressure plate. When the facing becomes overheated, clutch slippage can result, followed by more heat. Overheated clutch discs do not have a long service life.

There are two types of facings. Moulded friction facings can withstand greater pressure-plate loading force without damage. Woven friction facings are used when additional cushioning action is needed for clutch engagement. Until recently, the material that was moulded or woven into facings was predominantly asbestos. Now, because of the hazards associated with asbestos, usually other materials, such as paper-base and ceramics, are being used instead. Particles of cotton, brass, rope, and wire are added to prolong the life of the clutch disc and provide torsional strength.

The amount of torque that can be transferred through the clutch disc mainly depends on how tightly the disc is squeezed, the facing's coefficient of friction, and the diameter of the driven disc. Basically, the coefficient of friction describes the amount of friction there is between two surfaces. The overall efficiency of a clutch assembly depends on its coefficient of friction. If a clutch's coefficient of friction is lower than desired, the clutch will slip. If the coefficient is higher than desired, the clutch will experience grabbing.

Grooves are cut across the face of the friction facings. This promotes clean disengagement of the driven disc from the flywheel and pressure plate; it also promotes better cooling. The facings are riveted to wave springs, also called marcel or **cushioning springs**, which cause the contact pressure on the facings to rise gradually as the springs flatten out when the clutch is engaged. These springs eliminate chatter by dampening the clutch engagement and also reduce the chance of the clutch disc sticking to the flywheel and pressure-plate surfaces when the clutch is disengaged. The wave springs and friction facings are fastened to the centre steel disc.

Retaining Plate

The retaining plate of a clutch disc connects the disc's hub to the disc's facing. When the clutch is engaged, the facing and hub are squeezed together, the same way the friction disc is squeezed against the flywheel. The marcel springs between the facing and the hub are compressed when the clutch is engaged and help to disengage the clutch when the pedal is depressed. The purpose of the marcel is to make engagement and disengagement as smooth as possible.

Pressure-Plate Assembly

A pressure-plate assembly is composed of a pressure ring (plate), a cover, pressure springs, and release levers. The purpose of the assembly **(Figure 38–7)** is twofold. First, it must provide the necessary clamping force to squeeze the clutch disc onto the flywheel with sufficient force to create enough static friction to transmit engine torque efficiently. Second, it must move away from the clutch disc so that the clutch disc can stop rotating when the clutch is disengaged, even though the flywheel and pressure plate continue to rotate.

The pressure ring is a flat, heavy iron ring that moves against the outside diameter of the clutch disc to squeeze it against the flywheel. The cover of a pressure-plate assembly is normally a stamped steel housing that serves as the mounting point for the pressure plate, pressure springs, and release levers. The cover bolts to the flywheel.

FIGURE 38–7

A clutch pressure plate.

Basically, there are two types of pressure plate assemblies: **coil spring pressure plate assembly** and **diaphragm spring pressure plate assembly**.

COIL-SPRING PRESSURE PLATE ASSEMBLY A coil-spring pressure plate assembly, shown in **Figure 38–8**, uses coil springs and release levers to move the pressure plate back and forth through its engagement and disengagement travel. The springs exert pressure to hold the pressure plate tightly against the clutch disc and flywheel. The release levers release the holding force of the springs. There are usually three release levers, placed 120° apart. Each one has two pivot points. One of these pivot points attaches the lever to a pedestal cast into the pressure plate, and the other attaches to a release lever

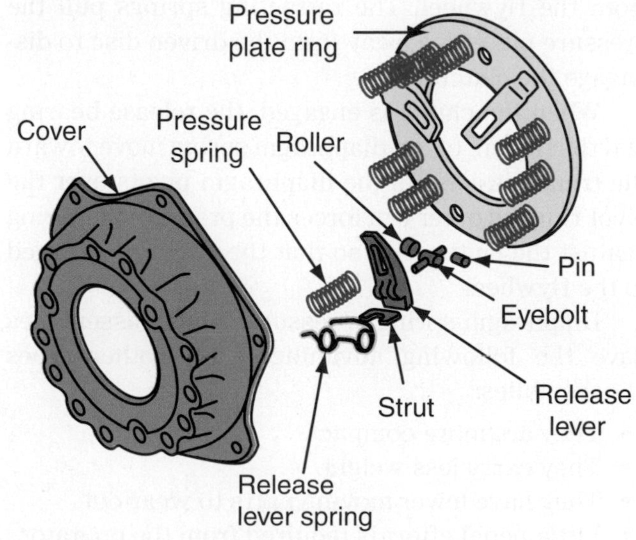

FIGURE 38–8

Parts of a coil-spring pressure plate.

yoke/keybolt that is bolted to the cover. The levers pivot on the pedestals and release lever yokes to move the pressure plate ring through its engagement and disengagement operations. To disengage the clutch, the release bearing pushes the inner ends of the release levers forward toward the flywheel. The release lever yokes act as fulcrums for the levers, and the outer ends of the release levers move backward, pulling the pressure plate ring away from the clutch disc. This action compresses the coil springs and disengages the clutch disc from the driving members.

When the clutch is engaged, the release bearing moves backward toward the transmission/transaxle. Without this force against the release levers, the coil springs are able to push the pressure plate ring and clutch disc against the flywheel with sufficient force to resist slipping.

DIAPHRAGM-SPRING PRESSURE-PLATE ASSEMBLY The diaphragm-spring pressure plate assembly relies on a cone-shaped diaphragm spring between the pressure plate ring and the pressure plate cover to move the pressure plate ring back and forth. The diaphragm spring (sometimes called a Belleville spring) is a single, thin sheet of spring steel that works in the same manner as the bottom of an oil can. The metal yields when pressure is applied to it. When the pressure is removed, the metal springs back to its original shape. The centre portion of the diaphragm spring is slit into numerous fingers that act as release levers **(Figure 38–9)**.

During clutch disengagement, these fingers are moved forward by the release bearing. The diaphragm spring pivots over the fulcrum ring (also called the pivot ring), and its outer rim moves away from the flywheel. The retracting springs pull the pressure plate ring away from the driven disc to disengage the clutch.

When the clutch is engaged, the release bearing and the fingers of the diaphragm spring move toward the transmission. As the diaphragm pivots over the pivot ring, its outer rim forces the pressure plate ring against the clutch disc so that the clutch is engaged to the flywheel.

Diaphragm-spring pressure plate assemblies have the following advantages over other types of assemblies:

- They are more compact.
- They carry less weight.
- They have fewer moving parts to wear out.
- Little pedal effort is required from the operator.
- They provide a balanced force around the pressure plate so that rotational unbalance is reduced.

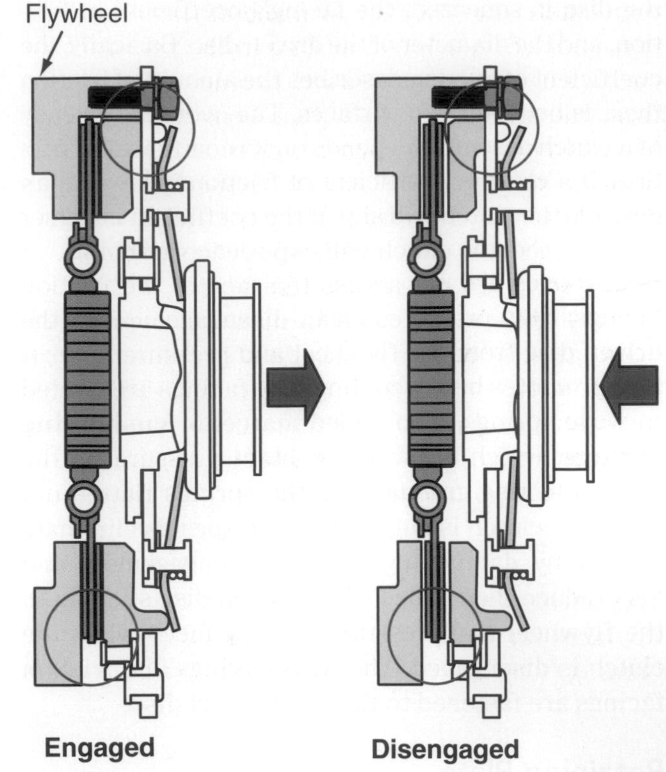

FIGURE 38–9

The action of a diaphragm spring–type pressure plate assembly.

- Clutch disc slippage is less likely to occur. Mileage builds because the force holding the clutch disc to the flywheel does not change throughout its service life.

Pilot Bushing/Bearing

A pilot bushing or bearing **(Figure 38–10)** is sometimes used to support the outer end of the transmission's input shaft. This pilot is normally pressed into a bore at the centre of the outer end of the engine's

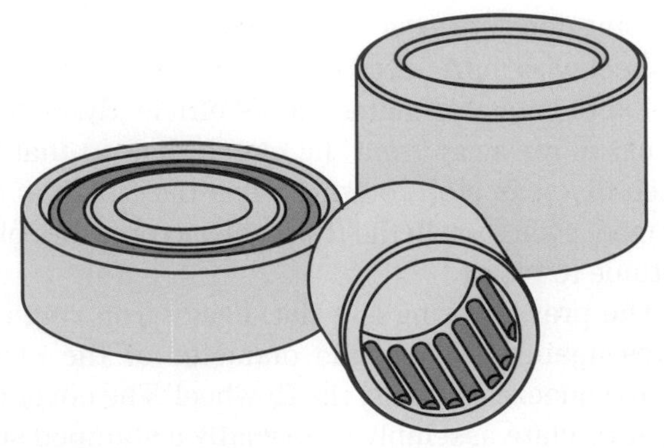

FIGURE 38–10

Different designs of pilot bushings and bearings used in today's clutch assemblies.

crankshaft. The transmission end of the input shaft is supported by a large bearing in the transmission case. Because the input shaft extends unsupported from the transmission, a pilot bushing is used to keep it in position. By supporting the shaft, the pilot bushing keeps the clutch disc centred in the pressure plate.

The pilot bushing can be made of sintered bronze saturated in oil during manufacturing. Pilot bearings are normally needle bearings but can be roller bearings or sealed bearing assemblies. Because the length of the input shaft is short and because the input shaft is supported internally by more than one bearing, there is no need for a pilot bearing or bushing with many transaxles.

Clutch Release Bearing

The **clutch release bearing**, also called a **throwout bearing**, is usually a sealed, prelubricated ball bearing **(Figure 38–11)**. Its function is to smoothly and quietly move the pressure plate release levers or diaphragm spring through the engagement and disengagement travel.

The release bearing is mounted on a hub, which slides on a hollow tube at the front of the transmission housing. This hollow tube, shown in **Figure 38–12**, is part of the transmission's front bearing retainer and fits over the transmission clutch (input) shaft **(Figure 38–13)**.

To disengage the clutch, the release bearing is moved forward on the transmission's front bearing retainer by the **clutch fork**. As the release bearing contacts the release levers or diaphragm spring of the pressure plate assembly, it begins to rotate with the rotating pressure plate assembly. As the release bearing continues forward, the clutch disc is disengaged from the pressure plate and flywheel.

To engage the clutch, the release bearing slides to the rear of the bearing retainer. The pressure plate ring moves forward and traps the clutch disc

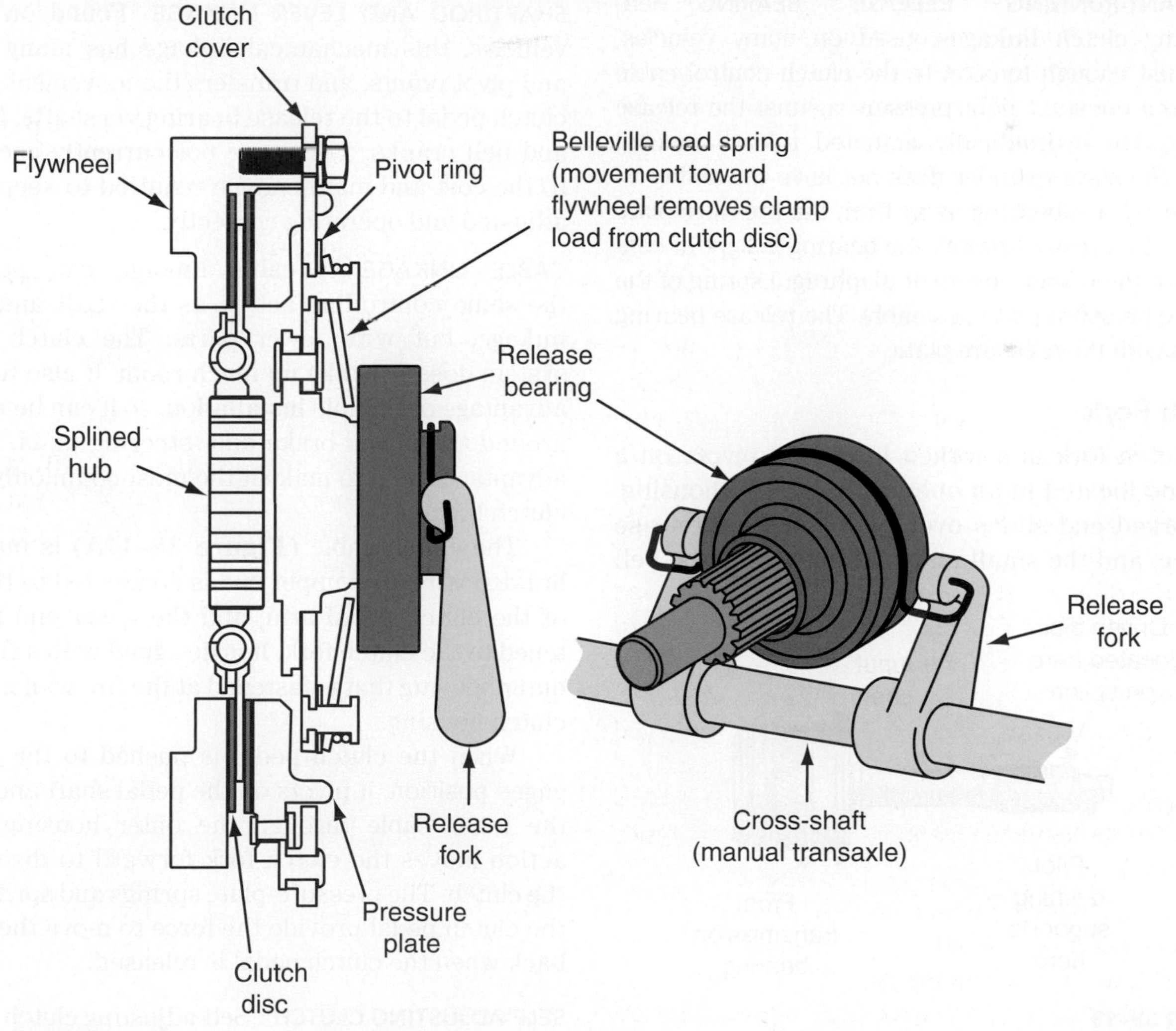

FIGURE 38–11
The clutch fork and throwout bearing location in the bell housing.

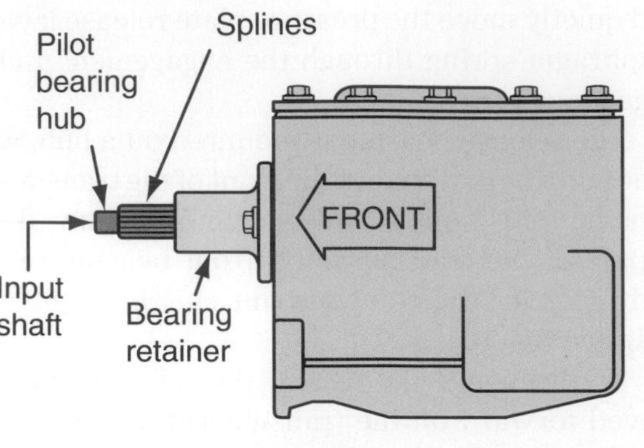

FIGURE 38–12

The release bearing slides on the hollow tube section of the transmission's front bearing retainer.

against the flywheel to transmit engine torque to the transmission input shaft. Depending on the release bearing design, with the clutch fully engaged, the release bearing may be stationary (non-constant running) or rotating (constant running).

CONSTANT-RUNNING RELEASE BEARING Self-adjusting clutch linkages, used on many vehicles, apply just enough tension to the clutch control cable to keep a constant light pressure against the release bearing. On hydraulically actuated release mechanisms, the slave cylinder does not have the ability to pull the release bearing away from the pressure-plate fingers. As a result, the release bearing is kept in contact with the release levers or diaphragm spring of the rotating pressure plate assembly. The release bearing rotates with the pressure plate.

Clutch Fork

The clutch fork is a forked lever that pivots on a ball stud located in an opening in the bell housing. The forked end slides over the hub of the release bearing, and the small end protrudes from the bell housing and connects to the clutch linkage and clutch pedal. The clutch fork moves the release bearing and hub back and forth during engagement and disengagement.

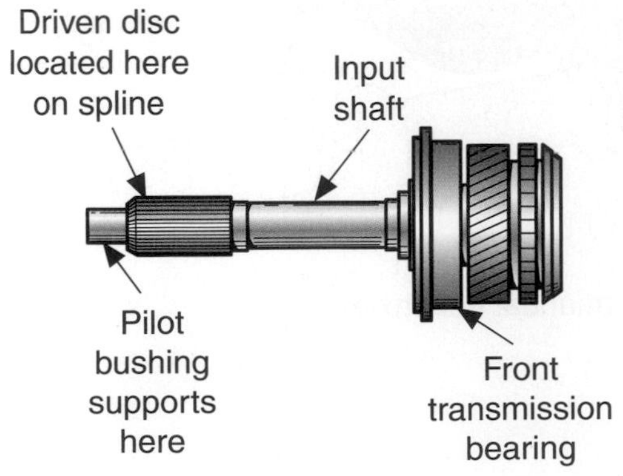

FIGURE 38–13

A clutch (input) shaft.

Clutch Linkage

The clutch linkage is a series of parts that connects the clutch pedal to the clutch fork. It is through the clutch linkage that the operator controls the engagement and disengagement of the clutch assembly smoothly, quietly, and with little effort.

On some vehicles when the clutch is engaged, the clutch linkage pulls the release bearing a small amount away from the release levers in the pressure plate. The slight amount the release bearing is moved away from the pressure plate defines the free play of the clutch pedal. If the vehicle has a self-adjusting mechanical linkage or hydraulic linkage, there is always slight contact between the release bearing and the pressure plate.

SHAFT/ROD AND LEVER LINKAGE Found on older vehicles, this mechanical linkage has many parts and pivot points, and transfers the movement of the clutch pedal to the release bearing via shafts, levers, and bell cranks. These are not currently used due to the cost and maintenance required to keep them adjusted and operating correctly.

CABLE LINKAGE A cable linkage can perform the same controlling action as the shaft and lever linkage, but with fewer parts. The clutch cable system does not take up much room. It also has the advantage of flexible installation, so it can be routed around the power-brake and -steering units. These advantages help to make it the most commonly used clutch linkage.

The clutch cable **(Figure 38–14A)** is made of braided wire. The upper end is connected to the top of the clutch pedal arm, and the lower end is fastened to the clutch fork. It is designed with a flexible outer housing that is fastened at the fire wall and the clutch housing.

When the clutch pedal is pushed to the disengaged position, it pivots on the pedal shaft and pulls the inner cable through the outer housing. This action moves the clutch fork forward to disengage the clutch. The pressure-plate springs and springs on the clutch pedal provide the force to move the cable back when the clutch pedal is released.

SELF-ADJUSTING CLUTCH Self-adjusting clutch mechanisms monitor clutch pedal play and automatically adjust it when necessary.

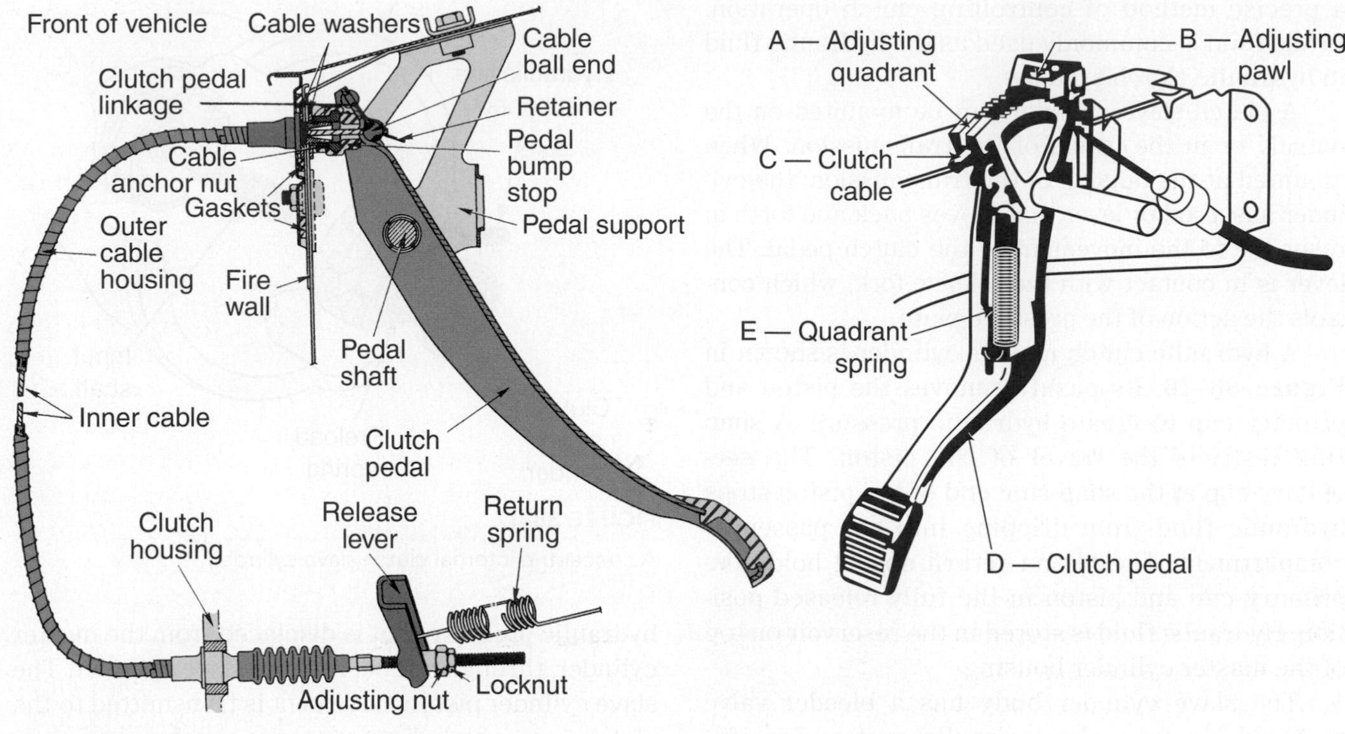

FIGURE 38–14

(A) A typical clutch cable system. (B) A self-adjusting clutch mechanism.

Usually, the self-adjusting clutch mechanism **(Figure 38–14B)** is a ratcheting mechanism located at the top of the clutch pedal behind the dash panel. The ratchet is designed with a pawl and toothed segment, and a pawl tension spring is used to keep the pawl in contact with the toothed segment. The pawl allows the toothed segment to move in only one direction in relation to the pawl.

The clutch cable is guided around and fastened to the toothed segment, which is free to rotate in one direction (backward) independently of the clutch pedal. The tension spring pulls the toothed segment backward.

When the clutch cable develops slack due to stretching and clutch disc wear, the cable is adjusted automatically when the clutch is released. The tension spring pulls the toothed segment backward and allows the pawl to ride over to the next tooth. This effectively shortens the cable. Actually, the cable is not really shortened, but the slack has been reeled in by the repositioning of the toothed segment. This self-adjusting action takes place automatically during the clutch's operational life.

HYDRAULIC CLUTCH LINKAGE Frequently, the clutch assembly is controlled by a hydraulic system **(Figure 38–15)**. In the hydraulic clutch linkage system, hydraulic (liquid) pressure transmits motion from one sealed cylinder to another through a hydraulic line. Like the cable linkage assembly, the hydraulic linkage is compact and flexible. Hydraulic linkages also allow engineers to place the release fork anywhere that gives them more flexibility in body design. In addition, the

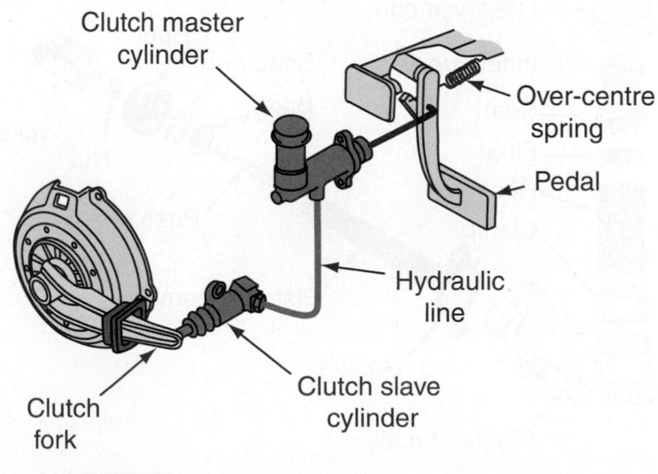

FIGURE 38–15

A typical hydraulic clutch linkage arrangement.

hydraulic pressure developed by the master cylinder decreases required pedal effort and provides a precise method of controlling clutch operation. Brake fluid is commonly used as the hydraulic fluid in hydraulic clutch systems.

A clutch slave cylinder can be mounted on the outside or in the inside of the transmission. When mounted at the outside of the transmission, the cylinder has a rod or lever that moves back and forth in response to the movement of the clutch pedal. The lever is in contact with the release fork, which controls the action of the pressure plate.

A hydraulic clutch master cylinder is shown in **Figure 38–16**. Its pushrod moves the piston and primary cup to create hydraulic pressure. A snap ring restricts the travel of the piston. The secondary cup at the snap-ring end of the piston stops hydraulic fluid from dripping into the passenger compartment. The piston return spring holds the primary cup and piston in the fully released position. Hydraulic fluid is stored in the reservoir on top of the master cylinder housing.

The slave cylinder body has a bleeder valve to bleed air from the hydraulic system for efficient clutch linkage operation. The cylinder body is threaded for a tube and fitting at the fluid entry port. Rubber seal rings are used to seal the hydraulic pressure between the piston and the slave cylinder walls. The piston retaining ring is used to restrict piston travel to a certain distance. Piston travel is transmitted by a pushrod to the clutch fork. The pushrod boot keeps contaminants out of the slave cylinder.

When the clutch pedal is depressed, the movement of the piston and primary cup develops hydraulic pressure that is displaced from the master cylinder, through a tube, into the slave cylinder. The slave cylinder piston movement is transmitted to the clutch fork, which disengages the clutch.

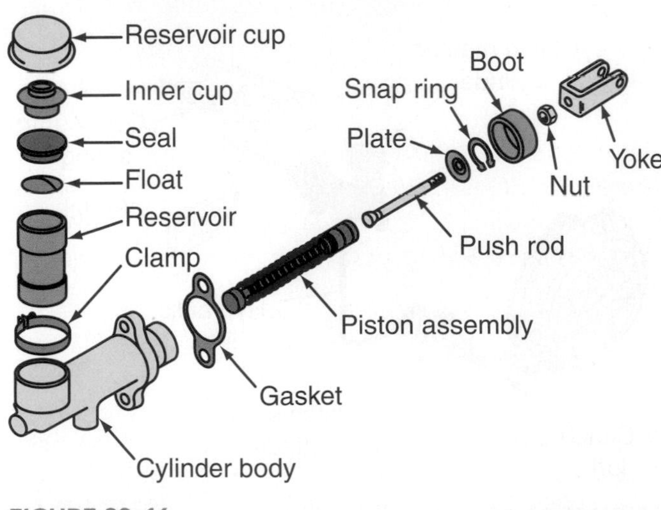

FIGURE 38–16

Parts of a hydraulic clutch master cylinder.

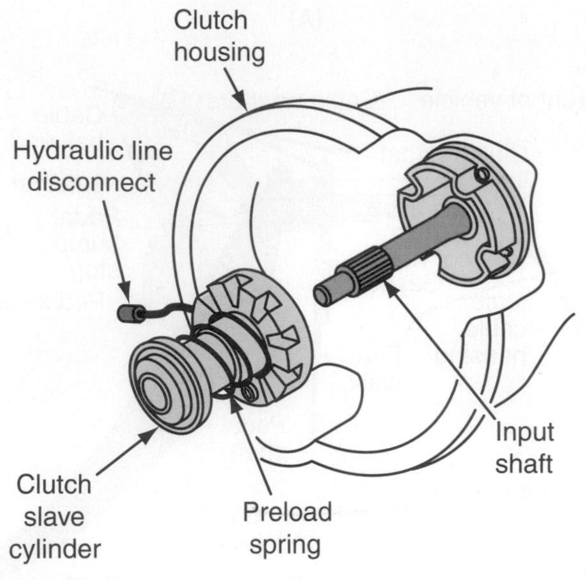

FIGURE 38–17

A concentric internal clutch slave cylinder.

When the clutch pedal is released, the primary cup and piston are forced back to the disengaged position by the master cylinder piston return spring. External springs move the slave cylinder pushrod and piston back to the engaged position. Fluid pressure returns through the hydraulic tubing to the master cylinder assembly. There is no hydraulic pressure in the system when the clutch assembly is in the engaged position.

CONCENTRIC SLAVE CYLINDERS A concentric (internal) slave cylinder is found on some cars and light trucks. These units are actually a combination of the slave cylinder and the clutch release bearing **(Figure 38–17)**. These slave cylinders may be made of plastic, aluminum, or cast iron. They also are part of a hydraulic clutch system and are controlled by pressurized mineral oil or brake fluid.

Having the slave cylinder directly behind the release bearing eliminates the need for actuating rods, levers, or cables. The movement of the release bearing is linear.

An internal slave cylinder is a doughnut-shaped unit that mounts to the front of the transmission, and the transmission's input shaft passes through it. The slave cylinder is either bolted to the transmission's front bearing cover or is held by a pressed pin **(Figure 38–18)**. A concentric slave cylinder reduces the weight of the clutch linkage assembly

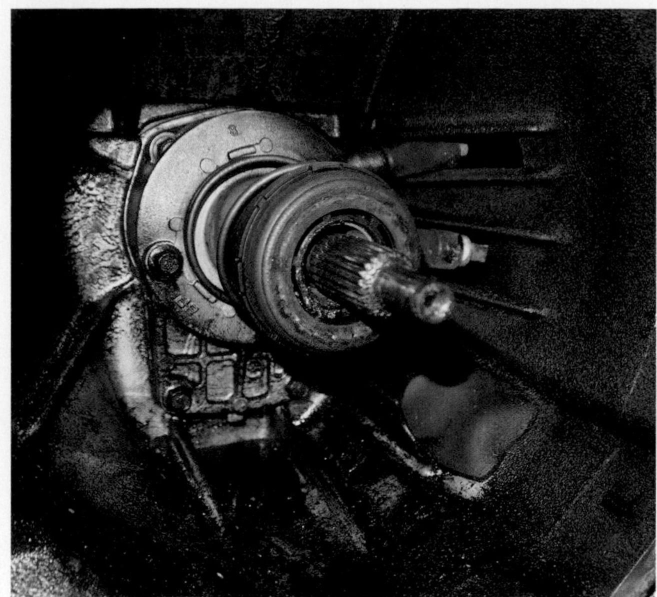

FIGURE 38–18

A concentric slave cylinder bolted to the transmission's front bearing cover.

and extends the assembly's service life because it has fewer parts. However, in most cases, the transmission must be removed to properly diagnose and service this type of slave cylinder.

CLUTCH PEDAL SWITCH Nearly all vehicles have a clutch pedal switch. This switch is used to let the PCM know the clutch pedal is depressed to allow the engine to start and to cancel cruise control operation **(Figure 38–19)**. The switch is a normally open switch that closes when the clutch pedal is fully depressed. If the switch is adjustable, it is important

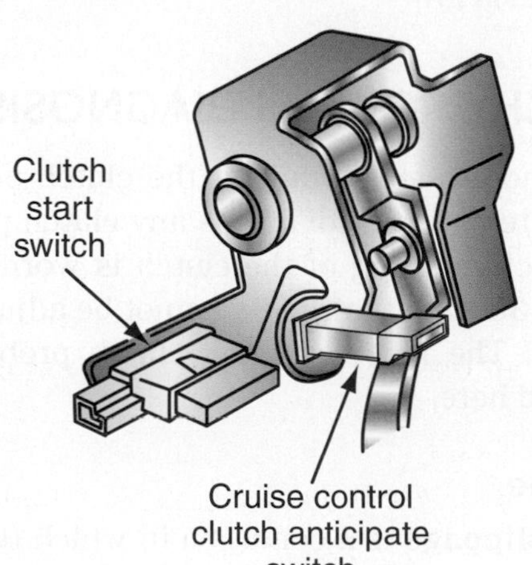

FIGURE 38–19

Various clutch switches are used to control engine starting and cruise control operation.

to make sure the adjustment is correct after any linkage adjustment has been made.

CLUTCH SERVICE SAFETY PRECAUTIONS

When servicing the clutch, exercise the following precautions:

- Always wear eye protection when working underneath a vehicle. Wear a mask with HEPA filtration to avoid breathing clutch disc–material dust.
- Remove asbestos dust only with a special, approved vacuum-collection system or an approved liquid-cleaning system.
- Never use compressed air or a brush to clean off asbestos dust.
- Follow all federal, provincial, and local laws when disposing of collected asbestos dust or liquid containing asbestos dust.
- Never work under a vehicle that is not raised on a hoist or supported by safety or jack stands.
- Use jack stands and special jacks to support the engine and transmission.
- Have a helper assist in removing the transmission.
- Be sure the work area is properly ventilated, or attach a ventilating hose to the vehicle's exhaust system when an engine is to be run indoors.
- Do not allow anyone to stand in front of or behind the automobile while the engine is running.
- Set the emergency brake securely, and place the gearshift in the NEUTRAL position when running the engine of a stationary vehicle.
- Avoid touching the hot engine and exhaust system parts. Whenever possible, let the vehicle cool down before beginning to work on it.

CLUTCH MAINTENANCE

All clutches require routine inspection. Older external, mechanical linkage systems may require lubrication and adjustment at regular intervals. These maintenance procedures are explained in this section.

USING SERVICE INFORMATION

Service information should include adjustment procedures and instructions for clutch removal, inspection, installation, and troubleshooting. It may also offer information to aid in clutch release-bearing distress analysis.

Clutch Fluid Level

The fluid for a hydraulic clutch system is checked by looking at the fluid's level in the master cylinder's reservoir. Most reservoirs have a mark to indicate the proper level. If there are no marks, check the service information. Normally, the desired level is 6 to 9 mm (¼ to ½ in.) from the top.

Clutch Linkage Adjustment

Except for systems with self-adjusting mechanisms, the release bearing should not touch the pressure-plate release levers when the clutch is engaged (pedal up). Clearance between these parts prevents premature wear on the clutch plate, pressure plate, and release bearing. As the clutch disc wears and becomes thinner, this clearance is reduced.

Clearance can be ensured by adjusting the clutch cable so that the pedal has a specified amount of play, or **free travel**. Free travel is the distance a clutch pedal moves when depressed before the release bearing contacts the clutch release lever or diaphragm spring of the pressure plate. Insufficient free travel can prevent the clutch from completely engaging, and excessive play may not allow the clutch to completely disengage.

To check pedal play, use a tape measure or ruler. Place the tape measure or ruler beside the clutch pedal and the end against the floor of the vehicle, and note the reading **(Figure 38–20)**. Then depress the clutch pedal just enough to take up the pedal play, and note the reading again. The difference between the two readings is the amount of pedal play.

To adjust clutch pedal play, refer to the manufacturer's service information for the correct procedure and adjustment points. Often, pedal play can be increased or decreased by turning a threaded fastener located either under the dash at the clutch pedal or where the cable attaches to the clutch fork.

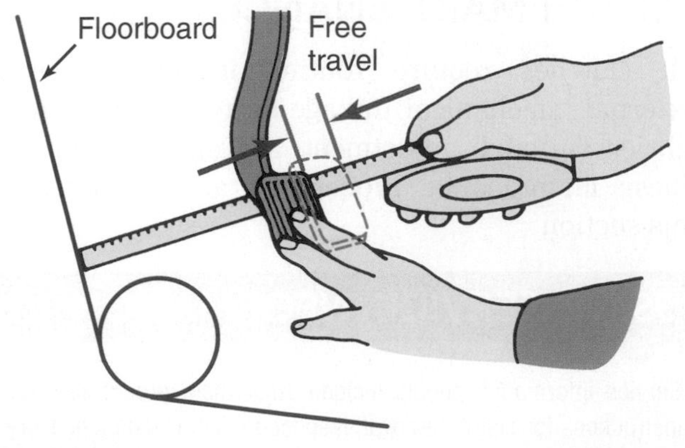

FIGURE 38–20
Checking clutch pedal play.

SHOP TALK

Usually, a properly adjusted clutch will have a predetermined amount of clutch pedal free play. Depending on the type of clutch, the free play may decrease as the clutch disc wears.

Clean the linkage with a shop towel and solvent, if necessary, before checking it and replacing any damaged or missing parts or cables. Check hydraulic linkage systems for leaks at the clutch master cylinder, hydraulic hose, and slave cylinder.

External Clutch Linkage Lubrication

Cables and older external clutch linkages should be lubricated at regular intervals, such as during chassis lubrication. Always use the type of lubricant recommended by the manufacturer. Many clutch linkages use the same chassis grease used for suspension parts and U-joints. Lubricate all the sliding surfaces and pivot points in the clutch linkage **(Figure 38–21)**. The linkage should move freely after lubrication.

On vehicles with hydraulic clutch linkage, check the clutch master cylinder reservoir fluid level. It should be approximately 6.35 mm (¼ in.) from the top of the reservoir. If it must be refilled, use approved brake fluid. Also, because the clutch master cylinder does not consume fluid, check for leaks in the master cylinder, connecting flexible line, and slave cylinder, if the fluid is low.

CLUTCH PROBLEM DIAGNOSIS

Check and attempt to adjust the clutch pedal play before attempting to diagnose any clutch problems. If the friction lining of the clutch is worn too thin **(Figure 38–22)**, the clutch cannot be adjusted successfully. The most common clutch problems are described here.

Slippage

Clutch slippage is a condition in which the engine rpm increases with the clutch engaged without generating any rpm increase to the transmission input shaft. It occurs when the clutch disc is not gripped firmly between the flywheel and the pressure plate. Instead, the clutch disc slips between these driving

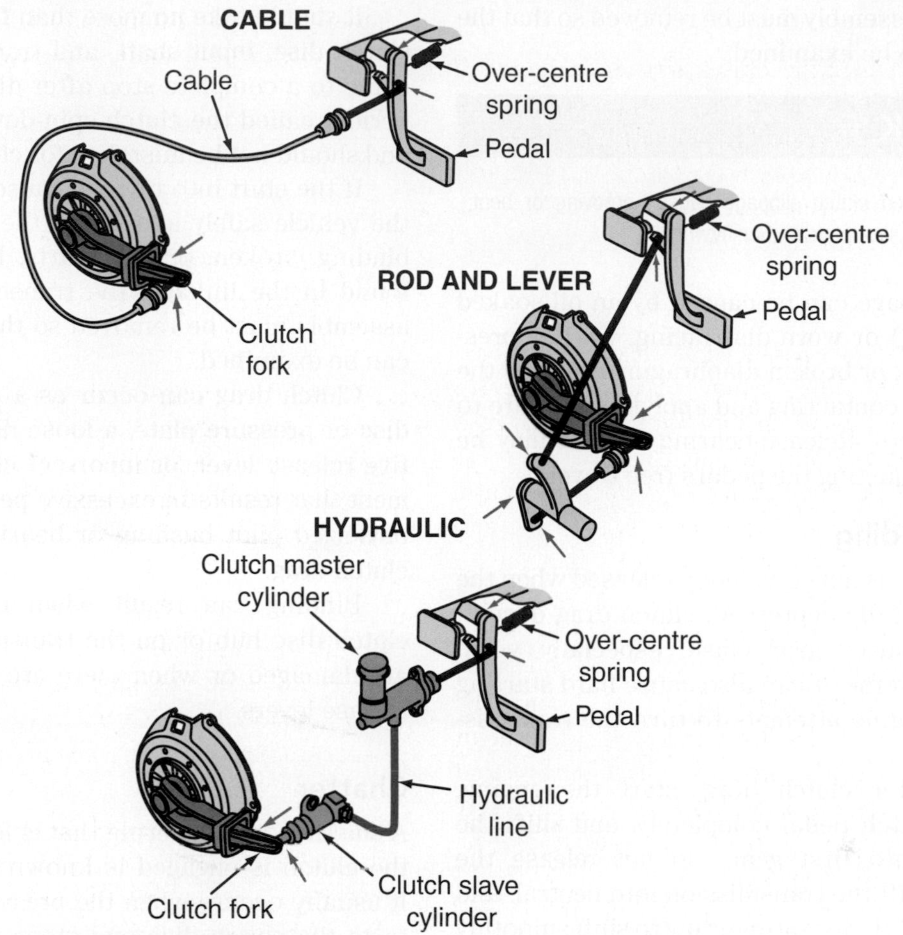

FIGURE 38–21

Clutch linkage lubrication points.

Nissan Canada

FIGURE 38–22

A severely worn clutch disc.

Schaeffler Group USA Inc.

members. Slippage can occur during initial acceleration or subsequent shifts, but it is usually most noticeable in higher gears.

One way to check for slippage is by driving the vehicle. Normal acceleration from a stop and several gear changes indicate whether the clutch is slipping.

Slippage also can be checked in the shop. Check the service information for the correct procedures. A general procedure for checking clutch slippage follows. Be sure to follow the safety precautions stated earlier.

With the parking brake on, disengage the clutch. Shift the transmission into third gear, and increase the engine speed to about 2000 rpm. Slowly release the clutch pedal until the clutch engages. The engine should stall immediately.

If it does not stall within a few seconds, the clutch is slipping. Safely raise the vehicle and check the clutch linkage for binding and broken or bent parts. If no linkage problems are found, the transmission and the clutch assembly must be removed so that the clutch parts can be examined.

SHOP TALK

Severe or prolonged clutch slippage causes grooving or heat damage to the pressure plate and/or flywheel.

Clutch slippage can be caused by an oil-soaked **(Figure 38–23)** or worn disc facing, warped pressure plate, weak or broken diaphragm spring, or the release bearing contacting and applying pressure to the release levers. Release-bearing contact may be corrected by adjusting the pedal's free travel.

Drag and Binding

If the clutch disc is not completely released when the clutch pedal is fully depressed, clutch drag occurs. Clutch drag causes gear clash, especially when shifting into reverse. It can also cause hard starting because the engine attempts to turn the transmission input shaft.

To check for clutch drag, start the engine, depress the clutch pedal completely, and shift the transmission into first gear. Do not release the clutch. Then shift the transmission into neutral, and wait five seconds before attempting to shift smoothly into reverse.

FIGURE 38–23
Grease and oil on the hub of this disc indicates that the disc may be contaminated with oil.
Schaeffler Group USA Inc.

It should take no more than five seconds for the clutch disc, input shaft, and transmission gears to come to a complete stop after disengagement. This period, called the clutch spin-down time, is normal and should not be mistaken for clutch drag.

If the shift into reverse causes gear clash, raise the vehicle safely and check the clutch linkage for binding, broken, or bent parts. If no problems are found in the linkage, the transmission and clutch assembly must be removed so that the clutch parts can be examined.

Clutch drag can occur as a result of a warped disc or pressure plate, a loose disc facing, a defective release lever, or incorrect clutch pedal adjustment that results in excessive pedal play. A binding or seized pilot bushing or bearing can also cause clutch drag.

Binding can result when the splines in the clutch disc hub or on the transmission input shaft are damaged or when there are problems with the release levers.

Chatter

A shaking or shuddering that is felt in the vehicle as the clutch is engaged is known as clutch **chatter**. It usually occurs when the pressure plate first contacts the clutch disc and stops when the clutch is fully engaged.

To check for clutch chatter, start the engine, depress the clutch completely, and shift the transmission into first gear. Increase engine speed to about 1500 rpm, and then slowly release the clutch pedal and check for chatter as the clutch begins to engage. Do not release the pedal completely, or the vehicle might jump and cause serious injury. As soon as the clutch is partially engaged, depress the clutch pedal immediately, and reduce engine speed to prevent damage to the clutch parts.

Usually, clutch chatter is caused by liquid leaking onto the clutch and contaminating its friction surfaces. This results in a mirror-like shine on the pressure plate or a glazed clutch facing. Oil and clutch hydraulic fluid leaks can occur at the engine rear main bearing seal, transmission input shaft seal, clutch slave cylinder, and hydraulic line. Other causes of clutch chatter include broken engine mounts, loose bell-housing bolts, and damaged clutch linkage.

During disassembly, check for a warped pressure plate or flywheel, hot spots on the flywheel, a burned or glazed disc facing, and worn input shaft splines. If the chattering is caused by an oil-soaked clutch disc and no other parts are damaged, then the

FIGURE 38–24

The marks on the surface of this pressure plate are caused by clutch chatter.

Schaeffler Group USA Inc.

FIGURE 38–25

This pressure plate shows that the release bearing is not evenly contacting the pressure plate.

Schaeffler Group USA Inc.

disc alone needs to be replaced. However, the cause of the oil leak must also be found and corrected.

Clutch chatter can also be caused by broken or weak torsional coil springs in the clutch disc **(Figure 38–24)** and by the failure to resurface the flywheel when a new clutch disc and/or pressure plate is installed. It is highly recommended that the flywheel be resurfaced every time a new clutch disc or pressure plate is installed. This will usually eliminate flywheel lateral runout, taper, grooving, and any hot spots that may exist.

Pedal Pulsation

Pedal pulsation is a rapid up-and-down movement of the clutch pedal as the clutch disengages or engages. This pedal movement usually is minor, but it can be felt through the clutch pedal. It is not accompanied by any noise. Pulsation begins when the release bearing makes contact with the release levers.

To check for pedal pulsation, start the engine, depress the clutch pedal slowly until the clutch just begins to disengage, and then stop briefly. Resume depressing the clutch pedal slowly until the pedal is depressed to a full stop.

On many vehicles, minor pulsation is considered normal. If pulsation is excessive, the clutch must be removed and disassembled for inspection.

Pedal pulsations can result from the misalignment of parts. Check for a misaligned bell housing or a bent flywheel. Inspect the clutch disc and pressure plate for warpage. Broken, bent, or warped release levers also create misalignment **(Figure 38–25)**.

CUSTOMER CARE

If you repair a vehicle with clutch slippage, tactfully inform the customer about the different poor-driving habits that can cause this problem. These habits include riding the clutch pedal and holding the vehicle on an incline by using the clutch as a brake.

Vibration

Clutch vibrations, unlike pedal pulsations, can be felt throughout the vehicle, and they occur at any clutch pedal position. These vibrations usually occur at normal engine operating speeds (more than 1500 rpm).

There are several possible sources of vibration that should be checked before disassembling the clutch to inspect it. Check the engine mounts and the crankshaft damper pulley. Look for any indication that engine parts are rubbing against the body or frame.

Accessories can also be a source of vibration. To check them, remove the drive belts one at a time.

Set the transmission in neutral and securely set the emergency brake. Start the engine and check for vibrations. Do not run the engine for more than one minute with the belts removed.

If the source of vibration is not discovered through these checks, the clutch parts should be examined. Be sure to check for loose flywheel bolts, excessive flywheel runout, and pressure-plate cover balance problems.

Noises

Many clutch noises come from bushings and bearings. Pilot bushing noises are squealing, howling, or trumpeting sounds that are most noticeable in cold weather. These bushing noises usually occur when the vehicle is stopped with the clutch pedal depressed and the transmission is in gear (as it would be at a stoplight). An rpm difference between the flywheel and transmission input shaft must occur to generate pilot bushing/bearing noise. Release bearing noise is a whirring, grating, or grinding sound that occurs when the clutch pedal is being depressed and stops when the pedal is fully released on a noncontact bearing. A constant-running release bearing is slightly different as the noise will lessen only when the clutch pedal is fully released. It is most noticeable when the transmission is in neutral, but it also can be heard when the transmission is in gear.

Hydraulic Clutch Diagnosis

Diagnostics of a hydraulic clutch system should begin with an inspection of the fluid. Check the fluid and reservoir for dirt and contamination. Foreign matter in the fluid will destroy the seals and wear grooves in the master and slave cylinders' bores.

A soft clutch pedal, excessive pedal travel, or a clutch that fails to release when the pedal is depressed can be caused by low fluid in the reservoir. To correct this problem, refill the reservoir to the correct level, and then bleed the system. This problem can also be caused by a faulty or damaged primary or secondary seal in the master cylinder. A leaking secondary seal will be evident by external leaks, whereas a primary seal leak will be internal. To correct either of these problems, replace or rebuild the master cylinder, and then refill and bleed the system. A slave cylinder with leaking inner hydraulic seals should be replaced and the system refilled with clean fluid and then bled.

If there is an extremely hard pedal, check the pedal mechanism and release fork for binding. If there is evidence of binding, repair and lubricate the assembly to ensure free movement. A hard pedal can also be caused by a blocked compensation port in the master cylinder. The port may be blocked by improper pushrod adjustments or because the piston is binding in the master cylinder bore. If the piston is binding, the master cylinder should be replaced or rebuilt, and the hydraulic system flushed, refilled, and bled. This problem may be also caused by swollen cup seals or contamination in the master or slave cylinders. If this is the problem, the master or slave cylinder should be replaced and the system flushed, refilled, and bled. Restricted hydraulic lines can also cause a hard pedal.

Restricted lines can also prevent the clutch from being totally engaged. The residual pressure will keep the release bearing in contact with the pressure plate. This problem will also cause wear on the pressure-plate fingers and the release bearing. Restricted lines should be replaced and the system flushed to remove all debris.

If the clutch does not fully engage when the pedal is released, check the pedal and release assemblies for binding or improper adjustment. A swollen primary cup in the master cylinder can also cause this problem. Swollen cups are caused by fluid contamination. This is typically the result of automatic transmission fluid (ATF) or power-steering fluid being added to the fluid reservoir, instead of DOT-3 brake fluid, which is the most commonly used fluid in a hydraulic-operated clutch system. If this is the case, the master and slave cylinders should be replaced and the system flushed.

CLUTCH SERVICE

A prerequisite for removing and replacing the clutch in a vehicle is removing the driveline or driveshafts and transmission or transaxle.

Removing the Clutch

After raising the vehicle on a hoist, clean excessive dirt, grease, or debris from around the clutch and transmission, and then disconnect and remove the clutch linkage. Cable systems need to be disconnected at the transmission.

On rear-wheel-drive automobiles, remove the driveline and the transmission. In some cases, the bell housing is removed with the transmission. It is a good practice to mark the driveshaft and yokes prior to driveshaft removal. This will reduce the chance of any balance and vibration issues. In some cases, the bell housing is removed with the transmission. In others, it is removed after the transmission is removed.

On front-wheel-drive vehicles with transaxles, any parts that interfere with transaxle removal must be removed first. These parts might include drive axles, parts of the engine, the brake and suspension system, or body parts. Check the service information for specific instructions.

The clutch assembly is accessible after the bell housing has been removed. Remember to use an approved vacuum-collection system or an approved liquid-cleaning system to remove clutch disc dust (which may contain asbestos) and dirt from the clutch assembly.

Photo Sequence 27 outlines the typical procedure for replacing a clutch disc and pressure plate. Always refer to the manufacturer's recommendations for bolt torque specifications prior to reinstalling the assembly.

While working on a clutch assembly, follow these guidelines:

- Check the bell housing, flywheel, and pressure plate for signs of oil leaks.
- Make sure the mating surfaces of the engine block and bell housing are clean. The smallest amount of dirt can cause a misalignment, which can cause premature wear of transmission shafts and bearings.
- Check the engine-to-bell-housing dowels and dowel bores. Replace or repair any damaged parts.
- Check both mounting surfaces of the bell housing for damage and runout (**Figure 38–26**).
- Check the flywheel for signs of burning or excessive wear. Check the runout of the flywheel with a dial indicator (**Figure 38–27**).

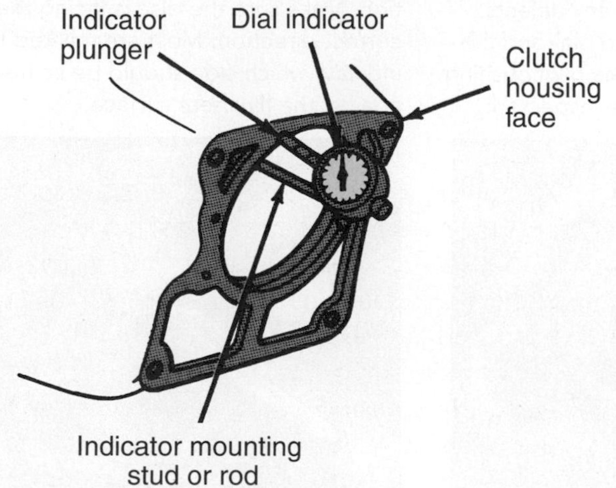

FIGURE 38–26

Use a dial indicator to check the runout of the clutch housing mounting surfaces.

Chrysler Group LLC

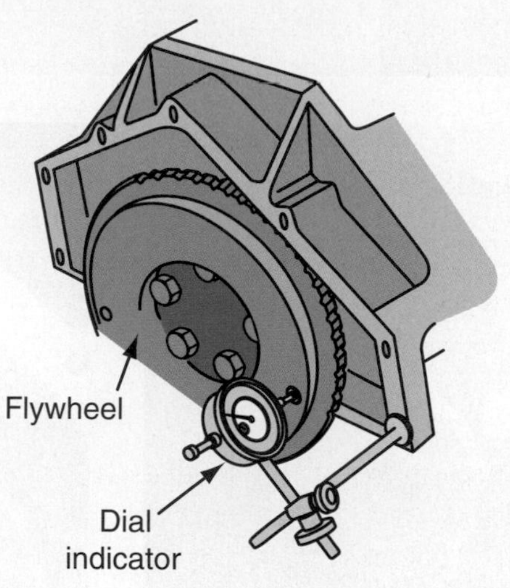

FIGURE 38–27

Measure the runout of the flywheel and crankshaft end play with a dial indicator.

- Check the teeth on the flywheel's ring gear; if there is damage, the ring gear or flywheel should be replaced.
- Use a clutch alignment tool during disassembly and reassembly (**Figure 38–28**). The tool will keep the disc centred on the pressure plate.
- Loosen and tighten the pressure-plate bolts according to the prescribed sequence.
- When installing a flywheel, make sure the bolts are tightened to specifications and in the prescribed sequence (normally, a star pattern).
- When measuring the lining thickness of a bonded clutch disc, measure the total thickness of the facing or lining. To measure the wear of a riveted lining, measure the material above the rivet heads.

FIGURE 38–28

An assortment of clutch alignment tools.

PHOTO SEQUENCE 27

INSTALLING AND ALIGNING A CLUTCH DISC

P27–1 The removal and replacement of a clutch assembly can be completed while the engine is in or out of the car. The clutch assembly is mounted to the flywheel that is mounted to the rear of the crankshaft.

P27–2 Before disassembling the clutch, make sure that alignment marks are present on the pressure plate and flywheel.

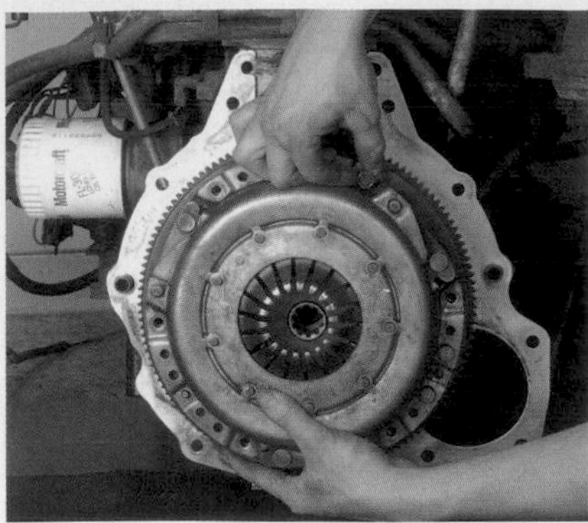

P27–3 The attaching bolts should be loosened before removing any of the bolts. With the bolts loosened, support the assembly with one hand while using the other to remove the bolts. The clutch disc will be free to fall as the pressure plate is separated from the flywheel. Keep it intact with the pressure plate.

P27–4 The surface of the pressure plate should be inspected for signs of burning, grooving, warpage, and cracks. Any faults normally indicate that the plate should be replaced.

P27–5 The surface of the flywheel should also be carefully inspected. Normally, the flywheel surface can be resurfaced to remove any defects. Also check crankshaft end play and the runout of the flywheel. The pilot bushing or bearing should also be inspected.

P27–6 The new clutch disc is placed into the pressure plate as the pressure plate is moved into its proper location. Make sure the disc is facing the correct direction. Most are marked to indicate which side should be seated against the flywheel surface.

P27–7 Install the pressure plate according to the alignment marks made during disassembly.

P27–8 Install the attaching bolts, but do not tighten. Then install the clutch alignment tool through the hub of the disc and the pilot bearing to centre the disc on the flywheel.

P27–9 With the disc aligned, tighten the attaching bolts according to the procedures outlined by the manufacturer, and check the release finger/lever height after tightening the bolts.

- Keep grease off the frictional surfaces of the clutch disc, flywheel, and pressure plate.
- Check the pressure-plate surface for warpage by laying a straightedge across the surface and inserting a feeler gauge between the surface and the straightedge. Compare the measurement against the specifications given in the service information.
- Check the release levers of the pressure plate for uneven wear or damage.
- Check the release bearing by turning it with your fingers and making sure it rotates freely.
- Check the clutch for damage.
- Check the pilot bushing or bearing for wear. Replace it if necessary (**Figure 38–29**).
- Lightly lubricate the input shaft and bearing retainer (**Figure 38–30**).
- Lubricate the clutch fork pivot points, the inside of the release-bearing hub, and the linkages.
- After the clutch assembly has been reinstalled, check the clutch pedal free travel.

FLYWHEEL INSPECTION The flywheel should have a smooth and flat surface. Check the flywheel for signs of overheating or excessive wear. If the surface of the flywheel appears to have slight grooves or shows signs of uneven wear, it should be resurfaced or replaced. If it has any evidence of cracks, it must be replaced. If the flywheel's surface appears

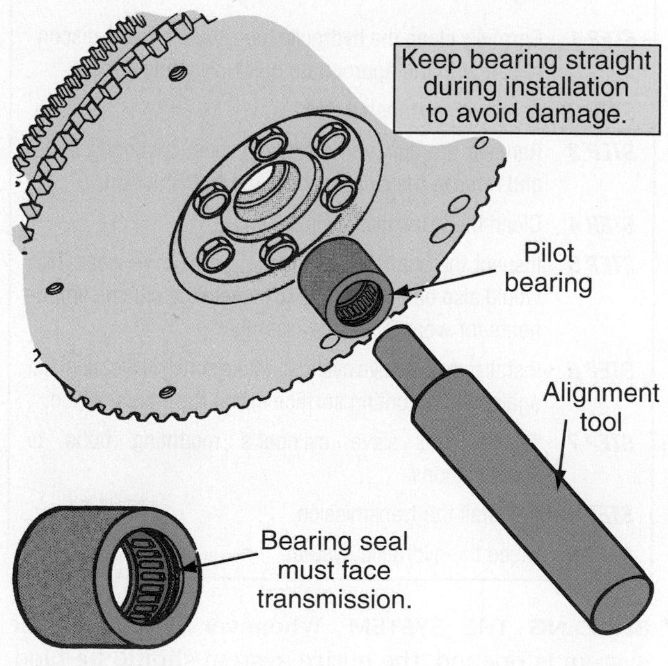
Keep bearing straight during installation to avoid damage.
Pilot bearing
Alignment tool
Bearing seal must face transmission.

FIGURE 38–29

A typical method for installing a pilot bearing.

Chrysler Group LLC

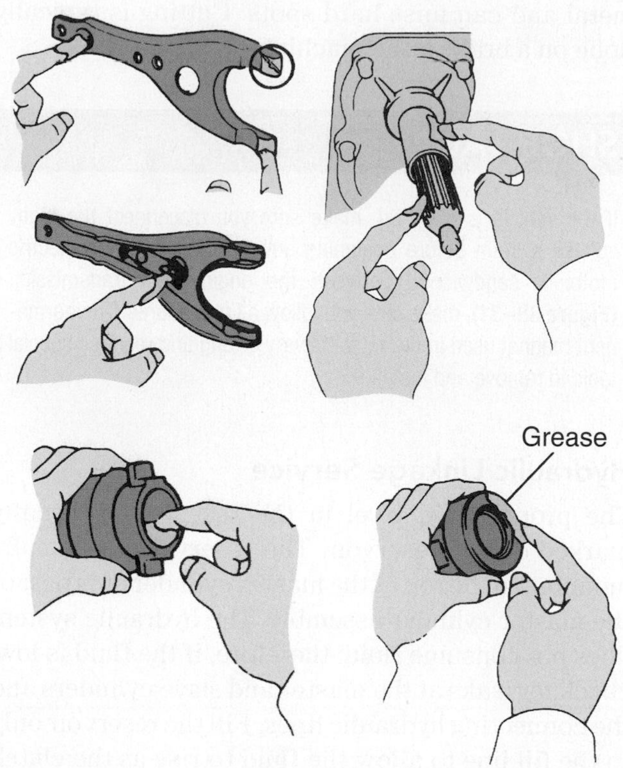
Grease

FIGURE 38–30

All contact points of the clutch release system should be lightly lubricated. Avoid applying too much grease.

to be fine, it does not need resurfacing or replacement. Check the teeth on the flywheel's ring gear; if there is damage, the ring gear or flywheel should be replaced.

Careful attention should be paid to make sure there are no signs of warpage or hard spots. Warpage occurs with overheating and can cause clutch chatter and vibrations. The flatness of the flywheel can be checked with a straightedge and a feeler gauge set. Typically, if the warpage is greater than 0.02 mm per centimeter (0.002 in. per inch) of the flywheel's diameter, the flywheel should be resurfaced or replaced. The overall runout of the flywheel should also be checked.

Hard spots are evident as raised areas on the surface that take on a bluish tint. Hard spots can cause clutch chattering. At times, the hard spots can be removed by resurfacing the flywheel; other times, flywheel replacement is the only cure.

FLYWHEEL SERVICE Resurfacing is done by grinding or cutting. Grinding normally relies on a dedicated grinder. Grinding is done as a wet process with silicone carbide stones, or a dry process with cubic boron nitride stones. Cutting the flywheel is not generally preferred because it can remove too much metal and can miss hard spots. Cutting is typically done on a brake lathe machine.

SHOP TALK

If the vehicle is a hybrid, make sure you disconnect the high-voltage system before beginning any work. Also, if the electric motor is sandwiched between the engine and transmission **(Figure 38–31)**, make sure you follow all procedures. The permanent magnet used in the motor is very strong and requires special tools to remove and install it.

Hydraulic Linkage Service

The proper fluid level in the reservoir is usually marked on the reservoir. The reservoir is normally mounted to the top of the master cylinder or is part of the master cylinder assembly. The hydraulic system does not consume fluid; therefore, if the fluid is low, check for leaks at the master and slave cylinders and the connecting hydraulic lines. Fill the reservoir only to the fill line to allow the fluid to rise as the clutch disc wears. Overfilling the system will cause slip and premature failure. Air can enter the system through the compensation and bleed ports if the fluid level in the reservoir is too low. The system must be bled to remove the trapped air.

Master cylinder problems are typically external or internal fluid leaks that require that the unit be replaced or rebuilt. Rebuild kits are available for most master cylinders. If a cast-iron master cylinder is rebuilt, the cylinder bore should be honed to remove any imperfections in the bore and new seals used. The bores of aluminum master cylinders should never be honed.

Figure 38–31
The electric motor for a hybrid vehicle with a clutch assembly attached to it.

Internal and external leaks are also typical problems for slave cylinders. Seldom are these cylinders rebuilt; rather, they are replaced. Replacing a slave cylinder is rather straightforward on most vehicles. Simply disconnect the hydraulic lines and unbolt the unit. If it appears that the piston of a slave cylinder is seized in its bore, check the movement of the release fork and lever at the clutch before replacing the slave cylinder. Leaks may also result from damaged or corroded hydraulic lines. These lines should be replaced, if damaged, with the same type of tube as was originally installed.

REPLACING A CONCENTRIC SLAVE CYLINDER When a concentric slave cylinder goes bad, fluid leakage and a soft or spongy pedal normally results. Because the slave cylinder is inside the transmission housing, the procedure for replacing it is different from that for replacing an external slave.

PROCEDURE

Replacing a Concentric Slave Cylinder

STEP 1 Carefully clean the hydraulic tube coupling, and disconnect it with the appropriate quick disconnect tool.

STEP 2 Remove the transmission.

STEP 3 Remove the concentric slave cylinder mounting bolts, and remove the cylinder from the transmission.

STEP 4 Clean the transmission input shaft.

STEP 5 Inspect the shaft for damage and excessive wear. This would also be a good time to inspect the clutch components for wear, before reassembly.

STEP 6 Install the new slave cylinder. Make sure it is installed flat against its mounting surface inside the transmission.

STEP 7 Tighten the slave cylinder's mounting bolts to specifications.

STEP 8 Reinstall the transmission.

STEP 9 Bleed the hydraulic system.

BLEEDING THE SYSTEM Whenever the hydraulic system is opened, the entire system should be bled. **Bleeding** may also be necessary if the system has run low on fluid, and air is trapped in the lines or cylinders. Bleeding can be accomplished through the use of a power bleeder (the same device used to bleed a brake system), a vacuum bleeder, or with the help of a co-worker. On most cars, it is impossible to pressurize the system and bleed the hydraulic lines at the same time; therefore, it is important that you have the proper equipment or someone to assist you. The typical procedure for bleeding the system follows:

PROCEDURE

Bleeding the Hydraulic System

STEP 1 Check the entire hydraulic circuit to make sure there are no leaks.

STEP 2 Check the clutch linkage for wear, and repair any defects before continuing.

STEP 3 Make sure all mounting points for the master and slave cylinders are solid and do not flex under the pressure of depressing the pedal.

STEP 4 Fill the master cylinder with the approved fluid.

STEP 5 Attach one end of a hose to the end of the bleeder screw and the other end into a catch can **(Figure 38–32)**. Loosen the bleed screw at the slave cylinder approximately one-half turn.

STEP 6 Fully depress the clutch pedal, and then move the pedal through three quick and short strokes. Allow the fluid and air to exit the system, and then immediately close the bleeder screw.

STEP 7 Release the pedal rapidly.

STEP 8 Recheck the fluid level in the master cylinder.

STEP 9 Repeat steps 3 and 4 until no air is evident in the fluid leaving the bleeder screw.

STEP 10 Close the bleeder screw immediately after the last downward movement of the pedal.

GRAVITY BLEEDING Often, the hydraulic system can be bled by allowing gravity to push the fluid through the system and move any trapped air with it. The process for doing this is rather simple. Begin by opening the bleeder screw on the slave cylinder; make sure there is a container below the bleeder to catch the dispelled fluid. Watch the flow of the fluid; when a constant flow without air bubbles is seen, tighten the bleeder and check the action of the pedal. This process may need to be repeated a few times. Make sure the reservoir is topped off between each procedure.

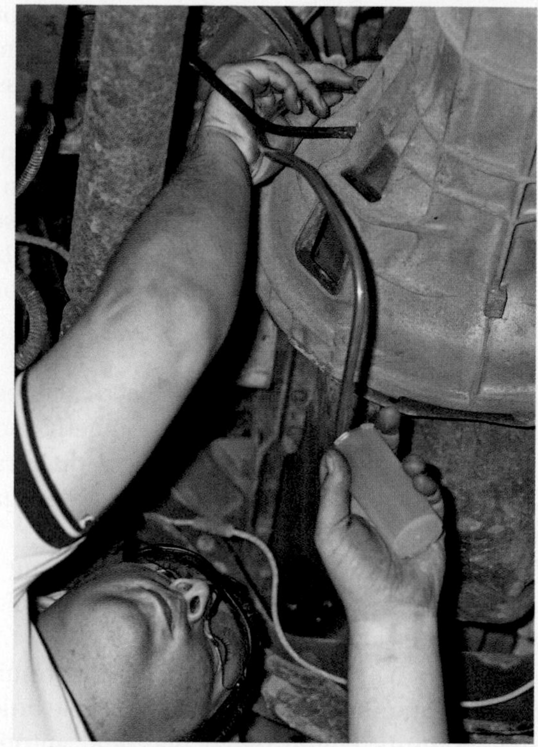

FIGURE 38–32

Attach a hose between the bleeder screw and a container before beginning to bleed the system.

KEY TERMS

Bell housing (p. 1154)
Bleeding (p. 1172)
Chatter (p. 1166)
Clutch disc (p. 1154)
Clutch fork (p. 1159)
Clutch release bearing (p. 1159)
Clutch slippage (p. 1164)
Coil spring pressure plate assembly (p. 1157)
Cushioning springs (p. 1157)
Diaphragm spring pressure plate assembly (p. 1157)
Flywheel (p. 1155)
Free travel (p. 1164)
Friction disc (p. 1154)
Pilot bushing (p. 1155)
Throwout bearing (p. 1159)

SUMMARY

- The clutch, located between the transmission and the engine, provides a mechanical coupling between the engine flywheel and the transmission input shaft. All manual transmissions and transaxles require a clutch.
- The flywheel, an important part of the engine, is also the main driving member of the clutch.

- The clutch disc receives the driving motion from the flywheel and pressure-plate assembly and transmits that motion to the transmission input shaft.
- The twofold purpose of the pressure plate assembly is to squeeze the clutch disc onto the flywheel and move away from the clutch disc so that the disc can stop rotating. There are basically two types of pressure plate assemblies: those with coil springs and those with a diaphragm spring.
- The clutch release bearing, also called a throwout bearing, smoothly and quietly moves the pressure-plate release levers or diaphragm spring through the engagement and disengagement processes.
- The clutch fork moves the release bearing and hub back and forth. It is controlled by the clutch pedal and linkage.
- Clutch linkage can be mechanical or hydraulic.
- The self-adjusting clutch is a clutch cable linkage that monitors clutch pedal play and automatically adjusts it when necessary.
- It is important that certain precautions are exercised when servicing the clutch. Clutch maintenance includes linkage adjustment and external clutch linkage lubrication.
- Slippage occurs when the clutch disc is not gripped firmly between the flywheel and the pressure plate. It can be caused by an oil-soaked or worn disc facing, warped pressure plate, weak diaphragm spring, or the release bearing contacting and applying pressure to the release levers.
- Clutch drag occurs if the clutch disc is not completely released when the clutch pedal is fully depressed. It can occur as a result of a warped disc or pressure plate, a loose disc facing, a defective release lever, or incorrect clutch pedal adjustment that results in excessive pedal play.
- Chatter is a shuddering felt in the vehicle when the pressure plate first contacts the clutch disc and stops when the clutch is fully engaged. Usually, chatter is caused when liquid contaminates the friction surfaces.
- Pedal pulsation is a rapid up-and-down movement of the clutch pedal as the clutch disengages or engages. It results from a misalignment of parts.
- Clutch vibrations can be felt throughout the vehicle, and they occur at any clutch pedal position. Sources of clutch vibrations include loose flywheel bolts, excessive flywheel runout, and pressure-plate cover balance problems.

REVIEW QUESTIONS

1. Name two types of friction facings.
2. *True or False?* If the fluid reservoir for a hydraulic clutch is low, the entire system should be checked for fluid leaks.
3. What is another name for the diaphragm spring?
4. Name three types of clutch linkages.
5. What is used to measure clutch pedal play?
6. Which of the following does the clutch friction disc always rotate with?
   a. the engine crankshaft
   b. the transmission input shaft
   c. the transmission output shaft
   d. the transmission main shaft
7. Which of the following statements about clutch disc torsional coil springs is correct?
   a. They cushion the clutch disc engagement rear to front.
   b. They produce the clamping force against the clutch disc.
   c. They absorb the engine's torque forces.
   d. They are located between the clutch disc friction facings.
8. How is contact between drive and driven members of a clutch maintained?
   a. mechanical clutch linkage pressure
   b. clutch pedal over-centre spring pressure
   c. static friction from the clutch disc surfaces
   d. spring pressure from the pressure plate
9. What type of fluid is used in a hydraulic clutch master cylinder?
   a. engine oil
   b. power-steering fluid
   c. brake fluid
   d. automatic transmission fluid
10. What does the clutch release bearing do?
    a. pushes against the pressure plate release fingers
    b. multiplies apply force on the clutch disc
    c. pulls the release fingers away from the pressure plate
    d. changes the direction of power flow
11. Which of the following clutch components is not a driving member?
    a. clutch cover **b.** clutch disc
    c. pressure plate **d.** flywheel

12. When the clutch is disengaged, where does the power flow stop?
   a. the transmission input shaft
   b. the driven disc hub
   c. the pressure plate and flywheel
   d. the torsion springs

13. Under what conditions would a bad pilot bearing be the most noticeable?
   a. under hard acceleration in first gear or reverse
   b. at steady cruising at highway speeds
   c. standing still, engine off, moving the clutch pedal up and down
   d. stopped at a red light in gear

14. What could be the result of insufficient clutch pedal clearance?
   a. gear clashing while shifting transmission
   b. a noisy front transmission bearing
   c. premature release bearing failure
   d. premature pilot bearing failure

15. What will happen as the clutch disc wears?
   a. transmission shifting becomes easier
   b. clutch pedal free play decreases
   c. clutch pedal free play increases
   d. clutch pedal free play is not affected

16. What can too much clutch pedal free play cause?
   a. hard gear shifting (clashing)
   b. fast release bearing wear
   c. release fork wear
   d. clutch slipping

17. Where are the wave springs or cushion springs located in a clutch disc?
   a. between the drive hub and disc assembly
   b. on the drive hub
   c. between the two clutch facings
   d. between the rear clutch facing and the pressure plate

18. What drives the clutch pressure plate?
   a. the clutch disc
   b. the flywheel through the clutch cover
   c. the bell housing
   d. the clutch shaft through the clutch levers

19. What is the purpose of the clutch disc cushion springs?
   a. cushion pressure plate torque on initial clutch engagement
   b. cushion pedal force to ensure easy engagement
   c. cushion the friction lining segments on clutch engagement
   d. cushion driveline shock and torque

20. What does the surface of the pressure plate ring contact?
   a. the transmission main shaft
   b. the release bearing
   c. the clutch disc
   d. the flywheel

CHAPTER 39 Manual Transmissions and Transaxles

LEARNING OUTCOMES

- Explain the design characteristics of the gears used in manual transmissions and transaxles.
- Explain the fundamentals of torque multiplication and overdrive.
- Describe the purpose, design, and operation of synchronizer assemblies.
- Describe the purpose, design, and operation of internal and remote gearshift linkages.
- Explain the operation and power flows produced in typical manual transmissions and transaxles.

The transmission or transaxle **(Figure 39–1)** is a vital link in the powertrain of any modern vehicle. The purpose of the transmission or transaxle is to use gears of various sizes to give the engine a mechanical advantage over the driving wheels. During normal operating conditions, power from the engine is transferred through the engaged clutch to the input shaft of the transmission or transaxle. Gears in the transmission or transaxle housing alter the torque and speed of this power input before passing it on to other components in the drivetrain. Without the mechanical advantage the gearing provides, an engine can generate only limited torque at low speeds. Without sufficient torque, moving a vehicle from a standing start would be impossible.

In any engine, the crankshaft always rotates in the same direction. If the engine transmitted its power directly to the drive axles, the wheels could be driven only in one direction. Instead, the transmission or transaxle provides the gearing needed to reverse direction so that the vehicle can be driven backward. There is also a neutral position that stops engine rotation and power from reaching the drive wheels.

FIGURE 39–1

A late-model transaxle.

TRANSMISSION VERSUS TRANSAXLE

Vehicles are propelled in one of three ways: by the rear wheels, by the front wheels, or by all four wheels. The type of drive system used determines whether a conventional transmission or a transaxle is used.

Vehicles propelled by the rear wheels normally use a transmission. Transmission gearing is located within an aluminum or iron casting called the transmission **case** assembly. The transmission case assembly is attached to the rear of the engine **(Figure 39–2)**, which is normally located in the front of the vehicle. A driveshaft links the output shaft of the transmission with the final drive gears and drive axles located in a separate housing at the rear of the vehicle. The differential splits the driveline power and redirects it to the two rear axle shafts, which then pass it on to the wheels. For many years,

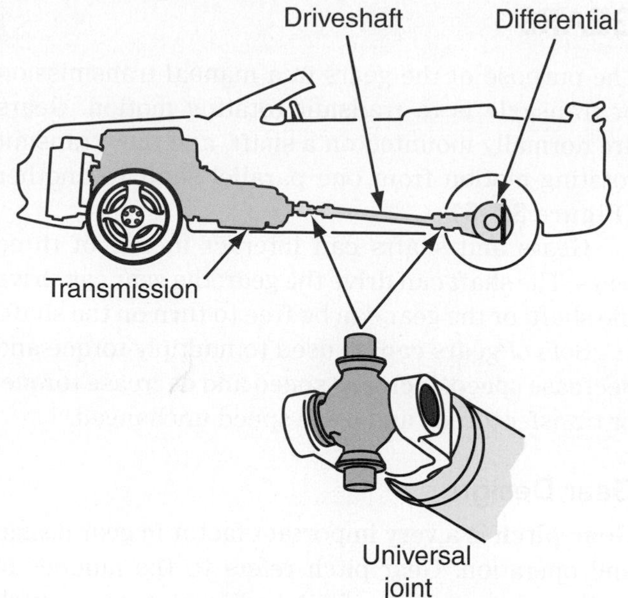

FIGURE 39–2

A transmission mounted to the rear of the engine connected to the drive axle by a driveshaft with universal joints.

rear-wheel-drive (RWD) systems were the conventional method of propelling a vehicle.

Cars with a mid- or rear-mounted engine send power to the rear wheels through a transaxle. Also, some front engine RWD vehicles have a shaft that connects the engine to a transaxle assembly at the rear axle.

Front-wheel-drive (FWD) vehicles are propelled by the front wheels. For this reason, they must use a drive design different from that of an RWD vehicle. The transaxle is the special power transfer unit commonly used on FWD vehicles. A transaxle combines the transmission gearing, final drive and differential, and drive axle connections into a single case aluminum housing located in front of the vehicle **(Figure 39–3)**. This design offers many advantages. One major advantage is the good traction on slippery roads due to the weight of the drivetrain components being directly over the driving axles of the vehicle. It is also more compact and lighter than the transmission of an RWD vehicle. Transverse engine and transaxle configurations also allow for lower hood lines, thereby improving the vehicle's aerodynamics.

Four-wheel-drive vehicles typically use a transmission and transfer case. The transfer case mounts on the side or back of the transmission. A chain or gear drive inside the transfer case receives power from the transmission and transfers it to two separate driveshafts. One driveshaft connects to the final drive on the front drive axle. The other driveshaft connects to the final drive on the rear drive axle.

Most manual transmissions and transaxles are **constant mesh, fully synchronized** units. Constant mesh means that whether or not the gear is locked to the output shaft, it is in mesh with its counter gear. All gears rotate in the transmission as long as the clutch is engaged. Fully synchronized means the

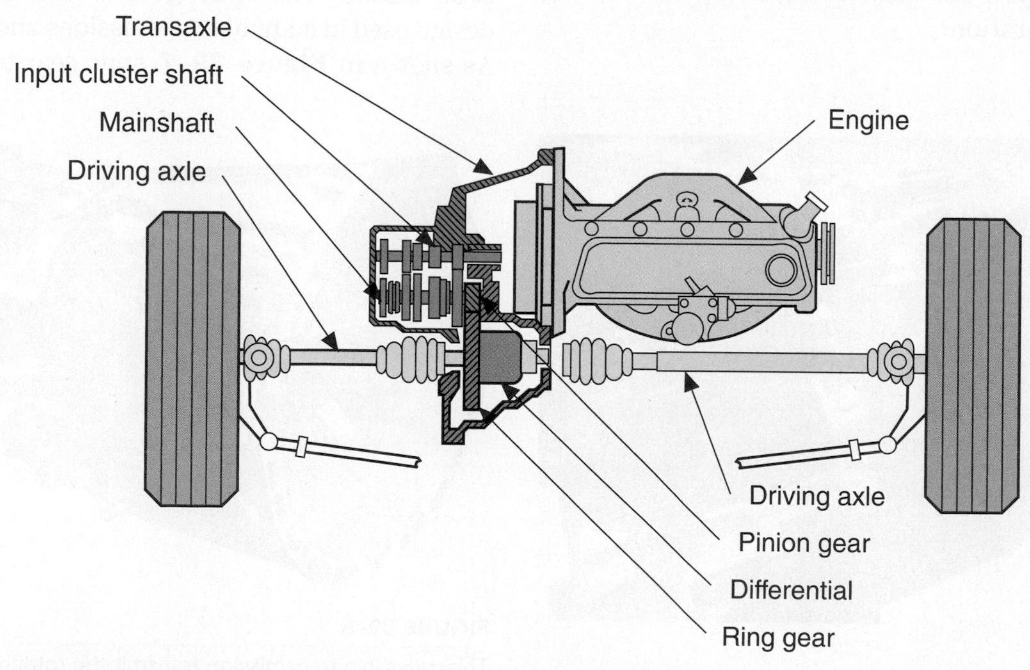

FIGURE 39–3

The location of typical front-wheel-drive powertrain components.

unit uses a mechanism of brass rings and clutches to bring rotating shafts and gears to the same speed before shifts occur. This promotes smooth shifting. In a vehicle equipped with a five-speed manual-shift transmission or transaxle, all five forward gears are synchronized. Reverse gearing may or may not be synchronized, depending on the type of transmission/transaxle.

Transmission Designs

All automotive transmissions/transaxles are equipped with a varied number of forward speed gears ranging between 3 and 7 **(Figure 39–4)**, a neutral gear, and one reverse speed. Transmissions can be divided into groupings based on the number of forward speed gears they have. The growing concern for improved gas mileage led to smaller engines with transmissions having higher numbers of forward gears. The additional gears allowed the smaller engines to perform better by matching the engine's torque curve with vehicle speeds and loads.

Six-speed transmissions and transaxles are now the most commonly used units. These transmissions can offer one or two overdrive gears. Overdrive reduces engine speed at a given vehicle speed, which increases top speed, improves fuel economy, and lowers engine noise. The addition of the two overdrive gears allows manufacturers to use lower final drive gears for acceleration. The fifth and sixth gears reduce the overall gear ratio and allow for slower engine speeds during highway operation.

GEARS

The purpose of the gears in a manual transmission or transaxle is to transmit rotating motion. Gears are normally mounted on a shaft, and they transmit rotating motion from one parallel shaft to another **(Figure 39–5)**.

Gears and shafts can interact in one of three ways: The shaft can drive the gear; the gear can drive the shaft; or the gear can be free to turn on the shaft.

Sets of gears can be used to multiply torque and decrease speed, increase speed and decrease torque, or transfer torque and leave speed unchanged.

Gear Design

Gear pitch is a very important factor in gear design and operation. Gear pitch refers to the number of teeth per given unit of pitch diameter. Gear pitch calculations are calculated in imperial measure (inches). Even gears that are manufactured to metric specifications are converted by dividing the millimetre measure by 25.4 to calculate the pitch. A simple way of determining gear pitch is to divide the number of teeth by the pitch diameter of the gear. For example, if a gear has 36 teeth and a 6-inch pitch diameter, it has a gear pitch of 6 **(Figure 39–6)**. The important fact to remember is that gears must have the same pitch to operate together. A five-pitch gear meshes only with another five-pitch gear, a six-pitch only with a six-pitch, and so on.

SPUR GEARS The **spur gear** is the simplest gear design used in manual transmissions and transaxles. As shown in **Figure 39–7**, spur gear teeth are cut

FIGURE 39–4

A seven-speed transmission from a current-model Corvette.

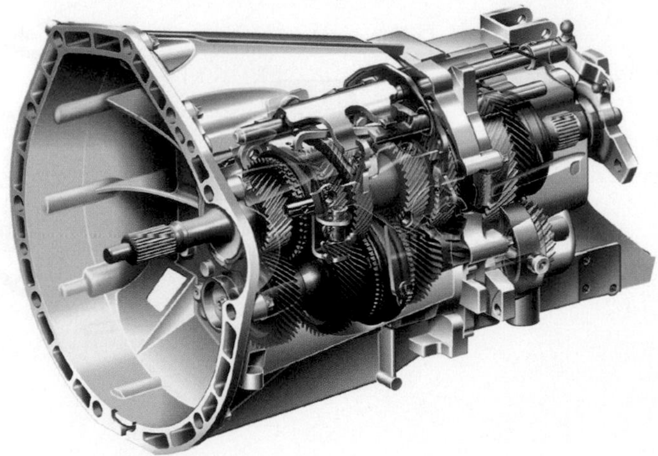

FIGURE 39–5

The gears in a transmission transmit the rotating power from the engine.

DaimlerChrysler Corporation

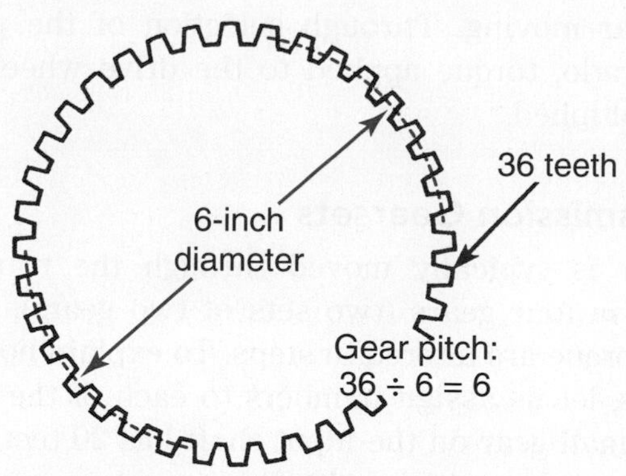

FIGURE 39–6

Determining gear pitch.

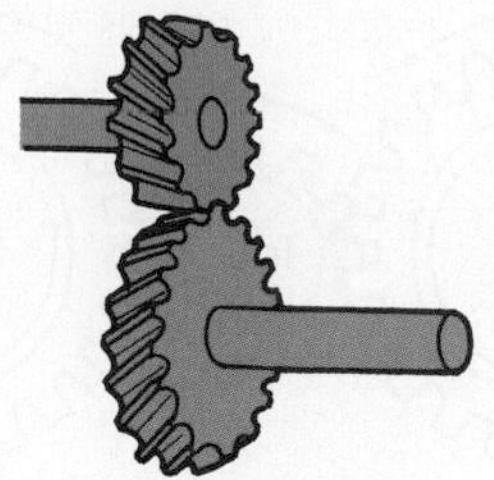

FIGURE 39–8

Helical gears have teeth cut at an angle to the gear's axis of rotation.

straight across the edge parallel to the gear's shaft. During operation, meshed spur gears have only one tooth in full contact at a time.

Its straight-tooth design is the spur gear's main advantage. It minimizes the chances of popping out of gear, an important consideration during acceleration/deceleration and reverse operation. For this reason, spur gears are often used for the reverse gear.

The spur gear's major drawback is the clicking noise that occurs as teeth contact one another. At higher speeds, this clicking becomes a constant whine. Quieter gears, such as the helical design, are often used to eliminate this gear whine problem.

SHOP TALK

When a small gear is meshed with a much larger gear, the small gear is often called a pinion or pinion gear, regardless of its tooth design.

HELICAL GEARS A **helical gear** has teeth that are cut at an angle or are spiral to the gear's axis of rotation **(Figure 39–8)**. This configuration allows two or more teeth to mesh at the same time, which distributes tooth load and produces a very strong gear. Helical gears also run more quietly than spur gears because they create a wiping action as they engage and disengage the teeth on another gear. One disadvantage is that helical teeth on a gear cause the gear to move fore or aft (axial thrust) on a shaft, depending on the direction of the angle of the gear teeth. This axial thrust must be absorbed by thrust washers and other transmission gears, shafts, or the transmission case.

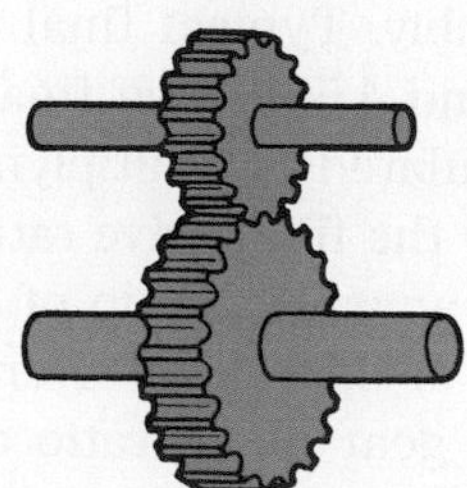

FIGURE 39–7

Spur gears have teeth cut straight across the gear edge parallel to the shaft.

Helical gears can be either right-handed or left-handed, depending on the direction the spiral appears to go when the gear is viewed face-on. When mounted on parallel shafts, one helical gear must be right-handed and the other left-handed. Two gears with the same direction spiral do not mesh in a parallel-mounted arrangement.

Spur and helical gears that have teeth cut around their outside diameter edge are called **external toothed gears**. When two external toothed gears are meshed together, one rotates in the opposite direction as the other **(Figure 39–9)**. If an external toothed gear is meshed with an internal toothed gear

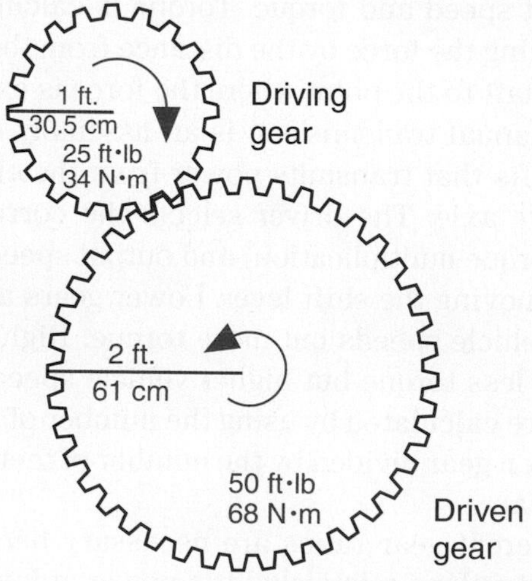

FIGURE 39–9

Externally meshed gears rotate in opposite directions.

FIGURE 39–10

An idler gear is used to transfer motion without changing rotational direction.

(one that has teeth around its inside diameter), both rotate in the same direction.

Idler Gears

An **idler gear** is a gear that is placed between a drive gear and a driven gear. Its purpose is to transfer motion from the drive gear to the driven gear without changing the direction of rotation. It can do this because all three gears have external teeth **(Figure 39–10)**.

Idler gears are used in reverse gear trains to change the directional rotation of the output shaft. In all forward gears, the input shaft and the output shaft turn in the same direction. In reverse, the output shaft turns in the opposite direction to the input shaft. This allows the vehicle drive wheels to turn backward.

BASIC GEAR THEORY

Gears are able to multiply and transfer torque to other rotating parts of the drivetrain. As gears with different numbers of teeth mesh, each rotates at a different speed and torque. Torque is calculated by multiplying the force by the distance from the centre of the shaft to the point where the force is exerted.

A manual transmission is an assembly of gears and shafts that transmits power from the engine to the drive axle. The driver selects the correct gear ratio (torque multiplication) and output speed potential by moving the shift lever. Lower gears allow for lower vehicle speeds but more torque. Higher gears provide less torque but higher vehicle speeds. Gear ratios are calculated by using the number of teeth on the driven gear divided by the number of teeth on the drive gear.

Different gear ratios are necessary because an engine develops relatively little power at low engine speeds. The engine must be turning at a fairly high speed before it can deliver enough power to get the car moving. Through selection of the proper gear ratio, torque applied to the drive wheels can be multiplied.

Transmission Gearsets

Power is typically moved through the transmission via four gears (two sets of two gears). Speed and torque are altered in steps. To explain how this works, let us assign numbers to each of the gears. The small gear on the input shaft has 20 teeth. The gear it meshes with has 40. This provides a gear ratio of 2:1. The output of this gearset moves along the shaft of the 40-tooth gear and rotates other gears. The gear involved with first gear has 15 teeth. This gear rotates with the same speed and with the same torque as the 40-tooth gear. However, the 15-tooth gear is meshed with a larger gear with 35 teeth. The gear ratio of the 15-tooth and the 35-tooth gearset is 2.33:1. However, the ratio of the entire gearset (both sets of two gears) is 4.67:1.

To calculate this gear ratio, divide the driven (output) gear of the first set by the drive (input) gear of the first set. Do the same for the second set of gears, and then multiply the answer from the first by the second. The result is equal to the gear ratio of the entire gearset. The mathematical formula for a compound gear ratio follows:

$$\frac{\text{driven}_1}{\text{drive}_1} \times \frac{\text{driven}_2}{\text{driven}_2} = \frac{40}{20} \times \frac{35}{15} = 4.67:1$$

Most of today's transmissions have at least one overdrive gear. Overdrive gears have ratios of less than 1:1. These ratios are achieved by using a large driving gear meshed with a smaller driven gear. Output speed is increased, and torque is reduced. The purpose of overdrive is to promote fuel economy and reduce operating noise while maintaining highway cruising speed.

The driveline's gear ratios are further increased by the gear ratio of the ring and pinion gears in the drive axle assembly. Typical final drive ratios are between 2.50:1 and 4.50:1. The final (overall) drive gear ratio is calculated by multiplying the transmission gear ratio by the final drive ratio. If a transmission is in first gear with a ratio of 3.63:1 and has a final drive ratio of 3.52:1, the overall gear ratio is 12.78:1. If fourth gear has a ratio of 1:1, using the same final drive ratio, the overall gear ratio is 3.52:1. The overall gear ratio is calculated by multiplying the ratio of the first set of gears by the ratio of the second set of gears ($3.63 \times 3.52 = 12.78$).

Reverse Gear Ratios

Reverse gear ratios involve two driving (driver) gears and two driven gears:

- The input gear is driver 1.
- The idler gear is driven 1.
- The idler gear is also driver 2.
- The output gear is driven 2.

If the input gear has 20 teeth, the idler gear has 28 teeth and the output gear has 48 teeth. However, because a single idler gear is used, the teeth of it are not used in the calculation of gear ratio. The idler gear merely transfers motion from one gear to another. The calculations for determining reverse gear ratio with a single idler gear follow:

$$\text{Reverse gear ratio} = \frac{\text{driven 2}}{\text{driver 1}}$$

$$= \frac{48}{20}$$

$$= 2.40$$

If the gearset uses two idler gears (one with 28 teeth and the other with 40 teeth), the gear ratio involves three driving gears and three driven gears:

- The input gear is driver 1.
- The no.1 idler gear is driven 1.
- The no.1 idler gear is also driver 2.
- The no.2 idler gear is driven 2.
- The no.2 idler gear is also driver 3.
- The output gear is driven 3.

The ratio of this gearset would be calculated as the compound gear ratio formula that follows:

$$\text{Reverse gear ratio} = \frac{\text{driven 1} \times \text{driven 2} \times \text{driven 3}}{\text{driver 1} \times \text{driver 2} \times \text{driver 3}}$$

$$= \frac{20 \times 40 \times 48}{20 \times 28 \times 40}$$

$$= \frac{53\,760}{22\,400}$$

$$= 2.40$$

As can be seen, idler gears do not affect the gear ratio.

TRANSMISSION/TRANSAXLE DESIGN

The internal components of a transmission or transaxle consist of a parallel set of metal shafts on which meshing gearsets of different ratios are mounted **(Figure 39–11)**. By moving the shift lever, gear ratios can be selected to generate different amounts of output torque and speed.

The gears are mounted or fixed to the shafts in a number of ways. They can be internally splined or

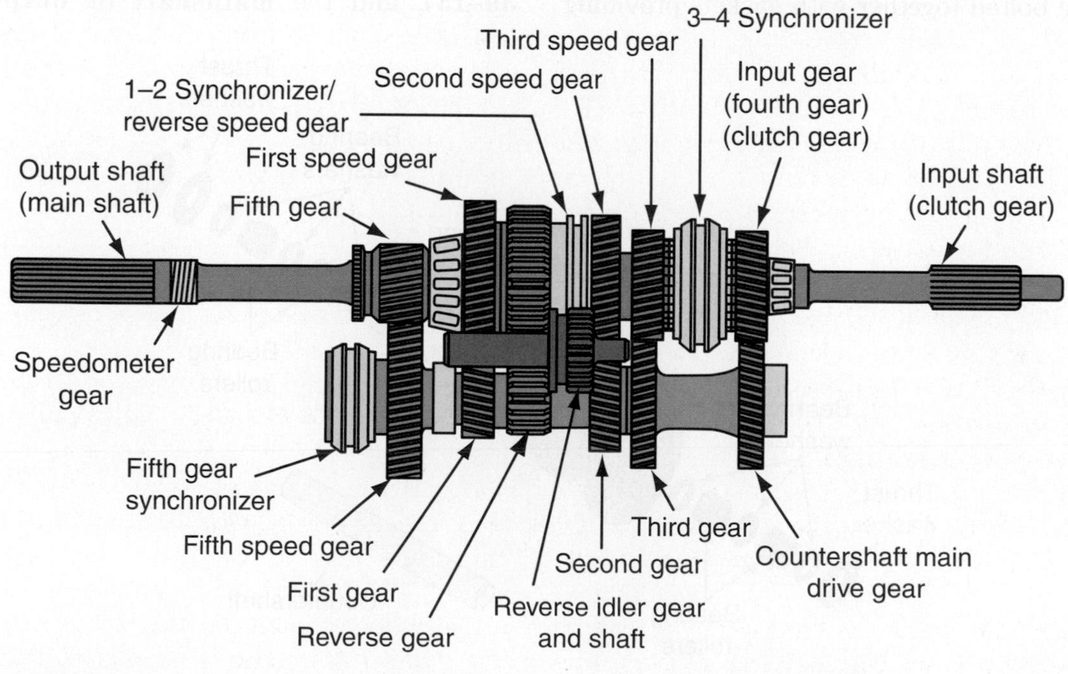

FIGURE 39–11

The arrangement of the gears and shafts in a typical five-speed transmission.

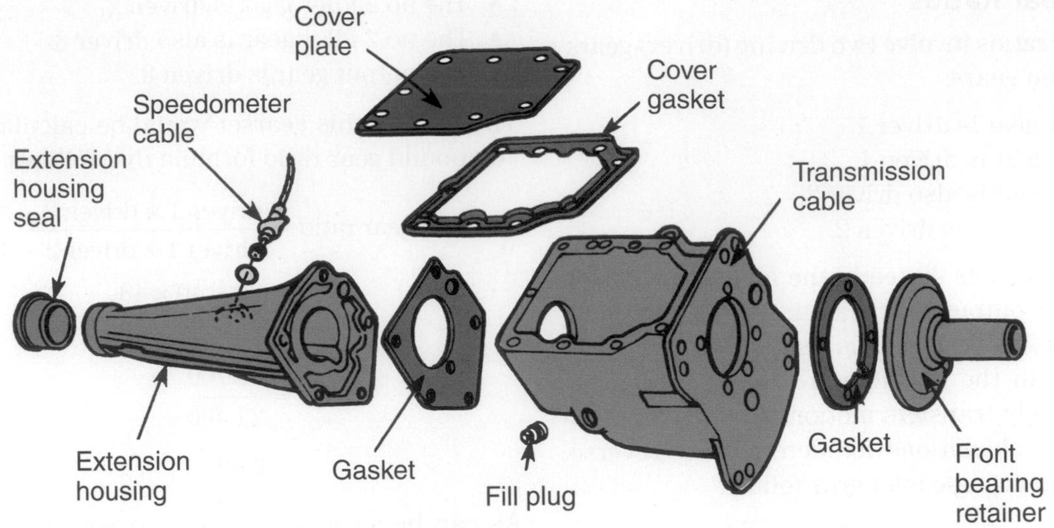

FIGURE 39–12

Typical manual transmission case components.

Ford Motor Company

keyed to a shaft. Gears can also be manufactured as an integral part of the shaft. Gears that must be able to freewheel around the shaft during certain speed ranges are mounted to the shaft using bushings or bearings, or directly onto the shaft.

The shafts and gears are contained in a transmission or transaxle case or housing. The components of this housing include the main case body, side or top cover plates, extension housings, and bearing retainers **(Figure 39–12)**. The metal components are bolted together with gaskets providing a leak-proof seal at all joints. The case is filled with lubricant to provide constant lubrication and cooling for the spinning gears and shafts.

Transmission Features

Although they operate in a similar fashion, the layout, components, and terminology used in transmissions and transaxles are not exactly the same.

Most transmissions have three specific shafts: the **input shaft**, the **countershaft (Figure 39–13)**, and the **mainshaft** or **output shaft**.

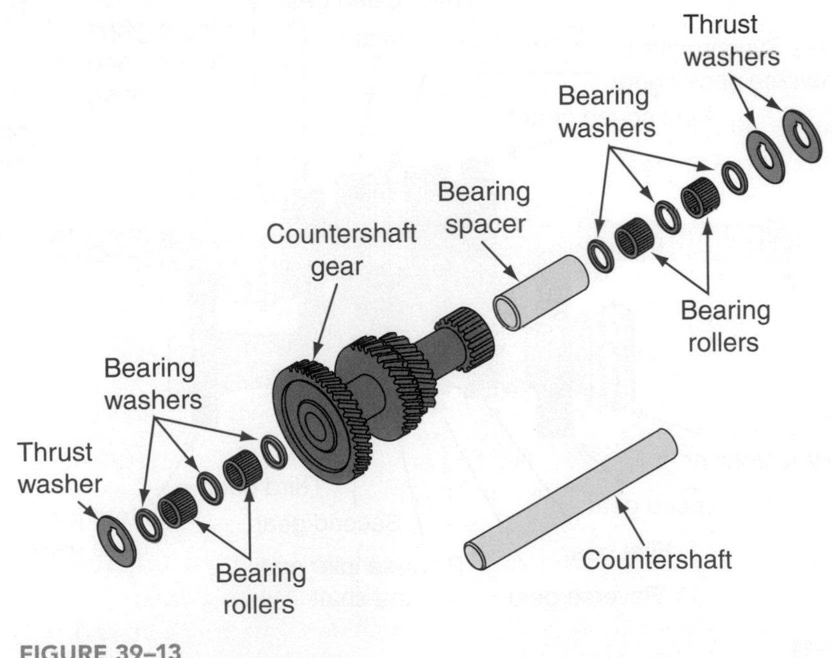

FIGURE 39–13

A typical countershaft assembly.

The clutch gear is an integral part of the transmission's input shaft and always rotates with the input shaft.

The countershaft is actually several gears machined out of a single piece of steel. The countershaft may also be called the **countergear** or **cluster gear**. The countergear can be solid (one piece) and mounted on front and rear bearings. Other designs have the countergear mounted via a shaft and roller bearings through its axis. The shaft-mounted countergear shaft is pinned or keyed into position and does not turn. Thrust washers control the amount of end play of the countergear in the transmission case.

The main gears on the mainshaft or output shaft transfer rotation from the countergears to the output shaft. The main gears are also called **speed gears**. They are mounted on the output shaft using bearings. Speed gears freewheel around the output shaft until they are locked to it by the engagement of their shift **synchronizer** unit.

Power flows from the transmission input shaft to the **clutch gear**. The clutch gear meshes with the large counter driven gear of the countergear assembly. This causes all of the countergears to rotate, and since the countergears are meshed with the main-shaft speed gears, the speed gears are forced to turn as well.

There can be no power output until one of the speed gears is locked to the mainshaft. This is done by activating a shift fork, which moves a synchronizer sleeve to lock the selected speed gear to the mainshaft. Power travels along the countergear until it reaches this selected speed gear. It then passes through this gear back to the mainshaft and out of the transmission to the driveline.

Transaxle Features

Transaxles use many of the design and operating principles found in transmissions. There are differences, however, due to the transaxle containing the final drive gears, differential, and drive axle connections. These changes results in unique internal operating considerations.

A transaxle typically has two separate shafts—an input shaft and an output shaft. The input shaft is the driving shaft. It is normally located above and parallel to the output shaft. The output shaft is the driven shaft. The transaxle's main (speed) gears freewheel around the output shaft unless they are locked to the shaft by their synchronizer assembly. The main speed gears are in constant mesh with the input shaft drive gears. The drive gears turn whenever the input shaft turns.

The names used to describe transaxle shafts vary between manufacturers. The service manuals or information systems for some vehicles refer to the input shaft as the mainshaft and the output as the driven pinion or driveshaft. Others call the input shaft and its gears the input gear cluster and refer to the output shaft as the mainshaft. For clarity, this text uses the terms input gear cluster for the input shaft and its drive gears, and **pinion shaft** for the output shaft.

A pinion gear is machined onto the end of the transaxle's pinion shaft. This pinion gear is in constant mesh with the final drive ring gear located in the lower portion of the transaxle housing. Because the pinion gear is part of the pinion shaft, it must rotate whenever the pinion shaft turns. With the pinion rotating, engine torque flows through the ring gear and differential gearing to the axle shafts and driving wheels.

Some transaxles have a third shaft designed to offset the power flow on the output shaft. Power is transferred from the output shaft to the third shaft using helical gears and by placing the third shaft in parallel with the output and input shafts. Other transaxles with a third shaft use an offset input shaft that receives the engine's power and transmits it to a mainshaft, which serves as an input shaft. The third shaft is only added to transaxles when an extremely compact transaxle is required **(Figure 39–14)**.

Unlike transmissions, most transaxle shaft and bearing preloads (or clearances) are designed into the housings. They are determined when the housings are bolted together. These housings usually have machined-smooth mating surfaces that require no sealant or anaerobic sealant in place of gaskets. Extreme care should be taken when handling the housings. Any burrs, nicks, or scratches must be carefully removed or repaired prior to housing assembly.

SYNCHRONIZERS

The synchronizer performs a number of jobs vital to transmission/transaxle operation. Its main job is to bring components that are rotating at different speeds to one synchronized speed. The second major job of the synchronizer is to actually lock these components together. The end result of these two functions is a clash-free shift. In some transmissions, a synchronizer can have another important job. When

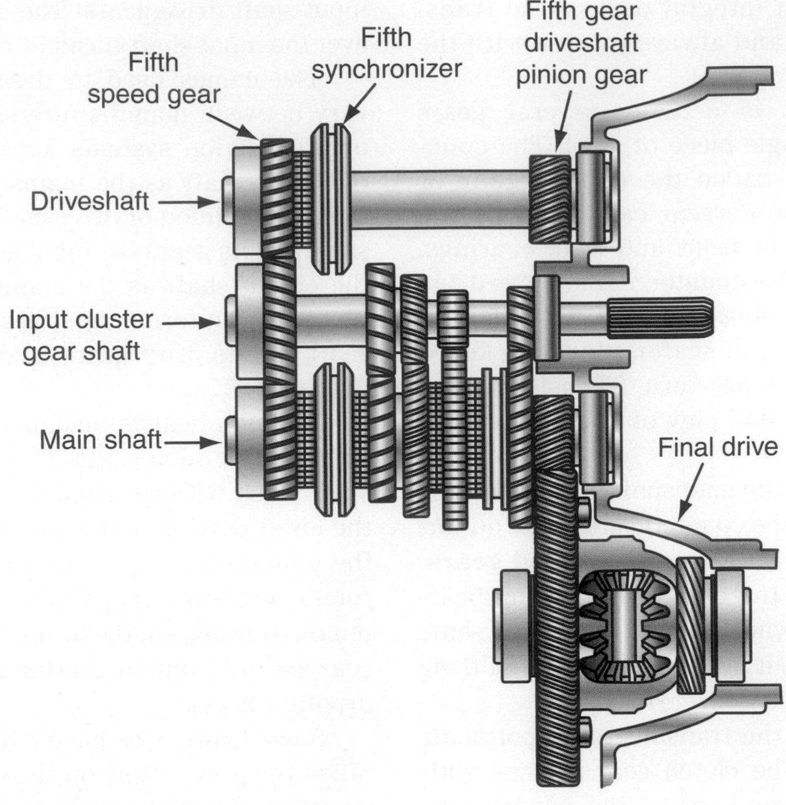

FIGURE 39–14

A transaxle with three gear shafts.

spur teeth are cut into the outer sleeve of the synchronizer, the sleeve can act as a reverse gear and assist in producing the correct direction of rotation for reverse operation.

In modern transmissions and transaxles, all forward gears are synchronized. In a transaxle, one synchronizer is placed between the first and second speed gears, the second synchronizer is placed between the third and fourth speed gears, and a third synchronizer will be needed for the fifth and sixth speed gears. Reverse gear is not normally fitted with a synchronizer. A synchronizer requires gear rotation to do its job, and reverse is selected with the vehicle at a stop.

Synchronizer Design

There are five primary components tied into synchronized shifting. The synchronizer sleeve has internal splines that slide over the external teeth of the synchronizer hub. The hub is fastened to the mainshaft. Spring-loaded keys (also referred to as dogs, struts, or insert plates) are positioned in the hub. Synchronizer or blocker rings have external splines and can move within the synchronizer sleeve. The outside of the synchronizer ring often has a cone-shaped surface to serve as a cone clutch. The other side of the cone clutch is machined to the side of a speed gear, which also has dog clutch teeth to allow the sleeve to fit over the speed gear. In a transmission or transaxle, there is usually a speed gear on both sides of the sleeve.

Figure 39–15 illustrates the most commonly used synchronizer—a **block synchronizer** or **cone synchronizer**. The synchronizer sleeve surrounds

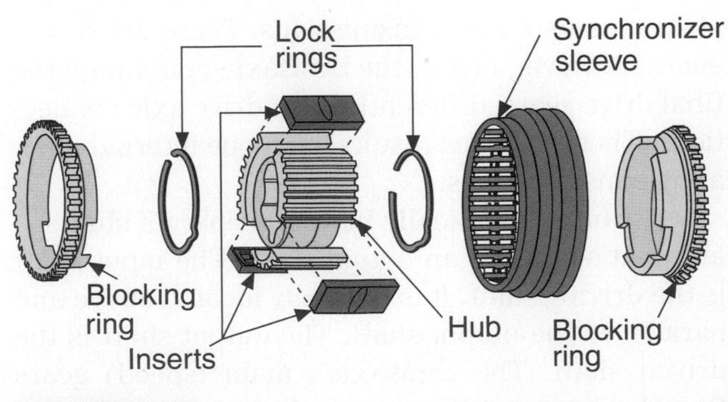

FIGURE 39–15

An exploded view of a blocking ring–type synchronizer assembly.

the synchronizer assembly and meshes with the external splines of the hub. The hub is splined to the transmission pinion shaft and is held in position by a snap ring. A few transmissions use pin-type synchronizers.

The synchronizer sleeve has a small internal groove and a large external groove in which the shift fork rests. Three slots are equally spaced around the outside of the hub. Inserts fit into these slots and are able to slide freely back and forth. These inserts, sometimes referred to as **shifter plates** or keys, are designed with a ridge in their outer surface. Insert springs hold the ridge in contact with the synchronizer sleeve internal groove.

The synchronizer sleeve is precisely machined to slide onto the hub smoothly. The sleeve and hub sometimes have alignment marks to ensure proper indexing of their splines when assembling to maintain smooth operation.

Brass, bronze, or powdered-iron synchronizing **blocking rings** are positioned at the front and rear of each synchronizer assembly. Some synchronizer assemblies use frictional material on the blocking rings to reduce slippage. Each blocking ring has three notches equally spaced to correspond with the three insert keys of the hub. Around the outside of each blocking ring is a set of bevelled clutching teeth, which is used for alignment during the shift sequence. The inside of the blocking ring is shaped like a cone. This coned surface is lined with many sharp grooves.

The cone of the blocking ring makes up only one-half of the total cone clutch. The second or matching half of the cone clutch is part of the speed gear to be synchronized. As shown in **Figure 39–16**, the shoulder of the speed gear is cone shaped to match the blocking ring. The shoulder also contains a ring of bevelled clutching teeth designed to align with the clutch/dog teeth on the blocking ring.

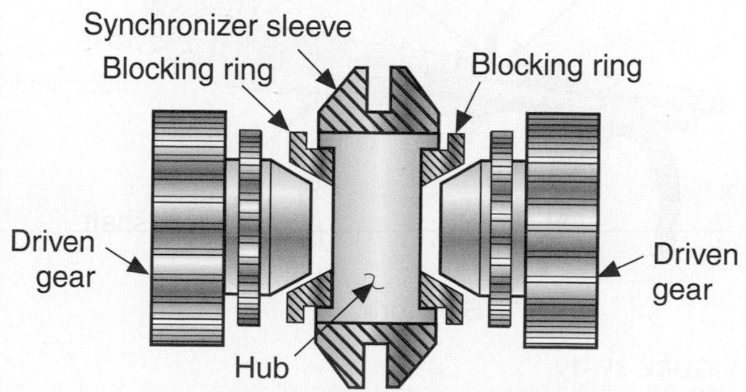

FIGURE 39–16

Gear shoulder and blocker ring mating surfaces.

Operation

When the transmission is in neutral, all of the synchronizers are in their neutral position and are not rotating with the pinion shaft. Gears on the mainshaft are meshed with their countershaft partners and are freewheeling around the pinion shaft at various speeds.

To shift the transmission into first gear, the clutch is disengaged and the gearshift lever is placed in first gear position. This forces the shift fork and the synchronizer sleeve toward the first speed gear on the mainshaft. As the sleeve moves, the inserts also move because the insert ridges lock the inserts to the internal groove of the sleeve.

The movement of the inserts forces the blocking ring's coned friction surface against the coned surface of the first speed gear shoulder. When the blocking ring and gear shoulder come into contact, the grooves on the blocking ring cone cut through the lubricant film on the first speed gear shoulder, and a metal-to-metal contact is made. The contact generates substantial friction and heat. This is one reason bronze or brass blocking rings are used. A nonferrous metal such as bronze or brass minimizes wear on the hardened-steel gear shoulder. This frictional coupling is not strong enough to transmit loads for long periods. As the components reach the same speed, the synchronizer sleeve can now slide over the external clutching teeth on the blocking ring and then over the clutching teeth on the first speed gear shoulder. This completes the engagement **(Figure 39–17)**. Power flow is now from the first speed gear, to the synchronizer sleeve, to the synchronizer hub, to the main output shaft, and out to the driveline.

To disengage the first speed gear from the pinion shaft and shift into second gear, the clutch must be disengaged. The shift fork moves away from the first speed gear and the sleeve disengages from the gear's clutch teeth. Because the first speed gear is no longer connected to the mainshaft, it is free to rotate but cannot transfer torque. As the transmission is shifted into second gear, the inserts again lock into the internal groove of the sleeve. As the sleeve moves forward, the forward blocking ring is forced by the inserts against the coned friction surface on the second speed gear

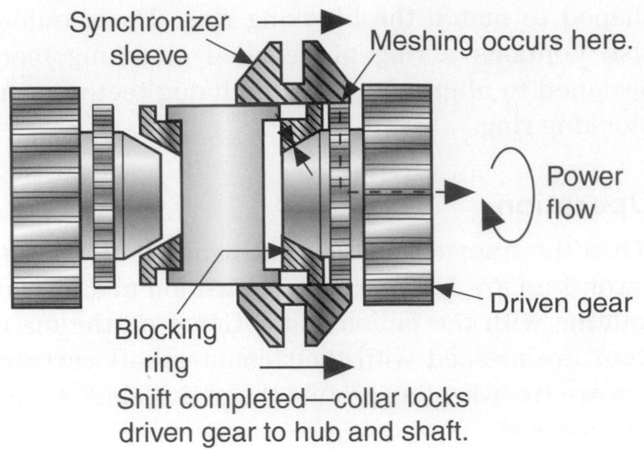

FIGURE 39–17

Gear shoulder and blocker ring mating surfaces when a gear is engaged.

©Cengage Learning 2015

FIGURE 39–18

A double-cone synchronizer.

Martin Restoule

shoulder. Once again, the grooves on the blocking ring cut through the lubricant on the gear shoulder to generate a frictional coupling that synchronizes the speed gear and shaft speeds. The shift fork can then continue to move the sleeve forward until it slides over the blocking ring and speed gear shoulder clutching teeth, locking them together. Power flow is now from the second speed gear, to the synchronizer sleeve, to the hub, and out through the mainshaft.

Advanced Synchronizer Designs

Many manufacturers are using multiple cone–type synchronizers in their transmissions. These transmissions are fitted with single-cone, double-cone, or triple-cone synchronizers. For example, first and second gears may have triple-cone synchronization, third and fourth may have double-cone, and fifth and sixth may have single-cone.

Double-cone synchronizers **(Figure 39–18)** have friction material on both sides of the synchronizer rings. The extra friction surfaces result in decreased shift effort and greater synchronizer durability. Triple-cone synchronizers provide a third surface of friction material.

With multiple-cone synchronizers, the size of the transmission can be reduced. The multiple-cone synchronizers offer a high synchronizer capacity in a smaller package. To obtain the same results in shifting, a single-cone synchronizer would need to have a larger diameter, which would increase the overall size and weight of the transmission.

GEARSHIFT MECHANISMS

Figure 39–19 illustrates a typical transmission shift linkage for a five-speed transmission. As you can see, there are three separate shift rails and forks. Each shift rail/shift fork is used to control the movement of a synchronizer, and each synchronizer is capable of engaging and locking two speed gears to the mainshaft. The shift rails transfer motion from the driver-controlled gearshift lever to the shift forks. The **shift forks (Figure 39–20)** are semicircular

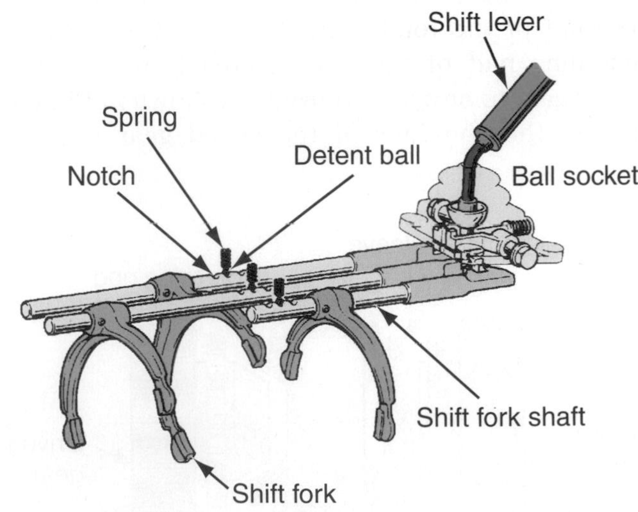

FIGURE 39–19

In a five-speed transmission, three separate shift rail/shift fork/synchronizer combinations control first/second, third/fourth, and fifth/reverse shifting.

©Cengage Learning 2015

FIGURE 39–20

The shift forks and rails are assembled in the cover of this transmission with a direct shift linkage.

castings connected to the shift rails with split pins. The shift fork rests in the groove in the synchronizer sleeve and surrounds about one-half of the sleeve circumference.

The gearshift lever is connected to the shift forks by means of a gearshift linkage. Linkage designs vary among manufacturers but can generally be classified as being direct or remote.

Gearshift Linkages

There are two basic designs of gearshift linkages: internal and external. Internal linkages are located at the side or top of the transmission. The control end of the shifter is mounted inside the transmission, as are all of the shift controls. Movement of the shifter moves a **shift rail** and shift fork toward the desired gear. This moves the synchronizer sleeve to lock the selected speed gear to the shaft. This type of linkage is often called a direct linkage, because the shifter is in direct contact with the internal gear-shifting mechanisms.

Shift rails are machined with **interlock** and **detent** notches. The interlock notches prevent the selection of more than one gear during shifting. When a shift rail is moved by the shifter, interlock pins hold the other shift rails in their neutral position **(Figure 39–21)**. The detent notches and matching spring-loaded pins or balls give the driver feedback as to when the synchronizer sleeve is adequately moved to lock the shift rail in place.

As the shift rail moves, a detent ball moves out of its detent notch and drops into the notch for the selected gear. At the same time, an interlock pin moves out of its interlock notch and into the other shift rails.

External linkages function in much the same way, except that rods, external to the transmission, are connected to levers that move the internal shift

The right interlock plate is moved by the 1-2 shift rail into the 3-4 shift rail slot.

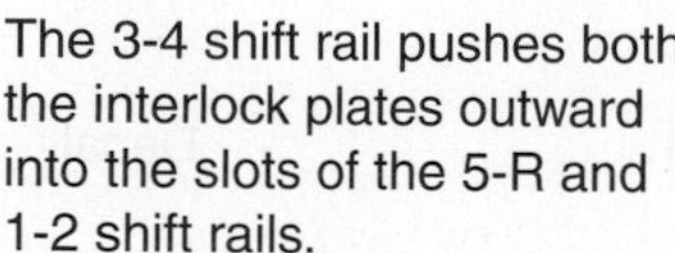

The 3-4 shift rail pushes both the interlock plates outward into the slots of the 5-R and 1-2 shift rails.

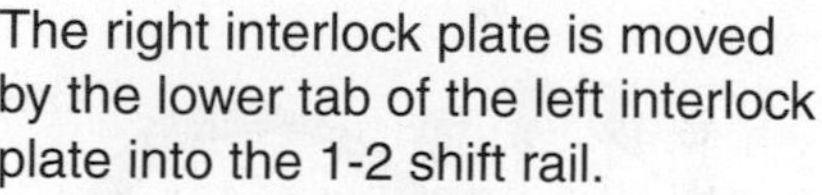

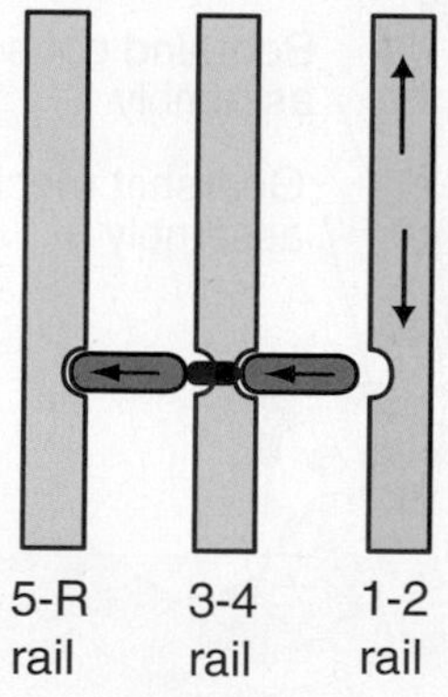

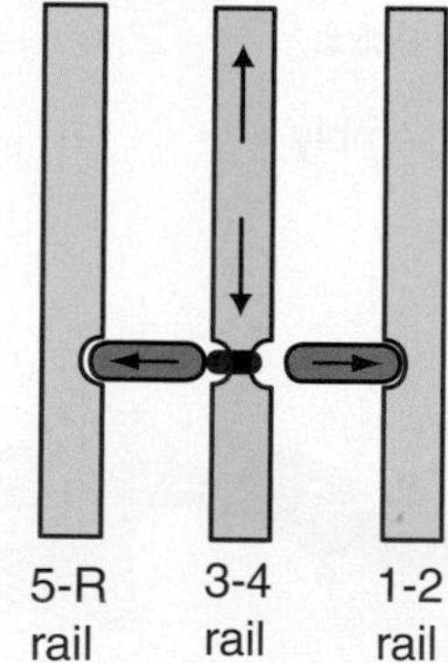

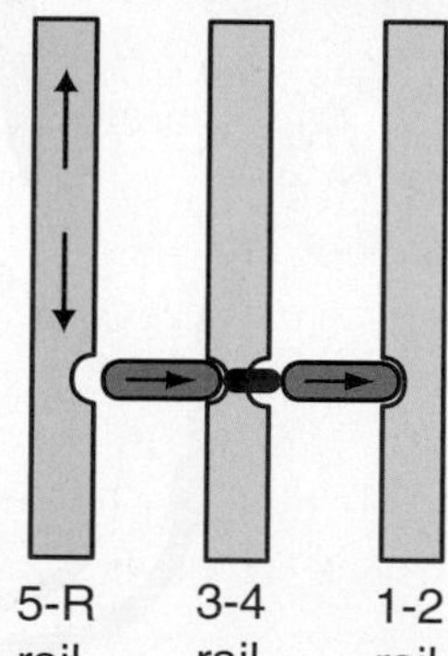

The left interlock plate is moved by the lower tab of the right interlock plate into the 5-R shift rail slot.

3-4 rail

The left interlock plate is moved by the 5-R shift rail into the 3-4 shift rail slot.

FIGURE 39–21

Interlock pins prevent the selection of one or more gears.

FIGURE 39–22

An external shifter assembly mounted to the transmission.

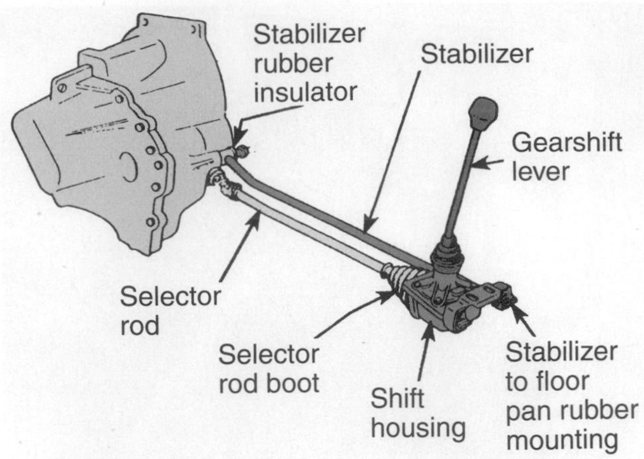

FIGURE 39–23

A remote gearshift showing linkage, selector rod, and stabilizer (stay bars).

Ford Motor Company

rails of the transmission **(Figure 39–22)**. Some transaxles are shifted by selector rods **(Figure 39–23)** or by cable **(Figure 39–24)**. These are remote design shift linkages.

TRANSMISSION POWER FLOW

The following sections describe the power-flow paths in a typical five-speed manual transmission.

Neutral

Neutral power flow is illustrated in **Figure 39–25**. The input shaft rotates at engine speed whenever the clutch is engaged. The clutch gear is mounted on the input shaft and rotates with it. The clutch gear meshes with the countergear, which rotates around the countershaft.

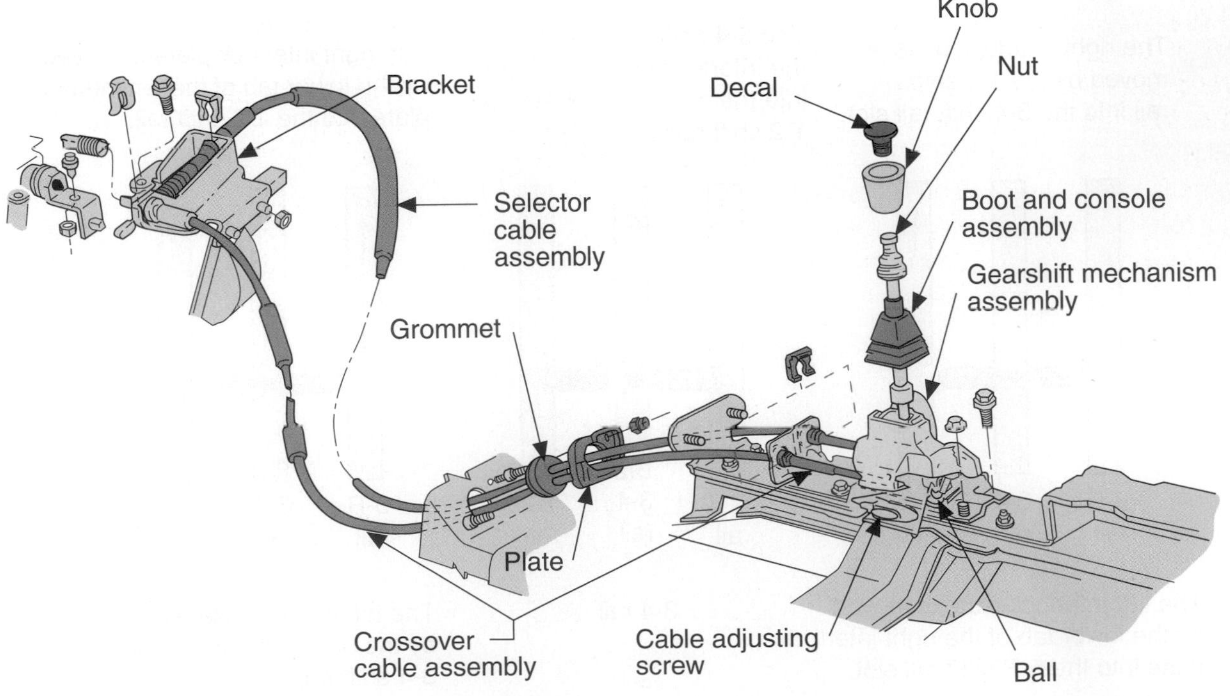

FIGURE 39–24

A cable-type external gearshift linkage used in a transaxle application.

Chrysler Group LLC

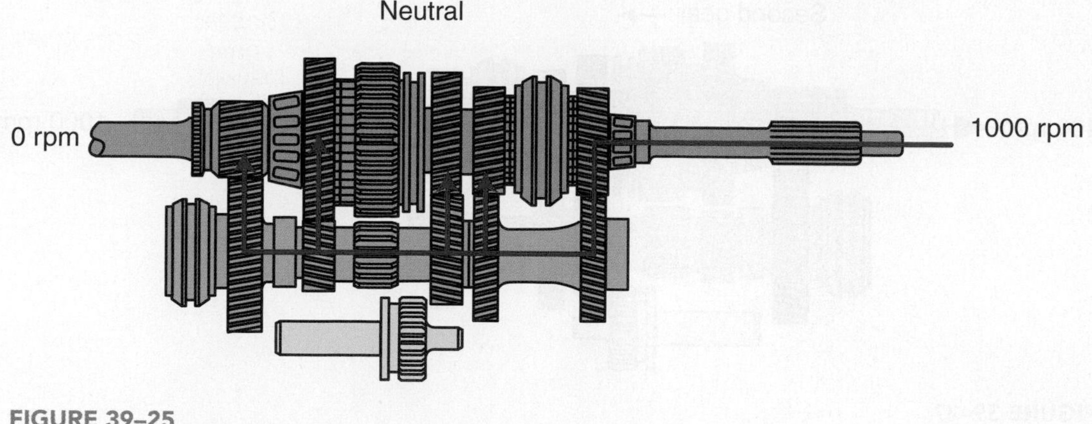

FIGURE 39–25

Transmission power flow in neutral.

The countergear transfers power to the speed gears on the mainshaft. However, because speed gears 1, 2, 3, and 4 are not locked to the mainshaft when the transmission is in neutral, they cannot transfer power to the mainshaft. The mainshaft does not turn, and there is no power output to the driveline.

All gear changes pass through the neutral gear position. When changing gears, one speed gear is disengaged, resulting in neutral, before the chosen gear is engaged. This is important to remember when diagnosing hard-to-shift problems.

First Gear

First gear power flow is illustrated in **Figure 39–26**. Power or torque flows through the input shaft and clutch gear to the countergear. The countergear rotates. The first gear on the cluster drives the first speed gear on the mainshaft. When the driver selects first gear, the first/second synchronizer sleeve moves to the rear to engage the first speed gear and lock it to the mainshaft. The first speed gear drives the main (output) shaft, which transfers power to the driveline. A typical first speed gear ratio is 3:1 (three full turns of the input shaft to one full turn of the output shaft). So if the engine torque entering the transmission is 298 N·m (220 ft·lb), it is multiplied three times to 895 N·m (660 ft·lb) by the time it is transferred to the driveline.

Second Gear

When the shift from first to second gear is made, the shift fork disengages the first/second synchronizer sleeve from the first speed gear and moves it until it locks the second speed gear to the mainshaft. Power flow is still through the input shaft and clutch gear to the countergear. However, now the second countergear on the cluster transfers power to the second speed gear locked on the mainshaft. Power flows from the second speed gear through the synchronizer to the mainshaft (output shaft) and driveline **(Figure 39–27)**.

In second gear, the need for vehicle speed and acceleration is greater than the need for maximum torque multiplication. To meet these needs, the second speed gear on the mainshaft is designed

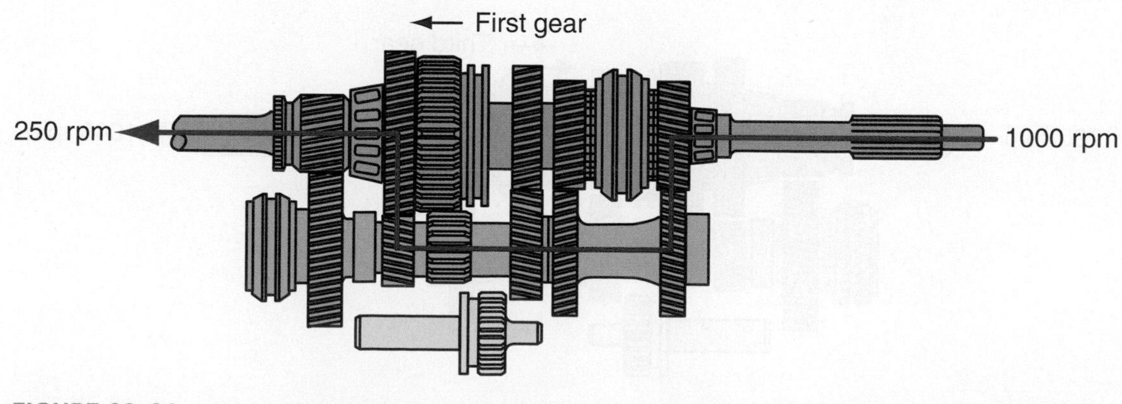

FIGURE 39–26

Transmission power flow in first gear.

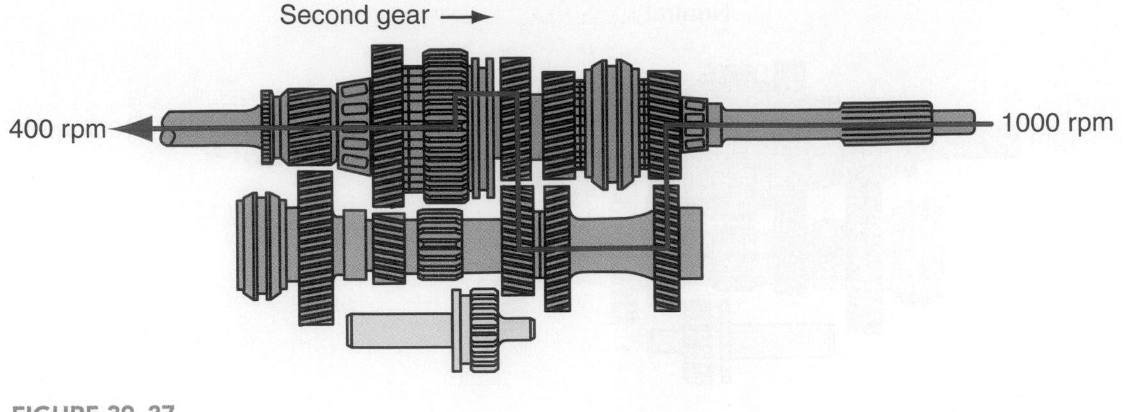

FIGURE 39–27

Transmission power flow in second gear.

slightly smaller than the first speed gear. This results in a typical gear ratio of 2.20:1, which reflects a drop in torque and an increase in speed.

Third Gear

When the shift from second to third gear is made, the shift fork returns the first/second synchronizer sleeve to its neutral position. A second shift fork slides the third/fourth synchronizer sleeve until it locks the third speed gear to the mainshaft. Power flow now goes through the third gear of the countergear to the third speed gear, through the synchronizer to the mainshaft and then to the driveline **(Figure 39–28)**.

Third gear permits a further decrease in torque and increase in speed. As you can see, the third speed gear is smaller than the second speed gear. This results in a typical gear ratio of 1.70:1.

Fourth Gear

In fourth gear, the third/fourth synchronizer sleeve is moved to lock the clutch gear on the input shaft to the mainshaft. This means power flow is directly from the input shaft to the mainshaft (output shaft) at a ratio of 1:1 **(Figure 39–29)**. This ratio results in maximum speed output and no torque multiplication. Fourth gear has no torque multiplication because it is used at cruising speeds to promote maximum fuel economy. The vehicle is normally downshifted to lower gears to take advantage of torque multiplication and acceleration when passing slower vehicles or climbing grades.

Fifth Gear

When fifth gear is selected, the fifth gear synchronizer sleeve locks the fifth speed gear to the mainshaft **(Figure 39–30)**. This causes a large gear on the countershaft to drive a smaller gear on the mainshaft, which results in an overdrive condition. Overdrive permits an engine speed reduction at higher vehicle speeds and a reduction in torque output in relation to input torque.

Reverse

In reverse gear, it is necessary to reverse the direction of the mainshaft (output shaft). This is done by introducing a reverse idler gear into the power flow path. The idler gear is located between the countershaft

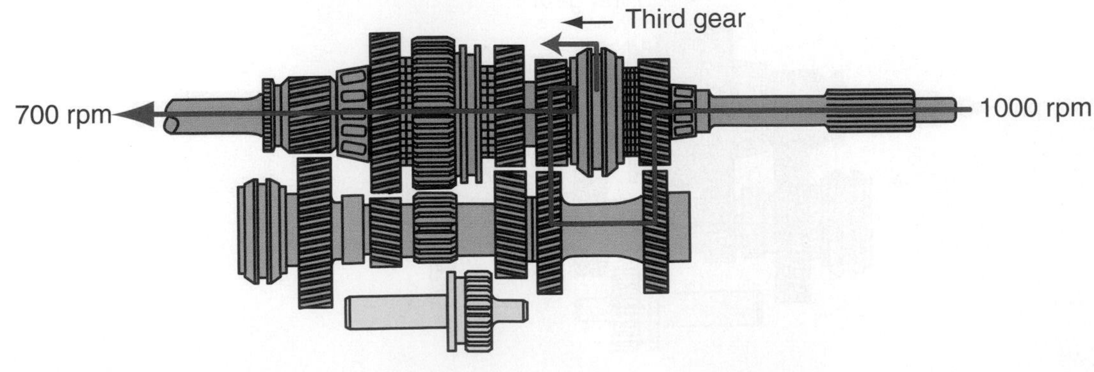

FIGURE 39–28

Transmission power flow in third gear.

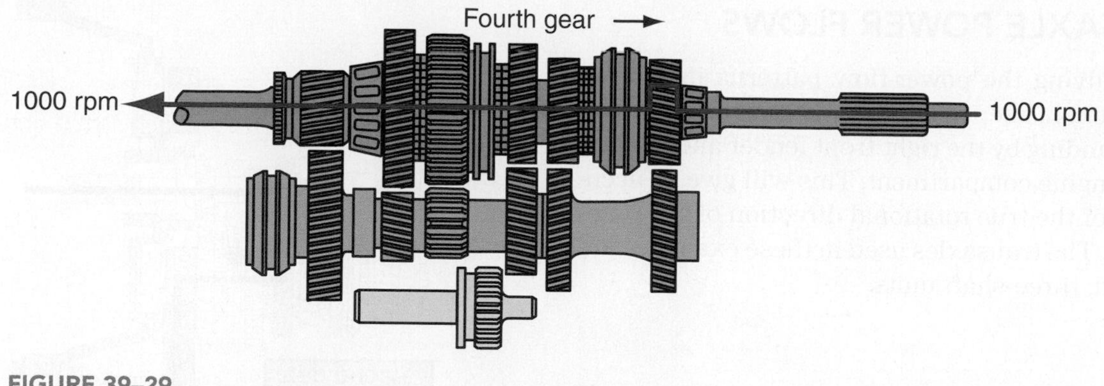

FIGURE 39–29

Transmission power flow in fourth gear.

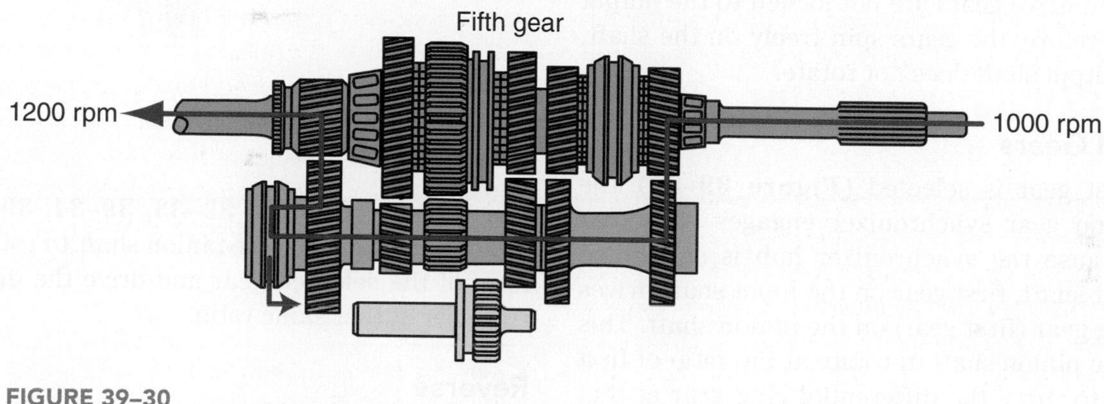

FIGURE 39–30

Transmission power flow in fifth gear.

reverse gear and the reverse gear on the mainshaft. The idler assembly is made of a short shaft independently mounted in the transmission case parallel to the countershaft. The idler gear may be mounted near the midpoint of the shaft.

The reverse gear is actually the external tooth sleeve of the first/second synchronizer.

When reverse gear is selected, both synchronizers are disengaged and in the neutral position. In the transmission shown in **Figure 39–31**, the shifting linkage moves the reverse idler gear into mesh with the first/second synchronizer sleeve. Power flows through the input shaft and clutch gear to the counter driven gear and countergear assembly. From the reverse gear on the counter/cluster shaft, torque then passes to the reverse idler gear, where it changes rotational direction. It then passes to the first/second synchronizer sleeve. Rotational direction is again reversed. From the sleeve, power passes to the mainshaft and driveline.

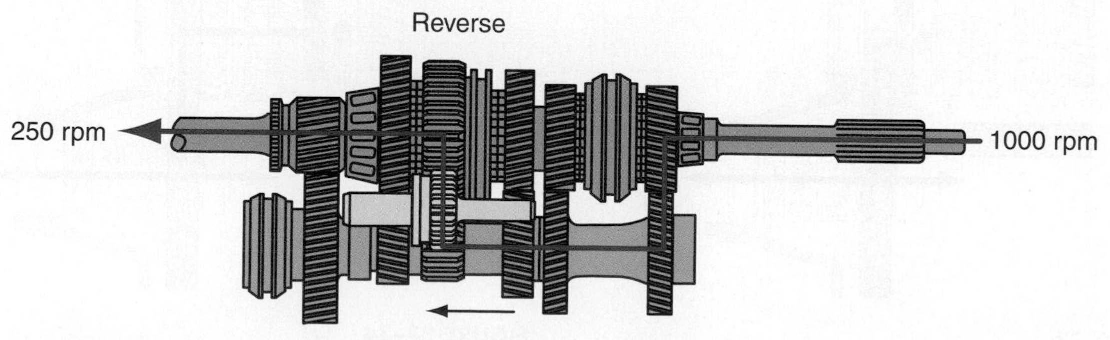

FIGURE 39–31

Transmission power flow in reverse gear.

TRANSAXLE POWER FLOWS

When studying the power-flow patterns in the following section, keep in mind that the views are based on you standing by the right front fender and looking into the engine compartment. This will give an accurate idea of the true rotational direction of the gears and shaft. The transaxles used in these examples are five-speed, three-shaft units.

Neutral

When a transaxle is in its position, no power is applied to the differential. Because the synchronizer collars are centred between their gear positions, the meshed drive gears are not locked to the output shaft. Therefore, the gears spin freely on the shaft, and the output shaft does not rotate.

Forward Gears

When first gear is selected **(Figure 39–32)**, the first/second gear synchronizer engages with first gear. Because the synchronizer hub is splined to the output shaft, first gear on the input shaft drives its mating gear (first gear) on the pinion shaft. This causes the pinion shaft to rotate at the ratio of first gear and to drive the differential ring gear at that same ratio.

As the other forward gears are selected, the appropriate shift fork moves to engage the selected synchronizer with the gear. Because the synchronizer's hub is splined to the pinion shaft, the desired gear on the input shaft drives its mating gear on the pinion shaft **(Figures 39–33, 39–34, 39–35,** and **39–36)**. This causes the pinion shaft to rotate at the ratio of the selected gear and drive the differential ring gear at that same ratio.

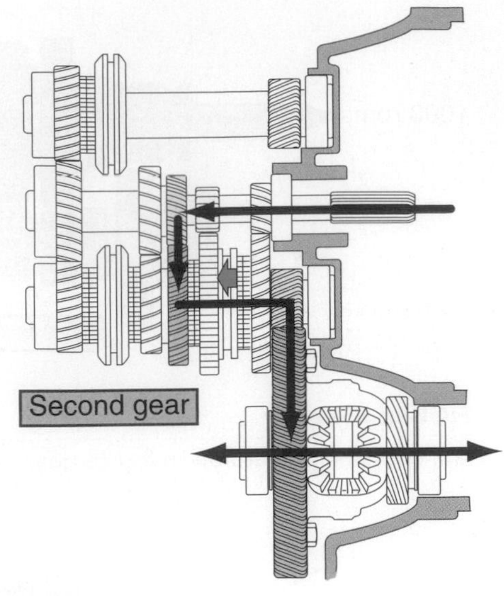

FIGURE 39–33

Transaxle power flow in second gear.

Reverse

When reverse gear is selected on transaxles that use a sliding reverse gear **(Figure 39–37)**, the shifting fork forces the gear into mesh with the input and pinion shafts. The addition of this third gear reverses the normal rotation of first gear and allows the car to change direction.

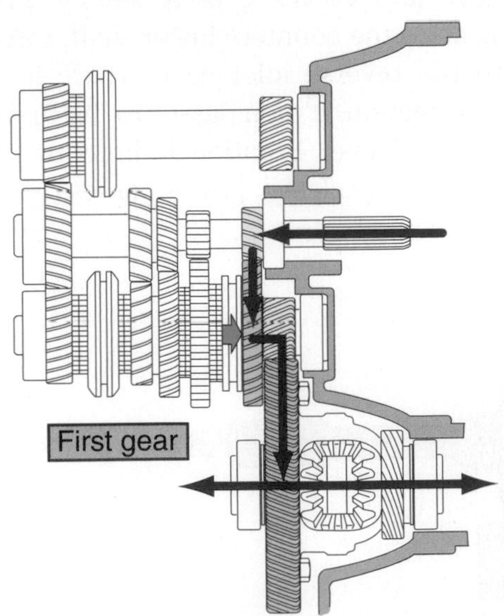

FIGURE 39–32

Transaxle power flow in first gear.

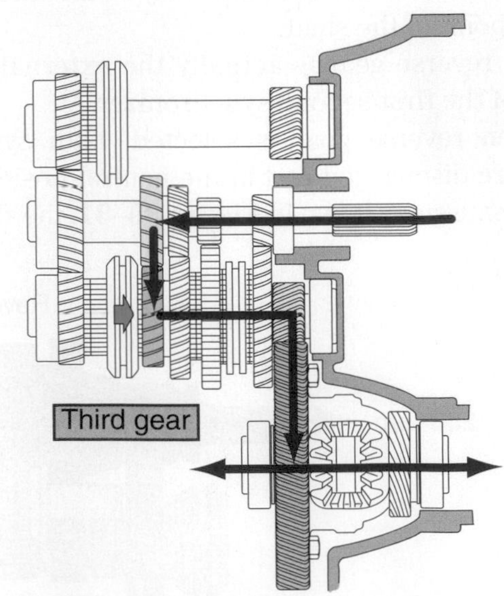

FIGURE 39–34

Transaxle power flow in third gear.

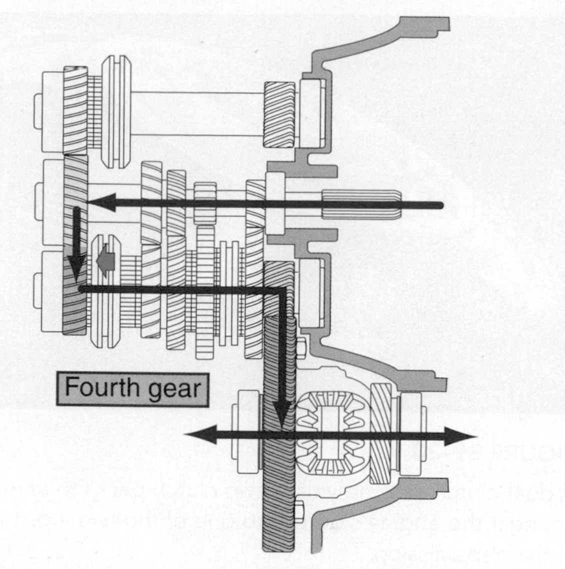

FIGURE 39–35

Transaxle power flow in fourth gear.

©Cengage Learning 2015

Differential Action

The final drive ring gear is driven by the transaxle's pinion shaft. The ring gear then transfers the power to the differential case. The case holds the ring gear with its mating pinion gear. The differential side gears are connected to the drive axles.

One major difference between the differential in an RWD car and the differential in a transaxle is direction of power flow. In an RWD differential, the power flow changes 90° between the drive pinion gear and

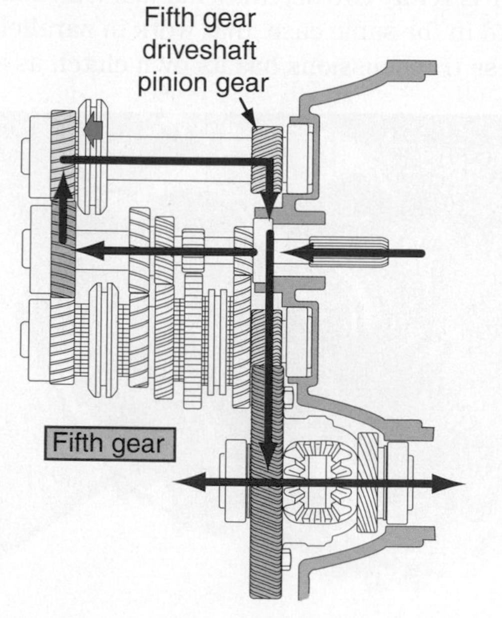

FIGURE 39–36

Transaxle power flow in fifth gear.

©Cengage Learning 2015

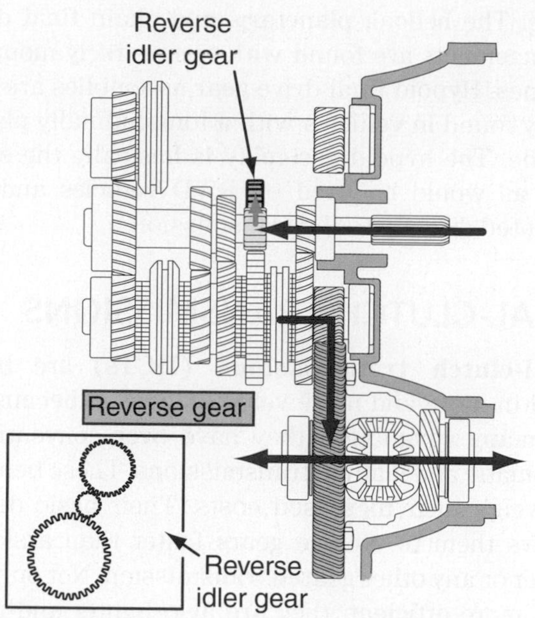

FIGURE 39–37

Transaxle power flow in reverse gear.

©Cengage Learning 2015

the ring gear. This change in direction is not needed with most FWD cars. The transverse engine position places the crankshaft so that it already is rotating in the correct direction. Therefore, the purpose of the differential is only to provide torque multiplication and divide the torque between the drive axle shafts so that they can rotate at different speeds.

Some transaxles need the 90° power flow change in the differential. These units are used in rear-engine with RWD applications, in longitudinally positioned engines with FWD, or some AWD vehicles.

FINAL DRIVE GEARS AND OVERALL RATIOS

All vehicles use a gearset to provide an additional gear reduction (torque increase) above and beyond what the transmission or transaxle gearing can produce. This is known as the **final drive gear**.

In a transmission-equipped vehicle, the final drive gearing is located in the rear axle housing. In a transaxle, however, the final reduction is produced by the final drive gears housed in the transaxle case.

A transaxle's final drive gears provide a way to transmit the transmission's output to the differential section of the transaxle. The pinion and ring gears and the differential assembly are normally located within the transaxle. There are four common configurations used as the final drives on FWD vehicles: helical gear, planetary gear, hypoid gear, and chain drive. The helical, planetary, and chain final drive arrangements are found with transversely mounted engines. Hypoid final drive gear assemblies are normally found in vehicles with a longitudinally placed engine. The hypoid assembly is basically the same unit as would be used on RWD vehicles and are mounted directly to the transmission.

DUAL-CLUTCH TRANSMISSIONS

Dual-clutch transmissions (DCTs) are being used in more and more vehicles. This is because of the many advantages they have over conventional automatic and manual transmissions. These benefits outweigh their increased costs. Their basic design allows them to change gears faster than a skilled driver or any other geared transmission. Not only are they more efficient, they are also lighter and more durable, and they require no regular maintenance. They also do not rely on a torque converter or planetary gearsets that can waste engine power. As a result, fuel consumption can decrease by 10 percent or more, and the driver will feel increased control over the engine's power output.

A dual-clutch transmission is essentially a fully automated, manual transmission with a computer-controlled clutch. DCTs have gear shafts fitted with helical and spur gears (just like most manual transmissions) to provide the various speed ratios. However, DCTs can operate in a fully automatic mode and drive like a very efficient automatic transmission. Current DCTs have six or seven forward speeds and are available from several manufacturers, and each may call the DCT by a different name and may have slightly different control mechanisms.

Input from the Engine

Dual-clutch transmissions use wet or dry clutch assemblies. These clutches are very similar; both have a compact multiplate design **(Figure 39–38)**. Some transmissions have a wet and dry clutch. Dry clutches are typically used in DCTs with low-power engines and in FWD vehicles **(Figure 39–39)**. They are smaller, lighter, more reliable, and less expensive to manufacture, and they can provide better fuel economy than wet clutch–equipped transmissions. This is due to their high clamping pressures that reduce the power lost as torque flows from the engine to the transmission and speed gears. DCTs have two separate clutch assemblies. The disadvantage of a dry clutch is that they wear quickly.

FIGURE 39–38

A dual-clutch assembly uses two clutch packs to alternately connect the engine's output to one of the two input shafts in the transmission.

BorgWarner

A wet clutch is commonly used with large, powerful engines. In a wet clutch assembly, the clutch discs are housed in a drum and are completely submerged in oil. The oil cools the discs, which allows the discs to have a relatively long service life. The cooling of the oil is also the reason why transmissions fitted to powerful engines have wet clutches. A typical wet clutch assembly has friction discs, friction plates, and a pressure plate within the drum. When engaged, hydraulic pressure forces the parts together and passes engine torque to the appropriate input shaft.

Operation

A DCT is really two separate manual transmissions, housed in the same case, that work in parallel. Each of these transmissions has its own clutch assembly.

FIGURE 39–39

A six-speed DCT with dry clutches for an FWD vehicle.

©Cengage Learning 2015

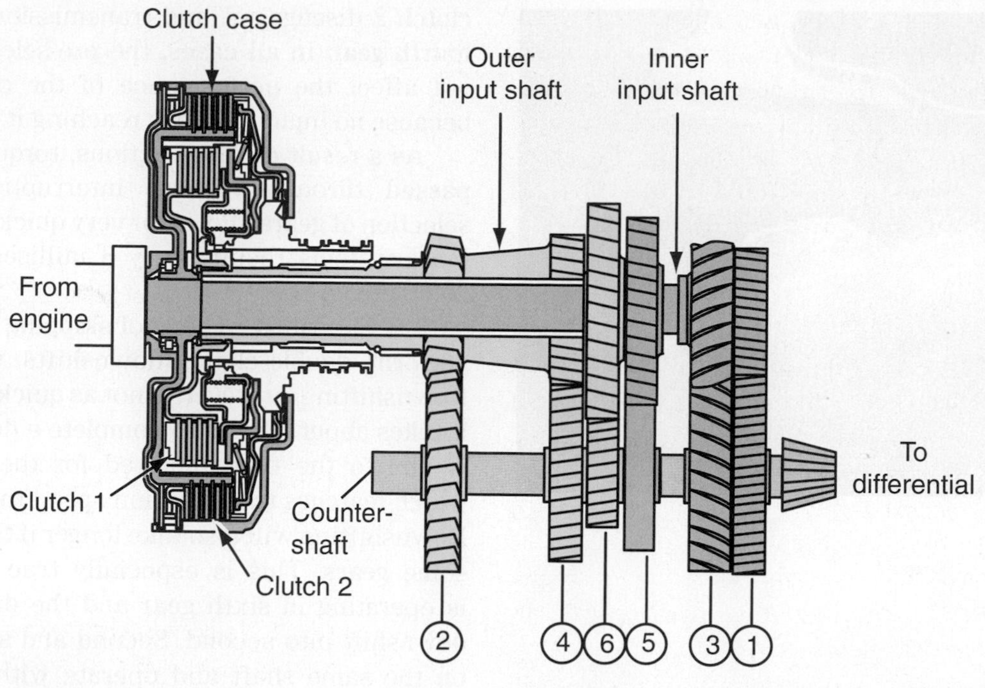

FIGURE 39–40

The basic layout for a six-speed DCT.

One of the basic transmission assemblies consists of the even-numbered gears, and the other contains the odd-numbered gears. For example, in a six-speed DCT, one clutch and transmission assembly is for first, third, and fifth gears; the other clutch and transmission is for second, fourth, and sixth gears. In order to do this, a two-part input shaft is used. Again, each part of the shaft works with the even or odd gears. The outer input shaft is hollow and has the inner shaft running through it **(Figure 39–40)**. Reverse gear is typically tied to the odd shaft.

A DCT has no clutch pedal. The clutches are engaged and disengaged by electrohydraulic actuators. The actual shifting of gears is completed by computer-controlled solenoids and hydraulics or electric motors and threaded shift rods. Dual-clutch transmissions can be shifted either manually or automatically. In the automatic mode, the computer selects the proper gear for the conditions. To operate the unit manually, the driver changes gears with buttons, paddles, or a shifter. Most dual-clutch transmissions have a traditional P-R-N-D shift pattern **(Figure 39–41)**. The transmission can shift automatically in either NORMAL (D) or SPORT (S) modes. While in the NORMAL mode, the DCT shifts into the higher gears early in order to minimize engine noise and maximize fuel economy. When the SPORT mode is selected, the transmission holds the lower gears longer for improved performance. The driver can also elect to operate the transmission manually by either sliding the shift lever to the side or pulling one of the paddles on the steering wheel **(Figure 39–42)**.

Instantaneous shifting is accomplished by complex electronic controls that change the power flow from one clutch and transmission assembly to the other when a change in gears is ordered

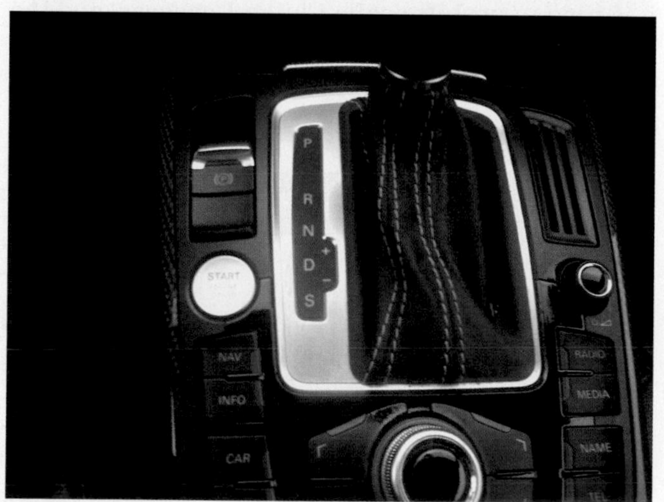

FIGURE 39–41

A shifter control for a DCT offering automatic and manually controlled shifting.

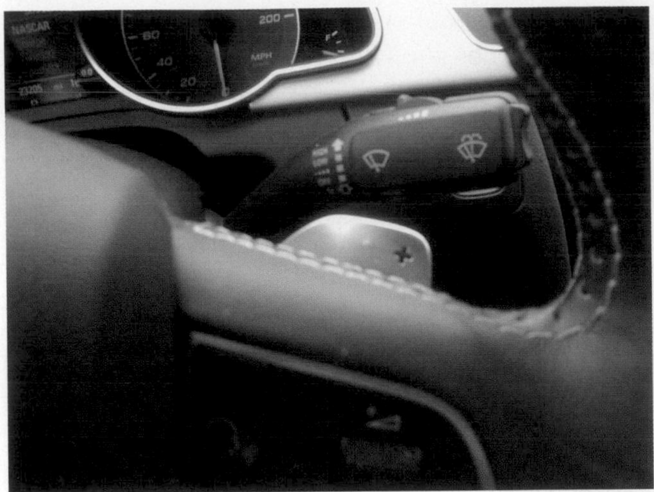
FIGURE 39–42

Shift paddles for a DCT.

©Cengage Learning 2015

(Figure 39–43). The electronic controls also have a predictive feature that pre-selects the next gear in the sequence.

To understand this concept, consider the power flow through a DCT. When the vehicle is at a standstill and the driver begins to accelerate, transmission 1 is in first gear while transmission 2 is in second. Clutch 1 engages, and the vehicle moves in first gear. When it is time to upshift, clutch 2 engages and clutch 1 disengages.

Because second gear was already engaged, the vehicle is now operating in second gear. As soon as clutch 1 is disengaged, transmission 1 immediately shifts into third gear. This sequence continues when it is time to upshift again: Clutch 1 engages, and the vehicle is operating in third gear. At the same time, clutch 2 disengages, and transmission 2 shifts into fourth gear. In all cases, the pre-selected gear will not affect the performance of the operating gear because no input torque is reaching it.

As a result of these actions, torque is smoothly passed through without interruption. The pre-selection of gears allows for very quick gear changes. Most systems require only 8 milliseconds (ms) to complete an upshift.

A DCT is also capable of skipping gears and will perform double-clutch downshifts when needed. Downshifting, however, is not as quick as upshifting. It takes about 600 ms to complete a downshift. This is due to the time required for the throttle blip, which matches transmission speed to engine speed. Downshifting will also take longer if the driver skips some gears. This is especially true if the vehicle is operating in sixth gear and the driver wants to downshift into second. Second and sixth gears are on the same shaft and operate with same clutch. To provide this change, the transmission must first downshift into fifth gear because it is on the other shaft. Then the transmission can engage second gear and the associated clutch. Typically, this sequence takes 900 ms; this is not a long time but is longer than upshifting.

Volkswagen/Audi Direct-Shift Gearbox (DSG)

Dual-clutch transmissions were first put into production by BorgWarner. This transmission was called a DualTronic unit. Volkswagen licensed the technology and uses BorgWarner clutches and control modules for their dual-clutch transmissions. Dual-clutch transmissions initially appeared in an

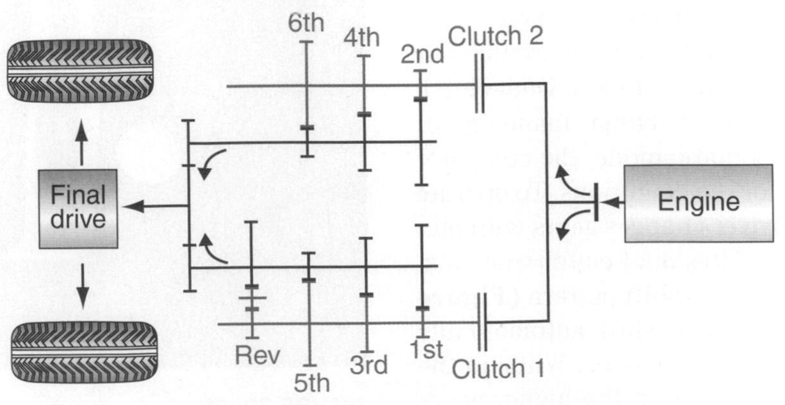

FIGURE 39–43

Power flow through a DCT.

©Cengage Learning 2015

Audi with the name **direct-shift gearbox (DSG)**, sometimes called a dual-shaft transmission. Since then, it has been available in several other Audi and Volkswagen models. The use of a DCT in the Bugatti Veyron EB 16.4 demonstrates that this transmission can handle large amounts of engine power. The Veyron's W16 engine is rated at 736.3 kW (987 peak horsepower).

Currently, the Volkswagen Group has more models available with a DCT than any other manufacturer. The most common is a six-speed transaxle with dry clutches. DSGs with seven speeds and wet clutches are also found in VW Group vehicles. The DSG's electronic control unit (ECU) is mounted to the transaxle. The ECU has adaptive learning, which allows it to learn how the vehicle is typically driven and will attempt to match the shift points to the driver. The ECU controls a hydro-mechanical unit. The control module is most complex component of a DSG transmission **(Figure 39–44)**. The transmission can be operated totally by the ECU, or the driver can manually shift gears with a shift lever or steering wheel–mounted paddles. The ECU controls the clutches, input and output shafts, transaxle cooling, gear selection and timing, and hydraulic pressures. In addition to the various DSG-specific sensors, the ECU relies on bus data to determine the current operating conditions.

Ford's PowerShift

The dual-clutch Ford **PowerShift** transmission is a six-speed transmission with dry clutches **(Figure 39–45)**. Ford has designed a dry clutch system that can handle engines with a broad torque range. Also, dry clutch systems are smaller, lighter, easier to service, more reliable, and cheaper than wet clutch

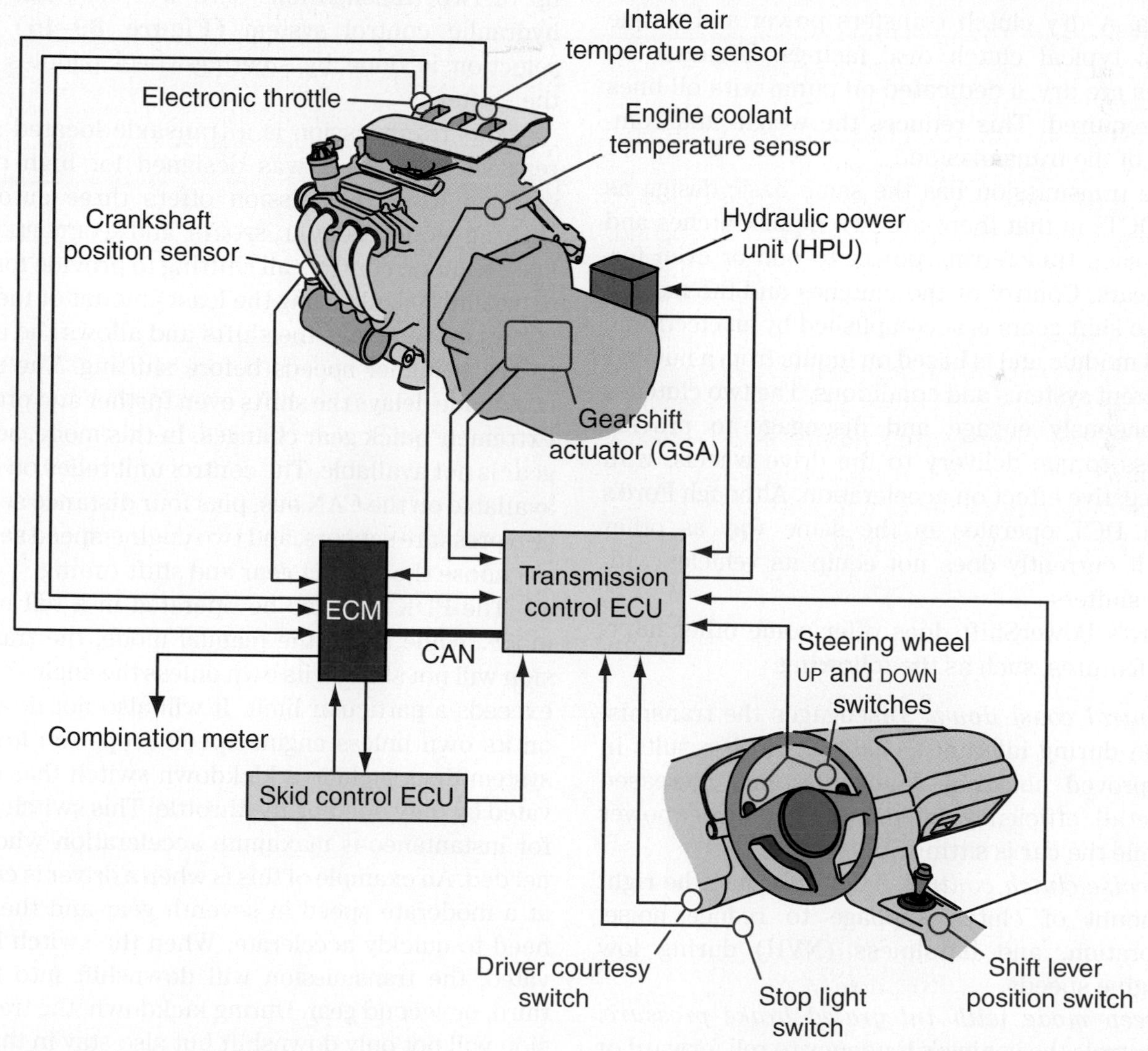

FIGURE 39–44

The main control components for a DSG.

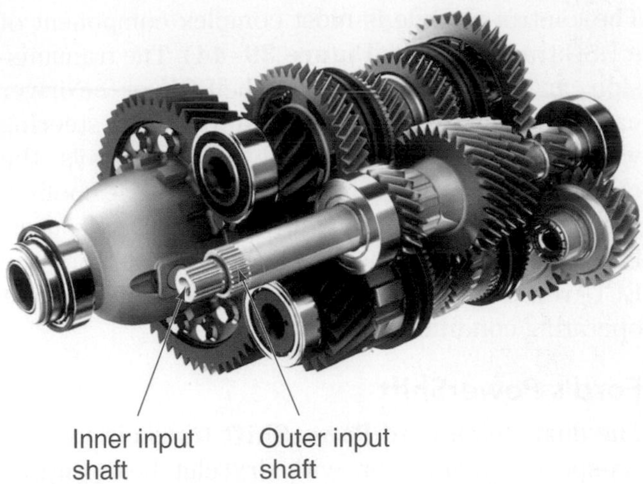

FIGURE 39–45

The gear train for a Ford PowerShift transaxle.
Ford Motor Company

systems. A dry clutch transfers power and torque through typical clutch disc facings. Because the clutches are dry, a dedicated oil pump with oil lines is not required. This reduces the weight and complexity of the transmission.

The transmission has the same basic design as other DCTs in that there are two input clutches and shafts, each transferring power to odd or even forward gears. Control of the clutches and the mechanisms to shift gears is accomplished by an electronic control module and is based on inputs from a number of different systems and conditions. The two clutches simultaneously engage and disengage to provide seamless torque delivery to the drive wheels, with little negative effect on acceleration. Although Ford's current DCT operates in the same way as other DCTs, it currently does not equip its vehicles with paddle shifters.

Ford's PowerShift does offer some other interesting features, such as the following:

- *Neutral coast down.* Disengages the transmission during idle and coasting, which results in improved coasting downshifts and increased overall efficiency by not using engine power while the car is sitting still.
- *Precise clutch control.* Allows for just the right amount of clutch slippage to reduce noise, vibration, and harshness (NVH) during low engine speeds.
- *Creep mode with integrated brake pressure.* Controls the vehicle's tendency to roll forward or backward when stopped and the transmission is in a forward or reverse gear.
- *Hill mode or launch assist.* Prevents the car from rolling back when it is stopped on a steep grade. The system's computer controls brake pressure and the engine to hold the car where it is.
- *SelectShift.* Allows the driver to change gears—up or down—by pressing a button on the shift handle.
- *Overspeed control.* Prevents the driver from downshifting too aggressively, which can cause engine damage by over-revving it.

Porsche Dual Klutch (PDK)

Porsche was one of the early developers of a dual-clutch transmission. Recently, the manufacturer introduced the **Porsche Dual Klutch (PDK)** transmission (the official name is *Porsche-Doppelkupplung*). This is a seven-speed dual-clutch transmission. Like other DCTs, the drive unit is made up of two transmissions, two wet clutches, and a hydraulic control system **(Figure 39–46)**. Gear selection is done by steering-wheel paddles or at the shifter.

This transmission is a transaxle located at the rear wheels, and it was designed for high output engines. The transmission offers three automatic shifting modes: NORMAL, SPORTS, and SPORTS PLUS. The NORMAL mode controls all shifting to provide for good driveability while using the least amount of fuel. The SPORTS mode delays the shifts and allows the engine to reach higher speeds before shifting. The SPORTS PLUS mode delays the shifts even further and provides extremely quick gear changes. In this mode, seventh gear is not available. The control unit relies on inputs available on the CAN bus, plus four distance sensors, two pressure sensors, and two engine-speed sensors, to choose the correct gear and shift timing.

The PDK can also be operated in a full manual mode. While set in the manual mode, the transmission will not shift on its own unless the engine's speed exceeds a particular limit. It will also not downshift on its own unless engine speed drops too low. The system does include a kickdown switch that is activated by fully opening the throttle. This switch allows for instantaneous maximum acceleration when it is needed. An example of this is when a driver is cruising at a moderate speed in seventh gear and there is a need to quickly accelerate. When the switch is activated, the transmission will downshift into fourth, third, or second gear. During kickdown, the transmission will not only downshift but also stay in that gear until the engine redlines; then it will upshift one gear at a time as long as the throttle is fully depressed.

FIGURE 39–46

The power flow through the two input shafts of this PDK.

Cars equipped with the optional Sports Chrono Plus package have a launch-control program that aggressively applies (dumps) the clutch automatically once 6500 rpm has been reached, after the brake pedal is depressed, the throttle is opened completely, and the brake pedal is released. This allows the car to accelerate under full power without wheel slip or an interruption of traction.

ELECTRICAL SYSTEMS

Although manual transmissions are not electrically operated or controlled, a few accessories of the car are controlled or linked to the transmission. The transmission may also be fitted with sensors that give vital information to the computer that controls other car systems. There are a few transmissions that have their shifting controlled or limited by electronics.

Reverse Lamp Switch

All vehicles sold in North America after 1971 have been required to have backup (reverse) lights. Backup lights illuminate the area behind the vehicle and warn other drivers and pedestrians that the vehicle is moving in reverse. Most manual transmissions are equipped with a separate switch located on the transmission **(Figure 39–47)** but can be mounted to the shift linkage away from the transmission. If the switch is mounted in the transmission, the shifting fork closes the switch and

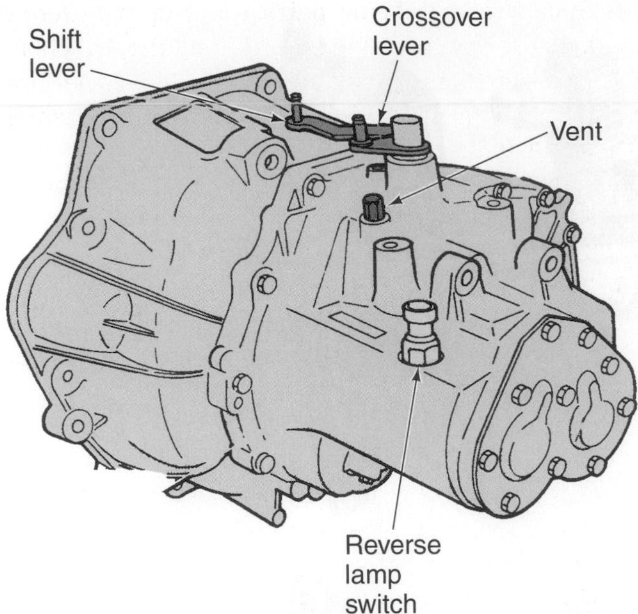

FIGURE 39–47
The typical location of a backup light switch in a transaxle.
Chrysler Group LLC

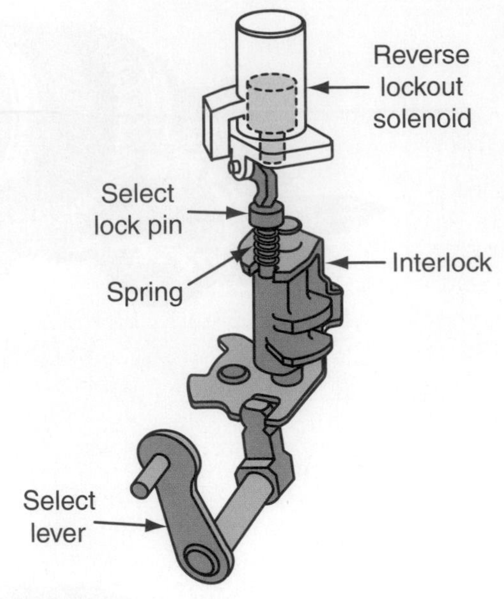

FIGURE 39–48
A PCM-controlled reverse lockout assembly.

completes the electrical circuit whenever the transmission is shifted into reverse gear. If the switch is mounted on the linkage, the switch is closed directly by the linkage.

Vehicle Speed Sensor

Most late-model transmissions and transaxles are fitted with a vehicle speed sensor (VSS). This sensor sends an electrical signal to the vehicle's PCM. This signal represents the speed of the transmission's output shaft. The PCM then calculates the speed of the vehicle. This information is used for many systems, such as cruise control, fuel and spark management, and instrumentation.

Reverse Lockout System

Some vehicles electronically prevent the transmission from being shifted into reverse when the vehicle is moving. These systems are typically called reverse lockout systems **(Figure 39–48)**. Normally, when the vehicle is moving at a speed of 20 km/h (12 mph) or more, a signal from the PCM energizes the reverse lockout solenoid. The solenoid pushes a cam and lock pin down. This action makes it impossible for the transmission to be shifted into reverse. When the vehicle is sitting still or travelling at lower speeds, the solenoid is not energized. A return spring holds the lock pin away from the interlock assembly, and the shifter is free to move into reverse.

Shift Blocking

Some six-speed transmissions have a feature called shift blocking that prevents the driver from shifting into second or third gears from first gear when the engine's coolant temperature is below a specified degree; the speed of the car is between 20 and 30 km/h (12 and 22 mph); and the throttle is opened less than 35 percent. Shift blocking helps improve fuel economy. These transmissions are also equipped with reverse lockout, as are some others. This feature prevents the engagement of reverse whenever the vehicle is moving forward.

Shift blocking is controlled by the PCM. A "skip shift" solenoid is used to block off the shift pattern from first gear to second and third gears. The driver moves the gearshift from its up position to a lower position, as if shifting into second gear, but fourth gear is selected.

CUSTOMER CARE

Just because the technician gets a little dirty in the course of a repair does not mean the vehicle should too. Treat every car that enters the shop with the utmost care and consideration. Scratches from belt buckles or tools and grease smears on the steering wheel, upholstery, or carpeting are inexcusable, and a sure way of losing business. Always use fender, seat, and floor covers when the job requires them. Check your hands for cleanliness before driving a vehicle or operating the windows and dash controls.

KEY TERMS

Block synchronizer (p. 1184)
Blocking rings (p. 1185)
Case (p. 1176)
Cluster gear (p. 1183)
Clutch gear (p. 1183)
Cone synchronizer (p. 1184)
Constant mesh (p. 1177)
Countergear (p. 1183)
Countershaft (p. 1182)
Detent (p. 1187)
Direct-shift gearbox (DSG) (p. 1197)
Dual-clutch transmissions (DCTs) (p. 1194)
External toothed gears (p. 1179)
Final drive gear (p. 1193)
Fully synchronized (p. 1177)
Gear pitch (p. 1178)
Helical gear (p. 1179)
Idler gear (p. 1180)
Input shaft (p. 1182)
Interlock (p. 1187)
Mainshaft (p. 1182)
Output shaft (p. 1182)
Pinion shaft (p. 1183)
Porsche Dual Klutch (PDK) (p. 1198)
PowerShift (p. 1197)
Shift forks (p. 1186)
Shift rail (p. 1187)
Shifter plates (p. 1185)
Speed gears (p. 1183)
Spur gear (p. 1178)
Synchronizer (p. 1183)

SUMMARY

- A transmission or transaxle uses meshed gears of various sizes to give the engine a mechanical advantage over its driving wheels.
- Transaxles contain the gear train plus the final drive gearing needed to produce the final drive gear ratios. Transaxles are commonly used on front-wheel-drive vehicles.
- Transmissions are normally used on rear-wheel-drive vehicles.
- Gears in the transmission/transaxle transmit power and motion from an input shaft to an output shaft. These shafts are mounted parallel to one another.
- Spur gears have straight-cut teeth, whereas helical gears have teeth cut at an angle. Helical gears run without creating gear whine.
- When a small gear drives a larger gear, output speed decreases but torque (power) increases.
- When a large gear drives a smaller gear, output speed increases but torque (power) decreases.
- When two external toothed gears are meshed and turning, the driven gear rotates in the opposite direction of the driving gear.
- Synchronizers bring parts rotating at different speeds to the same speed for smooth clash-free shifting. The synchronizer also locks and unlocks the driven (speed) gears to the transmission/transaxle output shaft.
- Idler gears are used to reverse the rotational direction of the output shaft for operating the vehicle in reverse.
- In typical five-speed transmission shift linkage, there are three separate shift rails and forks. Each shift rail/shift fork is used to control the movement of a synchronizer.
- Gear ratios indicate the number of times the input drive gear is turning for every turn of the output driven gear. Ratios are calculated by dividing the number of teeth on the driven gear by the number of teeth on the drive gear. You can also use the rpm speeds of meshed gears to calculate gear ratios.
- A gear ratio of less than one indicates an overdrive condition. This means the driven gear is turning faster than the drive gear. Speed is high, but output torque is low.
- All vehicles use a gearset in the final drive to provide additional gear reduction (torque increase) above and beyond what the transmission or transaxle gearing can produce.

REVIEW QUESTIONS

1. What determines whether a conventional transmission or a transaxle is used?

2. Explain the relationship between output speed and torque.

3. *True or False?* A reverse idler gear changes the direction of torque flow to the opposite direction of engine rotation.

4. Define final drive gear.

5. Explain the role of shift rails and shift forks in the operation of transmissions and transaxles.

6. How is a torque increase achieved in a standard transmission?

a. The input must turn slower than the output.
b. Both shafts must turn in the same direction.
c. The output must turn slower than the input.
d. Both the input and output must turn at the same speed.

7. What is the term for the calculation of the number of gear teeth per unit of measure?

a. ratio **b.** pitch
c. size **d.** load

8. Which of the following gear ratios shows an overdrive condition?
   a. 2.15:1 **b.** 1:1
   c. 0.85:1 **d.** none of the above
9. Which type of gear develops the problem of gear whine at higher speeds?
   a. spur gear **b.** helical gear
   c. spiral bevel **d.** herringbone
10. What occurs when a small gear is used to drive a larger gear?
   a. Torque output will be increased.
   b. Torque output will be decreased.
   c. Output speed will be increased.
   d. Input speed will be decreased.
11. What will a driven gear do when an idler gear is placed between it and a driving gear?
   a. rotates in the same direction as the driving gear
   b. rotates in the opposite direction of the driving gear
   c. remains stationary
   d. causes the driven gear to rotate faster
12. What component is used to ensure that the mainshaft (output shaft) and main (speed) gear to be locked to it are rotating at the same speed?
   a. synchronizer **b.** shift linkage
   c. shift fork **d.** transfer case
13. What do the shift forks in a synchronized transmission control or move?
   a. the synchronizer hub
   b. the synchronizer splines
   c. the synchronizer keys
   d. the synchronizer sleeve
14. In a transaxle, what does the pinion gear on the pinion shaft mesh with?
   a. the reverse idler gear
   b. the ring gear
   c. the countershaft drive gear
   d. the input gear
15. What mechanism prevents two gears from engaging at the same time?
   a. shift detent mechanism
   b. shift interlock mechanism
   c. shift synchronizer mechanism
   d. reverse idler gear shift lever
16. Where are the constant mesh speed gears located?
   a. a press fit on the mainshaft
   b. interference fit on the countershaft
   c. mounted on bearings on the mainshaft
   d. mounted on bearings on the input shaft
17. Which of the following gear ratios generates the highest torque or power output?
   a. 0.85:1 **b.** 2.67:1
   c. 5.23:1 **d.** 11.12:1
18. What occurs in a transmission (with constant mesh gearing) in neutral with the engine running and the clutch engaged?
   a. All forward gears in the transmission revolve.
   b. Only the mainshaft gears revolve.
   c. The main drive gear and the countershaft gears revolve.
   d. No gears revolve.
19. How is the ratio calculated for the reverse gearset?
   a. drive gear divided by idler, times idler divided by driven gear
   b. idler divided by drive gear, times driven gear divided by idler
   c. drive gear divided by driven gear
   d. driven gear divided by drive gear
20. In a transmission, what drives the countergear?
   a. the mainshaft
   b. the pinion shaft
   c. the clutch gear
   d. the idler gear

CHAPTER 40
Manual Transmission/ Transaxle Service

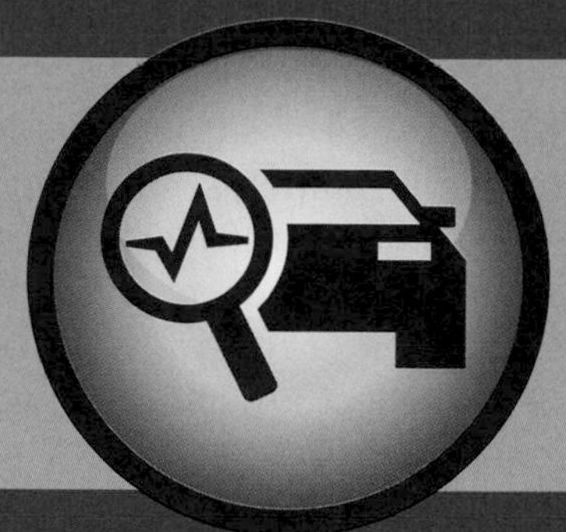

LEARNING OUTCOMES

- Perform a visual inspection of transmission/transaxle components for signs of damage or wear.
- Check transmission oil level correctly, detect signs of contaminated oil, and change oil as needed.
- Describe the steps taken to remove and install transmissions/transaxles, including the equipment and safety precautions used.
- Identify common transmission problems and their probable causes and solutions.
- Describe the basic steps and precautions taken during transmission/ transaxle disassembly, cleaning, inspection, and reassembly procedures.

When properly operated and maintained, a manual transmission/transaxle normally lasts the life of the vehicle without a major breakdown. All units are designed so that the internal parts operate in a bath of lubricant circulated by the motion of the gears and shafts. Some units also use a pump to circulate lubricant to critical wear areas that require more lubrication than the natural circulation provides.

Maintaining good internal lubrication is the key to long transmission/transaxle life. If the amount of lubricant falls below minimum levels, or if the lubricant becomes too dirty, problems result.

SHOP TALK

Whenever you are diagnosing or repairing a transaxle or transmission, make sure you refer first to the appropriate service information before you begin.

Prior to beginning any diagnosis, service, or repair work, be sure you know exactly which transmission you are working on. This will ensure that you are following the correct procedures and specifications and are installing the correct parts. Proper identification can be difficult because transmissions cannot be accurately identified by the way they look. The only positive way to identify the exact design of the transmission is by its identification numbers.

Transmission identification numbers are found either as numbers stamped on the case or on a metal tag held by a bolt head. Use service information to decipher the identification number. Most identification numbers include the model, gear ratios, manufacturer, and assembly date **(Figure 40–1)**. Whenever you work with a transmission with a metal ID tag, make sure the tag is put back on the transmission so that the next technician will be able to properly identify the transmission.

If the transmission does not have an ID tag, the transmission must be identified by comparing it with those in the vehicle's service information.

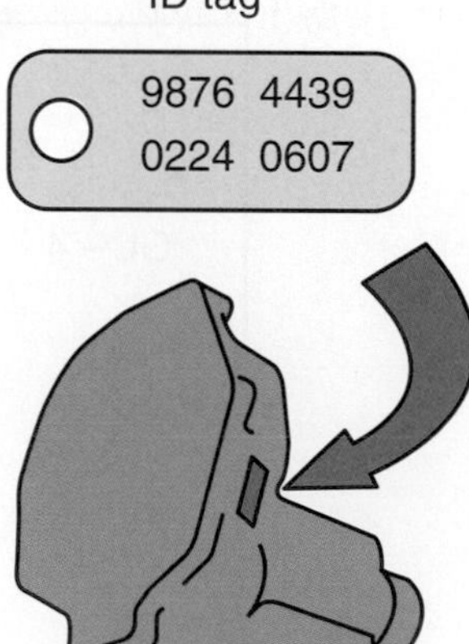

FIGURE 40–1
A typical transmission ID tag.

LUBRICANT CHECK

The transmission/transaxle gear lubricant level should be checked at the intervals specified in the service information system. Normally, these range from every 12 000 to 48 000 km (7500 to 30 000 mi.). For service convenience, many units are now designed with a dipstick **(Figure 40–2)** and filler tube accessible from beneath the hood. Check the lubricant with the engine off and the vehicle resting in a level position. If the engine has been running, wait two to three minutes before checking the gear lubricant level.

Some vehicles have no dipstick. Instead, the vehicle must be placed on a lift, and the lubricant level checked through the **fill plug** opening on the side of the unit. Clean the area around the plug before loosening and removing it. Lubricant should be level with, or not more than 12.7 mm (½ in.) below, the fill hole. Add the proper grade of lubricant as needed using a filler pump.

Manual transmission/transaxle lubricants in use today include single- and multiple-viscosity gear oils, engine oils, special hydraulic fluids, and automatic transmission fluid. Always refer to the service information to determine the correct lubricant and viscosity range for the vehicle and operation conditions **(Figure 40–3)**.

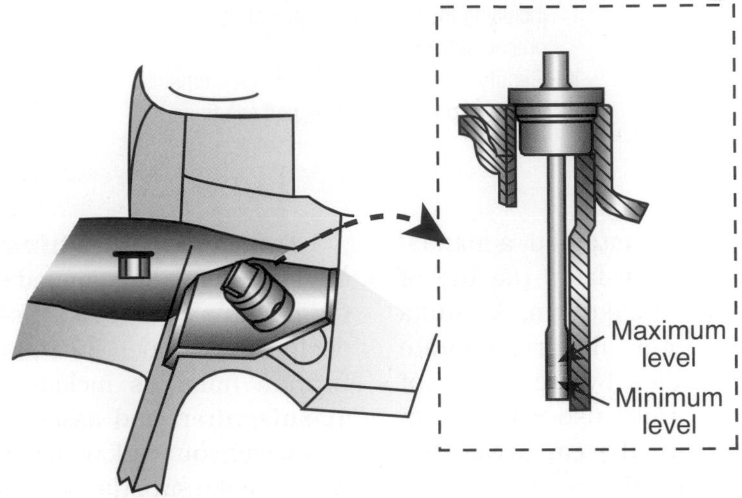

FIGURE 40–2

An example of a dipstick found on some manual transaxles.

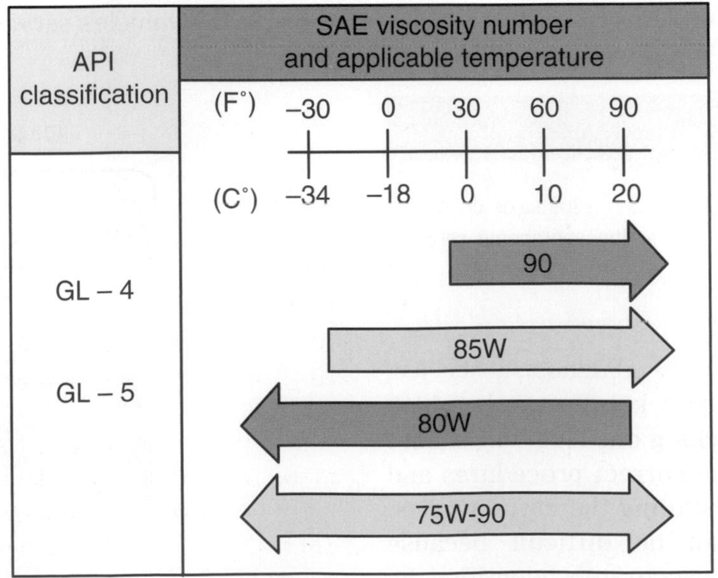

FIGURE 40–3

Typical transmission/transaxle gear oil classification and viscosity range data.

Dual-Clutch Transmission Maintenance

Most dual-clutch transmissions (DCTs) require no special maintenance. Although DCTs are maintenance-free, the fluid should be changed at some point. How soon depends on the manner in which the car has been driven. However, the required lubricant varies with the type of DCT. Always check the service information to see what lubricant should be used in the transmission, as well as when service is required.

Typically, a DCT with dry clutches requires the same gear oil used in manual transmissions. If the DCT is equipped with wet clutches, a special blend of automatic transmission fluid and gear oil must be used. You should also be aware that some DCTs require two separate types of fluids. These DCTs have dual sumps, one to lubricate the transmission's bearings, shafts, and gears, and the other to supply hydraulic pressure for the operation of the wet clutches.

Lubricant Leaks

Normally, the location and cause of a transmission fluid leak can be quickly identified by a visual inspection **(Figure 40–4)**. The following are common causes for fluid leakage:

- An excessive amount of lubricant in the transmission or transaxle
- The use of the wrong type of fluid, which will foam excessively and leave through the vent
- A loose or broken input shaft bearing retainer
- A damaged input shaft bearing retainer O-ring and/or lip seal
- Loose or missing case bolts
- A cracked case or case with a porosity problem

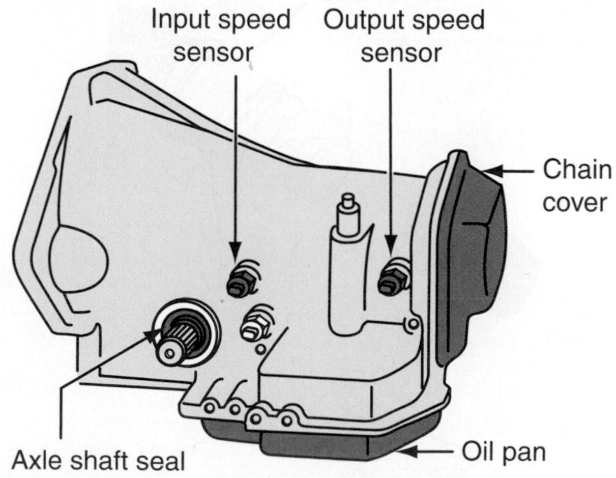

FIGURE 40–4

Possible sources of fluid leaks.

- A leaking shift lever seal
- Gaskets or seals damaged or missing
- A loose drain plug

Fluid leaks from the seal of the extension housing can be corrected with the transmission in the car. Often, the cause for the leakage is a worn extension housing bushing, which supports the sliding yoke. When the driveshaft is installed, the clearance between the sliding yoke and the bushing should be minimal. If the clearance is satisfactory, a new oil seal will correct the leak. If the clearance is excessive, the repair requires that a new seal and a new bushing be installed. If the seal is faulty, the transmission vent should be checked for blockage. If the vent is plugged, the lubricant will be under high pressure when the transmission is hot, and this pressure can cause seal leakage.

A lubricant leak at the speedometer cable can be corrected by replacing the O-ring seal. A lubricant leak stemming from the mating surfaces of the extension housing and the transmission case may be caused by loose bolts. To correct this problem, tighten the bolts to the specified torque.

Fluid Changes

The manufacturers of most transmissions do not recommend that the fluid be changed at any scheduled time. Older transmissions typically had 32 000 km (20 000 mi.) fluid-change intervals. When the vehicle has been operated under severe conditions, such as towing trailers or in high heat or dusty road conditions, the fluid may need to be periodically changed. Check the service information for the manufacturer's recommendations.

To change the transmission fluid, drive the vehicle to warm the fluid. Then raise the vehicle on a hoist. Locate the lubricant drain plug in the bottom of the transmission case or extension housing. Make sure the car is level so that all of the fluid can drain out. Remove the drain plug with a catch pan positioned below the hole, and let the lubricant drain into the pan. Let the transmission drain completely. The fluid is normally very thick, and it takes some time to drain it all out.

Inspect the drained fluid for gold-coloured metallic and other particles. The gold-coloured particles come from the brass blocking rings of the synchronizers. Metal shavings are typically from the wearing of gears. After the fluid has drained out, take a small magnet and insert it into the drain hole, and then sweep it around the inside to remove all metal particles. Because brass is not magnetic, it will not show on the magnet. An excess of iron or brass shavings indicates severe wear in the transmission.

Before refilling the transmission, reinstall the drain plug with a new washer. Some manufacturers recommend that sealer be used on the plug; check the service information. Tighten the drain plug to the recommended torque. Remove the filler plug, which is normally located above the drain plug. Check the service information to identify the location of the filler plug and the proper type and quantity of fluid for that transmission. Fill the transmission case until the fluid just starts to run out the filler hole or until it is at the bottom of the bore. Reinstall the plug with a new washer. You should check the case's vent to make sure it is not blocked with dirt. If the case is not properly vented, the fluid can easily break down, and the pressure buildup can cause leaks. Make sure you fill the transmission with the correct type and amount of fluid. Too much or too little fluid can destroy a transmission.

IN-VEHICLE SERVICE

Much service and maintenance work can be done to transmissions while they are in the car. Only when a complete overhaul or clutch replacement is necessary does the transmission need to be removed from the car. The following are procedures for common service operations: the replacement of a rear oil seal and bushing, linkage adjustments, and replacement of the backup light switch and the speedometer cable retainer and drive gear on older transmissions.

Rear Oil Seal and Bushing Replacement

Procedures for the replacement of the rear oil seal and bushing on a transmission vary little with each car model. Typically, to replace the rear bushing and seal, follow these steps:

PROCEDURE

Replacement of Rear Oil Seal and Bushing

STEP 1 Index and remove the driveshaft.

STEP 2 Remove the old seal from the extension housing.

STEP 3 Pull the bushing from the housing **(Figure 40–5)**.

STEP 4 Drive a new bushing into the extension housing.

STEP 5 Lubricate the lip of the seal, and then install the new seal in the extension housing **(Figure 40–6)**.

STEP 6 Install the driveshaft.

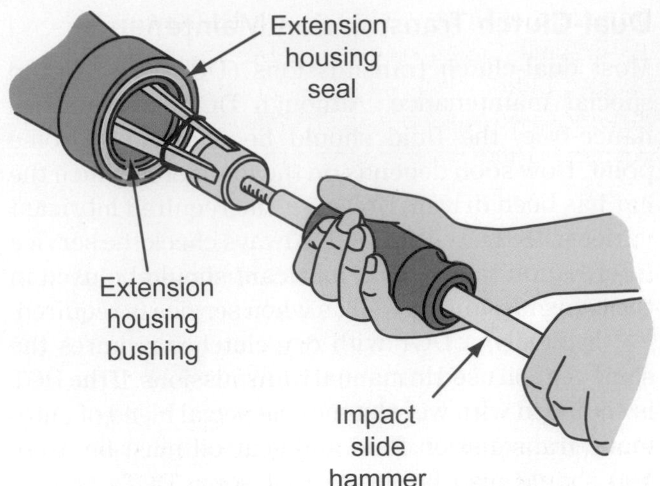

FIGURE 40–5

Removing the extension housing's seal and bushing.

Linkage Adjustment

Transmissions with internal linkage have no provision for adjustments. However, external linkages can be adjusted. Linkages are adjusted at the factory, but worn parts may make adjustments necessary. Also, after a transmission has been reassembled, adjustments may be necessary.

To begin the adjustment procedure, raise the car and support it on jack stands. Then follow the procedure given in the service information.

Backup Light Switch Service

To replace the backup light switch, disconnect the electrical lead to the switch. Put the transmission into reverse gear and remove the switch. Never shift the

FIGURE 40–6

Drive the new seal into place with a hammer and seal driver.

transmission until a new switch has been installed. To prevent fluid leaks, wrap the threads of a new backup light switch with Teflon tape in a clockwise direction before installing it. Tighten the switch to the correct torque, and reconnect the electrical wire to it.

Speedometer Drive Gear Service

On older vehicles that have cable-driven speedometers, begin to remove the speedometer cable retainer and drive gear by cleaning off the top of the speedometer cable retainer **(Figure 40–7)**. Then remove the hold-down screw that keeps the retainer in its bore. Carefully pull up on the speedometer cable, pulling the speedometer retainer and drive gear assembly from its bore. Unscrew the speedometer cable from the retainer.

To reinstall the retainer, lightly grease the O-ring on the retainer, and gently tap the retainer and gear assembly into its bore while lining the groove in the retainer with the screw hole in the side of the clutch housing case. Install the hold-down screw, and tighten it in place.

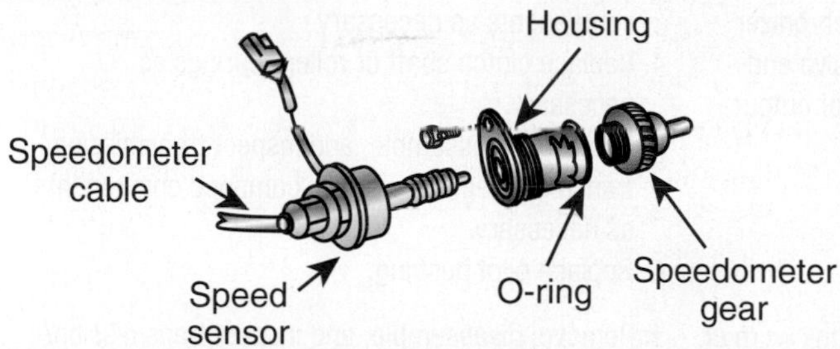

FIGURE 40–7

Oil leaks at the speed sensor can be caused by loose bolts or a bad seal.

DIAGNOSING PROBLEMS

Proper diagnosis involves locating the exact source of the problem. Many problems that seem transmission/transaxle-related may actually be caused by problems in the clutch driveline or differential. Check these areas along with the transmission/transaxle, particularly if you are considering removing the transmission/transaxle for service.

Table 40–1 is a troubleshooting chart for common transmission and transaxle problems.

Remember to begin all diagnostics with an interview of the customer. Then verify the customer's complaint or concern. Also, after repairs are made, make sure you verify the repair.

Visual Inspection

Visually inspect the transmission/transaxle at regular intervals. Perform the following checks:

1. Check for lubricant leaks at all gaskets and seals.
2. Check the case body for signs of porosity that show up as leakage or seepage of lubricant.
3. Push up and down on the unit. Watch the transmission mounts to see if the rubber separates from the metal plate. If the case moves up but not down, the mounts require replacement.
4. Move the clutch and shift linkages around, and check for loose or missing components. Cable linkages should have no kinks or sharp bends, and all movement should be smooth.
5. Check the transaxle drive axle boots for cracks, deformation, or damage.
6. Thoroughly inspect the constant velocity joints on transaxle drive axles.

TABLE 40–1 TRANSMISSION/TRANSAXLE TROUBLESHOOTING CHART

PROBLEM	POSSIBLE CAUSE	REMEDY
Gear clash when shifting from one gear to another	1. Clutch adjustment incorrect 2. Clutch linkage or cable binding 3. Clutch housing misalignment 4. Lubricant level low or incorrect lubricant 5. Gearshift components or synchronizer	1. Adjust clutch. 2. Lubricate or repair as necessary. 3. Check runout at rear face of clutch housing. Correct runout. 4. Drain and refill transmission/transaxle, and check for lubricant leaks if level was low. Repair as necessary. 5. Remove, disassemble, and inspect blocking rings for wear or damage. Replace worn or damaged components as necessary.
Clicking noise in any one gear range	1. Damaged teeth on input or intermediate shaft gears (transaxles), or damaged teeth on the countergear, cluster gear assembly, or output-shaft gears (transmissions)	1. Remove, disassemble, and inspect unit. Replace worn or damaged components as necessary.

(continued)

TABLE 40–1 (*continued*)

PROBLEM	POSSIBLE CAUSE	REMEDY
Does not shift into one gear	1. Gearshift internal linkage or shift rail assembly worn, damaged, or incorrectly assembled 2. Shift rail detent plunger worn, spring broken, or plug loose 3. Gearshift lever worn or damaged 4. Synchronizer sleeves or hubs damaged or worn	1. Remove, disassemble, and inspect transmission/transaxle cover assembly. Repair or replace components as necessary. 2. Tighten plug or replace worn or damaged components as necessary. 3. Replace gearshift lever. 4. Remove, disassemble, and inspect unit. Replace worn or damaged components.
Locked in one gear—cannot be shifted out of that gear	1. Shift rails worn or broken, shifter fork bent, setscrew loose, centre detent plug missing or worn 2. Gearshift lever broken or worn, shift mechanism in cover incorrectly assembled, or broken, worn, or damaged gear train components	1. Inspect and replace worn or damaged components. 2. Disassemble transmission/transaxle. Replace damaged parts of assembly correctly.
Slips out of gear	1. Clutch housing misaligned 2. Gearshift offset lever nylon insert worn or lever attachment nut loose 3. Gearshift mechanisms, shift forks, shift rail, detent plugs, springs, or 4. Clutch shaft or roller bearings worn or damaged 5. Gear side teeth worn or tapered, synchronizer assemblies worn or damaged, excessive end play caused by worn thrust washers or output shaft gears 6. Pilot bushing worn	1. Check runout at rear face of clutch housing. 2. Remove gearshift lever and check for loose offset lever nut or worn insert. Repair or replace as necessary. 3. Remove, disassemble, and inspect transmission cover assembly. Replace worn or damaged components as necessary. 4. Replace clutch shaft or roller bearings as necessary. 5. Remove, disassemble, and inspect transmission/transaxle. Replace worn or damaged components as necessary. 6. Replace pilot bushing.
Vehicle moving—rough growling noise isolated in transmission/transaxle and heard in all gears	1. Intermediate shaft front or rear bearings worn or damaged (transaxle) or output shaft rear bearing worn or damaged (transmission)	1. Remove, disassemble, and inspect transmission/transaxle. Replace damaged components as necessary.
Rough growling noise when engine operating with transmission/transaxle in neutral	1. Input shaft front or rear bearings worn or damaged (transaxle) or input shaft bearing, countergear, or countershaft bearings worn or damaged (transmission)	1. Remove, disassemble, and inspect transmission/transaxle. Replace damaged components as necessary.
Vehicle moving—rough growling noise in transmission—noise heard in all gears except direct drive	1. Output shaft pilot roller bearings	1. Remove, disassemble, and inspect transmission. Replace damaged components as needed.
Transmission/transaxle difficult to shift	1. Clutch adjustment incorrect 2. Clutch linkage binding 3. Shift rail binding 4. Internal bind in transmission/transaxle caused by shift forks, selector plates, or synchronizer assemblies 5. Clutch housing misalignment 6. Incorrect lubricant	1. Adjust clutch. 2. Lubricate or repair as necessary. 3. Check for mispositioned roll pin, loose cover bolts, worn shift rail bores, worn shift rail, distorted oil seal, or extension housing not aligned with case. Repair as necessary. 4. Remove, disassemble, and inspect unit. Replace worn or damaged components as necessary. 5. Check runout at rear of clutch housing. Correct runout. 6. Drain and refill.

Transmission Noise

Most manual transmission/transaxle complaints centre around noise in the unit. Once again, be certain the noise is not coming from other components in the drivetrain. Unusual noises may also be a sign of trouble in the engine or transmission mounting system. Improperly aligned engines, improperly torqued mounting bolts, damaged or missing rubber mounts, cracked brackets, or even a stone rattling around inside the engine compartment can create noises that appear to be transmission/transaxle-related.

SHOP TALK

If during the test drive you hear a noise you suspect is coming from inside the transmission/transaxle, bring the vehicle to a stop. Disengage the clutch. If the noise stops with the engine at idle and the clutch disengaged, the noise is probably inside the unit.

Once you have eliminated all other possible sources of noise, concentrate on the transmission/transaxle unit. Noises from inside the transmission/transaxle may indicate worn or damaged bearings, gear teeth, or synchronizers. A noise that changes or disappears in different gears can indicate a specific problem area in the transmission.

CAUTION!

When the transmission/transaxle is in gear and the engine is running, the driving wheels and related parts will turn. Avoid touching moving parts. Severe physical injury can result from contact with spinning drive axles and wheels.

The type of noise detected will help identify the problem.

ROUGH, GROWLING NOISE This noise can be a sign of several problems in a transaxle or transmission, depending on when it occurs. If the noise occurs when the transaxle is in neutral and the engine is running, the problem may be the input shaft roller bearings. The input shaft is supported on either end by tapered roller bearings, and these are the only bearings in operation when the transaxle is in neutral. In its early stages, the problem should not cause operational difficulties; however, left uncorrected, it grows worse until the bearing race or rolling element fractures. Solving the problem involves transaxle disassembly and bearing replacement.

When the vehicle is moving, both the input shaft and mainshaft (pinion shaft) are turning in the transaxle. If the noise occurs in forward and reverse gears, but not in neutral, the output or pinion shaft bearings are the likely failed component. Replacement is the solution.

In transmissions, the problem can also be bearing related. If the rough growling noise occurs when the engine is running, the clutch is engaged, and the transmission is in neutral, the front input shaft bearing is likely at fault. Rough growling when the vehicle is moving in all gears indicates faulty countergear bearings or countershaft-to-cluster assembly needle bearings. If the problem occurs in all gears except direct drive, the bearing at the rear of the transmission input shaft may be at fault. This bearing supports the pilot journal at the front of the transmission output shaft. In all forward gears except direct drive, the input shaft and output shaft turn at two different speeds. In reverse, the two shafts turn in opposite directions. In direct drive, the two shafts are locked together, and this bearing does not turn. Disassembly, inspection, and replacement of damaged parts is needed.

If the rough growling noise occurs when the engine is running, the clutch is disengaged, and a gear is selected in the transmission, the input shaft's pilot bearing is likely at fault.

CLICKING OR KNOCKING NOISE Normally, the helical gears used in modern transmissions/transaxles are quiet because the gear teeth are constantly in contact. (When spur-cut gear teeth are found in the reverse gearing, clicking or a certain amount of **gear whine** is normal, particularly when backing up at faster speeds.)

Clicking or whine in forward-gear ranges may indicate worn helical gear teeth. This problem may not require immediate attention. However, chipped or broken teeth are dangerous because the loose parts can cause severe damage in other areas of the transmission/transaxle. Broken parts are usually indicated by a rhythmical knocking sound, even at low speeds. Complete disassembly, inspection, and replacement of damaged parts is the solution to this problem.

Gear Clash

Gear clash is indicated by a grinding noise during shifting. The noise is the result of one gearset remaining partly engaged while another gearset attempts to turn the mainshaft. Gear clash can be caused by incorrect clutch adjustment or binding of clutch or gearshift linkage. Damaged, worn, or defective synchronizer blocking rings can cause gear clash, as can use of an improper gear lubricant.

Hard Shifting

If the shift lever is difficult to move from one gear to another, check the clutch linkage adjustment, linkage pivots, and/or cable condition. Hard shifting may also be caused by damage inside the transmission/transaxle, or by a lubricant that is too thick. Common hard-shifting causes include badly worn bearings and damaged clutch gears, control rods, shift rails, shift forks, and synchronizers.

Jumping Out of Gear

If the car jumps out of gear into neutral, particularly when decelerating or going down hills, first check the shift lever and internal gearshift linkage. Excessive clearance between gears and the input shaft or badly worn bearings can cause jumping out of gear. Other internal transmission/transaxle parts to inspect are the clutch pilot bearing, gear teeth, shift forks, shift rails, and springs or detents.

Locked in Gear

If a transmission or transaxle locks in one gear and cannot be shifted, check the gearshift lever linkage for misadjustment or damage. Low lubricant level can also cause needle bearings, gears, and synchronizers to seize and lock up the transmission.

If these checks do not identify the problem, the transmission or transaxle must be removed from the vehicle and disassembled. After disassembly, inspect the countergear, clutch shaft, reverse idler, shift rails, shift forks, and springs or detents for damage. Also, check for worn support bearings.

If the problem seems to be in the clutch assembly, make sure the transmission/transaxle is out of gear, set the parking brake, and start the engine. Increase the engine speed to about 1500 to 2000 rpm, and gradually apply the clutch until the engine torque causes tension at the drivetrain mounts. Watch the torque reaction of the engine. If the engine's reaction to the torque appears to be excessive, broken or worn drivetrain mounts may be the cause and not the clutch.

The engine mounts on front-wheel-drive (FWD) cars are important to the operation of the clutch and transaxle **(Figure 40–8)**. Any engine movement may change the effective length of the shift and clutch control cables and therefore may affect the engagement of the clutch and/or gears. To check the condition of the transaxle mounts, pull up and push down on the transaxle case while watching the mount. If the mount's rubber separates from the metal plate or if the case moves up but not down, replace the mount. If there is movement between the metal plate and its attaching point on the frame, tighten the attaching bolts to an appropriate torque.

If it is necessary to replace the transaxle mount, make sure you follow the procedure for maintaining the alignment of the driveline. Some manufacturers recommend that a holding fixture or special bolt be used to keep the unit in its proper location. A broken clutch cable may be caused by worn mounts and improper cable routing. Inspect all clutch and transaxle linkages and cables for kinks or stretching. Often, transaxle problems can be corrected by replacing or repairing the clutch or gearshift cables and linkage.

Shift Linkage

Check the shift linkage for smooth gear changes and full travel. If the linkage cannot move enough to fully engage a gear, the transmission/transaxle will jump out of gear while it is under a load. Some FWD cars have experienced the problem of jumping out of second or fourth gear. Two causes have been identified with this problem: The upshift light interferes with the shifter, or there are improper shifter-to-shifter boot clearances. Both conditions prevent the transaxle's shift forks from moving enough to fully engage the synchronizer sleeve to its mating gears. If correcting these problems does not solve the complaint, the cause may be the engine mounts or an internal problem in the transaxle.

USING SERVICE INFORMATION

Service information is absolutely necessary when performing any type of transmission/transaxle disassembly work. Not only will the material clearly illustrate all components and their disassembly procedure, it also lists many vital specifications, such as shaft and gear thrust (side) clearances, synchronizer ring and cone clearances, and bolt torque values. Special service tools, such as transmission service stands, oil seal presses, bearing replacers, shaft removers, pullers, and installing tools, are also illustrated and explained.

TRANSMISSION/TRANSAXLE REMOVAL

Removing the transmission from a rear-wheel-drive (RWD) vehicle is generally more straightforward than removing one from an FWD model, as there is typically one cross member, one driveshaft, and easy access to the cables, wiring, and bell housing bolts. Transmissions in FWD cars, because of their limited space, can be more difficult to remove as you

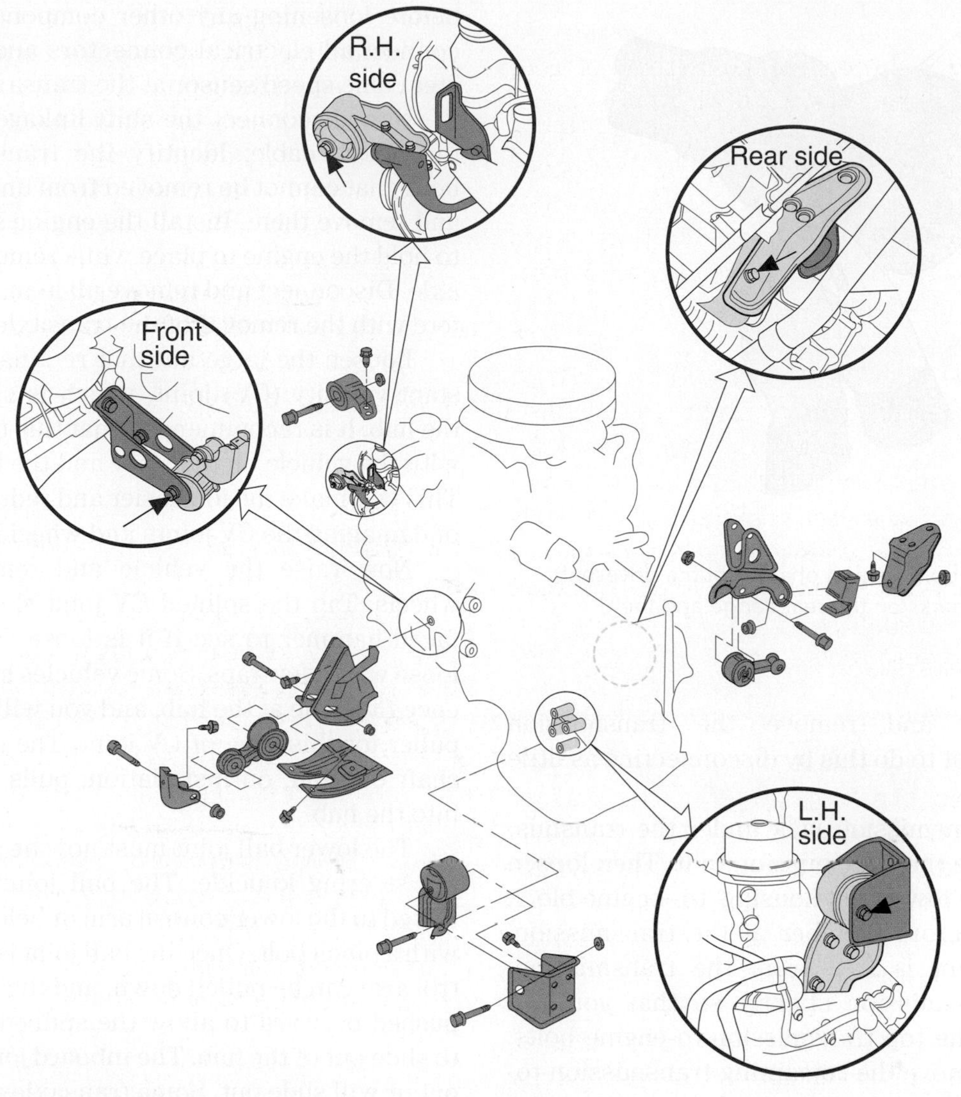

FIGURE 40–8

Typical engine and transaxle mounts.

Nissan Canada

may need to disassemble or remove large assemblies such as engine cradles, suspension components, brake components, splash shields, or other pieces that would not usually affect RWD transmission removal. The engine may also need to be supported with fixtures while removing the transmission.

GO TO Chapter 10 for a detailed discussion of how to remove an engine and/or transmission from a vehicle.

RWD Vehicles

The correct procedure for removing a transmission varies with each year, make, and model of vehicle; always refer to the service information for the correct procedure. Normally, the procedure begins with placing the vehicle on a hoist.

Once the vehicle is in position, disconnect the negative battery cable and place it away from the battery. Carefully check under the hood to identify anything that may interfere with transmission removal. Then raise the vehicle and disconnect the parts of the exhaust system that may get in the way. Disconnect all electrical connections and the speedometer cable/speed sensor at the transmission. Make sure you place these away from the transmission so that they are not damaged during transmission removal or installation.

Place a drain pan under the transmission, and drain the transmission's fluid. Then move the drain pan to the rear of the transmission. Before removing the driveshaft, use chalk to show the alignment of the rear U-joint and the pinion flange **(Figure 40–9)**; then remove the driveshaft.

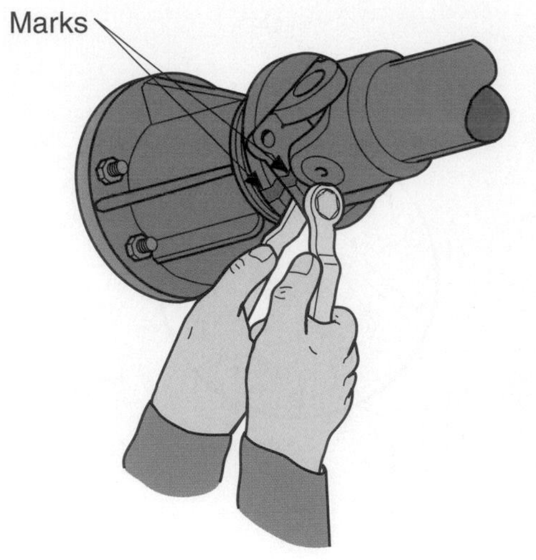

FIGURE 40–9

To ensure proper balance and phasing of the driveshaft, make alignment marks on the rear flange and the rear yoke.

Reprinted with permission from Toyota.

Disconnect and remove the transmission linkage. It is best to do this by disconnecting as little as possible.

Place a transmission jack under the transmission and secure the transmission to it. Then loosen and remove the lower bell housing–to–engine block bolts and the cross member at the transmission. After the mount is free from the transmission, lower the transmission slightly so that you can easily access the top transmission-to-engine bolts. Loosen and remove the remaining transmission-to-engine bolts.

Slowly and carefully move the transmission away from the engine until the input shaft is out of the clutch assembly. Then slowly lower the transmission. Once the transmission is out of the vehicle, carefully move it to the work area, and mount it to a stand or bench.

On some cars, the engine and transmission must be removed as a unit. The assembly is lifted with an engine hoist or lowered underneath the car.

FWD Vehicles

On some vehicles, the recommended procedure may include removing the engine with the transaxle. Always refer to the service information before proceeding to remove the transaxle. Identify any special tool needs and precautions recommended by the manufacturer. You will waste much time and energy if you do not check the manual or information system first.

Begin removal by placing the vehicle on a lift. Working under the hood, disconnect the battery before loosening any other components. Then disconnect all electrical connectors and the speedometer cable/speed sensor at the transaxle.

Now disconnect the shift linkage or cables and the clutch cable. Identify the transaxle-to-engine bolts that cannot be removed from under the vehicle, and remove them. Install the engine support fixture to hold the engine in place while removing the transaxle. Disconnect and remove all items that will interfere with the removal of the transaxle.

Loosen the large nut that retains the outer constant velocity (CV) joint, which is splined shaft to the hub. It is recommended that this nut be loosened with the vehicle on the floor and the brakes applied. This will make the job easier and reduces the chance of damaging the CV joints and wheel bearings.

Now raise the vehicle and remove the front wheels. Tap the splined CV joint shaft with a soft-faced hammer to see if it is loose. Most will come loose with a few taps. Some vehicles have an interference fit spline at the hub, and you will need a special puller for this type of CV joint. The tool pushes the shaft out and, on installation, pulls the shaft back into the hub.

The lower ball joint must now be separated from the steering knuckle. The ball joint will either be bolted to the lower control arm or held in the knuckle with a pinch bolt. Once the ball joint is loose, the control arm can be pulled down, and the knuckle can be pushed outward to allow the splined CV joint shaft to slide out of the hub. The inboard joint can be pried out or will slide out. Some transaxles have retaining clips that must be removed before the inner joint can be removed. Pull the drive axles out of the transaxle **(Figure 40–10)**.

While removing the axles, make sure the brake lines and hoses are not stressed. Suspend them with wire to relieve the weight on the hoses and to keep them out of the way.

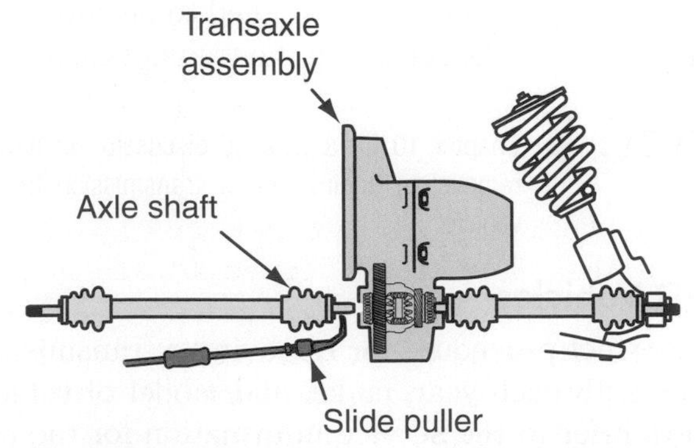

FIGURE 40–10

The inboard joints are typically pulled out of the transaxle.

On some cars, the inner CV joints have flange-type mountings. These must be unbolted for removal of the shafts. In some cases, the flange-mounted driveshafts may be left attached to the wheel and hub assembly and only unbolted at the transmission flange. The free end of the shafts should be supported and placed out of the way.

Now the remaining shift linkages and electrical connections should be disconnected. The exhaust system may also have to be lowered or partially removed.

Remove the starter. The starter's wiring may be left connected, or you may remove the starter from the vehicle to get it totally out of the way.

With the transmission jack supporting the transmission, remove the transaxle mounts. If the car is equipped with an engine cradle that will separate, remove the half of the cradle that allows for transaxle removal. Then remove all remaining transaxle-to-engine bolts. Slide the transaxle away from the engine. Make sure the input shaft is out of the clutch assembly before lowering the transmission.

CLEANING AND INSPECTION

Disassembly and overhaul procedures can vary greatly between transmission/transaxle models, so always follow the exact steps outlined in the service information.

Clean the transmission/transaxle with a steam cleaner, degreaser, or cleaning solvent. As you begin to disassemble the unit, pay close attention to the condition of its parts. Using a dial indicator, measure and record the end play of the input and main shafts **(Figure 40–11)**. This information will be needed during the reassembly of the unit for selecting the appropriate selective shims and washers.

SHOP TALK

Before disassembling a transmission, observe the effort it takes to rotate the input shaft through all the forward gears and reverse. Extreme effort in any or all gears may indicate an end play or preload problem.

Remove the bell housing from the transmission case, extension housing **(Figure 40–12)**, and the side or top cover. The seal and bushing should be removed from the extension housing (tail shaft) prior to cleaning. With the housing and cover removed, the gears, synchronizers, and shafts are exposed, and the shift forks can be removed **(Figure 40–13)**.

FIGURE 40–11

Use a dial indicator to measure the end play of the shafts before disassembling the unit.

SHOP TALK

It is good practice to lay the parts on a clean rag as you remove them, and to keep them in order to aid you during reassembly.

Each transmission design has its own specific service procedures. Photo Sequence 28 guides you through the disassembly of a typical transaxle. Always refer to your service information prior to overhauling a transmission or transaxle.

In some cases, the countershaft must be removed before the input and mainshaft **(Figure 40–14)**. In other cases, the mainshaft is removed with the extension housing. It may be removed through the

FIGURE 40–12

Slide the extension housing off the output shaft.

FIGURE 40–13

Remove the top cover and the shift linkage.

©Cengage Learning 2015

shift cover opening. To avoid difficulty in disassembly, follow the recommended sequence. A **gear puller** or hydraulic press is often needed to remove gears and synchronizer assemblies from transmission/transaxle pinion shafts.

Bearing removal and installation procedures require that the force applied to remove or install the bearing should always be placed on the tight bearing race. In some cases, the inner race is tight on the shaft, whereas in others it is the outer race that is tight in its bore. Removal or installation force should be applied to the tight race. Serious damage to the bearing can result if this practice is not followed.

FIGURE 40–14

Lift the output shaft from the case, and then remove the countershaft's rear bearing cup from the case.

©Cengage Learning 2015

Use a soft-faced hammer or a brass drift and ball-peen hammer if tapping is required. Never use excessive force or hammering.

During assembly of the transmission, never attempt to force parts into place by tightening the front bearing retainer bolts or extension housing bolts. All parts must be fully in place before tightening any bolts. Check for free rotation and shifting. New gaskets and seals should always be used.

The following are some general cleaning and inspection guidelines that result in quality workmanship and service:

1. Wash all parts, except sealed ball bearings and seals, in solvent. Brush or scrape all dirt from the parts. Remove all traces of old gasket. Wash roller bearings in solvent; dry them with a clean cloth. Never use compressed air to spin the bearings.
2. Inspect the front of the transmission case for nicks or burrs that could affect its alignment with the flywheel housing. Remove all nicks and burrs with a fine stone (cast-iron casing) or fine file (aluminum casing).
3. Replace any cover that is bent or distorted. If there are vent holes in the case, make certain they are open.
4. Inspect the seal and bushing in the extension housing. Measure the inside diameter of the bushing, and compare that to specifications. Replace bushings if they are worn or damaged.
5. Inspect ball bearings by holding the outer ring stationary and rotating the inner ring several times. Inspect the raceway of the inner ring from both sides for pits and spalling. Light particle indentation is acceptable wear, but all other types of wear merit replacement of the bearing assembly. Next, hold the inner ring stationary, and rotate the outer ring. Examine the outer ring raceway for wear, and replace as needed.
6. Examine the external surfaces of all bearings. Replace the bearings if there are radial cracks on the front and rear faces of the outer or inner rings, cracks on the outside diameter or outer ring, or deformation or cracks in the ball cage.
7. Lubricate the cleaned bearing raceways with a light coat of oil. Hold the bearing by the inner ring in a vertical position. Spin the outer ring several times by hand. If roughness or vibration is felt, or the outer ring stops abruptly, replace the bearing.

DISASSEMBLING A TYPICAL TRANSAXLE

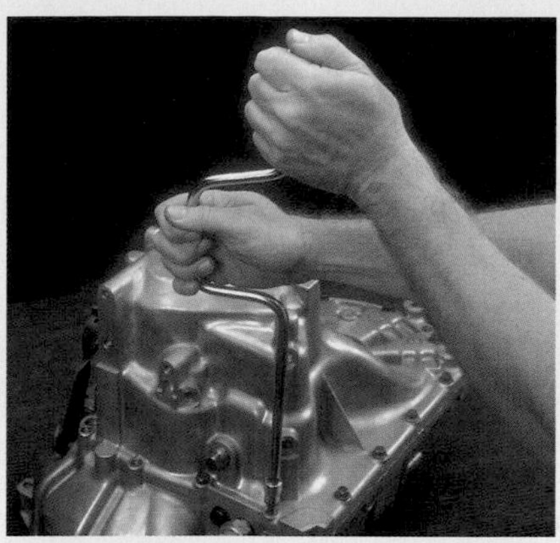

P28–1 Place the transaxle into a suitable work stand. Remove the reverse idler shaft retaining bolt and detent plunger retaining screw. Then loosen and remove all transaxle case-to-clutch housing attaching bolts.

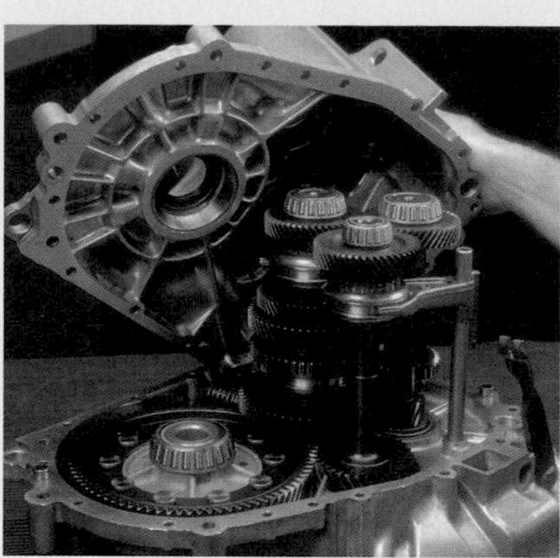

P28–2 Separate the housing from the case. If the housing is difficult to loosen, tap it with a soft mallet.

P28–3 Remove the C-clip retaining ring from the fifth-gear shift-relay lever pivot pin.

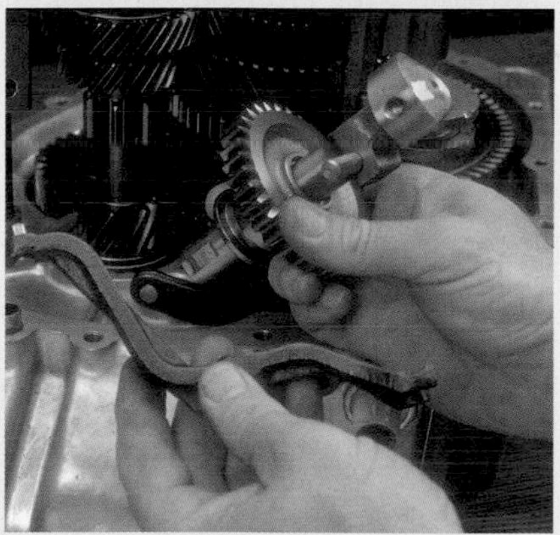

P28–4 Remove the fifth-gear shift-relay lever, reverse idler shaft, and reverse idler gear from the case.

P28–5 Use a punch to drive the roll pin from the shift-lever shaft.

P28–6 Remove the shift-lever shaft by gently pulling on it.

P28–7 Remove the relay lever and bias spring assembly.

P28–8 Grasp the input and main shafts and lift them as an assembly from the case. Note the position of the shift forks as an aid when reinstalling them.

P28–9 Remove the fifth-gear shaft assembly and the fifth-gear shift fork assembly.

DISASSEMBLING A TYPICAL TRANSAXLE *(continued)*

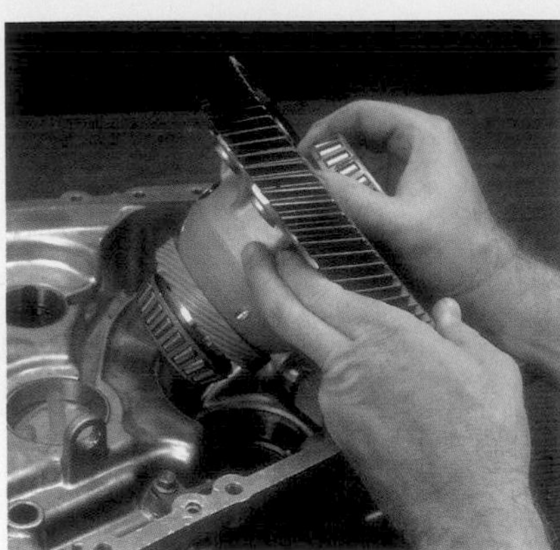

P28–10 Remove the differential assembly from the case.

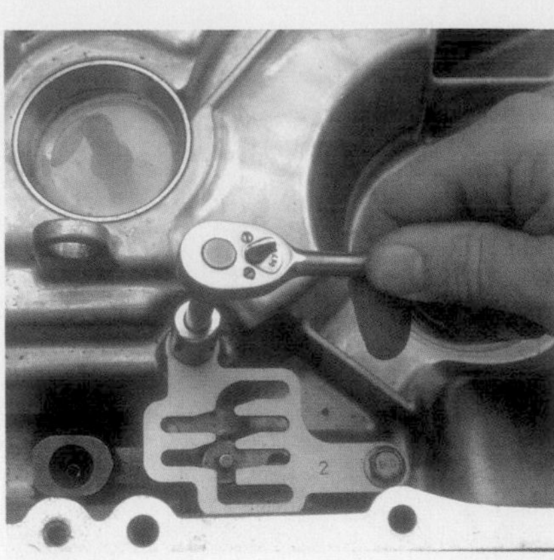

P28–11 Remove the bolts for the shift gate support bracket, and then remove the assembly.

P28–12 Carefully separate the shift rail and forks from the gears on the mainshaft.

P28–13 Remove the bearing at the fourth-speed gear end of the mainshaft.

P28–14 Slide the fourth-speed gear from the shaft.

P28–15 Remove the synchronizer blocking ring from the assembly.

P28–16 Remove the third- and fourth-gear synchronizer retaining ring.

P28–17 Lift the assembly off the shaft. Then remove the remaining gears and synchronizer assembly as a unit.

P28–18 Separate the synchronizer's hub, sleeve, and keys, noting their relative positions and scribing their location on the hub and the sleeve prior to separation.

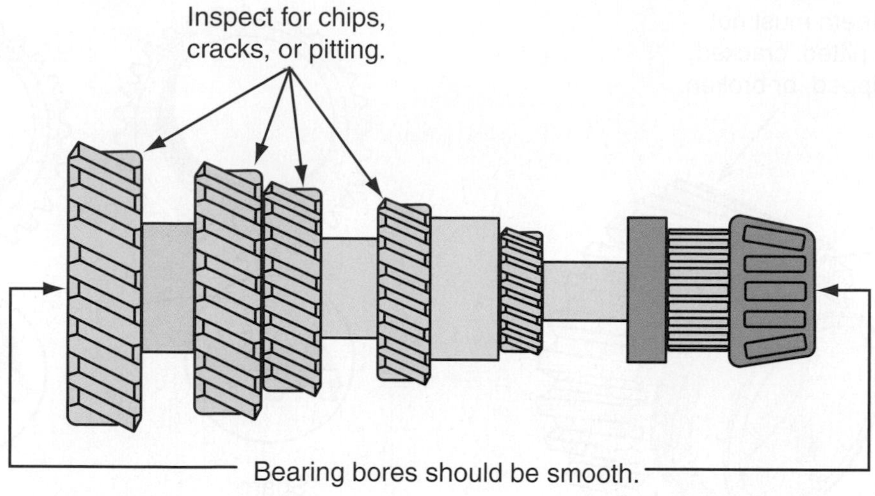
Inspect for chips,
cracks, or pitting.
Bearing bores should be smooth.

FIGURE 40–15

Carefully inspect the countergear assembly.

8. Replace any roller bearings that are broken, worn, or rough. Inspect their respective races. Replace them as needed.
9. Replace the counter (cluster) gear if its gear teeth are chipped, broken, or excessively worn **(Figure 40–15)**. Replace the countershaft if the shaft is bent, scored, or worn. Also, inspect the bore for the countershaft. If the bore is excessively worn or damaged, the needle bearings will not seat properly against the shaft.
10. Replace the reverse idler gear or sliding gear if its teeth are chipped, worn, or broken. Replace the idler gear shaft if it is bent, worn, or scored.
11. Replace the input shaft and gear if its splines are damaged or if the teeth are chipped, worn, or damaged **(Figure 40–16)**. If the roller bearing surface in the bore of the gear is worn or rough, or if the cone surface is damaged, replace the gear and the gear rollers.
12. Replace all main or speed gears that are chipped, broken, or worn. **(Figure 40–17)**.
13. Check the synchronizer sleeves for free movement on their hubs. Alignment marks (if present) should be properly indexed **(Figure 40–18)**.
14. Inspect the synchronizer blocking rings for widened index slots, rounded clutch teeth, and smooth internal surfaces. Remember that the blocking rings must have machined grooves on their internal surfaces to cut through lubricant **(Figure 40–19)**. Units with worn, flat grooves must be replaced. Also, check the clearance between the blocker ring and speed-gear dog teeth against service information specifications **(Figure 40–20)**.

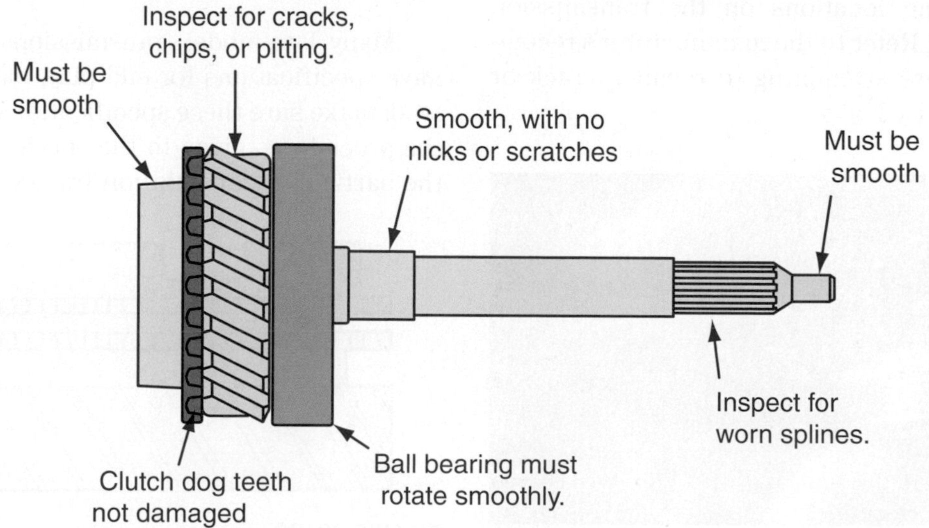
Inspect for cracks,
chips, or pitting.
Must be
smooth
Smooth, with no
nicks or scratches
Must be
smooth
Inspect for
worn splines.
Ball bearing must
rotate smoothly.
Clutch dog teeth
not damaged

FIGURE 40–16

The input shaft, including the splines, should also be carefully inspected.

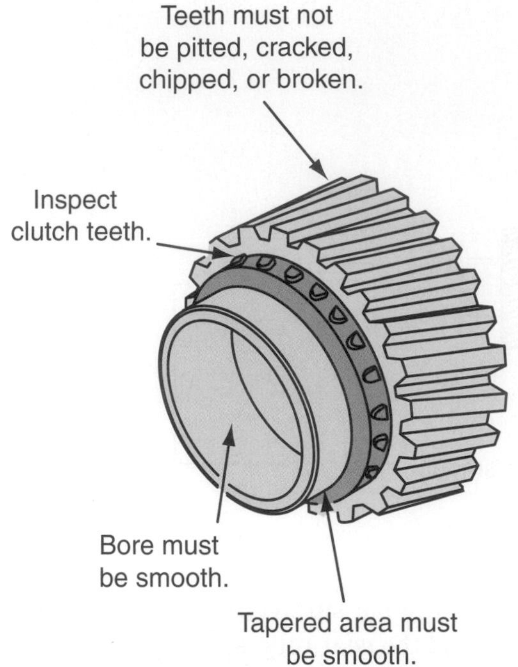
Teeth must not be pitted, cracked, chipped, or broken.
Inspect clutch teeth.
Bore must be smooth.
Tapered area must be smooth.

FIGURE 40–17

Every gear should be checked.

15. On older transmissions/transaxles, replace the speedometer drive gear if its teeth are damaged. Install the correct replacement part.
16. Replace the output shaft if there is any sign of wear or runout or if any of the splines are damaged.

Aluminum Case Repair

Normally, the case is replaced if it is cracked or damaged. However, some manufacturers recommend the use of an epoxy-based sealer on some types of leaks in some locations on the transmission **(Figure 40–21)**. Refer to the manufacturer's recommendations before attempting to repair a crack or correct for porosity leaks.

FIGURE 40–18

The movement of each synchronizer unit should be checked.

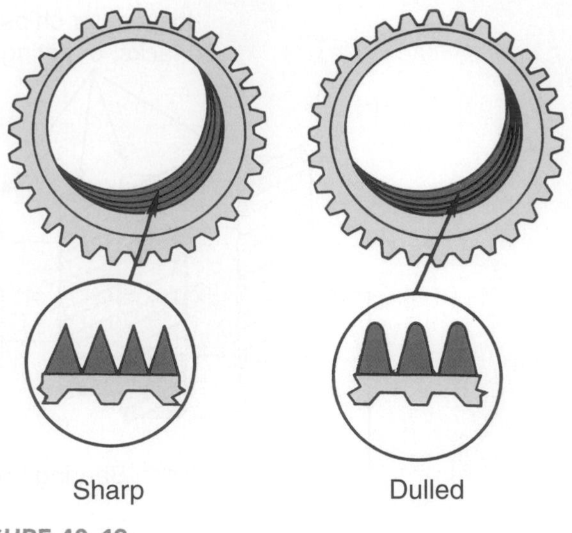
Sharp
Dulled

FIGURE 40–19

Grooves on the internal surface of the synchronizer blocker ring must be sharp.

If a threaded area in an aluminum housing is damaged, helicoil-type service kits can be used to insert new threads in the bore. Some threads should never be repaired; check the service information to identify which ones can be repaired.

After all parts are inspected and the defective parts replaced, you can begin to reassemble the transmission/transaxle. While you are doing so, coat all parts with gear lube.

SHOP TALK

If the transmission/transaxle is fitted with paper-type blocking rings, soak them in automatic transmission fluid prior to installing them.

Many late-model transmissions and transaxles have specifications for end play, backlash, and preload; make sure these specifications are met. Follow the procedures given in the service information for the particular transmission/transaxle being worked

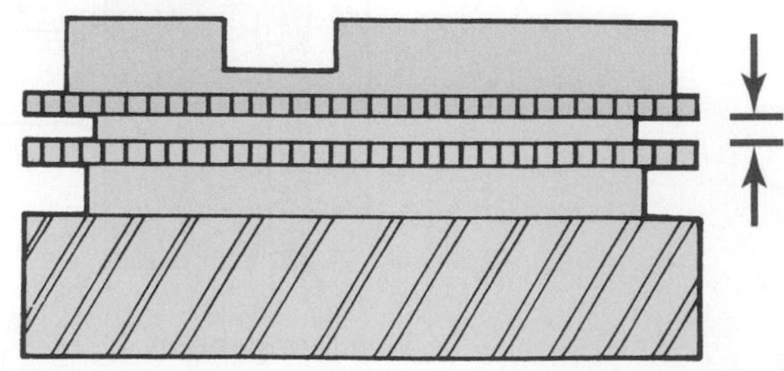

FIGURE 40–20

The clearance between the synchronizer blocker ring and the gear's clutching dog teeth must meet specifications.

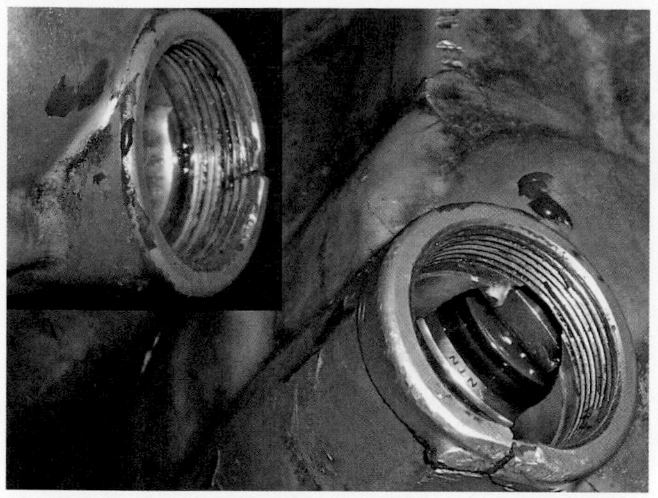

FIGURE 40–21

(A) A crack around the filler plug opening. (B) The crack was repaired with epoxy.

on **(Figure 40–22)**. For most transmissions, there are specifications for the end play and preload of the input shaft, the countershaft, and the differential. These are usually set by shims under the bearing cups. Reuse the original shims, if possible.

Specific repair and assembly instructions will vary from transaxle to transaxle and from transmission to transmission. Therefore, before beginning to reassemble the unit, gather the specific information about the unit you are working on. Photo Sequence 29 shows the reassembly of a typical transaxle.

DISASSEMBLY AND REASSEMBLY OF THE DIFFERENTIAL CASE

Although it is a part of the transaxle, the differential is often kept together while making a repair to the transmission part of the transaxle **(Figure 40–23)**. The differential case normally can be removed as soon as the transaxle case has been separated. It may be the source of the problem and be the only part of the transaxle that needs service. Therefore, the disassembly and reassembly of the differential is set aside from the procedures listed for the transaxle **(Figure 40–24)**.

Begin the disassembly by separating the ring gear from the differential case. Then remove the pinion shaft lock bolt. Remove the pinion shaft, and then remove the gears and thrust washers from the case. If the differential side bearings are to be replaced, use a puller to remove the bearings. Use the correct installer for reinstallation of the side bearings.

Clean and inspect all parts. Replace any damaged or worn parts. Install the gears and thrust washers

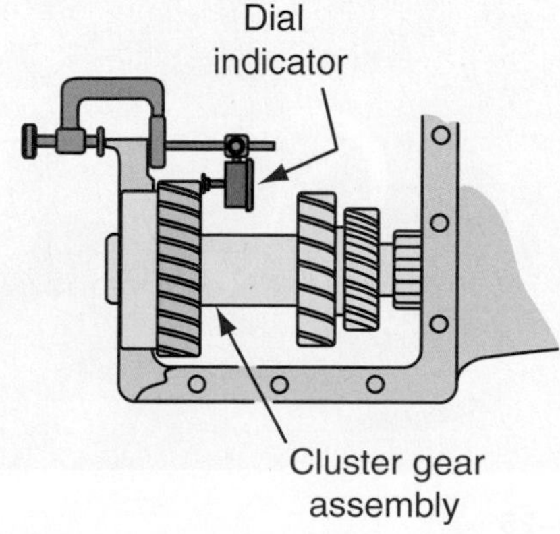

FIGURE 40–22

A typical setup for checking countershaft end play.

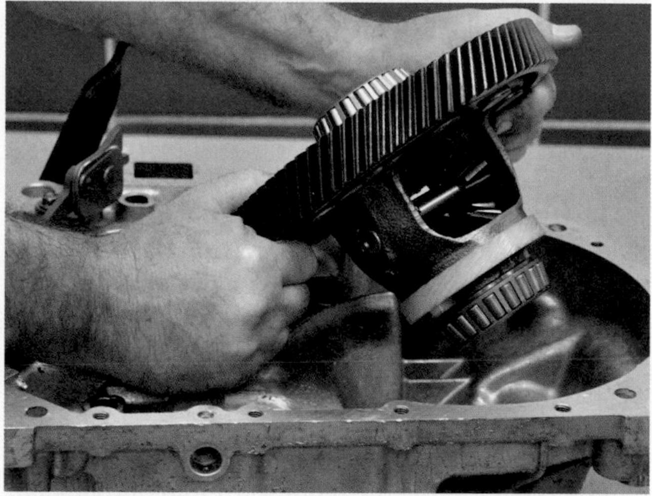

FIGURE 40–23

Once the transaxle case is opened, the differential assembly can be removed.

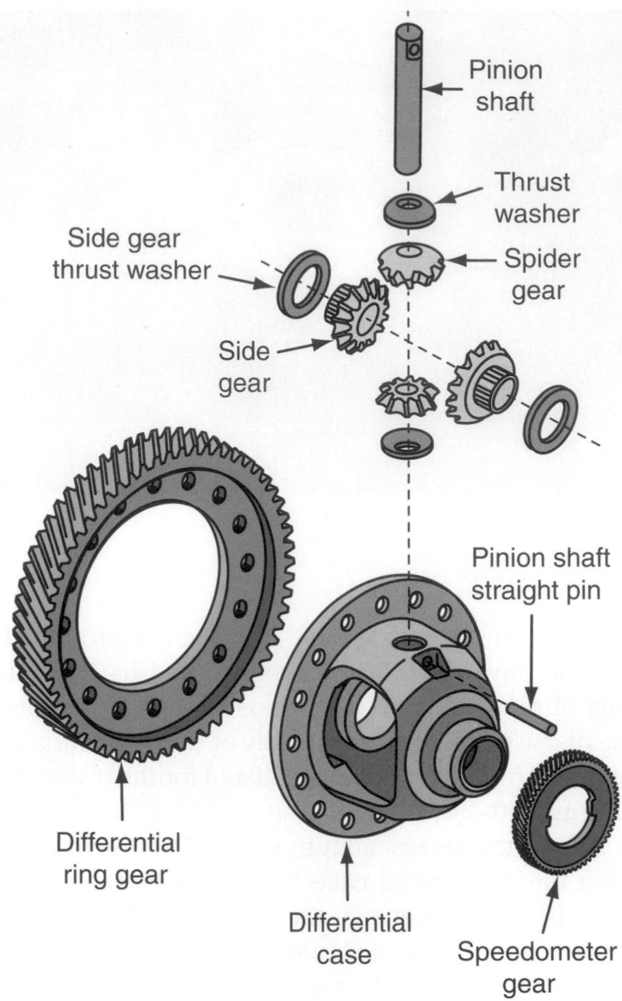

FIGURE 40–24

The differential assembly for a transaxle.

into the case, and install the pinion shaft and lock bolt. Tighten the bolt to the specified torque. Attach the ring gear to the differential case, and tighten to the specified torque.

Shim Selection

While you are disassembling the differential or transaxle, make sure to keep all shims and bearing races together, and identify them for reinstallation in their original location. Carefully inspect the bearings for wear and/or damage and determine if a bearing should be replaced. Replacement tapered roller bearings will be available with a nominal thickness service shim. A nominal thickness service shim will handle re-shimming the input shaft and output shaft bearings during normal repair.

When it is necessary to replace a bearing, race, or housing, refer to the manufacturer's recommendation for nominal shim thickness. If only other parts of the differential or transaxle are replaced, reuse the original shims. When repairs require the use of a service shim, discard the original shim. Never use the original shim together with the service shim. The shims must be installed only under the bearing cups at the transaxle case end of both the input and output shafts.

REASSEMBLY/REINSTALLATION OF TRANSMISSION/TRANSAXLE

Transmission/transaxle reassembly and reinstallation procedures are basically the reverse of disassembly. Once again, refer to the service information for any special procedures. New parts are installed as needed, and new gaskets and seals are always used.

Serviceable gears are pressed onto the mainshaft using special press equipment. Separate needle bearings should be held in place with petroleum jelly so that shafts can be inserted into place. During reassembly, measure all shaft end play. Adjust them to specifications with shims, spacers, or snap rings of different thicknesses **(Figure 40–25)**. In addition to checking end play, some manufacturers suggest that the torque required to rotate an assembly be checked. This is simply done with a torque wrench. All fasteners should be tightened to the manufacturer's torque specifications.

Soft-faced mallets can be used to tap shafts and other parts into place. After reassembly, secure the transmission to a transmission jack with safety chains and raise it into place. Before the transmission is reinstalled, inspect and service the clutch as necessary.

FIGURE 40–25

End play is controlled by shims.

REASSEMBLING A TYPICAL TRANSAXLE

PHOTO SEQUENCE 29

P29–1 Lightly oil the parts of the synchronizer.

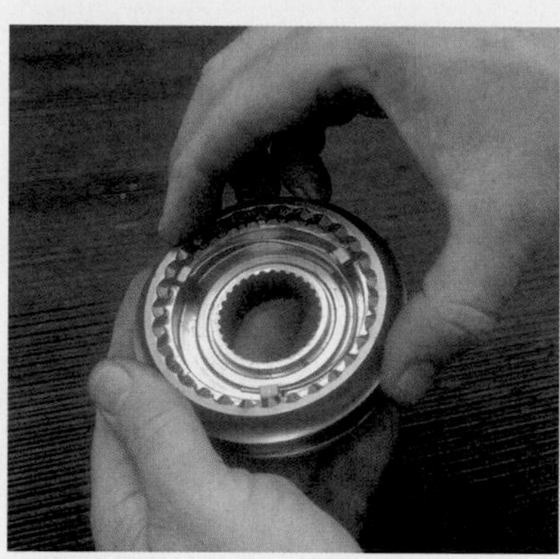

P29–2 Assemble the synchronizer assemblies, being careful to align the index marks made during disassembly.

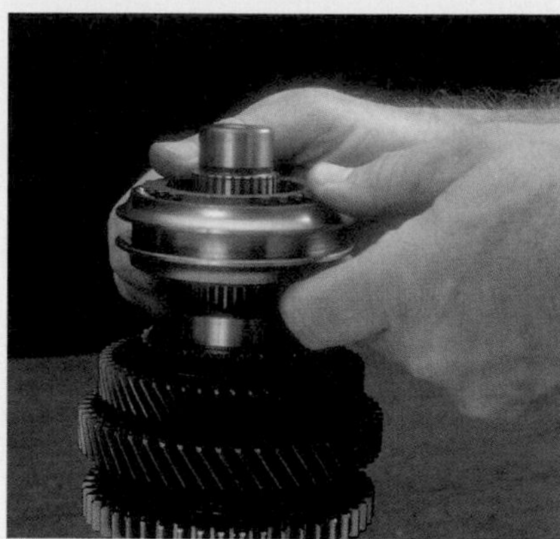

P29–3 Install the synchronizer assemblies onto the mainshaft.

P29–4 Install the fourth-speed gear and its bearing. The bearing may need to be lightly pressed into position.

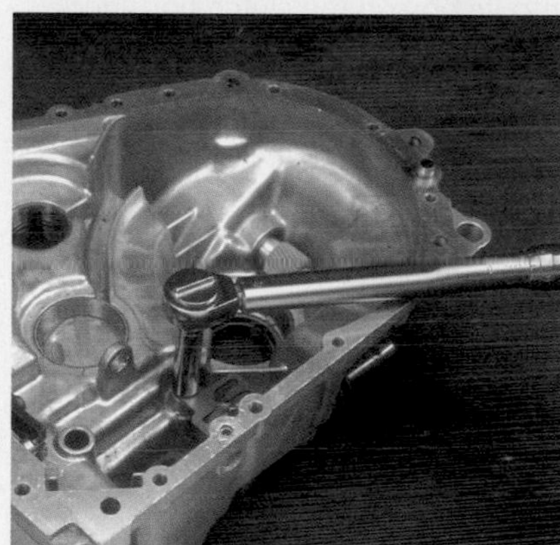

P29–5 Install and tighten the shift gate.

P29–6 Install the differential assembly into the transaxle case.

P29–7 Install the fifth-gear shaft assembly and fork shaft into the case.

P29–8 Place the mainshaft control-shaft assembly on the mainshaft so that the shift forks engage in their respective slots in the synchronizer sleeves. Then install the mainshaft assembly.

P29–9 Properly position the shift-relay lever and bias springs.

REASSEMBLING A TYPICAL TRANSAXLE *(continued)*

P29–10 Install the inhibitor spring and ball in the fifth and reverse-inhibitor shaft hole.

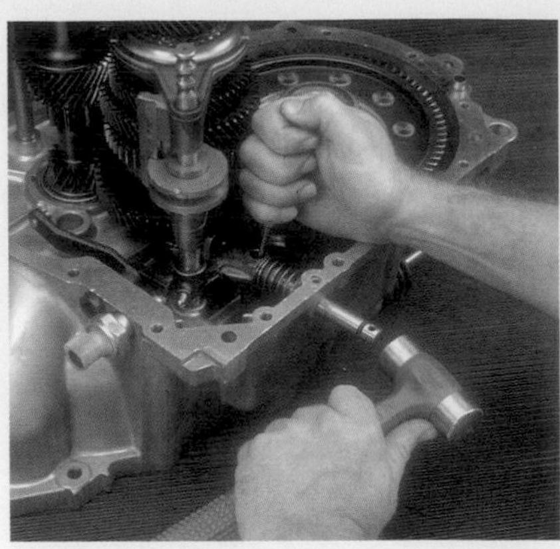

P29–11 Depress the inhibitor ball and spring using a drift, and slide the pivot shift-lever shaft through the shift-relay lever. Then tap the shaft into its bore in the clutch housing.

P29–12 Install the reverse idler gear shaft and gear into the appropriate bore in the case.

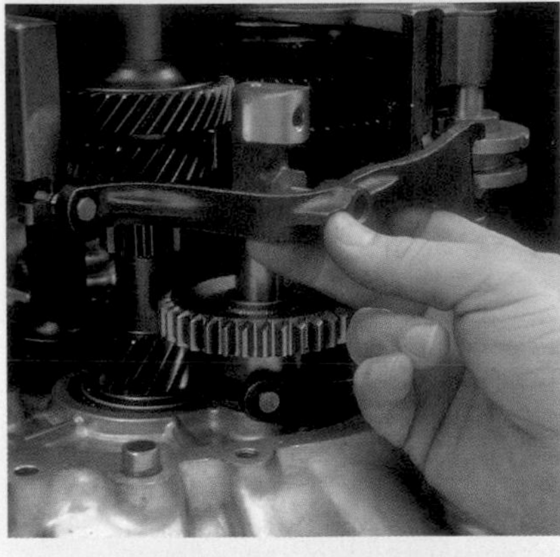

P29–13 Install the fifth-gear relay shaft, and align it with the fifth-gear fork slot and interlock plate.

P29–14 Install the retaining clip onto the fifth-gear relay shaft.

P29–15 Apply a thin bead of anaerobic sealant on the case's mating surface for the clutch housing.

P29–16 Install the clutch housing to the case. Be careful that the mainshaft, shift-control shaft, and fifth-gear shaft align with the bores in the case.

P29–17 After the housing and case are fit snugly together, tighten the attaching bolts to the specified torque.

P29–18 Install and tighten the interlock plate retaining bolt.

Installing the Transmission/Transaxle

SHOP TALK

Always check for free rotation of the transmission in all gears before installing in a vehicle. If the shafts do not rotate freely, identify the cause and correct it.

After the unit is together, install the clutch assembly and a new release or throwout bearing before installing the transmission/transaxle. Generally, installation is the reverse procedure as removal. When installing the transmission, never let the transmission hang by its input shaft. Use a transmission jack to hold the transmission while you guide it into place. The input shaft should be lightly coated with grease prior to installation to aid installation and serve as a lubricant for the pilot bearing. Avoid putting too much grease on the shaft, because the excess may fly off and get on the clutch disc, causing it to slip and/or burn.

Most transmissions are dowelled to the engine or bell housing. Use the dowels to locate and support the transmission during installation. Tighten the mounting bolts evenly, making sure nothing is caught between the housings. Then lightly coat the driveshaft's slip joint, and carefully insert it into the extension housing to prevent possible damage to the rear oil seal. Reattach and adjust the shift linkage, and fill the transmission with the proper fluid.

WARNING!

If the transmission/transaxle does not fit snugly against the engine block, or if you cannot move the transmission into place, do not force it. Pull the transmission back and lower it. Inspect the input shaft splines for dirt or damage. Also, check the mating surfaces for dirt or obstructions. If you try to force the transmission into place by tightening the bolts, you may break the case.

KEY TERMS

Fill plug (p. 1204)	Gear puller (p. 1214)
Gear clash (p. 1209)	Gear whine (p. 1209)

SUMMARY

- Proper lubrication is vital to long transmission/transaxle life. The transmission gear lubricant must be checked and changed at manufacturer's suggested intervals.
- Metal particles or shavings in the gear lubricant indicate extensive internal wear or damage.
- The first step in diagnosing transmission/transaxle problems is to confirm that the problem exists inside the transmission/transaxle. Clutch and driveline problems may often appear to be transmission/transaxle problems.
- The initial visual inspection should include checks for lubricant leakage at gaskets and seals, transmission mount inspection, clutch and gearshift linkage checks, and drive axle and CV joint inspection.
- Rough growling noise inside the transmission/transaxle housing is an indication of bearing problems.
- A clicking noise may indicate excessive gear tooth wear. Rhythmical knocking is a sign of loose or broken internal components.
- Hard shifting can be caused by shift linkage problems, improper lubricant, or worn internal components, such as bearings, gears, shift forks, or synchronizers.
- Jumping out of gear can be caused by misaligned drivetrain mounts, a worn or poorly adjusted shift linkage, excessive clearance between gears, or badly worn bearings.
- Low lubricant levels, poorly adjusted shift linkages, or damaged internal components can result in transmission/transaxle lock-up.
- Always follow service information recommendations for removing the transmission/transaxle from the vehicle and disassembling it.
- Use recommended bearing pullers, gear pullers, and press equipment to remove and install gears and synchronizers on shafts.
- Clean and inspect all parts carefully, replacing worn or damaged components. Never force components in place during reassembly. Follow all clearance specifications listed in the service information.
- Always use new snap rings, gaskets, and seals during reassembly.

REVIEW QUESTIONS

1. After draining gear oil from a transaxle, the technician notices the oil has shiny, metallic particles in it. What does this indicate?
2. List at least five separate checks that should be made during the visual inspection of transmission/transaxle components.

3. List at least three causes of noise that are not transmission/transaxle-related but may appear to be.
4. What tool is often needed to remove gears and synchronizer assemblies from the transmission mainshaft?
5. When removing or installing bearings, where should force be applied?
6. What could cause a standard transmission to jump out of gear?
   a. The clutch does not fully disengage.
   b. The wrong lubricant is being used.
   c. There is a faulty interlock mechanism.
   d. There are weak or broken detent springs.
7. What could allow a synchronizer to not slow or to speed up a speed gear?
   a. worn clutch dog teeth on the speed gear
   b. worn or dulled blocker ring oil cutting grooves
   c. worn or damaged speed gear teeth
   d. worn or damaged countergear teeth
8. A rough, growling noise is heard from a transaxle while it is in neutral with the engine running. If the vehicle is stationary and the clutch is engaged, what problem area would this noise likely indicate?
   a. transaxle input shaft bearings
   b. transaxle mainshaft (intermediate shaft) bearings
   c. first/second synchronizer assembly
   d. pinion and ring gear interaction
9. What could a clicking noise during transmission/transaxle operation be an indication of?
   a. worn mainshaft (input shaft) bearings
   b. faulty synchronizer operation
   c. failed oil seals
   d. worn, broken, or chipped gear teeth
10. What could low lubricant levels most likely cause?
    a. gear clash
    b. hard shifting
    c. gear lock-up or seizure
    d. gear jump-out
11. What could using a lubricant that is thicker than what is specified in the service information lead to?
    a. gear jump-out **b.** hard shifting
    c. gear lock-up **d.** gear slippage
12. What could cause a rough growling noise in a transmission/transaxle that is heard when operating in all gears?
    a. worn or damaged output shaft bearings
    b. worn or damaged input shaft bearings
    c. worn or damaged countershaft bearings
    d. worn or damaged reverse idler gear bearings
13. What is the correct procedure for removing a bearing from a transmission shaft?
    a. Use a puller on the outer race.
    b. Use a puller on the inner race.
    c. Use a ball-peen hammer on the outer race.
    d. Use a ball-peen hammer on the inner race.
14. What could cause a rough growling noise in a transmission that is heard when operating in all gears except direct drive?
    a. worn or damaged speed gear teeth
    b. worn or damaged countergear teeth
    c. worn or damaged output shaft pilot roller bearings
    d. worn or damaged output shaft bearings
15. A poorly adjusted shift linkage can cause which of the following problems?
    a. gear clash **b.** hard shifting
    c. gear jump-out **d.** all of the above
16. Which of the following can be reused when rebuilding a transmission?
    a. snap rings
    b. gaskets
    c. seals
    d. synchronizer assemblies
17. What could cause gear clashing when selecting only one particular gear?
    a. worn synchronizer blocker ring
    b. improper lubricant
    c. worn clutch
    d. chipped gear teeth
18. What position should the gear selector be in for backup light switch replacement?
    a. neutral **b.** reverse
    c. the lowest gear **d.** the highest gear
19. Which of the following could cause a rhythmic knocking noise?
    a. worn helical gears
    b. worn or damaged output shaft bearings
    c. worn synchronizer blocker ring
    d. chipped or broken gear teeth
20. When inspecting a synchronizer blocker ring, which of the following would require blocker ring replacement?
    a. narrow index slots
    b. straight edges and corners on the dog (clutching) teeth
    c. sharp grooves on the internal surface
    d. no space between the blocker ring and speed gear dog teeth

CHAPTER 41
Drive Axles and Differentials

LEARNING OUTCOMES

- Name and describe the components of a front-wheel-drive axle.
- Describe the operation of a front-wheel-drive axle.
- Diagnose problems in CV joints.
- Perform maintenance on CV joints.
- Explain the difference between CV joints and universal joints.
- Name and describe the components of a rear-wheel-drive axle.
- Describe the operation of a rear-wheel-drive axle.
- Explain the function and operation of a differential and drive axles.
- Describe the various drive axle designs, including complete, integral carrier, removable carrier, and limited slip.
- Describe the three common types of driving axles.
- Explain the function of the main driving gears, drive pinion gear, and ring gear.
- Describe the operation of hunting, nonhunting, and partial nonhunting gears.
- Describe the different types of axle shafts and axle shaft bearings.

The drive axle assembly transmits torque from the engine and transmission to drive the vehicle's wheels. The drive axle changes the direction of the power flow, multiplies torque, and allows different speeds between the two drive wheels. Drive axles are used for both front-wheel-drive and rear-wheel-drive vehicles.

FRONT-WHEEL-DRIVE (FWD) AXLES

Front-wheel-drive (FWD) axles, also called axle shafts, typically transfer engine torque from the transaxle's differential to the front wheels. One of the most important components of FWD axles is the constant velocity (CV) joint. These joints are used to transfer uniform torque at a constant speed, while operating through a wide range of angles.

On front- or four-wheel-drive cars, operating angles of as much as 40° are common **(Figure 41–1)**. The drive axles must transmit power from the engine to the front wheels that must drive, steer, and cope with the severe angles caused by the up-and-down movement of the vehicle's suspension. To accomplish this, these cars must have a compact joint that ensures the driven shaft is rotated at a constant velocity, regardless of angle. CV joints also allow the distance between the transaxle case and

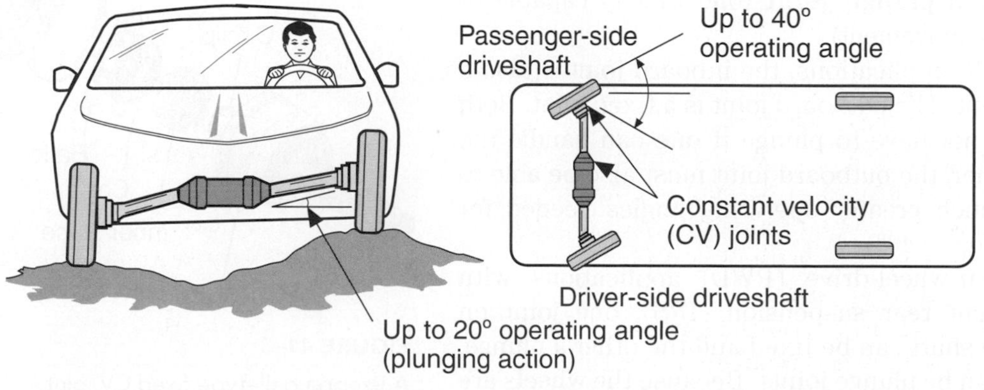

FIGURE 41–1

FWD drive axle (axle shaft) angles.

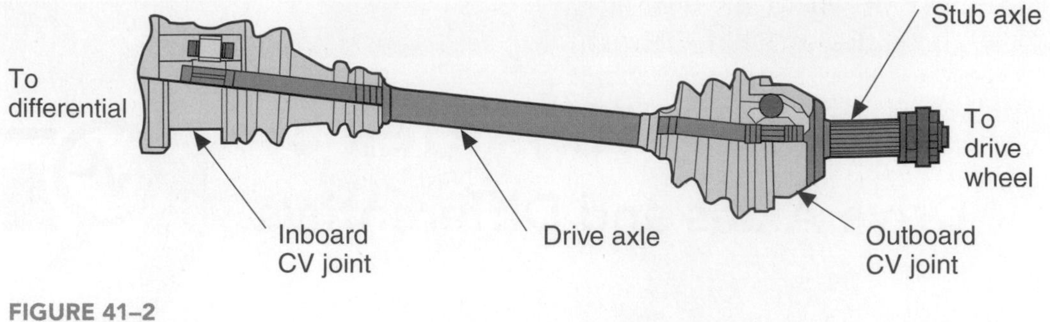

FIGURE 41–2

A typical FWD drive axle assembly.

MAHLE

wheels to change as the wheel travels up and down (plunging action).

TYPES OF CV JOINTS

Constant velocity joints come in a variety of styles. The different types of joints can be referred to by position (inboard or outboard), by function (fixed or plunge), or by design (ball type or tripod).

Inboard and Outboard Joints

On FWD vehicles, two CV joints are used on each half shaft **(Figure 41–2)**. The joint nearer the transaxle is the inner or **inboard joint**, and the one nearer the wheel is the outer or **outboard joint**. In an RWD vehicle with independent rear suspension, the joint nearer the carrier housing can also be referred to as the inboard joint. The one closer to the wheel is the outboard joint.

Fixed and Plunge Joints

CV joints are either a **fixed joint** (meaning it does not plunge in and out to compensate for changes in length) or a **plunge joint** (one that is capable of in-and-out movement).

In FWD applications, the inboard joint is also a plunge joint. The outboard joint is a fixed joint. Both joints do not have to plunge if one can handle the job. Further, the outboard joint must also be able to handle much greater operating angles needed for steering (up to 40°).

In rear-wheel-drive (RWD) applications with independent rear suspension (IRS), one joint on each axle shaft can be fixed and the other a plunge, or both can be plunge joints. Because the wheels are not used for steering, the operating angles are not as great. Therefore, plunge joints can be used at either or both ends of the axle shafts.

Ball-Type Joints

There are two basic varieties of CV joints: the **ball-type joints** and **tripod-type joints**. Both types are used as either inboard or outboard joints, and both are available in fixed or plunge designs.

FIXED BALL–TYPE CV JOINTS The **Rzeppa joint**, or fixed ball–type joint, consists of an inner ball race, six balls, a cage to position the balls, and an outer housing **(Figure 41–3)**. Tracks machined in the inner race and outer housing allow the joint to flex. The inner race and outer housing form a ball-and-socket arrangement. The six balls serve both as bearings between the races and the means of transferring torque from one to the other.

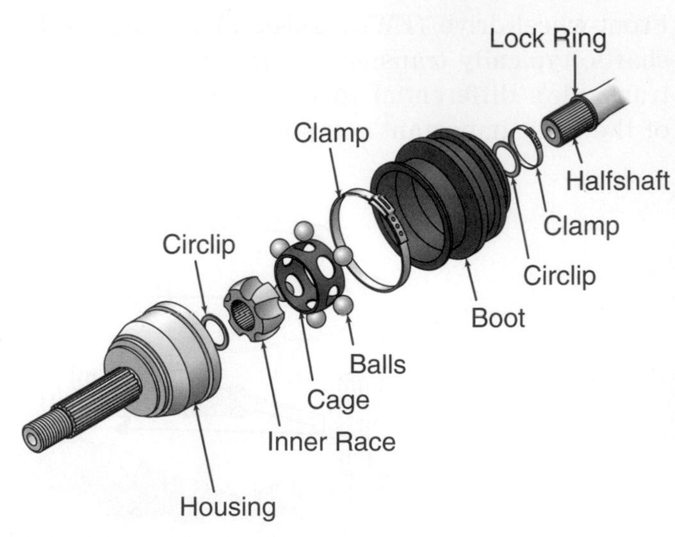

FIGURE 41–3

A Rzeppa ball–type fixed CV joint.

© AA1 Car Auto Diagnosis & Repair Help, www.AA1Car.com

If viewed from the side, the balls within the joint always bisect the angle formed by the shafts on either side of the joint regardless of the operating angle. This reduces the effective operating angle of the joint by half and virtually eliminates all vibration problems. The input speed to the joint is always equal to the output velocity of the joint—thus the description *constant velocity*. The cage helps to maintain this alignment by holding the six balls snugly in its windows. If the cage windows become worn or deformed over time, the resulting play between ball and window typically results in a clicking noise when turning. It is important to note that opposing balls in a Rzeppa CV joint always work together as a pair. Heavy wear in the tracks of one ball almost always results in identical wear in the tracks of the opposing ball.

PLUNGING BALL–TYPE JOINTS There are two basic styles of plunging ball–type joints: the **double-offset joint** and the **cross-groove joint**. This is a more compact design with a flat, doughnut-shaped outer housing and angled grooves.

The double-offset joint **(Figure 41–4)** uses a cylindrical outer housing with straight grooves and is typically used in applications that require higher operating angles (up to 25°) and greater plunge depth (up to 60 mm [2.4 in.]). This type of joint can be found at the inboard position on some front-wheel-drive half shafts as well as on the propeller shaft of some four-wheel-drive shafts.

The cross-groove joint **(Figure 41–5)** has a much flatter design than any other plunge joint. It is used as the inboard joint on FWD half shafts or at either end of an RWD independent rear suspension axle shaft.

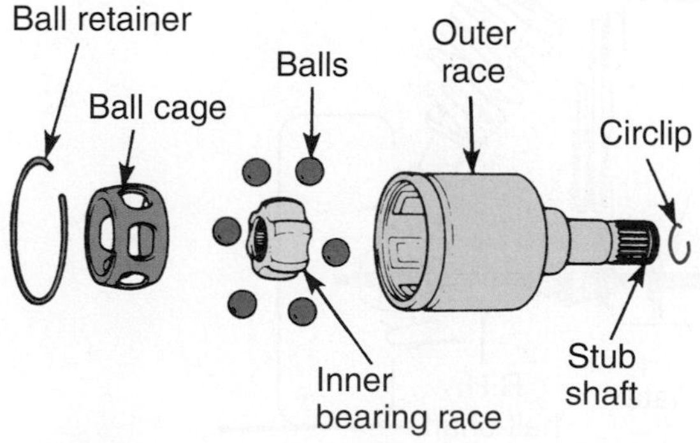

FIGURE 41–4

A double-offset CV joint.

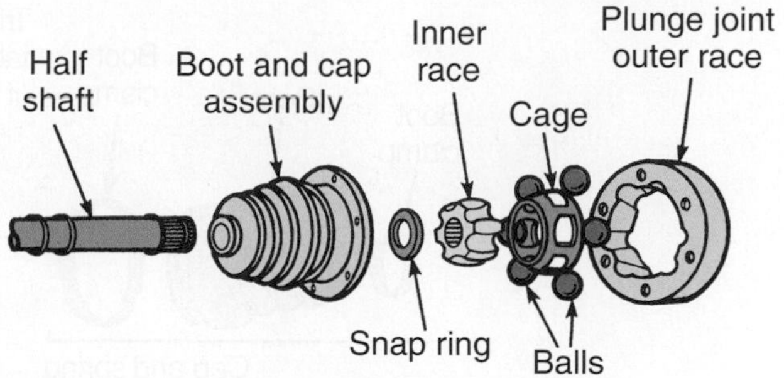

FIGURE 41–5

A cross-groove joint.

The feature that makes this joint unique is its ability to handle a fair amount of plunge (up to 46 mm [1.8 in.]) in a relatively short distance. The inner and outer races share the plunging motion equally, so less overall depth is needed for a given amount of plunge. The cross groove can handle operating angles up to 22°.

Tripod CV Joints

As with ball-type CV joints, tripod joints come in two varieties: plunge and fixed.

TRIPOD PLUNGING JOINTS Tripod plunging joints **(Figure 41–6)** consist of a central drive part or tripod (also known as a "spider"). This has three trunnions fitted with spherical rollers on needle bearings and an outer housing (sometimes called a "tulip" because of its three-lobed, flowerlike appearance). On some tripod joints, the outer housing is closed, meaning the roller tracks are totally enclosed within it. On others, the tulip is open, and the roller tracks are machined out of the housing. Tripod joints are most commonly used as FWD inboard plunge joints.

FIXED TRIPOD JOINTS The fixed tripod joint is sometimes used as the outboard joint in FWD applications. In this design, the trunnion is mounted in the outer housing, and the three roller bearings turn against an open tulip on the input shaft. A steel locking spider holds the joint together.

The fixed tripod joint has a much greater angular capability. The only major difference from a service standpoint is that the fixed tripod joint cannot be removed from the axle shaft or disassembled because of the way it is manufactured. The complete joint and shaft assembly must be replaced if the joint goes bad.

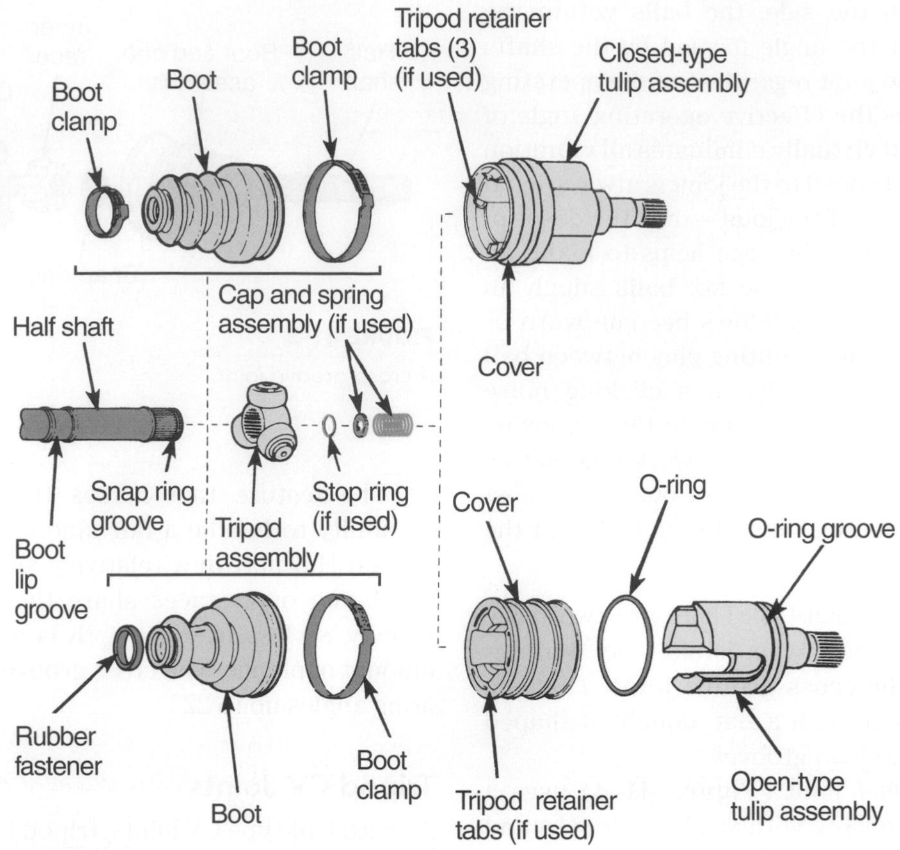

FIGURE 41–6

Inner tripod plunge-type joints: closed housing and open housing.

FRONT-WHEEL-DRIVE APPLICATIONS

FWD half shafts can be solid or tubular, be of equal **(Figure 41–7)** or unequal length **(Figure 41–8)**, and come with or without damper weights. Equal-length shafts are used in some vehicles to help reduce torque steer (the tendency to steer to one side as engine power is applied). In these applications, an intermediate shaft is used as a link from the transaxle to one of the half shafts. This intermediate shaft can use an ordinary Cardan universal joint (described later in this chapter) to a yoke at the transaxle. At the outer end is a support bracket

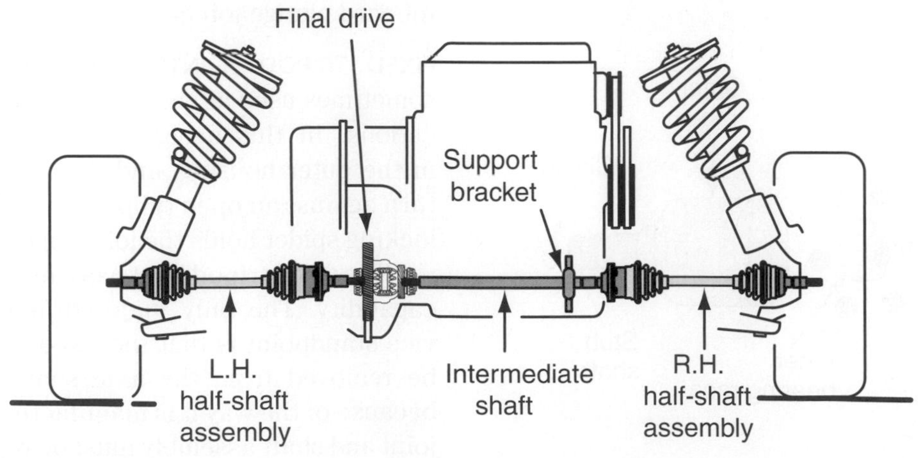

FIGURE 41–7

Equal-length FWD half shafts with an intermediate shaft.

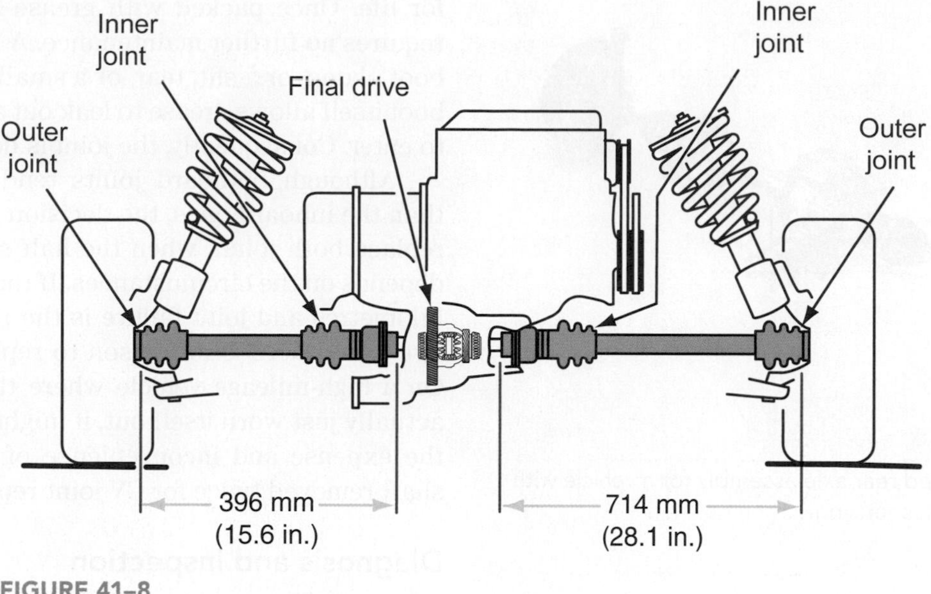

FIGURE 41–8

Unequal-length FWD half shafts.

and bearing assembly. Looseness in the bearing or bracket can create vibrations. These items should be included in any inspection of the drivetrain components. The small damper weight, called a **torsional damper**, that is sometimes attached to one half shaft serves to dampen harmonic vibrations in the drivetrain and to stabilize the shaft as it spins, not to balance the shaft **(Figure 41–9)**.

Regardless of the application, outer joints typically wear faster than inner joints because of the increased range of operating angles to which they are subjected. Inner joint angles may change only 10° to 20° as the suspension travels through jounce and rebound. Outer joints can undergo changes of up to 40°, in addition to jounce and rebound, as the wheels are steered. That, combined with more flexing of the outer boots, is why outer joints have a higher failure rate. On average, nine outer CV joints are replaced for every inner CV joint. That does not mean the technician should overlook the inner joints. They wear too. Every time the suspension travels through jounce and rebound, the inner joints must plunge in and out to accommodate the different arcs between the axle shafts and suspension. Tripod inner joints tend to develop unique wear patterns on each of the three rollers and their respective tracks in the housing, which can lead to noise and vibration problems.

Other Applications

CV joints are also found on the front axles of many 4WD vehicles and on vehicles with rear independent suspension systems **(Figure 41–10)**. Their use in these designs offers the same benefits as when they are used for front-wheel drive.

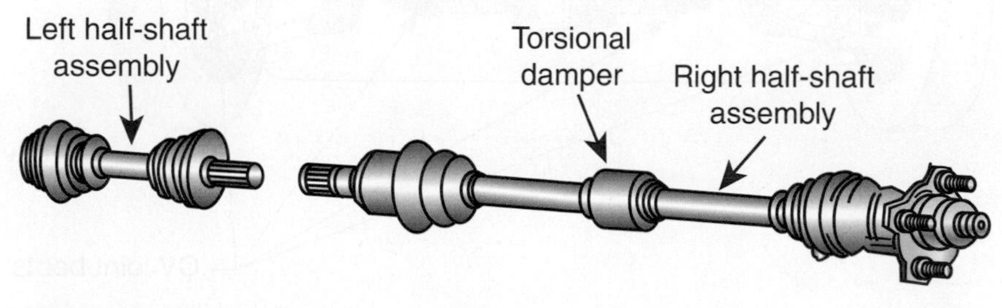

FIGURE 41–9

Some long drive axles are fitted with a torsional damper.

FIGURE 41–10

A CV joint–equipped rear axle assembly for a vehicle with independent rear suspension.

Dana Corporation

CV JOINT SERVICE

With proper service, CV joints can have a long life, despite having to perform extremely difficult jobs in hostile environments. They must endure extreme heat and cold and survive the shock of hitting potholes at high speeds. Fortunately, high-torque loads during low-speed turns and many thousands of high-speed kilometres normally do not bother the CV joint. It is relatively trouble-free unless damage to the boot or joint goes unnoticed.

All CV joints are encased in a protective rubber (neoprene, natural, or silicone) or thermoplastic (Hycrel) boot **(Figure 41–11)**. The job of the boot is to retain grease and to keep dirt and water out. The importance of the boot cannot be overemphasized because without its protection the joint does not survive. For all practical purposes, a CV joint is lubed for life. Once packed with grease and installed, it requires no further maintenance. A loose or missing boot clamp, or a slit, tear, or a small puncture in the boot itself allows grease to leak out and water or dirt to enter. Consequently, the joint is destroyed.

Although outboard joints tend to wear faster than the inboard ones, the decision as to whether to replace both joints when the half shaft is removed depends on the circumstances. If the vehicle has low kilometres and joint failure is the result of a defective boot, there is no reason to replace both joints. On a high-mileage vehicle where the bad joint has actually just worn itself out, it might be wise to save the expense and inconvenience of having the half shaft removed twice for CV joint replacement.

Diagnosis and Inspection

Any noise in the engine, drive axle, steering, or suspension is a good reason for a thorough inspection of the vehicle. A road test on a smooth surface is a good place to begin. The test should include driving at average highway speeds, some sharp turns, acceleration, and coasting. Look and listen for the following signs:

- A popping or clicking noise when turning indicates a possible worn or damaged outer joint **(Figure 41–12)**. To help identify the exact cause, put the vehicle in reverse and back up in a circle. If the noise gets louder, the outer joints should be replaced.
- A clunk during accelerating, decelerating, or putting an automatic transaxle into drive can be caused by excessive play in the inner joint on FWD vehicles. A clunking noise when putting an automatic transmission into gear or when

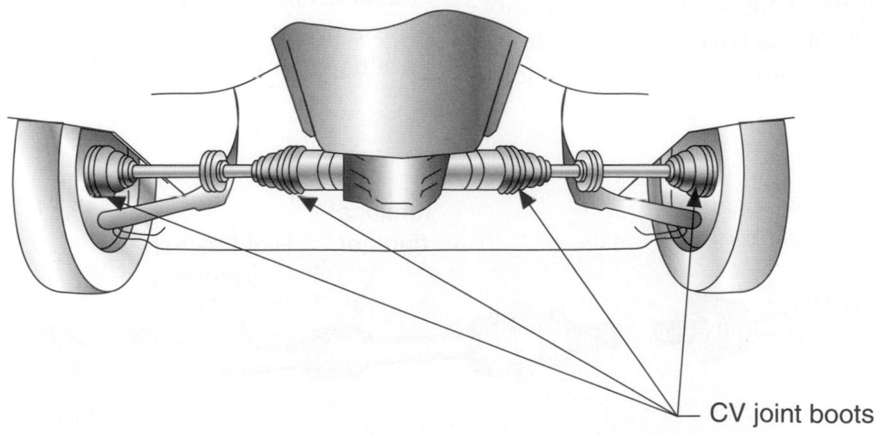

FIGURE 41–11

Location of the CV joint boots.

©Cengage Learning 2015

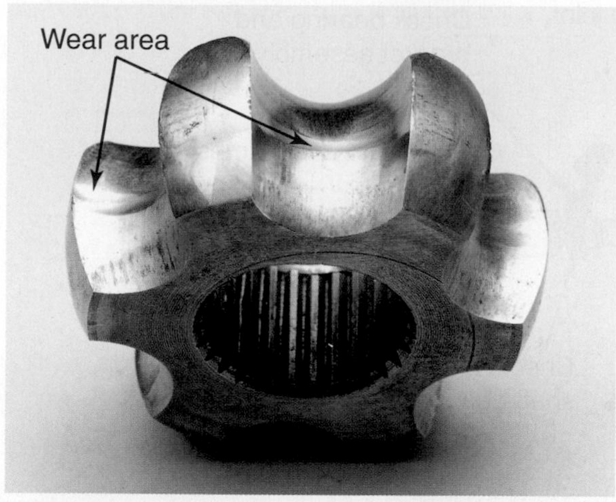

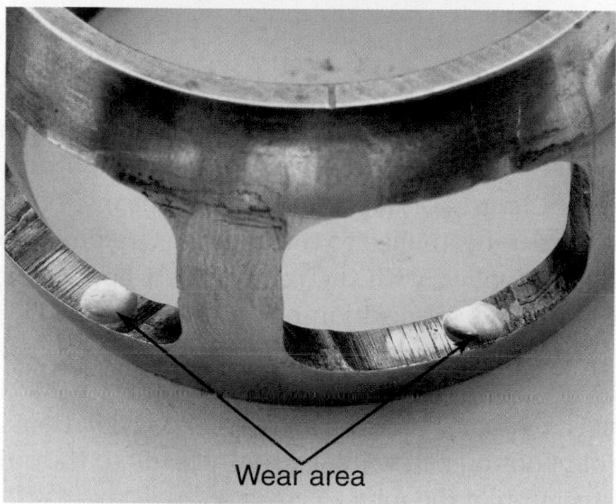

FIGURE 41–12

A worn cage or race can cause a clicking sound during a turn.
Federal-Mogul Corporation

starting out from a stop usually indicates excessive play in an inner or outer joint. Be warned, though, that the same kind of noise can also be produced by excessive backlash in the final drive gears and transmission. Alternately accelerating and decelerating in reverse while driving straight can reveal worn inner plunge joints. A bad joint clunks or shudders.

- A humming or growling noise is sometimes due to inadequate lubrication of either the inner or the outer CV joint. It is more often due to worn or damaged wheel bearings, a bad intermediate shaft bearing on equal-length half-shaft transaxles, or worn shaft bearings within the transmission.
- A shudder or vibration when accelerating is often caused by excessive play in either the inboard or outboard joint but more likely it is the inboard plunge joint. These vibrations can also be caused by a bad intermediate shaft bearing on transaxles with equal-length half shafts. On FWD vehicles with transverse-mounted engines, this kind of vibration can also be caused by loose or deteriorated engine/transaxle mounts. Be sure to inspect the rubber bushings in the engine's upper torque strap to rule out this possibility. A vibration or shudder that increases with speed or comes and goes at a certain speed may be the result of excessive play in an inner or outer joint. A bent axle shaft can cause the same problem. Note, however, that some shudder could also be inherent to the vehicle.
- A cyclic vibration that comes and goes between 72 and 100 km (45 and 60 mph) may lead the technician to think there is a wheel that is out of balance. However, as a rule, an out-of-balance wheel produces a continuous vibration. A more likely cause is a bad inner tripod CV joint. The vibration occurs because one of the three roller tracks has become dimpled or rough. Every time the tripod roller on the bad track hits the rough spot, it creates a little jerk in the driveline, which the driver feels as a cyclic vibration.
- A noise heard while driving straight ahead that ceases while turning means the problem is usually not a defective outer CV joint, but a bad front wheel bearing. Turning changes the load on the bearing, which may make it become quieter than before.
- A vibration that increases with speed is rarely due to CV joint problems or FWD half-shaft imbalance. An out-of-balance tire or wheel, an out-of-round tire or wheel, or a bent rim are the most likely causes. It is possible that a bent half shaft, as the result of collision or towing damage, could cause the vibration. A missing damper weight could also be the culprit.

Begin CV joint inspection **(Figure 41–13)** by checking the condition of the boots. Splits, cracks, tears, punctures, or thin spots caused by rubbing call for immediate boot replacement. If the boot appears rotted, this indicates improper greasing or excessive heat, and it should be replaced. Squeeze all boots. If any air escapes, replace the boot.

If the inner boot appears to be collapsed or deformed, venting it (allowing air to enter) might solve the problem. Place a round-tipped rod between the boot and driveshaft. This equalizes the outside and inside air and allows the boot to return to its normal shape.

Make sure that all boot clamps are tight. Missing or loose clamps should be replaced. If the boot

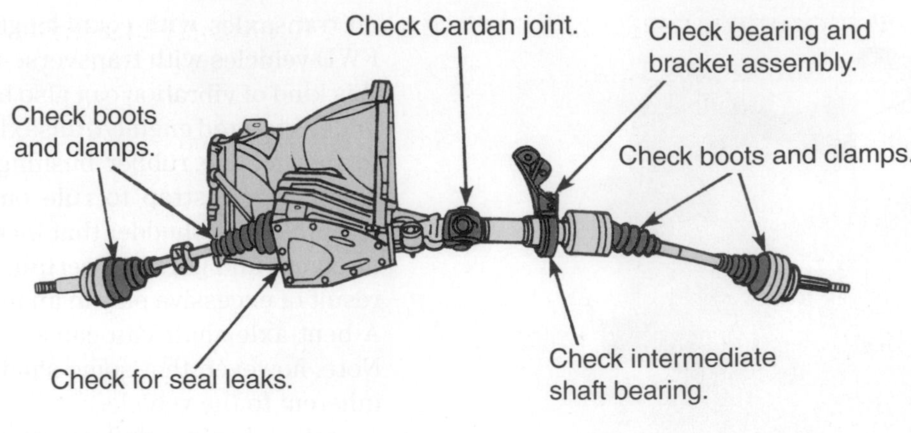
Check Cardan joint.
Check bearing and bracket assembly.
Check boots and clamps.
Check boots and clamps.
Check for seal leaks.
Check intermediate shaft bearing.

FIGURE 41–13

Inspection points for an FWD vehicle.

appears loose, slide it back and inspect the grease inside for possible contamination. A milky or foamy appearance indicates water contamination. A gritty feeling when rubbed between the fingers indicates dirt. In most cases, a water- or dirt-contaminated joint should be replaced.

The drive axles should be checked for signs of contact or rubbing against the chassis. Rubbing can be a symptom of a weak or broken spring or engine mount, as well as chassis misalignment. On FWD transaxles with equal-length half shafts, inspect the intermediate shaft U-joint, bearing, and support bracket for looseness by rocking the wheel back and forth and watching for any movement. Oil leakage around the inner CV joints indicates a faulty transaxle shaft seal. To replace the seal, the half shaft must be removed.

Obtaining CV Repair Parts

To repair a drive axle, a complete shaft should be installed. Most aftermarket parts suppliers offer a complete line of original equipment axle shafts for FWD vehicles. These shafts come fully assembled and ready for installation. This repair method eliminates the need to tear down and rebuild an old shaft.

If only the CV joints need service, a CV joint service kit should be installed. Joint service kits typically include a CV joint, boot, boot clamps and seals, special grease for lubrication (various joints require different amounts of grease; the correct quantity is packed in each kit), retaining rings, and all other attachment parts.

When only the boot is damaged, part manufacturers also produce a line of complete boot sets for each application, including new clamps and the appropriate type and amount of grease for the joint. CV joints require a special high-temperature, high-pressure grease. Substituting any other type of grease may lead to premature failure of the joint. Be sure to use all the grease supplied in the joint or boot kit. The same rule applies to the clamps. Use only those clamps supplied with the replacement boot. Follow the directions for positioning and securing them.

Old boots should never be reused when replacing a CV joint. In most cases, failure of the old joint is caused by some deterioration of the old boot. Reusing an old boot on a new joint usually leads to the quick destruction of the joint.

Photo Sequence 30 shows the procedure for removing a typical drive axle and replacing a CV joint boot. Always refer to the service information for the exact service procedure. The diagnosis and service chart shown in **Table 41–1** gives an idea of the types of front-wheel drivetrain problems that can occur.

CV Joint Service Guidelines

The following are some guidelines to follow when servicing CV joints:

- Always support the control arm when doing on-the-car balancing of the front wheels to avoid high-speed operation at a steep half-shaft angle. Off-the-car balancing might be a wiser choice.
- Do not use half shafts as lift points for raising a car.
- Use a plastic or metal shield over rubber boots to protect them from accidental tool damage when performing other wheel, brake, suspension, or steering system maintenance **(Figure 41–14)**.
- Clean the axle boots with only soap and water and avoid contact with gasoline, oil, or degreaser compounds.

TABLE 41–1 PROBLEM DIAGNOSIS AND SERVICE FOR FWD DRIVELINES

PROBLEM	POSSIBLE CAUSE	CORRECTIVE REMEDY
Vibrations in steering wheel at highway speeds	Front-wheel balance	Front-wheel unbalance is felt in the steering wheel. Front wheels must be balanced.
Vibrations throughout vehicle	Worn inner CV joints	Worn parts of the inner CV joint not operating smoothly.
Vibrations throughout vehicle low speed	Bent axle shaft	Axle shaft does not operate on centre of axis; thus, a vibration develops.
Vibrations during acceleration	Worn or damaged outer or inner CV joints Fatigued front springs	CV joints not operating smoothly due to damage or wear on parts. Sagged front springs are causing the inner CV joint to operate at too great an angle, causing vibrations.
Grease dripping on ground	Ripped or torn CV joint boots	Front-wheel-drive CV joints are immersed in lubricant or sprayed on chassis parts. If the CV joint boot has a rip or is torn, lubricant leaks out. Condition must be corrected as soon as possible.
Clicking or snapping noise heard when turning curves and corners	Worn or damaged outer CV joint Bent axle shaft	Worn parts are clicking and noisy as loading and unloading on CV joint takes place. Irregular rotation of the axle shaft causing a snapping, clicking noise.

- Never jerk or pull on the axle shaft when removing it from a vehicle with tripod inner joints. Doing so may pull the joint apart, allowing the needle bearings to fall out of the roller. Pull on the outer housing, and support the outer end of the shaft until the shaft is completely out.
- Always install new hub nuts, and torque them to specifications. This is absolutely necessary to properly preload the wheel bearings. Do not guess. The specifications can vary from 101 to 318 N·m (75 to 235 ft·lb). Most axle hub nuts are staked in place after they have been tightened **(Figure 41–15)**. Others have a castellated nut that is secured with a cotter pin.
- Never use an impact wrench to loosen or tighten axle hub nuts. Doing so may damage the wheel bearings as well as the CV joints.
- On vehicles with antilock brakes, use care to protect the wheel speed sensor and tone ring on the outer CV joint housings. If misaligned or damaged during joint replacement, it can cause wheel speed sensor problems.
- Always recheck the alignment after replacing CV joints. Marking the camber bolts is not enough, because camber can be off as much as three-quarters of a degree due to differences between the size of the camber bolts and their holes.

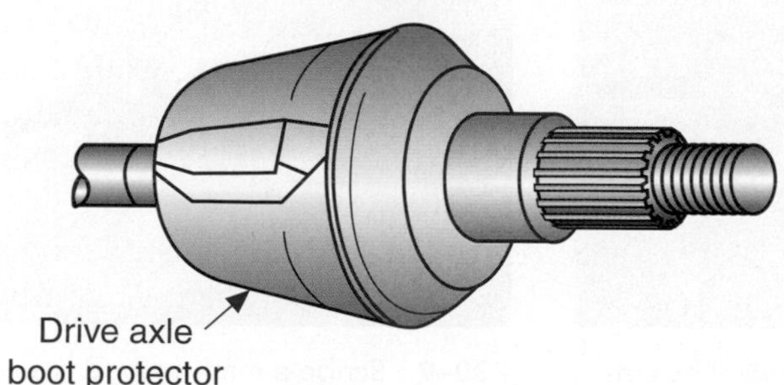

FIGURE 41–14

A typical axle boot protector.

©Cengage Learning 2015

FIGURE 41–15

Most axle hub nuts are staked after they are tightened to lock them in place.

Federal-Mogul Corporation

PHOTO SEQUENCE 30

REMOVING AND REPLACING A CV JOINT BOOT

P30–1 Removing the axle from the car begins with the removal of the wheel cover and wheel hub cover. The hub nut should be loosened before raising the car and removing the wheel.

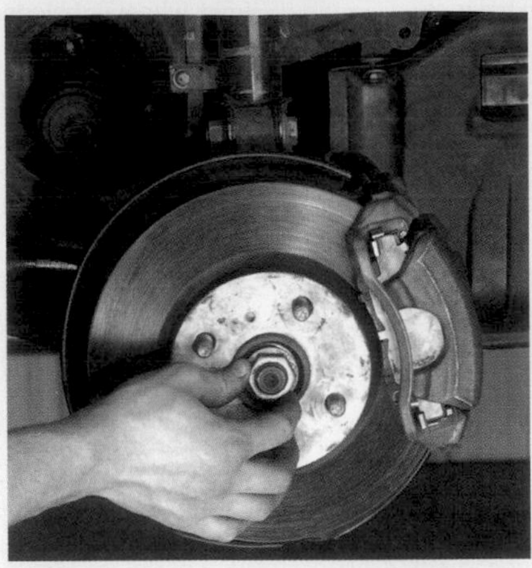

P30–2 After the car is raised and the wheel is removed, the hub nut can be unscrewed from the axle shaft.

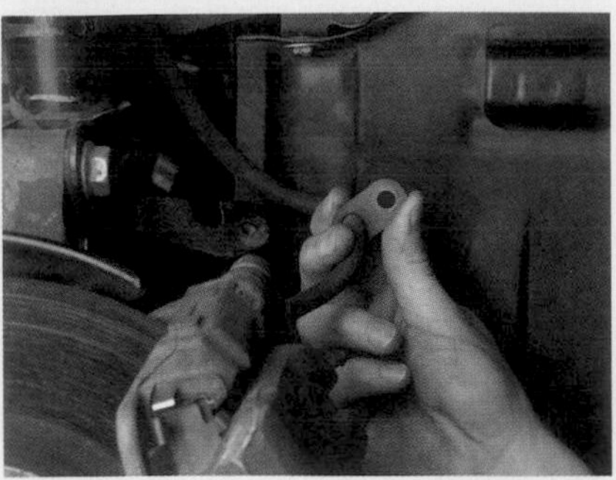

P30–3 The brake line holding clamp must be loosened from the suspension.

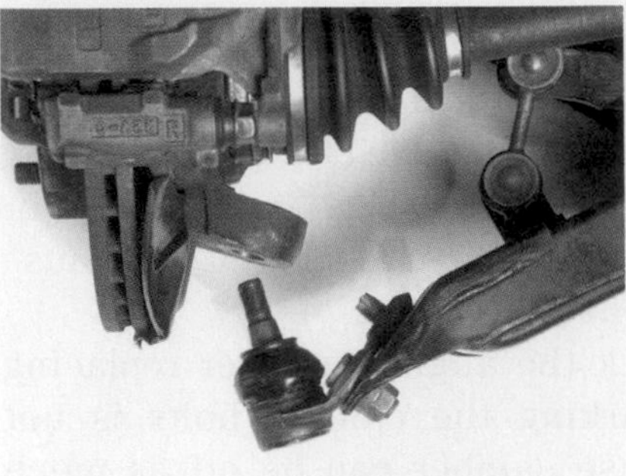

P30–4 The ball joint must be separated from the steering knuckle assembly. To do this, first remove the ball joint retaining bolt. Then pry down on the control arm until the ball joint is free.

P30–5 The inboard joint can be pulled free from the transaxle.

P30–6 A special tool is normally needed to separate the axle shaft from the hub, allowing the axle to be removed from the car.

P30–7 The axle shaft should be mounted in a soft-jawed vise for work on the joint. Pieces of wood on either side of the axle work well to secure the axle without damaging it.

P30–8 Begin boot removal by cutting and discarding the boot clamps.

P30–9 Scribe a mark around the axle to indicate the boot's position on the shaft. Then move the boot off the joint.

P30–10 Remove the circlip, and separate the joint from the shaft.

P30–11 Slide the old boot off the shaft.

P30–12 Wipe the axle shaft clean, and install the new boot onto the shaft.

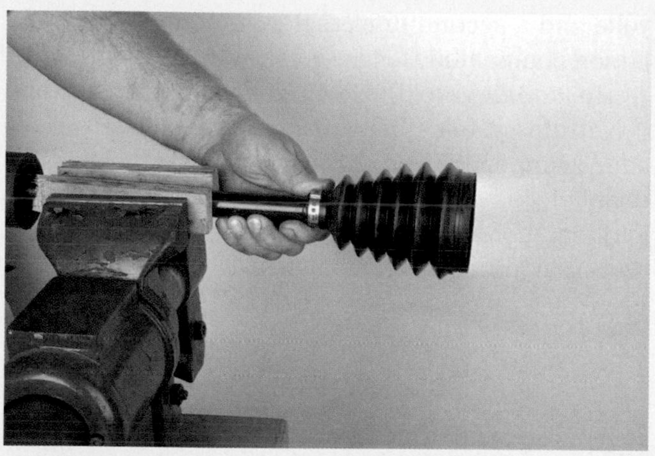

P30–13 Place the boot into its proper location on the shaft, and install a new clamp.

P30–14 Using a new circlip, reinstall the joint on the shaft. Pack joint grease into the joint and boot. The entire packet of grease that comes with a new boot needs to be forced into the boot and joint.

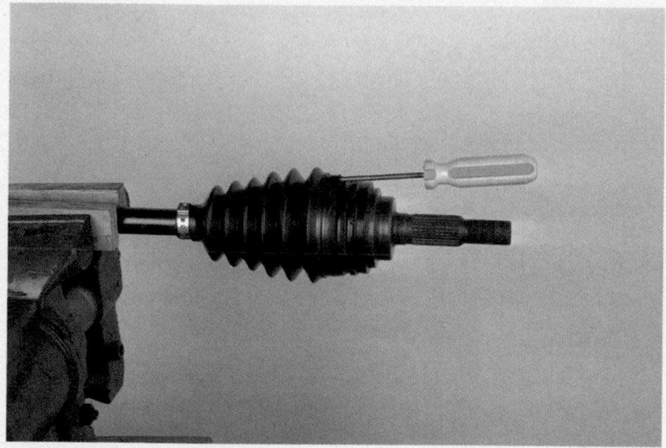

P30–15 Pull the boot over the joint and into its proper position. Use a dull screwdriver to lift an edge of the boot up to equalize the pressure inside the boot with the outside air.

P30–16 Install the new large boot clamp, and reinstall the axle into the car. Torque the hub nut after the wheels have been reinstalled and the car is sitting on the ground.

REAR-WHEEL DRIVESHAFTS

A driveshaft must smoothly transfer torque while rotating, changing length, and moving up and down. All different designs of driveshafts attempt to ensure a vibration-free transfer of the engine's power from the transmission to the rear axle. This goal is complicated by the fact that the engine and transmission are bolted solidly to the frame of the car, whereas the rear axle is mounted on springs. As the rear wheels go over bumps in the road or changes in the road's surface, the springs compress or expand, changing the angle of the driveshafts between the transmission and the rear axle, as well as the distance between the two. To allow for these changes, the Hotchkiss-type driveline is fitted with one or more U-joints to permit variations in the angle of the drive, and a slip joint that permits the effective length of the driveline to change.

SHOP TALK

When a vehicle is intentionally raised or lowered, the length of the driveshaft should be changed to allow for normal travel of the slip yoke on the output shaft.

FIGURE 41–17

A centre hanger bearing assembly.

Starting at the front or transmission end of an RWD shaft, there is a slip yoke, universal joint, driveshaft yoke, and driveshaft **(Figure 41–16)**. At the rear or rear axle end, there is another driveshaft yoke and a second universal joint connected to the pinion companion flange or yoke.

In addition to these basic components, some drivetrain systems use a centre hanger bearing for added support **(Figure 41–17)**. Some vehicles use a double U-joint arrangement, called a double-Cardan joint or a constant velocity U-joint, to help minimize universal joint operating angles.

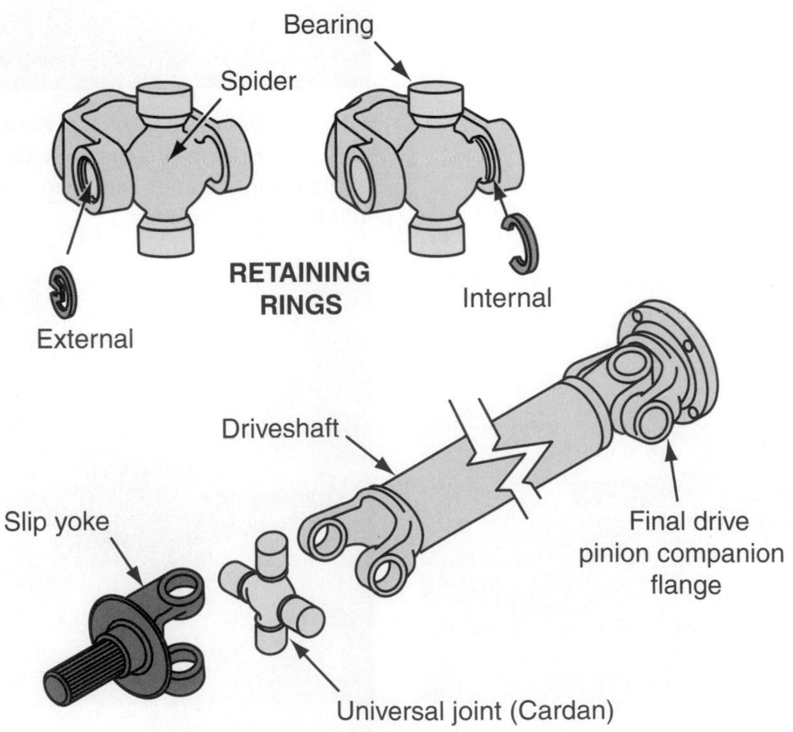

FIGURE 41–16

A driveshaft assembly with exploded view of U-joints.

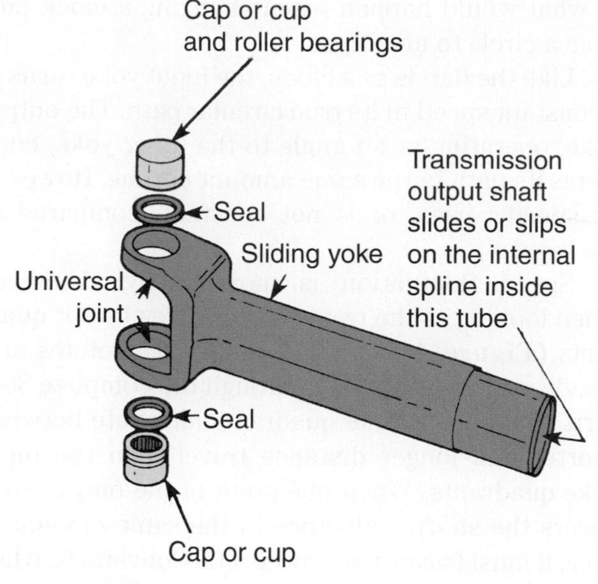

FIGURE 41–18

A typical slip or sliding yoke.

Slip Yoke

The most common sliding or **slip yoke (Figure 41–18)** has an internally splined, externally machined bore that lets the yoke rotate at transmission output shaft speed and slide at the same time (hence the name slip yoke). While the need for rotation is obvious, without the linear flexibility, the driveshaft would bend like a bow the first time the suspension jounced.

Driveshaft and Yokes

The driveshaft is nothing more than an extension of the transmission output shaft. The driveshaft, which is usually made from seamless steel tubing, transfers engine torque from the transmission to the rear driving axle. The yokes, which are either welded or pressed onto the shaft, provide a means of connecting two or more shafts together. At the present time, a limited number of vehicles are equipped with fibre-composite—reinforced fibreglass, graphite, and aluminum—driveshafts. The advantages of using these materials are weight reduction, torsional strength, fatigue resistance, easier and better balancing, and reduced interference from shock loading and torsional problems. Some driveshafts are fitted with a torsional damper to reduce torsional vibrations.

The driveshaft, like any other rigid tube, has a natural vibration frequency. If one end were held tightly, it would vibrate at its own frequency when deflected and released. It reaches its natural frequency at its critical speed. Critical driveshaft speed depends on the diameter of the tube and its length. Diameters are as large as possible and shafts as short as possible to keep the critical speed frequency above the driving speed range. It should be remembered that because the driveshaft generally turns three to four times faster than the tires, proper driveshaft balance is required for vibration-free operation.

OPERATION OF U-JOINTS

The U-joint allows two rotating shafts to operate at a slight angle to each other. A French mathematician named Cardan developed the original joint in the 16th century. In 1902, Clarence Spicer modified Cardan's invention for the purpose of transmitting engine torque to an automobile's rear wheels.

The universal joint is basically a double-hinged joint consisting of two Y-shaped yokes, one on the driving, or input, shaft and the other on the driven, or output, shaft, plus a cross-shaped unit called the cross **(Figure 41–19)**. A yoke is used to connect the U-joints together. The four trunnions of the cross are fitted with bearings in the ends of the two shaft yokes. The input shaft's yoke causes the cross to rotate, and the two other trunnions of the cross cause the output shaft to rotate. When the two shafts are at an angle to each other, the bearings allow the yokes to swing around on their trunnions with each revolution. This action allows two shafts, at a slight angle to each other, to rotate together.

Universal joints allow the driveshaft to transmit power to the rear axle through varying angles that are controlled by the travel of the rear suspension. Because power is transmitted on an angle, U-joints do not rotate at a constant velocity, nor are they vibration-free.

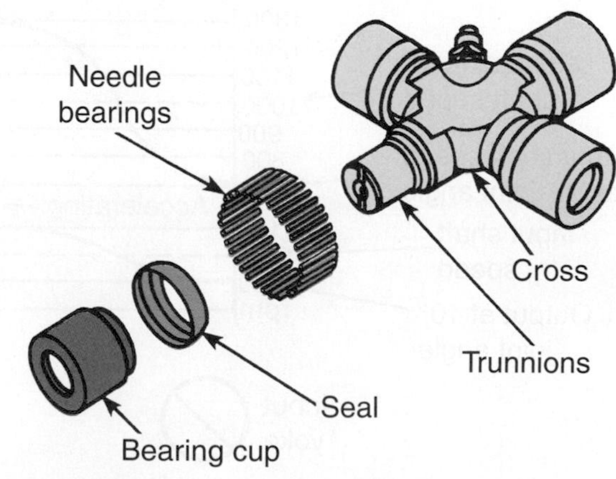

FIGURE 41–19

A Cardan joint.

Speed Variations (Fluctuations)

Although simple in appearance, the universal joint is more intricate than it seems, because its natural action is to speed up and slow down twice in each revolution while operating at an angle. The amount that the speed changes varies according to the steepness of the U-joint's angle.

U-joint **operating angle** is determined by taking the difference between the transmission installation angle and the driveshaft installation angle. When the universal joint is operating at an angle, the driven yoke speeds up and slows down twice during each driveshaft revolution.

These four speed changes are not normally visible during rotation. But they may be understood more easily after examining the action of a U-joint. A universal joint is a coupling between two shafts not in direct alignment, usually with changing relative positions. It would be logical to assume that the entire unit simply rotates. This is true only for the universal joint's input yoke.

The output yoke's circular path looks like an ellipse because it can be viewed at an angle instead of straight on. This effect can be obtained when a coin is rotated by the fingers. The height of the coin stays the same, even though the sides seem to get closer together.

This illusion might seem to be merely a visual effect, but it is more than that. The U-joint rigidly locks the circular action of the input yoke to the elliptical action of the output yoke. The result is similar to what would happen when changing a clock face from a circle to an ellipse.

Like the hands of a clock, the input yoke turns at a constant speed in its true circular path. The output yoke, operating at an angle to the other yoke, completes its path in the same amount of time. However, its speed varies, or is not constant, compared to the input.

Speed fluctuation is more easily visualized when looking at the travel of the yokes by 90° quadrants **(Figure 41–20)**. The input yoke rotates at a steady or constant speed through the complete 360° turn. The output yoke quadrants alternate between shorter and longer distance travel than the input yoke quadrants. When one point of the output yoke covers the shorter distance in the same amount of time, it must travel at a slower rate. Conversely, when travelling the longer distance (but only 90°) in the same amount of time, it must move faster.

Because the average speed of the output yoke through the four 90° quadrants (360°) equals the constant speed of the input yoke during the same revolution, it is possible for the two mating yokes to travel at different speeds. The output yoke is falling behind and catching up constantly. The resulting acceleration and deceleration produces a fluctuating torque and torsional vibrations characteristic of all Cardan U-joints. The steeper the U-joint angle, the greater the fluctuations in speed will be. Conversely, the smaller the angle, the speed will change less.

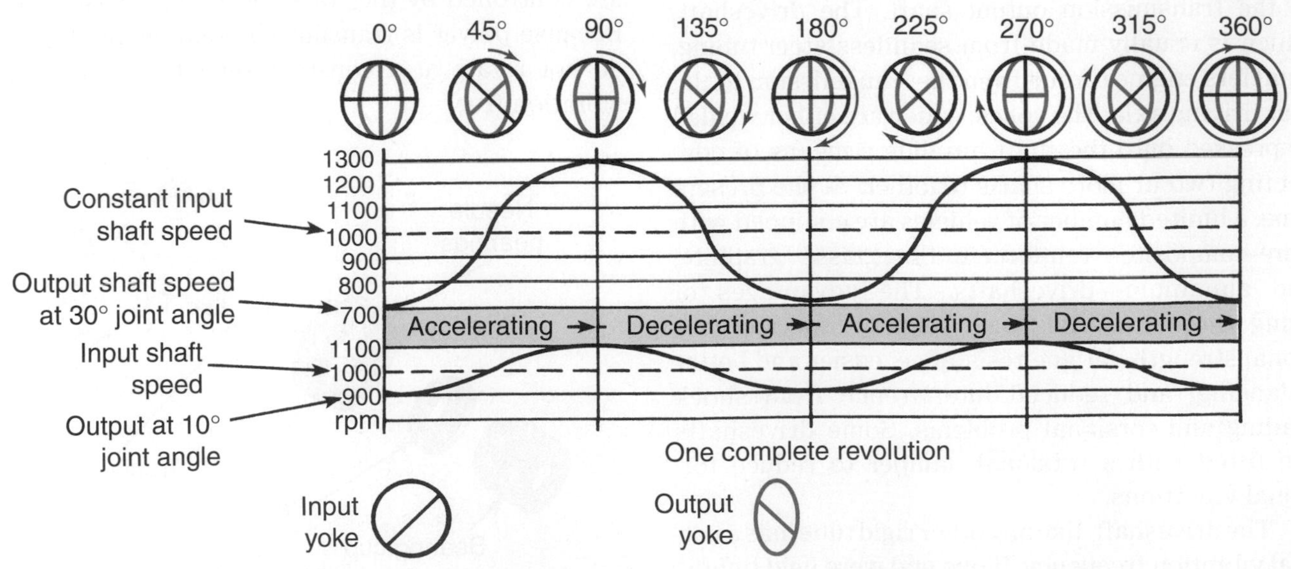

FIGURE 41–20

A graph showing typical driveshaft yoke speed fluctuations.

Phasing of Universal Joints

The torsional vibrations set up by the fluctuations in velocity are transferred down the driveshaft to the next U-joint. At this joint, similar speed fluctuation occurs. Because these speed variations take place at equal and opposite angles to the first joint, they cancel out each other. To provide for this cancelling effect, driveshafts should have at least two U-joints, and their operating angles must be equal to each other. Speed fluctuations can be cancelled if the driven yoke has the same point of rotation, or same plane, as the driving yoke. When the yokes are in the same plane, the joints are said to be *in phase*.

On a two-piece driveshaft, you may encounter problems if you are not careful. The centre U-joint must be disassembled to replace the centre hanger (support) bearing. The centre driving yoke is splined to the front driveshaft. If the yoke's position on the driveshaft is not indicated in some manner, the yoke could be installed in a position that is out of phase. Manufacturers use different methods of indexing the yoke to the shaft. Some use aligning arrows; others machine a master spline that is wider than the others. The yoke and shaft cannot be reassembled until the master spline is aligned properly. When there are no indexing marks, you should index the yoke to the driveshaft before disassembling the U-joint. This saves time and frustration during reassembly.

Cancelling Angles

Vibrations can be reduced by using cancelling angles **(Figure 41–21)**. Carefully examine the illustration, and note that the operating angle at the front of the driveshaft is offset by the one at the rear of the driveshaft. When the front universal joint accelerates, causing a vibration, the rear universal joint decelerates, causing a vibration. The vibrations created by the two joints oppose and dampen the vibrations from one to the other. The use of cancelling angles provides a smoother driveshaft operation.

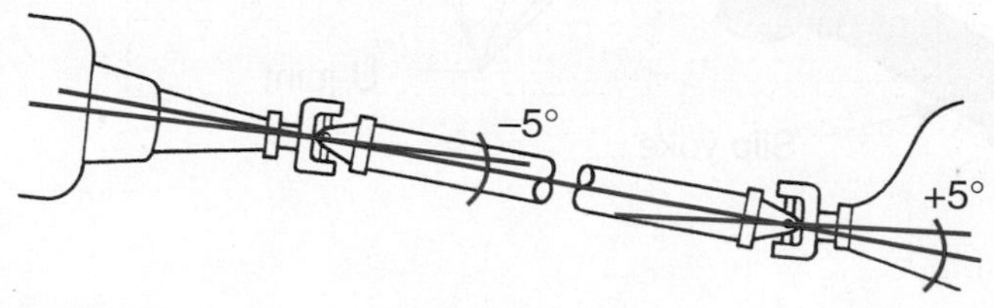

FIGURE 41–21

When a driveshaft's joints are in phase and have cancelling angles, inherent vibrations are reduced.

TYPES OF U-JOINTS

There are three common designs of U-joints: single U-joints retained by either an inside or outside retaining ring, coupled U-joints, and U-joints held in the yoke by U-bolts or lock plates.

Single Universal Joints

The single Cardan/Spicer universal joint is also known as the cross or four-point joint. These two names aptly describe the single Cardan, as the joint itself forms a cross, with four machined trunnions, or points, equally spaced around the centre of the axis. Needle bearings used to reduce friction and provide smoother operation are set in bearing cups. The trunnions of the cross fit into the cup assemblies, and the cup assemblies fit snugly into the driving and driven universal joint yokes. U-joint movement takes place between the trunnions, needle bearings, and bearing cups. There should be no movement between the bearing cup and its bore in the universal joint yoke. The bearings are normally held in place by retaining rings that drop into grooves in the yoke's bearing bores. The bearing caps allow free movement between the trunnion and yoke. The needle bearing caps may also be pressed into the yokes, bolted to the yokes, or held in place with U-bolts or metal straps.

There are other styles of single U-joints. The method used to retain the bearing caps is the major difference between these designs. The Spicer style **(Figure 41–22A)** uses an outside retaining ring that fits into a groove machined in the outer end of the yoke. The bearing cups for this style are machined to accommodate the retaining ring.

The Mechanics, or Detroit/Saginaw, style **(Figure 41–22B)** uses an inside retaining ring or C-clip that fits into a groove machined in the bearing cup on the side closer to the grease seal. When installed, the clip rests against the machined inside portion of the yoke. Some manufacturers use nylon injected into the retaining ring grooves to retain the cups to the yokes.

The Cleveland style is an attempt to combine different joint styles to have more applications from one joint. The bearing cups for this U-joint are machined to accommodate either Spicer- or Mechanics-style snap rings. If a replacement U-joint comes with both style clips, use the clips that pertain to your application.

Double-Cardan Universal Joint

A **double-Cardan U-joint** is used with split driveshafts and consists of two Cardan universal joints closely connected by a centring ball socket and a

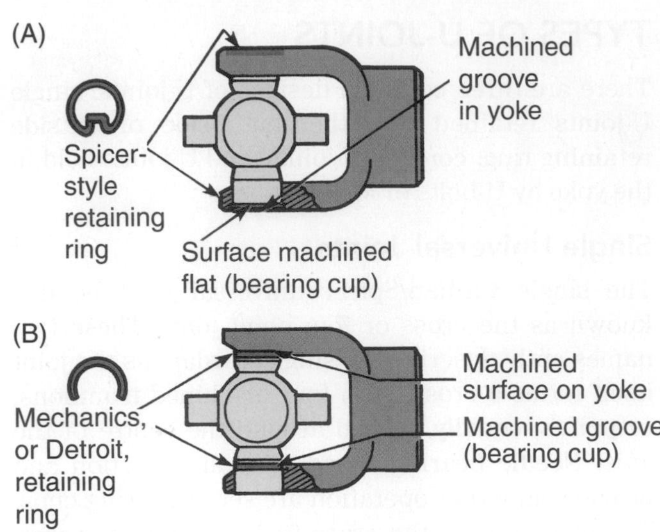

FIGURE 41–22

(A) A Spicer-style U-joint and (B) a Mechanics, or Detroit-style, U-joint.

centre yoke, which functions as a ball-and-socket. The ball-and-socket splits the angle of the two shafts between two U-joints **(Figure 41–23)**. Because of the centring socket yoke, the total operating angle is divided equally between the two joints. Because the two joints operate at the same angle, the normal fluctuations that result from the use of a single U-joint are cancelled out. The acceleration and deceleration of one joint is cancelled by the equal and opposite action of the other.

The double-Cardan joint is classified as a constant velocity universal joint. It is most often found in front-engine RWD luxury-type vehicles, SUVs, and pickup trucks.

DIAGNOSIS OF DRIVESHAFT AND U-JOINT PROBLEMS

A failed U-joint or damaged driveshaft can exhibit a variety of symptoms. A clunk that is heard when the transmission is shifted into gear is the most obvious. You can also encounter unusual noise, roughness, or vibration.

To help differentiate a potential driveline problem from other common sources of noise or vibration, it

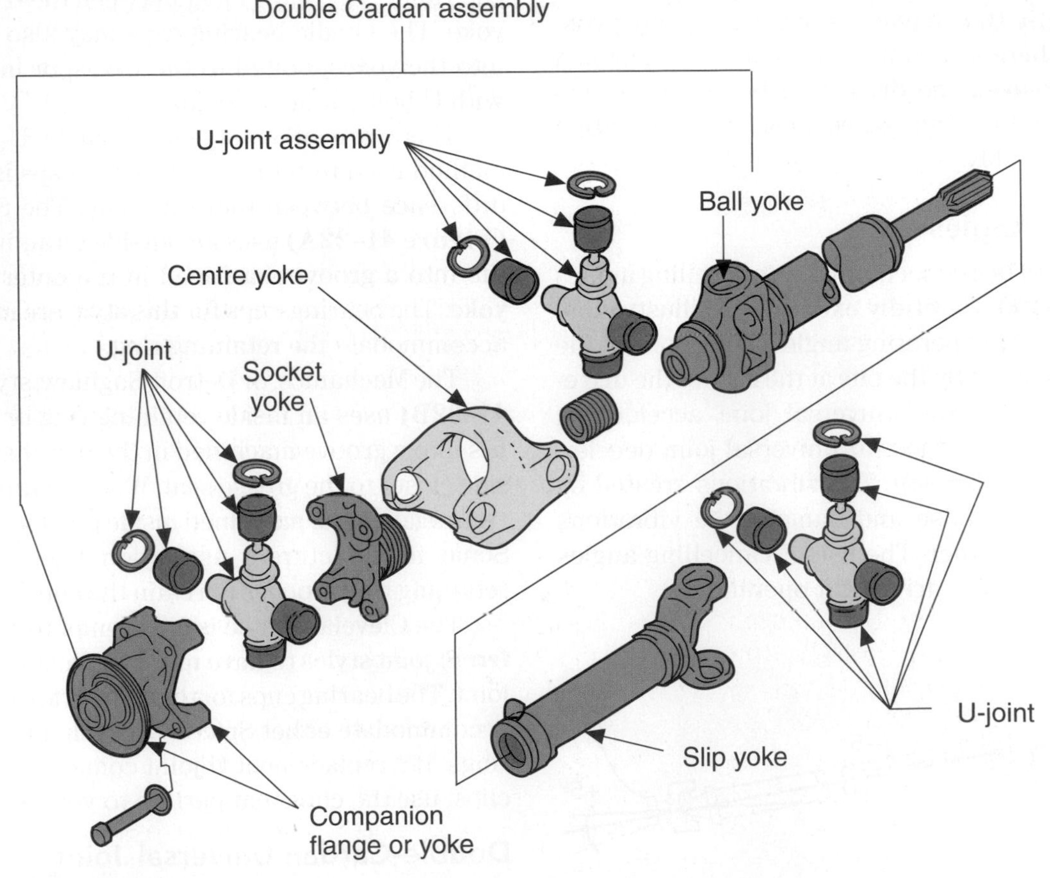

FIGURE 41–23

A double-Cardan joint.

Ford Motor Company

is important to note the speed and driving conditions at which the problem occurs. Unfortunately, it is often very difficult to accurately pinpoint driveline problems with only a road test.

Visual Inspection

Shift the transmission into NEUTRAL and release the parking brake. Raise and support the vehicle on a hoist or jack stands so that the wheels are free to rotate. The driveshaft should be kept at an angle equal to or close to the curb-weighted position. The first problem most likely encountered is an undercar fluid leak. If a lot of lube is escaping from the pinion shaft seal, a driveline noise could be caused by a bad pinion bearing. To confirm the problem, start the engine, put the transmission in gear, and listen at the drive axle housing. If the bearing is noisy, it must be replaced. If the bearing sounds fine but the pinion seal is still leaking, suggest an on-the-car seal replacement.

A leaking transmission extension housing seal may help to pinpoint the cause of a driveline vibration. A worn extension housing bushing allows for excessive driveshaft yoke movement, which can cause a vibration. If excessive yoke movement at the transmission is present, the extension housing bushing and seal must be replaced.

The replacement procedures for both the final drive pinion bearing and seal as well as the transmission extension housing bushing and seal are covered later in this chapter.

FIGURE 41–24

The entire length of the driveshaft should be inspected.

If both of these seals pass the test, continue the driveline examination by checking the entire length of the driveshaft for excess undercoating, dents, missing weights, or other damage that could cause an imbalance and result in a vibration **(Figure 41–24)**. Look at each joint; if the joint has a red dust on it, this could mean it is dry and lacks lubrication. Also check the seating of the joints in their yokes.

Rotate the driveshaft slowly by hand, and feel for binding or looseness at the U-joints and slip splines. Each joint can be further checked by gripping the input and output yokes and attempting to twist them back and forth in opposite directions, and then holding one stationary as you try to move the other vertically and side to side. Also grip the slip yoke at the rear of the transmission and try to move it vertically and side to side. If there is excessive movement, the slip yoke should be replaced.

Inspect the U-joints' grease seals, located at the bottom of the bearing caps, for signs of rust, leakage, or contamination of the lubricant. U-joints can be quickly checked for wear or damage by grasping the yoke and the driveshaft. Carefully watch the two parts as you turn them in opposite directions; there should be no noticeable movement. Then attempt to move the shaft up and down in the yoke. If any movement is possible, the joint is worn. Naturally, the amount of movement indicates the amount of wear.

Lastly, if the driveshaft has a centre bearing, carefully inspect it. Look for looseness, a broken rubber mounting, and damage due to excessive heat. Then rotate the driveshaft by hand. If the centre bearing shows signs of roughness or is noisy, install a new driveshaft assembly.

When a U-joint is damaged or excessively worn, it must be replaced. Photo Sequence 31 covers the typical procedure for removing a U-joint from a driveshaft. After a replacement joint is obtained, it needs to be installed. Photo Sequence 32 covers the reassembly of a common U-joint.

PHOTO SEQUENCE 31

DISASSEMBLING A SINGLE UNIVERSAL JOINT

P31–1 Clamp the slip yoke in a vise, and support the outer end of the driveshaft.

P31–2 Remove the retaining rings on the tops of the bearing cups. Make index marks in the yoke so that the joint can be assembled with the correct phasing.

P31–3 Select a socket that has an inside diameter large enough for the bearing cup to fit into; usually, a 32 mm socket works.

P31–4 Select a second socket that can slide into the shaft's bearing cup bore—usually, a 14 mm socket.

P31–5 Place the large socket against one vise jaw. Position the driveshaft yoke so that the socket is around a bearing cup.

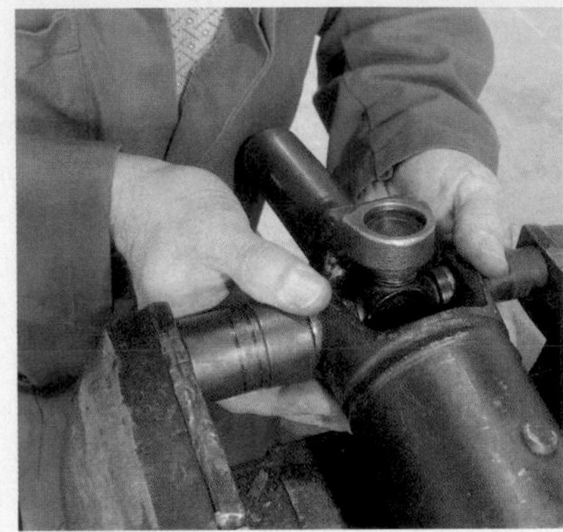

P31–6 Position the other socket to the centre of the bearing cup opposite to the one in line with the large socket.

P31–7 Carefully tighten the vise to press the bearing cup out of the yoke and into the large socket. Reposition the sockets at the opposite ends to press out the other bearing cup.

P31–8 Separate the joint by turning the shaft over in the vise and driving the spider and remaining bearing cup out through the yoke with the two sockets.

P31–9 Following the same procedure above, use the two sockets to remove the bearing cups and joint from the slip yoke.

REASSEMBLING A SINGLE UNIVERSAL JOINT

PHOTO SEQUENCE
32

P32–1 Clean any dirt from the yoke and the retaining ring grooves.

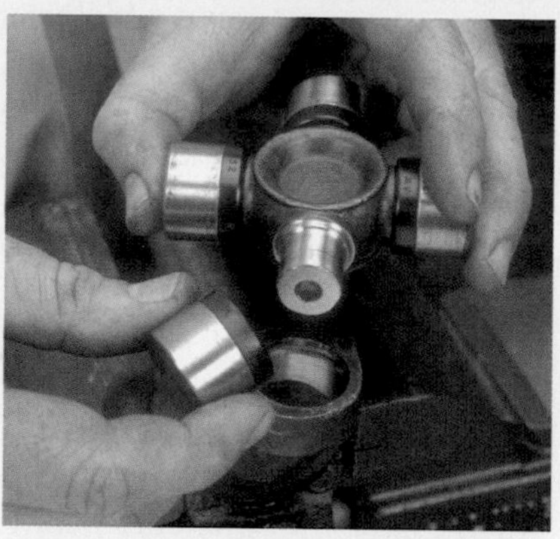

P32–2 Carefully remove the bearing cups from the new U-joint. Lubricate the needle bearings with petroleum jelly or chassis grease to prevent bearings from falling.

P32–3 Place the new spider inside the yoke, and push it to one side.

P32–4 Start one cup into the yoke's bore and over the spider's trunnion.

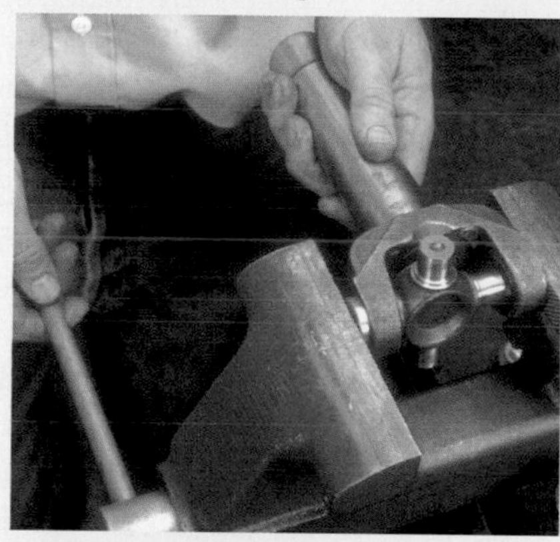

P32–5 Carefully place the assembly in a vise, and press the cup as far as possible through the yoke bore. Use a socket to allow the trunnion at the opposite end to go past the yoke bore.

P32–6 Remove the shaft from the vise, and push the spider toward the other side of the yoke.

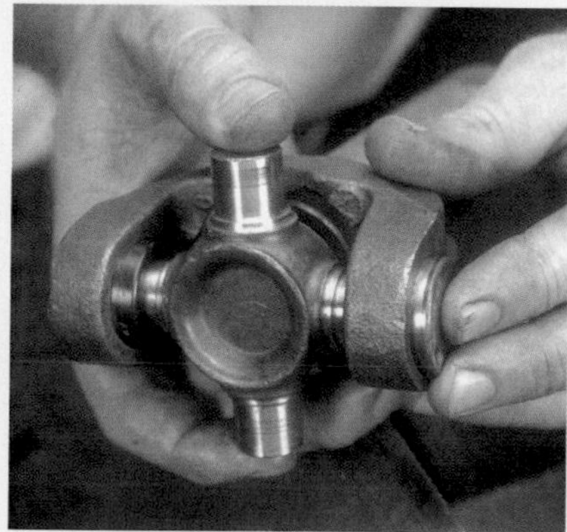

P32–7 Start a cup into the yoke's bore and over the trunnion.

P32–8 Place the shaft in the vise, and tighten the jaws to press the bearing cup into the bore and over the trunnion. Then install the snap rings. Make sure they are seated in their grooves.

P32–9 Position the joint's spider in the driveshaft yoke, and install the two remaining bearing cups.

Noise and Vibration Diagnosis

A failed U-joint or damaged driveshaft can exhibit a variety of symptoms. Most often, abnormal vibrations are felt or abnormal noises are heard. To help differentiate a potential drivetrain problem from other common sources of noise or vibration, it is important to note the speed and driving conditions at which the problem occurs.

As a general guide, a worn U-joint is noticeable during acceleration or deceleration and is less speed sensitive than an unbalanced tire (occurring in the 50 to 100 km/h [30 to 60 mph] range) or a bad wheel bearing (noticeable at higher speeds).

SHOP TALK

When diagnosing driveline noise, if a chirping sound that increases with speed is heard, suspect a dry U-joint. The chirping typically occurs with a frequency two to four times faster than the speed of the wheels.

Before assuming the driveline is the source of a noise or vibration, make sure the tires or exhaust are not the cause. To make sure the noises are not caused by tire tread patterns and/or wear, drive the car on various types of road surfaces (asphalt, concrete, and packed dirt). If the noise changes with the road surfaces, it means the tires are the cause of the noise.

Another way to isolate tire noises is to coast at speeds less than 50 km/h (30 mph). If the noise is still heard, the tires are probably the cause. Driveline and differential noises are less noticeable at these speeds. Accelerate and compare the sounds to those made while coasting. Driveline and differential noises change. Tire noise remains constant.

At times, the exhaust's noise may sound like gear whine or wheel bearing rumble. Tires, especially snow tires, can have a high-pitched tread whine or roar, which is similar to gear noise.

Sometimes, it is difficult to distinguish between axle-bearing noises and noises coming from the differential. Differential noises often change with the driving mode, whereas axle-bearing noises are usually constant. The sound of the bearing noise usually increases in speed and loudness as vehicle speed increases.

Often, noises and vibration appear to be coming from the axle, when they are actually caused by other problems. After the road test, the chassis should be carefully checked on a hoist. Exhaust system components positioned too close to the frame or underbody may be the cause of the noise. Loose or bent wheels, bad tire treads, worn U-joints, and damaged engine mounts all are capable of creating noise and/or vibration that appear to be coming from the axle.

By describing the noise or vibration, you can usually identify the most probable problem areas.

Driveshaft Runout

A vibration that occurs during all modes of driving, especially at 55 to 75 km/h (35 to 45 mph), is most often caused by a faulty universal joint, a bent drive pinion shaft, or a damaged pinion flange. All of these will cause the driveshaft to have excessive runout.

The runout of the driveshaft can be checked by putting the transmission in NEUTRAL and releasing the parking brake. Raise and securely support the vehicle on a hoist or jack stands so that the wheels can freely rotate. Clean any dirt or rust from around the driveshaft at its centre and 76 mm (3 in.) from each end.

Mount a dial indicator to the vehicle's underbody, and position the needle so that it points directly toward the centre of the driveshaft **(Figure 41–25)**. Rotate the driveshaft until the dial indicator displays

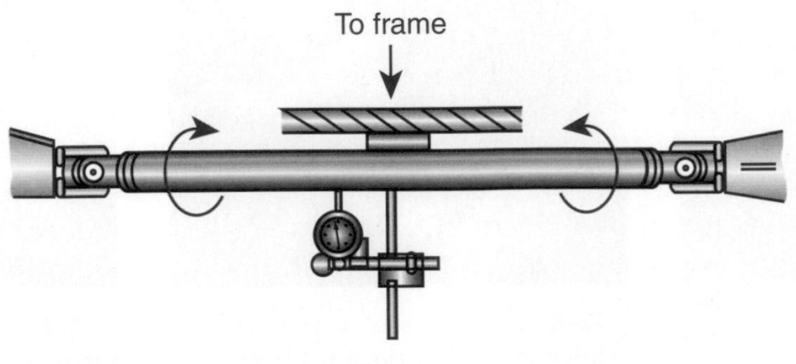

FIGURE 41–25

Checking the runout of a driveshaft.

its lowest reading; adjust the dial to zero, and rotate the driveshaft to the highest reading. Record that reading and the location of the high point, then repeat the procedure at the top and bottom of the driveshaft. If the runout is more than 1 mm (0.04 in.), install a new driveshaft.

Driveshaft Angles

If the runout is fine, check the phasing of the joints and their angle. This problem can be caused by worn and sagging rear springs or excessive load on the vehicle. If the vehicle is weighted down, the suspension will sag, which changes the operating angle of the rear universal joint. This problem can also be caused by worn or defective transmission and/or engine mounts, worn or loose wheel bearings, or unbalanced tire/wheel assemblies.

The angle at which the transmission is mounted to the frame does not change during vehicle operation. However, the angle at which the rear axle pinion is mounted to the frame changes constantly while the suspension is responding to the road's surface. When the vehicle is stationary, the driveshaft is set to a specific angle—the installation angle.

In order for the angles of the universal joints to have a cancelling effect, their yokes must be on the same plane. When they are, the universal joints are in phase.

Normally, the driveshaft's operating angle will be correct if the car is at its correct ride height and the rear axle housing and transmission have not moved on their mounts. Typically, the maximum allowable driveshaft angle is 3° or 4° from the car's horizontal. Normally, if the angles are wrong, the rear axle has moved in its mounting.

This check is extremely important for vehicles that have lift kits to increase ground clearance. This is a common modification for off-the-road vehicles, and those doing the modifications often overlook cancelling angles.

MEASURING THE ANGLES To check the driveshaft's operating angle, use an inclinometer. This instrument, when attached to the driveshaft, displays the angle of the driveshaft at any point.

Before checking the angles of the U-joints, make sure the vehicle is empty and has a full or near-full fuel tank. Raise the car and rotate the driveshaft so that the bearing caps in the rear axle's pinion flange and the transmission's slip yoke are straight up and down. Then clean the bearing caps of each joint. Install the inclinometer with the magnet on the downward-facing bearing cap. Using the adjusting knob on the tool, centre the weighted cord on its scale **(Figure 41–26)**.

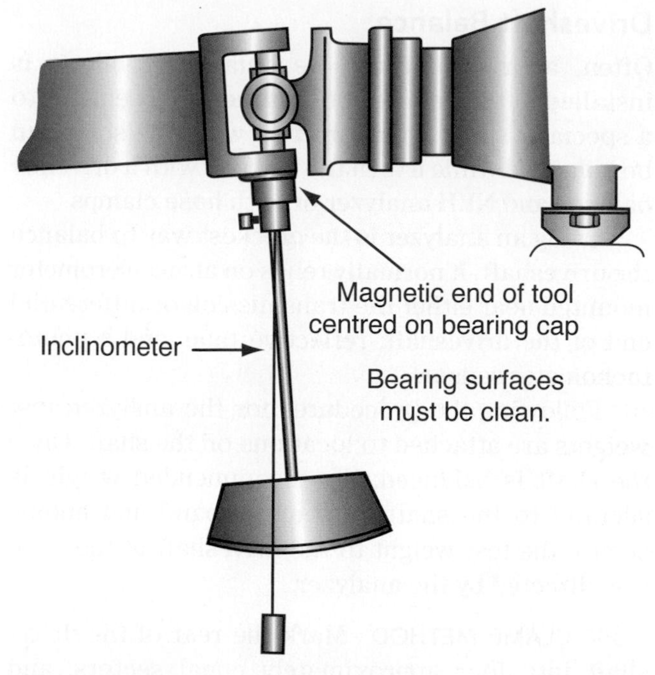

FIGURE 41–26
An inclinometer mounted to a U-joint.

Remove the inclinometer. Make sure not to bump the adjustment knob. Now rotate the driveshaft 90°, attach the inclinometer to the bearing cap now facing down, and then record the reading. Repeat this process until the driveshaft has completed one revolution, and then check the angle of the rear joint.

Compare rear and front angle readings. The difference is the operating angle. Your findings from this test should be compared to specifications.

ADJUSTING THE ANGLES If the installation angles of the two joints are not equal, normally, the rear angle is adjusted so that it equals the front angle. On vehicles with leaf springs, the rear angle can be changed by rotating the rear axle assembly on the spring pads or by installing tapered shims between the springs and the spring pads.

If the vehicle has rear coil springs, the joint angle can be changed by inserting wedge-shaped shims between the rear axle assembly and rear control arms, by adjusting an eccentric washer at the control arm, or by replacing the control arm with one of a different length.

If it is necessary to change the angle of the front U-joint, shims should be installed between the transmission's extension housing and the transmission mount.

Driveshaft Balance

Often, a driveshaft can be balanced while it is installed in the vehicle. Other times, it is sent out to a specialty shop. There are two ways the shaft can be balanced while it is in the vehicle: with a driveline balance and NVH analyzer or with hose clamps.

Using an analyzer is the quickest way to balance the driveshaft. It normally relies on an accelerometer mounted near either the transmission or differential end of the driveshaft, reflective tape, and a phototachometer sensor.

Following the procedures for the analyzer, test weights are attached to locations on the shaft. Once the shaft is balanced, the recommended weight is secured to the shaft by a metal band and epoxy; secure the test weight to the driveshaft at the position directed by the analyzer.

HOSE CLAMP METHOD Mark the rear of the driveshaft into four approximately equal sectors, and number the marks 1 through 4. Install a gear clamp on the driveshaft with its head at position 1 and another at position 2. Check for vibration at road speed.

Recheck with the clamp at each of the other positions to find the position that shows minimum vibration. If two adjacent positions show equal improvement, position the clamp head between them.

If the vibration persists, add another clamp at the same position, and recheck for vibration.

If there is no improvement, rotate the clamps in opposite directions at equal distances from the best position determined previously **(Figure 41–27)**. Separate the clamp heads about 13 mm (1/2 in.), and recheck for vibration at the road speed.

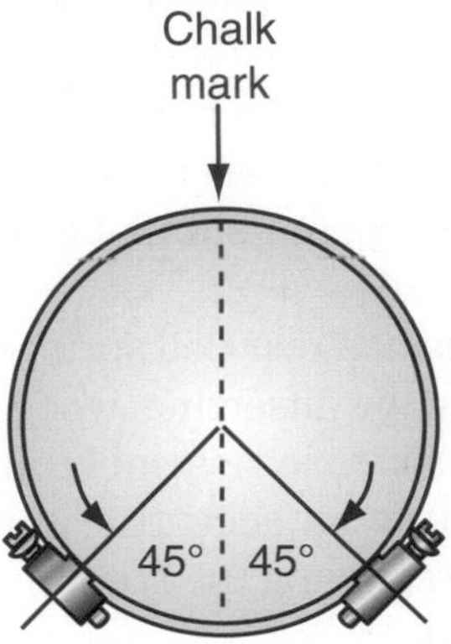

FIGURE 41–27

While balancing the shaft with hose clamps, reposition them in equal amounts.

FIGURE 41–28

A zerk fitting is used to lubricate a U-joint.

Dana Holding Corporation

Maintenance

The driveshaft itself requires no maintenance except for an occasional cleaning off of any dirt buildup on the shaft. Dirt and any other buildup on the shaft will affect its balance. Factory-installed U-joints are normally sealed and also require no periodic lubrication. However, replacement U-joints equipped with a zerk (grease) fitting **(Figure 41–28)** should be lubricated on a regular basis.

FINAL DRIVES AND DRIVE AXLES

The final drive is a geared mechanism located between the driving axles of a vehicle. The differential gearing of the final drive rotates the driving axles at different speeds when the vehicle is turning a corner **(Figure 41–29)**. It also allows both axles to turn at the same speed when the vehicle is moving in a straight line. The drive axle assembly directs driveline torque to the vehicle's drive wheels. The gear ratio of the drive axle's ring and pinion gears is used to increase torque. The differential serves to establish equal forces between the drive wheels and allows the drive wheels to turn at different speeds when the vehicle changes direction.

On an FWD car or truck, the differential is normally an integral part of the transaxle assembly located at the front of the vehicle. Transaxle design and operation depends on whether the engine is mounted transversely or longitudinally. With a transversely mounted engine, the crankshaft centreline and drive axle are on two parallel planes. With a longitudinally mounted power plant, the differential must change the power flow 90°.

On RWD vehicles, the differential is located in the rear axle housing or carrier. The driveshaft connects

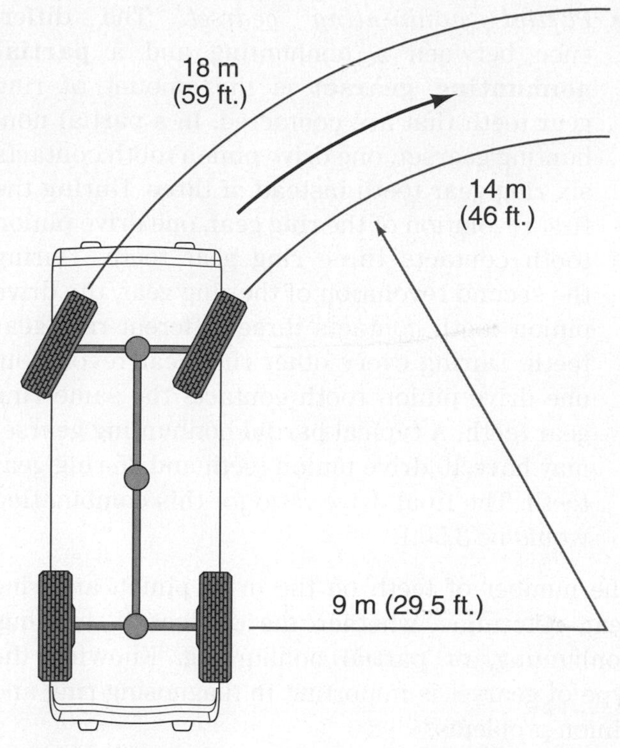

FIGURE 41–29
Travel of wheels when a vehicle is turning a corner.

the transmission with the rear axle gearing. Four-wheel-drive vehicles have differentials on both their front and rear axles.

The differential allows the drive wheels to rotate at different speeds when negotiating a turn or curve in the road and redirects the engine torque from the driveshaft to the rear drive axle shafts.

The final drive gears in the drive axle assembly are also sized to provide a gear reduction, or a torque multiplication. Axles with a low (numerically high) gear ratio allow for fast acceleration and good pulling power. Axles with high gear ratios allow the engine to run slower at any given speed, resulting in better fuel conservation.

Final Drive Components

The components of commonly used final drive units are shown in **Figure 41–30**. There are several other basic design arrangements. However, the most commonly used design has pinion/ring gears and a **pinion shaft**. The latter is normally a hypoid gear machined into an input (pinion) shaft. The shaft is mounted in the front end of the carrier and supported by two or three bearings. An overhung pinion gear is supported by two tapered bearings spaced far enough apart to provide the needed leverage to rotate the ring gear and drive axles **(Figure 41–31)**. A straddle-mounted pinion gear rests on three bearings: Two tapered bearings on the front support the input shaft, and one roller bearing is fitted over a short shaft extending from the rear end of the **pinion gear**.

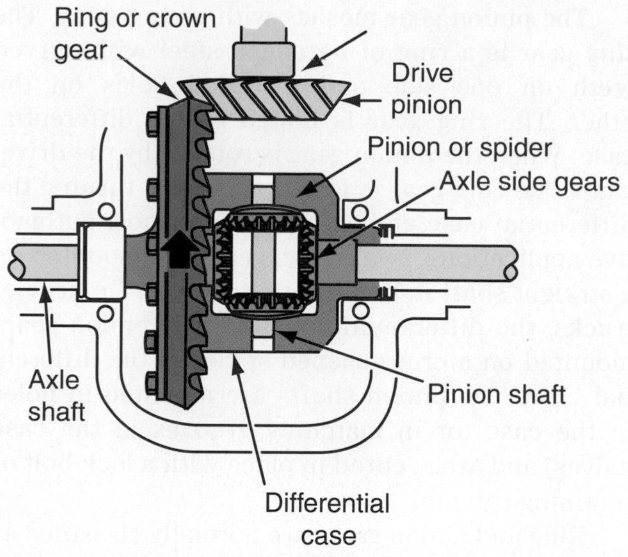

FIGURE 41–30
The components of a typical final drive unit.

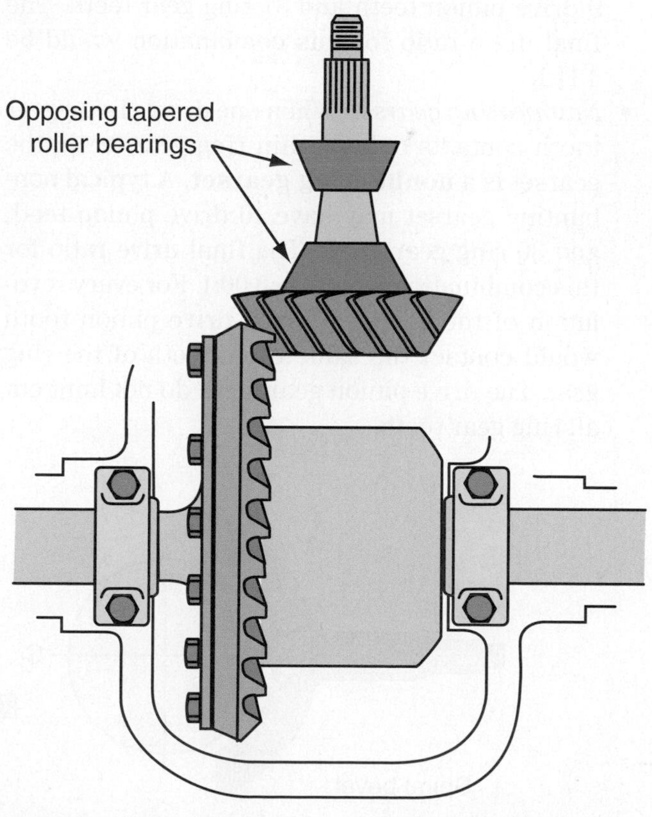

FIGURE 41–31
The pinion drive is supported by bearings.

The pinion gear meshes with a **ring gear**. The ring gear is a ring of hardened steel with curved teeth on one side and threaded holes on the other. The ring gear is bolted to the differential case. When the pinion gear is rotated by the driveshaft, the ring gear is forced to rotate, turning the differential case and axle shafts. In most automotive applications, two pinion gears are mounted on a straight shaft in the differential case. On heavier trucks, the differential contains four pinion gears mounted on a cross-shaped spider in the differential case. The pinion shafts are mounted in holes in the case (or in matching grooves in the case halves) and are secured in place with a lock bolt or retaining roll pin.

Ring and pinion gears are normally classified as hunting, nonhunting, or partial nonhunting gears. Each type of gearset has its own requirements for a satisfactory gear tooth contact pattern. These classifications are based on the number of teeth on the pinion and ring gears.

- *Hunting gearset.* When one drive pinion gear tooth contacts every ring gear tooth after several revolutions, the gearset is a **hunting gearset**. In other words, the drive pinion hunts out each ring gear tooth. A typical hunting gearset may have 9 drive pinion teeth and 37 ring gear teeth. The final drive ratio for this combination would be 4.11:1.
- *Nonhunting gearset.* When one drive pinion gear tooth contacts only certain ring gear teeth, the gearset is a **nonhunting gearset**. A typical nonhunting gearset may have 10 drive pinion teeth and 30 ring gear teeth. The final drive ratio for this combination would be 3.00:1. For every revolution of the ring gear, each drive pinion tooth would contact the same three teeth of the ring gear. The drive pinion gear teeth do not hunt out all ring gear teeth.
- *Partial nonhunting gearset.* The difference between a nonhunting and a **partial nonhunting gearset** is the amount of ring gear teeth that are contacted. In a partial nonhunting gearset, one drive pinion tooth contacts six ring gear teeth instead of three. During the first revolution of the ring gear, one drive pinion tooth contacts three ring gear teeth. During the second revolution of the ring gear, the drive pinion tooth contacts three different ring gear teeth. During every other ring gear revolution, one drive pinion tooth contacts the same ring gear teeth. A typical partial nonhunting gearset may have 10 drive pinion teeth and 35 ring gear teeth. The final drive ratio for this combination would be 3.50:1.

The number of teeth on the drive pinion and ring gear determine whether the gearset is hunting, nonhunting, or partial nonhunting. Knowing the type of gearset is important in diagnosing ring and pinion problems.

A **hypoid gear** contacts more than one tooth at a time. The hypoid gear also makes contact with a sliding motion. This sliding action, however, is smoother than that of the spiral gear, resulting in quieter operation. The biggest difference is that, in a hypoid gear, the centrelines of the ring and pinion gears do not match. Using hypoid gears, the drive pinion gear is placed lower in the carrier. The drive pinion meshes with the ring gear at a point below its centreline **(Figure 41–32)**.

The sliding effect of two hypoid gears meshing tends to wipe lubricant from the face of the gears **(Figure 41–33)**, resulting in eventual damage. Drive axles require the use of extreme pressure-type lubricants. The additives in this type of lubricant allow the lubricant to withstand the wiping action of the gear teeth without separating from the gear face.

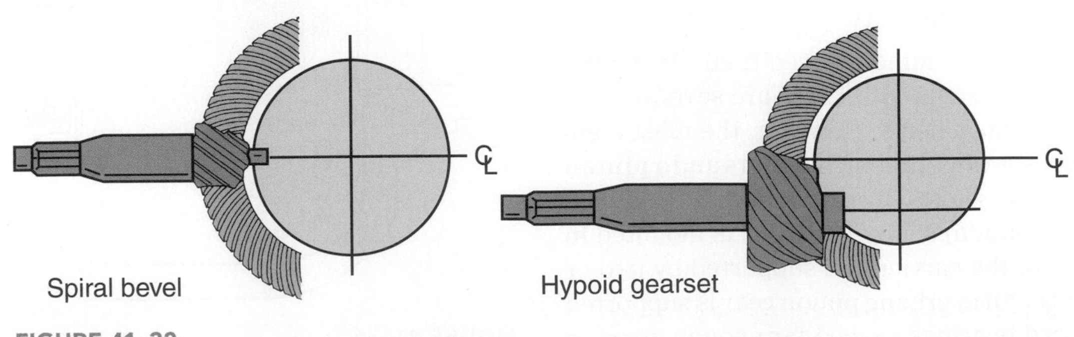

FIGURE 41–32

In a hypoid gearset, the drive pinion meshes with the ring gear at a point below its centreline.

FIGURE 41–33

The flow of oil in a hypoid gearset as it spins.

Dana Corporation

The differential also contains two side gears. The inside bore of the side gears is splined and mates with splines on the ends of the axles shafts. The differential pinion gears and side gears are in constant mesh. The pinion gears are mounted on a pinion gear shaft, which is mounted in the differential case. As the case turns with the ring gear, the pinion shaft and gears also turn. The pinion gears deliver torque to the side gears.

When the pinion and ring gears are manufactured, the faces of the gear teeth are machined to provide smooth mating surfaces. The pinion gear and ring gears are always matched to provide a good mesh **(Figure 41–34)**. Pinion gears and the ring gears should always be installed as a set. Otherwise, the mismatched gearset might operate noisily. A matched gearset code is etched in each drive pinion and ring gearset.

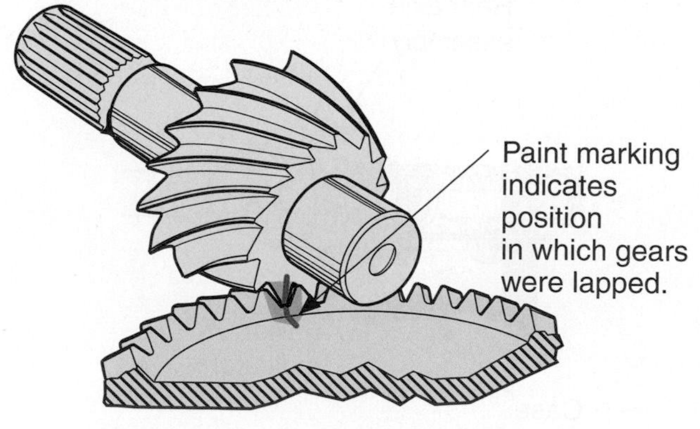
Paint marking
indicates
position
in which gears
were lapped.

FIGURE 41–34

Index marks on a nonhunting or partial nonhunting ring and pinion gearset.

Ford Motor Company

Drive Axle Housing and Casing

The differential and final drive gears in a rear-drive vehicle are housed in the rear axle housing, or carrier housing. The axle housing also contains the two drive axle shafts. Two types of axle housings are found on modern automobiles: the removable carrier and the integral carrier. The removable carrier axle housing is open on the front side. Because it resembles a banjo, it is often called a banjo housing. The backside of the housing is closed to seal out dirt and contaminants and keep in the lubricant. The differential is mounted in a carrier assembly that can be removed as a unit from the axle housing **(Figure 41–35)**. Removable carrier axle housings are most commonly used today on trucks, other heavy-duty vehicles, and some foreign vehicles.

The **integral housing** is most often found on late-model cars and light trucks **(Figure 41–36)**. A cast-iron carrier forms the centre of the axle housing. Steel axle tubes are pressed into both sides of the carrier to form the housing. The housing and carrier have a removable rear cover that allows access to the differential assembly **(Figure 41–37)**. Because the carrier is not removable, the differential components must be removed and serviced separately. For many operations, a case spreader **(Figure 41–38)** must be used to remove the components. In addition to providing a mounting place for the differential, the axle housing also contains brackets for mounting suspension components such as control arms, leaf springs, and coil springs.

Some vehicles have an ABS speed sensor attached to the carrier housing for rear-wheel lock-up prevention during braking.

FIGURE 41–35

A typical removable-carrier axle housing.

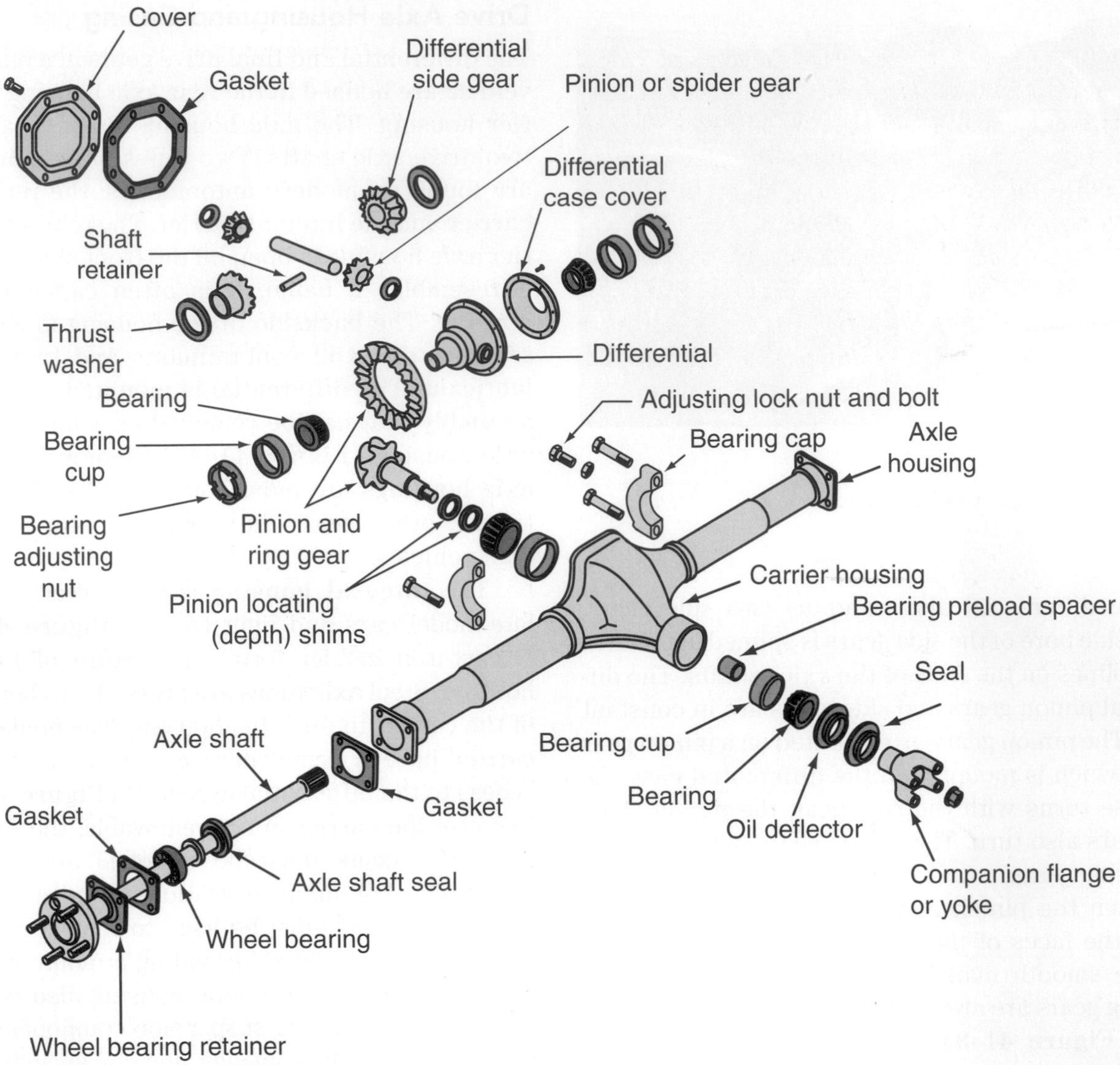

FIGURE 41–36

An exploded view of integral-carrier axle housing.

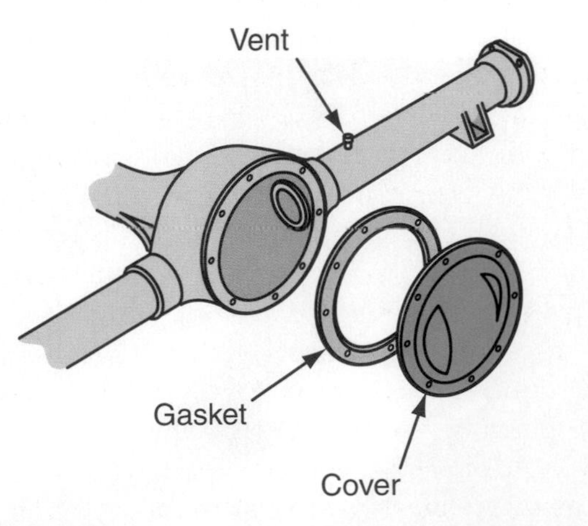

FIGURE 41–37

An integral drive axle housing.

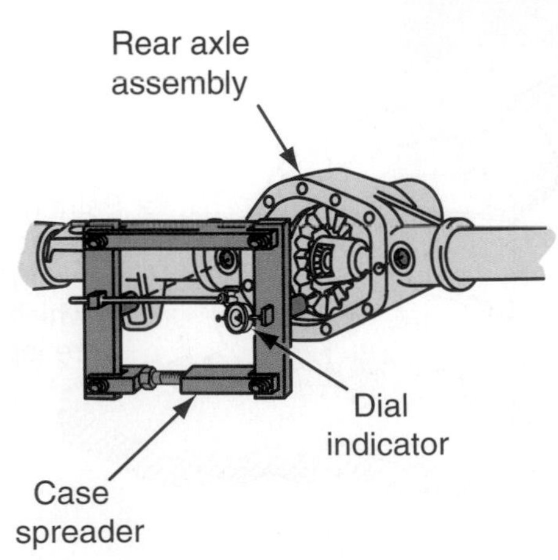

FIGURE 41–38

A case spreader.

Differential Operation

The amount of power delivered to each driving wheel by the differential is expressed as a percentage. When the vehicle moves straight ahead, each driving wheel rotates at 100 percent of the differential case speed. When the vehicle is turning, the inside wheel might be getting 90 percent of the differential case speed. At the same time, the outside wheel might be getting 110 percent of the differential case speed.

Power flow through the axle begins at the drive pinion yoke, or companion flange **(Figure 41–39)**.

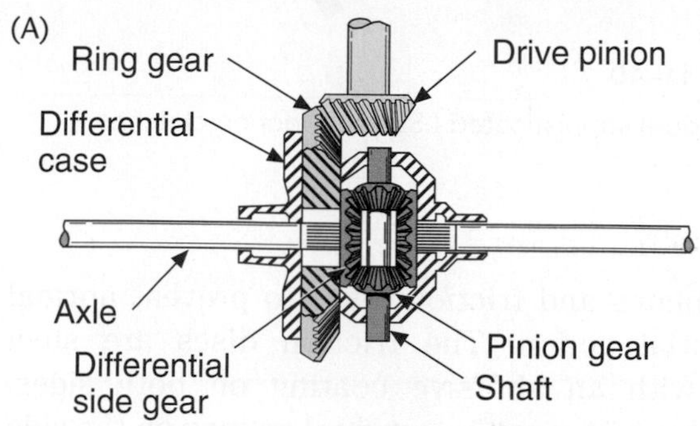

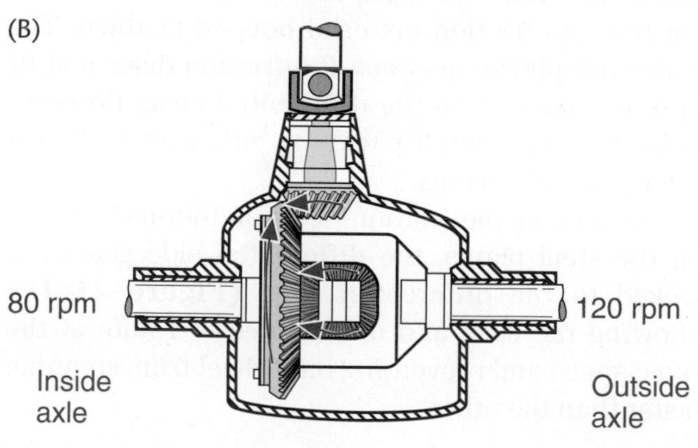

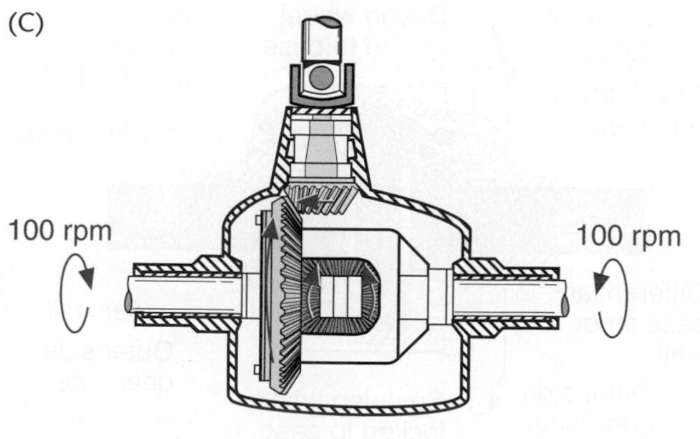

FIGURE 41–39

(A) Basic differential components, (B) differential action while the vehicle is turning left, and (C) differential action while the vehicle is moving straight.

The companion flange accepts torque from the rear U-joint. The companion flange is attached to the drive pinion gear, which transfers torque to the ring gear. As the ring gear turns, it turns the differential case and the pinion shaft. The differential pinion gears transfer torque to the side gears to turn the driving axle shafts. The differential pinion gears determine how much torque goes to each driving axle, depending on the resistance an axle shaft or wheel has while turning. The pinion gears can move with the differential housing or case, and they can rotate on the pinion shaft.

When driveshaft torque is applied to the input shaft and drive pinion, the shaft rotates in a direction that is perpendicular to the vehicle's drive axles. When this rotary motion is transferred to the ring gear, the torque flow changes direction and becomes parallel to the axle shafts and wheels. Because the ring gear is bolted to the differential case, the case must rotate with the ring gear. The pinion gear shaft mounted in the differential case must also rotate with the case and the ring gear. The pinions turn end over end. Pinions do not rotate on the pinion shaft when both driving wheels are turning at the same speed. They rotate end over end as the differential case rotates. Because the pinions are meshed with both side gears, the side gears rotate and turn the axle shafts. The ring gear, differential gears, and axle shafts turn together without variation in speed as long as the vehicle is moving in a straight line.

When a vehicle turns into a curve or negotiates a turn, the wheels on the outside of the curve must travel a greater distance than the wheels on the inside of the curve. The outer wheels must then rotate faster than the inside wheels. This would be impossible if the axle shafts were locked solidly to the ring gear. However, the differential allows the outer wheels and axle shaft to increase in speed and the inner wheels and axle to slow down, thus preventing the skidding and rapid tire wear that would otherwise occur. The differential action also makes the vehicle much easier to control while turning.

For example, when a car makes a sharp right-hand turn, the left-side wheels, axle shaft, and side gear must rotate faster than the right-side wheels, axle shaft, and side gear. The left side of the axle must speed up, and the right side must slow down. This is possible because the pinions to which the side gears are meshed are free to rotate on the pinion shaft. The increased speed of the left-side wheels causes the side gear to rotate faster than the differential case. This causes the pinions to rotate and walk around the slowing-down side gear. As the pinions turn to allow the left-side gear to increase speed, a reverse action—known as a reverse walking effect—is produced on the right-side gear. It slows down an amount that is inversely proportional to the increase in the left-side gear.

LIMITED-SLIP DIFFERENTIALS

Driveline torque is evenly divided between the two rear drive axle shafts by the differential. As long as the tires grip the road, providing a resistance to turning, the drivetrain forces the vehicle forward. When one tire encounters a slippery spot on the road, it loses traction, resistance to rotation drops, and the wheel begins to spin. Because resistance has dropped, the torque delivered to both drive wheels changes. The wheel with good traction is no longer driven. If the vehicle is stationary in this situation, only the wheel over the slippery spot rotates. When this is occurring, the differential case is driving the differential pinion gears around the stationary side gear.

This situation places stress on the differential gears. When the wheel spins because of traction loss, the speed of some of the differential gears increases greatly, while others remain idle. The amount of heat developed increases rapidly; the lube film breaks down; metal-to-metal contact occurs; and the parts are damaged. If spinout is allowed to continue long enough, the axle could break. The final drive and differential gears can also be damaged from prolonged spinning of one wheel. This is especially true if the spinning wheel suddenly has traction. The shock of the sudden traction can cause severe damage to the drive axle assembly.

To overcome these problems, differential manufacturers have developed the **limited-slip differential (LSD)**. Limited-slip differentials are manufactured under such names as sure-grip, no-spin, positraction, or equal-lock. Some vehicles use a viscous clutch in their limited-slip drive axles. These units are predominantly used in 4WD vehicles and are discussed in Chapter 45. Also, many late-model vehicles use electronic controls, rather than gearing and clutches, to send torque to the best drive wheel.

Clutch-Based Units

Many LSDs use friction material to transfer the torque applied to a slipping wheel to the one with traction. Those that use a clutch pack **(Figure 41–40)** have two sets (one for each side gear) of clutch plates and friction discs to prevent normal differential action. The friction discs are steel plates with an abrasive coating on both sides. These discs fit over the external splines on the side gears' hub. The clutch plates are also made of steel but have no friction material bonded to them. The plates are placed between the friction discs and fit into internal lugs in the differential case. Pressure is kept on the clutch packs by either an S-shaped spring or coil springs.

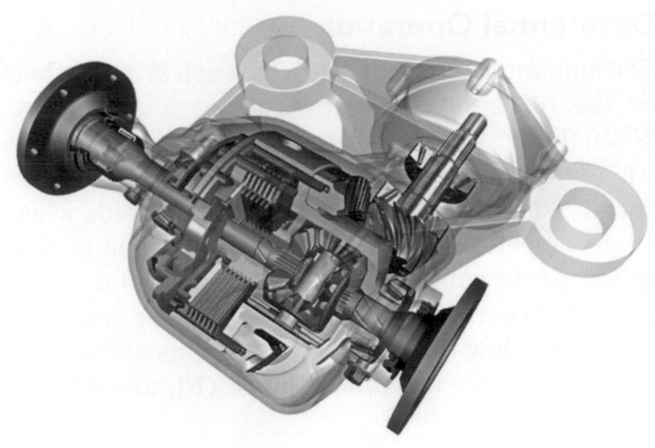

FIGURE 41–40

A late-model sophisticated LSD with friction clutches.

Dana Corporation

As long as the friction discs maintain their grip on the steel plates, the differential side gears are locked to the differential case **(Figure 41–41)**, allowing the case and drive axles to rotate at the same speed and preventing one wheel from spinning faster than the other.

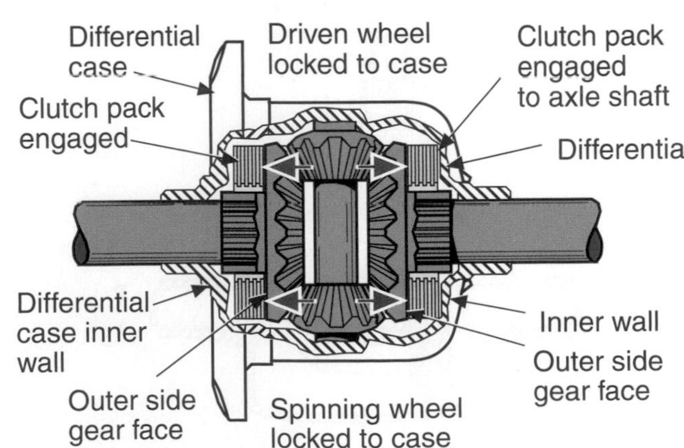

FIGURE 41–41

Action of the clutches in a limited-slip differential.

An older LSD uses two cone-shaped parts to lock the side gears to the differential case. The cones are located between the side gears and the case and are splined to the side gear hubs. The exterior surface of the cones is coated with a friction material that grabs the inside surface of the case. Four to six coil springs mounted in thrust plates between the side gears maintain a preload on the cones. When the cones are forced against the case, the axles rotate with the differential case.

The clutch plates and cones are designed to slip when a predetermined amount of torque is applied to them, which allows the vehicle to have differential action when it is turning a corner.

Gear-Based Units

Manufacturers are using a wide range of LSD designs other than the typical clutch type. These designs were born out of the need to improve vehicle stability and tire traction. Many are gear-based and are often called torque-bias or torque-sensing (Torsen) units. The basis of these units is a parallel-axis helical gearset **(Figure 41–42)**. The Torsen differential multiplies the torque available from the wheel that is starting to spin or lose traction and sends it to the slower turning wheel with the better traction. This action is initiated by the resistance between the sets of gears in mesh.

Helical-geared LSDs respond very quickly to changes in traction. They also do not bind in turns and do not lose their effectiveness with wear as clutch-based units can.

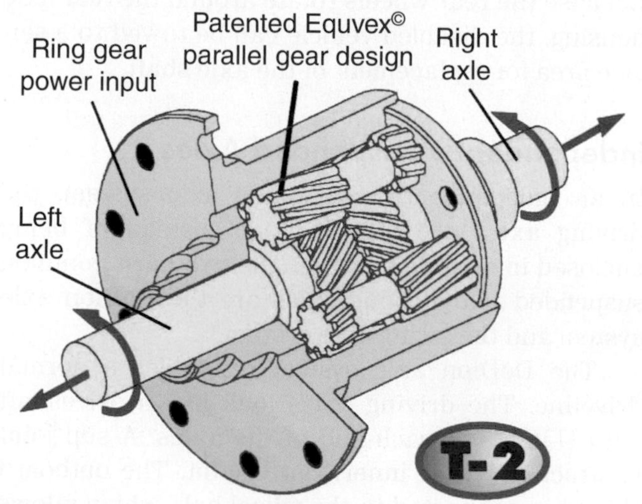

FIGURE 41–42

A Torsen torque-sensitive LSD.

Dana Corporation

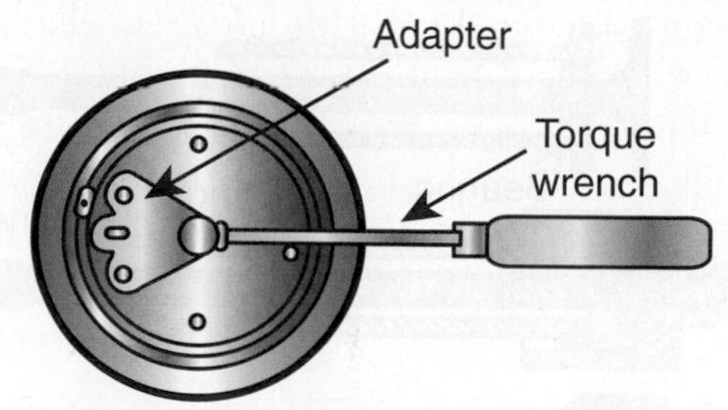

FIGURE 41–43

Checking the breakaway torque of a limited-slip differential.

Limited-Slip Differential Diagnostics

A limited-slip differential can be checked for proper operation without removing the unit from the axle housing. Put the transmission in NEUTRAL, and set one rear wheel on the floor with the other rear wheel raised off the floor. The vehicle's service information will give specifications for the required breakaway torque. Breakaway torque is the amount of torque required to start the rotation of the wheel that is raised off the floor. Breakaway torque is measured with a torque wrench **(Figure 41–43)**.

The initial breakaway torque reading may be higher than the torque required for continuous turning, but this is normal. The axle shaft should turn with even pressure throughout, without slipping or binding. If the torque reading is less than the specified amount, the differential needs to be checked.

AXLE SHAFTS

The purpose of an axle shaft is to transfer driving torque from the differential assembly to the vehicle's driving wheels. There are two types of axles: dead and live (or drive). A **dead axle** does not drive a vehicle. It merely supports the vehicle load and provides a mounting place for the wheels. The rear axle of an FWD vehicle is a dead axle, as are the axles used on trailers.

A **live axle** is one that drives the vehicle. Drive axles transfer torque from the differential to each driving wheel. Depending on the design, rear axles can also help carry the weight of the vehicle or even act as part of the suspension. Three types of driving axle shafts are commonly used **(Figure 41–44)**: semi-floating, three-quarter floating, and full-floating.

All three use axle shafts that are splined to the differential side gears. At the wheel ends, the axles can be attached in any one of a number of

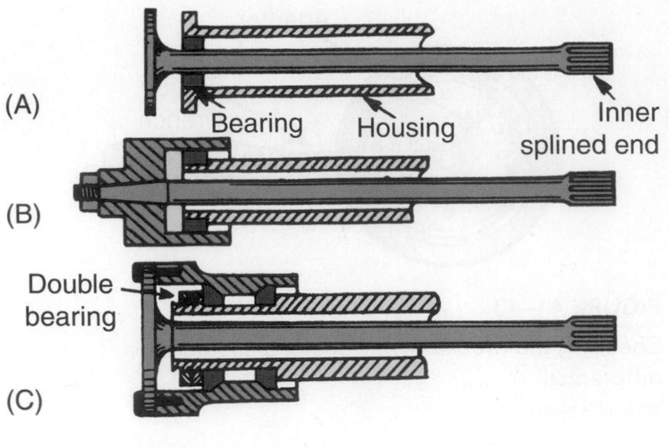

FIGURE 41–44

The types of rear axle shafts: (A) semi-floating; (B) three-quarter floating; and (C) full-floating.

ways. This attachment defines the type of axle it is and the manner in which the shafts are supported by bearings.

Semi-Floating Axle Shafts

Semi-floating axles help to support the weight of the vehicle. The axles are supported by bearings located in the axle housing. An axle shaft bearing supports the vehicle's weight and reduces rotational friction. The inner ends of the axle shafts are splined to the axle side gears. The axle shafts transmit only driving torque and are not acted upon by other forces. Therefore, the axle shafts are said to be floating.

The driving wheels are bolted to the outer ends of the axle shafts. The outer axle bearings are located between the axle shaft and axle housing. This type of axle has a bearing pressed into the end of the axle housing. This bearing supports the axle shaft. The axle shaft is held in place with either a bearing retainer bolted to a flange on the end of the axle housing or by a C-shaped washer that fits into grooves machined in the splined end of the shaft. A flange on the wheel end of the shaft is used to attach the wheel.

When semi-floating axles are used to drive the vehicle, the axle shafts push on the shaft bearings as they rotate. This places a driving force on the axle housing, springs, and vehicle chassis, moving the vehicle forward. The axle shaft faces the bending stresses associated with turning corners and curves, skidding, and bent or wobbling wheels, as well as the weight of the vehicle. In the semi-floating axle arrangement with a C-shaped washer-type retainer, if the axle shaft breaks, the driving wheel comes away from or out of the axle housing.

Three-Quarter Floating Axle Shafts

The wheel bearing on a **three-quarter floating axle** is on the outside of the axle housing instead of inside the housing as in the semi-floating axle. The wheel hubs are bolted to the end of the axle shaft and are supported by the bearing. In this arrangement, the axle shaft only supports 25 percent of the vehicle's weight. The weight is transferred through the wheel hub and bearing to the axle housing. Three-quarter floating axles are found on older vehicles and some trucks.

Full-Floating Axle Shafts

Most medium- and heavy-duty vehicles use a **full-floating axle shaft**. This design is similar to the three-quarter floating axle shaft except that two bearings rather than one are used to support the wheel hub. These are slid over the outside of the axle housing and carry all of the stresses caused by torque loading and turning. The wheel hubs are bolted to flanges on the outer end of each axle shaft.

In operation, the axle shaft transmits only the driving torque. The driving torque from the axle shaft rotates the axle flange, wheel hub, and rear driving wheel. The wheel hub forces its bearings against the axle housing to move the vehicle. The stresses caused by turning, skidding, and bent or wobbling wheels are taken by the axle housing through the wheel bearings. If a full-floating axle shaft should break, it can be removed from the axle housing. Because the rear wheels rotate around the rear axle housing, the disabled vehicle can be towed to a service area for replacement of the axle shaft.

Independently Suspended Axles

In an independently suspended axle system, the driving axles are usually open instead of being enclosed in an axle housing. The two most common suspended rear driving axles are the DeDion axle system and the swing axle system.

The DeDion axle system resembles a normal driveline. The driving axles look like a driveshaft with U-joints at each end of the axles. A slip joint is attached to the innermost U-joint. The outboard U-joint is connected to the wheel hub, which allows the driving axle to move up and down as it rotates.

On vehicles that use a swing axle, the driving axle shafts can be open or enclosed. An axle fits into the differential by way of a ball-and-socket system. The ball-and-socket system allows the axle to pivot up and down. As the axle pivots, the driving wheel swings up and down. This system best describes the drive axles of an FWD vehicle.

Axle Shaft Bearings

The axle shaft bearing supports the vehicle's weight and reduces rotational friction. In an axle mount, radial and thrust loads are always present on the axle shaft bearing when the vehicle is moving. Radial bearing loads act at 90° to the axle shaft's centre of axis. **Radial loading** is always present whether or not the vehicle is moving.

Thrust loading acts on the axle bearing parallel with the centre of axis. It is present on the driving wheels, axle shafts, and axle bearings when the vehicle turns corners or curves.

There are three designs of axle shaft bearings used in semi-floating axles: ball-type bearing, straight roller bearing, and tapered roller bearing.

The bearing load of primary concern is axle shaft end thrust. When a vehicle moves around a corner, centrifugal force acts on the vehicle body, causing it to lean to the outside of the curve. The vehicle's chassis does not lean because of the tires' contact with the road's surface. As the body leans outward, a thrust load is placed on the axle shaft and axle bearing. Each type of axle shaft bearing handles end thrust differently.

Normally, the way the axles are held in the housing is quite obvious after the rear wheels and brake assemblies have been removed. If the axle shaft is held in by a retainer and three or four bolts, it is not necessary to remove the differential cover to remove the axle. Most ball-type and tapered roller bearing supported axle shafts are retained in this manner **(Figure 41–45)**. To remove the axle, remove the bolts that hold the retainer to the backing plate, and then pull the axle out. Normally, the axle shaft slides out without the aid of a puller. Sometimes a puller is required.

A straight roller bearing–supported axle shaft does not use a retainer to secure it. Rather, a C-shaped washer is used to retain the axle shaft **(Figure 41–46)**. This C-shaped washer is located inside the differential, and the differential cover must be removed to gain access to it. To remove this type of axle, first remove the wheel, brake drum, and carrier housing cover. Then remove the differential pinion shaft retaining bolt and differential pinion shaft. Now push the axle shaft in and remove the C-shaped washer. The axle can now be pulled out of the housing.

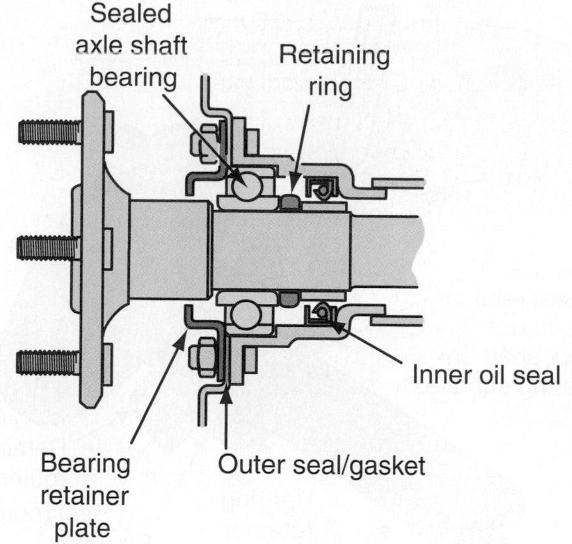

FIGURE 41–45

The location of an axle bearing retainer.

Ball bearings are lubricated with grease packed in the bearing at the factory. An inner seal, designed to keep the gear oil from the bearing, rides on the axle shaft just in front of the retaining ring. This type of bearing also has an outer seal/gasket to prevent grease from spraying onto the rear brakes. Ball-type axle bearings are pressed on and off the axle shaft. The retainer ring is made of soft metal and is pressed

FIGURE 41–46

To remove the axle shafts from the differential in an integral housing, a C-lock must normally be removed.

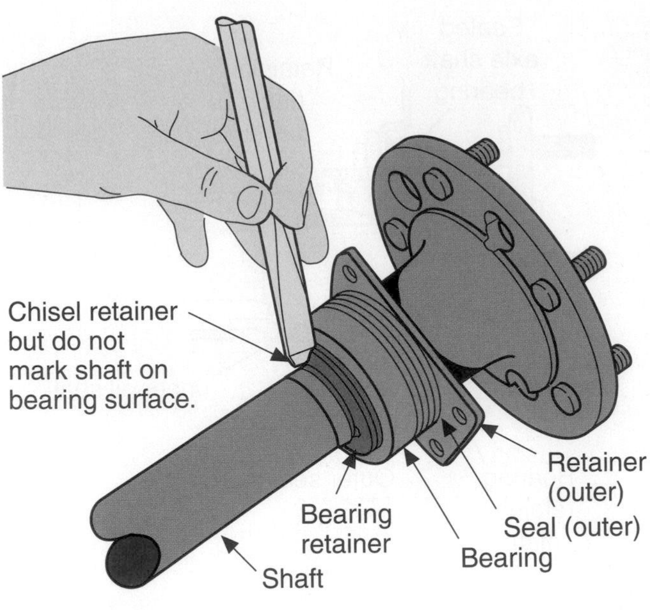

FIGURE 41–47

Freeing the retainer ring from an axle shaft.

onto the shaft against the wheel bearing. Never use a torch to remove the ring. Rather, drill into it or notch it in several places with a cold chisel to break the seal **(Figure 41–47)**. The ring can then be slid off the shaft easily. Heat should not be used to remove the ring because it can take the temper out of the shaft and thereby weaken it. Likewise, a torch should never be used to remove a bearing from an axle shaft.

Straight roller axle bearings are lubricated by the gear oil in the axle housing. Therefore, only a seal to protect the brakes is necessary with these bearings. These bearings are typically pressed into the axle housing and not onto the axle. To remove them, the axle must first be removed and then the bearing pulled out of the housing. With the axle out, inspect the area where it rides on the bearing for pits or scores. If pits or score marks are present, replace the axle shaft.

Tapered roller axle bearings are not lubricated by gear oil. They are sealed and lubricated with wheel grease. This type of bearing uses two seals and must be pressed on and off the axle shaft using a press. After the bearing is pressed onto the shaft, it must be packed with wheel-bearing grease. After packing the bearing, install the axle in the housing. Shaft end play must be checked. Use a dial indicator, and adjust the end play to the specifications given in the service information. If the end play is not within specifications, change the size of the bearing shim.

The installation of new axle shaft seals is recommended whenever the axle shafts have been removed. Some axle seals are identified as being either right or left side. When installing new seals, make sure to install the correct seal in each side. Check the seals or markings of right or left or for colour coding.

USING SERVICE INFORMATION

The driveline can create some especially difficult diagnostic problems. The driveline easily picks up vibrations and noises from other parts of the vehicle. A test drive is the best way to begin diagnosis. Most service information contains a checklist that helps with identifying the cause of a noise or vibration.

Wheel Studs

Whenever the wheels are removed, the wheel studs should be inspected for damage. If damage is evident, the stud(s) should be replaced. The studs are typically damaged when the wheel lugs are under or over torque or were damaged when the lug nuts cross-threaded the stud when they were installed. If a stud has minor distortions, run a die over it to correct the thread.

To replace a stud, the old one needs to be pressed out of the hub. The new stud must be pressed into the hub. This is easily done with a hydraulic press when the axle shaft has been removed from the housing. However, it can also be done with the axle in place. This is accomplished by placing strong washers over the stud after it has been inserted into its bore. Then the lug nut is attached to the stud, flat side down, and tightened. Tightening the lug nut will pull and seat the stud into its bore. Make sure the head of the stud is fully seated in the flange or hub.

SERVICING THE FINAL DRIVE ASSEMBLY

Before removing a final drive unit for service, make sure it needs to be serviced. Typically, problems with the differential and drive axles are first noticed as a leak or noise. As the problem worsens, vibrations or a clunking noise might be felt during certain operating conditions. Diagnosis of the problem should begin with a road test in which the vehicle is taken through the different modes of operation.

Before the test drive, check out the test-drive chart **(Figure 41–48)**. The four modes of driving given in most service information should be checked out for driving-axle and differential problems. For the drive mode, accelerate the vehicle. The throttle

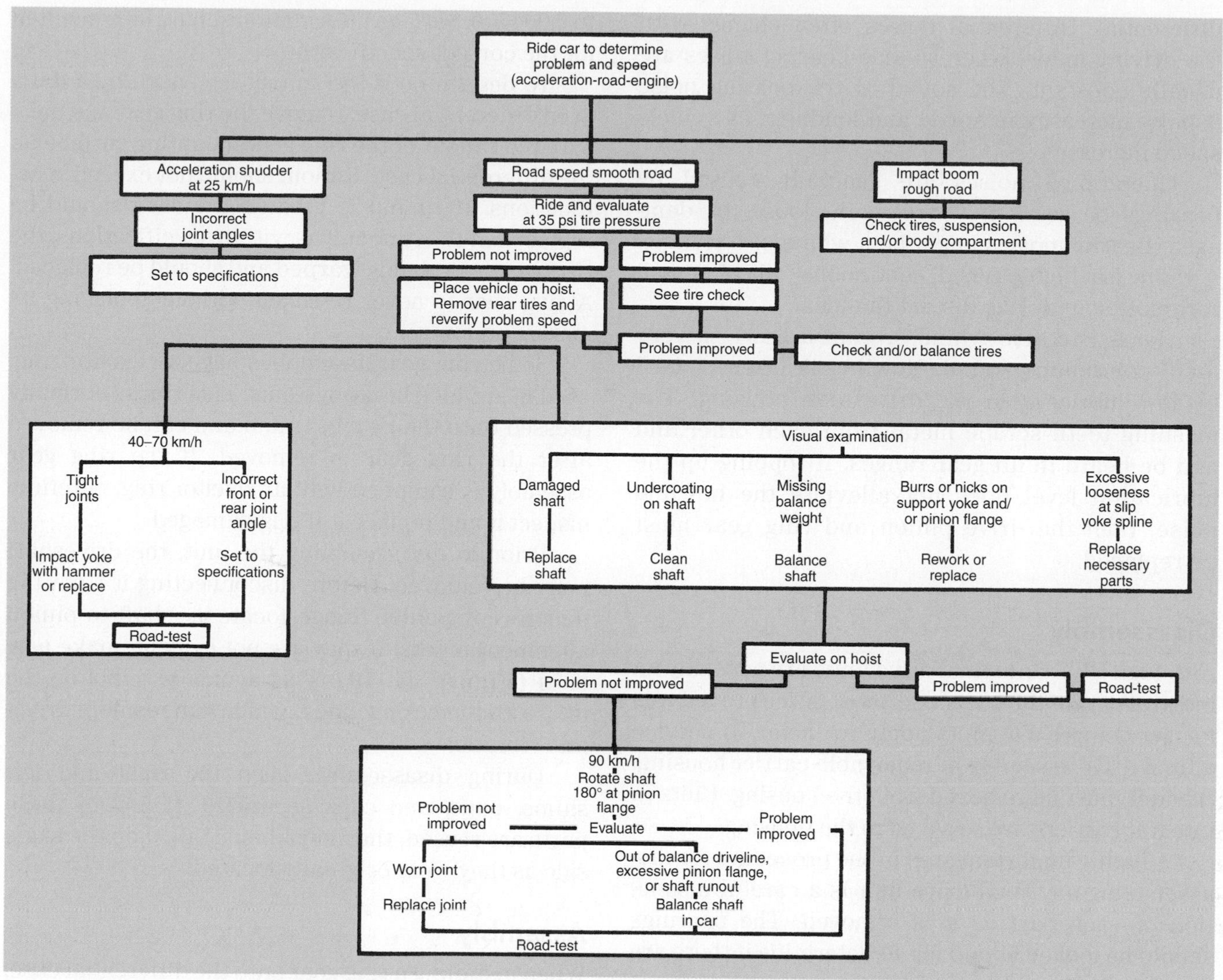

FIGURE 41–48

A typical test-drive troubleshooting chart.

must be depressed enough to apply sufficient engine torque. In the cruise mode, vehicle speed must be constant, which means that the throttle must be applied at all times. The speed must be held at a predetermined rpm on a level road. For the coast mode, take the foot off the throttle. Let the vehicle coast down from a specific speed. The float mode is a controlled deceleration. Back off the throttle gradually and continually. Do not brake or accelerate during this mode.

CAUTION!

Remember that driving safely is always important! Hard cornering or sudden braking should be avoided. Driving conditions other than normal can worsen a problem or create a new one. Carelessness during the test drive can result in an accident. Remember, you are responsible for someone else's vehicle.

Basic Diagnosis

It is common for the source of a noise to be tires, not the final drive unit. To make sure the noises are not caused by tire tread patterns and/or wear, drive the car on various types of road surfaces (asphalt, concrete, and packed dirt). If the noise changes with the road surfaces, it means the tires are the cause of the noise.

Another way to isolate tire noises is to coast at speeds less than 48 km/h (30 mph). If the noise is still heard, the tires are probably the cause. Drive axle and differential noises are less noticeable at these speeds. Accelerate and compare the sounds to those made while coasting. Drive axle and differential noises change. Tire noise remains constant.

Sometimes, it is difficult to distinguish between axle-bearing noises and noises coming from the differential. Differential noises often change with the driving mode, whereas axle-bearing noises are usually constant. The sound of the bearing noise usually increases in speed and loudness as vehicle speed increases.

Operational noises are generally caused by bearings or gears that are worn, loose, or damaged. Bearing noises might be a whine or a rumble. A whine is a high-pitched, continuous "whee" sound. A rumble sounds like distant thunder.

Gears can also whine or emit a howl—a very loud, continuous sound. Howling is often caused by low lubricant in the drive axle housing. The meshing teeth scrape metal from each other and can be heard in all gear ranges. If topping up the lubrication level does not alleviate the howling noise, then the drive pinion and ring gear must be replaced.

Disassembly

Although FWD axle final drive units are normally an integral part of the transaxle, most of the procedures for servicing RWD units apply to them. To service a final drive assembly in removable carrier housing, the unit must be removed from the housing. Units in integral carriers are serviced in the housing.

A highly important step in the procedure for disassembling any final drive unit is a careful inspection of each part as it is removed. The bearings should be looked at and felt to determine if there are any defects or evidence of damage.

After the ring and pinion gears have been inspected and before they have been removed from the assembly, check the side play. Using a screwdriver, attempt to move the differential case assembly laterally. Any movement is evidence of side play. Side play normally indicates that as the result of loose bearing cones on the differential case hubs, the differential case must be replaced.

Prior to disassembling the assembly, measure the runout of the ring gear. Excessive runout can be caused by a warped gear, worn differential side bearings, warped differential case, or particles trapped between the gear and the case. Runout is checked with a dial indicator mounted on the carrier assembly. The plunger on the indicator should be set at a right angle to the gear. With the dial indicator in position and its dial set to zero, rotate the ring gear and note the highest and lowest readings. The difference between these two readings indicates the total runout of the ring gear. Normally, the maximum permissible runout range is 0.0508 to 0.1016 mm (0.002 to 0.004 in.). Service information should be consulted for the correct specification.

To determine if the runout is caused by a damaged differential case, remove the ring gear and measure the runout of the ring gear mounting surface on the differential case. Runout should not exceed specifications. If runout is greater, the case should be replaced. If the runout was within specifications, the ring gear is probably warped and should be replaced. A ring gear is never replaced without replacing its mating pinion gear.

Some ring gear assemblies have an excitor ring, used in antilock brake systems. This ring is normally pressed onto the ring gear hub and can be removed after the ring gear is removed. If the ring gear assembly is equipped with an excitor ring, carefully inspect it and replace it if it is damaged.

Prior to disassembling the unit, the driveshaft must be removed. Before disconnecting it from the pinion's companion flange, locate the shaft-to-pinion alignment marks. If they are not evident, make new ones **(Figure 41–49)**. This avoids assembling the unit with the wrong index, which can result in driveline vibration.

During disassembly, keep the right and left shims, cups, and caps separated. If any of these parts are reused, they must be installed on the same side as they were originally located.

Assembly

When installing a ring gear onto the differential case, make sure the bolt holes are aligned before pressing the gear in place. While pressing the gear, pressure should be evenly applied to the gear. Likewise, when tightening the bolts, always tighten them in steps

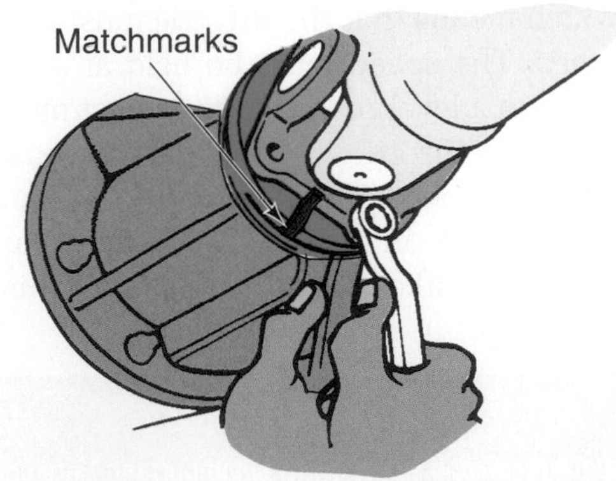

FIGURE 41–49

Make sure to index the rear yoke to the companion flange before removing the driveshaft.

©Cengage Learning 2015

and to the specified torque. These steps reduce the chances of distorting the gear.

Examine the gears to locate any timing marks on the gearset that indicate where the gears were lapped by the manufacturer. Normally, one tooth of the pinion gear is grooved and painted, and the ring gear has a notch between two painted teeth. If the paint marks are not evident, locate the notches. Proper timing of the gears is set by placing the grooved pinion tooth between the two marked ring gear teeth. Some gearsets have no timing marks. These gears are hunting gears and do not need to be timed. Nonhunting and partial nonhunting gears must be timed.

Whenever the ring and pinion gears or the pinion or differential case bearings are replaced, pinion gear depth, pinion bearing preload, and the ring and pinion gear tooth patterns and backlash must be checked and adjusted. This holds true for all types of differentials except most FWD differentials that use helical-cut gears, where checking tooth patterns is not necessary. Nearly all other final drive units use hypoid gears that must be properly adjusted to ensure a quiet operation.

Pinion gear depth is adjusted with shims placed behind the pinion bearing **(Figure 41–50)** or in the housing. The thickness of the drive pinion rear bearing shim controls the depth of the mesh between the pinion and ring gear. To determine and set pinion depth, a special tool is normally used to select the proper pinion shim **(Figure 41–51)**. Always follow the procedures in the service information when setting up the tool and determining the proper shim.

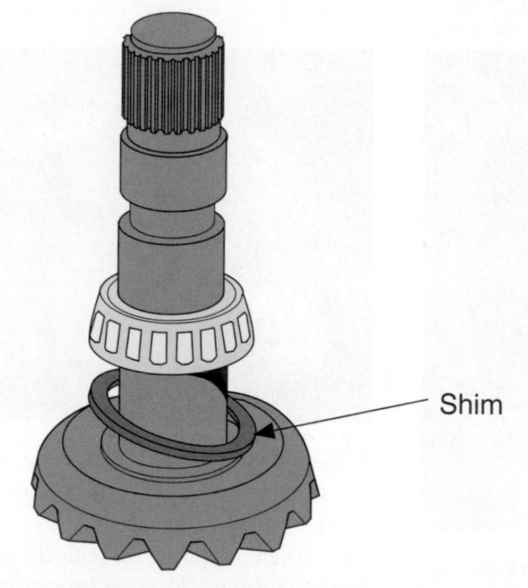

FIGURE 41–50

The typical placement of a pinion gear depth shim.

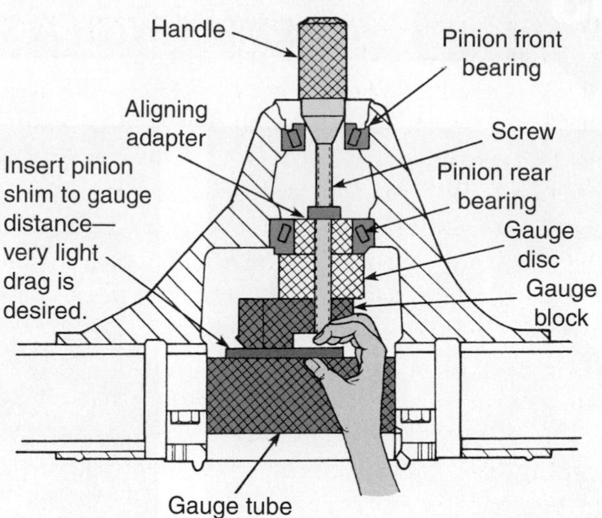

FIGURE 41–51

A special tool for measuring proper pinion gear depth.

Pinion bearing **preload** is set by tightening the pinion nut until the desired number of rotational kilogram-centimetres is required to turn the shaft. Tightening the nut crushes the collapsible pinion spacer, which maintains the desired preload. Never overtighten and then loosen the pinion nut to reach the desired torque reading. Tightening and loosening the pinion nut damages the collapsible spacer. It must then be replaced. For the exact procedures and specifications for bearing preload, refer to the service information. Incorrect bearing preload can cause rear axle noise. Some cases use shims to set pinion bearing preload.

It is recommended that a new pinion seal be installed whenever the pinion shaft is removed from the carrier housing. To install a new seal, thoroughly lubricate it, and press it in place with an appropriate seal driver.

Backlash of the gearset is adjusted at the same time as the side-bearing preload. Side-bearing preload limits the amount the differential is able to move laterally in the axle housing. Adjusting backlash sets the depth of the mesh between the ring and pinion gear teeth. Both of these are adjusted by shim thickness or by the adjustments made by the side-bearing adjusting nuts. Photo Sequence 33 goes through the typical procedure for measuring and adjusting backlash and side-bearing preload on a gearset that uses shims for adjustment. Photo Sequence 34 covers the same steps for a unit that has adjusting nuts.

A typical procedure for measuring and adjusting backlash and preload involves rocking the ring gear

PHOTO SEQUENCE 33

MEASURING AND ADJUSTING BACKLASH AND SIDE-BEARING PRELOAD ON A FINAL DRIVE ASSEMBLY WITH A SHIM PACK

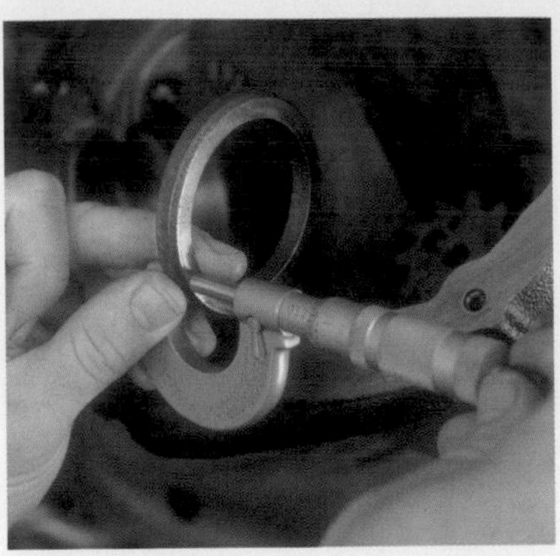

P33–1 Measure the thickness of the original side-bearing preload shim and spacer.

P33–2 Install the differential case into the carrier housing.

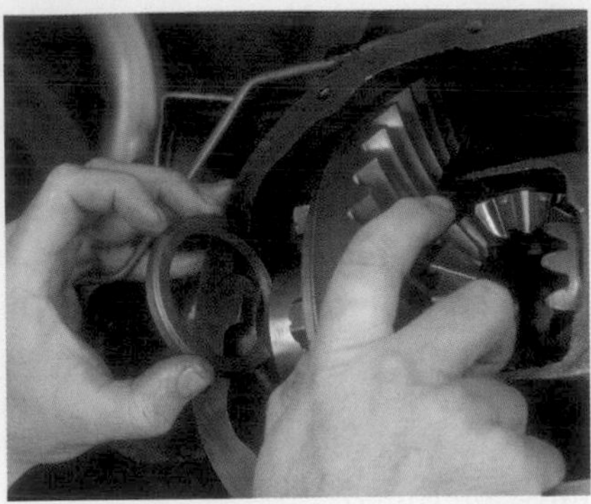

P33–3 Install service spacers that are the same thickness as the original preload shims between each bearing cup and the housing.

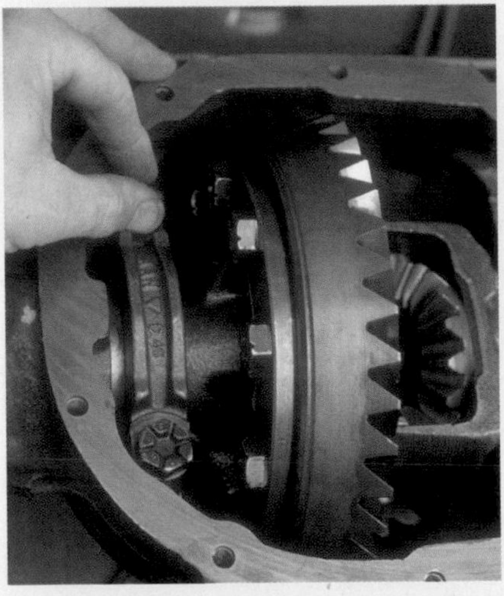

P33–4 Install the bearing caps, and finger-tighten the bolts.

P33–5 Mount a dial indicator to the housing so that the button of the indicator touches the face of the ring gear. Using two screwdrivers, pry between the shims and the housing. Pry to one side and set the dial indicator to zero, and then pry to the opposite side and record the reading.

P33–6 Select two shims with a combined thickness to that of the original shims plus the indicator reading, and then install them.

P33–7 Using the proper tool, drive the shims into position until they are fully seated.

P33–8 Install and tighten the bearing caps to specifications. Mount a dial indicator to check and set backlash. Ring gear radial runout should also be checked.

P33–9 Check the backlash by rocking the ring gear and noting the movement on the dial indicator. Adjust the shim pack to allow for the specified backlash. Recheck the backlash at four points equally spaced around the ring gear.

MEASURING AND ADJUSTING BACKLASH AND SIDE-BEARING PRELOAD ON A FINAL DRIVE ASSEMBLY WITH ADJUSTING NUTS

PHOTO SEQUENCE **34**

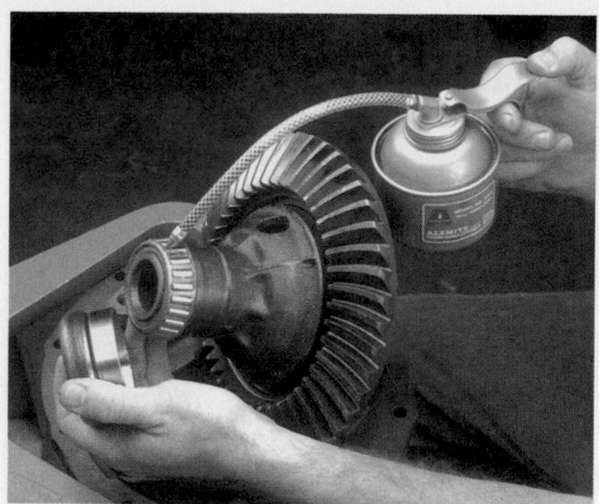

P34–1 Lubricate the differential bearings, cups, and adjusters.

P34–2 Install the differential case into the carrier housing.

P34–3 Install the bearing cups and adjusting nuts onto the carrier housing.

P34–4 Snugly tighten the top-bearing cup bolts, and finger-tighten the adjusting nuts.

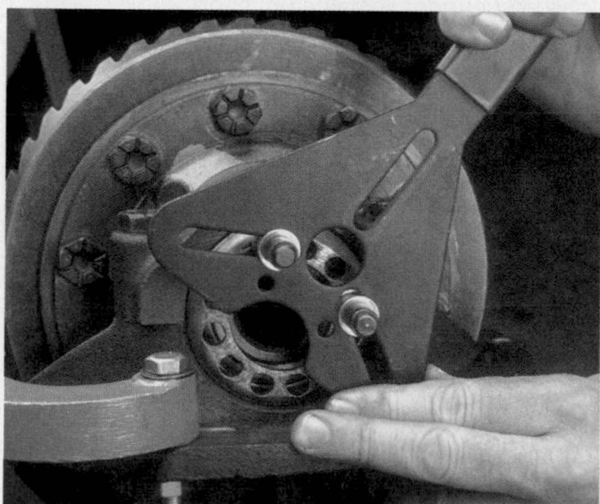

P34–5 Turn each adjuster until bearing free play is eliminated, with little or no backlash present between the ring and pinion gears.

P34–6 Seat the bearings by rotating the pinion several times each time the adjusters are moved.

PHOTO SEQUENCE 34

MEASURING AND ADJUSTING BACKLASH AND SIDE-BEARING PRELOAD ON A FINAL DRIVE ASSEMBLY WITH ADJUSTING NUTS *(continued)*

P34–7 Install a dial indicator, and position the plunger against the drive side of the ring gear. Set the dial to zero. Using two screwdrivers, pry between the differential case and the housing. Observe the reading.

P34–8 Determine how much the preload needs to be adjusted, and set the preload by turning the right adjusting nut.

P34–9 Check the backlash by rocking the ring gear and noting the movement on the dial indicator.

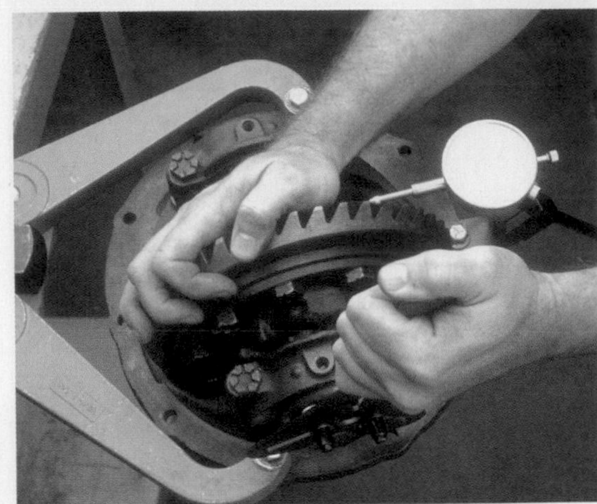

P34–10 Adjust the backlash by turning both adjusting nuts the same amount so that the pre-load adjustment remains unchanged.

P34–11 Install the locks on the adjusting nuts.

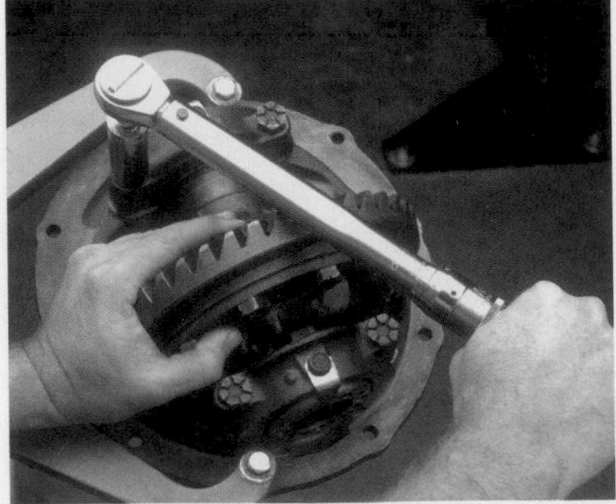

P34–12 Tighten the bearing cup bolts to the specified torque.

and measuring its movement with a dial indicator. Compare measured backlash with the specifications. Make the necessary adjustments. Then recheck the backlash at four points equally spaced around the ring gear. Normally, backlash should be less than 0.1778 mm (0.007 in.).

The pattern of gear teeth determines how quietly two meshed gears run. The pattern also describes where on the faces of the teeth the two gears mesh. The pattern should be checked during teardown for gear noise diagnosis, after adjusting backlash and side-bearing preload, or after replacing the drive pinion and setting up the pinion bearing preload. The terms commonly used to describe the possible patterns on a ring gear and the necessary corrections are shown in **Figure 41–52**.

To check the gear tooth pattern, paint several ring gear teeth with non-drying Prussian blue, ferric oxide, or red or white lead-marking compound **(Figure 41–53)**. White marking compound is preferred by many technicians because it tends to be more visible than the others are. Use the pinion gear yoke or companion flange to rotate the ring gear. Use a pry bar and place it between the differential case and the carrier housing to create some resistance to the ring gear rotation. (This will preload the ring gear while it is rotating and will simulate vehicle load.) Rotate the ring gear so that the painted teeth contact the pinion gear. Move it in both directions enough to get a clearly defined pattern. Examine the pattern on the ring gear, and make the necessary corrections.

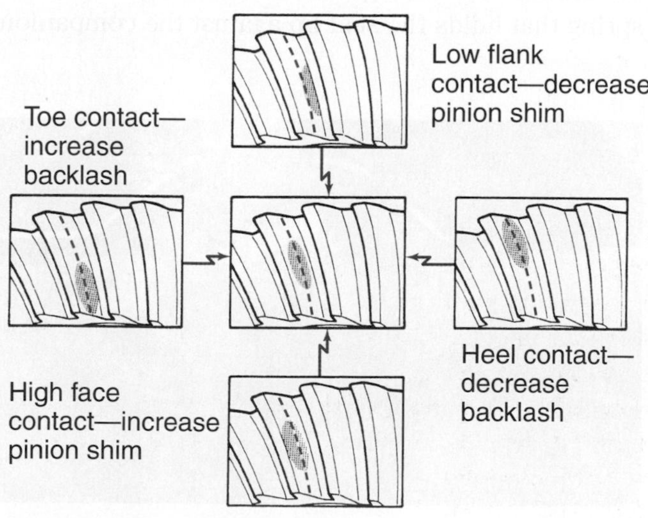

FIGURE 41–52

Commonly used terms for describing the possible patterns on a ring gear with the recommended corrections.

Ford Motor Company

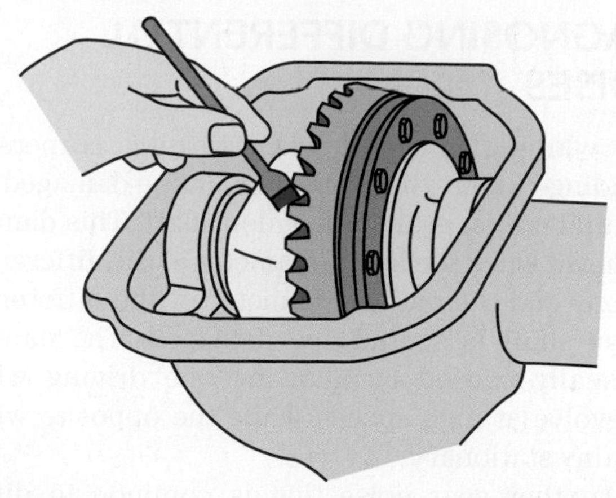

FIGURE 41–53

To check gear tooth patterns, several teeth of the ring gear are coated with a marking compound, and the pinion gear is rotated with the ring gear. The resultant pattern shown on the teeth determines how the gearset ought to be adjusted.

Most new gearsets purchased today come with a pattern prerolled on the teeth. This pattern provides the quietest operation for that gearset. Never wipe this pattern off or cover it up. When checking the pattern on a new gearset, only coat half of the ring gear with the marking compound, and compare the pattern with the prerolled pattern.

Maintenance

Maintenance includes inspecting the level of and changing the gear lubricant, and lubricating the U-joints if they are equipped with zerk or grease fittings. Most modern U-joints are of the extended life design, meaning they are sealed and require no periodic lubrication. However, it is wise to inspect the joints for hidden grease plugs or fittings.

Proper lubrication is necessary for drive axle durability. Different applications require different gear lubes. The American Petroleum Institute (API) has established a rating system for the various gear lubes available. In general, rear axles use either SAE 80- or 90-weight gear oil for lubrication, meeting API GL-4 or GL-5 specifications. With limited-slip axles, it is very important that the proper gear lube be used. Most often, a special friction modifier fluid should be added to the fluid. If the wrong lubricant is used, damage to the clutch packs and grabbing or chattering on turns will result. If this condition exists, try draining the oil and refilling with the proper gear lube before servicing it.

DIAGNOSING DIFFERENTIAL NOISES

If a whining is heard when turning corners or rounding curves, the problem might be damaged differential pinion gears and pinion shaft. This damage is caused when the inside diameter of the differential pinions and the outside diameter of the differential pinion shaft is scored and damaged. The damage is usually caused by allowing one driving wheel to revolve at high speeds while the opposite wheel remains stationary.

Another gear noise that is common in differentials is the chuckle. A chuckle is a low "heh-heh" sound that occurs when gears are worn to the point where there is excessive clearance between the pinion gear and the ring gear. Chuckle sounds occur most often in the decelerating mode, particularly below 65 km/h (40 mph). As the vehicle decelerates, the chuckle also slows and can be heard all the way to a stop.

A knock or clunk is caused by excessive wear or loose or broken parts. A knock is a repetitious rapping sound that occurs during all phases of driving but is most noticeable during acceleration and deceleration, when the gears are loaded.

A clunk is a sharp, loud noise caused by one part hitting another. Unlike a knock, a clunk can be felt as well as heard. Clunks are generally caused by loose parts striking each other.

Limited-slip clutch packs or cones that need servicing might be heard as a chatter or a rapid clicking noise that creates a vibration in the vehicle. Chattering is usually noticed when rounding a corner. A change of differential lubricant and adding friction modifier to the fluid sometimes corrects this problem. After draining the oil, replace it with the manufacturer's suggested friction modifier and lubricant. Road-test the vehicle again.

To make sure that the noise heard during the test drive is coming from the differential, stop the vehicle, and shift the transmission into NEUTRAL. Run the engine at various rpm levels. If the noise is heard during this procedure, it is caused by a problem somewhere other than in the differential.

Vibration Problems

Often, the source of vibration is a bent axle or axle flange, or improper mounting of the wheel to the flange. To check the runout of the flange, position a dial indicator against the outer flange surface of the axle. Apply slight pressure to the centre of the axle to remove the end play in the axle, and then zero the indicator. Slowly rotate the axle one complete revolution, and observe the readings on the indicator. The total amount of indicator movement is the total amount of axle flange lateral runout. Compare the measured runout with specifications.

Driveline Fluid Leaks

To find the exact source of the leak, carefully inspect the driveline for wet spots. Thoroughly clean the area around the leak so that the exact source can be found.

If the extension housing seal is leaking, it can be easily replaced. However, before replacing the seal, check the extension housing bushing, and replace it with the seal if it is worn. That is the most likely reason the seal went bad in the first place. Once the yoke is removed, an internal expanding bearing/bushing puller makes short work of bushing replacement. Before pushing the slip yoke back in after the new seal is installed, make sure the machined surface of the bore is free of scratches, nicks, and grooves that could damage the seal. For that added margin of safety, a little transmission lube or petroleum jelly on the lip of the seal helps the parts slide in easily.

An improperly installed or damaged drive pinion seal will allow the lubricant to leak past the outer edge of the seal **(Figure 41–54)**. Any damage to the seal's bore, such as dings, dents, and gouges, will distort the seal casing and allow leakage. Also, the spring that holds the seal lip against the companion

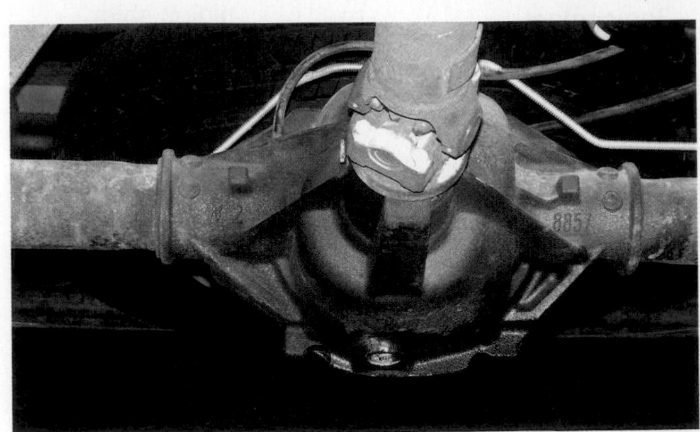

FIGURE 41–54

The housing is wet because of a leaking pinion seal.

flange may be knocked out and allow leakage past the seal's lip.

If a lot of lube is escaping from the pinion shaft seal, the drivetrain noise could be caused by a bad pinion bearing. To confirm the problem, start the engine, put the transmission in gear, and listen at the carrier. If the bearing is noisy, it is necessary to make one of those difficult judgment calls. If the bearing sounds fine but the pinion seal is still leaking, suggest an on-the-car seal replacement.

On some vehicles, seal replacement is a simple procedure that involves removing the pinion flange and replacing the seal. Others are a little more complex because the pinion shaft is retained with a nut that must be removed to gain access to the seal. These units require special tools to loosen and tighten the pinion nut **(Figure 41–55)**, which allows for the removal of the flange. Always refer to the service information for the correct procedure. On many units, there is a **collapsible spacer** behind the pinion nut. Whenever the nut is loosened, a new spacer should be installed before torquing the nut.

If the seal needs to be replaced, check the runout of the flange. A damaged seal can be caused by excessive runout. To do this, install a dial indicator with its base on the carrier and its plunger on the flange **(Figure 41–56)**. Rotate the wheels while observing the movement of the indicator. Any reading indicates some runout. Compare the reading to specifications. Also inspect the surface of the flange that rides in the seal. During reassembly, make sure the outer surface of the flange is lubricated before pushing it into the seal.

FIGURE 41–55

The tools required to tighten the flange nut.

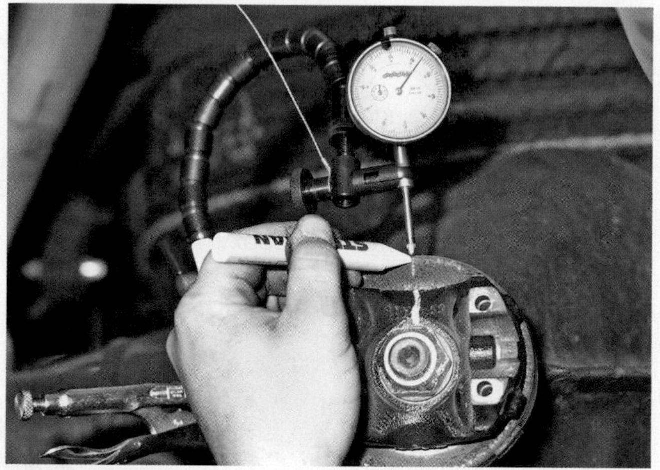

FIGURE 41–56

The setup for checking the runout of a companion flange.

It is also possible for oil to leak past the threads of the drive pinion nut or the pinion retaining bolts. These leaks can be stopped by removing the nut or bolts, applying thread or gasket sealer on the threads, and torquing the nuts or bolts to specifications.

There may be leakage at the retaining nuts for the carrier or cover. One way to correct this problem is to install copper washers under the nuts. This is recommended whenever the carrier is removed, whether or not copper washers were installed originally. If the assembly was originally equipped with steel washers, replace them with copper. Always make sure there is a copper washer under the axle ID tag.

Most gasket leaks result from poor installation, loose retaining bolts, or a damaged mating surface. Most late-model vehicles do not use a gasket on the housing cover; rather, silicone sealer is used. The old sealer should be cleaned off before applying a new coat to the surface. If a housing cover is leaking, inspect the surface for imperfections, such as cracks or nicks. File the surface true, and install a new gasket. If the surface cannot be trued, apply some gasket sealer to the surface before installing the new gasket. Always follow the correct tightening sequence and torque specifications when tightening an axle housing cover.

At times, lubricant will leak through the pores of the housing. There are two recommended ways to repair these leaks, other than replacing the housing. If the porous area is small, force some metallic body filler into the area. After the filler has set, seal the

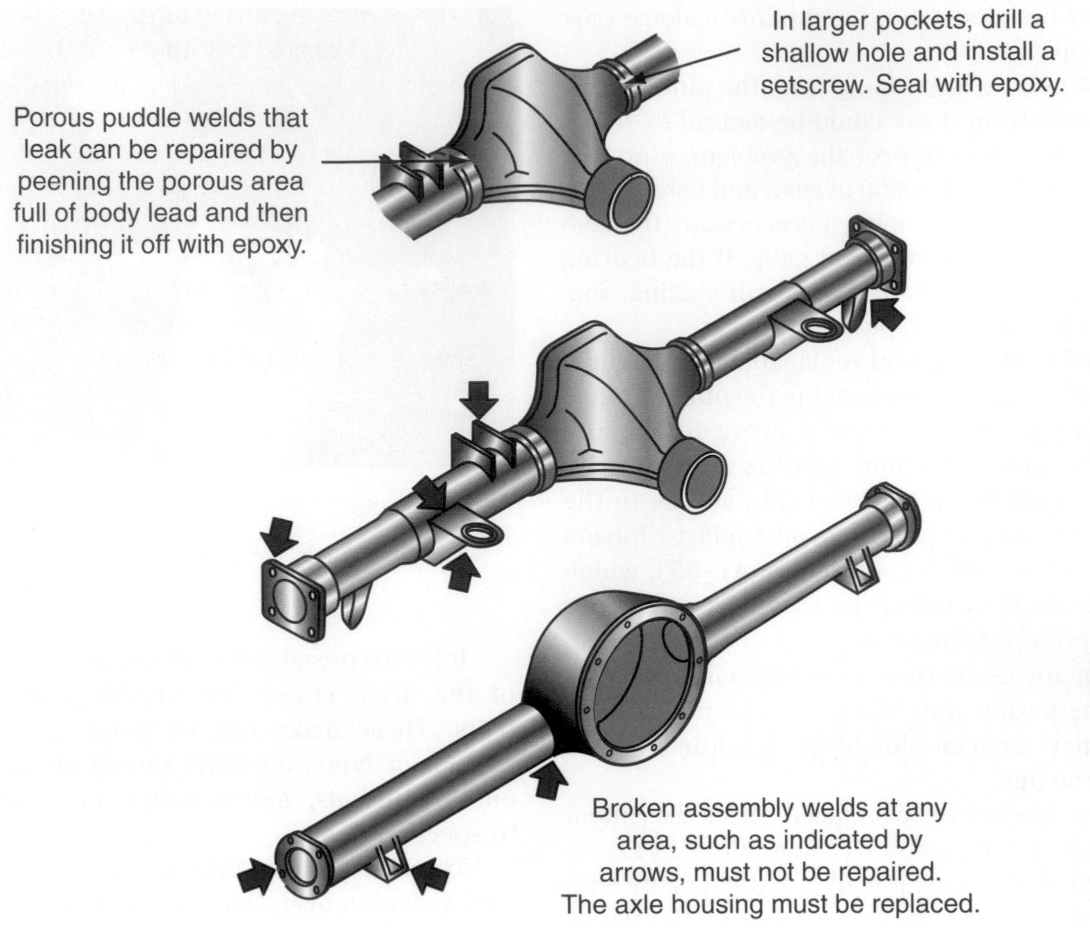
In larger pockets, drill a shallow hole and install a setscrew. Seal with epoxy.
Porous puddle welds that leak can be repaired by peening the porous area full of body lead and then finishing it off with epoxy.
Broken assembly welds at any area, such as indicated by arrows, must not be repaired. The axle housing must be replaced.

FIGURE 41–57

Basic guidelines for correcting axle housing leaks.

area with an epoxy-type sealer. If the area is rather large, drill a hole and tap an appropriately sized setscrew into the hole, and then cover the area with an epoxy sealer. Minor weld leaks can also be sealed with epoxy sealer. However, if a weld is broken, the housing should be replaced **(Figure 41–57)**.

If the wrong vent was installed in the axle housing or if there is an excessive amount of lubricant or oil turbulence in the axle, lubricant may leak through the axle vent hose. The cause of this type of leakage will only be found by a careful inspection. Check for a crimped or broken vent hose. Replace the hose it is damaged. If the cause of the leakage is an overfilled axle assembly, drain the housing and refill the unit with the specified amount and type of lubricant.

Some differential units are fitted with an ABS sensor. Lubricant can leak from around a damaged O-ring. To correct this problem, remove the sensor and replace the O-ring.

KEY TERMS

Backlash (p. 1259)
Ball-type joints (p. 1226)
Collapsible spacer (p. 1265)
Cross-groove joint (p. 1227)
Dead axle (p. 1253)
Double-Cardan U-joint (p. 1239)
Double-offset joint (p. 1227)
Fixed joint (p. 1226)
Full-floating axle shaft (p. 1254)
Hunting gearset (p. 1248)
Hypoid gear (p. 1248)
Inboard joint (p. 1226)
Integral housing (p. 1249)
Limited-slip differential (LSD) (p. 1252)
Live axle (p. 1253)
Nonhunting gearset (p. 1248)
Operating angle (p. 1238)
Outboard joint (p. 1226)
Partial nonhunting gearset (p. 1248)
Pinion gear (p. 1247)
Pinion shaft (p. 1247)
Plunge joint (p. 1226)
Preload (p. 1259)
Radial loading (p. 1255)
Ring gear (p. 1248)

Rzeppa joint (p. 1226)
Semi-floating axles (p. 1254)
Slip yoke (p. 1237)
Three-quarter floating axle (p. 1254)
Thrust loading (p. 1255)
Torsional damper (p. 1229)
Tripod-type joints (p. 1226)

SUMMARY

- FWD axles generally transfer engine torque from the transaxle to the front wheels.
- Constant velocity (CV) joints provide the necessary transfer of uniform torque and a constant speed while operating through a wide range of angles.
- In FWD drivetrains, two CV joints are used on each half shaft. The different types of joints can be referred to by position (inboard or outboard), by function (fixed or plunge), or by design (ball-type or tripod).
- Front-wheel-drive half shafts can be solid or tubular, of equal or unequal length, and with or without damper weights.
- Most problems with front-wheel-drive systems are noted by noise and vibration.
- Lubricant is the most important key to a long life for the CV joint.
- A U-joint is a flexible coupling located at each end of the driveshaft between the transmission and the pinion flange on the drive axle assembly.
- A U-joint allows two rotating shafts to operate at a slight angle to each other; this is important to RWD vehicles.
- A failed U-joint or damaged driveshaft can exhibit a variety of symptoms. A clunk that is heard when the transmission is shifted into gear is the most obvious. You can also encounter unusual noise, roughness, or vibrations.
- A differential is a geared mechanism located between the driving axle shafts of a vehicle. Its job is to direct power flow to the driving axle shafts. Differentials are used in all types of powertrains.
- The differential performs several functions. It allows the drive wheels to rotate at different speeds when negotiating a turn or curve in the road, and the differential drive gears redirect the engine torque from the driveshaft to the rear drive axles.
- The final drive and differential of an RWD vehicle is housed in the axle housing, or carrier housing.
- The purpose of the axle shaft is to transfer driving torque from the differential and final drive assembly to the vehicle's driving wheels.
- The two common types of driving axle shafts used are semi-floating and full-floating. The three-quarter floating axle type is rarely used.
- Axle shaft bearings may support the vehicle's weight but always reduce rotational friction.
- Problems with the differential and drive axle shafts are usually first noticed as a leak or noise. As the problem progresses, vibrations or a clunking noise might be felt in various modes of operation.

REVIEW QUESTIONS

1. Name the three ways in which CV joints can be classified.
2. What type of axle housing resembles a banjo?
3. What type of axle merely supports the vehicle load and provides a mounting place for the wheels?
4. What type of floating axle has one wheel bearing per wheel on the outside of the axle housing?
5. How are problems with the differential and drive axles usually first noticed?
6. What type of vehicle does not typically use constant velocity joints?
   a. conventional front-wheel-drive vehicles
   b. front-wheel-drive vehicles with independent front suspension
   c. conventional rear-wheel-drive vehicles
   d. rear-wheel-drive vehicles with independent rear suspension
7. In front-wheel drivetrains, what is the CV joint nearer to the transaxle called?
   a. fixed joint
   b. inboard joint
   c. outboard joint
   d. Rzeppa joint
8. What is a CV joint that is capable of in-and-out movement called?
   a. plunge joint **b.** fixed joint
   c. outboard joint **d.** Rzeppa joint
9. What component in the final drive assembly are the axle shafts splined to?
   a. pinion gears **b.** side gears
   c. differential case **d.** crown gear

10. What operating conditions typically require the use of the double-offset joint?
    a. higher operating angles and greater plunge depth
    b. lower operating angles and lower plunge depth
    c. higher operating angles and lower plunge depth
    d. lower operating angles and greater plunge depth
11. Which type of joint has a flatter design than any other?
    a. double-offset b. disc
    c. cross groove d. fixed tripod
12. Which of these is the best way to determine which CV joint is faulty?
    a. a squeeze test
    b. a runout test
    c. a visual inspection
    d. a road test
13. What would cause a final drive noise that is heard only when cornering?
    a. an incorrect tooth contact pattern
    b. a broken crown gear tooth
    c. worn differential case gears
    d. a bent axle housing
14. What driveshaft component provides a means of connecting two or more shafts together?
    a. pinion flange b. U-joint
    c. yoke d. biscuit
15. Which type of driving axle supports the weight of the vehicle?
    a. semi-floating
    b. three-quarter floating
    c. full-floating
    d. non-floating
16. When performing a final drive gear tooth pattern inspection, what would be done to correct a pattern that was too close to the heel of the tooth?
    a. Increase the backlash between the ring and pinion gears.
    b. Decrease the backlash between the ring and pinion gears.
    c. Increase the pinion gear depth.
    d. Decrease the pinion gear depth.
17. How many bearings are required to support an overhung mounted pinion?
    a. one b. two
    c. three d. four
18. In a limited-slip differential, what are the friction discs and clutch plates located between?
    a. the two side gears
    b. the two spider gears
    c. the side gears and the differential case
    d. the spider and crown gears
19. What occurs in a differential during cornering?
    a. The inside axle side gear turns faster than the outer axle side gear.
    b. The case turns at the same speed as the driveshaft.
    c. The crown gear will slow down slightly.
    d. The pinion/spider gears rotate on the pinion shaft.
20. What does the term *hypoid* final drive indicate?
    a. A spur bevel pinion and ring gear set is used.
    b. The centreline of the drive pinion is below the ring/crown gear centreline.
    c. The ring/crown gear and drive pinion gear centrelines are on the same plane.
    d. A spiral bevel pinion and ring gear set is used.

CHAPTER 42
Automatic Transmissions and Transaxles

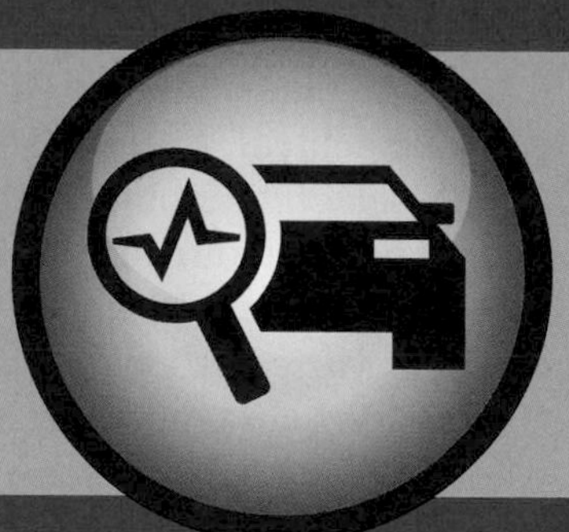

LEARNING OUTCOMES

- Explain the basic design and operation of standard and lock-up torque converters.
- Describe the design and operation of a simple planetary gearset and Simpson gear train.
- Name the major types of planetary gear controls used on automatic transmissions, and explain their basic operating principles.
- Describe the construction and operation of common Simpson gear train–based transmissions and transaxles.
- Describe the construction and operation of common Ravigneaux gear train–based transmissions.
- Describe the construction and operation of transaxles that use planetary gearsets in tandem.
- Describe the construction and operation of automatic transmissions that use helical gears in constant mesh.
- Describe the construction and operation of CVTs.
- Describe the design and operation of the hydraulic controls and valves used in modern transmissions and transaxles.
- Explain the role of the following components of the transmission control system: pressure regulator valve, throttle valve, governor assembly, manual valve, shift valves, and kick-down valve.
- Identify the various pressures in the transmission, state their purpose, and tell how they influence the operation of the transmission.

WE SUPPORT THE CANADIAN INTERPROVINCIAL STANDARDS PROGRAM

Many rear-wheel-drive (RWD) and four-wheel-drive (FWD) vehicles are equipped with automatic transmissions **(Figure 42–1)**. Automatic transaxles, which combine an automatic transmission and final drive assembly in a single unit, are used on FWD, all-wheel-drive (AWD), and some RWD vehicles **(Figure 42–2)**.

An automatic transmission or transaxle selects gear ratios according to engine speed, powertrain load, vehicle speed, and other operating factors. Little effort is needed on the part of the driver, because both upshifts and downshifts occur automatically. A driver-operated clutch is not needed to change gears, and the vehicle can be brought to

FIGURE 42–1
An eight-speed automatic transmission.
©Cengage Learning 2015

FIGURE 42–2
A cutaway of an automatic transaxle.
Chrysler Group LLC

a stop without shifting to neutral. This is a great convenience, particularly in stop-and-go traffic. The driver can also manually select a lower forward gear, reverse, neutral, or park. Depending on the forward range selected, some transmissions can provide engine braking during deceleration.

The number of available forward gears in current vehicles varies from four to nine. Some have zero fixed ratios and use a constantly variable design, where the ratio changes according to conditions. Most early automatic transmissions were three- or four-speed units. Today, the most common are five- and six-speed units. Nearly all automatics also have a lock-up torque converter. Some automatic transmissions are fitted with a transfer case that sends torque to additional drive axles to allow for four-wheel or all-wheel drive **(Figure 42–3)**.

Until recently, all automatic transmissions were controlled by hydraulics that mainly responded to the operating conditions of the engine. However, most systems now feature computer-controlled operation of the torque converter and transmission. Based on input data supplied by electronic sensors and switches, the computer sets the torque converter's operating mode and controls the transmission's hydraulic system.

Today's transmissions have at least one overdrive gear to reduce fuel consumption, lower emission levels, and reduce noise while the vehicle is cruising. The transmissions have a lock-up torque converter that eliminates loss of power through the torque converter. A select group of newer cars have eight or nine speeds. With a greater number of speeds and smaller ratio steps, the engine can remain in its higher efficiency ranges longer, with reduced engine speed and fuel consumption.

FIGURE 42–3

An automatic transmission fitted with a transfer case for all-wheel drive.

TORQUE CONVERTER

Automatic transmissions use a fluid coupling known as a torque converter to transfer engine torque from the engine to the transmission.

The torque converter operates through hydraulic force provided by automatic transmission fluid (ATF), often simply called transmission oil. The torque converter changes or multiplies the twisting motion of the engine crankshaft and directs it to the transmission.

The torque converter automatically engages and disengages power from the engine to the transmission in relation to engine rpm. With the engine running at the correct idle speed, there is not enough fluid flow for power transfer through the torque converter. As engine speed is increased, the added fluid flow creates sufficient force to transmit engine power through the torque converter assembly to the transmission.

Design

Nearly all torque converters, or T/Cs, are one-piece, welded units that can be repaired only by specialty shops. They are located between the engine and transmission and are sealed, doughnut-shaped units **(Figure 42–4)** that are always filled with automatic transmission fluid. All of the vital parts of a torque converter are housed within its shell **(Figure 42–5)**.

A flexplate, or drive plate, is used to mount the torque converter to the crankshaft. The flexplate transfers the rotation of the crankshaft to the shell of the torque converter. This design allows the plate

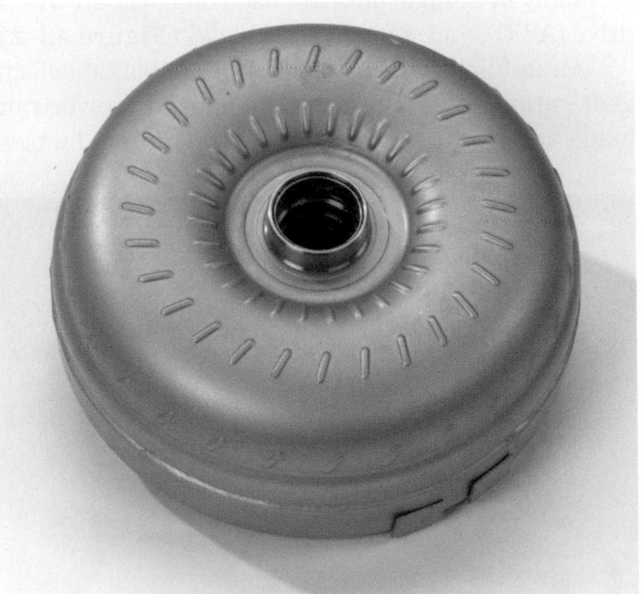

FIGURE 42–4

A torque converter.

Transtar Industries Inc.

FIGURE 42–5

A cutaway of a modern torque converter.

Transtar Industries Inc.

to flex in response to the slight change in torque converter size as pressure builds in it. It is bolted to a flange on the rear of the crankshaft and to mounting pads on the front of the torque converter shell. A heavy flywheel is not needed because the mass of the torque converter and flexplate work like a flywheel to smooth out the intermittent power strokes of the engine. The flexplate or torque converter also contains the ring gear for the starting motor.

Components

A standard torque converter consists of three elements **(Figure 42–6)**: the pump assembly, often called an impeller; the stator assembly; and the turbine.

The **impeller** assembly is the input (drive) member. It receives power from the engine. The **turbine** is the output (driven) member. It is splined

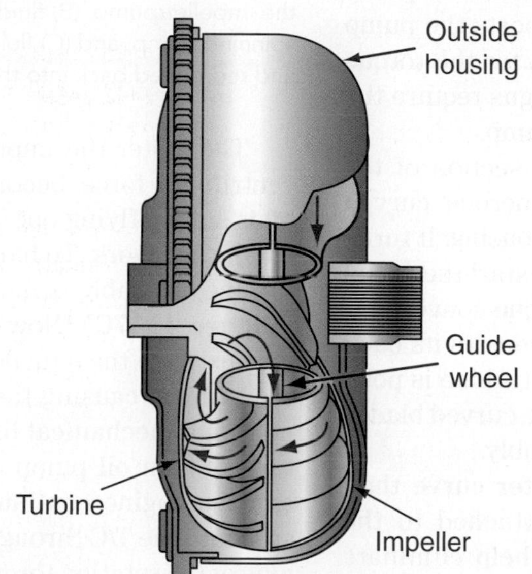

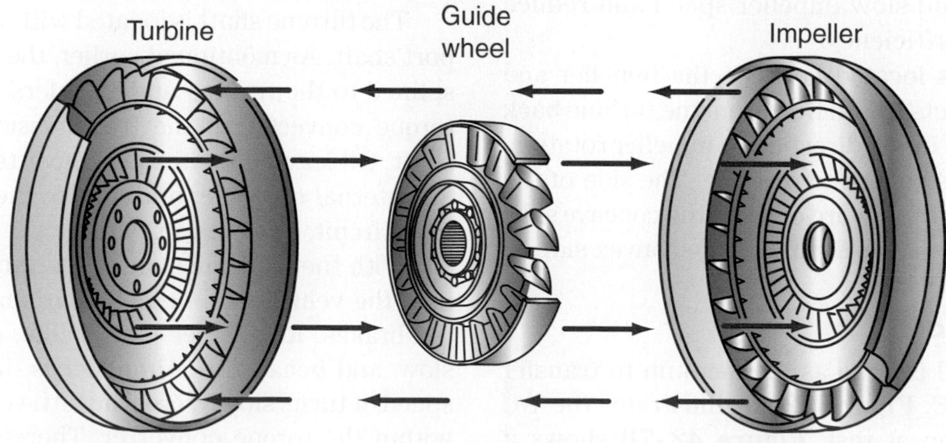

FIGURE 42–6

A torque converter's major internal parts are its impeller, turbine, and stator.

©Cengage Learning 2015

to the transmission's turbine shaft. The **stator** assembly is the reaction member or torque multiplier. The stator is supported on a one-way clutch, which operates as an overrunning clutch. This one-way clutch allows the stator to rotate freely in the direction of converter rotation and lock up in the opposite direction.

The exterior of the torque converter shell is shaped like two bowls standing on end, facing each other. To support the weight of the torque converter, a short stubby shaft projects forward from the front of the torque converter shell and fits into a pocket at the rear of the crankshaft. At the rear of many torque converter shells is a hollow hub with notches or flats at one end, ground 180° apart. This hub is called the pump drive hub. These notches or flats drive the transmission pump assembly, creating fluid flow as soon as the engine's crankshaft begins to rotate. At the front of the transmission within the pump housing is a pump bushing that supports the pump drive hub and provides rear support for the torque converter. Some other transaxle designs require the use of a separate shaft to drive the pump.

The impeller forms one internal section of the torque converter shell and has numerous curved blades that rotate as a unit with the housing. It turns at engine speed, acting like a pump to start the transmission oil circulating within the torque converter.

Whereas the impeller is positioned with its back facing the transmission housing, the turbine is positioned with its back to the engine. The curved blades of the turbine face the impeller assembly.

The turbine blades have a greater curve than the impeller blades. Guide rings attached to the impeller and turbine in the centre help eliminate oil turbulence between the turbine and impeller blades that would slow impeller speed and reduce the converter's efficiency.

The stator is located between the impeller and turbine. It redirects the oil flow from the turbine back into the impeller in the direction of impeller rotation, with minimal loss of speed or force. The side of the stator blade with the inward curve is the concave side. The side with an outward curve is the convex side.

Basic Operation

Transmission oil is used as the medium to transfer energy in the T/C. **Figure 42–7A** illustrates the T/C impeller or pump at rest. **Figure 42–7B** shows it being driven. As the pump impeller rotates, centrifugal force throws the oil outward and upward due to the curved shape of the impeller housing.

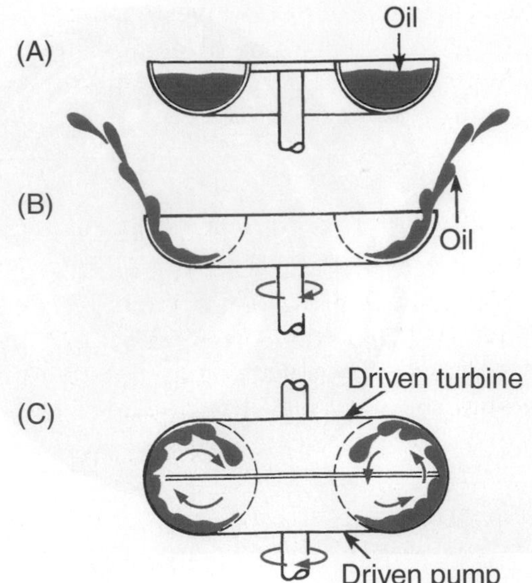

FIGURE 42–7

Fluid travel inside the torque converter: (A) fluid at rest in the impeller/pump, (B) fluid thrown up and outward by the spinning pump, and (C) fluid flow harnessed by the turbine and redirected back into the pump.

The faster the impeller rotates, the greater the centrifugal force becomes. In **Figure 42–7B**, the oil is simply flying out of the housing and is not producing any work. To harness some of this energy, the turbine assembly is mounted on top of the impeller **(Figure 42–7C)**. Now the oil thrown outward and upward from the impeller strikes the curved vanes of the turbine, causing the turbine to rotate. (There is no direct mechanical link between the impeller and turbine.) An oil pump driven by the converter hub and the engine continually delivers oil under pressure into the T/C through a hollow shaft at the centre axis of the rotating torque converter assembly. A seal prevents the loss of fluid from the system.

The turbine shaft is located within the stator support shaft. As mentioned earlier, the turbine shaft is splined to the turbine and transfers power from the torque converter to the transmission's main input shaft. Oil leaving the torque converter is directed to an external oil cooler and then to the transmission's lube circuit, oil sump, or pan.

With the transmission in gear and the engine at idle, the vehicle can be held stationary by applying the brakes. Remember that at idle, engine speed is slow, and because the impeller is driven by engine speed, it turns slowly, creating little centrifugal force within the torque converter. Therefore, little or no power is transferred to the transmission.

When the throttle is opened, engine speed, impeller speed, and the amount of centrifugal force generated in the torque converter increase dramatically. Oil is then directed against the turbine blades, which transfer power to the turbine shaft and transmission.

Types of Oil Flow

Two types of oil flow take place inside the torque converter: rotary and vortex flow **(Figure 42–8)**. **Rotary oil flow** is the oil flow around the circumference of the torque converter caused by the rotation of the torque converter on its axis. **Vortex oil flow** is the oil flow occurring from the impeller to the turbine and back to the impeller, at a 90° angle from engine rotation.

Figure 42–9 also shows the oil flow pattern as the speed of the turbine approaches the speed of the impeller. This is known as the **coupling point**. The turbine and the impeller are running at almost the same speed. They cannot run at exactly the same speed due to slippage between them. The only way they can turn at exactly the same speed is by using a lock-up clutch to mechanically tie them together.

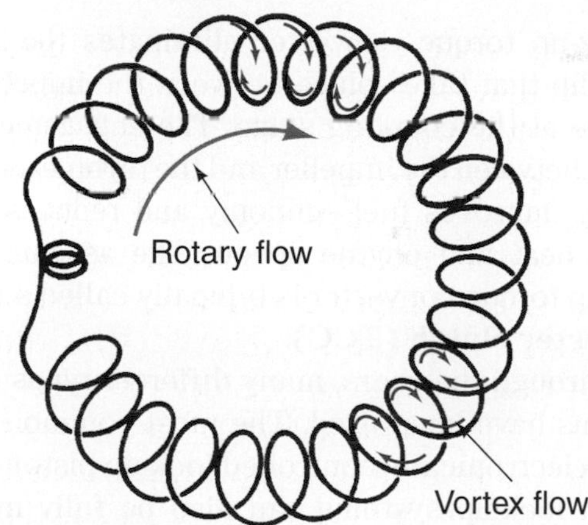

FIGURE 42–8

The difference between rotary and vortex flow. Note that vortex flow spirals its way around the converter.

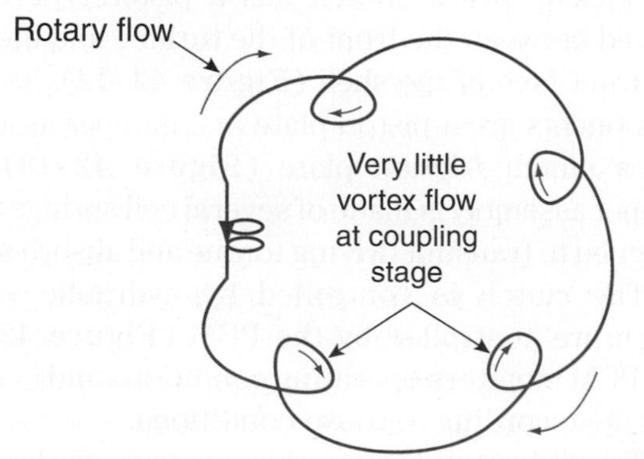

FIGURE 42–9

Rotary flow is at its greatest at the coupling stage.

The stator mounts through its splined centre hub to a mating stator shaft. The stator turns clockwise when the impeller and turbine reach the coupling stage.

The stator redirects the oil leaving the turbine back to the impeller, which helps the impeller rotate more efficiently **(Figure 42–10)**. Torque converter multiplication can only occur when the impeller is rotating faster than the turbine.

A stator is either a rotating or fixed type. Rotating stators are more efficient at higher speeds because there is less slippage when the impeller and turbine reach the coupling stage.

Overrunning Clutch

An **overrunning clutch** keeps the stator assembly from rotating when driven in one direction and permits overrunning (rotation) when turned in the opposite direction. Rotating stators generally use a

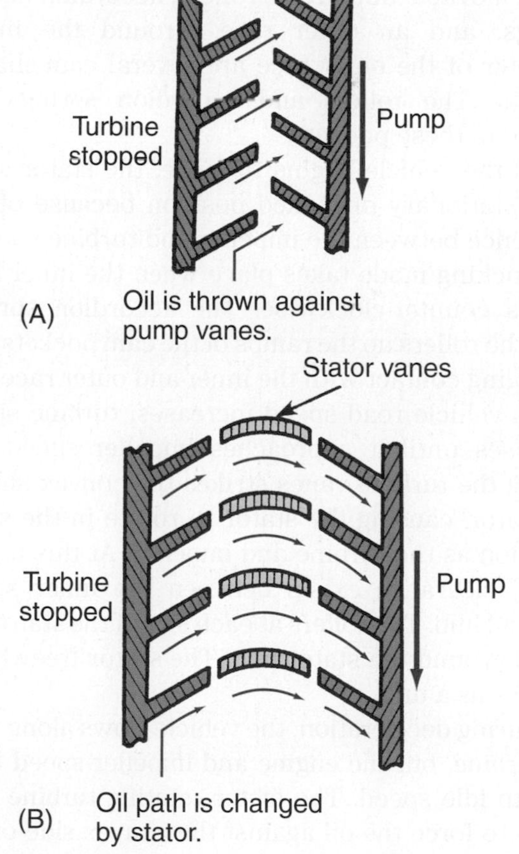

FIGURE 42–10

(A) Without a stator, fluid leaving the turbine works against the direction in which the impeller or pump is rotating. (B) With a stator in its locked (non-coupling) mode, fluid is directed to help push the impeller in its rotating direction.

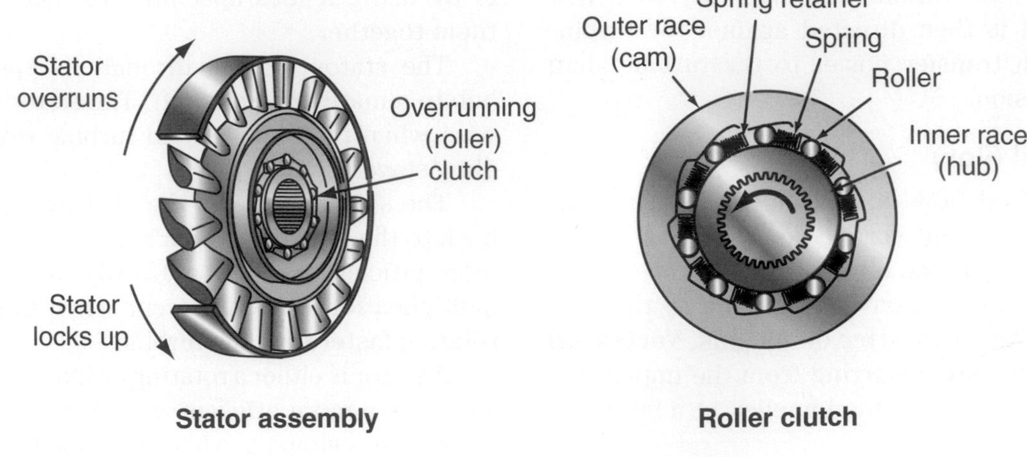

FIGURE 42–11

A typical roller-type overrunning clutch in a stator assembly.

roller-type overrunning clutch that allows the stator to freewheel (rotate) when the speed of the turbine and impeller reach the coupling point.

The roller clutch **(Figure 42–11)** is designed with a movable inner race, rollers, accordion (apply) springs, and an outer race. Around the inside diameter of the outer race are several cam-shaped pockets. The rollers and accordion springs are located in these pockets.

As the vehicle begins to move, the stator stays in its stationary or locked position because of the difference between the impeller and turbine speeds. This locking mode takes place when the inner race rotates counter-clockwise. The accordion springs force the rollers up the ramps of the cam pockets into a wedging contact with the inner and outer races.

As vehicle road speed increases, turbine speed increases until it approaches impeller speed. Oil exiting the turbine vanes strikes the convex side of the stator, causing the stator to rotate in the same direction as the turbine and impeller. At this higher speed, clearance exists between the inner stator race and hub. The rollers at each slot of the stator are pulled around the stator hub. The stator freewheels or turns as a unit.

During deceleration, the vehicle slows along with the turbine, but the engine and impeller speed drop back to idle speed. The faster rotating turbine continues to force the oil against the convex side of the stator vanes, allowing the stator to continue rotating.

As the vehicle continues to slow to a stop, the flow will again change direction back to the traditional vortex flow. The rollers will again jam between the inner race and hub, locking the stator in position. In this stationary position, the stator will again redirect the oil exiting the turbine so that torque is again multiplied.

LOCK-UP TORQUE CONVERTER

A lock-up torque converter eliminates the 10 percent slip that takes place between the impeller and turbine at the coupling stage. The engagement of a clutch between the impeller and the turbine assembly greatly improves fuel economy and reduces operational heat and engine speed. The assembly of a lock-up torque converter is typically called a **torque converter clutch (TCC)**.

Through the years, many different types of TCC systems have been used. The most common design is the electronically controlled lock-up piston clutch. Clutch lock-up systems can also be fully mechanical, centrifugally controlled, or dependent on a viscous coupling.

Lock-Up Piston Clutch

The lock-up piston clutch has a piston-type clutch located between the front of the turbine and the interior front face of the shell **(Figure 42–12)**. Its main components are a piston plate and damper assembly and a clutch friction plate **(Figure 42–13)**. The damper assembly is made of several coil springs and is designed to transmit driving torque and absorb shock.

The clutch is controlled by hydraulic valves, which are controlled by the PCM **(Figure 42–14)**. The PCM monitors operating conditions and controls lock-up according to those conditions.

To understand how this system works, consider this example. To provide for clutch control, Chrysler adds a three-valve module to its standard

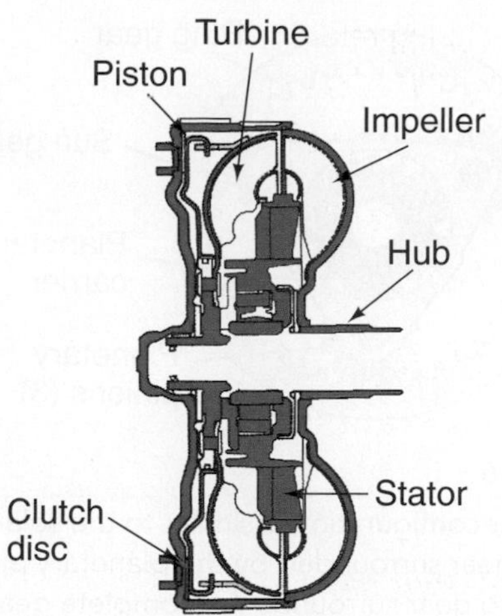

FIGURE 42–12

A piston-type converter lock-up clutch assembly.

Chrysler Group LLC

FIGURE 42–13

The friction surfaces of a torque converter clutch.

Martin Restoule

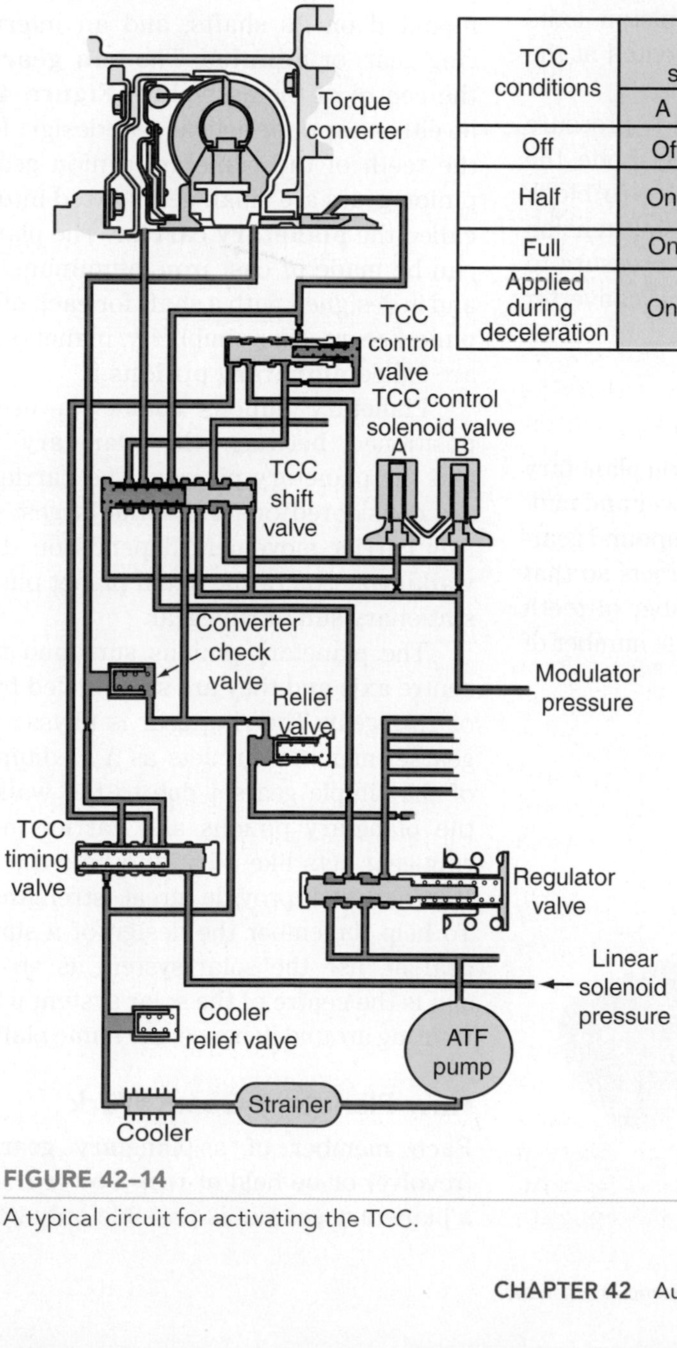

TCC conditions	TCC control solenoid valve		Linear solenoid pressure
	A	B	
Off	Off	Off	High
Half	On	Duty operation Off ↔ On	Low
Full	On	On	High
Applied during deceleration	On	Duty operation Off ↔ On	Low

FIGURE 42–14

A typical circuit for activating the TCC.

transmission valve body. The three valves are the lock-up valve, fail-safe valve, and switch valve. The lock-up valve actually controls the clutch. The fail-safe valve prevents lock-up until the transmission is in third gear. The switch valve directs fluid through the turbine shaft to fill the torque converter and changes its flow for lock-up operation.

When the converter is not locked, fluid enters the converter and moves to the front side of the piston, keeping it away from the shell or cover. Fluid flow continues around the piston to the rear side and exits between the neck of the torque converter and the stator support shaft.

During the lock-up mode, the switch valve moves and reverses the fluid path. This causes the fluid to move to the rear of the piston, pushing it forward to apply the clutch to the shell and allowing for lock-up. Fluid from the front side of the piston exits through the turbine shaft that is now vented at the switch valve.

During acceleration, system fluid pressure increases. If the converter is in its lock-up mode, the higher pressure moves the fail-safe valve to block fluid pressure to the lock-up valve. Spring tension moves the switch valve, directing fluid pressure to the front side of the piston. The torque converter then returns to its non–lock-up mode.

PLANETARY GEARS

Nearly all automatic transmissions rely on planetary gearsets **(Figure 42–15)** to transfer power and multiply engine torque to the drive axle. Compound gearsets combine two simple planetary gearsets so that load can be spread over a greater number of teeth for strength, and also to obtain the largest number of gear ratios possible in a compact area.

FIGURE 42–15

A single planetary gearset.

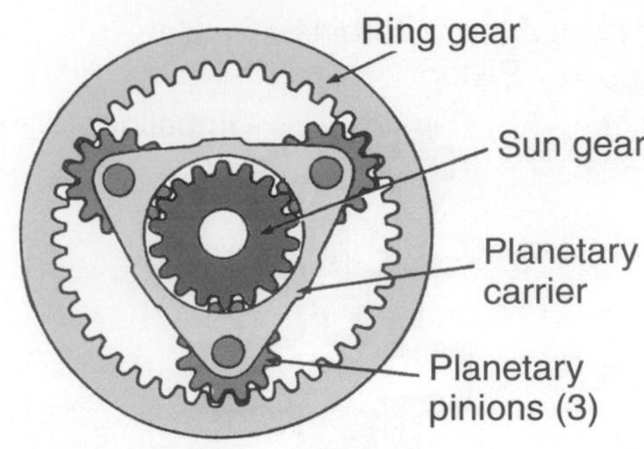

FIGURE 42–16

Planetary gear configuration is similar to the solar system, with the sun gear surrounded by the planetary pinion gears. The ring gear surrounds the complete gearset.

A simple planetary gearset consists of three parts: a sun gear; a carrier with planetary pinions mounted on its shafts; and an internally toothed ring gear, or **annulus**. The **sun gear** is located in the centre of the assembly **(Figure 42–16)**. It can be either a spur or helical gear design. It meshes with the teeth of the planetary pinion gears. Planetary pinion gears are small gears fitted into a framework called the **planetary carrier**. The planetary carrier can be made of cast iron, aluminum, or steel plate and is designed with a shaft for each of the planetary pinion gears. (For simplicity, planetary pinion gears are called **planetary pinions**.)

Planetary pinions rotate on needle bearings positioned between the planetary carrier shaft and the planetary pinions. The carrier and pinions are considered one unit—the largest gear member. Any carrier movement depends on the rotation or walking action of the small planet pinions around a stationary sun or ring gear.

The planetary pinions surround the sun gear's centre axis and they are surrounded by the annulus, or ring gear. The ring gear is physically the largest gearset member but acts as a medium-size member of the simple gearset due to the walking action of the planetary pinions and carrier movement. The ring gear acts like a band to hold the entire gearset together and provide great strength to the unit. To help remember the design of a simple planetary gearset, use the solar system as an example. The sun is the centre of the solar system with the planets rotating around it; hence, the name planetary gearset.

How Planetary Gears Work

Each member of a planetary gearset can spin (revolve) or be held at rest. Power transfer through a planetary gearset is possible only when one of the members is held at rest or if two of the members are locked together.

Any one of the three members can be used as the driving or input member. At the same time, another member might be kept from rotating and thus becomes the held or stationary member. The third member then becomes the driven, or output, member. Depending on which member is the driver, which is held, and which is driven, either a torque increase (underdrive) or a speed increase (overdrive) is produced by the planetary gearset. Output direction can also be reversed through various combinations.

Table 42–1 summarizes the basic laws of simple planetary gears. It indicates the resultant speed, torque, and direction of the various combinations available. Also, remember that when an external-to-external gear tooth set is in mesh, there is a change in the direction of rotation at the output. When an external gear tooth is in mesh with an internal gear, the output rotation for both gears is the same.

MAXIMUM FORWARD REDUCTION With the ring gear held and the sun gear (the input) turning clockwise, the sun gear rotates the planetary pinions counter-clockwise on their shafts. The small sun gear (driving) rotates several times, driving the large planetary carrier (the output) one complete revolution, resulting in the most gear reduction or the maximum torque multiplication that can be achieved in one planetary gearset. Input speed is high, but output speed is low.

MINIMUM FORWARD REDUCTION In this combination, the sun gear is held and the ring gear (input) rotates clockwise. The ring gear drives the planetary pinions clockwise, forcing them to walk around the stationary sun gear (held). The planetary pinions drive the planetary carrier (output) in the same direction as the ring gear—forward. This results in more than one turn of the input as compared to one complete revolution of the output. The result is torque multiplication. The planetary gearset is operating in a forward reduction with the mid-size ring gear driving the large planetary carrier. Therefore, the combination produces minimum forward reduction.

MAXIMUM OVERDRIVE With the ring gear held and the planetary carrier (input) rotating clockwise, the three planetary pinion shafts push against the inside diameter of the planetary pinions. The pinions are forced to walk around the inside of the ring gear, driving the sun gear (output) clockwise. In this combination, the large planetary carrier is rotating less than one turn and driving the smaller sun gear at a speed greater than the input speed. The result is overdrive with maximum speed increase.

MINIMUM OVERDRIVE In this combination, the sun gear is held and the carrier rotates (input) clockwise. As the carrier rotates, the pinion shafts push against the inside diameter of the pinions, and they are forced to walk around the held sun gear. This drives the ring gear (output) faster, and the speed increases. The carrier turning less than one turn causes the pinions to drive the ring gear one complete revolution in the same direction as the planetary carrier, and a minimum overdrive occurs.

SLOW REVERSE Here the sun gear (input) is driving the ring gear (output), with the planetary carrier held stationary. The planetary pinions, driven by the sun gear, rotate counter-clockwise on their shafts. While the sun gear is driving, the planetary pinions are used as idler gears to drive the ring gear counter-clockwise. This means the input and output shafts

TABLE 42–1 LAWS OF SIMPLE PLANETARY GEAR OPERATION

SUN GEAR	CARRIER	RING GEAR	SPEED	TORQUE	DIRECTION
1. Input	Output	Held	Maximum reduction	Increase	Same as input
2. Held	Output	Input	Minimum reduction	Increase	Same as input
3. Output	Input	Held	Maximum increase	Reduction	Same as input
4. Held	Input	Output	Minimum increase	Reduction	Same as input
5. Input	Held	Output	Reduction	Increase	Reverse of input
6. Output	Held	Input	Increase	Reduction	Reverse of input
7. When any two members are driven together, speed and direction are the same as input. Direct 1:1 drive occurs.					
8. When no member is held or locked together, output cannot occur. The result is a neutral condition.					

are operating in the opposite, or reverse, direction to provide a reverse power flow. Because the driving sun gear is small and the driven ring gear is mid-size, the result is slow reverse.

FAST REVERSE For fast reverse, the carrier is held, but the sun gear and ring gear reverse roles, with the ring gear (input) now being the driving member and the sun gear (output) driven. As the ring gear rotates counter-clockwise, the pinions rotate counter-clockwise as well, while the sun gear turns clockwise. In this combination, the input ring gear uses the planetary pinions to drive the output sun gear. The sun gear rotates in reverse to the input ring gear, providing fast reverse.

DIRECT DRIVE In the direct drive combination, both the ring gear and the sun gear are input members. They turn clockwise, turning at the same speed. The internal teeth of the clockwise-turning ring gear try to rotate the planetary pinions clockwise as well. But the sun gear, which rotates clockwise, tries to drive the planetary pinions counter-clockwise. These opposing forces lock the planetary pinions against rotation so that the entire planetary gearset rotates as one complete unit, providing direct drive. Whenever two members of the gearset are locked together, direct drive results.

NEUTRAL OPERATION When no member is held or locked, a neutral condition exists.

The following are helpful tips for remembering the basics of simple planetary gearset operation:

- When the planetary carrier is the drive (input) member, the gearset produces an overdrive condition. Speed increases, torque decreases.
- When the planetary carrier is the driven (output) member, the gearset produces a forward underdrive direction. Speed decreases, torque increases.
- When the planetary carrier is stationary (held), the gearset produces a reverse.

COMPOUND PLANETARY GEARSETS

A limited number of gear ratios are available from a single planetary gearset once it is inside the transmission. To increase the number of available gear ratios, additional gearsets are added. The typical automatic transmission with three or four forward speeds has at least two planetary gearsets.

In automatic transmissions, two or more planetary gearsets **(Figure 42–17)** are connected together to provide the various gear ratios needed to efficiently move a vehicle. There are two common designs of compound gearsets: the Simpson gearset, in which two planetary gearsets share a common sun gear; and the Ravigneaux gearset, which has two sun gears, two sets of planet gears, and a common ring gear. Many transmissions are fitted with additional single planetary gearsets connected in tandem to provide additional forward-speed ratios.

Simpson Gear Train

The **Simpson gear train** is an arrangement of two separate planetary gearsets with a common sun gear, two ring gears, and two planetary pinion carriers. A Simpson gear train is the most commonly used compound planetary gearset and is used to provide three forward gears. One-half of the compound set, or one planetary unit, is referred to as the front planetary, and the other planetary unit is the rear planetary **(Figure 42–18)**. The two planetary units do not need to be the same size or have the same number of teeth on their gears. The size and number of gear teeth determine the actual gear ratios obtained by the compound planetary gear assembly.

Gear ratios and direction of rotation are the result of applying torque to one member of either planetary unit, holding at least one member of the gearset, and using another member as the output. For the most part, each automobile manufacturer uses different parts of the planetary assemblies as inputs, outputs, and reaction members. The role of the planetary members also varies with the different transmission models from the same manufacturer. There are also many different apply devices used in the various transmission designs.

A Simpson gearset can provide the following gear ranges: neutral, first reduction gear, second reduction gear, direct drive, and reverse. The typical power flow through a Simpson gear train when it is in neutral has engine torque being delivered to the transmission's input shaft by the torque converter's turbine. No planetary gearset member is locked to that shaft; therefore, engine torque enters the transmission but goes nowhere else.

When the transmission is shifted into first gear **(Figure 42–19)**, engine torque is again delivered into the transmission by the input shaft. The input shaft is now locked to the front planetary ring gear that turns clockwise with the shaft. The front ring gear drives the front planetary gears, also in a clockwise direction. The front planetary gears drive the sun gear in a counter-clockwise direction. The rear

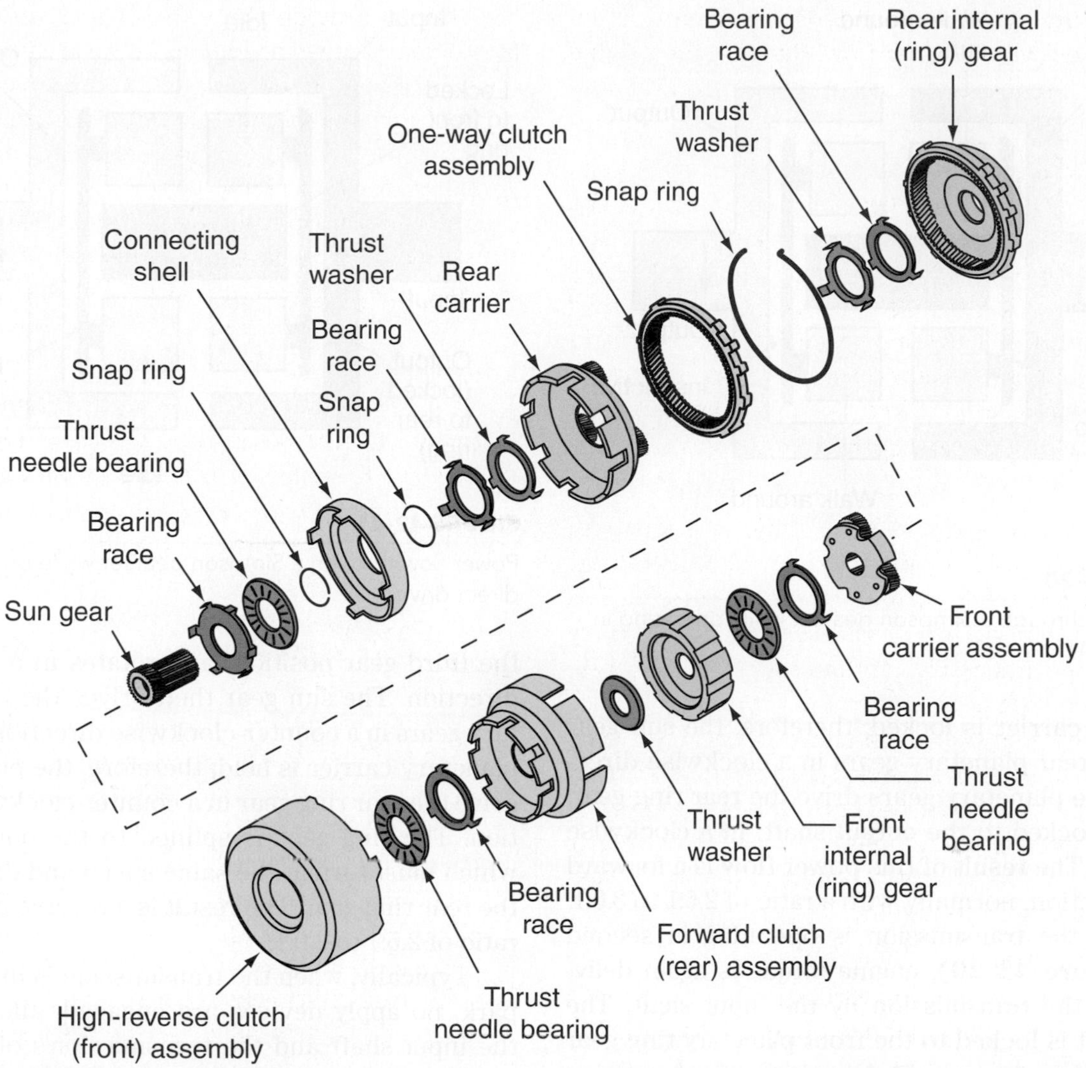

FIGURE 42–17

A Simpson planetary gearset.

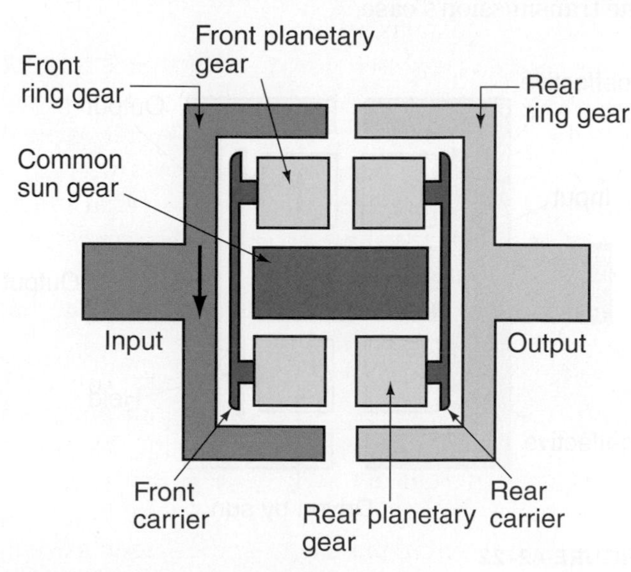

FIGURE 42–18

Components of a Simpson gearset.

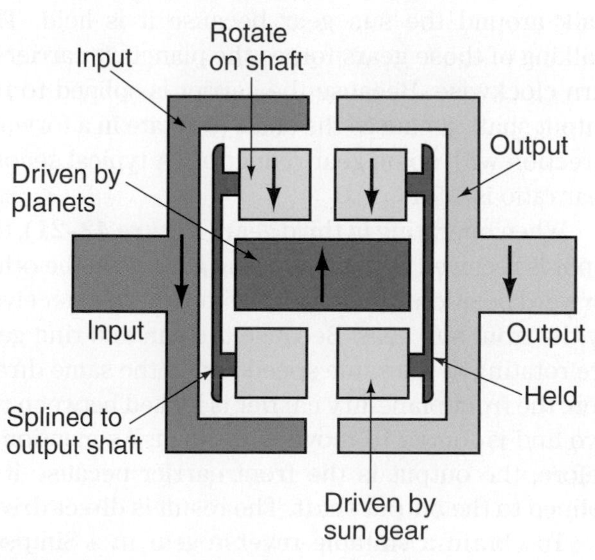

FIGURE 42–19

Power flow through a Simpson gearset while operating in first gear.

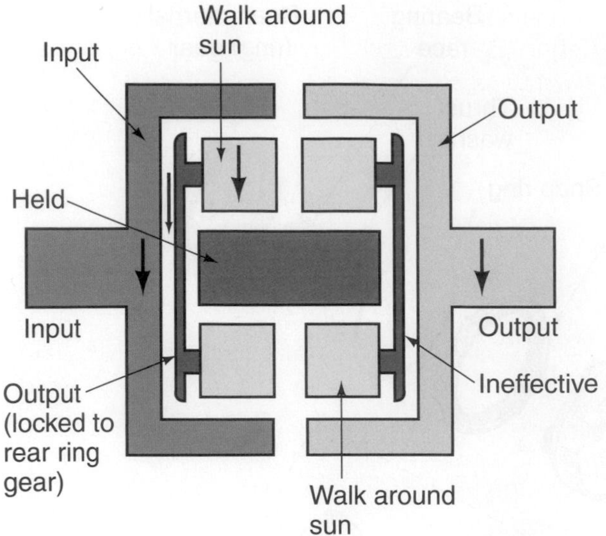

FIGURE 42–20

Power flow through a Simpson gearset while operating in second gear.

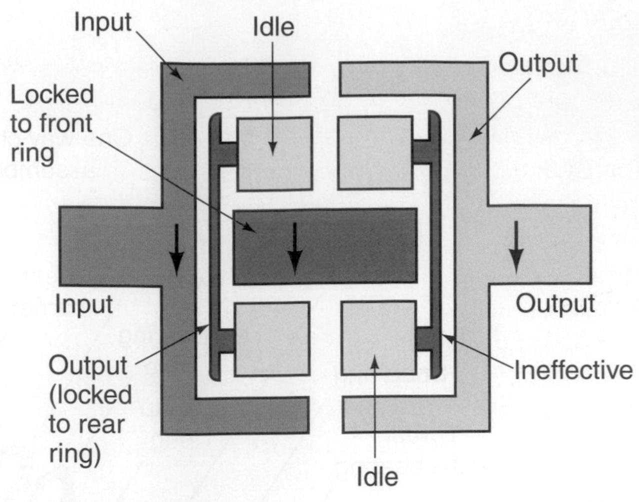

FIGURE 42–21

Power flow through a Simpson gearset while operating in direct drive.

planetary carrier is locked; therefore, the sun gear spins the rear planetary gears in a clockwise direction. These planetary gears drive the rear ring gear, which is locked to the output shaft, in a clockwise direction. The result of this power flow is a forward gear reduction, normally with a ratio of 2.5:1 to 3.0:1.

When the transmission is operating in second gear **(Figure 42–20)**, engine torque is again delivered into the transmission by the input shaft. The input shaft is locked to the front planetary ring gear that turns clockwise with the shaft. The front ring gear drives the front planetary pinion gears, also in a clockwise direction. The front planetary pinion gears walk around the sun gear because it is held. The walking of those gears forces the planetary carrier to turn clockwise. Because the carrier is splined to the output shaft, it causes the shaft to rotate in a forward direction with some gear reduction. A typical second gear ratio is 1.5:1.

When operating in third gear **(Figure 42–21)**, the input is received by the front ring gear, as in the other forward positions. However, the input is also received by the front sun gear. Because the sun and ring gear are rotating at the same speed and in the same direction, the front planetary carrier is locked between the two and is forced to move with them. As mentioned before, the output is the front carrier because it is splined to the output shaft. The result is direct drive.

To obtain a suitable reverse gear in a Simpson gear train, there must be a gear reduction, but in the opposite direction as the input torque **(Figure 42–22)**. The input is received by the sun gear, as in the third gear position, and rotates in a clockwise direction. The sun gear then drives the rear planetary gears in a counter-clockwise direction. The rear planetary carrier is held; therefore, the planet gears drive the rear ring gear in a counter-clockwise direction. The ring gear is splined to the output shaft, which must turn at the same speed and direction as the rear ring gear. The result is a reverse gear with a ratio of 2.5:1 to 2.0:1.

Typically, when the transmission is in neutral or park, no apply devices are engaged, allowing only the input shaft and the transmission's oil pump to turn with the engine. In park, a pawl is mechanically engaged to a parking gear that is splined to the transmission's output shaft, locking the drive wheels to the transmission's case.

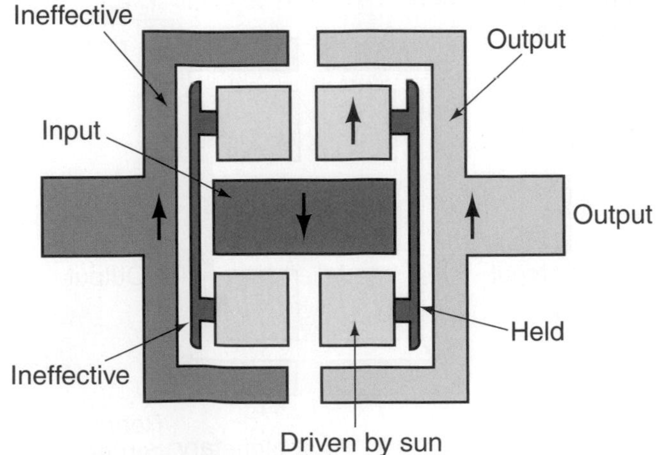

FIGURE 42–22

Power flow through a Simpson gearset while operating in reverse.

Ravigneaux Gear Train

The **Ravigneaux gear train**, like the Simpson gear train, provides forward gears with a reduction, direct drive, overdrive, and a reverse operating range. The Ravigneaux offers some advantages over a Simpson gear train. It is very compact. It can carry large amounts of torque because of the great amount of tooth contact.

The Ravigneaux gear train is designed to use two sun gears, one small and one large **(Figure 42–23)**. This type of gear train also has two sets of planetary pinion gears: three long pinions and three short pinions. The planetary pinion gears rotate on their own shafts that are fastened to a common planetary carrier. A single ring gear surrounds the complete assembly.

The small sun gear is meshed with the short planetary pinion gears. These short pinions act as idler gears to drive the long planetary pinion gears. The long planetary pinion gears mesh with the large sun gear and the ring gear.

Typically, when the gear selector is in neutral position, engine torque, through the converter turbine shaft, drives the forward clutch drum. Because the forward clutch is not applied, the power is not transmitted through the gear train, and there is no power output.

When the transmission is operating in first gear **(Figure 42–24)**, engine torque drives the forward clutch drum. The forward clutch is applied and drives the forward sun gear clockwise. The planetary carrier is prevented from rotating counter-clockwise by the one-way clutch; therefore, the forward sun gear drives the short planetary gears counter-clockwise. The direction of rotation is reversed as the short planetary gears drive the long planetary gears, which drive the ring gear and output shaft in a clockwise direction but at a lower speed than the input. This results in a gear reduction and a ratio of approximately 2.5:1.

In second gear **(Figure 42–25)** operation, the intermediate clutch is applied and locks the outer race of the one-way clutch. This prevents the reverse clutch drum, shell, and reverse sun gear from turning counter-clockwise. The forward clutch is also applied and locks the input to the forward sun gear, and it rotates in a clockwise direction. The forward sun gear drives the short planetary gears counter-clockwise. The direction of rotation is reversed as the short planetary gears drive the long planetary gears, which walk around the stationary reverse sun gear. This walking drives the ring gear and output shaft in a clockwise direction and at a reduction with a gear ratio of approximately 1.5:1.

During third gear **(Figure 42–26)** operation, there are two inputs into the planetary gear train. As in other forward gears, the turbine shaft of the

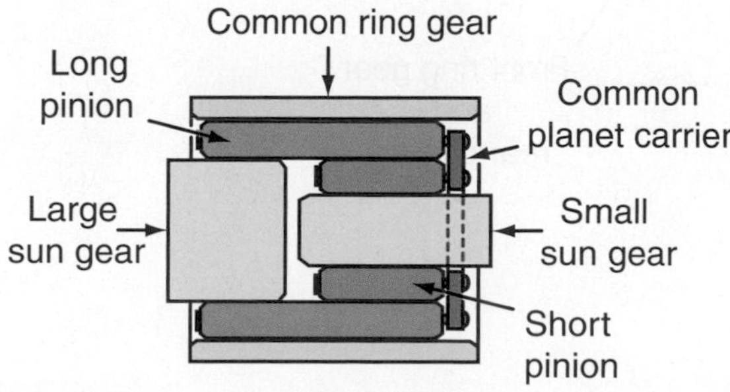

FIGURE 42–23
The parts of a Ravigneaux gearset.

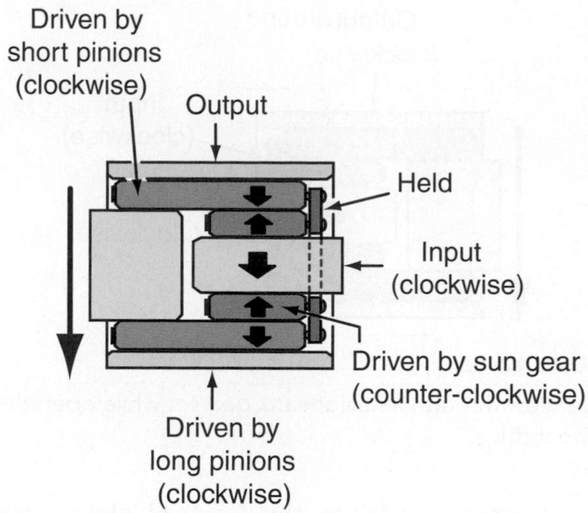

FIGURE 42–24
Power flow through a Ravigneaux gearset while operating in first gear.

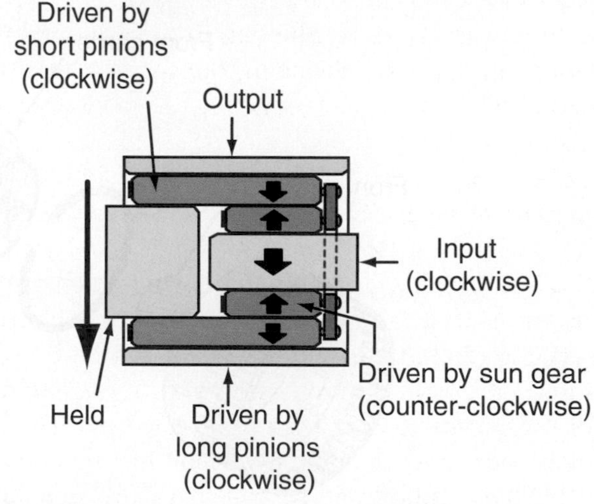

FIGURE 42–25
Power flow through a Ravigneaux gearset while operating in second gear.

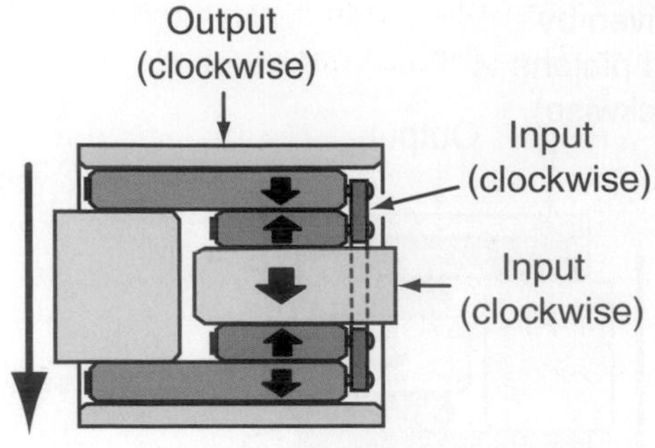

FIGURE 42–26

Power flow through a Ravigneaux gearset while operating in direct drive.

torque converter drives the forward clutch drum. The forward clutch is applied and drives the forward sun gear in a clockwise direction. Input is also received by the direct clutch that is driven by the torque converter cover. The direct clutch is applied and drives the planetary carrier. Because two members of the gear train are being driven at the same time, the planetary carrier and the forward sun gear rotate as a unit. The long planetary gears transfer the torque, in a clockwise direction, through the gearset to the ring gear and output shaft. This results in direct drive.

To operate in overdrive or fourth gear, input is received only by the direct clutch from the torque converter cover. The direct clutch is applied and drives the planetary carrier in a clockwise direction. The long planetary gears walk around the stationary reverse sun gear in a clockwise direction and drive the ring gear and output shaft. This results in an overdrive condition with an approximate ratio of 0.75:1.

During reverse gear operation, input is received through the torque converter's turbine shaft to the reverse clutch. The reverse clutch, when applied, connects the turbine shaft to the reverse sun gear. The low/reverse band is applied and holds the planetary gear carrier. The clockwise rotation of the reverse sun gear drives the long planetary gears in a counter-clockwise direction. The long planetary gears then drive the ring gear and output shaft in a counter-clockwise direction with a speed reduction and a gear ratio of approximately 2.0:1.

Planetary Gearsets in Tandem

Rather than relying on the use of a compound gearset, some automatic transmissions use two simple planetary units in series **(Figure 42–27)**. In this type of arrangement, gearset members are not shared; instead, the holding devices are used to lock different members of the planetary units together.

Although the gear train is based on two simple planetary gearsets operating in tandem, the

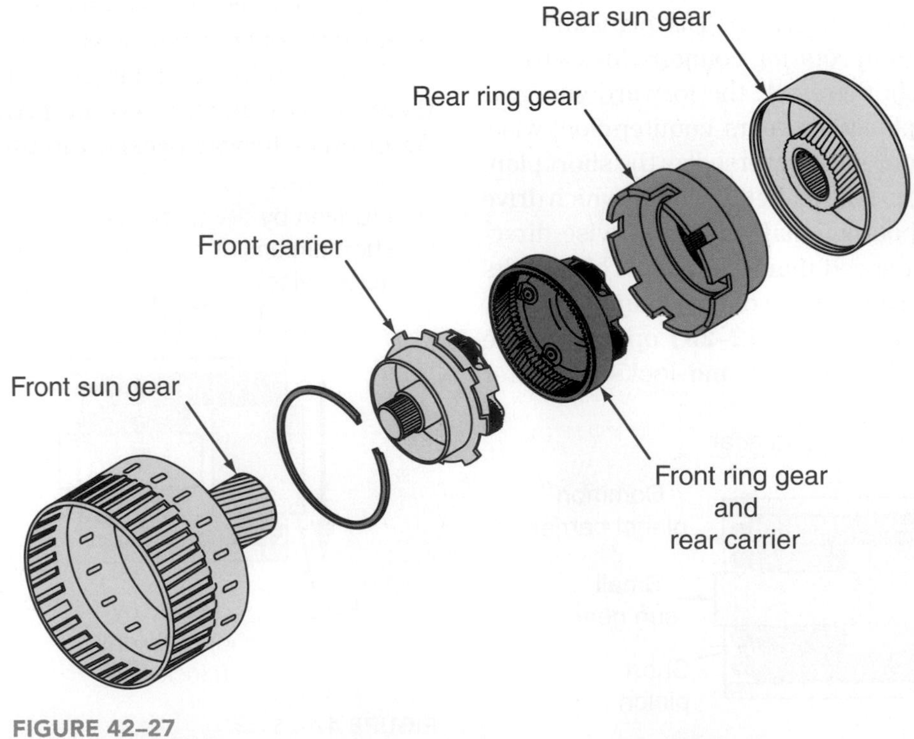

FIGURE 42–27

Two planetary units with the ring gear of one gearset connected to the planet carrier of the other.

combination of the two planetary units does function much like a compound unit. The two tandem units do not share a common member; rather, certain members are locked together or are integral with each other. The front planetary carrier is locked to the rear ring gear, and the front ring gear is locked to the rear planetary carrier. A transaxle may have additional planetary gearsets, including one that is used as the final drive gearset.

Lepelletier System

Some late-model six-, seven-, and eight-speed transmissions use the **Lepelletier system**. The Lepelletier system connects a simple planetary gearset to a Ravigneaux gearset. This design has been around for many years but has been difficult to control. Today's electronic technologies have made it practical. With this design, transmissions can be made with additional forward speeds without an increase in size and weight. In fact, most six-speed transmissions are more compact and are lighter than nearly all four- or five-speed transmissions.

In this arrangement, the ring gear of the simple gearset serves as the main input to the gearsets, and the input can also be connected to the carrier of the Ravigneaux gearset at the same time **(Figure 42–28)**. As engine torque passes through different input gears, it drives a variety of combinations of gears in the simple and the Ravigneaux gearsets. These combinations result in the various forward-speed ratios. The ring gear of the Ravigneaux gearset is the output member for the transmission.

In some models, the input shaft is always connected to the ring of the simple planetary gear and, in addition, can be connected to the carrier and large sun gear of the Ravigneaux gear. This allows for additional gear combinations. The ring gear of the Ravigneaux gearset still serves as the output.

POWER FLOW The power flow through a Lepelletier gearset is based on splitting the input into two ratios: the ratio of the single set and the ratio of the second. The overall ratio is the combination of the two. Here is an example of the power flow for a typical six-speed unit.

- *Drive—first gear.* For first gear, the rear carrier is held. This transfers torque from the sun gear in the rear planetary assembly to the ring gear **(Figure 42–29)**. The transmission is now in first gear.
- *Drive—second gear.* When shifting into second gear, the large sun gear of the rear planet is held. The sun gear now transfers torque through the short and long planetary pinions to the ring gear, which is the output.
- *Drive—third gear.* Third gear is achieved by locking the input shaft and both of the sun gears in the compound gearset together. This forces both of the rear planetary assemblies to lock and drive the output ring gear.
- *Drive—fourth gear.* For fourth gear, the input is transferred from the single gearset's carrier to the large sun gear and the planetary carriers in the compound gear set **(Figure 42–30)**. The ring gear is then driven with a slight reduction.
- *Drive—fifth gear.* In fifth gear, the input moves from the single planetary unit's carrier to the small sun gear and the planetary carrier in the compound gearset. The ring gear is then driven at an overdrive.
- *Drive—sixth gear.* Sixth speed is made by holding the front sun gear in the compound

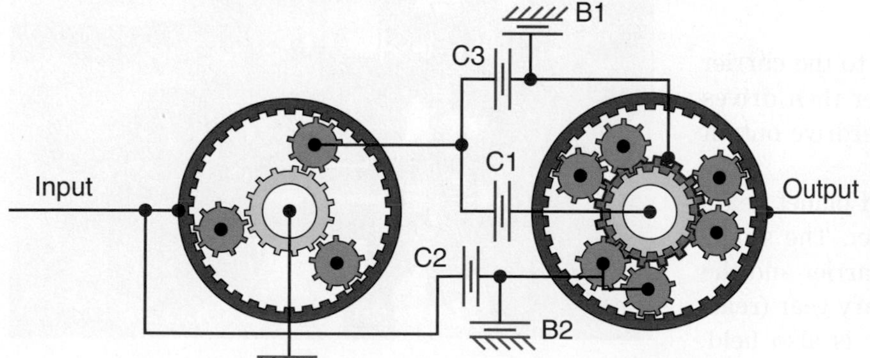

Range	C1	C2	C3	B1	B2
1	X				
Manual 1	X				X
2	X			X	
3	X		X		
4	X	X			
5		X	X		
6		X		X	
Reverse			X		X

FIGURE 42–28

A six-speed Lepelletier transmission based on a planetary gear and a Ravigneaux gearset.

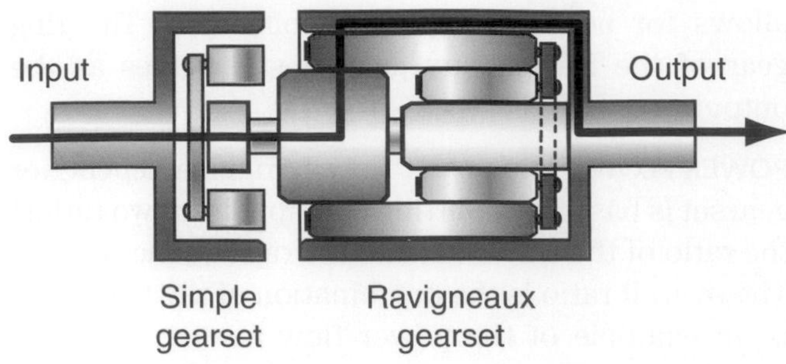

FIGURE 42–29

Power flow in first gear for a typical Lepelletier gearset.

©Cengage Learning 2015

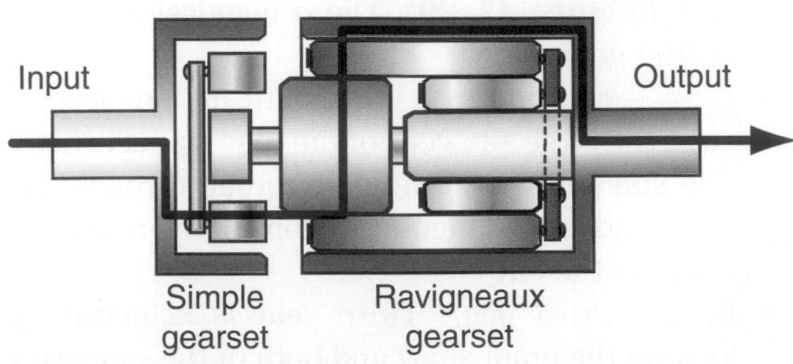

FIGURE 42–30

Power flow in fourth gear for a typical Lepelletier gearset.

©Cengage Learning 2015

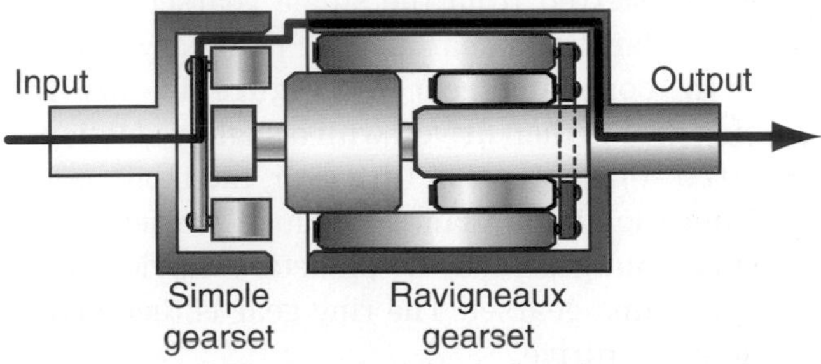

FIGURE 42–31

Power flow in sixth gear for a typical Lepelletier gearset.

©Cengage Learning 2015

gearset. The input shaft is locked to the carrier of the compound gear. The carrier then drives the ring gear to provide for an overdrive output **(Figure 42–31)**.

- *Reverse.* For reverse, the ring and planet gears of simple planetary rotate together. The single sun gear is held and rotates the carrier and the sun gear of the compound planetary gear (rear) assembly. The rear planet carrier is also held. The long planets in the compound gear rotate with the rear sun gear. This causes the ring gear to rotate backward and drives the output shaft in reverse.

HONDA'S NONPLANETARY-BASED TRANSMISSION

The Honda nonplanetary-based transaxles are used in many Honda and Acura cars. Certain Saturn automatic transaxles are also a similar design. These transaxles are unique in that they use constant-mesh helical and square-cut gears **(Figure 42–32)** in a manner similar to that of a manual transmission.

These transaxles have a mainshaft and countershaft on which the gears ride. To provide the forward gear ranges, different pairs of gears are locked to the shafts by hydraulically controlled clutches **(Figure 42–33)**. Reverse gear is obtained through the use of a shift fork that slides the reverse idler gear into position.

Power flow through these transaxles is similar to that through a manual transaxle. The action of the clutches is much the same as the action of the synchronizer assemblies in a manual transaxle. Older four-speed Hondas used four multiple-disc clutches, a sliding reverse idler gear, and a one-way clutch to control the gears.

CONTINUOUSLY VARIABLE TRANSMISSIONS (CVT)

Another unconventional transmission design, the **continuously variable transmission (CVT)**, is a transmission with no fixed forward speeds. The gear ratio varies with engine speed and temperature. These transmissions are, however, fitted with a one-speed reverse gear. Some CVT transaxles do not have a torque converter; rather, they use a manual transmission–type flywheel with a start clutch. The

FIGURE 42–32

Honda automatic transaxles use constant-mesh helical gears instead of planetary gearsets.

Martin Restoule

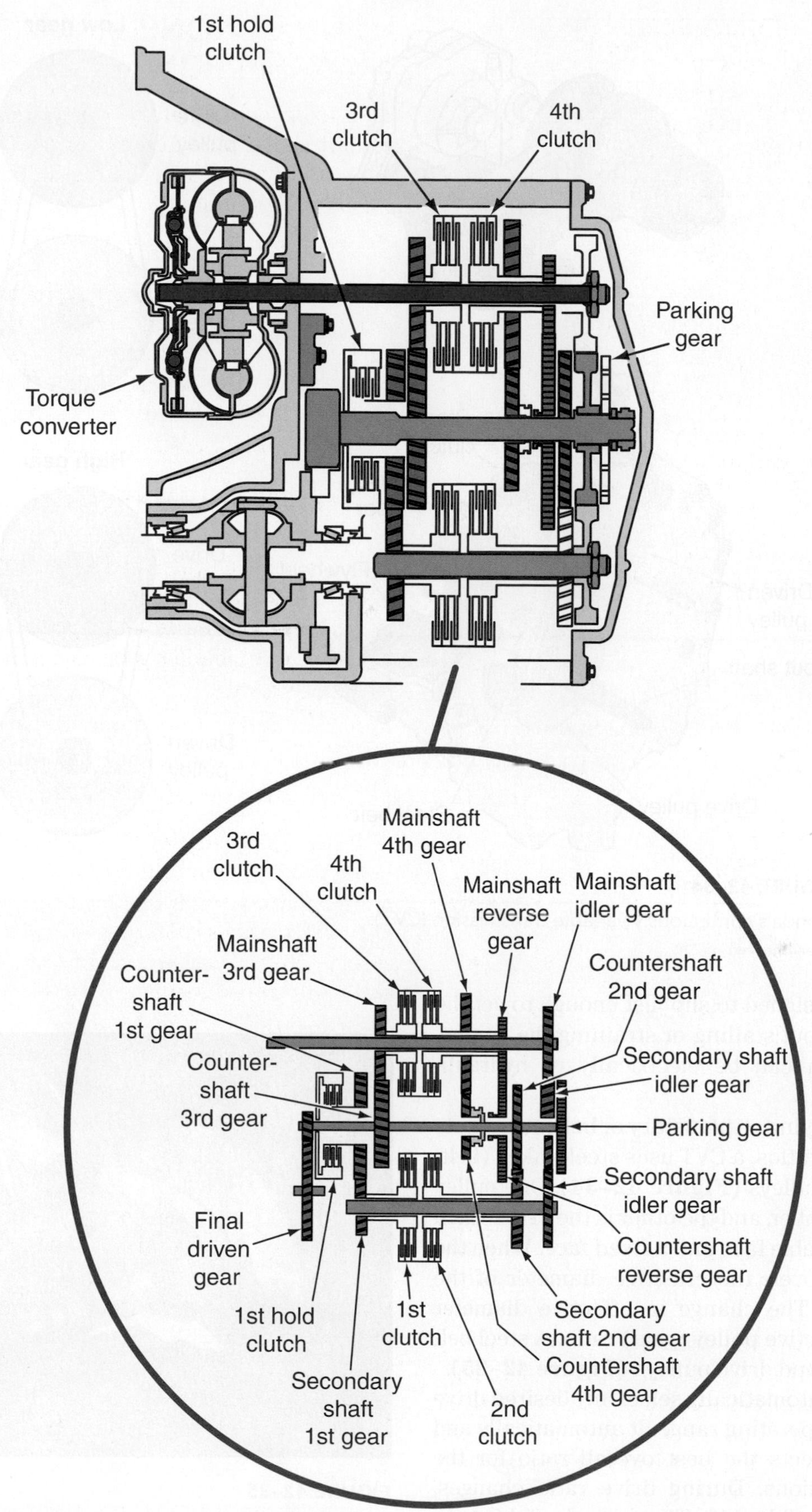

FIGURE 42–33

Arrangement of gears and reaction devices in a typical nonplanetary gearset transaxle.

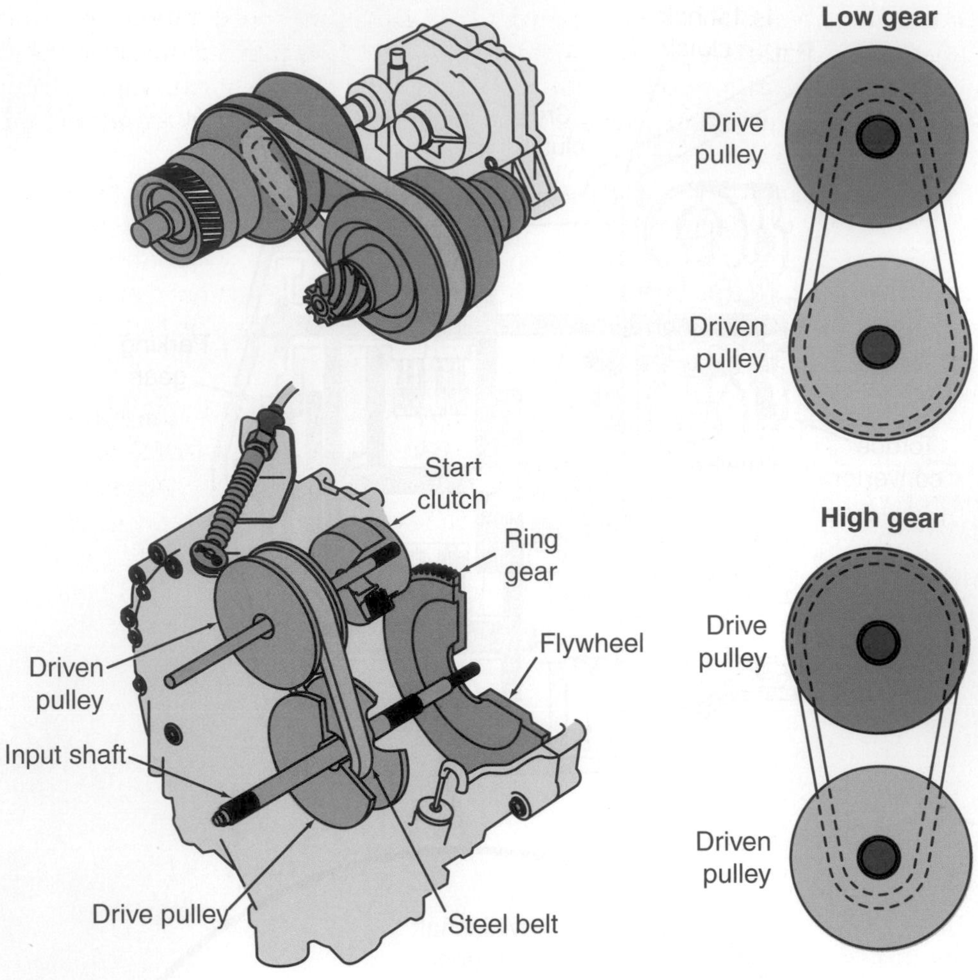

FIGURE 42–34

Honda's continuously variable transmission (CVT).

©Cengage Learning 2015

start clutch is designed to slip just enough to get the car moving without stalling or straining the engine. The start clutch can be electrically or hydraulically controlled.

Instead of relying on planetary or helical gearsets to provide drive ratios, a CVT uses steel-linked (belt-like) chain and pulleys **(Figure 42–34)**. One pulley is the driven member, and the other is the drive. Each pulley has a movable face and a fixed face. When the movable face moves, the effective diameter of the pulley changes. The change in effective diameter changes the effective pulley (gear) ratio. A steel belt links the driven and drive pulleys **(Figure 42–35)**.

A CVT can automatically select any desired drive ratio within its operating range. It automatically and continuously selects the best overall ratio for the operating conditions. During drive ratio changes, there is no perceptible shift. The controls of this type of transmission attempt to keep the engine operating at its most efficient speed. This decreases fuel

FIGURE 42–35

CVTs use pulleys that change size and are connected by a steel belt.

Nissan Canada

consumption and exhaust emissions. During maximum acceleration, the drive ratio is adjusted to maintain peak engine horsepower. At a constant vehicle speed, the drive ratio is set to obtain maximum fuel mileage while maintaining good driveability.

To achieve a low pulley ratio, high hydraulic pressure works on the movable face of the driven pulley to make it larger. In response to this high pressure, the pressure on the drive pulley is reduced. Because the steel belt (chain) links the two pulleys and proper belt tension is critical, the drive pulley reduces just enough to keep the proper tension on the steel belt (chain). The increase of pressure at the driven pulley is proportional to the decrease of pressure at the drive pulley. The opposite is true for high pulley ratios. Low pressure causes the driven pulley to decrease in size, whereas high pressure increases the size of the drive pulley.

Different speed ratios are available any time the vehicle is moving. Because the size of the drive and driven pulleys can vary greatly, vehicle loads and speeds can be changed without changing the engine's speed.

Many late-model CVTs are equipped with a feature that simulates the activity of a manual shifting automatic transmission. These transmissions have five or six predetermined areas that the pulleys stop in, thereby giving the feel and shift effect of distinct shifts.

CVT CONTROLS The control system for a typical CVT consists of a TCM, various sensors, linear solenoids, and an inhibitor solenoid. Input from the various sensors determines which drive ratio will command **(Figure 42–36)**. Activating the shift control solenoid changes the shift control valve pressure,

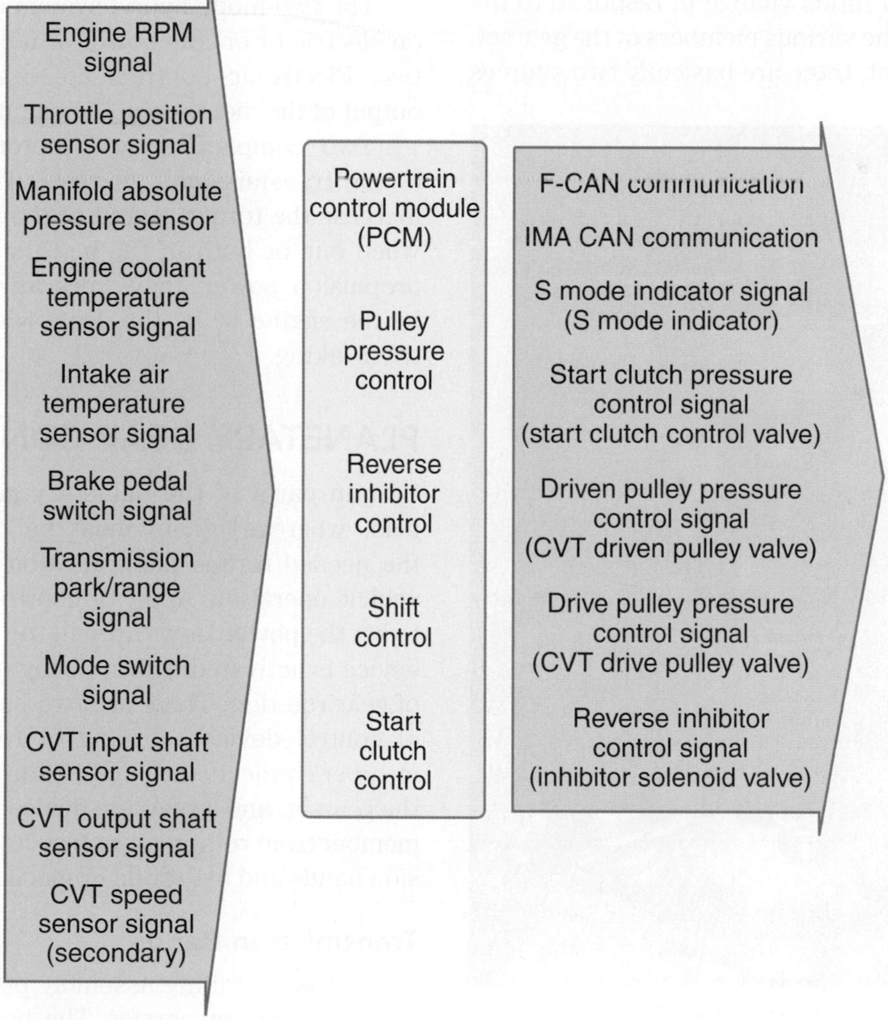

FIGURE 42–36

The input, processing, and control systems for the electronic control of a Honda CVT.

causing the shift valve to move. This changes the pressures applied to the driven and drive pulleys, which changes the effective pulley ratio.

Planetary Gear–Based CVTs

Hybrid electric vehicles from Toyota and Ford rely on a planetary gearset to provide for a CVT. The transaxle contains two electric motor/generators, a differential, and a simple planetary gearset. The engine and the motor/generators are connected directly to the planetary gear unit **(Figure 42–37)**. The planetary gearset is called the power split device because it can transfer power between the engine, motor/generators, drive wheels, and nearly any combination of these. The power split device splits power from the engine to different paths: to drive one of the motor/generators or to drive the car's wheels, or both. The other motor/generator can drive the wheels, assist the engine in driving the wheels, or be driven by the wheels. The speed ratios change in response to the torque applied to the various members of the gearset. In this arrangement, there are basically two sources of torque: the engine and an electric traction motor. Both rotate in the same direction, but not at the same speed. Therefore, one can assist the rotation of the other or slow down the rotation of the other, or they can work together.

FIGURE 42–37

In this hybrid CVT, the planetary gearset is located between the two electric motors.

Two-Mode Hybrid System

GM, BMW, and Chrysler together developed a two-mode full-hybrid system. The **two-mode full hybrid system** is another planetary gear–based CVT. The system fits into a standard transmission housing and is basically two planetary gearsets coupled to two electric motors, which are electronically controlled. This combination results in a CVT with motor/generators for hybrid operation. The system has two distinct modes of operation. It operates in the first mode during low speed and low load conditions and in the second mode while cruising at highway speeds.

The two-mode hybrid system can operate solely on electric or engine power or a combination of the two. Electronic controls are used to control the output of the motors and the engine.

Two compact electric motors are connected to the transmission's gearsets. The gears work to increase the torque output of the motors. Typically, when one or both of the motors are not providing propulsion power, they work as generators driven by the engine or by the drive wheels for regenerative braking.

PLANETARY GEAR CONTROLS

Certain parts of the planetary gear train must be held, whereas others must be driven to provide the needed torque multiplication and direction for vehicle operation. Apply or control devices are used to set the power flow through the gearset. A control device is activated hydraulically or by the direction of gear rotation. There are two basic classifications of control devices: driving or reaction. A driving member connects the engine's torque to a member of the gearset, and a reaction device prevents a gearset member from rotating. Control devices are transmission bands and hydraulic or mechanical clutches.

Transmission Bands

A **band** is a braking assembly positioned around a rotating drum or carrier. The band brings a drum to a stop by wrapping itself around the drum and holding it. The band is hydraulically applied by a servo assembly. Connected to the drum is a member of the planetary gear train. The purpose of a band is to hold a member of the planetary gearset by holding the drum and connecting planetary gear member stationary. Bands provide excellent holding characteristics and require a minimum amount of space within the transmission housing.

When a band closes around a rotating drum, a wedging action takes place to stop the drum from rotating. The wedging motion is known as self-energizing action. A typical band is designed to be larger in diameter than the drum it surrounds. This design promotes self-disengagement of the band from the drum when servo apply force is decreased to less than servo release spring tension. A friction material is bonded to the inside diameter of the band.

Typically, if the band will be holding a low-speed drum, the lining material of a band is a semimetallic compound. If the band is designed to hold a high-speed drum, it will have a paper-based lining.

Band lugs are either spot welded or cast as a part of the band assembly. The purpose of the lugs is to connect the band with the servo through the actuating (apply) linkage and the band anchor (reaction) at the opposite end. The band's steel strap is designed with slots or holes to release fluid trapped between the drum and the applying band.

SHOP TALK

A holding planetary control unit is also called a brake or reaction unit because it holds a gear train member stationary, reacting to rotation.

The bands used in automatic transmissions are rigid, flexible, single-wrap, or double-wrap types. Steel single-wrap bands **(Figure 42–38A)** are used to hold gear train components driven by high-output engines. Self-energizing action is low because of the rigidity of the band's design. Thinner steel bands are not able to provide a high degree of holding power, but because of the flexibility of design, self-energizing action is stronger and provides more apply force.

The double-wrap band is a circular external contracting band normally designed with two or three segments **(Figure 42–38B)**. As the band closes, the segments align themselves around the drum and provide a cushion. The steel body of the double-wrap band may be thin or thick steel strapping material. Modern automatic transmissions use thin single- or

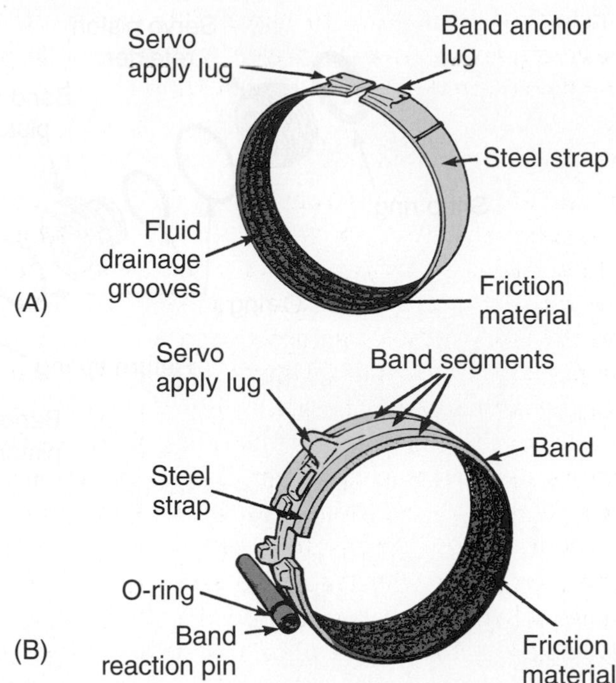

FIGURE 42–38

(A) Typical single-wrap and (B) double-wrap transmission band designs.

Chrysler Group LLC

double-wrap bands for increased efficiency. Double-wrap bands made with heavy thick steel strapping are required for high output engines.

Transmission Servos

The **servo** assembly converts hydraulic pressure into a mechanical force that applies a band to hold a drum stationary. Simple and compound servos are used to engage bands in modern transmissions.

SIMPLE SERVO In a simple servo **(Figure 42–39)**, the servo piston fits into the servo cylinder and is held in the released position by a coil spring. The piston is sealed with a rubber ring, which keeps fluid pressure confined to the apply side of the servo piston.

In Figure 42–39 (but not on all servo designs), the piston pushrod is drilled through the centre, which permits fluid pressure to be directed to the apply side of the servo piston. The pushrod moves the band apply strut, which is seated in the band apply lug, to tighten the band. At the opposite end of the band is the anchor strut and adjustment screw. They receive the engagement force of a band.

To apply a band, fluid pressure is directed down the servo pushrod to the apply side of the servo piston. The servo piston moves against the servo coil spring and develops servo apply force. This force is applied to the band lug through the apply lever and

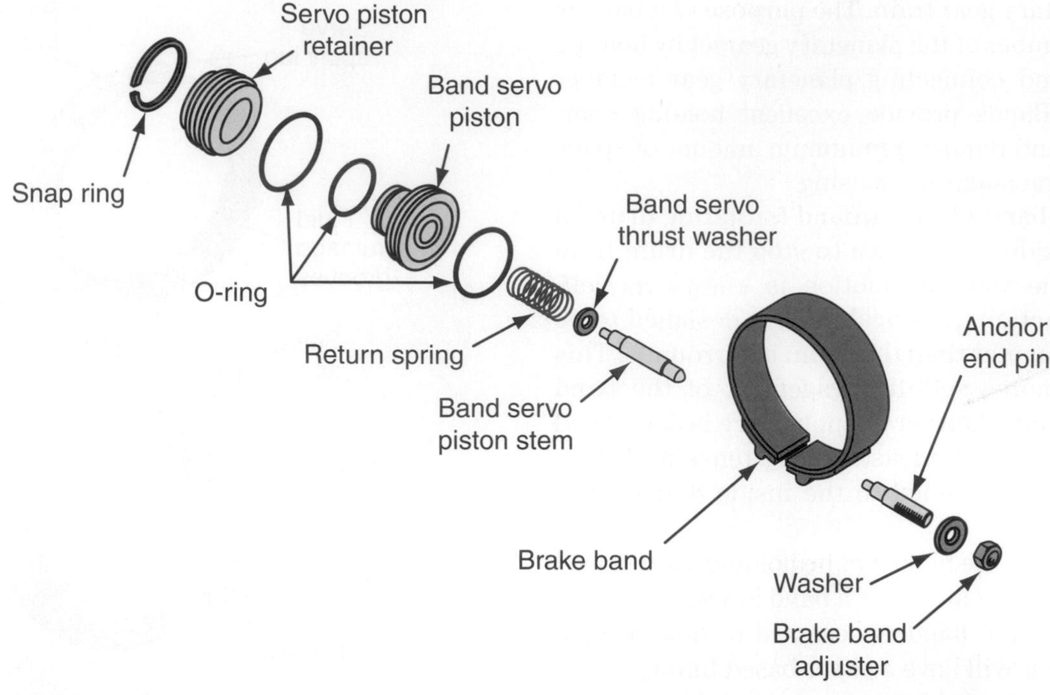

FIGURE 42–39

A typical band and servo assembly.

strut. The band tightens around the rotating drum. The rotating drum comes to a stop and is held stationary by the band.

When servo apply force is released, the servo coil spring forces the servo piston to move up the cylinder. With the servo apply force removed, the band springs free and permits drum rotation.

COMPOUND SERVO A compound servo **(Figure 42–40)** has a cylinder that is cast as part of the transmission housing. If the servo is located near the front of the transmission, it uses seal rings capable of withstanding the heat generated by the torque converter and engine.

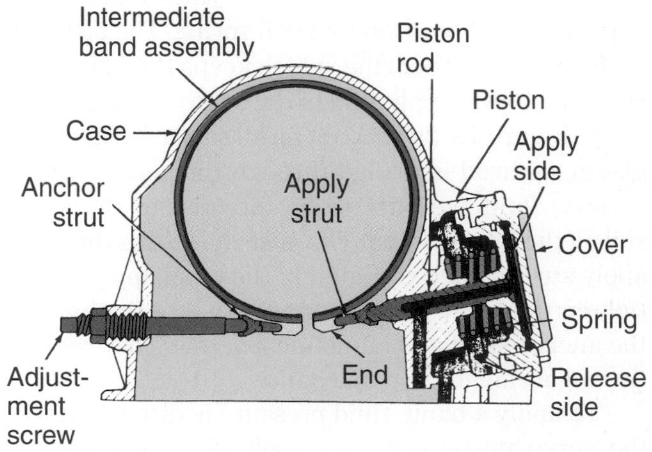

FIGURE 42–40

A typical compound servo design.

Ford Motor Company

When the compound servo is applied, fluid pressure flows through the hollow piston pushrod to the apply side of the servo piston. The piston compresses the servo coil spring and forces the pushrod to move one end of the band toward the adjusting screw and anchor. The band tightens around the rotating drum and brings it to a stop. The apply of the compound servo piston is much like the simple servo, but there the similarity ends.

The release side of the compound servo piston usually has a larger diameter, which means a greater surface area. Fluid pressure is applied to the release side of the servo piston when the band is to be released. With the greater release side surface area, the release force is greater than the apply force. This action, along with the servo release spring force, creates a more positive or faster release of the band than the simple servo, which relies on only spring force. This design is used to ensure that a band is released before another reaction member is applied.

TRANSMISSION CLUTCHES

In contrast to a band, which can only hold a planetary gear member, transmission clutches are capable of both holding and driving members.

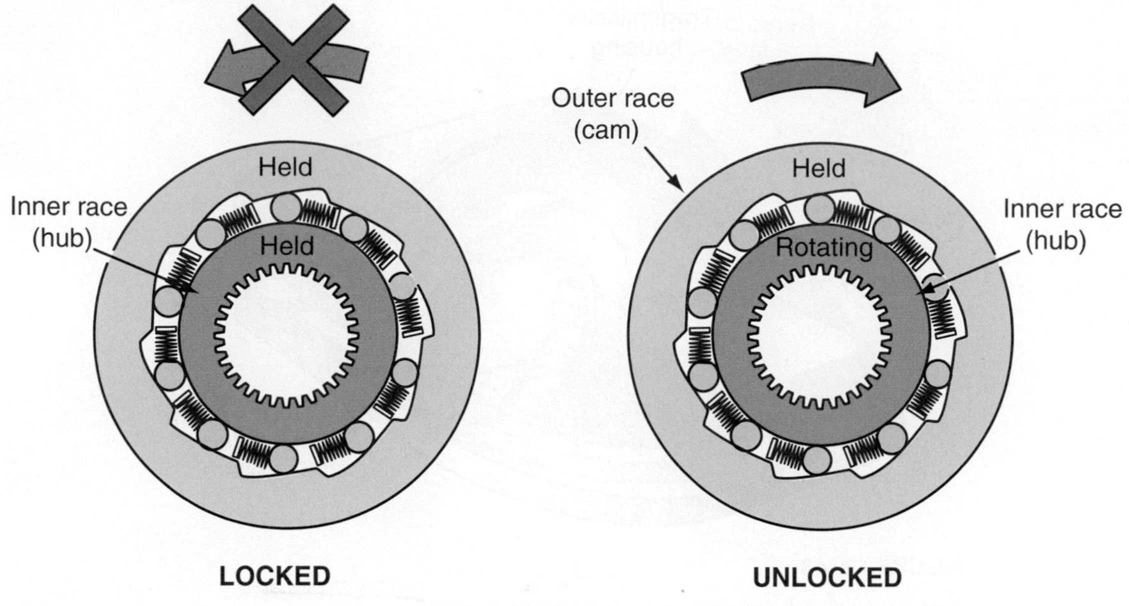

FIGURE 42–41

The action of a one-way roller clutch.

One-Way Clutches

In an automatic transmission operation, both sprag and roller overrunning clutches are used to hold or drive members of the planetary gearset. These clutches operate mechanically. An overrunning clutch allows rotation in only one direction and operates at all times. One-way overrunning clutches can be either roller-type or sprag-type clutches.

In a roller-type **(Figure 42–41)**, roller bearings are held in place by springs to separate the inner and outer race of the clutch assembly. Around the inside of the outer race are several cam-shaped indentations. The rollers and springs are located in these pockets. Rotation of one race in one direction locks the rollers between the two races, causing both to rotate together and prevent both from moving. When the race is rotated in the opposite direction, the roller bearings move into the pockets and are not locked, and the races are free to rotate independently.

A one-way sprag clutch **(Figure 42–42)** consists of a hub and drum separated by figure eight–shaped metal pieces called sprags. The sprags are shaped so that they lock between the races when a race is turned in one direction only. One end of the sprag is longer than the distance between the two races. Springs hold the sprags at the correct angle and maintain the sprags' contact with the races, thereby allowing for instantaneous engagement. When a race rotates in one direction, the sprags lift and allow the races to move independently. When a race is moved in the opposite direction, the sprags straighten and lock the two races together.

Sprag and roller clutches can be used to hold a member of the gearset by locking the inner race to the outer race, which is held by the transmission housing **(Figure 42–43)**. Both types also are effective as long as the engine powers the transmission. When the transmission is in a low gear and is coasting, the

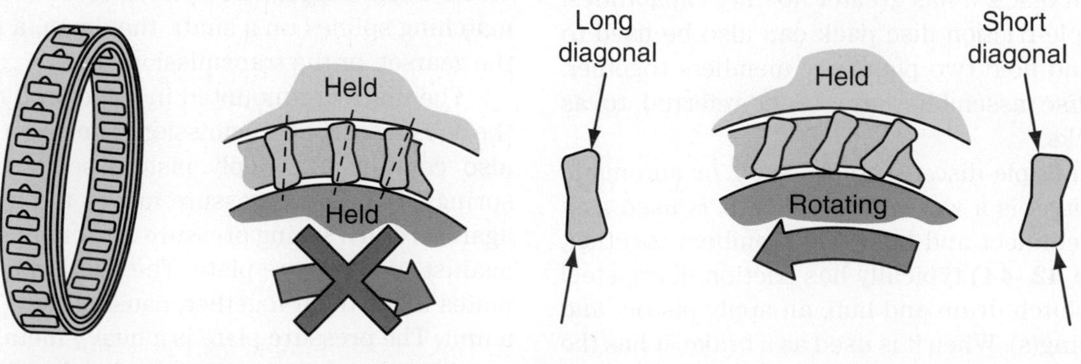

FIGURE 42–42

The action of a sprag-type one-way clutch.

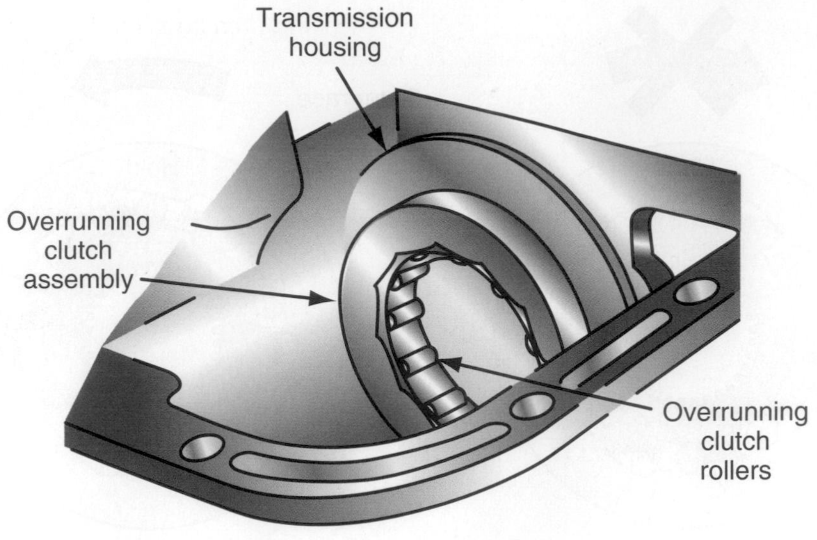

FIGURE 42–43

A one-way clutch secured in a transmission housing.

©Cengage Learning 2015

drive wheels rotate the transmission's output shaft with more power than is present on the input shaft. This allows the sprags or rollers to unwedge and begin freewheeling. This means they rotate without affecting the input or output of the gearset. When one-way clutches are freewheeling, they are off or ineffective. Freewheeling normally takes place when the clutch is rotating in a counter-clockwise direction.

Multiple-Friction Disc Clutch and Brake Assemblies

The **multiple-disc clutch assembly** can be used to drive or hold a member of the planetary gearset. The assemblies can serve as a brake by locking a gearset member to the transmission housing to prevent it from rotating. They can also be used to lock one race of a one-way clutch. As a brake, a multiple-disc pack serves the same purpose as a band; however, because there is much more surface area on the friction discs, it has greater holding capabilities. The multiple-friction disc pack can also be used to connect and hold two planetary members together. Multiple-disc assemblies are often referred to as clutch packs.

The multiple-disc assembly used in automatic transmissions is a wet clutch. When it is used as a clutch to connect and hold two members together, it **(Figure 42–44)** typically has friction discs, steel plates, a clutch drum and hub, an apply piston, and return spring(s). When it is used as a brake, it has the same basic parts but uses the transmission housing instead of a clutch drum.

The clutch pack has several plates lined with friction material and steel separator discs that are placed alternately inside a clutch drum. The friction plates are lined with rough frictional material on their faces, whereas the steel discs have smooth faces without friction material. A friction plate has friction material bonded to both sides of a steel plate **(Figure 42–45)**. Paper cellulose is the most commonly used friction material because it offers good holding power without the high frictional wear of metallic materials. The friction plates often have grooves cut in them to help keep them cool, thereby increasing their effectiveness and durability. Friction discs are always mounted between two steel plates.

The steel plates provide a smooth surface for the friction discs to engage. Plates can be installed steel-to-steel to create a specific clearance for the clutch pack. The set of friction or steel plates has splines on its inner edges, whereas the other set is splined on its outer edges. The splines of each set fit into matching splines on a shaft, the drum, a member of the gearset, or the transmission case.

The discs are mounted in the clutch drum or in the housing of the transmission. The drum or housing also contains the apply piston, seals, and return springs. Hydraulic pressure moves the apply piston against return spring pressure and clamps the plates against the pressure plate. The friction between the plates locks them together, causing them to turn as a unit. The pressure plate is a heavy metal plate that provides the clamping surface for the plates and is installed at one, or both, ends of the pack. The seals

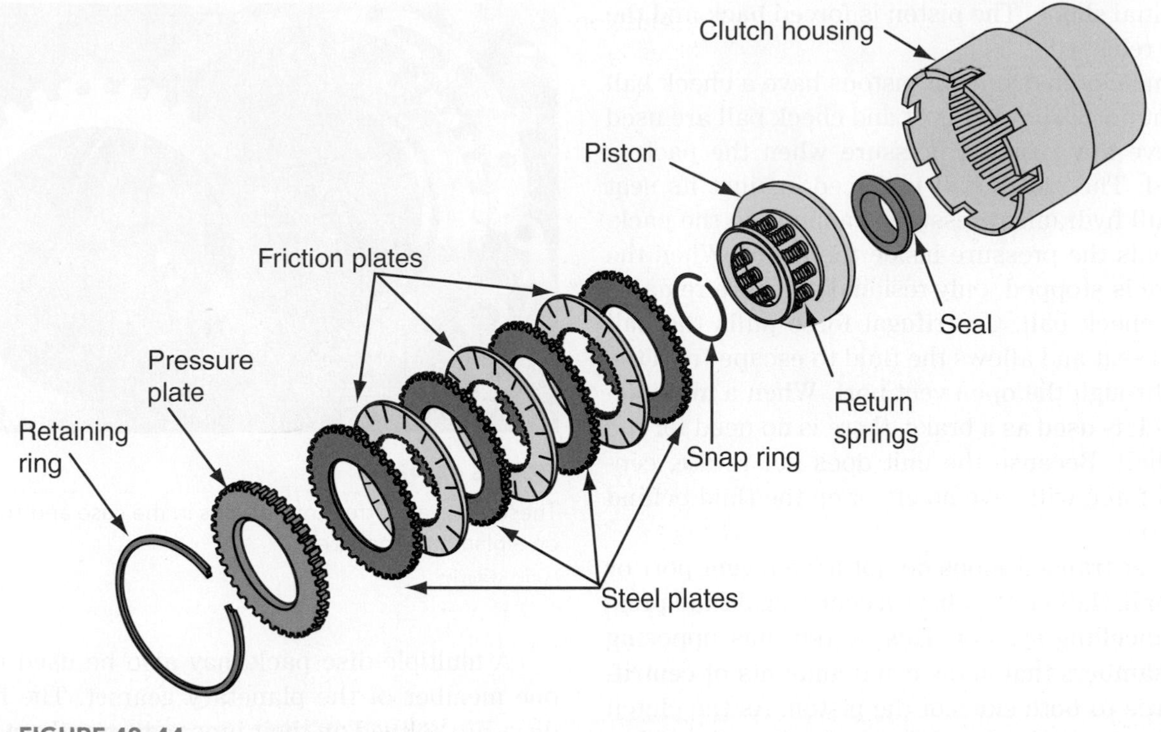

FIGURE 42–44
A multiple-disc clutch assembly.

hold in the hydraulic pressure when the clutch pack is applied. In a typical pack, the apply piston is held in place by the return springs and a spring retainer secured by a snap ring.

The apply piston is retracted by one large coil spring, several small springs **(Figure 42–46)**, or a single Belleville spring. The type and number of return springs used in the pack is determined by the pressure needed to release the piston quickly enough to prevent dragging. However, the amount of spring tension is limited to minimize the resistance to moving the piston.

FIGURE 42–45
The friction plates have friction material bonded to both sides.
Martin Restoule

A Belleville spring acts to improve the clamping force of the assembly and as a piston return spring. The spring is locked into a groove inside the drum by a snap ring. As the piston moves to apply the pack, it moves the inner ends of the Belleville spring's fingers into contact with the pressure plate to apply the assembly. The spring's fingers act as levers against the pressure plate and increase the application force of the pack. When hydraulic pressure to the piston is stopped, the spring relaxes and returns to

FIGURE 42–46
The apply piston can be retracted by several small springs.
Martin Restoule

its original shape. The piston is forced back and the pack is released.

Some clutch drums or pistons have a check ball and vent port. The vent port and check ball are used to relieve any residual pressure when the pack is released. The check ball is forced against its seat when full hydraulic pressure is applied to the pack. This holds the pressure inside the drum. When the pressure is stopped, only residual pressure remains on the check ball. Centrifugal force pulls the ball from its seat and allows the fluid to escape from the drum through the open vent port. When a multiple-disc pack is used as a brake, there is no need for the check ball. Because the unit does not rotate, centrifugal force will have no effect on the fluid behind the piston.

Newer transmissions do not have a vent port or check ball. Rather, they have a centrifugal fluid pressure cancelling system. This system has opposing fluid chambers that apply equal amounts of centrifugal force to both sides of the piston. As the clutch drum rotates, fluid in the cancelling pressure chamber counters the pressure built up inside the drum pressure chamber. This cancels out the effects of centrifugal force on piston movement and leads to smoother shifts.

Multiple-disc packs can function as holding or driving devices, depending on what they are splined to. When a pack locks two members of a planetary gearset together, it is a driving device **(Figure 42–47)**. To apply the clutch, hydraulic pressure is routed to the side of the piston opposite the return springs. When the pressure overcomes the tension of the return spring(s), the springs compress, and the piston squeezes the friction discs and steel plates tightly together so that they rotate as a unit.

FIGURE 42–47

These discs fit into the clutch drum.

Martin Restoule

FIGURE 42–48

These discs are fitted into splines in the case and splines on a planetary member.

Martin Restoule

A multiple-disc pack may also be used to hold one member of the planetary gearset. The friction discs are splined on their inner edges and are fit into matching splines on the outside of a drum. The steel discs are splined on their outer edges and fit into matching splines machined into the transmission case **(Figure 42–48)**. When the pack is applied, the gearset member cannot rotate as it is locked to the transmission case.

Several factors contribute to the effectiveness of a clutch pack:

- The type of frictional material used on the friction discs
- The composition of the steel plates and their surface finish
- Sufficient fluid flow to cool the clutch pack
- Condition and type of transmission fluid
- Proper grooving of the lining on the friction plates to aid in the cooling process
- Proper clutch plate clearances
- The force used to apply the clutch

BEARINGS, BUSHINGS, AND THRUST WASHERS

When a component slides over or rotates around another part, the surfaces that contact each other are called bearing surfaces. A gear rotating on a fixed shaft can have more than one bearing surface; it is supported and held in place by the shaft in a radial direction. Also, the gear tends to move along the shaft in an axial direction as it rotates and is therefore held in place by some other components.

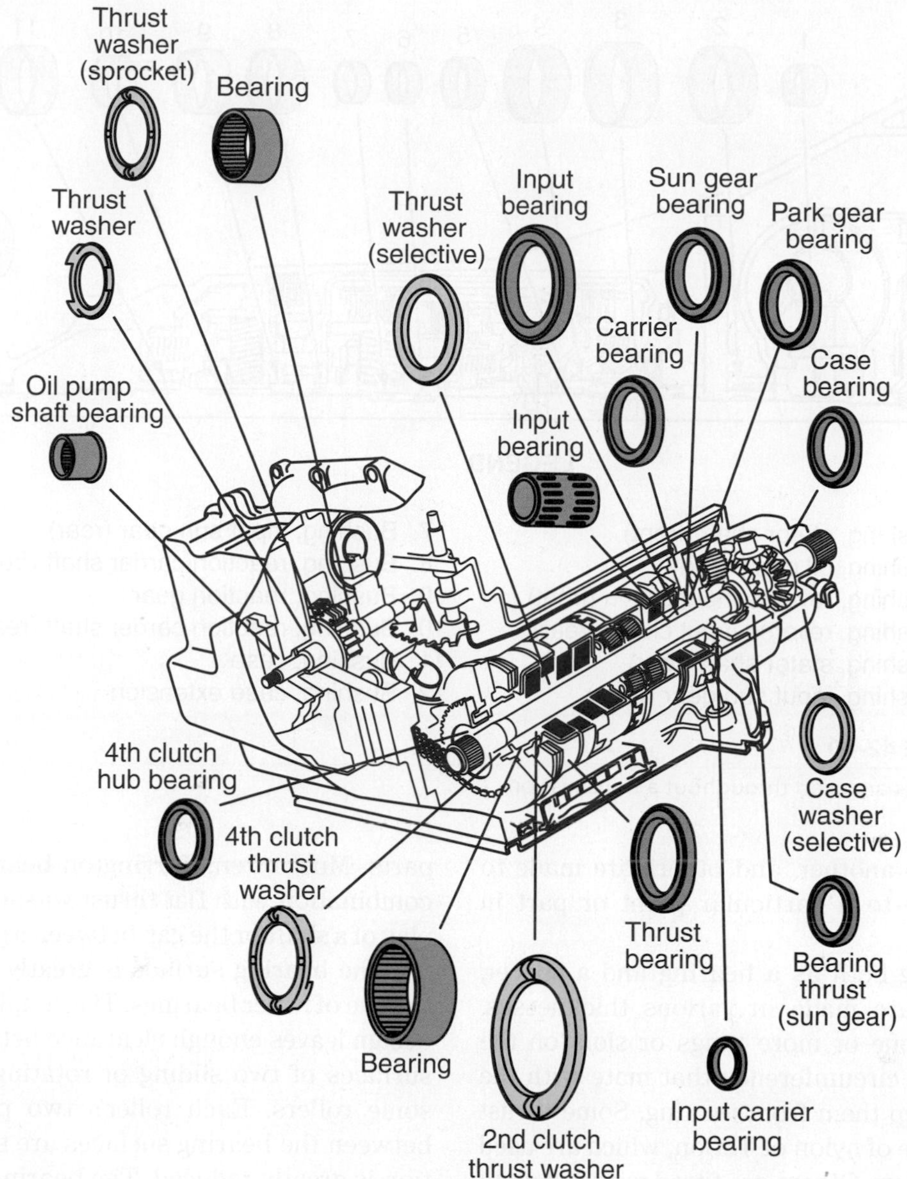

FIGURE 42–49

The location of various bearings and thrust washers in a typical transaxle.

The surfaces between the sides of the gear and the other parts are bearing surfaces.

A bearing is a device placed between two bearing surfaces to reduce friction and wear. Most bearings have surfaces that either slide or roll against each other. In automatic transmissions, sliding bearings are used where one or more of the following conditions prevail: low rotating speeds, very large bearing surfaces compared to the surfaces present, and low use applications. Rolling bearings are used in high-speed applications, high load with relatively small bearing surfaces, and high use.

Transmissions use sliding bearings composed of a relatively soft bronze alloy. Many are made from steel with the bearing surface bonded or fused to the steel. Those that take radial loads are called bushings, and those that take axial loads are called thrust washers **(Figure 42–49)**. The bearing's surface usually runs against a harder surface such as steel to produce minimum friction and heat wear characteristics.

Bushings are cylindrically shaped and usually held in place by press fit. Because bushings are typically made of a soft metal, they act like a bearing and support many of the transmission's rotating parts **(Figure 42–50)**. They are also used to precisely guide the movement of various valves in the transmission's valve body. Bushings can also be used to control fluid flow; some restrict the flow

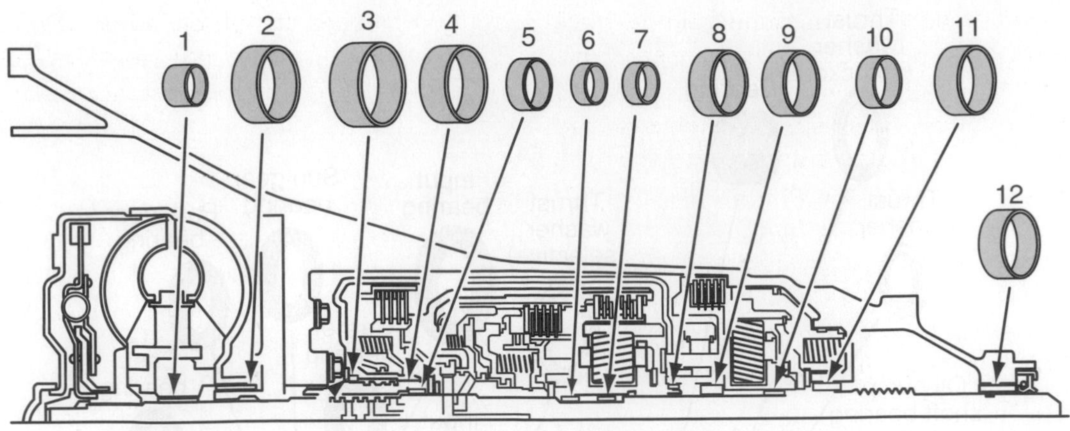

FIGURE 42–50

Bushings are used throughout a transmission.

from one part to another, and others are made to direct fluid flow to a particular point or part in the transmission.

Often serving both as a bearing and a spacer, thrust washers are made in various thicknesses. They may have one or more tangs or slots on the inside or outside circumference that mate with the shaft bore to keep them from turning. Some thrust washers are made of nylon or Teflon, which are used when the load is low. Others are fitted with rollers to reduce friction and wear.

Thrust washers normally control free axial movement or end play. Because some end play is necessary in all transmissions because of heat expansion, proper end play is often accomplished through selective thrust washers. These thrust washers are inserted between various parts of the transmission. Whenever end play is set, it must be set to manufacturer's specifications. Thrust washers work by filling the gap between two objects and become the primary wear item because they are made of softer materials than the parts they protect. Normally, thrust washers are made of copper- or babbitt-faced soft steel, bronze, nylon, or plastic.

Torrington bearings (Figure 42–51) are thrust washers fitted with roller bearings. These thrust bearings are primarily used to limit end play but also to reduce the friction between two rotating parts. Most often, Torrington bearings are used in combination with flat thrust washers to control end play of a shaft or the gap between a gear and its drum.

The bearing surface is greatly reduced through the use of roller bearings. The simplest roller-bearing design leaves enough clearance between the bearing surfaces of two sliding or rotating parts to accept some rollers. Each roller's two points of contact between the bearing surfaces are so small that friction is greatly reduced. The bearing surface is more like a line than an area.

If the roller length to diameter is about 5:1 or more, the roller is called a needle, and such a bearing

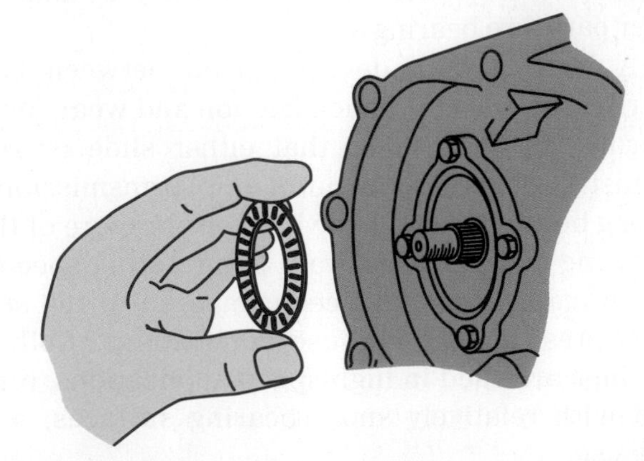

FIGURE 42–51

A Torrington-type thrust bearing.

is called a needle bearing. Sometimes, the needles are loose or they can be held in place by a steel cylinder or by rings at each end. Often, the latter are drilled to accept pins at the ends of each needle that act as axles. These small assemblies help save the agony of losing one or more loose needles and the delay caused by searching for them.

Many other roller bearings are designed as assemblies. The assemblies consist of an inner and outer race, the rollers, and a cage. There are roller bearings designed for radial loads and others designed for axial loads.

A tapered roller bearing is designed to accept both radial and axial loads. Its rollers turn on an angle to the bearing assembly's axis rather than parallel to it. The rollers are also slightly tapered to fit the angle of the inner and outer races. The bearing assembly consists of an inner race, the rollers, the cage, and the outer race. Tapered roller bearings are normally used in pairs and are rarely used in automatic transmissions. They are commonly used in final drive units.

The heaviest radial loads in automatic transmissions are carried by either roller or ball bearings. Ball bearings are constructed similarly to roller bearings, except that the races are grooved to accept the balls. The groove radius is slightly larger than the ball radius, which reduces the bearing surface area more than the roller bearing does. A ball bearing can also withstand light axial loads. Lip seals are sometimes built into ball bearings to retain lubricants.

SNAP RINGS

Many different sizes and types of snap rings are used in today's transmissions. External and internal snap rings are used as retaining devices throughout the transmission. Internal snap rings are used to hold servo assemblies and clutch assemblies together. In fact, snap rings are also available in several thicknesses and may be used to adjust the clearance in multiple-disc clutches. Some snap rings for clutch packs are waved to smooth clutch application. External snap rings are used to hold gear and clutch assemblies to their shafts.

GASKETS AND SEALS

The gaskets and seals of an automatic transmission help to contain the fluid within the transmission and prevent the fluid from leaking out of the various hydraulic circuits. Different types of seals are used in automatic transmissions; they can be made of rubber, metal, or Teflon **(Figure 42–52)**. Transmission gaskets are made of rubber, cork, paper, synthetic materials, metal, or plastic.

Gaskets

Gaskets are used to seal two parts together or to provide a passage for fluid flow from one part of the transmission to another. Gaskets are easily divided into two separate groups, hard and soft, depending on their application. Hard gaskets are used whenever the surfaces to be sealed are smooth. This type of gasket is usually made of paper. A common application of a hard gasket is the gasket used to seal the valve body and oil pump against the transmission case. Hard gaskets are also often used to direct fluid flow or to seal off some passages between the valve body and the separator plate.

Gaskets that are used when the sealing surfaces are irregular or in places where the surface may distort when the component is tightened into place are called soft gaskets. A typical location of a soft gasket is the oil pan gasket that seals the oil pan to the transmission case. Oil pan gaskets are typically a composition-type gasket made with rubber and cork. However, some late-model transmissions use an RTV sealant instead of a gasket to seal the oil pan.

Seals

As valves and transmission shafts move within the transmission, it is essential that the fluid and pressure be contained within its bore. Any leakage would decrease the pressure and result in poor transmission operation. Seals are used to prevent leakage around valves, shafts, and other moving parts. Rubber, metal, or Teflon materials are used throughout a transmission to provide for static and dynamic sealing. Both static and dynamic seals can provide for positive and non-positive sealing. A definition of each of the different basic classifications of seals follows:

- *Static.* A seal used between two parts that do not move in relationship to each other.
- *Dynamic.* A seal used between two parts that do move in relationship to each other. This movement is either a rotating or reciprocating (up and down) motion.
- *Positive.* A seal that prevents all fluid leakage between two parts.
- *Non-positive.* A seal that allows a controlled amount of fluid leakage. This leakage is typically used to lubricate a moving part.

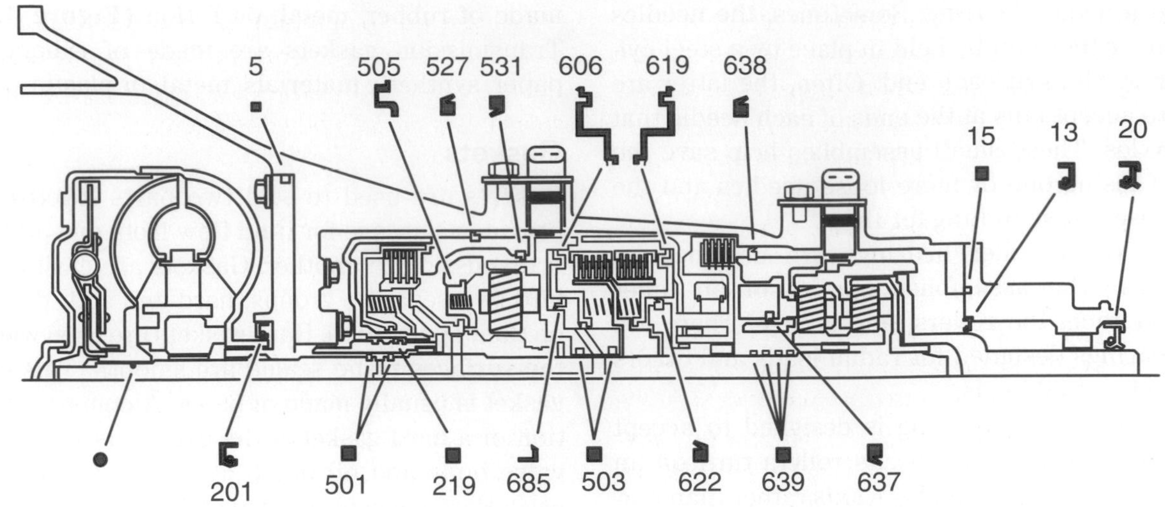

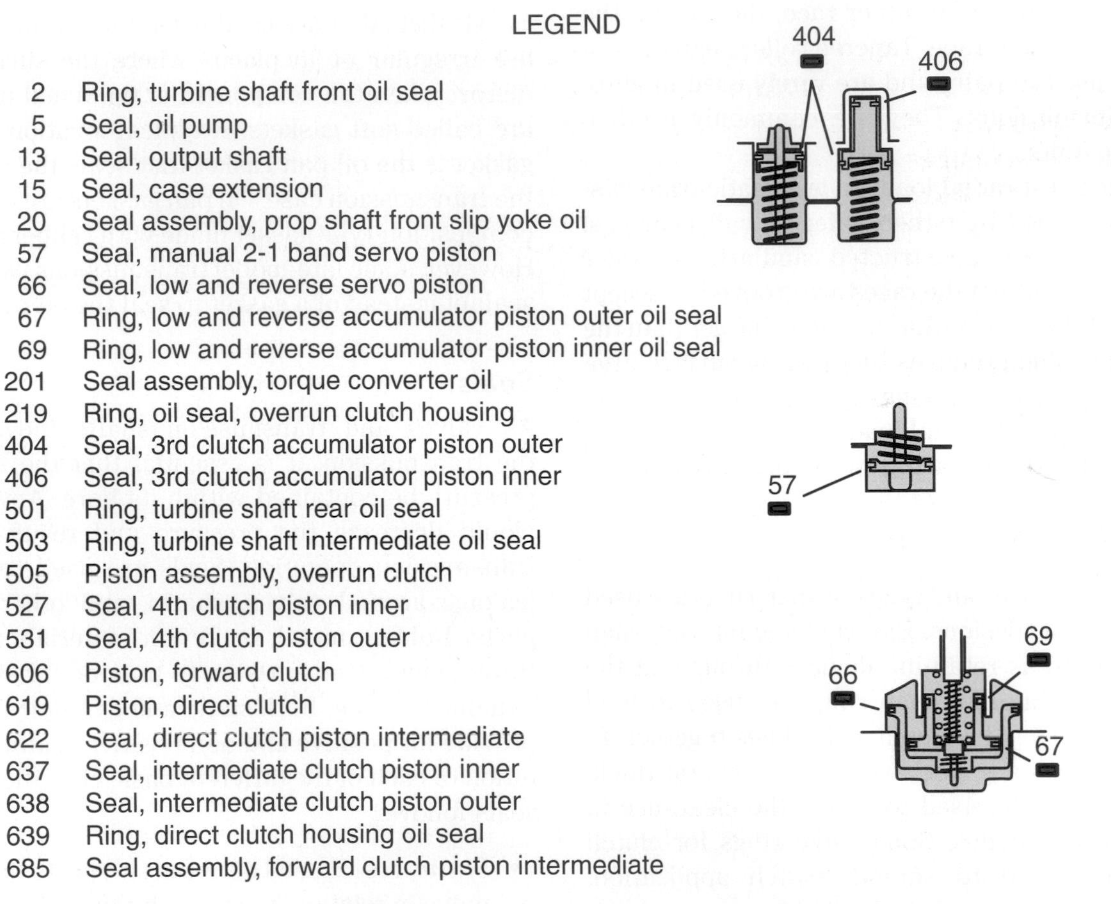

LEGEND
2 Ring, turbine shaft front oil seal
5 Seal, oil pump
13 Seal, output shaft
15 Seal, case extension
20 Seal assembly, prop shaft front slip yoke oil
57 Seal, manual 2-1 band servo piston
66 Seal, low and reverse servo piston
67 Ring, low and reverse accumulator piston outer oil seal
69 Ring, low and reverse accumulator piston inner oil seal
201 Seal assembly, torque converter oil
219 Ring, oil seal, overrun clutch housing
404 Seal, 3rd clutch accumulator piston outer
406 Seal, 3rd clutch accumulator piston inner
501 Ring, turbine shaft rear oil seal
503 Ring, turbine shaft intermediate oil seal
505 Piston assembly, overrun clutch
527 Seal, 4th clutch piston inner
531 Seal, 4th clutch piston outer
606 Piston, forward clutch
619 Piston, direct clutch
622 Seal, direct clutch piston intermediate
637 Seal, intermediate clutch piston inner
638 Seal, intermediate clutch piston outer
639 Ring, direct clutch housing oil seal
685 Seal assembly, forward clutch piston intermediate
404
406
57
66
69
67

FIGURE 42–52

The location of various seals and gaskets in a typical transmission.

Three major types of rubber seals are used in automatic transmissions: the O-ring, the lip seal, and the square-cut seal. Rubber seals are made from synthetic rubber rather than natural rubber.

O-rings are round seals with a circular cross section. Normally, an O-ring is installed in a groove cut into the inside diameter of one of the parts to be sealed. When the other part is inserted into the bore and through the O-ring, the O-ring is compressed between the inner part and the groove. This pressure distorts the O-ring and forms a tight seal between the two parts.

O-rings can be used as dynamic seals but are most commonly used as static seals. An O-ring can be used as a dynamic seal when the parts have relatively low amounts of axial movement. If there is a considerable amount of axial movement, the O-ring will quickly be damaged as it rolls within its groove. O-rings are never used to seal a shaft or part that has rotational movement.

Lip seals are used to seal parts that have axial or rotational movement. They are round to fit around a shaft, but the entire seal does not serve as a seal; rather, the sealing part is a flexible lip. The flexible lip is normally made of synthetic rubber and shaped so that it is flexed when it is installed to apply pressure at the sharp edge of the lip. Lip seals are used around input and output shafts to keep fluid in the housing and dirt out. Some seals are double-lipped.

When the lip is around the outside diameter of the seal, it is used as a piston seal **(Figure 42–53)**. Piston seals are designed to seal against high pressures, and the seal is positioned so that the lip faces the source of the pressurized fluid. The lip is pressed firmly against the cylinder wall as the fluid pushes against the lip; this forms a tight seal. The lip then relaxes its seal when the pressure on it is reduced or exhausted.

Lip seals are also commonly used as shaft seals. When used to seal a rotating shaft, the lip of the seal is around the inside diameter of the seal, and the outer diameter is bonded to the inside of a metal housing. The outer metal housing is pressed into a bore. To help maintain good sealing pressure on the rotating shaft, a garter spring is fitted behind the lip. This toroidal spring pushes on the lip to provide for uniform contact on the shaft. Shaft seals are not designed to contain pressurized fluid; rather, they are designed to prevent fluid from leaking over the shaft and out of the housing. The tension of the spring and of the lip is designed to allow an oil film of about 0.00254 mm (0.0001 in.). This oil film serves as a lubricant for the lip. If the tolerances increase, fluid will be able to leak past the shaft, and if the tolerances are too small, excessive shaft and seal wear will result.

A **square-cut seal** is similar to an O-ring; however, a square-cut seal can withstand more axial movement than an O-ring can. Square-cut seals have a rectangular or square cross-section. They are designed this way to prevent the seal from rolling in its groove when there are large amounts of axial movement. Added sealing comes from the distortion of the seal during axial movement. As the shaft inside the seal moves, the outer edge of the seal moves more than the inner edge causing the diameter of the sealing edge to increase, which creates a tighter seal.

Metal Sealing Rings

There are some parts of the transmission that do not require a positive seal and in which some leakage is acceptable. These components are sealed with ring seals that fit into a groove on a shaft **(Figure 42–54)**. The outside diameter of the ring seals slide against the walls of the bore into which the shaft is inserted. Most ring seals in a transmission are placed near pressurized fluid outlets on rotating shafts to help retain pressure. Ring seals are made of cast iron, nylon, or Teflon.

Three types of metal seals are used in automatic transmissions: **butt-end seals**, open-end seals, and hook-end seals. In appearance, butt-end and open-end seals are much the same; however, when an

FIGURE 42–53

A typical lip seal installed on a piston.

Martin Restoule

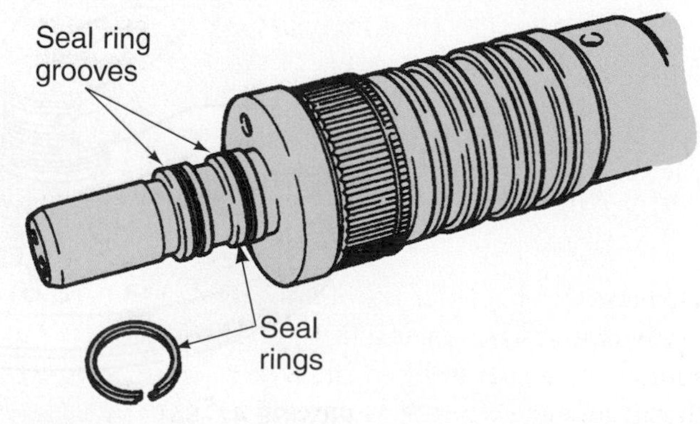

FIGURE 42–54

Metal sealing rings are fit into grooves on a shaft.

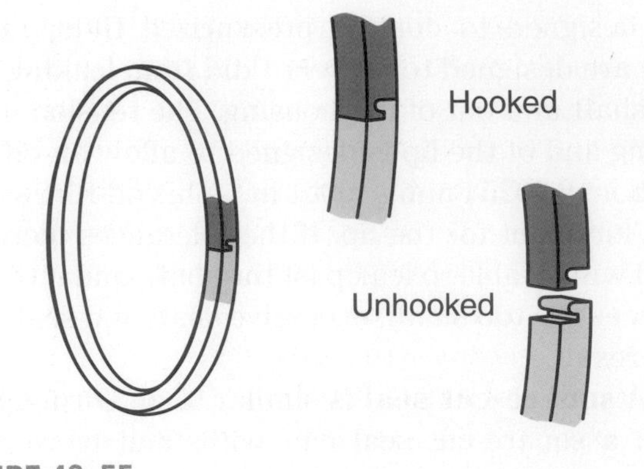

FIGURE 42–55
Hook-end sealing rings.

open-end seal is installed, there is a gap between the ends of the seal. **Hook-end seals (Figure 42–55)** have small hooks at their ends that are locked together during installation to provide better sealing than the open-end or butt-end seals.

Teflon Seals

Some transmissions use Teflon seals instead of metal seals. Teflon provides for a softer sealing surface, which results in less wear on the surface that it rides on and therefore a longer-lasting seal. Teflon seals are similar in appearance to metal seals, except for the hook-end type. The ends of locking-end Teflon seals are cut at an angle **(Figure 42–56)**, and the locking hooks are somewhat staggered. These seals are often called scarf-cut seals.

Many late-model transmissions are equipped with solid one-piece Teflon seals. Although the one-piece seal requires some special tools for installation, they provide for a nearly positive seal. These Teflon rings form a better seal than other metal sealing rings.

General Motors uses a different type of synthetic seal on some late-model transmissions. The material used in these seals is Vespel, which is a flexible but highly durable plasticlike material.

FINAL DRIVES AND DIFFERENTIALS

The last set of gears in the drivetrain is the final drive. In most RWD cars, the final drive is located in the rear axle housing. On most FWD cars, the final drive is located within the transaxle. Some FWD cars with longitudinally mounted engines locate the differential and final drive in a separate case that bolts to the transmission. AWD and 4WD vehicles have a final drive unit in the front and rear drive axles.

There are four common configurations used as the final drives on FWD vehicles: helical gear, planetary gear, hypoid gear, and chain drive. The helical, planetary, and chain final drive arrangements are found with transversely mounted engines. Hypoid final drive gear assemblies are normally found in

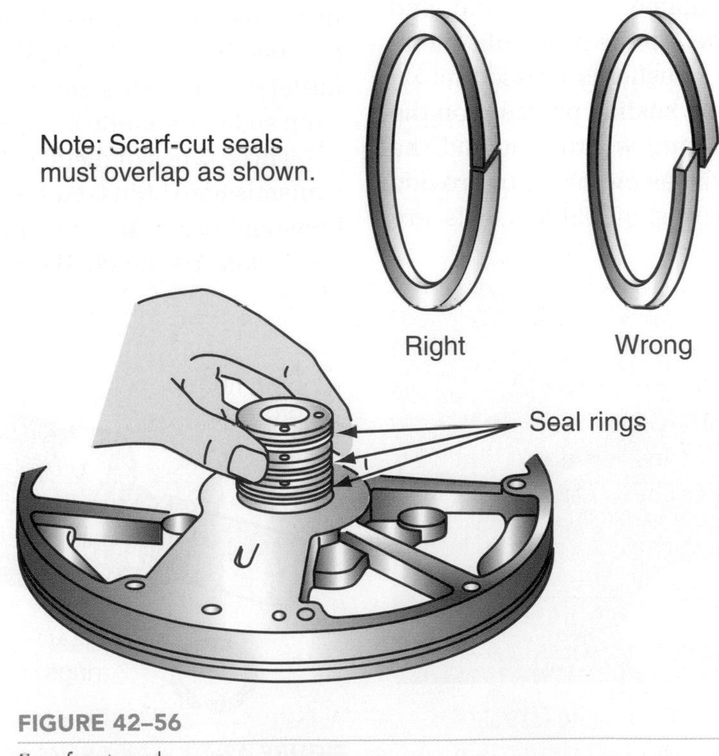

FIGURE 42–56
Scarf-cut seals.

vehicles with a longitudinally placed engine. The hypoid assembly is basically the same unit as would be used on RWD vehicles and is mounted directly to the transmission.

Some transaxles route power from the transmission through two helical-cut gears to a transfer shaft. A helical-cut pinion gear attached to the opposite end of the transfer shaft drives the differential ring gear and carrier. The differential assembly then drives the axles and wheels.

Rather than use helical-cut or spur gears in the final drive assembly, some transaxles use a simple planetary gearset for its final drive. The sun gear of this planetary unit is driven by the final drive sun gear shaft, which is splined to the front carrier and rear ring gear of the transmission's gearset. The final drive sun gear meshes with the final drive planetary pinion gears, which rotate on their shafts in the planetary carrier. The pinion gears mesh with the ring gear, which is splined to the transaxle case. The planetary carrier is part of the differential case, which contains typical differential gearing, two pinion gears, and two side gears.

The ring gear of a planetary final drive assembly has lugs around its outside diameter that fit into grooves machined inside the transaxle housing. These lugs and grooves hold the ring gear stationary. The transmission's output is connected to the planetary gearset's sun gear. In operation, the transmission's output drives the sun gear that, in turn, drives the planetary pinion gears. The pinion gears walk around the inside of the stationary ring gear. The rotating planetary pinion gears drive the planetary carrier and differential case. This combination provides maximum torque multiplication from a simple planetary gearset.

Chain-drive final drive assemblies use a multiple-link chain to connect a drive sprocket, connected to the transmission's output shaft, to a driven sprocket that is connected to the differential's pinion shaft. This design allows for remote positioning of the differential within the transaxle housing. Final drive gear ratios are determined by the size of the driven sprocket compared to the drive sprocket.

HYDRAULIC SYSTEM

A hydraulic system uses a liquid to perform work. In an automatic transmission, this liquid is automatic transmission fluid (ATF). ATF is one of the most complex fluids produced by the petroleum industry for the automobile.

ATF is special oil designed for transmission operation. Transmissions are equipped with a fluid cooler that prevents the overheating of the fluid, which can result in damage to the transmission. The transmission's pump is the source of all fluid flow and provides a constant supply of fluid under pressure to operate, lubricate, and cool the transmission. **Pressure regulating valves** change the fluid's pressure to control the shift quality and, in some transmissions, the shift points of the transmission. **Flow-directing valves** direct the pressurized fluid to the appropriate apply device to cause a change in gear ratios. The hydraulic system also keeps the torque converter filled with fluid.

The reservoir for ATF is the transmission's oil pan. Fluid is drawn from the pan and returned to it. The pressure source is the oil pump. The valve body contains control valving to regulate or restrict the pressure and flow of fluid within the transmission. The output devices for the hydraulic system are the servos or clutches operated by hydraulic pressure.

Hydraulic Principles

An automatic transmission uses ATF fluid pressure to change gears automatically through the use of various pressure regulators and control valves.

Basic hydraulic theory is explained in Chapter 7 of this book. Hydraulic circuits are direct applications of Pascal's law, which states, "Pressure exerted on a confined liquid or fluid is transmitted undiminished and equally in all directions and acts with equal force on all areas."

Fluids are perfect conductors of pressure. Therefore, when a piston in a cylinder moves and displaces fluid, that fluid is distributed equally within the circuit.

To form a complete, working hydraulic system, the following elements are needed: a fluid reservoir, a pressure source, control valving, and output devices.

The automatic transmission reservoir is the transmission oil pan. Transmission fluid is drawn from the pan and returned to it. The pressure source in the system is the oil pump. The valve body contains control valving to regulate or restrict the pressure and flow of fluid within the transmission. Output devices are the servos or clutches operated by hydraulic pressure.

Application of Hydraulics in Transmissions

A common hydraulic system within an automatic transmission is the servo assembly, which is used to control the application of a band. The band must

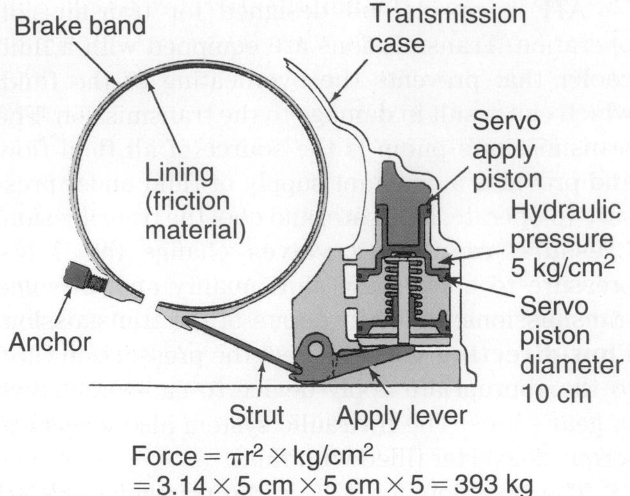

FIGURE 42–57

Calculating the output force developed by a servo assembly.
Chrysler Group LLC

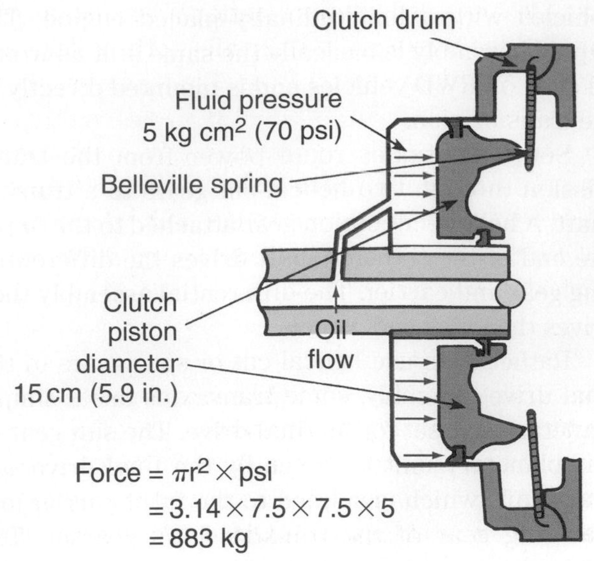

FIGURE 42–58

Using hydraulics to increase work in a multiple-disc clutch.
Chrysler Group LLC

tightly hold the drum or planetary carrier it surrounds when it is applied. The holding capacity of the band is determined by the construction of the band and the pressure applied to it. This pressure or holding force is the result of the action of a servo. The servo multiplies the force through hydraulic action.

If a servo has a diameter of 10 cm (3.94 in.) and has a pressure of 5.0 kg/cm^2 (71 psi) applied to it, the apply force of the servo is 393 kg (865 lb.) **(Figure 42–57)**. The force exerted by the servo is further increased by its lever-type linkage and the self-energizing action of the band. The total force applied by the band stops and holds the rotating drum connected to a planetary gearset member.

A multiple-disc assembly is also used to stop and hold gearset members. This assembly also uses hydraulics to increase its holding force. If the fluid pressure applied to the clutch assembly is 5 kg/cm^2 (71 psi) and the diameter of the clutch piston is 15 cm (5.9 in.), the force applying the clutch pack is 883 kg (1946.7 lb.). If the clutch assembly uses a **Belleville spring** or piston spring **(Figure 42–58)**, which adds a mechanical advantage of 1.25, the total force available to engage the clutch will be 883 kg (1946.7 lb.) multiplied by 1.25, or 1104 kg (2433.4 lb.).

Functions of ATF

The ATF circulating through the transmission and torque converter and over the parts of the transmission cools the transmission. The heated fluid typically moves to a transmission fluid cooler, where the heat is removed. As the fluid lubricates and cools the transmission, it also cleans the parts. The dirt is carried by the fluid to a filter, where the dirt is removed.

Another critical job of ATF is its role in shifting gears. ATF moves under pressure throughout the transmission and causes various valves to move. The pressure of the ATF changes with changes in engine speed and load.

ATF is also used to operate the various apply devices (clutches and bands) in the transmission. At the appropriate time, a switching valve opens and sends pressurized fluid to the apply device that engages or disengages a gear. The valving and hydraulic circuits are contained in the valve body.

Reservoir

A fluid reservoir stores fluid and provides a constant source of fluid for the system. In an automatic transmission, the reservoir is the pan, typically located at the bottom of the transmission case. ATF is forced out of the pan by atmospheric pressure and into the pump, and is then returned to it after it has circulated through the selected circuits. A transmission dipstick placed within a filler tube is typically used to check the level of the fluid and to add ATF to the transmission. Other transmissions have a side plug on the pan or the transmission to check and replenish fluid level.

Venting

All reservoirs must have an air vent that allows atmospheric pressure to force the fluid into the pump when the pump creates a low pressure at its inlet port. The pans of many automatic transmissions vent through the handle of the dipstick; others rely on a vent in the transmission case. Transmissions must also be vented to allow for the exhaust of built-up air pressure that results from heat and the moving components inside the transmission. The movement of these parts can force air into the ATF, which would not allow it to increase in pressure, cool, or lubricate the transmission properly.

Transmission Coolers

The removal of heat from ATF is extremely important to the durability of the transmission. Excessive heat causes the fluid to break down. Once broken down, ATF no longer lubricates well and has poor resistance to oxidation. Oxidized ATF may damage transmission seals. When a transmission is operated for some time with overheated ATF, varnish is formed inside the transmission. Varnish buildup on valves can cause them to stick or move slowly. The result is poor shifting and glazed or burned friction surfaces. Continued operation can lead to the need for a complete rebuilding of the transmission.

It is important to note that ATF is designed to operate at 80°C (176°F). At this temperature, the fluid should remain effective for 160 000 km (100 000 mi.). However, when the operating temperature increases, the useful life of the fluid quickly decreases. An 11.1°C (52°F) increase in operating temperature will decrease the life of ATF by one-half!

Transmission housings are fitted with ATF cooler lines **(Figure 42–59A)** that direct the hot fluid from the torque converter to the transmission cooler, normally located in the vehicle's radiator. The heat of the fluid is reduced by the cooler, and the cool ATF returns to the transmission. In some transmissions, the cooled fluid flows directly to the transmission's bushings, bearings, and gears. Then the fluid is circulated through the rest of the transmission. The cooled fluid in other transmissions is returned to the oil pan, where it is drawn into the pump and circulated throughout the transmission.

Some vehicles, such as those designed for heavy-duty use, are equipped with an auxiliary fluid cooler **(Figure 42–59B)** in addition to the one in the radiator. This cooler removes additional amounts of heat from the fluid before it is sent back to the transmission.

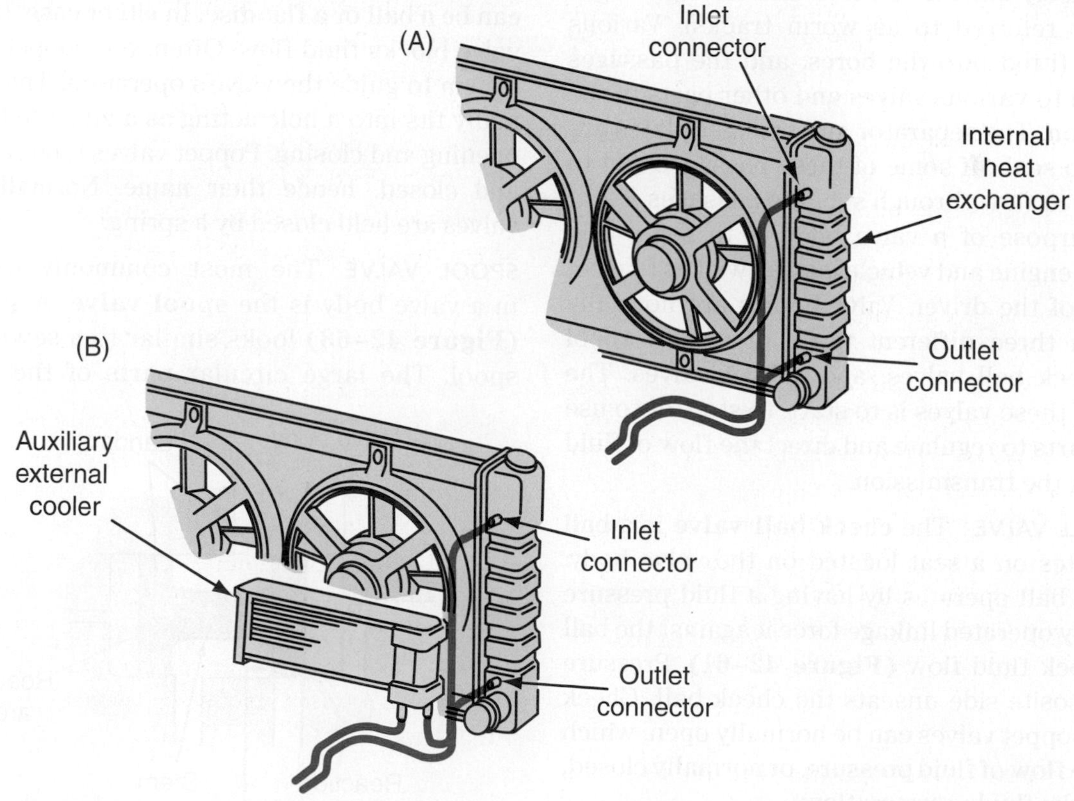

FIGURE 42–59

(A) A transmission cooler (heat exchanger) located in a radiator. (B) An auxiliary cooler added to the normal cooler circuit.

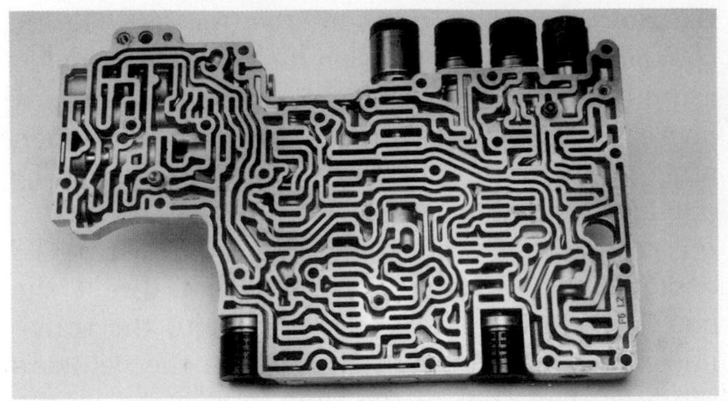

FIGURE 42–60
A typical valve body.

Valve Body

For efficient transmission operation, the bands and multiple-disc packs must be released and applied at the proper time. The **valve body** assembly **(Figure 42–60)** is responsible for the control and distribution of pressurized fluid throughout the transmission. This assembly is made of two or three main parts: a valve body, separator plate, and transfer plate. These parts are bolted as a single unit to the transmission housing. The valve body is machined from aluminum or iron and has many precisely machined bores and fluid passages (commonly referred to as worm tracks). Various valves are fitted into the bores, and the passages direct fluid to various valves and other parts of the transmission. The separator and transfer plates are designed to seal off some of these passages and to allow fluid to flow through specific passages.

The purpose of a valve body is to sense and respond to engine and vehicle load as well as to meet the needs of the driver. Valve bodies are normally fitted with three different types of valves: spool valves, check ball valves, and poppet valves. The purpose of these valves is to start, to stop, or to use movable parts to regulate and direct the flow of fluid throughout the transmission.

CHECK BALL VALVE The **check ball valve** is a ball that operates on a seat located on the valve body. The check ball operates by having a fluid pressure or manually operated linkage force it against the ball seat to block fluid flow **(Figure 42–61)**. Pressure on the opposite side unseats the check ball. Check balls and poppet valves can be normally open, which allows free flow of fluid pressure, or normally closed, which blocks fluid pressure flow.

At times, the check ball has two seats to check and direct fluid flow from two directions, being seated and unseated by pressures from either source.

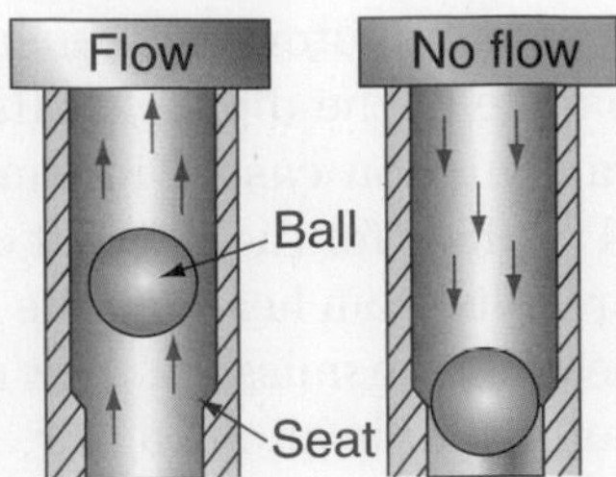

FIGURE 42–61
The operation of a check ball valve.

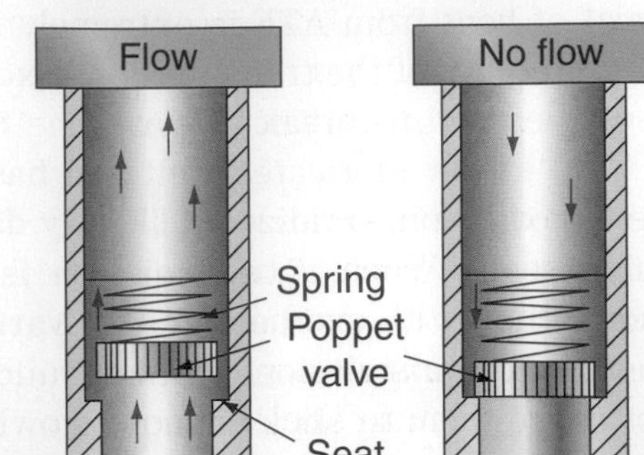

FIGURE 42–62
Typical poppet valve operation.

POPPET VALVE A **poppet valve (Figure 42–62)** can be a ball or a flat disc. In either case, the poppet valve blocks fluid flow. Often, the poppet valve has a stem to guide the valve's operation. The stem normally fits into a hole acting as a guide to the valve's opening and closing. Poppet valves tend to pop open and closed, hence their name. Normally, poppet valves are held closed by a spring.

SPOOL VALVE The most commonly used valve in a valve body is the **spool valve**. A spool valve **(Figure 42–63)** looks similar to a sewing thread spool. The large circular parts of the valve are

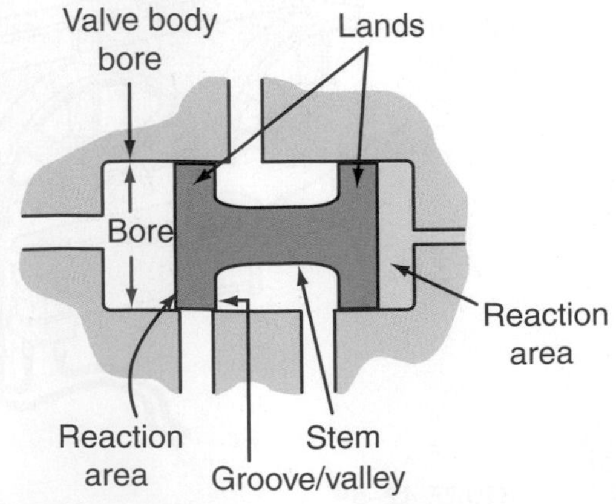

FIGURE 42–63
Components of a spool valve assembly.

called the lands. There is a minimum of two lands per valve. Each land of the assembly is connected by a stem. The space between the lands and stem is called the valley. Valleys form a fluid pressure chamber between the spools and valve body bore. Fluid flow can be directed into other passages, depending on the spool valve and valve body design.

Precisely machined around the periphery of each valve, the land is the part of the valve that rides on a thin film of fluid in a bore of the valve body. The land must be treated very carefully because any damage, even a small score or scratch, can impair smooth valve operation. As the spool valve moves, the land covers (closes) or uncovers (opens) ports in the valve body.

The reaction area, also known as the face, is the space at the outside of the lands at the end of the valve. Forces acting against the reaction area that cause the valve to move include spring tension, fluid pressure, or mechanical linkage.

Oil Pump

The source of fluid flow through the transmission is the oil pump **(Figure 42–64)**. Three types of oil pumps are commonly used in automatic transmissions: the gear type **(Figure 42–65A)**, rotor type, and vane type **(Figure 42–65B)**. Oil pumps are driven by the pump drive hub of the torque converter or oil pump shaft. Therefore, whenever the torque converter cover is rotating, the oil pump is driven. The oil pump creates fluid flow throughout the transmission.

FIGURE 42–64

An oil pump.

Transtar Industries Inc.

Pressure Regulator Valve

Transmission pumps are capable of creating excessive amounts of pressure that may cause damage; therefore, the transmission is equipped with a pressure regulator valve, normally located in the valve body. Pressure regulator valves are typically spool valves that toggle back and forth in their bores to open and close an exhaust passage. By opening the exhaust passage, the valve decreases the pressure of the fluid. As soon as the pressure decreases to a predetermined amount, the spool valve moves to close off the exhaust port and pressure again begins

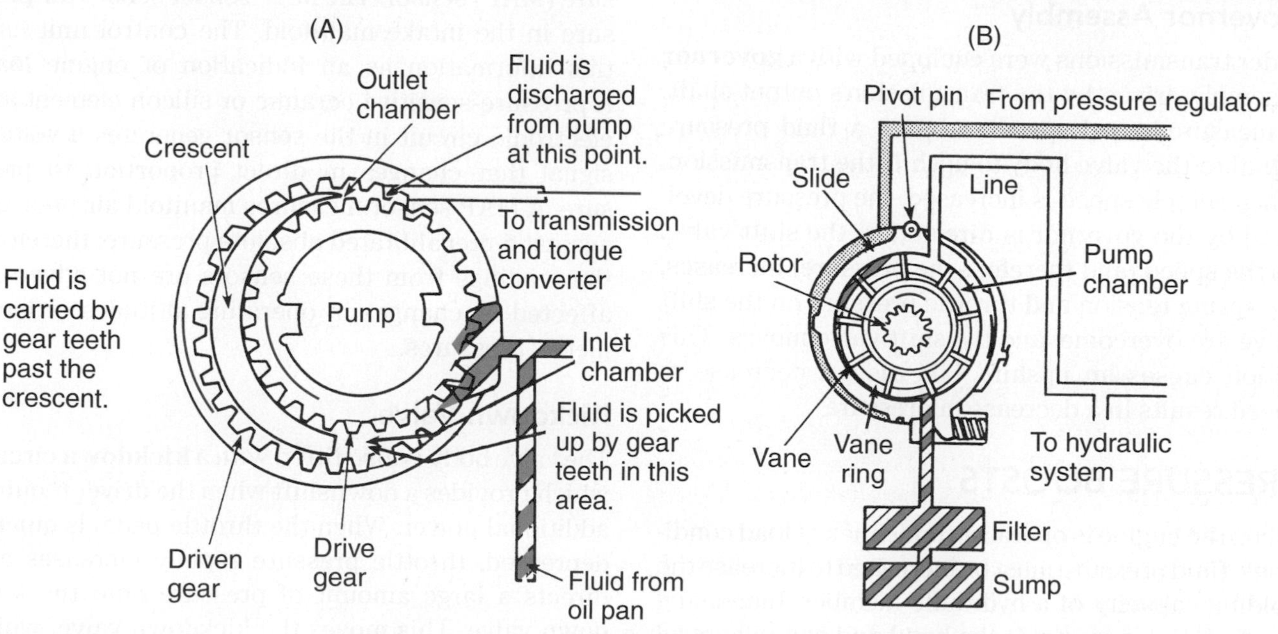

FIGURE 42–65

Operation of (A) a gear-type pump and (B) a vane-type pump.

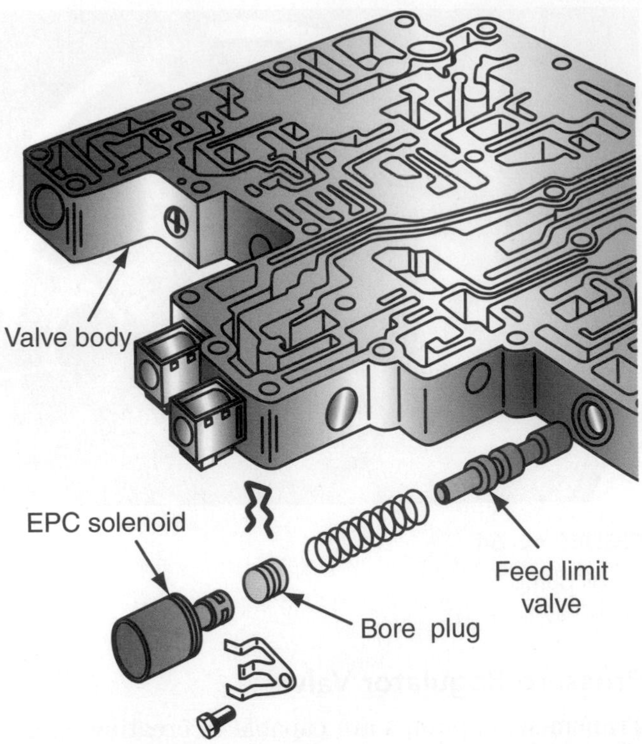

FIGURE 42–66

Normally, the EPC solenoid is inserted into the valve body.

to build. The action of the spool valve regulates the fluid pressure.

Many late-model transmissions use an electronic pressure control (EPC or PC) solenoid to regulate system pressure **(Figure 42–66)**.

Governor Assembly

Older transmissions were equipped with a **governor** assembly driven by the transmission's output shaft. It measured road speed and sent a fluid pressure signal to the valve body to upshift the transmission. When vehicle speed is increased, the pressure developed by the governor is directed to the shift valve. As the speed (and therefore the pressure) increases, the spring tension and throttle pressure on the shift valve are overcome, and the shift valve moves. This action causes an upshift. Likewise, a decrease in speed results in a decrease in pressure.

PRESSURE BOOSTS

When the engine is operating under heavy load conditions, fluid pressure must be increased to increase the holding capacity of a hydraulic member. Increasing the fluid pressure holds the band and clutch control units tighter to reduce the chance of slipping while under heavy load. This is accomplished by sending pressurized fluid to one side of the pressure regulator's spool valve. This pressure works against the spool valve's normal movement to open the exhaust port and allows pressure to build to a higher point than normal.

Engine load can be monitored electronically by various electronic sensors (primarily the TP and MAP sensors) that send information to an electronic control unit, which in turn controls the pressure at the valve body. Load can also be monitored by throttle pressure. Throttle pedal movement moves a **throttle valve** in the valve body via a throttle cable. When the throttle plate is opened, the throttle valve opens and applies pressure to the pressure regulator. This delays the opening of the pressure regulator valve, which allows for an increase in pressure. When the driver lets off the throttle pedal, the pressure regulator valve is free to move, and normal pressure is maintained.

Many early transmissions were equipped with a **vacuum modulator**, which uses engine vacuum to change transmission pressure. The vacuum modulator allows for an increase in pressure when vacuum is low and decreases it when vacuum is high.

MAP Sensor

Engine load can be monitored electronically through the use of various electronic sensors that send information to an electronic control unit, which in turn controls the pressure at the valve body. The most commonly used sensor is the manifold absolute pressure (MAP) sensor. The MAP sensor senses air pressure in the intake manifold. The control unit uses this information as an indication of engine load. A pressure-sensitive ceramic or silicon element and electronic circuit in the sensor generates a voltage signal that changes in direct proportion to pressure. A MAP sensor measures manifold air pressure against a precalibrated absolute pressure; therefore, the readings from these sensors are not adversely affected by changes in operating altitudes or barometric pressures.

Kickdown Valve

The valve body is also fitted with a **kickdown** circuit, which provides a downshift when the driver requires additional power. When the throttle pedal is quickly depressed, throttle pressure rapidly increases and directs a large amount of pressure onto the kickdown valve. This moves the kickdown valve, which opens a port and allows mainline pressure to flow against the shift valve. The spring tension on the shift valve, the kickdown pressure, and throttle pressure will push on the end of the shift valve, causing it to move to the downshift position and forcing a quick downshift.

SHIFT QUALITY

All transmissions are designed to change gears at the correct time according to engine and vehicle speed, load, and driver intent. However, transmissions are also designed to provide for positive change of gear ratios without jarring the driver or passengers. If a band or clutch is applied too quickly, a harsh shift will occur. **Shift feel** is controlled by the pressure at which each hydraulic member is applied or released, the rate at which each is pressurized or exhausted, and the relative timing of the apply and release of the members.

To improve shift feel during gear changes, a band is often released while a multiple-disc pack is being applied. The timing of these two actions must be just right or both components will be released or applied at the same time, which would cause engine flare-up or clutch and band slippage. Several other methods are used to smooth gear changes and improve shift feel.

Multiple friction-disc packs sometimes contain a wavy spring-steel separator plate that helps smooth the application of the clutch. Shift feel can also be smoothed out by using a restricting **orifice** or an **accumulator** piston in the band or clutch apply circuit. A restricting orifice or check ball in the passage to the apply piston restricts fluid flow and slows the pressure increase at the piston by limiting the quantity of fluid that can pass in a given time. Manufacturers have also applied electronics to get the desired shift feel. One of the most common techniques is the pulsing (turning on and off) of the shift solenoids, which prevents the immediate engagement of a gear by allowing some slippage.

Accumulators

Shift quality with a brake or clutch is dependent upon how quickly it is engaged by hydraulic pressure and the amount of pressure exerted on the piston. Some apply circuits use an accumulator to slow down application rates without decreasing the holding force of the apply device.

An accumulator **(Figure 42–67)** works like a shock absorber and cushions the application of servos and disc packs. An accumulator cushions sudden increases in hydraulic pressure by temporarily diverting some of the apply fluid into a parallel circuit or chamber. This allows the pressure to gradually increase and provides for smooth engagement of a brake or clutch.

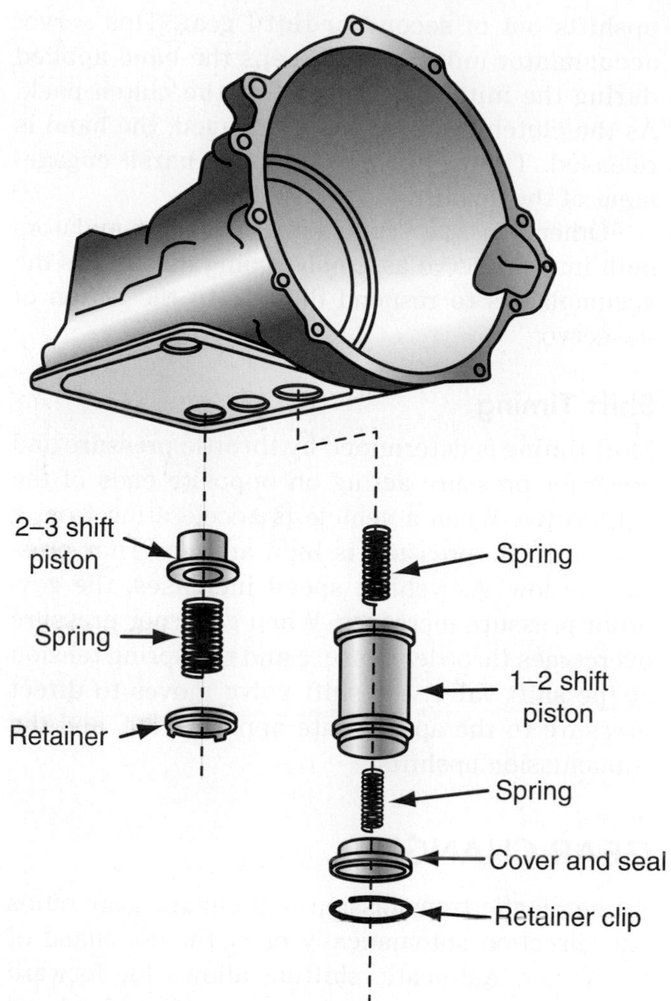

FIGURE 42–67

An accumulator assembly is used to control shift feel.

An accumulator is basically a large diameter piston located in a bore and held in position by a heavy, calibrated spring that acts against hydraulic pressure. Accumulators are placed in the hydraulic circuit between the shift valve and the holding device.

Some transmission designs do not rely on an accumulator for shift quality; rather, they have a restrictive orifice in the line to the servo or multiple-disc pack's piston. This restriction decreases the amount of initial apply pressure but will eventually allow for full pressure to act on the piston.

Several transmissions use servo units that also work as an accumulator. These units are typically used with the intermediate band and for upshifts out of second or third gear. This servo/accumulator unit actually keeps the band applied during the initial engagement of the clutch pack. As the clutch becomes more engaged, the band is released. This action prevents the harsh engagement of the upshift.

Other designs have multiple accumulators built into the servo assembly. Doing this allows the accumulators to respond directly to the action of the servo.

Shift Timing

Shift timing is determined by throttle pressure and governor pressure acting on opposite ends of the shift valve. When a vehicle is accelerating from a stop, throttle pressure is high and governor pressure is low. As vehicle speed increases, the governor pressure increases. When governor pressure overcomes throttle pressure and the spring tension at the shift valve, the shift valve moves to direct pressure to the appropriate apply device, and the transmission upshifts.

GEAR CHANGES

An automatic transmission will change gear ratios and direction automatically or at the command of the driver. Automatic shifting allows for forward gear ratio changes in response to engine speed and load—and in the case of electronically controlled units, in response to the PCM's interpretation of operating conditions. The transmission will not automatically move into park, reverse, or neutral. The driver makes these gear selections. The driver can also select certain forward gears. The forward gears available for manual selection vary with each transmission model.

A discussion on how gear changes are made is a summary of the operation of the transmission's components. Keep in mind that the way a particular transmission operates depends on its construction and components. This discussion is based on a common four-speed design using a Simpson gearset and an add-on overdrive planetary unit **(Figure 42–68)**. There are no electronic controls mentioned in this discussion because it important for you to understand how the gearset responds to the various hydraulic components in a transmission before you can understand how electronics improve the performance of a transmission. The clutch, brake, and band application for this typical transmission is shown in **Figure 42–69**. Refer to both figures as you read the following discussion.

Park/Neutral

When park is selected, the shift linkage moves the park pawl or lever into the park gear, located on the output shaft or final drive shaft **(Figure 42–70)**. The shift linkage also moves the manual shift valve to block the fluid passages to the clutches and bands. No member of the gearset receives input torque and no member is held; therefore, there is no output. Fluid

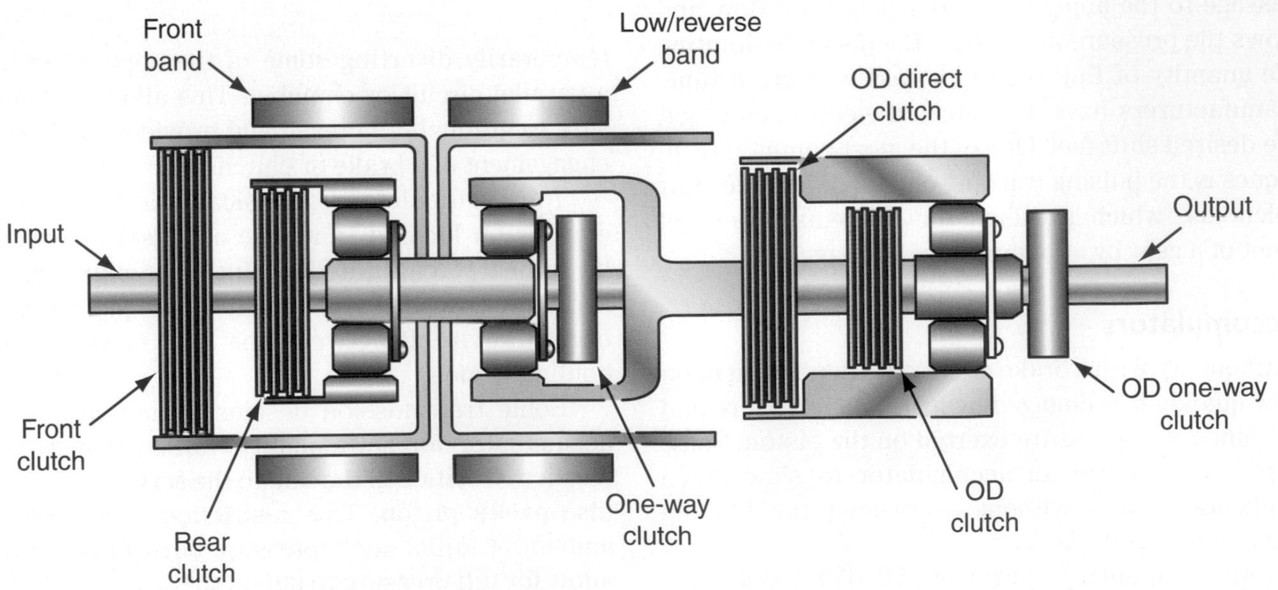

FIGURE 42–68

A typical four-speed transmission using a Simpson gearset and an add-on overdrive planetary unit.

TRANSMISSION CLUTCHES AND BANDS						OVERDRIVE CLUTCHES		
Lever Position	Front Clutch	Rear Clutch	One-way Clutch	Front Band	Low/Rev Band	OD Clutch	Direct Clutch	One-way Clutch
P—Park								
R—Reverse		X			X		X	
N—Neutral								
D—Drive								
First		X	X				X	X
Second		X		X			X	X
Third	X	X					X	X
Fourth	X	X				X		
2—Range second		X	X	X			X	X
1—Range first		X	X		X		X	X

FIGURE 42–69

The clutch, brake, and band application for the transmission shown in Figure 42–68.

flows to feed the torque converter and to lubricate and cool the transmission. In the neutral position, all of the same conditions of the park mode exist, except the park lever is not engaged to the park gear.

Reverse

When reverse is selected, the manual shift valve is moved, and fluid is sent to the front clutch and the low/reverse band. The pressure of the fluid is greater than mainline because boost pressure is added to the circuit at the pressure regulator valve. This added pressure provides good clamping and prevents clutch and band slippage.

The front clutch locks into the drive shell, and the sun gear becomes the input. The low/reverse band holds the rear planetary carrier. The sun gear rotates in a clockwise direction and drives the rear planetary pinions in a counter-clockwise direction. The pinions then cause the rear ring gear to rotate in a counter-clockwise direction. Because the rear ring gear is splined to the output shaft, the output shaft rotates in the opposite direction as the input shaft.

While the transmission is in reverse, the separate overdrive planetary gearset is in direct drive. This is accomplished by the engagement of the overdrive direct clutch that locks the sun and ring gears together. This clutch also provides for engine braking in reverse and the forward reduction gears.

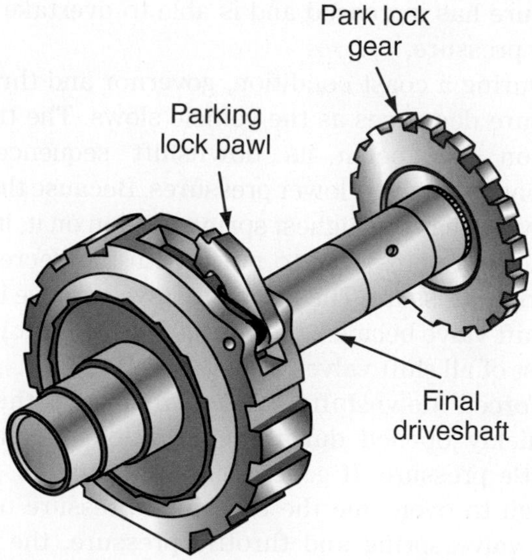

FIGURE 42–70

A park gear and shaft assembly.

Drive Range

When the shift linkage is moved to the DRIVE position, the manual shift valve allows fluid flow to engage the rear clutch. Fluid pressure is also directed to the shift valve. When the vehicle is stopped and in first gear, the various gear shift valves are closed by spring tension.

Input is transmitted through the rear clutch to the front ring gear in a clockwise direction. The front carrier is splined to the output shaft and is held by the weight of the vehicle at the drive wheels. The front ring gear drives the front planetary pinions in the same direction. The planets, in turn, drive the common sun gear in the opposite, or counter-clockwise, direction. The rear carrier is held by the one-way clutch as the planetary pinions attempt to rotate the rear ring gear; they walk the rear carrier in a counter-clockwise direction. The ring gear is splined to the output shaft, and the output is in the same direction as the input but at greater torque. This is first gear.

While the transmission is in the forward gears, the separate overdrive planetary gearset is in direct drive. The direct drive mode keeps transmission output connected to the input of the transmission. This unit also uses a one-way clutch that locks the shaft connecting the two planetary units to the output shaft.

This mode of operation will continue as long as the one-way clutch is locked or until the 1-2 shift valve forces an upshift. The clutch will remain locked until engine torque is released from the planetary unit or when the inertia of the vehicle is coasting and driving the planetary unit by the output shaft. The rear carrier rotates in a clockwise direction with the output shaft. This releases the one-way clutch. Because there is no longer a reactionary member, the transmission is effectively in neutral.

As the vehicle accelerates in first gear, the automatic movement of the shift valve controls progressive shifts through the other forward gears. As the vehicle moves forward, throttle pressure builds and helps the shift valve spring keep the valve in position. As output shaft and vehicle speed increases, governor pressure builds. Once governor pressure is great enough to overcome the combined pressure of the spring and throttle pressure, the 1-2 shift valve is pushed open. Fluid is directed to the front band, which holds the driving shell and the sun gear.

The rear clutch is still engaged and drives the front ring gear and front planetary pinions in a clockwise direction. The front planets drive the front carrier and the output shaft with some reduction. This is second gear. The rear planetary unit is ineffective because the rear carrier freewheels in the one-way clutch.

As the output shaft speed increases, governor pressure continues to build. Much more pressure is needed to move the 2-3 shift valve than was required to move the 1-2 shift valve. This is because it is held by a heavier spring tension and throttle pressure. Once governor pressure is great enough to overcome the pressure holding the 2-3 shift valve, the shift valve moves and opens the fluid circuit to the front clutch. At this time, fluid flow to the front band is stopped, and the sun gear is released. The rear clutch remains applied, and input is applied to the front ring gear through the rear clutch and to the sun gear through the front clutch. The front carrier is locked between these two input members and drives the output shaft at the same speed as the input. This is third gear.

To move the 3-4 shift valve requires more pressure than the 1-2 and 2-3 shift valves. Again, this is due to high spring tensions. Once vehicle and output shaft is great enough to increase governor pressure enough to move the 3-4 shift valve, fluid is sent to the overdrive clutch. When this happens, the 2-3 shift valve remains open and direct drive output is available to the overdrive planetary.

When pressure is first applied to the overdrive clutch, the overdrive clutch piston compresses the spring for the direct clutch and releases it. Then the pressure engages the overdrive clutch and locks the sun gear to the transmission housing.

Automatic Downshifting

Upshifts occur because governor pressure builds up enough to overcome the spring tension and the throttle pressure on a shift valve. Downshifting occurs when governor pressure decreases to the point where it can no longer overcome those pressures. Downshifting will also occur when throttle pressure has increased and is able to overtake governor pressure.

During a coast condition, governor and throttle pressure decreases as the vehicle slows. The transmission will begin its downshift sequence by responding to these lower pressures. Because the 3-4 shift valve has the highest spring tension on it, it will be the first shift valve to move with the decreased governor pressure. The last shift valve to move is the 1-2 shift valve because the spring tension on it is the lowest of all shift valves.

Forced downshifts occur when the throttle is quickly opened during acceleration, increasing throttle pressure. If governor pressure is not great enough to overcome the combined pressure of the shift valve spring and throttle pressure, the shift valve will close. This drops the transmission into the lower gear until governor pressure builds up enough to overtake the pressure on the shift valve.

Manual Low

When the driver selects a gear other than DRIVE, the manual shift valve controls the action of the other shift valves. By selecting a gear, the driver tells the transmission to stay in that gear or start off in that gear. The latter is often used when initiating movement when the vehicle is on ice or other slippery surfaces.

Basically, the movement of the gear selector into a manual gear inhibits the shift valve. When the driver selects DRIVE 2 while the transmission is in third gear, the transmission may downshift regardless of vehicle and engine speed. The manual shift valve cuts off line pressure to the 2-3 shift valve, and the transmission drops to second gear.

KEY TERMS

Accumulator (p. 1307)
Annulus (p. 1276)
Band (p. 1288)
Belleville spring (p. 1302)
Butt-end seals (p. 1299)
Check ball valve (p. 1304)
Continuously variable transmission (CVT) (p. 1284)
Coupling point (p. 1273)
Flow-directing valves (p. 1301)
Governor (p. 1306)
Hook-end seals (p. 1300)
Impeller (p. 1271)
Kickdown (p. 1306)
Lepelletier system (p. 1283)
Lip seals (p. 1299)
Multiple-disc clutch assembly (p. 1292)
Orifice (p. 1307)
Overrunning clutch (p. 1273)
Planetary carrier (p. 1276)
Planetary pinions (p. 1276)
Poppet valve (p. 1304)
Pressure regulating valves (p. 1301)
Ravigneaux gear train (p. 1281)
Rotary oil flow (p. 1273)
Servo (p. 1289)
Shift feel (p. 1307)
Simpson gear train (p. 1278)
Spool valve (p. 1304)
Square-cut seal (p. 1299)
Stator (p. 1272)
Sun gear (p. 1276)
Throttle valve (p. 1306)
Torque converter clutch (TCC) (p. 1274)
Torrington bearings (p. 1296)
Turbine (p. 1271)
Two-mode full hybrid system (p. 1288)
Vacuum modulator (p. 1306)
Valve body (p. 1304)
Vortex oil flow (p. 1273)

SUMMARY

- The torque converter is a fluid clutch used to transfer engine torque from the engine to the transmission. It automatically engages and disengages power transfer from the engine to the transmission in relation to engine rpm. It consists of three elements: the impeller (input), turbine (output), and stator (torque multiplier).
- Two types of oil flow take place inside the torque converter: rotary and vortex flow.
- An overrunning clutch keeps the stator assembly from rotating in one direction and permits overrunning when turned in the opposite direction.
- A lock-up torque converter eliminates the 10 percent slip that takes place between the impeller and turbine at the coupling stage of operation.
- Planetary gearsets transfer power and alter the engine's torque. Compound gearsets combine two simple planetary gearsets so that load can be spread over a greater number of teeth for strength and also so the largest number of gear ratios possible can be obtained in a compact area. A simple planetary gearset consists of a sun gear, a carrier with planetary pinions mounted to it, and an internally toothed ring gear.
- Planetary gear controls include transmission bands, servos, and clutches. A band is a braking assembly positioned around a drum. There are two types: single-wrap and double-wrap. Simple and compound servos are used to engage bands. Transmission clutches, either overrunning or multiple-disc, are capable of both holding and driving members.
- There are two common designs of compound gearsets: the Simpson gearset, in which two planetary gearsets share a common sun gear, and the Ravigneaux gearset, which has two sun gears, two sets of planet gears, and a common ring gear.
- Tandem gearsets are becoming more popular. These can feature two or more simple gearsets with various members joined together to provide additional forward gears. These can also be combinations of simple and compound gearsets, such as the Lepelletier system that uses a simple planetary gearset combined with a Ravigneaux gearset.
- Most Honda transaxles do not use a planetary gearset; rather, constant-mesh helical and square-cut gears are used in a manner similar to that of a manual transmission.
- The operation of most CVTs is based on a steel belt (chain) linking two variable pulleys.
- An automatic transmission uses ATF pressure to control the action of the planetary gearsets. This fluid pressure is regulated and directed to change gears automatically through the use of

various pressure regulators and control valves. To form a complete working hydraulic system, the following elements are needed: fluid reservoir (transmission oil pan), pressure source (oil pump), control valving (valve control body), and output devices (servos and clutches).

- The transmission's pump is driven by the torque converter drive hub at engine speed. The purpose of the pump is to create fluid flow and pressure in the system. Excessive pump pressure is controlled by the pressure regulator valve. Three common types of oil pumps are installed in automatic transmissions: the gear, rotor, and vane.
- The valve body is the control centre of the automatic transmission. It is made of two or three main parts. Internally, the valve body has many fluid passages called worm tracks.
- The purpose of a valve is to start, stop, or to direct and regulate fluid flow. Generally, in most valve bodies used in automatic transmissions, three types of valves are used: check ball; poppet; and, most commonly, spool.
- To prevent stalling, the automatic transmission pump has a pressure regulator valve normally located in the valve body. It maintains a basic fluid pressure. The valve's movement to the exhaust position is controlled by calibrated coil spring tension. The three stages of pressure regulation are charging the torque converter, exhausting fluid pressure, and establishing a balanced position. There are times when fluid pressure must be increased above its baseline pressure to accomplish these stages.
- The purpose of the governor assembly is to sense vehicle road speed and send a fluid pressure signal to the transmission valve body to upshift the transmission. Throttle pressure delays the transmission upshift and forces the downshift.
- Bearings that take radial loads are called bushings, and those that take axial loads are called thrust washers.
- The gaskets and seals of an automatic transmission help to contain the fluid within the transmission and prevent the fluid from leaking out of the various hydraulic circuits. Different types of seals are used in automatic transmissions; they can be made of rubber, metal, or Teflon.
- Three major types of rubber seals are used in automatic transmissions: the O-ring, the lip seal, and the square-cut seal.
- Three types of sealing rings are used in automatic transmissions: butt-end seals, open-end seals, and hook-end seals.
- There are four common configurations used as the final drives on FWD vehicles: helical gear, planetary gear, hypoid gear, and chain drive.

REVIEW QUESTIONS

1. Explain the difference between rotary and vortex fluid flow in a torque converter.
2. What component keeps the stator assembly from rotating when driven in one direction and permits rotation when turned in the opposite direction?
3. When a transmission is described as having two planetary gearsets in tandem, what does this mean?
4. The four common configurations used as the final drives on FWD vehicles are the ______ gear, ______ gear, ______ gear, and ______.
5. Describe the fluid flow during non-lock-up and lock-up operation of the piston clutch design of a lock-up torque converter.
6. What must occur in a simple planetary gearset to achieve a direct drive?
   a. The sun gear must be the input member.
   b. The ring gear must be the input member.
   c. The planetary carrier must be the input member.
   d. Two gearset members must be input members.
7. What must occur in a simple planetary gearset to achieve a slow overdrive?
   a. The sun gear must be the input member.
   b. The ring gear must be the input member.
   c. The planetary carrier must be the input member.
   d. The ring gear must be held.
8. What would be the result of holding the planetary carrier of a simple planetary gearset?
   a. reverse
   b. direct drive
   c. fast overdrive
   d. forward reduction
9. What allows the stator to operate correctly during vortex and rotary flow?
   a. the torque converter clutch
   b. an overrunning clutch
   c. governor pressure
   d. throttle pressure

10. What can throttle pressure be used for in automatic transmissions?
   a. to reduce governor pressure
   b. to increase governor pressure
   c. to reduce mainline pressure
   d. to increase mainline pressure
11. What affects shift valves to cause a downshift?
   a. governor pressure overcoming throttle pressure
   b. engine vacuum pressure overcoming governor pressure
   c. throttle pressure overcoming governor pressure
   d. vehicle speed overcoming governor pressure
12. What is the drive member of a torque converter?
   a. the stator
   b. the impeller
   c. the turbine
   d. the governor
13. What type of fluid flow is present in the torque converter during maximum torque multiplication?
   a. viscous flow
   b. vortex flow
   c. rotary flow
   d. radial flow
14. What makes a compound planetary gearset a Ravigneaux gearset?
   a. a common sun gear
   b. a common ring gear
   c. a common planetary carrier
   d. a one-piece front sun gear and rear ring gear assembly
15. What component is capable of driving a planetary gearset member in an automatic transmission?
   a. a band
   b. an overrunning clutch
   c. an input shaft–mounted multiple-disc clutch
   d. a case-mounted multiple-disc clutch
16. Which of the following is capable of controlling radial loads?
   a. Torrington washers
   b. nylon thrust washers
   c. steel-backed thrust washers
   d. bushings
17. In the Chrysler piston-type converter lock-up clutch assembly, what directs the oil flow to and from the lock up piston?
   a. lock-up valve
   b. shift solenoid
   c. fail-safe valve
   d. switch valve
18. What makes a compound planetary gearset a Simpson gearset?
   a. a common sun gear
   b. a common ring gear
   c. a common planetary carrier
   d. a one-piece front sun gear and rear ring gear assembly
19. Some CVT transaxles do not have a torque converter. What is used in its place?
   a. a flywheel with a start clutch
   b. a flywheel with a regular clutch assembly
   c. a viscous coupling
   d. a multiple-disc clutch
20. What torque converter component is responsible for torque multiplication?
   a. the impeller
   b. the stator
   c. the turbine
   d. the torque converter lock-up clutch

CHAPTER 43
Electronic Automatic Transmissions

LEARNING OUTCOMES

- Explain the advantages of using electronic controls for transmission shifting.
- Briefly describe what determines the shift characteristics of each selector lever position.
- Identify the input and output devices in a typical electronic control system, and briefly describe the function of each.
- Diagnose electronic control systems, and determine needed repairs.
- Conduct preliminary checks on the EAT systems, and determine needed repairs or service.
- Perform converter clutch system tests, and determine needed repairs or service.
- Inspect, test, and replace electrical/electronic sensors.
- Inspect, test, bypass, and replace actuators.

Electronic transmission control provides automatic gear changes when certain operating conditions are met. Through the use of electronics, transmissions have better shift timing and quality. As a result, the transmissions contribute to improved fuel economy, lower exhaust emission levels, and improved driver comfort. Although these transmissions function in the same way as earlier hydraulically based transmissions, a computer determines their shift points and the shift quality. The computer uses inputs from different sensors and matches the information to a predetermined program.

Today's electronic transmissions also have self-diagnostic capabilities and can accurately control mainline pressure, apply pressures, torque converter clutch (TCC) operation, and engine torque during shifts. Electronically controlled transmissions typically do not have governors or throttle pressure devices. The control unit monitors and controls the action of shift solenoids **(Figure 43–1)**. The solenoids do not directly control the transmission's clutches and bands. The clutches and bands are engaged or disengaged in the same way as hydraulically controlled units. The solenoids simply control the fluid pressures and do not perform a mechanical function.

In an electronic automatic transmission (EAT) system, there is a central processing unit, inputs, and outputs **(Figure 43–2)**. Often, the central processing unit is a separate computer designated for transmission control. The computer receives information from various inputs and controls two to five (perhaps even more!) solenoids that control hydraulic pressure and fluid flow to the apply devices and to the clutch of the torque converter. This computer may be the **transmission control module (TCM)**, the body control module (BCM), or the powertrain control module (PCM). Most manufacturers mount the TCM on the valve body inside the transmission. Doing this reduces the number of required electrical connectors and the amount of wiring. The TCM is tied into the serial (CAN) bus and shares information with the PCM.

TRANSMISSION CONTROL MODULE

A TCM is programmed to provide gear changes at the optimum time as well as shift timing, shift feel, and the operation of the torque converter clutch. To determine the best operating strategy for the transmission, the TCM uses inputs from some engine-related and driver-controlled sensors. It also receives input signals from specific sensors connected to the transmission. By monitoring all of these inputs, the TCM can determine when to shift gears or when to apply or release the torque converter clutch.

The decision to shift or not to shift is based on the shift schedules and logic. A **shift schedule** contains the actual shift points to be used by the computer according to the input data it receives from the sensors. Shift schedule logic chooses the appropriate gear and then determines the correct shift schedule

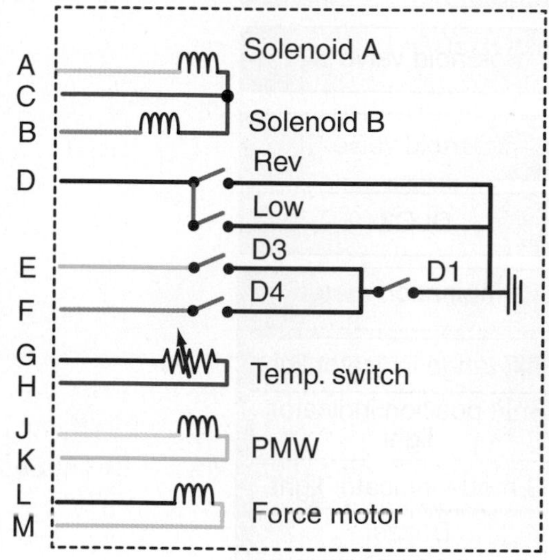

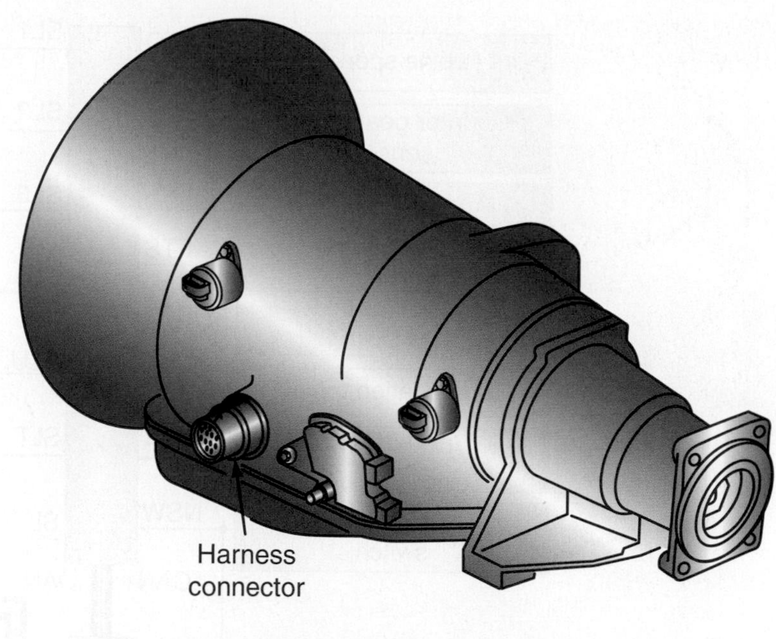

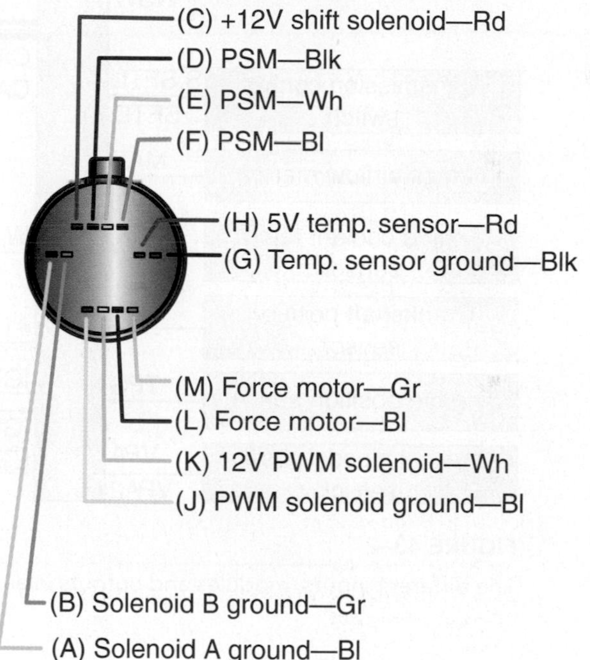

FIGURE 43–1

The key components of a typical EAT.

or pattern that should be followed. Each possible engine/transmission combination for a vehicle has a different set of shift schedules. The shift schedules set the conditions required for a change in gears. The computer frequently reviews the input information and can make quick adjustments to the schedule, if needed and *as* needed. To control shift quality, the TCM works with the PCM to momentarily alter engine output. This is accomplished by changing the ignition timing during a shift. The reduction in engine torque during shifting allows for smooth gear changes.

The electronic control systems used by the manufacturers differ with the transmission models and the engines to which they are attached. The components in each system and their operation also vary with the different transmissions. However, all operate in a similar fashion and use basically the same parts.

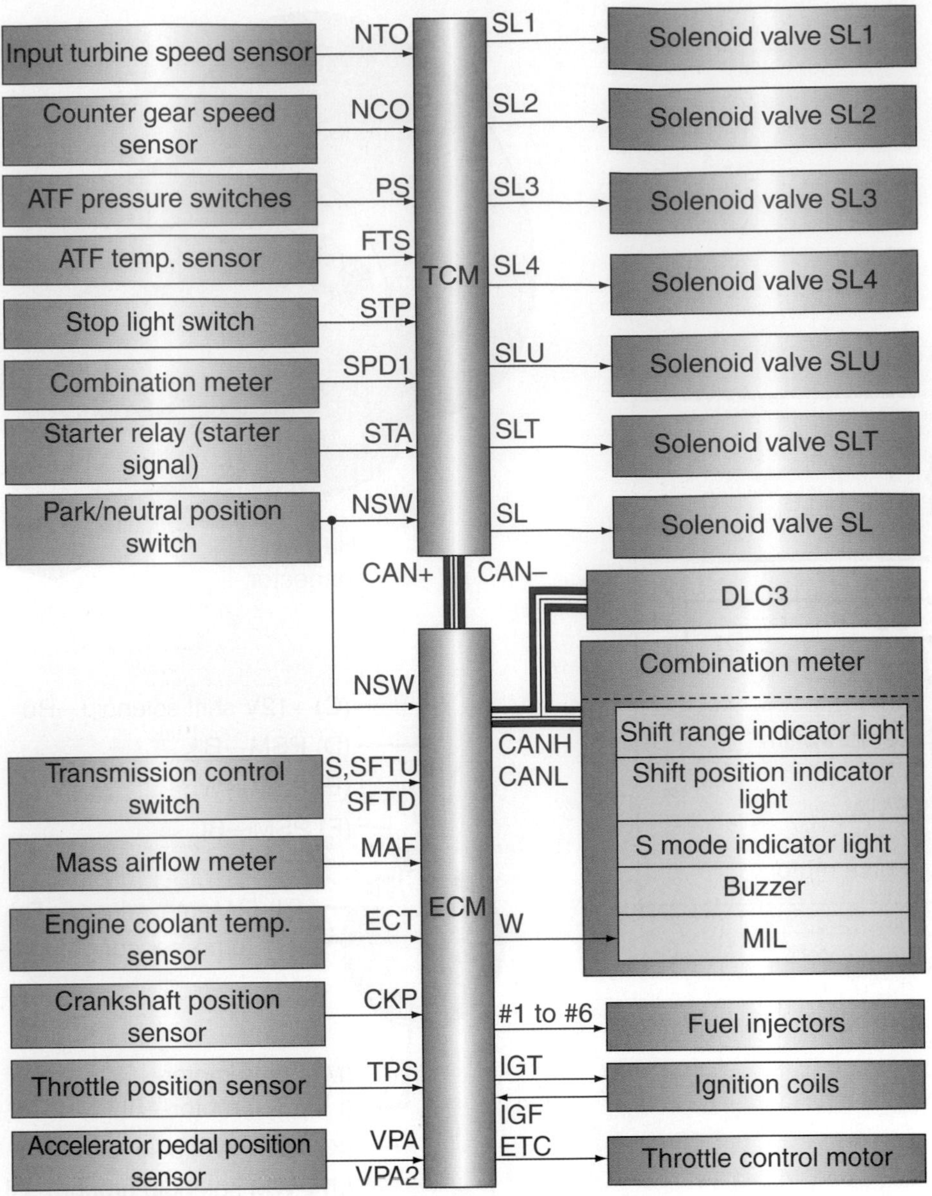

FIGURE 43–2

The different inputs, modules and outputs in an EAT circuit.

Inputs

The computer may receive information from two different sources: directly from a sensor or through the CAN communication bus that connects all of the vehicle's computer systems **(Figure 43–3)**. Normal engine-related inputs are used by the TCM to determine the best shift points. Many of these inputs are available at the common data bus. Other information, such as engine and body identification, the ECM's target idle speed, and speed control operation, are not the result of monitoring by sensors; rather, these have been calculated or determined by the ECM and made available on the bus.

Typical data inputs used by the TCM include the following:

- Throttle position (TP)
- Engine load
- Engine speed
- Gear range
- Vehicle speed
- Manifold absolute pressure (MAP)
- Mass airflow (MAF)
- Intake air temperature (IAT)
- Barometric pressure (BARO)
- Engine coolant temperature (ECT)
- ATF temperature

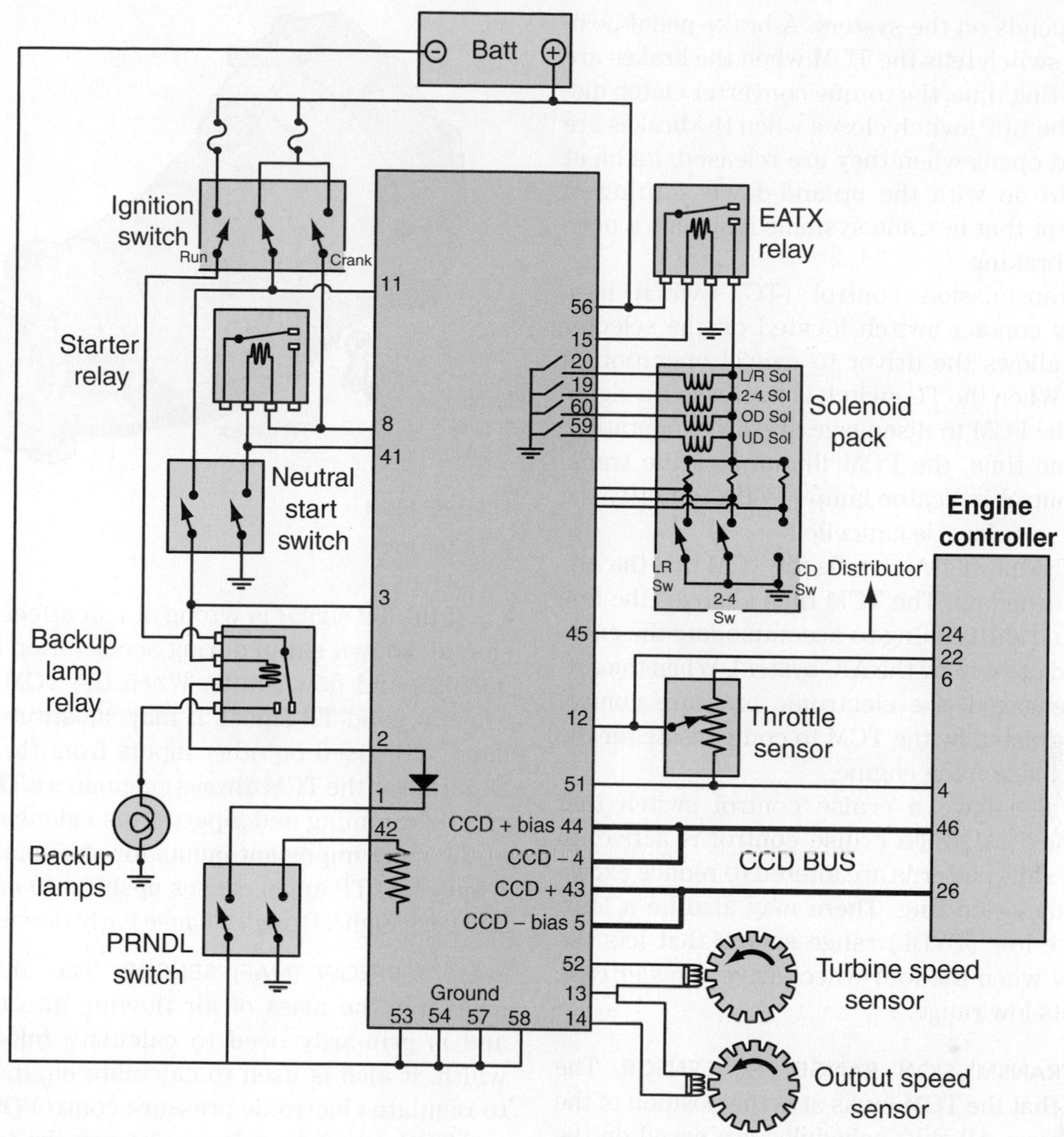

FIGURE 43–3

The electric circuit for a typical electronically controlled transmission. *Note:* The CCD BUS is the data source for other inputs in this multiplexed circuit.

- Crankshaft position (CKP) sensors
- Line pressure
- Input shaft speed
- Brake application
- A/C ON–OFF
- Cruise control ON–OFF

These provide the TCM with information about the operating condition of the engine. Through these, the TCM is able to control shifting and torque converter clutch operation according to the temperature, speed, and load of the engine.

Inputs that are tied directly to the TCM are typically not available on the bus circuit. Many of these sensors produce an analog signal that must be changed to a digital signal before the TCM can respond. This conversion is handled by an analog-to-digital (A/D) converter, the PCM, or a digital radio adapter controller (DRAC). These typically convert an analog AC signal to a digital 5-volt square wave.

The TCM constantly monitors what is happening inside the transmission with various speed and gear range sensors that tell it if the gears are shifting correctly and at what speeds. Once the TCM commands a gear change, solenoids are energized to hydraulically control the engagement of the proper clutch or band.

ON–OFF SWITCHES Several simple switches are used in the control of an EAT. The number and purpose of each depends on the system. A brake pedal position (BPP) switch tells the TCM when the brakes are applied. At that time, the torque converter clutch disengages. The BPP switch closes when the brakes are applied and opens when they are released. Its input has little to do with the up-and-down shifting of gears, except that in some systems it signals a need for engine braking.

The transmission control (TC) switch is a momentary contact switch located on the selector lever that allows the driver to cancel operation of overdrive. When the TC switch is depressed, a signal is sent to the PCM to disengage overdrive operation. At the same time, the PCM illuminates the transmission control indicator lamp (TCIL) to notify the driver that overdrive is cancelled.

An A/C request switch tells the TCM that the A/C has been turned on. The TCM then changes the line pressure and shift timing to accommodate the extra engine load created by the A/C system. When the A/C clutch is engaged, the electronic pressure control (EPC) is adjusted by the TCM to compensate for the additional load on the engine.

EATs also have a cruise control switch that informs the TCM when cruise control is active. In this mode, shift patterns are altered to reduce excessive and harsh shifting. There may also be a four-wheel-drive low (4WDL) range switch that lets the TCM know when the four-wheel-drive transfer case gear is in its low range.

DIGITAL TRANSMISSION RANGE (TR) SENSOR The first input that the TCM looks at is the position of the gearshift lever. All shift schedules are based on the gear selected by the driver. These schedules are coded by the position of the gear selector and the current gear range, and use throttle angle and vehicle speed as primary determining factors. A digital transmission range (TR) sensor informs the TCM of the gear selected by the driver. This sensor normally also contains the neutral safety switch and the reverse light switch. A TR sensor is typically a multiple pole–type on–off switch **(Figure 43–4)**. The digital TR sensor may be located on the outside of the transmission at the manual lever or it may be part of the TCM.

THROTTLE POSITION (TP) SENSOR The TP sensor sends a voltage signal to the TCM in response to throttle position. Not only is this signal used to inform the TCM of the driver's intent, it is also used in place of the hydraulic throttle pressure linkage. The TP sensor is very important to the operation of an EAT and is used for shift scheduling, EPC, and TCC control.

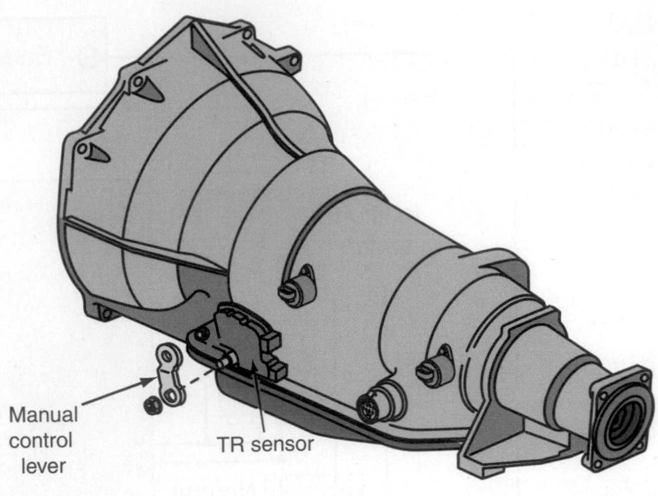

FIGURE 43–4

A TR sensor.

If the TP signal is wrong, it can affect transmission kickdown shifts during acceleration as well as upshifts and downshifts. When the TCM is unable to get a good TP signal, it may substitute a "calculated" TP based on other inputs from the data bus. Regardless, the TCM always calculates a TP based on its programming and inputs. This calculation is one of the most important inputs for shift pattern logic. A very low TP angle causes upshifts to occur early, and very high TP angles cause early downshifts.

MASS AIRFLOW (MAF) SENSOR The MAF sensor measures the mass of air flowing into the engine and is primarily used to calculate injector pulse width. It also is used to calculate engine load and to regulate electronic pressure control (EPC), shift and TCC scheduling. Depending on the type of fuel injection system used (speed density or airflow), engine load may be determined by the TP sensor, MAP sensor, and/or a vane airflow (VAF) sensor or MAF sensor.

Signals from a BARO sensor are used to adjust line pressures according to changes in altitude. This sensor input may not be used; its use depends on the type of intake air monitoring system the vehicle is equipped with. On those vehicles using the BARO sensor as an input, the sensor may be integrated in the PCM or it may be mounted externally.

TEMPERATURE SENSORS Shift schedules are also influenced by the engine's temperature. Sometimes, the engine's temperature is raised by delaying gear shifts. When the engine is overheating, shifts will occur sooner. The computer may also engage the converter clutch in second or third gear if the coolant temperature rises. Engine temperatures are often tied to ATF temperatures in the computer and are critical to the operation of a transmission.

The IAT sensor provides the fuel injection system with mixture temperature information. The IAT sensor is used both as a density corrector for airflow calculation and to proportion cold enrichment fuel flow. It is installed in the air cleaner outlet tube. The IAT sensor also is used in determining line pressures.

The TCM may also use the signal from the IAT to calculate the temperature of the battery. The TCM then uses this temperature to estimate transmission fluid temperature.

ENGINE SPEED To ensure that the transmission shifts at the correct time, the TCM must receive an engine speed input. This can be done through the ignition module, engine control module, serial data bus, or through a direct connection from a dedicated circuit between the CKP sensor and TCM. With the direct connection, all time delays at the bus circuit are eliminated, and the TCM always knows the current engine speed. This input is used to determine shift timing, wide-open throttle (WOT) shift control, TCC control, and EPC pressure. Also, to prevent the engine from running at too high of a speed, the TCM will order an upshift.

TRANSMISSION FLUID TEMPERATURE (TFT) SENSOR The transmission fluid temperature (TFT) sensor is normally located on the valve body. Signals from the sensor tell the TCM what the temperature of the ATF is. The sensor is a thermistor whose output signal varies with ATF temperature. The TCM uses this signal to control shift timing, shift feel, and TCC engagement. When the signal indicates normal operating temperature, the transmission is shifted normally; however, when the fluid is too cold or too hot, shifting is altered.

When the fluid temperature is too high, the TCM will operate the transmission in a way to allow it to cool. Shifts will occur sooner than normal, and the torque converter will apply in all forward ranges except first. When the fluid is too cold, shifting is delayed to help warm the fluid. Fluid temperature is used, along with other inputs, to control TCC clutch engagement. When the fluid is cold, the TCM prevents TCC engagement until the fluid reaches a specific temperature.

Some transmissions operate with distinct shift schedules based on fluid temperature. **Figure 43–5** shows the operating characteristics of a Chrysler 45RFE transmission when the ATF is at different temperatures.

If the TFT sensor fails or the TCM determines temperature signals are incorrect, the TCM will look at engine temperature to estimate the temperature of the ATF.

TRANSMISSION PRESSURE SWITCHES Various transmission pressure switches **(Figure 43–6)** can be used to keep the TCM informed as to which hydraulic circuits are pressurized and which clutches and brakes are applied. These input signals can serve as verification to other inputs and as self-monitoring or feedback signals. The most commonly used pressure switch is the transmission fluid pressure (TFP) switch, which monitors fluid pressure to determine when a clutch or band is applied or released.

VOLTAGE-GENERATING SENSORS A variety of speed sensors, most commonly Hall-effect or PM generator

TEMPERATURE RANGE	GEAR OPERATION	TCC OPERATION
Below −27°C (−16°F)	Operates only in park, neutral, reverse, 1st, and 3rd gears.	No
Between −24°C (−12°F) and 12°C (10°F)	Upshifts from 2nd to 3rd and 3rd to 4th are delayed and downshifts from 4th to 3rd and 3rd to 2nd are early.	No
Between 12°C (10°F) and 2°C (36°F)	All shifts are delayed.	No
Between 4°C (40°F) and 27°C (80°F)	Normal shifting.	No
Between 27°C (80°F) and 116°C (240°F)	Normal shifting.	Yes, normal
More than 116°C (240°F), or if the engine is overheating	2nd to 3rd and 3rd to 4th shifts are delayed.	Yes, but longer
Above 127°C (260°F)	2nd to 3rd and 3rd to 4th shifts are greatly delayed.	Yes, nearly full time

FIGURE 43–5

The operation of a Chrysler 45RFE transmission with the ATF at different temperatures.

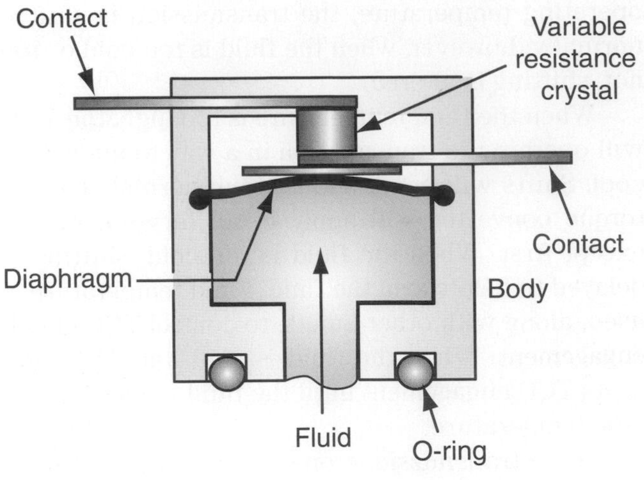

FIGURE 43–6

A pressure switch.

sensors, are used in today's EATs. These sensors serve as the governor signal and monitor the operation of the transmission. Most TCMs receive a signal from a vehicle speed sensor (VSS) and use this to determine the correct time to shift. Some EATs have an output shaft speed (OSS) sensor. This sensor may be used with the VSS or to provide a vehicle speed reference for the TCM. When a vehicle has both sensors, the OSS signal is used as a verification signal for the VSS, or vice versa. The OSS is typically used for control of torque converter clutch operation, shift timing, and fluid pressure control.

Some transmissions have an input speed sensor (ISS) or a turbine speed (TSS) sensor. This sensor is identical to the OSS, and its signal is used to calculate converter turbine speed **(Figure 43–7)**. It is also used, along with an engine speed input, to determine the amount of T/C slip by providing the TCM with the difference between engine speed and transmission input shaft speed. These inputs make it possible to accurately adjust the TCC apply pressure. Four-wheel-drive vehicles may have an additional speed sensor in the transfer case.

Outputs

Common outputs used with EATs are indicator lamps and solenoids. The indicator lamps can show the selected gear range in the instrument panel. There may also be a malfunction indicator lamp (MIL) designated for the transmission, or this warning light may be incorporated into the circuit for the engine's MIL. Shift, pressure control, and TCC solenoids are used in all modern EATs. All of these are controlled by the TCM. The solenoids are typically

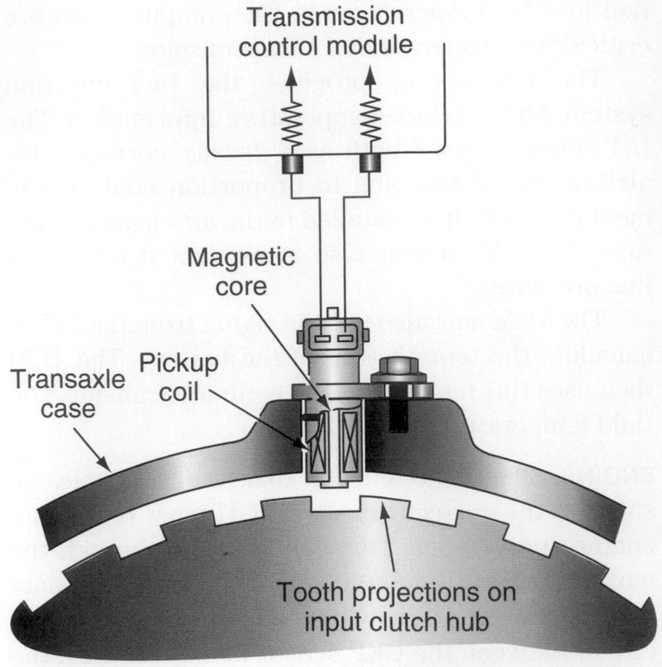

FIGURE 43–7

A voltage-generating speed sensor.

located inside the transmission and are mounted to the valve body.

SHIFT SOLENOIDS Shift solenoids **(Figure 43–8)** are used to regulate shift timing by controlling the delivery of fluid to the manual shift valve. These solenoids are on–off solenoids that are normally off and in the open position. When open, line pressure is present at the manual or selected shift valve. When the shift solenoids are energized, they block line pressure and allow pressure to exhaust from around the shift valve. This allows the valve to move.

The number and purpose of each depends on the model of transmission. A typical four-speed unit has two shift solenoids. The two solenoids offer four

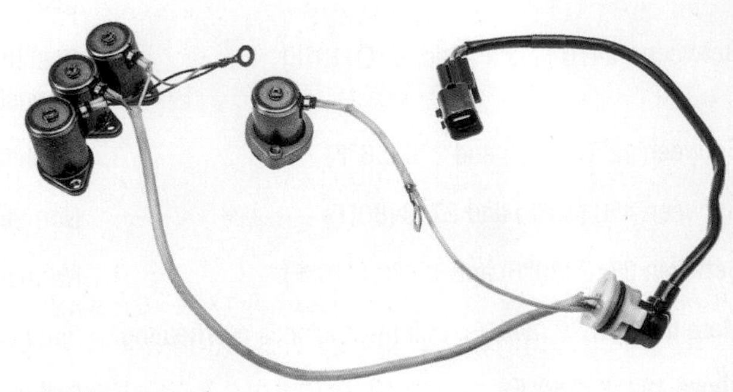

FIGURE 43–8

The solenoids for a transmission and their wiring harness.

LEVER POSITION	COMMANDED GEAR	SHIFT SOLENOID A	SHIFT SOLENOID B	TCC SOLENOID
P/R/N	1	ON	OFF	Disabled hydraulically
D	1	ON	OFF	Disabled hydraulically
D	2	OFF	OFF	Controlled electronically
D	3	OFF	ON	Controlled electronically
D	4	ON	ON	Controlled electronically
O/D switched off				
1	1	ON	OFF	Disabled hydraulically
2	2	OFF	OFF	Controlled electronically
3	3	OFF	ON	Controlled electronically
Manual 2	2	OFF	OFF	Controlled electronically
Manual 1	1	ON	OFF	Disabled hydraulically
Manual 1	2	OFF	OFF	Controlled electronically

FIGURE 43–9

The action of the solenoids in a typical four-speed transmission.

possible on–off combinations to control fluid to the various shift valves. This provides the engagement of the four forward gears **(Figure 43–9)**. Transmissions with additional forward gears rely on additional shift solenoids. The on–off combinations of these provide the additional gears. It is important to note that the TCM will stop current flow to a shift solenoid if it detects a fault in the solenoid or its circuit.

PRESSURE CONTROL SOLENOIDS The pressure control solenoid replaces the conventional TV cable setup to provide changes in line pressure in response to engine running conditions and engine load. The action of the solenoid controls the operating hydraulic pressures to the clutches and brakes to provide smooth and precise shifting. The pressure control solenoid is installed on the valve body.

In most cases, the pressure control solenoid is called an **electronic pressure control (EPC) solenoid** and is duty cycle–controlled by the TCM. Most of these solenoids are **variable force solenoids (VFSs)** or pulse-width modulated (PWM) solenoids and contain a spool valve or plunger and a spring. To control fluid pressure, the PCM sends a varying signal to the solenoid. This changes the amount the solenoid will cause the spool valve to move. When the solenoid is off, spring tension keeps the valve in place to maintain maximum pressure. When the solenoid is energized, it moves the spool valve. The movement of the valve uncovers the exhaust port around the valve, thereby causing a decrease in pressure **(Figure 43–10)**.

TCC SOLENOID The TCC solenoid is used to control the application, modulation, and release of the TCC. The operation of the converter clutch is also totally controlled by the TCM. The exception to this is when the clutch is hydraulically disabled to prevent engagement regardless of the commands from the computer. The converter clutch is hydraulically applied and electrically controlled through a PWM solenoid controlled by the TCM. When the solenoid is off, TCC signal fluid exhausts, and the converter clutch remains released. Once the solenoid is energized, TCC signal fluid passes to the TCC regulator valve and the clutch, engaging it. Modulating the pressure to the clutch allows for smooth engagement and disengagement and also allows for partial engagement of the clutch.

Whenever the TCM detects a problem in any of the solenoids in a transmission, it will disable the TCC solenoid.

Adaptive Controls

Most late-model EATs have systems that allow the TCM to change transmission behaviour based on the current condition of the transmission and engine, current operating conditions, and the habits of the driver. When systems are capable of doing this, they have **adaptive learning** capabilities. Adaptive

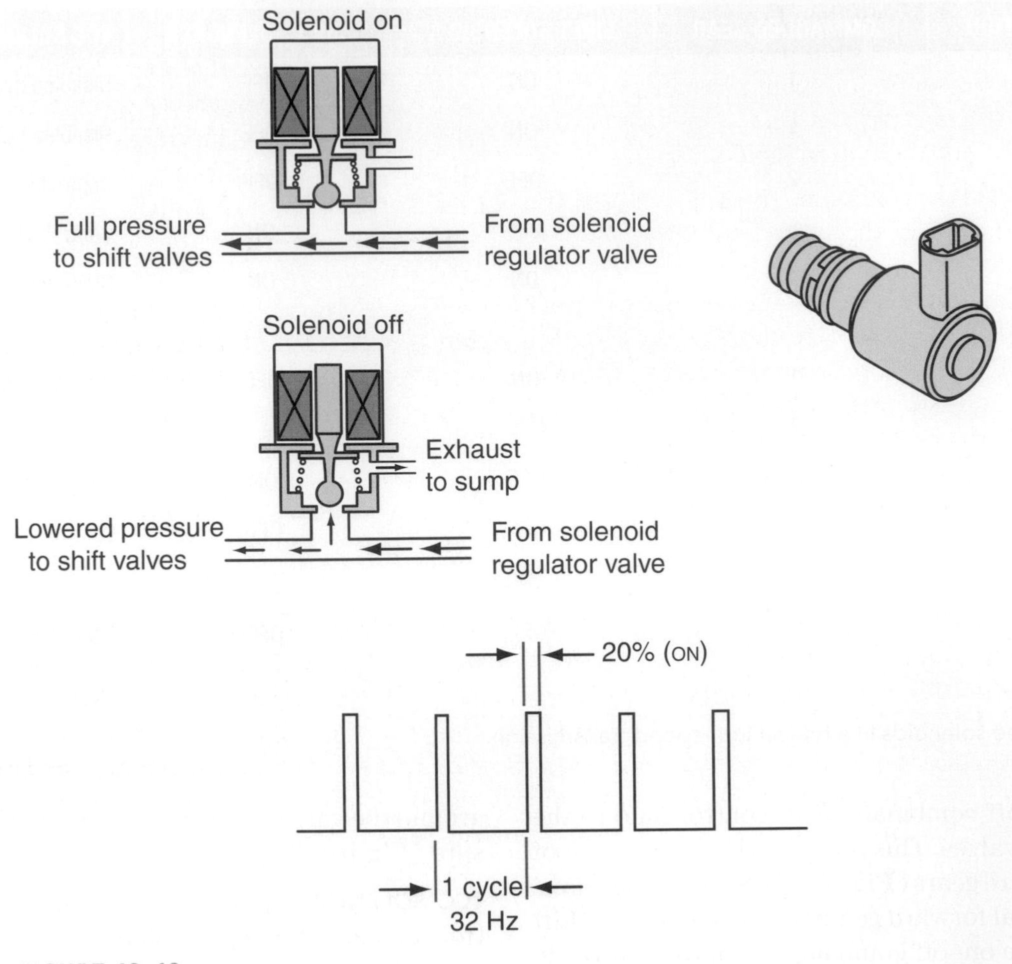

FIGURE 43–10

(Top) A typical PWM solenoid. (Bottom) The signal representing the control or ordered duty cycle from the computer.

learning provides consistent quality shifting and increases the durability of the transmission.

These transmissions have a line pressure control system that compensates for the normal wear of a transmission. As parts wear or conditions change, the time required to apply a clutch or brake changes. The TCM monitors the input and output speeds of the transmission during commanded shifts to determine if a shift is occurring too fast (harsh) or too slow (soft), and adjusts the line pressure to maintain the desired shift feel. Adaptive learning takes place as the TCM reads input and output speeds more than 140 times per second.

The system may also monitor the condition of the engine and compensate for any changes in the engine's performance. It also monitors and memorizes the typical driving style of the driver and the operating conditions of the vehicle. With this information, the computer adjusts the timing of shifts and converter clutch engagement to provide good shifting at the appropriate time.

SHOP TALK

EATs with adaptive learning may display harsh and/or abrupt shifting when they are new. This is normal because the TCM has not yet received enough information to learn about the vehicle or driver. This shifting behaviour may also occur after the vehicle's battery has been disconnected for a long period.

Limp-In Mode

When the TCM detects a serious transmission problem or a problem in its circuit (or in some cases, an engine control or data bus problem), it may switch to a default, limp-in, or fail-safe mode. Limp-in may also be initiated if the TCM loses its battery power feed. This mode allows for limited driving capabilities and is designed to prevent further transmission damage while allowing the driver to drive with decreased power and efficiency to a service facility for repairs.

The capabilities of a transmission while it is in limp-in mode depend on the extent of the fault, the manufacturer, and the model of transmission. When the TCM moves into the limp-in mode, a DTC is set, and the transmission will only operate in this mode until the problem is corrected. Examples of operating characteristics during limp-in include the following:

- The transmission locks in third gear when the gear selector is in the DRIVE position or second gear when it is in a lower position.
- The transmission will remain in whatever gear it was in but will shift into third or second gear and stay there as soon as the vehicle slows down.
- The transmission will only use first and third gears while in the DRIVE position.
- The transmission will operate only in PARK, NEUTRAL, REVERSE, and second gears, and will not upshift or downshift.

Operational Modes

With electronic controls, automatic transmissions can be programmed to operate in different modes. The desired mode is selected by the driver. The mode selection switch can be located on the centre console or the instrument panel. Most transmissions with this feature have two selective modes, usually called normal and power. During the normal mode, the transmission operates according to the shift schedule and logic set for normal or regular operation. In the power mode, the TCM uses different logic and shift schedules to provide for better acceleration and performance with heavy loads. Usually, this means delaying upshifts.

If three modes are available, the third mode is called the auto mode. The auto mode is a mixture between the normal and power modes. While in this mode, the TCM will control the shifts in a normal way. However, if the throttle is quickly opened, the shift pattern will switch to the power mode.

Some late-model vehicles have a TOW/HAUL switch. This operating mode delays upshifts and increases operating pressures, thus preventing the transmission's one-way clutches from freewheeling when the vehicle is operating with a heavy load. The mode also provides quicker downshifts during deceleration, which allows for more engine braking. The torque converter's lock-up clutch is applied sooner and in more gears than normal during acceleration. This helps to keep the temperature of the fluid in the torque converter down. Also, the clutch remains engaged for a longer period of time during deceleration to improve engine braking. In this mode, the transmission is able to shift into overdrive when the vehicle's load is overcome.

Manual Shifting

One of the features of electronically controlled transmissions is the availability of manual shift controls. Basically, these systems allow the driver to manually upshift and downshift the transmission at will, much like a manual transmission. Unlike a manual transmission, though, the driver does not need to depress a clutch pedal, nor is there a clutch assembly on the flywheel. The driver simply moves the gear selector or hits a button, and the transmission changes gears. If the driver does not change gears and engine speed is high, the transmission shifts on its own. If the driver elects to let the transmission shift automatically, a switch disconnects the manual control and the transmission operates automatically.

CVT Controls

The electronic control system for Honda's CVT consists of a TCM, various sensors, linear solenoids, and an inhibitor solenoid. Pulley ratios are always controlled by the control system. Input from the various sensors determines which linear solenoid the TCM will activate. Activating the shift control solenoid changes the shift control valve pressure, causing the shift valve to move. This changes the pressure applied to the driven and drive pulleys, which change the effective pulley ratio. Activating the start clutch control solenoid moves the start clutch valve. This valve allows or disallows pressure to the start clutch assembly. When pressure is applied to the clutch, power is transmitted from the pulleys to the final drive gearset **(Figure 43–11)**.

The start clutch allows for smooth starting. Because this transaxle does not have a torque converter, the start clutch is designed to slip just enough to get the car moving without stalling or straining the engine. The slippage is controlled by the hydraulic pressure applied to the start clutch. To compensate for engine loads, the TCM monitors the engine's vacuum and compares it to the measured vacuum of the engine while the transaxle was in PARK or NEUTRAL.

The TCM controls the pulley ratios to reduce engine speed and maintain ideal engine temperatures during acceleration. If the car is continuously driven at full throttle acceleration, the TCM causes an increase in pulley ratio. This reduces engine speed and maintains normal engine temperature while not adversely affecting acceleration. After the car has been driven at a lower speed or not accelerated for a while, the TCM lowers the pulley ratio. When the gear selector is placed into REVERSE, the TCM sends a signal to the PCM. The PCM then turns off the

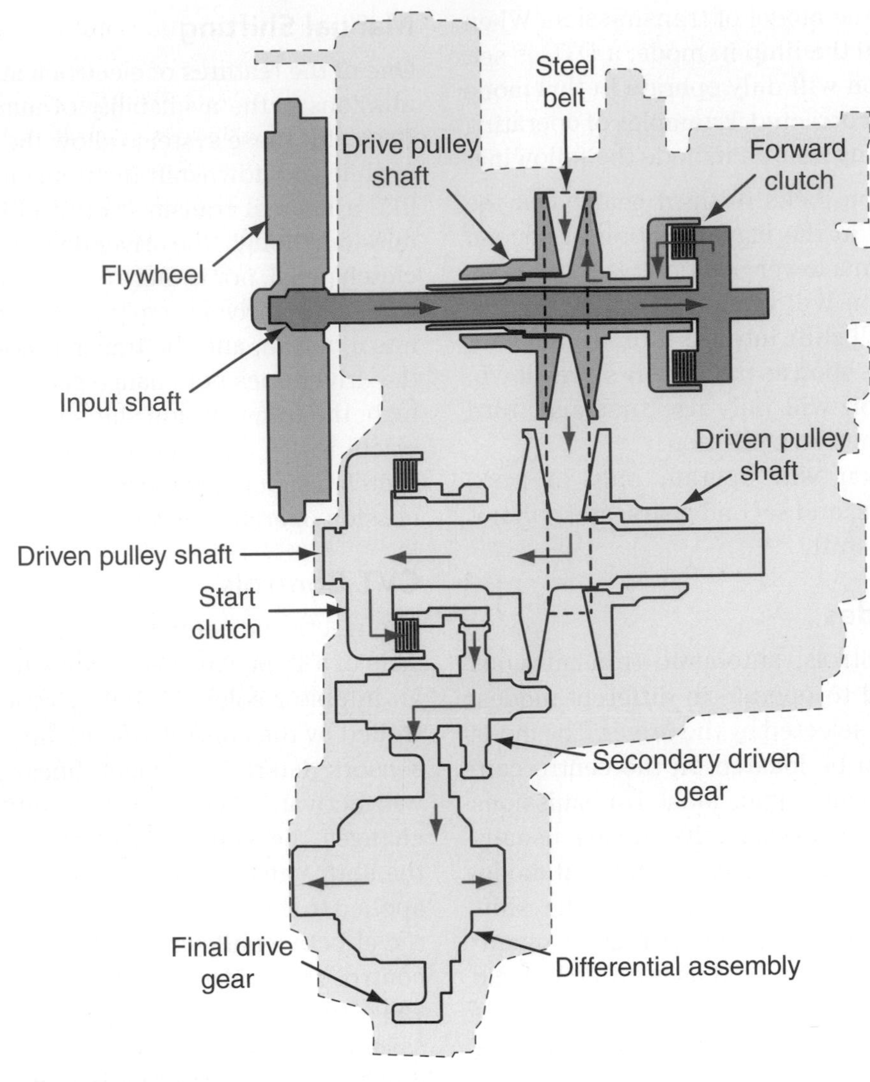

FIGURE 43–11

The electronic control system for a Honda CVT.

car's air conditioning and causes a slight increase in engine speed.

Audi's stepless Multitronic CVT is based on what it refers to as a variator. The variator is made of vanadium-hardened steel that is encased in oil and offers more durability than belt-driven CVTs. A manual gear selection mode is available, and six simulated gear ratios can be selected. In the automatic mode, Multitronic calculates the optimum gear ratio with the aid of a dynamic regulating program, according to engine load, the driver's preferences, and driving conditions.

HYBRID TRANSMISSIONS

Perhaps the most complex EATs are those used in most hybrid vehicles. The transmissions are fitted with electric motors that not only help propel the vehicle but also provide a constantly variable drive ratio. These CVTs do not rely on belts and pulleys; rather, the electric motors change the drive gear ratios. It is important to note that some hybrid vehicles rely on conventional CVTs and manual or automatic transmissions. This section covers the common nontraditional hybrid transmissions.

GO TO Chapter 36 for explanations on common hybrid vehicle systems.

Honda Hybrid Models

A modified version of an automatic, manual, or CVT transaxle is used in Honda hybrid vehicles. The transaxle is more compact, so it can fit behind the electric motor mounted at the rear of the engine **(Figure 43–12)** and occupy the same amount of

FIGURE 43–12

The IMA motor in Honda hybrids is mounted between the engine and the transaxle.

space as the transaxle does in a non-hybrid vehicle. These transaxles operate in the same way as other Honda units. The automatic transaxle is fitted with an integrated electric oil pump and different gear ratios that provide for better acceleration, fuel economy, and regenerative braking.

Toyota and Lexus Hybrids

The power split device used in Toyota and Lexus hybrids **(Figure 43–13)** operates as a continuously variable transaxle, although it does not use the belts and pulleys normally associated with CVTs. The variability of this transaxle depends on the action of a motor/generator, referred to as MG1, and the torque supplied by another motor/generator, referred to as MG2, and/or the engine.

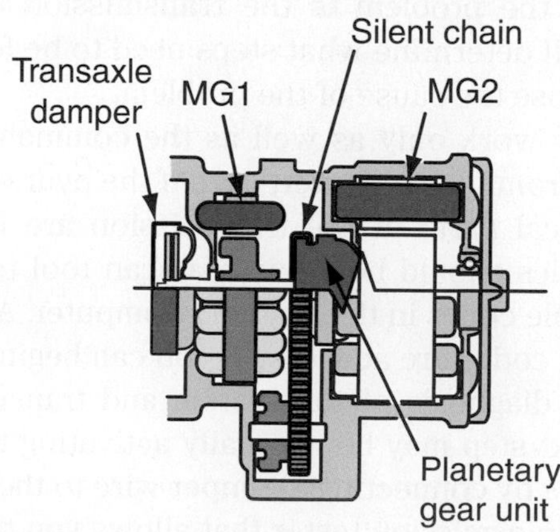

FIGURE 43–13

A typical power split transaxle for a Toyota hybrid.

This transaxle does not have a torque converter or clutch. Rather, a damper is used to cushion engine vibration and the power surges that result from the sudden engagement of power to the transaxle.

The engine and the two electric motors are connected to the planetary gearset. The gearset transfers power between the engine, MG1, MG2, and/or the drive wheels. The system splits power from the engine to different paths: to drive MG1 or drive the car's wheels, or both. MG2 can drive the wheels or be driven by them.

In the planetary gearset, the sun gear is attached to MG1. The ring gear is connected to MG2 and the final drive unit in the transaxle. The planetary carrier is connected to the engine's output shaft **(Figure 43–14)**. The key to understanding how this system splits power is to realize that when there are two sources of input power, they rotate in the same direction, but not at the same speed. Therefore, one can assist the rotation of the other, slow down the rotation of the other, or work together. Also, keep in mind that the rotational speed of MG2 largely depends on the power generated by MG1. Therefore, MG1 basically controls the continuously variable transmission function of the transaxle.

Ford Motor Company Hybrids

Ford's hybrids are equipped with an electronically controlled continuously variable transmission (eCVT). Based on a simple planetary gearset like Toyota's, however, Ford's transaxle is different in construction. In a Ford transaxle, the traction

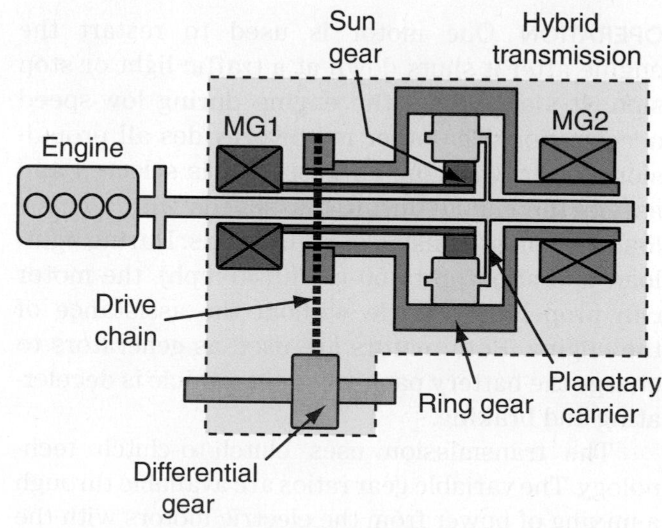

FIGURE 43–14

The arrangement of the electric motors and planetary gears in a power split transmission.

motor is not directly connected to the ring gear of the gearset. Rather, it is connected to the transfer gear assembly.

The effective gear ratios are determined by the speed of the motor/generator, engine, and traction motor. This means that these determine the torque that moves to the final drive unit in the transaxle. These power plants are controlled by the VSC through the TCM. Based on commands from the VSC and information from a variety of inputs, the TCM calculates the amount of torque required for the current operating conditions. A motor/generator control unit then sends commands to the inverter. The inverter, in turn, sends phased AC to the stator of the motors. The timing of the phased AC is critical to the operation of the motors, as is the amount of voltage applied to each stator winding.

Two-Mode Transmissions

The two-mode full-hybrid transmission relies on advanced hybrid, transmission, and electronic technologies to improve fuel economy and overall vehicle performance. It is claimed that the fuel consumption of a full-size truck or SUV is decreased by at least 25 percent when it is equipped with this hybrid system.

The system fits into a standard transmission housing and is basically three planetary gearsets coupled to two electronically controlled electric motors, which are powered by a 300-volt battery pack. This combination results in four forward speeds plus continuously variable gear ratios at low speeds and motor/generators for hybrid operation **(Figure 43–15)**.

OPERATION One motor is used to restart the engine after it shuts down at a traffic light or stop sign. It also assists the engine during low-speed acceleration. The other motor provides all propulsion power when the REVERSE gear is selected and assists the engine during low speeds with a heavy load and when cruising at high speeds. During light-load operation up to 50 km/h (30 mph), the motor can propel the vehicle without the assistance of the engine. Both motors are used as generators to charge the battery pack when the vehicle is decelerating and braking.

The transmission uses clutch-to-clutch technology. The variable gear ratios are available through a mixing of power from the electric motors with the engine's power. When the motors are not providing power, the transmission operates like a conventional four-speed automatic.

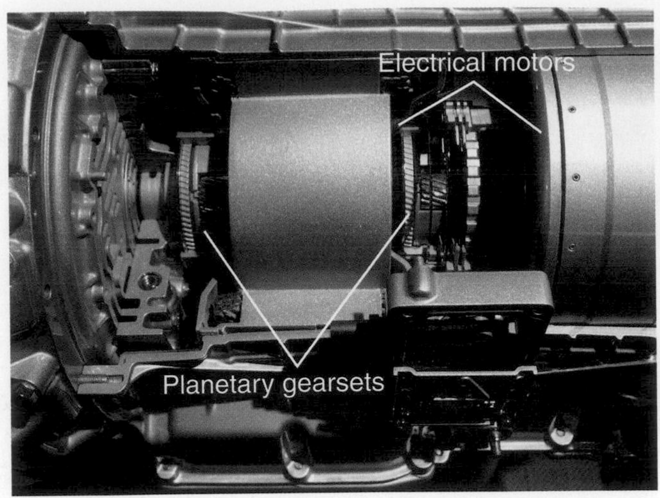

FIGURE 43–15

A planetary gearset is positioned in front of the two electric motors.

The hybrid system has two distinct modes of operation. It operates in the first mode during low-speed and low-load conditions, and the second mode is used while cruising at highway speeds. The system can operate solely on electric or engine power or by a combination of the two. Typically, when one or both of the motors are not providing propulsion power, they are working as generators driven by the engine or by the drive wheels for regenerative braking.

BASIC EAT TESTING

One of the first tasks during diagnosis of an EAT is to determine if the problem is caused by the transmission or by electronics. To determine this, the transmission must be observed to see if it responds to commands given by the computer. Identifying whether the problem is the transmission or electrical will determine what steps need to be followed to diagnose the cause of the problem.

EATs work only as well as the commands they receive from the computer, even if the hydraulic and mechanical parts of the transmission are fine. All diagnostics should begin with a scan tool to check for trouble codes in the system's computer. After the received codes are addressed, you can begin a more detailed diagnosis of the system and transmission. Your next step may be manually activating the shift solenoids by connecting a jumper wire to them or by using a transmission tester that allows you to manually activate the solenoids. Prior to doing this, study the wiring diagram for the solenoids to determine if the computer activates them by supplying voltage to them or by completing the ground circuit **(see Figure 43–3)**. In addition, you need to know in which gear certain solenoids are activated. The best way to diagnose an electronically controlled transmission is to approach solving the problem in a logical way. The recommended procedure for troubleshooting an EAT involves seven distinct steps that should be followed according to the order given:

PROCEDURE

Troubleshooting an EAT

STEP 1 Verify the customer's complaint. Pay attention to the conditions that exist when the problem occurs.

STEP 2 Check for any related symptoms, such as engine overheating, a lit MIL, and other driveability problems.

STEP 3 Conduct preliminary inspections and checks.

STEP 4 Check all service information for information that may apply to the complaint, including service bulletins, symptom charts, and recall notices.

STEP 5 Interpret and respond to all diagnostic codes.

STEP 6 Follow the diagnostic routines given by the manufacturer to define and isolate the cause of the problem.

STEP 7 Fix the problem and verify the repair.

Because many EAT problems are caused by the basics, it is wise to conduct all of the preliminary checks required for a nonelectronically controlled transmission. Also, thoroughly inspect the electronic system. This inspection should include a check of the MIL and the retrieval of diagnostic codes. Doing this will also allow you to pull engine-related codes as well as transmission codes. Whenever diagnosing a transmission, remember that an engine problem can and will cause the transmission to act abnormally.

Scan Tool Checks

Using a scan tool is one of the first steps of EAT diagnostics. Prior to retrieving the DTCs, pay attention to the MIL. The MIL is basically an engine malfunction light, but if the TCM detects a problem, it will send a request over the data bus to the PCM to turn on the MIL lamp. Therefore, if the MIL is lit, the engine or transmission can have a problem. Remember, the MIL does not light after all DTCs are set; however, it will illuminate if the fault is emission related.

GO TO ▶ Chapter 32 for details on connecting a scan tool and retrieving DTCs.

When diagnosing an EAT, make sure there are no engine-related codes that could affect the operation of the transmission. If there is an engine problem, fix it before continuing with your diagnosis.

DTCs that relate to transmission faults can be caused by engine or transmission input and/or output devices. These codes may seem to indicate a problem with an input or output circuit but may actually be caused by an internal transmission problem. Remember, codes are set by out-of-range values. Therefore, when the TCM is receiving a too low or high input signal, the cause is not necessarily a bad sensor. The sensor can be fine and a mechanical or hydraulic transmission problem can be causing the abnormal signals. Not only can internal transmission problems cause codes to be set, but so can basic electrical problems. Problems such as loose connections, broken wires, corrosion, and poor grounds will affect the signals.

If the TCM is unable to communicate with the PCM, there is a data bus problem. These problems will normally result in poor operation as well as the inability to retrieve DTCs from the TCM. The PCM constantly monitors the data bus and if it is unable to establish communication, it will order a data bus DTC.

Although the first steps in diagnosis include retrieving DTCs, there are problems that will not be evident by a code. These problems are solved with further testing, symptom charts, or pure logic. This logic must be based on an understanding of the transmission and its controls. It is possible to pinpoint the exact cause of a transmission problem: Monitor the serial data with a scan tool **(Figure 43–16)**. The serial data stream allows you to monitor system activity during operation. Comparing the observed values to the manufacturer's specifications will greatly help in diagnostics. However, it is possible that the data displayed by a scan tool are not the actual values. Most computer systems will disregard inputs that are well out of range and rely on a default value held in its memory. These default values are hard to recognize and do little for diagnostics; this is why the use of basic electrical troubleshooting equipment, such as wiring diagrams, diagnostic charts, DMMs, lab scopes, and special transmission testers, to check the system is common.

PID NAME	DESCRIPTION OF PID
EPC	Commanded electronic pressure control pressure—in psi
GEAR	Commanded gear—not actual
LINEDSD	Commanded line pressure—in psi
OSS	Input from output shaft speed sensor—in rpm
RPM	Input from engine speed sensor—in rpm
SSA	Commanded state of shift solenoid no. 1—ON or OFF
SSB	Commanded state of shift solenoid no. 2—ON or OFF
SS1F	Shift solenoid no. 1 circuit fault—YES or NO
SS2F	Shift solenoid no. 2 circuit fault—YES or NO
TCCACT	Slippage of torque converter clutch—in rpm
TCCCMD	Commanded state of torque converter clutch solenoid—in %
TCCF	Torque converter clutch solenoid circuit fault—YES or NO
TFT	Transmission fluid temperature—in voltage or degrees
TP	Throttle position—in voltage
TR	Transmission range sensor—by position
TRANRAT	Actual transmission gear ration—by position
TSS	Input from turbine shaft speed sensor—in rpm

FIGURE 43–16

Common transmission PIDs for Ford products.

USING SERVICE INFORMATION

At times, the manufacturer will make new software available that will correct common customer complaints. Normally, these are concerns that have no obvious physical or electronic cause. Flashing the TCM may take care of the concern. Always check the latest TSBs from the manufacturer when diagnosing a transmission concern.

Preliminary EAT Checks

Critical to proper diagnosis of EAT and TCC control systems is a road test. The road test should be conducted in the same way as one for a nonelectronic transmission except that a scan tool should also be connected to the circuit to monitor engine and transmission operation.

During the road test, the vehicle should be driven in the normal manner. Pay close attention to all gear changes. Also, the various computer inputs should be monitored and the readings recorded for future reference **(Figure 43–17)**. Some scan tools have the capability of printing out a report of the test drive. Critical information from the inputs includes engine speed, TP sensor, engine coolant and transmission fluid temperatures **(Figure 43–18)**, operating gear, and the time it took to shift gears. If the scan tool does not have the ability to give a summary of the road test, you should record this same information after each gear or change in operating condition.

Often, accurately defining the problem and locating related information in TSBs and other materials can identify the cause of the problem. When a

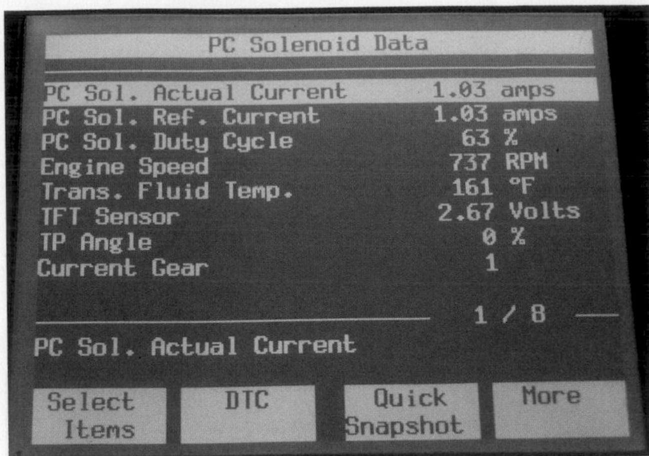

FIGURE 43–17

An example of some of the transmission-related data available on a scan tool.

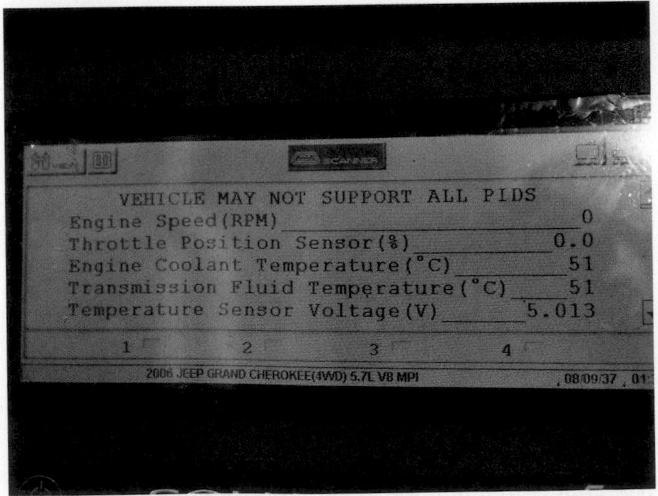

FIGURE 43–18

A display of other transmission-specific data.

manufacturer recognizes common occurrences of a problem, a bulletin will be issued regarding the fix of the problem. Also, for many DTCs and symptoms, the manufacturers give a simple diagnostic chart or path for identifying the cause of the problem. These are designed to be followed step by step and will lead to a conclusion if you follow the path matched to the symptom. Check all available information before moving on in your diagnostics.

Sometimes, the symptom will not match any of those described in the service information. This does not mean that it is time to guess; rather, it is time to clearly identify what is working right. By eliminating those circuits and components that are working correctly from a list of possible causes, you can identify what may be causing the problem and what should be tested further.

Common problems that affect shift timing and quality as well as the timing and quality of TCC engagement are incorrect battery voltage, a blown fuse, poor connections, a defective TP sensor or VSS, defective solenoids, crossed wires to the solenoids or sensors, corrosion at an electrical terminal, or faulty installation of an accessory.

Often, computer-controlled transmissions will start off in the wrong gear. This can happen for several reasons, such as internal transmission problems or external control system problems. Internal transmission problems can be faulty solenoids or stuck valves. External problems can be the result of a complete loss of power or ground to the control circuit, or the transmission is operating in its fail-safe mode. Typically, the default gear is simply the gear that is applied when the shift solenoids are off.

Electronic Defaults

While diagnosing a problem, always refer to the appropriate service information to identify the normal "default" operation of the transmission. You could spend time tracing the wrong problem by not recognizing that the transmission is operating in default.

Whenever the computer sees a potential problem that may increase wear and/or damage to the transmission, the system also defaults to limp-in mode. Minor slipping can be sensed by the computer through its input and output sensors. This slipping will cause premature wear and may cause the computer to move into its default mode; some systems may increase fluid pressure to compensate for the problem. A totally burnt clutch assembly will cause limp-in operation, as will some internal pressure leaks that may not be apparent until pressure tests are run.

Guidelines for Diagnosing EATs

The following guidelines will help diagnose EATs:

1. Make sure the battery has at least 12.6 volts before troubleshooting the transmission.
2. Check all fuses, and identify the cause of any blown fuses.
3. Check the physical condition of all sensors and the wiring going to them **(Figure 43–19)**.
4. Compare the wiring to all suspected components with the wire colours given in the service information.
5. When testing electronic circuits, always use a high-impedance testlight or DMM.
6. If an output device is not working properly, check the power circuit to it.
7. If an input device is not sending the correct signal back to the computer, check the reference voltage it is receiving and the voltage it is sending back to the computer.
8. Compare the voltages in and out of a sensor with the voltages the computer is sending out and receiving.
9. Before replacing a computer, check the solenoid isolation diodes according to the procedures outlined in the service information.
10. Make sure the computer wiring harnesses do not run parallel with any high-current wires or harnesses. The magnetic field created by the high current may induce a voltage in the computer harness. Take necessary precautions to prevent the possibility of static discharge while working with electronic systems.
11. While checking individual components, always check the voltage drop of the ground circuits. This becomes more and more important as cars are made of less material that conducts electricity well.
12. Make sure the ignition is off when you disconnect or connect an electronic component.
13. Check all sensors in cold and hot conditions.
14. Check all wire terminals and connections for tightness and cleanliness.
15. Use a TV-tuner cleaning spray or comparable product to clean all connectors and terminals.
16. Use dielectric grease at all connections to prevent future corrosion.
17. If you must break through the insulation of a wire to take an electrical measurement, make sure you tightly tape over or shrink-wrap the area after you are finished testing.

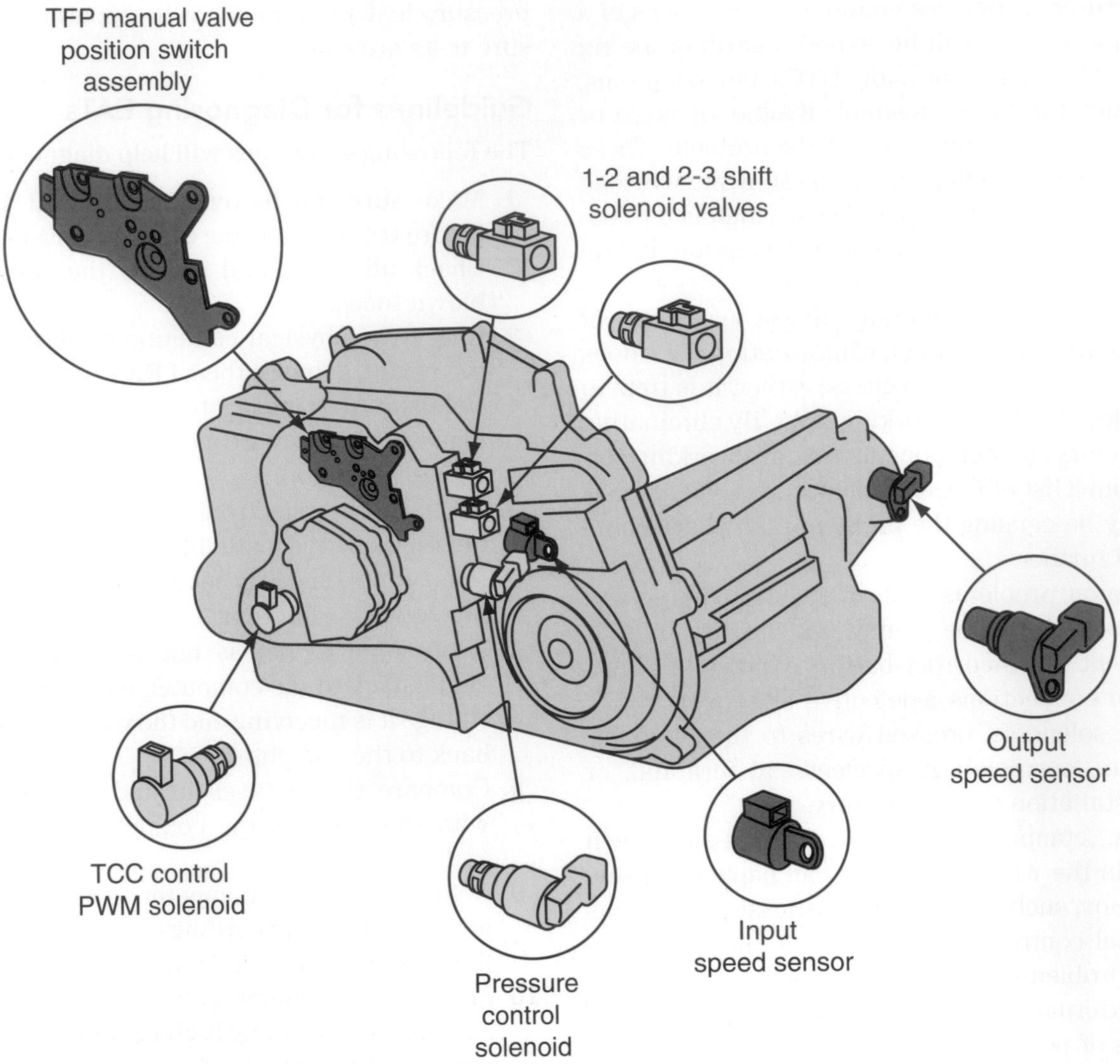

TFP manual valve position switch assembly

1-2 and 2-3 shift solenoid valves

Output speed sensor

TCC control PWM solenoid

Input speed sensor

Pressure control solenoid

FIGURE 43–19

An electronically controlled transaxle.

WARNING!

Static electricity can destroy an electronic part or render it useless. When handling any electronic part, do whatever is possible to reduce the chances of electrostatic buildup on your body and the inadvertent discharge to the electronic part.

CONVERTER CLUTCH CONTROL DIAGNOSTICS

To properly diagnose converter clutch problems, you must know when the TCC should engage and disengage and understand the function of the various controls involved with the system **(Figure 43–20)**. Although the actual controls for a TCC vary with the different models of transmissions, they all have certain operating conditions that must be met before the clutch can be engaged.

Diagnosis of a TCC circuit should be conducted in the same way as any other computer system. The computer will recognize problems within the system and store trouble codes that reflect the problem area of the circuit. A road test should be conducted to see if the problem is related to the TCC **(Figure 43–21)**.

On early electronically controlled systems with a TCC solenoid, the clutch is typically applied when oil flow through the torque converter is reversed. This change can be observed with a pressure gauge. Connect a pressure gauge to the fluid line from the transmission to the cooler. Position the gauge so that it is easily seen from the driver's seat. Then raise the vehicle on a hoist with the drive wheels off the ground and able to spin freely. Operate the

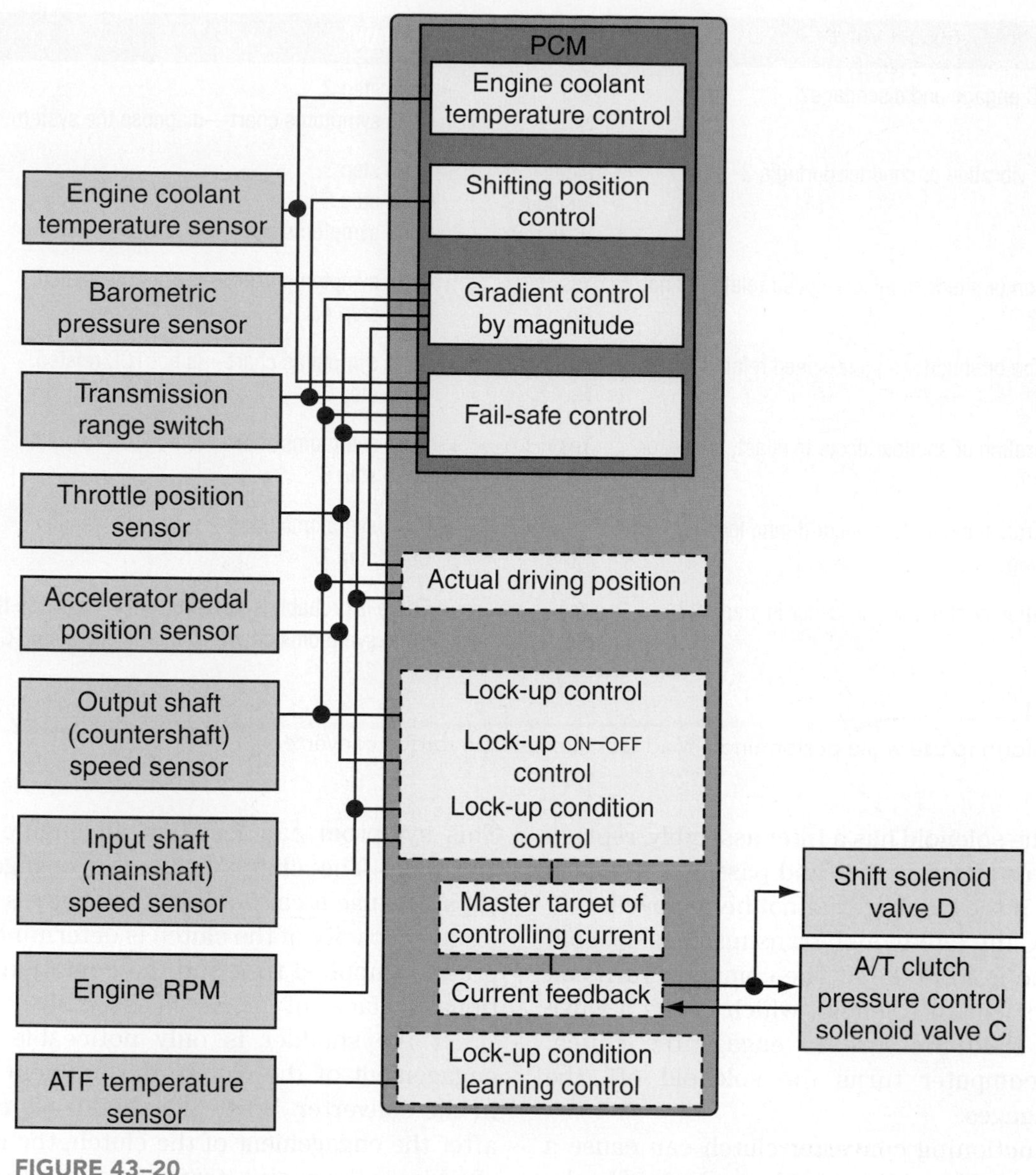

FIGURE 43–20

The electronic control system for a TCC.

vehicle until the transmission shifts into high gear. Then maintain a speed of approximately 90 km/h (55 mph). Once the speed is maintained, watch the pressure gauge.

If the pressure decreases 0.35 to 0.70 kg/cm^2 (5 to 10 psi), the converter clutch was applied. With this action, you should feel the engagement of the clutch as well as a drop in engine speed. If the pressure changed but the clutch did not engage, the problem may be inside the converter or at the end of the input shaft. If the input shaft end is worn or the O-ring at the end is cut or worn, there will be pressure loss at the converter clutch. This loss in pressure will prevent full engagement of the clutch. If the pressure did not change and the clutch did not engage, suspect a faulty clutch valve or control solenoid or a fault in the solenoid control circuit.

If the clutch does not engage, check for power to the solenoid. If power is available, make sure the ground of the circuit is good. If there is power available and the ground is good, check the voltage drop across the solenoid. The solenoids should drop very close to source voltage. If less than that is measured, check the voltage drop across the power and ground sides of the circuit. If the voltage drop testing results are good, remove the solenoid and test it with an ohmmeter. Suspect clutch material, dirt, or other material plugging up the solenoid valve passages if the solenoid checks out fine with the ohmmeter. Attempt to flush the valve with clean ATF if blockage

STEP	FINDINGS	REMEDY
1. Does the TCC engage and disengage?	Yes No	Go to step 2. Go to symptoms chart—diagnose the system.
2. Describe the vibration or shudder during a 3-4 or 4-3 shift.	Light Medium Heavy	Go to step 3. Go to step 3. Go to symptoms chart—is not TCC related.
3. Is the vibration or shudder vehicle-speed related or not gear related?	Yes No	Go to symptoms chart—is not TCC related. Go to step 4.
4. Is the vibration or shudder engine-speed related or not gear related?	Yes No	Go to symptoms chart—is not TCC related. Go to step 5.
5. Does the vibration or shudder occur in coast, cruise, or reverse gear?	Yes No	Go to symptoms chart—is not TCC related. Go to step 6.
6. Does the vibration or shudder occur during long periods of light braking?	Yes No	Go to symptoms chart—is not TCC related. Go to step 7.
7. Did the vibration or shudder only occur in step #2?	Yes No	There is probably a TCC problem—diagnose the system. Go to symptoms chart—the problem is not TCC related.

FIGURE 43–21

An evaluation form to use while performing a road test to check the torque converter.

is found. If the solenoid has a filter assembly, replace the filter after cleaning the fluid passages. Replace the solenoid if the blockage cannot be removed.

The TCC in late-model transmissions is controlled by the PCM or TCM. The computer turns on the converter clutch solenoid, which opens a valve and allows fluid pressure to engage the clutch. When the computer turns the solenoid off, the clutch disengages.

A malfunctioning converter clutch can cause a wide variety of driveability problems. Normally, the application of the clutch should feel like a smooth engagement into another gear. It should not feel harsh, nor should there be any noises related to the application of the clutch.

If the clutch engages at the wrong time, a sensor or switch in the circuit is probably the cause. If clutch engagement occurs at the wrong speed, check all speed-related sensors. A faulty temperature sensor may cause the clutch not to engage. If the sensor is not reading the correct temperature, the PCM may never realize that the temperature is suitable for engagement. Checking the appropriate sensors can be done with a scan tool, DMM, and/or lab scope.

Engagement Quality

If the clutch engages prematurely or is not applied with full pressure, a shudder or vibration results from the rapid grabbing and slipping of the clutch. This symptom can feel like an engine misfire or vibration. The clutch begins to engage and then slips because it cannot hold the engine's torque. The torque capacity of the clutch is determined by the oil pressure applied to it and the condition of the frictional surfaces of the clutch assembly.

If the shudder is only noticeable during the engagement of the clutch, the problem is typically in the converter. When the shudder is only evident after the engagement of the clutch, the cause of the shudder is the engine, transmission, or another component of the driveline. If the shudder is caused by the clutch, the converter must be replaced to correct the problem.

A faulty clutch solenoid or its return spring may cause low apply pressure. The valve controlled by the solenoid is normally held in position by a coil-type return spring. If the spring loses tension, the clutch will engage too soon. Because there is insufficient pressure to hold the clutch, shudder occurs as the clutch begins to grab and then slip. If the solenoid valve and/or return spring are faulty, they should be replaced, as should the torque converter. If the TCC fails to release, it can cause the engine to jerk and stall when the vehicle is stopping.

An out-of-round torque converter prevents full clutch engagement, which will also cause shudder, as will contaminated clutch frictional material. The frictional material can become contaminated by metal particles circulating through the torque converter and collecting on the clutch. Broken or worn clutch dampener springs will also cause shudder.

DETAILED TESTING OF INPUTS

There are many different designs of sensors that are part of the control system for an EAT. The transmission will not work properly if it receives bad information from its sensors or from the CAN bus. The transmission may shift at the wrong speeds, shift harshly, or operate only in the limp-in mode.

Some sensors are nothing more than a switch that completes a circuit. Others are complex devices that react to chemical reactions and generate their own voltage during certain conditions. If the preliminary tests pointed to a possible problem in an input circuit, the circuit should be tested. Make sure to check all suspect circuits for resistance problems; conduct voltage drop tests on those circuits. Often, the manufacturers will give specific testing procedures for their sensors; always follow them.

GO TO ▶ Chapter 25 for details on specific tests of input devices.

Testing Switches

Many different switches are used as inputs or control devices for EATs. Most of the switches are either mechanically or hydraulically controlled. These switches can be easily checked with an ohmmeter. With the meter connected across the switch's leads, there should be continuity or low resistance when the switch is closed, and there should be infinite resistance across the switch when it is open. A testlight can also be used. When the switch is closed, power should be present at both sides of the switch. When the switch is open, power should be present at only one side.

Pressure switches can be checked by applying air pressure to the part that would normally be exposed to fluid pressure **(Figure 43–22)**. When applying air pressure to these switches, check them for leaks. Although a malfunctioning electrical switch will probably not cause a shifting problem, it will if it leaks. If the switch leaks off the applied pressure in a hydraulic circuit to a holding device, the holding member may not be able function properly. When possible, you should check pressure switches when they are installed and controlled by the vehicle.

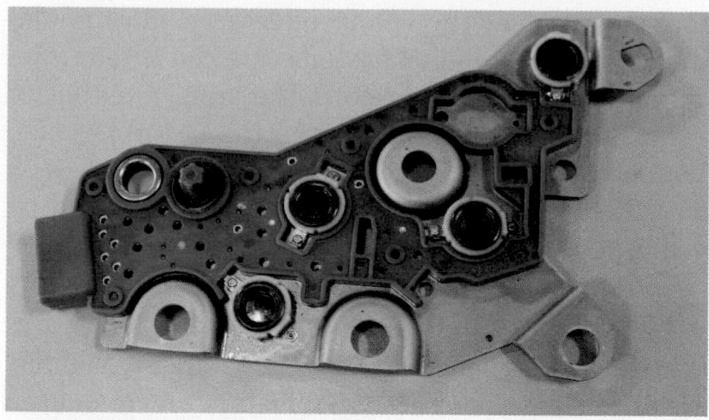

FIGURE 43–22

A typical transmission switch assembly.

Martin Restoule

Throttle Position (TP) Sensor

Another type of switch is a potentiometer. Rather than open and close a circuit, a potentiometer controls the circuit by varying its resistance in response to something. A TP sensor is a potentiometer. A bad TP sensor can cause the following problems: no upshifts, quick upshifts, delayed shifts, line-pressure problems with transmissions that have a line pressure control solenoid, and erratic converter clutch engagement.

SHOP TALK

A faulty TP sensor may not cause a DTC to be set. The diagnostic capabilities of the PCM must be able to determine if the sensor is working correctly or not. If it does not have this capability, it may not set a code. The PCM must be able to look at the input from the TP sensor and compare it to other inputs, such as engine speed, MAP inputs, and airflow. If the PCM determines that the TP signal does not reflect a true value based on engine speed and load inputs, it will set a code.

A TP sensor can be checked with an ohmmeter or a voltmeter. If checked with an ohmmeter, you should be able to watch the resistance of the sensor change as the throttle is opened and closed. Often, there will be a resistance specification given in the service information. Compare your reading to this.

With a voltmeter, you will be able to measure the reference voltage and the output voltage. Both of these should be within specified amounts. If the reference voltage is lower than normal, check the voltage drop across the reference voltage circuit from the computer to the TP sensor. Replace the TP sensor if it is defective.

Testing with a lab scope is a good way to watch the sweep of the resistor. The waveform is a DC signal that moves up as the voltage increases. Most potentiometers in computer systems are fed a reference voltage of 5 volts. Therefore, the voltage output of these sensors will typically range from 0.5 to 4.5 volts. The change in voltage should be smooth. Look for glitches in the signal. These can be caused by changes in resistance or an intermittent open.

Mass Airflow Sensor

When an MAF sensor fails or sends faulty signals, the engine runs roughly and tends to stall as soon as you put the transmission into gear. This sensor can be checked with a multimeter set to the Hz frequency range. Check the service information for specific values. Normally, at idle, 30 Hz is measured and the frequency will increase as the throttle opens.

A scan tool may also be used to test this sensor; most have a test mode that monitors MAF sensors. The output of some MAFs can be observed with a DMM; their output is variable DC voltage. While diagnosing these systems, keep in mind that cold air is denser than warm air.

Temperature Sensors

Temperature sensors can be checked with an ohmmeter. To do so, disconnect the sensor. In most cases, the sensor can be checked at room temperature **(Figure 43–23)**. Determine the temperature of the sensor, and measure the resistance across it. Compare your reading to the chart of normal resistance for that temperature, which is given in the service information.

Thermistor activity can be monitored with a lab scope. With the scope connected to the thermistor or temperature sensor, run the engine and watch the waveform. As the temperature increases, there should be a smooth increase or decrease in voltage. Look for glitches in the signal. These can be caused by changes in resistance or an intermittent open.

Speed Sensors

Speed sensors negate the need for hydraulic signals from a governor. When this sensor fails or sends faulty readings, it can cause complaints similar to those caused by a bad TP sensor. The most common complaints are no overdrive, no converter clutch engagement, and no upshifts.

The operation of a PM generator–type speed sensor can be tested with a DMM set to measure AC voltage. Raise the vehicle on a lift. Allow the wheels to be suspended so that they can rotate freely. Connect the meter to the speed sensor. Start the engine, and put the transmission in gear. Slowly increase the engine's speed until the vehicle is at approximately 30 km/h (19 mph), and then measure the voltage at the speed sensor. Slowly increase the engine's speed and observe the voltmeter. The voltage should increase smoothly and precisely with an increase in speed.

Magnetic pulse generator speed sensors can be tested with a lab scope. Connect the lab scope's leads across the sensor's terminals. The expected pattern is an AC signal, which should be a perfect sine wave when the speed is constant. When the speed is changing, the AC signal should change in amplitude and frequency. If the readings are not steady and do not smoothly change with speed, suspect a faulty connector, wiring harness, or sensor.

A speed sensor can also be tested when it is out of the vehicle. Connect an ohmmeter across the sensor's terminals. The desired resistance readings across the sensor will vary with every individual sensor; however, you should expect to have continuity across the leads. If there is no continuity, the sensor is open and should be replaced. Reposition the leads of the meter so that one lead is on the sensor's case and the other to a terminal. There should be no continuity in this position. If there is any measurable amount of resistance, the sensor is shorted.

Test connections	Specified value
E2 - THO1	90Ω to 156kΩ
E2 - ground	156kΩ or higher
THO1 - ground	156kΩ or higher

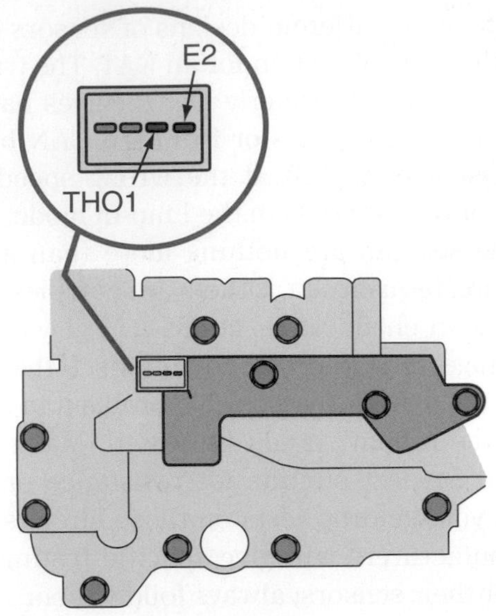

FIGURE 43–23

Checking an ATF temperature sensor.

DETAILED TESTING OF ACTUATORS

If you were unable to identify the cause of a transmission problem through the previous checks, you should continue your diagnostics with testing the solenoids. This will allow you to determine if the shifting problem is the solenoids or their control circuit, or if it is a hydraulic or mechanical problem.

GO TO Chapter 32 for details on specific tests of output devices, including solenoids.

Before continuing, however, you must first determine if the solenoids are case grounded and fed voltage by the computer, or if they always have power applied to them and the computer merely supplies the ground. While looking in the service information to find this, also find the section that tells you which solenoids are on and which are off for each of the different gears.

To begin this test, you should collect the tools and/or equipment necessary to manually activate the solenoids. Switch panels that connect into the solenoid assembly are available and allow the technician to switch gears by depressing or flicking a switch.

SHOP TALK

A solenoid tester is easily made. Get a wiring harness for the transmission you want to test. Connect the harness to simple switches. Follow the solenoid/gear pattern when doing this. To change gears, all you will need to do is turn off one switch and turn on the next.

With the tester, the solenoids will be energized in the correct pattern; observe the action of the solenoids. To totally test the transmission, you should shift gears under light, half, and full throttle. If the transmission shifts fine with the movement of the switches, you know that the transmission is fine. Any shifting problem must therefore be caused by something electrical. If the transmission did not respond to the switch movements, the problem is probably in the transmission.

At times, a solenoid will work fine during light throttle operation but may not allow the valve to exhaust enough fluid when pressure increases. To verify that the valve is not exhausting, activate the solenoid with the vehicle under load. If the valve cannot exhaust, the transmission will downshift. Restricted solenoids are a common cause of rough shifting under heavy loads or full throttle but good shifting under light throttle.

Testing Actuators with a Lab Scope

You will be able to watch the actuator's electrical activity by observing its action on a lab scope. Normally, if there is a mechanical fault, this will affect its electrical activity as well. Some actuators are controlled by pulse-width-modulated signals **(Figure 43–24)**. These devices are controlled by varying the pulse width, signal frequency, and/or voltage levels. By watching the control signal, you can see the turning on and off of the solenoid **(Figure 43–25)**. All waveforms should be checked for amplitude, time, and shape. You should also observe changes to the pulse width as operating conditions change. A bad waveform will have noise, glitches, or rounded corners. You should be able to see evidence that the actuator immediately turns off and on according to the commands of the computer.

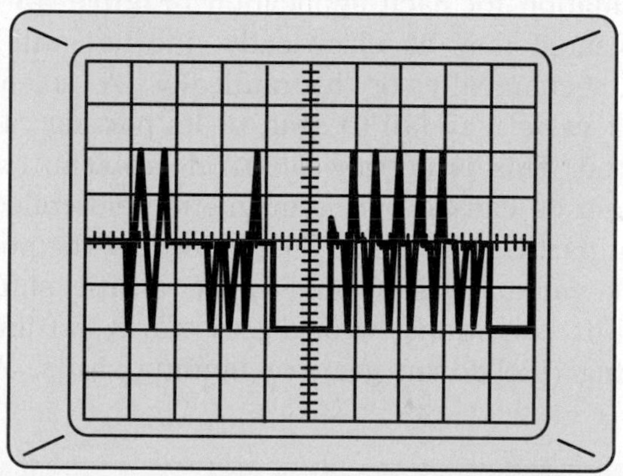

FIGURE 43–24

A typical control signal for a pulse-width-modulated solenoid.

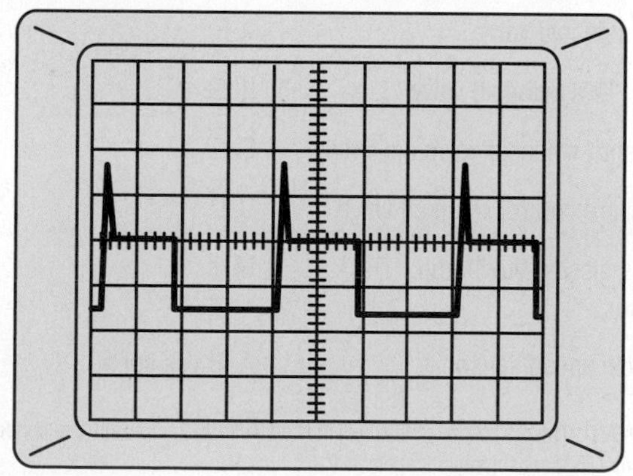

FIGURE 43–25

The activity of a solenoid as it cycles on and off.

Testing Actuators with an Ohmmeter

Solenoids can be checked for circuit resistance and shorts to ground. This can typically be done without removing the oil pan. The test can be conducted at the transmission case connector. Individual solenoids can be checked with an ohmmeter by identifying the proper pins in the connector. Remember, lower than normal resistance indicates a short, whereas higher than normal resistance indicates a problem of high resistance. If you get an infinite reading across the solenoid, the solenoid windings are open. The ohmmeter can also be used to check for shorts to ground. Simply connect one lead of the ohmmeter to one end of the solenoid windings and the other lead to ground **(Figure 43–26)**. The reading should be infinite. If there is any measurable resistance, the winding is shorted to ground.

Solenoids can also be tested on a bench. Resistance values are typically given in the service information for each application **(Figure 43–27)**. A solenoid may be electrically fine but still may fail mechanically or hydraulically. A solenoid's check valve may fail to seat, or its porting can be plugged. This is not an electrical problem; rather, it could be caused by the magnetic field collecting metal particles in the ATF and clogging the port or check valve. These would cause erratic shifting, no shift conditions, wrong gear starts, no limited passing (kickdown) gear, or binding shifts. When a solenoid affected in this way is activated, it will make a slow, dull thud. A good solenoid tends to snap when activated.

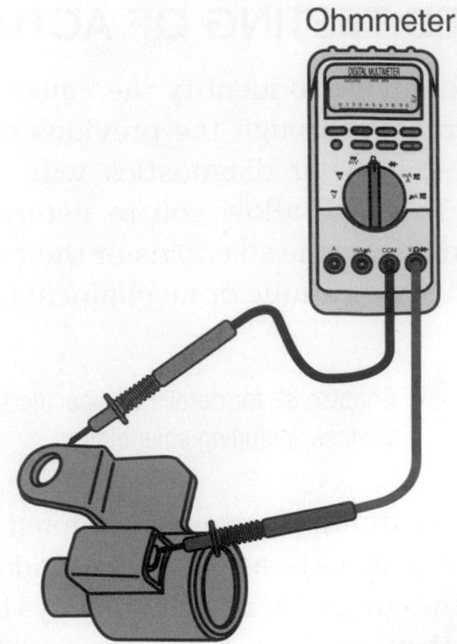

FIGURE 43–26

The meter connections for checking a solenoid for an open or short as well as resistance value.

CAUTION!

When servicing the transmission of a Honda hybrid, be careful of the high-voltage motor that is sandwiched between the engine and the transmission. Always disconnect the high voltage before working on or near the transmission.

COMPONENTS	PASS THRU PINS	RESISTANCE AT 20°C	RESISTANCE AT 100°C	RESISTANCE TO GROUND (CASE)
1-2 shift solenoid valve	A, E	19–24Ω	24–31Ω	Greater than 250MΩ
2-3 shift solenoid valve	B, E	19–24Ω	24–31Ω	Greater than 250MΩ
TCC solenoid valve	T, E	21–24Ω	26–33Ω	Greater than 250MΩ
TCC PWM solenoid valve	U, E	10–11Ω	13–15Ω	Greater than 250MΩ
3-2 shift solenoid valve assembly	S, E	20–24Ω	29–32Ω	Greater than 250MΩ
Pressure control solenoid valve	C, D	3–5Ω	4–7Ω	Greater than 250MΩ
Transmission fluid temp. (TFT) sensor*	M, L	3088–3942Ω	159–198Ω	Greater than 10MΩ
Vehicle speed sensor	A, B Vss conn	1420Ω @ 25°C	2140Ω @ 150°C	Greater than 10MΩ

*IMPORTANT: The resistance of this device is necessarily temperature dependent and will therefore vary far more than that of any other device. Refer to the specific transmission fluid temp (TFT) sensor specifications.

FIGURE 43–27

The appropriate service information lists the resistance values and test points for various transmission solenoids and sensors.

KEY TERMS

Adaptive learning (p. 1321)
Electronic pressure control (EPC) solenoid (p. 1321)
Shift schedule (p. 1314)
Transmission control module (TCM) (p. 1314)
Variable force solenoids (VFSs) (p. 1321)

SUMMARY

- A TCM is a separate computer designated for transmission operation or is part of the PCM. The PCM can be a multifunction computer that controls all engine and transmission operations.
- A shift schedule contains the actual shift points to be used by the computer according to the input data it receives from the sensors. Its logic chooses the proper shift schedule for the current conditions of the transmission.
- In most electronically controlled transmission systems, the hydraulically operated clutches are controlled by the transaxle controller.
- The typical output devices are solenoids and motors, which cause something mechanical or hydraulic to change.
- The computer may receive information from two different sources: directly from a sensor or through a twisted-pair bus circuit, which connects all of the vehicle computer systems.
- Fluid flow to the apply devices is directly controlled by the solenoids.
- Pressure switches give input to the transmission computer; they are all located within the solenoid assembly.
- Normally, the shift solenoids receive voltage through the ignition switch and are grounded through the PCM.
- The TCM is programmed to adjust its operating parameters in response to changes within the system, such as component wear. As component wear and shift overlap times increase, the TCM adjusts the line pressure to maintain proper shift timing calibrations. This is called adaptive learning.
- The pulse-width-modulated solenoid is a normally closed valve installed in the valve body. It controls the position of the TCC apply valve.
- If the TCM loses source voltage, the transmission will enter into limp-in mode. The transmission will also enter into default if the computer senses a transmission failure. While in the default mode, the transmission will operate only in PARK, NEUTRAL, REVERSE, and one forward gear. This allows the vehicle to be operated, although its efficiency and performance are hurt.
- The basic shift logic of the computer allows the releasing apply device to slip slightly during the engagement of the next engaging apply device.
- The EPC solenoid provides changes in pressure in response to engine load.
- Toyota, Ford, and Nissan hybrid systems rely on electric motors to determine the overall gear ratio in their CVTs.
- The two-mode hybrid system fits into a standard automatic transmission housing and is comprised of two planetary gearsets coupled to two electric motor/generators. This arrangement allows for two distinct modes of hybrid drive operation.
- If a computer's input signals are correct and its output signals are incorrect, the computer must be replaced.
- Input devices are critical to the operation of an EAT and should be checked with a scan tool, DMM, or lab scope.
- Solenoid valves can be checked by measuring their resistance or by applying a current to it and listening and feeling for its movement. They can also be checked with a lab scope or DMM.

REVIEW QUESTIONS

1. What happens if the TCM determines that the signals from the TFT sensor are incorrect?
2. How can air pressure be used to check an electrical switch?
3. Some transmissions receive information through multiplexing. How does this work?
4. List five things that could cause incorrect shift times and poor shifting quality.
5. Most late-model transmission control systems have adaptive learning. What does this mean?
6. Which of the following is *not* a voltage-generating type sensor?
   a. vehicle speed sensor
   b. OSS
   c. MAP
   d. ISS
7. A glitch appears in the waveform of a vehicle speed sensor. Which of the following is *not* a probable cause of the problem?
   a. a loose connector
   b. a damaged wire
   c. a poorly mounted sensor
   d. a damaged magnet in the sensor

8. In a Toyota hybrid CVT, which of the following parts effectively controls the overall gear ratio of the transaxle?
   a. MG1 **b.** MG2
   c. reduction unit **d.** transaxle damper
9. Which of the following would probably not be caused by a plugged filter screen at a solenoid?
   a. erratic shifting
   b. no shift conditions
   c. a solenoid that will not energize
   d. wrong gear starts
10. What solenoid replaces the conventional throttle valve cable setup in electronically controlled automatic transmissions?
    a. electronic pressure control (EPC) solenoid
    b. shift solenoid
    c. torque converter clutch (TCC) solenoid
    d. pulse-width-modulated (PWM) solenoid
11. What component in an electronically controlled automatic transmission replaces the governor valve?
    a. the throttle position sensor (TPS)
    b. the manifold absolute pressure sensor (MAP)
    c. the vehicle speed sensor (VSS)
    d. the transmission fluid temperature sensor (TFT)
12. What would be noticed when a good shift solenoid is activated?
    a. It will make a slow, dull thud sound.
    b. It will make a quick snap sound.
    c. A lab scope waveform will have rounded corners.
    d. An ohmmeter will read infinite across the solenoid.
13. Which of the following sensors could cause the engine to stall when a transmission range is selected?
    a. throttle position
    b. mass airflow
    c. transmission temperature
    d. vehicle speed
14. What occurs when the brake pedal position (BPP) switch tells the TCM that the brakes are applied?
    a. The electronic pressure control (EPC) adjusts transmission pressure.
    b. The torque converter clutch (TCC) disengages.
    c. The PCM illuminates the transmission control indicator lamp (TCIL).
    d. A signal is sent to the PCM to disengage overdrive.
15. What does the TCM alter when the fluid temperature is too low?
    a. make shifts sooner
    b. apply the torque converter clutch (TCC) sooner
    c. apply the TCC in all forward ranges except first
    d. prevent TCC engagement
16. What should be the first step in diagnosing an electronic automatic transmission?
    a. Check all fuses and identify the cause of any blown fuses.
    b. Check the physical condition of all sensors.
    c. Perform a road test.
    d. Check the wiring going to all sensors and actuators.
17. Which of the following is true of a shift solenoid?
    a. Shift solenoids receive voltage through the ignition switch.
    b. Shift solenoids are grounded through the chassis ground.
    c. Shift solenoids receive voltage through the PCM.
    d. Shift solenoids are grounded through the PCM.
18. What determines the default gear obtained when the transmission is operating in its fail-safe mode?
    a. the one occurring when the shift solenoids are all de-energized
    b. the one occurring when the shift solenoids are all energized
    c. the one occurring when the EPC solenoid is at its maximum duty cycle
    d. the one occurring when the EPC solenoid is fully de-energized
19. For adaptive learning to take place, how often does the TCM read sensor input and actuator output speeds?
    a. 7 times per second **b.** 14 times per second
    c. 70 times per second **d.** 140 times per second
20. Why does the TCM work with the PCM to momentarily change the ignition timing during a shift?
    a. to reduce engine torque to allow for smooth shifts
    b. to increase engine torque to allow for harsher shifts
    c. to increase engine rpm to produce a stronger shift
    d. to decrease engine rpm to produce a softer shift

CHAPTER 44
Automatic Transmission and Transaxle Service

LEARNING OUTCOMES

- Listen to the driver's complaint, road-test the vehicle, and then determine the needed repairs.
- Diagnose unusual fluid usage, level, and condition problems.
- Replace automatic transmission fluid and filters.
- Diagnose noise and vibration problems.
- Diagnose hydraulic and vacuum control systems.
- Perform oil pressure tests, and determine needed repairs.
- Inspect and adjust external linkages.
- Describe the basic steps for overhauling a transmission.

Because of the many similarities between a transmission and a transaxle, most diagnostic and service procedures are similar. Therefore, all references to a transmission apply equally to a transaxle unless otherwise noted. Whenever you are diagnosing or repairing a transaxle or transmission, make sure you refer first to the appropriate service information before you begin.

Transmissions are strong and typically trouble-free units that require little maintenance. Normal maintenance usually includes fluid checks, scheduled linkage adjustments, and oil and filter changes.

Nearly all current automatic transmissions have electronic controls that work with the hydraulic and mechanical systems of the transmission to provide reliable and efficient operation.

IDENTIFICATION

Before beginning any service or repair work, be sure you know exactly which transmission you are working on to ensure that you are following the correct procedures and specifications and are installing the correct parts. Proper identification can be difficult because transmissions cannot be accurately identified by the way they look. The only exception to this is the shape of the oil pan, which can sometimes be used for identification; however, this is not foolproof.

The only positive way to identify the exact design of the transmission is by its identification numbers. **Transmission identification numbers** are found on stickers on the transmission **(Figure 44–1)**, as stamped numbers in the case, or on a metal tag held by a bolt head. Use service information to decipher the identification number. Most identification numbers include the model, manufacturer, and assembly date. Whenever you work with a transmission with a metal ID tag, make sure the tag is put back on the transmission so that the next technician can properly identify the transmission. Most late-model transmissions have labels with bar codes that can be scanned for transmission identification.

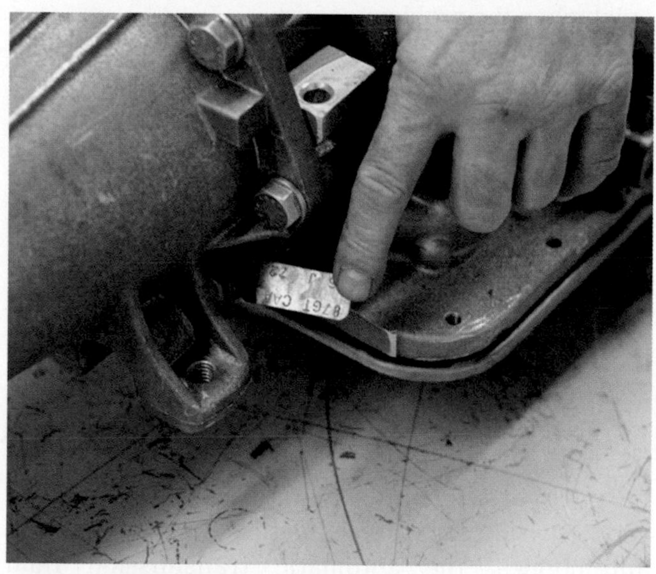

FIGURE 44–1

Make sure you properly identify the transmission before performing any service.

DIAGNOSTICS

Automatic transmission problems are usually caused by poor engine performance, problems in the hydraulic system, abuse resulting in overheating, mechanical malfunctions, electronic failures, and/or improper adjustments. Diagnosis of these problems should begin by checking the condition of the fluid and its level, conducting a thorough visual inspection, checking the various linkage adjustments, and scanning the transmission control module (TCM) for DTCs.

Engine performance can affect torque converter clutch (TCC) operation. If the engine is running too poorly to maintain a constant speed, the converter clutch will engage and disengage at higher speeds. The customer complaint may be that the vehicle vibrates.

If the vehicle has an engine performance problem, the cause should be found and corrected before any conclusions on the transmission are made. A quick way to identify whether the engine is causing shifting problems is to connect a vacuum gauge to the engine and take a reading while the engine is running. The gauge should be connected to intake manifold vacuum. A normal vacuum gauge reading is steady and at 432 mm Hg (17 in. Hg). The rougher the engine runs, the more the gauge readings will fluctuate.

In order to properly diagnose a problem, you must totally understand the customer's concern or complaint. Make sure you know the conditions that exist when the problem occurs.

Fluid Check

GO TO Chapter 8 for additional details on checking the fluid level in an automatic transmission.

Diagnosis should begin with a fluid level check. Make sure the vehicle is on a level surface when checking the level. Many late-model transmissions do not have a dipstick, and the fluid level is checked in the same way as a manual transmission.

If the transmission has a dipstick, locate it **(Figure 44–2)**, remove it, and wipe all dirt off the protective disc and the dipstick handle. On most automobiles, the automatic transmission fluid (ATF) level can be checked accurately only when the transmission is at operating temperature and the engine is running. Remove the dipstick and wipe it clean with a lint-free white cloth or paper towel.

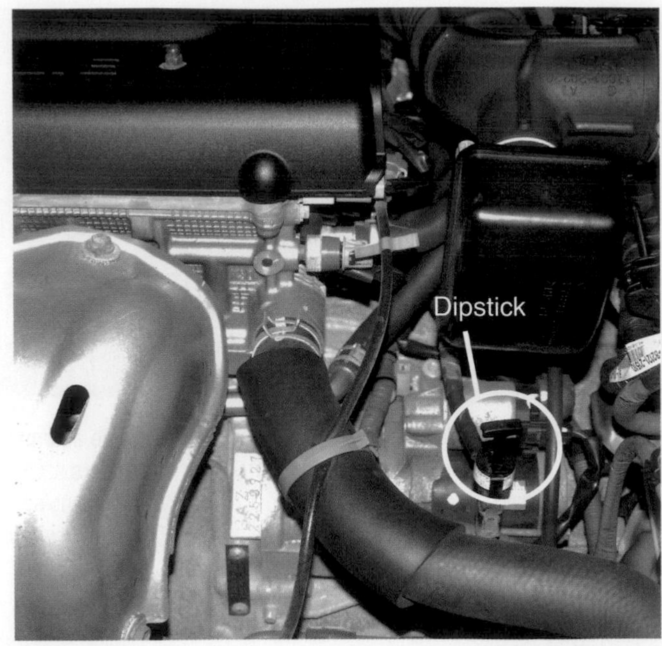

FIGURE 44–2

Typical location of a transmission dipstick in a transaxle.

Reinsert the dipstick, remove it again, and note the reading. Markings on a dipstick indicate ADD levels and, on some models, FULL levels for cool, warm, or hot fluid.

Low fluid levels can cause a variety of problems. Air can be drawn into the oil pump's inlet circuit and mix with the fluid. This will result in aerated fluid, which causes slow pressure buildup, and low pressures, which cause slippage between shifts. Air in the pressure regulator valve will cause a buzzing noise when the valve tries to regulate pump pressure. If the fluid level is low, the problem could be external fluid leaks. Check the transmission case, oil pan, and cooler lines for signs of leaks.

Excessively high fluid levels can also cause **aeration**. As the planetary gears rotate in high fluid levels, air can be forced into the fluid. Aerated fluid can foam, overheat, and oxidize. All of these problems can interfere with normal valve, clutch, and servo operation. Foaming may be evident by fluid leakage from the transmission's vent.

The condition of the fluid should be checked while checking its level. Examine the fluid carefully. The normal colour of most ATF is red. If the fluid has a dark brownish or blackish colour and/or a burned odour, the fluid has been overheated. A milky colour indicates that engine coolant has been leaking into the transmission's cooler in the radiator. If there is any question about the condition of the fluid, drain out a sample for closer inspection.

Synthetic ATF is normally a darker red than petroleum-based fluid. Synthetic fluids tend to look and smell burnt after normal use; therefore, the appearance and smell of these fluids is not a good indicator of the fluid's condition.

After checking the ATF level and colour, wipe the dipstick on absorbent white paper, and look at the stain left by the fluid. Dark particles are normally band and/or clutch material, whereas silvery metal particles are normally caused by the wearing of the transmission's metal parts. If the dipstick cannot be wiped clean, it is probably covered with varnish, which results from fluid oxidation. Varnish will cause the spool valves to stick, causing shifting malfunction. Varnish or other heavy deposits indicate the need to change the transmission's fluid and filter.

Contaminated fluid can sometimes be felt better than seen. Place a few drops of fluid between two fingers and rub them together. If the fluid feels dirty or gritty, it is contaminated with burned friction material.

TRANSMISSIONS WITHOUT A DIPSTICK Many late-model transmissions do not have a dipstick, and the fluid level is checked in a way similar to a manual transmission. Photo Sequence 35 covers the typical procedure for checking the transmission fluid level on transmissions without a dipstick. The dipstick and filler tube were removed from these transmissions to prevent overfilling. Research has found that many transmission failures were caused by overfilling and/or using the wrong fluid. Without a dipstick, it is difficult to check the fluid level and condition. These transmissions have a vent/fill cap typically located on the side of the transmission. Some also have a drain plug in the bottom of the pan. In addition, these transmissions are fitted with a fluid level sensor that will inform the driver when the fluid level is dangerously low.

Recommended Applications

Several ratings or types of ATF are available **(Figure 44–3)**; each type is designed for a specific application. The different classifications of transmission fluid have resulted from the inclusion of new or different additives that enhance the operation of the different transmission designs. Each automobile manufacturer specifies the proper type of ATF that should be used in its transmissions. Both the design of the transmission and the desired shift characteristics are considered when a specific ATF is chosen.

FIGURE 44–3

Examples of the different fluids used in today's transmissions.

Martin Restoule

To reduce wear and friction inside a transmission, most commonly used transmission fluids are mixed with friction modifiers. Transmission fluids with these additives allow for the use of lower clutch and band application pressures, which, in turn, provide for a very smooth-feeling shift. Transmission fluids without a friction modifier, such as types F and G, tend to have a firmer shift because higher clutch and band application pressures are required to avoid excessive slippage during gear changes.

Some manufacturers recommend the use of other types of fluid but normally list Dexron-VI as a secondary choice. Always use the exact automatic transmission fluid type specified by the manufacturer. The dipsticks of some transmissions indicate the type of fluid that should be used in the transmission.

Fluid Changes

The transmission's fluid and filter should be changed whenever there is an indication of oxidation or contamination. Periodic fluid and filter changes are also part of the preventive program for most vehicles. The mileage interval recommended depends on the type of transmission.

Change the fluid only when the engine and transmission are at normal operating temperatures. Photo Sequence 36 shows a typical procedure for changing a transmission's fluid and filter. On most transmissions, you must remove the oil pan to drain the fluid. Some transmission pans include a drain plug. A filter or screen is normally attached to the bottom of the valve body. Filters are made of paper or fabric and are held in place by screws, clips, or bolts, or by the

PHOTO SEQUENCE 35

CHECKING TRANSMISSION FLUID LEVEL ON A VEHICLE WITHOUT A DIPSTICK

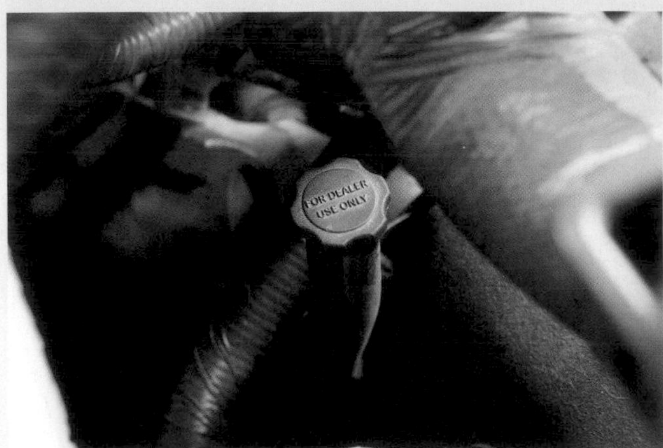

P35–1 Examine the transmission and determine how the fluid level is checked. If there is not a dipstick, often a service tag will indicate that special procedures are required.

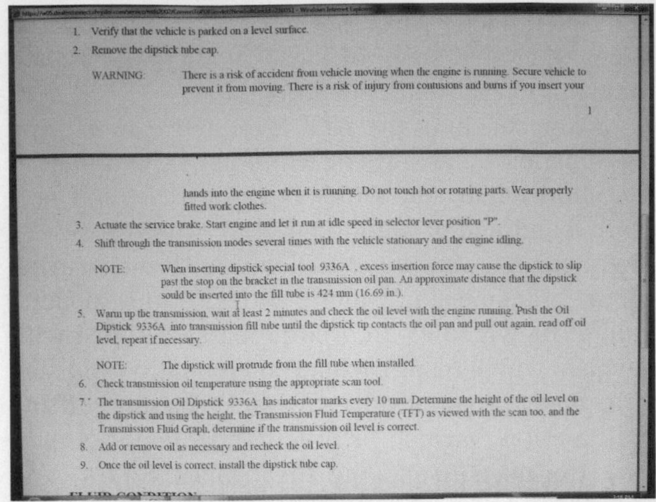

1. Verify that the vehicle is parked on a level surface.
2. Remove the dipstick tube cap.

WARNING: There is a risk of accident from vehicle moving when the engine is running. Secure vehicle to prevent it from moving. There is a risk of injury from contusions and burns if you insert your hands into the engine when it is running. Do not touch hot or rotating parts. Wear properly fitted work clothes.

3. Actuate the service brake. Start engine and let it run at idle speed in selector lever position "P".
4. Shift through the transmission modes several times with the vehicle stationary and the engine idling.

NOTE: When inserting dipstick special tool 9336A , excess insertion force may cause the dipstick to slip past the stop on the bracket in the transmission oil pan. An approximate distance that the dipstick sould be inserted into the fill tube is 424 mm (16.69 in.).

5. Warm up the transmission, wait at least 2 minutes and check the oil level with the engine running. Push the Oil Dipstick 9336A into transmission fill tube until the dipstick tip contacts the oil pan and pull out again, read off oil level, repeat if necessary.

NOTE: The dipstick will protrude from the fill tube when installed.

6. Check transmission oil temperature using the appropriate scan tool.
7. The transmission Oil Dipstick 9336A has indicator marks every 10 mm. Determine the height of the oil level on the dipstick and using the height, the Transmission Fluid Temperature (TFT) as viewed with the scan too, and the Transmission Fluid Graph, determine if the transmission oil level is correct.
8. Add or remove oil as necessary and recheck the oil level.
9. Once the oil level is correct, install the dipstick tube cap.

P35–2 Using the service information, determine how to check the transmission fluid level.

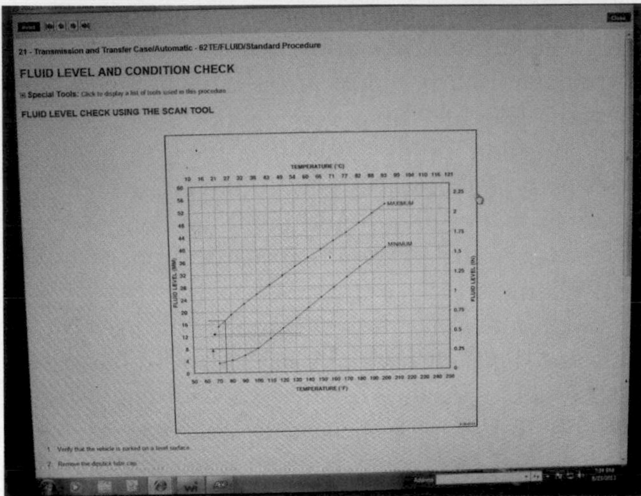

P35–3 Many vehicles require that the fluid level be checked with the fluid within a specific temperature range. This is to ensure that the fluid level is accurately determined, as the fluid level reading changes as the fluid heats up.

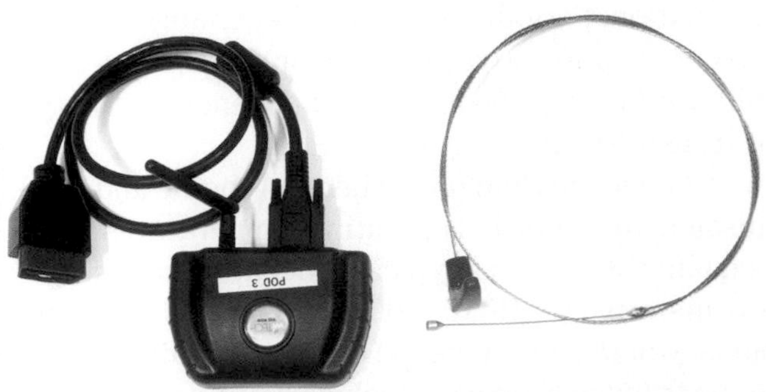

P35–4 For this vehicle, a scan tool and a transmission fluid level tool are required to accurately determine the fluid level.

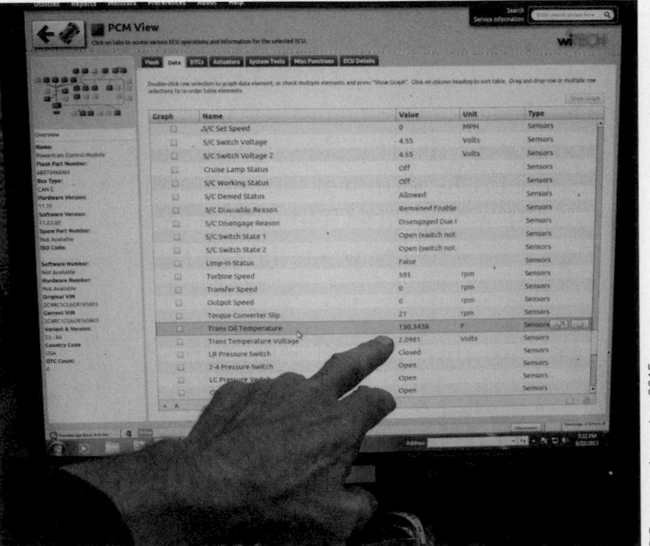

P35–5 To check transmission fluid temperature, connect a scan tool, and locate the transmission fluid temperature PID. Once the fluid is within the correct range, continue with checking the fluid level.

P35–6 Locate the fluid check plug in the transmission. Remove the plug, using the appropriate tools.

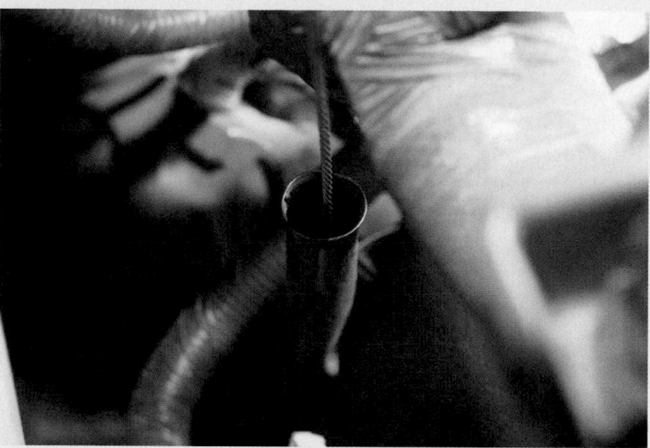

P35–7 Install the fluid tool in the transmission.

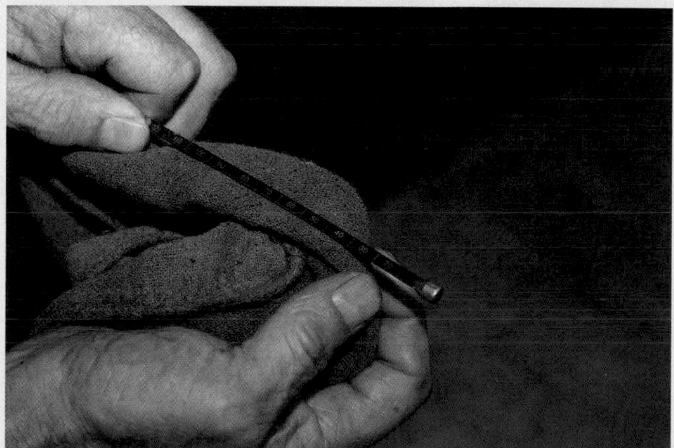

P35–8 Remove the tool and note the fluid level on the tool.

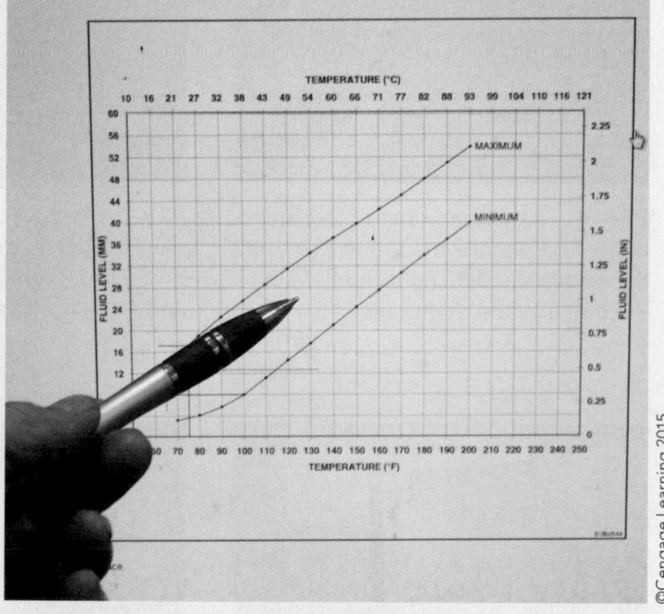

P35–9 Using the temperature range chart, the fluid level as measured at 128°F (52°C) indicates the fluid level is acceptable.

PHOTO SEQUENCE 36

CHANGING AUTOMATIC TRANSMISSION FLUID AND FILTER

P36–1 Raise the vehicle to a good working height and so it is safely positioned on the lift.

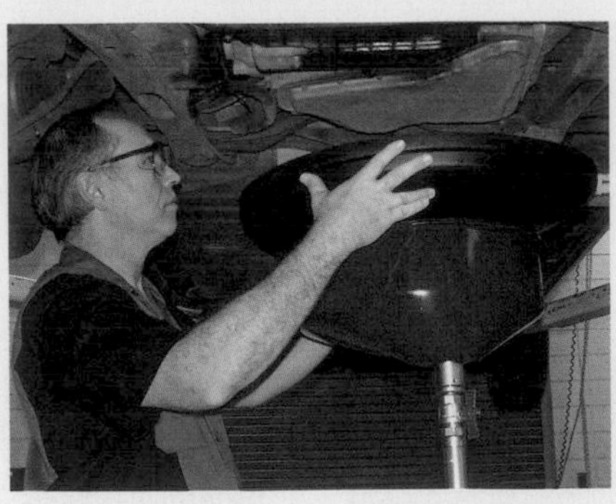

P36–2 Place a large-diameter oil drain pan under the transmission pan.

P36–3 Loosen all the pan bolts except three at one end. This will allow some fluid to drain out.

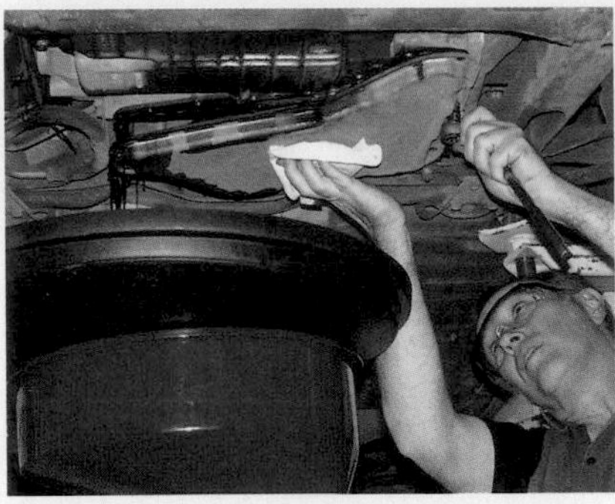

P36–4 Support the pan with one hand, and remove the remaining bolts to remove the pan. Pour the fluid in the pan into the drain pan.

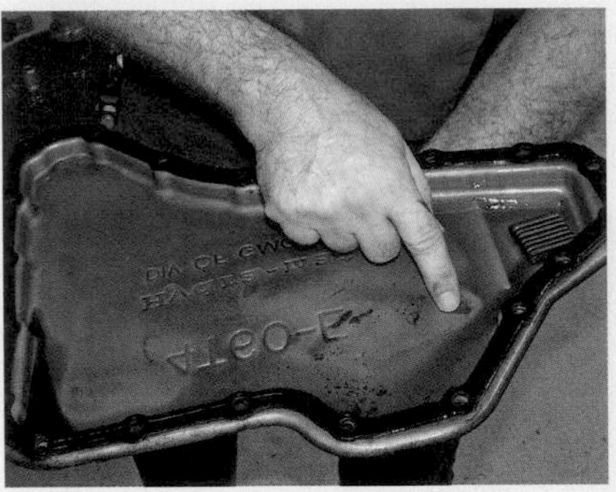

P36–5 Inspect the residue in the pan for indications of transmission problems. Then remove the old pan, and clean it with a lint-free rag.

P36–6 Unbolt the filter from the transmission.

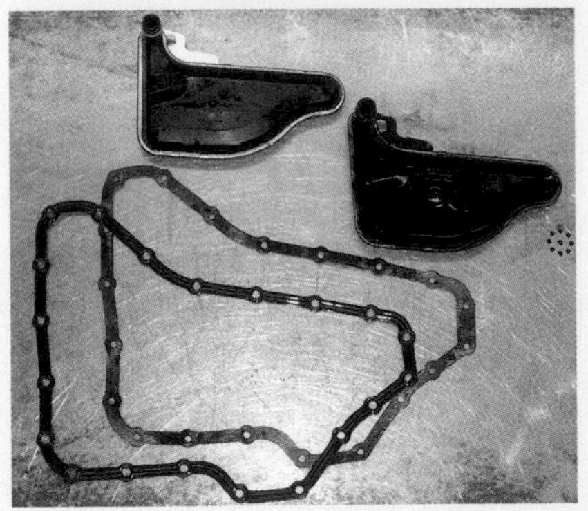

P36–7 Compare the replacement gasket and filter with the old ones to make sure the replacements are the right ones for this application.

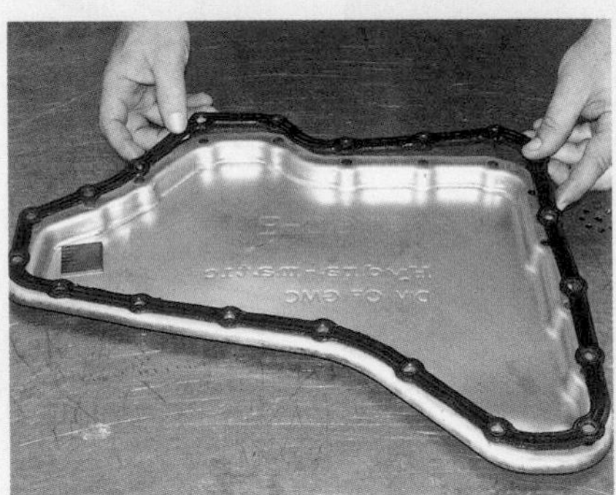

P36–8 Install the new filter and tighten the attaching bolts to specifications. Then lay the new gasket over the sealing surface of the pan.

P36–9 Install the pan onto the transmission. Install and tighten the bolts to specifications. Then lower the vehicle and pour new fluid into the transmission. Run the engine to circulate the new fluid, and then turn it off and raise the vehicle. Check for fluid leaks.

FIGURE 44–4

Transmission fluid filters are attached to the transmission case by screws, bolts, retaining clips, and/or the pickup tube.

pressure on the pickup tube seal **(Figure 44–4)**. Filters should be replaced, not cleaned.

Some late-model Saturn and a few other transmissions do not have a filter in the pan. Rather, they are fitted with a spin-on filter **(Figure 44–5)**. This filter looks like an engine filter. They are serviced in the same way as an engine oil filter. Other transmissions may have an in-line filter in the fluid lines leading to the cooler **(Figure 44–6)**.

After draining the fluid, carefully remove the pan. Check the bottom of the pan for deposits and metal particles. Slight contamination—blackish deposits from clutches and bands—is normal. Other contaminants should be of concern. Clean the oil pan and its magnet **(Figure 44–7)**.

Remove the filter and inspect it. Use a magnet to determine if metal particles are steel or aluminum. Steel particles indicate severe internal transmission wear or damage. If the metal particles are aluminum, they may be part of the torque-converter stator. Some torque converters use phenolic plastic stators; therefore, aluminum particles found in these transmissions must be from the transmission itself.

Remove any traces of the old pan gasket on the case and oil pan. Then, install a new filter and gasket, and tighten the retaining bolts to the specified torque. If the filter is sealed with an O-ring, make sure it is properly installed. Then reinstall the

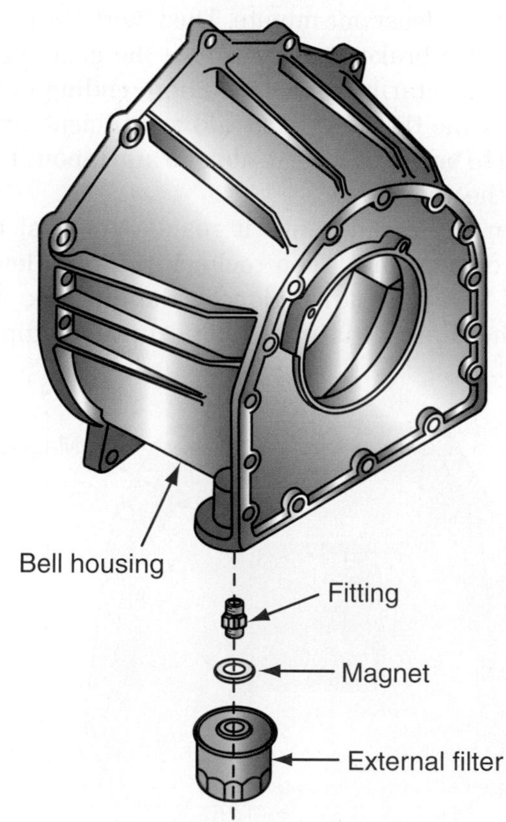

FIGURE 44–5

An engine oil filter–style transmission filter.

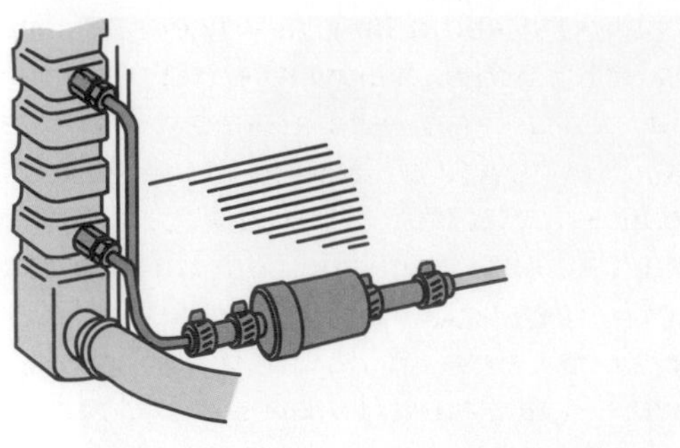

FIGURE 44–6

An in-line filter for a fluid cooler.

pan, using the gasket or sealant recommended by the manufacturer. Tighten the pan retaining bolts to the specified torque. Make sure you check the specifications. The required torque is often given in kilogram-centimetres (inch-pounds). You can easily break the bolts or damage something if you tighten the bolts to foot-pounds.

Pour a little less than the required amount of fluid into the transmission through the filler tube or fill point. Always use the recommended type of ATF. The wrong fluid will alter the shifting characteristics of the transmission. Start the engine and allow it to idle for at least one minute. Then, with the parking and service brakes applied, move the gear selector lever momentarily to each position, ending in PARK. Recheck the fluid level and add a sufficient amount of fluid to bring the level to about 3 mm (about 1/8 in.) below the ADD mark.

Run the engine until it reaches normal operating temperature. Then recheck the fluid level; it should be in the HOT region on the dipstick. Make sure the dipstick is fully seated into the dipstick tube opening to prevent dirt from entering into the transmission.

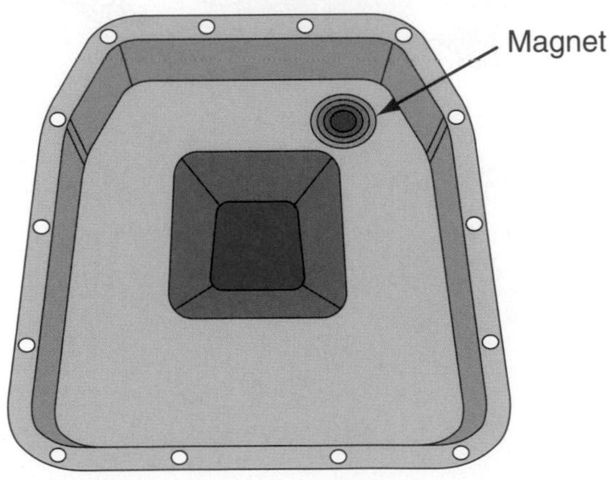

FIGURE 44–7

The magnet in the oil pan should be cleaned, and the material it gathered should be analyzed.

Parking Pawl

Any time you have the oil pan off, you should inspect all of the exposed parts, especially the parking pawl assembly. This component is typically not hydraulically activated; rather, the gearshift linkage moves the pawl into position to lock the output shaft of the transmission. Unless the customer's complaint indicates a problem with the parking mechanism, no test will detect a problem here.

Check the pawl assembly for excessive wear and other damage. Also, check to see how firmly the pawl is in place when the gear selector is shifted into the PARK mode. If the pawl can be moved out easily, it should be repaired or replaced.

VISUAL INSPECTION

Diagnosis of transmission problems should continue with conducting a thorough visual inspection, checking the various linkage adjustments, retrieving all DTCs, and checking basic engine operation. Also check the voltage of the battery as improper voltage can affect the performance of the transmission.

Fluid Leaks

Check all drivetrain parts for looseness and leaks. If the transmission fluid was low or there was no fluid, raise the vehicle and carefully inspect the transmission for signs of leakage. Leaks are often caused by defective gaskets or seals. Common sources of leaks are the oil pan seal, rear cover and final drive cover (on transaxles), extension housing, speedometer gear assembly, and electrical switches mounted into the housing **(Figure 44–8)**. The housing may have a porosity problem, allowing fluid to seep through the metal. Case porosity may be repaired with an epoxy sealer.

Oil Pan

A common source of fluid leakage is between the oil pan and the transmission housing. If fluid is present around the rim of the pan, retorquing the pan bolts may correct the problem. If it does not correct the problem, the pan must be removed and a new gasket installed. Make sure the sealing surface of the pan is flat and capable of providing a seal before reinstalling it.

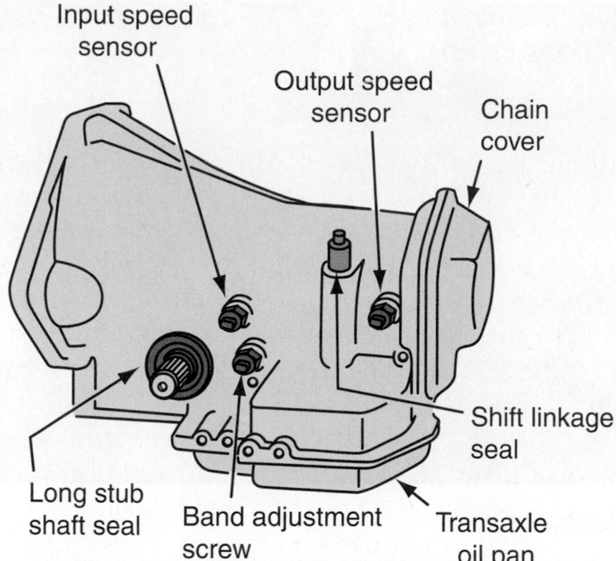

FIGURE 44–8

Possible sources of fluid leaks on this transaxle.

Torque Converter

Torque converter problems can be caused by a leaking converter **(Figure 44–9)**. This type of problem may be the cause of customer complaints of slippage and a lack of power. To check the converter for leaks, remove the converter access cover, and examine the area around the torque converter shell. An engine oil leak may be falsely diagnosed as a converter leak. The colour of engine oil is different from that of transmission fluid and may help identify the true source of the leak. However, if the oil or fluid has absorbed much dirt, both will look the same. An engine leak typically leaves an oil film on the front of the converter shell, whereas a converter leak will cause the entire shell to be wet. If the transmission's oil pump seal is leaking, only the back side of the shell will be wet. If the converter is leaking or damaged, it should be replaced.

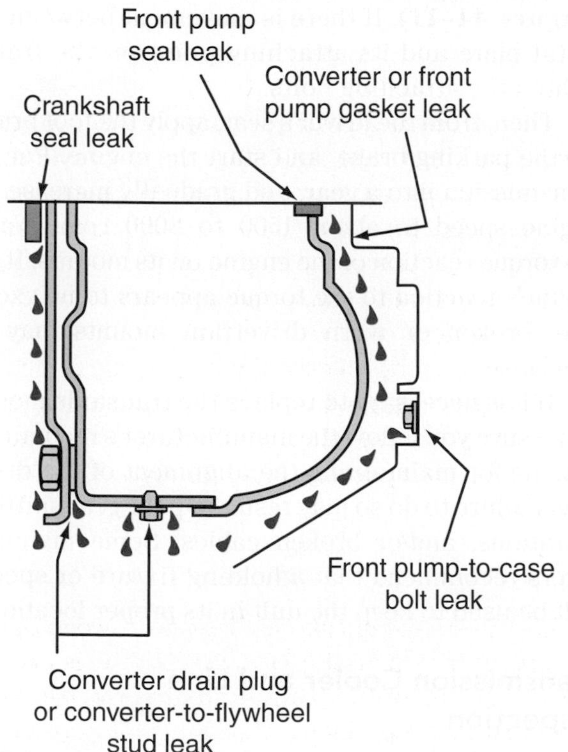

FIGURE 44–9

By determining the direction of fluid travel, the cause of a fluid leak around the torque converter can be identified.

Extension Housing

An oil leak stemming from the mating surfaces of the extension housing and the transmission case may be caused by loose bolts. To correct this problem, tighten the bolts to the specified torque. Also check for signs of leakage at the rear of the extension housing. Fluid leaks from the seal of the extension housing can be corrected with the transmission in the car. Often, the cause for the leakage is a worn extension housing bushing, which supports the sliding yoke of the driveshaft. When the driveshaft is installed, the clearance between the sliding yoke and the bushing should be minimal. If the clearance is satisfactory, a new oil seal will correct the leak. If the clearance is excessive, the repair requires that a new seal and a new bushing be installed. If the seal is faulty, the transmission vent should be checked for blockage.

Speed Sensor

The vehicle's speedometer can be purely electronic, which requires no mechanical hookup to the transmission. It can also be driven via a cable or off the output shaft. If the transmission is equipped with a vehicle speed sensor (VSS), the bore and sensor can be a source of leaks. The sensor is retained in the bore with a retaining nut or bolt. An oil leak at the speedometer cable or VSS can be corrected by replacing the O-ring seal. Always lubricate the O-ring and sensor prior to installation.

Electrical Connections

Check all electrical connections to the transmission. Faulty connectors or wires can cause harsh or delayed and missed shifts. On transaxles, the connectors can normally be inspected through the engine compartment, whereas they can only

FIGURE 44–10

Carefully disconnect all connectors and inspect them.

FIGURE 44–11

All transmission and engine mounts should be carefully checked for cracks and damage.

be seen from under the vehicle on longitudinally mounted transmissions. To check the connectors, release the locking tabs, and disconnect them one at a time from the transmission. Carefully examine them for signs of corrosion, distortion, moisture, and transmission fluid **(Figure 44–10)**. A connector or wiring harness may deteriorate if ATF reaches it. Also check the connector at the transmission. Using a small mirror and flashlight may help you get a good look at the inside of the connectors. Inspect the entire transmission wiring harness for tears and other damage. Road debris can damage the wiring and connectors mounted underneath the vehicle.

Because the operation of the engine and transmission are integrated through the control computer, a faulty engine sensor or connector may affect the operation of the engine and the transmission. The various sensors and their locations can be identified by referring to the appropriate service information. As stated in Chapter 43, the engine control sensors that most likely cause shifting problems are the throttle position (TP), manifold absolute pressure (MAP), and VSS.

Checking Transmission and Transaxle Mounts

The engine and transmission mounts on front-wheel-drive (FWD) cars are important to the operation of the transaxle. Any engine movement may change the effective length of the shift and throttle cables or wiring harnesses and therefore may affect the engagement of the gears. Delayed or missed shifts may result from linkage changes as the engine pivots on its mounts.

Visually inspect the mounts for looseness and cracks. With a pry bar, pull up and push down on the transaxle case while watching the mount **(Figure 44–11)**. If there is movement between the metal plate and its attaching point on the frame, tighten the attaching bolts.

Then, from the driver's seat, apply the foot brake, set the parking brake, and start the engine. Put the transmission into a gear, and gradually increase the engine speed to about 1500 to 2000 rpm. Watch the torque reaction of the engine on its mounts. If the engine's reaction to the torque appears to be excessive, broken or worn drivetrain mounts may be the cause.

If it is necessary to replace the transaxle mount, make sure you follow the manufacturer's recommendations for maintaining the alignment of the driveline. Failure to do so may result in poor gear shifting, vibrations, and/or broken cables. Some manufacturers recommend that a holding fixture or special bolt be used to keep the unit in its proper location.

Transmission Cooler and Line Inspection

Transmission coolers are a possible source of fluid leaks. The efficiency of the coolers is also critical to the operation and longevity of the transmission.

Follow these steps when inspecting the transmission cooler and associated lines and fittings:

1. Check the engine's cooling system. The transmission cooler cannot be efficient if the engine's cooling system is defective. Repair all engine cooling system problems before continuing to check the transmission cooler.
2. Inspect the fluid lines and fittings between the cooler and transmission. Check these for looseness, damage, signs of leakage, and wear. Replace any damaged lines and fittings.
3. Inspect the engine's coolant for traces of transmission fluid. If ATF is present in the coolant, the transmission cooler leaks.
4. Check the transmission's fluid for signs of engine coolant. Water or coolant will cause the fluid to appear milky with a pink tint. This milky appearance is also an indication that the transmission cooler leaks and is allowing engine coolant to enter into the transmission fluid.

The cooler can be checked for leaks by disconnecting and plugging the transmission to cooler lines at the radiator. Then remove the radiator cap to relieve any pressure in the system. Tightly plug one of the ATF line fittings at the radiator. Using the shop air supply with a pressure regulator, apply 345 to 485 kPa (50 to 70 psi) of air pressure into the cooler at the other cooler line fitting **(Figure 44–12)**. Look into the radiator. If bubbles are observed, the cooler leaks.

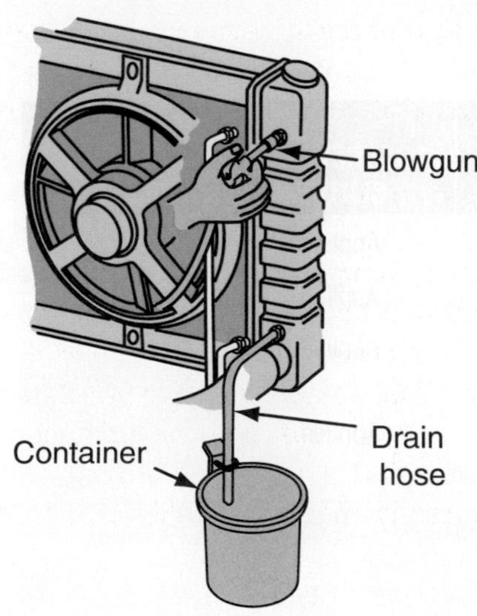

FIGURE 44–12

Shop air can be used to check a transmission cooler for leaks and restrictions.

ROAD-TESTING THE VEHICLE

Critical to proper diagnosis of an automatic transmission is a road test. If the vehicle has an EAT, connect a scan tool with recording capabilities, if possible, before taking the drive. Also find the appropriate chart that shows the activity of the various solenoids for the transmission **(Figure 44–13)**. Knowing their activity will help you determine if one or more of them are causing a shifting problem.

All transmission concerns should be verified by attempting to duplicate the customer's concern. Knowing the exact conditions that cause the symptom will allow you to accurately diagnose the cause. Diagnosis becomes easy if you think about what is happening in the transmission when the problem occurs. If there is a shifting problem, think about the parts that are being engaged and disengaged.

Also, before beginning your road test, find and duplicate, from service information, the chart **(Figure 44–14)** that shows the band and clutch application for different gear selector positions. Using these charts will greatly simplify your diagnosis of automatic transmission problems. It is also wise to have a notebook or piece of paper to jot down notes about the operation of the transmission.

Begin the road test with a drive at normal speeds to warm the engine and transmission. If a problem appears only when starting and/or when the engine and transmission are cold, record this symptom on the chart or in your notebook.

During the road test, the transmission should be operated in all possible modes and its operation noted. Check for proper gear engagement as the selector lever is moved to each gear position, including PARK. There should be no hesitation or roughness as the gears are engaging. Check for proper operation in all forward ranges, especially the upshifts and converter clutch engagement during light throttle operation. These shifts should be smooth and positive and occur at the correct speeds. These same shifts should feel firmer under medium to heavy throttle pressures. Transmissions equipped with a torque converter clutch should be brought to the specified apply speed and their engagement noted. Again, record the operation of the transmission in these different modes in your notebook or on the diagnostic chart. Also, the

		SOLENOID					
MANUAL VALVE LEVER POSITION	**GEAR**	**NA UD**	**NV OD**	**NV 4TH CLUTCH**	**NV 2ND CLUTCH**	**NV L/R**	**NA MS**
R	R						^x^
P/N	P/N					{x}	
OD	D1					{x}	x
OD	D2				x		x
OD	*D2*			x			x
OD	D3		x				x
OD	*D3*						
OD	D4	x	x	x			x
OD	*D4*	x		x			
OD L/I	3rd						
M2 L/I	2nd						

{x} = effective *Dx* = operating in

^x^ = On only if shift to reverse is above 13 km/h (8 mph)

FIGURE 44–13

A solenoid application chart for a typical electronic transmission.

various computer inputs should be monitored and the readings recorded.

Force the transmission to kick down, and pay attention to the quality of this shift. Manual downshifts should also be made at a variety of speeds. The reaction of the transmission should be noted, as should all abnormal noises as well as the gears and speeds at which they occur.

SHOP TALK

Always refer to the service information to identify the particulars of the transmission you are diagnosing. Also check for any technical service bulletins that may be related to the customer's complaint.

RANGE	1-2-3-4 CLUTCH	3-5 REVERSE CLUTCH	4-5-6 CLUTCH	2-6 CLUTCH	LOW/REVERSE CLUTCH	LOW SPRAG CLUTCH
Park					Applied	
Reverse		Applied			Applied	
Neutral					Applied	
1st Braking	Applied				Applied	
1st	Applied					
2nd	Applied			Applied		
3rd	Applied	Applied				
4th			Applied			
5th		Applied	Applied			
6th			Applied	Applied		

FIGURE 44–14

A clutch application chart for a typical six-speed transmission.

After the road test, check the transmission for signs of leakage. Any new leaks and their probable cause should be noted. Then compare your written notes from the road test to the information given in the service information to identify the cause of the malfunction. This information usually has a diagnostic chart to aid you in this process.

Diagnosis of Noise and Vibration Problems

Many noise and vibration problems that appear to be transmission-related may be caused by problems in the engine, driveshaft, universal or CV joints, wheel bearings, wheel/tire imbalance, or other conditions. Problems in those areas can lead customers—and, unfortunately, some technicians—to mistakenly suspect that the problems are caused by the transmission or torque converter. The entire driveline should be checked before assuming that the noise is transmission-related.

Common transmission noise and vibrations and their typical cause are shown in **Figure 44–15**. As can be seen, they can be caused by faulty bearings, damaged gears, worn or damaged clutches and bands, or a bad oil pump as well as contaminated fluid or improper fluid levels.

Most vibration problems are caused by an unbalanced torque converter assembly, loose converter mounting bolts, a loose or cracked flexplate, or a faulty output shaft. The key to determining the cause

Problem	Probable Cause(s)
Racheting noise	Damaged, weak, or misassembled return spring for the parking pawl
Engine speed–sensitive whine	Faulty torque converter Faulty pump
Popping noise	Pump cavitation—bubbles in the ATF Damaged fluid filter or filter seal
Buzz or high-frequency rattle, whine, or growl	Cooling system problem Stretched drive chain Broken teeth on drive and/or driven sprockets Nicked or scored drive and/or driven sprocket bearing surfaces Pitted or damaged bearing sufaces
Final drive hum	Worn final drive gear assembly Worn or pitted differential gears Damaged or worn differential gear thrust washers
Noise in forward gears	Worn or damaged final drive gears
Noise in specific gears	Worn or damaged components pertaining to that gear
Vibration	Out-of-balance torque converter Faulty torque converter Misaligned transmission or engine Worn or damaged output shaft bushing Out-of-balance input shaft Worn or damaged input shaft bushing

FIGURE 44–15
A basic symptom-based diagnostic chart.

of the vibration is to pay particular attention to the vibration in relationship to engine and vehicle speed. If the vibration changes with a change in engine speed, the cause of the problem is most probably the torque converter. If the vibration changes with vehicle speed, the cause is probably the output shaft or the driveline connected to it. The latter can be a bad extension housing bushing or universal joint, which would become worse at higher speeds.

To determine if the problem is caused by the transmission or the driveline, put the transmission in gear and apply the foot brakes. If the noise is no longer evident, the problem can be in the driveline or the output of the transmission. If the noise is still present, the problem must be in the transmission or torque converter.

Noise problems are also best diagnosed by paying a great deal of attention to the speed and the conditions at which the noise occurs. If the noise is engine speed–related and is present in all gears, including PARK and NEUTRAL, the oil pump is the most probable source because it rotates whenever the engine is running. However, if the noise is engine speed–related and is present in all gears except PARK and NEUTRAL, the most probable sources of the noise are those parts that rotate in all gears, such as the drive chain, the input shaft, and torque converter.

Noises that only occur when a particular gear is operating must be related to those parts responsible for providing that gear, such as a brake or clutch. Often, the exact cause of noise and vibration problems can only be identified through a careful inspection of a disassembled transmission.

CHECKING THE TORQUE CONVERTER

Many transmission problems are related to the operation of the torque converter. Normally, torque converter problems will cause abnormal noises, poor acceleration in all gears, normal acceleration but poor high-speed performance, or transmission overheating.

If the vehicle lacks power during acceleration, it has a restricted exhaust, or the torque converter's one-way stator clutch is slipping. To determine which of these problems is causing the power loss, test for a restricted exhaust first. Other possible causes of this problem include a restricted air or fuel filter and a defective fuel pump.

If there is no evidence of a restricted exhaust, it can be assumed that the torque converter's stator clutch is slipping and not allowing any torque multiplication to take place in the converter. To correct this problem, the torque converter should be replaced.

If the engine's speed flares up during acceleration in DRIVE and does not have normal acceleration, the clutches or bands in the transmission are slipping. This symptom is similar to the slipping of a clutch in a manual transmission. Often, this problem is mistakenly blamed on the torque converter.

Complaints of thumping or grinding noises are often thought to be caused by the torque converter (T/C), when they are actually caused by bad thrust washers or damaged gears and bearings in the transmission. Non-transmission problems, such as bad CV joints and wheel bearings, also cause these noises.

A faulty engine dampener can feel like a bad T/C, allowing vibrations that seem to be related to the torque converter. These vibrations can be caused by an out-of-balance torque converter. If the T/C is not properly balanced, the vibration will increase at higher engine speeds and while the transmission is in gear.

Checking the TCC

During the road test, check to see if the TCC engages and disengages. Engagement should be smooth and free of abnormal vibrations and noise. If the clutch prematurely engages or is not being applied by full pressure, a shudder or vibration results from the rapid grabbing and slipping of the clutch. If the shudder is noticeable only during clutch engagement, the problem is typically the converter. When the shudder is evident after the engagement of the clutch, the engine, transmission, or another component of the driveline may be the cause.

A shudder when the TCC is being engaged or disengaged can be caused by an out-of-round torque converter or contaminated clutch frictional material. An out-of-round converter prevents full clutch engagement.

If the clutch does not engage, check the TCC solenoid circuit for power and excessive voltage drops. The solenoids should drop very close to source voltage. If the solenoid is electrically sound, suspect clutch material, dirt, or other material plugging the solenoid valve passages. If blockage is found, attempt to flush the valve with clean ATF. If the solenoid has a filter assembly **(Figure 44–16)**, replace the filter after cleaning the fluid passages. If the blockage cannot be removed, replace the solenoid **(Figure 44–17)**.

Other conditions that prevent the TCC from applying include converter case leaks, internal seal

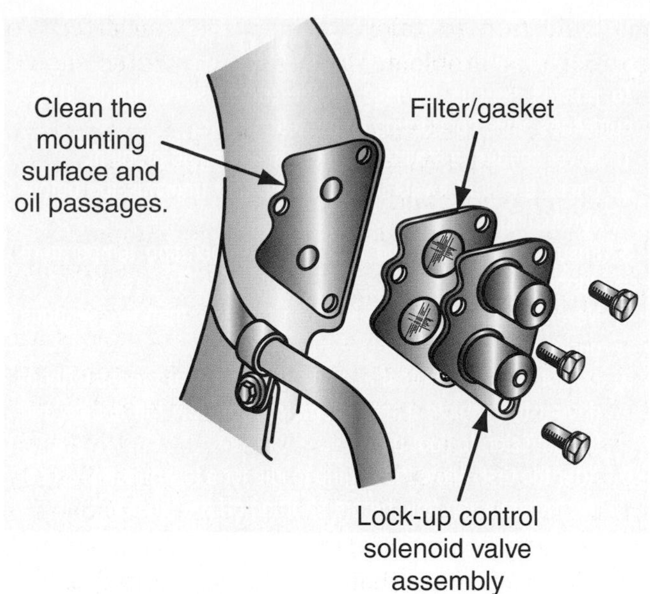

FIGURE 44–16

Some torque converter clutches have a replaceable filter.

FIGURE 44–17

A TCC solenoid.

or piston damage, or defects with the lock-up plate's friction material.

If the TCC stays applied and tends to stall the engine at a stop, the problem may be a damaged piston plate or the plate is stuck to the T/C cover. These problems can be caused by overheating the converter. Look at the converter for signs of excessive heat. This problem can also be caused by the lack of end play clearance inside the converter. In either case, the T/C should be replaced.

If the concern with the TCC is electrical, all related inputs and outputs should be tested with the appropriate meter. A scan tool should be used to retrieve all DTCs **(Figure 44–18)**. The scan tool

DTC	Description	Condition	Symptom
P0741	TCC slippage detected	The PCM noticed an excessive amount of slippage during normal operation.	TCC slippage, erratic operation, or no TCC operation. Flashing transmission MIL.
P0743	TCC solenoid circuit failure during on-board diagnostic	The TCC solenoid circuit does not provide the specified voltage drop across the solenoid. The circuit is open or shorted or the PCM driver failed during the diagnostic test.	With a short circuit, the engine will stall in second gear at low idle speeds with brake applied. With an open circuit, the TCC will not engage. This may cause the MIL to flash.
P0740	TCC electrical failure	The TCC solenoid circuit does not provide the specified voltage drop across the solenoid. The circuit is open or shorted or the PCM driver failed during the diagnostic test.	With a short circuit, the engine will stall in second gear at low idle speeds with brake applied. With an open circuit, the TCC will not engage. This may cause the MIL to flash.
P1740	TCC malfunction	A mechanical failure of the solenoid was detected.	If the solenoid is stuck in the ON position, the engine will stall in second gear at low idle speeds with brake applied. If the solenoid stays in the OFF position, the TCC will not engage. This may cause the MIL to flash.
P1742	TCC solenoid failed on	The TCC solenoid has failed, which was caused by an electric, mechanical, or hydraulic problem.	The transmission will exhibit harsh shifts.
P2758	TCC solenoid circuit failure, stuck ON		The TCC will never engage and there will be no adaptive or self-learning strategies.

FIGURE 44–18

A sample of some TCC-related DTCs.

should also be used to monitor the serial data while the vehicle is taken on a road test.

Most systems have the provision for operating outputs from the scan tool. This feature allows you to engage and disengage the TCC so that its action can be felt and heard.

Stall Test

To test the torque converter, many technicians perform a stall test. The stall test checks the holding capacity of the converter's stator overrunning clutch assembly as well as the clutches and bands in the transmission. Some manufacturers do not recommend the stall test. Rather, diagnosis should be based on the symptoms of the problem.

SHOP TALK

Stall testing places extreme stress on the transmission and should only be conducted if recommended by the manufacturer.

CAUTION!

If a stall test is not correctly conducted, the converter and/or transmission can be damaged.

To conduct a stall test, connect a tachometer to the engine, and position it so that it can be easily read from the driver's seat. Set the parking brake, raise the hood, and place blocks in front of the vehicle's non-driving tires. Conduct the test outdoors, if possible, especially if it is a cold day. If the test is conducted indoors, place a large fan in front of the vehicle to keep the engine cool. With the engine running, press and hold the brake pedal. Then move the gear selector to the DRIVE position, and press the throttle pedal to the floor. Hold the throttle down for two seconds, note the tachometer reading, and then immediately let off the throttle pedal and allow the engine to idle. Compare the measured stall speed to specifications.

WARNING!

Make sure no one is around the engine or the front of the vehicle while a stall test is being conducted. A lot of stress is put on the engine, transmission, and brakes during the test. If something lets go, somebody could get seriously hurt.

CAUTION!

To prevent serious damage to the transmission, follow these guidelines while conducting a stall test:

1. Never conduct a stall test if there is an engine problem.
2. Check the fluid levels in the engine and transmission before conducting the test.
3. Ensure that the engine is at normal operating temperature during the test.
4. Never hold the throttle wide open for more than five seconds during the test.
5. Do not perform the test in more than two gear ranges without driving the vehicle a few kilometres to allow the engine and transmission to cool down.
6. After the test, allow the engine to idle for a few minutes to cool the transmission fluid before shutting off the ignition.

The engine will reach a specific speed if the torque converter and transmission are functioning properly. If the tachometer indicates a speed above or below specifications, a possible problem exists in the transmission or torque converter. If a torque converter is suspected as being faulty, it should be removed and the one-way clutch checked on the bench.

A restricted exhaust or slipping stator clutch is indicated if the stall speed is below the specifications. If the stator's one-way clutch is not holding, ATF leaving the turbine of the converter works against the rotation of the impeller and slows down the engine. Both of these problems cause poor acceleration. If the stall speed is only slightly below normal, the engine is probably not producing enough power and should be diagnosed and repaired.

If the vehicle has poor acceleration but had good results from the stall test, suspect a seized one-way clutch. Excessively hot ATF in the transmission is a good indication that the clutch is seized. However, other problems can cause these same symptoms, so be careful during your diagnosis.

If the stall speed is above specifications, the bands or clutches in the transmission may be slipping and not holding properly.

A stall test will generate a lot of noise, most of which is normal. If you hear any metallic noises during the test, however, diagnose the source of these noises. Operate the vehicle at low speeds on a hoist with the drive wheels free to rotate. If the noises are still present, the source of the noises is probably the torque converter.

TC-Related Cooler Problems

Vehicles equipped with a torque converter clutch (TCC) may stall when the transmission is shifted into REVERSE gear. The cause of this problem may be plugged transmission cooler lines, or the cooler itself may be plugged. Fluid normally flows from the torque converter through the transmission cooler. If the cooler passages are blocked, fluid is unable to exhaust from the torque converter, and the TCC piston remains engaged. When the clutch is engaged, there is no vortex flow in the converter, and therefore little torque multiplication is taking place in the converter.

To verify that the transmission cooler is plugged, disconnect the cooler return line from the radiator or cooler. Connect a short piece of hose to the outlet of the cooler, and allow the other end of the hose to rest inside an empty container. Start the engine, and measure the amount of fluid that flows into the container after 20 seconds. Normally, approximately 1 litre (L) of fluid should flow into the container. If less than that filled the container, a plugged cooler is indicated.

To correct a plugged transmission cooler, disconnect the cooler lines at the transmission and the radiator. Blow air through the cooler, one end at a time, then through the cooler lines. The air will clear large pieces of debris from the transmission cooler. Always use low air pressure, no more than 345 kPa (50 psi). Higher pressures may damage the cooler. If there is little airflow through the cooler, the radiator or external cooler must be removed and flushed or replaced.

DIAGNOSING HYDRAULIC AND VACUUM CONTROL SYSTEMS

The best way to identify the exact cause of the problem is to use the results of the road test, logic, and the oil circuit charts for the transmission being worked on. Before doing this, however, always check all sources for information about the symptom first. Also make sure to check the basics: trouble codes in the computer, fluid level and condition, leaks, and mechanical and electrical connections.

The basic oil flow is the same for all transmissions. The oil pump creates the fluid flow used throughout the transmission. Fluid from the pump always goes to the pressure regulating valve and torque converter. From there, the fluid is directed to the manual shift valve. When the gear selector is moved, the manual valve directs the fluid to other valves and to the apply devices. By following the flow of the fluid on the oil circuit chart, you can identify which valves and apply devices should be operating in each particular gear selector position. Through a process of elimination, you can identify the most probable cause of the problem.

In most cases, the transmission or transaxle is removed to repair or replace the items causing the problem. However, some transmissions allow for a limited amount of service to the apply devices and control valves.

Mechanical and/or vacuum controls can also contribute to shifting problems. The condition and adjustment of the various linkages and cables should be checked whenever there is a shifting problem. If all checks indicate that the problem is either an apply device or in the valving, an air pressure test can help identify the exact problem. Air pressure tests are also performed during disassembly to locate leaking seals and during reassembly to check the operation of the clutches and servos.

Pressure Tests

If you cannot identify the cause of a transmission problem from your inspection or road test, a pressure test should be conducted. This test measures the fluid pressure of the different transmission circuits during the various operating gears and gear selector positions **(Figure 44–19)**. The number of hydraulic circuits that can be tested varies with the different makes and models of transmissions. However, most transmissions are equipped with pressure taps, which allow the pressure test equipment to be connected to the transmission's hydraulic circuits **(Figure 44–20)**.

Before conducting a pressure test on an electronic automatic transmission, check and correct all trouble codes retrieved from the system. Also make sure the transmission fluid level and condition is okay and that the shift linkage is in good order and properly adjusted.

The test is best conducted with three pressure gauges, but two will work. Two of the gauges should read up to 2760 kPa (400 psi) and the other to 690 kPa (100 psi). The two 2760 kPa (400 psi) gauges are usually used to check mainline and an individual circuit, such as mainline and direct or forward circuits. If a circuit is 100 kPa (15 psi) lower than mainline pressure when they are both tested at exactly the same time, a leak is indicated. A 690 kPa (100 psi) gauge may be used be used on TV and governor circuits.

Gear Selector Position	Actual Gear	PRESSURE TAPS					
		Underdrive Clutch	Overdrive Clutch	Reverse Clutch	Torque Converter Clutch Off	2/4 Clutch	Low/Reverse Clutch
Park* 0 km/hr	Park	0–2	0–5	0–2	60–110	0–2	115–145
Reverse* 0 km/hr	Reverse	0–2	0–7	165–235	50–100	0–2	165–235
Neutral* 0 km/hr	Neutral	0–2	0–5	0–2	60–110	0–2	115–145
L# 30 km/hr	First	110–145	0–5	0–2	60–110	0–2	115–145
3# 45 km/hr	Second	110–145	0–5	0–2	60–110	115–145	0–2
3# 65 km/hr	Direct	75–95	75–95	0–2	60–90	0–2	0–2
OD# 45 km/hr	Overdrive	0–2	75–95	0–2	60–90	75–95	0–2
OD# 75 km/hr	Overdrive with TCC	0–2	75–95	0–2	0–5	75–95	0–2

*With vehicle stationary

FIGURE 44–19

A typical oil pressure chart for an engine running at approximately 1500 rpm.

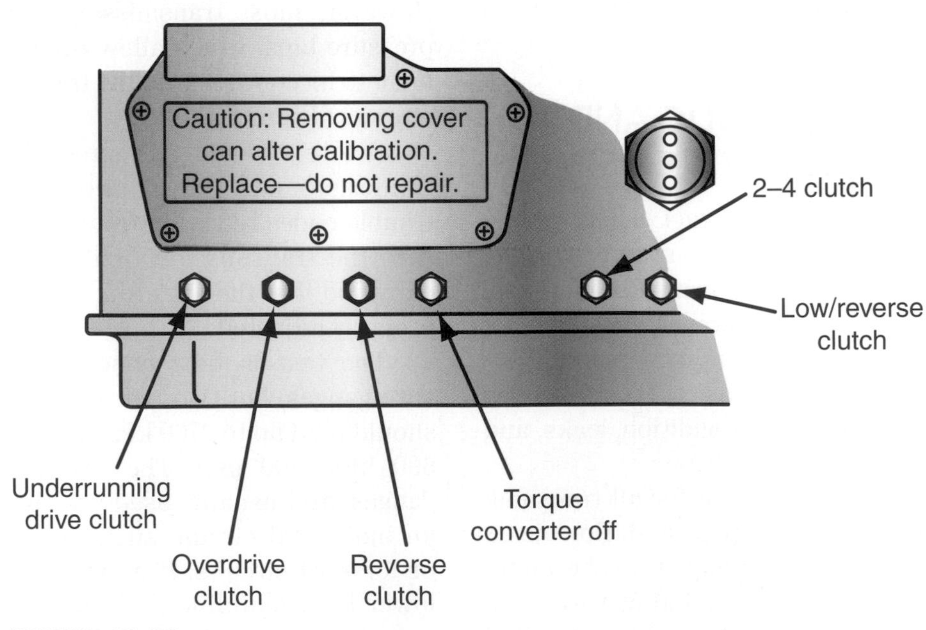

FIGURE 44–20

Pressure taps on the outside of a typical transaxle.

The pressure gauges are connected to the pressure taps in the transmission housing and routed so that the gauges can be seen by the driver. The vehicle is then road-tested and the gauge readings observed during the following operational modes: slow idle, fast idle, and wide-open throttle (WOT).

During the road test, observe the starting pressures and the steadiness of the increases that occur with slight increases in load. The pressure drops as the transmission shifts from one gear to another also should be noted. The pressure should not drop more than 100 kPa (15 psi) between shifts.

Any pressure reading not within the specifications indicates a problem **(Figure 44–21)**. Typically, when the fluid pressures are low, there is an internal leak, clogged filter, low oil pump output, or faulty pressure regulator valve. If the pressure increased at the wrong time or the pressure was not high enough, sticking valves or leaking seals are indicated. If the pressure drop during shifts was greater than approximately 100 kPa (15 psi), an internal leak at a servo or clutch seal is indicated. Always check the manufacturer's specifications for maximum drop-off.

On transmissions equipped with an electronic pressure control (EPC) solenoid, if the line pressure is not within specifications, the EPC pressure needs to be checked. To do this, connect the pressure gauge to the EPC tap. Start the engine and check EPC pressure, then compare it to specifications. If the pressure is not within specifications **(Figure 44–22)**, follow the procedures for testing the EPC. If the pressure is okay, there is a mainline pressure problem.

If the pressure tests suggest a governor problem, it should be removed, disassembled, cleaned, and

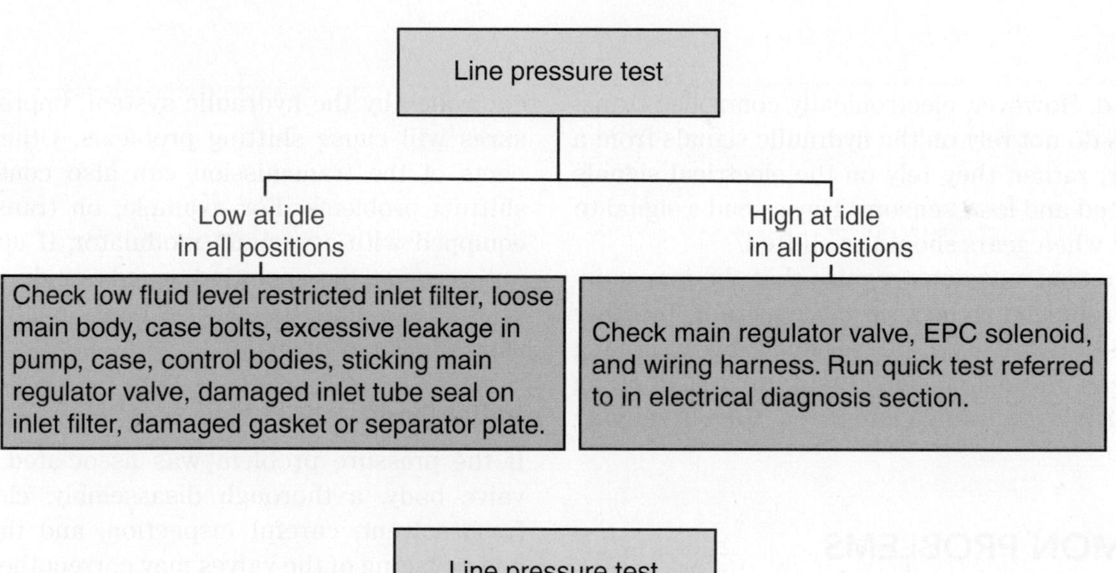

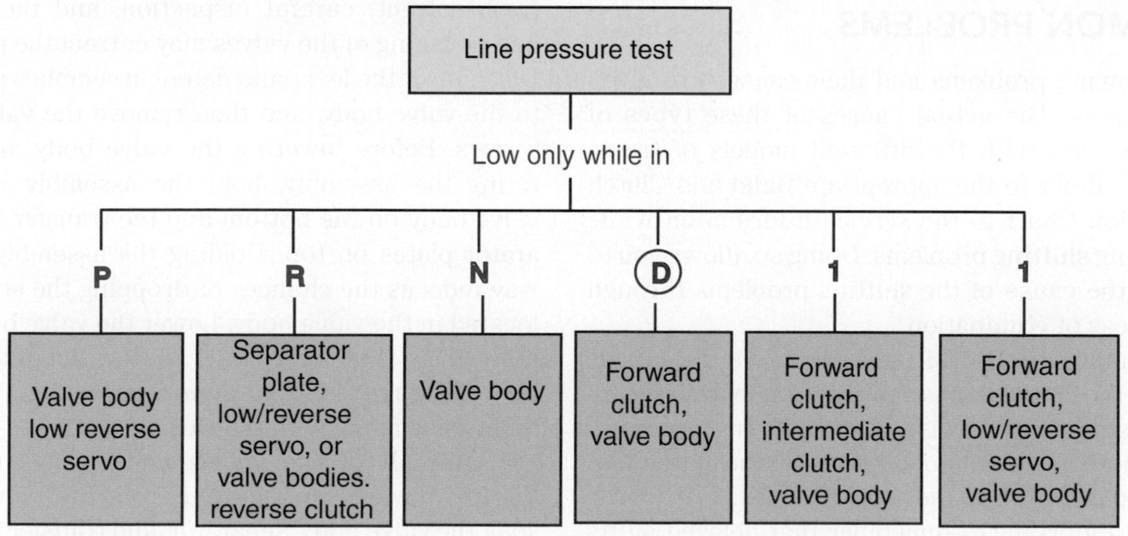

FIGURE 44–21

A troubleshooting chart for abnormal line pressure.

TRANSMISSION PRESSURE WITH TP AT 1.5 VOLTS AND VEHICLE SPEED AT 8 KM/H (5 MPH)					
GEAR	EPC TAP	LINE PRESSURE TAP	FORWARD CLUTCH TAP	INTERMEDIATE CLUTCH TAP	DIRECT CLUTCH TAP
1	276–345 kPa (40–50 psi)	689–814 kPa (100–118 psi)	620–745 kPa (90–108 psi)	641–779 kPa (93–113 psi)	0–34 kPa (0–5 psi)
2	310–345 kPa (45–50 psi)	731–869 kPa (106–126 psi)	662–800 kPa (96–116 psi)	689–827 kPa (100–120 psi)	655–800 kPa (95–116 psi)
3	341–310 kPa (35–45 psi)	620–758 kPa (90–110 psi)	0–34 kPa (0–5 psi)	586–724 kPa (85–105 psi)	551–689 kPa (80–100 psi)

FIGURE 44–22

A pressure chart for a transmission equipped with an EPC solenoid.

inspected. However, electronically controlled transmissions do not rely on the hydraulic signals from a governor; rather, they rely on the electrical signals from speed and load sensors. These send a signal to the TCM when gears should be shifted.

Some transmissions require that the transmission be removed to service the governor. In other transmissions, it can be serviced by removing the extension housing or oil pan, or by detaching an external retaining clamp and then removing the unit.

COMMON PROBLEMS

The following problems and their causes are given as examples. The actual causes of these types of problems vary with the different models of transmissions. Refer to the appropriate Band and Clutch Application Chart in the service information while diagnosing shifting problems. Doing so allows you to identify the cause of the shifting problems through the process of elimination.

Normally, if the shift for all forward gears is delayed, the clutch that is applied in all forward gears may be slipping. Likewise, if the slipping occurs in one or more gears but not all, suspect the clutch that is applied only during those gear ranges.

It is important to remember that delayed shifts or slippage may also be caused by leaking hydraulic circuits or sticking spool valves in the valve body. Because the application of bands and clutches is controlled by the hydraulic system, improper pressures will cause shifting problems. Other components of the transmission can also contribute to shifting problems. For example, on transmissions equipped with a vacuum modulator, if upshifts do not occur at the specified speeds or do not occur at all, the modulator may be faulty or the vacuum supply line may be leaking.

Valve Body

If the pressure problem was associated with the valve body, a thorough disassembly, cleaning in fresh solvent, careful inspection, and the freeing and polishing of the valves may correct the problem. Disconnect the lever and detent assemblies attached to the valve body, and then remove the valve body screws. Before lowering the valve body and separating the assembly, hold the assembly with the valve body on the bottom and the transfer and separator plates on top. Holding the assembly in this way reduces the chances of dropping the steel balls located in the valve body. Lower the valve body, and note where these steel balls are located in the valve body **(Figure 44–23)**, then remove them and set them aside along with the various screws.

After all of the valves and springs **(Figure 44–24)** have been removed from the valve body, soak the valve body, separator, and transfer plates in mineral spirits for a few minutes. Thoroughly clean all parts, and make sure all passages within the valve body are clear and free of debris. Carefully blow-dry

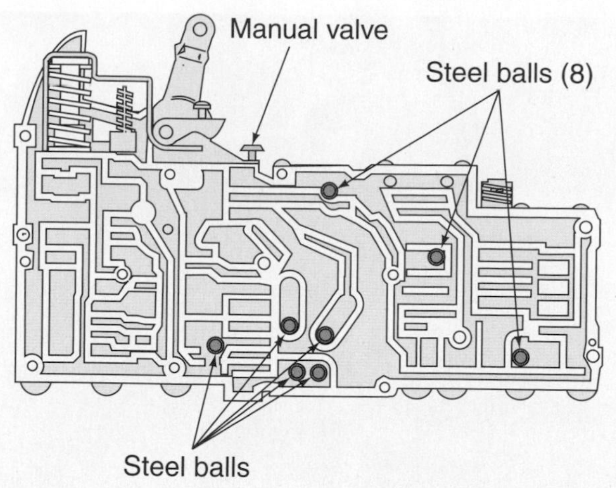

FIGURE 44–23

The location of the manual shift valve and check balls in a typical valve body.

Chrysler Group LLC

each part individually with dry compressed air. Never wipe the parts of a valve body with a wiping rag or paper towel. Lint from either will collect in the valve body passages and cause shifting problems. As the parts of the valve body are dried, place them in a clean container.

Examine each valve for nicks, burrs, and scratches. Check that each valve properly fits into its respective bore. If a valve cannot be cleaned enough to move freely in its bore, the valve body is normally replaced. Individual valve body parts are available, as well as bore reamers. Care must be taken when rebuilding valve bodies.

FIGURE 44–24

Examples of a spool valve and return spring in a valve body.

©Cengage Learning 2015

During reassembly of the valve body, lube the valves with fresh ATF. Check the valve body gasket (if used) to make sure it is the correct one by laying it over the separator plate and holding it up to the light. No oil holes should be blocked. Then install the bolts to hold valve body sections together and the valve body to the case. Tighten the bolts to the torque specifications to prevent valve body warpage and possible leakover.

Servo Assemblies

On some transmissions, the servo assemblies are serviceable with the transmission in the vehicle **(Figure 44–25)**. Others require the complete disassembly of the transmission. Internal leaks at the servo or clutch seal will cause excessive pressure drops during gear changes.

When removing the servo, check both the inner and outer parts of the seal for wet oil that means leakage. When removing the seal, inspect the sealing surface, or lips, before washing **(Figure 44–26)**. Look for unusual wear, warping, cuts and gouges, or particles embedded in the seal.

The servo piston, spring, piston rod, and guide should be cleaned and dried. Then check the sealing rings to make sure they are able to turn freely in the groove of the piston ring. These seal rings are not typically replaced unless they are damaged, so carefully inspect them. Check the servo piston for cracks, burrs, scores, and wear. Inspect the servo cylinder

FIGURE 44–25

To remove the servo in some transmissions, the retaining snap ring must be removed.

©Cengage Learning 2015

FIGURE 44–26

When removing the servo assembly, carefully inspect all parts.

FIGURE 44–27

An example of the location of an external band adjusting screw.

for scores or other damage. Move the piston rod through the piston rod guide, and check for freedom of movement. If all of the parts are in good condition, the servo assembly can be reassembled.

Lubricate the sealing ring with ATF, and carefully install it on the piston rod. Lubricate and install the piston rod guide with its snap ring into the servo piston. Then install the servo piston assembly, return spring, and piston guide into the servo cylinder. Some servos are fitted with rubber lip seals that should be replaced. Lubricate and install the new lip seal. On lip seals, make sure the spring is seated around the lip and that the lip is not damaged.

Band Adjustment

On some transmissions, slippage during shifting can be corrected by adjusting the holding bands. To help identify if a band adjustment will correct the problem, refer to the written results of your road test. Compare your results with the Clutch and Band Application Chart in the service information. If slippage occurs when there is a gear change that requires the holding by a band, the problem may be corrected by tightening the band.

On some vehicles, the bands can be adjusted externally with a torque wrench. On others, the transmission fluid must be drained and the oil pan removed. Locate the band-adjusting nut **(Figure 44–27)**, and then clean off all dirt on and around the nut. Now loosen the band-adjusting bolt locknut and back it off approximately five turns. Use a calibrated pound-inch torque wrench to tighten the adjusting bolt to the specified torque. Then back off the adjusting screw the specified number of turns, and tighten the adjusting bolt locknut while holding the adjusting stem stationary. Reinstall the oil pan with a new gasket, and refill the transmission with fluid. If the transmission problem still exists, an oil pressure test or transmission teardown must be done.

WARNING!

Do not excessively back off the adjusting stem as the anchor block may fall out of place. It will then be necessary to remove and disassemble the transmission to fit it back in place.

LINKAGES

Many transmission problems are caused by improper adjustment of the linkages. All transmissions have either a cable or a rod-type gear selector linkage. Older transmissions also have a throttle valve linkage or an electric switch connected to the throttle to control forced downshifts.

Transmission Range Switch

The transmission range (TR) switch or sensor sends gear shifter location to the TCM and PCM. A faulty switch can cause the transmission to not upshift, not stay in gear, delay gear engagement, and allow the engine to start in positions other than PARK and NEUTRAL.

To check the TR sensor, connect a scan tool, and prepare it to read the transmission data. Move the gear selector through its various positions, and see if the scan tool is displaying the selected location. Depending on the type of sensor, you may need a digital multimeter (DMM) to check the resistance through each position of the switch. Always refer to the manufacturer's service procedures for proper testing information.

The TR sensor will not work properly if the gear shift linkage is damaged or out of adjustment.

Gear Selector Linkage

Most transmissions have a shift cable. The cable connects the gear shifter to the transmission's shift lever. The shift lever is connected to the manual shift valve. This valve directs fluid through the valve body according to the position of the shift lever.

Some automatic transmissions have no linkage; rather, the gear shift works as an input device for the TCM or PCM.

A worn or misadjusted gear selection linkage will affect transmission operation. The transmission's manual shift valve must completely engage the selected gear **(Figure 44–28)**. Partial manual shift valve engagement will not allow the proper amount of fluid pressure to reach the rest of the valve body. If the linkage is misadjusted, poor gear engagement, slipping, and excessive wear can result. The gear selector linkage should be adjusted so that the manual shift valve detent position in the transmission matches the selector level detent and position indicator.

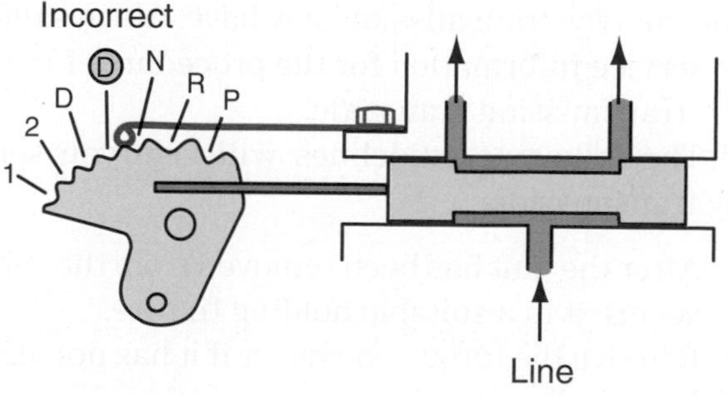

FIGURE 44–28

Incorrect linkage adjustments may cause the manual shift valve to be positioned improperly in its bore and cause slipping during gear changes.

To check the adjustment of the linkage, move the shift lever from PARK to the lowest drive gear. Detents should be felt at each of these positions. If the detent cannot be felt, the linkage needs to be adjusted. While moving the shift lever, pay attention to the gear position indicator. On some older vehicles, the indicator will move with an adjustment of the linkage, but the pointer may need to be adjusted so that it shows the exact gear after the linkage has been adjusted.

REPLACING, REBUILDING, AND INSTALLING A TRANSMISSION

Obviously, in order to rebuild or service a transmission, it must be removed. The procedure for this is very similar to removing a manual transmission or transaxle. The exact procedure for removing a transmission will vary with each year, make, and model of vehicle. Always refer to the service information for the correct procedure.

GO TO ▶ Chapter 40 for details on removing and installing a transmission.

CAUTION!

Always begin by disconnecting the battery ground cable. This is a safety-related precaution to help avoid any electrical surprises when removing starters or wiring harnesses. It is also possible to send voltage spikes, which may kill the PCM if wiring is disconnected when the battery is still connected.

Before removing the transmission, remove the torque converter access cover. Check for loose torque converter bolts. Then rotate the engine and watch the movement of the T/C. If it wobbles, this may be caused by a damaged flexplate or converter.

Place an index mark on the converter and the flexplate to ensure the two will be properly mated during installation. Using a flywheel turning tool, rotate the flywheel until some of the converter-to-flexplate bolts are exposed. Loosen and remove the bolts. Once the bolts are removed, slide the converter back into the transmission.

Once the vehicle is in position, disconnect all transmission linkages connected to the engine. Also remove the ATF dipstick. Plug the cooler line fittings on the housing and the lines. Then proceed to remove the transmission.

Inspecting the Torque Converter

After the transmission has been removed, do the following:

- Inspect the flexplate for evidence of cracking or other damage.
- Check the condition of the starter ring gear teeth, and make sure the gear is firmly attached (welded) to the flexplate.
- Inspect the drive studs or lugs used to attach the converter to the flexplate.
- Check the shoulder area around the lugs and studs for cracked welds or other damage.
- Inspect the converter attaching bolts or nuts, and replace them if they are damaged.
- Check the T/C for ballooning; this is caused by excessive pressure. If the converter is ballooned, it should be replaced and the cause of the high pressure corrected.
- Check the converter's balance weights to make sure they are still firmly attached to the unit.
- Check the pilot of the converter for wear and other damage.
- Check the area around the pilot for cracks.
- Check the drive hub of the torque converter for wear and other damage **(Figure 44–29)**.
- Check for excessive runout of the flexplate and the converter's hub with a dial indicator.

FIGURE 44–29

Check the oil pump and input shaft mating areas of the torque converter.

In general, a torque converter should be replaced if there is fluid leakage from its seams or welds, loose drive studs, worn drive stud shoulders, stripped drive stud threads, heavily grooved hub, or excessive hub runout. The following additional tests of the T/C can be conducted to determine its condition:

- Stator one-way clutch check
- Internal interference check
- End play check
- Converter leakage tests

Summary of T/C Checks

These conditions suggest that the converter should be replaced:

1. If there is a stator clutch failure
2. If there is internal interference
3. If the transmission's front pump is badly damaged
4. If the converter hub is severely damaged or scored
5. If there are signs of external fluid leaks
6. If the drive studs or lugs are damaged or loose
7. If there are signs of overheating
8. If heavy amounts of metal were found in the fluid
9. If any damage is evident that the converter is no longer balanced

Although some specialty shops will rebuild a converter, nearly all technicians replace the converter when any of the above conditions exists. This is especially true of converters with a clutch. Rebuilding a converter is not a normal task for an automobile technician. This procedure requires special equipment and knowledge; do not attempt to repair a faulty converter.

TRANSMISSION OVERHAUL

The exact procedures for rebuilding a transmission depend on the specific transmission as well as the problems the transmission may have. Always refer to the service information for the procedures for a specific transmission/transaxle.

The following guidelines will help you service any transmission:

- After the unit has been removed from the vehicle, secure it in a suitable holding fixture.
- Remove the torque converter if it has not already been removed.
- Check and record the end play of the input shaft before unbolting and removing the oil pan and gasket **(Figure 44–30)**.

FIGURE 44–30

Before and after taking apart a transmission, check the end play of the input shaft.

- Remove the oil pan, oil filter, and all components readily accessible at this point.
- If the transaxle is fitted with a drive chain, inspect it for side play and stretch **(Figure 44–31)**.
- If the unit is a transaxle, remove the differential assembly.

GO TO ▶ Chapter 41 for information on checking and servicing final drive units in a transaxle.

FIGURE 44–31

While removing the drive chain from a transaxle, make sure you check it for damage and stretch.

- Before removing and disassembling the oil pump, mark the gears with machinist bluing ink or paint first. This ensures proper reassembly.
- Measure the clearances within the oil pump.
- Some oil pumps have a wear plate. Carefully inspect the plate for damage and distortion. The plate should not be scored or nicked, or have grooves cut into it. The thickness of the plate at the wear area should be measured and compared to specifications.
- Carefully check the bushing of the oil pump and replace it if it is damaged or worn. The best way to determine if it is worn is to measure its inside diameter. If the measurement exceeds specifications, the bushing should be replaced.
- To replace the bushing, remove all gears from the pump's body. Support the body so that both ends of the bushing's bore are away from the bench. Install the removal tool into the bore, and press or drive the bushing out of the bore. Clean the bore, and then press or drive the bushing into the bore.
- Check all pumps, valve bodies, and cases for warpage, and ensure that they are flat-filed to take off any high spots or burrs prior to reassembly.
- When removing the check balls and springs from the valve body, note their exact location, and count them as they are removed.
- Check each spring for signs of distortion.
- Examine each valve for nicks, burrs, and scratches **(Figure 44–32)**.
- Check the oil passages in the upper and lower valve bodies for varnish deposits, scratches, or other damage that could restrict the movement of the valves. Make sure all fluid drain back openings are clear and have no varnish or dirt buildup.
- Band servos and accumulators are basically pistons with seals in a bore held in position by springs and retaining bolts or snap rings. Some pistons have cast-iron seal rings that may not need replacement. Always replace rubber and elastomer seals.
- To disassemble an accumulator, depress the piston to remove the accumulator cover snap ring. If the piston assembly is retained by a cover, unbolt the cover. Depending on the transmission, a special tool may be required to keep the spring compressed so that the cover can be removed.
- Note the direction of the piston during the teardown, it is possible to install some pistons

FIGURE 44–32

When removing valves from the valve body, check them for damage.

upside down. It is common for manufacturers to mate servo piston assemblies with accumulators. A servo is disassembled in a similar way as an accumulator. Some have two or more pistons and four or five fluid passages and fluid control orifices or check balls inside them. Make sure you do not mix the parts. Also note the exact location of the check balls.

- Inspect the outside area of the servo/accumulator seal. If it is wet, determine if the oil is leaking out or if it is merely a lubricating film of oil. Check both the inner and outer parts of the seal for wet oil, which means leakage. When removing the seal, inspect the sealing surface, or lips.
- The accumulator or servo's piston, spring, piston rod, and guide should be cleaned and dried. Check the piston for cracks, burrs, scores, and wear. The seal groove should be free of nicks or any imperfection that might pinch or bind the seal. Clean up any problems with a small file. The piston ring should rotate freely in its groove. If it does not, clean and inspect the grooves. Replace the piston ring during reassembly.
- Check the condition of the servo/accumulator pins. Look for signs of wear and damage. Also check the fit of the pins and pistons in the case. The bores in the case should not allow the pins and pistons to wobble. If they do, they are worn or the bores in the case are worn.
- Inspect the servo or accumulator spring for cracks. Also check the area where the spring rests against the case or piston. The spring may wear a groove, so make sure the piston or case material has not worn too thin.
- Inspect the servo cylinder and other parts for wear, scores, or other damage. Move the piston rod through the piston rod guide, and check for freedom of movement. Replace all other components as necessary, and then reassemble the servo assembly.
- When removing and disassembling a multiple-disc assembly, use the correct puller and spring compressor.
- Keep every part absolutely clean, and air-dry all parts. Only use lint-free rags to wipe parts off; lint can collect and damage the transmission.
- Check all of the threaded holes and related bolts and screws for damaged threads.
- Check the end play of each shaft as it is being assembled.
- Make sure that the correct size thrust washers are used throughout the transmission **(Figure 44–33)**.
- Soak all friction materials in clean ATF for at least 30 minutes prior to installing them.

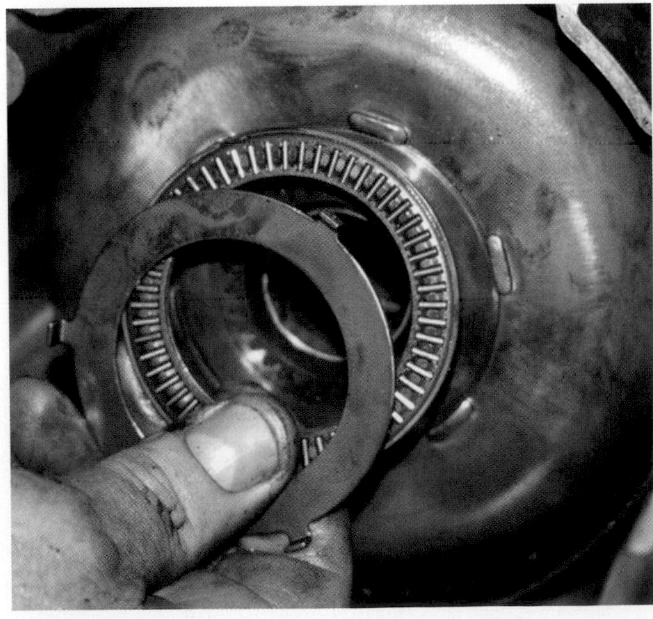

FIGURE 44–33

Make sure all thrust washers and bearings are the correct thickness before assembling the transmission.

- Coat all thrust washers, bearings, and bushings with transjel (assembly lubricant) before putting them into position.
- Typically, to remove and install a one-way clutch, its retainer must be removed. Rotate the clutch clockwise until it lines up with the roller clutch cam ramps. Then remove the roller clutch from the inside of the roller clutch cam.
- Inspect one-way clutches for wear or damage, spline damage, surface finish damage, and damaged retainers and grooves. All smooth surfaces should be checked for any imperfections.
- Replace all rollers and races that show any type of damage or surface irregularities. Check the folded springs for cracks, broken ends, or flattening out. Replace all distorted or otherwise damaged springs.
- When reassembling a roller clutch, make sure the rollers and springs are facing the correct direction. If they are reversed, the clutch will not lock in either direction.
- To reassemble the roller clutch assembly, press the clutch assembly onto the housing.
- Once the one-way clutches are ready for installation, verify that they overrun in the proper direction. Most one-way clutches have some marking that indicates which direction the clutch should be set. During installation, use a new retaining snap ring.
- Solid sealing rings are commonly used in transmissions. These rings are made of a Teflon-based material and are never reused. To remove them, pry them out of their groove and carefully cut through the seal. Installing a new seal requires two tools: an installation tool and a resizing tool. To install a Teflon seal, warm the new seal in hot water to soften the material; this will make installation easier. Lubricate the new seal and tool. Slide the seal over the tool, and seat it into its groove on the hub or shaft.
- Inspect all bushings that may be inserted into the planetary gearsets. These are commonly found in sun gears. Measure the inside diameter, and compare that dimension to specifications. If the diameter exceeds those specifications, replace the gear assembly.
- When tightening any fastener that directly or indirectly involves a rotating shaft, rotate the shaft during and after tightening to ensure that the part does not bind.
- Always use new gaskets and seals throughout the transmission.

FIGURE 44–34

Make sure all of the electrical connectors are in place and tight.

- Make sure the differential and final drive unit is properly set up before installing it into the transaxle.
- Make sure that all electrical wiring to the solenoids and sensors **(Figure 44–34)** are connected and that all the harness clamps hold the wiring where it should.
- Air-test the entire transmission **(Figure 44–35)** before installing the oil filter and oil pan.
- Install a new oil filter and pan gasket. Tighten the oil pan bolts to specifications.
- Always flush out the transmission cooling system before using a rebuilt or new transmission. The cooling system is a good place for debris to collect. Some manufacturers require replacement of the cooler rather than flushing it.

Installation

Transmission installation is generally the reverse of the removal procedure. A quick check of the following will greatly simplify your installation and reduce the chances of destroying something during installation:

- Make sure that the block alignment pins (dowels) are in their appropriate bores and are in good shape, and that the alignment holes in the bell housing are not damaged.
- Make sure that the pilot hole in the crankshaft is smooth and not out of round **(Figure 44–36)**. This will allow the converter to move in and out of the flexplate.

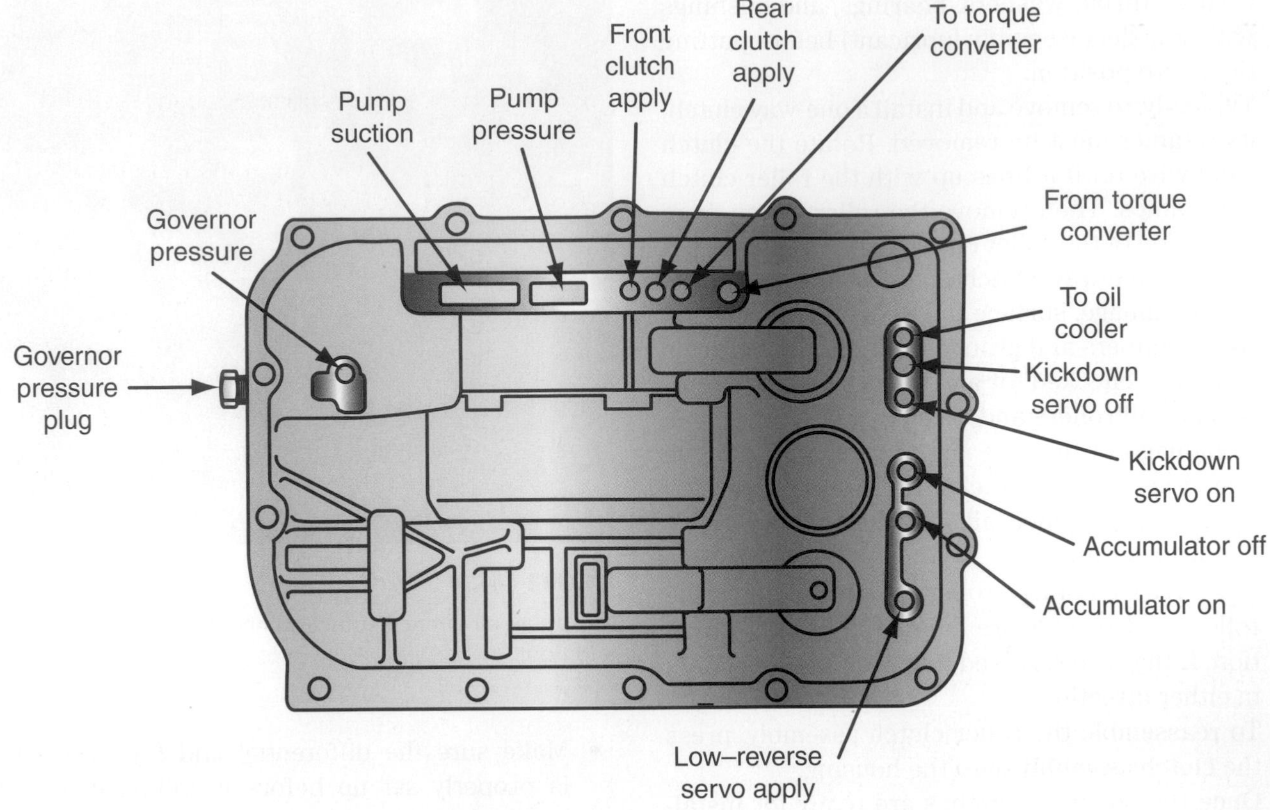

FIGURE 44–35

Air-testing points in a typical transmission.

- Make sure that the pilot hub of the converter is smooth and that you cover it with a light coating of chassis lubricant to prevent chafing or rust.
- Make sure that the converter's drive hub is smooth and that you coat it with trans gel or equivalent lubrication.

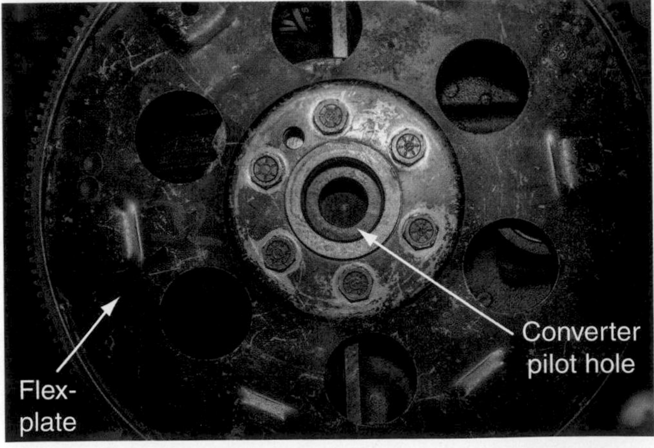

FIGURE 44–36

A pilot hole in a flexplate.

- Secure all wiring harnesses out of the way to prevent their being pinched between the bell housing and engine block.
- Flush out the converter. It is recommended that clutch-type converters be replaced, because it is not possible to tell how much debris is in the unit.
- Always perform an end play check, and check the overall height before reinstalling a torque converter or installing a fresh unit out of the box.
- Pour 1 L of the recommended fluid into the converter before mounting the converter to the transmission. This will ensure that all parts in the converter have some lubrication before start-up.

SHOP TALK

If the transmission is an EAT, check the service information before taking it on its initial road test. Most EATs require that a "Learning Procedure" be followed. This includes a variety of driving conditions. Because you need to road-test and teach the transmission, you may as well do both at the same time.

KEY TERMS

Aeration (p. 1340)

Transmission identification numbers (p. 1339)

SUMMARY

- The ATF level should be checked at regular mileage and time intervals. Typically, when the fluid is checked, the vehicle should be level and running, and the transmission should be at operating temperature.
- Both low fluid levels and high fluid levels can cause aeration of the fluid, which in turn can cause a number of transmission problems.
- Uncontaminated ATF is typically red in colour and has no dark or metallic particles suspended in it.
- The fluid should be changed when the engine and transmission or transaxle are at normal operating temperatures. After draining the fluid, the pan should be inspected and the filter replaced.
- If ATF is leaking from the pump seal, the transmission must be removed from the vehicle so that the seal can be replaced. Other worn or defective gaskets or seals can be replaced without removing the transmission. Case porosity may be repaired using an epoxy-type sealer.
- Slippage during shifting can indicate the need for band adjustment.
- Improper shift points can be caused by a malfunction in the governor or governor drive gear system or a misadjusted throttle linkage.
- The road test gives the technician the opportunity to check the transaxle or transmission operation for slipping, harshness, incorrect upshift speeds, and incorrect downshift.
- Accurate diagnosis depends on knowing what planetary controls are applied in a particular gear range.
- A pressure test checks hydraulic pressures in the transmission by using gauges attached to the transmission. Pressure readings reveal possible problems in the oil pump, governor, and throttle circuits.
- Proper adjustment of the gear selector or manual linkage is important to have the manual valve fluid inlet and outlets properly aligned in the valve body. If the manual valve does not align with the inlet and outlets, line pressure could be lost to an open circuit.

REVIEW QUESTIONS

1. What is the most probable cause of a low fluid level?
2. What does milky-coloured ATF indicate?
3. What do varnish or gum deposits on the dipstick indicate?
4. How can air pressure be used to check an electrical switch?
5. What should you do if a valve does not move freely in its bore in the valve body?
6. What is the typical maximum pressure drop allowed during shifts?
   a. 35 kPa (5 psi)
   b. 69 kPa (10 psi)
   c. 104 kPa (15 psi)
   d. 138 kPa (20 psi)
7. What would dark particles found in the automatic transmission indicate?
   a. aeration
   b. varnish buildup
   c. worn clutch friction discs or bands
   d. worn clutch steel disc
8. Which of the following is the most likely cause for a shudder during the engagement of a lock-up torque converter?
   a. a bad converter
   b. worn or damaged CV or U-joints
   c. a worn front planetary gearset
   d. a loose flexplate
9. When should the ATF level be checked on most vehicles?
   a. when the engine is cool
   b. when the engine is at operating temperature and the engine is off
   c. when the engine is at operating temperature and the engine is on
   d. It does not matter.
10. How should new friction discs be installed when rebuilding a clutch pack?
    a. dry
    b. soaked in new automatic transmission fluid
    c. coated in petroleum jelly
    d. soaked in engine oil
11. What would be the most likely cause of a transmission vibration?
    a. a worn transmission band
    b. an unbalanced torque converter
    c. worn oil pump housing
    d. burnt clutch discs

12. How does road-testing an electronic automatic transmission vary from road-testing a non-electronic transmission?
   a. The vehicle should be driven in a normal manner.
   b. The pressures and gear change condition should be noted.
   c. A scan tool should be connected to monitor transmission operation.
   d. The vehicle should be driven in a way to duplicate the complaint.
13. What would be the most likely cause of slippage in only one transmission operating gear?
   a. a failed clutch pack
   b. a faulty governor valve
   c. a worn transmission oil pump
   d. a failed torque converter stator one-way clutch
14. What would higher than specified engine rpm during a stall test indicate?
   a. Clutches and bands are not releasing.
   b. The stator one-way clutch will not release.
   c. The stator one-way clutch will not lock or hold.
   d. A clutch or band is slipping.
15. Which of the following automatic transmission fluid leaks describe porosity problems?
   a. fluid seeping through the transmission case
   b. fluid leaks at electrical switches
   c. fluid leaks at gasket mating surfaces
   d. fluid leaks at the output shaft seal
16. Which of the following could cause a lower than specified oil pressure reading during all gears in an electronically controlled automatic transmission under higher loads or engine rpm only?
   a. a worn transmission oil pump
   b. a sticking electronic pressure control solenoid
   c. a faulty vehicle speed sensor
   d. a sticking governor valve
17. When should the transmission fluid and filter be changed?
   a. at the periodic change interval only, with the transmission cold
   b. at the periodic change interval only, with the transmission at operating temperature
   c. whenever oxidation or contamination is present with the transmission cold
   d. whenever oxidation or contamination is present with the transmission at operating temperature
18. What would a low stall speed during a stall test indicate?
   a. a slipping stator one-way clutch
   b. a slipping clutch pack or band
   c. low governor pressure
   d. a failed torque converter clutch
18. When conducting a pressure test, what circuit can a 690 kPa (100 psi) gauge be connected to?
   a. mainline pressure circuit
   b. reverse gear circuit
   c. forward gear circuits
   d. governor circuit
20. What could cause delayed or no upshifts in a non-electronically controlled automatic transmission?
   a. low throttle pressure
   b. a worn transmission oil pump
   c. a faulty vacuum modulator
   d. a failed torque converter clutch

CHAPTER 45
Four- and All-Wheel Drive

LEARNING OUTCOMES

- Identify the advantages of four- and all-wheel drive.
- Name the major components of a conventional four-wheel-drive system.
- Name the components of a transfer case.
- State the difference between the transfer, open, and limited-slip differentials.
- State the major purpose of locking/unlocking hubs.
- Name the five shift lever positions on a typical four-wheel-drive vehicle.
- Understand the difference between four- and all-wheel drive.
- Know the purpose of a viscous clutch in all-wheel drive.

With the popularity of four-wheel-drive (4WD) and all-wheel-drive (AWD) sport utility vehicles (SUVs), pickup trucks, and crossover vehicles, the need for technicians who can diagnose and service four-wheel-drive systems has drastically increased. These vehicles have topped the Canadian sales charts because they offer utility, comfort, and confidence whether driving on urban streets or rural roads. Although all-wheel-drive passenger cars are available, most prospective buyers for all-wheel-drive and four-wheel-drive vehicles are opting for truck-based SUVs and pickups.

Four-wheel-drive (4WD) and **all-wheel-drive (AWD)** systems can dramatically increase a vehicle's traction and handling ability in rain, snow, and off-road driving. Considering that the vehicle's only contact with the road is the small areas of the tires, driving and handling is vastly improved if the work load is spread out evenly among four wheels rather than two.

Factors such as the side forces created by cornering and wind gusts have less effect on vehicles with four driving wheels. The increased traction also makes it possible to apply greater amounts of energy through the drive system. Vehicles with 4WD and AWD can maintain control while transmitting levels of power that would cause two wheels to spin either on take-off or while rounding a curve. The improved traction of 4WD and AWD systems allows the use of tires that are narrower than those used on similar 2WD vehicles. These narrow tires tend to cut through snow and water rather than hydroplane over them. Of course, wider and larger tires are often used in off-the-road adventures **(Figure 45–1)**.

Both 4WD and AWD systems add initial cost and weight. With most passenger cars, the weight problem is minor. A typical 4WD system adds approximately 77 kg (170 lb.) to a passenger car. An AWD system adds even less weight. The additional weight in larger 4WD trucks can be as much as 180 kg (400 lb.) or more.

The systems also add initial cost to the vehicle. Vehicles equipped with 4WD and AWD require special service and maintenance not needed on 2WD vehicles. However, the slight disadvantages of 4WD

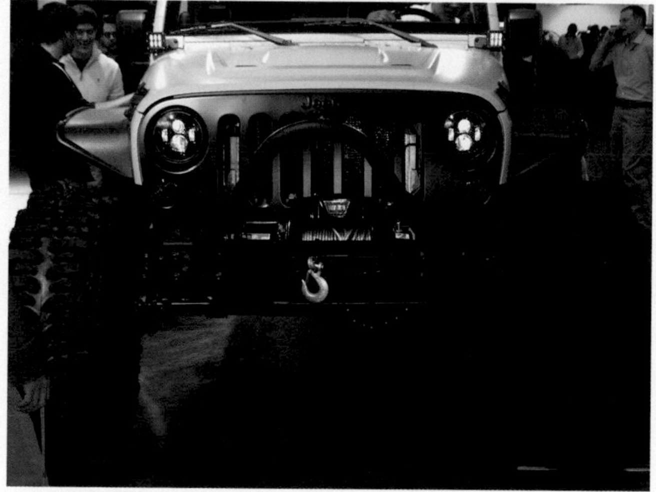

FIGURE 45–1

This Jeep is equipped for serious off-road operation.

and AWD are heavily outweighed by the traction and performance these systems offer.

TYPES OF FOUR-WHEEL DRIVES

Because of the many names manufacturers give their drive systems, it is often difficult to clearly define the difference between 4WD and AWD. Both have the ability to send torque to all four wheels. To do this, these vehicles have a transfer case and a differential unit at the front and rear axles **(Figure 45–2)**. A **transfer case** splits the power from the transmission between the front and rear axles. For clarity, the primary difference between the two is that the transfer case in a 4WD vehicle offers two speed ratios: high and low. These systems are mostly found in pickups and large SUVs. An AWD vehicle does not offer this. AWD vehicles are smaller SUVs or passenger cars.

Although most 4WD trucks, crossovers, and SUVs are design variations of most basic rear-wheel-drive (RWD) vehicles, most passenger cars with 4WD or AWD were developed from a front-wheel-drive (FWD) base vehicle **(Figure 45–3)**.

Four-wheel-drive cars normally differ from heavier-duty 4WD trucks and SUVs in several other ways. First, there is no separate transfer case; any gearing needed to transfer power to the rear driveline is usually contained in the transaxle housing or small bolt-on extension housing **(Figure 45–4)**. The reason some cars are equipped with AWD is to improve their handling, although the ability to drive all four wheels does help during inclement weather conditions.

Four-wheel-drive vehicles do not drive all of the wheels all of the time. This is because they have open differentials at the front and rear drive axles, and in the centre differential, if equipped. However, some have an open centre differential. Limited-slip (LSD) and locking differentials or couplings with torque management capability increase the ability to actually transfer torque to all four wheels.

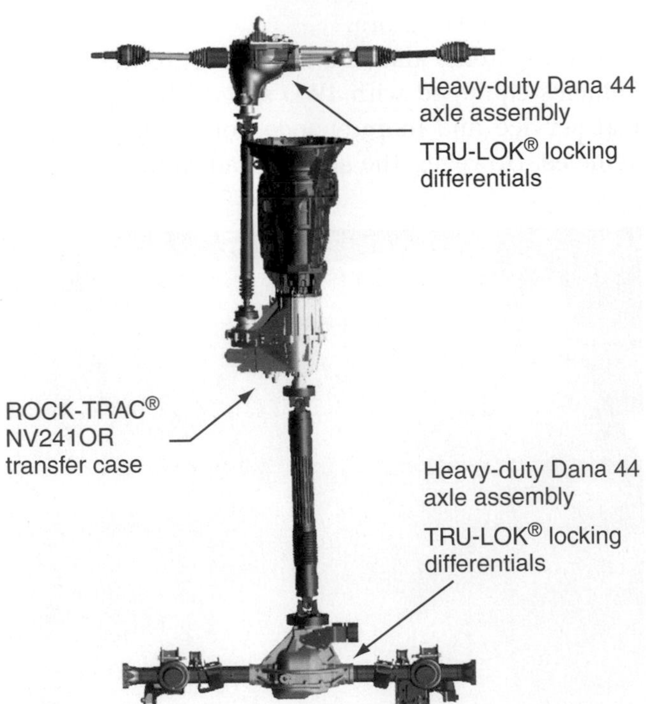

FIGURE 45–2

The drivetrain layout for a part-time, heavy-duty 4WD vehicle.

Chrysler Group LLC

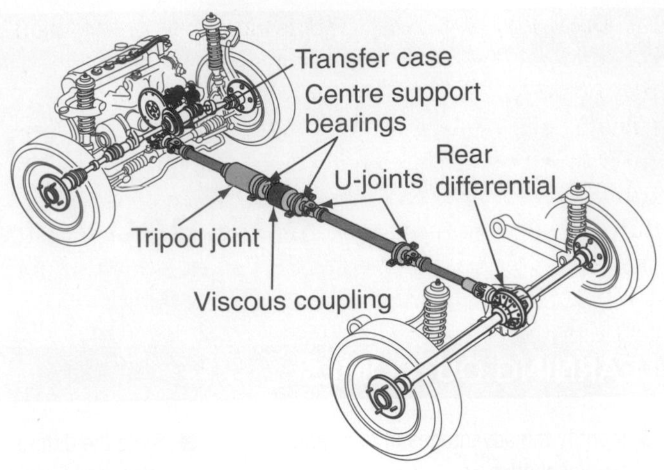

FIGURE 45–3

The layout of an AWD vehicle based on an FWD vehicle.

FIGURE 45–4

A nine-speed transaxle with a bolt-on transfer case to provide AWD.

Basically, part-time 4WD is designed to be used off-road. This system makes a mechanical connection between the front and rear driveshafts when the transfer case is shifted into 4WD. Full-time 4WD or AWD transfer cases transfer constant power to the front and rear axles and include a centre or interaxle differential between the front and rear output shafts. Full-time 4WD can also be called "anytime 4WD" or AWD, although this is not entirely a correct designation.

Part-Time 4WD

Part-time systems are most often found on 4WD pickups and older SUVs. These are basically RWD vehicles fitted with a two-speed transfer case and axle connects/disconnects. These 4WD systems operate in two-wheel drive (2WD) until the driver manually selects 4WD.

When 4WD is selected, more traction is available to allow the vehicle to either carry heavier loads and/or to travel in adverse terrain conditions. The system was designed just for those purposes and not for use on dry or smooth pavement. When operating in 4WD, the front and rear drive axles rotate with same torque, and there is no allowance for axle speed differences. When in 4WD and turning on dry surfaces, the tires scrub against the road surface since no difference in axle speed is available; as a result, the vehicle does not handle well. However, when the vehicle is cornering on a slippery surface, the tires easily skid or spin across the surface to allow for the required axle speed differences.

It must be kept in mind that the distances the tires must travel are different when the vehicle is making a turn **(Figure 45–5)**. The outside front wheel will travel further than the outside rear wheel and the result is called **driveline windup**. This is the main reason for using new full-time transfer case designs with clutch packs or viscous couplings to permit full-time use on dry pavement. Driveline windup can cause handling problems (jerking steering), particularly when rounding turns on dry pavement.

The driveshafts from the transfer case shafts connect to differentials at the front and rear drive axles. Universal joints are used to connect the driveshafts to the differential and the transfer case. The drive axles are either connected directly to the hub of the wheels or are connected to the hubs by U-joints. Universal joints are also normally used to connect the front axles to the wheel hubs on heavy-duty trucks. Light-duty vehicles and 4WD passenger

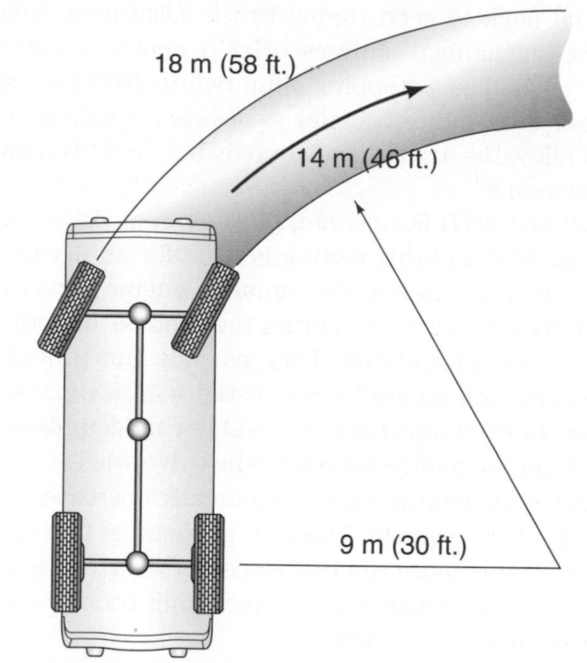

FIGURE 45–5

The distance the wheels travel while a vehicle is making a turn.

©Cengage Learning 2015

cars generally use half shafts and CV joints in their front drive axle assembly.

Nearly all part-time 4WD systems have a transfer case that not only allows for 2WD or 4WD but also offers two different speeds ratios: high and low. This feature allows the driver to control how engine torque will be distributed to the drive wheels. The selector switch **(Figure 45–6)** or mechanical shifter that controls the transfer case normally has 2WD, 4WD high, and 4WD low positions. The shift control either physically moves a gear in the transfer case or activates an electrically operated solenoid or

FIGURE 45–6

A 4WD-mode selector switch.

Martin Restoule

clutch pack to send torque to the front axle. Also, most mechanical shifters offer a neutral position that is used as a stopping point before 4WD low can be selected. Other transfer cases are a single speed and allow the driver to select only between 2WD and 4WD modes.

When 2WD is selected, only the rear drive axle is engaged, and this setting is used for all dry-road driving. 4WD high mode transfers engine power to both axles without affecting the amount of torque sent to the drive wheels. This gear selection provides 4WD traction on surfaces covered with ice, packed snow, or even loose gravel. 4WD low mode provides great torque multiplication to the drive wheels. This high torque multiplication drastically decreases the speed of the vehicle. The sole purpose of this gear is to provide more tractive force at the drive wheels to overcome rough terrain, including rocks, snow, gravel, or steep inclines.

Shifting

Some 4WD vehicles use a vacuum motor or mechanical linkage to move a splined sleeve to connect or disconnect the front drive axle **(Figure 45–7)**. With this system, locking hubs are not needed. When 2WD is selected, one axle is disconnected from the front differential. As a result, all engine torque moves to the side of the differential with the axle disconnected. This is due to normal differential action. When the vehicle is shifted into 4WD, the shift collar connects the two sections of the axle shaft together.

Other axle disconnects are operated electrically. An electric motor can be used to connect and disconnect the axle **(Figure 45–8)**. This system allows for a smooth transition from two- to four-wheel drive. General Motors uses a system whereby selecting 4WD on the selector switch energizes a heating element in the axle disconnect. The heating element heats a gas, causing the plunger to operate the shift mechanism.

Most late-model vehicles with a mechanical or electronic 4WD control allow the driver to switch between 2WD and 4WD while the vehicle is moving, or "on the fly." These systems work with the vehicle's PCM and integrated wheel ends (IWE) solenoids. The driver can typically switch between 2WD and 4WD at speeds up to 90 km/h (55 mph). However, switching from 4WD high to 4WD low requires that

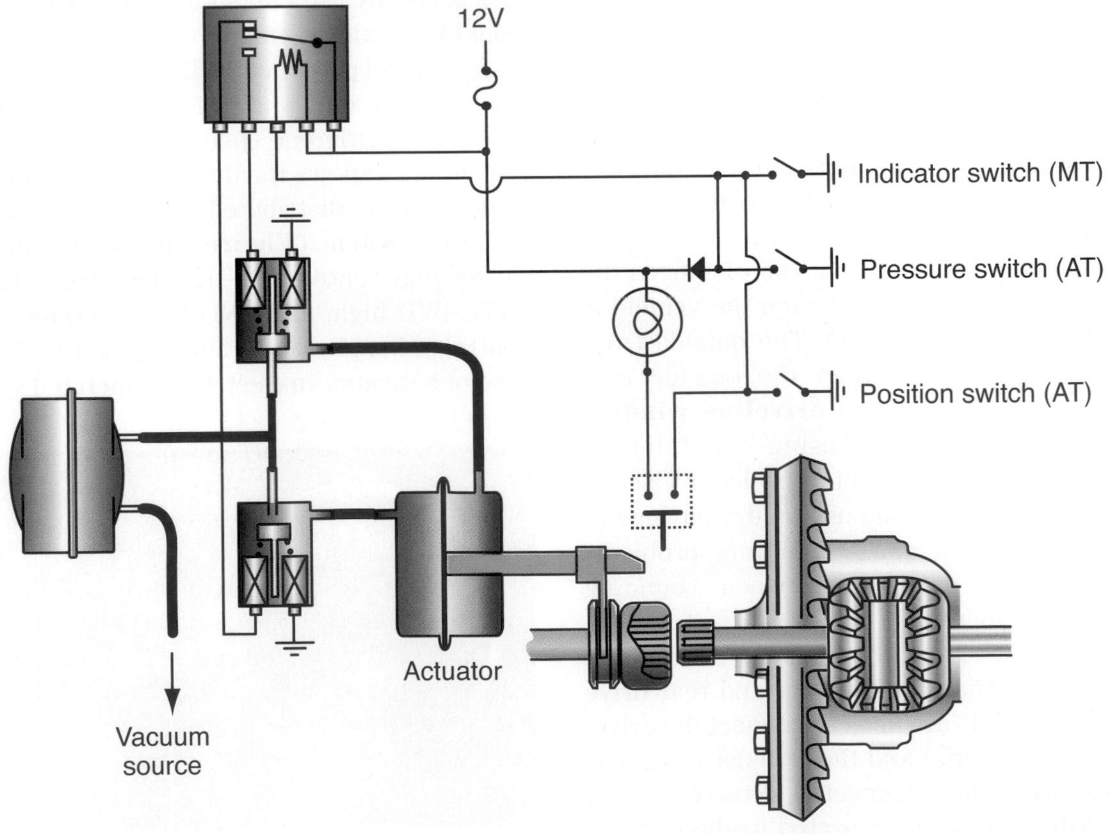

FIGURE 45–7

A vacuum-operated axle disconnect system.

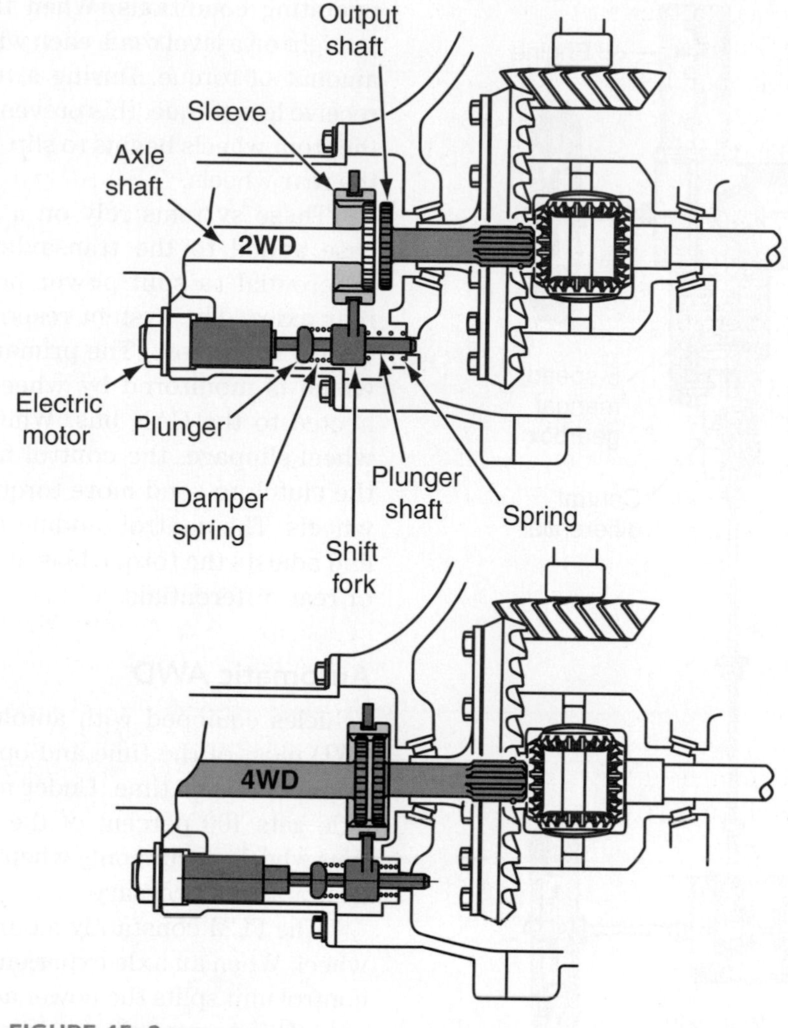

FIGURE 45–8

An electric motor disconnects the axles in this system.

the vehicle be travelling at less than 5 km/h (3 mph), the brake pedal be pressed, and the transmission be in NEUTRAL. The transfer case is fitted with an electromagnetic clutch that is used to synchronize the front driveline with the transmission's output. When the manual shift lever is moved into 4WD, the electromagnetic clutch is energized. This causes a shifting collar to engage the transfer case's main shaft hub to the drive sprocket for the front drive shaft. The front axle integrated wheel ends (IWEs) are engaged, and then the electromagnetic clutch is disengaged. In electronic shifter transfer cases, the clutch ultimately controls a shift motor.

FWD-BASED SYSTEMS Some FWD vehicles fitted for 4WD use a compact transfer case bolted to the transaxle. A driveshaft then carries the power to the rear differential. The driver can switch from 2WD to 4WD by pressing a switch that activates a solenoid vacuum valve or electric motor.

Full-Time 4WD

Like part-time 4WD, full-time 4WD provides a vehicle with more traction, but it makes 4WD more practical for everyday use, as it can be used full time on all surfaces, including dry pavement. These systems cannot be selected out of 4WD, and output torque is spread equally to the drive wheels if the road conditions permit this. To prevent tire scrub during cornering, the system may have an additional differential incorporated into the transfer case or between the front and rear drive axles. The extra differential, called a centre or **interaxle differential**, allows for a speed difference between the front and rear axles **(Figure 45–9)**.

Full-Time AWD

Full-time AWD is similar to full-time 4WD in that it powers all four wheels at all times. However, that 4WD low setting is not available. Because it does

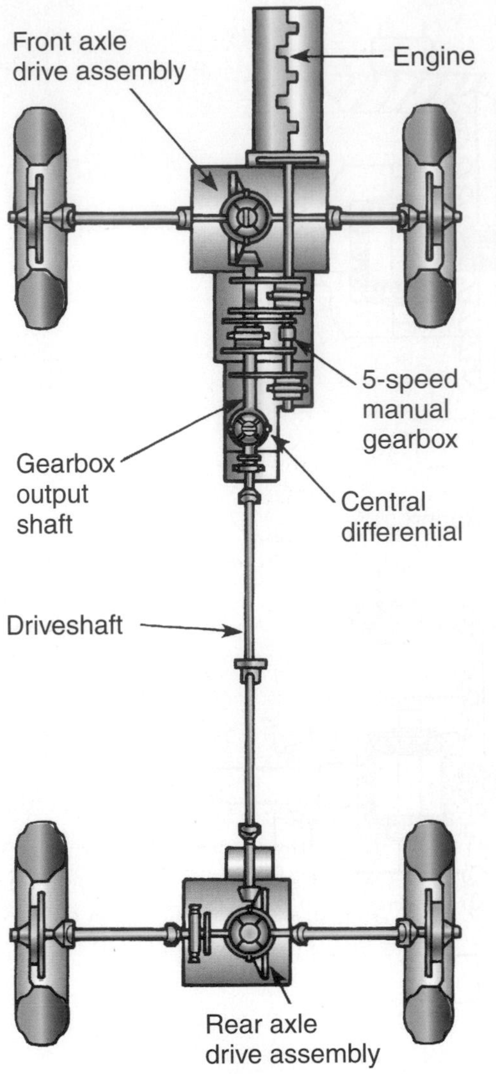

FIGURE 45–9

The placement of a centre differential for a full-time 4WD system.

not have this low gear, vehicles with AWD are not designed for off-road excursions, but can be used on all other surfaces, including dry pavement.

Most passenger cars and smaller SUVs equipped with AWD are based on FWD designs. These modified FWD systems consist of a transaxle and differential to drive the front wheels, plus some type of mechanism for connecting the transaxle to a rear driveline. They may not have a separate transfer case; rather, a viscous clutch, centre differential **(Figure 45–10)**, or electromagnetic transfer clutch is used to transfer power from the transaxle to a rear driveline and rear axle assembly.

In these systems, all torque is normally sent to the front wheels, but that can change with changing operating conditions. When the vehicle is moving straight on a level road, each wheel receives the same amount of torque. During a turn, the front wheels receive less torque; this prevents wheel slip. If one of the front wheels begins to slip, more torque is sent to the rear wheels.

These systems rely on a single-speed transfer case added to the transaxle and/or an interaxle differential to split power between the front and rear axles. The system responds to inputs from a variety of sensors. The primary input is wheel slip, which is monitored by wheel speed sensors connected to the CAN bus. When the system detects wheel slippage, the control module fully activates the clutch to send more torque to the front or rear wheels. The control module monitors these inputs and adjusts the torque bias at the centre, front, and/or rear differentials.

Automatic AWD

Vehicles equipped with automatic AWD operate in 2WD most of the time and operate in 4WD for only brief periods of time. Under normal conditions, one axle gets 100 percent of the torque, and power to four wheels occurs only when the conditions make it desirable or necessary.

The PCM constantly monitors the speed of each wheel. When an axle experiences some slippage, the control unit splits the power accordingly to the other axle. This power split can be accomplished hydraulically, mechanically, or electrically, depending on the system. As soon as the axle is no longer slipping and all four wheels are rotating at the same speed, all torque is sent to that axle.

Normally, to split the power between axles, a multiple disc clutch is used. This clutch serves as an interaxle differential and permits a speed difference between the front and rear drive axles. Sensors monitor front and rear axle speeds, engine speed, and load on the engine and driveline **(Figure 45–11)**. An electronic control unit (ECU) receives information from the sensors and controls a solenoid that operates on a duty cycle to control the fluid flow that engages the transfer clutch. The duty solenoid pulses, cycling on and off very rapidly, which develops a controlled slip condition. As a result, the transfer clutch operates like an interaxle differential and allows for a power split from 95 percent FWD and 5 percent RWD to 50 percent FWD and 50 percent RWD. This power split takes place so rapidly that the driver is unaware of the traction problem.

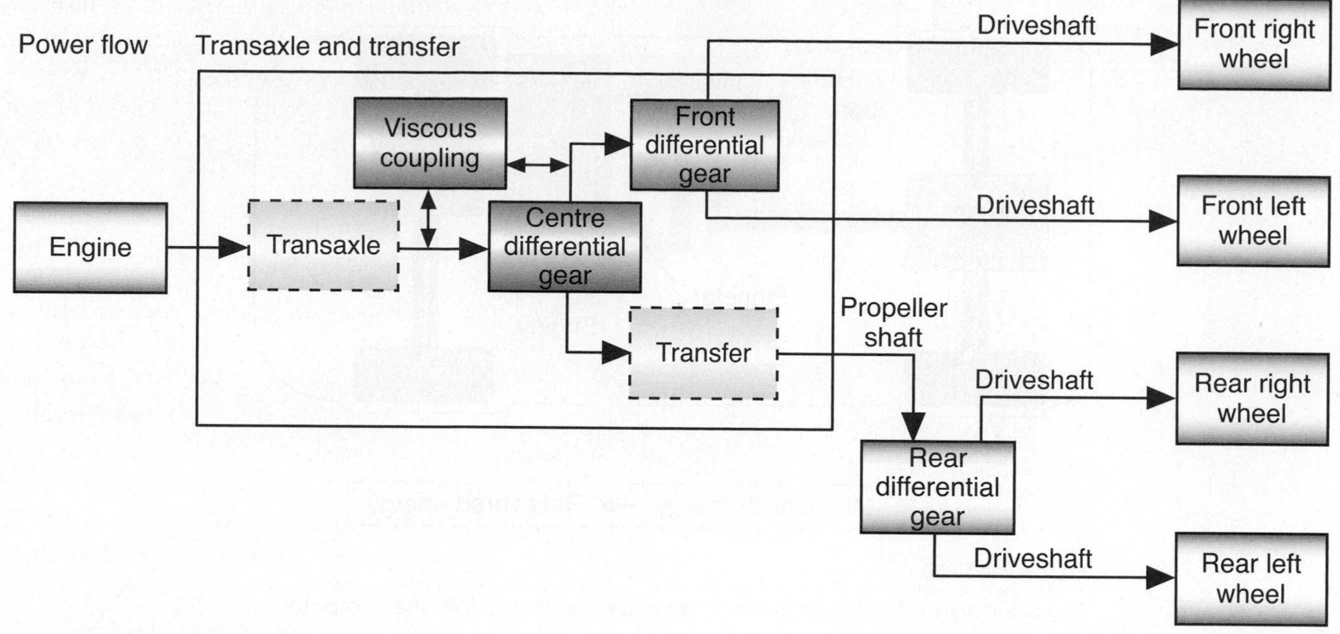

FIGURE 45–10

The power flow through a viscous clutch–type centre differential.

4WD Hybrids

On Toyota and Lexus hybrid SUVs, the front transaxle assembly has been modified to include a speed reduction unit. This unit is a planetary gearset coupled to the power-split planetary gearset. Also, at the rear axle, an additional motor/generator (MGR) is placed in its own transaxle assembly to rotate the rear drive wheels. Unlike conventional 4WD vehicles, there is no physical connection between the front and rear axles **(Figure 45–12)**. The aluminum

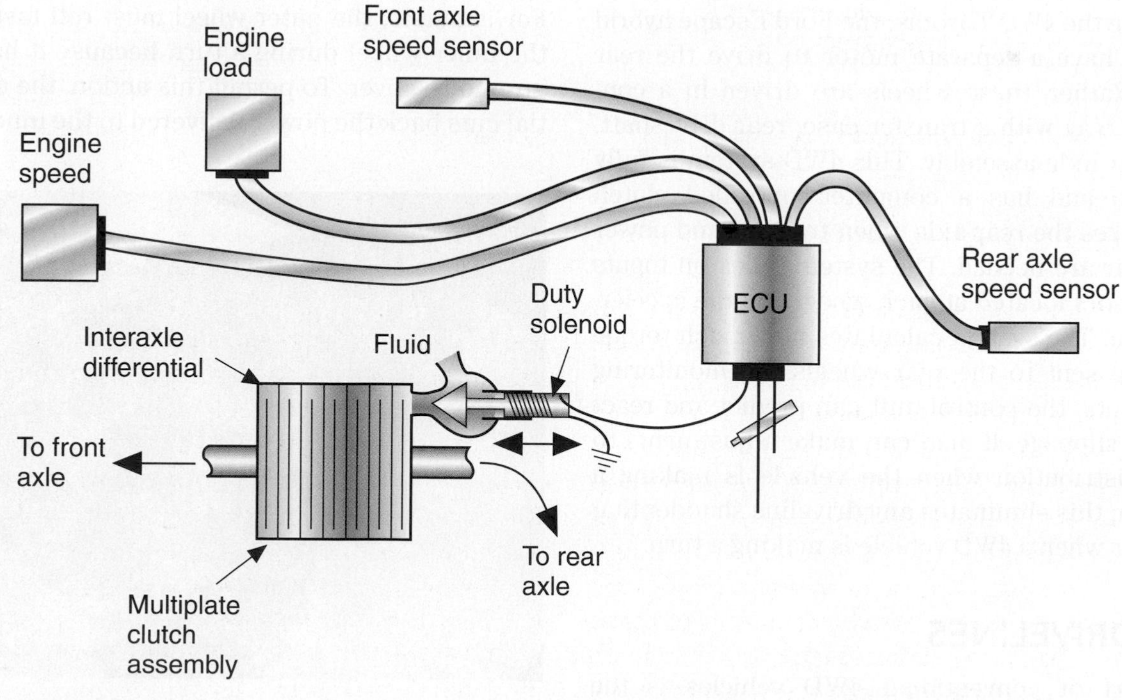

FIGURE 45–11

A simple schematic for an electronically controlled AWD system.

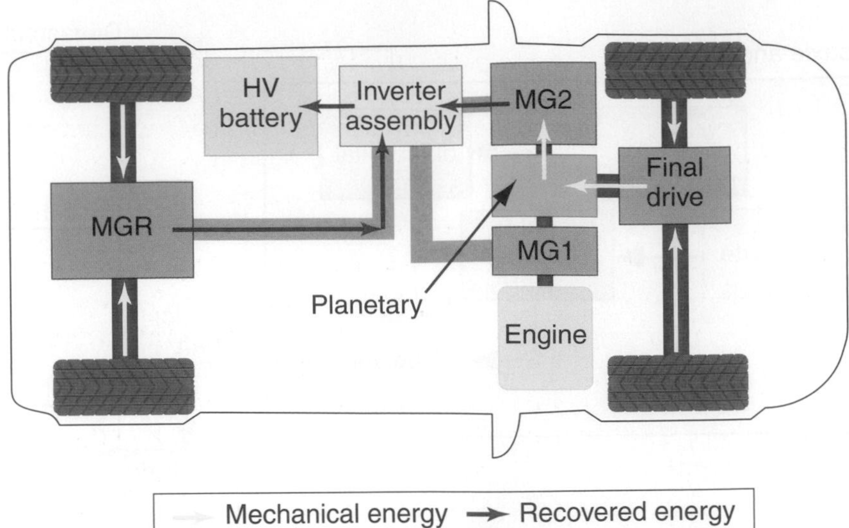

FIGURE 45–12

The basic layout for a hybrid vehicle equipped with an additional motor to drive the rear wheels.

©Cengage Learning 2015

housing of the rear transaxle contains the motor/generator (MGR), a counterdrive gear, a counter driven gear, and a differential. The unit has three shafts: MGR and the counterdrive gear are located on the mainshaft (MGR drives the counterdrive gear), the counter driven gear and the differential drive pinion gear are located on the second shaft, and the third shaft holds the differential.

Unlike the 4WD Toyotas, the Ford Escape hybrid does not have a separate motor to drive the rear wheels. Rather, these wheels are driven in a conventional way with a transfer case, rear driveshaft, and a rear axle assembly. This 4WD system is fully automatic and has a computer-controlled clutch that engages the rear axle when traction and power at the rear are needed. The system relies on inputs from sensors located at each wheel and the accelerator pedal. The system calculates how much torque should be sent to the rear wheels. By monitoring these inputs, the control unit can predict and react to wheel slippage. It also can make adjustments to torque distribution when the vehicle is making a tight turn; this eliminates any driveline shudder that can occur when a 4WD vehicle is making a turn.

4WD DRIVELINES

The heart of conventional 4WD vehicles is the transfer case, which may be integrated into the transmission **(Figure 45–13)** or mounted to the back of the transmission **(Figure 45–14)**. A chain or gear drive within the case receives the power from the transmission and transfers it to the front and rear axles through two separate driveshafts.

The driveline from the transfer case shafts runs to differentials at the front and rear axles. As on two-wheel-drive vehicles, these axle differentials are used to compensate for road and driving conditions by adjusting the rpm to opposing wheels. For example, the outer wheel must roll faster than the inner wheel during a turn because it has more ground to cover. To permit this action, the differential cuts back the power delivered to the inner wheel

FIGURE 45–13

The transfer case is integral with the transmission.

©Cengage Learning 2015

FIGURE 45–14

The transfer case is bolted to the rear of this transmission.

and boosts the amount of power delivered to the outer wheel.

U-joints are used to couple the driveline shafts with the differentials and transfer cases on all these vehicles. U-joints can also be used on some vehicles to connect the rear axle and wheels. Normally, however, rear axles are simply bolted to the wheel hubs.

The coupling between front wheels and axles is normally done with U-joints or CV (constant velocity) joints. Generally, half axles or half shafts with CV joints are found on 4WD passenger cars. They can also be found on a number of passenger vans and on mini-pickups and trucks.

To provide independent front suspension, some vehicles have one half shaft and one solid axle for the front drive axle **(Figure 45–15)**. The half shaft is able to move independently of the solid axle, thereby giving the vehicle the ride characteristics desired from an independent front suspension.

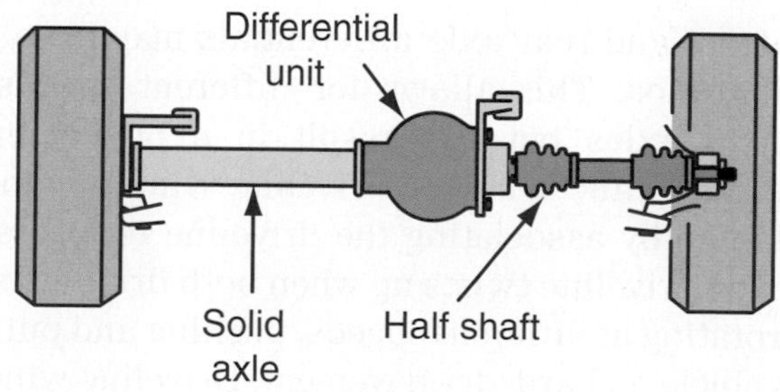

FIGURE 45–15

A front suspension for a front drive axle that relies on one solid axle and a half shaft to provide independent suspension.

Transfer Case

In a 4WD vehicle, as mentioned earlier, the transfer case delivers power to both the front and rear assemblies. Two driveshafts normally operate from the transfer case, one to each drive axle.

A transfer case **(Figure 45–16)** is constructed much like a transmission. Some transfer cases rely on a chain and sprockets, and others use gears to rotate the driveshaft to the front or rear axle.

It uses shift forks to select the operating mode, plus splines, gears, shims, bearings, and other components found in manual and automatic transmissions. The outer case of the unit is made of cast iron, magnesium, or aluminum. It is filled with lubricant (oil) that cuts friction on all moving parts.

CUSTOMER CARE

It is very important to remind your customers that the fluid level in a transfer case must be checked at recommended time intervals. (Check the owner's manual.) Remind them to make sure only the correct fluid is used.

Power Transfer (Take-Off) Units

On 4WD systems adapted from front-wheel-drive systems, a separate front differential and driveline are not needed. The front wheels are driven by the transaxle differential of the base model. A **power transfer (take-off) unit (PTU)** is added to the transaxle to transmit power to the rear wheels in four-wheel drive. This take-off gearing is housed in

FIGURE 45–16

This transfer case receives the output from the transmission via a chain drive before power is sent to the rear wheels.

FIGURE 45–17

A PTU bolted to a transaxle.

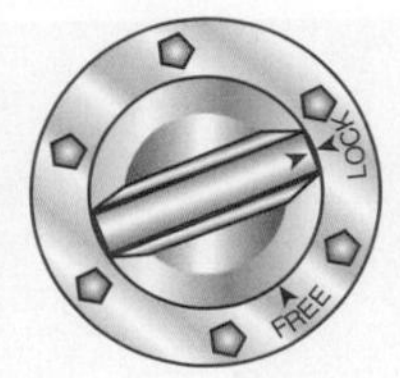

FIGURE 45–18

The knob positions for a manual locking hub.

or bolted to the transaxle housing **(Figure 45–17)**. It is simply a gearset, driven by the transaxle's final drive, that transfers the output torque at a 90° angle and rotates the rear driveshaft at the same speed as the front wheels.

Locking Hubs

A transfer case can only disconnect power to one of the driveshafts, either to the front or rear. Although the driveshafts are disconnected from the engine, the wheels of the disconnected axle will still drive the driveshaft and the gears and axle shafts. This causes unnecessary wear on the system and wastes fuel. Some 4WD vehicles are equipped with wheel hubs that engage or disengage the axle shafts from the engine's power. These hubs can be gear and ratchet assemblies that manually lock or unlock the axles power flow; other hubs operate automatically.

On some 4WD vehicles, the front axles are engaged by locking wheel hubs. The hubs are designed to stop the rotation of the front axles and differential when 4WD is not selected. **Locking hubs** can be either manual or automatic. Newer systems have automatic hubs that engage when the driver switches into 4WD. Older vehicles had manual hubs that required the driver to turn a knob on the front wheels. These hubs connect the front wheels to the front drive axles when they are in the locked position.

Manual locking hubs require that a lever or knob be turned by hand to the 2WD or 4WD position **(Figure 45–18)**. Automatic locking hubs can be locked by shifting to 4WD and moving forward slowly **(Figure 45–19)**. They are unlocked by slowly backing up the vehicle. On certain late-model 4WD systems, a front axle lock is used in place of individual locking hubs.

A locking hub is a type of clutch that engages or disengages the outer ends of the front axle shafts from the wheel hub. When the hub is in the locked position, the ring of the clutch is set onto the splines of the axle shaft. When the hub is in the unlocked position, spring pressure forces the clutch ring away from the axle shaft, thereby disconnecting the wheel hub and the axle.

Although automatic hubs are more convenient for the driver, they do have a disadvantage. Many self-locking hub designs are unlocked when the vehicle is moved in reverse. Therefore, if the vehicle is stuck and needs to back out of the trouble spot, only RWD will be available to move it. Other automatic hubs unlock immediately when 4WD is disengaged without the need to back up. On these systems, the hubs are automatically locked, regardless of the direction the vehicle is moving.

Locking hubs are not needed with full-time 4WD. The wheels and hubs are always engaged with the axle shafts. The interaxle differential or transfer case prevents damage and undue wear of the parts of the powertrain.

Interaxle (Centre) Differentials

The front and rear axle differentials may have different ratios. This allows for different speeds at the two axles but can result in a phenomenon called driveline windup. Driveline windup can be explained by associating the driveline to a torsion bar. The driveline twists up when both driving axles are rotating at different speeds, pushing and pulling the vehicle on hard, dry pavement. Driveline windup can cause handling problems, particularly when rounding turns on dry pavement. This is because the front axle wheels must travel farther than the rear axle wheels when rounding a curve. On wet or slippery roads, the front and rear wheels slide enough

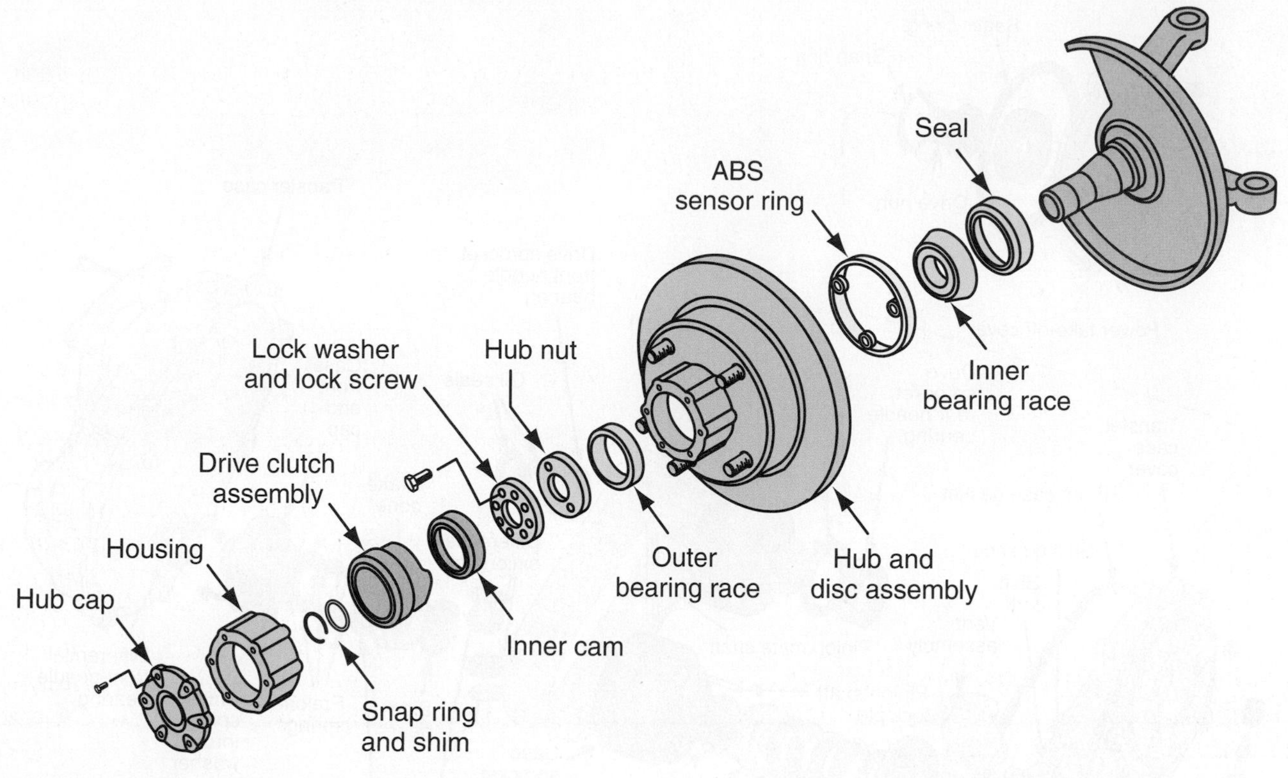

FIGURE 45–19

The parts of an automatic locking front hub.

to prevent damage to the driveline components. However, this is not the case on dry surfaces.

Full-time 4WD and AWD systems require a device that allows the front and rear axles to rotate at different speeds. These units work to eliminate driveline windup. The most common setup has an interaxle (centre) differential **(Figure 45–20)** located in the transfer case. The front and rear drivelines are connected to the interaxle differential. Just as a drive axle differential allows for different left and right drive axle shaft speeds, the interaxle differential allows for different front and rear driveline shaft speeds.

Although the interaxle differential solves the problem of driveline windup during turns, it also lowers performance in poor traction conditions because the interaxle differential tends to deliver more power to the wheels with the least traction. The result is increased slippage, the exact opposite of what is desired.

To counteract this problem, some interaxle differentials are designed much like a limited-slip differential. They use a multiple-disc clutch pack to maintain a predetermined amount of torque transfer before the differential action begins to take effect **(Figure 45–21)**.

Electromagnetic systems use an electromagnetic coil to engage a multiplate clutch to transfer torque to the rear axle. If the front wheels are turning faster than the rear wheels, the PCM applies the clutch to drive the rear wheels. Some systems use a Torsen or Haldex unit in the centre differential. A Torsen, or *tor*que *sen*sing unit, is a mechanical differential capable of transferring power from a wheel with little or no traction to the other wheel that has traction. Other systems, such as the one shown in Figure 45–20, use a cone braking system rather than a clutch pack, but the end result is the same. Power is supplied to both axles regardless of the traction encountered.

Most systems also give the driver the option of locking the interaxle differential in certain operating modes **(Figure 45–22)**. This eliminates the differential action altogether. However, the interaxle differential should only be locked when driving in slippery conditions and only activated at low speeds.

Viscous Clutch

Some systems use a **viscous clutch** in the transfer case, outside the transfer case **(Figure 45–23)**, or as a separate unit to split the power to the front and rear axles. A viscous clutch is used to drive the axle with low tractive effort, taking the place of the

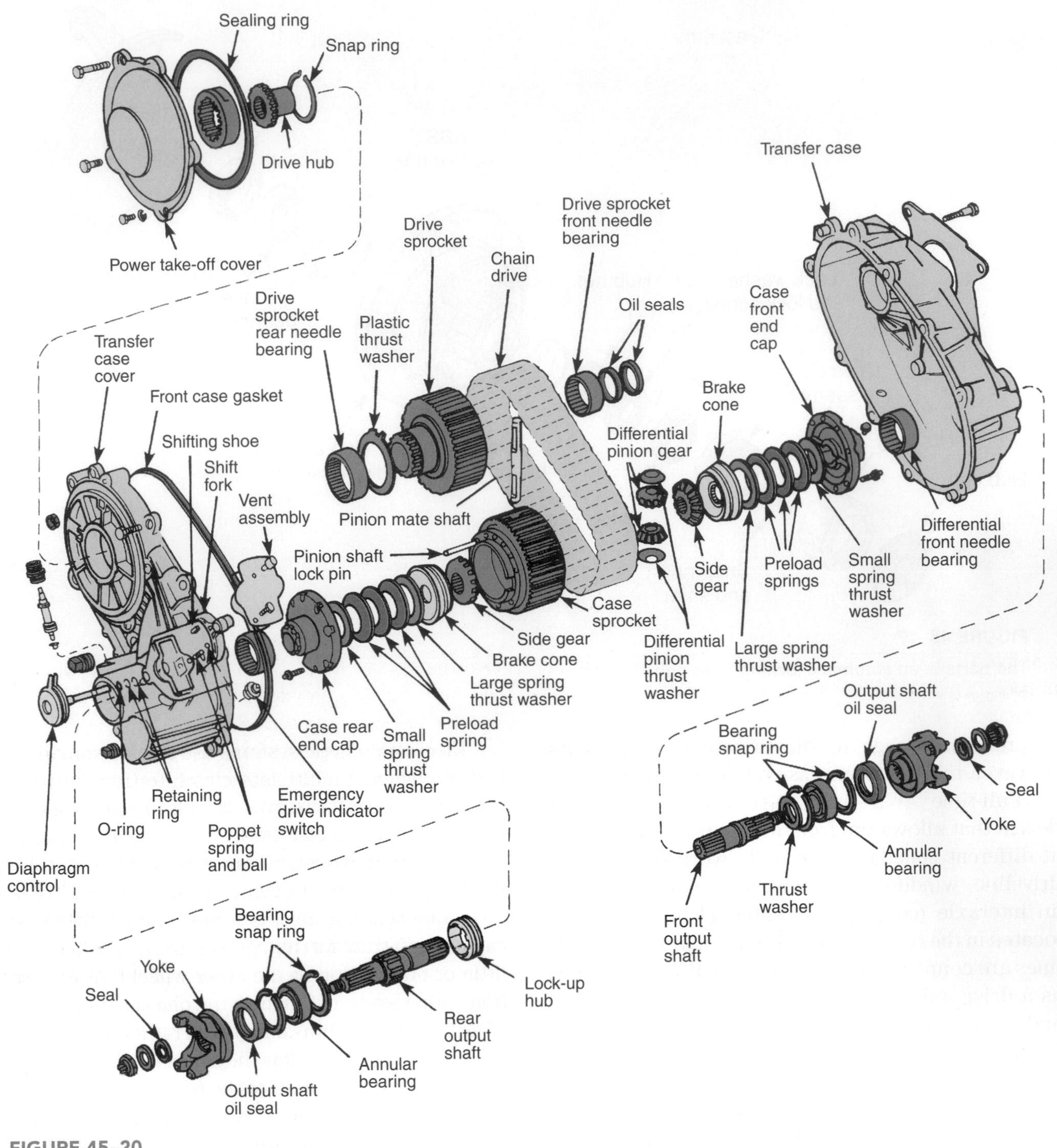

FIGURE 45–20

Four-wheel-drive transfer case with integral differential and cone brakes for limited slip.

interaxle differential. In existence for several years, the viscous clutch is installed to improve mobility under difficult driving conditions. Viscous clutches operate automatically while constantly transmitting power to the axle assembly as soon as it becomes necessary to improve driving wheel traction. This action is also known as biasing driving torque to the axle with tractive effort. The viscous clutch assembly is designed similarly to a multiple-disc clutch with alternating driven and driving plates.

The viscous clutch pack fits inside a drum that is completely sealed. The clutch pack is made up of alternating steel and friction plates. One set of steel plates is splined internally to the clutch assembly hub. The second set of clutch plates is splined externally to the clutch drum. The clutch housing is filled with a

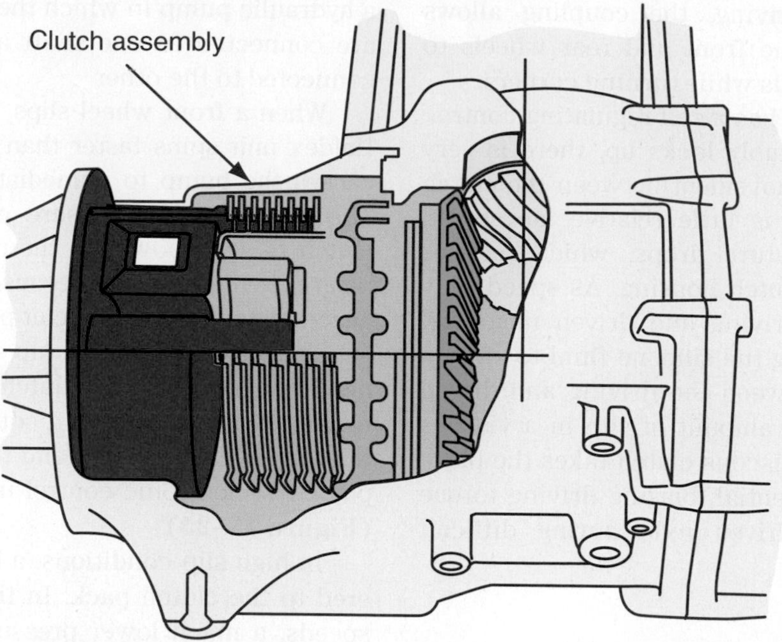

FIGURE 45–21

A mechanical clutch assembly for a centre differential.

small amount of air and special silicone fluid with the purpose of transmitting force from the driving plates to the driven plates.

When clutch pack rotates slowly, little torque is transferred through the coupling. As the speed increases, so does the amount of transferred torque. When a difference of 8 percent exists between the low traction axle and the axle with tractive effort, the silicone fluid is sheared by the clutch plates. The shearing action causes heat to build within the housing very rapidly, which results in the silicone fluid stiffening. The stiffening action causes a locking action between the clutch plates to take place within approximately 1/10 second. A locking action results from the stiff silicone fluid, making it very hard for the plates to shear. The stiff silicone fluid transfers power from the driving to the driven plates. The driving shaft is then connected to the driven shaft through the clutch plates and stiff silicone fluid.

FIGURE 45–22

The centre differential can be controlled by these selectors.

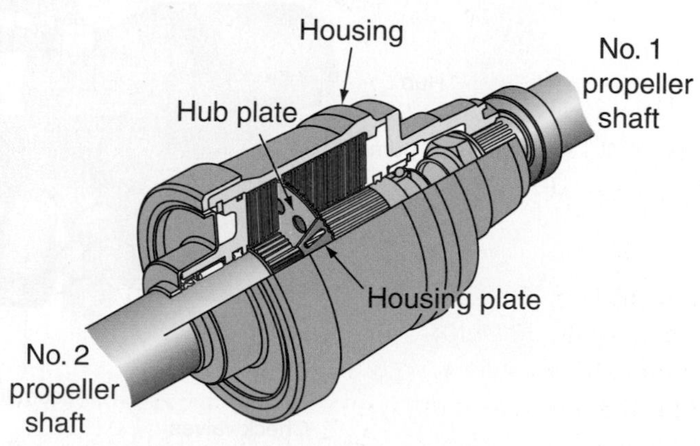

FIGURE 45–23

A viscous clutch is a sealed assembly with disc clutches.

During normal driving, the coupling allows enough slippage for the front and rear wheels to travel at different speeds while turning corners.

The viscous clutch has a self-regulating control. When the clutch assembly locks up, there is very little, if any, relative movement between the clutch plates. Because there is little relative movement, silicone fluid temperature drops, which reduces pressure within the clutch housing. As speed fluctuates between the driving and driven members, heat increases, causing the silicone fluid to stiffen. Speed differences between the driving and driven members regulate the amount of slip in a viscous clutch driveline. The viscous clutch takes the place of the interaxle differential, biasing driving torque to the normally undriven axle during difficult driving conditions.

Haldex Clutch

Some AWD vehicles are equipped with a **Haldex clutch (Figure 45–24)**, which serves as a centre differential. This clutch unit distributes the drive force variably between two axles. In a typical application, the Haldex unit mounts in front of the rear differential and receives torque from the front axle.

The Haldex unit has three main parts: the hydraulic pump driven by the slip between the axles or wheels, a wet multidisc clutch, and an electronically controlled valve. The unit is much like a hydraulic pump in which the housing and a piston are connected to one shaft and a piston actuator connected to the other.

When a front wheel slips, the input shaft to the Haldex unit spins faster than its output shaft. This causes the pump to immediately generate oil flow. The oil flow and pressure engages the multidisc clutch to send power to the rear wheels. This happens extremely quickly because an electric pump and accumulator keep the circuit primed.

The oil from the pump flows to the clutch's piston to compress the clutch pack. The oil returns to the reservoir through a controllable valve, which adjusts the oil pressure and the force on the clutch pack. An electronic control unit controls the valve **(Figure 45–25)**.

In high slip conditions, a high pressure is delivered to the clutch pack: In tight curves or at high speeds, a much lower pressure is provided. When there is no difference in speed between the front and rear axles, the pump does not supply pressure to the clutch pack.

VOLKSWAGEN 4MOTION The Volkswagen 4MOTION AWD system uses a Haldex coupling as a centre differential. The coupling is mounted in front and is part of the rear axle housing. The input shaft of the Haldex centre differential is driven by a driveshaft from the transaxle. In a Haldex, the input shaft is totally separated from the output shaft, which is

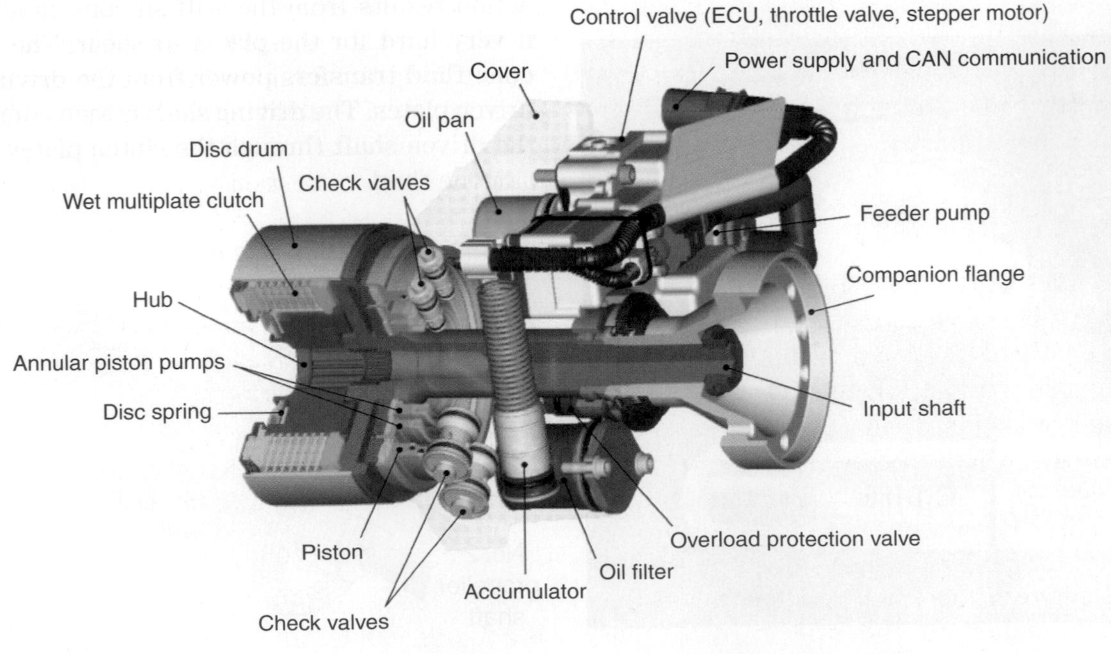

FIGURE 45–24

A Haldex clutch assembly.

Haldex Traction Systems

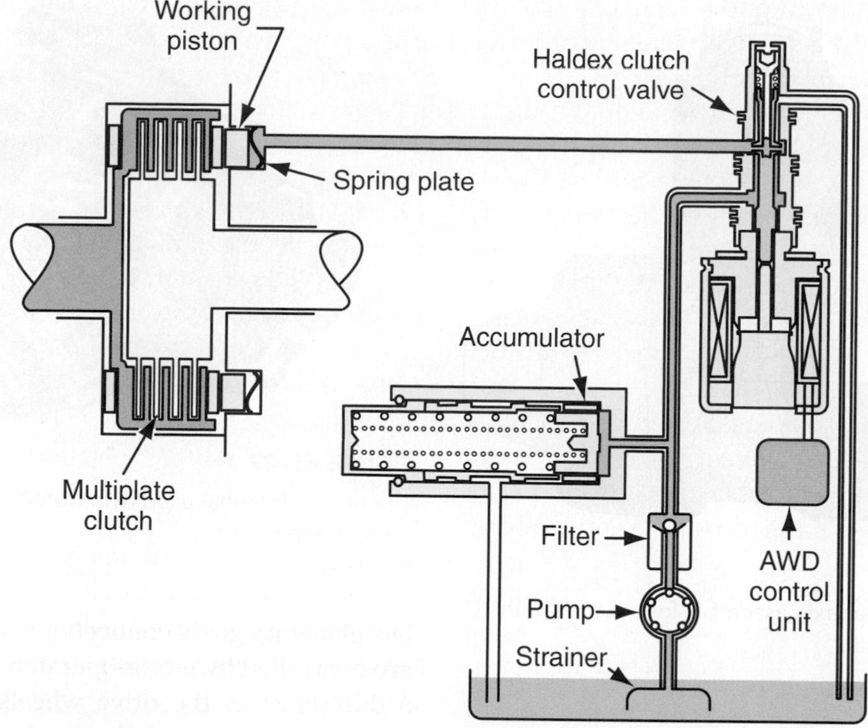

FIGURE 45–25

The hydraulic circuit for a typical Haldex clutch.

connected to the rear final drive gears. Therefore, power is sent to the rear wheels only when the Haldex clutch is engaged.

The Haldex unit is controlled by a PCM that receives inputs from a variety of sensors. This means the system can respond to other driving conditions and not just wheel slip. When there is no slip, understeer, or oversteer, the vehicle operates as a 2WD vehicle. It distributes power to the rear axle only when it is needed.

TORQUE VECTORING

Recently, AWD vehicles have been equipped with electronically controlled torque vectoring systems. These are advanced traction and stability controls. They are capable of transferring torque not only from front to rear but also from side to side. The benefits of these systems include improved overall handling, excellent cornering performance, and improved vehicle stability and safety.

A basic torque vectoring system can vary the torque between the front and rear wheels, in response to conditions. The system also monitors each wheel independently and will increase or decrease the amount of torque sent to each. To improve cornering, the system will automatically add torque to the outside rear wheel to allow the vehicle to turn more quickly.

Through the vehicle's CAN system, the systems monitor vehicle speed, individual wheel speeds, operating gear, braking force, steering angle, yaw rate, and other inputs.

Some AWD systems rely on a passive system in which the differential mechanically reacts to conditions to provide stability. Other current AWD systems control wheel slip and stability by braking the wheel with little or no traction or by reducing the power from the engine; these are referred to as active brake systems. True torque vectoring is achieved with special differentials that distribute power to the wheel or wheels that have traction; these are called active differential systems. Basic torque vectoring differentials can be found on some FWD.

GO TO ▶ Chapter 53 for more information on traction and stability control systems.

Active Differential Systems

Honda's Super Handling-All Wheel Drive (SH-AWD) is a full-time, fully automatic AWD traction and handling system with an active rear differential. The system does not have a centre differential or a limited-slip differential at either axle.

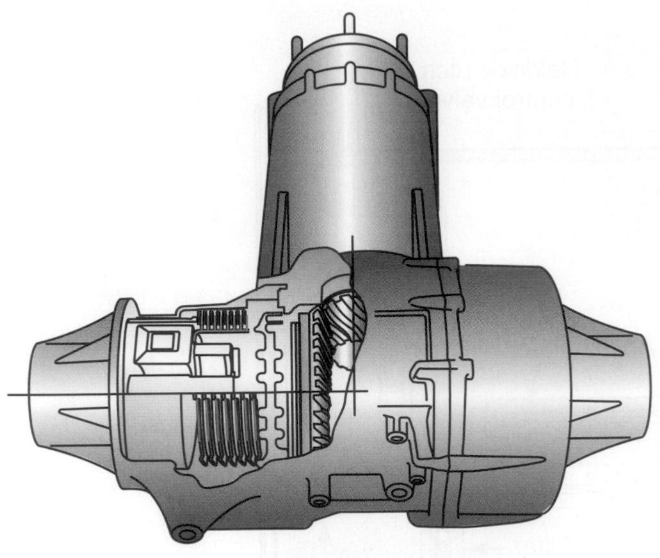

FIGURE 45–26

The differential and rear axle assembly for a Honda with SH-AWD.

©Cengage Learning 2015

FIGURE 45–27

The rear differential used with BMW's active torque vectoring system.

BMW Group

The active differential is in the rear axle. It relies on planetary gears to increase the rotation speed, creating a speed difference between the input and output, and two electromagnetic clutches to transfer driving torque to the rear axles **(Figure 45–26)**. Each clutch is tied to an individual drive wheel and is electronically controlled. The system monitors many inputs that are available on the CAN bus to automatically add torque to the wheel with the most traction and the outside rear wheel during corners. Before the clutches are activated, wheel slip must be detected.

The system, which normally distributes torque 90 percent up front and 10 percent in the rear, can respond to conditions by allowing for a 50/50 split during acceleration or hard cornering. The system can then send some or all of that 50 percent going to the rear axle directly to the outside tire to make the vehicle bend into a corner more sharply.

TORQUE VECTORING SYSTEMS BY ZF Late-model Audi Quattros and AWD BMWs are equipped with torque vectoring systems from ZF or Ricardo. Both of these systems are similar and rely on an active differential. These torque vectoring systems use wet clutches and planetary gearsets in the front, centre, and/or rear differentials **(Figure 45–27)**. They are controlled by electrical, electromechanical, or electrohydraulic systems.

One difference between the Audi system and BMW's is that Audi uses a wet clutch in the centre differential that is mechanically controlled by the worm gears in its Torsen differential. The rear differential has planetary gears connected to each drive axle that are controlled by motor-operated clutches to provide a difference in the drive wheels' rotational speed. Normally, the front differential is open to allow for normal steering. The PCM controls the operation of the centre and rear clutches of both systems. The system is tied to the stability control system, the active steering system, and the ABS. It helps steer the vehicle by sending torque to either of the rear wheels, which not only prevents wheel slip but also eliminates oversteer and understeer. When the possibility of understeer is detected, power to the front wheels is reduced. When oversteering is detected, more power is sent to the front wheels.

The systems are also designed to anticipate slippage by monitoring many inputs on the CAN bus, and they can provide yaw torque when the vehicle is cornering. The yaw torque forces the outside rear wheel to push forward while the inside rear wheel pulls back a bit. This is done by slightly altering the torque at the rear wheels. The overall effect is the same as the rear steering option offered on some vehicles.

MITSUBISHI S-AWC Mitsubishi's Super All Wheel Control (S-AWC) is a full-time 4WD system used in the Lancer Evolution. The system is integrated with the vehicle's active centre differential (ACD), active yaw control (AYC), active stability control (ASC), and sports ABS components. This provides for regulation of torque and braking force at each wheel. The system is best described as an active differential (centre and rear) with active braking at all wheels.

The ACD is an electronically controlled hydraulic multiplate clutch that limits the action of the centre differential gears. The ACD regulates the torque split between the front and rear drive axles. The AYC acts like a limited-slip differential by reducing rear-wheel slip to improve traction. It controls rear-wheel torque to limit the yaw of the vehicle to improve the vehicle's cornering performance. The ASC regulates engine power and the braking force at each wheel. This system improves vehicle stability and improves traction during acceleration. The system relies on two ECUs: one controls the ACD and AYC, and the other controls ASC and ABS. The ECUs communicate to each other via the CAN bus.

Passive AWD Control Systems

Audi AWD vehicles carry the designation of "Quattro." Its basic system has evolved through the years. The early systems relied on Torsen differentials in the centre and rear differentials. This system was capable of varying the torque received by any wheel of the vehicle. As time moved on, the basic system gained more electronic controls and is currently using a torque vectoring system by ZF, which is discussed earlier in this section.

Audi's Quattro permanent all-wheel drive system transfers torque from the front to the rear and side to side, as needed. The centre differential compensates for the speed differences between the front and rear axles and distributes engine power between the front and rear wheels. The system automatically regulates the distribution of power within milliseconds. This action is based on engine speed and torque, wheel spreads, and longitudinal and lateral acceleration rates.

AWD Control Braking Systems

The Mercedes-Benz Torque Vectoring Brake system can apply braking to only the inside rear wheel during cornering. If the system detects understeer, it generates a quick burst of yaw torque. This allows the driver to have more control of the steering, and the car can go through the corner with improved control.

This active brake torque vectoring system is integrated with Mercedes' stability control system and is always engaged. At the slightest hint of wheel slip, the system begins to brake that wheel or starts to decrease power to it.

PORSCHE TORQUE VECTORING (PTV) Although there are many AWD variations available, the 911 is unique in that it is a rear-wheel drive with the engine mounted to the rear as well. A special assembly is used to transfer power from the rear to the front axle **(Figure 45–28)**. Torque vectoring is not currently available for the front and rear wheels.

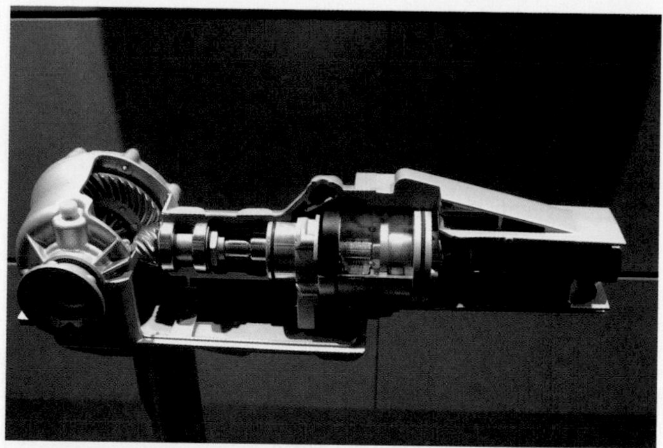

FIGURE 45–28

The assembly used to transfer engine power from the rear to the front axle in a Porsche 911.

The PTV system is a rear-wheel-only active braking system first used in the 2010 Porsche 911 Turbo. The PTV can apply the brakes individually to stabilize and improve traction while the car is making a turn. The system relies on CAN inputs, such as steering angle, vehicle speed, throttle position, and yaw rate. PTV applies the brake on the inside rear wheel to minimize understeer while entering corners.

FWD TORQUE VECTORING Ford introduced torque vectoring in its 2012 FWD Focus. Basically, this torque vectoring system works like an electronic limited-slip differential. The system relies on inputs from various sensors to apply an appropriate amount of brake force to the wheel with the least amount of traction.

DIAGNOSING FOUR-WHEEL-DRIVE AND ALL-WHEEL-DRIVE SYSTEMS

Components of 4WD vehicles can be serviced in basically the same manner as the identical components of a 2WD vehicle. Before doing any servicing, however, it is necessary to give the undercar a complete inspection. Pay particular attention to the steering dampers, steering linkage, wheel bearing, ball joints, coil springs, and radius arm bushings.

Current 4WD and AWD systems are controlled by the PCM or a separate control module. Like all computer-monitored and computer-controlled systems, faults have to be identified by DTCs.

PROCEDURE

Diagnosing Late-Model 4WD Systems

STEP 1 Verify the customer's complaint.

STEP 2 Visually inspect for obvious signs of mechanical or electrical damage. Check the condition of the following (some systems will not have all of these):

- Half shafts
- Locking hubs
- Driveshaft and U-joints
- Shift lever and/or linkage
- Mode switch
- Shift motor
- Electromagnetic clutch
- Vacuum and fluid lines
- Axle disconnect units
- Matching tire sizes
- Transfer case
- Transfer case linkages
- System fuses
- Wiring harnesses and connectors

STEP 3 If the cause was not visually evident, connect the scan tool to the DLC.

STEP 4 Note the DTCs retrieved.

STEP 5 Clear the DTC.

STEP 6 Conduct the self-test for the 4WD control module.

STEP 7 If the DTCs retrieved are related to the concern, interpret the codes and follow the specific pinpoint tests for those codes **(Figure 45–29)**.

STEP 8 If the DTCs are not related to the concern or there are no DTCs, go to the manufacturer's symptom chart **(Figure 45–30)**.

STEP 9 Identify and repair the cause of the problem.

STEP 10 Verify the repair.

Basic System Diagnosis

Computer-controlled 4WD systems are diagnosed just like other computer-controlled systems. Make sure the scan tool can communicate with the vehicle by completing the test provided by the scan tool. If the system has a separate self-diagnostic routine for the 4WD system, run that, and retrieve and record all DTCs. Always follow the procedures outlined by the manufacturer when connecting a scan tool to the system. Also always use the manufacturer's specifications when interpreting data.

COMPONENT TESTING Individual inputs should be tested according to the procedures recommended by the manufacturer. The speed sensors are typically Hall-effect or magnetic pulse units. These are best checked with a lab scope. Often, the manufacturer will recommend testing the sensors with a voltmeter or ohmmeter. Normally, these tests involve taking readings across a multiple-pin connector and at specific terminals.

The selector switch can be checked by connecting an ohmmeter across the terminals of the switch. As the switch is moved to various positions, the circuit through the switch should be either open or closed, depending on the position. By referring to the wiring diagram, you can easily identify what should happen in the different switch positions. If the switch does not function as it should, it must be replaced.

When the transfer case shift motor is suspected as being faulty, begin diagnosis with a careful inspection. Then follow the manufacturer's procedures for testing it. Often, testing involves checking the motor for opens, shorts, and high resistance. If the motor is found to be defective, it must be replaced. Where placing the motor, make sure it is mounted correctly and its retaining bolts are tightened to specifications.

If the system uses a magnetic clutch, it can be checked with an ohmmeter. Noise from the clutch can be caused by a faulty clutch or a clutch that needs adjustment.

SHIFT-ON-THE-FLY SYSTEMS Shift-on-the-fly systems switch between 2WD and 4WD electrically by a switch on the instrument panel. Many of these systems use electrically controlled vacuum motors to connect and disconnect the wheels from the axle and an electrical motor or electromagnetic clutch at the transfer case to transfer power to the additional axle. The 4WD indicator light will illuminate when 4WD is engaged. If there is a problem with the system or if 4WD cannot be engaged, the indicator lamp will blink to notify the driver of a problem. Diagnosis begins with watching the frequency of the blinking lamp. Check the service information for an interpretation of the blinking lamp.

If the vehicle will not engage into 4WD, there may be a faulty transfer case actuator motor or faulty vacuum solenoid assembly. If the motor in the transfer case is a possible cause, remove it and check its operation. Check the transfer case, and repair or replace any faulty components if the motor is not the cause of the problem. If no problem is found in the transfer case, it is likely that the electronic control unit is faulty.

Check for engine vacuum at the actuator and/or solenoid before checking anything else if the vacuum actuator is suspect. If no vacuum is present when the

DTC	DESCRIPTION
B1317	Battery Voltage High
B1318	Battery Voltage Low
C1979	IWE Solenoid Circuit Failure
C1980	IWE Solenoid Short to Battery
P1812	4-Wheel-Drive Mode Select Circuit Failure
P1815	4-Wheel-Drive Mode Select Short Circuit to Ground
P1820	Transfer Case Clockwise Shift Relay Coil Circuit Failure
P1822	Transfer Case Clockwise Shift Relay Coil Short Circuit to Battery
P1824	4-Wheel-Drive Clutch Relay Circuit Failure
P1828	Transfer Case Counter-Clockwise Shift Relay Coil Circuit Failure
P1849	Transmission Transfer Case Contact Plate A Short Circuit to Ground
P1853	Transmission Transfer Case Contact Plate B Short Circuit to Ground
P1857	Transmission Transfer Case Contact Plate C Short Circuit to Ground
P1861	Transmission Transfer Case Contact Plate D Short Circuit to Ground
P1867	Transmission Transfer Case Contact Plate General Circuit Failure
P1891	Transmission Transfer Case Contact Plate Ground Return Open Circuit
U1900	CAN Communication BUS Fault—Received Error
U2051	One or More Calibration Files Missing/Corrupt

FIGURE 45–29

A 4WD DTC chart.

CONDITION	POSSIBLE CAUSE
No communication with the control module (PCM)	Scan tool Data link connector (DLC) Control module (PCM) Circuitry
The 4WD indicators do not operate correctly or do not operate	Indicator lamp Circuitry Control module (PCM) Ignition switch
The vehicle does not shift between modes correctly	Mode select switch Transfer case Transfer case clutch Control module (PCM) Circuitry Locking hubs Ignition switch Transfer case shift lever Mode indication switch Ignition switch

FIGURE 45–30

A typical symptom chart.

CONDITION	POSSIBLE CAUSE
4WD does not engage correctly at speed	Transfer case clutch coil Control module (PCM) Locking hubs Ignition switch
The front axle does not engage or disengage correctly or makes noise in 2WD under heavy throttle	Mode select switch Locking hubs Vacuum leaks Control module (PCM) Front half shaft Ignition switch
The transfer case jumps out of gear	Transfer case Vacuum solenoid vent Mode select switch Transfer case Shift lever
Driveline windup while moving straight	Tire inflation pressure Tire and wheel size Tire wear Axle ratio
Grinding noise during 4WD engagement, especially at high speeds	The front half shafts are not turning at the same speed as the rear axle shafts
Flashing 4WD indicators	Loss of CAN communication between 4WD control module and instrument cluster Loss of high-speed (HS-CAN) communication between 4WD control module and PCM Ignition switch
The transfer case makes noise	Tire inflation pressure Unmatched tire and wheel size Tire tread wear Internal components Fluid level
The vehicle binds in turns, resists turning, or pulsates or shudders while moving straight	Unmatched tire sizes Unequal amounts of tire wear Unequal tire inflation pressures Unmatched front and rear axle ratios

FIGURE 45–30 ***(continued)***

engine is running, check the transfer position switch. A voltage signal should be sent to the vacuum solenoid when 4WD is selected. If no signal is being sent, then the switch or switch circuit is faulty. However, if the solenoid is receiving a voltage signal, suspect a faulty solenoid assembly. Check the vacuum circuit for leaks if there are no electrical problems.

Vacuum solenoids can be checked by connecting jumper leads from the battery to the terminals of the solenoid. Remove the vacuum solenoids, and connect the battery to the terminals of each solenoid. Connect a handheld vacuum pump to the inlet port of the solenoid and a vacuum gauge to the port for axle engagement. With the battery connected to the solenoid, there should be vacuum at the outlet port to the axle. Move the vacuum gauge to the other outlet port. Vacuum should not be present there. Now disconnect the battery from the solenoid. Recheck the vacuum at the outlet ports. A good solenoid will have vacuum only at the disengagement port when vacuum is applied and the solenoid is not connected to the battery.

ON-DEMAND SYSTEMS On-demand systems operate at the discretion of the PCM or override controls selected by the driver. The division of power between the front and rear axles is controlled by the output clutch at the transfer case. The activity of the clutch is controlled by regulating its duty cycle. During normal operation, the duty cycle is low. This allows for a slight speed difference between the front and rear driveshafts, which normally occurs when the vehicle is moving through a curve. When slip is detected at the rear wheels, the duty cycle to the clutch is increased until the difference between the front and rear drive axles is reduced.

To vary the torque split, a computer monitors many things, especially the rotational speeds of the front and rear drive axles. Some systems rely on wheel speed sensors for this, whereas others have additional speed sensors on the front and rear output shafts from the transfer case. Many of the inputs to the electronic module for 4WD are shared and work with other systems, and they are available on CAN. The best place to start when diagnosing these systems is the service information. Go to the section on the system, and identify the components involved in the various modes of operation.

AWD SYSTEMS Most AWD systems have a viscous coupling. A viscous coupling can be checked with a torque bias test. This check measures the torque required to rotate one front wheel when it is raised off the ground while the others are on the ground. The transfer case should be in 4WD and the transmission in NEUTRAL; compare the measured torque to specifications. A worn coupling is indicated by low readings.

Problems with AWD systems normally occur with the controls that provide for braking efficiency and AWD in reverse gear. Some systems use a vacuum solenoid setup to provide AWD in reverse. The solenoids are used to bypass the overrunning clutch. If AWD is not available in reverse gear, suspect the vacuum solenoids, their controls, or the dog-clutch assembly.

An overrunning clutch is used to prevent any feedback of front-wheel braking torque to the rear wheels. This allows the brake system to control braking as if it were a 2WD vehicle. The controls for this feature vary with the manufacturer, and you should always follow the troubleshooting charts provided by the manufacturer of the vehicle.

Axle Hub Diagnosis

Front hubs may make a ratcheting sound when water or dirt has entered the hub and contaminated the lubricant. This prevents free movement of the components in the hub. A ratcheting sound from an automatic locking hub may indicate that the hub on the opposite side of the axle is not disengaging.

Locking hubs can be quickly checked by rotating the brake drum or rotor slightly and turning the hub selector into the lock position. A click should be heard when the hub engages the axle, and the axle should now turn with the hub. Next, with the hub still turning, turn the selector to the free position. The axle should now be free of the hub. If both of these events do not happen, the hub assembly needs to be repaired or replaced.

Hubs can also be checked by raising the front wheels and spinning the front tires. Next, engage 4WD, and spin the tire. If the hub is locked, both the tire and the axle will spin. Release 4WD and spin the tire again to make sure the hub releases. If the hub does not lock or stays locked, the hub needs to be serviced or replaced.

Wheel Bearings

To check the adjustment of the wheel bearings, raise the front of the vehicle. Grasp each front tire at the front and rear, and push the wheel inward and outward. If any free play is noticed between the hub, rotor, and front spindle, adjust the wheel bearings.

Wheel bearings should be disassembled and serviced any time the hubs have been submerged in water. During normal operation, the bearings get warm, and when they are quickly cooled off by the splash of water, their lubricant breaks down, and the bearings can be destroyed. Always replace the bearings if they are worn or damaged.

Another frequent cause of wheel bearing failure is the use of oversized tires mounted on wheels with substantial offset. These switch the load from being equally placed on both the inner and outer wheel bearings to being exaggerated onto mostly the outer bearing. This extreme constant load, not to mention shock loads during rough road operation against mainly one bearing, reduces that bearing's lifespan.

SERVICING 4WD VEHICLES

The U-joints, slip joints, or CV joints on a 4WD drivetrain must be lubed on a regular basis. To service the driveshafts on a four-wheel-drive vehicle, use the general instructions given for a cross universal joint. A four-wheel drive simply uses two driveshafts instead of one.

Maintenance

It is also important to make sure the transfer case and differentials have the correct fluid and are filled to the correct level. To check the differential fluid, locate the plug. Most manufacturers require the use of either a 3/8-inch drive ratchet or a wrench to remove the plug. Be careful not to damage it or its threads. The fluid level should be up to the bottom of the fill plug hole. If fluid is needed, refer to the owner's manual or service information for the correct lubricant.

Some AWD vehicles, such as the Pilot and CR-V from Honda, require special lubricants for the viscous clutches and require the fluid to be replaced per the maintenance schedule. Failure to maintain the fluid can result in noise and damage to the clutch. Make sure your customers who own these types of vehicles are familiar with the special maintenance needs so that services can be performed before a problem arises.

Most transfer cases and PTUs have a plug in the side of the case for checking the oil level. The oil should be at the bottom of this opening. Some transfer cases use ATF and others use gear oil; always use the recommended oil.

4WD and AWD Tires

Because 4WD and AWD vehicles can direct torque to the front and rear and side-to-side tires, maintaining proper rolling circumference of the tires is critical. Most manufacturers recommend tire circumference be maintained within 6.35 to 12.70 mm (¼ to ½ in.) of each other. If the difference in tire circumference is excessive, the speed differences may result in the 4WD or AWD system attempting to manage torque and traction all the time. This could result in rapid wear and damage to viscous clutches and other components.

In the event the owner of a 4WD or AWD vehicle has a tire blowout, replacing all four tires may be the only way to maintain the correct tire circumferences. Replacing one tire while leaving the remaining three worn may cause enough variation in rotational speeds to affect the 4WD or AWD system. And depending on the system, even replacing both tires on one axle while leaving the other two old tires installed on the other axle may cause problems.

GO TO Chapter 41 for a detailed look at servicing joints and driveshafts.

Servicing the Transfer Case

As with all automobile servicing procedures, be sure to check the manufacturer's service information for specific transfer case repair and overhaul procedures. It gives details for the particular make and model of transfer case to be worked on.

> **CAUTION!**
> When removing and working on a transfer case, be sure to support it on a transmission jack or safety stands. The unit is heavy and, if dropped, could cause part damage and/or personal injury.

To remove the transfer case, raise and support the vehicle. Remove any skid plates and brace rods that block access to the transfer case or that are attached to it. Disconnect and remove all driveline or propeller shaft assemblies. Be sure to mark the parts and their relative positions on their yokes so that the proper driveline balance can be maintained when reassembled. Disconnect the linkage to the transfer case shift lever. Also disconnect wires to switches for 4WD dash indicator lights, if used. Support the transfer case, using a transmission jack, and remove the fasteners that secure it to the transmission. Slide the transfer case off the rear of the transmission, and remove it from the vehicle.

Once the transfer case has been removed from the vehicle and safely supported, take off the case cover and disconnect any electrical connections. Visually inspect for any oil leaks. Then carefully loosen and drive out the pins that hold the shift forks in place. Remove the front output shafts and chain drive or gearsets from the case. Keep in mind that some cases use chain drives, whereas others use spur or helical cut gearsets to transfer torque from the transaxle or transmission to the output shafts. Planetary gearsets provide the necessary gear reductions in some transfer cases. Photo Sequence 37 shows the procedure for disassembling and assembling a warner 13-56 transfer case, which is a commonly used unit.

Clean and carefully inspect all parts for damage and wear. Check the slack in the chain drive by following the procedure given in the service information. Replace any defective parts. It may be necessary to measure the shaft assembly end play. If excessive, new snap rings and shims may be used to correct the situation.

When reassembling the transfer case, the procedure is essentially the reverse of the removal. Be sure to use new gaskets between the covers when reassembling the unit. Photo Sequence 37 shows the procedure for reassembling a Warner 13-56 transfer case.

TYPICAL PROCEDURE FOR DISASSEMBLING AND ASSEMBLING A WARNER 13-56 TRANSFER CASE

PHOTO SEQUENCE 37

P37–1 If not previously drained, remove the drain plug and allow the oil to drain, then reinstall the plug. Loosen the flange nuts. Remove the two output shaft, yoke nuts, washers, rubber seals, and output yokes from the case.

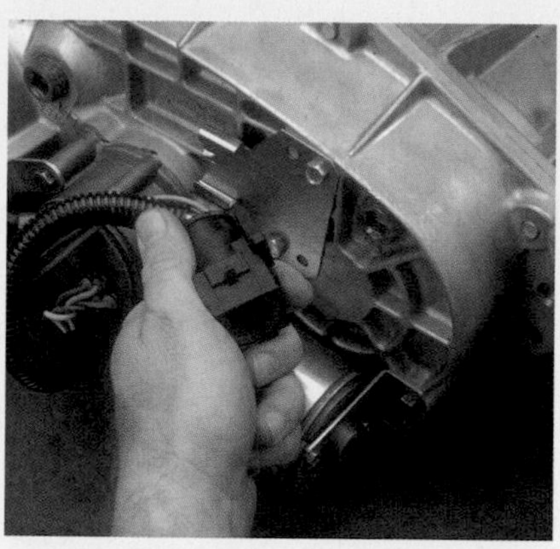

P37–2 Remove the four-wheel-drive indicator switch from the cover.

P37–3 Remove the wires from the electronic shift harness connector.

P37–4 Remove the speed sensor retaining the bracket screw, bracket, and sensor.

P37–5 Remove the bolts securing the electric shift motor and remove the motor.

P37–6 Note the location of the triangular shaft in the case and the triangular slot in the electric motor.

P37–7 Loosen and remove the front case to rear case retaining bolts. Separate the two halves by prying between the pry bosses on the case.

P37–8 Remove the shift rail for the electric motor.

P37–9 Pull the clutch coil off the mainshaft.

TYPICAL PROCEDURE FOR DISASSEMBLING AND ASSEMBLING A WARNER 13-56 TRANSFER CASE *(continued)*

P37–10 Pull the 2WD/4WD shift fork and lock-up assembly off the mainshaft.

P37–11 Remove the chain, driven sprocket, and drive sprocket as a unit.

P37–12 Remove the mainshaft with the oil pump assembly.

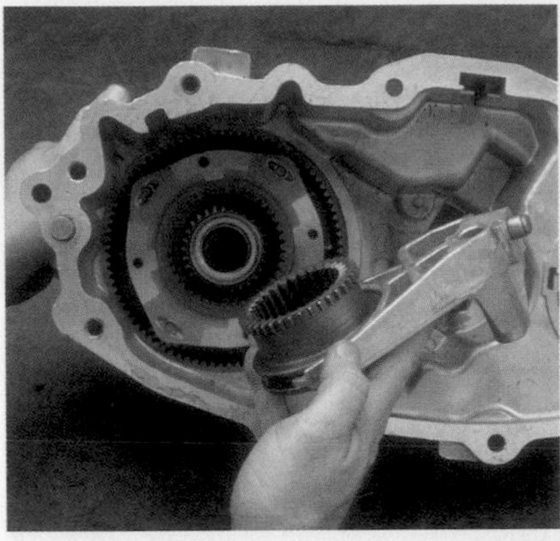
P37–13 Slip the high-low range shift fork out of the inside track of the shift cam. Remove the high-low shift collar from the shift fork.

P37–14 Unbolt and remove the planetary gear mounting plate from the case. With snap ring pliers, pull the planetary gearset out of the mounting plate.

P37–15 To begin reassembly, install the input shaft and front output shaft bearings into the case.

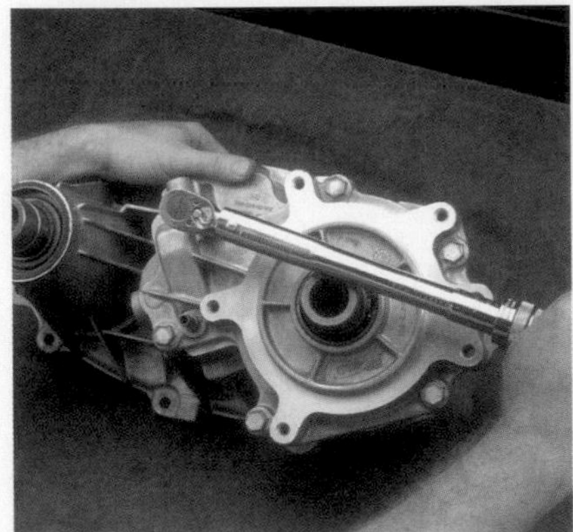
P37–16 Apply a thin bead of sealer around the ring gear housing. Install the input shaft with planetary gearset, and tighten retaining bolts to specifications.

P37–17 Install the high–low shift collar into the shift fork and install the assembly into the case.

P37–18 Install the mainshaft with oil pump assembly into the case.

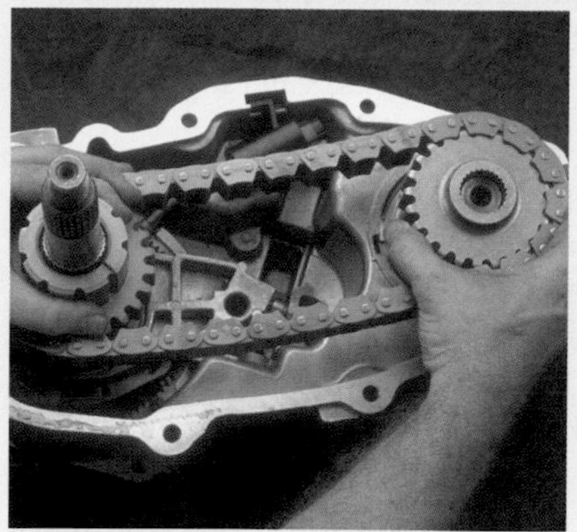

P37–19 Install the drive and driven sprockets and chain into position in the case.

P37–20 Install the shift rails.

P37–21 Install the 2WD/4WD shift fork and lock-up assembly onto the mainshaft.

P37–22 Install the clutch coil onto the mainshaft. Clean the case mating surfaces and apply a thin bead of sealer.

P37–23 Position the shafts and tighten the case halves together. Tighten attaching bolts to specifications.

P37–24 Apply a thin bead of sealer to the mating surface of the electric shift motor. Align the triangular shaft with the motor's triangular slot.

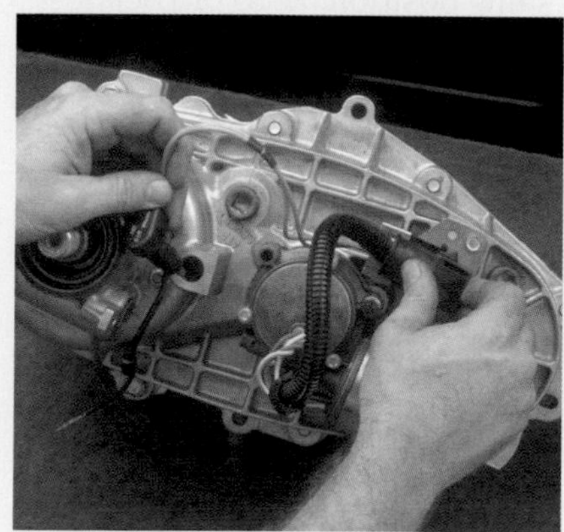

P37–25 Install the motor over the shaft, and wiggle the motor to make sure it is fully seated on the shaft. Tighten the motor's retaining bolts to specifications.

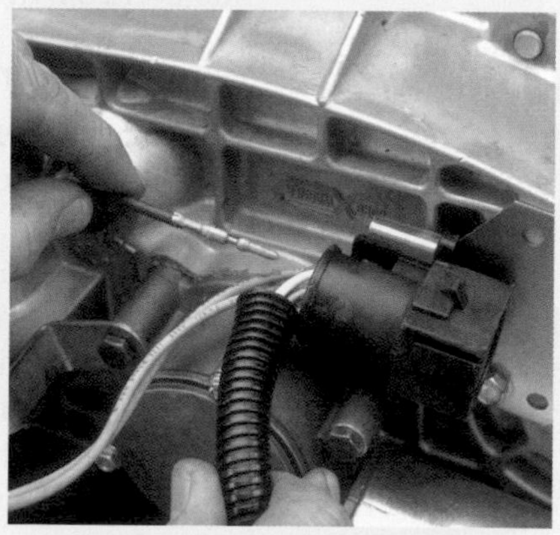

P37–26 Reinstall the wires into the connector, and connect all electric sensors.

P37–27 Install the companion flanges' seal, washer, and nut. Then tighten the nut to specifications.

Front Axles and Hubs

The axles and locking hubs must be removed before the front differential and final drive assembly can be removed from the axle housing. Removing these components requires special procedures, and each type of 4WD system requires its own specific procedure.

GO TO ▶ Chapter 41 for details on servicing a final drive unit in a transaxle.

CUSTOMER CARE

It is very important to remind your customers that the fluid level in a transfer case must be checked at recommended time intervals. (Check the service information.) It is wise to show your customers where the transfer case fill plug is located and how to remove it. Also tell them that the lubricant should be almost even with the fill hole. Always refer to the service information for recommended transfer case lubricants. Many transfer cases require extreme pressure (EP) lubricants, as used in differentials and in some manual transmissions.

Removal of the axle shafts begins with the removal of the hubs. Normally, the procedures for replacing manual hubs are somewhat different from those for replacing automatic hubs. Locking hubs are not serviceable; therefore, service of the hubs consists of merely replacing them. Also, the procedure for the replacement of the hubs varies with the type of hub and the manufacturer. The following are general procedures for the removal and installation of both manual and automatic locking hubs. Always follow the specific procedures for the type of hub you are servicing.

MANUAL LOCKING HUB To remove a manual hub, set the hub's control knob to the FREE or unlocked position. Then remove the cover bolts and the outer snap ring. If hubs are equipped with shims, remove them as well. Then remove the drive flange or gear from the axle. Now remove the bolts that attach the hub body to the hub. The hub assembly can now be pulled from the axle shaft. After the hub assembly has been removed, inspect the splines of the axle shaft for nicks and burrs.

To install a manual hub, separate the base and handle units of the hub lock assembly. Apply a light coat of grease to the axle shaft splines and the hub lock base. Also apply a light coating of lubricant to the O-ring. Make sure the gasket surface is smooth and clean, and then place a new hub gasket onto the hub. Set the control handle to the unlocked position, and install the hub assembly and the snap ring onto the axle shaft. Tighten the attaching bolts to the specified torque. Install the hub cover with a new gasket onto the hub, and tighten the remaining attaching bolts to specifications. Check the control handle for ease of operation.

AUTOMATIC LOCKING HUB There are basically two designs of automatic locking hubs: internally and externally retained. The procedure for servicing automatic hubs depends on how they are retained. Remove the bolts from the outer cap assembly, and pull the cap off the body of the hub. Then remove the axle shaft bolt, lock washer, and axle shaft stop. In the wheel hub, there is a groove that retains a lock ring; remove it and slide the hub assembly off the wheel's hub. Then loosen the set screws in the spindle locknut until their heads are flush with the face of the locknut. Remove the spindle locknut.

To install an automatic hub, adjust the wheel bearings, and install the spindle locknut and tighten it to specifications. Firmly tighten the locking set screws. Apply multipurpose grease to the inner splines of the hub; do not pack the hub with grease. Slide the assembly into the wheel hub. Push firmly on the body until it seats in the hub. Then install the lock ring in the groove of the wheel hub. Place the lock washer and axle shaft stop onto the axle bolt. Install the bolt, and tighten it to specifications. Apply a small amount of lube to the seal on the cap assembly. Do not grease the cap assembly. Install the cap assembly over the body assembly and into the wheel hub. Install the attaching bolts, and tighten them to specifications. Firmly turn the control from stop to stop to check the assembly. Then set both hub controls in the same position: AUTO or LOCK.

Wheel Bearings

The front-wheel bearings on a 4WD vehicle should be disassembled, cleaned, inspected, and lubricated on a regular basis. Upon reassembly, it is very important that the bearings be properly adjusted. The procedure for doing this is similar to other wheel bearings. However, because of the load on the bearings, the adjustment is more critical.

To adjust the wheel bearings, begin by removing the snap ring at the end of the spindle, and then the axle shaft spacer, needle thrust washer, bearing spacer, outer wheel bearing locknut, and bearing washer. Then loosen the inner bearing locknut, and tighten it fully to seat the bearings. The wheel bearing locknuts on most 4WD vehicles require the use of a four- or six-pronged spanner wrench. Spin the brake rotor on the spindle, then loosen the inner bearing locknut approximately 1/4 turn. Install the outer bearing locknut, and tighten it to the specified amount, usually 95 to 122 N·m (70 to 90 lb·ft). Assemble the remaining parts of the hub assembly, making sure that the snap ring is fully seated in its groove on the axle.

Most of these bearing assemblies previously mentioned are becoming obsolete. Most new 4WD vehicles use hub bearing assemblies due to their reduced maintenance and ease of replacement.

KEY TERMS

All-wheel drive (AWD) (p. 1369)
Driveline windup (p. 1371)
Four-wheel drive (4WD) (p. 1369)
Haldex clutch (p. 1382)
Interaxle differential (p. 1373)
Locking hubs (p. 1378)
Power transfer (take-off) unit (PTU) (p. 1377)
Transfer case (p. 1370)
Viscous clutch (p. 1379)

SUMMARY

- The importance of excellent traction and the benefits of four- and all-wheel-drive systems become readily apparent in snow or heavy rain.
- The heart of most conventional 4WD systems is the transfer case.
- The interaxle is placed in the transfer case to operate in the same fashion as the differentials at the axles. The only difference is that the transfer differential controls the speed of the drivelines instead of the axles.
- Many vehicles require that the front hubs be in a locked condition to operate as 4WD vehicles. This lock-up may be made either manually or automatically.
- On 4WD vehicles with on-the-fly shifting, it is not necessary to come to a complete stop when changing operational modes.
- Components of 4WD vehicles can be serviced in basically the same manner as the identical components on a 2WD vehicle.
- All-wheel-drive vehicles may use a viscous clutch, rather than a transfer case, to drive the axle with low tractive effort.
- Some AWD vehicles have a third differential, called an interaxle differential, instead of a transfer case.
- All-wheel-drive passenger cars are typically based on FWD platforms.

REVIEW QUESTIONS

1. What is the main advantage of all-wheel drive?
2. Name the three main driveline components that are added to a 2WD vehicle to make it a 4WD vehicle.
3. What are the primary differences between a full-time and a part-time 4WD system?
4. What is the purpose of the inter-axle differential?
5. Describe the differences between manual and self-locking hubs.
6. What is tractive effort defined as?
   a. driving force where a driving gear meshes with a driven gear
   b. place where the clutch disc makes contact with the flywheel and pressure plate
   c. driving force in which the wheel contacts the road's surface
   d. driving force at the wheel
7. What effect do the plates of a viscous clutch have on the fluid when they rotate at different speeds?
   a. They separate the fluid, allowing easier movement.
   b. They cool down the fluid, allowing it to thicken.
   c. They pump the fluid from the clutch assembly.
   d. They heat up the fluid rapidly.
8. What shift position should be selected when operating a four-wheel-drive SUV on slippery winter roads?
   a. four-wheel-drive high range
   b. four-wheel-drive low range
   c. two-wheel-drive high range
   d. two-wheel-drive low range
9. What type of vehicle uses a centre differential instead of a transfer case?
   a. a four-wheel-drive sport utility vehicle (SUV)
   b. a four-wheel-drive pickup truck
   c. a four-wheel-drive passenger car
   d. an all-wheel-drive passenger car
10. How are viscous clutches applied?
    a. vacuum-operated shift mechanism
    b. electric motor–controlled shift collars
    c. automatically by the silicone fluid
    d. mechanical shift linkage

11. Which of the following shifts can be performed on the fly?
  a. from four-low to four-high
  b. from four-high to four-low
  c. from two-high to four-low
  d. from two-high to four-high
12. What drive mode(s) can all-wheel-drive vehicles operate in?
  a. four-wheel drive only
  b. two-wheel drive or four-wheel drive
  c. two-wheel drive only
  d. four-wheel-drive high or four-wheel-drive low
13. What does the transfer clutch in the all-wheel-drive automatic transaxle take the place of?
  a. transmission b. reduction gears
  c. torque converter d. interaxle differential
14. In a typical all-wheel-drive automatic transaxle, what controls the transfer clutch?
  a. transmission control unit operating the pressure regulator valve
  b. engine computer controlling line pressure to the various clutches in the transmission
  c. transmission control unit and the duty solenoid
  d. front axle speed sensor and the vehicle speed sensor
15. Which of the following components is replaced as an assembly?
  a. the transfer case
  b. the viscous clutch
  c. the interaxle differential
  d. all-wheel-drive transaxles
16. What happens to the silicone fluid in the viscous clutch when it is heated?
  a. It becomes a very thin fluid.
  b. It boils to a vapour.
  c. It thickens to a solid mass.
  d. It stiffens.
17. What is used in a typical 4WD pickup truck to prevent driveline windup during turns?
  a. constant velocity joints in the driveshafts
  b. a Haldex clutch
  c. an interaxle differential in the transfer case
  d. automatic locking wheel hubs
18. Which of the following is an advantage of four-wheel drive?
  a. It weighs less.
  b. It requires less maintenance.
  c. The tires cost less.
  d. It offers improved traction for off-road driving.
19. What is the purpose of the viscous clutch?
  a. to engage engine torque to the transmission
  b. to improve mobility when driving conditions become difficult
  c. to drive both rear driving wheels when a vehicle is stuck
  d. to drive the rear axle assembly
20. Which transfer case shift position will provide the greatest torque to the wheels?
  a. four-wheel-drive high
  b. four-wheel-drive low
  c. two-wheel-drive high
  d. No position will provide any difference in torque output.

CHAPTER 46
Tires and Wheels

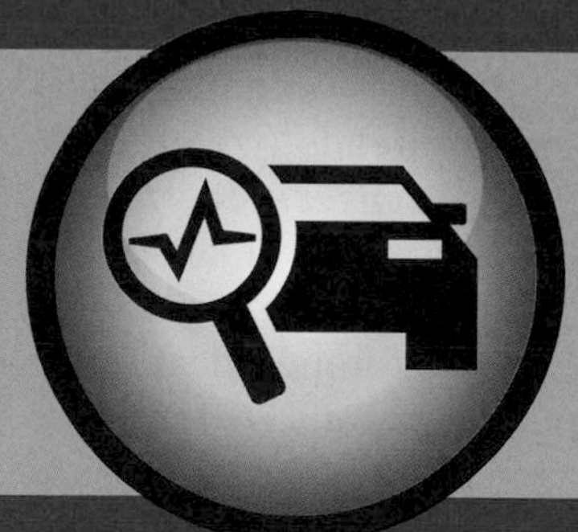

LEARNING OUTCOMES

- Describe basic wheel and hub design.
- Recognize the basic parts of a tubeless tire.
- Explain the tire ratings and designations in use today.
- Describe why certain factors affect tire performance, including inflation pressure, tire rotation, and tread wear.
- Remove and install a wheel and tire assembly.
- Dismount and remount a tire.
- Repair a damaged tire.
- Describe the differences between static balance and dynamic balance.
- Balance wheels both on and off a vehicle.
- Describe the three popular types of wheel hub bearings.
- Service tapered roller bearings.

Tire and wheel assemblies provide the only connection between the road and the vehicle. Tire design has improved dramatically during the past few years. Modern tires require increased attention to achieve their full potential of extended service and correct ride control. Tire wear that is uneven or premature is usually a good indicator of steering and suspension system problems. Tires, therefore, become not only a good diagnostic aid to a technician, but can also be clear evidence to the customer that there is need for service.

WHEELS

Wheels are made of either stamped or pressed steel discs riveted or welded together. They are also available in the form of aluminum or magnesium rims that are die-cast or forged **(Figure 46–1)**. Magnesium wheels are commonly referred to as mag wheels, although they are usually made of an aluminum alloy. Aluminum wheels are lighter in weight when compared with the stamped steel type. This weight savings is important because the wheels and tires on a vehicle are unsprung weight. This means the weight is not supported by the vehicle's springs.

Near the centre of the wheel are mounting holes that are tapered to fit tapered mounting nuts (**lug nuts** or bolts) that centre the wheel over the hub. The rim has a hole for the tire's **valve stem** and a **drop centre** area designed to allow for easy tire removal and installation. **Wheel offset** is the vertical distance between the centreline of the rim and the mounting face of the wheel. The offset is considered positive if the centreline of the rim is inboard of the mounting face, and negative if outboard of the mounting face. The amount and type of offset is critical because changing the wheel offset changes the front suspension loading as well as the scrub radius.

GO TO  Chapter 49 for details on scrub radius and wheel alignment.

The wheel is bolted to a **hub**, either by lug bolts that pass through the wheel and thread into the hub, or by studs that protrude from the hub. In the

FIGURE 46–1

An alloy wheel on a late-model car.

case of studs, special lug nuts are required. A few vehicles have left-hand threads (which turn counterclockwise to tighten) on the driver's side and right-hand threads (which turn clockwise to tighten) on the passenger's side. All other vehicles use right-hand threads on both sides. Wheel studs should never be lubricated with oil or antiseize compounds. Lug nut or bolt torque specifications are intended for dry, clean threads. Aftermarket wheels may have additional lug holes so that the wheels can fit a variety of stud patterns. Installing aftermarket wheels may require the use of a centerbore spacer in the wheel. The spacer is used to match the centerbore of the wheel to fit the hub of the vehicle.

Wheel size is designated by rim width and rim diameter **(Figure 46–2)**. Rim width is determined by measuring across the rim between the flanges. Rim diameter is measured across the bead seating areas from the top to the bottom of the wheel. Wheel bolt pattern is determined by the distance between the studs. Rims have safety ridges near their lips. In the event of a tire blowout, these ridges tend to keep the tire from moving into the dropped centre and from coming off the wheel.

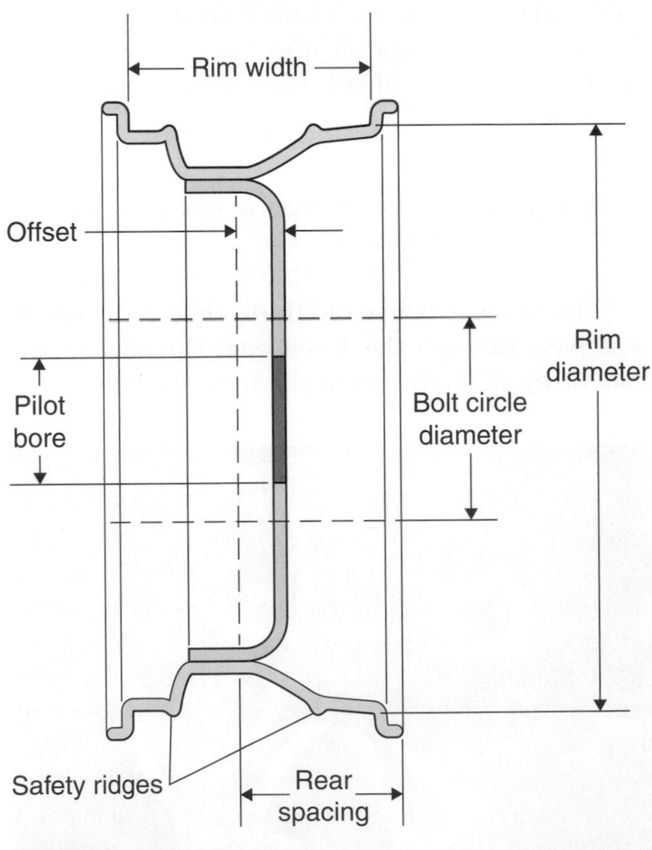

FIGURE 46–2

Wheel dimensions are important when replacing tires.

Based on James E. Duffy, Modern Automotive Technology (The Goodheart-Willcox Company); from Goodyear Tire & Rubber Company.

Replacement wheels must be equal to the original equipment wheels in load capacity, diameter, width, offset, and mounting configuration. An incorrect wheel can affect wheel and bearing life, ground and tire clearance, or speedometer and odometer calibrations. A wrong size wheel can also affect the antilock brake and traction control system.

Some performance-oriented cars are equipped with different-size wheels at the front and rear. It is important that the specific-size wheel be at the specific axle. The wider or larger wheels are designed to use different-size tires. For example, a Crossfire has 8 × 18 front wheels that are fitted with 225/40 ZR 18 tires. The rear is equipped with 10 × 19 wheels that carry 255/35 ZR 19 tires.

TIRES

The primary purpose of tires is to provide traction. Tires also help the suspension absorb road shocks, but this is a side benefit. They must perform under a variety of conditions. The road might be wet or dry or paved with asphalt, concrete, or gravel, or there might be no road at all. The car might be travelling slowly on a straight road, or moving quickly through curves or over hills. All of these conditions call for special requirements that must be present, at least to some degree, in all tires.

In addition to providing good traction, tires are also designed to carry the weight of the vehicle, to withstand side thrust over varying speeds and conditions, and to transfer braking and driving torque to the road. As a tire rolls on the road, static friction is created between the tire and the road. All of this has to occur on a contact patch that may be as small as 225 cm^2 (35 in^2) per tire. This friction gives the tire its traction. Although good traction is desirable, it must be limited. Too much traction means there is too much friction. Too much friction means there is a lot of rolling resistance. Rolling resistance wastes engine power and fuel; therefore, it must be kept to a minimal level. This dilemma is a major concern in the design of today's tires.

Tube and Tubeless Tires

Early vehicle tires were solid rubber. These were replaced with pneumatic tires, which are filled with air.

There are two basic types of pneumatic tires: those that use inner tubes and those that do not. The latter are called tubeless tires and are about the only type used on passenger cars today. A tubeless tire has a soft inner lining that keeps air from leaking between the tire and rim. On some tires, this inner lining can form a seal around a nail or other object that punctures the tread. A self-sealing tire holds in air even after the object is removed. The key to this sealing is a lining of sticky rubber compound on the inside of the tread area that will seal a hole up to 4.77 mm (3⁄16 in.).

A tubeless tire air valve has a central core that is spring loaded to allow air to pass inward only, unless the pin is depressed. If the core becomes defective, it can be unscrewed and replaced. The airtight cap on the end of the valve prevents dust and moisture from entering and affecting valve operation. A tubeless tire is mounted on a special rim that retains air between the rim and the tire casing when the tire is inflated.

Figure 46–3 shows a cutaway view of a typical tubeless tire. The basic parts are shown. The cord body or casing consists of layers of rubber-impregnated cords, called **plies**, that are bonded into a solid unit. Typically, tires are made of four plies. Bias tires were actually constructed of many plies, up to 10 or 12 layers to increase tire strength. Modern tires do not require the many layers to increase the tire's load-carrying capacity. The **bead** is the portion of the tire that helps keep it in contact with the rim of the wheel. It also provides the air seal on tubeless tires. The bead is constructed of a heavy band of steel wire wrapped into the inner circumference of the tire's ply structure. The **tread**, or crown, is the portion of the tire that comes in contact with the road surface. It is a pattern of grooves and ribs that provides traction. The grooves are designed to drain off water, and the ribs grip the road surface. Tread thickness varies with tire quality. On some tires, small cuts, called **sipes**, are moulded into the ribs of the tread. These sipes open as the tire flexes on the road, offering additional gripping action, especially on wet road surfaces. The **sidewalls** are the sides of the tire's body. They are constructed of thinner material than the tread, to offer greater flexibility.

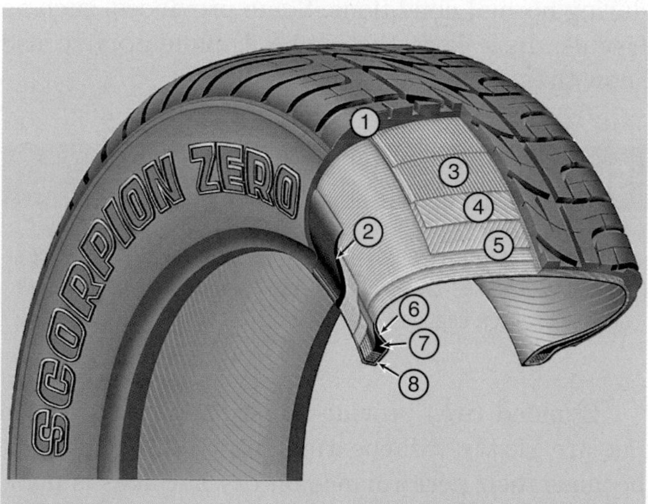

FIGURE 46–3

A typical tubeless tire.

Courtesy of Pirelli Tyre S.p.A.

The tire body and belt material can be made of rayon, nylon, polyester, fibreglass, steel, aramid, or Kevlar. Each has its advantages and disadvantages. For instance, rayon and cord tires are low in cost and give a good ride, but do not have the inherent strength needed to cope with long high-speed runs or extended periods of abusive use on rough roads. Nylon-cord tires generally give a slightly harder ride than rayon—especially for the first few kilometres after the car has been parked—but offer greater toughness and resistance to road damage. Polyester and fibreglass tires offer many of the best qualities of rayon and nylon, but without the disadvantages. They run as smoothly as rayon tires but are much tougher. They are almost as tough as nylon but give a much smoother ride. Steel is tougher than fibreglass or polyester, but it gives a slightly rougher ride because the steel cord does not give under impact, as do fabric plies. Aramid and Kevlar cords are lighter than steel cords and, kilogram for kilogram, stronger than steel.

Types of Tire Construction

There are two basic types of tire construction in use today **(Figure 46–4)**. The oldest design currently in use is the **belted bias ply**, which can be found on some heavy equipment, trailers, and vintage cars. It has a body of fabric plies that run alternately at opposite angles to form a crisscross design. The angle varies from 30° to 38° with the centreline of the tire and has an effect on high-speed stability, ride harshness, and handling. Generally speaking, the lower the cord angle, the better the high-speed stability, but also the harsher the ride.

In addition, there are two or more belts running the circumference of the tire under the tread. This construction gives strength to the sidewall and

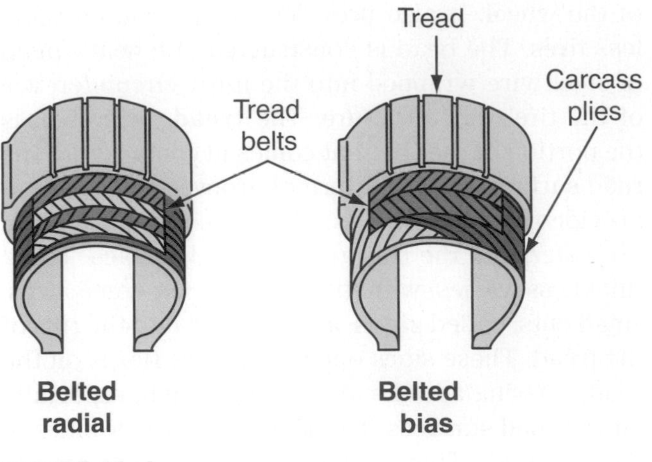

FIGURE 46–4

The construction of the two basic types of tires.

greater stability to the tread. The belts reduce tread motion during contact with the road, thus improving tread life. Plies and belts of various combinations of rayon, nylon, polyester, fibreglass, and steel are used with belted bias construction.

Radial ply tires have body cords that extend from bead to bead at an angle of about 90°—"radial" to the circumferential centreline of the tire—plus two or more layers of relatively inflexible belts under the tread. The construction of various combinations of rayon, nylon, fibreglass, and steel gives greater strength to the tread area and flexibility to the sidewall. The belts restrict tread motion during contact with the road, thus improving tread life and traction. Radial ply tires also offer greater fuel economy, increased skid resistance, and more positive braking.

One of the newer tire designs has steel reinforced sidewalls that allow the tire to maintain shape with no air pressure. The so-called run-flat tires give the driver a limited amount of travel after a tire has been punctured. Because they were designed to be filled with air, their ride and handling characteristics are much better when they have air than when they do not. However, the tires will not leave the driver stranded or forced to put on a spare tire when the tire is punctured.

Specialty Tires

Specialty tires reflect the advances made in tire development. These tires are designed for specific road conditions or applications. All-season tires are designed to perform well on all types of road conditions, but they are not excellent performers on all road surfaces. To provide traction in the snow and mud, at least 25 percent of the tread area is void

FIGURE 46–5

A directional all-season radial tire.

(Figure 46–5). This leaves open areas for the snow or mud to move into as the tire rotates. The open spaces also give the tires some bite as they move. The remaining tread area is designed to provide good traction on normal surfaces.

Tires can also be designed for heavy snow. These are commonly called snow tires. The tread area is much more aggressive, with larger voids than all-season tires. Because of the decreased contact with the road, these tires do not provide good traction during normal conditions. Some provinces, such as Quebec, have laws that make it mandatory to use snow tires during the winter months.

SHOP TALK

Snow tires are designed to increase traction in the snow. They should be installed on all wheels and replaced with normal tires once the season changes.

Studded tires provide superior traction on ice but are slowly disappearing from the tire market because their performance on dry surfaces is poor. Although these tires are still popular in Quebec during the winter, many provinces have outlawed their use because they damage the road and can be a safety hazard on dry surfaces. Because the studs have more contact on the road than the rubber, it is easy for the tire to slide on the road during cornering and stopping. The studs offer much less friction than the rubber tire tread.

Tires can be designed to be great on dry surfaces or wet surfaces. However, it is nearly impossible to have a tire that performs extremely well on both. Tires designed for dry and smooth roads do not really need a tread pattern. They can be "slicks." The smooth surface would give the tires maximum grip on a smooth, dry surface. However, when a slick hits a wet spot, there is no traction. The tire simply slides on the water.

Tires designed for wet surfaces have tread designs that pump the road's water behind and to the side of the tire. Moving the water is the only way a tire can grip the road. When too much water separates the tire from the road, hydroplaning takes place. This causes the tire to lift off the road and rotate on a layer of water. Traction is reduced, which can create an unsafe condition. Needless to say, when a tire has many directed and open channels for water in its tread, less rubber meets the road.

Some cars are equipped with summer performance tires. This type of tire maximizes handling and road-holding, but only in warm weather. In most cases, these tires should not be used in snow or when temperatures are 4.4°C (40°F) or below. Use of the tires below the recommended temperatures can cause reduced traction as the tires may not develop enough heat to become "sticky."

Run-Flat Tires

There are several types of run-flat tires available. Vehicles equipped with these have no spare tire or jack in the luggage compartment. Run-flats can be divided into three categories: self-sealing, self-supporting, and auxiliary supported systems. Each of these uses different ways to allow a vehicle to be driven after a tire is punctured.

SELF-SEALING Self-sealing tires are designed to quickly and permanently seal most tread-area punctures. The tires are constructed like other tires, but there is an additional lining inside the tire under the tread area. The lining is coated with a sealant that can permanently seal most punctures up to 4.76 mm (3/16 in.) in diameter. The lining seals the area around the puncture and can fill in the hole once the object is removed from the tire **(Figure 46–6)**. There is no provision for reinflating the tire, so the tire will need to be inflated after the repair.

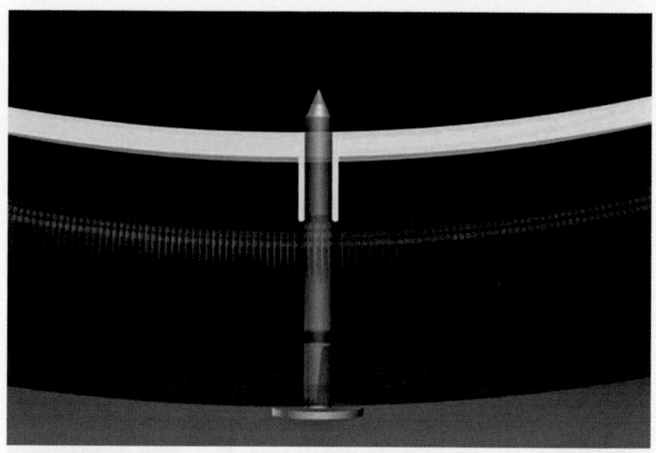

FIGURE 46–6

The action of a self-sealing tire.

Used with permission of Ric Snyder.

SELF-SUPPORTING When a self-supporting tire loses all of its air pressure, it is able to temporarily carry the weight of the vehicle and allow the vehicle to be driven. This is a result of the tire's construction. These tires have reinforced sidewalls and special beads **(Figure 46–7)**. The first "run-flat" tire available on a regular production vehicle was a self-supporting tire that was offered as an option on the 1994 Chevrolet Corvette. Today, many tire manufacturers offer self-supporting tires, and some even come as standard equipment on some vehicles. Typically, a self-supporting tire can be driven for 80 km (50 mi.) at speeds up to 90 km/h (55 mph) after it has lost air pressure.

AUXILIARY SUPPORTED RUN-FLAT SYSTEMS Auxiliary supported systems are much different from other run-flat tires. They are systems that have special tires and wheels. The basis of these systems is a

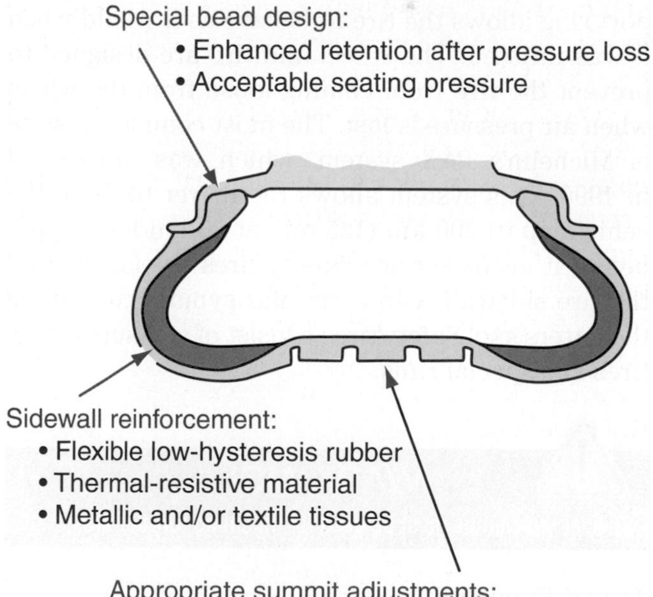
Special bead design:
• Enhanced retention after pressure loss
• Acceptable seating pressure
Sidewall reinforcement:
• Flexible low-hysteresis rubber
• Thermal-resistive material
• Metallic and/or textile tissues
Appropriate summit adjustments:
• Maintain inflated performance (comfort and handling like standard tires)

FIGURE 46–7

Features of a run-flat tire.

FIGURE 46–8

A cutaway of a run-flat tire with an insert for support in case the tire goes very flat.

The Goodyear Tire Rubber Company

solid supporting ring that allows a flat tire's tread to rest on a support ring attached to the wheel when the tire loses pressure **(Figure 46–8)**. This support ring allows the tire to behave as it would when it was inflated. The wheel and tire are designed to prevent the tire from coming loose from the wheel when air pressure is lost. The most common system is Michelin's PAX system, which was introduced in 1996. This system allows the driver to drive the vehicle up to 200 km (125 mi.) at 90 km/h (55 mph) before it needs service. Safety tires are labelled on the tire sidewall with a circular symbol containing the letters RSC. Safety tires consist of self-supporting tires and special rims.

WARNING!

Special rim clamps must be used for mounting run-flat tires.

Tread Designs

The real purpose of a tire is to get a grip on the road. The ideal tire is one that wears little, holds the road well to provide sure handling and braking, and provides a cushion from road shock. The ideal tire should also provide maximum grip on dry roads, wet roads, and snow and ice, and operate quietly at any speed. This is a tall order, so tire manufacturers compromise on one or two of these qualities for the sake of excelling at another. A tire's tread design dictates what the tire will excel at.

There are basically three categories of tread patterns: directional, nondirectional, and symmetric and asymmetric.

A directional tire is mounted so that it revolves in a particular direction. These tires have an arrow on the sidewalls that show the designed direction of travel. A directional tire offers good performance only when it is rotating in the direction in which it was designed to rotate **(Figure 46–9A)**. Directional tread patterns are used with all-season tires to more effectively channel water from under the tire and improve wet traction. Many sports car tires also use a directional tread design. A nondirectional tire has the same handling qualities in either direction of rotation. The most common and basic tread designs are nondirectional. A symmetric tire has the same tread pattern on both sides of the tire. An asymmetrical tire has a tread design that is different from one side to the other **(Figure 46–9B)**. Asymmetrical tires are typically designed to provide good grip when travelling straight (the inside half) and good grip in turns (the outside half of the tread). Some asymmetric tires are also directional tires.

The number and size of the blocks, sipes, and grooves on a tire's tread not only determine how much rubber contacts the road and how much water can be displaced but also how quiet the tire will be during travel. The more aggressive the tread, the more noise it will make. This statement is especially true if the tire's tread is made of a hard compound. Softer tires typically make less noise but wear more quickly. Soft tires and tread designs with more sipes also adhere to the road better.

Channels are cut into a tire's tread to allow water to move away from the tire's direction of travel. Obviously, the deeper the channel, the more water the tire can move. The disadvantage of these channels is decreased road contact.

SHOP TALK

Whenever a customer wants a better handling tire, make sure he or she knows a better gripping tire may make more noise and not wear as long as other tires. Knowing what design of tire will meet a customer's needs is a science. Always consult with a tire specialist before recommending one tire or another.

(A)

(B)

FIGURE 46–9

There are many different tread designs available for today's vehicles. (A) This is a symmetrical tread design. (B) This is an asymmetrical tread design.

Spare Tires

Nearly all vehicles are equipped with a spare tire to be used in case one of the vehicle's tires loses air and goes flat. A spare tire can be a tire that matches the tires on the vehicle or can be a compact spare. Compact spares are designed to reduce weight and storage space but still provide the driver with a tire in the case of an emergency. Compact tires are typically one of three types: high-pressure mini-spare, space-saver spare, and lightweight skin spare.

A high-pressure mini-spare tire is a temporary tire **(Figure 46–10)**. It should not be used for extended mileage or for speeds above 80 km/h

FIGURE 46–10

Notice the warnings affixed to this temporary spare tire.

(50 mph). These tires typically require an inflation pressure of 415 kPa (60 psi). A space-saver spare must be blown up with a compressor that operates from the cigarette lighter/power outlet or a built-in air compressor. A skin spare is a normal bias ply-type tire with a reduced tread depth.

CUSTOMER CARE

Make sure you warn your customer that a mini, space-saver, or similar type compact spare should be used only as a temporary tire. It should never be used as a regular tire. Any continuous load use of a temporary spare will result in tire failure, loss of vehicle control, and possible injury to the vehicle's occupants.

The Ratings and Designations

The construction of a tire depends on its application. Needless to say, there are many different tires. These differences are based on not only size but also their construction to meet intended driving conditions. There are also standards that tire manufacturers must meet to ensure that the tire will be safe, not wear rapidly, and offer good road isolation for the passengers in the vehicle. The uniqueness of each tire is represented by information given on the sidewall of every tire produced. In fact, everything you need to know about a tire is imprinted on the tire **(Figure 46–11)**.

Tire Size

The best way to describe and explain the information given on the sidewall of a tire is to look at an example. Look at the tire size designation of P215/65 R15 89H and see what it tells.

FIGURE 46–11

There is a lot of information about a tire on its sidewall.

The Goodyear Tire Company

On a **P**215/65 R15 89H tire, the *P* represents the application of the tire; in this case, P is for passenger car. If the tire had an *LT* designation, the tire would be for a light truck.

The *215* in P**215**/65 R15 89H represents the width of the tire measured in millimetres from sidewall to sidewall. This tire width is 215 mm. This is also called section width, and it varies with the wheel (rim) on which the tire is mounted. A wide rim increases the section width, whereas a narrow one decreases it. The measurement given on the tire was taken with the tire on a specific rim.

The *65* in P215/**65** R15 89H indicates the **aspect ratio**, or **profile (series)**, of the tire **(Figure 46–12)**. A tire's aspect ratio is the relationship of its cross-sectional height (from tread to bead) to its cross-sectional width (from sidewall to sidewall). In our example, the tire's height is equal to 65 percent of its width (the width equals 215 mm × 65 percent, or 140 mm). The aspect ratio determines a tire's performance characteristics. Higher aspect ratios provide a softer ride because they will deflect more over irregular surfaces and under heavy loads. Shorter sidewall heights demand stiffer sidewalls. Therefore, tires with a low aspect ratio have a harsher ride. However, they provide a larger contact area with the road and therefore better traction. Common aspect ratios are 75, 70, 65, 60, 55, 50, 40, 35, and even 30, which has become popular.

For a tire rated as P215/65 **R**15 89H, the *R* represents the basic ply construction of the tire. This letter can be an *R* for radial construction, a *B* for belted-bias construction, or a *D* for bias ply (bias means the plies are set diagonally or at a slant).

The diameter of the wheel is indicated by the 15 after the R. The diameter of the wheel for this tire is 15 inches. Wheel diameter is the height of the wheel from one end to the other.

Following the size notation is the load and speed ratings. These are expressed by a number and a letter; in this case, the ratings are given as 89H. The *89* is the load index and the *H* is the speed rating.

The maximum load rating lists the maximum amount of weight the tire can carry at the recommended tire pressure. For bias tires, the load rating and the number of tread and sidewall plies are proportional. In most cases, the more plies a tire has, the more weight it can support. For modern radial tires, the higher the tire's load index number, the greater its load-carrying capacity. The load ratings for passenger car and light truck tires range from 70

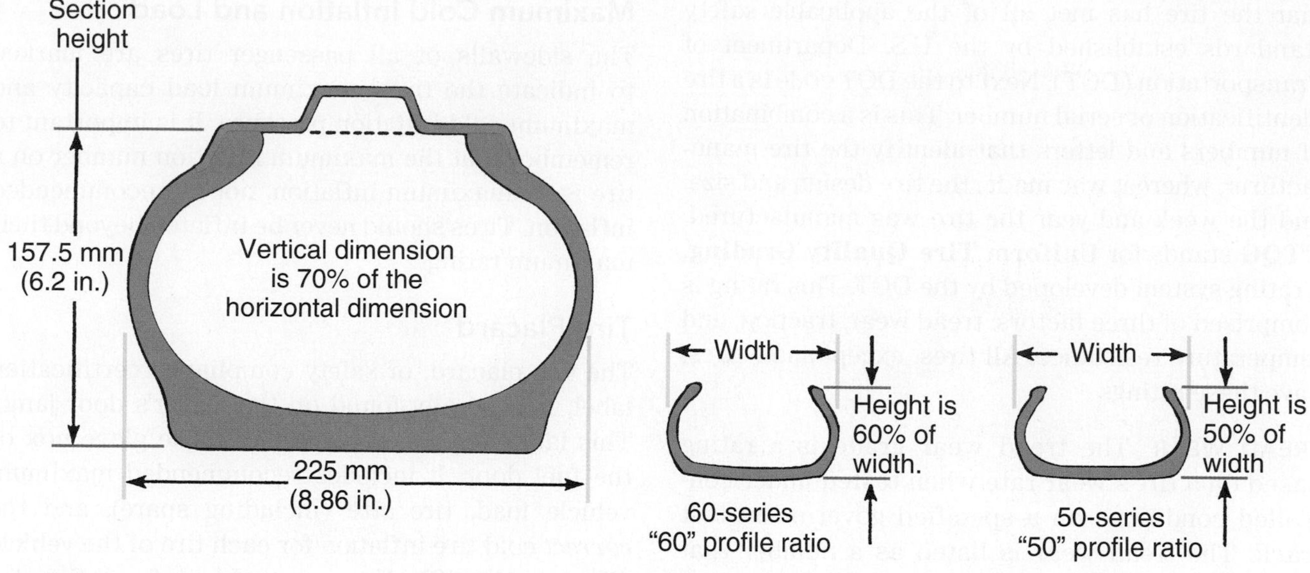

FIGURE 46–12

The aspect ratio (profile) of a tire is its cross-sectional height compared to its cross-sectional width expressed in a percentage.

to 110. Following are some examples of load ratings and the weight they represent:

71 345 kg (761 lb.)
79 437 kg (963 lb.)
89 580 kg (1279 lb.)
99 775 kg (1709 lb.)
109 1030 kg (2271 lb.)

So in our example, the tire can carry 580 kg (1279 lb.).

The speed rating indicates the maximum speed at which the tire should be used. In this case, the *H* means the tire has been tested to be safe at speeds up to 210 km/h (130 mph). The speed rating of a tire is really nothing more than an expression of how well the tire will withstand the temperatures of high speed. This does not necessarily mean that a high-speed-rated tire will perform better at low speeds than a lower-rated tire.

Table 46–1 lists the various letters used to designate the speed rating of a tire and the maximum speed at which the tire was designed to safely operate. Driving a vehicle at speeds greater than the speed rating of the tires is risky. The heat generated can cause the tire to come apart. If this happens at high speeds, it will be close to impossible for the driver to maintain control of the vehicle.

Although high-performance tires withstand heat better than normal tires, they still wear and must be replaced. In some European countries, the replacement tire must have, by law, the same speed rating as the original equipment (OE) tire. In Canada, a tire cannot be replaced with a tire that has a lower weight rating than what is recommended by the original equipment manufacturer (OEM).

TABLE 46–1 SPEED RATINGS

SYMBOL	MAX. SPEED
Q	160 km/h (99 mph)
S	180 km/h (112 mph)
T	190 km/h (118 mph)
U	200 km/h (124 mph)
H	210 km/h (130 mph)
V	240 km/h (149 mph)
Z	Above 240 km/h (149+ mph)
Z–W	270 km/h (168 mph)
Z–Y	300 km/h (186 mph)

All metric system measurements are now being given along with a standard translation. A typical metric tire shows its width in millimetres, its inflation pressure in kilopascals (kPa), and its load capacity in kilograms (kg). One kilopascal equals 6.895 psi. A typical all-metric radial size is 190/65R-390. It fits a 390 mm diameter wheel.

Other Information

The sidewall of a tire also has a DOT safety code, tire identification or serial number, UTQG ratings, and maximum inflation values. The DOT code indicates that the tire has met all of the applicable safety standards established by the U.S. Department of Transportation (DOT). Next to the DOT code is a tire identification or serial number. This is a combination of numbers and letters that identify the tire manufacturer, where it was made, the tire design and size, and the week and year the tire was manufactured. **UTQG** stands for **Uniform Tire Quality Grading**, a rating system developed by the DOT. This rating is comprised of three factors: tread wear, traction, and temperature resistance. All tires, except snow tires, have these ratings.

TREAD WEAR The tread wear grade is a rating based on a tire's wear rate when tested under controlled conditions on a specified government test track. The tread wear is listed as a number that ranges from 100 to 500. A rating of 100 means the tire will wear easily and quickly with a projected tread life of 20 000 km (about 12 500 mi.). A rating of 500 means the tire is harder and will resist wear with a projected tread life of 100 000 km (about 62 000 mi.). These ratings should be used to compare the anticipated wear of tires from the same manufacturer and not to compare wear between manufacturers.

TRACTION Tire traction ratings are based on a tire's ability to stop on wet concrete and asphalt. It is not an indication of how well a tire will handle. The traction rating is given as AA, A, B, or C. A tire rated as C will provide less traction than one rated with an A.

TEMPERATURE RESISTANCE This rating is an indication of how well a tire will dissipate heat and how it works when it is heated. The temperature rating applies only to a properly inflated tire that is not overloaded. Heat builds up when a tire is underinflated or overloaded. Temperature also increases with excessive speeds. Temperature resistancc rating is given as A, B, or C. A rating of C means the tire is acceptable. A tire with an A temperature rating will be able to withstand high temperatures better than one rated B or C.

ADDITIONAL RATINGS Some tires carry additional markings related to their intended service. An M&S or M+S designation means the tire has been rated by the manufacturer as suitable for mud and snow use. This is the designation for many all-season tires. Some tires may also have a mountain/snowflake symbol, which indicates that these tires are suitable for severe snow conditions; these are commonly called winter tires.

Maximum Cold Inflation and Load

The sidewalls of all passenger tires are marked to indicate the tires' maximum load capacity and maximum cold inflation pressure. It is important to remember that the maximum inflation number on a tire is its maximum inflation, not its recommended inflation. Tires should never be inflated beyond their maximum rating.

Tire Placard

The tire placard, or safety compliance certification label, is generally found on the driver's door jamb. This label may also be located in the glove box or the fuel door. It includes recommended maximum vehicle load, tire size (including spare), and the *correct* cold tire inflation for each tire of the vehicle **(Figure 46–13)**. Never use this information for other cars.

As a general rule, tires should be replaced with the same size and design or an approved optional size as recommended by the auto or tire manufacturer. Also, always follow the manufacturer's recommendations for tire type, inflation pressures, and rotation patterns. Never replace a tire with a smaller size than recommended on the tire placard.

SHOP TALK

Tires with a larger or smaller diameter than originally installed will affect the operation of the antilock brake system and the accuracy of the speedometer. It might be necessary to recalibrate the ABS computer and change the speedometer drive gears when tire size has been changed. Check the vehicle's service information for details.

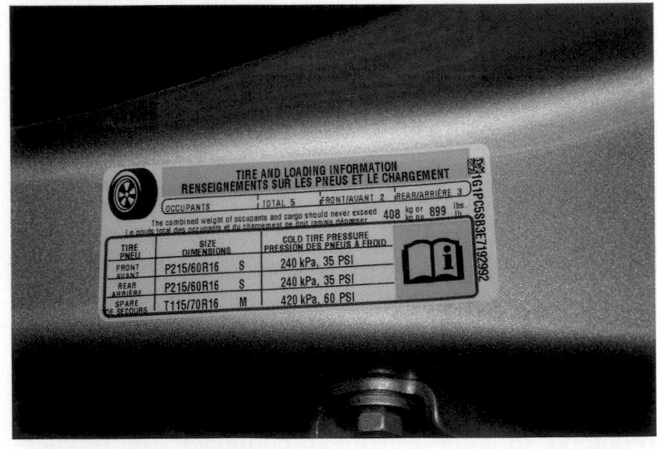

FIGURE 46–13

A tire placard on a doorjamb.

Martin Restoule

Tire Care

To maximize tire performance, inspect for signs of improper inflation and uneven wear, which can indicate a need for balancing, rotation, or wheel alignment. Tires should also be checked frequently for cuts, stone bruises, abrasions, and blisters, and for objects that might have become embedded in the tread. More frequent inspections are recommended when rapid or extreme temperature changes occur, or where road surfaces are rough or occasionally littered with debris.

To clean tires, use a mild soap and water solution only. Rinse thoroughly with clear water. Do not use any caustic solutions or abrasive materials. Never use steel wool or wire brushes. Avoid gasoline, paint thinner, and similar materials having a mineral oil base. These materials will cause premature drying of the tire's rubber. As the rubber dries, it gets harder and the tire will lose some of its performance characteristics. Tires will also dry out with age. For this reason and the fact that snow tires have aggressive tread designs and use softer rubber compounds, they do not normally last beyond three years.

INFLATION PRESSURE A properly inflated tire gives the best tire life, riding comfort, handling stability, and fuel economy for normal driving conditions. Too little air pressure can result in tire squeal, hard steering, excessive tire heat, abnormal tire wear, and increased fuel consumption by as much as 10 percent. An excessively low tire can cause severe and permanent damage to the sidewalls if driven on for even a short period of time. An underinflated tire shows maximum wear on the outside edges of the tread. There is little or no wear in the centre.

Conversely, an overinflated tire shows its wear in the centre of the tread and little wear on the outside edges. A tire inflated higher than recommended can cause a hard ride, tire bruising, and rapid wear at the centre of the tire **(Figure 46–14)**. Always use the recommended inflation pressure on the tire placard and not the maximum inflation marked on the tire sidewall.

It is not unusual for a vehicle to specify different inflation pressures for the front and rear tires. Many inflation pressures listed for imported vehicles are given in kilopascals (kPa) rather than psi. **Table 46–2** converts kPa to psi.

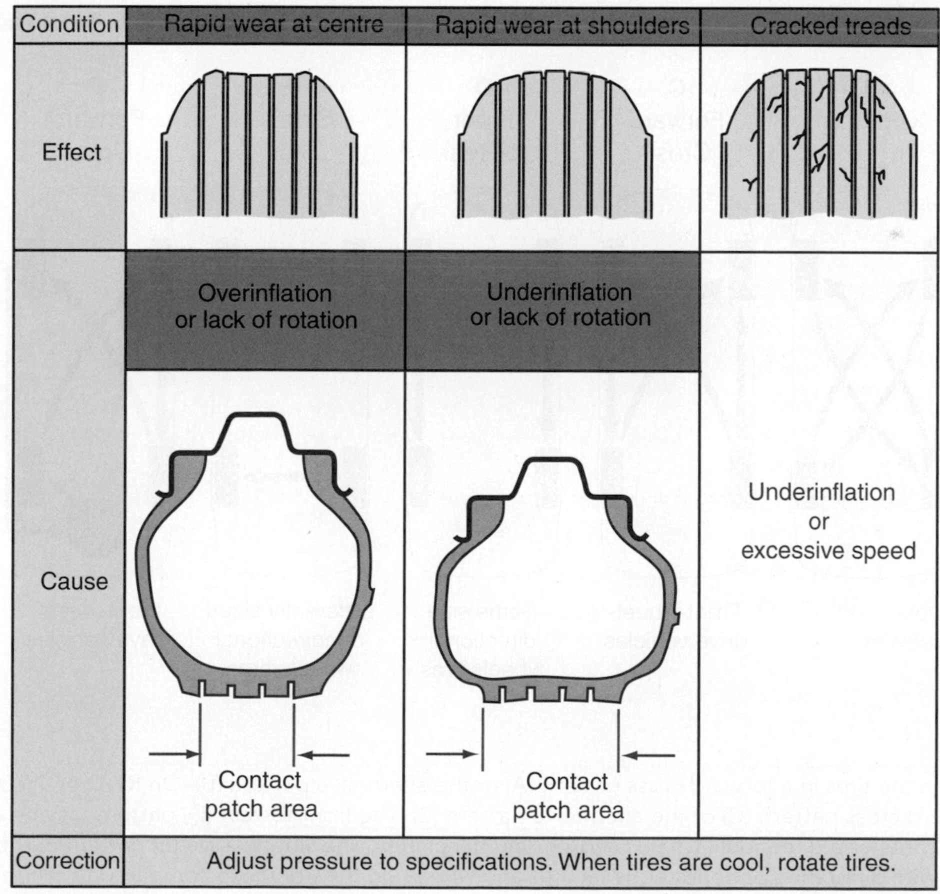

FIGURE 46–14

Effects of inflation on tread contact.

TABLE 46–2 INFLATION PRESSURE CONVERSION (KILOPASCALS TO PSI)

kPa	psi	kPa	psi
140	20	215	31
145	21	220	32
155	22	230	33
160	23	235	34
165	24	240	35
170	25	250	36
180	26	275	40
185	27	310	45
190	28	345	50
200	29	380	55
205	30	415	60

Conversion: 6.9 kPa = 1 psi.

Only a few older vehicles were equipped with tire inflation monitoring systems. Some current vehicles actually use part of the vehicle's anti-lock brake system to monitor tire pressures. The wheel speeds are closely monitored and constantly compared to each other. If the speed of one wheel varies from the others during non-braking periods, the ABS system will determine a low tire inflation condition and illuminate a warning light on the dash. The more common systems have air pressure sensors strapped around the drop centre of each wheel, or special sensor-equipped valve cores. The sensors monitor the pressure in the tire. When the pressure is below or above a specified range, the vehicle's computer causes a warning light on the dash to illuminate. This alerts the driver to the problem. Tire inflation monitoring systems are now mandatory on all new vehicles as of the 2007 model year.

TIRE ROTATION To equalize tire wear, most car and tire manufacturers recommend that the tires be rotated. Remember that front and rear tires perform different jobs and can wear differently, depending on driving habits and the type of vehicle. To equalize wear, it is recommended that tires be rotated as illustrated in **Figure 46–15**. Most radial tires should

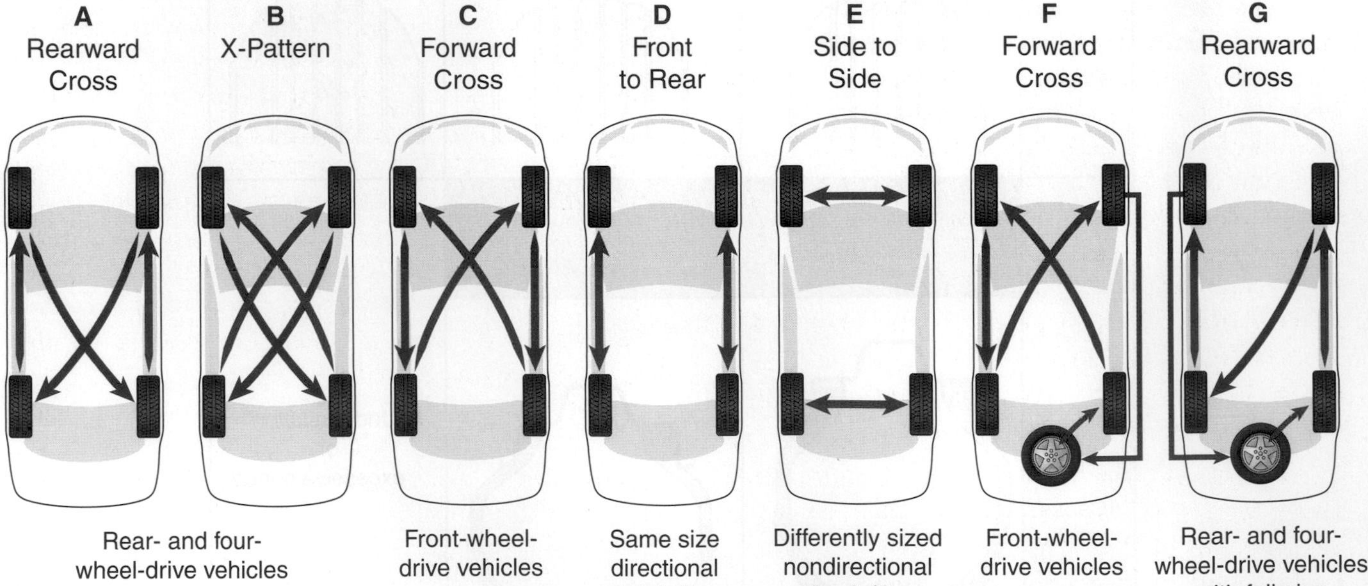

FIGURE 46–15

On FWD cars, rotate the tires in a forward cross pattern (A) or the alternative pattern (B). On RWD or 4WD vehicles, rotate the tires in a rearward cross pattern (C) or the alternative pattern (B). The front-to-rear (D) pattern may be used for vehicles equipped with the same-size directional wheels and/or directional tires. The side-to-side (E) pattern may be used for vehicles equipped with different-size nondirectional tires and wheels on the front axle compared to the rear axle. On FWD cars with a full-size matching spare, rotate the tires in a forward cross pattern (F). On RWD or 4WD vehicles with a full-size matching spare, rotate the tires in a rearward cross pattern (G).

Based on "Tire Rotation Instructions", tirerack.com

be initially rotated at 12 000 km (7500 mi.) and then at least every 24 000 km (15 000 mi.) thereafter. It is important that directional tires are kept rotating in the correct direction after rotating them. This means the tires may need to be dismounted from the wheel, flipped, and reinstalled on the rim before being put on the other side of the car. Many auto shops keep a record of tire rotation periods so that they can notify their customers when it should be done next.

SHOP TALK

Although the rotation patterns shown in Figure 46–15 can serve as the general rule, always refer to the service information. Most vehicles equipped with radial tires should have their tires rotated in a crisscross pattern. The correct rotation method depends on the design of the vehicle's suspension and steering systems.

For best winter driving results, a set of four snow tires should be installed. Although this is not the most common practice, traction on both the drive and steer axles is greatly improved. When installing only two snow tires, these should be installed on the drive axle, with the two best remaining tires installed on the other axle. When the snow tires are removed, rotate the best tires to the drive axle and the stored tires to the other axle. Snow tires should be mounted on a separate set of rims to reduce the chance of damage to the tire bead from mounting and dismounting before and after each winter season. Do not rotate studded tires. Always remount them in their original positions.

When storing tires, lay them flat on a clean, dry, oil-free floor. Keep them away from ozone, which comes from the electrical sparking frequently produced by electric motors. Store them in the dark. Direct sunlight is hard on tires.

TREAD WEAR Most tires used today have built-in tread wear indicators to show when they need replacement. These indicators appear as 12 mm (½ in.) wide bands when the tire tread depth wears to 1.5 mm (2⁄32 in.). When the indicators appear in two or more adjacent grooves at three locations around the tire, or when cord or fabric is exposed, tire replacement is necessary.

If the tires do not have tread wear indicators, a tread depth indicator **(Figure 46–16)** quickly shows in millimetres and/or 32nds of an inch how much tire tread is left. When only 1.5 mm (2⁄32 in.) is left, it is time to replace a tire.

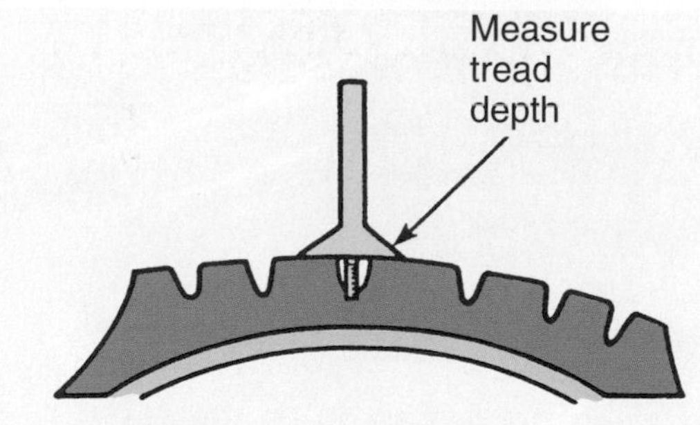

FIGURE 46–16
Checking tread depth.

Proper wheel alignment allows the tires to roll straight without excessive tread wear. The wheels can go out of alignment from striking raised objects or potholes. Misalignment subjects the tires to uneven and/or irregular wear.

GO TO Chapter 49 for more information about wheel alignment.

Tire Pressure Monitor (TPM)

The U.S. DOT National Highway Traffic Safety Administration (NHTSA) has developed a federal motor vehicle safety standard that requires the installation of a **tire pressure monitoring (TPM) system** on all 2007 and newer passenger cars, trucks, multipurpose passenger vehicles, and buses with a gross vehicle weight rating of 4500 kg (10 000 lb.) or less, except those vehicles with dual wheels on an axle. This safety standard is not mandatory in Canada. Many post-2008 domestic and foreign vehicles purchased in Canada are equipped with a TPM system due to North American base model runs or consumer selection of this option. The monitoring systems must illuminate a warning light to warn the driver of an under-inflated tire if one or more tires is at least 25 percent below the recommended cold-inflation pressure. As a result of this standard, two basic TPM designs are being used.

The most commonly used system is referred to as a direct system. In this system, an air pressure sensor is strapped around the drop centre of each wheel **(Figure 46–17A)**, or the sensor is attached to a special tire valve **(Figure 46–17B)**. The pressure sensor measures the tire's inflation pressure and relays this information to the vehicle via radio waves. These signals are picked up by separate body-mounted antennas for each wheel. A central electronic control unit processes the signals from the four wheels and reports any variations to the system.

(A)

(B)

FIGURE 46–17

(A) A TPMS strapped to the inside of the wheel. (B) A tire pressure monitor and air valve assembly.

The TPM checks the inflation pressures in all four tires at frequent periodic intervals. The TPM's sensors keep track of the tire pressures both when the vehicle is moving and when it is stationary. When the TPM detects changes in any tire's inflation pressure, it responds by triggering a warning light on the instrument panel.

A typical direct TPM system has the following components:

- *Tire pressure warning transmitter and air valve.* This is a single unit with a built-in battery that measures tire pressure and temperature and transmits a signal and ID number for that particular tire.

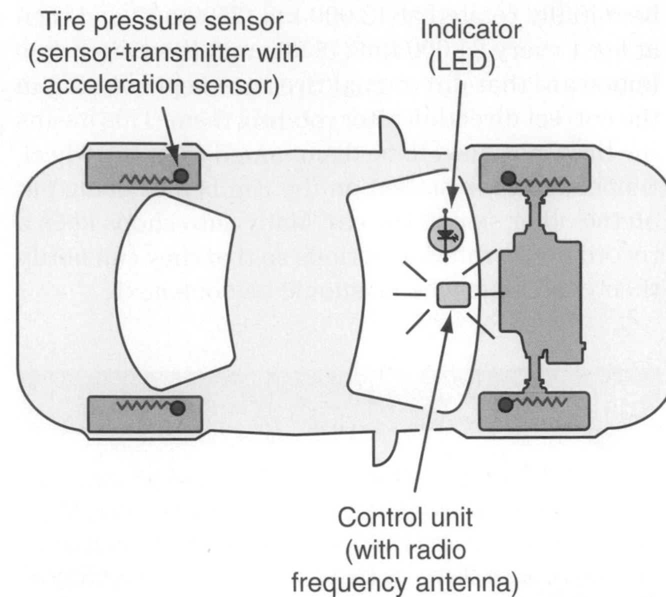

FIGURE 46–18

Basic components for a direct TPM system.

- *Tire pressure warning antenna and receiver.* This unit **(Figure 46–18)** receives and transmits the signals from the transmitters to the tire pressure warning control unit.
- *Tire pressure warning control unit.* This unit receives the signal from the receiver. If the measured air pressure is equal to or lower than a specified value, this unit transmits a signal, causing the air pressure warning light to illuminate.
- *Tire pressure warning light.* Located in the instrument cluster, this unit informs the driver of low tire pressure or a problem in the system.
- *Tire pressure warning reset switch.* Some models have a tire pressure warning reset switch. This unit is used after sensor, tire, or wheel replacement. It is used to allow the control unit to relearn the system.

In an attempt to meet the TPM standard without the cost of a direct system, some manufacturers use an indirect system. These systems do not use pressure sensors; rather, they rely on the inputs from wheel speed sensors. These signals have been used for ABS or other systems. With indirect TPM, the PCM is reprogrammed to use those signals to identify when a tire has lost air pressure. Indirect systems are also used on some older vehicles with run-flat tires. This was necessary because run-flat tires without air appear normal. The driver needs to be alerted to the loss of air.

The input signals from the wheel speed sensors are used to compare the rotational speeds of the four tires. When a tire loses or gains air pressure, it will roll at a slightly different number of revolutions per mile than the other three tires. When the computer senses this difference, a warning lamp will light.

Indirect systems are not as effective as direct systems, however. These systems cannot tell the driver which tire has low pressure. They also are not capable of informing the driver when all four tires are losing air pressure. This commonly happens when the outside temperature drops.

WARNING SYSTEM The TPM warning system can vary from a warning lamp to a graphic display that shows which tire is low on pressure. Some systems allow the driver to monitor the current air pressure of each tire.

If the PCM detects a problem with the TPM system, a warning lamp or message will appear in the instrument cluster. After tires have been rotated or replaced, the system may need to relearn tire position. During this time, a message or warning lamp will be illuminated. Once the relearning process is completed, the message will turn off.

If the system is working correctly, the TPM lamp and the low tire pressure lamp should illuminate for about two seconds when the ignition is turned on. If the lamps do not turn off, there is a problem in the system. The system will typically set a DTC when there is a problem with the system or when it detects low pressure in any of the four tires.

Testing a TPM System

The TPM system in most vehicles is tied directly to the PCM, therefore, faults cause DTCs to set. TPM system data can often be monitored and DTCs retrieved with a scan tool **(Figure 46–19)**. Special tools are required to accurately test and locate the problem tire(s). A TPM sensor tool is a wireless tool that may be used with a scan tool for diagnosing sensors and allowing the system to relearn when a part has been replaced. Although trouble codes can identify a system problem, the DTCs do not indicate the exact location of the troubled tire. The TPM sensor tool is used to find the tire or sensor that is causing the problem.

The TPM sensor tester is used to reset the system, which may be needed after tires are rotated, after tires or wheels are replaced, after repairs are made to the system, and when the vehicle's battery was low or replaced. The tester activates the sensors, and the transmitted data from them can be observed. Most systems require placing the tool near the sensor and then activating the sensor via the tool **(Figure 46–20)**. The sensor then broadcasts information such as the sensor ID, tire pressure and temperature, and battery information. If a sensor fails to communicate, the internal battery may be dead, or the sensor is faulty. Any problem with the sensor requires sensor replacement.

Kia 1993-2010

Data Display

Number Of Tread Warnings	3
Warning #1 Status	HISTORIC
Warning #1 Type	INFLATION
Warning #1 Sensor ID	8072900E
Warning #1 Pressure(PSI)	23.5
Warning #1 Temperature(°F)	32
Warning #2 Status	HISTORIC
Warning #2 Type	INFLATION
Warning #2 Sensor ID	8072900E
Warning #2 Pressure(PSI)	23.5
Warning #2 Temperature(°F)	34

1 2 3 4

FIGURE 46–19

TPM data displayed on a scan tool.

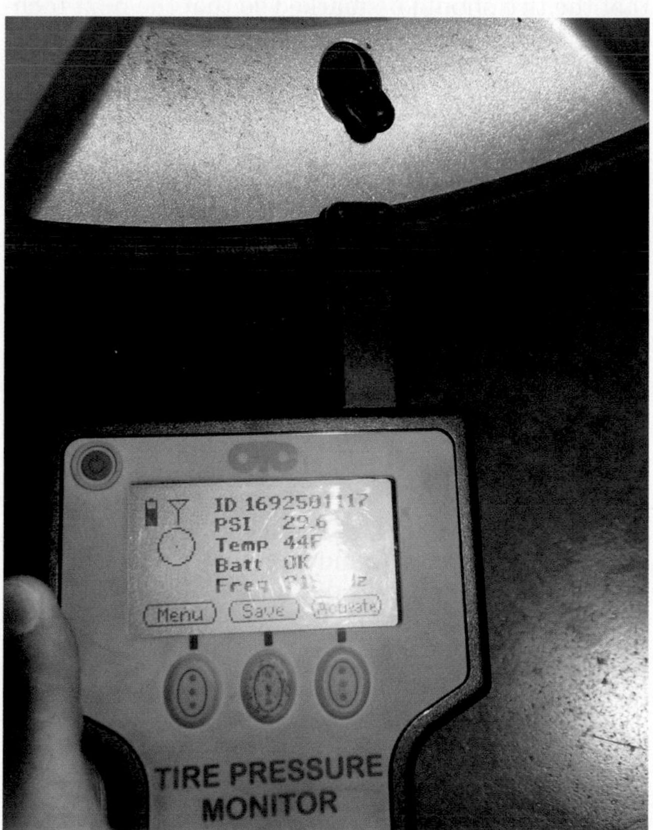

FIGURE 46–20

A TPM tester.

TIRE REPAIR

The most common tire problem besides wear is a puncture. When properly repaired, the tire can be put back in service without the fear of an air leak recurring. Punctures in the tread area are the only ones that should be repaired or even attempted to be repaired. Never attempt to service punctures in the tire's shoulders or sidewalls unless otherwise stated by the tire manufacturer. In addition, do not service any tire that has sustained the following damage:

- Bulges or blisters
- Ply separation
- Broken or cracked beads
- Fabric cracks or cuts
- Wear to the fabric or visible wear indicators
- Punctures larger than 6 mm (¼ in.) in diameter

Some car owners attempt to seal punctures with tire sealants. These sealants are injected into the tire through the valve stem. Sometimes the chemicals in the sealant do a great job sealing the hole; other times they fail. The sealants should never be used and will not work on sidewall punctures. Some of the sealants are very flammable and carry a warning that the tire should be marked so that the next technician knows the sealant has been used.

WARNING!

Tire sealants injected through the valve stem can produce wheel rust and tire imbalance.

To locate a puncture in a tire, inflate it to the maximum inflation pressure indicated on its sidewall. Then submerge the tire/wheel assembly in a tank of water or sponge it with a soapy water solution. Bubbles will identify the location of any air leakage.

Mark the location of the leak with a crayon so that it can be easily found once the tire is removed from the wheel. Also use the crayon to mark the location of the valve stem so that original tire and wheel balance can be maintained after the tire is put back on the wheel.

The proper procedure for dismounting and remounting a tire is illustrated in Photo Sequence 38. Do not use hand tools or tire irons alone to change a tire because they might damage the beads or wheel rim. When mounting or dismounting tires on vehicles using aluminum or wire spoke wheels, it is recommended that the tire changer manufacturer be contacted about the accessories that are required to protect the wheel's finish.

WARNING!

Once a tire puncture has been repaired, many tire manufacturers state that the tire no longer retains its speed rating. This is because the puncture compromises the tire and could ultimately fail under the stress of high-speed driving. Because of this, the work order must be noted and the customer informed that the speed rating is no longer applicable to the repaired tire.

TPM Sensors

If the wheel/tire assembly has a direct TPM system, the sensor can be removed after the air pressure has been released from the tire. This should be done before the tire is removed from the rim. Unbolt the air valve assembly and allow it to drop into the tire. Before servicing the tire or wheel, remove the sensor. After tire and/or wheel repairs have been made, install the sensor with a new rubber O-ring or seal and aluminum retaining nut. The retaining nut must be torqued to specifications.

SHOP TALK

Some air valves for TPM systems are made of brass and have an aluminum valve stem. Over time, the valves will experience galvanic corrosion and will seize within the brass valve core. If the tire you are working on shows these problems, replace this unit with a nickel-plated valve core.

Repair Methods

Once the tire is off the wheel and the cause of the puncture is removed, the tire can be permanently serviced from the inside using a combination service plug and vulcanized patch. Although the service kit's instructions should always be followed, there are some general guidelines that help make a good, permanent patch of the puncture. The following methods are the most common methods used to repair a tire.

PLUG REPAIR The head-type plug **(Figure 46–21)** is commonly used. A plug that is slightly larger than the size of the puncture is inserted into the hole from the inside of the tire with an insertion tool. Before doing this, insert the plug into the eye of the tool and coat the hole, plug, and tool with vulcanizing fluid.

While holding and stretching the long end of the plug, insert it into the hole. The plug must extend above both the tread and inner liner surface. If the

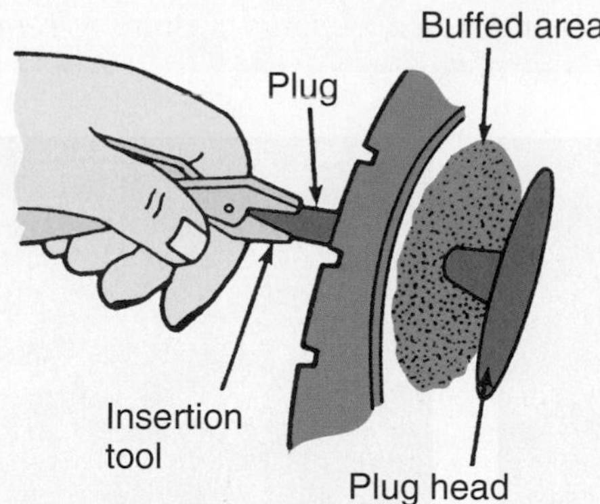

FIGURE 46–21

A plug for a radial tire.

plug pops through, throw it away and insert a new plug. Once the plug is in place, remove the tool, and trim off the plug 0.7 mm (1/32 in.) above the inner surface. Be careful not to pull on the plug while cutting it.

COLD PATCH REPAIR When using a cold patch, carefully remove the backing from the patch. Spread vulcanizing fluid on the punctured area. Let it dry, and then centre the patch base over the punctured area. Run a stitching tool over the patch to help bind it to the tire.

WARNING!

When repairing radial tires, use only a patch specially approved for that application. These special patches have arrows that must be lined up parallel to the radial plies.

HOT PATCH REPAIR A hot tire patch application is similar to a cold patch. The difference is that the hot patch is clamped over the puncture, and heat is applied to the patch to make it adhere.

Installation of Tire/Wheel Assembly on the Vehicle

A wheel should be carefully inspected each time a tire is to be mounted on it. The major causes of wheel failure are improper maintenance, overloading, age, and accidents, including pothole damage. Wheels must be replaced when they are bent, dented, or heavily rusted; have leaks or elongated bolt holes; and have excessive lateral or radial runout. Wheels with lateral or radial runout greater than specifications can cause high-speed vibrations. Wobble or shimmy caused by a damaged wheel eventually damages the wheel bearings. Stones wedged between the wheel and disc brake rotor or drum can unbalance the wheel. Also check the lug nuts to be sure that they are set according to the torque given in the vehicle's service information. Loose lug nuts can cause shimmy and vibration and can also distort the stud holes in the wheels. Other wheel mounting problems are caused by improperly positioning the wheel on the wheel hub or by using an improper tightening sequence (pattern).

SHOP TALK

When mounting new tires, always install new valve stems. The life of tire rubber is close to the life of the valve stem rubber. Most stems are the snap-in type. These are installed from inside the wheel with a pulling tool. Make sure that the stem is properly seated. Another style of stem has a retaining nut that must be removed when pulling off the old stem. Be sure to completely tighten the new nut.

Before reinstalling a tire/wheel assembly on a vehicle, inspect the wheel bearings as described later in this chapter, then clean the axle/rotor flange and wheel bore with a wire brush or steel wool. Coat the axle pilot flange with disc brake caliper slide grease or an equivalent. Place the wheel on the hub. Install the locking wheelcover pedestal (if used) and lug nuts, and tighten them alternately to draw the wheel evenly against the hub. They should be tightened to a specified torque and sequence **(Figure 46–22)** to avoid distortion. Many tire technicians snug up the lug nuts. Then when the car is lowered to the floor, they use a torque wrench for the final tightening.

WARNING!

Overtorquing and uneven torquing of the lug nuts is the most common cause of disc brake rotor distortion. Also, an overtorqued lug distorts the threads of the lug and could lead to premature failure.

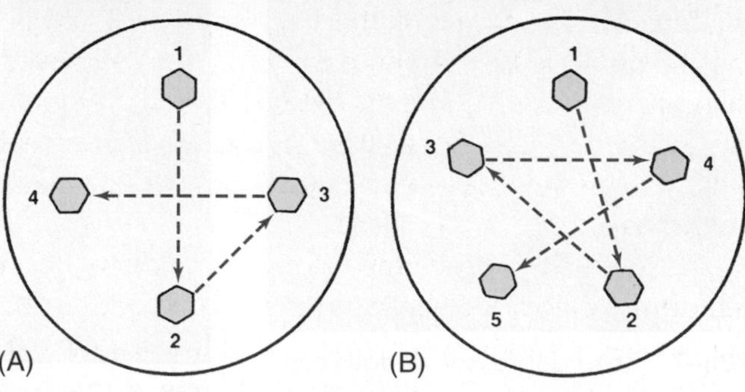

FIGURE 46–22

The lug nut tightening sequence for (A) a four-lug wheel and (B) a five-lug wheel.

PHOTO SEQUENCE **38**

DISMOUNTING AND MOUNTING A TIRE ON A WHEEL ASSEMBLY

P38–1 Dismounting the tire from the wheel begins with releasing the air, removing the valve stem core, mounting the tire on the tire machine, and unseating the tire from its rim. The machine's bead-loosening roller does the unseating. The technician positions the upper and lower rollers through a control pad.

P38–2 Once both sides of the tire are unseated, lubricate the tire beads, and place the machine's tool head under the upper bead.

P38–3 Rotate the tire and wheel assembly while allowing the tool head to pull the upper bead over the rim for dismounting.

P38–4 Use the tool head again to pull the lower bead over the rim and rotate the wheel assembly to dismount the tire from the rim.

P38–5 After the tire is totally free from the rim, remove the tire. Inspect the wheel for rust, corrosion, or damage. If necessary, prepare the wheel for mounting the tire by using a wire brush to remove all dirt and rust from the sealing surface. Apply rubber compound to the bead area of the tire.

P38–6 A combination of the tool head easing the tire bead over the rim and the bead press arm pushing down on the tire will mount each bead onto the rim when the rim is rotated.

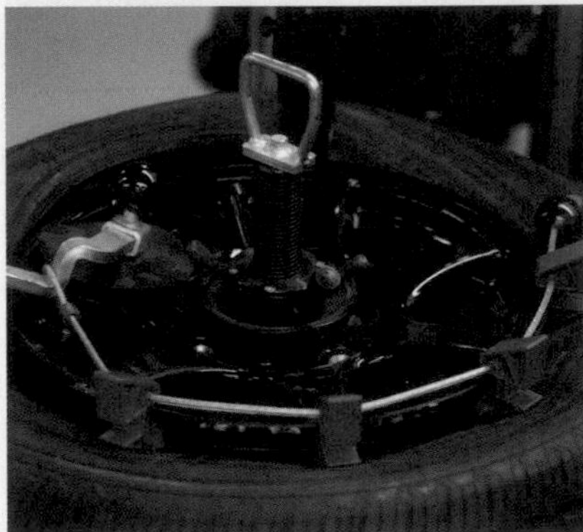

P38–7 Bead depressor wedges can be used when mounting a tire with stiff side walls. These wedges hold the bead down in the drop centre section, allowing easier bead mounting.

P38–8 The air blast attachment can be used to supply a large burst of compressed air during inflation to assist in seating the beads.

P38–9 Reinstall the valve stem core and inflate the tire to the recommended inflation.

Be sure the wheels and hub are clean. To clean aluminum wheels, use a mild soap and water solution and rinse thoroughly with clear water. Do not use a steel wool abrasive cleaner or strong detergent containing high alkaline or caustic agents because they might damage the protective coating and cause discoloration. Once the vehicle is on the ground, check and adjust the air pressure in all tires.

Inflation

After the tire is mounted to the wheel, inflate the tires to the recommended pressure. Also check to see if any air is leaking from the beads or the point of repair.

NITROGEN TIRE INFLATION Many tire experts recommend that tires be filled with nitrogen rather than compressed air. Other experts say there is no need to do this if owners watch the air pressure in their vehicle's tires. The idea behind using nitrogen is simple: Nitrogen molecules are larger than air molecules. Therefore, nitrogen is less likely to leak out of a tire. Those in favour of using nitrogen claim that nitrogen-filled tires stay inflated about three times longer than air-filled tires. Nitrogen also helps keep tires cooler while travelling on the highway. This means the air pressure stays more constant and is less likely to leak out. The supposed result of these advantages is safer and longer-lasting tires. The idea of using nitrogen in tires is not new. Race cars, commercial airliners, and trucks have used nitrogen-filled tires for many years.

Nitrogen inflation requires a nitrogen filling station. Typically, a fill/purge cycle fills the tire, purges the air, and refills the tire with nitrogen until the desired amount of nitrogen is reached. Once a tire is filled with nitrogen, a green valve stem cap is installed to alert others to refill the tire with nitrogen only **(Figure 46–23)**.

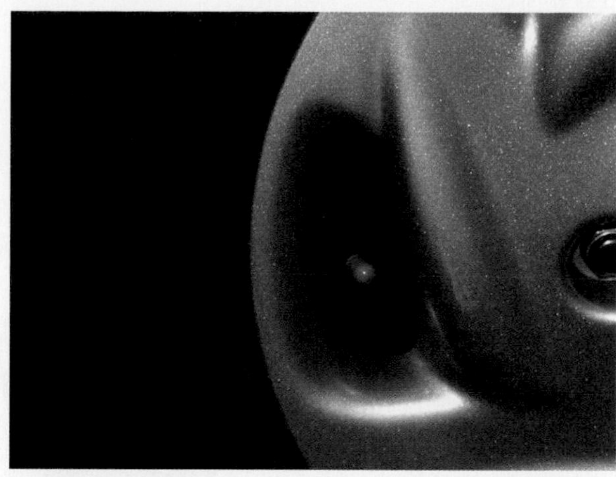

FIGURE 46–23

A nitrogen-filled tire.

TIRE/WHEEL RUNOUT

A tire that is off centre is said to run out; that is, it has **radial runout** or eccentricity. One that wobbles side to side is said to have **lateral runout**. If a tire with some built-in runout is mismatched with a wheel's runout, the resulting total runout can exceed the ability of the balance weights to correct the problem. For this reason, part of a wheel balance check should be a check for excessive runout. Sometimes, tires or wheels can be remounted to lessen or correct runout problems.

To avoid false readings caused by temporary flat spots in the tires, check runout only after the vehicle has been driven. Visually inspect the tire for abnormal bulges or distortions. The extent of runout should be measured with a dial indicator. All measurements should be made on the vehicle with the tires inflated to recommended load inflation pressures and with the wheel bearing adjusted to specification.

Measure tire radial runout at the centre and outside ribs of the tread face. Measure tire lateral runout just above the buffing rib on the sidewall **(Figure 46–24)**. Mark the high points of lateral and radial runout for future references. On radial tires, radial runout must not exceed 2.05 mm (0.081 in.), and lateral runout must not exceed 2.5 mm (0.099 in.).

If total radial or lateral runout of the tire exceeds specified limits, it is necessary to check wheel runout to determine whether the wheel or tire is at fault. Wheel radial runout is measured at the wheel rim just inside the wheel cover retaining nibs. Wheel lateral runout is measured at the wheel rim bead flange just inside the curved lip of the flange. Wheel radial runout should not exceed 0.88 mm (0.035 in.), and wheel lateral runout should not exceed 1.0 mm (0.040 in.). Mark the high points of radial and lateral runout for future reference.

If total tire runout, either lateral or radial, exceeds the specified limit, but wheel runout is within the specified limit, it might be possible to reduce runout to an acceptable level. This is done by changing the position of the tire on the wheel so that the previously marked high points are 180° apart. Many computer wheel balancing machines

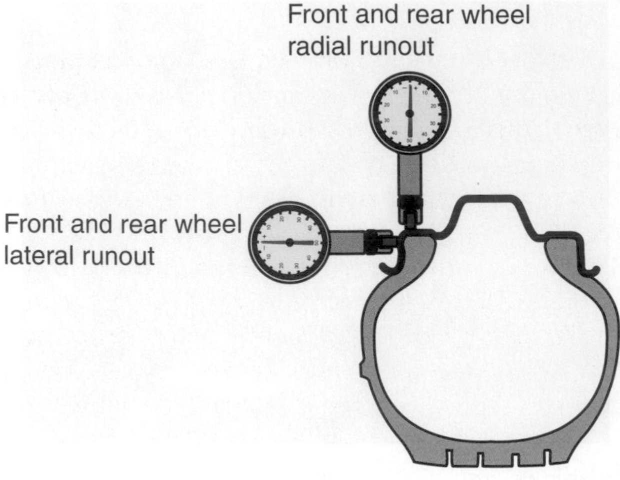

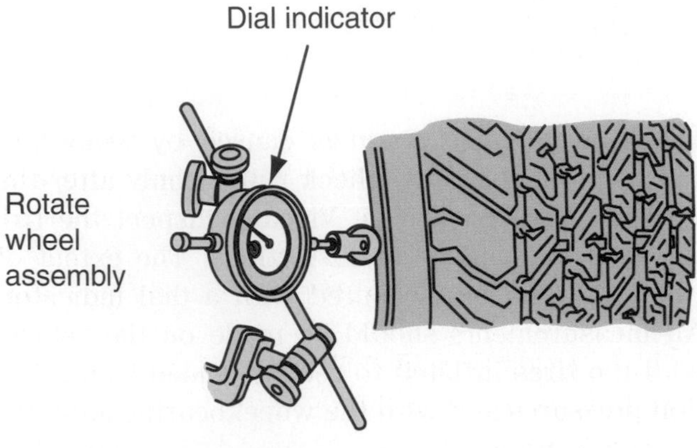

FIGURE 46–24

Checking wheel runout.

©Cengage Learning 2015

can detect wheel and tire runout and determine the best way to remount the tire on the wheel to reduce total runout.

Tire Pull

It is not uncommon for a tire to cause a pull to one side when driving. This is caused by conicity, meaning the tire has a cone shape. As the tire rotates, it tries to go in the direction of the narrow end of the cone shape, pulling the vehicle either left or right. Unfortunately, this problem occurs during tire construction and cannot be fixed. Depending on the severity of the conicity problem, the tire, if installed on the rear of the vehicle, may not cause a pull until it is moved to the front of the vehicle during a tire rotation. If this occurs, the customer may return to the shop complaining that the car pulls to one side since the tires were rotated. To check if conicity is the cause, switch the front tires from side to side, and test-drive the vehicle. If the pull goes away or pulls in the other direction, conicity is the problem.

TIRE REPLACEMENT

Tires should be replaced when they are worn or heavily damaged. In addition, because modern tires last much longer than those even 20 years ago, tire age needs to be considered. Tires, especially those that remain outside all year, are subject to dry rot and cracking. Over time, this weakens the tire and can lead to tire failure. Although there is not a current replacement age requirement in the Canada, many tire and rubber manufacturers and recommend tire replacement based on age alone at six to ten years. Tires that reach six years of service should be very carefully inspected for wear, damage, and rot. Once a tire reaches ten years, it should be replaced.

Replacement tires should match the tires that were on the vehicle originally. This is the preferred choice unless a change in appearance or handling is desired. Vehicle manufacturers know how important the right tire is. They spend a great deal of time developing a suspension system that works in the way they believe it should. Part of that development is tire design. However, the tire choice by manufacturers is a compromise—a compromise between characteristics. Manufacturers choose the characteristics they believe owners of that model vehicle want. If the owner is happy with the vehicle, recommend that the replacement tires match the ones the vehicle was originally sold with. When replacing tires, installing tires that do not meet the requirements of the original equipment (OE) could place you in legal trouble. If a tire does not meet the OE specifications for the vehicle, such as for load capacity or speed rating, and the replacement tire fails, you could face legal consequences for installing a substandard tire.

Replacing One Tire

In some cases, only one tire needs to be replaced. The usual causes of this are that the tire was damaged due to an accident, a road hazard, or vandalism. Replacing one tire is recommended only if the other tires have a satisfactory amount of tread left. Make sure the replacement tire is the same brand, type, size, and speed rating as the other tires. If the replacement tire is different from the rest, the vehicle can exhibit unsafe handling problems. Also, the replacement tire should have a similar tread design as the remaining tires. This helps reduce noise and other issues that can be caused by dissimilar tread designs. The replacement tire should be mounted on the rear axle, and the tire (of the remaining three) with the most tread depth should be mounted on the opposite side of the axle.

Replacing Two Tires

If there is a need to replace two tires, and the other two have good treads, the replacement pair should be mounted on the rear axle (even on FWD vehicles). The reason for this practice is to allow the driver to maintain control by preventing the rear axle from not being able to follow the front axle due to much better steering traction. This is usually due to hydroplaning, which produces dangerous oversteer when it takes place on the rear axle. The replacement tires should match the remaining pair of tires as closely as possible.

Four-Wheel-Drive and All-Wheel-Drive Vehicles

Because of the increased popularity of small SUVs and other vehicles using all-wheel drive, replacing tires as a set of four is now recommended. These all-wheel-drive systems use a centre differential or viscous clutch to apply power to a set of wheels when there is a speed difference between the front and rear wheels. If only a pair of tires is replaced, this leaves a set of worn tires on one axle. The worn tires will have a slightly smaller diameter than the new tires on the other axle. Even this small amount of size difference can cause a speed difference between the front and rear tires. The difference in size creates a difference in wheel rotation speeds, which can damage the differential or viscous clutch over prolonged use.

Changing Tire and/or Wheel Size

An owner may want more emphasis on handling or fuel economy and may desire a different type of tire and/or wheel. There are several other factors that may dictate a change in tire size, and the customer may come to you for advice. Perhaps one of the most important considerations is that the tire must be able to carry the weight of the vehicle. The load-carrying capacity of a tire must be the same as or higher than the OE tire. Changing tires from one aspect ratio to another also changes the sectional width, which relates to the load-carrying capacity of the tire.

Most tire width changes affect the overall diameter of the tire. A change in the tire's outside diameter will cause a change in the overall gear ratio and will affect the accuracy of the speedometer and odometer. A change in tire diameter or aspect ratio may also affect overall driveability. This is due to false readings from the vehicle or wheel speed sensors. On passenger cars and mini-vans, a 3 percent or less (20 mm [3/4 in.] or less) change in tire diameter is acceptable. Most SUVs and pickups can handle a change of as much as 15 percent. The overall diameter of a tire can be calculated.

A vehicle may have the ability to change the tire size that is programmed into the on-board computer system. Typically, the option to change the tire size is to one of the sizes offered as an option from the vehicle manufacturer.

PLUS SIZING A way to change the contact area of a tire without seriously affecting its overall diameter is using the plus-sizing system. This system is based on the overall diameters of a combination of different-size tires and wheels. The system requires much research but it is the best way to achieve the desired results. For example, the customer wants to have a wider tread area **(Figure 46–25)**. The OEM tire was a 195/75-14 tire mounted on a 14 × 6 inch wheel. The overall diameter of this assembly is 647.7 mm (25.5 in.). There are three available wheel/tire

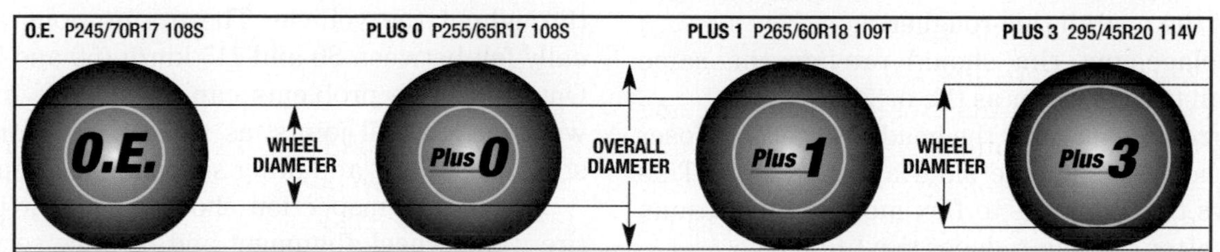

FIGURE 46–25

The effects of plus sizing.

Used with permission of Toyo Tires Canada. Found at http://toyotires.com/tires-101/plus-sizing-tires

combinations that closely match that diameter. If a 205/65-15 tire is mounted on a 15 × 7 inch wheel, the width increases by 10 mm (0.39 in.), whereas the diameter stays the same. If a 16 × 7.5 wheel is used along with a 225/55-16 tire, the width increases by 30 mm (1.18 in.) and the overall diameter increases by only 5.1 mm (0.2 in.). Going even wider, if a 235/45-17 tire is mounted on a 17 × 8 wheel, the tire's width increases by 40 mm (1.57 in.) and the overall diameter decreases by 5.1 mm (0.2 in.). Although the latter two changes do not match the OEM diameter, they are certainly within the 3 percent rule.

ADDITIONAL POINTS Here are some additional important points to consider:

- Handling improvement typically comes from more tire contact on the road, and fuel economy increases with less.
- Increasing tread width and tire contact can affect steering effort as more tire is in contact with the road surface.
- Tires of different sizes, constructions, and wear may affect handling, stability, and fuel economy.
- Too wide a tire may rub against the body or suspension.
- Radial tires should never be mixed with another type of tire on the same vehicle.
- All tires on the vehicle should be the same size, construction, tread design, and speed rating unless the vehicle was otherwise equipped by the OEM.
- Tires should be replaced with ones of the same or higher speed rating. Speed ratings should not be downgrades from original equipment ratings.
- A hard tread will provide long wear and low rolling resistance but will also have poorer traction.
- An aggressive tread pattern may provide resistance to hydroplaning or better traction in the snow, but they are noisier on dry surfaces.
- A tire with stiff sidewalls will increase high-speed stability and improve handling, but it will make the overall ride rougher.
- A replacement rim should provide the same overall tire diameter as the original.
- A narrower rim pulls the beads of the tire closer together, causing the sidewalls to curve. This allows the sidewalls to flex more, which results in a softer ride but reduces tire life.
- A wide rim increases the distance between the beads, which stiffens the sidewalls and provides a harsher ride and shorter tire life. However, it will improve the handling of the vehicle.
- A replacement tire should have, at minimum, the qualities of the tires installed when the vehicle was new. Replacing tires with ones with a reduced load capacity, traction, or temperature rating is not recommended.

CALCULATING TIRE DIMENSIONS When replacing tires with other than the original tire size, you may need to calculate the dimensions of a desired replacement tire. Like everything else, there are formulas to make these calculations.

To determine the sectional height of a tire, multiply its aspect ratio by the sectional width.

$$\text{Width} \times \text{Aspect ratio} = \text{Sectional height}$$

To determine the overall diameter of a tire, multiply the sectional height by two (this is called the combined sectional height because there are two), and then add the diameter of the wheel.

$$\text{Combined sectional height (Sectional height} \times 2\text{)} + \text{Wheel diameter} = \text{Tire diameter}$$

TIRE/WHEEL ASSEMBLY SERVICE

For most tire/wheel service, the assembly must first be removed from the vehicle. The wheel and the tire must be separated whenever tires are replaced or repaired. The wheel and tire mount on the hub flange, which may be on the end of an axle or part of a brake rotor or drum. The wheel is held against the hub and drum or rotor by the wheel nuts or bolts. Most cars and trucks use lug nuts that thread onto studs. However, many European cars use wheel bolts that thread into holes in the hub flange to secure the wheels in place.

Tire/Wheel Balance

An out-of-balance condition can cause steering wheel shimmy and vibrations that can be felt throughout the vehicle. These vibrations are typically felt between 80 and 115 km/h (50 and 70 mph). Out-of-balance problems can also cause increased wear on the ball joints, as well as deterioration of shock absorbers and other suspension components.

Should an inspection show uneven or irregular tire wear, wheel alignment and balance service is a must. Wheel balancing distributes weights along the wheel rim, which counteract heavy spots in the wheels and tires and allow them to roll smoothly without vibration. The wheel weights are adhered to the wheel or are clipped over the edge of the wheel's rim. There are two types of wheel imbalance: static and dynamic.

STATIC BALANCE **Static balance** is the equal distribution of weight around the wheel. Wheels that are statically unbalanced cause a bouncing action called **wheel tramp** that gets worse with speed. Even a slight imbalance will magnify with speed. A 28 g (1 oz.) imbalance will create an imbalance force of 4.5 kg (10 lb.) at 110 km/h (70 mph). This condition eventually causes uneven tire wear. As the name implies, static balance means balancing a wheel at rest. This is done by adding a compensating weight. A statically unbalanced wheel tends to rotate by itself until the heavy portion is down. A bubble balancer is used to statically balance a tire and wheel. When it is placed on the balancer, any imbalance moves the bubble off centre.

Many equipment manufacturers recommend static balancing a wheel at equal distances from the centre of the light area. Balance weights are normally hammered on with their holding tabs between the tire bead and rim **(Figure 46–26)**. Wheel weights are not normally hammered onto alloy or mag wheels; rather, special tape weights are adhered to the wheels to balance them.

FIGURE 46–26

A typical wheel weight attached to a wheel.

DYNAMIC BALANCE **Dynamic balance** is the equal distribution of weight on each side of the centreline. When the balanced tire spins, there is no tendency for the assembly to move from side to side. Wheels that are dynamically unbalanced can cause **wheel shimmy** and a wear pattern **(Figure 46–27)**. Dynamic balance, simply stated, means balancing a wheel in motion. Once a wheel starts to rotate and is in motion, the static weights try to reach the true plane of rotation of the wheel because of the action of centrifugal force. In an attempt to reach the true plane of rotation when there is an imbalance, the static weights force the spindle to one side.

At 180° of wheel rotation, static weights kick the spindle in the opposite direction. The resultant side thrusts cause the wheel assembly to wobble or wiggle. When the imbalance is severe enough, as already mentioned, it causes vibration and front-wheel shimmy. These vibrations and shimmies

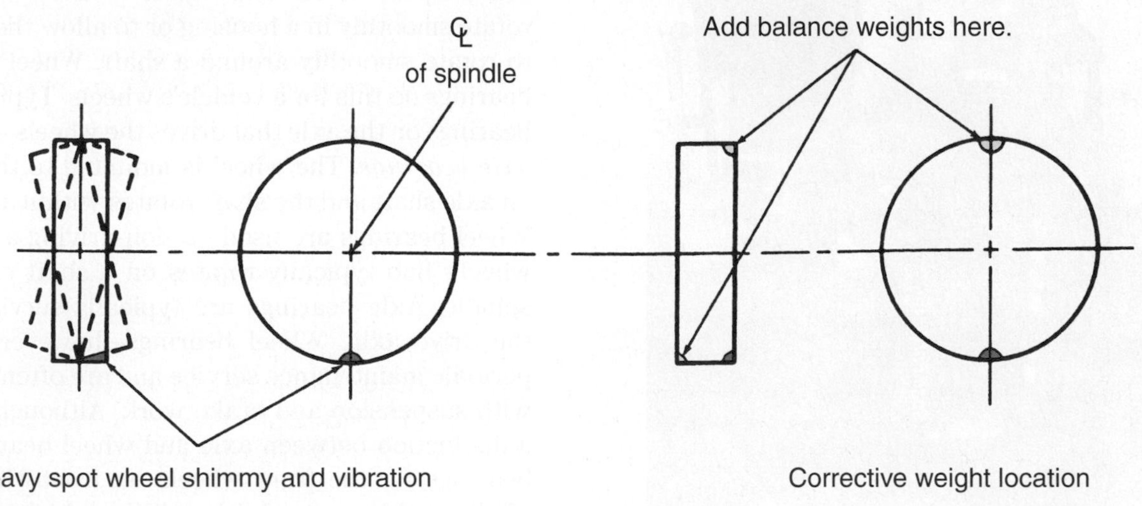

FIGURE 46–27

Dynamic wheel balancing calls for a weight to be attached to the wheel to compensate for a heavy spot on the wheel.

Chrysler Group LLC

tend to become more noticeable at different speeds. A customer may complain that a harsh vibration is present from 85 to 90 km/h (50 to 55 mph) and again even harsher at 110 km/h (70 mph).

To correct dynamic unbalance, equal weights are placed 180° opposite each other, one on the inside of the wheel and one on the outside, at the point of unbalance. This corrects the couple action or wiggle of the wheel assembly. Also, note that dynamic balance is obtained, whereas static balance remains unaffected.

Wheel Balancing

The most commonly used dynamic wheel balancer requires that the tire/wheel assembly be taken off and mounted on the balancer's spindle **(Figure 46–28)**. The machine spins the entire assembly and determines the correct placement of weights to correct any static or dynamic imbalance. The results are shown in the balancer's display, which indicates where and how much weight to apply to the wheel. Newer balancers can help hide adhesive weights behind wheel spokes to give the wheel a cleaner appearance.

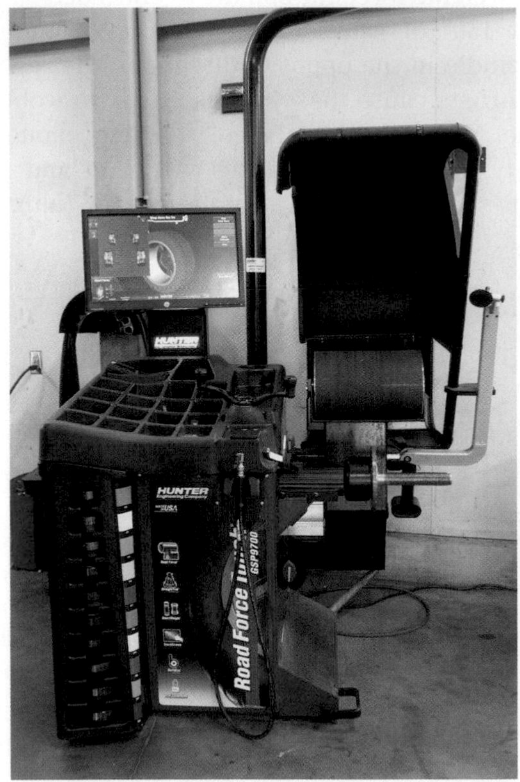

FIGURE 46–28

A computerized road force tire balancer.

Martin Restoule

There are several electronic dynamic/static balancer units available that will permit balancing while the wheel and tire are on the car. A switch on the console sets the machine for either static or dynamic balancing or both. When the wheel balancing assembly is mounted for static balancing, it rotates until the heavy spot falls to the bottom. Weights are added to balance the assembly.

In the dynamic balance mode, the wheel assembly is rotated at low or high speed, depending on the type of balancer. Observing the balance scale, the operator reads out the amount of weight that has to be added and the location where the weights should be placed.

Many newer wheel balance machines also check for wheel and tire runout. These machines may use cameras or lasers to monitor wheel and tire movement while spinning the tire and checking the balance.

ROAD FORCE MEASUREMENTS The balancer shown in Figure 46–28 is designed to also eliminate all causes of vibration due to the wheel and tire assembly. Like many other current wheel balancers, this unit can simulate a road test with a load roller. This roller applies a heavy force on the tire while it is rotating on the balancer. The roller also measures the deflection of the tire as it rolls under pressure. The machine then makes recommendations for the service required to remove all runout and ensure vibration-free operation.

WHEEL BEARINGS

The purpose of all bearings is to allow a shaft to rotate smoothly in a housing or to allow the housing to rotate smoothly around a shaft. Wheel and axle bearings do this for a vehicle's wheels. Typically, the bearings on the axle that drives the wheels are called *axle bearings*. The wheel is mounted to the hub of an axle shaft and the shaft rotates within a housing. Wheel bearings are used on non-driving axles. The wheel's hub typically rotates on a shaft called the spindle. Axle bearings are typically serviced with the drive axle. Wheel bearings, however, require periodic maintenance service and are often serviced with suspension and brake work. Although there is a distinction between axle and wheel bearings, the bearings for the front wheels on front-wheel-drive (FWD) and four-wheel-drive (4WD) vehicles are commonly called *wheel bearings*. Regardless of what they are called, bad bearings will cause handling and tire wear problems.

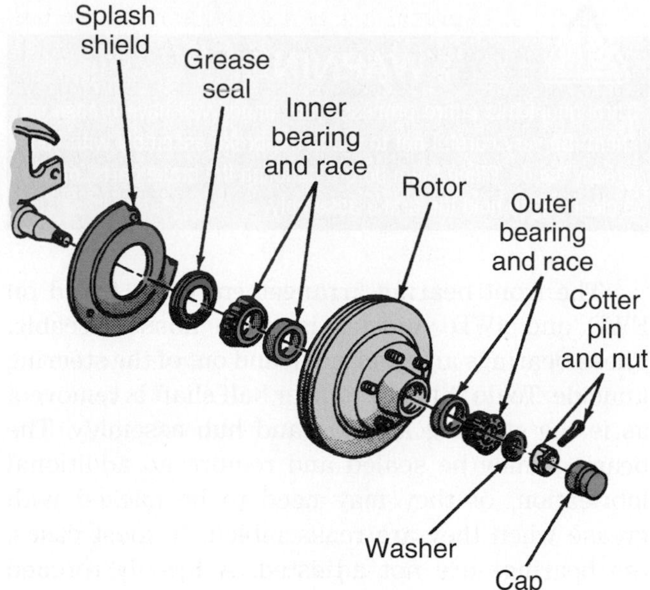

FIGURE 46–29

An exploded view of a typical front wheel bearing assembly for an RWD vehicle.

Chrysler Group LLC

Front Wheel Hubs

Often, the front wheel hub bearing assembly for driven and non-driven wheels is actually two tapered bearings **(Figure 46–29)** facing each other. Each of the bearings rides in its own race. Most front wheel bearings are sealed units and are lubricated for life. They are replaced and serviced as an assembly. Others are serviceable and require periodic lubrication and adjustment.

With tapered roller bearings, except when making slight adjustments to the bearings, the bearing assembly must be removed for all service work. This is done with the vehicle on lifts and the wheel assembly removed. In the centre of the hub is a dust (grease) cap. Using slip-joint pliers or a special dust cap removal tool **(Figure 46–30)**, wiggle the cap out of its recess in the hub. Now remove the cotter pin and nut lock from the end of the spindle. Loosen the spindle nut while supporting the brake assembly and hub. On many vehicles, you will need to remove the brake caliper to remove the brake disc and hub. Once the hub is free to come off the spindle, remove the spindle nut and the washer behind the nut. Move the hub slightly forward, and then push it back. This should free the outer bearing so that you can remove it. Now remove the hub assembly.

A grease seal located on the back of the hub normally keeps the inner bearing from falling out when the hub is removed. To remove the bearing assembly, the grease must be removed first. In most cases, all you need to do to remove the seal is pry and pop it out of the hub. The inner bearing should then fall out. Keep the outer bearing and inner bearing separated if you plan on reusing them.

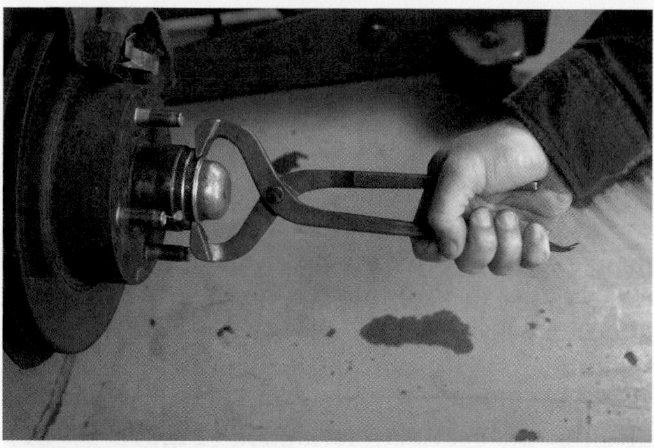

FIGURE 46–30

A special tool for removing a dust cap. If one is not available, use slip-joint pliers.

Wipe the bearings and races, and use a parts cleaner to clean them. While doing this, pay close attention to the condition and movement of the bearings. The bearings need to rotate smoothly. Also visually inspect the bearings and races after they have been cleaned. Any noticeable damage means they should be replaced. Also inspect the spindle. If it is damaged or excessively worn, the steering knuckle assembly should be replaced.

Whenever a bearing is replaced, its race must also be replaced. Races are pressed in and out of the hub. Typically, the old race can be driven out with a large drift and a brass hammer. Once the race has been removed, wipe all grease from the inside of the hub. The new race should be installed with the proper brass driver to prevent race damage.

During assembly, the bearings and hub assembly must be thoroughly and carefully lubricated. Care must be taken not to get grease on the brake disc or on any part that will directly contact the disc. Always use the recommended grease on this assembly (usually NLGI no.2). The grease must be able to withstand much heat and friction. If the wrong grease is used, it may not offer the correct protection, or it may liquefy from the heat and leak out of the seals.

The bearings should be packed with grease. It is important that the grease is forced into and around all of the rollers in the bearing. Merely coating the outside of the bearing with grease will not do the job. A bearing packer does the best job at packing in the grease. If one is not available, force grease into the bearing with your palm.

Install the greased inner bearing into the hub. Install a new grease seal into the hub. To avoid damaging the seal, use the correct size driver to press the seal into the hub. Lubricate the spindle, and then slip the hub over the spindle. Install the outer bearing, washer, and spindle nut. The spindle nut should be adjusted to the exact specifications given in the service information **(Figure 46–31)**. Often, it is tightened until the hub cannot rotate, and then it is loosened about one-half turn before it is set to the specified free play. The initial tightening seats the bearings into their races. Once the spindle nut is tightened, install the nut lock, and use a new cotter pin to retain the lock.

The bearings require free play, and the adjustment can be checked with a dial indicator. Mount the base of the indicator as close as possible to the centre of the hub. Locate the tip of the indicator's plunger on the tip of the spindle. Set the indicator to zero. Firmly grasp the brake disc and move it in and out. The total movement shown on the indicator is the amount of free play at the bearing. Compare your reading to the specifications, and make adjustments as necessary. The most common free play specification is 0.03 to 0.13 mm (0.001 to 0.005 in.).

WARNING!

Throughout this entire process, your hands will have grease on them. Be very careful not to touch the brake assembly with your greasy hands. Clean them before handling the brake parts, or use a clean rag to hold the brake assembly.

The front bearing arrangement often found on FWD and 4WD vehicles is often nonserviceable. These bearings are pressed in and out of the steering knuckle. To do this, the axle or half shaft is removed, as is the steering knuckle and hub assembly. The bearings may be sealed and require no additional lubrication, or they may need to be packed with grease when they are reassembled. In most cases, the bearings are not adjusted. A heavily torqued axle nut is used to hold the assembly in place on the axle. This nut is typically replaced after it has been removed and is staked in place after it is tightened.

SHOP TALK

Because an axle nut is heavily torqued, it is wise to loosen the nut before the vehicle is raised for service. The weight of the vehicle on the tires will stop the hub from rotating while the nut is being loosened. The same holds true for final torquing. Tighten the nut as much as possible with the vehicle raised, and then lower the vehicle and torque the nut to specifications.

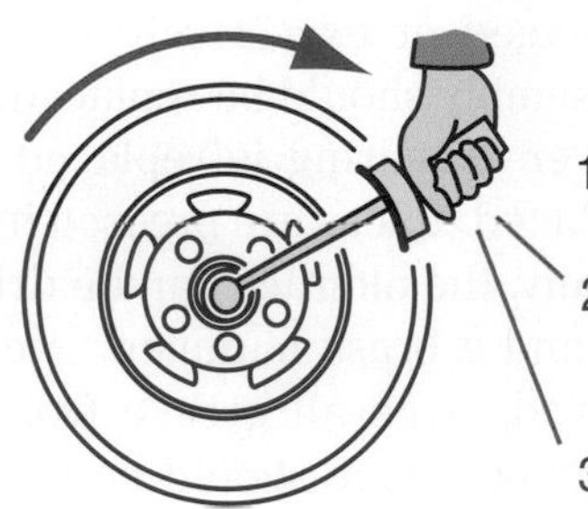

1. Hand spin the wheel.
2. Tighten the nut to 16 N • m (12 ft•lb) to fully seat the bearings and to overcome any burrs on threads.
3. Back off the nut until just loose.
4. Snug up the nut by hand.
5. Loosen the nut until a hole in the spindle lines up with a slot in the nut. Insert cotter pin.
6. When the bearing is properly adjusted, there will be from 0.03 to 0.13 mm (0.001 to 0.005 in.) end play.
Bend the ends of cotter pin legs flat against nut. Cut off extra length.

FIGURE 46–31

A typical procedure for adjusting wheel bearings.

Rear Hubs

The rear bearings on an FWD vehicle are serviced in the same way as the non-driving front wheel bearings. Most rear-weel-drive (RWD) axle bearings are of the straight roller bearing design, in which the drive axle tube serves as the bearing race. Some rear-wheel axle bearings are of the ball or tapered roller bearing type.

Wheel Bearing Grease Specification

The grease for wheel bearings should be smooth-textured, consist of soaps and oils, and be free of filler and abrasives. Recommended are lithium complex (or equivalent) soaps or solvent-refined petroleum oils. Additives could be used to inhibit corrosion and oxidation. The grease should be non-corrosive to bearing parts, with no chance of separating during storage or use.

Always make sure you use the grease recommended by the manufacturer. Greases are classified by the National Lubricating Grease Institute (NLGI) to indicate their application. NLGI no.2 grease is the grease classification for wheel bearings. Failure to maintain proper lubrication might result in bearing damage, causing a wheel to lock **(Figure 46–32)**. To lubricate a bearing, force grease around the outside of the bearing and between the rollers, cone, and cage. Also pack grease into the wheel hub. The depth of the grease should be level with the outside diameter of the cup. Different brands of grease should never be mixed. The different compositions can separate and liquify and then will not provide proper lubrication.

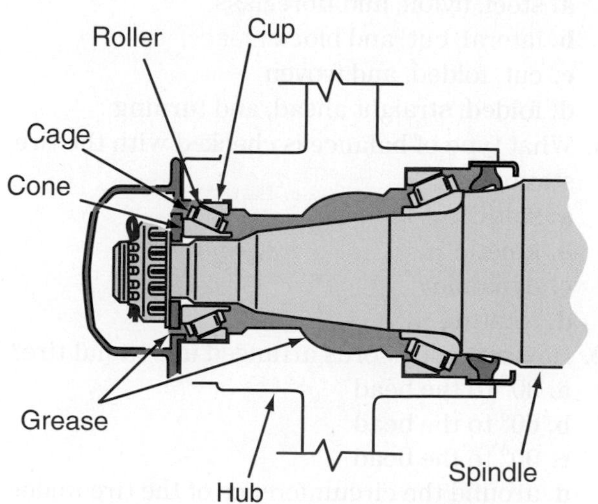

FIGURE 46–32
Wheel bearing lubrication.

GO TO Chapter 8 for a detailed discussion and chart of NLGI lubricants.

Bearing Troubleshooting

Wheel bearings are designed for longevity. Their life expectancy, based on metal fatigue, can usually be calculated if general operating conditions are known. Bearing failures not caused by normal material fatigue are called permanent failures. The causes can range from improper lubrication or incorrect mounting, to poor condition of the shaft housing or bearing surfaces.

When servicing, replacing, or installing wheel bearings, always follow the procedure given in the service information.

KEY TERMS

Aspect ratio (p. 1404)
Bead (p. 1399)
Belted bias ply (p. 1399)
Drop centre (p. 1397)
Dynamic balance (p. 1419)
Hub (p. 1397)
Lateral runout (p. 1415)
Lug nuts (p. 1397)
Plies (p. 1399)
Profile (series) (p. 1404)
Radial ply (p. 1400)
Radial runout (p. 1415)
Sidewalls (p. 1399)
Sipes (p. 1399)
Static balance (p. 1419)
Tire pressure monitoring (TPM) system (p. 1409)
Tread (p. 1399)
Uniform Tire Quality Grading (UTQG) (p. 1406)
Valve stem (p. 1397)
Wheel offset (p. 1397)
Wheel shimmy (p. 1419)
Wheel tramp (p. 1419)

SUMMARY

- Wheels are made of either stamped or pressed steel discs riveted or welded into a circular shape or are die-cast or forged aluminum or magnesium rims.
- The primary purpose of tires is to provide traction. They are also designed to carry the weight of the vehicle, withstand side thrust over varying speeds and conditions, transfer braking and driving torque to the road, and absorb much of the rock shock from surface irregularities.
- Pneumatic tires are of two types: those that use inner tubes and those that do not. The latter are called tubeless tires and are about the only type used on passenger cars today.

- There are two types of tire construction in use today: belted bias ply (specialty trailer and construction equipment) and radial ply.
- Tires are rated by their profile, ratio, size, load range, and UTQG.
- Tire construction affects both dimensions and ride characteristics, creating differences that can seriously affect vehicle handling.
- An ideal tire is one that wears little, holds the road well to provide sure handling and braking, and provides a cushion from road shock. It should also provide maximum grip on dry roads, wet roads, and snow and ice, and operate quietly at any speed.
- There are basically three categories of tread patterns: directional, nondirectional, and symmetric and asymmetric. A directional tire is mounted so that it revolves in a particular direction and will perform well only when it is rotating in that direction. A nondirectional tire has the same handling qualities in either direction of rotation. A symmetrical tire has the same tread pattern on both sides of the tire. An asymmetrical tire has a tread design that is different from one side to the other.
- The number and size of the blocks, sipes, and grooves on a tire's tread determine how much rubber contacts the road, how much water can be displaced, and how quiet the tire will be during travel.
- To maximize tire performance, inspect for signs of improper inflation and uneven wear, which can indicate a need for balancing, rotation, or alignment. Tires should also be checked frequently for cuts, bruises, abrasions, and blisters, and for stones or other objects that might have become imbedded in the tread.
- A properly inflated tire gives the best tire life, riding comfort, handling stability, and even fuel economy during normal driving conditions.
- To equalize tire wear, most car and tire manufacturers recommend that the tires be rotated. It must be remembered that front and rear tires perform different jobs and can wear differently, depending on driving habits and the type of vehicle.
- Most tires used today have built-in tread wear indicators to show when tires need replacement.
- There are three popular methods of tire repair: head-type plug, cold patch repair, and hot patch repair.
- There are two types of wheel balancing: static balance and dynamic balance.
- The bearings on an axle that drives the wheels are called axle bearings. Wheel bearings are used on non-driving axles. Axle bearings are typically serviced with the drive axle. Tapered roller wheel bearings require periodic maintenance service and are often serviced with suspension and brake work.
- Bad axle and wheel bearings will cause handling and tire wear problems.
- The front wheel hubs on ball or tapered roller bearings are lubricated by wheel bearing grease.
- Rear wheels are bolted to integral or detachable hubs.

REVIEW QUESTIONS

1. How is a modern tire wheel constructed? What materials are used?
2. Define dynamic and static wheel balance.
3. Describe the procedure for removing a tire from and replacing it on a wheel rim.
4. Describe the procedure for using a plug to seal a puncture in a tire.
5. What can cause an increase in a tire's rolling resistance and why is this important?
6. When describing a tire, what is "series" another name for?
   - **a.** size
   - **b.** load range
   - **c.** bead
   - **d.** profile
7. What are the three belting systems available for today's tires?
   - **a.** steel, nylon, and fibreglass
   - **b.** lateral, cut, and block
   - **c.** cut, folded, and woven
   - **d.** folded, straight ahead, and turning
8. What type of balance is checked with the tire stationary?
   - **a.** static
   - **b.** kinetic
   - **c.** dynamic
   - **d.** rotating
9. How are body cords arranged in a radial tire?
   - **a.** 30° to the bead
   - **b.** 60° to the bead
   - **c.** 90° to the bead
   - **d.** around the circumference of the tire under the tread

10. Where does an overinflated tire show its wear?
   a. in the centre of the tread
   b. on both outside edges
   c. on the inner edge
   d. on the outer edge
11. Which of the following is a major cause of wheel failure?
   a. overloading
   b. age
   c. improper maintenance
   d. all of the above
12. How is tire aspect ratio found?
   a. multiplying the tire height by its width
   b. dividing the tire section height by the section width
   c. dividing the tire section width by the rim size
   d. multiplying the tread width by the rim size
13. What is the primary reason for the grooves in tire tread?
   a. identify one tire from another
   b. trap sand and dirt in them to provide better traction
   c. allow water to run out from between the tire and the road
   d. allow the tire to hydroplane
14. Where can most of the information about a car's tires be found?
   a. on the engine block
   b. on the tire placard
   c. on the undercarriage of the vehicle
   d. on the steering wheel
15. What is the term used for a tire that wobbles from side to side?
   a. radial runout
   b. lateral runout
   c. eccentric runout
   d. concentric runout
16. What is rapid up-and-down movement of the wheel called?
   a. wheel tramp
   b. wheel misalignment
   c. wheel kick-up
   d. wheel shimmy
17. Which of the following statements is true?
   a. A cold patch should be vulcanized to the inner surface of the carcass.
   b. A tread puncture that is less than 6.35 mm (¼ in.) in diameter is repairable.
   c. Wheel balancing is required only when new tires are installed.
   d. For best results, punctures should be repaired from outside the tire.
18. When adjusting wheel bearings, how much end play should be present?
   a. zero end play
   b. 0.03–0.13 mm (0.001–0.005 in.) free play
   c. 0.13–0.25 mm (0.005–0.010 in.) free play
   d. 0.25–0.38 mm (0.010–0.015 in.) free play
19. What does the Uniform Tire Quality Grading system (UTQG) include?
   a. tread design, maximum inflation, and load capability
   b. tread wear capability, tread design, and load capability
   c. traction, tread wear capability, and temperature resistance
   d. maximum inflation, tread wear capability, and temperature resistance
20. What section of a wheel allows for easy tire removal and installation?
   a. the drop centre
   b. the wheel offset
   c. the safety ridges
   d. the pilot bore

CHAPTER 47
Suspension Systems

LEARNING OUTCOMES

- Name the different types of springs and how they operate.
- Name the advantages of ball joint suspensions.
- Explain the important differences between sprung and unsprung weight with regard to suspension control devices.
- Identify the functions of shock absorbers and struts, and describe their basic construction.
- Identify the components of a MacPherson strut system, and describe its functions.
- Identify the functions of bushings and stabilizers.
- Perform a general front suspension inspection.
- Check chassis height measurements to specifications.
- Identify the three basic types of rear suspensions, and know their effects on traction and tire wear.
- Identify the various types of springs, their functions, and their locations in the rear axle housing.
- Describe the advantages and operation of the three basic electronically controlled suspension systems: level control, adaptive, and active.
- Explain the function of electronic suspension components, including air compressors, sensors, control modules, air shocks, electronic shock absorbers, and electronic struts.
- Explain the basic towing, lifting, jacking, and service precautions that must be followed when servicing air springs and other electronic suspension components.

Like the rest of the systems on cars and light trucks, the suspension system has greatly changed through the years. Not only has technology brought about these changes, so has the quest for great-handling and comfortable vehicles. Suspension systems for the front and rear of a vehicle can get quite complex.

As a vehicle moves, the suspension and tires must react to the current driving conditions. Specifically, the suspension system does the following:

- Supports the weight of the vehicle
- Keeps the tires in contact with the road
- Controls the direction of the vehicle's travel
- Attempts to maintain the correct vehicle ride height
- Maintains proper wheel alignment
- Reduces the effect of shock forces as the vehicle travels on an irregular surface

FRAMES

To provide a rigid structural foundation for the vehicle body and a solid anchorage for the suspension system, a frame of some type is essential. There are two basic frames in common use today.

Conventional Frame Construction

In the conventional body-over-frame construction, the frame is the vehicle's foundation. The body and all major parts of a vehicle are attached to the frame **(Figure 47–1)**. It must provide the support and strength needed by the assemblies and parts attached to it. In other words, the frame is an independent, separate component because it is not

FIGURE 47–1

An example of body-over-frame construction.

FIGURE 47–2

Unibody construction is the most common frame design today.

©Cengage Learning 2015

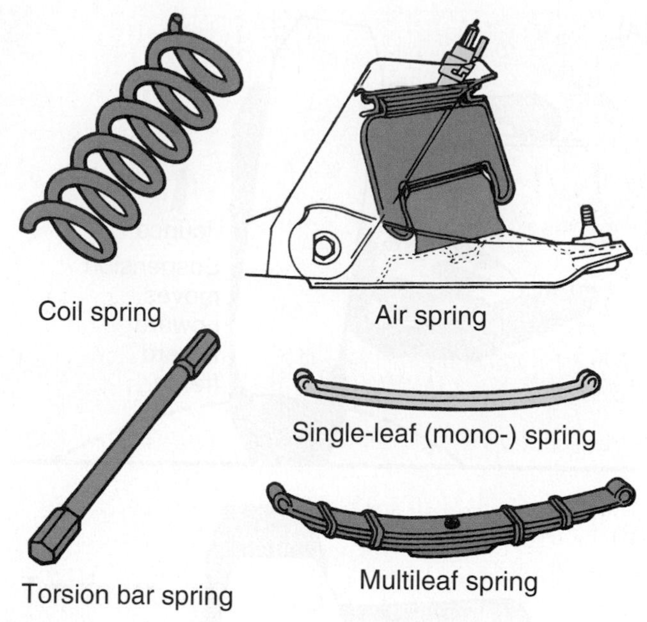

FIGURE 47–3

Various types of automotive springs.

welded to any of the major units of the body shell. The major frame types are perimeter, "X" type, and ladder, which is the most common.

Unibody Construction

Unibody construction has no separate frame. The body is constructed in such a manner that the body parts themselves supply the rigidity and strength required to maintain the structural integrity of the car **(Figure 47–2)**. The unibody design significantly lowers the base weight of the car, and that in turn increases gas mileage capabilities. The two major types are unibody and unibody with a bolt-on subframe.

SUSPENSION SYSTEM COMPONENTS

Most automotive suspension systems have the same basic components and operate similarly. Their differences are found in the method in which the basic components are arranged.

Springs

The spring is the core of nearly all suspension systems. It is the component that absorbs shock forces while maintaining correct riding height. If the spring is worn or damaged, the other suspension elements shift out of their proper positions and are subject to increased wear. The increased effect of shock impairs the vehicle's handling. If one spring is considered worn or damaged and replacement is required, both springs on the axle must be replaced.

Various types of springs are used in suspension systems—coil, torsion bar, leaf (both mono- and multileaf types), and air springs **(Figure 47–3)**. Springs are rubber mounted to reduce road shock and noise.

Automotive springs are generally classified by the amount of deflection exhibited under a specific load. This is referred to as the **spring rate**. According to Hooke's law of physics, a force (weight) applied to a spring causes it to compress in direct proportion to the force applied. When that force is removed, the spring returns to its original position, if not overloaded. Remember that a heavy vehicle requires stiffer springs than a lightweight car.

The springs take care of two fundamental vertical actions: jounce and rebound. **Jounce** occurs when a wheel hits a bump and moves up **(Figure 47–4A)**. When this happens, the suspension system acts to pull in the top of the wheel, maintaining an equal distance between the two front wheels and preventing a sideways scrubbing action as the wheel moves up and down. **Rebound** occurs when the wheel hits a dip or hole and moves downward **(Figure 47–4B)**. In this case, the suspension system acts to move the wheel in at both the top and bottom equally, while maintaining an equal distance between the wheels.

The spring goes back and forth from jounce to rebound. Each time, jounce and rebound become smaller and smaller. This is caused by friction of the spring's molecular structure and the suspension pivot joints. A **shock absorber** is added to each suspension to dampen and stop the motion of the spring after

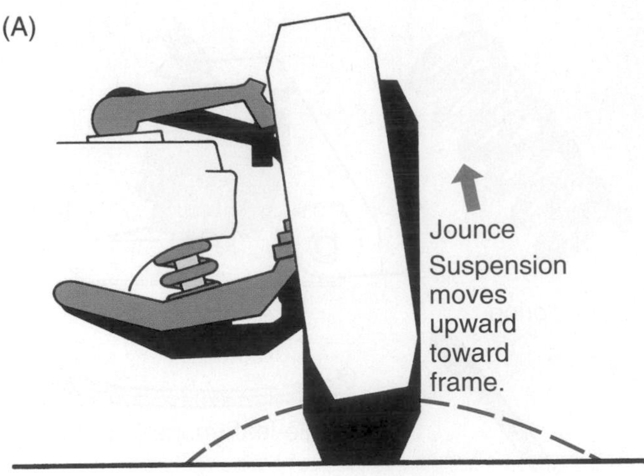
(A)
Jounce
Suspension moves upward toward frame.

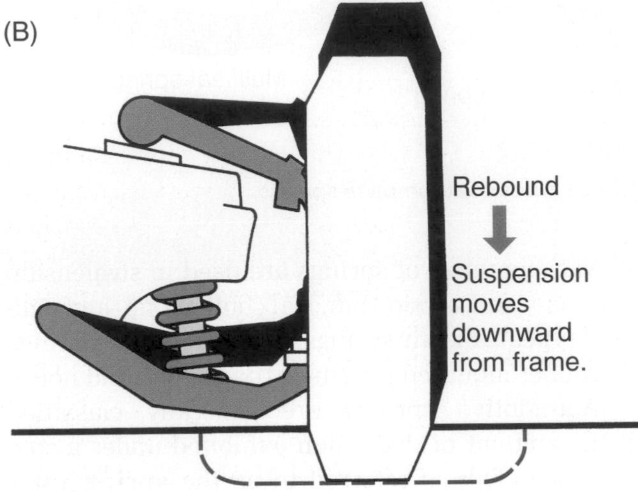
(B)
Rebound
Suspension moves downward from frame.

FIGURE 47–4

(A) Upward and (B) downward suspension movement.

jounce. The shock absorber accomplishes this by converting kinetic energy (energy from motion) to heat.

All of the vehicle's weight supported by the suspension system is known as **sprung weight**. The weight of those components not supported by the springs is known as **unsprung weight**. The vehicle's body, frame, engine, transmission, and all of its components are considered sprung weight. Undercar parts classified as unsprung weight include the steering knuckles and rear axle assemblies (but not always the final drive). Keep in mind that, in general, the lower the ratio of unsprung weight to sprung weight, the better the vehicle's ride will be.

COIL SPRINGS Two basic designs of coil springs are used: linear rate and variable rate **(Figure 47–5)**. **Linear rate** springs characteristically have one basic shape and a consistent wire diameter. All linear springs are wound from a steel rod into a cylindrical shape with even spacing between the coils. As the load is increased, the spring is compressed, and the coils twist (deflect). As the load is removed, the coils flex (unwind) back to the normal position. The amount of load necessary to deflect the spring 25.4 mm (1 in.) is the spring rate. On linear rate springs, this is a constant rate, no matter how much the spring is compressed. For example, 112 kg (250 lb.) compress the spring 25.4 mm (1 in.), and 340 kg (750 lb.) compress the spring 76.2 mm (3 in.). Spring rates for linear rate springs are normally calculated between 20 and 60 percent of the total spring deflection.

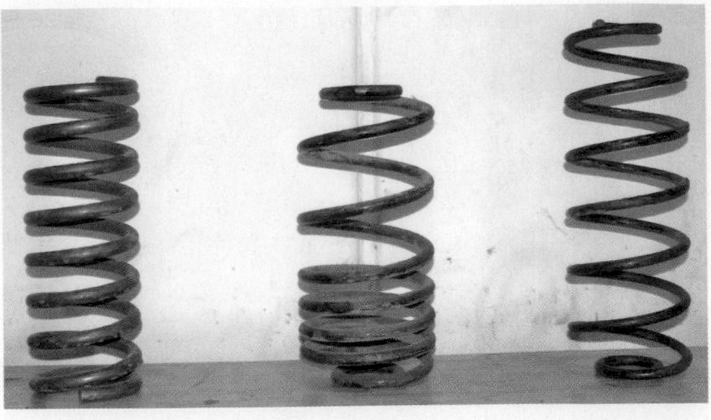

FIGURE 47–5

The different designs of coil springs.

Variable rate spring designs are characterized by a combination of wire sizes and shapes. The most commonly used variable rate springs have a consistent wire diameter, are wound in a cylindrical shape, and have unequally spaced coils. This type of spring is called a progressive rate coil spring or cargo coil when used for heavy-duty purposes.

The design of the coil spacing gives the spring three functional ranges of coils: inactive, transitional, and active. Inactive coils are usually the end coils and introduce force into the spring. Transitional coils become inactive as they are compressed to their point of maximum load-bearing capacity. Active coils work throughout the entire range of spring loading. Theoretically, in this type of design, at stationary loads, the inactive coils are supporting all of the vehicle's weight. As the loads are increased, the transitional coils take over until they reach maximum capacity. Finally, the active coils carry the remaining overload. This allows for automatic load adjustment while maintaining vehicle height.

Another common variable rate design uses tapered wire to achieve this same type of progressive rate action. In this design, the active coils have a large wire diameter, and the inactive coils have a small wire diameter.

Later designs of variable rate springs deviate from the old cylindrical shape. These include the truncated cone, the double cone, and the barrel spring. The major advantage of these designs is the ability of the coils to nest, or bottom out, within each other without touching, which lessens the amount of space needed to store the springs in the vehicle.

Unlike a linear rate spring, a variable rate spring has no predictable standard spring rate. Instead, it has an average spring rate based on the load of a predetermined spring deflection. This makes it impossible to compare a linear rate spring to a variable rate spring. Variable rate springs, however, handle a load of up to 30 percent over standard rate springs in some applications.

Servicing Coil Springs. A technician must know how to check and replace coil springs, select the proper replacement, and recommend the proper size and type of spring to the customer.

The first step in coil spring selection is to check for the original equipment (OE) part number. This is usually on a tag wrapped around the coil. In many instances, this tag falls off before replacement is necessary. If a set of aftermarket springs has been installed, the part number might be stamped on the end of the coil. Next, check to see what type of ends the coil springs have. There are three types of ends used in automotive applications: full wire open, tapered wire closed, and pigtail. Springs with full wire open ends are cut straight off and sometimes flattened or ground into a D or square shape. Tapered wire closed ends are wound to ensure squareness and ground into a taper at the ends. Pigtail ends are wound into a smaller diameter at the ends.

The final step is to check the application in the catalogue. To do this, it is necessary to know the make, year, model, body style, and engine size (number of cylinders), and whether the vehicle is equipped with air conditioning. In some cases, it is also good to know the type of transmission, seating capacity, and other specifics that add extra weight to the vehicle. In most catalogues, springs are listed by vehicle application in two sections: front and rear.

LEAF SPRINGS Although leaf springs were the first type of suspension spring used on automobiles, today they are generally found only on light-duty trucks, vans, and some passenger cars. There are three basic types of leaf springs: multiple leaf, monoleaf, and fibre composite.

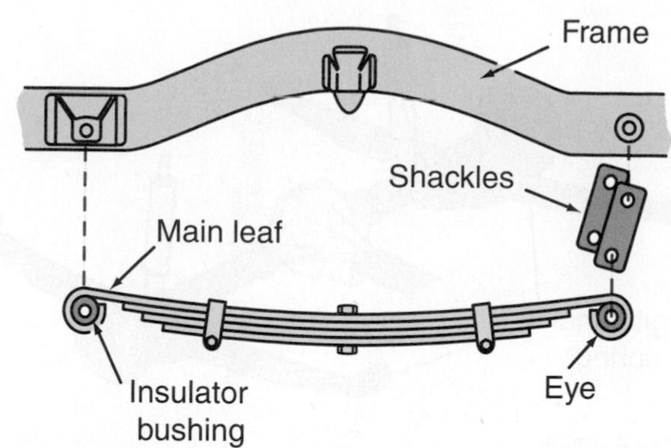

FIGURE 47–6

A leaf spring and its related hardware.

Multiple-Leaf Springs. **Multiple-leaf** springs consist of a series of flat steel leaves that are bundled together and held with clips or by a bolt placed slightly ahead of the centre of the bundle **(Figure 47–6)**. One leaf, called the **main leaf**, runs the entire length of the spring. The next leaf is a little shorter and attaches to the main leaf. The next leaf is shorter yet and attaches to the second leaf, and so on. This system allows almost any number of leaves to be used to support the vehicle's weight **(Figure 47–7)**. It also gives a progressively stiffer spring. The spring easily flexes over small distances for minor bumps. The farther the spring is deflected, the stiffer it gets. The more leaves and the thicker and shorter the leaves, the stronger the spring. It must be remembered that as the spring flexes, the ends of the leaves slide over one another. This sliding could be a source of noise and can also produce friction. These problems are reduced by interleaves of zinc and plastic placed between the spring's leaves. As the multiple leaves slide, friction produces a harsh ride as the spring flexes. This friction also dampens the spring motion.

FIGURE 47–7

To provide more ground clearance, a lift kit that included 10 leaf rear springs was installed on this pickup.

Superlift Suspension Systems

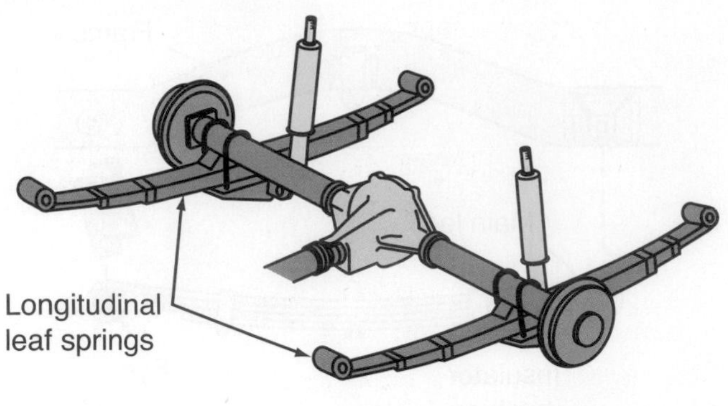

FIGURE 47–8

These leaf springs support and locate the drive axle.

Multiple-leaf springs have a curve in them. This curve, if doubled, forms an ellipse. Thus, leaf springs are sometimes called semielliptical or quarter-elliptical. The semi or quarter refers to how much of the ellipse the spring actually describes. The vast majority of leaf springs are semielliptical.

Leaf springs are commonly mounted at right angles to the axle **(Figure 47–8)**. In addition to absorbing blows from road forces, they also serve as suspension locators by fixing the position of the suspension with respect to the front and rear of the vehicle. A centring pin is frequently used to ensure that the axle is correctly located. If a spring is broken or misplaced, the axle might be mislocated and the alignment impaired. Some vehicles have a transversely mounted leaf spring. The centre of the spring is mounted to the vehicle's chassis, and the outer ends are fastened to the ends of the axle housing or wheel spindles.

The front eye of the main leaf at either end of the axle is attached to a bracket on the frame of the vehicle with a bolt and bushing connection. The rear eye of the main leaf is secured to the frame with a shackle, which permits some fore and aft movement **(Figure 47–9)** in response to physical forces of acceleration, deceleration, and braking.

Monoleaf Springs. **Monoleaf** or single-leaf springs are usually the tapered plate type with a heavy or thick centre section tapering off at both ends. This provides a variable spring rate for a smooth ride and good load-carrying ability. In addition, single-leaf springs do not have the noise and static friction characteristic of multiple-leaf springs.

Fibre Composite Springs. Although most leaf springs are still made of steel, **fibre composite** types are increasing in popularity **(Figure 47–10)**. Some automotive people call them plastic springs in

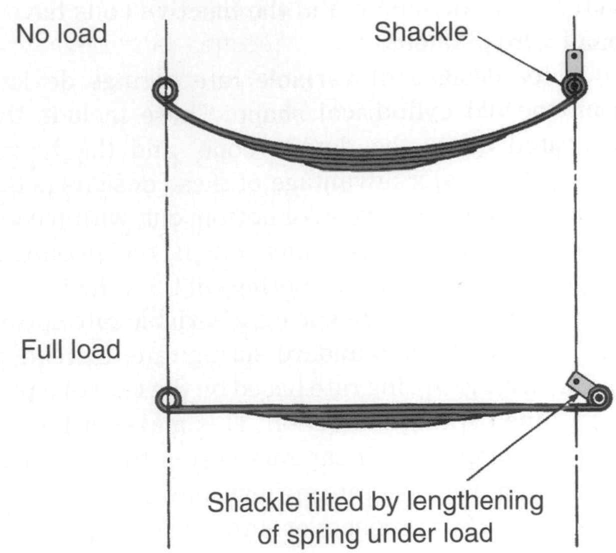

FIGURE 47–9

The action of a leaf spring as it compresses.

spite of the fact that the springs contain no plastic at all. They are made of fibreglass, laminated and bonded together by tough polyester resins. The long strands of fibreglass are saturated with resin and bundled together by wrapping (a process called filament winding) or squeezed together under pressure (compression moulding).

Fibre composite leaf springs are incredibly lightweight and possess some unique ride-control characteristics. Conventional monoleaf steel springs are real heavyweights, tipping the scale at anywhere from 11 to 20 kg (25 to 45 lb.) apiece. Some multiple-leaf springs can weigh almost twice as much. A fibre composite leaf spring is a featherweight by comparison, weighing a mere 3.6 to 4.5 kg (8 to 10 lb.). As every performance enthusiast knows, springs are dead weight. Reducing the weight of the suspension reduces not only the overall weight of the vehicle but also the sprung mass of the suspension itself. This reduces the spring effort and amount of shock control

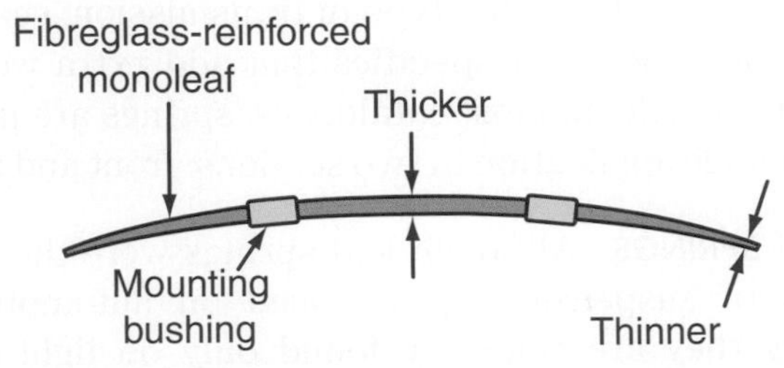

FIGURE 47–10

The construction of a fibreglass-reinforced monoleaf spring.

that is required to keep the wheels in contact with the road. The result is a smoother riding, better handling, and faster responding suspension, which is exactly the sort of thing every performance enthusiast wants.

AIR SPRINGS Another type of spring, an air spring, is used in an air-operated microprocessor-controlled system that replaces the conventional coil springs with air springs to provide a comfortable ride and automatic front and rear load levelling. This system, fully described later in this chapter, uses four air springs to carry the vehicle's weight. The air springs are located in the same positions where coil springs are usually found. Each spring consists of a reinforced rubber bag pressurized with air. The bottom of each air bag is attached to an inverted pistonlike mount that reduces the interior volume of the air bag during jounce **(Figure 47–11)**. This has the effect of increasing air pressure inside the spring as it is compressed, making it progressively stiffer. A vehicle equipped with an electronic air suspension system is able to provide a comfortable street ride, about a third softer than conventional coil springs. At the same time, its variable spring rate helps absorb bumps and protect against bottoming.

Torsion Bar Suspension System

Torsion bars serve the same function as coil springs. In fact, they are often described as straightened-out coil springs. Instead of compressing like coil springs, a torsion bar twists and straightens out on the recoil. That is, as the bar twists, it resists up-and-down movement. One end of the bar—made of heat-treated alloy spring steel—is attached to the vehicle frame. The other end is attached to the lower control arm **(Figure 47–12)**. When the wheel moves up and down, the lower control arm is raised and lowered. This twists the torsion bar, which causes it to absorb road shocks. The bar's natural resistance to twisting quickly restores it to its original position, returning the wheel to the road.

FIGURE 47–11

A rear suspension setup with air springs.

BMW Group

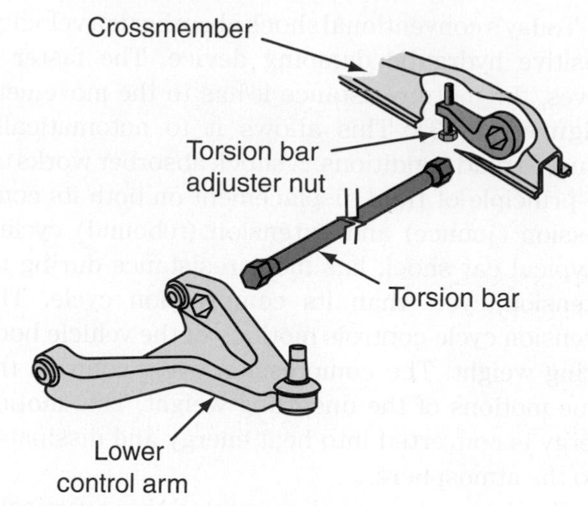

FIGURE 47–12

A torsion bar setup.

When torsion bars are manufactured, they are pre-stressed to give them fatigue strength. Because of directional pre-stressing, torsion bars are directional. The torsion bar is marked either right or left to identify on which side it is to be used.

Because the torsion bar is connected to the lower control arm, the lower ball joint is the load carrier. A shock absorber is connected between the lower control arm and the frame to damp the twisting motion of the torsion bar.

Many late-model pickups and sport utility vehicles (SUVs) use torsion bars in their front suspensions. They are primarily used in this type of vehicle because they can be mounted low and out of the way of the driveline components. Torsion bars, like coil springs, must be replaced in pairs.

Shock Absorbers

Shock absorbers damp or control motion in a vehicle. If unrestrained, springs continue expanding and contracting after a blow until all the energy is absorbed. Not only would this lead to a rough and unstable—perhaps uncontrollable—ride after consecutive shocks, it would also create a great deal of wear on the suspension and steering systems. Shock absorbers prevent this. Despite their name, they actually dampen spring movement instead of absorbing shock. As a matter of fact, in England and almost everywhere else but North America, shock absorbers are referred to as **dampers**.

Today's conventional shock absorber is a velocity-sensitive hydraulic damping device. The faster it moves, the more resistance it has to the movement **(Figure 47–13)**. This allows it to automatically adjust to road conditions. A shock absorber works on the principle of fluid displacement on both its compression (jounce) and extension (rebound) cycles. A typical car shock has more resistance during its extension cycle than its compression cycle. The extension cycle controls motions of the vehicle body spring weight. The compression cycle controls the same motions of the unsprung weight. This motion energy is converted into heat energy and dissipated into the atmosphere.

Shock absorbers can be mounted either vertically or at an angle. Angle mounting of shock absorbers improves vehicle stability and dampens accelerating and braking torque.

Conventional hydraulic shocks are available in two styles: single tube and double tube. The vast majority of domestic shocks are double tubed. Although they are a little heavier and run hotter than the single-tubed type, they are easier to make. The double-tube shock has an outer tube that completely covers the inner tube. The area between the tubes is the oil reservoir. A compression valve at the bottom of the inner tube allows oil to flow between the two tubes. The piston moves up and down inside the inner tube.

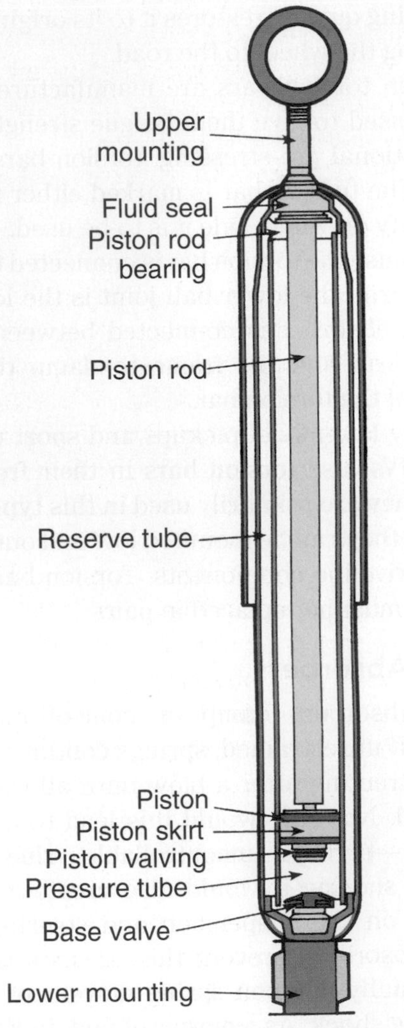

FIGURE 47–13

A cross-section of a conventional shock absorber.

In a single **monoshock**, there is a second floating piston near the bottom of the tube. When the fluid volume increases or decreases, the second piston moves up and down, compressing the reservoir. The fluid does not move back and forth between a reservoir and the main chamber. There are no other valves in a single-tube shock besides those in the main piston. The second piston prevents the oil from splashing around too much and getting air bubbles in it. Air in the shock oil is detrimental. Air, unlike oil, is compressible and slips past the piston easily. When this happens, the result is a shock that offers poor vehicle control on bumpy roads.

In addition to these conventional hydraulic shocks, there are a number of others the technician may encounter.

GAS-CHARGED SHOCK ABSORBERS On rough roads, the passage of fluid from chamber to chamber becomes so rapid that foaming can occur. Foaming is simply the mixing of the fluid with any available air. Because aeration can cause a skip in the shock's action, engineers have sought methods of eliminating it. One is the spiral groove reservoir, the shape of which breaks up bubbles. Another is a gas-filled cell or bag (usually nitrogen) that seals air out of the reservoir so that the shock fluid is isolated from contacting air or gas.

A gas-charged shock absorber **(Figure 47–14)** operates on the same hydraulic fluid principle as conventional shocks. It uses a piston and oil chamber similar to other shock absorbers. Instead of a double tube with a reserve chamber, it has a dividing piston that separates the oil chamber from the gas chamber. The oil chamber contains a special hydraulic oil, and the gas chamber contains nitrogen gas under pressure equal to approximately 25 times atmospheric pressure.

As the piston rod moves downward in the shock absorber, oil is displaced, just as it is in a double-tube shock. This oil displacement causes the divided piston to press on the gas chamber. The gas is compressed, and the chamber reduces in size. When the piston rod returns, the gas pressure returns the

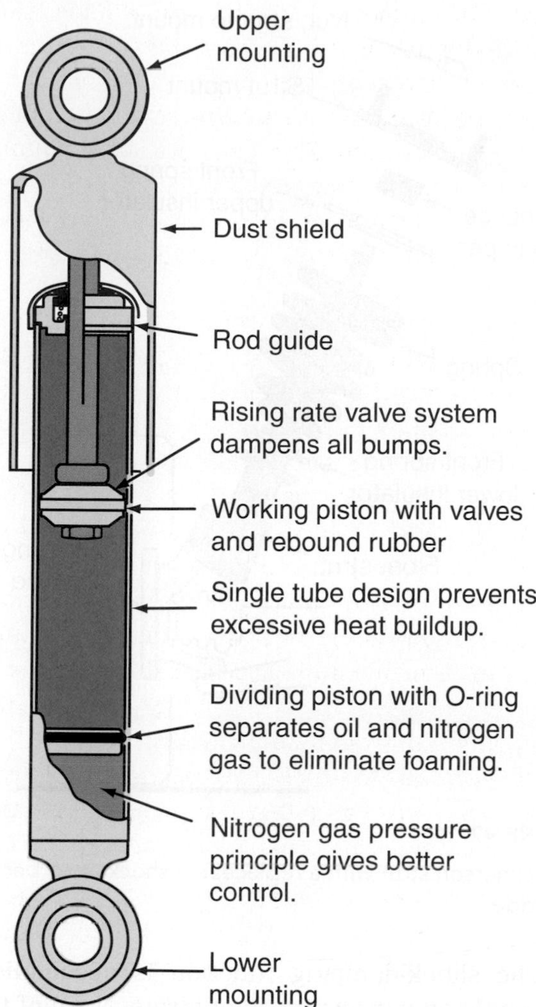
Upper mounting
Dust shield
Rod guide
Rising rate valve system dampens all bumps.
Working piston with valves and rebound rubber
Single tube design prevents excessive heat buildup.
Dividing piston with O-ring separates oil and nitrogen gas to eliminate foaming.
Nitrogen gas pressure principle gives better control.
Lower mounting

FIGURE 47–14

Gas-pressure-damped shocks operate like conventional oil-filled shocks. Gas is used to keep oil pressurized, which reduces oil foaming and increases efficiency under severe conditions.

dividing piston to its starting position. Whenever the static pressure of the oil column is held at approximately 700 to 2500 kPa (100 to 360 psi) (depending on the design), the pressure decreases behind the piston and therefore cannot be high enough for the gas to escape to the oil column. As a result, a gas-filled shock absorber operates without aeration.

SHOP TALK

Some high-pressure gas-charged shocks are monotube shocks with fluid and gas in separate chambers. The gas is charged to 2500 kPa (360 psi). Its basic design does not allow the valving range needed for a more responsive ride over a broad range of road conditions. The high-pressure gas charge can provide a harsh ride under normal driving conditions and is usually found on small trucks.

AIR SHOCK SYSTEMS There are two basic adjustable air shock systems: manual fill and automatic load levelling. The manual fill system can be ordered on new vehicles or can be installed on almost any vehicle manufactured without it.

There are several different types of manual fill air shock systems available. One common manual fill air shock system uses a high-speed, direct current (DC) motor to transfer a command signal that is manually selected from the driver's seat. In another manual air system, the units are inflated through air valves mounted at the rear of the vehicle. Air lines run between the shocks and the valve. A tire air pressure pump is used to fill the shocks to bring the rear of the vehicle to the desired height.

The automatic load-levelling air shock absorber system uses many of the same suspension components as the manual system, with the addition of suspension height sensors or height sensing valves, which control the operation of the high-speed DC motor–driven air compressor. When the vehicle is weighted down with either extra passengers or heavy loads, the compressor will automatically turn on and add more air to the shock absorbers to raise the vehicle to the correct ride height and conversely allow air to bleed off when the load is removed.

SHOCK ABSORBER RATIO Most shock absorbers are valved to offer roughly equal resistance to suspension movement upward (jounce) and downward (rebound). The proportion of a shock absorber's ability to resist these movements is indicated by a numerical formula. The first number indicates jounce resistance. The second indicates rebound resistance. For example, passenger cars with normal suspension requirements use shock absorbers valued at 50/50 (50 percent jounce/50 percent rebound). Drag racers, on the other hand, use shocks valued at about 90/10. Small vehicles, because of their light weight and soft springs, require more control in both jounce and rebound in the shock absorbers. Damping rates within the shock absorbers are controlled by the size of the piston, the size of the orifices, and the closing force of the valves.

It is important to keep in mind that the shock absorber ratio only describes what percentage of the shock absorber's total control is compression and what percentage is extension. Two shocks with the same ratio can differ greatly in their control capacity. This is one reason the technician must be sure correct replacement shocks are installed on the vehicle.

MACPHERSON STRUT SUSPENSION COMPONENTS

The **MacPherson strut** suspension was developed by Earle MacPherson and is dramatically different in appearance from the traditional independent front suspension **(Figure 47–15)**, but similar components operate in the same way to meet suspension demands.

The MacPherson strut suspension's most distinctive feature is the combination of the main elements into a single assembly. It typically includes the spring, upper suspension locator, and shock absorber. It is mounted vertically between the top arm of the steering knuckle and the inner fender panel.

Struts have taken two forms: a concentric coil spring around the strut itself **(Figure 47–16)** and a spring located between the lower control arm and the frame **(Figure 47–17)**. The location of the spring on the lower control arm, not on the strut as in a conventional MacPherson strut system, allows for larger engine compartments and larger engines without transferring the extra weight and road shock through the strut to the chassis. This system is called modified strut suspension.

Struts

The core element of this type of suspension is the strut. With its cylindrical shape and protruding piston rod, it looks quite similar to the conventional shock absorber. In fact, the strut provides the damping function of the shock absorber, in addition to serving to locate the spring and to fix the position of the suspension.

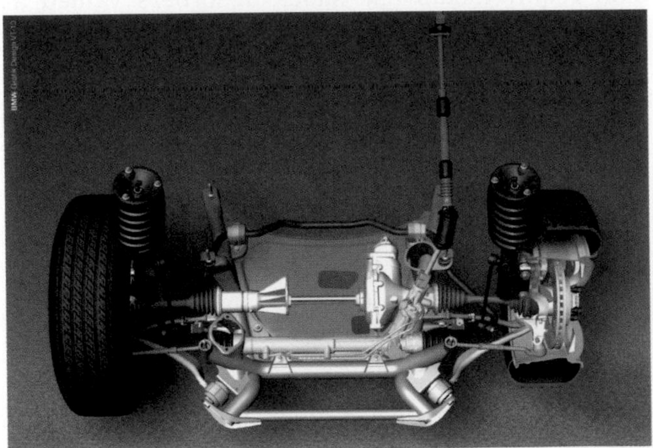

FIGURE 47–15

A complete MacPherson strut front suspension.

BMW of North America, Inc.

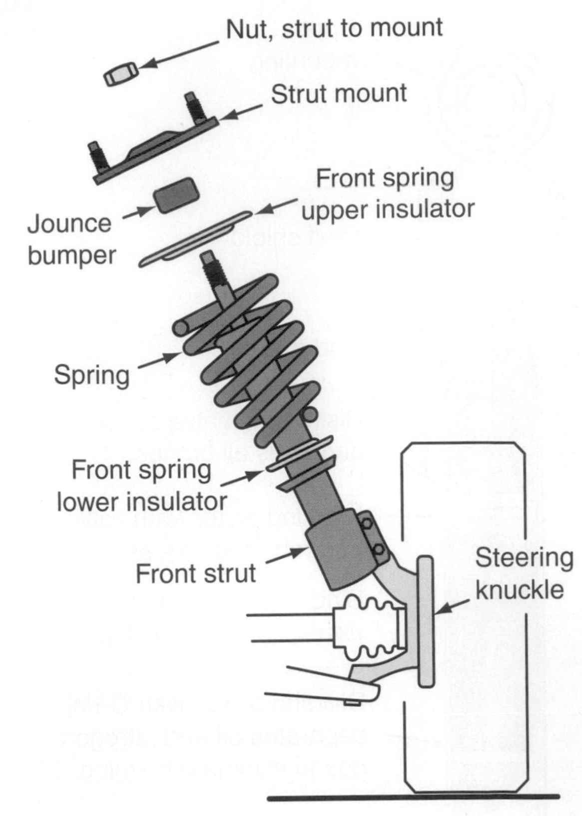

FIGURE 47–16

A MacPherson strut with a replaceable shock absorber cartridge.

The shock-damping function is accomplished differently on various types of struts. None of them uses a separate shock absorber, as the traditional front suspension does. Some versions are designed so that the damper can be independently serviced.

Struts fall into two broad categories: sealed and serviceable units. A sealed strut is designed so that the top closure of the strut assembly is permanently sealed. There is no access to the shock absorber cartridge inside the strut housing and no means of replacing the cartridge. Therefore, it is necessary to replace the entire strut unit. A serviceable strut is designed so that the cartridge inside the housing, which provides the shock-absorbing function, can be replaced with a new cartridge. Serviceable struts use a threaded body nut in place of a sealed cap to retain the cartridge.

The shock absorber device inside a serviceable strut is generally wet. This means the shock absorber contains oil that contacts and lubricates the inner wall of the strut body. The oil is sealed inside the strut by the body nut, O-ring, and piston rod seal. Servicing a wet strut with the equivalent components involves a thorough cleaning of the inside of the strut body, absolute cleanliness, and

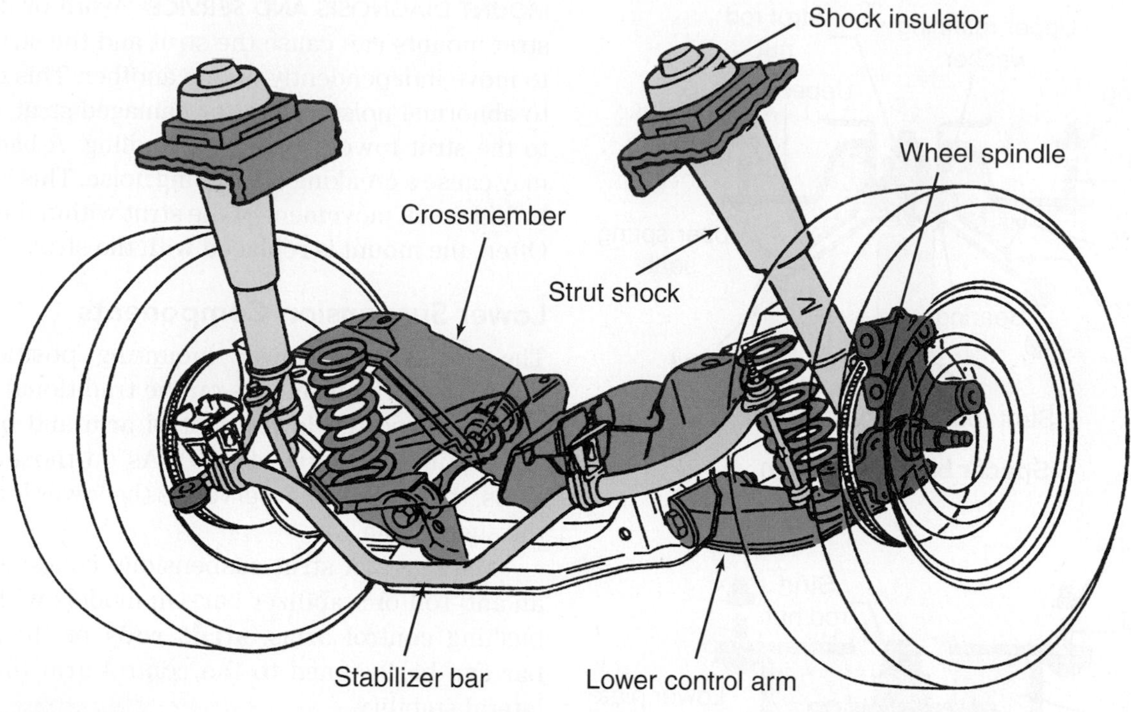

FIGURE 47–17

A modified MacPherson suspension has the spring mounted separately from the strut.
Chrysler Group LLC

great care in reassembly (including replenishing the strut with oil).

Cartridge inserts were developed to simplify servicing wet struts. The insert is a factory-sealed replacement for the strut shock absorber. The replacement cartridge is simply substituted for the original shock absorber cartridge and retained with the body nut.

Most OE domestic struts are serviced by replacement of the entire unit. There is no strut cartridge to replace. Sealed OE units can also be serviced by replacement with an aftermarket unit that permits future servicing by cartridge replacement.

The use of the strut reduces suspension space and weight requirements. By mounting the bottom of the strut assembly to the steering knuckle, the upper control arm and ball joint of the traditional suspension are eliminated. In place of the ball joint, the upper mount, which is bolted to the fender panel, is the load-carrying member on MacPherson suspensions.

Strut Mounts

A MacPherson strut has a mount between the top of the strut and the chassis where the strut is supported. These mounts are designed to dampen vibrations as well as secure the strut in position. Often, the mounts include a bearing, although some use a bushing, or a combination of both. Bearings are most commonly found in front suspensions because the suspension turns or pivots between left and right. Some strut suspension use bushings, although they can also just have a single bushing. Every application is different; however, most applications fall into one of three groupings **(Figure 47–18)**.

SPACER BUSHING The spacer bushing design is used mostly by Volkswagen, Toyota, Mazda, and Mitsubishi, and it was used in early Chrysler vehicles. This design has a bearing centred in the mount and a separate inner bushing. The bearing is pressed into the strut mount. The bearing, bushing, and upper plate support the strut rod. If the bushing is cracked or torn, or if the bearing is binding or seized, the strut mount must be replaced.

INNER PLATE The inner plate design is used by GM and Ford. It has a rubber-encased inner plate placed between upper and lower surface plates. The plate prevents the strut piston rod from pushing through the upper or lower surface plate if the inner plate fails. The bearing is at the bottom of the strut mount and is not serviceable. If the bearing is bad, the mount must be replaced.

CENTRE SLEEVE The centre sleeve design is widely used by Chrysler. It has a centre sleeve moulded to a rubber bushing. The stem of the strut passes through the sleeve. The bearing is not a part of the strut mount;

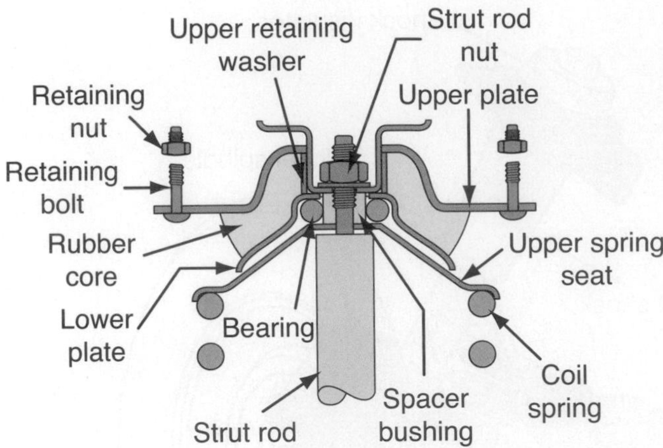

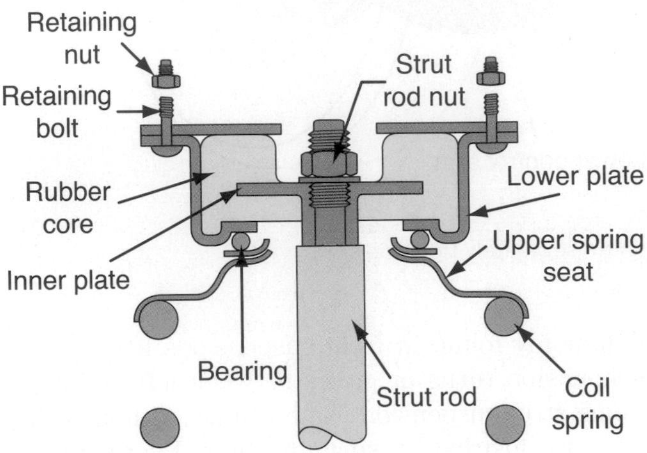

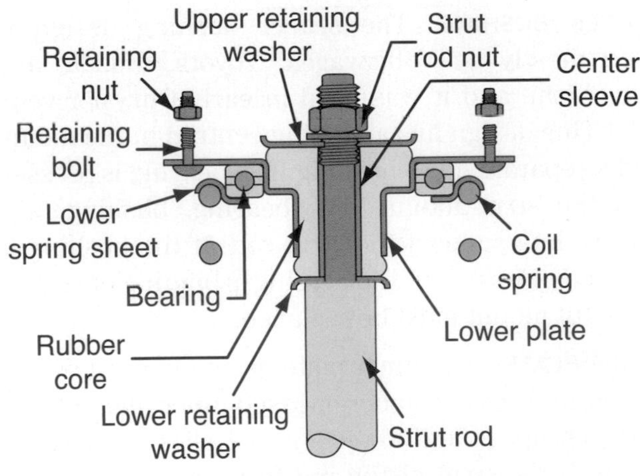

FIGURE 47–18

The different types of upper mounts for a strut.

rather, it is a separate unit. To prevent the strut rod from pushing through the mount, upper and lower retainer washers are used. If there are cracks, tears, or other damage to the bushing, the mount should be replaced.

MOUNT DIAGNOSIS AND SERVICE Worn or damaged strut mounts can cause the strut and the strut tower to move independently of one another. This can lead to abnormal noise, a bent or damaged strut, damage to the strut tower, and poor handling. A bad mount may cause a creaking or popping noise. This is caused by too much movement of the strut within the mount. Often, the mount is replaced with the strut.

Lower Suspension Components

The suspension's lower mounting position continues to be the frame, as on the traditional suspension, because the lower control arm and ball joint are retained **(Figure 47–19)**. As on those suspensions, the control arm serves as the lower locator for the suspension.

MacPherson strut suspensions continue to use an anti-roll or stabilizer bar. On models with single-bushing control arms, **strut rods** or the anti-roll bar can be fastened to the control arm to provide lateral stability.

The lower **ball joint** is a friction or steering ball joint and is used to stabilize the steering and to retard shimmy. The only exception is on modified

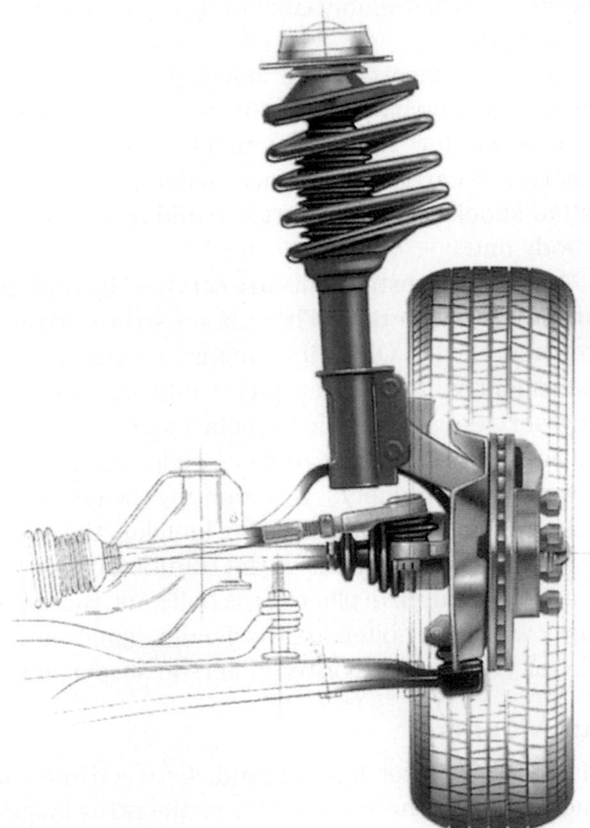

FIGURE 47–19

A front MacPherson strut assembly.

Ford Motor Company

MacPherson suspensions. In this design, the ball joint becomes the load bearer; the upper mount becomes the steering component.

Springs

Coil springs are used on all strut suspensions. A mounting plate welded to the strut serves as the lower spring seat. The upper seat is bolted to the strut piston rod. A bearing or rubber bushing in the upper mount permits the spring and strut to turn with the motion of the wheel as it is steered.

INDEPENDENT FRONT SUSPENSION

Front-suspension systems are fairly complex. They have somewhat contradictory jobs. They must keep the wheels rigidly positioned and at the same time allow them to steer right and left. In addition, because of weight transfer during braking, the front suspension system absorbs most of the braking torque. While accomplishing this, it must provide good ride and stability characteristics.

Short-Long Arm Suspension

The unequal length control arm or **short-long arm (SLA)** suspension system has been common on domestic-made vehicles for many years **(Figure 47–20)**. Each wheel is independently connected to the frame by a steering knuckle, ball joint assemblies, and short upper and longer lower control arms. Because the upper arm pivots in a shorter arc, the top of the wheel moves in and out slightly, but the tire's road contact remains constant **(Figure 47–21)**.

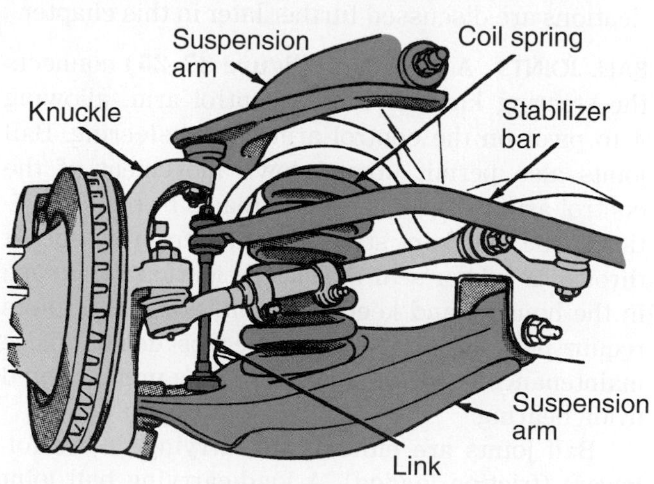

FIGURE 47–20

A typical SLA front suspension.

Chrysler Group LLC

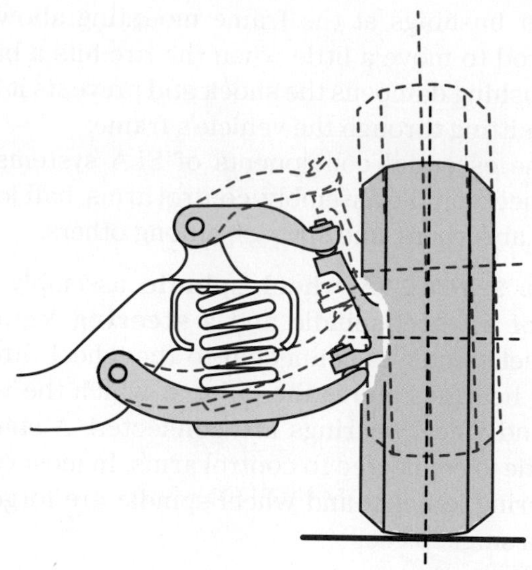

FIGURE 47–21

The movement of the wheel as a short-long arm suspension system moves up and down.

One design of an SLA uses a narrow lower control arm, shaped like an "I" **(Figure 47–22)**. A strut rod is used to hold the control arm in place. The strut rod is attached to the control arm close to the steering knuckle and to the frame in front of the wheel assembly to prevent fore and aft movement.

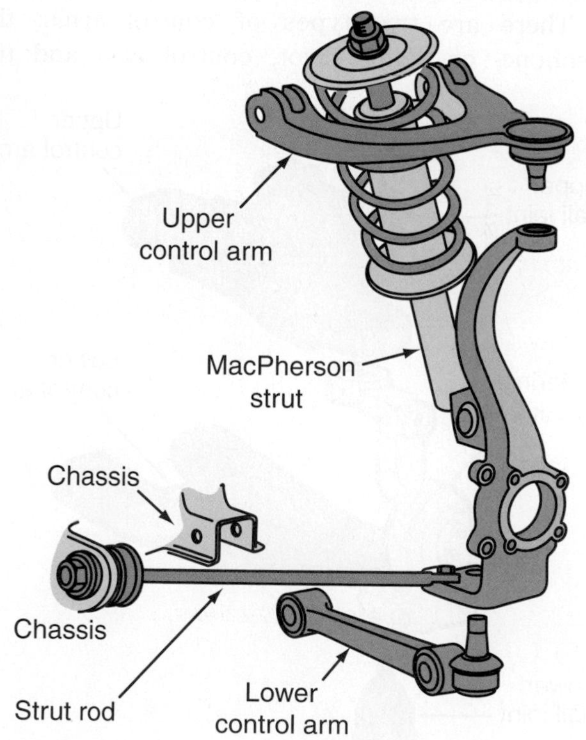

FIGURE 47–22

An FWD wishbone suspension system with a narrow lower control arm mounted on a single pivot.

Rubber bushings at the frame mounting allow the strut rod to move a little when the tire hits a bump. The bushing dampens the shock and prevents it from transmitting through the vehicle's frame.

The essential components of SLA systems are the wheel spindle assembly, control arms, ball joints, shock absorbers, and springs, among others.

WHEEL SPINDLE A **wheel spindle** assembly consists of a wheel spindle and a **steering knuckle**. A wheel spindle is connected to the wheel through wheel bearings and is the point at which the wheel hub and wheel bearings are connected. A steering knuckle is connected to control arms. In most cases, a steering knuckle and wheel spindle are forged to form a single piece.

CONTROL ARMS The upper and lower **control arms** on the traditional **independent front suspension (IFS)** function primarily as locators. They fix the position of the system and its components relative to the vehicle and are attached to the frame with bushings that permit the wheel assemblies to move up and down separately in response to irregularities in the road surface. The outer ends are connected to the wheel assembly with ball joints **(Figure 47–23)** inserted through each arm into the steering knuckle.

There are two types of control arms: the wishbone, or double-pivot, control arm and the single-pivot, or single-bushing, control arm **(Figure 47–24)**. The wishbone offers greater lateral stability than the single-pivot arm, which is lighter and requires less space than the wishbone but also requires modifications in suspension design to compensate for the reduced lateral stability. Those modifications are discussed further later in this chapter.

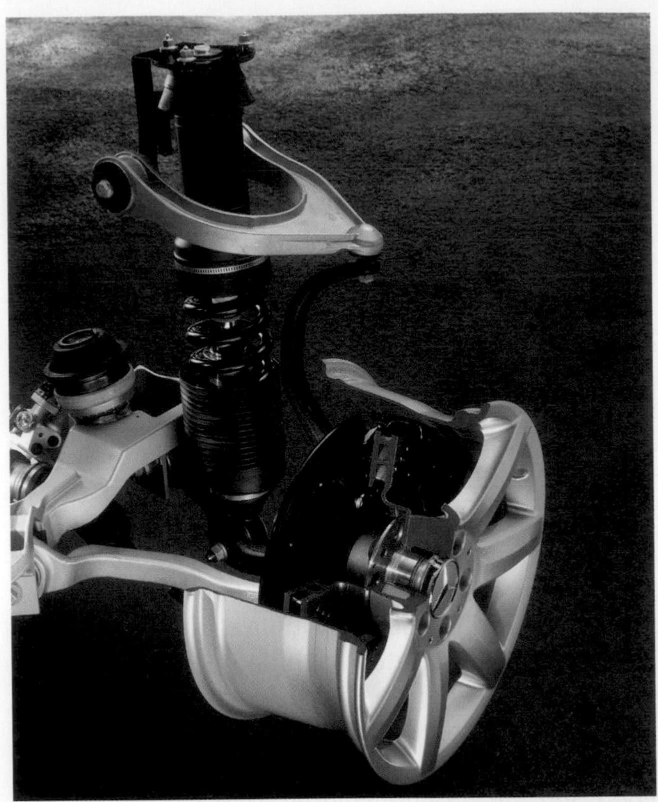

FIGURE 47–24

A front suspension system with an upper control arm shaped like a wishbone, which is what this type of suspension is called.

DaimlerChrysler Corporation

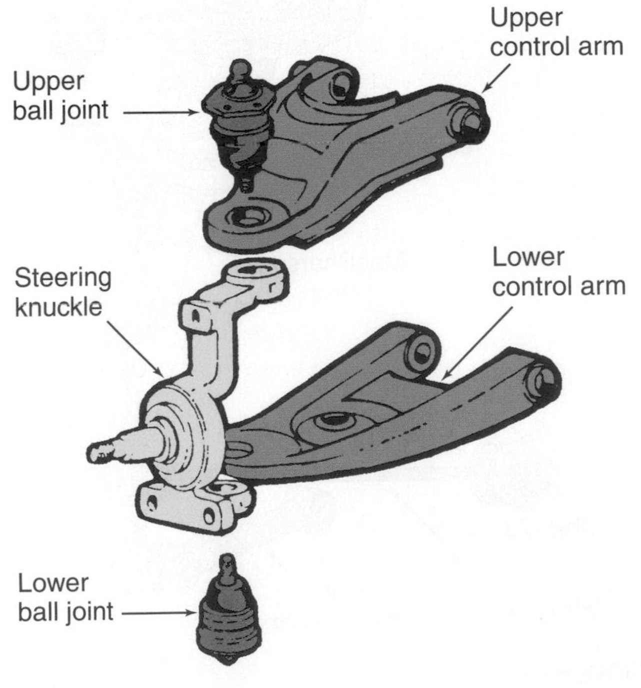

FIGURE 47–23

Ball joint locations.

Federal-Mogul Corporation

BALL JOINTS A ball joint **(Figure 47–25)** connects the steering knuckle to the control arm, allowing it to pivot on the control arm during steering. Ball joints also permit up-and-down movement of the control arm as the suspension reacts to road conditions. The ball joint stud protrudes from its socket through a rubber seal that keeps lubricating grease in the housing and keeps dirt out. Some ball joints require periodic lubrication, while most do not. These maintenance-free ball joints move in a prelubricated nylon bearing.

Ball joints are either load carrying or are followers (friction loaded). A load-carrying ball joint supports the car's weight and is generally in the control arm that holds or seats the spring. Load-carrying joints can be called tension-loaded or

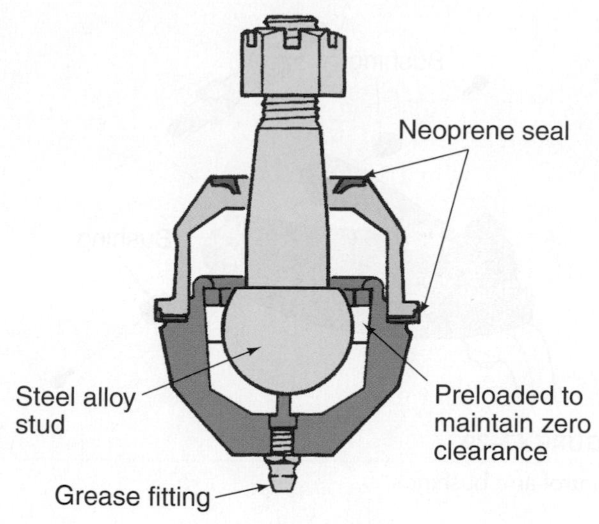

FIGURE 47–25

A typical ball joint.

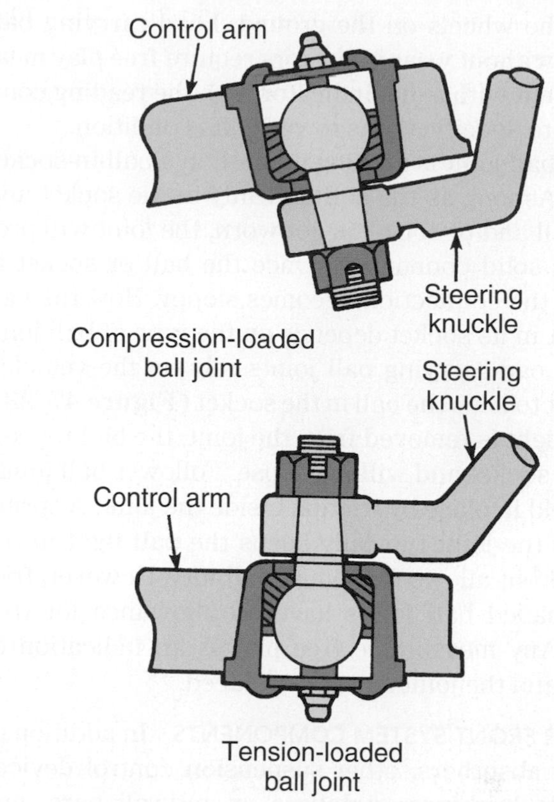

FIGURE 47–26

Two basic types of load-carrying ball joints.

compression-loaded ball joints **(Figure 47–26)**. The correct term depends on whether the force of the load tends to push the ball into the socket (compression) or pull it out of the socket (tension).

Follower ball joints are often called friction-loaded ball joints. A follower ball joint mounts on the control arm that does not provide a seat for the spring. The follower does not support vehicle weight and does not get the same stress as the load carrier.

Depending on the location of the suspension system's spring, either the upper or lower ball joint will be the load-carrying joint. In a MacPherson strut suspension, there is usually only one ball joint on each side and it is typically a follower. In modified strut suspensions, the ball joint is a load-carrying joint because the spring is positioned between the frame crossmember and the lower control arm.

Some ball joints have wear indicators. As the joint wears, the grease fitting of the joint recedes into the housing. When the shoulder of the fitting is flush with the housing, the joint needs to be replaced **(Figure 47–27)**. This type of joint must be checked

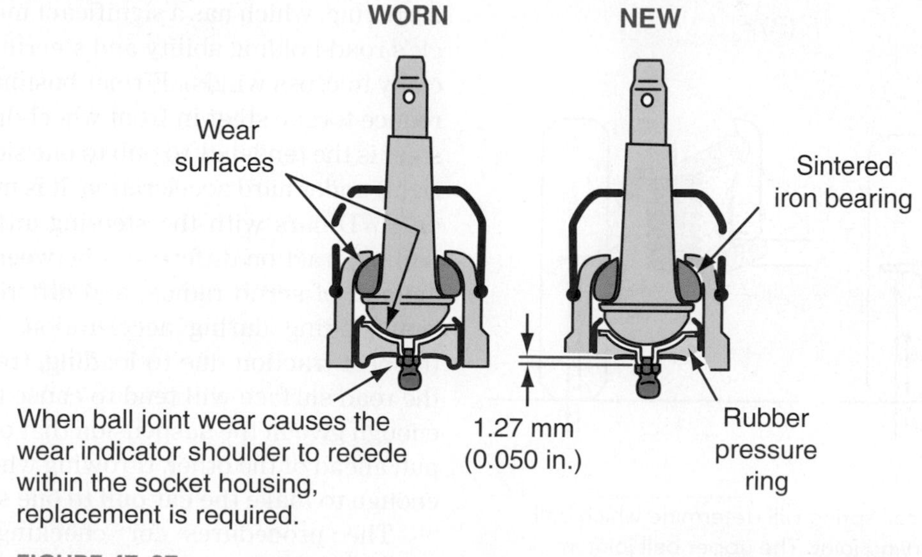

FIGURE 47–27

A wear indicator on a ball joint.

with the wheels on the ground. Load-carrying ball joints without wear indicators require free play measurement with a dial indicator with the reading compared to specifications to verify its condition.

A ball joint is nothing more than a ball-in-socket joint. As long as the ball is firmly in the socket and the ball and/or socket is not worn, the joint will provide a solid connection. Once the ball or socket is worn, the connection becomes sloppy. How the ball is kept in its socket depends on the type of ball joint it is. Load-carrying ball joints rely on the vehicle's weight to keep the ball in the socket **(Figure 47–28)**. As weight is removed from the joint, the ball relaxes in the socket and will feel loose. Follower ball joints are held in place by friction inside the joint. A spring inside the joint typically keeps the ball tight in the socket but allows for some flexibility. However, friction-loaded ball joints have no allowance for free play. Any measurable free play is an indication of wear and the joint must be replaced.

OTHER FRONT SYSTEM COMPONENTS In addition to shock absorbers, other suspension control devices include bushings, stabilizer or anti-roll bars, and strut rods. In the design of these suspension control devices, the difference between sprung and unsprung weight is important.

Bushings. Rubber or polyurethane bushings are found on many suspension components, such as the control arms **(Figure 47–29)**, radius arms, and strut rods. They make good suspension system pivots, minimize the number of lubrication points, and allow for slight assembly misalignments. Bushings help to absorb road shock, allow some movement, and reduce noise entering the vehicle.

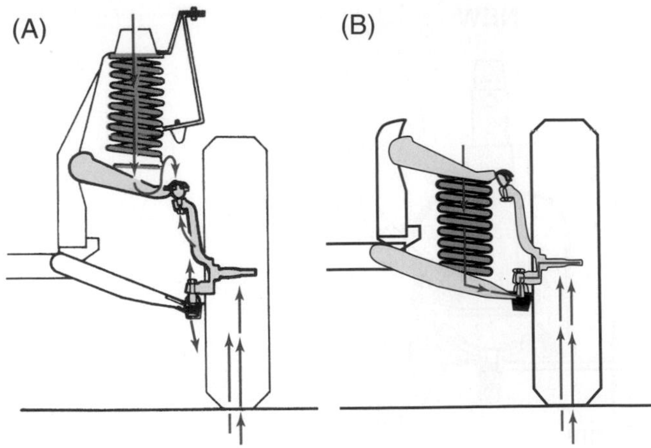

FIGURE 47–28

The location of the coil spring will determine which ball joint is the load-carrying joint. The upper ball joint in (A) and the lower ball joint in (B) are load-carrying.

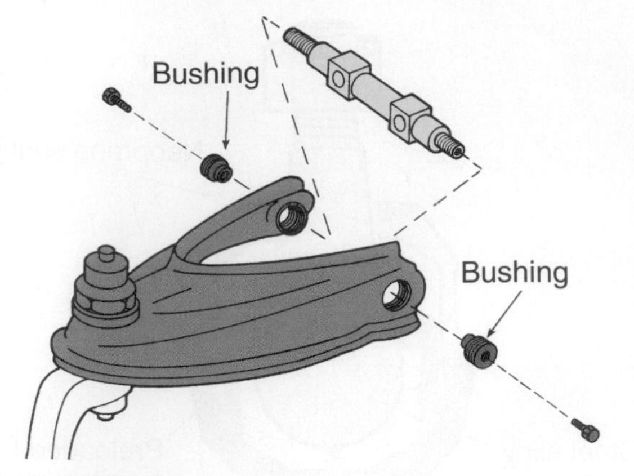

FIGURE 47–29

Control arm bushings.

Suspension bushings can deteriorate fairly rapidly, causing tire wear. They are a common cause of misalignment. Replacement bushings come in two basic varieties: stock and performance. The latter are usually made of a high-grade polyurethane material and are sold primarily as an upgrade to improve handling response and ride control. Do not use oil or grease when servicing the chassis or replacing any rubber or urethane suspension bushings. This can cause swelling and rapid deterioration of the bushings.

The harder urethane bushings eliminate unwanted compliance or give in the suspension. When hard cornering overtaxes stock bushings and causes them to deflect excessively, undesirable chamber changes occur in the front wheels that result in excessive outer shoulder wear on the tires. Compliance also slows down the action of the anti-roll bar when cornering, which has a significant impact on the vehicle's road-holding ability and steering stability (especially in cross winds). Firmer bushings can also help reduce torque steer in front-wheel-drive cars. Torque steer is the tendency to pull to one side (usually to the right) under hard acceleration. It is more of a problem on FWD cars with the steering and driving wheels pulling, traction differences between the wheels, the amount of scrub radius, and differences in suspension loading during acceleration. The wheel with the best traction due to loading, tread condition, or the road surface will tend to cause the pull. There is enough give in the suspension that one wheel tries to pull ahead of the other, throwing wheel alignment off enough to make the car pull to one side.

The procedures for checking bushings and the methods for replacing them are given in service information.

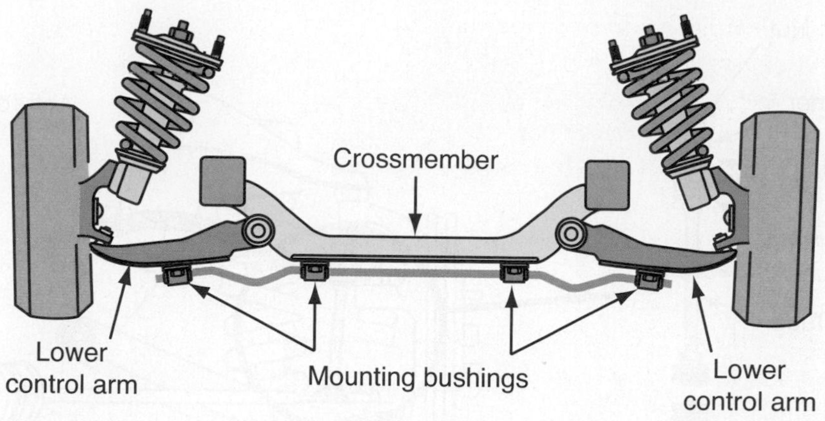

FIGURE 47–30

The typical location of an anti-roll bar.

Stabilizers. A variety of devices are used with the basic suspension components to provide additional stability. One of the most common is the **anti-roll bar**. This bar is commonly referred to as a sway bar, **anti-sway bar**, **roll bar**, or stabilizer bar. Its actual purpose is to provide a more level ride and to prevent body roll when cornering. This is a metal rod running between the opposite lower control arms **(Figure 47–30)**. As the suspension at one wheel responds to the road surface, the anti-roll bar transfers a similar movement to the suspension at the other wheel. For example, if the right wheel is drawn down by a dip in the road surface, the anti-roll bar is drawn with it creating a downward draw on the left wheel as well. In this way, a more level ride is produced. Roll during cornering is also reduced.

If both wheels go into a jounce, the anti-roll bar simply rotates in its insulator bushings. It is a different matter when only one wheel goes into jounce. The bar twists, just like a torsion bar, to lift the frame and the opposite suspension arm. This action reduces body roll.

The anti-roll bar can be a one-piece, U-shaped rod fastened directly into the control arms with rubber bushings, or it can be attached to each control arm by a separate anti-roll bar link **(Figure 47–31)**. The anti-roll bar link may be comprised of a bolt, nut, washer, and rubber bushings. Many modern anti-roll bar links use a steel or plastic rod and ball-socket joints to connect the anti-roll bar to the control arm or strut assembly. The anti-roll bar is also mounted to the frame in the centre with rubber bushings **(Figure 47–32)**. If the anti-roll bar is too small, it has little effect on stability.

On suspensions that use single-housing lower control arms instead of wishbone types, the anti-roll

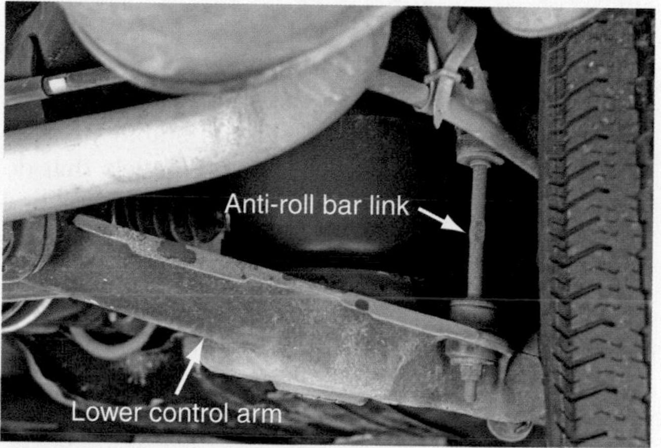

FIGURE 47–31

An anti-roll bar link connects the anti-roll bar to the lower control arm.

FIGURE 47–32

The centre section of the anti-roll bar rides in bushings.

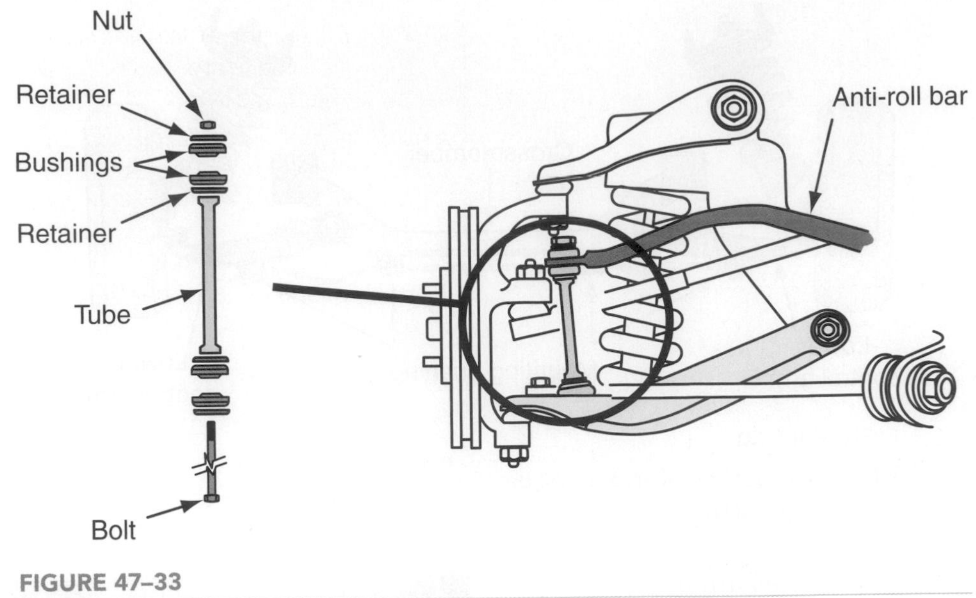

FIGURE 47–33

The strut bar assembly for attaching the anti-roll bar to the control arm.

bar can also be used to add lateral stability to the control arm. Strut rods are used on models that do not use the anti-roll bar like this. They are attached to the arm and frame with bushings, allowing the arm a limited amount of forward and backward movement **(Figure 47–33)**. Strut rods are directly affected by braking forces and road shocks, and their failure can quickly lead to failure of the entire suspension system.

Four-Link or Multilink Front Suspension

A four-link front suspension, also called a multilink suspension, fixes the wheel with four rod-type control arms and the tie rod **(Figure 47–34)**. The suspension strut supports the vehicle weight against the body via the load-bearing link. By separating wheel attachment and suspension elements, this suspension optimizes ride quality and movement. The influence of drive forces on the steering system is minimal.

Four-Wheel-Drive Front Suspensions

Most modern four-wheel-drive (4WD) trucks use an independent front suspension. The suspension may use short-long arms and a torsion bar or may use a multilink arrangement **(Figure 47–35)**. Mounting the front differential solidly to the frame and using two short drive axles allows for an independent suspension that improves ride quality compared to older, dependent suspensions. Older 4WD trucks and some new heavy-duty versions use a live axle. Although very strong, using a live axle in the front suspension reduces ride quality and handling.

BASIC FRONT-SUSPENSION DIAGNOSIS

Diagnosis of suspension problems should follow a logical sequence. The following procedure can be used on most vehicles; however, it is also best to follow the sequence given by the manufacturer for a specific vehicle.

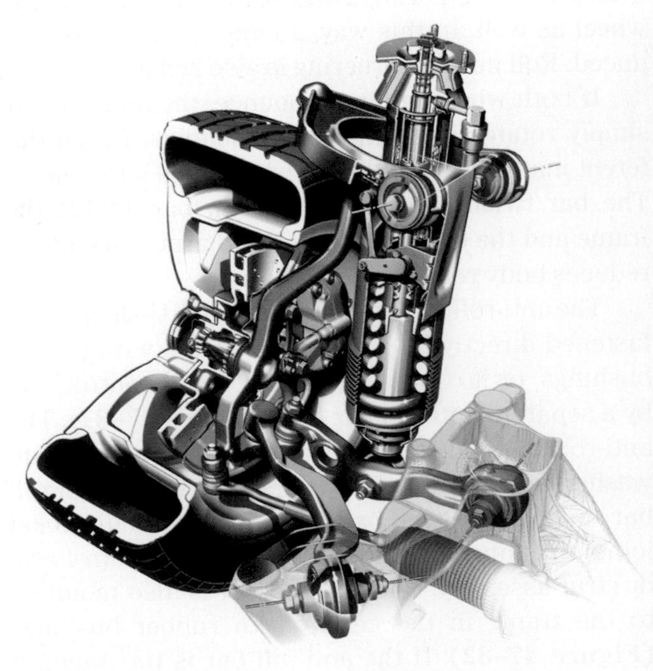

FIGURE 47–34

A four-link front suspension system.

DaimlerChrysler Corporation

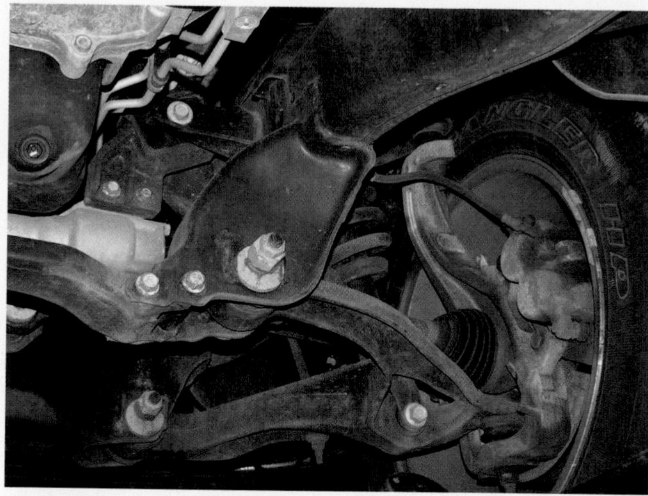

FIGURE 47–35

An independent 4WD front suspension.

PROCEDURE

Diagnosing a Suspension System

STEP 1 Take the vehicle on a road test, and verify the customer's concern.

STEP 2 Inspect the tires. Check their condition and air pressure. Also make sure that the tires and wheels are the correct size.

STEP 3 Inspect the chassis and underbody. Remove any excessive accumulation of mud, dirt, or road deposits. Then do the following:

- Inspect all parts to identify any aftermarket modifications that may have been made.
- Check vehicle attitude for evidence of overloading or sagging. Be sure the chassis height is correct.
- Raise the vehicle off the floor. Grasp the upper and lower surfaces of the tire, and shake each front wheel to check for worn wheel bearings.
- Look for loose or damaged front and rear suspension parts.
- Check loose, damaged, or missing suspension bolts.
- Check the ball joints for looseness and wear.
- Check the condition of the struts' upper mounts.
- Check the shock absorbers and struts for signs of fluid leakage **(Figure 47–36)** and damage.
- Check all of the mountings for the shocks and struts.
- Check all suspension bushings for looseness, splits, cracks, misplacement, and noises.
- Check the steering mounts, linkages, and all connections for looseness, binding, or damage.
- Check for damaged or sagging springs.
- Check the drive axles for damage and looseness.

STEP 4 If the cause of the customer's concern was found, make the repair as necessary, and verify that the repair fixed the problem.

STEP 5 If the cause of the concern was not found, refer to the symptom chart given in the service information, and conduct all applicable checks. Then make the repair as necessary, and verify that the repair fixed the problem.

SHOP TALK

Before visual inspection or suspension height measurements can be performed, the vehicle must be on a level surface. Tires must be at recommended pressures; the gas tank must be full; and there should be no passenger or luggage compartment load. Beginning at the rear bumper, jounce the car up and down several times. Proceed to the front bumper and repeat, releasing during the same cycle as rear jounce.

Shock Absorber or Strut Bounce Test

With the vehicle on the ground, a quick check, called the bounce test, can be performed to check the operation of the shocks and struts. When the bounce test is performed, the bumper is pushed two or three times downward with considerable weight applied on each corner of the vehicle. The bumper is released after each push, and the vehicle should oscillate about 1½ cycles and then settle. One free upward bounce should stop the vertical chassis movement if the shock absorber or strut provides proper spring control. If the vehicle's bumper does more than 1½ free upward bounces, the shock absorber or strut is defective.

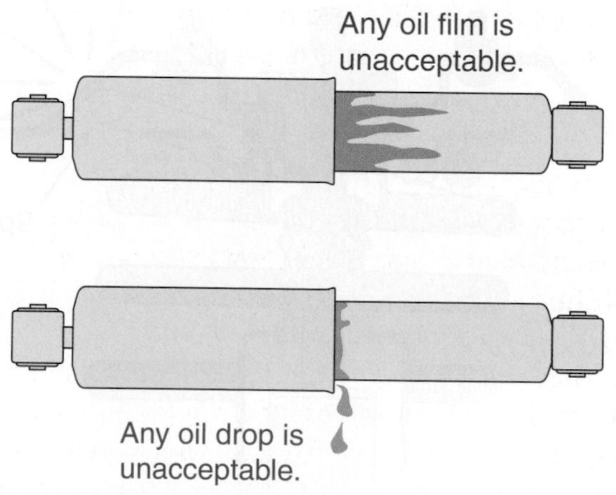
Any oil film is unacceptable.
Any oil drop is unacceptable.

FIGURE 47–36

Check the shock absorbers for signs of fluid leakage.

Excessive Body Roll

If a vehicle leans or rolls excessively during corners, the anti-roll bar links are likely broken. Broken links can also cause noise over bumps and while turning.

Noises

Abnormal noises from the suspension system can be caused by a number of problems. Tire noise varies with road surface conditions, whereas differential noise is not affected when various road surfaces are encountered. Uneven tread surfaces may cause tire noises that seem to originate elsewhere in the vehicle. These noises may be confused with differential noise. Differential noise usually varies with acceleration and deceleration, whereas tire noise remains more constant in relation to these forces. Tire noise is most pronounced on smooth asphalt road surfaces at speeds of 24 to 72 km/h (15 to 45 mph).

Rattling on road irregularities can be caused by worn shock absorber bushings or grommets, worn spring insulators, a broken coil spring or broken spring insulators, worn control arm bushings, worn stabilizer bar bushings, worn strut rod grommets, worn leaf spring shackles and bushings, and worn torsion bars, anchors, and bushings. Dry or worn control arm bushings may cause a squeaking noise on irregular road surfaces.

Chatter while cornering can be caused by worn upper strut mounts. Front strut noise on sharp turns or during suspension jounce may be caused by interference between the coil spring and the strut tower **(Figure 47–37)** or between the coil spring and the upper mount. Worn upper strut mounts or bearings can also cause binding and poor steering return.

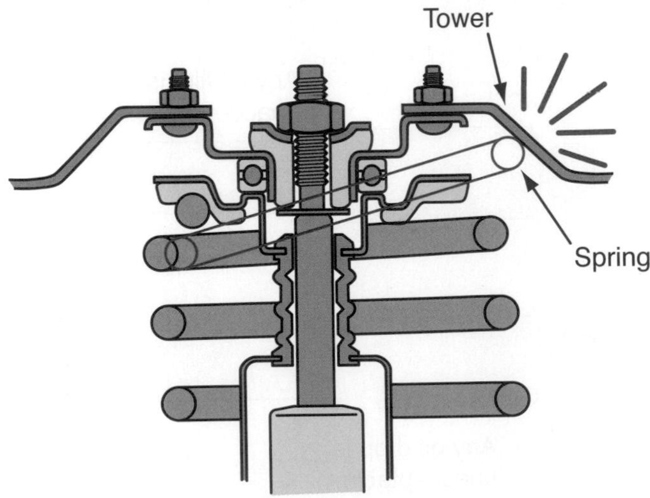

FIGURE 47–37

Coil spring to strut tower interference.

Knocking noises can be caused by broken anti-roll bar links, worn and loose ball joints, worn control arm, strut rod, or radius arm bushings. Severely worn struts or shocks can also cause knocking sounds, especially over very bumpy roads.

Chassis Height Specifications

A quick overall visual inspection detects any obvious sag from front to rear or from side to side. Under the car, at the level of the two ends of the control arms, check for out-of-level, damaged, or worn rubber bumpers, or shiny or worn spring coils. All indicate weak coil springs. A quick inspection of the tires to ensure that they are all of the same size and condition should also be made.

A more accurate inspection reveals less obvious problems by measuring heights at specific points on each side of the suspension system.

USING SERVICE INFORMATION

For the most accurate measurement of chassis height, use the service information to check against the manufacturer's recommendations for the specific model. Photo Sequence 39 shows the typical procedure for checking vehicle ride height. Be careful. The measurement points vary from one model to another, even if manufactured by the same company. When coil spring wear is suspected, it might be necessary to load the vehicle to the manufacturer's suggested capacities and measure at the designated points.

FRONT-SUSPENSION COMPONENT SERVICING

Each major component of the suspension system needs to be carefully checked. Each has its own procedure for doing this as well as the service procedures. The only maintenance required on suspension systems is a periodic chassis lubrication. If the owner has failed to have this done, many problems can result. Very few vehicles require any periodic chassis lubrication anymore, as nearly all suspension and steering components are sealed and greased for life. However, if a vehicle has had components replaced, such as ball joints or steering sockets, these parts will likely need to be greased.

GO TO Chapter 8 for details on how to lubricate the chassis.

MEASURING FRONT AND REAR CURB RIDING HEIGHT

PHOTO SEQUENCE 39

P39–1 Position the car on a level shop floor or alignment rack.

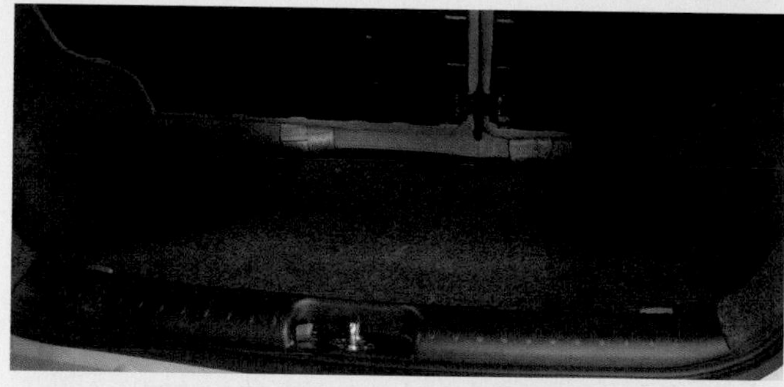

P39–2 Check the trunk or hatch for extra weight.

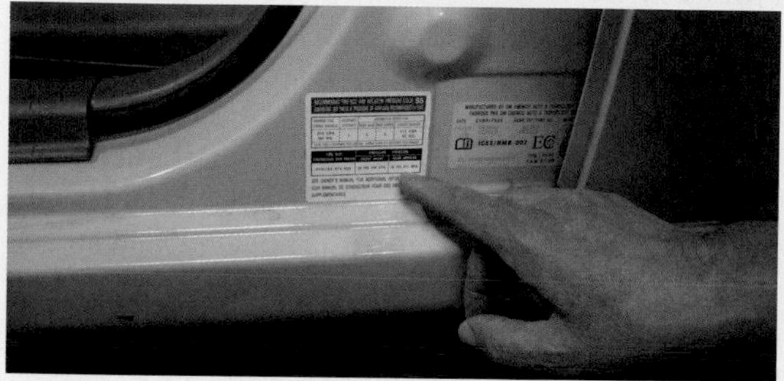

P39–3 Locate the tire placard, and identify the tire size and inflation pressure.

P39–4 Check the tires for the recommended pressure.

P39–5 Inflate as required.

P39–6 Find the vehicle's specified curb riding height measurement locations from a service information system or manual.

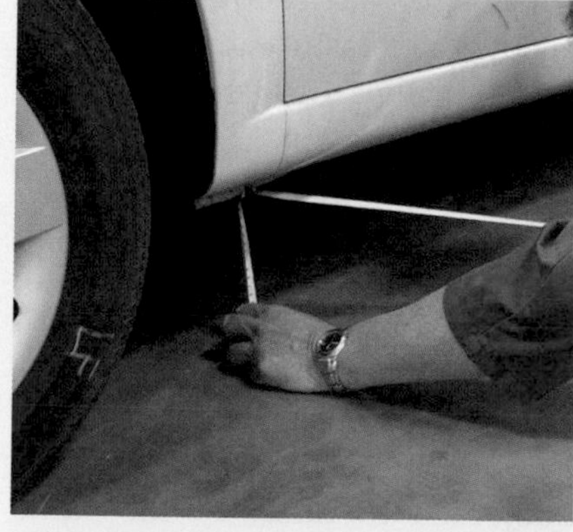

P39–7 Measure and record the front right and left curb riding height.

P39–8 Measure and record the rear right and left curb riding height.

P39–9 Compare the measurement results to the specified curb riding height.

Coil Springs

The never-ending twisting and untwisting of the coil spring (or the torsion bar) lead to inevitable loss of elasticity and spring sag. Coil springs, then, require replacement because they sag in service. A sagged coil spring upsets vehicle trim height, resulting in upset wheel alignment angles, steering angles, headlight aiming, braking distribution, riding quality, tire tread life, shock life, and U-joint life.

Coil springs also break. Downsized cars are often forced to carry the same loads as their larger counterparts, mostly because people and their hauling needs did not downsize along with the cars.

> **CAUTION!**
>
> The coil spring exerts a tremendous force on the control arm. Before you disconnect either control arm from the knuckle for any service operation, contain the spring with a spring compressor to prevent it from flying out and causing injury.

REMOVING A SPRING To remove a coil spring, raise and support the vehicle by its frame. Let the control arm hang free. Remove wheels, shock absorbers, and stabilizer links. Disconnect the outer tie-rod ends from their respective arms. Remove the brake caliper and support it with a wire or bungee cord so that it does not hang on the brake hose.

Unload the ball joints with a roll-around floor jack. Jack under the lower control arm from the opposite side of the vehicle. This allows the jack to roll back when the control arm is lowered. Position the jack as close to the lower ball joint as possible for maximum leverage against the spring.

The spring is ready for the installation of the spring compressor **(Figure 47–38)**. There are many different types of spring compressors. One type uses a threaded compression rod that fits through two plates, an upper and lower ball nut, a thrust washer, and a forcing nut. The two plates are positioned at either end of the spring. The compression rod fits through the plates with a ball nut at either end. The upper ball nut is pinned to the rod. The thrust washer and forcing nut are threaded onto the end of the rod. Turning the forcing nut draws the two plates together and compresses the spring.

Removing a torsion bar often requires a special tool used to unload the bar. Begin by lifting and supporting the vehicle by the frame. Remove any covers or plates covering the torsion bar adjuster. Measure and record or mark the adjuster bolt before loosening it. This measurement will be used when installing the bar and setting the tension. Install the torsion bar tension relief tool, and tighten the tool until the torsion bar adjuster is loose. Remove the adjuster, and loosen the tool until tension is off the bar. Remove the tool, and then remove the torsion bar from the control arm.

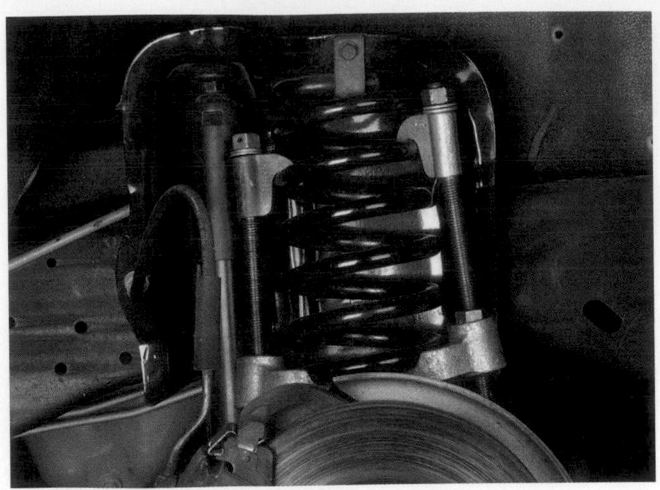

FIGURE 47–38
A coil spring compressor is used to compress the spring before disconnecting some suspension parts.

Install the new torsion bar, and apply tension using the tool. Tighten until the new adjustment bolt can be installed, and set the bolt at the depth measured on the old bolt before removal. Loosen and remove the tool. Once the torsion bar is installed, set the vehicle back on the ground; bounce it several times to settle the suspension; and measure the ride height. If ride height is incorrect, adjust the torsion bar until the height is within specifications. Once ride height is correct, check the wheel alignment.

In some cases, it is necessary to break the tapers of both upper and lower ball joints so that the steering knuckle can be moved to one side **(Figure 47–39)**. If the vehicle is equipped with a strut rod, this must be disconnected at the lower control arm. Push the control arm down until the spring can be removed. If necessary, a pry bar can be used to remove the spring from its lower seat. Remove the spring and compressor.

If the same spring is to be reinstalled, leave the compressor in position. If a new spring is to be used, slowly release the pressure on the tool by backing off the forcing nut. Compress the new spring prior to installing it.

Torsion Bars

Torsion bars are subjected to many of the same conditions affecting coil springs. Periodic adjustment of the torsion bars is necessary to maintain the

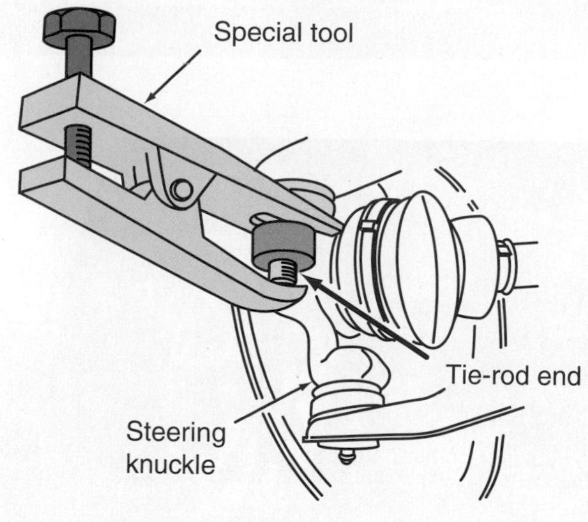

FIGURE 47–39

When this tool is expanded, it can force the tie-rod end or ball joint stud out of the steering knuckle taper.

proper height. Replacement is sometimes necessary because of breakage. It should be noted that the bars are not interchangeable from side to side, and they should be replaced in pairs.

Height inspection and measurements for vehicles with torsion bar suspensions are the same for coil springs. However, sagging can usually be corrected by adjusting the bars. Always follow the manufacturer's recommended procedures for adjusting torsion bars.

Ball Joints

Begin your inspection of a ball joint by checking to see if the ball joint has a wear indicator on it. If it does, check the placement of the grease fitting. If it is recessed, the ball joint is worn and should be replaced. On some vehicles, it is recommended that you check to see if the grease fitting can wiggle in the ball joint. If it does, the ball joint should be replaced. Always check the service information when checking ball joints.

Look carefully at the joint's boot. A damaged boot or joint seal will allow lubricant to leak out and allow dirt to enter and contaminate the lubricant. If the boot is damaged, the ball joint should be replaced.

If no boot damage is evident, gently squeeze the boot. If the boot is filled with grease, it will feel somewhat firm. If the joint has a grease fitting and appears not to be filled with grease, use a grease gun, and refill the joint. Fill the joint until fresh grease is seen flowing out of the boot's vent. If too much grease is forced into the joint or it is forced in too quickly, the boot can unseat or tear.

Ball joints should be checked for excessive wear. As stated before, load-carrying joints will have some play when the weight of the vehicle is taken off them. Follower joints should never have play. To check a load-carrying joint, it must be unloaded.

When the coil spring is on the lower control arm, raise the vehicle by jacking under the control arm as close to the ball joint as possible. This gives the maximum amount of leverage against the spring. The ball joint is unloaded when the upper strike out bumper is not in contact with the control arm or frame. A quick check for looseness can be made by using a pry bar between the tire and the ground. To find out if the ball joint is loose beyond manufacturer's specifications, use an accurate measuring device. The following checking procedures demonstrate the use of a dial indicator. The dial indicator is a precision instrument and should be handled carefully to prevent damage. The mounting procedure for the checking tool might vary depending on the style of ball joint used on the vehicle. Manufacturer's tolerances can be axial (vertical), radial (horizontal), or both. To conduct these checks, follow these procedures.

TYPICAL RADIAL CHECK For a radial check, attach a dial indicator to the control arm of the ball joint being checked. Position and adjust the plunger of the dial indicator against the side of the steering knuckle close to the ball joint being checked. Slip the dial ring to the zero marking. Move the wheel in and out, and note the amount of ball joint radial looseness registered on the dial **(Figure 47–40)**. The procedure for checking the radial movement of a lower ball joint on a MacPherson strut front suspension is shown in Photo Sequence 40.

FIGURE 47–40

A typical mounting of a dial indicator for a radial check.

Federal-Mogul Corporation

PHOTO SEQUENCE 40

MEASURING THE LOWER BALL JOINT RADIAL MOVEMENT ON A MACPHERSON STRUT FRONT SUSPENSION

P40–1 Position the car correctly on a chassis hoist.

P40–2 Grasp the front tire at the top and bottom, and rock the tire inward and outward while a co-worker visually checks for movement in the front wheel bearing. If there is movement, adjust or replace the wheel bearing.

P40–3 Position a dial indicator against an inner surface of the steering knuckle close to the ball joint. Preload and zero the dial indicator.

P40–4 Grasp the bottom of the tire and push outward.

P40–5 With the tire held outward, read the dial indicator.

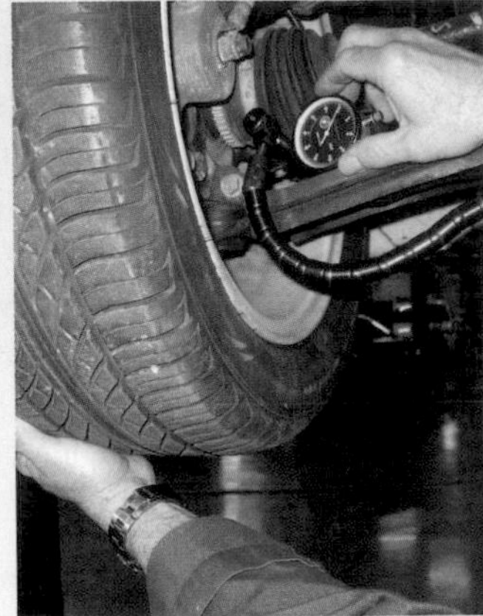

P40–6 Pull the bottom of the tire inward, and be sure the dial indicator reading is zero. Adjust the dial indicator as required.

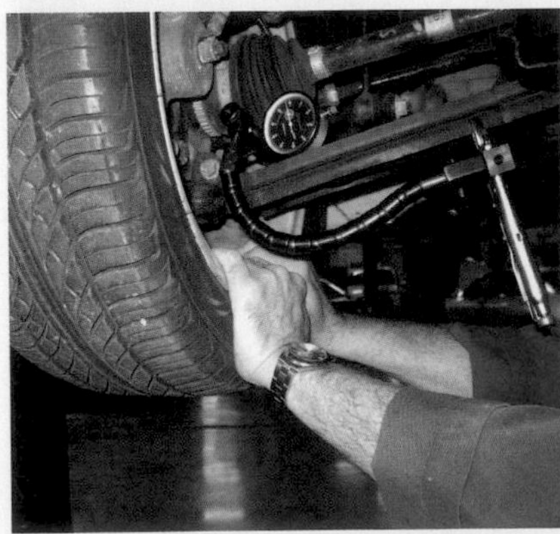

P40–7 Grasp the bottom of the tire and push outward.

P40–8 With the tire held in this position, read the dial indicator.

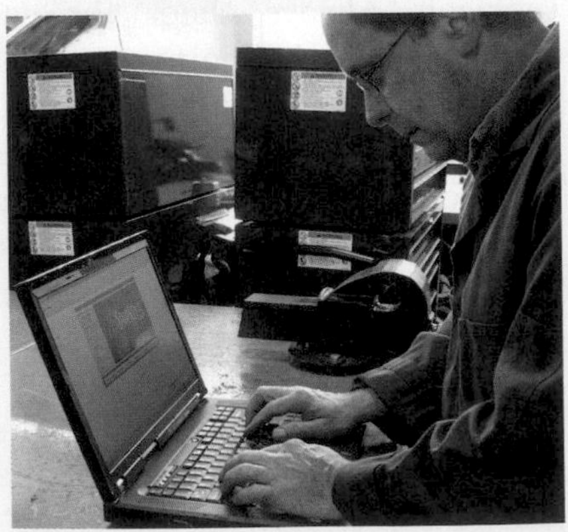

P40–9 Compare the measurement results to the ball joint specifications.

TYPICAL AXIAL CHECK For an axial check, first fasten the dial indicator to the control arm, and then clean off the flat on the spindle next to the ball joint stud nut. Position the dial indicator plunger on the flat of the spindle, and depress the plunger approximately 0.9 mm (0.35 in.). Turn the lever to tighten the indicator in place. Pry the bar between the floor and tire. Record the reading **(Figure 47–41)**.

If the ball joint looseness reading on the dial indicator exceeds manufacturer's specifications, the ball joint should be replaced.

When the load-carrying ball joints are on the upper control arm (spring mounted on the upper arm), place a steel wedge between the bottom of the upper control arm and the frame. Next, raise the vehicle by its frame to unload the ball joints and hold them in their normal position. Pry up on the tire to check for play in the load-carrying joint. To determine the condition of the non-load-carrying (or follower) ball joint, vigorously push and pull on the tire while watching the ball joint for signs of movement. Refer to the manufacturer's specifications for tolerances.

INSPECTION OF WEAR INDICATORS Wear indicator–type ball joints must remain loaded to check for wear. The vehicle should be checked with the suspension at curb height. The most common type has a small diameter boss, which protrudes from the centre of the lower housing. As wear occurs internally, this boss recedes very gradually into the housing (review Figure 47–27). When it is flush with the housing, the ball joint should be replaced. To remove and install a ball joint, follow the procedure given in the service information.

FIGURE 47–41

A typical mounting of a dial indicator for an axial check.

Federal-Mogul Corporation

Replacing Ball Joints

Ball joints are mounted to the control arm in one of four basic ways: rivets, bolts, press-fit, and threaded. To replace a ball joint, safely raise and support the vehicle. Depending on the vehicle and spring location, you may need to support the control arm. In most cases, remove the brake caliper, and support it using a wire or bungee cord. Next, remove the ball joint to the steering knuckle nut, and separate the knuckle from the joint. Once the knuckle is out of the way, you will have access to the ball joint.

Many ball joints are riveted into place at the factory. Replacing these joints required drilling or cutting out the rivets. Once the rivets are removed, the new ball joint is bolted in the control arm. Bolted-in ball joints are removed by simply removing the bolts and nuts securing it to the arm. Once the new joint is installed, torque the bolts and nuts to specifications.

The most common method is press-fit. Some manufacturers require you to replace the entire control arm assembly if a ball joint is to be replaced. In these cases, the ball joint and control arm are made as a single assembly, and individual parts are not available.

Press-fit ball joints are typically removed and installed using a special ball joint press. While pressing the ball joint out of or into the control arm, make sure you do not damage the arm.

A few vehicles use threaded ball joints. In this style, external threads are formed in the outside of the ball joint housing. The joint threads into the control arm. To replace, remove the steering knuckle and unscrew the ball joint. Install the new joint, and torque it to specifications.

Once the new ball joints are installed, reinstall the steering knuckle, and torque the ball joint nut to specifications. If a castle nut is used and the hole in the ball joint stud does not align with the hole in the nut, tighten the nut until the two align. Install a new cotter pin, and bend one or both prongs so that the pin cannot back out of the joint.

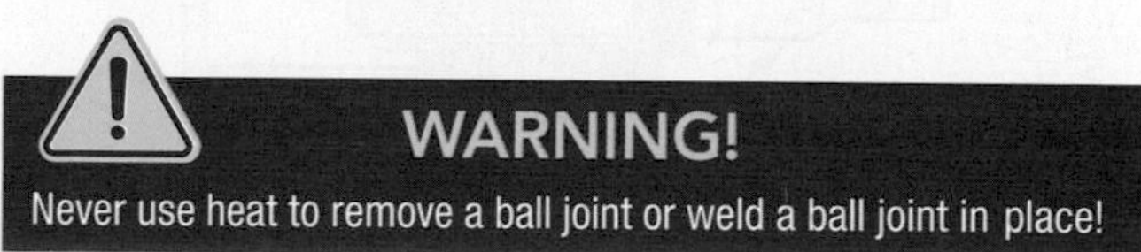

WARNING!
Never use heat to remove a ball joint or weld a ball joint in place!

Control Arm Bushings

If the bushings, which attach the arms to the frame, are not in good condition, precise wheel alignment settings cannot be maintained.

Visually inspect each rubber bushing for signs of distortion, movement, off-centre condition, and presence of heavy cracking. Check metal bushings for noise and loose seals.

To remove the control arm bushings, raise the vehicle and support the frame on safety jack stands. Remove the wheel assembly. Install a spring compressor on the coil spring.

Disconnect the ball joint studs from the steering knuckle as described previously. Remove the bolts attaching the control arm assembly to the frame and remove the control arm.

Bushings are pressed in and out of their bores by using special tools. A special tool is installed over the bushing **(Figure 47–42)** after the correct size adapter for the tool has been selected. Tightening the tool pushes the bushing out of the control arm. The same process is used to press a new bushing into the control arm. As the special tool is tightened, the bushing moves into its bore. When installing new bushings, make sure they stay straight while they are being pressed in.

Once the new bushings are started into the control arm, measure and mark the centre between mounting holes and centre the control arm. Now, alternately press in the bushings on each side, keeping the reference marks aligned. This ensures the shaft is not off centre, causing binding. End cap nuts or bolts should not be torqued until the vehicle is at curb height and the suspension has been bounced and allowed to settle out.

Reattach the control arm, and tighten the bolts to specifications, and then install the coil spring into position. Install the ball joint studs into the control arms. Remove the coil spring compressor. Install the wheel assembly, and lower the car. Road-test the car, retighten all bolts, and set wheel alignment.

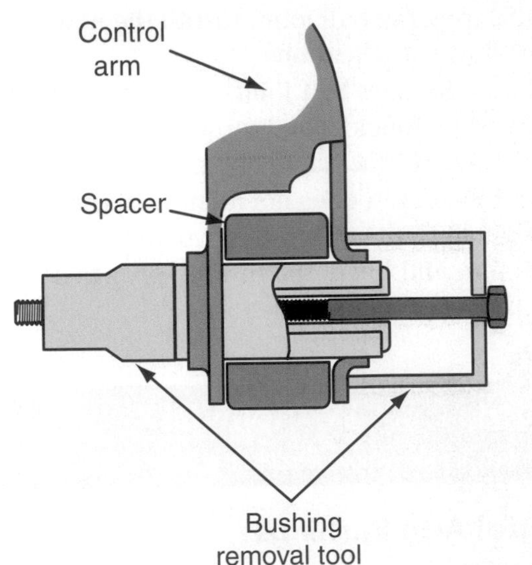

FIGURE 47–42

Removing a control-arm bushing.

Strut Rod Bushings

Except in the case of accidental damage, the strut rod itself is rarely replaced. Rather, it is the bushing that wears, deteriorates, and needs replacement.

Anti-Roll Bar Bushings

These bushings anchor the anti-roll bar securely to the vehicle frame and the control arms on each side. The condition of the bushings affects the performance of the bar. Visual inspection of mounting bushings indicates if the bushings are worn, have taken a permanent set, or are possibly missing. Also check the anti-roll bar links. Any damage indicates that they should be replaced.

Anti-roll bar links on older vehicles typically use a long bolt, washers, bushings, a spacer, and a nut. The link is assembled between the control arm and anti-roll bar **(Figure 47–43)**. Over time, the links rust and

FIGURE 47–43

A broken anti-roll bar link.

FIGURE 47–44

An example of an anti-roll bar link.

break. Install an anti-roll bar link kit, making sure the the bushings and washers are installed in the correct positions. Tighten the bolt and nut to specifications to prevent overtightening and collapsing the bushings.

Most vehicles now use plastic or metal links with ball socket joints on the ends. The ball sockets connect the link to the anti-roll bar and control arm or strut **(Figure 47–44)**. This type of link typically does not come apart without destroying the ball socket. This means that if the link needs to be removed, such as when replacing a strut, the link will likely need to be replaced also as the ball sockets do not usually separate without damage.

Shock Absorbers

A shock absorber that is functioning properly ensures vehicle stability, handling, and rideability. Most motorists fail to notice gradual changes in the operation of their cars as a result of worn shock absorbers. Some common indications of shock absorber failure follow:

- Steering and handling are more difficult.
- Braking is not smooth.
- Bouncing is excessive after bumps and stops.
- Tire wear patterns are unusual, especially cupping.
- Springs are bottoming out.

WARNING!

Gas-pressurized shock absorbers will extend on their own and can create a dangerous situation. Never apply heat or a flame to the shock absorbers during removal; the heat will cause the gas to expand, and the shock will quickly extend. Before disposing of used gas-pressurized shocks, refer to the manufacturer's service information about releasing the gas pressure.

Shock absorbers should be inspected for loose mounting bolts and worn mounting bushings. If these components are loose, they will rattle, and replacement of the bushings and bolts is necessary. The upper mounts of struts should also be carefully checked.

In some shock absorbers, the bushing is permanently mounted in the shock, and the complete unit must be replaced if the bushing is worn. When the mounting bushings are worn, the shock absorber will not provide proper spring control.

Vibrations set up by a worn shock absorber can cause premature wear in many of the undercar systems. They can cause wear in the front and rear component parts of the suspension system, the linkage component parts of the steering system, and the U-joints and motor or transmission mounts of the driveline. Also, vibrations can cause unnatural wear patterns on the tires.

When replacing shocks, refer to the manufacturer's service information. Some rear shocks are mounted into the passenger compartment or trunk, requiring extra caution during replacement. For most front shocks, remove the shock's lower mounting bolt(s), and then remove the upper mount. If the shock mount uses bushings, install the new bushings on the shock to prevent metal-on-metal contact. Install the shock, and torque all fasteners to specs.

Rear shock replacement may require using a jack to support the rear axle or suspension, but the replacement procedure is basically the same as for front shock replacement. If the old shocks are gas pressurized, refer to the manufacturer's procedures for releasing the gas before disposing of the shock.

A shock absorber can be bench-tested. First, turn it up in the same direction it occupies in the vehicle. Then extend it fully. Next, turn it upside down and fully compress it. Repeat this operation several times. Install a new shock absorber if a lag or skip occurs near mid-stroke of the shaft's change in travel direction, or if the shaft seizes at any point in its travel, except at the ends. Also, install a new shock absorber if noise, other than a switch or click, is encountered when the stroke is reversed rapidly, if there are any leaks, or if action remains erratic after purging air.

MacPherson Strut Suspension

The MacPherson strut suspension system is based on a triangular design. The strut shaft is a structural member that does away with the upper control arm bushings and the upper ball joint. Because this shaft is also the shock absorber shaft, it receives a tremendous amount of force vertically and horizontally. Therefore, this assembly should be inspected very closely for leakage, bent shaft, and poor damping.

To remove and replace the MacPherson strut, proceed as shown in Photo Sequence 41.

During the disassembly of the strut, make sure you check the strut pivot bearing **(Figure 47–45)**. Move the bearing with your hand. If the bearing is hard to move or seems to bind, it must be replaced. When replacing the bearing, make sure the correct side is up. Manufacturers normally mark the up side with paint or some other marking. Make sure you check all rubber insulators for deterioration and other damage, and replace them if necessary. Also make sure you mark the eccentric camber bolts before loosening them. Returning the bolts in the same location will help maintain the correct camber angle after reassembly.

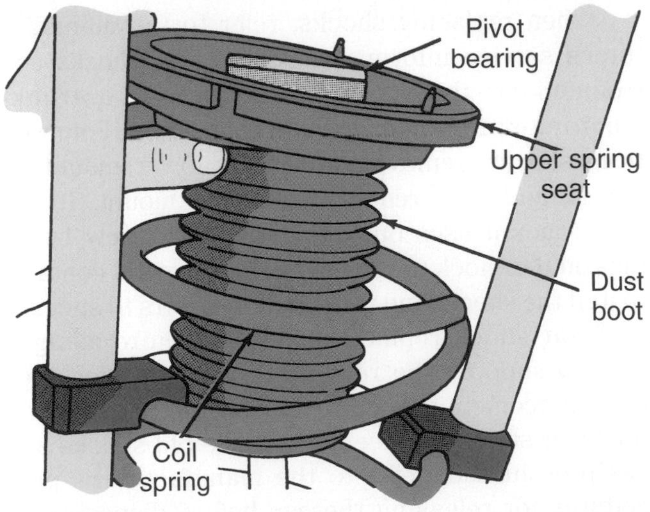

FIGURE 47–45

Check the strut's pivot bearing for free movement.

Chrysler Group LLC

REAR SUSPENSION SYSTEMS

There are three basic types of rear suspensions: live axle, semi-independent, and independent. There are distinct designs of each, but the types of components and the principles involved are the same as on front-suspension systems described earlier in this chapter. Live-axle suspensions are found on rear-wheel-drive (RWD) trucks, vans, and many four-wheel-drive (4WD) passenger cars. Semi-independent systems are used on front-wheel-drive (FWD) vehicles. Independent suspensions can be found on both RWD and FWD vehicles as well as 4WD cars.

Live-Axle Rear Suspension Systems

This traditional rear suspension system consists of springs used in conjunction with a live axle (one in which the final drive, axles, wheel bearings, and brakes act as a unit). The springs are either of leaf or coil type.

LEAF SPRING LIVE-AXLE SYSTEM Two springs—either multiple-leaf or monoleaf—are mounted at right angles to the axle and, along with the shock absorbers, are positioned below the rear axle housing. The front of the two springs is attached to brackets on the vehicle's frame by a bolt and bushing inserted through the eyes of the springs. Although the bushing allows the spring to move, it isolates the rest of the vehicle from noisy road vibrations.

The centre of each leaf spring is connected to the rear axle housing with U-bolts. Rubber bumpers are located between the rear axle housing and frame or unit body to dampen severe shocks. The rear eye pivot bushings are held to the frame with shackles, which attach to the springs by a bolt and bushing **(Figure 47–46)**.

The advantage to this live-axle suspension system is that the leaf springs act as excellent axle locators. The springs allow for the up-and-down movement of the axle and wheels, but at the same time, the springs prevent them from wandering. Therefore, control arms are not required with leaf springs.

There are some disadvantages to the live-axle suspension system. First, this design has a large amount of unsprung weight. Another drawback is the instability caused by the use of a solid axle.

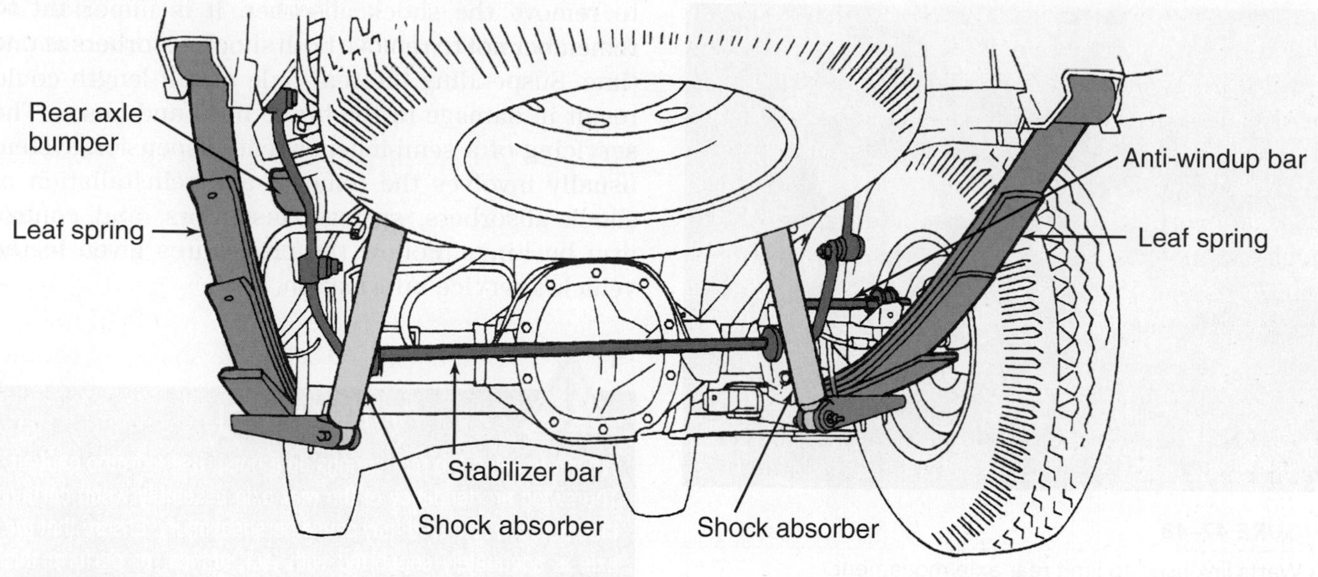

FIGURE 47–46

A typical rear suspension with a live axle.

Ford Motor Company

Because both rear wheels are connected to the same axle, movement up or down by one wheel affects the other. Consequently, poor traction results because both wheels are pushed out of alignment with the road. Under severe acceleration, this type of suspension is subject to **axle tramp**, or wheel hop, a rapid up-and-down jumping of the rear axle due to the torque absorption of the leaf springs. This condition can break spring mounts and shock absorbers and cause premature wear of wheel bearings. Axle tramp is reduced by mounting shock absorbers on the opposing sides (front and back) of the axle. Some heavy-duty vehicles have two-stage springs that allow the vehicle to ride comfortably with both a light or heavy load.

COIL SPRING LIVE-AXLE SYSTEM Some vehicles use two coil springs at the rear with a live rear axle. Because coil springs can only support weight and have little axle-locating capabilities, such vehicles need forward and lateral control arms or links. This type of suspension is called the link-type rigid axle.

The coil springs, located between the brackets on the axle housing and the vehicle body or frame, are held in place by the weight of the vehicle and sometimes by the shock absorbers **(Figure 47–47)**. The control arms are usually made of channelled steel and mounted with rubber bushings. Accelerating, driving, and braking torque are transmitted through three or four control arms, depending on the design. Two forward links are always used, but either one or two lateral links can be found on individual models. **Trailing arms** mount to the underside of the axle and run forward at a 90° angle to the axle to brackets on the car frame. Rubber bushings are used at mounting locations to permit up-and-down movement of the arms and to reduce noise and the effect of shock.

Some rear axle assemblies are connected to the body by two lower control arms and a tracking bar. A single torque arm is used in place of upper control arms and is rigidly mounted to the rear axle housing at the rear and through a rubber bushing to the transmission at the front. A few manufacturers are using Watts links in their rear suspensions **(Figure 47–48)**. A Watts link is used to limit rear axle motion from side to side.

LIVE-AXLE SUSPENSION SYSTEM SERVICING Typical service to both coil-spring and leaf-spring systems include the replacement of shock absorbers or

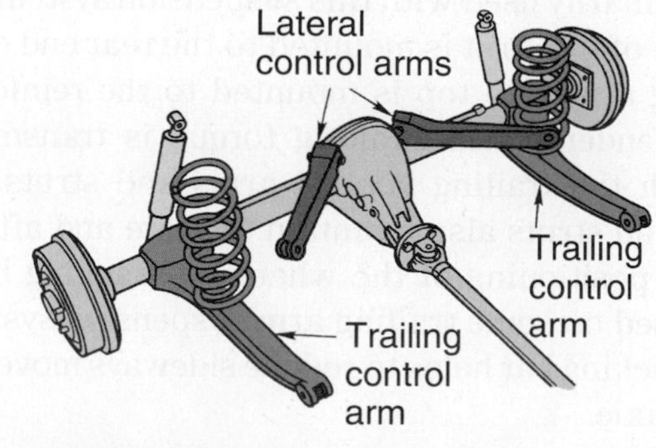

FIGURE 47–47

A typical live-axle suspension with coil springs.

FIGURE 47–48

A Watts link used to limit rear axle movement.

springs. Bushings, shackles, or control arms do not need replacement frequently. Always follow the procedures outlined in the service information whenever servicing the rear suspension.

Semi-Independent Rear Suspension Systems

A semi-independent suspension system is used on many front-wheel-drive models. On some, the suspension position is fixed by an axle beam, or crossmember, running between two trailing arms. Although there is a solid connection between the two halves of the suspension because of the axle beam, the beam twists as the wheel assemblies move up and down. The twisting action not only permits semi-independent suspension movement, but it also acts as a stabilizer. Frequently, a separate shock and spring trailing arm system is also used. In either an integrated or separate shock system, each rear wheel is independently suspended by a coil spring.

A coil spring and shock absorber–strut assembly are ordinarily used with this suspension system. The bottom of the strut is mounted to the rear end of the trailing arm. The top is mounted to the reinforced inner fender panel. Braking torque is transmitted through the trailing control arms and struts. The arms and struts also maintain the fore and aft, and lateral positioning of the wheels. A tracking bar is also used on some trailing arm suspension systems. The tracking bar helps to reduce sideways movement of the axle.

SEMI-INDEPENDENT SUSPENSION SYSTEM SERVICING

As in most rear-system servicing, the first step is to remove the shock absorber. It is important to remember not to remove both shock absorbers at one time. Suspending the rear axle at full length could result in damage to the brake lines and hoses. The servicing of a semi-independent suspension system usually involves the removal and reinstallation of shock absorbers, springs, insulators, and control arm bushings. Follow the procedures given in the vehicle's service information.

WARNING!

When removing the rear springs on a hoist, use extreme caution. The swing arc tendency of the rear axle assembly when certain fasteners are removed may cause personal injury. Ensure that the axle is adequately supported with jack stands, or perform this operation on the floor if necessary.

Independent Rear Suspension Systems

Independent suspensions can be found in large numbers on both FWD and RWD vehicles. The introduction of independent rear suspensions was brought about by the same concerns for improved traction ride that prompted the introduction of independent front suspensions. If the wheels can move separately on the road, traction and ride is improved.

Independent coil-spring rear suspensions can have several control arm arrangements. For example, A-shaped control arms are sometimes employed. When the wide bottom of a control arm is toward the front of the car, and the point turns in to meet the upright, they are called trailing arms **(Figure 47–49)**. When the entire A-shaped control arms are

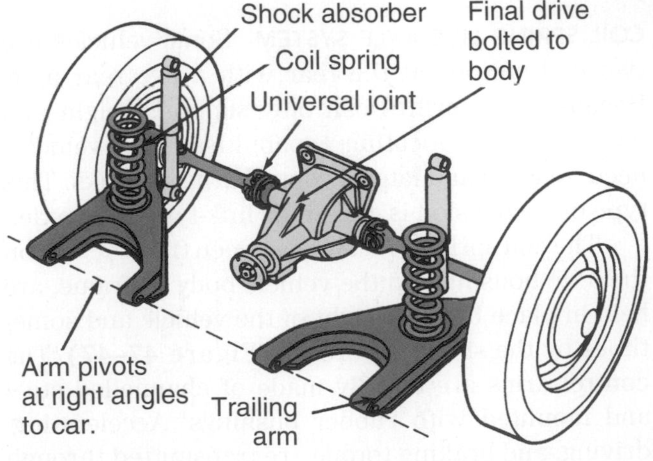

FIGURE 47–49

Trailing arms are often used with independent rear suspensions.

REMOVING AND REPLACING A MACPHERSON STRUT

P41–1 The top of the strut assembly is mounted directly to the chassis of the car.

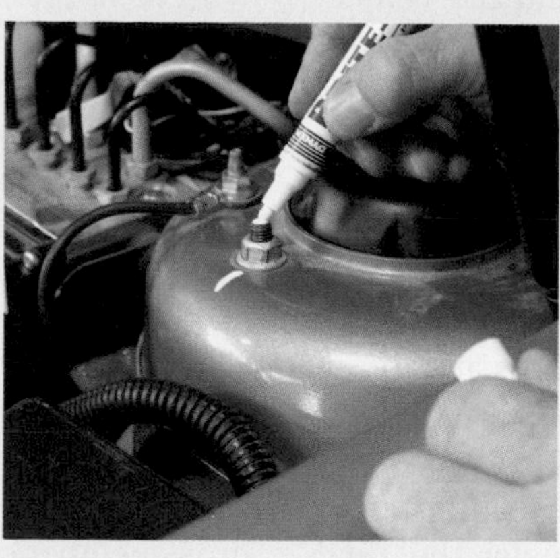

P41–2 Prior to loosening the strut chassis bolts, scribe alignment marks on the strut bolts and the chassis.

P41–3 With the top strut bolts or nuts removed, raise the car to a working height. It is important that the car be supported on its frame and not on its suspension components.

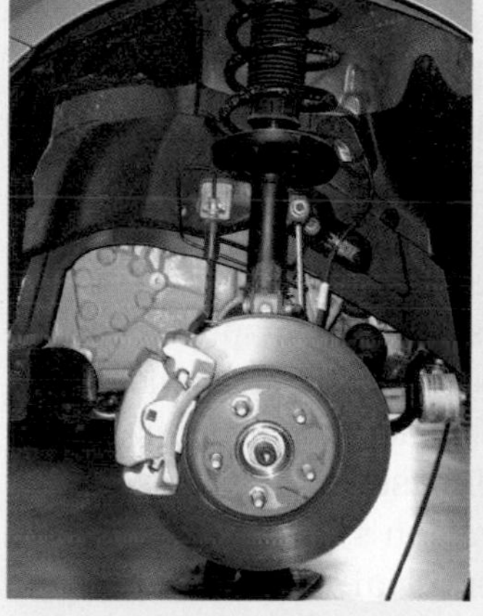

P41–4 Remove the wheel assembly. The strut is accessible from the wheel well after the wheel is removed.

P41–5 Remove the bolt that fastens the brake line or hose to the strut assembly.

P41–6 Remove the strut's two steering knuckle bolts.

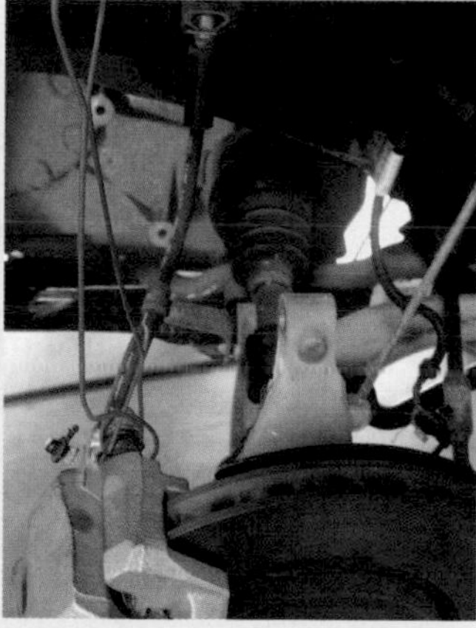

P41–7 Support the steering knuckle with wire, and remove the strut assembly from the car.

P41–8 Install the strut assembly into the proper type of spring compressor. Then compress the spring until it is possible to safely loosen the retaining bolts.

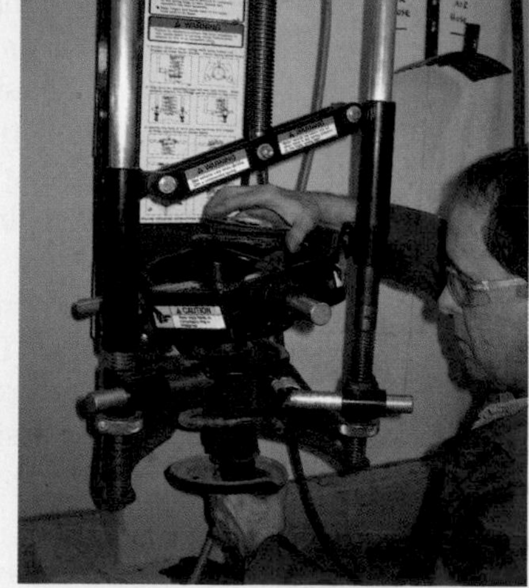

P41–9 Remove the old strut assembly from the spring, and install the new strut. Compress the spring to allow for reassembly, and tighten the retaining bolts.

P41–10 Reinstall the strut assembly into the car. Make sure all bolts are properly tightened and in the correct locations.

mounted at an angle, they are known as semitrailing control arms or multilink suspensions. Coil springs are used between the control arm and the vehicle body. The control arms pivot on a crossmember and are attached at the other end to a spindle. A shock absorber is attached to the spindle or control arm.

Some vehicles use a rear suspension system that uses a lower control arm and open driving axles. A crossmember supports the control arms, while the tops of the shock absorbers are mounted to the body. The springs are set in seats at the bottom and top of the crossmember.

A few cars use only lower control arms but substitute a wishbone-shaped subframe for the upper control arms. Two torque arms transfer the final drive torque to the subframe. In fact, many cars are now featuring rear **double-wishbone** suspension. Torque loads create bushing and control arm deflection during braking, cornering, acceleration, and deceleration. It is interesting to note this rear suspension system allows for a small amount of toe-in change to enhance straight line stability. The toe-in change during cornering leads to quicker and more responsive turning. The rear suspension system can also be tuned to ensure minimal dive under braking and minimal squat under acceleration.

Currently, struts are replacing conventional shock absorbers in independent rear suspension systems. One of the latest strut rear suspension designs used by car manufacturers is shown in **Figure 47–50**.

On this type of system, the spindle is used to secure the strut, the outer ends of two of the four control arms, the rear ends of the tie rods, and a rear wheel. The control arms contain bushings of different sizes at their outer ends. The ends with the smaller bushings attach near the body centreline. The ends with the larger bushings attach at the spindle. (When replacing control arms, it is mandatory that offsets at their outer ends and the flanges on the arms face in the direction prescribed in the manufacturer's service information.) This system is also called the nonmodified MacPherson strut system.

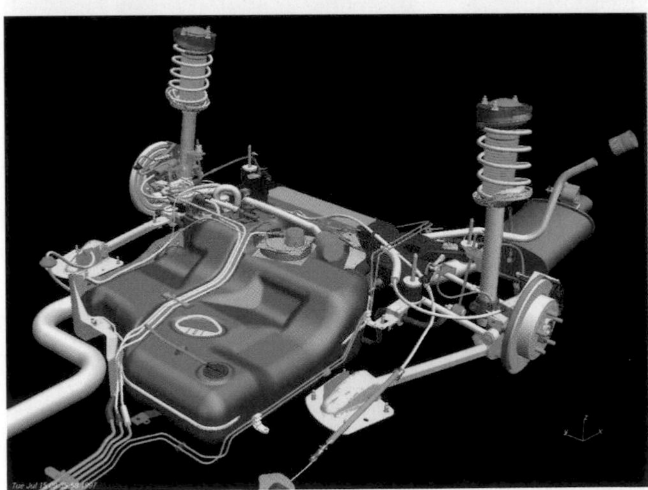

FIGURE 47–50

A strut-based rear suspension system.

Chrysler Group LLC

The modified MacPherson strut rear suspension is common for vehicles with front-wheel drive. The major components on each side of the vehicle are a modified MacPherson shock strut, lower control arm, tie rod, and wheel spindle. A coil spring mounts between the lower control arm and the body crossmember/side rail. The spindle, in addition to supporting the rear wheel, is used as an attaching location for the outer end of the control arm and the rear end of the tie rod. The inner end of the control arm attaches to the crossmember. The forward end of the tie rod attaches to the side rail.

Another rear strut design uses a **Chapman strut** that is similar to the modified MacPherson strut. The name is from its inventor, Colin Chapman of Lotus. The difference between the two struts is that the MacPherson strut requires a lower suspension arm, and the Chapman strut does not. The Chapman strut utilizes the drive shaft for the lower suspension arm. Chapman struts can only be used on drive axles.

Rear leaf-spring suspension systems are used on many vehicles with conventional rear drives. These leaf springs are generally mounted longitudinally in the same manner as described earlier in this chapter for live-axle systems. A few leaf-spring systems, however, employ springs mounted transversely. Both multiple-leaf and monoleaf, or single-leaf, can be used. The transverse-leaf spring is mounted to the final drive housing rather than to the vehicle frame as in the longitudinal installation. The transversely mounted spring's eyes are connected to the wheel spindle assemblies.

Rear shock absorbers or shock struts have the same service limitations as those used on the front of the vehicle. They cannot be adjusted, refilled, or repaired. The procedure for inspecting rear shocks or shock struts is similar to previously mentioned front-end parts inspection. Repetition is not necessary.

Multilink Rear Suspension

A multilink rear suspension uses several control arms to guide the wheel **(Figure 47–51)**. Different models feature different types of multilink rear suspensions that satisfy the varying demands of vehicle

FIGURE 47–51

A multilink rear suspension.

Chrysler Group LLC

dynamics, ride comfort, and space requirements. These include the double-wishbone rear suspension, trailing-link double-wishbone rear suspension, and trapezoidal-link rear suspension.

In the double-wishbone suspension, the wheel is guided by two triangulated lateral control arms (the wishbones) and a tie rod. The suspension strut is attached to the lower wishbone to provide vertical support.

The trailing-link double-wishbone suspension has a trailing link that also carries the wheel and upper and lower wishbones. The spring is located on the trailing link ahead of the centre of the wheel; the shock absorber is behind it.

The trapezoidal-link rear suspension permits excellent performance, handling, and comfort. The rear wheel is fixed by an upper lateral control arm and a trapezoidal lower link with a tie rod behind it. For reduced weight, the trapezoidal link and upper control are hollow aluminum castings.

Servicing Independent Suspension Systems

Most of the servicing techniques for rear independent suspension systems—except coil, control arm, and strut removal and installation—are similar to other front and rear suspension parts. They have been covered earlier in this chapter. Of course, check the service information for all inspection and repair techniques of the vehicle's independent rear system.

SERVICING REAR COIL SPRING Raise the vehicle on a frame contact hoist, or position jack stands under the frame forward of the rear axle assembly. This allows the shock absorbers to fully extend. Place a floor jack under the centre of the rear axle housing, and support the weight of the rear axle, but do not lift the vehicle off the jack stands. Disconnect the lower end of the shock absorber. Then lower the floor jack until all of the coil spring force is relieved. If a coil spring positioner is used, remove it from the centre of the coil spring. The coil spring can usually be removed from the vehicle at this time by lifting it from its spring seat. If the springs are to be used again, mark or tag each one so that it can be returned to its original location. When a replacement is needed, always replace coil springs in pairs. This ensures equal height.

To install a spring, place the insulator on top of the coil spring, and position the spring on the spring seat. The end of the top coil must be positioned to line up with the recess in the spring seat. Jack up the rear axle housing so that the spring is properly seated at the lower end and the shock absorbers line up. Reconnect the shock absorbers.

There are some definite advantages to working on one spring at a time. First of all, the assembled side of the vehicle helps support the disassembled side. It also keeps the parts aligned and eliminates the possibility of putting the parts on the wrong side of the vehicle.

SERVICING REAR CONTROL ARMS To remove the upper rear control arms from the vehicle, remove the bolts passing through the control arms at the frame and at the axle ends. Usually, the rear coil spring does not have to be removed for this. Service one side of the vehicle at a time. This simplifies realigning the parts during assembly. On a serviceable control arm, replace the control arm bushings by removing the defective bushing with an appropriate puller. Properly position the new bushing, and press it into place in the same manner as is done on front suspensions. Position the repaired control arm on the vehicle, and loosely install the bolts. Repeat the service on the other control arm, if necessary. Properly torque the nuts and bolts once the vehicle's entire weight is on the springs again.

The coil springs must be removed to service the lower rear control arms. Again, one side of the vehicle should be serviced at a time. Once the vehicle is properly supported and the springs are dismantled, remove the nuts and bolts that pass through the control arm. Remove the control arm from the vehicle, and service it in the same way as the upper control arm.

Check the service information to see if there is an adjustment for the driveline working angle. If none is specified, torque the control arm bolts to specification while the full vehicle weight is on the rear axle. This sets neutral bushing tension at normal curb height.

When there is a driveline working angle adjustment, adjust the angle before torquing the control arm bolts. After the rear suspension has been serviced, always check the working angle of the universal joints on the driveshaft. This minimizes the possibility of driveline vibration.

Some independent rear suspension systems have ball joints that perform a function similar to the front ball joints, and they should be inspected in the same way.

Although very few cars have rear wheels that steer, some independent suspension systems have components that would normally be seen only on vehicles with four-wheel steering. Components such as tie-rod ends may appear to serve the same purpose as if they were on the front suspension. They are used to adjust the angle of the wheels for stable straight ahead performance. These suspension and steering items are covered in greater detail in Chapters 48 and 49 of this book.

ELECTRONICALLY CONTROLLED SUSPENSIONS

All of the suspension systems covered up to this point are known as passive systems. Vehicle height and damping depend on fixed nonadjustable coil springs, shock absorbers, or MacPherson struts. When weight is added, the vehicle lowers as the springs are compressed. Air-adjustable shock absorbers may provide some amount of flexibility in ride height and ride firmness, but there is no way to vary this setting during operation. Passive systems can be set to provide a soft, firm, or compromise ride. Vehicle body motion and tire traction vary due to road conditions and turning and braking forces. Passive systems have no way of adjusting to these changes.

Advances in electronic sensor and computer control technology have led to a new generation of suspension systems. The simplest systems are level control systems that use electronic height sensors to control an air compressor linked to air-adjustable shock absorbers.

More advanced adaptive suspensions are capable of altering shock damping and ride height continuously. Electronic sensors **(Figure 47–52)** provide input data to a computer. The computer adjusts air spring and shock damping settings to match road and driving conditions.

The most advanced computer-controlled suspension systems are true active suspensions. These systems are hydraulically, rather than air, controlled. They use high-pressure hydraulic actuators to carry the vehicle's weight rather than conventional springs or air springs.

The unique feature of an active suspension is that it can be programmed to respond almost perfectly to various operating conditions. For example, by raising the height of the outside actuators and lowering the inside actuators when going around a curve, the vehicle can be made to lean into a curve, much like a motorcycle. Active systems using hydraulic actuators are presently used on a limited number of high-performance vehicles. Most manufacturers are introducing various adaptive suspension systems that rely on pneumatically actuated air springs and dampers.

Some late-model pickups and SUVs offer air suspension systems. These systems are added to existing leaf-spring suspensions. The air spring is positioned between the centre of the leaf spring and the frame of the truck. The air spring serves as an adjustable and additional spring at each end of the axle.

Adaptive Suspensions

Adaptive suspensions use electronic shock absorbers with variable valving. In some cases, variable air spring rates are used to adapt the vehicle's ride characteristics to the prevailing road conditions or driver demands.

Electronic sensors monitor factors such as vehicle height, vehicle speed, steering angle, braking force, door position, shock damping status, engine vacuum, throttle position, and ignition switching. A computer is used to analyze this input and switch the suspension into a preset operating mode that matches existing conditions. Some systems are fully automatic. Others allow the driver to select the ride mode.

At present, adaptive suspensions are less costly and complicated than hydraulically controlled active suspensions. However, they do have some limitations. Although they can reduce body roll, adaptive suspensions cannot eliminate it like true active systems. Adaptive systems also experience a slight delay in their reaction time, although some systems can change shock valving in as little as 150 microseconds.

SYSTEM COMPONENTS There are many different designs and components used by the manufacturers to accomplish the same task. Some systems use adjustable shocks, whereas others use air springs at each side of the axles. The air spring membrane is similar to a tire in construction. A solenoid valve and filter assembly allows clean air to be added or

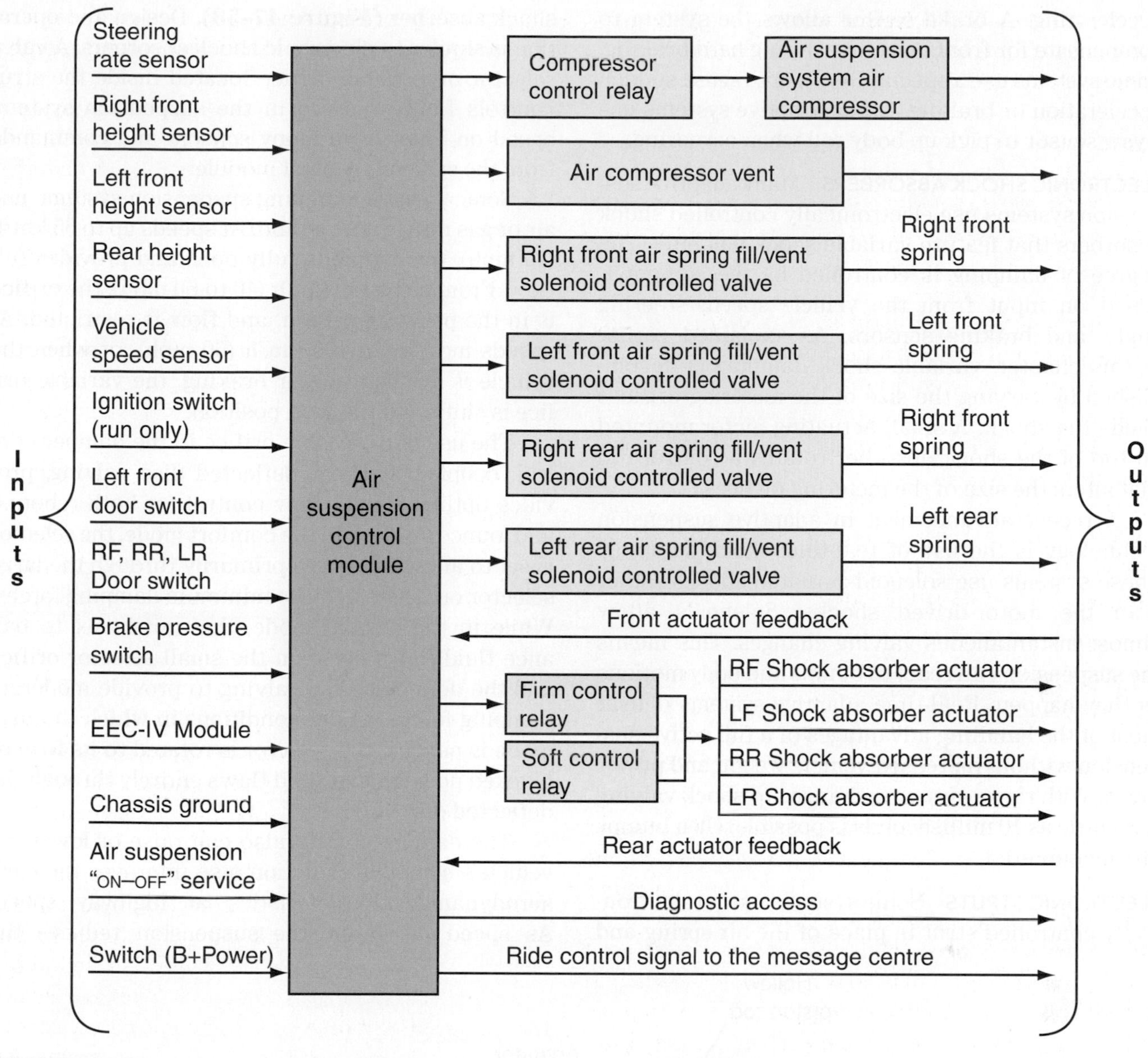

FIGURE 47–52

The various inputs and outputs for an electronic suspension system.

released from the air spring. Adding or removing air changes the ride height of the vehicle.

The airflow to the springs is controlled by the interaction of the air compressor, system sensors, computer control module, and solenoid valves. All of the air-operated parts of the system are connected by nylon tubing.

COMPRESSOR The compressor supplies the air pressure for operating the entire system. It is often a positive displacement single piston pump powered by a 12-volt DC motor. A regenerative air dryer is attached to the compressor output to remove moisture from the air before it is delivered to the air springs. The compressor is operated through the use of an electric relay controlled by the computer module.

SENSORS Vehicle height sensors can be rotary Hall-effect sensors that enable the computer to more accurately measure ride height as well as to compensate for road variations. This prevents the vehicle from bottoming out when crossing over railroad tracks or similar road irregularities.

Advanced systems also read the steering angle by using a photo diode and shutter location inside the steering column. This allows the system to firm up the suspension when the vehicle is turning. The system also reads engine vacuum or throttle position to stiffen the suspension when the vehicle is accelerating. A brake sensor allows the system to compensate for front nose dive during hard braking. Some systems use a special G-sensor to sense sudden acceleration or braking. Other adaptive systems use a yaw sensor to pick up body roll when cornering.

ELECTRONIC SHOCK ABSORBERS Many adaptive suspension systems use electronically controlled shock absorbers that feature variable shock damping. The degree of damping is controlled by the computer, based on input from the vehicle speed, steering angle, and braking sensors. As explained earlier in this chapter, variable shock damping is accomplished by varying the size of the metering orifices inside the shock. A small actuating motor mounted on top of the shock absorber rotates a control rod that alters the size of the metering orifices.

A recent advancement in adaptive suspension technology is the use of real-time shock damping. These systems use solenoid-actuated shocks rather than the motor-driven shocks. Solenoids allow almost instantaneous valving changes. This means the suspension can react to bumps and body motions as they happen. Real-time adaptive systems deliver most of the handling advantages of a full active suspension without increased vehicle weight and power drain. With these systems, changes to shock valving in as little as 10 milliseconds is possible when bumps are encountered.

ELECTRONIC STRUTS Some systems use an electronically controlled strut in place of the air spring and shock absorber **(Figure 47–53)**. Design and operation is similar to electronic shock absorbers. A valve selector or variable orifice located inside the strut controls fluid pressure in the suspension system, based on input from many sensors and commands from the system's control module.

Some variable damping suspension systems use air or gas rather than a fluid. At speeds up to 65 km/h (40 mph), the orifice is fully open and provides full flow. From 65 to 100 km/h (40 to 60 mph), the orifice is in the normal position, and flow is restricted. At speeds more than 100 km/h (60 mph), or when the vehicle is accelerating or braking, the variable orifice is shifted to the firm position.

The use of a variable orifice in the damper control, coupled with the deflected disc valving, provides optimum fluid flow control for both rebound and jounce strokes. In the comfort mode, the selector is set to allow fluid flow primarily through the large selector orifice to achieve minimum damping forces. While in the normal mode, the unit is set to balance fluid flows between the small selector orifice and the deflected disc valving to provide moderate damping forces. Under conditions in which the firm mode is needed, the selector is rotated to its firm or blocked position and fluid flows entirely through the deflected disc valving.

The damper control also can raise or lower the vehicle's height. This action also improves the car's aerodynamic characteristics at highway speed. As speed increases, the suspension reduces the

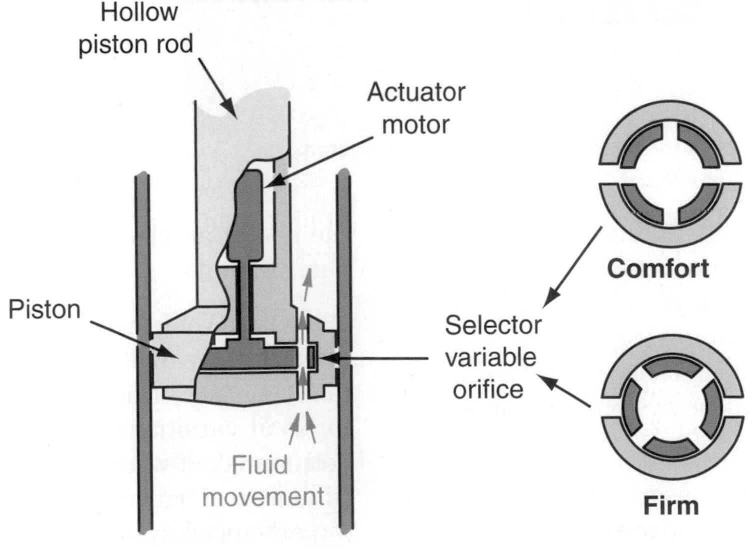

FIGURE 47–53

A computer-command ride strut. Four electronically controlled struts are used on many adaptive suspension systems. Based on input from the computer, the valving selector shifts the variable bypass orifice to a comfort, normal, or firm setting.

vehicle's height, and the front end angles downward. This action tends to reduce wind resistance for greater stability and better gas mileage. As the vehicle slows, the suspension brings the body up to its normal height and level position.

COMPUTER CONTROL MODULE A microcomputer-based module controls the air compressor motor (through a relay), the compressor vent solenoid, and the four air spring solenoids. The computer module also controls operation of electronic shock absorber actuating motors and electronic strut valving selectors. The control module receives input from all system sensors.

The computer module also has the capability of performing diagnostic tests on the system. It has a preprogrammed routine for properly fitting air springs after servicing. The module also controls the dash-mounted system warning light.

Electrical power to operate the basic air suspension system is distributed by the main body wiring harness. Each wiring harness involved has a special function in the typical air suspension system.

WARNING!

The compressor relay, compressor vent solenoid, and all air spring solenoids have internal diodes for electrical noise suppression and are polarity sensitive. Care must be taken when servicing these components not to switch the battery feed and ground circuits, or component damage will result. When charging the battery, the ignition switch must be in the OFF position if the air suspension switch is on, or damage to the air compressor relay or motor may occur. However, use of a battery charger while performing the diagnostic test or air spring fill option is acceptable. Set to a rate to maintain, but not damage, the vehicle battery.

ELECTRONIC LEVELLING CONTROL Adaptive suspension systems are capable of adjusting the suspension system during operation. Less complicated electronic level-control systems are used on many large and mid-size vehicles.

These systems do not use a computer module. In most cases, height sensors are the only types of sensors used. These height sensors sense when passenger weight or cargo is added to or removed from the vehicle **(Figure 47–54)**. The height sensors control two basic circuits. The compressor relay coil grounds circuits that activate the compressor. The exhaust solenoid coil grounds circuits that vent air from the system.

To prevent falsely actuating the compressor relay or exhaust solenoid circuits during normal ride motions, the sensor circuitry provides an 8- to 15-second delay before either circuit can be completed.

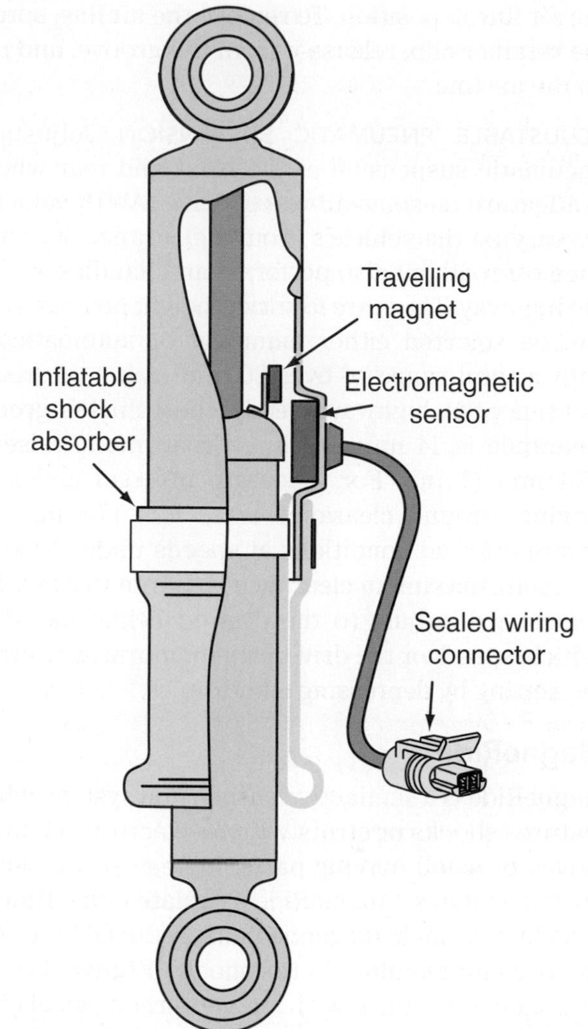

FIGURE 47–54

A load-sensing shock absorber.

In addition, the typical sensor electronically limits compressor run time or exhaust solenoid energized time to a maximum of approximately 3½ minutes. This time-limit function is necessary to prevent continuous compressor operation in a case of a solenoid malfunction. Turning the ignition off and on resets the electronic timer circuit to renew the 3½-minute maximum run time. The height sensor is mounted to the frame crossmember in the rear. The sensor actuator arm is attached to the rear upper control arm by a link. The link should be attached to the metal arm when making any trim adjustment.

When the air line is attached to the shock absorber fittings or compressor dryer fitting, the retainer clip snaps into a groove in the fitting, locking the air line in position. To remove the air line, spread the retainer clip, release it from the groove, and pull on the air line.

ADJUSTABLE PNEUMATIC SUSPENSION Adjustable pneumatic suspension at the front and rear wheels is a feature on some all-wheel-drive (AWD) vehicles. By varying the vehicle's ground clearance, it can be used off-road but also performs and handles well on the highway. There are four ride-height positions that can be selected either manually or automatically, with a total range of over 20 mm (8 in.) of ground clearance. At highway speeds, the vehicle's ground clearance is 14 mm (5.6 in.). Urban mode raises it 25.4 mm (1 in.). For moderate off-road and local driving, ground clearance is 19 mm (7.6 in.). For severe off-road conditions at speeds under 40 km/h (25 mph), maximum clearance is 21 mm (8.2 in.). The vehicle will adjust to the desired height based on vehicle speed, or the driver can temporarily override the setting by depressing a button.

MagneRide

MagneRide is a semiactive suspension system, which features shocks or struts with no electromechanical valves or small moving parts. Instead of valve-controlled orifices, MagneRide regulates the flow of fluid by a variable magnetic field produced by a small electric coil mounted in the shock **(Figure 47–55)**. The shocks are filled with magneto-rheological (MR) fluid. MR fluid consists of magnetically soft particles, such as iron, suspended in synthetic hydrocarbon fluid.

FIGURE 47–55

A magneto-rheological fluid-based strut.

The action of the shock forces the MR fluid through a magnetized opening in each shock. When the shock is in its off state, the fluid is not magnetized and flows freely through the orifice. When current is sent to the coil, the fluid becomes magnetized, and its viscosity changes instantly **(Figure 47–56)**.

The material changes from a fluid state to a semisolid state that is directly proportional to the magnetic field applied to it. With little or no electrical current, the iron particles are randomly distributed, and the fluid passes freely through the piston orifice. When a strong electrical current is applied to the coil, the resulting magnetic field aligns the iron particles so that the fluid stiffens and the flow is

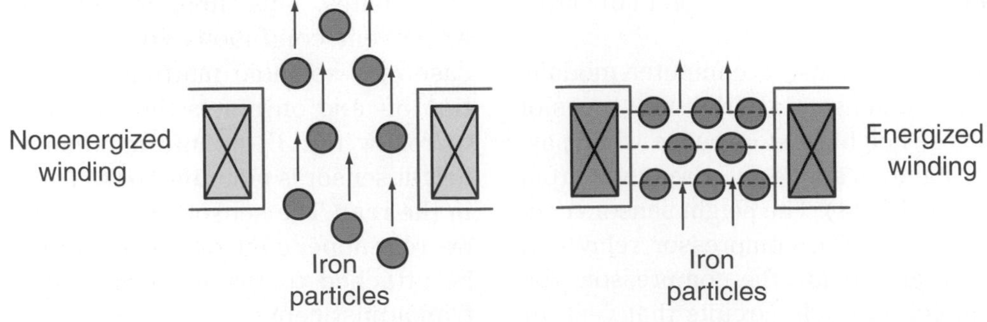

FIGURE 47–56

The iron particles in the MR fluid align themselves when they pass through the magnetic field, causing the fluid to stiffen.

resisted. This condition causes heavy damping. The resulting damping force is proportional to the viscosity, which is proportional to the strength of the magnetic field.

Sensors monitoring wheel position, lateral acceleration, vehicle speed, steering wheel angle, and brake pedal angle are inputs to the control module that sends current to the coil in the shocks. This system provides extremely quick response time, typically about 5 milliseconds, and the fluid is capable of reacting 1000 times per second.

SERVICING ELECTRONIC SUSPENSION COMPONENTS

Most electronic suspension servicing requires the removal of the component from the system, replacing or repairing it, and then reinstalling the component back into the system. Procedures for individual component replacement are covered in the vehicle's service information. Serviceable components include the air compressor, charger, mounting brackets, height sensors, air springs, air lines and connections, gas struts, strut mounts, control arm components, shock absorbers, and stabilizer bars.

Failure to keep the following procedures in mind might result in a sudden failure of the air spring or suspension system. Suspension fasteners are important attaching parts. They could affect performance of vital components and systems or could result in major service expenses. They must be replaced with fasteners of the same part number or with an equivalent part, if replacement becomes necessary. Do not use a replacement part of lesser quality or substitute design. Torque values must be used as specified during assembly to ensure proper retention of parts. New fasteners must be used in the place of the old ones whenever they are loosened or removed and when new component parts are installed.

WARNING!

Do not remove an air spring under any circumstances when there is pressure in the air spring. Do not remove any components supporting an air spring without either exhausting the air or providing support for the air spring. Power to the air system must be shut off by turning the air suspension switch (in the luggage compartment) off or by disconnecting the battery when servicing any air suspension components. Most air suspension systems are equipped with a warning light. The light comes on if there is a problem or when servicing the system.

CUSTOMER CARE

Because the technician is seldom present when a vehicle requires towing, it is important to advise the customer of proper procedures so that the tow operator does not damage the electronic suspension system. You must also know the proper hoist lifting and jacking restrictions. Although it is necessary to check the service information for specific instructions, the following are the basics for electronic suspension. When towing, it must be remembered that when the ignition is off, the automatic levelling suspension may or may not be still on. Before lifting the vehicle, be sure the ignition switch is turned off and the trunk switch deactivated. When lifting a vehicle equipped with an electronic suspension system, service information should be consulted to determine if the manufacturer has specific procedures for lifting or hoisting. All warnings advising of the correct hoisting procedures should be followed to prevent damage to the system components. Most vehicles can be lifted on any hoist when the system is deactivated.

Diagnosis

A scan tool and/or a special electronic tester is used to diagnose most electronic suspension systems. These can only retrieve diagnostic trouble codes (DTCs); they may also be able to activate various actuators in the system. The exact procedures and available data from the vehicle's computer will vary with manufacturer and the system found on the vehicle. Always refer to the correct service information when diagnosing electronic systems.

Diagnosis should begin with gathering as much information as possible from the customer. Make sure you know exactly what the concern is and the conditions at the time the malfunction occurred. Verify the customer's concern by attempting to duplicate these conditions during a road test.

Check the voltage at the battery. If the voltage is below 11 volts, recharge or replace the battery before continuing with diagnostics. Check the fuses, connectors, and wiring harnesses for the suspension system, and repair them as necessary. Start the engine and allow it to warm up. Then connect the scan tool or tester. Retrieve all DTCs from the system. If the scan tool or tester is unable to communicate with the vehicle's computer, diagnose the cause of this before proceeding.

Check the DTC charts to see if the DTCs match the symptoms exhibited by the vehicle. If they do, proceed to follow the troubleshooting chart related to each DTC. If the DTCs do not match the symptoms, clear them and recheck for trouble codes after your diagnosis of the suspension system has been completed. If there are no DTCs related to the problem,

PID	DESCRIPTION	EXPECTED VALUES
4X4_HIGH	4X4 High Input	IN, OUT
4X4_LOW	4X4 Low Input	IN, OUT
AS_COMP	Compressor Relay Status	ON---, ONO--, ON-B-,ON--G, OFF---, OFFO--, OFF-B-, OFF--G
AS_GATE	Front Gate Solenoid Status	ON---, ONO--, ON-B-,ON--G, OFF---, OFFO--, OFF-B-, OFF--G
AS_VENT	Vent Solenoid Status	ON---, ONO--, ON-B-,ON--G, OFF---, OFFO--, OFF-B-, OFF--G
BOO_ARC	Brake Pedal Position Switch Input	ON, OFF
CCNTARC	Number of Continuous DTCs Counted by the ARC Module	one count per bit
DR_OPEN	Door Ajar Input	OPEN, CLOSED
F_FILL	Front Fill Solenoid Status	ON---, ONO--, ON-B-,ON--G, OFF---, OFFO--, OFF-B-, OFF--G
FHGTSEN	Front Height Sensor	#, ## VDC
HGTSENS	Height Sensor	ON, OFF
IGN_RUN	Detection of Ignition Switch in the RUN Position	RUN, not RUN
LFSHK_E	Left Front Shock Encoder Status	SOFT, FIRM
LRSHK_E	Left Rear Shock Encoder Status	SOFT, FIRM
OFFROAD	Vehicle Off-Road Status	ON, OFF
OPSTRAT	Operational Strategy	ARC
PCM_ACC	Acceleration Signal from the Powertrain Control Module (PCM)	YES, NO
R_FILL	Rear Fill Solenoid Status	ON---, ONO--, ON-B-,ON--G, OFF---, OFFO--, OFF-B-, OFF--G
RASGATE	Rear Gate Solenoid Status	ON---, ONO--, ON-B-,ON--G, OFF---, OFFO--, OFF-B-, OFF--G
RFSHK_E	Right Front Shock Encoder Status	SOFT, FIRM
RHGTSEN	Rear Height Sensor	#, ## VDC
RRSHK_E	Right Rear Shock Encoder Status	SOFT, FIRM
STEER_A	Steering Rotation Sensor A	LOW, HIGH

FIGURE 47–57

Typical PIDs for an electronic suspension system.

record the PIDs, and compare them to specifications **(Figure 47–57)**. If the cause of the problem is still not evident, refer to the symptoms chart provided by the manufacturer. Check those areas identified as possible causes of the problem. If the tester has the capability of activating the system's actuators, do this now.

In many cases, the tester can cause action at one wheel at a time. The suspension of one wheel can be manually energized, and the vehicle will tilt. The actuator can then be de-energized, and the vehicle will return to its original position. This procedure should be completed at each wheel. If the suspension does not respond correctly to these commands, the corner of the suspension that did not should be thoroughly checked. If all corners responded as they should, the problem is most likely not caused by something in the electronic system, and diagnosis should continue with checks of the conventional suspension parts.

Once the cause is identified, it should be replaced or repaired, and then the repair ought to be confirmed. Drive the vehicle under the same conditions during which the concern was present. Also recheck the system for any new DTCs.

Vehicle Alignment

Aligning a vehicle with an electronic suspension system is essentially the same as the aligning procedure described in Chapter 49—with one notable exception—curb height.

Curb height is an important dimension because it affects the other alignment angles. Caster is the most obvious one that is affected, but front camber and toe can also be included. Curb height is especially critical when checking rear camber and toe on independent rear suspensions. With electronic suspension, the ride can vary depending on various circumstances. The only way to guarantee the suspension at curb height is to preset it.

ACTIVE SUSPENSIONS

Some of the advanced adaptive suspension systems may be called **active suspensions**. In this text, active suspensions refer to those controlled by double-acting hydraulic cylinders or solenoids (usually called actuators) that are mounted at each wheel. Each actuator maintains a sort of hydraulic equilibrium with the others to carry the vehicle's weight, while maintaining the desired body attitude. At the same time, each actuator serves as its own shock absorber, eliminating the need for yet another traditional suspension component.

In other words, each hydraulic actuator acts as both a spring (with variable-rate damping characteristics) and a variable-rate shock absorber. This is accomplished in an active suspension system by varying the hydraulic pressure within each cylinder and the rate at which it increases or decreases. By bleeding or adding hydraulic pressure from the individual actuators, each wheel can react independently to changing road conditions.

The components that make such a system possible are the actuator control valves, various sensors, and the chassis computer **(Figure 47–58)**. Feeding information to a computer are a number of specialized sensors. Each actuator has a linear displacement sensor and an acceleration sensor to keep the computer informed about the actuator's relative position. This enables the computer to track the extension and compression of each actuator and to know when each wheel is undergoing jounce or rebound. There are also load sensors and hub acceleration sensors in each wheel to measure how heavily each wheel is loaded.

A steering angle sensor is used to signal the computer when the vehicle is turning. To monitor body motions, a roll sensor, lateral acceleration and G-sensors are used. The computer also monitors speed of the pump and the hydraulic pressure within the system.

Once it has all the necessary inputs, the computer can then regulate the flow of hydraulic pressure within each individual actuator according to any number of variables and its own built-in program. Another nice feature of a suspension such as this is that it can be programmed to behave in a variety of unique and currently impossible ways: leaning or rolling into turns, for example, or even raising a flat tire on command in order to change the tire without using a separate jack.

When the wheel of an active suspension hits a bump, the sensors detect the sudden upward deflection of the wheel. The computer recognizes the change as a bump and instantly opens a control valve to bleed pressure from the hydraulic actuator. The rate at which pressure is bled from the actuator determines the cushioning of the bump and the relative harshness or softness of the ride. The rate can be varied at any point during jounce or rebound to produce a variable spring rate effect. In other words, the feel of the suspension can be programmed to respond in an almost indefinite variety of ways. Once the bump has been absorbed by the actuator, pressure is forced back into it to keep the wheel in contact with the road and to maintain the suspension's desired ride height.

During hard braking with a conventional suspension system, there is a tendency for a vehicle to make a dive. The weight of the vehicle seemingly pushes the front of the car downward and the back upward. During hard braking, the active suspension increases hydraulic pressure in the front actuators and reduces hydraulic pressure in the rear actuators. These actions minimize dive to keep the vehicle level and make it easier for the driver to control. After braking, valves operate to equalize hydraulic pressures in front and rear actuators and level the vehicle again.

Frequently, when a driver depresses the accelerator quickly during hard acceleration, the front end of the vehicle tends to lift up, while the rear end lowers. The action is known as **squat**. With an active suspension system, squat is controlled by the operating valve's solenoids, which increase the hydraulic pressure in rear-wheel actuators and reduce hydraulic pressure in front-wheel actuators. When the vehicle is no longer accelerating quickly, the control system operates valves to equalize hydraulic pressures and level the vehicle. Thus, an active suspension changes the height of the front, rear, or either side of the vehicle to counteract tilting, rolling, and leaning. These active attitude control functions improve vehicle stability and increase tire traction and driver control.

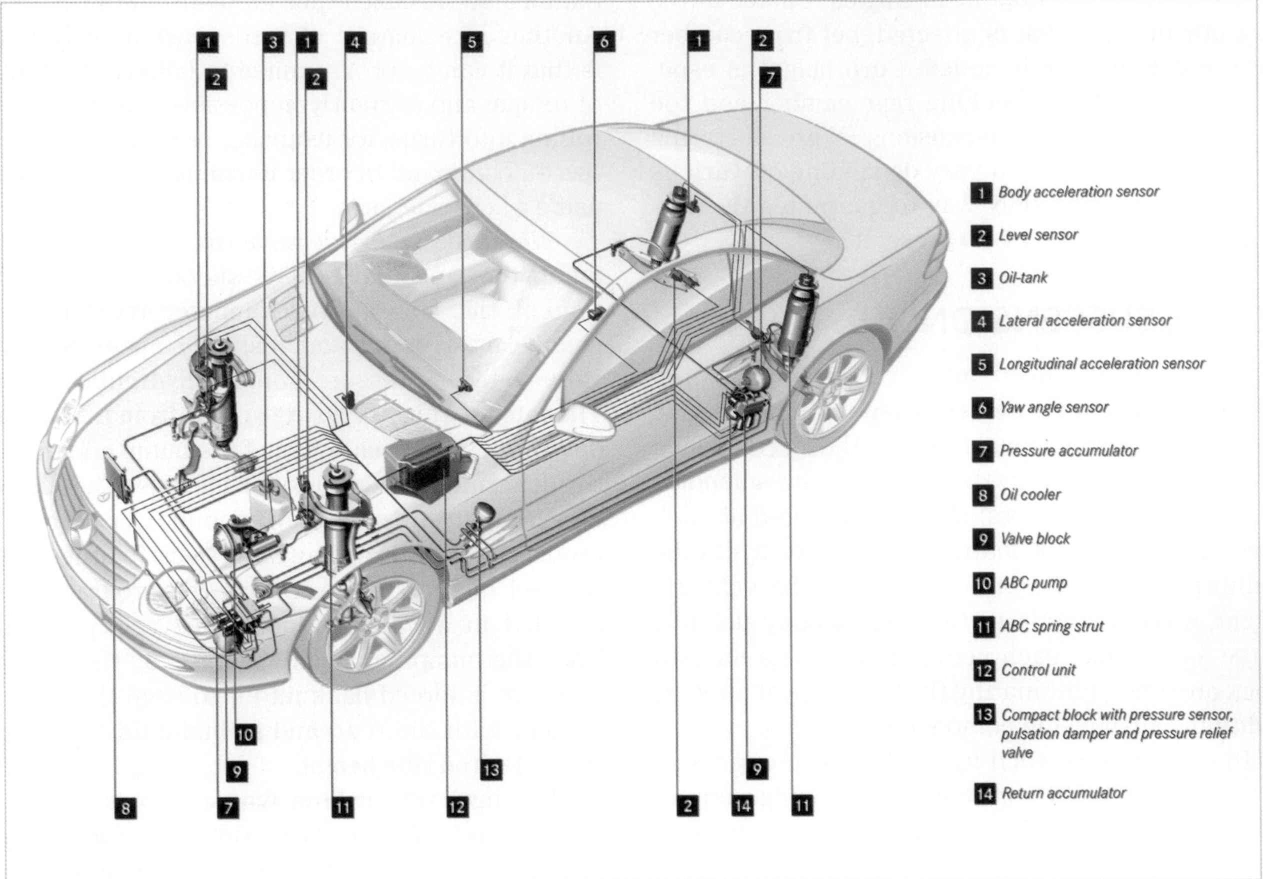

FIGURE 47–58

An active suspension system.

DaimlerChrysler Corporation

The power required for a totally active system is only 3 to 5 horsepower (about the same as a typical power steering pump). Power consumption is lowest when the system is least active, as when driving on a smooth road. Rough roads and hard manoeuvres, on the other hand, put more of a demand on the system. The hydraulic pump works harder and thus requires more power.

Power consumption can be reduced by going with a semiactive suspension that uses small springs with the hydraulic actuators. The springs help to support the vehicle's weight, which reduces the load on the actuators. Smaller actuators that require less hydraulic power can then be used, which reduces the bulk and weight of the system. The addition of springs also adds a certain margin of safety to the system to keep it from going flat should the hydraulics spring a leak.

Although not as widely used as electronic levelling or adaptive suspension systems, hydraulic active suspensions are sure to become more common.

Chassis Lubrication

Grease fittings, called zerk fittings, are threaded into the part that should be lubricated. These fittings have a one-way spring-loaded check valve that allows grease into the joint but prevents it from leaking out. To lubricate the chassis of a vehicle, safely raise the vehicle and lock the lift, or set the vehicle on safety stands. Locate all of the lubrication points, and wipe the fittings clean with a shop towel. If the lubrication points do not have zerk fittings but have plugs, remove the plugs and install new zerk fittings, or use a special lubrication adapter.

Carefully look at the joints and determine if the joint boots are sealed. If the boots are good, push the grease gun nozzle straight into the fitting, and pump grease slowly into the joint. If the joint has a sealed boot, put just enough grease into the joint to cause the boot to expand slightly. If the boot is not sealed, put in enough grease to push the old grease out. Then wipe off the old grease. Repeat this at all lubrication points, and wipe all excess grease off the joints and fittings.

KEY TERMS

Active suspensions (p. 1465)
Adaptive suspensions (p. 1458)
Anti-roll bar (p. 1441)
Anti-sway bar (p. 1441)
Axle tramp (p. 1453)
Ball joint (p. 1436)
Chapman strut (p. 1456)
Control arms (p. 1438)
Dampers (p. 1431)
Double-wishbone (p. 1456)
Fibre composite (p. 1430)
Independent front suspension (IFS) (p. 1438)
Jounce (p. 1427)
Linear rate (p. 1428)
MacPherson strut (p. 1434)
Main leaf (p. 1429)
Monoleaf (p. 1430)
Monoshock (p. 1432)
Multiple-leaf (p. 1429)
Rebound (p. 1427)
Roll bar (p. 1441)
Shock absorber (p. 1427)
Short-long arm (SLA) (p. 1437)
Spring rate (p. 1427)
Sprung weight (p. 1428)
Squat (p. 1465)
Steering knuckle (p. 1438)
Strut rods (p. 1436)
Torsion bars (p. 1431)
Trailing arms (p. 1453)
Unsprung weight (p. 1428)
Variable rate (p. 1428)
Wheel spindle (p. 1438)

SUMMARY

- Four types of springs are used in suspension systems: coil, leaf, torsion bar, and air.
- Springs take care of two fundamental wheel actions: jounce and rebound.
- Common coil spring materials include carbon steel, carbon boron, steel, and alloy steels. Alloy steels, such as those containing chromium and silicon, improve the coil's resistance to relaxation. Most coil springs are manufactured by either a cold-coiling or a hot-coiling process. Hot coiling usually requires hardening and tempering along with short peening to increase the fatigue strength of the base material.
- Two basic designs of coil springs are used in vehicles: linear rate and variable rate.
- Leaf springs are made of steel or a fibre composite.
- In torsion suspension, the bar may either run from front to rear or side to side across the chassis.
- Air springs are generally only used in microprocessor-controlled suspension systems.
- Shock absorbers damp or control motion in a vehicle. A conventional shock absorber is a velocity-sensitive hydraulic damping device. The faster it moves, the more resistance it has to the movement.
- Shock absorbers can be mounted either vertically or at an angle. Angle mounting of shock absorbers improves vehicle stability and dampens accelerating and braking torque.
- There are two basic adjustable air shock systems: the manual fill type and the automatic or electronic load-leveling type.
- MacPherson struts provide the damping function of a shock absorber. In addition, they serve to locate the spring and to fix the position of the suspension.
- Domestic struts have taken two forms: a concentric coil spring around the strut itself and a spring located between the lower control arm and the frame.
- MacPherson suspensions use anti-roll or stabilizer bars. Coil springs are used on all strut suspensions.
- Independent front suspension (IFS) must keep the wheels rigidly positioned and at the same time allow them to steer right and left. In addition, because of weight transfer during braking, the front suspension system absorbs most of the braking torque. When accomplishing this, it must provide good ride and stability characteristics.
- The unequal length arm or short-long arm (SLA) suspension system is most commonly used on domestic vehicles.
- Live axle is the traditional rear suspension system and consists of springs used in conjunction with a live axle (one in which the differential axle, wheel bearings, and brakes act as a unit). The springs are either of leaf or coil type.
- Semi-independent suspension is used on many front-wheel-drive models.
- Three strut designs are frequently used in IFS systems: the conventional MacPherson strut, the modified MacPherson strut, and the Chapman strut.
- The two basic types of computer suspension systems are adaptive and active.
- Electronically controlled suspensions can be either simple load-levelling systems, adaptive systems, or fully active systems. Adaptive and active suspension systems are computer controlled. Most load-levelling systems do not use a computer.
- Adaptive suspensions can alter vehicle ride height and shock absorber damping while the

vehicle is in motion. Such systems use air springs and electronic shock absorbers or struts.

- Active suspensions are hydraulically operated actuators to control up-and-down and side-to-side movement. They can be programmed to respond to certain road conditions and turning forces.

REVIEW QUESTIONS

1. How does a stabilizer bar work?
2. Explain the difference between sprung and unsprung weight.
3. What is the principle of the air spring?
4. Explain the action of the conventional shock absorber on both compression (jounce) and rebound strokes.
5. Describe the action of the independent front-wheel suspension system.
6. What is the core of any suspension system?
   a. wheel spindle assembly
   b. spring
   c. ball joints
   d. control arm
7. What occurs when a wheel hits a dip or hole and moves downward?
   a. jounce **b.** free length
   c. deflection **d.** rebound
8. Which of the following is part of the sprung weight of a vehicle?
   a. steering linkage **b.** tires
   c. engine **d.** steering knuckle
9. What occurs when a wheel hits a dip or hole and moves upward?
   a. jounce **b.** free length
   c. deflection **d.** rebound
10. Which of the following ball joints may have an allowance for free play?
    a. load carrying **b.** follower
    c. friction loaded **d.** all upper ball joints
11. What is the most common shock absorber ratio used on passenger cars?
    a. 80/20 **b.** 70/30
    c. 60/40 **d.** 50/50
12. What stabilizes the up-and-down motion of the vehicle when travelling on a bumpy road?
    a. the upper control arm
    b. the torsion bar
    c. the shock absorber
    d. the coil spring
13. What is the purpose of a vehicle's coil spring?
    a. to support the weight of the vehicle
    b. to provide axle location
    c. to stabilize the up-and-down motion
    d. to stabilize rebound only
14. What type of ball joint is used on the lower control arm of an SLA suspension when the coil spring is placed between the lower control arm and the frame, and the control arm is beneath the knuckle?
    a. a friction-loaded ball joint
    b. a follower ball joint
    c. a compression loaded–load carrying
    d. a tension loaded–load carrying
15. Which of the following statements about torsion bars is correct?
    a. Torsion bars are interchangeable from side to side.
    b. Torsion bars can be replaced individually.
    c. Torsion bars are pre-stressed for specific sides of the vehicle.
    d. Torsion bars are nonadjustable.
16. What are the two SLA systems in common use today?
    a. coil spring and strut suspension
    b. coil spring and torsion bar suspension
    c. coil spring and single control arm suspension
    d. single and double control arm suspension
17. What component is used to transfer suspension movement from one side to the other during cornering?
    a. a torsion bar **b.** an anti-roll bar
    c. strut rods **d.** shock absorbers
18. How full should the fuel tank be to correctly check the vehicle's ride height?
    a. full **b.** ¾ full
    c. ½ full **d.** ¼ full
19. What vehicles commonly use a modified MacPherson strut rear suspension?
    a. front-wheel-drive vehicles
    b. rear-wheel-drive vehicles
    c. pickup trucks
    d. station wagons
20. Which suspension system can actually change the ride height during operation?
    a. adaptive suspensions
    b. active suspensions
    c. air shock suspensions
    d. automatic load-levelling air suspension

CHAPTER 48
Steering Systems

LEARNING OUTCOMES

- Describe the similarities and differences between parallelogram, worm and roller, and rack and pinion steering linkage systems.
- Identify the typical manual-steering system components and their functions.
- Identify basic types of steering linkage systems.
- Identify the components in a parallelogram steering linkage arrangement and describe the function of each.
- Identify the components in a manual rack and pinion steering arrangement and describe the function of each.
- Describe the function and operation of a manual-steering gearbox and the steering column.
- Identify the components of the various power steering systems.
- Describe the function and operation of the various power steering system designs.
- Describe the service to the various power steering designs.
- Perform general power steering system checks.
- Inspect and service steering linkage components.
- Inspect and service power steering pumps.
- Describe the common four-wheel steering systems.

The purpose of the steering system is to turn the front wheels. In some cases, it also turns the rear wheels. The wheels constantly change direction—when switching lanes, when rounding sharp turns, and when avoiding roadway obstacles. Few older vehicles had some sort of power assist to turn the wheels; today, all vehicles have power-assist steering. This point is important because, in most cases, the power-assist systems work with manual-steering systems. Therefore, it is important to look at manual systems and then power-assist systems to understand how today's steering systems work **(Figure 48–1)**.

FIGURE 48–1

Allowing a vehicle to do something other than go straight is the job of the steering system.

Martin Restoule

MANUAL-STEERING SYSTEMS

The steering system is composed of three major subsystems: the steering linkage, steering gear, and steering column and wheel. As the steering wheel is turned by the driver, the steering gear transfers this motion to the steering linkage. The steering linkage turns the wheels to control the vehicle's direction. Although there are many variations to this system, these three major assemblies are in all steering systems.

Steering Linkage

The term **steering linkage** is applied to the system of pivots and connecting parts placed between the steering gear and the steering arms that are attached to the front or rear wheels that control the direction of vehicle travel. The steering linkage transfers the motion of the steering gear output shaft to the steering arms, turning the wheels to manoeuvre the vehicle.

The type of front-wheel suspension (independent wheel suspension as compared with a solid front axle) greatly influences steering geometry. Most passenger cars and many light trucks and recreational vehicles have independent front-wheel suspension systems. Therefore, a steering linkage arrangement that tolerates relatively large wheel movement must be used.

Parallelogram Steering Linkage

A parallelogram type of steering linkage arrangement was at one time the most common type used on passenger cars. It is used with the recirculating ball steering gear and can be classified into two distinct configurations: parallelogram steering linkage placed behind the front-wheel suspension **(Figure 48–2A)** and parallelogram steering linkage placed ahead of the front-wheel suspension **(Figure 48–2B)**. This type of steering linkage is most often used where motor and chassis components would interfere with normal operation of the steering linkage.

These designs are the basic steering systems used in conjunction with short-long arm independent front-wheel suspensions. This type of linkage also provides good steering and suspension geometry. However, road vibrations and impact forces are transmitted to the linkage from the tires, causing wear and looseness in the system, which permits intermittent changes in the toe setting of the front wheels, allowing further tire wear.

In a parallelogram steering linkage, the tie rods have ball socket assemblies at each end. One end is attached to the wheel's steering arm and the other end to the centre link.

The components in a parallelogram steering linkage arrangement are the pitman arm, idler arm, links, and tie rods.

PITMAN ARM The **pitman arm (Figure 48–3)** connects the linkage to the steering column through a steering gear located at the base of the column. It transmits the rotary motion it receives from the gear to the linkage, causing the linkage to move left or right to turn the wheels in the appropriate direction. It also serves to maintain the height of the centre link. This ensures that the tie rods are able to be parallel to the control arm movement and avoid unsteady toe settings or **bump steer**. *Toe*, a critical alignment factor, is a term that defines how well the tires point to the direction of the vehicle.

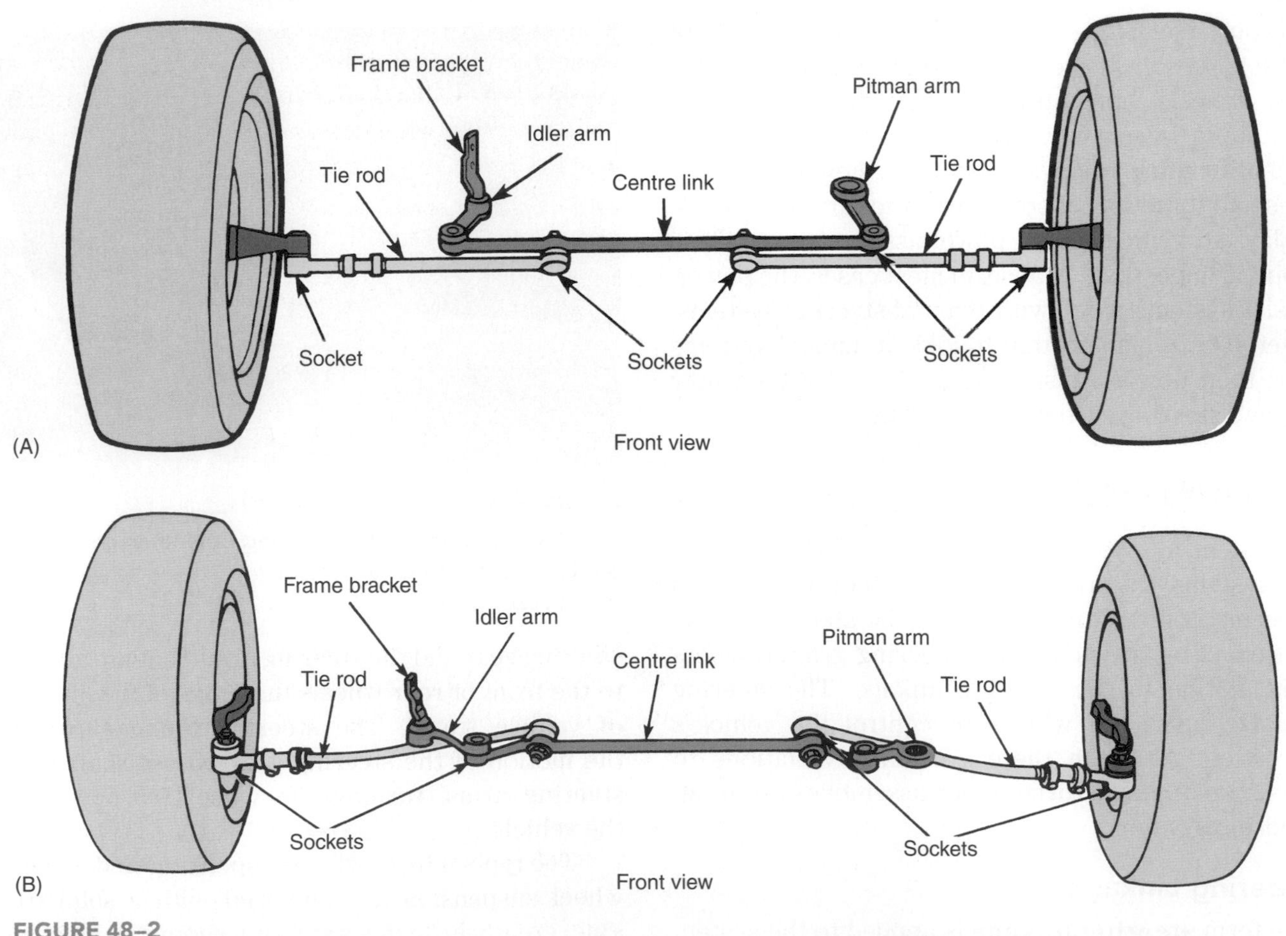

FIGURE 48–2

A parallelogram steering system mounts (A) behind the front suspension and (B) ahead of the front suspension.

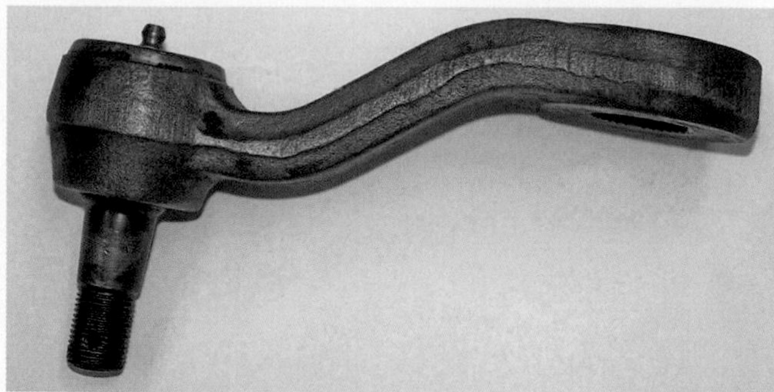

FIGURE 48–3

A typical pitman arm. It connects the steering column to the centre link.

There are two basic types of pitman arms: wear and nonwear. Service needs differ, depending on which type of arm is used. Nonwear arms have tapered holes at their centre link ends and normally need to be replaced only if they have been damaged in an accident or have been mounted with excessive tolerance. Wear arms have studs at the centre link end and are subject to deterioration from normal operation. These arms must be inspected periodically to determine whether they are still serviceable.

IDLER ARM The **idler arm** or idler arm assembly **(Figure 48–4)** is normally attached on the opposite side of the centre link from the pitman arm and to the car frame, supporting the centre link at the correct height. A pivot built into the arm or assembly permits sideways movement of the linkage. On some linkages, such as those on a few light-duty trucks, two idler arms are used.

Idler arms normally wear more than pitman arms because of this pivot function, with wear usually showing up at the swivel point of the arm or assembly. Worn bushings or stud assemblies on idler arms permit excessive vertical movement in the idler arms.

LINKS Links, depending on the design application, can be referred to as **centre links**, **drag links**, or

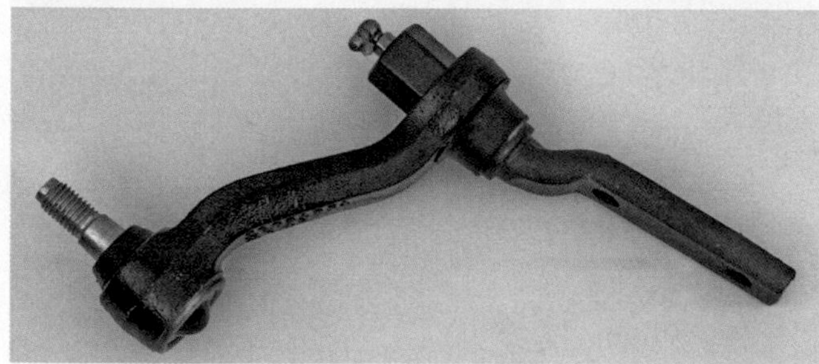

FIGURE 48–4

A typical idler arm. It supports the centre link and is mounted to the frame.

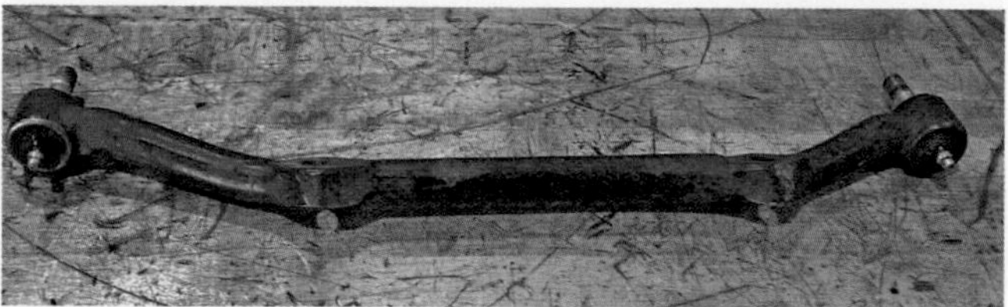

FIGURE 48–5

A typical centre link.

steering links **(Figure 48–5)**. Their purpose is to control sideways linkage movement, which changes the wheel's direction. Because they usually are also mounting locations for tie rods, they are very important for maintaining correct toe settings. If they are not mounted at the correct height, toe is unstable, and toe change or bump steer is produced. Centre links and drag links can be used either alone or in conjunction with each other, depending on the particular steering design. On some vehicles, a steering damper may be attached to the centre link and to the frame.

There are several common designs of centre links. Like pitman arms, they can be broadly characterized as either wear or nonwear. Centre links with stud or bushing ends are likely to become unserviceable from the effects of normal operation and should be inspected periodically. Links with open tapers are nonwear and usually need to be replaced only if they have been damaged in an accident or through excessive tolerance at the mounting position of the idler or pitman arms.

SHOP TALK

A linkage component can be a wear or nonwear component. If the part has a ball joint socket, it is considered a wear component. If the part only has the hole for the ball joint stud of another component to connect to, it is considered a nonwear component. If a centre link is nonwear, the pitman arm is normally a wear arm. If the link is wear, the pitman arm is usually nonwear. Idler arms or assemblies are always subject to wear.

TIE RODS **Tie rods** and tie-rod assemblies make the final connections between the steering linkage and steering knuckles. In a parallelogram steering linkage, the tie rods have ball socket assemblies at each end. One end is attached to the steering arm and the other end to the centre link. Tie-rod assemblies consist of inner tie-rod ends, which are connected to the opposite sides of the centre link; outer tie-rod ends, which connect to the steering knuckles; and adjusting sleeves or bolts, which join the inner and

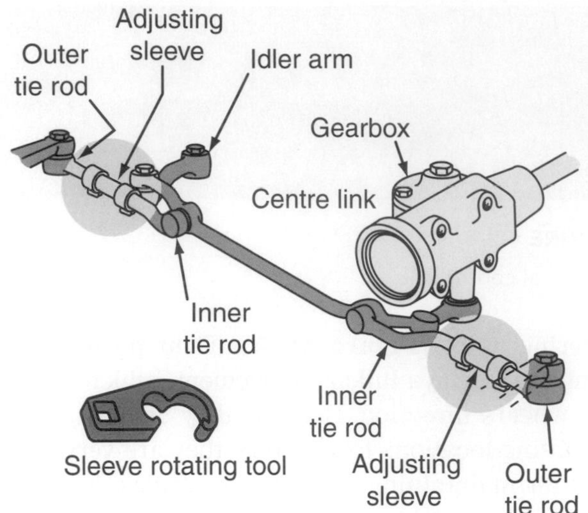

FIGURE 48–6

A tie-rod assembly.

FIGURE 48–7

A cross-steer linkage system on a 4WD truck.

©Cengage Learning 2015

outer tie-rod ends, permitting the tie-rod length to be adjusted for correct toe settings **(Figure 48–6)**.

Tie rods are subject to wear and damage, particularly if the rubber or plastic dust boots covering the ball stud have been damaged or are missing. Contaminants such as dirt and moisture can enter and cause rapid part failure. A special bonded ball stud, in which no boot is used, is available for use on certain light-duty two-wheel-drive and four-wheel-drive (4WD) trucks. An elastomer bushing bonded to the stud ball provides strong shock absorption and steering return in downsized vehicles.

WARNING!

Never apply heat to any part of the steering linkage while servicing it. If any parts must be heated in order to remove them, they must be replaced and not reinstalled.

Cross-Steer and Haltenberger Linkages

Vehicles with solid front axles, front live axles, or twin I-beam suspensions typically use either a cross-steer or Haltenberger linkage **(Figure 48–7)**. These linkage arrangements use long tie rods, often called drag links, and a Pitman arm, instead of parallelogram linkage. These linkages allow for the movement of the front suspension without causing bump steer.

Rack and Pinion Steering Linkage

Rack and pinion is lighter in weight and has fewer components than parallelogram steering **(Figure 48–8)**. Tie rods are used in the same fashion on both systems, but the resemblance stops there. Steering input is received from a pinion gear attached to the steering column. This gear moves a toothed rack that is attached to the tie rods.

In the rack and pinion steering arrangement, there is no pitman arm, idler arm assembly, or centre link. The **rack** performs the task of the centre link. Its movement pushes and pulls the tie rods to change the wheel's direction. The tie rods are the only steering linkage parts used in a rack and pinion system.

Most rack and pinion constructions **(Figure 48–9)** are composed of a tube in which the steering rack can slide. The rack is a rod with gear teeth cut along its length (spur or helical). The inner tie rods attach either at the ends of the rack gear or bolt into its centre. When the tie rods thread onto the ends of the rack gear, it is called an end take-off rack. When the tie rods bolt to the centre of the rack gear, it is called a centre take-off rack.

FIGURE 48–8

A rack and pinion steering system.

©Cengage Learning 2015

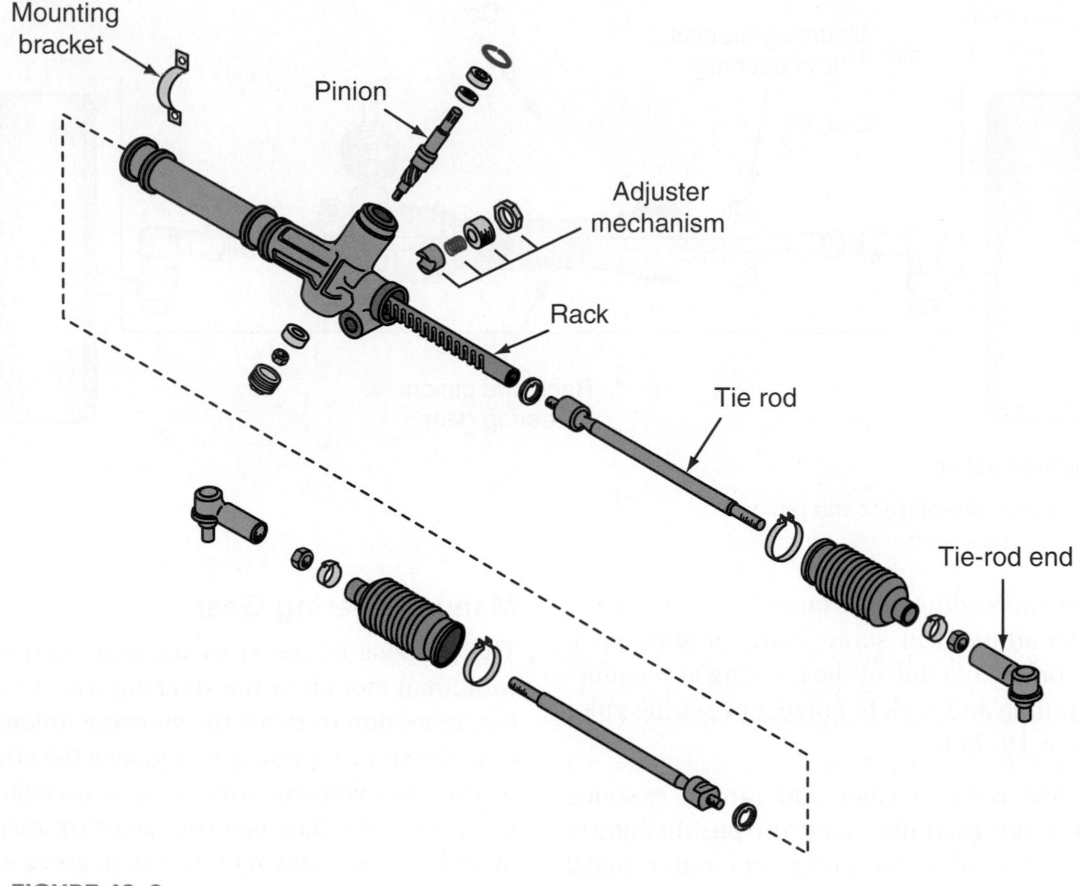

FIGURE 48–9

A disassembled view of a manual rack and pinion steering gear.

The rack meshes with the teeth of a small pinion gear. The pinion gear is at the end of the steering column. The two inner tie-rod ends, which are attached to the rack gear, are covered by rubber bellows boots that protect the rack from contamination. The inner tie rods connect to outer tie-rod ends, which connect to the steering arms. The rack and pinion housing is fastened to the vehicle at two or three points. Typically, the rack is mounted using a rubber bushing. Like the parallelogram linkage, it can be mounted in front of or behind the suspension.

In some cases, the rack and pinion steering gear on unibody cars is bolted directly to a body panel, like a cowl. This is common with centre take-off racks **(Figure 48–10)**. When this is done, the body panel must hold the steering gear in its correct location. The unibody structure must maintain the proper relationship of the steering and suspension parts to each other. Along with other advantages, the rack and pinion steering system combined with the MacPherson strut suspension system is found in most front-wheel-drive unibody vehicles because of their weight- and space-saving feature.

The driver gets a greater feeling of the road with rack and pinion because there are fewer friction points. This means a higher probability of car owners with steering complaints. Fewer friction points can reduce the system's total ability to isolate and dampen vibrations.

RACK The rack is a toothed bar contained in a metal housing. The rack maintains the correct height of the steering components so that the tie-rod movement is able to parallel control arm movement.

The rack is similar to the parallelogram centre link in that its sideways movement in the housing is what pulls or pushes the tie rods to change wheel directions.

PINION The pinion is a toothed or worm gear mounted at the base of the steering column assembly where it is moved by the steering wheel. The pinion gear meshes with the teeth in the rack so that the rack is propelled sideways in response to the turning of the pinion.

YOKE ADJUSTMENT The rack-to-pinion lash, or preload, affects steering harshness, feedback, and

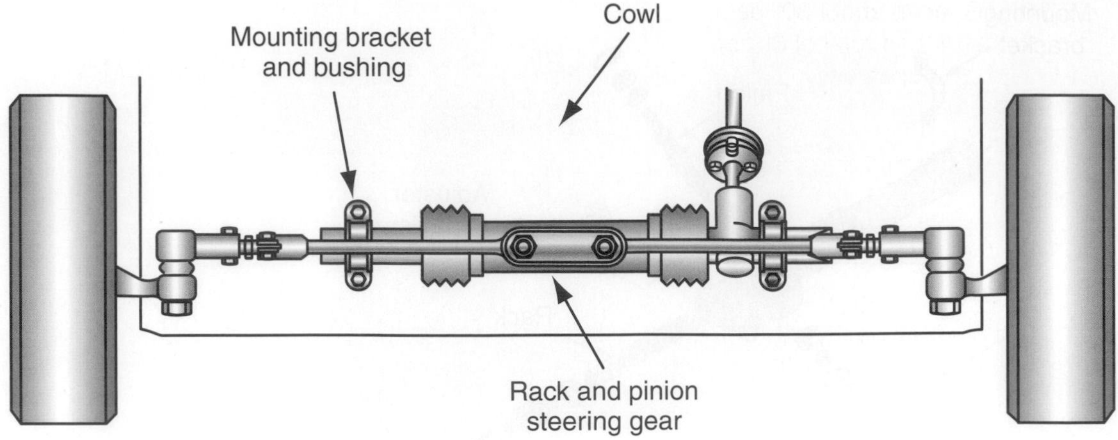

FIGURE 48–10

A centre take-off rack and pinion.

noise. It is set according to the manufacturer's specifications. An adjustment screw, plug, or shim pack are located on the outside of the housing at the junction of the pinion and rack to correct or set the **yoke lash (Figure 48–11)**.

TIE RODS Tie rods in rack and pinion systems are very similar to those used on parallelogram systems. They consist of inner and outer ends. Typically, the rod of the inner tie-rod threads into the outer tie rod. This allows for changing the length of the tie-rod assembly, which changes wheel position and toe angle. The inner tie rod ends on rack and pinion units are usually spring-loaded ball sockets that screw onto the rack ends **(Figure 48–12)**. They are preloaded and protected against contaminant entry by rubber bellows or boots. There is no free play allowed in any of the tie-rod ball sockets.

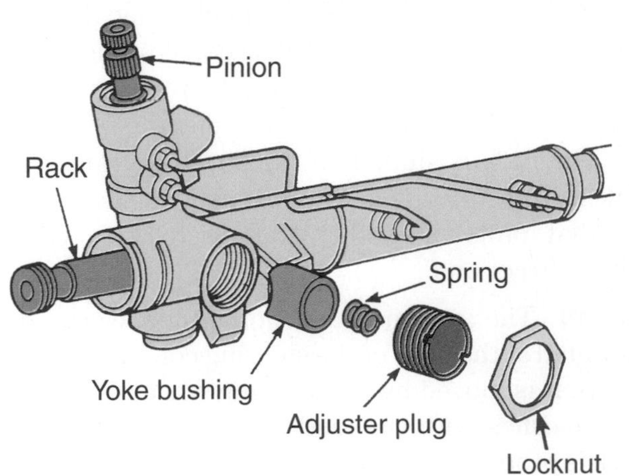

FIGURE 48–11

The rack preload (yoke lash) is adjusted by a screw, plug, or shim pack.

Manual-Steering Gear

The purpose of the steering gear is to change the rotational motion of the steering wheel to a reciprocating motion to move the steering linkage. In addition, the steering gear also reduces the effort needed to turn the wheels. This is done by using gears of different sizes. The steering gearbox ratio is determined by the total number of degrees of steering wheel rotation divided by the total degrees of movement of the front tires **(Figure 48–13)**. Gearboxes

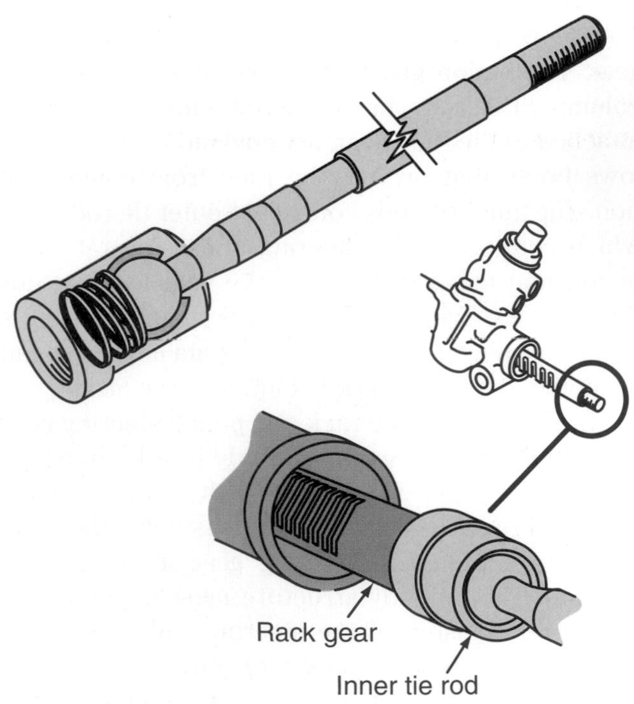

FIGURE 48–12

The inner tie rod is a spring-loaded ball socket in a rack and pinion steering box.

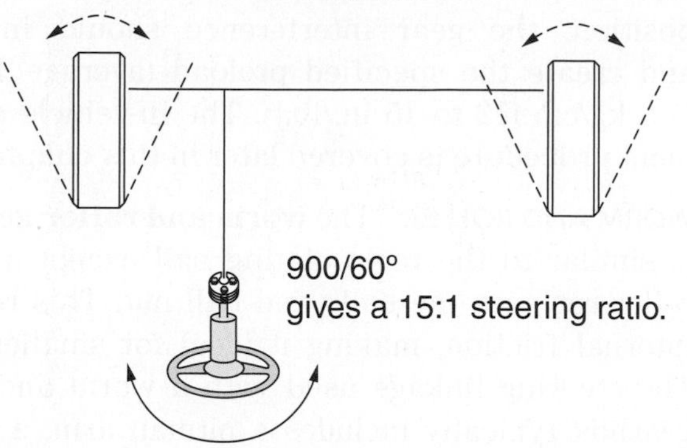

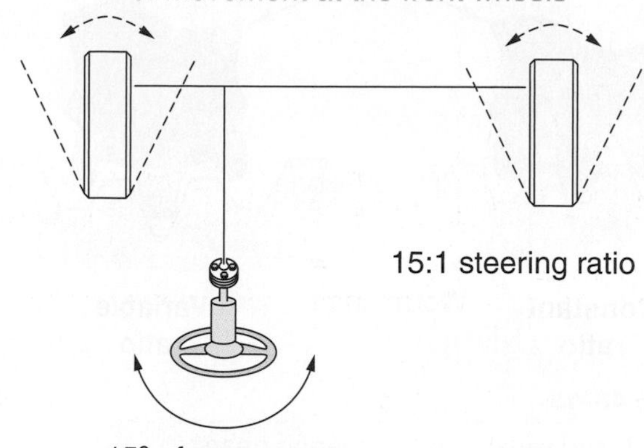

FIGURE 48–13

Steering gearbox ratio.

©Cengage Learning 2015

with high numerical ratios, such as 24:1, provide easier but less precise steering. Lower ratios, such as 15:1, provide better driver feedback and feel but at increased effort.

There are three styles currently in use: the recirculating ball, the worm and roller, and the rack and pinion. The latter gear assembly incorporates the already described rack and pinion linkage system and steering gear as a single unit.

The recirculating ball, as shown in **Figure 48–14**, is generally found in light-duty trucks. A sector shaft is supported by needle bearings in the housing and a bushing in the sector cover. A ball nut is used that has threads that mate to the threads of the wormshaft via continuous rows or ball bearings between the two. Ball bearings recirculate through two outside loops, referred to as ball return guide tubes. The ball nut has gear teeth cut on one face that mesh with gear teeth on the sector shaft. As the steering wheel is rotated, the wormshaft rotates, causing the ball nut to move up or down the wormshaft. Because the gear teeth on the ball nut are meshed with the gear teeth on the sector shaft, the movement of the nut causes the sector shaft to rotate and swing the pitman arm.

The design of two separate circuits of balls results in an almost friction-free operation of the ball nut and the wormshaft. When the steering wheel is turned, the ball bearings roll in the ball thread grooves of the wormshaft and ball nut. When the ball bearings reach the end of their respective circuit, they enter the guide tubes and are returned to the other end of the circuits.

The teeth on the sector shaft and the ball nut are designed so that an interference fit exists between the two when the front wheels are straight ahead. This interference fit eliminates gear tooth lash for a positive feel when driving straight ahead. Proper mesh engagement between the sector and ball nut is obtained by an adjusting screw that moves the sector shaft axially.

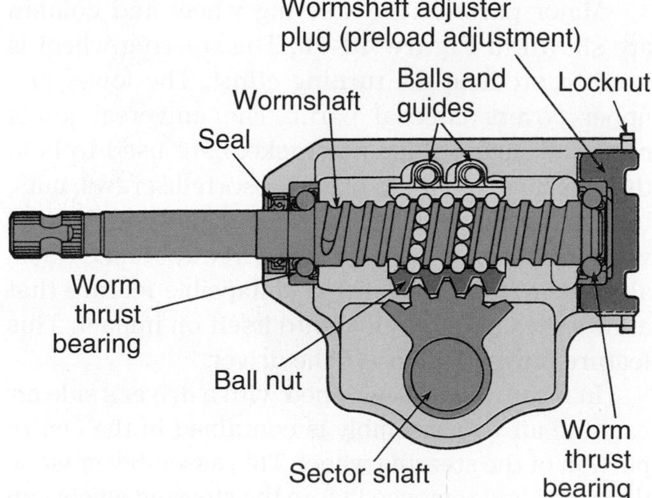

FIGURE 48–14

A top cutaway of a manual recirculating ball steering gear.

CAUTION!

Proper adjustment between the worm gear and the pitman gear is important for adequate steering response. Refer to the vehicle service information for specific adjustment procedures.

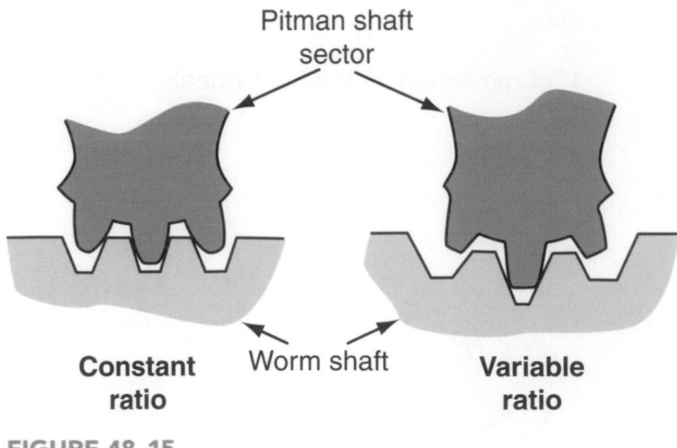

FIGURE 48–15

Constant- and variable-ratio steering gears.

The number of input turns per output turn of the steering gearbox is called the gearbox ratio. Steering gears can have a constant or a variable ratio. The sector teeth in a constant-ratio unit are identical in size and shape, whereas the sector of a variable-ratio unit has larger centre teeth **(Figure 48–15)**. This makes the steering faster in turns than in a straight direction.

The adjustments on this gearbox are ideally performed off-vehicle. With the location of the steering gear and with the steering column and linkage attached, access and drag measurements may be affected. The three typical adjustments are wormshaft bearing preload, sector shaft end play, and sector over-centre preload.

With only the wormshaft installed in the steering gear, the worm thrust bearing adjuster can be turned to provide proper preloading of the wormshaft thrust bearings. The amount of preload can be determined by rotating the wormshaft with a kilogram/centimetre or inch/pound torque wrench. Although steering gears may vary, the average torque reading is 5.75 to 8 kg/cm (5 to 7 in./lb.). Worm bearing preload eliminates wormshaft end play and is necessary to prevent steering free play and vehicle wander. The sector shaft end play and sector over-centre preload are obtained together. By turning the sector lash adjustment screw, the sector shaft will move axially in or out of mesh with the ball nut. The sector shaft end play adjustment should be performed with the steering gear turned off-centre. Turn the adjusting screw until the specified lash (average 0.05 mm [0.002 in.]) is obtained. This clearance allows for smooth operation between the sector shaft gear and the ball nut. To check the sector over-centre preload, a kilogram/centimetre or inch/pound torque wrench can be used to rotate the wormshaft through the steering gear's centre position. Through the centre position, the gear interference should increase and create the specified preload (average 13.8 to 17.3 kg/cm [12 to 15 in./lb.]). The in-vehicle adjustment procedure is covered later in this chapter.

WORM AND ROLLER The **worm and roller gearbox** is similar to the recirculating ball except a single roller replaces the balls and ball nut. This reduces internal friction, making it ideal for smaller cars. The steering linkage used with a worm and roller gearbox typically includes a pitman arm, a centre link, an idler arm, and two inner and outer tie-rod assemblies. The function of these components is the same as in the parallelogram steering linkage described earlier in this chapter.

In operation, the steering shaft rotates the worm gear. It, in turn, engages the roller, causing the roller shaft to turn. The shaft moves the pitman arm left or right to steer the vehicle. It must be noted that the steering gear does not cause the vehicle to pull to one side, nor does it cause road wheel shimmy.

Steering Wheel and Column

The purpose of the steering wheel and column is to produce the necessary force to turn the steering gear. The exact type of steering wheel and column depends on the year and the car manufacturer. The steering column, also called a steering shaft, relays the movement of the steering wheel to the steering gear. An electric motor, used to provide power steering assist, may be part of the steering column.

Major parts of the steering wheel and column are shown in **Figure 48–16**. The steering wheel is used to produce the turning effort. The lower and upper covers conceal parts. The universal joints rotate at angles. Support brackets are used to hold the steering column in place. Assorted screws, nuts, bolt pins, and seals are used to make the steering wheel and column perform correctly. Since 1968, all steering columns have a collapsible feature that allows the column to fold into itself on impact. This feature prevents injury to the driver.

In most vehicles equipped with a driver's side air bag, the air bag assembly is contained in the centre portion of the steering wheel. This assembly must be disarmed and removed before the steering wheel can be removed.

Differences in steering wheel and column designs include fixed column, telescoping column,

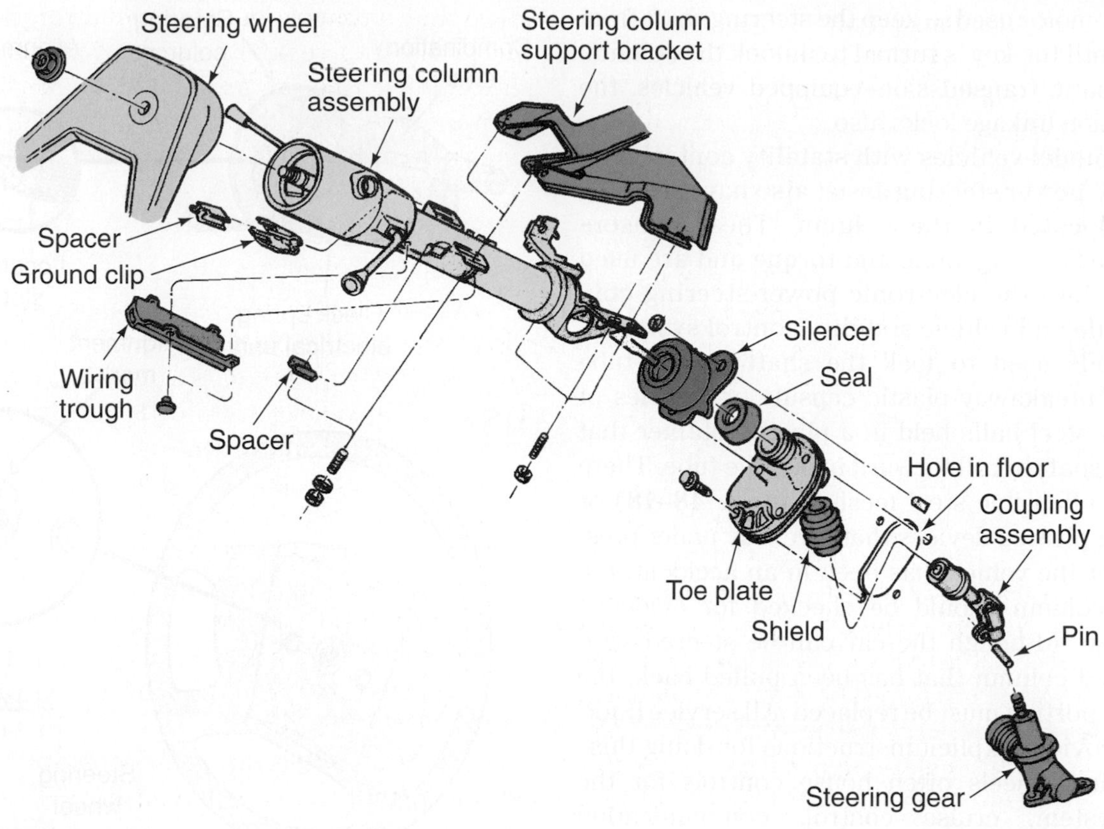

FIGURE 48–16

Typical steering column components. The steering wheel is splined to the shaft that extends through the column and down to the steering gearbox.

Chrysler Group LLC

WARNING!

Always follow the manufacturer's service procedures to disarm an air bag before beginning to remove an air bag assembly. Failure to do this may result in accidental air bag deployment. (See Chapter 24 of this textbook.) Air bag service precautions vary depending on the vehicle.

tilt column, manual transmission, floor shift, and automatic transmission column shift. The tilt columns **(Figure 48–17)** may feature preset detents that have at least five driving positions (two up, two down, and a centre position). Many tilt systems allow setting column position anywhere between a fixed upper and lower limit. A feature on newer cars and trucks is power tilt and telescoping steering columns. Two small electric motors are used to move the column and wheel into the driver's desired position. One motor is used for tilt, and the other motor is for the telescoping function. Powered columns are common on vehicles with programmable memory or personality presets for different drivers.

Most columns contain a multifunction switch, which operates the emergency warning flasher control, turn signal switch, lights (high–low beams), horn, and windshield wipers and washers, as well as an antitheft device that locks the steering system in the column. The antitheft device may be a rod, cable,

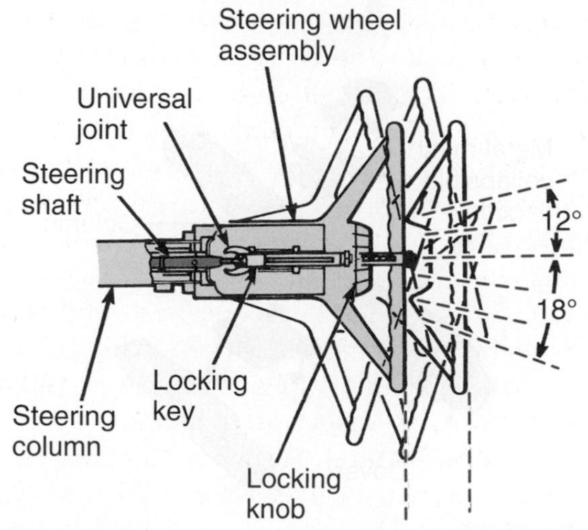

FIGURE 48–17

Tilt steering column operation.

or electric motor used to keep the steering shaft from turning until the key is turned to unlock the column. On automatic transmission–equipped vehicles, the transmission linkage locks also.

Late-model vehicles with stability control and/or electric power steering assist also have steering sensors located in the column. These sensors determine steering angle and torque and are used as inputs into the electronic power-steering control module and vehicle stability control system.

Methods used to lock the shaft to the tube include a breakaway plastic capsule or a series of inserts or steel balls held in a plastic retainer that allow the shaft to roll forward inside the tube. There are also collapsible steel mesh **(Figure 48–18)** or accordion-pleated devices that give way under pressure. After the vehicle has been in an accident, the steering column should be checked for evidence of collapse. Although the car can be steered with a collapsed column that has been pulled back, the collapsed portion must be replaced. All service information provides explicit instructions for doing this.

Steering wheels often house controls for the audio system, cruise control, communication system, and other systems. The steering wheel is usually held in place on the steering column by either a bolt or nut **(Figure 48–19)**. When the blocked tooth on the steering gear input shaft is in the 12 o'clock position, the front wheels should be in the straight-ahead position and the steering wheel spokes in their normal position. If the spokes are not in their normal position, they can be adjusted by changing the toe adjustment. This adjustment can be made only when the steering wheel indexing mark is aligned with the steering column indexing marks. As a rule, indexing teeth or mating flats on the wheel hub and steering shaft prevent misindexing of these components. The alignment of the notches on the steering wheel hub and steering shaft confirm correct orientation.

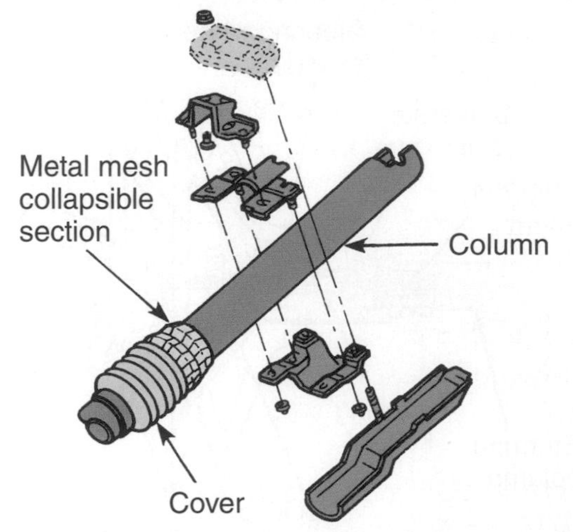

FIGURE 48–18

A collapsing mesh steering column.

Ford Motor Company

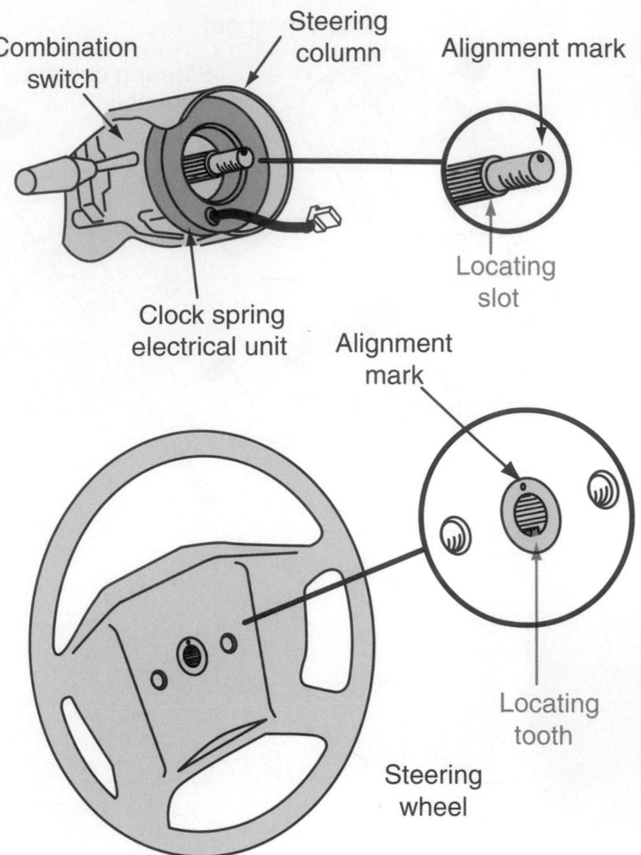

FIGURE 48–19

The steering wheel is splined to the steering column and held in place by a nut.

Steering Damper

The purpose of a steering damper is simply to reduce the amount of road shock that is transmitted up through the steering column. Steering dampers are found mostly on 4WD vehicles, especially those fitted with large tires. The damper serves the same function as a shock absorber but is mounted horizontally to the steering linkage—one end to the centre link and the other to the frame.

POWER-STEERING SYSTEMS

The power-steering unit is designed to reduce the amount of effort required to turn the steering wheel. It also reduces driver fatigue on long drives and makes it easier to steer the vehicle at slow road speeds, particularly during parking.

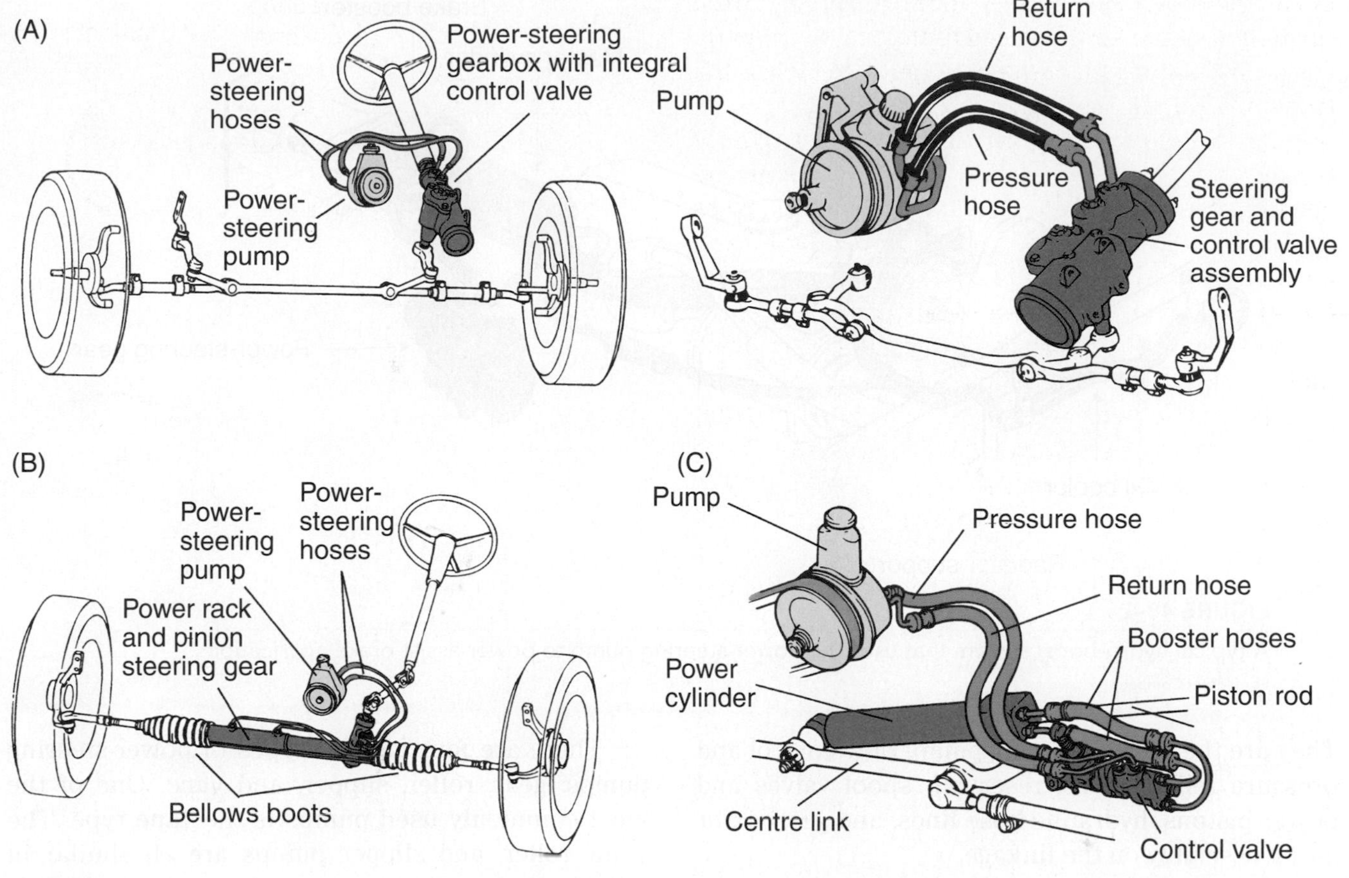

FIGURE 48–20

Three types of power-steering systems: (A) integral piston linkage, (B) rack and pinion, and (C) external piston linkage.
Federal-Mogul Corporation

Power steering can be broken down into two design arrangements: conventional and nonconventional (or electronically controlled). In the conventional arrangement, hydraulic power is used to assist the driver. In the nonconventional arrangement, an electric motor and electronic controls provide power assistance in steering.

There are several power-steering systems in use on passenger cars and light-duty trucks. The most common ones are the integral piston and power-assisted rack and pinion system **(Figure 48–20)**.

Integral Piston System

The **integral piston** system is the most common conventional power-steering system in use today. It consists of a power-steering pump and reservoir, power-steering pressure and return hose, and steering gear. The power cylinder and the control valve are in the same housing as the steering gear.

On some late-model cars and light trucks, instead of the conventional vacuum-assist brake booster, the hydraulic fluid from the power-steering pump is also used to actuate the brake booster. This brake system is called the hydro-boost system **(Figure 48–21)**.

Power-Assisted Rack and Pinion System

The power-assisted rack and pinion system is similar to the integral system because the power cylinder and the control valve are in the same housing. The rack housing acts as the cylinder, and the power piston is part of the rack. Control valve location is in the pinion housing. Turning the steering wheel moves the valve, directing pressure to either end of the rack piston. The system utilizes a pressure hose from the pump to the control valve housing and a return line to the pump reservoir. This type of steering system is common in most vehicles today.

Components

Several of the manual-steering parts described earlier in this chapter, such as the steering linkage, are used in conventional power-steering systems. The components that have been added for power steering provide the hydraulic power that drives the system.

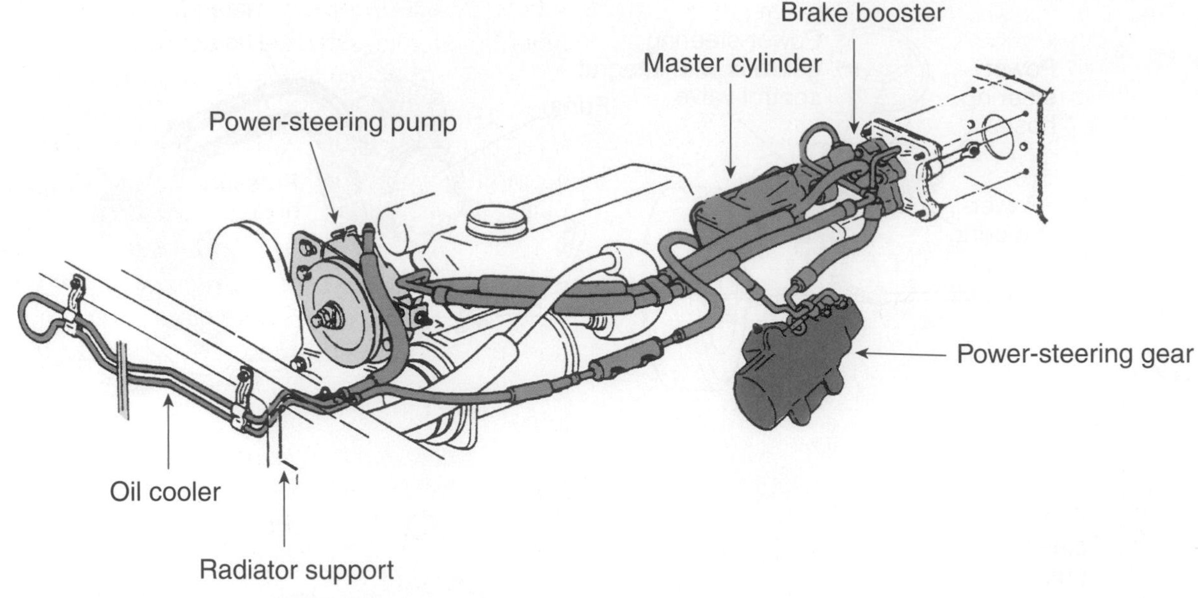

FIGURE 48–21

A typical hydro-boost system that uses the power-steering pump to power-assist brake applications.

Ford Motor Company

They are the power-steering pump, flow control and pressure relief valves, reservoir, spool valves and power pistons, hydraulic hose lines, and gearbox or assist assembly on the linkage.

POWER-STEERING PUMP The power-steering pump is used to develop hydraulic flow, which provides the force needed to operate the steering gear. The pump is belt driven from the engine crankshaft, providing flow any time the engine is running. It is usually mounted near the front of the engine **(Figure 48–22)**. The pump assembly includes a reservoir and an internal flow control valve. The drive pulley is normally pressed onto the pump's shaft.

FIGURE 48–22

A power-steering pump.

Visteon Corporation

There are four general types of power-steering pumps: gear, roller, slipper, and vane. One of the most commonly used pumps is the vane type. The vane, roller, and slipper pumps are all similar in construction and are all positive displacement–balanced pumps. The vane pump consists of a slotted centre rotor, a number of flat vanes that are fitted into the slotted rotor, and an outer pump cam or cam ring that is slightly elliptical in shape to form two pumping chambers. As the rotor turns, the vanes follow the contour of the cam ring by centrifugal force and hydraulic pressure against the inner end and the flat side of each vane. The area between each pair of vanes forms individual pumping chambers that increase and decrease in relation to the distance between the rotor and cam ring. As the chamber's size increases, fluid from the reservoir flows in. The fluid is carried around the cam ring to the section where the chamber size decreases, and the fluid is forced out of the chamber to the power-steering pressure line. This occurs between each pair of vanes on both sides of the rotor, which balances the force against the rotor. These pumps also move a certain volume of fluid for each turn of the pump, whether turning at low or high rpm, and its volume at engine idle is sufficient to produce maximum steering assist **(Figure 48–23).** Hydraulic fluid for the power-steering pump is stored in a reservoir. Fluid is routed to and from the pump by hoses and lines. Excessive pressure is controlled by a relief valve.

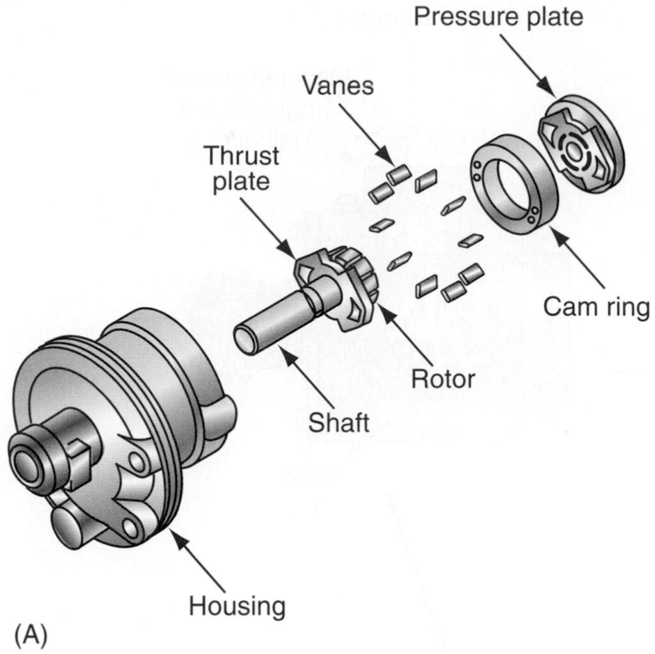

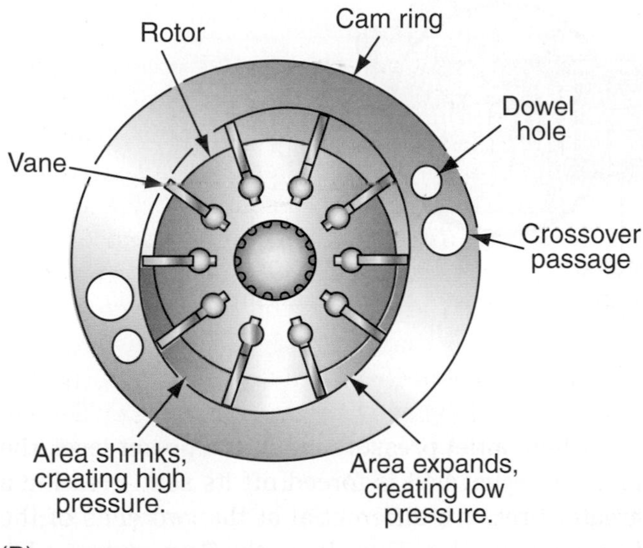

FIGURESS 48–23

(A) A vane-type power-steering pump. (B) The basic operation of a vane-type pump.

POWER-STEERING PUMP DRIVE BELTS Many power-steering pumps are driven by a belt that connects the crankshaft pulley to the power-steering pump pulley. The belt may also drive other components, such as the water pump. The sides of a V-belt are the friction surfaces that drive the power-steering pump. If the sides of the belt are worn and the lower edge of the belt is contacting the bottom of the pulley, the belt will slip. The power-steering pump pulley, crankshaft pulley, and any other pulleys driven by the V-belt must be properly aligned. If these pulleys are misaligned, excessive belt wear occurs.

A ribbed V-belt is used on many vehicles, and this belt may be used to drive all the belt-driven components. Most ribbed V-belts have a spring-loaded automatic belt tensioner that eliminates periodic belt tension adjustments. Because the ribbed V-belt may be used to drive all of the belt-driven components, these components are placed on the same vertical plane, which saves a considerable amount of underhood space. The smooth backside of the ribbed V-belt may also be used to drive one of the components. Regardless of the type of belt, the belt tension is critical. A power-steering pump will never develop full pressure if the belt is slipping.

ELECTRIC POWER STEERING Many vehicles use a 12- or 42-volt electric motor mounted to or in the steering gear **(Figure 48–24)**. The motor replaces the conventional pump and its belts and hoses. These systems are discussed in more detail later in this chapter.

FLOW CONTROL AND PRESSURE RELIEF VALVES A flow control/pressure relief valve controls the volume output from the pump. This valve is necessary because of the variations in engine rpm and the need for consistent steering ability in all ranges, from idle to highway speeds. It is positioned in a chamber that is exposed to pump outlet pressure at one end and supply hose pressure at the other. A spring is used at the supply pressure end to help maintain a balance.

As the fluid leaves the pump rotor, it passes the end of the flow control valve and is forced through an orifice that causes a slight drop in pressure. This reduced pressure, aided by the springs, holds the flow control valve in the closed position. With the engine idling and no steering effort at the steering gear, the pressure within the system can range from 210 to 1050 kPa (30 to 150 psi). This pressure is from the restriction of the entire pump flow being circulated through the lines, the steering gear directional control valve, and back to the power-steering pump reservoir.

When engine speed increases, the pump can deliver more flow than is required to operate the system. Because the outlet orifice restricts the amount of fluid leaving the pump, the difference in pressure at the two ends of the valve becomes greater until pump outlet pressure overcomes the combined force of supply line pressure and spring force. The valve is pushed down against the spring, opening a passage that returns the excess flow back to the inlet side of the pump.

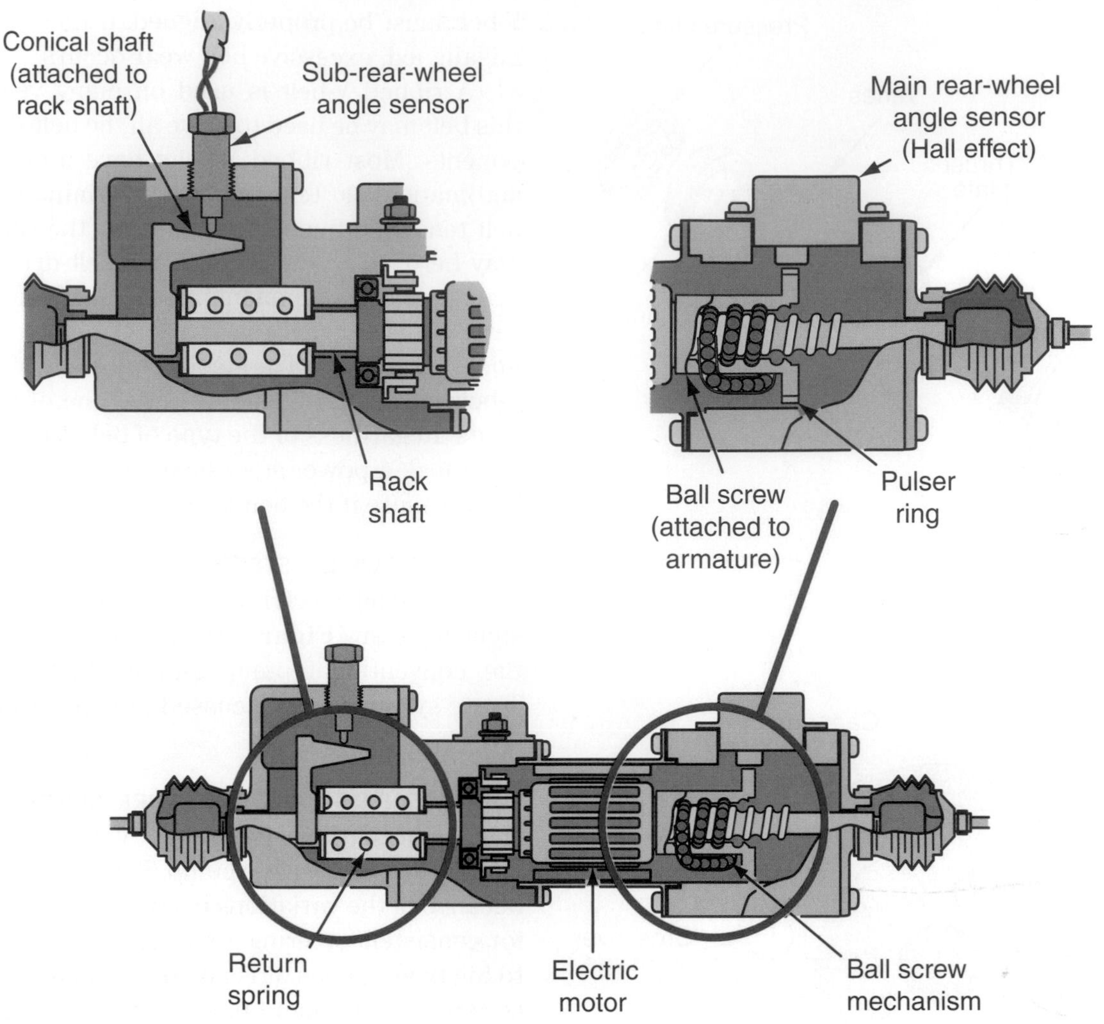

FIGURE 48–24

An electric/electronic rack and pinion system.

When there is steering effort at the steering gear, the directional control valve is directing partial pump flow toward the steering gear power piston and restricting the remaining flow back to the pump. Depending on the amount of restriction at the steering gear, the pressure within the system can range from 4150 to 5500 kPa (600 to 800 psi). This increase in pressure provides the extra force against the power piston that provides the steering assist.

A spring and ball contained inside the flow control valve are used to relieve pump outlet pressure. This is done to protect the system from damage due to excessive pressure when the steering wheel is held against the stops or the wheels are wedged against a curb. Because the directional control valve is directing maximum flow against the power piston with very minimal flow returning to the pump, the pump would continue to build pressure until a hose ruptured or the pump destroyed itself.

When outlet pressure reaches a preset level, the pressure relief ball is forced off its seat, creating a greater pressure differential at the two ends of the flow control valve. This allows the flow control valve to open wider, permitting more pump volume to flow back to the pump inlet, and pressure is held at a safe level.

When the pressure relief valve is operating, the fluid flow through the valve can generate a squealing noise. The sharp increase in pressure, usually 8300 to 10 350 kPa (1200 to 1500 psi), will create a rapid temperature increase of the power-steering fluid and possibly cause damage to the power-steering pump and other system components. The pump should not operate in this mode for more than a couple of seconds.

Some vehicles electronically control the fluid flow from the pump **(Figure 48–25)**. These variable-effort systems can decrease fluid flow to increase steering firmness and improve steering feel based on

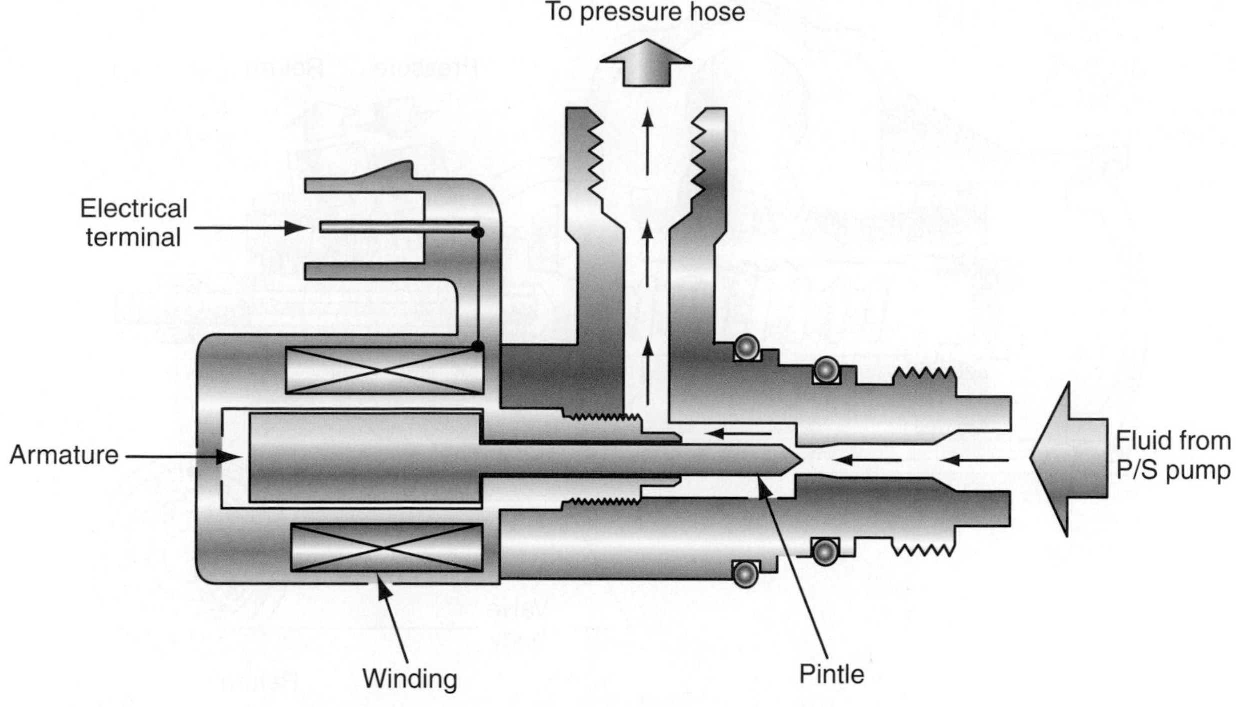

FIGURE 48–25

Electrohydraulic variable-effort steering.

vehicle speed. When vehicle speed is reduced, flow is increased to allow easier steering.

POWER-STEERING GEARBOX A power-steering gearbox is basically the same as a manual **recirculating ball gearbox** with the addition of a hydraulic assist. A power-steering gearbox is filled with hydraulic fluid and uses a control valve.

The integral power-steering system has the directional control valve (DCV) and a power piston integrated with the recirculating ball gearbox. The DCV directs the oil pressure to the left or right power chamber to steer the vehicle. The DCV is actuated by a torsion bar that is located between the input (stub) shaft and the wormshaft **(Figure 48–26)**. The

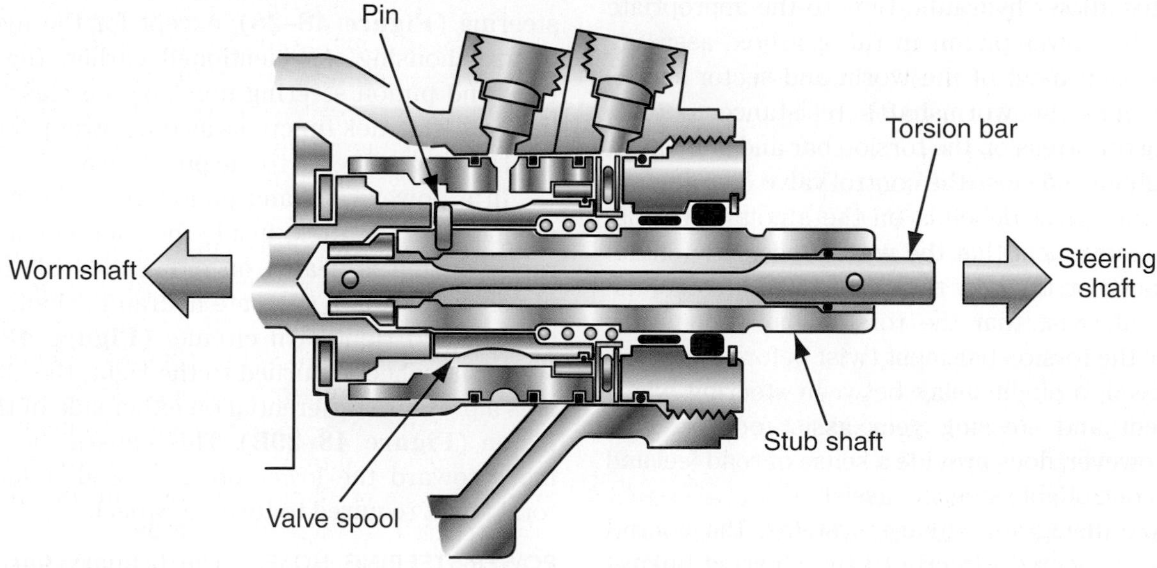

FIGURE 48–26

A torsion bar moves the spool valve to direct the oil flow to the piston.

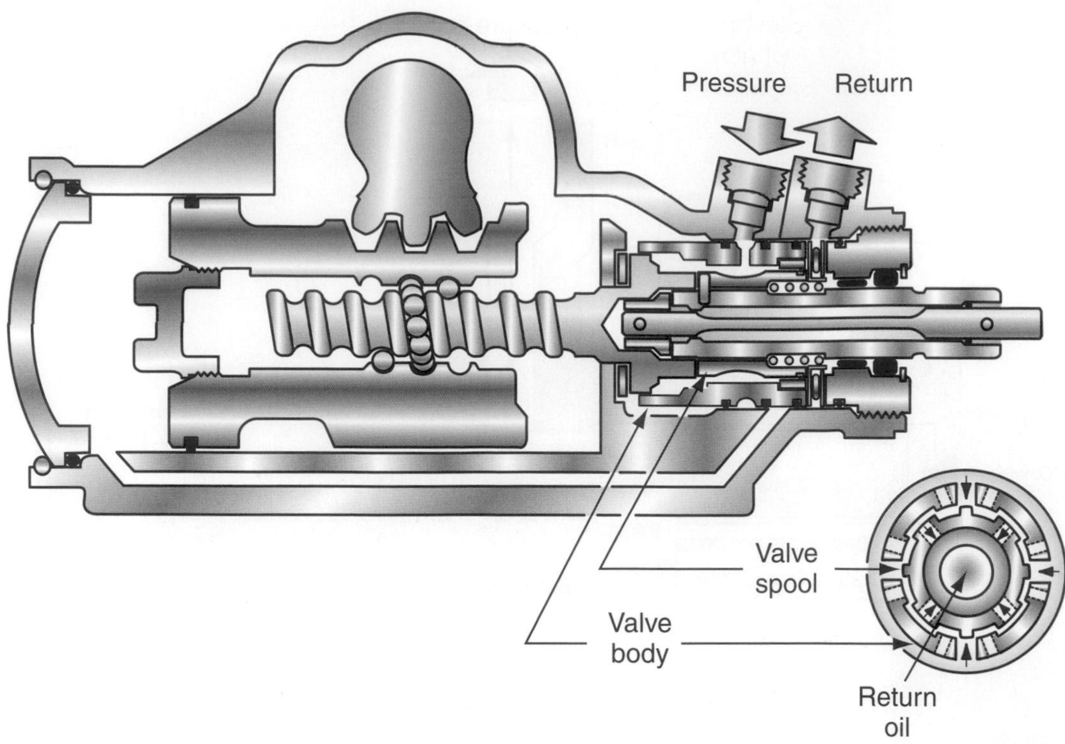

FIGURE 48–27

The spool valve with the wheels straight ahead.

directional control valve is comprised of an inner and outer spool valve. The inner spool is connected to the input shaft and one end of the torsion bar, and the outer spool is connected to the wormshaft and the other end of the torsion bar. When steering effort is placed on the input shaft through to the wormshaft, the torsion bar will twist slightly, opening the DCV. This allows hydraulic flow to the appropriate side of the power piston in the gearbox, assisting with the movement of the worm and sector shafts. This reduces the wormshaft's resistance to turn, relieving the stress on the torsion bar and allowing it to straighten and close the control valve. The amount of steering assist depends on the amount of resistance to movement that the worm and sector shafts have placed on them by the wheels and, in turn, the amount of twist that the torsion bar encounters. Because the torsion bar must twist before any assist is provided, a slight delay between steering wheel movement and steering gear assist occurs. This delay, however, does provide a sense of road feel and a more controllable steering assist.

In parallelogram linkage systems, the control valve is connected directly to the steering linkage through the pitman arm on the steering gear. Any movement of the steering wheel and the pitman arm compresses the centring spring and moves the valve spool. This opens and closes a series of ports, directing fluid under pressure from the pump to one side or the other of the power cylinder piston **(Figure 48–27)**.

POWER-ASSISTED RACK AND PINION STEERING Power-assisted rack and pinion components are basically the same as for manual rack and pinion steering **(Figure 48–28)**, except for the hydraulic control housing. As mentioned earlier, the power rack and pinion steering unit may be classified as integral. The rack functions as the power piston, and the DCV is connected to the pinion gear.

In a power rack and pinion gear, the piston is mounted on the rack, inside the rack housing. The rack housing is sealed on either side of the rack piston to form two separate hydraulic chambers for the left and right turn circuits **(Figure 48–29A)**. When the wheel is turned to the right, the DCV creates a pressure differential on either side of the rack piston **(Figure 48–29B)**. This causes the rack to move toward the lower pressure and reduces the total effort required to turn the wheels.

POWER-STEERING HOSES The primary purpose of power-steering hoses is to transmit power (fluid under pressure) from the pump to the steering gearbox, and to return the fluid ultimately to the

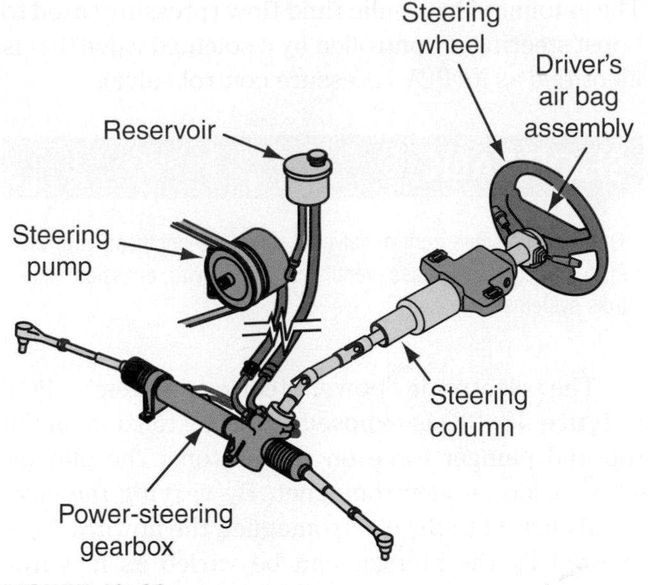

FIGURE 48–28

A complete power-assisted rack and pinion steering system.

©Cengage Learning 2015

pump reservoir. Hoses also, through material and construction, function as additional reservoirs and act as sound and vibration dampers.

Hoses are generally a reinforced synthetic rubber (neoprene) material coupled to metal tubing at the connecting points. The pressure side must be able to handle pressures up to 10 350 kPa (1500 psi). For that reason, wherever there is a metal tubing to a rubber connection, the connection is crimped. Pressure hoses are also subject to surges in pressure and pulsations from the pump. The reinforced construction permits the hose to expand slightly and absorb changes in pressure.

Two internal diameters of hose **(Figure 48–30)** may be used on the pressure side; the larger diameter or pressure hose is at the pump end. It acts as a reservoir and as an accumulator absorbing pulsations. The smaller diameter or return hose reduces the effects of kickback from the gear itself. By restricting fluid flow, it also maintains constant back pressure on the pump, which reduces pump noise. If the hose is of one diameter, the gearbox is performing the damping functions internally.

Because of working fluid temperature and adjacent engine temperatures, these hoses must be able to withstand temperatures up to 150°C (300°F). Due to various weather conditions, they must also tolerate subzero temperatures. Hose material is specially formulated to resist breakdown or deterioration due to oil or temperature conditions.

CAUTION!

Hoses must be carefully routed away from engine manifolds. Power-steering fluid is very flammable. If it comes in contact with hot engine parts, it could start an underhood fire.

(A)

(B)

FIGURE 48–29

(A) A power-steering piston and cylinder for a rack and pinion steering gear. (B) The assembly that connects the steering wheel to the rack and pinion assembly.

©Cengage Learning 2015

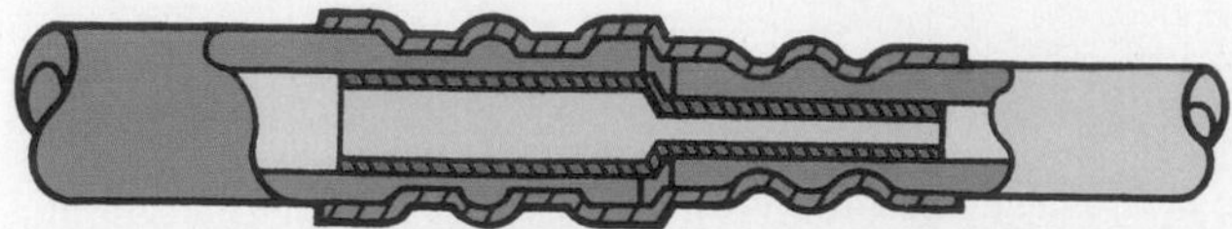

FIGURE 48–30

Power-steering hoses may have internal diameters.

VARIABLE-EFFORT (ELECTRONICALLY CONTROLLED) POWER-STEERING SYSTEMS

The object of power steering is to make steering easier at low speeds, especially while parking. However, higher steering efforts are desirable at higher speeds in order to provide improved down-the-road feel. The electronically controlled power-steering (EPS) systems **(Figure 48–31)** provide both of these benefits. The hydraulic boost of these systems is tapered off by electronic control as road speed increases. Thus, these systems require well under 0.45 kg (1 lb.) of steering effort at low road speeds and 1.36 kg (3 lb.) plus of steering effort at higher road speeds to enable the driver to maintain control of the steering wheel for improved high-speed handling.

A rotary valve (DCV) electronic power-steering system consists of the power-steering gearbox, power-steering oil pump, pressure hose, and return hose.

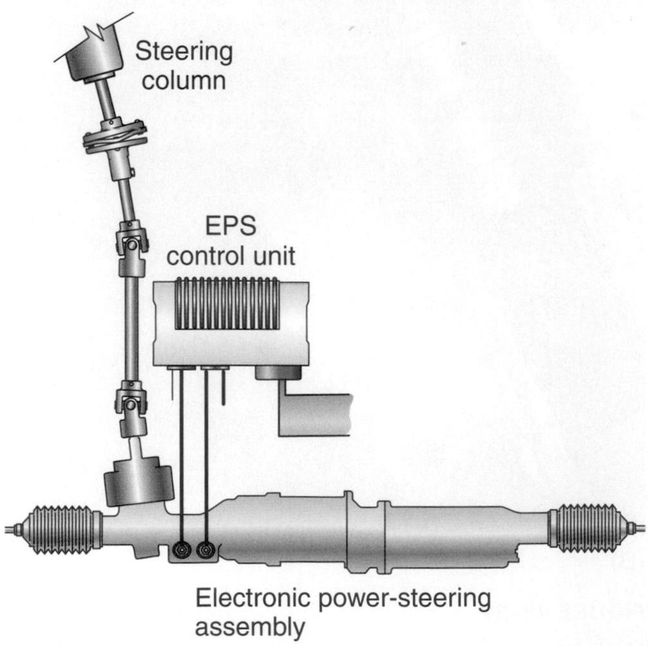

FIGURE 48–31

A variable-assist power-steering system.

The amount of hydraulic fluid flow (pressure) used to boost steering is controlled by a solenoid valve that is identified as its PCV (pressure control valve).

SHOP TALK

The PCV (pressure control valve) is not to be confused with the PCV (positive crankcase ventilator) used with emission controls systems.

The electronic power-steering system's PCV **(Figure 48–32)** is exposed to spring tension on the top and plunger force on the bottom. The plunger slips inside an electromagnet. By varying the electrical current to the electromagnet, the upward force exerted by the plunger can be varied as it works against the opposing spring. Current flow to the electromagnet is variable with vehicle road speed and therefore provides steering to match the vehicle's road speed.

General Motors' variable-effort steering (VES) system relies on an input signal from the vehicle speed sensor to the VES controller to control the amount of power assist. The controller, in turn, supplies a pulse-width-modulated voltage to the actuator solenoid in the power-steering pump. The controller also provides a ground connection for the solenoid.

When the vehicle is operating at low speeds, the controller supplies a signal to cycle the solenoid faster so that it allows high pump pressure. This provides for maximum power assist during cornering and parking. As the vehicle's speed increases, the solenoid cycles less, and the pump provides a lower amount of assist. This gives the driver better road feel during high speeds.

Magnasteer

In the mid-1990s, General Motors introduced magnetic speed variable assist, called Magnasteer. These systems use a special rack and pinion gearbox that contains an electromagnet inside the spool valve. By varying current flow through the coil, assist is increased or decreased. No current flow through the coil maintains a default amount of hydraulic assist.

Active Steering

Active steering improves vehicle stability by turning the wheels more or less sharply than commanded by the turn of the steering wheel during some situations.

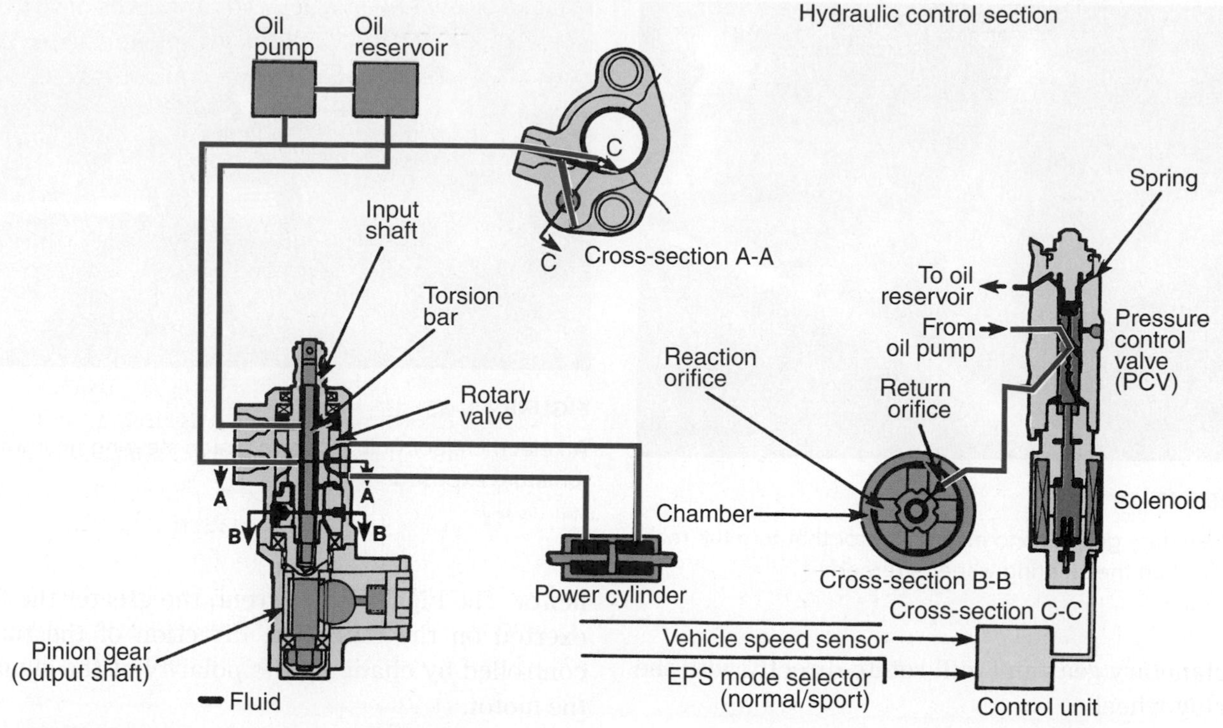

FIGURE 48–32

An outline of electronic power-steering components. The EPS PCV is exposed to spring tension and plunger force.

Mitsubishi Motor Sales of America, Inc.

Through inputs and computer programming, this system can adjust the steering to respond quickly to the threat of skidding **(Figure 48–33)**. The system also allows for a variable steering ratio dependent on vehicle speed.

Current active steering systems are not true steer-by-wire systems. There is still a mechanical connection between the steering wheel and vehicle's wheels **(Figure 48–34)**. The systems have an overriding drive built into the steering column. This drive is controlled by an electric motor, which is controlled by the system's computer. The computer determines whether the steering angle needs to be changed and by how much. If the system fails,

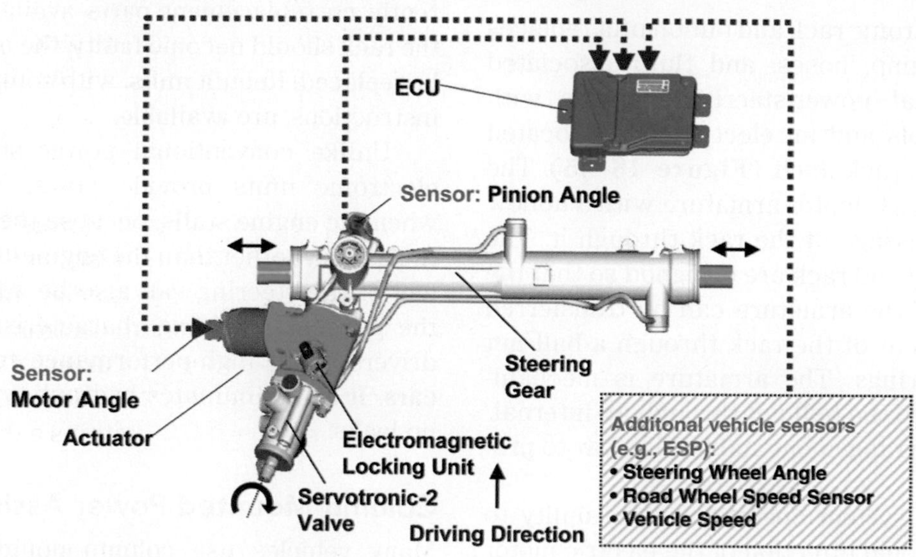

FIGURE 48–33

The main components and circuits of an active steering system.

Robert Bosch Corp.

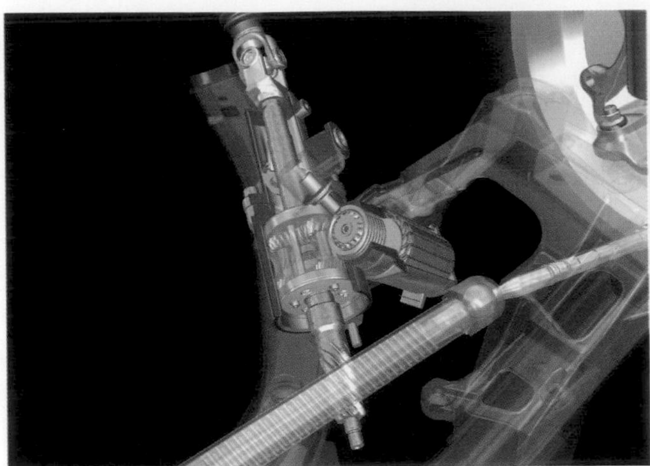

FIGURE 48–34

The planetary gearset and electric motor that turn the road wheels when the steering wheel is turned.

BMW Group

FIGURE 48–35

An electric/electronic rack and pinion steering gear with its armature exposed.

Martin Restoule

the planetary gear unit will rotate directly with the steering wheel.

General Service

When servicing, as with any power-steering system, the first step is to look for fluid leaks, damaged components, a slipping drive belt, and so on. Only after these things have been checked and no problems have been found should the electronics be suspected as the problem. Check the appropriate service information for the correct troubleshooting procedures of the electronics.

Electric/Electronic Rack and Pinion System

The electric/electronic rack and pinion unit replaces the hydraulic pump, hoses, and fluid associated with conventional power-steering systems with electronic controls and an electric motor located concentric to the rack itself **(Figure 48–35)**. The design features a DC motor armature with a hollow shaft to allow passage of the rack through it. The outboard housing and rack are designed so that the rotary motion of the armature can be transferred to linear movement of the rack through a ball nut with thrust bearings. The armature is mechanically connected to the ball nut through an internal/external spline arrangement and torsion bar to provide road feel.

The basis of system operation is its ability to change the rotational direction of the electric motor while being able to deliver the necessary amount of current to meet torque requirements at the same time. The system can deliver up to 75 amps to the motor. The higher the current, the greater the force exerted on the rack. The direction of the turn is controlled by changing the polarity of the signal to the motor.

The field assembly houses permanent ceramic magnets while providing structural integrity for the gear system. In essence, the electronic/electric rack design allows for a direct power source to the rack and steering linkage. The system monitors steering wheel movement through a sensor mounted on the input shaft of the rack and pinion steering gear. After receiving directional and load information from the sensor, an electronic controller activates the motor to provide power assistance.

These units are readily retrofitted to conventionally equipped vehicles. As for servicing, there are currently no replacement parts available; therefore, if the rack should become faulty, the entire unit should be replaced. Rebuilt units, with complete installation instructions, are available.

Unlike conventional power steering, electric/electronic units provide power assistance even when the engine stalls, because the power source is the battery rather than the engine-driven pump. The feel of the steering can also be adjusted to match the particular driving characteristics of cars and drivers, from high-performance to luxury-touring cars. It also eliminates hydraulic oil, which means no leaks.

Column-Mounted Power Assist

Many vehicles use column-mounted power-assist systems. An electric motor is mounted to the steering column and uses a worm gear to directly drive the steering shaft **(Figure 48–36)**. Steering input is

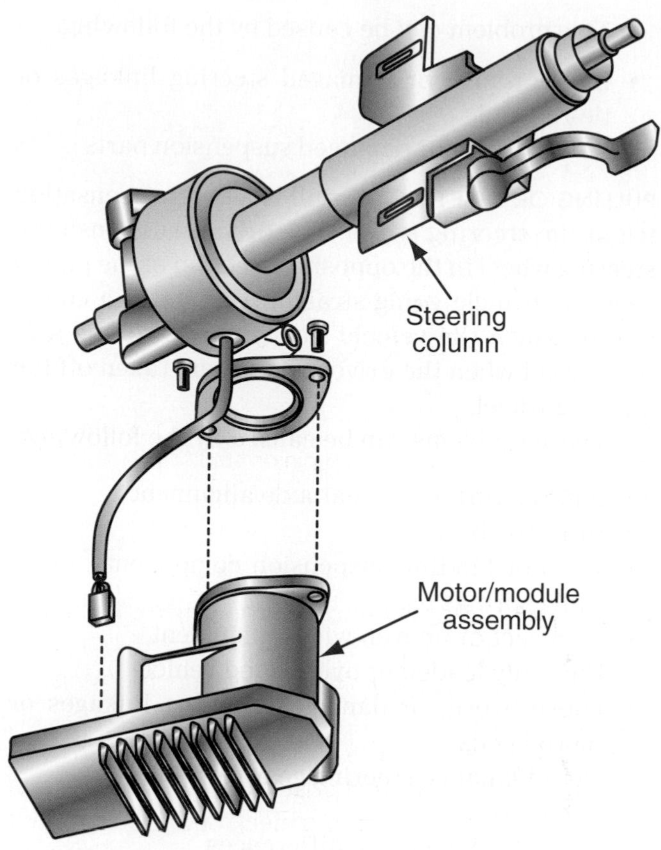

FIGURE 48–36

A column-mounted power-assist motor.

detected by one or more torque sensors. As the steering shaft turns, the sensors detect the change in movement and torque applied to the steering shaft. Input from other sensors, such as vehicle speed and others, are used to determine the amount of assist provided by the motor.

Steer-by-Wire System

Completely steer-by-wire systems **(Figure 48–37)** are not found on any production vehicles today. They are being tested and have appeared on many concept cars. The 2014 Infinity Q50 Hybrid has what is called "Infinity Adaptive Steering," which is a steer-by-wire system with a complete mechanical backup system.

Steer-by-wire systems do not use a steering column or shaft to connect the steering wheel to the steering gear. The system is totally electronic. The turning of the steering wheel is monitored by a sensor. The sensor sends an input signal to a controller. The controller, in turn, sends commands to an electric motor in the steering gear. The commands from the controller are also based on inputs from a variety of other inputs, such as vehicle speed.

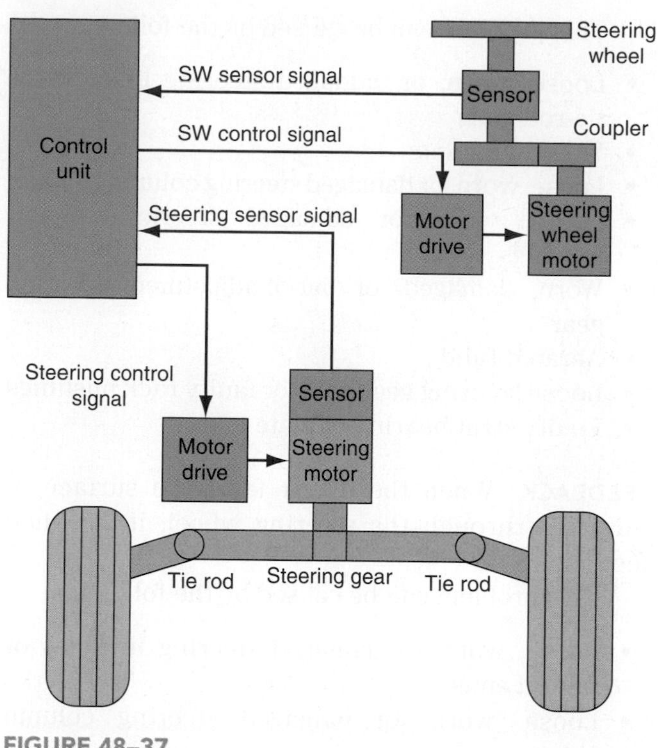

FIGURE 48–37

The basic layout for a steer-by-wire system.

These systems also have a small motor attached to the mount for the steering wheel. This motor is controlled by a steering controller. This motor provides the correct steering feel for the current conditions. The driver needs this feel to maintain control of the vehicle.

Steer-by-wire systems allow total customization of steering performance and can provide a constantly variable steering ratio. The absence of a steering column opens up space in the vehicle's interior and engine compartment. The systems are also lighter than conventional steering systems.

STEERING SYSTEM DIAGNOSIS

It is important to realize that many steering complaints are caused by problems in areas other than the steering system. A good diagnosis is one that finds the exact cause of the customer's complaint. Although customers may describe the problem in different ways, the most common complaints and their typical causes are discussed next.

Common Complaints

EXCESSIVE STEERING WHEEL PLAY Excessive play in the steering wheel is apparent when there is too much steering wheel movement before the wheels begin to turn. A small amount of play is normal.

This problem can be caused by the following:

- Loose, worn, or damaged steering linkages or tie-rod ends
- Worn ball joints
- Loose, worn, or damaged steering column U-joints
- Loose, worn, or damaged steering column bearings
- Worn, damaged, or out-of-adjustment steering gear
- Aerated fluid
- Loose steering gear bolts or faulty rack bushings
- Faulty strut bearing or plate

FEEDBACK When the driver feels the surface of the road through the steering wheel, it is called feedback.

This problem can be caused by the following:

- Loose, worn, or damaged steering linkages or tie-rod ends
- Loose, worn, or damaged steering column U-joints
- Loose or damaged steering gear mounting bolts
- Damaged or worn steering column bearings
- Loose suspension bushings, fasteners, or ball joints

HARD STEERING Obviously, a complaint of hard steering results when extra effort is needed to turn the steering wheel. This problem may be simply an absence of power assist. Hard-steering problems can occur whenever the steering wheel is turned or just when it has been turned close to its limit.

This problem can be caused by the following:

- A faulty power-steering pump
- Damaged or faulty steering column bearings
- Seized steering column U-joints
- Steering gear set too tight or is binding
- Stuck flow control valve
- Inadequately inflated tires
- Restricted power-steering lines or hoses

Internal leaks around the flow control valve seals in rack and pinion units are a common cause of hard steering, especially on cold starts. Once the fluid warms, the hard steering may improve.

NIBBLE This feeling is similar to a shimmy. Nibble results from the interaction of the tires with the road's surface. The customer's complaint may describe the nibble problem as slight rotational oscillations of the steering wheel.

This problem can be caused by the following:

- Loose, worn, or damaged steering linkages or tie-rod ends
- Loose, worn, or damaged suspension parts

PULLING OR DRIFTING A pull is a tugging sensation felt at the steering wheel. The driver must push the steering wheel in the opposite direction of the pull to keep the vehicle going straight. Drifting is a condition in which the vehicle slowly moves to one side of the road when the driver's hands are taken off the steering wheel.

These problems can be caused by the following:

- Improper frame or rear axle alignment
- Dragging brakes
- Worn or binding suspension components, especially springs
- Incorrect or uneven wheel alignment
- Unevenly loaded or overloaded vehicle
- Loose, worn, or damaged steering linkages or tie-rod ends
- Out-of-balance steering gear valve
- Torque steer
- Tire inflation or size differences
- Binding strut bearing

It must be noted that the steering gear does not cause the vehicle to pull to one side, nor does it cause road wheel shimmy.

SHIMMY When the wheels shimmy, the driver will feel large, consistent, rotational oscillations at the steering wheel. These motions are caused by the lateral movement of the tires.

This problem can be caused by the following:

- Loose, worn, or damaged steering linkages or tie-rod ends
- Loose, worn, or damaged suspension parts
- Out-of-balance tires
- Excessive wheel runout
- Tire belt separation or different tread designs on same axle
- Loose wheel bearings

STICKING STEERING OR POOR RETURN Poor returnability and sticky steering describe the steering wheel's resistance to return to centre after a turn.

This problem can be caused by the following:

- Binding steering column U-joints
- Loose, worn, or damaged steering linkages or tie-rod ends
- Steering gear set too tight or is binding

- Loose, damaged, or worn suspension parts
- Incorrect wheel alignment
- Binding steering column bearings

WANDERING When a vehicle wanders, the driver must constantly turn the steering wheel to the left and right to keep the vehicle going straight on a level road.

This problem can be caused by the following:

- Loose or worn suspension components
- Incorrect or uneven wheel alignment
- Unevenly loaded or overloaded vehicle
- Loose or damaged steering gear or rack bolts
- Loose steering column U-joint bolts
- Loose, worn, or binding steering linkages or tie-rod ends
- Improper steering gear preload adjustment
- Leaking rack pistons

Diagnosing

As with the diagnosis of any problem, your diagnosis should begin by trying to duplicate the customer's complaint. For steering problems, this is done on a road test. Make sure you drive carefully and cautiously, especially because the vehicle has a control problem. It is very important that during the road test the vehicle be driven under conditions similar to the owner's normal driving. Although it may be somewhat inconvenient to seek out a particular type of road surface, this may be the only way to verify the condition. Take plenty of time because it could save hours of service time in the final diagnosis. Before going on the road test, do a thorough safety inspection of the vehicle that includes the tires.

Once the road test has been completed and it has been determined that there is an abnormal condition, certain tests are necessary to pinpoint the exact cause. In some cases, partial disassembly of the system may be required to complete a test. Do not short-cut test steps to save time. Doing so can alter the results, leading to an inaccurate diagnosis and unnecessary replacement of parts. It is also vital that specifications be confirmed in the service information. Guessing gives incorrect results. Continue your diagnosis with a thorough visual inspection.

Power-Steering Fluid

The power-steering fluid can be checked either hot or cold. Fluid level will vary with temperature, however, and a more accurate check is done when the engine is warm. The reservoir cap has a dipstick **(Figure 48–38)** typically marked HOT and COLD on opposite sides of the dipstick. Make sure to check the level on the right side of the dipstick. If necessary, add fluid to correct the level. Some manufacturers recommend a specific fluid for use in a power-steering system. However, most recommend a specific type of automatic transmission fluid (ATF). Always check the service information before installing fluid to the system. Using an incorrect fluid can cause damage to the system and a loss of assist.

FIGURE 48–38
Check the fluid level with the dipstick attached to the cap.

Noise Diagnosis

Often, customers complain of abnormal noises or vibrations coming from the steering system. Pay attention to these during the road test. There can be many causes for these; some are not the steering system. The cause of these noises is best identified by paying close attention to where the noise is coming from. Some noises may be caused by tires or interference between the steering wheel and the steering column covers. Others can result from a faulty power-steering pump or system. **Figure 48–39** features a list of common noises and the possible problem areas.

A handy tool for identifying the exact cause of a noise is called the ChassisEar™ **(Figure 48–40)**. This is a wireless electronic device that can identify

SYMPTOM	POSSIBLE CAUSE
Drive belt squeal or chirp when moving the steering wheel from stop to stop	Loose or worn drive belt
Noise during a cold start	Blockage in the power-steering fluid reservoir Air in the steering hydraulic system
Steering grunt, growl, or shudder when turning into or out of a turn at low speeds	Air in the steering hydraulic system Restricted power-steering hoses
Steering system clunk	Air in the steering hydraulic system Steering column U-joints Steering gear
Steering gear squeak	Incorrect fluid in system Steering gear rotary seal Steering column components
Power-steering hiss or whistle	Steering column shaft or binding or misaligned Damaged or worn steering gear input shaft and valve Power-steering pump low relief pressure Restricted power-steering lines
Power-steering pump moan when the steering wheel is rotated to the stop position	Low fluid Air in the hydraulic system Power-steering fluid reservoir or screen is blocked or damaged Power-steering pump brackets loose or misaligned Bad steering gear isolators
Power-steering pump whine noise	Aerated fluid Damaged power-steering pump
Rattle, chuckle, or knocking noise or roughness felt in the steering wheel when the vehicle is driven over rough surfaces	Steering column shaft/coupling joints damaged or worn Loose, damaged, or worn tie-rod ends Steering gear insulators or mounting bolts loose or damaged Steering column shaft/coupling bolts are loose Steering column damaged or worn Loose suspension bushings, bolts, or ball joints
Steering column rattle	Loose bolts or attaching brackets Loose, worn or insufficiently lubricated column bearings Steering shaft insulators damaged or worn Steering column shaft/coupling compressed or extended Clearance in the intermediate shaft
Steering column squeak, cracks, or grinds	Poorly lubricated steering shaft bushings Loose or misaligned steering column shrouds Steering wheel rubbing against steering column shrouds Upper or lower bearing sleeves out of position

FIGURE 48–39

Common causes of steering noises.

the source of a noise during a road test. The unit relies on inductive pickups that clamp on or near the component you want to listen to. The device also has a control unit with adjustable volume and headphones. The value of this tool is that it simply is hard to identify the source of a noise while you are driving.

Power-Steering Pressure Checks

Many steering problems are caused by incorrect pressures in the power-steering system. Most late-model vehicles have a power-steering pressure sensor that can be monitored with a scan tool. In others, you must connect a power-steering pressure gauge to the system.

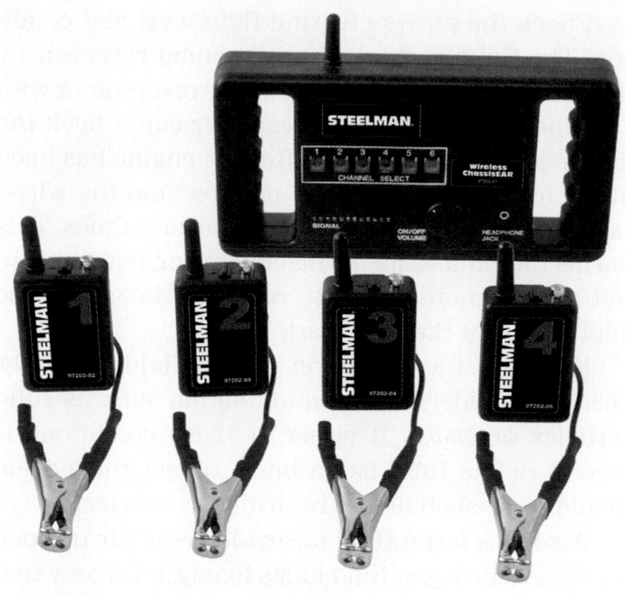

FIGURE 48–40

The wireless ChassisEar™ tool is valuable for identifying the source of steering and suspension noises.

JS Products, Inc.

PROCEDURE

Checking the Pressure in a Power-Steering System Using a Scan Tool

STEP 1 Place a thermometer in the power-steering fluid reservoir.

STEP 2 Connect the scan tool.

STEP 3 Set the tool to monitor the power-steering pressure sensor PID.

STEP 4 Start the engine.

STEP 5 With the engine at idle, raise the power-steering fluid temperature to the specified amount by rotating the steering wheel fully to the left and right several times.

STEP 6 With the steering wheel in the straight-ahead position and the engine speed at idle, record the pressure reading.

STEP 7 Compare the readings to specifications. If the pressure reading is higher than specifications, check the power-steering lines and hoses for restrictions.

STEP 8 Turn the steering wheel to the left and right stops. Record the pressure readings at each stop. The pressure reading at both stops should be almost the same. If the pressure reaches the maximum pump pressure at one stop but not the other, install a new steering gear. If the pressure does not reach the maximum pump pressure at either side, install a new pump.

Older vehicles may use a pressure switch that only detects low or high pressure. Using a scan tool, the switch may show LOW or OFF under light power-steering loads. When the wheels are turned and power-steering load increases, the switch may show HIGH or ON. This signals the PCM to boost idle speed to prevent the engine from stalling under low-speed, high power-steering demands.

TESTING SYSTEMS WITH A PRESSURE TESTER With the engine off, disconnect the pressure hose at the pump. Install the pressure gauge between the pump and the steering gear, and bleed the system.

Run the engine for about two minutes, and then stop the engine and add fluid to the power-steering pump, if necessary. Restart the engine, allow it to idle, and observe the pressure reading. The readings should be about 200 to 1040 kPa (30 to 150 psi). If the pressure is low, the pump may be faulty. If the pressure is too high, the problem may be restricted hoses.

Continue the testing by closing the shutoff valve of the tester and observing the pressure reading **(Figure 48–41)**. Never keep the valve closed for

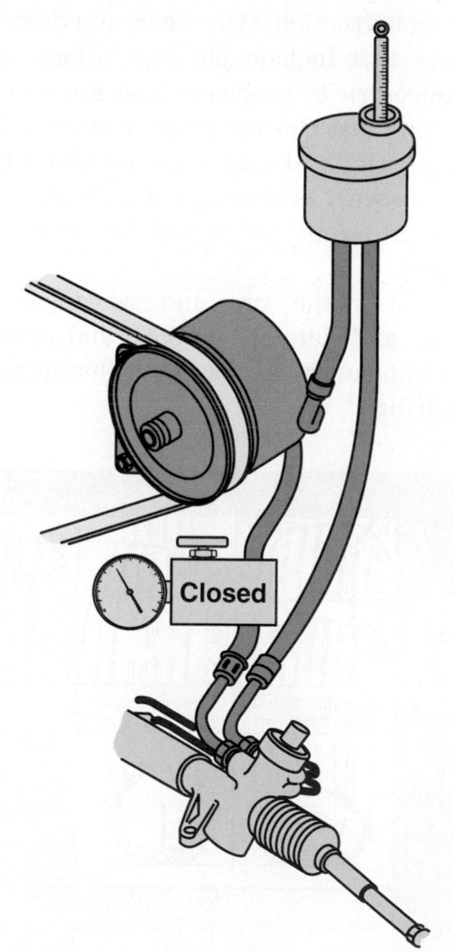

FIGURE 48–41

The final step when checking power-steering pump pressure is to close the tester's valve and observe the pressure in the system. Never keep the valve closed for more than five seconds.

more than five seconds. When the valve is closed, the pressure should increase. If the pressure is too high, a faulty pressure relief valve is suggested. If the pressure is too low, the pump may be bad. Always refer to the manufacturer's specifications for pressures.

VISUAL INSPECTION

Often as you go over the vehicle's systems, you will run across something that appears to be faulty, worn, or damaged. At that point, you may wish to check out the component further before continuing your inspection. These checks and some services to the components are given with the details of the visual inspection. Before actually checking or servicing a component, check with the service information to make sure you are doing it correctly.

Begin your visual inspection of the steering system by inspecting the tires. Check for correct pressure, construction, size, wear, and damage, and for defects that include ply separations, sidewall knots, concentricity problems, and force problems. Keep in mind that tire wear patterns are good indicators of steering and suspension problems **(Figure 48–42)**. Tire wear is also a great indicator of wheel alignment problems, which will be covered in the next chapter.

Also check the tire and wheel assemblies for radial and lateral runout, and static and dynamic imbalance. Check the adjustment of the wheel bearings.

Check the power-steering fluid level and condition. The fluid is checked at the pump reservoir by observing the fluid level through the reservoir or with a dipstick attached to the reservoir cap. Check the fluid level in the reservoir after the engine has been run at idle for two or three minutes and the wheel has been cycled from lock to lock several times. This warms the fluid to its normal operating temperature and gives a more accurate reading. Make sure the fluid level is at the FULL mark.

Examine the condition of the fluid carefully. Check for evidence of contamination such as solid particles or water. If either of these conditions is present or the fluid has a burnt odour, the system should be flushed before returning to service.

Also check the fluid for evidence of air trapped in the system. If the fluid looks foamy, it is likely that air is in the system. To verify this, run the engine until it reaches normal operating temperature. Then turn the steering wheel to the left and to the right several times without hitting the stops. If there is air in the system, bubbles will appear in the fluid reservoir. To remove the air, the system must be bled. The method of bleeding depends on the type of power-steering system. Follow the procedures given in the service information. The typical procedure involves connecting a vacuum pump to the cap's opening in the reservoir. While the engine is running, vacuum is applied to the reservoir and maintained for about five minutes. The vacuum is then released and the reservoir refilled with fluid. Vacuum is again applied to the reservoir, and the steering wheel is cycled

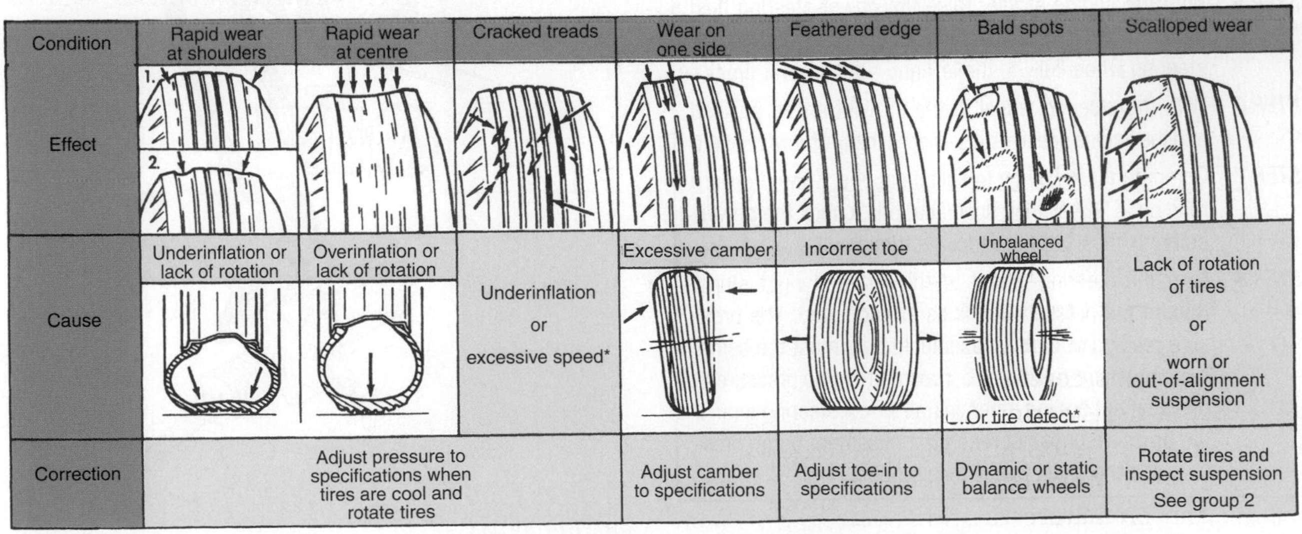

Condition	Rapid wear at shoulders	Rapid wear at centre	Cracked treads	Wear on one side	Feathered edge	Bald spots	Scalloped wear
Effect	1. 2.						
Cause	Underinflation or lack of rotation	Overinflation or lack of rotation	Underinflation or excessive speed*	Excessive camber	Incorrect toe	Unbalanced wheel ... Or tire defect*	Lack of rotation of tires or worn or out-of-alignment suspension
Correction	Adjust pressure to specifications when tires are cool and rotate tires			Adjust camber to specifications	Adjust toe-in to specifications	Dynamic or static balance wheels	Rotate tires and inspect suspension See group 2

*Have tire inspected for further use.

FIGURE 48–42

Tire wear patterns.

Chrysler Group LLC

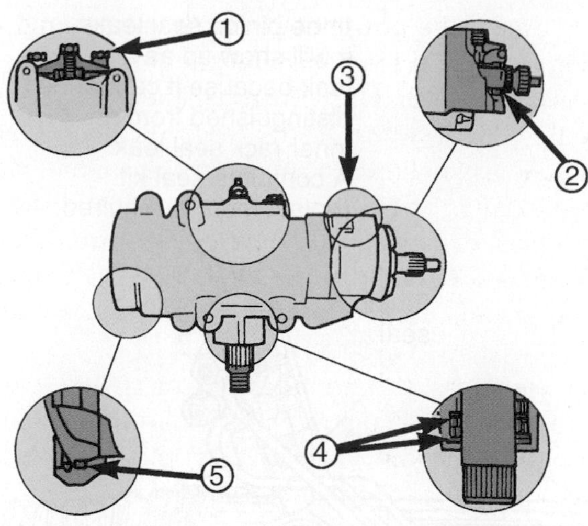

1. Side cover leak—torque side cover bolts to specification. Replace the side cover seal if the leakage persists.
2. Adjuster plug seal—replace the adjuster plug seals.
3. Pressure line fitting—torque the hose fitting nut to specifications. If leakage persists, replace the seal.
4. Pitman shaft seals—replace the seals.
5. Top cover seal—replace the seal.

FIGURE 48–43

Points for fluid leaks at a steering gear.

Chrysler Group LLC

from stop to stop every 30 seconds for at least five minutes. After this period of time, the vacuum is released and the reservoir cap installed. Then the system is checked for leaks.

With the ignition turned off, wipe off the outside of the power-steering pump, pressure hose, return hose, fluid cooler, and steering gear **(Figure 48–43)**. Start the engine and turn the steering wheel several times from stop to stop. Check for leaks **(Figure 48–44)**. Fluid leakage will not only cause abnormal noises but may result in unequal and abnormal steering efforts. If no signs of leakage are apparent, repeat the wheel cycling process and inspection several more times. Hoses should also be carefully inspected for swelling and cracks. Always replace power-steering hoses with an exact replacement hose. Never attempt to patch or seal a leak in a hose or the hose's fittings.

On all systems, carefully check all of the mechanical parts of the steering and suspension system. Many suspension parts affect the operation of the steering system. Worn ball joints can cause erratic steering and premature tire wear. Bad suspension bushings will allow excessive wheel movement that can affect braking, handling, and wheel alignment. If any part is found to be defective, it should be replaced.

Power-Steering Pump Belt

Power-steering belt condition and tension are extremely important for satisfactory power-steering pump operation. A loose belt causes low pump pressure and hard steering. A loose, dry, or worn belt may cause squealing and chirping noises, especially during engine acceleration and cornering. The power-steering pump belt should be checked for tension, cracks, oil soaking, worn or glazed edges, tears, and splits. If any of these conditions are present, the belt should be replaced.

Belt tension can be checked by measuring the belt deflection. Press on the belt with the engine stopped to measure the belt deflection, which should be 1.25 cm per 30 cm (0.5 in. per 1 ft.) of free span. The belt tension may also be checked with a belt tension gauge placed over the belt. The tension on the gauge should equal the vehicle manufacturer's specifications.

To adjust the tension of the belt, loosen the power-steering pump bracket or tension adjusting bolt and the power-steering pump mounting bolts. Pry against the pump ear and hub with a pry bar to tighten the belt. Some pump brackets have a half-inch square opening in which a breaker bar may be installed to move the pump and tighten the belt. Hold the pump in the desired position, and tighten the bracket or tension adjusting bolt. Once tightened, recheck the belt tension with the tension gauge. If the belt does not have the specified tension, readjust it, and tighten the tension adjusting bolt and the mounting bolts to the specified torque.

Some power-steering pumps have a ribbed V-belt, which has an automatic tensioning pulley; therefore, a tension adjustment is not required. The belt, however, should be checked to make sure it is installed properly on each pulley in the belt drive system and that it is in good condition.

Electric Power-Steering Systems

Many vehicles, including all hybrid and electric vehicles, have electric power steering. These systems do not have a power-steering pump and require no fluid services. Because these systems are electronically controlled, diagnosis is performed using a scan tool. A warning lamp on the dash comes on when a problem is detected **(Figure 48–45)**. A DTC will also be set. In most cases, when the warning lamp is lit, there will be no power assist available. However, the system will allow for manual steering.

The system is constantly (from key on to key off) monitored by the control module. If a problem

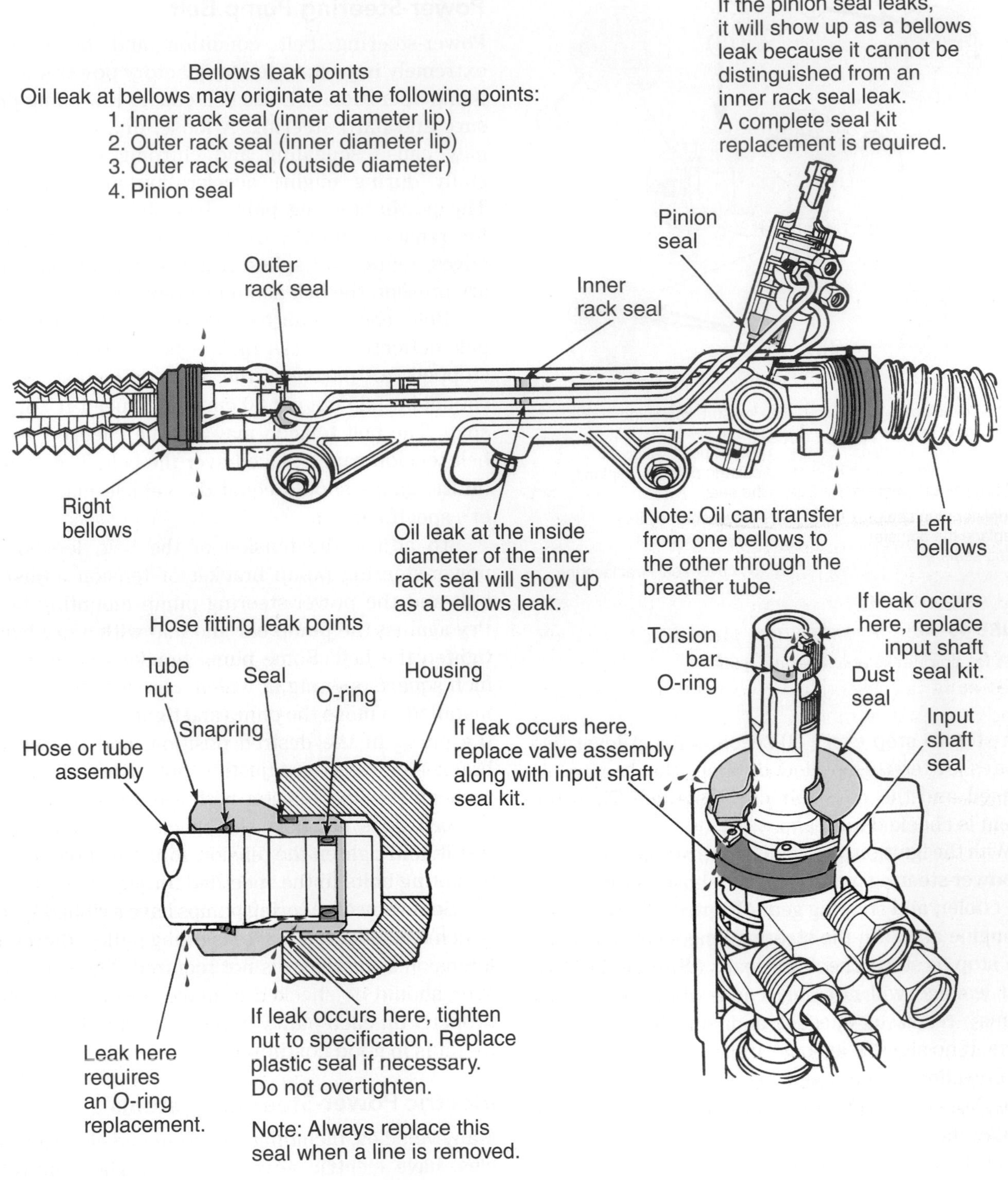
If the pinion seal leaks, it will show up as a bellows leak because it cannot be distinguished from an inner rack seal leak. A complete seal kit replacement is required.
Bellows leak points
Oil leak at bellows may originate at the following points:
1. Inner rack seal (inner diameter lip)
2. Outer rack seal (inner diameter lip)
3. Outer rack seal (outside diameter)
4. Pinion seal
Pinion seal
Outer rack seal
Inner rack seal
Right bellows
Oil leak at the inside diameter of the inner rack seal will show up as a bellows leak.
Note: Oil can transfer from one bellows to the other through the breather tube.
Left bellows
Hose fitting leak points
Tube nut
Seal
Housing
O-ring
Snapring
Hose or tube assembly
If leak occurs here, replace valve assembly along with input shaft seal kit.
Leak here requires an O-ring replacement.
If leak occurs here, tighten nut to specification. Replace plastic seal if necessary. Do not overtighten.
Note: Always replace this seal when a line is removed.
Torsion bar O-ring
If leak occurs here, replace input shaft seal kit.
Dust seal
Input shaft seal

FIGURE 48–44

Possible leakage points on power-steering systems.

is detected, the EPS lamp may stay lit after the ignition is turned on, or it may come on while it is being driven. To identify the reason for the illumination of the lamp, interview the customer to find out when and where it first came on. Try to duplicate the situation during a road test, and then retrieve any and all DTCs. If the problem cannot be duplicated, do a careful inspection of all associated wiring and connectors.

Steering problems in an EPS can be caused by the same things as a conventional steering system. However, power assist is provided by electronics, not hydraulics. Therefore, power-steering problems are related to the electronic/electrical components.

FIGURE 48–45

An electric power-steering fault light.

©Cengage Learning 2015

If the driver complains about how hard it is to turn the steering wheel, the problem can be caused by the power-steering motor, the speed sensor, the power-steering control unit, or electrical circuit problems.

If the steering effort does not change with vehicle speed or is greater when turning left or right, suspect a faulty power-steering motor, speed sensor, or power-steering control unit. A defective power-steering motor may be evident by a high-pitched sound that occurs when the vehicle is stopped and the steering wheel is turned slowly.

On some models, if the power-steering motor gets too hot, the control module will decrease the current flow to it. This can cause increased steering effort. The systems do this to prevent permanent damage to the motor. As soon as the motor's temperature drops, normal current flow resumes. Therefore, a temporary loss of full assist may be normal.

Pitman Arm

Because of its function, the pitman arm is the most heavily stressed point in the system. To inspect the pitman arm, grasp it and vigorously shake it to detect any looseness. Check the socket to reveal any damage or looseness. Either condition must be corrected by replacing the worn part. Their removal normally requires the use of a special puller **(Figure 48–46)**.

FIGURE 48–46

A pitman arm puller.

Federal-Mogul Corporation

Idler Arm

A worn or damaged idler arm can cause steering instability, uneven tire wear, front-end shimmy, hard steering, excessive play in steering, or poor returnability. Because an idler arm is the weakest link in a parallelogram steering system, it wears more quickly than the rest of the system.

The procedure is simple for checking an idler arm for looseness or wear. The suspension should be normally loaded on the ground or on an alignment rack. When raised by a frame contact hoist, the vehicle's steering linkage is allowed to hang, and proper testing cannot be done. Check the idler arm ends for worn sockets or deteriorated bushings. Grasp the centre link firmly with your hand at the idler arm end. Push up with approximately an 11.5 kg (25 lb.) load. Pull down with the same load. The allowable movement of the idler arm and support assembly in one direction is 3 mm, (1/8 in.) for a total acceptable movement of 6 mm (1/4 in.). The load can be accurately measured by using a dial indicator or pull-spring scale located as near the centre link end of the idler arm as possible. Keep in mind that the test forces should not exceed 11.5 kg (25 lb.), as even a new idler arm might be forced to show movement due to steel flexing when excessive pressure is applied. It is also necessary that a scale or ruler be rested against the frame and used to determine the amount of movement. Observers tend to overestimate the actual movement when a scale is not used. The idler arm should always be replaced if it fails this test. Jerking the right front wheel and tire assembly back and forth (causing an up-and-down movement in the idler arm) is not an acceptable method of checking, as there is no control on the amount of force being applied.

Centre Link

Worn or bent centre links can cause front-end shimmy, vehicle pull to one side, or change in the toe setting, causing excessive tire wear.

When inspecting the centre link, look closely to ensure it has not been bent or damaged. Grasp the centre link firmly, and try moving it in all directions. Any movement, or sign of damage, is reason for replacement. Tapered openings seldom wear but should be checked for enlargement caused by a loose connection. If necessary, replace the centre link.

Tie-Rod Assembly

Worn tie-rod ends result in incorrect toe-in settings, scalloped and scuffed tires, wheel shimmy, understeering, or front-end noise and tire squeal on turns.

Tie-rod end and centre link inspections are similar. Grasp the tie-rod end firmly. Push vertically with the stud, and inspect for movement at the joint with the steering knuckle. Any movement over 3 mm (⅛ in.) or observation of damaged or missing parts, such as seals, is sufficient evidence that replacement is necessary. Another way to detect looseness in a tie rod is to perform a dry park test. A technician lightly rotates the steering wheel back and forth without turning the wheels while another technician underneath checks for any component looseness. Adjusting sleeves resemble a piece of internally threaded pipe. They have a slot or separation that runs either their entire length or just part way. Adjusting sleeves also have two crimping or squeezing clamps located at each end to lock the toe adjustment. Badly rusted, worn, or damaged adjusting sleeves should be replaced.

An additional check of the tie rods can be made by rotating each tie-rod end to feel for roughness or binding, which could indicate that the socket has probably rusted internally. A special puller is often required to separate a tie-rod end from the steering knuckle **(Figure 48–47)**.

FIGURE 48–47

A tie-rod end-separating tool.

Federal-Mogul Corporation

Steering Damper

The steering dampers found in some steering linkage designs are generally nonadjustable, nonrefillable, and not repairable. At each inspection interval, inspect the mountings, and check the assembly for damage (such as being bent) and fluid leaks. A light film of fluid is evidence of fluid leakage. A leaking or damaged damper should be replaced. A bad steering damper may cause wheel shimmy, even though the rest of the suspension and steering system is fine.

Dry Park Check

An excellent overall check for worn or loose conventional steering components is the **dry park check**. With the full weight of the vehicle on the wheels, have an assistant rock the steering wheel back and forth. Start your inspection from one side to the other side. Note any looseness in tie rod, centre link, idler arm, or pitman arm sockets **(Figure 48–48)**.

Turning Effort

If an owner's concern indicates excessive turning effort, a pull scale should be used to read the actual force required to turn the wheel **(Figure 48–49)**. Compare the test results to the specifications in the service information. If the effort exceeds the maximum, carefully inspect the entire steering system before performing a pressure test.

Tie-Rod Articulation Effort

The effort required to move the tie rod or its inner ball socket should be checked with a pull scale if excessive steering effort or looseness is noted during

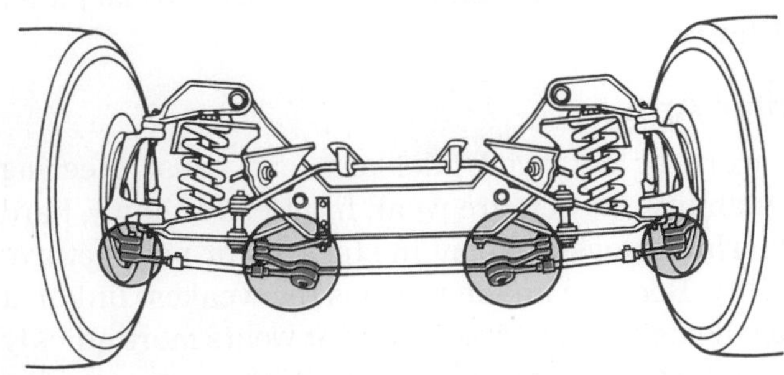

FIGURE 48–48

Circled areas indicate where a dry park check of steering linkage should be made.

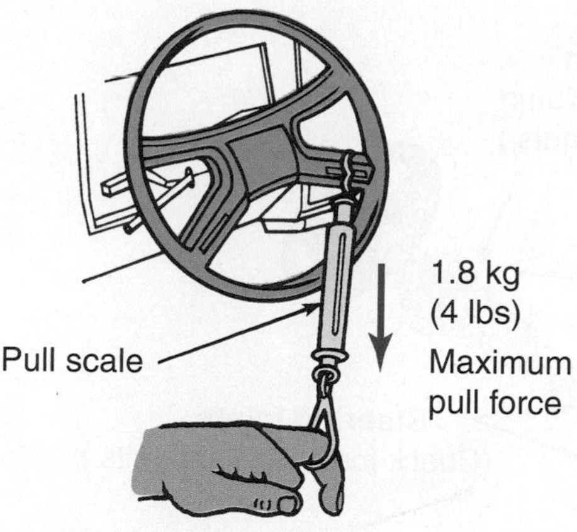

FIGURE 48–49

A pull scale is used to measure steering effort.

Reprinted with permission.

the road test. If the effort is not within the specified limits, the tie rod must be replaced.

Rack and Pinion Steering

A rack and pinion system has no idler or pitman arms and no centre link. Instead, they are replaced with a rack. This reduces the number of wear points on rack and pinion systems to the four tie-rod ends and the rack mounts. Power rack and pinion assemblies should be carefully checked for leaks. Leaking rack end seals will allow fluid to accumulate in the bellows. Leaks can also be caused by faulty pinion seals and fluid transfer lines. If leaks cause the pump to run out of fluid, the pump will be damaged.

In order to solve customer complaints, a very thorough inspection of the entire system is needed. Everything, including ball joints, tires, outer tie rods, bellows boots, inner tie rods, rack-mounting bushings, mounting bolts, steering couplings, and gearbox adjustment, must be checked. Rack and pinion steering inspection must be very thorough because of the system's sensitivity.

PROCEDURE

Rack and Pinion Steering Inspection

STEP 1 Check all working components **(Figure 48–50)** of the systems. Inspect the flexible steering coupling or the universal joints for wear or looseness. If any play is found, recommend replacement. Universal joints can also seize or bind. They should be checked closely.

STEP 2 Grasp the pinion gear shaft at the flexible steering coupling, and try to move it in and out of the gear. If there is movement, the pinion bearing preload might need adjustment. If there is no adjustment, internal components have to be replaced.

STEP 3 Carefully inspect the rack housing. In most cases, the rack and pinion steering assemblies are mounted in rubber bushings. As the vehicle gets older, these mounting bushings deteriorate from heat, age, and oil leakage from the engine. When this happens, the housing moves within its mounting and causes loose and erratic steering. Also be alert for excessive movement of the rack housing. Stiffness in steering can be caused by a bent rack assembly, a tight yoke bearing adjustment, a loose power-steering belt, a weak pump, internal leaks in the power-steering system, and damaged CV joints in front-wheel-drive vehicles.

STEP 4 Check the inner tie-rod socket assemblies located inside the bellows. The most foolproof way of checking these sockets is to loosen the inner bellows clamp and pull the bellows back, giving a clear view of the socket. During the dry park check, observe any looseness. The inner tie-rod socket can also be checked by squeezing the bellows boot until the inner socket can be felt. Push and pull on the tire. If looseness is found in the tie rod, it should be replaced. On some vehicles, the boot might be made of hard plastic. For this type of boot, lock the steering wheel, and push and pull on the tire. Watch for in-and-out movement of the tie rod. If movement is observed, replace the inner tie rod.

One fact to keep in mind is that the condition of the bellows boot determines the life of the inner socket. The bellows boot protects the rack from contamination. It might also contain fluid that helps keep the rack lubricated. If any cracks, splits, or leaks exist, the boot should be replaced. Also be sure that clamps for the bellows are in their proper place and fastened tightly.

STEP 5 Inspect the outer tie-rod ends. In addition to the dry park check, grab each end and rotate to feel for any roughness that would indicate internal rusting. Be sure to check for bent or damaged forgings and studs, split or deteriorated seals, and damaged, out-of-round, or loose tapers. If any of these conditions exist, the parts should be replaced.

STEERING SYSTEM SERVICING

When a steering system component is found to be faulty, it is replaced. Most often, part replacement is quite straightforward, but you should always refer to the service information before proceeding. At times, diagnosis will indicate a need to adjust the steering gear or inspect and repair the steering column.

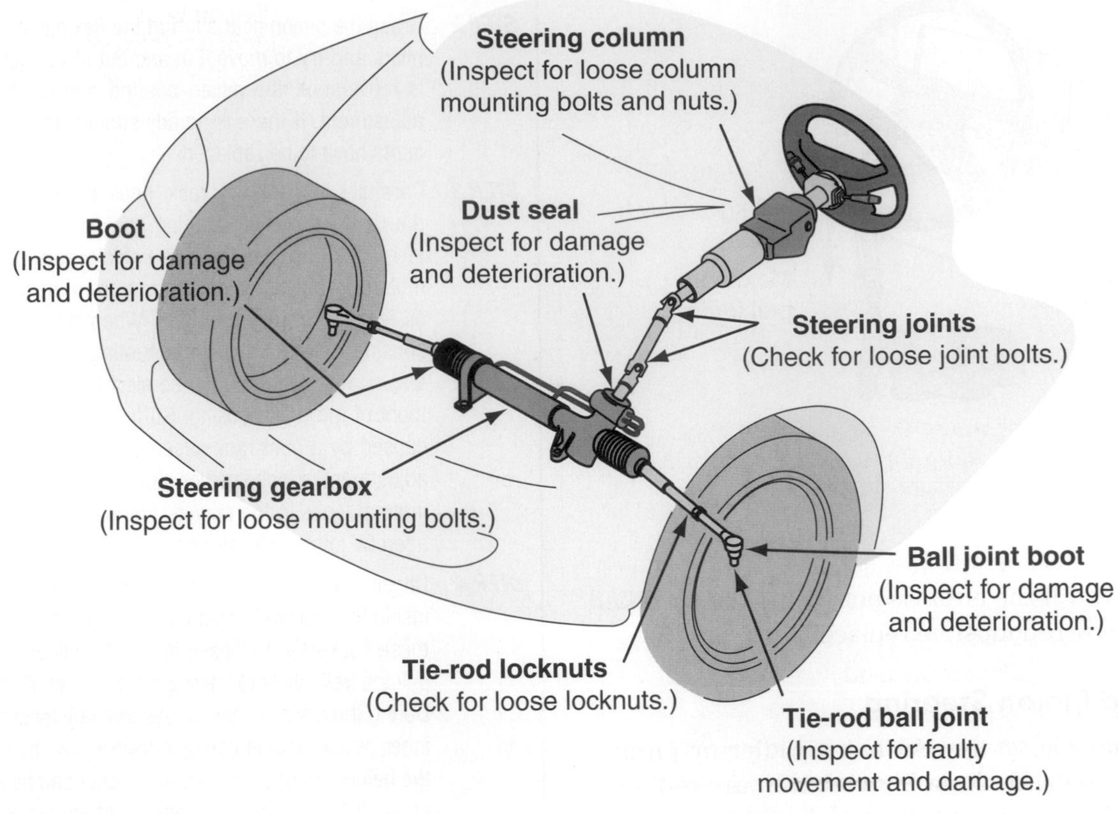

FIGURE 48–50

All steering parts should be carefully checked during diagnosis.

Steering Linkage

Eventually, steering linkage components wear and require replacement. Removing tie rods and the centre link is straightforward. First, remove the nut securing the stud in place. This may require removing and discarding the cotter pin. Next, break the taper between the stud and component it is connect to. This is often done using a tie-rod puller or pickle fork. Once the part is removed, install the new component, and torque the nut to specifications. If a castle nut and cotter pin are used, and the nut does not align with the hole in the stud, tighten the nut until it aligns, and then install a new cotter pin. Bend one or both of the free ends to prevent the pin from working loose and falling out. To replace an idler arm, remove it from the centre link, and then unbolt the arm from the frame. Install the new arm, and torque the bolts and nuts to specifications. To replace a pitman arm, first remove the retaining nut and washer. Next, install a pitman arm puller over the arm, and tighten the shaft against the pitman shaft. Pitman arms tend to be tight, so don't be surprised if it takes a bit of work to get it loose from the gearbox. Once off, install the new pitman arm, washer, and nut. Tighten the nut to specifications.

Before assembling any steering linkage parts, thoroughly check all tapered holes for out-of-roundness and wear. Thoroughly clean all bores that the stud tapers mount in. On new and reused parts, firmly install the tapered stud into its tapered hole. The stud must seat firmly without rocking. Only thread should protrude from the hole. If the parts do not meet these requirements, the mating part is worn and must be replaced, or the correct parts are not being used. Always follow the manufacturer's stud and mounting bolt torque specifications when installing chassis parts.

Rack and Pinion Inner Tie Rods

For end take-off rack and pinion units, removing the inner tie rods begins by removing the outer tie rods from the steering knuckle. Measure and note the position of the tie rod and jamb nut on the inner tie rod. This will make setting the toe a little quicker if it is already close to where it was before the tie rod was replaced. Loosen the jam nut securing the inner and out tie rods tight, and then unthread the outer tie

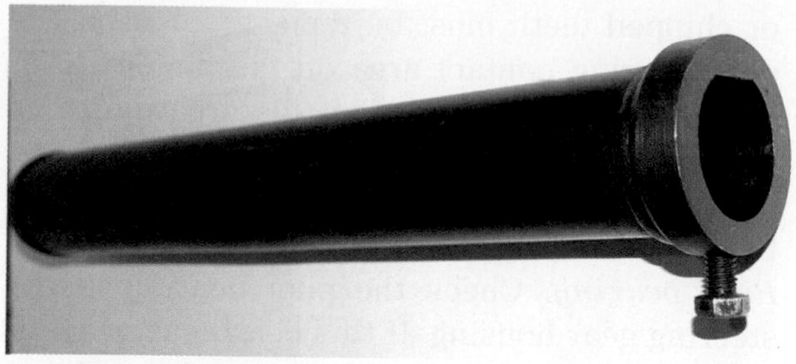

FIGURE 48–51

An inner tie-rod service tool.

rod from the inner. Remove the jam nut and then the clamps holding the bellows to the inner tie rod and the rack housing. Pull the bellows off the rack, and inspect it for cracks or other damage.

Check the service information for tie-rod removal. Some tie rods have set screws or staked washers that must be removed first. Also, some manufacturers want the rack gear to be held from moving when loosening the tie rod. Removing the inner tie rod often requires a special tool **(Figure 48–51)**. Place the tool over the tie rod, and make sure it fits snugly. Attach a ratchet to the tie-rod socket, and remove the tie rod. Make sure the replacement tie rod matches the old tie rod, and install it on the rack gear. Tighten the inner tie rod to specifications, and reinstall any set screws or other retention devices as outlined in the service information. Reinstall and re-clamp the bellows in place. Install the jamb nut and outer tie rod. Tighten the outer tie-rod nut to specifications, and install a new cotter pin. If the outer tie rod uses a nylon friction nut to secure it to the steering knuckle, do not reuse the nut. Install a new nut and torque it to specs. Once the tie rod is installed, the toe will need to be checked and adjusted.

Steering Gear Adjustments

Before any adjustments are made or servicing procedures performed to the steering gear, a careful check should be made of front-end alignment, shock absorbers, wheel balance, and tire pressure for possible steering system problems.

Before adjusting or servicing a manual steering gear, the technician must disconnect the battery ground cable. Raise the vehicle with the front wheels in the straight-ahead position. Remove the pitman arm nut. Mark the relationship of the pitman shaft. Remove the pitman arm with a pitman arm puller. Loosen the steering gear adjuster plug lock nut, and back the adjuster plug off one-quarter turn **(Figure 48–52)**. Remove the horn shroud or

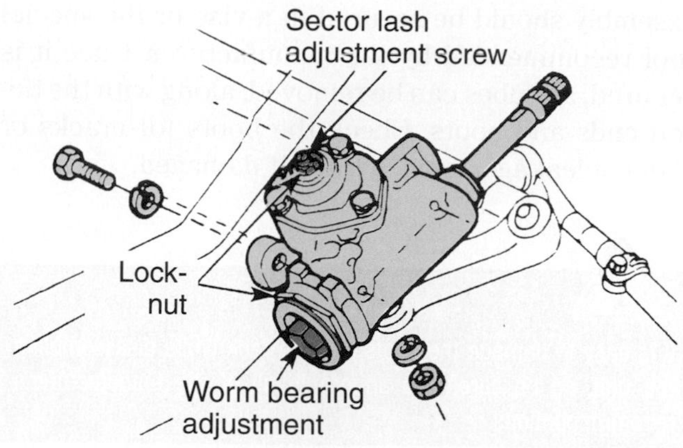

FIGURE 48–52

Typical steering gear adjustment points.

button cap. Turn the steering wheel gently in one direction until stopped by the gear, and then turn back one-half turn. Measure and record bearing drag by applying a torque wrench with a socket on the steering wheel nut and rotating through a 90° arc. Check the service information for the correct amount of drag.

Once these steps are taken, the steering gear is ready for adjusting or servicing as per instructions in the vehicle's service information.

Rack and Pinion Service

When removing a rack and pinion steering gear, make sure the front wheels of the vehicle are in the straight-ahead position. The exact procedure for doing this will vary with the model and make of the vehicle; always refer to the procedures outlined by the manufacturer. A typical procedure includes removing the wheels and disconnecting the outer tie rods from the steering knuckles. Next, remove hydraulic pressure and return lines at the rack. Do not remove the transfer lines. Disconnect any electrical connectors from the rack. Locate the clamp that secures the steering shaft to the pinion shaft. The clamp is typically covered by a protective boot that may be secured to the body of vehicle where the column passes through the firewall. Remove the clamp bolt, and slide the clamp off the pinion shaft. Remove the rack mounting bolts and the rack. The rack may have to pass through the wheel opening in the body, or in some cases the cradle may have to be lowered or the engine and/or transaxle must be removed to remove the rack.

Typically, a faulty rack is replaced, and internal repairs are not performed. However, if it is necessary to disassemble and inspect the assembly, the rack assembly should be secured in a vise or the special tool recommended by the manufacturer. Once it is secured, all tubes can be removed, along with the tie-rod ends and boots. Check the boots for cracks or signs of leakage; replace them if damaged.

WARNING!

It is easy to damage the boots and the rack while disassembling the unit, so be careful to adhere to all of the manufacturer's recommendations.

Make an index mark on the outer tie-rod end, jam nut, and tie rod **(Figure 48–53)**. Remove the rack guide spring cap, compression spring, and rack guide subassembly. Then remove the O-ring from the rack guide spring cap and the dust cover. Now remove the control valve assembly and subassembly. In some cases, the subassembly is removed by pressing it away from the main assembly. Once the subassembly is separated, remove the snap rings, seals, and spacers from it. The cylinder and "stoppers" in the rack are removed next, along with the steering rack bushing. When doing this, be careful not to drop the bushing or damage the inside of the steering gear housing.

Check the following:

- *Inner tie-rod sockets.* If the inner tie-rod ends were found to be loose during the inspection or if they require too much effort to move, replace them.
- *Pinion and bearing assembly.* If the pinion bearing is loose on the shaft, replace the pinion and bearing assembly. A pinion shaft with worn or chipped teeth must be replaced. Inspect the pilot bearing contact area on the pinion shaft. Wear, pitting, or scoring in this area indicates that a new pinion shaft is required. If the pinion bearing is bad, replace the pinion shaft and bearing assembly.
- *Pilot bearing.* Check the pilot bearing in the steering gear housing. If this bearing is worn or scored, replace the pilot bearing and the pinion shaft and bearing assembly.
- *Rack bushing.* If the rack bushing is worn, bushing replacement is necessary. Remove the bushing retaining ring prior to bushing removal. Position the puller fingers behind the bushing, and operate the slide hammer on the puller to remove the bushing. If a puller is not available, an appropriate bushing driver or socket may be used to drive the bushing out of the housing.
- *Mounting bushings.* If the mounting bushings are loose, replace the bushings. Always replace the bushings in pairs. If the bushings are in satisfactory condition, do not disturb them.

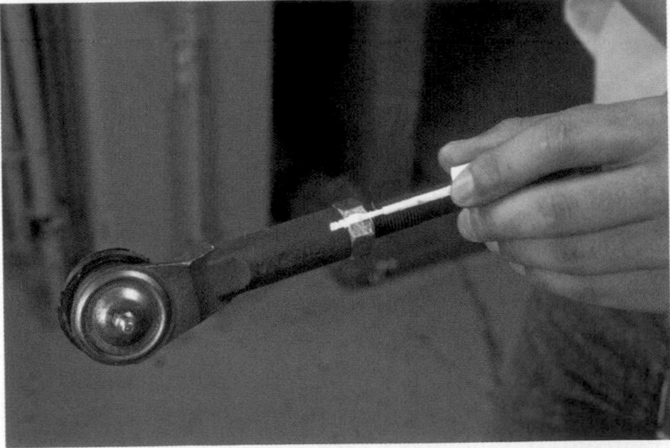

FIGURE 48–53

Before disassembling the steering gear, make an index mark on the outer tie-rod end, jam nut, and tie rod.

When installing new parts to the rack assembly or when reassembling the unit, all moving parts should be lubricated with power-steering fluid, molybdenum disulfide lithium-based grease, MP grease, or silicon grease. Also install new bellows, clamps, bushings, and seals. There are certain parts that need to be preloaded; again, follow the manufacturer's recommendations. Reassembly is the reverse of the disassembly procedure.

Installing a new or rebuilt rack and pinion steering gear follows the opposite procedure as when removing it. Make sure everything is tightened to the torque specified by the manufacturer, with the wheels in a straight-ahead position. Also, make sure to check the fluid level when finished and before the vehicle is driven. The vehicle will also need to have its wheels aligned.

Steering Columns

To perform service procedures on the steering column upper end components, it is not necessary to remove the column from the vehicle. The steering wheel, horn components, directional signal switch, ignition switch, and lock cylinder can be removed with the column remaining in the vehicle.

To determine if the energy-absorbing steering column components are functioning as designed, or if repairs are required, a close inspection should be made. An inspection is called for in all cases where damage is evident or whenever the vehicle is being repaired due to a front-end collision. If damage is evident, the affected parts must be replaced. Because of the differences in the steering column styles and various components, consult the service information for more explicit inspection and servicing procedures.

Steering Wheels

At times, a customer may complain about an uncentred steering wheel. When the steering wheel is in its centred position, the front wheels should be pointing straight ahead. If the wheels are not in the straight-ahead position, this can be corrected by adjusting the toe of the vehicle. However, this adjustment should only be made if the steering wheel index mark is aligned with the steering column index marks. As a rule, indexing teeth or mating flats on the wheel hub and steering shaft prevent misindexing of these components. One way to verify an incorrectly positioned steering wheel is to turn it from stop to stop and count the number of turns it took. Then take that number and divide it by two; the result represents the centre or straight-ahead position. Now turn the steering wheel to a stop and then turn it the number of turns that represents the centre. Look at the front wheels; if they are not in the straight-ahead position, either the steering was installed wrong or the wheels need to have their toe adjusted. On some 4WD vehicles, those with certain cross-link steering linkages, the steering wheel must be removed and recentred once the front wheel toe has been set. Refer to the manufacturer's wheel alignment procedures.

At times, the steering wheel must be removed, such as when servicing the multifunctional switch, horn, or the steering column covers. The steering wheel is very unlikely to cause a steering problem.

PROCEDURE

Removing and Reinstalling a Steering Wheel

STEP 1 Place the front wheels facing straight ahead.

STEP 2 Disable the supplemental restraint system (SRS).

STEP 3 Remove the steering pad and air bag.

STEP 4 Remove the steering wheel assembly set nut.

STEP 5 Put alignment marks on the steering wheel and the steering main shaft.

STEP 6 Disconnect the connectors from the spiral cable.

STEP 7 Using the correct puller, remove the steering wheel assembly **(Figure 48–54)**.

To install:

STEP 1 Slip the steering wheel over the main shaft.

STEP 2 Align the alignment marks on the steering wheel assembly and steering main shaft.

STEP 3 Install the steering wheel assembly set nut. Torque the nut to specifications.

STEP 4 Connect the connectors to the clockspring (spiral cable) subassembly.

STEP 5 Connect the cable to the negative battery terminal.

STEP 6 Check the SRS warning light.

WARNING!

Set the parking brake before removing the steering column. Also remove the battery cable from the negative terminal. Remember that special precautions must be observed before beginning disassembly and during assembly to ensure correctly fitting together the steering column shaft and steering gear shaft connections.

POWER-STEERING SYSTEM SERVICING

Vehicles with power-steering systems have the same type of steering linkage as manual steering. The power-steering linkage is checked and serviced as previously described. Actually, the only difference is the servicing of the hydraulic components such as the hoses, pump, and power-steering gear. One of the common procedures that is recommended by manufacturers as part of a preventive maintenance program is flushing the hydraulic system.

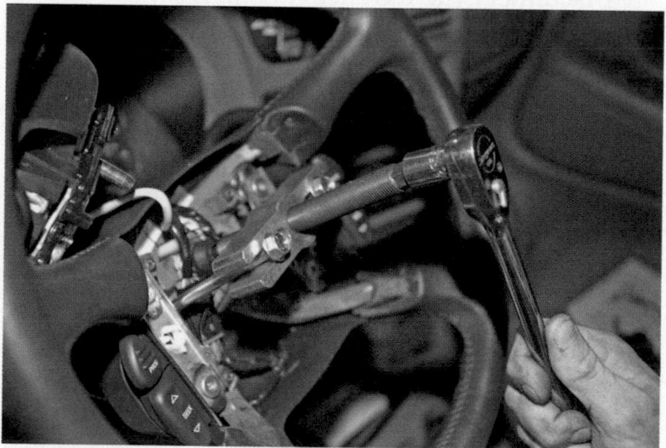

FIGURE 48–54

Use the correct puller to remove the steering wheel assembly.

WARNING!

The centre of the steering wheel contains the driver's-side air bag. Failure to disarm it can cause serious personal injury.

GO TO Chapter 24 for specific precautions that should be followed when working around or near an air bag.

Flushing the System

The reason for flushing the system should be obvious. Hydraulic fluid becomes contaminated by moisture and dirt. Flushing removes the old fluid with its contaminants, and new fluid is added to the system. It is also wise to flush the system after you have replaced or repaired a part in the system. Before beginning to flush the system, you should disable the engine's ignition. Then disconnect the power-steering return hose and plug the reservoir. Attach an extension hose between the power-steering return hose and an empty container. Raise the vehicle's front wheels off the ground. Fill the reservoir with the correct type of fluid.

WARNING!

Never mix oil types. Any mixture or any unapproved oil can lead to seal deterioration and leaks.

Turn the steering wheel from stop to stop while cranking the engine until the fluid leaving the return hose is clean. Never crank the engine for more than five seconds at a time. Add fluid to the reservoir to make sure it does not empty. Once the fluid is clear, fill the reservoir to its FULL mark and lower the vehicle.

Disconnect the extension hose from the power-steering return hose, and reconnect the return hose to the reservoir. Check the fluid level again, and add fluid as necessary. Now enable the ignition system. Start the engine and turn the steering wheel from stop to stop. If the power-steering system is noisy and bubbles are forming in the fluid, the system must be purged of air.

Flushing the system is also commonly performed with the engine running at idle speed. With the return hose placed in a suitable container, start the engine, and turn the steering wheel lock to lock while adding new fluid. Once the fluid coming out of the return hose is new and clean, shut the engine off. Reconnect the return hose, and refill the system. Start the engine, and bleed any air from the system.

Bleeding the System

Often, the procedure for bleeding the power-steering system is called "purging" the system because it moves air that may be trapped in the fluid. Purging the system must be done after the replacement of any part of the power-steering hydraulic system or when there is a problem that indicates there may be aerated fluid, such as a whining noise.

The method for bleeding the system depends on the type of power steering the vehicle is equipped with. Follow the procedures given in the service information. If the system is not purged correctly or if air is allowed to remain in the system, the power-steering pump can fail prematurely. What follows is a typical procedure.

PROCEDURE

Bleeding a Power-Steering Hydraulic System

STEP 1 Remove the reservoir cap.

STEP 2 Tightly install the adapter of the vacuum pump to the reservoir opening **(Figure 48–55)**.

STEP 3 Start the engine.

STEP 4 Connect the vacuum pump.

STEP 5 Apply 68 to 85 kPa (20 to 25 in. Hg) of vacuum.

Note: If the vehicle has a hydro-boost system, depress the brake pedal two times.

STEP 6 Fully cycle the steering wheel from stop to stop 10 times.

STEP 7 Turn off the engine.

STEP 8 Release the vacuum, and remove the adapter from the reservoir.

STEP 9 Fill the reservoir with the correct fluid.

STEP 10 Tightly install the adapter of the vacuum pump to the reservoir opening.

STEP 11 Start the engine.

STEP 12 Apply 68 to 85 kPa (20 to 25 in. Hg) of vacuum.

Note: If the vehicle has a hydro-boost system, depress the brake pedal two times.

STEP 13 Turn off the engine.

STEP 14 Release the vacuum, and remove the adapter from the reservoir.

STEP 15 Fill the reservoir with the correct fluid.

STEP 16 Check the system for signs of leaks. Make repairs as necessary.

STEP 17 Install the reservoir cap.

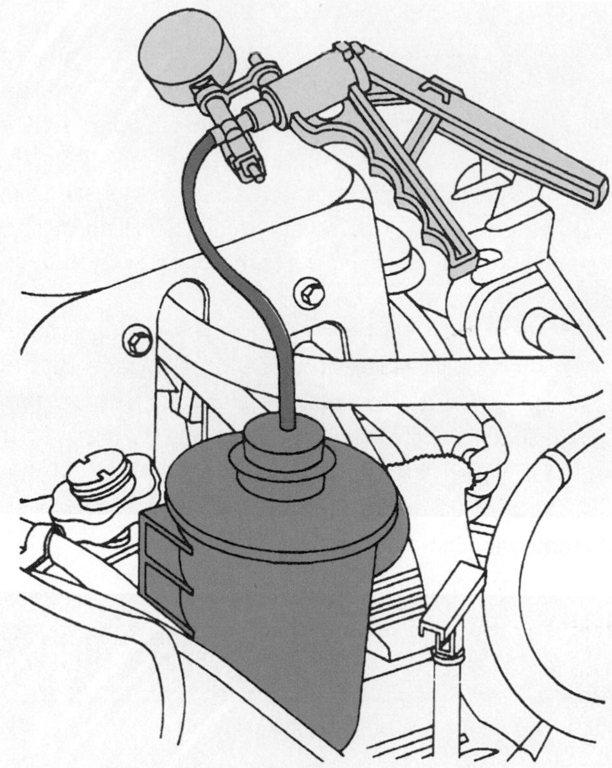

FIGURE 48–55

To bleed the system, a vacuum pump is connected to the fluid reservoir.

Ford Motor Company

Electric Power-Steering Systems

When working on a vehicle with EPS, remember that the electric motor adds the assist. It must be positioned properly in order to provide assist equally to both sides. Therefore, it is important that the front wheels are in their straight-ahead position whenever any part of the steering linkage is removed or installed. If it is necessary to disconnect the steering column shaft from the steering gear, mark the alignment of the two before disconnecting it. When reattaching the shaft, make sure the marks are aligned.

Some systems use a torque sensor to help determine how much power assist to provide. A torsion bar links the input shaft from the steering column to the pinion in the steering gear. The torsion bar twists in response to the movement of the steering wheel. The torque sensor detects the twist of the torsion bar and converts the torque applied to the torsion bar into an electrical signal. The torque sensor must be recalibrated to its zero point whenever the steering gear, steering wheel, steering column, or steering control module have been removed or replaced. This calibration must also be completed if there is a difference in steering effort when turning right or left. Follow the manufacturer's directions for calibrating this sensor. If the sensor is faulty, steering effort in both directions may increase, or the steering wheel may not return properly.

CAUTION!

The power-steering unit on some hybrid vehicles is powered by high voltage. Make sure you disarm the high-voltage circuit before working on or near all high-voltage systems.

Hoses and Lines

Hoses and lines should also be carefully inspected for leaks, dents, sharp bends, cracks, and swelling. Always replace the power-steering hoses with an exact replacement hose. Never attempt to patch or seal a leak in a hose or the hose's fittings. Lines and hoses must not rub against other components; this can wear a hole in the line or hose. Many high-pressure lines are made of high-pressure steel-braided hose with moulded steel fittings on each end.

PROCEDURE

Replacing a Power-Steering Hose

STEP 1 With the engine stopped, remove the return hose at the power-steering gear, and allow the fluid to drain into a drain pan.

STEP 2 Loosen and remove all hose fittings from the pump and steering gear.

STEP 3 Remove all hose-to-chassis clips.

STEP 4 Remove the hoses from the chassis, and cap the pump and steering gear fittings.

STEP 5 If O-rings are used on the hose ends, install new O-rings. Some lines have gaskets. The old gasket must be pried out of the fitting before installing the new lines. Lubricate the O-rings with power-steering fluid.

STEP 6 Install the new hose by reversing the steps for removal. Make sure to tighten all fittings to the specified torque. Be sure all hose-to-chassis clips are in place. Do not position hoses where they rub on other components.

STEP 7 Fill the pump reservoir to the FULL mark with the manufacturer's recommended fluid. Bleed air from the power-steering system; then check the fluid level in the reservoir and add fluid as required.

Power-Steering Pump

To replace a power-steering pump, follow the manufacturer's service procedures. General replacement procedures include locating the pump and removing the drive belt. Next, determine if the pulley must be removed to remove the pump from its bracket. If so, remove the pulley using the appropriate puller. Remove the return hose and drain the power-steering fluid into an oil pan, and then remove the pressure hose. Next, unbolt the pump from the bracket, and remove the pump. Compare the old pump and the replacement pump. Make sure the replacement pump has the same size fitting and mounting points. Install the new pump, and torque the fasteners to specs. Reconnect the hoses, and reinstall the pulley and drive belt. Fill the system with the correct power-steering fluid, and bleed the air from the system. Start the engine, check for leaks, and ensure proper power-steering operation.

Although power-steering pumps are not typically rebuilt by technicians, some parts are replaceable. The actual parts that can be replaced depend on the make of the pump. The common parts that can be replaced are discussed next.

POWER-STEERING PUMP PULLEY REPLACEMENT If the pulley wobbles while it is rotating, it is undoubtedly bent and should be replaced. Worn pulley grooves and/or cracks also indicate that the pulley should be replaced. A pulley that is loose on the pump's shaft must be replaced. Never hammer on the pump's driveshaft during pulley removal or installation. This will damage the internal parts of the pump.

If the pulley is pressed onto the pump's shaft, a special puller **(Figure 48–56A)** is required to remove it, and a pulley installation tool is used to install the pulley **(Figure 48–56B)**. When replacing a pump, the replacement pump usually does not come with a pulley. You will need to remove the pulley from the old pump and reinstall it on the new pump. To remove the pulley, install the puller over the lip of the pulley, and tighten the puller bolt against the pump's driveshaft. Hold the body of the puller with a wrench and tighten the puller bolt. This will pull the pulley from the pump shaft. To reinstall, align the pulley onto the pump shaft and thread the installation bolt into the driveshaft. Tighten the bolt and hold it from turning with a wrench or socket. Using a wrench, tighten the nut against the pulley. This will push the pulley onto the driveshaft. Once the pulley is seated in place, spin the pulley to make sure it does not wobble. A bent pulley should be replaced because it will cause premature belt wear.

If the power-steering pump pulley is retained with a nut, mount the pump in a vise. Always tighten the vise on one of the pump's mounting bolt surfaces. Do not tighten the vise with excessive force. Use a special holding tool to keep the pulley from turning, and loosen the nut with a box-end wrench. Remove the nut, pulley, and woodruff key. Inspect the pulley, shaft, and woodruff key for wear. Replace all worn components.

(A) Removal

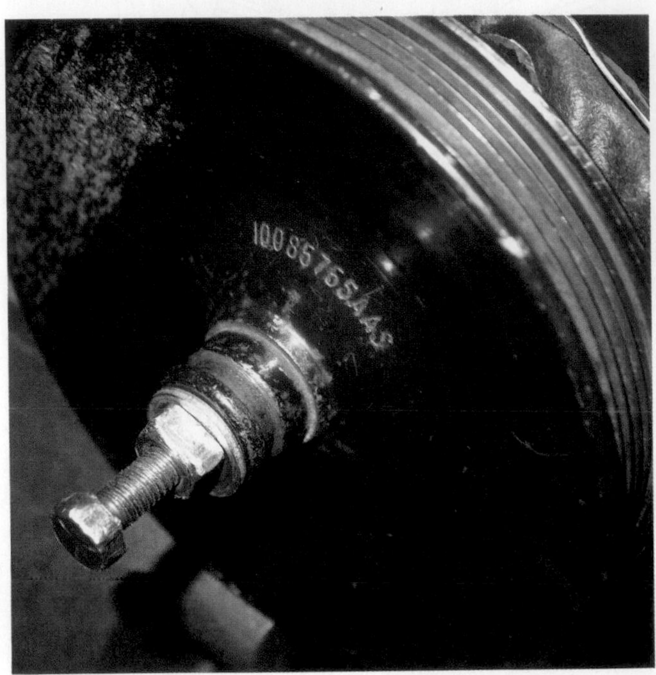

(B) Installation

FIGURE 48–56

Removing and installing a press-fit pulley.

REMOVE AND REPLACE THE FLOW CONTROL VALVE AND END COVER To replace the flow control valve and end plate, remove the retaining ring with a slotted screwdriver and punch. Then remove the flow control valve, end cover, spring, and magnet. Check the flow control valve for burrs. Remove minor burrs with crocus cloth, and clean the flow control valve in solvent. Damaged or worn flow control valves must be replaced. Inspect the end cover sealing surface for damage. Also check the pump's driveshaft for corrosion and damage. Remove any corrosion with crocus cloth. Clean all parts and lubricate the end cover with power-steering fluid. Make sure to clean the magnet with a shop towel. To reassemble, install the flow control valve, end cover, retaining ring, and related components.

REMOVE AND REPLACE THE PRESSURE RELIEF VALVE Follow this procedure to service the pressure relief valve:

1. Wrap a shop towel around the land end of the flow control valve and clamp this end in a soft-jawed vise. Be very careful not to damage the valve lands.
2. Remove the hex-head ball seat. Clean the components in solvent. A worn or damaged pressure relief ball, spring, guide, or seat must be replaced.
3. Reinstall the new or cleaned components, and then install the ball seat.

FOUR-WHEEL STEERING SYSTEMS

A few manufacturers have offered four-wheel steering systems in which the rear wheels also help to turn the car by electrical, hydraulic, or mechanical means. Although they certainly are not very common, you should be aware of how they work.

Production-built cars tend to understeer or, in a few instances, oversteer. If a car could automatically compensate for an understeer/oversteer problem, the driver would enjoy nearly neutral steering under varying operating conditions. **Four-wheel steering (4WS)** is a serious effort on the part of automotive design engineers to provide near-neutral steering with the following advantages:

- The vehicle's cornering behaviour becomes more stable and controllable at high speeds as well as on wet or slippery road surfaces **(Figure 48–57)**.
- The vehicle's response to steering input becomes quicker and more precise throughout the vehicle's entire speed range.
- The vehicle's straight-line stability at high speeds is improved. Negative effects of road irregularities and crosswinds on the vehicle's stability are minimized.
- Stability in lane changing at high speeds is improved. High-speed slalom-type operations become easier. The vehicle is less likely to go into a spin even in situations in which the driver must make a sudden and relatively large change of direction.
- By steering the rear wheels in the direction opposite the front wheels at low speeds, the vehicle's turning circle is greatly reduced. Therefore, vehicle manoeuvring on narrow roads and during parking becomes easier.

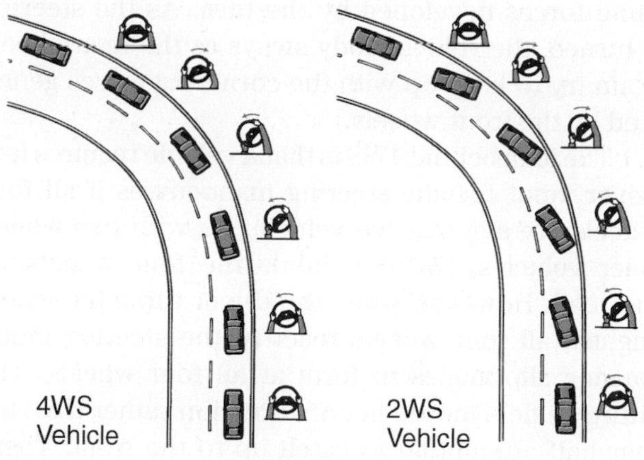

FIGURE 48–57

A comparison of 2WS and 4WS vehicle behaviour during cornering.

To understand the advantages of 4WS, it is wise to review the dynamics of typical steering manoeuvres with a conventional front-steered vehicle. The tires are subject to the forces of grip, momentum, and steering input when making a movement other than straight-ahead driving. These forces compete with each other during steering manoeuvres. With a front-steered vehicle, the rear end is always trying to catch up to the directional changes of the front wheels. This causes the vehicle to sway. As a normal part of operating a vehicle, the driver learns to adjust to these forces without thinking about them.

When turning, the driver is putting into motion a complex series of forces. Each of these must be balanced against the others. The tires are subjected to road grip and slip angle. Grip holds the car's wheels to the road, and momentum moves the car straight ahead. Steering input causes the front wheels to turn. The car momentarily resists the turning motion, causing a tire slip angle to form. Once the vehicle begins to respond to the steering input, cornering forces are generated. The vehicle sways as the rear wheels attempt to keep up with the cornering forces already generated by the front tires. This is referred to as rear-end lag because there is a time delay between steering input and vehicle reaction. When the front wheels are turned back to a straight-ahead position, the vehicle must again try to adjust by reversing the same forces developed by the turn. As the steering is turned, the vehicle body sways as the rear wheels again try to keep up with the cornering forces generated by the front wheels.

The idea behind 4WS is that a vehicle requires less driver input for any steering manoeuvres if all four wheels are steering the vehicle. As with two-wheel-steer vehicles, tire grip holds the four wheels on the road. However, when the driver turns the wheel slightly, all four wheels react to the steering input, causing slip angles to form at all four wheels. The entire vehicle moves in one direction rather than the rear half attempting to catch up to the front. There is also less sway when the wheels are turned back to a straight-ahead position. The vehicle responds more quickly to steering input because rear-wheel lag is eliminated.

Because each 4WS system is unique in its construction and repair needs, the vehicle's service information must be followed for proper diagnosis, repair, and alignment of a four-wheel system.

Mechanical 4WS

In a straight-mechanical type of 4WS, two steering gears are used—one for the front and the other for the rear wheels. A steel shaft connects the two steering gearboxes and terminates at an eccentric shaft that is fitted with an offset pin **(Figure 48–58)**. This pin engages a second offset pin that fits into a planetary gear.

The planetary gear meshes with the matching teeth of an internal gear that is secured in a fixed position to the gearbox housing. This means that the planetary gear can rotate, but the internal gear cannot. The eccentric pin of the planetary gear fits into a hole in a slider for the steering gear.

A 120° turn of the steering wheel rotates the planetary gear to move the slider in the same direction that the front wheels are headed. Proportionately, the rear wheels turn about 1.5° to 10° of the steering wheel. Further rotation of the steering wheel, past the 120° point, causes the rear wheels to start straightening out due to the double-crank action (two eccentric pins) and rotation of the planetary gear. Turning the steering wheel to a greater angle, about 230°, finds the rear wheels in a neutral position regarding the front wheels. Further rotation of the steering wheel, at about 45°, results in the rear wheels going counter-phase with regard to the front wheels. About 5.3° maximum counter-phase rear steering is possible.

Mechanical 4WS is steering-angle sensitive. It is not sensitive to vehicle road speed.

Hydraulic 4WS

The hydraulically operated 4WS system shown in **Figure 48–59** is a simple design, both in components and operation. The rear wheels turn only in the same direction as the front wheels. They also turn no more than 1½°. The system only activates at speeds above 50km/h (30 mph) and does not operate when the vehicle moves in reverse.

A two-way hydraulic cylinder mounted on the rear stub frame turns the wheels. Fluid for this cylinder is supplied by a rear steering pump that is driven by the differential. The pump only operates when the front wheels are turning. A tank in the engine compartment supplies the rear steering pump with fluid.

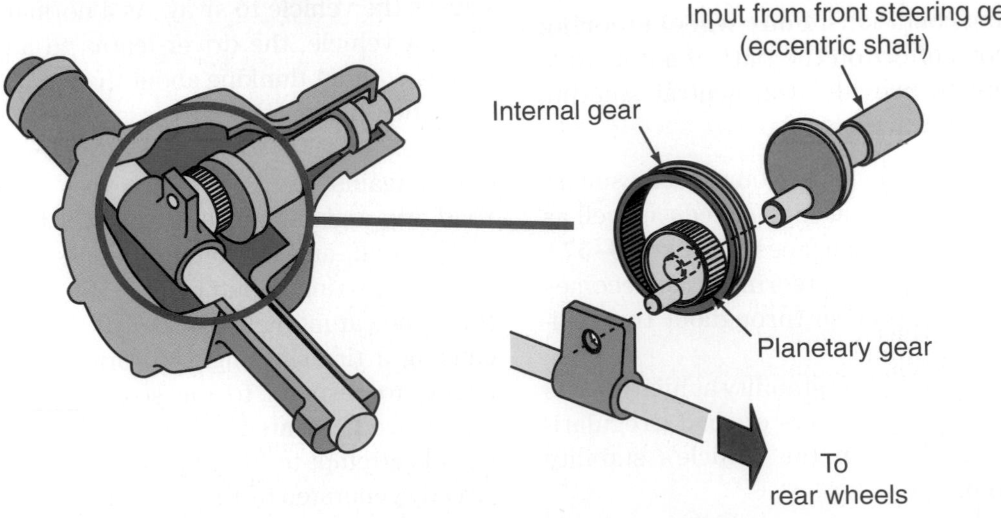

FIGURE 48–58

Inside a rear-steering gearbox is a simple planetary gear setup.

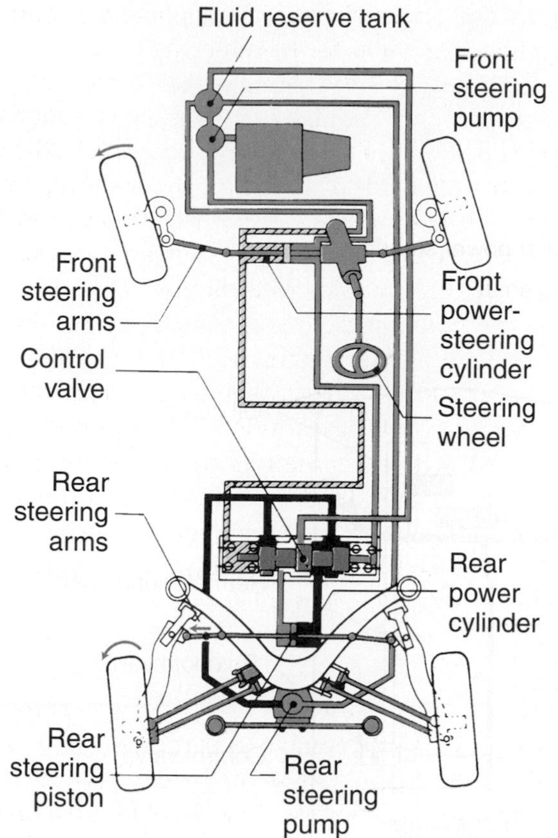

FIGURE 48–59

A simple hydraulic 4WS system.

Mitsubishi Motor Sales of America, Inc.

When the steering wheel is turned, the front steering pump sends fluid under pressure to the rotary valve or directional control valve (DCV) in the front rack and pinion unit. This forces fluid into the front power cylinder, and the front wheels turn in the direction steered. The fluid pressure varies with the turning of the steering wheel. The faster and farther the steering wheel is turned, the greater the fluid pressure.

The fluid is also fed under the same pressure to the control valve where it opens a spool valve in the control valve housing. As the spool valve moves, it allows fluid from the rear steering pump to move through and operate the rear power cylinder. The higher the pressure on the spool, the farther it moves. The farther it moves, the more fluid it allows through to move the rear wheels. As mentioned earlier, this system limits rear-wheel movement to 1½° in either the left or right direction.

Electrohydraulic 4WS

Several 4WS systems combine computer electronic controls with hydraulics to make the system sensitive to both steering angle and road speeds. In this design, a speed sensor and steering wheel angle sensor feed information to the electronic control unit (ECU). By processing the information received, the ECU commands the hydraulic system to steer the rear wheels. At low road speed, the rear wheels of this system are not considered a dynamic factor in the steering process.

At moderate road speeds, the rear wheels are steered momentarily counter-phase, through neutral, then in phase with the front wheels. At high road speeds, the rear wheels turn only in phase with the front wheels. The ECU must know not only road speed, but also how much and how quickly the steering wheel is turned. These three factors—road speed, the amount of steering wheel turn, and the quickness of the steering wheel turn—are interpreted by the ECU to maintain continuous and desired steering angle of the rear wheels.

Another electrohydraulic 4WS system is shown in **Figure 48–60**. The basic working elements of the design are the control unit, a stepper motor, a swing arm, a set of bevelled gears, a control rod, and a control valve with an output rod. Two electronic sensors tell the ECU how fast the car is going.

The yoke is a major mechanical component of this electrohydraulic design. The position of the control yoke varies with vehicle road speed. For example, at speeds below 53 km/h (33 mph), the yoke is in its downward position, which results in the rear wheels steering in the counter-phase (opposite front wheels) direction. As road speeds approach and exceed 35 km/h (22 mph), the control yoke swings up through a neutral (horizontal) position to an up position. In the neutral position, the rear wheels steer in phase with the front wheels.

The stepper motor moves the control yoke. A swing arm is attached to the control yoke. The position of the yoke determines the arc of the swing rod. The arc of the swing arm is transmitted through a control arm that passes through a large bevel gear. Stepper motor action eventually causes a push-or-pull movement of its output shaft to steer the rear wheels up to a maximum of 5° in either direction.

The electronically controlled 4WS system regulates the angle and direction of the rear wheels in response to speed and driver's steering. This speed-sensing system optimizes the vehicle's dynamic characteristics at any speed, thereby producing enhanced stability and, within certain parameters, agility.

The actual 4WS system consists of a rack and pinion front steering that is hydraulically powered by a main twin-tandem pump. The system also has

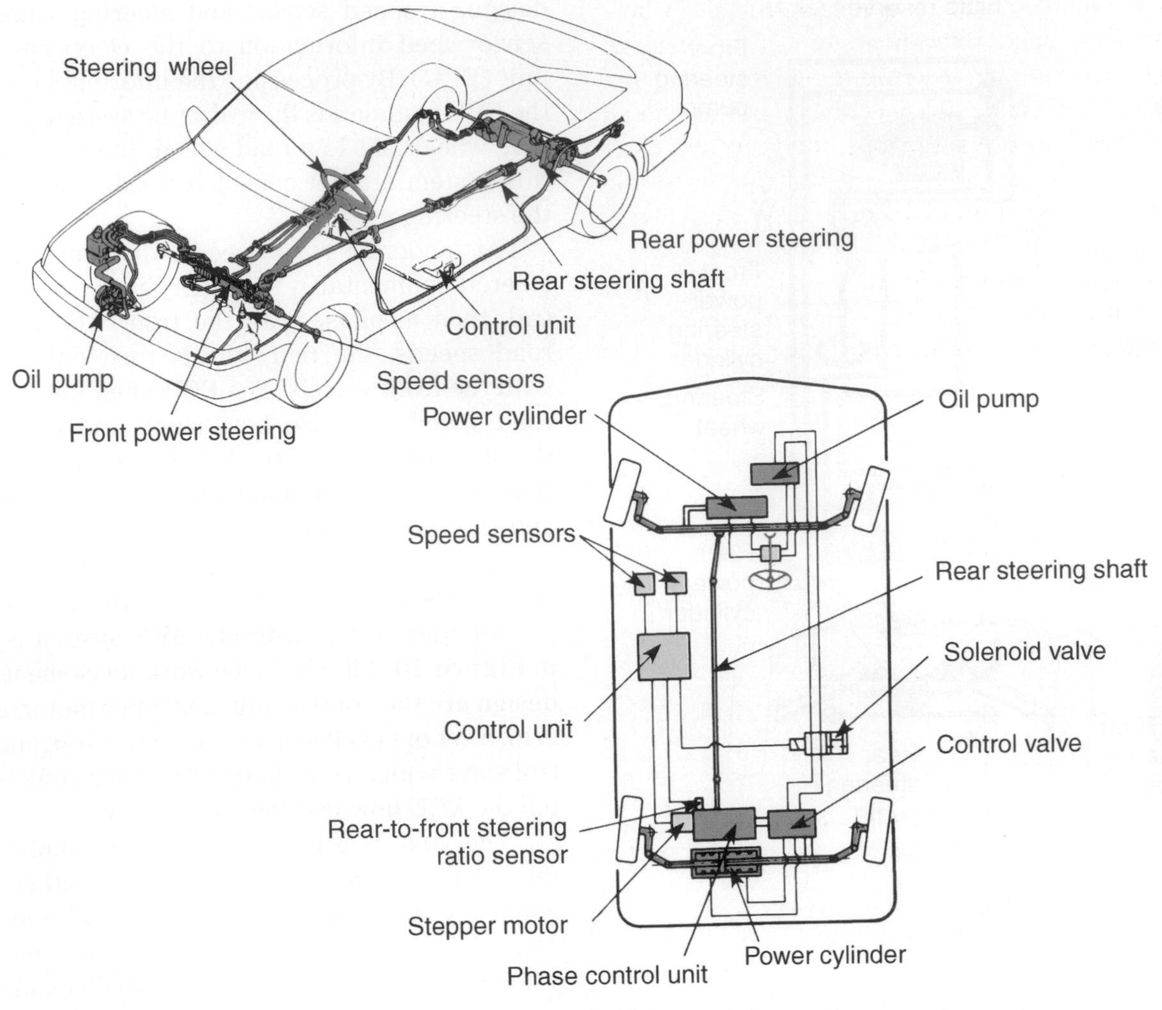

FIGURE 48–60

An electronically and hydraulically controlled 4WS system using a stepper motor and control yoke.

Mazda Motor of America, Inc.

a rear-steering mechanism, hydraulically powered by the main pump. The rear-steering shaft extends from the rack bar of the front-steering assembly to the rear-steering-phase control unit.

The rear steering is comprised of the input end of the rear-steering shaft, vehicle speed sensors, and steering-phase control unit (deciding direction and degree), a power cylinder, and an output rod. A centring lock spring is incorporated that locks the rear system in a neutral (straight-ahead) position in the event of hydraulic failure. Additionally, a solenoid valve that disengages the hydraulic boost (thereby activating the centring lock spring in case of an electrical failure) is included.

All 4WS systems have fail-safe measures. For example, with the electrohydraulic setup, the system automatically counteracts possible causes of failure, both electronic and hydraulic, and converts the entire steering system to a conventional two-wheel steering type. Specifically, if a hydraulic defect should reduce pressure level (by a movement malfunction or a broken driving belt), the rear-wheel steering mechanism is automatically locked in a neutral position, activating a low-level warning light.

An electrical failure would be detected by a self-diagnostic circuit integrated in the 4WS control unit. The control unit stimulates a solenoid valve, which neutralizes hydraulic pressure, thereby alternating the system to two-wheel steering. The failure would be indicated by the system's warning light in the main instrument display.

On any 4WS system, there must be near-perfect compliance between the position of the steering wheel, the position of the front wheels, and the position of the rear wheels. It is usually recommended that the car be driven about 6 m (20 ft.) in a dead-straight line. Then the position of the front/rear wheels is checked with respect to steering wheel position. The base reference point is a strip of masking tape on the steering wheel hub and the steering column. When the wheel is positioned dead centre, draw a line down the tape. Run the car a short distance straight ahead to see if the reference line holds. If not, corrections are needed, such as repositioning the steering wheel.

Even severe imbalance of a rear wheel on a speed-sensitive 4WS system can cause problems and make basic troubleshooting a bit frustrating.

Quadrasteer

Quadrasteer is a 4WS system that improves low-speed manoeuvrability (decreasing the turning radius), high-speed stability, and trailering capabilities for full-size pickups, vans, and SUVs. The system combines normal front-wheel steering with an electrically powered and electronically controlled rear-wheel steering system. Besides the mechanical part, the system uses wheel position and vehicle speed sensors and a central control module **(Figure 48–61)**.

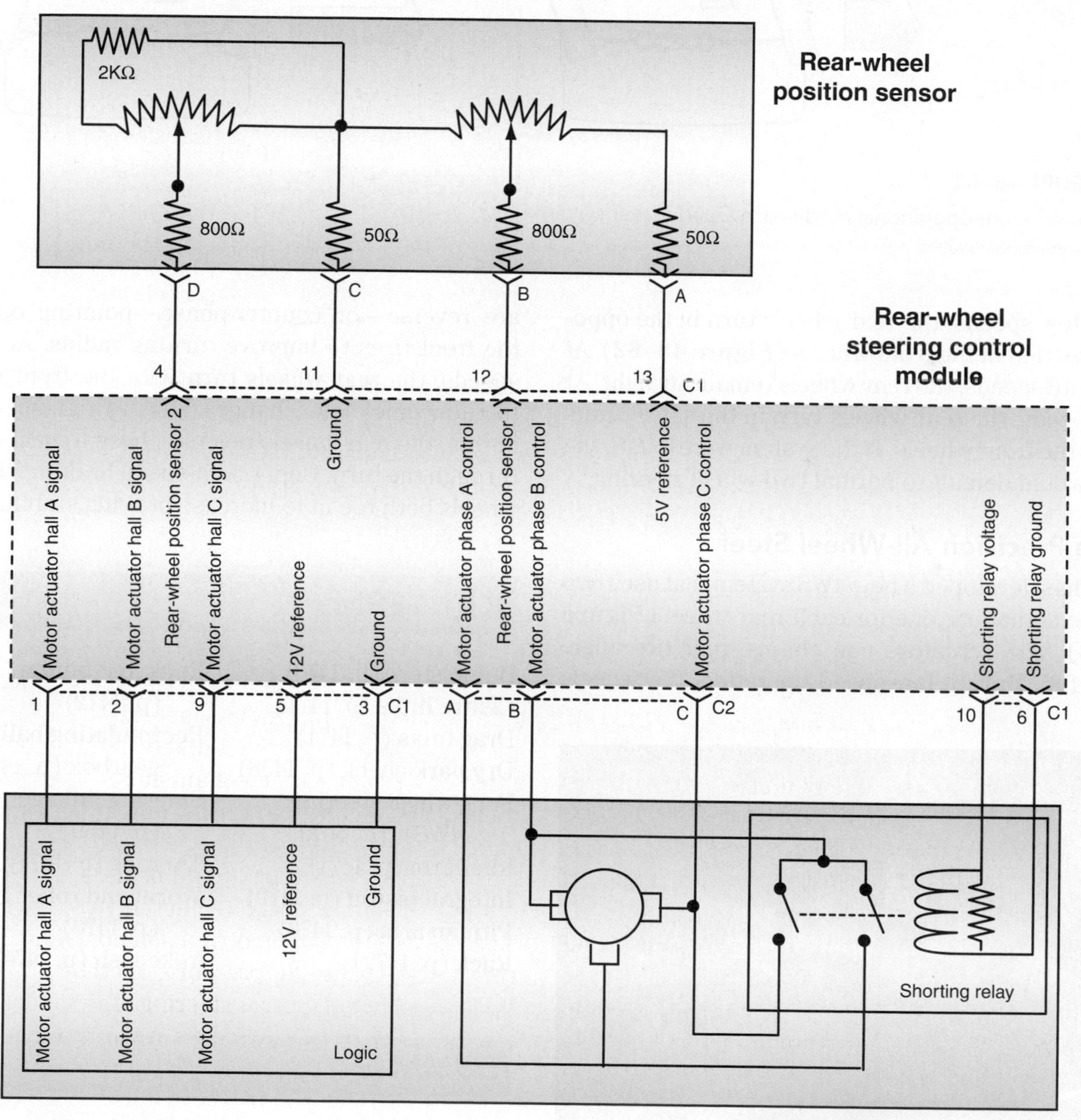

FIGURE 48–61

The components controlling the four-wheel steering system.

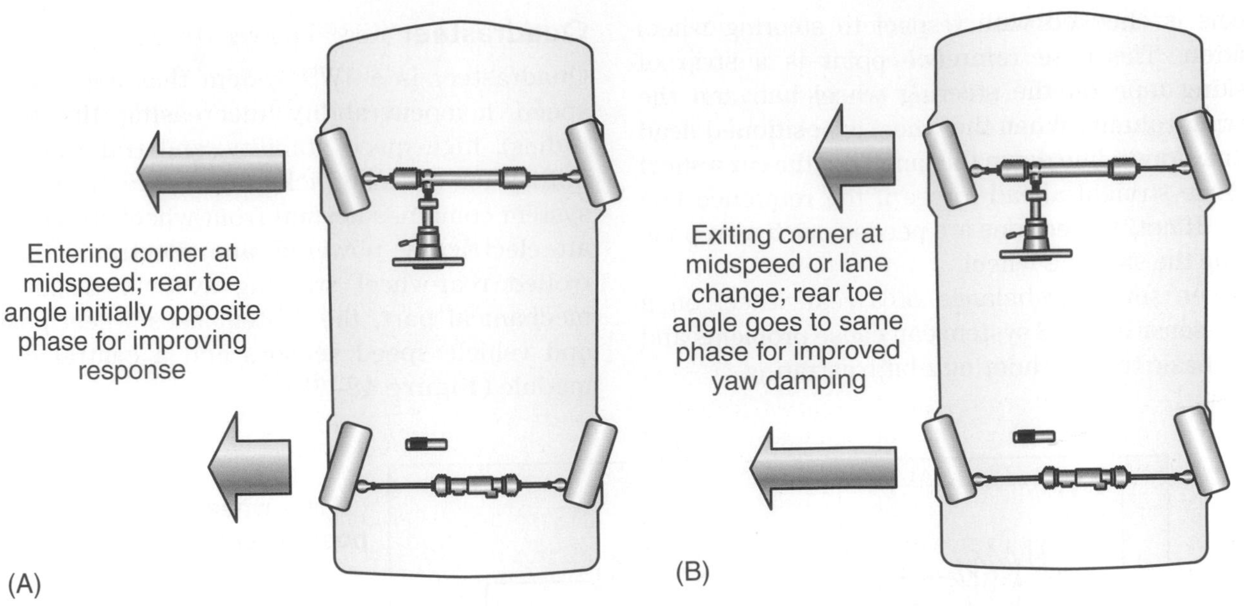

FIGURE 48–62

The different operational modes of a Quadrasteer system.

At low speeds, the rear wheels turn in the opposite direction of the front wheels **(Figure 48–62)**. At moderate speeds, the rear wheels remain straight. At high speeds, the rear wheels turn in the same direction as the front wheels. If the system were to fail, the truck would default to normal two-wheel steering.

Acura Precision All-Wheel Steer

Acura has developed a new 4WS system that uses two electric actuators, one for each rear wheel **(Figure 48–63)**. The actuators can change rear tire angle up to 1.8°. During low-speed operation, the wheels are reverse—or counter-phase—pointing opposite the front tires to improve turning radius. At higher speeds, the rear wheels turn with the front wheels to allow quick lane changes. During high-speed cornering, the rear wheels reverse-phase to help the car through the turn. Under high-speed braking, the rear wheels both toe in to increase braking stability.

FIGURE 48–63

Acura's 4WS system uses two electric actuators, one for each rear wheel.

KEY TERMS

Bump steer (p. 1470)
Centre links (p. 1471)
Drag links (p. 1471)
Dry park check (p. 1498)
Four-wheel steering (4WS) (p. 1507)
Idler arm (p. 1471)
Integral piston (p. 1479)
Pitman arm (p. 1470)
Rack (p. 1472)
Rack and pinion (p. 1472)
Recirculating ball gearbox (p. 1483)
Steering linkage (p. 1469)
Tie rods (p. 1471)
Worm and roller gearbox (p. 1476)
Yoke lash (p. 1474)

SUMMARY

- The components of a manual-steering system include the steering linkage, steering gear, and the steering column and wheel.
- The term steering linkage is applied to the system of pivots and connecting parts placed between

the steering gear and the steering arms that are attached to the front wheels, controlling the direction of vehicle travel. The steering linkage transfers the motion of the steering gear output shaft to the steering arms, turning the wheels to manoeuvre the vehicle.

- Basic components of a parallelogram steering linkage system include the pitman arm; the idler arm; links; tie rods; and, in some designs, a steering damper.
- In rack and pinion steering linkage, steering input is received from a pinion gear attached to the steering column. This gear moves a toothed rack that is attached to the tie rods that move the wheels.
- There are two types of manual steering gears in use today: recirculating ball and rack and pinion.
- The steering wheel and column produce the necessary force to turn the steering gear.
- The power-steering unit is designed to reduce the amount of effort required to turn the steering wheel. It also reduces driver fatigue on long drives and makes it easier to steer the vehicle at slow road speeds, particularly during parking.
- There are several power-steering systems in use on passenger cars and light-duty trucks. The most common ones are the integral, nonintegral, hydro-boost, and power-assisted rack and pinion systems.
- The major components of a conventional power-steering system are the steering linkage, power-steering pump, flow control and pressure relief valves, reservoir, spool valves and power pistons, hydraulic hose lines, and gearbox or assist assembly on the linkage.
- Electronic rack and pinion systems replace the hydraulic pump, hoses, and fluid associated with conventional power-steering systems with electronic controls and an electric motor located concentric to the rack itself.
- Four-wheel-steering (4WS) advantages include cornering capability, steering response, straight-line stability, lane changing, and low-speed manoeuvrability.

REVIEW QUESTIONS

1. Describe how rack and pinion steering and parallelogram steering systems operate.
2. A power-steering hose transmits fluid under pressure from the ______ to the ______.
3. What is an integral power-steering system?
4. Define the term *gearbox ratio.*
5. What are the basic features of all 4WS systems?
6. What is the primary purpose of power-steering hoses?
   a. to lubricate the pump
   b. to relieve pressure
   c. to transmit power through fluid under pressure
   d. to allow the system to be filled
7. What component is the power piston part of in an integral recirculating ball power-steering gear?
   a. the pitman arm
   b. the ball nut
   c. the sector shaft
   d. the wormshaft
8. What controls the amount of assist in most power-steering systems?
   a. the flow control valve
   b. the pressure relief valve
   c. the torsion bar in the directional control valve
   d. the power-steering pump
9. Which steering gearbox adjustment should be performed first?
   a. sector over centre preload
   b. wormshaft bearing preload
   c. sector gear lash
   d. steering effort
10. When is hydraulic fluid pressure in the power-steering system the greatest?
    a. during no steering assist operation with the engine idling
    b. during no steering assist operation with the engine at high rpm
    c. during steering assist while driving
    d. during steering assist with the wheels held against the steering stops
11. What type of vehicle usually uses a recirculating ball type of steering gear?
    a. a front-wheel-drive car
    b. a full-sized rear-wheel-drive car
    c. a front-wheel-drive minivan
    d. an all-wheel-drive car
12. In a rack and pinion steering system, what protects the rack from contamination?
    a. the inner tie-rod socket
    b. the outer tie-rod socket
    c. grommets
    d. the bellows boot

13. During a power-steering pressure test, what could cause the pressure to be lower than specified when the gauge shutoff valve is closed?
a. a worn power-steering pump
b. a faulty pressure relief valve
c. excessive sector over-centre preload
d. a seized tie-rod adjustment sleeve

14. Which one of the following is a probable cause of wheel shimmy?
a. binding steering column U-joints
b. loose wheel bearings
c. out-of-balance tires
d. loose, worn, or damaged suspension parts

15. What is the main job of the idler arm?
a. to support the left side of the centre link
b. to support the right side of the centre link
c. to support the pitman arm
d. to keep both ends of the steering system level

16. Which of the following could cause a vehicle to wander?
a. a binding steering column U-joint
b. insufficient steering gear preload
c. excessive steering gear preload
d. a worn power-steering pump

17. Which component in the integral power rack and pinion steering gear has the power piston attached to it?
a. the pinion gear
b. the rack gear
c. the tie-rod assembly
d. the input shaft

18. In a mechanical four-wheel steering system, what position will the rear wheels move to when the steering wheel is turned past 230°?
a. parallel
b. same-phase
c. counter-phase
d. neutral position

19. Which of the following is true of rack and pinion steering?
a. It is lighter in weight and has fewer components than parallelogram steering.
b. It does not provide as much feel for the road as parallelogram steering.
c. It does not use tie rods in the same fashion as parallelogram steering.
d. It is not as direct as a recirculating ball steering gear.

20. What connects the steering column to the wheels?
a. the steering shaft
b. the steering linkage
c. the gearbox
d. the pitman arm

CHAPTER 49
Wheel Alignment

LEARNING OUTCOMES

- Explain the benefits of accurate wheel alignment.
- Explain the importance of correct wheel alignment angles.
- Describe the different functions of camber and caster with regard to the vehicle's suspension.
- Identify the purposes of steering axis inclination.
- Explain why toe is the most critical tire wear factor of all the alignment angles.
- Identify the purposes of turning radius or toe-out.
- Explain the condition known as tracking.
- Perform a pre-alignment inspection.
- Describe the various types of equipment that can be used to align the wheels of a vehicle.
- Describe how alignment angles can be changed on a vehicle.
- Understand the importance of rear-wheel alignment.
- Know the difference between two-wheel and four-wheel alignment procedures.
- Identify and understand steering sensor calibration requirements.

A vehicle's wheels, tires, suspension system, and steering system are all designed to work together to provide safe, stable, and reliable handling. The end result of all of these systems working correctly together is wheel alignment. During a wheel alignment, the angles of the wheels are measured and adjusted. The outcome is to place the tires at their ideal position to provide maximum performance, ride quality, and tire life. These angles are adjusted by changing the position of various steering and suspension parts. The desired angles are those set by the vehicle's manufacturer.

ALIGNMENT FUNCTION

Correct wheel alignment allows the wheels to roll without scuffing, dragging, or slipping on different types of road surfaces. Proper alignment of both the front and rear wheels ensures greater safety, easier steering, longer tire life, reduction in fuel consumption, and less strain on the steering and suspension systems.

The alignment of the wheels should be checked whenever new tires or steering and suspension parts are installed. Also, the wheels should be aligned whenever the tires are wearing abnormally. A vehicle's wheels become out of alignment due to worn suspension parts, a change in ride height, or driving hard into a pothole or curb. All of these can affect the angle of the wheels.

A wheel alignment restores the geometry of the suspension to the angles that were determined to properly locate the vehicle's weight on the tires and to facilitate steering.

Types of Wheel Alignment

For many years, and until the 1980s, two-wheel alignments were performed. This is because most vehicles were RWD and had live rear axles that did not allow for four-wheel alignment. Two-wheel alignment was common before suspension and steering systems became more complex. Today, four-wheel alignments are performed. This is true even if only the front-wheel alignment can be adjusted.

Four-wheel alignment measures the angles at the four wheels. On some vehicles, adjustments are made only to the front wheels. This is primarily due to the fact that there is no way to make adjustments to the rear wheels. However, by adjusting the front wheels so that they are rotating in the same direction as the rear wheels, the vehicle will tend to move straight. Many vehicles have provisions for adjusting the rear wheels. When this is the case, the rear wheels are adjusted first, and then the fronts are aligned to the vehicle's centreline.

Road Crown

Most roads are not designed to be flat. They are paved to slope at a slight angle. This allows water to flow off the road's surface rather than allowing it to accumulate. The angle is called **road crown** and can cause a vehicle to tend to pull toward the right of the road. To compensate for this, different angles may be set at each side of the vehicle.

ALIGNMENT GEOMETRY

The proper alignment of a suspension/steering system centres on the accuracy of the following angles.

Caster

Caster is the angle of the steering axis of a wheel from the vertical, as viewed from the side of the vehicle. The forward or rearward tilt from the vertical line **(Figure 49–1)** is caster. Caster is the first angle adjusted during an alignment. Tilting the steering axis forward is negative caster. Tilting backward is positive caster.

Caster is designed to provide steering stability. The caster angle for each wheel on an axle should be equal or very close to equal. Caster can be used to compensate for road crown by setting the right wheel caster slightly more positive than the left. Excessively unequal caster angles cause the vehicle to steer toward the side with less caster. Too much negative caster can cause the vehicle to have sensitive steering at high speeds. The vehicle might wander as a result of negative caster. Caster is not related to tire wear unless the caster setting is excessive.

Caster is affected by worn or loose strut rod and control arm bushings. Caster adjustments are not possible on some strut suspension systems. Where they are provided, they can be made at the top or bottom mount of the strut assembly. When adjusting caster, the side-to-side reading should be within 0.5°.

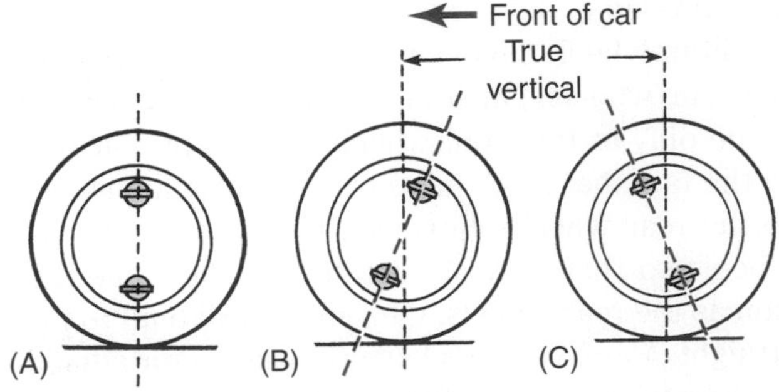

FIGURE 49–1

Three types of caster are (A) zero, (B) positive, and (C) negative.

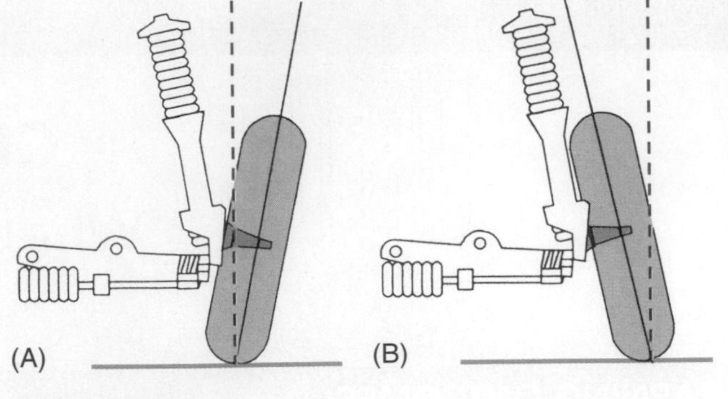

FIGURE 49–2

(A) Positive and (B) negative camber.

Camber

Camber is the angle represented by the tilt of either the front or rear wheels inward or outward from the vertical as viewed from the front of the car **(Figure 49–2)**. Camber is designed into the vehicle to compensate for road crown, passenger weight, and vehicle weight. Camber is usually set equally for each wheel. Equal camber means each wheel is tilted outward or inward the same amount. Unequal camber causes the vehicle to steer toward the side that is more positive. The typical maximum side-to-side camber difference is 0.5°. Excessive camber causes tire wear.

Camber angle changes, through the travel of the suspension system, are controlled by pivots. Camber is affected by worn or loose ball joints, control arm bushings, and wheel bearings. Anything that changes chassis height also affects camber. Camber is adjustable on most vehicles. Some manufacturers prefer to include a camber adjustment at the spindle assembly. Camber adjustments are also provided on some strut suspension systems at the top mounting of the strut. Very little adjustment of camber (or caster) is required on strut suspensions if the tower and lower control arm positions are in their proper place. If serious camber error has occurred and the suspension mounting positions have not been damaged, it is an indication of bent suspension parts. In this case, diagnostic angle and dimensional checks should be made on the suspension parts. Damaged parts should be replaced.

Toe

Toe is the distance comparison between the leading edge and trailing edge of the front tires at spindle height. If the leading edge distance is less, then there

Positive toe (toe-in)

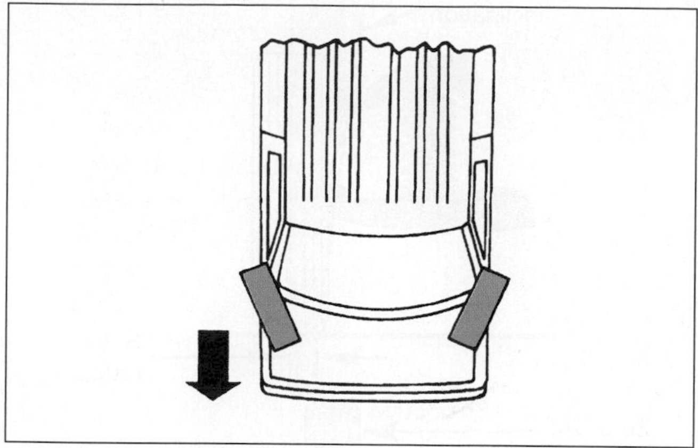

Negative toe (toe-out)

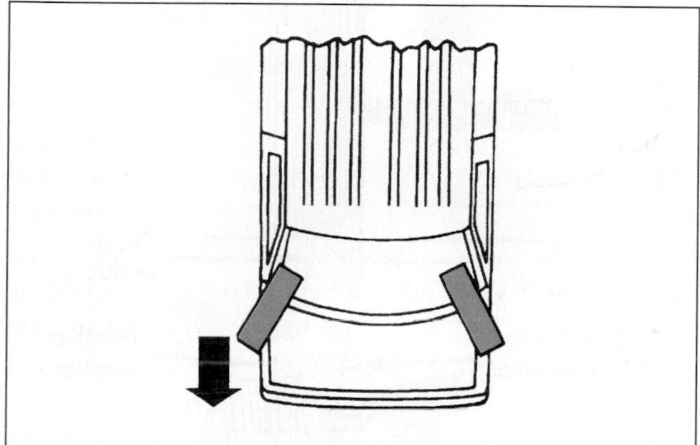

FIGURE 49–3

The position of the wheels during toe-in and toe-out.

Ford Motor Company

is positive toe, or toe-in. If it is greater, there is negative toe, or toe-out **(Figure 49–3)**. Actually, toe is critical as a tire-wearing angle. Wheels that do not track straight ahead have to drag as they travel forward. Excessive toe measurements (in or out) cause a sawtooth edge on the tread surface from dragging the tire sideways. Excessive toe-in will cause tire wear on the outside edge of the tire. Toe-out causes wear on the inside edge.

Toe adjustments are made at the tie rod. They must be made evenly on both sides of the car. If the toe settings are not made equal with the steering wheel centred, the steering wheel will be off centre when driving straight ahead. Equal toe with an off-centre wheel may cause wander due to the steering box being turned from the straight-ahead position and not at the sector over-centre preload position. Apart from the tire wear factor, unequal toe can also cause a vehicle to wander due to the track of the tires being placed at different angles. Toe is the last adjustment made in an alignment.

Toe is affected by worn steering linkage components, worn springs, ball joints, or anything that changes camber. Toe changes caused by loose or worn parts often cause bump steer. Bump steer occurs when the vehicle darts left or right after hitting a bump in the road. The change in suspension and/or steering geometry changes the toe, causing the wheels to move left or right.

Toe will change with vehicle speed. As the vehicle moves, friction forces the tires to move straight ahead or have zero toe. However, aerodynamic forces on the vehicle cause a change in its riding height. This will also change the toe as well as camber. Therefore, most toe specifications anticipate these changes and are set to provide zero toe at highway speeds.

Thrust Line Alignment

A main consideration in any alignment is to make sure the vehicle runs straight down the road, with the rear tires tracking directly behind the front tires when the steering wheel is in the straight-ahead position. The geometric centreline of the vehicle should parallel the road direction. This is the case when rear toe is parallel to the vehicle's geometric centreline in the straight-ahead position. If rear toe does not parallel the vehicle centreline, a thrust direction to the left or right is created **(Figure 49–4)**. This difference of rear toe from the geometric centreline is called the **thrust angle**.

Any time the centreline of the front axle is not parallel to the rear axle, handling will be affected. This is because the vehicle will tend to travel according to the angle of the rear axle. This causes the vehicle to pull in the opposite direction as the thrust angle. If the thrust line is to the right, the vehicle will pull to the left; when the thrust line is to the left, the vehicle will pull to the right.

This problem can cause tire wear and poor directional stability on ice, snow, or wet pavement. It

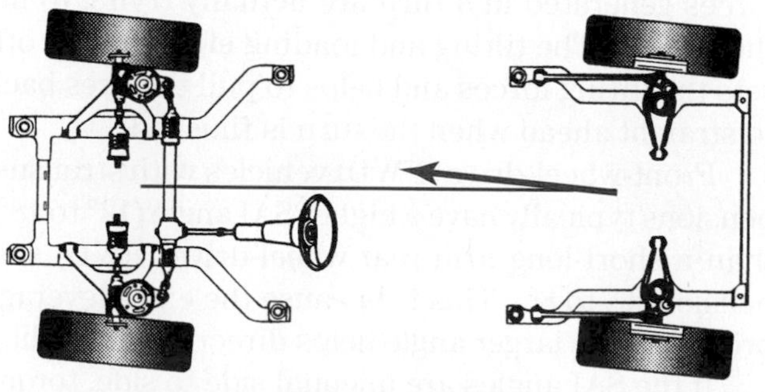

FIGURE 49–4

The thrust line, or driving direction, of the rear wheels.

Hunter Engineering Company

can also make a vehicle pull during braking or hard acceleration. Also, if the thrust angle is not zero, the steering wheel will not be centred. Nonparallel axles, or thrust line deviations, are usually caused by the shifting of the rear axle on its spring supports, rear-wheel misalignment, or damage from an accident.

To correct this problem, begin by setting individual rear-wheel toe equally in reference to the geometric centreline. Four-wheel-alignment machines check individual toe on each wheel. Once the rear wheels are in alignment with the geometric centreline, set the individual front toe in reference to the thrust angle. Following this procedure assures that the steering wheel is straight ahead for straight-ahead travel. If you set the front toe to the vehicle geometric centreline, ignoring the rear toe angle, a cocked steering wheel results.

Steering Axis Inclination (SAI)

Steering axis inclination (SAI) locates the vehicle weight to the inside or outside of the vertical centreline of the tire. The SAI is the angle between true vertical and a line drawn between the steering pivots as viewed from the front of the vehicle **(Figure 49–5)**. It is an engineering angle designed to project the weight of the vehicle to the road surface for stability. The SAI helps the vehicle's steering system return to straight ahead after a turn.

If the vehicle has 0 (zero) SAI, the upper and lower ball joints (or strut pivot points) would be located directly over one another. Problems associated with this simple relationship include tire scrub in turns, lack of control, and increased effort during turn recovery. If the SAI is tilted, a triangle is formed between ball joints and spindle. An arc is then formed when turning. There is a high point at straight-ahead position and a drop downward turning to each side. This motion travels through the control arms to the springs and, finally, to the weight of the vehicle. The forces generated in a turn are actually trying to lift the vehicle. The tilting and loading effect of SAI offsets the lifting forces and helps to pull the tires back to straight ahead when the turn is finished.

Front-wheel-drive (FWD) vehicles with strut suspensions typically have a higher SAI angle (12° to 18°) than a short-long arm rear-wheel-drive (RWD) suspension (6° to 8°). This is because the extra leverage provided by a larger angle helps directional stability.

If the SAI angles are unequal side to side, torque steer, brake pull, and bump steer (jerking from side to side) can occur even if static camber angles are within specifications.

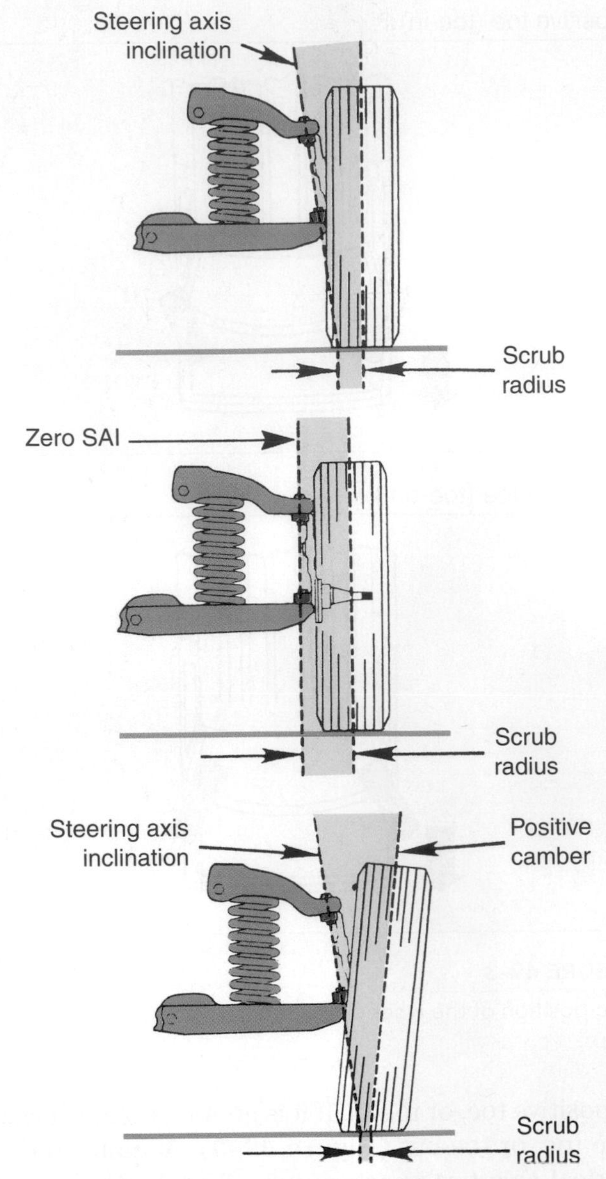

FIGURE 49–5

The effects of steering axis inclination changes.

Checking the SAI angle can help locate various problems that affect wheel alignment. For example, an SAI angle that varies from side to side may indicate an out-of-position upper strut tower, a bowed lower control arm, or a shifted centre crossmember.

On a short-long arm suspension, SAI is the angle between true vertical and a line drawn from the upper ball joint through the lower ball joint. In a strut-equipped vehicle, this line is drawn through the centre of the strut's upper mount down through the centre of the lower ball joint.

INCLUDED ANGLE When the camber angle is added to the SAI angle, the sum of the two is called the **included angle (Figure 49–6)**. This angle is not

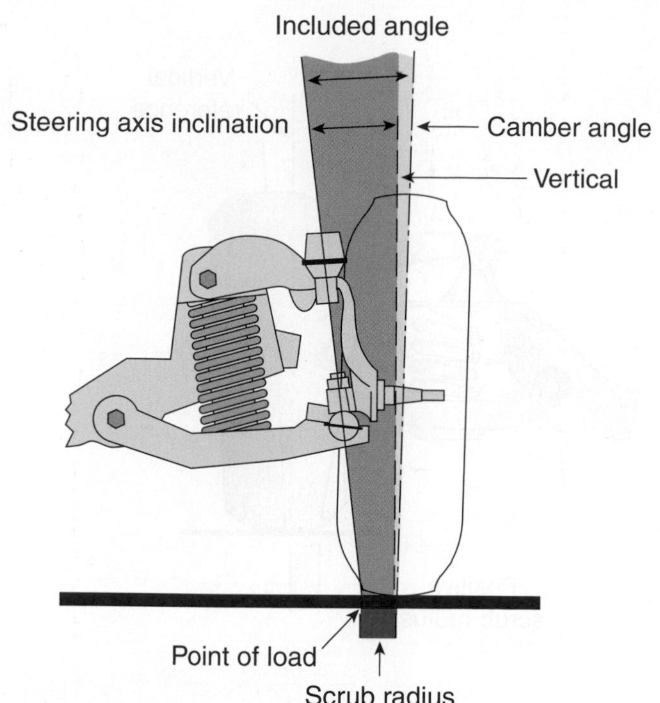

FIGURE 49–6

The included angle is the sum of the camber and SAI angles.

measured by an alignment machine; it is simply obtained by adding the camber and SAI on one side of the vehicle together. The included angle must be the same on each side of the vehicle, even if the camber on each side is different. If it is not, the vehicle will pull.

Comparing the SAI, included, and camber angles can help identify damaged or worn components. For example, if the SAI reading is correct but the camber and included angles are less than specifications, the steering knuckle or strut tower may be bent. **Table 49–1** summarizes the various angle combinations used to troubleshoot short-long arm, strut, and twin I-beam suspension system alignment problems.

SCRUB RADIUS **Scrub radius** is the distance between the centre of the tire and where the SAI angle intersects the ground (**Figure 49–7**). The scrub radius must be equal on both sides of the vehicle. Otherwise, the vehicle will pull to one side. Scrub radius is not adjustable or measured; rather, it is observed. Scrub radius is part of the suspension's design. There is positive scrub when the tire's contact patch is outside the SAI angle and negative when the patch is inside the angle. Most FWD vehicles have a negative scrub radius. This is done to reduce torque steer. To the contrary, most SLA suspensions have a positive scrub radius. If a vehicle pulls after it has been properly aligned, look for offset wheels or a problem that would affect SAI.

Turning Radius (Ackerman's Principle)

Turning radius is the amount of toe-out present in turns (**Figure 49–8**). The measurement of the turning radius is also called "toe-out on turns,"

TABLE 49–1 **ALIGNMENT ANGLE DIAGNOSTIC CHART**

SUSPENSION SYSTEM	SAI	CAMBER	INCLUDED ANGLE	PROBABLE CAUSE
Short-Long Arm Suspension	1. Correct 2. Less 3. Greater 4. Less	1. Less 2. Greater 3. Less 4. Greater	1. Less 2. Correct 3. Correct 4. Greater	1. Bent knuckle 2. Bent lower control arm 3. Bent upper control arm 4. Bent knuckle
MacPherson Strut Suspension	1. Correct 2. Correct 3. Less 4. Greater 5. Greater 6. Less 7. Less	1. Less 2. Greater 3. Greater 4. Less 5. Greater 6. Greater 7. Less	1. Less 2. Greater 3. Correct 4. Correct 5. Greater 6. Greater 7. Less	1. Bent knuckle and/or bent strut 2. Bent knuckle and/or bent strut 3. Bent control arm or strut tower (out at top) 4. Strut tower (in at top) 5. Strut tower (in at top) and spindle and/or bent strut 6. Bent control arm or strut tower (out at top) plus bent knuckle and/or bent strut 7. Strut tower (out at top) and knuckle and/or strut bent or bent control arm
Twin I-Beam Suspension	1. Correct 2. Greater 3. Less 4. Less	1. Greater 2. Less 3. Greater 4. Greater	1. Greater 2. Correct 3. Correct 4. Greater	1. Bent knuckle 2. Bent I-beam 3. Bent I-beam 4. Bent knuckle

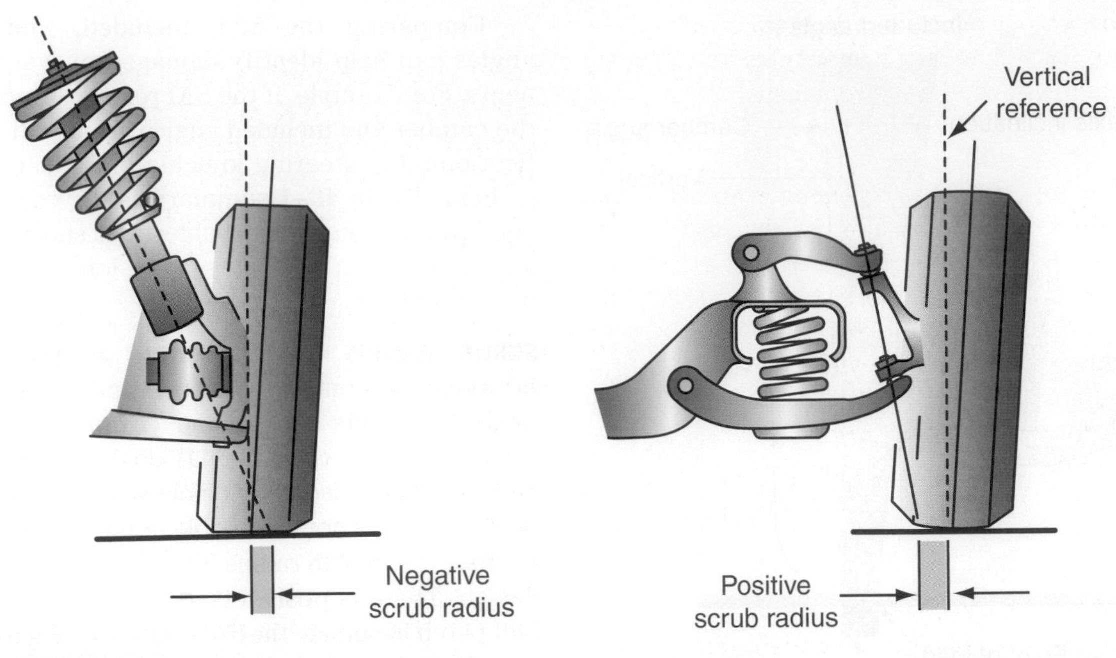

FIGURE 49–7

Scrub radius is affected by wheel size and offset.

"Ackerman's angle," or "turning angle." As a car corners, the inside tire must travel in a smaller radius circle than the outside tire. This is accomplished by designing the steering geometry to turn the inside wheel sharper than the outside wheel. Ackerman's principle creates this steering difference by having the steering arms aimed from the centre of the steering axis to the centre of the rear axle. The inner wheel turns to a greater angle because the angled steering arm is travelling through its arc while the outer wheels arm is travelling across the base of its circle. The result can be seen as toe-out in turns.

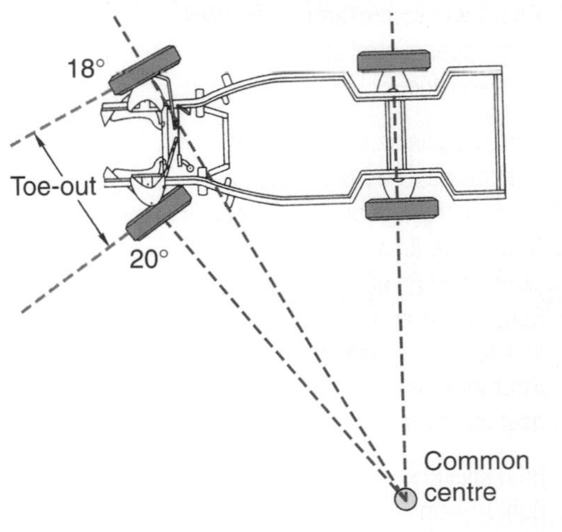

FIGURE 49–8

Turning radius is affected by toe-out in turns.

This eliminates tire scrubbing on the road surface by keeping the tires pointed in the direction they have to move. The usual angle difference from side to side between the front wheels during a turn is 2°.

When a car has a steering problem, the first diagnostic check should be a visual inspection of the entire vehicle for anything obvious: bent wheels, misalignment of the cradle, and so on. If there is nothing obviously wrong with the car, make a series of diagnostic checks without disassembling the vehicle. One of the most useful checks that can be made with a minimum of equipment is a jounce-rebound check.

This jounce-rebound check determines if there is misalignment in the rack and pinion gear. For a quick check, unlock the steering wheel and see if it moves during the jounce or rebound. For a more careful check, use a pointer and a piece of chalk. Use the chalk to make a reference mark on the tire tread and place the pointer on the same line as the chalk mark. Jounce and rebound the suspension system a few times while someone watches the chalk mark and the pointer. If the mark on the wheel moves unequally in and out on both sides of the car, chances are there is a steering arm or gear out of alignment. If the mark does not move or moves equally in and out on both sides of the car, the steering arm and gear are probably all right. Each wheel or side should be checked.

Turning radius is not an adjustable angle. If the angle is not correct, steering or suspension parts are damaged and will need to be replaced.

Thrust Line

All vehicles are built around a geometric centreline that runs through the centre of the chassis from the back to the front. The **thrust line** is the direction the rear axle would travel if unaffected by the front wheels. This condition is also called **tracking**. An ideal alignment has all four wheels parallel to the centreline, making the thrust line parallel to the centreline. However, the rear-wheel thrust line of a vehicle might not always be parallel to the actual centreline of the vehicle, so the angle of the thrust line must be checked first.

Rear-wheel-drive vehicles rarely need thrust line adjustment unless they have been in an accident or experienced severe usage. Independent rear suspension can have offset thrust angles from unequal rear toe adjustments. An offset thrust angle affects handling by pulling in the direction away from the thrust line, and it can cause tire wear similar in pattern to that of incorrect toe settings. As a general rule, minor variations between the thrust line and centreline are not noticeable and do not cause handling problems as long as the front wheels are aligned parallel to the thrust line.

Correct tracking refers to a situation with all suspension and wheels in their correct location and condition, and aligned so that the rear wheels follow directly behind the front wheels while moving in a straight line **(Figure 49–9)**. For this to occur, all wheels must be parallel to one another, and axle and spindle lines must be at 90° angles to the vehicle centreline. Simply stated, all four wheels should form a perfect rectangle.

Load Distribution

Load distribution refers to the load placed on each wheel. Every vehicle is engineered to operate at a designed curb height (also called **trim height**). At this height, each wheel must carry the correct amount of weight. Excessive loading to the front, rear, or one side of the vehicle changes the curb height, upsetting vehicle balance and steering geometry.

Incorrect alignment, sagging springs, and bent suspension parts can also change this condition, upsetting geometry and placing excessive load on only one or two wheels.

All of these elements—springs, shocks, suspension, and geometry—are engineered to work together as a balanced team to provide safe and comfortable riding and handling. Quite naturally, if one wheel is running under a different condition of weight load and steering geometry, the vehicle does not ride and handle as it is capable of doing.

PREALIGNMENT INSPECTION

Before beginning to align a vehicle, make sure you know why the vehicle needs to be aligned or why the customer has requested an alignment. Often, the symptoms will lead to identifying the reason the wheels need to be aligned. Follow the customer interview with a test drive. While driving the car, check to see that the steering wheel is straight. Feel for vibrations in the steering wheel as well as in the floor or seats. Notice any pulling or abnormal handling problems, such as hard steering, tire squeal while cornering, or mechanical pops or clunks. This helps find problems that must be corrected before proceeding with the alignment.

An extremely important part of a wheel alignment is the prealignment inspection. During this

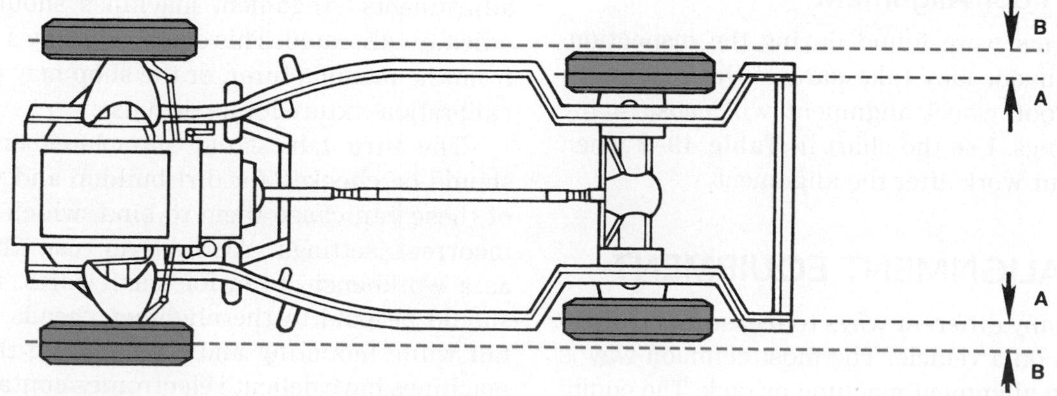

FIGURE 49–9

When a car is tracking correctly, its rear wheels are the same distance from the front wheels on both sides.

inspection, if any parts are found to be defective, they should be replaced before proceeding with the alignment. The inspection should include a careful look at the tires, wheels, suspension system, and steering system. The procedures for inspecting these are detailed in the previous three chapters. A typical alignment inspection report is shown in **Figure 49–10**.

It is important that all abnormal loads be removed before taking any measurements. Obviously, added weight will affect the vehicle's ride height and therefore the alignment angles. If the vehicle is normally used to carry heavy objects, such as toolboxes, leave them in the vehicle. In some cases, the vehicle manufacturer will specify that a certain amount of weight be added to the vehicle during the alignment. This is to better reflect the actual driving conditions the vehicle will experience.

Ride Height

Check the vehicle's ride height. Every vehicle is designed to ride at a specific curb height. Curb height specifications and the specific measuring points are given in the service information. Proper alignment is impossible if the ride height is incorrect. This is especially true for camber because as the height of a vehicle changes, so does the camber of the wheels. The procedure for checking ride height is found in Chapter 47.

Some current alignment machines have adapters that allow the unit to measure ride height. This is done without a tape measure; the machine's sensors are installed on the fenders, and the measurements are instantly transmitted to the alignment machine **(Figure 49–11)**.

Results of Poor Alignment

If no problems were found during the inspection, a good alignment may take care of the customer's concerns. Poor wheel alignment will cause many different things. Use the chart in **Table 49–2** when checking your work after the alignment.

TABLE 49–2 **EFFECTS OF INCORRECT WHEEL ALIGNMENT**

PROBLEM	EFFECT
Incorrect camber setting	Tire wear Ball joint/wheel bearing wear Pull to side of most positive/least negative camber
Too much positive caster	Hard steering Excessive road shock Wheel shimmy
Too much negative caster	Wander Weave Instability at high speeds
Unequal caster	Pull to side of most negative/least positive caster
Incorrect SAI	Instability Poor steering returnability Pull to side of lesser inclination Hard steering
Incorrect toe setting	Tire wear
Incorrect turning radius	Tire wear Squeal in turns

WHEEL ALIGNMENT EQUIPMENT

There are many different ways to measure the alignment angles on a vehicle. The most common way is the use of an alignment machine or rack. The equipment used for checking alignment angles has evolved from string and measuring tapes to computerized machines. Today, computerized machines are most commonly used **(Figure 49–12)**. A typical computerized system gives information on a standard computer monitor to guide the technician step by step through the alignment process.

Alignment Machine Care

The alignment machine is a precise piece of equipment; therefore, it needs to be taken care of. Failure to do so will lead to incorrect measurements and adjustments. Alignment machines should be periodically calibrated. This can be done by a technician from the manufacturer, or the shop may purchase a calibration fixture for its own use.

The turn tables and slip plates on the rack should be checked for dirt buildup and wear. Both of these can cause them to bind, which will cause incorrect settings. Also, never use the console as a workbench, even for small parts. Extra care should be paid to the alignment heads. They may fall while mounting and dismounting them. Older machines have delicate electronics contained in the heads; they can be easily damaged. If the heads are dropped, they should be recalibrated before they are used again.

PREALIGNMENT INSPECTION CHECKLIST

Owner ______________________ Phone ______________ Date ___________

Address ______________________ VIN ______________________

Make ________ Model ________ Year ______ Lic. number __________ Mileage __________

1. Road test results	Yes	No	Right	Left
Above 50 km/h				
Below 50 km/h				
Bump steer				
When braking				
Steering wheel movement				
Stopping from 3-5 km/h (Front)				
Vehicle steers hard				
Strg wheel returnability normal				
Strg wheel position				
Vibration	Yes	No	Frnt	Rear

2. Tire pressure Specs Frnt ____ Rear ____

Record pressure found

RF ______ LF ________ RR ________ LR ______

3. Chassis height Specs Frnt ____ Rear ____

Record height found

RF ______ LF ________ RR ________ LR ______

	Yes	No
Springs sagged		
Torsion bars adjusted		

4. Rubber bushings	OK
Upper control arm	
Lower control arm	
Sway bar/stabilizer link	
Strut rod	
Rear bushing	

5. Shock absorbers/struts	Frnt	Rear

6. Steering linkage	Frnt OK	Rear OK
Tie-rod ends		
Idler arm		
Centre link		
Sector shaft		
Pitman arm		
Gearbox/rack adjustment		
Gearbox/rack mounting		

7. Ball joints		OK
Load bearings		
Specs	Readings	
Right _____ Left _____	Right _____ Left _____	
Follower		
Upper strut bearing mount		
Rear		

8. Power steering	OK
Belt tension	
Fluid level	
Leaks/hose fittings	
Spool valve centred	

9. Tires/wheels	OK
Wheel runout	
Condition	
Equal tread depth	
Wheel bearing	
10. Brakes operating properly	

11. Alignment	Spec		Initial reading		Adjusted reading	
	R	L	R	L	R	L
Camber						
Caster						
Toe						

Bump steer	Toe change right wheel		Toe change left wheel	
	Amount	Direction	Amount	Direction
Chassis down 3"				
Chassis up 3"				

	Spec		Initial reading		Adjusted reading	
	R	L	R	L	R	L
Toe-out on turns						
SAI						
Rear camber						
Rear total toe						
Rear indiv. toe						
Wheel balance						
Radial tire pull						

FIGURE 49–10

A checklist for a prealignment inspection and wheel alignment.

FIGURE 49–11

The sensors used to measure ride height with a wheel aligner.

Hunter Engineering Company

FIGURE 49–12

A computerized wheel alignment machine.

Hunter Engineering Company

The heads on newer alignment machines **(Figure 49–13)** do not contain electronic parts. Rather, the heads serve as targets **(Figure 49–14)** for cameras mounted above the console **(Figure 49–15)**. Light from LEDs on the alignment machine is reflected back and picked up by the digital cameras. Damaged or dirty targets lose their reflectivity and may not be picked up by the cameras.

Other alignment equipment often used are turning radius gauges, caster-camber gauges, optical toe gauges, and trammel bar gauges (also known as tram gauges).

Turning Radius Gauges

Turning radius gauges measure how many degrees the front wheels are turned. They are commonly used to measure camber, caster, and toe-out on turns. Turning radius gauges (sometimes called turn tables) may be portable but are commonly found as part of an alignment rack. To use these gauges, the front wheels are centred on the gauge plates. Then the locking pins are removed to allow the plate to turn with the tires. As the tires are turned, a pointer will indicate how many degrees the tires have turned. To check toe-out on turns, turn one of the tires to 20°. Then look at the gauge on the other tire.

FIGURE 49–13

A wheel alignment machine that relies on high-imaging cameras and targets to gather measurements.

FIGURE 49–14

These targets are in the machine shown in Figure 49–13 in place of the traditional wheel heads.

Caster-Camber Gauges

A caster-camber gauge is used with the turning radius plate to check caster and camber. This gauge is often referred to as a bubble gauge. The gauge is normally attached to the wheel hub with a magnet. Make sure the vehicle is on a level surface, and then jounce the front bumper several times to stabilize the suspension. Now look at the bubble gauges to read camber and compare yours with the specifications. Do both front wheels. Apply the brakes and hold the brake pedal down with a brake pedal lock. Turn one front tire 20° out, and adjust the bubble gauge to read zero. Now turn the wheel 20° in, and take the reading from the bubble gauge. This is the caster reading for that wheel. Compare it to specifications.

FIGURE 49–15

These high-imaging cameras are positioned on a beam above the console of the machine shown in Figure 49–13.

Now measure caster on the other front tire. If any reading is outside the specifications, the angles need to be adjusted.

Optical Toe Gauges

Checking wheel toe is also done with the tires sitting on the turning radius plates. Make sure the tires are straight ahead and the plates are at 0° before removing the locking pins on the plate. Then install a steering wheel clamp to prevent the steering wheel from moving. Then install the brake pedal lock. Remove the hubcaps and grease caps from the front wheels. Jounce the front suspension several times to allow it to stabilize. Note the location of the steering gear. If the steering linkage is at the rear of the front wheels, push outwardly on the rear of the tires. If the linkage is to the front, push on the front of the tires. Mount the optical toe gauges onto the front-wheel hubs. Use the level indicators on the gauges to make sure the gauges are level. Then read wheel toe on the gauges and compare them to specifications.

Trammel Bar Gauges

A trammel bar gauge is a purely mechanical way to check toe. It is used to measure the distance between the centre of the two front tires. First, a distance measurement is taken at the rear of the tires (centreline to centreline); then the same measurement is taken at the front of the tires. The difference in the measurements is the amount of toe. Compare this reading with specifications.

Miscellaneous Tools

Figure 49–16 shows an assortment of the special tools required for wheel alignment and other steering and suspension system work.

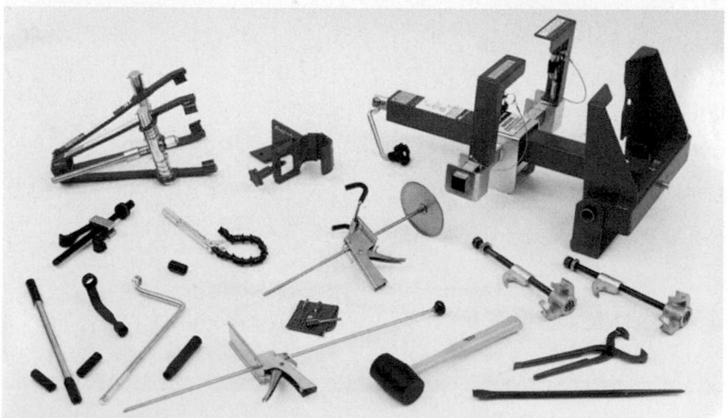

FIGURE 49–16

An assortment of steering and suspension tools.

Snap-on Tools Company

ALIGNMENT MACHINES

Most shops use an alignment machine to check all of the alignment angles. Normally, an alignment rack is part of the alignment machine's package. The rack is best described as a limited-purpose vehicle hoist (lift) equipped with turning radius plates. There are many varieties of alignment machines that have been used through the years. Some are equipped to measure alignment angles at all four wheels of the vehicle, and others measure the angles at only two wheels. Some alignment machines simply display the angle readings, whereas others display the readings plus give advice on how to correct the angles **(Figure 49–17)**.

Four-wheel alignment (or **total wheel alignment**), whether front or rear drive, solid axle or independent rear suspension, sets the alignment angles on all four wheels so that they are positioned straight ahead with the steering wheel centred. The wheels must also be parallel to one another and perpendicular to a common centreline. More than 85 percent of all new vehicles require that all four wheels are aligned.

To accomplish this, the readings for all four wheels must be determined and rear toe adjusted where possible to bring the rear axle or wheels into square with the chassis. The front toe setting can then be adjusted to compensate for any rear alignment deviation that might persist.

Four-wheel alignment also includes checking and adjusting rear-wheel camber as well as toe, and doing all the traditional checks of front camber, toe, caster, toe-out on turns, and steering axis inclination.

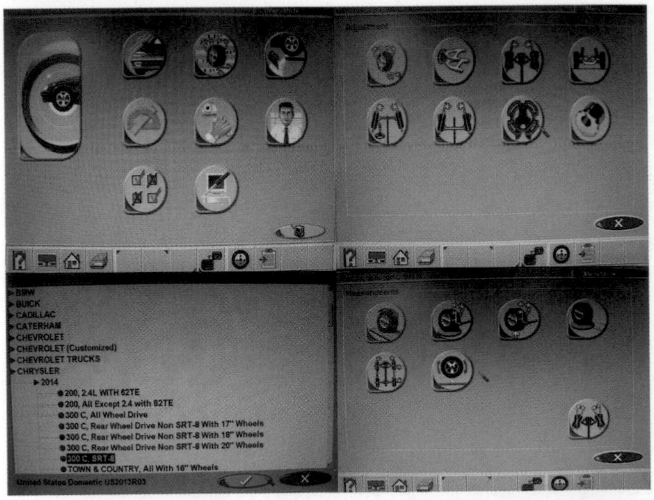

FIGURE 49–17

Examples of the screens available on the latest alignment machines.

Hunter Engineering Company

The most important thing a four-wheel alignment job tells the technician is whether the rear axle or rear wheels are square with respect to the front wheels and chassis.

With two-wheel alignment equipment that aligns the front wheels to the geometric centreline, the assumption is made that the rear wheels are square with respect to the centreline. If that is true, the alignment job produces satisfactory results. If not, steering and tracking might be a problem.

Rear-wheel alignment can be checked on a two-wheel machine by simply backing the vehicle onto the rack. The rear wheels can also be checked for square by measuring the wheelbase on both sides with a track bar to make sure it is equal. The obvious drawback to this approach is that four-wheel alignment involves several extra steps. There is also the question of aligning the wheels to the centreline or thrust line.

The best approach in terms of both accuracy and completeness for wheel alignment is to reference the front wheels to the rear wheels. With this approach, individual rear toe is measured so that the thrust line can be determined and the front wheels adjusted to the centreline. It also eliminates the need to back the vehicle onto the alignment machine because heads are provided for the rear wheels.

The actual procedure for total four-wheel alignment, as with two-wheel alignment, begins with a thorough inspection of the tires and suspension. Do not forget to check the ride height based on the normal vehicle load.

Once the alignment heads are installed and compensated for wheel runout, read front and rear camber and front and rear toe. The thrust angle created by the rear wheels can then be determined.

Most alignment machines compensate for wheel setback. Some show how much setback is present and where. Nearly all machines will also show SAI. Steering axis inclination, an angle frequently neglected on quick alignment jobs, is important to include because it can help pinpoint damaged parts. If the SAI does not match specifications, compare both sides. Problems can be caused by bent struts, control arm spindles, or steering knuckles. A mislocated strut tower or control arm anchor can also throw off the reading.

ADJUSTING WHEEL ALIGNMENT

All wheel alignment angles are interrelated. Regardless of the make of a car or the type of suspension, the same adjustment order—caster, camber, toe—should be followed, as far as the car permits such adjustments to be made. Some MacPherson suspensions do not provide for caster or camber adjustments. Additionally, adjustment methods vary from model to model and, occasionally, even in different model years.

After the vehicle is properly placed on the rack and turn tables, vehicle information is keyed into the machine, and the wheel units (heads) are installed. On some machines, the wheel units or heads must be compensated for wheel runout. When compensation is complete, alignment measurements are instantly displayed. Also displayed are the specifications for that vehicle. In addition to the normal alignment specifications, the CRT may display asymmetric tolerances, different left- and right-side specifications, and cross-specifications (the difference allowed between the left and right sides). Graphics and text on the screen show the technician where and how to make adjustments **(Figure 49–18)**. As the adjustments are made on the vehicle, the technician can observe the centre block slide toward the target. When the block aligns with the target, the adjustment is within half the specified tolerance. A typical procedure for checking the alignment of all four wheels with a computerized alignment machine is shown in Photo Sequence 42.

Specifications

All angles and measurements should be set to the manufacturer's specifications. These specifications usually list a preferred angle for camber, caster, and toe. They are also listed with a minimum and maximum specification, sometimes listed as a plus or minus **(Figure 49–19)**. When making adjustments, you should attempt to achieve the preferred angle. If this is not possible, make sure the measurements are within the minimum and maximum range. If you cannot make an adjustment within that range, something is wrong, and the suspension and frame should be carefully checked.

In most cases, the bolts or nuts need to be loosened to make the necessary adjustments. Make sure that all of these are retightened to the required torque.

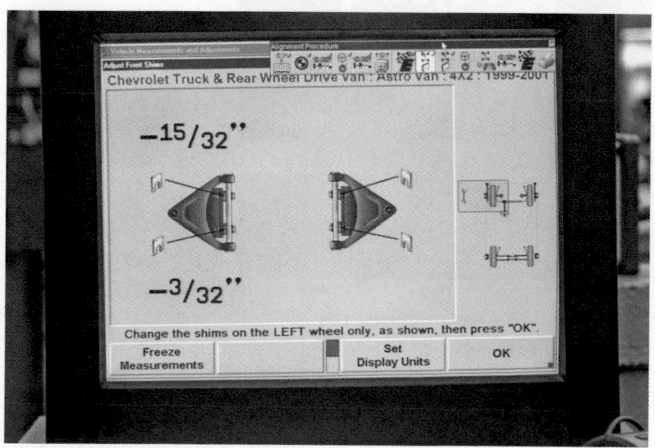

FIGURE 49–18

On many wheel alignment machines, the screen shows how and where to make adjustments.

Toe-in (total): 0 ± 0.08 in. (0 ± 2 mm)
Wheel turning angle—Inside wheel: 38.37° ± 2°
Outside wheel reference: 33.55°
Camber: −0.67° ± 0.75°
Right–left difference: 0.75° or less
Caster: 3.00° ± 0.75°
Right–left difference: 0.75° or less
Steering axis inclination: 12.25° ± 0.75°
Right–left difference: 0.75° or less
Rear toe-in (total): 0.16 ± 0.08 in. (−4 ± 2 mm)
Rear camber: −1.30° ± 0.75°
Right–left difference: 0.75° or less

FIGURE 49–19

An example of alignment specifications.

Caster/Camber Adjustment

Caster affects steering stability and steering wheel returnability. Zero (0) caster is present when the upper ball joint or top strut bearing and lower ball joint are in the same plane as viewed from the side of the vehicle. Positive caster exists when the upper ball joint or top strut bearing is toward the rear of the vehicle in relationship to the lower ball joint. When the upper ball joint or top strut bearing is toward the front of the vehicle in relationship to the lower ball joint, negative caster is present. If the caster at both wheels is not equal, the vehicle will tend to drift toward the side with the lowest positive caster.

Camber is the inward or outward tilt of the top of the wheel. Adjusting camber centres the vehicle's weight on the tire. Proper camber adjustment minimizes tire wear. Zero (0) camber is present when the wheel is at a perfectly vertical position. The tires have positive camber when the top of the tire is tilted out, or away from the engine. When the top of the tire is tilted in, there is negative camber. Incorrect camber will cause excessive stress and wear on suspension parts. Too much negative camber will cause wear on

PHOTO SEQUENCE 42

TYPICAL PROCEDURE FOR PERFORMING FOUR-WHEEL ALIGNMENT WITH A COMPUTER WHEEL ALIGNER

P42–1 Position the vehicle on the alignment rack.

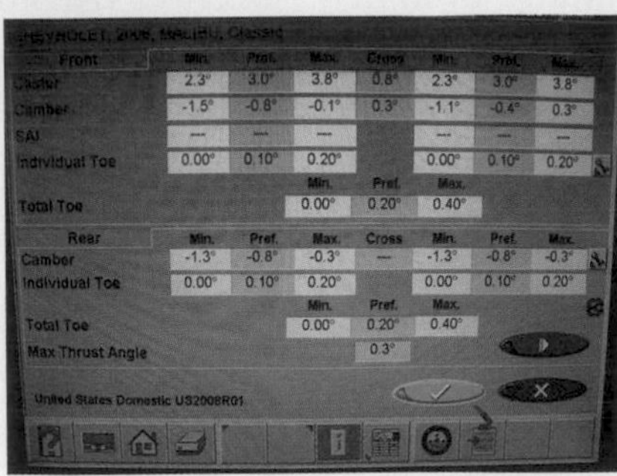

P42–2 Start the alignment machine, and retrieve the vehicle specifications.

P42–3 Install the wheel heads or reflectors on the wheels.

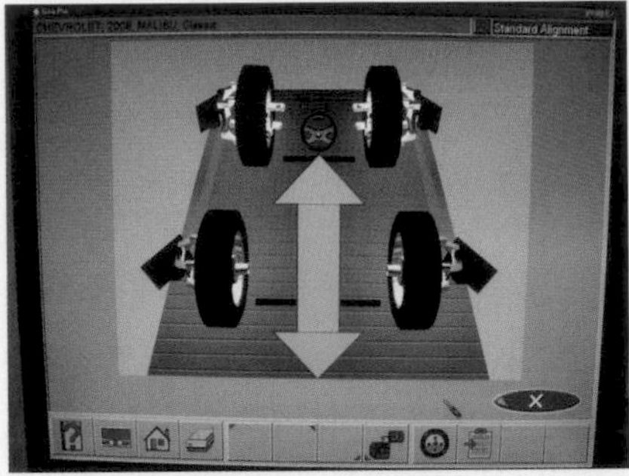

P42–4 Proceed through the alignment machine steps, beginning with performing wheel runout compensation.

P42–5 While following the on-screen directions, roll the vehicle rearward.

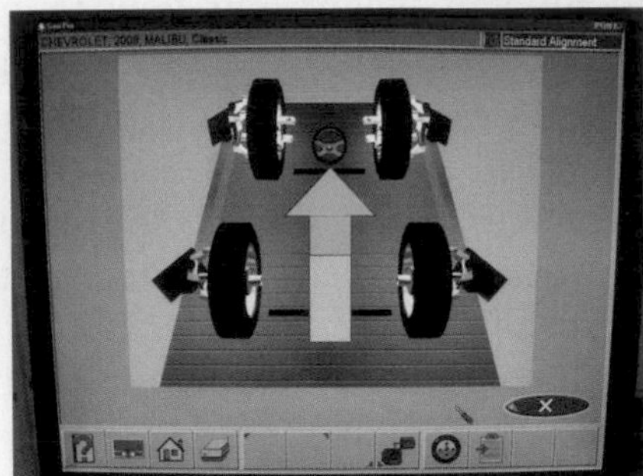

P42–6 Now roll the vehicle forward.

P42–7 Continue rolling the vehicle forward until instructed to stop.

P42–8 At this time, chock the rear wheels to prevent the vehicle from unintentionally rolling during the alignment.

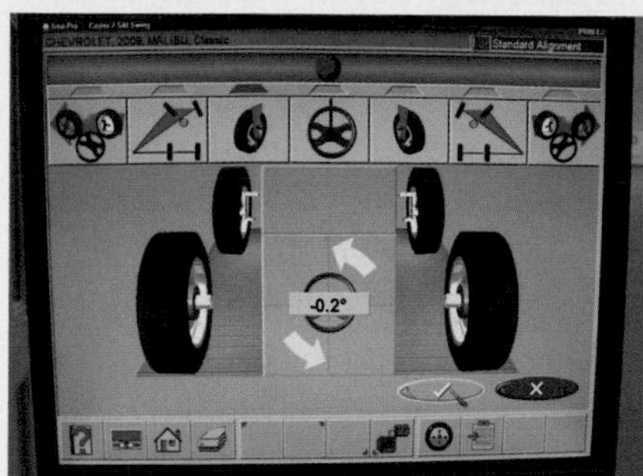

P42–9 While following the on-screen directions, perform a caster swing by steering the wheels to the left.

P42–10 Grasp the wheel firmly and move it left while watching the aligner screen for correct positioning.

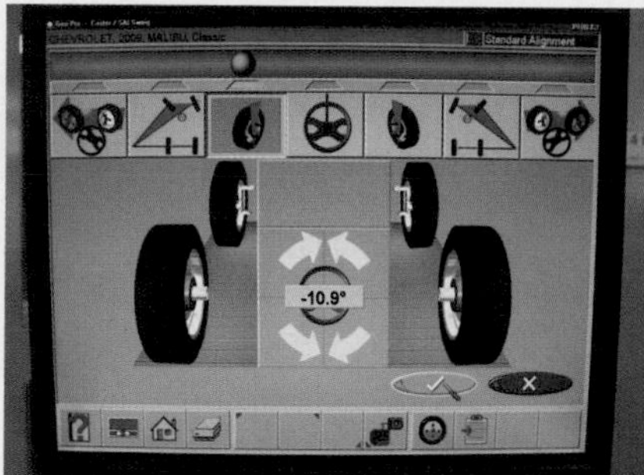

P42–11 The aligner screen will indicate the correct positioning of the wheels for measurement.

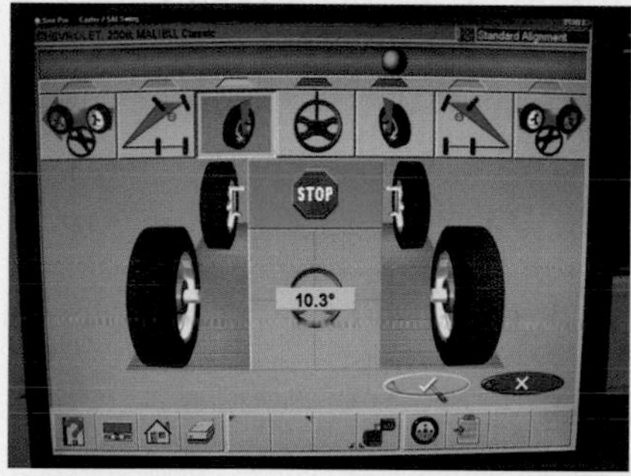

P42–12 Now turn the wheels to the right to complete the caster measurement.

P42–13 Again, turn the wheel to the right while watching the aligner screen.

P42–14 Once your caster readings are complete, return the wheels to the centre position.

P42–15 Proceed to the measurement screen to analyze the readings.

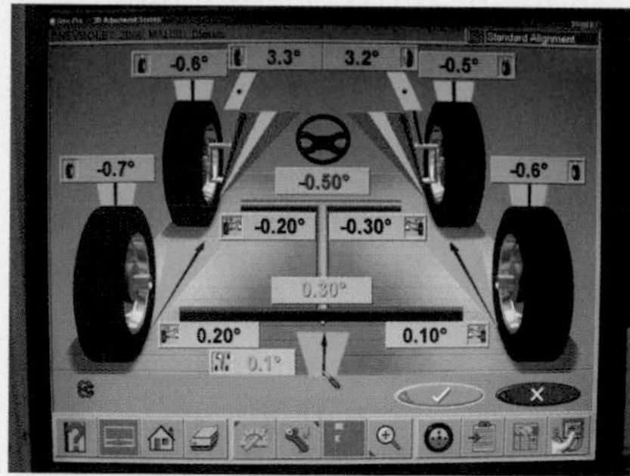

P42–16 Perform the adjustments required to bring all readings into the "green," or acceptable, range.

the inside tread of the tire, whereas too much positive camber will cause tire wear on the outside tread. If camber is not the same on both wheels, the vehicle will pull toward the side with the most positive camber.

SHOP TALK

Sometimes, it is impossible to adjust the camber to specifications. When this happens, suspect a poorly seated wheel hub and bearings, a bent frame, weak springs, or a bent control arm.

ADJUSTING FOR ROAD CROWN To compensate for road crown, most alignment specifications allow for slightly more negative caster or slightly more positive camber on the left side of the vehicle. Road crown exists at the centre of most roads to allow for water drainage from the road. This is true of two-lane roads and four- and multi-lane highways. Because most vehicles are driven on two-lane roads and the right lane of most multi-lane highways (except when passing), the road crown compensation is always to the left to prevent the vehicle from pulling to the right.

SHOP TALK

When the vehicle has been aligned to compensate for road crown, it will pull to the left when it is travelling on a flat road or on a road that is angled to the left.

Typically, specifications call for slightly more negative camber (approximately ¼°) on the right side of the vehicle. Or the caster on the left should be set with slightly more negative caster (approximately ¼°).

Adjusting Caster and Camber

Several methods are used to adjust caster and camber; always check the service information to determine how the alignment angles can be changed. Most current wheel alignment machines feature illustrations, pictures, or videos of what should happen and where.

Camber on nearly all front suspensions is adjustable, although installing special bolts or making modifications to the strut or knuckle may be required. On strut-equipped vehicles, camber can be adjusted by moving the top of the strut mount or by adjusting an eccentric bolt located where the strut attaches to the steering knuckle. Caster is adjustable on some strut-equipped vehicles. This is done by again moving the top of the strut. Typically, there is no provision for adjusting caster. Vehicles that have no simple means for adjusting camber or caster require the installation of special kits from the aftermarket to obtain the correct angles. These kits basically contain an adjustable strut mount that permits the top of the strut to move front to back and left to right.

On other vehicles, the upper or lower control arm is used to adjust camber and caster. This is done by adding or removing shims between the control arm and the frame or by rotating an eccentric shaft or eccentric washers. Two bolts attach the control arm to the frame. An equal amount of shims is placed behind or in front of both bolts to correct camber. To gain more negative camber, the lower control arm must be moved outward or the upper arm moved inward. The opposite is true for gaining positive camber.

Caster is adjustable on all vehicles with an upper and a lower control arm. This is done by rotating an eccentric bushing at one of the pivot points for the control arm or adding or subtracting shims between a control arm and the frame.

Rear-wheel camber may be adjustable. When it is, camber is adjusted by eccentric bushings at the control arm pivot point or by an eccentric on the lower mounting of the strut. Solid rear axles seldom have a provision for camber adjustment.

SHOP TALK

A feature of some alignment machines makes the job of adjusting camber and caster much easier and quicker. The feature has many names; the most common one is "jack and hold." This allows for caster and camber adjustments with the wheels off the ground. The advantage here is simply more room to work, and the vehicle's weight is off the adjusting points. Although camber and caster will change when the wheel is lifted, the machine will display the reading before the wheel was raised. This allows you to make adjustments and monitor them just as if the vehicle was sitting flat. Of course, all measurements should be verified after the wheels are back on the ground.

SHIMS Many cars use shims for adjusting caster and camber **(Figure 49–20)**. The shims can be located between the control arm pivot shaft and the inside of the frame. Both caster and camber can be adjusted in one operation requiring the loosening of the shim bolts just once. Caster is changed by adding or subtracting shims from one end of the pivot shaft only. Then, camber is adjusted by adding or subtracting an equal amount of shims from the front and rear bolts. This procedure allows camber to change without affecting the caster setting.

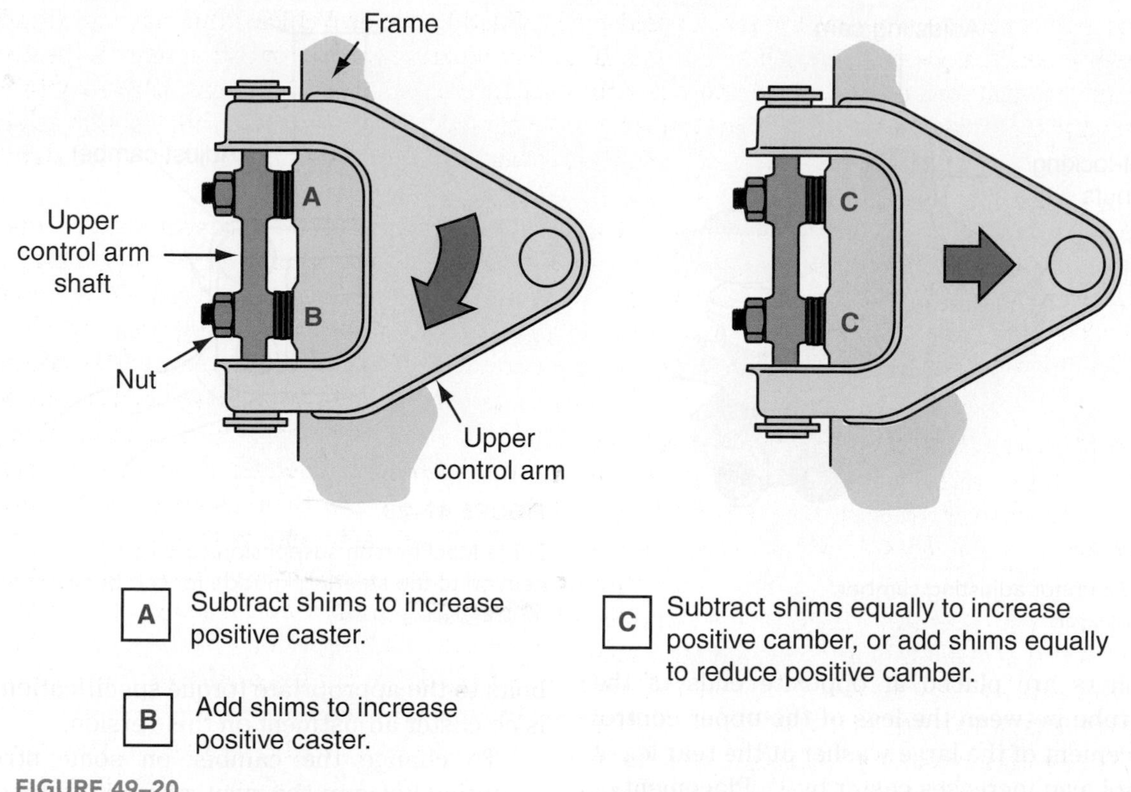
Frame
Upper control arm shaft
Nut
Upper control arm
A
B
C
C
A Subtract shims to increase positive caster.
B Add shims to increase positive caster.
C Subtract shims equally to increase positive camber, or add shims equally to reduce positive camber.

FIGURE 49–20

Adding and subtracting shims between the control arm and the frame will change caster and camber.

Some cars use shims located between the control arm pivot shaft and the outside of the frame. The adjustment procedure is the same as just described. Always look at the shim arrangements to determine the desired direction of change before loosening the bolts.

ECCENTRICS AND SHIMS Eccentrics and shims are used on some vehicles to adjust caster and camber. In some designs, an eccentric bolt and cam on the upper control arm adjust both caster and camber. To adjust, the nuts on the upper control arm are loosened first. Then, one eccentric bolt at a time is turned to set caster. Both bolts are turned equally to set camber.

The **eccentric bolt and cam** assembly **(Figure 49–21)** can be located on the inner pivot of the lower or upper control arm. Unlike other designs, camber is adjusted first. Some car models have a camber eccentric between the steering knuckle and the upper control arm. The camber eccentric is rotated to set camber **(Figure 49–22)**. Caster is set with an adjustable strut rod.

SLOTTED FRAME The slotted frame adjustment has slotted holes under the control arm inner shaft that allow the shaft to be repositioned to the correct caster and camber settings. Caster and camber adjusting tools help in making adjustments. One end of the shaft is moved for caster adjustment. Both ends of the shaft are moved for camber adjustment. Turning a nut on one end of the rod changes its length and adjusts caster. Camber is set by an eccentric at the inner end of the lower control arm, or by a camber eccentric in the steering knuckle of the upper support arm, as described earlier.

ROTATING BALL JOINT AND WASHERS In this design, camber is increased by disconnecting the upper ball joint, rotating it 180°, and reconnecting. This positions the flat of the ball joint flange inboard and increases camber approximately 1°. Caster angle is changed with a kit containing two washers, one 3 mm (0.12 in.) thick and one 9 mm (0.36 in.) thick.

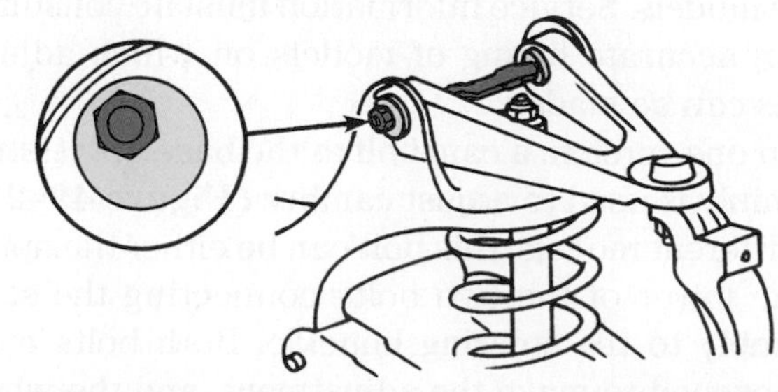

FIGURE 49–21

Eccentric bolt and cam, shown on an upper control arm.

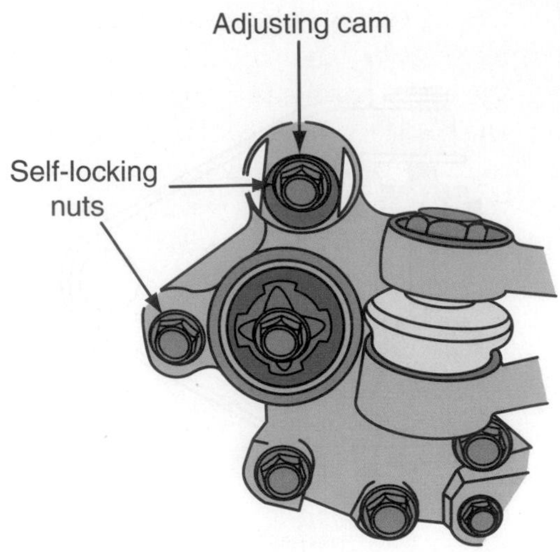

FIGURE 49–22

Graduated cam for adjusting camber.

©Cengage Learning 2015

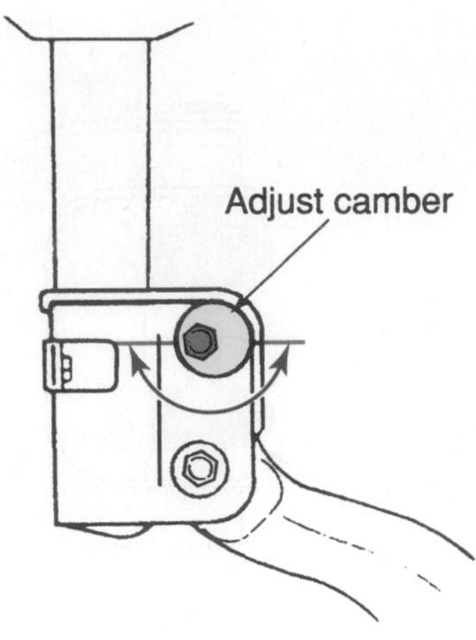

FIGURE 49–23

Some MacPherson suspensions use cam bolts at the connection to the steering knuckle for camber adjustments.

Hunter Engineering Company

The washers are placed at opposite ends of the locating tube between the legs of the upper control arm. Placement of the large washer at the rear leg of the control arm increases caster by 1°. Placement of the large washer at the front leg of the control arm decreases caster by 1°.

BALL JOINT STUD BUSHINGS Some suspension systems have an eccentric bushing at the top of the steering knuckle. This bushing can be used to adjust camber and caster. The bore for the ball joint stud through the bushing is off centre. Rotating the bushing moves the wheel's geometry. If the correct alignment cannot be attained by rotating the bushing, replacement bushings are available. You will need to order a bushing that changes the caster and/or camber specific amounts.

MacPherson Suspension Adjustments

Caster/camber adjustments are made only on certain models with MacPherson suspensions. There are two general OEM procedures for doing this, although aftermarket kit adapters are available for some models. Service information must be consulted for an accurate listing of models on which adjustments can be made.

In one version, a cam bolt at the base of the strut assembly is used to adjust camber **(Figure 49–23)**. On different models, this bolt can be either the upper or the lower of the two bolts connecting the strut assembly to the steering knuckle. Both bolts must be loosened to make the adjustment, and the wheel assembly must be centred. Turn the cam bolt to reach the correct alignment, and then retighten the bolts to the appropriate torque specifications. There is no caster adjustment on this version.

To change the camber on some struts, the mounting holes in the strut where it attaches to the steering knuckle are enlarged. Extending the holes allows the strut to move relative to the knuckle, changing the camber.

In the other form of kit adapter, both caster and camber are adjustable at the strut upper mount. Slots in the mounting plate permit the strut assembly to be shifted to reach the alignment specifications. To adjust caster, loosen the three locknuts on the mounting studs and relocate the plate. Do not remove the nuts **(Figure 49–24)**. Loosen the centre locknut, and slide it toward or away from the engine as needed to adjust camber correctly.

Although caster cannot be adjusted on many MacPherson strut front suspensions, camber can be adjusted. The camber is such that, although it is locked in place with a pop rivet, it can be adjusted by removing the rivet from the camber plate and loosening the three nuts that hold the plate to the body apron **(Figure 49–25)**. Camber is changed by moving the top of the shock strut to the position in which the desired camber setting is achieved. The nuts are then tightened to specifications. (The pop rivet is OEM only, and it is not necessary to install a new pop rivet.)

Rear-Wheel Camber Adjustments

Like front camber, rear camber affects both tire wear and handling. The ideal situation for many cars is to have zero running toe and camber on all four

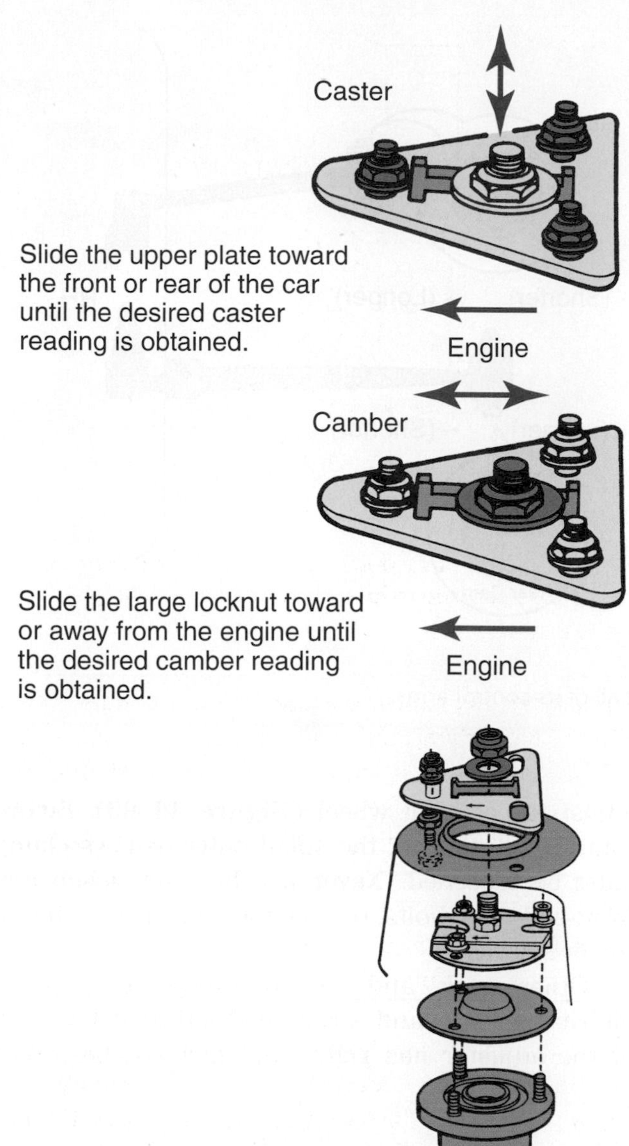
Caster
Slide the upper plate toward the front or rear of the car until the desired caster reading is obtained.
Engine
Camber
Slide the large locknut toward or away from the engine until the desired camber reading is obtained.
Engine

FIGURE 49–24

Caster and camber adjustment of locknuts.

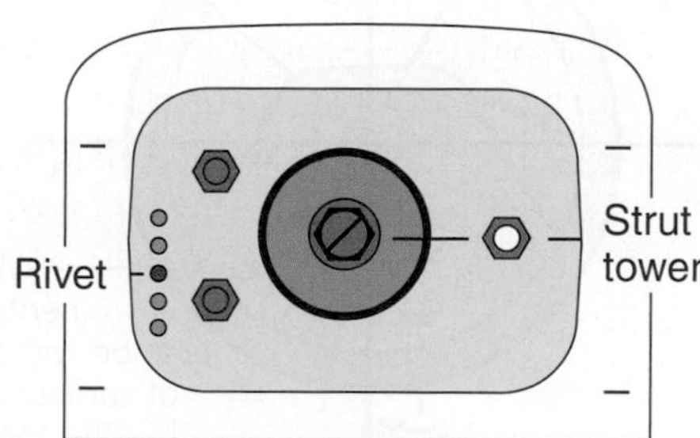
Rivet
Strut
tower

FIGURE 49–25

An upper strut mount with a camber plate. Note the location of the rivet and the attaching bolts.

Hunter Engineering Company

wheels when travelling at highway speed. It is generally accepted that the effect of airflow on the vehicle, and average passenger and cargo loads including fuel, will change the vehicle height enough to change camber from the static adjustment reading to zero when at highway speed. It is also accepted that the push or pull on the tires will make the tires actually track at zero toe at highway speed. This will keep the tread in full contact with the road for optimum traction and handling.

Camber is not a static angle. It changes as the suspension moves up and down. Camber also changes as the vehicle is loaded, and the suspension sags under the weight.

To compensate for loading, some vehicles with independent rear suspension often call for a slight amount of positive camber. Other vehicles, particularly sports cars, tend to use negative camber, which increases the handling abilities of the car. A collapsed or mislocated strut tower, bent strut, collapsed upper control arm bushing, bent upper control arm, sagging spring, or an overloaded suspension can cause the rear wheels to have negative camber. A bent spindle or strut or bowed lower control arm can cause too much positive camber. Even rigid rear axle housings in rear-wheel-drive vehicles can become bowed by excessive torque, severe overloading, or road damage.

CAUTION!

Before lifting the rear of an FWD vehicle, make sure you are using the correct lift point. Never jack on rear control arm rods or rear tie rods.The weight of the vehicle may cause the components to bend and result in misalignment of the rear wheels. Always lift the vehicle at the recommended lifting points.

Besides wearing the tires unevenly across the tread, uneven side-to-side camber (as when one wheel leans in and the other does not) creates a steering pull just like it does when the camber readings on the front wheels do not match. It is like leaning on a bicycle. A vehicle always pulls toward a wheel with the most positive camber. If the mismatch is at the rear wheels, the rear axle pulls toward the side with the greatest amount of positive camber. If the rear axle pulls to the right, the front of the car drifts to the left, and the result is a steering pull even though the front wheels may be perfectly aligned.

The methods used to adjust rear suspensions vary. On some semi-independent suspensions, camber and toe are adjusted by inserting different sizes of shims between the rear spindle and the

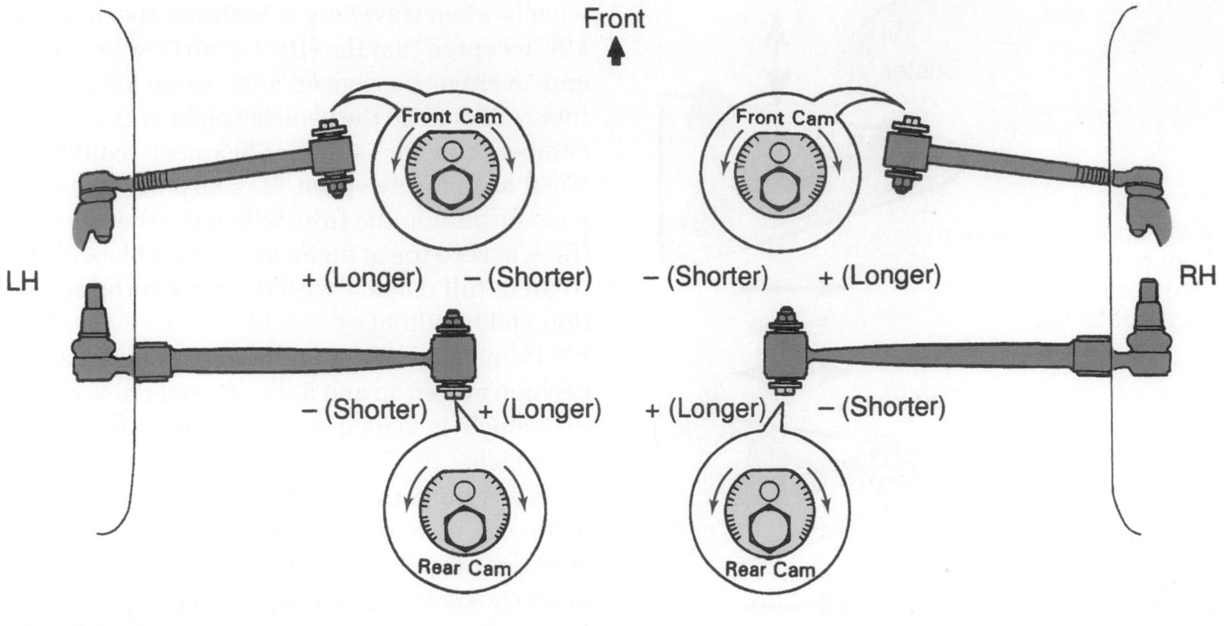

FIGURE 49–26

An independent rear suspension may have eccentric cams at all of its control arms.

Reprinted with permission.

spindle mounting. The shim thickness is changed between the top or bottom of the spindle to adjust camber. Many shims are now available that are round but have different thicknesses through their diameters. Most alignment computers can display exactly how to install a rear shim to set camber and toe at the same time.

On others, a camber adjustment can be made by installing a wedge spacer between the top of the knuckle and the strut. Still others have eccentric bolts and cams at the mounting points for the control and/or trailing arms **(Figure 49–26)**.

Remember that rear camber is adjusted first. Once the rear camber is set, rear toe is adjusted. Once the rear wheels are properly aligned, the thrust line should be parallel with the vehicle centreline and the front-wheel alignment checked and adjusted.

Toe Adjustment

Toe is the last alignment angle to be set. The same procedure is followed on all vehicles, except those with bonded ball stud sockets. Correct toe will minimize tire wear and rolling friction.

To adjust toe, start by being sure the steering wheel is centred **(Figure 49–27)** when the front wheels point straight ahead. Then loosen the retaining bolts on the tie-rod adjusting sleeves. Turn the sleeves to move the tie-rod ends **(Figure 49–28)**.

On many rack and pinion systems, the tie-rod locknut must be loosened and the tie rod rotated to adjust toe at each wheel **(Figure 49–29)**. Before rotating the tie rod, the small outer bellows clamp must be loosened. Never use heat to loosen any front-end nuts, bolts, or components such as the tie rod or sleeves.

Other rack and pinion tie-rod ends have internal threads and a threaded adjuster. One end of the adjuster has right-hand threads; the other

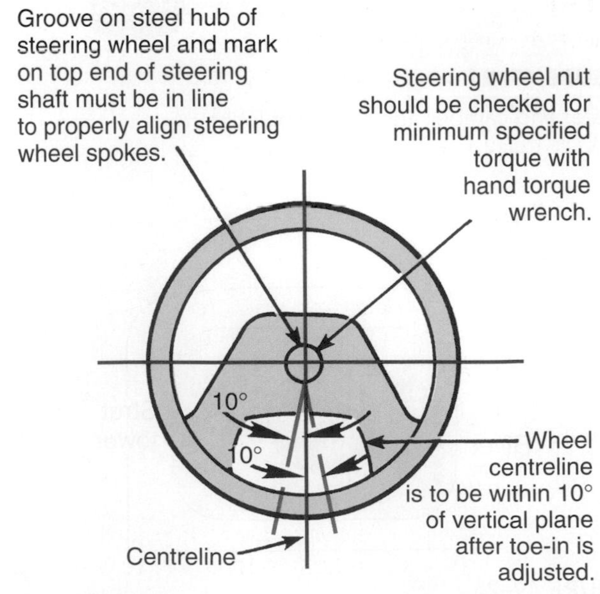

FIGURE 49–27

A typical acceptable steering wheel position—measured from a normal spoke angle.

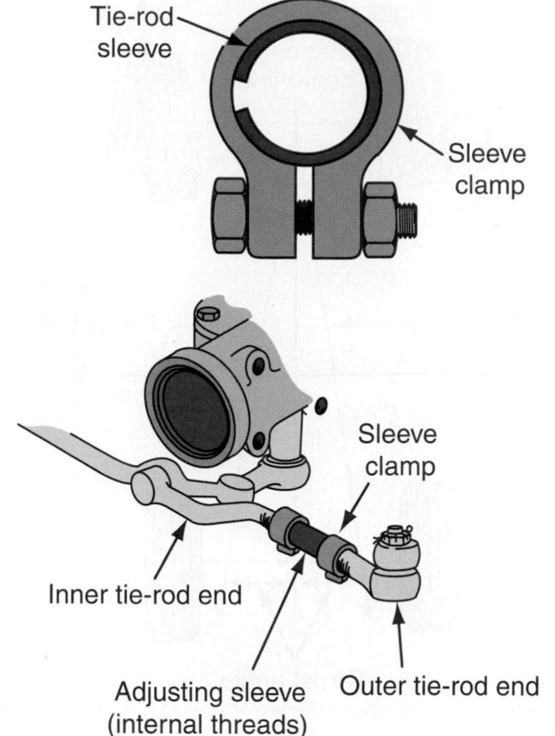

FIGURE 49–28

To adjust toe, the sleeve clamps on the tie-rod assembly are loosened and the sleeve is rotated.

has left-hand threads. As the adjuster is turned, it changes the overall length of the tie rod, thereby changing toe.

An ideal toe condition is both wheels exactly straight ahead while driving, which would minimize tire wear. This, however, is not possible because of the many factors affecting alignment. As a result of these numerous conditions dealing with both tire wear and handling, all suspensions are designed with a slight toe-in or toe-out.

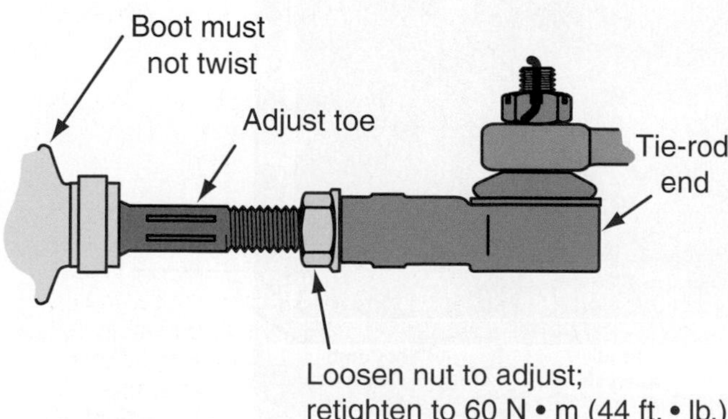

FIGURE 49–29

Rotating the tie rod to adjust the toe on a rack and pinion steering gear.

Any misalignment of the steering linkage pivot point or control arm pivot point (such as the centre link or rack and pinion out of place) causes the condition known as **toe-change**. Toe-change involves turning the wheels from their straight-ahead position as the suspension moves up and down.

The change might be only one wheel, both wheels in the same direction, or both wheels in the opposite direction. Regardless of the condition, any change of one or more wheels is a toe-change condition. The results are tire wear and a hard-to-handle vehicle. The poor handling effects can get to the point that the vehicle is dangerous to drive.

Toe change is not a specification; it is a condition in which the toe setting constantly varies. It must be determined by equipment or a method that measures individual wheel toe at all suspension heights. There must be a change in suspension heights for any changes to occur.

Lightweight FWD vehicles can be affected greatly by toe-change. With these vehicles, the front wheels are no longer being pushed. They actually pull the vehicle forward, and as a result, if the wheels are not maintaining a straight-ahead position, they affect directional control. Adverse road conditions, such as wet or icy conditions, can also increase the handling effects created by toe-change in an FWD car.

REAR TOE Rear toe, like front toe, is a critical tire wear angle. If toed-in or toed-out, the rear tires scuff just like the front ones. Either condition can also contribute to steering instability as well as reduced braking effectiveness. Keep this in mind with antilock brake systems.

Like camber, rear toe is not a static alignment angle. It changes as the suspension goes through jounce and rebound. It also changes in response to rolling resistance and the application of engine torque. With FWD vehicles, the front wheels tend to toe in under power while the rear wheels toe out in response to rolling resistance and suspension compliance. With RWD vehicles, the opposite happens: The front wheels toe out while the rear wheels on an independent suspension try to toe in as they push the vehicle ahead.

If rear toe is not within specifications, it affects tire wear and steering stability just as much as front toe. A total toe reading that is within specifications does not necessarily mean the wheels are properly aligned—especially when it comes to rear toe measurements. If one rear wheel is toed in while the other is toed-out by an equal amount, total toe would be within specifications. However, the vehicle would have a steering pull because the rear wheels would not be parallel to centre. The trust line will also show as off in this case.

Remember that the ideal situation is to have all four wheels at zero running toe when the car is travelling down the road. This is especially true with antilock brakes, where improper toe can affect brakes; such a condition can affect brake balance when braking on slick or wet surfaces, causing the antilock brakes to cycle on and off to prevent a skid. Without antilock brakes, this condition may upset traction enough to cause an uncontrollable skid.

Rear toe may be adjusted by adjusting a tie rod, similar to those found in the front steering linkage; by eccentric bolts; or by moving slotted control arms.

THRUST LINE If both rear wheels are square to one another and the rest of the vehicle, the thrust line is perpendicular to the rear axle and coincides with the vehicle's centreline. But if one or both rear wheels are toed in or toed out, or one is set back slightly with respect to the other, the thrust line is thrown off centre. This creates a thrust angle that causes a steering pull in the opposite direction. For example, a thrust line that is off centre to the right makes the car pull left **(Figure 49–30)**.

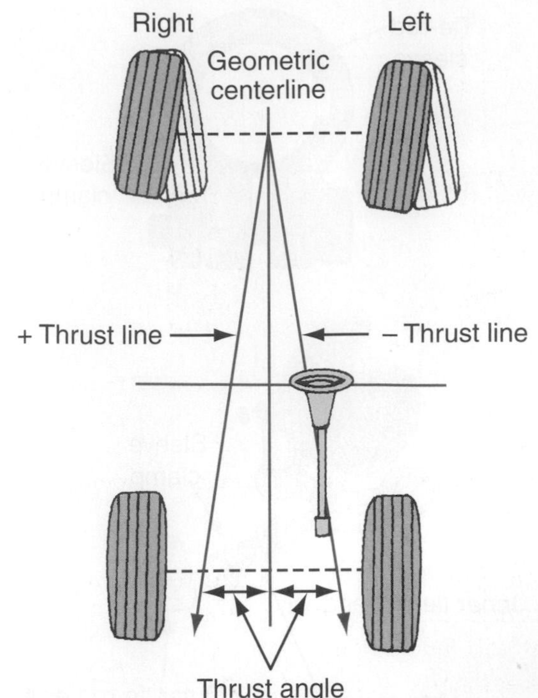

FIGURE 49–30

The thrust angle is the angle between the geometric centreline of the vehicle and the thrust line.

The only way to eliminate the problem is to eliminate the thrust angle. The thrust line can be recentred by realigning rear toe. On most FWD applications, that can be easily done by using the factory-provided toe adjustments, by placing toe/camber shims between the rear spindles and axle, or by using eccentric bushing kits. With RWD vehicles that have a solid rear axle, changing rear toe is not as easy **(Figure 49–31)**. Sometimes, the floorpan or frame rails are misaligned from the factory or from collision damage. Short of pulling the chassis on a collision bench to restore the correct control arm or spring-mount geometry, the only other options are to

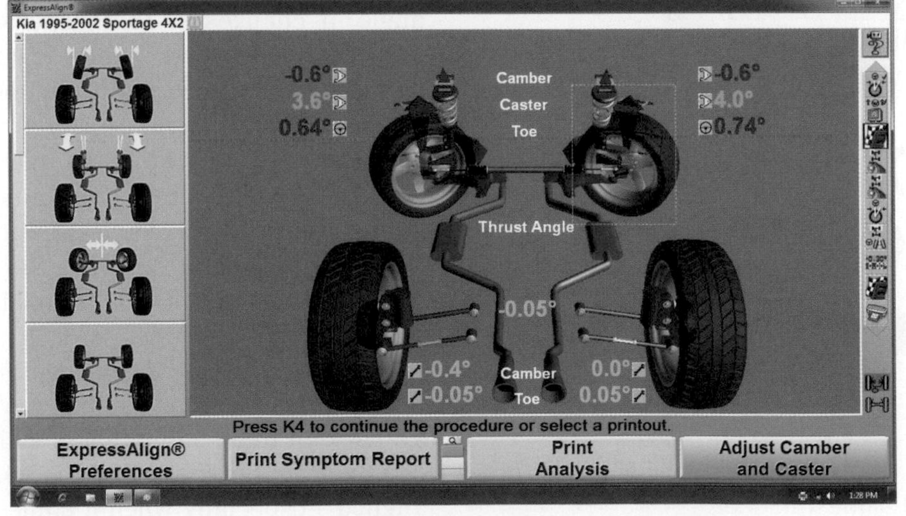

FIGURE 49–31

Thrust angle and a vehicle with nonadjustable rear toe.

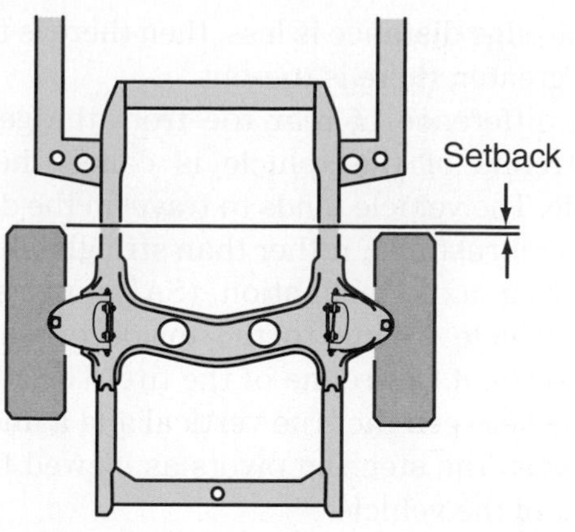

FIGURE 49–32

Setback is a condition when one wheel on an axle is set behind the other.

try some type of offset trailing arm bushing with the coil springs or to reposition the spring shackles or U-bolts with leaf springs.

If rear toe cannot be easily changed, the next best alternative is to align the front wheels to the rear axle thrust line rather than the vehicle centreline. This is called a thrust alignment. Doing this puts the steering wheel back on centre and eliminates the steering pull—but it does not eliminate dog tracking. Dog tracking occurs when the rear wheels do not follow directly behind the front wheels when the vehicle is moving forward.

Setback

Setback is a condition when one wheel on an axle is set behind the other **(Figure 49–32)**. This means the distance between the centres of the tires on one side of the vehicle will be different than on the other side. Like the thrust angle, setback will cause an off-centred steering wheel. Most alignment machines measure setback. Excessive setback is evident by a reading of more than 6.35 mm (¼ in.). Setback is typically caused by a bent suspension part, problems with the upper strut mount, or a misaligned lower control arm. It can also be caused by a misaligned front cradle. If the setback is slight, the difference will be compensated during a good four-wheel alignment.

FOUR-WHEEL-DRIVE VEHICLE ALIGNMENT

With FWD and full-time four-wheel-drive (4WD) vehicles, the front wheels are also driving wheels. As the front wheels pull the vehicle, the wheels tend to toe in when torque is applied. To offset this tendency, the front wheels usually need less static toe-in to produce zero running toe. In fact, the preferred toe alignment specifications in this instance can be zero to slightly toed out (1.5 mm [1⁄16 in.] toe-out).

It is important to note that when the front wheels of a part-time 4WD system are freewheeling, they behave the same as the front wheels in an RWD vehicle. That is, they roll rather than pull. The wheels tend to toe out, so the static toe setting would have to toe in to achieve zero running toe when driving in the two-wheel mode.

The tires suffer in proportion to toe misalignment. For a tire that is only 3 mm (⅛ in.) off (¼°), the tire is scrubbed sideways 3.6 m (12 ft.) for every 1.6 km (1 mi.) travelled. That may not sound like much, but 3.6 m (12 ft.) of sideways scrub every 1.6 km (1 mi.) can cut a tire's life in half.

If rapid tire wear seems to be the problem, look for the telltale feathered wear pattern **(Figure 49–33)**. If the wheels are running toe-in, the feathered wear pattern leaves sharp edges on the inside edges of the tread. If the wheels are running toe-out, the sharp edges are toward the outside of the tread. It is usually easier to feel the feathered wear pattern than to see it. To tell which way the wear pattern runs, rub your fingers sideways across the tread.

On most 4WD vehicles, caster is not adjustable. Aftermarket companies do provide caster adjustment kits for some pickups. These kits may contain shims or eccentric cam and bolt. For some pickups, the aftermarket cam kit will also provide for camber adjustments.

On other 4WD vehicles, camber is adjusted by installing adjustment shims between the spindle and the steering knuckle or by installing and/or adjusting

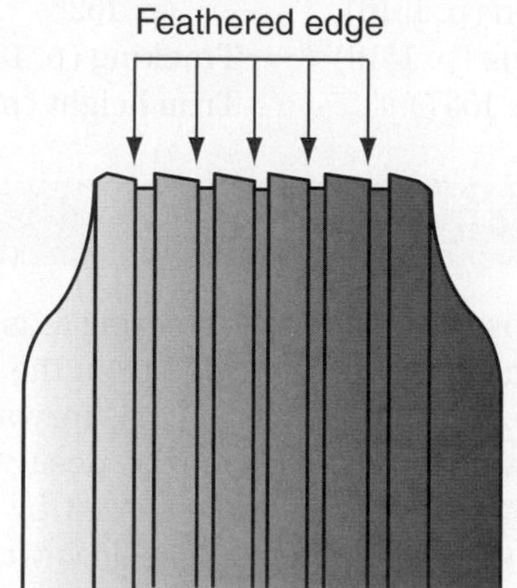

FIGURE 49–33

Incorrect toe can cause a feathered wear pattern.

an eccentric bushing at the upper ball joint. Most aftermarket parts manufacturers have camber adjustment shims available in various thicknesses and diameters. Never stack the shims. Only one shim per side should be used.

Steering Angle Sensor Calibration

Cars and light trucks that have electric power-steering assist and/or vehicle stability control use sensors in the steering column or rack to measure steering shaft angle and torque. When performing an alignment or after work is performed on the steering system, these sensors may require calibration. This includes when any steering component has been removed and reinstalled or replaced.

Sensor recalibration procedures vary from relatively simple to requiring a factory scan tool to access the system. Refer to the manufacturer's service information for procedures specific to the vehicle being aligned. Newer alignment machines can alert you to if a vehicle needs calibration. The alignment machine may also have the ability to recalibrate the sensor by connecting to the DLC. Some vehicles do not require connecting to the on-board computer system.

KEY TERMS

Camber (p. 1516)
Caster (p. 1516)
Eccentric bolt and cam (p. 1531)
Four-wheel alignment (p. 1526)
Included angle (p. 1518)
Road crown (p. 1516)
Scrub radius (p. 1519)
Setback (p. 1537)
Steering axis inclination (SAI) (p. 1518)
Thrust angle (p. 1517)
Thrust line (p. 1521)
Toe (p. 1516)
Toe-change (p. 1535)
Total wheel alignment (p. 1526)
Tracking (p. 1521)
Trim height (p. 1521)

SUMMARY

- Caster is the angle of the steering axis of a wheel from the vertical, as viewed from the side of the vehicle. Tilting the steering axis forward is negative caster. Tilting backward is positive caster.
- Camber is the angle represented by the tilt of either the front or rear wheels inward or outward from the vertical, as viewed from the front of the car.
- Toe is the distance comparison between the leading edge and trailing edge of the front tires. If the edge distance is less, then there is toe-in. If it is greater, there is toe-out.
- The difference of rear toe from the geometric centreline of the vehicle is called the thrust angle. The vehicle tends to travel in the direction of the thrust line, rather than straight ahead.
- Steering axis inclination (SAI) angles locate the vehicle weight to the inside or outside of the vertical centreline of the tire. The SAI is the angle between the true vertical and a line drawn between the steering pivots as viewed from the front of the vehicle.
- Turning radius or cornering angle is the amount of toe-out present on turns.
- In correct tracking, all suspensions and wheels are in their correct locations and conditions and are aligned so that the rear wheels follow directly behind the front wheels while moving in a straight line.
- It is important to remember that approximately 85 percent of today's vehicles not only undergo front-end alignment but require rear-wheel alignment as well.
- The primary objective of four-wheel or total-wheel alignment, whether front or rear drive, solid axle or independent rear suspension, is to align all four wheels so that the vehicle drives and tracks straight with the steering wheel centred. To accomplish this, the wheels must be parallel to one another and perpendicular to a common centreline.

REVIEW QUESTIONS

1. Define positive camber.
2. What tire wear pattern will result from excessive negative camber?
3. What can be caused by excessive negative caster?
4. Describe the difference between toe-in and toe-out.
5. Describe thrust angle.
6. What does Ackerman's principle refer to?
   a. allows the inside wheel to turn in a tighter radius than the outside
   b. allows the outer wheel to turn in a tighter radius than the inside
   c. allows the vehicle to make tighter turns than conventional steering
   d. allows both wheels to turn exactly the same amount

7. What direction must you move the steering axis to make the caster more positive?
   a. forward **b.** rearward
   c. inward **d.** outward
8. In what direction must you move the steering axis to make the camber more positive?
   a. forward **b.** rearward
   c. inward **d.** outward
9. Which of the following is the correct definition of SAI?
   a. forward tilt of the top of the steering knuckle
   b. inward tilt of the spindle steering arm
   c. inward tilt of the top of the ball joint, kingpin, or MacPherson strut
   d. outward tilt of the top of the ball joint, kingpin, or MacPherson strut
10. Which of the following is the correct adjustment sequence for a four-wheel alignment?
   a. Adjust the front wheels first, then the rear wheels.
   b. Adjust the rear wheels first, then the front wheels.
   c. Adjust both right wheels first, then the left wheels.
   d. Adjust both left wheels first, then the right wheels.
11. Which of the following angles when out of adjustment will produce the most tire wear?
   a. toe **b.** camber
   c. caster **d.** thrust line
12. What alignment angle can be checked with a trammel bar gauge?
   a. steering axis inclination
   b. caster
   c. camber
   d. toe
13. Once the alignment machine heads are mounted on the wheels, what must be performed next?
   a. the caster sweep
   b. adjust toe to specifications
   c. set runout compensation
   d. note the camber angle
14. What is the first step in adjusting front-wheel toe?
   a. Adjust the right tie-rod collar to correctly position the wheel.
   b. Adjust the left tie-rod collar to correctly position the wheel.
   c. Adjust both tie-rod collars equally to correctly position both wheels.
   d. Centre the steering wheel.
15. What is the first alignment angle to be adjusted when performing a front-wheel alignment?
   a. camber **b.** caster
   c. toe **d.** steering axis inclination
16. What will unequal caster angles cause?
   a. a pull to the side with the most positive caster
   b. a wandering condition
   c. a wheel shimmy condition
   d. a pull to the side with the least positive caster
17. Which alignment angle will produce feather-edged tire wear?
   a. incorrect caster
   b. incorrect camber
   c. incorrect toe
   d. incorrect steering axis inclination
18. Which of the following angles is not adjustable during a wheel alignment?
   a. toe-in
   b. toe-out
   c. camber
   d. steering axis inclination
19. When adjusting a vehicle to counter the effect of road crown, what adjustment would you make?
   a. Adjust the left front wheel with slightly less positive caster.
   b. Adjust the right front wheel with slightly less positive caster.
   c. Adjust the right front wheel with slightly more positive camber.
   d. Adjust both front wheels with slight toe-out.
20. What two alignment angles are added together to form the included angle?
   a. caster and steering axis inclination
   b. camber and steering axis inclination
   c. steering axis inclination and the turning radius
   d. the turning radius and the thrust line

CHAPTER 50
Brake Systems

LEARNING OUTCOMES

- Explain the basic principles of braking, including kinetic and static friction, friction materials, application pressure, and heat dissipation.
- Describe the components of a hydraulic brake system and their operation, including brake lines and hoses, master cylinders, system control valves, and safety switches.
- Perform both manual and pressure bleeding of the hydraulic system.
- Briefly describe the operation of drum and disc brakes.
- Inspect and service hydraulic system components.
- Describe the operation and components of both vacuum-assist and hydraulic-assist braking units.

WE SUPPORT THE CANADIAN INTERPROVINCIAL STANDARDS PROGRAM

It is commonly believed that the purpose of a brake system is to slow or halt the motion of a vehicle. However, that is really not true. The friction of the tires against the road is what slows down and stops a vehicle. The brake system slows or stops the rotation of the wheels. This is a minor point but one that extends the responsibility for braking to the tires as well as the brake system.

The brake system converts the momentum of the vehicle into heat by slowing and stopping the vehicle's wheels. This is done by causing friction at the wheels. The application of the friction units is controlled by a hydraulic system. This chapter looks at the basics of all brake systems and gives a detailed look at the hydraulic systems required to stop a vehicle.

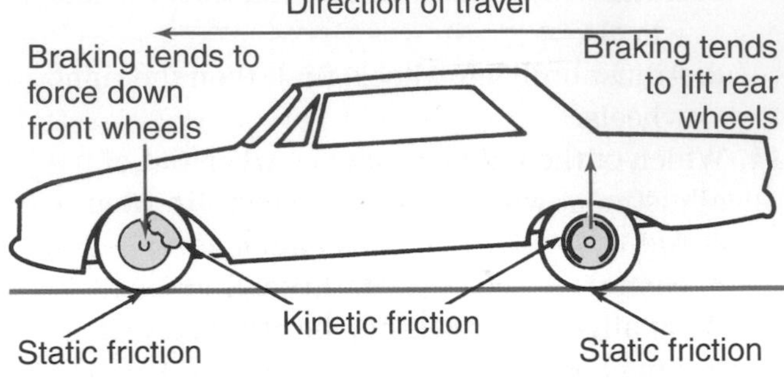

FIGURE 50–1

Braking action creates kinetic friction in the brakes and static friction between the tire and road to slow the vehicle. When the brakes are applied, the vehicle's weight is transferred to the front wheels and is unloaded on the rear wheels.

FRICTION

There are two basic types of friction that explain how brake systems work: **kinetic**, or moving, and **static**, or stationary, **friction (Figure 50–1)**. The amount of friction, or resistance to movement, depends on the type of materials in contact, the smoothness of their rubbing surfaces, and the pressure holding them together (often gravity or weight). Friction always converts moving, or kinetic, energy into heat. The greater the friction between two moving surfaces, the greater the amount of heat produced.

The faster a vehicle travels and/or the more the vehicle weighs will determine the amount of the kinetic energy (KE) developed. Kinetic energy will determine the amount of braking power that must be available to stop the vehicle. This is why it is harder to stop a car that is travelling fast compared to a slow-moving car. All of this energy must be converted into heat by the brakes. If the weight of a vehicle is doubled, the kinetic energy and the amount of heat generated by the brakes will also double. If the speed of the vehicle is doubled, the kinetic energy will increase by four times and so must the brake effectiveness. The formula for calculating the kinetic energy developed by a vehicle is as follows.

$$KE = \frac{1}{2}M \times m/s^2$$

M = the mass or weight of the vehicle in kilograms

m/s = vehicle speed in metres per second (1 km = 0.278 m/s)

Kinetic energy is measured in joules (metric) and can be converted to foot-pounds for imperial measurement (1 J = 0.738 ft·lb).

As the brakes on a moving automobile are applied, pads or shoes are pressed against rotating parts of the vehicle—either rotors (discs) or drums. The kinetic energy, or momentum, of the vehicle is then converted into heat energy by the kinetic friction of rubbing surfaces, and the car or truck slows down.

When the vehicle comes to a stop, it is held in place by static friction. The friction between the surfaces of the brakes and between the tires and the road resists any movement.

Factors Governing Braking

Four basic factors determine the braking power of a system. The first three factors govern the generation of friction: pressure, coefficient of friction, and frictional contact surface. The fourth factor is a result of friction. It is heat or, more precisely, heat dissipation.

An additional factor influences how well a vehicle will stop when the brakes are applied, that being weight transfer. When the brakes are applied while the vehicle is moving forward, the weight of the vehicle shifts forward. This causes the front of the vehicle to drop, or "nose dive." It also means that the front brakes will need the most stopping power. If the vehicle is overloaded or if the front suspension is weak, more weight will be thrown forward and the brakes will need to work harder.

PRESSURE The amount of friction generated between moving surfaces contacting one another depends in part on the pressure exerted on the surfaces. For example, if you slowly increase the downward pressure on the palm of your hand as you move it across a tabletop, you feel a gradual increase in friction.

In a brake system, hydraulic systems provide application pressure. Hydraulic force is used to move brake pads or brake shoes against spinning rotors or drums mounted to the wheels. The amount of pressure is determined by the pressure on the brake pedal and the design of the brake system.

COEFFICIENT OF FRICTION The amount of friction between two surfaces is expressed as a **coefficient of friction (COF)**. The coefficient of friction is determined by dividing the force required to pull an object across a surface by the weight of the object **(Figure 50–2)**. For example, if it requires a force of 100 kg (220 lb.) to slide a 100 kg (220 lb.) block of rubber across a concrete floor, the coefficient of friction is 100 ÷ 100, or 1. To pull a 100 kg (220 lb.) block of ice across the same floor might require only 2 kg (4.4 lb.) of force. The coefficient of friction then would be only 0.02. As it applies to automotive brakes, the COF expresses the frictional relationship between pads and rotors, or shoes and drums, and is carefully engineered to ensure optimum performance. Therefore, when replacing pads or shoes, it is important to use replacement parts with similar COF. If, for example, the COF is too high, the brakes are too sticky to stop the car smoothly. Premature wheel lock-up or grabbing would result. If the coefficient is too low, the friction material tends to slide over the machined surface of the drum or rotor rather than slowing it down. Most automotive friction materials are thus engineered with a COF of between 0.25 and 0.55, depending on their intended application. Because the COF of friction material changes at different temperatures, the COF is measured cold and hot. The friction material will usually have a letter code stamped on its edge, with each letter representing a COF range when cold and when hot—for example, **FF** or **FG**.

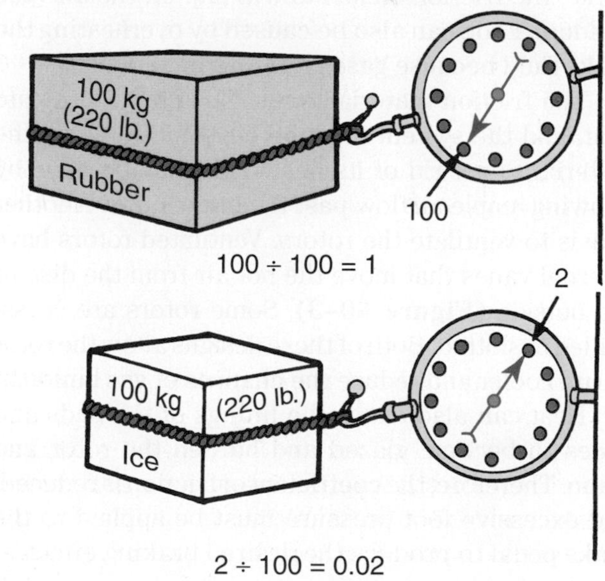

FIGURE 50–2

The coefficient of friction is equal to the kilograms of pull divided by the weight of the object.

FRICTIONAL CONTACT SURFACE The third factor is the amount of surface, or area, that is in contact. Simply put, bigger brakes stop a car more quickly than smaller brakes used on the same car. For the most part, the vehicle's weight and potential speed determines the size of the friction surface areas. Also, the greater the surface areas of the wheel brake units, the faster heat can be dissipated.

HEAT DISSIPATION Any braking system must be able to effectively handle the heat created by friction within the system. The tremendous heat created by the rubbing brake surfaces must be conducted away from the pad and rotor (or shoe and drum) and be absorbed by the air. The greater the surface areas of the brake components, the faster the heat can be dissipated. Thus, the weight and potential speed of the vehicle determine the size of the braking mechanism and the friction surface area of the pad or shoe. Brakes that do not effectively dissipate heat experience brake fade during hard, continuous braking.

Brake fade is a condition where the stopping power of the brakes has been drastically reduced. This is commonly caused by excessive heat buildup. With brake fade, the brake pedal seems normal, but there is reduced stopping ability. Brake fade may become worse as heat builds up; this may be due to outgassing. As the shoes or pads become extremely hot, they can generate a gas. This gas can become an air bearing between the frictional material and the rotor or drum. Rather than clamp on the wheel brake, the friction elements will slip on the air (gas buildup). Fade can also be caused by overheating the brake fluid because gases form in the fluid.

The friction materials must be able to dissipate heat, and the system must be designed to allow the material to get rid of its heat. This may be done by allowing ample airflow past the brake units. Another way is to ventilate the rotors. Ventilated rotors have internal vanes that move the hot air from the disc to the outside **(Figure 50–3)**. Some rotors are cross-drilled or slotted. Both of these designs allow the rotor to run cooler and reduce the chances of gas buildup.

Heat can also cause the linings of the pads and shoes to become glazed and harden the rotor and drum. Therefore, the coefficient of friction is reduced, and excessive foot pressure must be applied to the brake pedal to produce the desired braking effect.

Brake Lining Friction Materials

The friction materials used on brake pads and shoes are called brake linings. Brake linings are either riveted or bonded to the backing of the pad or shoe.

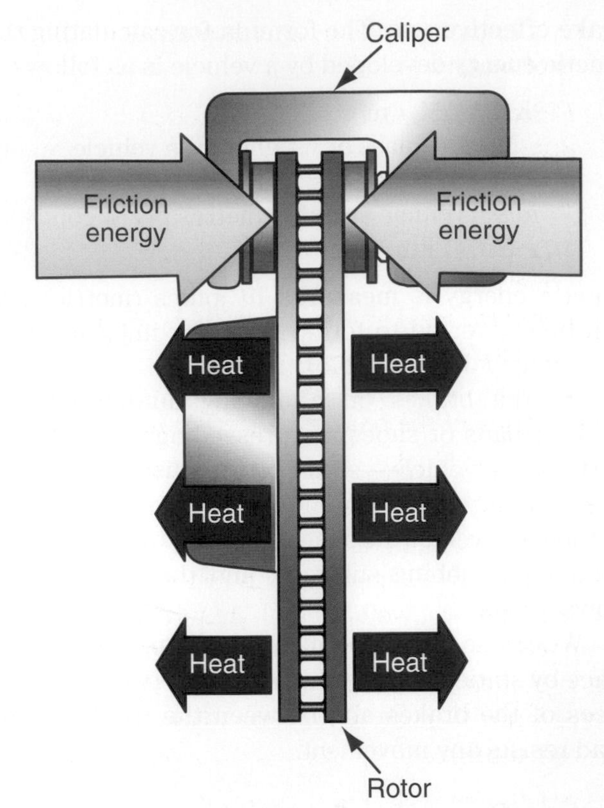

FIGURE 50–3

Heat generated during braking is dissipated as air moves through the rotor.

Some newer brake pads are integrally moulded. These can be identified by looking at the backing of the pad. Integrally moulded pad assemblies will have holes that are partially or totally filled with the lining material.

For many decades, asbestos was the standard brake lining material. It offered good friction qualities, long wear, and low noise. New materials, such as composite/organic, ceramics, and carbon fibres, are being used because of the health hazards of breathing asbestos dust. Some countries around the world have banned the use of asbestos in brake components for use on new vehicles and in the large aftermarket parts industry. North American bans have been successfully challenged in the courts and have been repealed. Even with the repeal, many manufacturers have eliminated asbestos from their shoes and pads mostly for liability reasons.

NONASBESTOS ORGANIC **Nonasbestos organic (NAO) linings** are installed on many vehicles by the OEM. Organic linings are made of nonmetallic fibres bonded together to form a composite material.

Today's organic brake linings contain the following types of materials:

- *Friction materials and friction modifiers.* Some common examples of these are graphite, powdered metals, and even nut shells.
- *Fillers.* Fillers are secondary materials added for noise reduction, heat transfer, and other purposes.
- *Binders.* Binders are glues that hold the other materials together.
- *Curing agents.* These accelerate the chemical reaction of the binders and other materials.

Organic linings have a high COF, and they are economical and quiet, wear slowly, and are only mildly abrasive to rotors and drums. However, organic linings fade more quickly than other materials and do not operate well at high temperatures. High-temperature organic linings are available for high-performance use, but they do not work as well at low temperatures. They also wear faster than regular organic linings.

METALLIC LININGS Fully metallic materials were used for many years in racing. **Metallic lining** is made of powdered metal that is formed into blocks by heat and pressure. These materials provide excellent resistance to brake fade but require high brake pedal pressure; they create the most wear on rotors and drums. Metallic linings work very poorly until they are fully warmed. Improved high-temperature organic linings and semimetallic materials have made metallic linings almost obsolete for late-model automotive use. Metallic linings are also extremely noisy, which is something that must be considered by customers when choosing the type of brake lining to install on their vehicle.

SEMIMETALLIC LININGS Semimetallic materials are made of a mixture of organic or synthetic fibres and certain metals moulded together. Semimetallics are harder and more fade resistant than organic materials but require higher brake pedal effort.

Most **semimetallic linings** contain about 50 percent iron and steel fibres. Copper also has been used in some semimetallic linings and, in smaller amounts, in organic linings. Concerns about copper contamination of the nation's water systems has led to its reduced use in brake linings, however.

Semimetallic linings operate best above 93.3°C (200°F) and actually must be warmed up to bring them to full efficiency. Consequently, they are typically less efficient than organic linings at low temperatures.

Semimetallic linings were sometimes used on older heavy or high-performance vehicles with four-wheel drum brakes. The lighter braking loads on rear brakes, particularly on FWD cars, may never heat semimetallic linings to their required operating efficiency. Semimetallic linings also have a lower static COF than organic linings, which makes them less efficient with parking brakes.

SYNTHETIC LININGS The goals of improved braking performance and the disadvantages of the other lining materials have led to the development of **synthetic lining** materials. They are classified as synthetic because they are made of nonorganic, nonmetallic, and nonasbestos materials. Two types of synthetic materials are commonly used as brake linings for drum brakes: fibreglass and **aramid fibres**.

Fibreglass was introduced as a brake lining material to help eliminate asbestos. Like asbestos, it has good heat resistance, good COF, and excellent structural strength. The disadvantages of fibreglass are its higher cost and its reduced friction at very high temperatures. Overall, fibreglass linings perform similarly to organic linings and are used primarily in rear drum brakes.

Aramid fibres are a family of synthetic materials that are five times stronger than steel, pound for pound, but weigh little more than half what an equal volume of fibreglass weighs. Friction materials made with aramid fibres are made similarly to organic and fibreglass linings. Aramid fibres have a COF similar to semimetallic linings when cold and close to that of organic linings when hot. Overall, the performance of aramid linings is somewhere between organic and semimetallic materials but with much better wear resistance and longevity than organic materials.

CERAMIC Ceramic linings are found on many FWD vehicles because they have high heat resistance. Most ceramic pads are made of a ceramic material mixed with copper fibres. These pads are quiet and produce little dust, making them popular as both OE and as aftermarket replacement parts.

CARBON-METALLIC/CERAMIC Carbon-metallic pads are often used on high-performance cars due to their good COF and high heat resistance. Carbon linings are also able to withstand very high temperatures without causing brake fade. A few aftermarket companies offer linings made of carbon, Kevlar, and various other materials. These also have ceramic heat shields that reduce the amount of heat that can transfer from the linings to the rest of the brake system. Some high-performance cars are fitted with carbon-ceramic pads. These pads are comprised of a ceramic composite of carbon fibre reinforced with silicon carbide. These pads offer excellent braking performance and are extremely lightweight. They also provide a consistent COF through a wide range of temperatures and weather conditions.

PRINCIPLES OF HYDRAULIC BRAKE SYSTEMS

A hydraulic system **(Figure 50–4)** uses a brake fluid to transfer force from the brake pedal to the pads or shoes. This transfer of force is reliable and consistent because liquids are not compressible. That is, force applied to a liquid in a closed system pressurizes the liquid. This fluid pressure is transmitted by that liquid equally to every other part of that system. When an input force of 50 kg (110 lb.) is placed on a 5 cm^2 (0.775 in.^{2}) master cylinder piston, a pressure of 10 kg/cm^2 (142 psi or 981 kPa) will be placed on the fluid by the master cylinder. You could measure this same pressure anywhere in the lines and inside any of the hydraulic components.

The force can be increased at output (that is, at the wheel) by increasing the size of the wheel's piston, though piston travel decreases. The force at output can be decreased by decreasing the size of the wheel piston, but the piston travel increases.

Thus, to double the output force of the 50 kg (110 lb.) at the master cylinder to 100 kg (220 lb.) at the wheels, simply use a wheel cylinder with a piston surface area of 10 cm^2 (1.55 in.^{2}). To triple the output force at the wheels to 150 kg (330 lb.), use a wheel cylinder piston with an area of 15 cm^2 (2.325 in.^{2}) **(Figure 50–5)**. No matter what the fluid pressure is, the output force can be increased with a larger piston, though piston travel decreases proportionately. In actual practice, however, fluid movement in an automotive hydraulic brake system

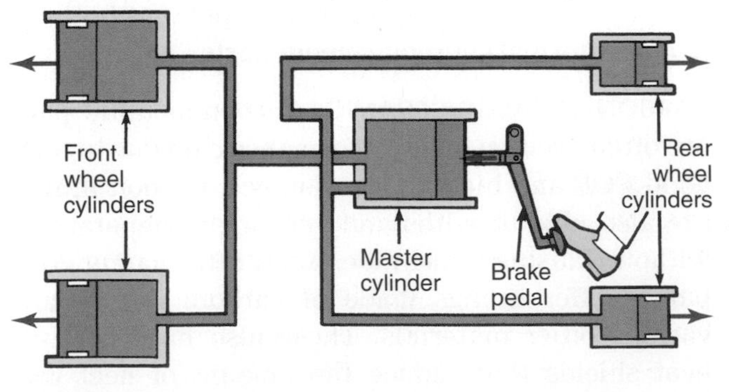

FIGURE 50–4

A schematic of a basic automotive hydraulic brake system.

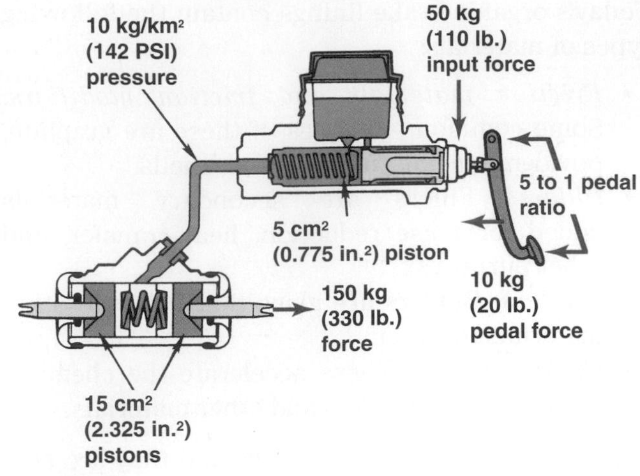

FIGURE 50–5

Output force increases with piston size.

is very slight. In an emergency, when the pedal goes all the way to the floor, the volume of fluid displaced amounts to only about 20 cc. About 15 cc goes to the front discs and 5 cc goes to the rear drums. Even under these conditions, the wheel cylinder and caliper pistons move only slightly.

Of course, the hydraulic system does not stop the car all by itself. In fact, it really just transmits the action of the driver's foot on the brake pedal out to the wheels. In the wheels, sets of friction pads are forced against rotors or drums to slow their turning and bring the car to a stop. Mechanical force (the driver stepping on the brake pedal) is changed into hydraulic pressure, which is changed back into mechanical force (brake shoes and disc pads contacting the drums and rotors). The amount of force acting on the friction pads and shoes is equal to the system pressure multiplied by the area of the output piston affected. A force of 10 kg (22 lb.) applied to the brake pedal multiplied by a brake pedal ratio (mechanical advantage) of 5:1 would equal a 50 kg (110 lb.) input force. With this force pushing against a master cylinder piston with an area of 5 cm^2 (0.775 in.^{2}), the system pressure would be 10 kg/cm^2 (142 psi or 981 kPa).

Dual Braking Systems

Since 1967, federal law has required that all cars be equipped with two separate brake systems. If one circuit fails, the other provides enough braking power to safely stop the car.

The dual system differs from the single system by employing a tandem master cylinder, which is essentially two master cylinders formed by installing two separate pistons and fluid reservoirs into one

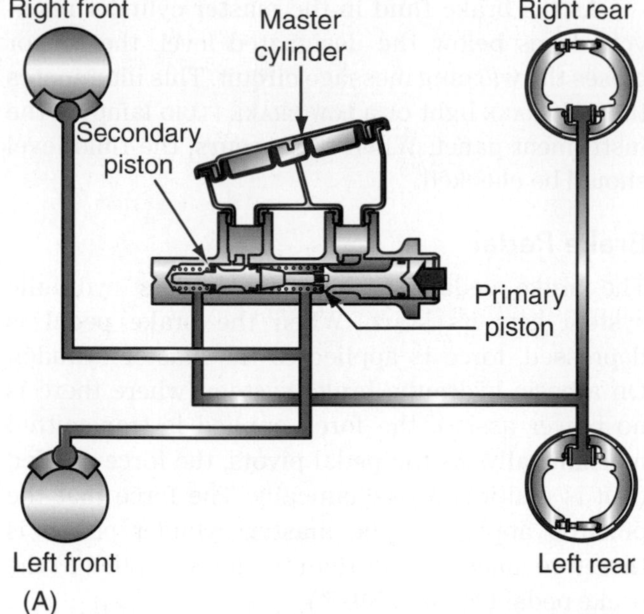

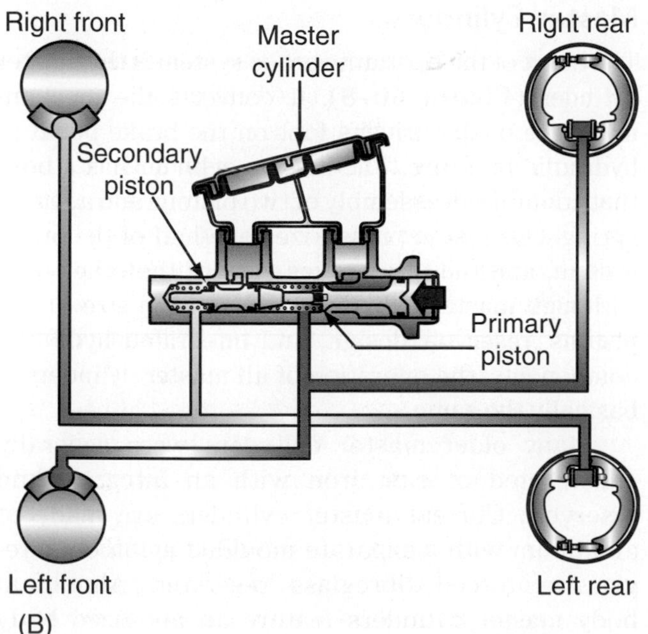

FIGURE 50–6

(A) A front/rear split of the braking system. (B) A diagonal split. The master cylinder is also split to allow pressure only to its designated wheel units.

cylinder bore. Each piston applies hydraulic pressure to two wheels **(Figure 50–6)**.

FRONT/REAR SPLIT SYSTEM In early dual systems, the hydraulic circuits were separated front and rear. Both front wheels were on one hydraulic circuit and both rear wheels on another. If a failure occurred in one system, the other system was still available to stop the vehicle. However, the front brakes do approximately 70 percent of the braking work. A failure in the front brake system would leave only 20 to 40 percent braking power. This problem was somewhat reduced with the development of diagonally split systems.

DIAGONALLY SPLIT SYSTEM The **diagonally split system** is now used on most passenger cars. In 1999, the Canadian Motor Vehicle Safety Standards (CMVSS) was updated to mandate a 50 percent brake capacity when a failure occurs in either one of the two brake circuits. The diagonally split system meets the mandatory minimum allowable split. This system operates on the same principles as the front and rear split system. It uses primary and secondary master cylinders that move simultaneously to exert hydraulic pressure on their respective systems.

The hydraulic brake lines on this system, however, have been diagonally split front to rear (left front to right rear and right front to left rear). The circuit split commonly occurs within the master cylinder but may split externally at a proportioning valve or pressure differential switch.

In the event of a system failure, the remaining good system would do all the braking on one front wheel and the opposite rear wheel, thus maintaining 50 percent of the total braking force.

HYDRAULIC BRAKE SYSTEM COMPONENTS

The following sections describe the major components of a hydraulic brake system, including power-assisted systems and antilock braking systems.

Brake Fluid

Brake fluid is the lifeblood of any hydraulic brake system and is what makes the system operate properly.

Modern brake fluid is specially blended to enable it to perform a variety of functions. Brake fluid must be able to flow freely at extremely high temperatures (2608°C [5008°F]) and at very low temperatures (–758°C [–1048°F]). Brake fluid must also serve as a lubricant to the many parts with which it comes into contact to ensure smooth and even operation. In addition, brake fluid must fight corrosion and rust in the brake lines and various assemblies and components in services. Another important property of brake fluid is that it must resist evaporation.

Most brake fluids are hygroscopic; that is, they readily absorb water. This is why brake fluid should always be kept in a sealed container and should be exposed to outside air only for limited periods of time. The fluid's hygroscopic qualities also make it necessary to flush the brake system every two years.

Some of the earliest brake fluids had chemicals in them that ate away at the rubber components in the brake system (that is, cups and seals). Modern brake fluid must be compatible with rubber to avoid damage to the cups and seals in the system. Brake fluid must provide a controlled amount of swell to the brake system cups and seals. There must be just enough swell to form a good seal. However, the swell cannot be too great. If it is, drag and poor brake response occur. Every can of brake fluid carries the identification letters of SAE and **DOT**. These letters (and corresponding numbers) indicate the nature, blend, and performance characteristics of that particular brand of brake fluid.

There are three basic types or classifications of hydraulic brake fluids. DOT 3 is a conventional brake fluid with a minimum dry **equilibrium reflux boiling point (ERBP)** of 205°C (401°F) and a minimum wet ERBP (3 percent moisture contamination) of 140°C (284°F). It is generally recommended for most ABS systems and some power brake setups. DOT 4 is a conventional brake fluid with a minimum dry ERBP of 230°C (446°F) and a minimum wet ERBP of 180°C (356°F). It is the most commonly used brake fluid for conventional brake systems. DOT 5 is a unique silicone-based brake fluid with a minimum dry ERBP of 260°C (500°F) and a minimum wet ERBP of 180°C (356°F). In the last few years, DOT 5 has lost its demand by brake servicing experts. DOT 5.1 is a nonsilicone-based polyglycol fluid and is amber in colour. This is a severe-duty fluid that has the same boiling point as DOT 5. However, remember that it is best to follow the vehicle manufacturer's fluid type and maintenance recommendations.

WARNING!

Use only approved brake fluid in a brake system. Any other lubricant, such as power-steering fluid, automatic transmission fluid, or engine oil, which has a petroleum base, must never be used in the brake system. Petroleum-based fluids attack the rubber components in the brake system, such as the piston cups and seals, and cause them to swell and disintegrate.

Some vehicles have brake fluid level sensors that provide the driver with an early warning message that the brake fluid in the master cylinder reservoir has dropped below the normal level.

As the brake fluid in the master cylinder reservoir drops below the designated level, the sensor closes the warning message circuit. This illuminates the red BRAKE light or a LOW BRAKE FLUID lamp on the instrument panel. When this occurs, the fluid level should be checked.

Brake Pedal

The brake pedal is where the brake's hydraulic system gets its start. When the brake pedal is depressed, force is applied to the master cylinder. On a basic hydraulic brake system (where there is no power assist), the force applied is transmitted mechanically. As the pedal pivots, the force applied to it is multiplied mechanically. The force that the pushrod applies to the master cylinder piston is therefore much greater than the force applied to the brake pedal **(Figure 50–7)**.

Master Cylinders

The heart of the hydraulic brake system is the master cylinder **(Figure 50–8)**. It converts the mechanical force of the driver's foot on the brake pedal to hydraulic pressure. The master cylinder has a bore that contains an assembly of two pistons and a return spring. One piston pressurizes one-half of the brake system, and the other takes care of the other half. Although master cylinders differ in the size of the pistons, reservoir design, and integrated hydraulic components, the operation of all master cylinders is basically the same.

Many older master cylinders were generally constructed of cast iron with an integral fluid reservoir. Current master cylinders are made of aluminum with a separate moulded nylon or fibreglass-reinforced fibreglass reservoir. Aluminum body master cylinders feature an anodized body **(Figure 50–9)** to protect against corrosion and to extend bore life.

CAUTION!

It is recommended that aluminum master cylinders not be rebuilt if pitting or scoring of the cylinder bore is evident. A new unit should be installed.

MASTER CYLINDER OPERATION

It is important that you understand how a master cylinder works. This understanding will be of great help when diagnosing a brake system. The basic principle of operation is a simple one. The brake pedal,

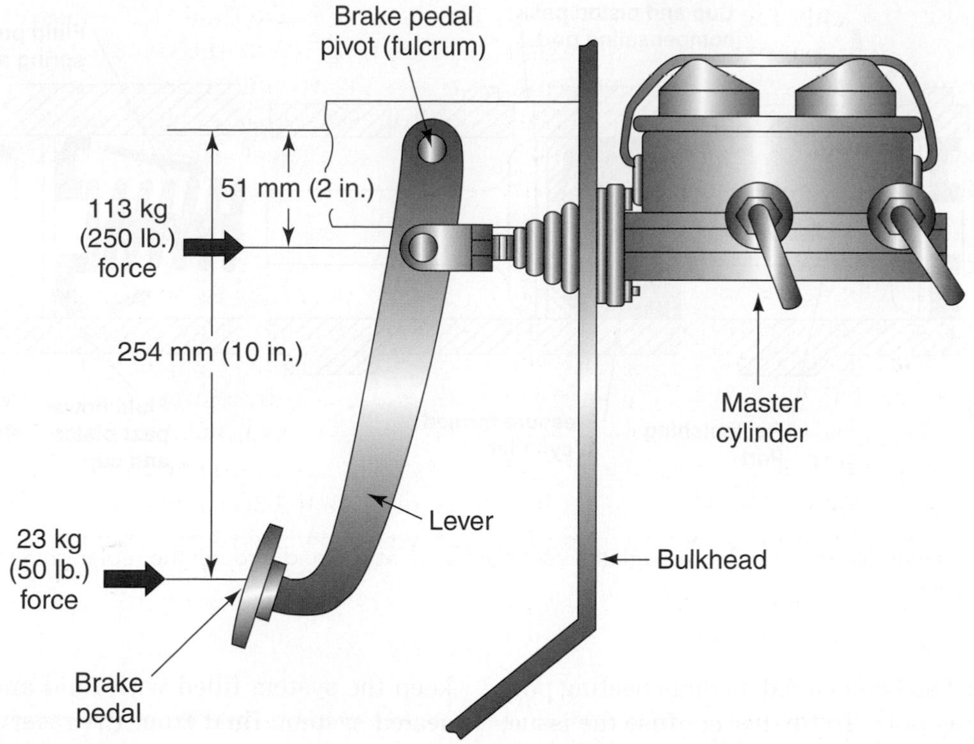

FIGURE 50–7

The brake pedal uses leverage to increase the force applied to the master cylinder.

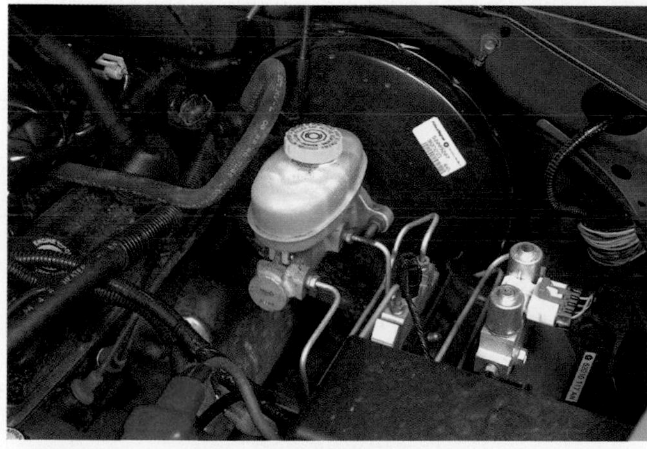

FIGURE 50–8

A brake master cylinder.

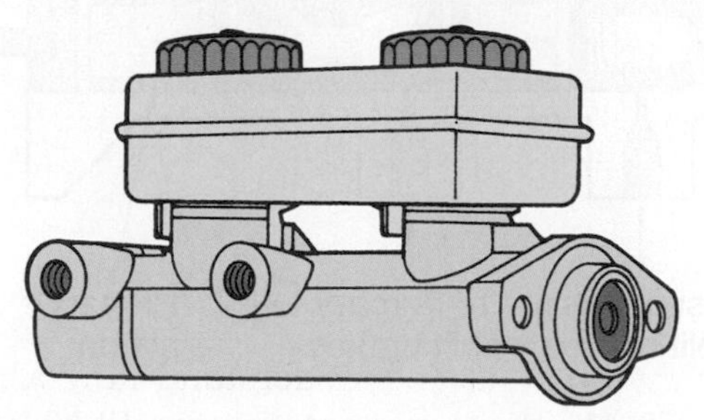

FIGURE 50–9

A typical aluminum/composite dual master cylinder.

connected to the master cylinder's piston assembly by a pushrod, controls the movement of the pistons. As the pedal is depressed, the piston assembly moves. The piston exerts pressure on the fluid, which flows out under pressure through the fluid outlet ports into the rest of the hydraulic system. When the brake pedal is released, the return spring in the cylinder forces the piston assembly back to its original position.

Because the master cylinder is a hydraulic device, the system must remain sealed and full of fluid. To do this, each piston has a primary and a secondary cup. The primary cup compresses the fluid when the pedal is depressed and keeps the brake fluid ahead of the pistons when they are put under pressure. The movement of the piston and cup shut off the fluid supply from the reservoir and create a sealed system from the primary cup forward. To prevent the fluid in the reservoir from leaking out of the cylinder when the primary cup has moved forward, the primary piston has a secondary seal. The secondary seal only blocks nonpressurized fluid.

MASTER CYLINDER PORTS Different names have been used for the ports in the master cylinder. This text refers to the forward port as the vent port and the rearward port as the replenishing port. These are the names established by SAE Standard J1153.

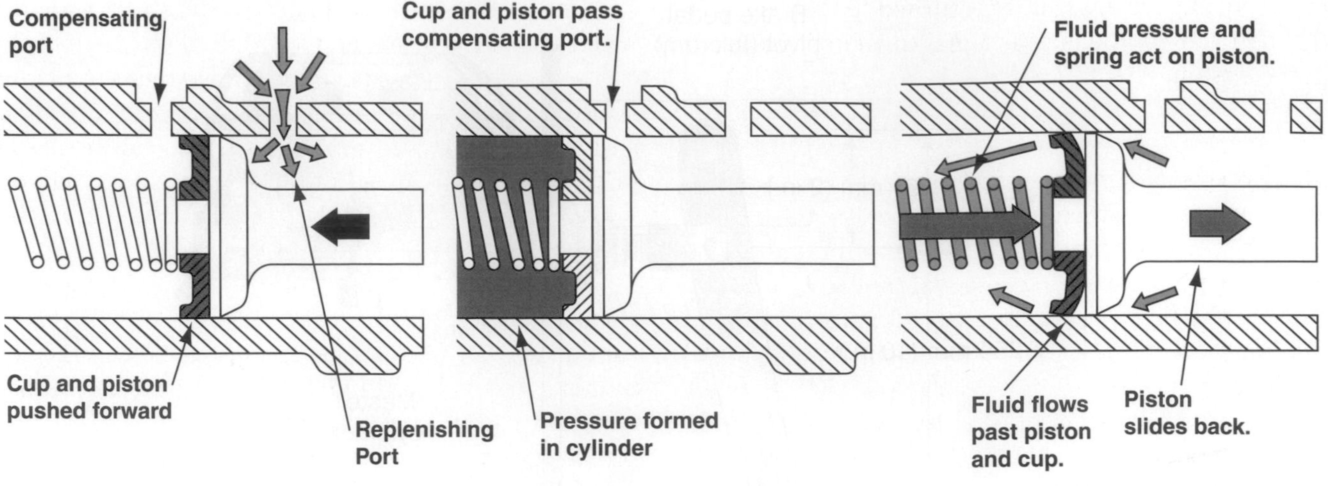
Compensating port
Cup and piston pass compensating port.
Fluid pressure and spring act on piston.
Cup and piston pushed forward
Replenishing Port
Pressure formed in cylinder
Fluid flows past piston and cup.
Piston slides back.

FIGURE 50–10

Fluid from the reservoir fills the cylinder through the vent port when at rest and through the replenishing port during brake release.

The **vent port** has been called a compensating port or a replenishing port. To further confuse the issue, the **replenishing port** has been described by many manufacturers as the compensating port, as well as the vent port, the bypass port or hole, the filler port, or the intake port. The vent ports and replenishing ports let fluid pass between each pressure chamber and its fluid reservoir during operation. The names of these ports are not important as long as you understand their purposes and operations.

When the brake pedal is not depressed, fluid from the reservoir keeps the cylinder filled. The fluid enters the cylinder through vent (intake) ports or compensating and replenishing ports that connect the reservoir to the piston chamber.

Each piston has a return spring in front of it. This holds the primary piston cup slightly behind the vent port **(Figure 50–10)**, which allows the brake fluid stored in the reservoir to enter the master cylinder bore. In addition to keeping the compensating ports uncovered, the return springs also help to return the brake pedal when the force has been removed from it. As the brakes are applied **(Figure 50–11)**, the stiffer primary piston spring pushes the secondary piston and spring slightly. Then the cup at the front end of the secondary piston passes and closes off the vent port on the secondary side of the master cylinder.

When the piston moves quickly back to its resting place, the brake fluid cannot return through the lines fast enough to avoid creating a low-pressure condition ahead of it. The fluid must reach the low-pressure area in time for another stroke of the cylinder. To keep the system filled with fluid and to maintain a sealed system, fluid from the reservoir then enters the cylinder through the replenishing port to fill the void. Fluid also enters the system by flowing past the primary cups. During the return stroke, the edges of the cup pull away from the bore enough to allow fluid to pass around the piston assembly to the area of low pressure. This also allows extra fluid to be trapped ahead of the piston during rapid pedal application, to pump up the brakes.

Finally, the pads and shoes return to position. That is, once the brake pedal is released and the master cylinder piston has returned to the rest position, shoe return springs (in the drum brakes) and piston seals (in the disc brakes) cause these pistons to retract. When all of the brakes are fully

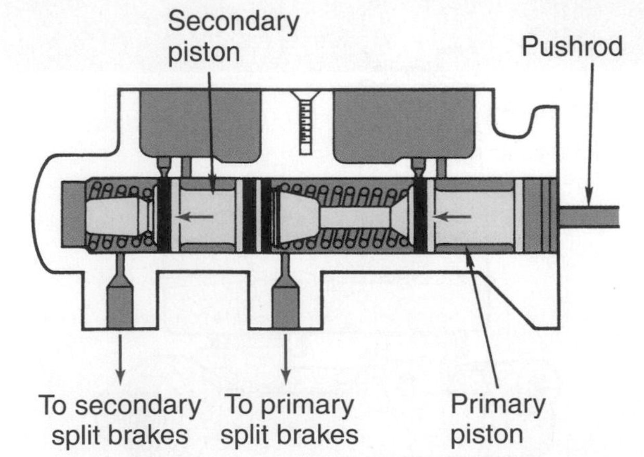
Secondary piston
Pushrod
To secondary split brakes
To primary split brakes
Primary piston

FIGURE 50–11

The position of the pistons in a master cylinder when the brakes are applied.

released, any excess fluid is returned to the reservoir through the compensating port, to relieve pressure in the system.

Master Cylinder Components

The two pistons in the master cylinder are not rigidly connected. Each piston has a return spring, with the primary piston spring between the two pistons. Stepping on the brake pedal moves a pushrod, causing the first or primary piston to move forward. The fluid ahead of it cannot be compressed, so the secondary piston moves. As the pistons progress deeper into the cylinder bore, the brake fluid that is put under pressure transmits this force through both systems to friction pads at the wheels. A retaining ring fits into a groove near the end of the bore and holds the piston inside the cylinder.

Extra brake fluid is stored in two separate reservoirs. The reservoirs are made of either cast iron or specially blended plastic and are designed to function independently as a protection against lost brake fluid **(Figure 50–12)**. The cap or cover for the reservoir keeps an airtight seal on the master cylinder while letting it breathe. It also keeps moisture from entering the system.

Split Hydraulic Systems

Most late-model vehicles have a diagonally split hydraulic system. If there is a hydraulic failure in the brake lines served by the master cylinder's secondary piston, both pistons will move forward when the brakes are applied, but there is nothing to resist piston travel except the secondary piston spring. This lets the primary piston build up only a small amount of pressure until the secondary piston bottoms in the cylinder bore. Then the primary piston will build enough hydraulic pressure to operate the brakes served by this half of the system.

In case of a hydraulic failure in the brake system served by the primary piston, the piston will move forward when the brakes are applied but will not build up hydraulic pressure. Very little force is transferred to the secondary piston through the primary piston spring until the piston extension screw comes in contact with the secondary piston. Then, pushrod force is transmitted directly to the secondary piston, and enough pressure is built up to operate its brakes.

Fast-Fill and Quick Take-Up Master Cylinders

Several manufacturers use fast-fill, or quick take-up, master cylinders. These cylinders fill the hydraulic system quickly to take up the slack in the caliper pistons of low-drag disc brakes. Low-drag calipers retract the pistons and pads farther from the rotor than traditional calipers. This reduces friction and brake drag and improves fuel mileage.

If a conventional master cylinder were used with low-drag calipers, excessive pedal travel would be needed on the first stroke to fill the lines and

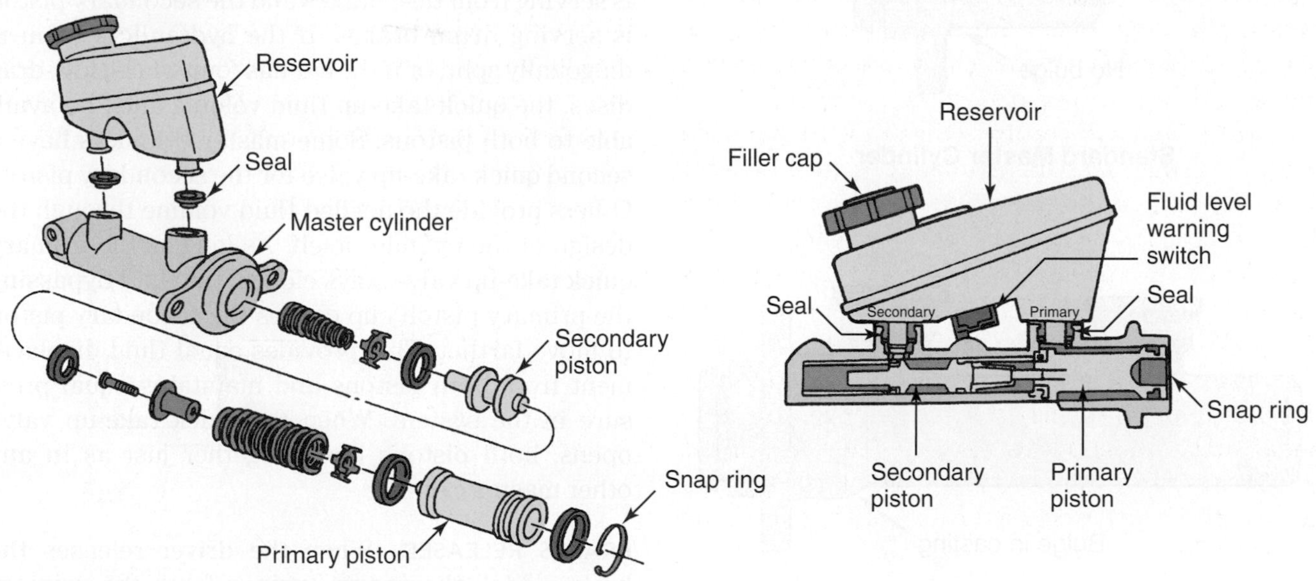

FIGURE 50–12

Although the fluid reservoir looks like it is one reservoir, it is actually two separate reservoirs, one for each half of the hydraulic system.

Ford Motor Company

calipers with fluid and take up the slack in the pads. To overcome this, fast-fill, or **quick take-up**, master cylinders provide a large volume of fluid on the first stroke of the brake pedal.

You can recognize a fast-fill, or quick take-up, master cylinder by the bulge, or larger diameter, on the outside of the casting **(Figure 50–13)**. The cylinder has a larger diameter bore for the rear primary piston than the front primary piston. Inside the cylinder, a fast-fill, or quick take-up, valve replaces the conventional vent and replenishing ports for the primary piston. Some master cylinders for four-wheel disc brakes also have a quick take-up valve for the secondary piston.

The quick take-up valve contains a spring-loaded check ball that has a small bypass groove cut in the edge of its seat. The outer circumference of the valve is sealed to the cylinder body with a lip seal. Several holes around the edge of the hole let fluid bypass the lip seal under certain conditions. Some valves (those more often called "fast-fill" valves) are pressed into the cylinder body and sealed tightly by an O-ring. A rubber flapper-type check valve under the fast-fill valve performs the same functions as a lip seal of a quick take-up valve.

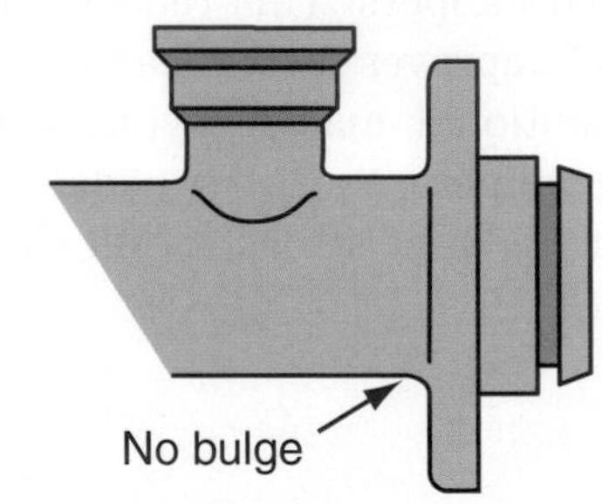

Standard Master Cylinder

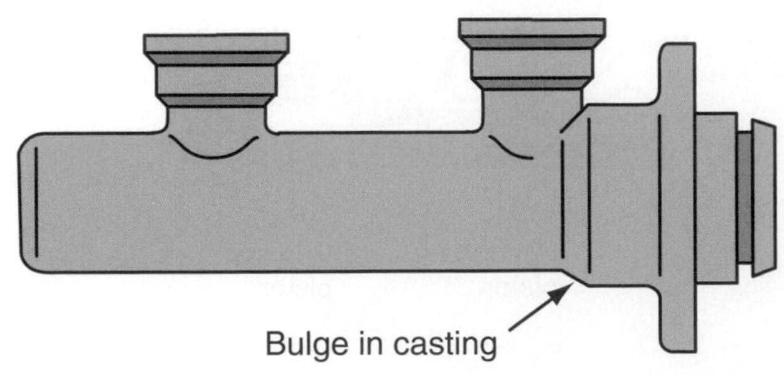

Quick Take-Up Master Cylinder

FIGURE 50–13

You can recognize a quick take-up master cylinder by the bulge or step in the casting.

BRAKES NOT APPLIED When the brakes are off, both master cylinder pistons are retracted, and all compensating/vent and replenishing ports are open. However, fluid to both ports of the primary piston must flow through the groove in the check ball seat.

BRAKES APPLIED As the brakes are applied, the primary piston moves forward in its bore. Remember that the diameter of the primary chamber is larger than the diameter of the secondary. As the secondary piston moves forward, the volume is reduced. This causes hydraulic pressure to instantly rise in the low-pressure chamber. The higher pressure forces the large volume of fluid into the low-pressure chamber past the cup seal of the primary piston. This provides the extra volume of fluid to take up the slack in the caliper pistons.

The lip seal of the quick take-up valve keeps fluid from flowing from the low-pressure chamber back to the reservoir. Initially a small amount of fluid bypasses the check ball through the bypass groove, but this is not enough to affect quick take-up operation.

As brake application continues, pressure in the low-pressure chamber rises to about 480 to 690 kPa (70 to 100 psi). The check ball in the quick take-up valve then opens to let excess fluid return to the reservoir. Pressure in both chambers of the primary piston equalize, and the piston moves forward to actuate the secondary piston.

All of the actions apply to the primary piston if it is serving front disc brakes and the secondary piston is serving drum brakes. If the hydraulic system is diagonally split, or if the car has four-wheel, low-drag discs, the quick take-up fluid volume must be available to both pistons. Some master cylinders have a second quick take-up valve for the secondary piston. Others provide the needed fluid volume through the design of the cylinder itself. As long as the primary quick take-up valve stays closed, the fluid bypassing the primary piston cup causes the secondary piston to move farther. This provides equal fluid displacement from both pistons and maintains equal pressure in the system. When the quick take-up valve opens, both pistons move together just as in any other master cylinder.

BRAKES RELEASED When the driver releases the brake pedal, the return springs force the primary and secondary pistons to move back. Pressure drops in the high-pressure chambers, and fluid bypasses the piston cup seals from the low-pressure chambers. Low pressure is created in the low-pressure chamber, which allows atmospheric pressure to move past the lip seal of the quick take-up valve. Fluid from the reservoir then flows through both the vent and replenishing ports to equalize pressure in the pressure chambers and valley areas.

On the return stroke, fluid flows to the secondary piston through the replenishing port unless the secondary piston also has a quick take-up valve. If a secondary quick take-up valve is installed, it works in the same way as a primary quick take-up valve.

Central-Valve Master Cylinders

Some antilock brake systems (ABS) use master cylinders that have central check valves in the tops of the pistons. These valves are designed to prevent seal damage and pedal vibration. If the master cylinder provides pressure during antilock operation and the system also has a motor-driven pump, the master cylinder's pistons may shift back and forth rapidly during antilock operation. This will cause excessive pedal vibration and—more importantly—wear on the piston cups where they pass over the vent ports.

When the brakes are released, fluid flows from the replenishing ports to the low-pressure chambers, through the open central check valves, and into the high-pressure chambers. As the brakes are applied, the central valves close to hold fluid in the high-pressure chambers. When the brakes are released again, the check valves open to let fluid flow back through the pistons to the low-pressure chambers and the reservoir.

The central check valves provide supplementary fluid passages to let fluid move rapidly back and forth between the high- and low-pressure chambers during antilock operation. This is not much different in principle from non-ABS fluid flow, but the extra passages reduce piston and pedal vibration and cup seal wear.

HYDRAULIC TUBES AND HOSES

Steel tubing and flexible synthetic rubber hosing serve as the arteries and veins of the hydraulic brake system. These brake lines transmit brake fluid pressure (the blood) from the master cylinder (the heart) to the wheel cylinders and calipers (the muscles and working parts) of the drum and disc brakes.

Fluid transfer from the driver-actuated master cylinder is usually routed through one or more valves and then into the steel tubing and hoses **(Figure 50–14)**. The design of the brake lines offers quick fluid transfer response with very little friction

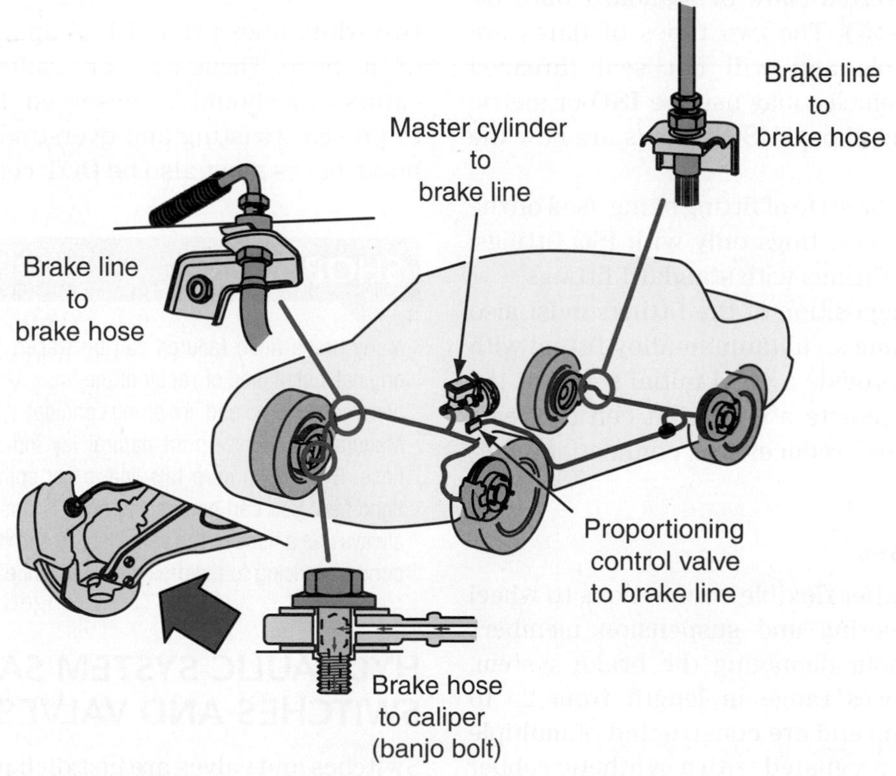

FIGURE 50–14

A typical layout of the hoses and tubes for the brake system.

loss. Engineering and installing the brake lines so that they do not wrap around sharp curves is very important in maintaining this good fluid transfer.

Brake Line Tubing

Most brake line tubing consists of copper-fused double-wall seamless steel tubing in diameters ranging from 3 to 9 mm (⅛ to ⅜ in.). Some OEM brake tubing is manufactured with soft steel strips, sheathed with copper. The strips are rolled into a double-wall assembly and then bonded in a furnace at extremely high temperatures. Corrosion protection is often added by tin-plating the tubing. Some newer lines, referred to as 90/10 brake line, are of a copper and nickel composition for increased strength.

Fittings

Assorted fittings are used to connect steel tubing to junction blocks or other tubing sections. The most common fitting on older vehicles is the double or inverted flare style. Double flaring is important to maintain the strength and safety of the system. Single-flare or sleeve compression fittings may not hold up in the rigorous operating environment of a standard vehicle brake system.

Fittings are constructed of steel or brass and use either the inverted flare or standard flare fitting **(Figure 50–15)**. The two types of flares are not interchangeable and will not seal threaded together. Newer vehicles may use the **ISO** or metric bubble flare fitting, though ISO fittings are now the most common.

Never change the style of fitting being used on the vehicle. Replace ISO fittings only with ISO fittings. Replace standard fittings with standard fittings.

The metal composition of the fittings must also match exactly. Using an aluminum-alloy fitting with steel tubing may provide a good initial seal, but the dissimilar metals create a corrosion cell that eats away the metal and reduces the connection's service life.

Brake Line Hoses

Brake line hoses offer flexible connections to wheel units so that steering and suspension members can operate without damaging the brake system. Typical brake hoses range in length from 25 to 76 mm (10 to 30 in.) and are constructed of multiple layers of fabric impregnated with a synthetic rubber **(Figure 50–16)**. Brake hose material must offer high heat resistance and withstand harsh operating conditions. All brake hoses or flex lines now have two white lines printed 180° apart along the length of the hose. These lines are called natural lay indicators and should be observed during installation to prevent twisting and overstressing the hose. All brake hoses must also be DOT certified.

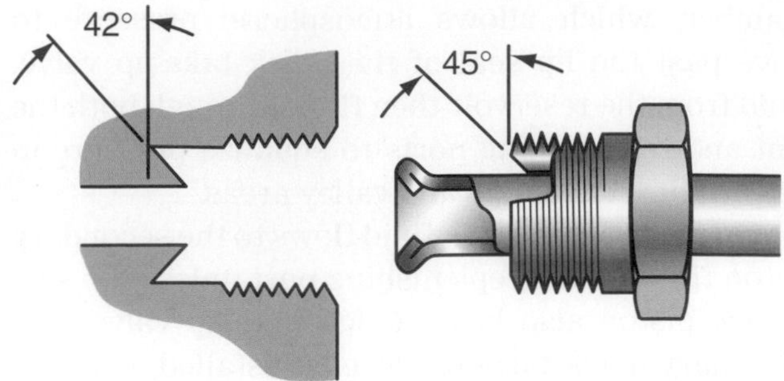

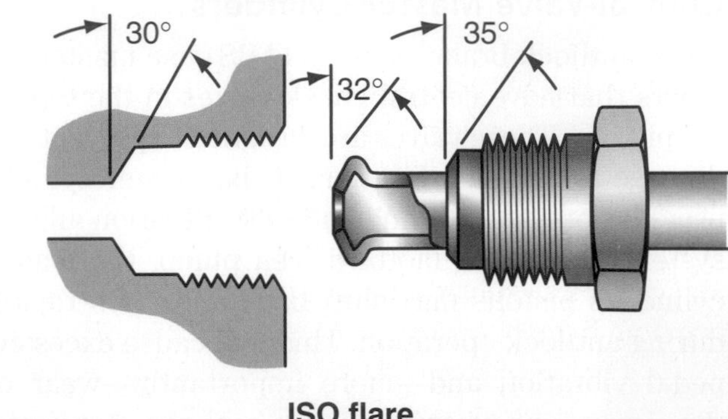

FIGURE 50–15

The two types of flares—double and ISO, or bubble—are not interchangeable.

> **SHOP TALK**
>
> Many brake hose failures can be traced to errors made in the original installation or repair of the hose. Hoses twisted into place become stressed and are prime candidates for leaks and bursting. Manufacturers now print natural lay indicators or lines on the hose. By making sure this line is not spiralled after fittings are tightened, you can ensure the hose is not overly stressed. Also, always use a hose of the same length and diameter as the original during servicing to maintain brake balance at all wheels.

HYDRAULIC SYSTEM SAFETY SWITCHES AND VALVES

Switches and valves are installed in the brake system hydraulic lines to act as warning devices to pressure control devices. Though the adoption of four-wheel disc brakes and antilock brake systems has

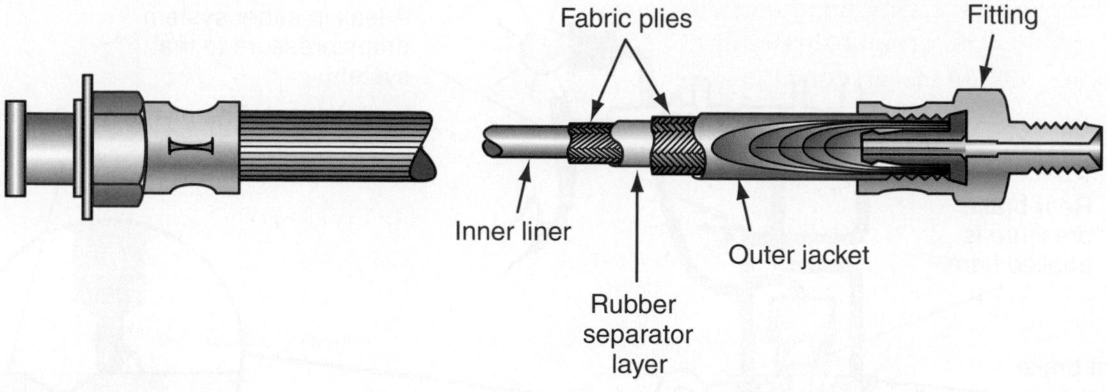

FIGURE 50–16

Flexible brake hoses are constructed of layers of reinforcement materials.

decreased the use of some valves in the hydraulic system, many vehicles still use one or more valves for pressure control or to illuminate the BRAKE warning light.

Pressure Differential (Warning Light) Switches

A pressure differential valve, in the event of a hydraulic leak, is used to shut off one hydraulic circuit and operate a warning light switch. Its main purpose is to tell the driver if pressure is lost in either of the two hydraulic systems. Because each brake hydraulic system functions independently, it is possible the driver might not notice immediately that pressure and braking are lost. When a pressure loss occurs, brake pedal travel increases, and pedal feel typically becomes soft, spongy, and much less firm. This results in a more-than-usual effort needed for braking. Should the driver not notice the extra effort needed, the warning light is actuated by the hydraulic system safety switch.

Under normal conditions, the hydraulic pressure on each side of the pressure differential valve piston is balanced. The piston is located at its centre point, so the spring-loaded warning switch plunger fits into the tapered groove of the piston. This leaves the contacts of the warning switch open. The brake warning light stays off.

If there is a leak in the front or rear braking system, the hydraulic pressure in the two systems is unequal. For example, if there is a leak in the system supplying the front brakes, there is lower pressure in the front system when the brake pedal is applied. The hydraulic pressure in the rear system then pushes the piston toward the front side, where the pressure is lower. As the piston moves, the plunger is pushed out **(Figure 50–17)**. This closes the switch and illuminates the brake warning light.

Although all brake warning light switches serve the same function, there are three common variations in the design of these switches. These variations include switches with centring springs, without centring springs, and with centring springs and two pistons.

Metering and Proportioning Valves

Metering and proportioning valves are used to balance the braking characteristics of disc and drum brakes.

The braking response of the disc brakes is immediate when the brake pedal is applied. It is directly proportionate to the effort applied at the pedal. Drum brake response is delayed while rear brake hydraulic pressure moves the wheel cylinder pistons to overcome the force of their return springs and force the brake shoes to contact the drum. Once applied, their actions may be self-energizing and tend to multiply the pedal effort. This reduces the need for increasing pressure to the rear brakes.

METERING VALVE A **metering valve** in the front brake line holds off pressure going from the master cylinder to the front disc calipers. This delay allows pressure to build up in the rear drums first. When the rear brakes begin to take hold, the hydraulic pressure builds to the level needed to open the metering valve **(Figure 50–18)**. When the metering valve opens, line pressure is high enough to operate the front discs. This process provides for better balance of the front and rear brakes. It also prevents lock-up of the front brakes by keeping pressure from them until the rear brakes have started to operate. The

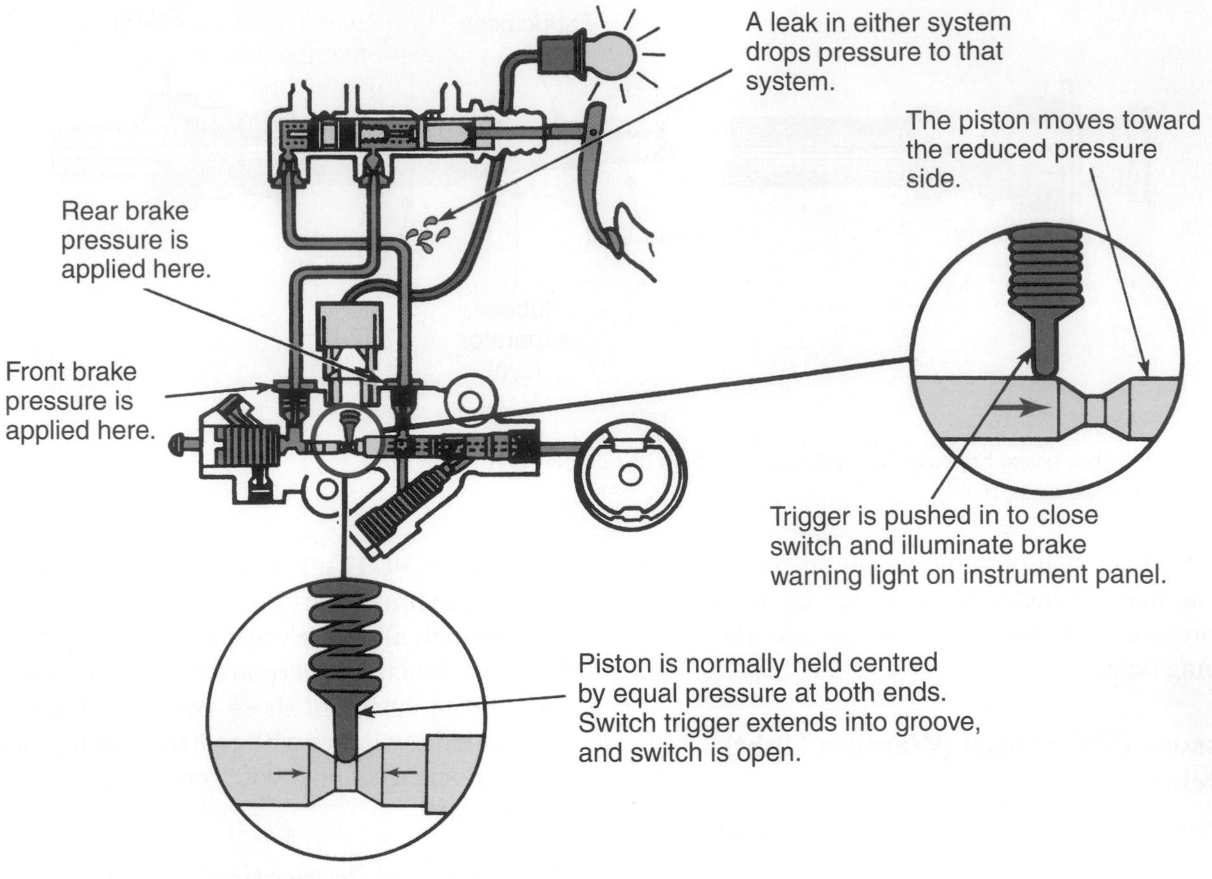
A leak in either system drops pressure to that system.
The piston moves toward the reduced pressure side.
Rear brake pressure is applied here.
Front brake pressure is applied here.
Trigger is pushed in to close switch and illuminate brake warning light on instrument panel.
Piston is normally held centred by equal pressure at both ends. Switch trigger extends into groove, and switch is open.

FIGURE 50–17

A pressure differential valve under normal conditions.

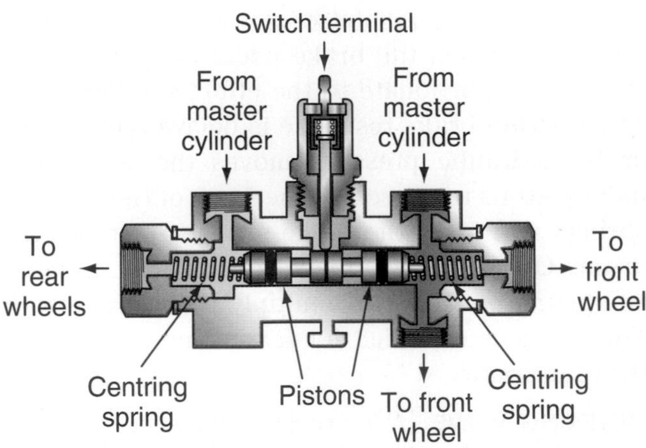
Switch terminal
From master cylinder
From master cylinder
To rear wheels
To front wheel
Centring spring
Pistons
To front wheel
Centring spring

FIGURE 50–18

The pressure differential valve can turn on the brake warning light if a drop in pressure occurs in the hydraulic system.

metering valve has the most effect at the start of each brake operation and all during light-braking conditions.

PROPORTIONING VALVE The self-energizing action of the delayed-response rear drum brakes can cause them to lock the rear wheels at a lower hydraulic pressure than the front brakes. The **proportioning valve** (balance valve) **(Figure 50–19)** is used to control rear brake pressures, particularly during hard stops. When the pressure to the rear brakes reaches a specified level, the proportioning valve overcomes the force of its spring-loaded piston, stopping the flow of fluid to the rear brakes. By doing so, it regulates rear brake system pressure and adjusts for the difference in pressure between front and rear brake systems. This keeps front and rear braking forces in balance.

HEIGHT-SENSING PROPORTIONAL VALVE The height-sensing proportional valve provides two different brake balance modes to the rear brakes, based on vehicle load. This is accomplished by turning the valve on or off. When the vehicle is not loaded, hydraulic pressure is reduced to the rear brakes. When the vehicle is carrying a full load, the actuator lever moves up to change the valve's setting. The valve now allows full hydraulic pressure to the rear brakes. The valve contains a plunger, cam, torsional clutch spring, and actuator shaft **(Figure 50–20)**.

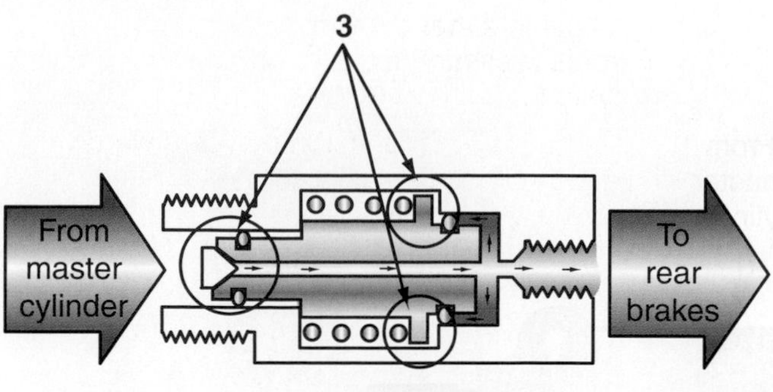

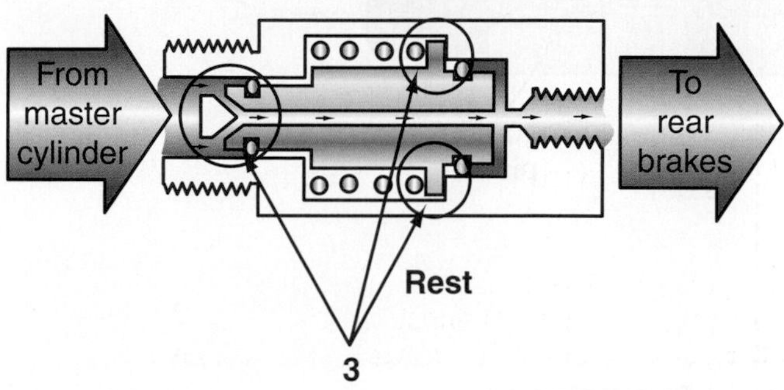

FIGURE 50–19

A proportioning valve.

The valve is mounted to the frame above the rear axle and has an actuator lever connected by a link to the lower shock absorber bracket. The valve is turned on and off as the axle-to-frame height changes due to load in the vehicle. The torsional clutch spring attached to the valve shaft is used as an override. Once the valve is positioned during braking, the spring prevents the valve from changing position if the vehicle goes over a bump or moves off the road.

Height-sensing proportional valves are replaced when defective and are not adjustable.

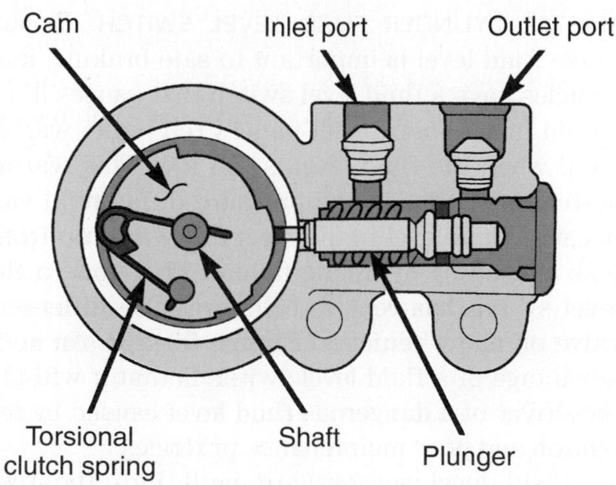

FIGURE 50–20

A height-sensing proportional valve.

SHOP TALK

Distribution and weight on some vehicles, such as vans, is such that a rear-to-front weight transfer does not present a similar problem. For this reason, standard proportioning valves are not required on all van models.

COMBINATION VALVES Most newer cars without ABS, have a **combination valve (Figure 50–21)** in their hydraulic system. This valve is simply a single unit that combines the metering and proportioning valves with the pressure differential valve. Combination valves are described as three-function or two-function valves, depending on the number of functions they perform in the hydraulic system.

Three-Function Valve. This type of valve performs the functions of the metering valve, brake warning light switch, and proportioning valve.

Two-Function Valves. There are two variations of the two-function combination valve. One variation does the proportioning valve and brake warning light switch functions. The other performs the metering valve and brake warning light switch functions.

If any one of its several operations fail, the entire combination valve must be replaced because these units are not repairable.

Warning Lights

A wide variety of electrical and electronic components are found in a brake system, especially with ABS. These include the warning lamp switch operation of a pressure differential valve and the electrical switches to operate the parking brake warning light, hydraulic failure warning lamp, and the sensors that indicate low brake fluid level and brake pad wear **(Figure 50–22)**.

FAILURE WARNING LAMP SWITCH The pressure differential valve has a hydraulically operated switch that controls the brake failure warning lamp on the instrument panel. Each side of the pressure differential valve is connected to half of the hydraulic system (one chamber of the master cylinder). Each master cylinder piston provides pressure to a separate hydraulic system. If one of the circuits fails,

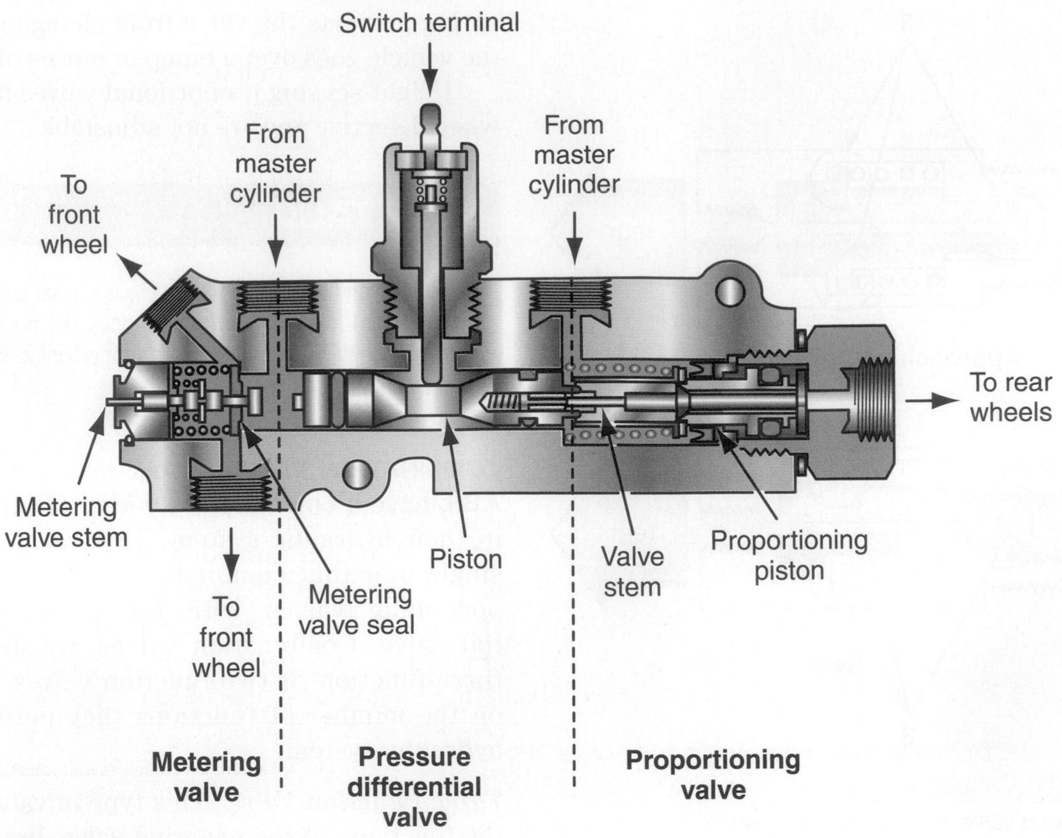

FIGURE 50–21

A typical combination valve.

©Cengage Learning 2015

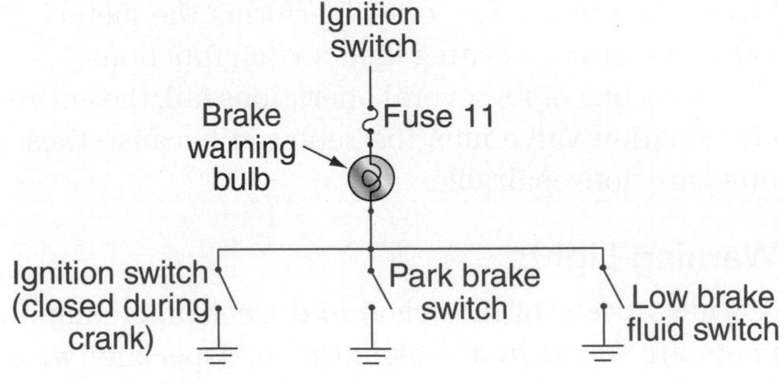

FIGURE 50–22

An example of a brake warning light circuit.

©Cengage Learning 2015

the brake pedal travel will increase, and more brake pedal effort will be required to stop the car. The driver might not notice a problem, however, but the lamp on the instrument panel will provide a warning in case of hydraulic failure.

Failure in one half of the hydraulic system causes a pressure loss on one side of the pressure differential valve. Pressure on the other side moves the valve's plunger into contact with the switch terminal. This closes the circuit, and the warning lamp is illuminated. All pressure differential valves work in this basic way but differ in the details of the shape of the piston and the use of centring springs.

The pressure differential valve on most late-model vehicles is part of a combination valve or built into the body of the master cylinder.

MASTER CYLINDER FLUID LEVEL SWITCH Because brake fluid level is important to safe braking, many vehicles have a fluid level switch that causes illumination of the instrument panel's red brake warning lamp when the fluid level is too low. This warning system is similar to the pressure differential valve because fluid level in the reservoir will drop from a leak caused by hydraulic failure. Therefore, a fluid level switch has replaced the pressure differential valve on many vehicles **(Figure 50–23)**. An added advantage of a fluid level switch is that it will alert the driver of a dangerous fluid level caused by inattention and poor maintenance practices.

Fluid level sensors are built into the reservoir body or cap. One type has a float with a pair of switch contacts on a rod that extends above the float. If the fluid level drops too low, the float will

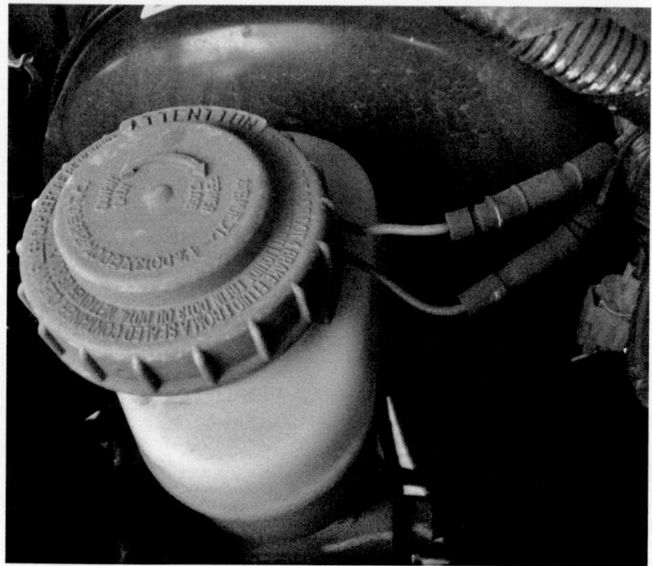

FIGURE 50–23

A reservoir cap with a fluid level sensor built in.

drop and cause the rod-mounted contacts to touch a set of fixed contacts and close the lamp circuit. Another type of switch uses a magnet in a movable float. If the float drops to a predetermined level, the floating magnet will activate the reed switch, pulling the switch contacts together to close the lamp circuit. The contacts typically provide a ground path for the brake warning lamp. The switch itself is not serviceable. If there is problem with the switch or float assembly, the entire reservoir needs to be replaced.

PARKING BRAKE SWITCH The parking brake should only be applied to hold a vehicle stationary. If the parking brake is even partially applied while the vehicle is moving, it will produce enough heat to glaze friction materials, expand drum dimensions, and increase pedal travel. On rear disc brake systems with integral parking brake actuators, driving with the parking brake applied will distort the brake rotors and reduce pad life.

A normally closed, single-pole, single-throw switch is used to ground the circuit of the red brake warning lamp in the instrument cluster. This switch is located within the parking brake handle or pedal assembly and is designed to turn on the light whenever the parking brake is applied.

Some vehicles with daytime running lights (DRL) use the parking brake switch to complete a circuit that prevents the headlights from coming on if the parking brake is applied when the engine is started. When the parking brake is released, the DRLs operate normally.

Stop Lamps

Stop lamps are included in the right and left tail lamp assemblies. Vehicles built since 1986 also have a centre high-mounted stop lamp (CHMSL).

Brake stop lamp switches are operated hydraulically or mechanically. Hydraulic switches were used on older vehicles and were installed in the master cylinder's high-pressure chamber; they were activated by system pressure. A mechanical switch is mounted on the bracket for the brake pedal and activated by the movement of the pedal lever.

Mechanical switches are found on today's vehicles because they can be adjusted to illuminate the stop lamps with the slightest pedal movement. Stop lamp switches may be single-function or multifunction units. Single-function switches have only one set of switch contacts that control electric current to the stop lamps at the rear of the vehicle. Multifunction switches have one set of switch contacts for the stop lamps and at least one additional set of contacts for the torque converter clutch, cruise control, or ABS. Some multifunction switches have contacts for all of these functions.

The brake lamp switch contacts are often connected to the brake lamps through the turn signal and hazard flasher switch. If a vehicle has antilock brakes, there is a connection or a separate switch for the ABS control unit to sense when the brakes are being applied. Wiring diagrams are essential for accurate identification of brake pedal switches and their functions.

DRUM AND DISC BRAKE ASSEMBLIES

Although drum and disc brakes are explained in great detail in later chapters, a brief explanation of their components and operating principles is essential at this point.

Drum Brakes

A drum brake assembly consists of a cast-iron drum, which is bolted to and rotates with the vehicle's wheel, and a fixed backing plate to which the shoes, wheel cylinders, automatic adjusters, and linkages are attached **(Figure 50–24)**. Additionally, there might be some extra hardware for parking brakes. The shoes are surfaced with frictional linings, which contact the inside of the drum when the brakes are applied. The shoes are forced outward by pistons located inside the wheel cylinder. The pistons are

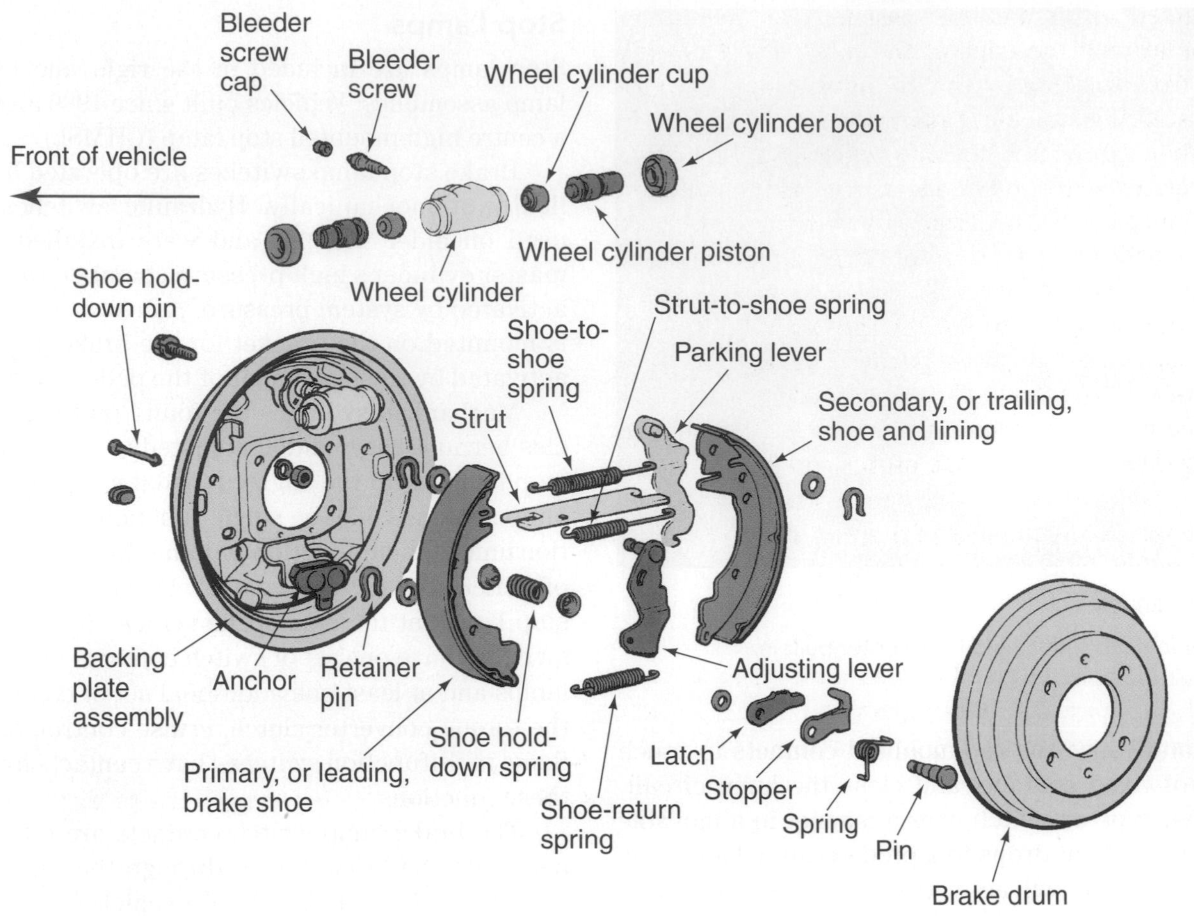

FIGURE 50–24

A typical drum brake assembly.

actuated by hydraulic pressure. As the drum rubs against the shoes, the energy of the moving drum is transformed into heat. This heat energy is passed into the atmosphere. When the brake pedal is released, hydraulic pressure drops, and the pistons are pushed back to their unapplied position by return springs.

Disc Brakes

Disc brakes resemble the brakes on a bicycle: The friction elements are in the form of pads, which are squeezed or clamped about the edge of a rotating wheel. With automotive disc brakes, this wheel is a separate unit, called a **rotor**, inboard of the vehicle wheel **(Figure 50–25)**. The rotor is typically made of cast iron. Because the pads clamp against both sides of it, both sides are machined smooth. Usually the two surfaces are separated by a finned centre section for better cooling (such rotors are called **ventilated rotors**). The pads are attached to a metal backing and held in place inside the body of the caliper. As with the drum brakes, the pads are actuated by pistons and hydraulic pressure. The pistons are

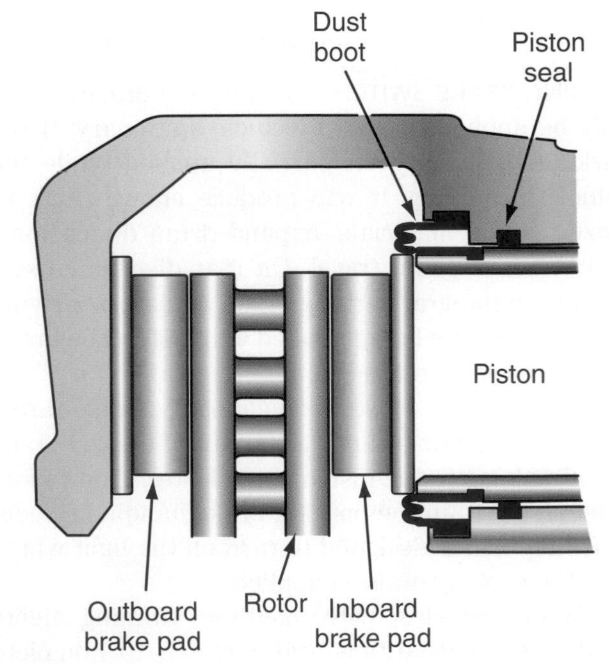

FIGURE 50–25

A view of how the disc brake caliper, pads, and rotor are arranged.

contained within a caliper assembly, a housing that wraps around the edge of the rotor. The caliper is kept from rotating by way of bolts holding it to the car's steering knuckle framework.

The **caliper** is a housing containing the pistons and related seals and boots, as well as the cylinders and fluid passages necessary to force the friction linings or pads against the rotor. The caliper resembles a hand in the way it wraps around the edge of the rotor. It is attached to the steering knuckle. To keep the pads close to the rotor during operation, a unique type of seal is used to allow the piston to be pushed out the necessary amount; then the piston retracts just enough to pull the pad off the rotor.

Unlike shoes in a drum brake, the pads act perpendicular to the rotation of the disc when the brakes are applied. This effect is different from that produced in a brake drum, where frictional drag actually pulls the shoe into the drum. Disc brakes are said to be nonenergized and so require more force to achieve the same braking effort. For this reason, they are ordinarily used in conjunction with a power brake unit.

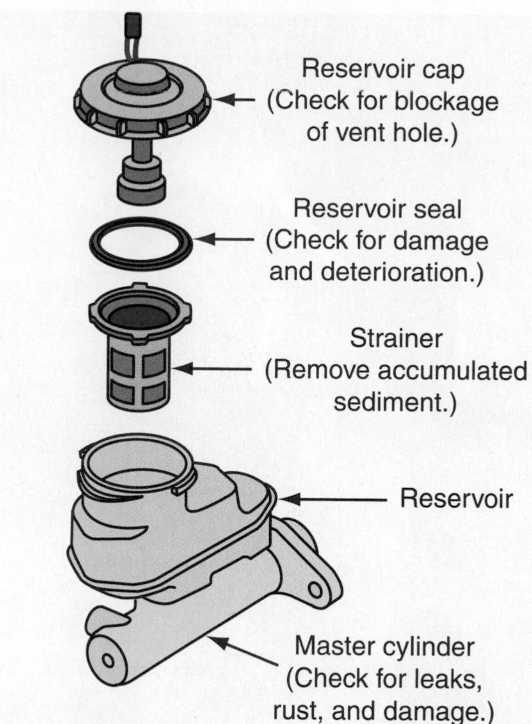

FIGURE 50–26

When inspecting a master cylinder, make sure to carefully check these items.

HYDRAULIC SYSTEM SERVICE

Hydraulic system service is relatively uncomplicated, but it is vital to the vehicle's safe operation.

Brake Fluid Inspection

The master cylinder is usually located under the hood and near the firewall on the driver's side.

WARNING!

To check brake fluid level, you should observe the fluid through the reservoir and not remove the cap unless necessary. This helps prevent moisture from getting to the fluid. If the master cylinder cover must be removed, clean the cover before removal to avoid dropping dirt into the reservoir.

Remove the cover and check the gasket, or diaphragm **(Figure 50–26)**. Inspect the cover for damage or plugged vent holes. Clean the vent holes, if necessary.

Check the brake fluid level in the master cylinder. The plastic reservoir may have fluid level marks. Do not overfill a reservoir. If fluid must be added, a leak probably has developed, or the shoes and/or pads are worn. Check the system carefully to locate the leak.

To check for contaminated fluid, place a small amount of brake fluid in a clear glass jar. If the fluid is dirty or separates into layers, it is contaminated. Contaminated fluid must be replaced, and the system should be flushed. The fluid's colour does not necessarily indicate the level of contamination. Clear fluid can still contain more than 3 percent moisture.

Test strips are also available. A strip of treated paper is dipped into the reservoir **(Figure 50–27)**. The paper will change colours, corresponding to the condition of the fluid. The resulting colour is then matched to the colour chart that accompanies the test strips. Special brake fluid testers are also available. These measure the fluid's boiling point, which is an indication of moisture content **(Figure 50–28)**.

Contaminated brake fluid can damage rubber parts and cause leaks. When replacing contaminated brake fluid, it is necessary to flush and refill the brake system with new fluid. Always follow the manufacturer's recommendations.

If the brake fluid level is low and there are no signs of external leaks, it is likely that the brake pads are nearing their wear limit. As the disc brake pad linings and rotors wear, the caliper pistons extend further out of the caliper bore. This increases the

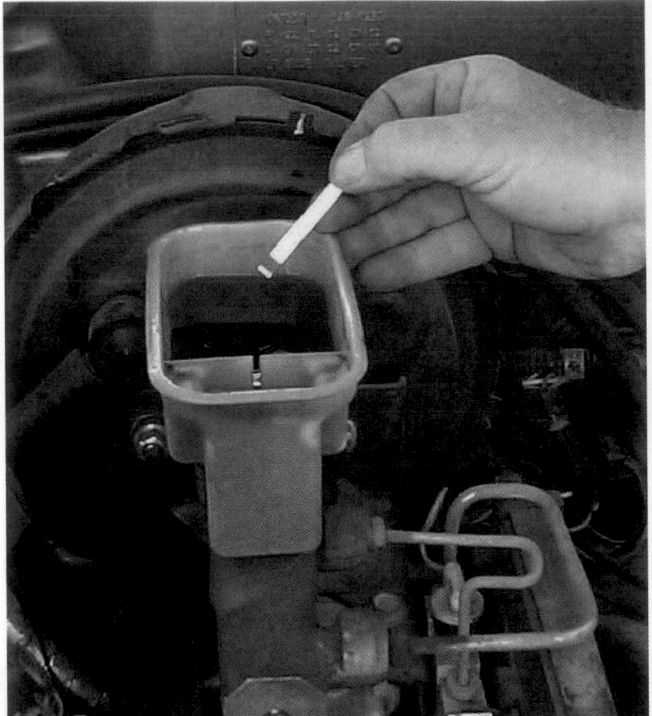

FIGURE 50–27

A test strip can be used to check for copper contaminants in the brake fluid.

Phoenix Systems, LLC

FIGURE 50–28

Using a moisture tester to check brake fluid condition.

volume of fluid inside the calipers and reduces the fluid level in the master cylinder reservoir. If the fluid level is low, possibly causing the BRAKE warning light to illuminate, and no leaks are present, recommend to the customer to have the entire brake system inspected to check for worn linings.

Master Cylinder Inspection

Master cylinder problems are quite common, but not always readily evident. However, there are times when the master cylinder is suspect but the problem lies elsewhere. Accurate and logical troubleshooting is the only way to truly determine if the master cylinder is working properly. Although brake pedal response and reservoir fluid levels are strong indicators of problems with the master cylinder or hydraulic system, other tests can be performed to help pinpoint the problem.

Check the master cylinder housing for cracks and damage. Look for drops of brake fluid around and behind the master cylinder. If a reservoir chamber is cracked, it may be completely empty, and the surrounding area may be dry. This is because the fluid drained very quickly and has had time to evaporate or wash away. But with only one half of the brake system operational, the brake warning lamp should be lit, and a test drive should reveal the loss of braking power.

Refill the master cylinder reservoir section that is empty, and apply the brakes several times. Wait five to ten minutes, and check for leakage or fluid level drop in the reservoir.

Hydraulic brake system leaks can be internal or external. Most internal leaks are actually fluid bypassing the cups in the master cylinder. If the cups lose their ability to seal the pistons, brake fluid leaks past the cups, and the pistons cannot develop system pressure **(Figure 50–29)**.

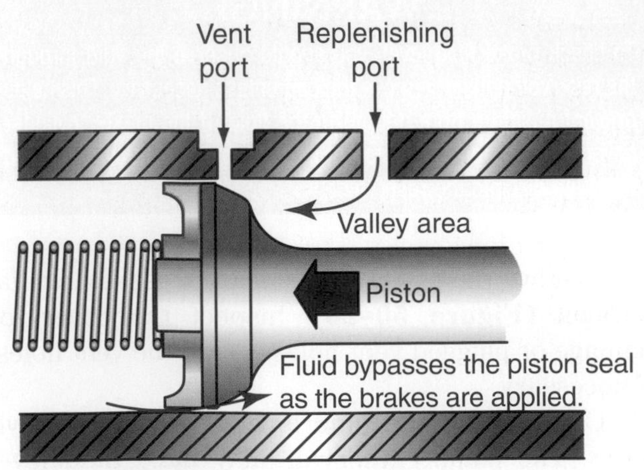

FIGURE 50–29

Fluid can leak around or bypass the piston seals in the master cylinder, causing a low pedal and poor braking.

Internal and external rubber parts wear with use or can deteriorate with age or fluid contamination. Moisture or dirt in the hydraulic system can cause corrosion or deposits to form in the bore, resulting in the wear of the cylinder bore or its parts. Although internal leaks do not cause a loss of brake fluid, they can result in a loss of brake performance. Internal leakage will cause sinking pedal complaints and can be hard to pinpoint.

If the primary piston cup seal is leaking, the fluid will bypass the seal and move between the vent and replenishing ports for that reservoir or, in some cases, between reservoirs. If there are no signs of external leakage, but the BRAKE warning lamp is lit, the master cylinder may have an internal leak. To check for an internal leak in the master cylinder, remove the master cylinder cover, and be sure the reservoirs are at least half full. Watch the fluid levels in the reservoirs while a helper slowly presses the brake pedal and then quickly releases it. If the fluid level rises slightly under steady pressure, the piston cups are probably leaking. Fluid level rising in one reservoir and falling in the other as the brake pedal is pressed and released also can indicate that fluid is bypassing the piston cups. Replace or rebuild the master cylinder if there is evidence of leakage.

Another quick test for internal leakage is to hold pressure on the brake pedal for about one minute. If the pedal drops but no sign of external leakage exists, fluid is probably bypassing the piston cups.

QUICK TAKE-UP VALVE CHECKS Remember, the quick take-up valve is used to provide a high volume of fluid on the first pedal stroke. This action takes up the slack in the low-drag caliper pistons. No direct test method exists for a quick take-up valve, but excessive pedal travel on the first stroke may indicate that fluid is bypassing the valve. If this symptom exists, check for a damaged or unseated valve. If the pedal returns slowly when the brakes are released, the quick take-up valve may be clogged so that fluid flow from the cylinder to the reservoir is delayed.

Air Entrapment Test

Poor pedal feel or action may be caused by trapped air in the system. Trapped air can be the result of a worn or defective master cylinder or other parts of the system. Air will also enter the system if there is a fluid leak. If there are no signs of leaks, check the system for trapped air.

> **CAUTION!**
> This test may result in brake fluid bubbling or spraying out of the master cylinder reservoir. Wear safety goggles. Cover the master cylinder reservoirs with clear plastic wrap or other suitable cover to keep brake fluid off the vehicle's paint.

To check for entrapped air, remove the cover of the master cylinder, and make sure the reservoirs are filled to the proper level. Hold the cover and gasket against the reservoir top, but do not secure it with its clamp or screws. Then have an assistant pump the brake pedal 10 to 20 times rapidly and maintain pressure after the last pedal application. Remove the cover and have the assistant quickly release pedal pressure. Watch for a squirt of brake fluid from the reservoirs. If air is compressed in the system, it will force fluid back through the compensating ports faster than normal and cause fluid to squirt in the reservoir. If a fluid squirt appears in one side of the reservoir but not the other, that side of the split hydraulic system contains the trapped air. If there is air trapped in the system, bleed the system and recheck.

System Flushing

Currently, more than a dozen manufacturers specify periodic brake fluid changes for some, or all, of their models built during the past 12 years. Change intervals vary from as often as every 12 months or 25 000 km (15 000 mi.) to as infrequently as every 100 000 km (60 000 mi.). All brake systems accumulate sludge over some period of time. Flushing the system can remove this sludge and any moisture, but once you have disturbed the sludge, you want to be sure you get it *all* out of the system. Stirring up sludge from the master cylinder reservoir may cause it to get into ABS valves and pumps if you do not get it all out of the system.

Brake hoses for disc brakes usually enter the caliper near the top of the caliper body. The bleeder valve is also located at the top of the caliper bore. If sludge accumulates in the caliper bore, it collects at the bottom. A quick, superficial bleeding of the caliper will not flush out the sludge and all of the old fluid. To flush a caliper thoroughly, pump several ounces of fluid through it. On some vehicles, during brake pad replacement, you may want to remove the caliper from its mounts and retract the piston to force out all the old fluid. Then reinstall it and thoroughly flush it with fresh fluid.

Flushing should be done at each bleeder screw in the same manner as bleeding. Open the bleeder screw approximately 1½ turns, and force fluid through the system until the fluid emerges clear and uncontaminated. Do this at each bleeder screw in the system. After all lines have been flushed, bleed the system, using one of the common bleeding procedures and the proper bleeding sequence. All contaminated fluid should be drawn out of the master cylinder reservoir before bleeding. Make sure you dispose of the old brake fluid in the proper manner.

BRAKE LINE INSPECTION Check all tubing, hoses, and connections, from under the hood to the wheels, for leaks and damage. Wheels and tires should also be inspected for signs of brake fluid leaks. Check all hoses for flexibility, bulges, and cracks. Check parking brake linkage, cable, and connections for seizure, damage, and wear. Replace parts where necessary.

BRAKE PEDAL INSPECTION Depress and release the brake pedal several times (engine running for power brakes). Check for friction and noise. Pedal movement should be smooth, with no squeaks from the pedal or brakes. The pedal should return quickly when it is released.

When operating the engine, be sure the transmission lever is in NEUTRAL or PARK. Be sure the area is properly ventilated for the exhaust to escape.

Apply heavy foot pressure to the brake pedal (engine running for power brakes). Check for a spongy pedal and pedal reserve. Spongy pedal action is springy. Pedal action should feel firm. **Pedal reserve** is the distance between the brake pedal and the floor after the pedal has been depressed fully. The pedal should not go lower than 25 or 50 mm (1 or 2 in.) above the floor.

With the engine off, hold light foot pressure on the pedal for about 15 seconds. There should be no pedal movement during this time. Pedal movement indicates a leak. Repeat the procedure using heavy pedal pressure (engine running for power brakes).

If there is pedal movement but the fluid level is not low, the master cylinder has internal leakage. It must be rebuilt or replaced. If the fluid level is low, there is an external leak somewhere in the brake system. The leak must be repaired.

Depress the pedal and check for proper stoplight operation.

To check power brake operation, depress and release the pedal several times while the engine is stopped. This eliminates vacuum from the system. Hold the brake down with moderate foot pressure, and start the engine. If the power unit is operating properly, the brake pedal moves downward when the engine is started.

MASTER CYLINDER REBUILDING A master cylinder is rebuilt to replace leaking seals or gaskets. If a more serious problem exists, the master cylinder should be replaced.

To remove a master cylinder, disconnect the brake lines at the master cylinder. Install plugs in the brake lines and master cylinder to prevent dirt from entering. Remove the nuts that attach the master cylinder to the firewall power brake unit, and remove the cylinder.

Remove the cover and seal. Drain the master cylinder, and carefully mount it in a vise. Remove the piston assembly and seals according to the manufacturer's instructions. New pistons, pushrods, and seals are usually included in rebuilding kits.

Clean master cylinder parts only with brake fluid, brake cleaning solvent, or alcohol. Do not use a solvent containing mineral oil, such as gasoline. Mineral oil is very harmful to rubber seals.

Inspect the master cylinder. Damage, cracks, porous leaks, and worn piston bores mean the master cylinder must be replaced. Check very carefully for pitting or roughness in the bore. If any are present, the cylinder must be replaced. Only older cast-iron master cylinders can have minor bore wear corrected by honing. If honing is allowed, only brake fluid should be used as a honing lubricant.

Reassemble, install, and bleed the master cylinder according to the manufacturer's directions. Photo Sequence 43 is a typical procedure for bench-bleeding a master cylinder.

WARNING!

Brake fluid will remove paint. Always use fender covers to protect the vehicle's finish, and take extra care not to spill brake fluid.

Hydraulic System Bleeding

Fluids cannot be compressed, whereas gases are compressible. Any air in the brake hydraulic system is compressed as the pressure increases. This action reduces the amount of force that can be transmitted by the fluid; therefore, it is very important to keep all air out of the hydraulic system. To do this, air must be bled from brakes. This procedure is called **bleeding** the brake system.

Bleeding is a process of forcing fluid through the brake lines and out through a bleeder valve or

TYPICAL PROCEDURE FOR BENCH-BLEEDING A MASTER CYLINDER

PHOTO SEQUENCE 43

P43–1 Mount the master cylinder firmly in a vise, being careful not to apply excessive pressure to the casting. Position the master cylinder so that the bore is horizontal.

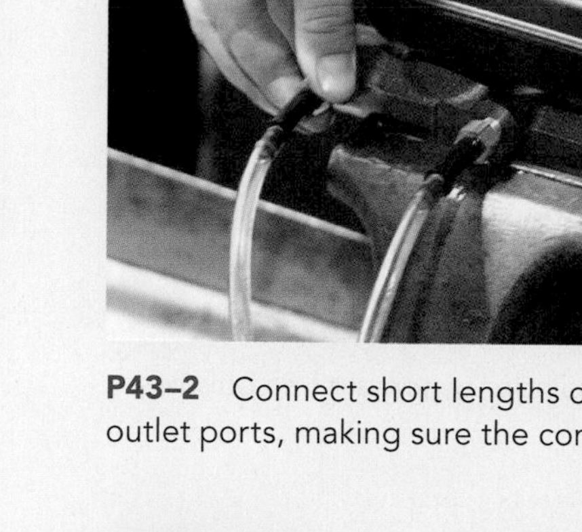

P43–2 Connect short lengths of tubing to the outlet ports, making sure the connections are tight.

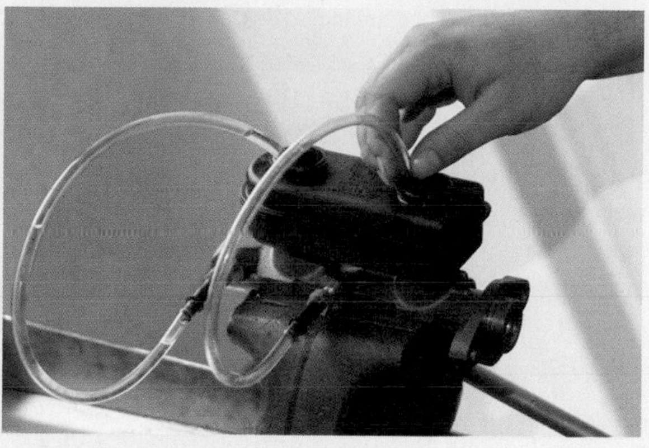

P43–3 Bend the tubing lines so that the ends are in each chamber of the master cylinder reservoir.

P43–4 Fill the reservoirs with fresh brake fluid until the level is above the ends of the tubes.

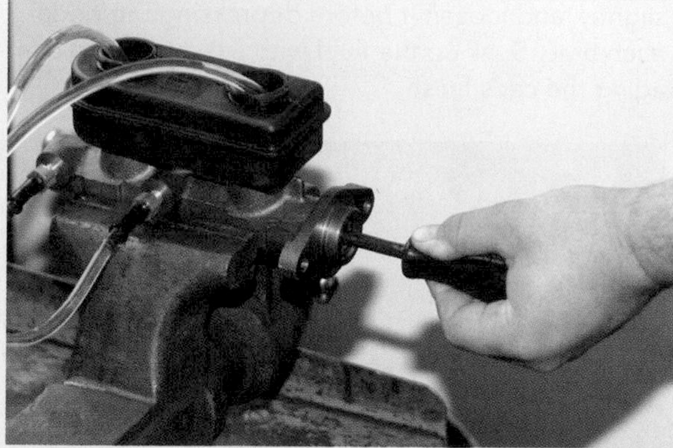

P43–5 Using a wooden dowel or the blunt end of a drift or punch, slowly push on the master cylinder pistons until both are completely bottomed out in their bore.

P43–6 Watch for bubbles to appear at the tube ends immersed in the fluid. Slowly release the cylinder piston, and allow it to return to its original position. On quick take-up master cylinders, wait 15 seconds before pushing in the piston again. On other units, repeat the stroke as soon as the piston returns to its original position. Slow piston return is normal for some master cylinders.

TYPICAL PROCEDURE FOR BENCH-BLEEDING A MASTER CYLINDER *(continued)*

P43–7 Pump the cylinder piston until no bubbles appear in the fluid.

P43–8 Remove the tubes from the outlet ports, and plug the openings with temporary plugs or your fingers. Keep the ports covered until you install the master cylinder on the vehicle.

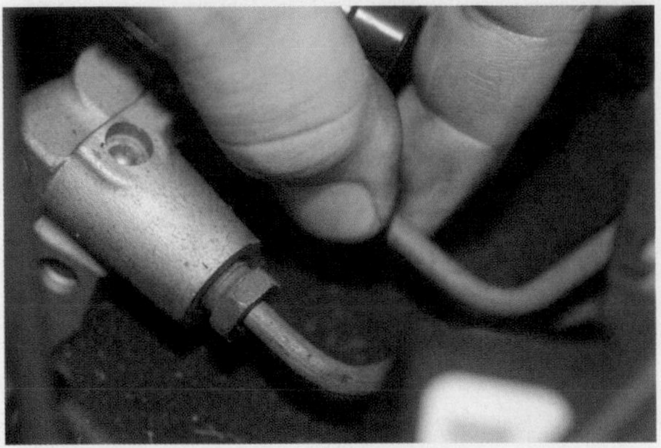

P43–9 Install the master cylinder on the vehicle. Attach the lines, but do not tighten the tube connections.

P43–10 Slowly depress the pedal several times to force out any air that might be trapped in the connections. Before releasing the pedal, tighten the nut slightly, and loosen it before depressing the pedal each time. Soak up the fluid with a rag to avoid damaging the car's finish.

P43–11 When there are no air bubbles in the fluid, tighten the connections to the manufacturer's specifications. Make sure the master cylinder reservoirs are adequately filled with brake fluid.

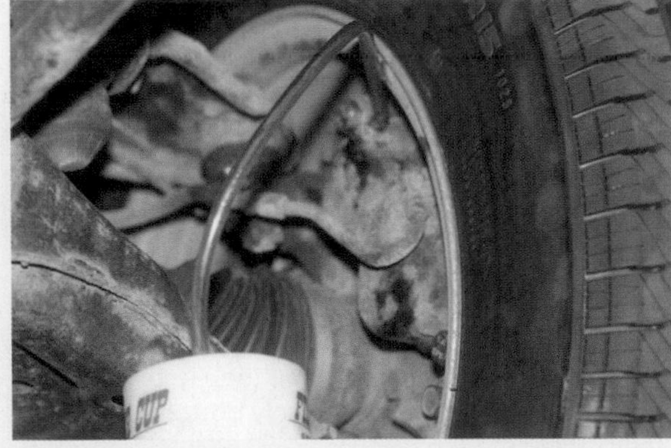

P43–12 After reinstalling the master cylinder, bleed the entire brake system on the vehicle.

FIGURE 50–30

The wheel brake units are fitted with bleeder screws.

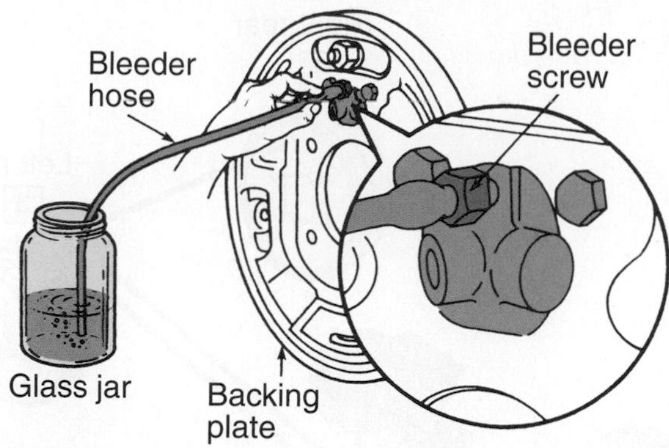

FIGURE 50–31

While bleeding the brake system, releasing the fluid into a glass jar will help to determine when the system is free of air.

bleeder screw **(Figure 50–30)**. The fluid eliminates any air that might be in the system. Bleeder screws and valves are fastened to the wheel cylinders or calipers. Before attempting to open, the bleeder should be cleaned. A drain hose then is connected from the bleeder to a glass jar **(Figure 50–31)**.

CUSTOMER CARE

Remind your customers to use only approved and recommended brake fluid in a brake system. Any other lubricant, such as power-steering fluid, automatic transmission fluid, or engine oil, which has a petroleum base, must never be used in the brake system. Petroleum-based fluids attack the rubber components in the brake system, such as the piston cups and seals, and cause them to swell and disintegrate. If you open a brake fluid reservoir and find the diaphragm is swelled out of shape and is larger than it should be, you can be sure someone has previously put some kind of petroleum-based fluid in it. Such a discovery is costly to the vehicle's owner as all rubber components in the entire brake system will need to be replaced.

Four types of brake-bleeding procedures are used: manual, gravity, vacuum, and pressure bleeding. On some antilock brake systems, a scan tool is required to bleed the brakes. Always follow the manufacturer's recommendations when bleeding brakes. The sequence in which bleeding is performed can be critical. When bleeding a power brake system, remove the vacuum line from the power unit, and plug the unit. To remove vacuum from the vacuum assist unit, the engine must be off; then pump the brake pedal several times.

WARNING!

Always use fresh brake fluid when bleeding the system. Do not use fluid from an open container. Do not use fluid that has been drained. Drained fluid may be contaminated and can damage the system.

BLEEDING SEQUENCE All manufacturers recommend a specific sequence to follow when bleeding a vehicle's brakes. These recommendations can be found in service information and should be followed. If the manufacturer's recommendations are not available, the following sequence will work on most vehicles. This sequence starts at the wheel farthest from the master cylinder and works to the wheel closest:

1. Master cylinder
2. Combination valve or proportioning valve (if fitted with bleeder screws)
3. Right rear
4. Left rear
5. Right front
6. Left front
7. Height-sensing proportioning valve (if there is a bleeder screw)

Remember to stop after bleeding each wheel and refilling the fluid in the master cylinder. Failure to perform this step may result in running out of fluid and pulling air into the system.

This sequence is based on the principles of starting at the highest point in the system and working downward, then starting at the wheel farthest from the master cylinder and working to the closest. A couple of more general rules also are worth remembering.

If the brake system is split between the front and rear wheels, the rear wheels (which are farthest

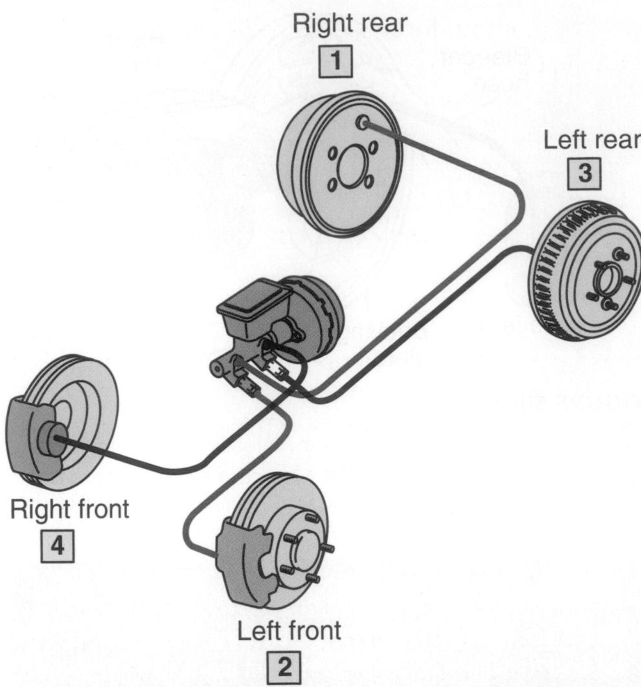

FIGURE 50–32
The recommended bleeding sequence for a diagonally split brake system.

from the master cylinder) usually are bled first. If the brake system is split diagonally, the most common sequence is: RR-LF-LR-RF **(Figure 50–32)**. This sequence also applies to most systems with a quick take-up master cylinder. If you bleed a quick take-up system in any other sequence, you may chase air throughout the system. Exceptions to the general rules exist, however. Chrysler, for example, recommends bleeding both rear brakes before the front brakes, regardless of how the hydraulic system is split.

MANUAL BLEEDING A manual bleeding procedure requires two people. One person operates the bleeder; the other, the brake pedal. Bleed only one wheel at a time.

WARNING!

Be sure the bleeder hose is below the surface of the liquid in the jar at all times. Do not allow the master cylinder to run out of fluid at any time. If these precautions are not followed, air can enter the system, and it must be bled again. The master cylinder cover must be kept in place.

Place the bleeder hose and jar in position. Have a helper pump the brake pedal several times and then hold it down with moderate pressure. Slowly open the bleeder valve. After fluid/air has stopped flowing, close the bleeder valve. Have the helper slowly release the pedal. Repeat this procedure until fluid that flows from the bleeder is clear and free of bubbles.

Discard all used brake fluid. Fill the master cylinder reservoir after bleeding each wheel brake. Check the brakes for proper operation.

WARNING!

Clean the master cylinder and cover before adding fluid. This is important for preventing dirt from entering the reservoir.

PRESSURE BLEEDING A pressure bleeding procedure can be done by one person. Pressure bleeding equipment uses pressurized fluid that flows through a special adapter fitted into the master cylinder **(Figure 50–33)**.

The use of pressure bleeding equipment varies with different automobiles and different equipment makers. Always follow the automobile manufacturer's recommendations when using pressure bleeding equipment.

On automobiles with metering valves, the valve must be held open during pressure bleeding. A special tool is used to hold open the metering section of a combination valve.

USING SERVICE INFORMATION

Consult the service information to be sure the proper bleeding sequence is followed. If a vehicle requiring a special sequence is bled in the conventional manner, air might be chased throughout the system.

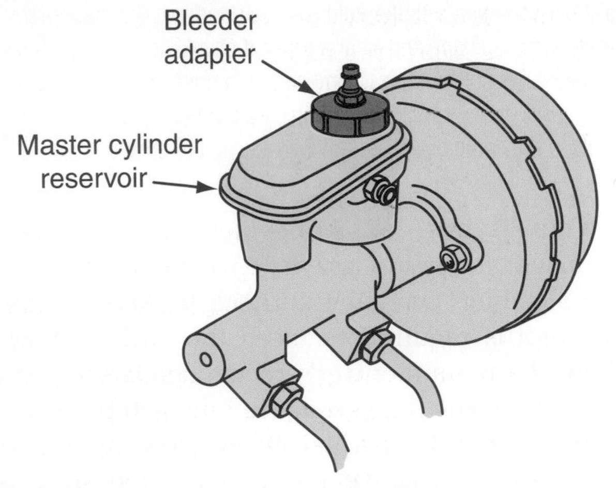

FIGURE 50–33
An adapter that installs onto the reservoir's fill opening is required to pressure bleed the brake system.

Open the bleeder valves one at a time until clear, air-free fluid is flowing. Progress from the wheel cylinder farthest from the master cylinder to the cylinder closest.

Do not exceed recommended pressure while bleeding the brakes. Always release air pressure after bleeding. Clean and fill the master cylinder after pressure bleeding. Check the brakes for proper operation. Be sure to remove the special tool used to hold the metering valve.

Power Brakes

Power brakes are nothing more than a standard hydraulic brake system with a booster unit located between the brake pedal and the master cylinder to help activate the brakes.

Two basic types of power-assist mechanisms are used. The first is **vacuum assist**. These systems use engine vacuum, or sometimes vacuum pressure developed by an external vacuum pump, to help apply the brakes. The second type of power assist is **hydraulic assist**. It is normally found on larger or diesel engine–powered vehicles. This system uses hydraulic pressure developed by the power-steering pump or other external pump to help apply the brakes.

Both vacuum and hydraulic assist act to multiply the force exerted on the master cylinder pistons by the driver. This increases the hydraulic pressure delivered to the wheel cylinders or calipers while decreasing driver foot pressure.

Vacuum-Assist Power Brakes

All vacuum-assisted units are similar in design. They generate application energy by opposing engine vacuum to atmospheric pressure. A piston and cylinder, flexible diaphragm, or bellows use this energy to provide braking assistance.

All modern vacuum-assist units are vacuum-suspended systems. This means the diaphragm inside the unit is balanced using engine vacuum until the brake pedal is depressed. Applying the brake allows atmospheric pressure to unbalance the diaphragm and allows it to move generating application pressure.

Atmospheric pressure is normally between 0.98 and 1.05 kg/cm^2 (14 and 15 psi). If the diameter of the diaphragm is 30.5 cm (12 in.), the area of the diaphragm is about 730 cm^2 (113 in.^{2}). Because the vacuum to the booster is typically 50.8 cm Hg of mercury (20 in. Hg) or about 0.70 kg/cm^2 (10 psi), which is the pressure difference on the diaphragm, the resulting output force would be 511 kg (0.70 × 730) or 1130 lb. (10 × 113).

Vacuum boosters may be single diaphragm or tandem diaphragm. The unit consists of three basic elements combined into a single power unit **(Figure 50–34)**.

The three basic elements of the single diaphragm follow:

1. A vacuum power section that includes a front and rear shell, a power diaphragm, a return spring, and a pushrod.
2. A control valve built as an integral part of the power diaphragm and connected through a valve rod to the brake pedal. It controls the degree of brake application or release in accordance with the pressure applied to the brake pedal.
3. A hydraulic master cylinder, attached to the vacuum power section that contains all the elements of the conventional brake master cylinder except for the pushrod. It supplies fluid under pressure to the pressure applied by the brake booster.

OPERATION The vacuum booster has three operating modes: applied, hold, and release. The applied mode is used when the brakes are applied; the valve rod and plunger move to the left in the power diaphragm. This action closes the control valve's vacuum port and opens the atmospheric port to admit air through the valve at the rear diaphragm chamber. With vacuum in the rear chamber, a force develops that moves the power diaphragm, hydraulic pushrod, and hydraulic piston or pistons to close the compensating port or ports and force fluid under pressure through the residual check valve or valves and lines into the front and rear brake assemblies.

As pressure develops in the master cylinder, a counterforce acts through the hydraulic pushrod and reaction disc against the vacuum power diaphragm and valve plunger. This force tends to close the atmospheric port, trapping whatever pressure is present in the rear chamber, which creates a holding condition or hold mode. Because this force is in opposition to the force applied to the brake pedal by the operator, it gives the operator a feel for the amount of brake applied.

If the input force is reduced, the counterforce from the hydraulic pressure against the master cylinder pistons will push the control valve plunger further rearward, opening the vacuum port and allowing both the front and rear chambers open to engine vacuum. This position releases pressure

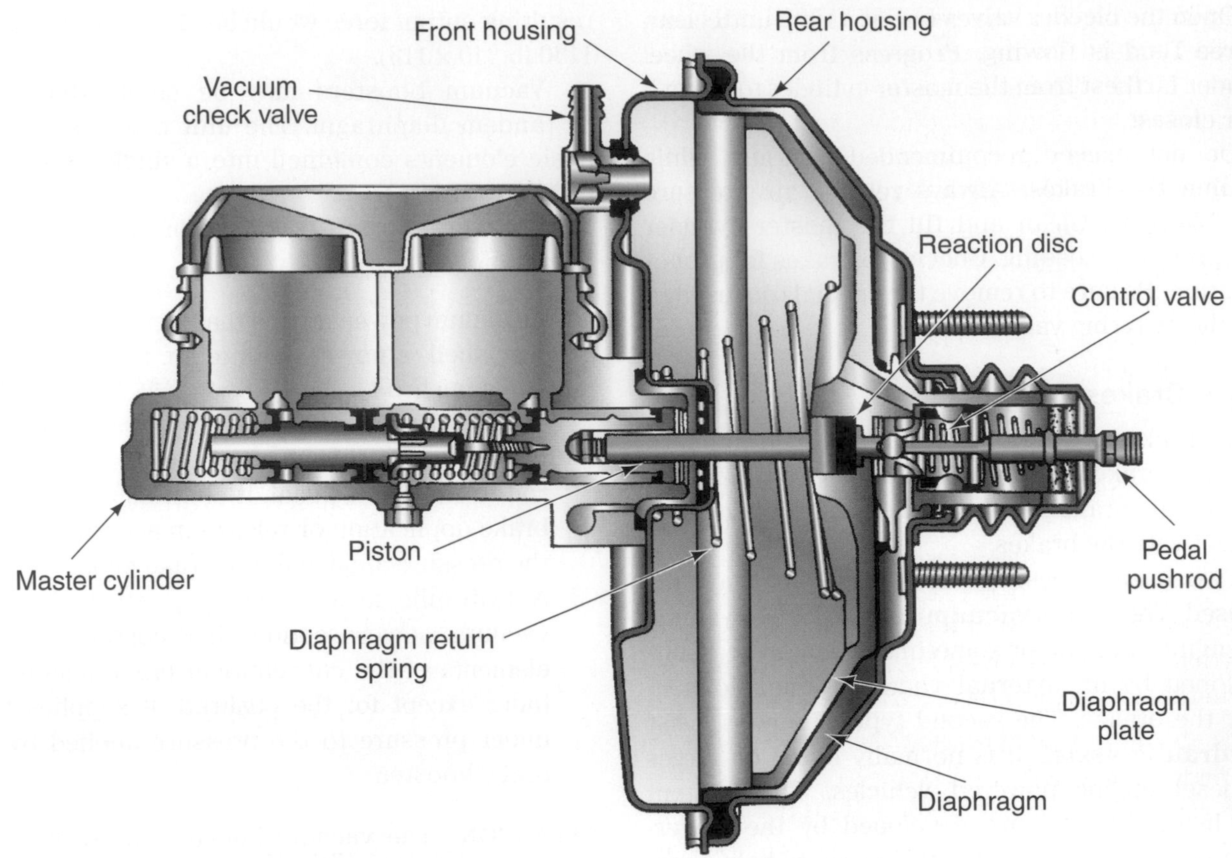

FIGURE 50–34

A typical vacuum brake booster.

against the master cylinder and therefore is the release position or mode.

SERVICING VACUUM-ASSIST BOOSTER UNITS The fact that a vehicle's brakes still operate when the vacuum-assist unit fails indicates that the hydraulic brake system and the vacuum-assist system are two separate systems. This means you should always check for faults in the hydraulic system first. If it checks out satisfactorily, start inspecting the vacuum-assist circuit.

For a fast check of vacuum-assist operation, press the brake pedal firmly, and then start the engine. The pedal should fall away slightly, and less pressure should be needed to maintain the pedal in any position. This also checks the vacuum check valve's operation. The pedal should be easy to press at least once or twice and then get very firm. If the check valve is not working, the pedal will be very hard to press the first time after the engine is off.

PRESSURE CHECK Another simple check can be made by installing a suitable pressure gauge in the brake hydraulic system. Take a reading with the engine off and the power unit not operating. Maintain the same pedal height, start the engine, and read the gauge. There should be a substantial pressure increase if the vacuum-assist booster is operating correctly.

PEDAL TRAVEL **Pedal travel** and total travel are critical on vacuum-assisted vehicles. Pedal travel should be kept strictly to specifications listed in the vehicle's service information.

VACUUM READING If the power unit is not giving sufficient assistance, take a manifold vacuum reading. If manifold vacuum level is below specifications, tune the engine and retest the unit. Loose or damaged vacuum lines and clogged air intake filters reduce braking assistance. Most units have a check valve that retains some vacuum in the system when the engine is off. A vacuum gauge check of this valve indicates if it is restricted or stays open.

RELEASE PROBLEMS Failure of the brakes to release is often caused by a tight or misaligned connection between the power unit and the brake linkage. Broken pistons, diaphragms, bellows, or return springs can also cause this problem.

To help pinpoint the problem, loosen the connection between the master cylinder and the brake booster. If the brakes release, the problem is caused by internal binding in the vacuum unit. If the brakes do not release, look for a crimped or restricted brake line or similar problem in the hydraulic system.

HARD PEDAL Power brakes that have a **hard pedal** may have collapsed or leaking vacuum lines of insufficient manifold vacuum. Punctured diaphragms or bellows and leaky piston seals all lead to weak power unit operation and hard pedal. A steady hiss when the brake is held down indicates a vacuum leak that causes poor operation.

GRABBING BRAKES First, look for common causes of brake grab, such as greasy linings or scored rotors or drums. If the trouble appears to be in the power unit, check for a damaged reaction control. The reaction control is made up of a diaphragm, spring, and valve that tend to resist pedal action. It is put into the system to give the driver more brake pedal feel.

CHECK OF INTERNAL BINDING Release problems, hard pedal, and dragging (slow-releasing) brakes can all be caused by internal binding. To test a vacuum unit for internal binding, place the transmission/transaxle in neutral and start the engine. Increase engine speed to 1500 rpm, close the throttle, and completely depress the brake pedal. Slowly release the brake pedal and stop the engine. Remove the vacuum check valve and hose from the vacuum assist unit. Observe for backward movement of the brake pedal. If the brake pedal moves backward, there is internal binding and the unit should be replaced.

PUSHROD ADJUSTMENT

Proper adjustment of the master cylinder pushrod is necessary to ensure proper operation of the power brake system. A pushrod that is too long causes the master cylinder piston to close off the replenishing port, preventing hydraulic pressure from being released and resulting in brake drag. A pushrod that is too short causes excessive brake pedal travel and causes groaning noises to come from the booster when the brakes are applied. A properly adjusted pushrod that remains assembled to the booster with which it was matched during production should not require service adjustment. However, if the booster, master cylinder, or pushrod are replaced, the pushrod might require adjustment.

Two methods can be used to check for proper pushrod length and installation: the gauge method and the air method.

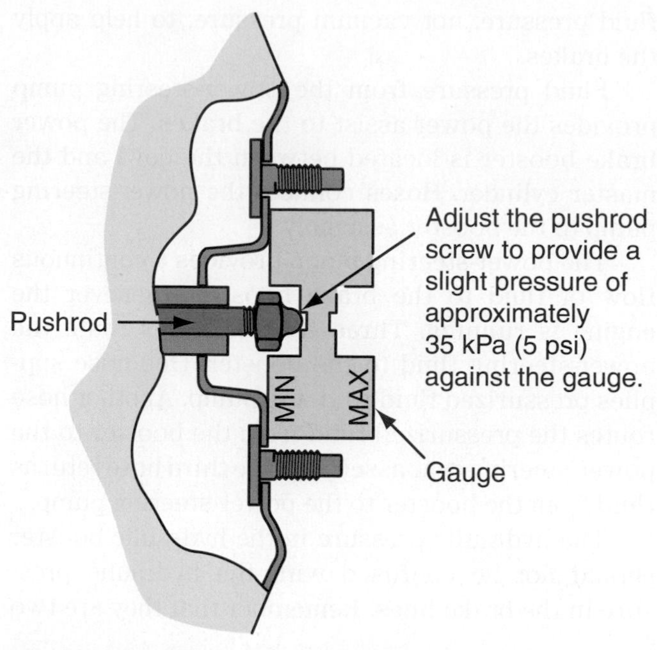

Adjust the pushrod screw to provide a slight pressure of approximately 35 kPa (5 psi) against the gauge.
Pushrod
MIN
MAX
Gauge

FIGURE 50–35

A gauge for measuring pushrod length.

Gauge Method

In most vacuum power units, the master cylinder pushrod length is fixed, and length is usually checked only after the unit has been overhauled or replaced. A typical adjustment using the gauge method is shown in **Figure 50–35**.

Air Method

The air-testing method uses compressed air applied to the hydraulic outlet of the master cylinder. Air pressure is regulated to a value of approximately 35 kPa (5 psi) to prevent brake fluid spraying from the master cylinder.

If air passes through the compensating port, which is the smaller of the two holes in the bottom of the master cylinder reservoir, the adjustment is satisfactory. If air does not flow through the compensating port, adjust the pushrod as required, either by means of the adjustment screw (if provided) or by adding shims between the master cylinder and power unit shell until the air flows freely.

HYDRAULIC BRAKE BOOSTERS

Decreases in engine size, increased use of diesel engines, plus the continued use of engine vacuum to operate other engine systems, such as emission control devices, led to the development of hydraulic-assist power brakes. These systems use fluid pressure, not vacuum pressure, to help apply the brakes.

Fluid pressure from the power-steering pump provides the power assist to the brakes. The power brake booster is located between the cowl and the master cylinder. Hoses connect the power-steering pump to the booster assembly.

The power-steering pump provides a continuous flow of fluid to the brake booster whenever the engine is running. Three flexible hoses route the power-steering fluid to the booster. One hose supplies pressurized fluid from the pump. Another hose routes the pressurized fluid from the booster to the power-steering gear assembly. The third hose returns fluid from the booster to the power-steering pump.

The hydraulic pressure in the hydraulic booster should not be confused with the hydraulic pressure in the brake lines. Remember that they are two separate systems and require two different types of fluid: power-steering fluid for the pump and brake fluid for the brake system. Never put power-steering fluid in the brake reservoir. If the brake fluid becomes contaminated by power-steering fluid, the rubber components will be damaged. This may require the replacement of all parts that contain rubber, and the steel lines of the brake system will need to be flushed.

Some systems have a nitrogen-charged pneumatic accumulator on the booster to provide reserve power-assist pressure. If power-steering pump pressure is not available due to belt failure or similar problems, the accumulator pressure is used to provide brake assist.

The booster assembly **(Figure 50–36)** consists of an open-centre spool valve and sleeve assembly, a lever assembly, an input rod assembly, a power

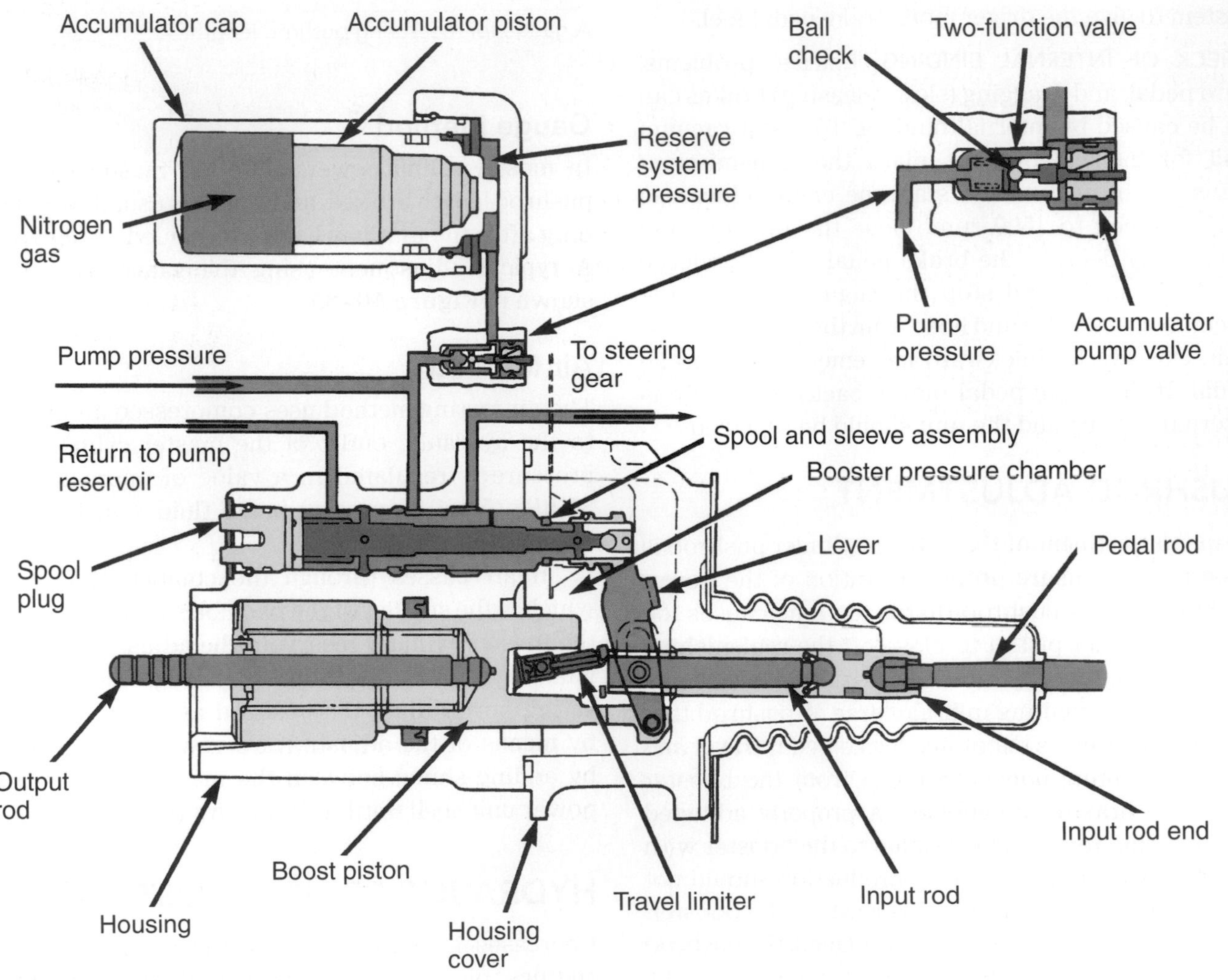

FIGURE 50–36

A hydraulic brake booster.

Chrysler Group LLC

piston, an output pushrod, and the accumulator. The booster assembly is mounted on the vehicle in much the same manner as a vacuum booster. The pedal rod is connected at the booster input rod end.

Power-steering fluid flow in the booster unit is controlled by a hollow-centre spool valve. The spool valve has lands, annular grooves, and drilled passages. These mate with grooves and lands in the valve bore. The flow pattern of the fluid depends on the alignment of the valve in the bore.

Operation

When the brake pedal is depressed, the pedal's pushrod moves the master cylinder's primary piston forward. This causes the lever assembly of the booster to move a sleeve forward to close off the holes leading to the open centre of the spool valve. A small additional lever movement moves the spool valve into the spool valve bore. The spool valve then diverts some hydraulic fluid into the cavity behind the booster piston, building up hydraulic pressure that moves the piston and a pushrod forward. The output pushrod moves the primary and secondary master cylinder pistons that apply pressure to the brake system. When the brake pedal is released, the spool and sleeve assemblies return to their normal positions. Excess fluid behind the piston returns to the power-steering pump reservoir through the return line. After the brakes have been released, pressurized fluid from the power-steering pump flows into the booster through the open centre of the spool valve and back to the power-steering pump.

There have been many variations of and names given to hydraulic brake booster systems. The most common names are the hydro-boost system produced by Bendix and the Powermaster system produced by General Motors. The Powermaster system is a little different from the rest in that it uses a self-contained hydraulic booster that is built directly onto the master cylinder. Instead of relying on the power-steering pump for hydraulic pressure, as is done in the other systems, the Powermaster has its own vane pump and electric motor to provide the hydraulic pressure required for booster operation.

Any investigation of a hydraulic boost complaint should begin with an inspection of the power-steering pump belt, fluid level, and hose condition and connections. Hydraulic boost systems do not work properly if they are not supplied with a continuous supply of clean, bubble-free power-steering fluid at the proper pressure.

WARNING!

Always depressurize the accumulator of any hydraulic boost system before disconnecting any brake lines or hoses. This is usually done by turning the engine off and depressing and releasing the brake pedal up to 20 times.

Basic Operational Test

The basic operational test of these systems is as follows. With the engine off, pump the brake pedal numerous times to bleed off the residual hydraulic pressure that is stored in the accumulator. Hold firm pressure on the brake pedal, and start the engine. The brake pedal should move downward and then push up against the foot.

Accumulator Test

To be sure the **accumulator** is performing properly, rotate the steering wheel with the engine running, until it stops, and hold it in that position for no more than five seconds. Return the steering wheel to the centre position, and shut off the engine. Pump the brake pedal. You should feel two to three power-assisted strokes. Now repeat the steps. That pressurizes the accumulator. Wait one hour and then pump the brake pedal. There should be two or three power-assisted strokes. If the system does not perform as just described, the accumulator is leaking and should be replaced.

Noise Troubleshooting

The booster is also part of another major subsystem of the vehicle, the power-steering system. Problems or malfunctions in the steering system may affect brake-assist operation. The following are some common troubleshooting tips.

Moan or low-frequency hum usually accompanied by a vibration in the pedal or steering column might be encountered during parking or other very low-speed manoeuvres. This can be caused by a low fluid level in the power-steering pump, or by air in the power-steering fluid due to holding the pump at relief pressure (steering wheel held all the way in one direction) for an excessive amount of time (more than five seconds). Check the fluid level and add fluid, if necessary. Allow the system to sit for one hour with the cap removed to eliminate the air. If the condition persists, it might be a sign of excessive pump wear. Check the pump according to the vehicle manufacturer's recommended procedure.

At or near power runout (brake pedal near fully depressed position), a high-speed fluid noise (like a faucet can make) might occur. This is a normal condition and will not be heard except in emergency braking conditions.

Whenever the accumulator pressure is used, a slight hiss is noticed. It is the sound of the hydraulic fluid escaping through the accumulator valve and is completely normal.

After the accumulator has been emptied and the engine is started again, another hissing sound might be heard during the first brake application or the first steering manoeuvre. This sound is caused by the fluid rushing through the accumulator charging orifice. It is normal and will be heard only once after the accumulator is emptied. However, if this sound continues, even though no apparent accumulator pressure assist was made, it could indicate that the accumulator is not holding pressure. Check for this possibility, using the accumulator test discussed previously.

After bleeding, a gulping sound might be present during brake applications, as noted in the bleeding instructions. This sound is normal and should disappear with normal driving and braking.

Diagnosis and testing of the Powermaster unit requires the use of a special adapter and test gauge or aftermarket equivalents. The Powermaster pressure switch is removed, and the adapter and test gauge are installed in its port. The unit can then be energized, and the switch's high-pressure cutoff and low-pressure turn-on points observed and checked against specifications. Follow the service information instructions for the connection and operation of the test gauge and all system test procedures.

ELECTRIC PARKING BRAKES

Electrically operated parking brakes are becoming more common and are replacing mechanical systems. These systems operate as a conventional hydraulic brake for normal braking and as an electric brake for parking. With electric parking brakes, there is no need for a parking brake lever or pedal. This frees up space in the interior.

Electric parking brakes are seen as the first step toward brake-by-wire systems. Two different techniques are currently being used by manufacturers. Some systems have an electric motor mounted on the rear brake calipers, and others use an undercar motor to pull on the parking brake cables.

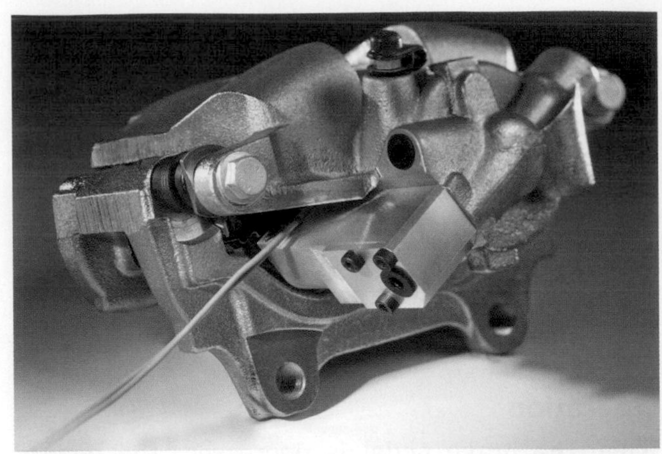

FIGURE 50–37

This caliper is fitted with an electronically controlled motor that controls the application of the parking brake.

Robert Bosch Corp.

When the caliper is fitted with a motor **(Figure 50–37)**, there is no need for parking brake cables and linkages. The motor is controlled by the powertrain control module (PCM) or a specific park brake module. The system interfaces with the vehicle's controller area network (CAN) for continuous monitoring and feedback. This allows the system to do many things besides apply the parking brake, such as the following:

- Provide some control during emergency braking
- Help stop the car if the hydraulic system fails
- Automatically release the parking brakes when the throttle is opened
- Automatically engage when the ignition is turned off
- Automatically engage when the driver's door is opened
- Keep the vehicle from rolling backward when stopped on a hill, by applying the rear brakes until the driver operates the clutch or throttle pedals

KEY TERMS

Accumulator (p. 1571)
Aramid fibres (p. 1543)
Bleeding (p. 1562)
Brake fade (p. 1542)
Caliper (p. 1559)
Coefficient of friction (COF) (p. 1541)
Combination valve (p. 1555)
Diagonally split system (p. 1545)
DOT (p. 1546)
Equilibrium reflux boiling point (ERBP) (p. 1546)
Friction (p. 1540)
Hard pedal (p. 1569)
Hydraulic assist (p. 1567)

ISO (p. 1552)
Kinetic friction (p. 1540)
Metallic lining (p. 1543)
Metering valve (p. 1553)
Nonasbestos organic (NAO) linings (p. 1542)
Pedal reserve (p. 1562)
Pedal travel (p. 1568)
Proportioning valve (p. 1554)
Quick take-up (p. 1550)
Replenishing port (p. 1548)
Rotor (p. 1558)
Semimetallic linings (p. 1543)
Static friction (p. 1540)
Synthetic lining (p. 1543)
Vacuum assist (p. 1567)
Ventilated rotors (p. 1558)
Vent port (p. 1548)

SUMMARY

- The four factors that determine a vehicle's braking power are pressure, which is provided by the hydraulic system; coefficient of friction, which represents the frictional relationship between pads and rotors, or shoes and drums, and is engineered to ensure optimum performance; frictional contact surface, meaning that bigger brakes stop a car more quickly than smaller brakes; and heat dissipation, which is necessary to prevent brake fade.
- With asbestos banned as a brake lining material, in many countries today, full metallics and semimetallics are used, as well as other nonasbestos substances.
- Since 1967, all cars have been required to have two separate brake systems. The dual brake system uses a tandem master cylinder, which is two master cylinders formed by installing two separated pistons and fluid reservoirs in one cylinder bore.
- Brake fluid is the lifeblood of any hydraulic brake system. DOT 4 is the type most commonly used on conventional systems. DOT 3 is recommended for most ABS and some power brake systems.
- The brake lines transmit brake fluid pressure from the master cylinder to the wheel cylinders and calipers of drum and disc brakes. Brake hoses offer flexible connections to wheel units and must offer high heat resistance.
- A pressure differential valve is used in all dual brake systems to operate a warning light switch that alerts the driver if pressure is lost in either hydraulic system.
- The metering valve, located in the front brake line, provides for better balance of the front and rear brakes while also preventing lock-up of the front brakes. The proportioning valve controls rear brake pressure, particularly during hard stops.
- Hydraulic brake system service is vital to safe vehicle operation. It includes brake fluid, brake line, and brake pedal inspection.
- Bleeding removes air from the hydraulic system.
- Power brakes can be either vacuum assist or hydraulic assist. Vacuum-assisted units use engine vacuum or vacuum developed by an external pump to help apply the brakes. Hydraulic-assisted units use fluid pressure.

REVIEW QUESTIONS

1. Explain how vacuum is used to provide a power assist.
2. Explain why modern hydraulic braking systems are dual designs and why this is important.
3. Define the different types of power brake designs, and tell how they operate.
4. Describe the purpose of a pressure differential valve.
5. Describe the functions of the hydraulic system combination valve.
6. When not skidding, what type of friction is generated between the tire and the road?
   a. static **b.** kinetic
   c. heat **d.** coefficient
7. What is removed from the hydraulic brake system during bleeding?
   a. water
   b. contaminated brake fluid
   c. air
   d. dirt
8. What is the purpose of the master cylinder?
   a. to generate the hydraulic pressure needed to apply the brake assemblies
   b. to automatically pump the brakes during panic stops
   c. to apply braking power when wheel slippage occurs
   d. to delay pressure to the front brakes until the rear brakes apply
9. Which of the following bleeding methods requires two people to perform?
   a. manual **b.** gravity
   c. vacuum **d.** pressure
10. What is the purpose of the white lines printed along the length of brake hoses?
    a. to identify the type of brake hose
    b. to identify the manufacturer of the brake hose
    c. to help prevent brake hose twisting during installation
    d. to determine the strength of the brake hose

11. What hydraulic component is used to help balance the braking characteristics in a disc and drum brake system?
   a. the metering valve
   b. the proportioning valve
   c. the pressure differential valve
   d. the residual pressure check valve
12. Which type of brake requires greater application force and is commonly used with power-boost units?
   a. rear drum brakes
   b. front drum brakes
   c. disc brakes
   d. parking brakes
13. Which of the following best describes the function of the compensating port?
   a. allows the brake fluid in the system to operate at atmospheric pressure when the pedal is released
   b. allows the brake fluid in the system to operate at atmospheric pressure when the pedal is depressed
   c. balances pressure between the primary and the secondary circuits
   d. prevents atmospheric pressure from reaching the brake fluid in the system
14. In brake fluid, what percentage of moisture contamination is considered excessive?
   a. 2% b. 3%
   c. 4% d. 5%
15. What hydraulic component is used to prevent the front disc brakes from applying before the rear drum brakes overcome return spring pressure and contact the drums?
   a. the metering valve
   b. the proportioning valve
   c. the pressure differential valve
   d. the residual pressure check valve
16. What is the source of brake system boost when using a hydraulic brake booster?
   a. engine vacuum pressure
   b. engine compression pressure
   c. electric/hydraulic motor pressure
   d. power-steering pump pressure
17. Which of the following could cause swelling and deterioration of internal brake system rubber components?
   a. using DOT 3 brake fluid in a system that recommends DOT 4
   b. using DOT 4 brake fluid in a system that recommends DOT 3
   c. air in the hydraulic system
   d. using power-steering fluid in the hydraulic system
18. A force of 100 kg (220.5 lb.) is placed on a 5 cm^2 (0.775 $in.^2$) master cylinder piston. What would the output force be from a 35 cm^2 (5.425 $in.^2$) caliper piston?
   a. 500 kg (1102 lb.)
   b. 700 kg (1543 lb.)
   c. 35 000 kg (77 162 lb.)
   d. 175 000 kg (385 809 lb.)
19. If the speed of the vehicle is doubled, how is the braking force required to stop the vehicle affected?
   a. The braking force required is cut in half.
   b. The braking force required remains the same.
   c. The braking force required is also doubled.
   d. The braking force required must be increased four times.
20. Which of the following could cause an extremely hard brake pedal?
   a. air in the system
   b. excessively worn brake pads
   c. use of the wrong fluid
   d. a leaking diaphragm in the vacuum power booster

CHAPTER 51
Drum Brakes

LEARNING OUTCOMES

- Explain how drum brakes operate.
- Identify the major components of a typical drum brake, and describe their functions.
- Explain the difference between duo-servo and nonservo drum brakes.
- Perform a cleaning and inspection of a drum brake assembly.
- Recognize conditions that adversely affect the performance of drums, shoes, linings, and related hardware.
- Reassemble a drum brake after servicing.
- Explain how typical drum parking brakes operate.
- Adjust a typical drum parking brake.

For many years, drum brakes **(Figure 51–1)** were used on all four wheels on virtually every vehicle on the road. Today, disc brakes have replaced drum brakes on the front wheels of most vehicles and some models are equipped with both front and rear disc brakes. One reason for their continued use is that drum brakes can easily handle the 20 to 40 percent of total braking load placed on the rear wheels. Another is that drum brakes can also be built with a simple parking brake mechanism.

DRUM BRAKE OPERATION

Drum brake operation is fairly simple. The most important feature contributing to the effectiveness of the braking force supplied by the drum brake is the brake shoe pressure or force directed against the drum **(Figure 51–2)**. With the vehicle moving in either the forward or reverse direction with the brakes on, the applied force of the brake shoe pressing against the brake drum increasingly multiplies itself because the brake's anchor pin acts as a brake shoe stop and prohibits the brake shoe from its tendency to follow the movement of the rotating drum. The result is a wedging action between the brake shoe and brake drum. The wedging action combined with the applied brake force creates a self-multiplied brake force called self-energizing.

FIGURE 51–1
A drum brake assembly.

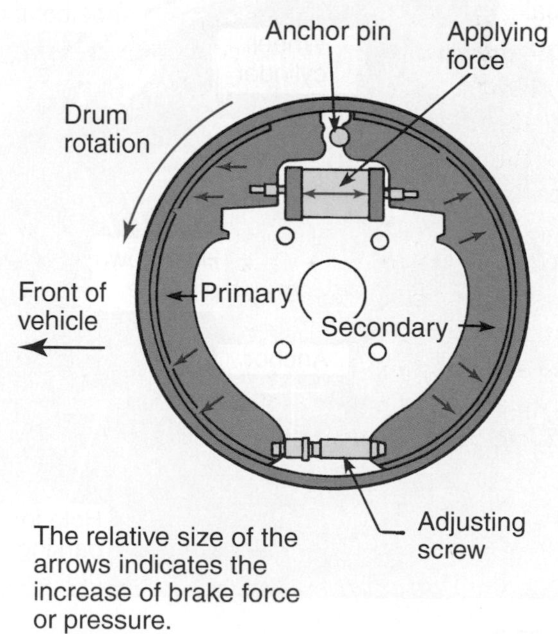

FIGURE 51–2
The wheel cylinder pushes the primary and secondary shoes against the inside surface of the rotating brake drum.

DRUM BRAKE COMPONENTS

The **backing plate** provides a foundation for the brake shoes and associated hardware **(Figure 51–3)**. The plate is secured and bolted to the axle flange or spindle. The wheel cylinder, under hydraulic pressure, forces the brake's shoes against the drum. There are also two linked brake shoes attached to the backing plate. Brake shoes are the backbone of a drum brake. They must support the lining and carry it into the drum so that the pressure is distributed across the lining surface during brake application. Shoe return springs and shoe hold-down parts maintain the correct shoe position and clearance. Some drum brakes are self-adjusting. Others require manual adjustment mechanisms. Brake drums provide the rubbing surface area for the linings. Drums must withstand high pressures without excessive flexing and must also dissipate large quantities of heat generated during brake application. Finally, the rear drum brakes on most vehicles include the parking brakes.

Wheel Cylinders

Wheel cylinders convert hydraulic pressure from the master cylinder into a mechanical force at the brakes **(Figure 51–4)**. The wheel cylinder bore is filled with fluid. When the brake pedal is depressed, additional brake fluid is forced into the cylinder. The additional fluid moves the cups and pistons outward. This piston movement forces the brake shoes outward to the contact drum and thus applies the brakes. Piston stops prevent the fluid leakage or air from getting into the system when the pistons move to the end of their bores.

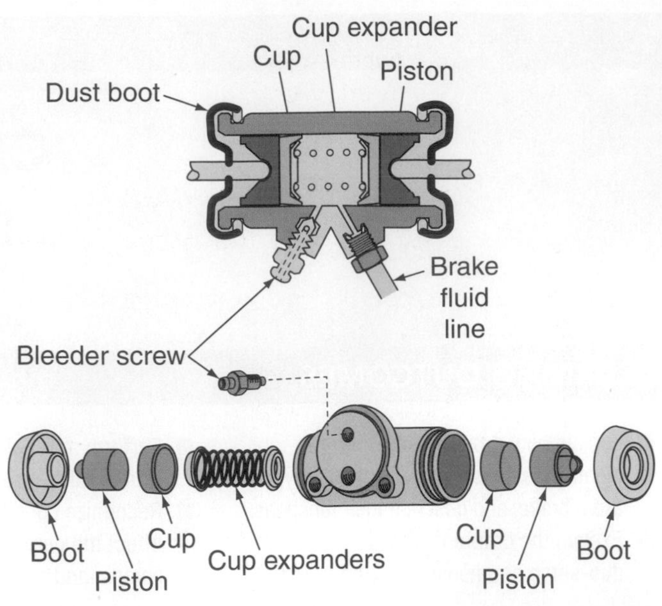

FIGURE 51–4

An exploded view of a typical wheel cylinder.

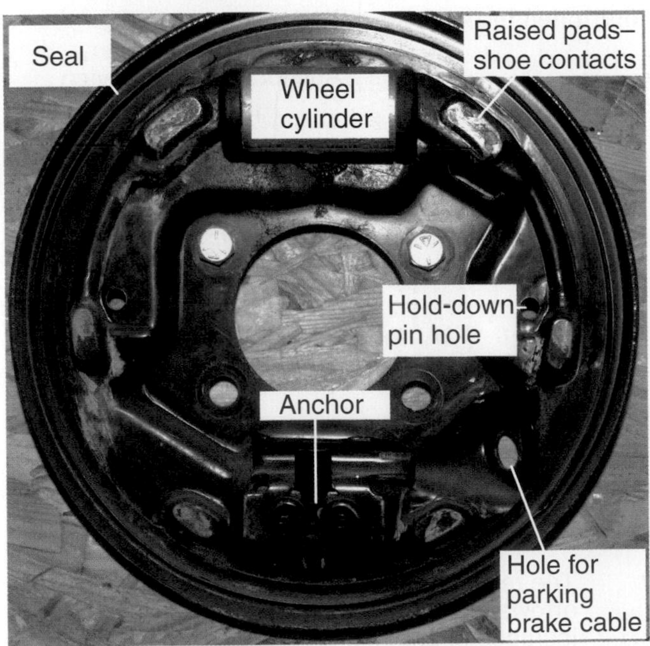

FIGURE 51–3

The backing plate provides a foundation for the brake shoes and associated hardware.

Brake Shoes and Linings

In the same brake shoe sizes, there can be differences in **web** thickness, shape of web cutouts, and positions of any reinforcements **(Figure 51–5)**.

The shoe rim is welded to the web to provide a stable surface for the lining. The web thickness might differ to provide the stiffness or flexibility needed for a specific application. Many shoes have nibs or indented places along the edge of the rim. These nibs rest against shoe support ledges on the backing plate and keep the shoe from hanging up.

Each drum in the drum braking system contains a set of shoes. The **primary shoe** (or leading shoe) is the first shoe in the direction of forward wheel rotation and is usually the one that is toward the front of the vehicle. The friction between the primary shoe and the brake drum forces the primary shoe to shift slightly in the direction that the drum is turning. (An **anchor pin** permits just limited movement.) The shifting of the primary shoe forces it against the bottom of the secondary shoe, which causes the secondary shoe to contact

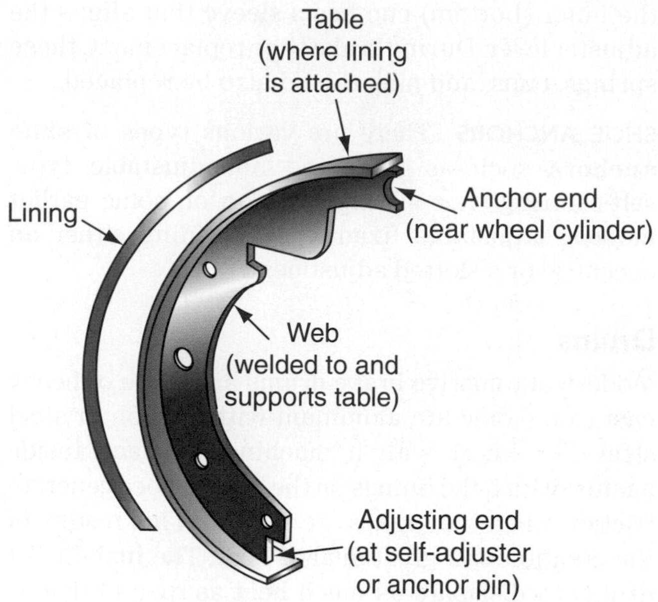

FIGURE 51–5

A typical brake shoe with the lining detached.

the drum. The **secondary shoe** (or trailing shoe) is the one that is toward the rear of the vehicle. It comes into contact as a result of the self-energizing movement and pressure from the primary shoe and wheel cylinder piston and increases the braking action.

The brake shoe lining provides friction against the drum to stop the car. It contains heat-resistant fibres. The lining is moulded with a high-temperature synthetic bonding agent.

The two general methods of attaching the lining to the shoe are riveting and bonding. Regardless of the method of attachment, brake shoes are usually held in a position by spring tension. They are either held against the anchor by the shoe return springs or against the support plate pads by shoe **hold-down springs**. The shoe webs are linked together at the end opposite the anchor by an adjuster and a spring. The adjuster holds them apart. The spring holds them against the adjuster ends.

Mechanical Components

In the unapplied position, the shoes are held against the anchor pin by the return springs. The shoes are held to the backing plate by hold-down springs or spring clips. Opposite the anchor pin, a star wheel adjuster links the shoe webs and provides a threaded adjustment that permits the shoes to be expanded or contracted. The shoes are held against the star wheel by a spring.

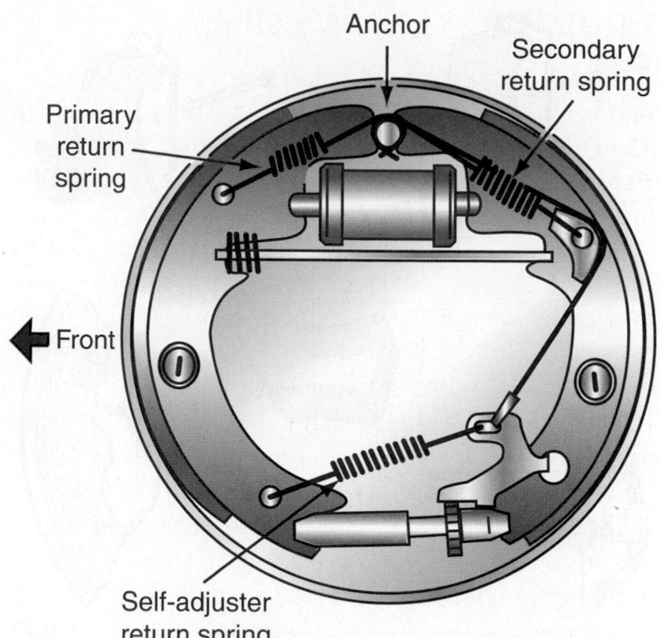

FIGURE 51–6

Typical return spring installation in a duo-servo system.

SHOE RETURN SPRINGS **Return springs** can be separately hooked into a link or a guide (Figure 51–3) or strung between the shoes. Springs are normally installed on the anchor with servo-type brakes, as shown in **Figure 51–6**. In nonservo brake assemblies, return springs typically connect each shoe together, as shown in **Figure 51–7**.

Although shoe brake springs look the same, they are usually not interchangeable. Sometimes to help distinguish between them, they are colour-coded.

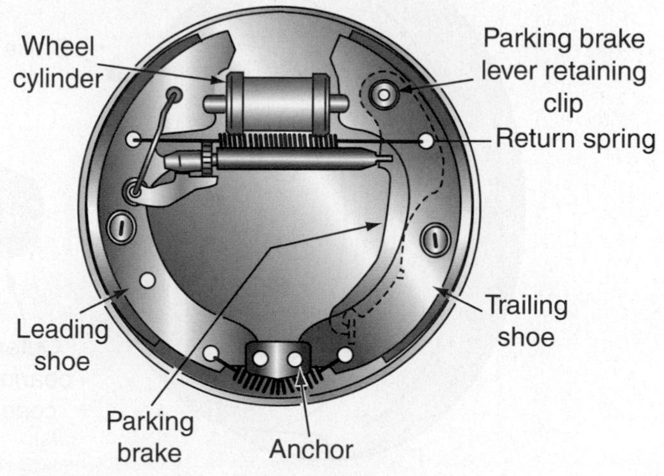

FIGURE 51–7

In nonservo systems, the return springs typically connect each brake shoe together.

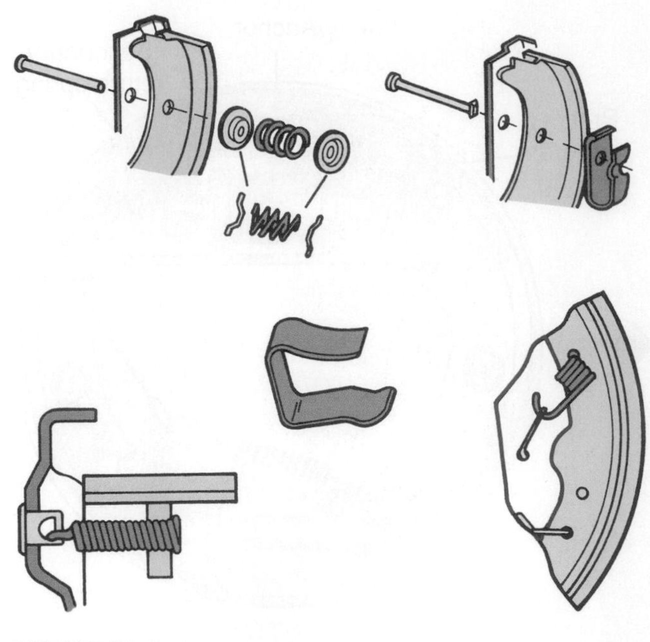

FIGURE 51–8

Types of brake shoe hold-downs.

Pay close attention to the colours and the way they are hooked up. It is a recommended practice to replace these springs when replacing the brake shoes.

SHOE HOLD-DOWNS Various shoe hold-downs are illustrated in **Figure 51–8**. To unlock or lock the straight pin hold-downs, depress the locking cup and coil spring or the spring clip, and rotate the pin or lock 90°. On General Motors' lever adjusters, the inner (bottom) cup has a sleeve that aligns the adjuster lever. During brake shoe replacement, these springs, cups, and pins should also be replaced.

SHOE ANCHORS There are various types of **shoe anchors** such as the fixed nonadjustable type, self-centring shoe sliding type, or, on some earlier models, adjustable fixed-type providing either an eccentric or a slotted adjustment.

Drums

Modern automotive brake drums are made of heavy cast iron (some are aluminum with an iron or steel sleeve or liner) with a machined surface inside against which the linings on the brake shoes generate friction when the brakes are applied. This results in the creation of a great deal of heat. The inability of drums to dissipate as much heat as disc brakes is one of the main reasons discs have replaced drums at the front of all late-model cars and light trucks, and at the rear of some sports and luxury cars.

Sometimes, the rear drums of front-wheel-drive (FWD) cars and larger trucks are integral with the hub and cannot be removed without disassembling the wheel bearing. These are called fixed drums as they are held or fixed in place by other components **(Figure 51–9)**. The rear drums of other FWD and most rear-wheel-drive (RWD) cars are held in place by the wheel lugs so that they can be removed without tampering with the wheel bearings. These are called floating drums.

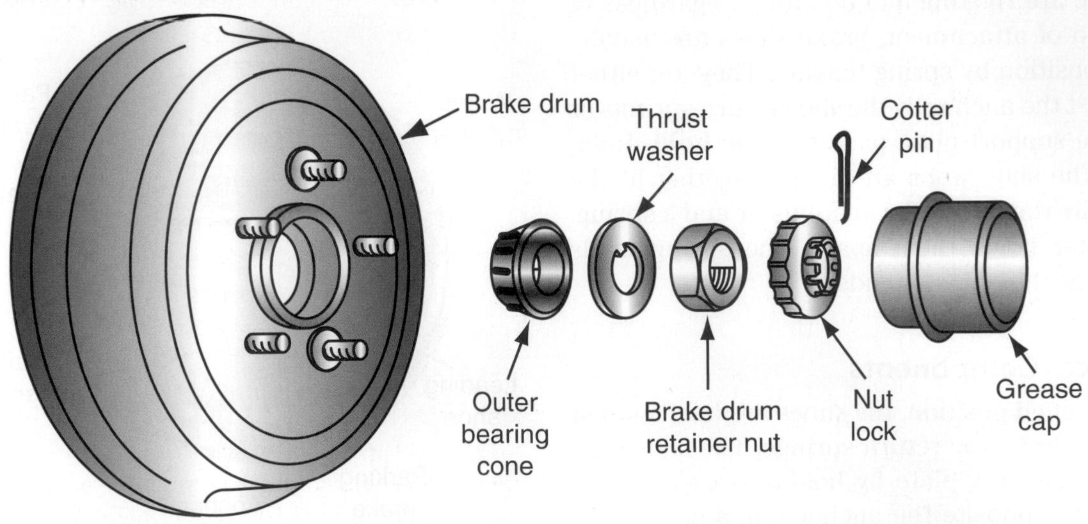

FIGURE 51–9

To remove a fixed drum assembly, remove the retaining nut and slide the assembly off with its bearings.

DRUM BRAKE DESIGNS

There are two brake designs in common use. They are **duo-servo drum brakes** (also called self-energizing) and **nonservo drum brakes** (also called leading-trailing).

Most large North American cars use the duo-servo design of brake. However, the nonservo type has become popular as the size of cars has become smaller. Because the smaller cars are lighter, this type of brake helps reduce rear brake lock-up without reducing braking ability.

Duo-Servo Drum Brakes

The name duo-servo drum brake is derived from the fact that the **self-energizing force** is transferred from one shoe to the other with the wheel rotating in either direction. Both the primary (front) and secondary (rear) brake shoes are actuated by a double-piston wheel cylinder **(Figure 51–10)**. The upper end of each shoe is held against a single anchor by a heavy coil return spring. An adjusting screw assembly and spring connect the lower ends of the shoes. There is no lower shoe anchor.

The wheel cylinder is mounted on the backing plate at the top of the brake. When the brakes are applied, hydraulic pressure behind the wheel cylinder cups forces both pistons outward, causing the brakes to be applied.

When the brake shoes contact the rotating drum in either direction of rotation, they tend to move with the drum until one shoe contacts the anchor, and the other shoe is stopped by the star wheel adjuster link **(Figure 51–11)**. With forward rotation, frictional forces between the lining and the drum of the primary shoe result in a force acting on the adjuster link to apply the secondary shoe. This adjuster link force into the secondary shoe is many times greater than the wheel cylinder output force acting on the primary shoe. The force of the adjuster link into the secondary shoe is again multiplied by the frictional forces between the secondary lining and rotating drum, and all of the resultant force is taken on the anchor pin. In normal forward braking, the friction developed by the secondary lining is greater than the primary lining. Therefore, the secondary brake lining is usually thicker and has more surface area. The roles of the primary and secondary linings are reversed in braking the vehicle when backing up.

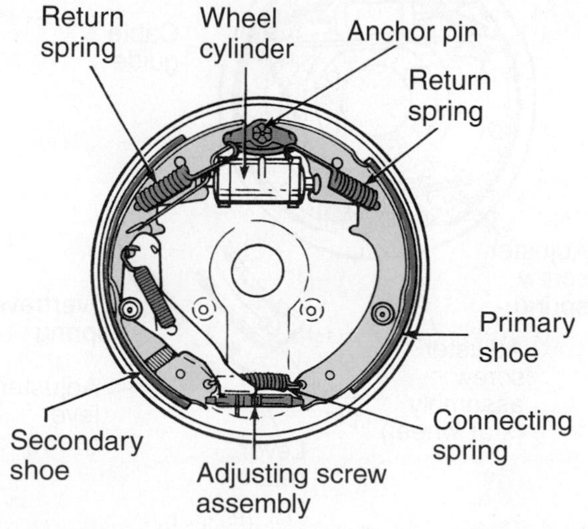

FIGURE 51–10

A typical duo-servo drum brake.

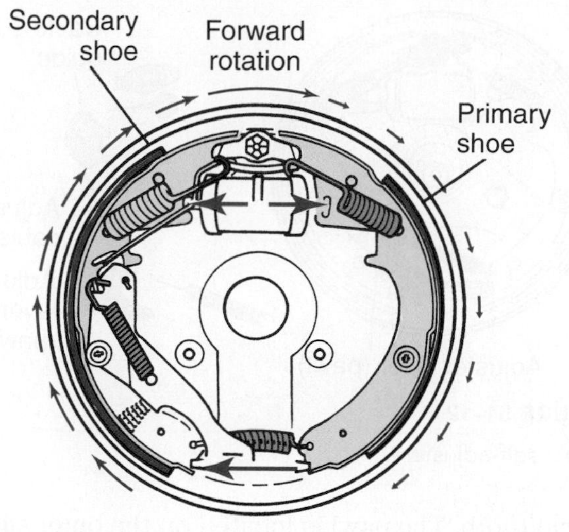

FIGURE 51–11

Duo-servo braking forces.

Automatically Adjusted Servo Brakes

Since the early 1960s, automatic drum brake adjusters have been used on all North American and most import vehicles. There are several variations of automatic adjusters used with servo brakes. The more common types available follow.

BASIC CABLE A typical automatic adjusting system is shown in **Figure 51–12**. Adjusters, whether cable, crank, or lever, are installed on one shoe and operated whenever the shoe moves away from its anchor. The upper link, or cable eye, is attached to the anchor. As the shoe moves, the cable pulls over a guide mounted on the shoe web (the crank or lever pivots on the shoe web) and operates a lever (pawl), which is attached to the shoe so that it engages a star

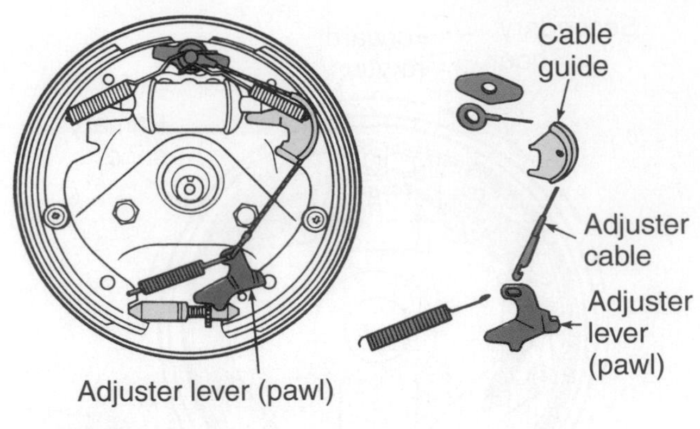

FIGURE 51–12

Cable self-adjusters.

wheel tooth. The pawl is located on the outer side of the star wheel and, on different styles, slightly above or below the wheel centreline so that it serves as a ratchet lock, which prevents the adjustment from backing off. However, whenever the lining wears enough to permit sufficient shoe movement, brake application pulls the pawl high enough to engage the next tooth. As the brake is released, the adjuster spring returns the pawl, thus advancing the star wheel one notch.

On most vehicles, the adjuster system is installed on the secondary shoe and operates when the brakes are applied as the vehicle is backing up. On a few models, it is located on the primary shoe and operates when the brakes are applied as the vehicle is moving forward. Left-hand and right-hand threaded star wheels are used on opposite sides of the car, so parts should be kept separated. If the wrong star wheel thread is installed, the system will not adjust the shoes tighter to the drum, but loosen them away from the drum.

Another system uses a cable and pawl, with the left brake having right-hand threads and the right brake, left-hand threads. The first cable guide is usually retained on the shoe web by the secondary shoe return spring, and the lever-pawl engages a hole in the shoe web. The adjuster operates in either direction of vehicle movement.

CABLE WITH OVERTRAVEL SPRING A system with an upstroke pawl advance is shown in **Figure 51–13**. The left brake has left-hand threads, and the right brake has right-hand threads. The lever (pawl) is installed on a web pin with an additional pawl return mousetrap spring. The cable is hooked to the lever (pawl) by means of an **overtravel spring** installed in the cable hook. The overtravel spring dampens movements and prevents unnecessary adjustment should sudden hard braking cause excessive drum deflection and shoe movement.

LEVER WITH OVERRIDE The system illustrated in **Figure 51–14** uses a downstroke pawl advance. The left brake has right-hand threads, and the right brake has left-hand threads.

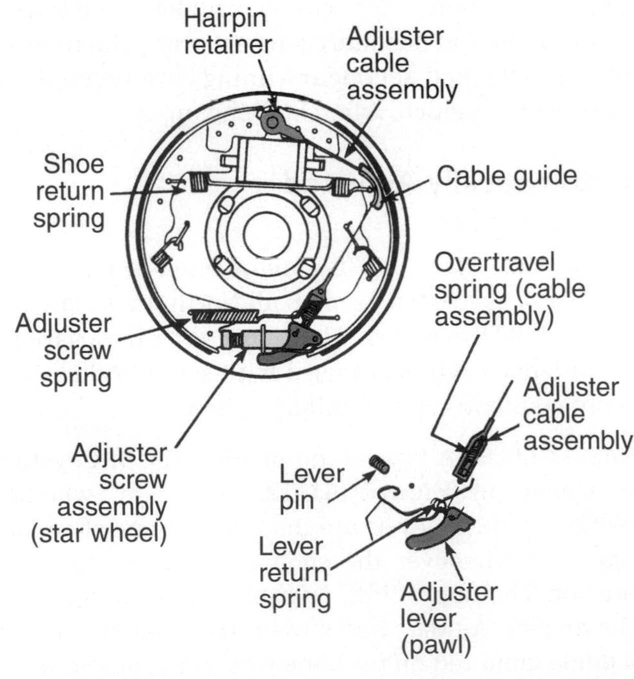

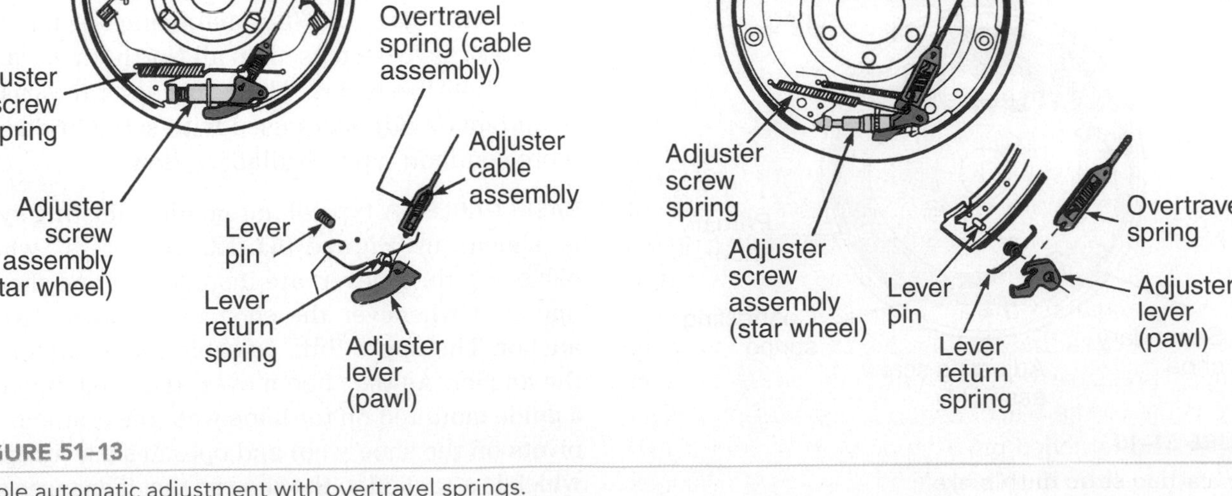

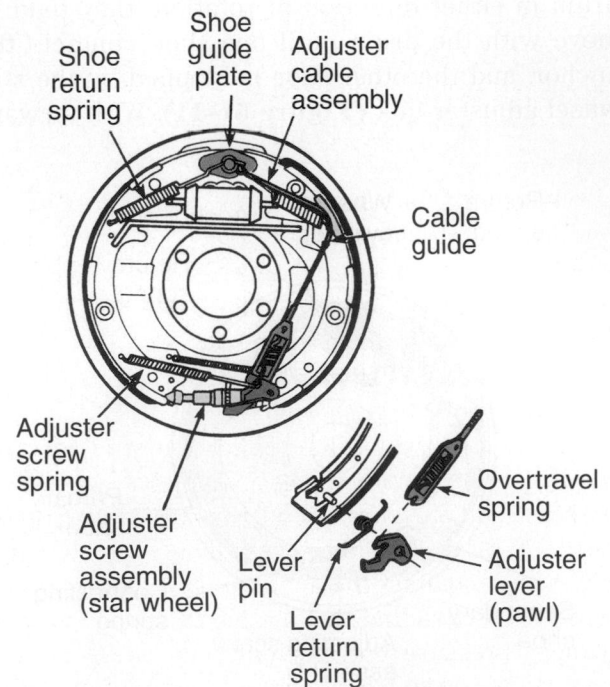

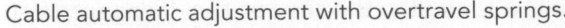

FIGURE 51–13

Cable automatic adjustment with overtravel springs.

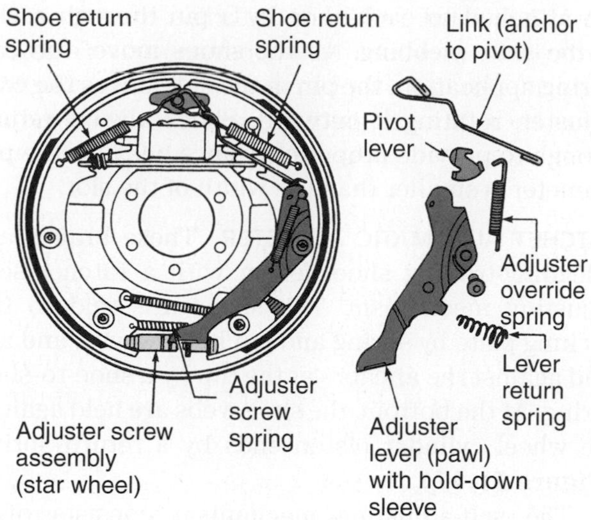

FIGURE 51–14

An adjusting lever with a pivot and override spring.

The lever (pawl) is mounted on a shoe hold-down, pivoting on a cup sleeve. It has a separate lever-pawl return spring located between the lever and the shoe table. A pivot lever and an override spring assembled to the upper end of the main lever dampen movement, preventing unnecessary adjustment in the event of excessive drum deflection.

LEVER AND PAWL The system illustrated in **Figure 51–15** uses a downstroke pawl advance. The left brake has right-hand threads, and the right brake has left-hand threads. The lever is mounted on a shoe hold-down, pivoting on a cup sleeve, and engages the pawl. A separate pawl return spring is located between the pawl and the shoe.

Nonservo Drum Brakes

The nonservo (or as it is better known today, the leading-trailing shoe) drum brake is often used on small cars. The basic difference between this type and the duo-servo brake is that both brake shoes are held against a fixed anchor at the bottom by a retaining spring **(Figure 51–16)**. Nonservo brakes have no servo action.

On a forward brake application, the forward (leading) shoe friction forces are developed by wheel cylinder fluid pressure forcing the lining into contact with the rotating brake drum. The shoe's friction forces work against the anchor pin at the bottom of the shoe. The trailing shoe is also actuated by wheel cylinder pressure but can only support a friction force equal to the wheel cylinder piston forces. The trailing shoe anchor pin supports no friction load. The leading shoe in this brake is energized and does

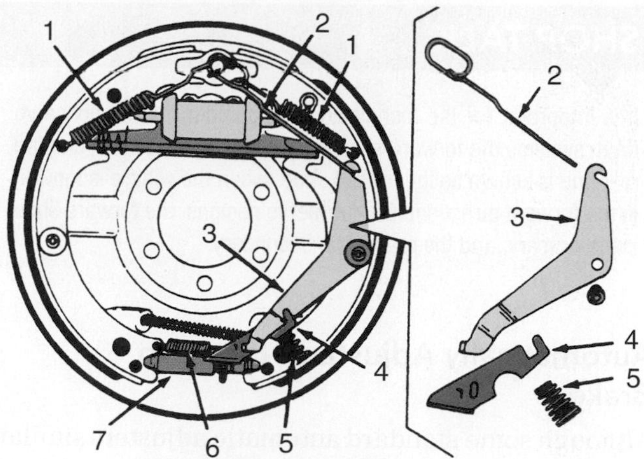

FIGURE 51–15

Lever and pawl automatic adjustment.

Federal-Mogul Corporation

most of the braking in comparison to the nonenergized trailing shoe. In reverse braking, the leading and trailing brake shoes switch functions.

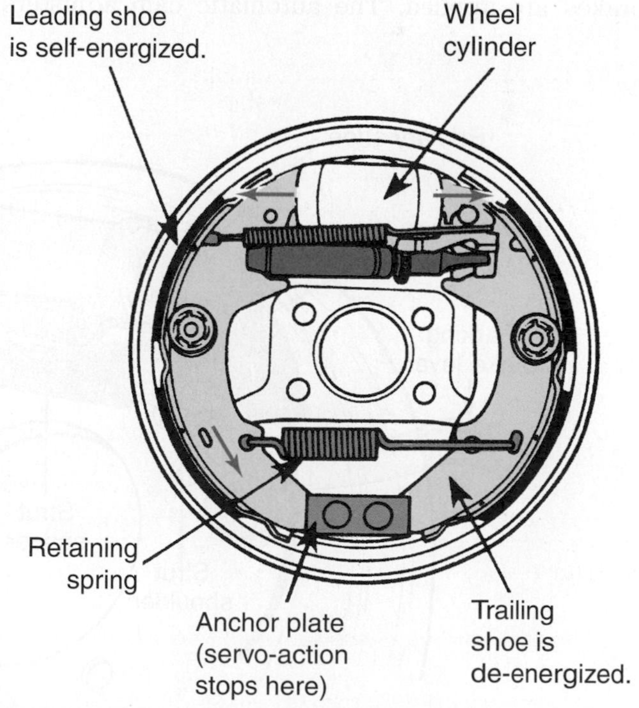

FIGURE 51–16

A typical nonservo drum brake.

SHOP TALK

It is important for the technician to remember that on nonservo drum systems the forward shoe is called the leading shoe, and the rear one is known as the trailing shoe (when the vehicle is moving in the forward direction). On duo-servo designs, the forward shoe is the primary, and the rear is the secondary.

Automatically Adjusted Nonservo Brakes

Although some standard automatic adjusters similar to the ones already discussed are employed on small cars, some of the automatic adjuster mechanisms are unique and varied, using expanding struts between the shoes, or special ratchet adjusting mechanisms. Among the more common of these designs are automatic cam, ratchet automatic, and semiautomatic adjusters.

AUTOMATIC CAM ADJUSTERS This rear nonservo drum brake is for use with front disc brakes and has one forward acting (leading) and one reverse acting (trailing) shoe. Shoes rest against the wheel cylinder pistons at the top and are held against the anchor plate by a shoe-to-shoe pull-back spring. The anchor plate and retaining plate are riveted to the backing plate. Adjustment of the brake shoes takes place automatically as needed when the brakes are applied. The automatic cam adjusters are attached to each shoe by a pin through a slot in the shoe webbing. As the shoes move outward during application, the pin in the slot moves the cam adjuster, rotating it outward. Shoes always return enough to provide proper clearance because the pin diameter is smaller than the width of the slot.

RATCHET AUTOMATIC ADJUSTER These brakes are a leading-trailing shoe design with a ratchet self-adjusting mechanism. The shoes are held to the backing plate by spring and pin hold-downs, and are held against the anchors at the top by a shoe-to-shoe spring. At the bottom, the shoe webs are held against the wheel cylinder piston ends by a return spring **(Figure 51–17)**.

The self-adjusting mechanism consists of a spacer strut and a pair of toothed ratchets attached to the secondary brake shoe. The parking brake actuating lever is pivoted on the spacer strut.

The self-adjusting mechanism automatically senses the correct lining-to-drum clearance. As the linings wear, the clearance is adjusted by increasing the effective length of the spacer strut. This strut has projections to engage the inner edge of the secondary shoe via the hand brake lever and the inner edge of the large ratchet on the secondary shoe. As wear on the linings increases, the movement of the shoes to bring them in contact with the drums becomes greater than the gap. The spacer strut, bearing on the shoe web, is moved together with the primary shoe to close the gap. Further movement

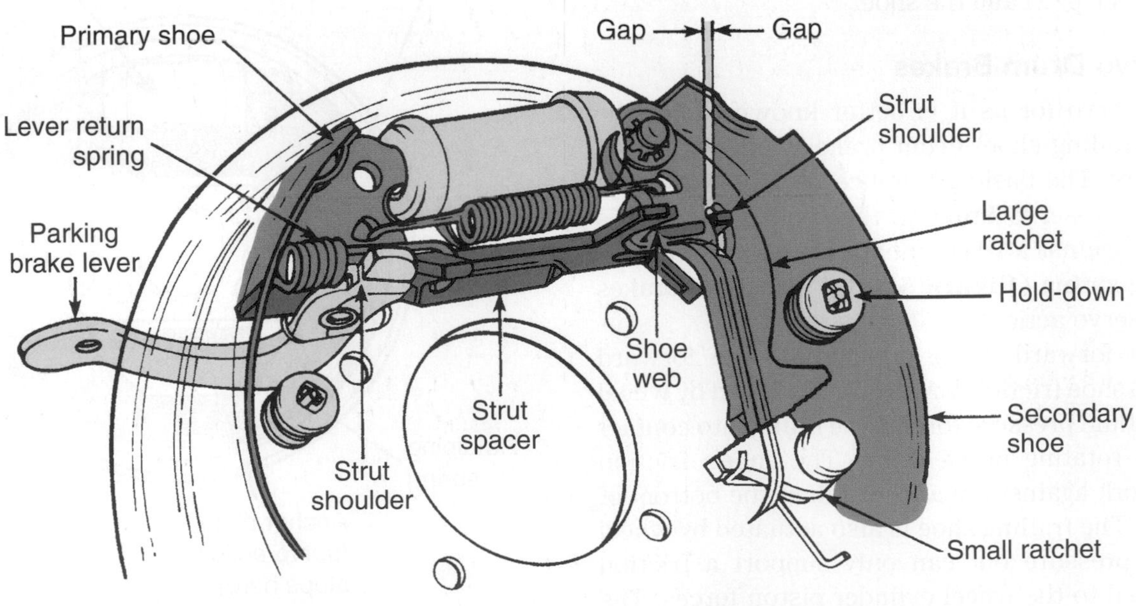

FIGURE 51–17
A typical nonservo self-adjusting mechanism.

causes the large ratchet behind the secondary shoe to rotate inward against the spring-loaded small ratchet, and the serrations on the mating edges maintain this new setting until further wear on the shoe results in another adjustment. On releasing brake pedal pressure, the return springs cause the shoes to move into contact with the shoulders of the spacer strut/hand brake actuating lever. This restores the clearance between the linings and the drum proportionate to the gap.

Inspection and Service

The first rule of quality brake service is to perform a complete job. For example, perform an inspection of the entire brake system, not just the front or rear brakes. Also, if new linings are installed without regard to the condition of the hydraulic system, the presence of a leaking wheel cylinder quickly ruins the new linings. Braking power and safety are also compromised.

Problems such as spongy pedal, excessive pedal travel, pedal pulsation, poor braking ability, brake drag, lock, or pulling to one side, as well as braking noises can be caused by trouble in the hydraulic system or the mechanical components of the brake assembly. To aid in doing a complete inspection and diagnosis, a form like the one shown in **Figure 51–18** is very helpful. Working with such a form helps the technician avoid missing any brake test and components that may cause problems.

Brake Noise

All customer complaints related to brake performance must be carefully considered. The number-one customer complaint is brake noise. Noise is often the first indication of wear or problems within the braking system, particularly in the mechanical components. Rattles, clicking, grinding, and hammering from the wheels when the brake is in the unapplied position should be carefully investigated. Be sure the noise is not caused by the bearings or various suspension parts. If the noise is coming from the brake assembly, it is most likely caused by worn, damaged, or missing brake hardware, or the poor fastening or mounting of brake components. Grinding noises usually occur when a stone or other object becomes trapped between the lining material and the rotor or drum.

When the brakes are applied, a clicking noise usually indicates play or hardware failure in the attachment of the pad or shoe. On recent systems, the noise could be caused by the lining tracking cutting tool marks in the rotor or drum. A nondirectional finish on rotors eliminates this, and so does a less pointed tip on the cutting tool used to refinish drums.

Grinding noises on application can mean metal-to-metal contact, either from badly worn pads or shoes, or from a serious misalignment of the caliper, rotor, wheel cylinder, or backing plate. Wheel cylinders and calipers that are frozen due to internal corrosion can also cause grinding or squealing noises.

Other noise problems and their solutions are covered later in this chapter.

ROAD-TESTING BRAKES

Road-testing allows the brake technician to evaluate brake performance under actual driving conditions. Whenever practical, perform the road test before beginning any work on the brake system. In every case, road-test the vehicle after any brake work to make sure the brake system is working safely and properly.

WARNING!

Before test-driving any car, first check the fluid level in the master cylinder. Depress the brake pedal to be sure there is adequate pedal reserve. Make a series of low-speed stops to be sure the brakes are safe for road-testing. Always make a preliminary inspection of the brake system in the shop before taking the vehicle on the road.

Brakes should be road-tested on a dry, clean, reasonably smooth, and level roadway. A true test of brake performance cannot be made if the roadway is wet, greasy, or covered with loose dirt. Testing is also adversely affected if the roadway is crowned so as to throw the weight of the vehicle toward the wheels on one side, or if the roadway is so rough that wheels tend to bounce.

Test brakes at different speeds with both light and heavy pedal pressure. Avoid locking the wheels and sliding the tires on the roadway. There are external conditions that affect brake road-test performance. Tires having unequal contact and grip on the road cause unequal braking. Tires must be equally inflated and the tread pattern of right and left tires must be approximately equal. When the vehicle has unequal loading, the most heavily loaded wheels require more braking power than others, and a heavily loaded vehicle requires more braking effort.

PRE-BRAKE-JOB INSPECTION CHECKLIST

Owner ______________ LAST ______________ FIRST ______________ Phone ______________ Date ___|___|___

Address ______________ License No. ______________

Make ______ Model ______ Kilometres ______ Serial No. ______ Year ______

Special Key for Hubcaps/Wheels Location ______________ Owner Use Parking Brake Yes ☐ No ☐

4 Drum ☐ 4 Disc ☐ Disc/Drum ☐ P/B No ☐ Yes ☐ Vacuum ☐ Hydro ☐ ABS ☐

Owner Comments ______________

1. CHECKS BEFORE ROAD TEST	Safe	Unsafe
Stoplight Operation		
Brake Warning Light Operation		
Master Cylinder Checks		
Fluid Level		
Fluid Contamination		
Under Hood Fluid Leaks		
Under Dash Fluid Leaks (No Power)		
Bypassing		
BRAKE PEDAL HEIGHT AND FEEL		

Check One		Check One	
Low		Spongy	
Med		Firm	
High			

Power Brake Unit Checks

VACUUM	Safe	Unsafe	HYDRO	Safe	Unsafe
Vacuum Unit			Hydro Unit		
Engine Vacuum			P/S Fluid		
Vacuum Hose			P/S Belt Tension		
Unit Check Valve			P/S Belt Condition		
Reserve Braking			P/S Fluid Leaks		
			Reserve Braking		

Tire Pressure Specs	Front	Rear
Record Pressure Found		

RF ______ LF ______ RR ______ LR ______

Tire Condition

RF ______ LF ______ RR ______ LR ______

2. ROAD TEST	Yes	No	RF	LF	RR	LR
Brake Pull						
Brake Clunk						
Brake Scraping						
Brake Squeal						
Brake Grabby						
Brakes Lock Prematurely						
Wheel Bearing Noise						
Vehicle Vibrates						

STEERING WHEEL MOVEMENT WHEN STOPPING FROM 3–5 KM/H YES/NO/RGT/LFT

Does ABS Work	YES	NO
Pedal Pulsation when Braking	YES	NO
Steering Wheel Oscillation when Braking	YES	NO
No Stopping Power	YES	NO
Warning Light Comes on when Braking	YES	NO
Difference in Pedal Height after Cornering	YES	NO
Nose Dive	YES	NO

3. In Shop Checks On Hoist	Yes	No	RF	LF	RR	LR
Brake Drag						
Intermittent Brake Drag						
Brake Pedal Linkage Binding						
Wheel Bearing Looseness						
Missing or Broken Wheel Fasteners						
Suspension Looseness						

Mark Wheels and Remove

Caliper/Piston Stuck RF LF RR LR

	Front					Rear				
	Right			Left		Right			Left	
Mark Drums and Remove	Spec	Safe	Unsafe	Safe	Unsafe	Spec	Safe	Unsafe	Safe	Unsafe
Measure Rotor Thickness or Drum Diameter										
Measure Rotor Thickness Variation										
Measure Rotor Runout										
Lining Thickness										
Tubes and Hoses										
Fluid Leaks										
Broken Bleeders										
Leaky Seals										
Self-Adjuster Operation										
									Safe	Unsafe
Parking Brake Cables and Linkage										

FIGURE 51–18

A sample of a pre-brake-job inspection checklist.

Hennessy Industries, Inc.

A loose front wheel bearing permits the drum and wheel to tilt and have spotty contact with the brake linings, causing erratic brake action. Misalignment of the front end may cause the brakes to pull to one side. Faulty shock absorbers that do not prevent the car from bouncing on quick stops can give the erroneous impression that the brakes are too severe.

DRUM BRAKE INSPECTION

Before inspecting drum brakes, place the vehicle in neutral, release the parking brake, and raise the vehicle on the hoist. Once the vehicle is raised, mark the wheel-to-drum and drum-to-axle positions so that the components can be accurately reassembled. To access the drum brake assembly, remove the lug nuts and pull the wheel off the hub.

> **CAUTION!**
>
> When servicing wheel brake parts, do not create dust by cleaning with a dry brush or with compressed air. Current domestic brake linings for the most part do not contain asbestos fibres; however, some foreign aftermarket brake linings may contain asbestos because of the cost factor and the lack of industry restrictions. The dust from any lining fibre material (asbestos or not) and lining resins, along with road dust, should not be inhaled. To clean away brake dust from drum brake assemblies, use an approved brake dust vacuum, or rinse the assemblies thoroughly with water to prevent the dust from becoming airborne. When using special equipment, always follow the equipment manufacturer's operating instructions.

Drum Removal

Drum removal procedures are different for fixed and floating drums. In all cases, however, you may need to back off the parking brake adjustment or manually retract the self-adjusters **(Figure 51–19)** to have enough shoe-to-drum clearance to remove the drum. Wear on the friction surface of the drum creates a ridge at the edge of the drum's rim. As the self-adjusters move the shoes outward to take up clearance, the shoe diameter becomes larger than the ridge diameter. If the adjuster is not retracted, the drum's ridge may jam on the shoes and prevent drum removal. Trying to force the drum over the shoes may damage brake parts. To retract the shoes, reach through the adjusting slot with a thin screwdriver (or similar tool) and carefully push the self-adjusting lever away from the star wheel a maximum of 1.59 mm (1/16 in.). While holding the lever back, insert a brake adjusting tool into the slot, and turn the star wheel in the proper direction until the brake drum can be removed. On vehicles that have the adjusting slot in the drum rather than in the backing plate, reach through the slot with a thin wire hook, and pull the adjuster lever away from the star wheel.

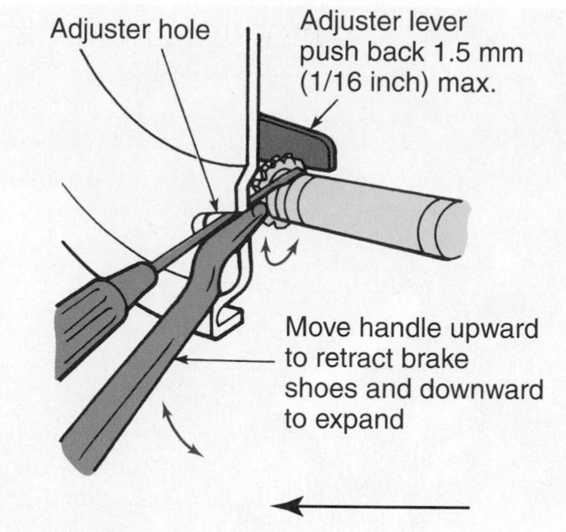

FIGURE 51–19

Backing off the self-adjusters in order to remove the brake drum.

Before you remove a drum, mark it "L" or "R" for left or right so that it gets reinstalled on the same side of the vehicle from which it was removed.

Brake drums that are made as a one-piece unit with the wheel hub are common as rear drums on FWD cars and on the front wheels of older vehicles with four-wheel drum brakes. The hub contains the wheel bearings and is held onto the spindle by a single large nut. This nut also is used to adjust the wheel bearings. To remove this type of drum, remove the dust cap from the centre of the hub. Then remove the cotter pin from the castellated nut or nut lock on the spindle. Next, remove the spindle nut and washer. Pull the drum outward to slide it off the spindle.

On some 4WD trucks, the rear drums are held in place by the rear axle and wheel bearings. First remove the bolts securing the axle to the hub, and then remove the axle. In most cases, a special socket is required to remove the retaining nut and remove the bearings. Once the bearings are removed, the drum can be removed.

Floating drums do not have a built-in hub. In most cases, the drums are held in place by studs on the axle flange and the wheel and lug nuts. On many floating drums, push nuts or speed nuts are used during vehicle assembly to hold the drum onto two

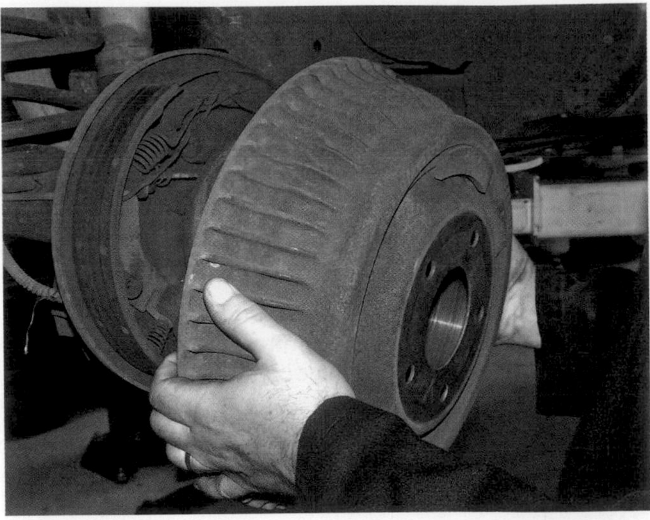

FIGURE 51–20

Pull the drum away from the axle flange or hub, being careful not to drop it.

FIGURE 51–21

Faulty axle seals allow axle fluid to leak and contaminate the brake linings.

or three studs. Typically, the push nuts do not need to be reinstalled after service. However, on some vehicles, the push nuts are used to hold the drum squarely against the axle or hub flange.

Floating drums are pulled off the hub or axle flange **(Figure 51–20)**. If the brake drum is rusted or corroded to the axle flange and cannot be removed, lightly tap the axle flange to the drum mounting surface with a ball-peen hammer. Penetrating oil may help in loosening a stuck drum. If the drum is stuck to its flange, use a large scribe or centre punch to score around the joint at the drum and flange and break the surface tension. Remember that if the drum is worn, the brake shoe adjustment has to be backed off for the drums to clear the brake shoes. Do not force the drum or distort it. Do not pry against the drum at the backing plate as this will damage and possibly bend the backing plate. Once loose, do not allow the drum to drop.

Once the drum is removed, inspect the brake assembly for signs of fluid leakage from the wheel cylinder and, on RWD vehicles, from the rear axle seals. Faulty axle seals, shown in **Figure 51–21**, allow differential lubricant to leak out and contaminate the rear brake components. When this occurs, the shoes must be replaced, as the fluid cannot be removed from the lining material.

FIXED DRUMS After the drum is removed, inspect the grease in the hub and on the bearings. If the grease is dirty or dried out and hard, it is a clue to possible bearing damage. Also, inspect the rear axle gaskets and wheel seals for leaks. Replace worn components as needed. Set the drum and all bearing parts aside for cleaning and close inspection. If the grease seems to be in good condition, place the drum on a bench with the open side down. Cover the outer bearing opening with a shop cloth to keep dirt out.

Drum Inspection

One of the most important safety inspections to be made is that of the brake drum. First, visually inspect the brake shoes, as installed on the vehicle. Often, their condition can reveal defects in the drums. If the linings on one wheel are worn more than the others, it might indicate a rough drum. Uneven wear from side to side on any one set of shoes can be caused by a tapered drum. If some linings are worn badly at the toe or heel, it might indicate an out-of-round drum.

Thoroughly clean the drums with a water-dampened cloth or a water-based solution. Equipment for washing brake parts is commercially available **(Figure 51–22)**. Wet cleaning methods must be used to prevent brake dust or fibres from becoming airborne. If the drums have been exposed to leaking oil or grease, thoroughly clean them with a non-oil-based solvent after washing to remove dust and dirt. It is important to determine the source of the oil or grease leak and correct the problem before reinstalling the drums.

Brake drums act as a heat sink. They absorb heat and dissipate it into the air. As drums wear from normal use or are machined, their cooling surface area is reduced and their operating temperatures

FIGURE 51–22

Before disassembling the brakes, use an OSHA-approved washer to make sure all asbestos dust is removed from the parts.

Martin Restoule

increase. Structural strength also reduces. This leads to distortion, which causes some of the drum conditions shown in **Figure 51–23**.

SCORED DRUM SURFACE The most common cause of a scored drum surface is buildup of brake dust and dirt between the brake lining and drum. A glazed brake lining, hardened by high heat, or in some cases, by very hard inferior grade brake lining, can also groove the drum surface. Excessive lining wear that exposes the rivet head or shoe steel will score the drum surface. If the grooves are not too deep, the drum can be turned; otherwise, replacement of both drums on the axle is required.

BELL-MOUTHED DRUM A bell-mouthed drum is caused by a distortion due to extreme heat and braking pressure. It occurs mostly on wide drums and is caused by poor support at the outside of the drum. Full drum-to-lining contact cannot be achieved and fading can be expected. Drums must be turned or replaced.

CONCAVE DRUM A concave drum is the result of an excessive wear pattern in the centre area of the drum brake surface. Extreme braking pressure can distort the shoe platform so that braking pressure is concentrated at the centre of the drum.

CONVEX DRUM This wear pattern is greater at the closed end of the drum. It is the result of excessive heat or an oversized drum, which allows the open end of the drum to distort.

HARD SPOTS ON THE DRUM This condition in the cast-iron surface, sometimes called chisel spots or islands of steel, results from a change in metallurgy caused by braking heat. Chatter, pulling, rapid wear, hard pedal, and noise can occur. These spots can be removed by grinding only, not typical machining with a cutting bit. However, only the raised surfaces are removed, and they can reappear when heat is applied. If this condition reappears, the drum must be replaced. With this consideration, most shops recommend replacement rather than grinding.

THREADED DRUM SURFACE An extremely sharp or chipped tool bit or a lathe that turns too fast can result in a threaded drum surface. This condition can cause a snapping sound during brake application as the shoes ride outward on the thread and then snap back. To avoid this, recondition drums using a rounded tool and proper lathe speed. Check the edge of the drum surface around the mounting flange side for tool marks indicating a previous machining. If the drum has been machined, it might have worn too thin for use. Check the diameter.

HEAT CHECKS **Heat checks** are visible, unlike hard spots that do not appear until the machining of the drum **(Figure 51–24)**. Extreme operating temperatures are the major cause. The drum might also show a bluish-gold tint, which is a sign of high temperatures. Hardened carbide lathe bits or special grinding attachments are available through lathe manufacturers to service these conditions. Excessive damage by heat checks or hard spots requires drum replacement.

CRACKED DRUM Cracks in the cast-iron drum are caused by excessive stress. They can be anywhere but usually are in the vicinity of the bolt circle or at the outside of the flange. Fine cracks in the drums

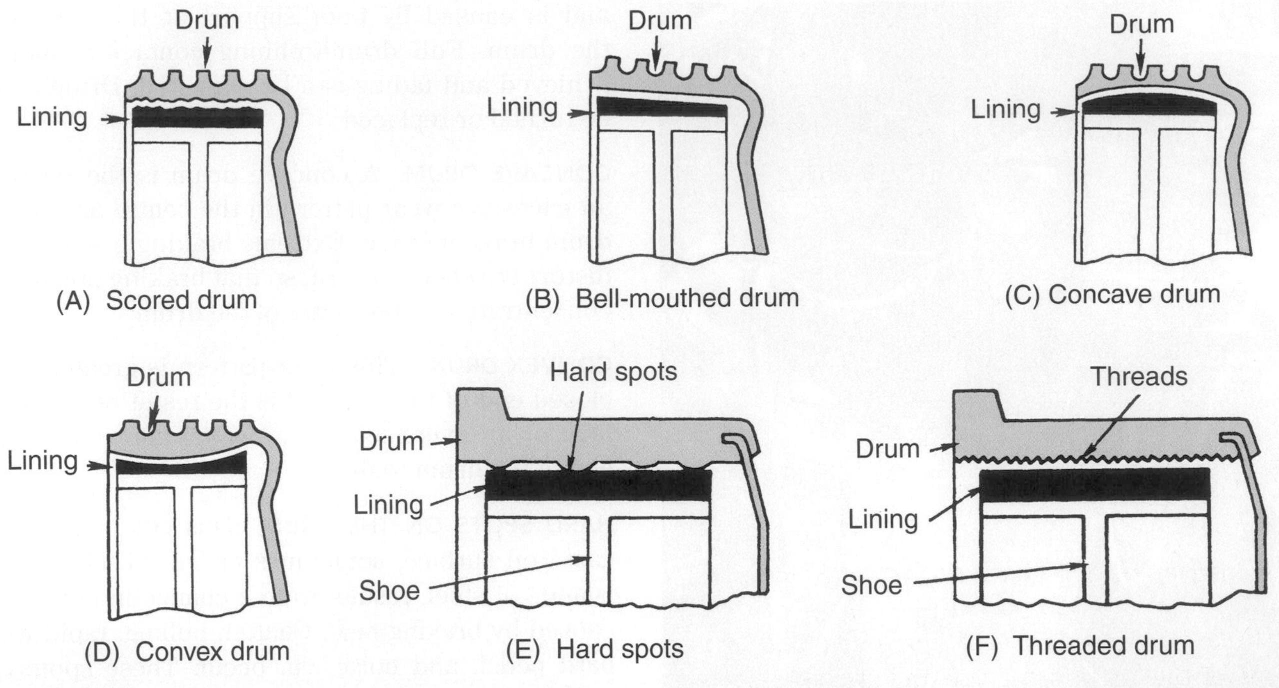

FIGURE 51–23

Drum wear conditions.

are often hard to see and, unfortunately, often do not show up until after machining. Nevertheless, should any cracks appear, no matter how small, the drum must be replaced.

OUT-OF-ROUND DRUMS Drums with eccentric distortion might appear fine to the eye but can cause pulling, grabbing, and pedal vibration or pulsation. An out-of-round or egg-shaped condition **(Figure 51–25)** is often caused by heating and cooling during normal brake operation. Out-of-round drums can be detected before the drum is removed by adjusting the brake to a light drag and feeling the rotation of the drum by hand. After removing the drum, gauge it to determine the amount of eccentric distortion. Drums with this defect should be machined or replaced.

Drum Measurements

Measure every drum with a drum micrometer **(Figure 51–26)**, even if the drum passed a visual inspection, to make sure that it is within the safe oversize limits. If the drum is within safe limits, even though the surface appears smooth, it should be turned to ensure a true drum surface and to remove any possible contamination in the surface from previous brake linings, road dust, and so forth. Remember that if too much metal is removed from a drum, unsafe conditions can result.

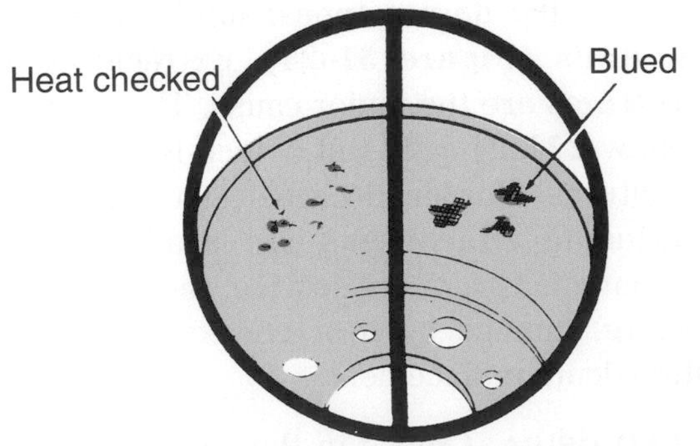

FIGURE 51–24

An example of a heat-checked and overheated brake drum.

Federal-Mogul Corporation

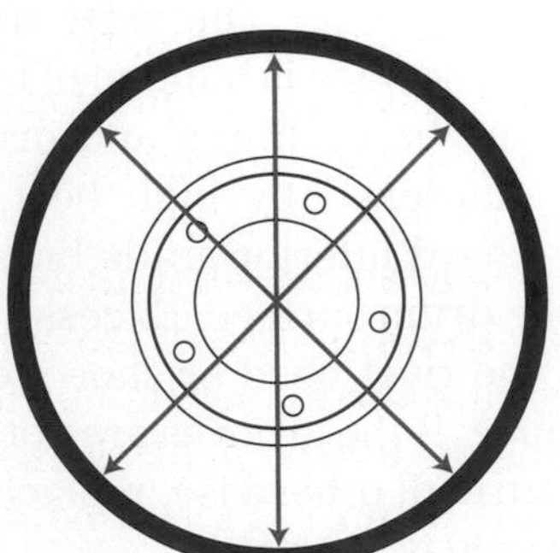

FIGURE 51–25

Measure the inside diameter of the drum in several spots to determine out-of-roundness.

FIGURE 51–26
Measuring the inside diameter with a drum micrometer.

Take measurements at the open and closed edges of the friction surface and at right angles to each other. Drums with taper or out-of-roundness exceeding 0.152 mm (0.006 in.) are unfit for service and should be turned or replaced. If the maximum diameter reading (measured from the bottom of any grooves that might be present) exceeds the new drum diameter by more than 1.5 mm (0.060 in.), the drum cannot be reworked. If the drums are smooth and true but exceed the new diameter by 2.2 mm (0.090 in.) or more, they must be replaced.

If the drums are true, smooth up any slight scores by polishing with fine emery cloth. If deep scores or grooves are present that cannot be removed by this method, the drum must be turned or replaced.

Drum Refinishing

Brake drums can be refinished by either turning or grinding on a **brake lathe (Figure 51–27)**.

Only enough metal should be removed to obtain a true, smooth friction surface. When one drum must be machined to remove defects, the other drum on the same axle set must also be machined in the same manner and to the same diameter so that braking is equal.

Brake drums are stamped with a discard dimension **(Figure 51–28)**. This is the allowable wear dimension and not the allowable machining dimension. There must be 0.762 mm (0.030 in.) left for wear after turning the drums. Some provinces have laws about measuring the limits of a brake drum.

Machining or grinding brake drums increases the inside diameter of the drum and changes the lining-to-drum fit. When remachining a drum, follow the equipment instructions for the specific tool you are using.

FIGURE 51–27
Brake drums can be resurfaced by grinding or turning them on a brake lathe.
Hennessy Industries, Inc.

Cleaning Newly Refaced Drums

The friction surface of a newly refaced drum contains millions of tiny metal particles. These particles not only remain free on the surface, but they also always lodge themselves in the open pores of the newly machined surface. If the metal particles are allowed to remain in the drum, they become embedded in the brake lining. Once the brake lining gets contaminated in this manner, it acts as a fine grinding stone and scores the drum.

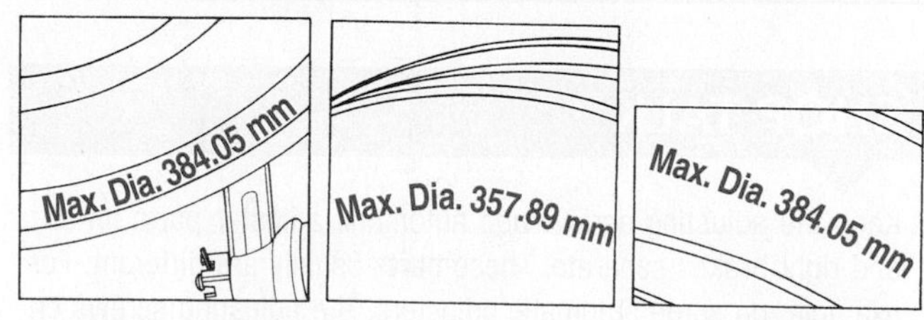

FIGURE 51–28
The drum's discard diameter is stamped on the drum.

PROCEDURE

Mechanical Component Service of Duo-Servo Drum Brakes

STEP 1 Clean the brake dust from the brake assembly, using the appropriate cleaning equipment.

STEP 2 If required, install wheel cylinder clamps on the wheel cylinders to prevent fluid leakage or air from getting into the system while the shoes are removed. Some brakes have wheel cylinder stops; therefore, wheel cylinder clamps are not required. Regardless of whether the clamps are needed, do not press down on the brake pedal after shoe return springs have been removed. To prevent this, block up the brake pedal so that it cannot be depressed.

STEP 3 Remove the brake shoe return springs. Use a brake spring removal and installation tool to unhook the springs from the anchor pin or anchor plate **(Figure 51–29)**.

STEP 4 Remove the shoe retaining or hold-down cups and springs. Special tools are available **(Figure 51–30)**, but the hold-down springs can be removed by using pliers to compress the spring and rotating the cup with relation to the pin.

STEP 5 Self-adjuster parts can now be removed. Lift off the actuating link, lever and pivot assembly, sleeve (through lever), and return spring. No advantage is gained by disassembling the lever and pivot assembly unless one of the parts is damaged.

STEP 6 Spread the shoes slightly to free the parking brake strut and remove the strut with its spring. Disconnect the parking brake lever from the secondary shoe. It can be attached with a retaining clip, bolt, or simply hooked into the shoe.

STEP 7 Slip the anchor plate off the pin. No advantage is gained by removing the plate if it is bolted on or riveted. Spread the anchor ends of the shoes, and disengage them from the wheel cylinder links, if used. Remove the shoes connected at the bottom by the adjusting screw and spring, as an assembly.

STEP 8 Overlap the anchor end of the shoes to relieve spring tension. Unhook the adjusting screw spring, and remove the adjusting screw assembly.

SHOP TALK

Keep the adjusting screws and automatic adjuster parts for left and right brakes separate. These parts usually are different. For example, on some automatic adjusters, the adjusting screws on the right brakes have left-hand threads, and the adjusting screws on the left brakes have right-hand threads.

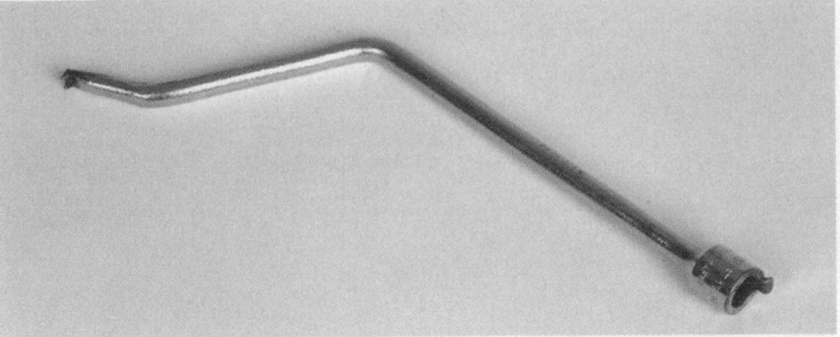

FIGURE 51–29

A brake spring tool.

FIGURE 51–30

A hold-down spring tool.

PROCEDURE

Disassembling Nonservo or Leading-Trailing Brakes

STEP 1 Install the wheel cylinder clamp. Then unhook the adjuster spring from the parking brake strut and reverse shoe.

STEP 2 Unhook the upper shoe-to-shoe spring from the shoes and unhook the antinoise spring from the spring bracket.

STEP 3 Remove the parking brake strut, and disengage the shoe webs from the flat, clamp shoe hold-down clips.

STEP 4 Unhook the lower shoe-to-shoe spring, and remove the forward shoe. Disconnect the parking brake cable and then remove the reserve shoe.

STEP 5 Remove the shoe hold-down clips from the backing plate.

STEP 6 Press off the C-shaped retainers from the pins, and remove the parking brake lever, automatic adjuster lever, and adjuster latch.

STEP 7 Remove the parking brake lever.

SHOP TALK

Mark the shoe positions if shoes and linings are to be reused. When disassembling an unfamiliar brake assembly, refer to the manufacturer's manual or service information system for the correct disassembly and assembly procedures.

PROCEDURE

Cleaning and Inspecting Brake Parts

STEP 1 Clean the backing plates, struts, levers, and other metal parts to be reused using a water-dampened cloth or a water-based solution. Equipment is commercially available to perform washing functions of brake parts. Wet cleaning methods must be used to prevent brake dust and fibres from becoming airborne.

STEP 2 Carefully examine the raised shoe pads on the backing plate to make sure they are free from corrosion or other surface defects that might prevent the shoes from sliding freely **(Figure 51–31)**. Use fine emery cloth to remove surface defects, if necessary. Clean them thoroughly.

STEP 3 Check to make sure that the backing plates are not cracked or bent. If so, they must be replaced. Make sure backing plate bolts and bolted-on anchor pins are torqued to specifications.

STEP 4 If replacement of the wheel cylinder is needed, it should be done at this time. To determine wheel cylinder condition, carefully inspect the boots. If they are cut, torn, heat-cracked, or show evidence of leakage, the wheel cylinders should be replaced. If more than a drop of fluid spills out, leakage is excessive and indicates that replacement is necessary.

STEP 5 Disassemble the adjusting screw assembly **(Figure 51–32)**, and clean the parts in a suitable solvent. Make sure the adjusting screw threads into the pivot nut over its complete length without sticking or binding. Check that none of the adjusting screw teeth are damaged. Lubricate the adjusting screw threads with brake lubricant.

STEP 6 Examine the shoe anchor, support plate, and small parts for signs of looseness, wear, or damage that could cause faulty shoe alignment. Check springs for spread or collapsed coils, twisted or nicked shanks, and severe discoloration **(Figure 51–33)**. Operate star wheel automatic adjusters by prying the shoe lightly away from its anchor or by pulling the cable to make sure the adjuster advances easily, one notch at a time. Adjuster cables tend to stretch, and star wheels and pawls become blunted after a long period of use. For rear axle parking brakes, pull on the cable and shoe linkage to make sure no binding condition is present that could cause the shoes to drag when the parking brake is released.

WARNING!

Do not use ordinary grease to lubricate drum brake parts. It does not hold up under high temperatures.

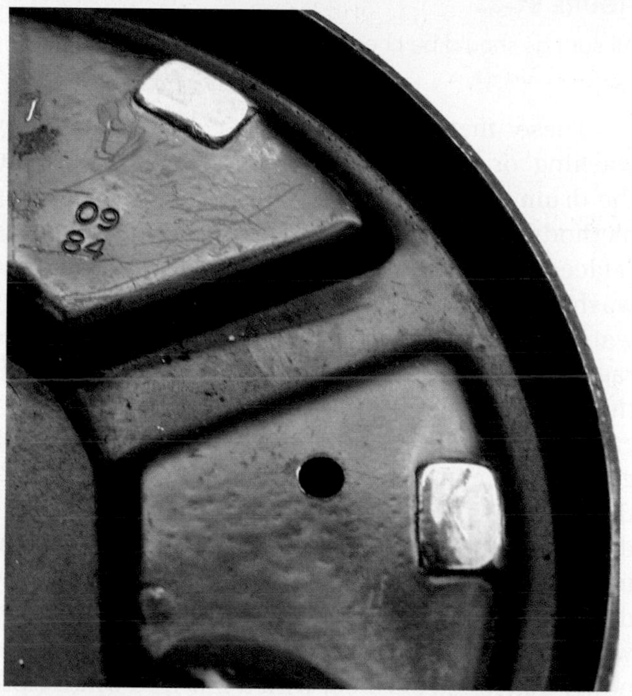

FIGURE 51–31
Carefully examine the raised shoe support pads on the backing plate.

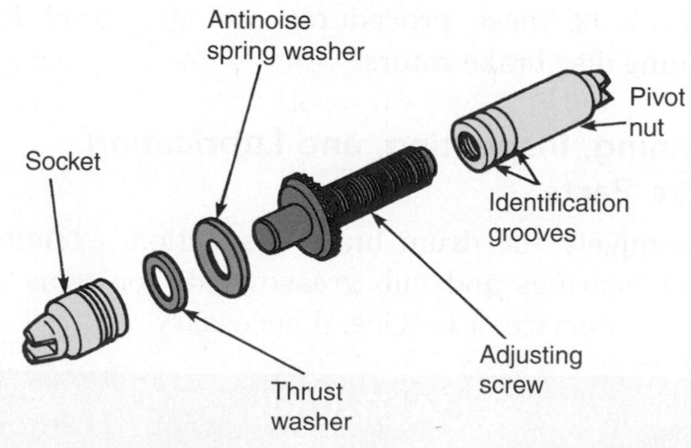

FIGURE 51–32
An exploded view of a brake adjuster assembly.

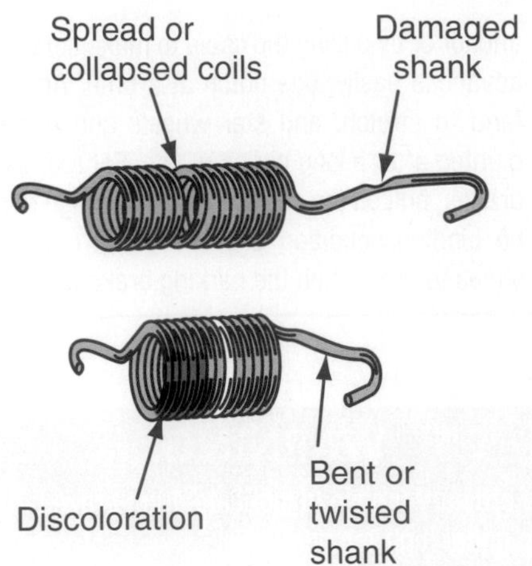

FIGURE 51–33

All springs should be checked for distortion and damage.

These metal particles must be removed by washing or cleaning the drum. Do not blow out the drum with air pressure. Either of the following methods is recommended for cleaning a newly refaced brake drum. The first method involves washing the brake drum thoroughly with hot soapy water or plain hot water and wiping with a lint-free rag. Then use the air pressure to thoroughly dry it. If the front hub and drums are being cleaned, be very careful to avoid contaminating the wheel bearing grease. Or completely remove all the old grease, then regrease and repack the wheel bearing after the drum has been cleaned and dried. The wheel bearings and the grease seals must be removed from the drum before cleaning. The second method involves wiping the inside of the brake drum (especially the newly machined surface) with a lint-free white cloth dipped in one of the many available brake cleaning solvents that do not leave a residue. This operation should be repeated until dirt is no longer apparent on the wiping cloth. Allow the drum to dry before reinstalling it on the vehicle.

Both of these procedures are also good for cleaning disc brake rotors.

Cleaning, Inspecting, and Lubricating Brake Parts

To complete the drum brake inspection, examine wheel bearings and hub grease seals for signs of damage. Service or replace, if necessary.

CAUTION!

When servicing wheel brake parts, do not create dust by cleaning with a dry brush or compressed air.

SHOP TALK

Some brake technicians check brake spring tension by the drop method. This method is not overly scientific, and the results are not always correct. Drop the brake spring on a clean concrete floor. If it bounces with a chunky sound, it is good. If the bounced spring gives off a tinny sound, it is tired and should be replaced. Many technicians simply replace the springs as part of the brake service. This ensures the brake hardware will be in good condition.

BRAKE SHOES AND LININGS

Lining materials influence braking operation. The use of a lining with a friction value that is too high can result in a severe grabbing condition. A friction value that is too low can make stopping difficult because of a hard pedal.

Overheating a lining accelerates wear and can result in dangerous lining heat fade—a friction-reducing condition that hardens the pedal and lengthens the stopping distance. Continual overheating eventually pushes the lining beyond the point of recovery into a permanent fade condition. In addition to fade, overheating can cause squeal.

Overheating is indicated by a lining that is charred or has a glass-hard glazed surface, or if severe, random cracking of the surface is present.

CAUTION!

Automotive friction lining materials may contain substantial amounts of asbestos. To protect yourself and co-workers, all brake linings should be treated as if they do contain asbestos. Studies indicate that exposure to excessive amounts of asbestos dust can be a potential health hazard. It is important that anyone handling brake linings understands this and takes the necessary precautions to avoid injury.

Inspect the linings for uneven wear, embedded foreign material, loose rivets, and to see if they are oil soaked. If linings are oil-soaked, replace them.

If linings at any wheel show a spotty wear pattern or an uneven contact with the brake drum, it is an indication that the linings are not centred in the drums. Linings should be circle ground to provide better contact with the drum.

Brake Shoe Replacement

Brake linings that are worn to within 1.59 mm (2/32 in.) of a rivet head or that have been contaminated with brake fluid, grease, or oil must be replaced **(Figure 51–34)**. Failure to replace worn linings

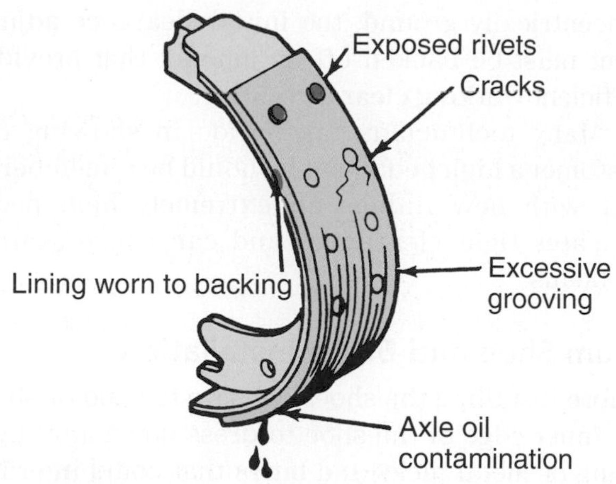

FIGURE 51–34

Potential brake shoe problems.

FIGURE 51–35

Identification codes, called automotive friction material edge codes, are printed on the edges of drum brake linings.

results in a scored drum. When it is necessary to replace brake shoes, they must also be replaced on the wheel on the opposite side of the vehicle. Inspect brake shoes for distortion, cracks, or looseness. If these conditions exist, the shoe must be discarded.

Do not let brake fluid, oil, or grease touch the brake lining. If a brake lining kit is used to replace the linings, follow the instructions in the kit, and install all the parts provided.

The two general methods of attaching the linings to the brake shoes are bonding and riveting. The **bonded linings** are fastened (glued) with a special adhesive to the shoe, clamped in place, and then cured in an oven. Instead of using an adhesive, some linings are riveted to the shoe. Riveted linings allow for better heat transfer than bonded linings.

Selecting Replacement Linings

Identification codes, called the automotive friction material edge codes, are printed on the edges of drum brake linings **(Figure 51–35)** and disc brake pads. The letters and numbers identify the manufacturer of the lining material and the material used, and the last two letters identify the cold and hot coefficients of friction (COF).

These codes do not address lining quality or its hardness. From a service standpoint, the COF codes are the most important and are coded as follows:

C not over 0.15
D over 0.15, but not over 0.25
E over 0.25, but not over 0.35
F over 0.35, but not over 0.45
G over 0.45, but not over 0.55
H over 0.55

It is also important to use the recommended friction material when replacing brake shoes. The incorrect type of friction material can affect the stopping characteristics of the car.

Hard and soft are terms applied to linings within a general category of material. Thus, any particular organic lining may be considered as a hard or a soft organic material. Overall, organic linings are considered softer than semimetallic linings, and semimetallic linings are considered softer than fully metallic linings. A hard lining usually has a low COF but resists fade better and lasts longer than a soft lining. A soft lining has a higher COF but fades sooner and wears faster than a hard lining. A soft lining is less abrasive on drum surfaces and operates more quietly than a hard lining. It also is common to use linings with a lower COF on the rear brakes than on the front to minimize rear brake lock-up.

Sizing New Linings

Modern brake shoes are usually supplied with what is known as cam, offset, contour, or eccentric shape, which is ground in at the factory. That is, the full thickness of the lining is present at the middle of the shoe but is ground down slightly at the heel and toe. The diameter of the circle that the shoes make is slightly smaller than that of the drum. This compensates for the minor tolerance variations of drums and brake mountings and promotes proper wearing-in of the linings to match the drum.

SHOP TALK

On duo-servo shoe designs, the forward shoe is the primary, and the rear is the secondary. The secondary shoe lining is longer.

Lining Adjustment

New **eccentric-ground** linings tolerate a closer new lining clearance adjustment than concentric ground linings. With manual adjusters, the shoes should be expanded into the drums until the linings are at the point of drag, but not dragging heavily against the drum. With star wheel automatic adjusters, a drum/shoe gauge **(Figure 51–36)** provides a convenient means of making the preliminary adjustment. This type of gauge, when set at actual drum diameter, automatically provides the working clearance of the shoes **(Figure 51–37)**. If new linings have been concentrically ground, the initial clearance adjustment must be backed off an amount that provides sufficient working clearance.

Many technicians take pride in showing the customer a high pedal, but it should be remembered that with new linings an extremely high pedal indicates tight clearances and can cause seating problems.

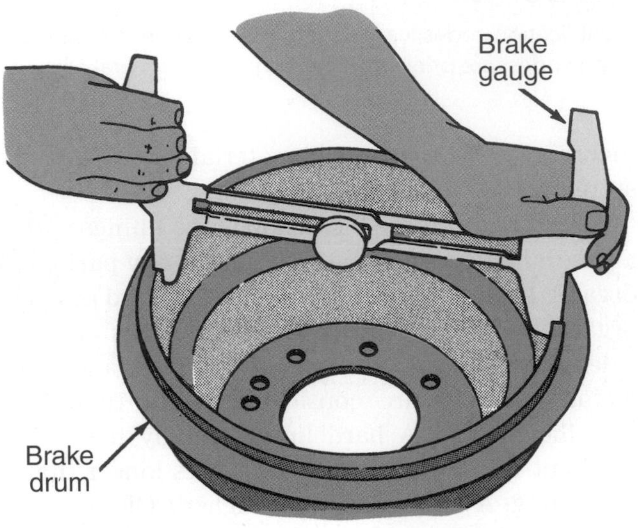

FIGURE 51–36

Using a brake gauge to measure the diameter of a brake drum.

Chrysler Group LLC

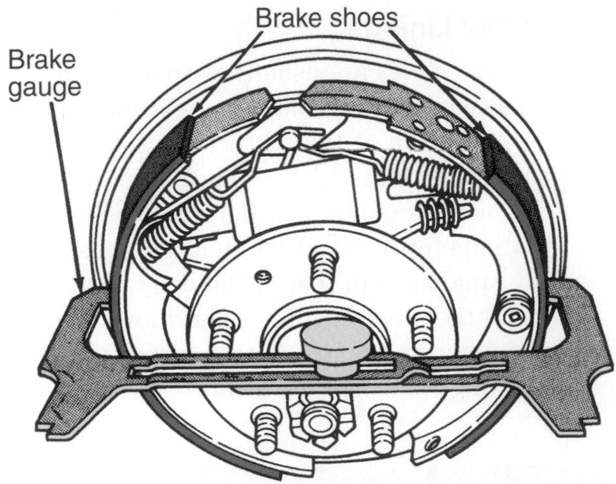

FIGURE 51–37

Using the other side of the brake gauge to set the brake shoes.

Chrysler Group LLC

Drum Shoe and Brake Installation

Before installing the shoes, be sure to sand or stone the inner edge of the shoe to dress down any slight lining or metal nicks and burrs that could interfere with the sliding on the support pads.

A support (backing) plate must be tight on its mount and not bent. Stone the shoe support pads brightly, and dress down any burrs or grooves that could cause the shoes to bind or hang up.

Using an approved lubricant, lightly coat the support pads **(Figure 51–38)** and the threads of servo star wheel adjusters. On rear axle parking brakes, lubricate any point of potential binding in the linkage and the cable. Do not lubricate nonservo brake adjusters other than to free a frozen adjuster with penetrating oil.

Reassemble the brakes in the reverse order of disassembly. Make sure all parts are in their proper locations and that both brake shoes are properly positioned in either end of the adjuster. Also, both brake shoes should correctly engage the wheel cylinder pushrods and parking brake links. They should be centred on the backing plate. Parking brake links and levers should be in place on the rear brakes. With all of the parts in place, try the fit of the brake

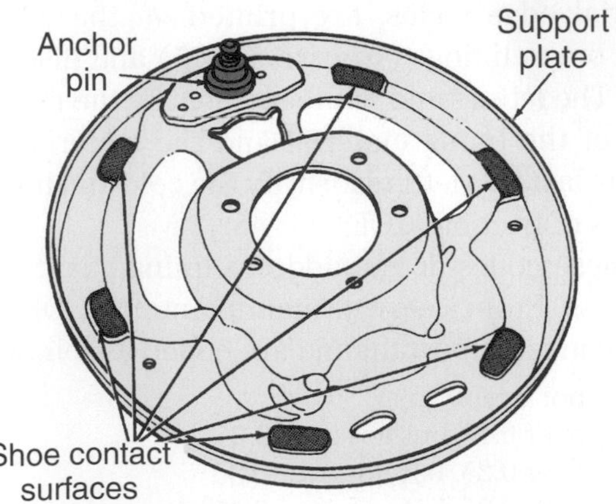

FIGURE 51–38

The areas or pads where the brake shoe will rub or contact the backing plate.

Chrysler Group LLC

drum over the new shoes. If not slightly snug, pull it off, and turn the star wheel until a slight drag is felt when sliding on the drum. A brake preset gauge makes this job easy and final brake adjustment simple. Then install the brake drum, wheel bearings, spindle nuts, cotter pins, dust caps, and wheel/tire assemblies, and make the final brake adjustments as specified in individual instructions in the vehicle's service information. Torque the spindle and lug nuts to specifications.

WHEEL CYLINDER INSPECTION AND SERVICING

Wheel cylinders might need replacement when the brake shoes are replaced or when they begin to leak.

Inspecting and Replacing Wheel Cylinders

Wheel cylinder leaks reveal themselves in several ways: (1) Fluid can be found when the dust boot is peeled back **(Figure 51–39)**; (2) the cylinder, linings, and backing plate, or the inside of a tire might be wet; or (3) there might be a drop in the level of fluid in the master cylinder reservoir.

Such leaks can cause the brakes to grab or fail and should be immediately corrected. Any amount of fluid present when the dust boot is pulled back is a sign of a leaking wheel cylinder. Even a small amount of fluid seepage dampening the interior of the boot warrants wheel cylinder replacement.

FIGURE 51–39

Pull back the wheel cylinder's dust boot to check for internal wheel cylinder leaks.

WARNING!

Hydraulic system parts should not be allowed to come in contact with oil or grease. They should not be handled with greasy hands. Even a trace of any petroleum-based product is sufficient to cause damage to the rubber parts.

Cylinder binding can be caused by rust deposits, by swollen cups due to fluid contamination, or by a cup wedged into an excessive piston clearance. If the clearance between the pistons and the bore wall exceeds allowable values, a condition called heel drag might exist. It can result in rapid cup wear and can cause the piston to retract very slowly when the brakes are released.

Evidence of a scored, pitted, or corroded cylinder bore is a ring of hard, crystal-like substance. This substance is sometimes noticed in the cylinder bore in which the piston rests after the brakes are released.

Light roughness or deposits can be removed with crocus cloth or an approved cylinder hone. While honing lightly, brake fluid can be used as a lubricant. After honing, the cylinder should be washed with soap and water and lubricated with brake fluid. If the bore cannot be cleaned up readily, the cylinder must be replaced.

Care must be taken when installing new or reconditioned wheel cylinders on cars equipped with wheel cylinder piston stops. The rubber dust boots and the pistons must be squeezed into the cylinder before it is tightened to the backing plate. If this is not done, the pistons jam against the stops, causing hydraulic fluid leaks and erratic brake performance.

DRUM PARKING BRAKES

The parking brake keeps a vehicle from rolling while it is parked. It is important to remember that the parking brake is not part of the vehicle's hydraulic braking system. It works mechanically, using a lever assembly connected through a cable system to the rear drum service brakes.

Types of Parking Brake Systems

Parking brakes can be either hand or foot operated. In general, downsized cars and light trucks use hand-operated self-adjusting lever systems **(Figure 51–40)**. Full-size vehicles normally use a foot-operated parking brake pedal **(Figure 51–41A)**. The pedal or lever assembly is designed to latch into

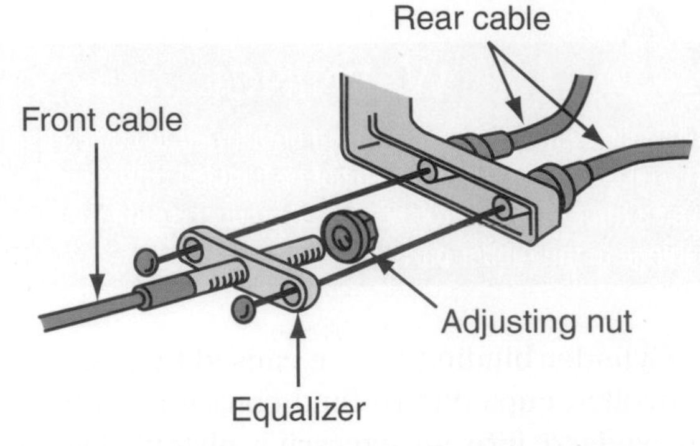

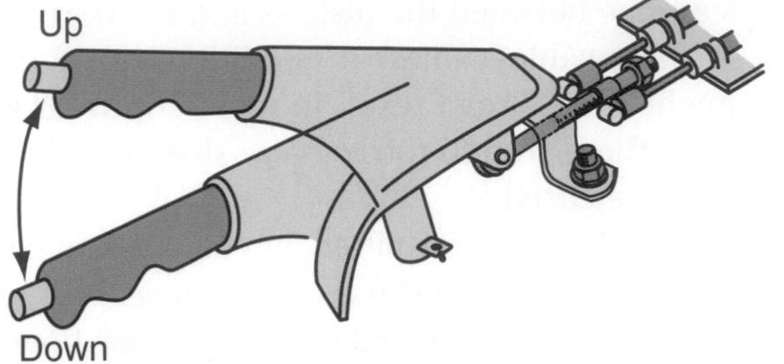

FIGURE 51–40

A typical setup for a centre-mounted hand-operated parking brake.

an applied position and is released by pulling a brake release handle or pushing a release button.

On some vehicles, a vacuum power unit **(Figure 51–41B)** is connected by a rod to the upper end of the release lever. The vacuum motor is actuated to release the parking brake whenever the engine is running and the transmission is in forward driving gear. The lower end of the release lever extends down for alternate manual release in the event of vacuum power failure or for optional manual release at any time. Hoses connect the power unit and the engine manifold to a vacuum release valve on the steering column.

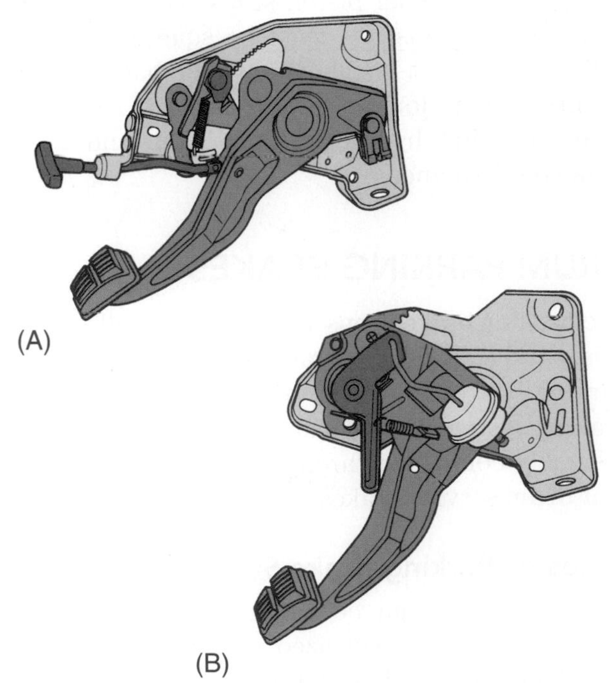

FIGURE 51–41

Typical pedal-operated parking brakes: (A) mechanical release and (B) vacuum release.

The starting point of a typical parking brake cable and lever system is the foot pedal or hand lever. This assembly is a variable-ratio lever mechanism that converts input effort of the operator and pedal/lever travel into output force with less travel. Tensile force from the front cable is transmitted through the car's brake cable system to the rear brakes. This tension pulls the flexible steel cables attached to each of the rear brakes. It operates the internal lever and strut mechanism of each rear brake, expanding the brake shoes against the drum. Springs return the shoes to the unapplied position when the parking brake pedal is released and tensile forces in the cable system are relaxed.

An electric switch, triggered when the brake pedal is applied, lights the brake indicator in the instrument panel when the ignition is turned on. The light goes out when either the pedal or control is released or the ignition is turned off. Some newer vehicles are equipped with electronic parking brake systems.

The cable/lever routing system in a typical parking brake arrangement **(Figure 51–42)** employs a two-lever setup to multiply the physical effort of the operator. First is the pedal assembly or hand grip. When moved, it multiplies the operator's effect and pulls the front cable. The front cable, in turn, pulls the equalizer.

PROCEDURE

Replacing a Wheel Cylinder

STEP 1 Because brake hoses are an important link in the hydraulic system, it is recommended they be replaced when a new cylinder is to be installed or when the old cylinder is to be reconditioned. Remove the brake shoe assemblies from the backing plate before proceeding. The smallest amount of brake fluid contaminates the friction surface of the brake lining.

STEP 2 Use the appropriate tubing wrench, and disconnect the hydraulic line where it enters the wheel cylinder. Care must be exercised in removing this steel line. It might break or bend at this point and be difficult to install once new wheel cylinders are mounted to the backing plate.

STEP 3 Remove the plates, shims, and bolts that hold the wheel cylinder to the backing plate. Some later-designed wheel cylinders are held to the backing plate with a retaining ring that can be removed with two small picks.

STEP 4 Remove the wheel cylinder from the backing plate, and clean the area with a proper cleaning solvent.

STEP 5 Install the new wheel cylinder. Care must be taken when installing wheel cylinders on cars equipped with wheel cylinder piston stops. The rubber dust boots and pistons must be squeezed into the cylinder before it is tightened to the backing plate. If this is not done, the pistons will jam against the stops, causing fluid leaks and erratic brake performance.

STEP 6 Thread the brake line into the cylinder before attaching the wheel cylinder to the backing plate. Once the cylinder's mounting bolts are tightened to specifications, tighten the brake line. Then reassemble the brake unit and bleed the system.

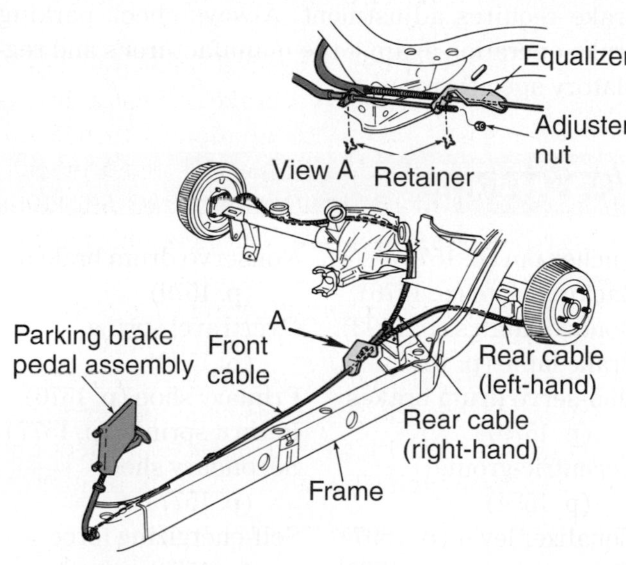

FIGURE 51–42

Typical parking brake routing to a cable equalizer and the rear drum brakes.

The **equalizer lever** splits the effort of the pedal assembly, or hand grip, equally to both rear cables. The equalizer functions by allowing the rear brake cables to slip slightly to balance out small differences in cable length or adjustment. The rear park brake cables pull on a second set of levers, the park brake levers inside the drum brake assembly. These levers multiply the apply force from the cables and place this force directly on the secondary shoe and through a strut bar to the primary shoe. This force pushes both shoes into contact with the drum to hold the drum stationary.

USING SERVICE INFORMATION

Service information lists the standard brake drum inside diameter along with the discard dimension. It also states the standard and minimum lining thickness. Illustrations should be used to accurately identify all components plus the disassembly/reassembly procedure. Tightening torques for backing plate nuts and other components should always be followed.

INTEGRAL PARKING BRAKES

Integral parking brakes are for vehicles with rear-wheel drum brakes. **Figure 51–43** shows a typical integral parking brake. When the parking brake pedal is applied, the cables and equalizer exert a balanced pull on the parking brake levers of both rear brakes. The levers and the parking brake struts move the shoes outward against the brake drums. The shoes are held in this position until the parking brake pedal is released.

The rear cable enters each rear brake through a conduit **(Figure 51–44)**. The cable end engages the lower end of the parking brake lever. This lever is hinged to the web of the secondary shoe and linked with the primary shoe by means of a strut. The lever and strut expand both shoes away from the anchor and wheel cylinder and into contact with the drum

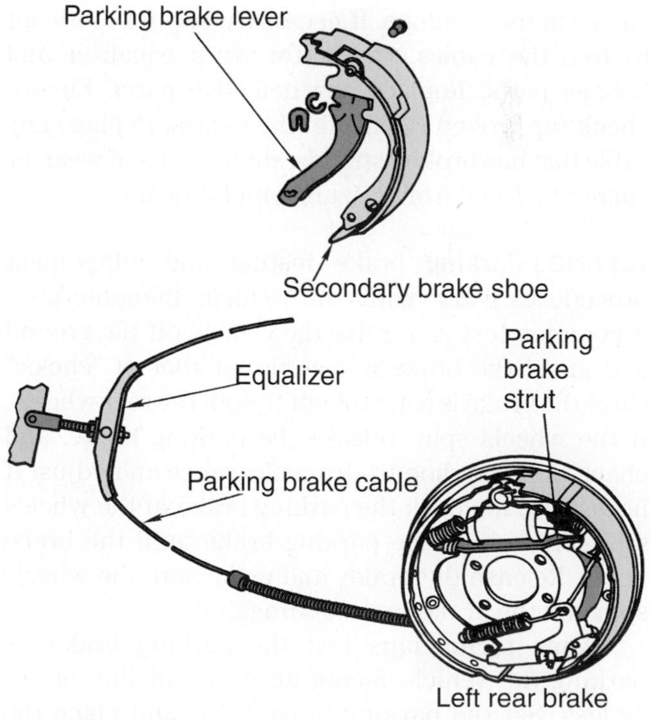

FIGURE 51–43

Integral parking brake components.

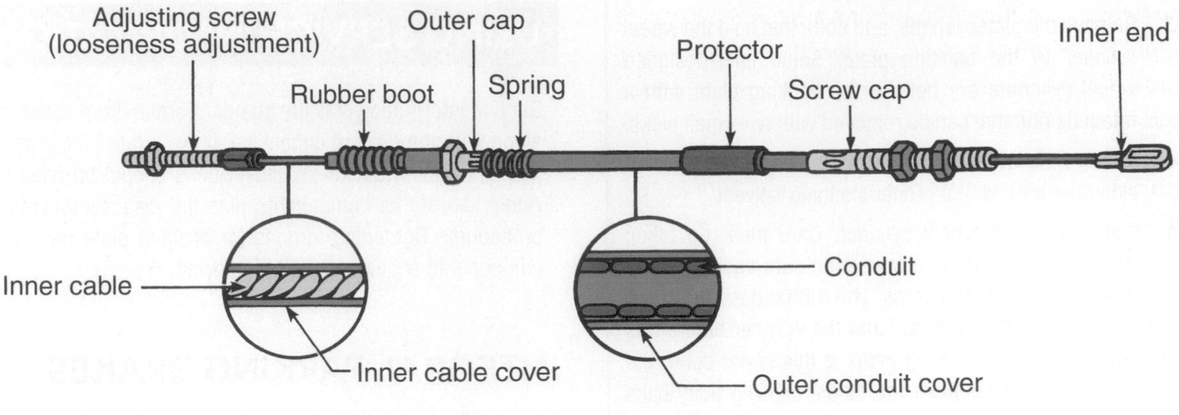

FIGURE 51–44

Rear cable and conduit details.

as the cable and lever are drawn forward. The shoe return springs reposition the shoes when the cable is slacked.

To remove and replace the rear brake shoes, it might be necessary to relieve the parking brake cable tension by backing off the adjusting check nuts at the equalizer. Count the turns backed off in order to restore the nuts to their original position.

Adjusting and Replacing Parking Brakes

Regular wheel brake service should be completed before adjusting the parking brake. Then check the parking brake for free movement of the parking brake cables in the conduits. If necessary, apply a lubricant to free the cables. Check for worn equalizer and linkage parts. Replace any defective parts. Finally, check for broken strands in the cables. Replace any cable that has broken strands, shows signs of wear, or cannot be freed from seizure with lubricant.

TESTING Parking brake testing and adjustment procedures vary with the vehicle manufacturer. A common test is to raise the vehicle off the ground and apply the brake a specific number of "clicks." Once the brake is set, attempt to spin the rear wheels. If the wheels spin, release the parking brake, and check the rear shoe-to-drum clearance and adjust if necessary. Recheck the parking brake. If the wheels still spin, adjust the parking brake until the brake holds. Release the brake, and make sure the wheels spin and the brakes are not dragging.

Some technicians test the parking brake by parking the vehicle facing up on an incline of 30° or less. Set the parking brake fully, and place the transmission in neutral. The vehicle should hold steady. Reverse the vehicle position so that it is facing down the incline, and repeat the test. If the vehicle creeps or rolls in either case, the parking brake requires adjustment. Always check parking brake operation against the manufacturer's and regulatory specifications.

KEY TERMS

Anchor pin (p. 1576)
Backing plate (p. 1576)
Bonded linings (p. 1593)
Brake lathe (p. 1589)
Duo-servo drum brakes (p. 1579)
Eccentric-ground (p. 1594)
Equalizer lever (p. 1597)
Floating drums (p. 1585)
Heat checks (p. 1587)
Hold-down spring (p. 1577)
Nonservo drum brakes (p. 1579)
Overtravel spring (p. 1580)
Primary shoe (p. 1576)
Return springs (p. 1577)
Secondary shoe (p. 1577)
Self-energizing force (p. 1579)
Shoe anchors (p. 1578)
Web (p. 1576)

SUMMARY

- Drum brakes are still used on the rear wheels of many cars and light trucks.
- The drum is mounted to the wheel hub. When the brakes are applied, a wheel cylinder uses hydraulic pressure to force two brake shoes against the inside surface of the drum. The resulting friction between the shoe's lining and drum slows the drum and wheel.

- The brake's anchor pin acts as a brake shoe stop, keeping the shoes from following the rotating drum. This creates a wedging action that multiplies braking force.
- The shoes and wheel cylinder are mounted on a backing plate. Hardware such as shoe return springs, hold-down parts, and linkages are also mounted on the backing plate.
- The primary or leading shoe is toward the front of the vehicle, and the secondary or trailing shoe is toward the rear of the vehicle.
- Brake lining can be attached to the shoes by riveting or a special adhesive bonding process.
- Brake drums act as a heat sink to dissipate the heat of braking friction. Drums can be refinished on a brake lathe provided the inside diameter is not increased above the specified limit before the discard dimension.
- Servicing brakes requires performing a complete system inspection. Partial replacement of worn or damaged parts does not solve the braking problems and may ruin the new parts installed.
- When servicing brakes, care should be taken to avoid generating harmful brake dust.
- Wheel cylinders should be replaced if they show any signs of hydraulic fluid leakage or component wear.
- Drum brakes allow for the use of a simple parking brake mechanism that can be activated with a hand lever or foot pedal. This is a mechanical system, completely separate from the service brake hydraulic system.

REVIEW QUESTIONS

1. Name the two methods of attaching brake lining materials to the brake shoes.
2. Explain how drum brakes create a self-multiplying brake force.
3. List at least five separate types of wear and distortion to look for when inspecting brake drums.
4. What is the job of wheel cylinder stops?
5. Explain the operation of an integral drum brake parking brake.
6. In a typical drum brake, which component provides a foundation for the brake shoes and associated hardware?
   a. wheel cylinder **b.** drum
   c. backing plate **d.** lining
7. How is the trailing shoe of a leading-trailing brake assembly applied?
   a. wheel cylinder hydraulic force and self-energization
   b. wheel cylinder hydraulic force plus servo application
   c. wheel cylinder hydraulic force only
   d. servo application only
8. Which of the following statements about drum brakes is correct?
   a. Web thickness varies only with changes in shoe size.
   b. Friction lining material can only be riveted to the brake shoe.
   c. The secondary shoe usually has the smaller brake lining.
   d. The primary shoe is the first shoe from the wheel cylinder in the direction of forward rotation.
9. When should brake linings be replaced?
   a. when linings are worn to within 0.79 mm (1/32 in.) of a rivet head
   b. when linings are smooth and not rough textured
   c. when self-adjustment does not occur
   d. when the drum surface appears glazed
10. In the unapplied position, how are the drum brake shoes held against the anchor pin?
    a. hold-down springs
    b. star wheel adjuster
    c. shoe hold-down
    d. return springs
11. What are duo-servo drum brakes also known as?
    a. leading-trailing brakes
    b. self-energizing brakes
    c. nonservo brakes
    d. dual anchor brakes
12. Which of the following statements about drum brakes equipped with star wheel automatic adjusters is correct?
    a. The adjustment mechanism is located on the primary shoe.
    b. They are set to adjust when the vehicle is moving forward.
    c. The adjustment mechanism is located on the secondary shoe.
    d. They automatically adjust when the brakes are not applied.

13. What is the function of the wheel cylinder?
   a. to maintain the correct brake shoe to drum clearance
   b. to prevent the shoe from rotating with the brake drum
   c. to convert hydraulic pressure to mechanical force against the brake shoe
   d. to apply the brake shoes when the park brake is applied
14. How should backing plates, struts, levers, and other metal brake parts be cleaned?
   a. wet-cleaned using water or a water-based solution
   b. wet-cleaned using an alcohol-based solvent
   c. wet-cleaned using gasoline
   d. dry-cleaned only
15. What can brake dust and dirt between the lining and the drum cause?
   a. concave/barrel-shaped drum
   b. convex/tapered drum
   c. threaded drum surface
   d. scored drum surface
16. What does the diameter stamped on the brake drum refer to?
   a. the diameter of the drum when new
   b. the diameter that the drum should wear to before machining
   c. the diameter that the drum can be machined to
   d. the maximum (discard) diameter
17. What can cause a drum to become convex shaped?
   a. dirt between the drum and lining
   b. machining with a dull cutting bit
   c. overheating the drum
   d. linings contaminated by a leaking wheel cylinder
18. What park brake component ensures that both drum assemblies receive the same application force?
   a. the park brake lever
   b. the front park brake cable
   c. the equalizer
   d. the rear park brake cables
19. What is the maximum brake drum out-of-roundness allowed before machining or drum replacement is required?
   a. 0.075 mm (0.003 in.)
   b. 0.150 mm (0.006 in.)
   c. 0.225 mm (0.009 in.)
   d. 0.300 mm (0.012 in.)
20. Which of the following components do not require same axle replacement (on both sides) when performing drum brake servicing?
   a. wheel cylinder
   b. brake drum
   c. brake springs and hardware
   d. brake linings

CHAPTER 52
Disc Brakes

LEARNING OUTCOMES

- List the advantages of disc brakes.
- List disc brake components, and describe their functions.
- Explain the difference between the three types of calipers commonly used on disc brakes.
- Describe the two types of parking brake systems used with disc brakes.
- Describe the causes of common disc brake problems.
- Explain what precautions should be taken when servicing disc brake systems.
- Describe the general procedure involved in replacing disc brake pads.
- List and describe five typical disc brake rotor problems.

Disc brakes resemble the brakes on a bicycle. The friction elements are in the form of pads, which are squeezed or clamped about the edge of a rotating wheel. With automotive disc brakes, this wheel is a separate unit mounted to the wheel hub, called a **rotor (Figure 52–1)**. The rotor is typically made of cast iron. However, much research is being done with composite materials in the design of rotors. Because the pads clamp against both sides of a rotor, both sides are machined smooth. Usually, the two surfaces are separated by a finned centre section for better cooling. Such rotors are called ventilated rotors. The pads are attached to metal shoes, which are actuated by pistons contained within a **caliper** assembly, and a housing that wraps around the edge of the rotor. The caliper is mounted to the steering knuckle to stop it from rotating.

FIGURE 52–1

A typical disc brake assembly.

The caliper contains the pistons and related seals, springs, and boots as well as the cylinder(s) and fluid passages necessary to force the pads against the rotor. The caliper resembles a hand in the way it wraps around the edge of the rotor.

Unlike shoes in a drum brake, the pads act perpendicularly to the rotation of the disc when the brakes are applied. In a brake drum, frictional drag actually pulls the shoe tighter into the drum. This does not happen with disc brakes; therefore, they require more force to achieve the same braking effort. For this reason, they are typically used with a power brake unit.

Disc brakes offer a number of major advantages over drum brakes. Disc brakes are more resistant to heat fade during high-speed brake stops or repeated stops because the design of the disc brake rotor exposes more surface to the air and thus dissipates heat more efficiently. They are also resistant to water fade because the rotation of the rotor tends to throw off moisture and the squeeze of the sharp edges of the pads clears the surface of water. Disc brakes perform more straight-line stops because they do not produce any self-energization action. Due to their clamping action, disc brakes are less apt to pull. Finally, disc brakes automatically adjust as pads wear.

DISC BRAKE COMPONENTS AND THEIR FUNCTIONS

The disc brakes used today are typically of two basic designs: **fixed caliper (Figure 52–2)** or **floating caliper**. There is also a **sliding caliper**, but its design is very similar to the floating caliper **(Figure 52–3)**. The only difference is that sliding calipers slide on surfaces that have been machined smooth for this purpose, and floating calipers slide on special pins or bolts. The disc brake, regardless of its design, consists of a hub and rotor assembly, a caliper assembly, and the brake pads.

Rotors

The disc brake rotor has two main parts: the hub and the braking surface. The hub is where the wheel is mounted and may contain the wheel bearings. The braking surface is the machined surface on both sides of the rotor. It is carefully machined to provide a friction surface for the brake pads. The entire rotor is usually made of cast iron, which provides an excellent friction surface.

FIGURE 52–2

A fixed caliper assembly.

©Cengage Learning 2015

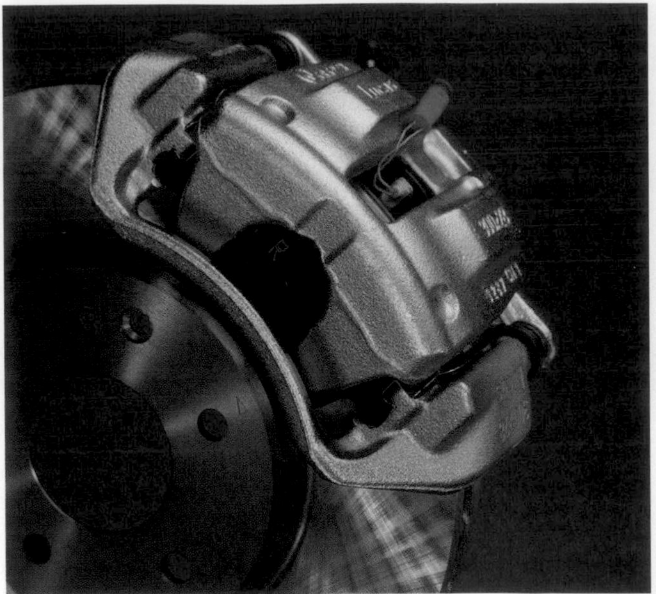

FIGURE 52–3

A floating caliper assembly.

DaimlerChrysler Corporation

The size of the rotor braking surface is determined by the diameter of the rotor. Large and high-performance cars, which require more braking energy, have large rotors. Smaller, lighter cars can use smaller rotors. Generally, manufacturers want to keep parts as small and light as possible while maintaining efficient braking ability.

The rotor is protected from water and dirt due to road splash by a **splash shield** bolted to the steering knuckle **(Figure 52–4)**. The outboard side is shielded by the vehicle's wheel. The splash shield

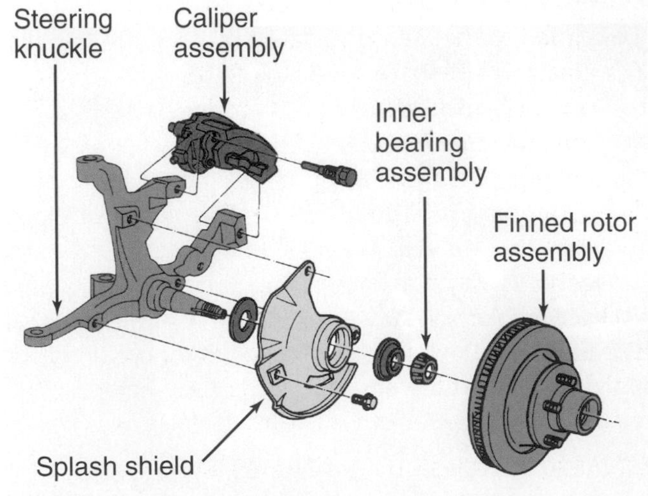

FIGURE 52–4

A steering knuckle, a splash shield, bearings, and a finned rotor for a disc brake assembly.

Ford Motor Company

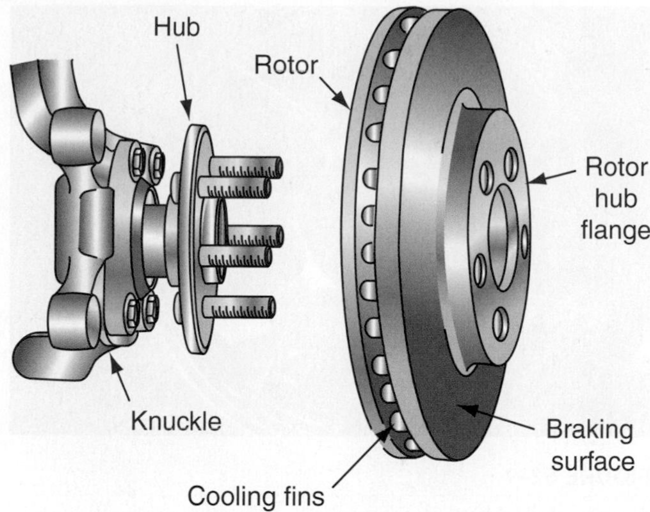

FIGURE 52–5

A rotor that is cast as a separate part and fastened to the hub is called a floating rotor.

and wheel also are important in directing air over the rotor to aid cooling.

FIXED AND FLOATING ROTORS Rotors are classified by their hub design. A **fixed rotor** has the hub and the rotor cast as a single unit. The rotor illustrated in Figure 52–4 is an example of a fixed rotor. **Floating rotors** and hubs are made as two separate parts. The hub is a conventional casting and is mounted on wheel bearings or on the axle. The wheel studs are mounted in the hub and pass through the rotor centre section **(Figure 52–5)**. One advantage of this design is that the rotor is less expensive and can be replaced easily and economically.

COMPOSITE ROTORS The need to reduce vehicle weight led to the development of **composite rotors**. Composite rotors are made of different materials, usually cast iron and steel, to reduce weight. The friction surfaces and the hubs are cast iron, but the supporting parts of the rotor are made of lighter steel stampings **(Figure 52–6)**. The steel and iron sections are bonded to each under heat and high pressure to form a one-piece finished assembly. Composite rotors may be fixed or floating rotors. Because the friction surfaces of composite rotors are cast iron, the wear standards are generally the same as they are for other rotors.

CERAMIC ROTORS In the late 1990s, Porsche first offered a carbon-ceramic brake option on the 911 GT2 and then on its 911 Turbo in 2000. Today, ceramic brakes are an option on all Ferraris and most Lamborghinis, Porsches, Bentleys, and some Corvettes **(Figure 52–7)**. Soon they will be standard or optional equipment on many more vehicles.

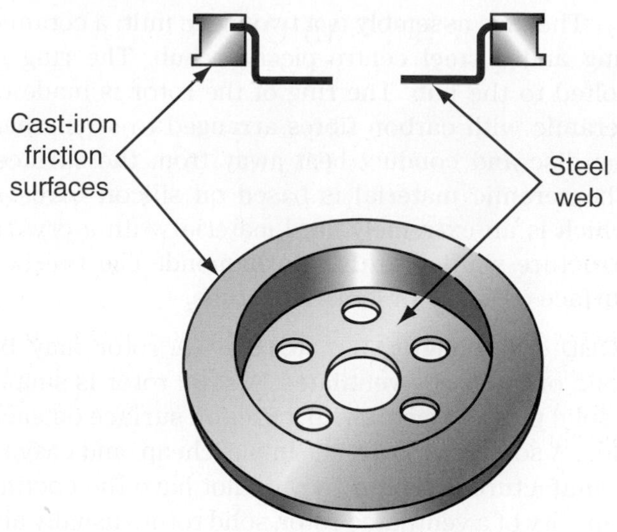

FIGURE 52–6

Composite rotors are made of different materials, usually cast iron and steel.

Ceramic brakes are costly but weigh about one-half of a conventional rotor. This means they allow for lower unsprung weight that helps ride quality and handling and improves fuel economy. They also last four times longer than steel discs. Brake pads also last about three times longer. The brake pads designed to be used with ceramic discs contain a ceramic powder mixed with metal wires or particles. The pads have heat shields to prevent the heat from travelling through the system.

Ceramic brakes have excellent fade resistance and stopping power. Also, the vehicle's wheels stay cleaner because no black brake dust is released.

FIGURE 52–7

A carbon-ceramic brake assembly.

The disc assembly is a two-piece unit: a ceramic ring and a steel centre piece or hub. The ring is bolted to the hub. The ring of the rotor is made of ceramic with carbon fibres arranged to strengthen the disc and conduct heat away from the surface. The ceramic material is based on silicon carbide, which is an extremely hard material with a crystal structure similar to that of diamond. The finished surface of the rotor looks like stone.

SOLID AND VENTILATED ROTORS A rotor may be solid or it may be ventilated. A solid rotor is simply a solid piece of metal with a friction surface on each side. A solid rotor is light, simple, cheap, and easy to manufacture. Because they do not have the cooling capacity of a ventilated rotor, solid rotors usually are used on small cars of moderate performance and the rear brakes of performance-oriented vehicles.

A ventilated rotor has cooling fins cast between the braking surfaces to increase the cooling area of the rotor. When the wheel is in motion, the rotation of these fins in the rotor also increases air circulation and brake cooling **(Figure 52–8)**. Although ventilated rotors are larger and heavier than solid rotors, these disadvantages are more than offset by their better cooling ability and heat dissipation.

Some ventilated rotors have cooling fins that are curved or formed at an angle to the hub centre. These fins increase the centrifugal force on the rotor airflow and increase the air volume that removes heat. Such rotors are called unidirectional rotors because the fins only work properly when the rotor rotates in one direction. Therefore, unidirectional rotors cannot be interchanged from the right side to the left side on the car.

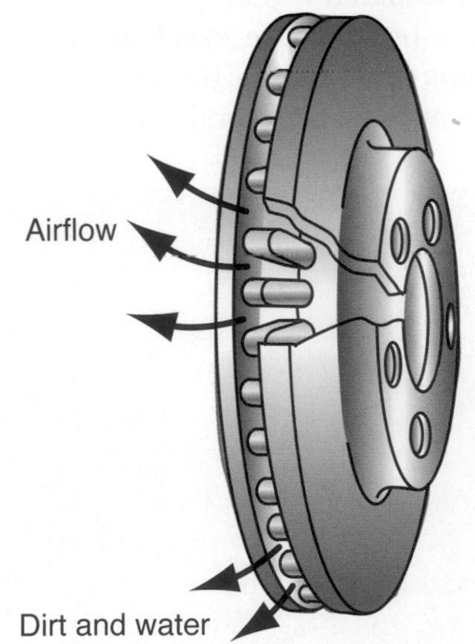

FIGURE 52–8

A ventilated rotor has cooling fins cast between the braking surfaces to increase the cooling area of the rotor.

FIGURE 52–9

A cross-drilled brake rotor.

DRILLED VERSUS SLOTTED ROTORS Many high-performance vehicles are fitted with cross-drilled rotors **(Figure 52–9)**. The idea behind having holes through the rotor is simply to allow heat, gases, and dirt to escape. In addition, the edges of the holes give a place for the pads to grab. They, however, also decrease the overall surface area of the rotor, which reduces the thermal capacity of the discs, and the discs have a poor service life. The latest trend is to cut a series of tangential slots or channels into the surface. These slots do the same thing as the holes without the disadvantages.

Rotor Hubs and Wheel Bearings

Tapered roller bearings, which are installed in the wheel hubs, were common on the front wheels of rear-wheel-drive (RWD) and the rear wheels of front-wheel-drive (FWD) vehicles. Normally, late-model vehicles do not use tapered roller bearings; however, there are many vehicles still on the road that have them. The tapered roller bearing has two main parts: the inner bearing cone and the outer bearing cup. The bearing cone is an assembly that contains steel tapered rollers. The rollers ride on an inner cone-shaped race and are held together by a bearing cage. The bearing fits into the outer cup, or race, which is pressed into the hub. This provides two surfaces, an inner cone and outer cup, for the rollers to ride on. The bearings are held in place with a thrust washer, nut, locknut, and cotter pin. A dust cap fits over the assembly to keep dirt out and lubricant in. A seal on the inboard side prevents lubricant from escaping at this end.

Caliper Assembly

A brake caliper converts hydraulic pressure into mechanical force. The caliper housing is usually a one-piece construction of cast iron or aluminum and has an inspection hole in the top to allow for lining wear inspection. The housing contains the cylinder bore(s). In the cylinder bore is a groove that seats a square-cut seal. This groove is tapered toward the bottom of the bore to increase the compression on the edge of the seal that is nearest hydraulic pressure. The top of the cylinder bore is also grooved as a seat for the dust boot. A fluid inlet hole is machined into the bottom of the cylinder bore, and a bleeder valve is located near the top of the casting **(Figure 52–10)**.

A caliper can contain one, two, four, or six cylinder bores and pistons that provide uniform pressure distribution against the brake's friction pads. The pistons are relatively large in diameter and short in stroke to provide high pressure on the friction pad assemblies with a minimum of fluid displacement.

Basically, the hydraulics of disc brakes are the same as for drum brakes, in that the master cylinder piston forces the brake fluid into the wheel cylinders and against the wheel pistons.

The disc brake piston is made of steel, aluminum, or fibreglass-reinforced **phenolic resin**. Steel pistons are usually nickel-chrome plated for improved durability and smoothness. The top of the pistons is grooved to accept the **dust boot** that seats in a groove at the top of the cylinder bore and also in a groove in the piston. The dust boot prevents moisture and road contamination from entering the bore.

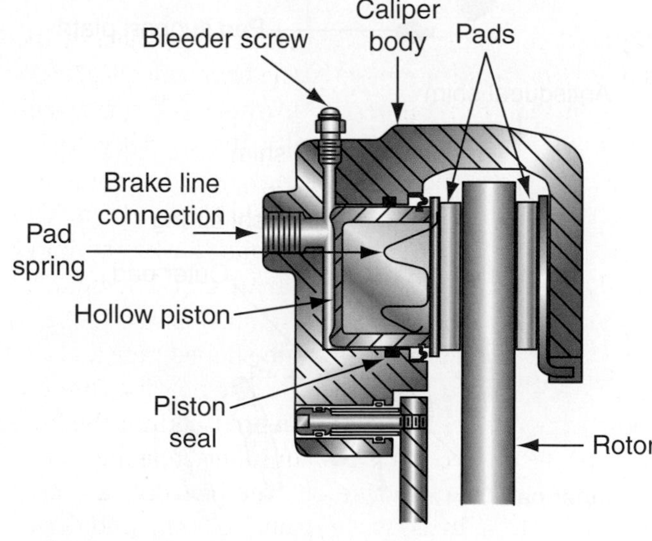

FIGURE 52–10

Cross-section of a typical caliper.

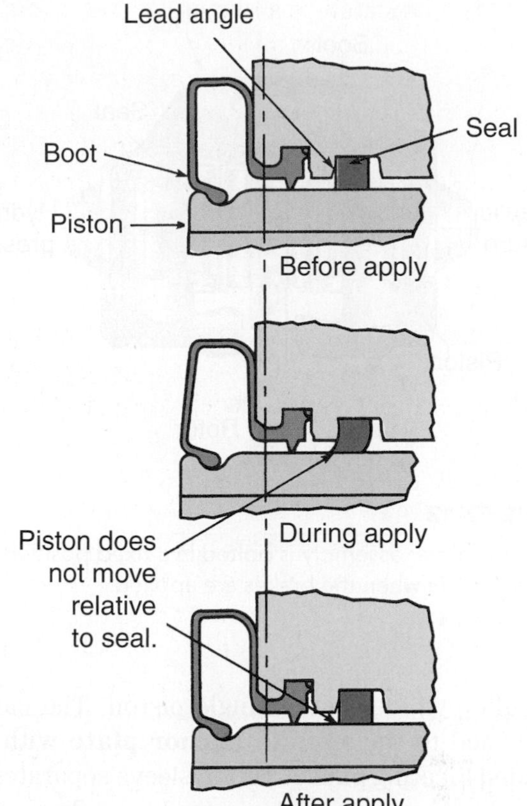

FIGURE 52–11

Action of the piston's hydraulic seal in the caliper's cylinder.

A piston hydraulic (square-cut) seal prevents fluid leakage between the cylinder bore wall and the piston. This rubber sealing ring also acts as a retracting mechanism for the piston when hydraulic pressure is released, causing the piston to return in its bore **(Figure 52–11)**. When hydraulic pressure is diminished, the seal functions as a return spring to retract the piston.

In addition, as the disc brake pads wear, the seal allows the piston to move farther out to adjust automatically for the wear, without allowing fluid to leak. Because the brake pads need to retract only slightly after they have been applied, the piston moves back only slightly into its bore. The additional brake fluid in the caliper bore keeps the piston out and ready to clamp the surface of the rotor.

FIXED CALIPER DISC BRAKES Fixed caliper disc brakes have a caliper assembly that is bolted in a fixed position and does not move when the brakes are applied. The pistons in both sides of the caliper come inward to force the pads against the rotor **(Figure 52–12)**.

FLOATING CALIPER DISC BRAKES A typical floating caliper disc brake is a one-piece casting that has one

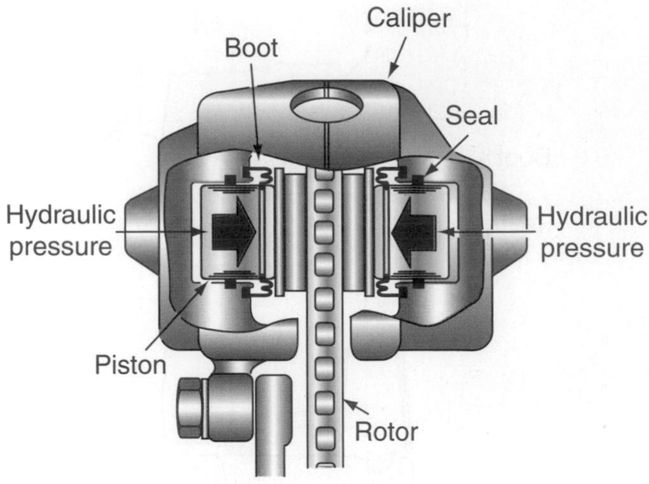

FIGURE 52–12

The fixed caliper assembly is bolted in a fixed position and does not move when the brakes are applied.

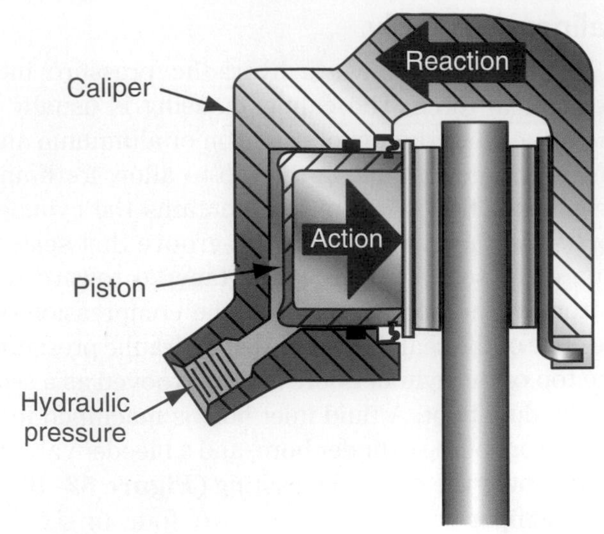

FIGURE 52–13

Operation of a floating caliper.

hydraulic cylinder and a single piston. The caliper is attached to the **spindle anchor plate** with two threaded locating pins. A Teflon sleeve separates the caliper housing from each pin, and the caliper slides back and forth on the pins as the brakes are actuated. When the brakes are applied, hydraulic pressure builds in the cylinder behind the piston and seal. Because hydraulic pressure exerts equal force in all directions, the piston moves evenly out of its bore.

The piston presses the inboard pad against the rotor. As the pad contacts the revolving rotor, greater resistance to outward movement is increased, while hydraulic pressure also pushes the caliper away from the piston. This action forces the outboard pad against the rotor **(Figure 52–13)**. This allows one piston to apply both pads with equal pressure.

SLIDING CALIPER DISC BRAKES With a sliding caliper assembly, the caliper slides or moves sideways when the brakes are applied. As mentioned previously, in operation, this older brake design is almost identical to the floating type. But unlike the floating caliper, the sliding caliper does not float on pins or bolts attached to the anchor plate. It has angular machined surfaces at each end that slide in mating machined surfaces on the anchor plate. This is where the caliper slides back and forth.

Some sliding calipers use a support key to locate and support the caliper in the anchor plate **(Figure 52–14)**. The caliper support key is inserted between the caliper and the anchor plate. A worn support key may cause tapered brake pad wear. Always inspect the support keys when replacing

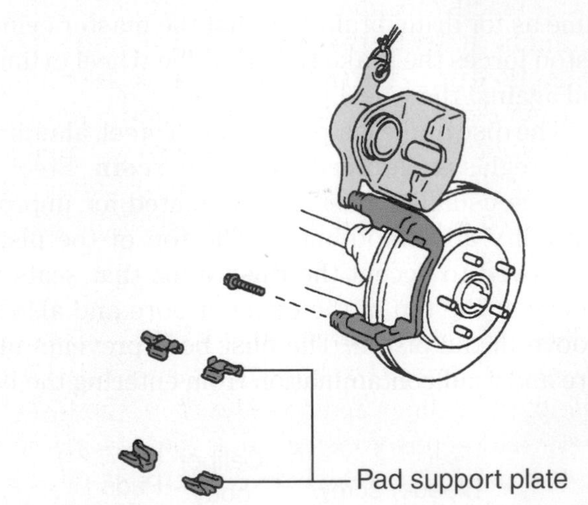

FIGURE 52–14

A sliding caliper with support plates (keys).

FIGURE 52–15
An electric brake caliper.
BWI North America

brake pads. Also make sure they are lubricated when reassembling the unit.

ELECTRIC CALIPERS A recent development is the use of electrically operated calipers **(Figure 52–15)**. Using electric calipers eliminates the need to run hydraulic lines to the wheels and results in better brake control. This system is computer controlled and can evenly distribute braking power at the four wheels. At the rear wheels, this system can also be used as the parking brake, thereby eliminating the need to have cables at the rear wheels.

Brake Pad Assembly

Brake pads are metal plates with the linings either riveted or bonded to them. Pads are placed at each side of the caliper and straddle the rotor. The inner brake pad is positioned against the piston and may or may not be interchangeable with the outer brake pad, which is located against the outer caliper housing. The linings are made of semimetallic or other nonasbestos material.

DISC PAD WEAR SENSORS Some brake pads have wear-sensing indicators. The three most common designs of wear sensors are audible, visual, and tactile.

Audible sensors are thin, spring steel tabs that are riveted to or installed onto the edge of the pad's backing plate and are bent to contact the rotor when the lining wears down to a point that replacement is necessary. At that point, the sensor causes a high-pitched squeal whenever the wheel is turning, except when the brakes are applied. Then the noise goes away. The noise gives a warning to the driver that brake service is needed and perhaps saves the rotor from destruction **(Figure 52–16)**. The tab is generally located on the leading edge of the inner or outer brake pad, depending on the vehicle.

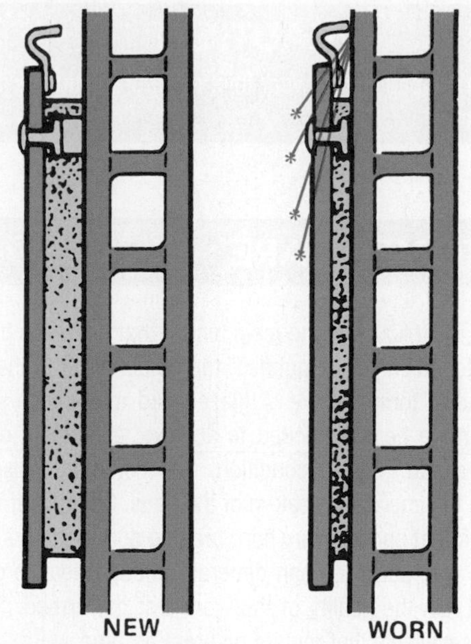

FIGURE 52–16
Operation of the wear indicator.
Federal-Mogul Corporation

Some vehicles have systems with an electronic wear indicator in the disc brake pads. As the pad wears to a predetermined point, a warning light in the instrument panel is illuminated by the wear sensors. In some systems, a small pellet is contained in the brake pad's friction material. The pellets are wired in series or in parallel to the red brake warning lamp circuit **(Figure 52–17)** and complete the lamp circuit when the pellets contact the rotor.

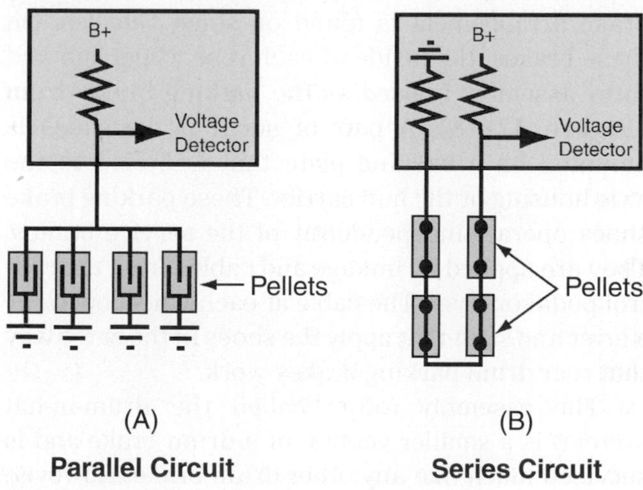

FIGURE 52–17
Electronic pad wear sensors can be wired in parallel or in series with the warning lamp. The pellets are embedded in the brake pads.

CAUTION!

As disc brake pad linings wear thin, more brake fluid is needed in the system.

CUSTOMER CARE

Excessive heat liquefies the resin binder that holds the brake pad material together. Once liquefied, the binder rises to the surface of the pad to form a glaze. A glazed pad may cause squealing because more heat is needed to achieve an amount of friction equal to a pad in good condition. One common cause of pad glazing is the improper break-in of the pads. Remember to inform customers that unnecessary hard braking during the first 320 km (200 mi.) of a pad's life can generate enough heat to glaze the pads and ruin the quality of the work just performed. There are pads now available that require no break-in time.

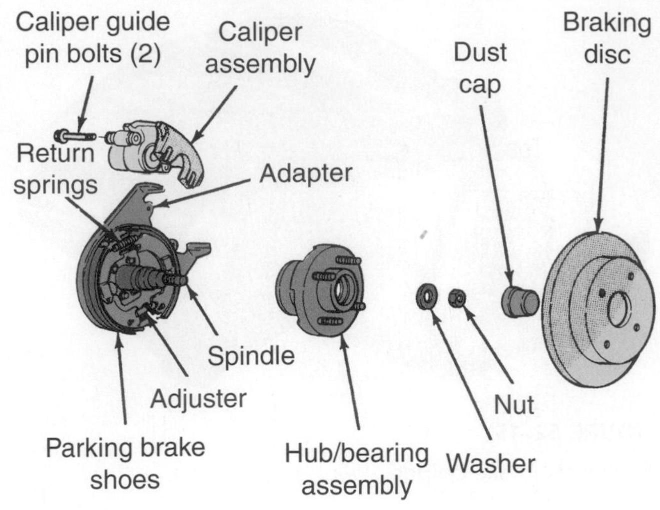

FIGURE 52–18

A typical parking brake that uses a drum built into the centre of a rotor.

Chrysler Group LLC

REAR-WHEEL DISC BRAKES

Rear-wheel disc brake calipers may be fixed, floating, or sliding, and all of these designs work in the same way as when they are used at the front wheels. The only difference between a front and rear disc brake caliper is that the rear disc brake caliper needs a parking brake. Four-wheel disc brake installations must have some way to apply the rear brakes when the parking brake is set.

Rear Disc/Drum (Auxiliary Drum) Parking Brake

The rear disc/drum or auxiliary drum parking brake arrangement is found on some vehicles. On these brakes, the inside of each rear wheel hub and rotor assembly is used as the parking brake drum **(Figure 52–18)**. A pair of small brake shoes is mounted on a backing plate that is bolted to the axle housing or the hub carrier. These parking brake shoes operate independently of the service brakes. They are applied by linkage and cables from the control pedal or lever. The cable at each wheel operates a lever and strut that apply the shoes in the same way that rear drum parking brakes work.

The assembly (often called the drum-in-hat system) is a smaller version of a drum brake and is serviced much like any other drum brake. However, they do not have self-adjusters. The parking brakes must be adjusted manually with star wheels that are accessible through the backing plate or through the outboard surface of the drum.

Late-model GM trucks use an expandable, single metal band covered with friction material inside the parking brake drum that is machined on the inside of the rear brake disc.

Caliper-Actuated Parking Brakes

Most floating or sliding caliper rear disc brakes mechanically apply the calipers' pistons to lock the pads against the rotors for parking. All caliper-actuated parking brakes have a lever that protrudes from the inboard side of the caliper. These levers are operated by linkage and cables from the parking brake control pedal or lever.

The two most common types of caliper-actuated parking brakes are the screw-and-nut type and the ball-and-ramp type. A few imported cars have a third type that uses an eccentric shaft and a rod to apply the caliper piston. An eccentric acts like a cam. One portion of the shaft is oval shaped. As the shaft rotates, the high part of the oval pushes the operating rod out to apply the brakes.

General Motors' floating caliper rear disc brakes are the most common example of the screw-and-nut parking brake mechanism **(Figure 52–19)**. The caliper lever is attached to an actuator screw inside the caliper that is threaded into a large nut. The nut, in turn, is splined to the inside of a large cone that fits inside the caliper piston. When the parking brake is applied, the caliper lever rotates the actuator screw. Because the nut is splined to the inside of the cone, it cannot rotate, so it forces the cone outward against the inside

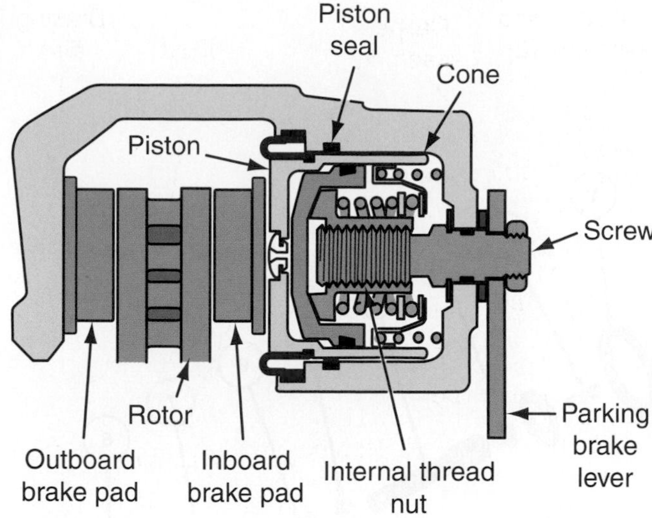

FIGURE 52–19

A GM screw-and-nut parking brake mechanism for a disc brake.

of the piston, forcing it outward. Similarly, the piston cannot rotate because it is keyed to the brake pad, which is fixed in the caliper. The piston then applies the inboard brake pad, and the caliper slides as it does for service brake operation and forces the outboard pad against the rotor. An adjuster spring inside the nut and cone rotates the nut outward when the parking brakes are released to provide self-adjustment. Rotation of the nut takes up clearance as the brake pads wear.

Ford's floating caliper rear disc brakes are the most common example of the ball-and-ramp parking brake mechanism **(Figure 52–20)**. The

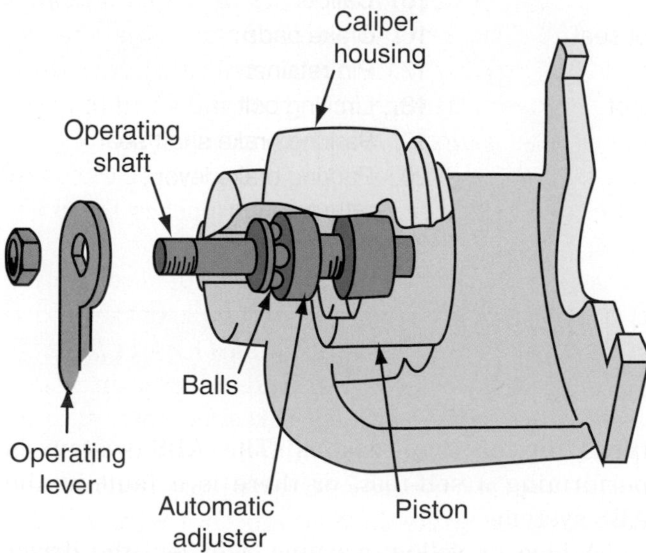

FIGURE 52–20

A Ford ball-and-ramp parking mechanism for a disc brake.

caliper lever is attached to a shaft inside the caliper that has a small plate on the other end. Another plate is attached to a thrust screw inside the caliper piston. The two plates face each other, and three steel balls separate them. When the parking brake is applied, the caliper lever rotates the shaft and plate. Ramps in the surface of the plate force the balls outward against similar ramps in the other plate. As the plates move farther apart, the thrust screw forces the piston outward. The thrust screw cannot rotate because it is keyed to the caliper. The piston then applies the inboard brake pad, and the caliper slides as it does for service brake operation and forces the outboard pad against the rotor. When the piston moves away from the thrust screw, an adjuster nut inside the piston rotates on the screw to take up clearance and provide self-adjustment. A drive ring on the nut keeps it from rotating backward.

Another way to tighten the pads against the rotor when the parking brake is applied is to use a threaded, spring-loaded pushrod **(Figure 52–21)**. As the parking brakes are applied, a mechanism rotates or unscrews the pushrod, which in turn pushes the piston out.

DISC BRAKE DIAGNOSIS

Many problems experienced on vehicles with disc brakes are the same as those evident with drum brake systems. Some problems occur only with disc brakes. Before covering the typical complaints, it is important to remind you to get as much information as possible about the complaint from the customer, and then road-test the vehicle to verify the complaint. A complete inspection of the rotor, caliper, and pads should be done any time you are working on the brakes.

What follows is a brief discussion of common complaints and their typical causes.

Warning Lights

Today's vehicles are normally equipped with more than one brake warning light on the instrument panel. Regardless of what warning light is lit, it is an indication of warning to the driver. You need to understand what would cause the different lights to illuminate in order to take care of the problem. Keep in mind that a vehicle may have one, two, or all of these lights **(Figure 52–22)**.

The red warning light indicates there is a problem in the regular brake system, such as low

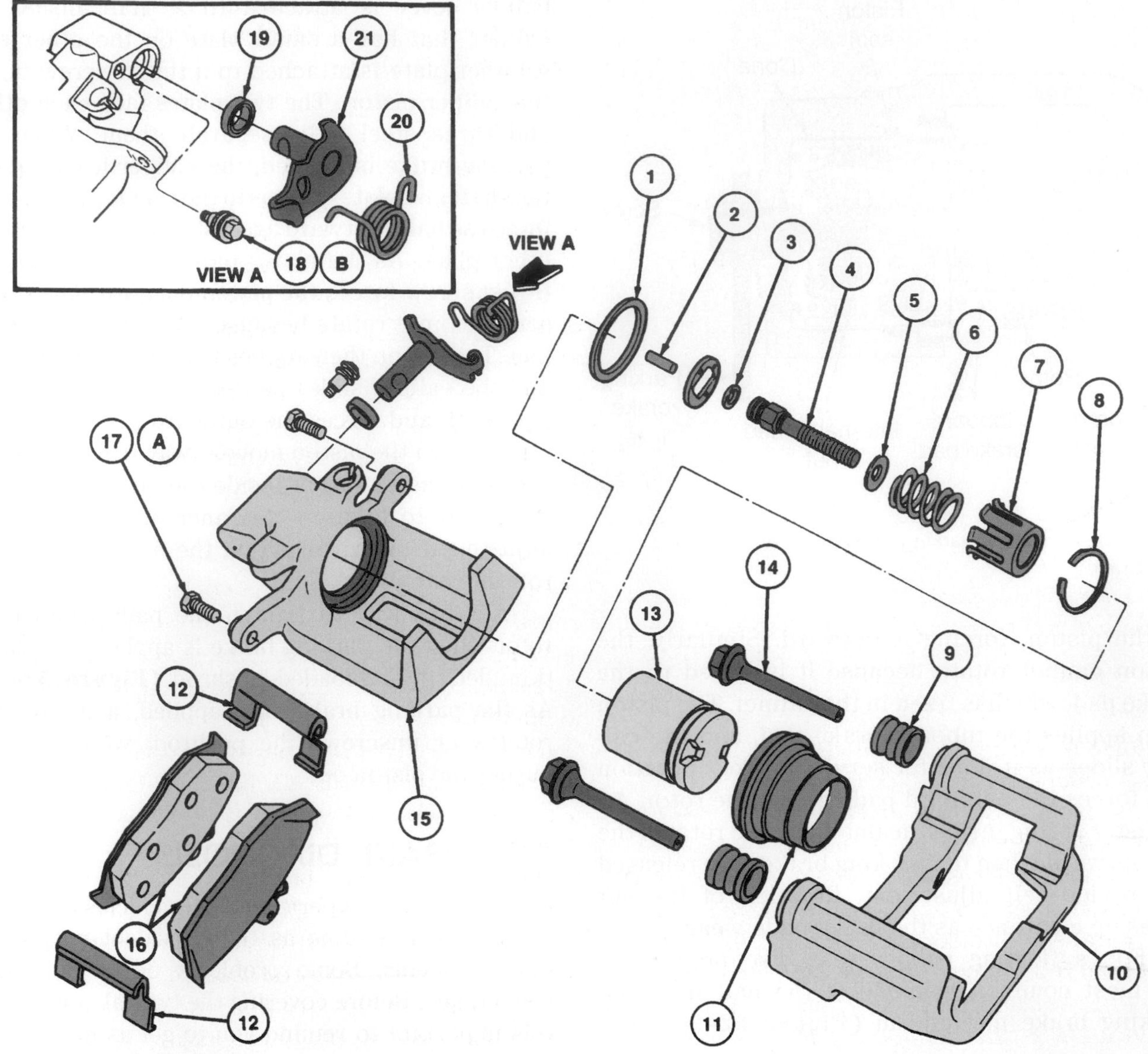

1. Piston seal
2. Pin
3. O-ring
4. Pushrod
5. Flat washer
6. Spring
7. Spring cage
8. Snap ring
9. Slider pin boot seal
10. Rear support
11. Piston dust boot
12. Antirattle clip
13. Piston
14. Locating pin
15. Caliper
16. Brake pads
17. Pin retainer
18. Limiting bolt
19. Parking brake shaft seal
20. Parking brake lever return spring
21. Parking brake lever

FIGURE 52–21

A rear caliper with a threaded drive for the piston.

Ford Motor Company

brake fluid levels or that the parking brake is on. A low fluid light may be present in addition to the red brake warning light. Whenever the fluid is low, you should suspect a leak or very worn brake pads.

The yellow or amber brake warning light is tied into the antilock brake system. This light turns on for two reasons: The ABS system is performing a self-test, or there is a fault in the ABS system.

A blue or yellow warning light lets the driver know the wheels are slipping because of poor road conditions.

FIGURE 52–22

Typical brake warning lamps.

Pulsating Pedal

Customers will feel a vibration or shudder in the brake pedal and/or the steering wheel when the brakes are applied if a brake rotor is warped. A warped brake rotor is one that no longer has parallel friction surfaces or has side-to-side movement. If this symptom exists, check the rotors for runout and parallelism, discussed later in this chapter. A warped rotor may need to be replaced and can be caused by an improper tightening sequence of the wheel lug nuts. In fact, uneven lug nut torque can cause vibrations or shudders at the brake pedal. A brake rotor that has its friction surfaces out of parallel (parallelism) can cause brake pedal pulsations as the pads follow the uneven rotor surfaces. This condition is more commonly found in older ventilated rotors with rusted webs. Under heavy brake pad application force, these corroded webs can collapse slightly, creating a rotor of varying thickness. An out-of-parallel rotor should be replaced. You should be aware that pedal pulsation is normal on vehicles with ABS when the antilock brake system is working.

Spongy Pedal

With a spongy pedal, the customer will probably feel the need to pump the brake pedal to get good stopping ability. The complaint may also be described as a soft pedal. This problem is caused by air in the hydraulic system. Although bleeding the system may remove the air, you should always question how the air got in there. Check for leaks and for proper master cylinder operation.

Hard Pedal

The driver's complaint of a hard pedal normally indicates a problem with the power brake booster. However, it can also be caused by a restricted brake line or hose. Carefully check the lines and hoses for damage. Feel the brake hoses. If they seem to have lost their rigidity, the hose may have collapsed on the inside and this is causing the restriction. Make sure the brake hoses have not been twisted. Incorrect caliper installation can twist and restrict the hose. A hard pedal can also be caused by frozen caliper or wheel cylinder pistons.

Dragging Brakes

Dragging brakes make the vehicle feel as if it has lost or is losing power as it drives down the road. The problem also wastes a lot of fuel and generates destructive amounts of heat that can cause serious brake damage and brake failure. While trying to find the cause of this problem, check the parking brake first. Make sure it is off. Check the rear wheels to make sure the parking brakes are released when they should be. If the problem is not in the parking brakes, check for restricted brake hoses, keeping pressure applied to the calipers. Inspect the calipers and wheel cylinders for sticky or seized caliper or wheel cylinder pistons. A collapsed brake hose (flex line) may also allow fluid flow to a piston during application while preventing fluid return during brake release.

Seized caliper pistons can be found by using a pry bar or C-clamp to push the pistons back into the calipers. Position the pry bar to pull the caliper outward, forcing the piston backward into its bore. If the piston does not retract, loosen the bleeder valve and retry. If the caliper still does not move, the piston is likely seized. If the piston retracts, the brake hose is likely the cause of the problem.

Grabbing Brakes

When the brakes seem to be overly sensitive to pedal pressure, they are grabbing. Normally, this problem is caused by contaminated brake linings. If the linings are covered or saturated with oil, find the source of the oil and repair it. Then replace the pads, and refinish or replace the rotor. This is more commonly found in rear drum brake systems where leaking wheel cylinder cups or axle seals can readily contaminate the linings.

Noise

If the customer's complaint is noisy brakes, verify during the road test that the problem is in the brakes. If the noise is caused by the brakes, pay attention to the type of noise, and let that lead you to the source of the problem. Remember that some brake pads have wear sensors that are designed to make a high-pitched squeal when the pads are worn. Other causes could be the rotor rubbing against the splash shield or something that has become wedged between the rotor and another part of the vehicle. Noise may also be caused by failure to install all of the hardware when placing a caliper or brake pads in service.

Other sources of noise include the pads themselves. Depending on the lining materials, some pads are more prone to making noise. These noises can be a light wire brush against metal sound, grinding, or high-pitched squeal, depending on the pad.

Pulling

When a vehicle drifts or pulls to one side while cruising or when braking, the cause could be in the brake system or in the steering and suspension system. Check the inflation of the tires and the tires' tread condition, and verify that the tires on each axle are the same size. Check the operation of the brakes. If only one front wheel is actually doing the braking, the vehicle will seem to stumble or pivot on that one wheel. A range of brake system problems can cause pulling, including seized caliper slider pins, frozen pistons, a restricted brake hose, or friction material–related causes such as worn, contaminated, or missing lining segments. If no problems are found in the brake system, suspect an alignment or suspension problem such as worn control arm bushings.

SERVICE GUIDELINES

The following general service guidelines apply to all disc brake systems and should always be followed:

- Be sure the vehicle is properly centred and secured on stands or a hoist.
- If the vehicle has antilock brakes, depressurize the system according to the procedures given in the service information.
- Before any service is performed, carefully check the following:
  - Tires for excessive wear or improper inflation
  - Wheels for bent or warped rims
  - Wheel bearings for looseness or wear
  - Suspension components to see if they are worn or broken
  - Brake fluid level
  - Master cylinder, brake lines or hoses, and each wheel for leaks
  - Parking brake operation
- Before you remove a brake hydraulic part, use a pedal depressor to slightly depress the brake pedal. This closes the master cylinder's ports and prevents fluid from draining. It also makes the bleeding process easier after the system is reassembled.
- Before beginning brake work, remove about two-thirds of the brake fluid from the master cylinder's reservoir. If this is not done, the fluid could overflow and spill when the pistons are forced back into the caliper bore. You also can open the caliper bleeder screw and run a hose to a container to catch the fluid that is expelled. This prevents dirty brake fluid from being forced back into the ABS control unit.
- If the bleeder screws are frozen tight with corrosion, it is sometimes possible to free them by tapping on the caliper body at the bleeder with a hammer. If the bleeder screws cannot be loosened, the entire caliper is probably in bad shape, so it is best to replace the entire unit.
- During servicing, grease, oil, brake fluid, or any other foreign material must be kept off the brake linings, caliper, surfaces of the disc, and external surfaces of the hub. Handle the brake disc and caliper in such a way as to avoid deformation of the disc and nicking or scratching of the brake linings.
- When a hydraulic hose is disconnected, plug it to prevent any foreign material from entering.
- Never permit the caliper assembly to hang with its weight on the brake hose. Support it on the suspension or hang it away from the assembly **(Figure 52–23)**.
- Inspect the caliper for leaks. If leakage is present, the caliper must be overhauled or replaced.
- When using compressed air to remove caliper pistons, avoid high pressures. A safe pressure to use is 207 kPa (30 psi).
- Clean the brake components in either denatured alcohol or clean brake fluid. Do not use mineral-based cleaning solvent such as gasoline, kerosene, carbon tetrachloride, acetone, or paint thinner to clean the caliper. It causes rubber parts to become soft and swollen in an extremely short time.

FIGURE 52–23

Never permit the caliper assembly to hang with its weight on the brake hose. Support it on the suspension or hang it away from the assembly.

- Lubricate any moving member such as the caliper housing or mounting bracket to ensure a free-moving action. Use only recommended lubricant.
- Obtain a firm brake pedal after servicing the brakes and before moving the vehicle. Be sure to road-test the vehicle.
- Always torque all brake fasteners to specifications. Torque the lug nuts in the correct sequence when installing a wheel onto the vehicle. Never use an impact gun to tighten the lug nuts. Warpage of the rotor could result if an impact gun is used.

WARNING!

A propane or oxy-acetylene torch should never be used to loosen bleeder screws or brake line fittings at the caliper. The heat can raise the temperature of the caliper and brake fluid enough to damage rubber seals and rupture brake hoses. Personal injury could occur.

GENERAL CALIPER INSPECTION AND SERVICING

Frequently, caliper service involves only the removal and installation of the brake pads. However, because the new pads are thicker than the worn-out set they replace, they locate the piston farther back in the bore where dirt and corrosion might cause the seals to leak. For this reason, it is often good practice to carefully inspect the calipers whenever installing new pads. Of course, it is also good practice to true-up or replace the rotors when replacing brake pads.

When bench-working a caliper assembly, use a vise that is equipped with protector jaws. Excessive vise pressure causes bore and piston distortion.

Caliper Removal

To be able to replace brake pads, service the rotor, or replace the caliper, the caliper must be removed. The procedure for doing this varies according to caliper design. Always follow the specific procedures given in the service information. Use the following as an example of these procedures:

1. Remove about two-thirds of the brake fluid from the master cylinder.
2. Raise the vehicle and remove the wheel and tire assembly.
3. On a sliding or floating caliper, install a C-clamp with the solid end of the clamp on the caliper housing and the screw end on the metal portion of the outboard brake pad. Tighten the clamp until the piston bottoms in the caliper bore **(Figure 52–24)**, and then remove the clamp.

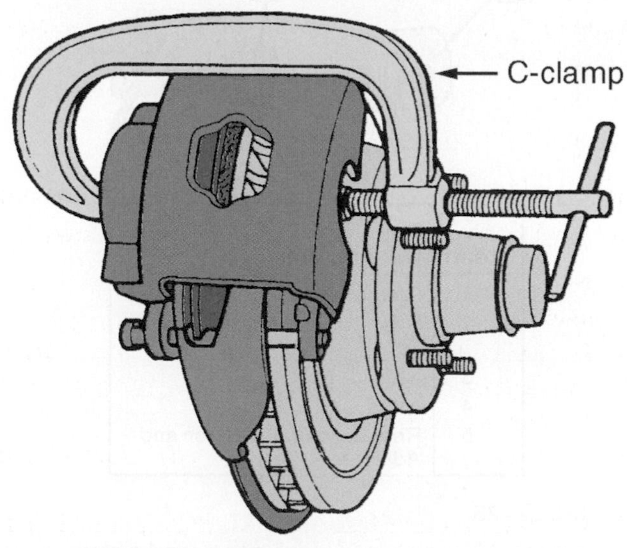

FIGURE 52–24

Bottoming the piston in the caliper's bore.

Bottoming the piston allows room for the brake pad to slide over the ridge of rust that accumulates on the edge of the rotor.

4. On threaded-type rear calipers, the piston must be rotated to depress it, which requires a special tool **(Figure 52–25)**. This is discussed later in this chapter.
   If only the brake pads are to be replaced, do not disconnect the brake hose. If rebuilding or replacing the caliper, disconnect the brake hose from the caliper, remove the copper gasket or washer, and cap the end of the brake hose.
5. Remove the two mounting brackets to the steering knuckle bolts. Support the caliper when removing the second bolt to prevent the caliper from falling.
6. On a sliding caliper, remove the top bolts, retainer clip, and **antirattle springs (Figure 52–26)**. On a floating caliper, remove the two special pins that hold the caliper to the anchor plate **(Figure 52–27)**. On a fixed caliper, remove the bolts holding it to the steering knuckle. On all three types, get the caliper off by prying it straight up and lifting it clear of the rotor.

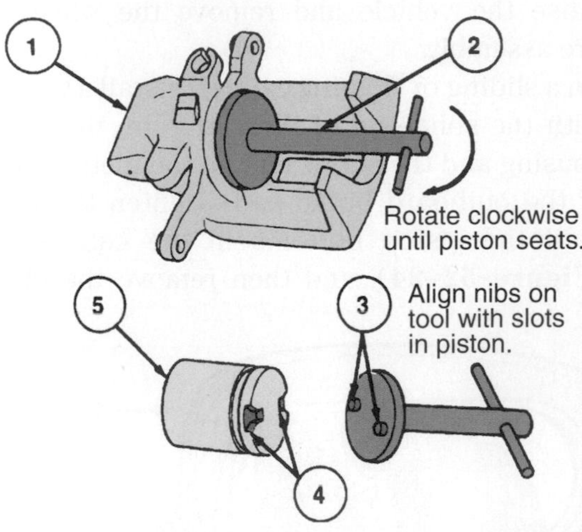

Item	Description
1	Caliper Housing
2	Rear Caliper Piston Adjuster Tool
3	Nibs
4	Slots
5	Rear Disc Brake Piston and Adjuster

FIGURE 52–25

A special tool is required to move a threaded piston into its bore.

Ford Motor Company

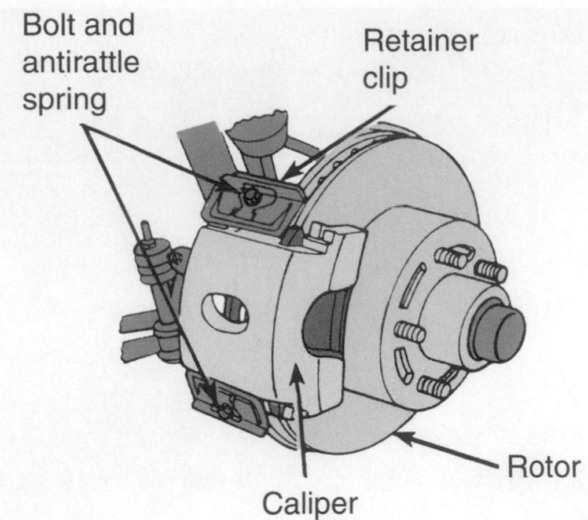

FIGURE 52–26

Sliding caliper removal.

Brake Pad Removal

Disc brake linings should be checked periodically or whenever the wheels are removed. Some calipers have inspection holes in the caliper body. If they do not, the brake pads can be visually inspected from the outer ends of the caliper.

If you are not sure the pads are worn enough to warrant replacement, measure them at the thinnest part of the pad. Compare this measurement to the minimum brake pad lining thickness listed in the vehicle's service information, and replace the pads if needed.

When a pad on one side of the rotor has worn more than on the other side, the condition is called uneven wear. Uneven pad wear often means the caliper is sticking and not giving equal pressure to both pads. On a sliding caliper, the problem could be caused by poor lubrication or deformation of the

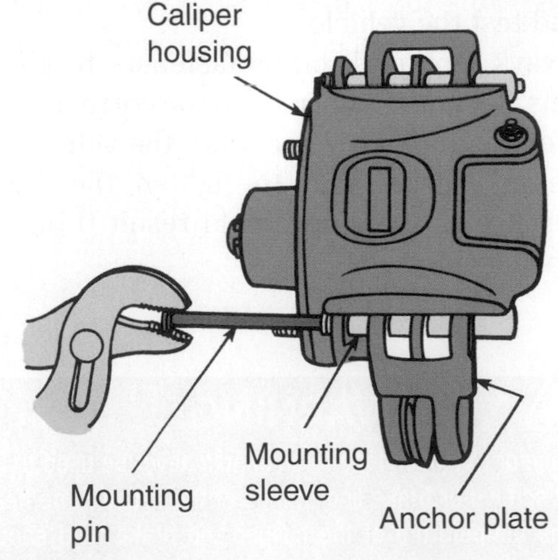

FIGURE 52–27

Floating caliper removal.

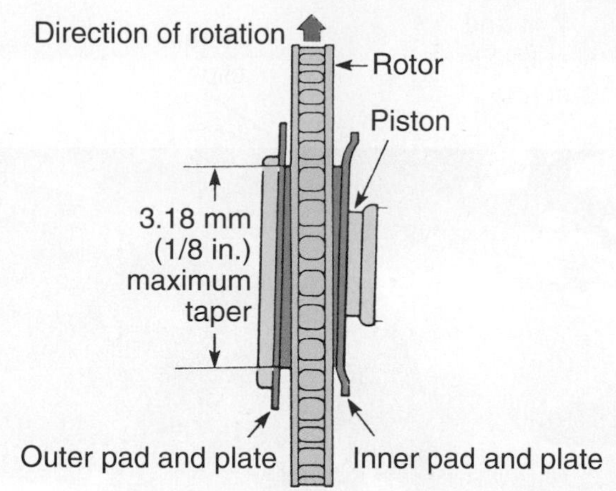

FIGURE 52–28

Normal pad wear pattern.

machined sliding areas on the caliper and/or anchor plate. Rust buildup on the pad guides is a common cause of the pads not sliding properly. A slightly tapered wear pattern on the pads of certain models is caused by caliper twist during braking. It is normal if it does not exceed 3 mm (⅛ in.) taper from one end of the pad to the other **(Figure 52–28)**.

Sliding or floating calipers are usually lifted off the rotor for pad replacement. However, some designs allow the caliper to be tilted out of the way, allowing the pads to be removed and replaced without fully removing the caliper. Even if caliper removal is not necessary, it is a good practice to remove it to allow for closer inspection and cleaning of the components.

Fixed calipers might have pads that can be replaced by removing the retaining pins or clips instead of having to lift off the entire caliper. Brake pads may be held in position by retaining pins, guide pins, or a support key. Note the position of the shims, antirattle clips, keys, bushings, or pins during disassembly. A typical procedure for replacing brake pads is outlined in Photo Sequence 44.

If only the pads are going to be replaced, lift the caliper off the rotor, and hang it up by a wire. Remove the outer pad and inner pad. Remove the old sleeves and bushings, and install new ones. Replace rusty pins on a floating caliper to provide for free movement. You may need to transfer shoe retainers, which can be clips or springs, onto the new pads.

SHOP TALK

Most often, calipers are replaced rather than rebuilt. The old caliper is sent back to the manufacturer as a core. The following overview for rebuilding is intended only to give you an understanding of what may be done. Always refer to the appropriate service information when rebuilding a caliper.

Caliper Disassembly

If the caliper must be rebuilt, it should be taken to the workbench for servicing. Drain any brake fluid from the caliper by way of bleeder screws. Remove the bleeder valve protector, if so equipped.

On a floating caliper, examine the mounting pins for rust that could limit travel. Most manufacturers recommend that these pins and their bushings be replaced each time the caliper is removed. This is a good idea because the pins are inexpensive and a good insurance against costly comebacks. On a fixed caliper, check the pistons for sticking, and rebuild the caliper if this problem is found.

To disassemble the caliper, the piston and dust boot must first be removed. Place the caliper face down on a workbench **(Figure 52–29)**. Insert the used outer pad or a block of wood into the caliper. Place a folded shop towel on the face of the lining to cushion the piston. Apply low air pressure (*never more than* 207 kPa [30 psi]) to the fluid inlet port of the caliper to force the piston from the caliper housing.

CAUTION!

Wear safety glasses while disassembling the caliper to protect your eyes from spraying brake fluid.

WARNING!

Be careful to apply air pressure very gradually. Be sure there are enough cloths to catch the piston when it comes out of the bore. Never place your fingers in front of the piston for any reason when applying compressed air. Personal injury could occur if the piston is popped out of the bore.

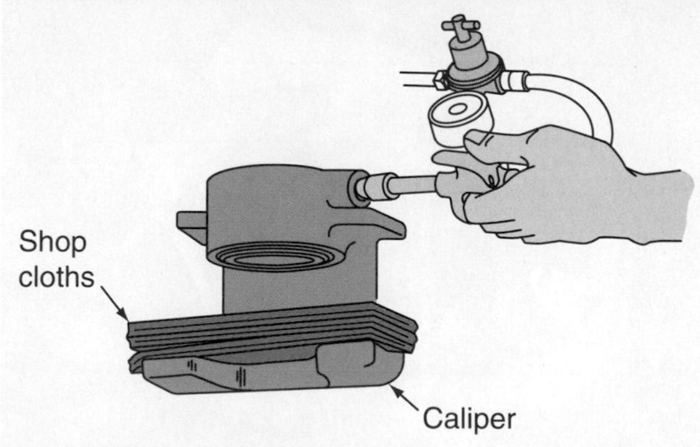

FIGURE 52–29

Using air to remove a piston.

REMOVING AND REPLACING BRAKE PADS

P44–1 Front brake pad replacement begins with removing brake fluid from the master cylinder reservoir.

P44–2 Raise the car. Make sure it is safely positioned on the lift. Remove its wheel assemblies.

P44–3 Inspect the brake assembly. Look for signs of fluid leaks, broken or cracked lines, or a damaged brake rotor. If any problem is found, correct it before installing the new brake pads.

P44–4 Loosen and remove the bolts from the caliper locator pins.

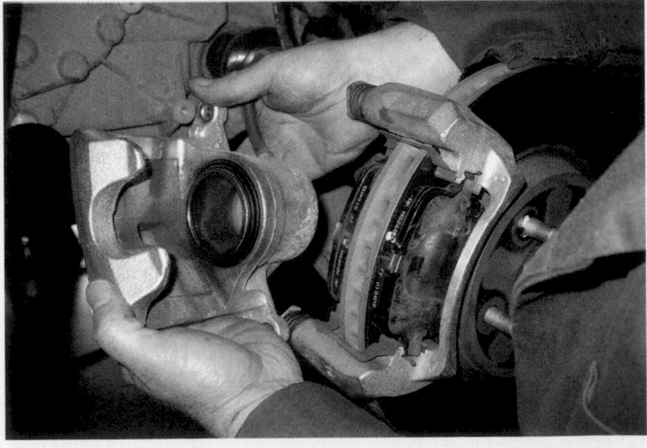

P44–5 Lift and rotate the caliper assembly from the rotor.

P44–6 Remove the brake pads from the caliper mounting adapter.

P44–7 Fasten a piece of wire to the car's frame, and support the caliper with the wire.

P44–8 Check the condition of the caliper locating pins and dust boots.

P44–9 Place a piece of wood over the caliper's piston, and install a C-clamp over the wood and caliper. Tighten the clamp to force the piston back into its bore.

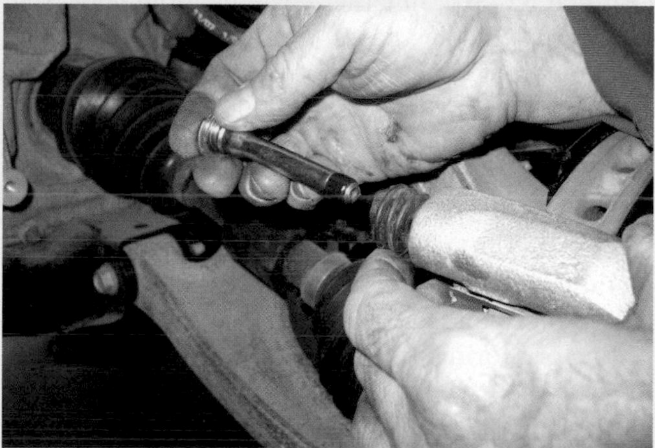

P44–10 Remove the clamp, and install new caliper locating pins and boots, if necessary.

P44–11 Install the new brake pads into the caliper mounting adapter.

P44–12 Set the caliper over the pads and rotor. After the caliper is positioned correctly, install and torque the caliper mount bolts.

If a piston is frozen, release air pressure and tap the piston into its bore with a soft-faced hammer or mallet. Reapply air pressure. Frozen phenolic (plastic) pistons can be broken into pieces with a chisel and hammer. Be careful not to damage the cylinder bore while doing this. Internal expanding pliers are sometimes used to remove pistons from caliper bores.

Inspect phenolic pistons for cracks, chips, or gouges. Replace the piston if any of these conditions are evident. If the plated surface of a steel piston is worn, pitted, scored, or corroded, it also should be replaced.

Dust boots vary in design, depending on the type of piston and seal, but they all fit into one groove in the piston and another groove in the cylinder. One type comes out with the piston and peels off. Another type stays in place, and the piston comes out through the boot and then is removed from the cylinder **(Figure 52–30)**. In either case, peel the boot from its groove. In some cases, it might be necessary to pry it out, but be careful not to scratch the cylinder bore while doing so. The old boot can be discarded as it must be replaced along with the seal.

Remove the piston seal from the cylinder bore by prying it out with a wooden or plastic tool **(Figure 52–31)**. Do not use a screwdriver or other metal tool. Any of these could nick the metal in the caliper bore and cause a leak. Inspect the bore for pitting or scoring. A bore that shows light scratches or corrosion can usually be cleaned with crocus cloth. However, a bore that has deep scratches or scoring normally indicates that the caliper should be replaced. In some cases, the cylinder can be honed. Check the service information before doing this. If there is no mention of honing the bore, the manufacturer probably does not recommend it. Black stains on the bore walls are caused by piston seals. They do no harm.

When using a hone, be sure to install the hone baffle before honing the bore. The baffle protects the hone stones from damage. Use extreme care in cleaning the caliper after honing. Remove all dust and grit by flushing the caliper with alcohol. Wipe it dry with a clean lint-free cloth, and then clean the caliper a second time in the same manner.

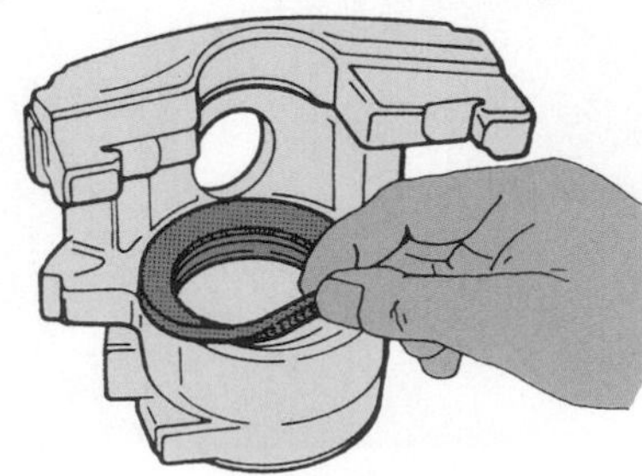

FIGURE 52–30

Peeling off the dust boot.

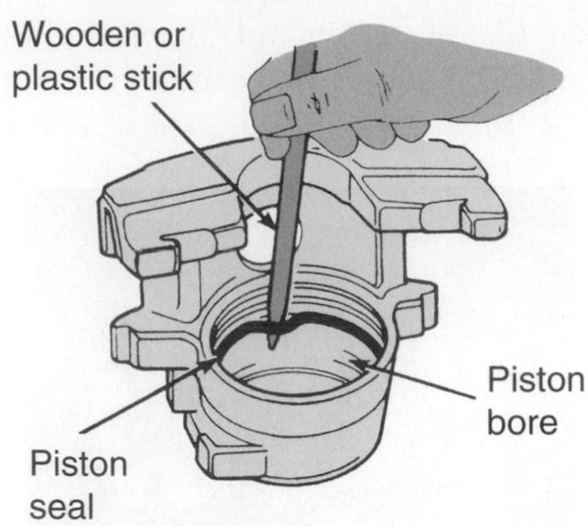

FIGURE 52–31

Removing a piston seal with a wooden or plastic stick.

Loaded Calipers

There is now a trend toward installing loaded calipers **(Figure 52–32)** rather than overhauling calipers in the shop. Loaded calipers are completely assembled with friction pads and mounting hardware included. Besides the convenience and the savings of installation time, preassembled calipers also reduce the odds of errors during caliper overhaul.

Mistakes frequently made when replacing calipers include forgetting to bend pad-locating tabs

FIGURE 52–32

A pair of loaded brake calipers.

Bendix Brakes by Allied Signal/Honeywell

that prevent pad vibration and noise, leaving off antirattle clips and pad insulators, and reusing corroded caliper mounting hardware that can cause a floating caliper to bind up and wear the pads unevenly. Installing a loaded caliper ensures that all parts requiring replacement are replaced.

SHOP TALK

Avoid mismatching friction materials from side to side. When one caliper is bad, both calipers should be replaced using the same friction material on both sides.

Caliper Reassembly

Before assembling the caliper, clean the phenolic piston (if so equipped) and all metal parts to be reused with soap and water or clean denatured alcohol. Then clean out and dry the grooves and passageways with compressed air. Make sure that the caliper bore and component parts are thoroughly clean.

To replace a typical piston seal, dust boot, and piston, first lubricate the new piston seal with clean brake fluid or assembly lubricant (usually supplied with the caliper rebuild kit). Make sure the seal is not distorted. Insert it into the groove in the cylinder bore so that it does not become twisted or rolled. Install a new dust boot by setting the flange squarely in the outer groove of the caliper bore. Next, coat the piston with brake fluid or assembly lubricant, and install it in the cylinder bore. Be sure to use a wood block or other flat stock when installing the piston back into the piston bore. Never apply a C-clamp directly to a phenolic piston, and be sure the pistons are not cocked. Spread the dust boot over the piston as it is installed. Seat the dust boot in the piston groove.

With some types of boot/piston arrangements, the procedure of installation is slightly different from that already described. That is, the new dust boot is pulled over the end of the piston **(Figure 52–33)**. Lubricate the piston with brake fluid before installing it in the caliper. Then by hand, slip the piston carefully into the cylinder bore, pushing it straight so that the piston seal is not damaged during installation. Use an installation tool or wooden block to seat the new dust boot **(Figure 52–34)**.

Another point to keep in mind is that some caliper designs have a slot cut in the face of the pistons that must align with an **antisqueal shim**. Make sure that the piston and shim align. It might be necessary

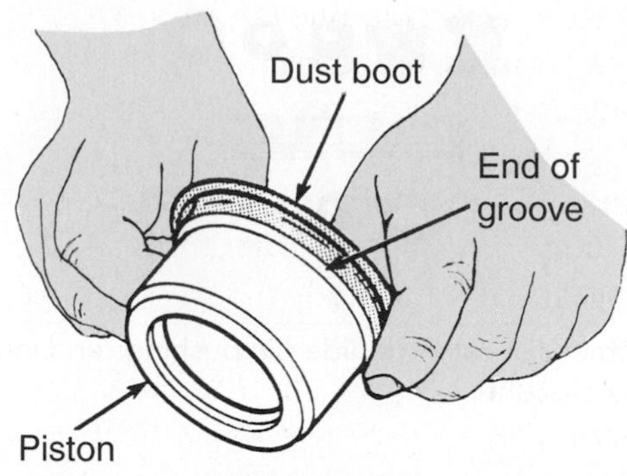

FIGURE 52–33

Some installation procedures require the dust boot to be pulled over the end of the piston.

to turn the piston to achieve proper alignment. To complete the caliper assembly job, install the bleeder screw.

CAUTION!

On fixed calipers, bridge bolts are used to hold the two caliper halves together. These are high-tensile bolts ordered only by a specific part number. They require accurate torque tightness to prevent leakage. Do not attempt to use standard bolts in place of bridge bolts.

Brake Pad Installation

It is a good practice to replace disc brake hardware **(Figure 52–35)** when replacing disc brake pads. Replacement of the hardware ensures proper caliper movement and brake pad retention. It also aids in preventing brake noise and uneven brake pad wear.

One of the major causes of premature brake wear is rust. It causes improper slider and piston operation that leads to uneven pad wear. Tests have

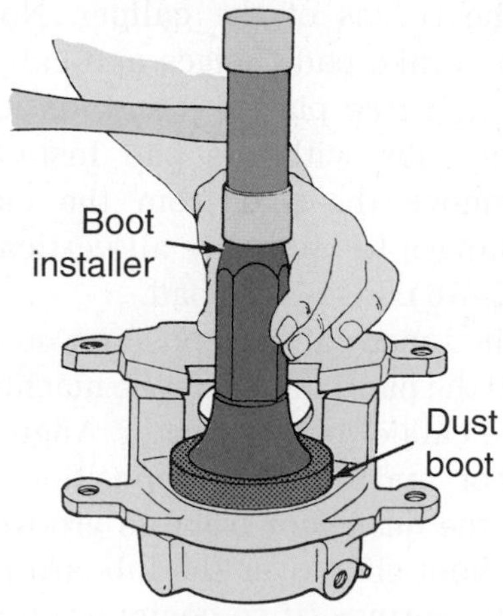

FIGURE 52–34

Seating a dust boot with a boot installer.

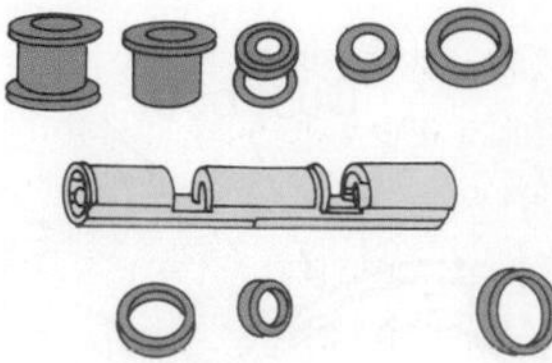

FIGURE 52–35

An assortment of caliper guide pin bushings and insulators.

Federal-Mogul Corporation

shown that when only the pads are replaced, the new pads can wear out in half the kilometres they should if the calipers or slides are corroded. Therefore, if the calipers are corroded, replace them. You should always replace the pad guides. If the pads slide on a bracket that does not use replaceable guides, thoroughly clean the contact areas, and lubricate as recommended in the service information.

FIXED CALIPER BRAKE PADS The designs of fixed caliper disc brakes vary slightly. Generally, to replace the pads, insert new pads and plates in the caliper, with the metal plates against the end of the pistons. Be sure that the plates are properly seated in the caliper. Spread the pads apart, and slide the caliper into position on the rotor. With some pads, mounting bolts are used to hold them in place. These bolts are usually tightened 108 to 122 N·m (80 to 90 lb·ft). On some fixed disc brakes, the pads are held in place by retaining clips and/or retaining pins. Reinstall the antirattle spring/clips and other hardware (if so equipped).

SLIDING CALIPER BRAKE PADS Push the piston carefully back into the bore until it bottoms. Lightly lubricate the sliding surfaces of the caliper and the caliper anchor with brake fluid. Slide a new outer pad into the recess of the caliper. No free play between the brake pad flanges and caliper fingers should exist. If free play is found, do not attempt to bend the tabs with the pad installed in the caliper. Remove the pad from the caliper, and bend the flanges to eliminate all vertical free play **(Figure 52–36)**. Install the pad.

Place the inner pad into position on the caliper anchor with the pad's flange on the machined sliding area. Fit the caliper over the rotor. Align the caliper to the anchor, and slide it into position. Be careful not to pull the dust boot from its groove when the piston and boot slide over the inboard pad. Install the antirattle springs (if so equipped) on top of the retainer plate, and tighten the retaining screws to specification.

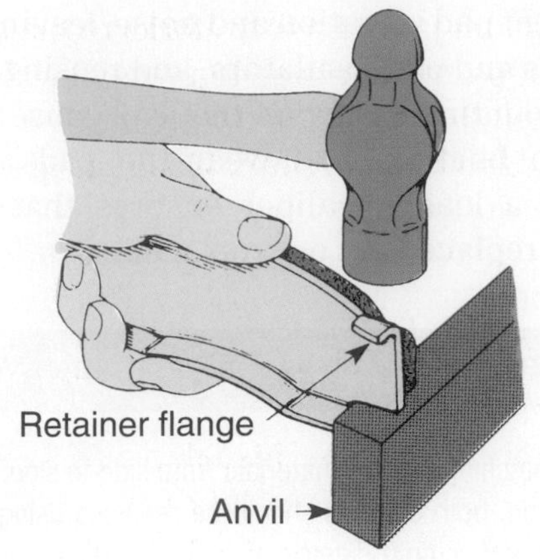

FIGURE 52–36

Bend the retaining flange if there is excessive free play.

On some calipers, especially those used as rear brakes, there is a notch or groove in the piston and a tab on the rear of the inner pad. During installation of the pad, the tab must fit into the groove in the piston **(Figure 52–37)**.

FLOATING CALIPER BRAKE PADS For floating or pin caliper disc brakes, compress the flanges of the outer bushing in the caliper fingers, and work them into position in the hole from the outer side of the caliper. Compress the flanges of the inner guide pin bushings and install them.

Slide the new pad and lining assemblies into position in the adapter and caliper. Be sure that the metal portion of the pad is fully recessed in the caliper and adapter and that the proper pad is on the outer side of the caliper.

Hold the outer pad and carefully slide the caliper into position on the anchor and over the disc. Align the guide pin holes of the anchor with those of the

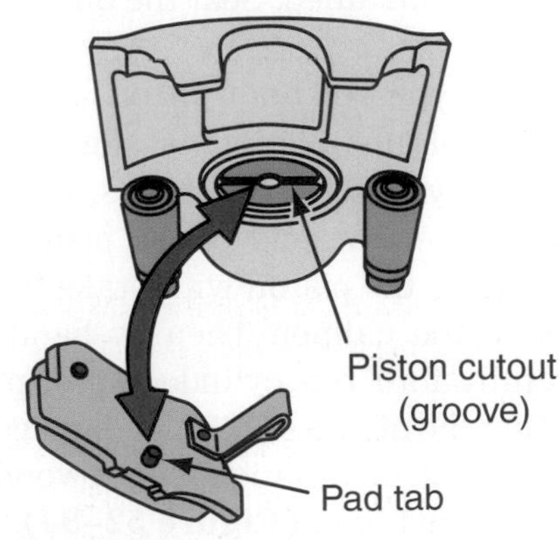

FIGURE 52–37

On some rear brake assemblies, there is a tab on the back of the pad that must line up with the groove in the piston.

inner and outer pads. Lightly lubricate the guide pins with a non-petroleum-based grease, and install the pins through the bushings, caliper, anchor, and inner and outer pads into the outer bushings in the caliper and antirattle spring.

When installing any type of caliper, follow these guidelines:

- Make sure the correct caliper is installed on the correct anchor plate.
- Lubricate the rubber insulators (if so equipped) with the recommended lubricant.
- After the caliper assembly is in its mounting brackets, connect the brake hose to the caliper. If copper washers or gaskets are used, be sure to use new ones—the old ones might have taken a set and might not form a tight seal if reused. Torque the caliper hose bolt to specifications.
- Fill the master cylinder reservoirs, and bleed the hydraulic system.
- Check for fluid leaks under maximum pedal pressure.
- Lower the vehicle and road-test it.

Rear Disc Brake Calipers

Rear disc brakes calipers with some type of parking brake mechanism have different inspection and overhaul procedures than front brake calipers. To overhaul a typical rear-wheel caliper when there is no auxiliary drum parking brake, follow the steps in the following Procedure box:

PROCEDURE

Overhauling a Typical Rear-Wheel Caliper with No Auxiliary Drum Parking Brake

STEP 1 Unbolt the caliper.

STEP 2 Disconnect the parking brake cable from the lever on the caliper.

STEP 3 Disconnect the brake hose from the caliper, remove the caliper mounting bolts, and lift the caliper off its support.

STEP 4 Remove the brake pads and all pad shims and retainers.

STEP 5 Remove the piston from its bore; this procedure will vary based on the type of caliper being serviced. When the piston is free, remove the piston boot.

STEP 6 Carefully inspect the piston for wear. Replace the piston if it is worn or damaged in any way.

STEP 7 Remove the piston seal from the caliper, using the tip of a screwdriver or a wooden or plastic scraper and being careful not to scratch the bore.

STEP 8 Service of the remaining internal components will vary depending on the caliper. Refer to the manufacturer's service information.

STEP 9 Coat the new piston seal and piston boot with silicone grease, and install them in the caliper.

STEP 10 Coat the outside of the piston with brake fluid, and install it on the adjusting bolt while rotating it clockwise with the locknut wrench.

STEP 11 Install the new brake pads, pad shims, retainers, and springs onto the caliper bracket.

STEP 12 Reinstall the caliper and the splash shield, and tighten the caliper bolts to torque specifications.

STEP 13 Reconnect the brake hose to the caliper with new sealing washers and tighten the banjo bolt to specifications.

STEP 14 Then reconnect the parking brake cable to the arm on the caliper, and reinstall the caliper shield.

STEP 15 Top off the master cylinder reservoir, and bleed the brake system. Adjust the parking brake as needed. Before making adjustments, be sure the parking brake arm on the caliper touches the pin.

ROTOR INSPECTION AND SERVICING

The rotors should be inspected whenever brake pads are replaced and when the wheels are removed for other services. They should be carefully checked to determine if they can be reused or machined or if they should be replaced. When inspecting the rotor, make sure you look at the sensor wheel for the wheel speed sensor **(Figure 52–38)**.

If a good look at the surface is impossible because of dirt, clean the surfaces with a shop cloth dampened in brake-cleaning solvent or alcohol. If the surface is rusted, remove it with medium-grit sandpaper or emery cloth, and then clean it with brake cleaner or alcohol.

Most brake rotors have a discard thickness dimension cast into them. If you cannot find this dimension on the rotor or if it is hard to read, check service information for thickness specifications. Rotor discard thickness dimensions are given in two or three decimal points (hundredths of a millimetre or hundredths or thousandths of an inch), such as 24.75 mm, or 1.25 inches, 1.375 inches, 0.750 inch. If you resurface the rotor, it must be 0.38 to 0.76 mm (0.015 to 0.030 in.) thicker than the discard

FIGURE 52–38

Check the wheel speed sensor's rotor when servicing a rotor.

dimension after machining to allow for wear. If a rotor is already below the minimal thickness spec, replace it. It is always wise to replace both rotors on the same axle.

New rotors come with a protective coating on the friction surfaces. To remove this coating, use brake cleaner or the solvent recommended by the manufacturer.

SHOP TALK

Cross-drilled or slotted brake rotors may not be able to be machined. Therefore, if the rotor is scored or otherwise damaged, it should be replaced.

CAUTION!

Never turn the rotor on one side of the vehicle and not the other.

Lateral Runout

Excessive **lateral runout** is the wobbling of a rotor from side to side when it rotates **(Figure 52–39)**. This wobble knocks the pads farther back than normal, causing the pedal to pulse and vibrate during braking. Over time, runout can cause excessive parallelism. This occurs as the rotor rubs against the pads as it rotates, wearing away at the rotor and causing it to become thinner in certain sections. Chatter can also result. Lateral runout also causes excessive pedal travel because the pistons have farther to travel to reach the rotor. If runout exceeds specifications, the rotor must be turned or replaced.

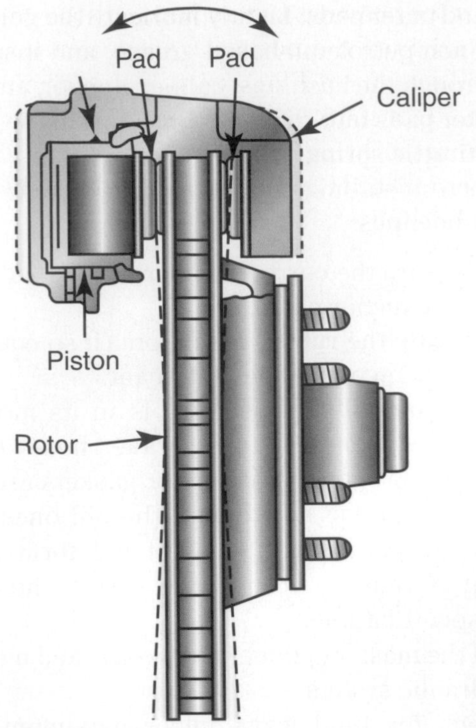

FIGURE 52–39

Excessive lateral runout causes the rotor to wobble from side to side when it rotates.

For the best braking performance, lateral runout should be less than 0.08 mm (0.003 in.) for most vehicles. Some manufacturers, however, specify runout limits as small as 0.05 mm (0.002 in.) or as great as 0.20 mm (0.008 in.).

Runout measurements are taken only on the outboard surface of the rotor, using a dial indicator and suitable mounting adapters **(Figure 52–40)**. If the rotor is mounted on adjustable wheel bearings,

FIGURE 52–40

Checking the lateral runout of a brake rotor.

readjust the bearings to remove bearing end play. Do not overtighten the bearings. On rotors bolted solidly to the axles of FWD vehicles, bearing end play is not a factor in rotor runout measurement. If there is excessive bearing end play, the bearing assembly must be replaced. Bearing end play is best checked with a dial indicator.

Clamp the dial indicator support to the steering knuckle or other suspension part that will hold it securely as you turn the rotor. Position the dial indicator so that its tip contacts the rotor at 90°. Place the indicator tip on the friction surface about 25 mm (1 in.) in from the outer edge of the rotor. Do not place the dial indicator on a dirty, rusted, grooved, or scored area. Rotate the rotor until the lowest reading appears on the dial indicator, and then set the indicator to zero. Turn the rotor through one complete revolution, and compare the lowest to the highest reading. This is the maximum runout of the rotor.

Thickness and Parallelism

To measure rotor thickness, place a brake disc micrometer about 35 mm (1 in.) in from the outer edge of the rotor, and measure the thickness **(Figure 52–41)**. Compare the measurement to specifications. Repeat the measurement at about eight points equidistant (45°) around the surface of the rotor and compare each measurement to specifications. Take all measurements at the same distance from the edge so that rotor taper does not affect the measurements. If the rotor is thinner than the minimum thickness at any point or if thickness variations exceed limits, it must be replaced. Also check the service information for an allowable thickness variation. Many manufacturers hold tolerances on thickness variations as close as 0.013 mm (0.0005 in.).

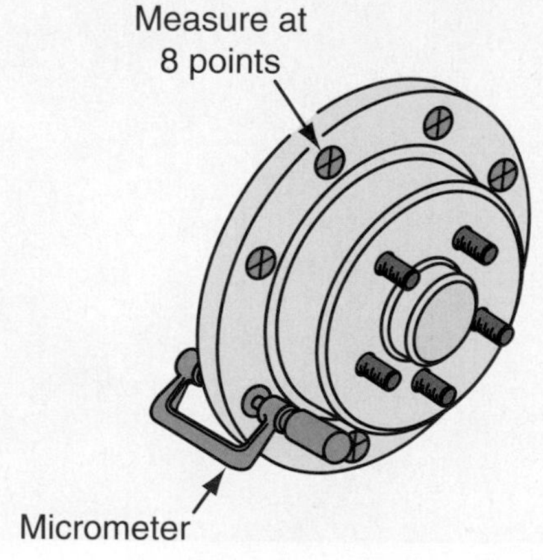

FIGURE 52–41

To check a rotor's parallelism, measure the thickness of the rotor at eight different spots.

Rotor **parallelism** refers to thickness variations in the rotor from one measurement point to another around the rotor surface. If the rotor is out of parallel, it can cause excessive pedal travel; front end vibration; pedal pulsation; chatter; and, on occasion, grabbing of the brakes. The rotor then must be resurfaced or replaced.

Additional Checks

The following are some of the typical rotor conditions that warrant disc replacement or machining.

GROOVES AND SCORING Inspect both rotor surfaces for scoring and grooving. Scoring or small grooves up to 0.25 mm (0.010 in.) deep are usually acceptable for proper braking performance. Scoring can be caused by linings that are worn through to the rivets or backing plate or by friction material that is harsh or unkind to the mating surface. Rust, road dirt, and other contamination could also cause rotor scoring. Any rotor having score marks more than 0.38 mm (0.015 in.) should be refinished or replaced.

If the rotor is deeply grooved, it must be thick enough to allow the grooves to be completely removed without machining the rotor to less than its minimum thickness. Measure rotor thickness at the bottom of the deepest groove. If rotor thickness at the bottom of the deepest grooves is at or near the discard dimension, replace the rotor.

CRACKS Check the rotor thoroughly for cracks or broken edges. Replace any rotor that is cracked or chipped, but do not mistake small surface checks in the rotor for structural cracks. Surface checks will normally disappear when a rotor is resurfaced. Structural cracks, however, will be more visible when surrounded by a freshly turned rotor surface.

BLUING OR HEAT CHECKING Inspect the rotor surfaces for heat checking and hard spots **(Figure 52–42)**. Heat checking appears as many small interlaced cracks on the surface. Heat checking lowers the heat dissipation ability and friction coefficient of the rotor surface. Heat checking does not disappear with resurfacing. Therefore, a rotor with heat checks should be replaced.

Hard spots appear as round, shiny, bluish areas on the friction surface. Hard spots on the surface of a rotor usually result from a change in the metallurgy caused by brake heat. Pulling, rapid wear, hard pedal, and noise occur. These spots

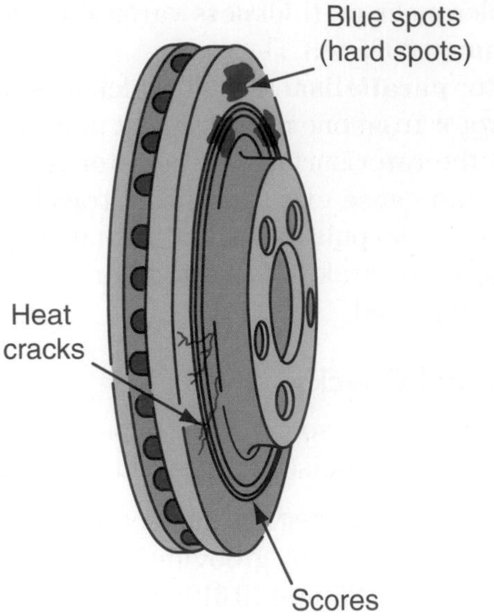

FIGURE 52–42

Some of the typical conditions that should be looked for on a brake rotor.

can be removed by machining. However, only the raised surfaces are removed, and they could reappear when heat is again encountered. Instead of machining the spots again if they reappear, the rotor should be replaced.

RUST If the vehicle has not been driven for a period of time, the discs will rust in the area not covered by the lining and cause noise and chatter. This also can result in excessive wear and scoring of the discs and pads. Wear ridges on the discs can cause temporary improper pad contact if the ridges are not removed before the installation of new pads. Rusted rotors should be cleaned before any measurements are taken.

Inspect the fins of vented rotors for cracks and rust. Rust near the fins can cause the rotor to expand and lead to rotor thickness variations and excessive runout problems. Machining the rotor may remove runout and thickness variations, but rotor expansion due to rust may cause these problems to reappear soon. Rusted rotors should be replaced.

Rotor Service

If the thickness of the rotor is below or close to the minimal allowable thickness or is badly distorted, it must be replaced. If the thickness is greater than the minimum specifications, it can be trued and smoothed with a brake lathe. Rotors that have minor imperfections or are slightly unparallel can be turned true and smooth with a brake lathe.

Removing a Rotor

To remove a rotor, raise the vehicle and remove the wheel. Then remove the caliper from the rotor, and suspend it with wire from the suspension of the vehicle. Before you remove a rotor, mark it "L" or "R" for left or right so that it gets reinstalled on the same side of the vehicle from which it was removed. If the rotors have not been off before, make an index mark on the rotor to a wheel stud or hub so that the rotor is reinstalled in the same position on the hub.

If the rotor is a floating rotor, remove it from the hub by pulling it off the hub studs. If you cannot pull the rotor off by hand, apply penetrating oil on the front and rear rotor-to-hub mating surfaces. Strike the rotor between the studs using a ball-peen hammer. If this does not free the rotor, attach a three-jaw puller to the rotor and pull it off.

Whenever you separate a floating rotor from the hub flange, clean any rust or dirt from the mating surfaces of the hub and rotor. Neglecting to clean rust and dirt from the rotor and hub mounting surfaces before installing the rotor will result in increased rotor lateral runout, leading to premature brake pulsation and other problems.

If the rotor and hub are a one-piece assembly, remove the outer wheel bearing, and lift the rotor and hub off the spindle. Some rotors, such as that shown in **Figure 52–43**, are bolted to the hub and bearing assembly and can be removed only by removing the hub unit. Once removed, the rotor is unbolted from the hub.

FIGURE 52–43

Some rotors are bolted to the hub and bearing assembly and can be removed only by removing the entire assembly.

Brake Lathes

A brake lathe cuts metal away to achieve the desired surface finish. There are basically two types of brake disc lathes used by the industry. The first one, a bench brake lathe, has the capability of resurfacing brake drums and brake discs after they have been removed from the vehicle. The lathe rotates the disc as cutting tools work their way across the braking surface of the disc. The second type is an on-vehicle brake lathe. This type of brake lathe is a time saver because the rotor does not need to be removed from the vehicle. Special fixtures are used to straddle the rotor so that the cutting tools can precisely cut both sides of the rotor. An electric motor is used to rotate the disc and hub assembly during cutting.

Whenever you refinish a rotor, remove the least amount of metal possible to achieve the proper finish. This helps to ensure the longest service life from the rotor. Never turn the rotor on one side of the vehicle without turning the rotor on the other side. Left- and right-side rotors should be the same thickness, generally within 0.05 and 0.08 mm (0.002 and 0.003 in.). Similarly, equal amounts of metal should be cut off both surfaces of a rotor.

BENCH LATHES On a bench, off-vehicle lathe, the rotor is mounted on the lathe's arbor and turned at a controlled speed while a cutting bit passes across the rotor surface to remove a few hundredths of a millimetre (few thousandths of an inch) of metal **(Figure 52–44)**. The lathe turns the rotor perpendicularly to the cutting bits so that the entire rotor surface is refinished. Most rotor cutting assemblies have two cutting bits. The rotor mounts between the bits and is pinched between them. As the cut is made, the same amount of surface material should be cut from both sides of the rotor.

FIGURE 52–44

A typical brake lathe.

Hunter Engineering Company

ON-VEHICLE BRAKE LATHES The advantage of an on-vehicle lathe **(Figure 52–45)** is that the rotor does not need to be removed. On-vehicle lathes also are ideal for rotors with excessive runout problems.

To install the lathe, remove the wheel and then remove the caliper. If any end play is present in an adjustable tapered roller bearing, carefully tighten the adjusting nut by hand just enough to remove the end play before installing the lathe. After turning the rotor, readjust the bearing. To hold a two-piece floating rotor to its hub, reinstall the wheel nuts with flat washers or adapters against the rotor. Carefully follow the manufacturer's mounting instructions and attach the lathe to the rotor. Some on-vehicle lathes mount on the caliper support; others are supported on a separate stand.

Some on-vehicle lathes use the vehicle's power to rotate the rotor; others use an electric motor. If the engine is used to turn the rotor, the lathe can be

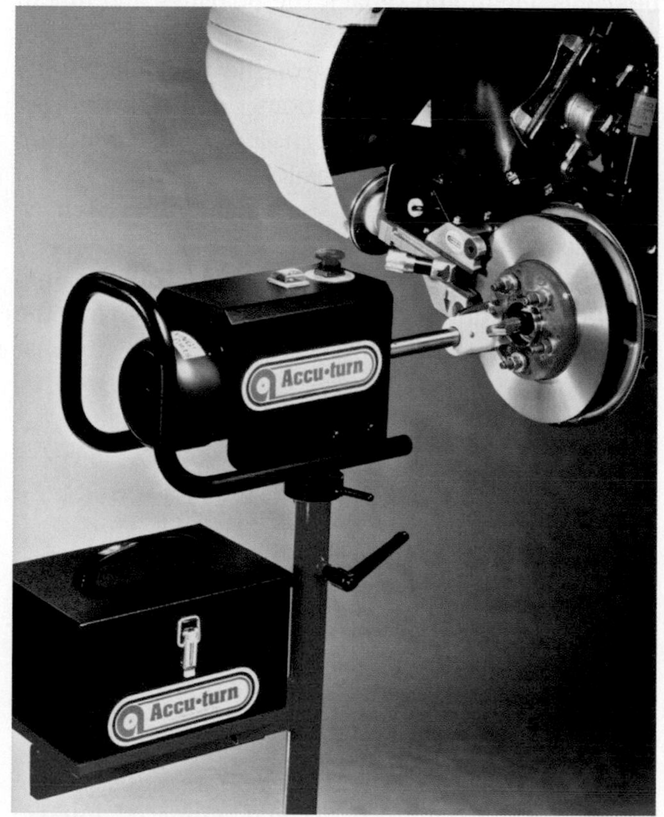

FIGURE 52–45

An on-vehicle brake lathe.

Accu Industries

used only on drive wheels. However, this presents a problem because the differential gearing in the transaxle transmits the power to the opposite wheel, not to the rotor to be resurfaced. To prevent that opposite wheel from turning, that wheel can be lowered to the floor or the brake on the opposite side can be applied. This is done by removing and plugging the brake hose to the caliper on the rotor you are refinishing. Apply the brakes; this will lock the other wheel.

Self-powered on-vehicle lathes are more popular. These have the advantage of being able to machine on non-driving wheels, and rotor speed can be controlled more exactly.

An on-car lathe may be mounted on the brake caliper support or on its own stand and indexed to the hub and the wheel studs. Each lathe has its own operating instructions, which you must follow carefully.

Installing a Rotor

If the rotor is a two-piece floating rotor, make sure all mounting surfaces are clean. Apply a small amount of antiseize compound to the pilot diameter of the disc brake rotor before installing the rotor on the hub. Reinstall the caliper. If the rotor is a fixed, one-piece assembly with the hub that contains the wheel bearings, clean and repack the bearings and install the rotor.

Install the wheel and tire on the rotor, and torque the wheel nuts to specifications, following the recommended tightening pattern. Failure to tighten in the correct pattern may result in increased lateral runout, brake roughness, or pulsation as well as damage to the wheels.

After lowering the vehicle to the ground, pump the brake pedal several times before moving the vehicle. This positions the brake linings against the rotor.

SHOP TALK

Special cutters are required to resurface composite rotors. Make sure you have the correct machine and cutting tools before attempting to true up a composite brake rotor.

KEY TERMS

Antirattle spring (p. 1614)
Antisqueal shim (p. 1619)
Audible sensor (p. 1607)
Caliper (p. 1601)
Composite rotor (p. 1603)
Dust boot (p. 1605)
Fixed caliper (p. 1602)
Fixed rotor (p. 1603)
Floating caliper (p. 1602)
Floating rotor (p. 1603)
Lateral runout (p. 1622)
Parallelism (p. 1623)
Phenolic resin (piston) (p. 1605)
Rotor (p. 1601)
Sliding caliper (p. 1602)
Spindle anchor plate (p. 1606)
Splash shield (p. 1602)

SUMMARY

- Disc brakes offer four major advantages over drum brakes: resistance to heat fade, resistance to water fade, increased straight-line stopping ability, and automatic adjustment.
- The typical rotor is attached to and rotates with the wheel hub assembly. Heavier vehicles generally use ventilated rotors. Splash shields protect the rotors and pads from road moisture and dirt.
- The caliper assembly includes cylinder bores and pistons, dust boots, and piston hydraulic seals.
- Brake pads are placed in each side of the caliper and together straddle the rotor. Some brake pads have wear sensors.
- Fixed caliper disc brakes do not move when the brakes are applied. Floating caliper disc brakes slide back and forth on pins or bolts. Sliding calipers slide on surfaces that have been machined smooth for this purpose.
- In a rear disc brake system, the inside of each rear wheel hub and rotor assembly is used as the parking brake drum.
- Rear disc parking brakes have a mechanism that forces the pads against the rotor mechanically.
- The general procedures involved in a complete caliper overhaul include tasks such as caliper removal, brake pad removal, caliper disassembly, caliper assembly, brake pad installation, and caliper installation.
- The first step in proper caliper service is to remove the caliper assembly from the vehicle.
- Disc brake pads should be checked periodically or whenever the wheels are removed. They should be replaced if they fail to exceed minimum lining thickness as listed in the service information.
- To disassemble the caliper, the piston and dust boot must first be removed. Compressed air is used to pop the piston out of the bore.
- Before assembling the caliper, all metal parts and the phenolic piston are cleaned with soap and water or denatured alcohol. The grooves and passageways of the caliper are cleaned out and dried with compressed air.
- It is a good practice to replace disc brake hardware when replacing disc brake pads.

- Disc brake rotor conditions that must be corrected include lateral runout, lack of parallelism, scoring, blueing or heat checking, and rusty rotors.

REVIEW QUESTIONS

1. Name four major advantages of disc brakes over drum brakes.
2. Name the three major assemblies that make up a disc brake.
3. Name the three types of calipers used on disc brakes.
4. What type of brake uses the inside of each rear wheel hub and rotor assembly as a parking brake drum?
5. How many cylinder bores and pistons can a caliper assembly contain?
6. What returns the caliper piston away from the rotor during brake release?
   a. a piston return spring
   b. the caliper square cut seal
   c. the caliper dust boot
   d. the brake fluid flow back to the master cylinder
7. What type of rotor is usually used on larger vehicles?
   a. solid rotors **b.** two-piece rotor
   c. ventilated rotors **d.** dual-sided rotors
8. What allows the brake caliper to be self-adjusting?
   a. the master cylinder compensating port
   b. the caliper dust boot
   c. the shape of the brake pads
   d. the caliper square-cut seal
9. What channels the flow of air over the exposed rotor surfaces?
   a. webs **b.** hub assembly
   c. splash shield **d.** caliper assembly
10. Where is the audible wear sensor tab located on the brake pad?
    a. at the leading edge **b.** at the trailing edge
    c. on the outer edge **d.** on the inner edge
11. What prevents moisture from entering the cylinder bore?
    a. phenolic piston **b.** drag caliper
    c. splash shield **d.** dust boot
12. What is the recommended maximum air pressure used to force the piston from the caliper?
    a. 207 kPa (30 psi) b. 414 kPa (60 psi)
    c. 621 kPa (90 psi) **d.** 828 kPa (120 psi)
13. What is the maximum pad taper allowed before replacement is required?
    a. 0.79 mm (1/32 in.)
    b. 1.59 mm (1/16 in.)
    c. 2.38 mm (3/32 in.)
    d. 3.18 mm (1/8 in.)
14. What could cause only the left front disc brake assembly to drag?
    a. the master cylinder piston blocking the compensating port
    b. insufficient clearance between the power booster apply pin and the master cylinder primary piston
    c. a restricted brake hose
    d. worn brake pads
15. Which term refers to variations in thickness of the rotor?
    a. torque **b.** lateral runout
    c. parallelism **d.** pedal pulsation
16. What causes the formation of hard spots in the rotor?
    a. manufacturing defects
    b. oil-contaminated brake pads
    c. overheating
    d. dirt between the pad and rotor
17. Which of the following directional rotor statements is correct?
    a. The rotor fins must point forward.
    b. The rotor fins are straight.
    c. These rotors are interchangeable from right to left.
    d. The rotor fins must point rearward.
18. Which of the following is likely to cause a pulsating brake pedal?
    a. a damaged brake flex hose
    b. worn brake pad linings
    c. excessive lateral runout
    d. parallel rotors
19. When installing a piston into a caliper, what type of lubricant should be used?
    a. petroleum jelly
    b. brake fluid
    c. silicone lubricant
    d. no lubricant; install dry
20. What tool(s) should be used to measure rotor lateral runout?
    a. a dial indicator
    b. a vernier caliper
    c. a micrometer
    d. a straightedge and feeler gauge

CHAPTER 53
Antilock Brake, Traction Control, and Stability Control Systems

LEARNING OUTCOMES

- Explain how antilock brake systems work to bring a vehicle to a controlled stop.
- Describe the differences between an integrated and a nonintegrated antilock brake system.
- Briefly describe the major components of a two-wheel antilock brake system.
- Briefly describe the major components of a four-wheel antilock brake system.
- Describe the operation of the major components of an antilock brake system.
- Describe the operation of the major components of automatic traction and stability control systems.
- Explain the best procedure for finding ABS faults.
- List the precautions that should be followed whenever working on an antilock brake system.

Antilock brake systems (ABS) and traction and stability control systems are rapidly gaining popularity. ABS is now standard equipment on most vehicles. These systems add yet another group of electronically controlled systems to the increasingly complex modern vehicle.

ANTILOCK BRAKES

Modern **antilock brake systems (Figure 53–1)** can be thought of as electronic/hydraulic pumping of the brakes for stopping under panic conditions. Good drivers have always pumped the brake pedal during panic stops to avoid wheel lock-up and the loss of steering control. Antilock brake systems simply get the pumping job done much faster and in a much more precise manner than the fastest human foot. Keep in mind that a tire on the verge of slipping produces more friction with respect to the road than one that is locked and skidding. Once a tire loses its grip, friction is reduced, control is compromised, and the vehicle takes longer to stop.

FIGURE 53–1

A common four-wheel antilock brake system.

DaimlerChrysler Corporation

Pressure Modulation

When the driver quickly and firmly applies the brakes and holds the pedal down, the brakes of a vehicle not equipped with ABS will almost immediately lock the wheels. The vehicle slides rather than rolls to a stop. During this time, the driver also has a very difficult time keeping the vehicle under control, and the vehicle can skid out of control. The skidding and lack of control was caused by the locking of the wheels. If the driver was able to release the brake pedal just before the wheels locked up and then reapply the brakes, the skidding could be avoided.

This release and apply of the brake pedal is exactly what an antilock system does. When the brake pedal is pumped or pulsed, pressure is quickly applied and released at the wheels. This is called **pressure modulation**. Pressure modulation works to prevent wheel locking. Antilock brake systems can modulate the pressure to the brakes as often as 20 times per second. By modulating the pressure to the brakes, friction between the tires and the road is maintained, and the vehicle is able to come to a controllable stop.

The only time reduced friction aids in braking is when a tire is in snow or on loose gravel. A locked tire allows a small wedge of snow or gravel to build up ahead of it, which allows it to stop in a shorter distance than a rolling tire.

Steering is another important consideration. As long as a tire does not slip, it goes only in the direction in which it is turned. But once it skids, it has little or no directional stability. One of the big advantages of ABS, therefore, is the ability to keep control of the vehicle under all conditions.

Slip Rate

The manoeuvrability of the vehicle is reduced if the front wheels are locked, and the stability of the vehicle is reduced if the rear wheels are locked. Antilock brake systems precisely control the slip rate **(Figure 53–2)** of the wheels to ensure maximum grip force from the tires, and it thereby ensures the manoeuvrability and stability of the vehicle. An ABS control module calculates the slip rate of the wheels based on the vehicle speed and the speed of the wheels, and then it controls the brake fluid pressure to attain the target slip rate.

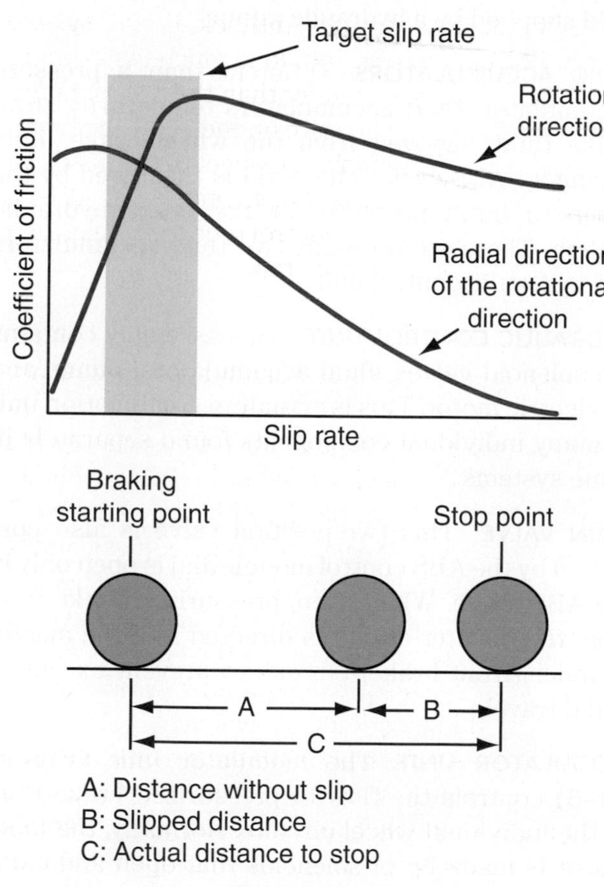

$$\textbf{Slip rate} = \frac{B}{C} = \frac{\text{Vehicle speed} - \text{Wheel speed}}{\text{Vehicle speed}}$$

FIGURE 53–2

Defining slip rate.

Although ABS prevents complete wheel lock-up, it allows some wheel slip in order to achieve the best braking possible. During ABS operation, the target slip rate can be from 10 to 40 percent. Zero (0 percent) slip means the wheel is rolling freely, whereas 100 percent means the wheel is fully locked. A slip rate of 25 percent means the velocity of a wheel is 25 percent less than that of a free-rolling wheel at the same vehicle speed. Many things are considered when determining the target slip rate for a particular vehicle. For some, the range is very low—5 to 10 percent—and on others, it is high—20 to 40 percent.

CUSTOMER CARE

Remind your customers that pumping the brake pedal while stopping will prevent the ABS from activating. They should always keep firm steady pressure on the brake pedal during braking.

Pedal Feel

The brake pedal on a vehicle equipped with ABS has a different feel than that of a conventional braking system. When the ABS is activated, a small bump followed by rapid pedal pulsations will continue until the vehicle comes to a stop or the ABS turns off. These pulsations are the result of the modulation of pressure to the brakes and are felt more on some systems than on others. This is due to the use of damping valves in some modulation units. If pedal feel is of concern during diagnosis of a brake problem, compare the brake pedal feel with that of a similar vehicle with a normal operating antilock brake system. With ABS, the brake pedal effort and pedal feel during normal braking are similar to that of a conventional power brake system.

ABS COMPONENTS

Many different designs of antilock brake systems are found on today's vehicles. These designs vary in their basic layout, operation, and components. There are also variations based on the type of power assist used. Systems that combine ABS and power-assist functions are called integral systems. The ABS components that may be found on a vehicle can be divided into two categories: hydraulic and electrical/electronic components. Keep in mind that no one system uses all of the parts discussed here. Normal or conventional brake parts are part of the overall brake system but are not in the following discussion.

Hydraulic Components

ACCUMULATOR An accumulator is used to store hydraulic fluid to maintain high pressure in the brake system and to provide residual pressure for power-assisted braking. Normally, the accumulator is charged with nitrogen gas **(Figure 53–3)** and is an integral part of the modulator unit. This unit is typically found on vehicles with a hydraulically assisted brake system. Depending on the system, the accumulator may have to be fully depressurized to accurately check the reservoir fluid level. This is usually performed by pumping the brake pedal, with the key off, 20 to 50 times, depending on the model.

ANTILOCK HYDRAULIC CONTROL VALVE ASSEMBLY This assembly controls the release and application of brake system pressure to the wheel brake assemblies. It may be of the integral type, meaning this unit is combined with the power booster and master cylinder units into one assembly **(Figure 53–4)**. The nonintegral type is mounted externally from the master cylinder/power booster unit and is located between the master cylinder and wheel brake assemblies. Both types generally contain solenoid valves that control the releasing, the holding, and the applying of brake system pressure. The number of solenoid valves will vary depending on the number of channels in the particular antilock system. Each channel requires a pair of solenoid valves to perform the antilock braking functions.

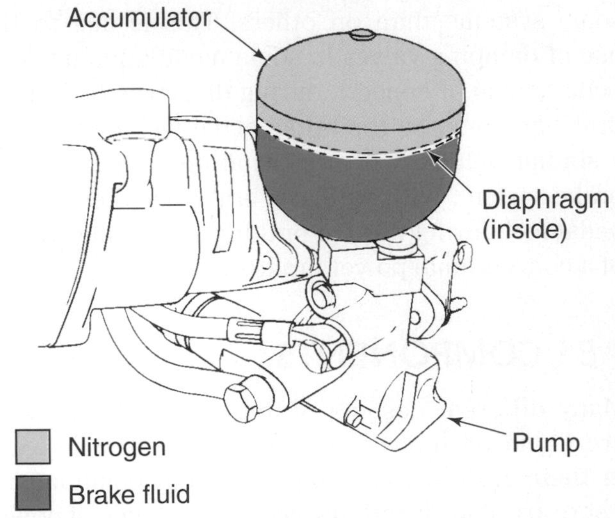

FIGURE 53–3

Pressure in an accumulator.

Federal-Mogul Corporation

BOOSTER PUMP The booster pump is an assembly of an electric motor and pump. The booster pump is used to provide pressurized hydraulic fluid for the ABS. The pump's motor is controlled by the system's control unit. A pressure switch determines the high-pressure pump cut-out point and the low-pressure pump cut-in point. The booster pump is also called the electric pump and motor assembly.

BOOSTER/MASTERCYLINDERASSEMBLY The booster/master cylinder assembly **(Figure 53–5)**, sometimes referred to as the hydraulic unit, contains the valves and pistons needed to modulate hydraulic pressure in the wheel circuits during ABS operation. Power brake assist is provided by pressurized brake fluid supplied by a hydraulic pump.

FLUID ACCUMULATORS Different than a pressure accumulator, fluid accumulators temporarily store brake fluid removed from the wheel brake units during an ABS cycle. This fluid is then used by the pump to build pressure for the brake hydraulic system. There are normally two fluid accumulators in a hydraulic control unit.

HYDRAULIC CONTROL UNIT This assembly contains the solenoid valves, fluid accumulators, pump, and an electric motor. This is actually a combination unit of many individual components found separately in some systems.

MAIN VALVE This two-position valve is also controlled by the ABS control module and is open only in the ABS mode. When open, pressurized brake fluid from the booster circuit is directed into the master cylinder (front brake) circuits to prevent excessive pedal travel.

MODULATOR UNIT The modulator unit **(Figure 53–6)** controls the flow of pressurized brake fluid to the individual wheel circuits. Normally, the modulator is made up of solenoids that open and close valves, several valves that control the flow of fluid to the wheel brake units, and electrical relays that activate or deactivate the solenoids through the commands of the control module. This unit may also be called the hydraulic actuator, hydraulic power unit, or the electrohydraulic control valve.

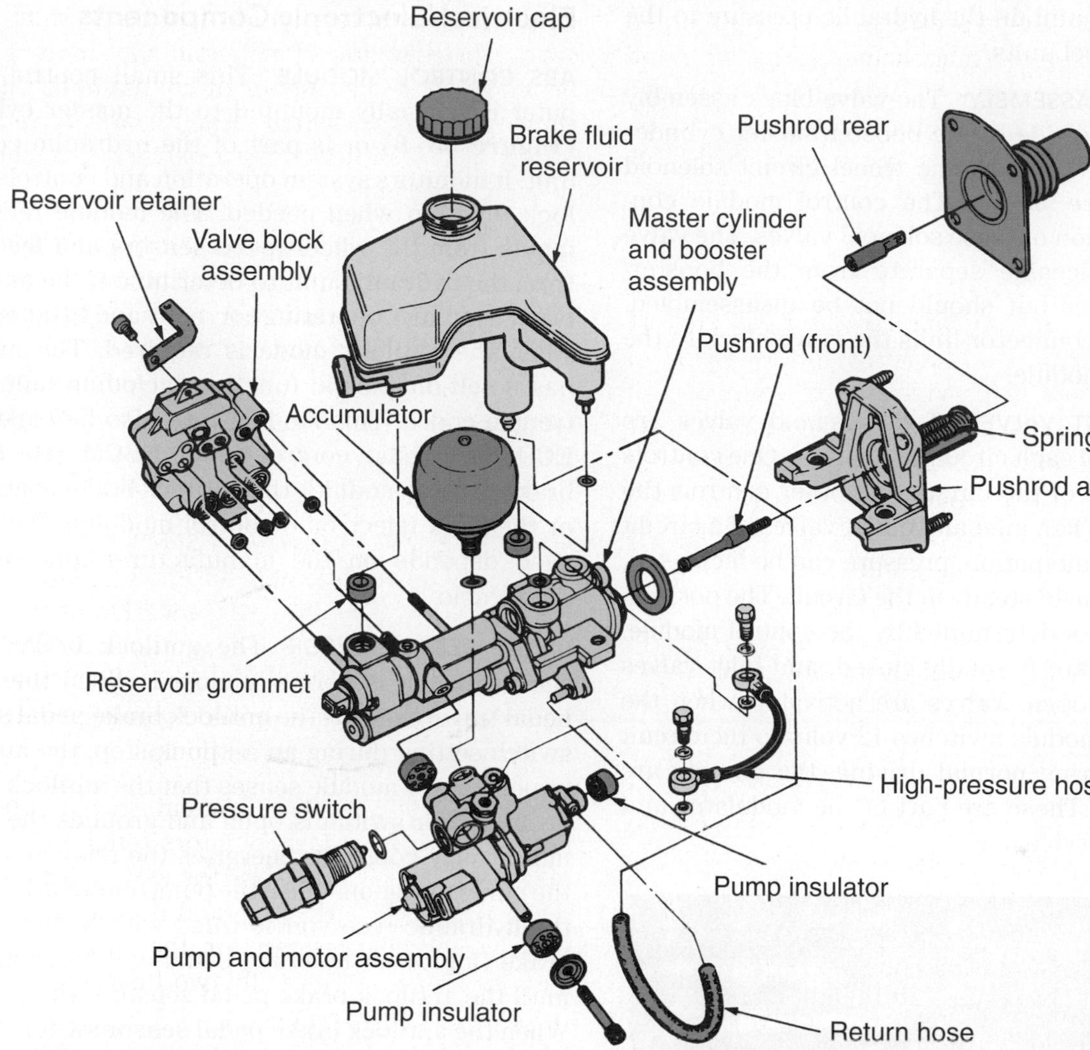

Reservoir cap
Brake fluid reservoir
Pushrod rear
Reservoir retainer
Master cylinder and booster assembly
Valve block assembly
Pushrod (front)
Accumulator
Spring
Pushrod assembly
Reservoir grommet
Pressure switch
High-pressure hose
Pump insulator
Pump and motor assembly
Pump insulator
Return hose

FIGURE 53–4

A hydraulic unit for a Teves antilock brake system.

Federal-Mogul Corporation

Reservoir assembly
Solenoid valve block
Electric motor and pump

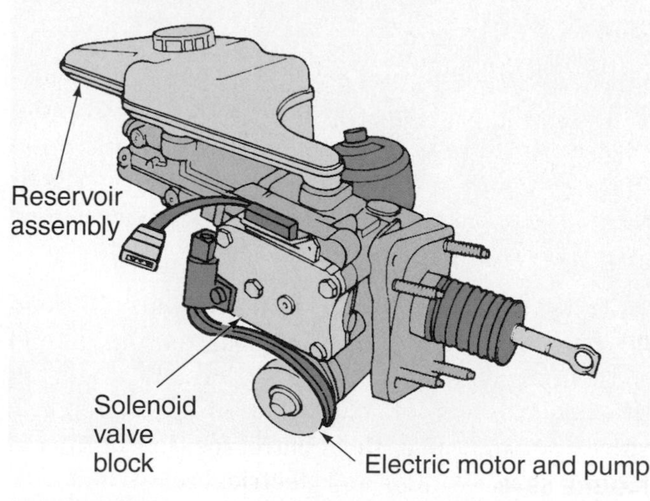

FIGURE 53–5

A master brake cylinder in a typical system uses an electric pump for power boost.

Ford Motor Company

SOLENOID VALVES The solenoid valves are located in the modulator unit and are electrically operated by signals from the control module. The control module switches the solenoids on or off to increase,

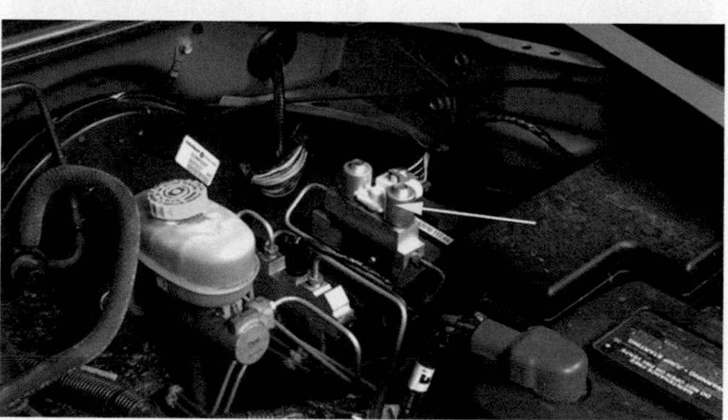

FIGURE 53–6

A rear-wheel ABS modulator assembly.

decrease, or maintain the hydraulic pressure to the individual wheel units.

VALVE BLOCK ASSEMBLY The valve block assembly attaches to the side of the booster/master cylinder and contains the hydraulic wheel circuit solenoid valves **(Figure 53–7)**. The control module controls the position of these solenoid valves. The valve block is serviceable separate from the booster/master cylinder but should not be disassembled. An electrical connector links the valve block to the ABS control module.

WHEEL CIRCUIT VALVES Two solenoid valves are used to control each circuit or channel. One controls the inlet valve of the circuit; the other controls the outlet valve. When inlet and outlet valves of a circuit are used in combination, pressure can be increased, decreased, or held steady in the circuit. The position of each valve is determined by the control module. Outlet valves are normally closed, and inlet valves are normally open. Valves are activated when the ABS control module switches 12 volts to the circuit solenoids. During normal driving, the circuits are not activated. These are part of the modulator unit and are not serviceable.

FIGURE 53–7

The valve block assembly attaches to the side of the booster/master cylinder and contains the hydraulic wheel circuit solenoid valves.

Electrical/Electronic Components

ABS CONTROL MODULE This small control computer is normally mounted to the master cylinder **(Figure 53–8)** or is part of the hydraulic control unit. It monitors system operation and controls antilock function when needed. The module relies on inputs from the wheel speed sensors and feedback from the hydraulic unit to determine if the antilock brake system is operating correctly and to determine when the antilock mode is required. The module has a self-diagnostic function including numerous trouble codes. This module may also be called the ECU (electronic control unit), EBCM (electronic brake control module), the antilock brake controller, or the ECM (electronic control module). The name used depends on the manufacturer and year of the vehicle.

BRAKE PEDAL SENSOR The antilock brake pedal sensor switch is normally closed. When the brake pedal travel exceeds the antilock brake pedal sensor switch setting during an antilock stop, the antilock brake control module senses that the antilock brake pedal sensor switch is open and grounds the pump motor relay coil. This energizes the relay and turns the pump motor on. When the pump motor is running, the hydraulic reservoir is filled with high-pressure brake fluid, and the brake pedal will be pushed up until the antilock brake pedal sensor switch closes. When the antilock brake pedal sensor switch closes, the pump motor is turned off, and the brake pedal

FIGURE 53–8

The ABS control module is normally mounted on the wheel housing, mounted to the master cylinder, or is part of the hydraulic control unit.

will drop some with each ABS control cycle until the antilock brake pedal sensor switch opens and the pump motor is turned on again. This minimizes pedal feedback during ABS cycling.

DATA LINK CONNECTOR (DLC) The DLC provides access and/or control of vehicle information, operating conditions, and diagnostic information.

DIAGNOSTIC TROUBLE CODE (DTC) These trouble codes are numeric identifiers for fault conditions identified by the ABS's internal diagnostic system.

INDICATOR LIGHTS Most ABS-equipped vehicles are fitted with two different warning lights. One of the warning lights is tied directly to the ABS, whereas the other light is part of the base brake system. All vehicles have a red warning light. This is lit when there is a problem with the brake system or when the parking brake is on. An amber warning light indicates a fault in the ABS. Both lamps will illuminate if there is a major problem in the base system, causing the ABS to be inhibited **(Figure 53–9)**.

LATERAL ACCELERATION SENSOR Used on vehicles with stability control, this switch monitors the sideward movement of the vehicle while it is turning a corner. This information is sent to the control module to ensure proper braking during turns.

PRESSURE SWITCH This switch controls pump motor operation and the low pressure warning light circuit. The pressure switch grounds the pump motor relay coil circuit, activating the pump when

FIGURE 53–9

All vehicles have a red warning light. This lamp lights when the brake fluid level is low, when there is a problem with the brake system, or when the park brake is on. An amber warning lamp lights when there is a fault in the ABS.

accumulator pressure drops below 14 000 kPa (2030 psi). The switch cuts off the motor when the pressure reaches 18 000 kPa (2610 psi). The pressure switch also contains switches to activate the dash-mounted warning light if accumulator pressure drops below 10 343 kPa (1500 psi). This unit is typically found on vehicles with a hydraulically assisted brake system.

PRESSURE DIFFERENTIAL SWITCH The pressure differential switch is located in the modulator unit. This switch sends a signal to the control module whenever there is an undesirable difference in hydraulic pressures within the brake system.

RELAYS Relays are electromagnetic devices used to control a high-current circuit with a low-current switching circuit. In ABS, relays are used to switch motors and solenoids. A low-current signal from the control module energizes the relays that complete the electrical circuit for the motor or solenoid.

TOOTHED RING The toothed ring can be located on an axle shaft, differential gear, or a wheel's hub. This ring is used in conjunction with the wheel speed sensor. The ring has a number of teeth around its circumference. The number of teeth varies by manufacturer and vehicle model. As the ring rotates and each tooth passes by the wheel speed sensor, an AC voltage signal is generated between the sensor and the tooth. As the tooth moves away from the sensor, the signal is broken until the next tooth comes close to the sensor. The end result is a pulsing signal that is sent to the control module. The control module translates the signal into wheel speed. The toothed ring may also be called the reluctor, tone ring, or gear pulser.

WHEEL SPEED SENSOR The wheel speed sensors **(Figure 53–10)** are mounted near the different toothed rings. Two types of sensors are in use: passive sensors, which generate AC voltage signals, and active sensors, which generate digital DC signals. With a passive sensor design, the ring's teeth rotate past the sensor, and AC voltage is generated. As the teeth move away from the sensor, the signal is broken until the next tooth comes close to the sensor. The end result is a pulsing signal that is sent to the control module **(Figure 53–11)**. The control module translates the signal into wheel speed. The sensor is normally a small coil of wire with a permanent magnet (PM) in its centre.

Active sensors may have two or three wires and are either Hall-effect or magnetoresistive sensors

FIGURE 53–10

A wheel speed sensor on an FWD drive axle.

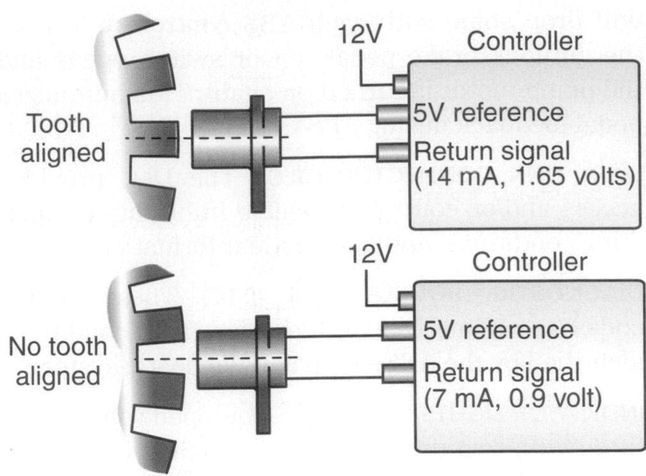

FIGURE 53–12

Active sensors have two or three wires and are either Hall-effect or magnetoresistive sensors that produce a square-wave digital signal.

that produce a square-wave digital signal **(Figure 53–12)**. Active sensors are now commonly used because, unlike PM sensors, digital sensors can be used down to zero km/h and can determine in which direction the wheel is turning. This is necessary for electronic stability system operation as well as vehicles with hill-assist take-off.

Multiplexing

The electronic circuit and wiring of an ABS is tied into the vehicle's CAN network **(Figure 53–13)**. This allows the ABS control unit to communicate with other control modules and share input devices with them. Every computer in the vehicle has access to all of the data in the CAN network, but individual computers use only the data that apply to their function.

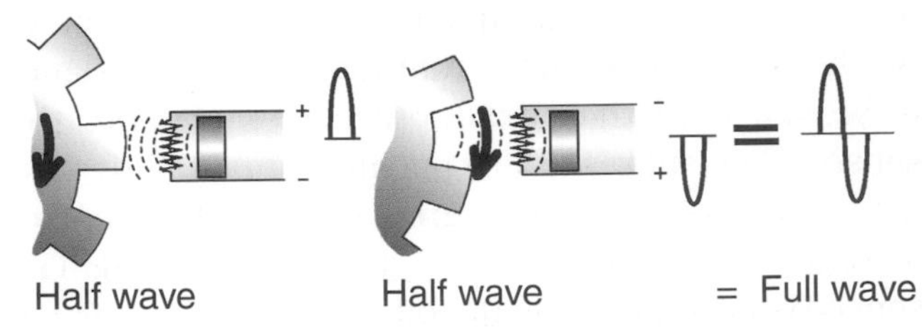

FIGURE 53–11

The output signal from a wheel speed sensor.

CAN communications is especially important when ABS is modified with additional features, such as traction and stability control.

Basic Operation

The control unit processes inputs and controls the operation of **isolation/dump valves** in the hydraulic modulator unit. The isolation/dump valves block off or isolate the master cylinder from certain wheel brakes. As long as the brakes are applied and the vehicle is moving, the master cylinder remains isolated, so additional fluid cannot be directed to those brakes. At the same time, the dump valve opens and allows a very small amount of fluid from that brake circuit to enter an accumulator **(Figure 53–14)**. This reduces the hydraulic pressure at the brake, and it is slightly released to allow the wheels to turn. If the wheels speed up too much, the dump valve reverses, and the accumulator forces a small amount of fluid back into that brake circuit. This constant dump/recharge is what causes the pulsation of the brake pedal during a panic or ABS stop. Most systems have a dedicated isolation/dump valve for each wheel.

TYPES OF ANTILOCK BRAKE SYSTEMS

The ABSs found on today's vehicles are manufactured by one of many different companies. Each manufacturer has a unique way to accomplish the same thing—vehicle control during braking. When

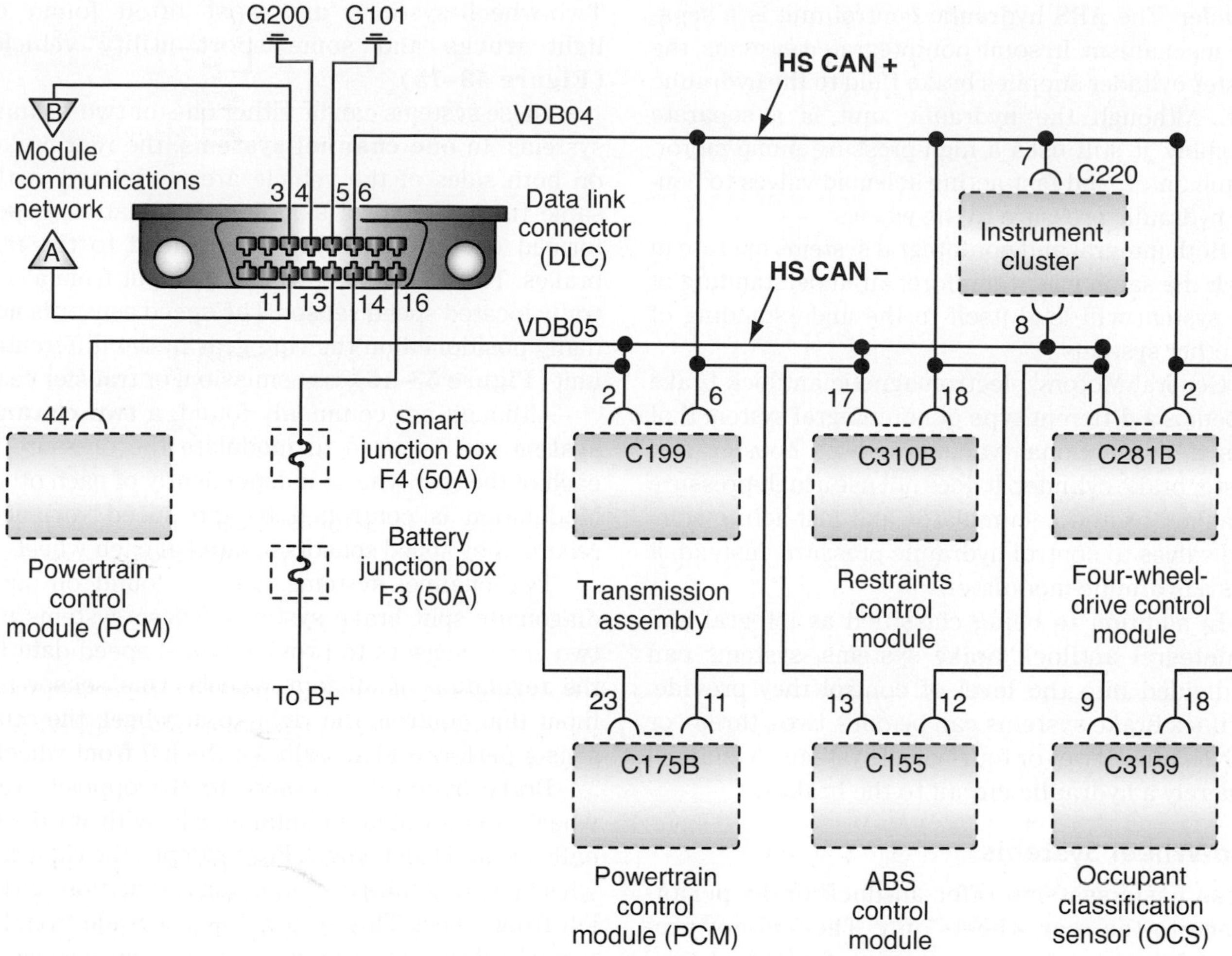

FIGURE 53–13

The electronic circuit and wiring of an ABS control unit is tied to the CAN network. This allows it to communicate with other control modules and share input devices with them.

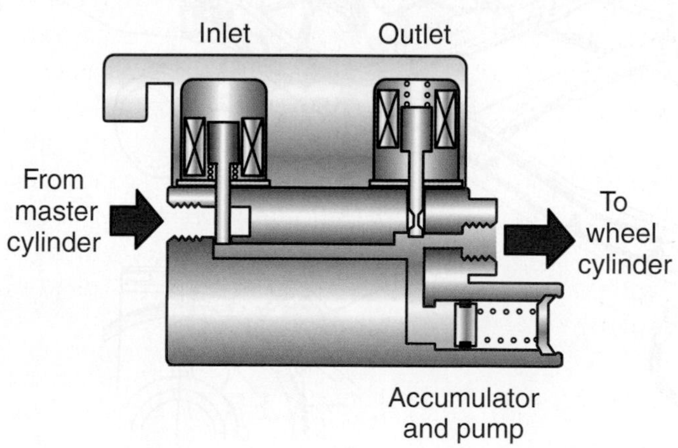

FIGURE 53–14

The control unit processes inputs and controls the operation of the isolation/dump valves to block off or isolate the master cylinder from the brakes.

working with ABS, it is important that you identify the exact system you are working with and follow the specific service procedures for that system. Keep in mind that there have been nearly 50 different ABSs used by the industry in recent years.

The exact manner in which hydraulic pressure is controlled depends on the exact ABS design. The great majority of the earlier antilock brake systems were integrated, or **integral antilock brake systems**. They combine the master cylinder, hydraulic booster, and ABS hydraulic circuitry into a single hydraulic assembly. These systems are becoming more prevalent because of the increasing number of hybrid and electric vehicles on the market. This is because integral systems use electric pumps instead of vacuum boosters to supply assist.

Nearly all of today's non-hybrid vehicles have **nonintegral antilock brake systems**. They use a conventional vacuum-assist booster and master cylinder. The ABS hydraulic control unit is a separate mechanism. In some nonintegrated systems, the master cylinder supplies brake fluid to the hydraulic unit. Although the hydraulic unit is a separate assembly, it still uses a high-pressure pump/motor, accumulator, and fast-acting solenoid valves to control hydraulic pressure to the wheels.

Both integral and nonintegral systems operate in much the same way; therefore, an understanding of one system will lend itself to the understanding of the other systems.

General Motors' electromagnetic antilock brake system is a different type of nonintegral system that uses a conventional vacuum power booster and master brake cylinder. It does not use a high-pressure pump/motor and accumulator and fast-acting solenoid valves to control hydraulic pressure. Instead, it uses a hydraulic modulator.

In addition to being classified as integral and nonintegral antilock brake systems, systems can be divided into the level of control they provide. Antilock brake systems can be one-, two-, three-, or four-channel, two- or four-wheel systems. A channel is merely a hydraulic circuit to the brakes.

Two-Wheel Systems

These basic systems offer antilock brake performance to the rear wheels only. They do not provide antilock performance to the steering wheels. Two-wheel systems are most often found on light trucks and some sport utility vehicles **(Figure 53–15)**.

These systems can be either one- or two-channel systems. In **one-channel systems**, the rear brakes on both sides of the vehicle are modulated at the same time to control skidding. Modulation is performed on the single brake line feed to the rear brakes. These systems rely on the input from a centrally located speed sensor. The speed sensor is normally positioned on the ring gear in the differential unit **(Figure 53–16)**, transmission, or transfer case.

Although not commonly found, a **two-channel system** can be used to modulate the pressure to each of the rear wheels independently of each other. Modulation is controlled by the speed variances recorded by speed sensors located at each wheel.

Two-channel systems may be found on some diagonally split brake systems. These systems use two speed sensors to provide wheel speed data for the regulation of all four wheels. One sensor has input that controls the right front wheel; the other sensor performs identically for the left front wheel.

Brake hydraulic pressure to the opposite rear wheel is controlled simultaneously with its diagonally located front wheel. For example, the right rear wheel receives the same pumping instructions as the left front wheel. This system is an upgrade from the two-wheel system as it does provide steering control.

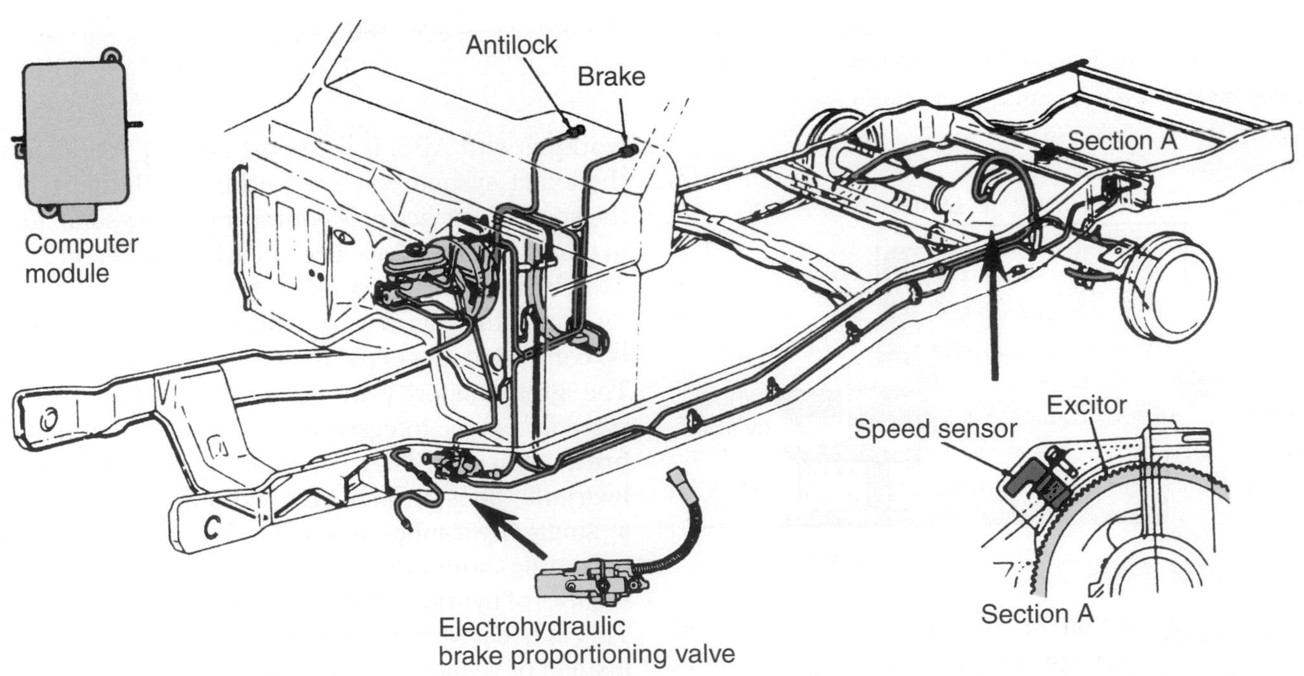

FIGURE 53–15

The main components of a rear-wheel ABS.

Federal-Mogul Corporation

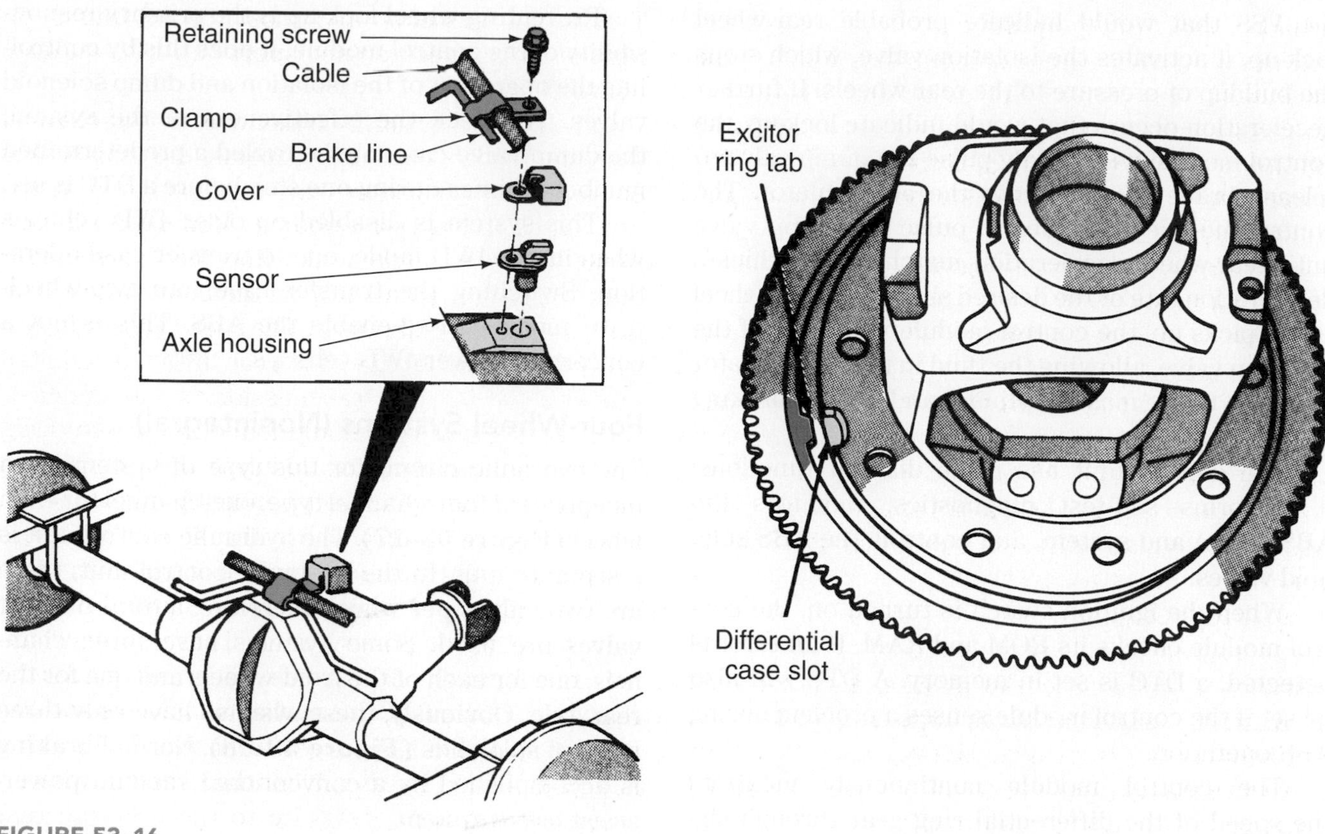

FIGURE 53–16

A speed sensor for both rear wheels located on the differential unit.

Chrysler Group LLC

However, it can have shortcomings under certain operating conditions.

Full (Four-Wheel) Systems

Some hydraulic systems that are split from front to rear use a **three-channel system** and are called four-wheel antilock brake systems. These systems have individual hydraulic circuits to each of the two front wheels, and a single circuit to the two rear wheels.

The most effective and most common ABS available is a **four-channel system**, in which sensors monitor each of the four wheels. With this continuous information, the ABS control module ensures each wheel receives the exact braking force it needs to maintain both antilock and steering control.

ABS OPERATION

The exact operation of an antilock brake system depends on its design and manufacturer. It would take many pages to try to explain the operation of each, and as soon as you read the explanations, there would be two or more new systems that would have to be explained. The exact operation of any system can be easily understood if you understand the basic operation of a few. The primary difference in operation between them all is based on the components used by the system. Therefore, the following systems were chosen as examples of how certain systems operate with the components they have.

Two-Wheel Systems (Nonintegral)

These systems are used to prevent rear-wheel lockup on pickup trucks and SUVs, especially under light payload conditions. They consist of a standard power brake system, an electronic control unit (control module), and an isolation/dump valve assembly. The valve assembly is attached to the master cylinder at the rear brake line. Both rear-wheel brake assemblies are controlled by the valve assembly under ABS conditions.

Under normal braking, pressure will pass through the valve assembly. The control module receives a signal from the brake switch when brakes are applied and begins to monitor the vehicle speed sensor (VSS) signal at speeds over 8 km/h (5 mph). If the control module detects a deceleration rate from the VSS that would indicate probable rear-wheel lock-up, it activates the isolation valve, which stops the buildup of pressure to the rear wheels. If further deceleration occurs that would indicate lock-up, the control module will rapidly pulse the dump valve to release brake pressure into the accumulator. The control module continues to pulse the dump valve until rear-wheel deceleration matches the vehicle's deceleration rate or the desired slip rate. When wheel speed picks up, the control module will turn off the isolation valve, allowing the fluid in the accumulator to return to the master cylinder and normal braking control to resume.

The control unit has three distinct functions: It performs self-test diagnostics, monitors the ABS action and system, and controls the ABS solenoid valves.

When the ignition switch is turned on, the control module checks its ROM and RAM. If an error is detected, a DTC is set in memory. A DTC will also be set if the control module senses a problem during ABS operation.

The control module continuously monitors the speed of the differential ring gear through signals from the rear-wheel speed sensor. The control module also receives signals from the brake light switch, brake warning light switch, reset switch, and four-wheel-drive (4WD) switch.

Preventing wheel lock-up is the primary responsibility of the control module. It does this by controlling the operation of the isolation and dump solenoid valves. To check the effectiveness of the system, the dump valve can only be cycled a predetermined number of times during one stop before a DTC is set.

This system is disabled on older 4WD vehicles when in the 4WD mode, due to transfer case operation. Switching the transfer case into two-wheel-drive mode will re-enable the ABS. This is not a concern on newer 4WD vehicles.

Four-Wheel Systems (Nonintegral)

The hydraulic circuit for this type of system is an independent four-channel type, one-channel for each wheel **(Figure 53–17)**. The hydraulic control unit is a separate unit. In the hydraulic control unit, there are two valves per wheel; therefore, a total of eight valves are used. Some systems have three channels, one for each of the front wheels and one for the rear axle. Obviously, these systems have only three pairs of solenoids **(Figure 53–18)**. Normal braking is accomplished by a conventional vacuum power-assist brake system.

The system prevents wheel lock-up during an emergency stop by modulating brake pressure. It allows the driver to maintain steering control and stop the vehicle in the shortest possible distance

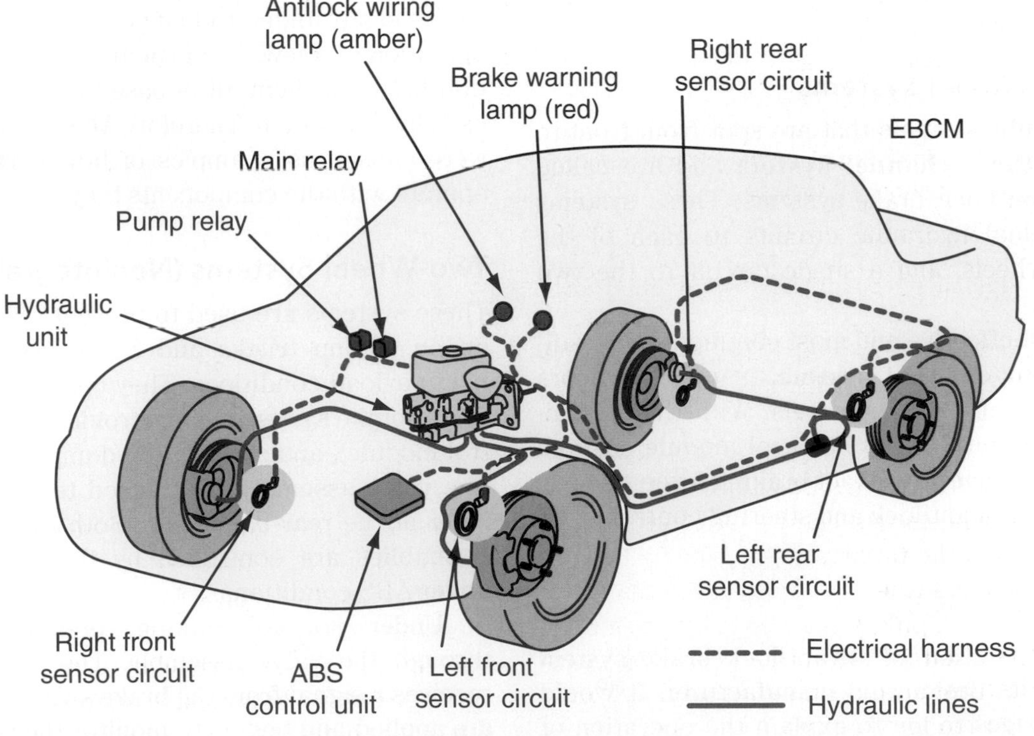

FIGURE 53–17

The basic electrical and hydraulic components of a four-wheel antilock brake system.

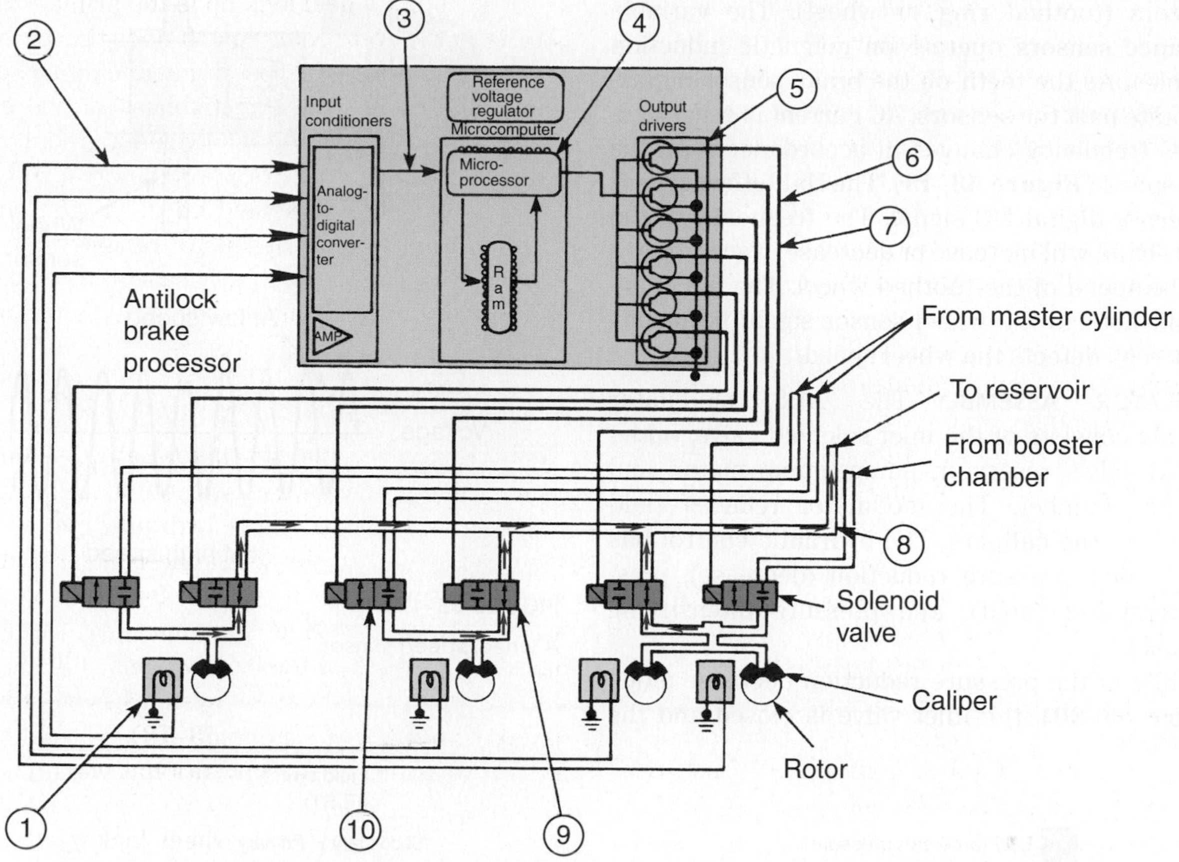

1. Sensors detect wheel rotation.
2. Sensors relay analog signal, indicating impending locking condition.
3. Analog signal converted to digital signal.
4. Microprocessor compares input with information in RAM and determines potential brake lock-up.
5. Output drivers close.
6. Actuator ground circuits close.
7. Current flows to solenoids.
8. Hydraulic pressure to brake is reduced.
9. Normally closed outlet valves open.
10. Normally open inlet valves closed.

FIGURE 53–18

An ABS operation—potential brake lock condition.

under most conditions. During ABS operation, the driver will sense a pulsation in the brake pedal and a clicking sound.

OPERATION The ABS control module calculates the slip rate of the wheels and controls the brake fluid pressure to attain the target slip rate. If the control module senses that a wheel is about to lock based on input sensor data, it pulses the normally open inlet solenoid valve closed for that circuit. This prevents any more fluid from entering that circuit. The ABS control module then looks at the sensor signal from the affected wheel again. If that wheel is still decelerating faster than the other three wheels, it opens the normally closed outlet solenoid valve for that circuit. This dumps any pressure trapped between the closed inlet valve and the brake back to the master cylinder reservoir. Once the affected wheel returns to the same speed as the other wheels, the control module returns the valves to their normal condition, allowing fluid flow to the affected brake.

Based on the inputs from the vehicle speed and wheel speed sensors, the control module calculates the slip rate of each wheel and transmits a control signal to the modulator unit solenoid valve when the slip rate is high.

Wheel speed at each wheel is measured by variable-reluctance or Hall-effect sensors and sensor indicators (toothed ring or wheel). The variable reluctance sensors operate on magnetic induction principles. As the teeth on the brake sensor indicators rotate past the sensors, AC current is generated. The AC frequency changes in accordance with the wheel speed **(Figure 53–19)**. The Hall-effect sensor produces a digital DC signal. The frequency of the digital signal will increase or decrease in accordance with the speed of the toothed wheel. The ABS control unit detects the wheel sensor signal frequency and thereby detects the wheel speed.

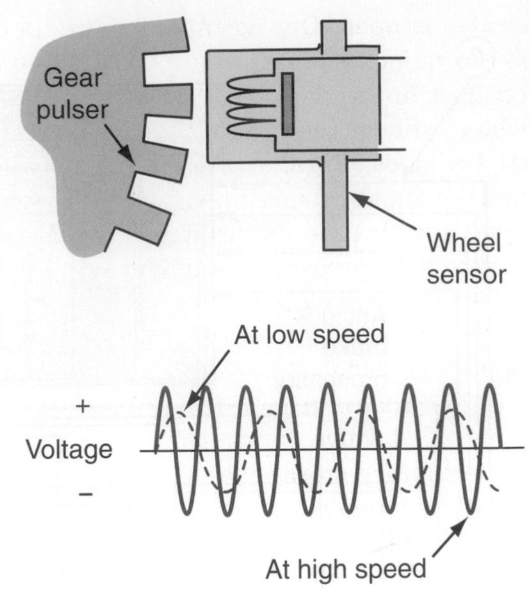

FIGURE 53–19
A wheel speed sensor.

MODULATOR ASSEMBLY The ABS modulator assembly consists of the inlet solenoid valve, outlet solenoid valve, reservoir, pump, pump motor, and damping chamber. The modulator reduces fluid pressure to the calipers. The hydraulic control has three modes: pressure reduction (decrease), pressure retaining (hold), and pressure intensifying (increase).

While in the pressure reduction decrease mode **(Figure 53–20)**, the inlet valve is closed and the

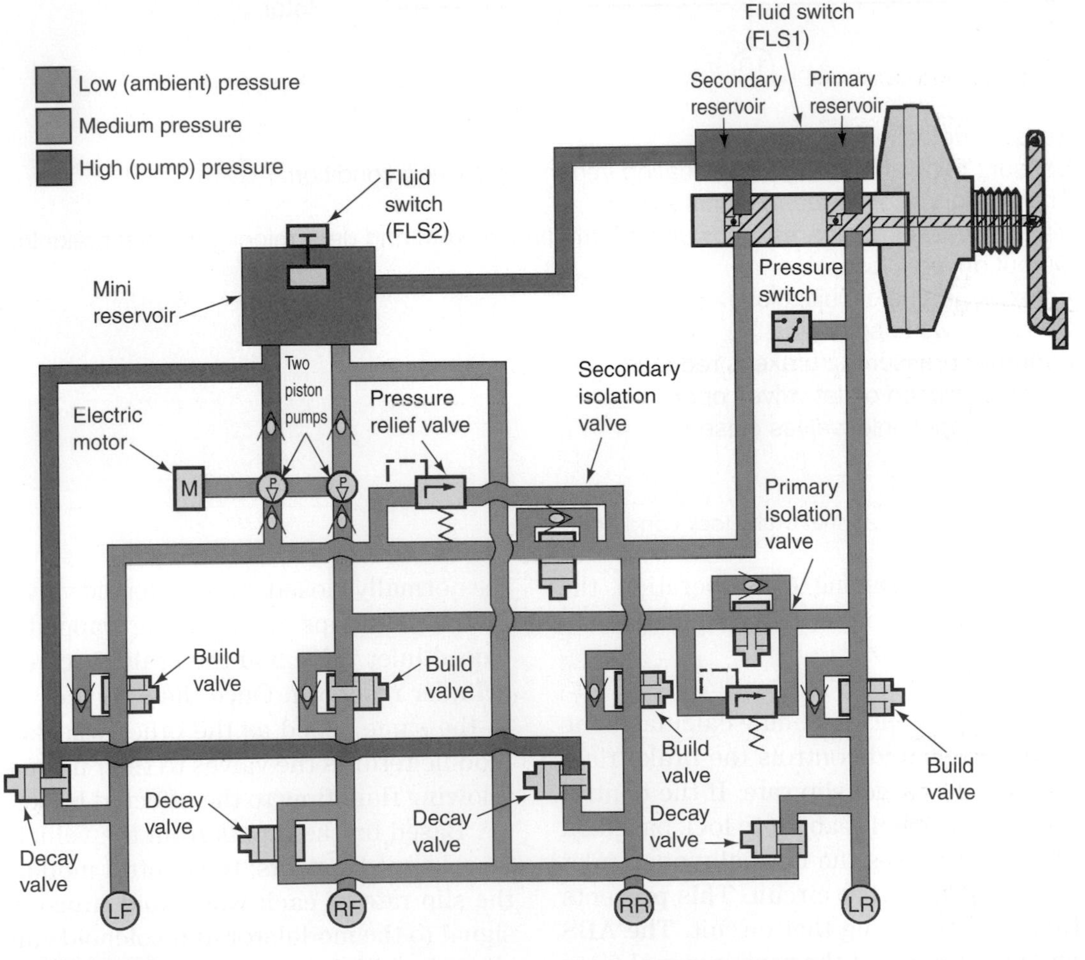

FIGURE 53–20
Antilock braking with pressure reduction.
Chrysler Group LLC

outlet valve is open. During this mode, fluid pressure to the caliper is blocked, and the existing fluid in the caliper flows through the outlet valve back to the master cylinder reservoir. During the pressure-intensifying mode **(Figure 53–21)**, the inlet valve is open and the outlet valve is closed. Pressurized fluid is pumped to the caliper. To keep the pressure at the caliper during the pressure-retaining mode, the inlet and outlet valves are closed.

The pump/motor assembly provides the extra fluid required during an ABS stop, and the pump is supplied fluid that is released to the accumulators when the outlet valve is open. The accumulators provide temporary storage for the fluid during an ABS stop. The pump is also used to drain the accumulator circuits after the ABS stop is complete. The pump is run by an electric motor controlled by a relay that is controlled by the ABS control module. The pump is continuously on during an ABS stop and remains on for about five seconds after the stop is complete.

Keep in mind that the activity of the solenoid valves changes rapidly, several times each second. This means the fluid under pressure must be redirected quickly. This is the primary job of the pump.

SELF-DIAGNOSIS The control module monitors the electromechanical components of the system. A malfunction of the system will cause the control module to shut off or inhibit the system. However, normal power-assisted braking remains. Malfunctions are indicated by a warning indicator in the instrument cluster. The system is self-monitoring. When the ignition switch is placed in the RUN position, the ABS control module will perform a preliminary self-check on its electrical system, indicated by a second illumination of the amber ABS indicator in the instrument cluster. During vehicle operation and during normal and antilock braking, the control module monitors all electrical ABS functions and some hydraulic functions. With most malfunctions of the ABS, the amber ABS indicator is illuminated and a DTC recorded.

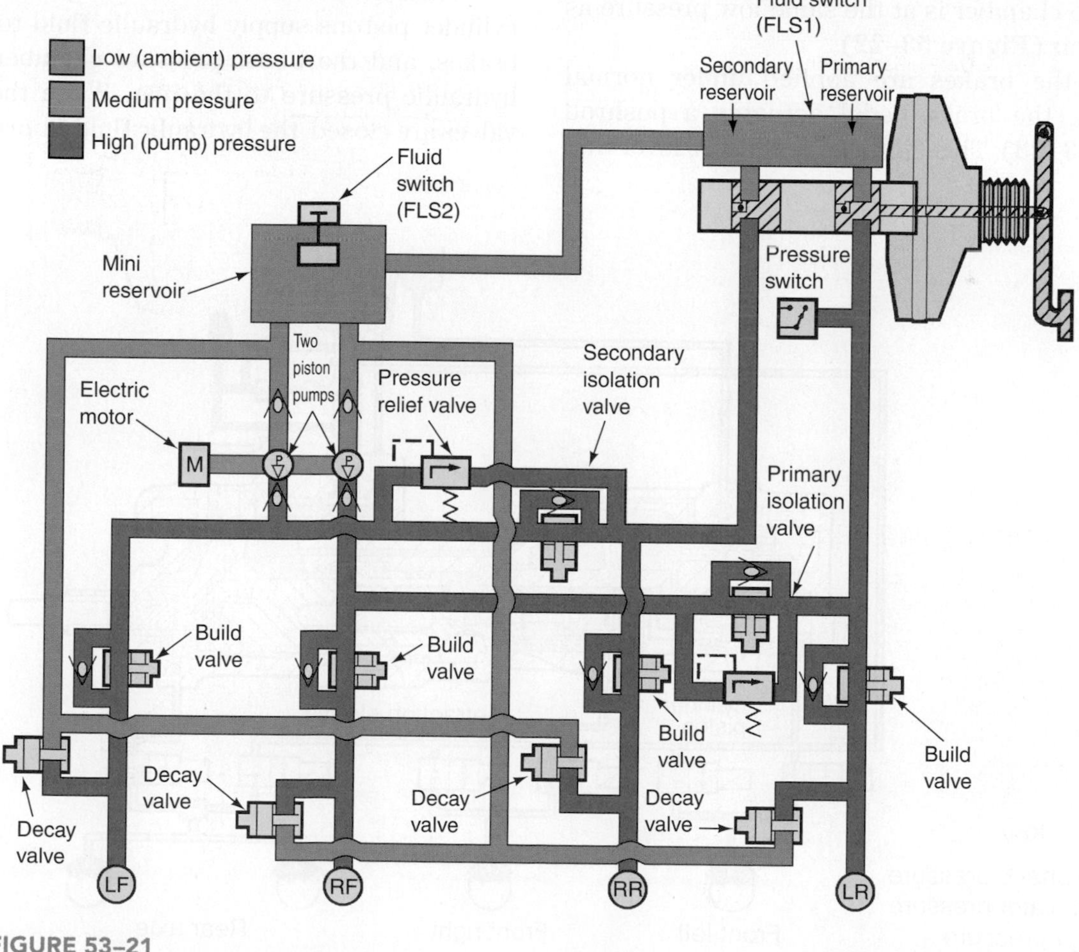

FIGURE 53–21

Antilock braking with pressure increase.

Chrysler Group LLC

Four-Wheel Systems (Integral)

When a wheel speed sensor signals the ABS control module that a high rate of deceleration is taking place at its wheel (that is, that the wheel is likely to lock and skid), the control module initially signals the hydraulic unit to keep hydraulic pressure at the wheel constant. If the wheel continues to decelerate, the control module then signals the circuit valve solenoids to reduce hydraulic pressure to the affected wheel. This reduces braking at the wheel, reducing the risk of lock-up.

The wheel accelerates again as a result of the reduced braking pressure. Once a specific limit has been reached, the control module registers the fact that the wheel is not being sufficiently braked. The wheel is decelerated again by increasing the pressure that was initially reduced. The control cycles may be executed every second, depending upon the road conditions.

In operation, when the brakes are released, the piston in the master cylinder is retracted. The booster chamber is vented to the reservoir, and the fluid in the chamber is at the same low pressure as the reservoir **(Figure 53–22)**.

When the brakes are applied under normal conditions, the brake pedal actuates a pushrod **(Figure 53–23)**. This operates a scissors lever that moves a spool valve. When the spool valve moves, it closes the port from the booster chamber to the reservoir and partially opens the port from the accumulator in proportion to the pressure on the brake pedal. This allows hydraulic fluid under pressure from the accumulator to enter the booster chamber. As hydraulic pressure enters, it pushes the booster piston forward, providing hydraulic assist to the mechanical thrust from the pushrod.

When the control module determines the wheels are locking up, it opens a valve that supplies one chamber between the two master cylinder pistons and another chamber between the retraction sleeve and the first master cylinder piston **(Figure 53–24)**. The hydraulic pressure on the retraction sleeve retracts the pushrod, pushing back the brake pedal. In effect, the hydraulic pressure to the front wheels is now supplied by the accumulator, not by the brake pedal action. The control module also opens and closes the solenoid valves to cycle the brakes on the wheels that have been locking up.

When the solenoid valves are open, the master cylinder pistons supply hydraulic fluid to the front brakes, and the boost pressure chamber provides hydraulic pressure to the rear. When the solenoid valves are closed, the hydraulic fluid from the master

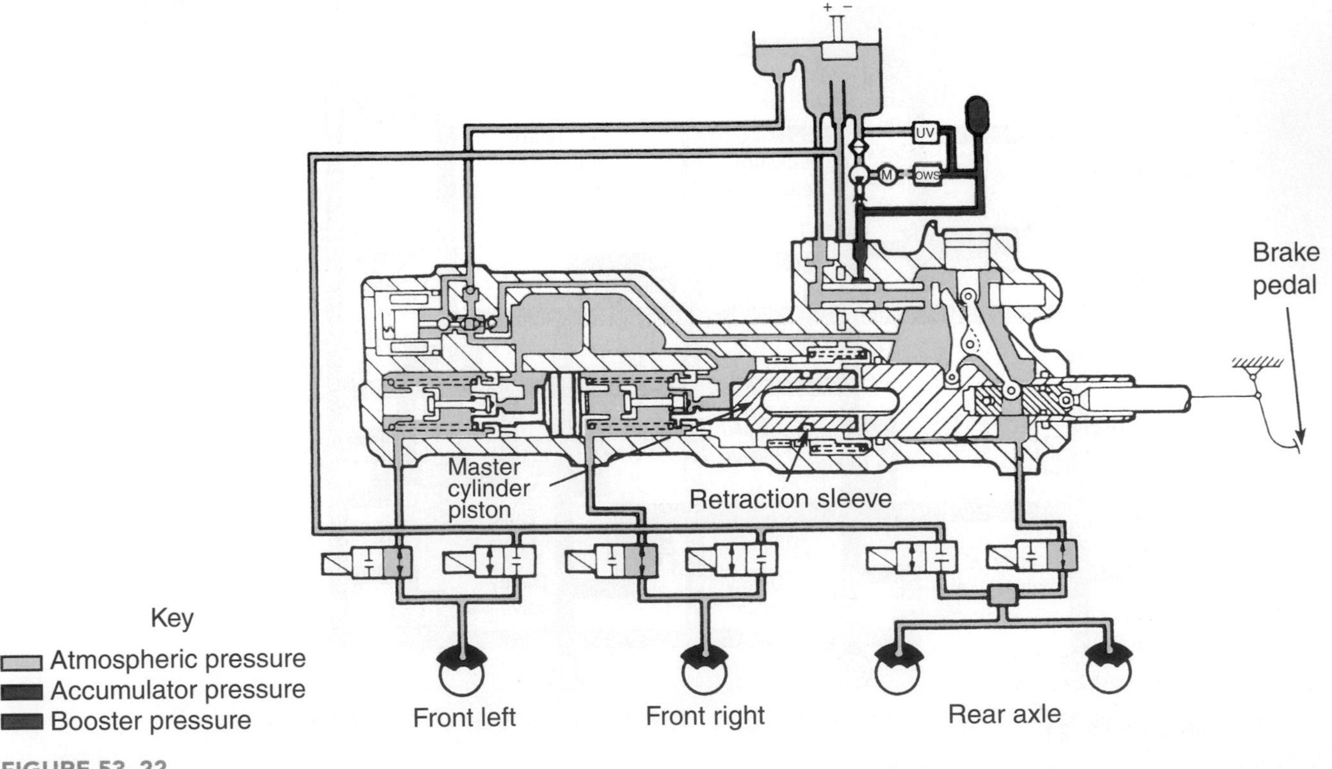

FIGURE 53–22

ABS operation—no braking.

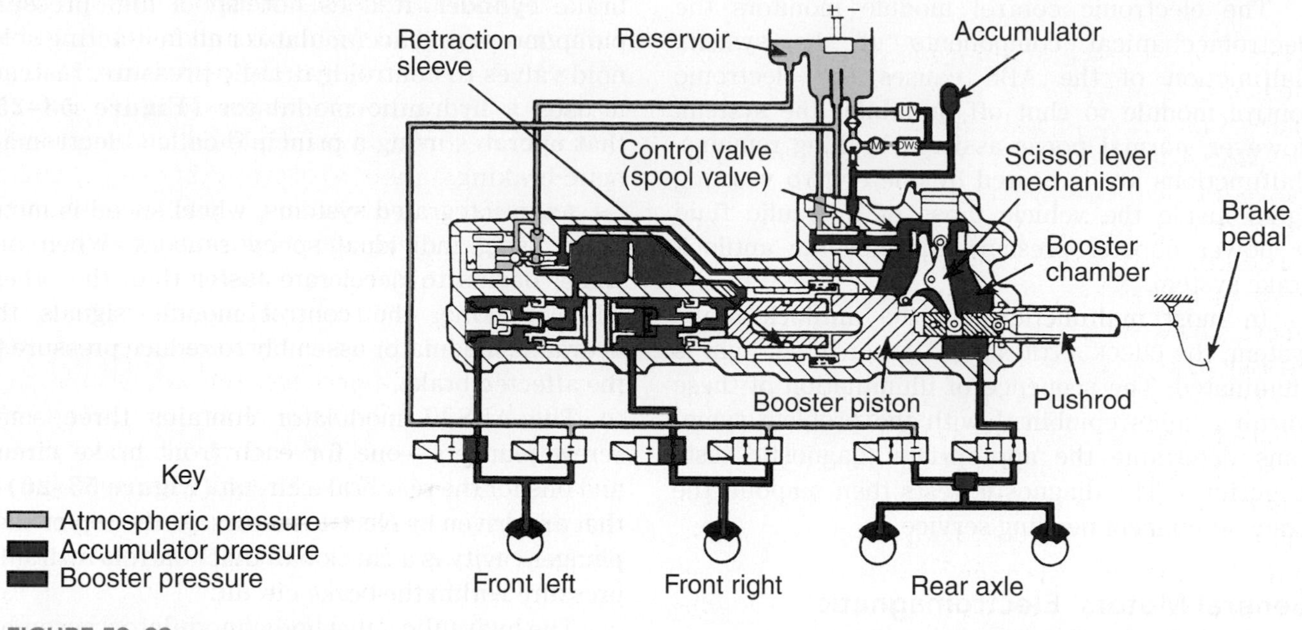

FIGURE 53–23

Normal braking with an antilock system.

cylinder pistons and booster pressure chamber is cut off. The hydraulic fluid is returned from the brakes to the reservoir.

A few systems are equipped with a lateral acceleration switch to change the system's programming when hard braking while cornering. The switch may be no more than a left-hand and right-hand mercury switch located in a common housing that responds to forces generated by left-hand/right-hand turns and cornering movements. Late-model lateral acceleration sensors use a Hall-effect switch.

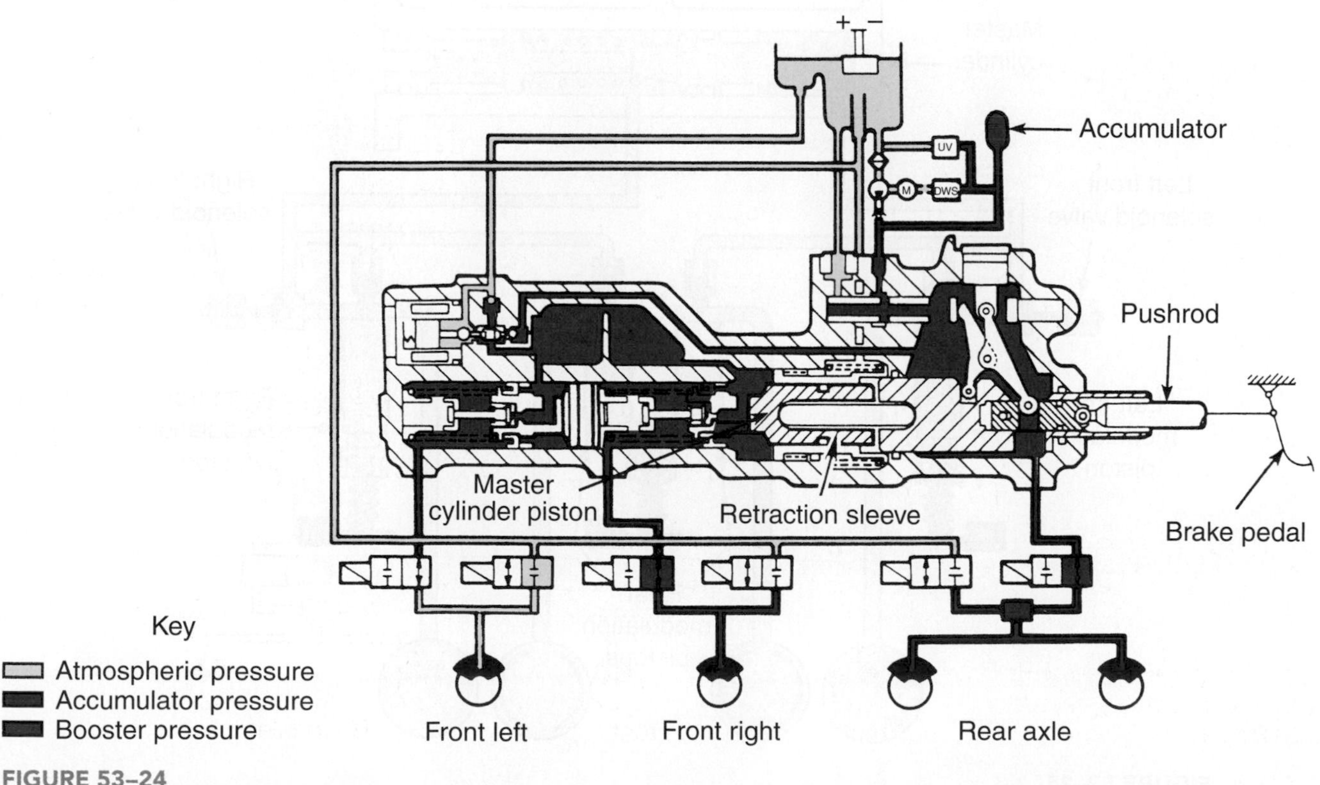

FIGURE 53–24

An antilock braking system in operation.

The electronic control module monitors the electromechanical components of the system. Malfunction of the ABS causes the electronic control module to shut off or inhibit the system. However, normal power-assisted braking remains. Malfunctions are indicated by one or two warning lights inside the vehicle. Loss of hydraulic fluid or power booster pressure disables the antilock brake system.

In most malfunctions in the antilock brake system, the CHECK ANTILOCK BRAKES or brake light is illuminated. The sequence of illumination of these warning lights combined with the problem symptoms determine the appropriate diagnostic tests to perform. The diagnostic tests then pinpoint the exact component needing service.

General Motors' Electromagnetic Antilock Brake Systems

Beginning in 1991, General Motors began equipping certain small and mid-size vehicles with an antilock braking system called the ABS-VI. This system is a nonintegral or add-on system that uses a conventional vacuum power booster and master brake cylinder. It does not use a high-pressure pump/motor and accumulator and fast-acting solenoid valves to control hydraulic pressure. Instead, it uses a hydraulic modulator **(Figure 53–25)** that operates using a principle called electromagnetic braking.

As in integrated systems, wheel speed is monitored using individual speed sensors. When one wheel begins to decelerate faster than the others while braking, the control module signals the hydraulic modulator assembly to reduce pressure to the affected brake.

The ABS-VI modulator contains three small screw plungers—one for each front brake circuit and one for the rear brake circuit **(Figure 53–26)**—that are driven by electric motors. At the top of each plunger cavity is a check ball that controls hydraulic pressure within the brake circuit.

The hydraulic valve body (modulator) assembly and motor pack is mounted to the master cylinder. The hydraulic brake circuit for each front wheel is controlled by a motor, gear-driven ball screw, solenoid, piston, and check valve. The rear-wheel circuit is controlled by check balls and a single motor; therefore, both rear brakes are modulated together.

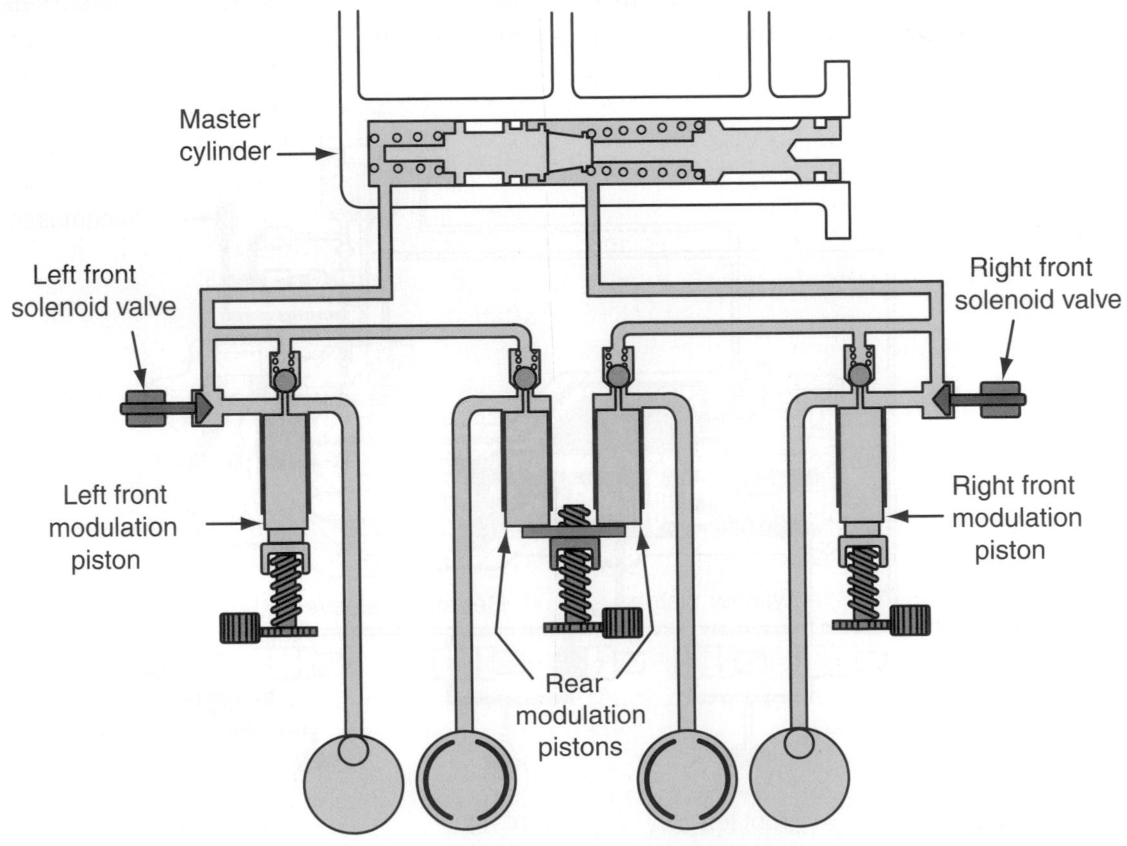

FIGURE 53–25

The hydraulic layout for GM's ABS-VI.

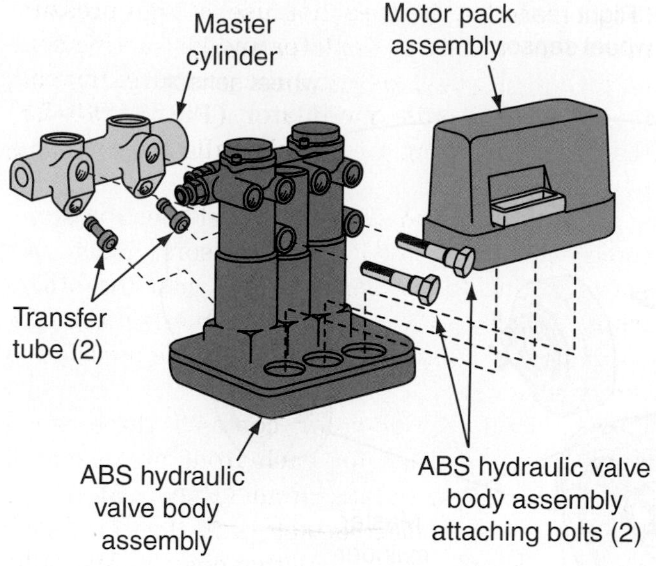

FIGURE 53–26

The hydraulic modulator is an add-on component to the conventional master cylinder. The electric motors can be replaced separately, as can the two solenoids.

The motors are high-speed, bidirectional motors that quickly and precisely position the ball screws. Each motor has a brake that allows its ball screw to maintain its position against hydraulic pressures. The front motors have an electromagnetic brake (EMB), and the rear motor uses an expansion spring brake (ESB).

During normal braking conditions, each plunger is all the way up. The check ball at the top is unseated, and the bypass solenoid is normally open, which allows brake pressure from the master cylinder and vacuum power booster to apply the brakes during normal stopping.

During panic stops, the ABS mode operates. In this situation, brake pressure must be reduced to prevent wheel lock-up. This is done by closing the normally open solenoid to isolate the circuit and then turning the plunger down to reduce braking pressure. As the plunger turns down, it increases the volume within the brake circuit. This causes a drop in pressure that keeps the wheel from locking. The amount of pressure applied is controlled by running the plunger up and down as required. To decrease pressure further, the plunger is run down. To reapply pressure, the plunger moves back up.

The system can cycle the brakes seven times per second. Because the system does not have a high-pressure pump and accumulator, it cannot increase brake pressure above what can be provided by the master cylinder and vacuum-assist booster.

Other Brake System Controls

A few automobiles are now equipped with an electronic brake-assist system that can recognize emergency braking and automatically apply full-power brake force for shorter stopping distances. This system is activated only in emergency braking situations and does not affect normal brake operation.

The system recognizes emergency braking by the speed at which it was depressed, and the brakes are automatically applied under full power. The system is driver adaptive, as it learns the driver's braking habits by using sensors to monitor every movement of the brake pedal. When the sensors detect an emergency stop, an electronic valve at the power brake booster is turned on. This supplies full braking power to the wheels. The ABS prevents the wheels from locking in spite of the full-power braking force, and it may even modulate more pressure to the wheels that are not slipping than is applied by the master cylinder.

AUTOMATIC TRACTION CONTROL

Automakers use the technology and hardware of antilock braking systems to control tire traction and vehicle stability. As explained earlier, an ABS pumps the brakes when a braking wheel attempts to go into a locked condition. **Automatic traction control (ATC)** systems apply the brakes when a drive wheel attempts to spin and lose traction **(Figure 53–27)**.

The system works best when one drive wheel is working on a good traction surface and the other is not. The system also works well when the vehicle is accelerating on slippery road surfaces, especially when climbing hills. Automatic traction control will only operate when the vehicle is accelerating and the driver is not depressing the brake pedal.

ATC is most helpful on four-wheel-drive or all-wheel-drive vehicles in which loss of traction at one wheel could hamper driver control. It is also desirable on high-powered front-wheel-drive (FWD) vehicles for the same reason. Often, if traction control is fitted to an FWD vehicle, the modified ABS is a three-channel system because ATC is not needed at the rear wheels. On rear-wheel-drive (RWD) and 4WD vehicles, the system is based on a four-channel ABS.

During road operation, the ATC system uses an electronic control module to monitor the

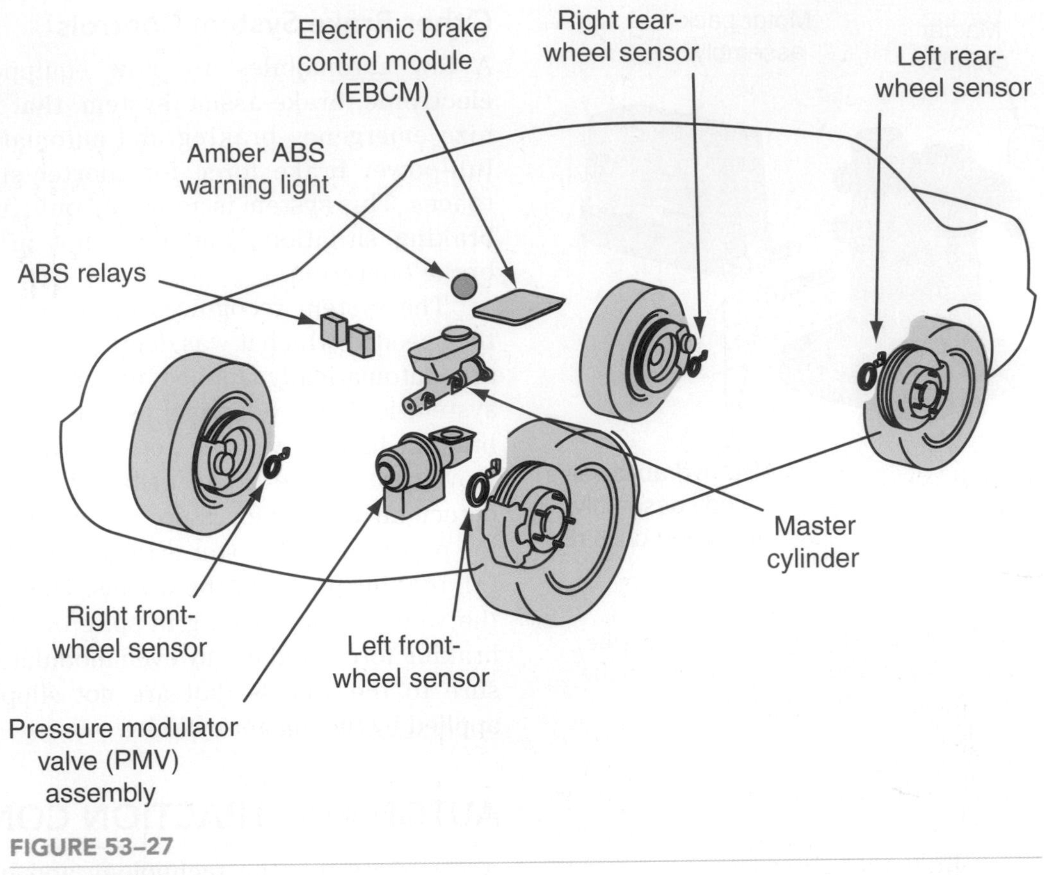

FIGURE 53–27

A typical ATC system.

wheel-speed sensors. If a wheel enters a loss-of-traction situation, the module applies braking force to the wheel in trouble. Loss of traction is identified by comparing the vehicle's speed to the speed of the wheel. If there is a loss of traction, the speed of the wheel will be greater than expected for the particular vehicle speed. Wheel spin is normally limited to a 10 percent slippage. Some traction control systems use separated hydraulic valve units and control modules for the ABS and ATC, whereas others integrate both systems into one hydraulic control unit and a single control module. The pulse rings and wheel speed sensors remain unchanged from the ABS to the ATC.

Some ATC systems function only at low road speeds, from 8 to 40 km/h (5 to 25 mph). These systems are designed to reduce wheel slip and maintain traction at the drive wheels when the road is wet or snow covered. The control module monitors wheel speed. If during acceleration the module detects drive wheel slip and the brakes are not applied, the control module enters into the traction control mode. The inlet and outlet solenoid valves are pulsed and allow the brake to be quickly applied and released. The pump/motor assembly is turned on and supplies pressurized fluid to the slipping wheel's brake.

Engine Controls

More advanced systems work at higher speeds and integrate some engine control functions into the control loop. Most ATC systems rely on inputs available on the CAN bus and compare front-wheel speeds to rear-wheel speeds to determine if drive wheels lose traction.

When drive wheel slip is detected while the brake is not applied, the electronic brake control module (EBCM) will enter into the traction control mode. At that time, the PCM will initiate an engine torque reduction routine to slow down the drive wheels. The following shows how the PCM reduces torque to the drive wheels:

- By retarding spark timing
- By decreasing the opening of the throttle plate
- By reducing or cutting off fuel injection pulses to one or more cylinders
- By increasing exhaust gas recirculation (EGR) flow
- By momentarily upshifting the transmission to a higher gear

If the engine torque reduction does not eliminate drive wheel slip, the EBCM will gradually apply the brakes at the driving wheels. The master cylinder's isolation valve closes to isolate the cylinder from the rest of the hydraulic system. Then the prime valve opens to allow the pump to accumulate brake fluid and build hydraulic pressure. The drive wheel inlet and outlet solenoid valves then open and close and pass through stages of pressure hold, pressure increase, and pressure decrease.

Driver Controls and Indicators

Most traction control systems (TCSs) have two warning lights. However, some vehicles display the status of the system with messages on the instrument cluster. Normally, an amber lamp will illuminate or a service message will appear as any of the ABS-disabling DTCs is set. When this occurs, the TCS is automatically disabled by the control unit. When the system is actively controlling wheel spin, a green lamp lights or a message is displayed saying the system is active.

There may also be a manual cutoff switch so that the driver can turn off the TCS. This will cause the amber lamp to light or a message to be displayed, stating that the system is off.

AUTOMATIC STABILITY CONTROL

Various stability control systems are found on today's vehicles. Like TCSs, stability controls are based on and linked to the ABS **(Figure 53–28)**. On some vehicles, the stability control system is also linked to the electronic suspension system **(Figure 53–29)**. Most often, the stability control system is called an **electronic stability control (ESC)** system, although many other names are used. ESC helps prevent skids, swerves, and rollover accidents. Basically, the system applies the brakes at one or more wheels to help correct the steering. In some cases, power to the drive wheels is also reduced.

It is important to remember that a vehicle's tendency to roll is influenced by its height, its track width, and the stiffness of its suspension. ESC cannot override a car's physical limits nor can it increase traction. If the vehicle is pushed beyond its traction limits, ESC may not be able to correct the vehicle's movement. ESC simply helps the driver maintain control using the available traction.

ESC systems can control the vehicle during acceleration, braking, and coasting. If the brakes are

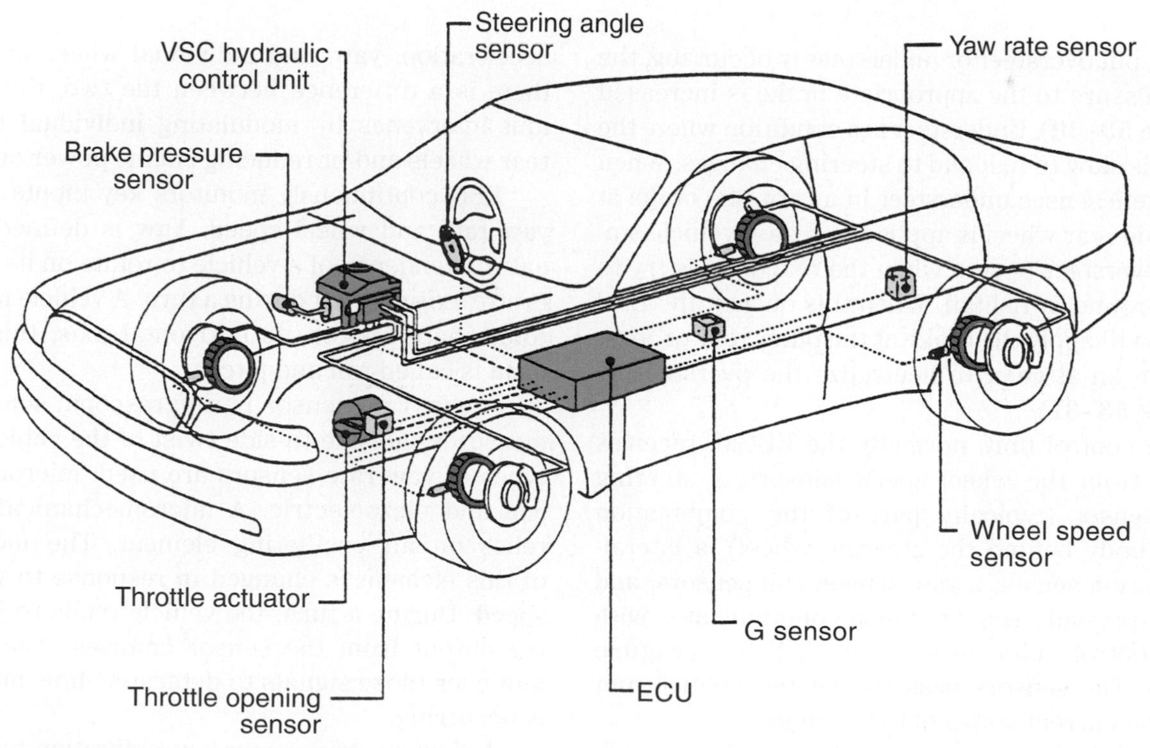

FIGURE 53–28

The components of a typical vehicle stability control system.

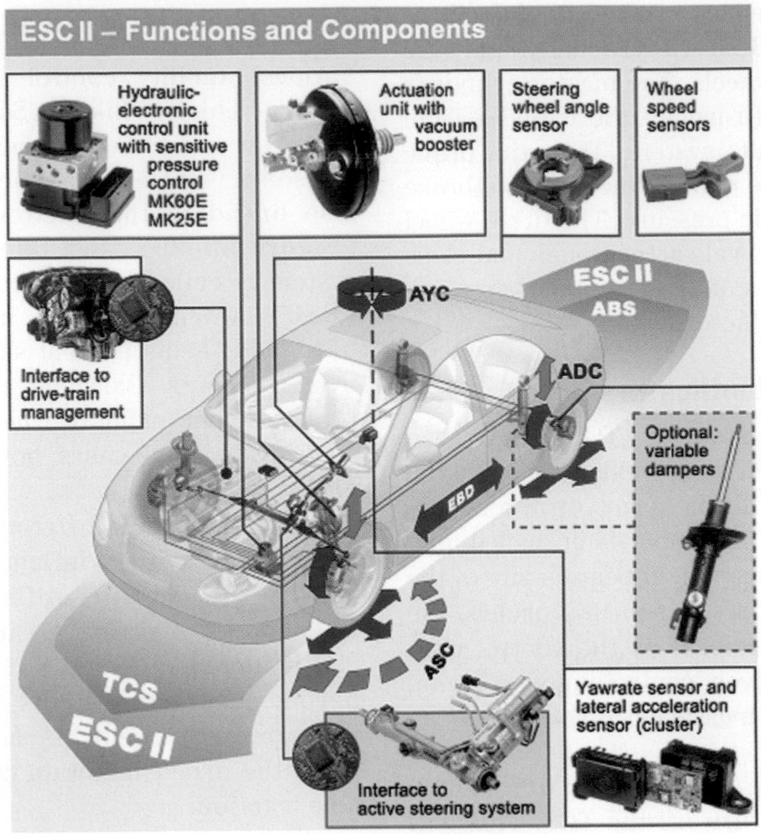

FIGURE 53–29

On some stability control systems, the suspension is also adjusted to provide for safe handling.

Continental Automotive Systems

applied, but oversteer or understeer is occurring, the fluid pressure to the appropriate brake is increased **(Figure 53–30)**. Understeer is a condition where the vehicle is slow to respond to steering changes. When the system senses understeer in a turn, the brake at the inside rear wheel is applied to regain vehicle stability. Oversteer occurs when the rear wheels try to swing around or fishtail. When this occurs, the ESC system will apply the brake at the outer rear or front wheel in an attempt to neutralize the oversteering **(Figure 53–31)**.

The control unit, normally the EBCM, receives signals from the wheel speed sensors, a steering angle sensor (typically part of the combination switch body behind the steering wheel), a lateral-acceleration sensor, a yaw sensor, roll sensors, and a brake-pressure sensor. It also communicates with other control units through the CAN bus **(Figure 53–32)**. The sensors basically let the control unit know the current status of the vehicle.

The ESC control unit compares the driver's intended direction (by monitoring steering angle) to the vehicle's actual direction (by measuring lateral acceleration, yaw, and individual wheel speeds). If there is a difference between the two, the control unit intervenes by modulating individual front or rear wheels and/or reducing engine power output.

ESC continuously monitors key inputs such as yaw rate and wheel speed. **Yaw** is defined as the natural tendency of a vehicle to rotate on its vertical centre axis or twist during a turn. A vehicle may also rotate naturally on its horizontal axis; this movement is called roll and pitch.

A yaw rate sensor is a gyroscopic sensor that measures the side-to-side twist of the vehicle. Two types of yaw rate sensors are used: micromechanical and piezoelectric. A micromechanical sensor relies on an oscillating element. The movement of this element is changed in response to yaw and speed. During a turn, the vehicle tends to yaw and the output from the sensor changes. The control unit uses those signals to determine how much yaw is occurring.

A piezoelectric sensor has a vibration-type gyroscope shaped like a tuning fork **(Figure 53–33)**. The device is divided into two sections: upper and lower.

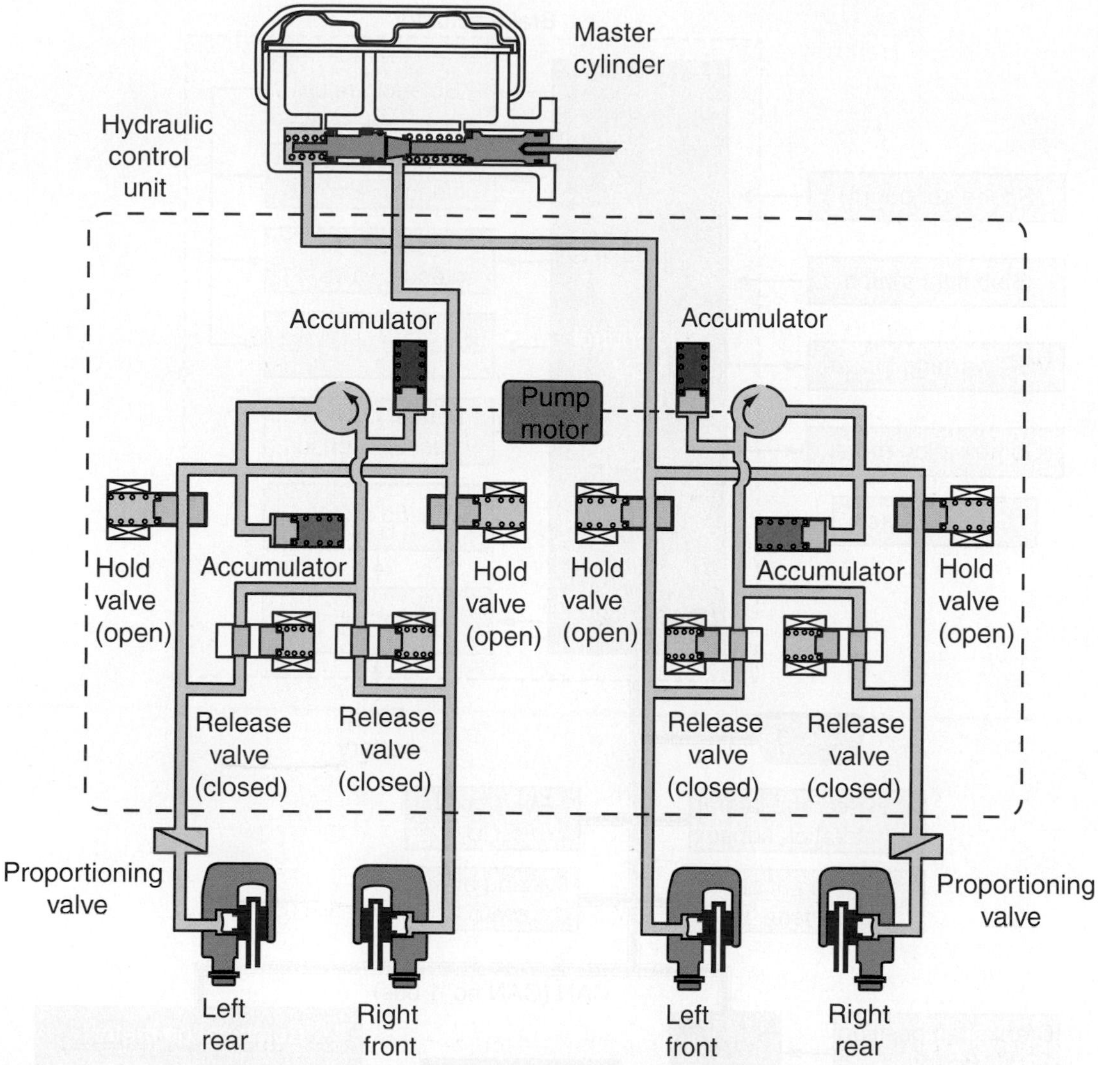

FIGURE 53–30

The hydraulic layout of a typical integrated antilock brake traction control and stability control system.

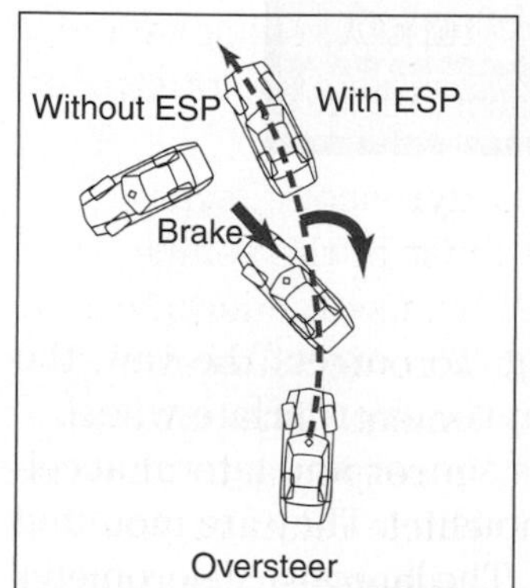

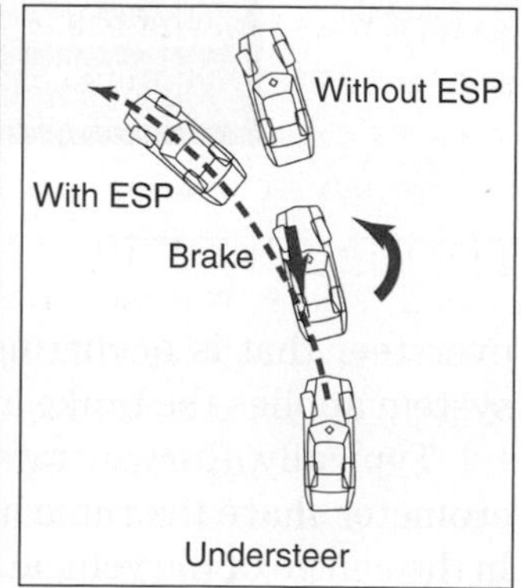

FIGURE 53–31

The effects of oversteer and understeer and how they are affected by electronic stability control (ESC) systems.

DaimlerChrysler Corporation

Both sections have piezoelectric elements attached to them. As current flows through the piezoelectric materials, the sections oscillate from one side to the other. When the vehicle is making a turn, the movement of the vehicle causes the upper elements to move away from the lower elements. This action generates an AC voltage signal that represents the vehicle's speed and yaw rate. In a right turn, the signal voltage increases and decreases when the vehicle is turning right. The output signals range from 0.25 to 4.75 volts. When the vehicle has 0 yaw, the output signal will be 2.5 volts.

The control unit looks at the actual yaw rate and compares it to the calculated desired rate. It responds to the difference between the two. This difference represents the amount of understeer or

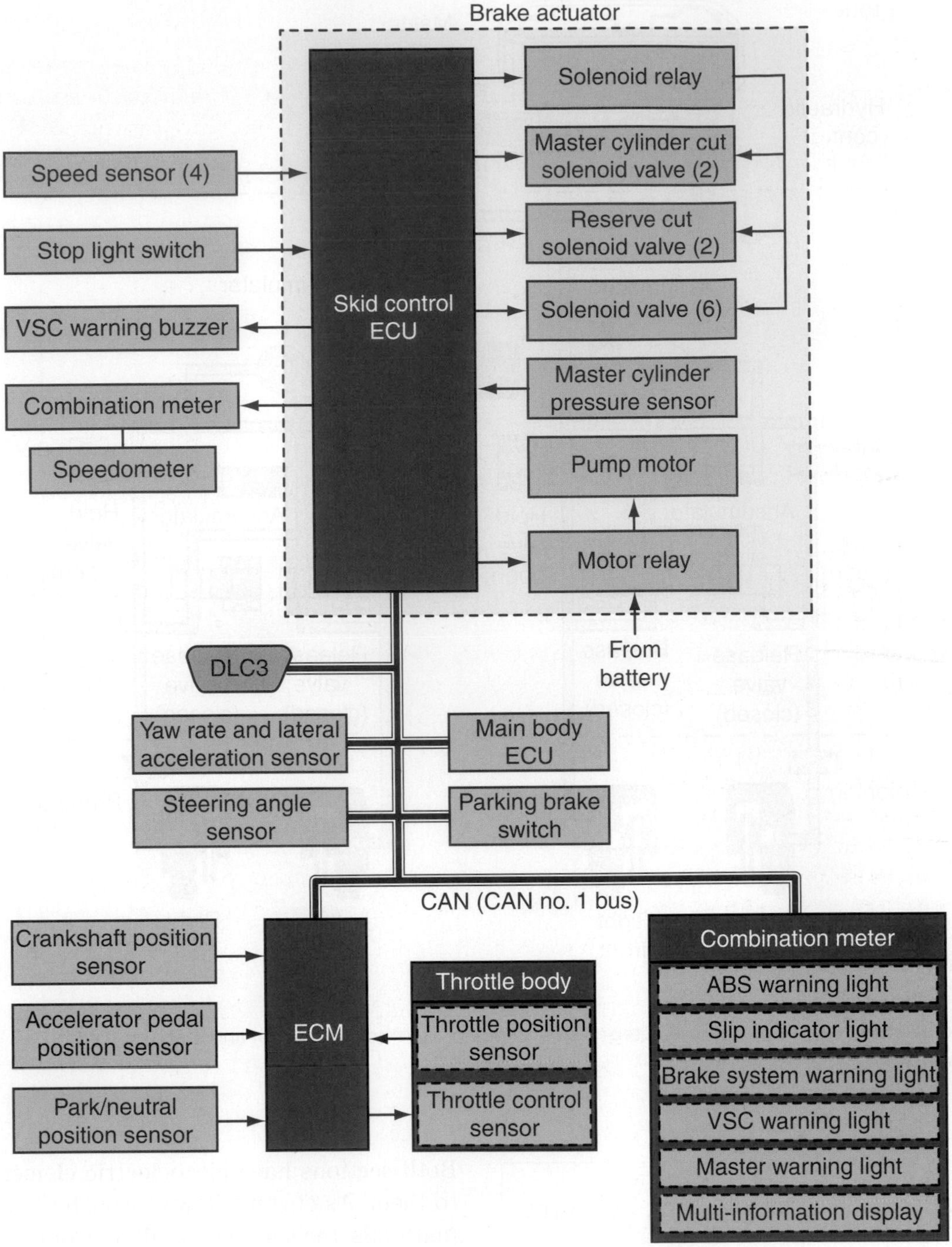

Brake actuator
Skid control ECU
Speed sensor (4)
Stop light switch
VSC warning buzzer
Combination meter
Speedometer
Solenoid relay
Master cylinder cut solenoid valve (2)
Reserve cut solenoid valve (2)
Solenoid valve (6)
Master cylinder pressure sensor
Pump motor
Motor relay
From battery
DLC3
Yaw rate and lateral acceleration sensor
Steering angle sensor
Main body ECU
Parking brake switch
CAN (CAN no. 1 bus)
Crankshaft position sensor
Accelerator pedal position sensor
Park/neutral position sensor
ECM
Throttle body
Throttle position sensor
Throttle control sensor
Combination meter
ABS warning light
Slip indicator light
Brake system warning light
VSC warning light
Master warning light
Multi-information display

FIGURE 53–32

A typical system diagram for an ESC system.

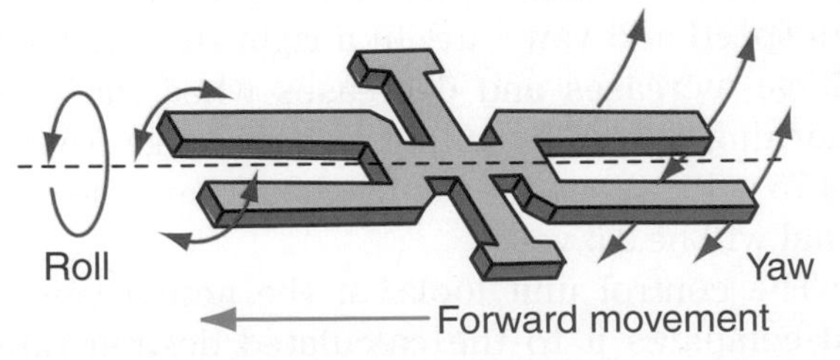

FIGURE 53–33

The basis of a yaw rate sensor.

oversteer that is occurring. To correct the yaw, the system applies the brake at the appropriate wheel.

Typically, the yaw rate sensor and lateral accelerometer share the same housing. They are mounted in the centre of the vehicle. The lateral accelerometer monitors acceleration, deceleration, and cornering forces. These sensors are commonly Hall-effect or piezoelectric units. Semiconductor materials are placed on a plate and are set 45° away from the centreline of the vehicle. The plate is supported by four

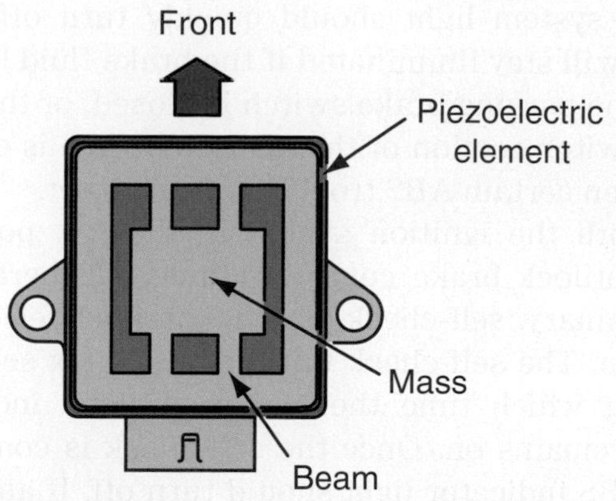

FIGURE 53–34

An acceleration/deceleration sensor. The semiconductor material rests on a plate that is supported by four beams that flex as thrust is applied in any direction.

beams **(Figure 53–34)**. The beams are designed to be able to flex in response to the movement of the vehicle. The amount of flex determines the output signal from the sensor. The signal can range from 0.25 to 4.75 volts, depending on the G-forces the vehicle is experiencing.

G-force is a measurement of a vehicle's acceleration. The term is based on the normal acceleration rate due to gravity (32.174 ft/s^2 or 9.80665 m/s^2). A G-force value of 1 means the acceleration rate of the vehicle is the same as the acceleration rate of gravity. A value of less than 1 means the rate is lower than that of gravity. Values greater than 1 indicate that the forces acting on the vehicle are greater than the normal acceleration rate of gravity. Most lateral accelerometer sensors have an operating range of –1.5 to +1.5 g. Zero lateral acceleration provides a 2.5-volt output signal. The voltage of the signal increases with an increase in lateral acceleration.

Stability Control System Indicators

ESC systems use an indicator light or the message centre on the dash to tell the driver when the system is active (that is, it has detected and corrected yaw rates). A lamp or the message centre will also inform the driver when the system is turned off. If the control unit detects a problem in the system, it may shut down the system and alert the driver of needed service. Also, many ESC systems have an OFF switch so that the driver can disable ESC—for example, when stuck in mud or snow. However, ESC defaults to ON when the ignition is restarted. When the system has been turned off by the driver, a warning lamp or message is displayed.

ANTILOCK BRAKE SYSTEM SERVICE

Most of the service done to the brakes of an antilock brake system are identical to those in a conventional brake system. There are, however, some important differences. Always refer to the recommended procedures in the appropriate service information before attempting to service the brakes on an ABS-equipped vehicle. It may be necessary to depressurize the accumulator to prevent personal injury from high-pressure fluid.

Safety Precautions

- When replacing brake lines and/or hoses, always use lines and hoses designed for and specifically labelled for use on ABS vehicles.
- Do not use silicone brake fluids in ABS vehicles unless otherwise specified (such as Corvette). Use only the brake fluid type recommended by the manufacturer (normally, DOT 3 or DOT 4).
- Never begin to bleed the hydraulic brake system on a vehicle equipped with ABS until you have checked the service information for the proper procedure.
- Never open a bleeder screw or loosen a hydraulic brake line or hose while the ABS is pressurized.
- Disconnect all vehicle computers, including the ABS control unit, before doing any welding on the vehicle.
- Never disconnect or reconnect electrical connectors while the ignition switch is on.
- Never install any antennas, signal boosters, or amplifiers close to the ABS control unit or any other control module or computer.
- Check the wheel speed sensor to sensor ring air gap after any parts of the wheel speed circuit have been replaced.
- Keep the wheel sensor clean. Never cover the sensor with grease unless the manufacturer specifies doing so; in those cases, use only the recommended type.
- When replacing speed (toothed) rings, never beat on them with a hammer. These rings should be pressed onto their flange. Hammering may result in a loss of polarization or magnetization.

Relieving Accumulator Pressure

Some services require that brake tubing or hoses be disconnected. Many ABS use hydraulic pressures as high as 19 300 kPa (2800 psi) and an accumulator to store this pressurized fluid. Before disconnecting any lines or fittings in many systems, the accumulator must be fully depressurized. A common method of depressurizing the ABS follows:

1. Turn the ignition switch to the OFF position.
2. Pump the brake pedal between 25 and 50 times.
3. The pedal should be noticeably harder when the accumulator is discharged.

Procedures for depressurizing the system vary with the design of the system. Always refer to the service information.

DIAGNOSIS AND TESTING

Always follow the vehicle manufacturer's procedures when diagnosing an ABS. In general, ABS diagnostics require three to five different types of testing that must be performed in the specified order listed in the service information. Types of testing may include the following:

1. Prediagnostic inspections and test drive
2. Warning light symptom troubleshooting
3. On-board ABS control module testing (trouble code reading)
4. Individual trouble code or component troubleshooting

CAUTION!

Following the wrong sequence or bypassing steps may lead to unnecessary replacement of parts or incorrect resolution of the symptom. The information and procedures given in this chapter are typical of the various antilock systems on the market. For specific instructions, consult the vehicle's service information.

Prediagnostic Inspection

Before undertaking any actual checks, take just a few minutes to talk with the customer about his or her ABS complaint. The customer is a very good source of information, especially when diagnosing intermittent problems. Make sure you find out what symptoms are present and under what conditions they occur.

All ABS have some sort of self-test. This test is activated each time the ignition switch is turned on. You should begin all diagnosis with this simple test.

WARNING LAMPS Place the ignition switch in the START position while observing both the red brake system light and amber ABS indicator lights. Both lights should turn on. Start the vehicle. The red brake system light should quickly turn off. This lamp will stay illuminated if the brake fluid level is low, the parking brake switch is closed, or the bulb test switch section of the ignition switch is closed, or when certain ABS trouble codes are set.

With the ignition switch in the RUN position, the antilock brake control module will perform a preliminary self-check on the antilock electrical system. The self-check takes three to six seconds, during which time the amber antilock indicator light remains on. Once the self-check is complete, the ABS indicator light should turn off. If any malfunction is detected during this test, the amber lamp will either flash or light continuously to alert the driver of the problem. In some systems, a flashing ABS indicator lamp indicates that the control unit detected a problem but has not suspended ABS operation. However, a flashing ABS indicator lamp is a signal that repairs must be made to the system as soon as possible.

A solid ABS indicator lamp indicates that a problem has been detected that affects the operation of ABS. No antilock braking will be available, but normal, base brake operation will remain. In order to regain ABS braking ability, the ABS system must be serviced.

If both brake indicators stay on and the parking brake is fully released, the front-to-rear braking distribution system may be shut down due to a pressure loss.

If the lamp does not light when the ignition is turned on, the computer probably will not go into its self-test mode. The problem may be as simple as a burned-out bulb, or it may be a problem with the computer itself. To identify the cause for the bulb not illuminating, begin by checking the bulb. If the bulb is good, you will need to make some voltage tests at a diagnostic connector. Follow the testing procedures given by the manufacturer. Nearly all diagnostic connectors have a ground terminal that is used for one or more test modes. Use a voltmeter or ohmmeter to check the continuity between the diagnostic ground terminal and the battery negative terminal. High ground resistance or an open circuit can keep the computer out of the self-test mode and may be a clue to other system problems.

Various other terminals on the diagnostic connector may have other levels of voltage applied to them at different times. Some may have battery (system) voltage under certain conditions, whereas others may have 5 volts, 7 volts, or a variable voltage applied to them.

Visual Inspection

The prediagnosis inspection consists of a quick visual check of system components. Problems can often be spotted during this inspection, which can eliminate the need to conduct other more time-consuming procedures. This inspection should include the following:

1. Check the master cylinder fluid level.
2. Inspect all brake hoses, lines, and fittings for signs of damage, deterioration, and leakage. Inspect the hydraulic modulator unit for any leaks or wiring damage.
3. Inspect the brake components at all four wheels. Make sure that no brake drag exists and that all brakes react normally when they are applied.
4. Inspect for worn or damaged wheel bearings that may allow a wheel to wobble. Many vehicles have the wheel speed sensor (WSS) built into the wheel bearing assembly, and a worn wheel bearing can affect ABS operation.
5. Check the alignment and operation of the outer CV joints.
6. Make sure the tires meet the legal tread-depth requirements and that they are the correct size.
7. Inspect all electrical connections for signs of corrosion, damage, fraying, and disconnection.
8. Inspect the wheel speed sensors and their wiring. Check the air gaps between the sensor and ring, and make sure these gaps are within the specified range. Also check the mounting of the sensors and the condition of the toothed ring and wiring to the sensor.

SHOP TALK

Remember that faulty base brake system components may cause the ABS to shut down. Do not condemn the ABS too quickly.

Test Drive

After the visual inspection is completed, test-drive the vehicle to evaluate the performance of the entire brake system. Begin the test drive with a feel of the brake pedal while the vehicle is sitting still. Then accelerate to a speed of about 30 km/h (20 mph). Bring the vehicle to a stop using normal braking procedures. Look for any signs of swerving or improper operation. Next, accelerate the vehicle to about 40 km/h (25 mph), and apply the brakes with firm and constant pressure. You should feel the pedal pulsate if the antilock brake system is working properly.

During the test drive, both brake warning lights should remain off. If either of the lights turns on, take note of the condition that may have caused it. After you have stopped the vehicle, place the gear selector into PARK or NEUTRAL, and observe the warning lights. They should both be off.

If the control module detects a problem with the system, the amber ABS indicator light will either flash or light continuously to alert the driver of the problem. In some systems, a flashing ABS indicator light indicates that the control unit detected a problem but has not suspended ABS operation. However, a flashing ABS indicator light is a signal that repairs must be made to the system as soon as possible.

A solid ABS indicator light indicates that a problem has been detected that affects the operation of ABS. No antilock braking will be available, but normal, base brake performance will remain. In order to regain ABS braking ability, the ABS must be serviced.

The red brake warning light will be illuminated when the brake fluid level is low, a hydraulic system is leaking, the parking brake switch is closed, the bulb test switch section of the ignition switch is closed, or when certain ABS trouble codes are set.

Intermittent problems can often be identified during the test drive. If the scan tool is equipped with a "snapshot" feature, use it to capture system performance during normal acceleration, stopping, and turning manoeuvres. If this does not reproduce the malfunction, perform an antilock stop on a low coefficient surface such as gravel, from approximately 50 to 80 km/h (30 to 50 mph) while triggering the snapshot mode on the scan tool.

Self-Diagnostics

The control module monitors the electromechanical components of the system. A malfunction of the system will cause the control module to shut off or inhibit the system. However, normal power-assisted braking remains. Malfunctions are indicated by a warning indicator in the instrument cluster. The system is self-monitoring. When the ignition switch is placed in the RUN position, the ABS control module will perform a preliminary self-check on its electrical system indicated by a second illumination of the amber ABS indicator in the instrument cluster. During vehicle operation, the control module monitors all electrical ABS functions and some hydraulic functions during normal and antilock braking. With most malfunctions of the ABS, the amber ABS indicator will be illuminated and a DTC recorded.

The electronic control system of most ABSs includes sophisticated on-board diagnostics that, when accessed with the proper scan tool, can identify the source of a problem within the system. Each of the DTCs represents a specific possible problem in the system **(Figure 53–35)**. The service information contains a detailed step-by-step troubleshooting chart for each DTC. Each system has its own self-diagnostic capabilities. For example, an ABS-VI control module has five separate diagnostic modes. Data available for troubleshooting include wheel speed sensor readings, vehicle speed, battery

DTC	DEFINITION
C1211	ABS indicator signal circuit high
C1214	System relay contact or coil circuit open
C1217	Brake pressure modulator valve (BPMV) pump motor control circuit shorted
C1218	Pump motor voltage
C1221–C1224	Wheel speed sensor input signal is zero
C1225–C1228	Excessive wheel speed sensor variation
C1232–C1235	Wheel speed sensor circuit open or shorted
C1236	Low system voltage
C1237	High system voltage
C1238	Drive wheel brake rotor excessive temperature
C1242	BPMV pump motor shorted to ground
C1243	BPMV pump motor stalled
C1245	Tire inflation monitor
C1246	Brake lining wear circuit open
C1248	Dynamic rear proportioning control system
C1251	Road sensing suspension (RSS) indicated malfunction
C1252	LF normal force malfunction
C1253	RF normal force malfunction
C1254	Checksum error
C1255	Electronic brake traction control motor (EBTCM) internal malfunction (ABS/TCS disabled)
C1256	EBTCM internal malfunction
C1261	LF inlet solenoid valve malfunction
C1262	LF outlet solenoid malfunction
C1263	RF inlet solenoid valve malfunction
C1264	RF outlet solenoid malfunction
C1265	LR inlet solenoid valve malfunction
C1266	LR outlet solenoid malfunction
C1266	RR inlet solenoid valve malfunction
C1267	RR outlet solenoid malfunction

FIGURE 53–35

Examples of some of the DTCs that can be set by an ABS, ATC, or ESC system.

DTC	DEFINITION
C1271	LF TCS master cylinder isolation valve malfunction
C1272	LF TCS prime valve malfunction
C1273	RF TCS master cylinder isolation valve malfunction
C1274	RF TCS prime valve malfunction
C1276	Delivered torque signal circuit malfunction
C1277	Requested torque signal circuit malfunction
C1278	TCS temporarily inhibited by PCM
C1281	Steering sensor uncorrelated malfunction
C1282	Yaw rate sensor bias circuit malfunction
C1283	Excessive time to centre steering
C1284	Lateral accelerometer self-test malfunction
C1285	Lateral accelerometer circuit malfunction
C1286	Steering sensor bias malfunction
C1287	Steering sensor rate malfunction
C1288	Steering sensor circuit malfunction
C1291	Open brake switch contacts during deceleration
C1293	DTC C1291 set in previous ignition cycle
C1294	Brake light switch circuit always active
C1295	Brake light switch circuit open
C1298	Class 2 serial data link malfunction
P1571	Requested torque signal circuit malfunction
P1644	Delivered torque signal voltage invalid
P1689	Delivered torque signal voltage invalid

FIGURE 53–35 ***(continued)***

voltage, individual motor and solenoid command status, warning light status, and brake switch status. Numerous trouble codes are programmed into the control module to help pinpoint problems. Other diagnostic modes store past trouble codes. These data can help technicians determine if an earlier fault code, such as an intermittent wheel speed sensor, is linked to the present problem, such as a completely failed wheel sensor. Another mode enables testing of individual system components.

Testers and Scanning Tools

Different vehicle manufacturers provide ABS test and scan tools with varying capabilities. Some testers are used simply to access the digital trouble codes. Others may also provide functional test modes for checking wheel sensor circuits, pump operation, solenoid testing, and so forth.

On some vehicles, the amber ABS and red brake lights flash out the digital trouble codes. As you can see, it is important to research the capabilities and proper use of the test equipment the vehicle manufacturer provides. Misuse of test equipment can be dangerous. For example, connecting test equipment during a test drive that is not designed for this use may lead to loss of braking ability.

Once all system malfunctions have been corrected, clear the ABS's DTCs. Codes cannot be erased until all codes have been retrieved, all faults have been corrected, and the vehicle has been driven above a set speed (usually 30 to 40 km/h [18 to 25 mph]). It may be necessary to disconnect a fuse for several seconds to clear the codes on some systems. After service work is performed on the ABS, repeat the previous test procedure to confirm that all codes have been erased.

Testing Components with ABS Scan Tools

ABS scan tools and testers can often be used to monitor and/or trigger input and output signals in the ABS. This allows you to confirm the presence of a suspected problem with an input sensor, switch, or output solenoid in the system. You can also check that the repair has been successful before driving the vehicle. Manual control of components and automated functional tests are also available when using many diagnostic testers. Details of typical functional tests follow.

Solenoid Leak Test

To test for solenoid leaks, follow the procedure set forth below.

PROCEDURE

Checking for Solenoid Leaks in the Hydraulic Modulator

STEP 1 Disconnect the inspection connector from the connector cover, and connect the inspection connector to the tester.

STEP 2 Remove the modulator reservoir filter, and then fill the reservoir to the MAX level with fresh fluid.

STEP 3 Bleed the high-pressure fluid from the maintenance bleeder connection. Often, this procedure requires the use of a special tool.

STEP 4 Start the engine and release the parking brake.

STEP 5 Set the tester to the proper mode, and press the start-test button.

STEP 6 While the ABS pump is running, place your finger over the top of the solenoid return tube in the modulator reservoir. If you can feel brake fluid coming from the return tube, one of the solenoids is leaking. Go to step 7. If you cannot feel brake fluid coming from the return tube, the solenoids are okay. Reinstall the modular reservoir filter and refill the reservoir to the MAX level.

STEP 7 Bleed the high-pressure fluid from the maintenance bleeder, and then run through steps 3 and 6 with the tester. Repeat this procedure three or four times.

STEP 8 Repeat steps 5 and 6. If the solenoid leakage has stopped, reinstall the modulator reservoir filter and refill the reservoir to the MAX level. If one of the solenoids is leaking, the entire hydraulic modulator may require replacement. In some cases, it is possible to remove, inspect, and replace individual solenoids.

WARNING!

Certain components of the ABS are not intended to be serviced individually. Do not attempt to remove or disconnect these components. Only those components with approved removal and installation procedures in the manufacturer's service information should be serviced.

Testing Components with a Lab Scope

Like most electrical/electronic systems, antilock brake traction control and stability control system components can be tested with a lab scope. A lab scope offers one distinct advantage over many other testing tools: You can watch the component's activity over time. For example, all antilock-based systems rely on the cycling of solenoid valves. When looking at a solenoid on the lab scope, you should see a change in the waveform as soon as ABS operation begins. When a wheel tries to slip, the ABS control module should begin pulsing the solenoid to that wheel.

A critical input to the antilock brake, traction control, and stability control systems is from the wheel sensors. These too can be monitored on a lab scope **(Figure 53–36)**. As the wheel begins to spin, the waveform of the sensor's output should begin to oscillate above and below 0 volts. The oscillations should get taller as speed increases. If the wheel's speed is kept constant, the waveform should also stay constant.

Component Replacement

A typical antilock braking system consists of a conventional hydraulic brake system (the base system) plus a number of antilock components. The base brake system consists of a vacuum power booster, master cylinder, front disc brakes, rear drum or disc brakes, interconnecting hydraulic tubing and hoses, a low fluid sensor, and a red brake system warning light.

Antilock components are added to this base system to provide antilocking braking ability. Most ABSs use the same operational principles, but the major components may be configured and/or named differently.

The electrical components of the ABS are generally very stable. Common electrical system failures

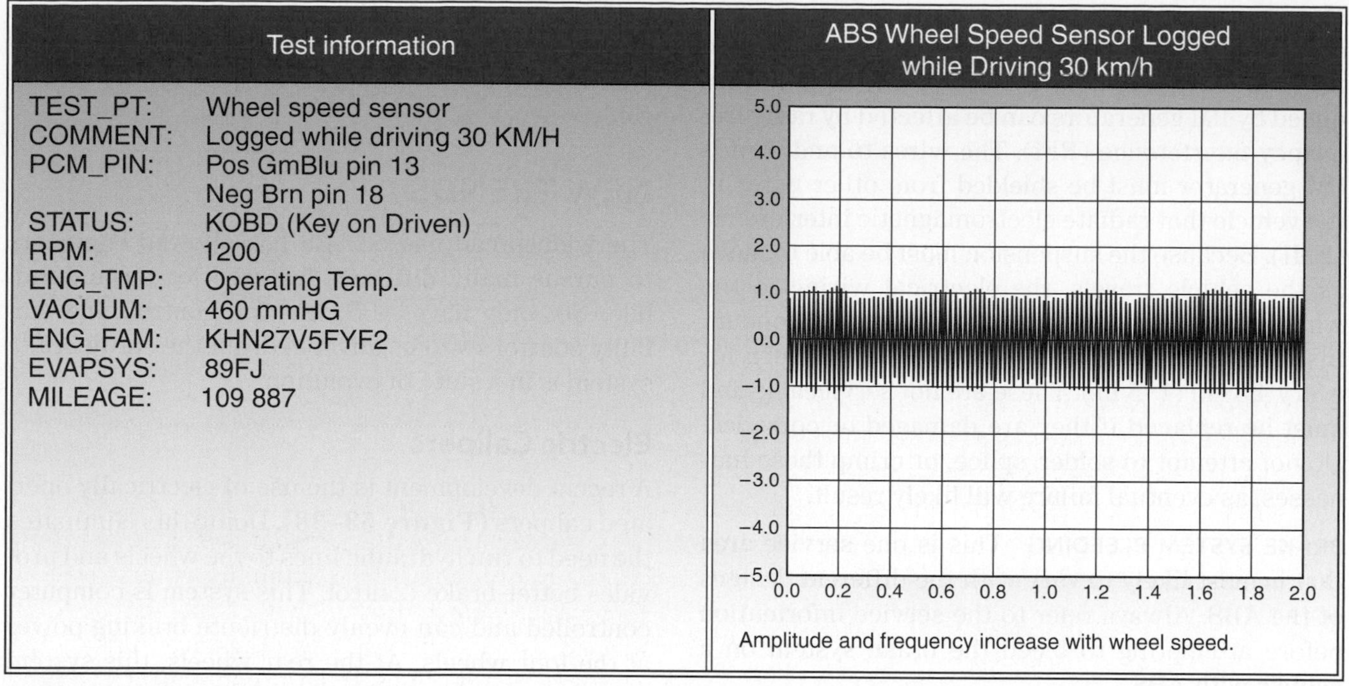

FIGURE 53–36

The waveform from a wheel speed sensor.

are usually caused by poor or broken connections. Other common faults can be caused by a malfunction of the wheel speed sensors, pump and motor assembly, or the hydraulic module assembly.

Many of the components of ABSs are simply remove-and-replace items. On some systems, wheel speed sensors must be adjusted. Normal brake repairs—such as replacing brake pads, caliper replacement, rotor machining or replacement, brake hose replacement, master cylinder or power booster replacement, or parking brake repair—can all be performed as usual. In other words, brake service on an ABS-equipped vehicle is similar to service on a conventional system with a few exceptions.

Always refer to the service information for the correct adjustment and replacement procedures for any and all brake parts in an ABS.

WHEEL SPEED SENSOR SERVICE Visually inspect the wheel speed sensor pulsers for chipped or damaged teeth. Use a non-steel feeler gauge (plastic or brass) to measure the air gap between the sensor and rotor **(Figure 53–37)** all the way around while rotating the driveshaft, wheel, or rear hub bearing unit by hand. If there is a specification on this gap, make certain it is within the required specification (typically 0.4 to 1.0 mm, or 0.02 to 0.04 in.). If the gap exceeds specifications, the problem is likely a distorted knuckle that should be replaced.

Sensors are replaced by simply disconnecting the wiring at the sensor and unbolting the fasteners. Be careful not to twist the wiring cables or harness when installing the sensors.

Many vehicles have a jumper harness made of highly flexible twisted-pair wiring between each wheel speed sensor and the main wiring harness.

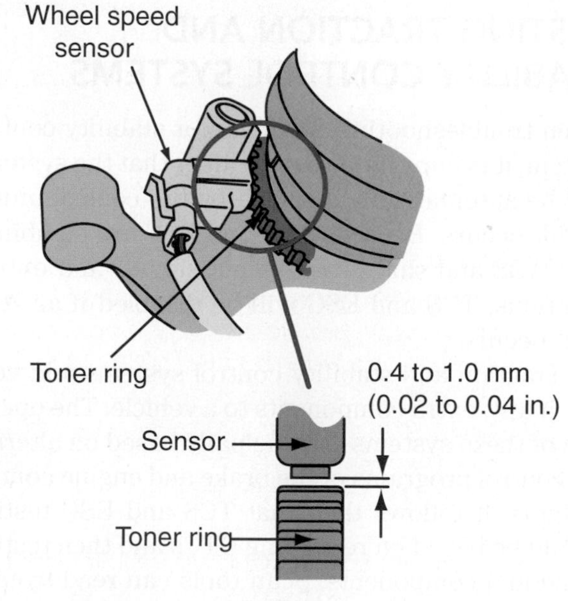

FIGURE 53–37

Wheel sensor gap measurement.

Many wheel speed sensors are PM generators that have AC voltage outputs. Radio signals are also AC-modulated signals; therefore, the signal produced by PM generators can be affected by radio frequency interference (RFI). The wires to and from a PM generator must be shielded from other wires in the vehicle that radiate electromagnetic interference (EMI). Because the suspension must be able to move as the vehicle travels, the electrical wiring to the wheel speed sensors cannot be shielded in conduit. Instead, the wires are twisted at least one turn for every 4.5 cm (1.75 in.). These are not serviceable and must be replaced if they are damaged or corroded. Do not attempt to solder, splice, or crimp these harnesses, as eventual failure will likely result.

BRAKE SYSTEM BLEEDING This is one service area that is most likely to vary with the different designs of the ABS. Always refer to the service information before attempting to bleed the brake system on a vehicle with ABS.

Some systems require that the accumulator be depressurized and the power source to the ABS control module disconnected. After these have been done, the system can be bled like a conventional brake system. On other systems, the bleeder screws are opened while the system is turned on, and they have one procedure for the front brakes and another for the rear brakes.

Failure to follow the correct procedure can result in personal injury, destruction of ABS components, or the trapping of more air in the system.

TESTING TRACTION AND STABILITY CONTROL SYSTEMS

When troubleshooting a traction or stability control system, it is important to remember that the systems will be automatically disabled by the control unit if a fault occurs. Because TCS and ESC are combined with ABS and share many input signals and output functions, TCS and ESC will be disabled if an ABS fault occurs.

Traction and stability control systems add very few, if any, extra components to a vehicle. The operation of these systems is principally based on altering the control programs of the brake and engine control systems. It follows then that TCS and ESC testing should be based on retrieving DTCs and then testing individual components. Scan tools can read trouble codes and operating system data. The scan tool will also identify any communications errors that may be present. Proper communications is important for both traction and stability control systems. As with other electronic tests, you should refer to the manufacturer's troubleshooting procedures for the vehicle you are servicing.

NEW TRENDS

The widespread use of ABS has allowed engineers to pursue many different features for an automobile. Not only have ABS, traction control, and stability control systems advanced, but the entire brake system is in a state of evolution.

Electric Calipers

A recent development is the use of electrically operated calipers **(Figure 53–38)**. Doing this eliminates the need to run hydraulic lines to the wheels and provides better brake control. This system is computer controlled and can evenly distribute braking power at the four wheels. At the rear wheels, this system can also be used as the parking brake, thereby eliminating the need to have cables at the rear wheels.

Brake-by-Wire

In a brake-by-wire system, the braking command of the brake pedal is electronically processed. A control module receives sensor input from the brake pedal and activates the master cylinder. For safety purposes, the normal linkage from the pedal to the master cylinder may still exist. The advantage of brake-by-wire is simply that the pressure on the pedal can be immediately and directly relayed to the master cylinder and wheel brakes. Brake-by-wire is currently used with hybrid electric vehicles (HEVs)

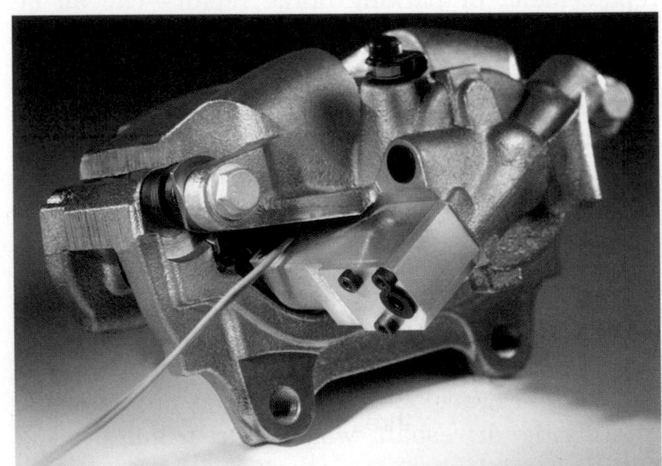

FIGURE 53–38

This caliper is fitted with an electronically controlled solenoid that controls the application of the parking brake.

Robert Bosch Corp.

so that braking energy is captured and used to recharge the battery. In a true brake-by-wire system, the hydraulic system has been totally eliminated; these systems are currently being analyzed.

Electronic Wedge Brake (EWB)

Much development is being done with electronic wedge brakes **(Figure 53–39)**. This system is totally electrically operated and electronically controlled. There is no hydraulic system. This development opens the door for brake-by-wire systems. One of the obstacles to brake-by-wire systems has been providing enough clamping power on the rotor to safely stop the vehicle. The electronic wedge brake uses metal wedges with a series of interlocking triangular teeth between the caliper and the disc.

The wedges are designed to react to the speed of the wheels; as speed increases, so does the clamping pressure of the brakes. When the pad is pushed against the rotor, the momentum of the rotating disc draws the pads farther up the interlocking series of wedges, applying greater braking pressure and increasing braking efficiency. This allows for the use of a low-power 12-volt electric motor to initially push the brake pad toward the rotor. The motor turns roller screws that press the wedge and pads against the rotor.

The system relies on the vehicle's battery for power but also has a built-in backup battery. This is important because one of the fears of brake-by-wire systems is the loss of power. Because the system uses a low-power motor, a small backup battery can power the brakes if needed. Also, as a safety feature, each wheel brake in the system is independent of the others. This means that if one brake fails, there are three others that can be used to stop the vehicle. The system can also quickly retract the pads when the brake pedal is released; this reduces brake drag and allows for ABS cycling to occur up to six times faster.

It is claimed that EWB systems greatly shorten stopping distances, require much less energy than hydraulic brakes, allow for much better control in ATC and ESC systems, are light, and are more reliable and durable than hydraulic brake systems.

Several suppliers, such as Siemens VDO, Continental AG, Robert Bosch LLC, Delphi Corporation, and TRW Automotive, also are working on electromechanical brake programs.

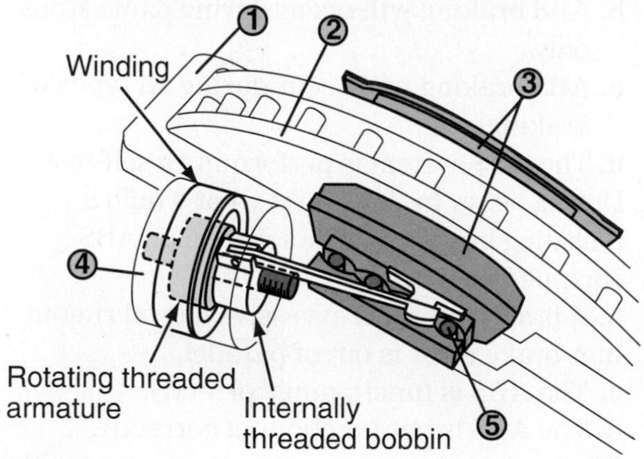

Winding
1
2
3
4
5
Rotating threaded armature
Internally threaded bobbin

FIGURE 53–39

The layout of an electronic wedge brake. The brake caliper (1) spans the brake disc and (2) on two sides. The brake disc is braked by a pad (3) that is moved by an electric motor, (4) via several rollers, and (5) along wedge-shaped faces.

Wet Brakes

Some vehicles are equipped with rain-sensing windshield wipers. These systems have been modified by a few manufacturers (notably BMW) to include removal of water from the brake rotors. When the windshield rain sensor sends a signal to the ABS, the brakes are lightly applied to wipe water off the discs and pads.

Hold Feature

Some vehicles with manual transmissions have a brake hold feature. The purpose of this system is to apply the brakes while the vehicle is stopped on a hill. This prevents the vehicle from rolling. To enable the system, a driver presses a button while the brakes are applied. This locks the brakes until the throttle pedal is depressed.

KEY TERMS

Antilock brake systems (p. 1628)
Automatic traction control (ATC) (p. 1645)
Electronic stability control (ESC) (p. 1647)
Four-channel system (p. 1637)
Integral antilock brake systems (p. 1635)
Isolation/dump valves (p. 1634)
Nonintegral antilock brake systems (p. 1635)
One-channel systems (p. 1636)
Pressure modulation (p. 1628)
Three-channel system (p. 1637)
Two-channel system (p. 1636)
Yaw (p. 1648)

SUMMARY

- Brake fluid is the lifeblood of any hydraulic brake system. DOT 3 is recommended for most antilock brake systems and some power brake systems.
- Modern antilock brake systems provide electronic/hydraulic pumping of the brakes for stopping under panic conditions.
- In addition to being classified as integral and nonintegral, antilock brake systems can be divided into the level of control they provide. They can be one-, two-, three-, or four-channel, two- or four-wheel systems.
- Pressure modulation works to prevent wheel locking. Antilock brake systems can modulate the pressure to the brakes as often as 20 times per second.
- Integrated antilock brake systems combine the master cylinder, hydraulic booster, and hydraulic circuitry into a single assembly. The great majority of antilock brake systems are of this type.
- On a nonintegrated ABS, the master cylinder and hydraulic valve unit are separate assemblies, and a vacuum boost is used. In some nonintegrated systems, the master cylinder supplies brake fluid to the hydraulic unit.
- Automatic traction control (ATC) is a system that applies the brakes when a drive wheel attempts to spin and lose traction.
- Automatic stability systems correct oversteer and understeer by applying one wheel brake.
- Malfunction of the ABS causes the electronic control module to shut off or inhibit the system. However, normal power-assisted braking remains. Malfunctions are indicated by one or two warning lights inside the vehicle.
- Loss of hydraulic fluid or power booster pressure disables the antilock brake system.

REVIEW QUESTIONS

1. Describe the ways antilock brake systems can be classified.
2. What is the primary difference between an automatic traction control system and a stability control system?
3. Explain the role of the wheel sensor(s) in the ABS.
4. Define the difference between an integrated and nonintegrated antilock braking system.
5. Briefly describe the proper steps and testing needed to accurately diagnose antilock braking systems.
6. In a one-channel antilock brake system, what wheel(s) are affected?
   a. the right front only
   b. both front wheels
   c. the right rear only
   d. both rear wheels
7. What does a magnetic pulse type of wheel speed sensor produce?
   a. varying AC frequency
   b. varying DC frequency
   c. digital AC frequency
   d. digital DC frequency
8. What occurs at the wheel cylinder or caliper when the inlet wheel circuit valve is open and the outlet valve is closed?
   a. Brake pressure can increase.
   b. Brake pressure is held.
   c. Brake pressure is released.
   d. The fluid accumulator is increasing pressure.
9. How often can some antilock brake systems modulate pressure to the brakes?
   a. 10 times per second
   b. 20 times per second
   c. 30 times per second
   d. 40 times per second
10. What occurs when the amber ABS indicator light is constantly illuminated?
    a. No ABS operating will occur during braking.
    b. ABS braking will occur during panic stops only.
    c. ABS braking will occur during all types of braking.
    d. The ABS system is performing a self-test.
11. During heavy braking only, what would a pulsating brake pedal indicate on an ABS-equipped vehicle?
    a. A brake rotor has excessive lateral runout.
    b. A brake rotor is out of parallel.
    c. The ABS is functioning correctly.
    d. The ABS is not functioning correctly.
12. What is an accumulator usually charged with?
    a. oxygen **b.** nitrogen
    c. hydrogen **d.** argon

13. What fluid is recommended for most antilock brake systems?
    a. power-steering fluid
    b. DOT 3 brake fluid
    c. DOT 4 brake fluid
    d. DOT 5 brake fluid
14. How are the channels of a three-channel antilock brake system arranged?
    a. right front wheel, left front wheel, and both rear wheels together
    b. right rear wheel, left rear wheel, and both front wheels together
    c. right front and left rear, left front and right rear
    d. both front wheels together and both rear wheels together
15. What will occur at the brakes when both wheel circuit valves are closed?
    a. A pressure increase will occur.
    b. A pressure decrease will occur.
    c. A pressure hold will occur.
    d. The accumulator will boost pressure.
16. What is the name for the gas-filled pressure chamber that is part of the antilock braking system's pump and motor assembly?
    a. control module
    b. accumulator
    c. sensor
    d. reservoir
17. What occurs when the automatic traction control system is operating?
    a. The brakes will be modulated on the non-spinning drive wheel.
    b. The brakes will be modulated on the spinning drive wheel.
    c. The brakes will be modulated on both drive wheels.
    d. The brakes will be modulated on all wheels equally.
18. Where would the speed sensors be located in a four-channel antilock brake system?
    a. at each non-drive wheel and on the final drive ring gear of the drive axle
    b. at each drive wheel only
    c. at each non-drive wheel only
    d. at all wheels
19. What must be performed before removing a high-pressure accumulator-equipped master cylinder on an ABS-equipped vehicle?
    a. With the key off, open all bleeder screws to relieve the pressure.
    b. With the key on, open all bleeder screws to relieve the pressure.
    c. Pump the brake pedal 25 to 50 times with the key off.
    d. Pump the brake pedal 25 to 50 times with the key on.
20. What tool should be used to measure a wheel-speed sensor gap?
    a. a dial indicator
    b. a vernier caliper
    c. a steel feeler gauge
    d. a brass feeler gauge

CHAPTER 54
Heating and Air Conditioning

LEARNING OUTCOMES

- Identify the purpose of a ventilation system.
- Identify the common parts of a heating system.
- Compare the vacuum and mechanical controls of a heating system.
- Diagnose temperature control problems in the heater/ventilation system.
- Remove, inspect, and reinstall the heater control valve(s) and heater core.
- Describe how an automotive air-conditioning system operates.
- Describe how refrigerants are used in the air-conditioning system.
- Locate, identify, and describe the function of the various air-conditioning components.
- Describe the operation of the types of air-conditioning control systems.

The first automobiles had few features that provided for driver and passenger comfort. In winter, heavy coats and blankets were all that allowed passengers to survive behind a woefully inadequate windscreen. In summer, the breeze generated by 25 km/h (15 mph) travel was the only thing that cooled the passengers.

Today, ventilation, heating, and air-conditioning (A/C) systems are very important elements for providing passenger comfort. Ventilation and heating systems are standard equipment on all passenger vehicles, and air conditioning is standard on most and available for nearly all. These systems move heat to warm or cool the passengers.

GO TO ▶ Chapter 7 for a detailed discussion of heat and heat transfer.

The large number of vehicles with air conditioning, plus recent changes in the methods used to cool a vehicle and to service the systems, make a basic knowledge of air-conditioning systems a must for automotive technicians.

VENTILATION SYSTEM

The ventilation system on most vehicles is designed to supply outside air to the passenger compartment through upper or lower vents or both. Several systems are used to vent air into the passenger compartment, the most common of which is the flow-through system **(Figure 54–1)**. In this arrangement, a supply of outside air, called ram air, flows into the car when it is moving. When the car is not moving, a steady flow of outside air can be produced by the heater fan. In operation, ram air is forced through an inlet grille. The pressurized air then circulates throughout the passenger and trunk compartment. From there, the air is forced outside the vehicle through an exhaust area that is usually located in the rear door sill or in the trunk area.

On certain older vehicles, air is admitted by opening or closing two vent knobs under the dashboard. The left knob controls air through the left inlet. The right knob controls air through the right inlet. The air is still considered ram air and is circulated through the passenger compartment.

Rather than using ram air (especially if the vehicle is stopped), a ventilation fan can be used. It is accessible either from under the dashboard or from inside the engine compartment. A blower motor and fan assembly is placed inside the blower housing. As the fan, often called a squirrel cage, rotates, it produces a strong suction on the intake. A pressure is also created on the output. When the fan motor is energized by using the temperature controls on the dashboard, air is moved through the passenger compartment.

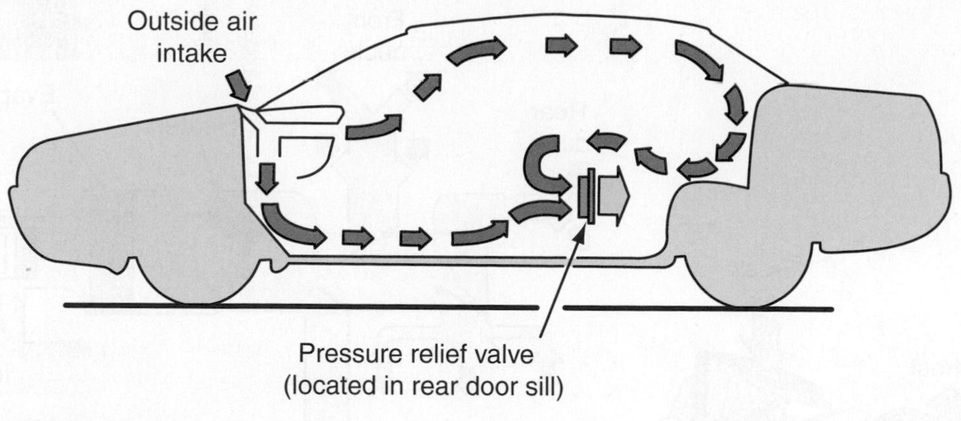

FIGURE 54–1

A flow-through ventilation system.

AUTOMOTIVE HEATING SYSTEMS

The heating system's primary job is to provide a comfortable passenger compartment temperature and to keep car windows clear of fog or frost. The heating system works together with the engine's cooling system to maintain proper temperatures inside the car.

The primary components of the automotive heating system are the heater core, heater control valve, blower motor and fan, and heater and defroster ducts **(Figure 54–2)**.

The engine's cooling system is the source of heat for the passenger compartment heating system. Hot coolant from the engine is transferred by a heater hose to the heater control valve and then to the heater core inlet **(Figure 54–3)**. As the coolant circulates through the core, heat is transferred from the coolant to the tubes and fins of the core. Air is drawn into the heater housing and blown through the core by the blower motor and fan. The air then picks up the heat from the surfaces of the core and transfers it into the passenger compartment of the car. After releasing its heat, the coolant then flows out through the heater core outlet, where it is returned to the engine to be recirculated by the water pump.

Heater Core

The **heater core** is generally designed and constructed much like a miniature radiator **(Figure 54–4)**. It features an inlet and outlet tube and a tube and fin core to facilitate coolant flow between them.

Although all heater cores basically function in the same manner, several variations in design and materials are used by different automakers to achieve the same results. Most have an aluminum core with plastic tanks.

Heater Control Valve

The **heater control valve** (sometimes called the water flow valve) controls the flow of coolant into the heater core from the engine. In a closed position, the valve allows no flow of hot coolant to the heater core, keeping it cool. In an open position, the valve allows heated coolant to circulate through the heater core, maximizing heater efficiency. Heater control valves are operated in several ways: by cable, thermostat, or vacuum, or electronically.

Cable-operated valves are controlled directly from the heater control lever on the dashboard **(Figure 54–5)**. Thermostatically controlled valves feature a liquid-filled capillary tube located in the discharge airstream off the heater core. This tube senses air temperature, and the valve modulates the flow of coolant to maintain a constant temperature, regardless of engine speed or temperature.

Many vehicles use electrically controlled heater control valves. The valve may use a stepper motor to block the flow of coolant or use the motor to move a lever and block the flow, as in non-electronic valves.

Most vehicles equipped with air conditioning have heater valves that are vacuum operated. These valves are normally located in the heater hose line or mounted directly in the engine block. When a vacuum signal reaches the valve, a diaphragm inside the valve is raised, either opening or closing the valve against an opposing spring. When the temperature selection on the dashboard is changed,

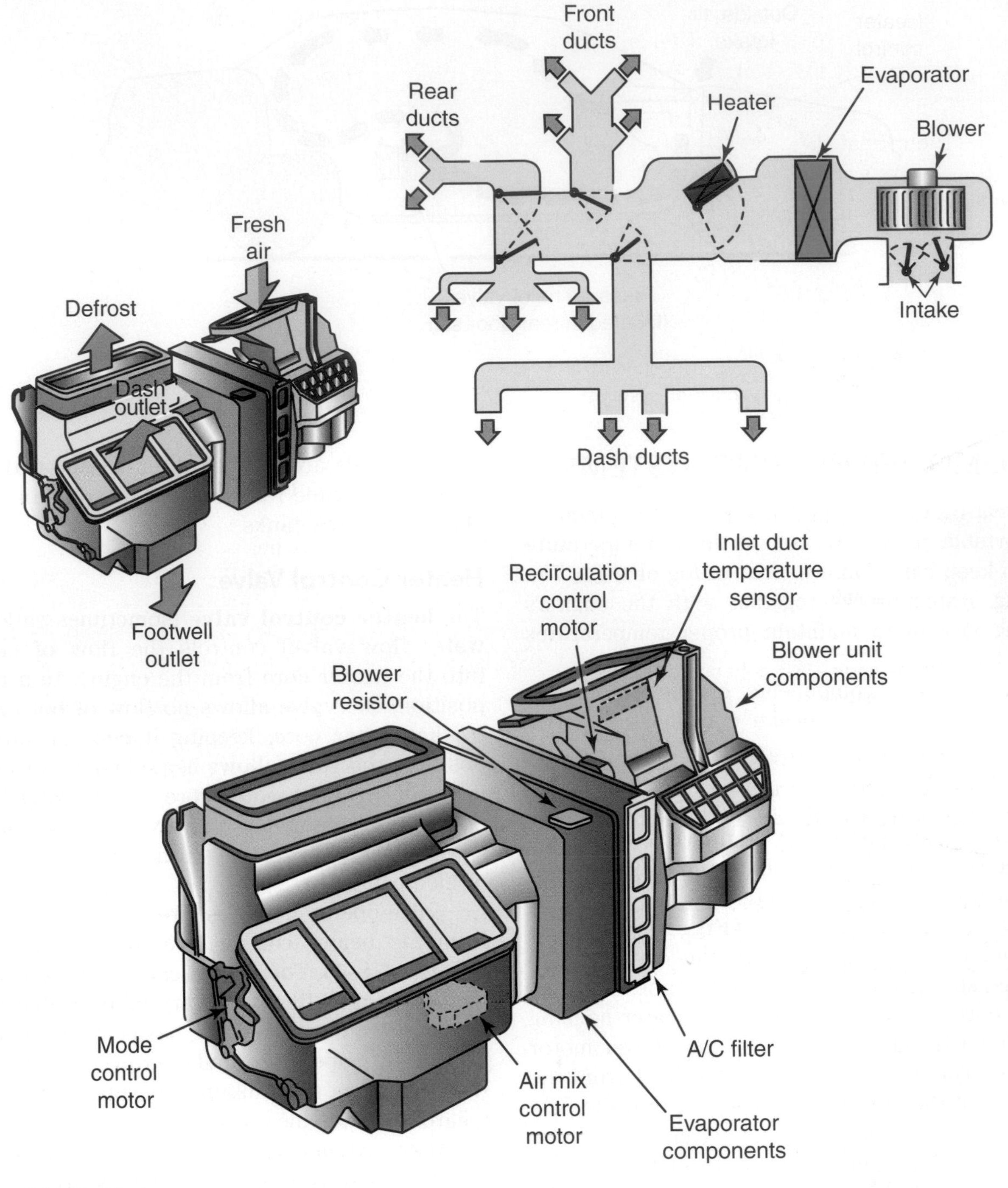

FIGURE 54–2

A typical HVAC airflow and components for heating and air-conditioning systems.

vacuum to the valve is vented, and the valve returns to its original position. Vacuum-actuated heater control valves are either normally open or normally closed designs.

On late-model vehicles, heater control valves are typically made of plastic for corrosion resistance and light weight. These valves feature few internal working parts and no external working parts. With the reduced weight of these valves, external mounting brackets are not required.

The engine thermostat, which helps regulate the coolant temperature in the cooling system, plays a large part in the heating system. A malfunctioning thermostat can cause the engine to overheat or not reach normal operating temperature, or it can be the cause of poor heater performance.

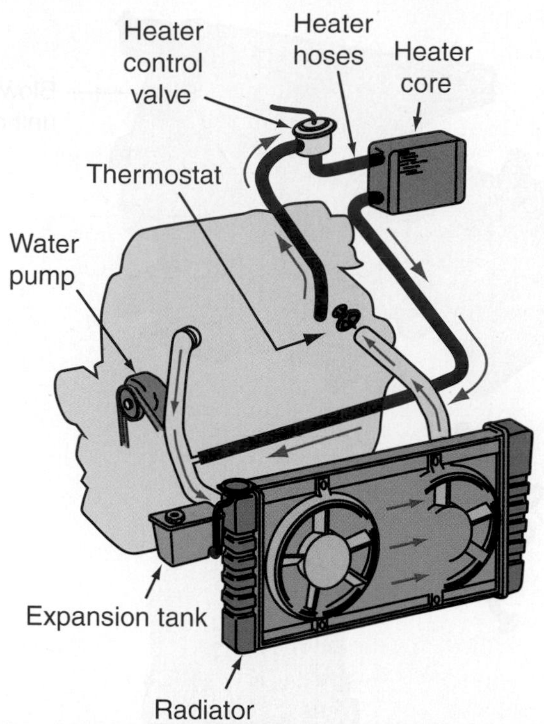

FIGURE 54–3

Hot coolant from the engine is sent from the upper portion of the engine to the heater core.

SYSTEMS WITHOUT A CONTROL VALVE Some systems do not have a heater control valve; rather, heat inside the passenger compartment is controlled by changing the airflow over the heater core. In these systems, hot coolant always flows through the heater core. When heat is needed, air is allowed to flow through the heater core. The amount of heat is determined by the amount of air allowed to pass through the core.

FIGURE 54–4

A heater core.

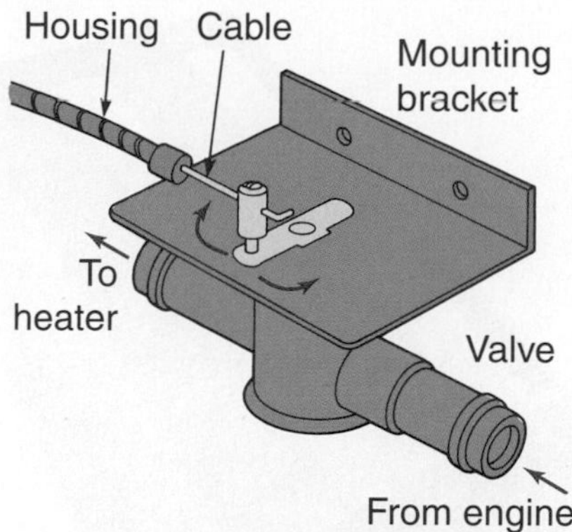

FIGURE 54–5

A cable-operated heater valve used to control the flow of water or coolant through the heater core.

Airflow is controlled by a **blend door** inside the blower housing. The blend door directs airflow across either the heater core or air-conditioning evaporator. This allows the driver to change the temperature of the air from the system. The movement of this door can be controlled by a cable or vacuum, or electrically with a motor.

PTC HEATERS Today's engines are designed to reduce the amount of waste heat they emit; therefore, the amount of heat available for the heating system may be limited. This is especially true with hybrid vehicles and not all-electric vehicles with the stop–start feature. Also, in conventional vehicles, interior heat is not available until the coolant warms up and transfers heat through the heater core.

Most models have an auxiliary positive temperature coefficient (PTC) unit **(Figure 54–6)**. These heaters are able to provide warm air before the engine's coolant warms up. PTC heaters are electronically controlled and work independently of the cooling system.

When current passes through a PTC element, heat is produced, and the air that passes through the heater core is warmed. Some vehicles have an additional PTC element installed in an air duct from the blower housing; this helps to increase the air temperature in the ducts. PTC elements react quickly to changes in current. The elements are small ceramic stones. These are typically barium titanate polycrystalline ceramics doped with metals. Most PTC elements are self-regulating, which means there is no need for a thermostat to control its operation. As the element heats, its resistance increases. Once a predetermined temperature is reached, the internal resistance increases

FIGURE 54–6

An electronic PTC cabin heater.

BorgWarner

significantly, reducing current flow and preventing further increase in temperature.

Blower Motor

The blower motor is usually located in the heater housing assembly. It ensures that air is circulated through the system **(Figure 54–7)**. Its speed is controlled by a multiposition switch in the control panel that works in conjunction with a resistor block that is usually mounted on the heater housing, with its resistor elements exposed to the airflow inside the heater housing to help keep them cool.

On some vehicles, when the engine is running, the blower motor is in constant operation at low speed. On automatic temperature control (ATC) systems, the blower motor is activated only when the engine reaches a predetermined minimum temperature. The blower motor circuit is protected by a fuse located in the fuse panel. The fuse rating is usually 20 to 30 amps.

The blower motor resistor block is used to control the blower motor speed. The typical resistor block is composed of three or four wire resistors in series with the blower motor, which control its voltage and current. The speed of the motor is determined by the control panel switch, which puts the resistors in series. Increasing the resistance in the system slows the blower speed.

Vehicles with automatic climate control typically use pulse-width modulation to control blower speed.

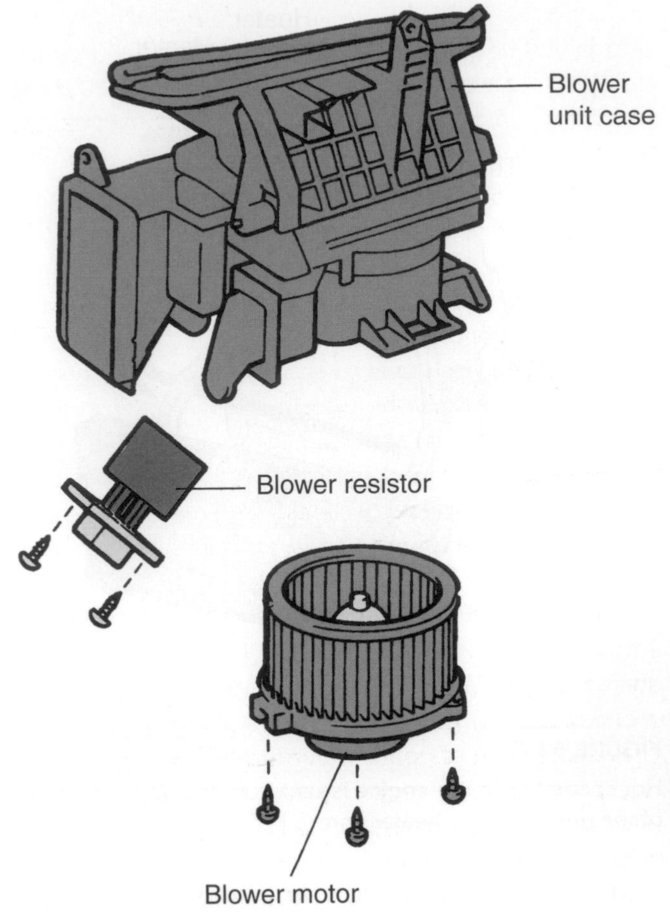

FIGURE 54–7

A blower motor assembly.

Reprinted with permission.

Instead of a fixed resistor block and fixed blower speeds, a control module pulses the blower motor power circuit on and off. By varying the on-time of the motor, fan speed is increased or decreased as needed.

Some late-model vehicles, including the Toyota Prius and some Mercedes-Benz models, use solar cells to power the blower motor to circulate air through the passenger compartment when the vehicle is off. The sunload and ambient temperature sensors provide input to the climate control module, which can vary blower speeds as needed. Circulating air through the cabin helps reduce the temperature and the load on the A/C system once the vehicle is restarted.

Heater and Defroster Duct Hoses

Transferring heated air from the heater core to the passenger compartment heater and defroster outlets is the job of the heater and defroster ducts. The ducts are typically part of a large plastic or steel shell that connects to the necessary inside and outside vents.

This ductwork also has mounting points for the evaporator and heater core assemblies. Contained inside the duct are also the doors required to direct air to the floor, dash, and/or windshield. Sometimes the duct is connected directly to the vents; other times, hoses are used.

RECIRC/FRESH MODE Many vehicles have a recirc/fresh mode. This feature allows the driver to recirculate (recirc) the air inside the vehicle. This mode is normally selected when outside air that is entering the vehicle has an unpleasant odour. Outside or fresh air normally enters the interior through the normal heating, ventilation, and A/C systems. While operating in the recirc mode, the system uses the air from the interior for those systems. A recirculation door closes off the outside airflow and opens the intake for interior air.

Some late-model vehicles use sensors to detect smog and pollution present in the incoming air. On some models, the sensor also is able to detect carbon monoxide, hydrocarbons, and oxides of nitrogen. Upon detection of contaminates, airflow is automatically switched to recirculation mode. The recirculated air passes through a dust and pollen filter and charcoal filter to remove any odours.

Heaters and A/C systems in Hybrid Vehicles

In a hybrid electric vehicle, the engine can be used to supply the heat, so heating and defrosting systems are similar to those used in conventional vehicles. However, some hybrids have additional electrical heaters. These keep the passenger compartment warm when the engine is off.

Other systems have a 12- or 42-volt auxiliary electric water pump. This pump is used to circulate coolant through the engine to maintain heater performance during an idle stop. A heater control module operates the pump based on an idle stop signal from the powertrain control module (PCM). If the engine's temperature drops below a predetermined level during an idle stop, the engine will restart. The system also has a temperature sensor at the heater core. This sensor monitors the activity of the electric water pump. If it senses that the heater core is losing heat quickly, the control module will assume that the pump has failed and will start the engine and set a diagnostic trouble code (DTC).

The cooling systems used in some hybrids feature coolant storage tanks. Hot coolant is stored in a container **(Figure 54–8)** where it can remain hot for quite a long period. The hot coolant is circulated through the engine immediately after start-up. The fluid also may circulate through the engine many hours after it is shut off. The stored coolant can provide heat for the passenger compartment and allows the engine to warm up quickly, thereby reducing emission levels during start-up.

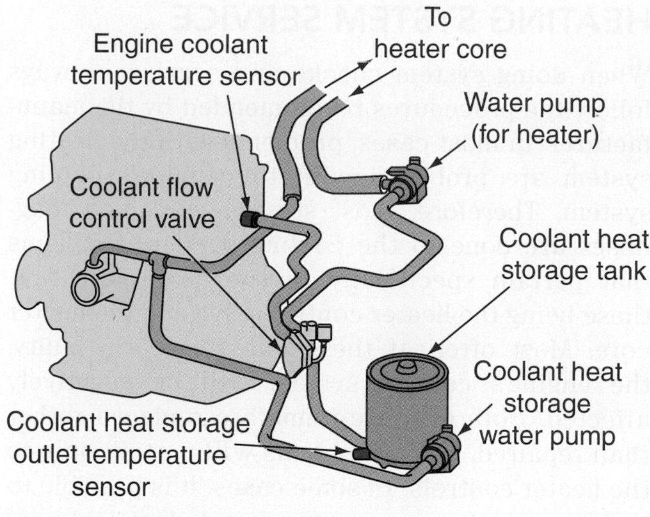

FIGURE 54–8

The coolant heat storage tank is a large, vacuum-insulated container that is capable of storing hot coolant for a long period.

The heating and A/C system is also used on some hybrid vehicles to maintain high-voltage (HV) battery temperature. Because temperature is important for battery efficiency, a network of heating and cooling lines may be attached to the battery box. Even if separate liquid heating and A/C lines are not used, there is a ventilation system for the HV battery. In many hybrids, such as the Prius, Honda IMA hybrids, and others, a battery ventilation fan and ductwork system is used to draw cabin air into and through the battery box. Air is then discharged outside the vehicle. It is important that the inlet for the battery cooling system not be blocked or restricted, as this can increase battery temperature and set battery temperature DTCs.

Vehicles that use Li-Ion batteries, such as the Volt, have dedicated heating and A/C lines for the battery box. In addition to circulating coolant through the inverter and drive motor assemblies, the Volt uses a separate A/C evaporator, called a battery chiller, to remove heat from the battery charger and pack. Depending on operating conditions, the liquid cooling system can be used to either warm the battery pack or remove heat. As greater heat dissipation is required, the A/C compressor is turned on, and the A/C system cools the battery.

HEATING SYSTEM SERVICE

When doing system checks and services, always follow the procedures recommended by the manufacturer. In most cases, problems with the heating system are problems with the engine's cooling system. Therefore, most service work and diagnoses are done to the cooling system. Problems that pertain specifically to the heater are few, these being the heater control valve and the heater core. Most often, if these two items are faulty, the engine's cooling system will be negatively affected. Both of these items are replaced rather than repaired. Some problems will pertain only to the heater controls. In some cases, it is possible to make repairs to vacuum hose and electrical connections without removing the heater assembly. If it is necessary to remove the heater assembly, the cooling system must be drained before removing the heater core.

GO TO ▶ Chapter 14 for details on diagnosing and servicing engine cooling systems.

Basic Heater Inspection and Checks

When there is a problem of insufficient heat, begin your diagnosis with a visual inspection and a check of the coolant level and condition. If the level is correct, turn the heater controls on, and run the engine until it reaches normal operating temperature. If the engine does not reach operating temperature, typically around 105°C (220°F), suspect a faulty thermostat or fan clutch. Then measure the temperature of the upper radiator hose. The temperature can be measured with a pyrometer or thermocouple and a DMM. If these are not available, carefully touch the hose. You should not be able to hold the hose long because of the heat. While doing this, make sure you stay clear of the area around the cooling fan. A spinning fan can chop off your hand. If the temperature of the hose is not within specifications, suspect a faulty thermostat.

If the hose is the correct temperature, check the temperature of the two heater hoses. They should both be hot. If only one of the hoses is hot, suspect the problem to be the heater control valve or a plugged heater core.

Leaking heater cores are often easy to diagnose because of the presence of one or more of three symptoms: coolant leaking onto the floorboards, steam coming from the vents, or a sweet smell inside the passenger compartment. Any of these conditions should prompt a complete inspection of the cooling system.

Most maintenance schedules recommend the coolant be changed every two years. This is an important maintenance item because additives that prevent corrosion will deplete over time, causing damage to aluminum and steel components. Sediment may also build up in low-flow areas, creating blockages.

Heater Core Service

Like the radiator, heater core tanks, tubes, and fins can become clogged over time by rust, scale, and mineral deposits circulated by the coolant. Debris carried through the system can actually act as a sandblaster and eat away at the core from the inside, causing leaks. Feel the heater inlet and outlet hoses while the engine is idling and warm, with the heater temperature control on hot. If the hose downstream of the heater valve does not feel hot, the valve is not opening.

If the heater core appears to be plugged, the inlet hose may feel hot up to the core, but the outlet hose remains cool. Reverse-flushing the core with a power flusher may open up the blockage, but usually the core has to be removed for cleaning or replacement. Air pockets in the heater core can also interfere with proper coolant circulation. Air pockets form when the coolant level is low or when the cooling system is not properly filled after draining.

When the heater core leaks and must be repaired or replaced, it is often a very difficult and time-consuming job, primarily because of the core's location deep within the bulkhead of the car. For this reason, always leak-test a replacement heater core before installation. Also flush the cooling system and replace the coolant.

PTC HEATERS PTC (positive temperature coefficient) heaters can be checked by measuring the resistance of the heating element. If the resistance does not meet specifications **(Figure 54–9)**, the assembly should be replaced.

Heater Control Valve Service

When there is a problem with the control valve, it is typically caused by the controls or the valve itself. When the valve is bad, it should be replaced. If the heater control valve or its controls prevent the valve from opening all the way, heating will be reduced in the passenger compartment. If the valve does not completely close, the compartment will not cool properly when the A/C is turned on.

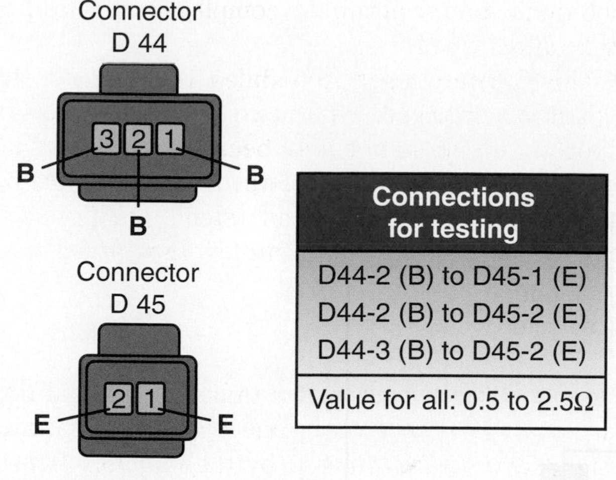

FIGURE 54–9

Resistance testing points at the terminals on a PTC heater.

With cable-operated control valves, check the cable for sticking, slipping (loose mounting bracket), or misadjustment. Vacuum-operated control valves should be checked for complete travel (fully open to fully closed) with a vacuum pump. The vacuum motor should hold a constant vacuum and not bleed off. When the control panel is in the appropriate position, vacuum should be present at the control valve. Most control valves use vacuum to close off the flow of coolant to the heater core, although some use vacuum to allow coolant flow.

THERMOSTAT SERVICE The thermostat, which helps regulate the coolant temperature in the cooling system, plays a large part in the heating system. A malfunctioning thermostat can cause the engine to overheat or not reach normal operating temperature, or it can be the cause of poor heater performance.

Late-model vehicles can set a DTC if the thermostat is not allowing the engine to reach normal operating temperature. A thermostat that is stuck open or is opening too soon can often be diagnosed by watching the flow of coolant in the radiator during engine warm-up or by monitoring coolant temperature. Most engines have thermostats that begin to open around 80°C (176°F) and are fully open at 90°C (194°F). If the coolant temperature remains low, not even reaching 90°C (194°F), suspect a defective thermostat.

Replacing a thermostat varies greatly from engine to engine, so refer to the manufacturer's service information for details. Make sure the new thermostat is installed properly, with the wax pellet oriented toward the engine. Also make sure the thermostat seats and seals properly in its housing before tightening the housing bolts to specs. Once the thermostat has been replaced, refill the cooling system, using the specified coolant, and make sure all air is bled from system.

BLOWER MOTOR SERVICE If the blower motor does not operate, use a digital multimeter (DMM) to make sure there is voltage on both sides of the fuse. Then check to see if there is voltage and ground at the motor. Some systems supply constant power and vary the ground, whereas others ground the motor and vary the voltage **(Figure 54–10)**. If the system uses a PWM motor, connect a scan tool, and check for DTCs while you try to control the blower with the scan tool. You may need to connect a meter or scope to the fan wiring to determine if the fan control circuit is operating.

For non-PWM blower motors, if the blower motor is getting voltage and ground, the problem is a burned-out blower motor. If voltage is present but ground is not, ground the motor to check if it operates. If the motor works, the problem is likely an open resistor or switch. In situations where no voltage is available at the motor, backtrack to check for an open resistor, switch, or relay. Check for proper relay operation and for burned or corroded connections at the blower relays or in the bulkhead connectors. Carefully inspect all wiring and connectors in the fan circuit. A faulty connection can overheat, melting the plastic and damaging the connector and terminals.

As the blower motor ages and/or debris accumulates in the blower housing, blower motor speed can be reduced. This can cause the motor to increase its current draw. Excessive current draw by the motor can then cause burned resistors and damage connections. Many technicians measure blower motor current draw whenever replacing a blower resistor, and compare the reading to specs or against similar vehicles. If the motor is drawing more current than specified, it should be replaced.

Heater and Defroster Duct Service

If the blower motor runs but no air comes out of the ducts, the problem is either a stuck or inoperative airflow control valve or blend door, or a plugged cabin filter. This can also affect the operation of the defrosters. These doors may be cable, electrically, or vacuum operated. To further diagnose the problem, change the position of the temperature selector knob, sliding it from hot to cold. If you do not hear the sound of doors opening and closing, it means the control cables have slipped loose from the dash switch or door arm, which can sometimes occur,

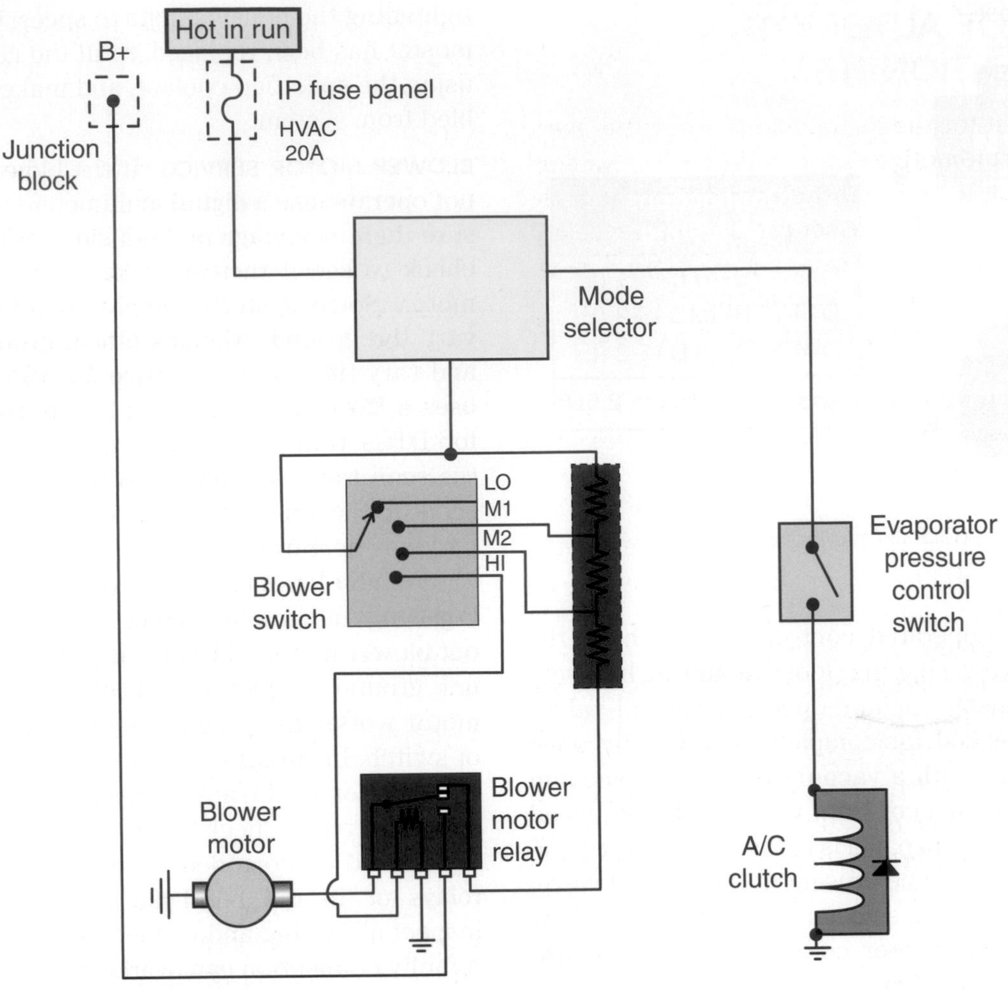

FIGURE 54–10

An example of a power-controlled blower motor circuit.

rendering the door inoperative. If this is the case, you will feel little or no resistance when sliding the temperature control knob. A kinked or rusted cable can also prevent a door from working. If this is the case, you will feel resistance when trying to move the control knob. In either case, it is necessary to get under the dash, find the cable, and then replace, reroute, or reconnect it. Doors can also be jammed by objects that have fallen down the defroster ducts. Remove the obstruction from the plenum by fishing through the heater outlet with a coat hanger or magnet, or remove the plenum. Be careful when fishing for the obstruction; some heater cores are easily pierced.

Most current HVAC systems use electric servomotors to move the various doors. Each motor is fitted with a potentiometer that sends a signal to the A/C control unit regarding the actual position of the door. If the actual position is not the same as the commanded position, most systems will set a DTC, indicating that the door is not working properly. Before proceeding with detailed diagnostics of the electrical system, make sure there are no linkage problems or that the door is not physically stuck. Normally, if the problem is electrical, the A/C control unit, servomotor, or wiring harness is faulty.

With vacuum-controlled doors, the most common causes of failure are leaky or loose vacuum hoses or defective diaphragms in the vacuum motors that move the doors. Check the vacuum by starting the engine and disconnecting the hose that goes to one of the door's vacuum motor. If you feel a vacuum or hear a hissing sound when trying different temperature settings, the vacuum source is good. Apply vacuum to the motor with a handheld pump; if it moves and holds vacuum for about one minute and then bleeds off, the problem is a bad vacuum motor. If it does not, check for leaky vacuum hose connections, a defective temperature control switch, or a leaky vacuum reservoir or check valve under the dash or in the engine compartment.

THEORY OF AUTOMOTIVE AIR CONDITIONING

It is important for the technician to know the basic theory of automotive air-conditioning systems. Understanding how a system moves heat from the confined space of the passenger compartment and dissipates it into the atmosphere **(Figure 54–11)** will assist the technician in analyzing system failures and in performing the required maintenance and service.

All air-conditioning systems are based on four fundamental laws of nature, as discussed below.

Heat Flow

An air-conditioning system is designed to pump heat from one point to another. All materials or substances above absolute zero (–273°C [–459°F]) have heat in them. Below this temperature, called absolute zero, no heat energy remains. Heat always flows from a warmer object to a colder one. For example, if one object is at 0°C (32°F) and another object is 27°C (80°F), heat flows from the warmer object (27°C [80°F]) to the colder one (0°C [32°F]). The greater the temperature difference between the objects, the greater the amount of heat flow.

Heat Absorption

Objects can be in one of three forms: solid, liquid, or gas. When objects change from one state to another, large amounts of heat can be transferred. For example, when water temperature goes below 0°C (32°F), water changes from a liquid to a solid (ice). If the temperature of water is raised to 100°C (212°F), the liquid turns into a gas (steam). But an interesting thing occurs when water, or any matter, changes from a solid to a liquid and then from a liquid to a gas. Additional heat is necessary to change the state of the substance, even though this heat does not register on a thermometer. For example, ice at 0°C (32°F) requires heat to change into water, which will also be at 0°C (32°F). Additional heat raises the temperature of the water until it reaches the boiling point of 100°C (212°F). More heat is required to change water into steam. But if the temperature of the steam were measured, it would also be 100°C (212°F). The amount of heat necessary to change the state of a substance is called **latent heat**—or hidden heat—because it cannot be measured with a thermometer. This hidden heat is the basic principle behind all air-conditioning systems.

Heat from sun and outside air
Condenser
Evaporator
Heat
Heat
Compressor
Road heat

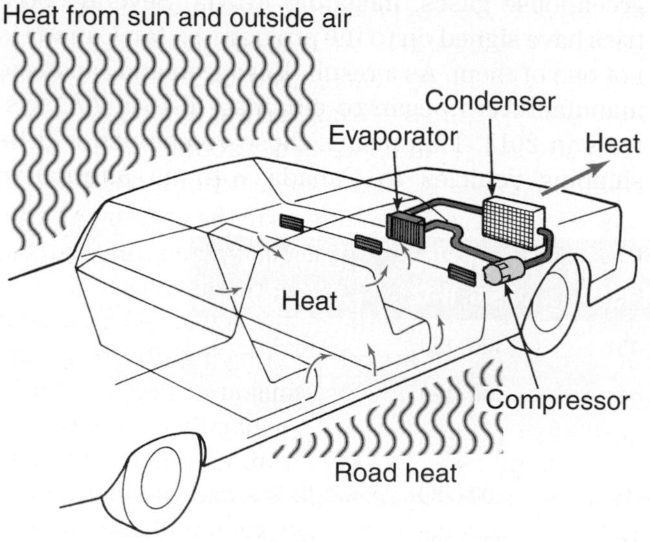

FIGURE 54–11

Heat flow from inside the car to outside.

Everco Industries, Inc.

GO TO ▶ Chapter 7 for a discussion of heat and how it is measured.

Pressure and Boiling Points

Pressure also plays an important part in air conditioning. Pressure on a substance, such as a liquid, changes its boiling point. The greater the pressure on a liquid, the higher the boiling point. If pressure is placed on a vapour, the vapour condenses at a higher than normal temperature. In addition, as the pressure on a substance is reduced, the boiling point can also be reduced. For example, the boiling point of water is 100°C (212°F). The boiling point can be increased by increasing the pressure on the fluid. This is because as the pressure increases, the energy required for the water molecules to expand and turn to a gas also increases. It can also be decreased by reducing the pressure or placing the fluid in a vacuum. By reducing the pressure, the water molecules can change state at a lower temperature. If the pressure is reduced enough, water can boil at room temperature.

Relative Humidity

The amount of moisture that air can hold is directly related to the temperature of the air. The warmer the air, the more moisture it can hold. Therefore, lowering the temperature of the air extracts the moisture from the air and lowers the air's relative humidity. What we call relative humidity is the amount of moisture in the air compared to the amount the air can actually hold at that temperature. If the air cannot hold any more moisture, it condenses back into water. This is what causes dew and condensation on surfaces. For many people, low relative humidity is typically more comfortable than high humidity.

REFRIGERANTS

Types of Refrigerant

The substance used to remove heat from the inside of an air-conditioned vehicle is called the **refrigerant**. On older automotive air-conditioning systems, the refrigerant was Refrigerant-12 (commonly referred to as R-12 and Freon). R-12 is dichlorodifluoromethane (CCl_2F_2). An amendment to the Canadian Environmental Protection Act (CEPA) Ozone-depleting Substances Regulation and international agreements mandated the phase-out of R-12. This came as a result of research that found the earth's ozone layer was being deteriorated by the chemicals found in R-12. The ozone layer is the earth's outermost shield of protection. This delicate layer protects against harmful effects of the sun's ultraviolet rays. The thinning of the ozone layer has become a worldwide concern. The ozone depletion is caused in part by release of **chlorofluorocarbons (CFCs)** into the atmosphere. R-12 is in the chemical family of CFCs. Because air-conditioning systems with R-12 are susceptible to leaks, further damage to the ozone layer could be avoided by not using R-12 in air-conditioning units.

R-134a

Of the many chemicals that could be used in place of R-12, the automobile manufacturers decided to use R-134a, which contains no chlorine. This refrigerant may also be referred to as SUVA. R-134a is tetrafluoroethane (CH_3CF_3) and considered a **hydrofluorocarbon (HFC)** that causes less damage to the ozone layer when released into the atmosphere.

Although R-134a air conditioners operate in the same way and with the same basic components as R-12 systems, the two refrigerants are not interchangeable. R-134a systems operate at higher pressures and are designed to handle these pressures **(Figure 54–12)**. R-134a systems also require different service techniques and equipment. All R-134a systems are identified by an underhood decal **(Figure 54–13)** and by the hoses and fittings used in the system.

To keep systems operating that were originally equipped with R-12, there are retrofit kits available for a changeover to R-134a. Although R-134a is less likely to have an adverse effect on the ozone layer, it still has the capability of contributing to the greenhouse effect when released into the atmosphere. Therefore, the recovery and recycling of R-12 and R-134a is mandatory by law.

Legislation has drastically affected the life of an air-conditioning technician. In 1987, many countries worldwide agreed to the Montreal Protocol agreement. This agreement was responsible for the phase-out of R-12 refrigerant. The production and importation phase-out for all CFCs, including R-12, was scheduled for January 1, 1996. Laws have also been passed that dictate the certification of equipment and technicians. The laws also define how older vehicles can be updated, through a conversion to R-134a or the use of an alternative refrigerant.

The Kyoto Protocol, a worldwide environmental program, has set standards for the reduced use of greenhouse gases, including R-134a. Several countries have signed on to the program; so far, Canada is not one of them. As a result, European and Japanese manufacturers began to eliminate R-134a A/C systems in 2013. This means those companies will be shipping vehicles to Canada with an alternative

AMBIENT TEMPERATURE	HIGH-SIDE PSIG, R-134a	LOW-SIDE PSIG, R-134a	HIGH-SIDE PSIG, R-12	LOW-SIDE PSIG, R-12
15.6°C (60°F)	120–170	7–15	120–150	5–15
21.1°C (70°F)	150–250	8–16	140–180	8–16
26.7°C (80°F)	190–280	10–20	160–250	10–18
32.2°C (90°F)	220–330	15–25	200–280	12–25
37.8°C (100°F)	250–350	20–30	220–300	15–30

FIGURE 54–12
A chart comparing the pressures of R-134a and R-12 at various temperatures.

FIGURE 54–13

A typical decal informing the technician that the system uses R-134a.

refrigerant, and Canadian service facilities will need to be equipped to service R-134a systems and its replacement, HFO-1234yf. Full adoption of 1234yf is proposed to occur by 2017.

HFO-1234yf

Currently, the replacement for R-134a will be HFO-1234yf. HFO means that the refrigerant is a hydrofluoroolefin-based chemical. The refrigerant 1234yf differs from R-134a in that it has a global warming potential (GWP) of 4 compared to that of 1430 for R-134a. This provides a major reduction in the impact of the refrigerant on global warming if accidently released into the atmosphere. Another benefit to changing to 1234yf is that it can be used in existing mobile air-conditioning systems with very little change to the system, as operating pressures and temperatures are very similar to R-134a.

One issue with 1234yf that has caused some debate and possible delays in its adoption as the next refrigerant has arisen from Daimler. In 2012, Daimler announced it would not use 1234yf after crash tests of its vehicles equipped with 1234yf revealed flammability issues. A potential for the refrigerant to ignite exists when the refrigerant is released in a hot engine compartment. Because of this concern, other refrigerants are being considered by Daimler and Volkswagen.

Alternative Refrigerants

There are different refrigerants that can be used in automotive A/C systems; these are typically called alternative refrigerants. The U.S. Environmental Protection Agency (EPA) has a list of approved alternative refrigerants, which have been accepted by Transport Canada and Environment Canada; however, the OEMs say that only R-134a should be used when retrofitting an R-12 system.

Many brand-name hydrocarbon (HC) refrigerant blends are sold in Canada. The main components of these products are butane and propane. Because they have no ozone-depleting potential, they are not regulated under most provincial regulations; therefore, there are no restrictions on their use in Canada, nor requirements that the users be certified.

The Canadian Vehicle Manufacturers' Association (CVMA) and the Association of International Automobile Manufacturers of Canada (AIAMC) do not support or encourage the use of HC blends due to their safety concerns as refrigerants in vehicle air-conditioning systems.

Because their flammability risk has not been assessed, the EPA considers the use of these refrigerants unacceptable in vehicle air-conditioning systems. The use of HC refrigerants are banned in a number of states.

The safety concerns have not been adequately addressed with regard to the manufacturing, use, servicing, and disposal of hydrocarbon refrigerants.

Finding an alternative to R-134a is the focus of much research. The replacement cannot contribute significantly to global warming or to ozone layer depletion. There are always a number of new refrigerants being tested; some of these may eventually be found in automobiles. A rating scale has been developed to specify a refrigerant's global warming potential (GWP). R-134a has a GWP of 1430. An alternative refrigerant is R-152a, which has a GWP level of 124. R-152a is a hydrofluorocarbon variant.

CO_2 Systems

To meet the standards set up by the Kyoto Protocol, auto manufacturers need to find an alternative to R-134a. BMW and other European manufacturers have chosen CO_2 as a possible refrigerant for their A/C systems in the future. CO_2 is nontoxic when used as a refrigerant. It is known as R-744 and has a GWP of 1. The extremely low GWP means there is little environmental concern with its use. In fact, this refrigerant will not need to be recovered and recycled. If CO_2 leaks from an A/C system, the effect on the environment is very small. CO_2 is abundantly available in our air; therefore, it does not need to be manufactured.

It is claimed that CO_2-based systems will be up to 25 percent more energy efficient than the best of today's R-134a systems. In addition, CO_2 can be used for heat pump systems. Heat pump systems can supply cool and warm air and can be used in hybrid vehicles. Also, CO_2 systems need a smaller amount of gas than conventional systems, which means the size of the system is substantially less than a conventional system.

The operation of a CO_2 system is the same as that of any other A/C system, but the components are different due to the higher pressures. CO_2 systems operate at pressures that are nearly 10 times that of an R-134a system. CO_2 also has a critical temperature that is much lower than R-134a. Critical temperature is the temperature above which a substance cannot exist in the liquid state regardless of the pressure. Therefore, these systems need an internal heat exchanger (IHX) and accumulator. The materials used for hoses and gaskets must also be slightly different. The IHX is located between the condenser and the evaporator. The accumulator is part of the IHX **(Figure 54–14)**.

Basic Operation of an Air-Conditioning System

Refrigerants are used to carry heat from the inside of the vehicle to the outside of the vehicle. Automotive refrigerants have a low boiling point (the point at which **evaporation** occurs). For example, at any temperature above –26.3°C (–26.3°F), liquid R-134a can boil and become a vapour. As a refrigerant changes state, it absorbs large amounts of heat. Because the heat that it absorbs is from the inside of the vehicle, passengers are cooler.

To understand how a refrigerant is used to cool the interior of a vehicle, the effects of pressure and temperature on it must first be understood. If the pressure of the refrigerant is high, so is its temperature; if the pressure is low, so is its temperature. Therefore, the temperature of the refrigerant can be changed by changing its pressure. **Figure 54–15** compares the pressures of R-12 and R-134a at various temperatures. Both refrigerants have an increase in pressure with an increase in temperature, and vice versa.

To absorb heat, the temperature and pressure of the refrigerant is kept low. To dissipate heat, the temperature and pressure are high. Remember that the heat from something hot always moves to something colder. The refrigerant absorbs heat from the inside of the vehicle and dissipates it to the cooler outside air. As the refrigerant absorbs heat, it changes from a liquid to a vapour. This is called evaporation. As it dissipates heat, it changes from a vapour to a liquid. The change from a vapour to a

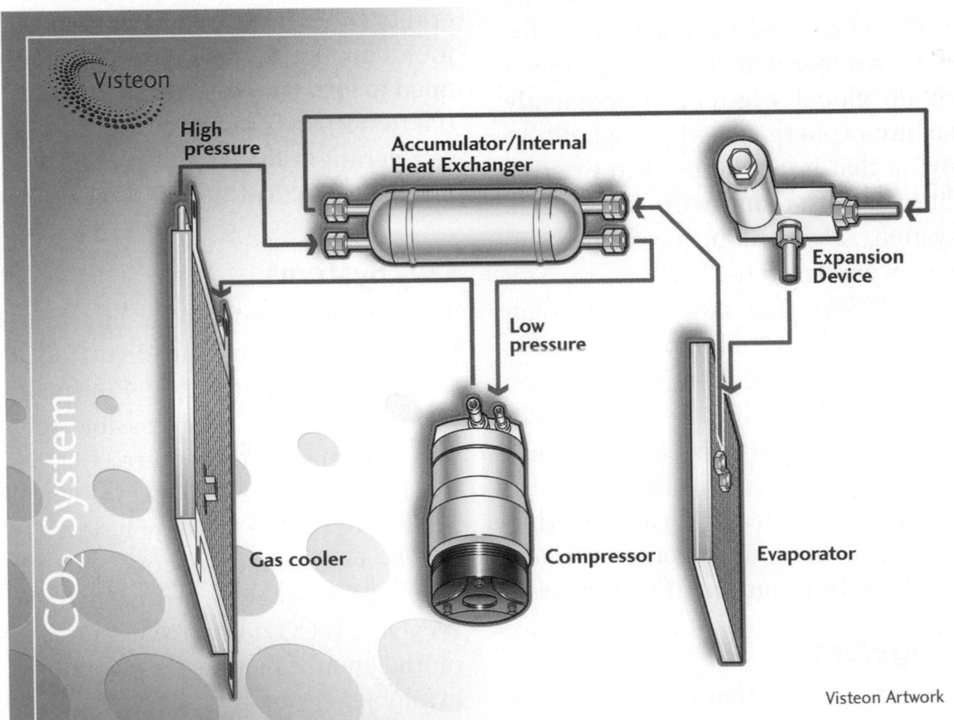

FIGURE 54–14

The layout of a CO_2 A/C system.

Visteon Corporation

TEMPERATURE °F	TEMPERATURE °C	PRESSURE Psig	PRESSURE kPa
10	−12.2	12.0	82.7
20	−6.67	18.4	127
30	−1.11	25.3	174
40	4.44	35.0	241
50	10.0	45.4	313
60	15.6	57.4	396
70	21.1	17.1	490
80	26.7	86.7	598
90	32.2	104.3	719
100	37.8	124.1	856
110	43.3	146.3	1009
120	48.9	171.1	1180
130	54.4	198.7	1370
140	60.0	229.2	1580
150	65.6	262.8	1812

FIGURE 54–15

A simple chart showing the relationship of temperature and pressure of R-134a.

liquid is called **condensation**. These two changes of state—evaporation and condensation—occur continuously as the refrigerant circulates through the air-conditioning system.

THE AIR-CONDITIONING SYSTEM AND ITS COMPONENTS

An automotive air-conditioning system is a closed, pressurized system. It consists of a compressor, condenser, receiver/dryer or accumulator, expansion valve or orifice tube, and an evaporator.

In a basic air-conditioning system, the heat is absorbed and transferred in the following steps **(Figure 54–16)**:

1. Refrigerant leaves the compressor as a high-pressure, high-temperature vapour.
2. By removing heat via the condenser, the vapour becomes a high-pressure, high-temperature liquid.
3. In some systems, at this point, a receiver/dryer is used to remove any vapour that did not condense in the condenser along with moisture and contaminants. Excess refrigerant in the system is also stored in the receiver/dryer until it is needed. These are found in systems that use a thermostatic expansion valve.
4. A thermostatic expansion valve or a fixed orifice tube controls the flow of liquid refrigerant into the evaporator.
5. Heat is absorbed from the air inside the passenger compartment by the low-pressure liquid refrigerant entering the evaporator. The low-pressure liquid refrigerant's boiling point is lower than the temperature of the air around the evaporator, and the refrigerant uses this available heat to evaporate to a vapour.
6. In some systems, at this point, an accumulator is used to remove any liquid that did not evaporate in the evaporator along with moisture and contaminants. Excess refrigerant in the system is also stored in the accumulator until it is needed. These are found in systems that use a fixed orifice tube.
7. The refrigerant returns to the compressor as a low-pressure, low-temperature vapour.

To understand the operation of the five major components, remember that an air-conditioning system is divided into two sides: the high side and

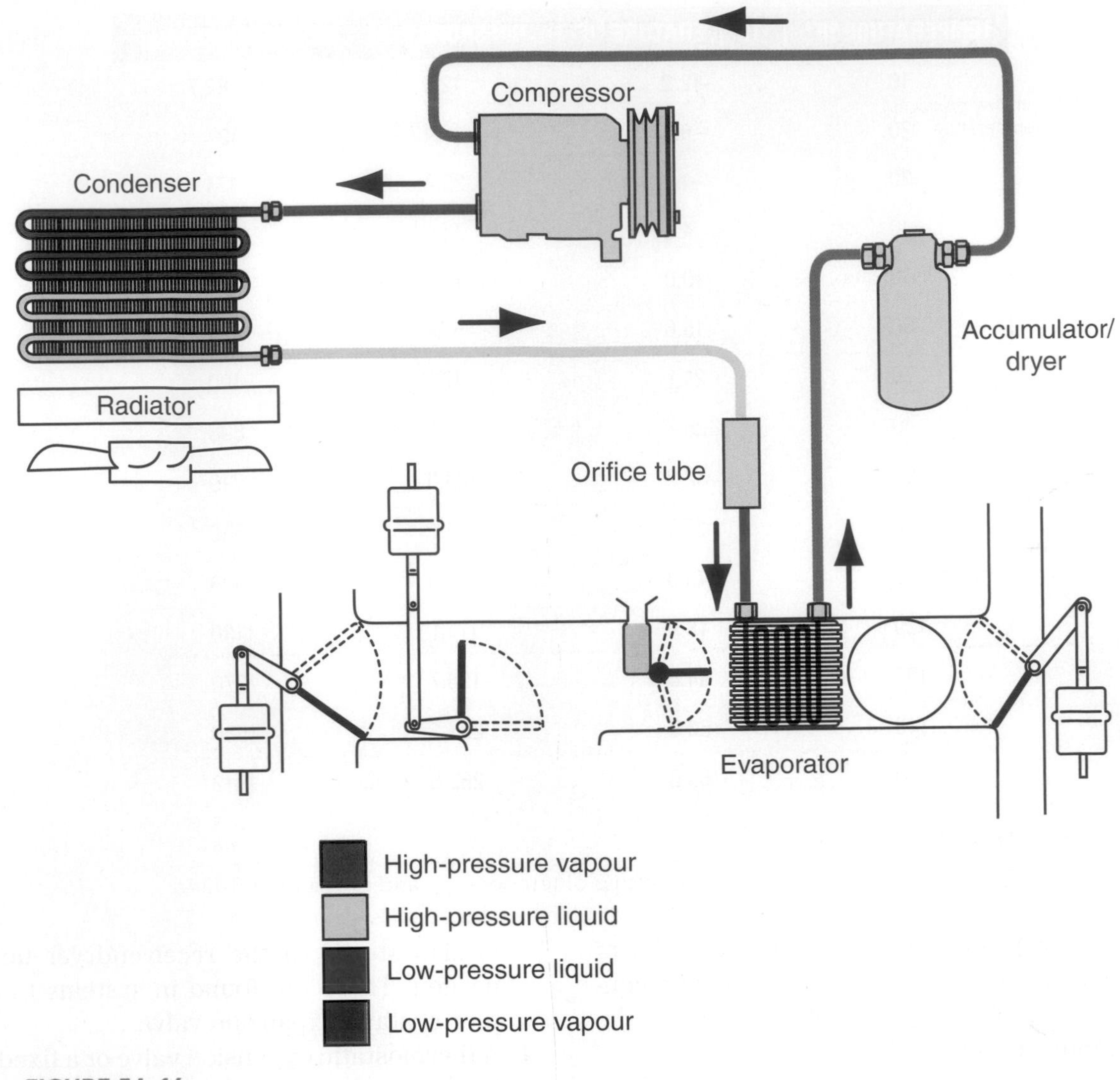

FIGURE 54–16
The basic refrigerant flow cycle.

the low side **(Figure 54–17)**. **High side** refers to the side of the system that is under high pressure and high temperature. **Low side** refers to the low-pressure, low-temperature side of the system. Each side can also be divided into a liquid and vapour side.

The Compressor

The compressor is the heart of the automotive air-conditioning system. It separates the high-pressure and low-pressure sides of the system. The primary purpose of the unit is to draw the low-pressure and low-temperature vapour from the **evaporator** and compress this vapour into high-temperature, high-pressure vapour. This action results in the refrigerant having a higher temperature than surrounding air, and enables the **condenser** to condense the vapour back to a liquid. The secondary purpose of the compressor is to circulate or pump the refrigerant through the A/C system under the different pressures required for proper operation. The compressor is located on the engine and is driven by the engine's crankshaft via a drive belt. Hybrid and electric vehicles typically have a three-phase high-voltage motor in the compressor for when the engine is off.

Although there are numerous types of compressors in use today, they are usually one of three types, as described below.

PISTON COMPRESSOR This type of compressor **(Figure 54–18)** can have its pistons arranged in an in-line, axial, radial, or V design. It is designed to have an intake stroke and a compression stroke for each cylinder. On the intake stroke, the refrigerant from the low side of the system (evaporator) is drawn into the compressor. The intake of refrigerant occurs through intake reed valves **(Figure 54–19)**.

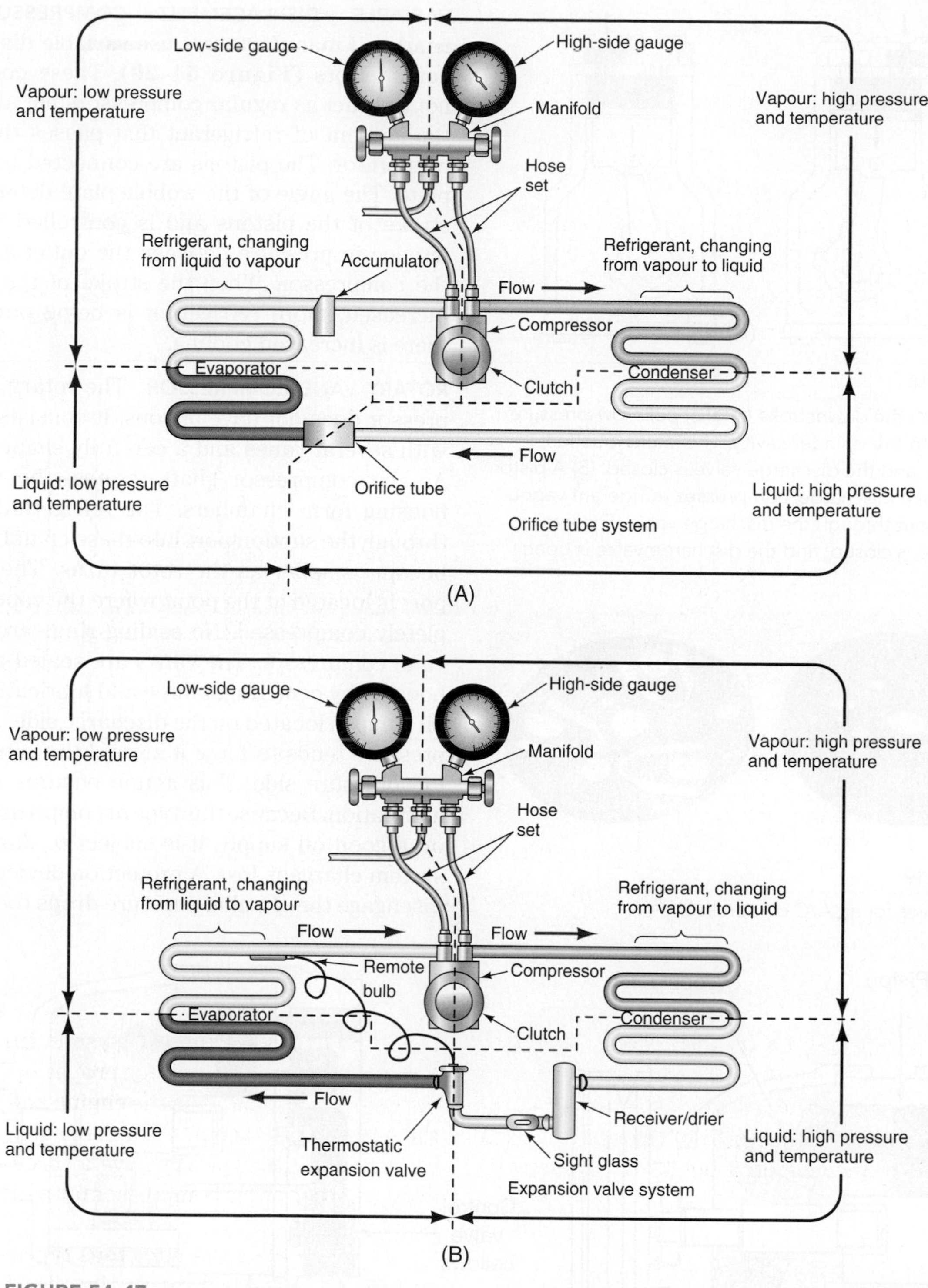

FIGURE 54–17

How the two systems are divided into high- and low-pressure sections: (A) orifice tube system and (B) expansion valve system.

These one-way valves control the flow of refrigerant vapours into the cylinder. During the compression stroke, the vaporous refrigerant is compressed. This increases both the pressure and the temperature of the heat-carrying refrigerant. The outlet or **discharge-side** reed valves then open to allow the refrigerant to move to the condenser. The outlet reed valves are the beginning of the high side of the system. Reed valves are made of spring steel, which can be weakened or broken if improper charging procedures are used, such as liquid charging with the engine running.

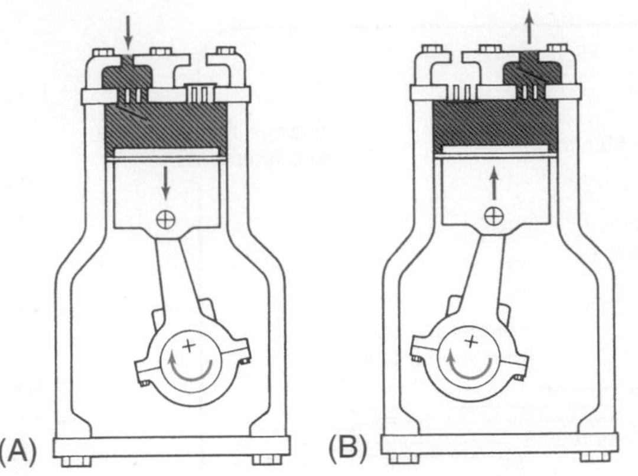

FIGURE 54–18

(A) A piston on the downstroke (intake) pulls low-pressure refrigerant into the cylinder cavity. The intake (suction) valve is open, and the discharge valve is closed. (B) A piston on the upstroke (discharge) compresses refrigerant vapour and forces it out through the discharge valve. The intake (suction) valve is closed, and the discharge valve is open.

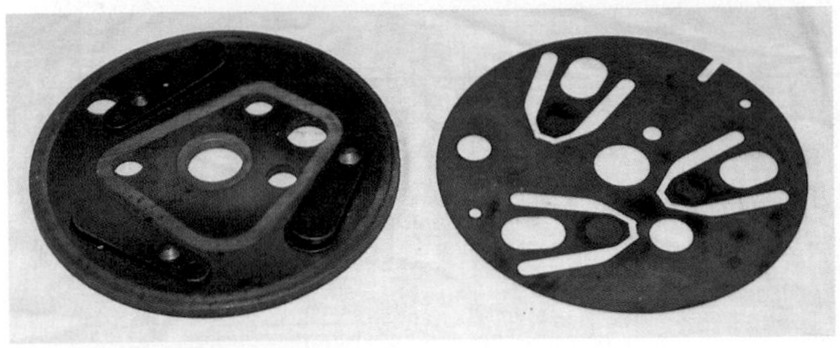

FIGURE 54–19

The reed valves for an A/C compressor.

VARIABLE DISPLACEMENT COMPRESSOR Today, nearly all manufacturers use variable displacement compressors **(Figure 54–20)**. These compressors not only act as regular compressors but also control the amount of refrigerant that passes through the evaporator. The pistons are connected to a wobble plate. The angle of the wobble plate determines the stroke of the pistons and is controlled by the difference in pressure between the outlet and inlet of the compressor. When the stroke of the pistons is increased, more refrigerant is being pumped, and there is increased cooling.

ROTARY VANE COMPRESSOR The rotary vane compressor does not have pistons. It consists of a rotor with several vanes and a carefully shaped housing. As the compressor shaft rotates, the vanes and housing form chambers. The refrigerant is drawn through the suction port into these chambers, which become smaller as the rotor turns. The discharge port is located at the point where the vapour is completely compressed. No sealing rings are used in a vane compressor. The vanes are sealed against the housing by centrifugal force and lubricating oil. The oil sump is located on the discharge side, so the high pressure tends to force it around the vanes into the low-pressure side. This action ensures continuous lubrication. Because this type of compressor depends on a good oil supply, it is subject to damage if the system charge is lost. A protection device is used to disengage the clutch if pressure drops too low.

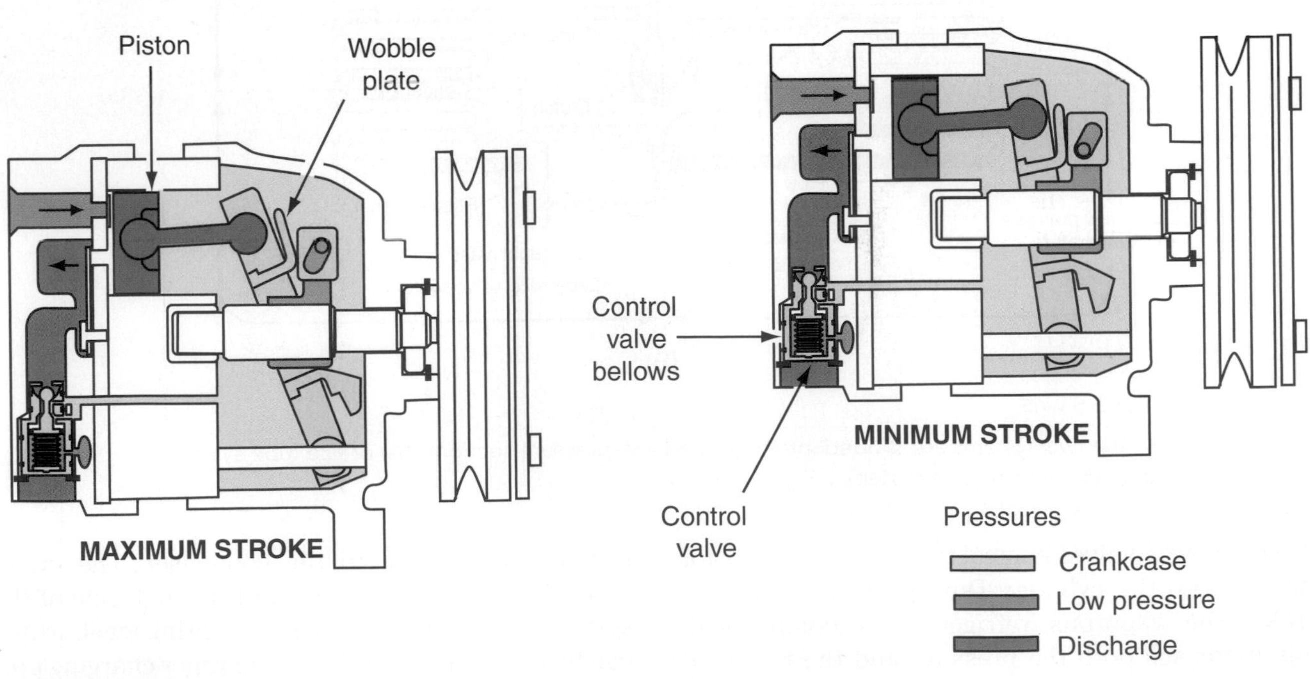

FIGURE 54–20

A V5 variable displacement compressor.

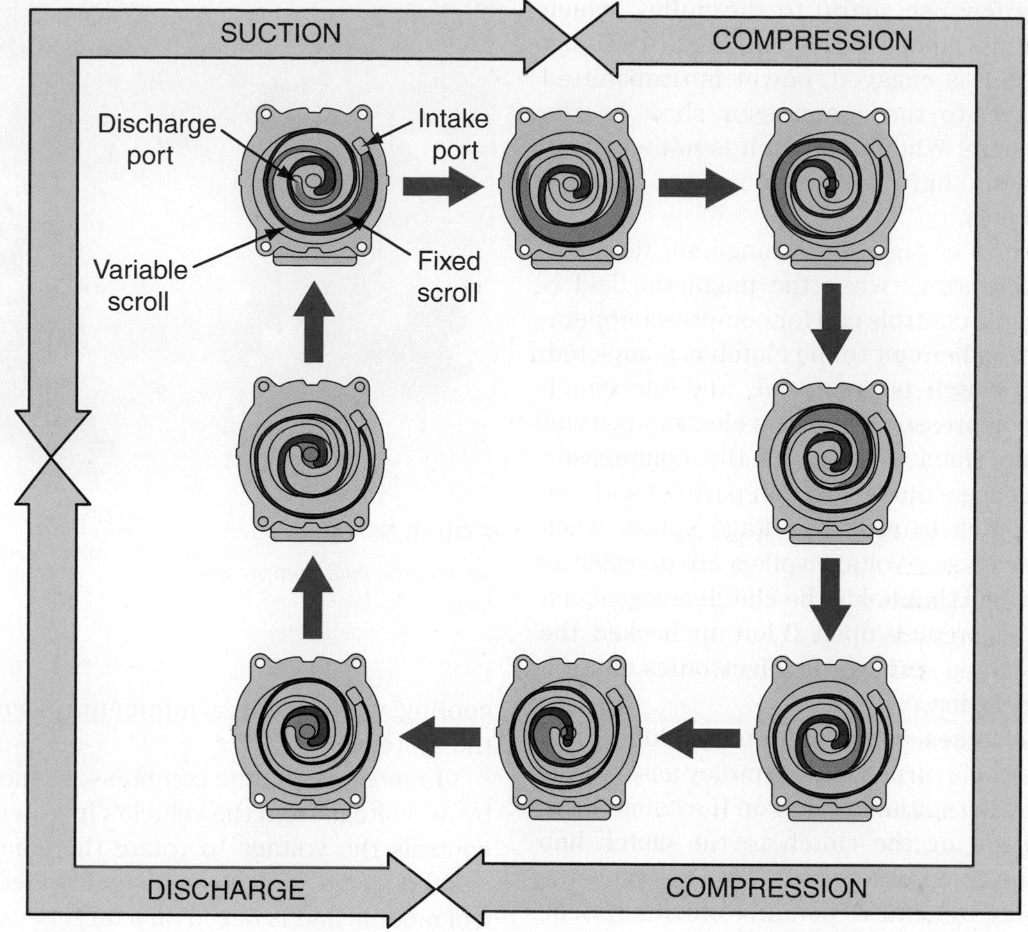

FIGURE 54–21

A scroll compressor passes through three distinct phases: suction, compression, and discharge.

SCROLL-TYPE COMPRESSOR The scroll-type compressor has a movable scroll and a fixed or nonmovable scroll that provide an eccentric-like motion. As the compressor's crankshaft rotates, the movable scroll forces the refrigerant against the fixed scroll and toward the centre of the compressor. This motion pressurizes the refrigerant. The action of a scroll-type compressor can be compared to that of a tornado. The pressure of air moving in a circular pattern increases as it moves toward the centre of the circle. A delivery port is positioned at the centre of the compressor and allows the high-pressure refrigerant to flow into the air-conditioning system **(Figure 54–21)**. This type of compressor operates more smoothly than other designs. Some hybrids use electrically driven compressors that do not require the engine to operate to run the air-conditioning system.

Compressor Clutches

Every compressor is equipped with an electromagnetic clutch as part of the compressor pulley assembly **(Figure 54–22)**. It is designed to engage the pulley to the compressor shaft when the clutch coil is energized. The purpose of the clutch is to transmit power from the engine to the compressor and to provide a means of engaging and disengaging the compressor crankshaft from engine operation.

The clutch is driven by power from the engine's crankshaft, which is transmitted through one or

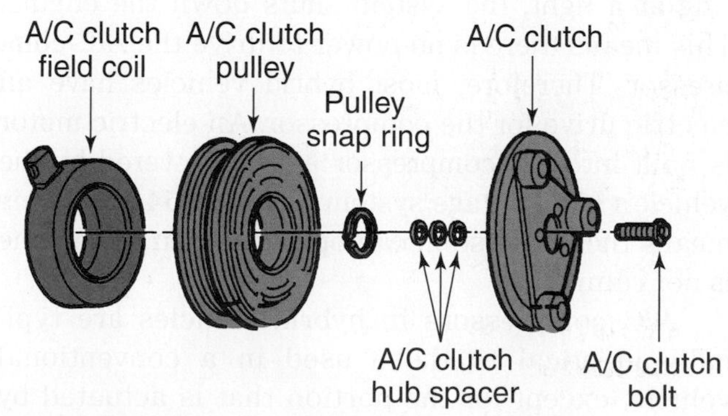

FIGURE 54–22

Parts of a compressor clutch assembly.

Ford Motor Company

more belts (a few use gears) to the pulley, which is in operation whenever the engine is running. When the clutch is engaged, power is transmitted from the pulley to the compressor shaft by the clutch drive plate. When the clutch is not engaged, the compressor shaft does not rotate, and the pulley freewheels.

The clutch is engaged by a magnetic field and disengaged by springs when the magnetic field is broken. When the controls call for compressor operation, the electrical circuit to the clutch is completed; the magnetic clutch is energized; and the clutch engages the compressor. When the electrical circuit is opened, the clutch disengages the compressor. Many systems use a diode wired in parallel with the compressor clutch to absorb voltage spikes when the clutch disengages. Voltage spikes are possible as the magnetic field that holds the clutch engaged collapses once the circuit is open. If left unchecked, the voltage spike could damage the electronics controlling clutch operation.

Modern clutches use a stationary clutch coil. When the system is turned on, the pulley assembly is magnetized by the stationary coil on the compressor body, thus engaging the clutch to the clutch hub attached to the compressor shaft. This activates the air-conditioning system. Depending on the system, the magnetic clutch is usually pressure controlled to cycle the operation of the compressor (depending on system temperature or pressure). In some system designs, the clutch might operate continually when the system is turned on. With stationary coil design, service is not usually necessary except for an occasional check on the electrical connections.

Electric Drive Compressors

One of the features of hybrid vehicles is the idle–stop or stop–start system. When the vehicle is sitting at a light, the system shuts down the engine. This means there is no power to drive the A/C compressor. Therefore, most hybrid vehicles have an electric drive for the compressor. An electric motor is built into the compressor and is powered by the vehicle's high-voltage system **(Figure 54–23)**. This means the A/C system can operate when the engine is not running.

A/C compressors in hybrid vehicles are typically identical to those used in a conventional vehicle, except for the portion that is actuated by the electric motor. Because the compressor is energized by electricity, the control module can control its speed. This enables the system to provide ideal cooling while using a minimum of energy to run the system.

FIGURE 54–23

An electric A/C compressor.

DENSO Corporation

In most cases, the compressor's motor is driven by AC voltage from the vehicle's inverter. The system controls the voltage to rotate the compressor at a desired speed. This speed is calculated by the control module and is based on a target evaporator temperature and the actual evaporator temperature.

The Accord hybrid is also equipped with a dual-scroll "hybrid" A/C compressor **(Figure 54–24)**. The A/C system uses two compressors built into a single housing; one compressor is driven by the engine, and the other is driven by an electric motor powered by the high-voltage battery. A driver circuit in the control module provides switched high voltage to the motor. The action of the driver is determined by inputs from the climate control settings and the CAN bus. The mechanically driven side uses a normal electric clutch controlled by the climate control module. The electrically driven side uses a high-voltage, brushless, three-phase motor driven by a controller. When full cooling is needed, the A/C unit relies on both power sources to provide maximum cooling. During normal cooling, the A/C is powered by either the belt-driven compressor or the electric motor-driven compressor.

Refrigerant Oils

Normally, the only source of lubrication for a compressor is the oil mixed with the refrigerant. Because of the loads and speeds at which the compressor operates, proper lubrication is a must for long

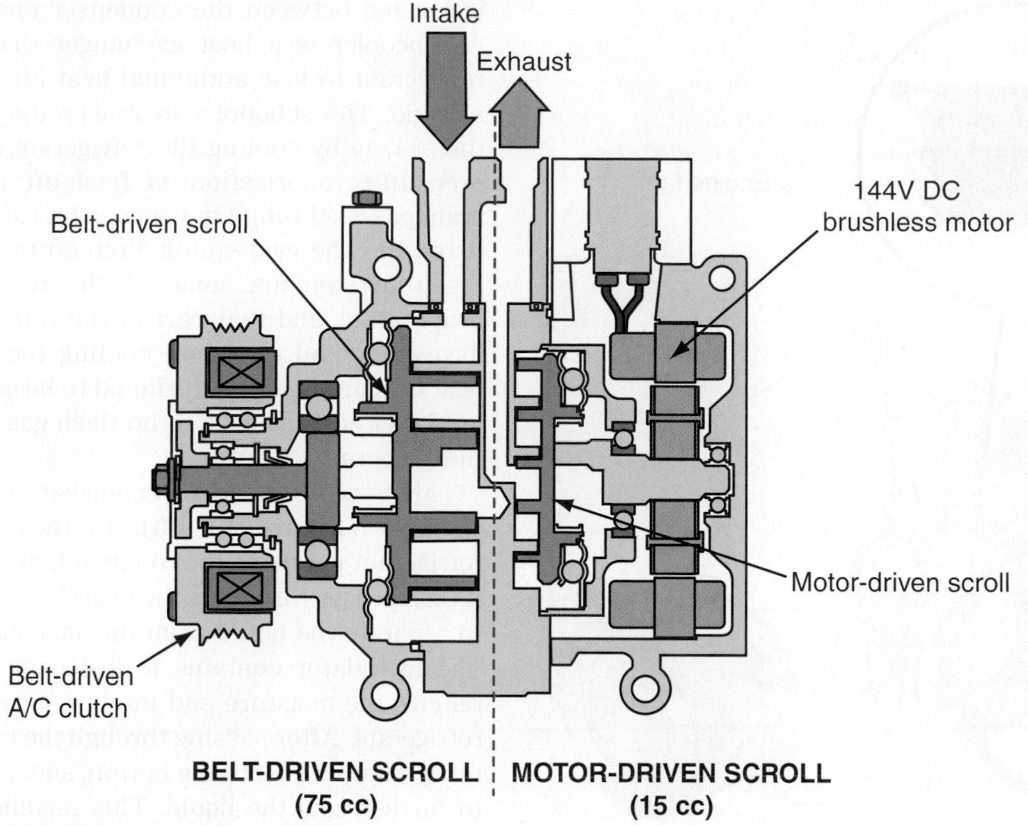

FIGURE 54–24
A dual-scroll A/C compressor powered by a drive belt and/or an electric motor.

compressor life. The refrigerant oil required by the system depends on a number of things, but it is primarily dictated by the refrigerant used in the system. R-12 systems use a mineral oil. Mineral oil, however, cannot be used with R-134a. R-134a systems require that a synthetic oil, **polyalkalene glycol (PAG)**, be used. There are a number of different blends of PAG oil. Always use the one recommended by the vehicle manufacturer or compressor manufacturer. Failure to use the correct oil will cause damage to the compressor.

WARNING!

Hybrid vehicles often have an electrically driven compressor. Because the electric motor is inside the compressor case and is in contact with the oil in the compressor, only the specified oil should be used in the compressor. This oil has electrical insulating qualities that protect you from dangerous electrical shocks. Also, if you use the wrong oil, the A/C unit will be contaminated, and this may result in a need to replace the compressor, condenser, evaporator, and/or all of the refrigerant lines.

Condenser

The condenser **(Figure 54–25)** consists of coiled refrigerant tubing mounted in a series of thin cooling fins to provide maximum heat transfer in a minimum amount of space. The condenser is normally mounted just in front of the vehicle's radiator. It receives the full flow of ram air from the movement of the vehicle or airflow from the radiator fan or dedicated condenser fan when the vehicle is standing still.

The purpose of the condenser is to condense or liquefy the high-pressure, high-temperature vapour coming from the compressor. To do so, it must give up its heat. The condenser receives very hot (normally, 93°C to 204°C [200°F to 400°F]), high-pressure refrigerant vapour from the compressor through its discharge hose. The refrigerant vapour enters the inlet at the top of the condenser, and as the hot vapour passes down through the condenser coils, heat (following its natural tendencies) moves from the hot refrigerant into the cooler air as it flows across the condenser coils and fins. This process causes a large quantity of heat to be transferred to the outside air and the refrigerant to change from a high-pressure

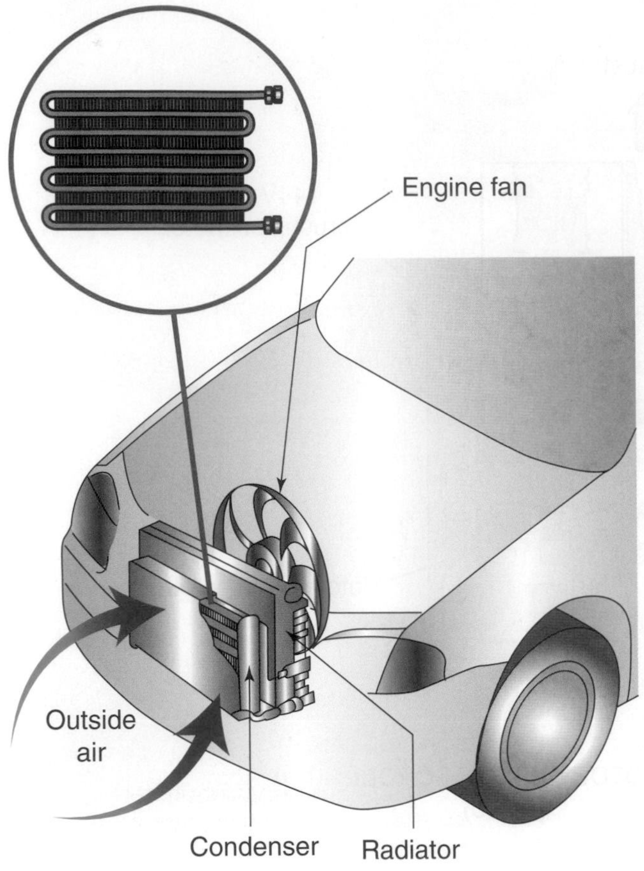

FIGURE 54–25

A typical condenser.

hot vapour to a high-pressure warm liquid. This high-pressure warm liquid flows from the outlet at the bottom of the condenser, through a line, to the receiver/dryer or to the refrigerant metering device if an accumulator instead of a dryer is used.

In an air-conditioning system operating under an average heat load, the condenser has a combination of hot refrigerant vapour in the upper two-thirds of its coils. The lower third of the coils contains the warm liquid refrigerant, which has condensed. This high-pressure liquid refrigerant flows from the condenser and on toward the evaporator. In effect, the condenser is a true heat exchanger.

Subcoolers

Some vehicles have a subcooler built into the condenser or have a separate subcooler. Subcooling is a process by which sensible heat is removed from liquid refrigerant, resulting in lower refrigerant temperatures. Sensible heat is heat that can be felt and measured and represents the energy required to change the temperature of a substance without causing a change in its state. The separate subcooler is located between the condenser and evaporator. A subcooler is a heat exchanger that allows the refrigerant to lose additional heat after it becomes a liquid. The subcooler increases the efficiency of the system by cooling the refrigerant and prevents premature vaporization, or flash-off, as the refrigerant passes through the expansion valve and before it reaches the evaporator. Premature flash-off can result in stopping some of the refrigerant from evaporating, and that part of the refrigerant would have no useful effect on cooling the vehicle. The heat exchange causes the liquid to be subcooled to a level that ensures little or no flash gas on its way to the evaporator.

Many vehicles have a condenser with a subcool chamber **(Figure 54–26)**. In these condensers, refrigerant enters at the top as a high-pressure gas. It then passes through a receiver/dryer or modulator to separate the liquid from the gaseous refrigerant. The modulator contains a desiccant and filter to remove the moisture and foreign material from the refrigerant. After passing through the modulator, the refrigerant flows into the bottom subcooler chamber to further cool the liquid. This results in complete liquidization of the refrigerant and better A/C performance. These condensers have thin tubes and low fin height, which improve the heat exchange rate. Because of this two-step approach, the refrigerant sent to the evaporator is almost completely liquefied.

Receiver/Dryer

Used on many systems, the **receiver/dryer** is a storage tank for the liquid refrigerant from the condenser, which flows into the upper portion of the receiver tank containing a bag of **desiccant** (moisture-absorbing material such as silica alumina or silica gel). As the refrigerant flows through an opening in the lower portion of the receiver, it is filtered through a mesh screen attached to a baffle

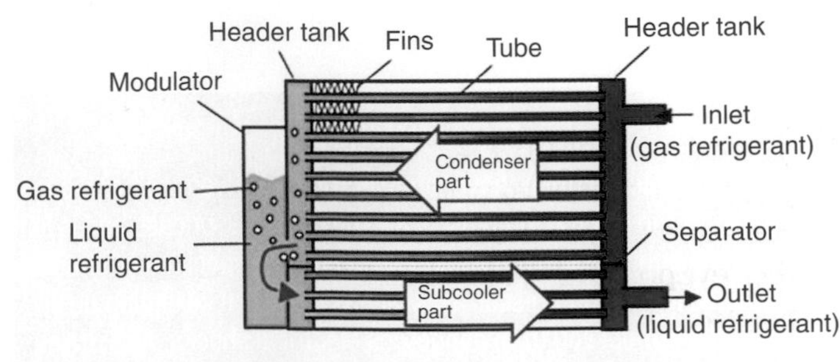

FIGURE 54–26

A condenser with a built-in subcooler.

DENSO Corporation

at the bottom of the receiver. The desiccant in this unit absorbs any moisture present that might enter the system during assembly. These features of the unit prevent obstruction to the valves or damage to the compressor.

Depending on the manufacturer, the receiver/dryer may be known by such names as filter or dehydrator. Regardless of its name, the function is the same. Included in many receiver/dryers are additional features such as high-pressure fittings, a pressure relief valve, and a sight glass for determining the state and condition of the refrigerant in the system.

The receiver/dryer is often neglected when the air-conditioning system is serviced or repaired. Failure to replace it can lead to poor system performance or replacement part failure. It is recommended that the receiver/dryer and/or its desiccant be changed whenever a component is replaced, when the system has lost the refrigerant charge, or when the system has been open to the atmosphere for any length of time.

WARNING!

The desiccant draws moisture like a magnet and can become contaminated in less than five minutes if it is exposed to the atmosphere. Keep it sealed. Once opened, put it under a vacuum quickly.

Accumulator

Many systems are not equipped with a receiver/dryer; rather, they use an accumulator to accomplish the same thing **(Figure 54–27)**. The accumulator is connected into the low side at the outlet of the evaporator. It also contains a desiccant and is designed to store, filter, and dry excess refrigerant. If liquid refrigerant flows out of the evaporator, it will be collected by and stored in the accumulator. The main purpose of an accumulator is to prevent liquid from entering the compressor.

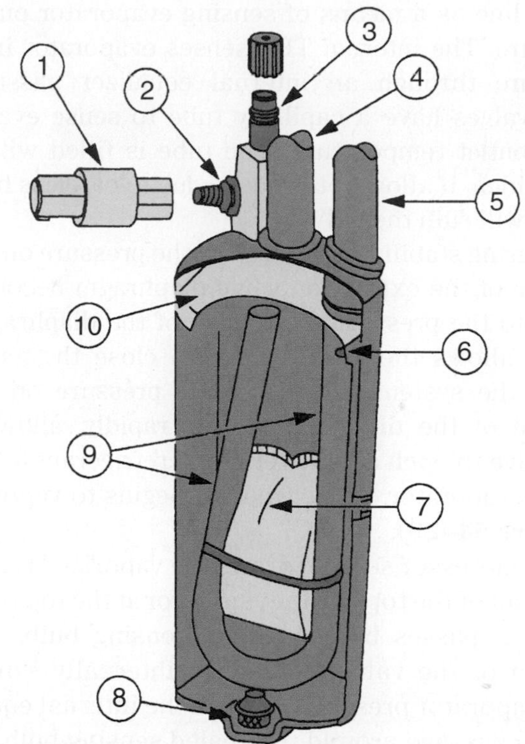

1. Cycling switch
2. O-ring seal
3. Low-side port
4. From evaporator
5. To compressor
6. Antisiphon hole
7. Desiccant bag
8. Oil return filter
9. Vapour return tube
10. Accumulator/dryer dome

FIGURE 54–27

An accumulator/dryer.

Ford Motor Company

Thermostatic Expansion Valve/Orifice Tube

The refrigerant flow to the evaporator must be controlled to obtain maximum cooling while ensuring complete evaporation of the liquid refrigerant within the evaporator. This is accomplished by a thermostatic expansion valve (TEV or TXV) or a fixed **orifice tube (Figure 54–28)**.

The TEV is mounted at the inlet to the evaporator and separates the high-pressure side of the system from the low-pressure side. The TEV regulates refrigerant flow to the evaporator to prevent evaporator flooding or starving. In operation, the TEV regulates the refrigerant flow to the evaporator by balancing the inlet flow to the outlet temperature.

Both externally and internally equalized TEVs are used in air-conditioning systems. The only difference between the two valves is that the external TEV uses an equalizer line connected to the evaporator

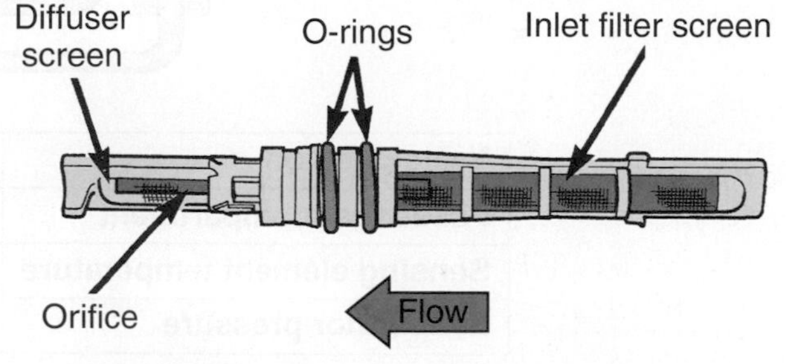

FIGURE 54–28

A typical orifice expansion tube.

Chrysler Group LLC

outlet line as a means of sensing evaporator outlet pressure. The internal TEV senses evaporator inlet pressure through an internal equalizer passage. Both valves have a capillary tube to sense evaporator outlet temperature. The tube is filled with a gas, which, if allowed to escape due to careless handling, will ruin the TEV.

During stabilized conditions, the pressure on the bottom of the expansion valve diaphragm becomes equal to the pressure on the top of the diaphragm, which allows the valve spring to close the valve. When the system is started, the pressure on the bottom of the diaphragm drops rapidly, allowing the valve to open and meter liquid refrigerant to the lower evaporator tubes, where it begins to vaporize **(Figure 54–29)**.

Compressor suction draws the vaporized refrigerant out of the top of the evaporator at the top tube, where it passes by the sealed sensing bulb. The bottom of the valve diaphragm internally senses the evaporator pressure through the internal equalization passage around the sealed sensing bulb. As evaporator pressure is increased, the diaphragm flexes upward, pulling the pushrod away from the ball seat of the expansion valve. The expansion valve spring forces the ball onto the tapered seat, and the liquid refrigerant flow is reduced.

As the pressure is reduced due to restricted refrigerant flow, the diaphragm flexes downward again, opening the expansion valve to provide the required controlled pressure and refrigerant flow condition. As the cool refrigerant passes by the body of the sensing bulb, the gas above the diaphragm contracts and allows the expansion valve spring to close the expansion valve. When heat from the passenger compartment is absorbed by the refrigerant, it causes the gas to expand. The pushrod again forces the expansion valve to open, allowing more refrigerant to flow so that more heat can be absorbed.

Like the thermostatic expansion valve, the orifice tube is the dividing point between the high- and low-pressure parts of the system. However, its metering or flow rate control does not depend on comparing evaporator pressure and temperature. It is a fixed orifice. The flow rate is determined by pressure difference across the orifice and by subcooling. Subcooling is additional cooling of the refrigerant (cooling below the saturation or boiling point of a liquid) in the bottom of the condenser after it has changed from vapour to liquid. The flow rate

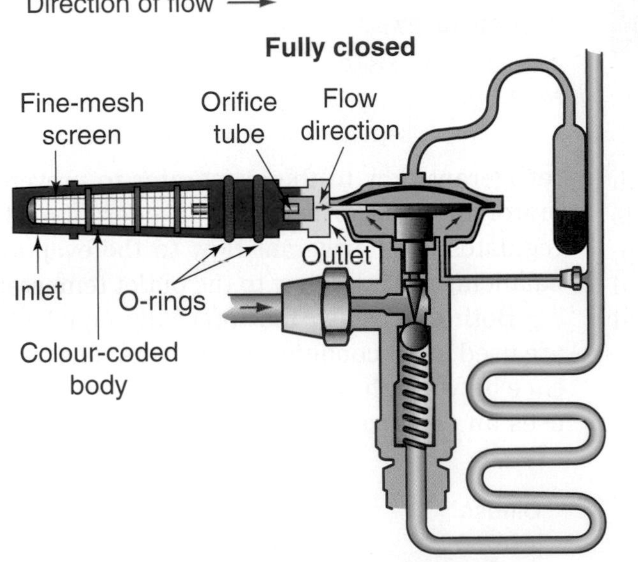

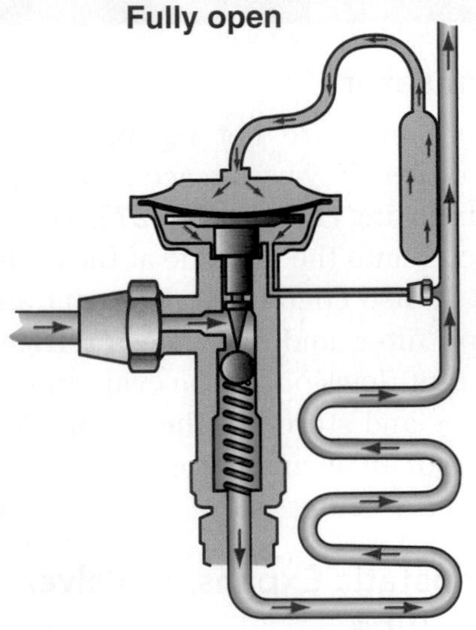

Valve position	Closed	Open
Passenger compartment	Desired temperature reached	Heat soaked
Sensing element temperature	Low	High
Evaporator pressure	High	Low

FIGURE 54–29

How a TXV regulates refrigerant flow through the evaporator.

through the orifice is more sensitive to subcooling than to pressure difference.

Evaporator

The evaporator, like the condenser, consists of a refrigerant coil mounted in a series of thin cooling fins **(Figure 54–30)**. It provides a maximum amount of heat transfer in a minimum amount of space. The evaporator is usually located beneath the dashboard or instrument panel.

On receiving the low-pressure, low-temperature liquid refrigerant from the thermostatic expansion valve or orifice tube in the form of an atomized (or droplet) spray, the evaporator serves as a boiler or vaporizer. This regulated flow of refrigerant boils almost immediately. Heat from the core surface is lost to the boiling and vaporizing refrigerant, which is cooler than the core, thereby cooling the core. The air passing over the evaporator loses its heat to the cooler surface of the core, thereby cooling the air inside the car. As the process of heat loss from air to the evaporator core surface is taking place, any moisture (humidity) in the air condenses on the outside of the evaporator core and is drained off as water. A drain tube or hole in the bottom of the evaporator housing leads the water outside the vehicle. This dehumidification of air is an added feature of the air-conditioning system that adds to passenger comfort. It can also be used as a means of controlling fogging of the vehicle windows. Under certain conditions, however, too much moisture can accumulate on the evaporator coils. An example would be when humidity is extremely high and the maximum cooling mode is selected or when a TEV cannot monitor evaporator temperature correctly. The evaporator temperature might become so low that moisture would freeze on the evaporator coils before it could drain off.

Through the metering, or controlling, action of the thermostatic expansion valve or orifice tube, greater or lesser amounts of refrigerant are provided in the evaporator to adequately cool the car under all heat load conditions. If too much refrigerant is allowed to enter, the evaporator floods. This results in poor cooling due to the higher pressure (and temperature) of the refrigerant. The refrigerant can neither boil away rapidly nor vaporize. Conversely, if too little refrigerant is metered, the evaporator starves. Poor cooling again results because the refrigerant boils away or vaporizes too quickly before passing through the evaporator.

The temperature of the refrigerant vapour at the evaporator outlet will be approximately 2.2°C to 8.9°C (4°F to 16°F) higher than the temperature of the liquid refrigerant at the evaporator inlet. This temperature differential is the superheat that ensures that the vapour will not contain any droplets of liquid refrigerant that would be harmful to the compressor.

Refrigerant Lines

All the major components of the system have inlet and outlet connections that accommodate O-ring fittings. The refrigerant lines that connect between these units are made up of an appropriate length of hose with O-ring fittings at each end as required **(Figure 54–31)**.

There are three major refrigerant lines: suction, liquid, and discharge. Suction lines are located between the outlet side of the evaporator and the inlet side, or suction side, of the compressor. They carry the low-pressure, low-temperature refrigerant vapour to the compressor, where it is again recycled through the system. Suction lines are always distinguished from the discharge lines by touch and size. They are cold to the touch. The suction line is larger in diameter than the liquid line because refrigerant

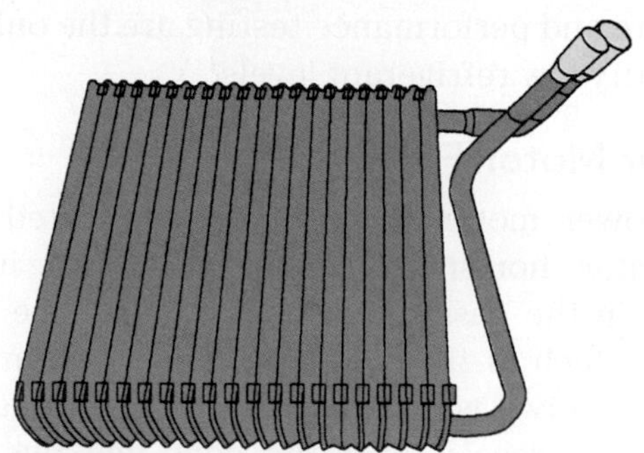

FIGURE 54–30

A typical evaporator.

Ford Motor Company

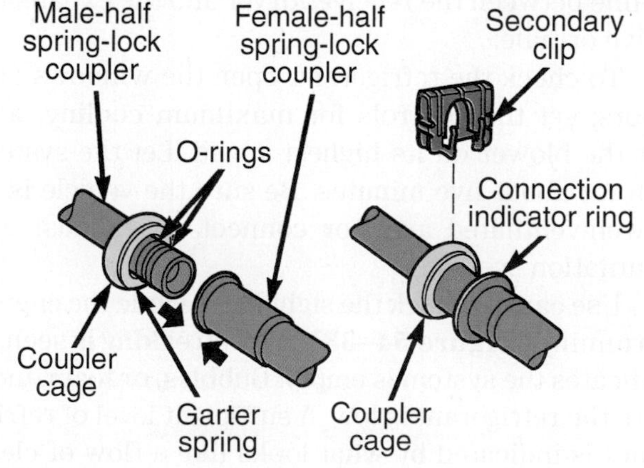

FIGURE 54–31

A spring-lock line coupler with O-rings.

Chrysler Group LLC

in a vapour state takes up more room than refrigerant in a liquid state.

Beginning at the discharge outlet on the compressor, the discharge or high-pressure line connects the compressor to the condenser. The liquid lines connect the condenser to the receiver/dryer and the receiver/dryer to the inlet side of the expansion valve. Through these lines, the refrigerant travels in its path from a gas state (compressor outlet) to a liquid state (condenser outlet) and then to the inlet side of the expansion valve, where it vaporizes on entry to the evaporator. Discharge and liquid lines are always very warm to the touch and easily distinguishable from the sucion lines.

Aluminum tubing is often used to connect air-conditioning components where flexibility is not required. Where the line is subjected to vibrations, special rubber hoses are used. Typically, the compressor outlet and inlet lines are rubber hoses with aluminum ends and fittings.

R-134a systems are required to be fitted with quick-disconnect fittings throughout the system **(Figure 54–32)**. These systems also have hoses specially made for R-134a that have an additional layer of foil that serves as a barrier to prevent the refrigerant from escaping through the pores of the hose. Some late-model R-12 systems also use these barrier hoses to prevent the loss of refrigerant through the walls of the hoses.

(A)

(B)

FIGURE 54–32

(A) A quick-disconnect fitting on an A/C line. (B) Connecting a pressure gauge to the fitting.

Sight Glass

The **sight glass** allows the technician to see the flow of refrigerant in the lines. A sight glass is normally found on systems using R-12 and a thermal expansion valve. It is not commonly found on R-134a systems. It can be located on the receiver/dryer or in-line between the receiver/dryer and the expansion valve or tube.

To check the refrigerant, open the windows and doors; set the controls for maximum cooling; and set the blower on its highest speed. Let the system run for about five minutes. Be sure the vehicle is in a well-ventilated area, or connect an exhaust gas ventilation system.

Use care to check the sight glass while the engine is running **(Figure 54–33)**. If oil streaking is seen, it indicates the system is empty. Bubbles, or foam, indicate the refrigerant is low. A sufficient level of refrigerant is indicated by what looks like a flow of clear water, with no bubbles. A clouded sight glass is an indication of desiccant contamination, with subsequent infiltration and circulation through the system.

Accumulator systems do not have a receiver/dryer to separate the gas from the liquid as it flows from the condenser. The liquid line will always have a certain amount of bubbles in it. Therefore, it would be useless to have a sight glass in these systems. Pressure and performance testing are the only ways to identify low refrigerant levels.

Blower Motor/Fan

The blower motor/fan assembly is located in the evaporator housing. Its purpose is to increase airflow in the passenger compartment. The blower motor, which is the same as those used in heater systems, draws warm air from the passenger compartment, forces it over the coils and fins of the evaporator, and blows the cooled, cleaned, and dehumidified air into the passenger compartment. The blower motor is controlled by a fan switch.

FIGURE 54–33
The location of a typical sight glass.

During cold weather, the blower motor/fan provides the airflow, and the heater core provides the heat for the passenger compartment.

AIR-CONDITIONING SYSTEMS AND CONTROLS

There are two basic types of automotive air-conditioning systems. They are classified according to the method used in obtaining temperature control and are known as cycling clutch systems or evaporator pressure or temperature control systems.

Evaporator Pressure Control System

Evaporator controls maintain a back pressure in the evaporator. Because of the refrigerant temperature/pressure relationship, the effect is to regulate evaporator temperature. The temperature is controlled to a point that provides effective air cooling but prevents the freezing of moisture that condenses on the evaporator.

In this type of system, the compressor operates continually when dash controls are in the air-conditioning or defrost modes. Evaporator outlet air temperature is automatically controlled by an evaporator pressure control valve. This type of valve throttles the flow of refrigerant out of the evaporator as required to establish a minimum evaporator pressure and thereby prevent freezing of condensation on the evaporator core.

Cycling Clutch System

In every **cycling clutch** system, the compressor is run intermittently by means of controlling the application and release of its clutch through a thermostatic or pressure switch. The thermostatic switch senses the evaporator's outlet air temperature through a capillary tube that is part of the switch assembly. With a high sensing temperature, the thermostatic switch is closed, and the compressor clutch is energized. As the evaporator outlet temperature drops to a preset level, the thermostatic switch opens the circuit to the compressor clutch. The compressor then ceases to operate until such time as the evaporator temperature rises above the switch setting. From this on-and-off operation is derived the term cycling clutch. In effect, the thermostatic switch is calibrated to allow the lowest possible evaporator outlet temperature that would prevent the freezing of condensation that might form on the evaporator.

Variations of the cycling clutch system include a system with a thermostatic expansion valve and a system with an orifice tube.

CYCLING CLUTCH SYSTEM WITH THERMOSTATIC EXPANSION VALVE Some factory installations utilize a cycling clutch system that incorporates a TEV and receiver/dryer, as do some add-on units. The evaporator and control components are either in the engine compartment or an integral part of the cowl. In such cases there is a common blower and duct work for both heating and air-conditioning purposes. Also in these installations, the thermostatic switch has no temperature control knob and is usually mounted on the evaporator or its case. Temperature control is accomplished by using fresh or recirculating air and by reheating the cooled air in the heater core. The clutch cycles only to prevent evaporator icing.

CYCLING CLUTCH SYSTEM WITH ORIFICE TUBE (CCOT) A typical **CCOT system** is illustrated in **Figure 54–34**. The system is factory installed and can use a thermostatic clutch cycling switch mounted on the evaporator case or a pressure cycling switch located on the accumulator. An expansion (orifice) tube is used in place of the TEV. Also, the system has an accumulator in the evaporator outlet. The accumulator is used primarily to separate vapour from liquid refrigerant before it enters the compressor. It also contains a drying agent, or desiccant, to remove moisture. The CCOT system has no receiver/dryer or sight glass. This system does have a special orifice, the oil bleed orifice, that allows refrigerant oil to return to the compressor rather than to collect in the accumulator.

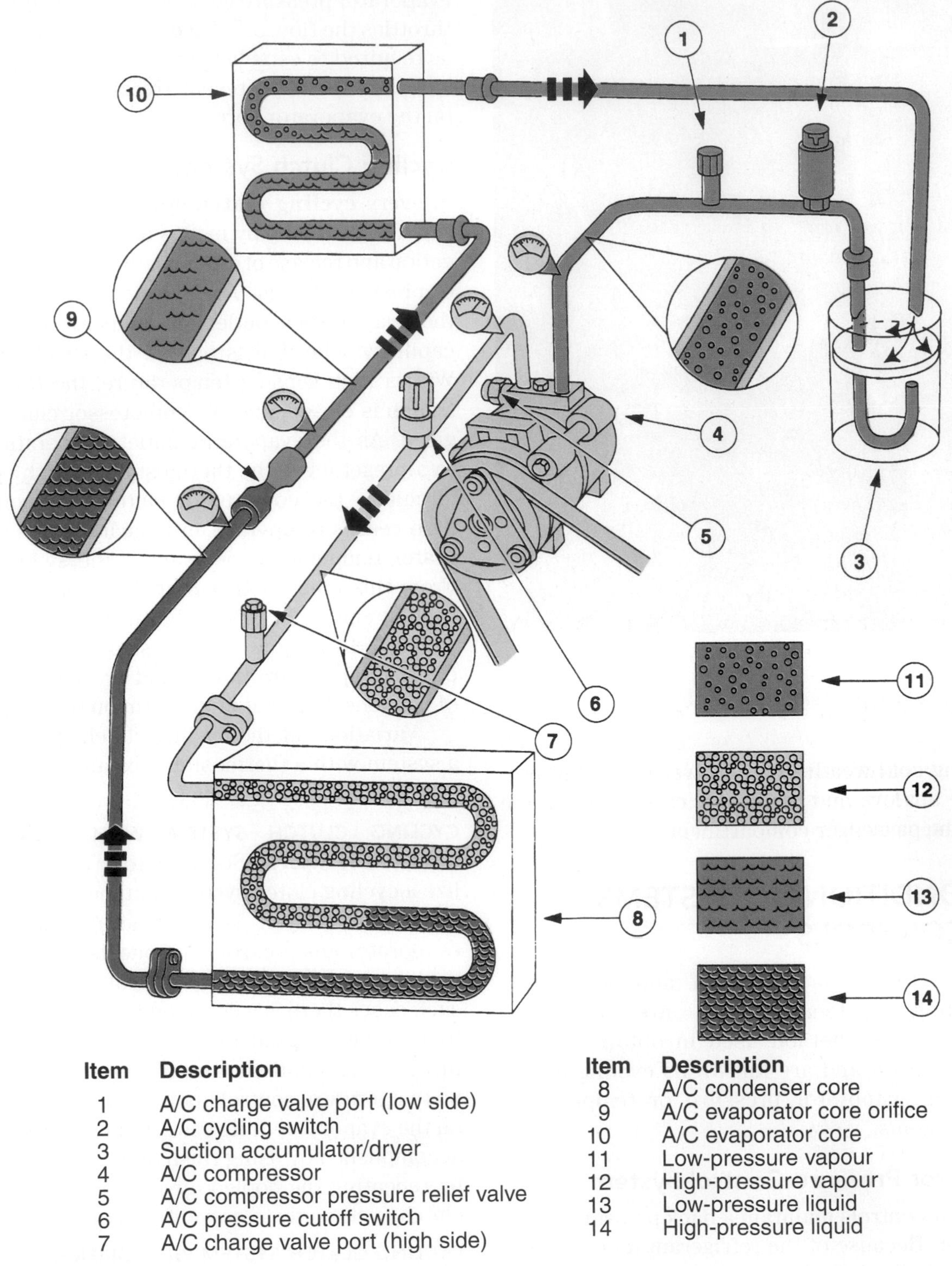

FIGURE 54–34

A typical cycling clutch system with an expansion (orifice) tube.

Ford Motor Company

Compressor Controls

Many controls are used to monitor and trigger the compressor during its operational cycle. Each of these represents the most common protective control devices designed to ensure safe and reliable operation of the compressor.

AMBIENT TEMPERATURE SWITCH This switch senses outside air temperature and is designed to prevent compressor clutch engagement when air conditioning is not required or when compressor operation might cause internal damage to seals and other parts.

The switch is in series with the compressor clutch electrical circuit and closes at about 2.7°C (37°F). At all lower temperatures, the switch is open, preventing clutch engagement.

On some vehicles, the ambient switch is located in the air inlet duct of air-conditioning systems regulated by evaporator pressure controls; other makes have it installed near the radiator. It is not required on systems with a thermostatic or pressure switch.

THERMOSTATIC SWITCH In cycling clutch systems, the thermostatic switch is placed in series with the compressor clutch circuit so that it can turn the clutch on or off. It has two purposes: It (1) de-energizes the clutch and (2) stops the compressor if the evaporator is at the freezing point **(Figure 54–35)**.

When the temperature of the evaporator approaches the freezing point (or the low setting of the switch), the thermostatic switch opens the circuit and disengages the compressor clutch. The compressor remains inoperative until the evaporator temperature rises to the preset temperature, at which time the switch closes, and compressor operation resumes.

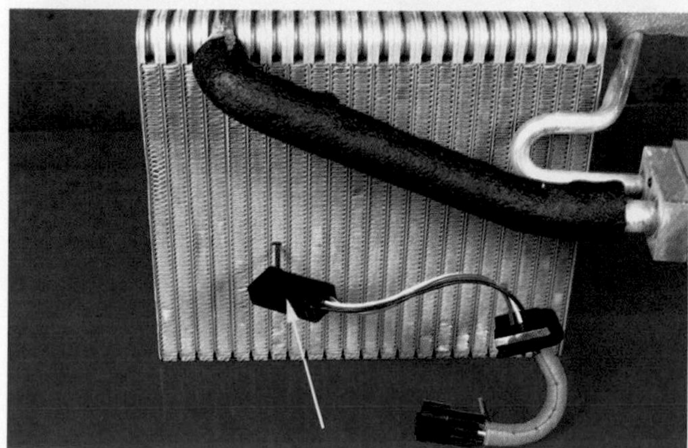

FIGURE 54–35

A thermostatic switch mounted onto an evaporator.

PRESSURE CYCLING SWITCH This switch is electrically connected in series with the compressor electromagnetic clutch. Like the thermostatic switch, the turning on and off of the pressure cycling switch controls the operation of the compressor.

LOW-PRESSURE CUTOFF, OR DISCHARGE PRESSURE, SWITCH This switch is located on the high side of the system and senses any low-pressure conditions. It is tied into the compressor clutch circuit, allowing it to immediately disengage the clutch when the pressure falls too low.

HIGH-PRESSURE CUT-OUT SWITCH This switch, normally located in the vicinity of the compressor or discharge (high side) muffler, is wired with the compressor clutch (in series). Designed to open (cut out) and disengage the clutch at 2400 to 2600 kPa (350 to 375 psi), it again closes and normally re-engages the clutch when pressure returns to 1700 kPa (250 psi)—or higher if the system uses R-134a.

HIGH-PRESSURE RELIEF VALVE A high-pressure relief valve is incorporated into many air-conditioning systems. This valve may be installed on the receiver/dryer, compressor, or elsewhere in the high side of the system. It is a high-pressure protection device that opens (normally at 3033 kPa [440 psi]) to bleed off excessive pressure that might occur in the system.

COMPRESSOR CONTROL VALVE This valve regulates the crankcase pressure in some compressors (commonly, General Motors' V5 compressor). It has a pressure-sensitive bellows exposed to the suction side that acts on a ball and pin valve, which is exposed to high-side pressure. The bellows also controls a bleed port that is also exposed to the low side. The control valve is continuously modulating—changing the displacement of the compressor according to pressure or temperature.

ELECTRONIC CYCLING CLUTCH SWITCH (ECCS) The ECCS prevents evaporator freeze-up by sending a signal to the engine control computer. The computer, in turn, cycles the compressor on and off by monitoring suction line temperature. If the temperature gets too low, the ECCS will open the input circuit to the computer, which causes the A/C clutch relay to open, disengaging the compressor clutch. Often, this switch is the thermostatic switch at the evaporator.

Solar (Sunload) Sensor

More than half of the heat that is in a vehicle's passenger compartment comes from solar radiation. Many A/C systems have a solar sensor, also called sunload sensor, that anticipates the amount

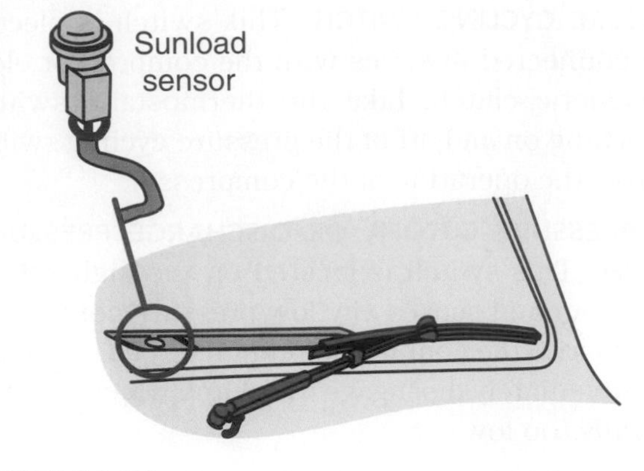

FIGURE 54–36

A solar (sunload) sensor.

of heat that will result from the sunlight. The solar sensor is usually located on top of the instrument panel **(Figure 54–36)**. The solar sensor is one of the main controls in many automatic climate control systems.

The solar sensor is a photo diode that receives a 5-volt reference signal from the control module. The diode normally blocks the flow of current in both directions, except in the presence of light. The signal voltage from the solar sensor varies with the amount of sunlight on the sensor. When the sunlight increases, the output voltage increases, and as the sunlight decreases, the output voltage decreases. Bright sunlight causes the vehicle's interior temperature to increase; therefore, the system adds more cool air to the passenger compartment to overcome the increased heat.

Some vehicles have a solar sensor that measures sunlight at two separate angles. This allows the system to react differently on the driver and passenger sides of the vehicle. Other dual-zone systems have left and right solar sensors.

Some late-model vehicles are using infrared sensors inside the passenger compartment to measure the temperature of the passengers. Input from the sensors is used to direct airflow and control compressor operation. Other vehicles monitor which seat belts are being used so that airflow can be routed where passengers are seated to maintain a more comfortable environment for them.

TEMPERATURE CONTROL SYSTEMS

Temperature control systems for air conditioners usually are connected with heater controls. Most heater and air-conditioning systems use the same plenum chamber for air distribution. Two types of air-conditioning controls are used: manual/semiautomatic and automatic.

Manual/Semiautomatic Temperature Controls

Air-conditioner manual/semiautomatic temperature controls (MTCs and SATCs) operate in a manner similar to heater controls. Depending on the control setting, doors are opened and closed to direct airflow. The amount of cooling is controlled manually through the use of control settings and blower speed **(Figure 54–37)**.

Automatic Temperature Control

An automatic or electronic temperature control system maintains a specific temperature automatically inside the passenger compartment. To maintain a selected temperature, heat sensors send signals to a computer unit that controls compressor, heater valve, blower, and plenum door operation. A typical electronic control system might contain a coolant temperature sensor, in-car temperature sensor **(Figure 54–38)**, outside temperature sensor,

FIGURE 54–37

Typical air-conditioning controls.

Martin Restoule

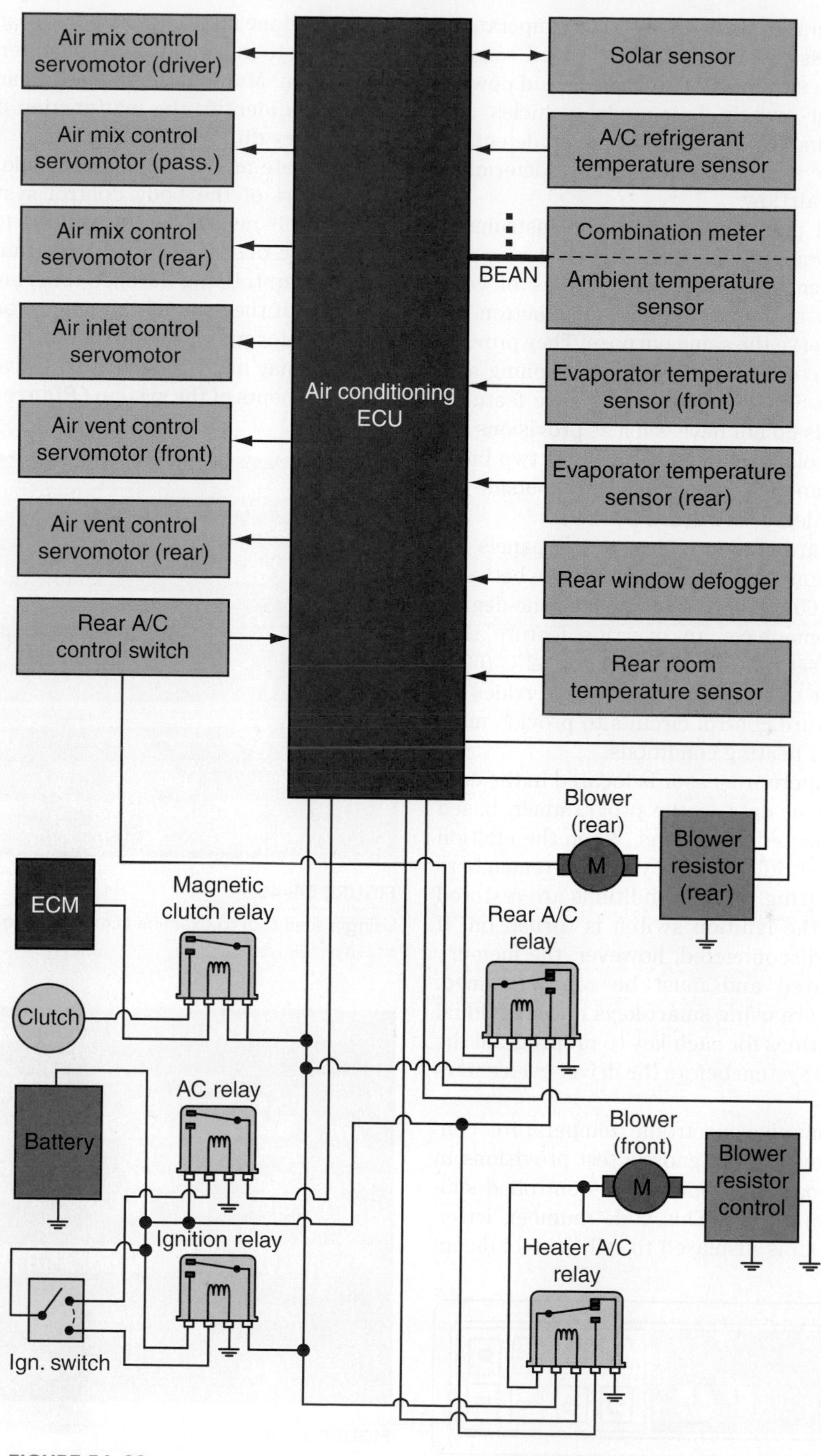

FIGURE 54–38

The layout of a computer-controlled automatic climate control system.

Chrysler Group LLC

high-side temperature switch, low-side temperature switch, low-pressure switch, vehicle speed sensor, throttle position sensor, sunload sensor, and power-steering cut-out switch. Late-model vehicles are also incorporating GPS data and occupant detection via seat belt use and infrared sensors to determine where to direct airflow.

The control panel is found in the instrument panel at a convenient location for both driver and front-seat passenger access. Three types of control panels may be found: manual, push-button, or touch pad. All serve the same purpose. They provide operator input control for the air-conditioning and heating system. Some control panels have features that other panels do not have, such as provisions for dual zone controls for climate control for two independent zones and to display in-car and outside air temperature in degrees **(Figure 54–39)**.

Provisions are made on the control panels for operator selection of an in-car temperature between 18°C and 29°C (65°F and 85°F) in half- or one-degree increments. Some have an override feature that provides for a setting of either 15°C or 32°C (60°F or 90°F). Either of these two settings overrides all in-car temperature control circuits to provide maximum cooling or heating conditions.

Usually, a microprocessor is located in the control head to input data to the programmer, based on operator-selected conditions. When the ignition switch is turned off, a memory circuit remembers the previous setting. These conditions are restored the next time the ignition switch is turned on. If the battery is disconnected, however, the memory circuit is cleared and must be reprogrammed. Some vehicles are using smart-keys and individual personality settings for each key to preprogram the climate control system before the driver even enters the vehicle.

Many automotive electronic temperature control systems have self-diagnostic test provisions in which an on-board microprocessor-controlled subsystem displays a code. This code (number, letter, or alphanumeric) is displayed to tell the technician the cause of the malfunction. Some systems also display a code to indicate which computer detected the malfunction. Manufacturers' specifications must be followed to identify the malfunction display codes because they differ from car to car.

FIGURE 54–39

An ATC switch and message display assembly.

Most late-model vehicles include the climate control part of the body control system **(Figure 54–40)**. This means that in addition to DTCs, scan data can be observed to help determine the cause of a complaint. Using the scan tool can allow you to determine if the system's inputs are being received and the information acted upon. In addition, the scan tool may have the ability to actively command the components of the system **(Figure 54–41)**. This

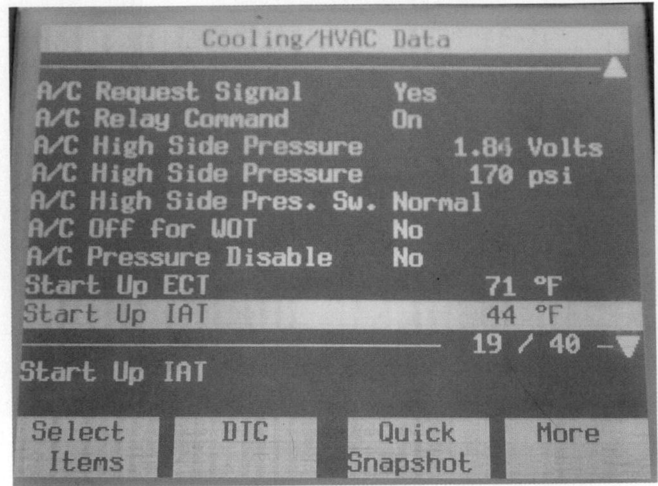

FIGURE 54–40

Using a scan tool to examine climate control data.

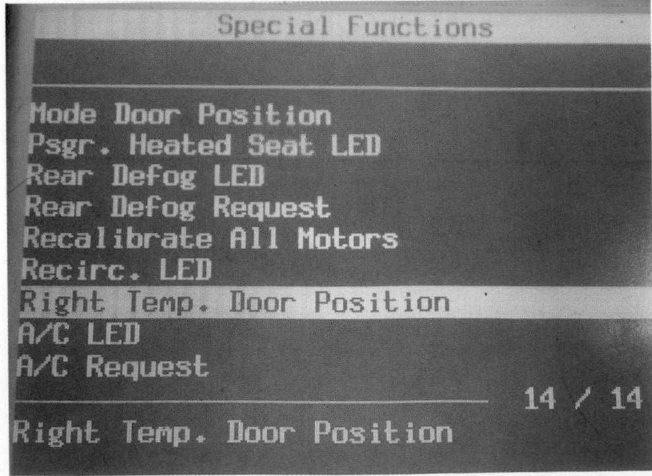

FIGURE 54–41

An example of active commands for a climate control system.

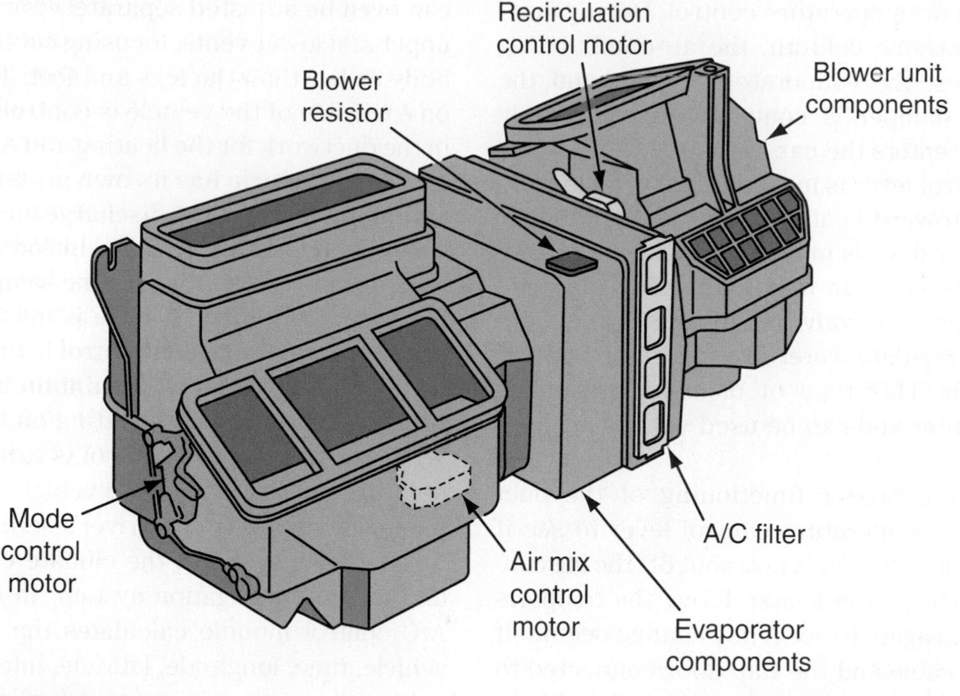

FIGURE 54–42

Typical heater/air-conditioner ducts.

can be very useful when testing stepper motors, blower motors, and the air-conditioning compressor clutch.

Case and Duct Systems

A typical automotive heater/air-conditioner case and duct system is shown in **Figure 54–42**. The purpose of the system is twofold: It is used to house the heater core and the air-conditioner evaporator, and to direct the selected supply air through these components into the passenger compartment of the vehicle. The supply air selected can be either fresh (outside) or recirculated air, depending on the system mode. After the air is heated or cooled, it is delivered to the floor outlet, dash panel outlets **(Figure 54–43)**, or the defrost outlets.

In domestic vehicles, there are two basic duct systems employed. In the stacked-core-reheat system, the basic control is in the water valve. For maximum air, the water valve is completely closed. All air enters the vehicle compartment through the heater core.

The access door, which is activated by a cable, controls only fresh or recirculated air. Recirculated air is used during maximum cold operation. The air-conditioning unit is not operative, and the evaporator will not be cold. The evaporator-only is used in the maximum air or maximum cold position. As the control level inside the car is moved, it controls the water valve by means of a vacuum or a cable to control the amount of hot water entering the heater core and the temperature of the air at the unit outlet.

The blend-air-reheat system is found on General Motors' and Ford vehicles and some truck units with factory-controlled heater system units. During heater-only operation, the air-conditioning unit is shut off, and the evaporator performs no function in

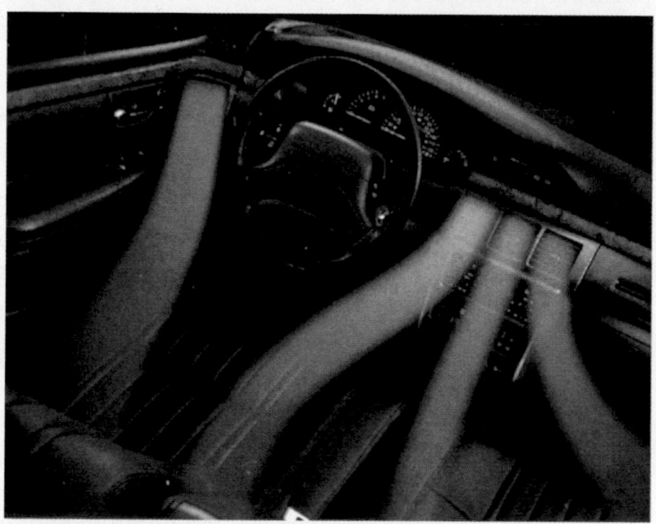

FIGURE 54–43

Air moving out from the dash vents to create comfort zones for the driver and passengers.

DaimlerChrysler Corporation

air distribution or temperature control. During maximum air or extreme cold air, the air-conditioning system operates; the evaporator is cold; and the blend-air-door damper is completely closed. Only conditioned air enters the car.

As the control lever is moved in the vehicle from maximum air toward heat with the air conditioner on, the blend-air door is moving. In maximum cold, it is completely shut. On maximum hot, it is completely open. The water valve on this unit is a vacuum on–off unit to regulate water flow. Normal position would be open. This type of blend-air system is extremely popular and can be used with or without a water valve.

To check the proper functioning of the duct work, move the temperature control lever to see if any change occurs. If it does not, shut off the air conditioner and turn on the heater. Move the temperature control arm again to see if any change occurs. If not, check the cable and the flap door connected to the temperature control lever. You might be able to reach under the dash to reconnect the cable or free a stuck flap.

If no substantial airflow is coming out of the registers, check the fuses in the blower circuit. Remove the fan switch and test it. Check the blower motor operation by following the recommended service procedures found in the manufacturer's service information.

Dual-Zone and Multiple-Zone Systems

Some climate control systems offer separate temperature settings for the driver's seat and passenger's seat **(Figure 54–44)** as well as for rear-seat passengers. On some BMW models, the temperature can even be adjusted separately for the driver's side upper and lower vents, focusing cooling on the upper body rather than the legs and feet. The temperature on each side of the vehicle is controlled by the doors in the ductwork for the heating and A/C system. Each side of the vehicle has its own air temperature actuator motor, controlling discharge air from the blower case. Each side also has an inside air temperature sensor and a solar sensor. The temperature setting for either side of the vehicle is made at the control panel. If the passenger's control is turned off, the climate control system will maintain both sides of the vehicle according to the setting on the driver's side. When the passenger's control is turned on, the temperature on that side of the vehicle can be adjusted independently from the driver's setting.

On some systems, the climate control system is tied into the navigation system. In these cases, the A/C control module calculates the direction of the vehicle, time, longitude, latitude, intensity of the sunlight, ambient temperature, and other information to determine the ideal amount of heat or cooling for each side of the vehicle. This system automatically controls the temperature in response to all of these inputs.

The automatic climate control also may be linked to the heated and ventilated seats. As the driver or passengers select their desired temperatures, the climate control system may heat, cool, or use a combination of heating and cooling based on temperature and humidity levels.

Rear Systems

Some vehicles, typically larger vans and SUVs, have a separate rear A/C system to provide comfort and temperature control for the passengers in the rear

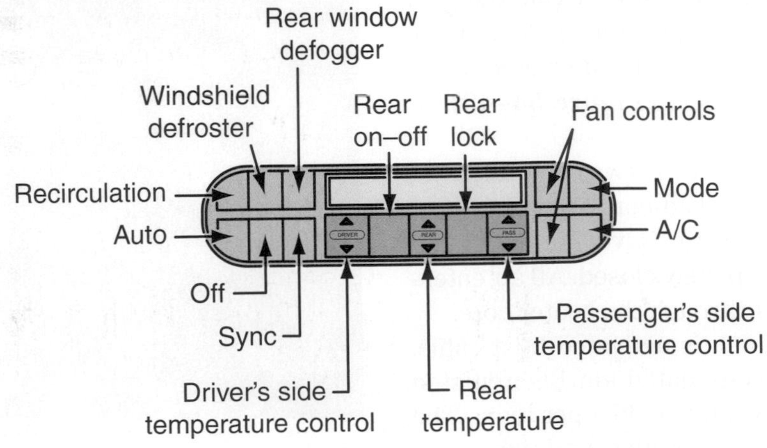

FIGURE 54–44

The control panel for a vehicle with dual-zone and a rear heating and A/C system.

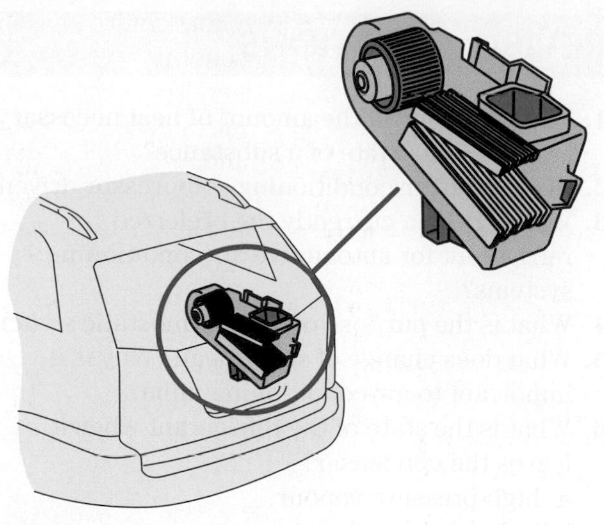

FIGURE 54–45

A rear A/C unit.

of the vehicle. These vehicles have a separate evaporator mounted in the rear of the vehicle **(Figure 54–45)**. A single conventional compressor is used to move refrigerant through the front and rear systems. There are also independent controls for the front and rear of the vehicle. Refrigerant lines connect the rear system to the compressor.

KEY TERMS

Blend door (p. 1665)
CCOT system (p. 1687)
Chlorofluorocarbons (CFCs) (p. 1672)
Condensation (p. 1675)
Condenser (p. 1676)
Cycling clutch (p. 1687)
Desiccant (p. 1682)
Discharge-side (p. 1677)
Evaporation (p. 1674)
Evaporator (p. 1676)
Heater control valve (p. 1663)
Heater core (p. 1663)
High side (p. 1676)
Hydrofluorocarbon (HFC) (p. 1672)
Latent heat (p. 1671)
Low side (p. 1676)
Orifice tube (p. 1683)
Polyalkalene glycol (PAG) (p. 1681)
Receiver/dryer (p. 1682)
Refrigerant (p. 1672)
Sight glass (p. 1686)

SUMMARY

- Ventilation, heating, and air conditioning provide for the comfort of the vehicle's passengers.
- The ventilation system on most vehicles is designed to supply outside air to the passenger compartment through upper or lower vents or both. Several systems are used to vent air into the passenger compartment. The most common is the flow-through system. In this arrangement, a supply of outside air, called ram air, flows into the car when it is moving.
- Automotive heating systems have been designed to work with the cooling system to maintain proper temperature inside the car. The heating system's primary job is to provide a comfortable passenger compartment temperature and to keep car windows clear of fog or frost.
- The main components of an automotive heating system are the heater control valve, heater core, blower motor and fan, and heater and defroster ducts.
- All air-conditioning systems are based on four fundamental laws of nature: heat flow, heat absorption, pressure and boiling points, and relative humidity.
- The major components of an air-conditioning system are compressor, condenser, receiver/dryer or accumulator, expansion valve or orifice tube, and evaporator.
- The compressor is the heart of an automotive air-conditioning system. It separates the high-pressure and low-pressure sides of the system. The primary purpose of the unit is to draw the low-pressure vapour from the evaporator and compress this vapour into high-temperature, high-pressure vapour. This action results in the refrigerant having a higher temperature than surrounding air, enabling the condenser to condense the vapour back to liquid.
- The secondary purpose of the compressor is to circulate or pump the refrigerant through the A/C system under the different pressures required for proper operation. The compressor is located in the engine compartment.
- The condenser consists of a refrigerant coil tube mounted in a series of thin cooling fins to provide maximum heat transfer in a minimum amount of space. The condenser is normally mounted just in front of the vehicle's radiator.
- The receiver/dryer is a storage tank for the liquid refrigerant.
- The refrigerant flow to the evaporator must be controlled to obtain maximum cooling while ensuring complete evaporation of the liquid refrigerant within the evaporator. This is accomplished by a thermostatic expansion valve or a fixed orifice tube.

- The evaporator, like the condenser, consists of a refrigerant coil mounted in a series of thin cooling fins. The evaporator is usually located beneath the dashboard or instrument panel.
- An amendment to the Canadian Environmental Protection Act (CEPA) Ozone-depleting Substances Regulation and international agreements have mandated that R-12 be phased out as the refrigerant of choice. This came as a result of research that found the earth's ozone layer was being deteriorated by the chemicals found in R-12.
- Of the many chemicals that could have been used in place of R-12, the automobile manufacturers have decided to use R-134a, a hydrofluorocarbon (HFC) that causes less damage to the ozone layer when released to the atmosphere.
- Although R-134a air conditioners operate in the same way and with the same basic components as R-12 systems, the two refrigerants are not interchangeable. Because it is less efficient than R-12, R-134a operates at higher pressures to make up for the loss of performance and requires new service techniques and system component designs. Basically, the higher system pressures of R-134a mean the system must be designed for those higher pressures.
- The production and importation phase-out for all CFCs, including R-12, took place on January 1, 1996. To keep systems operating that were originally equipped with R-12, there are retrofit kits available for a changeover to R-134a. The recovery and recycling of R-134a became mandatory on November 15, 1995.
- There are two basic types of automotive air-conditioning systems. They are classified according to the method used in obtaining temperature control and are known as cycling clutch systems or evaporator pressure control systems.
- Evaporator controls maintain a back pressure in the evaporator. Because of the refrigerant temperature/pressure relationship, the effect is to regulate evaporator temperature.
- Temperature control systems for air conditioners are usually connected with heater controls. Most heater and air-conditioner systems use the same plenum chamber for air distribution.
- Two types of air-conditioner controls are used: manual/semiautomatic and automatic.

REVIEW QUESTIONS

1. What do you call the amount of heat necessary to change the state of a substance?
2. How is the air-conditioning compressor driven?
3. Why is R-134a currently the preferred refrigerant for automotive air-conditioning systems?
4. What is the purpose of the thermostatic switch?
5. What does change of state mean? Why is it important to air-conditioning units?
6. What is the state of the refrigerant when it leaves the condenser?
   a. high-pressure vapour
   b. high-pressure liquid
   c. low-pressure vapour
   d. low-pressure liquid
7. What causes condensed water to leak from the air-conditioning system?
   a. moisture from the passenger compartment air settling on and draining from the heater core
   b. a chemical reaction from the refrigerant boiling in the heater housing
   c. moisture from the passenger compartment air settling on and draining from the evaporator
   d. moisture from the passenger compartment air settling on and draining from the condenser
8. What component in the fixed orifice tube system removes moisture from the refrigerant?
   a. the accumulator
   b. the compressor
   c. the receiver/dryer
   d. the thermostatic expansion valve
9. What is the state of the refrigerant when it leaves the evaporator?
   a. high-pressure vapour
   b. high-pressure liquid
   c. low-pressure vapour
   d. low-pressure liquid
10. Which of the following statements is true?
    a. Refrigerant leaves the compressor as a high-pressure, high-temperature liquid.
    b. Refrigerant leaves the condenser as a low-pressure, low-temperature liquid.
    c. Refrigerant returns to the compressor as a low-pressure, high-temperature vapour.
    d. Refrigerant leaves the compressor as a high-pressure, high-temperature vapour.

11. Which of the following air-conditioning component statements is correct?
 a. The condenser is usually located just in front of the radiator.
 b. The receiver/dryer is located in the low side.
 c. The evaporator is usually located just in front of the radiator.
 d. The accumulator is located in the high side.
12. Why are vehicle manufacturers looking for a replacement for R-134a?
 a. R-134a contributes to the greenhouse effect.
 b. R-134a contains chlorine.
 c. R-134a has a high ozone-depletion potential.
 d. R-134a contains hydrocarbons and is flammable.
13. What is the reason for the extreme heat transfer at the evaporator surface?
 a. the pressurized refrigerant circulating through the evaporator
 b. the sensible heat required to reduce the temperature of the refrigerant
 c. the latent heat required to evaporate the liquid refrigerant
 d. the sensible heat required to evaporate the liquid refrigerant
14. Which of the following components separate the air-conditioning system's high and low sides?
 a. the compressor and the condenser
 b. the compressor and the thermostatic expansion valve
 c. the compressor and the evaporator
 d. the condenser and the evaporator
15. What could cause no airflow from a heater housing when the fan selector is in the low-speed position only?
 a. a blocked blend door
 b. a burnt-out blower motor
 c. a blocked mode door
 d. a burnt-out resistor in the resistor block
16. What type of compressor clutch is used on most modern compressors?
 a. the stationary coil type
 b. the rotating coil type
 c. the permanent magnet type
 d. the centrifugal force type
17. Why must R-134a refrigerant be recovered and recycled?
 a. It is a chlorofluorocarbon.
 b. It is extremely harmful to the ozone layer.
 c. It contributes to the greenhouse effect.
 d. It forms dichlorodifluoromethane gas.
18. Which of the following oils is used in an original equipment R-134a system?
 a. mineral oil
 b. CCOT
 c. ester oil
 d. PAG
19. Which of the following components can prevent hot coolant from reaching the heater housing?
 a. the heater core
 b. the water control valve
 c. the receiver/dryer
 d. the engine thermostat
20. What is the purpose of the accumulator?
 a. to ensure only vapour reaches the compressor
 b. to ensure only liquid reaches the compressor
 c. to ensure only liquid reaches the condenser
 d. to ensure only vapour reaches the condenser

CHAPTER 55
Heating and Air-Conditioning Diagnosis and Service

LEARNING OUTCOMES

- Understand the special handling procedures for automotive refrigerants.
- Explain the concerns and precautions regarding retrofitting an air-conditioning (A/C) system.
- Describe how to connect a manifold gauge set to a system.
- Describe methods used to check refrigerant leaks.
- Use approved methods and equipment to discharge, reclaim/recycle, evacuate, and recharge an automotive air-conditioning system.
- Perform a performance test on an A/C system.
- Interpret pressure readings as an aid to diagnose A/C problems.
- Diagnose and repair A/C control systems.

Air-conditioning service, in many ways, is different than service to other parts of the vehicle. Although there are few parts in the system **(Figure 55–1)**, each component has a specific purpose and service procedure. That in and of itself is not a big deal. The challenge with air conditioning is how it operates. The system operates on changes of refrigerant pressure. Many things can cause the pressure to change; some are part of the system, some are part of the environment, and some are faults or bad components in the system.

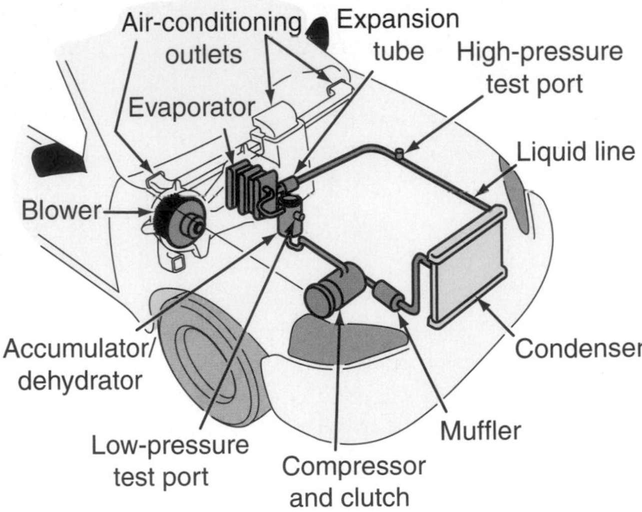

FIGURE 55–1

A late-model air-conditioning system.

MAINTENANCE PRECAUTIONS

Air-conditioning systems are extremely sensitive to moisture and dirt. Therefore, clean working conditions are extremely important. The smallest particle of foreign matter in an air-conditioning system contaminates the refrigerant, causing rust, ice, or damage to the compressor. For this reason, all replacement parts are sold in vacuum-sealed containers and should not be opened until they are ready to be installed in the system. If, for any reason, a part has been removed from its container for any length of time, the part must be completely flushed using only recommended flush solvent to remove any dust or moisture that might have accumulated during storage. When the system has been open for any length of time, the entire system must be purged completely, and a new accumulator must be installed, because the element of the existing unit will have become saturated and unable to remove any moisture from the system once the system is recharged. The following general practices should be observed to ensure chemical stability in the system.

SHOP TALK

It is important to remember that just one drop of water added to the refrigerant will start chemical changes that can result in corrosion and eventual breakdown of the chemicals in the system. Hydrochloric acid is the result of R-12 mixed with water. When R-134a and water mix, hydrofluoric acid is formed. This acid is even more corrosive than hydrochloric acid. With this being said, the smallest amount of moist air in either of these refrigerant systems might start reactions that can cause malfunctions.

Whenever it becomes necessary to disconnect a refrigerant connection, wipe away any dirt or oil at and near the connection to eliminate the possibility of dirt entering the system. Both sides of the connection should be immediately capped or plugged to prevent the entrance of dirt, foreign material, and moisture. It must be remembered that all air contains moisture. Air that enters any part of the system carries moisture with it, and the exposed surface collects the moisture quickly. In an older system that has been retrofitted, check for green-coloured Teflon or neoprene O-rings. These O-rings have replaced the older rubber ones that were more susceptible to moisture entry.

Keep tools clean and dry, including the gauge set and replacement parts. Be careful not to overtighten any connection. Overtightening results in line and flare seat distortion and a system leak.

When adding oil, the container and the transfer tube through which the oil will flow should be exceptionally clean and dry. Refrigerant oil is as moisture-free as it is possible to make it. Therefore, it quickly absorbs any moisture it contacts. For this reason, the oil container should not be opened until ready for use and then it should be capped immediately after use.

When it is necessary to open a system, have everything needed ready and handy so as little time as possible is required to perform the operation. Do not leave the system open any longer than is necessary.

Cleanliness is especially important when servicing compressors because of the very close tolerances used in these units. Consequently, repairs to the compressor itself should not be attempted unless all proper tools are at hand and a spotless work area is provided.

Any time the system has been opened and sealed again, it must be properly evacuated. Keep in mind that the complete and positive sealing of the entire system is vitally important and that this sealed condition is absolutely necessary to retain the chemicals and keep them in a pure and proper condition.

REFRIGERANT SAFETY PRECAUTIONS

Follow these tips when working with or around refrigerants:

- Always work in a well-ventilated and clean area. Refrigerant (R-12 and R-134a) is colourless and invisible as a gas. Refrigerant is heavier than oxygen and will displace oxygen in a confined area. Avoid breathing the refrigerant vapours. Exposure to refrigerant may irritate your eyes, nose, and throat.
- Refrigerant evaporates quickly when it is exposed to the atmosphere. Never rub your eyes or skin if you have been in contact with refrigerant. It will freeze anything it contacts. If liquid refrigerant gets in your eyes or on your skin, it can cause frostbite. Immediately flush the exposed areas with cool water for 15 minutes and seek medical help. Also check the material safety data sheets (MSDSs) for the refrigerant, to identify other safety-related procedures.
- An A/C system's high pressure could cause severe injury to your eyes and/or skin if a hose were to burst. Always wear eye protection when working around the A/C system and refrigerant. It is also advisable to wear protective gloves and clothing.
- Never use R-134a in combination with compressed air for leak-testing. Pressurized R-134a in the presence of oxygen may form a combustible mixture. Never introduce compressed air into R-134a containers (empty or full ones), A/C systems, or A/C service equipment.
- Be careful when handling refrigerant containers. Never drop, strike, puncture, or burn the containers. Always use Transport Canada, CSA-approved refrigerant containers.
- Never expose A/C system components to high temperatures. Heat will cause the refrigerant's pressure to increase. Never expose refrigerant to an open flame. When R-12 comes into contact with an open flame, **dichlorodifluoromethane gas** is formed. A poisonous gas called **phosgene gas** (mustard gas) can also be created in small quantities during the combustion and breakdown of the refrigerant. If phosgene gas is breathed, it can cause severe respiratory irritation. This gas

also occurs when an open flame leak detector is used. These fumes, like phosgene fumes, have an acrid (bitter) smell. If the refrigerant needs to be heated during service, the bottom of the refrigerant container should be placed in warm water (less than 52°C [125°F]).

- Never overfill refrigerant containers. The filling level of the container should never exceed 60 percent of the container's gross weight rating. Always store refrigerant containers in temperatures below 52°C (125°F), and keep them out of direct sunlight.
- Refrigerant comes in 14 and 23 kg (30 and 50 lb.) cylinders. Remember that these drums are under considerable pressure and should be handled following precautions. Keep the drums in an upright position. Make sure that valves and safety plugs are protected by metal caps when the drums are not in use. Avoid dropping the drums. Handle them carefully. When transporting refrigerant, do not place containers in the vehicle's passenger compartment.
- R-12 should be stored in white containers, whereas R-134a should be stored in light blue containers. R-12 and R-134a should never be mixed. Their oils and desiccants are not compatible. If the two refrigerants are mixed, contamination will occur and may result in A/C system failure. Separate service equipment, including recovery/recycling machines and service gauges, should be used for the different refrigerants. Always read and follow the instructions from the equipment manufacturer when servicing A/C systems.
- To prevent cross-contamination, identify whether the A/C system being worked on uses R-12 or R-134a. Check the fittings in the system; all R-134a based-systems use ½-inch 16 ACME threaded fittings and quick-disconnect service couplings. Most R-134a systems can be identified by underhood labels that clearly state that R-134a is used. Most manufacturers identify the type of refrigerant used by labelling the compressor. Also look for a label with the words, CAUTION—SYSTEM TO BE SERVICED BY QUALIFIED PERSONNEL. This label or plate can be found under the hood, near a component of the system, and will tell you what kind of refrigerant is used, as well as the type of refrigerant oil. When servicing an A/C system, determine the type of refrigerant used before beginning or use an identifier first.
- R-134a can also be identified by viewing the sight glass, if the system has one. The appearance of R-134a will be milky due to the mixture of refrigerant and the refrigerant oil.
- Before storing refrigerant in a tank, evacuate the tank before filling it. Tank pressure should never exceed the maximum allowable pressure as indicated on the tank.

Special Precautions for Hybrid Vehicles

Hybrid vehicles have high-voltage systems. Most often, the A/C compressor is powered by high voltage. Careless handling of some components can lead to serious injury, including death. Always follow and adhere to the precautions given by the manufacturer. These precautions are clearly labelled in the manufacturers' service information. All service procedures should be followed exactly as defined by the manufacturer. Being careless and/or not following the procedures can cause serious injury and can cause the battery to explode! Following is a list of common-sense items to consider when working on a hybrid vehicle:

- Before doing any service on a hybrid vehicle, refer to the service information for that specific vehicle. All hybrids have similar operation but have different systems and components; this is true for vehicles made by the same manufacturer.
- All high-voltage wires and harnesses are wrapped in orange-coloured insulation. Respect the colour and stay away from it unless the system is depowered.
- Warning and/or caution labels are attached to all high-voltage parts. Be careful not to touch these cables and parts without the correct protective gear, such as safety gloves.
- Make sure that the high-voltage system is shut down and isolated from the vehicle before working near or with any high-voltage component.
- When working on or near the high-voltage system, even when it is depowered, always use insulated tools.
- Never leave tools or loose parts under the hood or close to the battery pack. These can easily cause a short.
- Never wear anything metallic, such as rings, necklaces, watches, and earrings, when working on a hybrid vehicle.

GO TO ▶ Chapter 36 for details on isolating the high-voltage system in a hybrid vehicle.

GUIDELINES FOR CONVERTING (RETROFITTING) R-12 SYSTEMS TO R-134A

The following guidelines should be followed when converting an older A/C system to R-134a. These guidelines should allow you to provide the customer with a cool vehicle and also meet current legislative mandates. After January 1, 2002, Environment Canada, through the Canadian Environmental Protection Act's (CEPA) Environmental Code of Practice, banned the refilling of any air-conditioning system with a chlorofluorocarbon (CFC) refrigerant (R-12). The guidelines are listed in order and reflect the necessary steps for making this conversion.

1. Visually inspect all air-conditioning and heater system components.
2. Use a refrigerant identifier to make sure the system contains only R-12.
3. Check the system for leaks.
4. Run a performance test, and record the temperature and pressure readings.
5. Remove all R-12 from the system with a recycling machine.
6. If the system uses a compressor with an oil sump, remove the compressor, and drain all the oil from it. Measure the amount of oil drained out.
7. Remove and inspect the expansion valve or the orifice tube. Replace it if necessary. If either is contaminated, flush the condenser.
8. Remove the receiver/dryer or accumulator, drain it, and measure the amount of oil in it.
9. Before converting, make all necessary system changes, such as hoses, gaskets, and seals.
10. Install R-134a-compatible oil into the system; put in the same amount you took out. (PAG oil is recommended.)
11. Install a new filter/dryer or accumulator with the correct desiccant.
12. Permanently install conversion fittings, using a thread-locking chemical.
13. Install a high-pressure cutoff switch if the system does not have one.
14. Install conversion labels, and remove the R-12 label.
15. Connect R-134a system evacuation equipment, and evacuate for at least 30 minutes.
16. Recharge the system to approximately 80 percent of the original R-12 charge.
17. Run a performance check, and compare readings with those taken before the conversion.
18. Leak-check the system.

GO TO Chapter 6 for a complete description of the various tools used to diagnose and service A/C systems.

Refrigerants

Always use the same type of refrigerant and refrigerant oil that the system was designed for. If the system is being converted to accept an alternative refrigerant, make sure all old refrigerant and oil is removed from the system. Also make sure the vehicle is clearly labelled **(Figure 55–2)** as to the type of refrigerant it was converted to use. There are a number of alternate refrigerants available other than R-134a, but these may or may not be ozone-depleting substances (ODS) or a blend containing some ODS, the use of which in Canada is prohibited. Environment Canada does not approve or disapprove of any of these non-ODS refrigerants. The U.S. Environmental Protection Agency (EPA) does provide information on, and technical evaluations of, many of these alternate, top-up or drop-in refrigerants. These may be used in many provinces, but the systems may have to be altered to prevent connection with R-12 or R-134a equipment. The air-conditioning system may also have to be modified to contain and effectively use these refrigerants. There are several disadvantages to using these alternative refrigerants. One is based on the future. If the system has been converted to use an alternative refrigerant and the vehicle needs service in the future, will the alternative refrigerant be available at that shop or at all? Another disadvantage is that

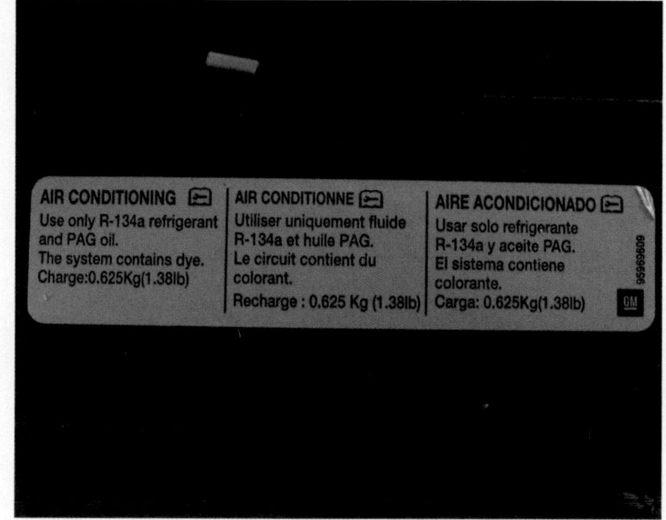

FIGURE 55–2

Air-conditioning systems should be labelled as to what refrigerant should be used in them.

many air-conditioning parts manufacturers will not warrant their parts if something other than R-12 or R-134a is used.

WARNING!

Under no circumstances, because of environmental and health concerns, should refrigerants be mixed. Never add one type of refrigerant to a system that has or has had another type of refrigerant in it.

INITIAL SYSTEM CHECKS

The first step in A/C diagnosis, as in all automotive diagnostic work, is to get the customer's story. Is the problem no cold air; the air from the ducts never gets cold enough; or the unit cools only at certain times of the day? The complaint could also be that the system does not work at all or that the air blows out of the wrong ducts. An accurate description of the customer's problem helps pinpoint whether the problem is refrigerant, mechanical, vacuum, or electrical related. It also reduces diagnosis time and, most importantly, satisfies the customer.

Because of the many construction and operational variations that exist, there is no uniform or standard diagnostic procedure applicable to all automotive A/C systems except the performance test. For complete specific diagnostic information on a given air conditioner, check the manufacturer's service information.

A quick verification of the customer's concern can be done by starting the engine and allowing it to warm up. Move the temperature control to its full heat position. Check the amount of airflow and heat in each of the fan positions. The amount of air coming out the system should change with a change in blower speed. Also pay attention to all smells and noises. Turn the temperature control to the defrost mode. Make sure there is adequate airflow coming out of the defroster vents. Now turn the temperature control to cool or to the A/C position. Pay attention to how the engine reacts when the A/C compressor clutch engages. Also pay attention to any abnormal noises or smells. The air leaving the vents should be cool. While doing these checks, attempt to duplicate the customer's concerns and the conditions that exist when the problem occurs. If the system does not respond as expected, the system should be inspected and performance tested.

Inspection

A visual check of the A/C and cooling systems can result in an immediate diagnosis. Begin by checking the condition of the compressor's drive belt. Check the tension of the belt with a belt tension gauge. (If the compressor is driven by a serpentine belt, check the belt tensioner markings.) Carefully look at the compressor clutch for signs of oil leakage and belt slippage. Belt slippage may be evident by large amounts of black dust around the clutch plate. The presence of oil on or around the clutch normally indicates a bad compressor seal.

When the system is low on refrigerant charge, the performance of the system suffers. Low charge is typically the result of system leaks. The area of leakage can often be identified with a thorough inspection. Because refrigerant oil leaks out with the refrigerant, there will be an oily film at the point of leakage **(Figure 55–3)**. This film will collect dirt, and a buildup of dirt around a fitting is a good indication of a leak. Check the following refrigerant hoses and fittings for signs of leakage or damage and bends or distortion:

- Compressor to the condenser
- Condenser to the evaporator
- Evaporator to the compressor

Although leaks from the metal lines are not very common, leaks from the rubber hoses and from around connections and fittings are. Make sure to check around electrical switches and any place a hose or line is secured or can contact another component.

Visually check the condenser and the area between the condenser and radiator for a buildup of

FIGURE 55–3

The oily film by this fitting is evidence of a refrigerant leak.

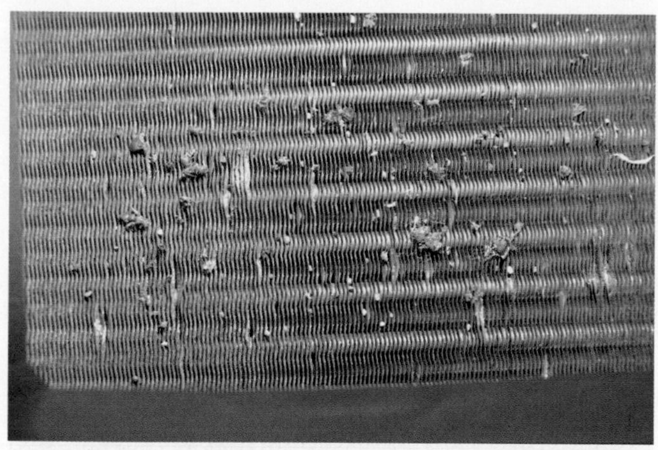

FIGURE 55–4

The condenser should be checked for damage and a buildup of debris that could cause a restriction to airflow.

dirt, leaves, and other debris **(Figure 55–4)**. Also check the fins of the condenser for damage. If the airflow through the condenser is blocked, clean the condenser and the area around it. If the fins are damaged, they may be able to be straightened. In most cases, a damaged condenser is replaced.

A/C systems rely on many switches; these can be the cause of several operating problems. Look at the wiring and connectors to all sensors and switches carefully. Make sure all connections are secure and clean. Sometimes, a bad connection or corrosion can cause intermittent problems.

In a well-ventilated area, start the engine, place the transmission in PARK (if the vehicle has a manual transmission, put it in NEUTRAL), set the parking brake, and turn on the A/C system to MAX cooling. On most vehicles, the electric engine cooling fan will turn on when the A/C system is turned on. Check to see if the fan(s) came on.

Listen to the system and record any unusual sounds. Be sure to listen to the system while the compressor clutch cycles on and off. You should hear a click or a slight change in engine speed when the clutch engages. Of course, if the compressor runs at all times, the clutch will not engage or disengage with a change of the A/C controls.

SIMPLE CHECKS Feel the air coming out of the vents. Cold or slightly cool air means the system has some cooling ability and the compressor is working. If the air is not cool, this means the compressor is not working, or the system is not capable of providing conditioned air. If the air is warm or hot, there is probably a problem with the air duct system or the controls.

Move the temperature control from hot to cool. Feel the air from the vents. If the air temperature does not change or changes little, there is a problem with the blend door or doors in the air distribution case.

Operate the blower control through all of its speed positions; the amount of air should change. If not, there is a problem with the switch or the resistor block for the blower.

Carefully feel the discharge line from the compressor to the condenser. The line should be hot, and the temperature should be the same along its full length. Any change is a sign of restriction, and the line should be flushed or replaced. Check the condenser by feeling up and down the face or along the return bends. There should be a gradual change from hot to warm as you move from the top to the bottom. Any abrupt change indicates a restriction, and the condenser has to be flushed or replaced.

If the system has a receiver/dryer, check the inlet and outlet lines. They should have the same temperature. Any temperature difference or frost on the lines or receiver tank are signs of a restriction. The receiver/dryer must be replaced. Also feel the liquid line from the receiver/dryer to the expansion valve. The entire length of the line should be warm. Typically, the formation of frost on the outside of a line or component means there is a restriction to refrigerant flow. A restriction can cause the refrigerant to stay longer in the condenser. This is called condenser flooding and will cause a starved evaporator.

The expansion valve should be free of frost, and there should be a sharp temperature difference between its inlet and outlet. On vehicles equipped with an orifice tube, feel the liquid line from the condenser outlet to the evaporator inlet. A restriction is indicated by any temperature change in the liquid line before the orifice tube. Flush the line or replace the orifice tube if restricted.

Carefully place your hand on the inlet line of the evaporator **(Figure 55–5)**. Do the same on the outlet. Both tubes should be colder than ambient temperature and both should be about the same temperature. If the inlet is significantly colder than the outlet, the system is probably low on refrigerant. If both lines are not colder than ambient temperature, the system needs further diagnosis.

The suction line to the compressor should be cool to the touch from the evaporator to the compressor. If it is covered with thick frost, this might indicate that the expansion valve is flooding the evaporator. The accumulator should also be cool to the touch.

FIGURE 55–5

Feeling the temperature of the pressure lines in an A/C system is part of a basic visual inspection of the system.

SIGHT GLASS Generally, only older R-12 systems used a sight glass that allowed a technician to see the flow of refrigerant in the lines. A sight glass is normally found on systems using a thermal expansion valve. It is not commonly found on R-134a systems because air or moisture in an R-134a system does not show in a sight glass as it did in R-12 systems. It can be located on the receiver/dryer, between the receiver/dryer and the expansion valve or tube, or in the liquid line.

For R-12 systems, check the sight glass while the engine is running **(Figure 55–6)**. If oil streaking is seen, this indicates that the system is empty. Bubbles, or foam, indicate that the refrigerant is low. A sufficient level of refrigerant is indicated by what looks like a flow of clear water with no bubbles. A clouded sight glass is an indication of desiccant contamination with subsequent infiltration and circulation through the system.

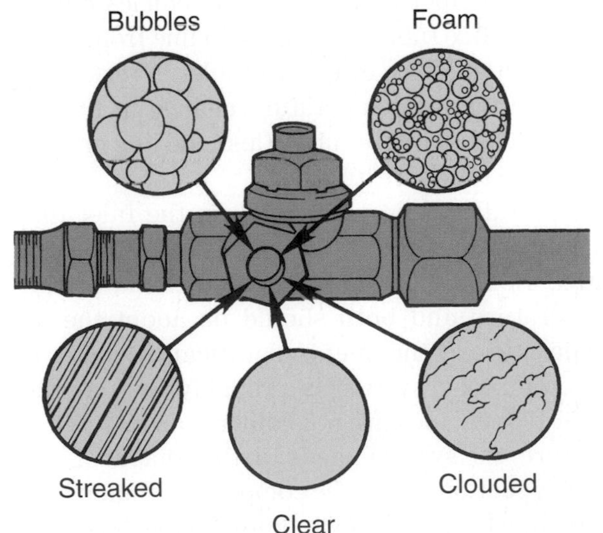

FIGURE 55–6

The appearance of the refrigerant in the sight glass can give an indication of system problems.

Accumulator systems do not have a receiver/dryer to separate the gas from the liquid as it flows from the condenser. The liquid line will always have a certain amount of bubbles in it. R-134a systems can also be checked by viewing the sight glass if the system has one. The normal appearance of R-134a tends to be milky. This is why R-134a systems and systems with an accumulator normally do not have a sight glass. Pressure, line temperatures, and performance testing are the only ways to identify low-refrigerant levels.

DIAGNOSIS

Often, diagnosis of an A/C system can be done by observing the operation of the system, including listening for noises. When abnormal noises are heard, they can lead to the problem area.

Noise

Often, a customer's concern is the noise emitted by the A/C system. The following discussion covers the common abnormal noises and their causes.

CLUTCH NOISES The clutch normally makes a clicking noise when it engages and disengages. This noise will become louder as the clutch wears. Once the clutch is severely worn, it will make a squealing noise when it is engaged. This noise can also result from oil on the clutch. A very loud screech or squeal can indicate a seized compressor. A bent drive pulley also can cause a growling or rubbing noise.

HOSE NOISE The change in pressures in the suction/discharge hoses can set up vibrations that cause sounds to appear from the inside of the vehicle. The noise is typically caused by a hose contacting another hose or part in the engine. Check the routing of the hoses to make sure they are not in contact. Also look for abrasions on the hoses. These can be caused by the contact.

COMPRESSOR NOISE If the mounts for the compressor are loose or damaged, there will be a rattling or groaning noise. The noise is normally random. At times, loose mounts will also cause premature wear of the drive belt. Worn bearings and/or internal damage to the compressor can cause whining and growling noises when the compressor is engaged.

HISSING OR WHISTLING This is a common and normal noise. When the A/C is turned off, a high-pitched whistle may be heard. This sound is caused by the pressures that are equalizing in the system. The high pressure moves to the low-pressure side through the metering device. This movement can result in a whistle or hiss.

Odours

A somewhat common concern of customers is a smell emitting from the A/C system. This mouldy and musty smell is due to a buildup of moisture on the evaporator or the cabin filter. Moisture buildup is caused by the evaporator doing its job. As the temperature inside the vehicle is lowered, moisture is separated from the air; the moisture should be able to drain, but sometimes the drain is plugged or dislocated. When a customer complains about the smell, check the evaporator drain. Also, the cabin air filter needs to be periodically changed; if it has not been changed, there will be odours that offend the passengers. Products are available to disinfect the evaporator and ductwork. These kill any odour-causing bacteria. Application of these products may require access to the evaporator, requiring disassembly of the HVAC system.

CABIN FILTERS Cabin air filters are designed to capture soot, dirt, pollen, and other pollutants that enter a vehicle through its heating, A/C, and defrost systems. Today, more than 80 percent of all new vehicles sold in North America have a cabin air filter or a slot where one can be installed. Cabin air filters are typically located behind the glove compartment, under the dash, and under the hood **(Figure 55–7)**. Some cars have the filter in the HVAC case between the blower motor and evaporator core. A cabin filter is a critical part of the HVAC system. If the filter becomes dirty or clogged, less air will be able to pass through the filter. This will adversely affect the operation of the HVAC system.

FIGURE 55–7

A typical cabin air filter and housing.

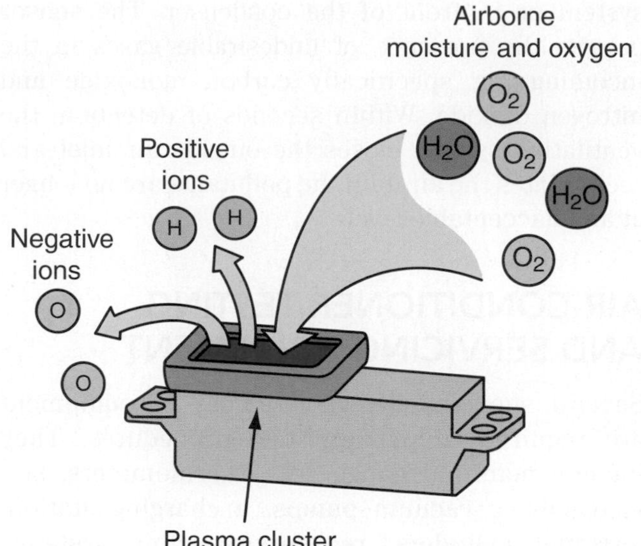

FIGURE 55–8

Many Toyotas have an ion generator called the Plasmacluster generator that improves interior air quality.

Many Toyotas have an ion generator called the Plasmacluster generator **(Figure 55–8)** that improves interior air quality. It uses a high-voltage device that emits a slight sound. With these systems, dust will accumulate at the driver-side air vent. Clean it with a cloth; never spray solvent into the air vent.

Replacement of the filters is part of a vehicle's preventive maintenance program; normally, a filter should be replaced every 19 000 to 24 000 km (12 000 to 15 000 mi.), or at least once a year. This interval really depends on where the vehicle is typically driven. The procedure for replacing the filters varies with make and model; always check with the manufacturer.

There are two basic types of cabin air filters: particle-trapping, and one that has an additional charcoal layer for odour absorption. Some particulate filters have a section that traps larger particles and another that is electrostatically charged to attract and hold smaller particles.

A charcoal type will hold a lot of odourant, and under some conditions it may release trapped odours. This is why these filters should be replaced on a regular basis. There are two alternatives to the charcoal filter: a filter element impregnated with baking soda and a filter with a biocidal cartridge. Both prevent the growth of bacteria, mould, mildew, algae, and yeast.

Some vehicles are also equipped with an air-quality sensor in the main inlet duct of the HVAC system or in front of the condenser. The sensor detects the presence of undesirable gases in the incoming air, specifically carbon monoxide and nitrogen dioxide. Within seconds of detection, the ventilation system closes the outside air inlet and recirculates the air until the pollutants are no longer at an unacceptable level.

AIR-CONDITIONER TESTING AND SERVICING EQUIPMENT

Several specially designed pieces of equipment are required to perform test procedures. They include manifold gauge sets, thermometers, service valves, vacuum pumps, a charging station, charging cylinders, recovery/recycling systems, and leak-detecting devices.

> **CAUTION!**
> Always wear safety goggles when working on or around air-conditioning systems.

Manifold Gauge Set

The **manifold gauge set** shown in **Figure 55–9** is one of the most valuable air-conditioning tools. It is used when discharging, charging, and evacuating, and for diagnosing trouble in the system. With the legislation on handling refrigerants, all gauge sets are required to have a valve device to close off the end of the hose so that the fitting not in use is automatically shut.

FIGURE 55–9

A manifold gauge set for R-134a.

The low-pressure gauge is graduated into kilopascals from 10 to 850 kPa in 10 kPa increments, and pounds per square inch (psi) of pressure from 1 to 120 psi in 1 psi increments, with a cushion to 2400 kPa or 250 psi. In the opposite direction, the vacuum side is graduated in millimetres of mercury (mm Hg) from 0 to 760 mm Hg, and inches of mercury (in. Hg) from 0 to 30 in. Hg. This is the gauge that should always be used in checking pressure on the low-pressure side of the system. The gauge in Figure 55–10 is graduated from 0 to 3400 kPa in 50 kPa increments and from 0 to 500 psi in 10 psi increments. This is the high-pressure gauge that is used for checking pressure on the high-pressure side of the system.

The centre manifold fitting is common to both the low and the high side and is for evacuating or adding refrigerant to the system. When this fitting is not being used, it should be capped. A test hose connected to the fitting directly under the low-side gauge is used to connect the low side of the test manifold to the low side of the system. A similar connection is found on the high side.

The gauge manifold is designed to control refrigerant flow. When the manifold test set is connected into the system, pressure is registered on both gauges at all times. During all tests, both the low- and high-side hand valves are in the closed position (turned inward until the valve is seated).

Refrigerant flows around the valve stem to the respective gauges and registers the system's low-side pressure on the low-side gauge and the system's high-side pressure on the high-side gauge. The hand valves isolate the low and high side from the central portion of the manifold. When the gauges are first connected to the gauge fittings with the refrigeration system charged, the gauge lines should always be purged. Purging is done by cracking each valve on the gauge set to allow the pressure of the refrigerant in the refrigeration system to force the air to escape through the centre gauge line. Failure to purge lines can result in air or other contaminants entering the refrigeration system.

Because R-134a is not interchangeable with the older R-12, separate sets of hoses, gauges, and other equipment are required to service vehicles. All equipment used to service R-134a and R-12 systems must meet SAE standard J1991. The service hoses on the manifold gauge set must have manual or automatic backflow valves at the service port connector ends. This prevents the refrigerant from being released into the atmosphere during connection and disconnection. Manifold gauge sets for R-134a can be identified by one or all of the following: a label reading FOR USE WITH R-134A, a label reading HFC-134 or R-134A, and/or a light blue colour on the face of the gauges.

WARNING!

Do not open the high-side hand valve with the system operating and a refrigerant can connected to the centre hose. The refrigerant will flow out of the system under high pressure into the can. High-side pressure is between 1030 and 2070 kPa (150 and 300 psi) and will cause the refrigerant tank to burst. The only occasion for opening both hand valves at the same time would be when evacuating the system or when reclaiming refrigerant with the system off.

For identification purposes, R-134a service hoses must have a black stripe along their length and be clearly labelled SAE J2196/R-134A. The low-pressure hose is blue with a black stripe. The high-pressure hose is red with a black stripe, and the centre service hose is yellow with a black stripe. Service hoses for one type of refrigerant will not easily connect into the wrong system, as the fittings for an R-134a system are different than those used in an R-12 system.

Purity Test

When you are not sure of the refrigerant used in a system or if you suspect a mixing of refrigerants has occurred, you should run a purity test and/or use a refrigerant identifier **(Figure 55–10)**. This should also be performed on any vehicle that has had previous system service. Equipment is available to do both of these. Knowing what refrigerant is in the system, or what condition it is in, will help you determine what steps you need to take to properly service the system.

Service Valves

System service valves, which incorporate a service gauge port for manifold gauge set connections, are provided on the low- and high-sides of most air-conditioning systems **(Figure 55–11)**. When making gauge connections, purge the gauge lines first by

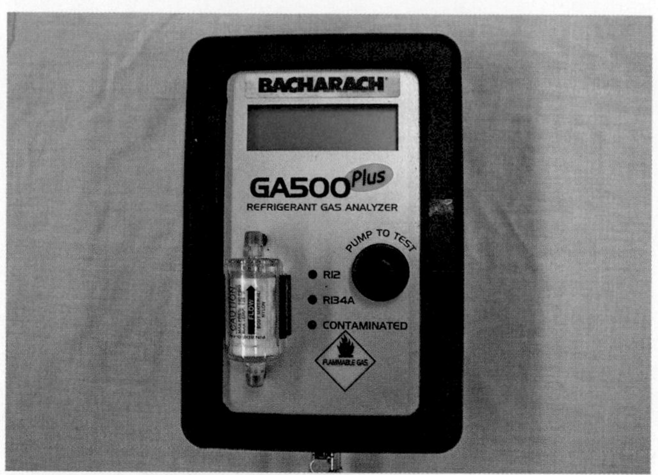

FIGURE 55–10

A refrigerant identifier.

cracking the charging valve and allowing a small amount of refrigerant to flow through the lines, and then connect the lines immediately.

Two basic types of service valves are used: stem and Schrader.

STEM SERVICE VALVE The **stem valve** is sometimes used on old two-cylinder reciprocating-piston compressors. Access to the high-pressure and low-pressure sides of the system is provided through service valves mounted on the compressor head. The low-pressure valve is mounted at the inlet to the compressor, and the high-pressure valve is mounted at the compressor outlet. Both of these valves can be used to shut off the rest of the air-conditioning system from the compressor when the compressor is being serviced. These valves have a stem under

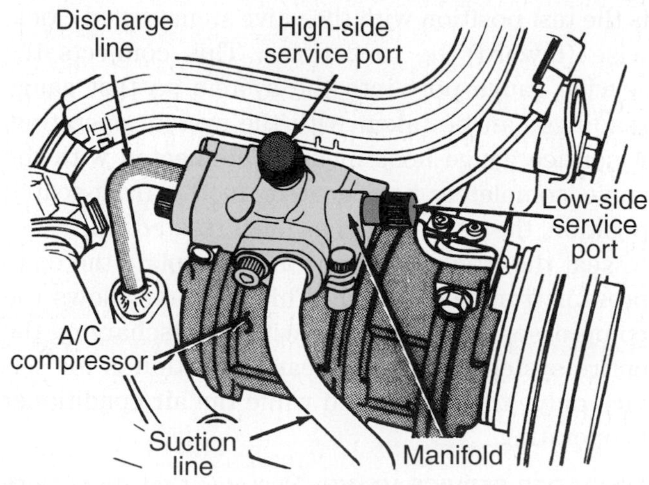

FIGURE 55–11

A/C service port locations.

Chrysler Group LLC

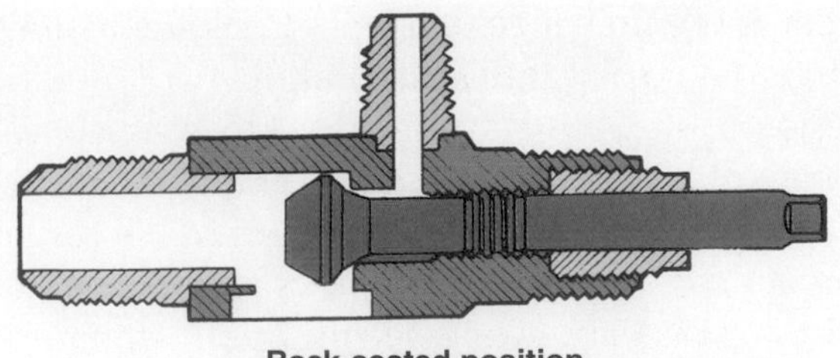

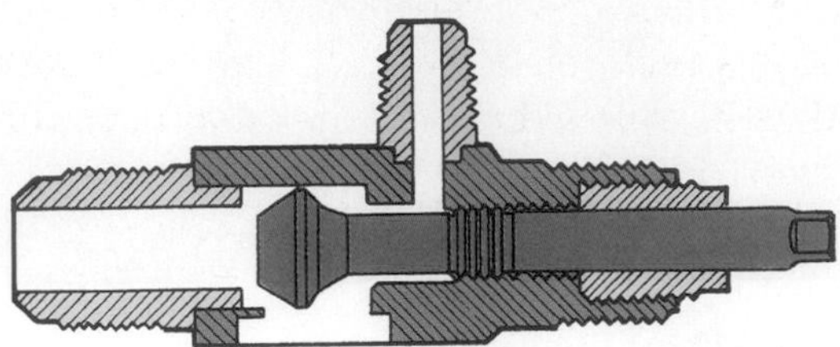

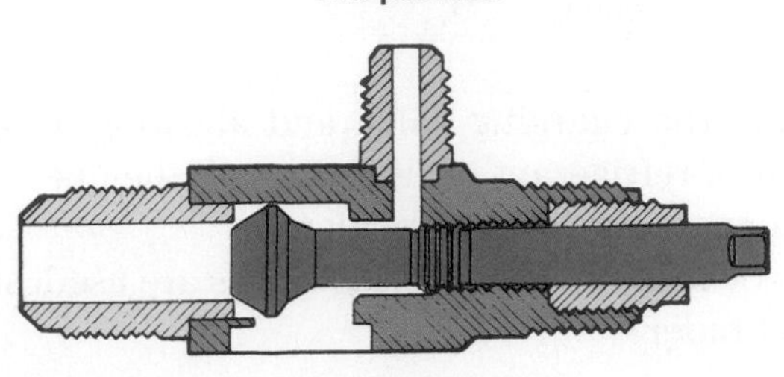

FIGURE 55–12

Stem service valve positions.

a cap **(Figure 55–12)** with the hose connection directly opposite.

A special wrench is used to open the valve to one of three positions. The back-seated position is the normal operating position with the valve stem rotated counter-clockwise to seat the rear valve face and seal off the service gauge port. The mid-position is the test position with the valve stem turned clockwise (inward) 1½ to 2 turns. This connects the service gauge port into the system so that gauge readings can be taken with the system operating. A service gauge hose must be connected with the valve completely back-seated. In the front-seated position, the valve stem has been rotated clockwise to seat the front valve face and to isolate the compressor from the system. This position allows the compressor to be serviced without discharging the entire system. The front-seated position is for service only. It is never used while the air conditioner is operating.

SCHRADER SERVICE VALVE Systems that do not utilize stem service have **Schrader service valves (Figure 55–13)** in both the high and low portions of the system for test purposes. Closely resembling a tire valve, Schrader valves are usually located in the high-pressure line (from compressor to condenser) and in the low-pressure line (from evaporator to condenser) to permit checking of the high side and low side of the system. All test hoses have a Schrader core depressor in them. As the hose is threaded into the service port, the pin in the centre of the valve is depressed, allowing refrigerant to flow to the manifold gauge set. When the hose is removed, the valve closes automatically.

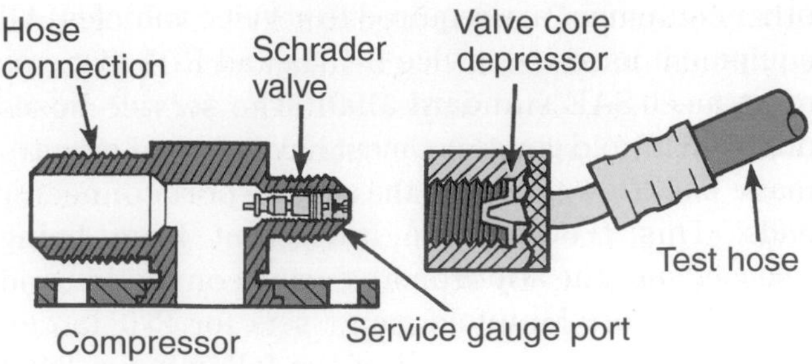

FIGURE 55–13

Schrader service valves.

After disconnecting gauge lines, check the valve areas with soapy water to be sure the service valves are correctly seated and the Schrader valves are not leaking.

Quick-Connect Fittings

R-134a systems use two different-size quick-connect fittings, the low-pressure fitting being the smaller of the two. This is to prevent connecting the service hoses incorrectly. The service hose fittings may have a manual valve that threads through the fitting to open the service port connection. This manual valve is used to reduce the amount of refrigerant released when the service hoses are disconnected from the system.

Leak-Testing a System

Testing the refrigerant system for leaks is one of the most important phases of troubleshooting. Over a period of time, all air conditioners lose or leak some refrigerant. In systems that are in good condition, R-134a losses of up to ¼ kg (0.5 lb.) per year are considered normal. Higher loss rates signal a need to locate and repair the leaks.

Leaks are most often found at the compressor crankshaft seal area, hose connections, and the various fittings and joints in the system. Refrigerant can be lost through hose permeation. Leaks can also be traced to pinholes in the condenser and evaporator caused by acid that forms when water and refrigerant mix. Because oil and refrigerant leak out together, oily spots on hoses, fittings, and components mean there is a leak. Any suspected leak should be confirmed by using any one of these methods of detection. Once a refrigerant leak is found, most provinces require that the vehicle be tagged stating the specifics of the leak. By law, all leaks must be repaired before any refrigerant can be added to the system.

VISUAL INSPECTION As mentioned before, the presence of oil around the fitting of an air-conditioning line or hose is an indication of a refrigerant leak. Carefully check all system connections. This method of leak detection is the easiest to conduct but also the least effective.

STOP LEAK PRODUCTS Products that can be added to seal leaks in the system are sold. Many OEMs do not recommend their use. Sealants may cause buildups in the system, causing unwanted restrictions. They may also cause compressor clutch problems if there is a leak around the compressor shaft. These manufacturers also will void the warranty on the system if a sealant has been installed. Always check with the manufacturer before adding a sealant to the system.

ELECTRONIC LEAK DETECTOR This is the preferred method of leak detection; it is safe, effective, and can be used with all types of refrigerants. The hand-held battery-operated electronic leak detector **(Figure 55–14)** contains a test probe that is moved about 2.5 cm (1 in.) in areas of suspected leaks. (Remember that some refrigerant vapour is heavier than air, thus the probe should be positioned below the test point.) An alarm or a buzzer on the detector indicates the presence of a leak. On some models, a light flashes to establish the leak.

FIGURE 55–14

An electronic leak detector.

SPX Corporation, Aftermarket Tool and Equipment Group

Electronic (halogen) leak detectors are common because of their accuracy and ability to detect several types of refrigerants. This type of detector must be capable of measuring a leak as small as 4 grams (0.15 oz.) per year. When selecting a leak detector, make sure the unit meets the SAE J2791 performance standard for refrigerant leak detection. J2791 sets standards for leak-detection capabilities as well as how quickly and from what distance the leak can be found.

When using an electronic leak detector, keep the following in mind:

- Many leak detectors have the sensing unit inside the tool and not at the tip.
- When checking the system, move slowly, about an inch per second, to allow the detector to get a sample of the air.
- Normally, small leaks can be found only when the A/C and engine are turned off and there are no cooling fans operating.
- Low-side leaks may be easier to find when the system is turned off.
- High-side leaks may be easier to find when the system is running.

Leak-testing guidelines are also listed in the Environmental Code of Practice for Elimination of Fluorocarbon Emissions from Refrigeration and Air Conditioning Systems. Before this test is performed, there cannot be any oil deposits from large leaks visible and no sign of physical damage. After the test is performed, all test refrigerant must be recovered if a leak is found.

Fluorescent Dye Leak Detection

A common method of leak detection is using a fluorescent dye. The dye can be injected into the system for testing. Many new vehicles have R-134a fluorescent dye installed in the refrigerant system when the vehicle is produced. The fluorescent dye mixes and flows with the PAG oil throughout the A/C system. In these systems, additional dye should not added unless more than 50 percent of the refrigerant oil has been lost because of bad fittings, hose rupture, or other major leakage. There is no need to add dye to the system after the system has been flushed. Part of the flushing process is the replacement of the accumulator or receiver/dryer. New accumulators or receiver/dryers for these systems have a fluorescent dye element in the desiccant bag. The element will

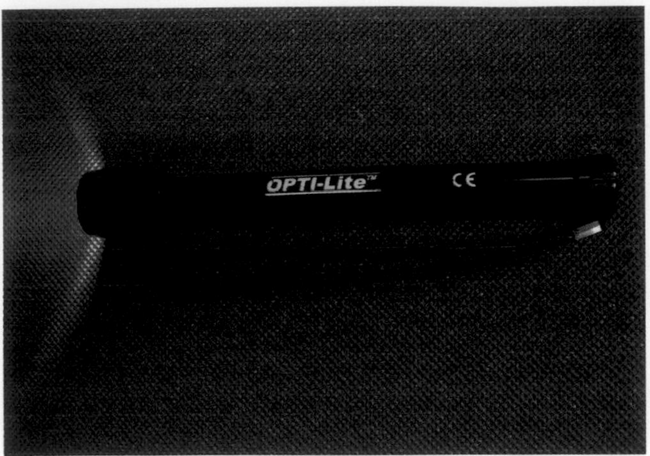

FIGURE 55–15

A fluorescent dye is added to the system before checking for leaks with the black light.

Tracer Products

dissolve while the system runs, adding the correct amount of dye to the system.

Using a UV lamp, and in some cases special glasses or goggles, system leaks will be apparent by a luminous yellow-green **(Figure 55–15)** trace. When checking for leaks, inspect all components, lines, and fittings of the refrigerant system. This includes the evaporator drain tube. The latter is done to check the evaporator for leaks. If a leak is found, continue checking the rest of the system because there may be more than one leak. It is important to remember that PAG oil is water soluble. Therefore, condensation on the evaporator or the refrigerant lines may wash the oil and fluorescent dye away from where the leak is.

If a leak is found, recover the refrigerant. Then correct the problem. Next, evacuate and charge the system. Thoroughly clean all traces of fluorescent dye from any area where leaks were found. Run the system for a short time and recheck the system for leaks to verify the repair.

LEAK DETECTOR Fluid leaks can also be located by applying leak detector fluid around areas to be tested. There are dyes available for R-134a and older R-12 systems. Make sure you use the correct one for the refrigerant you are looking for. If a leak is present, it will form clusters of bubbles around the source. It should be noted that an average-sized air bubble that forms in one second indicates a leak of approximately 1420 grams per year (50 oz./yr.). A very small leak will cause a white foam to form around the leak source within several seconds to a minute. Adequate lighting over the entire surface being tested is necessary for an accurate diagnosis.

ADDING FLUORESCENT DYE Dye may need to be added to a system that was produced with dye in the system or added to other systems for leak detection. This can be done by injecting it into the system or by pouring it directly into a removed component. It is important to note that all fluorescent dyes are not compatible with PAG oil. Before adding dye to the system, make sure it is the right type. Make sure the right quantity of dye is installed; never overcharge the system with dye. Also, the dye needs time to properly circulate through the system before it can be useful. This can be anywhere from 15 minutes to several days, depending on the size of the leak.

There are two basic ways to inject dye into the system. One uses an A/C charging station and a manifold gauge set. This should only be used when the system is not fully charged. A small tank of dye and injector is attached to the centre hose on the gauge set. The dye is added with some refrigerant to the low side of the system.

When the system is fully charged, a special kit must be used. The kit contains a reservoir for the dye and is connected between the high-pressure and low-pressure service valves. With the A/C turned on, the high-side valve is opened and remains open until the dye has left the reservoir. This method does not add refrigerant to the system.

Regardless of which type of leak detector is used, the system should have a minimum of 480 kPa (70 psi) pressure to accurately detect a leak. Be sure to check the entire system. If a leak is found at a connection, tighten the connection carefully and recheck. If the leak is still apparent, **discharge** the system. Replace the damaged components, evacuate, charge the system with nitrogen (approximately 1725 kPa [250 psi]), and inspect the entire system for leaks with soapy water. Look for bubble formation or foaming. If the system is free of any further leaks, the system can be drained and evacuated and charged with refrigerant. The system should again be checked for leaks after it has been completely recharged.

SERVICE PROCEDURES

All refrigerant must be discharged from the system before repair or replacement of any component (except for compressors with stem service valves). This procedure is accomplished through the use of a manifold gauge set and a refrigerant reclaiming unit that makes it possible to control the rate of refrigerant discharge, preventing any loss of oil from the system.

Today, the refrigerant must be reclaimed and recycled. This means the refrigerant cannot be released into the atmosphere. According to the Canadian Environmental Protection Act's (CEPA) Environmental Code of Practice for Elimination of Fluorocarbon Emissions from Refrigeration and Air Conditioning Systems, no one repairing or servicing a motor vehicle's air-conditioning system can do so without the proper refrigerant recycling equipment. It further states that no one should do this service unless that person is properly trained and certified.

Certification

Before you can purchase refrigerant, you must be certified by an Environment Canada or provincially approved program. Actually, you should be certified to work on the refrigerant part of any air-conditioning system. To become certified, a technician must have and use approved refrigerant recycling equipment and must pass an exam on refrigerant recovery and recycling. The tests for this certification are administered in some provinces by environment ministries or an approved agent.

Recycling collects old refrigerant from the system being serviced and cleans it up and prepares it to be used in the future.

The Society of Automotive Engineers (SAE) has set standards for recycled R-134a to make sure the refrigerant provides proper system performance and longevity. J1991 and J2099 are purity standards for recycled refrigerant and specify limits, in ppm (parts per million) by weight. These limits are placed on three potential contaminants. There can be no more than 15 ppm of moisture in recycled R-134a. The limit for the amount of refrigerant oil is 500 ppm for R-134a. The additional contaminant looked at by the SAE was air or noncondensable gases. There should be no more than 150 ppm in R-134a.

Refrigerant Recovery

To minimize the amount of refrigerant released to the atmosphere when A/C systems are serviced, always follow these steps:

1. The recycling equipment **(Figure 55–16)** must have shutoff valves within 300 mm (12 in.) of the hoses' service ends. With the valves closed, connect the hoses to the vehicle's air-conditioning service fittings.
2. Always follow the equipment manufacturer's recommended procedures for use. Recover the refrigerant from the vehicle, and continue the process until the vehicle's system shows vacuum instead of pressure. Turn off the recovery/recycling unit for at least five minutes. If the system still has pressure, repeat the recovery process to remove any remaining refrigerant. Continue until the A/C system holds a stable vacuum for two minutes.
3. Close the valves in the recovery/recycling unit's service lines, and disconnect them from the system's service fittings. On recovery/recycling stations with automatic shutoff valves, make sure they work properly.
4. You may now make repairs and/or replace parts in the system.

FIGURE 55–16

A refrigerant recovery, recycling, and recharging machine that is capable of doing all refrigerant services.

Martin Restoule

Recycling

All recycled refrigerant must be safely stored in Transport Canada–regulated, CSA-approved containers. Containers specifically made for R-134a should be so marked. Before any container of recycled refrigerant can be used, it must be checked for noncondensable gases.

There are currently two types of refrigerant recovery machines, the single pass and the multi-pass. Both have the ability to draw the refrigerant from the vehicle, filter and separate the oil from it, remove moisture and air from it, and store the refrigerant until it is reused.

In a single-pass system, the refrigerant goes through each stage before being stored. In multi-pass systems, the refrigerant may go through all stages or some of the stages before being stored. Either system is acceptable if it has the CSA-approved label.

Because the A/C service equipment may recycle in addition to recover and recharge, periodic maintenance, such as filter changes, may be necessary to keep the unit operating properly. Before using the A/C equipment, you should read and understand its operation and ensure that it is in proper working condition. Refer to the manual that came with the equipment to determine what type of maintenance is required and how often.

Compressor Oil Level Checks

Generally, compressor oil level is checked only where there is evidence of a major loss of system oil that could be caused by a broken refrigerant hose, a severe hose fitting leak, a badly leaking compressor seal, or collision damage to the system's components.

When replacing refrigerant oil, it is important to use the specific type and quantity of oil recommended by the compressor manufacturer. If there is a surplus of oil in the system, too much oil circulates with the refrigerant, causing the cooling capacity of the system to be reduced. Too little oil results in poor lubrication of the compressor. When there has been excessive leakage or it is necessary to replace a component of the refrigeration system, certain procedures must be followed to ensure that the total oil charge in the system is correct after leak repair or the new part is installed on the car.

When the compressor is operated, oil gradually leaves the compressor and is circulated through the system with the refrigerant. Eventually, a balanced condition is reached in which a certain amount of oil is retained in the compressor, and a certain amount is continually circulated. If a component of the system is replaced after the system has been operated, some oil goes with it. To maintain the original total oil charge, it is necessary to compensate for this by adding oil to the new replacement part. Because of the differences in compressor designs, be sure to follow the manufacturer's instructions when adding refrigerant oil to their unit.

R-12–based systems use mineral oil, whereas R-134a systems use synthetic polyalkylene glycol (PAG) oils. Using a mineral oil with R-134a will result in A/C compressor failure because of poor lubrication. Use only the oil specified for the system.

Performance Testing

Performance testing provides a measure of operating efficiency for an air-conditioning system. A manifold pressure gauge set is used to determine both high and low pressures in the refrigeration system. The desired pressure readings will vary according to temperature. Use temperature/pressure charts as a guide to determine the proper pressures. At the same time, a thermometer is used to determine air discharge temperature into the passenger compartment **(Figure 55–17)**.

Before making this test, it should be established that the air distribution (air door) portion of a factory-installed system is functioning properly. This ensures that all air passing through the evaporator is being routed directly to the air outlet nozzles.

Operating pressures vary with humidity as well as with outside air temperature. Accordingly, on more humid days, operating pressures will be on the high side of the range indicated in the service information performance charts. On less humid days, the operating pressures will read toward the lower side. If operating pressures are found to be within the normal range, the refrigeration portion of the

FIGURE 55–17

A thermometer is used to determine air discharge temperature in the passenger compartment.

air-conditioning system is functioning properly. This can be further confirmed with a check of evaporator outlet air temperatures.

Always refer to the pressure charts given in the service information when basing your diagnosis of the system on system pressures. Although the specifications will vary, here are some guidelines to help you interpret abnormal readings:

- If the high-side pressure is too high, suspect air in the system, too much refrigerant in the system, a restriction in the high side of the system, and poor airflow across the condenser.
- If the high-side pressure is too low, suspect a low refrigerant level or defective compressor.
- If the low-side pressure is higher than normal, suspect refrigerant overcharge, a defective compressor, or a faulty metering device.
- If the low-side pressure is lower than normal, suspect a faulty metering device, poor airflow across the evaporator, a restriction in the low side of the system, or a system undercharged with refrigerant.

Evaporator outlet air temperature also varies according to outside (ambient) air and humidity conditions. Further variations can be found, depending on whether the system is controlled by a cycling clutch compressor or an evaporator pressure control valve. Because of these variations, it is difficult to pinpoint what evaporator outlet air temperature should be on all applications. In general, with low-side air temperatures (21°C [70°F]) and humidity (20 percent), the evaporator outlet air temperature should be in the 2°C to 4°C (35°F to 40°F) range. On the other extreme of 27°C (80°F) outside air temperatures and 90 percent humidity condition, the evaporator air outlet temperature might be in the 13°C to 15°C (55°F to 60°F) range.

PROCEDURE

Performance-Testing the Air-Conditioning System

STEP 1 Turn on the air-conditioning recycling/charging station prior to system testing to allow it to evaluate the refrigerant in the hose and tank.

STEP 2 Connect the manifold gauge set to the respective high- and low-pressure fittings. These fittings are found in various locations within the high- and low-pressure sides of the system.

STEP 3 Close all of the doors and windows of the vehicle.

STEP 4 Adjust the air-conditioning controls to maximum cooling and high blower position.

STEP 5 Idle the engine in NEUTRAL or PARK with the brake on. For the best results, place a high-volume fan in front of the radiator grille to ensure an adequate supply of airflow across the condenser.

STEP 6 Increase engine speed to 1500 to 2000 rpm.

STEP 7 Measure the temperature at the evaporator air outlet grille or air duct nozzle (should be 1.6°C to 4.4°C [35 to 40°F]).

STEP 8 Read the high and low pressures, and compare them to the normal range of the operating pressure given in the service information.

Because it is impractical to provide a specific performance chart for all the different types of air-conditioning systems, it is desirable to develop an experience factor for determining the correlation that can be anticipated between operating pressures and outlet air temperatures on the various systems. For example, feel the discharge line from the compressor to the condenser. The discharge line should be the same temperature along its full length. Any temperature change is a sign of restriction, and the line should be flushed or replaced. Perform this test carefully because the discharge line will be hot.

Other tests can be performed with the engine running:

- Check the condenser by feeling up and down the face or along the return bends for a temperature change. There should be a gradual change from hot to warm as you go from the top to the bottom. Any abrupt change indicates a restriction, and the condenser has to be flushed or replaced.
- If the system has a receiver/dryer, check it. The inlet and outlet lines should be the same temperature. Any temperature difference or frost on the lines or receiver tank are signs of a restriction. The receiver/dryer must be replaced.
- If the system has a sight glass, check it as previously described.
- Feel the liquid line from the receiver/dryer to the expansion valve. The line should be warm for its entire length.
- The expansion valve should be free of frost, and there should be a sharp temperature difference between its inlet and outlet.
- The suction line to the compressor should be cool to the touch from the evaporator to the

compressor. If it is covered with thick frost, this might indicate that the expansion valve is flooding the evaporator.

- On vehicles equipped with the orifice tube system, feel the liquid line from the condenser outlet to the evaporator inlet. A restriction is indicated by any temperature change in the liquid line before the crimp dimples the orifice tube in the evaporator inlet. Flush the liquid line or replace the orifice tube if restricted.
- The accumulator as well as the suction line must be cool to the touch from the evaporator outlet to the compressor.
- By combining the results of both the hands-on checks and an interpretation of pressure gauge readings, the technician has a good indication that some unit in the system is malfunctioning and that further diagnosis is needed.

GENERAL SERVICE

Because vehicle air-conditioning systems contain various switches, there will be times when one of them malfunctions. Sometimes, this is not due to the switch itself but is caused instead by a bad connection or corrosion that has built up on the switch leads, leading to a broken contact. The switch can sometimes function intermittently because of this buildup. It is important to check the switches from time to time and clean off the leads. Check that each switch is making good contact and is free of dirt and corrosion.

Clutch Service

The compressor clutch is engaged by a magnetic field and disengaged by springs when the magnetic field is broken. When the controls call for compressor operation, the electrical circuit to the clutch is completed, the magnetic clutch is energized, and the clutch engages the compressor. When the electrical circuit is opened, the clutch disengages the compressor. The clutch assembly should be carefully inspected for discoloration, peeling, or other damage. If there is damage, replace the clutch assembly. Also check the play and drag of the compressor pulley bearing by rotating the pulley by hand. Replace the clutch assembly if it is noisy or has excessive play or drag. The field coil for the clutch can be checked with an ohmmeter **(Figure 55–18)**. The exact testing points and acceptable resistance readings are given in the vehicle's service information. If resistance is not within specifications, replace the field coil. Also check the clutch clamping diode. If the clutch is not operating, check the electrical connections to it. Make sure they are secure and corrosion-free. Then check for power to the clutch with a testlight or digital multimeter (DMM). If there is no power, locate and repair the problem. If there is power to the clutch, check the ground circuit with a DMM. If there is power and a good ground, the clutch is defective and must be replaced.

CLUTCH CLEARANCE Nearly all clutch assemblies have a clearance spec for the distance between the clutch and the pressure plate. This clearance

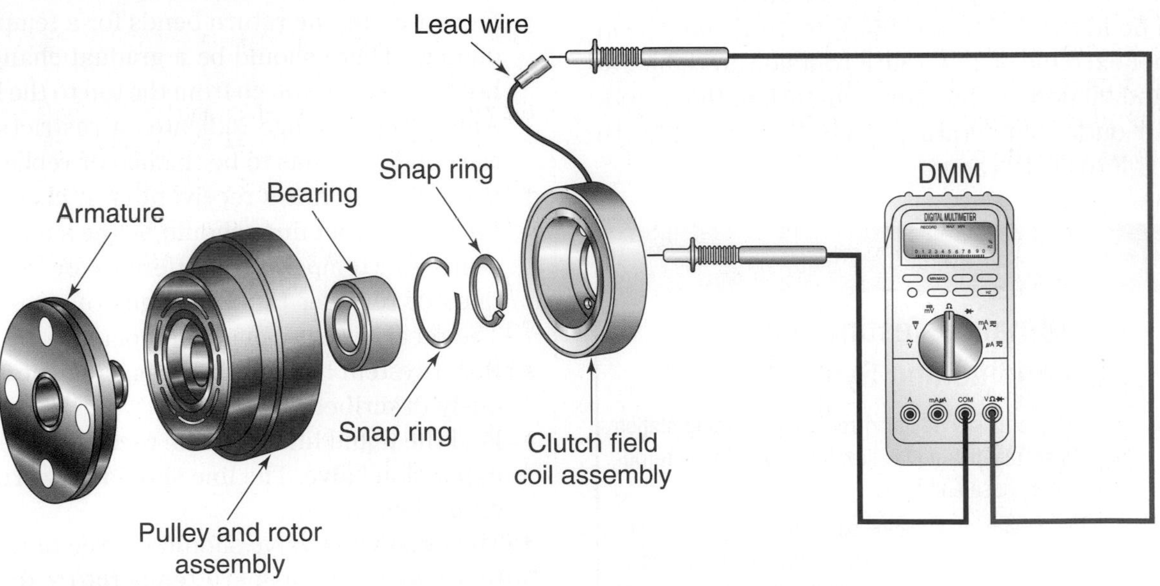

FIGURE 55–18

Testing a compressor clutch with an ohmmeter.

is measured with a feeler gauge. If the clearance is too great, the clutch may slip and cause a scraping or squealing noise. If the clearance is insufficient, the compressor may run when not electrically activated, and the clutch may chatter at all times. As the clutch assembly wears, the clearance increases and therefore should be checked and adjusted whenever symptoms suggest doing so. Always follow the specific procedures given by the manufacturer for measuring and correcting the gap.

Measure the clearance between the rotor pulley and the armature plate **(Figure 55–19)**. Do this at more than one location. If the clearance is not within the specified limits, remove the armature plate, and add or remove shims as needed to increase or decrease the clearance. The shims are available in different thicknesses, and it is recommended that no more three shims be installed to correct the clearance.

Some clutches are press-fit to the compressor shaft. The gap of these clutches is adjusted by tightening the retaining bolt. With a feeler gauge placed between the clutch plate and pulley, tighten the bolt until the feeler gauge of the specified size has a slight drag.

CLUTCH AND PULLEY REPLACEMENT On some vehicles, the clutch can be serviced on the vehicle. On others, the compressor assembly must be removed to service the clutch. Refer to the service information on your specific vehicle to determine what must be done.

Normally, to remove the clutch assembly, the magnetic clutch hub must be held while the retaining bolt is loosened. After the bolt is removed, the clutch hub and shims can be removed. This will give access to a snap ring around the compressor shaft. Using snap ring pliers, expand and remove the snap ring, and then remove the rotor of the magnetic clutch. Be careful not to damage the compressor seals while doing this. Now remove the snap ring for the clutch stator and then the stator.

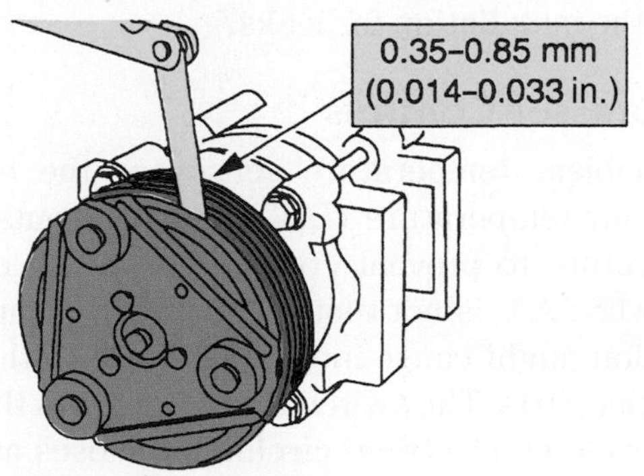

FIGURE 55–19

Checking the gap of a compressor clutch assembly with a brass or plastic feeler gauge.

Ford Motor Company

When installing a new clutch assembly, begin by aligning the protrusion on the stator with the notch on the compressor. Then install a new snap ring with its chamfered side facing up. Install the clutch rotor with a new snap ring. Now install the clutch shims and hub. Hold the hub while tightening the centre bolt to specifications. While installing the clutch assembly, make sure all parts are kept clean and free of oil or grease.

CLUTCH PULLEY BEARING REPLACEMENT The clutch pulley can be a source of unwanted noise and can be replaced on most systems. To replace the bearing, remove the pulley. Use a bearing driver or press to push the bearing out of the front of the pulley. To install a new bearing, align the pulley, and press it into place.

Compressor Service

To remove and install a compressor, follow these steps:

PROCEDURE

Removing and Installing a Compressor

STEP 1 Disconnect the negative battery cable.

STEP 2 Identify and disconnect all electrical connections to the compressor.

STEP 3 Discharge the system using a recovery/recycling machine.

STEP 4 Disconnect the refrigerant lines at the compressor. Immediately cap or seal the ends of the lines or hoses.

STEP 5 Remove the drive belt for the compressor.

STEP 6 Loosen and remove the compressor mounting brackets. Note the location of the bolts because they are typically a different length.

STEP 7 Remove the compressor.

STEP 8 To install a new compressor, add the specified amount of oil to the new compressor, and reverse the removal procedure. Make sure the drive belt is tightened to the proper tension.

SHAFT SEAL REPLACEMENT The seal at the compressor shaft is a common source of leaks. To replace the seal, discharge the system, and then remove the clutch assembly. Now remove the internal snap ring that holds the seal in place. Install the special seal remover/installer tool against the face of the seal. Twist it to expand the jaws of the tool on the seal. With a gentle twisting and pulling motion, remove the seal. Then remove and discard the O-ring seal that is located between the housing and the shaft seal.

Lubricate and install a new O-ring into its groove. Coat the new shaft seal with refrigerant oil. Then place the seal into the jaws of the seal remover/installer tool. Install a seal protector over the threads of the compressor's shaft **(Figure 55–20)**. With a gentle twisting motion, slide the seal over the protector and into the groove in the compressor. Release the installer from the seal. Then reinstall the snap ring and clutch assembly. The system will need to be evacuated before it is recharged.

FLUSHING Compressor failure causes foreign material to pass into the system. The condenser must be flushed and the receiver/dryer or accumulator replaced. Filter screens are sometimes located in the suction side of the compressor and in the receiver/dryer. The compressor inlet screen should be replaced whenever the compressor is replaced. These screens confine foreign material to the compressor, condenser, receiver/dryer, and connecting hoses. If a screen becomes clogged, it will block the flow of refrigerant. Use only recommended flushing solvents. Never use CFCs or methylchloroform for flushing. Some manufacturers recommend replacing the clogged components and installing a liquid line (in-line) filter just ahead of the expansion valve or orifice tube instead of flushing the system. If flushing is recommended by the manufacturer, use only the recommended flushing agent, and follow the specified procedures. After the system has been flushed, be sure to oil all components that require it.

FIGURE 55–20

Install the protector over the threads of the compressor shaft. Place the new seal on the installer, and install the seal with a gentle twisting motion.

FIGURE 55–21

An in-line filter.

IN-LINE FILTERS The most effective way to collect debris that may be in the system is to install an in-line filter **(Figure 55–21)**. These filters should be installed whenever the compressor seizes or has severe damage and needs to be replaced. A filter should also be installed if the expansion valve or orifice tube is plugged with debris. Filters can contain an orifice, and these should be installed in a different location than filters without an orifice. If the filter does not have an orifice, it should be installed in the liquid line between the condenser outlet and the evaporator inlet. If the filter has an orifice, it is installed between the low-pressure side of the system beyond the expansion tube. If the filter has a built-in orifice, remove the original expansion tube. In either case, the filter should be inserted at a point where the refrigerant line is straight.

To install the filter, cut the line at the desired filter location. Make sure the cut is smooth and straight. Slide the ferrules, cones, and seals for the filter onto each end of the cut line. Lubricate the seals, and insert the filter and tighten the ferrules. Make sure the lines are properly sealed, and then evacuate and charge the system. When finished, check the installation for leaks.

Compressor Controls

The ambient temperature sensor or the engine's intake air temperature (IAT) measures outside air temperature to prevent compressor clutch engagement when A/C is not required or when compressor operation might cause internal damage to the seals and other parts. The switch is in series with the compressor clutch electrical circuit and closes at about 2.8°C (37°F). At all lower temperatures, the switch is open, preventing clutch engagement.

In cycling clutch systems, a thermostatic switch is placed in series with the compressor clutch circuit so that it can turn the clutch on or off. It de-energizes the clutch and stops the compressor if the evaporator is at the freezing point. When the temperature of the evaporator approaches the freezing point, the thermostatic switch opens the circuit and disengages the compressor clutch. The compressor remains inoperative until the evaporator temperature rises to the preset temperature, at which time the switch closes and compressor operation resumes.

A pressure cycling switch is electrically connected in series with the compressor electromagnetic clutch. Like the thermostatic switch, the turning on and off of the pressure cycling switch controls the operation of the compressor.

The low-pressure cutoff, or discharge pressure, switch is located on the low side of the system and senses any low-pressure conditions. It is tied into the compressor clutch circuit, allowing it to immediately disengage the clutch when the pressure falls too low.

The electronic cycling clutch switch (ECCS) prevents evaporator freeze-up by sending a signal to the engine control computer. The computer, in turn, cycles the compressor on and off by monitoring suction line temperature. If the temperature gets too low, the ECCS will open the input circuit to the computer, causing the A/C clutch relay to open, which disengages the compressor clutch. Often, this switch is the thermostatic switch at the evaporator.

The high-pressure relief valve is used to keep system pressures from reaching a point that may cause compressor lock-up or other component damage because of excessively high pressures. When system pressures exceed a predetermined point, the pressure relief valve opens, reducing the system's pressure.

The A/C high-pressure switch is used for additional A/C system pressure control. It is normally closed, and its high-pressure contacts open at a predetermined A/C pressure. This results in the A/C turning off, preventing the A/C pressure from rising to a level that would open the A/C high-pressure relief valve. The A/C pressure transducer sensor is located in the high-pressure (discharge) side of the A/C system. The A/C pressure transducer sensor provides a voltage signal to the powertrain control module (PCM) that is proportional to the A/C pressure. The PCM uses this information for A/C clutch control, fan control, and idle speed control.

The evaporator temperature sensor measures the temperature of the cool air immediately after the evaporator. This input to the A/C control module is used to detect evaporator freeze-up.

PROCEDURE

Checking A/C Control Devices

STEP 1 Using the service information, locate the various pressure and temperature sensors in the system that control the operation of the compressor.

STEP 2 Check the service information to determine if each of these is normally closed or normally open.

STEP 3 Checking one sensor at a time, complete the following steps:

(a) Disconnect the wires at the sensor or switch.

(b) Connect the DMM across the switch terminals **(Figure 55–22)**.

(c) Set the meter for resistance or continuity checks. If the switch is normally closed, the reading should be zero ohms. If the switch is normally open, you should get an infinite reading. Any reading other than these is an indication of a faulty switch.

STEP 4 If a switch needs to be replaced, the system may need to be discharged. Check the service information before proceeding. Some systems have a Schrader-type disconnect below the switch. Switch replacement in these systems does not require system evacuation.

STEP 5 To remove the defective switch, loosen it and remove it.

STEP 6 To install the new switch, carefully thread it into place and tighten it. Then reconnect the wires leading to the switches.

Hoses and Fittings

Although total hose replacement is the preferred way to correct for a hose leak, there are several accepted ways to repair refrigerant hoses and fittings. Using insert barb fittings and a length of replacement hose, you can fabricate an acceptable replacement for an

FIGURE 55–22
All sensors and switches can be checked with an ohmmeter.

original equipment hose. Insert barb fittings can also be used to replace bad original fittings or to replace a section of a hose.

PROCEDURE

Fabricating a Replacement A/C Pressure Hose

STEP 1 Measure and mark the required length of replacement high-pressure hose.

STEP 2 Using a razor blade, cut the hose to the desired length. Make sure the cut is square and flat.

STEP 3 Apply clean refrigerant oil to the inside of the hose.

STEP 4 Install the correct ferrule onto the end of the hose.

STEP 5 Carefully inspect the new fitting to make sure it is free of nicks and other damage.

STEP 6 Coat the fitting with refrigerant oil, and insert it into the hose.

STEP 7 Using the crimping tool designed for the ferrule, crimp the ferrule **(Figure 55–23)**.

When using an insert barb fitting **(Figure 55–24)** to replace a bad original fitting, the original fitting must be removed without damaging too much of the hose. Remove the hose from the vehicle. Set the hose in a vise while cutting through the fitting's ferrule with a hacksaw. Make this cut in the direction of the hose, not through the hose. With the ferrule cut, use pliers to peel or pull the ferrule off. Cut the hose at a point just beyond the insert of the original fitting. Make sure the cut is square and flat. Apply clean refrigerant oil to the inside of the hose and to the fitting's insert. Insert the fitting into the hose. Position the hose clamp over the barb closest to the fitting, before tightening it.

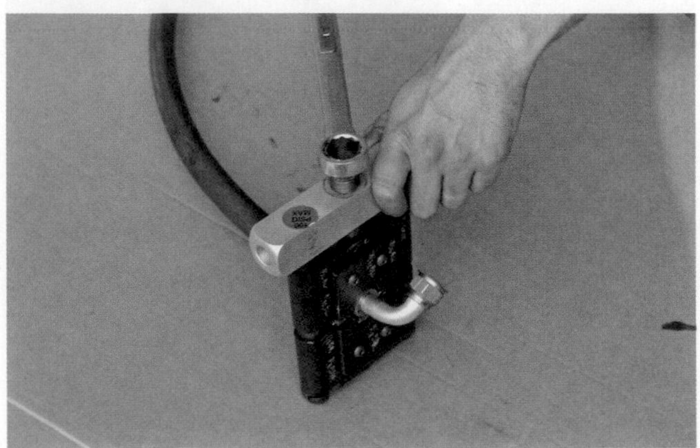

FIGURE 55–23
The ferrule is crimped with a special crimping tool.

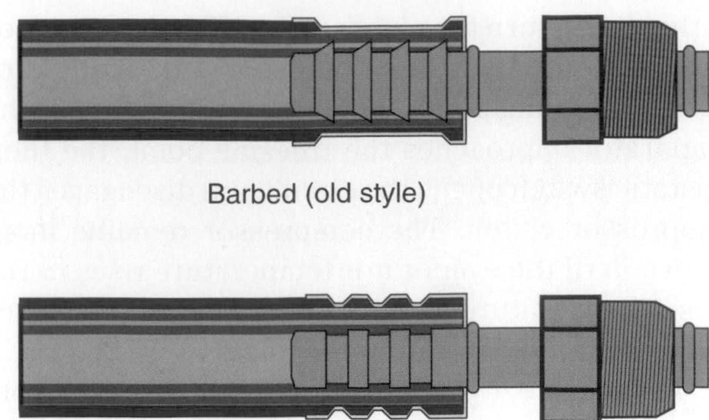

FIGURE 55–24
A comparison of a barbed fitting and the new-style beadlock fitting.

If the hose or line has spring lock fittings, a special tool is required to separate the sections. Put the tool over the coupling. Close the tool and push the lines into the tool to release the female fitting from the garter spring of the coupling. Then pull the male and female fittings apart. Remove the tool. When joining two fittings with a spring lock coupling, lubricate new O-rings and put them in their proper location on the male fitting. Insert the male fitting into the female fitting. Firmly push them together until they are secured by the spring lock coupling.

RIGID LINE REPAIR Like pressure hoses, solid or rigid refrigeration lines are replaced, rather than repaired, if they leak. However, repairs can be made with special collars **(Figure 55–25)** inserted into the line to correct a kink or leak. The damaged section of the tube is cleanly cut out. The collar or fitting is inserted into the line after sealant has been applied to it. Using a special tool, the collar is pressed into the ends of the existing line **(Figure 55–26)**. The line is then crimped around the insert. This provides for a permanent seal.

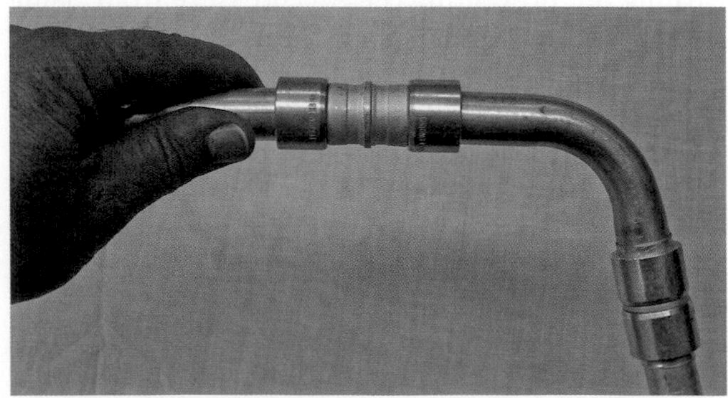

FIGURE 55–25
Repairs to rigid lines can be made with special collars.

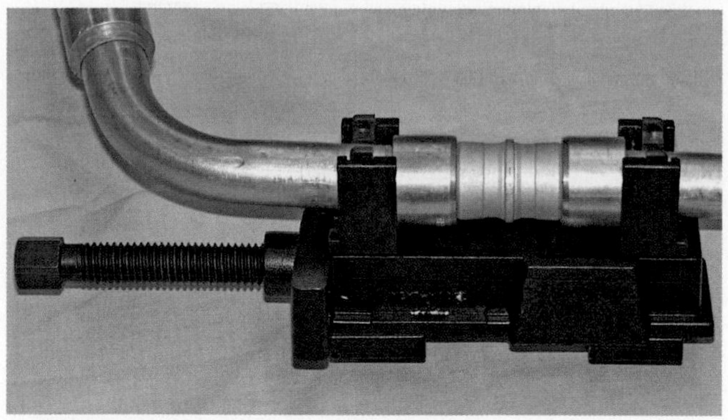

FIGURE 55–26

A designated tool is used to press the collar into the tube.

Receiver-Dryer/Accumulator

The receiver-dryer/accumulator is often neglected when the A/C system is serviced or repaired. Failure to replace it can lead to poor system performance or replacement part failure. It is recommended that the receiver-dryer/accumulator and/or its desiccant be changed whenever a component is replaced, the system has lost the refrigerant charge, or the system has been open to the atmosphere for any length of time.

The desiccant draws moisture like a magnet, and it can become contaminated in less than five minutes if it is exposed to the atmosphere. Keep it sealed. To replace a receiver-dryer/accumulator, disconnect the electrical connector to the low-pressure switch, if there is one. Then disconnect the inlet and outlet hoses or lines at the receiver-dryer/accumulator. Remove the mounting bolts and brackets from the receiver-dryer/accumulator, and lift it out. The procedure for installation is the reverse of removal.

ACCUMULATOR Most late-model systems do not have a receiver/dryer; rather, they use an accumulator. The accumulator is connected to the low side at the outlet of the evaporator. The accumulator also contains a desiccant and is designed to store excess refrigerant and to filter and dry the refrigerant. If liquid refrigerant flows out of the evaporator, it will be collected by and stored in the accumulator. An accumulator (also known as an accumulator/dryer) should be replaced if there is excessive moisture or debris in the A/C system or when the accumulator leaks. To replace an accumulator, disconnect the lines at the accumulator's inlet and outlet fittings. Then loosen and/or remove the accumulator mounting bolts or screws. Lift the accumulator out. The procedure for installation is the reverse of removal. As soon as the accumulator is installed, immediately evacuate the system.

Evaporator

The evaporator and its controls often need service. If the evaporator is clogged, leaking, or damaged, it should be replaced.

PROCEDURE

Removing and Installing an Evaporator

STEP 1 Recover the refrigerant using approved recovery/recycling equipment.

STEP 2 Disconnect the negative battery cable if specified in the service information.

STEP 3 If the evaporator and heater core are a combined unit, drain the engine's coolant.

STEP 4 Disconnect and label all electrical connectors, cables, and vacuum hoses that are connected to the evaporator.

STEP 5 Disconnect the refrigerant hoses at the evaporator, and plug or cap the hose ends to prevent dirt and moisture from entering the system.

STEP 6 Unbolt and remove the evaporator.

STEP 7 Drain the oil from the evaporator into a graduated container. Record the amount of oil removed.

STEP 8 Check the oil for dirt. If the oil is contaminated, replace the accumulator or receiver/dryer.

STEP 9 Add the above amount of new refrigerant oil or the amount specified in the service information to the new evaporator.

STEP 10 Coat the new O-rings for the evaporator and the line fittings on the evaporator with clean refrigerant oil.

STEP 11 Install the new evaporator and tighten the fittings. Also install a new receiver-dryer/accumulator or its desiccant bag.

STEP 12 Immediately evacuate the system.

STEP 13 Reinstall and reconnect all parts, wires, cables, and vacuum hoses that were disconnected during removal. Add coolant if needed.

STEP 14 Connect the negative battery cable.

STEP 15 Evacuate and recharge the system.

STEP 16 Perform a leak test, and correct any problems.

EVAPORATOR WATER DRAIN Because the evaporator is also responsible for controlling the humidity in the vehicle, a check of its drain is necessary. This is especially true if the customer had concerns about odour or water on the carpet. To check the drain, raise the vehicle on a lift. Locate the evaporator case drain tube. Place the drain pan under the evaporator case. Disconnect the tube from the case. If no water comes out, carefully clean out the drain hole and tube at the evaporator case. You may need to insert a rod through the tube to clean it out; do this carefully because the tube is accordion shaped, and one of its bends may feel like a restriction. If no water comes out while trying to clear the tube, the entire evaporator case must be cleared. If water comes out of the case, the outside drain tube should be cleared by inserting the air nozzle into an end of the tube. Low air pressure should remove any restrictions in the tube.

Odour Control

Because the evaporator becomes wet with condensed water vapour, odour from mildew buildup in the evaporator case or on the evaporator itself is a common concern. To reduce odours, manufacturers and the aftermarket supply disinfecting kits to clean the evaporator. When performing a cleaning of the evaporator, refer to the manufacturer's service information for the proper disinfectant and procedures. In general, you will need to apply the disinfectant directly to the evaporator and allow the chemicals to work on the surface, and then rinse with clean water.

CAUTION!

Always wear safety goggles and protective gloves when handling disinfectants and cleaning the system.

EXPANSION DEVICES Basically, there are two types of expansion devices. On late-model vehicles, the orifice tube is the most common. To replace an orifice tube assembly, always use the tools and procedures recommended by the manufacturer, including proper refrigerant recovery. Begin by disconnecting the inlet line at the evaporator. Then pour a small amount of clean refrigerant oil into the orifice tube. Insert the orifice removal tool. Turn the handle of the tool just enough for it to engage onto the tabs of the orifice tube. Hold the handle of the tool in position while turning the tool's outer sleeve clockwise to remove the orifice tube **(Figure 55–27)**. Coat the new orifice tube with clean refrigerant oil. Place it into the evaporator line inlet. Push it in until it stops. Install a new O-ring on the refrigerant line **(Figure 55–28)**, and reconnect it.

To replace a thermostatic expansion valve, disconnect the inlet and outlet lines at the thermostatic expansion valve (TXV). Then remove whatever is used to keep the remote sensing bulb secure. Loosen and/or remove the mounting clamp for the TXV, and remove the TXV from the evaporator. The procedure for installation is the reverse of removal.

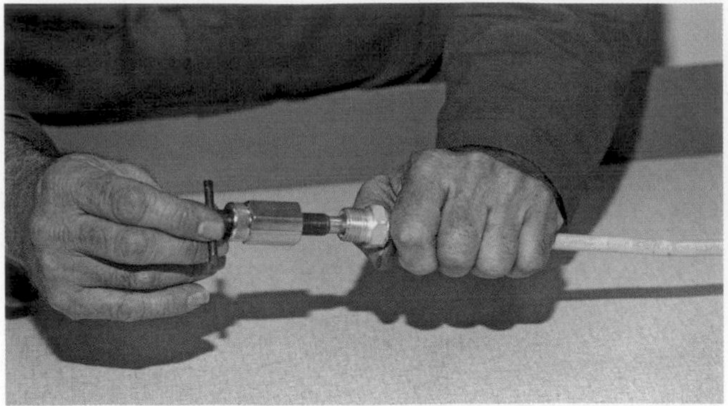

FIGURE 55–27
The tool designed to remove and replace an orifice tube.

Condenser

The condenser is quite reliable and only needs replacement when there are leaks due to damage. On many vehicles, the radiator, cooling fan, and/or shroud must be removed to gain access to the condenser. If the radiator must be removed, drain the coolant. Disconnect and label all electrical connectors that are connected to the condenser and all those that may be in the way. Disconnect the refrigerant hoses at the condenser, and plug or cap the hose ends to prevent dirt and moisture from entering the system. Unbolt and remove the condenser. If the receiver-dryer/accumulator mounts to the condenser, disconnect it and remove it with the condenser. Drain and measure the oil from the

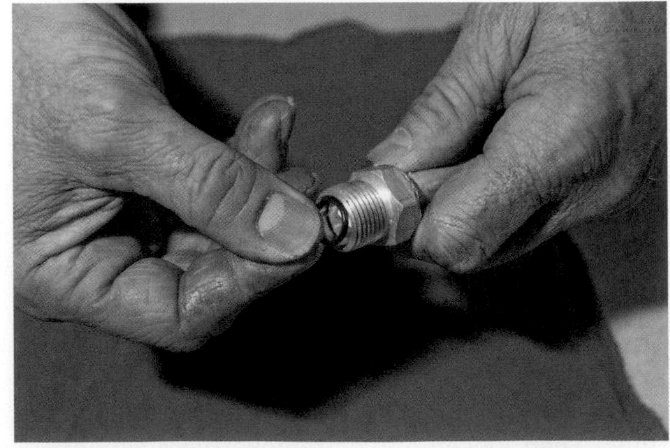

FIGURE 55–28
Always install a new O-ring when replacing an orifice tube.

condenser into a graduated container. Check the oil for dirt and debris. If the oil is contaminated with moisture, replace the accumulator or receiver/dryer. Add the above amount of new refrigerant oil or the amount specified in the service information to the new condenser. Install the new condenser, and loosely tighten the mounting bolts. Coat the new O-rings for the condenser and the line fittings on the condenser with clean refrigerant oil. Tighten the fittings, and then tighten the mounting bolts. Then immediately evacuate the system. Reinstall and reconnect all parts, wires, cables, and vacuum hoses that were disconnected or removed during removal. Connect the negative battery cable. Check for coolant and refrigerant leaks.

RECHARGING THE SYSTEM

All of the refrigerant in the system must be recovered prior to evacuation. Evacuation is the name given to the process that pulls all traces of air and moisture from the system. This is done by creating a vacuum in the system **(Figure 55–29)**. A vacuum pump is connected to the system to do this. The vacuum pump should remain on and connected to the system for at least 30 minutes after 660 to 740 mm Hg (26 to 29 in. Hg) is reached.

> **CAUTION!**
> Always wear safety goggles when working with refrigerant containers or servicing A/C systems.

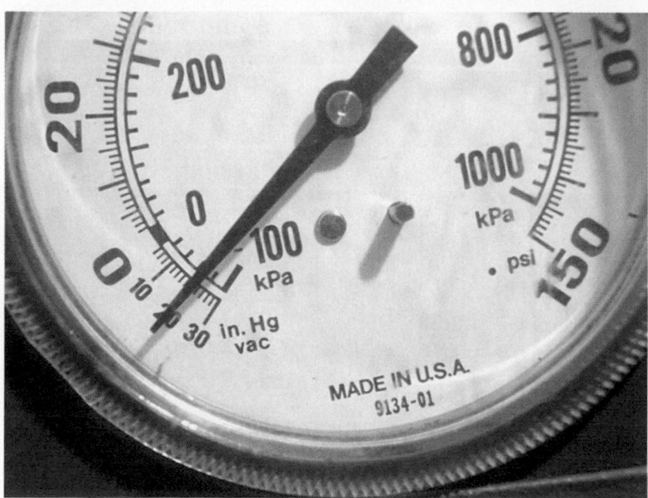

FIGURE 55–29

The reading on a low-pressure gauge with a vacuum pump connected to an A/C system through a gauge set.

Any air or moisture that is left inside an A/C system reduces the system's efficiency and eventually leads to major problems, such as compressor failure.

Air causes excessive pressure within the system, restricting the refrigerant's ability to change its state from gas to liquid within the refrigeration cycle, which drastically reduces its ability to absorb and transfer heat. Moisture, on the other hand, can cause freeze-up at the orifice tube or expansion valve, which restricts refrigerant flow or blocks it completely. Both of these problems result in intermittent cooling or no cooling at all. Moisture mixed with R-134a forms hydrofluoric acid, causing internal corrosion, which is especially dangerous to the compressor. The vacuum pump reduces system pressure, vaporizes the moisture, and then exhausts it with the air.

The main responsibility of the vacuum pump is to remove the contaminating air and moisture from the system. The vacuum pump reduces system pressure in order to vaporize the moisture and then exhausts the vapour along with all remaining air. The pump's ability to clean the system is directly related to its ability to reduce pressure—create a vacuum—low enough to boil off all the contaminating moisture. As pressure in the system reaches 737 mm Hg (29 in. Hg), the boiling point of any moisture in the system approaches room temperature. As pressure is further reduced, the moisture will boil at even lower temperatures. Between 737 mm Hg and 757 mm Hg (29 in. Hg and 29.8 in. Hg), the boiling point drops to about 0°C (32°F).

An electronic thermistor vacuum gauge is designed to work with the vacuum pump to measure the last, most critical 25 mm Hg (1 in. Hg) of mercury during evacuation. It constantly monitors and visually indicates the vacuum level so that you know when the system has a full vacuum and will be moisture-free. After the system is evacuated, it can be recharged. If the system will not pull down to a good vacuum, there is probably a leak somewhere in the system. Photo Sequence 45 goes through a common way to evacuate and recharge a system.

Charging

Refilling the system with refrigerant is called charging the system. Refrigerant is added through the system's service ports **(Figure 55–30)**. Depending on the method and source of new refrigerant, refrigerant is introduced through the low or high side of the system. When the system is operating, refrigerant is charged through the low side. On

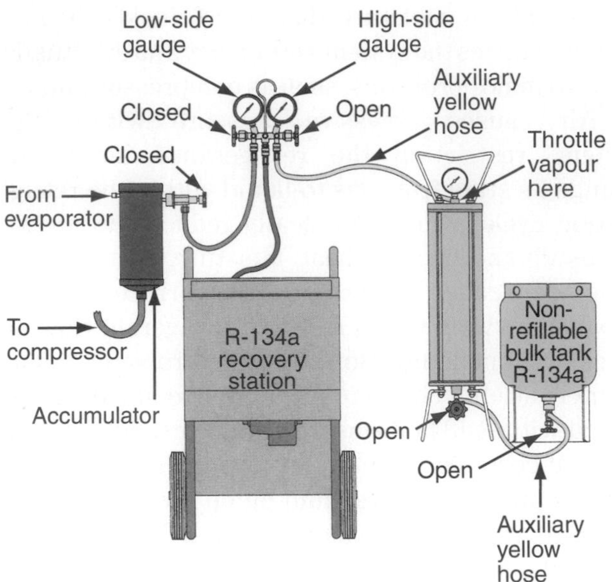

FIGURE 55–30

The connections for filling a system with refrigerant using a charging cylinder.

SPX Corporation, Aftermarket Tool and Equipment Group

some vehicles with the system off, charging is done through the high side. Always refer to the service information for the correct procedure.

Because late-model vehicles require less refrigerant than in the past, modern recharging equipment must be more accurate in determining the charge amount. New charging stations must meet the SAE J2788 standard, which requires an accuracy of 15 CCs (0.5 oz.). This helps prevent overcharging, degrading system performance, and causing damage to the system.

Always use the same type of refrigerant and refrigerant oil for the system. The importance of the correct charge cannot be stressed enough. The efficient operation of the A/C system greatly depends on the correct amount of refrigerant in the system. A low charge results in inadequate cooling under high heat loads due to a lack of reserve refrigerant, and can cause the clutch cycling switch to cycle faster than normal. An overcharge can cause inadequate cooling because of a high liquid refrigerant level in the condenser. Refrigerant controls will not operate properly, and compressor damage can result. In general, an overcharge of refrigerant will cause higher than normal gauge readings and noisy compressor operation.

The charging cylinder is designed to meter out a desired amount of a specific refrigerant by weight. Compensation for temperature variations is accomplished by reading the pressure on the gauge of the cylinder and dialling the plastic shroud. The calibrated chart on the shroud contains corresponding pressure readings for the refrigerant being used.

When charging an A/C system with refrigerant, the pressure in the system often reaches a point at which it is equal to the pressure in the cylinder from which the system is being charged. To get more refrigerant into the system to complete the charge, heat must be applied to the cylinder.

The A/C system may only be charged through the high side with the system off **(Figure 55–31)**. Inverting the refrigerant container disperses liquid refrigerant. As a general practice, the system is vapour-charged through the low side while the system is running. If the system is charged through the high side (with the system off), the compressor should be turned a few times by hand afterward to ensure there is no liquid refrigerant on top of the piston. If liquid refrigerant enters the compressor, this component may be damaged. The vehicle should always be charged from the low-pressure side if the engine is on. If the engine is off, it can be charged from the high-pressure side.

Never open the high-side hand valve with the system operating and a refrigerant can connected to the centre hose. The refrigerant will flow out of the system under high pressure into the can. High-side pressure is between 1030 and 2070 kPa (150 and 300 psi) and will cause the refrigerant tank to burst. The only occasion for opening both hand valves at the same time would be when evacuating the system or when reclaiming refrigerant with the system off.

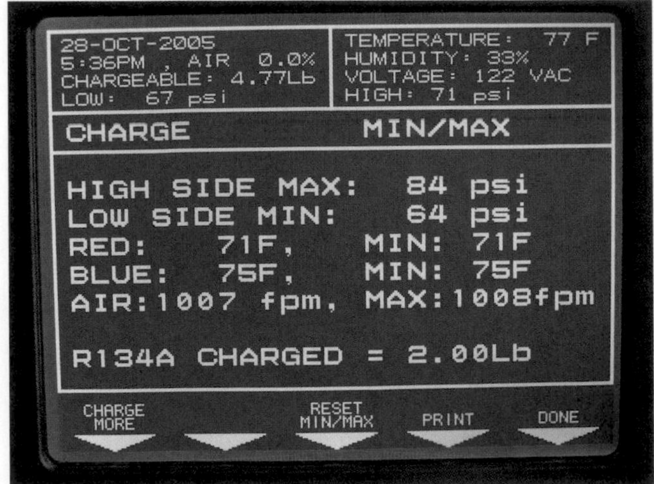

FIGURE 55–31

The recharging equipment will display how much refrigerant was added to the system.

EVACUATING AND RECHARGING AN A/C SYSTEM WITH RECYCLING AND CHARGING STATIONS

PHOTO SEQUENCE 45

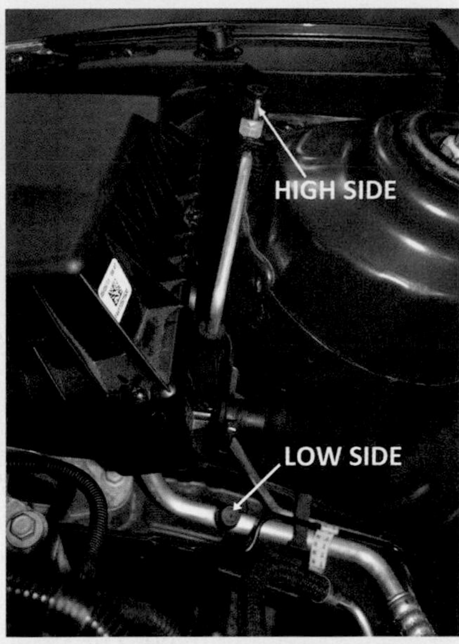

P45–1 Physically locate the pressure fittings of the system you will be working with. Consulting service information will assist in proper identification.

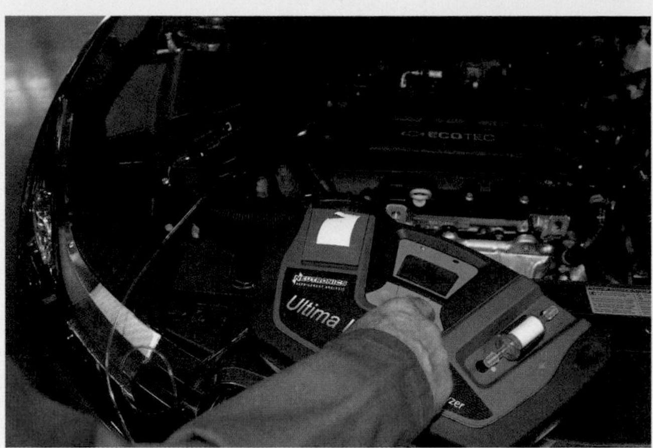

P45–2 Prior to connecting the refrigerant recovery station to the vehicle, a refrigerant identifier should be used to verify the type used in the system.

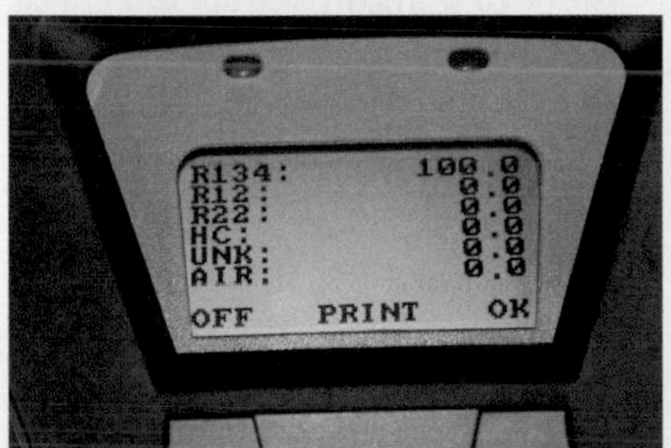

P45–3 Only uncontaminated refrigerant should be drawn into the refrigerant recovery equipment.

P45–4 Connect the recovery station to the vehicle's service ports.

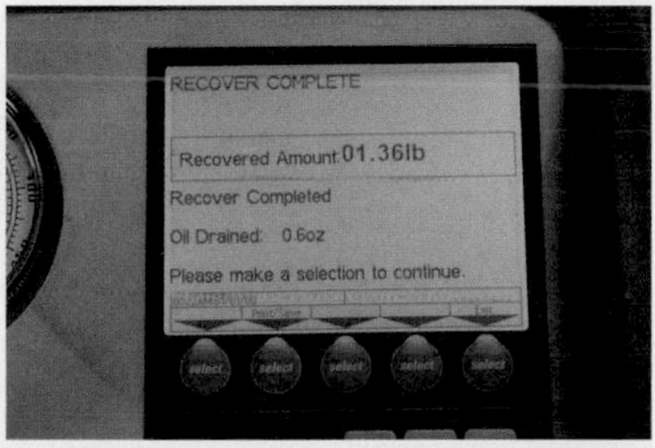

P45–5 Recover all the refrigerant, and note the amount of oil recovered for replacement later.

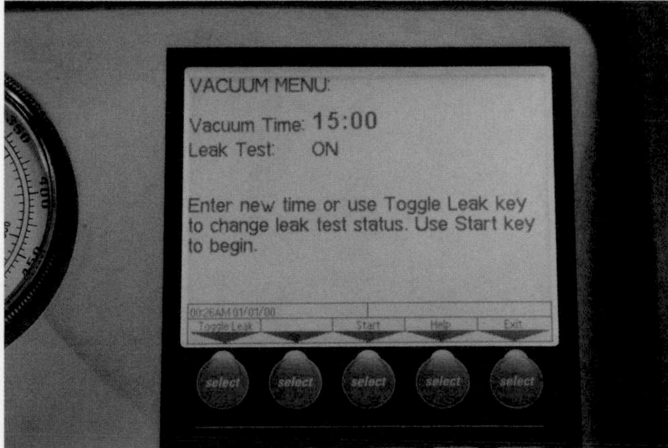

P45–6 Vacuum the system for at least 15 minutes before performing a leak test.

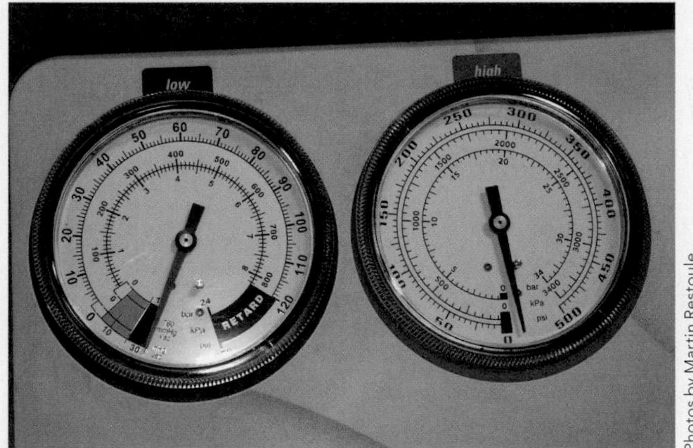

P45–7 Watch the recovery station's low- and high-side gauges to ensure sufficient vacuum is obtained. The system should be vacuumed for a minimum of 30 minutes to ensure that all moisture is removed from the system.

Photos by Martin Restoule

EVACUATING AND RECHARGING AN A/C SYSTEM WITH RECYCLING AND CHARGING STATIONS *(continued)*

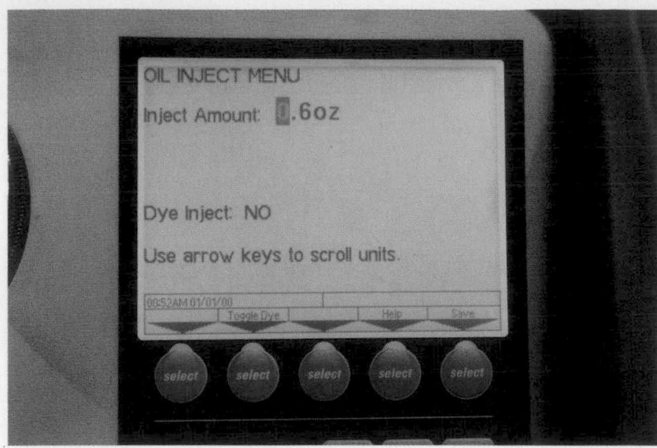

P45–8 Prior to charging, inject the amount of oil that was removed during the recovery process.

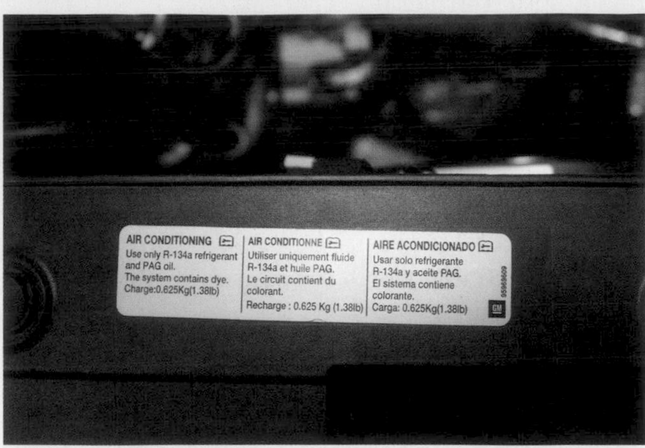

P45–9 Consult the service information or the underhood refrigerant label to determine the amount of refrigerant required to recharge the system.

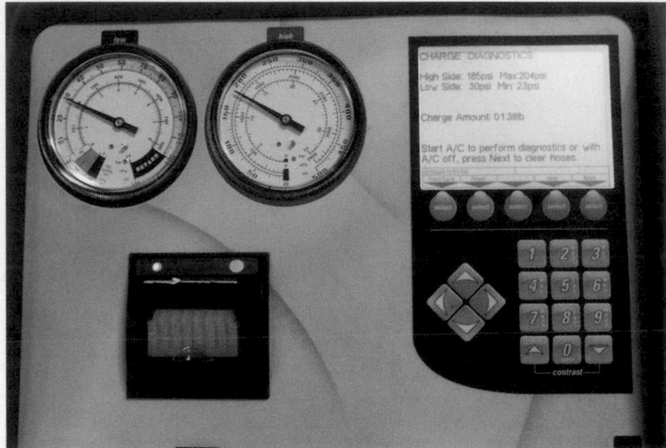

P45–10 Once the system is recharged, verify correct system operation while the machine is still connected to the vehicle's service ports.

P45–11 Be sure to replace the service port caps once the recycling/charging station has been removed. These caps are considered a primary refrigerant seal.

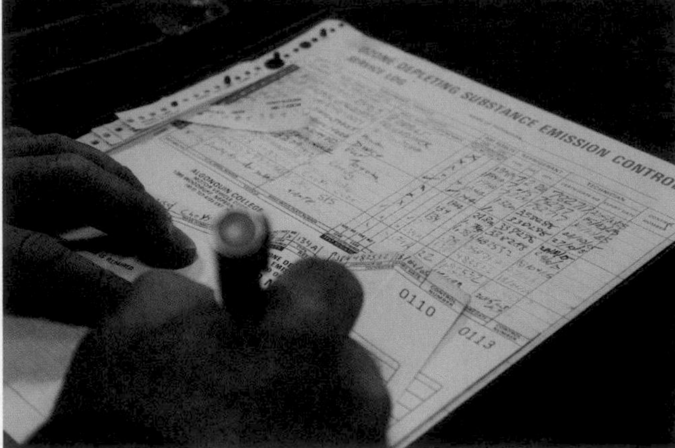

P45–12 Complete a substance emission control form/tag. This form contains system condition information and must be dated and signed by a certified A/C service technician.

Photos by Martin Restoule

P45–13 The tag must be securely attached to the A/C system, readily visible, accessible, and close to a service port.

Charging Cylinder

With an increase in temperature in any cylinder filled with refrigerant, there is a corresponding increase in pressure and a change in the volume of liquid refrigerant in the cylinder. To measure an accurate charge according to the weight of a cylinder, it is absolutely necessary to compensate for liquid volume variations caused by temperature variations. These temperature variations are directly related to pressure variations, and accurate measurements by weight can be calibrated in relation to pressure.

The charging cylinder is designed to meter out a desired amount of a specific refrigerant by weight. Compensation for temperature variations is accomplished by reading the pressure on the gauge of the cylinder. A calibrated chart on the machine contains the corresponding pressure readings for the refrigerant being used.

When charging an A/C system with refrigerant, the pressure in the system often reaches a point at which it is equal to the pressure in the cylinder from which the system is being charged. To get more refrigerant into the system to complete the charge, the cylinder can be placed on a heating plate.

TESTING FOR NONCONDENSABLE GASES Whenever using recycled refrigerant, it should be checked for noncondensable gases. Install a calibrated pressure gauge on the container and measure the pressure in the container. Measure the temperature of the air 10 cm (4 in.) away from the container's surface. Compare the measured pressure and temperature to the manufacturer's pressure/temperature chart. Determine if the recycled refrigerant has excessive noncondensable gases. If the refrigerant has excessive noncondensable gases, it should be recycled again and then retested before it is used.

CLIMATE CONTROL SYSTEMS

When diagnosing climate control systems, it is essential to have accurate information about the system before beginning. This information can be obtained from a number of sources: manufacturer service information, manufacturer service information websites, CD- or Internet-based computerized information systems, technical service bulletins, and aftermarket printed manuals.

Self-Diagnosis

Most climate control systems are computer controlled and have a self-diagnostic feature. An indicator lamp on the control panel may be used to flash codes. Manufacturers' specifications must be followed to identify the malfunction display codes, because they differ from car to car. To check the system, refer to the service information and determine if the automatic climate control system relies on the PCM, BCM, or a separate computer for control. Modern systems are part of CAN; therefore, DTCs and data are available through a scan tool. Typically, system pressures and input and output data are also available. In addition, active command of system operation, such as blend doors and blower speed, is often possible. By using active commands or bidirectional control, you can determine if a component and circuit are functioning.

If the unit is part of the PCM/BCM system, connect the scan tool and retrieve any codes that may be present. If the unit is controlled by its own computer system, use your service information and the procedure for retrieving trouble codes. If the system is a semiautomatic system, check the service information for the proper diagnostic procedures for checking evaporator and heating controls. Test the components or subsystems identified by the trouble codes.

PROCEDURE

Accessing DTCs from a System Using a Scan Tool

STEP 1 Using the service information, determine if and how to retrieve diagnostic information from the body control module (BCM).

STEP 2 Connect the scan tool.

STEP 3 Turn the ignition on.

STEP 4 Program the scan tool for that vehicle.

STEP 5 From the menu on the scan tool, select the BCM.

STEP 6 Retrieve and record all DTCs.

STEP 7 To observe the activity of the inputs and outputs, select DATA. Pay attention to any data that are outside its normal operating range, such as the ambient temperature, evaporator temperature, and low- and high-side pressures.

STEP 8 Diagnose any abnormal operating perimeters.

STEP 9 Turn the ignition off.

STEP 10 Disconnect the scan tool.

PROCEDURE

Accessing DTCs from a System Using the Control Panel

STEP 1 Refer to the service information for the correct procedure to enter into the HVAC's self-diagnostic mode. Normally, this is done by depressing and holding in buttons on the panel.

STEP 2 Determine what display on the panel should be observed to retrieve the DTCs and data.

STEP 3 Turn the ignition on.

STEP 4 Press and hold the designated buttons for the designated time. Make sure you hold the buttons down for the required period.

STEP 5 Record all DTCs that were displayed, and describe what is indicated by each.

STEP 6 If there are no DTCs detected and the system is still not operating properly, check the sensor input to the HVAC control unit. *Note:* The procedure for checking sensor inputs will vary with the model of vehicle.

STEP 7 Turn the ignition switch off.

STEP 8 Depress and hold the designated buttons.

STEP 9 Start the engine and release the buttons after the engine starts.

STEP 10 The control panel should begin to display the data for the sensors. Compare all readings to specifications.

STEP 11 Diagnose and correct all problems noted.

STEP 12 Cancel the self-diagnostic function.

STEP 13 After completing all repairs to the system, run the self-diagnostic test again to make sure there are no other problems.

The DTCs from the system can lead a technician to the following problems:

- An open or short in the air mix control motor circuit
- A problem with the air mix control linkage, door, or motor circuit
- An open or short in the mode control circuit
- A problem with the mode control linkage, doors, or motor circuit
- A problem in the blower motor circuit
- A problem with the HVAC control unit
- An open or short in the evaporator temperature sensor circuit
- A problem with temperature and sunload sensors

Some automatic temperature control systems allow for testing through the control panel. By pressing and holding two buttons, such as the AUTO and RECIRCULATE buttons, at the same time, the control panel will enter a self-diagnostic mode. During this mode, the system will move through different output modes, such as full heating to full cooling. Once completed, trouble codes are displayed on the control panel. Because each system is different, you will need to refer to the manufacturer's service information for procedures specific to the vehicle.

Blower Motor

The blower motor is usually located in the heater housing assembly. It ensures that air is circulated through the system. Its speed is controlled by a multiposition switch in the control panel. The switch works in connection with a resistor block that is usually located on the heater housing. Vehicles with automatic climate control typically use a blower speed control module, which uses pulse-width modulation (PWM) to control motor speed.

On some vehicles, when the engine is running, the blower motor is in constant operation at low speed. On automatic temperature control (ATC) systems, the blower motor is activated only when the engine reaches a predetermined temperature. The blower motor circuit is protected by a fuse located in the fuse panel. The fuse rating is usually 20 to 30 amperes.

The blower motor resistor block is used to control the blower motor speed. The typical resistor block is composed of three or four resistors in series with the blower motor, which control its voltage and current. The speed of the motor is determined by the control panel switch, which puts the resistors in series. Increasing the resistance in the system slows the blower speed.

If the blower does not operate, use a testlight to make sure there is voltage on both sides of the fuse. Check the wiring diagram, and identify whether the blower circuit is controlled by a ground side switch or an insulated (power side) switch. If the ground side of the circuit is controlled, check to see if voltage is arriving at the motor. On cars where the blower motor is behind the inner fender shell, hunt out the wiring and check for voltage. If the blower motor is getting voltage, the problem is either a burned-out blower motor or an open ground in the motor circuit. In situations where no voltage is available at the motor, backtrack to check for an open. Check also for burned or corroded connections in the blower relays and connectors. An open blower motor ground would cause the motor not to run at all. If the correct amount of voltage is available to the motor and the ground is good, suspect a bad motor.

If the blower works at some speeds but not all, check the voltage to the blower motor at the various switch positions. If the voltage does not change when a new position is selected, check the circuit from the switch to the resistor block. If there was zero voltage in a switch position, check for an open in the resistor block.

If the motor runs when it should not, there is probably a short in the circuit. If a ground-side switch controls the circuit, check for a short to ground in the control circuit. The exact problem can be isolated by disconnecting portions of the circuit until the motor stops. The short is in that part of the circuit that was disconnected last. If the circuit is controlled by an insulated switch, check for a wire-to-wire short. Check other circuits of the vehicle to identify what circuit is involved in this problem. That circuit will also experience a lack of control, or when that circuit is turned off, the blower motor will turn off. The exact problem can be isolated by disconnecting portions of the circuit until the motor stops. The short is in that part of the circuit that was disconnected last.

To diagnose PWM blower motor control systems, begin by setting the system so that the blower should be operating. If the blower does not work at all, refer to the manufacturer's service information to determine how to access the self-diagnostics for the ATC system. In general, diagnosis of a blower motor concern is similar to a non-ATC system. Determine if the motor is power or ground circuit controlled, and test for power and ground at the motor. Verify which constant, either power or ground, is present. If the blower control circuit is intact, the motor is faulty.

If the PWM signal is not present at the motor, you will need to determine the cause. Using either the self-diagnostic function or a scan tool, determine whether a blower motor request is received and if the control module is attempting to control blower motor speed. If the inputs are present and correct, and the blower is commanded on but does not operate, look for an open in the wiring between the blower control and the motor.

Doors and Ductwork

To check the proper functioning of the ductwork, move the temperature control lever to see if any change occurs. If it does not, shut off the air conditioner and turn on the heater. Move the temperature control arm again to see if any change occurs. If not, check the cable and the flap door connected to the temperature control lever. You might be able to reach under the dash to reconnect the cable or free a stuck flap.

If no substantial airflow is coming out of the registers, check the fuses in the blower circuit. Remove the fan switch and test it. Check the blower motor by hot-wiring it directly to the battery with jumper wires.

Controls

Most systems use electric motors to control the blend and mode doors; however, some use vacuum. To check these systems, identify and inspect the hoses and components for vacuum motors and doors. Look for disconnected or broken hoses, broken connectors, misrouted vacuum lines, and loose or disconnected electrical connectors at vacuum switches and controls. Identify the vacuum source for the various doors and switches of the duct system, and make sure vacuum is available when the engine is running. Disconnect that vacuum source. Connect a vacuum gauge to the inlet for the defroster door motor(s), and record the vacuum available there when the master control switch is in the following positions: MAX, NORM, BILEVEL, VENT, HEAT, BLEND, DEFROST, and OFF. If they do not move with the engine's vacuum, use a hand-operated vacuum pump to check the component's activity and its ability to hold a vacuum. Use the vacuum pump to check all one-way check valves. Also check all mechanical or cable linkages and controls. Make sure the cables are properly attached to the levers of the control switch. If the cable(s) are equipped with an automatic adjuster, make sure the adjusting mechanism operates freely and is not damaged.

CONTROL PANEL At times, the control panel is the root of problems **(Figure 55–32)**. Inspect the connectors to the control unit, looking for damage and corrosion. For systems with mechanical switches, use an ohmmeter to check for continuity through the blower motor switch in all of its positions. Compare your results with the wiring diagram for the switch. Check the resistance across the temperature control switch. If the mode selector switch is electrical, check for continuity across the terminals as you move the switch through the various mode selections. If the mode selector switch is a vacuum switch, apply a vacuum to the inlet of the switch and feel for a vacuum at the various tube connectors on the switch while you move the selector through its various positions. Compare your results to the service information.

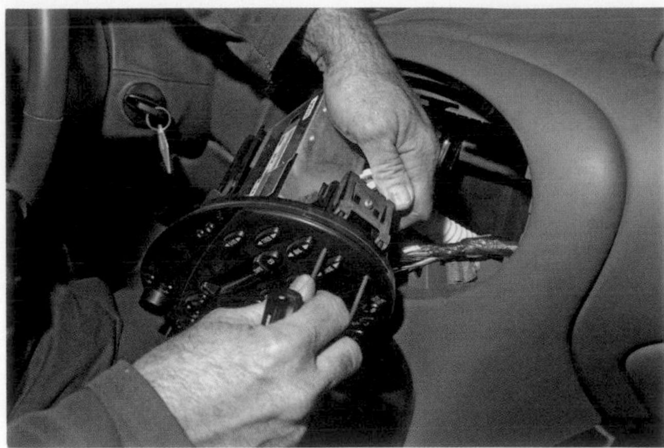

FIGURE 55–32

The control panel must be removed to test it and what it controls.

For systems with ATC or electronic controls, you will need to connect a scan tool to monitor the operation of the input switches. ATC panels are not repaired; rather, they are replaced as a unit. Care needs to be taken when handling and replacing these units. Like the other electronic circuits on a vehicle, the control panel may be part of the BCM circuit. Follow all precautions to eliminate static and voltage spikes at the BCM.

Solar Sensor

The solar sensor detects sunlight. It controls the climate control system on many vehicles. The output voltage from the sensor varies with the amount of sunlight. When the sunlight increases, the output voltage increases. As the sunlight decreases, the output voltage decreases. The sensor is monitored by the A/C control module. DTCs will indicate a perceived problem. The systems typically look for an open or short in the solar sensor circuit.

HEATING SYSTEM SERVICE

When performing any checks and service procedures, always follow the procedures recommended in the manufacturer's service information. In most cases, problems with the heating system are problems with the engine's cooling system. Therefore, most service work and diagnosis is done to the cooling system. Problems that pertain specifically to the heater are few: the heater control valve and the heater. Most often, if these two items are faulty, the engine's cooling system will be negatively affected. Both of these items are replaced rather than repaired. In some cases, it is possible to make repairs to vacuum hoses and electrical connections without removing the heater assembly. If it is necessary to remove the heater assembly, the cooling system must be drained before removing the heater core.

Basic Heater Inspection and Checks

When there is a problem of insufficient heat, begin your diagnosis with a visual inspection and a check of the coolant level. If the level is correct, turn the heater controls on; run the engine until it reaches normal operating temperature; and then measure the temperature of the upper radiator hose. The temperature can be measured with a pyrometer. If one is not available, gently touch the hose. You should not be able to hold the hose long because of the heat. While doing this, make sure you stay clear of the area around the cooling fan. A spinning fan can chop off your hand. If the temperature of the hose is not within specifications, suspect a faulty thermostat.

If the hose was the correct temperature, check the temperature of the two heater hoses. They should both be hot. If only one of the hoses is hot, suspect the heater control valve or a plugged heater core.

KEY TERMS

Dichlorodifluoromethane gas (p. 1699)
Discharge (p. 1710)
Manifold gauge set (p. 1706)
Phosgene gas (p. 1699)
Schrader service valves (p. 1708)
Stem valve (p. 1707)

SUMMARY

- Air-conditioning systems are extremely sensitive to moisture and dirt. Therefore, clean working conditions are extremely important. The smallest particle of foreign matter in an air-conditioning system contaminates the refrigerant, causing rust, ice, or damage to the compressor.
- Air-conditioner testing and servicing equipment includes a manifold gauge set, service valves, vacuum pumps, a charging station, a charging cylinder, recovery/recycling systems, and leak-detecting devices.
- To put air-conditioning systems in their best condition, the following service procedures should be performed: system discharging, system flushing, compressor oil level checks, evacuation, and system charging.

- Refrigerant evaporates quickly when it is exposed to the atmosphere. Never rub your eyes or skin if you have come in contact with refrigerant. It will freeze anything it contacts. If liquid refrigerant gets in your eyes or on your skin, it can cause frostbite. Immediately flush the exposed areas with cool water for 15 minutes and seek medical help.
- Never expose A/C system components to high temperatures. Heat will cause the refrigerant's pressure to increase. Never expose refrigerant to an open flame. A poisonous gas, dichlorodifluoromethane gas (similar to phosgene gas), will result. If breathed, this gas can cause severe respiratory irritation.
- Different refrigerants should never be mixed. Their oils and desiccants are not compatible. If the two refrigerants are mixed, contamination will occur and may result in A/C system failure. Separate service equipment, including recovery/recycling machines and service gauges, should be used for the different refrigerants.
- The manifold gauge set is used when discharging, charging, evacuating, and for diagnosing trouble in the system. Because R-134a is not interchangeable with R-12, separate sets of hoses, gauges, and other equipment are required to service vehicles.
- Performance testing provides a measure of air-conditioning system operating efficiency.
- A manifold pressure gauge set is used to determine both high and low pressures in the refrigeration system. The desired pressure readings will vary according to temperature.
- Using an electronic leak detector is the preferred method of leak detection.
- Fluorescent tracer systems use a fluorescent dye and a black light to locate the source of leaks.
- All refrigerant must be discharged from the system before repair or replacement of any component, except for compressors with stem service valves.
- Before you can purchase R-134a or work with refrigerant, you must be certified by an Environment Canada–approved program.
- Recycling collects old refrigerant from the system being serviced, cleans it up, and prepares it to be used in the future.
- When replacing refrigerant oil, it is important to use the specific type and quantity of oil recommended by the compressor manufacturer.
- Older R-12–based systems used mineral oil, whereas R-134a systems use synthetic polyalkylene glycol (PAG) oils. Using a mineral oil with R-134a will result in A/C compressor failure because of poor lubrication. Use only the oil specified for the system.
- While evacuating the system, the vacuum pump should remain on and connected to the system for at least 30 minutes after 660 to 740 mm Hg (26 to 29 in. Hg) is reached.
- When there is a problem of insufficient heat, begin your diagnosis with a visual inspection and a check of the coolant level. Then measure the temperature of the upper radiator hose.

REVIEW QUESTIONS

1. What should you do if you suspect that an alternative refrigerant was used in the A/C system, and it is not performing properly?
2. Define vacuum.
3. What can a manifold gauge set be used for?
4. What test instrument is commonly used to check the coil of a compressor clutch?
5. Why should refrigerants not be vented to the atmosphere?
6. Which of the following best defines the term evacuation?
   a. a process in which the refrigerant in a system is released
   b. a condition that exists when a compressor runs until there is no refrigerant left in the system
   c. the process that pulls all traces of air, moisture, and refrigerant from the system
   d. a process that uses air to force moisture and dirt out of the system
7. Which of the following statements about orifice tube systems is correct?
   a. The suction line must be warm to the touch from the evaporator outlet to the compressor.
   b. The liquid line from the condenser outlet to the evaporator inlet should be cool.
   c. The evaporator should feel very hot.
   d. The accumulator must be cool to the touch.
8. How is a new compressor shaft seal installed?
   a. tapped into place with a deep socket
   b. tapped into place with a brass punch
   c. tapped into place with a seal installer tool
   d. hand-twisted into place with a seal installer tool

9. To charge an air-conditioning system while it is running, where should the refrigerant be added?
   a. the high side
   b. the low side
   c. both the high and low sides
   d. either the high or the low side
10. When using a manifold gauge set to evacuate an R-134a air-conditioning system, which hose should be connected to the vacuum pump?
   a. the red high-pressure hose
   b. the centre yellow hose
   c. the blue low-pressure hose
   d. the centre black hose
11. Which of the following could cause a heater to not provide heated air?
   a. the water flow valve stuck in the closed position
   b. the water flow valve stuck in the open position
   c. no vacuum to a normally open water flow valve
   d. a faulty mode door vacuum motor
12. What refrigerant oil should be used in an R-134a air-conditioning system?
   a. mineral-based
   b. polyalkalene glycol (PAG) oil
   c. ester oil
   d. vacuum pump oil
13. What would an abrupt temperature change at the centre of the condenser indicate?
   a. The condenser is working efficiently.
   b. The airflow through the condenser is insufficient.
   c. The orifice tube is severely restricted.
   d. The condenser is restricted at the temperature change point.
14. What does the presence of oil around an air-conditioning line fitting indicate?
   a. The fitting is leaking oil.
   b. The fitting was lubricated during assembly.
   c. The fitting is leaking refrigerant.
   d. The system contains excessive refrigerant oil.
15. When an air-conditioning line is damaged and losing all refrigerant, other than the damaged line, what else should be replaced?
   a. the evaporator
   b. the condenser
   c. the accumulator
   d. the compressor
16. When performing a hand-feel air-conditioning system operational check, where should the two temperature change points be?
   a. at the compressor and the orifice tube
   b. at the compressor and the condenser
   c. at the condenser and the orifice tube
   d. at the evaporator and the orifice tube
17. What would a thick frost buildup on the suction line to the compressor indicate?
   a. The thermostatic expansion valve is starving the evaporator.
   b. The thermostatic expansion valve is flooding the evaporator.
   c. The accumulator is not bleeding enough oil into the system.
   d. The compressor is weak and should be replaced.
18. What is the preferred method of refrigerant leak detection?
   a. using soapy water
   b. using a halide torch
   c. using fluorescent dye
   d. using an electronic leak detector
19. While evacuating an air-conditioning system and a vacuum of 740 mm Hg (29 in. Hg) is reached, what is the minimum time that the vacuum pump should remain on?
   a. 15 minutes   b. 30 minutes
   c. 45 minutes   d. 60 minutes
20. Which of the following refrigerant statements is true?
   a. Empty refrigerant containers can be filled with compressed air.
   b. R-134a refrigerant will not cause damage to eyes or skin.
   c. Refillable refrigerant containers can be 100 percent filled.
   d. R-134a comes in light blue containers.

APPENDIX A

DECIMAL AND METRIC EQUIVALENTS

METRIC (MM)	DECIMAL (IN.)	FRACTIONS	METRIC (MM)	DECIMAL (IN.)	FRACTIONS
0.397	0.015625	1/64	13.097	0.515625	33/64
0.794	0.03125	1/32	13.494	0.53125	17/32
1.191	0.046875	3/64	13.891	0.546875	35/64
1.588	0.0625	1/16	14.288	0.5625	9/16
1.984	0.078125	5/64	14.684	0.578125	37/64
2.381	0.09375	3/32	15.081	0.59375	19/32
2.778	0.109375	7/64	15.478	0.609375	39/64
3.175	0.125	1/8	15.875	0.625	5/8
3.572	0.140625	9/64	16.272	0.640625	41/64
3.969	0.15625	5/32	16.669	0.65625	21/32
4.366	0.171875	11/64	17.066	0.671875	43/64
4.763	0.1875	3/16	17.463	0.6875	11/16
5.159	0.203125	13/64	17.859	0.703125	45/64
5.556	0.21875	7/32	18.256	0.71875	23/32
5.953	0.234275	15/64	18.653	0.734375	47/64
6.35	0.250	1/4	19.05	0.750	3/4
6.747	0.265625	17/64	19.447	0.765625	49/64
7.144	0.28125	9/32	19.844	0.78125	25/32
7.54	0.296875	19/64	20.241	0.796875	51/64
7.938	0.3125	5/16	20.638	0.8125	13/16
8.334	0.328125	21/64	21.034	0.828125	53/64
8.731	0.34375	11/32	21.431	0.84375	27/32
9.128	0.359375	23/64	21.828	0.859375	55/64
9.525	0.375	3/8	22.225	0.875	7/8
9.922	0.390625	25/64	22.622	0.890625	57/64
10.319	0.40625	13/32	23.019	0.90625	29/32
10.716	0.421875	27/64	23.416	0.921875	59/64
11.113	0.4375	7/16	23.813	0.9375	15/16
11.509	0.453125	29/64	24.209	0.953125	61/64
11.906	0.46875	15/32	24.606	0.96875	31/32
12.303	0.484375	31/64	25.003	0.984375	63/64
12.7	0.500	1/2	25.4	1.00	1

APPENDIX B

GENERAL TORQUE SPECIFICATIONS

The values in this chart should be used only when manufacturer's specifications are *not* available. Also, the values are valid only when SAE 10 oil is used to lubricate the threads of the bolt.

BOLT DIAMETER (IN MILLIMETRES)	TORQUE: KG CM* KG M PROPERTY CLASS:									
	4.6	4.8	5.6	5.8	6.6	6.8	6.9	8.8	10.9	12.9
6	49*	63*	61*	79*	74*	95*	103*	126*	172*	206*
8	119*	153*	148*	178*	178*	230*	250*	306*	417*	500*
10	235*	303*	294*	379*	353*	455*	495*	606*	8.2	10
12	411*	529*	427*	662*	616*	7.9	8.6	10.5	14	17
14	654*	8.4	8.2	10.5	10	12	13	17	23	27
16	10	13	12	16	15	20	21	26	36	43
18	14	18	17	23	21	27	30	36	49	59
22	27	35	34	44	41	52	57	70	95	114

BOLT DIAMETER (IN INCHES)	TORQUE (POUNDS-FEET)		
	SAE 2	SAE 5	SAE 8
1/4	7	10	14
5/16	14	21	30
3/8	24	37	52
7/16	39	60	84
1/2	59	90	128
9/16	85	130	184
5/8	117	180	255
3/4	205	320	450
7/8	200	515	730
1	300	775	1090

B

C

D

E

F

G

H

N

O

P

Q

R

S

T

W

X

Y

Z